GRAY'S ANATOMY

CHURCHILL LIVINGSTONE
Medical Division of Longman Group Limited

© Longman Group Ltd. 1980

Exclusive distribution in United States of America granted to
W. B. Saunders Company, Philadelphia.

ISBN 0 7216 9128 5
Library of Congress Catalog Card Number 80–52146

Thirty-fifth edition 1973
Thirty-sixth edition 1980

Printed in Great Britain by Jarrold and Sons Ltd, Norwich

GRAY'S ANATOMY

36th Edition

EDITED BY

PETER L. WILLIAMS & ROGER WARWICK

D.Sc., M.A., M.B., B.Chir.   B.Sc., Ph.D., M.D.

PROFESSORS OF ANATOMY

GUY'S HOSPITAL MEDICAL SCHOOL, UNIVERSITY OF LONDON

Associate Editors

MARY DYSON, B.SC., PH.D. & LAWRENCE H. BANNISTER, B.SC., PH.D.

With the Assistance of our Colleagues

RICHARD E. M. MOORE, D.F.A., M.M.A.A., F.R.S.A.—*Illustrations*
JENNIFER HALSTEAD, M.M.A.A.—*Illustrations*
MICHAEL C. E. HUTCHINSON, M.B., B.S., B.D.S.—*Special Preparations*
JEFFREY W. OSBORN, B.D.S., Ph.D., F.D.S.R.C.S.—*Dental Anatomy*
SUSAN M. STANDRING, B.Sc., Ph.D.—*Bibliography*
E. LOWELL REES, M.B., B.S.—*Index*

W. B. SAUNDERS COMPANY

PHILADELPHIA

CHURCHILL LIVINGSTONE

EDINBURGH LONDON MELBOURNE AND NEW YORK 1980

PREVIOUS EDITIONS AND EDITORS

First Edition 1858
Second Edition 1860

By HENRY GRAY, F.R.S., F.R.C.S.
St. George's Hospital

Third Edition 1863
Fourth Edition 1865
Fifth Edition 1869
Sixth Edition 1872
Seventh Edition 1875
Eighth Edition 1877
Ninth Edition 1880

By TIMOTHY HOLMES, M.A., F.R.C.S.
St. George's Hospital

Tenth Edition 1883
Eleventh Edition 1887
Twelfth Edition 1890
Thirteenth Edition 1893
Fourteenth Edition 1897

By T. PICKERING PICK, F.R.C.S.
St. George's Hospital

Fifteenth Edition 1901
Sixteenth Edition 1905

By T. PICKERING PICK, F.R.C.S.
St. George's Hospital
and R. HOWDEN, M.A., M.B., C.M., D.Sc., LL.D.
University of Durham

Seventeenth Edition 1909
Eighteenth Edition 1913
Nineteenth Edition 1916
Twentieth Edition 1918
Twenty-first Edition 1920
Twenty-second Edition 1923
Twenty-third Edition 1926

By R. HOWDEN, M.A., M.B. C.M., D.Sc., LL.D.
University of Durham

Twenty-fourth Edition 1930
Twenty-fifth Edition 1932
Twenty-sixth Edition 1935
Twenty-seventh Edition 1938

By T. B. JOHNSTON, C.B.E., M.D.
Guy's Hospital Medical School
University of London

Twenty-eighth Edition 1942
New Impression 1944
New Impression 1945
Twenty-ninth Edition 1946
New Impression 1947
Thirtieth Edition 1949 (Completely Reset)
New Impression 1950
Thirty-first Edition 1954
New Impression 1956

By T. B. JOHNSTON, C.B.E., M.D.
and J. WHILLIS, M.D., M.S.
Guy's Hospital Medical School
University of London

Thirty-second (Centenary) Edition 1958
New Impression 1960

By T. B. JOHNSTON, C.B.E., M.D.
University of London
D. V. DAVIES, M.A.(Cantab.), M.B., B.S.
University of London
and F. DAVIES, M.D., D.Sc.(Lond), F.R.C.S.
University of Sheffield

Thirty-third Edition 1962
New Impression 1964

By D. V. DAVIES,
D.Sc.(Lond.), M.A.(Cantab.), M.B., B.S., F.R.C.S.
University of London
and F. DAVIES, M.D., D.Sc.(Lond.), F.R.C.S.
University of Sheffield

Thirty-fourth Edition 1967

By D. V. DAVIES,
D.Sc.(Lond.), M.A.(Cantab.), M.B., B.S., F.R.C.S.
St. Thomas's Hospital Medical School
University of London

Thirty-fifth Edition 1973 (Completely Reset)
New Impressions 1975, 1978

By ROGER WARWICK, B.Sc., Ph.D., M.D.
Professor of Anatomy
Guy's Hospital Medical School
University of London
and PETER L. WILLIAMS, D.Sc., M.A., M.B., B.Chir.
Professor of Anatomy
Guy's Hospital Medical School
University of London

Thirty-sixth Edition 1980 (Completely Reset)

By PETER L. WILLIAMS, D.Sc., M.A., M.B., B.Chir.
Professor of Anatomy
Guy's Hospital Medical School
University of London
and ROGER WARWICK, B.Sc., Ph.D., M.D.
Professor of Anatomy
Guy's Hospital Medical School
University of London

R. Howden 1901–1926

Timothy Holmes 1863–1880

T. Pickering Pick 1883–1905

T. B. Johnston 1930–1958

Henry Gray, 1858–1860, seen here in the Dissecting Room
of St. George's Hospital 1860

J. Whillis 1942–1954

D. V. Davies 1958–1967

Francis Davies 1958–1962

PREFACE

TO THE 35TH EDITION

When our predecessor, Professor David Vaughan Davies, began his preface to the last edition, his first words were to regret the death of his co-editor (Professor Francis Davies) and to acknowledge, with characteristic generosity, the latter's editorial virtues. It is now our equally saddening task to record the death of an editor and friend for whom we shared, with so many others throughout and beyond the confines of human anatomy and medicine, a deep respect both as a scientist and teacher, and as a man of many accomplishments outside his profession. During a period of fifteen years he brought to this editorship the same unyielding standards of scholarship which he applied so outstandingly in his own discipline and school. For both of us his death had also immediate personal associations, for one (P.L.W.) had enjoyed twenty-five years as his student, colleague and companion, and for a number of years acted as his indexer, while the other (R.W.)—frequently a co-examiner—was expecting him on the second day of an examination of our own students when he was so suddenly stricken. We both regard it as an honour to carry forward his editorial obligations.

At the time of his death Professor Davies had completed a small number of revision notes, and these and his collection of references have been valuable to us. However, the decision had already been taken to modernize the format of this volume, with a complete resetting of the text; and it was therefore opportune to undertake a more extensive revision than would have been otherwise possible. The original role of *Gray's Anatomy* as a treasury of 'descriptive and applied' systematic human topography has become amplified through more than a century of usefulness by the early addition of histology and embryology, and by the gradual development of introductory sections to the various 'systems'. To many readers, both graduate and undergraduate, this elder textbook has, however, represented *in excelsis* the field of naked-eye or *dissectional* anatomy. We have neither disturbed nor curtailed this aspect of the volume: rather have we added to it—by the correction of errors, the addition of a host of new observations, the inclusion of hundreds of new references to cover details beyond the scope of even a large textbook, and by reinstatement of variations in respect of many structures. On the other hand we have endeavoured to reduce prolixities of language, as far as the shortened period of revision remaining to us has permitted; no page has escaped such attention, frequently extensive in degree. A considerable saving of space has been thus effected—to offset large additions of new writing and to keep this new edition within a single volume. To rewrite the whole text would have been an impossibly lengthy task, and there has inevitably resulted an increase in size. Subsequently we hope and intend to render all of the established text in a simpler and more succinct style, but to do this throughout with no loss of factual detail is a massive undertaking, and for our limited success in this regard we ask our readers' forbearance. Moreover, we were convinced that other tasks were more urgent and important.

Particularly in this century, the conviction has increased amongst anatomists that isolated observation and description is not enough, and that an experimental approach to problems of structure is as necessary as in other biological sciences. In addition, the greater advances in technique—especially in the study of finer detail, in living and developing organs, tissues and cells—have enlarged the scope of anatomy far beyond the parent stem of macroscopic structure. These advances have engendered a spate of new specialities, such as histology, cytology, ultrastructure, embryology, neurology, electromyography, kinesiology, ergonomics, and so on, to a degree dependent only upon the choice of individual minds and the canalization of techniques. The expanding scope of structural knowledge and the exacting demands of more elaborate techniques do indeed dictate such specialization; but all such knowledge remains a continuum—except insofar as extensive gaps of ignorance and uncertainty persist. Unfortunately, and perhaps particularly in the medical sphere, the compartmentalization of anatomy into several disciplines or subjects—with attendant titles, individual chairs, and even separated departments—tends towards disintegration. To study some such region as a limb, in all its proportions, activities, and even evolution, and then its major structures—bones, joints, muscles, vessels and so forth—and to proceed to the microscopic, ultrastructural and ultimately biochemical details of its tissues and cells, appears to us a continuous process, and most desirably so to a balanced education for medicine, however elementary the standard. Unfortunately, these different levels of organization and function are perforce usually considered in separate laboratories, departments, lectures and books.

The defects of this compartmentalization are widely recognized, and have resulted in much effort to 'integrate' teaching. In this persuasion we have re-arranged certain contents of this volume, and in particular have transferred most of the existing section of histology to the appropriate systems. Hence, in *Myology* will be found not only a systematic description of the muscles of the human body, but also of muscle as a tissue. Moreover, there are certain general considerations such as, in this instance, the variable form and the mode of action of muscles and accessory structures—tendons, aponeuroses, bursae, and the like—which have already in recent editions been set out in introductory sections. It is precisely in these generalized aspects of human anatomy that the greatest interest often lies, attracting the major volume of research. We have concentrated special attention upon these sections, which are the more difficult to keep in accord with current progress in research. For this reason, and because the *significance* of structural data is more apparent in generalizations, we have found it necessary to rewrite all such introductory sections and to extend them—to a marked degree in most instances.

The accelerating tempo of enquiry in all respects of structure has a dual effect: there is not only a continuous correction and, more especially, accumulation of data, but also in the main a sharpening awareness of defects in knowledge and deficiences in its interpretation. It is our belief that ignorance and uncertainty should be more prominently stated in textbooks than they usually are. We have tried to imbue this new edition with rather more of this attitude, not merely in introductory sections, but throughout the systematic text. Where the actions of a muscle are not

ix

convincingly demonstrable by direct observation or experiment, we believe uncertainty should be admitted; and where, for example, intricate central nervous organization has been investigated in another animal—perhaps quite remote from man—we consider that the need for caution in extrapolation to mankind should, in a textbook of human anatomy, be clearly appreciated.

It would be burdensome to categorize the changes and additions in this edition; approximately seven hundred pages of completely new writing have been contributed, covering all systems and a wide spectrum of topics. Doubtless our efforts have been uneven; we have, for example, not been able to revise some aspects of cardiac anatomy as thoroughly as we wished, but these and other defects will be remedied in the next edition. While hoping that many will welcome the substantial changes and additions in histology (and especially ultrastructure), in the sections on the teeth (completely rewritten), joints, muscles, lymphatic system, nervous system, special senses and endocrine organs, we also hope that our readers will freely and constructively criticize our mistakes, excesses and deficiencies. To keep abreast of all the new work reported is a formidable undertaking, and we trust that no one will hesitate to inform us of our shortcomings in this or any other regard.

In one particular this edition is unique; for the first time since the first edition, in 1858, this, the *direct* descendant of the original 'Gray', will be available in the United States side by side with the American scion, which has also persisted through almost an equal period, having begun in 1859 as a reprint of the first edition, but having diverged markedly in its evolution from the direct British descendant. The American parallel version ceased to refer to the British senior heir as long ago as 1896, though preserving the name of Henry Gray on its title-page. With subsequent editions, and particularly this, the divergence has become so great that the two books cannot be regarded as parallel versions; each has its own character.

Our editorial labours have been lightened by much help from others. Doctor Lawrence H. Bannister has contributed the new section on cytology and many of the histological and ultrastructural passages in the sections devoted to myology, neurology, and splanchnology. Doctor Jeffrey W. Osborn has recast the section on dental anatomy. Doctor Susan M. Standring has prepared the bibliography, which appears as a new feature at the end of the text, and has meticulously supervised all reference material. Doctor E. Lowell Rees has not only undertaken the complex revision of an expanded index, but has also provided many special dissections, histological preparations and advice. However, the majority of the revision remains the work of the editors, who are wholly responsible for any errors of judgment, incorrect terminology, omissions, misquotations, or lack of clarity throughout the volume. With the complete resetting of the text, index, illustration captions and tabbing, and the addition of a new bibliography, it is inevitable that, despite prolonged and repeated proof reading, some typographical errors will have eluded us; for these we apologize.

In accord with the comprehensive textual changes in this edition, the illustrations have also received much attention. More than 200 of the 1,305 figures of the 34th edition have been removed, with the addition of over 600 new items; thus, in this 35th edition almost a third of the illustrations are new. Moreover, new blocks have been prepared for all illustrations retained—from the original artwork wherever possible. Apart from those radiographs and reproductions from external sources, acknowledged below, all new illustrations have been prepared in our Medical Centre. We are much indebted to Doctor Aszal Riaz and Doctor John D. Dow (Department of Diagnostic Radiology) and to Mr. Kenneth Twinn and Mrs. Joy Taylor (Department of Physics) for much help with radiographs. In our own department Doctors E. Lowell Rees and Michael C. E. Hutchinson have produced many special dissections and other preparations, assisted in this by Doctors Andrew M. Seal and William J. Owen. Most of these have been photographed by Mr. Kevin Fitzpatrick, our photographer, to whom we are much indebted. Mr. Derek Lovell and Mr. David Ristow have provided skilled technical assistance in respect of electron microscopy and histology. Many

other workers on our staff, past and present, have also helped with illustration material, including Doctors Mary Dyson, Murray Brookes, Karen Hiiemae, Wimal Jayaratnam and Kenneth J. W. Taylor; Doctors David R. Turner and Roy O. Weller (now in Pathology), Doctor David N. Landon (now in the Department of Neurobiology, National Hospital for Nervous Diseases, London), Doctor Eric W. Baxter, Mr. Eric C. Tatchell and Miss Hilary Phillip (Department of Biology) and Doctors J. P. Black and P. Barkhan (Department of Haematology) have all afforded us generous aid with preparations for photography. We also gratefully record the expert help of our School's Librarian, Miss Jean M. Farmer and her willing staff.

Mr. S. W. Woods had already prepared six new illustrations (chiefly in embryology), with the same high standards with which he embellished several previous editions while serving Professor Davies. Most of the new artwork, however—amounting to about 210 items—has been carried out by our colleague, Mr. Richard E. M. Moore, D.F.A.Lond., M.M.A.A., member of l'Association Internationale pour l'Etude de la Mosaique Antique, and Fellow of the Royal Society of Arts. His combination of meticulous draughtsmanship, a most unusual ability to comprehend the scientific intent of projected illustrations, and his extraordinary patience and stamina throughout two and a half years of continuous effort, have made our collaboration most fruitful and enjoyable.

Many other authorities in various fields have allowed us to reproduce, copy or adapt illustrations from papers and monographs, or have special materials available for photography. It is a pleasure to acknowledge the generosity of Professor Janos Szentágothai (University of Budapest), Doctor Elizabeth Crosby (University of Michigan), Doctor W. J. W. Sharrard (University of Sheffield), Doctor Michael J. Hogan, Doctor Jorge J. Alvarado and Mrs. Joan E. Weddell (University of California), Doctor Alan M. Laties (University of Pennsylvania Medical School), Mr. Emanuel Rosen (Royal Eye Hospital, Manchester), Doctor N. A. Locket (Institute of Opththalmology, London), Professor Alf Brodal (University of Oslo), Professor William J. Hamilton (Professor Emeritus, University of London), Professor Peter M. Daniel (Institute of Psychiatry, University of London), Professor J. André-Balisaux (University of Brussels), Doctors Keith E. Webster and A. Robert Lieberman (University College, London), Professor Don Fawcett (Harvard University Medical School), Professor N. Cauna (University of Pittsburgh), Professor Yves Clermont (McGill University), Doctor Max Levene and Mr. Emrys Turner (St. Helier Hospital, Surrey), Doctor Charles Levene (Sir William Dunn Department of Cellular Pathology, University of Cambridge), Doctor R. C. Edwards (University of Cambridge), Doctor A. F. Holstein (University of Hamburg), Doctor L. M. Franks (Imperial Cancer Research Fund, London), Doctor Don H. Tompsett (Royal College of Surgeons of England), Doctor M. A. Sleigh (University of Bristol), Professor Sir John O. Eccles (Laboratory of Neurobiology, State University of New York), Doctor Berta Scharrer (Albert Einstein College of Medicine, New York), Professor Paul D. Maclean (National Institute of Mental Health, Maryland), Doctor Walle J. H. Nauta (Massachusetts Institute of Technology), Doctor Webb Haymaker (Ames Research Center, N.A.S.A., California), Doctor Bror Rexed (Socialstyrelsen, Stockholm), Professor James M. Sprague (University of Pennsylvania), Doctor Ray S. Snider (University of Rochester, New York), Professor Clinton N. Woolsey (University of Wisconsin), Doctor Julia Fourman (University of Leeds), Professor David B. Moffat (University College, Cardiff), Doctor Alexander Barry (University of California), Professor Viktor Hamburger (Washington University, Missouri), Professor Setsuya Fujita (Prefectural University Medical School, Kyoto), Doctor Douglas R. Anderson (University of Miami), Professor Koji Uchizono (University of Tokyo).

We have had the benefit of special advice from many of the authorities acknowledged above, and in addition from Professor M. A. MacConaill (University College, Cork), Professor Jack J. Pritchard and the late Doctor James H. Scott (University of Belfast), Doctor F. Torrent Guasp (University of Barcelona), Professor John Z. Young (University College, London), Mr. D.

G. Wilson Clyne, Professor Patrick D. Wall (University College, London), Professor J. V. Basmajian (Emory University, Atlanta), and Doctor Keith Jones (Keeper of the Jodrell Laboratory, Royal Botanic Gardens, Kew, Surrey). Among colleagues in other departments in our centre we are especially indebted to Doctor R. T. Grant (late of the Department of Experimental Medicine) and Professor Paul E. Polani and his staff (Department of Paediatric Research) for much help respectively in regard to arteriovenous anastomoses and genetics. Similarly we have had the advantage of advice from many other colleagues in their own specialities, including Doctor Sidney Liebowitz (Immunological Pathology), Doctor John R. Henderson (Physiology) and Doctor David Watts (Biochemistry).

Although we have striven to acknowledge with punctilio the many publishers of scientific journals and books who have allowed us, with customary generosity, to use copyright material, we trust that any neglect in this respect of which we may have been guilty will be forgiven in the same generous spirit.

Throughout the ardours and pressures of this ambitious revision we have enjoyed the most cordial relationship with our publishers and printers, who have both given us the greatest freedom and encouragement. In particular we wish to mention Mr. John A. Rivers of Churchill, and Mr. William G. Henderson, Mr. Gerald J. Hooton and Mr. Alfred S. Knightley, of Churchill Livingstone, who have been our companions in many anxious, sometimes convivial, and always protracted discussions.

We are most grateful to our departmental secretaries, Miss Margaret Collins and Mrs Patricia Elson, for much sporadic and demanding help, and to our official secretarial assistant, Mrs. Irene Williams, who has patiently translated innumerable notes and drafts into immaculate typescript for the printers.

It is customary to eulogize the patience of wives—faint praise which is scarcely *galante*. Far from tolerating our preoccupation, our wives have supported us unfailingly with a true and critical interest and sympathy in our labours. To them and all our friends and colleagues, who have helped more than they know to sustain our enthusiasm, we remain profoundly grateful.

1973   P.L.W. and R.W.

CONTENTS

HENRY GRAY, F.R.S., F.R.C.S.

Since readers of *Gray's Anatomy* will be interested to learn something of the original author, Henry Gray, the following information as to his career has been extracted from an article which appeared in the *St. George's Hospital Gazette* of 21 May 1908.

Gray, whose father was private messenger to George IV, and also to William IV, was born in 1827, but of his childhood and early education nothing is known.

On 6 May 1845, he entered as a perpetual student at St. George's Hospital, London, and he is described by those who knew him as 'a most painstaking and methodical worker, and one who learnt his anatomy by the slow but invaluable method of making dissections for himself'.

While still a student he secured, in 1848, the triennial prize of the Royal College of Surgeons for an essay entitled, 'The origin, connexions and distribution of the nerves to the human eye and its appendages, illustrated by comparative dissections of the eye in other vertebrate animals'.

At the early age of twenty-five he was, in 1852, elected a Fellow of the Royal Society, and in the following year he obtained the Astley Cooper prize of three hundred guineas for a dissertation 'On the structure and use of the spleen'.

He held successively the posts of demonstrator of anatomy, curator of the museum, and Lecturer on anatomy at St. George's Hospital, and was in 1861 a candidate for the post of assistant surgeon. Unfortunately he was struck down by an attack of confluent smallpox, which he contracted while looking after a nephew who was suffering from that disease, and died at the early age of thirty-four. A career of great promise was thus untimely cut short. Writing on 15 June 1861, Sir Benjamin Brodie said, 'His death, just as he was on the point of obtaining the reward of his labours . . . is a great loss to the Hospital and School.'

In 1858 Gray published the first edition of his *Anatomy*, which covered 750 pages and contained 363 figures. He had the good fortune to secure the help of his friend, Dr. H. Vandyke Carter, a skilled draughtsman and formerly a demonstrator of anatomy at St. George's Hospital. Carter made the drawings from which the engravings were executed, and the success of the book was, in the first instance, undoubtedly due in no small measure to the excellence of its illustrations. This edition was dedicated to Sir Benjamin Collins Brodie, Bart., F.R.S., D.C.L. A second edition was prepared by Gray and published in 1860.

The portrait of Gray published in the present section is a reproduction of one which appeared in the *St. George's Hospital Gazette* of 21 May 1908, where the original is described as being 'a very faded photograph taken by Mr. Henry Pollock, second son of the late Lord Chief Baron Sir Frederick Pollock, and one of the earliest members of the photographic society of London'.

INTRODUCTION

To perform an 'anatomy' was to make a 'dissection'; the two words are no longer synonymous. *Dissection* has remained a technique; *Anatomy* has become a field of study—a corpus of observations, still dependent upon technique, but capable of rational correlation among themselves and with other biological studies. Most narrowly, anatomy may be the investigation of biological structure—in plants or animals—with no other motive than description of form. Even so, such *topographical anatomy* has not remained insulated from technological progress; the usefulness of direct visual dissection persists, but its relatively crude results have become incalculably augmented by the advent of light microscopy, micro-dissection, electron microscopy, histochemistry, radiology, autoradiography, and many other techniques. The application of these, with ever-growing modifications and extensions, has revealed great new fields of discovery. Some, such as *histology* and *cytology*, the study of tissues and cells, are true extensions of the parent discipline; others—*electron microscopy, histochemistry* and *auto-radiography*—are merely techniques capable of providing particular types of data.

The theme of growth and differentiation, both in individual development or *ontogeny* and in that of the species or kind—*phylogeny*—has led to the particular studies of *embryology*, *comparative anatomy* and *morphology*. Embryology, the study of individual development, also embraces problems of gameto-genesis, fertilization and embryonic nutrition, and in the investigation of these in relation to mankind, *comparative embryology* has proved invaluable.

Studies of growth, whether upon the epochal time-scale of evolution or the rapid cycles of ontogeny, emphasize the mutability of structures; and the dynamic nature of all *living* structures entails an inescapable relation between form and function. It is, of course, possible to consider form in isolation, an exercise of most limited value, though the data of pure description may have particular applications. Such *applied anatomy* is usually concerned with human structural observations which are useful in medicine, especially in surgical technique, but also in clinical diagnosis. Descriptive anatomy, however, has a far more extensive application in relation to function. Few biological structures can be regarded as functionless, and no biological function is known to occur outside living fabric, which includes, it must be noted, everything from the whole creature to its molecular structure. In human considerations of nature it is possible to divorce structure from function (though not the reverse); but few structuralists are disinterested in function. The mere fact that topography can be described upon a *regional* or *systematic* basis necessarily entails functional considerations, for while the former is of particular vocational interest in medicine, *systematic anatomy* is based upon recognition of function. The *locomotor system* embraces structures directly concerned with movement—skeletal elements, articulations and ligaments, and muscles, the study of which may be formalized as *osteology, arthrology* and *myology*. Similarly, *neurology*, which treats of the nervous sytem, including its sensory organs, and *angiology*, the study of cardiovascular arrangements of organs and tissues, are correlated as much by

function as by structure. It is equally clear that the *respiratory, alimentary, urogenital* and *endocrine systems* (though often grouped under the unilluminating anatomical term *splanchnology*), are clearly functional as well as anatomical fields of study.

All such 'systems' are investigated at macroscopic, histological, cytological, ultrastructural and biochemical levels. Furthermore, experimentation upon structures—classically illustrated by early work on the circulation, reflex behaviour and endocrine influences—has formed the vanguard of anatomical research, with increasing momentum in this century. *Experimental anatomy, experimental cytology* and *experimental embryology*, have contributed greatly to the advancement of knowledge in the field of human anatomy. Throughout the following sections, where relevant, numerous allusions will be made to the results of such experimental studies.

Descriptive anatomy obviously demands an internationally acceptable repertoire of names for structures, and there is also a need for an agreed convention upon terms for their spatial relationships (*see* accompanying figure). For this purpose the human body is assumed to be in its usual bipedal or erect position with the arms pendent and eyes and hands facing forwards. This position is open to certain objections; for example, the arms are rotated laterally at the shoulder joints and the forearms are fully supinated and are thus not in their usual position of rest (*see* p. 350). Moreover, comparisons of human anatomy with that of other animals, which are mostly quadrupedal in habit, are confronted with some difficulties. Nevertheless, the erect 'anatomical position' does provide an unambiguous system of correlation for man. In this posture the *median plane* divides the body vertically into right and left halves which are approximately symmetrical, apart from certain visceral details. The superficial contours of this plane form *anterior* and *posterior median lines* on the surface of the body. The median plane is also frequently called the *sagittal plane* (after the cranial suture of that name), but this term is also sometimes applied to any vertical plane parallel to the median, and hence the latter may be termed *paramedian* or *parasagittal planes*. To avoid confusion and to maintain simplicity, it is perhaps best to reserve the term *median* for the single *median plane* defined above, and to regard all other planes parallel to this as *sagittal*. Vertical planes at right angles to the median plane are usually described as *coronal planes*, after the coronal suture (p. 298). To complete the three-dimensional reference grid, *horizontal planes* are those which traverse the body at right angles to both the median and coronal planes.

The adjectives *anterior* and *posterior* are applied to the front or back surfaces of the body, including the limbs. Synonyms for these are *ventral* and *dorsal*, which, since they can be applied equally to quadrupeds, are sometimes preferable. All these terms are in fact used more extensively to specify the aspects or surfaces of individual structures *within* the body, and often to denote their relative positions. Thus—the heart is posterior (or dorsal) to the sternum, the posterior surface of which is close to the anterior (or ventral) aspect of the heart. Similarly, *superior* (or *cranial*) and *inferior* (or *caudal*) are adjectives qualifying the positions of structures in the vertical sense. Of the two venae cavae one is

INTRODUCTION

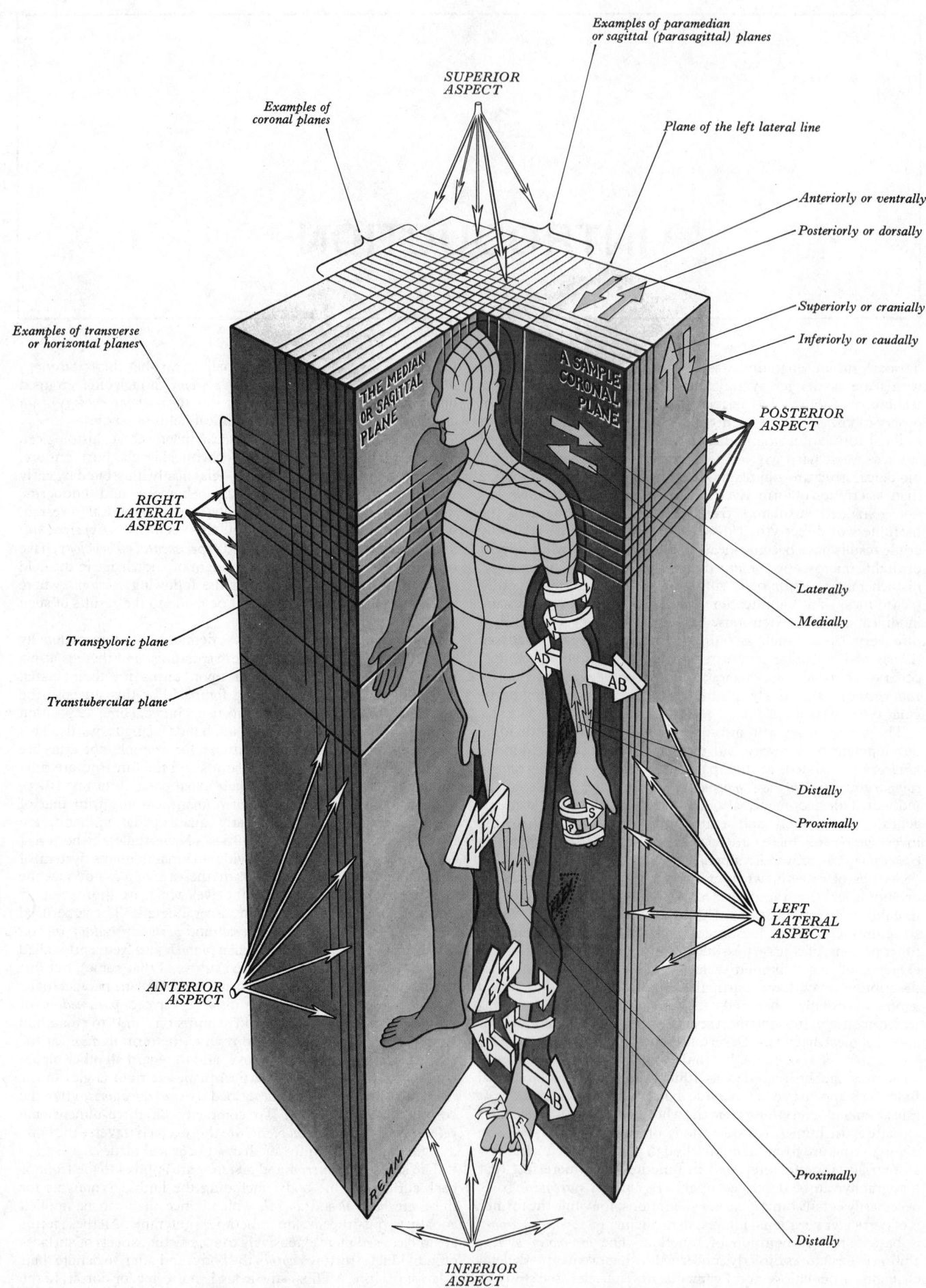

The terminology widely used in descriptive anatomy is illustrated in the figure above. The abbreviations on the solid arrows: AD—adduction, AB—abduction, FLEX—flexion (of the thigh at the hip joint), EXT—extension (of the leg at the knee joint), M and L—medial and lateral rotation, P and S—pronation and supination, I and E—inversion and eversion.

superior, cranial (headward) with respect to the other which is *inferior, caudal* (tailward) in position. (Incidentally, the *superior* and *inferior* venae cavae are *anterior* and *posterior* in the quadruped, but the terms cranial and caudal, which obviate this confusion, have not been adopted in this instance in mankind.) In describing structures within the cranium, the term *cranial* is obviously unsatisfactory. Particularly in describing positional terms to cerebral structures it is more appropriate to employ the term *rostral*, which indicates that a particular entity is nearer the *rostrum* (beak or nose). Thus, the cerebrum is *rostral* to the cerebellum, which is *caudal* to the former. To define the relation of structures to the median plane the terms *medial* and *lateral* are employed: the heart is *medial* to the lungs, which are *lateral* to it, and so on.

Any number of oblique planes can be imagined, and likewise the spatial relations of structures are not always so orthogonally simple as anterior, posterior, superior, medial, and so forth. Combined terms are therefore sometimes used for intermediate positional arrangements, such as anterolateral (ventrolateral), postero-inferior (dorsocaudal), etc., and these are self-explanatory. In the limbs certain variant terms are current—and since these do not involve reference to the 'anatomical position', they have some value in obviating confusion. Thus, structures which are superior, and hence nearer the limb root, are dubbed *proximal*; those relatively inferior in position are clearly more *distal*. Anterior and posterior aspects or regions may be described respectively as *flexor* or *extensor* in upper limbs, but the terms do not correspond in the lower limbs, which have undergone a contrasting form of morphological rotation (pp. 150, 350), such that the primitively extensor or dorsal aspect is now anterior. In the forearm, the terms *radial* and *ulnar* are occasional synonyms for lateral and medial, as are *fibular* (peroneal) and *tibial* in the lower limb. *Palmar* and *plantar* are variants for the flexor surface of the hand and foot. Finally, *superficial* and *deep* specify distance from the surface of the body; the somewhat similar terms, *external* and *internal*, are usually applied to the walls of hollow structures such as the head, thorax and abdomen and various viscera, including vessels and ducts.

It will be noted that throughout the text, the units of linear measurement are those of the Système Internationale (S.I.). These include the *micrometre* (micron, μm) $1~\mu m = 1 \times 10^{-6}$ metre; $1000\mu m = 1$ mm), and the *nanometre* (nm) $1 nm = 1 \times 10^{-9}$ metre; 1000 nm $= 1~\mu$m). Accordingly the use of Ångstrom unit ($\text{Å} = 1 \times 10^{-10}$ metre; $10\text{Å} = 1$ nm) has been discontinued.

We have, in the main, continued the policy of adherence to *Nomina Anatomica* (3rd ed., by G. A. G. Mitchell, Excerpta Medica Foundation, 1968). Familiar variants and some eponyms have also been included where deemed advisable. We have also attempted to follow the proposed *Nomina Histologica* and *Nomina Embryologica*, prepared by the subcommittee of the International Anatomical Nomenclature Committee and presented to the Eleventh International Congress of Anatomists held in Leningrad in August 1970, at a plenary session at which these two drafts were approved. Unfortunately, both contain many common terms at variance with each other, important omissions, and some terms which have aroused belated dissatisfaction. Moreover,

some ultrastructural details were overlooked, often obliging us to follow current practice, itself confused by synonyms and vernacular jargon. Therefore, we have regarded the recommendations of *Nomina Histologica* and *Nomina Embryologica* as less obligatory than *Nomina Anatomica*, which has been exposed to much more critical revision. Nevertheless, the latter does not meet all contingencies, especially in the central nervous system; it also still retains some intrinsically unsatisfactory terms. Notwithstanding, we have continued to adhere to most of these, though we have preferred not to handicap a whole section with the title of 'Syndesmology', for which 'Arthrology' has priority in time and in clarity of communication. We have also disregarded official disapproval of the hyphen, employing it frequently, though not perhaps consistently, to separate vowels likely to be compounded as diphthongs. (Since these remarks were written a 4th Edition of *Nomina Anatomica* (Ed. Roger Warwick, 1977) has appeared from the same publisher, and some of the difficulties referred to have been resolved. We have also had the advantage of consulting revisional proposals for a 5th Edition, which will be published shortly after the Twelfth International Congress of Anatomists, which is to take place in Mexico in 1980. The 4th Edition contains *Nomina Histologica* and *Nomina Embriologica*, which have therefore now been available for assessment for three years. It is to be hoped that active workers, especially in the fields of Cytology and Histology, will collaborate adequately with the I.A.N.C. to elaborate acceptable compromises between the jargon of the former and the sometimes impractical Latinity of the latter.) Even in Europe and the Americas, most university students, however, fall short of even 'a little Latin and no Greek', and this takes no account of the large numbers of schools outside these continents. Consequently, it must be assumed that a large majority of undergraduates and younger postgraduates find words of Greek and Latin derivation unfamiliar, and that this difficulty will increase in the future. Even the Latinist may find *bulbourethral* bothersome; *ou* is a diphthong not only in English but in other European languages; similarly, *sacro-iliac* aids in both pronunciation and understanding. The diaeresis is often frowned upon these days: but while *cooperate* perhaps no longer requires hyphen or diaeresis, *spermatozoön* and *oöcyte* are awkward words. We believe that an increasing number of readers need this help if unfamiliar words are to be pronounced with confidence, to the betterment of international communication. Officially, all anatomical terms are expressed entirely in Latin; but there is little objection to translating *flexor digitorum superficialis* into 'the superficial flexor of the digits', and this kind of vernacularization is practised in many countries, especially in Europe. Human anatomy, however, is a worldwide science, and the Latin terms are recognizable everywhere. A little effort with the first few pages of an elementary Latin grammar will quickly unveil the mysteries of masculine and feminine plurals, genitives and so on—a small personal concession to international understanding, in a world which must surely welcome unequivocal terms, even if based on a 'dead' language—perhaps all the more acceptable because no longer a contender in petty national rivalries.

1
CYTOLOGY

Introduction

All living organisms show distinctive patterns of organization with respect to time and space. Temporally, these involve orderly sequences of physical and chemical events which are conventionally described in terms of movement, metabolism, growth, differentiation, reproduction, reactivity to external change, and evolutionary progression (see also pp. 72–78). Spatially, these sequences are not haphazard, but arranged in a complex structural framework which determines their direction and their coordination. The integration of these phenomena, and their dependency upon environmental energy sources, constitutes what we call 'life' in the biological sense.

The microscopic study of living things concerns such intimate organization of their life processes, and since their ultimate structure is on a molecular scale, various special means have to be used to analyse them.

The *biological molecules* are arranged in complex aggregations playing specific roles in the living process; these aggregations are the *organelles*; multiples of different types of organelles are further combined in specific membrane-bound units, or *cells*. The cell is an important unit, since it is the smallest aggregation to show all the major features of living organisms mentioned above. In some simple organisms, e.g. bacteria, protozoa, the cell is also the whole organism, and capable of an independent existence. In more complex organisms, multiples of cells are grouped together, with spatial differentiation of various regions to perform particular roles, such as digestion, reproduction, and so forth. These may have the form of simple co-operative layers of cells, or *tissues* (a term introduced by Bichat (1771–1802) for the different groups of cells, muscular, nervous, and so on, in post-mortem man). Multiples of tissue layers are further grouped together to carry out more complicated co-operative actions; these constitute the *organs* of the body. Finally, the whole consortium of organs, co-ordinated and unified by specialized communication systems, is capable of the activities which characterize human life. To analyse adequately the biological basis of such activities, it is necessary to understand its parts, and the subjects of *cytology* and *histology* have as their aim the clarification of these at the microscopic level.

Advances in cytology, as in all scientific pursuits, have awaited technical progress (Singer 1931; Taton 1966). The early microscopes of the sixteenth and seventeenth centuries only allowed examination of large cells, such as those of protozoa and plants, and the English microscopist Robert Hooke (1665) was the first to use the term *'cell'*, which was applied by him to the compartments of cork wood. With the refinement of the optical microscope by Abbé and Leitz, and the introduction of staining and sectioning techniques in the late nineteenth century, the field was set for the rapid expansion of cytology and histology as a serious discipline. The further introduction of bright field, phase contrast, and interference microscopy in the 1930–50 period allowed direct detailed observation of living cells. By this time histology had lost some of its initial impetus, but the growth of electron microscopy, increasing resolution by three orders of magnitude, together with the development in biochemistry of cell-fractionation techniques, and of methods of biophysical analysis of molecules, allowed observation and interpretation at the molecular as well as the cellular level.

The concept of the cell is a convenient starting point; each cell comprises a discrete unit enclosed by a membrane which is interposed between it and its micro-environment, and envelops the living material or *protoplasm*, included in which is one or more *nuclei*. The protoplasm is a heterogeneous aqueous phase in which is the chemical machinery for metabolic processes, and the material of heredity, which specifies the character of the cell from one generation to the next. In some very primitive cells (*prokaryotes*) such as bacteria and actinomycetes, the hereditary and metabolic materials are not separated from each other; in more complex cells (*eukaryotes*) the hereditary instructions are almost entirely sequestered in a special membrane-bound region, the *nucleus* (*karyon*), which is distinct from the remainder of the cell, the *cytoplasm*.

The protoplasm of cells consists chemically of large and small organic molecules, and inorganic ions in aqueous solution (DeRobertis *et al.* 1975); water comprises about 70 per cent of the total cell volume. Of the large organic molecules, the most abundant are those of carbohydrates, lipids, and proteins which provide important structural materials and metabolic machinery in the form of enzymes. Nucleic acids are also important in directing the activities of the cell. The cytoplasm of each cell is relatively unstable in its composition, and must be held ionically and osmotically within a narrow range for the effective functioning of its metabolic apparatus. However, each cell is also in a constant dynamic interchange with its external environment (p. 72), including other cells, and the continuous expenditure of metabolic energy is needed for a cell to maintain its steady state. If this is lost, the cell dies.

Most mammalian cells lie within the size range of 5–50 μm in diameter. Although many cells possess only one nucleus, some, formed either by fusion of uninucleate cells (*syncytia*), or by nuclear division without corresponding cytoplasmic division (*plasmodia*), are multinucleate. The latter may achieve a much larger volume than a uninucleate cell, although the *ratio* of nuclear to cytoplasmic volume is similar.

One great advantage of cellularity is that diffusion of materials between and within the living units is relatively rapid, so that control systems can operate rapidly within a mass of cells, and also that gaseous, nutritive, and excretory exchange processes can keep pace with the high demands of active cells. In this regard it should be noted that as a cell increases in size, its surface area (available for diffusion) increases by the square of the diameter, whereas the volume of protoplasm increases by the cube; this relationship puts a limit on the maximum size a cell can attain (D'Arcy Thompson 1942; see also p. 78). Preservation of cellularity also allows the emergence of different cell types joined together into functionally distinct tissues.

In the living state most individual cells are greyish in appearance when examined by transmitted light, and each is bounded by a deformable elastic membrane. Physical measurements of protoplasmic viscosity (DeRobertis *et al.* 1975) indicate that it is a heterogeneous material which can be highly viscous (the *gel* state), or relatively non-viscous (the *sol* state), or both states may coexist and interchange. Metabolic processes can profoundly alter these physical characteristics.

Motility is also a characteristic of most cells. This may take the form of intracellular streaming, with the movement of materials within the cell, or may produce movement of the whole cell by the progressive formation of finger-like projections (*pseudopodia*), or other extensions, of the cell surface (p. 35). Cell movements are also involved in the multiplication of cells by division to form two—*binary fission*—and in the uptake of extracellular materials such as solid particles (*phagocytosis*) or fluids (*pinocytosis*).

Cells in many situations form aggregates by reciprocal adhesion, and the cell surfaces behave as though sticky. Small differences in ionic composition can alter this aggregation, e.g. the presence of calcium ions is necessary for cell adhesion. When similar motile cells growing in nutrient media outside the body (i.e. *in vitro*) come into contact, movement normally ceases, a phenomenon termed *contact inhibition* (p. 91). Such a process is believed to govern the aggregation of cells into stable tissues in the body. These activities of the cell will be examined further in subsequent sections of this chapter.

1.1 A three-dimensional reconstruction of some of the principal architectural features of an absorptive cell lying in the simple columnar epithelium of the small intestine. Part of the cell is cut away to expose the nuclear envelope (green); nuclear contents are the nucleolus (red) and chromosomes (black). Outside in the cytoplasm lie the endoplasmic reticulum (yellow) with ribosomes (red) in clusters; the Golgi apparatus (pink) is shown in a supranuclear position, various cytoplasmic vesicles including lysosomes, microtubules (blue), microfilaments, and a centriole pair (grey) are also shown. The apical surface of the cell is covered with microvilli supported by microfilaments which are inserted into a filamentous terminal web. Junctional complexes are seen at the lateral borders of cells apically. A lamina basalis (purple) forms the boundary of the epithelium basally, and lies in close relation to the underlying reticulin and collagen.

CELL STRUCTURE

Electron microscopy has had a great impact on our understanding of cell structure; although no two types of cell are identical, there is a common pattern of organization, so that some generalizations can be made. In describing their internal structure, it is convenient to divide the cell into cytoplasmic and nuclear divisions (Bloom and Fawcett 1975; Ham 1974; Junquiera *et al.* 1975).

In the cytoplasm several distinct systems of organelles are discernible, comprising the external and internal cytoplasmic membranes, membrane-bound bodies such as mitochondria, lysosomes, lipid vacuoles and the Golgi complex; ribosomes, the microtubules, microfilaments, centrioles, cilia and flagella, and microvilli; also the proteins in suspension, termed collectively the *cytosol*.

The nucleus is made up of the nuclear membranes with nuclear pores, the nucleoplasm (nuclear sap), chromosomes (chromatin), nucleoli, and intranuclear bodies of various kinds. These structures will now be described in some detail (**1.1**, 2, 3, 4, 5).

Cytoplasm

THE MEMBRANE SYSTEMS OF THE CELL

With the advent of electron microscopy, it was confirmed that cells are bounded by a distinct membrane, and internally are permeated by membrane-lined vacuoles and channels. Both external and internal membranes (*cytomembranes*) have many

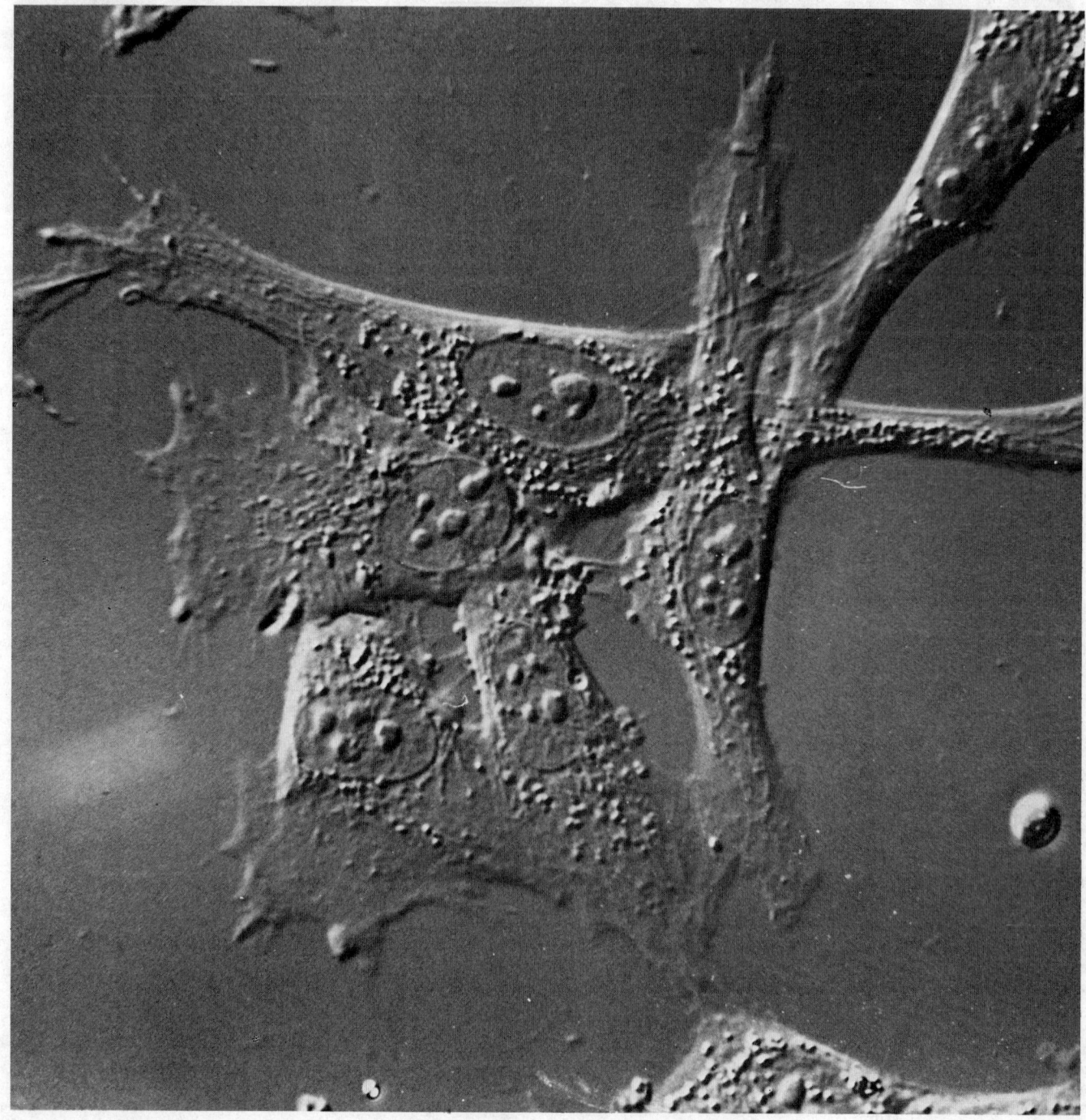

1.2 A high-power micrograph of living cells (fibroblasts in tissue culture), viewed by Normarski interference microscopy. Kindly provided by the Paediatric Research Unit, Guy's Hospital Medical School. Magnification × 2,000.

features in common. They are all composed chemically of phospholipids and proteins, usually in an approximately 3:2 ratio, with a small amount of carbohydrate. The amount of lipid in an external membrane is enough to give a layer two molecules thick over the cell surface, so it was proposed as long ago as 1925 by Gorter and Grendel that cell membranes are bilayers of lipid, with the hydrophobic ends of each lipid molecule pointing towards the interior of the membrane, and the hydrophilic ends pointing outwards. Later, Davson and Danielli (1935) suggested that the protein might be situated on both sides of the lipid to form a protein/lipid sandwich, but more recently a 'fluid mosaic' model has been proposed (Singer and Nicolson 1972), in which the proteins are envisaged as embedded, or floating, in the lipid bilayer (**1.6A**). Some proteins, by virtue of extensive hydrophobic portions of their polypeptide chains, appear deeply embedded in the lipid, spanning its entire width (*intrinsic proteins*) whilst others are only superficially embedded and can be more easily detached by mild treatment (*extrinsic proteins*). Carbohydrates in the form of oligosaccharides and polysaccharides are attached either to proteins (glycoproteins) or to lipids (glycolipids), projecting outwards from the surface of the membrane (Nicolson and Poste 1976).

These features, deduced partly from biochemical and biophysical data, can be correlated with appearances in the electron microscope. Membranes when suitably fixed and stained with heavy metals show, in section, two densely stained layers separated by an electron-translucent zone, the total thickness being about 8 nm (the classical *'unit membrane'* of Robertson 1959), probably reflecting the binding of stain by the 'heads' of the phospholipid molecules. Freeze-fractured and/or etched specimens (Branton 1971; Stolinsky and Breathnach 1975), in which the deeply frozen sample is cleaved to expose membrane interiors, and of which a metal-shadowed carbon replica is then made and viewed in the electron microscope (**1.8, 9**), have also demonstrated a bilaminar structure in membranes. The cleavage planes usually pass along the midline of each membrane where the hydrophobic 'tails' of the phospholipids meet. This method has also demonstrated protein 'particles' embedded in the lipid layers; these particles are in the 5–15 nm range and may represent large protein molecules or assemblies of several smaller protein molecules. It is interesting that the particles are distributed asymmetrically between the two half-membranes, usually adhering more to one face than the other. In plasma membranes (which form the cell surface), the *inner* or protoplasmic half-membrane carries most of the particles, exposed at its externally-facing (P or A) surface, whilst the internally-facing (E or B) surface of the *external* half-membrane usually shows pits into which the particles fit (**1.8, 9**). Not all the proteins of membranes are *visible* as particles, however; some are either too small or not compact enough to appear in this form, and have, as yet, only been demonstrated biochemically.

Biophysical measurements have shown the phospholipid bilayer to be a highly fluid structure, allowing diffusion *along the plane* of the membrane at rates as high as 2 μm/sec. Thus some proteins are able to move freely along the membrane plane unless there are special attachment devices within the cell preventing this, and aggregates of proteins are thus able to form (*see* p. 7). Some internal membranes possess much higher amounts of protein than the external cell membrane, for example, the inner mitochondrial membrane which is rich in enzyme activity, and the fluidity of such membranes is correspondingly much reduced.

The functions of cell membranes are many; they form boundaries which selectively limit diffusion, separate different phases inside the cell—dividing, in general terms, those regions within the channel system (*vacuoplasm*) from those outside it (*hyaloplasm*). Membranes actively control the passage of electrolytes and small organic molecules, generate bioelectric potentials, and they act as surfaces for the attachment of enzymes and other metabolic systems, often associated with the movement of reaction products from one side of the membrane to the other (*vectorial metabolism*). Membranes also serve as sites for the reception of external stimuli including hormones and other chemical agents, and for the recognition and attachment of other cells. Lastly, they can themselves act as points of attachment of

intracellular structures, thus providing the basis for locomotor activity and for cytoskeletal stability.

The membranous elements of a cell can, on occasion, fuse with each other and so form a potentially continuous system. However, there are severe restrictions on the fusion of different types of membrane so that each maintains its unique chemical and functional features distinct from those of others. How the cell is able to prevent the indiscriminate mixing of membranes when they do fuse, for example, in the passage of metabolites within vacuoles from one organelle to another, is not known. So although the membrane systems of the cell can be viewed as a single entity, they are also highly distinctive and localized in their activities.

Cell membranes are synthesized by the granular endoplasmic reticulum (p. 12), usually in collaboration with the Golgi apparatus (Poste and Nicolson 1977).

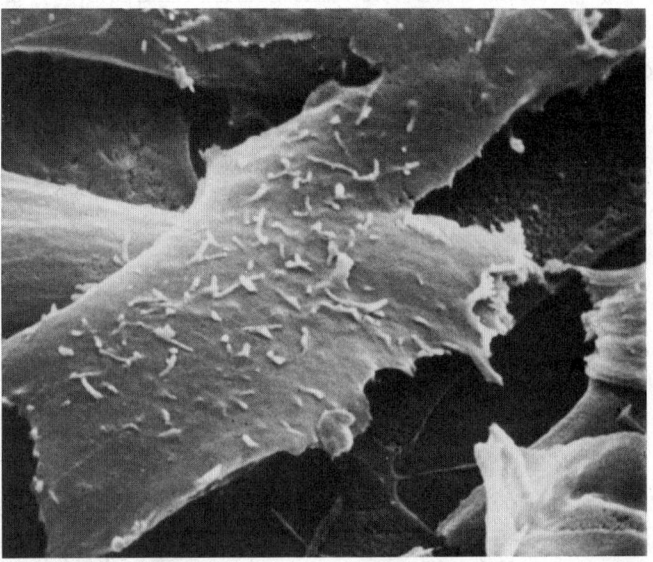

1.3 Flattened fibroblasts growing in tissue culture, viewed with the scanning electron microscope. Note the numerous fine lateral projections and the presence of a few microvilli on the upper surfaces of the cells. Magnification × 4,000.

PLASMA MEMBRANE (PLASMALEMMA OR CELL MEMBRANE)

This constitutes the external boundary of the cell, and differs from the other membranes in that it bears a diffuse carbohydrate-rich coat, the *cell coat*, or *glycocalyx*, on its external surface. The cell coat varies in composition from cell to cell, but usually contains much sialic acid; it confers on the cell surface a net negative electrostatic charge. Disturbances in the cell coat and surface charge are associated with neoplastic transformations, and may play a part in the failure of some malignant cells to adhere to one another, with their consequent metastasis to other sites. Many tissue and blood *antigens* are also located in the coat in its glycoproteins and glycolipids.

The plasma membrane is capable of active transport of small molecules without any apparent structural change, but the uptake of larger molecules is associated with the invagination and rounding up of the plasma membrane to form small vacuoles termed *endocytic vesicles* (**1.1, 4**), which are transported to other regions within the cell. The reverse process—the extrusion of organic molecules—is achieved by *exocytic vesicles* which fuse with the plasma membrane and release their contents to the exterior.

Although the plasma membrane carries out the functions common to all membranes, it has especial importance as the boundary of the whole cell, in co-ordinating many of its activities, by mediating changes in the cell's environment to the cell interior, and in maintaining the cell's shape and coherence (De Pierre and Karnovsky 1973). The plasma membrane, therefore, acts as a

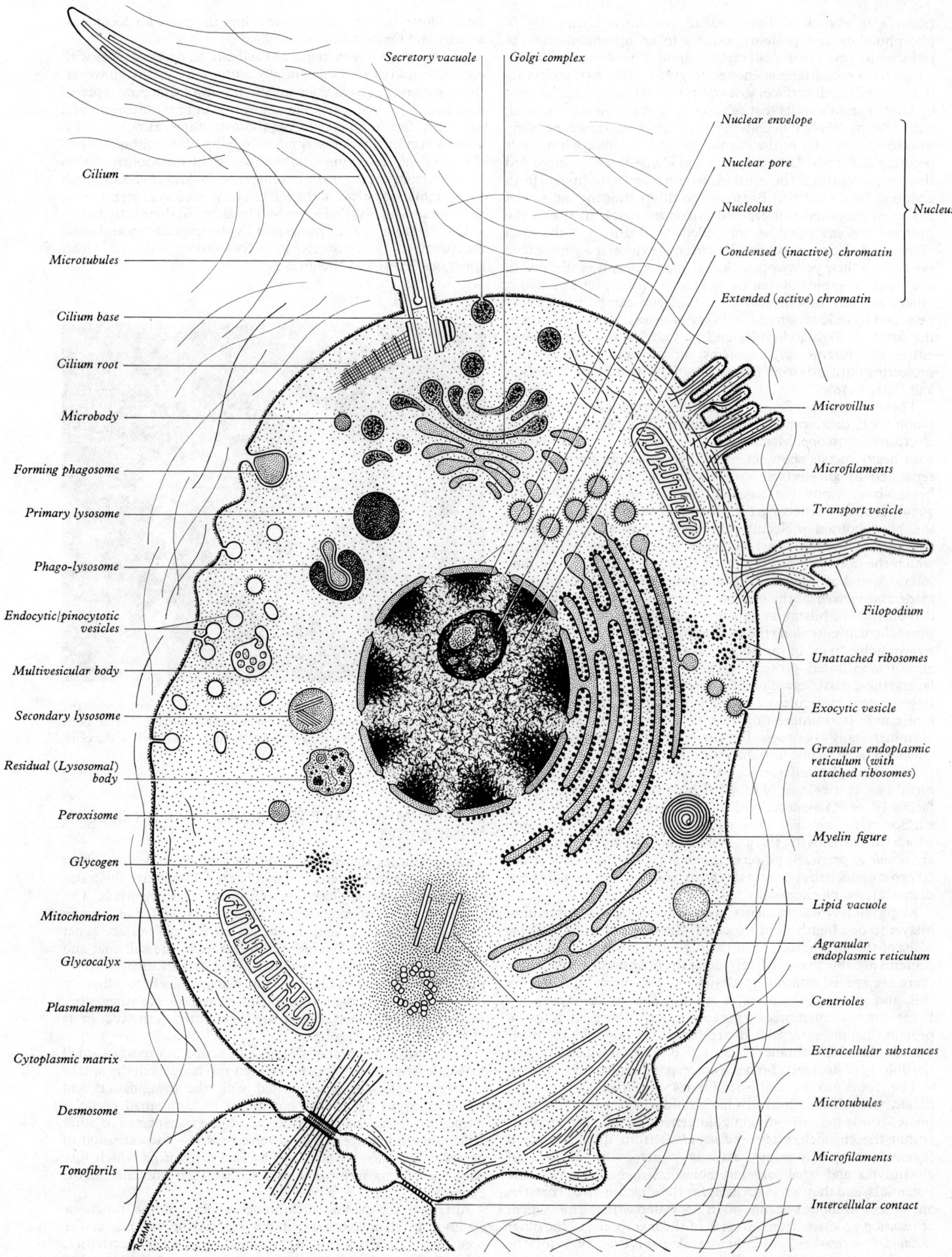

Secretory vacuole   Golgi complex

Nuclear envelope

Nuclear pore

Nucleolus   } Nucleus

Condensed (inactive) chromatin

Extended (active) chromatin

Cilium

Microtubules

Cilium base

Cilium root

Microbody

Forming phagosome

Primary lysosome

Phago-lysosome

Endocytic/pinocytotic
vesicles

Multivesicular body

Secondary lysosome

Residual (Lysosomal)
body

Peroxisome

Glycogen

Mitochondrion

Glycocalyx

Plasmalemma

Cytoplasmic matrix

Desmosome

Tonofibrils

Microvillus

Microfilaments

Transport vesicle

Filopodium

Unattached ribosomes

Exocytic vesicle

Granular endoplasmic
reticulum (with
attached ribosomes)

Myelin figure

Lipid vacuole

Agranular
endoplasmic reticulum

Centrioles

Extracellular substances

Microtubules

Microfilaments

Intercellular contact

1.4   A composite diagram showing the principal structures found within
tissue cells. Only a proportion of the features illustrated will be present in
any specific cell type.

sensory surface, and possesses a wide variety of special receptor molecules, some responding only to a narrow range of stimuli, as for example the receptors for specific steroid hormones, whilst others are perhaps activated by more general factors such as the contact with other cells or inorganic surfaces. Stimulation of the cell surface may result in changes in the bioelectric transmembrane potential which accompany fluxes of inorganic ions; this is most striking in the *excitable* plasma membranes of nerve and muscle cells in which the 'resting' voltage can change transiently from as much as 100 mV (negative on the inside) to one of 50 mV (positive on the inside) when suitably stimulated, a phenomenon which results from the opening and subsequent closure of channels selectively permeable to sodium and potassium and other ions (*see* pp. 9, 824).

Stimulation of receptors at the cell surface also often results in the activation of a chemical 'second order messenger' which may cause profound changes in the metabolism or motility of the whole cell (Weissmann and Claiborne 1975). Adenylate cyclase, an enzyme associated with the plasma membrane of probably all nucleated cells, is an important element in this process; activation of this enzyme results in changes in concentrations of cyclic AMP (cyclic adenosine monophosphate) within the cell, in turn leading to alterations in, for example, metabolic pathways, DNA synthesis, gene expression, protein synthesis, actin and myosin interactions, and many other intracellular events. Cyclic GMP (cyclic guanidine monophosphate) is likewise controlled by similar enzyme systems, and may have effects antagonistic to those of cyclic AMP. Some hormones and neutrotransmitters have been shown to act on the cell in this manner, and it is probable that many types of intercellular communication use the same or similar systems (p. 830).

Maintenance of cell shape by the plasma membrane is largely brought about by filamentous proteins, including microfilaments, which are attached to it; the same structures may also be responsible for changes in shape such as those associated with cell motility (for a fuller discussion *see* p. 35).

Although these considerations apply to the plasma membranes of all cells, those of specialized cells are often highly developed in some particular respect. Thus, sensory membranes derived from the plasma membrane, for example those of the retinal photoreceptors, have numerous photoreceptive proteins embedded in their surfaces; the acetylcholine-receptive sole plate membranes of skeletal muscle cells likewise are studded with protein particles capable of binding the transmitter. Other specialized cell membranes, such as that of the erythrocyte, have important metabolic and cell shape-determining components. This area of research is indeed of great potential, since we may learn much about cellular functions in general by first understanding the cell surface which is responsible for co-ordinating so many cellular activities. We may also gain a deeper understanding of the actions of drugs on cells since the cell membrane is undoubtedly the primary target of a wide variety of chemotherapeutic agents, anaesthetics, and other substances of medical importance.

As mentioned above, *the cell coat* forms an integral part of the plasma membrane, projecting 2–20 nm, or perhaps more from the lipoprotein elements considered above. It is composed of the carbohydrate portions of glycoproteins and glycolipids embedded in the plasma membrane (**1.6A**), often consisting of highly branched oligosaccharides and polysaccharides, the terminal residues of which are usually negatively charged sialic acids such as n-acetyl neuraminic acid, but are also often rich in galactose residues. These and other carbohydrates can be readily demonstrated with plant-derived chemical probes termed *lectins* (e.g. concanavalin A, wheat germ agglutinin, phytohaemagglutinin) which bind specifically to particular carbohydrate groups. By conjugating lectins with fluorescent molecules, or with an electron microscopic tracer such as ferritin or horseradish peroxidase, the surface carbohydrates can readily be visualized (**1.6B**) and even measured (Nicolson 1974, 1976). Since the glycoproteins and glycolipids are usually free to move in the plane of the membrane, addition of lectins can cause the aggregation of these carbohydrate-rich membrane molecules which become cross-linked to form raft-like groups or 'patches'. If these are

further aggregated by motile activities of the cell, they merge to form a 'cap' at one pole of the cell (p. 36). In other cells (e.g. erythrocytes) the carbohydrate-rich molecules are prevented from wandering or forming patches by internal anchoring proteins.

The plasma membrane itself, like other cytomembranes, is in a constant state of flux, the whole surface being regularly renewed by addition from exocytic vesicles, by subtraction and degradation in the lysosomal system of the cell, or by the loss of components into the surrounding micro-environment.

INTERCELLULAR CONTACTS

The plasma membrane is, of course, the surface which establishes contact with other cells and promotes various kinds of cellular interaction. Structurally we can distinguish two main classes of contact: (1) those which are not marked by the formation of obvious cytological surface specializations in the areas of contact, and (2) those in which relatively long-lasting junctional features are cytologically well defined, and which establish and maintain the intercellular contacts. In the first category (Moores and Partridge 1974), which is by far the most common, cells adhere to one another, usually at any part of their surfaces, enabling them, for example, to form loose aggregations whilst permitting some degree of relative movement. A distance of about 20 nm separates the membrane surfaces, a distance probably determined by the interplay of *electrostatic repulsion* and *adhesive forces*, and perhaps by the thicknesses of the cell coats on the adjacent membrane surfaces. It is not clear what type of adhesive forces operate in this type of contact: it has been suggested that attraction between oppositely polarized surface groups may be important, that divalent cations (e.g. Ca^{++}) may form bridges between adjacent surfaces, or that other *non-specific attractive forces* may operate (Curtis 1973). Recently, there has been increasing evidence that at least in some cases, more *specific interactions* may occur between cell surfaces, with complex sites for mutual recognition and adhesion analogous to those between antibodies and antigens. Such specific adhesion is evidenced by the tendency of cells from the same tissues or embryonic germ layers to reaggregate after they have been experimentally separated, and in the migration of cells and growth of cell extensions (e.g. in the developing nervous system) to make contact with specific types of 'target' cell (Cox 1974). It must be added that the degree of adhesion varies considerably in different cell varieties; epithelial cells normally adhere to one another strongly, but macrophages—often highly motile cells—may move freely over other cells. Reduction of the normal adhesive properties of cells occurs in malignant neoplasms which therefore rapidly spread locally, or form secondary colonies (metastases) elsewhere.

Specialized junctional structures are regions of mutually adherent cells where there are distinctive structures visible with the electron microscope (**1.7, 8**). A number of different types have been demonstrated (Cox 1974; McNutt and Weinstein 1973), including those at which the cell membranes are separated by a 20 nm gap or more, and those where the membranes either approach closely or are actually in contact. The former include the *macula adherens* and *zonula adherens*, the latter the *zonula occludens* and the 'communicating junction' or *junctio communicans*.

The *macula adherens* (*desmosome*) is a plaque-like structure (**1.1, 7, 8**) found at any region of the cell surface, where the plasma membrane is coated on its inner aspect with a layer of dense protein into which intracytoplasmic tonofibrils are inserted. The gap between the two cells is bridged by thin looping filaments which interlock, traversed by broad bands of densely staining material. This structure apparently forms anchorage points between cells, particularly where there is a requirement for strong cohesion as in the stratum spinosum of the epidermis, where they are responsible for the post-fixation prickly appearances of the cells. Single-sided or hemi-desmosomes are also found between epidermal (and other) cells and the underlying basement membrane.

The *zonula adherens* (**1.1, 7, 8**) is a continuous belt formed

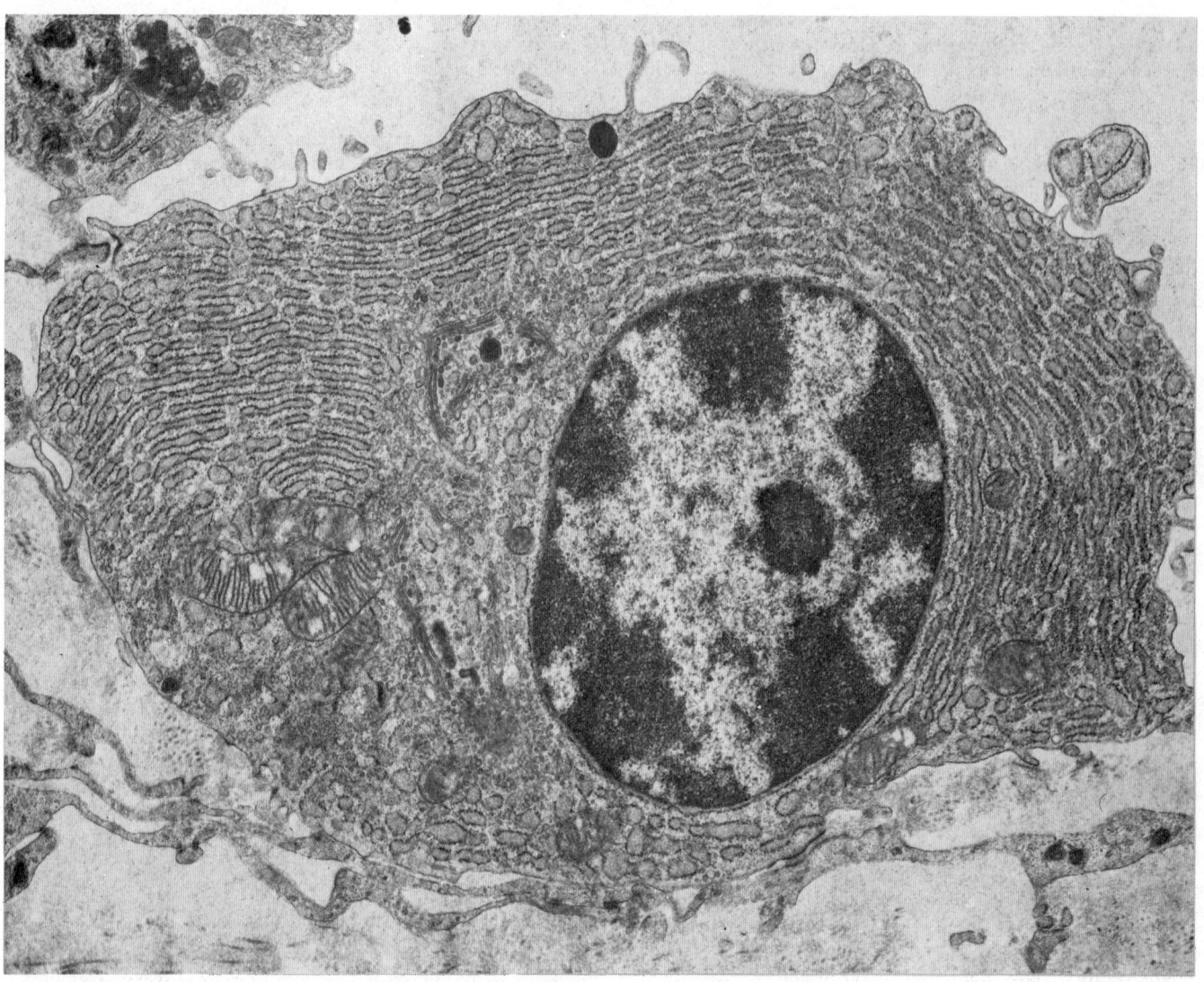

1.5 An electron micrograph of a protein-synthesizing cell in section showing abundant rough endoplasmic reticulum, Golgi apparatus, mitochondria, lysosomes, and a cell membrane with surface protrusions (filopodia). Magnification × 12,000.

around the apices of epithelial cells, or a zone between smooth muscle cells, and as a 'single-sided' junction between muscle cells and their collagenous sheaths. As in the desmosomes, there are cytoplasmic densities beneath the membrane surfaces in these structures, and filaments (usually microfilaments) are often inserted on the cytoplasmic side. No intercellular filaments span the gap between the cell surfaces, however, but it is possible that an adhesive, non-stainable material intervenes at this point. They probably serve anchoring functions like the desmosome.

The *zonula occludens* (tight or occluding junction) is an annular junction (1.1, 7, 8) present around the apices of epithelial cells, between endothelial cells, and is found in other sites where there is a barrier to diffusion through the intercellular space. This is exemplified by the failure of colloidal tracers such as ferritin and horseradish peroxidase to move from the lumen of the intestine into the spaces between the epithelial cells or even deeper into the mucosa. At a zonula occludens the membranes of the adjacent cells are in actual contact, obliterating the gap between them and so creating a barrier to the movement of molecules. The efficacy of this barrier does, however, vary in different tissues, so that relatively 'leaky' tight junctions that allow the slow diffusion of large molecules have been demonstrated in the kidney, blood vessels and in other structures (Friend and Gilula 1972). Freeze-etching has shown the contacts between the membranes to lie along branching and anastomosing lines which, in the case of 'leaky' tight junctions, are said to be discontinuous. The lines take

the form of ridges which are probably formed by the incorporation of fibrils or chains of particles within the membranes, distorting and stiffening them at the points of contact. The tight junction is of great importance, for example, in preventing the leakage of toxic substances from the lumina of viscera into the surrounding tissues, and in retaining the colloids within the bloodstream.

Communicating junctions or *maculae communicantes* -('gap' *junctions, electrical junctions, nexuses*) are similar to tight junctions in sectioned specimens except that the apposed membranes are separated by an apparent gap about 3 nm wide, which is traversed by numerous dense beads arranged in hexagonal arrays on the two membrane surfaces (1.8). Such junctions form limited attachment plaques (Goodenough and Revel 1970) rather than the continuous belts which the tight junctions form, thereby allowing the free passage of substances along the cleft between the cells which they join. They are present in numerous tissues including the liver, epidermis, connective tissues, between embryonic cells, between cardiac muscle cells, and between smooth muscle cells. In the central nervous system (Staehelin 1974) they join adjacent cells belonging to ependyma, to neuroglia, and they form electrical synapses between neurons in some regions of mammalian brains, and more commonly in those of lower vertebrates and invertebrates (*see* p. 829).

The importance of the communicating junction is that it forms channels for the diffusion of ions and larger particles—up to a

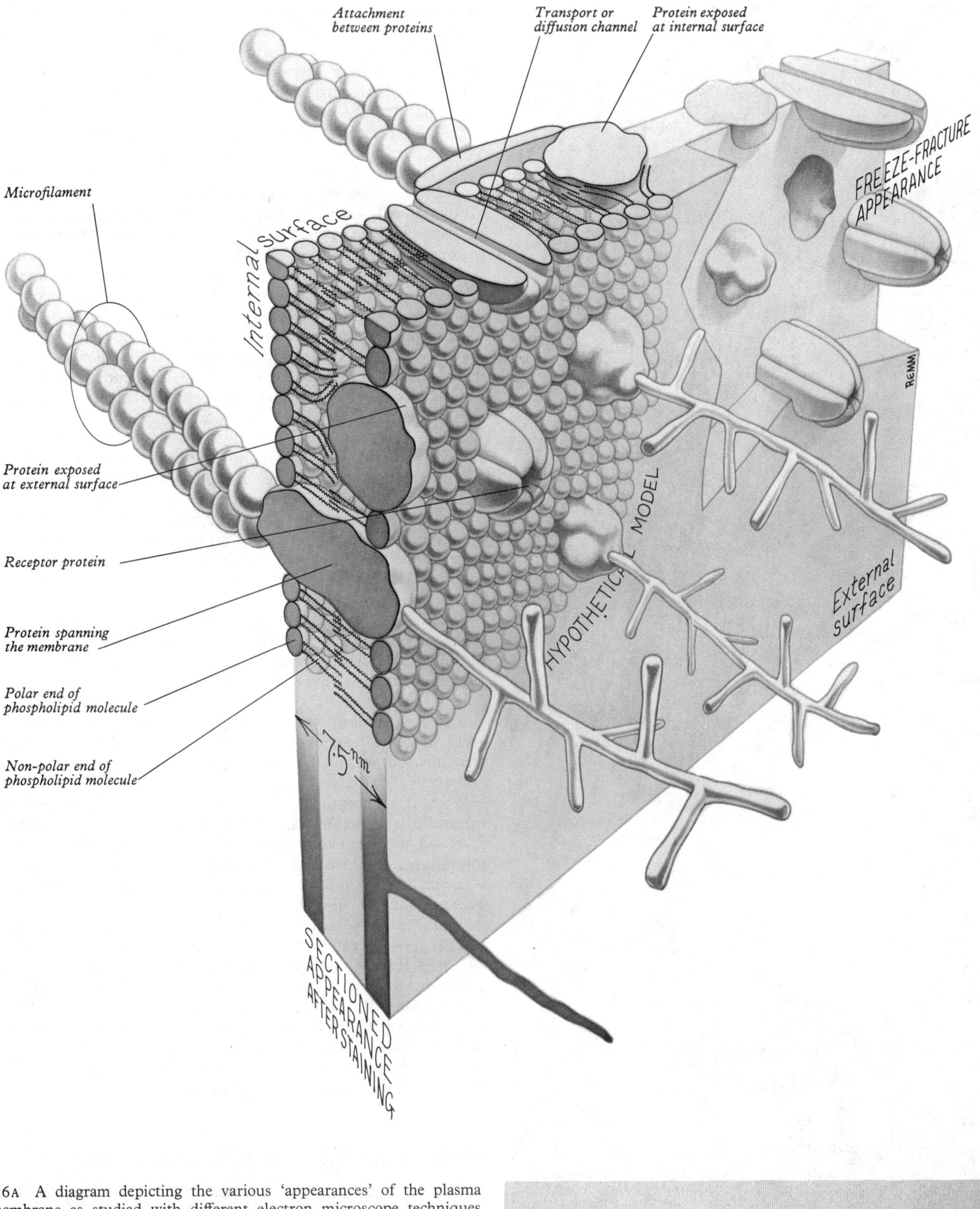

Attachment
between proteins

Transport or
diffusion channel

Protein exposed
at internal surface

FREEZE-FRACTURE
APPEARANCE

Microfilament

Internal surface

Protein exposed
at external surface

Receptor protein

Protein spanning
the membrane

Polar end of
phospholipid molecule

Non-polar end of
phospholipid molecule

HYPOTHETICAL MODEL

External surface

7.5 nm

SECTIONED
APPEARANCE
AFTER STAINING

1.6A A diagram depicting the various 'appearances' of the plasma membrane as studied with different electron microscope techniques including sectioning and freeze-fracturing, together with current interpretations of the results of these, and other biophysical methods. In this diagram, membrane proteins (green) are either confined to a single leaflet of the lipid bilayer, or they span both layers. Branched carbohydrate chains (grey) are shown attached to some transmembrane proteins on the external surface of the membrane, and a double helical actin filament (pink) is attached to the inside (left). The hydrophobic tails of the lipid molecules are shown as thin black lines, and their hydrophilic heads as spheres—blue for the outer surface of the membrane and yellow for the inner surface. These various shapes are, of course, only schematic, as the various protein, lipid, and carbohydrate molecules of which cell membranes are constituted, have differing detailed shapes.

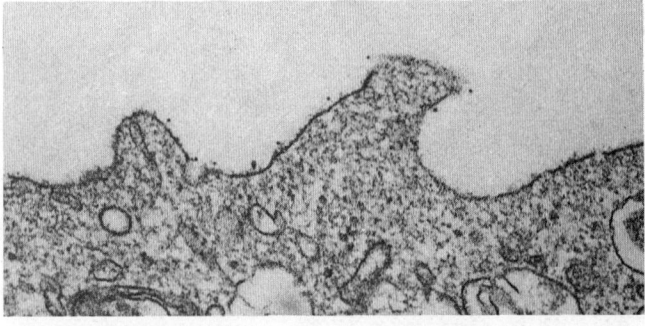

1.6B An electron micrograph of the surface of a fibroblast in which the presence of a surface carbohydrate has been made visible using concanavalin A conjugated to ferritin, which appears as dark particles. Magnification × 60,000.

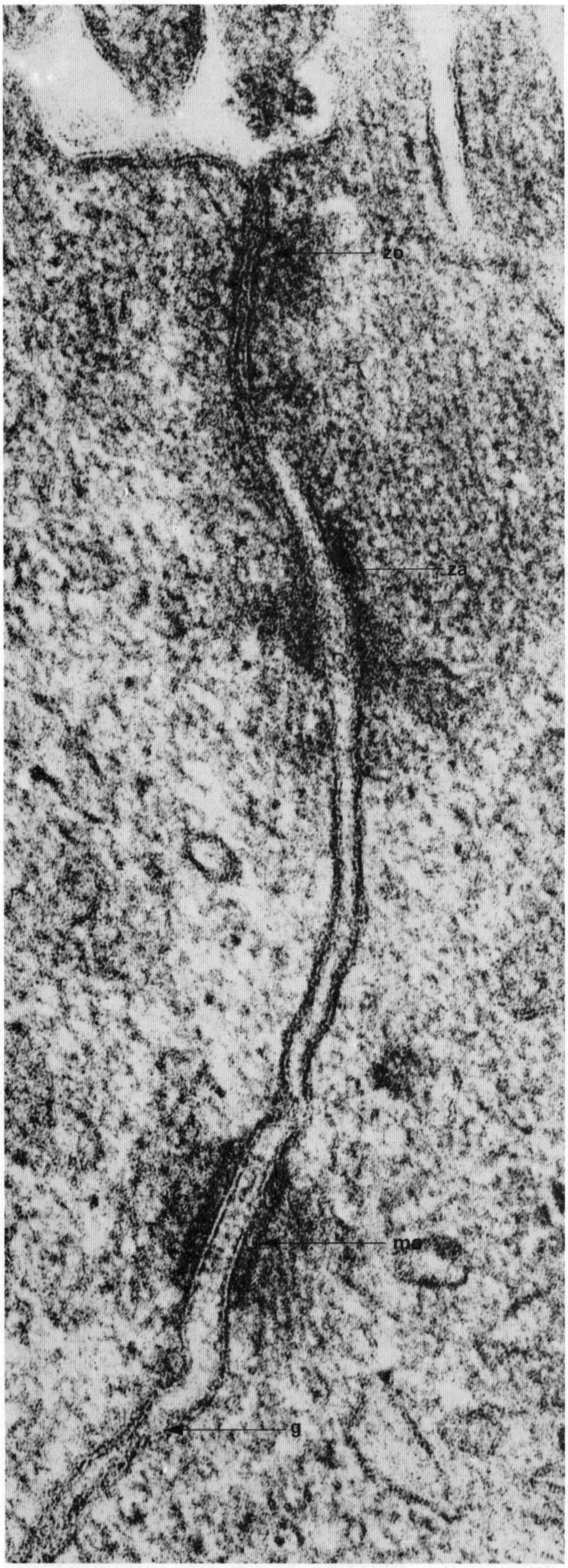

1.7 A high-power electron micrograph of a junctional complex between two epithelial cells showing a zonula occludens (tight junction) (*zo*), a zonula adherens (*za*), a macula adherens (desmosome) (*ma*), and a normal intercellular gap (*g*). Magnification × 130,000.

molecular size of about 4,000 daltons—*from one cell to the next*. Thus, where excitable cells (muscle cells, nerve cells) are concerned, one cell can directly invade its neighbour by electrotonic current flow without the intervention of a chemical transmitter, in contrast to the usual mode of synaptic transmission. In other cells their significance is not entirely clear, although experimentally they have been shown to be permeable to various dyes and to form low resistance pathways to the flow of ionic current. It appears probable that the communicating junction permits metabolic cooperation between adjacent cells or groups of cells. Thus, during embryonic life such junctions are perhaps involved in the establishment of pattern, and in the co-ordinated differentiation of, for example, the whole blastula, within and between germ layers, or of more localized tissues by allowing regulatory substances involved in gene blocking, or gene repression and de-repression, to diffuse freely or establish morphogenetic gradients (*see* p. 85). Communicating junctions may also be involved in the control of cell division since in damaged tissues they disappear, and the cells undergo reparative mitotic divisions until junctions reappear, when regeneration ceases (Bennett 1973; Fusijawa *et al.* 1976).

Freeze-fracturing and -etching studies (Staehelin 1974) have shown hexagonal arrays of membrane particles on both sides of communicating junctions, each particle composed of subunits probably surrounding a central channel (**1.**8) which, when located opposite a similar unit in an adjacent cell, forms a small communicating passage between the two. Such particles may also exist in smaller numbers, or singly, elsewhere on the cell, so that intercellular communication may not be restricted to special junctional *areas*.

Junctional complexes are present at the apex of epithelial cells (Farquhar and Palade 1963), where a combination of a zonula occludens, zonula adherens and macula adherens (desmosome) is a regular feature (**1.**1, 7, 8). This array of junctions corresponds in part to the *terminal bar* of light microscopy.

Synapses and *neuromuscular junctions* are specialized junctional areas involved respectively in interneuronal and nerve-muscle transmission, and will be described under these headings in a later chapter.

With the exception of the latter two specializations, all intercellular junctions are rather labile, and can be readily resorbed or reformed in suitable circumstances, permitting slow migration and division of cells, and in some cases the passage of other cells along intercellular clefts, as for example, leucocytes between endothelial cells.

Topographical specializations of the plasma membrane include fine finger-like extensions—microvilli, filopodia, microspikes, cilia and flagella; and less regular, often rounded or leaf-like extensions—'blebs', pseudopodia, and ruffled membranes. Indentations of the membrane include small endocytic and pinocytotic vesicles associated with the uptake of materials by the cell, and small exocytic vesicles and larger secretory vesicles involved in the outward transport of materials. These various structures will be discussed in detail later.

ENDOPLASMIC RETICULUM (ERGASTOPLASM)

This term is used to describe the system of interconnecting membrane-lined channels (Palade 1975) within the cytoplasm of most cells (**1.**1, 2, 10A, B, C). These channels are either cisternae (flattened sacs), or tubules, or else they may be vesicular. The membranes divide the cytoplasm into two major compartments, that inside the channel system, and that outside. The former constitutes the space in which secretory products are stored, or transported to the Golgi complex and cell exterior; the latter is made up of the colloidal proteins such as enzymes, carbohydrates and small molecules, together with the ribosomes and ribonucleic acid. As a whole, the extra-channel material is termed *hyaloplasm* or *cytosol*, and that within the channels, the *vacuoloplasm*.

Structurally, the channel system can be divided into *granular* or *rough endoplasmic reticulum*, to the external surface of which ribosomes are attached, and *agranular* or *smooth endoplasmic reticulum* which lacks ribosomes. When cells are fractionated by

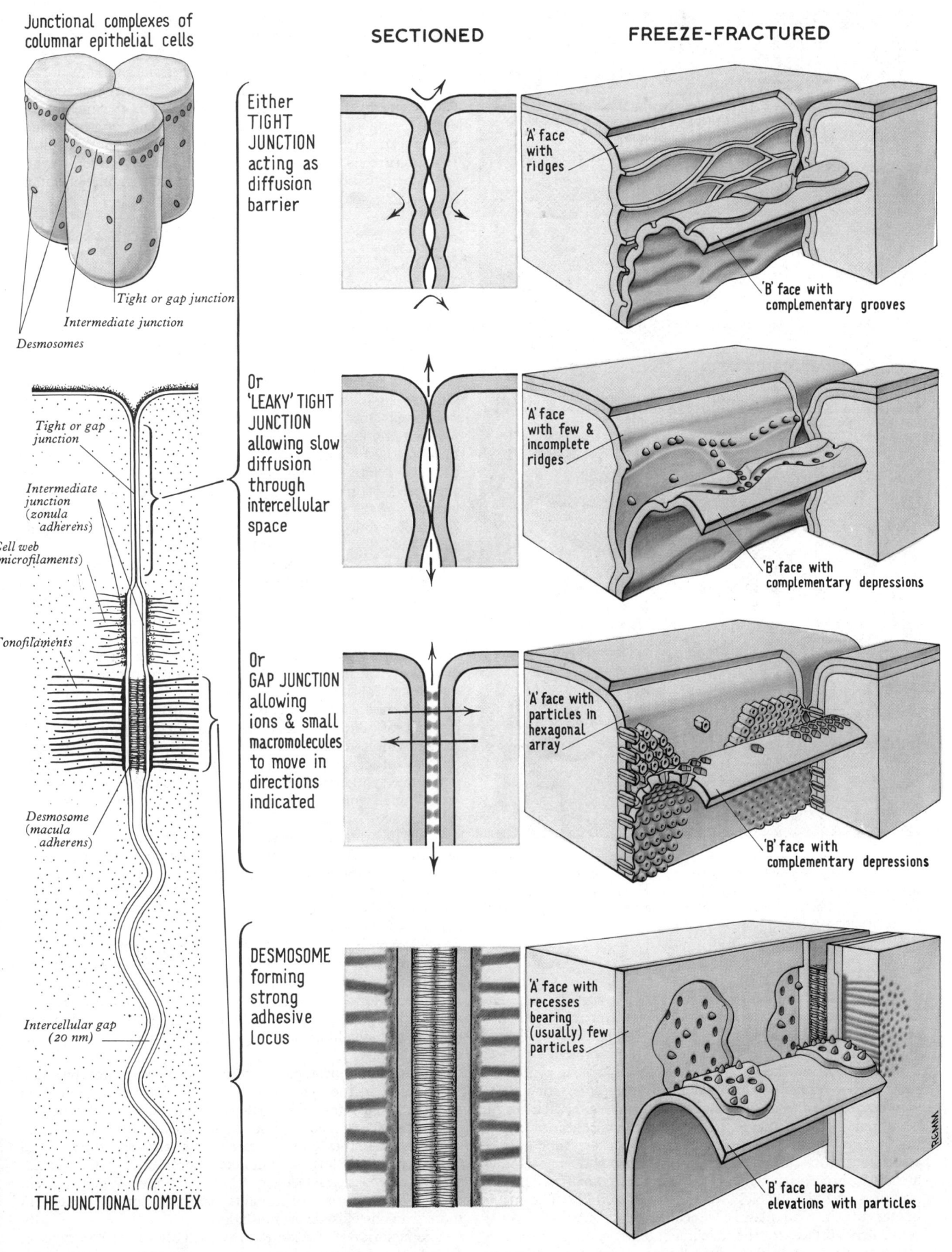

Junctional complexes of columnar epithelial cells

Tight or gap junction

Intermediate junction

Desmosomes

THE JUNCTIONAL COMPLEX

Tight or gap junction

Intermediate junction (zonula adherens)

Cell web (microfilaments)

Tonofilaments

Desmosome (macula adherens)

Intercellular gap (20 nm)

SECTIONED   **FREEZE-FRACTURED**

Either TIGHT JUNCTION acting as diffusion barrier

'A' face with ridges

'B' face with complementary grooves

Or 'LEAKY' TIGHT JUNCTION allowing slow diffusion through intercellular space

'A' face with few & incomplete ridges

'B' face with complementary depressions

Or GAP JUNCTION allowing ions & small macromolecules to move in directions indicated

'A' face with particles in hexagonal array

'B' face with complementary depressions

DESMOSOME forming strong adhesive locus

'A' face with recesses bearing (usually) few particles

'B' face bears elevations with particles

1.8   Schemata of various junctions commonly occurring between cells, showing current interpretations of the electron microscopic appearances seen after sectioning, and with freeze-fracturing techniques. For clarity, the freeze-fracturing data are presented as though it were possible to separate and fold back the two leaflets of the plasma membrane, to expose the intramembranous protein particles inserted on their two faces. The *A* face=the *P* fracture face, and the *B*=the *E* fracture face in current terminology.

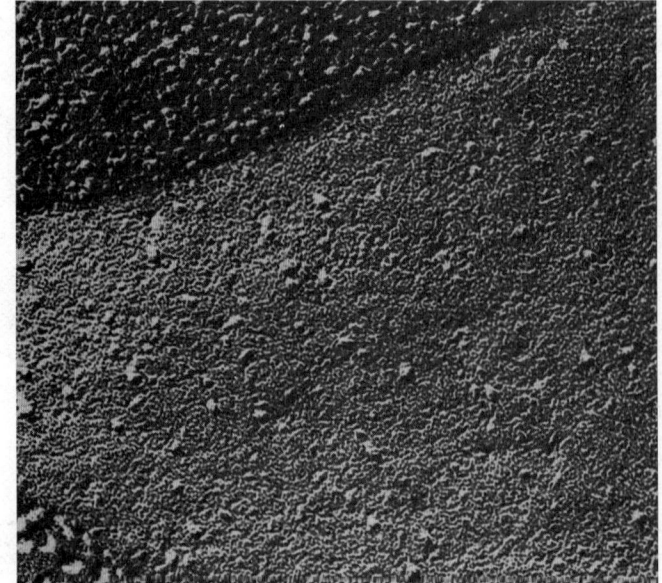

A

B

1.9A and B Electron micrographs of the plasma membrane of an erythrocyte prepared by freeze-fracturing. In A the external (E) fracture face, bearing few particles and some pit-like depressions is shown; in B the internal (P) fracture face is visible, showing numerous particles. Magnification × 80,000.

disruption and centrifugation, the rough and smooth endoplasmic reticula break up into vesicles collectively termed *microsomes*.

Granular endoplasmic reticulum carries out protein synthesis by means of its attached ribosomes, and is also able to synthesize some carbohydrates (p. 246). Generally ribosomes attached to membranes synthesize proteins which thereafter remain in membrane-bound bodies such as lysosomes or else are secreted to the exterior of the cell. Agranular endoplasmic reticulum is associated with carbohydrate metabolism, and many other metabolic processes, including respiration, detoxification, and synthesis of lipids, cholesterol and other steroids. The membranes of the endoplasmic reticulum as a whole serve as convenient surfaces for the attachment of many enzyme systems which are thus accessible to the substrates in solution within the cell. They also act co-operatively with the Golgi apparatus to elaborate new cell membranes, the protein, carbohydrate and lipid components being added in different regions.

Specialized types of endoplasmic reticulum are present in various cell types. In striated muscle cells the smooth

endoplasmic reticulum (*sarcoplasmic reticulum*) is essential to the binding of calcium ions, which are liberated to initiate contraction on appropriate stimulation (p. 512).

In embryonic cells, endoplasmic reticulum although present is scant, and ribosome groups are mostly unattached within the hyaloplasm. With differentiation, the membranes of the cytoplasm usually increase greatly and ribosomes may become attached to form the rough endoplasmic reticulum. In mature cells, synthesis of some proteins starts only when attachment to membranes occurs, a step which is under hormonal regulation in some systems.

RIBOSOMES

Ribosomes (ribonucleoprotein particles) are relatively small bodies about 15 nm across (1.10A, B), composed of protein and (ribosomal) RNA, which are responsible for the synthesis of proteins from amino acids (Palade 1955; Siekevitz and Palade 1960). Each ribosome is made of two subunits (Nanninga 1973; Wittmann 1976), one slightly larger than the other, and sedimenting in the centrifuge at different rates (in nucleated cells mainly at 60S and 40S, where S = the Svedberg unit of sedimentation rate, a function of density, shape, etc.; the whole ribosome sediments at 80S). The subunits can be further dissociated into a number of protein subunits, about 30 in all, composed of *structural proteins* determining the shape and integrity of the ribosome, and *functional proteins* governing its synthetic activity. Two small RNA strands (28S and 7S), highly convoluted, lie in the 60S subunit, and one in the 40S subunit (18S), such RNA molecules all being derived from the nucleolus (see below). The 60S and 40S subunits are usually separate from each other when not engaged in protein synthesis, and can easily be dissociated in the laboratory by ionic manipulation.

Ribosomes may exist as solitary, relatively inactive *monosomes*, or as multiples (*polyribosomes* or *polysomes*) attached to, and in the process of, translating messenger RNA during protein synthesis. Polysomes can be *attached* to membranes, so constituting the granular endoplasmic reticulum (see above) or may lie *free* in the hyaloplasm. Attached polysomes synthesize proteins which remain within vesicles (e.g. lysosomal enzymes), or are secreted by way of exocytic vacuoles (e.g. procollagen); unattached polysomes synthesize proteins for use outside the channel system, including enzymes of the hyaloplasm, structural proteins of the cell (e.g. actin, tubulin) and haemoglobin in erythroblasts.

Since in a mature polysome all the attachment sites of the messenger RNA are occupied as the ribosomes move along it, synthesizing protein according to its instructions, the number of ribosomes in a polysome is a measure of the length of the messenger RNA molecule and therefore of the size of the protein being made.

The two major subunits have separate roles in protein synthesis. The smaller subunit is the site of attachment and translation of the messenger RNA; the larger subunit is responsible for the release of the newly constructed protein and, where appropriate, attachment to the endoplasmic reticulum and thereafter channelling the protein through its membranes into the cisternal cavity.

The subunit proteins of the ribosomes themselves are synthesized in the cytoplasm by other ribosomes. They then apparently pass into the nucleus where they are attached to ribosomal RNA from the nucleolus to form the two major subunits; these then pass separately back into the cytoplasm and only associate to form a complete ribosome when they attach themselves to a messenger RNA molecule. When protein synthesis is over, the two subunits dissociate, but may be used in more than one episode of translation.

The majority of ribosomes in the cell are of the 80S type, as already stated, having a molecular weight of about 4×10^6 daltons; however, those inside mitochondria are smaller (about 55S). The ribosomes of fungi and prokaryote organisms such as bacteria are also smaller than those of the nucleated cells of animals (and plants), which may perhaps reflect the less complex regulatory mechanisms of these relatively primitive species.

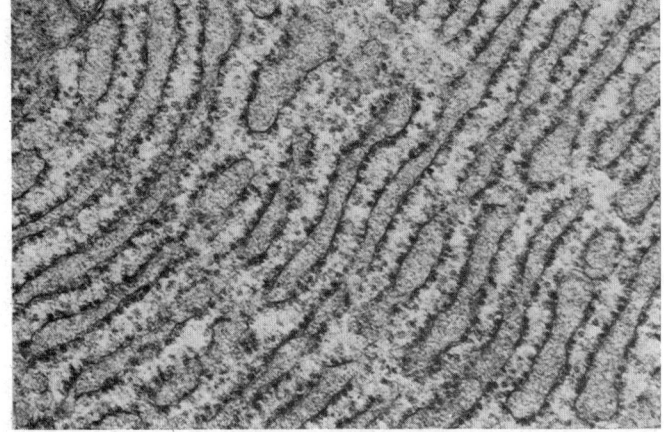

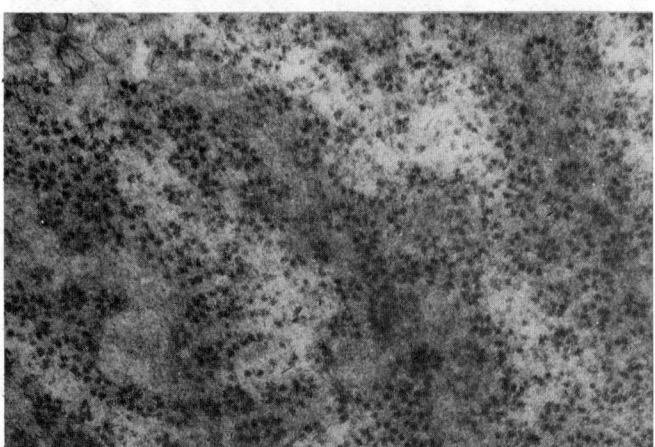

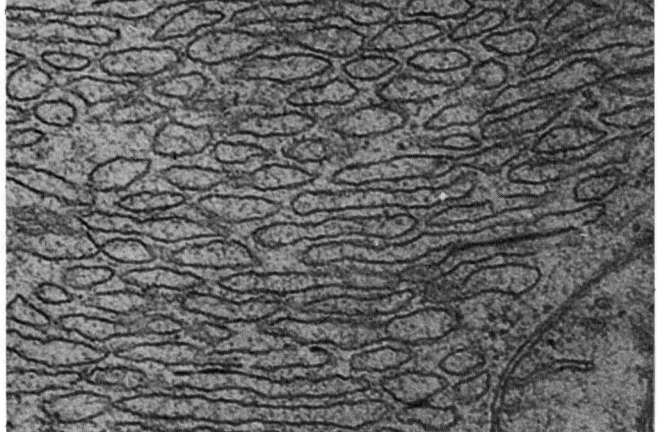

inner membrane is highly pleated to form incomplete transverse partitions or tubular invaginations termed *cristae mitochondriales*, or simply 'cristae'. This folding creates a relatively large surface area. The cristae are more numerous and complex in cells with a high metabolic rate than in relatively inactive ones; in heart muscle, for instance, the cristae are numerous and show complex pleats.

In the inner lumen of the mitochondrion lies the finely granular mitochondrial *matrix* which has a variable density. Thus it has been demonstrated that the shape and size of mitochondria can vary with activity, being smaller and with a denser matrix when inactive (Hackenbrock 1972). Within the matrix are many soluble enzymes and also a variety of inclusions have been described, including calcium salts, organic crystals and glycogen. There are also some ribosomes and the nucleic acids RNA and DNA (Nass 1969). The ribosomes and nucleic acids are quite distinct in their physical and chemical properties from those of the rest of the cell; the ribosomes are slightly smaller than those outside mitochondria and they sediment in the ultracentrifuge at a different rate. The DNA thread is joined at each end to form a ring (55S), and has a proportion of nitrogen bases different from that of chromosomal DNA. These components are vital to the continuity of mitochondria, which are able to multiply by division during interphase (Racher 1975); experiments with fungi indicate that some of the proteins of mitochondrial inner membranes are specified by their own nucleic acids, an example, therefore, of *cytoplasmic inheritance*. The ribosomes and nucleic acids are, interestingly, similar to those of bacteria, and speculations have been made on the possible origin of mitochondria in the far distant past as symbiotic bacteria which lost their ability to lead an independent existence.

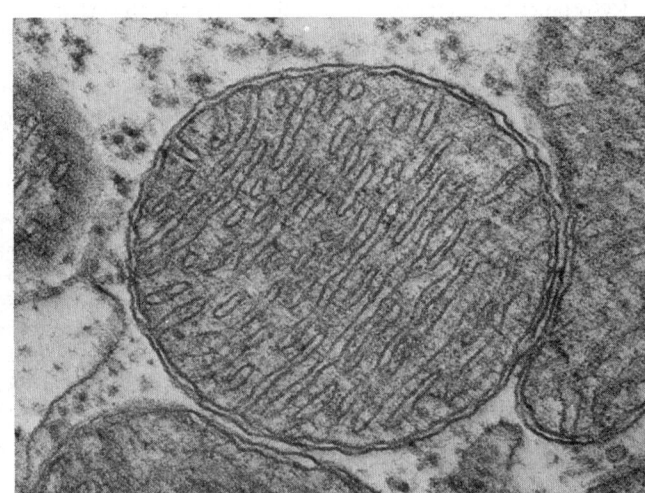

1.10A–C Ultrastructural details of the endoplasmic reticulum: (A) rough endoplasmic reticulum; (B) polysome groups attached to obliquely sectioned cisternae showing 'rosette' configuration; (C) smooth endoplasmic reticulum. (B) kindly provided by Dr. D. R. Turner of Guy's Hospital Medical School. Magnification × 30,000.

1.11 Electron micrograph showing details of mitochondrial structure: the outer membrane, the inner membrane folded to form cristae, and their related intra-mitochondrial spaces are visible. Magnification × 30,000.

MITOCHONDRIA

Mitochondria are membrane-bound organelles of great metabolic significance (Slater 1972). Mitochondria were first observed with the light microscope as thread-like, spherical, or ellipsoidal bodies present in the cytoplasm of most cells and particularly those with a high metabolic rate such as the secretory cells of the exocrine glands. In living cells, when viewed by bright field, phase contrast or interference microscopy, they have been seen to move within the cell, change size and shape, and to divide. In size most mitochondria range from 0·5–2·0 μm wide by 3–4 μm long.

With the electron microscope (1.1, 11) each mitochondrion is seen to be made up of an outer and an inner membrane, the two being separated by a variable gap. The outer membrane is smooth and sometimes in continuity with the endoplasmic reticulum; the

The functions of mitochondria have been extensively examined by cell fractionation techniques, and they have been shown to be the principal site of a number of enzyme systems, particularly those concerned with the oxidative phosphorylation which is associated with the tricarboxylic acid (Kreb's) cycle and the cytochrome electron transport sequences of respiration. They are the chief sites where chemical energy is derived from the breakdown of organic compounds during respiration, to form high-energy organic phosphate compounds (particularly adenosine triphosphate, ATP, and guanosine triphosphate, GTP). These compounds pass to other parts of the cell where they take part in energy-consuming reactions. The various enzyme systems of the Kreb's cycle are situated in the mitochondrial matrix, while those of the cytochrome system and oxidative phosphorylation are localized chiefly in the inner mitochondrial membrane. Some of these ATPases form *enzyme assemblies* which, when mitochondria are hypotonically disrupted and negatively stained, become visible as minute spheres supported by stalks projecting from the inner membrane. These are termed *elementary* or *stalked particles*. Mitochondria are also concerned with other transformations, particularly fatty acid metabolism (Racher 1975) and with calcium concentration.

It is interesting that mitochondria are distributed within the cell according to regional energy requirements, e.g. near the bases of cilia in certain epithelia, and between the folds of the membrane at the base of the cells of proximal convoluted renal tubules, where considerable active transport occurs.

Estimates of mitochondrial numbers in cells also show variability; in mammalian hepatic cells they average about 800 per cell; in lymphocytes only a few may be present.

LYSOSOMES

Lysosomes are membrane-bound, spheroidal, or ellipsoidal bodies ranging in size from 0·08–0·8 μm, which contain hydrolases capable of degrading a wide variety of substances (DeDuve 1965; Dingle *et al.* 1969a & b, 1973, 1975, 1976; Holtzman 1976). They are present in all cells except erythrocytes, and play a fundamental role in the degradative activities of cells, often being particularly numerous in those with a rapid metabolism, and also in phagocytes. So far, more than 40 lysosomal enzymes have been shown by biochemical analysis of organelles separated from cells by centrifugation methods, including many varieties of proteases, lipases, carbohydrases, esterases and nucleases. Cytochemically, the enzyme acid phosphatase (β-glycerylphosphatase) has been widely used as a marker of lysosomes for light and electron microscopy, although it is not invariably present.

Lysosomes have a rather complex life history, each stage being structurally distinct, and so they may take several forms even in one cell. Their contents are first formed in the granular endoplasmic reticulum where some carbohydrate is added to the enzymes, then passed to the Golgi complex for packaging in vesicular form and final processing; some lysosomes, however, may apparently be formed without the involvement of the Golgi complex. When first released as *primary lysosomes* they are relatively small but may sometimes fuse to form larger primary lysosomes up to 0·8 μm across, typically with dense granular interiors (**1.12A**). Primary lysosomes can be used either to degrade materials taken into the cell by phagocytosis, pinocytosis and endocytosis (*heterophagy*) or to degrade worn out, damaged or 'unwanted' organelles within the cell (*autophagy*). In the case of heterophagy, phagocytic vacuoles (*phagosomes*) containing objects such as bacteria, fuse with one or many primary lysosomes to form *secondary lysosomes* (*phagolysosomes*) which typically contain signs of digestion—for instance, membranous whorls, fibrillar débris, etc. (**1.12A**). Smaller pinocytic and endocytic vesicles, containing fluids or small particles fuse with primary lysosomes to form *multivesicular bodies* (**1.12B**)—another form of secondary lysosome. In both heterophagy and autophagy, lysosomes become smaller and denser as digestion proceeds, and eventually the residual, undigested, insoluble contents may be ejected from the cell by exocytosis, or they may remain within the cell—an example of 'storage excretion'. In non-dividing cells such as neurons, *residual bodies* of this type (**1.12C**), which may fuse to form much larger aggregates, are retained throughout the life span of the cell, so that in old age they are conspicuous in the form of the lipid-rich pigment lipofuscin ('*senility pigment*') within the nervous system.

In some cells, lysosomes are a predominant feature, for instance, in neutrophil leucocytes, where the granules are primary lysosomes employed in the killing of phagocytosed bacteria. Lysosomal enzymes may also possibly be released externally to damage neighbouring foreign organisms, or pathologically, as in certain types of arthritis, damaging the surrounding tissues. Osteoclasts may also erode bone by releasing similar enzymes on to the neighbouring bone surface (*see* p. 261).

Lysosomal membranes, normally impermeable to their enclosed enzymes, may under certain circumstances allow the enzymes to leak. Ionizing radiations, some carcinogens, silica

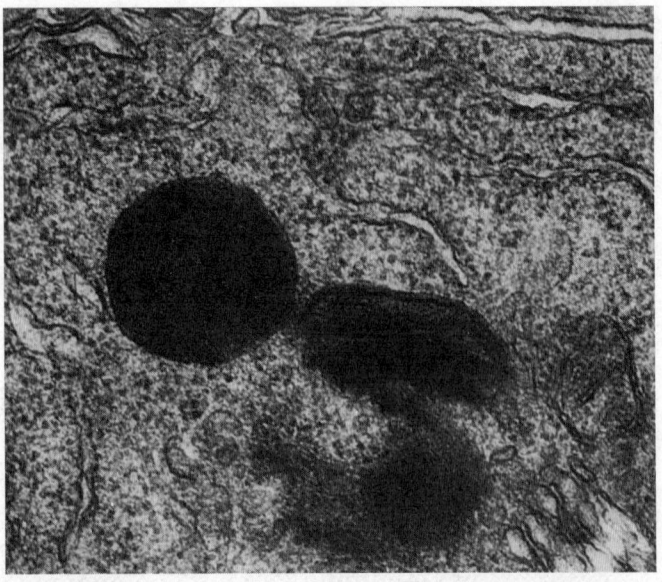

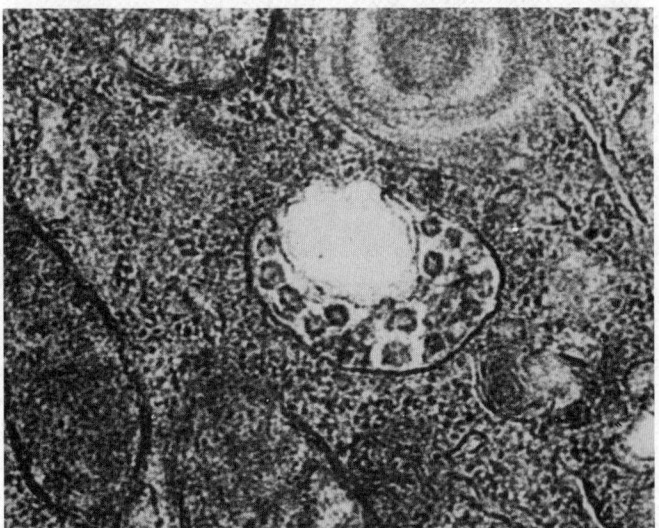

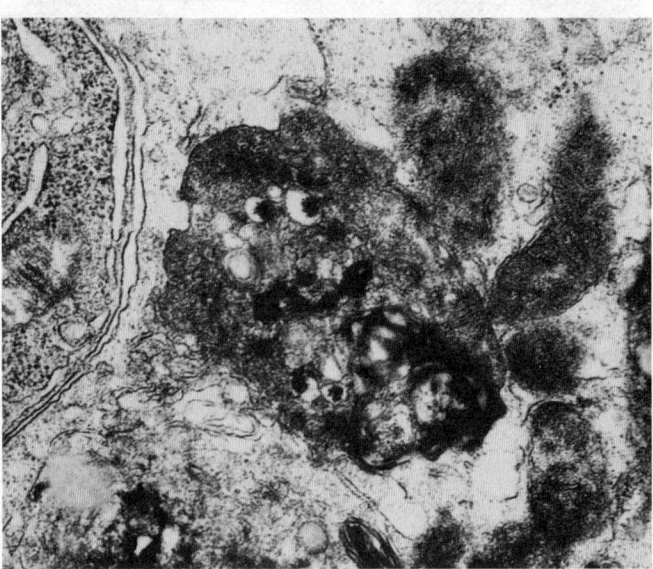

1.12A–C Electron micrographs of lysosomes in various stages: (A) a group of lysosomes in an olfactory receptor cell showing small primary lysosomes (left), and larger secondary lysosomes containing lamellar débris; (B) a multivesicular body with which endocytic vesicles are in the process of fusing; (C) a residual body, the end stage of lysosomal hydrolysis of engulfed cellular organelles. All magnifications × 30,000.

particles, asbestos dust, anoxia, heat, and a wide variety of drugs have been suggested as causing such effects, with consequent cellular damage, cell death or, if the genes are affected, the transformation of the cell into a neoplastic one. Conversely, some drugs, for example cortisone, can stabilize lysosomal membranes, and may therefore inhibit their fusion with phagocytic vesicles. Lysosomal leakage may also be of great natural significance in causing the *programmed death* of redundant cells during the embryonic period, or at least in facilitating their removal once dead (see also p. 79).

Because of their involvement in breakdown processes in metabolism, defects in lysosomal enzymes can lead to profound morphogenetic changes, and a range of genetic disorders (*lysosomal storage diseases*) has also been described. Examples of the latter are Tay-Sachs' disease in which a faulty carbohydrase leads to the accumulation of glycolipid in the central nervous system, with consequent death at an early age, and Hurler's syndrome, where failure to metabolize certain glycosamino-glycans (mucosubstances) causes abnormalities in the formation of connective tissues.

PEROXISOMES

Peroxisomes (*microbodies*) are membrane-bound vacuoles (**1.4**) about 0·6 μm across, often with dense cores or crystalline interiors (DeDuve 1973). They are concerned with catabolic reactions which release hydrogen peroxide, formed by the action of their contained enzymes—peroxidase, D-amino acid oxidase, and urate oxidase. Catalase, another major enzyme of peroxisomes, destroys hydrogen peroxide released by the others, preventing this toxic material from accumulating. Peroxisomes are probably present in all nucleated cells, and are particularly numerous in hepatocytes where they have been most extensively studied.

Other *membrane-bound bodies* are found in many cells, such as vacuoles containing lipids, mucopolysaccharides, and various secretory or storage products. The complex 'granules' of pancreatic islet β-cells, and the neurosecretory bodies of the neurohypophysis are two such examples.

GOLGI COMPLEX (GOLGI APPARATUS OR DICTYOSOME)

Beginning with Camillo Golgi, the Italian cytologist, optical microscopists of the nineteenth and early twentieth centuries recognized a distinct region of cytoplasm near the nucleus, particularly prominent in secretory cells, which stained heavily with metallic salts. This was considered an artefact by many, but with the advent of electron microscopy it was authenticated as a genuine cellular organelle.

The Golgi complex consists of one or more zones of smooth endoplasmic reticulum (**1.1, 4, 13**) arranged as stacks of flattened cisternae; membranous channels interconnect the cisternae, and small vesicles are found clustered at their borders. In glandular cells with an apical secretory zone, the Golgi complex is positioned between the secretory surface and the nucleus; in fibroblasts, where the secretory activity is more general, two or more groups of flattened vesicles, and in liver cells up to 50, may form a more widely distributed complex.

The Golgi complex is often cup-shaped, with the convex side nearest the nucleus. Around the perimeter of the concave side, larger *condensing vacuoles* are usually present. Biochemical and autoradiographic studies have shown that the organelle is a metabolically active region where secretory products and lysosomal enzymes are accumulated by fusion of *transport vesicles*. After moving through the cisternae of the Golgi complex, enzymes and secretory products are budded off either to be retained in the cytoplasm or extruded as *secretory vacuoles* (Jamieson and Palade 1967). In passing through the complex the secretory materials are altered; in mucus-secreting cells, for instance, the proteins synthesized in the rough endoplasmic reticulum are condensed in the condensing vacuoles into a smaller space, probably by subtraction of water, and carbohydrates rich in sulphur are added. In other cells, too, much of the carbohydrate

moiety of their secretions is added at the Golgi complex, whose membranes serve as a site for attachment of the enzymes associated with carbohydrate synthesis, such as glycosyltrans-ferase. The enzyme thiamine pyrophosphatase is also a valuable histochemical marker of Golgi complex activity.

The Golgi complex is also active in primary lysosome production (*vide supra*) which may in some cases take place in a special region, the Golgi-endoplasmic reticulum-lysosome complex (*G.E.R.L.*)—*see* Novikoff (1973)—and in the synthesis of cell membranes and external cell coats.

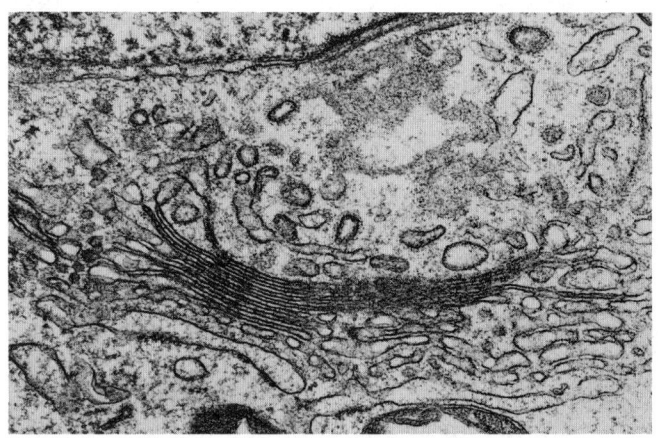

1.13 Electron micrograph of a Golgi apparatus in a spermatocyte showing the central multilamellar cisternae, and clusters of peripheral vesicles. Magnification × 15,000.

ANNULATED LAMELLAE AND MYELIN FIGURES

Annulated lamellae are arrays of parallel cisternae which are sometimes found in the cytoplasm and occasionally the nuclei of cells undergoing rapid protein synthesis (Kessel 1968), for example oöcytes, some embryonic and some neoplastic cells. Characteristically the membranes fuse intermittently, where they are punctuated by structures closely resembling nuclear pores (p. 21), reflecting their origin from the outer nuclear membrane. It has been suggested that they may serve as attachment sites for stored RNA, but their functions are otherwise obscure.

Myelin figures constitute regions of tightly packed membranous whorls, frequent within neuronal cytoplasm, and also commonly found in degenerating cells. They may be signs of breakdown of parts of the endoplasmic reticulum, and are often associated with lysosomal activity.

Other inclusions not bounded by membranes, and lying within the hyaloplasm, include clusters of *glycogen granules*, *ferritin*, and a variety of inorganic crystals; the hyaloplasm of erythrocytes is predominantly made up of haemoglobin.

TRANSPORT VESICLES

These include membrane-bound vacuoles, usually less than 0·1 μm in diameter, which are involved in the movement of fluids and particles into, out of, and within the cell (Kanaseki and Kadota 1969; Allison 1973). The process of uptake is broadly termed *endocytosis*, and the passage of materials in vesicles to the cell surface and their release, *exocytosis*. In the case of endocytosis the terminology is somewhat confused, but some authors distinguish between the ingestion of large particles in *phagocytic vesicles*, the bulk uptake of fluids in relatively large, often complex, *pinocytotic vesicles* which is an energy-dependent process, and the formation of small, rounded, *endocytic vesicles* which can still occur in cells deprived of energy sources, and may depend on physical rearrangements of plasma membrane components. Such *micropi-nocytotic vesicles* often possess a bristly coat on their exteriors, composed of a series of cross-linking pegs of a protein ('clathrin') which may enable them to round up initially, and may also be involved in their transport around the cell as 'coated vesicles'. Not all small internal vesicles are coated, however, so the functions of such coats are far from obvious.

Exocytic vesicles, carrying materials out of cells, vary greatly in size, and may only be distinguished from endocytic vesicles experimentally by using extracellular tracers, which will enter the latter but not the former (*see* pp. 35, 36). Vesicles of these various types are present in all cells, but are particularly numerous where rapid exchange of materials between the cell and its environment is occurring, for example, endothelial cells, synaptic terminals of neurons, and in secretory cells.

MICROTUBULES AND MICROFILAMENTS

Microtubules (**1.14A, B**, and *see* Stephens and Edds 1976) are elongated cylinders about 24 nm in diameter, of varying length (up to some 70 μm in spermatozoan flagella), and made of protein (*see* Soifer 1975). They are frequently observed components of most cell types, but are particularly abundant in neurons, leucocytes, blood platelets, and in the mitotic spindles of dividing cells. They also form part of the structure of cilia, flagella and centrioles (*vide infra*). Helical microtubules occur in Schwann cells and oligodendrocytes (pp. 835, 843). Massed parallel bundles of interconnected microtubules are found in the pillar cells supporting the sensory hair cells of the cochlea. Microtubules normally show few acute bends, indicating some degree of stiffness.

The microtubules are composed of two proteins, *tubulins a and b*, arranged as dimers (Fujiwara and Tilney 1975). Each protein molecule is about 5 nm across, and microtubules in transverse section are made up of a ring of 12 or 13 such globular subunits which are arranged in long columns of alternating tubulins a and b, forming the walls of a cylinder. Each column is slightly shifted on its neighbour, so that a spiral pattern of subunits is also generated (Amos and Klug 1974). Microtubules are formed by the simple polymerization of the tubulin dimers, and there is in some cells an equilibrium between polymerized and unpolymerized tubulin, so that some types of microtubule may break down and reform rapidly, although there is much variation in their stability. Consequently, chemicals which bind to tubulin cause the equilibrium to shift, so bringing about the depolymerization of microtubules; the drugs colchicine, colchemide, vinblastin and griseofulvin are well known as anti-microtubule agents acting in this manner.

Formation of microtubules can take place by addition to preexisting microtubules, or formation *de novo* in, as yet, poorly understood special cytoplasmic centres with a microtubule-organizing capacity such as centrioles, around which spindle microtubules polymerize during cell division (Pickett-Heaps 1975). In other instances microtubule-organizing centres have no obvious structural basis, thus centrioles are not always essential to microtubule formation.

Microtubules are often linked to adjacent structures by filamentous extensions, such connexions having been observed in relation to cell membranes, other microtubules, microfilaments, and transport vesicles. In some of these situations their role is obviously to give mechanical support; in others they appear to be able to cause movement of adjacent structures. Their role in cell motility will be discussed later (p. 35).

MICROFILAMENTS

Microfilaments (**1.4, 14A**) are thinner (6–8 nm) than microtubules and appear solid in cross-section; they are prominent components of all cells (except perhaps mature erythrocytes) and are intimately involved in cell motility in general (Wessels *et al.* 1971). Microfilaments are composed of *actin*, of which the 6 nm subunits are arranged in two entwined helices. A number of different varieties of actin are known (Goldman *et al.* 1976), some of them forming long relatively stable microfilaments, whilst others form shorter and more transient ones. Some types are primarily associated with cell plasma membranes; others lie deeper within the hyaloplasm (Pollack *et al.* 1975). Muscle cells

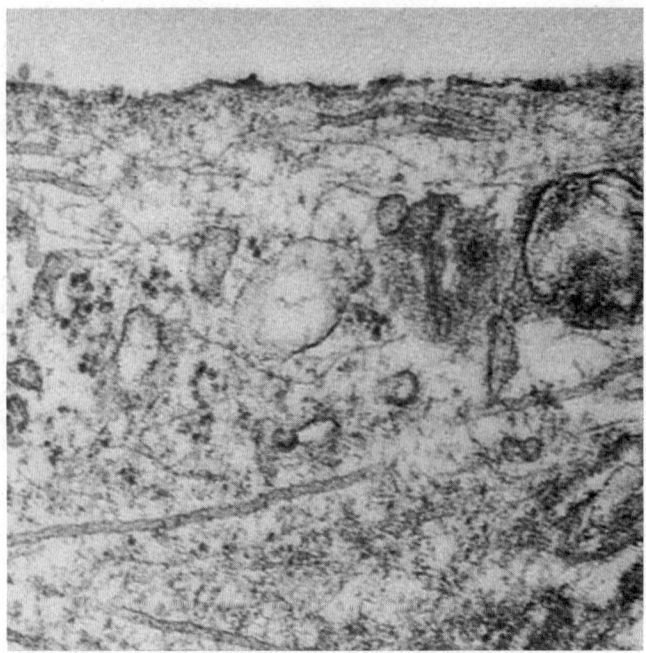

1.14A Electron micrograph of microtubules and microfilaments in a tangential section through a fibroblast. Magnification × 50,000.

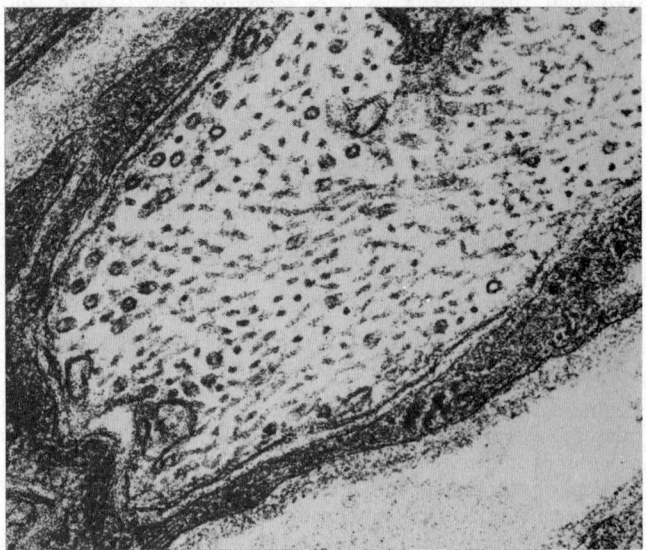

1.14B Microtubules and neurofilaments in a transverse/oblique section of part of a small nerve fibre. The microtubules have a circular cross-section, whereas the smaller neurofilaments appear solid. Magnification × 50,000.

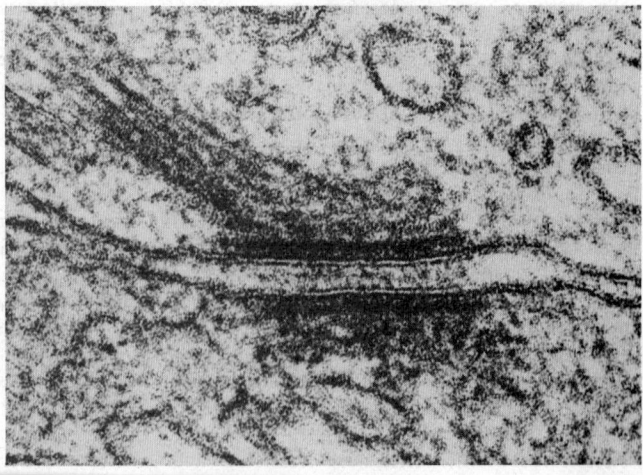

1.14C Tonofilaments associated with a desmosome (macula adherens) between two epithelial cells. Magnification × 100,000.

contain particularly robust forms of microfilament (p. 507). In all cases, actin is able to bind to the protein *myosin* when an energy source (ATP) is available, and when organized in an appropriate manner. This phenomenon leads to various types of shearing movement which can produce cell motility of different kinds, including muscular contraction (p. 510). The presence of actin microfilaments can be detected electron microscopically by incubating cells, made permeable to proteins, with heavy meromyosin which binds to actin and decorates the filaments with arrowhead formations. The drug cytochalasin B causes depolymerization of some actin microfilaments (*see* Spudich 1972), and is therefore a valuable experimental tool in research into their functions. Often associated with actin microfilaments are a number of other proteins (e.g. troponin, tropomyosin B, actinin) which assist in various motile activities of these organelles (Lazarides 1976).

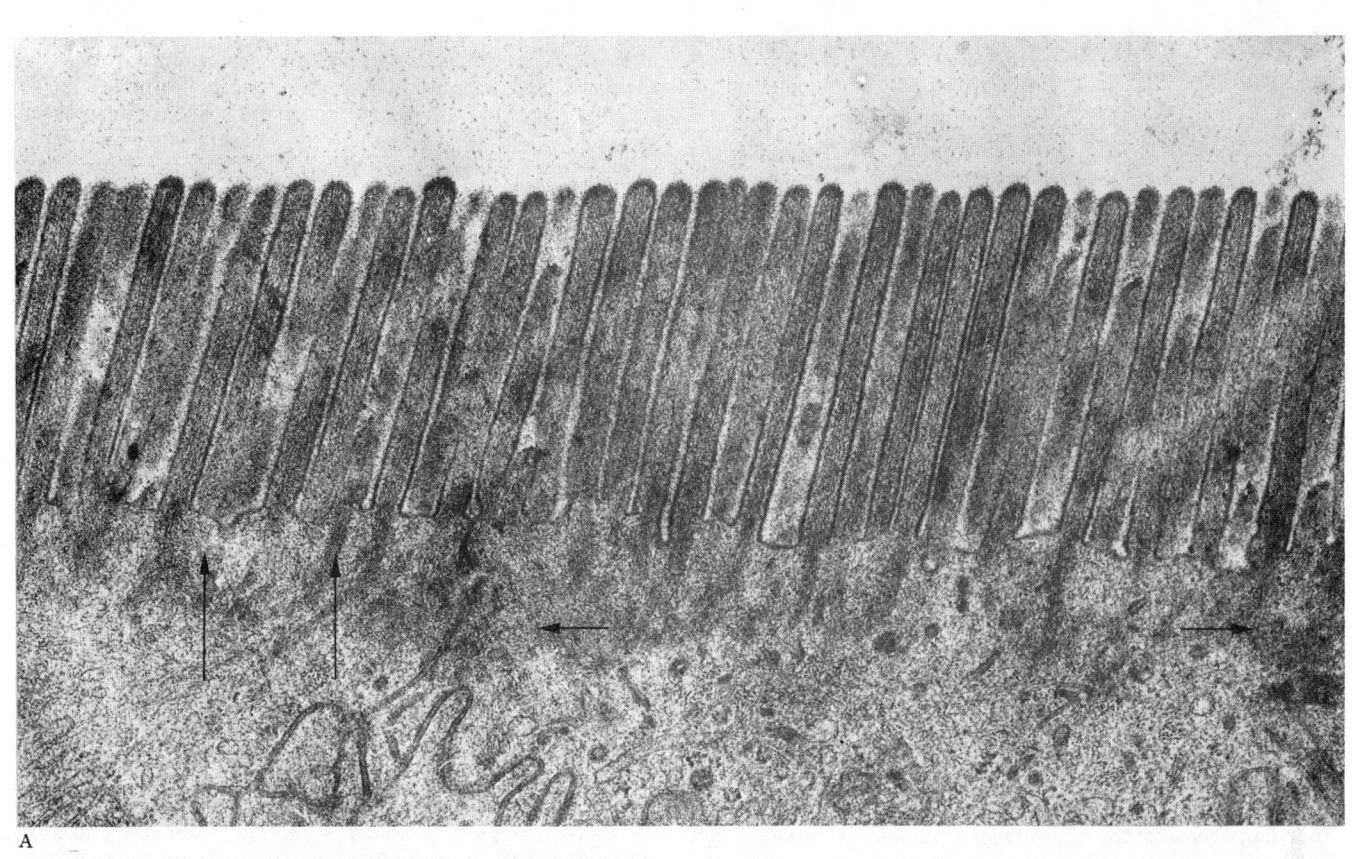

A

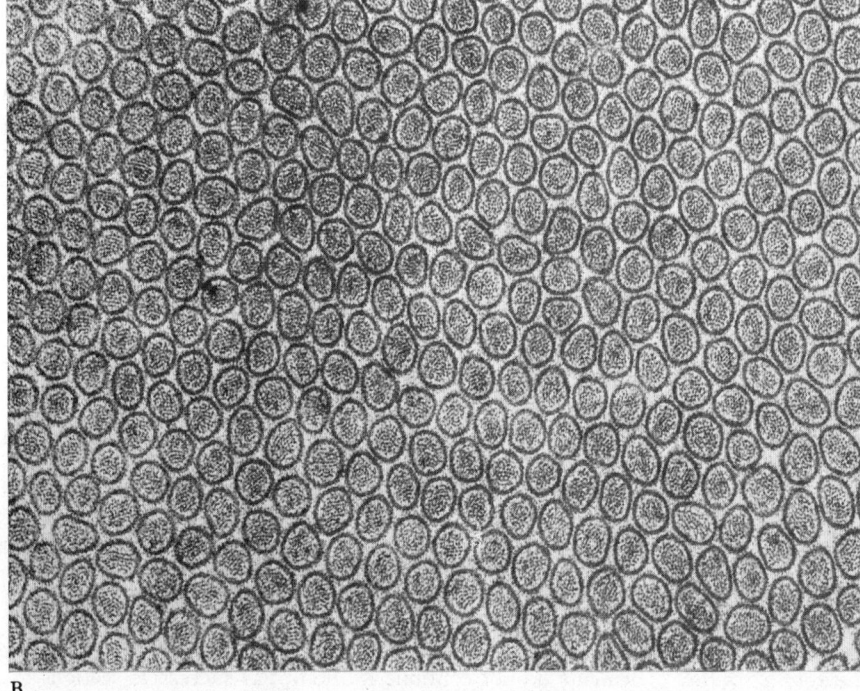

B

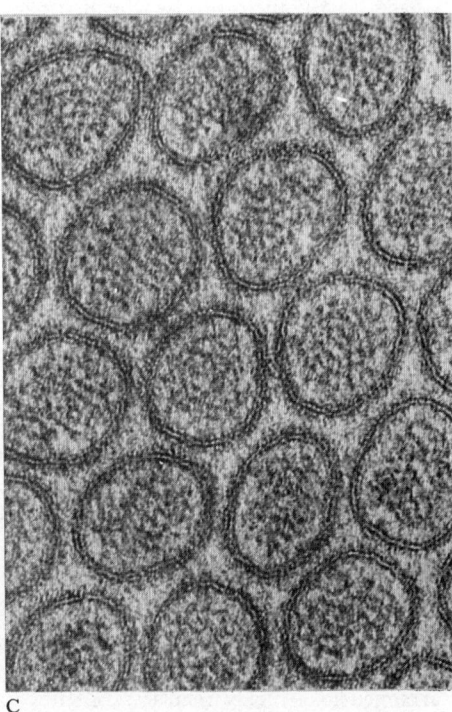

C

1.15A–C Electron micrograph of microvilli from the striated border of an absorptive columnar epithelial cell of the small intestine (compare with **1.1**): (A) vertical section showing microvilli with supporting microfilaments (long arrows) and terminal web (short arrows); (B) horizontal section through cell surface cutting each microvillus transversely. Insert (C) shows the details of a few microvilli from (B): note the bilayer structure of the plasma membrane and the fuzzy external cell coat. Magnification A and B × 20,000; C × 110,000.

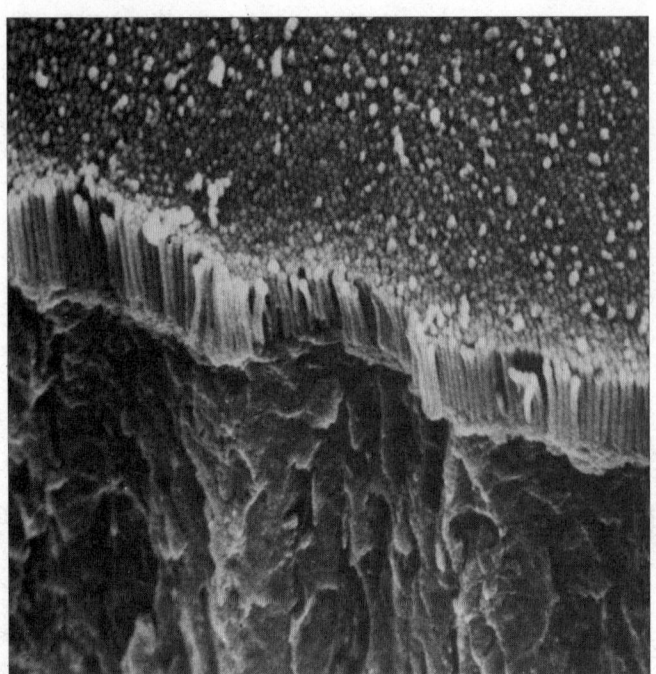

1.16   Scanning electron micrograph of a striated border from the small intestine showing numerous microvilli. The epithelium has been cut vertically to allow the microvilli to be viewed from the side as well as from above. Magnification ×6,000.

OTHER CELLULAR FILAMENTS

These include protein filaments of various sizes found in different cell types (Goldman and Knipe 1972). The most widespread are *tonofibrils* (*tonofilaments*) or *intermediate filaments*, often present in large numbers where structural strength is needed (**1.14C**), as in epidermal prickle cells. They tend to form extensive bundles running for long distances in the cytoplasm, and to be inserted into desmosomes at the cell surface. Tonofibrils are about 10 nm in diameter and often appear tubular in cross-section (Blose and Chako 1976); their precise chemistry has not, as yet, been satisfactorily analysed.

Keratin filaments are similar to tonofibrils in appearance; they are proteinaceous structures confined to the mature (and dead) keratinocytes of the epidermis (pp. 41, 1218).

Neurofilaments (**1.14B**) are characteristic elements of neurons, both in their cell somata and neurite extensions (axons and dendrites); similar filaments are present in some neuroglial cell processes. They are about 12 nm across, are sometimes hollow in section, and may extend for large distances, perhaps throughout an entire axon. Neurofilaments are proteinaceous; they do not, however, appear to contain tubulin or actin (Shelanski *et al.* 1971). They may be important in providing mechanical support and, incidentally, are of value to the histologist since they can be impregnated with silver salts using specialized staining techniques, thus demonstrating aspects of the histological structure of nervous tissue. For further comment *see* p. 820.

Intranuclear filaments have also been reported in a variety of cell types; the significance of these structures is not clear, although they may have a developmental (or sometimes pathological) connotation in particular instances.

MICROVILLI

These are finger-like extensions of cell surfaces (**1.15A–C**, 16) from less than 0·1 μm in diameter and up to about 5 μm long. When arranged in a highly regular series, they constitute a *striated border*, as at the absorptive surfaces of the epithelial cells of the small intestine (Limbrick and Finean 1970); when they are less regular, as in the gall-bladder epithelium, the term *brush border* is applied.

Internally they are supported by bundles of actin microfila-

ments, which are implanted within the apical cytoplasm of the cell amongst other transversely running filaments to create a meshwork, the *terminal web*. The microfilaments are attached to the membrane surface of each microvillus by finer lateral filaments along the margins, and are inserted apically into a dense mass of the protein α-actinin (Mooseker and Tilney 1975).

Microvilli create a highly folded cell surface, and are found at sites of active absorption where extensive surface areas are necessary, such as in the proximal and distal convoluted tubules of the kidney. In the small intestine absorptive cells bear microvilli which have a thick cell coat, probably of great importance in providing a site for adsorption of digestive enzymes. The supporting filaments within the microvilli are possibly associated with the transport of absorbed materials from the cell surface to the cell interior. Several enzyme systems associated with active transport across membranes have been localized within striated borders.

Irregular microvilli, or *filopodia*, are also found on the surfaces of many types of cell, particularly free macrophages and fibroblasts. Again, these may be associated with transport processes, particularly phagocytosis, and with cell motility.

Large, regular microvilli (*stereocilia*) also occur at some sensory surfaces, for example, on taste-buds, cochlear and vestibular receptor cells, and in the epididymis.

Similar, though more transient structures, found on developing or motile cells include stiff elongate 'microspikes' and *ruffled membranes* (*see* p. 36; also **1.37**).

CILIA AND FLAGELLA

Seen with the light microscope these are hair-like (cilium= eyelash) and whip-like (flagellum=whip) projections of the cell surface capable of movement, creating currents in the overlying fluid, or movements of a cell to which they are attached. Cilia are found on many lining layers, particularly the epithelia of most of the respiratory tract, and parts of the male and female reproductive tracts, the ependyma lining the central canal of the spinal cord and ventricles of the brain, and the mesothelia of the peritoneal and pleural cavities. They also occur at olfactory receptor endings and, modified, as portions of the rods and cones of the retina; cilia are, in addition, present on many dividing tissue cells. Many cilia may be present on a single cell, as in the bronchial epithelium, or only one or two as with some mesothelial cells. Each male gamete possesses a single flagellum about 70 μm long.

With the electron microscope (**1.17A–D**, 18) each cilium or flagellum is seen to consist of a *shaft*, constituting most of its length, with a diameter of about 0·25 μm, a tapering *tip*, and, within the surface cytoplasm of the cell, a *kinetosome* (*basal body*, *basal granule*, or *blepharoplast*) about 1 μm long (Gibbons and Grimstone 1960). The whole structure is bounded, except at its base, by a tubular extension of the plasma membrane. In freeze-fracture preparations chain-like groups of characteristic membrane particles, collectively termed the *ciliary necklace*, are seen to surround the proximal end of the cilium (Gilula and Satir 1972). These particles may be involved in the control of ciliary beating. Within this lies a cylinder of nine double microtubules, surrounding a central pair of single microtubules.

At the *base* of the cilium each microtubule doublet, the two parts of which are designated the A and B subfibres, are twisted through 40 degrees, and another microtubule, the C subfibre, is added. The central pair sometimes end immediately above the cell surface in a dense sphere or *axosome*; beneath this lies a transverse partition or *basal plate*. Associated with the kinetosome are often one or more cross-banded filamentous *rootlets* and a plate-like *basal foot*, which are probably concerned with anchoring the cilium into the cytoplasm.

At the *apical end* of the structure, the various microtubular elements do not continue to the tip; as successive sections are followed, first the A subfibres terminate, then the B subfibres, and finally the central pair.

Within the shaft (**1.17E**) lie several structures associated with the microtubules, such as radial spokes which extend inwards from the outer microtubules towards the central pair. The outer

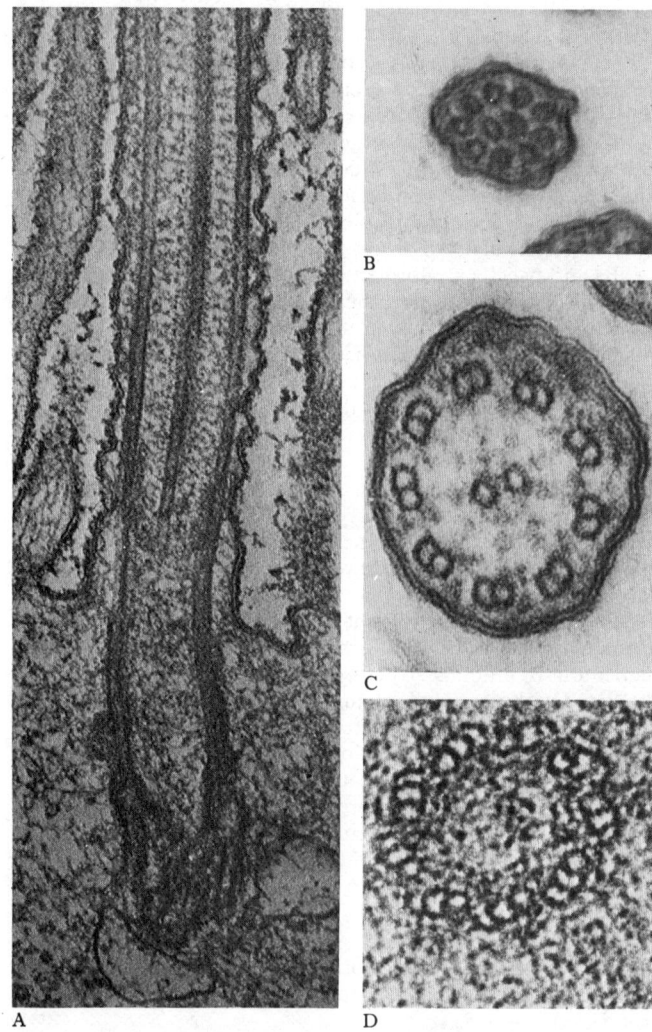

A   D

1.17A–D Electron micrographs of cilia cut (A) longitudinally, and (B, C, D) through the tip, shaft and base respectively. Magnification A×50,000; B–D ×110,000.

doublet microtubules bear tangentially positioned *arms*; these are attached to the A subfibre of one doublet and point towards the B subfibre of the neighbouring doublet. A helical filament encircles the central pair of microtubules, which are also joined together by ladder-like spokes. Because of the 9+2 pattern of tubules, the cilium (or flagellum) possesses a plane of symmetry which passes perpendicular to a line joining the central pair. The direction of this plane also corresponds closely to the plane of bending of the cilium.

The movements of cilia and flagella (*see* Sleigh 1974, 1977), although somewhat different in detail, are similar in broad outline (1.19A–C).

Flagella cause movement by the passage of rapid successive waves of bending, from the attached to the free end, to generate a resultant motile force in rather the same manner as an eel whilst swimming. In human spermatozoa there is a helical component in this motion, although the precise form is complex from a hydrodynamic point of view (*see* e.g. Brokaw 1975).

In cilia, the beating is planar, but is asymmetrical. In the first phase, the *effective stroke*, the cilium remains stiff except at the base where it bends to produce an oar-like stroke. This is followed by the *recovery stroke* during which the cilium behaves as though flaccid, the wave of bending passing from the base to the tip so that the cilium returns to its initial position ready for the next cycle. The activity of adjacent cilia is usually coupled so that the bending of one cilium is rapidly followed by the bending of the next and so on, to produce long travelling or *metachronal* waves which pass over the tissue surface in the same direction as the effective stroke, in the cilia of vertebrates. This coupling is known to be caused by the viscous interaction of adjacent cilia, although the rate of beating may be affected by extraneous factors (Sleigh 1974).

The mechanism of beating is still not clear; structural observations indicate that when a cilium bends, the microtubules do not change in length but slide upon one another (Satir 1974); biochemical studies have demonstrated in the 'arms' of peripheral doublets a protein 'dynein', with an enzyme site for splitting ATP. An interaction of adjacent doublets involving these arms, to cause mutual sliding, may underly the active beating of cilia and flagella (Gibbons and Rowe 1965). Re-arrangement of the spoke-

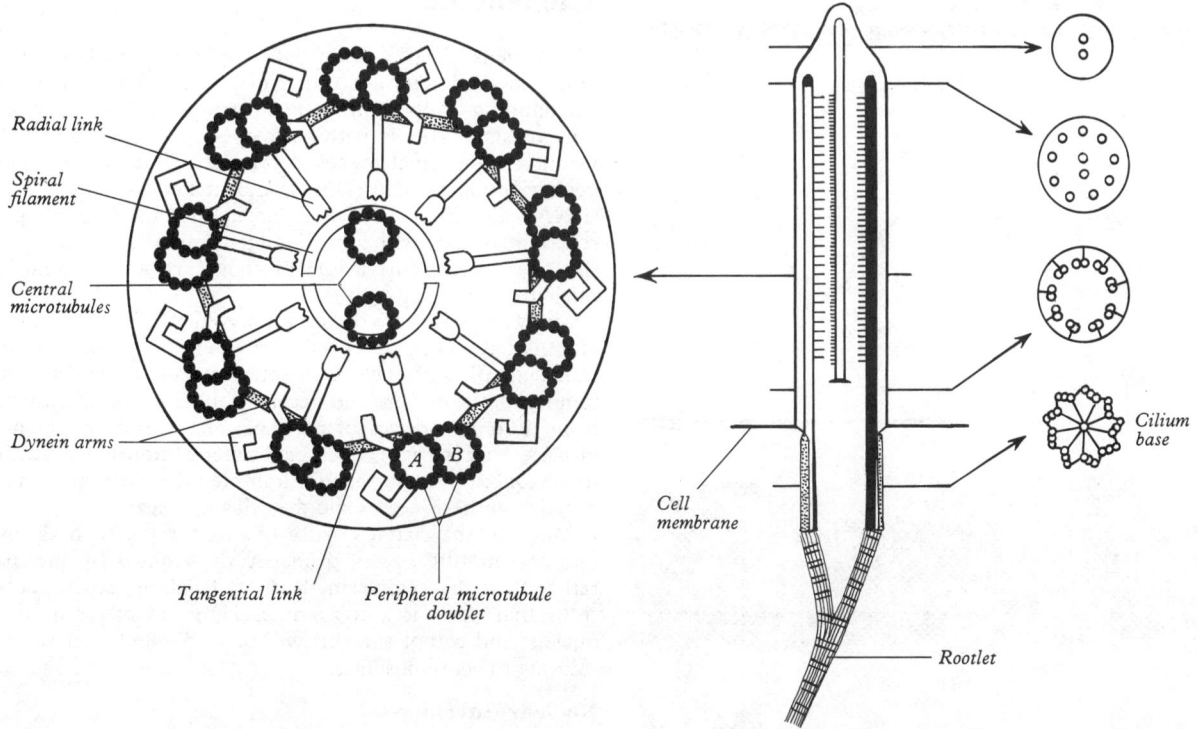

1.17E Diagram of the internal structure of a cilium in longitudinal section (centre), and showing the disposition of microtubules at various levels (right), and the detailed cross-sectional appearance of the shaft of the cilium (left). In the latter, hook-like dynein arms extend from the

peripheral doublets and radial links with expanded heads project towards the central pair of microtubules. A filamentous spiral encircles the central pair. Adjacent peripheral doublet microtubules are also connected by tangential links. Redrawn from Sleigh (1977).

1.18 Scanning electron micrograph of the ciliated surface in the trachea. Magnification × 10,000.

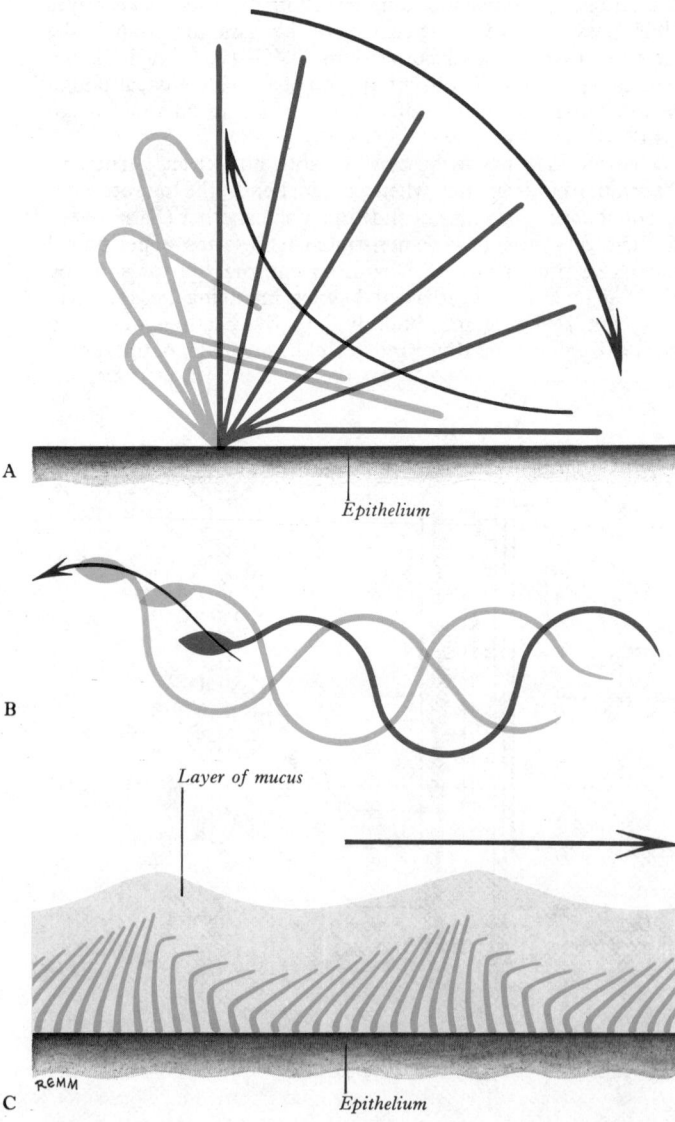

A

Epithelium

B

Layer of mucus

C REMM

Epithelium

1.19A–C Diagram of (A) ciliary and (B) flagellar action. In (A) the effective stroke is shown in red and the recovery stroke in blue. In (B) successive movements of the sperm tail are indicated in red, blue, and green; (C) represents a number of cilia in the respiratory tract showing the metachronal wave and the direction of mucus movement.

like elements may also modify the form of the ciliary bending (Warner and Satir 1974).

The cilia of olfactory receptors possess the typical '9+2' arrangement of subfibres; the cilium segment of retinal rods and cones, however, lacks the central pair. Many tissues containing rapidly dividing cells such as the adrenal cortex also show a small number of cilia which protrude from the cell surfaces into the intercellular spaces. The function of these is unknown, and they may represent an unavoidable but mechanically functionless correlate of cell division processes.

Cilia and flagella are formed by the polymerization of tubulin on centrioles which are synthesized deep in the cell, and then move to lie immediately beneath the cell membrane before the cilia begin to sprout (Sorokin 1968; Staprans and Dirksen 1974). Once initial sprouting has occurred, they grow by the addition of tubulin and other materials to the distal end.

CENTRIOLES

Centrioles are cylindrical bodies about 1 μm long by 0·25 μm in diameter (Fulton 1971), identical in structure with the bases of cilia (*vide supra*) but lying free within the cytoplasm (1.1, 20). At least two centrioles are found in all cells capable of division. Usually they form a pair lying at right angles to each other, within a somewhat dense region of cytoplasm, the *centrosome*. Various filamentous or granular structures (*pericentriolar bodies*) are often associated with the centrioles in their vicinity.

Centrioles, or their peri-centriolar cytoplasm, are concerned in some way with the synthesis of microtubules, as in the generation of the spindle and aster microtubules during cell division, the sprouting of cilia, and the provision of axonal and dendritic microtubules in developing neurons. Prior to cell division a new centriole is formed near each old one, and the resulting pairs pass during cell division to the two new cells (Rattner and Phillips 1973)

The nature of this organizing capacity is not known; in flowering plants spindle microtubules are apparently formed in the absence of centrioles, so they may not be essential to this process.

The Nucleus

The nucleus of each cell is usually a spherical or ellipsoidal body, often indented, some 4–10 μm in diameter and separated from the surrounding cytoplasm by a membranous nuclear envelope. The nucleus stains deeply with basic dyes such as haematoxylin by virtue of the nucleic acids present in its chromosomes—deoxyribonucleic acid (DNA), and nucleoli—ribonucleic acids (RNA). The DNA is also responsible for the strongly positive reaction found with the Feulgen method. Basic proteins (histones), and some acidic proteins, are also present in the nucleus.

Many investigations of the part played by the nucleus in the life of cells have been carried out, involving experiments such as transplantation of nuclei between different species of protozoa, removal of the nucleus and, recently, the fusion of dissimilar cells from different species of metazoa, after prior treatment with viruses (Harris 1970). Such experiments indicate that although the cytoplasm can survive for a limited time without a nucleus, protein synthesis ceases and the cell soon dies.

Many of the characteristics of intact cells, such as division rate and motility, seem primarily determined by the nucleus rather than the cytoplasm. Isotopic labelling experiments also show that there is a constant exchange of materials between nucleus and cytoplasm; this will be elaborated later when considering the chromosomes.

Nuclear Envelope

This is a flattened sac lined on both sides by the membranes which constitute the innermost two layers of the cytomembrane system, and they are separated by a distance of about 20 nm (1.1, 4, 5). The outer membrane is part of the endoplasmic reticulum proper and

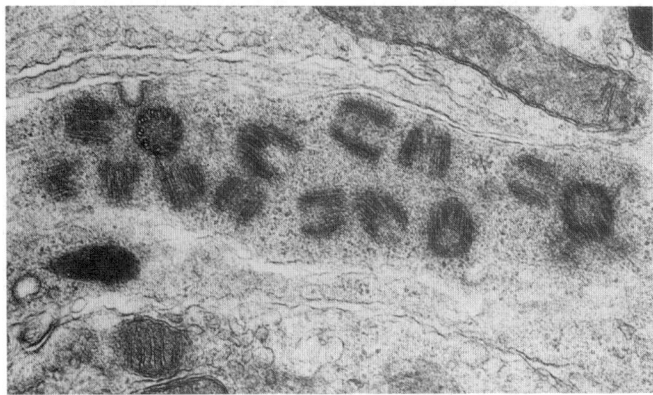

I.20 Transmission electron micrograph of a group of centrioles in a developing nerve cell dendrite. Magnification × 25,000.

Chromosomes

Within the nucleus most of the nucleic acid is deoxyribonucleic acid (DNA) which is associated with protein to form elongated threads or chromosomes. Between cell divisions (interphase) the chromosomes are usually highly extended filaments which form a diffuse network collectively termed *chromatin*. This may be rather dispersed and poorly staining *euchromatin*, or clumped and densely staining *heterochromatin* (I.22). These two states are associated, respectively, with high and low degrees of synthetic activity. Nuclei with a largely euchromatic appearance are known as *open-face* nuclei, and are larger in volume than heterochromatic, *closed-face* nuclei. In moribund cells, the nuclei are often small and stain densely (*pyknotic nuclei*). Pathological nuclei (for example in neoplastic cells) may be highly irregular in outline, and take on a variety of appearances (*pleomorphic nuclei*).

often shows attached ribosomes; products of protein synthesis may accumulate between the two membranes.

At points of fusion between the membranes, the envelope is punctured by circular holes, the *nuclear pores*, each about 50 nm in diameter (I.I, 21 A, B). Surrounding each pore is a 120 nm ring of dense proteinaceous material composed of eight sets of granules or fibrils, and a diaphragm which, in some circumstances, may restrict the permeability of the pore. This marginal ring is the *annulus* and together with the pore forms the *pore complex* (Wischnitzer 1973). Electrical measurements in some cells, though not in all, suggest that the pores are not always patent. The pores appear to serve chiefly as a passage for diffusion of materials into and out of the nucleus, particularly the large molecules of ribonucleic acid and various proteins.

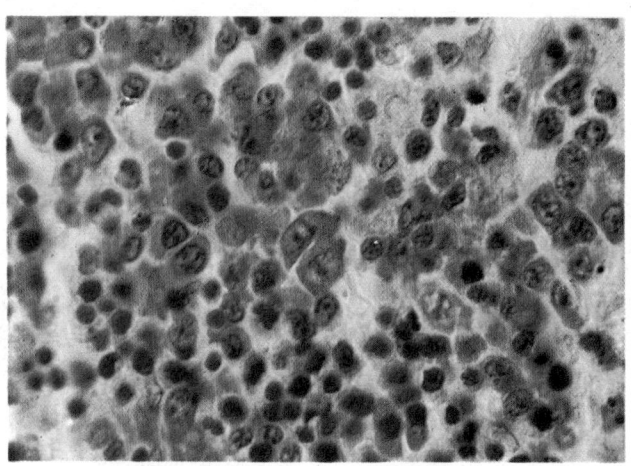

I.22 Photomicrograph showing the configuration of chromatin in 'closed-face', heterochromatic nuclei of lymphocytes (small cells), and 'open-face', euchromatic nuclei of larger reticular cells. The section is stained with methyl green-pyronin stain which also gives the characteristic red colour to the nucleolus and cytoplasm of the larger cells, denoting the presence of much RNA.

In most somatic cells there is a mixture of the two states of chromatin; heterochromatin typically lies close against the inside of the nuclear envelope, leaving gaps at the nuclear pores, whilst the euchromatin occupies the more central region of the nucleus.

During cell division (*vide infra*) chromosomes shorten to a small fraction of their interphase length; each cell is seen to contain a fixed number of chromosomes characteristic for that species of organism. In the somatic cells of man this number is 46, which is the *diploid* number; in the gametes and the germinal cells giving rise immediately to them, the number is half this (the *haploid* number—p. 30). When fertilization takes place, the fusion of two haploid sets, one from each gamete nucleus, restores the diploid number. The number of sets is termed the *ploidy*. If more than two sets are present the cell is said to be *polyploid*.

Each chromosome in a haploid set is unique in size and shape; in the diploid set there are two of each type of chromosome (*homologous* pairs), one from the ovum and the other from the spermatozoön. However, two paired chromosomes are not always identical; these are the *sex chromosomes* (*see* Mittwoch 1967), which are distinct (in man) from the remaining 22 pairs of chromosomes (*autosomes*). In the female (the *homogametic sex*) the sex chromosomes are identical, each being termed an X *chromosome* so giving 44 autosomes + 2 X sex chromosomes. In the male (the *heterogametic sex*) there is one X and an unequal Y *chromosome* giving a diploid complement of 44 autosomes + X and Y sex chromosomes.

During gametogenesis the homologous chromosomes separate, one passing to each gamete. This ensures that in the male, there are equal numbers of gametes possessing either an X or a Y sex chromosome, whereas in the female the ova all contain an X

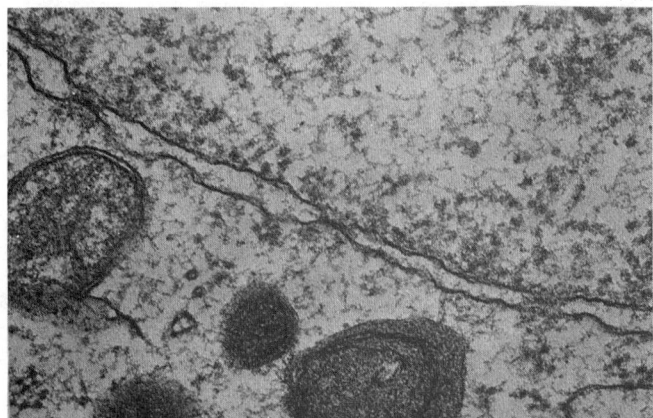

A

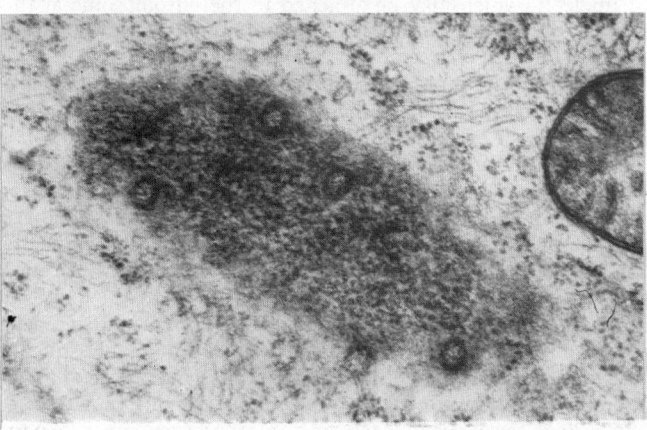

B

I.21 A, B Electron micrograph of the nuclear envelope: (A) in transverse section to show inner and outer membranes, and nuclear pores; (B) in tangential section to show pore complexes (dense rings). Micrograph (B) kindly provided by Dr. M. Dyson of Guy's Hospital Medical School. Magnification × 25,000.

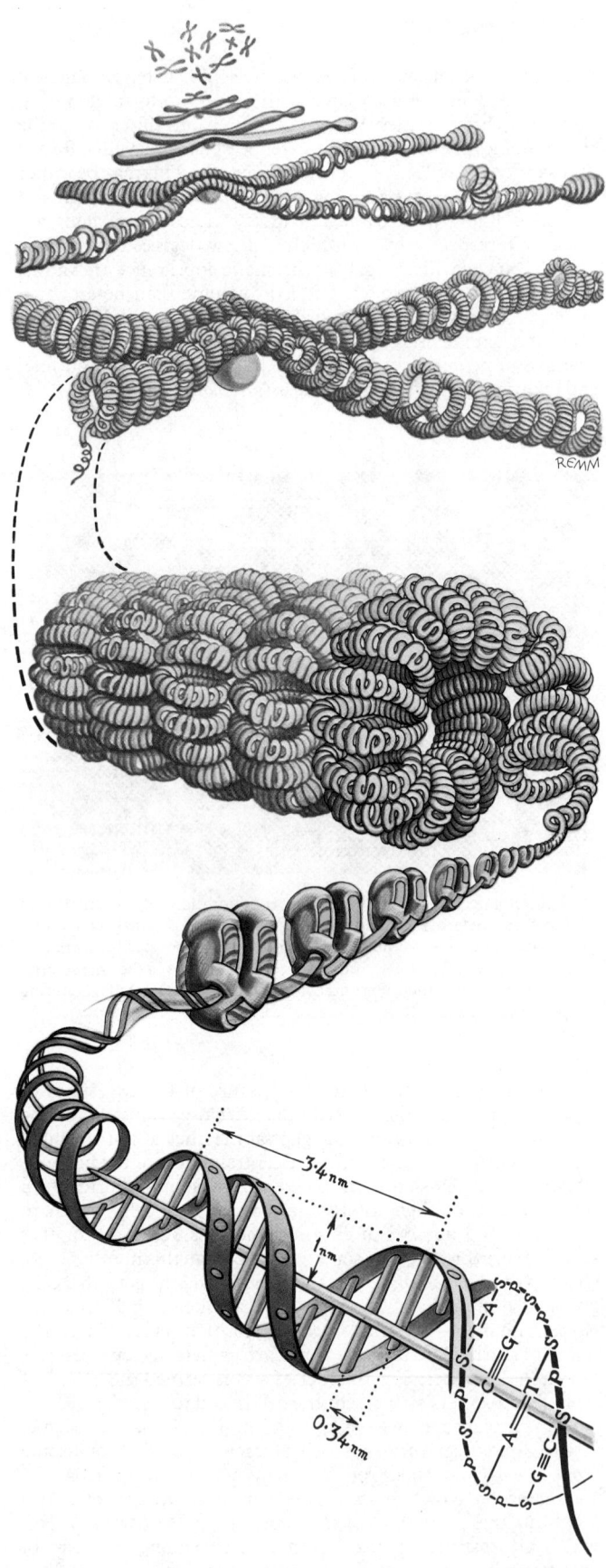

chromosome. At fertilization there is therefore an equal chance of a male (XY) or a female (XX) zygote being formed, although at birth there is a slight predominance of females because of a higher prenatal mortality amongst male embryos.

CHROMOSOMAL STRUCTURE

In early cell division when the chromosomes are quite short and thick, each pair of homologous chromosomes has a characteristic common basic structure (1.23; and *see* Stubblefield 1973). All chromosomes thus consist of two parallel and identical filaments, *chromatids* (1.24, 25), joined together at a narrowed region, the *primary constriction*, within which is a pale-staining region, the

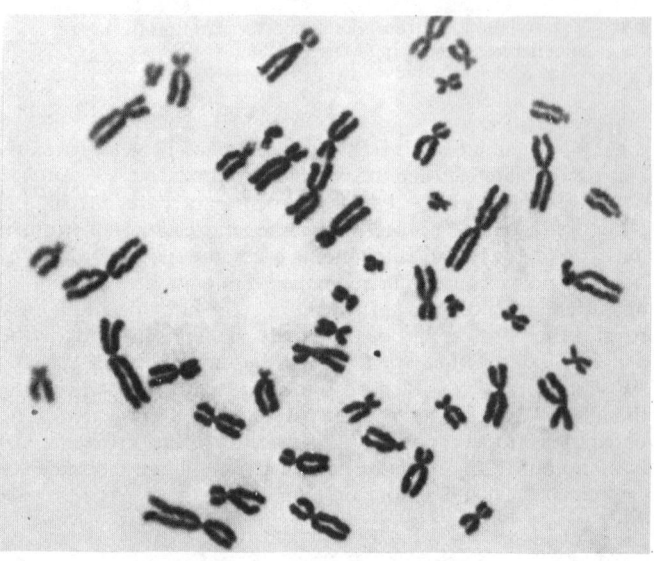

1.24 A spread preparation of metaphase chromosomes of a human fibroblast. Kindly provided by the Paediatric Research Unit, Guy's Hospital Medical School.

centromere or *kinetochore*, which is attached to the spindle fibres (*fusal microtubules*) during cell division. The free ends of the chromatids are known as *telomeres*.

Each chromatid is divided by the centromere into two *arms*. The centromere may be near the middle of the chromosome (*metacentric*) or near one end (*acrocentric*). In some mammals the centromere may be placed at the extreme end (the *telocentric* condition). In certain chromosomes there is another narrowing (the *secondary constriction*) near one end of each chromatid, frequently dividing off terminal *satellite bodies*. These constrictions are believed to be often associated with the organization of nucleoli, which may be attached to the chromosome at this point in interphase. The classification of human chromosomes on the basis of such characteristics will be considered later (p. 33).

Each complete diploid set of chromosomes contains the cell's hereditary instructions, or *genome*, and each cell in the body, apart from the germ cells, has an identical genetic complement. The chromosomal threads bearing these instructions are termed the *chromonemata* (singular: *chromonema*). It is known that each chromosome bears a linear sequence of *genes*, each situated at a characteristic position (*gene locus*), which together determine the inheritable characters of the organism carrying it, the theoretical units of inheritance being termed the *genes*.

During interphase, each chromosome retains its integrity and so the sequence of genes is also preserved. Although the sequence is highly stable, permanent, inheritable changes (*mutations*) brought about by external forces such as ionizing radiation, or exposure to certain chemicals, can occur in the genome. When the affected region is restricted these are called *gene mutations*. When whole arms or relatively large segments are involved, they are termed *chromosomal mutations*. Such mutations are some of the principal factors responsible for the inheritable *variations* which all populations show (but see also meiotic cell division p. 30).

1.23 A diagram showing a possible model for the organization of DNA and protein in chromsomes during early metaphase in mitosis. Each chromosome consists of two chromatids united at the primary constriction where the centromere is also sited. In the exposed region of the DNA molecule, the chemical groups of the 'backbone' of each helix— S (sugar) and P (phosphate), and the bases—A (adenine), T (thymine), C (cytosine) and G (guanine), are represented. Several nucleosomes are also depicted (see text).

Within members of such a population, homologous chromosomes may determine a number of alternative characters since they may carry some mutant genes; such groups of alternative genes are termed *alleles* (see also p. 101).

In any genome homologous chromosomes may possess, at a particular locus, identical alleles (the *homozygous* condition) or non-identical ones (the *heterozygous* condition). Since, in the formation of gametes, the homologous chromosomes are segregated into haploid sets, the alleles which they bear are also segregated. The various types of recombination of these which occur at fertilization, and their expression in the resulting organism, are the substance of the basic laws of inheritance, first propounded by Gregor Mendel (see also pp. 101–102).

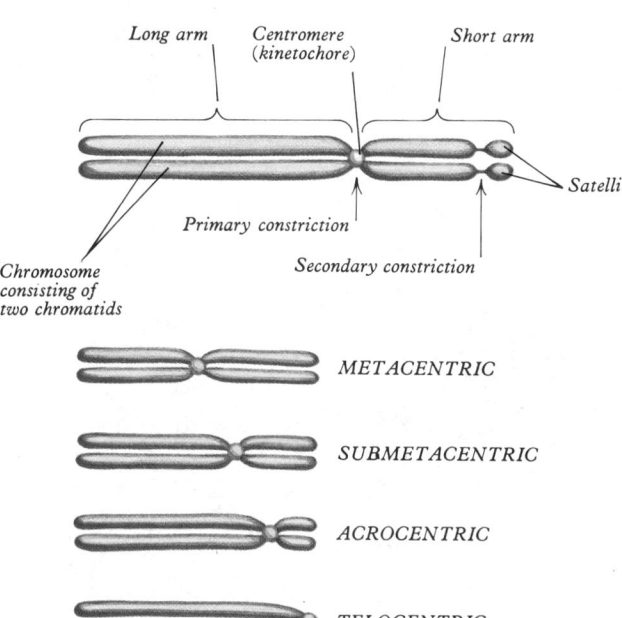

1.25 A diagram depicting the major structural features of chromosomes seen in mitotic metaphase; the terminology of the different chromosomal regions is given, together with the various major categories of chromosome, classified according to differences in the position of the primary constriction.

COMPOSITION AND ACTIVITIES OF CHROMOSOMES

Chromosomes are made of fine filaments of the nucleic acid DNA (**1.23, 26**), each associated with varieties of nucleoprotein. In man, each diploid somatic cell nucleus contains about $5\cdot6 \times 10^{-12}$ g of DNA (Rudkin 1967). The identification of the latter as the hereditary material has a long history, dating from the characterization of nucleic acids by Miescher in the 1860s, the recognition of the part played by nuclei in fertilization and the cytological events occurring in mitotic and meiotic cell divisions, and their correlation with the laws of inheritance propounded by Mendel in 1865, and then rediscovered in 1900. Further milestones were the analysis of the chemistry and distribution of the nucleic acids within cells, the investigation of viruses, which are largely composed of either RNA or DNA, and the final characterization of DNA by X-ray diffraction crystallography. The latter led to the 'double helix' model of DNA structure (Watson and Crick 1953), which opened a new window on the molecular organization of living systems. Since that time the nature of the genetic store and its mode of action have been intensively studied.

The nucleic acids DNA and RNA act as bearers of coded information which can be translated to direct the activities of each cell, and therefore the structure and metabolism of the whole organism (**1.26**). DNA is the central molecule, since its information is handed on unaltered from parent cell to its descendants, throughout life, and, by way of the germ cells, to subsequent generations. RNA is the transient intermediary, taking coded information from the DNA of the chromosomes and translating it as the structure of the enzymes and other proteins which determine the morphology and chemistry of the cell. This flow of information has been termed 'the central dogma' of molecular biology, that is, DNA→RNA→protein. At each step in this flow there is an expenditure of chemical energy, and the mediation of previously existing enzymes.

All three of these compounds (DNA, RNA and proteins) have one characteristic in common: they are all composed of molecules arranged as a linear sequence of subunits. Proteins are long chain polymers of 20 types of amino acid arranged in a variety of sequences. They vary in length, each arrangement determining the precise nature of a particular protein; for example, the way in which its chain folds into more complex shapes, such as the macromolecules of globular proteins, and its catalytic activity, in the case of enzymes. Likewise, DNA is a long linear molecule, and there is a direct correspondence between the arrangement of its chemical groups and those of the protein for which it codes. The transfer of information from DNA to RNA by the synthesis of the latter alongside the former is termed *transcription*, and the final decoding of the RNA message into the amino acid sequence constituting a protein is called *translation*. The other activity, the exact duplication of the DNA code to provide each daughter cell with identical gene sequences, is called *replication*.

THE CELL CYCLE

Synthetic activity in the cell is not uniform at all times in its life cycle, since, for example, appreciable protein or DNA synthesis does not occur during mitosis. Likewise in interphase, DNA synthesis is restricted mainly to a middle period of the cell cycle, whereas protein synthesis occurs throughout (Mitchison 1971). Accordingly, we can divide interphase (**1.27**) into a *G1* (*first gap*), *S* (*synthetic*), and *G2* (*second gap*) stage. DNA replication occurs in the S phase, and the manufacture of proteins to be used in mitosis takes place in G2. The duration of the cell cycle and its different stages varies greatly between cell types. G1 may last from a few hours to many years, or it may be absent, as in the rapidly dividing blastomeres of the early embryo. The S stage usually occupies about 7 hours, and the G2 stage up to 5 hours. Mitosis in human cells is complete in approximately 1 hour.

DNA STRUCTURE

The Watson and Crick model of DNA structure, now widely confirmed, proposes that DNA consists of two parallel molecular chains running in opposite directions in the form of a double helix, $2\cdot5$ nm in diameter, repeating every $3\cdot4$ nm. Each chain is composed of a 'backbone' of pentose sugar (deoxyribose) groups linked by phosphoric acid bridges, in such a way that they confer a particular directionality. Each pentose group bears a nitrogen-carbon ring base, which is directed sideways towards a corresponding group on the opposing chain, and is linked to that base by hydrogen bonds. There are four such bases: the purines, adenine and guanine, and the pyrimidines, cytosine and thymine. These bases are dissimilar in shape; the space between the two chains, and the type of bonding available, dictates that adenine on one chain always pairs with thymine on the other, and guanine with cytosine. The rigid complementary nature of this pairing is the basis of the genetic code. This is written in terms of three-letter 'words', each made up of three bases in a row, corresponding to a single amino acid in the final protein. Multiples of such triplets encode a whole single polypeptide, and such a sequence corresponds to a *structural gene*, or *cistron*, of classical genetics. Some triplets do not code for amino acids, but are, in effect, instructions to start or terminate the formation of amino acid chains, rather like the punctuation in a sentence. Often a number of functionally related proteins, for example, an interdependent system of enzymes, are encoded by a series of genes which may be controlled as a unit, a *supergene* or *polycistron*.

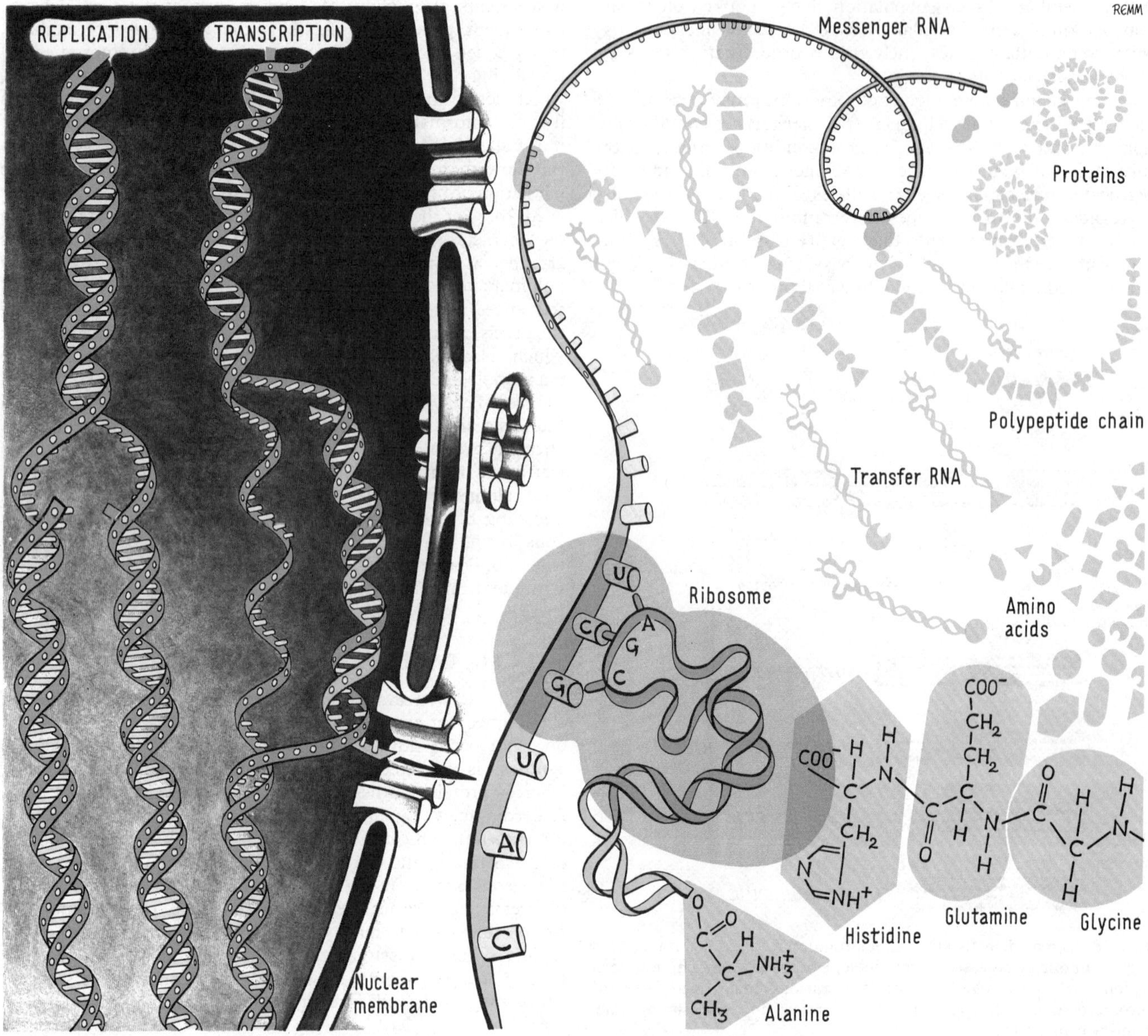

1.26 A diagram of the processes occurring during DNA replication and protein synthesis. The processes of DNA replication (on left), and transcription (middle, right) are indicated; the abbreviations are the same as those used in **1.23**, with the addition of U=uridine.

ORGANIZATION OF CHROMOSOMAL DNA

In all eukaryote chromosomes the DNA appears to be present in the form of a single continuous thread (prior, that is, to the S phase, when the DNA is being duplicated). From the known amount of DNA in each human cell it can be calculated that each nucleus contains a total length of about 170 cm, and a single chromosome is as much as 7 cm if fully extended (DuPraw 1968). In life, most of the DNA exists in a highly folded state, seen at its most extreme in the head of the mature spermatozoön (p. 95). In general it seems that the DNA which is active in transcription is most extended (*euchromatic*) and that the inactive DNA regions are most folded (*heterochromatic*), a difference which is seen very clearly in the giant polytene chromosomes of the salivary glands of the fruit fly *Drosophila*, where the active zones form extended loops ('*puffs*') when suitably stimulated to synthesize RNA (Berendes 1973). As might be expected, the degree of DNA folding varies with the phases of the cell cycle (Golomb and Bahr 1974), in parallel with the rate of protein synthesis; it is most folded during cell division (at metaphase) and, in an active cell, least folded in interphase. However, some DNA remains apparently permanently folded and heterochromatic; this includes the DNA which at some phase of development was active, but later becomes inactive (*facultative heterochromatin—*

see Brown 1966), such as the inactive X chromosome in human females. Other chromosomal regions which appear permanently heterochromatic (*constitutive heterochromatin*), are the primary and secondary constrictions of some, and perhaps all, chromosomes (Hsu and Arrighi 1971). The significance of constitutive heterochromatin is not clear although many suggestions have been made, for example, that it may have some 'intrinsic' role in maintaining chromosome stability.

Biochemical analyses of DNA have shown that a small proportion of the total consists of highly *repetitive sequences of bases*, rich in guanosine and cytosine, and therefore differing from other types of DNA in buoyant density; this is termed *satellite DNA*. Repetitive DNA of this kind is known to occur at the secondary constrictions of certain chromosomes where ribosomal RNA is synthesized, and at the primary (pericentrosomal) constrictions, where their functions are not known, so that at least in these two situations, constitutive heterochromatin seems to be identical with satellite DNA (Jones 1970). It has been suggested that repetitive DNA may represent parts of the chromosome where many identical genes coding for 'intrinsic' molecules (e.g. regulatory proteins, transfer RNAs, ribosomal RNAs) might be present. Repetitive DNA varies widely in different, even closely related species, so there is no obvious significance in the amounts present.

The exact manner in which the chromosomal DNA is folded has been much debated. It has been suggested that heterochromatin (e.g. in metaphase chromosomes) is thrown into tight loops (DuPraw 1971) or into supercoiled helices (1.23). On a more minute scale, it has recently been shown that the DNA is intermittently wound around groups of protein (*see below*).

Association of DNA and Nucleoproteins

Nucleoproteins form about 80 per cent of the dry weight of chromatin, in which they are closely associated with DNA. These proteins include in particular the basic histones and protamines, and also acidic proteins and enzymes (Elgin and Weintraub 1975). Most of the protein is aggregated in spheroidal particles about 10 nm across, strung like beads along the DNA thread (Olins *et al.* 1976), which is coiled around each particle to form a complex, the *nucleosome* or *v*-body. Each nucleosome is composed largely of four histones (H2A, H2B, H3, H4) which form two symmetrical groups (Weintraub *et al.* 1976). It has also been suggested that the nucleosomes may be able to move along the DNA, or to become detached when the DNA is perhaps active in transcription.

Enzymes associated with DNA metabolism and protein synthesis are also attached to the DNA; some of these (e.g. RNA polymerase) can be detected structurally as dense granules about 5 nm across adhering to the DNA thread (Miller and Beatty 1969).

1.27 Diagram of the major events occurring during the cell cycle of a growing and mitotically active cell to show the major stages: *G1* or 'first gap' phase; *S* or 'synthetic' phase; *G2* or 'second gap' phase; *M* or 'mitotic' phase.

TRANSCRIPTION

In transcription (*see* Watson 1975) the two helical strands of DNA are able to separate locally to expose a particular cistron on one of them, the *master strand* (the other strand being complementary). Upon this template, another strand of RNA is constructed with the aid of RNA polymerase and ATP. RNA is similar to DNA in composition, except that it is single stranded, contains ribose instead of deoxyribose, and contains the base uracil in place of thymine. The same type of base pairing now takes place so that a chain of RNA is formed alongside the DNA master strand. The DNA triplets are thus transcribed into complementary triplets, or *codons*, of RNA. When the complete cistron has been transcribed in this way, the codon-bearing RNA separates and moves to the cytoplasm, and is therefore known as *messenger* RNA (mRNA). At some point the mRNA attaches to the small subunit of a ribosome which, with other chemical factors, acts as a complex enzyme. By means of another class of nucleic acid called *transfer* RNA (tRNA), specific cytoplasmic amino acids are recognized and matched with the triplet codons of the mRNA. There exist a large number of specific tRNA molecules in the cytoplasm, each capable of 'recognizing', on the one hand, a single mRNA codon and, on the other, a corresponding amino acid. The recognition of the codons again occurs by complementary base-pairing and the relevant part of the tRNA molecule involved is called the tRNA *anticodon*. Recognition of the specific amino acid is less well understood; it may not involve the amino acid directly but, rather, an intermediary specific enzyme involved in its linkage to the growing polypeptide chain.

Each ribosome moves along the mRNA strand, 'reading' its code, so that a corresponding sequence of amino acids is formed, each being linked to the next by a peptide bond. Several ribosomes may be reading the mRNA strand at a given time, so forming the characteristic polysome rosettes or helices. When decoding is complete, the ribosome detaches itself and both mRNA and protein are released.

Since each nucleus contains the whole complement of genetic information for an individual, the expression of this must be subject to complex interlocking control mechanisms for any directed action to occur. These permit only a small fraction of the total to be transcribed at any moment, depending on the demands made on the cell, and its role in the economy of the total organism.

Jacob and Monod (1961) proposed a twofold mechanism of gene control on the basis of bacterial experiments, involving a hierarchial system of three gene types. In the first type of control, *regulation by induction*, each group of structural genes is closely linked to a controlling *operator gene*, so forming a unit or *operon*.

The operator gene is normally repressed by a protein, the *repressor*, produced by another gene, the *regulator gene*, which is not necessarily on the same chromosome, and so prevents RNA polymerase from transcribing the structural genes. A metabolite of the cell which induces protein transcription does so by combining with the repressor, which then cannot repress the operator gene, so allowing transcription. Alternatively, a metabolite may inhibit protein synthesis—*regulation by repression*. This second scheme is similar to the first, but proposes that the repressor can only inhibit the operator gene after combining with the inducer substance, which thus prevents transcription by the structural genes.

In higher organisms, the regulatory mechanisms, though basically similar to those of prokaryotes, show important differences. In multicellular animals all the cells of the body, while carrying identical genes, express them in a wide variety of ways, so that in any given cell the great majority of genes are *inactive* and only those few which are relevant to the maintenance of all cells, and to the specialist functions of the particular cell, are expressed. Experimental evidence indicates that the DNA is maintained in the repressed state by its close association with *histones*, the major basic nucleoproteins of chromatin (*see* Elgin and Weintraub 1975, and also p. 85). There are only a few classes of histones and these are very similar in a very wide range of animals, so that they are unlikely to be related *specifically* to any particular pattern of genes, unlike the repressor substances found in bacteria (*see* above, and p. 85). Genes in eukaryotes appear to be activated by another class of *acidic, non-histone proteins* which interact with specific genes (Paul 1972). Many types of such proteins are known, and these vary widely from cell to cell and between different species. They probably operate by first disengaging the histones from their close association with a particular gene sequence, and then by attaching to the DNA to promote the co-attachment of a molecule of RNA polymerase which then proceeds to transcribe the gene or group of genes. Such activities have been studied extensively in the interaction of hormones with specific 'target' cells. In some instances, as in the action of progesterone (a steroid hormone of low molecular weight) on the uterine epithelium, the hormone passes into the cytoplasm of the cell where it combines with specific receptor proteins and then, in this complexed form, moves to the nucleus to combine further with the appropriate region of DNA through its associated histones, resulting in its transcription, and the formation of proteins for cell growth (O'Malley and Schrader 1976; King and Mainwaring 1974). In other examples a hormone of too large a molecular size to pass through the plasma membrane (e.g. the polypeptide adrenocorticotrophic hormone, ACTH) may instead stimulate an enzyme, adenyl cyclase, present in the plasma membrane, to release an intracellular 'second order messenger' which, by way of a rather complex train of events, eventually activates the relevant genes in a manner similar to that of other hormones (Stein *et al.* 1975). The sensitivity of a target cell to its stimulatory hormone presumably depends on its ability to synthesize the appropriate receptor molecules, and upon the relative susceptibility of its gene sequences to derepression. The establishment of such patterns of gene regulation in a particular cell is poorly understood at present, although this process, *cellular differentiation*, is known to be dependent on many factors both inside and outside the differentiating cell.

CELLULAR DIFFERENTIATION

As development of an organism proceeds, its cells pass through a series of changes in gene expression, reflected in alterations of cell structure and behaviour (see also pp 83–92). Initially, all cells possess rather similar properties, but as embryogenesis gathers momentum, they begin to diversify, first separating into broad categories (e.g. the principal germ layers, etc.) and then into narrower categories (tissue and subtypes of tissues) until finally they mature into the 'end cells' of their particular lineage (Gurdon 1973). Some cells which are capable of giving rise to others throughout life (*stem cells*) never proceed to the ultimate point of this progression, and retain some embryonic characteristics (p. 60), but in all cases there is a *sequential pattern of gene expression* which changes and limits the cell to a particular specialized range of activities. We can detect such changes as alterations in cell structure and biochemical properties, particularly in the types of proteins which are synthesized. At the level of the gene, differentiation of this kind must be based on a change in the pattern of repression and activation of the DNA sequences transcribing the specific proteins of that stage of development; in the lifetime of a particular cell lineage, many such changes will take place (see, for example, the development of haemal cells—p. 63). It appears that the selection of a pattern of gene activity may occur some while before its expression in protein synthesis. Thus, a cell may be *committed* to a particular line of specialization without manifesting its commitment until later; once 'switched' in this way, cells are not usually able to revert to an earlier stage of development, so that an irreversible repression of some gene sequences must have occurred. *Stem cells* may remain permanently at a level of partial differentiation, although some of their offspring will be committed to full differentiation. In general, as the degree of differentiation progresses, cell division becomes less frequent (e.g. erythroblasts—p. 66), although some structurally specialized cells such as Schwann cells (p. 843) and certain glandular cells, when suitably stimulated, may undergo repeated mitotic divisions.

What constitutes the appropriate stimulus for a cell line to differentiate at successive stages in development is one of the most fascinating of biological problems, and one which also lies at the heart of many pathological transformations including carcinogenesis. In the embryo, differentiation depends upon a wide variety of circumstances (*see* pp. 83–92), although in most cases chemical interactions between cells or between different regions of the same cell are thought to be of primary significance. The initiation of a particular pathway of cellular development may, however, also depend upon complex, competitive interactions between cells, or upon the position which a cell occupies within a cell group (*see* discussion on 'positional information' on pp. 85, 87). In addition, differentiation may also depend upon some temporal sequence such as the number of previous cell divisions (pp. 81, 84). Even in mature tissues in which cell turnover occurs, similar mechanisms appear to ensure the final differentiation to a functional end cell. In some cases this is linked to the presence of a physiological stimulus as, for example, in the case of the 'B' lymphocytes which respond to exposure to an appropriate antigen by differentiating into plasma cells which secrete a neutralizing antibody. In other cases, particularly where a cell is part of a highly organized system, more subtle mechanisms must be involved, for example, the regional differentiation of keratinocytes to form undulant epidermal friction ridges in hairless skin.

REPLICATION

Replication involves the separation of the two intertwined DNA chains (1.26) and the manufacture of complementary chains alongside each of the originals which, during this process, act as templates so that the sequence of bases is copied exactly. Thus, in each of the two resulting double helices, one chain of the original double helix is conserved, so that replication is said to be *semi-conservative* (Meselson and Stahl 1958; Taylor 1958). During replication DNA is synthesized by a DNA polymerase from nucleotide precursors, so that if radioactive thymidine, a precursor of the specific DNA base thymine, is added to a cell, it is incorporated into the chromosome and can be detected autoradiographically. Using this method it has been shown that DNA synthesis occurs during the S phase simultaneously along many short stretches of DNA, beginning at a series of *initiation points* where DNA polymerase molecules can attach and then move along the chromonema to synthesize new DNA chains. Such regions of synthesis spread in both directions away from the initiation points, forming *replication forks* which can also be visualized by autoradiography (Cairns 1966; Taylor and Hozier 1976), until the whole chromosome is replicated.

Each type of chromosome has a characteristic pattern of DNA synthesis; permanently heterochromatic regions replicate later than the rest, and the inactive heterochromatic X chromosome in females replicates later than other chromosomes in the set (German 1964).

While the majority of the chromosomal DNA synthesis is confined to the S phase, there is also a constant low level synthesis of DNA which is concerned with the *repair* of minor damage to the chromosomes resulting, for example, from ultraviolet light impinging on epidermal cells. A variety of enzymes are needed for such reparative activities, including endonucleases which recognize a damaged DNA strand and begin to cut it, exonucleases which excise the damaged region, a DNA polymerase which synthesizes new DNA alongside the unaffected chain, and a ligase which joins the new portion to the cut ends of the old. In diseases where repair mechanisms are faulty, the body may be very sensitive to mutagens, such as in *xeroderma pigmentosum*, where the skin is easily damaged by ultraviolet light.

A small amount of DNA synthesis also accompanies the first meiotic prophase in gametogenesis (Stern and Hotta 1969), and this may be concerned with the exchange of genetic material between homologous chromosomes (p. 32).

The number of DNA threads within a single chromatid is not certain; the genetic evidence suggests one thread for each. After DNA synthesis the two chromonemata separate into the two distinct chromatids, so that each cell now possesses the tetraploid amount of DNA. At mitotic division (*vide infra*) the content in each cell is reduced to the diploid amount, to be doubled again in the following interphase.

It is interesting that wide variations of DNA content occur within the cells of different species of animal, even though closely related, suggesting that segments of DNA may be repeated many times. It is possible that in some cases many copies of each gene are present so that when activated, large-scale transcription can be initiated rapidly, as for example in the case of the nucleolar genes, of which there are multiple copies. Such a repetition of genes may also be a source of genetic variability.

NUCLEOLI

Each nucleus possesses a number of spheroidal nucleoli, dense bodies which take up basic dyes strongly; in man their number varies (one or several may occur), and collectively they are considered to consist of five pairs of nucleolar fragments related to particular chromosomes (*vide infra*). Nucleoli vary in size up to about 4 μm in diameter (1.28). Each nucleolus consists of a heterogeneous mass of fine granules and fibrils set in a less dense matrix, the whole agglomeration often containing cavities or

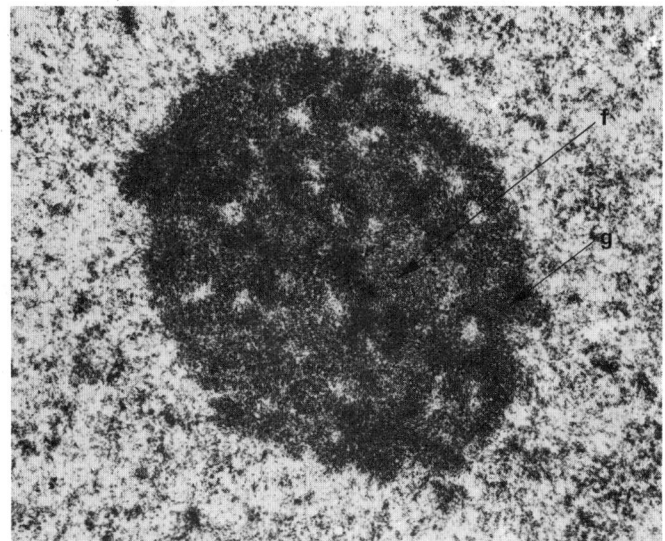

1.28 An electron micrograph of a nucleolus from a highly active cell, showing the pars filamentosa (f) enclosed in the darker pars granulosa (g). Magnification × 25,000.

channels continuous with the surrounding nucleoplasm, so that with the light microscope it may have the appearance of a folded thread (*nucleolonema*). With the electron microscope four structural elements can be detected: an inner region of dense filaments (*pars filamentosa*), an outer granular zone (*pars granulosa*), both of the former embedded in an apparently structureless matrix, the *pars amorpha*, whilst centrally is a series of somewhat less dense elongate DNA-rich regions, constituting the *pars chromosoma* (*see* Noel *et al.* 1971). These various parts can be correlated with the synthetic pathway of ribosomal RNA which is transcribed on the nucleolar organizing genes of certain chromosomes, situated in the pars chromosoma. The ribosomal RNA when first made is exceptionally long, forming the basis of the fibrils in the pars filamentosa, but is later broken into fragments, probably to form the pars granulosa, from which point the final ribosomal RNA passes to the cytoplasm through the nuclear pores. A considerable amount of phosphoprotein is also present in the nucleolus, together with various enzymes, and these may, at least in part, form the pars amorpha.

In man, *nucleolar organizing centres* are thought to be present on those chromosomes with *secondary constrictions*, and within a single nucleolus several such centres may be active. In some lower vertebrates there are large numbers of such organizing regions, and in oöcytes, hundreds of nucleoli are formed; regions of such chromosomes synthesizing nucleolar RNA form lateral loops giving the whole chromosome a bristly appearance (*lampbrush chromosomes*) and these, under the electron microscope, are seen to be festooned with RNA strands in the process of being transcribed (Mott and Callan 1975). As might be expected, cells with a high rate of protein synthesis and thus a large demand for ribosomes, have large nucleoli, for example, large neurons with extensive complex arborizations of neurites, active embryonic cells, and the cells of malignant neoplasms.

During cell division, when protein synthesis is suspended, the nucleoli disintegrate and their chromosomal portions are drawn back into the main mass of their parent chromosomes; at the end of cell division the nucleoli are rapidly reformed.

OTHER NUCLEAR INCLUSIONS

In addition to chromosomes and nucleoli, there are other less prominent structures present in all nuclei (Wischnitzer 1973). These include *granules*, some of which are associated with chromatin margins (*perichromatin granules*), about 40 nm across, and others lying between the chromatin masses (*interchromatin granules*) about 20 nm across, the significance of both types being obscure. *Perichromatin fibres* are also present at the margins of the chromatin.

In the prophase of the first meiotic division, long, fibrillar *synaptinemal complexes* form connexions between adjacent chromosomes (p. 32). Various other structures which have been reported in different types of cell include membranous vesicles and lamellae, bundles of filaments, crystals and further varieties of granule. Some of these may represent symbiotic C viruses which are common inhabitants of mammalian nuclei. The functions of the other inclusions are obscure. The nucleoplasm also contains numerous enzymes, nucleotide precursors and proteins which can only be demonstrated by cytochemical or biochemical means; these are, of course, involved in nuclear metabolism.

REPRODUCTION OF CELLS

During embryonic development most cells are undergoing repeated division as the body grows in size and complexity. As a particular cell matures it becomes differentiated with respect to its structure and function and may eventually lose the ability to divide, as do neurons, or it may persist as a stem cell capable of dividing throughout life, as do the blood-forming elements of the bone marrow.

The rate of cell division varies considerably in different tissues; in many epithelia subject to mechanical wear and tear, the replacement of damaged cells by division of stem cells may be rapid, as in the crypts between intestinal villi; it may also vary according to demand, as in healing of wounded skin, where cell proliferation rises to a peak and then drops again to the normal cell replacement level. The rate of cell division is therefore tightly coupled to the demand for growth and replacement; where this coupling is faulty, the tissues either fail to grow or replace their cells, or else they overgrow, giving rise to neoplasms.

The mechanism of normal control is poorly understood. During the earlier phases of embryonic life, the control appears to be local, involving the diffusion of metabolites from one cell group to another, so stimulating or inhibiting the division of cells. At later stages, general hormonal control is also involved. Several hormones are known to affect cell division rates either throughout the body (e.g. somatotrophic hormone), or local regions (e.g. progesterone), although the precise mechanism of these effects is not fully understood. Some hormones may act on cell division indirectly by affecting general metabolic rates (thyroid hormones) or protein synthesis (some corticosteroids and somatotrophic hormone). At least some of these effects may be mediated by the release within the cell of cyclic adenosine monophosphate (cyclic AMP) which inhibits the synthesis of new DNA during the S phase.

In adult tissues, local control of cell division is an important factor in such phenomena as wound healing. A class of compounds termed *chalones*, which inhibit cell division, has been isolated from normal tissue cells (see also pp. 78–81). A possible model for their action is that normal cells produce chalones which inhibit division of the cells in their locality; when some cells are damaged or destroyed, the concentration of chalones drops allowing cell division to occur, until the normal levels are restored, when division is again inhibited.

The Mechanism of Cell Division

Two distinct events occur in cell division, the division of the nucleus (*karyokinesis*) and that of the cytoplasm (*cytokinesis*); they are usually, but not always, coupled.

Nuclear division can be achieved in three ways; in the first, termed *amitotic* or *direct division*, nuclear material is randomly distributed to the resultant cells. This process, once thought to be common, is now known to be restricted to pathological

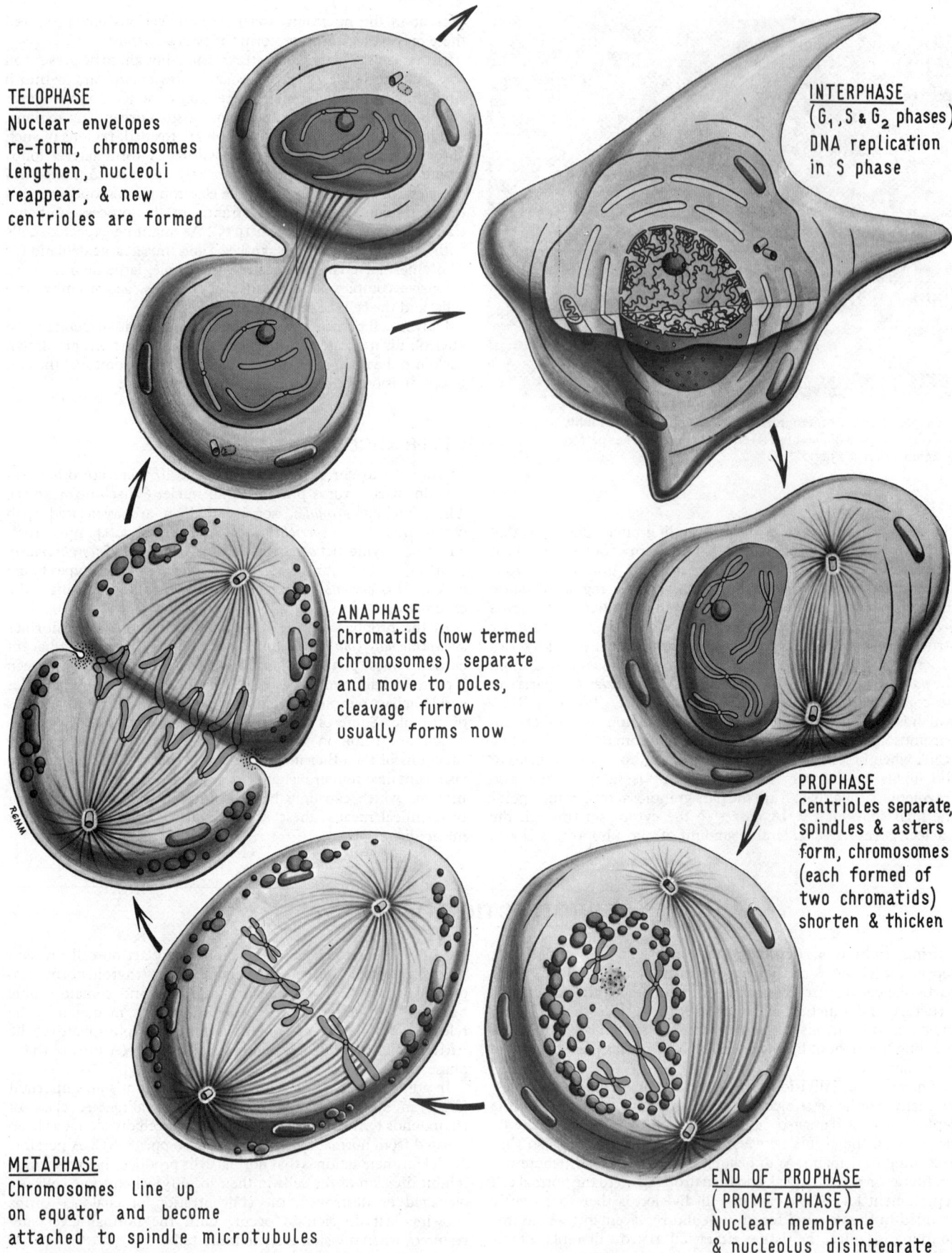

TELOPHASE
Nuclear envelopes
re-form, chromosomes
lengthen, nucleoli
reappear, & new
centrioles are formed

INTERPHASE
(G_1, S & G_2 phases)
DNA replication
in S phase

ANAPHASE
Chromatids (now termed
chromosomes) separate
and move to poles,
cleavage furrow
usually forms now

PROPHASE
Centrioles separate,
spindles & asters
form, chromosomes
(each formed of
two chromatids)
shorten & thicken

METAPHASE
Chromosomes line up
on equator and become
attached to spindle microtubules

END OF PROPHASE
(PROMETAPHASE)
Nuclear membrane
& nucleolus disintegrate

1.29 A diagram of the main stages of the mitotic cycle of a somatic cell.

conditions; in the other two types of nuclear division, complex chromosomal manoeuvres take place (*indirect division*).

Mitotic division (**1.29**) occurs in most somatic cells and results in the distribution of identical copies of the parent cell's genome to the resulting cells. In *meiosis*, which occurs in the divisions immediately preceding the production of gametes in the gonads,

the number of chromosomes is halved to the haploid number, so that when fertilization takes place the diploid number is restored; some exchange of genetic material also occurs between homologous chromosomes so that a *reassortment* of genes takes place (*see*, for example, Lewis and John 1970; Whitehouse 1973; White 1973). This leads to a wide range of genetic variability in a

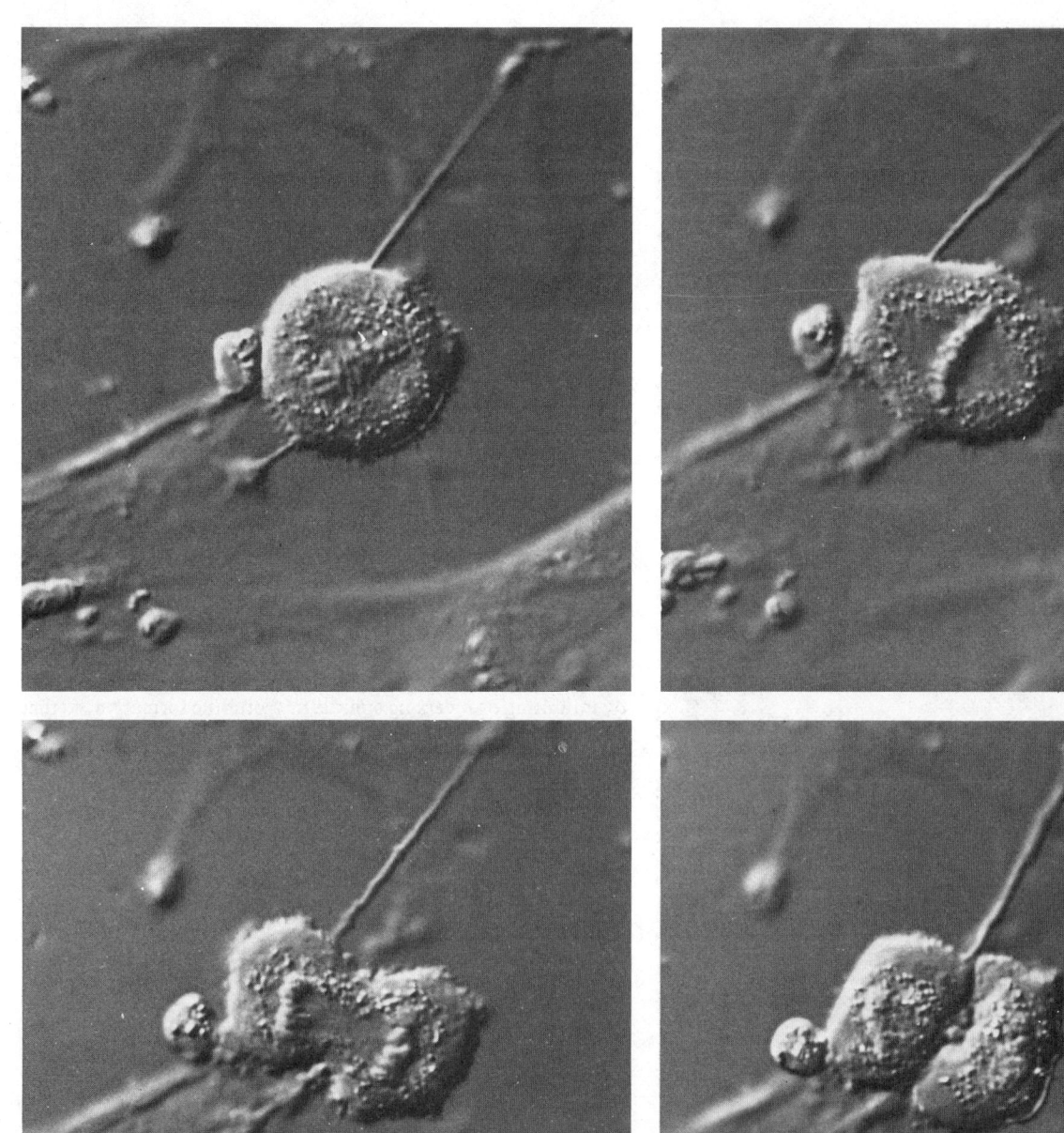

1.30   A series of micrographs taken at different stages of mitosis of a human fibroblast in tissue culture; compare with 1.29. Nomarski interference microscopy. Kindly provided by the Paediatric Research Unit, Guy's Hospital Medical School. Magnification × 1,500.

population, which is of great evolutionary advantage (see also p. 30).

Mitosis and meiosis resemble each other in many respects, differing chiefly in the behaviour of the chromosomes during early stages of cell division; meiosis is also considered to comprise two divisions in quick succession, the first of them, unlike mitosis (*meiosis I, heterotypical* division), the second, being more like mitosis (*meiosis II, homotypical* division).

Mitosis

As already stated, new DNA is synthesized during the S phase of interphase, so that in normal diploid cells the *amount* of DNA has doubled by the onset of mitosis to the tetraploid DNA value, although the chromosome *number* is still diploid. During mitosis, this amount is distributed equally between the two resulting cells, so that DNA quantity and chromosome number are now diploid in both. The nuclear changes which achieve this distribution are divided for descriptive purposes into four phases: prophase, metaphase, anaphase and telophase, (1.29, 30, 31).

Prophase

The strands of chromatin, which before mitosis are highly extended, begin to shorten, thicken, and resolve themselves into recognizable chromosomes, each made up of two chromatids joined at the centromere. Outside the nucleus the centrioles begin to separate and move apart, one to each pole of the cell; parallel microtubules are synthesized between them to create the *central spindle*, and others radiate out from the centrioles to form the *astral rays*, collectively termed asters. The spindle and the two asters together constitute the *achromatic figure*, or *diaster* (amphiaster).

Towards the end of prophase, the nucleoli disintegrate and disappear. The termination of prophase is marked by a sudden breakdown of the nuclear envelope to release the chromosomes. The membranes of the envelope disintegrate into small vesicles indistinguishable from those of the endoplasmic reticulum.

Metaphase

As the nuclear envelope disappears, the spindle microtubules invade the central region of the cell and the chromosomes move towards the equator of the spindle (a period known as *prometaphase*). Once they have arrived at this imaginary plane (the *metaphase* or *equatorial plate*) the chromosomes become attached by their centromeres to spindle microtubules and are so arranged in a star-like ring when viewed from either pole of the cell.

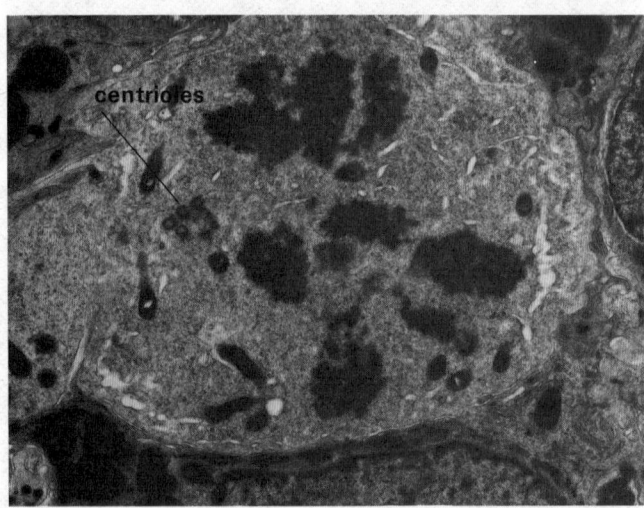

1.31   An electron micrograph of an epithelial cell in mitotic metaphase. The chromosomes are seen as blocks of dense material, and are sectioned approximately along the plane of the equator. Magnification × 10,000.

Anaphase

The centromere in metaphase is a double structure; it now separates into its two components, each carrying an attached chromatid, so that the original chromosome has, in effect, split into two new chromosomes. These move apart, one towards each pole of the cell. The mechanism of this movement is not yet understood, but it is known that the spindle microtubules are necessary for movement to occur. It has been suggested that the two halves of each centromere become attached to separate categories of microtubules, one from each pole of the cell (McIntosh *et al.* 1969). These microtubules would then move towards their respective poles, possibly by sliding interaction with other microtubules, pulling the new chromosomes along with them. Alternatively it has been proposed that the progressive disintegration and reassembly of microtubules might generate similar movements by addition of tubulin at the equator and removal at the poles (Inoué and Sato 1967).

Telophase

As a result of the events in anaphase the chromosomes become grouped at each end of the cell, both groups being diploid in number. The chromosomes now re-extend, and the nuclear envelope reappears, beginning as a series of membranous vesicles on the end surfaces of the chromosomes. The nucleoli also reappear.

Meanwhile, cytoplasmic division, which usually begins in early anaphase, is well started. This process is signalled at late metaphase by cytoplasmic movements involving the equal distribution of mitochondria and other organelles around the cell periphery. In anaphase an infolding of the cell equator begins and deepens as the *cleavage furrow* (Szollosi 1970). Small vacuoles form in the cytoplasm along the plane of cleavage, and eventually the furrow constricts the cell into two. Where the constriction meets the remains of the spindle a dense region of cytoplasm, the *mid-body*, is visible, but eventually the two new cells separate, each with its derived nucleus. The spindle remnant now disintegrates. When the cleavage furrow is active, a peripheral band of microfilaments has been observed in the constricting zone in some species; the contraction of this band is probably responsible for furrow formation. During telophase the filaments of the cleavage furrow eventually contract down on the remaining spindle microtubules to form the dense mid-body, which finally disappears, presumably by de-polymerization of its actin and tubulin constituents.

Failure of *disjunction* and '*lagging*' of chromatids may sometimes occur, so that two paired chromatids pass to the same pole. Of the two new cells, one will have more, and the other less, chromosomes than the diploid number. Exposure to ionizing radiation enhances the frequency of such events, and may, because of chromosomal damage, inhibit mitosis altogether. A typical symptom of *radiation sickness* is the failure of epithelia to replace lost cells, with consequent ulceration of the skin and mucous membranes.

Mitosis can also be disrupted by several chemical agents, particularly colchicine and its derivatives colcemide, podophyllin and podophyllotoxin, and by viniblastine. These compounds act by inhibiting or reversing spindle microtubule formation, so that mitosis is arrested in metaphase. This phenomenon is of great value in cytogenetic studies, since chromosomes are most easily examined in the metaphase condition.

Meiosis

During meiosis (1.32, 33A, B) there are two cell divisions (*see* Whitehouse 1973): in the interphase prior to the first division, DNA is replicated in the usual manner, resulting in the tetraploid amount of DNA, the number of chromosomes being diploid. During *meiosis I* the DNA is reduced to the diploid *amount* in each resultant cell, although the chromosome *number* is halved to the haploid value; in *meiosis II*, the DNA in each new cell formed is reduced to the haploid amount, the chromosome number remaining haploid.

Prophase I

This is a long and complex phase differing considerably from mitotic prophase. It is customarily divided into five, the leptotene, zygotene, pachytene, and diplotene stages and diakinesis (for details of structural changes *see* Comings and Okada 1972; Moens 1974; John 1976).

Leptotene stage—the chromosomes become visible as individual threads attached at one end to the nuclear envelope, and show characteristic beads (*chromomeres*) throughout their length.

Zygotene stage—the chromosomes are seen to have come together side by side in homologous pairs, a process which may already have occurred as early as the telophase of the previous mitotic division. The homologous chromosomes pair point for point progressively, beginning at the attachment point to the nuclear envelope (Moens 1974), so that corresponding regions lie in contact. This process is *synapsis*, *conjugation*, or *pairing*. Each chromosome pair is now termed a *bivalent*. In the case of the unequal X and Y sex chromosomes which during zygotene and pachytene are sequestered in a secluded zone of the nucleus, the *sex vesicle* (Solari and Tres 1967), only limited segments (the *pairing segments*) are homologous and these pair end to end (1.33); the remaining parts are *differential segments*. With the electron microscope, homologous chromosomes are seen to be held together by a highly structured fibrillar band, the *synaptinemal complex* (1.33C), which occupies the space (about 100 nm wide) between them (Gillies 1973).

Pachytene stage—spiralized shortening and thickening of

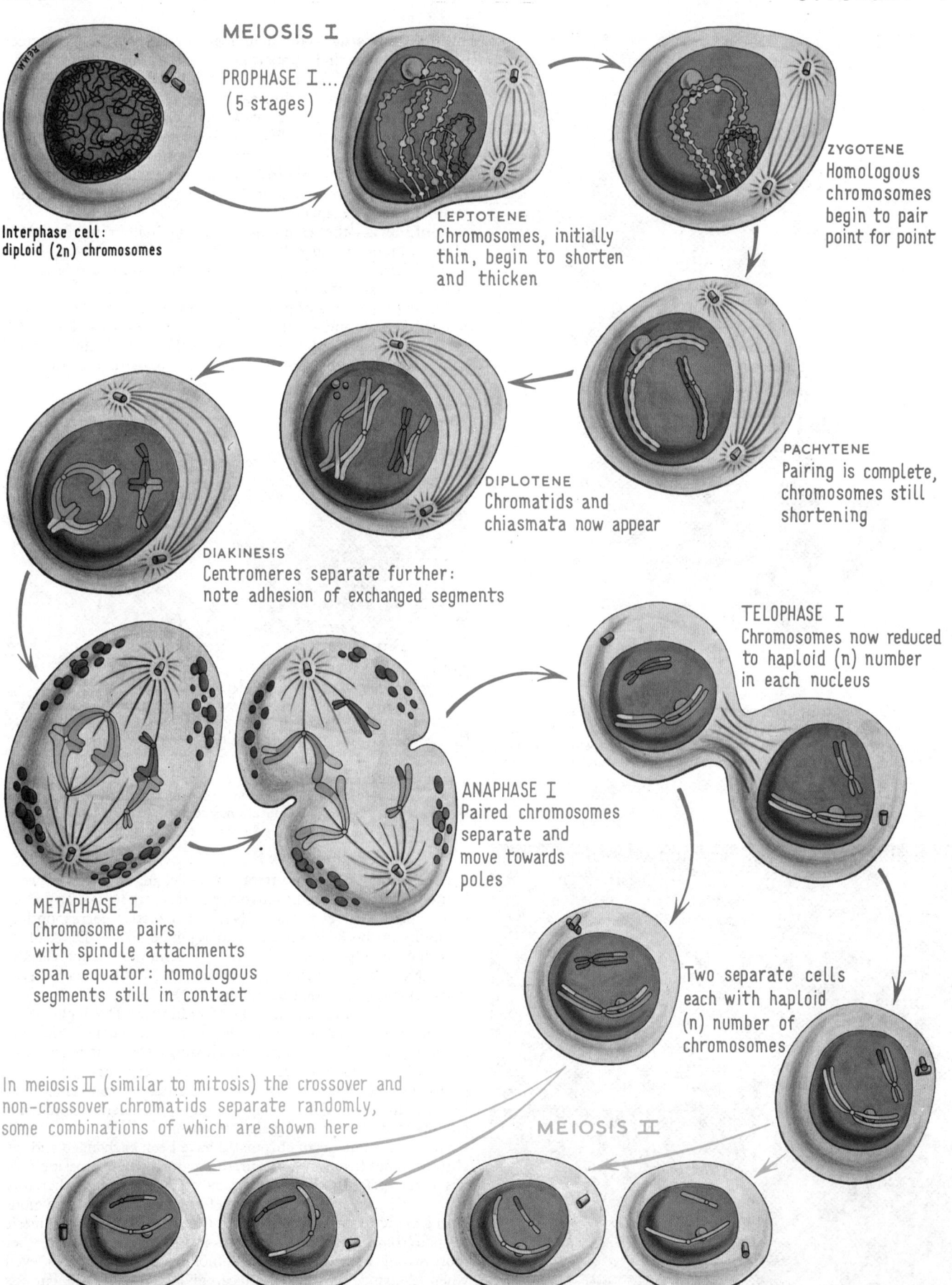

MEIOSIS I

PROPHASE I...
(5 stages)

LEPTOTENE
Chromosomes, initially
thin, begin to shorten
and thicken

Interphase cell:
diploid (2n) chromosomes

ZYGOTENE
Homologous
chromosomes
begin to pair
point for point

PACHYTENE
Pairing is complete,
chromosomes still
shortening

DIPLOTENE
Chromatids and
chiasmata now appear

DIAKINESIS
Centromeres separate further:
note adhesion of exchanged segments

TELOPHASE I
Chromosomes now reduced
to haploid (n) number
in each nucleus

ANAPHASE I
Paired chromosomes
separate and
move towards
poles

METAPHASE I
Chromosome pairs
with spindle attachments
span equator: homologous
segments still in contact

Two separate cells
each with haploid
(n) number of
chromosomes

In meiosis II (similar to mitosis) the crossover and
non-crossover chromatids separate randomly,
some combinations of which are shown here

MEIOSIS II

1.32 Chief stages in the meiotic cycle (male). For clarity only four
chromosomes out of the total 46 are shown.

each chromosome continues progressively, and each is now seen
to be formed of two chromatids joined at the centromere. Each
bivalent pair therefore consists of four chromatids and is called a
tetrad. During this stage, two chromatids, one from each bivalent

chromosome, become partially coiled around each other. Also
during this stage it is probable that exchange of DNA (*crossing
over* or *decussation*) by breaking and rejoining occurs, perhaps
facilitated by the synaptinemal complex.

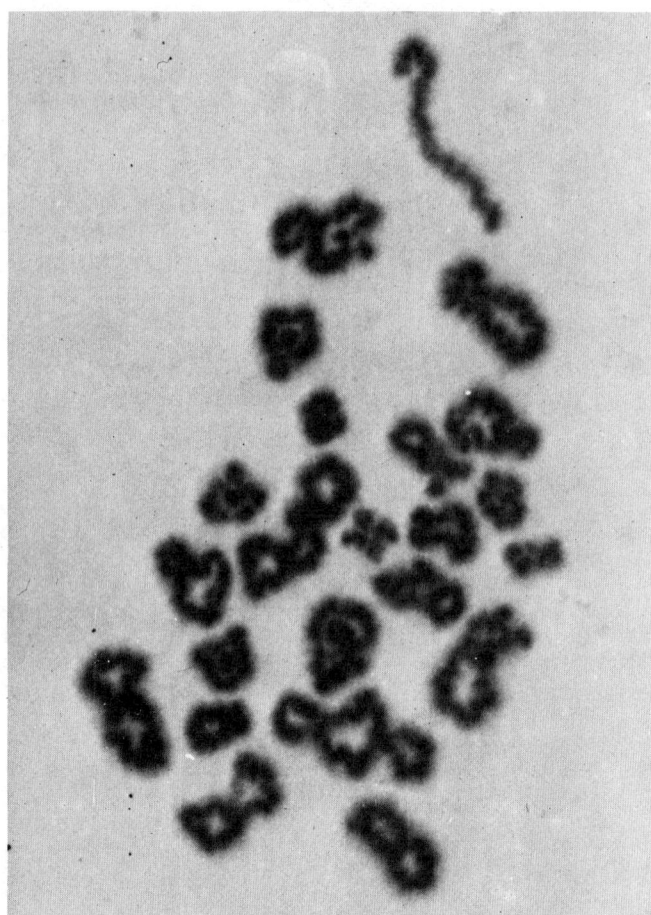

Diplotene stage—the homologous pairs, by this time much shortened, move slightly apart except at the points where crossing over has occurred (*chiasmata*). Sometimes, the chiasmata move towards the ends of the chromatids (*terminalization*); normally at least one chiasma forms between each pair of homologous chromosomes, and up to five have been observed (even up to ten in some species). In human meiosis, primary oöcytes become diplotene by the fifth month *in utero*, and remain in this state until the period prior to ovulation (some for decades, and even fifty years, in some instances).

Diakinesis—the remaining chiasmata finally resolve, and the chromosomes, which still form bivalents, become even shorter and thicker. The bivalent pairs move away from each other and become spread out against the nuclear envelope.

During the events of prophase the nucleoli have disappeared, and the spindle and asters have formed as in mitosis. The end of prophase is marked by dissolution of the nuclear envelope and movement of the bivalent chromosomes towards the equatorial plate (*prometaphase*).

1.33A A photomicrograph of chromosomes from a human spermatocyte in late prophase (diakinesis) of the first meiotic division showing bivalents beginning to separate. Note the paired sex chromosomes which are joined end to end. Kindly provided by the Paediatric Research Unit, Guy's Hospital Medical School.

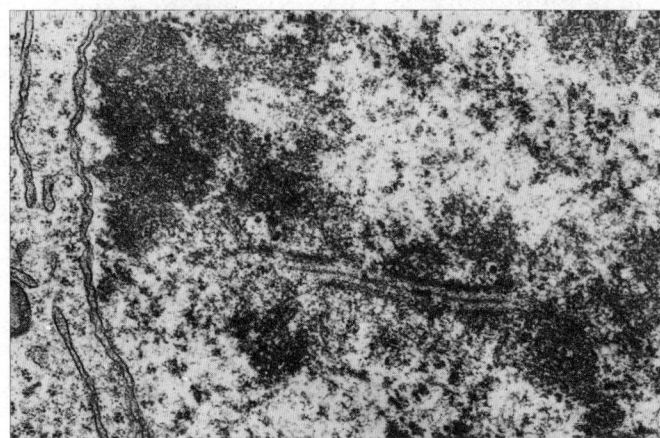

1.33C Detail of a synaptinemal complex in a cell similar to that shown in B. Note the masses of heterochromatin, and the fibrillar core of the synaptinemal complex. Magnification × 20,000.

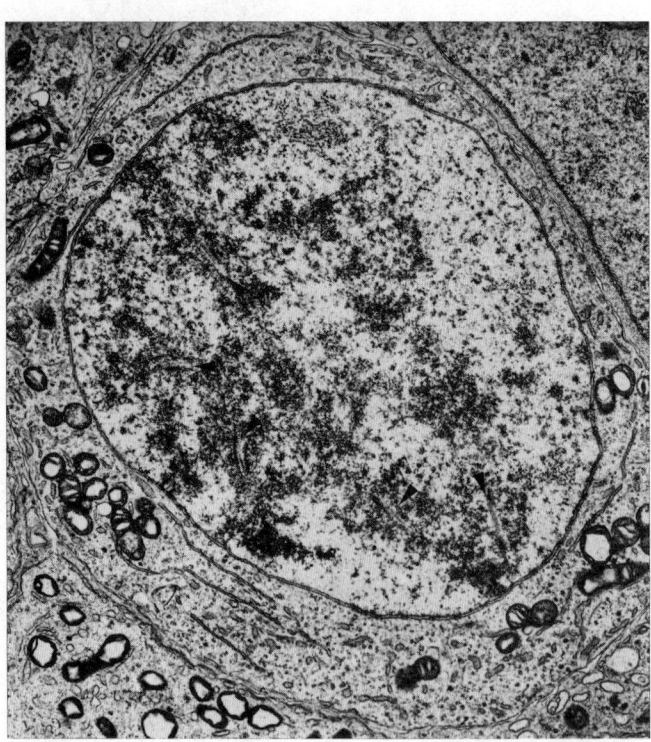

1.33B Electron micrograph of a primary spermatocyte (mouse) showing the nucleus in early prophase of the first meiotic division. Five synaptinemal complexes are visible (arrowheads). Magnification × 10,000.

Mechanism of decussation—the exchange of genes between homologous chromosomes involves some extraordinarily precise mechanism whereby the DNA, at exactly corresponding positions on both, is severed and rejoined to the DNA of its corresponding partner. How this is achieved is not certain, although it has been proposed that the DNA double chains of the two exchanging chromatid segments are exchanged one at a time, perhaps with some remodelling of redundant DNA chains by partial dissolution and resynthesis in the correct position (Holliday 1964; Whitehouse and Hastings 1965). It is probable that such exchanges occur during late zygotene and early pachytene (Whitehouse 1973), when there is a certain amount of DNA synthesis, and it is likely that the synaptinemal complex is an important mediator of genetic exchange.

Once the segments of chromatid have been exchanged and the chromosomes begin to separate, the regions where crossing over has taken place become visible as *chiasmata*; these therefore represent a *past event* and chiasma formation is *not* synonymous with genetic recombination. It is interesting that after diakinesis the paired chromatids of a given chromosome adhere to each other more strongly than to those of homologous chromosomes, even when exchange has taken place, giving rise to the curious configurations which chromosomes adopt in metaphase of meiosis I (1.32).

Metaphase I

This is similar to that of mitosis except that the bodies attaching to the spindle microtubules are bivalent and not single chromosomes. These become arranged so that the homologous pairs lie parallel to the equatorial plate with one member on either side.

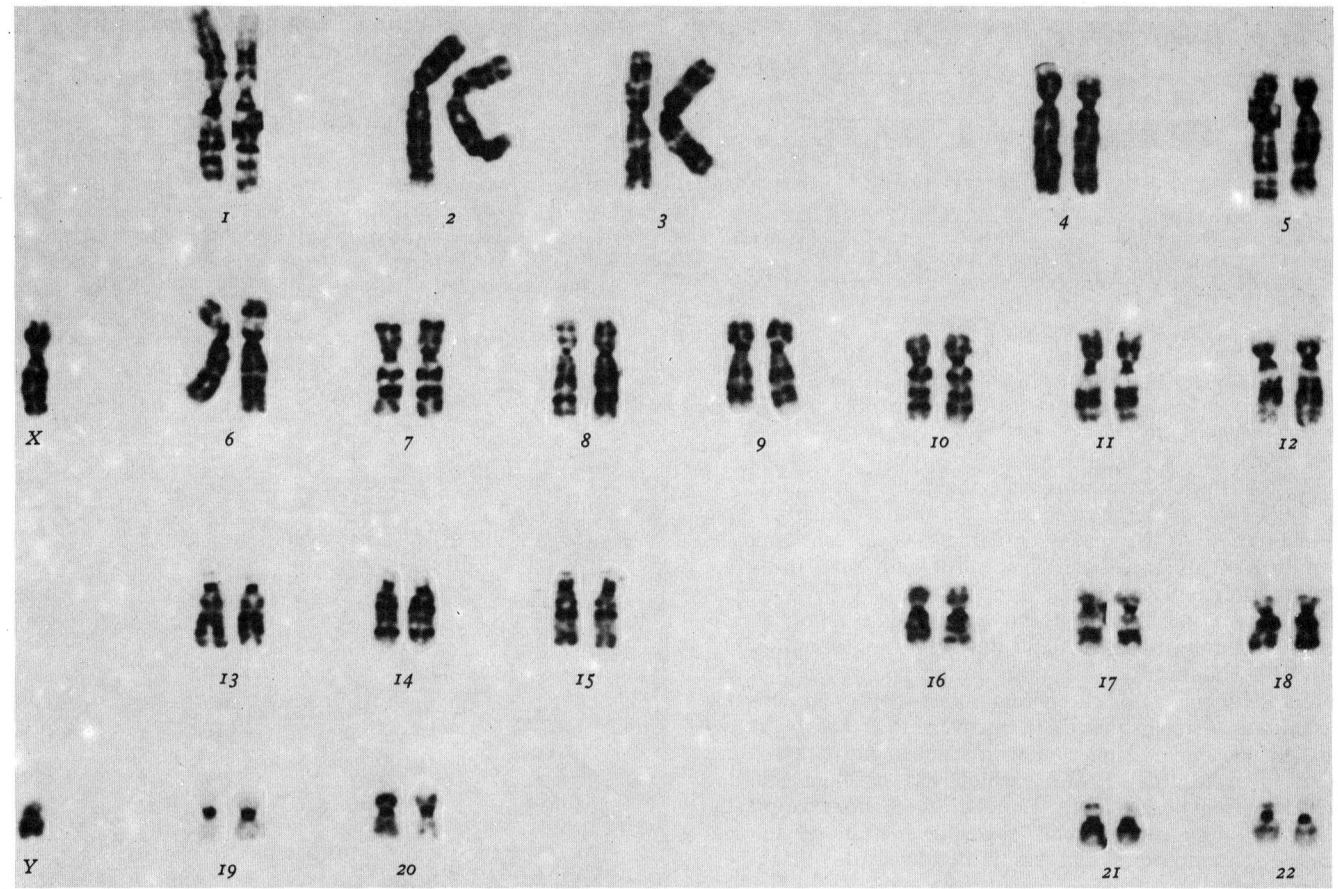

1.34 The human male karyotype prepared from a metaphase chromosome spread, stained by the Giemsa banding technique, to demonstrate the pattern of dye-binding used to characterize different chromosomes. The autosomes are shown on the right, and the sex chromosomes on the left. The numbering is that agreed by the Paris Convention (*see* text). Kindly provided by the Paediatric Research Unit, Guy's Hospital Medical School.

Anaphase and Telophase I

These also occur as in mitosis, with the exception that in anaphase, the centromeres do not split; thus, instead of the paired chromatids separating to move towards the poles, whole homologous chromosomes made up of two joined chromatids depart to opposite poles. Since positioning of bivalent pairs is random, there is a random assortment of maternal and paternal chromosomes in each telophase nucleus.

During meiosis I, cytoplasmic division occurs as in mitosis to produce two new cells.

Meiosis II

This commences after only a short interval during which no DNA synthesis occurs. This second meiotic division is more like a mitosis, with separation of chromatids during anaphase; but it should be noted that unlike mitosis, the separating chromatids are dissimilar genetically. Cytoplasmic division also proceeds and thus a total of *four* cells results from meiosis I and II. (However, during spermatogenesis there was some evidence that further divisions occur, but *see* p. 96.)

If cytoplasmic division fails during meiosis I, gametes with the diploid number of chromosomes result, and thus give, at fertilization, a triploid zygote.

Other abnormalities, some of them viable, are produced by *nondisjunction* of chromosomes, or by the *lagging* of individual chromosomes, during anaphase. These will be discussed later (p. 136). There is much evidence that with increasing maternal age, there is an increased frequency of such abnormalities.

CLASSIFICATION OF HUMAN CHROMOSOMES

Since a number of genetic abnormalities can be directly related to the chromosomal pattern, the characterization or *karyotyping* of chromosomes is of considerable diagnostic importance. Their identifying features are best seen during metaphase.

Living cells taken from blood samples, nasal swabs or other sources, can be cultured *in vitro* and caused to divide by treatment with mitotic stimulators such as phytohaemagglutinin. Mitosis is later interrupted at metaphase with spindle inhibitors (*see* p. 16), and the chromosomes are then dispersed by swelling the cells in hypotonic solutions followed by air drying, or squashing, to rupture the cells and flatten the chromosomes. These can then be treated by a variety of staining procedures to allow the identification of individual chromosomes by their size, shape, and overall distribution of stain. Methods of staining include *general techniques* to show the obvious landmarks, for example, length of arms, position of primary and secondary constrictions; and various *banding techniques* to demonstrate differential patterns of binding of the dye, which is characteristic for each chromosome type (**1.34, 35**). Since the invention and introduction of the original banding methods (Caspersson *et al.* 1968), several further procedures have been developed (*see* Caspersson and Zech 1973) such as fluorescence staining with quinacrine mustard and related compounds (**1.36A**), Giemsa staining after treatment with alkali, 'reverse-Giemsa' staining in which the light and dark areas are reversed, and the staining of constitutive heterochromatin (C-banding; see also p. 34).

Using these different methods, chromosomes can be classified according to their overall length, positions of primary and secondary constrictions, presence of satellites, the number and positions of the transverse bands, and the distribution of constitutive heterochromatin. A standard system of numbering the autosomal pairs of chromosomes in order of decreasing size, from 1 to 22, is adopted according to the conventions agreed in London (1960) and Chicago (1964). Before the introduction of banding, some chromosome types could not be easily distinguished from each other, and thus were grouped together. However, the more recent methods allow unambiguous identification, and a full system of classification agreed in Paris (1972) is now employed.

HUMAN KARYOTYPE CHROMOSOME NUMBER 2

HUMAN KARYOTYPE CHROMOSOME NUMBER 15

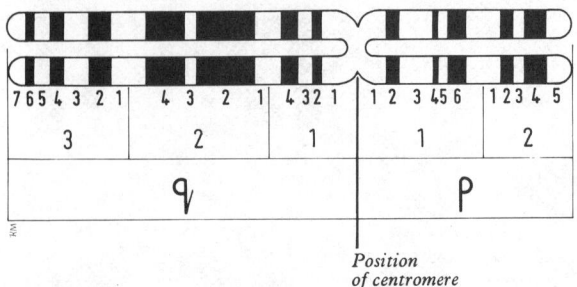

■ *Positive staining Giemsa and Quinacrine methods*

□ *Negative staining except for reverse Giemsa method*

▨ *Variable staining*

Position of centromere

1.35 Diagram of the nomenclature of the banding patterns in two human chromosome types, designated by the Paris Convention, for different staining techniques (for details *see* text).

These advances have also greatly helped the recognition of *abnormal chromosome patterns*, since the origins of extra whole chromosomes, or fragments of chromosomes, can often be identified in abnormal karyotypes. An additional means of identification exploits variations in the *time* of DNA synthesis during the S phase in different chromosome types; this is demonstrated by pulse-labelling the replicating chromosomes with ³H-thymidine at specific times during the S phase, followed by autoradiography of the chromosomes during the ensuing metaphase (Gianelli 1970). Alternatively they can be pulse-labelled in a similar manner but with the analogue of thymidine, bromo-deoxyuridine (BrDU), which has different staining properties from normal DNA (Pain *et al.* 1976). For example, the inactive X-chromosome replicates particularly late (p. 26), and extra X-chromosomes can easily be detected using these techniques.

Sex Chromatin

Interestingly, it was observed some years ago that the nuclei of female mammals are structurally different from those of males, in

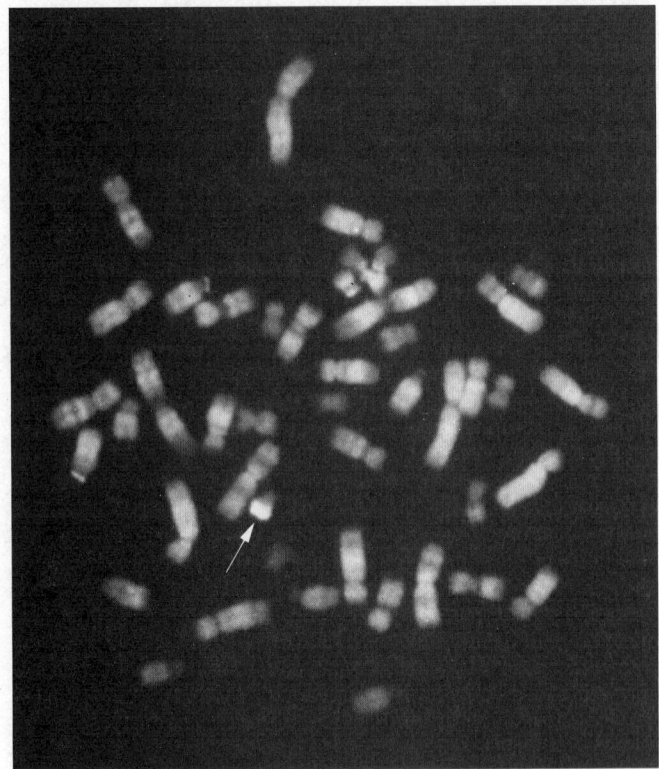

1.36A A chromosome spread from a normal human male cell, treated with a fluorescent dye (see text) and photographed using fluorescence microscopy. In addition to the terminal fluorescence of the long arm of the Y chromosome (arrow), bands which fluoresce less strongly are also present on the other chromosomes, providing a basis for their identification. Kindly supplied by the Paediatric Research Unit, Guy's Hospital Medical School. Magnification × 2,500.

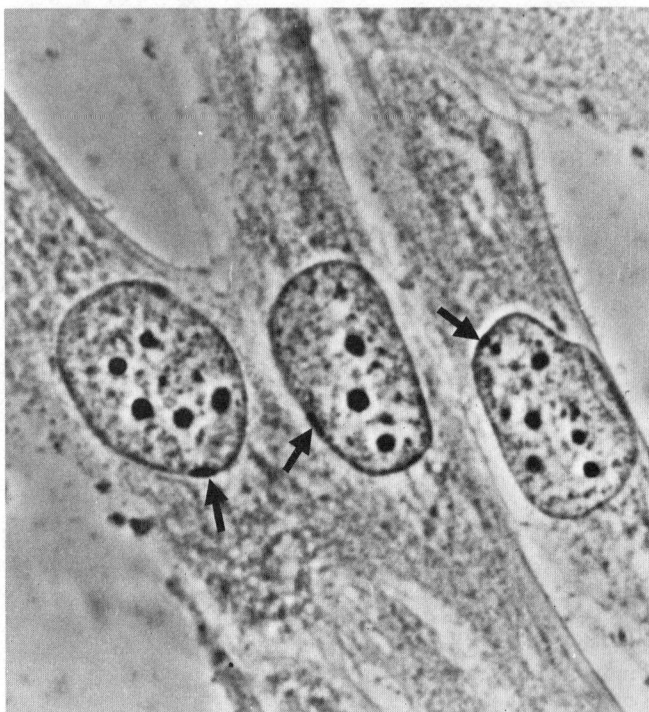

1.36B Fibroblasts from a human female in tissue culture, showing dense heterochromatic Barr bodies (arrows) lying close against the nuclear envelope. Kindly provided by the Paediatric Research Unit, Guy's Hospital Medical School. Magnification × 2,000.

that a heterochromatic body is usually seen close against the nuclear envelope in females (**1.36B**). This body, the *sex chromatin* or *Barr body* (Barr and Bertram 1949), may form a protrusion (drumstick) from the surface of polymorphic nuclei. It is now thought to be one of the X chromosomes which is synthetically inert, and thus heterochromatic, its euchromatic partner carrying out all necessary synthetic functions (Lyon 1962). The inert X chromosome also replicates later during interphase than its homologue.

Chromosome Abnormalities

As already stated, aberrations in mitotic and meiotic cell divisions result in chromosomal abnormalities. Extreme examples such as the formation of more than the doublet (diploid) set of chromosomes to form *polyploid genomes* (triploid, tetraploid, octoploid, etc.) are found in some organisms, but are lethal in advanced animal forms.

Failure in the equal segregation of chromosomes in anaphase can result in either more or less than the full number. If this takes place in germ cell development, the resulting offspring will also possess an altered number, either of autosomes or of sex chromosomes.

Where, instead of a homologous pair, three chromosomes are

present (*trisomy*), the additional chromosome may be identified by a variety of accessory techniques, including banding methods; the same applies to the presence of unpaired chromosomes (*monosomy*), where the number is reduced.

In mongolism, in which there are 47 chromosomes in the somatic cells, the additional number is one of the smallest of the set, known to be number 21. In a few cases of mongolism, the extra chromosome is attached to another chromosome of the set, as a result of a structural chromosome anomaly termed a *translocation*. Other abnormalities of chromosome structure include *ring forms* in which the two ends are joined together.

Anomalies of the sex chromosome complement are also well known. In the trisomy of Klinefelter's syndrome the cell contains an additional X chromosome (44 autosomes + XXY); these cases are male in character (phenotypically male), but the testes remain small, and spermatogenesis fails to occur. On the other hand, in Turner's syndrome there is only a single X chromosome (44 autosomes + XO); the subject is female in character, but the ovaries are rudimentary, and full oögenesis does not occur, although the embryonic ovary may appear normal with plentiful oögonia. Higher multiples of sex chromosomes (e.g. 44 autosomes + XXXXY), have also been described.

Chromosome abnormalities and triploidy, in man, are quite common (about 0·6 per cent of all conceptions), and occasionally are compatible with survival to birth. However, it is known that many instances of such abnormalities are lethal in embryonic life, thus accounting for their rarity among survivors at birth; 20–30 per cent of spontaneously aborted individuals are chromosomally abnormal.

CELL MOTILITY

All cells, except perhaps erythrocytes, exhibit some form of motility, ranging from the minute streaming of the cytoplasm, which is characteristic of all nucleated cells, to the powerful contractions of muscle cells. The types of motility vary widely; in most tissues there are constant modulations of cell shape and detailed surface topography, including the rounding of cells, particularly prior to mitosis, their flattening or elongation, e.g. in the developing neuroblasts of the neural tube, the formation of microvilli or of larger 'blebs' and pseudopodia at the cell surface (**1.37**). Such changes in shape are prominent in embryonic development and, later, in the repair of damaged tissues; many morphogenetic movements of cell layers are a result of the

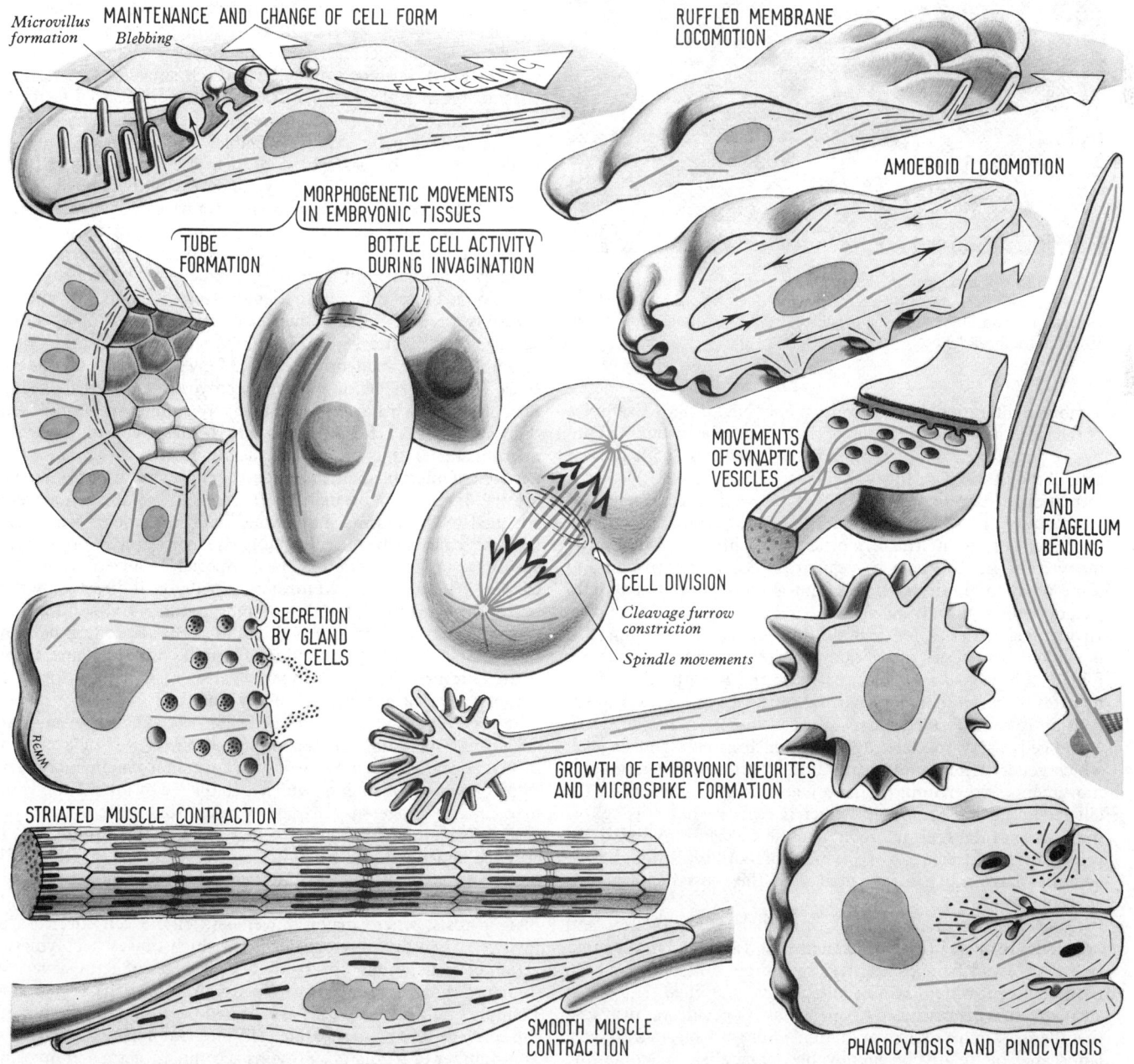

1.37 A diagram summarizing some major examples of cell motility associated with actin filaments (red) and microtubules (blue).

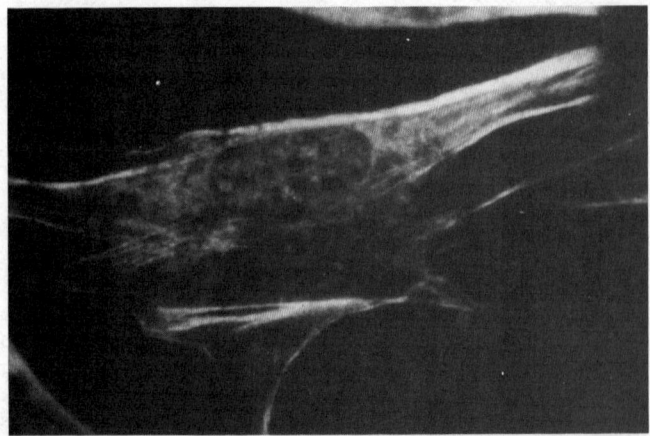

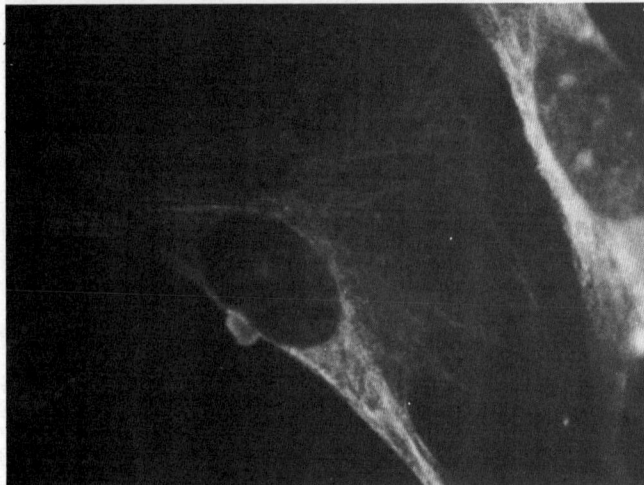

1.38A and B Fibroblasts in tissue culture stained by immunofluorescence techniques to demonstrate the distribution of: (A) actin filaments, and (B) microtubules. Kindly provided by Dr. L. Mallucci (Guy's Hospital Medical School).

combined alteration of their constituent cells (p. 86). Examples of the latter are—the rolling up of the neural tube, brought about by the transformation of approximately parallel-sided columnar cells of the neural plate into wedge-shaped elements (**2.5**), and the formation of cup-like depressions by the apical contraction of 'bottle cells' (**1.37**). Changes of cellular shape associated with the movement of the internal cytoplasm occur in the locomotion of many cells. In amoeboid movement, typical of neutrophil and some other migratory cells, the central cytoplasm streams into pseudopodia at the advancing edge, and is withdrawn at the rear. In other cell varieties, for example macrophages, the leading edge is marked by a series of delicate undulations, creating a highly folded, thin *ruffled membrane*, or, as in the growth cone of the neurites of nerve cells, long narrow processes or *microspikes* are constantly being formed and resorbed as the cone moves forward. Some cells carry specialized features such as cilia and flagella, which generate fluid movements at their surfaces by undulatory movements, either to move the mucus on the surface, as in the ciliated lining of the respiratory tract, or to move the entire cell, as in the case of the spermatozoön. Muscle cells and myoepithelial cells lie at the extreme end of this range of motility, although their activities have much in common with the movements of less specialized cells.

Cytoplasmic streaming within cells enables metabolites and organelles to move in a directed manner and at a much more rapid rate than could be accomplished by simple diffusion. Cells in tissue culture, which have been extensively studied, show that such streaming is restricted to certain fairly specific pathways, so that there is often a rapid flow of highly fluid cytoplasm through a landscape of relatively immovable organelles. Cytoplasmic streaming is particularly well developed in very large and extensively branched cells such as neurons, where metabolites are

conveyed for distances of, in some cases, a metre or more to the peripheral extremities of axons and dendrites, some materials then retracing the route back towards the cell soma. Speeds of up to 25 mm/day have been recorded for *axoplasmic flow* in peripheral nerves in the centrifugal direction.

Other internal cellular movements include the passage of secretory vacuoles towards the surface of glandular cells, with the subsequent release of the secretory product, the uptake of materials by the cell surface (endocytosis, phagocytosis, and pinocytosis: *see* p. 15) and their movement into the cytoplasm, and the movements of chromosomes during mitosis and meiosis.

The mechanisms of these various types of movement are far from being fully understood, but it is beginning to emerge that many of them have a fundamentally similar basis at the molecular level, particularly concerning the presence of the proteins actin and myosin, and in some cases, also tubulin (*see* Allison 1973). Motility is best understood in the highly organized cytoplasm of striated muscle fibres which lend themselves particularly well to structural, biochemical and biophysical analyses (p. 506). Contraction is now known to involve the energy-dependent sliding interaction of filaments of actin and myosin, both of which are polymers of smaller molecules, g-actin and tropomyosin A, respectively. These proteins are also present in a less organized form in all other nucleated cells, being most prominent in contractile units such as smooth muscle and myoepithelial cells, but are also easily detectable in endothelial cells, fibroblasts (**1.38A**), macrophages and many other cell types. They can also be demonstrated in blood platelets, which are also capable of contraction (p. 67). Actin is present in its polymerized form (*f-actin*) as microfilaments which are thought to exist in more than one state, some being relatively stable, whilst others are highly labile and may be assembled or dispersed in a few seconds (Wessels *et al.* 1971). The polymer is in equilibrium with an intracellular pool of actin monomer (g-actin), so that agents which disturb this equilibrium can either cause the formation, or the dismantling, of at least some classes of microfilament. Cell membranes often form sites of attachment for microfilaments. Myosin is more difficult to demonstrate structurally, and is probably present in monomeric or dimeric form, or as small aggregates in most cells, and it is only in muscle cells that larger assemblies are permanently present. Many movements of cells are caused by the interaction of actin and myosin, as stated above. In fibroblasts, which have been intensively studied, it is proposed that myosin may form sliding cross-links between adjacent microfilaments, thus generating shearing forces which cause the network of microfilaments to extend or contract as a whole, by the mutual interaction of its constituents. If the filaments were attached to the plasma membrane, such movements would be transmitted to the whole cell, giving rise to flattening or rounding according to the direction of shear. Special, localized extensions of the actin network could form microvilli, or, if directed by the presence of orientated bundles of microtubules (**1.38B**), to which the actin may also be attached, could perhaps produce certain types of locomotion. Contraction of bundles of actin filaments by sliding interaction, is probably also responsible for the cytoplasmic cleavage during cell division (p. 30).

It has further been proposed that certain very transient structures, including microspikes, and the waves of a ruffled membrane, are created by the rapid conversion of actin monomer into a network of filaments, which is stabilized by interaction with myosin, into a finger-like process or ridge. Such structures can be easily dismantled by depolymerization, and their presence or absence depends upon local fluxes of certain ions or other cell metabolites. Actin-myosin interactions also probably occur during some types of endocytosis, such as pinocytosis and phagocytosis, where major movements of the cell surface are involved. The drug, cytochalasin-B, which causes depolymerization of some, though probably not all, types of actin, severely impairs a wide range of cell movements (Allison 1973), so that, for example, multinucleate cells are formed because of the failure of the cleavage furrow to separate cells into two during mitosis.

A number of regulatory proteins accompany actin and myosin both in muscle and non-muscle cells, and these include tropomyosin B, troponin, and actinin; these can be demonstrated clearly, often in strikingly beautiful networks, by means of immunofluorescent labelling with antibodies evoked specifically against these proteins (Lazarides 1976).

Another major protein associated with movement is tubulin, which polymerizes to form microtubules (p. 16). These organelles are prominent, for example, during spindle extension in mitotic cells, and they may be involved in the movement of neurotransmitter vesicles in synaptic endings, in axoplasmic flow, and in the movements of cilia, flagella, and of some glandular cells. In some instances microtubules can be dispersed by colchicine and other drugs which bind to the cytoplasmic pool of dimers, and thus shift the equilibrium away from the polymeric form. Under these conditions motility is often affected (Stephens and Edds 1976). In cilia and flagella motility is achieved by the sliding interaction between adjacent microtubule pairs through the mediation of another protein, dynein (p. 18) which causes bending movements. In other sites it is probable that vesicles, and other organelles or colloids, may be able to slide along the outside of microtubules by means of interacting side arms, to produce directed movements and perhaps some types of cytoplasmic streaming (1.37). Other types of microtubule-induced movement may involve the assembly and dispersal of these structures, and certainly the growth of cilia and flagella is brought about in this way.

Finally, there are a number of other movements which probably operate by different mechanisms. These include the transport of various membrane-lined vesicles from one part of the endoplasmic reticulum to another, and the formation of minute endocytotic vesicles during micropinocytosis.

The patterns of control of cell movements are also of great significance, particularly in the study of morphogenesis, teratogenesis, and various pathological states in the adult. Ionic fluxes, perhaps associated, in some cases, with the release of such metabolites as cyclic adenosine monophosphate (cyclic AMP) within the cell are associated with many types of motile control

EPITHELIUM

The term *epithelium* is applied to the layer or layers of cells which immediately cover or line the body surfaces. Embryologically, epithelia are derived from all three major cell layers: those covering topologically external surfaces (the integument, alimentary tract and the distal parts of the urogenital tracts) are of ectodermal and endodermal origin. Those lining internal cavities and the proximal parts of the urogenital tract stem from mesoderm, and are variously known as *mesothelia* (lining the pericardial, pleural and peritoneal cavities), and *endothelia* (lining vascular channels), or again *epithelia* in the case of the urogenital passages.

Most glands are of epithelial origin, since they arise as diverticula of the body surfaces. The largest of these is the liver.

Epithelia in the broad sense function as selective barriers, capable of facilitating or preventing the passage of materials across the surfaces which they cover; they may also protect the underlying tissues against dehydration, chemical and mechanical damage; they may elaborate and secrete materials into the spaces which they bound; they may function as sensory surfaces. Indeed, many features of the central nervous system suggest that nervous tissue as a whole can be regarded as a modified epithelium.

Structure of Epithelia

Epithelia (1.39) are predominantly cellular, that is, they possess only a small proportion of extracellular material. The single or multiple layers of cells rest on a mucopolysaccharide *basement membrane* (see p. 52), and the 20 nm intercellular gaps between adjacent cells are occupied by glycoproteins. Junctional complexes (p. 10) are usually numerous. The shapes of the cells, which are typically polygonal, are in part determined by their array of internal organelles, and features of their microenvironment including the mechanical forces created by close-packing within a confined volume.

Epithelia also possess a marked ability to regenerate when injured, as seen most dramatically in the liver, which is able to replace excised portions rapidly. Constant cell replacement is also important at surfaces subject to wear and tear, and most epithelia covering external surfaces show a steady rate of cell division, which offsets the loss of cells caused by abrasion. Perhaps, as a correlate of this ability, tumours of epithelial origin (papillomata and carcinomata) are of common occurrence.

Usually, blood vessels are absent from epithelia, nutrition of which depends on diffusion from capillaries of neighbouring tissues. This puts a limit on the possible thickness of living cell layers. The epithelial cells, together with their supporting connective tissue, can often be removed surgically as one 'layer', collectively known as a *membrane*. Where the surface is moistened by mucous glands it is termed a *mucous membrane*, and where it is covered with a film of thin serum-like fluid, a *serous membrane*.

Classification of Epithelia (1.39)

They can be grouped into *unilayered* or *simple* epithelia—single layers of cells resting on a basement membrane; and *multilayered* epithelia—more than one cell layer. The latter can be subdivided into *replacing* or *stratified squamous* epithelia, in which superficial cells are constantly replaced from the basal layers, and *urothelium* (*transitional epithelium*), lining parts of the genito-urinary tract, and in which cells are only replaced when patches of cell death have followed injury.

Unilayered (Simple) Epithelia

This group may be further subdivided according to the shape of the constituent cells into squamous, cuboidal, and columnar types. The shape of the cells is largely related to the distribution of their internal constituents; where little cytoplasm is present, denoting low metabolic activity, the cells are squamous or low cuboidal. Highly active cells, such as those bearing cilia or possessing secretory functions, contain abundant mitochondria and endoplasmic reticulum, and are tall cuboidal, or columnar, in shape.

Various special structures such as cilia, microvilli (brush and striated borders), secretory vacuoles (in mucous and serous glandular cells) and a sensory apparatus may also characterize particular types of epithelial cell. *Myoepitheliocytes*, which are contractile, do not form complete sheets but occur as small isolated groups of cells.

SQUAMOUS (PAVEMENT) EPITHELIUM

This is composed of flattened, interlocking, polygonal cells (*squames*). The cytoplasm may in places be as little as $0 \cdot 1 \, \mu m$ thick, and the nucleus usually bulges into the overlying space (1.39, 40). This type of epithelium lines the alveoli of the lungs, renal corpuscles, the thin segments of the nephric tubules, various parts of the inner ear, and as a *mesothelium* it forms the surfaces of the pericardial, pleural and peritoneal cavities. Here, each cell possesses one or two cilia. As *endothelium* it lines the blood vascular and lymphatic channels (p. 622).

Squamous epithelium is often associated with passive movements of water or electrolytes, but may also be active in transport, as evidenced by the numerous pinocytotic vesicles often seen in such cells. The presence of *zonulae occludentes* between adjacent cells (*see* Staehelin 1975) ensures that materials pass primarily through the cells rather than between them, except where they are specially punctuated by holes, as in the fenestrated endothelia of the renal glomeruli.

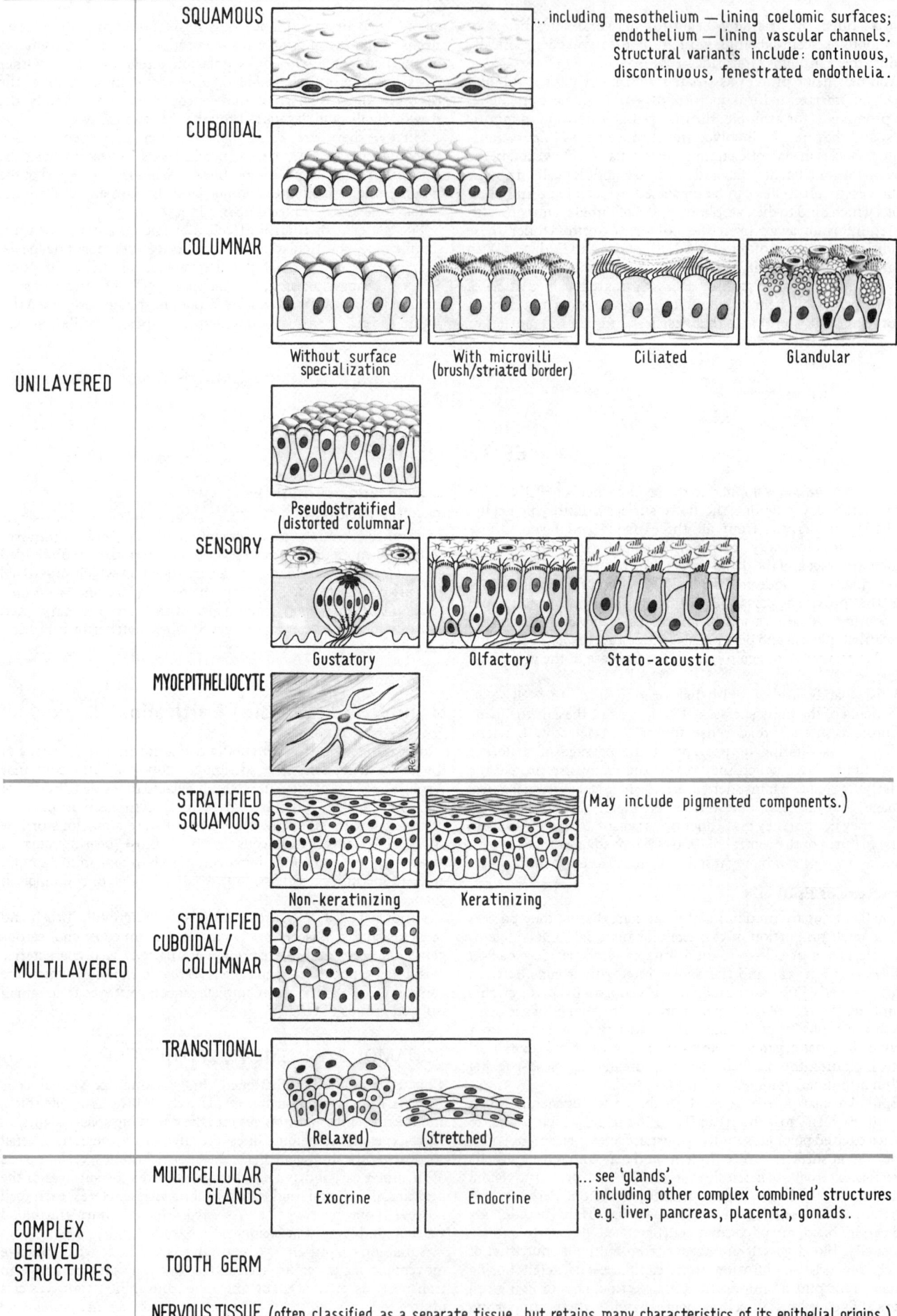

SQUAMOUS

... including mesothelium — lining coelomic surfaces; endothelium — lining vascular channels. Structural variants include: continuous, discontinuous, fenestrated endothelia.

CUBOIDAL

COLUMNAR

Without surface specialization

With microvilli (brush/striated border)

Ciliated

Glandular

UNILAYERED

Pseudostratified (distorted columnar)

SENSORY

Gustatory

Olfactory

Stato-acoustic

MYOEPITHELIOCYTE

STRATIFIED SQUAMOUS

(May include pigmented components.)

Non-keratinizing

Keratinizing

STRATIFIED CUBOIDAL/ COLUMNAR

MULTILAYERED

TRANSITIONAL

(Relaxed)

(Stretched)

COMPLEX DERIVED STRUCTURES

MULTICELLULAR GLANDS

Exocrine

Endocrine

... see 'glands', including other complex 'combined' structures e.g. liver, pancreas, placenta, gonads.

TOOTH GERM

NERVOUS TISSUE (often classified as a separate tissue, but retains many characteristics of its epithelial origins.)

1.39   A classification of the major types of epithelia and associated tissues described in the ensuing section.

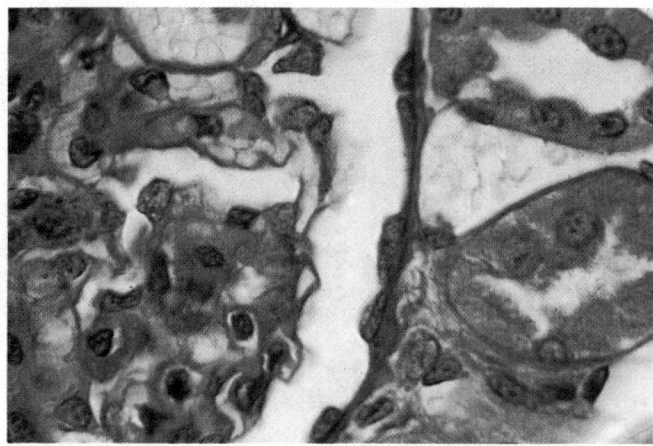

1.40 Part of a renal corpuscle showing vertically sectioned simple squamous epithelial cells in Bowman's capsule. Masson's trichrome stain.

CUBOIDAL AND COLUMNAR EPITHELIA

These consist of cylindrical cells set together to form a palisade-like layer (**1.41–48**). Each cell is polygonal in horizontal section; cuboidal cells are square in vertical section, whereas columnar cells are taller than their diameter. Commonly, microvilli are found on the free surface of such cells, providing a large absorptive area (p. 18), as in the epithelium of the small intestine (columnar cells with a striated border), the gall bladder (columnar cells with a brush border) and the proximal and distal convoluted tubules of the kidney (large cuboidal cells with brush borders). Ciliated columnar epithelium lines most of the respiratory tract (**1.44, 45, 46, 47**) as far as the respiratory bronchioles (excepting

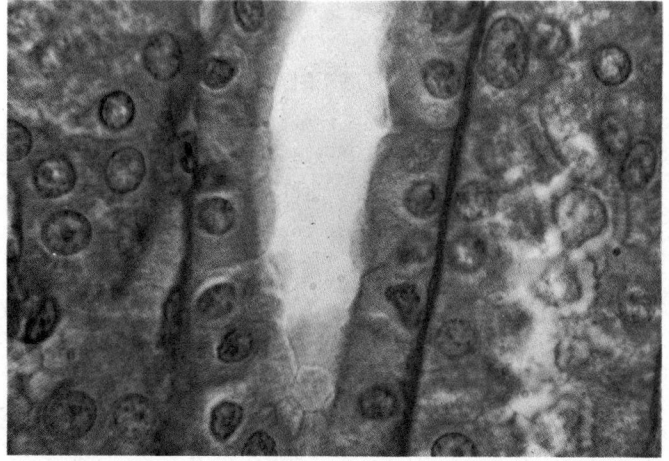

1.41 Simple cuboidal epithelium in renal collecting ducts. Masson's trichrome stain.

the lower pharynx and vocal folds), some of the tympanic cavity and auditory tube, the uterine tube and patches in the cavity of the infantile uterus, and the efferent ductules of the testis.

In the surfaces of the respiratory tract there are also mucous glands, and the cilia sweep a layer of viscous fluid and trapped dust particles, etc. from the lung towards the pharynx (the *rejection current*), so clearing the respiratory passages of inhaled particles. In the uterine tube, ciliary action assists the passage of ova from the peritoneal cavity to the uterus.

Some columnar cells are glandular, and their apical regions contain mucus or protein-carrying vacuoles. The mucin-secreting cells, and the protease-secreting chief cells of the gastric epithelium are typical examples. Often the mucous secretory cells are interspersed between non-secretory ones, and this allows the expansion of the apices of the secretory cells, giving a characteristic cell shape. The latter are known as *goblet*, *chalice* or *calciform* cells (**1.42, 43**), and they are particularly abundant in the intestinal epithelium.

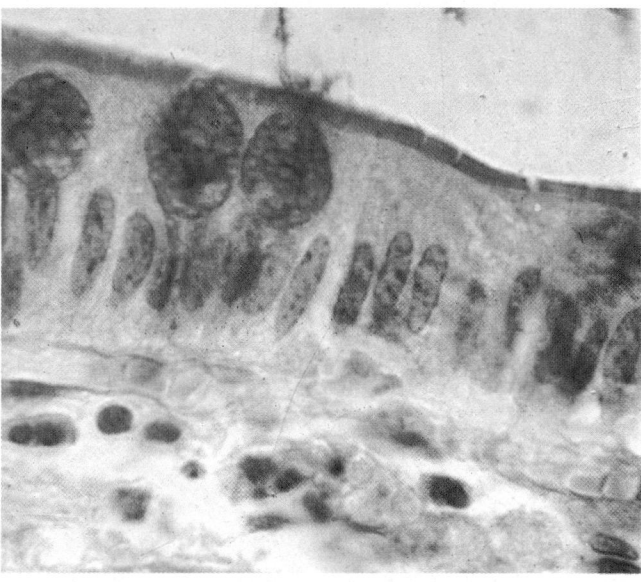

1.42 Simple columnar epithelium of the small intestine showing absorptive cells with a striated border and goblet cells. PAS stain.

Mucus is a viscous suspension of complex glycoproteins containing both sulphated and carboxylated glycosaminoglycans. Their long polymeric chains bind water and so protect surfaces against drying. They can also slide over one another with ease, providing good lubricating properties, and their negative charges may also assist in binding cations such as Na^+.

The synthesis of mucus has been followed in goblet cells by light and electron microscope autoradiography (Peterson and Leblond 1964). Protein synthesized in the rough endoplasmic reticulum is transported to the Golgi apparatus, where it is conjugated with sulphated carbohydrates to form the glyco-protein, *mucinogen*. This then passes out in small dense vesicles which swell as they approach the cell surface, and finally fuse with it, to release the mucus.

PSEUDOSTRATIFIED EPITHELIUM

This is a simple columnar epithelium in which the regular arrangement of the cells is such that their nuclei lie at different levels in a vertical section. The cells may be twisted with respect to each other so that only partial profiles of some cells appear in vertical section, giving the false appearance of more than one layer of cells. Columnar cells in many situations assume this appearance if subjected to lateral compression forces (**1.48**). In other cases some cells do not extend through the whole thickness of the epithelium, and may constitute a basal cell layer, often capable of cell division. It is also possible for migrating lymphocytes within a columnar epithelium to give the appearance of a pseudostratified epithelium because their nuclei are placed at different depths.

SENSORY EPITHELIA

These epithelia are restricted to the special sense organs of the olfactory, gustatory and vestibulo-cochlear receptor systems. All of them contain sensory cells surrounded by supportive, non-receptive cells. The olfactory receptors are modified neurons, their axons passing directly to the brain, but the other types are specialized epithelial cells synapsing with the terminals of afferent (and sometimes efferent) nerve fibres. They will be considered further in *Section 7 (Neurology)*.

MYOEPITHELIOCYTES

Also sometimes termed *basket cells*, these are dendritic in form, containing actin and myosin filaments, and are capable of contraction when suitably stimulated by a nervous or neuro-hormonal signal. They are present around the secretory portions

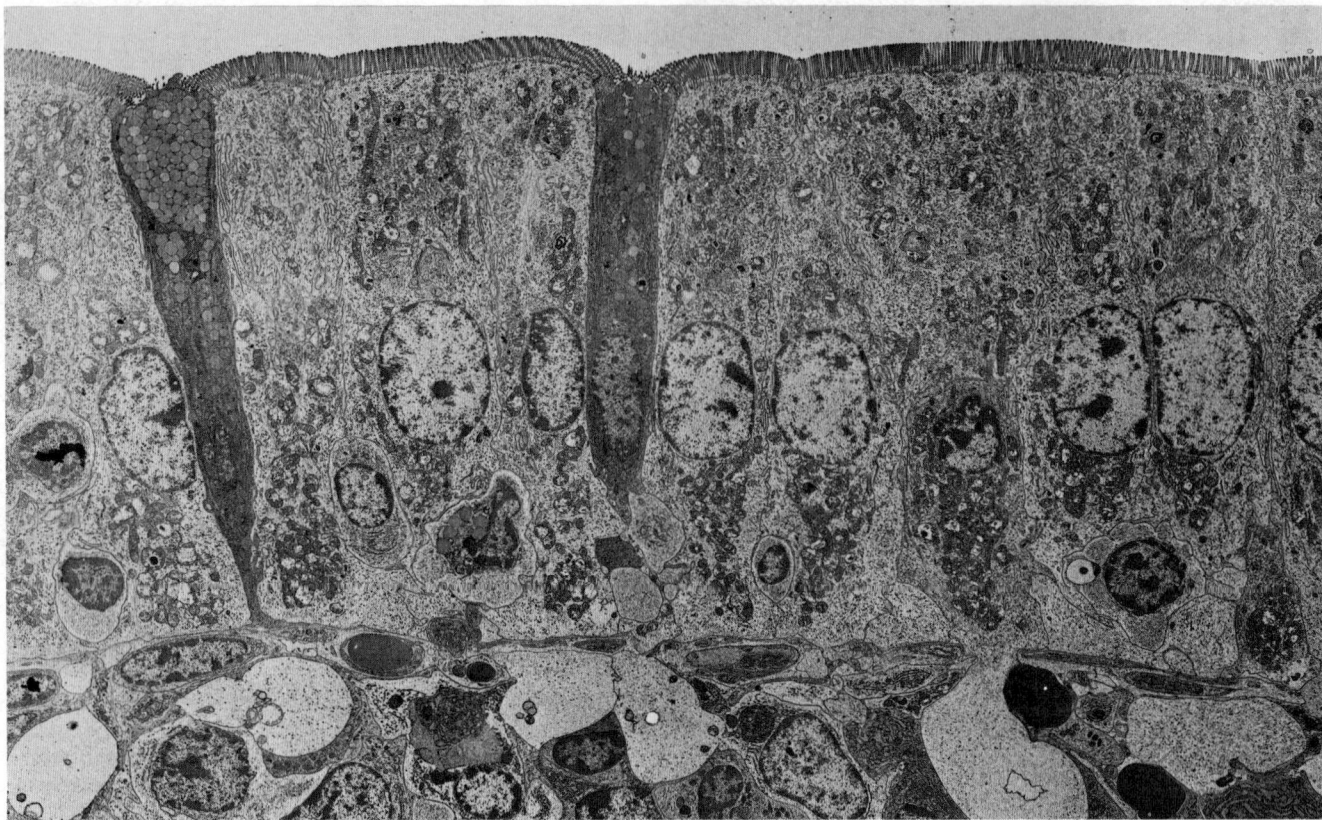

1.43 A low-power electron micrograph of a vertical section through simple columnar epithelium bearing microvilli; two goblet cells are also present. Note the presence of several small lymphocytes near the epithelial base. Small intestine. Magnification × 8,000. Kindly provided by Mr. Derrick Lovell, Guy's Hospital Medical School.

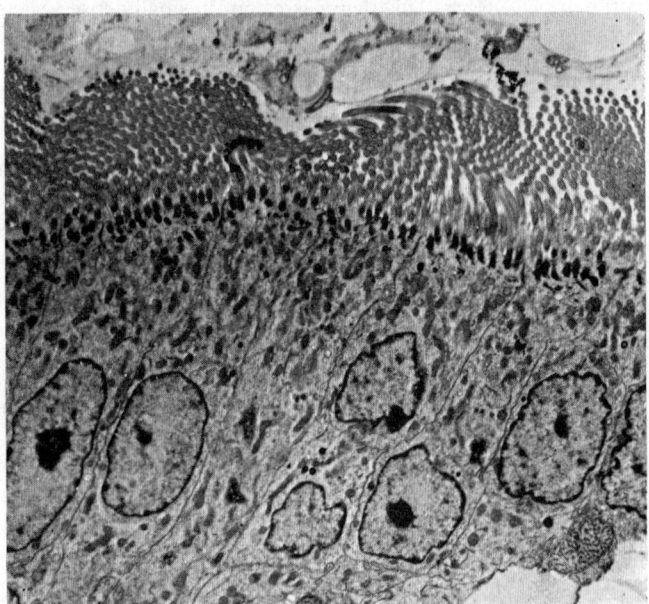

1.44 Low-power electron micrograph of simple ciliated columnar epithelial cells from the nasal mucosa. Magnification × 4,000.

and ducts of some glands, for example, mammary and salivary glands, where they assist the initial passage of secretion into the larger channels.

Multilayered Epithelia

These constitute surfaces where there is considerable wear and tear; their chief characteristic is the continued replacement of superficial cells by division of more deeply placed cells.

STRATIFIED SQUAMOUS EPITHELIUM

This type of epithelium consists of several layers of cells which vary greatly in shape. Those of the deepest layer are, for the most part, columnar and are placed vertically on the basement membrane. Superficial to these, the succeeding layers consist of polyhedral cells which become more and more compressed as they approach the surface; the most superficial cells are flattened scales (*squames*) which overlap one another and so present an imbricated appearance. The cells of the deepest layers proliferate and undergo a progressive change in structure as they pass towards the surface, where they are continually lost by abrasive wear and tear.

In stratified squamous epithelia the cells are closely bound to each other by desmosomes (p. 7), and their cytoplasm is pervaded by tonofibrils which act as an internal skeleton to strengthen the cell. Where surfaces are moist, the cells of the superficial stratum, though flattened and degenerating, still retain their nuclei. At dry surfaces they lay down the protein keratin and lose their nuclei. The keratin-filled squames effect waterproofing and afford mechanical protection. The first of these two types, termed *non-keratinizing* epithelium (**1.49**), is found in the mucous membranes of the mouth, lower pharynx, oesophagus, vagina, part of the cervix uteri, and in the conjunctiva and anterior surface of the cornea. The second type, *keratinizing* epithelium, forms the epidermis of the skin (**1.50**) and occurs in parts of the mouth. The degree of keratinization is dependent upon the amount of mechanical stress and dehydration experienced by the epithelium, and also on dietary factors. Diets deficient in vitamin A induce keratinization in many otherwise non-keratinizing epithelia, such as that of the cornea.

It has recently been shown that the 'parent' basal cell and its progeny (together forming an epidermal 'proliferative unit') have a very precise geometric relationship. Each mitosing basal cell forms a single vertical column of cells which ascends in a somewhat spiral manner because of the prismatic shape of its cellular elements (Potten *et al.* 1976).

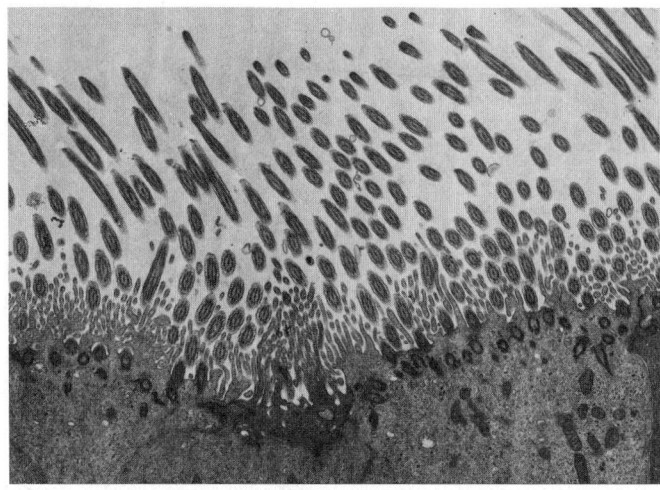

1.45A Detail of a group of cilia cut in various planes at the surface of a ciliated columnar epithelial cell from the upper respiratory tract. Note the presence of small microvilli interspersed between the cilia. Magnification ×7,500.

Epidermis

The epithelium (epidermis) of the skin (**1.50, 51**) shows several distinct layers of cells, the *keratinocytes*, arranged in two *zones* (Montagna and Parakkal 1974). The deeper of these, *zona germinativa*, consists of a single layer of columnar cells, the *stratum basale* (*stratum germinativum* or *Malpighian layer*), and a more superficial layer of variable thickness composed of polyhedral cells, the *stratum spinosum*. In the stratum basale the cells are placed perpendicularly on a basement membrane to which they are attached by half-desmosomes. Most of the mitoses are confined to this layer, and occasional pigment cells (*melanocytes*) of neural crest origin, which spread dendritic processes amongst the more superficial cells, are located here. The pigment produced protects the mitotic layer from adverse

effects of ultraviolet radiation. Pale staining *Langerhans* cells, of possible phagocytic function, are also situated in this layer.

In the *stratum spinosum* the cells have a complex highly folded surface, covered with desmosomes which are associated with cytoplasmic tonofibrils. In sections of paraffin-embedded skin, these cells shrink away from each other except at the desmosomes, so giving the appearance of cells studded with spines ('prickle' cells). Mitosis also occurs in the deeper layers of this stratum.

In the more superficial zone, *zona cornea*, there are three strata of cells. As the cells are pushed upwards from the stratum spinosum, they flatten and synthesize *keratohyalin granules* which stain deeply with basic dyes. These cells constitute the *stratum granulosum*. As they move outwards, these cells lose their nuclei and the keratohyalin granules fuse and mingle with the tonofibrils, changing to fibrous *keratin*, composed of fine filaments, and surrounded with lipid (*vide infra*). The cells first form a clear layer, the *stratum lucidum*, which is prominent only in the thick skin covering the palms and soles. The cells become compressed into opaque squames which are now full of the horny protein *keratin* and tonofibril remnants. These squames constitute the outer layer, the *stratum corneum*.

Keratohyalin is a protein which is rich in proline and sulphur-containing amino acids; it is a precursor of keratin, a highly insoluble and mechanically stable fibrous protein which provides skin with its waterproofing qualities. In skin, keratin adopts the α-configuration in which the molecules have a pleated or helical form. These can be cross-linked by disulphydryl bonds between adjacent cysteine groups, so giving great tensile strength. This type of configuration is particularly prominent in hair, which, as a derivative of the epidermis, is also highly keratinized. As final stabilization occurs in the stratum corneum, each keratin strand is covered with a layer of lipid, in the absence of which

1.46 A low-power scanning electron micrograph of the ciliated surface of the trachea. Magnification × 3,000.

keratinization may be defective. The epidermis will be further considered in *Section 7*, p. 1216.

Stratified Cuboidal and Columnar Epithelia

Two or more layers of cuboidal or columnar cells are typical of the walls of the larger ducts of some exocrine glands, such as the pancreas and salivary glands, presumably affording more strength than a single layer. In some regions, for example, some parts of the epiglottis, the covering stratified columnar epithelium is also ciliated.

1.45B A high-power view of cilia in a vertical section. Note particularly the details of the ciliary bases. Magnification ×40,000.

41

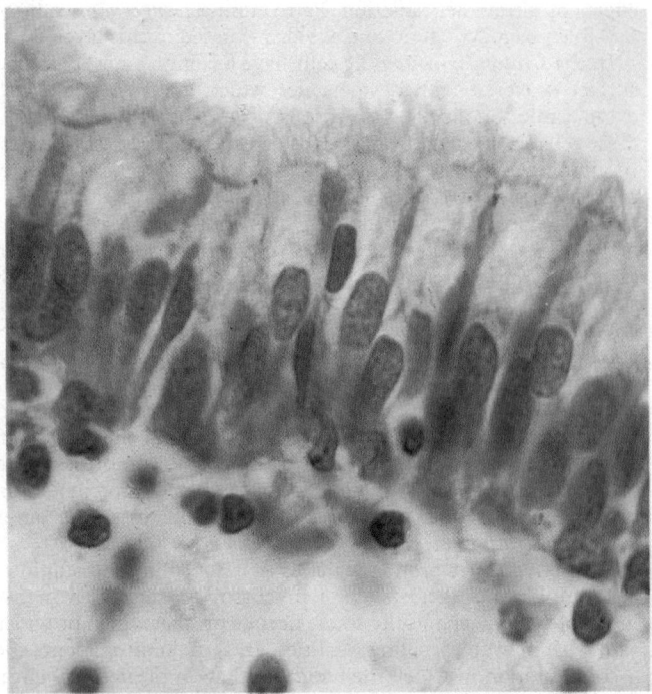

1.47 Simple ciliated columnar epithelium from the trachea of a rat. Magnification × 1,200.

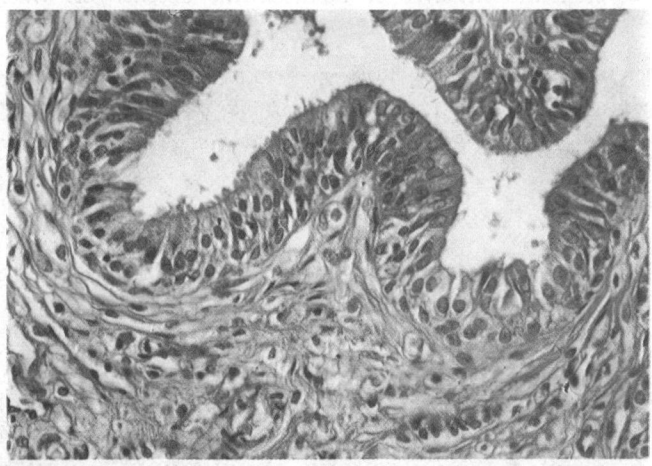

1.48 Pseudostratified epithelium from the male urethra. Haematoxylin and eosin stains. Magnification × 250.

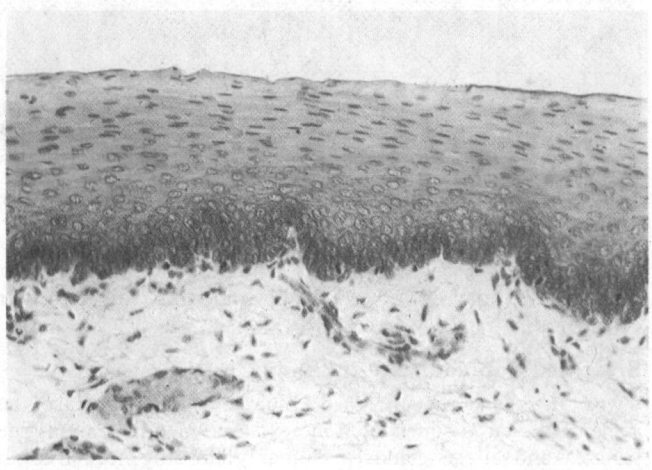

1.49 Non-keratinizing, stratified squamous epithelium from the human tongue. A vertical section stained with haematoxylin and eosin. Note the presence of nuclei in the surface cells. Magnification × 150.

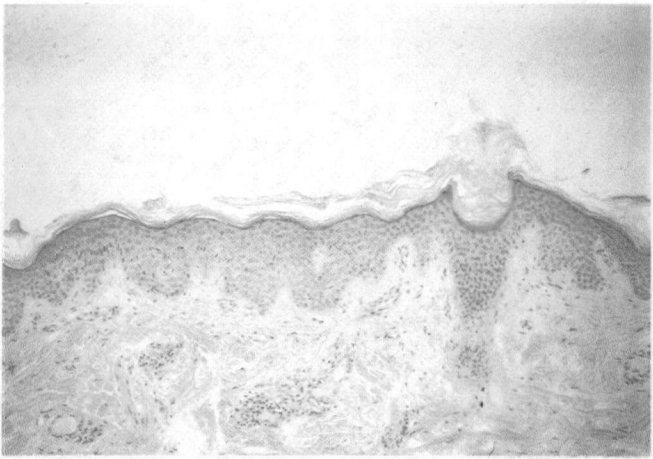

1.50 A vertical section of the epidermis from the thick skin of the human foot showing germinative, spinous, lucid, granular and keratinized layers. Mallory's trichrome stain. Magnification × 125.

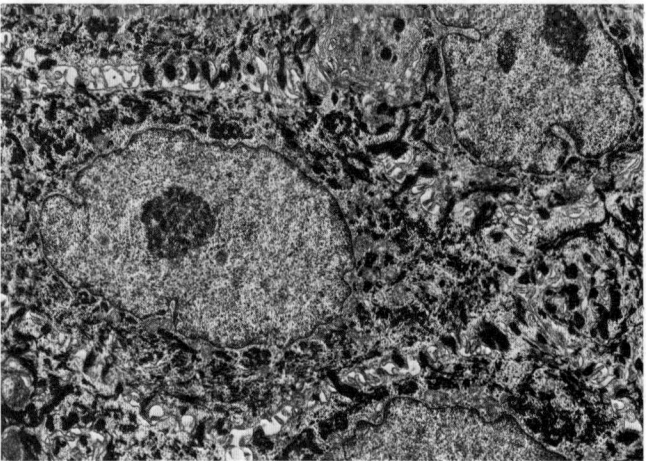

1.51 Electron micrograph of 'prickle' cells in the stratum spinosum of the human skin showing interdigitating surfaces, frequent desmosomes and prominent intracellular tonofibrils. Magnification × 4,000.

Urothelium (Transitional Epithelium)

This epithelium (**1.39, 52, 53A–C**) forms the characteristic lining of much of the urinary tract, extending from the ends of the collecting ducts of the kidneys, through the ureters and bladder to the distal end of the urethra. During development part of it is derived from mesoderm and part from ectoderm and endoderm. The epithelium is 4–6 cells thick, and lines organs which undergo considerable distension and contraction; it is therefore capable of being greatly stretched without losing its integrity. During such movements, the cells become flattened, or more rounded, without altering their positions relative to each other, since they are firmly connected by numerous desmosomes (*see* review by Hicks 1975). When relaxed, the basally situated cells are approximately cuboidal; they are uninucleate (diploid) and basophilic with many cytoplasmic ribosomes. More apically the cells progressively fuse to form larger, binucleate, or uninucleate but polyploid cells. The cells are largest at the surface, where they may even be octoploid, and are bounded over their luminal surfaces by a plasma membrane bearing plate-like aggregates of glycoprotein particles which are embedded in its lipid bilayer. These arrays confer a measure of stiffness on the membrane, so that when the epithelium is in the relaxed state, and the surface area of the cells is reduced, the glycoprotein-lipid plates are partially taken into the cytoplasm within vacuoles or diverticula, re-emerging on to the surface when its area increases once more with the accumulation of urine.

These unusual membranes, together with the tight junctions which surround the apices of the surface cells, form an effective barrier preventing urine, or its constituents, from passing into the epithelium or beyond into the adjacent tissues. The urothelium therefore creates a protective lining to the urinary system which prevents the rather toxic contents of the urine damaging surrounding structures (Hicks *et al.* 1974).

Normally, cell turnover is very slow in urothelium, and cell divisions, which are restricted to the basal layer, are infrequent. When damaged, however, the epithelium regenerates quite rapidly (Annis 1962).

Pigmented epithelial cells are found in various parts of the body. As layers of columnar to cuboidal cells, they form the external layer of the retina, and are present in the posterior epithelium of the iris. The pigment granules are very small in size, are crowded together within the cells, but do not invade the nuclei. *Melanocytes* of neural crest origin are different in shape, being *dendritic* rather than columnar or cuboidal. They are scattered intermittently, or form small pigmented patches, in the membranous labyrinth of the ear, in the deeper layers of the epidermis, and in hairs. Melanocytes synthesize the pigment

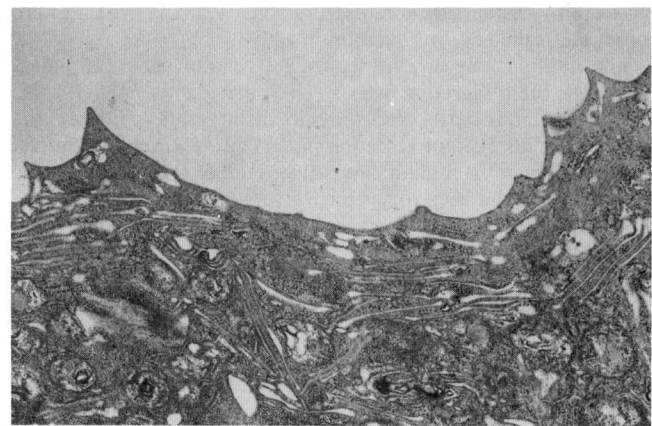

I.53A Transmission electron micrograph of the surface of the urothelium (transitional epithelium) lining the relaxed bladder. Note the angular profiles of the epithelial surface. Magnification × 15,000.

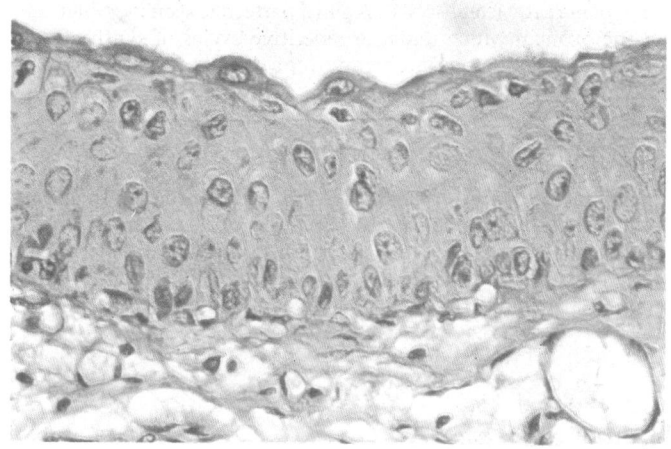

I.52 A vertical section through the surface of a ureter to show the urothelium lining its lumen, stained by Mallory's triple staining technique. Magnification × 600.

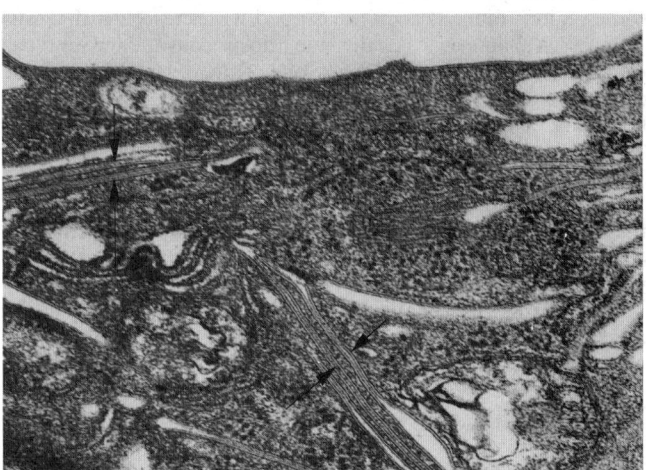

I.53B Detail of I.53A showing the plate-like modifications of the plasma membrane and its partial 'internalization' (arrows) in the relaxed state. Magnification × 30,000.

melanin, which is brown-black, and *phaeomelanin* of an orange hue. They will be further discussed in *Section 7* (p. 1219).

Glands

Glands are epithelial derivatives in which the component cells elaborate a secretion. They may lie in and discharge directly on to the epithelial surface from which they originate, for example goblet cells, or be depressed below the surface and discharge on to this by means of a duct. Such glands are termed *exocrine*. In other glands, which are termed *endocrine*, the cells have become detached from the epithelial surface, no duct exists (hence the term *ductless glands*) and the secretion is discharged into the blood and lymph streams.

Exocrine glands may be *unicellular* or *multicellular*. In the former case the secretory cell is incorporated in the epithelial surface on to which it secretes. In *multicellular glands* the cells are usually depressed below the surface, to which they remain connected by one or more ducts. When the glandular epithelium discharges directly into one duct it is called a *simple* gland; when the glandular epithelium discharges through minor into main ducts it is a *compound* gland. The secretory portion can be a *simple*, *straight*, *coiled* or *branched tube* of about the same diameter as the duct, a rounded sac termed an *acinus*, or a flask-shaped sac termed an *alveolus*. Thus there can be distinguished simple and compound tubular, acinar or alveolar glands. Intermediate types, such as tubulo-acinar and tubulo-alveolar, also occur. Glands

I.53C Scanning electron micrograph of the (relaxed) urothelial surface showing the plate-like arrangement of its plasma membrane. Magnification × 6,000.

with highly distended secretory portions (e.g. the prostate) are sometimes called *saccular* glands.

The method whereby the epithelial cells of exocrine glands discharge their secretion varies. In some glands, such as the sebaceous glands in the skin, the cells disintegrate to liberate the secretion. This type of secretion is termed *holocrine*. In others, such as the mammary gland, a small amount of cytoplasm may be released with the secretory vesicles (*apocrine* secretion). More frequently only the secretion is discharged; such secretions are termed *merocrine* or *epicrine*.

Glands are often classified by the type of secretion they produce; a common division of exocrine glands is made into *mucus-secreting* or *mucous* glands, where the cells possess frothy cytoplasm with basal flattened nuclei, and which stain with metachromatic stains and PAS methods, and *serous glands* where the cells have centrally placed nuclei, granular cytoplasm, and which synthesize and secrete proteins (e.g. lysozyme, a bactericide, or the digestive enzymes).

Some glands are entirely mucus-secreting (e.g. the sublingual salivary glands), whilst others are mainly serous (e.g. the parotid salivary glands). The submandibular gland is mixed, some lobules being predominantly mucous whilst others are serous. In some regions mucous acini are capped with crescents of serous cells (*serous demilunes*). The division into mucous and serous types is, however, a rather imprecise one, since even within these categories a wide range of secretory products exist, and such terms are of limited usefulness outside certain narrow contexts (e.g. salivary glands) as considered later in greater detail (p. 1276).

Exocrine glands are usually in part divided by connective tissue septa into *lobules*; small ducts draining the lobules pass to the septa and join to form the major duct or ducts. Glands are characterized by a rich vascular supply, which may be correlated with their high metabolic rate.

Endocrine glands are composed of groups of cells lying in close proximity to vascular channels; they mostly occur in clumps or cords of closely-knit cell groups supported by a network of fine reticulin fibres and associated cells; in the thyroid gland they form hollow balls of cells, or *follicles*, in the centres of which synthesized products can be stored. Endocrine glands have a particularly well developed circulation, and often the capillary vascular endothelium is *fenestrated* (p. 628), presumably facilitating the diffusion of large molecules of hormone into the blood.

The fine structure of gland cells is typified by the presence of an extensive endoplasmic reticulum. In protein-synthesizing cells such as goblet and pancreatic acinar cells, high concentrations of polyribosomes are attached to the rough endoplasmic reticulum, whereas in steroid-secreting cells (e.g. adrenal cortex) the elaborate endoplasmic reticulum is mainly smooth.

In salivary and mammary glands, the *release* of secretion is accelerated by the neurally elicited contraction of *myoepithelial cells* which surround alveoli and the initial segments of their ducts.

The patterns of synthesis and release of secretions by glandular cells varies considerably (Smith 1972). In some, synthesis and release occur continuously (but with quantitative fluctuations) as in the adrenal cortex. In others, synthetic products may be stored for long periods, only being released when suitably stimulated (e.g. pancreatic islet β-cells). A third pattern is seen in goblet cells of the bronchi which undergo repetitive cycles of synthesis and release without the intervention of any external triggering mechanism.

COMPLEX DERIVED STRUCTURES

These include those organs which are derived from epithelia and retain their highly cellular nature; they often possess typical epithelial features as secretory, absorptive and transport functions. Such structures include the liver, placenta, the early tooth germ, and the nervous system as a whole.

THE CONNECTIVE TISSUES

Introduction

The connective tissues may be defined as that group of elements, derived largely from embryonic mesoderm, where there is a considerable proportion of intercellular material secreted primarily by the cells themselves, which are consequently quite widely spaced. Many of the special properties of connective tissues are determined by the precise composition of the intercellular substances, and their classification is also based on its characteristics.

Connective tissues play several essential roles in the body (*see* Wagner and Smith 1967), both *structural*, since many of the extracellular elements possess special mechanical properties, and *defensive*, a role which has a cellular basis (p. 45). Connective tissues are conveniently divided into 'ordinary' types, which are distributed widely throughout the body, and special types, namely cartilage and bone—these are described elsewhere (pp. 238–66), although the same principles apply to both groups.

Connective tissue is thus formed of two components: *cells* and *extracellular matrix*. The matrix in turn is composed of *fibres* and an amorphous viscous *ground substance*. (Some authors apply the term 'matrix' to ground substance alone, but it will not be used in this manner here.) It should also be mentioned that a number of the cell types of connective tissue are also found in the circulating blood and lymph, and there is a dynamic equilibrium between the two.

The Cells of Connective Tissue

Unspecialized connective tissue consists of six principal types of cell and associated matrix—fibroblasts, macrophages, plasma cells, mast cells, fat cells and pigment cells (**1.54**). In addition, lymphocytes and neutrophil and eosinophil leucocytes form a variable population, which may increase dramatically in various pathological states.

Embryologically most of the cell types arise from relatively undifferentiated mesenchymal stem cells. Some of the latter are believed to remain in the tissues, providing a postnatal source of new cellular elements.

FIBROBLASTS

These are usually the most numerous cells. They are flattened and irregular in outline, with branching processes; in profile they appear fusiform or spindle-shaped (**1.55**). Their nuclei are relatively large, active or euchromatic (open-faced), and possess prominent nucleoli. The cytoplasm is clear, with a few fine granules and occasional fat droplets. In young and active cells the cytoplasm is abundant and basophilic because of the high concentration of rough endoplasmic reticulum (Hall and Jackson 1968). In old and inactive fibroblasts (often termed *fibrocytes*) the cytoplasm is sparse, the endoplasmic reticulum scanty, and the nucleus flattened and heterochromatic (close-faced). In active fibroblasts, mitochondria are abundant and the Golgi complex well represented.

Fibroblasts are concerned with the production of the extracellular materials (*vide infra*), and they are intimately associated with white collagen fibres, to which they often adhere. In some situations, such as glands, fine fibres of *reticulin* are produced instead of collagen. In other situations, such as the spleen, *reticular cells* may lay down reticulin as well as acting as phagocytes; the relationship between such cells, ordinary fibroblasts, and lymphopoietic stem cells (p. 770) is not yet clear.

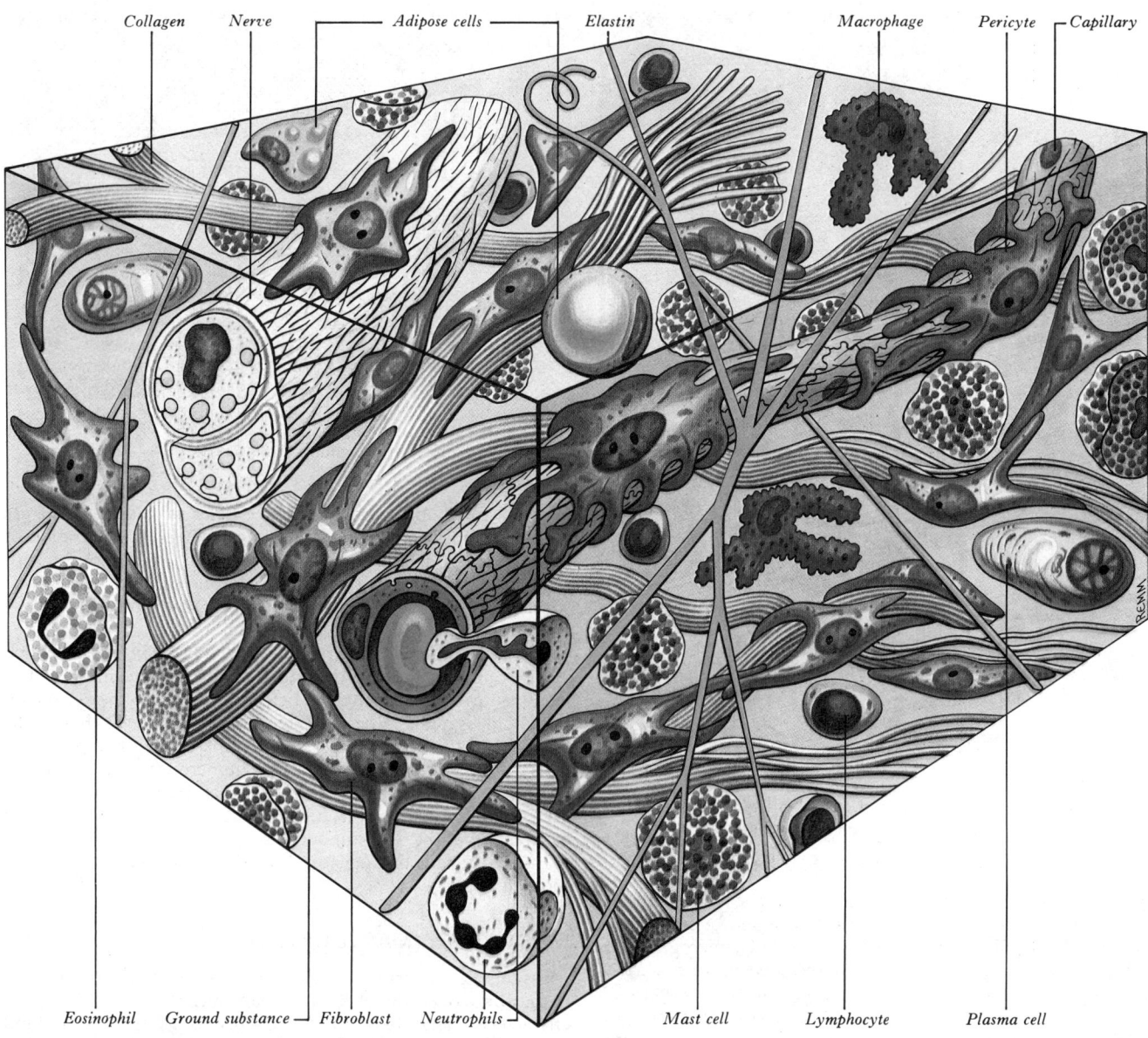

Collagen   Nerve   Adipose cells   Elastin   Macrophage   Pericyte   Capillary

Eosinophil   Ground substance   Fibroblast   Neutrophils   Mast cell   Lymphocyte   Plasma cell

1.54   A diagrammatic reconstruction of loose connective tissue showing the characteristic cell types, fibres and intercellular spaces.

Fibroblasts are particularly active during wound repair, and they multiply to produce large numbers of cells embedded in fibrovascular *granulation tissue* (*see* Ross 1968). In such situations, mesenchyme cells also may give rise to new generations of fibroblasts and other cell types, although it is thought by some that mesenchyme cells are absent, and that the fibroblasts themselves are capable of acting as stem cells. The contraction of wounds is at least in part caused by the shortening of specialised contractile fibroblasts (myofibroblasts) which arise in such areas (Gabbiani *et al.* 1973). Fibroblast activity is influenced by various factors such as steroid hormone levels, and dietary content. In vitamin C deficiency there is an impairment of collagen formation.

MACROPHAGES (HISTIOCYTES, CLASMATOCYTES)

These are also numerous in connective tissues. Their nuclei are smaller and more heterochromatic than those of fibroblasts, and are often indented on one side (**1.**56, 57). The cytoplasm contains variable numbers of granules and vacuoles. Macrophages may be either attached to the fibres of the matrix (*stationary* or 'fixed' macrophages), when their outlines are irregular and bear numerous filopodia, or they may be motile or *nomadic*, when the cell is more rounded.

Macrophages are, of course, phagocytic, forming part of the

reticuloendothelial (or *mononuclear phagocyte*) *system* (p. 765). They are capable of engulfing and digesting particulate organic materials such as bacteria and other foreign bodies, and also of disposing of damaged tissues prior to their regeneration.

Ultrastructurally, macrophages contain numerous lysosomes which are active in hydrolysing phagocytosed particles (p. 14). Inert materials such as small particles of carbon or metals may also be taken up. This quality is useful in demonstrating macrophages histologically, since their cytoplasm becomes filled with ingested particles if an experimental animal is previously injected with a suspension of India ink (**1.**57), trypan blue or lithium carmine (vital staining). Macrophages may also be separated magnetically from mixed cell samples by first treating them with iron carbamyl, which they ingest.

Macrophages are also involved in the immunological reactions of the body (p. 62). Antigens adhering to macrophages may (directly or after modification) be passed to, and thus stimulate, neighbouring immunologically competent cells (p. 265). They may selectively phagocytose particles previously coated by antibodies (*opsonins—see* p. 62) synthesized by lymphocytes, and they are themselves sites for *cytophilic antibody* attachment, which enables them to recognize and attack foreign substances (see also monocytes, p. 59).

Many properties of macrophages are similar to those of a number of specialized cell types in other sites, particularly   **45**

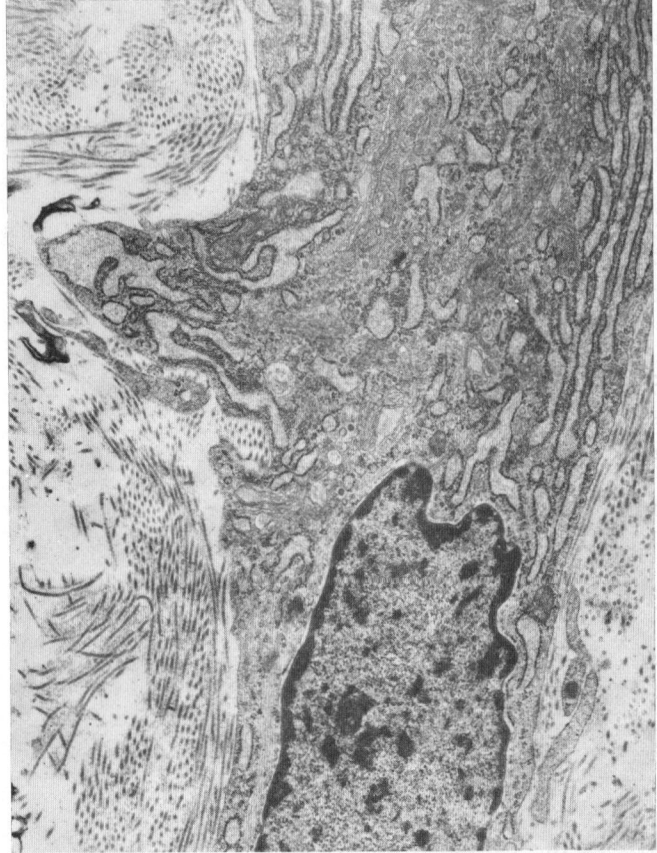

1.55 A transmission electron micrograph of a fibroblast. Note the abundant endoplasmic reticulum, and extensive Golgi complex. Some extracellular collagen fibres are also visible. Magnification × 15,000.

circulating monocytes, alveolar phagocytes in the lungs, reticulo-endothelial cells in the lymph nodes, spleen and bone marrow, endothelial (Kupffer) cells of the liver sinusoids, microglial cells of the brain and possibly osteoclasts of the bone marrow. Various cell-labelling experiments have shown that most, or perhaps all, the precursors of these arise in the bone marrow as monocytes which then pass in the circulation to their final destinations, where, however, they continue to divide.

All macrophages are capable of motility, when suitably stimulated. When grouped around a large foreign body, macrophages may also fuse together to form *syncytial giant cells* and *epithelioid cells*.

LYMPHOCYTES AND PLASMA CELLS

These cells are numerous in connective tissue only in pathological states. *Lymphocytes* may invade an area from adjacent lymphoid tissue or from the circulation. The majority are small cells (6–8 μm) with rounded, highly heterochromatic, often deeply indented nuclei (p. 20, and *see* 1.58, 68, 69). When appropriately stimulated they enlarge, developing numerous ribosomes. *Two* major *functional classes* of lymphocyte exist. The first type is derived from the bone marrow via various lymphoid tissues ('B' *lymphocytes*) and, when antigenically stimulated, synthesizes antibody in the extensive arrays of granular endoplasmic reticulum that it develops as it differentiates and matures into a *plasma cell* (1.58B, C) in lymph nodes and elsewhere. Mature plasma cells, which are incapable of cell division are oval or round in shape, and are virtually filled with rough endoplasmic reticulum except for a well marked Golgi apparatus. They produce large quantities of antibody which may be released locally, or widely dispersed via the circulation, or some may be temporarily stored in large vacuoles clearly visible with a light microscope as *Russell bodies*, which sometimes contain crystalline inclusions. (Some observers consider the presence of such bodies to indicate that the plasma cell in question is ageing and has a low

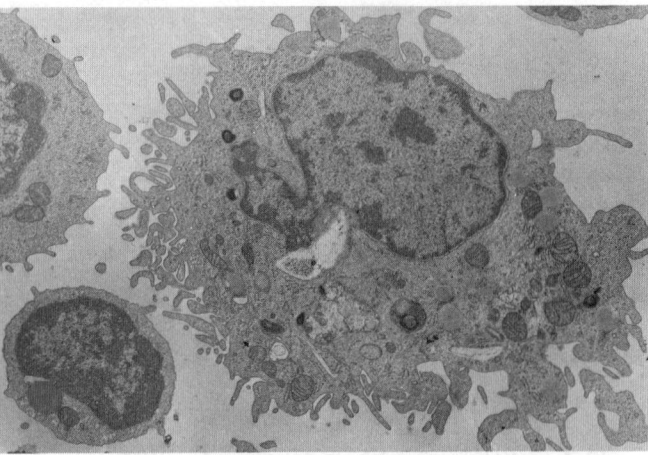

1.56 An electron micrograph of a macrophage (right) and a small lymphocyte (lower left). Magnification × 6,000.

level of functional activity.) The nuclei of plasma cells are spherical and have a characteristic 'cartwheel' configuration of heterochromatin. The prominent Golgi complex is also seen with a light microscope as a pale region to one side of the nucleus, whilst the remaining cytoplasm is deeply basophilic due to the abundant rough endoplasmic reticulum.

The second type is the thymus-derived *'T'-lymphocyte*, which stems from cell stocks initially formed by the bone marrow, but later sojourning in the thymus before passing into the peripheral lymphoid system. When antigenically stimulated, these cells enlarge and their cytoplasm becomes filled with 'free' polysome clusters. The functions of T cells are numerous and incompletely understood, but include the recognition and destruction of virus-infected cells, tumour cells, fungi, tissue and organ grafts, and the modulation of the 'B'-lymphocytes.

Further details of the natural history of lymphocytes may be found on pp. 59–63.

MAST CELLS (MASTOCYTES)

Mast cells occur particularly in loose connective tissues and often in the fibrous capsules of certain organs such as the liver. They are characteristically sited around blood vessels. Mast cells are round or oval, with many filopodia extending from the cell surface (1.59). The nucleus is centrally placed and relatively small, being surrounded by large numbers of prominent vesicles, a well developed Golgi apparatus, but scanty endoplasmic reticulum. The vesicles show a strongly positive reaction with the periodic

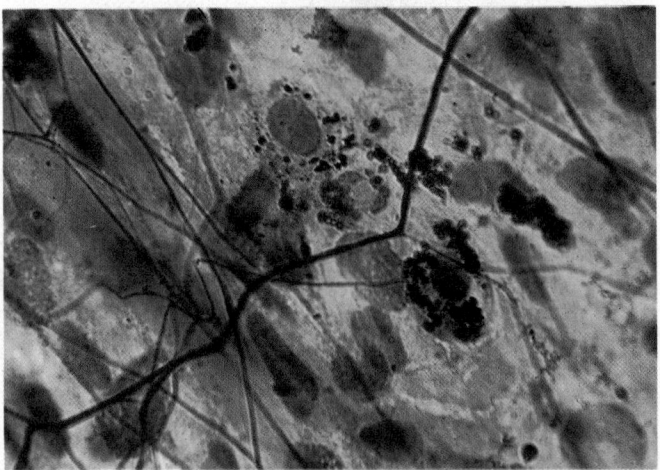

1.57 Loose connective tissue in the mesentery of a rabbit which had previously been injected intraperitoneally with India ink, showing fibroblasts and macrophages. The cytoplasm of the macrophages is full of phagocytosed particles, the collagen is stained pink and the elastin fibres black. Van Gieson's and Verhoeff's elastin stain. Magnification × 1,000.

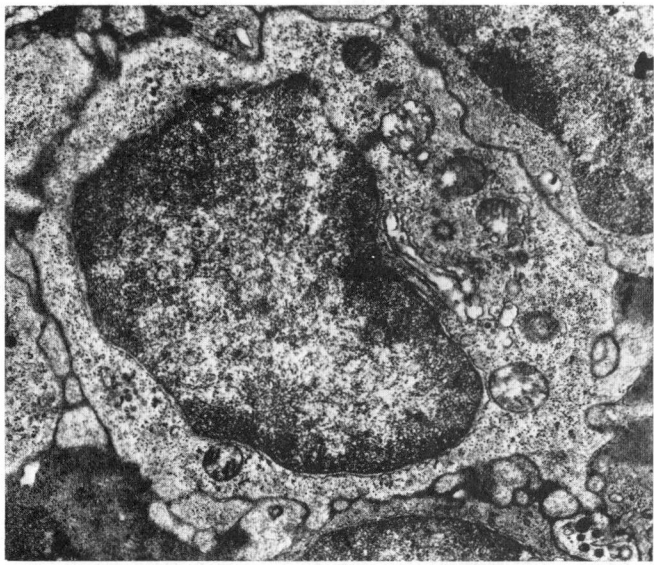

A

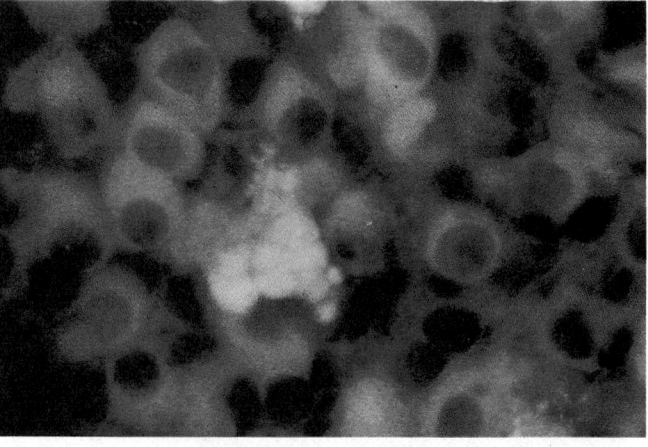

1.58C Plasma cells in a frozen section of the synovial membrane from a patient with rheumatoid arthritis. In this specimen the immunoglobulin being synthesized within the plasma cells has been demonstrated by immunofluorescence of rhodamine conjugated with rabbit anti-human IgG, viewed by ultraviolet microscopy. (Kindly provided by Dr. G. S. Panayi, Guy's Hospital Medical School.)

present in places a crystalline substructure. For these reasons they have sometimes been termed *compound granules*.

The available evidence points to the presence in mast cells of three active substances: *histamine* and *serotonin* (5-hydroxytryptamine)—agents in inflammatory changes—and *heparin*, a sulphated acid mucopolysaccharide which prevents clotting of plasma, and possibly of proteins which may leak in small quantities from the vascular channels. Mast cells may be disrupted to release their contents either by direct mechanical or chemical trauma, or sometimes following contact with particular antigens to which the body has previously been exposed (p. 62). The latter may result from interaction between the antigen and *cytophilic antibodies* of the IgE class associated with the mast cell plasma membrane, and may give rise to local responses (e.g. urticaria), or generalized ones (anaphylactic shock) following the release of large amounts of histamine and serotonin into the general circulation. They have thus been implicated in many of the phenomena occurring in inflammatory reactions, allergies and hypersensitivity states.

Mast cells closely resemble basophil leucocytes of the general circulation, and it is widely considered that they arise from the bone marrow and pass to the tissues as basophils, migrating through the capillary and venule walls to their final destination. However, there are minor differences between the basophil and the mast cell in terms of their cytochemistry, which suggest either a different lineage for the two, or perhaps more likely, that the basophil matures into a mast cell when it reaches its extravascular environment.

FAT CELLS (LIPOCYTES, ADIPOCYTES)

Fat cells occur singly or in groups in the meshes of many but not all connective tissues, being specially numerous in *adipose tissue* (**1.60**). When occurring singly the cells are oval or spherical in shape, but when mutually compressed they are polygonal. They vary in diameter, averaging about 50 μm. Each cell consists of a peripheral rim of cytoplasm, in which the nucleus is embedded, surrounding a single large central globule of fat. There is a slight accumulation of cytoplasm around the nucleus, which is oval in shape and appears compressed against the cell membrane by the fat droplet, as does the Golgi complex. In electron micrographs many microfilaments are also seen around the membrane-bound lipid vacuole. (Slavin 1972, Greenwood and Johnson 1977).

In sections not specially prepared to preserve fat this is usually dissolved out by the solvents used, particularly xylol or benzene; only the nucleus and the peripheral rim of cytoplasm surrounding a central empty space are left, so that the cell has a signet-ring

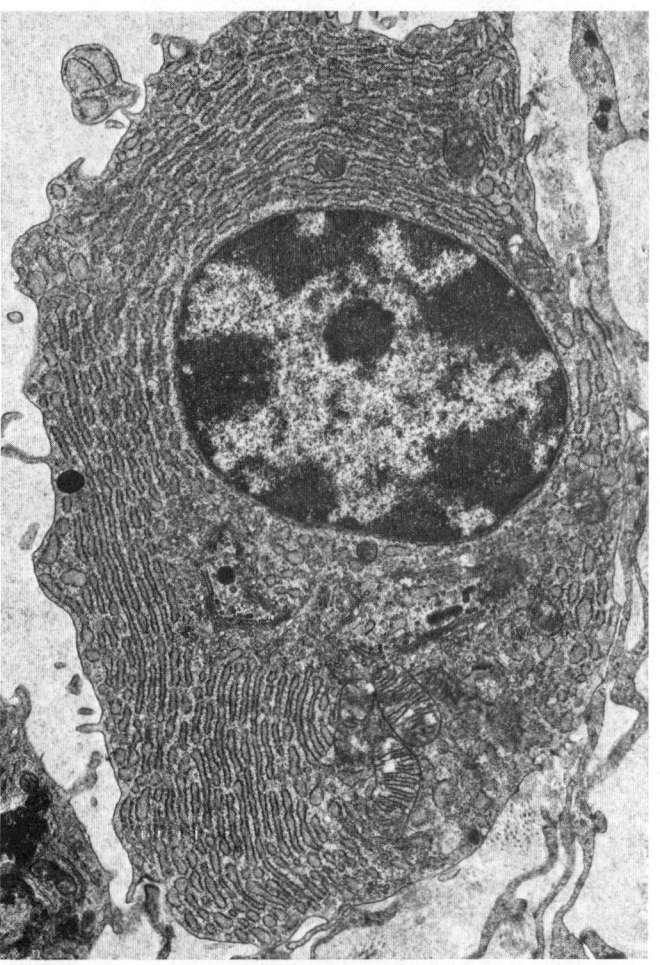

B

1.58A and B Electron micrographs of: (A) a small lymphocyte, (B) a plasma cell. Magnification ×6,000. (A kindly provided by Dr. D. R. Turner, Guy's Hospital Medical School.)

acid-Schiff (PAS) stain for carbohydrates, and with toluidine blue, methylene blue, azure A and alcian blue, they show strong *metachromatic* staining reactions (staining red), also indicating an acid mucopolysaccharide content. Ultrastructurally the vesicles (or 'granules') are seen to vary in size and shape (mean diameter about 0·5 μm), to be membrane-bound and to have a rather heterogeneous content, which varies with the species. In man the vesicles usually contain dense osmiophilic material which may be finely granular, lamellar, or in the form of membranous whorls; these variants may co-exist in the same vesicle, which may also

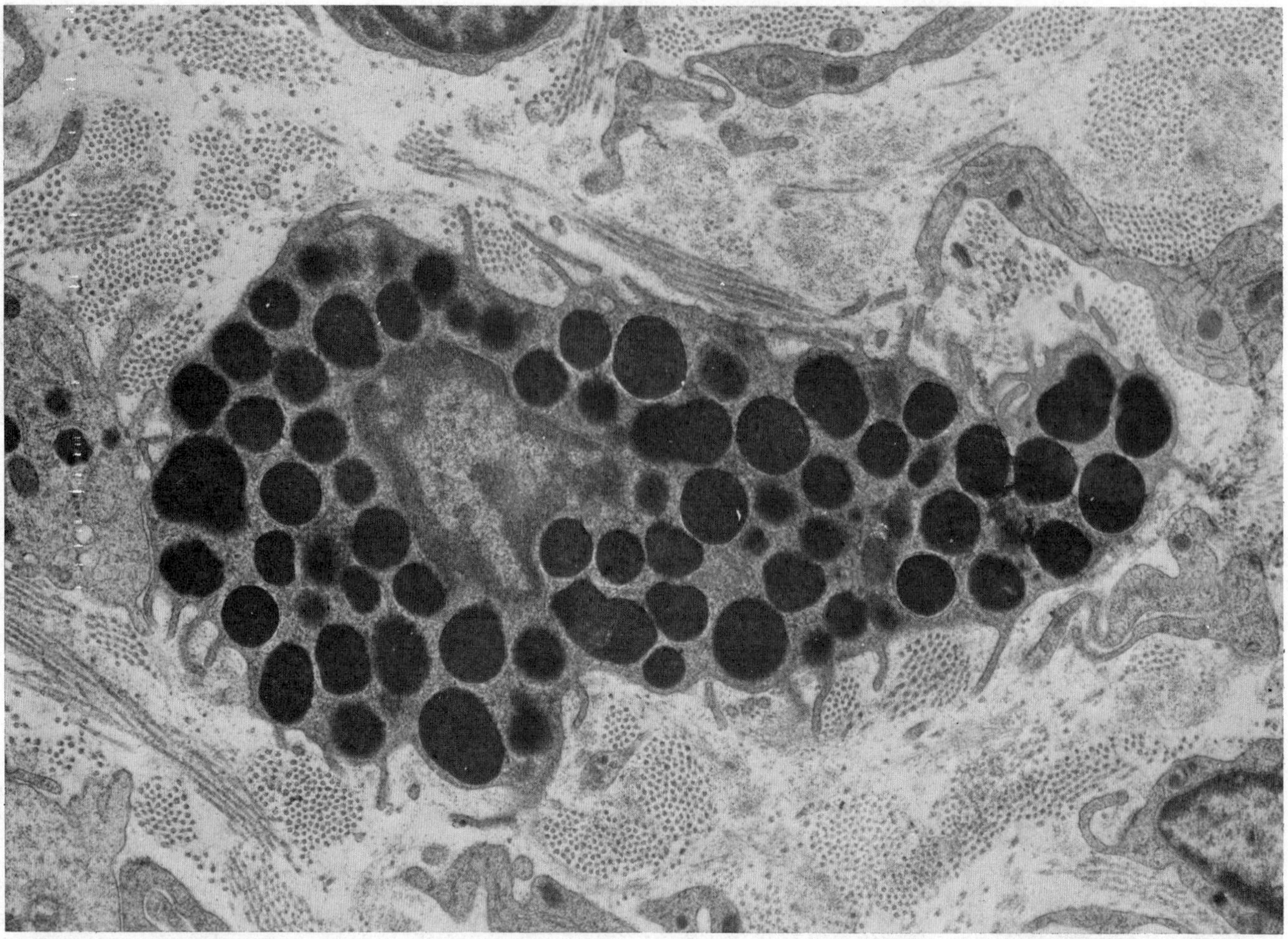

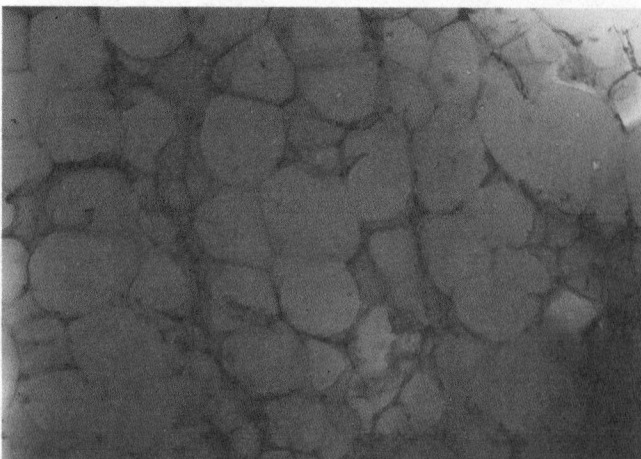

1.59 An electron micrograph of a mast cell showing the large densely staining membrane-bound cytoplasmic granules. Magnification × 6,000.

appearance. The fat consists of glycerol esters of oleic, palmitic and stearic acids.

Doubts exist as to whether fat cells are specifically and exclusively concerned with the storage, and perhaps the synthesis, of fat. Prior to the storage of fat within them they are stellate in shape and difficult to distinguish from fibroblasts and, when depleted of fat, they revert to this appearance. As fat accumulates the cells enlarge and become rounded, the fat first appearing as isolated small globules which later coalesce to form a single large droplet. Conversely, during depletion the single large globule diminishes in size and then breaks up into droplets while the cells become stellate in shape. Fat cells, however, seem to have a well defined distribution within the body and there is evidence that they are indeed specific cells.

The fat is fixed and stained by osmium tetroxide and specifically coloured by alcoholic solutions of certain dyes, notably Sudan III, Sudan black and Scharlach R, which are more soluble in fat than in the solvent; lipid is conveniently demonstrated in frozen or cryostat sections (1.60).

In some mammals, especially those which hibernate, certain deposits of fat, particularly in the interscapular region, are characterized by the presence of a large, glandular type of cell in which the fat is present as separate droplets and not as a single globule. These deposits of fat are often termed *brown fat* and they are concerned with heat production, mediated by mitochondria.

The mobilization of fat is under nervous or hormonal control, and noradrenalin released at sympathetic nerve endings in adipose tissue is particularly important in this respect (Napolitano 1965; Havel 1965).

Pigment cells (Chromatophores)

These occur in the corium of the skin, especially in dark races, and in the iris and choroid of the eye. They are frequently termed *chromatophores*. They are generally stellate cells with long processes and numerous dark brown or black granules, believed to be melanin, in their cytoplasm. Their origin is considered to be the embryonic neural crest.

Reticular Tissue

In many situations, as in exocrine and endocrine glands, a fine meshwork of reticulin fibres supports the cellular elements (*see* p. 51 and Carr 1970). The cells responsible for forming this are indistinguishable from fibroblasts, but are sometimes termed 'reticular cells'. This term is, however, also used for a variety of

1.60 Adipose tissue in frozen section, stained with Sudan red and haemalum. Magnification × 200.

other cells in the reticuloendothelial and haemopoietic tissues (p. 66), where reticulin is also laid down. The nomenclature and proposed relationships between these various elements is somewhat confused, and will not be further explored here.

The Matrix of Connective Tissue

This term is taken here to include all the extracellular materials of connective tissue, and may be divided into two components, the *fibrous elements* and the *ground substance*.

The fibrous elements consist of three types of fibre: collagen, reticulin and elastin.

COLLAGEN FIBRES

White collagen fibres are the most numerous and widely distributed fibrous elements in ordinary connective tissues. They are generally collected into bundles which pursue a straight or sinuous course (**1.54**) and within which the fibres are more or less longitudinally orientated, giving the bundle a faint longitudinal striation. Within the bundles, the fibres are bound by a small amount of amorphous cement substance, believed to be a mucoprotein. The amount of this material diminishes with age, and the constituent fibres in the bundle thicken.

In the fresh state, collagen fibres are soft and flexible; they have a high tensile strength, and are relatively inelastic and inextensible, with a Young's modulus of 10^{10} (Harkness 1961). They are homogeneous, transparent and show a form bire-fringence, indicating a longitudinal orientation of subunits. The individual fibres are 1–12 μm in diameter and do not branch, although they may, either singly or in small aggregates, pass from one bundle to another. They vary in length and may extend for considerable distances.

Collagen fibres are lightly eosinophilic; they also stain with acid aniline dyes such as aniline blue in Mallory's stain, and with acid fuchsin (red) in van Gieson's stain. They stain only weakly with silver impregnation.

Each fibre consists of bundles of finer parallel fibres, 0·3–0·5 μm in diameter, and these in turn consist of bundles of fibrils each 20–100 nm in diameter (**1.61**A, C); the latter are joined transversely by regular cross-connexions, probably glycoprotein in nature. These fibrils show a characteristic banding with a periodicity of 64 nm when viewed by electron microscopy. Within the major periodicity, there are several thinner cross striations arranged asymmetrically, so giving the fibril a polarity (**1.61**A, C).

The foregoing structural appearance is characteristic of fibrils of the protein collagen; when appropriately treated with dilute acids and alkalis, the collagen fibrils swell, and disintegrate into fibrillar units of *tropocollagen*. These units are 240 nm long by 1·4 nm wide, with a molecular weight of 350,000 daltons. *Gelatin*, obtained by boiling collagen, is composed largely of tropocollagen in suspension.

Each tropocollagen molecule can be further subdivided into three polypeptide chains (α-chains) arranged in a triple helix (*see* e.g. Hodge 1967; and **1.62**). These chains are rich in the amino acids glycine, which recurs at every third position, hydroxyproline and hydroxylysine, and are bonded by covalent

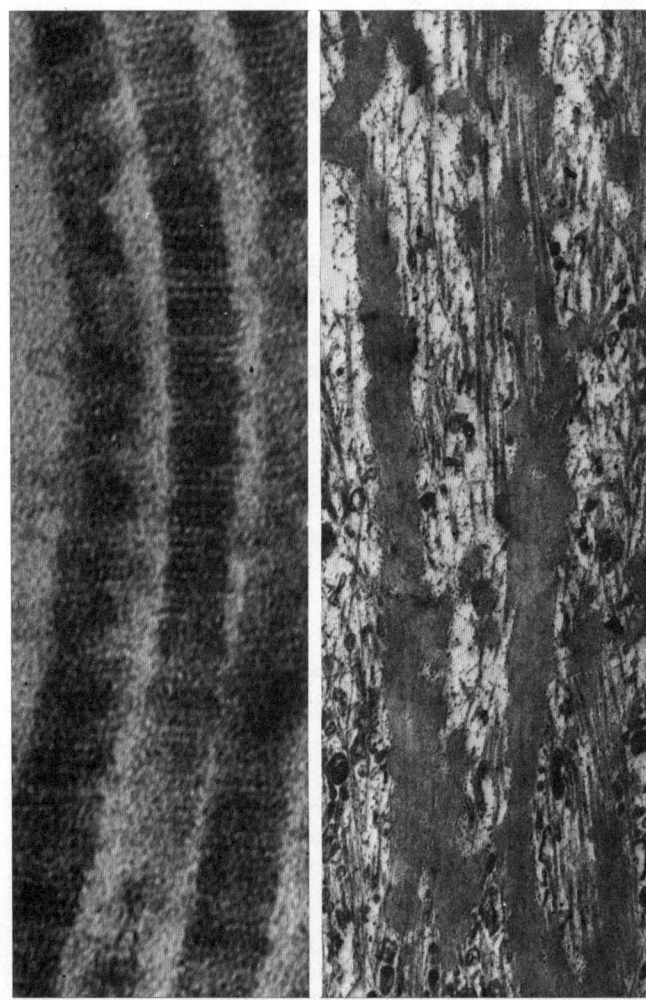

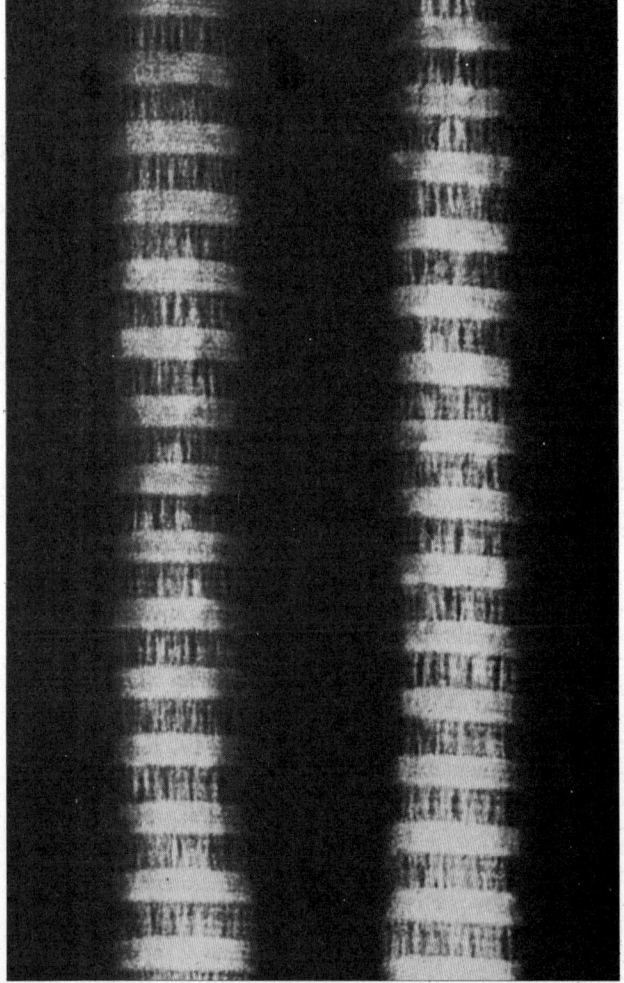

1.61A and B Electron micrographs showing (A) collagen fibres in longitudinal section from embryonic tissue in culture; the major 64 nm periodicity and finer intermediate bands are visible: (B) elastin fibres in longitudinal section, stained with phosphotungstic acid. Micrograph (A) kindly provided by Dr. C. Levine, and (B) by Dr. W. Jayaratnam.

1.61C Collagen fibrils viewed electron microscopically in negative contrast to show the characteristic cross-banding. Magnification × 140,000.

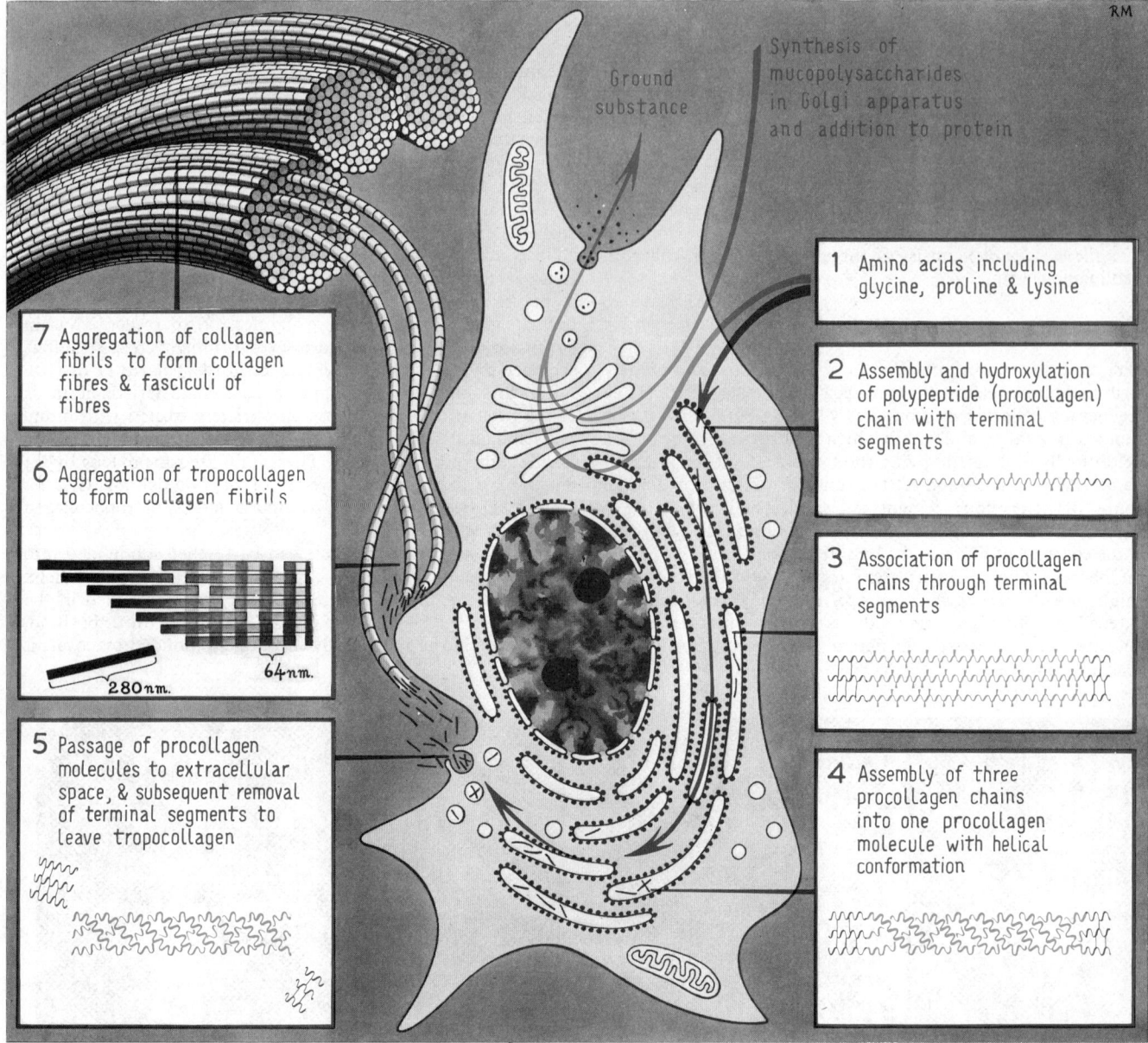

1.62 A schematic diagram showing the major steps in collagen synthesis by fibroblasts in connective tissue.

linkages at various sites along their length. Two major classes of chain, α_1 and α_2, differing slightly in their amino acid composition, are found in connective tissues; in tendons, for example, tropocollagen molecules are composed of two α_1 chains and one α_2. The lysine and hydroxylysine residues are also able to form strong cross-links between adjacent tropocollagen molecules, and it is the regular arrangement of parallel cross-linked molecules which gives collagen its great mechanical strength. Carbohydrate side-chains are also attached to lysine residues at various positions. The manner in which the tropocollagen molecules are arranged to give the 64 nm periodic bands has received much attention, and a general model has emerged.

The tropocollagen molecules lie parallel to each other in rows (**1.62**); each molecule, 280 nm long, is in echelon with respect to its lateral neighbour by 64 nm, a distance given by the arrangement of bonding between adjacent molecules. Each molecule thus stretches for four and a quarter period bands. Glycoproteins are important in the stabilization of this structure. The overlapping of these quarter-spaced molecules, when viewed in bulk, gives the characteristic pattern of bands, but only if the aggregations are in a definite geometric relation, possibly with groups of one tropocollagen molecule surrounded by six others as seen in cross-section, limiting the diameter of the fibril to about 100 nm. As

intimated above, variations in the amino acid compositions of the a-chains are present in different tissues, and at least four classes of collagen can be distinguished. *Type I* characterizes tendon, bone and much of the dermis; *Type II* pervades the matrix of cartilage; *Type II* is found throughout the cardiovascular system, gastrointestinal tract and in immature dermis; *Type IV* is present in basement membranes. (See also p. 246.)

The Biosynthesis of Collagen

The involvement of fibroblasts in the biosynthesis of collagen (**1.62**) has been studied biochemically and by autoradiographic electron microscopy. In many respects it appears to be similar to the synthesis of collagen by chondrocytes (p. 245). Amino acids are taken up by the cell and synthesized on the ribosomes of the rough endoplasmic reticulum (**1.10A–C**) to form long polypeptides, the *procollagen chains* (*see* review by Uitto and Lichtenstein 1976). These are longer than the final α-chains of tropocollagen because extra polypeptide segments (*extension peptides*), which are used to assemble the total molecule, are present at both ends, later to be removed. As chain synthesis proceeds and the polypeptide moves into the cisternae of the endoplasmic reticulum, various enzymes hydroxylate certain proline and lysine residues to hydroxyproline and hydroxylysine; ascorbic acid (vitamin C), molecular oxygen, and other factors are needed for this step, so that vitamin C deficiency and anoxia result in impaired collagen synthesis. Meanwhile, carbohydrate is attached to some hydroxylysine residues; then the three procollagen chains associate at one end by means of their extension peptides, intertwining as cross-links are formed, to give the triple helix of the *procollagen molecule*. In this form, the molecules are transferred via the Golgi apparatus, or perhaps directly, to the exterior in secretory vacuoles. Once outside, the extension peptides are removed by enzymes released by the cell, so forming tropocollagen which aggregates spontaneously (given the correct ionic environment) to create collagen fibres, cross-linked at the lysine and hydroxylysine residues. Tropocollagen molecules can be re-aggregated from acid solutions, in their characteristic pattern, by raising the *p*H to 7. The 64 nm banding pattern can, however, be altered by changing the glycosaminoglycan concentration.

The *direction* of fibre formation, on the other hand, appears to be dependent on the stresses acting in the tissue. The relation between stress on the one hand and the rate and direction of fibre formation on the other is uncertain, but may involve movement of fibroblasts along lines determined by piezo-electric currents consequent upon the deformation of pre-formed collagen fibres.

It is remarkable that in many situations collagen is laid down in a precise geometrical pattern, with successive layers regularly alternating in direction. This occurs in the cornea, where successive layers lie at 90 degrees to each other, helping to provide the cornea with its mechanical and unique optical properties. Regularity of pattern exists also in ligaments, tendons, aponeuroses and connective tissue capsules, etc.

Divergent views have been expressed concerning the role of collagen substructure as an essential determinant of the *pattern* of seeding and crystallite *orientation* during bone formation (*see* pp. 260–66).

RETICULIN FIBRES

Fine branching and anastomosing reticulin fibres form the supporting framework of many glands, the kidney, and the lympho-reticular tissues (lymph nodes, spleen, etc.); also in association with basement membranes, and often in the neighbourhood of collagen fibre bundles. Unlike collagen, reticulin fibres take up silver salts strongly, but do not stain strongly with acid fuchsin.

Ultrastructurally they show a periodic banding which is identical with that of collagen, display the same X-ray diffraction patterns, and exhibit a similar, though not identical, chemical composition (Windrum *et al.* 1955). Reticulin may also form an early framework on which the coarser collagen fibres are laid down during matrix synthesis. Fibroblasts are responsible for their production in most connective tissues, but perhaps other distinctive cells are involved in the various reticulo-endothelial tissues, for example the reticular cells of lymph nodes (p. 768).

ELASTIN FIBRES

These are less frequent than collagen fibres, and are, in contrast, yellowish in colour, and hence so are the tissues in which they abound. Elastin fibres (**1.61**B; **1.63**) branch and rejoin freely, and are usually thinner (1·0–0·2 μm) than collagen fibres, although on occasion they occur as thicker fibres, for example in the ligamenta flava, and sheets, as in the fenestrated elastic laminae of the aortic wall. Such fibres and laminae stretch easily, with an almost perfect recoil, Young's modulus of elasticity being 6×10^6 (Bergel 1961), although with advancing years they may calcify, losing elasticity.

Elastin fibres are highly resistant to hydrolysis in aqueous solutions of high or low *p*H, or boiling water, and to treatment with organic solvents. However, crude preparations of trypsin which contain the enzyme *elastase*, and bacterial elastases, are capable of dissociating the fibre structure.

Elastin fibres can be stained with orcein, with Weigert's

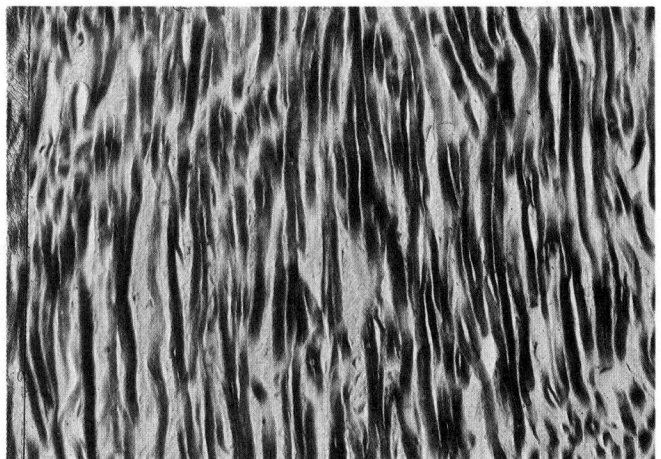

1.63 A longitudinal section through an elastic ligament of an ox showing elastin fibres (black) interspersed with collagen fibres. Verhoeff's and Van Gieson's stains. Magnification × 200.

resorcin-fuchsin and with Verhoeff's stain. The fibres display only weak or no birefringence when unstretched, indicating a lack of regular orientation, but become strongly birefringent when stretched, indicating the rearrangement of its molecules (strain birefringence).

The fibres are composed chiefly of the protein *elastin*, together with some glycoprotein, in fibrillar form (*oxytalan*), and traces of collagen, ground substance and lipid. Elastin is composed of globular subunits of *tropo-elastin* which have a molecular weight of 67,000 daltons and a diameter of 5·4 nm. These subunits are believed to be arranged regularly in three-dimensional sheets with no directionality, each subunit being linked to the next by bonds capable of deformation. Chemically, elastin contains a high content of the amino acid valine, and another named desmosine, which is apparently unique to elastin. In connective tissue, fibroblasts are responsible for elastin formation, although fibres and sheets of elastin can also be formed by smooth muscle cells.

For a detailed review of the biochemistry, ultrastructure, and biosynthesis of elastin, including an extensive bibliography, *see* Serafini-Fracassini and Smith (1974). For the pre-elastic role of the oxytalan fibrils during elastic fibre synthesis in auricular cartilage, *see* Bradamante and Švajger (1977). The latter consider that the glycoprotein microfibrils are synthesized first, and form a three-dimensional spatial framework, later to become impregnated by elastin.

THE GROUND SUBSTANCE

This forms the non-fibrous element of the matrix, in which cells and other components are embedded (**1.54**, **62**). In ordinary connective tissues it is a viscous gel containing a high proportion of water, bound largely to long-chain carbohydrate molecules and carbohydrate-protein complexes, and some tropocollagen (Balazs 1971).

Where the carbohydrate is a polysaccharide with one or more amino-sugar moieties it is termed a *glycosaminoglycan* (mucopolysaccharide); where the polysaccharide is in association with a protein it is known as a *glycoprotein* or *proteoglycan* depending on the proportion of each.

Various combinations of these molecules may occur at a particular site. Most of them have several features in common; the carbohydrate moiety is always composed of long, unbranched polymers of repeating disaccharides, each chain ranging from 6,000 to several million daltons in molecular weight. The disaccharide units consist of one hexosamine residue bearing an acetyl group, and a hexuronate or galactose residue. The disaccharides may also bear either a carboxyl or a sulphuric ester group, conferring a strong negative charge on the whole polymer which acts as a poly-anion, and therefore can be stained with cationic dyes such as Hale's colloidal iron, alcian blue, and various metachromatic stains.

The compounds present in all ordinary connective tissues vary in amount from one site to another. The principal types are: chondroitin 4-sulphate and chondroitin 6-sulphate (chondroitin sulphates A and C), hyaluronate, dermatan sulphate (chondroitin sulphate B), keratan sulphate, and heparin sulphate (see Montagna and Parakkal 1974). All of these, except hyaluronate, which is a huge randomly coiled molecule several million daltons in molecular weight, are normally conjugated with a protein core, from which they extend in rows, like bristles on a brush, being repelled from each other by their negative charges. Such arrangements usually confer a high viscosity on the ground substance because of the formation of complex three-dimensional networks capable of binding water strongly. This water component is of great importance in facilitating diffusion of metabolites, electrolytes and gases, between the capillaries and the cellular elements embedded within the ground substance. The presence of negative charges determines their ion-binding power, and they thus act as selective barriers to the passage of inorganic ions and charged molecules. All of these may be conjugated with proteins.

Because of its high viscosity the ground substance also forms a mechanical barrier; some bacteria secrete the enzyme hyaluronidase which decreases its viscosity, so facilitating the spread of the organisms in the connective tissues. The matrix is synthesized chiefly by fibroblasts within the granular endoplasmic reticulum and Golgi complex. Its production may be affected by hormones, as shown for instance in hypothyroidism, when much ground substance may be generated (myxoedema). The ground substance is difficult to demonstrate histologically, since it is highly soluble in water; it can, however, be retained in freeze-dried or freeze-substituted specimens, or after precipitation with cationic detergents such as cetyl pyridinium.

BASEMENT MEMBRANES

These are laminae of dense amorphous material which vary in thickness (Kefalides 1970), are associated with many types of cell embedded in or adjacent to connective tissue (e.g. Schwann cells, muscle cells, capillary endothelium, and epithelia in general). In some situations they may be particularly thick, and so, incidentally, easier to investigate, as in the glomerular membrane of the kidney, the lens capsule, the anterior limiting (Descemet's) membrane in the cornea and Reichert's membrane in the placenta of certain mammals.

Such membranes may possess a supporting function, and may be important in selectively altering ionic and molecular diffusion rates. Changes in their thickness are often associated with pathological conditions, as in the thickening of the glomerular membrane in glomerulonephritis and diabetes, although it is difficult to allot causes and effects in such cases.

Ultrastructurally, the basement membrane usually consists of two zones: a thin (50–80 nm) layer of fine (4 nm) fibrils embedded in an amorphous densely staining material, situated near the adjacent cell surface (and often referred to as the *basal lamina*), and a layer of fine reticulin or collagen fibres which merges into the neighbouring matrix. In many sites the basement membrane is continuous with the *glycocalyx* of the neighbouring cells (p. 5).

Basement membranes stain strongly with the PAS technique and with silver methenamine (Rambourg and Leblond 1967). Chemically they are composed chiefly of tropocollagen (80 per cent) with some 10 per cent glycoprotein (Kefalides 1971).

Although at one time basement membranes were thought to be formed as a condensation of the general connective tissue ground substance, recent immunological evidence shows that they are in part formed by the cells which they adjoin. For example, basement membranes of several types of epithelia have antigenic properties in common, which are, however, distinct from those of connective tissue; furthermore, endothelial basement membranes also have a distinct antigenic nature.

CLASSIFICATION OF CONNECTIVE TISSUE

The connective tissues differ considerably in appearance, consistency and composition in different regions according to the local functional requirements. These differences are related to the predominance or otherwise of one or other of the cell types, the concentration, arrangement and types of fibre, and the character of the ground substance. On these bases, *ordinary*

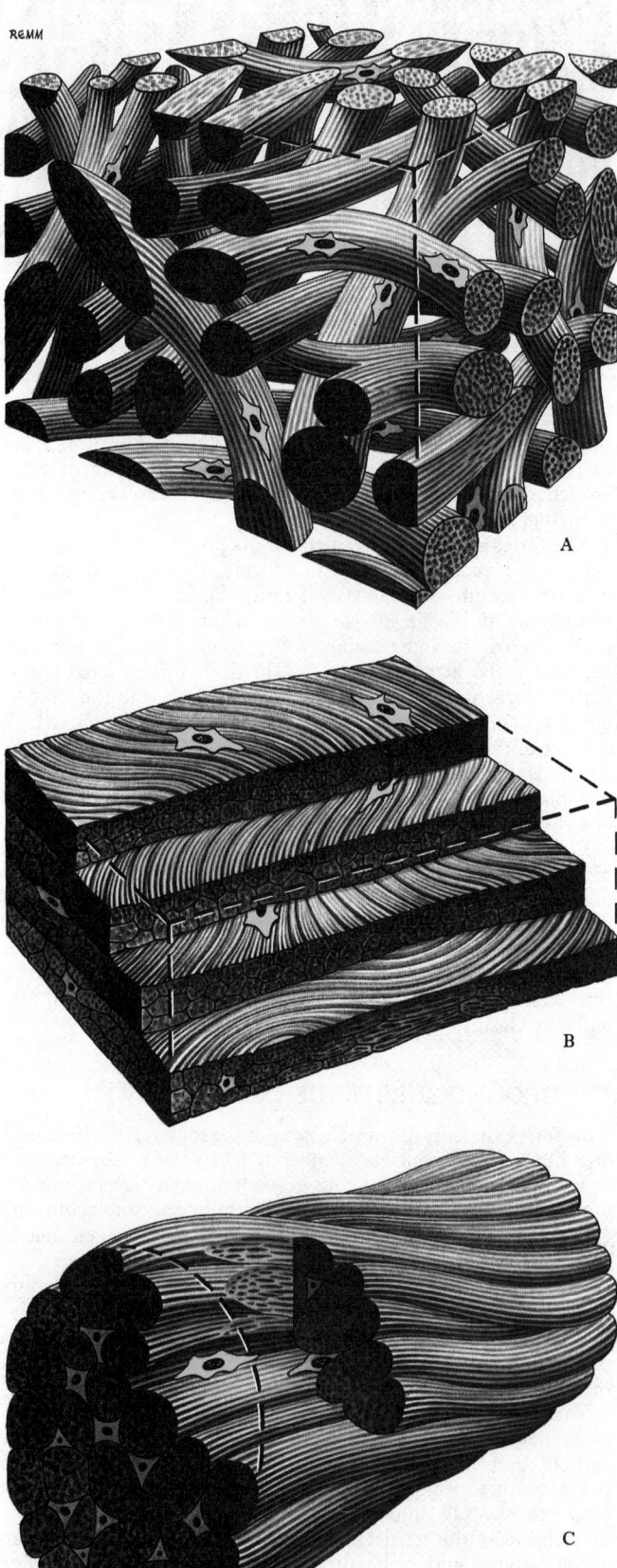

1.64A–C A diagram showing three types of arrangement of collagen fibres in: (A) dense irregular connective tissue; (B) a ligament; (C) a tendon.

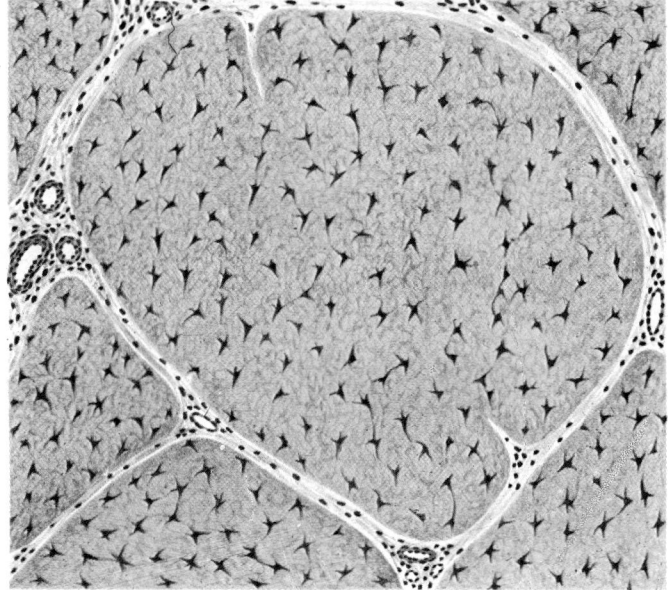

1.65 A transverse section of a tendon showing fibrocytes enclosed between bundles of collagen fibres.

connective tissues can be classified into *irregular* and *regular* types, distinguished by the absence or presence of a high degree of orientation in the fibrous elements.

Irregular Connective Tissue

This can be further subdivided into *loose, dense* and *adipose.*

Loose connective tissue (1.54, 57) is the most generalized form. It is extensively distributed and its chief use is to bind parts together, though allowing, by virtue of its extensibility and elasticity, a considerable amount of movement to take place. It occurs as the subcutaneous tissue in animals which possess a hairy coat, as the submucous coat in the digestive tract and as subserous tissue. In man it forms the subcutaneous tissue in regions where this is devoid of fat as in the eyelids, penis, the scrotum and labia. It is also found between muscles, vessels and nerves, forming investing sheaths for them and connecting them with surrounding structures. It is present in the interior of organs, binding together the lobes and lobules of the compound glands, the various coats of the hollow viscera and the constituent fibres of muscles and nerves.

Loose connective tissue consists of a meshwork of thin collagen and elastin fibres interlacing in all directions to give a measure of both elasticity and tensile strength. The large meshes contain the soft, pliable semifluid ground substance in which occur all varieties of connective tissue cells except reticular cells, scattered along the fibres or in the meshes. Occasional fat cells, usually in small groups, are seen.

Dense irregular connective tissue is found in regions which experience considerable mechanical stress and where protection is given to ensheathed organs. The matrix contains a high proportion of collagen fibres which form thick bundles (1.64A) interweaving in three dimensions, and giving considerable strength. Active fibroblasts are few in number and most are flattened with heterochromatic nuclei. The vascular supply, in accord with this, is limited.

Examples of this type of tissue may be found in the reticular layer of the dermis, the connective tissue sheaths of muscle and nerves, and the adventitia of large blood vessels. The capsules of various glands, the coverings of various organs such as the penis and testes, the sclera of the eye, and periostea and perichondria are all composed of dense irregular connective tissue.

Adipose tissue. A few fat cells occur in loose connective tissue in most parts of the body. However, in adipose tissue (1.60) these occur in great abundance and constitute the principal component. Fatty tissue occurs only in certain regions and this selective distribution suggests that the fat is deposited in genetically determined sites. It occurs in abundance in subcutaneous tissue, which is sometimes referred to as the *panniculus adiposus*, around the kidneys, in the mesenteries and omenta, in the marrow of bones and as localized pads in the synovial membrane of many joints. Its distribution in subcutaneous tissue shows characteristic age and sex differences.

Adipose tissue consists of fat cells embedded in a vascular areolar tissue which is usually divided into lobules by stronger fibrous septa carrying the larger blood vessels, whence each lobule receives an independent blood supply. Within the lobules the fat cells are round or, when mutually compressed, polygonal. Areolar tissue and septa both contain the other cellular components of fibrous tissue. These fat deposits serve not only as stores but in some situations have mechanical functions such as in the soles of the feet, palms of the hand and in synovial membranes. In these regions the connective tissue framework of the fatty tissue often contains large amounts of elastic tissue and in emaciation these deposits tend to be spared until a late stage. Elsewhere they help to conserve the body heat.

Regular Connective Tissue

This includes those highly fibrous tissues in which the fibres are regularly orientated with respect to one another, either to form sheets such as fasciae and aponeuroses, or thicker bundles as ligaments or tendons (1.64). The direction of the fibres within such structures is related to the stresses which they experience, but there is considerable interweaving of fibrous bundles, even within tendons, which increases their structural stability and perhaps elasticity.

The fibroblasts which secrete the fibres may eventually become trapped within the fibrous structure, where they are compressed, and present highly angular outlines. Cross-sections of tendons (1.64C, 1.65) show inactive fibroblasts with stellate profiles and small heterochromatic nuclei. Fibroblasts on the external surface may be active in continued fibre formation, and they afford a pool of cells from which repair after damage may stem (*see* McMinn 1969).

Regular connective tissue is predominantly collagenous, but elastic components also occur, as in the ligamenta flava of the vertebral laminae, and in the vocal folds. A smaller proportion of elastic elements is also present between the collagen lamellae of many other ligaments and fasciae. In other sites, the collagen fibres may form precise geometrical patterns, as in the cornea (*see* p. 1154).

Mucoid tissue is a fetal or embryonic type of connective tissue (1.66), found chiefly as a stage in the development of connective tissue from mesenchyme. It exists in the 'jelly of Wharton', which forms the bulk of the umbilical cord, and consists of a copious matrix, largely made up of hydrated 'mucosubstances' and a fine meshwork of collagen fibres, in which nucleated cells with branching processes (probably fibroblasts) are found (Boyd and Hamilton 1970). However, it has been demonstrated (Parry 1970)

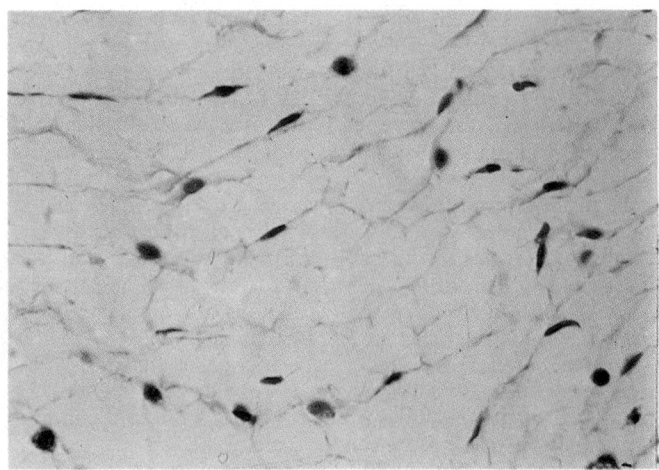

1.66 A section of fetal mesenchyme showing mucoid tissue sparsely populated with cells. Magnification × 500.

that the stellate appearance of the cells in Wharton's jelly probably results from fixation and staining of excised cords which have suffered a haemodynamic collapse. When this is avoided the cells are aligned and strap-like. Such findings may well apply to other situations in which mucoid tissue is found. Usually few fibres occur in typical mucoid tissue, though at birth the umbilical cord shows a considerable development of perivascular collagen fibres; after birth it is still to be seen in the pulp of a developing tooth. In the adult the vitreous body of the eye is a persistent form of mucoid tissue in which the fibres and cells are very few in number.

Pigmented connective tissue, such as occurs in the choroid and in the lamina fusca of the sclera of the eye, is composed of loose connective tissue, in which large numbers of pigment cells (melanocytes) are also present.

Vessels and Nerves of Connective Tissue

The *blood vessels* of connective tissue are very few—that is to say, few are supplied to the tissue itself, although many carrying blood to other structures may permeate one of its forms, viz. areolar tissue. In dense fibrous tissue the blood vessels usually run parallel to and between the longitudinal bundles, sending communicating branches across the bundles; in some of its forms, as in the periosteum and dura mater, they are fairly numerous. In yellow elastic tissue the blood vessels also run between the fibres. *Lymphatic vessels* are very numerous in most forms of connective tissue, especially in the loose tissue beneath the skin and the mucous and serous surfaces. They also occur abundantly in the sheaths of tendons, as well as in the tendons themselves. *Nerves* are found ending in dense connective tissues (p. 839); it is doubtful whether any nerves end in loose irregular tissue; at all events, they have not yet been demonstrated.

THE BLOOD

Introduction

All cells live within the body in a fluid environment upon the stability of which, in terms of physico-chemical characteristics, each cell is dependent for its normal functioning. This fluid environment, whilst being maintained at a steady state within certain limits, also allows for the diffusion of metabolites between cells and between the external and internal environments. In small animals with a relatively simple type of organization, unaided diffusion is rapid enough for nutrients, gases and waste materials to pass to and from the individual cells to satisfy their metabolic demands. In more complex animals diffusion is assisted by the muscular movements of the body, and in larger and yet more complex forms, particularly those with a coelom which interrupts the continuity of the tissues, a system of cell-lined tubes separates the body fluids into blood, confined within the tubes and able to circulate by muscular propulsion, and tissue fluid lying in the extracellular spaces outside the tubes. The coelom forms a third compartment which also possesses the function of a circulatory system in some groups of invertebrates. In the vertebrates a fourth division is added in the form of the medullary cavity of the central nervous system which contains the cerebrospinal fluid (p. 1050). A fifth compartment is provided by a system of further vessels which are cell-lined lymphatics, returning some of the tissue fluid to the blood circulation.

In animals with a true blood vascular system (p. 622) a high level of circulatory efficiency is possible. The system of cell-lined tubes also allows for the segregation within the circulation of some elements such as respiratory pigments able to store and carry oxygen and often carried within special cells as in vertebrates, various types of defensive cell, many of which can actively migrate into the tissues when required, so achieving rapid access to all parts of the body, and other large molecules such as polysaccharides and proteins, which may gain only limited entry to the tissues. With the development of a pumped circulation the hazard of leakage from damaged vessels is increased; various cellular and chemical *haemostatic* mechanisms ensure that such blood loss is limited.

In mammals most tissues are served by an extensive blood supply circulating by way of the heart through arteries, arterioles, capillaries, venules and veins (p. 622) The capillaries form the main site of exchange and diffusion between blood and tissues. Most cells lie within a few tens of microns of a capillary, active tissues receiving a richer and more intimate supply than less active ones. Some tissues, however, are not penetrated directly by capillaries, and diffusion gradients may be several hundreds of microns in length; such tissues are cartilage, some epithelia, and the connective tissue of the cornea and dentine.

Tissue Fluid and Lymph

The walls of the vascular system are lined by endothelial cells (p.

629) which allow the free diffusion of water, ions, gases in solution and small organic molecules. Large molecules are mostly retained within the blood, although some leakage does occur. Water and dissolved substances pass into the extravascular space which lies between the tissue cells, being termed *tissue fluid*. Because of the relatively high pressure at the arteriolar ends of the capillaries tissue fluid is constantly being formed; some of this passes back into the venous end of the capillary and the post-capillary venules, which are at relatively low pressure, but the remainder passes into the neighbouring lymphatic vessels as *lymph*. The lymphatic vessels eventually return the lymph to the blood circulation via the thoracic duct or the right lymphatic duct; *en route* to these ducts, the lymph is filtered by its passage through successive lymph nodes (p. 765), and whilst in them it picks up a population of lymphocytes and macrophages. Blind-ending lymphatics in the villi of the small intestine contain lymph which picks up great quantities of lipid (chyle) absorbed across the intestinal mucous membrane during alimentation. This is added to the lymph (and eventually to the blood) as minute lipid droplets (*chylomicrons*) each about 0·2–1·0 μm in diameter, which may be seen using light microscopy with dark field illumination. The lymphoid tissue also adds proteins in the form of antibodies (*immunoglobulins*) to the lymph and hence to the bloodstream.

Characteristics of Blood

Blood is an opaque turbid fluid with a viscosity somewhat greater than that of water (mean relative viscosity 4·75 at 18 °C), and a specific gravity of about 1·06 at 15 °C. When oxygenated, as in the systemic arteries, it is bright scarlet, and when deoxygenated, as in systemic veins, it is dark red to purple.

Blood is a heterogeneous fluid consisting of a clear liquid, *plasma*, and formed elements, *corpuscles*; because of this admixture it behaves hydrodynamically in a complex fashion, and belongs to that class of fluids termed non-Newtonian. This characteristic has important consequences in the physical study of blood flow in vessels (haemorrheology).

Plasma

This is a clear, slightly yellow fluid which contains many substances in solution or suspension; the *crystalloids* give a mean freezing-point depression of about 0·54 °C. Plasma is rich in sodium and chloride ions and also contains potassium, calcium, magnesium, phosphate, bicarbonate and many other ions, glucose, amino acids, etc. The *colloids* include the high molecular weight plasma proteins, composed chiefly of those associated with clotting, particularly prothrombin, the immunoglobulins, and complement proteins, involved in immunological defence (p. 61), and glycoproteins, polypeptides and steroids concerned with hormonal activities. Since most of the metabolic activities of the body are reflected in the composition of the plasma, routine chemical analysis of this fluid has become of great diagnostic importance and a considerable body of information on its chemistry is available.

The formation of clots by the precipitation of the protein fibrin from the plasma is initiated partly by the release of specific materials from damaged cells and blood platelets (p. 63) in the presence of calcium ions. If blood or plasma samples are allowed to stand, clot formation occurs to leave a clear yellowish fluid, the *serum*. Removal of the available calcium ions by means of citrate, various organic calcium chelators (EDTA, EGTA), and oxalate, prevents clot formation.

Blood as a Tissue

Blood has many affinities with connective tissue, as, for instance, in the mesenchymal origin of its cells, the free exchange of leucocytes with the connective tissues, and the relatively low cell:matrix ratio. Many of the plasma substances and some of the cells, however, arise from a variety of sources (e.g. the proteins associated with clotting are formed in the liver), and so blood is really a composite tissue pool.

The Formed Elements of Blood

Blood contains three groups of formed elements: red and white blood corpuscles, and platelets (*see* Wintrobe 1974, for a detailed description of these cells). Some structural aspects of these elements are visible in fresh blood, but many others are seen only in fixed and stained specimens. The examination of blood cells is of considerable clinical importance since their numbers, proportions of different cell types, and structure are valuable indicators of pathological changes in the body. Amongst other techniques, the Romanowsky methods of staining are particularly valuable and are widely used in clinical laboratories. These methods involve staining in aqueous solutions with methylene blue-eosin mixtures which colour both acidic dye binding and basic dye binding structures. The Giemsa and Leishman stains belong to this group. (It should be noted that throughout this section, the figures given for cell dimensions and numbers are approximate ranges only. The data provided by different authorities vary somewhat; further, the dimensions of some cells when measured in the fresh state are substantially smaller than when measured in a dried smear—with erythrocytes the converse applies. In particular the upper limit for the diameter range of the small lymphocyte given here—10 μm—is rather higher than that chosen by some workers, who hold that cells over 8 μm in diameter should not be so classified. The differences stem from methodological and site variations, and the somewhat arbitrary nature of the chosen dividing line.)

Red Blood Corpuscles

The red blood corpuscles (*erythrocytes* or *red blood cells*) (1.67, 68, 69) form the greater proportion of the blood cells (99 per cent of the total number), with a count of $4\cdot1–6\cdot0 \times 10^6/$mm^3 in adult males and $3\cdot9–5\cdot5 \times 10^6/$mm^3 in adult females. Each cell is a biconcave disc with a diameter of $6\cdot3–7\cdot9$ μm (mean $7\cdot1$ μm) and a rim thickness of $1\cdot9$ μm; in wet preparations the mean diameter is $8\cdot6$ μm. Erythrocytes lack nuclei and are pale red by transmitted light with paler centres because of their biconcavity. They show a tendency to adhere to one another by their rims to form loose piles of cells (*rouleaux*), a character probably determined by the properties of their cell coat. In normal blood a few assume a shrunken star-like, *crenated* form (1.67A), a shape which can be reproduced by placing normal biconcave erythrocytes in a hypertonic solution, which results in osmotic shrinkage. Such cells are called *echinocytes* (Bessis 1973). In hypotonic solutions erythrocytes take up water and become spherical and may eventually lyse to release their haemoglobin (*haemolysis*); they are then termed *red cell ghosts* (erthrocytic umbrae).

Erythrocytes are bounded by a plasma membrane and consist internally of a single protein, haemoglobin, apart from a few remnants from their initial development. The plasma membrane

of erythrocytes has received much attention because of the ease with which it can be obtained for analysis in quantity (Bretscher and Raff 1975). It is about 60 per cent lipid and glycolipid, and 40 per cent protein and glycoprotein. More than 15 classes of protein are present, including two major types. Firstly, the glycoprotein *glycophorin* (of about 50,000 daltons molecular weight) spans the membrane, and its negatively charged carbohydrate chains project from the outer surface of the cell, conferring most of the fixed charge on the cell exterior by virtue of its sialic acid groups. Secondly, the 'Band 2' protein which may bear some antigenic groups; the ABO antigens (p. 68) are all glycolipids (Race and Sanger 1975). Other proteins include several enzyme systems, some concerned with ionic regulation, others with the addition of lipid to the cell membrane from serum lipid (this is necessary because the cell does not possess its own synthetic apparatus).

The shape of the erythrocyte is largely determined by the protein *spectrin* (a name which reflects its biochemical preparation from red cell 'ghosts') which is attached to integral membrane

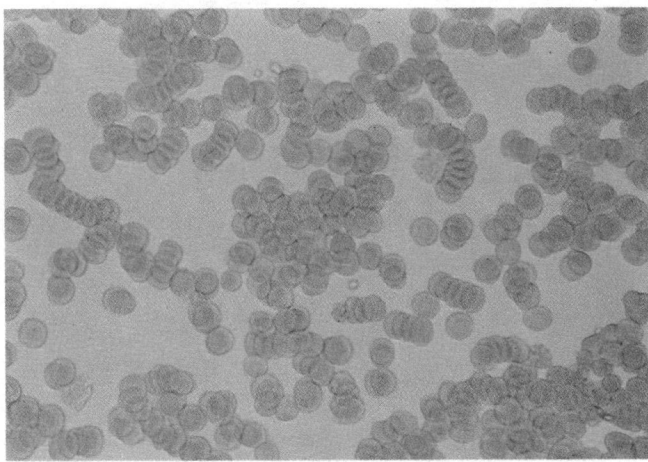

1.67A Fresh preparation of living erythrocytes showing rouleaux formation and red pigmentation. Magnification $\times$ 500.

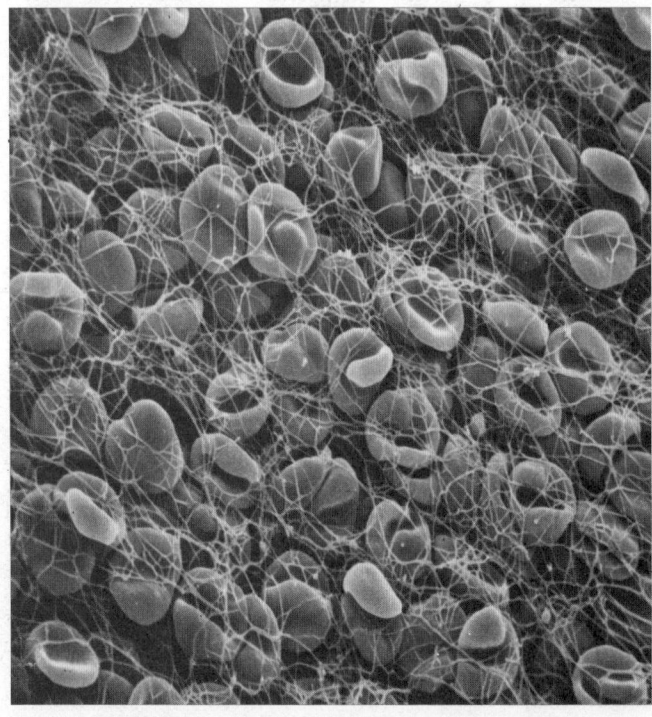

1.67B A scanning electron micrograph of erythrocytes, showing biconcave discoidal and other shapes; the fine filaments are fibrin resulting from clotting of the plasma after extravasation of blood. Magnification $\times$ 1,500. Photographed by Mr. Michael Crowder, Guy's Hospital.

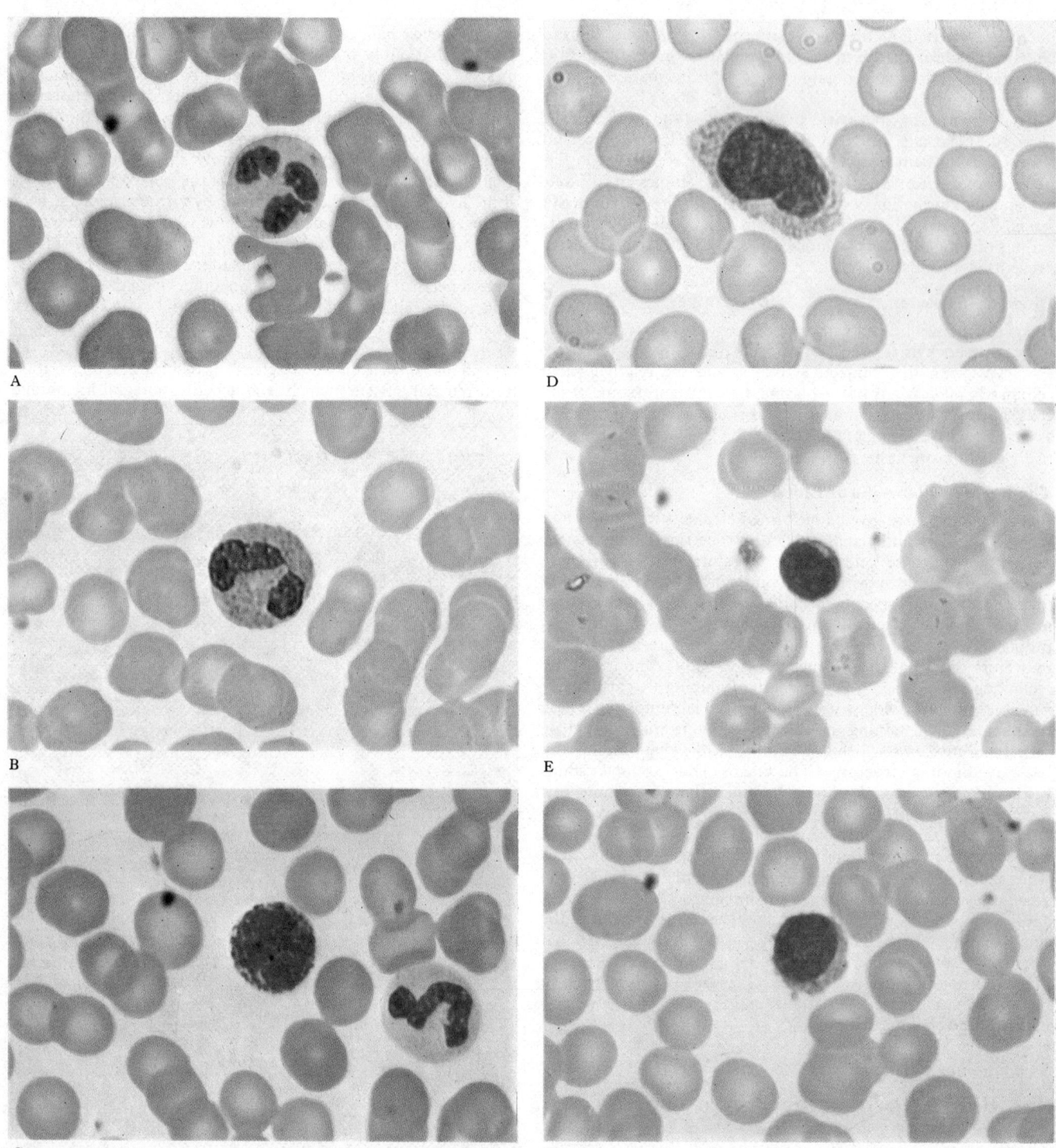

1.68A–F Blood cell types stained in smeared preparations by the Giemsa method. Erythrocytes are shown in all figures which also demonstrate other cell types: (A) neutrophil leucocyte and platelets; (B) eosinophil leucocyte; (C) basophil leucocyte with prominent densely staining cytoplasmic granules, and neutrophil leucocyte; (D) monocyte; (E) small lymphocyte and (F) medium-sized lymphocyte. Material kindly provided by Dr. J. P. Black of Guy's Hospital Haematology Dept.

proteins on the inner surface of the cell membrane, so forming a stabilizing network of attachments. This considerably stiffens the membrane, an effect which is aided by the large amount of cholesterol in the membrane itself. Red cells can thus regain their shape and dimensions after passing through the lumina of the finest ramules of the blood-vascular system; microscopic examination has shown that erythrocytes often pass through capillaries flattened face-first, buckling somewhat to a shield-like shape (Brånemark 1972) rather than rolling up, as might be expected.

Haemoglobin is a globular protein with a molecular weight of 67,000, consisting of *globulin* molecules bound to *haem*, an iron-containing *porphyrin* group (Perutz *et al.* 1960). Each molecule is made up of four subunits, each in turn consisting of a coiled polypeptide chain with a cleft holding a single *haem* group. In normal blood, four types of polypeptide chain can occur, namely: α, β, γ and δ. Each haemoglobin molecule contains two α-chains and two others, so that several combinations, and hence a number of different types of haemoglobin molecule are possible. Haemoglobin A (HbA) which is the major adult class, contains 2 α- and 2 β-chains. Haemoglobin A_2 (HbA2), a minor component in adults, is composed of 2 α- and 2 δ-chains. Haemoglobin F (HbF), found in fetal and early postnatal life consists of 2 α- and 2 γ-chains. In the pathological condition *thalassaemia* only one type of chain is synthesized, so that a molecule may contain 4 α-chains (α-thalassaemia) or, more commonly, 4 β-chains (β-thalassaemia)—Haemoglobin H (Wintrobe 1974).

Each polypeptide chain is determined by a separate gene; a number of variant haemoglobins are known in which only one or a few amino acid residues are abnormal, reflecting slight alterations in the corresponding genes. In the Haemoglobin S of sickle-cell disease a single alteration in the β-chains (valine substituted for glutamine) causes a major alteration in the behaviour of the red cell and its oxygen-carrying capacity which, however, may confer some protection against malarial infection, where the disease is endemic. Other common variants include Haemoglobins C and D (see Williams et al. 1977). The oxygen-binding power of haemoglobin is provided by the iron atoms of the haem groups, and these are always maintained in the ferrous (Fe^{++}) state by the presence of glutathione.

In addition to haemoglobin, erythrocytes possess a number of enzyme systems, notably those concerned with glycolysis and ionic transport, which together maintain low sodium levels within the cell against diffusion gradients, and thus create the appropriate conditions of pH and ionic strength for the normal functioning of haemoglobin. As intimated above, glutathione metabolism is also active. Although, of course, in the absence of a nucleus and ribosomes, no protein synthesis takes place in mature erythrocytes, lipid in the plasma membrane can, however, be replaced to some extent from circulating serum lipids, by the activity of membrane enzyme systems.

The iron-containing compound ferritin is also often present in newly formed erythrocytes, as are also persisting remnants of the apparatus of protein synthesis (ribosomes and other RNAs) from the stage of differentiation of the cell in the bone marrow. After Romanowsky staining, the residual RNA of young erythrocytes causes a slight bluish tinge; with the supravital stain brilliant cresyl blue it forms a reticulum, giving the name reticulocyte to this type of cell. Later in maturation such evidences of basophilia disappear. Other inclusions may be present in red cells, particularly in pathological conditions; amongst these are nuclear remnants (Howell-Jolly bodies) and altered haemoglobin (Heinz-Ehrlich bodies).

Life Span and Destruction

Erythrocytes which have been labelled radioactively or antigenically and then injected into the circulation, have been shown to last between 100 and 120 days before being destroyed (Berlin et al. 1959). As erythrocytes age they become increasingly fragile and their surface charge changes as their content of negatively charged membrane glycoproteins are progressively reduced (Marikovsky and Daron 1969). Perhaps as a result of the changes in charge they are eventually ingested by the macrophages of the spleen and liver sinusoids without previous lysis, and are then hydrolysed (Harris 1963). Here, the haemoglobin is broken into its globulin and porphyrin moieties; the globulin is then further degraded into its constituent amino acids which pass to the general amino-acid pool of the body. The iron is removed from the porphyrin and can be used either directly in the synthesis of new haemoglobin in the bone marrow, or stored in the liver as ferritin or haemosiderin; the remaining haem portion is converted in the liver to bilirubin and is then excreted in the bile.

Red cells are produced by the bone marrow (p. 66) and are destroyed at the rate of about 5×10^{11} cells a day.

Fetal erythrocytes differ markedly from those of adults, up to the fourth month of development, in that they are larger (10 μm), are nucleated, and contain a somewhat different type of haemoglobin (Hb-F). From this time they are progressively replaced by the adult type of cell.

The Leucocytes

Leucocytes (white blood corpuscles or cells) belong to at least five different categories, distinguishable by their size, nuclear shape and cytoplasmic inclusions (1.68, 69, 72). These types of cell fall into two main divisions, namely those with prominent stainable cytoplasmic granules, the granulocytes, and those without, the agranulocytes.

The granulocytes or granular leucocytes are all closely related developmentally; all possess irregular or multilobed nuclei and are often termed **polymorphonuclear leucocytes** for this reason. This group is comprised of three types of cell, the granules of which give different staining reactions with the Romanowsky dyes: they are acidophil (or eosinophil) leucocytes with granules which bind acidic dyes such as eosin, basophil leucocytes the granules of which bind basic dyes (methylene blue) strongly, and neutrophil leucocytes whose granules stain weakly with both elements by a different type of reaction (see Wintrobe 1974).

NEUTROPHIL LEUCOCYTES

Neutrophil polymorphonuclear leucocytes (neutrophils, neutrophiles, heterophile leucocytes, or 'polymorphs') form the largest proportion of the leucocytes (60–70 per cent, with a count of 3,000–6,000/mm³). In dried smears, where the cells have flattened, they have a circular profile with a diameter of 10–15 μm. In the living state the cells may be spherical whilst passively circulating, but can flatten and become actively motile on contact with a suitable surface.

Within the cytoplasm the numerous granules give a variety of colour shades ranging from violet to pink when stained with Romanowsky stains such as Wright's and May-Grünwald-Giemsa, which are commonly employed in haematology. Under the electron microscope, too, the granules are heterogeneous in size, shape, and content (see e.g. Zucker-Franklin 1967), but all of them are membrane-bound bodies containing hydrolytic enzymes, that is, they are various species of lysosome. Two major categories can be distinguished according to their developmental origin (Bainton et al. 1971). Firstly, non-specific or primary granules which are formed early in neutrophil genesis (p. 67); these are relatively large (0·5 μm) spheroidal bodies containing peroxidase, acid phosphatase, and several other enzymes. With light microscopy they stain strongly with neutral red and azure dyes. Secondly, specific or secondary granules, formed a little later, and which assume a wide range of shapes including spheres, ellipsoids and rods. These contain alkaline phosphatase, collagenase, and aminopeptidase, all of which are lacking in the primary granules. The secondary granules, however, lack peroxidase and acid phosphatase. Some enzymes such as lysozyme are present in both. The presence of these lysosomes correlates well with the phagocytic activity of neutrophils.

In mature neutrophil granulocytes the nucleus is characteristically multilobate with up to six segments joined by narrow nuclear strands (the segmented stage). Less mature cells have fewer lobes, the earliest to be released under normal conditions being juveniles (band, or stab cells) in which the nucleus is an unsegmented crescentic band. In certain clinical conditions, even earlier stages in neutrophil formation, with indented or rounded nuclei (metamyelocytes, or myelocytes—p. 64) may be released from the bone marrow. In mature cells the edges of the nuclear lobes are often irregular; in females about 3 per cent (range 1–17 per cent) of the nuclei of neutrophils show a conspicuous 'drumstick' formation which represents the sex chromatin of the inactive X chromosome (Barr body)—see also p. 34.

Mitochondria, a Golgi complex, a sparse endoplasmic reticulum, and glycogen are present in the cytoplasm. A conspicuous array of microtubules is often seen between the 'arms' of the nucleus, and it is interesting that locomotion of neutrophils is in the direction in which the free arms point.

Neutrophils form an important element in the defence systems of the body (see Wintrobe 1974); they can engulf microbes and particles in the circulation and, after squeezing between the endothelial cells lining capillaries or venules, can perform local phagocytosis in the extravascular tissues, wherever it is needed. The engulfing of foreign objects is followed by digestion through fusion of the phagocytic vacuole, first with primary granules (the acid phase of hydrolysis), then with the secondary granules (the alkaline phase), any undigested remains being finally either egested, or stored. Although phagocytosis is supported by anaerobic glycolysis, lysosome action requires oxygen, so that neutrophils that are active in defence have a high oxygen demand. Phagocytosis, or the release of granules, may be enhanced by the

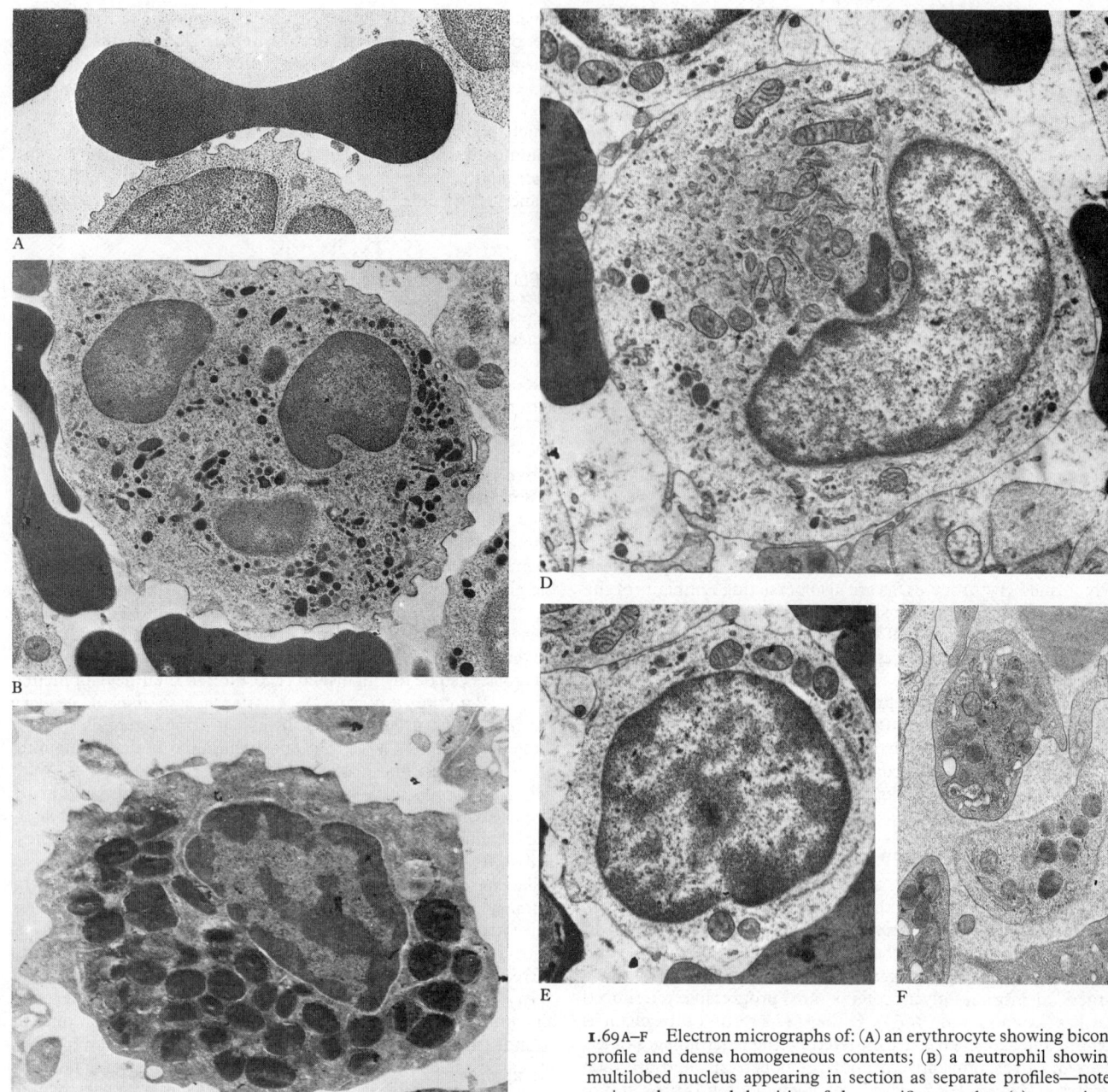

1.69A–F Electron micrographs of: (A) an erythrocyte showing biconcave profile and dense homogeneous contents; (B) a neutrophil showing its multilobed nucleus appearing in section as separate profiles—note the various shapes and densities of the specific granules; (C) an eosinophil, with crystalline inclusions in the specific granules; (D) a monocyte showing an indented nucleus, endoplasmic reticulum and lysosomes; (E) a small lymphocyte; (F) a group of platelets. See text for further details.

presence of antibodies attached to the surfaces of neutrophils, and which can bind specifically to target antigens, for example, a type of bacterium to which the body has previously been exposed. Opsonizing antibodies (*opsonins*—p. 62) coating the antigenic target may also promote phagocytosis. The antibodies in both cases are secreted by transforming lymphocytes (p. 61), and the neutrophil granulocyte is just one in a series of defensive cells which form an interrelated system for the elimination of foreign materials from the tissues (1.71).

After phagocytosis, the neutrophil's cytoplasmic granules gradually become used up, so that a marked reduction in their number (degranulation) occurs. Granules may also be discharged from the surface of the cell when it is suitably stimulated, to damage or kill neighbouring organisms or cells.

The numbers of neutrophil granulocytes in the blood vary considerably, often rising during episodes of acute bacterial infection, as they are released earlier in their formation by the bone marrow, and as granulocyte production rises. Once released into the blood they may circulate freely (the *circulatory pool*), or they may adhere to the walls of post-capillary venules and other

vessels (the *marginal pool*) to re-enter the circulation when suitably recruited, for example, by brief exercise, or by exposure to noradrenalin. However, neutrophil granulocytes do not remain long in circulation since, after a short time (half-life 7·5 hours), they may either be destroyed, or pass through the endothelial walls to the extravascular tissues, engaging in defence; alternatively, after entering various secretory ducts, such as the bronchi, salivary gland ducts, and the urinary tract, they are lost to the body.

EOSINOPHIL LEUCOCYTES

These cells (*acidophil leucocytes*) are similar in size, shape and motile capacity to neutrophils; they number 100–400/mm³. The granules of the cytoplasm are uniformly large (0·5 μm) and give the living cell a slightly yellowish colour. With Romanowsky stains they are uniformly orange to red. The nucleus has two prominent lobes connected by a thin strand. Ultrastructurally, the cytoplasm is seen to be packed with specific granules (1.69C) which are spherical or slightly ellipsoidal. Each of these bodies is bounded by a membrane and contains an amorphous material in which is embedded a prominent crystal; in man these crystals show a square lattice structure with a periodicity of about 4 nm. In other mammals the details of the granules vary somewhat. Several enzymes have been demonstrated cytochemically within the granules, including acid phosphatase, peroxidase, ribonucleases and cathepsin. They therefore appear to have a similarity to lysosomes. Mitochondria, endoplasmic reticulum and other cell organelles are similar to those of neutrophil granulocytes.

Eosinophil granulocytes are able to pass into the extravascular tissues from the circulation in the same way as neutrophils, when suitably stimulated. The total lifespan of the eosinophils is a few days, of which about 30 hours is spent in the circulation, and the remainder in the surrounding tissues.

The functions of the eosinophils are not yet fully understood, but since their numbers rise disproportionately to other leucocytes (an *eosinophilia*) in certain allergic disorders, and also in worm infestations, they seem to play a part in the immune system. They may be important in phagocytosing antibody-antigen complexes and fibrin, to concentrations of which they are attracted. They are also capable of engulfing bacteria and fungi, although this does not seem to be a major function. Recently, it has been shown that eosinophils can attach themselves to parasitic worms by cytophilic antibodies, and cause their destruction by liberating lysosomal enzymes on their surfaces.

BASOPHIL LEUCOCYTES

These are similar to neutrophils in size (10–15 μm) and shape, but form only a small proportion of the leucocytes (0·5–2 per cent) in normal blood, numbering from 25 to 200/mm³ (**1.68 c**).

They are distinguished from the other leucocytes by their large (0·5–1·5 μm) basophilic granules, which vary greatly in number (10–100 in each cell), but they are usually conspicuous. The nucleus is less irregular than in other granulocytes and often presents an S-shaped form. Ultrastructurally, the basophilic granules appear as membrane-bound vesicles with densely stained contents showing a variety of crystalline, granular and lamellar inclusions. With light microscopy basophil cells are periodic acid-Schiff (PAS) positive, and with some dyes such as Azure A, they are *metachromatic* (i.e. they stain a different colour from that of the dye), both reactions indicating the presence of polysaccharides. Heparin, histamine and serotonin have also been shown to be present. All these features indicate a close relation to, or identity with, mast cells of the connective tissues, and it is probable that basophils are mast cells derived from bone marrow, although, since some histochemical findings suggest slight differences, definitive evidence is lacking. In normal blood the turnover time of basophils is long (9–18 months in mice), but like other leucocytes they can migrate from the circulation to the extravascular tissues when antigenically stimulated; under these conditions they release histamine and other inflammatory agents from their granules (p. 47).

The agranular leucocytes are divided into two distinct groups, monocytes and lymphocytes, distinguishable on structural and functional grounds. Their nuclei are not lobulate, and both possess a moderately basophilic cytoplasm.

MONOCYTES

These are the largest of the agranular leucocytes (15–20 μm in diameter in smears) but they form only a small proportion of the leucocytes (2–8 per cent with a count of 100–700/mm³ of blood). The nucleus, which is euchromatic, is relatively large, and has a characteristic indentation on one side. The cytoplasm forms a wide rim around the nucleus, near the indentation of which lies a prominent Golgi complex and vesicles stainable with neutral red. Ultrastructurally (Cawley and Hayhoe 1973), many lysosomes are seen to be present, together with some peripheral rough endoplasmic reticulum. Mitochondria are quite abundant, reflecting the highly motile nature of the cell. Monocytes are actively phagocytic.

All of these characteristics are similar to those of macrophages of connective tissues, spleen, liver, lung, peritoneal cavity, and nervous tissue; this series of related cells is known as *mononuclear phagocytes* (corresponding approximately to the reticulo-endothelial system of former times) with which they are believed to be identical (p. 765), the circulating monocytes becoming macrophages in the extravascular tissues.

Lymphocytes

These cells (**1**.68, 69) are the second most numerous type of leucocyte, forming 20–30 per cent of all white cells and giving a total count of 1,500–2,700/mm³. Like the other leucocytes already described, they are also found in extravascular tissues; however, they differ in being formed in large numbers outside the bone marrow as well as within it and, accordingly, constitute a widely distributed *lymphoid system* which will be described in detail in *Section VI* p. 766.

Until relatively recently, lymphocytes were regarded as cells of little functional importance and, perhaps, representing degenerative stages of some other type of blood cell. The work of Miller, Yoffey, Gowans and others in the 1950s, however, showed that these cells play a vital part in the body's defences, and since that time they have come to be the focal point of modern immunology (*see* e.g. Roitt 1977).

The term 'lymphocyte' is customarily reserved for agranular leucocytes in the 5–15 μm diameter range, but they do not show the phagocytic activities characterizing monocytes, and the granular leucocytes. They are often classified as 'small' or 'large' although there is a continuous spectrum of sizes between these extremes, reflecting the ability of the small cells to transform into the large ones. Besides those cells engaged in defence, some of the small elements of similar appearance, are known to be circulating *stem cells* for other cell lineages of the haemal series, which can only be distinguished from the 'true' lymphocytes by methods of tissue culture.

Small lymphocytes, which are in the great majority in normal blood, have a rounded, densely-staining nucleus containing coarse strands of heterochromatin (a *leptochromatic nucleus: leptos*=threadlike) and one or more inconspicuous nucleoli. The cytoplasm forms a narrow rim around the nucleus and stains light blue with Romanowsky dyes. Under the electron microscope, few organelles appear in the cytoplasm, which contains only a small number of mitochondria, single, unattached ribosomes (monosomes), and occasional elements of agranular or granular endoplasmic reticulum; all of these features are signs of low metabolic activity. Small lymphocytes are, however, freely motile and can pass between endothelial cells to escape from (or enter) the vascular system, so that they are frequent in extravascular tissues where they may make extensive migrations, even passing into the body's secretions such as saliva.

Larger lymphocytes in the circulation, sometimes known as *lymphoblasts*, are a mixture, firstly, of *immature cells* capable of becoming small lymphocytes—a step which usually occurs in the lymphoid system—and secondly, of *transforming lymphocytes* which are formed when small lymphocytes become functionally active after antigenic stimulation. Both types of large lymphocyte are actively engaged in synthesizing protein, and hence possess a partly euchromatic nucleus with a prominent nucleolus, and a wide perinuclear zone of strongly basophilic cytoplasm. Ultrastructurally, polyribosome groups are numerous; in B-lymphocytes (*see* p. 61) these are chiefly attached to the endoplasmic reticulum, in which antibodies are synthesized prior to their secretion. In T-lymphocytes, and immature lymphoblasts, ribosomes are mostly unattached. The nuclei of both B- and T-lymphocytes, although more euchromatic than those of small lymphocytes, still retain a peripheral rim of heterochromatin, usually arranged in blocks not unlike the figures around a clock face, or the spokes of a wheel (the 'radkern' form), indicating that only a small part of the genome is active, as one might expect in a cell synthesizing a restricted range of specific proteins. In the most *immature* cells the nucleus is evenly euchromatic, reflecting the wide variety of proteins being made.

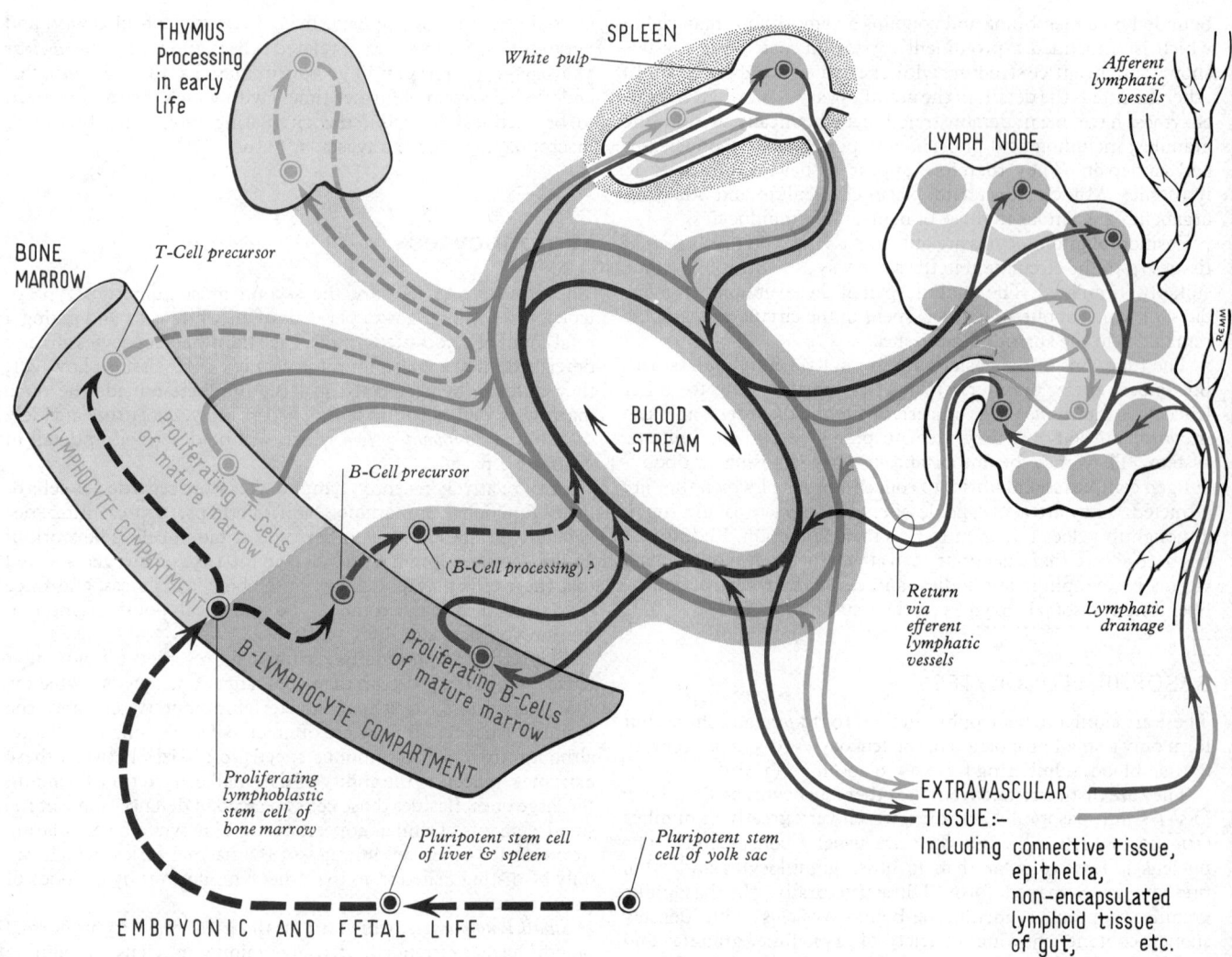

1.70 A diagram depicting current views of the origins and circulation of the two major classes of lymphocyte from the bone marrow to the peripheral lymphoid tissues. B-lymphocytes are shown in red, T-lymphocytes in blue.

The lifespan of lymphocytes varies from a few days to many years, and so we can distinguish between *short lived* and *long lived* lymphocytes. Those with long lives are almost certainly of great significance in *immunological 'memory'* (*see* p. 62).

Origins of Lymphocytes

Much of the knowledge of lymphocyte life history (**1.70, 71**) has come from experimental situations which include the tracing of radioactively or genetically labelled cells—the latter involving transfusion of lymphocytes with chromosomal abnormalities into normal but inbred strains of laboratory animals. Other experiments are based on the sensitivity of lymphocytes to ionizing radiation, the various components of the lymphocytic system being eliminated by selective irradiation, accompanied by appropriate transfusions or transplants of lymphopoietic tissue from bone marrow, lymph nodes, thymus and so forth. Thirdly, selective surgical removal of various lymphopoietic components and the removal of lymphocytes from the lymphatic channels has been carried out. In addition to these, a whole battery of immunological techniques has been brought to bear on the problem, including immunoassay, tissue culture methods, techniques for the localization of antibodies by their reactions with antigens or antigen-antibody complexes previously labelled with fluorescent compounds or electron-dense atoms (or enzymes) with subsequent examination by fluorescence and electron microscopy respectively.

The picture which has emerged is still far from clear; nevertheless tentative views can be expressed at this stage.

Lymphocytes originate in the embryo from mesenchymal cells of the yolk sac initially, and later in the liver and spleen. These primitive *stem cells* subsequently take up residence in the bone marrow, which becomes the only site of stem cell proliferation after birth. Similar considerations apply to the haemopoietic *colony-forming units* (*see* p. 66). Such stem cells divide unequally to throw off lymphoblasts which again divide and transform into small lymphocytes. Some of these pass in the blood circulation to the thymus where they migrate into the lobules and divide repeatedly (p. 780); the resulting small thymus-processed *T-lymphocytes* then re-enter the bloodstream and can return to the bone marrow or to the *peripheral lymphoid tissues* (tonsils, lymphoid tissue of the alimentary and respiratory tracts, lymph nodes, spleen), which they enter by migrating through the walls of the capillaries or post-capillary venules. Here, the cells are found in the sinusoids of the lymph nodes and in the loose superficial tissue of the lymph nodes (i.e. paracortical areas), and *peri-arteriolar white pulp* of the spleen, that is, in both cases, tissue neighbouring the *germinal centres* (p. 768). They can enter the general flow of lymph, returning to the bloodstream via the thoracic and right lymphatic ducts and the brachiocephalic veins, and so eventually back to the lymphoid tissues again. This *circulation of lymphocytes* (see also p. 773), first established by Gowans, is responsible for the large number of lymphocytes found in the blood (**1.70**). When stimulated antigenically such lymphocytes enlarge and multiply, and their progeny are capable of interacting with tissue cells (*vide infra*).

The *B-lymphocytes* do not pass through the thymus, but move directly via the circulation to the general lymphoid tissues, where they settle down. *En route*, they may undergo a phase of differentiation similar to that of the T-lymphocytes in the thymus, although it is not clear where this step might take place. Various sites have been suggested, including the intestinal lymphoid tissue, but as yet conclusive proof is not available. In birds, B-lymphocytes are derived from a specialized diverticulum of the cloaca, called the *bursa of Fabricius* (giving this type of lymphocyte its title—the *bursa equivalent*—or *B*-lymphocytes, although subsequently it appears that such a prefix can be even more appropriately applied to denote the bone marrow origin of the B-cell class in mammals). On antigenic stimulation they multiply to form the *germinal centres*; such lymphocytes can, while still within the lymphoid tissues, or after further migration, transform into the large pyroninophilic (i.e. RNA-rich) plasma cell series, which produce antibodies in their extensive rough endoplasmic reticulum.

FUNCTIONING OF THE LYMPHOCYTIC SYSTEM

The lymphocytic cells, together with the phagocytes of the reticulo-endothelial system, are responsible for the defensive reactions of the body, the former by antibody formation, the latter by phagocytosis. These defences are directed against foreign invasion by bacteria, viruses, fungi, protozoa, helminth worms, etc., or their metabolites (toxins), and against any unwanted or abnormal materials within the body such as proteins and effete, neoplastic, or virally transformed cells.

B-lymphocytes can, when antigenically stimulated, proliferate and their progeny transform to larger cells (*plasmocytes*) which synthesize and secrete *circulating antibodies*; the latter chemically 'recognize' and bind specifically to foreign chemical substances (*antigens*) so as to inactivate them, or cause their destruction. Antibodies may circulate freely in the body fluids (*soluble antibodies*) or may be secondarily attached to a variety of defensive cells (*cytophilic antibodies*) to enhance their activities, or to enable them to carry out a wider range of functions (**1**.70, 71).

Chemically, **antibodies** (or *immunoglobulins*) are proteins with a molecular weight of 150,000–950,000 daltons. Each antibody molecule consists of four polypeptide chains, two of them long (about 15 nm: the *heavy* chains) and two shorter (*light* chains) all joined together by sulphydryl links. The whole molecule is Y-shaped, two of the arms of the Y being able to swing around a central hinge. One end of the molecule (the *Fab* fraction) is highly variable in its amino-acid sequences, and since this end is responsible for specific binding to antigens, there is a vast number of possible antibody varieties, each type being able to bind a specific antigen in a manner closely analogous to enzyme-substrate binding. The other end of the molecule, the *Fc* portion, is much less variable and can attach to certain cells if they possess

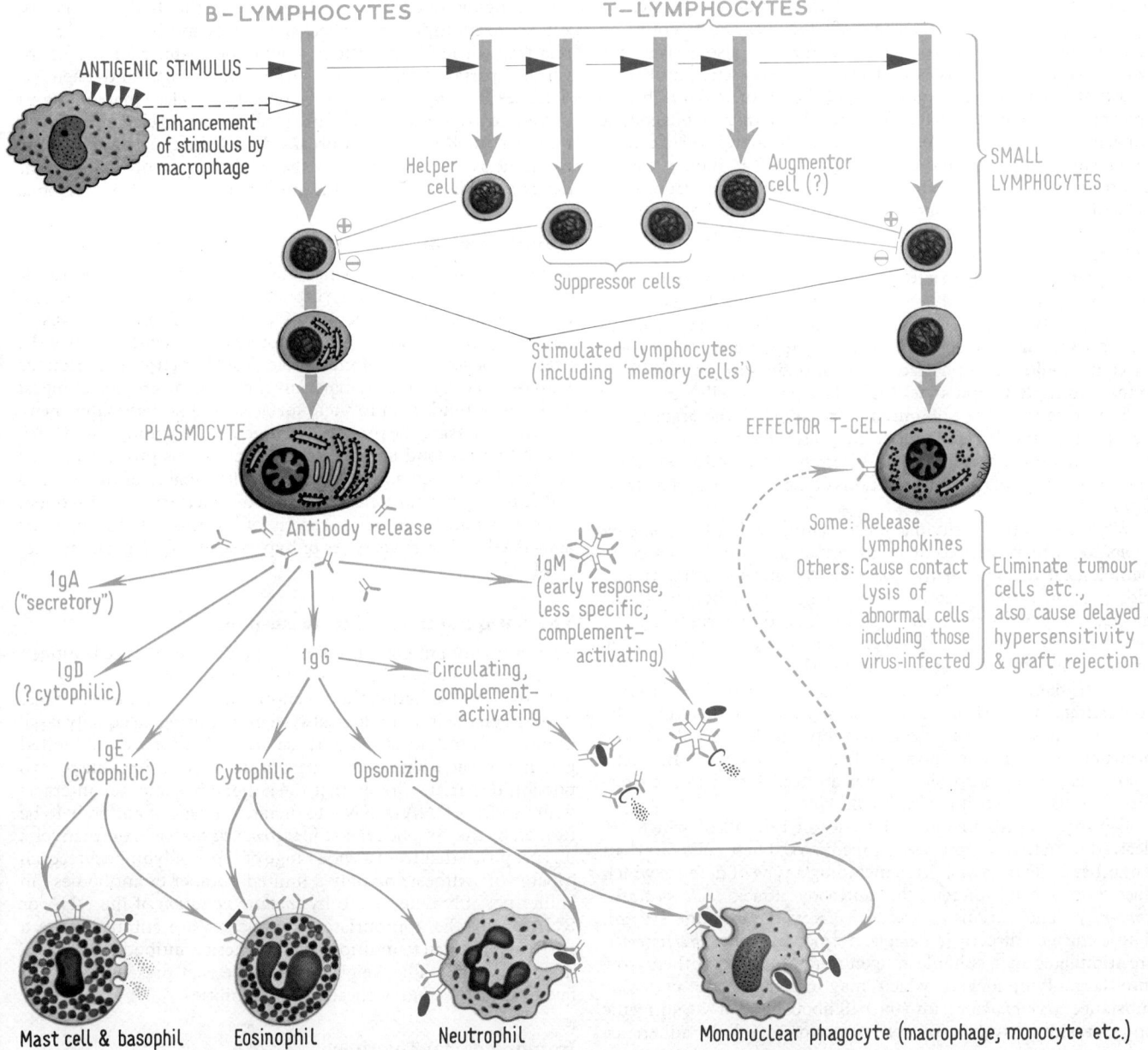

1.71 Schema of the origins and functions of the various subclasses of lymphocyte, including the role of antibodies (green) in inactivating antigens (red) and in initiating defensive activities in other classes of cell.

A further subclass of defensive 'killer cells', active against virally transformed cells, has been demonstrated to act cooperatively with effector T-cells.

specific *Fc* receptors on their surfaces, thus conferring antigen-binding properties upon them by virtue of the free *Fab* ends of the bound antibodies.

At present, five classes of antibodies are distinguishable in the blood plasma and elsewhere. These are: Immunoglobulin G (IgG), which forms the bulk of circulating antibodies; Immunoglobulin M (IgM), which is often formed early in the immune response and is usually present as a pentamer, the monomers joined at the *Fc* ends into a star-like aggregate; Immunoglobulin A (IgA), which is present in secretions of the body, particularly saliva and other fluids of the alimentary tract; Immunoglobulin E (IgE), which is a cytophilic antibody found on the surface of mast and perhaps other cells; and finally Immunoglobulin D (IgD) of uncertain significance, although believed to be important in the ontogeny of the immune system. All classes can be soluble antibodies, and IgG and IgE can also be cytophilic (*vide infra*).

The defensive functions of antibodies are numerous. They can *agglutinate* antigens by forming cross-links between them, so rendering them inactive as infective agents (e.g. viruses). After binding an antigen they may also co-bind *complement*, a protein complex in plasma, which punctures cell membranes, so causing the lysis of bacteria and other cells attacked by antibody. They may also bind to antigens after attachment by their 'safe' *Fc* ends to other defensive cells which are thus armed to detect, adhere to, and ingest or enzymatically damage foreign bodies; examples are macrophages and neutrophils both of which bear *Fc* receptors on their surfaces. Such cell-bound antibodies can also activate the complement complex to cause lysis of bacteria, etc. Antibodies may first coat an antigen (*opsonizing antibodies* or *opsonins*) before they attach to a macrophage, thereby stimulating its phagocytic capacity. Cytophilic antibodies may cause the activation of cells in other ways when they bind antigens. Thus, IgE antibodies trigger the release of histamine and other vasoactive agents from mast cells to which they are bound, when activated by antigens; this is seen in certain types of allergy, for example to the proteins of pollen.

B-lymphocytes, themselves, can be activated to divide and transform into antibody-secreting plasma cells by antigenic stimulation of cytophilic IgG bound to their surfaces, and T-lymphocytes are also activated in the same manner. In all these cases it should be emphasized, the antibodies are derived from transforming B-lymphocytes (including plasma cells) and come to be attached to other cells only after passing into the tissue fluids or blood. Some B-lymphocytes may also act to '*suppress*' the transformation of other types of B-lymphocyte, whilst yet others carry a '*memory*' of previous reactions of the immune system (*vide infra*).

When circulating antibodies bind to antigens they form *immune complexes* which, if present in abnormal quantities, may cause pathological damage to the vascular system and other tissues either, as in some types of glomerulonephritis, by interfering mechanically with permeability to fluids; or alternatively, by causing local activation of the complement system to attack cell membranes, thus causing vascular disease.

In pregnancy some IgG is transferred across the placenta, conferring a measure of *passive immunity* on the fetus; but in the case of Rhesus factor incompatibility, it brings about the destruction of Rhesus-positive fetal erythrocytes. In some mammals (oxen, sheep), antibodies are transferred in the first-formed milk (colostrum) after birth.

T-lymphocytes include subclasses of cell, all of which are derived from thymus-processed precursors. These cells carry out a number of distinct activities, including a type of defence which does not depend directly on antibody attack (*cell-mediated immunity*), and various *controlling roles* in the immune system. Those engaged directly in defence (termed by some, *effector cells*) are stimulated by a suitable antigen to reproduce and then grow into large lymphocytes which may either release cytotoxic substances (*lymphokines*) into the neighbourhood of the antigenic body, or, alternatively, cause its destruction by cellular adherence to it with the subsequent release of toxic materials at short range. They may also in some way 'arm' other cells (e.g. macrophages, or the recently described '*killer cells*') to carry out a similar aggressive, cytotoxic operation. Lymphokines are non-antibody substances with a molecular weight of from 20,000 to 80,000 daltons, and appear to consist of several types of molecule. In addition to their cytotoxicity they can also affect the migration of macrophages, neutrophils, and eosinophils, stimulate the mitosis of other lymphocytes, and also increase capillary permeability: all these activities of lymphokines enhance the effectiveness of defensive responses.

The type of immunity mediated by T-lymphocytes thus appears to be chiefly concerned with the elimination of foreign or abnormal eukaryote cells, including virus-infected cells, fungi, some protozoa (e.g. trypanosomes), neoplastic cells, and, unfortunately for medicine, foreign tissues in grafts (if their antigens are not sufficiently closely matched with those of the recipient). Cell-mediated immunity is also seen in such responses as delayed hypersensitivity.

Controlling functions of T-lymphocytes include '*helper*' actions to enhance the responses of B-lymphocytes and other T-lymphocytes, and '*suppressor*' effects which may inhibit the transformation of some classes of B- and T-cells. Finally, a proportion of T-lymphocytes are *memory* cells, able to respond rapidly to those antigens to which the body has previously reacted, as also in the case of some B-lymphocytes noted above.

Although the plethora of activities carried out by lymphocytes seems highly complex, it is to be expected that the potent and wide-ranging defensive mechanisms of the body should be subject to multiple strict checks, controls and balances. As yet, relatively little is understood about the manner in which the various parts of the whole system of cellular and chemical defences are *integrated*, but it is increasingly clear that they must be viewed as a *single system* of great efficiency and elegance. When, however, such integration breaks down, the effects may be far-reaching as, for example, in the wide variety of *auto-immune diseases*, and in neoplasia of the immune system, such as myeloma.

Immunological Memory

If after one antigenic response the body is again exposed to the same antigen, the second response is much more rapid and extensive, even after a period of years; this forms the basis of clinical immunization. This phenomenon evidently implies the presence of some type of immunological 'memory'. Experiments with isotopically labelled materials show that macrophages ingest antigens or bind them to their surfaces, and perhaps alter them, thereafter passing the processed antigen on to lymphocytes which may either respond immediately or, perhaps as progeny of such activated cells, remain as 'memory cells' capable of mounting a much more efficient defensive train of reactions on subsequent antigen stimulation. Over-reactivity of this system may be associated with various types of hypersensitivity, e.g. anaphylactic shock.

The Nature of the Antigenic Response

The antibody-antigen reaction is *highly specific*, each antigen requiring its own antibody for binding to occur.

In the life of an individual an enormous range of antigens must impinge on the immunological system, requiring an equally large number of antibodies. How these are coded for by the limited genome of each lymphocyte constitutes a severe problem; two possibilities exist. Firstly, that the antigen in some way interacts with the cell's DNA or RNA to dictate the type of antibody to be formed by the lymphocytes (the *instructive theory*). Alternatively, it is proposed that from a wide range of types of lymphocyte, each capable of synthesizing only a limited number of antibodies (an ability possibly determined by *somatic mutation* of the DNA or RNA), only the appropriate lymphocytes are stimulated by a particular antigen to multiply and synthesize antibody (the *clonal selection theory*). The weight of evidence is at present greatly in favour of the second of these two possibilities.

Immunological Tolerance

Since lymphocytes react to foreign antigens and not usually to the proteins and carbohydrates of the body itself, there must be some mechanism which ensures the distinguishing of *self* from *non-self*, that is an *immunological tolerance* to self. This may break down in auto-immune diseases such as Hashimoto's syndrome—self-immune destruction of the thyroid gland—and perhaps in disseminated sclerosis which causes demyelination of tracts in the central nervous system. Self-tolerance is achieved during fetal development in man, involving perhaps the action of the thymus (p. 782). Burnet has widened the theory of selective antigenic response (*vide supra*) to attempt an explanation for the mechanism of self-tolerance (the *clonal theory*), suggesting that those cell lines (*clones*) of lymphocytes which produce antibodies to the body's own tissues are suppressed in fetal life and are then no longer available to multiply at a later stage (*see* Burnet 1969; Edelman 1974). Self-tolerance may be extended to the antigens of a genetically identical or closely similar individual, so that grafts of tissue between monozygotic twins or inbred strains of animals may be accepted.

All tissues from within a particular species show antigenic classes similar to those seen in the erythrocytes (p. 68), and in man a number of *histocompatibility* tissue antigen groups have been recognized (p. 69). The success or failure of *isologous grafts* (i.e. grafts between genetically dissimilar members of the same species) depends largely on the degree of matching of these antigens between donor and recipient.

PLATELETS (THROMBOCYTES)

Blood platelets, also known as *thrombocytes*, (**1**.68, 69), are relatively small (2–4 μm), irregular or oval discs (Hovig 1968), present in large numbers (250,000–500,000/mm³) in blood. In fresh blood samples they rapidly stick to one another, and to all available surfaces, unless the blood is previously treated with citrate or other substances which reduce the availability of calcium.

In Romanowsky-stained preparations, platelets show an outer clear zone (*hyalomere*) and an inner basophilic granular region (*granulomere*). Ultrastructurally each platelet is seen to be anucleate, unlike similar cells in submammalian groups of vertebrates (for which the term thrombocyte is perhaps best reserved). Each platelet is surrounded by a plasma membrane with a thick cell coat; the latter is probably the basis of the adhesive properties of platelets. Beneath the surface is a band of about ten microtubules which runs around the perimeter of the cell and probably gives rise to its characteristic shape. Associated with the microtubules are microfilaments of an actin-like protein, *thrombosthenin*, responsible for platelet contraction (Zucker-Franklin 1969). Within the cytoplasm are mitochondria, some endoplasmic reticulum, glycogen, and membrane-bound vesicles of varying density. These include large lysosomes rich in acid phosphatase, large dense-cored vacuoles containing acid muco-substances, and small osmiophilic granules of 5-hydroxy-tryptamine (Day *et al.* 1969). Narrow, agranular endoplasmic reticulum, tubules, and a wider system of channels and vacuoles connected to the surface have also been described (*see* Cawley and Hayhoe 1973).

Platelets form an important element in the haemostatic mechanism by initiating thrombus formation due to their adhesion to each other (*agglutination*) and to damaged surfaces, and by the release of factors which assist in the formation of the fibrin clot. The contraction of individual platelets is also involved in clot retraction. Details of clot formation are given in many standard texts and will not be considered further here. The life span of platelets is 8–14 days (Heyssel 1961).

Haemopoiesis (1.72)

The earliest sign of blood *vessel* formation in the human embryo occurs when, during the early primitive streak stage of development, angioblastic tissue differentiates almost simul-taneously in various extraembryonic sites, namely, in the mesenchyme of the yolk sac wall, and in similar tissue of the connecting stalk and chorion (p. 180). The earliest formation of blood *cells* (**1**.72), however, appears to be confined to the wall of the secondary yolk sac, where they differentiate from deeply placed mesenchymal cells which lie next to the yolk sac endoderm. Whilst a mesodermal or endodermal origin for these cells is still debated, they are unquestionably mesenchymal in character, and differentiate into the primitive *stem cells* of the haemopoietic line, which give rise directly to fetal blood cells. Beginning in the second month, a number of intraembryonic sites of haemopoiesis appear and slowly replace the earlier sites. These intraembryonic sites *succeed, but overlap*, each other in time, each site gradually increasing in importance and then waning. Initially, the intraembryonic sites are broadly *intravascular*, but soon *extravascular* loci of haemopoiesis supervene. Rapidly the *liver* becomes the dominant organ of embryonic blood formation, its activities continuing until about the seventh month. Lagging somewhat behind the liver, the *spleen* is then added, its haemopoiesis continuing from the third to sixth months. From later in the third month, an additional source of blood cells emerges, namely the *bone marrow (myeloid tissue)* where *all blood cell types* are formed, and later the peripheral lymphoid tissues, where only lymphocytes are produced (lymphopoiesis). The *thymus* is an active lymphopoietic organ from this stage (whilst initially it also has some general haemopoietic functions). The myeloid and lymphoid tissues become the dominant source of supply by the seventh month, and shortly after birth all other sites have regressed completely (*see* Wintrobe 1974). Occasionally, clumps of tissue capable of total haemopoiesis are found outside the bone marrow (*extramedullary tissue*) in paravertebral sites; in pathological cases, where there is more demand for haemopoiesis, and especially during early childhood, complete haemopoiesis may also occur in the liver, spleen and lymph nodes, and infrequently, also in the kidneys, adrenals, adipose tissue, general connective tissue, and even in cartilage.

THE BONE MARROW

This is a soft pulpy tissue which is found not only in the cylindrical marrow cavities of the long bones but also in the spaces between the trabeculae of all bones and even in the larger Haversian canals. It differs in composition in different bones and at different ages, and occurs in two forms, *yellow* and *red marrow*.

During fetal life and at birth there is red marrow throughout the skeleton. After about the fifth year the red marrow is gradually replaced in the long bones by yellow marrow. The replacement commences earlier and is more advanced in the more distal bones. Further, in each bone the replacement, in general, proceeds from the distal to the proximal end, though some maintain that it commences in the centre of the shaft and extends in both directions, but more rapidly in the distal. By 20 to 25 years of age the red marrow persists only in the vertebrae, sternum, ribs, clavicles, scapulae, pelvis, cranial bones and in the proximal ends of the femora and humeri. In old age the marrow of the cranial bones undergoes degeneration and is then termed *gelatinous marrow*.

The yellow marrow consists of a basis of connective tissue, supporting numerous blood vessels and cells, most of which are fat cells, although a small population of typical red marrow cells persists.

The Red Marrow (1.74)

This consists of a framework of reticular connective tissue with argyrophilic reticulin fibres and attached phagocytic cells, containing in its meshes a variety of blood cells and their precursors, and a few fat cells. Small nodules of lymphoid tissue are also scattered through it. No lymph vessels have been demonstrated, however, in yellow or red marrow. The vascular supply of the bone marrow is derived from the nutrient artery to the bone and drained by the accompanying vein (p. 257). The artery ramifies in the bone marrow; its branches terminate in thin-walled arteries from which an extensive plexus of cell-lined sinusoids arises. These, in turn, drain into disproportionately large veins (Brookes and Harrison 1957). Many of these blood sinusoids are collapsed at any instant and are frequently but erroneously referred to as intersinusoidal capillaries. The marrow

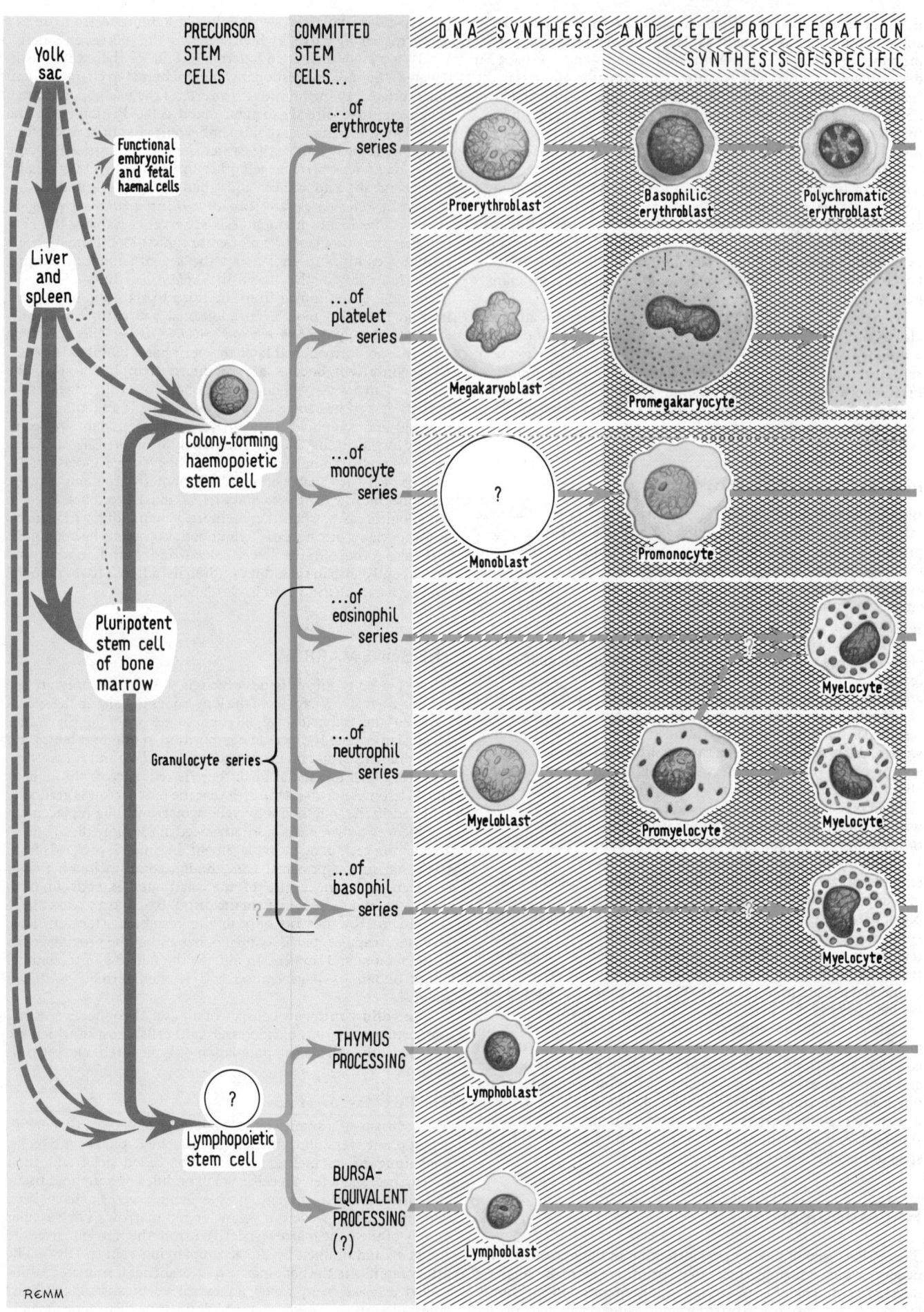

PRECURSOR STEM CELLS

COMMITTED STEM CELLS...

DNA SYNTHESIS AND CELL PROLIFERATION
SYNTHESIS OF SPECIFIC

Yolk sac

Functional embryonic and fetal haemal cells

Liver and spleen

Colony-forming haemopoietic stem cell

Pluripotent stem cell of bone marrow

Granulocyte series

...of erythrocyte series

Proerythroblast

Basophilic erythroblast

Polychromatic erythroblast

...of platelet series

Megakaryoblast

Promegakaryocyte

...of monocyte series

?

Monoblast

Promonocyte

...of eosinophil series

Myelocyte

...of neutrophil series

Myeloblast

Promyelocyte

Myelocyte

...of basophil series

Myelocyte

?

THYMUS PROCESSING

Lymphoblast

?

Lymphopoietic stem cell

BURSA-EQUIVALENT PROCESSING (?)

Lymphoblast

REMM

1.72 Haemopoiesis: a schema of the origins, developmental stages, and fates of the different classes of haemal cells. Where details of development are uncertain, putative routes are shown in broken lines. Developmental pathways of the prenatal period are indicated (left) in red, and different

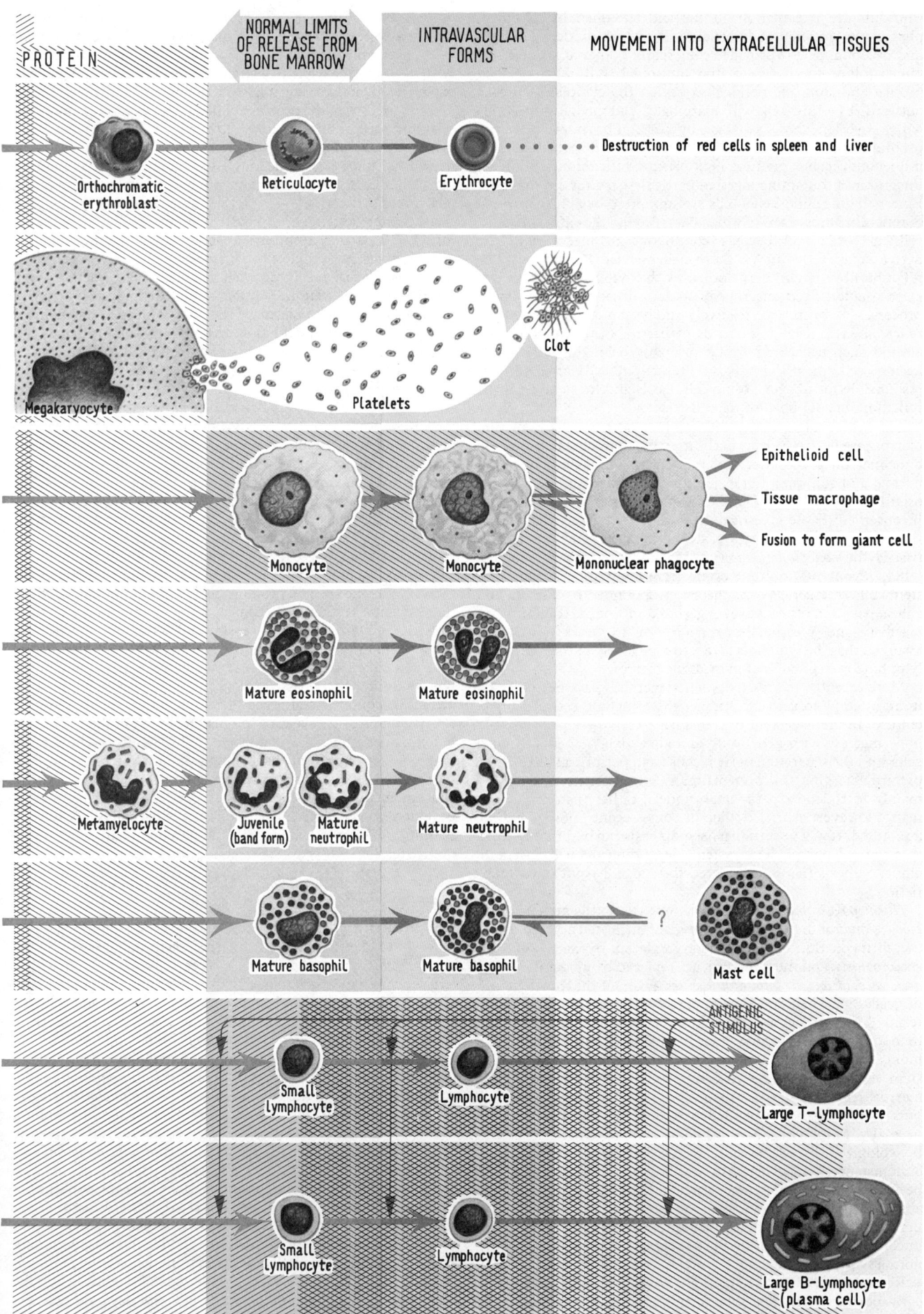

stages of cellular development are hatched diagonally according to the column headings. It should also be noted that there is evidence for macrophage proliferation in the extravascular tissues, a finding not indicated in this scheme.

sinusoids are irregular in outline and are lined by flattened, phagocytic, reticulo-endothelial cells, which under the light microscope appear to provide a complete endothelium. Their oval nuclei follow the curves of the sinusoidal wall, when seen in profile, and from the perinuclear region the cytoplasm spreads outwards as a thin veil, with marginal pseudopodial projections which overlap those of adjacent cells. The nucleus carries prominent nucleoli, and the cytoplasm is well provided with mitochondria, but granular endoplasmic reticulum is minimal. Phagosomes containing ingested particulate matter are common. Externally the endothelial cells are supported by a delicate mesh of reticulin fibres, a few of which pass through the extrasinusoidal cellular cords and become continuous in places with the perivascular reticulin of the smaller marrow arteries. The reticulin fibres are in part clothed by the cytoplasmic process of extravascular fixed macrophages, and, it is claimed, similar processes of primitive, relatively undifferentiated, *reticular* or *reticulum cells*. Islands or cords of haemopoietic tissue cluster around such cells, each group probably belonging to a single haemal cell line (Weiss 1977). Reticulum cells appear to be responsible for phagocytosing cell débris and for laying down a reticulin fibre support for the cells.

The initial stages of differentiation of haemal cells in mature marrow are far from clear. Early attempts to trace cell lineages in bone marrow preparations solely by histological examination of normal and abnormal tissue, led to much controversy about the origins and relationships of the various types of cell which could be observed. Some investigators considered it likely that all cell lines arose from a single type of stem cell (*haemocytoblast*) in adult tissues (the *monophyletic theory* of Maximow and others), whereas others favoured a multiple origin from many different types of stem cells (the *polyphyletic* theory, propounded by Ferrata and colleagues). Yet others have suggested compromise solutions (see Hardisty and Weatherall 1974). For many years it was also assumed that the lymphocytes arose exclusively in the peripheral lymphoid system and not in the bone marrow.

More recently, many experimental methods have been devised in an attempt to settle this issue; these include cell and organ culture, the replacement or transfusion of radiation-inactivated bone marrow with genetically or radioactively labelled haemal cells, and the separation of stem cells from peripheral blood. The picture emerging still presents many uncertainties, and applies mainly to rodent experimental models, rather than directly to man. However, a brief outline of some recent findings will be attempted. It will be seen that haemopoiesis can best be described as a limited polyphyletic system, depending upon the stage in embryogenesis that is considered the starting point of differentiation.

Haemopoiesis in embryonic life commences in the mesenchymal *blood islands* of the yolk sac, where giant nucleated erythroid cells are formed. But these are soon replaced by *nucleated fetal erythroblasts*, and these, in turn, are replaced by moderately large *anucleate, biconcave fetal erythrocytes* by about the fourth month of gestation. The adult type of erythrocytes are present by full term. Megakaryocytes, then granulocytes, lymphocytes and monocytes, appear in that order between the second and fourth months of gestation. During the earlier developmental stages, stem cells occurs in the yolk sac, liver, spleen and early bone marrow (p.180), and these tissues can give rise to *all haemal cell lines* if they are transplanted to adult spleens (i.e. they are haemally 'totipotent'). Meanwhile, the lymphoid organs start to be colonized by stem cells which, however, under normal circumstances only give rise to lymphocytes, and hence appear to be already committed to that particular line of cells. However, the ability of these lymphoid tissues to carry out full haemopoiesis under certain pathological conditions suggests that some uncommitted stem cells have been retained in these sites, but are normally suppressed.

In the *bone marrow* two types of *stem cell* have differentiated by early postnatal life; one type forming lymphocytes, whilst the other gives rise to erythrocytes, platelets, monocytes, neutrophils, eosinophils, and possibly basophils. The second of these stem cell types readily forms *haemal cell colonies* when transplanted to spleen cultures and each such cell is termed a *colony-forming unit*

(CFU). The pluripotent CFUs are not numerous (about 0·5 per cent of all haemopoietic cells in marrow), but they can circulate freely in the blood, where they appear as small lymphocyte-like cells, 7–10 μm across, with a large pale nucleus containing one or two nucleoli, and bearing a narrow rim of cytoplasm.

Next, we turn to the problem of the differentiation of such stem cells into the various cell lines of the blood. The early stages of these lines are difficult to distinguish since usually they cannot be recognized until well after they have begun to differentiate. In each cell line, however, a definite sequence of major stages can be recognized (or ascribed)—*see* **1**.72. These stages are: (1) the *final commitment* of a stem cell to a particular line of differentiation; (2) early cell *proliferation* to form a large pool of dividing cells; (3) *differentiation* as specific proteins characterizing the particular line are synthesized, accompanied by the gradual cessation of cell division; (4) final *maturation*, marked by the gradual closure of protein synthesis; (5) *release* of the cells from the bone marrow parenchyma into the circulation. In the case of some cell lines, cell division may continue outside the bone marrow (e.g. monocytes or macrophages), and maturation may also be completed after release (e.g. erythrocytes).

The earliest 'committed' stages of each cell line are outwardly similar to the CFU, but at the next proliferative stage, differences begin to emerge. In all, however, the initial picture is that of a rapidly dividing, relatively undifferentiated cell, in which the nucleus is large and euchromatic, with prominent nucleoli, the moderately basophilic cytoplasm contains unattached polyribosomes and the total cell size is large. As proliferation proceeds, the size of the cell usually diminishes as cell division outstrips cell growth. As the phase of final differentiation continues, the ribosomes become numerous, and the cytoplasmic basophilia increases, as the specific proteins begin to be synthesized, whilst the nucleus gradually becomes more heterochromatic and smaller, as DNA synthesis ceases, and ultimately the nucleus becomes multilobed or pyknotic as RNA synthesis terminates. The completed cell is then ready to be released, although the precise timing of its passage from the bone marrow varies with metabolic conditions and the 'demand' for more cells, so that relatively immature types of cell may be found in the circulating blood under abnormal conditions. This 'shift to the left' (a graphic convention of haematologists in which the cell is represented as maturing from the left to the right) is a useful concept in diagnostic haematology. These general features of development can be seen in several lines of haemal cells, and help to unify what appears at first sight to be a highly divergent series of progressions.

Erythropoiesis (1.72, 73)

The first identifiable cell of the erythroid series is the *proerythroblast* (*pronormoblast*), a large (14–20 μm) cell with a large euchromatic nucleus and moderately basophilic cytoplasm. The latter already contains small amounts of ferritin and contains some of the protein spectrin in its surface; both are characteristic of this cell line which must have been preceded by an erythropoietic stem cell sensitive to stimulation by erythropoietin (*see* p. 68). Proerythroblasts proliferate, and haemoglobin-RNA synthesis begins, as the smaller (12–17 μm) *basophilic* or *early erythroblast* (*basophilic normoblast*) appears, rich in ribosomes; shortly afterwards, haemoglobin synthesis commences so that the cytoplasm becomes partially eosinophilic (the *polychromatophilic*, or *intermediate erythroblast/normoblast*) which has a diameter of some 12–15 μm. As the cytoplasm becomes fully eosinophilic the cell is transformed into an *acidophilic erythroblast* (*orthochromatic* or *final erythroblast/normoblast*) which has a diameter of 8–12 μm. At this stage, most of the RNA is lost and the nucleus, becoming intensely pyknotic, is finally extruded from the cell, thus forming the anucleate *reticulocyte*. At this point the cell is released into the circulation, losing its residual RNA in a few days to become a mature *erythrocyte*. The whole process of erythropoiesis takes 5–9 days; after release, reticulocytes typically sojourn for up to 2 days in the marrow sinusoids, and then for an equal time in the spleen, perhaps because of their particularly adhesive cell coat.

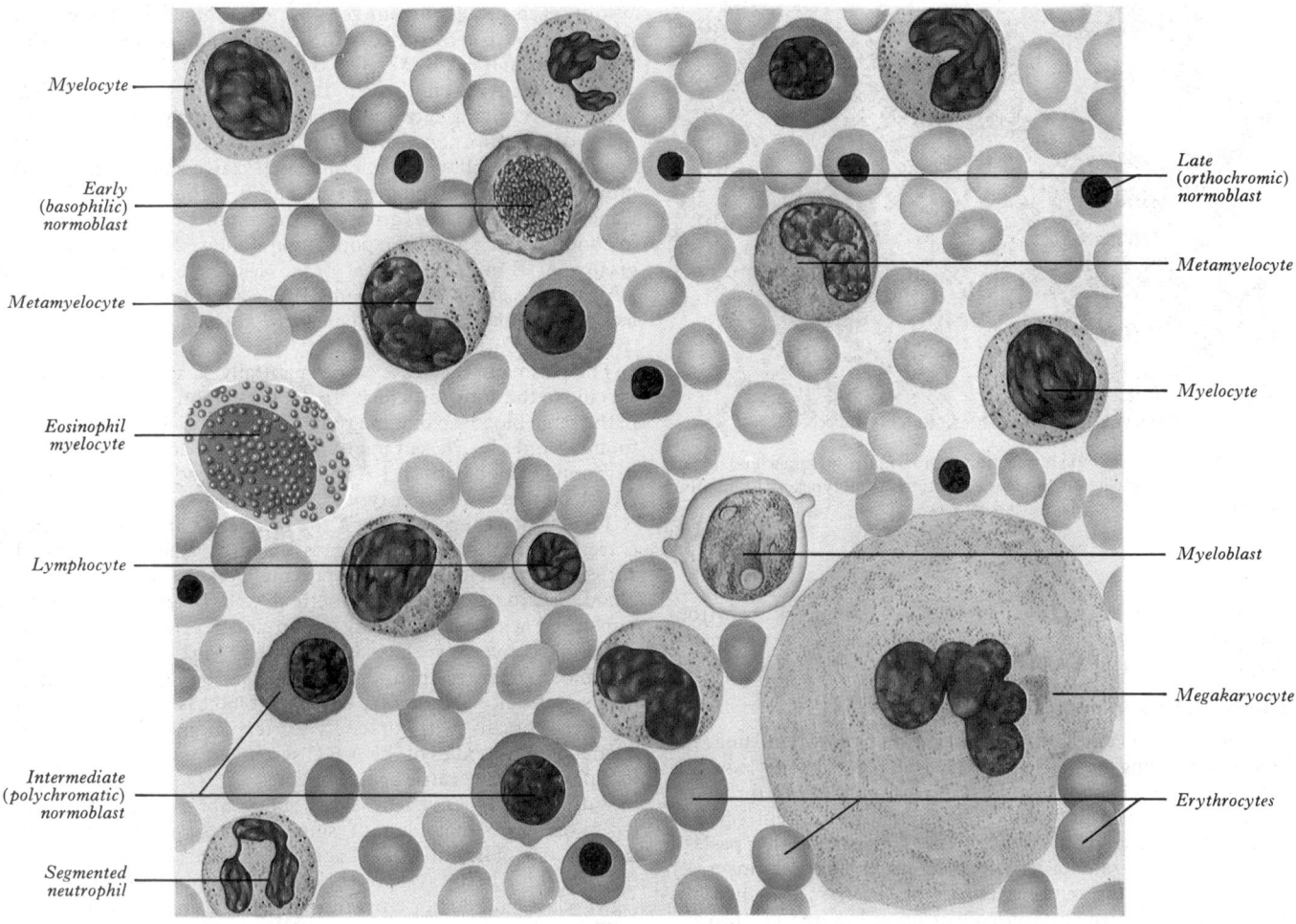

Myelocyte

Early
(basophilic)
normoblast

Metamyelocyte

Eosinophil
myelocyte

Lymphocyte

Intermediate
(polychromatic)
normoblast

Segmented
neutrophil

Late
(orthochromic)
normoblast

Metamyelocyte

Myelocyte

Myeloblast

Megakaryocyte

Erythrocytes

1.73   A smear preparation of normal human red bone marrow, obtained by sternal puncture. This is a composite figure, drawn from a number of normal smears prepared by the late Dr. R. L. Waterfield, Guy's Hospital, and stained with a modification of Leishman's stain.

The cell lineage of normal erythrocytes is often called the *normoblastic series* to distinguish it from abnormal erythroid lines; too few divisions of the early proliferative stages may give rise to abnormally large erythrocytes (*macrocytes*), whereas too many divisions, or insufficient early growth, can lead to *microcyte* formation. Disturbances in haemoglobin synthesis can also give rise to a variety of anaemias and other pathologies.

Granulocytopoesis (1.72, 73)

The details of this process are best known for the *neutrophil*, which will be described in some detail. Initially, the putative stem cells transform into the large (10–20 μm) *myeloblasts* which are similar in general size and appearance, though not in internal details, to the proerythroblast (*vide supra*). These proliferative cells differentiate into the larger *promyelocytes*, in which the first group of specific proteins is synthesized in the granular endoplasmic reticulum and Golgi apparatus, both of these organelles being quite prominent. The proteins are stored in large (0·3 μm) *primary* ('*non-specific*') *granules*, that are large lysosomes containing acid phosphatase and having azurophilic staining properties (p. 57). Next, in the smaller *myelocyte*—the last proliferative stage—the smaller *secondary* ('*specific*') *granules*, which contain a slightly different enzyme array, are formed in a similar manner. The nucleus is typically indented on one side in myelocytes (1.72, 73). Subsequently, in the *metamyelocyte* stage the cell size decreases further, the nucleus becomes heterochromatic and horse-shoe shaped, and protein synthesis practically ceases. Finally, as the neutrophil is released, the nucleus becomes heavily indented (the *juvenile* or '*stab*' form) and then partially divided into up to six lobes (the *segmented* or *mature neutrophil*).
Eosinophils pass through a similar sequence except that their nuclei never become as highly irregular as that of the neutrophil,

and only one set of lysosomal granules is synthesized. It is not known whether the eosinophil differentiates from the same myeloblast (or promyelocyte) stock as the neutrophil, or whether it is distinct from the colony forming unit (CFU) stage. In the case of *basophils*, it is not certain that they follow this general sequence at all; they may not even share the CFU as an ancestor.

Monocytes are also formed in the bone marrow, from stem cells of unknown structure; subsequently they pass through a proliferative *monoblast* stage, and then form differentiating *promonocytes* in which small lysosomes begin to be formed—these may be demonstrated by neutral red staining. After further divisions, monocytes are released into the general circulation, and then pass to perivascular and extravascular sites, which they populate as *mononuclear phagocytes* or *macrophages*.

Platelets, being fragments of cells, arise in a most unusual manner by the division of the cytoplasm of huge cells into many portions. The first detectable cell of this line is the highly *basophilic megakaryoblast* (15–50 μm); this is followed by a *promegakaryocyte* stage (20–80 μm) in which synthesis of granules begins; finally, the fully differentiated *megakaryocyte*—a giant cell (35–160 μm) with a large, dense, *polyploid*, *multilobate* nucleus—emerges. Once differentiation has commenced, mitoses proceed without cytoplasmic division, and the chromosomes are retained within a single polyploid nucleus containing 8n, 16n or 32n chromosomes, depending upon how many nuclear mitoses finally occur. Under the electron microscope, the cytoplasm is distinguished by numerous centrioles and spindle microtubules, both of which reflect the repetitive *mitotic* activity. Meanwhile, differentiation proceeds in the cytoplasm with the production of free polysomes, smooth endoplasmic reticulum, and fine basophilic granules. Cytomembranes within the cell fuse with one another, and, with invaginations of the plasma membrane and cell

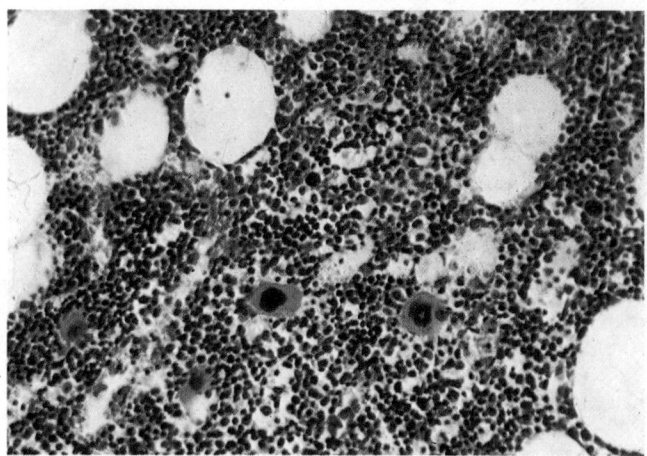

1.74 A photomicrograph of a section of bone marrow from a fetal human long bone. Note the heterogeneous collection of cell types including four large megakaryocytes. Magnification × 150.

coat to cut off portions of cytoplasm, which then break away from the parent cell to form platelets. The nucleus of the megakaryocyte eventually disintegrates.

Control of Haemopoiesis

The numbers of cells in the circulation are closely regulated, cell destruction being counterbalanced by cell replacement. How this system operates is known, at least in part, only for erythrocytes. Erythropoiesis is stimulated by a circulating protein *erythropoietin* synthesized by the tissues of the kidney and other parts of the body. The rate of erythropoietin synthesis is inversely proportional to the oxygen content of the tissues, hence low oxygen tensions, usually consequent upon lowered erythrocyte numbers, stimulate erythropoiesis, whereas high oxygen tensions cause the withdrawal of the stimulus. At high altitude the lowering of the partial pressure of oxygen in the atmosphere leads to a raised erythrocyte count.

Many other factors also affect the rate of haemopoiesis, for example thyroid hormones, somatotrophic hormone, androgenic steroids and other hormones. The numbers of cells in the blood show a *diurnal rhythm*, probably because of hormonal fluctuations. Infection, haemorrhage and other clinical disturbances also affect the pattern of cell production, as do cytotoxic chemicals and ionizing radiations, to which the dividing cells of the bone marrow are particularly susceptible.

Blood Groups

Early attempts at transfusion of blood led to the discovery that erythrocytes bear antigens on their surface (*see* Race and Sanger 1975) which can interact with naturally occurring antibodies in the plasma of other individuals, causing agglutination and lysis of the erythrocytes. Such antigens, which are not shared with all members of a particular species, are termed *iso-antigens*; other iso-antigens are found amongst cells of other tissues (*vide infra*). Erythrocyte antigens are known as *agglutinogens* and the corresponding antibodies as *agglutinins*.

Erythrocytes from an individual can bear several different types of antigen, each type belonging to an antigenic system in which a number of alternative antigens are possible in different persons. So far at least fifteen groups have been identified, which vary in their frequency of distribution amongst the various races of mankind, including the A B O, Rhesus, M N S, Lutheran, P, Kell, Lewis, Duffy, Kidd, Diego, Yt, Auberger, Ii, Xg and Dombrock systems. Clinically, only the A B O and Rhesus groups are of major importance. For a full description of blood groups in man, see the comprehensive account by Race and Sanger (1975).

All of these antigens are determined by genes carried by autosomes except Xg, which is borne by the X chromosome.

Within each group the antigens are determined by alleles and inheritance is in accordance with simple mendelian principles (p. 101). Thus, in the A B O system the genome may be homozygous and carry the A A complement, the blood group being A, the B B complement giving blood group B, or may carry neither (O O), the blood group consequently being O. In the heterozygous condition the following combinations can occur: A B (blood group A B), A O (blood group A), and B O (blood group B). In Caucasians and Negroes group O is the commonest, being present in about 50 per cent of the population, followed in frequency by groups A, B and AB in that order (Mourant 1975). In West Africans the Duffy determinant (Fy) is almost always absent, a lack which may confer resistance to *Plasmodium vivax* malaria (Miller and Carter 1976).

The plasma in each case carries naturally occurring antibodies specific to the antigens which are not present on the erythrocytes in the same blood, so that in group A blood, anti-B antibodies are found. Similarly, present in group B blood are anti-A antibodies, in group O blood both anti-A and anti-B antibodies, and in group AB blood there is neither type of antibody.

Transfusions succeed only if the recipient's antibodies do not correspond to the donor's antigens, and cross-matching of blood antigens is therefore vitally important. Persons with group AB blood, lacking antibodies to both A and B antigens, can be transfused with blood of any group and are termed *universal recipients*; conversely, those with group O, *universal donors*, can give blood to any recipient, the donor's antibodies being diluted to insignificance. Normally, however, blood is only transfused between persons with precisely corresponding groups, since anomalous antibodies of the A B O system are occasionally found in blood, and may cause agglutination.

Within the A B O system several subgroups exist (A_1, A_2, A_1B, etc.); the cross-matching of some of these is important in transfusions. The anti-A B O agglutinins, like all others (except those of the Rhesus system), belong to the immunoglobulin M (IgM) class, and do not cross the placenta during pregnancy.

The Rhesus antigen system, so-called because of its presence also in the erythrocytes of the Rhesus monkey, is determined by three sets of alleles, namely, Cc, Dd and Ee, the most important clinically being Dd. The commonest condition in Britain is CDe and about 83 per cent of the population is Rhesus-positive. Inheritance of the Rh factor obeys simple mendelian laws, and it is therefore possible for a Rhesus-negative mother to bear a Rhesus-positive child. Fetal Rh antigens can, under these circumstances, stimulate the production of anti-Rh antibodies by the mother, and since these belong to the immunoglobulin G group of antibodies they are able to cross the placental barrier and cause agglutination of fetal erythrocytes. In the first of such pregnancies little damage is usually caused, but in subsequent Rh-positive ones, massive destruction of fetal red cells may ensue, causing fetal or neonatal death. Treatment is by exchange transfusion of the neonate infant, or by desensitizing the mother after the first Rh-positive pregnancy with Rh-immune serum, which appears to destroy the fetal Rh antigen in the maternal circulation before the processes of immunological memory (p. 62) can be entrained (*see* Clarke 1975).

Of the other antigenic systems known, some of which are occasionally of clinical importance, many are restricted to individual genic groups or even families; they can be of great value to anthropologists when tracing demographic relationships, as of course are the major systems described above.

Other antigenic systems such as M N S (shared with other tissues in the body) can be used in medico-legal investigations to establish identity of blood, or in parental identification. These antigens can remain intact long after death, and have been detected even in mummified tissues from Egypt over 4,000 years old (Harrison *et al.* 1969).

The genetics of blood groups is complicated by gene linkage with other characters which may be of some clinical importance; duodenal ulcers show a higher incidence in those with group O blood than in the general population, for example.

Leucocytes also bear antigens, and about twelve such groups have so far been identified, ten of them belonging to the same complex system. These are similar to the *histocompatibility antigens* involved in graft rejection (*vide infra*), and they form the basis of possible *immunosuppressive* measures by means of *antilymphocytic sera*, which contain specific antibodies to the lymphocyte antigens.

Histocompatibility antigen system. Amongst important components constituting the cell membranes of all cells except erythrocytes, are factors which are responsible for determining the individuality of tissues from different persons, termed collectively the *histocompatibility* or HLA system (Bach and van Rood 1976). Various subdivisions of this system can be detected serologically, and four major subgroups, HLA-A, -B, -C, and -D, are known. Great variation of composition occurs within these subgroups, providing an enormous number of permutations amongst the human population. The HLA system has attracted much attention, lately, because of its importance in transplant surgery, and, in a quite different way, because of the association between certain of its subgroups and some types of disease. For example, the subtype HLA-B27 is present in nearly all cases of ankylosing spondylitis; the condition of haemochromatosis, a disturbance of iron metabolism, is strongly associated with HLA-A3, and rheumatoid arthritis with HL4-D4. Other conditions that display a statistical correlation with the HLA system include diabetes in juveniles, Hodgkin's disease, and multiple sclerosis. The reasons for these associations are not clear, but are thought to represent some type of loose *genetic linkage* between the HLA determinants and other alleles predisposing the body to a range of diseases. In mice, where this system has been widely investigated, genes that apparently control the responsiveness of the immune system to infection (the Immune related, or Ir genes) are situated between those genes that determine the major histocompatibility (H2) factors. In humans, where the HLA determinants are carried on the short arm of chromosome 6 (in a sequence A, C, B, D), genes coding for some of the serum complement factors, and those controlling antibody production, are also present in the same chromosomal region, and may therefore segregate with the HLA determinants during gamete formation.

Apart from these considerations, the HLA system is of great importance in the defence of the body against viruses and cancer. The HLA factors are integral proteins or glycoproteins of the plasma membranes of all nucleated cells (although the HLA-D subgroup appears to be specific to B-lymphocytes). Their main function is, apparently, to provide the means by which the immune system, particularly that associated with T-lymphocyte activity, can distinguish between normal cells and those which, like virally infected or neoplastic cells, bear alien or abnormal antigens, causing their subsequent elimination. Unfortunately, the same surveillance system applies to organ transplants and tissue grafts, for which procedures, recipient and donor histocompatibility subtypes must be matched as closely as possible to avoid immune attack and rejection. It is interesting that chemically the HLA factors have many features in common with antibodies, and it is feasible that they originated from a common cell-surface ancestor molecule, which provided a system of recognition between self and non-self, and which in the case of antibodies, later became adapted for specifically attacking foreign antigens (Cunningham 1977).

2

EMBRYOLOGY

Introduction

To distinguish between that which is living and that which is inanimate seems intuitively simple, and is a commonplace of everyday existence; but such a distinction continues to defy concise scientific definition. This question is obviously central to any study of living forms, and biological texts conventionally open by considering a series of attributes displayed by organisms, such as irritability or responses to environmental change, motility, ingestion followed by transformation (metabolism) of foodstuffs and excretion of end products, growth, reproduction, maturation and so forth.

However, the basic building blocks of living things, sub-atomic particles, atoms and molecules are common to all materials known to man; in this context the attributes listed above are seen to be the inevitable consequences of particular arrays or *organizations* of the building units.

THE DYNAMISM OF LIVING ORGANISMS

To date, our examination of both the physical universe and the biosphere has led to fundamental generalizations, firstly concerning the relativistic nature of all forms of observation, and secondly, the interrelation between matter and energy. Aspects of the latter are stated explicitly in the various *laws of thermodynamics*, which for our present purposes hold that the total energy content of the universe is constant, and that overall, the proportion of disorganized, randomized or dissipated energy (the *entropy* level) is steadily increasing. Living systems, however, appear to reverse this trend by their growth, reproduction and increasing structural complexity, that is to say, their entropy level is *decreasing*. This apparent paradox is only answered by considering life forms as interlocked structurally, functionally, and energetically with their environment, from which they are supplied by a constant flow of energy which is essential for their maintenance (Schrödinger 1967; Lehninger 1965; Young 1964). The ultimate source of this energy on earth is the structural degradation of materials in the sun with the emission of radiant energy partly as light. This is trapped by green plants and used in photosynthetic processes whereby simple substances, with the addition of energy, are transformed into complex molecules which are incorporated into the plant tissues and eventually provide a source of chemical energy, or food, for the animal kingdom. This absolute dependency upon supplies of environmental energy reflects the *dynamic* nature of life processes, which constantly undergo change. Further, the continued success of an organism in any habitat implies that its structural and functional organization is complementary to features of its environment, and they can only be meaningfully studied together, that is, as one system. Indeed, some observers hold that the various control systems of an organism necessarily embody a 'model' of their environment for effective responses to occur (Young 1971).

That living organisms are dynamic, of course, implies a changing state with the passage of time, but the time scale chosen for study may vary from millions of years to fractions of milliseconds. Examples of the latter include such rapid biochemical and biophysical interactions as the activation of visual pigments by photons, conformational changes in cell membranes with consequent changes in permeability, as during the transmission of a nervous impulse or the release of a neurochemical transmitter, and the formation and breaking of linkages between actin and myosin subunits during muscle contraction. A relatively short time scale is also involved in the multitude of feedback systems whereby the state of the internal and external environment is constantly monitored and appropriate adjustments are made to preserve the internal environment within the fairly narrow range of values necessary to the continued life of cells. Such a preservation of internal constancy, or *homeostasis* (p. 802), despite *short-term* fluctuations in the surroundings, is a central feature of the operation of all the principal organ systems, for example, the regulation of temperature, blood pressure, hydration, osmolarity, electrolyte and hydrogen ion concentrations, glucose and oxygen levels, etc.

To these may be added somewhat *slower homeostatic cycles* of endocrine secretory control systems, the varying rates of cell growth, death and replacement in various tissues, the behaviour patterns associated with mating, rearing of young, searching for and securing food, and also the ill-understood processes which lead to the laying down of short- and long-term memory traces.

REPRODUCTION AND EVOLUTION

Contrasting with these short-duration changes, all extant organisms have an ancestry which stretches back over hundreds of millions of years, a period which involved great changes in the natural environment. The latter were paralleled by a gradually increasing diversity of form and complexity of organisms, each more or less well adapted to life in some particular pocket of the changing environment. The *long-term homeostatic mechanism* which emerged and led to this diversification is intimately connected to another aspect of the dynamism of living organisms. Thus, in addition to securing an adequate flow of solar or chemical energy to drive its cellular machinery and continually replace its substance, each organism also passes through stages of *growth*, increasing *maturity* and, in complex forms, eventual *senescence* and *death*. Continuing life of the species is therefore dependent upon some mode of *reproduction*. Elementary life forms increase their mass and maturity until, at a critical point, they reproduce by simple equating division of their substance. Each successive organism has the same essential physico-chemical constitution as the parent, and part of its structure operates as an information store, that is, a chemical (or genetic) *memory system*, so that the offspring grow into identical replicas of the parent. Occasionally, however, apparently spontaneous changes occur in the organization of the genetic memory store, and such *mutations* lead to the emergence of organisms with altered characteristics. Often such *mutants* are less well fitted for survival under environmental stresses and they succumb, but sporadically a favourable mutation occurs. The transmission of this to succeeding generations leads to the establishment of a species with an enhanced probability of survival and, occasionally, the ability to populate new or changing environments.

Such **asexual** methods of reproduction are seen in many relatively simple extant forms of life, but the great diversity and complexity of the remainder followed the emergence of **sexual** methods of reproduction in which each offspring is the result of welding of half of the genetic memory of *two* parents. In essence each parent *segregates* certain specialized *reproductive cells* in which there is a *rearrangement* of the genetic material, and a *reduction* to half the quantity found in the general cells of the body (p. 27). Two reproductive cells, one from each parent, approach and *fuse*, and their genetic material intermingles in a specific manner (p. 99). The resulting offspring, whilst in part possessing certain characteristics which resemble those of one or other parent, and of course those of the species in general, also possesses unique features due to the new genetic combination and the previous genetic rearrangement. When the effects of occasional favourable mutations are added to the infinitely varied genetic combinations resulting from sexual reproduction over countless generations, the opportunities for the initiation of altered species are vastly increased. It is on such a wide range of variants that the forces of *natural selection* operate and result in the endless array of life forms which populate all corners of the earth's surface, the oceans and the skies.

Any organism, therefore, including mankind, is not only interlocked structurally and energetically with its environment, exhibiting transient homeostatic responses and rhythmic variations, but is also at some point in its individually unfolding life history (*ontogeny*). It is progressing towards senescence, and possesses a unique genetic apparatus, the result of countless generations of diversifying, sexually reproducing ancestors (*phylogeny*). All these grades of dynamic change should be considered when any study of the fabric and behaviour of mankind is undertaken.

The present section is concerned with certain aspects of the development or ontogeny of the human body. However, in a volume such as this, which deals primarily with the general

EARLY DEVELOPMENT OF CHORDATA

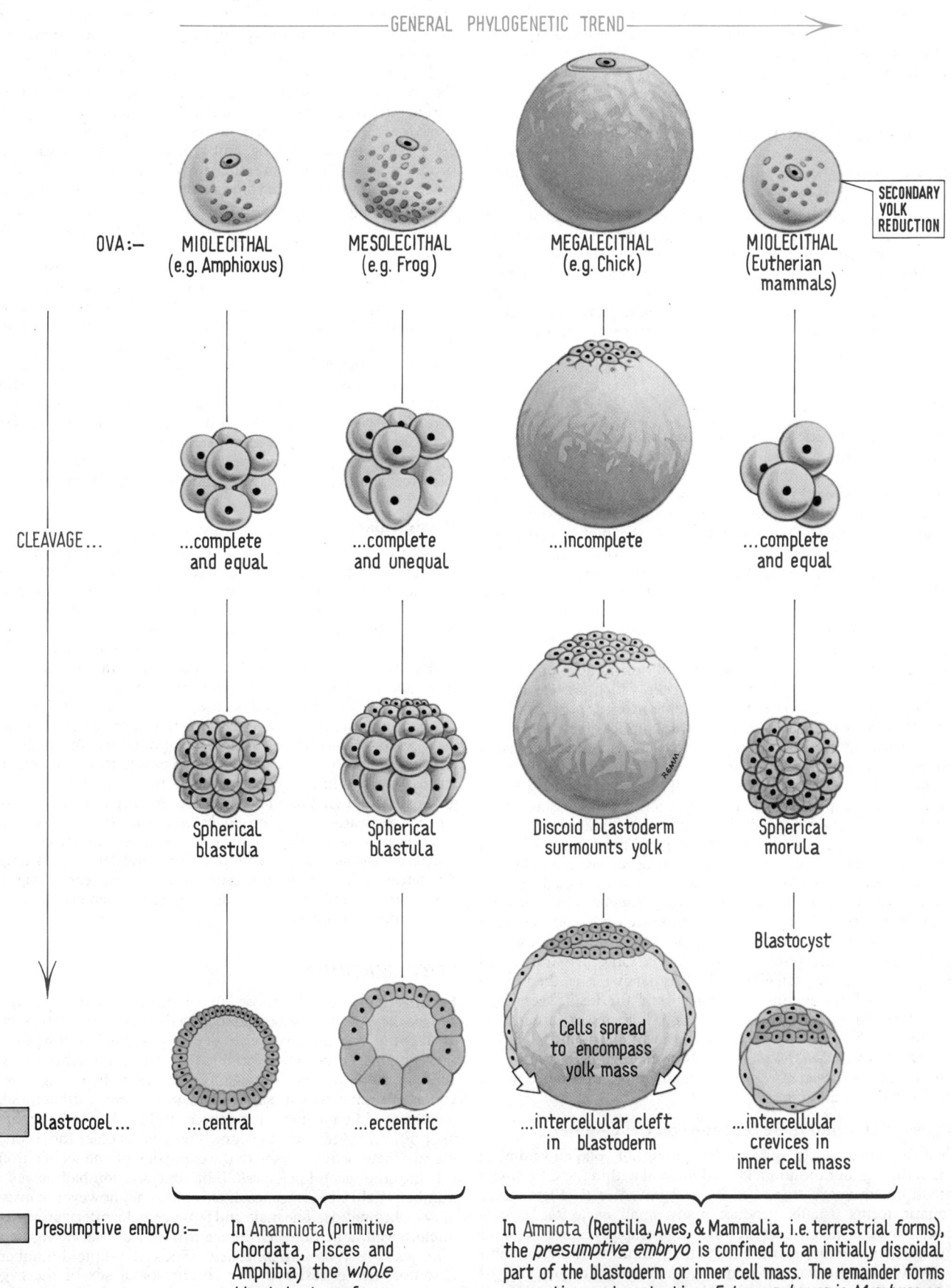

GENERAL PHYLOGENETIC TREND ⟶

OVA:— | MIOLECITHAL (e.g. Amphioxus) | MESOLECITHAL (e.g. Frog) | MEGALECITHAL (e.g. Chick) | MIOLECITHAL (Eutherian mammals)

SECONDARY YOLK REDUCTION

CLEAVAGE...

...complete and equal | ...complete and unequal | ...incomplete | ...complete and equal

Spherical blastula | Spherical blastula | Discoid blastoderm surmounts yolk | Spherical morula

Blastocyst

Cells spread to encompass yolk mass

Blastocoel... | ...central | ...eccentric | ...intercellular cleft in blastoderm | ...intercellular crevices in inner cell mass

Presumptive embryo:— | In Anamniota (primitive Chordata, Pisces and Amphibia) the *whole blastula* transforms into an *embryo*. | In Amniota (Reptilia, Aves, & Mammalia, i.e. terrestrial forms), the *presumptive embryo* is confined to an initially discoidal part of the blastoderm or inner cell mass. The remainder forms supportive and protective *Extra-embryonic Membranes* —amnion, chorion, yolk sac, allantois, & placental structures.

2.1 Variations in the patterns of cleavage in *Chordata* associated with differences in oöcyte size, yolk volume and distribution. Note that the overall phylogenetic trend is from left to right, and the progression of cleavage from above downwards. Note also the morphological *differences* and '*homologies*' between the various groups. Consult text for further comment.

anatomy of the body, a comprehensive treatment of human development cannot be entertained. What follows is a concise summary of the principal events, and those requiring further details should consult one of the excellent texts devoted to general developmental principles or specifically to human embryology (Patten 1953; Arey 1974; Hamilton *et al.* 1977).

Sexual Reproduction

Sexual reproduction can conveniently be divided into several chronological stages, namely *gametogenesis, fertilization, cleavage, gastrulation,* and *organogenesis,* whilst to the latter three stages the general term *embryogenesis* may be applied. This essentially reflects orderly temporal and spatial variations in *growth, morphogenetic movement,* and *patterned differentiation,* balanced against *regional degeneration* and *death* of cells and tissues. An abbreviated, preliminary description of these phases of development will be given before proceeding to a more specific account of human ontogeny

GAMETOGENESIS

The body of a multicellular organism which reproduces sexually can be regarded as consisting of general body (*somatic*) cells which, although exhibiting extreme ranges of variation in size, shape, and specific functional as well as morphological characteristics, all possess certain common features of their genetic apparatus. This contrasts with the highly distinctive differentiation process and genetic changes which occur within the *gonads* and result in the formation of mature *generative* (*germ*) cells or *gametes.* General somatic cells usually possess a full or *diploid number* of chromosomes, a half of which was originally derived from each parent. However, the *activity,* and *expression* of the many different gene loci on the chromosomes varies in different cell types in adult tissues, and during development as differentiation occurs. The somatic cells which are capable of dividing do so by the process of mitosis (p. 29). Each chromosome, having first made a faithful replica of itself and thus of its genes, divides longitudinally and each resultant cell is again equipped with the diploid number of chromosomes which are replicas of those in the parent cell. During gametogenesis, however, a complex series of changes termed *meiosis* are set in train (p. 30). Essentially, during this, the original chromosome number is reduced to a half, the *haploid number,* in which a *recombination* of the genetic material has occurred (the details of this process are considered on p. 99). The male gonad or *testis* produces many small motile gametes or *spermatozoa* in which the cytoplasmic machinery is much reduced; each consists of a nucleus bearing the genetic apparatus, the chromosomes, with a closely applied acrosome derived from a Golgi apparatus and succeeded by a long flagellum which is partly ensheathed by an energy-transforming mitochondrial sheath (for details see p. 95). In contrast the female gonad or *ovary* produces fewer, larger non-motile *ova,* their cytoplasm containing a variable quantity of food reserves (*yolk* or *deutoplasm*) and exhibiting regional variations of the cortex (*oöplasmic segregation*) which are of importance in the initial stages of differentiation (p. 76).

Types of Ova and their Influence on Development

Ova vary enormously in their size, amount of yolk, and number shed, throughout the animal kingdom, and within the *Chordata* a broad evolutionary trend can be recognized (**2.1**). Elementary aquatic forms usually produce many small *miolecithal* (yolk-poor) ova, which, after fertilization, rapidly develop into independent larvae, many of which succumb to predators and other inhospitable features of the environment. The larger ova of the amphibia are *mesolecithal* (with greater reserves of yolk), but again, are *wholly* transformed into aquatic larval stages which usually live and forage for food independently, and exhibit a long period of growth, differentiation and metamorphosis before the mature state is reached.

Colonization of dry land by forms not restricted by a periodic return to the water for reproductive purposes was, of course,

necessarily paralleled by drastic evolutionary modifications in their ontogeny. Reptiles, birds (and prototherian mammals) are characterized by the production of a small number of large, heavily yolked (*megalecithal*) eggs enclosed by membranes and a hard shell, which may be laid on land. They have no larval stage and much of the fertilized ovum does not contribute directly to the body of the embryo, but is concerned with the elaboration of various *extraembryonic membranes.* The latter provide mechanisms for the mobilization and transport of the massive yolk reserves to the developing embryo, deposition of waste end-products of metabolism, the provision of a local aquatic environment for the embryonic tissues, and to allow exchanges of carbon dioxide and oxygen across the porous shell (**2.2**).

Eutherian mammals, including man, although derived from such ancestors with heavily yolked eggs, again produce miolecithal eggs, but these are few in number, possess no hard shell, and development proceeds within the controlled environment of the female genital apparatus. Despite this drastic secondary reduction in yolk content, however, many features of mammalian development are only explicable by reference to such ancestry. Thus, although there are differences of detail between the various groups of mammals, they all develop the same array of extraembryonic membranes (yolk sac, allantois, amnion and chorion), with broadly similar functional associations. These membranes are, of course, modified in various ways for *viviparity,* that is, development *in utero,* with nutritive, respiratory, excretory and immunologic dependency upon the mother.

Recapitulation

Similarly, within the body of the mammalian embryo all the principal organ systems pass through stages which reflect fundamental patterns in the *ontogeny* of their ancestors, although these may be variously modified or abbreviated. Such patterns are particularly clear in the basic metamerism of the trunk, the progression from a simple tubular heart to a complex, folded, double pump with intervening septa, the common plan of the early nervous system and special sense organs, the craniocaudal sequence of nephric tubules, the plan of the early chondrocranium and the succession of gill arches in the early pharyngeal wall. Such *recapitulation* during development was originally, and too simply, interpreted as if an embryo passed through sequential stages which mimicked the *mature* form of its main ancestral groups. Although this view was later modified to include only structural features during *embryonic life* of the ancestors, emphasis is now placed on the essential similarity of the *morphogenetic mechanisms* common to all chordates. Accordingly, the latter closely resemble each other in the early stages of development, and intergroup differences only emerge gradually as development proceeds.

FERTILIZATION

Fertilization includes those mechanisms whereby a sperm approaches, becomes attached to, and then penetrates the surface of an ovum, and the early series of changes which follow. Briefly, the egg surface becomes modified to prevent further entry of sperms, the oöplasmic segregation of factors in the egg cortex which are important in subsequent processes of differentiation may be modified to a varying degree, and final completion of the meiotic cycle of the ovum occurs. There takes place the second of the two meiotic divisions with the expulsion of one set of chromatids into the second *polar cell.* The first division had halved the number of chromosomes, each chromosome, however, consisted of two chromatids. The male and female nuclear derivatives (pro-nuclei) swell, approach each other and then coalesce (*karyogamy*) to form a single *segmentation nucleus.* Thus, the diploid number of chromosomes is restored, the chromosomal sex of the zygote established, and a series of mitotic divisions which constitute *cleavage* are initiated.

Much evidence from comparative and experimental embryology (usually from non-mammalian forms) has accumulated concerning the forms of symmetry exhibited by ova, the mechanism of sperm attachment and entry, the site of sperm entry and its influence on the distribution of formative substances in the

FORMATION OF EXTRA-EMBRYONIC MEMBRANES

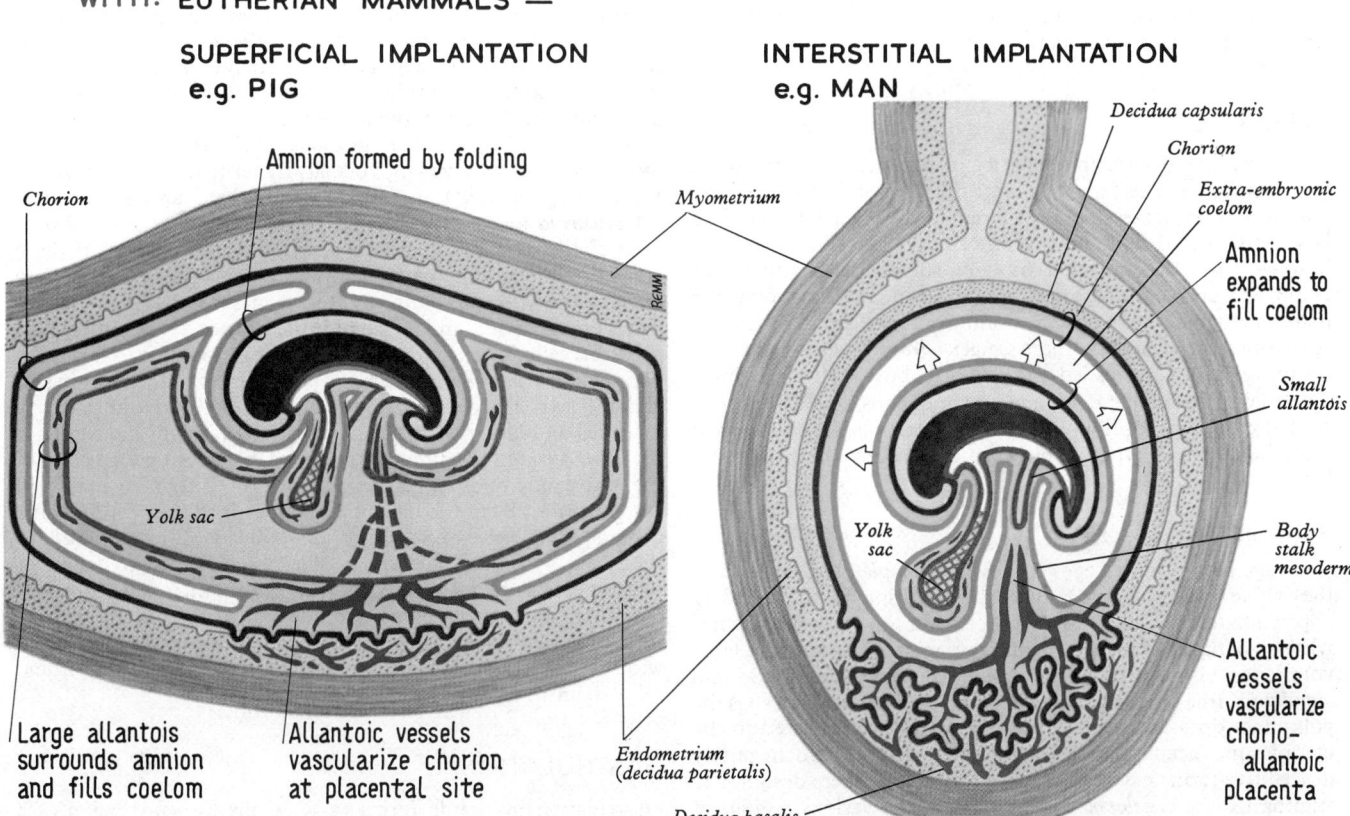

Somatopleure
Ectoderm
Somatic extra-embryonic mesoderm
Splanchnopleure
Splanchnic extra-embryonic mesoderm
Endoderm

Somatopleuric folds

Extra-embryonic coelom

Yolk

Folds meet defining Amnion and Chorion

Vascular allantois expanding into coelom

Yolk sac carrying vitelline vessels

COMPARE: BASIC PATTERN (EGG-LAYING AMNIOTE)

Chorio-allantois
Allantoic cavity
Yolk sac
Allantois (inner wall)

Expanded allantois almost fills extra-embryonic coelom; fusing with, and vascularizing overlying chorion

Amnion

Diminishing albumen

Diminishing yolk

WITH: EUTHERIAN MAMMALS —

SUPERFICIAL IMPLANTATION e.g. PIG

Amnion formed by folding

Chorion

Yolk sac

Large allantois surrounds amnion and fills coelom

Allantoic vessels vascularize chorion at placental site

INTERSTITIAL IMPLANTATION e.g. MAN

Myometrium

Decidua capsularis
Chorion
Extra-embryonic coelom
Amnion expands to fill coelom
Small allantois

Yolk sac

Body stalk mesoderm

Endometrium (decidua parietalis)

Decidua basalis

Allantoic vessels vascularize chorio-allantoic placenta

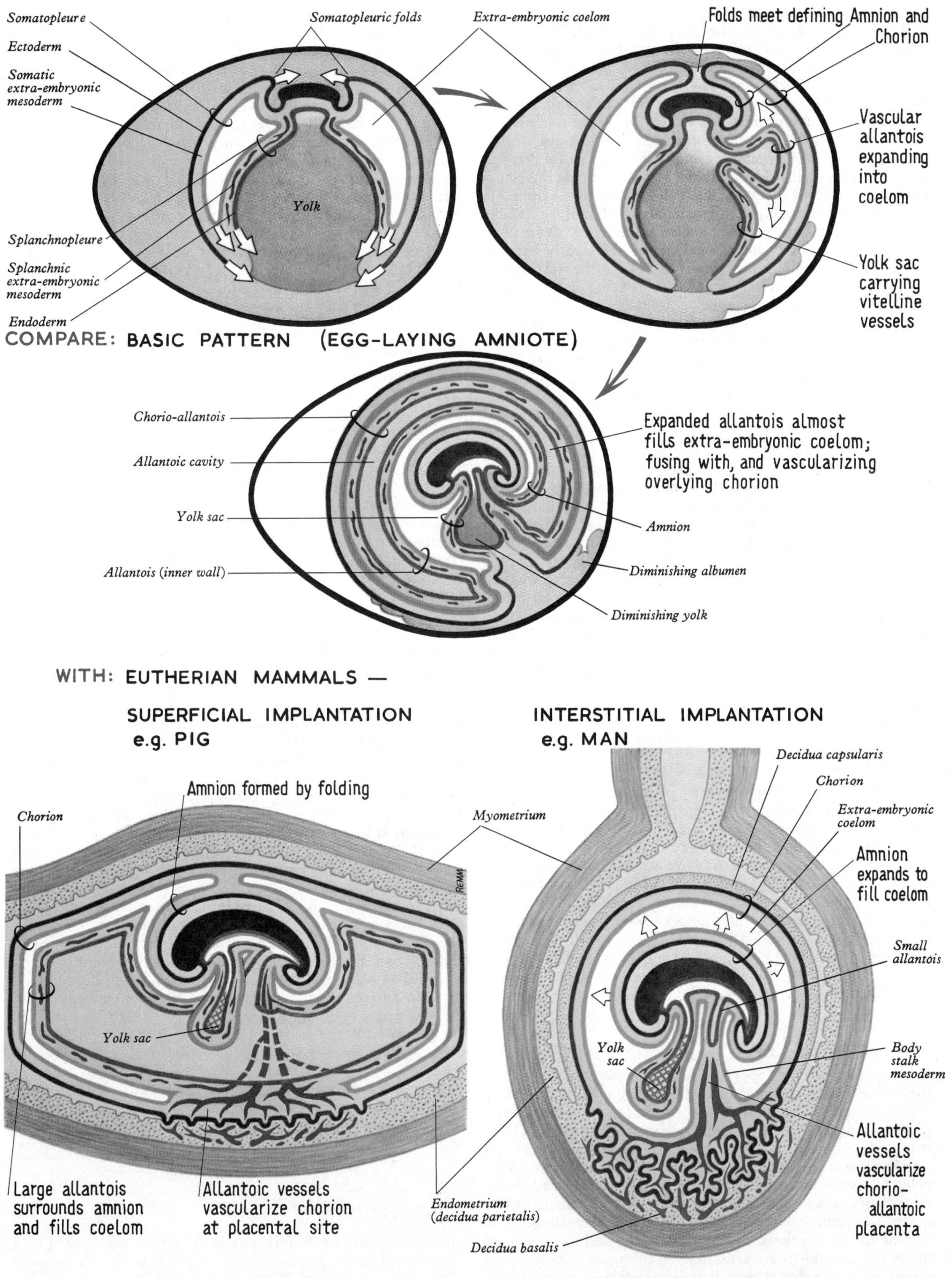

2.2 A generalized series of diagrams to allow comparison of the mode of formation of extraembryonic membranes in a megalecithal egg-laying amniote (e.g. chick) with that of a eutherian mammal exhibiting superficial implantation and a large allantois (e.g. pig), and a eutherian mammal showing interstitial implantation and a diminutive allantois (man). Note in each case the vascularization of the yolk sac, and the manner in which the expanded allantois (chick, pig) or its homologue in man, the body stalk mesoderm continuing from the small allantoic diverticulum, bear allantoic blood vessels which vascularize the overlying chorion. The latter is app.ied either to the shell, or to the uterine decidua.

egg cortex, and the relative fixity, or lability, of such substances. For some of the prominent earlier contributors to this important area of biology *see*, for example, Needham (1942); Waddington (1956, 1962); Brachet (1960, 1965); and Raven (1958, 1961, 1963, 1966). A comprehensive recent review lies outside the scope of the present volume, and only a few salient points can be mentioned. All ova are *polarized* with an axis extending between the largely cytoplasmic *animal pole*, near which the nucleus lies, and the more heavily yolked *vegetative pole*. Whilst ova of some species possess a fairly simple *radial symmetry* around this axis with experimentally demonstrable *gradients* of *developmental potential* extending between the poles, all others exhibit, in addition, further localizations of materials which confer a *bilateral symmetry* on the ovum. The time of appearance, degree and reversibility of this bilateral symmetry varies greatly between animal groups. Again, in occasional types of ova, the site of sperm entry is restricted by the egg coats to a small area on the surface, whereas, in the majority of others, entry may occur at any point. Further, in some, the bilateral symmetry is largely reorganized in relation to the point of sperm entry, whereas in others, it is little affected. There is little evidence that sperms are specifically attracted towards ova by any form of chemotropism, but in mammals it is often held that approach towards the egg surface, which at this time is surrounded by granulosa cells of the corona radiata and the zona pellucida (p. 93), is facilitated by hyaluronidase produced by the acrosome of the sperm. However, the degree of development of an acrosome and its concentration of hyaluronidase vary considerably—both are poorly developed in man. In some instances the egg cortex and its coats have been shown to contain proteins (*fertilizins*) which can activate and agglutinate sperms, whilst the sperms contain *antifertilizins* which, when extracted, can agglutinate ova; penetration is assumed to involve an interaction of the antigen-antibody type between the two. Finally, in some cases the deeper part of the acrosome has been shown to extrude an *acrosomal filament* which, when it contacts the egg surface, becomes engulfed by a *fertilization cone* of cytoplasm which protrudes from the surface of the egg. The degree to which these postulated mechanisms, symmetries and rearrangements apply to human ova and their fertilization remains uncertain.

CLEAVAGE

When compared with the majority of somatic cells (excepting large neurons and certain other large cells), recently fertilized ova are unusual in possessing but a single diploid nucleus within a large volume of cytoplasm. The mitotic divisions of cleavage are, however, not accompanied by any substantial synthesis of cytoplasm, and so the repeated replication and division of nuclear material, together with the partitioning of the existing cytoplasm, soon causes a return to the more usual *nucleo-cytoplasm ratio*. The resulting cells are called *blastomeres*, but their individual potentiality for further development varies considerably between animal groups; accordingly, a variety of terms have been used to describe cleavage 'types', which can only receive brief mention here.

Types of Cleavage (2.1)

Cleavage of miolecithal eggs is known as *complete* and *equal* since the whole zygote segments and the resulting blastomeres are approximately equal in size. In contrast, medialecithal eggs, whilst having a complete form of cleavage, it is *unequal* with large yolk-laden blastomeres at the vegetative pole which grade into small yolk-free ones at the animal pole. In heavily yolked eggs the yolk mass does not segment and divisions are restricted to the cytoplasmic accumulation at the animal pole. Such purely descriptive terms have been supplemented by others designed to summarize the *developmental potential* of the various regions of the zygote and the blastomeres derived from them. As has been intimated, experimental analysis of the *causal mechanisms* of early development has indicated that all zygotes exhibit some degree of segregation of cytoplasmic factors which, in their interactions with different blastomere nuclei, are of primary importance in the subsequent emergence of different cell varieties. However, the

degree, precision, and lability of this segregation varies greatly between animal groups. In some invertebrate ova it is highly ordered, and in such *'mosaic eggs'* cleavage often follows a fairly precise geometrical pattern (e.g. spirally cleaving forms), and is called *determinate*. The fate of each blastomere is relatively fixed, and removal of blastomeres results in the development of a partial embryo. In contrast, in so-called *'regulative eggs'* with *indeterminate cleavage* the segregation, although present, is much more diffuse and flexible. In such forms, a blastomere when separated from its fellows often proceeds to develop into a complete, although small, embryo. However, these descriptive terms unfortunately imply a distinction which is in reality only one of degree. Many intermediate types exist and even the most mosaic ova show a small regulative capacity, and conversely, the most regulative eggs are by no means equipotential systems, and only those regions which include presumptive chorda-mesoderm (*vide infra*) will continue to the formation of a complete embryo when separated from the remainder.

Cleavage in Chordata (2.1)

Cleavage of the miolecithal eggs of elementary chordates soon leads to the formation of a spherical *blastula*, i.e. a ball of cells with a central cavity or *blastocoel*, and the animal pole blastomeres are slightly smaller than those at the vegetative pole. The cleavage product of mesolecithal eggs is also a spherical blastula, but reflecting the large gradation in blastomere size from animal to vegetative pole, the blastocoel is eccentric and sited nearer the former. Cleavage in heavily yolked eggs results in a circular *blastodisc* at the animal pole, which is several blastomeres thick and separated from the underlying yolk by a subgerminal cavity containing products of yolk degradation. The blastocoel is probably represented by a small cleft which soon develops between the superficial and deeper parts of the blastodisc (i.e. it is not homologous with the subgerminal cavity). Finally, cleavage of the miolecithal eggs of mammals leads first to a solid ball of blastomeres, the *morula*, soon to be transformed into a hollow *blastocyst* with an eccentric cavity. The latter is surrounded on most aspects by a thin wall of cells (the early *trophoblast*), but at one point a localized group of cells, the *inner cell mass*, projects into the cavity. The mammalian blastocyst is to be carefully distinguished from the spherical blastula of the more elementary chordates. In the latter, the *whole blastula* undergoes a complex series of foldings, cell migrations and other *morphogenetic movements* and *interactions*, collectively termed *gastrulation*, and is then progressively transformed through early and late *gastrula*, then *neurula* stages, into the embryo. In contrast, only the *central cells* of the avian or reptilian blastodisc are engaged in these transformations. The peripheral blastomeres gradually grow to enclose the yolk, becoming arranged in several layers with an intervening cavity, and by complicated processes of differential growth, cellular changes and vascularization (2.2), form the various *extraembryonic membranes* (yolk sac, amnion, chorion and allantois) described more fully elsewhere. Similarly, in the early mammalian blastocyst, only a few *formative cells* in the centre of the *inner cell mass* are directly concerned in the future formation of the embryonic body. The remaining cells of the inner mass and the general walls of the blastocyst are destined to form the same general extraembryonic membranes already foreshadowed in their heavily yolked ancestors. The precocious segregation of blastomeres as extraembryonic structures in mammals contrasts with their later development in reptiles and birds, and the detailed fate of the various membranes and their role in *placentation* shows wide variation between the different mammalian orders (Mossman 1937; Amoroso 1952).

GASTRULATION

Gastrulation in simple forms (2.3) entails the conversion of a roughly spherical, hollow, single-walled *blastula* into a more elongated, double-walled *gastrula* containing a cavity, the *archenteron*, for the reception and assimilation of food, and with oral and blastoporic openings at opposite ends. In such forms, this is achieved by the infolding of one half of the blastula into the other, the orifice formed by the invaginating material being the

EARLY DEVELOPMENT OF THE ELEMENTARY CHORDATE

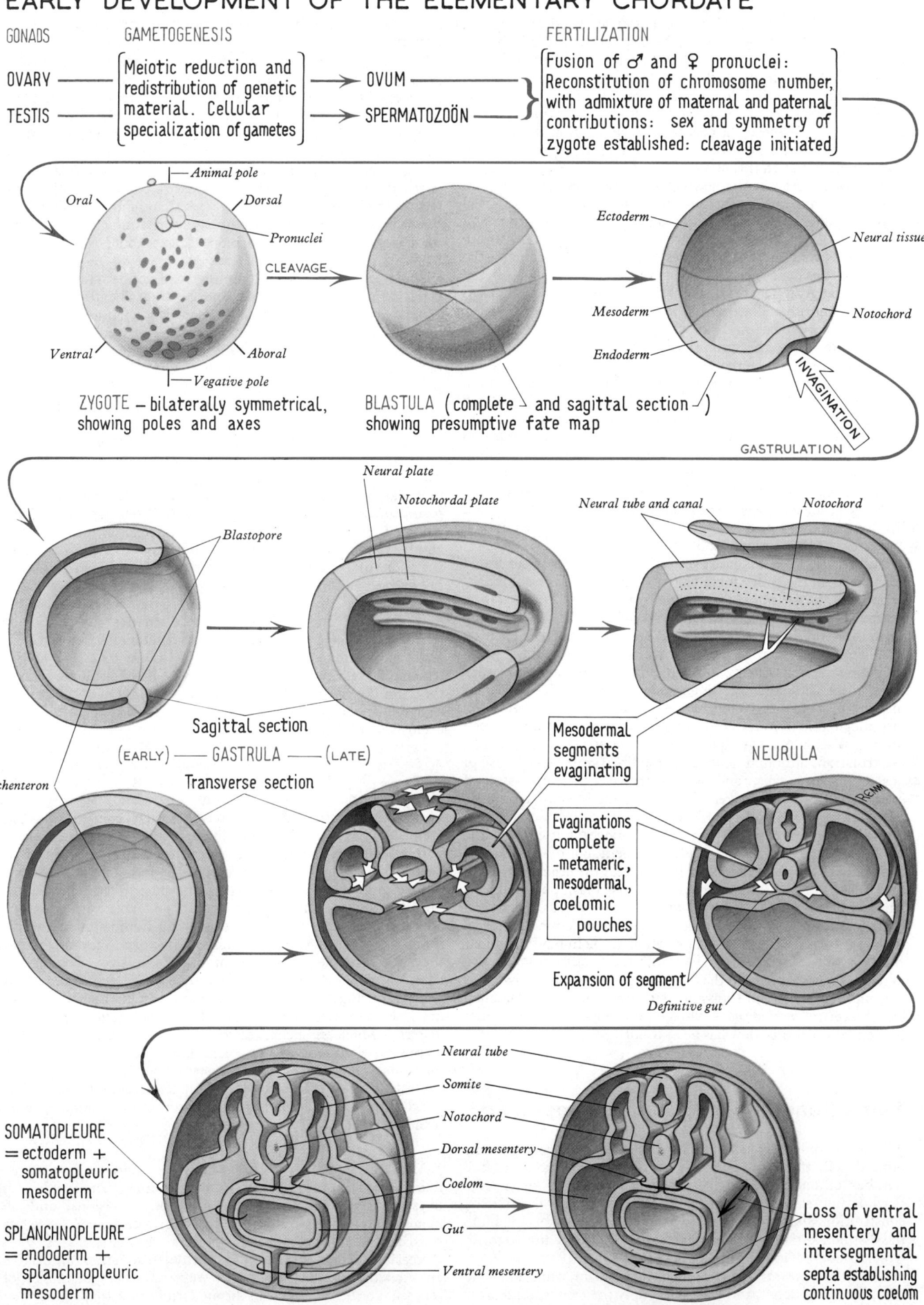

GONADS   GAMETOGENESIS

OVARY ———— ⎡Meiotic reduction and ⎤ ——→ OVUM ————
TESTIS ———— ⎢redistribution of genetic⎥
          ⎢material. Cellular   ⎥ ——→ SPERMATOZOÖN ————
          ⎣specialization of gametes⎦

FERTILIZATION

⎡Fusion of ♂ and ♀ pronuclei: ⎤
⎢Reconstitution of chromosome number,⎥
⎢with admixture of maternal and paternal⎥
⎢contributions: sex and symmetry of⎥
⎣zygote established: cleavage initiated⎦

Animal pole
Oral
Dorsal
Pronuclei
Ventral
Aboral
Vegative pole

ZYGOTE — bilaterally symmetrical,
showing poles and axes

CLEAVAGE

BLASTULA (complete and sagittal section)
showing presumptive fate map

Ectoderm
Neural tissue
Mesoderm
Notochord
Endoderm

INVAGINATION

GASTRULATION

Blastopore

Neural plate
Notochordal plate

Neural tube and canal
Notochord

Sagittal section

(EARLY) ——— GASTRULA ——— (LATE)

Mesodermal
segments
evaginating

NEURULA

Archenteron

Transverse section

Evaginations
complete
-metameric,
mesodermal,
coelomic
pouches

Expansion of segment

Definitive gut

Neural tube
Somite
Notochord
Dorsal mesentery
Coelom
Gut
Ventral mesentery

SOMATOPLEURE
= ectoderm +
somatopleuric
mesoderm

SPLANCHNOPLEURE
= endoderm +
splanchnopleuric
mesoderm

Loss of ventral
mesentery and
intersegmental
septa establishing
continuous coelom

2.3 Principal stages in the early development of an elementary chordate. The tabbing and illustrations proceed from left to right in a series of rows from above downwards. Note the poles and axes of the zygote, the presumptive fate map of the blastula, and the progression of gastrulation, neurulation, and coelom formation. Consult text for further details.

77

blastopore, and the point where the first invaginated cells contact the inner aspect of the opposite hemisphere eventually breaks down to form an *oral orifice*. The addition of a third, intermediate, mesodermal layer in the wall is usual in most groups of animals, and its origin is varied; in particular the fate of the walls of the primitive archenteron in the lower chordates is rather more complex (**2**.3). In these, the blastopore is conventionally regarded as presenting a dorsal, a ventral, and a pair of lateral *lips*. The material invaginated through the *dorsal lip* forms a midline strip in the roof of the archenteron which stretches from the blastopore to the margin of the future oral orifice; it underlies the external cellular wall and eventually folds off dorsally to form the *axial notochord*. Laterally placed strips, which flank the future notochord eventually form a series of paired outpouchings—the *metameric coelomic pouches*. The ventrolateral walls of the archenteron grow across ventral to the pouches and notochord to re-form a gut tube. The external wall of the gastrula dorsal to the mesodermal pouches and notochord, thickens and eventually folds in to form the primitive nervous system, whilst the remainder of the external wall forms the epidermis and its derivatives. In many forms the lateral lips of the blastopore gradually approach each other and fuse to form a *primitive streak* as they continue to bud off generations of mesodermal cells into the walls of the elongating gastrula. The dorsal lip continues to contribute notochordal cells and, with the fusion of the lateral lips, it rounds off and thus temporarily connects the archenteron with the caudal end of the developing nervous system (i.e. forming a transient *neurenteric canal*). In a similar way the ventral lip closes around a primitive cloaca, continuous with the gut tube and later with the urogenital ducts.

Accordingly, during gastrulation, originally widely separated cell masses in the walls of the blastula come into close secondary apposition, as the *presumptive organ rudiments* and tissue layers assume their definitive positions. *Presumptive fate maps* of the various regions of blastular walls have been produced by staining small regions of them differentially with vital dyes, and following the stained regions during subsequent development. Much work has also been directed towards an understanding of the mechanics of gastrulation itself and, as we shall see, the profound morphogenetic significance of the secondary association between formerly separated groups of blastomeres (**2**.4, 5, p. 81). Gastrulation, although necessarily modified in the blastodisc of heavily yolked eggs, and within the inner cell masses of mammalian blastocysts, remains an essentially similar process. Ectodermal and endodermal layers become defined, and within the former, a linear proliferative zone of uncommitted cells, the *primitive streak* persists and produces generations of mesodermal cells which pass between the two layers. Near the headward end of the streak a zone corresponding to the dorsal blastopore lip forms a notochordal process, which invaginates and breaks through to form a neurenteric canal, while caudal to the streak a cloacal membrane forms (cf. the transformation of the human bilaminar embryonic disc (pp. 110–114). The basic structural correspondence in these stages throughout the *Chordata* and in forms as diverse as *Amphioxus* and mankind is assumed to imply similar fundamental *morphogenetic mechanisms*—where experimental analysis has proved feasible, this view has been confirmed.

Some General Aspects of Development

The development of a large complex organism from a single diploid cell, the zygote, entails an increase in mass and cell number of many billions of times, and the emergence of a bewildering variety of cell types, each with a characteristic constitution, shape, specific activities, and variously welded into the distinctive tissue masses, sheets, tubes, liquids and so forth, which cooperate in the functioning of the whole.

The *developmental biology* of such an organism, while endlessly complicated in detail, and with as yet only the most rudimentary proposals concerning the control systems involved, can neverthe-less, in principle, be reduced to a limited number of basic processes, namely: temporo-spatial variations in *growth, pattern formation, morphogenetic movement, differentiation*, together with

degeneration and *death* of the cells and their aggregates (*see* for example pp. 79, 153).

The *description, definition* and *causal analysis* of these basic processes, and the degree to which they are independent or interdependent, still hold the centre of attention of modern developmental biologists, although many of the fundamental questions were first formulated, and some of the most dramatic experimental observations made, many years ago (*vide infra*). In the intervening years, as fresh experimental data have accumulated, the questions have often been rephrased, the definitions and terms used have been constantly revised and modified, and there has been a gradual change of emphasis concerning the conceptual frameworks currently in fashion (Locke 1966; Moscona and Monroy, to 1974).

Thus, in addition to attempting to describe and define the various forms, combinations, and differential rates of *positive growth* exhibited by organisms, there has been a more recent upsurge of interest in simultaneous *cell necrosis* and removal, and the factors which may control the regional differences in balance between these processes. Similarly, the detailed analysis of the *specific intracellular events* exhibited by cells undergoing *differentiation* has been paralleled by numerous hypotheses concerning the *spatial patterning* of differentiation. Concepts such as *oöplasmic segregation* of *morphogenetic factors*, *'organizer' phenomena*, *competence* of tissues, *induction*, *determination*, *morphogenetic gradients* and *developmental field phenomena*, and more latterly, ideas concerning *intercellular communication and positional information* available to cells in an aggregate, have emerged. In contrast, other authors have placed much greater emphasis on the probable morphogenetic significance of variations in the *cell cycle* (p. 23) with the establishment of *specific cell lineages* and the emergence of *cell clones* (*vide infra*). Finally, much interest has recently been aroused by the possible morphogenetic implications of the mathematical formulation of *catastrophe theory*. An immense scientific literature has accumulated concerning these fundamental fields of biological study, which of necessity can only be touched upon in a volume such as this, and the interested reader should consult any of the readily available reviews, monographs or original papers devoted exclusively to these topics.

GROWTH

Growth is a term widely used in everyday conversation and applied to both living and inanimate objects; it may, of course, within biology be applied to entire communities, individuals, or parts of an organism such as a whole body segment, an organ, tissue, group of cells or even subcellular features such as mitochondria, cell membrane and so on. This variety in the application of the term is paralleled by the number of definitions for it—evidence that none is wholly satisfactory. They range from the over-simplified 'increase in mass and size', to lengthy, complex, and sometimes incomprehensible descriptions. In general, during ontogeny, growth implies an increase in size and mass which results from synthesis of protoplasm and extracellular materials which are specific tissue components. Accordingly, any increases which follow the mere imbibition of fluid or the ingestion of food do not constitute true growth.

Growth of a tissue is classically considered to be either multiplicative, auxetic, accretionary or, quite often, a combination of these.

Multiplicative growth, as the name suggests, involves an increase in the number of cells (or nuclei and associated cytoplasm in syncytia) by a succession of mitotic divisions, and all mammalian tissues exhibit this form of growth at some stage. Its persistence, however, varies greatly between tissues with increasing differentiation and maturity. In some, multiplicative growth continues throughout life with continual replacement of senescent or dead cells, as in the epidermis, intestinal epithelium and the myeloid precursors of circulating erythrocytes. At the other extreme, cell division during embryonic life may lead to the full complement of a particular cell, e.g. oöcytes and neurons, and thereafter multiplicative growth ceases.

Much work has thus been directed at establishing the ranges of mitotic rates in many tissues throughout development into maturity and senescence, and towards an analysis of any cyclic (e.g. diurnal) variation in rate, and the extreme differences in potential for further growth during regeneration exhibited by different tissues. Despite the adequacy of such quantitative investigations in tissues of different age, sex and type, and the recognition of many general factors which influence such growth, e.g. genetic, nutritional (p. 135), endocrine, thermal, photic, mechanical, etc., little is known of the intimate processes which initiate, maintain, modify, and terminate, complex three-dimensional arrays of mitoses. In certain tissues (e.g. the epidermis and the growing nervous system) specific growth-promoting factors have been demonstrated (Levi Montalcini 1967). Recently, substances of widespread occurrence in many tissues termed *chalones* have received increasing attention (Bullough *et al.* 1967). They have been defined as internal secretions produced by a tissue which, by inhibition, control the mitotic activity of the cells of that tissue. It is considered that the chalone concentration in a tissue may be an important determinant of the balance which exists between mitotic rate, progressive cell differentiation and the assumption of a full functional role on the one hand, and the rate of cell ageing and progression towards senescence and death on the other. They are perhaps intermediaries between the general growth-influencing factors listed above and the differential activities of the genome (p. 85).

Auxetic growth implies an increase in the size of the individual cells in a tissue. It is particularly characteristic of certain invertebrate tissues such as the salivary glands of the *Diptera*, but is also well shown by some mammalian tissues. For example, there is a vast post-natal increase of both surface area and cytoplasmic volume in many neurons and glial cells (p. 821); the growing striated muscle fibre (p. 513), the oöcyte, the myelinating Schwann cells, and the smooth muscle cells of the pregnant uterus furnish further obvious examples. The majority of other tissues, however, show some auxetic growth but of much more limited degree, while in contrast, in some regions, continued multiplicative growth is accompanied by a *reduction* in cell volume (e.g. the granule cells of the cerebellar cortex, and the formation of small lymphocytes in lymphoid tissue).

It is widely held that the general *nucleocytoplasmic ratio* to which most of the body cells roughly approximate reflects the fixed quantity of DNA in their diploid nuclei which, in turn, imposes a rate limitation on the replacement of cytoplasmic proteins (each of which has a characteristic turnover rate). Thus, with continuing auxetic growth of a cell, its cytoplasmic volume eventually reaches a point beyond which the structural genes could not effectively replace the protein which is undergoing continual degradation. In most cases, growth ceases at this point, or nuclear replication with cell division occurs. The cases of auxetic growth cited above, however, often proceed far beyond the usual ratio of cytoplasmic volume to nuclear material and in these, various methods of providing auxiliary nuclear support have emerged. The large dipteran salivary gland cells develop 'giant' *polytene chromosomes* containing some multiple of the diploid DNA content. The striated muscle fibre and other 'giant' cells such as megakaryocytes are, of course, *multinucleate syncytia*. Finally, the enlarging oöcyte and neuron (possessing but a single haploid and a single diploid nucleus, respectively) have their surfaces clothed by numerous *satellite cells* (follicular or glial cells). Such satellite cells probably provide auxiliary metabolic and nuclear support for the enlarged central cell, that is, the two are functionally interlocked as a cytophysiological unit.

Accretionary growth denotes an increase in the amount of structural intercellular material between tissue, cells—bone and cartilage are the most commonly cited examples. Further, perhaps less obvious, examples are the other fibrous connective tissues, tendons, joint capsules, aponeuroses and fasciae (p. 522), the cornea, growing mesenchyme with its abundant intercellular matrix, and during the development of renal tissue and smooth muscle where the appearance of substantial basal laminae contributes significantly to the growing mass.

In practice, of course, these 'types' of growth as defined above are often welded together in various patterns, with differential growth rates and directions in different parts of the system; furthermore, they overlap with and merge into, the phases of cell differentiation, full functional activity, ageing and senescence. These *differential growth patterns* with either *random*, or *preferentially polarized* directions of mitotic division, together with alterations in cell size, shape, and surface consistency, are central features of embryonic development and are responsible for the moulding of tissues into specific shapes whether they be solid masses, hollow balls, tubes, sheets and so forth. Equally important in some regions, however, is a process of tissue regression, with degeneration, cell death and tissue removal. Where multiplicative (and sometimes accretionary) growth continues throughout the thickness of a tissue mass, it grows as a whole, expanding from within, and its growth is termed *interstitial*. Experimental markers placed within such a tissue during growth are later found to have become increasingly separated—excellent examples are embryonic limb bud ectoderm (p. 153) and the interior of young masses of hyaline cartilage (p. 251). Alternatively, in *appositional growth* new generations of cells and intercellular material are added to the *surface* of the tissue by the repeated division of the cells of a *cambial layer* which surrounds the tissue, for example periosteum and perichondrium (pp. 252, 260, 420).

Differential growth rates, as has been intimated, are an essential feature of many aspects of embryonic development. For example, if in an area of loose mesenchymatous tissue, a locus with a high proliferative rate appears, surrounded by 'shells' of tissue with a progressively lower rate, a mass of tightly packed cells which grades into the surrounding tissue is formed, i.e. a so-called *mesenchymal condensation*. Two adjacent loci in subepithelial mesenchyme with an intervening quiescent zone lead to neighbouring elevations of the epithelium, with an intermediate furrow (cf. the development of the facial processes, p. 146). In many situations, mesenchymal condensations appear and their *shape* foreshadows that of the structure, bone, cartilage or muscle mass, into which they subsequently differentiate. Many other examples of differential growth will be encountered as the development of the individual organ systems is considered. Particularly clear-cut examples are seen in descent of the gonads (p. 217), formation of the cauda equina (p. 161), neck elongation, septation of the heart (p. 184), and in the emergence of asymmetric arterial, venous and lymphatic systems from initially bilaterally symmetrical patterns.

MORPHOGENETIC CELL DEATH

Cell death and tissue removal are as much an integral part of embryogenesis as the positive forms of tissue growth discussed briefly in the foregoing chapters. As we have seen, it is currently held that mechanisms of *tissue homeostasis* exist, possibly operated by the local concentrations of tissue-specific chalones, whereby a balance is achieved between cell division, differentiation, functioning and senescence. Death of cells has been recognized as occurring as early as the blastocyst stage in rabbits (Daniel and Olson 1966), and has been demonstrated at many other stages of ontogeny throughout the vertebrates (Glucksmann 1951). An obvious example is the retrogression of the amphibian tail at metamorphosis, but similar degenerations occur during the development of the axial nervous system (Hughes 1968) and in association with the mammalian primitive streak. At a later stage, bone growth and remodelling in particular are accompanied by surface deposition in some regions and erosion and removal in others (*see* p. 266).

Particularly well studied have been the integrated patterns of growth and death of cells during the shaping of the tetrapod limb (Saunders and Fallon 1966). Massive zones of what has been dramatically termed *cataclysmic necrosis* sweep the length of the developing limb as its contours develop. The process is especially prominent in some localities, and it finally occurs in the interdigital clefts as the digital outlines become crystallized. Neighbouring mesenchymal cells become intensely phagocytic

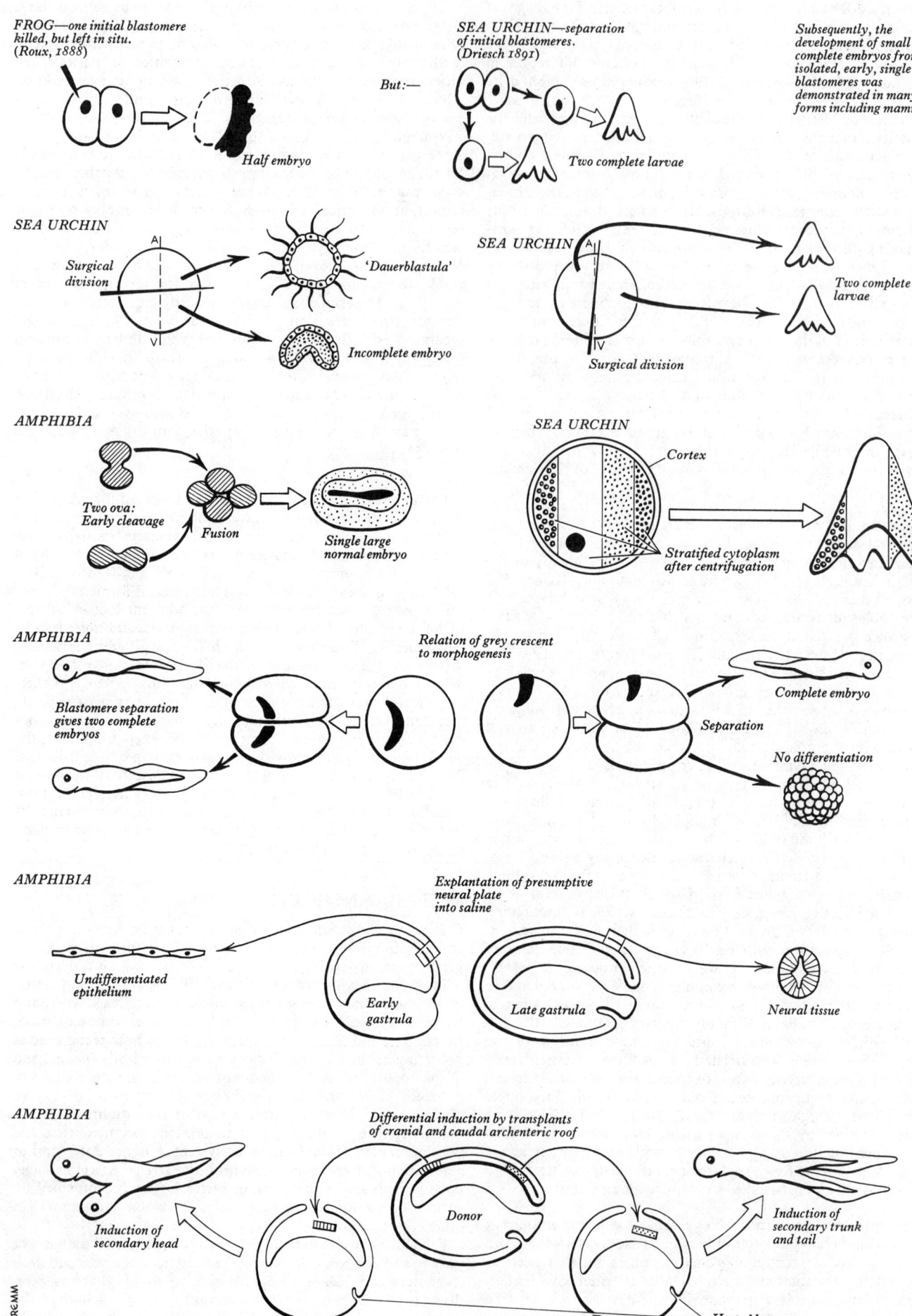

FROG—one initial blastomere killed, but left in situ. (Roux, 1888)

Half embryo

SEA URCHIN—separation of initial blastomeres. (Driesch 1891)

But:—

Two complete larvae

Subsequently, the development of small complete embryos from isolated, early, single blastomeres was demonstrated in many forms including mammals

SEA URCHIN

Surgical division

'Dauerblastula'

Incomplete embryo

SEA URCHIN

Two complete larvae

Surgical division

AMPHIBIA

Two ova: Early cleavage

Fusion

Single large normal embryo

SEA URCHIN

Cortex

Stratified cytoplasm after centrifugation

AMPHIBIA

Blastomere separation gives two complete embryos

Relation of grey crescent to morphogenesis

Separation

Complete embryo

No differentiation

AMPHIBIA

Explantation of presumptive neural plate into saline

Undifferentiated epithelium

Early gastrula

Late gastrula

Neural tissue

AMPHIBIA

Differential induction by transplants of cranial and caudal archenteric roof

Donor

Induction of secondary head

Induction of secondary trunk and tail

Host blastopore

REMM

2.4 A selection of the classical manoeuvres, which were milestones in the development of experimental embryology as a science. Consult text.

80

and engulf the dying cells and any resulting cellular débris. It should be emphasized that the *normal prospective fate* of such a cell mass is to degenerate and die, and since, in modern terminology, it has become fashionable to regard ontogeny as the unfolding of a *developmental programme*, ultimately encoded in the genome, such cells are described as succumbing to *programmed cell death*. Recently, numerous experiments have been devised in attempts to analyse the precise roles of such necrotic zones in normogenesis, and to elucidate what factors affect the enactment and timing of the so-called '*death clock*'. The *posterior necrotic zone* of the developing chick wing bud has, because of its predictable localization, timing, and susceptibility to microsurgical experiments, become a favourite object for intensive study (Saunders and Gasseling 1968), and its relationship to the establishment of *polarity* in the wing bud is becoming firmly recognized.

DIFFERENTIATION

The appearance of many highly distinctive cell types which become blended in various ways to form the functioning tissues, and the control mechanisms involved, has been for centuries, and remains today, one of the central problems in biology.

The differences between fully differentiated cells in terms of their size, shape, ultrastructural appearance, chemical constitution and functioning (e.g. mature muscle, nerve and glandular cells) are plain, no intermediaries exist, and there is no known experimental situation in which one form changes into another. As we shall see, however, the cells of early chordate embryos may, under suitable conditions, *ultimately* form any one of many different cell types, i.e. they are widely held by many developmental biologists to be *pluripotent*, and as development proceeds, there occurs a step by step reduction in their potential for forming a variety of cells, and finally, the fate of most cells is limited to one type of fully differentiated cell only. Even in the mature body, however, some cells retain some degree of flexibility and can still transform into a number of types—this is seen most powerfully in the ability to regenerate a whole body segment, e.g. limb or tail, shown by certain amphibia.

It may be mentioned here, however, and outlined further below (p. 83) that some groups of cell biologists contest the view just given of the early blastomeres as undifferentiated, totipotent or pluripotent cells with *immediate* access to a wide range of possible phenotypic fates, and advance instead hypotheses of much more strictly compartmentalized cell lineages during development.

Preformation and Epigenesis

For centuries, a controversy raged concerning the form taken by the differentiation which occurs during embryogenesis, whether there was a *preformation* in the early embryo, or whether *epigenesis* took place. The former view held that all the structures of the mature body, although so minute as to be invisible, were present in the zygote and simply enlarged and were thus revealed in subsequent development. The epigeneticists, however, held that the various organ primordia appearing during development were essentially new formations. In a modern context both views, somewhat modified, have their place. In one sense, *the total information content* of both the cytoplasm and the nuclear genetic apparatus of a zygote is a 'preformation' which through a series of complex interactions directs the 'epigenesis' of new structures in subsequent development.

The classical studies of *descriptive embryology* in which times of appearance, shape and size of the various embryonic layers and organ systems at different stages of development were meticulously described in serial histological sections, and similar *comparative studies* of different species, although providing an invaluable framework, could make no contribution to an *understanding* of differentiation. An analytical approach was necessary (**2**.4, 5), and the foundations of 'developmental mechanics' or an *experimental embryology*, which sought to understand the *causal interactions* of development, were laid by Wilhelm Roux late in the nineteenth century. His experiments on frog development were paralleled by other observations of fundamental importance by Hans Driesch on the early sea urchin

embryo. Then, passing into the present century, contributions of outstanding penetrance were made by such investigators as Hans Spemann, Ross G. Harrison, Sven Horstadius, J. Holtfreter, and many others. Only the most cursory mention of this field of biology can be contemplated, and those with wider interests should consult the original writings of these pioneers and, as an introduction to the equally important researches of modern development biologists, specific monographs (e.g. Spemann 1938; Willier *et al.* 1955; Hamburger 1960; Balinsky 1970; and the numerous text references up to p. 83).

Experimental Embryology

The final blow to rigid preformationist concepts was delivered when Hans Driesch showed that if the first two blastomeres of a sea urchin egg were separated, each could develop into a complete, although small, embryo (**2**.4). Thus, any idea of strict compartmentalism in the fertilized ovum had to be abandoned, since in normal development each of the blastomeres would have formed half the embryo, but when separated, each could *regulate* sufficiently to form a whole. On this basis Driesch supposed (mistakenly) that there existed *no segregation whatever* in the egg cytoplasm, and that it constituted a '*harmonious equipotential system*' in all directions. Since subsequent differentiation occurred, he regarded this as evidence to support a revival of the Aristotelian *entelechy*, holding that such a phenomenon could not be explained by the laws of physics and chemistry, and that a vital or spiritual 'internal perfecting principle' was present in all living forms.

However, as we have seen elsewhere (p. 76) much subsequent work showed that all ova and zygotes show some degree of *segregation of formative materials* in their cytoplasm, but its *degree* and *lability* varies between species. Ironically, had Driesch's separation been made in an equatorial plane rather than, as was the case, a meridional one, only partial embryos would have resulted!

The experimental methods employed by the earlier investigators (**2**.4, 5) included the preparation of '*presumptive fate maps*' of the blastula or blastodisc, *ablation* of regions of the blastula, *autotransplantation* of cell masses from one place to another in the same embryo, *explantation* of such masses or combinations of these into salt solutions and nutrient media, and finally, the powerful methods of *heteroplastic* and *xenoplastic grafting*, between different species. In the latter, the tissues of both the graft and the host are so distinctive that their relative contributions and interactions in any complex development that occurs after transplantation, can be followed.

The production of maps (**2**.3) showing the *presumptive fate* in normal development of different regions of the blastula or blastodisc, by the application of spots of vital dyes and following their subsequent history, was a necessary prerequisite of experimentation using microsurgical methods.

Self and Dependent Differentiation

From the earlier autotransplantation experiments, and later reinforced by explantation of blastula fragments into salt solutions (**2**.4), the first concepts of two distinct tissue states emerged, i.e. those that exhibited *self-differentiation* and the remainder in which *dependent differentiation* was the rule. In the former, as the name suggests, a fragment when transplanted to an apparently 'indifferent' region of the blastula, or, more critically, into salt solutions, proceeded to develop into the tissues it would have formed, if undisturbed. Regions with dependent differentiation, however, if transplanted early, failed to continue development and only did so if combined with some neighbouring mass of cells. If such a region, which at an early stage showed dependent differentiation only, was not transplanted until a later stage, it was often found to have become self-differentiating. Furthermore by combining such an early region with different regions of the embryo, development would often proceed, but the final tissue produced was changed (e.g. muscle rather than kidney).

From such work, therefore, it was realized that the *prospective significance* or *normal fate* of the various regions of the blastula did not reflect the *prospective potency* of these regions (i.e. the *various*

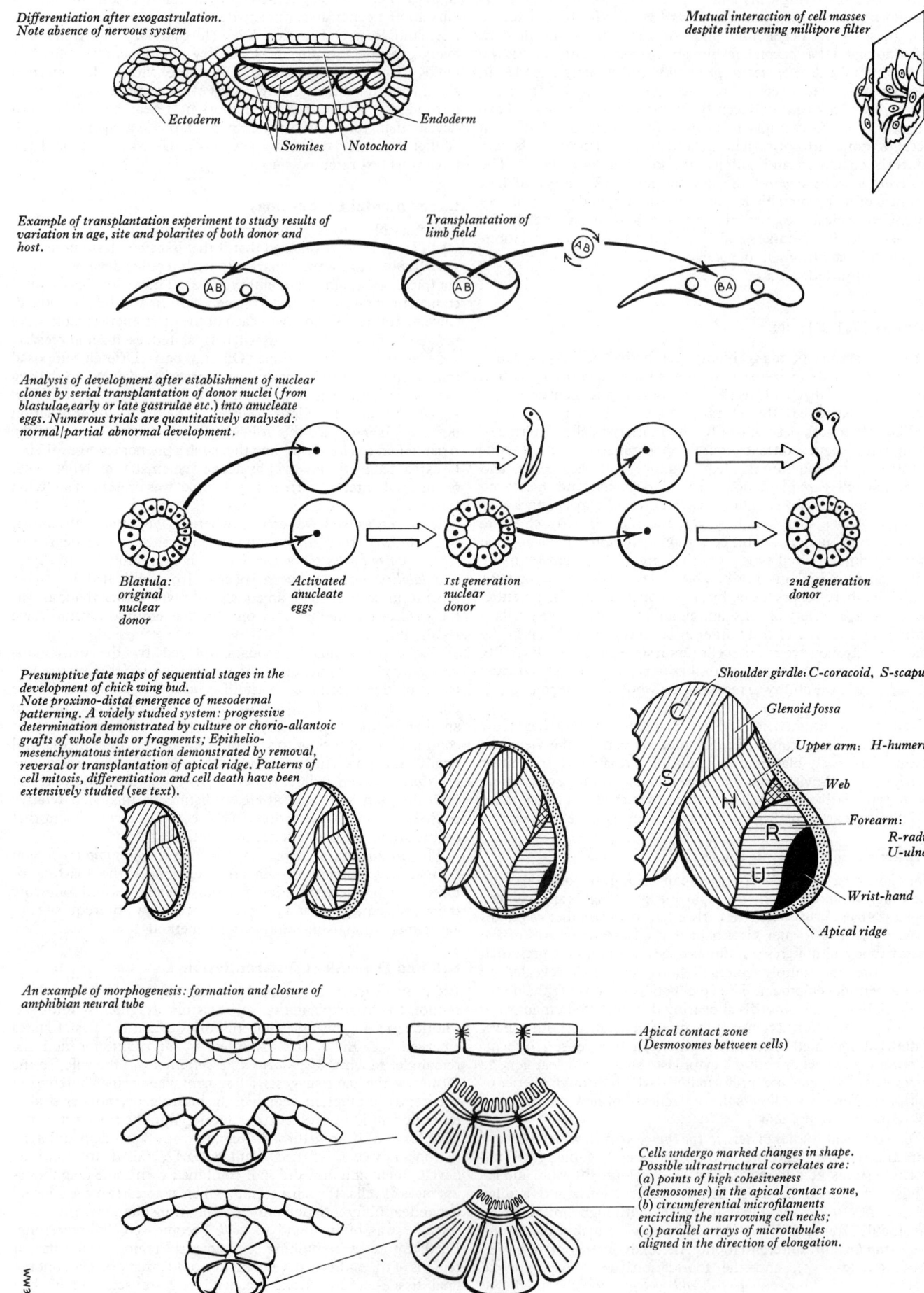

Differentiation after exogastrulation.
Note absence of nervous system

Ectoderm

Endoderm

Somites   Notochord

Mutual interaction of cell masses
despite intervening millipore filter

Example of transplantation experiment to study results of
variation in age, site and polarites of both donor and
host.

Transplantation of
limb field

Analysis of development after establishment of nuclear
clones by serial transplantation of donor nuclei (from
blastulae, early or late gastrulae etc.) into anucleate
eggs. Numerous trials are quantitatively analysed:
normal/partial abnormal development.

Blastula:
original
nuclear
donor

Activated
anucleate
eggs

1st generation
nuclear
donor

2nd generation
donor

Presumptive fate maps of sequential stages in the
development of chick wing bud.
Note proximo-distal emergence of mesodermal
patterning. A widely studied system: progressive
determination demonstrated by culture or chorio-allantoic
grafts of whole buds or fragments; Epithelio-
mesenchymatous interaction demonstrated by removal,
reversal or transplantation of apical ridge. Patterns of
cell mitosis, differentiation and cell death have been
extensively studied (see text).

Shoulder girdle: C-coracoid, S-scapula

Glenoid fossa

Upper arm: H-humerus

Web

Forearm:
R-radius,
U-ulna

Wrist-hand

Apical ridge

An example of morphogenesis: formation and closure of
amphibian neural tube

Apical contact zone
(Desmosomes between cells)

Cells undergo marked changes in shape.
Possible ultrastructural correlates are:
(a) points of high cohesiveness
(desmosomes) in the apical contact zone,
(b) circumferential microfilaments
encircling the narrowing cell necks;
(c) parallel arrays of microtubules
aligned in the direction of elongation.

REMM

2.5 A further selection of classical experimental embryological
procedures. Consult text.

fates possible under different experimental conditions). Further, it was found that from the earliest moments, certain restricted regions only exhibited *self-differentiation*, that is to say *determination* of their fate had occurred. The remaining regions with dependent differentiation, originally pluripotent suffered a stepwise restriction in their possible fates, i.e. an increasing fixity of determination occurred as development proceeded.

There followed a period in which the features of dependent differentiation were more clearly defined. Classical experiments showing the dependency of the development of the lens vesicle (later transforming into the crystalline lens of the eye), on the approach and close apposition of an evagination from the primitive nervous system (the optic vesicle), were performed, with many variations. These showed that, in the absence of an optic vesicle, no lens formation occurred, that many different regions of 'indifferent' ectoderm, if exposed to the influence of the vesicle could form lens tissue, and similarly, if the same ectoderm was used to replace that covering the hindbrain, it would develop into a membranous labyrinth. In this manner, the concept of *embryonic induction* developed. This held that for a restricted period of time a particular region of pluripotent tissue was *competent* to react to an inductive influence exerted by an adjacent tissue with different characteristics.

Primary Organization

Experiments by Hans Spemann involving the ablation, or the auto- or hetero-transplantation of the future dorsal lip of the blastopore region, further clarified the process. In the absence of a dorsal lip, continued development of an organized nature ceased. When, however, such a region was transplanted to the flank of another gastrula a dramatic series of changes ensued. The graft proceeded to proliferate, invaginate and develop into a notochord, surrounding skeletal structures and adjacent muscle masses, showing that extremely early determination had occurred, and it followed its normal course of self-differentiation. In addition, the surrounding host tissues displayed an equally dramatic series of responses to the inductive influences of the graft. The superjacent ectoderm developed into a remarkably complete nervous system, with recognizable subdivisions of brain and spinal cord, laterally placed host mesoderm formed musculature and kidney tissues, whilst on occasions, a subjacent gut tube was induced. Host and graft tissues often continued to interlock in the development of what was virtually a complete second embryo, attached to the abdomen of the host. Because of the early determination and self-differentiating nature of the tissues of the dorsal blastopore region, and their profound inductive influence on surrounding tissues which lead to the development of an embryonic axis and adjacent tissues, it was called the *primary organizer region*. Then followed a series of experiments which demonstrated regional differences in the inductive capacity of the tissues sequentially invaginated through the dorsal lip (**2.4**); thus, so-called head, trunk and tail 'organizer' regions were recognized. Subsequently, however, a wide variety of substances, including some inorganic compounds which do not exist in developing embryos, were shown to have inductive effects, and similar effects sometimes followed the microsurgical killing of a few blastomeres, or the introduction into the blastocoel of powdered glass. Interestingly, extracts of adult tissues produced various responses, either a *mesodermalizing* effect or a *regional neuralizing* effect (with the formation of predominantly forebrain, hindbrain or spinal structures)—*see* Toivonen and Nieuwkoop (1967). Thereafter, many subordinate inductive systems were shown to operate in coordinated sequences throughout chordate development. In some cases competent tissue was separated from its relevant inductor tissue by a 'millipore' filter which prevented the passage of cells (**2.5**); inductive effects still occurred, presumably because of the passage of chemical messengers between the two. (It should be mentioned, in this context, that since the advent of the scanning electron microscope, the full effectiveness of such filters has been subjected to a critical reappraisal.) Since the turn of the century much effort and ingenuity have been expended in attempts to isolate and identify specific intercellular chemical inducers, but despite many claims, none have been substantiated in the embryogenesis of metazoa (Grobstein 1967).

In contrast, considerable detail has now accumulated (Beckwith and Zipser 1970; Woods 1973) concerning the differential activation and repression of the genome in micro-organisms associated with the *induction of enzyme synthesis* (*vide infra*) and also concerning the two-stage inductive events which transform an uncommitted lymphoid stem cell to, firstly a committed cell (p. 59), and secondly to express this committal by the production of specific antibody (Feldman and Globerson 1974). In the latter case, the second stage inducer, the antigen, is, of course, often a chemically well-defined molecule. Whilst such studies are of great intrinsic interest, and may well have parallels with events in embryogenesis, direct extrapolation is at present premature, and it is perhaps unfortunate that the term induction has been applied to all these situations.

In recent years there has occurred a great change of emphasis when developmental biologists consider inductive phenomena. The dramatic responses demonstrated by Spemann following blastopore transplantation, coupled with his all-embracing term 'primary organizer', led for many years to the following view of early embryogenesis. Namely, that the majority of the early blastomeres were *undifferentiated*, *totipotent* cells, with highly labile control mechanisms, and with *numerous* alternative developmental paths *immediately* available to them. A *master organizer* region or regions, was assumed to produce complex organizer or inducer molecules of *high information content* which passed to, and *directed*, the subsequent developmental path of the reactive cells. Naturally, great interest focused on the possibility of isolating and characterizing physico-chemically these *in-structive molecules*. Gradually, however, as data have accumulated concerning other possible modes of interaction of cell aggregates (*vide infra*), the use of the term organizer has fallen into disuse, and the search for complex instructive molecules has slackened. It may well be that the inductive phenomena which undoubtedly occur during embryogenesis are mediated by slight variations in ubiquitous molecules of relatively simple chemical constitution (e.g. metabolites) of low, or absent, intrinsic information content, and that the developmental path followed is largely determined by the currently available metabolic options provided by the *cytoplasm* and *genome* of the reacting cells.

MODERN VIEWS OF DIFFERENTIATION

Differentiation has often been defined as the emergence of *irreversible*, *inheritable* differences between somatic cells, in contrast to the many *reversible* changes shown by cells in response to changed local environments, termed *modulations*. However, the apparent irreversibility seems to be a feature which is only especially prominent in cells of vertebrates in the relatively few experimental situations so far studied (and in this sense any such claim must remain empirical). Further, many marked changes in the cells of plants and higher invertebrates closely resemble processes of differentiation, but are in fact reversible (*see* p. 72). Perhaps the terms differentiation and modulation imply a rigid distinction between two states exhibited by cells, when in fact a range of states, with varying degrees of reversibility, may exist.

It has been held (Sager 1965) that from the earliest times the simplest, non-nuclear (*prokaryote*) units of life consisted of '. . . a stablized tripartite system: nucleic acids for replication, a photosynthetic or chemosynthetic system for energy conversion, and protein enzymes to catalyse the two processes'. To these essential ingredients common to all living forms, were later added the enzyme systems which accompanied a dependency on environmental oxygen, and then greater variety of cellular form paralleled the appearance of *eukaryote cells* with a well defined chromosome-bearing nucleus. It is in the *interactions* between these different phases of one cooperative system, the cell, that clues to the nature of differentiation are currently sought (Bell *et al.* 1965; De Reuck and Knight 1967; Bullough 1967; Toivonen and Nieuwkoop 1967; Grobstein 1967; Wolstenholme and Knight 1970; Harris 1970; Ashworth 1973; and subsequent text references).

As we have seen elsewhere (pp. 21–27), proteins are of fundamental significance in the construction and operation of all prokaryote and eukaryote cells. In great variety they are essential constituents of the cytoplasmic matrix, nucleoplasm, cytomembranes including the plasma membrane and walls of the membrane-bound organelles and, with nucleic acids, as the nucleoproteins of the genetic apparatus. Perhaps most significantly in this context, they form the basis of the innumerable enzyme arrays which catalyse all the steps in the structural and energetic transformations which constitute 'life'.

Thus, it has been stated quite simply that when cells with the same genome synthesize different proteins, differentiation has occurred (Jacob and Monod 1961, 1963). This widely held view implies, therefore, that an understanding of the processes of differentiation lies not in the details of the protein synthetic pathways as such, but rather, in the various control systems to which these are subjected. The latter are currently believed to involve varieties of chemical messengers which operate within and between cells.

proliferative cell cycles with amplification of cell numbers, the progeny having precisely the same synthetic capacity as the parent cell (**2.6B**); and *quantal cell cycles* which, at some point, make available for synthetic activities, regions of the genome in the subsequent cells which were *not available* in the mother cell. Such a view securely links the essential steps in differentiation to particular phases of a *critical series* of quantal mitoses, perhaps interspersed by proliferative cycles, and at each quantal cycle that parent cell is maximally *bipotential*. Thus, the division may be *symmetrical* giving derived cells with identical synthetic capacities, or *asymmetrical* in which their capacities differ (**2.6B**). In one particular type of asymmetrical division, one of the progeny retains the synthetic capacity of the parent cell (**2.6B**), that is, it persists as a *stem cell* (p. 27). A quantal cell division, therefore, is a decision point at which the progeny enter one of two new paths of differentiation. A series of such divisions would

THEORIES OF MITOSIS AND PROGRESSIVE DIFFERENTIATION

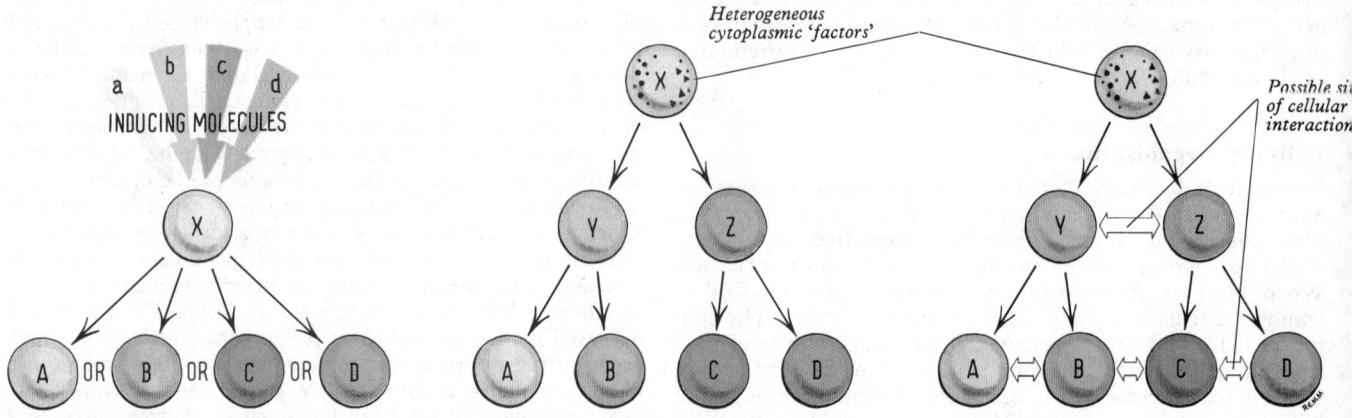

2.6A Theories of mitosis and progressive differentiation. The tabbing is further explained in the accompanying text.

The foregoing definition of differentiation, however, does not specify which gene products may be regarded as indices that differentiation has occurred, and disregards the phases of the cell cycle. On such a view, early 'pluripotent' embryonic cells have a *wide repertoire* of possible responses, any one of which would *immediately* follow a local change in the operative control system, during any phase of the particular cell cycle occurring at the time. Thereafter, the early, labile control systems would become progressively stablized in subsequent generations as the cells differentiate (**2.**6A).

TYPES OF MITOTIC DIVISION

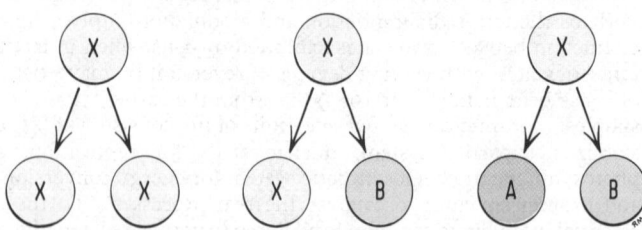

2.6B Varieties of mitotic cell division: these include a symmetric proliferative mitosis, and two types of asymmetric quantal mitosis. Consult text for discussion.

Other groups of cell biologists, in contrast, propose an alternative set of hypotheses (Holtzer *et al.* 1972, 1975; Reinert and Holtzer 1975, and **2.**6A). These authors class materials synthesized by cells as either '*essential molecules*', i.e. those synthesized by most cells and required for their viability, or '*luxury molecules*' which are 'those cell-unique molecules responsible for the state of differentiation of the cell that either synthesizes them or has inherited them'. Secondly, cell mitotic cycles are regarded as of two fundamentally different types—

clearly lead to diversification of cell types, each ultimate *clonal type* being the product of a precise and strictly compartmentalized *cell lineage*. This hypothesis, presented here in the barest outline, rejects the conventional view of the pluripotentiality of early blastomeres, holding that all cells are optimally differentiated for the stage of development they have reached, and further, relegates the inductive capacities of exogenous molecules to a relatively minor role (Holtzer *et al.* 1972, 1975; Reinert and Holtzer 1975). Such views are, however, not wholly in accord with those proposed by other groups in relation to the spatial patterning of differentiation (*vide infra*). The contrasting views concerning cell diversification, and the varieties of mitotic division are summarized in illustrations **2.**6A and B.

The Control of Gene Action

The general structure and role of nucleic acids in the form of nuclear and extranuclear deoxyribonucleic acids (DNAs), and various ribonucleic acids (RNAs), in relation to protein synthesis has been discussed elsewhere (p. 25). Mention was also made of the concept of the *operon* which emerged from elegant investigations on micro-organisms (Jacob and Monod 1961, 1963). Essentially, an operon is a functional grouping of DNA loci including several *structural genes*, an *operator segment* (*o*), a *promotor region* (*p*), and a *regulator gene* (*i*). Each structural gene carries the information necessary for the specification of a particular polypeptide (which forms part of either a structural or an enzymatic protein). In the case of the *lac operon* of *Escherichia coli*—the first to be thoroughly analysed by the study of the activities of mutant forms (Beckwith and Zipser 1970; Ashworth 1973)—the linear sequence of DNA loci has been demonstrated to be: $i\ p\ o\ z\ y\ a$ (z, y and a denoting structural genes which encode for the synthesis of β-galactosidase, a permease, and a transacetylase respectively). The prefix *lac* indicates that lactose is the physiologically important substrate for β-galactosidase. Subsequently, further examples of more complex operons have been analysed, with many features in common with the lac operon, but with alternative controlling mechanisms. A comparison of these studies is now leading to rapid advances in our knowledge of the main variations in gene activity in *prokaryotes*.

In general the activity of the structural genes is dependent upon the state of their associated *operator*; they are active only when it is 'free' and they become inactive when it is 'blocked' by combination with a specific chemical group. The neighbouring *regulator gene* produces a *repressor protein* which, depending on local conditions, may exist in an active form capable of binding with the operator, or in an inactive form when it cannot do so. In this manner the activity or otherwise of a particular group of structural genes is dependent upon the production of a *first order* chemical messenger or repressor protein. In turn, specific *second order* messengers (*effectors*) determine the state of the repressor molecules. Some repressors can only block the operator when combined with effectors termed *co-repressor* molecules. In some instances it has been shown that low molecular weight end products of certain *biosynthetic pathways* act as co-repressors, thereby providing a negative feedback system controlling the synthesis of enzymes essential to the biosynthetic path. Thus, heightened activity in the biosynthetic path leads ultimately to gene *repression*, whereas diminished activity leads to *derepression* of the relevant genes.

Other effectors, such as *enzyme inducers* act in a converse way— by combining with the repressor protein they so modify its structure that it is unable to combine with the operator, and consequently the related structural genes become active. The *promotor region*, which lies adjacent to the operator, is now known to be the site where binding of DNA-dependent RNA polymerases (*transcriptases*) occurs, when the operator is not blocked by repressor protein. These determine the *loci* for the *initiation of transcription*, and further, variations in a number of factors entering into the constitution of the transcriptases (e.g. the σ-factor) result in *modulation of the rate* of transcription.

Thus, through the balanced activities of *negative* (repressor/operator) and *positive* (transcriptase/promotor) controls, a great flexibility of gene action may be achieved varying from *on/off switches* to delicate *adjustments of transcription rates*.

Certain gene loci have also been considered to control the *mitotic* behaviour of cells; on this view, when the loci are active multiplicative growth persists, and when inactive, the cells are channelled into avenues of differentiation and ageing. The *tissue-specific chalones* mentioned previously have been proposed as co-repressors of such mitosis-initiating genes (Bullough 1967). The activities of the tissue chalones may be modified by the concentration of general circulating hormones and thus as intermediaries between the endocrine system and the state of the genome in the tissue cells. (In certain invertebrates a further class of substances called *pheromones* even pass between different individual members of a community, and affect the reactions of their tissues and also their behaviour patterns.)

However, as indicated previously, some cell biologists would not regard the foregoing as an adequate view of the role of mitosis in cell differentiation. Firstly, many would prefer to consider the delicate on-going quantitative controls of protein synthesis as more the province of the cell physiologist, and regard true differentiation to occur only at certain definite phases of *quantal cell cycles* (p. 25). As stated, in the latter, regions of the genome of derived cells become available for synthetic activities, which were *not available* in the parent cell. In this regard, mention must be made here of the elegant researches on the structure, distribution and possible roles of varieties of the basic proteins called *histones* (*see* reviews by Zubay 1968; Cook 1973; Wasseman 1973; Borun 1975). Five principal classes of histone are present in eukaryote cells, and these form strong, stable complexes with each other; and with DNA, thereby being important structural and functional components of the chromosomes (*see* the discussion on the *nucleosome*—p. 25, and **1**.22). It has been proposed that in combination with certain 'operator-like' regions of the chromosomes the histones provide a powerful *non-specific repression* of these regions and their associated structural genes. Such regions, it is thought, become *unblocked* by the cooperation of *non-histone components* only at specific times during chromosome replication in quantal cell cycles. It is possible that such non-histone

components, or their precursors, form the basis of the so-called *morphogenetic substances* which many workers have indirectly demonstrated to be differentially distributed in the cortex of the oöcyte, thereafter being unequally partitioned in the various blastomeres at cleavage and further, they may form the substrate for cell-cell interactions in later development. Experimental analysis of the components that may interact with the histone complexes is still in its infancy. Particular interest has recently been centred upon the structure and metabolism of certain '*acidic*' *non-histone chromosomal proteins* (Borun and Stein 1972; Stein and Borun·1972; Gerver and Humphrey 1973; Borun 1975) which may be responsible for the production of specific RNA transcripts and, accordingly, be closely related to aspects of the control of cellular differentiation. (Over 400 varieties of such proteins have recently been demonstrated in mammalian cells.)

Some authors have stressed that the genome of *prokaryotic cells* is permanently *unblocked* but can be repressed or derepressed (Tsanev and Sendov 1971; Tsanev 1975), while in metazoan *eukaryotic cells* the genome is initially *blocked* and during embryogenesis *sequential deblocking* different gene combinations occurs. The *functional state* of the fully differentiated cell is thereafter determined by repression and derepression 'within the limits of the deblocked regions of the genome'. Computer simulation of model systems of metazoan development indicates that the final stages of cellular differentiation can be reached only through 'a chain of events in which both processes of repression-derepression (realized by embryonic induction) and of blocking-deblocking (realized by the mitotic cycle) are successively involved'.

It should be noted that many of the control systems described above seem to be mediated by that ubiquitous regulator molecule cyclic adenosine monophosphate ($3'5'$cAMP) which therefore also appears to have a central role in cellular differentiation.

An abbreviated account such as the foregoing can do little justice to the numerous, elegant, and rapidly expanding researches in these fields. Nevertheless, enough has perhaps been said to emphasize that the earlier evidences stemmed largely from work with micro-organisms, and although considerable advances are now being made (Lewin 1976), the positive identification of the specific roles of intracellular and intercellular chemical messengers in advanced metazoan tissues is still awaited. However, as a working hypothesis it is considered that the patterns of growth, differentiation and functional expression in metazoa result from the varied activation and repression, coupled with sequential deblocking, of gene groups which stem from an interlocked hierarchy of chemical messengers—repressors, co-repressors, inducers, transcriptases, blocking/deblocking proteins, tissue chalones, and circulating hormones (and in some communities pheromones), which act in coordinated sequences in space and time, as integral parts of cell cycles. Many modifications of this scheme will undoubtedly prove necessary. Furthermore, the informational role of extrachromosomal nucleic acids and other self-replicating cytoplasmic substances (or *plasmagenes*) which may be of great importance in normal development have as yet been little investigated in the higher metazoa.

CONCEPTS RELATING TO PATTERN FORMATION

The events outlined in the previous sections dealt largely with hypotheses concerning the *intracellular* mechanisms controlling cell differentiation, neglecting two other fundamental aspects of development, namely *pattern formation* and *morphogenesis*. The latter, acting in concert with cell differentiation, are the means by which the total genetic information available to the zygote is ultimately expressed in the mature functioning phenotype.

Pattern formation concerns the processes whereby the individual members of a mass of cells, initially *apparently* homogeneous, undergo a number of different avenues of differentiation, which are precisely related to each other in an orderly manner in space and time. It is the mechanisms by which such temporo-spatial differences are initiated and maintained that currently engage the attention of many developmental biologists. The 'patterns' embraced by the term, of course, not only apply to regions of *regular* geometrical order such as the retina, crystalline lens, cerebellar cortex, etc., but all aspects of metazoan development including ordered but *asymmetric* structures such as the tetrapod limb, mammalian liver, etc. Thus, in the case of the tetrapod limb, for example, the intimate *intracellular* events leading to the emergence of chondroblasts, myoblasts, osteoblasts and so forth, will be identical in right and left sided limbs, fore-limbs and hind-limbs, and their counterparts in different species; nevertheless, the *patterned arrangement* of these elemental processes differ in each case.

Morphogenesis, or the assumption of form by the whole, or part, of a developing embryo is often for convenience treated as a separate field of study, involving differential rates of growth or degeneration of groups of cells, but particularly in early embryogenesis the relative *morphogenetic movements* of cells or cell aggregates. The movements involved in gastrulation (p. 76) or the wide dispersal of neural crest cells (p. 112) are particularly dramatic examples. Thus, the study involves an analysis of the *mechanisms* and *forces* responsible for cell movement, migration, changes in shape of individual cells and particularly the differential adhesiveness of their surfaces (*see* pp. 7, 92). However, despite the conveniences for research groups, these various elements in development are necessarily *interdependent*. Whether it is differential growth balanced against areas of degeneration, morphogenetic movements and migration, or cellular differentiation under consideration, it is necessary first to *specify* within a group of cells *which particular ones* shall divide, degenerate, migrate or differentiate, and this is the province of *pattern formation*. As an example, when a group of cells migrates to a distant location in normal development, mechanisms clearly exist to *locate* the group about to migrate, in relation to the remainder of the embryo. After an initial period of growth, the selected group enters initial stages of differentiation, which will lead to the development of those organelles, surface specializations and changes of cell shape, responsible for the mechanics of migration; further, these events must follow in correct *sequence* and *orientation*. Following migration, the group is now in a new cellular environment, and further periods of growth and differentiation follow, with crucial interactions between the migrated group and the cells forming its new environment, often affecting the particular avenues of differentiation followed by *both*.

It should be noted that pattern formation is an essential ingredient not only of normal embryogenesis but of processes of *regeneration*, good examples being the regeneration of new heads and tails in transected flatworms, new apical and basal regions in transected *Hydra*, or the replacement of a whole severed limb, or an appropriate part thereof, in adult *Amphibia*. In general, regeneration follows one of two main types; *morphallaxis* in which the whole reforms by rearrangement and differentiation of the existing tissues without further growth, or by *epimorphosis* in which there is new growth of blastemal tissue which subsequently matures into the full regenerate. Whatever control mechanisms are proposed for pattern formation in the embryo, they must be sufficiently flexible to also account for morphallaxis and epimorphosis.

It should be emphasized that the approach to the problems of pattern, form and differentiation contrasts sharply between different groups of workers. Some (Holtzer *et al.* 1975; Lederberg 1967) take the view that finer and finer probing of the genome and the detailed control of protein synthesis, leading to a reasonably complete understanding of the *intracellular* molecular events of cytodifferentiation would, in its turn, with little addition, provide an adequate explanation of changes in form and pattern. Others (Wolpert 1969, 1971), however, whilst conceding the importance of the molecular aspects of cytodifferentiation, hold that the laws and principles of control of pattern and form must initially be sought at the *cellular* and *intercellular* level, and that they will prove to be 'as general, universal, elegant and simple, as those that now apply to molecular genetics'. (For an extensive and critical account of pattern formation consult Wolpert, 1977.)

Since the closing decades of the last century, a series of concepts concerning pattern formation have slowly emerged, but with the passage of time these have frequently been rephrased and redefined, often resulting in some differences of definition and terminologic confusion between authors. These concepts were mentioned earlier (p. 81) and include ideas of polarity, axes, gradients and fields, and suggestions concerning the coordinates necessary for their specification.

In the physical sciences, the idea of a system possessing two regions of opposite tendencies or qualities, the *poles*, which can be considered to be joined by a linear *axis* is commonplace, for example in magnetism and electrostatics. At each point in space along and around the axis and poles, a force exists having a specific intensity and direction, i.e. a *vectorial force field* is present in three-dimensional space, and a *linear axial gradient* exists between the poles. Similar terms were adopted by developmental biologists, but whilst some authors (Wolpert 1971) stressed the analogy with the physical sciences, others (Waddington 1970) considered that, except for particular purposes, such approaches were grossly over-simplified. Descriptions of systems exhibiting forms of polarity, axes and gradients, abound throughout biology. Classical examples include the study of regeneration following the transection of *platyhelminths* or *Hydra* at different levels. The details will not be followed here, but in general the presence of an axial gradient concerning the form and frequency of successful regeneration was demonstrated, together with gradients of metabolic activity and susceptibility to toxic materials. Further, the *apical region* (head in flatworms and hypostome in *Hydra*) regenerated first and subsequently inhibited the formation of additional apical structures, and appeared to coordinate the regeneration of succeeding parts (a phenomenon termed *apical dominance*). Similar poles, axes and gradients have been described in many *oöcytes*, for example the animal-vegetative, oral-aboral and dorso-ventral axes (*see* p. 74 and **2**.3). The term axis has, however, been used in rather different ways in different situations. Thus, the animal-vegetative axis of oöcytes is often an obvious structural and cytophysiological feature, reflected by the eccentricity of the nucleus and the differential distribution of yolk platelets, other organelles, pigment granules, metabolic levels, etc. Further, in some forms which have been thoroughly analysed (Gustafson and Wolpert 1963, 1967), e.g. the sea urchin, there exists an experimentally demonstrable 'morphogenetic gradient' along this axis, although it is as yet uncertain what are the relevant 'morphogenetic factors' which are distributed along the axis, although tentative suggestions have been made (*see* p. 85), and of course the mechanisms controlling their distribution are completely unknown. However, it is clear that the animal-vegetative axis is determined early and once formed is extremely stable; in contrast, the dorso-ventral axis is determined much later, is more difficult to demonstrate and is more labile. Similar concepts of polarity and axes have often been applied, for example, to the developing tetrapod limb, and here again, the proximo-distal, cranio-caudal and dorso-ventral 'axes' are determined at different times and vary in their lability.

Such concepts are central to many of the hypotheses relating to pattern formation and may be considered in association with the two forms of development mentioned earlier (p. 76) termed *mosaic* and *regulative*. In a mosaic system, removal of a part results in a specific defect in the final embryo, whereas removal of part of a regulative system is followed by a total proportionate reorganization of the remaining portion, with the formation of a small but complete embryo. Further, division of a regulative system often results in duplication, with two small complete embryos. Similar considerations also apply not only to whole zygotes, but also to presumptive organ rudiments. Although no animal groups are wholly mosaic or regulative, some such as the annelids and molluscs show a high degree of mosaicism, whereas the sea urchin and chordates are typically regulative.

During oögenesis, and perhaps modified to some degree at fertilization, it has been demonstrated (Raven 1958, 1963, 1966; Davidson 1969) that morphogenetically important substances are localized (oöplasmic segregation) in the cortical zone in geometric patterns related to the principal axes of the oöcyte. It is assumed that the distribution gradients of such substances are differentially parcelled between the blastomeres during cleavage. In *mosaic systems* it is thought that the cytoplasmic parcelling which results from their highly ordered and specifically orientated forms of cleavage, is almost wholly responsible for directing subsequent pattern formation, morphogenesis, and cytodiffentiation, as a series of independent localized events, with relatively little intercellular communication. In contrast, it has been proposed that whilst the blastomeres of regulative eggs also necessarily differ qualitatively and quantitatively, the phenomenon of *regulation* exhibits *global features* wherein the mass of cells reorganizes *as a whole*, and there must exist avenues of *intercellular communication* of crucial significance in development. Such ideas led to the early formulation of the concept of the *morphogenetic field* (Huxley and DeBeer 1934; Child 1941) defined as 'a region throughout which some agency is at work in a co-ordinated way, resulting in the establishment of an equilibrium within the area of the field'. Subsequent years saw the rather sporadic (and occasionally inappropriate) use of the term embryonic field, and then for a period its use was largely abandoned by many developmental anatomists. Only recently has the phrase been subjected to a critical reappraisal and restatement, and as a result, two provocative but contrasting views have emerged.

The first concerns the idea of *positional information* being available to individual members of a mass of cells by virtue of the presence of *reference points* at its borders, the whole constituting a *morphogenetic field*. Essential to the idea is that *gradients* of activity (information) extend from the reference points throughout the cell mass by *cell–cell communication*: accordingly, the unique position of each cell is specified, and subsequently each differentiates appropriately, within the limits of the metabolic options provided at the time by the state of its genome. The picturesque model adopted to present this hypothesis is the so-called 'French Flag Problem' (2.7 A, B). In this, a line of cells with channels of intercellular communication between them, are considered to have three possibilities for molecular differentiation—blue, white or red, and they form a correctly proportioned French Flag whatever the number of cells in the line, and even if parts of the original line are removed. It is assumed that each cell is assigned a positional value by appropriate signals with respect to reference sources at the ends of the line. Reference to illustration 2.7 A, B will make it clear how such a proposal could successfully explain proportionate differentiation, not only with lines of different lengths, but also to include epimorphosis and morphallaxis. A number of model systems proposing possible mechanisms for the mode of operation of the signalling system have been examined. Briefly these include, most simply, a source of a metabolic product at one end of the line, a sink at the other, with the establishment of a linear diffusion gradient between them. More complex models have proposed signalling gradients of two or more substances, or the reversible operation of multiple sources and sinks depending upon local conditions (the so-called homeostatic model), whilst others have invoked active transport mechanisms, or the presence of paths of selective permeability. A most interesting alternative mechanism has been suggested (Goodwin and Cohen 1969; Cooke and Goodwin 1971) that a reference point may be provided by a *pacemaker cell* which emits two wave-like periodic propagations of biochemical activity which are transmitted from cell to cell throughout the morphogenetic field. Positional information would then be signalled by the *phase angle difference* between the arrival of the two waveforms at the cell in question. Recently, in a detailed analysis of amphibian gastrulation it was suggested that *two* such pacemaker regions were present, one at the animal pole and the other at the grey crescent. Their mutual positions change during gastrulation providing a dynamic *two-dimensional grid* of positional information in the mesodermal mantle, thereby specifying its subsequent pattern of cytodifferentiation.

Despite the fact that all the foregoing signalling mechanisms are relatively simple in principle, it has been stressed that such systems, when coupled with variations in the 'rules for interpretation of the signals' on the part of the genome of the reacting cells, could result in the production of highly complex, even asymmetric patterns. Nevertheless, although universality has been claimed for the theory throughout biology (Wolpert 1971, 1977), currently there is no direct evidence for the precise nature of the boundary zones (reference points or pacemaker cells), the detailed mechanisms of intercellular signalling, and the manner of cellular interpretation of the signals. As yet, therefore,

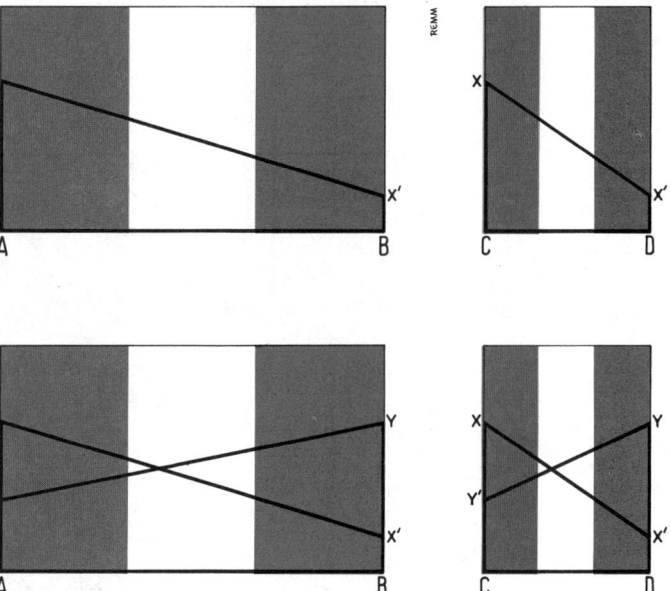

2.7A Diagram illustrating one hypothesis concerning positional information in relation to pattern formation, by reference to the so-called 'French Flag' problem. A–B and C–D indicate long and short rows of cells respectively. X, X' and Y, Y' the concentrations of morphogenetically active substances. Above: (X–X') shows the gradient of a single substance; below: (X–X' and Y–Y') the gradients of two substances. In each case a similar triple differentiation occurs. See text for discussion. (Modified from, and kindly provided by Professor L. Wolpert 1971, 1977.)

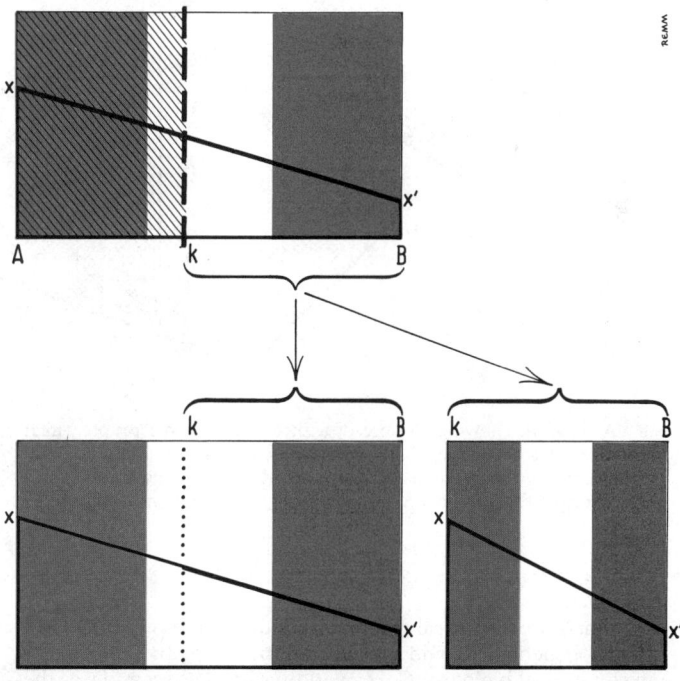

2.7B Diagram using the 'French Flag' analogy to illustrate application of the hypothesis of positional information advanced to explain regeneration, after ablation (cross-hatch), of part of an array of cells, either by morphallaxis, or by epimorphosis.

the hypothesis of a coordinate system which specifies positional information, whilst providing a powerful stimulus for new thought and experiment, must for the present remain the province of tentative theoretical biology.

During the period of emergence of theories of positional information in relation to morphogenetic fields, there has also occurred the convergence of ideas from two other sources. The

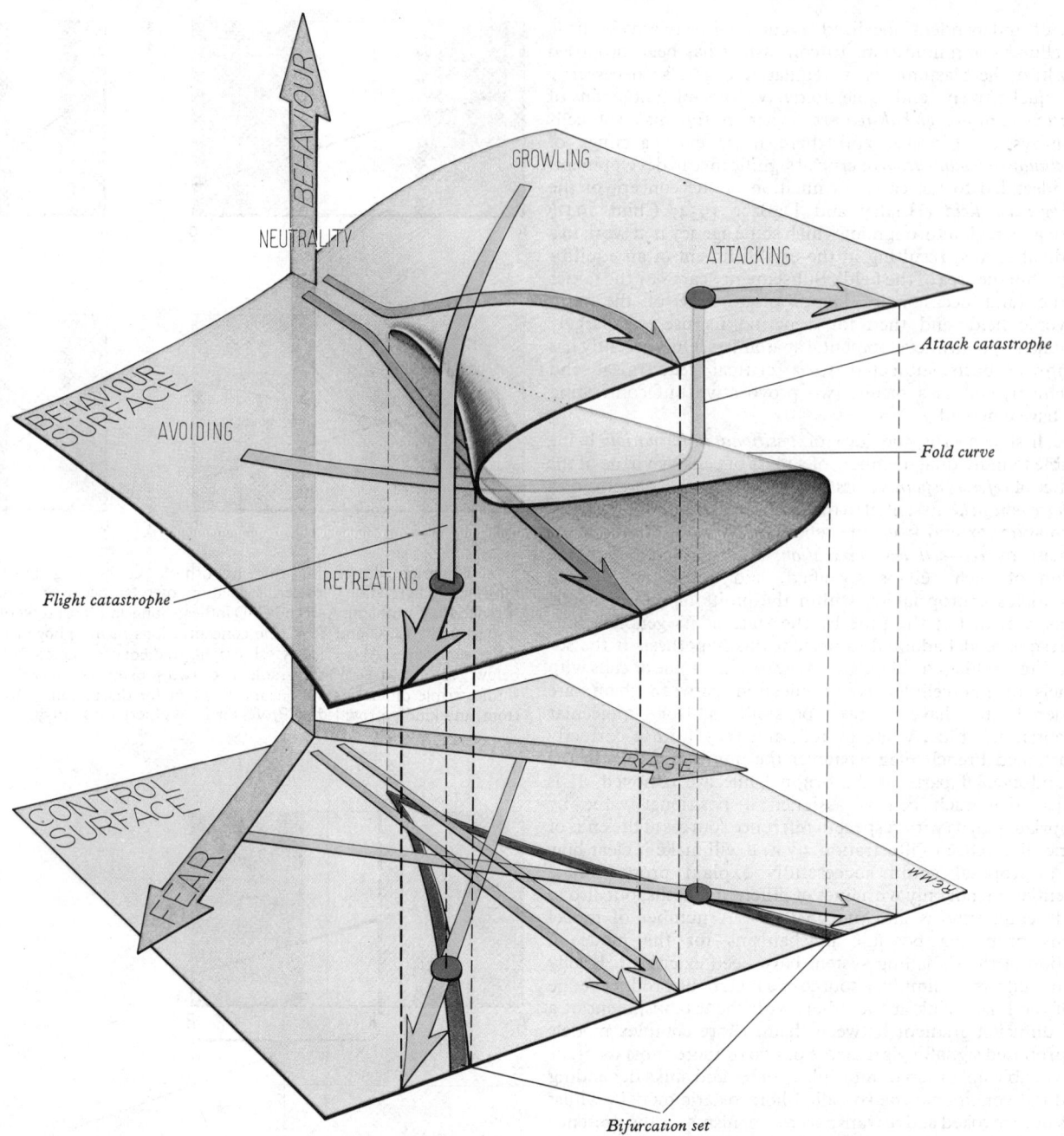

2.8   A diagram showing the use of a three-dimensional graph which embodies the single pleat of an elementary cusp-catastrophe model. In this, changes in the possible aggressive or submissive behavioural states resulting from the interactions of varying degrees of fear and rage in dogs have been illustrated. Consult adjoining text for a more detailed description, and explanation of the terms used. Redrawn, and slightly modified, by kind permission from the author, Professor E. C. Zeeman, and the *Scientific American*.

first constituted a critical reappraisal of the definition of the term 'morphogenetic field' and associated terminology by the eminent developmental biologist, C. H. Waddington. The second was the evolution of a new set of descriptive and explanatory mathematical models which have been grouped under the dramatic name of *catastrophe theory*.

During the attempts to redefine the term 'field' as applied to embryology, it was stressed that much confusion had arisen because of the original combination of the ideas of organization and induction (Waddington 1966), and reinforced by the use of terms such as 'organization centre' and 'organizer substance'. Thus it was originally supposed, as indicated above, that such a centre, during an act of embryonic induction, produced organizer substances which transmitted the instructions necessary for the subsequent complex organization of a neighbouring mass of competent cells. In order to avoid this confusion it was proposed (Waddington and Needham 1936; Waddington 1966) that the

term *evocation* should be used for 'induction that does not *transmit* organization'. *Individuation* was suggested as the name for organization of a mass of cells *per se*, whether it was preceded by an inductive *triggering mechanism* or not. In this sense it was proposed that a morphogenetic field was only appropriately named when applied to an *individuation field* and it was indicated that some authors had used the term too loosely, denoting merely the *locality* of a normal presumptive organ rudiment. It was also considered that to view a morphogenetic field as the manifestation of the presence of gradients of one, or at most a few, simple diffusable substances, was almost certainly a naïve oversimplification. Emphasis was placed on the fact that when attempts are made to describe a system in developmental biology, cognizance must be given not only to the three dimensions of *space*, but also to the essential dimension of *time*, to embrace the dynamic progressive changes involved. Further, at successive intervals it would be necessary in such a *multidimensional function space* to establish coordinates which allow a vectorial analysis of the *concentrations* and *directionality of interaction* of a large number of relevant chemical families. The mathematical concepts appropriate to a study of multidimensional space-surface transitions constitute *topology*. Those facets of topology which are of wide application in both the physical and biological sciences, including developmental biology, constitute *catastrophe theory* which has recently been invented, developed mathematically, and provocatively propounded by the distinguished French mathematician René Thom (1972, 1975). Hitherto, the great domains of the physical sciences, for example, the laws of motion and gravitation, the theory of electromagnetism, and the general theory of relativity, dealt with systems exhibiting gradual, smooth, continuous transitions which could be adequately described by the differential calculus invented by Newton and Leibniz. However, it has become clear that in many systems, both in physics and biology including the behavioural sciences, despite a gradual smooth increase or decrease of one or more parameters affecting the system, the response of the system although initially smooth and continuous, a point is reached when there occurs a sudden, dramatic, completely contrasting form of response. The science of topology is involved because the *resultant forces* in multidimensional space systems are best described by smooth, continuously curved *surfaces of equilibrium*. Abrupt changes in the form of the surface indicate the points at which the equilibrium breaks down, and at which equally abrupt changes in response occur—such points of discontinuity are mathematically termed *catastrophes*. They are not amenable to analysis by the differential calculus, and it is to embrace such phenomena that the new mathematical theory has been developed. The rigorous proof of the mathematical validity of the theory is highly complex (Thom 1972, 1975) and cannot concern us here, but a brief allusion to the manner of presentation of *results* of the proof are illustrated (**2**.8, 9).

Catastrophe theory has, at the time of writing, had its greatest impact on other branches of mathematics; applications in physics include shock-wave propagation, non-linear oscillations and the detailed analysis of caustics and fluid flow. It has been suggested, however, that in future years its potentially most powerful applications may lie in biology in relation to the behavioural sciences and in developmental studies. Illustration **2**.8 is redrawn from a paper devoted largely to behavioural science, because its construction and implications are relatively easy to understand. Space constraints in the present volume, however, preclude an adequate description and discussion of even this elemental model, and the interested reader should consult the original sources cited (Thom 1972, 1975; Zeeman 1975, 1976; Thom and Zeeman 1975). The illustration deals with aggressive behaviour in dogs and includes the following principal features. The assumption is made that aggressive behaviour is the resultant of two *measureable conflicting* factors, fear and rage, which are plotted as axes diverging from an origin on a horizontal *control surface*. The behaviour of the dog is plotted on a third, vertical axis, the most aggressive behaviour—growling leading to overt attack—being assigned the largest values, whilst submissive behaviour—avoidance reactions leading to frank retreat—being assigned progressively lower values. The total set of points on the vertical axis, plotted from every point on the control surface, together form the *behaviour surface*. It might be imagined that only *one* resultant behaviour would be predicted from each combination of fear and rage leading to a simple downward slope passing from attack, through growling, neutrality, avoidance and finally retreat. Over much of the behaviour surface there is, indeed, only one resultant behaviour point, but where rage and fear have roughly equal values there are *two* highly probable forms of behaviour, i.e. attack *or* retreat, but neutrality is the *least* probable form of behaviour. Application of catastrophe theory explains this *bimodality* in behaviour pattern. Expanding from a point vertically above the origin along the distant margin and then including the right and left regions, the behaviour surface forms a single continuous slope which is high to the right and low to the left. Centrally, however, starting at a *point of singularity*, and expanding forwards, a continuous double fold or *pleat* is formed. The upper and lower layers denote *two* alternative behaviours each with a *high probability* corresponding to a *single* point on the control surface (the intermediate layer of the pleat can be disregarded in terms of response because it represents the *least probable* form of behaviour). The actual behaviour pattern followed can only be predicted from a knowledge of the particular *starting point* in the behaviour surface and then tracing its route as the relevant parameters change with time. Thus, an enraged but only slightly fearful dog shows aggressive behaviour: as fear increases, behaviour changes only slowly (gentle slope) remaining aggressive, but the point on the upper layer of the behaviour surface moves centrally. Eventually the edge of the pleat is reached, and the behaviour plot *suddenly* drops vertically to the lower sheet denoting a complete contrast in behaviour which is now meek, submissive and ultimately leads to retreat. The projection of the edges of the pleat back on to the control surface comprises a cusp-shaped pair of divergent curved lines, the *bifurcation set*, and the whole system predicting sudden changes in response is therefore termed a *cusp-catastrophe model*. In summary, this model predicts a single behaviour pattern for most parameter combinations, but *bimodality exists within the bifurcation set*. The points at which jumps from the upper to the lower sheet occur, do not correspond with those points at which the converse direction of jumps occurs (*hysteresis*). It is also clear that starting at a point of neutrality, slight differences in the temporal change of parameters may lead to widely divergent responses (passing either on to the upper sheet or the lower sheet, the divergence occurring at the point of singularity). The forces which operate homeostatically within any particular system tending to maintain the behaviour point (or trajectory) in contact with either the upper or the lower sheet of the behaviour surface are collectively termed the *attractor* for that particular surface. The attractor effect is most powerful in unimodal regions which approach the horizontal and are of gentle slope. Its effect diminishes in the region of bimodality and increasing slope, and of course vanishes at the margin of the pleat where it is replaced by a new attractor.

When a maximum of four parameters are considered, Thom has shown mathematically that these can be satisfactorily described by one of seven *elementary catastrophe* models of which the cusp catastrophe is one of the simplest and, as yet, of widest application. The analysis of models embodying larger numbers of parameters and their possible applications, are currently engaging leading mathematicians in many centres.

In view of the recent wide acceptance of the conceptual relevance of catastrophe theory in such diverse fields as the physical and behavioural sciences, it is of the greatest interest that the first formal scientific paper devoted to the theory should have been preceded, by some years, by a brief statement in relation to *developmental biology* following conversations between Waddington, Thom and Zeeman (*see* Waddington 1966). Illustration **2**.9 summarizes some of the main points. For simplification, the developing three-dimensional embryo has been reduced to a pair of two-dimensional embryo maps (compare with the *control surfaces* described above) and these represent increasing developmental compexity at two successive points in time T1 and T2. From each point on the embryo maps, vectorial plots of the activities of relevant chemical groupings are made within multidimensional spaces termed *phenotype cubes*. These plots generate appropriate 'behaviour surfaces' the form of each indicating the homeostatic equilibria present. It will be noted that the almost horizontal featureless behaviour surface at T1, which embodies a single, powerful, effective, *attractor*, has during development by time T2 been subdivided into two different attractor surfaces, separated by a breakdown in homeostatic equilibrium and indicated by the pleat of an *elementary cusp catastrophe*. The term *chreod* (derived from Greek roots meaning 'it is necessary or fated' and 'a path') was introduced by Waddington to refer to equilibrium surfaces showing attractor phenomena in multidimensional space systems. The region of strong effective attractor action was termed by Thom the 'support of the chreod', and the surrounding zone where its effectiveness diminishes, and catastrophes occur was named 'the cone of the chreod'. Accordingly 'fields' in embryogenesis were portrayed as subregions in a multidimensional space-time system, each being chreodic in character with its particular attractor system, and with progressive development, divergence of chreods occurs by the

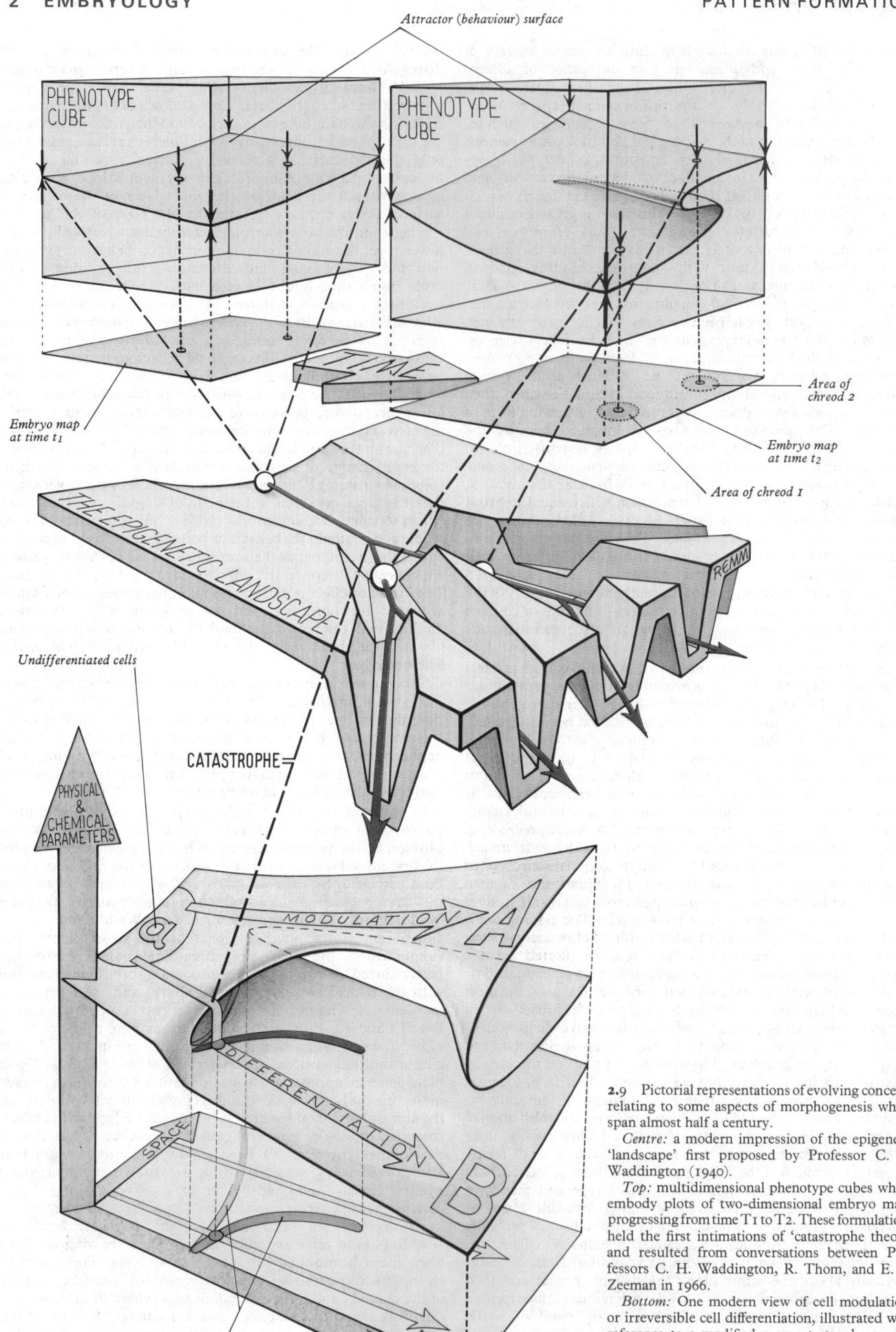

Attractor (behaviour) surface

PHENOTYPE CUBE

PHENOTYPE CUBE

Embryo map at time t1

TIME

Area of chreod 2

Embryo map at time t2

Area of chreod 1

THE EPIGENETIC LANDSCAPE

REMM

Undifferentiated cells

CATASTROPHE

PHYSICAL & CHEMICAL PARAMETERS

MODULATION

A

DIFFERENTIATION

B

Bifurcation set

SPACE

TIME

2.9 Pictorial representations of evolving concepts relating to some aspects of morphogenesis which span almost half a century.

Centre: a modern impression of the epigenetic 'landscape' first proposed by Professor C. H. Waddington (1940).

Top: multidimensional phenotype cubes which embody plots of two-dimensional embryo maps progressing from time T1 to T2. These formulations held the first intimations of 'catastrophe theory' and resulted from conversations between Professors C. H. Waddington, R. Thom, and E. C. Zeeman in 1966.

Bottom: One modern view of cell modulation, or irreversible cell differentiation, illustrated with reference to a modified cusp-catastrophe model. Derived from data provided by Professor E. C. Zeeman (1975, 1976). Consult text for an extended discussion of these diagrams.

interposition of catastrophe zones. In the central part of illustration 2.9 each attractor surface has been plotted down to a simplified contour impression of the picturesque concept of an *'epigenetic landscape'* which had been advanced by Waddington a quarter of a century earlier (Waddington 1940). This comprises a system of valleys, initially single but diverging into branch valleys with the passage of time. Each valley and its sloping walls constitutes a chreod. The homeostatic power of its attractor mechanisms is simulated by the steepness of the valley walls, and it is this feature that confers on a developmental field, or chreod, its regulative capacity (p. 81). The points of divergence of valleys are the points at which the homeostatic attraction breaks down with the occurrence of catastrophes. Such a system may be thought of as analogous to the emergence of, for example, a series of separate organ primordia during early development. The lowest part of illustration 2.9 displays a modified cusp-catastrophe model which applies catastrophe theory to two possible modes of change which may ensue when a group of 'undifferentiated' cells (*a*) are subjected to an array of physical and chemical parameters which change temporally and spatially. It will be noted that *modulation* of the cells to condition (*A*) is represented as a reversible process (remaining confined to the upper leaflet of the behaviour surface). In sharp contrast, *differentiation* of the cells to condition (*B*) can be considered as involving a sudden 'catastrophic' change (occurring, perhaps, when particular parameter combinations are present at the time of the *critical mitoses* discussed previously on p. 84). It should also be appreciated that in the model illustrated, the pleat in the behaviour surface is *incomplete*: i.e. there is no return path from *B* to *A* (which would involve a pleat edge). Thus it is assumed that, in this case, such a change (or dedifferentiation) back to condition *a* cannot occur.

It is hoped that the greatly abbreviated accounts given in the foregoing paragraphs relating to pattern formation and the concepts of positional information and catastrophe theory, will be sufficient to convey the general conceptual framework employed by modern developmental biologists. It must be stressed, however, that whilst they have provided powerful stimuli for reappraisal of many outstanding problems, precise laboratory verification and clarification is still awaited. Thus, whilst the *phenomenon* of embryonic induction is widely accepted as adequately demonstrated, its *mechanism* is still unknown. Similar, perhaps identical, questions are posed, but remain unanswered, by the *intercellular signalling* mechanisms implicit in the theory of positional information. In like manner, *mathematical models* based upon catastrophe theory have been constructed for the cardiac cycle, nerve impulse propagation, cell differentiation (2.9) and for amphibian gastrulation (Zeeman 1975, 1976), but as yet experimental validation is lacking (and indeed, occasional authors view with scepticism the over-enthusiastic acceptance and application of the theory in the presence of incomplete data—*see* Croll, 1976).

Nuclear Changes During Differentiation

Two main lines of investigation have attempted to demonstrate changes in the nuclei of tissues which have undergone differentiation—firstly, experiments using nuclear transplant-ation techniques, and, secondly, an examination of the phenomenon of chromosome 'puffing' shown by some species. Many kinds of transplantation experiment have been carried out (King and Briggs 1965; Gurdon 1967, 1968) but in principle, the nucleus is removed from an activated egg and replaced by one removed from a particular site in, for example, an early blastula, early or late gastrula, or more highly differentiated post-gastrulation tissue, and then any subsequent development which proceeds is analysed. In practice more reliable quantitative results can be achieved by the *serial transplantation* of donor nuclei before proceeding to a developmental analysis (2.5). Although the results show considerable variation, in general, the transplant-ation of early nuclei is often followed by the formation of a complete embryo, whereas nuclei from the later stages are sometimes only able to support the development of incomplete embryos or cell masses of a restricted tissue type. However, in contrast, it has been shown that nuclei from quite advanced

tissues such as amphibian neural plate cells, and ciliated gut epithelial cells of the tadpole can still support the formation of complete embryos, although the proportion of successful experiments is reduced, whilst nuclei from fully adult tissues fail to do so (Gurdon 1967, 1968). Whilst such results may reflect some stable changes occurring in nuclei as differentiation proceeds, much more work will be necessary before clear-cut interpretations are possible.

Direct visualization of differences between the chromosomes of various tissues, and at various times, has been provided by examination of the giant polytene chromosomes of the *Diptera* and the 'lampbrush' chromosomes of the newt. At certain points along their length such chromosomes exhibit hazy thickenings of their outlines ('*puffs*' or *Balbiani rings*). It is widely held that throughout inactive regions of chromosomes, the DNA threads are closely coiled and supercoiled, but at sites of activity the threads become unwound and spread laterally from the axis of the chromosome and form the basis of the puffs. Each puff may include a number of active operons at which transcription and synthesis of messenger RNA is occurring. In short, although many details of the process of puff formation remain to be elucidated, the puff pattern of all the functional nuclei of a particular cell type in a tissue are identical, but it differs significantly from the patterns in other tissues. The patterns also change as development proceeds; for example, there is a regularly repeating sequence of changes accompanying the 'moult' and 'intermoult' stages in the life of larvae. The latter system is particularly interesting since moulting is a response to a hormone, *ecdysone*, which has been isolated, and which affects the activity of the genome—details of the process are under intense analysis.

It is important to emphasize the fundamental question posed by such investigations, namely—during the growth, patterning and differentiation of complete multicellular organisms, is the *essential sequence* of instructive molecules in the genome of the different emergent cell types *altered* for example, by deletion of segments? Alternatively, is the overall sequence *preserved* and the *expression* of its various segments *modified* appropriately? The consensus supports the latter, with the reservation that the *degree of reversibility of expression control* varies widely with site and species.

Occasional examples of alteration of genomic sequences have been quoted, for example, the early segregation of gonocyte and somatic cell nuclei in *Dipterans*, with the apparent deletion of sex-determining loci in the latter; another case involves the *amplification* of repetitive gene sequences during oögenesis at the other extreme; it has proved possible, by careful tissue culture techniques, to grow a complete carrot plant from a single isolated parenchymal cell from a mature carrot.

Surface Interaction Between Cells

Further properties of developing cells which reflect some aspect of their state of differentiation, and have important implications in morphogenesis, concern the selective interactions between their surfaces when in contact. Such surface modifications are closely related to the formation of the cellular sheets, hollow balls, tubes, invaginations, solid masses and so on which characterize embryonic life. Following earlier tissue culture studies, (Weiss 1941, 1961; Abercrombie 1961; Abercrombie and Heaysman 1954, 1970) two important principles were enunciated. Many experiments showed that some cells when explanted on to a matrix with an heterogeneous, polarized substructure, often migrated along preferential paths determined by its architecture. This led to the concept of *contact guidance* of cells—particularly well shown in the preferred regeneration of nerve fibres along the surfaces of Schwann cells, or, during ontogeny, the growth of later generations of nerve fibres along the surfaces of the earlier pioneer fibres.

In contrast, when similar cells which were migrating freely in tissue culture, approached each other and made contact, all movement and often cell division in the plane of contact ceased, and this phenomenon of *contact inhibition* was proposed as an important morphogenetic factor. For a brief review of the mechanisms of cell movement *see* p. 35, and for more recent reviews on morphogenetic movements consult Trinkaus (1969);

Abercrombie and Heaysman (1970); Ingram (1969); Garrod (1973).

Since that time many experiments involving the *disaggregation* of tissue cells have been made (Townes and Holtfreter 1965). For example, if the cells of a neurula, or those of a developing kidney are disassociated to form random cell suspensions, they often subsequently reaggregate and similar cells reassociate so that a recognizable neurula, or a fragment of kidney tissue is reformed. Again, if cell suspensions from developing liver and kidney are intimately mixed, similar cells gradually reassociate and organize into kidney and liver tissue. In this manner, using different cell combinations, the potential of different cell types for mutual recognition, reorganization into tissues, sheet formation, invagination and so forth, have been analysed. An adequate interpretation of the phenomena would, of course, necessitate an intimate understanding of the physico-chemical properties and control systems of cell surfaces. Until such become available, the investigators in this active field of research are pursuing the quantitative variations in the *specific adhesiveness* between cells, and are proposing various models for further examination, such as the possession of specific *stereochemical surface configurations* which can only interlock with a complementary surface. For reviews consult Curtis (1967, 1970); Steinberg (1970); Karfunkel (1971); Baker and Schroeder (1967); Schroeder (1970); *see also* p. 7.

In the subsequent sections a brief account of human development is given and, as such, is necessarily and solely *descriptive embryology*. When reading such an account of the emergence of organ primordia and their sequential shape changes, the reader should be constantly aware of the many continuing fields of *experimental analysis* in other forms, which are daily adding to our comprehension of the *causal processes* of the almost incredible and beautiful phenomenon of development.

HUMAN DEVELOPMENT

The Female Gamete: The Ovum

The precise origin of the primordial germ cells, from which both ova (Shettles 1960; Austin 1961), and spermatozoa are derived, remains uncertain (p. 216); but those which reach the genital ridge in the female, the *oögonia*, proliferate and differentiate into *primary oöcytes*. The number of germ cells reaches a maximum estimated at about 6,000,000 in each ovary by the fifth month of intra-uterine development (Baker 1963, 1966). Subsequently, before birth, a widespread degeneration occurs, many oögonia disappearing, leaving a complement of about 2,000,000 primary oöcytes to last throughout reproductive life (Block 1953). However, the majority of these become atretic before puberty, at which time only some 40,000 oöcytes persist in each ovary (Pinkerton *et al.* 1961). A mere 400 or so of these will be shed at ovulation with the potential of fertilization during the reproductive years of the average woman. (Each primary oöcyte passes through a series of changes, becoming a *secondary oöcyte* just before ovulation, and mature *ovum* at, or just after, fertilization. The term 'ovum' is nevertheless used with a confusing vagueness for all three stages, a practice which is best avoided.) Immediately after formation, and still before birth has occurred, primary oöcytes enter the initial stages of prophase of the *first meiotic (reduction) division* in the process of *maturation*, which is not completed until shortly before ovulation. This reduction in the genetic apparatus of the cell is, of course, a central feature of maturation (p. 94). A primary oöcyte is distinguishable from other cells in the ovary by its large size, being about 35 μm in diameter, and by the fat granules in its cytoplasm. Its nucleus is relatively large, vesicular and usually eccentric, with a nucleolus also comparatively large. The Golgi apparatus is well developed and consists of parallel cisternae and numerous small vesicles; the latter also appear throughout the cytoplasm. Juxtanuclear annulate lamellae occur and there is a similar distribution of mitochondria, which are spherical or slightly elongated. At first the endoplasmic reticulum is vesicular and displays few ribosomes. As the oöcyte grows all these organelles increase in number, and usually in size, and they disperse from the juxtanuclear zone throughout the cytoplasm. Granular endoplasmic reticulum and free ribosomes, though not prominent, also increase. Lipid granules begin to appear, the granules corresponding to the yolk platelets of earlier vertebrates, though

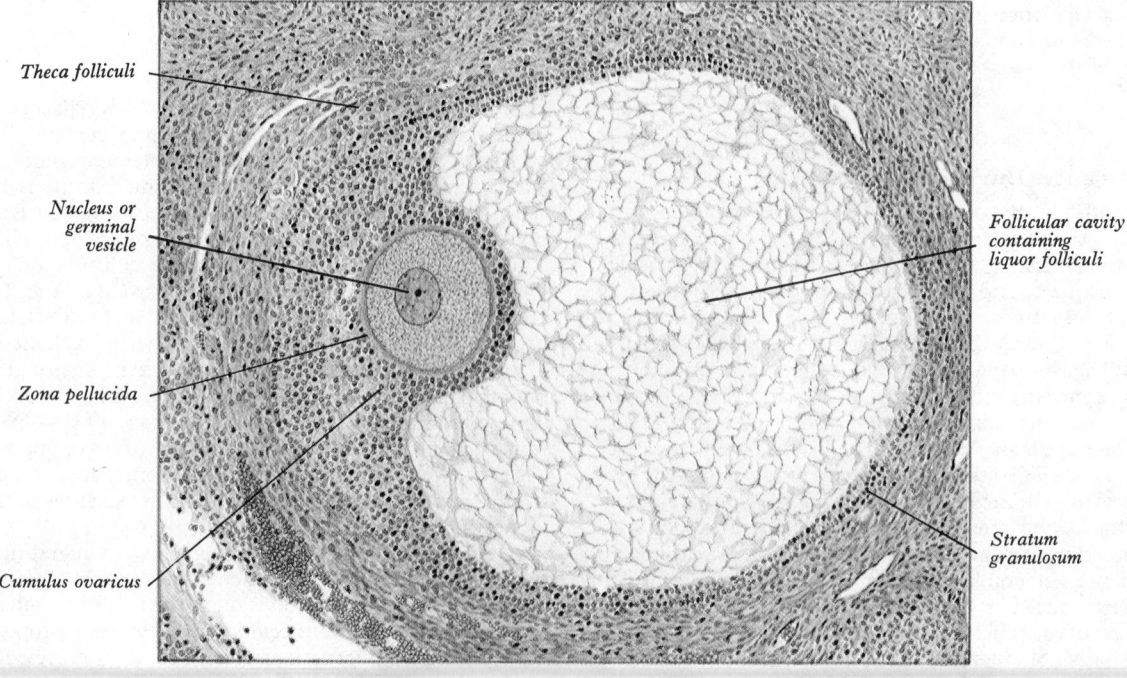

2.10 Ovarian follicle from a woman aged 28 years. Haematoxylin and eosin. Magnification × *c.* 90.

they are smaller and less frequent in primates. By the time the secondary or antral stage of follicular growth is reached the oöcyte has completed its increase in size. Usually each ovarian follicle contains one, but two or even more may be present. In some strains of certain animals polyovular follicles may be more frequent. The oöcyte is surrounded by a single layer of cubical cells, and at this stage electron microscopy shows that the relationship is one of close apposition, the adjacent plasma membranes being smooth and separated by 12·5–14·0 nm.

THE OVARIAN FOLLICLE

During the next stage of maturation considerable growth of the primary oöcyte and enlargement of the follicle occurs. By accumulation of cytoplasm and particularly of lipids, the oöcyte reaches a diameter of 117–142 μm—a volume increase of about a thousand times. This is fully achieved before follicular enlargement is completed. The follicular cells first multiply and form several layers; at, or even before, this stage electron microscopy reveals the accumulation of interrupted masses of amorphous material between the cells and the oöcyte. These soon fuse into a complete membrane around the oöcyte, seen with light microscopy as a PAS-positive, thick, and radially striated envelope, the *zona pellucida* (z. striata). The zona contains considerable amounts of carbohydrate and neutral glycoprotein as a structural component (Braden 1955). Variable amounts of sialic acid residues also occur and may contribute to its elasticity (Soupart and Noyes 1964; Soupart and Cleive 1965). Electron microscopy shows it to possess an intricate filamentous structure at the time of ovulation. The origin of the zonal membrane, whether from the oöcyte, follicular cells, or both, is still undecided (Hope 1965); an outer layer, consisting of acid mucopolysaccharide has, however, been attributed to the follicular cells (Wartenberg and Stegner 1960–1). Coincident with the development of the zona, the surface of the primary oöcyte presents numerous microvilli which project into the zona: less frequent processes from the follicular cells pass through the zona pellucida to make contact with the plasma membrane (oölemma) of the oöcyte, without actual cytoplasmic continuity. These intimate relationships may be concerned with transport of materials between the follicular cells and ovum (Shettles 1959, Chiquoine 1959; Sotelo and Porter 1959; Odor 1960).

The zona pellucida is semipermeable and is slowly lysed in alkaline media. It is best developed in placental mammals and persists until the blastocyst stage of development. It is probably the source of fertilizin (Bishop and Tyler 1956) in mammals, and its function in fertilization is discussed below (p. 99).

Dense extracellular granules, *Gall-Exner bodies*, develop between the maturing follicular cells. They are PAS-positive and display a complex ultrastructure. They may be derived from the Golgi apparatus, but are of uncertain significance. They occur in human follicles, but rarely in other mammals.

OVULATION

When the follicle has reached a multi-layered form, a fluid-filled antrum appears among its proliferating cells and gradually divides them into an internal stratum, the *cumulus ovaricus* (oöphorus), and an external layer, the *stratum granulosum*; but at one site the two strata maintain continuity (2.10). The oöcyte is in the cumulus, the granular stratum forming an envelope around the fluid and gradually thinning out as the latter, the *liquor folliculi*, accumulates. (Antral development has been analysed in detail in the rhesus monkey by Zamboni 1974.) The primary oöcyte retains its microvilli and its complex interrelationship with the follicular cells outside the zona pellucida (2.11) until shortly before ovulation, when formation of the first polar cell is initiated. At this point in time the circumvitelline space first begins to appear between the zona and what is now a secondary oöcyte. The processes of the follicular cells are also withdrawn from the zona, which therefore loses the striation evident under light microscopy and becomes more homogeneous. The follicular cells show an increasing abundance of mitochondria, granular endoplasmic reticulum, and free ribosomes, and their Golgi organelles become

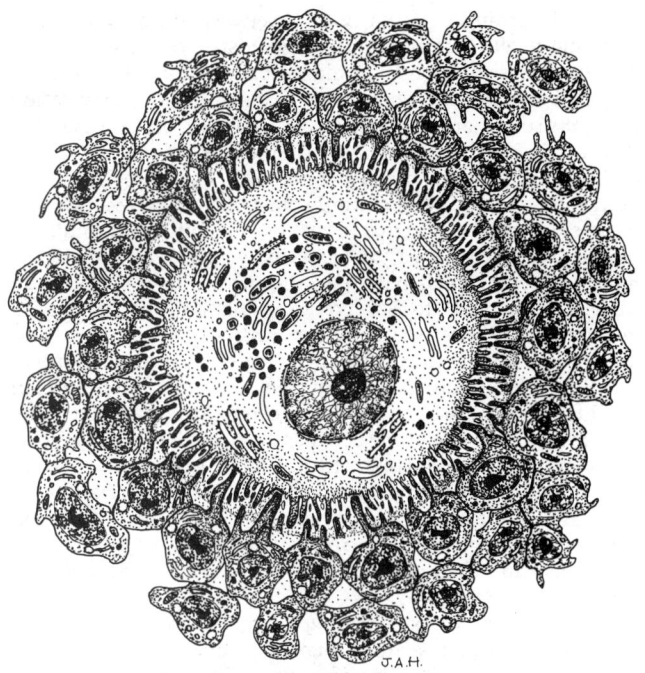

2.11A A drawing constructed from electron micrographs of a primary oöcyte surrounded by its zona pellucida, outside which are aggregated follicular cells of the corona radiata. Note extensive invasion of the zona by microvilli of the oöcyte interdigitating with cytoplasmic processes of the follicular cells. Where some of the latter approach the plasma membrane of the oöcyte, desmosomoid junctions are visible. Note the organelle-rich cytoplasm of the oöcyte. (Modified from Anderson and Beams 1960, with kind permission.)

more prominent, indicating an increasing rate of activity. The stratum granulosum becomes enveloped in a cellular and fibrous sheath derived from the ovarian stroma (2.10), termed the *theca folliculi* (p. 1424). By enlargement, and perhaps also some actual translation through the ovarian stoma, the ripening follicle projects from the surface of the ovary and finally ruptures, releasing the 'ovum' (2.11), surrounded by the cells of the cumulus, into the peritoneal cavity, from which it rapidly enters the uterine tube for slow transit to the uterus, aided by contractions of the fimbriae of the tube, and of the tube itself, together with ciliary movements of its epithelium (Austin 1963). It should be added here that the secondary oöcyte cannot be strictly regarded as an ovum until after completion of its division into ovum and second polar cell (3.12).

The follicular cells adherent to the liberated oöcyte are now termed the *corona radiata*, and they form two or three irregular

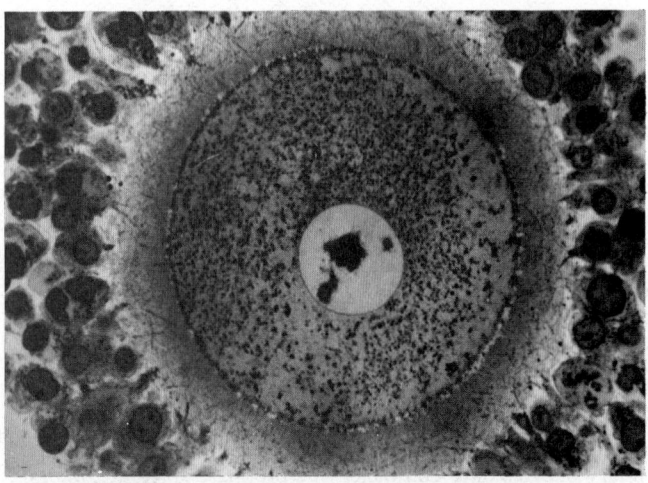

2.11B Human oöcyte photographed in transverse section. Follicular cells surround it, the zona pellucida intervening. Part of the antrum of the follicle is visible above. Epon-embedded, osmic-acid fixed, toluidine blue stained. Magnification × 550. (M. Baca and L. Zamboni, J. Ultrastr, Res., **19**, 1967.)

and loosely arranged layers (**2.11**). Between them is a matrix containing hyaluronic acid, and in some species there is strong evidence that seminal hyaluronidase brings about the disintegration of the corona which soon occurs. Unless fertilization occurs, the secondary oöcyte is discharged from the uterus in the débris of the next menstrual period; if it is fertilized, the zygote which results is retained and pregnancy begins. (For details of fertilization *see* p. 99.) Usually only one follicle matures fully and ruptures in each menstrual cycle in mankind, but multiple ovulation does occur (p. 138). Hence only a small fraction of the primary oöcytes which persist in the ovaries of the new-born female are destined to survive to ovulation, the remainder degenerating. Atresia occurs before birth, as stated above, and even at birth about half of the surviving oöcytes are abnormal. Several follicles usually develop normally to some extent in each reproductive cycle, but only one commonly completes the process, the others becoming degenerate, being invaded by blood vessels and connective tissue which ultimately replaces them as *atretic follicles* (eventually forming small *corpora albicantia*—p. 1425). An interesting laparoscopic technique for observing normal, and artificially controlled, ovulation and follicular development in primates other than man has been exploited by Dunkelow *et al.* (1973).

Although an oöcyte is exceedingly large compared with other general body cells, it resembles them in general structure, though some of its parts are given specific names. Thus, the cytoplasm is the *yolk* or *oöplasm*, the nucleus is termed the *germinal vesicle*, and the nucleolus the *germinal spot*. The cytoplasm is also often known as the *vitellus*, and hence the plasma membrane of the cell is the *vitelline membrane*. The yolk comprises two major components— (1) cytoplasm such as that of any other cells, termed *formative yolk*, and (2) the *nutritive yolk* or *deutoplasm*, which consists largely of fatty droplets containing lecithin. In mammalian ova the deutoplasm is small in amount and is only sufficient for the initial stages of embryonic development prior to implantation in the uterus. In contrast, as we have seen (p. 74), the eggs of birds contain a supply which nourishes the chick throughout its entire development up to hatching. The distribution of the deutoplasm varies in different animal groups; the abundant deutoplasm of certain ova tends to accumulate in one part of the cell, which is therefore termed *telolecithal*. In human ova the dispersion is rather more uniform or *isolecithal*. The formative yolk contains some of the organelles characteristic of cytoplasm, such as mitochondria, Golgi apparatus and centrosome; centrioles are sometimes visible in the neighbourhood of the nucleus, but they disappear when the female pronucleus is formed after fertilization (p. 100). (For a detailed analysis of the fine structure of the developing human oöcyte *in vitro* consult Zamboni *et al.* 1972.)

MEIOTIC DIVISION OF THE OÖCYTE

The series of changes which produce a fertilizable 'ovum' are known as maturation (Monroy and Tyler 1967), some features of which have already been mentioned above. The primary oöcyte's initial period of growth by accumulation of deutoplasm is followed by two successive cell divisions (**2.3**), the first being *heterotypical, reduction division,* or *meiosis,* during which the so-called *diploid* or full complement of chromosomes is halved, the second division being *homotypical* and involving no such reduction (p. 29). Unfortunately, both forms of division, intimately linked in the formation of mature ova (and spermatozoa) from their precursors, are, in current practice, grouped together under the term meiosis (which strictly implies a reduction or lessening), a custom which is misleading but probably now ineradicable.

The primary oöcyte, which is already in the prophase of the first meiotic (true reduction) division, contains the diploid number of chromosomes. It should be emphasized that most, if not all, the primary oöcytes have completed the prophase stage of the first meiotic division before the individual's birth (Manotaya and Potter 1963; Ohno and Smith 1964). Instead of proceeding to metaphase, they enter the so-called *dictyotene* stage, a resting state between prophase and metaphase, which is characterized by a reticular arrangement of chromatin and lasts until the first meiotic

division is resumed and completed shortly before ovulation. As this division is resumed, the nucleolus disappears, the nuclear membrane disintegrates, the chromosomes appear and become arranged in homologous or *bivalent* pairs at the equator of a spindle radially orientated at one pole of the oöcyte. The chromosomes separate, and one member of each pair passes centripetally into the central region of the cell, while the others move centrifugally, forming a projection at the upper pole of the oöcyte, which becomes separated off as the *first polar cell* or 'body'. Unlike that of the nucleus the division of the cytoplasm is highly *unequal*, the polar cell carrying with its equal chromosomal complement an exiguous share of the cytoplasm. The larger cell resulting from this reduction division is a *secondary oöcyte*, and it is not fully reconstituted before it divides again. Its haploid complement of chromosomes (22 autosomes + an X sex chromosome) is again rearranged around the equator of a spindle at the upper pole of the cell. The nuclear membrane is not reformed before this second maturation division commences. A similar division may occur in the first polar cell, but this is abortive, since all the polar cells degenerate. At the moment of spindle formation the secondary oöcyte is shed from its mature follicle to enter the uterine tube. The division is only completed after fertilization, as had long been suspected (Allen *et al.* 1930; Hamilton 1944, 1949).

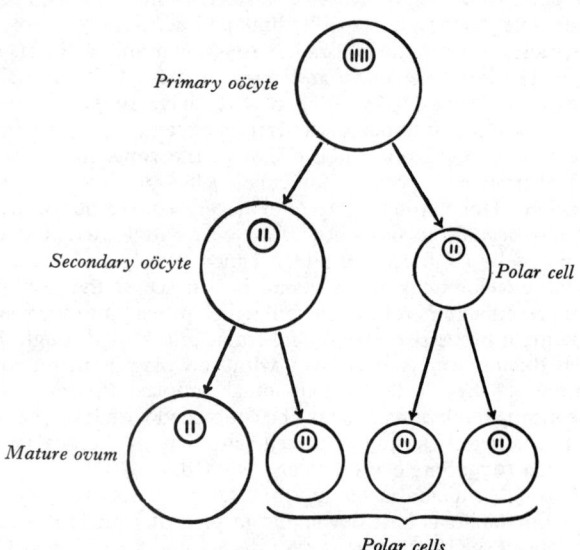

2.12 A diagram showing the reduction in number of the chromosomes during the maturation of the ovum. The first division is heterotypical the second homotypical.

If fertilization ensues, the haploid group of chromosomes split longitudinally, so that each of the resultant two cells receives an equal amount of genetic material, although the chromosome content has been halved. The larger cell is the ovum, which retains almost all the cytoplasm, a small amount only of which enters into the formation of the *second polar cell*. The first polar cell may not undergo division in every instance, but as many as three, theoretically the full complement, have been observed in association with fertilized ova within the zona pellucida (Shettles 1955).

The chromosomes of the fertilized ovum now lose their identity and form a reticulum near the centre of the cell and constitute its nucleus, at this stage termed the *female pronucleus*, around which a nuclear membrane reappears. The cytoplasm shrinks a little, so that a distinct circumvitelline space develops, in which the polar cells may be visible. At this stage the mature ovum is characterized by its chromosome content—half the typical number for the somatic cells of the species— and its large size, due to the great accumulation of deutoplasm. Its pronucleus is now ready for union with the male pronucleus, already in the cell, in the last act of fertilization, the formation of a zygote (p. 99). (The above remarks concentrate upon chromosomal and nuclear organization; for the equally essential organization of the *cytoplasm* of the oöcyte prior to fertilization and further development, *see* pp. 74–76.)

The Male Gamete: The Spermatozoön

Gametogenesis in the male exhibits marked similarities and differences in comparison with the development of ova. During the process of maturation there is the same reduction of chromosomes to the haploid number, but in the testis there is a continuous formation of spermatocytes and spermatozoa during the reproductive life of the individual, a difference linked with the enormous number of gametes which must be formed. In each ejaculate there are many times more spermatozoa than there are germ cells in both ovaries at their peak content before birth; whereas the latter is of the order of 10 to 12 million, a single ejaculation may contain 300 million spermatozoa (**2.**13), only one of which may fertilize an ovum. (The nomenclature of male gametes is not yet unified; spermatozoön, spermatoid, sperm and spermium are all used, but official agreement has not yet been satisfactorily established.)

THE MORPHOLOGY OF SPERMATOZOA

A spermatozoön, or sperm (Fawcett 1961, 1975; Piko 1969; Rothschild 1957), is a smaller cell than an ovum, highly specialized to reach the latter and to carry to it its own haploid chromosome complement. Its expanded *caput* or *head* contains little cytoplasm, and is connected by a short constricted section, the *cervix* or *neck*, to the *cauda* or *tail*. The latter is a flagellum of complex structure, usually divided into *middle*, *principal*, and *end parts* or *pieces*. Volumetrically the tail much exceeds the head, which varies greatly in different species (Rothschild 1957; Phillips 1975), being ovoid or piriform in man, somewhat flattened at the tip in lateral profile, with a maximum length of about 4 μm and a maximum diameter of 3 μm. The tail, about 45–50μm in length, displays a greater uniformity from species to species.

The head is a most extreme example of chromatin concentration, consisting essentially of a dense and uniform nucleus, with a distinct bilaminar nuclear membrane and a bilaminar *acrosomal cap* (head cap), the latter covering the anterior two-thirds of the nucleus and partly derived from the Golgi apparatus of the spermatid (p.97). The acrosomal cap is thin in the human spermatozoön, but in other species it is often large and more complex in shape. The acrosome has been shown to contain several enzymes including acid phosphatase and a protease (*acrosomase*), which are probably involved in penetration of the ovum. The nucleus and acrosome are enveloped in a continuous plasma membrane without intervening cytoplasm (Fawcett and Burgos 1956; Anberg 1957). The chromatin is stabilized by disulphide bonds, as if to protect its genetic content during the spermatozoön's journey (Fawcett 1975). So densely packed is the chromatin of the nucleus that it appears homogeneous even under electron microscopy. It has a strong affinity for basic stains, consisting of about 40 per cent (dry weight) deoxyribonucleic acid (Leuchtenberger *et al.* 1953) and a protein rich in arginine (Daoust and Clermont 1955). It is also resistant to physical stress, e.g. ultrasonication (Henle *et al.* 1938), and to mechanical shear (Mann 1949). Defects in condensation of nuclear material may be visible under light microscopy as relatively clear areas or *nuclear vacuoles*. Attempts to discover structural details in the nucleus in a variety of species, by polarization microscopy, X-ray diffraction, and freeze-fracturing techniques, have shown a lamellar structure which cannot yet be equated with chromosomal content. The human Y chromosome has been identified using fluorescence microscopy (Barlow and Vosa 1970; see also p. 34).

Between the head and the middle part or body of the spermatozoön is a slight constriction, the *neck*, about 0.3 μm long. In its central axis, close to a shallow recess in the base of the nucleus, is a well-formed centriole, corresponding to the *proximal* centriole of the spermatid from which the spermatozoön differentiated (p. 96). The axial filament complex (axoneme) is derived from the *distal* centriole, a funnel-shaped *connecting piece*, or *basal body* from which the outer fibrils of the tail extend (*vide infra*). (The *nuclear recess*, or *implantation fossa*, is the region of attachment of the complex filamentary structure of the tail. It is continuous with the postacrosomal part of the nuclear envelope concerned in fusion with the ovum and its nucleus.) A small

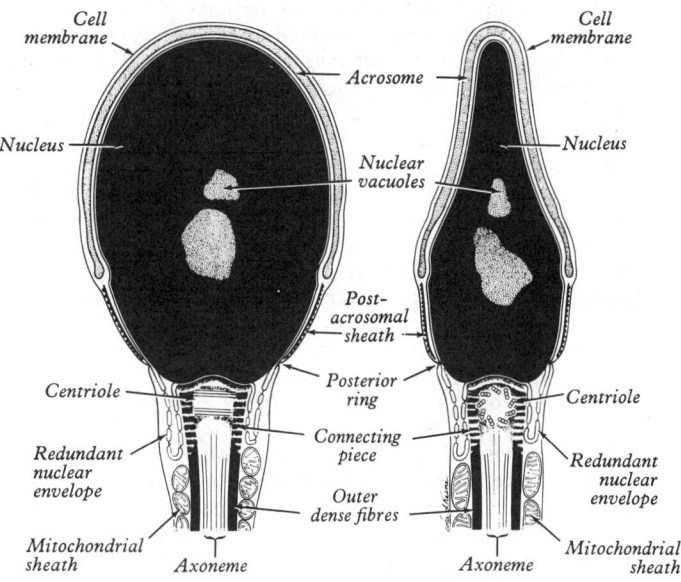

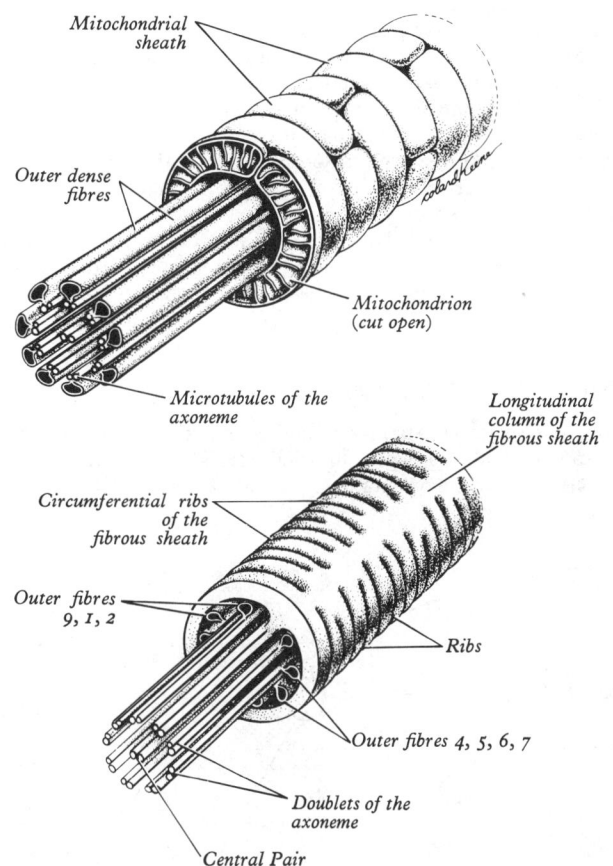

2.13 Diagrams of human spermatozoön, showing the head above, viewed in its major (left) and minor (right) diameters, the middle part below this, and the principal part or tail in the lowest diagram. (By courtesy of D. W. Fawcett, E. S. E. Hafez and C. V. Mosby Company.)

amount of cytoplasm exists in the neck, covered by a plasma membrane continuous with that of the head and tail.

The middle part or piece is a long cylinder, about 1 μm in diameter and 7 μm long. It consists of an *axial bundle of fibrils*, or *axoneme* (the axial 'filament' of light microscopy), surrounded by a *mitochondrial sheath* in which the mitochondria of the spermatid have become arranged in a helical manner (**2.**13), the whole being enveloped by cytoplasm and a plasma membrane, as in the neck. The axoneme consists of a central pair of fibrils within a symmetrical set of nine doublet fibrils, as in a typical cilium (p. 19), and outside this is a second ring of nine coarser fibres, less symmetrical in arrangement and unequal in size. These external fibres are also less regular in cross-sectional profile (*vide infra*), showing marked interspecific variations in size and regularity. It is no longer believed that they are contractile, although they demonstrate a striation in their surface layer. Their function is obscure. The mitochondrial helix exhibits ten to fourteen turns (Reed and Reed 1948), but like other structures in the spermatozoön. this sheath is subject to considerable variation in abnormal spermatozoa (Fujita *et al.* 1970). The number of mitochondria appears to be excessive in some species including *Homo sapiens*, when related to the energy requirements of the axonema. Their close relation to the external coarse fibres is suggestive, but as noted, these are apparently not contractile. At the caudal end of the middle part of the cell, immediately anterior to the tail, is an electron-dense body, the *annulus* (**2.**13). The mitochondria of the sheath are much compressed, but it is now certain that they retain their individuality (Fawcett and Ito 1965).

The principal part or tail of a spermatozoön is the motile part of the cell. Being about 40 μm long and 0·5 μm in diameter, it forms the greater part of the spermatozoön. The axial bundle of fibrils and the surrounding array of coarse fibres are continued uninterruptedly from the basal body, through the mitochondrial sheath and through the whole length of the tail except for its terminal 5 to 7 μm, in which the axial bundle alone persists, the coarse fibres ceasing before them. It is only in this terminal *end part* or *piece* that the tail has the typical structure of a flagellum; the coarse fibres are peculiar to mammalian spermatozoa, which also display other specializations. External to the fibres and fibrils, coarse and fine, is a circumferentially orientated dense *fibrous sheath*, whose individual elements branch and re-unite to form a tight reticulum. A small amount of cytoplasm and a plasma membrane complete the major elements in the structure of the tail. The finer details of the structure have been studied intensively in recent years in mammals such as the guinea-pig (Fawcett 1965), and while there is little doubt that the human spermatozoön is highly similar (Pedersen 1969), the following summary of these findings is based on appearances in the guinea-pig.

MOTILITY OF SPERMATOZOA

In cross-section the tail is oval and tapers caudally, and its central area is typical of a flagellum or cilium. The surrounding coarse fibres are obovate or petal-shaped and unequal in size, one being consistently the largest. This is given the number 1 and the rest are numbered from this in a clockwise manner. These fibres are separate into two unequal groups by slender *longitudinal* columns in the fibrous sheath which interrupt its circumferential fibres and extend inwards to meet the coarse fibres numbered 3 and 8. This divides the interior of the tail into *major* and *minor compartments*, containing respectively coarse fibres 4, 5, 6 and 7, and 9, 1 and 2. The plane through the two columns also passes through the central pair of the axial bundle of fibrils and can be used as a reference datum for other structural details of the spermatozoön. For example, the transverse diameter of the head has been considered to lie at right angles to the plane of the columns, but it has now been shown in the guinea-pig that the angle between the two planes is 20–30 degrees less than a right angle (Fawcett 1968). Such details may be instrumental in elucidating the motile activities of the tail. It is now generally accepted that the tail executes undulatory movements in one plane (Bishop 1958) (p. 20), but it has also been suggested that a helical component is superimposed upon this, there being perhaps two separable mechanisms, one involving flat waves travelling along the tail, the other associated with torsional activity (Gray 1958; Bishop 1962; Lindahl and Drevius 1964).

The latter variety of movement has been linked with the unequal size of the coarse fibrils, which are considered to act, like the fine fibrils, as contractile elements, though it has been suggested that the central pair act as axial stiffeners. The asymmetry of the spermatozoan head has also been invoked to explain supposed helical movement. However, it has to be admitted that the full details of the mechanisms of spermatozoal motility have yet to be worked out.

Some further details deserve mention. The dense outer fibres have been shown to exhibit an oblique or helical striation in replicas of dried whole mounts of rodent spermatozoa (Phillips and Olson 1974). The central axoneme (**2.**13) consisting of the usual ciliary pattern of nine double fibrils or 'doublets', has been intensively studied (*see* p. 19, and Fawcett 1975 for recent literature). Each fibril is a microtubule, itself constructed of a regular number of protofibrils. Protein bridges connect adjoining doublets at regular intervals, and radial links extend centrally to the central doublet of the axoneme (**2.**13).

Maturation of spermatozoa (**2.**14, 15) is a complex process which has received much attention in recent years. Spermatozoa show little independent motility while still in the male genital tract, though when removed from the epididymis they may even display circular swimming movements or even directive movements if taken from the cauda epididymis, near the beginning of the ductus deferens (Blandau and Rumery 1964). From the results of artificial insemination of rabbits with spermatozoa from the caput and cauda of the epididymis, it has been postulated that some form of maturation process takes place in this organ, during which the spermatozoön attains its specific pattern of motility (Gaddum 1968). Apart from these incomplete activities, spermatozoa are largely transported through the genital tract by ciliary action, fluid currents set up by localized secretion and absorption, and by muscular contractions. Associated with the maturation of spermatozoa in the genital tract in some mammals, is the extrusion of a small mass of cytoplasm, the *kinoplasmic* droplet or *residual body*, which migrates backwards along the surface of the head and middle part of the spermatozoön before disappearing. It consists of membrane-enclosed cytoplasm containing fine tubules and vesicles (Bloom and Nicander 1961; Guraya 1963). Human spermatozoa have not been shown to undergo any demonstrable structural changes during passage through the epididymis, but there is evidence of an increase in sulphide cross-linking between proteins (Bedford and Calvin 1974). Moreover, restorative surgery after vasectomy indicates that at least part of the human epididymis is essential for motility (Bedford *et al.* 1973).

As soon as they are ejaculated the spermatozoa display their full pattern of motility. The precise factors which trigger off these movements enabling human gametes to travel at a rate of 1·5 to 3 mm per minute, are not yet clear; but the other constituents of semen, derived from the epididymis and testis and from the seminal vesicle and prostate, are generally considered to exert an activating influence on spermatozoan motility. The motility

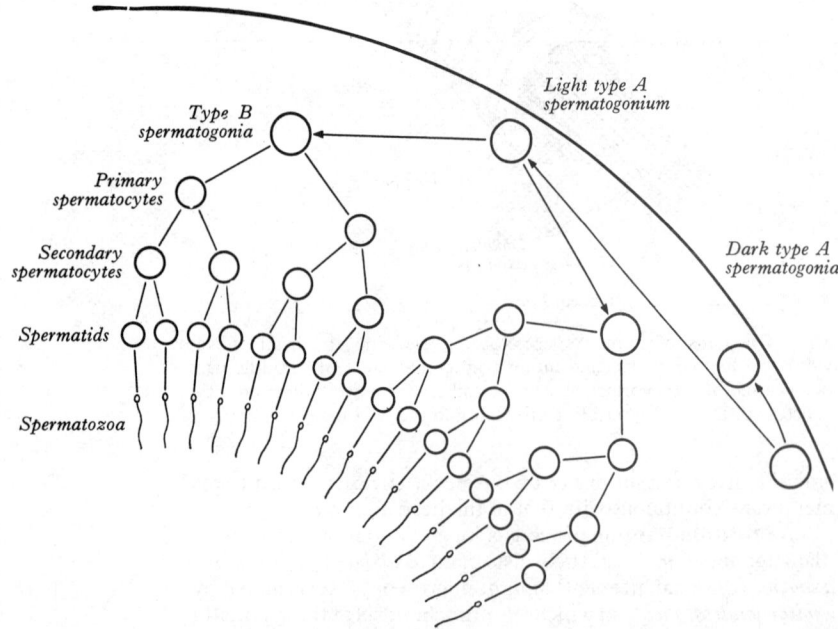

2.14 A diagram showing the stages in the maturation of the spermatozoön. The division of the primary spermatocyte is heterotypical; the remaining divisions are homotypical. Some observers have described a further division of the spermatids in the human testis—*see* text for further details and compare with illustration **8.**168.

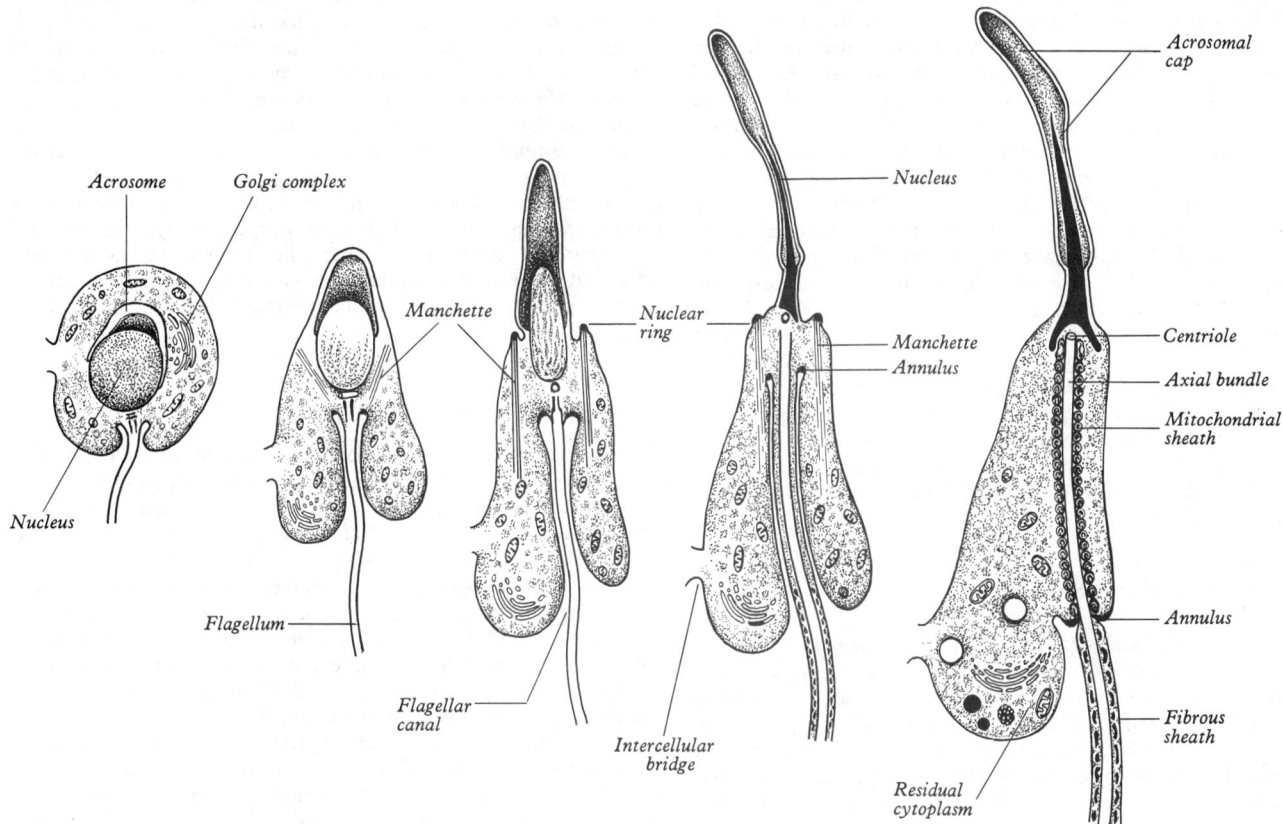

2.15 Differentiation of the spermatid (guinea pig), from left to right. Note nuclear condensation and elongation, appearance of the *manchette* and fibrous sheath, caudal movement of the annulus, and development of the mitochondrial sheath. (Modified after D. W. Fawcett in Bloom and Fawcett's *Manual of Histology*, Saunders. 1975.)

varies greatly in different species, disappearing in minutes in some fish, but usually persisting in mammals for hours and even days when introduced into the female genital tract. Exact figures for its persistence in the human female are uncertain, and are of doubtful value, since it is likely that, as in other mammals, human spermatozoa quickly lose their potency for fertilization, though still motile. They have been recovered in a motile state in human cervical mucus several days after insemination, and will survive in this condition for as long as seven days when implanted into such secretions *in vitro* (Paerloff and Steinberger 1964). These survival periods may, however, be of little significance, in view of the speed with which spermatozoa reach the infundibulum of the uterine tube and the brevity of their fertilizing power. Spermatozoa have been shown to reach their tubal destination in a manner of minutes after ejaculation in some mammals, and experiments on recently excised human uteri and tubes indicated a time of about 70 minutes (Brown 1944). The conclusion must be that factors other than their own motility are responsible for the transport of spermatozoa from their site of deposition in the vaginal fornix to the ovarian end of the uterine tube, and there is considerable evidence that contraction of the uterine and tubal musculature is responsible (Bickers 1960).

It is not usually recognized that a spermatozoön must be adaptable to a wide range of environments in its long journey from the seminiferous tubule to the uterine tube, encountering major changes in the electrolyte and non-electrolyte constituents in the fluids with which it is successively surrounded. Nevertheless, a collectively vast amount of observation and experiment has been recorded in connexion with the effects of the multitude of factors, both physical and chemical, in the natural media involved regarding the behaviour of these cells, and particularly their motility and fertility (Nelson 1967; Mann 1967). For example, the effects of respiratory gas tensions, reaction, various ions, antibodies, vitamins, hormones, inhibitory substances, temperature, different forms of radiation, and other factors have been studied in remarkable detail, for which monographs and original papers must be consulted. The effects of low temperatures in preserving spermatozoa, and perhaps prolonging their vitality, have attracted much research in connexion with artificial

insemination, both in stock-breeding and in infertile human marriage. Mammalian semen, including that of human beings (Parkes 1960), can be stored at temperatures of about $-70°C$ for weeks and even months, the motility and fertility of its suspended spermatozoa reappearing when the suspension is unfrozen. However, storage of human semen presents difficulties (Polge 1957).

Seminal plasma, the fluid component of *seminal fluid* or *semen*, contains a remarkable array of substances, including muco-proteins, a dozen or more identified proteolytic enzymes, the bases spermine, glycerylphosphorylcholine and ergothioneine, a group of organic acids called prostaglandins (which have pharmacodynamic actions on the uterus and smooth muscle in general), acids, such as citric, ascorbic, uric, lactic and pyruvic, and the sugars, sorbitol, inositol and fructose. The fructose, added to the fluid by the secretion of the seminal vesicle, is an essential substrate in the anaerobic glycolysis by which spermatozoa survive the low oxygen tensions existing in semen itself and in the female genital tract. Prostaglandins are now believed to play a role, perhaps by modulation of neurotransmitter release, in the contraction/relaxation activity of the non-striated muscle in the testicular capsule and interlobular septa adjacent to the seminiferous tubules (as suggested by von Euler 1936). Consult Ellis and Hargrove (1977) for literature.

CAPACITATION

After ejaculation into the female, the spermatozoa undergo the final step in their maturation, a process known as *capacitation*. It has been shown that spermatozoa are not able to fertilize ova until they have been within the genital tract of the female for a period of time, usually of hours but varying with the species (Austin 1951; Chang 1951; Austin and Watton 1952). The mechanism of capacitation, whereby the spermatozoön is activated to enter and fertilize the ovum is still uncertain. A confusing array of findings with regard to the interactions of the two gametes immediately prior to this event have been described, unfortunately in widely different vertebrates and invertebrates. It is highly probable that hyaluronidase hastens the separation of the corona radiata cells from the ovum, and thus facilitates the spermatozoön's approach to the zona pellucida. The origin of hyaluronidase from the acrosomal cap is associated with subsequent loss of the cap (Leuchtenberger and Schrader 1950), at least in part (Austin and Bishop 1958). 'Capacitated' spermatozoa observed in the zona pellucida or perivitelline space have invariably lost most of their acrosomal material (Leuchtenberger and Schrader 1950; Pikó and Tyler 1964), and it is clear that capacitation is some process of activation which precedes penetration. Antigenic 'coating' substances on the surface of mammalian spermatozoa, including those of man (Weil 1965), have been recorded and it is possible that an immunological reaction may be involved. Interaction between a *fertilizin*, derived from the ovum or elsewhere in the female genital tract, and a spermatozoan *anti-fertilizin* has been associated with capacitation, but the interrelationship between the various events is still *sub judice*, as the most recent reviews show (Mety and Monroy 1969). Capacitation may be regarded as the terminal event of maturation of the spermatozoön, prior to actual fertilization, for which it is a preparation. For a recent survey of current views *see* Chang and Hunter (1975).

Spermatogenesis

This is the complex series of changes by which spermatogonia are transformed into spermatozoa, similar in some general features—particularly in reduction division—to the evolution of ova from oögonia, but differing in the more profound morphological metamorphosis involved. Spermatogenesis may, for convenience, be divided into three phases. During the first, **spermatocytosis**, spermatogonia proliferate by mitotic division to replace themselves and to produce primary spermatocytes. In the second phase, **meiosis**, two successive maturation divisions, the first a true reduction division, as in the case of the oöcyte, produce *secondary spermatocytes* and then *spermatids*, all with the haploid number of chromosomes. In the third phase, **spermiogenesis** or **spermateliosis**, the spermatids become spermatozoa; it is during this period that the greatest visible transformation of structure occurs.

SPERMATOCYTOSIS

During embryonic, fetal, and perhaps also early post-natal life, the *primordial germ cells* (p. 216) in the tubules of the testes divide mitotically to produce spermatogonia (Everett 1963; Minty 1960), from which, at and subsequent to puberty, the development of spermatocytes and spermatozoa commences. It is the cyclic divisions of these cells, the details of which have attracted much attention in recent years, that form the starting place for production of the huge numbers of spermatozoa discharged into the seminal plasma to form the seminal fluid. The series of changes involved do not occur in a synchronous manner in all seminiferous tubules at the same time, though they do in considerable parts of an individual tube, with individual variations in different mammalian species, including mankind. As the cycle of change from spermatogonia to spermatids proceeds at any particular locus in a tubule, a succession of varying cell associations can be observed and measured; the process is termed the *cycle of the seminiferous epithelium* (Clermont and Leblond 1955) (p. 1414).

In man, three types of spermatogonia can be distinguished and are termed the *dark type A*, the *light type A*, and the *type B* (Clermont 1963). Spermatogonia are large rounded cells, the three types showing little difference in size or in their cytoplasm, but the A series, light and dark, are distinguishable by their nucleoli, which are eccentric and attached to the internal aspect of the nuclear membrane. The type B spermatogonia have a more constantly spherical nucleus, in which the nucleolus is central in position. The dark type A is distinguished from the pale type A by its dark nucleoplasm and a large pale-staining nuclear vacuole. This latter type is now considered, largely on morphological grounds, to be the progenitor or stem spermatogonium (Clermont 1963). Such cells, peripherally situated in the tubule, and often in pairs, divide mitotically at the beginning of a seminiferous cycle,

some to produce two further dark type A spermatogonia, thus replenishing the complement of stem cells, others into two type B spermatogonia. On theoretical grounds, it is probable that in man a larger series of spermatogonial divisions may occur, so that the ultimate spermatocyte progeny of a stem cell may in fact be more numerous than is here indicated (Clermont 1966). Each type B spermatogonium then divides again mitotically into two *resting primary spermatocytes* or preleptotene spermatocytes. The dark type A cells remain arranged along the basement membrane of the tubule, whereas the pale type A, and type B spermatocytes, and the spermatids derived from them, lie closer to the lumen, into which the free end-product, spermatozoa, will be discharged (**8**.168). These events constitute spermatocytosis, which is now followed by meiosis.

MEIOSIS OF SPERMATOCYTES

The primary spermatocytes soon enter the prophase of the *first maturation (reduction) division*, which is prolonged over several days through the successive stages of leptotene, zygotene, pachytene, diplotene and diakinesis (p. 30) (Clermont 1963, 1966). As the nuclear membrane now disappears in metaphase, the bivalent chromosomes are arranged on the equatorial plate, separating into two groups and moving to opposite poles in anaphase, followed in the usual manner by reformation of the nuclear membranes in telophase and division of the cell. These three phases occur much more rapidly than prophase, and during the whole process there is a considerable increase of nuclear and cytoplasmic material, bringing the primary spermatocyte back to a size comparable with that of the stem spermatogonium. The two *secondary spermatocytes* thus formed contain, of course, the haploid number of chromosomes, this *first* maturation division being the one which is strictly speaking *meiotic*. After a brief interphase each secondary spermatocyte now undergoes a *second maturation division*, which is by mitosis. The two resultant cells are spermatids, and with their formation, the phase of meiosis may be considered to end, being followed by their maturation into spermatozoa (spermiogenesis). Theoretically each primary spermatocyte may be expected to produce four spermatids, but in mankind the yield is less than this, presumably because some spermatocytes degenerate during maturation.

Criticism of the traditional view that spermatids do not divide has been expressed (Roosen-Runge 1952), and corroboration of this has come from electron microscope and other studies (Fawcett *et al.* 1959; Fawcett 1961). Electron microscopy has also shown that the division of the cell body (cytokinesis) in both spermatocytes and spermatids may be delayed, so that fine cytoplasmic bridges may remain interconnecting such cells even beyond the stage of the next nuclear division (Fawcett *et al.* 1959). These bridges, which may remain in the case of spermatids until a late phase in their transformation into spermatozoa, are short, devoid of spindle fibres or other remnants, and are enclosed in annular thickenings of the plasma membrane. They probably permit interchange of organelles, may be involved in synchronization of development, and may contribute to the mechanical stability of the spermatid-Sertoli cell complexes. Such cytoplasmic interconnexion may also help to explain the formation of multinucleated masses when the seminiferous epithelium is injured, as in making teased preparations. Except where connected together by bridges the developing spermatids are very closely associated with Sertoli cells, whose processes are insinuated between them.

During the differentiation of some rodent spermatids a fusiform conglomeration of microtubules, the 'spindle-shaped body' appears between the annulus and fibrous sheath. The same structure has been observed in human spermatids; its functional significance is uncertain but an association with the development of the fibrous sheath has been suggested. (Pedersen 1969; Wartenburg and Holstein 1975).

SPERMIOGENESIS

During *spermiogenesis* (spermateliosis), spermatids go through a complex series of changes to become spermatozoa (**2**.15) and this metamorphosis has been studied in particular by light micro-scopy, using preparations stained by the periodic-acid Schiff technique (PAS) (Leblond and Clermont 1952; Clermont 1963). Electron microscopy has confirmed these observations (Fawcett and Burgos 1956). In the newly formed spermatid the Golgi apparatus (idiosome-Golgi complex) is large but otherwise typical, consisting of flattened membrane-enclosed vesicles, usually stacked in a parallel array, together with rounded minute vesicles which are possibly nipped off from the flattened variety, the whole complex being juxtanuclear. A few electron-dense homogeneous *paracrosomal granules*, which are intensely PAS-positive, develop in separate Golgi complex vesicles, and the latter coalesce into a single large *acrosomal vesicle*, the separate granules fusing into a single spherical *acrosomal granule*. This vesicle, with its granule attached to its juxtanuclear wall, becomes adherent to the nuclear membrane over an area which will be anterior or 'leading' in the maturing spermatozoön. The granule flattens, but its central part bulges slightly into a shallow depression in the nucleus, which becomes progressively more ovoid (**2.15**). By successive absorption of further vesicles from the Golgi complex, the material in the acrosomal vesicle increases, and the vesicle expands as a bilaminar cap over the anterior two-thirds of the nucleus. Coincident with these changes, the spermatid elongates and the Golgi complex and associated cytoplasm migrate to the posterior part of the cell, bringing the external wall of the acrosomal vesicle into contact with the plasma membrane at the anterior aspect of the cell. The acrosomal granule now spreads out between the layers of the vesicle until it is uniformly distributed and no longer a localized structure. When the centrioles begin to separate (*vide infra*), microtubules develop forming an inverted conical array, *the manchette*, perinuclear in position, and expanding from the region of the acrosomal cap; its precise significance is not yet explained.

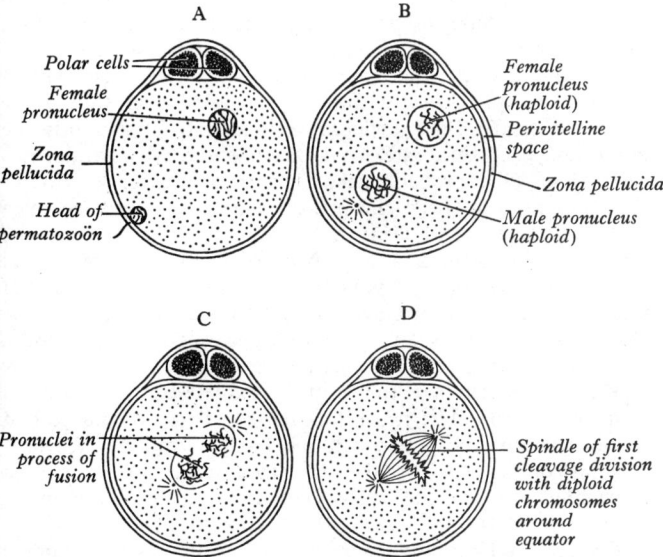

Polar cells
Female pronucleus
Zona pellucida
Head of spermatozoön

Female pronucleus (haploid)
Perivitelline space
Zona pellucida
Male pronucleus (haploid)

Pronuclei in process of fusion

Spindle of first cleavage division with diploid chromosomes around equator

2.16A–D The process of fertilization in a mammalian ovum. Diagrammatic. (After Sobotta.) The female pronucleus and second polar cell shown in A, are only formed after fertilization.

In the early spermatid the nucleus is of relatively low density, containing finely dispersed granules which aggregate into larger and denser masses as development proceeds. These finally agglomerate into a homogeneously dense mass, usually containing one or more regions of low electron-density and variable in size, position and shape—the *head vacuoles* (Fawcett and Burgos 1956). Correlated biochemical and ultrastructural studies indicate a considerable variation in the chromatin content of spermatozoa, this heterogeneity being more marked in mankind than other primates or rodents; it probably indicates a lower fertilizing power (Bedford *et al.* 1973).

The two centrioles are near the posterior aspect of the nucleus from an early stage. One remains unmodified, the other becomes modified into the basal body of the spermatozoön (*see* p.

95). The annulus arises close to the latter (the distal centriole), but its origin from it is doubtful. The axial fibrils begin to develop from the basal body, extending 'caudally' into the cytoplasm of the cell as it becomes progressively more elongated. Only the proximal part of the bundle of fibrils remains surrounded by cytoplasm, to form the definitive middle part of the spermatozoön. In this, the mitochondria of the spermatid assemble to form the helical sheath (Challice 1953). The detailed development of the fibrous sheath of the tail part is unknown. These changes complete what might be called the period of 'organogenesis' of the spermatid, whose further development into a spermatozoön is largely concerned with enlargement of the tail.

During the final maturation of spermatids into individual spermatozoa some of the cytoplasm is detached as a *residual body*. This contains some mitochondria, Golgi membranes and vesicles, RNA particles, lipid granules, but of course no nucleus. Residual bodies are prominent when spermatozoa are being released into their tubule. They are engulfed by Sertoli cells, which accounts for the increase in lipid content of these cells at this period (Lacy 1960).

As already stated, there is a close relation between developing spermatids and Sertoli sustentacular cells. The spermatogonia are external or basal to the Sertoli cells in the tubule, and the spermatocytes which develop from the former are embraced by Sertoli processes; the spermatids are even more deeply embedded in the supportive cells. (For further details *see* p. 1413.) These associations have been regarded as symbiotic, a single sustentacular cell being grouped with several spermatids. The Sertoli cells are phagocytic and absorb not only residual bodies but also degenerating germ cells. They have been attributed a metabolic role and may form, or at least transmit, hormones involved in the maturation of germ cells. Until released into the seminiferous tubule, spermatozoa are very firmly held by the Sertoli cells. Their release is sometimes termed *spermiation* and is followed by rapid translation of the spermatozoa to the epididymis.

Spermatogenesis is an orderly and complex sequence of events, with characteristic time constants and cell associations for each mammalian species. The details of these in the human testis will be discussed with that organ (p. 1411).

Fertilization—Union of the Gametes

It has become customary to speak of an ovum as being fertilized by a spermatozoön, and, in view of the apparent passivity of the former and the extreme specialization of the latter for motility, it is tempting to consider the ovum as being merely stimulated to further development, in which the part played by the sperm nucleus may be consequently overlooked. It is of interest to note that in some of the earliest organisms to propagate by sexual reproduction, such as primitive algae, the gametes are all alike, except presumably in their genetic content, whether chromosomal or cytoplasmic. The profound differences which have nevertheless evolved in the gametes of the great majority of plants and animals—vertebrate and invertebrate—appear to depend on a conflict between adaptation for carriage of nutriment and improvement in motility. The effectiveness of *syngamy*, the bringing together of two gametes to produce a new individual, has been furthered by the development of this marked dimorphism between them. This dimorphism has, in turn, entailed the evolution of equally profound differences between the individuals producing the two kinds of gametes, males and females, in regard to the organs concerned in bringing together these dissimilar gametes and ensuring the development of their fused product, the *zygote*, until able to undertake a separate existence. Strictly speaking, therefore, the expression 'fertilized ovum', insofar as it appears to assign a merely stimulatory role to the spermatozoön, is misleading and hence undesirable. However, it should be noted that almost all the cytoplasm of the zygote is derived from the ovum, and, as indicated elsewhere (pp. 74–76), its organization has a profound morphogenetic significance.

The occurrence of parthenogenesis in a very small fraction of animals, mainly invertebrates, does not invalidate the fact that *fusion* of two gametes, from separate parents, is the central feature of reproduction in most animals and plants (Tyler 1967). There is also abundant evidence that the critical element of this fusion is the conjunction in one cell of two haploid complements of the particular species' chromosomes, these being sufficiently identical to entail the development of a viable new unit in the species. Differences involving a number of genes, arising during the phenomena of crossover and segregation of the chromosomal material, or (very rarely) by sporadic mutation, are thus also able to express themselves in individual variation, with all its implications in evolutionary reactions between the species and its environment (p. 72).

Nevertheless, it would be equally misleading to regard the union of the ovum and spermatozoön as little more than restoration of the diploid number of chromosomes, because other events, of equal significance in the further development of the zygote, also occur. The ovum immediately resumes and completes its second maturation division; without fertilization this does not occur, and the ovum begins to degenerate within twenty-four hours (Allen *et al.* 1930; Hamilton 1949). A second polar cell is thus produced, and if the first has divided, as it sometimes does in human beings (Shettles 1955), three may be present in the circumvitelline space. In many mammals, and probably in mankind, what is apparently a small excess of nutritive yolk, or deutoplasm is extruded from the ovum into the same space after fertilization (*deutoplasmolysis*). This event is followed at once by segmentation, which is also dependent upon fertilization for its initiation.

ACTIVATION AND SEX DETERMINATION

The phenomena which ensue upon the entry of the spermatozoön can thus be conveniently divided into those described in the preceding paragraph, which together constitute *activation*, and those concerned in the intermingling of the separate hereditary influences of the two parents, and associated with the fusion of the nuclei of their gametes, a process described as *amphimixis*. The latter really falls within the scope of genetics (*see* p. 101), but a particular aspect—the determination of the sex of the zygote—may be mentioned here, because the actual determination is in a sense effected as soon as a spermatozoön has entered the ovum. In approximately equal numbers, spermatozoa of many animal groups, including mammals, contain either an X or a Y chromosome, whereas mature ova contain only an X chromosome. If both gametes contain an X chromosome the resultant individual is female; but when the spermatozoön contains a Y chromosome, the cells of the new individual will each contain an X and a Y chromosome, which is characteristic of the male. This may not, however, be the only factor involved in the determination of sex (p. 221).

UNION OF THE GAMETES

In mammals, generally only one spermatozoön pierces the vitelline membrane to enter the ovum, although numbers usually penetrate the corona radiata and even the zona pellucida. Enzymes of acrosomal origin cause dispersal of the cells of the corona, thus opening the route to the zona. Though some form of chemotaxis, as exists in the syngamy of plant gametes, has been claimed for some animals, no clear evidence for this has been recorded (Tyler and Bishop 1963). The mechanism of union of the ovum and spermatozoön has been studied extensively in all its stages. The primary reaction between the gametes appears to be of an immunological nature, which would account for the exclusive specificity of fertilization. Landsteiner (1899) and Metchnikoff (1899) first reported the antigenicity of spermatozoa, and a species recognition system, dependent upon an agent in the ovum, a 'fertilizin', was identified surprisingly early (Lillie 1913). The sequence of the reactions of this with spermatozoan antifertilizins (associated with the plasma membrane) have attracted much attention (Metz and Monroy 1967, Bedford 1977). Such antibody-antigen interaction appears not only to be essential to contact between the gametes but also to initiate the subsequent entry of the spermatozoön by the lytic action of enzymes derived from the acrosome

(*see* p. 95, and Austin and Bishop 1958; Strivastava *et al.* 1965). This *zona-lysin* enables the spermatozoön to pass through the zona pellucida into the circumvitelline space and to come into direct contact with the oöcyte. Penetration is followed immediately by permeability changes in the zona which effectively block the subsequent entry of other spermatozoa (Austin and Braden 1956). The *zona reaction*, as this phenomenon is called, is however, not a complete block to polyspermy; two or more spermatozoa may pass *simultaneously* through the zona, a not uncommon event in the case of some rodents, seemingly also a possibility in man. The result is one of the types of *triploid* zygotes that may arise. Further, there is evidence of a second line of defence, the *vitelline block*, associated with the plasma membrane of the ovum. Both blocking mechanisms may depend upon discharge of the contents of *cortical granules*—electron-dense particles subjacent to the plasma membrane—into the circumvitelline space (Szollosi and Ris 1961; Szollosi 1967). However, recent evidence derived from *in vitro* fertilization of *human* ova suggests that the zona pellucida is, under these conditions, a complete block to polyspermy (Soupart and Strong 1975). Electron microscope studies suggest that the actual entry of the spermatozoön into the ovum is effected by fusion of their plasma membranes, that of the former being left on the surface of the ovum (Szollosi and Ris 1961; Szollosi 1967). Once within the ovum the head and neck of the spermatozoön become detached from the middle part and tail, which soon disappear, their significance in fertilization, if any, and their ultimate disposal being obscure. For a review of the uncertainties and problems in regard to the mechanisms of fertilization consult Bedford (1975), and for ultrastructural details *see* Austin (1968). The occurrence of auto- and iso-immune anti-spermatozoal antibodies has been exhaustively surveyed by Hekman and Rümke (1976).

The second maturation division of the ovum is now completed, with the extrusion of a second polar cell into the circumvitelline space, which has become opened up by shrinkage of the oöplasm. Failure of extrusion of the second polar cell is another mechanism of formation of a *triploid* zygote. Meanwhile, the head of the spermatozoön swells to become the *male pronucleus*, being at this juncture morphologically indistinguishable from the *female pronucleus* (**2.16**). Both nuclei move to a central position in the ovum and each duplicates its DNA content. Meanwhile, two centrioles appear, probably derived from the anterior centriole of the spermatozoön. The chromosomes become organized from each pronucleus and arranged on the spindle between the centrioles, and since 23 are contributed from each, the typical diploid number for man is now re-established. The degree of fusion of the two pronuclei before the appearance of the chromosomes varies in different species, but usually the chromosome groups remain separate during prophase. They now split longitudinally and the resultant halves segregate as in an ordinary mitosis. The two sets, each consisting of the diploid number of chromosomes with the normal amount of DNA, retreat to opposite poles of the cell, on whose surface a deepening grove develops, as the zygote enters the next phase in its development, *cleavage* or *segmentation* (pp. 76, 103).

The chronological element in the events of the human fertilization are still only inexactly known. A particularly detailed study of the temporal march of events has been contributed in the pig (Hunter 1974).

PARTHENOGENESIS

It is generally agreed that a mature ovum contains within itself all the material potentiality to form a new being. Apart from the *natural parthenogenesis* which occurs in some invertebrates, this potentiality of the ovum to develop further without fertilization can be released experimentally by a variety of mechanical, chemical, and physico-chemical means, such as pricking, exposure to alteration of tonicity and reaction in the surrounding medium, treatment with various kinds of radiation, etc. (Beatty 1957, 1967). Such *artificial parthenogenesis* has been studied in a wide range of invertebrate and vertebrate animals, including mammals. Such offspring rarely survive beyond the embryonic stage, but viable young have occasionally been obtained in the case of rabbits, for example. The particular interest of these parthenogenetic phenomena in vertebrates is that they illustrate the fact that, apart from its normal necessary chromosomal contribution, the spermatozoön plays little part in embryogenesis. However, there is some evidence that the plane of bilateral symmetry is determined in many animals by the direction and place of entry of the spermatozoön. Nevertheless, both in this determination and the early polarization of the zygote, there must be a considerable flexibility, because the blastomeres resulting from its early segmentation are 'totipotent', in the sense that they may separate to form complete embryos, as in the formation of uniovular twins (but *see* p. 138).

Heredity and Human Genetics

The operation of genetic influences is so basic to reproduction, growth, differentiation, regeneration, and indeed to all life processes, that a brief account of heredity and its mechanisms must be included here. Great progress has been made in recent years, and not least in the medical field. Further details can be sought in the many excellent general and medical texts available. *See*: Clarke (1970); Whitehouse (1973); Thompson and Thompson (1973); Ford (1974); McKusick (1975); Fraser and Mayo (1975); Bodmer and Cavelli-Sforza (1976); Yuris (1977).

Before chromosomes were discovered, Gregor Mendel (1822–84) had postulated, as the basis of his explanation of the phenomena of heredity, the presence of certain factors which have subsequently been termed *genes* (Mendel 1866). The study of heredity has been increasingly intensified following the discovery of chromosomes, investigation of their behaviour during somatic and germinal cell divisions, and detailed mapping of them, coupled with studies of the transmission of inherited characteristics in complex organisms such as the fruit fly *(Drosophila)*.

More precise genetic analysis of fast-breeding and simple organisms, developments in molecular and chemical genetics concerning the identity of the genetic material, and investigation of the modes of transmission of genetic information from the genes in which it is encoded to the proteins fabricated in cell cytoplasm—all these have led to a clearer understanding of the processes of heredity. Consequently it is now established that the units of inheritance foreshadowed by Mendel (1866), and now called genes, do in fact exist, that they are contained by the chromosomes in all somatic cells, that each gene occupies a specific *locus* or position in its chromosome, and that the genes are transmitted from parents to offspring in the chromosomes of the ovum and spermatozoön.

AUTOSOMES AND SEX CHROMOSOMES

Each chromosome carries a number of genes arranged in a *linear series* along its length. The 46 chromosomes in the human species consist of 22 genetically homologous pairs of *autosomes* and one pair of *sex chromosomes*, one member of each pair being paternal and one maternal in origin. Corresponding genes, or *alleles* (allelomorphs), occupy identical loci in each of a chromosomal pair. The alleles occupying a pair of homologous loci may be identical or different, and thus exert the same, or a different, influence on the function (or functions) which the given gene controls. Although the two sex chromosomes are homologous in the female, both being X, there are X and Y chromosomes in the male which differ markedly in size (p. 34), Y being much the smaller. The X chromosome carries a number of genes which are not represented in Y and which, therefore, must be essential to ordinary life processes. The Y chromosome in man seems largely devoid of genes recognizable by their action on *somatic* tissues, but contains genetic factors responsible for the differentiation of the primordial indifferent embryonic *gonad* into a *testis*.

DOMINANT AND RECESSIVE GENES

At the molecular level each *structural gene* is concerned with determination of the amino-acid sequence in a particular polypeptide and therefore with the structure of the enzyme-proteins on which biochemical and morphological features of the cell depend (pp. 23, 85). (*Regulator genes* are those which modify the structural genes, hence the distinction—pp. 23, 25.) At the more general *phenotypic level* major genes control the development of specific characteristics. The two alleles sited at corresponding loci in a pair of homologous chromosomes may, as indicated above, be identical or different. If different, they may be in a *dominant-recessive* relationship with respect to the character which they influence. Thus, the expression of one allele may be *dominant* over the other member of the pair, which is therefore *recessive* in its expression. If in a given population there are just two different alleles at a particular locus—call them *A* and *a* to represent their dominant-recessive relation—different individuals in the population may carry a pair of identical alleles at the locus or a differing pair. In the former situation individuals are said to be *homozygous*, being either *AA* (homozygous dominant) or *aa* (homozygous recessive). In the other case the individual is *heterozygous—Aa*. If the effect of *A* is completely dominant over that of *a*, an *Aa* individual can be indistinguishable as regards the character involved from the *AA* genotype (*vide infra*). More complex situations exist, but the ABO blood group system is convenient to use as an example. Given the presence in a population of the three alleles at the ABO locus for the A, B, and O characters of the red blood cells, individuals can belong, on test, to groups A, B, AB, or O—as *phenotypes* (*vide infra*). Because of the dominant-recessive relationship which exists in this genetic arrangement, an individual of A phenotype may be either of *AA* or *AO* genotype. Similarly, an individual of group B may be genetically *BB* or *BO*. However, AB group individuals must be genetically *AB*, and group O must be *OO*. The allele for O thus acts recessively to those for either A or B group characteristics, while the alleles for the latter two traits act *co-dominantly*.

HEREDITARY CONDITIONS

In certain diseases, which are known to be hereditary, such as Friedreich's ataxia and fibrocystic disease of the pancreas, the autosomal gene concerned operates recessively; hence only those who carry the allele in question in 'double dose' (genotype *aa*) actually suffer from the disease. The frequency of such sufferers, and hence the incidence of this disease, is generally so low (but there are exceptions) that the majority of the population must be homozygous with respect to this particular gene, both alleles being of the 'normal' or 'wild' type, i.e. they specify a gene product which results in 'normality' of the character under consideration. In these circumstances the normal allele is widely distributed in the population and is therefore present in 'double dose' in most people. The abnormal allele is uncommon and is thus very rarely present in double dose. Persons carrying the abnormal alleles in double dose are, in the examples cited, sufferers from serious diseases. However, the frequency of heterozygotes is, as is always the case, greater than that of the abnormal homozygotes suffering from the disease. Since affected individuals rarely marry, the disease appears sporadically in the offspring of healthy parents, both of whom must be heterozygous in this respect, because it is practically only in these circumstances that the two abnormal alleles may come together in one individual's cells, the probability being 1 in 4. Careful inquiry into the family history in such instances generally adduces no evidence of affected forebears. Since the frequency of the gene responsible for Friedreich's ataxia is very low in the general population, a person who is known to carry it is more likely to mate with a person with the same allele if he or she marries a relative than if the spouse is unrelated. Furthermore, the closer the relationship, the greater the risk, which is naturally in ratio to the proportion of genes carried in common—that is, of genes derived from the common ancestor and therefore identical by descent. On this account intermarriage of the first cousins in families in which a recessive gene is responsible for a serious disease *segregates*, i.e. is transmitted, should always be discountenanced, when there is no way of distinguishing unaffected heterozygotes from homozygotes. Both are phenotypically healthy, but the former are genotypically capable of handing on the abnormal gene which their blood-related spouse may also carry. In some other hereditary diseases possible carriers can be positively identified, and therefore estimates of risk with regard to certain recessive pathological traits can be given precisely, even for close relationships. Other hereditary autosomal diseases, for example neurofibromatosis, are *dominantly* inherited because the abnormal gene acts dominantly over the normal, wild-type, allele. If one of the parents carries the abnormal allele in single dose he or she has a fifty-fifty chance of transmitting it, and with it the dominant affection, to any of his or her children. Yet other hereditary characteristics may be transmitted by genes on the *sex chromosomes*. An important feature of the genetic material of the chromosomes, the DNA (p. 23), is the fact that it can change in response to a variety of environmental influences. These special changes are called *mutations* and produce *mutant genes* which are in fact alleles of the original wild-type forms of the genes. A mutant gene may act dominantly and be the allele responsible for a dominant disease, e.g. neurofibromatosis. Alternatively the mutant gene may be recessive and transmitted silently for a number of generations.

GENETIC LINKAGE

Since each chromosome contains a large number of genes, neighbouring genes, specifically their alleles, in a linear series in the same chromosome will always tend to be inherited together; and the nearer they are in the chromosome the less often will the alleles of these genes be separated from each other and recombine following crossing-over in meiosis (p. 32). The characters associated with such alleles thus tend to appear together in the same individual, and their genes are said to be *linked*. *Linkage* refers specifically to the fact that the gene loci under consideration are on the same chromosome more or less close together. Some hereditary diseases, for example, Huntington's chorea, which is inherited as an autosomal dominant, may not display any recognizable signs or symptoms before marriageable age; this makes it difficult to advise members of affected families as to the desirability or otherwise of procreation. Definite knowledge of any other easily recognized characteristics with which such diseases could be proved to be closely linked, would allow a degree of prediction as to which members of certain affected families carry the gene. Unfortunately, present knowledge of instances of such linkage in human beings is scanty, and much additional data must be collated before such studies can make a significant contribution to genetic counselling in most conditions, although there are exceptions. (*See*, for example, histocompatibility antigens—p. 69.) Nevertheless, new experimental techniques involving *cell fusion* (especially human cells with rodent cells) are now making possible the analysis of human genes and, together with familial studies, are rapidly helping to build up 'maps' of human genes on man's chromosomes, similiar to the maps available for classical genetic organisms like *Drosophila* or the mouse. Further, mention should be made here of the intensive research currently being directed towards linkage analysis in various conditions, such as the juvenile 'rigid' form, or Westphal variant of Huntington's chorea (consult also, for example, Butterfield 1977; Caro 1977; and p. 136).

In mankind still the best demonstration of linkage concerns the X chromosome; while the Y chromosome seems to carry rather little in addition to genetic factors essential to determination of sex, the X chromosome carries genes concerned with other non-sexual characters. The *X-linked* genes, whose alleles determine characteristics said to be *sex-linked*, are concerned with haemophilia, the commoner forms of colour-blindness, a special blood group antigen (the Xg system), and a number of rare diseases of the skin, eyes and nervous system.

Haemophilia, in which clotting of shed blood is greatly slowed or abolished, leading to serious haemorrhage, is operated by a recessive sex-linked gene; it affects only males (with very rare exceptions), but is transmitted by females. The mother of a haemophiliac male must be a heterozygote carrier of the abnormal allele responsible for a deficient clotting factor. As the allele behaves recessively, the XX female, though not affected by the condition, may nevertheless be able to transmit it to her male offspring. However, at times, the mother is *not* a carrier because the abnormal allele has arisen as a *fresh mutation* of the clotting gene on the X chromosome received by the haemophiliac male. In this event subsequent males born in the family will not receive the haemophia allele. The father of a haemophiliac male is usually unaffected, because his X chromosome carries the normal allele for the elaboration of the normal clotting factor, and his Y chromosome does not contain genes concerned with blood clotting. When the mother is a carrier the female offspring will have an equal chance of being homozygous and normal (incapable of transmitting the disease) or heterozygous like their mother; the sons will have equal chances of being normal or haemophiliac. The mating of a normal male with a heterozygous female, and the possible varieties of their offspring may be represented as follows:

$$X^H Y^. \times X^H X^h$$
$$X^H X^H \quad X^H X^h \quad X^H Y^. \quad X^h Y^.$$

On the other hand, when a haemophiliac male ($X^h Y^.$) mates with a healthy woman who is a homozygote normal in this respect ($X^H X^H$), their sons will not be affected ($X^H Y^.$), but their daughters will all be heterozygotes ($X^H X^h$). The condition therefore skips this generation, but will reappear in half of the sons of these heterozygous females. The only possible combination which could produce a haemophiliac female is one between an affected male and a heterozygous female ($X^h Y^. \times X^H X^h$).

GENOTYPE AND PHENOTYPE

The *genetic constitution* of an organism is termed its *genotype*. Variations in this are due to chromosome segregation and to recombination of the genetic material in changed linear relationships as a result of crossing over during meiosis (p. 32) together with any *mutations* that may have occurred. *Chromosomal mutations* (involving structure or numbers) and *gene mutations* are sudden effects in the genetic material which alter its effects on the cell. Mutations are structural alterations in DNA brought about by a variety of environmental factors, such as ionizing radiation, certain chemical agents (e.g. nitrogen mustards), etc. Such variations in genetic constitution, particularly when including some form of mutation, are frequently harmful in their effects, ranging from non-viability of the initial zygote, embryo, or fetus, through serious to mild, and finally occult forms of abnormality or disease complex. However, it must also be emphasized that it is upon such variants that the forces of *natural selection* operate leading to the diversified forms of resulting from *organic evolution*.

The *expressed* constitution of the organism is its *phenotype*; this results not only from the totality of its genetic constitution, but also from interaction between this and the environment. Nature interacts with nurture, and defects in the latter may impede full expression of the genotype and alter the pattern of the genotype-controlled response. The phenotype may be the same even where the genotype is not, as exemplified by the operation of a dominant allele in homozygotes and heterozygotes. Conversely, in individuals with the same genotype (with respect to a given locus in a chromosome) their phenotypes may be different. This may be partly due to interactions between genes, partly to environmental effects. The influence of the latter can best be observed in comparisons of uniovular twins. Some agents, such as teratogenic drugs, may bring about changes in an organism leading to the production of abnormalities known to be dependent upon genetic influence in other instances. Such an abnormality, usually determined genetically, but artificially induced as a replica in an organism lacking the associated genetic mechanism, is termed *phenocopy*.

EARLY DEVELOPMENT OF THE HUMAN EMBRYO

CLEAVAGE

As we have seen (p. 76), cleavage is the process whereby a *unicellular* fertilized ovum with an exceptionally large ratio of cytoplasm to nucleus is transformed into a *multicellular* mass of *blastomeres*, each of which approximates to the ratio found in general somatic cells. For example, in the sea urchin, volumetric ratios are 550:1 at the onset of cleavage, reducing to 6:1 in individual blastomeres at its termination (Brachet 1950). Comparative studies show that cleavage is essentially a series of mitotic divisions in which little true growth occurs, but with a dramatic synthesis of nuclear DNA, and cytoplasmic factors important in subsequent development are differentially parcelled in the cytoplasm of the resultant blastomeres (p. 85).

As each generation divides, the nuclear population is, of course, doubled, preceded by replication of chromosomal DNA as in any other mitotic division (p. 29). The precursors for this replication are dispersed in the oöplasm of the zygote which is rich in various RNAs (p. 25); *cytoplasmic* DNA is contained in both mitochondria and yolk platelets, whilst the oöplasm also contains abundant low molecular weight nucleic acid precursors. The relative contributions from these different sources, to the processes of replication, transcription and translation remains uncertain. Similarly, the interesting question of whether cytoplasmic nucleic acids are wholly broken down into unitary molecules before they are incorporated into nuclear DNA, or perhaps are utilized in more substantial segments, sufficiently large to bear coded genetic information, remains unresolved. (For earlier comparative reviews on the synthesis and transformations of nucleic acids during early embryogenesis, consult Brachet and Quertier 1963; Brown 1966; Tyler 1967; Gurdon 1968; Brachet 1969.)

Synthesis of *new* ribosomal RNA is low during cleavage, but protein synthesis continues and this is particularly related to the formation of the microtubular asters of mitotic cell division, the synthesis of increasing quantities of cell membrane, and the formation of DNA polymerases. The foregoing remarks, until recently, applied mainly to observations on non-mammalia and, for technical reasons, mammalian data were less plentiful and lacked the same precision (*see* Jones-Seaton 1950; Dalcq 1954; Austin 1961, 1965, 1968). However, detailed analyses on mammalian oöcytes and during early embryogenesis are now accumulating (Neyfakh 1971; Herbert and Graham 1974; Church and Schultz 1974).

Relatively few specimens of the earliest stages of human development occurring in the normal environment of the female genital tract have been recovered and studied (Hertig *et al.* 1954, 1956)—these include one 2-cell stage, a few, possibly abnormal, up to the 12-cell stage, and one presumed normal 12-cell stage

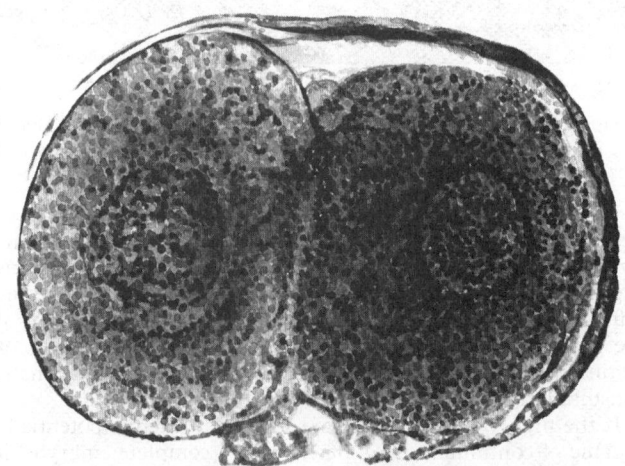

2.17   Early human cleavage. Two-blastomere stage recovered from the uterine tube. A polar cell is seen at each end of the cleavage plane. Magnification × *c.* 700. (A. T. Hertig, J. Rock, E. C. Adams and W. J. Mulligan, *Contr. Embryol.*, **35**, 1954.)

(*vide infra*). Recently a 7-cell stage has been recovered and analysed (Avenado *et al.* 1975).

Successful attempts at the *in vitro* fertilization of mature mammalian ova and their subsequent culture up to the blastocyst stage, have been made since the early 1930s (Lewis and Hartman 1933; Mulnard 1964, 1965). Early development was recorded using time-lapse cinematography, and over the last twenty-five years, similar techniques have been applied to human ova (Shettles 1953, 1955, 1958). Recently, techniques of *in vitro* fertilization of human ova have been greatly improved (Edwards *et al.* 1965, 1969). Preovulatory oöcytes, recovered by laparoscopy

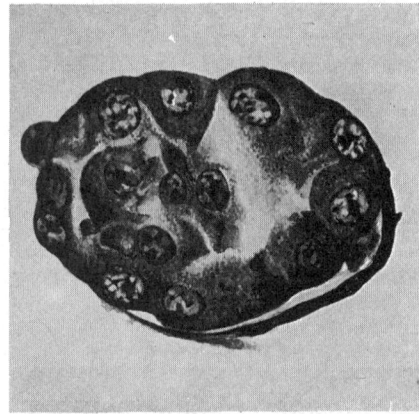 

2.18   Section of a 58-cell human blastocyst recovered from the uterine cavity showing the zona pellucida, trophoblast and inner cell mass. Magnification × *c.* 510. (A. T. Hertig, J. Rock, E. C. Adams and W. J. Mulligan, *Contr. Embryol.*, **35**, 1954.)

2.19   Section of a 107-cell human blastocyst recovered from the uterine cavity. The mural and polar trophoblastic cells and the inner cell mass can be distinguished. Magnification × *c.* 550. (A. T. Hertig, J. Rock, E. C. Adams and W. J. Mulligan, *Contr. Embryol.*, **35**, 1954.)

in women under the control of gonadotrophins (Steptoe and Edwards 1970), have been repeatedly fertilized and cultured to the 8- and 16-cell stages and beyond (**2.20**). Reintroduction of cultured morulae into the uterine cavities of women with pathological uterine tubes and in whom pregnancy would otherwise be impossible—Steptoe and Edwards (1976), although initially unsuccessful, resulted in the first full-term birth late in 1978. This clinical approach is now being vigorously pursued. However, it should be noted that studies of *in vitro* fertilization in various mammals other than man, stressed that such procedures are followed by an increased incidence of abnormal morphological development in rabbits (Fraser *et al.* 1975) and rats (Toyoda and Chang 1974), and of chromosomal abnormality in mice (Fraser *et al.* 1976; Barnes 1976).

The following account is based upon specimens from these various sources, supplemented where necessary by reference to other primates, especially the macaque monkey (Heuser and Streeter 1941).

Fertilization occurs in the lateral part of the uterine tube, probably within 24 hours of ovulation. After completion of the second maturation division the zygote divides into two blastomeres of approximately equal size (**2.17**). (Human cleavage may be *total*, *equal* and *indeterminate*, see p. 76). The human 2-cell stage recorded was recovered from the middle of the uterine tube 60 hours after a fertile mating (**2.8**), the menstrual pattern, however, indicating an age of 36 hours. By repeated division and subdivision of the blastomeres a mulberry-shaped mass of cells, hence termed the *morula*, is formed (**2.20**, 21). Intense activity of their cytoplasm is evidenced by the constant streaming of contained granules, and the individual blastomeres are not motionless, but restlessly alter their positions relative to one another and reorientate themselves within the zona pellucida. The human morula is believed to enter the uterus at

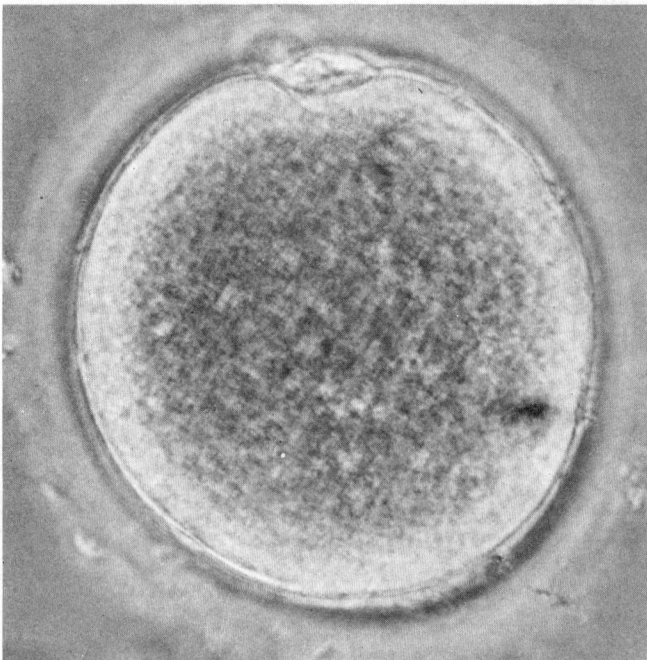

A An unfertilized ovum surrounded by the zona pellucida; the first polar cell can be seen.

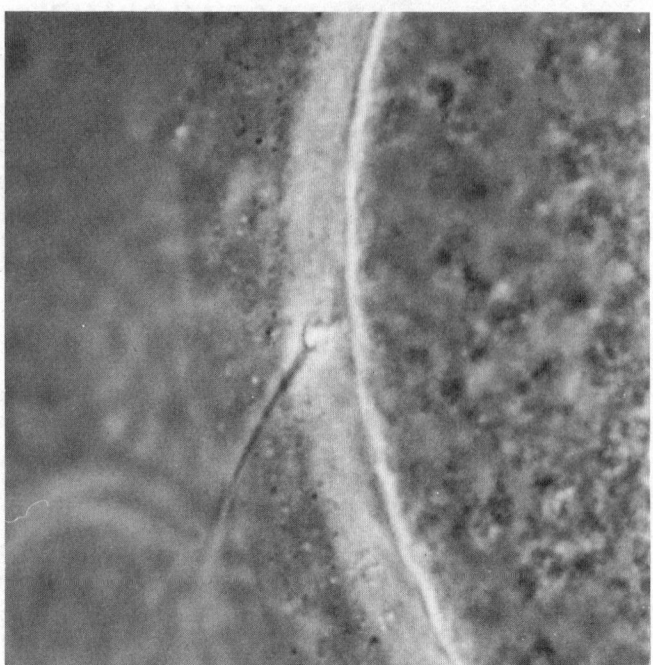

B High magnification of the surface of an unfertilized living human ovum; the head of a spermatozoön is visible in the perivitelline space and its tail is beating outside the zona pellucida.

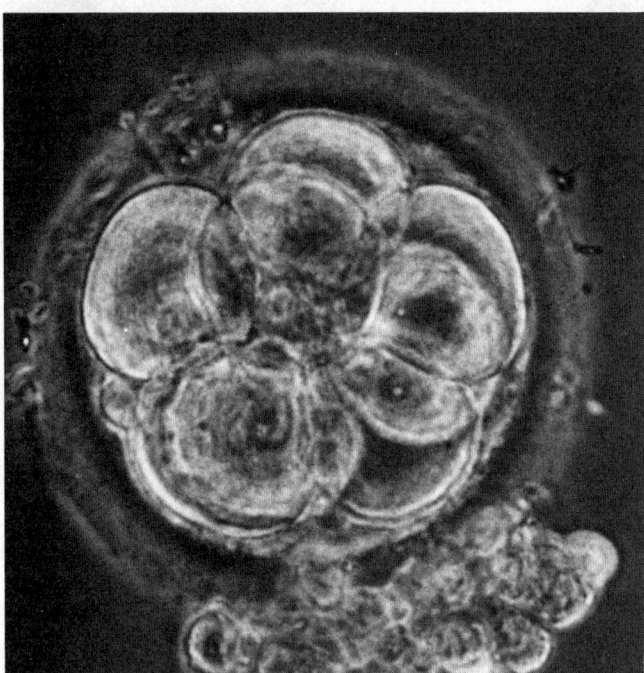

E 8-celled stage.

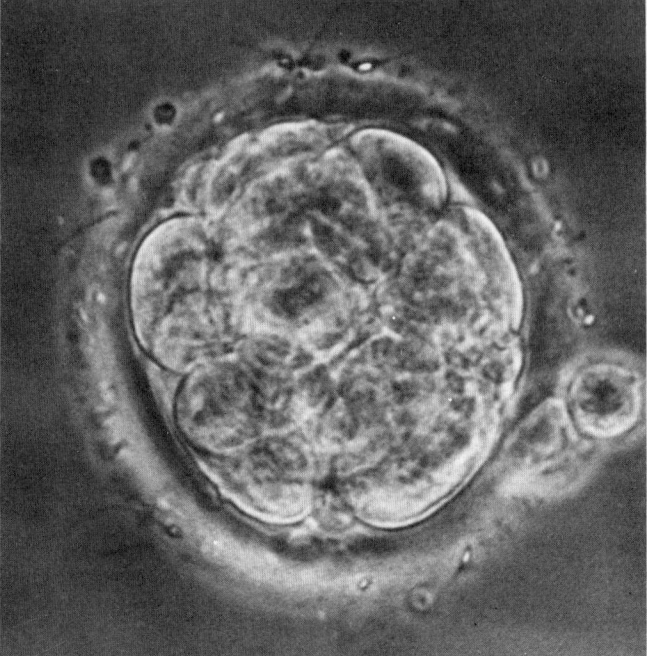

F 16-celled stage.

2.20A–H These are all photographs of living human specimens in tissue culture.

All these specimens were kindly provided by Dr. R. G. Edwards of the

Anatomy Department, Cambridge University, B, E, F and G are reproduced by courtesy of *Nature* and c by courtesy of the *Journal of Reproduction and Fertility.*

about the 8- to 12-cell stage and about 72 hours after fertilization. The four human morulae of 5–12 blastomeres which are recorded, however (all believed to have been abnormal), were recovered from the uterus on the fourth day after presumed ovulation and fertilization. The 12-cell stage, assumed to be normal, consisted of 11 small peripheral cells which surrounded a larger centrally placed one, indicating that during the earlier cleavage divisions two different groups of cells emerge, and the earliest visible signs of such differentiation has now been claimed even at the 7-blastomere stage (Avenado *et al.* 1975). Of these diverging cell lines one will form the embryo and the other will form the nourishing and protective membranes by which it is surrounded. The former is composed of centrally sited larger cells, few in number, which divide more slowly and retain to a high degree the

developmental potential of the fertilized ovum, termed the *formative cells*; the latter consists of smaller and more numerous superficially placed cells, which divide more rapidly as they differentiate into progenitor *trophoblastic cells*.(For a discussion of the possible *causal mechanisms* leading to the emergence of distinctive formative and trophoblastic cell lines in mammalia consult Herbert and Graham 1974.)

If the first two blastomeres are separated, each is potentially capable of continuing development into a complete embryo—it has been claimed that 25–30 per cent of instances of monozygotic twinning in man follow separation at this stage. It has been demonstrated experimentally in the mouse (Tarkowski 1961, 1963; Minty 1962, 1964, 1965) that if one of the first two blastomeres is killed, the remaining one can continue to develop

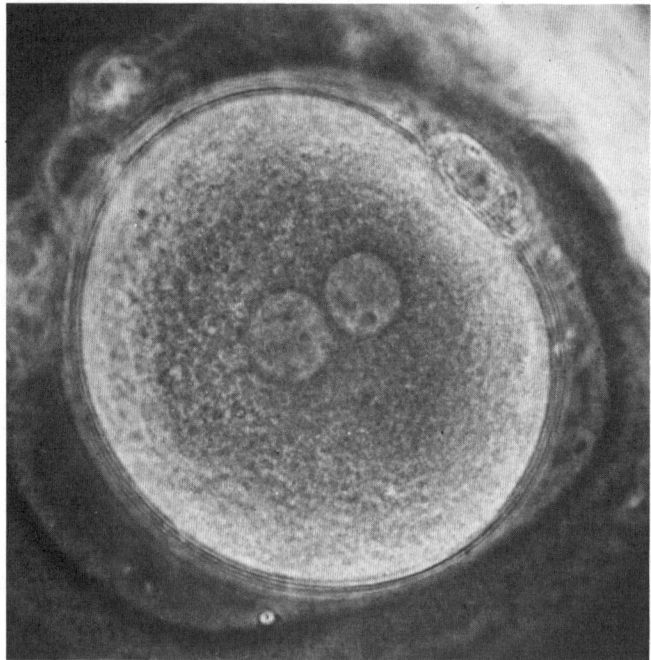

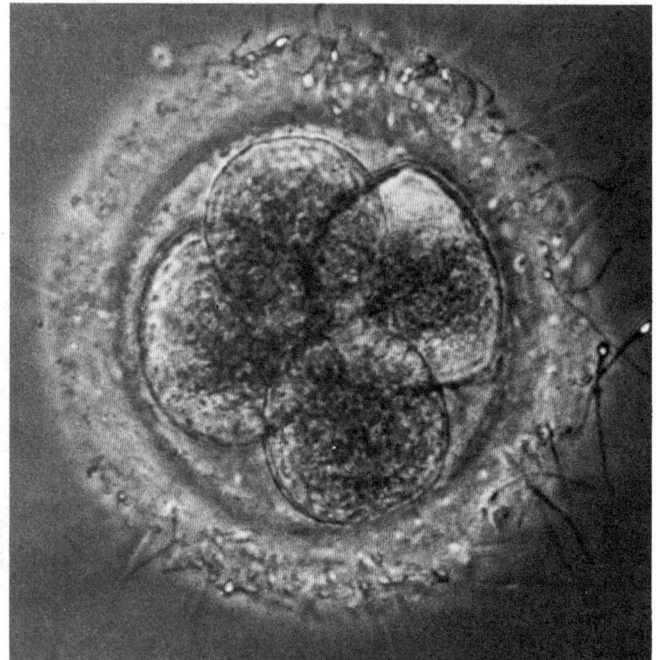

c After penetration by the spermatozoön, the male and female pronuclei can be seen near the centre of the ovum, whilst two polar cells lie beneath the zona pellucida.

D Human cleavage—4-celled stage.

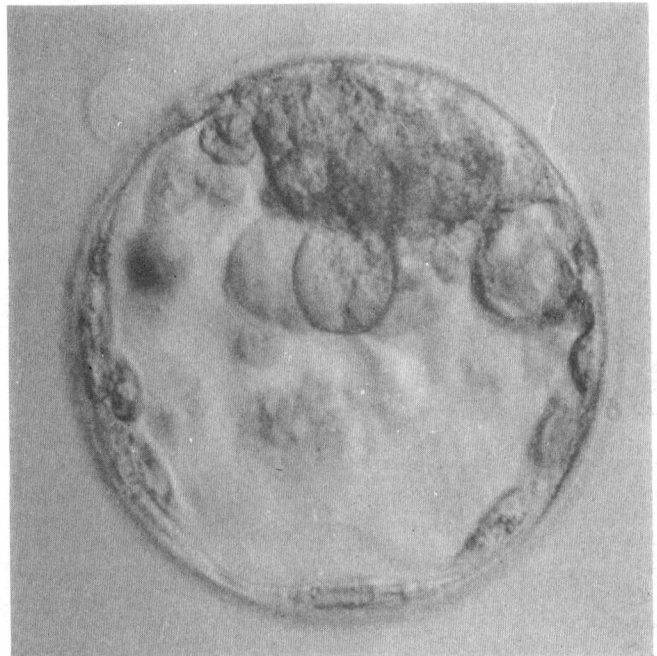

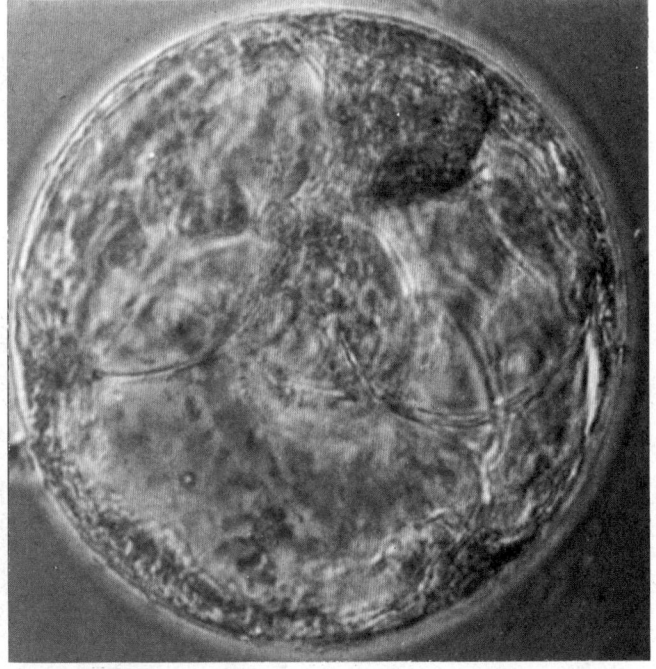

G An early living human blastocyst. Note the blastocyst cavity, small flattened mural trophoblastic cells, and the projecting clump of large cells constituting the inner cell mass.

H A blastocyst which has lost its zona pellucida and started to expand.

into a small but complete embryo. However, blastomeres isolated at the 4- or 8-cell stages often fail to continue normal development and are limited to the formation of simple *trophoblastic vesicles* which contain no inner cell mass and are therefore devoid of formative cells (Tarkowski 1965). Conversely, from the 2-cell to the morula stage, two cleaving zygotes may be fused, development proceeding to the formation of a single 'giant' embryo (2.4).

THE BLASTOCYST

Before the zona pellucida disappears, fluid, either secreted by the trophoblastic cells or derived from the uterine lumen, begins to accumulate within the morula. (Cell death of a number of deeply placed blastomeres has also been demonstrated in some

mammals, but as yet, evidence for this phenomenon is lacking in man.) The intercellular spaces enlarge and coalesce to form a single fluid-filled *blastocyst cavity*. (It should be noted that the latter is not directly homologous with the *blastocoele* of primitive chordates, p. 76.) This cavity is largely bounded by *mural trophoblast cells*, except at one place where a clump of cells, termed the *inner cell mass* (*embryoblast* or *formative mass*) (2.18, 19) projects into the cavity. This is widely considered to represent the residue of totipotential cells, some of which, the *embryogenic cells*, are destined to form the body of the embryo proper (*see*, however, p. 81). The segmenting zygote has now been converted into a *unilaminar blastocyst*. Of a total of some 60 cells composing the blastocyst at this stage, only about 5 are formative cells; the remainder constitute the flattened (mural) trophoblastic

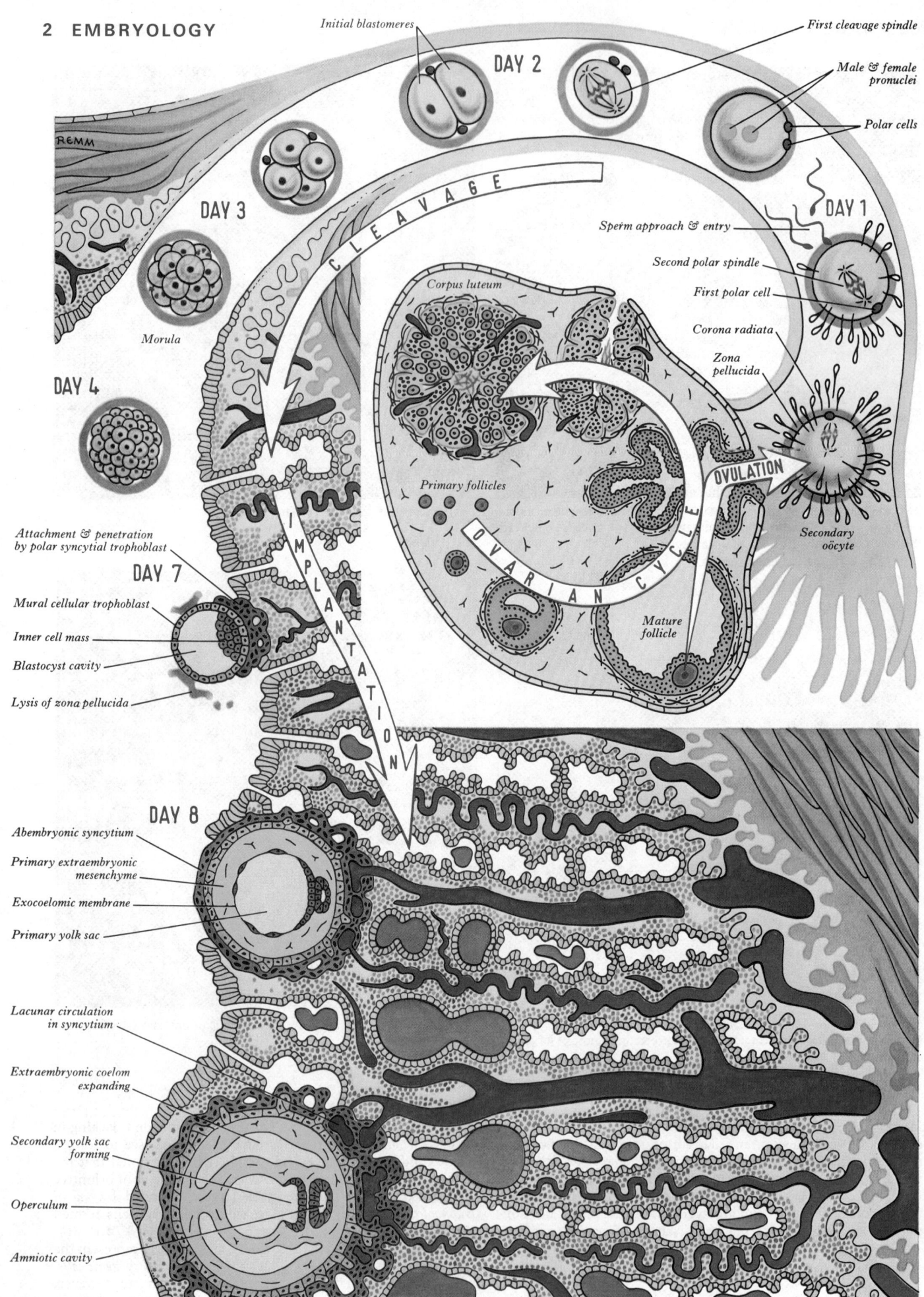

Initial blastomeres

First cleavage spindle

DAY 2

Male & female pronuclei

Polar cells

DAY 3

REMM

Sperm approach & entry

DAY 1

Second polar spindle

First polar cell

Corpus luteum

Corona radiata

Zona pellucida

Morula

Primary follicles

DAY 4

Secondary oöcyte

OVULATION

Mature follicle

Attachment & penetration by polar syncytial trophoblast

DAY 7

Mural cellular trophoblast

Inner cell mass

Blastocyst cavity

Lysis of zona pellucida

DAY 8

Abembryonic syncytium

Primary extraembryonic mesenchyme

Exocoelomic membrane

Primary yolk sac

Lacunar circulation in syncytium

Extraembryonic coelom expanding

Secondary yolk sac forming

Operculum

Amniotic cavity

CLEAVAGE

IMPLANTATION

OVARIAN CYCLE

2.21 A composite schema of the major events in the ovarian cycle, ovulation, fertilization, tubal transport and cleavage, differentiation of blastocyst, implantation, early embryogenesis, and incipient placentation.

epithelium forming the wall of the vesicle (Hertig *et al.* 1954). The unilaminar blastocyst is formed by the end of the fourth or the beginning of the fifth day after ovulation and lies free in the uterine cavity (**2.21**).

The zona pellucida disappears during the fifth day after ovulation by which stage the blastocyst is about 130–140 μm in diameter. The cells of the inner cell mass have multiplied and form an irregular clump insinuated into the wall of the blastocyst where it is flanked by trophoblast. One blastocyst described in detail at this stage (Hertig *et al.* 1954) consists of 107 cells (**2.19**), of which 69 are *mural trophoblast cells*. Of the remainder, 8 are large *embryonic cells* constituting the formative mass, whilst the remaining 30 are *polar trophoblast* cells and lie around its margins. The latter are so named because they soon spread to cover the external aspect of the formative mass (at the *embryonic pole* of the blastocyst). At the same time, the inner aspect of the mass becomes lined by a single layer of polyhedral cells, the *embryonic endoderm* (also variously termed at this early stage, the embryonic *hypoblast* or *endoblast*).

In the latter part of the sixth day after fertilization the polar trophoblast adheres to the uterine mucosa, exerting a histolytic action on its lining epithelium. As they engulf the degenerating uterine tissue, the trophoblast cells in this situation divide with great rapidity and their individual progeny fuse with each other (Boyd and Hamilton 1966) so that all trace of cellular definition is lost. Accordingly the formative mass becomes covered with a multinucleated mass of cytoplasm, the *syncytial trophoblast* (*syntrophoblast*), which continues to burrow into the stratum compactum of the uterine mucosa. As the blastocyst embeds more

deeply the remainder of its trophoblastic wall undergoes a similar change; but, at a slightly later stage, cell boundaries again appear on the inner or embryonic surface of the syncytial trophoblast. Thereafter the wall of the blastocyst consists of an inner cellular layer, the *cellular trophoblast* (*cytotrophoblast*), covered on its outer surface with syncytial trophoblast, which is thickest over the formative mass, i.e. at the area of deepest penetration or *embryonic pole*, and thinnest over the area most recently embedded or *abembryonic pole*.

The *site of implantation* is normally on the posterior wall of the uterus, nearer to the fundus than to the cervix, and may be in the median plane or to one or other side.

The youngest implanting human blastocyst hitherto recovered and described in detail (Hertig and Rock 1945) shows an early stage in the process. The polar trophoblast displays an extensive development of syncytium, which has destroyed a patch of uterine epithelium and underlying stroma (**2.22**). The blastocyst is not completely embedded and a portion of its wall, at the abembryonic pole still projects into the uterine lumen. The fertilization age is believed to be 7½ days (Hertig and Rock 1945).

In a slightly older specimen (Hertig and Rock 1945) (9½ days) the blastocyst is completely embedded. The syncytial trophoblast has undergone further proliferation and now covers the wall at the abembryonic pole as a thin layer. Irregular *lacunae* have developed in the syncytium which communicate with eroded maternal veins. The formative mass now consists externally of a thick plate of large, irregularly arranged cells, termed the *embryonic ectoderm* (*germ disc*, or *embryonic epiblast*) and internally of a single-layered sheet of *embryonic endoderm*, which inter-

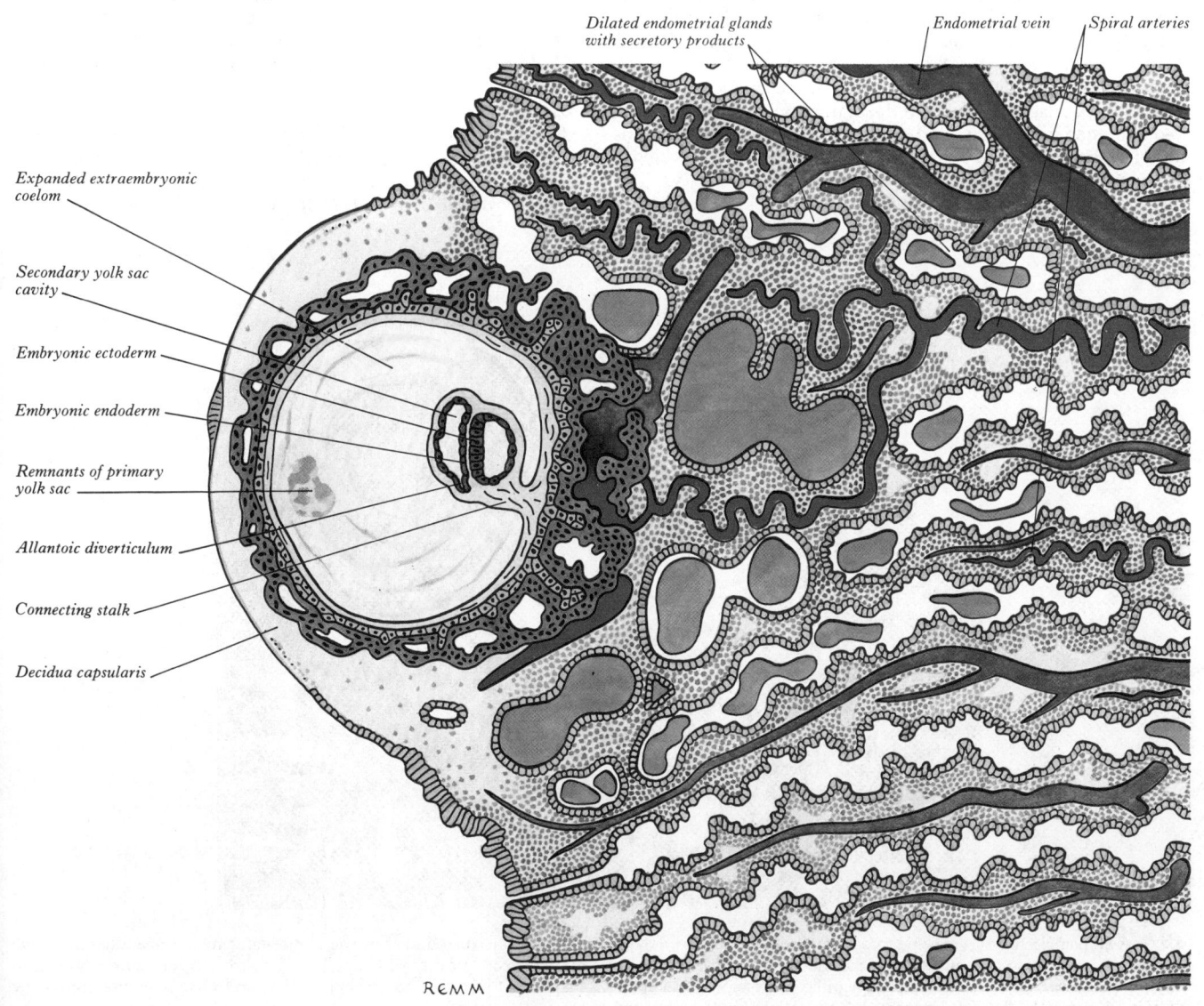

Dilated endometrial glands with secretory products

Endometrial vein   Spiral arteries

Expanded extraembryonic coelom

Secondary yolk sac cavity

Embryonic ectoderm

Embryonic endoderm

Remnants of primary yolk sac

Allantoic diverticulum

Connecting stalk

Decidua capsularis

REMM

2.21 (*continued*)   Further expansion and differentiation of the blastocyst.

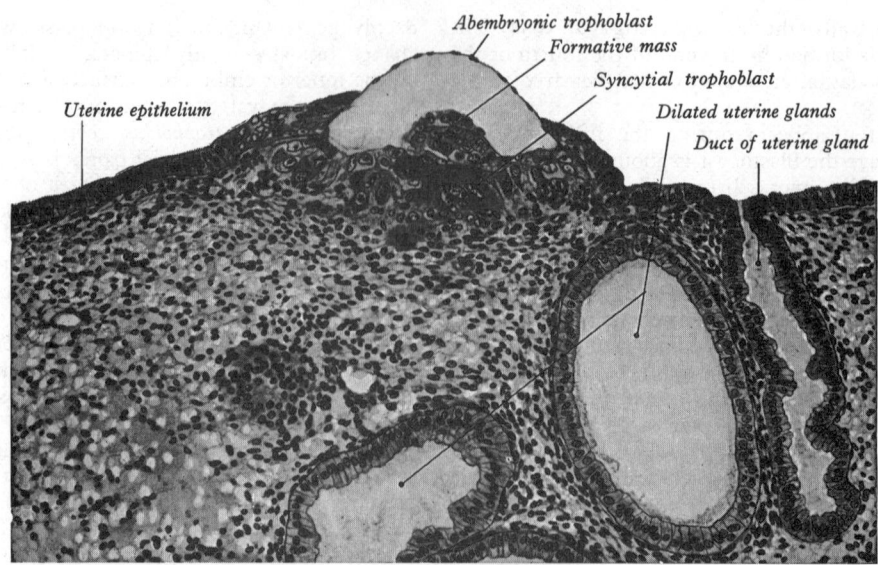

Uterine epithelium

Abembryonic trophoblast
Formative mass
Syncytial trophoblast
Dilated uterine glands
Duct of uterine gland

2.22 A human blastocyst (Carnegie, 8020), fertilization age 7–7½ days, in process of embedding in the uterine mucosa. (J. Rock and A. T. Hertig, *Am. J. Obst. Gynec.*, **44**, 1942, and **47**, 1944.) Magnification × *c.* 150.

In the actual specimen the abembryonic trophoblast had collapsed on the formative mass but, for purposes of clarity; it has been shown projecting into the uterine cavity. (Drawn from a photograph by Dr. A. T. Hertig.)

venes between the disc and the blastocyst cavity. On its external surface the germ disc is in process of separation from the trophoblast by the formation of a cavity, which will soon become the *cavity of the amnion* (**2.**23, 24, 25). Recently, re-examination of early rhesus monkey and human embryos has led to a modification of this view of amniogenesis (Luckett 1975). A *primordial amniotic cavity* develops by cavitation within the inner cell mass and initially has no direct relation with the overlying trophoblast. This occurs in 7-day human blastocysts. Subsequently the epiblastic roof thins and disappears forming a slightly cup-shaped germ disc which is surmounted by a transitory *tropho-epiblastic cavity*. The definitive amniotic epithelium forms by upfolding and pro-

liferation of the margins of the epiblastic disc, the process being completed by 9 days.

In the immediately succeeding stages the walls of the blastocyst become *bilaminar* as its cavity becomes lined with a thin layer of flattened cells, which are in continuity with the embryonic endoderm around its margins and enclose a cavity now termed the *primary yolk sac*. It is uncertain whether these are cells derived from the cytotrophoblast or whether they are formed by an extension from the endoderm (as in the rabbit, dog and some other mammals). Whatever their origin, a variety of terms have been applied to this tissue layer—*extraembryonic hypoblast* and later, *extraembryonic endoderm*, or the *exocoelomic (Heuser's)*

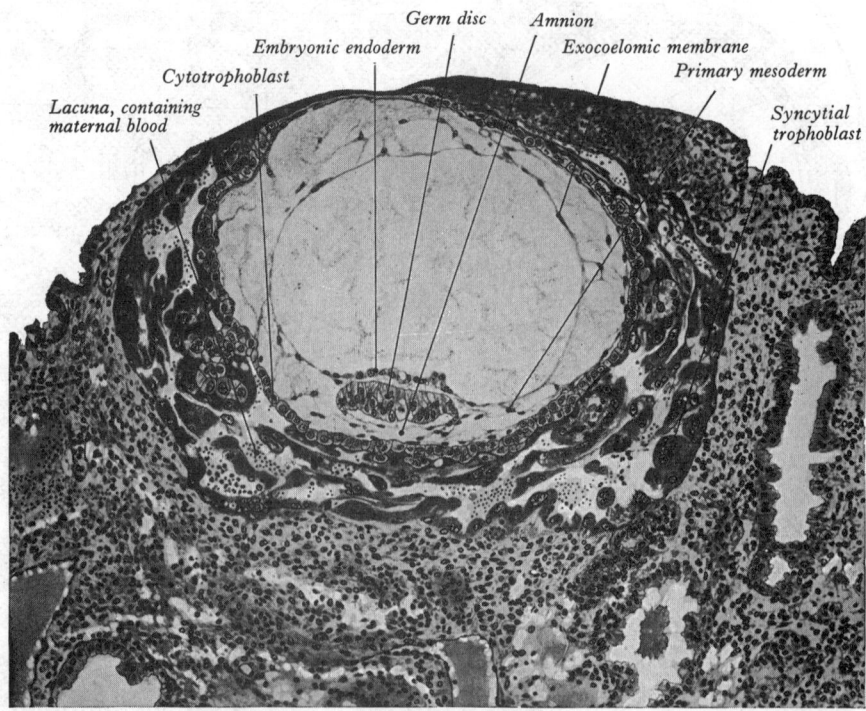

Lacuna, containing maternal blood
Cytotrophoblast
Embryonic endoderm
Germ disc
Amnion
Exocoelomic membrane
Primary mesoderm
Syncytial trophoblast

2.23 A human blastocyst (Carnegie, 7700), fertilization age 12–12½ days, embedded in the stratum compactum of the endometrium. (A. T. Hertig and J. Rock, *Contr. Embryol.*, **29**, 1941.) Magnification × *c.* 105. Compare with **2.**22 and note the development of the lacunae in the syncytial trophoblast, many containing maternal blood. Note that the primary yolk

sac, which is surrounded by the exocoelomic membrane, does not quite fill the blastocyst cavity. The cells of the germ disc are now columnar in shape and form an ectodermal plate. (Drawn from a photomicrograph by Dr A. T. Hertig.)

membrane. The primary yolk sac, formed in this way (**2.**24), appears to meet a functional need in the selective transport of the fluid which the trophoblast absorbs carrying cytolytic products from the uterine mucosa, making them available for the nourishment of the germ disc at this stage.

The cells of the germ disc, hitherto arranged in an irregular manner, now dispose themselves in the form of a plate of columnar cells. At the same time extraembryonic mesodermal cells delaminate from the inner surface of the cytotrophoblast and intervene between it and the primary yolk sac. The wall of the sac (now often termed the *exocoelomic membrane*) becomes retracted from the wall of the blastocyst, although it usually remains connected to it near the abembryonic pole by strands of mesoderm. Descriptions of several human blastocysts (Stieve 1936; Odgers 1937; Dible and West 1940; Hertig and Rock 1941, 1945; Davies 1944) which have reached this *trilaminar* stage and are already completely embedded in the uterine mucosa are now available.

The stage depicted in **2.**23 corresponds almost exactly with the one shown diagrammatically in **2.**24D. The trophoblast is thickest over the embryonic pole and over the sides of the blastocyst, but is exceedingly thin over the abembryonic pole, the last part to be embedded. It is lined over most of its extent by a single layer of cytotrophoblast and the syncytial trophoblast contains numerous lacunae. Maternal blood from invaded vessels occupies the lacunae and serves as a new source of nourishment.

Within the trophoblastic vesicle, the cells of the germ disc have now been arranged as a columnar epithelium, the *embryonic ectodermal plate* or *disc*. Externally they are separated from the cytotrophoblast by the amnion, its cavity and some extraembryonic mesoderm; internally they are in close contact with the embryonic endoderm, although a distinct basement membrane forms a sharp line of demarcation. The embryonic endoderm consists of a single layer of low cuboidal epithelial cells, lining the ventral surface of the germ disc and forming the roof of the primary yolk sac, which is lined elsewhere by the exocoelomic membrane, and which contains fluid and some coagulum resulting from the effects of histological fixation. The interval between the trophoblast on the one hand, and the primary yolk sac and the developing amnion, on the other, is occupied by fluid in which are found extraembryonic mesodermal cells delaminated or delaminating from the cytotrophoblast, together with a certain amount of coagulum. This extraembryonic mesoderm soon forms a loose reticulum, the *magma reticulare*, in which a series of clefts develop and begin to coalesce. As this proceeds the trophoblastic wall of the blastocyst with its lining of mesoderm becomes a separate membrane, the *chorion*, and the general cavity of the blastocyst can therefore now be termed the *chorionic cavity* or *extraembryonic coelom*.

The cavity of the amnion with its roof of epiblastic 'amniogenic' cells and extraembryonic mesoderm, and its floor of embryonic ectoderm, now constitutes the *amnio-embryonic vesicle*. At first mesoderm connects chorion to amnion over a wide area, but as the continued development and coalescence of clefts in the mesoderm extends the extraembryonic coelem, this attachment soon becomes circumscribed (**2.**24E) and condenses to form a *connecting stalk* (body stalk) which forms a permanent connexion between the embryo and the chorion, and is the pathway along which the blood vessels of the embryo later establish communication with those of the chorion. Subsequently it becomes converted into part of the *umbilical cord* (*vide infra*).

While the connecting stalk is being defined, the primary yolk sac suffers a reduction in size, the mechanism of which is obscure, although it seems probable that part of its wall becomes drawn out by the mesodermal strands which anchor it to the abembryonic chorion, and which then loses its connexion with the rest of the sac. The resultant smaller cavity is that of the *definitive* (or *secondary*) yolk sac (**2.**24E, 25), but is often referred to simply as *the yolk sac*. Some investigators (Heuser and Streeter 1941), however, maintain that the secondary yolk sac is a new formation by a rearrangement of the embryonic endodermal cells on the deep surface of the germ disc.

In the immediately succeeding developmental stage, of which many examples are available, the chorionic cavity contains two hollow vesicles (the amnion and the yolk sac), each covered with

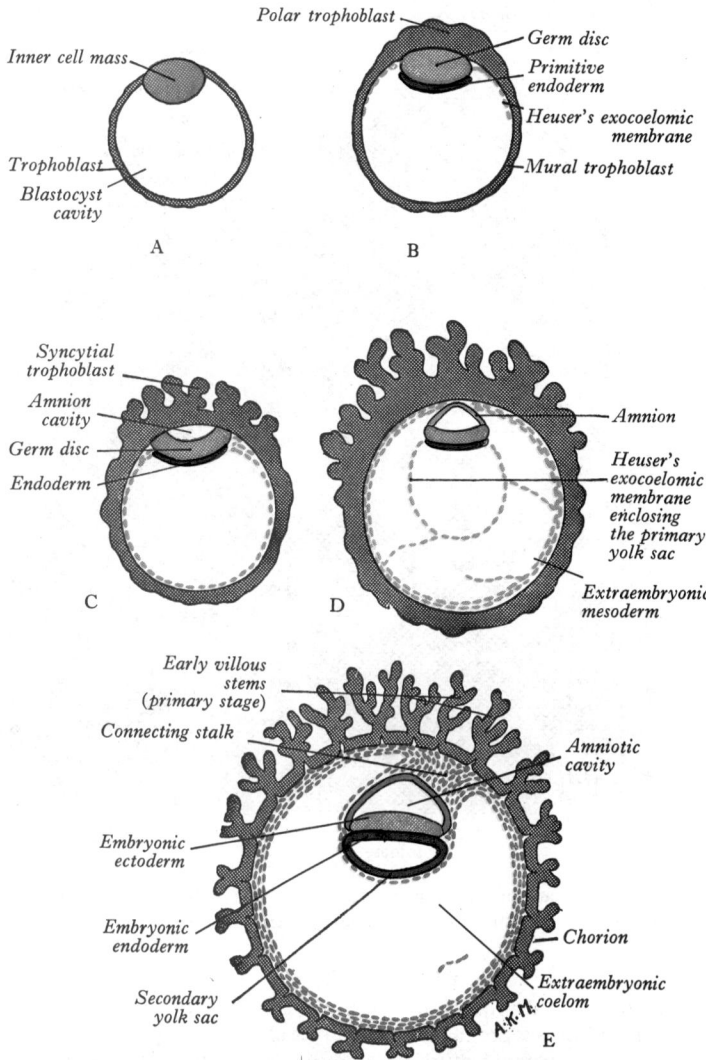

2.24A–E A Early blastocyst. B Differentiation of inner cell mass and development of primary mesoderm. C Formation of amniotic cavity. D Further development of amnion and primary yolk sac. E The primary yolk sac has become reduced in size to form the (secondary) yolk sac, which is covered externally with extraembryonic mesoderm. This is the condition found at the end of the second or the beginning of the third week. Compare with **2.**25. The trophoblastic shell and maternal tissues have been omitted.

extraembryonic mesoderm, by which they are connected to the chorionic wall (**2.**25, 33A). Only the cells where the two vesicles are in contact with each other contribute to the formation of the body of the embryo, and it is therefore this region which is termed the *bilaminar embryonic disc* (*embryonic shield* or *area*).

It is noteworthy that of the multitude of cells derived from the fertilized ovum *at this stage*, only a relatively small number take part in the formation of the embryo, while the vast majority form its covering and nourishing membranes, and certain other extraembryonic structures to be noted later. (Hypotheses concerning the causal mechanisms of this early differentiation have been critically examined in relation to development in the mouse—Herbert and Graham 1974.) The ectodermal cells of the germ disc are columnar and separated from the endoderm by a basement membrane. By the third week the ectoderm takes the form of three or four interlocking rows of cells continuous round its margins with the cells of the amnion which are flatter and more elongated, and form only a single stratum, which is covered externally with a layer of extraembryonic mesoderm. The endodermal cells of the disc are flattened, but those lining the rest of the yolk sac vary in shape, and patches of cubical or low columnar cells are found, especially on its caudal wall. Except in

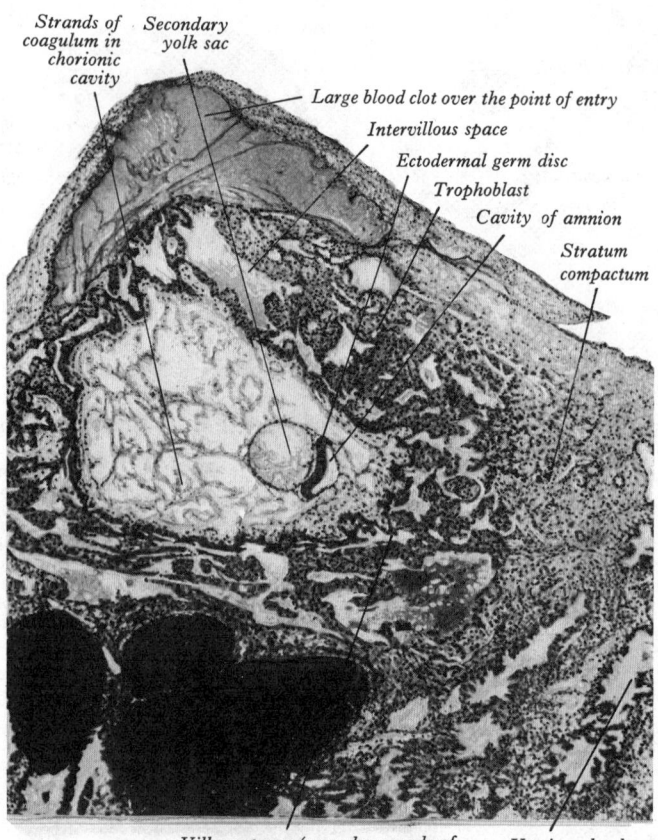

Strands of coagulum in chorionic cavity

Secondary yolk sac

Large blood clot over the point of entry

Intervillous space

Ectodermal germ disc

Trophoblast

Cavity of amnion

Stratum compactum

Villous stems (secondary grade of differentiation—see text)

Uterine gland

2.25 An advanced human blastocyst, embedded in the stratum compactum. Estimated age, 13·5 days. (Hertig and Rock 1942.)

two localized situations, to be noted later (p. 117), the endoderm never forms more than a single layer of cells.

An inconstant feature of the embryo at this stage is the presence of a mesodermal stalk connecting the amnion to the trophoblast which may contain an *amniotic duct*, extending from the amniotic cavity, to end blindly near the trophoblast. In reptiles, and in many mammals, the amnion arises by a process of folding around the periphery of the embryonic disc. The amniotic duct of the human embryo has been interpreted as the homologue of the point of closure of these folds (but it has also been suggested that its presence depends on the chance occurrence of a trophoblastic lacuna which breaks through into the amniotic cavity).

Concomitant with the changes which transform the zygote into a morula, the latter is slowly carried along the uterine tube by the action of the cilia of its lining epithelium and probably by gentle waves of muscular contraction which perhaps also generate currents in the tubal fluid. Passage along the tube covers a period of about three days, and the morula is then transformed into a blastocyst which remains free in the uterine cavity for a further three days before implantation. During this interval the uterine mucosa is preparing for the reception of the blastocyst (p. 126). The disappearance of the zona pellucida, after the formation of the blastocyst, allows the trophoblast cells to come into direct contact with the uterine mucosa and initiate implantation.

ECTOPIC IMPLANTATION

Implantation of the blastocyst normally occurs in the endometrium of the uterine body, more frequently on the posterior wall somewhere near the fundus, but may occur elsewhere in the uterus, or in an *extrauterine* or *ectopic* site. Implantation near the internal os results in the condition of *placenta praevia* with its attendant risk of severe antepartum haemorrhage (p. 136).

Alternatively, the zygote or segmenting morula may be arrested at any point during its migration through the uterine tube, and implant in its wall. Previous tubal inflammatory episodes may predispose to such tubal arrest, and it has also been suggested that congenital abnormalities of the tube, tubal tumours, transperitoneal migration of an ovum from one ovary to the opposite tube, delayed ovulation, and psychological stress which may cause tubal spasm, are additional predisposing factors (Woodruff and Pauerstein 1969).

Nidation in the intramural part of the tube often results in early abortion of the conceptus, whereas if it occurs elsewhere in the tube, development often proceeds for about two months and is then usually followed by tubal rupture with death of the embryo and severe intraperitoneal haemorrhage—a grave surgical emergency. However, slow rupture of the tube may occur, accompanied by a further implantation of the conceptus into any adjacent peritonealized surface (*secondary abdominal pregnancy*), which again ultimately leads to rupture of the surface with similar consequences.

Primary ovarian or *abdominal* pregnancies have also been described, in which it has been presumed that the fertilization occurred in the vicinity of the ovary; most cases, however, are probably of the secondary type following a slow tubal rupture or a slow extrusion of the conceptus through the abdominal ostium of the tube.

Apart from their important clinical implications, such conditions emphasize the interesting fact that the conceptus can implant successfully into tissues other than a normal progestational endometrium. Further, prolonged development can occur in such sites and is usually terminated by a mechanical or vascular accident, and not by a fundamental nutritive or endocrine insufficiency, or by an immune maternal response.

Differentiation of the Embryonic Area

Near the end of the second week, the embryonic area thus consists of an almost *circular, bilaminar disc* which separates the cavity of the amnion from that of the secondary yolk sac. As we have seen, the amniotic aspect of the area consists of a thick plate of columnar embryonic ectodermal cells, 2–4 cells thick, which is continuous around its margins with the flattened cells lining the amniotic cavity. A single layer of rather flattened polyhedral endodermal cells forms the roof of the secondary yolk sac; this roof is quite closely applied to the superjacent ectodermal plate, a basement membrane intervening, and it is continuous peripherally with the cuboidal or columnar cells lining the secondary yolk sac. At this stage no mesodermal cells are found between the ectoderm and endoderm of the embryonic area, but thin strata of extraembryonic mesoderm clothe the outer surfaces of the amnion and secondary yolk sac, and line the chorion (p. 124)—the space bounded by these various mesodermal layers constituting the *chorionic cavity* or *extraembryonic coelom.*

The walls of the amnion on the one hand, and that of the yolk sac on the other, may now be defined as extraembryonic *somatopleure* and *splanchnopleure* respectively. The former consisting of an epithelial layer (trophoblast or amniotic lining) associated with a layer of *somatopleuric extraembryonic mesoderm*, whilst the latter comprises the endodermal epithelial lining of the yolk sac together with *splanchnopleuric extraembryonic mesoderm.* At a *junctional zone* surrounding the margins of the embryonic area, where the walls of the amnion and yolk sac converge, the somatopleuric and splanchnopleuric layers of mesoderm are continuous.

At the early stage the embryonic area is radially symmetrical, with no specialized features distinguishing its future cephalic, caudal and lateral borders, but with the extensions of the extraembryonic coelom noted above, and the increasing separation of the amnion from the overlying chorion, the mesodermal connecting stalk becomes increasingly defined, and its embryonic attachment is soon limited to a zone which marks the future *caudal edge* of the embryonic area.

The first change within the embryonic area itself occurs near the opposite (*cephalic*) edge, where, in a localized region, the endoderm in the yolk sac roof thickens, forming an oval plate of large vesicular cells, the *prechordal plate*, which at the later stage contributes cells to the general head mesenchyme (p. 120) and which, together with the closely applied superjacent ectoderm, forms the transient *oropharyngeal membrane*. The attachment of the body stalk and the position of the prechordal plate have now defined the *plane of bilateral symmetry* of the embryonic area.

During the immediately succeeding stages the embryonic area changes from a circular *bilaminar* disc to first an oval, and then a pear-shaped *trilaminar* one as a third *mesodermal layer* becomes interposed between the embryonic ectoderm and endoderm. The long axis of the oval and pear lie in the future cephalo-caudal axis of the developing embryo, whilst the pear-shaped outline is narrow caudally and expanded at the cephalic end. These shape changes accompany the emergence of proliferative zones in the caudal midline of the primitive ectoderm of the embryonic area— a linear *primitive streak* with, at its cephalic end, a *primitive knot*. From the former generations of *intraembryonic mesodermal cells*, and from the latter an axial *head process* (ultimately forming the notochord), pass between the ectoderm and endoderm (*vide infra*).

Presumably (based on the evidence of comparative studies, p. 83), under the primary inductive influence of the chorda-mesoderm, the central regions of the ectoderm which overlies it initiates the formation of a *neural plate*, the primordium of the central nervous system (**2.26, 27**). The margins of the plate are surrounded by a strip of presumptive *neural crest* tissue (p. 115) which is, again, surrounded peripherally by the layer which will form the general body ectoderm. These various regions will now be considered further.

As noted elsewhere (p. 83) many experimental embryologists consider that the totipotency of the fertilized mammalian oöcyte, and initial blastomeres, suffers a stepwise reduction, as the determination and differentiation of the various embryonic regions proceeds. Alternative views have also been advanced—*see* p. 84.) Despite the relative paucity of experimental analyses, because of technical difficulties, on mammalia, it is widely held that by the blastocyst stage, the vast majority of the cells have already suffered a severe restriction in their developmental potency, being destined to form extraembryonic structures (p. 74), whilst the pluripotent cells are restricted to the embryonic area. It has been proposed (Streeter 1942) that, as the changes outlined above occur, these pluripotent cells become progressively circumscribed, first in the caudal part of the oval embryonic area, and then as a narrowing midline strip in the caudal end of the increasingly pear-shaped area, i.e. the site of formation of the *primitive streak* which, as development proceeds, becomes more obvious as a linear opacity when viewed from the amniotic aspect (For more recent *experimental* data on mammalia *see* Herbert and Graham 1974.)

In the early stages of development the living tissues are greyish and translucent, but any localized thickening due to cellular proliferation naturally interferes with the translucency and causes a localized opacity as in the case of the primitive streak, indicating that rapid multiplicative growth is occurring throughout its length. At its headward end a further area of exceptionally active growth forms a knob-like thickening which is termed the *primitive knot* or *Hensen's node* (**2.26**) and at this point the ectoderm is fused with the endoderm.

From the primitive knot a rod-like process of cells grows headwards in the midline separating the presumptive neural plate from the subjacent roof of the yolk sac until its cranial tip reaches the caudal margin of the prechordal plate. This is termed the *head process* (*notochordal process*), which is the forerunner of the skeletal axis of the body. This initially solid rod of cells now becomes canalized, and caudally the canal breaks through on to the ectodermal surface at the posterior end of the primitive knot. The endodermal cells lying ventral to the head process disappear, and the head process then becomes, briefly, a constituent part of the roof of the yolk sac in the median plane. The cells forming the floor of the canal of the head process break down so that the canal communicates freely with the yolk sac, and, at its caudal end, a temporary communication is established between the yolk sac and the amniotic cavity. This connexion, which pierces the embryonic area at the primitive knot, is the *neurenteric canal*. At this stage a transverse section across the embryonic area cranial to the primitive knot shows that, in the median plane, the roof of the

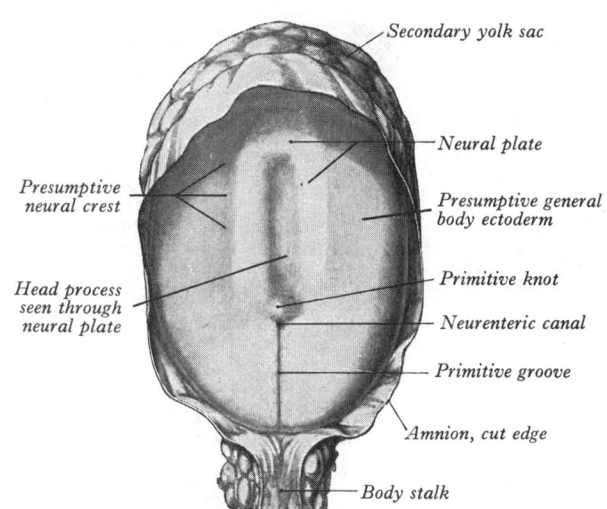

2.26 A human embryo, 1·16 mm long. The amnio-embryonic vesicle has been laid open widely, most of the amnion having been removed. The embryonic area is exposed in almost its whole extent and shows an early stage of differentiation. Estimated age, 19 days. (W. C. George, *Contr. Embryol.*, **30**, 1942.)

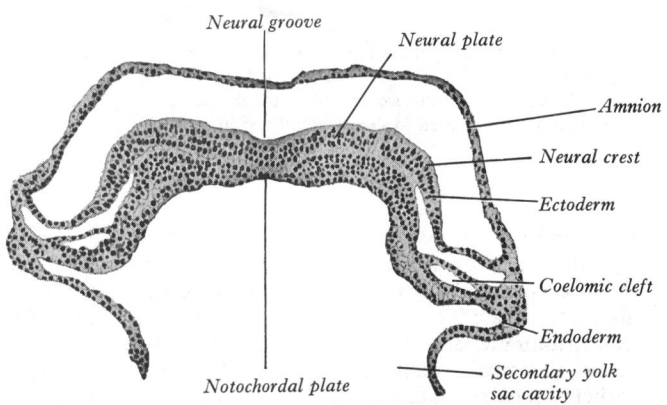

2.27 Transverse section of a human embryo, about 19 days old. (Embryo: Bryce-McIntrye, *Trans. Roy. Soc. Edin.*, **53**, 1924.) The canal of the head process has broken into the yolk sac at this level and the notochordal plate is incorporated in the roof of the gut. Observe that the coelomic clefts have appeared in the lateral plates of the mesoderm, and that they do not communicate with the extraembryonic coelom. The neural groove is shallow and the amnion is restricted to the dorsal and lateral aspects of the embryo.

yolk sac is, for a time, formed by the cells of the head process (**2.27**), and this intercalation, which forms the *chordal or notochordal plate*, extends forwards to include the region which will subsequently form the roof, or dorsal wall, of the pharynx. Later, these cells of the head process become separated from the endoderm and form the *notochord*, the roof of the yolk sac being repaired by the fusion of the adjoining endodermal cells. Subsequently the cells of the notochord develop around them a homogeneous sheath, and the continued proliferation of cells within the sheath results in the formation of a solid but flexible rod, which becomes surrounded by mesoderm to form the primitive or *blastemal vertebral column* (p. 140).

During the earlier part of this period another important change affects most of the embryonic area. From the sides of the primitive streak occurs an intensely active multiplicative cell growth. The cells formed spread laterally, forwards and backwards until they extend over the whole embryonic area (*with the exception of most of the median plane*). They insinuate themselves between the ectoderm and the underlying endoderm and produce a third layer, which constitutes the *intraembryonic mesoderm*. (For a review of experimental data of these stages of development in higher vertebrates consult Bellairs 1971.)

The primitive streak is the principal, but not the only source of intraembryonic mesoderm; it also receives accessions from the

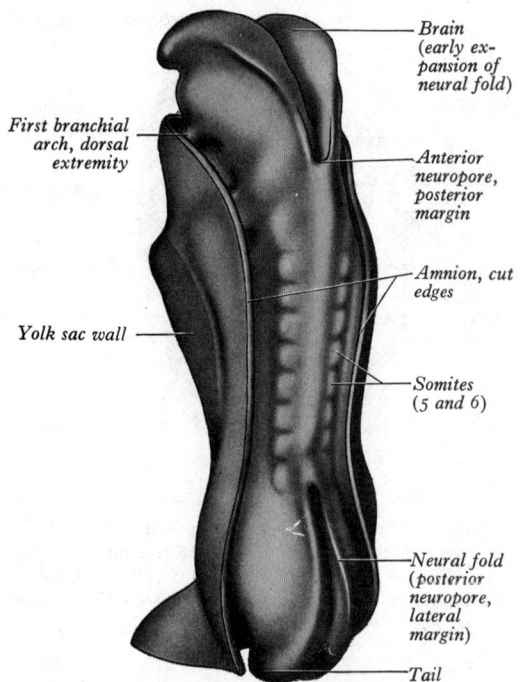

Brain
(early ex-
pansion of
neural fold)

First branchial
arch, dorsal
extremity

Anterior
neuropore,
posterior
margin

Amnion, cut
edges

Yolk sac wall

Somites
(5 and 6)

Neural fold
(posterior
neuropore,
lateral
margin)

Tail

2.28 A human embryo, 2·1 mm long, with nine somites. Viewed from the left lateral and dorsal aspects. (From a model by Eternod.) Nearly all the yolk sac has been cut away, and at the caudal end a portion of the amnion has been excised to show the tail region.

neural crest (p. 115) particularly in the head and branchial arches (Hörstadius 1950), and the latter also receive contributions from the thickened endoderm of the prechordal plate. As mentioned above, this plate forms an integral part of the future oropharyngeal membrane, and it is probably involved in inductive processes of cranial structures (termed a *head organizer region* by earlier workers) during further embryonic development. Similarly, in the midline at the opposite end of the embryonic area, a patch of endoderm thickens and becomes closely applied to the overlying ectoderm, between the caudal tip of the primitive streak and the attachment of the body stalk, to form the future *cloacal membrane*.

As the streams of intraembryonic mesodermal cells radiate from the margins of the primitive streak (2.31), they separate the ectoderm and endoderm of the embryonic area, and soon approach to become confluent with extraembryonic mesoderm around the margins of the area, i.e. at the *junctional zone* where the splanchnic and somatic strata of extraembryonic mesoderm merge. The streams passing in a cephalic direction flank the differentiating head process (later the notochord), skirt the margins of the prechordal plate, and then converge medially to fuse in the midline beyond its cephalic border. This transmedian mass is the *cardiogenic mesoderm* in which the heart and pericardium are to develop. Around the extreme cephalic margin of the embryonic area, the cardiogenic mesoderm fuses with the junctional zone of extraembryonic mesoderm mentioned above, and this region of the latter will eventually form the *septum transversum* and *primitive ventral mesentery* of the gut and their derivatives (p. 119).

The intraembryonic mesoderm which streams caudally from the primitive streak skirts the margins of the cloacal membrane and then converges to become continuous with the extraembryonic mesoderm of the connecting stalk.

The ectodermal thickening of the *neural plate* from which most of the central nervous system develops, corresponds precisely in length to the underlying developing notochord. Thus it extends from the cranial border of the primitive knot to the caudal border of the future oropharyngeal membrane, but it also extends laterally to cover the medial strips of intraembryonic mesoderm. As it grows, its margins become raised (2.27, 28, 29) as the *neural folds*, with a longitudinal, midline *neural grove* between them.

The neural folds become particularly prominent at the anterior end of the area and the walls of the groove become expanded (2.27). This enlargement is the first sign of brain formation. It is not a continuous enlargement, however, because two transverse constrictions indicate a division into three parts—the *prosencephalon* or *forebrain*, the *mesencephalon* or *midbrain*, and the *rhombencephalon* or *hindbrain*. As we have seen, between the cephalic margin of the forebrain and the cardiogenic mesoderm, the ectoderm and the endoderm are in contact over a small area forming the *oropharyngeal membrane* (2.33). As the neural groove deepens, its dorsal edges come into contact with each other and fuse to convert the groove into a sagittal slit-like canal (2.28, 29). It should be noted that as the edges of the groove approximate they carry with them the adjoining general body ectoderm and, when the process of fusion occurs, like fuses with like, i.e. ectoderm with ectoderm, and neural tissue with neural tissue to form a *neural tube*. The process is effected first in the hindbrain or upper cervical region (2.28) in the latter part of the third week, and extends both headwards and tailwards, until only a small opening is left at each end. These are the rostral and caudal *neuropores*; the former closes in the middle and the latter towards the end of the fourth week. The cells which lie in the line of fusion of the dorsal edges of the neural plate constitute the *neural crest* (2.29), the history of which will be considered in a subsequent section (p. 115).

From initiation, the walls of the neural groove and tube are bathed by the fluid in the amniotic cavity (*liquor amnii*) and presumably are, in part, dependent on it for their nourishment whilst the neuropores remain open. It is perhaps significant that closure of the neuropores coincides with the establishment of a blood vascular circulation for the neural tube.

The fusion of the dorsal lips of the neural plate in the region of the brain results in the formation of three zones termed, somewhat inappropriately, the *primary cerebral vesicles* (see p. 163). The walls of these vesicles become thickened and develop into the nervous tissues and neuroglia of the brain, and their cavities are modified to form the ventricles of the brain. In some important sites, however, the walls of the forebrain and hindbrain vesicles do not develop nervous tissue but remain thin and are modified to form *choroid plexuses* (p. 169). The remainder of the tube forms the spinal cord, its cavity persisting as the central canal (see also p. 157).

Meanwhile, throughout the extent of the notochord the intraembryonic mesoderm becomes arranged into three zones: (1) a thickened medial portion which lies immediately lateral to the grooved neural plate and notochord, the *paraxial mesoderm* (2.30), (2) a narrower *intermediate mesoderm*, situated lateral but directly continuous with the paraxial mesoderm, and (3) a flattened *lateral plate* which extends from the intermediate mesoderm to the periphery of the embryonic area where it is continuous with the extraembryonic mesoderm on the outer surfaces of the amnion and the yolk sac, i.e. at the *junctional zone* of mesoderm mentioned above.

THE EMBRYONIC LAYERS

The appearance of the intraembryonic mesoderm and its three major divisions completes the first stage of differentiation of the embryonic area. The embryo now consists of an outer protective layer, the ectoderm, an inner nutritive layer, the endoderm, and an intermediate layer, the intraembryonic mesoderm, which is available primarily as a muscle-forming layer. It is clear from their history and their early differentiation that these layers are of considerable phylogenetic and ontogenetic significance, but too much stress must not be laid on their independence from one another. The differentiation at this stage does not leave the constituent cells of the three layers with potencies so limited that complete divergence occurs in subsequent development. The developmental potential of the individual cell layers are reduced from the pluripotent condition found at an earlier period, but not to such an extent that the potencies of the cells of one layer are entirely different from those of the two remaining layers. Further, during subsequent development of the various organ systems,

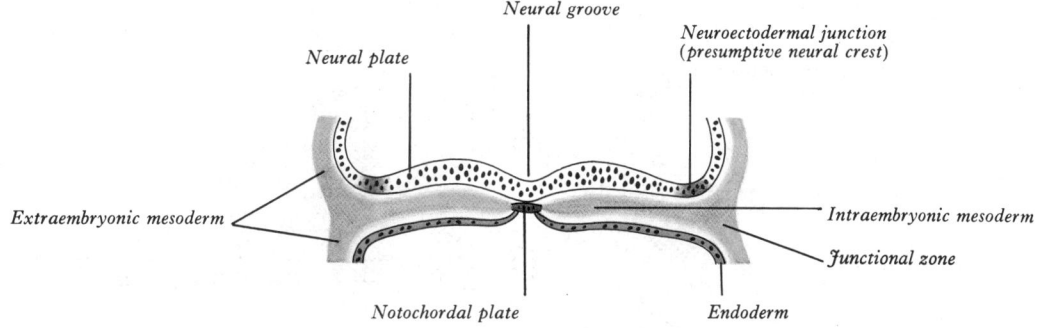

A

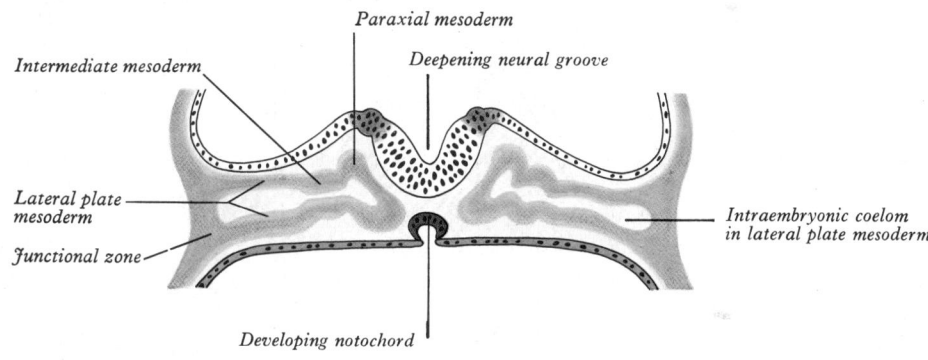

B

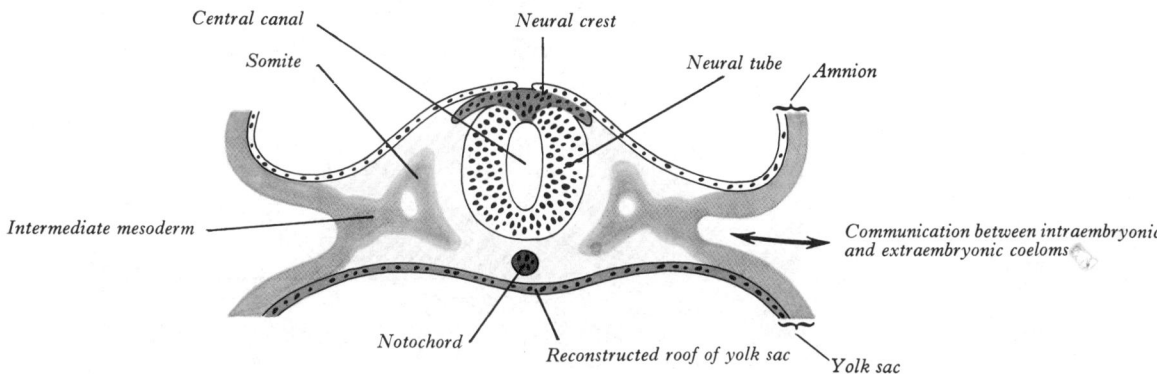

C

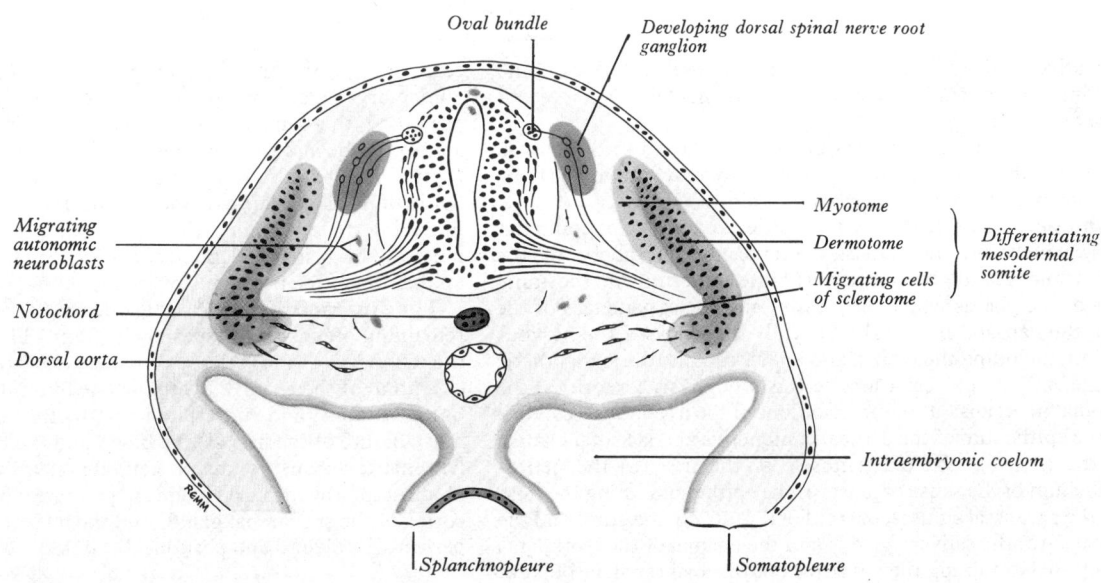

D

2.29A–D   A series of schematic diagrams showing important stages in the development and differentiation of the neural plate and tube, the neural crest, the notochord and the intraembryonic mesoderm and coelem. See text for a detailed description.

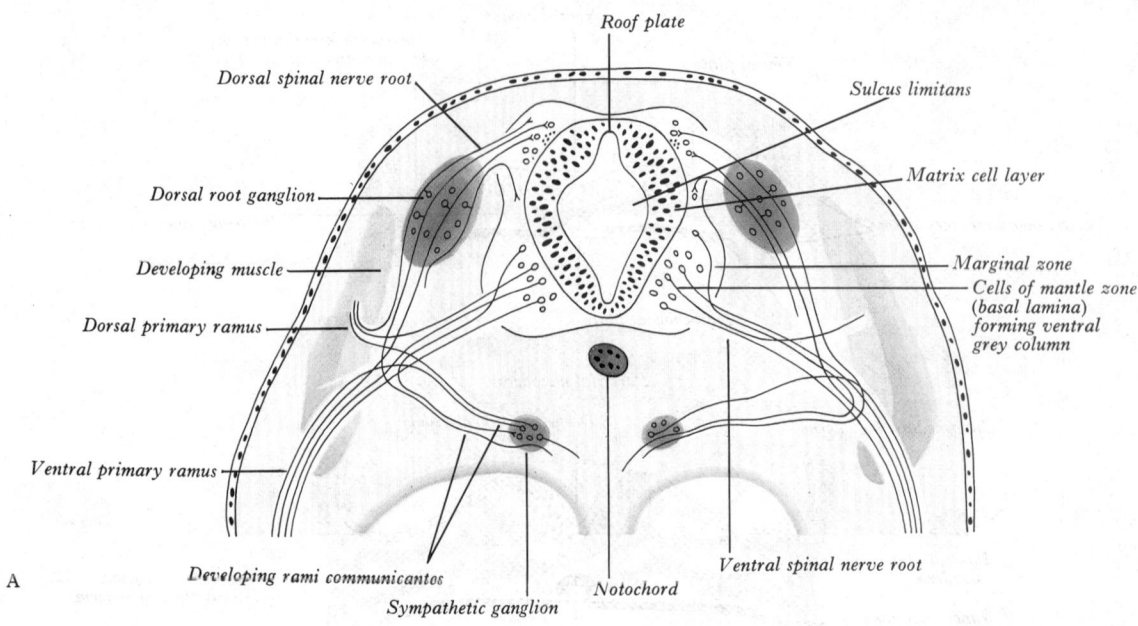

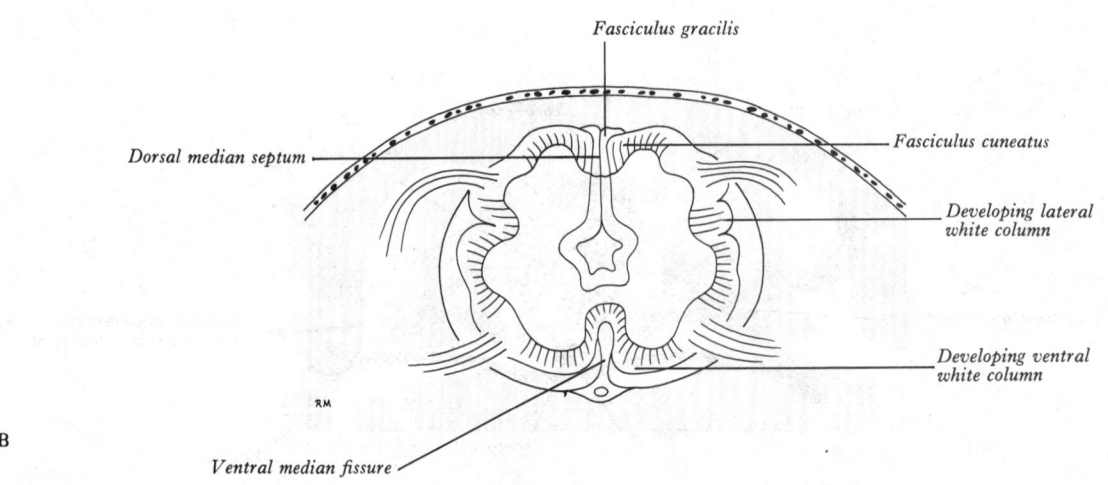

2.30A and B   Further stages in the development and differentiation of the neural tube. See text for further description and **2.**29 A–D for the earlier stages.

derivatives of the different layers are often closely interlocked and are interdependent in terms of fundamental *morphogenetic mechanisms* (pp. 85–91).

However, the contributions of the three layers in the human embryo to the formation of the different sytems and organs may, for convenience, but summarized as follows.

The primitive embryonic ectoderm consists of columnar cells, which become cubical towards the periphery of the embryonic area. It gives origin to: (1) the epidermis and the lining cells of the glands which open on it, and the appendages of the skin, the hair and nails; (2) practically the whole of the nervous system, including the cranial and spinal ganglia, the sympathetic ganglia and the posterior lobe of the hypophysis cerebri; (3) the chromaffin organs; (4) the anterior lobe of the hypophysis cerebri; (5) the epithelium of the cornea, conjunctiva and lacrimal glands; (6) the lens; (7) the plain muscle of the iris; (8) the neuro-epithelium of the sense organs; (9) the epithelium lining the nose and the paranasal sinuses, the roof of the mouth, the gums and the cheeks; (10) the salivary glands and the enamel of the teeth; (11) the epithelium lining the lower part of the anal canal and that of the terminal urethra.

The primitive embryonic endoderm consists at first of flattened cells, which subsequently become columnar. It gives origin to: (1) the epithelial lining of the whole of the alimentary canal, with the exception of those portions already ascribed to the ectoderm; (2) the lining cells of the glands which open into the alimentary canal, including the liver and the pancreas, but excluding the salivary glands; (3) the epithelium lining the auditory tube and the tympanic cavity; (4) the epithelium of the thyroid and parathyroid glands and the thymus; (5) the lining epithelium of the larynx, trachea and the smaller air passages, including the alveoli and the air saccules; (6) the epithelium of most of the urinary bladder and much of the urethra; (7) the epithelium of the prostate.

The intraembryonic mesoderm gives origin to the remaining organs and tissues of the body. These include: (1) all the connective and sclerous tissues; (2) the teeth with the exception of the enamel; (3) the whole musculature of the body, both striated and unstriated, with the exception of the musculature of the iris; (4) the blood and the blood vascular and lymphatic systems; (5) the urogenital system, with the exception of most of the urinary bladder, prostate and uretha; (6) the cortex of the suprarenal glands and the mesothelial linings of the pericardial, pleural and peritoneal cavities.

THE EMBRYONIC TISSUES

The cells of the embryo become arranged at this early stage into two fundamental types of tissue, *epithelial* and *mesenchymatous*, and, despite regional modifications introduced as the various tissues develop, these types persist in large measure throughout life.

Epithelial tissues are those in which the cells are closely packed, with narrow intercellular clefts containing minimal extracellular material, and subsequently the cells usually show inter-cell surface specializations such as desmosomes, tight junctions, gap junctions, etc. (pp. 7–10). Characteristically, they clothe internal and external surfaces as simple or compound cellular sheets which separate phases of differing composition (e.g. the external environment and the subepithelial tissue fluids; intravascular and extravascular fluids, and so forth). Traffic of materials in the intercellular clefts between cells is limited, and passage occurs across the cells and their limiting membranes, which function as energy-dependent selective barriers, enhancing the passage of some materials and impeding the passage of others.

The majority of the derivatives of the embryonic ectoderm and endoderm listed above retain their epithelial character throughout life. The third germ layer, mesoderm, also contributes many tissues of an epithelial type (although they may also be termed endothelia or mesothelia depending upon their topographic site). These include the visceral and parietal mesothelia of the coelomic cavities, many of the derivatives of the intermediate mesoderm such as the pro-, meso-, and meta-nephroi and their ducts, the paramesonephric (Müllerian) duct system, and the cortex of the suprarenal gland. Further, after first passing through a mesenchymatous stage (*vide infra*) many mesodermal cells contribute to the formation of the widespread endothelial cell layers which line the blood and lymphatic vascular systems. Still other mesothelia develop to line the various synovial cavities of the many joints, bursae and tendon sheaths, the anterior chamber of the eye and the periotic perilymphatic space of the internal eye. The myotome of mesodermal somite origin (p. 118) also maintains an epithelial arrangement throughout the early stages of its development.

Mesenchyme, a name first introduced almost a century ago (Hertwig 1881), forms the remaining tissue of the early embryo, occupying all the regions between the various epithelial layers described above. In contrast it consists of a loosely arranged tissue with wide extracellular spaces containing copious extracellular fluid which carries a variety of hydrated mucosubstances (p. 51), suspended in which are widely spaced, amoeboid, primitive *mesenchymal cells*. The latter usually appear stellate in fixed and stained preparations, they often show some degree of phagocytic ability, and they have retained a potentiality for developing into a wide array of cell types in different sites. Many mesenchymal cells are derived from various regions of the early intraembryonic mesoderm, e.g. from the lateral plate where they form the core between the limiting epithelia of the splanchnopleure and somatopleure, and from the dermotome and sclerotome of the mesodermal somites (p. 118).

Mesenchymal cells are, however, not exclusively of mesodermal origin; it is probable that the extraembryonic mesenchyme of the yolk sac and chorion receive contributions from the neighbouring endoderm and cytotrophoblast respectively. Within the embryonic body considerable accessions are also received, particularly in the head region from the neural crest and the endodermal prechordal plate.

In subsequent development mesenchymal cells differentiate into many different types of tissue including: the general and specialized connective tissues (cartilage, bone, dentine) and their many types of attendant cell (p. 258); smooth visceral, cardiac and, in some sites, striated muscle; the endothelium of the blood and lymphatic vascular systems; the various blood-forming sites, lymphoid tissue, bone marrow, spleen and the remaining elements of the reticulo-endothelial and macrophage systems; the tissues of joints including their synovial cells and those of bursae and tendon sheaths; finally the sheaths of nerves, muscles and the periostea of bones.

SUBREGIONS OF THE EMBRYONIC LAYERS

In addition to the general summaries of the derivatives of the primitive embryonic layers given above, it may also prove convenient here to indicate the principal *subregions* of these layers

and their main contributions to the developing embryo. (Inevitably this involves some repetition.)

The Embryonic Ectoderm

This may be divided into (a) general body ectoderm; (b) neural plate; (c) neural crest; (d) ectodermal placodes.

General body ectoderm gives rise to: (1) most of the cutaneous epidermal cells; the secretory, duct-lining, and myoepithelial cells of sweat, sebaceous, and mammary glands; the hairs and nails; (2) the epithelia of the cornea and conjunctiva, and the secretory and duct-lining cells of the lacrimal gland; (3) the respiratory nasal epithelium; the epithelia lining the paranasal sinuses, lips, cheeks, gums and palate; the secretory and duct-lining cells of the nasal, labial, palatine, general oral and salivary glands; (4) the embryonic enamel organ and dental enamel; (5) the epithelial lining of the external acoustic meatus and external epithelium of the tympanic membrane; (6) the epithelial lining of the lacrimal canaliculi, sac and nasolacrimal duct; (7) the epithelium of the lower anal canal and terminal urethra.

Neural plate cells give rise to: (1) all the neurons of the *central* nervous system; these include not only those neurons wholly confined to the central nervous system, but also those somatic motor and preganglionic efferent neurons with somata within the central system and axons which pass out into the peripheral nervous system; (2) the macroglial cells, i.e. varieties of astrocyte and oligodendrocyte; (3) generalized ependymal cells lining the ventricles, aqueduct and central canal of brain and spinal cord; (4) specialized ependyal cells, including tanycytes, those covering the choroid plexuses, and distinctive regions of the third ventricle such as the subcommissural organ; (5) cells belonging to the APUD series (the term is an acronym for cells showing characteristic amine handling properties—Amine Precursor Uptake and Decarboxylation—the evolution and modification of the concept is discussed more fully elsewhere—p. 1454, and at other appropriate points throughout the volume). These include pinealocytes, magnocellular and parvocellular peptide-secreting neurons of the hypothalamus, and certain adenohypophyseal cells; (6) all retinal cells, and the epithelia of the iris, ciliary body and processes.

Neural crest cells, first forming a strip of tissue flanking the neural plate, then *migrating* widely, often in well-defined streams, and subsequently forming local aggregations, or invading other tissues, give rise to: (1) the sensory neurons of the dorsal root ganglia of the spinal nerves; (2) many of the sensory neurons of the ganglia of the trigeminal, facial, vestibulocochlear, glosso-pharyngeal and vagal cranial nerves (other cranial nerves may have transient populations—*see* p. 174); (3) the principal postganglionic autonomic neurons of both sympathetic and parasympathetic moieties, including the enteric plexuses, ganglionated sympathetic chains, and other ganglionic aggregates associated with, for example, the cardiac, coeliac, mesenteric, renal and vesical plexuses; (4) many specialized peripheral sensory receptor cells (see also placodes below); Schwann cells, the perisomatic satellite cells of sensory ganglia, the 'glial' elements of autonomic ganglia and peripheral receptors; (5) the basal regions of the neurocranium, and viscerocranium (p. 141), in particular the trabeculae cranii and cartilages of the branchial arches, and possibly the osteogenic elements that become associated with these cartilages; (6) the dentine-producing osteoblasts of the tooth germs; probably admixed with mesenchyme from other sources, crest cells may contribute to the special visceral striated muscles derived from the branchial arches, and to the smooth muscle of the branchial vessels; (7) some, at least, of the meningeal pia-arachnoid; (8) cells belonging to the APUD series (*vide supra*, and p. 1454). These include: thyroid parafollicular (C) cells derived from ultimobranchial tissue, classical chromaffin tissue (phaeochromocytes of the adrenal medulla, paraganglia and para-aortic bodies, and other outlying masses—*see* **8**.204), varieties of small intensely fluorescent (SIF) cells associated with sympathetic ganglionic neurons, the glomus (type I) cells of the carotid body (and probably many other functionally similar sites throughout the body), and finally both dermal and epidermal melanoblasts, and other epidermal, non-melanoblastic dendritic cells; (9) contributions to connective tissues throughout the body. For a review of neural crest *see* Leikola (1976).

Ectodermal placodes are specialized patches of ectodermal cells (the named placodes being the olfactory, otic, dorsolateral, epibranchail, and lental) that invaginate, sometimes coming into close association with cells from other sources. They all have some characteristics in common with the neural crest and give rise to: (1) the receptor, and probably sustentacular cells of the olfactory part of the nasal mucosa; (2) the epithelial walls of the membranous labyrinth, and the specialized receptor cells of its utricular and saccular maculae, and the ampullary crests of its semicircular ducts; (3) *some* of the primary sensory neurons of the trigeminal, facial, vestibulocochlear, glossopharyngeal, and vagal cranial nerves (although their precise contribution to these in the *human* embryo needs further clarification); (4) the crystalline lens of the eye; (5) some members of the APUD series, in particular the parathyroid chief cells: certain authors prefer to regard the cells of the adenohypophysis as having a placodal origin through the medium of the ectodermal anlage of Rathke's pouch; and similar considerations may apply to the parafollicular cells of the thyroid, through the anlage of the ultimobranchial body (the latter may therefore imply the existence of endodermal 'placodes').

It may be noted here, that despite cautious suggestions made over a decade ago (Pearse 1966), only 6 of the 40 cell varieties now accepted as belonging to the APUD series have definitely been shown to stem from the neural crest. Thus the large number of varieties of enterochromaffin cells present in the mucosae of the stomach and intestines, in the pancreatic islets, and other varieties found in the bronchial mucosa, the lining epithelium of the urethra and prostatic parenchyma, have an uncertain origin. Their progenitors have been hypothecated as 'neuoendocrine-programmed ectoblastic cells', and a foregut endodermal origin has been postulated for the gastro-intestinal/pancreatic varieties (Pearse 1977). (For a further discussion of this topic and the introduction of the term 'paraneuron' *see* p. 1454, and refer to Fujita 1976.)

The Embryonic Mesoderm

This may be divided into, as noted above, *paraxial* mesoderm, *intermediate* mesoderm and the *lateral* plate. The paraxial mesoderm, as is detailed further below, segments into a longitudinal series of mesodermal somites, and each somite subdivides into a sclerotome, dermotome, and myotome. The intermediate mesoderm forming a longitudinal, cord-like, cylindrical mass, which is continuous medially with the somites, and laterally with the lateral plate, may best be considered as a cranio-caudal series of subregions—cervical, thoracic, lumbar, and sacrococcygeal. Clefts appearing in the cardiogenic mesoderm and lateral plate coalesce to form the intraembryonic coelom (*vide infra*). The latter separates the lateral plate into somatopleuric (body wall) and splanchnopleuric (enteric) layers, which line or cover the ectoderm and endoderm, and their derivatives, respectively. As it enlarges, the intermediate mesoderm projects into, and also forms a partial boundary for the coelom, whilst the *zones of continuity* between the somatopleuric and splanchnopleuric mesoderm, after much differential growth, excavation, repositioning, and in some places adhesion, become sculpted to form the definitive gastric, enteric, hepatic and splenic mesenteries. The epithelial or mesenchymatous nature of these various zones are discussed further below and in subsequent pages.

The paraxial mesoderm, through the medium of the somites, forms: (1) the vertebrae and their associated joints and ligaments, including much of the intervertebral discs; the ribs, costal cartilages, sternal plates (ultimately manubrium, sternebrae and xiphoid process), by migrations from the sclerotomes: cells from the latter probably contribute to the dura mater; (2) the dermotomic cells migrate ventrolaterally, admixing with neural crest cells and somatopleuric lateral plate mesoderm and, together with the costal elements, form the bulk of the tissues of the body wall (*vide infra*), (3) the cells of the myotomes, by migration, subdivision, rearrangement, and sometimes fusion of cell masses from neighbouring myotomes, ultimately differentiate into the striated musculature of the trunk (it appears that the limb muscles, and lingual muscles, although segmentally innervated, develop *in situ*, with little obvious migration from the myotomes); (4) perineural mesenchyme, of uncertain, but possibly myotomic origin, provides the microglia of the central nervous system.

The intermediate mesoderm through the intermediaries of the nephrogenic cord, genital ridge and overlying medial coelomic mesothelium forms: (1) the rudimentary pronephric, and fully functional meso- and meta-nephric renal corpuscles and nephric tubules; (2) the mesonephric (Wolffian) duct and its derivatives—the vesical trigone, ureter and renal calices and collecting tubules in both sexes; male testicular duct complexes (p.217), epididymis, ductus deferens, seminal vesicle, ejaculatory duct, and a small area of prostatic urethra; and merely rudimentary vesicular, tubular and duct systems are formed in the female (p. 217); (3) the paramesonephric (Müllerian) duct and its derivatives—the female uterine tube and uterus, the development of the vagina being more conjectural; vestigial testicular structures and prostatic utricle in the male; (4) in both sexes, the whole of the gonadal tissues with the exception of the sex cells (gonocytes) themselves; (5) from the medial coelomic bay, and encroaching on to the mesonephric ridge, the cortex of the adrenal gland; (6) the connective tissue framework, capsule, mesothelial covering and, where appropriate, the mesenteries of all the foregoing.

The lateral plate mesoderm becomes split, as noted above, by the appearance of the intraembryonic coelom, into somatic and splanchnic layers (p. 118), the somatic mesothelial lining of the coelom forming the serous parietal pericardium, pleura and peritoneum, whilst the splanchnic mesothelial lining forms the serous epicardium, pulmonary pleura, and the visceral peritoneum that wholly or partly clothes the various abdominal and pelvic organs. The *somatopleuric mesenchyme*, as mentioned above and elsewhere, receives accessions from the neighbouring neural crest, and is also admixed with invading cells from adjacent dermotomes, myotomes and (in the thoracic region) sclerotomes. Thereafter mesenchymatous condensations of heterogeneous origins appear, and growth and differentiation transform them into the osseous, cartilaginous, muscular, vascular, lymphatic and connective tissues of the body wall. As the appropriate spatio-temporal patterning of these tissues is occurring, they gradually receive the terminals of ingrowing bundles of axonal processes (and their accompanying Schwann cells) growing from the neuroblasts of the ventral lamina of the neural tube, and from the neuroblasts of the primordial spinal and autonomic ganglia. Similar tissues and transformations are involved in the formation of the limbs, the early limb buds consisting of an external covering of ectoderm and a mesenchymatous core of mixed origin.

The splanchnopleuric mesenchyme clothes the endodermal gut tube, including the endodermal cloaca, allanto-enteric diverticulum, and their various subdivisions and glandular derivatives. Its cells grow and differentiate into the non-striated muscle, connective, vascular, lymphatic, and adipose tissues of the walls of the stomach, small and large intestines, urinary bladder (excluding the trigone), and also the walls of (all but the finest terminal) biliary and pancreatic duct systems. Similar tissues are provided by splanchnopleuric mesenchymal cells in the walls of the tracheo-bronchial tree down to its finest pre-alveolar passages, but with the addition of cartilage as far as the commencement of the bronchiolar system. Most of these tissues receive the ingrowing terminals of post-ganglionic sympathetic axons, and preganglionic parasympathetic terminals from the vagus (foregut and midgut and their derivatives) and the pelvic parasympathetic nerves (hindgut). Corresponding regions are invaded by postganglionic parasympathetic neuroblasts derived, it is presumed, from the vagal and sacral neural crest respectively.

It should be noted that vascular venous differentiation in the splanchnopleuric mesenchyme caudal to the developing diaphragm, leads to numerous radicles draining virtually the whole of the subdiaphragmatic gut, liver, pancreas and spleen, and these gradually combine to form the hepatic portal vein. In the primitive pericardial region, similar mesenchyme, including its covering mesothelium, form all the non-nervous tissues of the heart. The spleen differentiates in the mesenchyme of the dorsal mesentery of the stomach (dorsal mesogastrium)

It should also be stressed that in the ventrolateral walls of the embryonic pharynx *no coelomic cavity* develops, so that in the branchial arches there is *unsplit* lateral plate mesenchyme that has received substantial additions from the cranial neural crest, and possibly also from adjacent sclerotomes and dermotomes. Thus, whilst it has become customary to loosely classify many branchial structures as 'visceral' or 'special visceral' it is by no means clear in mammals in general, and particularly in mankind, which of these sources of mesenchyme predominates in the anlage of any specific tissue derivative.

Finally it may be mentioned that the splanchnopleuric mesenchyme of the yolk sac, liver, and spleen are the earliest loci of red blood cell formation, preceding the appearance of progenitor cells in the red bone marrow.

The Embryonic Endoderm

The main derivatives of the embryonic endoderm have already been listed above, and will be further detailed in subsequent sections: to avoid excessive repetition, therefore, only some additional points or as yet unresolved questions will be mentioned here. Thus: (1) whilst in general it is usually stated that the endoderm forms the epithelium of the whole alimentary tract, and its associated glands, from stomatodeum to proctodeum, its precise extent is less certain cranially, and the origin of some of the epithelial cell types cannot be regarded as proven. The epithelia of the lips, gums, parotid glands, cheeks, dental enamel, and anterior part of the hard palate are accepted as definitely of *ectodermal* origin; similarly, most authorities accept that the lingual epithelium and that of the lingual glands are endodermal; the exact site of the transition line between the two is not yet established. It must be stressed that the line does not necessarily correspond to any particular outstanding topographic landmark such as the oropharyngeal isthmus or linguogingival sulcus, and authorities differ concerning the ectodermal or endodermal status they ascribe to the submandibular and sublingual salivary glands. In contrast, the generally accepted caudal limit of the hindgut endoderm is at the level of the anal 'valves'. (2) The endoderm of the remainder of the post-oral foregut forms the great variety of epithelia that characterize the various zones, recesses, diverticula and glands of the pharynx, and the general, glandular, and duct-lining epithelia of the oesophagus, stomach, and proximal duodenum. Some parts of the embryonic pharyngeal lining form endocrine glands that lose all connexion with the gut, whilst other regions become associated with specialized localizations of lymphoid tissue. (3) The origin of the hair cells of the lingual, palatal, pharyngeal and laryngeal taste buds (neural crest or 'neurally-programmed' foregut endoderm) is uncertain, but in addition to their sensory receptor function, they have been classified as members of the APUD neuroendocrine series, and dubbed paraneurons (Fujita 1976). (4) The lymphoid-associated endodermal derivatives constitute *Waldeyer's ring*: characteristically their epithelial surfaces are uneven (showing undulations, folds, crypts, fossulae, clefts or sinuses); they include the nasopharyngeal tonsil (adenoid), the (auditory) tubal tonsils, the palatine tonsils, and the lingual tonsil. (5) The pharyngeal epithelia range from respiratory, mucus-secreting, ciliated, columnar epithelium which lines the upper nasopharynx, the environs of the opening of the auditory tube, and the superior part of the dorsal surface of the soft palate, together with the larnygeal vestibule, to the non-keratinized stratified squamous epithelium which lines the rest of the pharynx and palate. (6) The epithelial lining of the auditory tube, tympanic cavity with its mesentery-like mucosal folds and internal lamina of the tympanic membrane, and of the tympanic antrum, is derived from endoderm (p. 1194) and ranges from tall columnar and ciliated, through cuboidal to simple squamous in different locations. (7) Other pharyngeal, endodermal, derivatives include the general mucous glandular and duct-lining cells; the main follicular and parafollicular cells of the thyroid (from different locations); parathyroid secretory cells (debate persists concerning their origin—Fujita 1976; Pearse 1977); and the cytoreticulum and concentric corpuscles of the thymus. (8) Little is to be added to the widely held view that the great majority of the epithelial cells, glandular and duct-lining cells of the oesophagus, stomach, jejunum, ileum, colon, rectum,

and upper anal canal are of endodermal origin: similar considerations applying to the hepatocytes, and duct-lining epithelia of the biliary duct system, and to the pancreatic acinar and duct-lining cells: again the majority of epithelial cells— ranging from tall columnar ciliated and mucus-producing, to an extremely attenuated simple squamous type—that line the whole of the tracheobronchial-alveolar respiratory tree, are endodermal in origin. A similar derivation is also accepted for most of the transitional epithelium lining the urinary bladder and urethra, and for the parenchymal secretory cells of the prostate and lesser urethral glands. In almost all these locations, however, as intimated above, are numerous cell types loosely termed *enterochromaffin cells*: characteristically these cells show properties of *receptor mechanisms* on their surfaces, and the production of *neuroendocrine secretions* and *neurotransmitter-like* substances. (Not all the cells postulated as enterochromaffin have, as yet, fulfilled all these criteria.) Particular interest is currently focused upon the pancreatic islet cells and the so-called neuro-insular complex (p. 1371). Further comments on these matters will be found under the appropriate organs in subsequent sections: see also pp. 1348, 1364. (9) Finally, reference must be made to the possible relationship between endodermal cells and the haemopoietic stem cells (pp. 60, 63), and to the origin of the primordial sex cells (p. 1412).

SEGMENTATION OF THE MESODERM

In the embryos of all vertebrates (**2**.3) the intraembryonic mesoderm becomes incompletely subdivided by a longitudinal groove into a *paraxial part* and a *lateral plate*, on each side of the midline (**2**.29). The mesoderm in the floor of the groove which connects these two parts is the *intermediate mesoderm*, most of which subsequently forms the *nephrogenic cord* (p. 210). Soon after the appearance of this longitudinal groove the paraxial mesoderm becomes subdivided into prismatic blocks by a series of transverse grooves. This process is termed the *segmentation of the mesoderm*, and the blocks of paraxial mesoderm so formed are known as *mesodermal somites* (*primitive segments* or *metameres*). Commencing at the end of the third or the beginning of the fourth week in the region of the hindbrain, the process extends in a *cranio-caudal direction*, additional somites being laid down as the embryo grows in length, until 42–45 pairs are present (p. 224). The most cranial four or possibly five somites lying alongside the hindbrain participate in the formation of the skull and are termed the *occipital somites*. However, other workers (de Beer 1937; Sensenig 1957) have suggested that originally nine somites were involved in the elaboration of part of the skull. The remainder flank the spinal cord and usually number 8 cervical, 12 thoracic, 5 lumbar, 5 sacral and 8–10 coccygeal. Variations in this sequence may occur, usually involving either an additional segment or a missing segment in either the thoracic, lumbar or sacral regions, a variation in one region often being accompanied by a compensatory variation in a neighbouring region. The first occipital pair are often asymmetric in their degree of development, and one or both may be partly or completely suppressed. Many workers hold that mesodermal masses which are serially homologous with the segmented paraxial mesoderm arise beside the cranial tip of the notochord in association with the development of the extrinsic ocular muscles (p. 152). (For experimental data concerning mesodermal segmentation in higher vertebrates consult Bellairs 1971.)

In the *human embryo* it is only the *paraxial* mesoderm alongside the notochord which is physically segmented, but in view of the obviously segmental arrangement of the nerves of the spinal cord and their areas of distribution to the various embryonic layers, it is reasonable to suppose that the segmentation of the other structures is present although obscured.

This basic *metameric segmentation* is a characteristic feature of the whole phylum *Chordata* and is particularly well seen in the primitive miolecithal forms in which, after gastrulation (p. 76 and **2**.3), the mesoderm arises as a series of evaginations from the archenteron, which form independent, paired, segmental, coelomic, mesodermal pouches. These expand, separating the ectoderm from the endoderm until they meet dorsal and ventral to

Clefts in lateral
plate mesoderm

Pericardial cavity

Septum transversum

Oropharyngeal
membrane

Cardiogenic
mesoderm

Head process

Coelomic
duct

Peritoneal
part of coelom
communicating
with extraembryonic
coelom

A   B   C

Primitive streak

Primitive knot

Notochord
seen through
floor of neural
groove

Primitive knot

Primitive streak

Cloacal membrane

Allanto-enteric diverticulum

Connecting stalk

2.31 A–C Diagrams to illustrate the formation of the intraembryonic coelom—the embryonic area is viewed from the dorsal aspect and the early stages of head and tail fold formation are not shown.

A The intraembryonic mesoderm (green) becomes continuous with extraembryonic mesoderm (blue) around the periphery of the embryonic area. At the headward end the cardiogenic mesoderm extends from side to side across the median plane, it is continuous with extraembryonic mesoderm which will form the septum transversum; the region of the developing neural plate (yellow) is devoid of mesoderm.

B The pericardial cavity (uncoloured) has appeared in the cardiogenic mesoderm, and a series of clefts (uncoloured) has appeared in the lateral plate mesoderm on each side. The oropharyngeal and cloacal membranes which are devoid of mesoderm are shown in yellow.

C The clefts have extended and become confluent with each other and with the pericardial cavity forming a coelomic duct on each side. In the caudal part of the area additional clefts have formed, coalesced and broken through into the pleuropericardial canals and into the extraembryonic coelom. Part of the coelomic duct will form the pleural coelom.

the gut as permanent dorsal and (often temporary) ventral mesenteries. It is quite evident that in these forms the metameric segmentation involves not only paraxial structures, but the intermediate mesoderm and both somatopleure and splanchnopleure (and the coelom itself).

THE SOMITE

A typical human *mesodermal somite* consists of tightly packed (epithelioid) cells and at first contains a transient central cavity termed the *myocoele* which is, however, soon obliterated by cells proliferated from its walls. The cells of the ventromedial part of the somite constitute the *sclerotome*. They become mesenchymatous and migrate medially to provide the tissue from which the axial skeleton is ultimately derived (p. 138). The cells of the dorsolateral part of the somite constitute the *dermomyotome*. Spindle-shaped cells proliferate from its margins to form a tightly packed cellular mass on its medial aspect, the *muscle plate* or *myotome* (p. 154). The remaining epithelially arranged cells constitute the *skin plate* or *dermatome* (**2.29D**). The myotome is the forerunner of much of the striated musculature of the body (p. 154), and after it has commenced differentiation, the cells of the dermatome lose their epithelioid character and become mesenchymatous. They spread to mix with the somatopleuric mesenchyme, and, probably with accessions from the neighbouring neural crest, lay down the foundation of the dermis (p. 155).

THE EARLY INTRAEMBRYONIC COELOM

In the initial stages of somite formation a midline cavity appears in the *cardiogenic mesoderm* and this cavity is the first indication of the *intraembryonic coelom*. As segmentation of the mesoderm proceeds, a number of clefts are also formed on each side in the lateral plate mesoderm. The clefts expand and gradually coalesce, and the lateral plate therefore becomes divided into a *somatic* and a *splanchnic* layer (**2.29, 31**). The somatic layer, with its covering of ectoderm, constitutes the *intraembryonic somatopleure*, whilst the splanchnic layer, with the underlying endoderm, constitutes the *intraembryonic splanchnopleure*.

The lateral extremities of the cavity in the cardiogenic

mesoderm extend caudally and link up with the coalescing clefts in the lateral plate, and as a result the intraembryonic coelom is formed as an inverted U-shaped tube, from which the pericardial, pleural and peritoneal cavities are subsequently developed. Around the periphery of the embryonic area the somatopleuric and splanchnopleuric mesoderms are continuous, at first both with each other and with the similar extraembryonic layers, at the *junctional zone* mentioned previously. Soon,

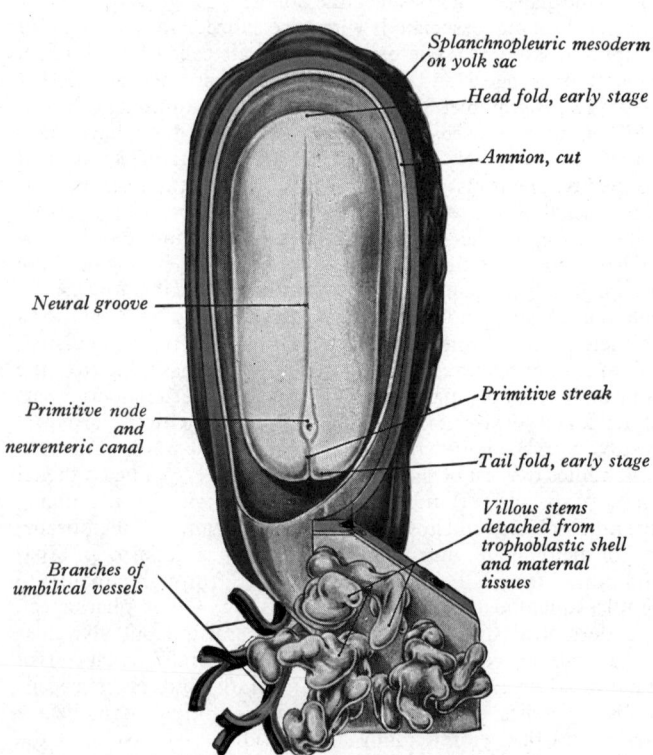

Splanchnopleuric mesoderm
on yolk sac

Head fold, early stage

Amnion, cut

Neural groove

Primitive node
and
neurenteric canal

Primitive streak

Tail fold, early stage

Villous stems
detached from
trophoblastic shell
and maternal
tissues

Branches of
umbilical vessels

2.32 Dorsal aspect of a model (by Eternod) of a human embryo, 1·3 mm in length. (After Graf Spee.) Compare with **2.34**.

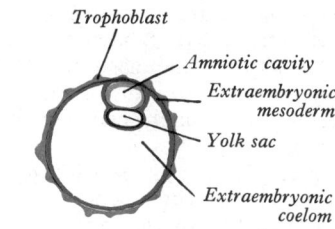

A

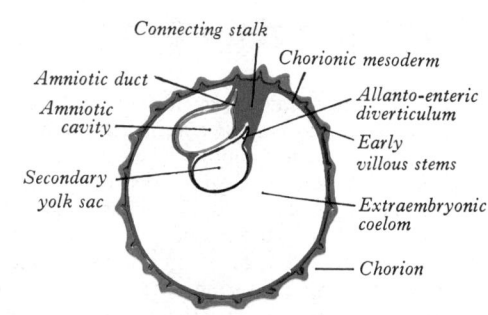

B

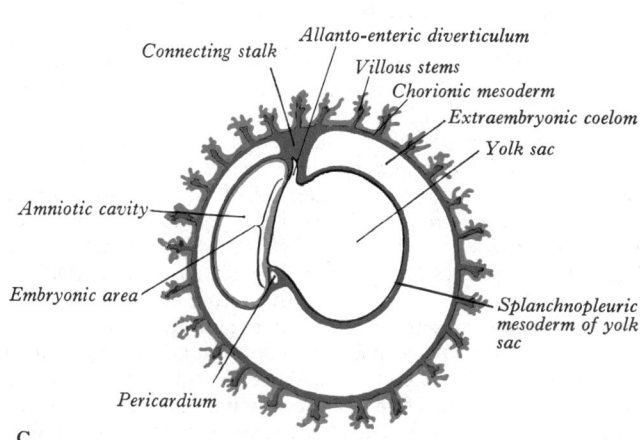

C

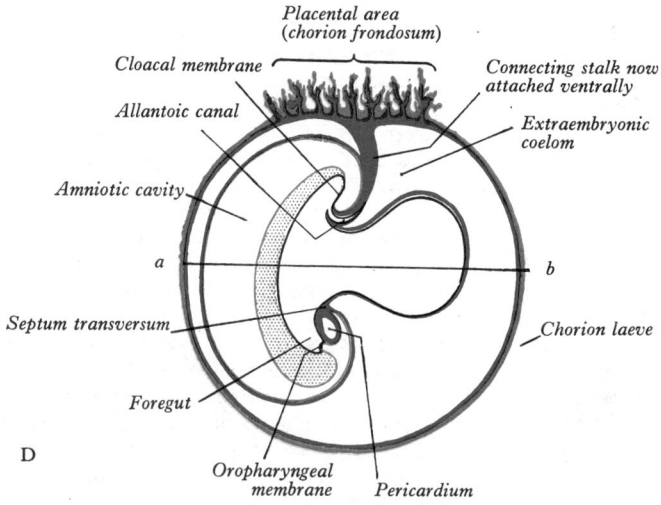

D

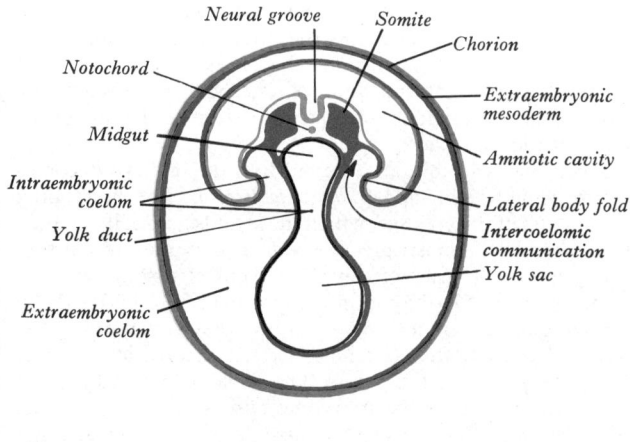

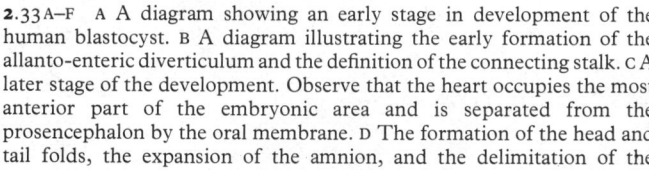

E

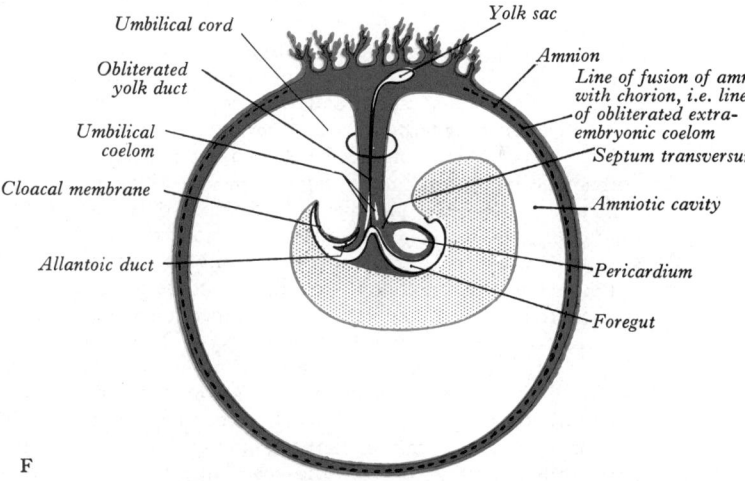

F

2.33A–F   A A diagram showing an early stage in development of the human blastocyst. B A diagram illustrating the early formation of the allanto-enteric diverticulum and the definition of the connecting stalk. C A later stage of the development. Observe that the heart occupies the most anterior part of the embryonic area and is separated from the prosencephalon by the oral membrane. D The formation of the head and tail folds, the expansion of the amnion, and the delimitation of the umbilicus. E A transverse section along the line *ab* in 2.33D. Observe that the intraembryonic coelom communicates freely with the extraembryonic coelom. F A later stage in the development of the umbilical cord.

It should be noted that whilst these diagrams, from an earlier edition, are useful for the general changes in disposition of the embryo and extraembryonic membranes, the modern view of development of the villous stems has been revised—consult text and 2.39.

however, the process of cavitation in the lateral plate extends beyond the embryonic area, and the continuity between these layers is broken, and the intraembryonic coelom is now thrown into free communication with the extraembryonic coelom (2.29 c, 31 c). This process, however, only occurs in the lateral regions; it does not affect the pericardial area or the regions immediately adjoining it on either side, where the median *primitive pericardial cavity* now opens into bilateral *pericardioperitoneal canals* (*coelomic ducts*) (2. 31 c). Accordingly, the junctional mesoderm around the cranial border of the embryo which will form the *septum transversum* and *primitive ventral mesentery* of the foregut (p. 200) remains intact.

The early formation of the coelom and its free communication with the extraembryonic coelom allows the fluid which fills the latter to gain access to the interior of the embryo. It therefore may act as a path for the passage of nutrients during the period which still has to elapse before the establishment of a blood vascular circulation (Streeter 1942). The walls of the coelom are formed of undifferentiated mesodermal cells (*mesoblasts*) which rapidly proliferate. From the mesoblasts of the somatopleure, together with those derived from the dermatome and some neural crest, the corium and subcutaneous tissues are formed. Splanchnopleuric mesenchymal cells, on the other hand, later differentiate into the muscles, blood vessels, lymphatics and connective tissues of the walls of the heart and gastro-intestinal tract.

It is not until a later stage that the mesoblasts which line the coelom itself become differentiated into the mesothelia which characterize the various serous membranes (pp. 37, 116).

The Formation of the Embryo

Hitherto, the embryonic area, initially bilaminar and later trilaminar, has been essentially a *flattened* disc which changes to an oval, and then a piriform outline. Towards the end of the third week, however, the embryo begins to assume its definitive shape. The immediate cause of this alteration is the difference in the rate at which adjoining areas are growing. The rate at the periphery fails to keep pace with the rate within the embryonic area, and, as the embryo is increasing more rapidly in its long axis, especially at its cranial end where the walls of the neural groove are expanding to form the forebrain, both extremities tend to project beyond the limits of the area (2.33). In this way a *head fold* is developed at the cranial extremity and a *tail fold* at the caudal extremity of the embryonic area. At the same time, and for similar reasons, right and left *lateral body folds* develop, and the extension of these four folds (2.33) gradually constricts off the embryo from the yolk sac and imparts its characteristic shape.

Resulting from head fold formation, the prosencephalon, which was hitherto separated from the cranial extremity of the embryonic area by the oropharyngeal membrane (p. 111) and pericardium, comes to lie at the extreme cranial tip of the embryo. This alteration in position of the forebrain is accompanied by a corresponding alteration in the relative positions of the membrane and pericardium (2.33 C–F). The former now lies on the *ventral surface* and forms the floor of a depression, which constitutes the primitive mouth or *stomatodeum*. Cranially the stomatodeum is bounded by the projecting forebrain, and caudally by the pericardium. The latter has not only altered its position relative to the cephalic extremity of the embryo but has also undergone a reversal of its surfaces, as will be examined below.

In addition to these positional alterations and reversals, head folding results in the inclusion within the embryo of a portion of the yolk sac termed the *foregut* (*proenteron*). The latter is now placed with the oropharyngeal membrane and pericardium on its ventral aspect, and with the hindbrain placed dorsally. It communicates at its caudal end with the *midgut* (*mesenteron*) through an opening termed the *anterior intestinal portal*.

The ventral bend of the head fold is associated with a correspondingly pronounced *midbrain flexure* which is concave ventrally and which underlies a projecting dorsal convexity (*midbrain prominence*) of the overlying ectoderm, seen to best advantage when the embryo is viewed from the side (e.g. 2.59).

Before tail fold formation, the caudal end of the embryonic area is anchored to the trophoblast by the connecting stalk, which is covered on one aspect by the amnion (2.33 C). The formation of the tail fold carries the connecting stalk round on to the ventral aspect of the embryo, so that it now assumes the permanent position of the umbilical cord. It will be remembered that the stalk is connected to the embryo at the caudal end of the primitive streak (but separated from it by the cloacal membrane, *vide infra*)

and, in consequence the primitive streak also extends round towards the ventral aspect of the embryo to the region which later lies immediately behind the anal orifice. Some, however, hold the view that the primitive streak is continued into the rudimentary tail and does not appear in the perineum or on the ventral surface of the embryo.

Just as a recess of the yolk sac is included within the head fold to form the foregut, so a corresponding part is included within the tail fold to form the *hindgut* (*metenteron*). But the similarity between these two included recesses goes further. A portion of the endoderm (earlier, the prechordal plate, p. 110) in the floor or ventral wall of the foregut is in direct contact with the ectoderm over an area which is termed the *oropharyngeal membrane*. This membrane soon disappears and the communication of the gut with the exterior through the mouth is thus established. In the region of the hindgut a similar relationship exists. As we have seen, even before the tail fold is defined, the ectoderm and endoderm are in contact with each other at the caudal end of the embryonic area, forming the *cloacal membrane*. As will be described later, this membrane subsequently breaks down in two places to form the *urogenital* and *anal orifices*.

Before the formation of the tail fold an endodermal diverticulum arises from the dorsocaudal portion of the yolk sac and grows into the mesoderm of the connecting stalk (2.33). This outgrowth constitutes the *allanto-enteric diverticulum* (2.35B). As the tail fold becomes defined the proximal part of the diverticulum becomes incorporated in the hindgut, and its distal portion persists as the *endodermal allantoic duct* (p. 123), which then communicates directly with the ventral surface of the hindgut. The region of the hindgut which lies caudal to this communication forms the *endodermal cloaca* (p. 202). As tail fold formation continues, the cloacal membrane progressively comes to lie in the ventral wall of the hindgut but it extends also on to the adjoining dorsal aspect of the connecting stalk, where it is associated with the allantoic portion of the allanto-enteric diverticulum (2.35C). Even before the latter is incorporated in the hindgut, the cloacal membrane becomes interrupted and shortened by the interposition of mesoderm between the endoderm of the diverticulum and the covering epithelium of the body stalk (Florian 1930).

Between the head fold and the tail fold the embryo becomes constricted off by right and left *lateral body folds*. The intervening dorsal portion of the yolk sac, which these folds gradually include within the embryo, constitutes the *midgut* (mesenteron). At first the midgut communicates freely on its ventral surface with the rest of the yolk sac, but the continued growth of the folds results in a narrowing of the connexion, which becomes drawn out as the *splanchnopleuric yolk stalk* (vitello-intestinal duct) which contains the *endodermal yolk duct* (2.33E,F). The remainder (distal part) of the yolk sac remains extraembryonic and is often termed the *umbilical vesicle*. The subsequent history of the duct and the vesicle will be dealt with later (pp. 122, 205).

THE NUTRITION OF THE EMBRYO

In the early stages of development the blastomeres derive their nourishment in part from the store laid up within the cell body of the primary oöcyte. Such stores are possibly maintained at a high concentration and are subsequently liberated in a more dilute form, for absorption, from the cavity of the blastocyst, and later the primary and secondary yolk sac cavities (p. 109). In addition, it is also assumed that the blastocyst derives nourishment from tubal and uterine secretions, and later, during the process of embedding, from products stemming from the lysed uterine tissues. There follows a period of about two weeks during which the embryonic disc is dependent on the nutriment it can obtain from the fluids which fill the cavities of the amnion, the coelom and the yolk sac. These fluids probably contain material absorbed by the trophoblast from the uterine tissues and the maternal blood, modified, perhaps, as they pass through the cellular walls of these various cavities. However, at an early stage in

development these sources of supply are cut off. The lumen of the neural tube is isolated by the closure of the neuropores, the extraembryonic coelom becomes greatly reduced (2.33D, F) and is later shut off from the intraembryonic coelom, and the obliteration of the yolk duct separates the yolk sac from the gut. It therefore becomes imperative that some other source should be rendered available at an early stage. This involves the maternal circulation coming into close, although indirect, apposition with the developing embryonic circulation.

The differentiating mesenchyme in which the embryonic vessels and erythrocytes develop is termed *angioblastic tissue*, and it is probably first formed from the deepest layer of mesenchyme which clothes the endoderm of the yolk sac early in the third week (p. 109).

Slightly later, angioblastic tissue can also be recognized in the connecting stalk and in the mesenchyme of the chorion, and it then appears also within the embryonic area. Tissue spaces form in the angioblastic tissue and the cells which line them differentiate into typical, flattened endothelial cells whilst adjoining spaces join to form capillary plexuses. While the yolk sac spaces are forming, small, localized groups of mesodermal cells project into them, soon to become cut off to form *blood islets*, the cells soon differentiating into embryonic erythrocytes (p. 180).

The vessels formed in the chorion soon establish an intimate relationship with the maternal circulation (p. 124). Vessels develop in the embryo as two longitudinal channels, which, at their headward ends, project into the dorsal wall of the pericardium. They are the rudimentary right and left dorsal aortae, and at their cranial ends, after curving ventrally in the lateral wall of the pharynx to reach the cranial end of the pericardium, they fuse, becoming continuous with the developing primitive tubular heart. At the caudal end of the embryo they traverse the connecting stalk as the rudimentary umbilical arteries and break up into capillaries in the chorion. The venules from the chorion converge on the stalk where they form the right and left umbilical veins, which run headwards in the somatopleure, close to the margin of the embryonic area, to reach the caudal end of the heart.

The pericardial cavity never communicates directly with the extraembryonic coelom, and at its cranial limit the somatopleure and splanchnopleure are continuous (**2.76**A). With the formation of the head fold the surfaces of the pericardium are reversed, and the original cranial aspect comes into intimate relation with the *ventral* wall of the foregut at the cranial border of the anterior intestinal portal (**2.33**C). As the caudal wall of the pericardium deepens dorso-ventrally, the mesenchyme between it, the gut and proximal yolk stalk forms a sheet, which is the *septum transversum*. Prior to reversal this mesoderm formed a U-shaped mass intervening between the pericardium and the extraembryonic coelom (p. 118). The septum transversum later plays an important part in the development of the diaphragm. At this stage it is bounded on its headward surface by the pericardium and on its caudal surface by the foregut; on its dorsal surface it is limited by the coelomic ducts, which connect the pericardium with the peritoneal cavity and on its lateral surface by the opening of the peritoneal cavity into the extraembryonic coelom. The umbilical and body wall veins, which run in the somatopleure, and the vitelline veins, which run in the splanchnopleure, meet one another in the junctional mesoderm of the septum transversum and so gain the venous end of the heart. Through these various channels the early embryonic circulation is established.

The Fetal Membranes and Placenta

THE ALLANTOIS

The allanto-enteric diverticulum (**2.33**B, 34, 35) arises early in the third week as a solid, endodermal outgrowth from the dorsocaudal part of the yolk sac and grows into the mesoderm of the connecting stalk. It soon becomes canalized and, when the hindgut is developed, the proximal (enteric) part of the diverticulum is incorporated in its ventral wall and the distal (allantoic) portion remains as the allantoic duct and is carried ventrally to open into the ventral aspect of the cloaca or terminal part of the hindgut (**2.33**, 35). The diverticulum is lined with endoderm and is surrounded by the mesoderm of the connecting stalk, in which the umbilical vessels develop at a slightly later stage.

In reptiles, birds and many mammals the allantoic diverticulum develops into a vesicle which continues expanding into the extraembryonic coelom and forms a vascular organ to which the term *allantois* should perhaps be restricted. In the bird it projects to the right side of the embryo, and gradually spreads over the dorsal surface of the embryo as a flattened sac between the amnion and the chorion (serosa), and ultimately also spreads to surround the yolk sac. Its outer wall becomes applied to, and fuses with, the chorion (forming an *allantochorion*) which lies immediately inside the shell membrane. Blood is carried to the allantoic sac by the two *allantoic* (or *umbilical*) *arteries*, which are continuous with the primitive aortae and, after circulating through the allantoic

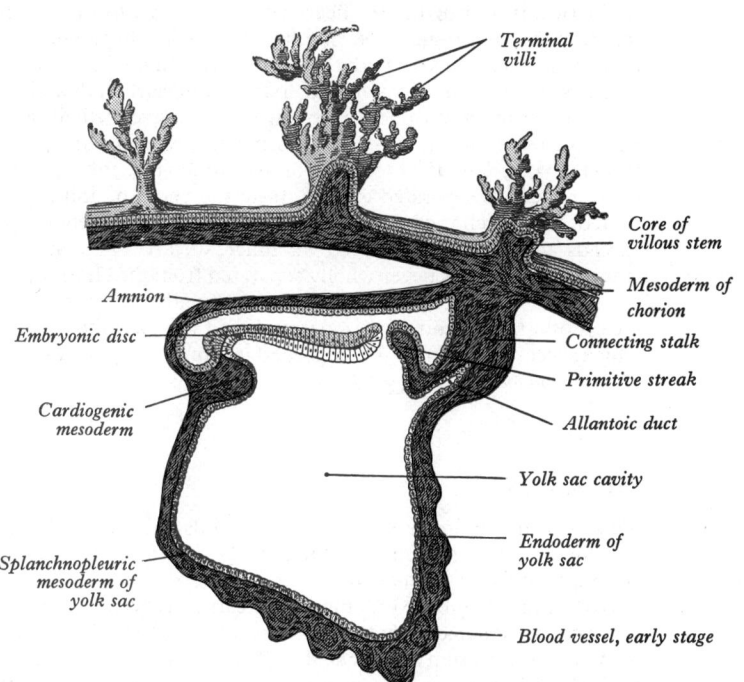

Labels on figure: Terminal villi; Core of villous stem; Mesoderm of chorion; Connecting stalk; Primitive streak; Allantoic duct; Yolk sac cavity; Endoderm of yolk sac; Blood vessel, early stage; Amnion; Embryonic disc; Cardiogenic mesoderm; Splanchnopleuric mesoderm of yolk sac

2.34 A sagittal section through the embryo which is represented in **2.32**. The cloacal membrane is not shown. (After Graf Spee.) Note that the concept of the chorionic villi as free-ending outgrowths from the chorion, as implied in this historic model, has since been discarded—*see* text for details.

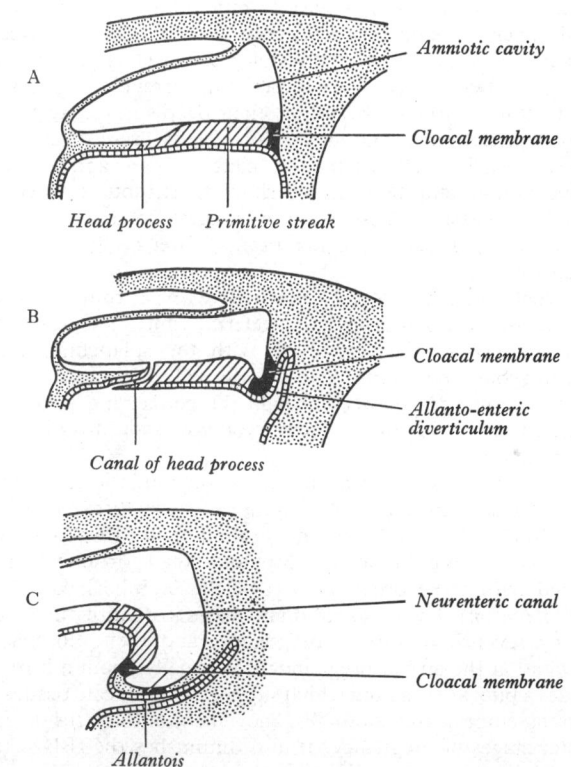

Labels on figure: Amniotic cavity; Cloacal membrane; Head process; Primitive streak; Cloacal membrane; Allanto-enteric diverticulum; Canal of head process; Neurenteric canal; Cloacal membrane; Allantois

2.35A–C Three stages in the development of the cloacal membrane. (Florian 1930.) A Prior to the formation of the tail fold and the allanto-enteric diverticulum. B The tail fold is indicated and the allanto-enteric diverticulum has formed. The cloacal membrane now extends in relation with the enteric part of the diverticulum on to the dorsal aspect of the body stalk. C The tail fold is clearly defined. Mesoderm has become interposed between the ectoderm and endoderm over the proximal part of the allanto-enteric diverticulum and the cloacal membrane is thus broken up into two parts, of which only the proximal one persists. The mesoderm is shown stippled.

capillaries, is returned to the heart by the two *umbilical veins*. In this way the chorio-allantoic circulation, which is of the utmost importance in connexion with the respiration and nutrition of the chick, is established. Oxygen is taken from, and carbon dioxide is given up to the atmosphere through the porous egg-shell, and nutritive materials are at the same time absorbed by the blood from the yolk. With the formation of the amnion the embryo is, in most mammals, separated entirely from the chorion, and is not united to the chorion again until the allantoic mesenchyme spreads to become applied to its inner surface. The human embryo, however, is never wholly separated from the chorion, its caudal end being from the first connected with the chorion by means of a thick band of mesoderm named the *connecting stalk*, which accordingly may be regarded as precociously formed *allantoic mesoderm*.

THE AMNION

This is a membranous sac which surrounds the embryo; it is developed in reptiles, birds and mammals (*Amniota*), but not in amphibia or fishes (*Anamniota*).

In the human embryo the amnion appears as a cavity within the inner cell mass adjacent to the overlying trophoblast. This cavity is roofed by a stratum of epithelial cells, and its floor is formed by the cells of the embryonic germ disc—continuity between the roof and floor being at the margin of the disc. (For details of the *epiblastic* origin of the roof cells *see* p. 109.) In the first half of pregnancy the epithelial cells are flattened and contain abundant glycogen; in later stages they become cuboidal or columnar over the placenta, the glycogen diminishes in amount and lipid globules appear in the cytoplasm (Goto 1959). Ultrastructurally, two types of amniotic epithelial cell have been described (Thomas 1965)—the 'Golgi type' and the 'fibrillar type'. The former is characterized by a pronounced Golgi apparatus, and a considerable content of rough endoplasmic reticulum and membrane-bound vesicles; the fibrillar cytoplasm of the latter contains few organelles. Possibly they are physiological variants of the same cell. Their free surfaces are beset with irregular microvilli embedded in a surface coat, whilst their deep surfaces rest on a basement membrane. The intracellular clefts present scattered desmosomes, but elsewhere the clefts widen and contain interlacing microvilli between which complex tubules penetrate the cell cytoplasm. These features suggest an active role in selective transport across the membrane (Lister 1968; Wynn and French 1968).

Externally the amnion is covered with a thin layer of somatopleuric extraembryonic mesoderm, which is continuous, at the margins of the disc, both with the splanchnopleuric extraembryonic mesoderm covering the yolk sac and with the intraembryonic mesoderm. Through the connecting stalk it is continuous also with the extraembryonic mesoderm lining the chorion (**2**.33).

Fluid, termed *liquor amnii*, occupies the amniotic cavity and increases steadily in volume, so that the sac gradually expands and encroaches on the extraembryonic coelom (**2**.33); this continues until the coelom is obliterated, except for a small portion which is included within the proximal part of the umbilical cord (the *umbilical coelom*). The liquor amnii increases in quantity up to the sixth or seventh month of pregnancy, and then diminishes somewhat; at the end of pregnancy it is usually about a litre. It provides a buoyant medium which supports the delicate tissues of the young embryo and allows free movement of the fetus during the later stages of pregnancy. It also diminishes the risk to the fetus of injury from without. It contains less than 2 per cent of solids, consisting of urea, inorganic salts, a small amount of protein, and frequently a trace of sugar.

The source of the amniotic liquid, whether fetal, maternal or both, is not finally settled. Its volume is regulated by a multiplicity of factors, both fetal and maternal, but it is suggested that normally it is predominantly fetal in origin in the normal (Davies 1960). In the early stages it resembles blood plasma in composition and is probably formed largely by transport across the amniotic membrane, but as pregnancy advances, it becomes

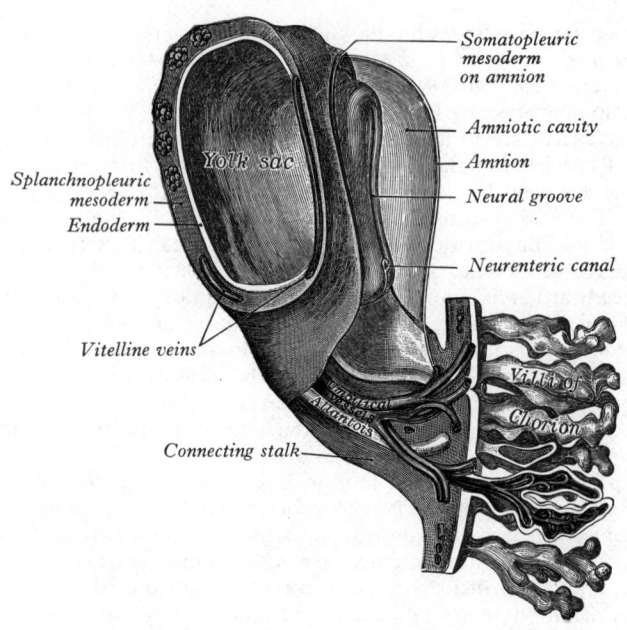

2.36 A human embryo, 1·3 mm long. (From a model by Eternod.) The left wall of the yolk sac and most of the amnion have been removed, and some of the primary mesoderm of the connecting stalk has been cut away to show the allantois and the umbilical vessels. A portion of the chorion and some of the villi related to the decidua basalis are shown. (No true villous stems or cytotrophoblastic shell were shown in this early model.) The embryonic area is seen from the dorsal and left lateral aspects. Compare with 2.32, which presents another view of the same model.

progressively more dilute, presumably by the addition of fetal urine. It has been shown experimentally that there is a considerable and rapid flux of water across the amniotic membrane. There is rapid exchange between the amniotic fluid and maternal and fetal circulations, probably by way of the placenta and fetal kidneys. By the end of the third month the expanding amnion has extensive contact with the chorion and only these thin membranes separate the amniotic fluid from the decidua parietalis, the vessels in which may provide another route for the exchange of water and dissolved substances (Plentl 1958). A volume of amniotic fluid in excess of two litres is generally considered to be abnormal and constitutes *hydramnios*. A deficiency is termed *oligamnios*. Both conditions may be associated with fetal abnormalities. For example, fetuses with agenesis of the kidneys or atresia of the lower urinary tract are often associated with oligamnios.

It has been suggested that fetal swallowing of amniotic fluid is a normal occurrence, and also that respiratory movements may possibly aspirate some fluid into the fetal lungs. In either case the fluid may be absorbed into the fetal circulation and then pass the placental barrier into the maternal circulation. Cases of oesophageal atresia, or anencephaly, in which swallowing is impossible, or impaired, are often associated with hydramnios.

THE CONNECTING STALK AND UMBILICAL CORD

The connecting stalk (**2**.33) is, as we have seen, a mass of precociously formed allantoic mesoderm, which at first connects the caudal end of the embryonic area with the chorion. Proximally it surrounds the short allanto-enteric diverticulum but it is traversed throughout its length by the umbilical (allantoic) vessels. At first its dorsal surface is covered with the amnion and its ventral surface is bounded by the extraembryonic coelom. As a result of the folding of the embryo and distension of the amnion, the connecting stalk comes to lie on the ventral surface of the embryo, and its mesoderm approaches that of the yolk sac and its stalk. With continued expansion of the amnion, the extraembryonic coelom is largely obliterated (**2**.33), and its only remaining part surrounds the elongating yolk stalk, and this part still communicates freely with the intraembryonic coelom. The mesoderm-covered surfaces of the head, tail and lateral body folds of the amnion now converge on the connecting stalk and yolk stalk (and their vessels), and the umbilical cord is formed as they meet one another (2.33) thus closing off the intraembryonic coelom.

The umbilical cord (2.37) thus consists of an outer covering of flattened amniotic epithelial cells, containing in its interior a mass of mesoderm of diverse origins (*vide infra*). Embedded in the latter are two endodermal tubes—the yolk and allantoic ducts—their associated vitelline and allantoic (umbilical) blood vessels, and near its fetal end, the remains of the extraembryonic coelom mentioned above.

The *mesodermal core* is derived from the somatopleuric extraembryonic mesoderm covering the amniotic folds, the splanchnopleuric extraembryonic mesoderm of the yolk stalk which carries the vitelline vessels and clothes the endodermal yolk duct, and similar allantoic mesoderm of the connecting stalk which clothes the allantoic duct and carries initially two umbilical arteries and two umbilical veins. These various mesoderms fuse and are gradually transformed into the viscid, mucoid connective tissue (*Wharton's jelly*) which characterizes the more mature cord. The tissue consists of widely spaced fibroblasts separated by an extensive intercellular space filled with a copious matrix consisting of a delicate three-dimensional meshwork of fine

Usually, the embryonic right umbilical vein disappears in the early months of pregnancy (and exceptionally only one artery may be present). The vessels of the umbilical cord are rarely straight but usually show a twisted conformation which may exist as either a right- or left-handed cylindrical helix. The number of turns involved may be relatively few or, at the other extreme, may even exceed 300. Their causation has been variously ascribed to unequal growth of the vessels, or to torsional forces imposed by fetal movements; their functional significance is obscure; perhaps their pulsations and contractions (*vide infra*) assist the venous return to the fetus in the umbilical vein. When fully developed the umbilical vessels, particularly the arteries, are provided with a strong muscular coat which contracts readily in response to mechanical stimuli. The outermost muscle bundles pursue an interlacing spiral course so that, when they contract, they produce shortening of the vessel and thickening of the media, with folding of the interna and considerable narrowing of the lumen. This action may account for the periodic sharp constrictions of contour—the so-called *valves of Hoboken* which often character-ize these vessels.

When fully developed, the umbilical cord is on average some 50 cm long and 1–2 cm in diameter, but the length is subject to great variation (20–120 cm). Obstetrical complications during labour may be imposed by the possession of an exceptionally short or long umbilical cord.

Cords may also exhibit *knots* which may be 'true' or 'false'. True knots presumably follow some exceptional form of fetal movement *in utero*, and may embarrass the circulatory flow in the umbilical vessels. False knots are sharp variations in contour which may accompany an abnormally pronounced looping of one of the umbilical vessels, or it may correspond to a local accumulation of a mass of Wharton's jelly.

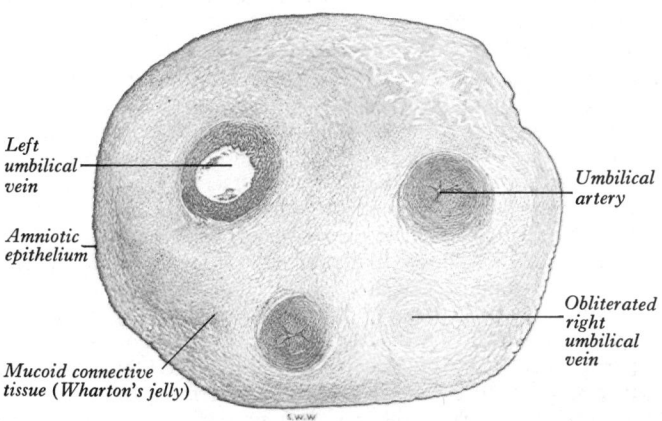

Left umbilical vein

Umbilical artery

Amniotic epithelium

Obliterated right umbilical vein

Mucoid connective tissue (Wharton's jelly)

2.37 Transverse section through a human umbilical cord. Stained with haematoxylin and eosin. Magnification × *c*. 8.

collagen fibres surrounded by a dilute ground substance containing a variety of hydrated mucopolysaccharides. In specimens which have been excised *before* fixation and staining, the fibroblasts present stellate profiles, as they do in other areas of mucoid tissue when similarly prepared. However, it has been shown that if the tissue is fixed *before* any haemodynamic collapse and consequent modification of the jelly has occurred, the cells are long and strap-like and present a regular orientation (Parry 1970).

The part of the extraembryonic coelom (the *umbilical coelom*) included in the base of the umbilical cord acts as a sac which receives the normal *umbilical hernia* of the midgut, developing in the embryo between the sixth and tenth weeks (p. 200). After the disappearance of this hernia the extraembryonic coelomic sac is normally obliterated.

The yolk sac becomes caught between the amnion and chorion as they fuse near the placental attachment of the cord (2.33, 38, 46)—it continues to grow slowly and is sometimes found at term in this site, as a small vesicle usually less than 5 mm in diameter. The yolk stalk and its contained endodermal duct and accompanying vessels gradually elongate with growth in length of the umbilical cord. The duct and vessels slowly degenerate and they have usually disappeared by mid-pregnancy.

The endodermal allantoic duct which is confined to the proximal end of the growing cord, also elongates and thins, but may persist as an interrupted series of epithelial strands until term. At the umbilicus the proximal strand is often continuous with the median intra-abdominal *urachus*, which in turn continues into the apex of the bladder (p. 218).

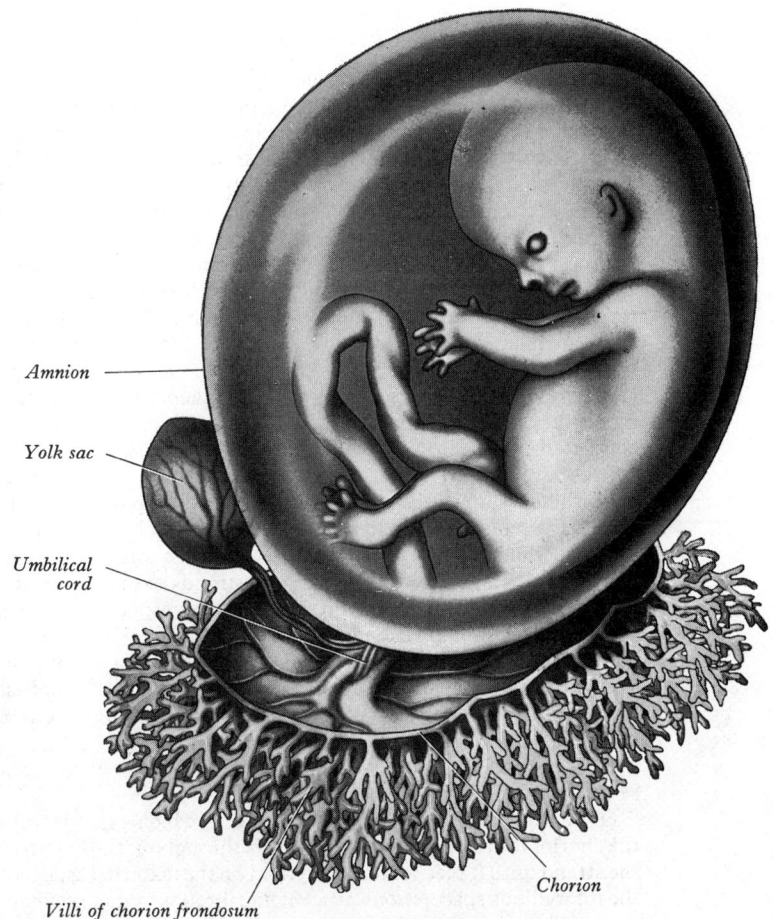

Amnion

Yolk sac

Umbilical cord

Villi of chorion frondosum

Chorion

2.38 A fetus of about eight weeks, enclosed in the amnion, magnified about 2½ diameters. A part of the chorion frondosum with its branching villous stems is shown in the lower part of the figure. The villous stems have been detached from the basal plate, which is not shown here.

For variations in the placental attachment of the cord, *see* p. 136, and for further details of its morphology and an extensive review of the relevant literature, *see* Boyd and Hamilton (1970).

IMPLANTATION

As already stated (p. 103), fertilization occurs in the lateral or ampullary part of the uterine tube, and is immediately followed by cleavage. The segmenting zygote is conveyed along the uterine tube to the cavity of the uterus by the action of the cilia of the epithelial lining of the tube aided by muscular tubal contractions, the journey occupying about three days. On reaching the uterine lumen the morula becomes a blastocyst but is still surrounded by the zona pellucida. The disappearance of the zona and the formation of polar syncytial trophoblast over the region of the formative mass are necessary before implantation of the ovum in the uterine mucosa can occur (*see* pp. 103, 105 for details). In the interval between ovulation and the arrival of the blastocyst in the uterine cavity, changes also occur in the uterine mucosa which prepare it for implantation. These progestational changes will be detailed later (p. 128), but when they have occurred, the syncytial trophoblast adheres to the uterine mucous membrane, destroys the epithelium over the area of contact, and excavates a cavity in the mucous membrane in which the blastocyst becomes embedded. In a conceptus described by Bryce and Teacher (1908) the point of entrance was visible as a small gap closed by a mass of fibrin and leucocytes; in another described by Peters (1899) the opening was covered with a mushroom-shaped mass of fibrin and blood clot, the narrow stalk of which plugged the aperture in the mucous membrane. It is believed that this *operculum* represents a portion of the syncytial trophoblast cut off by the decidua capsularis (Böving 1963).

THE TROPHOBLAST AND CHORION

The structure actively concerned in excavating the uterine mucous membrane is the syncytial trophoblast. This increases rapidly in thickness over the embryonic pole and then forms a progressively thinner layer over the rest of the wall towards the abembryonic pole, as the blastocyst becomes embedded. It invades and digests the uterine tissues, and attacks the walls of the uterine (maternal) blood vessels (*see* 2.23; p. 107, and consult Böving 1959, 1963). Lacunar spaces develop in the trophoblastic envelope and establish communications with each other. At an early period many of them contain maternal blood (2.23) derived from the dilated uterine capillaries and veins, the walls of which have been partially destroyed. As the conceptus grows and the lacunar spaces enlarge, becoming more confluent to form an early blood-filled *intervillous space*, their trophoblastic walls are converted at first into an *irregular* spongework, or *labyrinth*. With further growth, however, the main strands of trophoblastic syncytium assume a *radial* arrangement. Thus, from quite early stages, the intervillous space is *completely spanned* by these radial strands which undergo an orderly sequence of three main histological changes (*vide infra*). Such strands extend from the syncytial layer of the chorion (on the embryonic aspect of which is a layer of cytotrophoblast, lined by vascularized fetal mesoderm), across the intervillous space, to the layer of (peripheral) syncytium which is in close apposition to the excavated maternal tissues. Through spaces in the latter, extravasated maternal blood continues to enter the intervillous space.

The central core of each syncytial strand is now invaded sequentially by:

(1) A growing *column* of proliferating cells which extends from the chorionic cytotrophoblast and grows throughout the length of the strand until it reaches the syncytium on the maternal aspect of the intervillous space, *within which* it 'mushrooms' tangentially to meet and fuse with neighbouring outgrowths to form a spherical *cytotrophoblastic shell* around the conceptus (2.39).

(2) Each *cytotrophoblastic cell column* now develops, centrally, and throughout much of its length, a *core* of extraembryonic mesoderm. This never completely reaches the trophoblastic shell,

and it is uncertain how much of this mesoderm is derived by *inflexion* and *invasion* from the chorionic mesoderm, or whether it arises by differentiation of the central cytotrophoblastic cells.

(3) Capillaries now differentiate within the mesodermal core (2.39) and soon establish connexions with the radicles of the umbilical vessels in the general mesoderm of the chorion.

The radial trabeculae undergoing these transformations are termed the *villous stems* and they, and their subsequent side branches, may be regarded as passing through the three grades of *histological differentiation* (primary, secondary and tertiary) just described (2.39). The term 'villous stem' is used here in preference to the traditional primary, secondary and tertiary chorionic *villi* because it is evident that, at this stage, there are no true villi (i.e. finger-like projections with a free tip which grow across the intervillous space to reach the maternal tissues).

Each villous stem now consists, from its chorionic base and throughout much of its extent, of a vascularized mesodermal core, covered by a single (*Langhans'*) layer of cytotrophoblast, which is again ensheathed by a layer of syncytium. Near the maternal end, however, it contains no mesodermal core but is formed of a solid *cytotrophoblastic cell column* which is continuous peripherally with the trophoblastic shell, and is surrounded by a layer of syncytium.

With continued development two major changes now ensue (2.39); firstly, increasing complexity and subdivision of the maternal ends of the villous stems and, secondly, the development of free side branches. Expansion of the whole conceptus is accompanied by *radial growth* of the villous stems and, simultaneously, an integrated *tangential growth* with expansion of the trophoblastic shell. As these changes occur, the attached maternal extremity of each villous stem undergoes a series of longitudinal divisions. Eventually each stem forms a complex consisting of a single *trunk* (*truncus*) attached by its base to the chorion, and from which arise distally, second and third order branches (which have been named *rami* and *ramuli* respectively—Stieve 1926). Each of the latter remains attached at its maternal end to the trophoblastic shell.

Now, for the first time, 'true' villi with free tips make their appearance as outgrowths from the sides of the villous stems, and particularly from the rami and ramuli rather than from the truncus. These have been variously termed *free, terminal, absorption* or *fringing* villi, and it is predominantly across their walls that exchanges between the fetal and maternal circulations occur. Each terminal villus commences as a syncytial outgrowth which, as it continues to grow, is again invaded successively by cytotrophoblastic cells which then develop a core of fetal mesoderm; this is finally vascularized by fetal capillaries (i.e. each villus passes through primary, secondary and tertiary grades of histological differentiation). The terminal villi continue to form and branch, within the confines of the definitive placenta (*vide infra*) throughout gestation, projecting in all directions into the intervillous space, many of them making contact with those growing from adjacent villous stems. When this occurs some form of secondary adhesion or fusion may occur between their tips. Some earlier workers suggested that this may proceed to mesodermal or even vascular fusion between adjacent villi, but a recent survey of the available evidence strongly suggests that this does not occur to any appreciable extent and that any connexion between villi is limited to fibrinoid adhesion between their surfaces, or at most, fusion between their syncytial tips. As these changes proceed, the intervillous space, at first spanned by the early villous stems and their branches, and then increasingly permeated by growing free villi, is finally once again transformed into an exceedingly complex *labyrinth* of fine intercommunicating maternal vascular spaces.

The intervillous space, which contains the circulating maternal blood and is everywhere lined by syncytial trophoblast, is thus bounded:

(1) On its *fetal aspect* by a *chorionic plate*, consisting of syncytial, cytotrophoblastic and mesodermal layers of the chorion, the latter carrying the radicles of the umbilical vessels, and which is soon fused with the mesoderm of the expanding amnion.

(2) On its *maternal aspect* by a *basal plate* consisting of the peripheral syncytium that has been divided by the growth of the cytotrophoblastic shell into an inner layer enclosing the

NUTRITION THROUGHOUT GESTATION

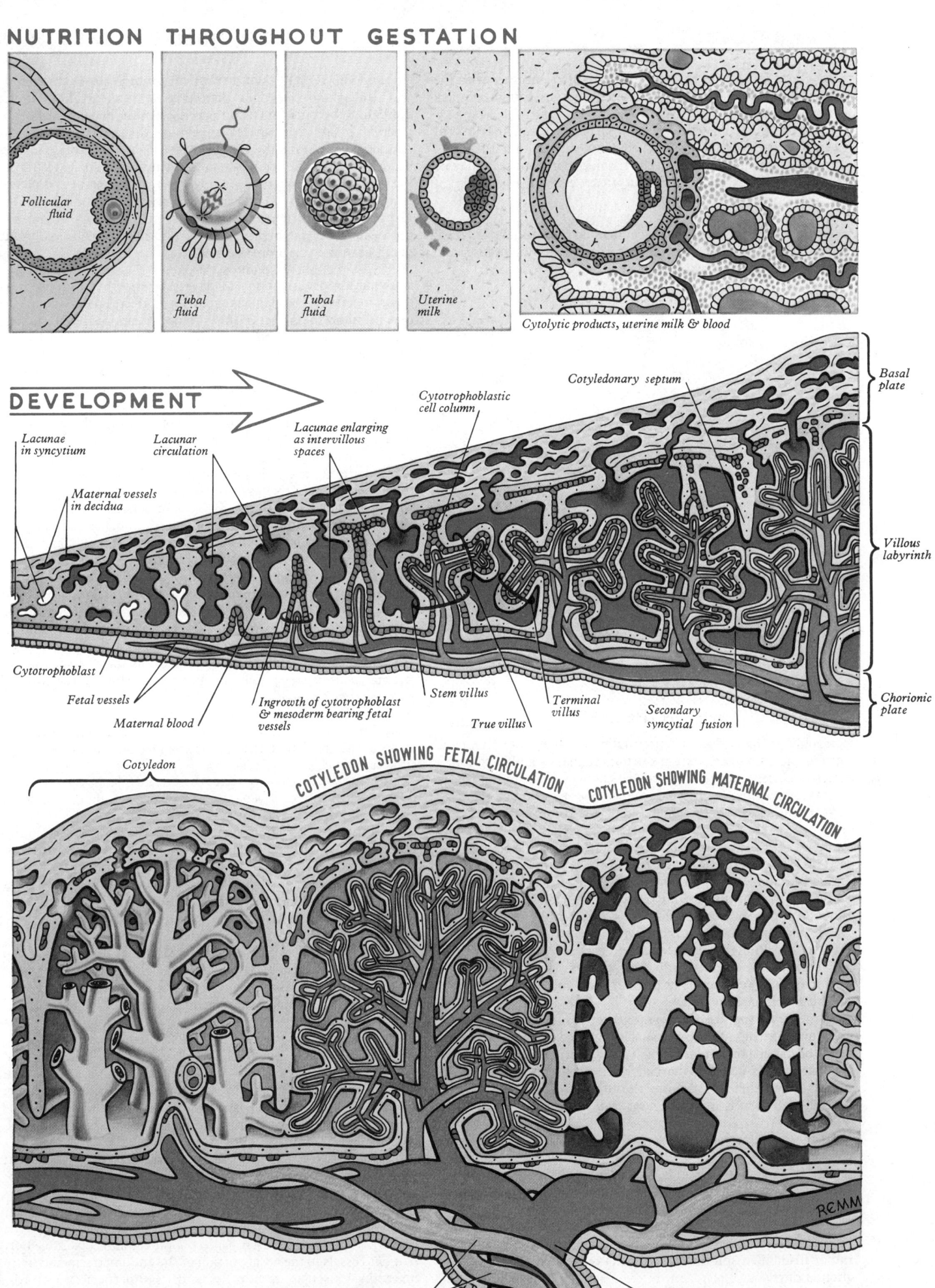

Follicular fluid

Tubal fluid

Tubal fluid

Uterine milk

Cytolytic products, uterine milk & blood

Basal plate

DEVELOPMENT

Cytotrophoblastic cell column

Cotyledonary septum

Lacunae in syncytium

Lacunar circulation

Lacunae enlarging as intervillous spaces

Maternal vessels in decidua

Villous labyrinth

Cytotrophoblast

Fetal vessels

Maternal blood

Ingrowth of cytotrophoblast & mesoderm bearing fetal vessels

Stem villus

True villus

Terminal villus

Secondary syncytial fusion

Chorionic plate

Cotyledon

COTYLEDON SHOWING FETAL CIRCULATION

COTYLEDON SHOWING MATERNAL CIRCULATION

Umbilical arteries

Umbilical vein

2.39 Nutrition of oöcyte, zygote, morula, free and embedded blastocyst, embryo and fetus, throughout gestation. Blastocystic and placental development proceed from left to right. Aspects of mature placental structure and circulation are shown below.

125

intervillous space, and external masses which, together with the adjacent, modified and excavated uterine tissues, form a complex *junctional zone*.

(3) *Crossing it* from chorionic to basal plates, the main trunks of the *villous stems* dividing into their rami, ramuli, and associated complex of free terminal villi; the trunk and its branches may be regarded as the essential *structural, functional* and *growth unit* of the developing placenta.

Further consideration of the placenta must now be deferred until the preparation of the uterine tissues for the implantation and development of the blastocyst has been briefly described.

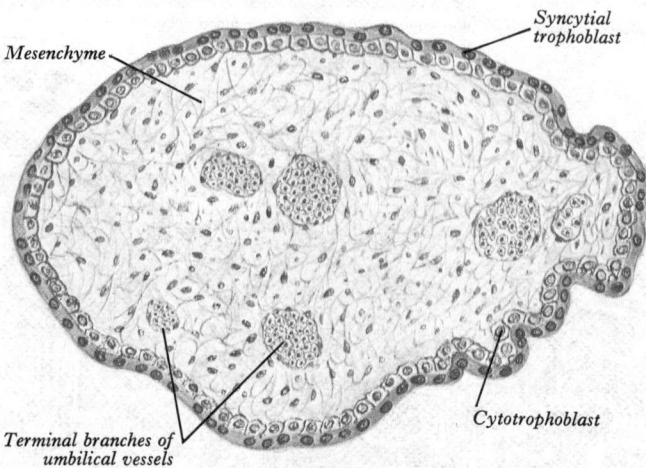

2.40 A transverse section of a terminal villus stained with haematoxylin and eosin.

CYCLICAL CHANGES IN THE UTERUS

Throughout the period of reproductive life (i.e. from about the 15th to the 45th year), except during pregnancy and lactation, a series of closely interrelated cyclical changes occur in the ovary, uterus and vagina. Each cycle extends over a period of about 28 days. In the *ovarian cycle*, which is described more fully elsewhere (pp. 93, 1424), one follicle usually reaches full maturity, ruptures and releases its secondary oöcyte during this period. The wall of the follicle is then transformed into an important endocrine gland, the *corpus luteum* (p. 1424). About ten days after ovulation the corpus luteum begins to regress, then ceases to function and is replaced by fibrous tissue.

The changes of the *uterine cycle (menstrual cycle)* chiefly involve the lining endometrium of the uterus and may, for convenience, be divided into four phases: (a) menstrual, (b) postmenstrual, (c) interval or proliferative, and (d) premenstrual or luteal (but see also below).

In the *menstrual (haemorrhagic) phase* the superficial part of the endometrium, next to the free surface, is shed piecemeal, leaving only the basal portion, adjacent to the uterine muscle (**2.42**A). Outwardly this phase is marked by a discharge of blood with necrotic epithelial débris from the uterus through the vagina. This discharge constitutes the *menstrual flow* and lasts a period of 3–6 days.

In the *postmenstrual (reparative) phase*, and even before the menstrual flow ceases, the epithelium from the persisting basal portions of the uterine glands grows luminally over the denuded surface of the endometrium. The endometrium is then 1–2 mm in thickness and lined by a low cuboidal epithelium. The glands are straight and narrow. The stroma is dense and contains small numbers of lymphocytes amongst its more general population of rather ill-defined spindle-shaped cells (**2.42**B). This phase lasts about 4 days.

During the *proliferative phase* which lasts about 10–12 days, there is a growth of the endometrium associated with the presence in the bloodstream of oestrogenic hormones (particularly *oestradiol*), an internal secretion produced by the ovary (**2.44**A).

The endometrium thickens to about 2–3 mm. Mitoses are present and the glands become distinctly tortuous. Their lining epithelium becomes tall columnar (**2.42**C). In the later part of the interval phase clear spaces appear in the basal parts of the epithelial cells lining the glands. These vacuoles increase in size, occupy more of the cytoplasm, and the nuclei are displaced from the basal regions of the cells towards the lumina of the glands (**2.43**). Secretion is exuded into the lumen, and for a while the glands may be distended with secretion containing mucin and glycogen which can be demonstrated both in the cells and in the gland lumina by appropriate histochemical methods. At the end of the interval phase no mitoses are seen.

Ovulation occurs about 14 days *before the onset* of the next menstrual flow. The changes occurring in the premenstrual phase depend upon the presence in the bloodstream of the hormones, *progesterone* and *oestradiol*, both of which are secreted by the corpus luteum (**2.44**A).

The *premenstrual phase* occurs during the 7 days immediately prior to the next menstrual flow. The mucous membrane becomes much thicker and at the end of the phase may be 7–8 mm deep. In their middle portions, the glands become more voluminous and their walls are folded upon themselves so that tuft-like processes project into the lumen. This gives the glandular wall a saw-toothed appearance in longitudinal section. Secretion is seen in the lumen of the active parts of the gland. The gland epithelium which was tall and columnar at the commencement of the phase becomes frayed and worn down on their luminal aspects, and the nuclei resume their original basal position. The basal parts of the glands adjacent to the uterine muscle take little or no part in these changes. Later in the premenstrual phase characteristic changes appear in the interglandular stroma. These changes are most marked in the superficial part of the endometrium and around the blood vessels. Here the stromal cells, hitherto not clearly defined, become enlarged and swollen and for the first time definite cell outlines can be clearly seen. A few mitoses may be seen in the stroma at this stage. These changes constitute the *stromal premenstrual decidual reaction*. Three strata can now be clearly recognized in the endometrium (**2.42**D):

(1) *Stratum compactum*, next to the free surface in which the necks of the gland are but slightly expanded and the stromal cells show a distinct decidual reaction.

(2) *Stratum spongiosum*, where the uterine glands are tortuous, dilated and ultimately separated from one another by a small amount of interglandular tissue.

(3) A thin *stratum basale*, next to the uterine muscle containing the tips of the uterine glands embedded in an unaltered stroma.

In the last days of the premenstrual phase lymphocytes appear in the endometrium in increasing numbers. They are found between and beneath the surface epithelial cells, amongst the stromal cells and around and between the gland cells. Towards the end of this period as regression of the corpus luteum occurs, those parts of the stroma showing a decidual reaction and the glandular epithelium both undergo degenerative changes and the endometrium often diminishes in thickness. These degenerative changes precede the phase of bleeding.

During *menstruation* blood escapes from the superficial vessels of the endometrium forming small haematomata beneath the surface epithelium which raise it up. Blood and necrotic endometrium then begin to appear in the uterine lumen. The shedding of the endometrium starts at the surface and extends into the deeper layers. The amount of tissue lost is variable, but usually the stratum compactum and most of the spongiosum are desquamated.

The endometrium is regenerated from the stratum basale and that part of the spongy layer which remains, the surface epithelium being reformed with remarkable rapidity.

The *vascular bed* of the endometrium undergoes significant changes during the menstrual cycle. The arteries to the endometrium arise from a myometrial plexus and consist of short *straight* vessels to the basal portion of the endometrium and more markedly muscular *spiral arteries* to its superficial two-thirds. The venous drainage consists of narrow perpendicular vessels which anastomose by cross branches and is common to both the superficial and basal layers of the endometrium. The arterial supply to the basal part of the endometrium remains unchanged during the menstrual cycle. The spiral arteries to the superficial strata, however, lengthen disproportionately, become increasingly coiled and their tips approach more closely to the uterine epithelium during the interval and, particularly, in the premenstrual phase of the menstrual cycle. This leads to a slowing of the circulation in the superficial strata with some vasodilation. Immediately before the menstrual flow these vessels begin to constrict intermittently causing stasis of the blood and anaemia of the superficial strata. During the periods of relaxation of the vessels blood escapes from the devitalized capillaries and veins, thus causing the *menstrual haemorrhage*.

The changes in the endometrium leading to the hypertrophy of

(progestational or *secretory)* phase, which includes the second half of the menstrual cycle, i.e. the premenstrual and menstrual stages. These two phases are of approximately equal duration, although there is a tendency for the pre-ovulatory phase to be more variable than the post-ovulatory phase. The former is largely under the control of oestrogens alone and the epithelial cells in this phase possess large microvilli, whilst their cytoplasm contains much rough endoplasmic reticulum, lipid globules, and lysosomes which become more numerous as the phase progresses. Histochemically the epithelium can be shown to contain large amounts of ribonucleoprotein and alkaline phosphatase, signifying the production and utilization of protein. In the pro-

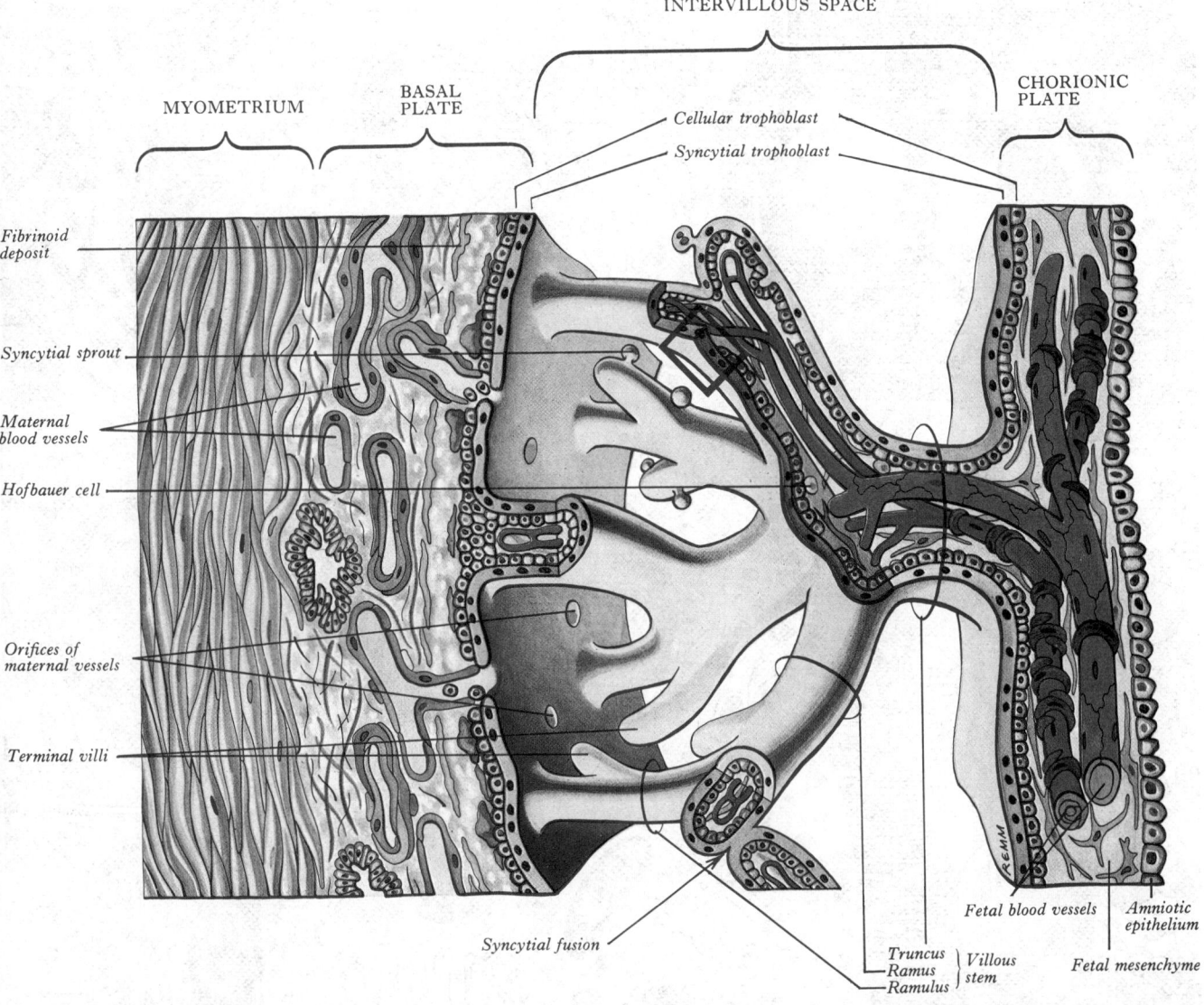

2.41A A schematic diagram to show the arrangement of the placental tissues. Note the chorionic and basal plates, and the intervillous space spanned by a villous stem and its divisions (truncus, rami and ramuli). The sectioned surfaces show the disposition of the fetal and maternal blood vessels, the amniotic epithelium, the cellular and syncytial layers of trophoblast and the complex junctional zone between the fetal and

maternal tissues in the basal plate containing deposits of fibrinoid material and isolated masses of peripheral syncytium. Note also the presence of surface syncytial sprouts, a stromal trophoblastic bud and Hofbauer cells associated with a terminal villus, and syncytial fusion occurring between the tips of two terminal villi. See text for further details. The region enclosed in the rectangle is shown greatly enlarged in **2.41B**.

the premenstrual phase are a preparation for the reception of the blastocyst and result from the action of oestrogen and progesterone (2.44A). If fertilization of the ovum does not occur, the corpus luteum undergoes degenerative changes. The breakdown of the endometrium follows this cessation of function and is apparently due to the loss of the stimulating action of the progesterone and oestradiol (*see* p. 1424).

The uterine cycle may thus, alternatively, be considered as consisting of a *pre-ovulatory (follicular)* phase, which includes the postmenstrual and interval stages, and a *post-ovulatory*

gestational phase, which is under the control of the oestrogens plus progesterone, the above activity declines and the surfaces of the cells become irregular with microvilli and cytoplasmic processes, large vesicles appear in the apical cytoplasm and more numerous lysosomes are present. Acid phosphatases, malic and succinic dehydrogenases, cytochrome oxidase and adenosine triphosphatase are demonstrable and large amounts of glycogen and mucopolysaccharide are present, possibly signifying the utilization of carbohydrate (McKay *et. al.* 1956; Nilsson 1962).

A form of menstruation frequently occurs in the absence of

ovulation, particularly around puberty and also in women approaching the menopause (*anovulatory cycle*). Instead of liberating its ovum, the ripe follicle fails to rupture and undergoes degeneration. This is accompanied by a rapid reduction in oestrogen secretion with consequent breakdown of the uterine mucosa in the interval phase and in the absence of the changes

evoked by the presence of progesterone in the bloodstream (this is termed an *oestrogen-withdrawal* bleeding—Corner 1938).

The Decidua

If fertilization and successful implantation occur, a hormone secreted by the trophoblast, *chorionic gonadotrophin*, prolongs the

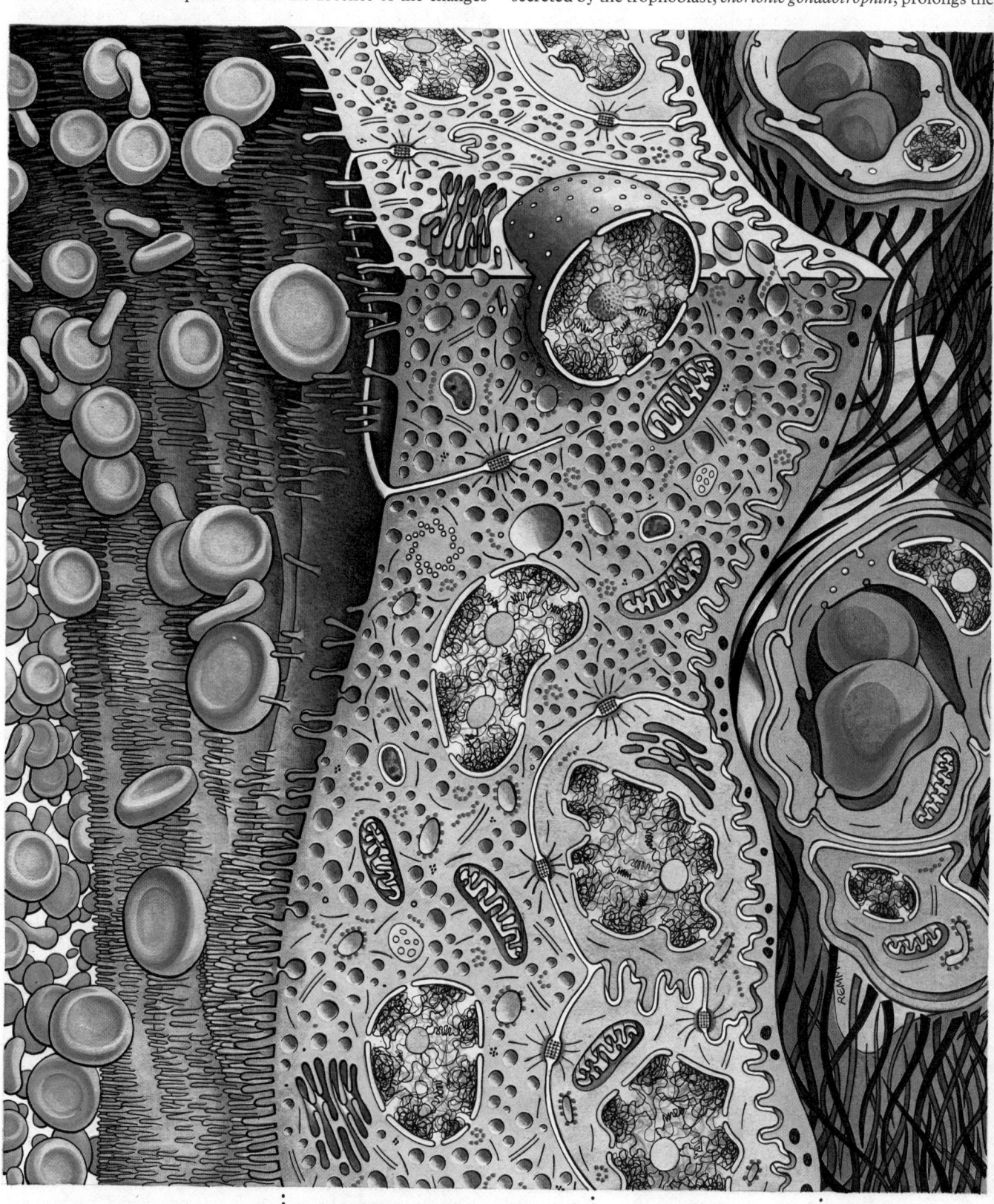

| MATERNAL BLOOD IN INTERVILLOUS SPACE | SYNCYTIAL TROPHOBLAST | CELLULAR TROPHOBLAST AND BASEMENT MEMBRANE | FETAL MESENCHYME AND BLOOD VESSELS |

2.41B A schematic diagram showing the detailed ultrastructural features of the tissues (enclosed in a rectangle in **2.41A**) which intervene between the maternal and fetal blood streams. Note the contrasting architecture of the syncytial and the cellular trophoblasts, and the

substantial basement membrane and delicate fetal mesenchyme which separate the trophoblast from the fetal vessels. For further description see text. (Based on data in Boyd and Hamilton 1970.)

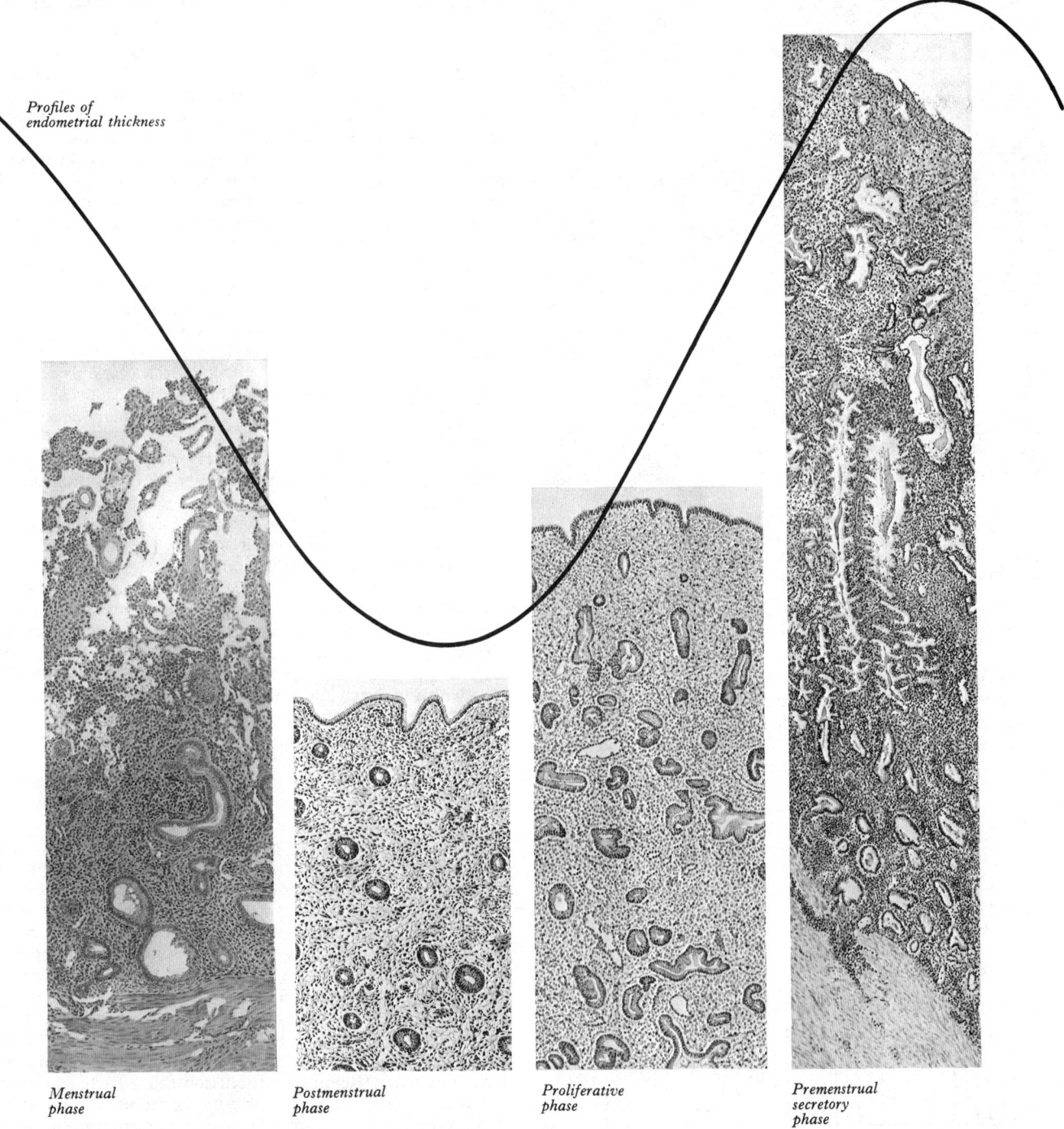

*Profiles of
endometrial thickness*

*Menstrual
phase*

*Postmenstrual
phase*

*Proliferative
phase*

*Premenstrual
secretory
phase*

2.42 A–D Stages in the human menstrual cycle—the sections of the endometrium have all been reproduced at approximately the same magnification. The black line is a rough indication of the changes in endometrial thickness throughout the cycle. (Specimens kindly provided by the Shattock Museum, St. Thomas's Hospital Medical School.)

life of the corpus luteum, which continues to secrete progesterone (*see* pp. 1424–1425 and **2**.44B). Menstruation does not occur and the vascular endometrium, now known as the *decidua of pregnancy*, thickens further to form a suitable nidus for the conceptus. (Chorionic gonadotrophin also appears early in the urine and its presence here is used as a basis for many of the tests for early pregnancy.) The interglandular tissue increases in quantity; it contains a number of leucocytes, and is crowded with large round, oval or polygonal *decidual cells*. These are stromal cells which have accumulated glycogen and lipid in their distended cytoplasm. They continue to be conspicuous in the early stages of gestation but tend to regress in the later months. Their precise significance is uncertain. Possibly they offer a nutritive pabulum which is engulfed by the syncytial trophoblast but, on the other hand, they are often regarded as a defensive mechanism to protect the endometrium from excessive destruction. These changes are

well advanced by the second month of pregnancy, when the three strata recognizable in the premenstrual phase, the stratum compactum, stratum spongiosum and stratum basale, are better differentiated and easily distinguished.

After the blastocyst is embedded, distinctive names are applied to different regions of the decidua. The part which covers the conceptus is the *decidua capsularis*; the part between the conceptus and the uterine muscular wall is named the *decidua basalis*, and it is here that the placenta is subsequently developed; the part which lines the remainder of the body of the uterus is known as the *decidua parietalis* (**2**.46).

Coincidentally with the growth of the embryo and the expansion of the cavity of the amnion (p. 131), the decidua capsularis is thinned and distended (**2**.45, 46) and the space between it and the decidua parietalis is gradually obliterated. By the beginning of the third month of pregnancy the decidua capsularis and decidua

129

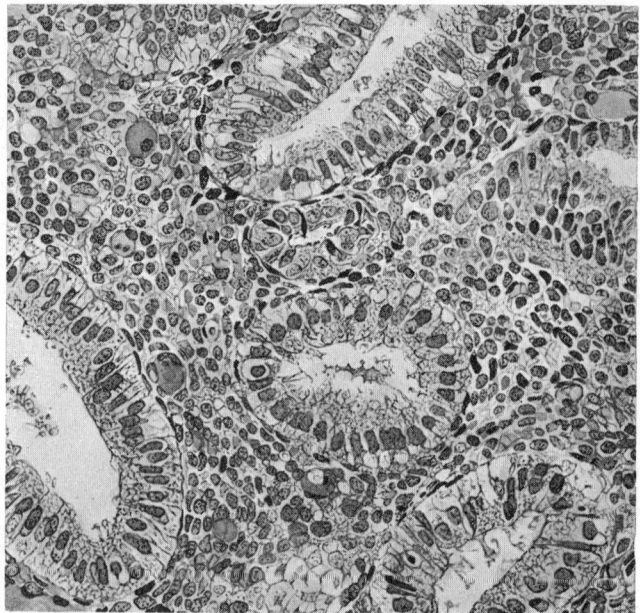

2.43 Section of human endometrium at about the seventeenth day of the menstrual cycle to show the accumulation of secretions in the basal parts of the epithelial cells lining the glands, resulting in displacement of the nuclei towards the lumen of the gland. Magnification × *c.* 300. Stained with haematoxylin and eosin. (Kindly lent by the Shattock Museum, St. Thomas's Hospital Medical School.)

parietalis are in contact; by the fifth month the decidua capsularis is greatly thinned, while during the succeeding months (**2.47**) it virtually disappears, and the decidua parietalis also largely atrophies, owing to the increased pressure. The glands of the stratum compactum are obliterated, and their epithelium is lost; in the stratum spongiosum the glands are compressed, they appear as oblique slit-like fissures, and their epithelium undergoes degeneration; in the limiting or boundary zone, however, the glandular epithelium retains a cuboidal form.

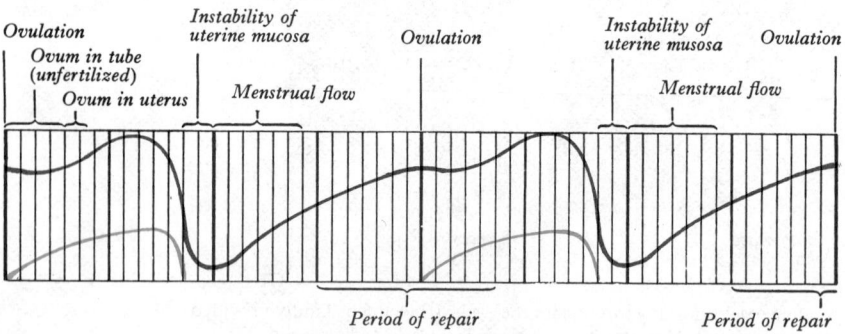

2.44A Graphic representation of the menstrual cycle in the absence of fertilization.

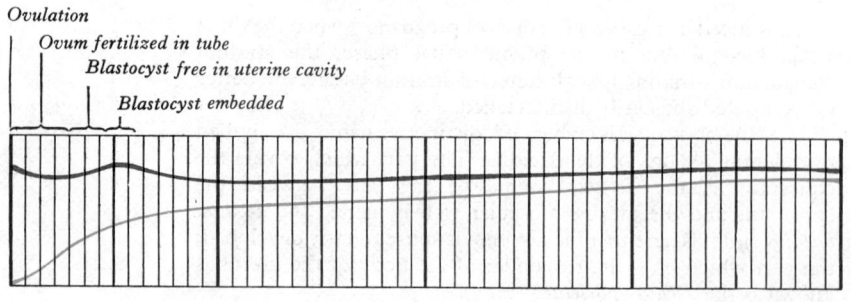

2.44B Graphic representation of the same period following fertilization, cleavage and embedding of the blastocyst. *Note:* Each vertical line represents a day, every seventh day being indicated by a heavy line. The *red* lines show the graphs of the secretion of *oestrogenic* substances, and the *blue* lines show the graphs of *progesterone* secretion.

Implantation, the Chorion and Placenta

As we have seen elsewhere (p. 103), during the first series of cleavage divisions of the fertilized ovum, two distinctive groups of blastomeres emerge; one forms the inner cell mass including the *embryogenic cells,* and the other consists of the more numerous, smaller, rather flattened, polyhedral cells of the primitive *blastocystic trophoblast.* The latter spreads to cover the inner cell mass and consequently *polar* and *mural* regions of this trophoblast may now be distinguished. Comparative studies have shown that such cells in the blastocysts of a number of different mammals possess many features in common. They are characterized by the presence of short, irregular microvilli on their external surfaces, numerous invaginations of the plasma membrane between the microvilli, with adjacent micropinocytotic vesicles in the cytoplasm, and junctional complexes occur between the cells at the outer end of the intercellular clefts (Enders and Shlafke 1965). Their cytoplasm carries many mitochondria, a variably developed Golgi complex, glycogen clusters, lipid inclusions, lysosomes, free ribosome rosettes, but scanty granular endoplasmic reticulum.

From the cells of this primitive layer all·the elements of the chorion are derived, including the varieties of syncytium and also its mesodermal lining.

The method of origin of *syncytial trophoblast* was long debated, many workers holding that it arose by endomitosis of the primitive cells without cell division, but there is now an increasing body of evidence (Boyd *et al.* 1968) that it arises by proliferation of the cellular layer, with subsequent fusion of primarily independent cells on its external surface. The cellular layer itself persists, somewhat modified, as the germinal *cytotrophoblast* which, by multiplicative growth, can add additional cells which fuse with the overlying syncytium, may contribute cells to the underlying extraembryonic mesoderm, and which, by its continued growth, is an important factor in the expansion of the volume of the whole conceptus.

Six or seven days after ovulation the endometrium is well into the development of its secretory phase (p. 126), and with the disappearance of the fibrillary zona pellucida, and the elaboration of polar synctium, the stage is set for the implantation of the blastocyst. From comparative studies (Böving 1959) the mechanisms of implantation have been described as *muscular,* involving blastocyst transport (and spacing in polyovular species), *adhesive,* whereby the blastocyst becomes attached to the uterine mucosa, and *invasive,* the means by which it enters the uterine tissues. Whilst the earliest stages of human blastocyst attachment and penetration have not been available for study, a number of mammals, including primates (Wislocki and Streeter 1938), have been subjected to close scrutiny using the light microscope, and more recently a number of ultrastructural studies have been made (Reinius 1967; Nilsson 1967; Tachi *et al.* 1970).

Based on these accounts, early implantation is initiated by a close approach of the trophoblastic plasma membrane to the tips of the microvilli and irregular surface protrusions of the uterine epithelial cells. At these points the trophoblast surface often presents patches of 'bristle-coated' membrane, with adjacent cytoplasmic plaques having a crystalline substructure, with numerous neighbouring lysosomes. The microvilli shorten and disappear, and for a period there is a close mutual adaptation between the contours of the trophoblast and the uterine epithelial cell surfaces. The syncytium now send finger-like projections between adjacent epithelial cells towards the underlying basement membrane, the two layers becoming closely interlocked by the formation of numerous tight junctions between them. The epithelial cells now degenerate and are engulfed by the trophoblast, resultant cellular débris becoming enclosed in phagosomes within the trophoblastic cytoplasm. The subepithelial basement membrane is soon penetrated, with the development of oedema, hyperaemia and cell degeneration in the neighbouring stroma.

After interstitial implantation of the human blastocyst, the process of invasion continues with erosion, degeneration and phagocytosis of maternal vascular endothelium, glandular epithelium with secretory products, and decidual cells, until the

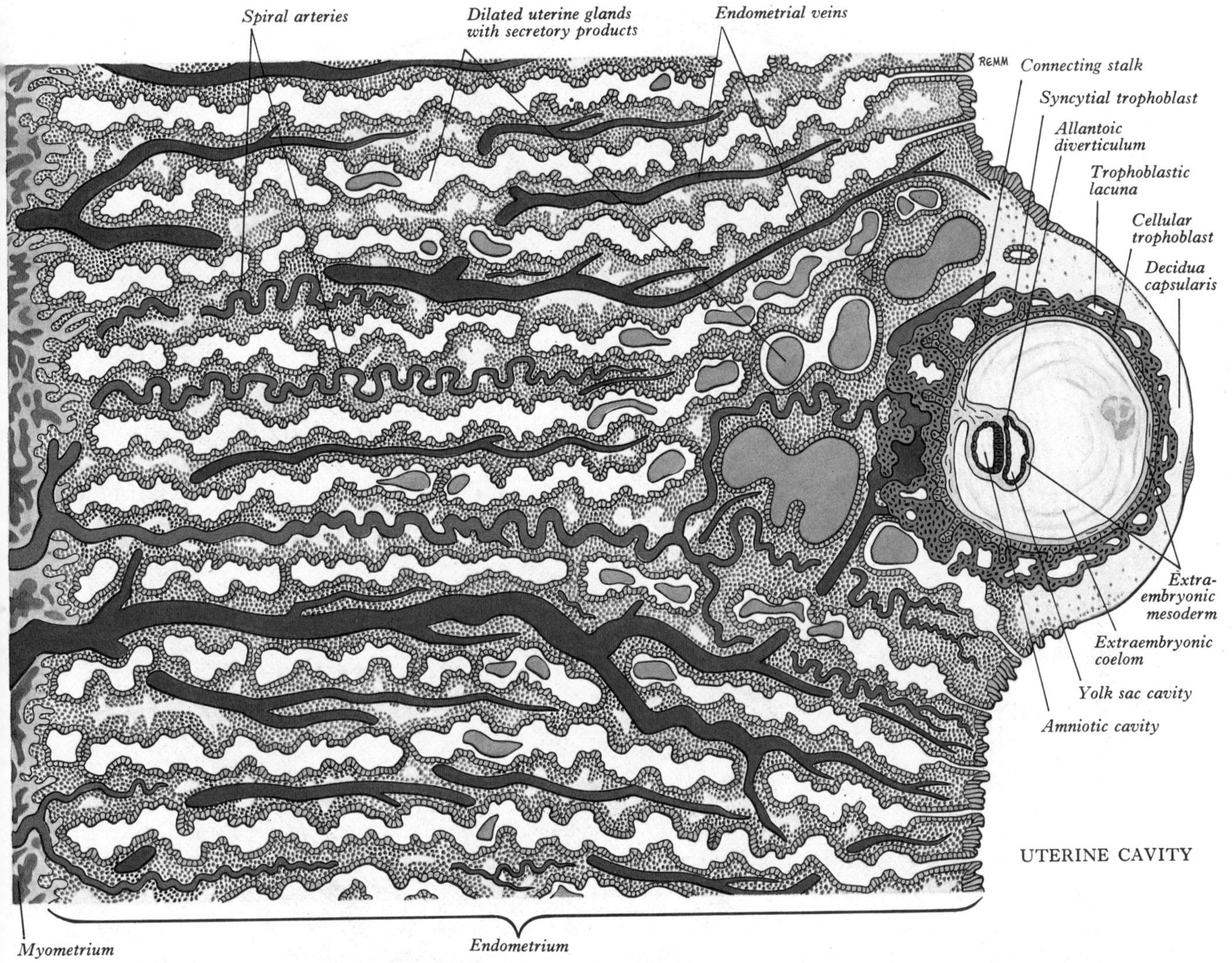

Spiral arteries *Dilated uterine glands with secretory products* *Endometrial veins*

REMM *Connecting stalk*

Syncytial trophoblast

Allantoic diverticulum

Trophoblastic lacuna

Cellular trophoblast

Decidua capsularis

Extra-embryonic mesoderm

Extraembryonic coelom

Yolk sac cavity

Amniotic cavity

UTERINE CAVITY

Myometrium   *Endometrium*

2.45 A diagram showing the general structure of the implanting blastocyst and its relationship to the tissues of the endometrium on the 15th day after fertilization. Note the arrangement and gradation in thickness of the syncytial trophoblast which has eroded the maternal tissues. Some of the deeper trophoblastic lacunae already contain maternal blood. Note also the active dilated uterine glands, the endometrial venous plexus, spiral arteries, and the stage of development of the cellular trophoblast, amnion, yolk sac, allantoic diverticulum, connecting stalk and the extraembryonic mesodermal layers lining the extraembryonic coelom. For further details see text.

blastocyst sinks into and occupies a ragged *implantation cavity* in the decidua.

The subsequent stages of development, which include the definition of the layers of the chorion (p. 124), the elaboration of syncytial trophoblast which is thick at the embryonic pole and thins towards the abembryonic pole and carries a lacunar circulation, and the principal stages in the development of the intervillous space, the villous stems and their branches, and the trophoblastic shell, have already been described (p. 124).

From the third week until about the second month of pregnancy the *entire chorion* is covered with villous stems which are thus continuous peripherally with the trophoblastic shell in close apposition with *both* the decidua capsularis and the decidua basalis. The latter, however, are stouter, longer and show a greater profusion of terminal villi. As the conceptus continues to expand, the decidua capsularis is progressively compressed and thinned, the circulation through it is gradually reduced and, accordingly, its related villous stems and villi slowly atrophy and disappear. This process starts at the abembryonic pole and by the end of the third month the abembryonic hemisphere of the conceptus is largely denuded. This continues until the whole chorion which is related to the capsularis is smooth (the *chorion laeve*); in contrast, the villous stems of the disc-shaped region of chorion related to the *decidua basalis* increase greatly in size and complexity (the *chorion frondosum*), and together the latter constitute the *definitive placental site*.

An exhaustive account of the growth, dimensional changes,

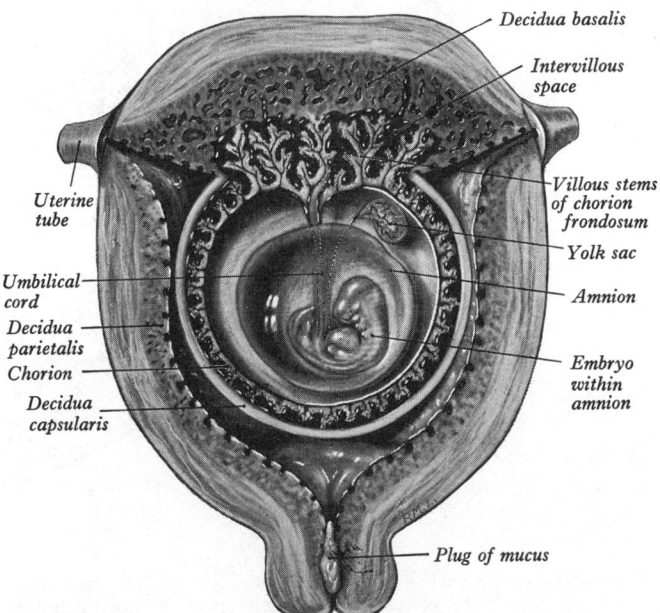

Decidua basalis

Intervillous space

Uterine tube

Villous stems of chorion frondosum

Yolk sac

Amnion

Umbilical cord

Decidua parietalis

Chorion

Decidua capsularis

Embryo within amnion

Plug of mucus

2.46 A plan of the gravid uterus in the second month. A placental site precisely in the uterine fundus as indicated in the plan is, however, rather unusual. (The dorsal, ventral or lateral wall of the corpus uteri is more usual.)

131

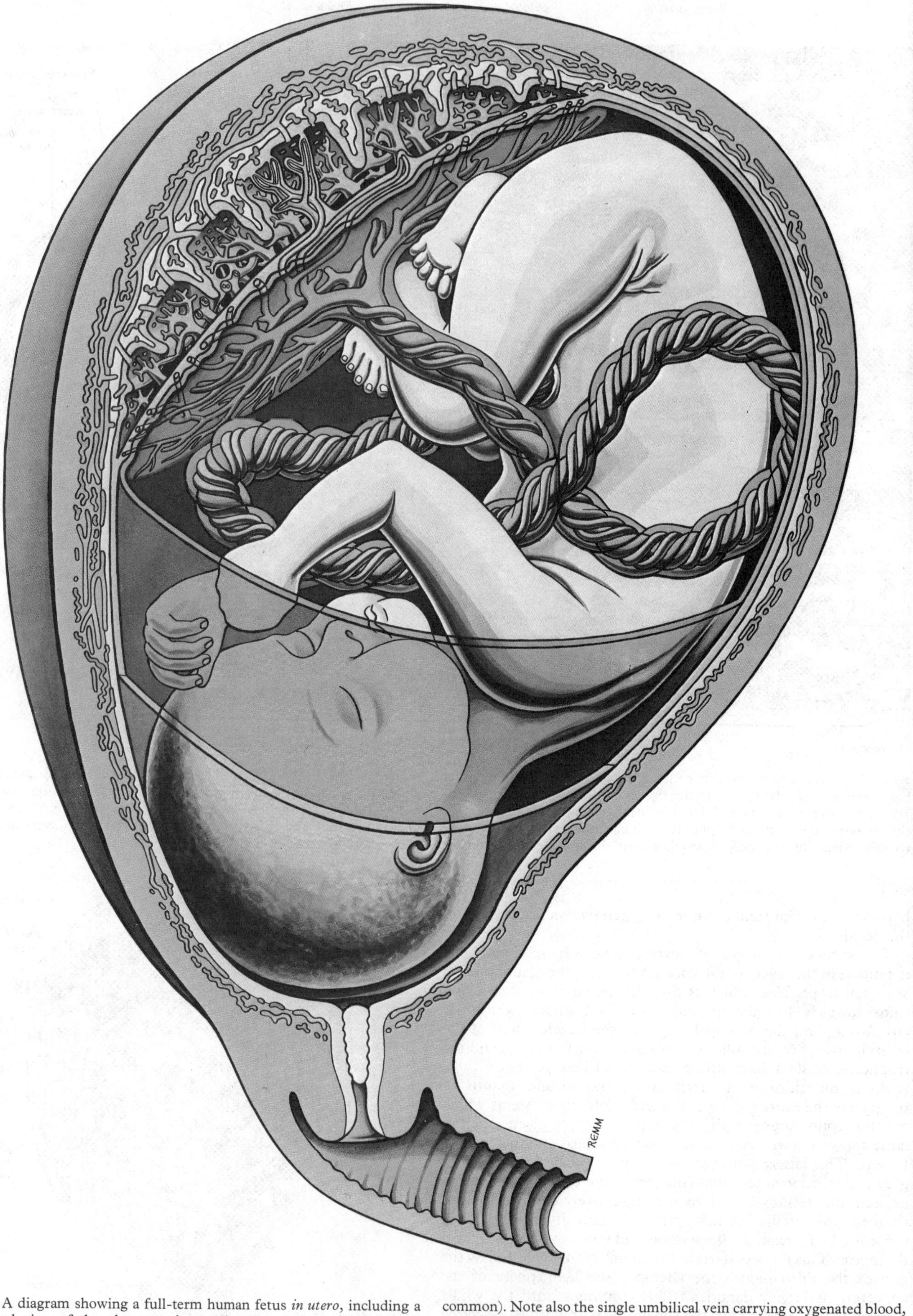

2.47 A diagram showing a full-term human fetus *in utero*, including a sectional view of the placenta, the amnion (mauve), chorion (green), uterine wall (orange), the umbilical cord and its contained vessels, the cervix with a plug of mucus in the cervical canal, and the rugose vaginal wall. Note the characteristic flexed posture of the fetus and its limbs, and the overall position within the uterus which the fetus commonly occupies (other positions, although less frequent, are, however, also quite common). Note also the single umbilical vein carrying oxygenated blood, the two umbilical arteries carrying deoxygenated blood, the arborization of these vessels in the chorionic plate (seen through the overlying amnion) and their branches which pass into the villous stems. The latter span the intervillous space and their trunci, rami, ramuli and terminal villi may be seen; incomplete placental septa project from the basal plate towards the chorionic plate. See text for further details.

vasculature and haemodynamics, cell varieties, ultrastructure and histochemistry of the placenta, and the physiological aspects of placental transfer and its status as a metabolic store and endocrine gland, lies beyond the scope of the present volume. What follows is necessarily an abbreviated account of selected topics, and the interested reader should consult the profusion of original papers devoted to these subjects (the classical volume by Professors J. D. Boyd and W. J. Hamilton (1970) provides an unrivalled source of information and an extensive bibliography). A more recent excellent survey of placental transfer mechanisms is to be found in Longo (1972).

Definition of the Human Placenta

The human placenta (Wislocki 1937; Hill 1932; Grosser 1936; Mossman 1937; Amoroso 1960) is defined as *discoidal* (in contrast to other shapes, e.g. *zonary, bidiscoidal, diffuse*, etc. seen in other forms). It is initially *labyrinthine* as the early villous stems are formed, but becomes secondarily *villous* with the development of generations of terminal villi, and finally, there is a partial return to a *labyrinthine* state as superficial adhesion or partial fusion occurs between the tips of a number of villi (*vide supra*). Maternal blood bathes the surfaces of the chorion which bound the intervillous space, and it is thus defined as *haemochorial*, distinguishing it from the different grades of fusion between the maternal and fetal tissues which exist in many other forms (*epitheliochorial, syndesmochorial, endotheliochorial*, and even an approach to a *haemoendothelial* condition but with the vascular endothelial basement membrane and a fine layer of perivascular connective tissue persisting). The chorion is vascularized by the allantoic blood vessels of the body stalk and the human placenta is termed *chorioallantoic* (whereas in some forms a *choriovitelline* placenta either exists alone or supplements the chorioallantoic variety). Finally, the human placenta is said to be *deciduate* because maternal tissue is shed with the placenta and membranes at term as part of the afterbirth (*vide infra*).

THE PLACENTA AT TERM

The expelled placenta (**2**.48), is a flattened discoidal mass with an approximately circular or oval outline, with an average volume of some 500 ml (range 200–950 ml), average weight about 500 gm (range 200–800 gm), average diameter 185 mm (range 150–200 mm), average thickness 23 mm (range 10–40 mm), and an average surface area of about 30,000 mm². It is, of course, thickest at its centre (the original embryonic pole) and rapidly diminishes in thickness towards the periphery where it is continuous with the chorion laeve.

Macroscopically, its *fetal* or *inner surface*, which is covered by the amnion, is smooth, shiny and transparent, and the mottled appearance of the subjacent chorion, to which it is closely applied, can be seen through it. The umbilical cord is usually attached near the centre of the fetal surface and the branches of the umbilical vessels radiate out under the amnion from this point, the veins being deeper and larger than the arteries. Beneath the amnion and close to the attachment of the cord, the remains of the yolk sac can sometimes be identified as a minute vesicle with a fine thread—a vestige of the yolk stalk—attached to it.

The *maternal surface* is finely granular and mapped into some fifteen to thirty lobes by a series of fissures or grooves. The lobes are often, somewhat loosely, termed *cotyledons* (but see also below), and the grooves themselves correspond to the bases of incomplete *placental septa* which become increasingly prominent from the third month onwards, and extend from the maternal aspect of the intervillous space (the basal plate) towards, but do not quite reach, the chorionic plate. The septa form initially as ingrowths of the cytotrophoblastic shell, with a covering of syncytium, but from the fourth month they develop a complex core derived from the maternal tissues which includes reticulin fibres and associated cells, decidual cells, remnants of glandular epithelium, occasional blood vessels, and deep to the cytotrophoblast, a stratum of fibrinoid material. Later cytotrophoblastic cells often penetrate into this stratum at various points with a resultant admixture of the fetal and maternal tissues, and in the

later months of pregnancy, patches of central degeneration occur in many septa, which, incidentally, increases the difficulty of histological interpretation.

The nature of the maternal surface of the expelled placenta is of course determined by the tissue plane of separation of the placenta at parturition.

Recent studies of the human placenta include morphometric analysis (Laga *et al.* 1973), surface architecture using scanning electron microscopy (Fox and Agrafojo-Blanco 1974), ultrastructural studies of angioarchitecture (Sheppard and Bonnar 1974), and the possible mechanisms whereby the maternal placental circulation is controlled (Bruce and Abdul-Karim 1974). The ultrastructure of biopsies taken from placental uterine beds, which were presumed to be normal, has been reviewed by Robertson and Warner (1974).

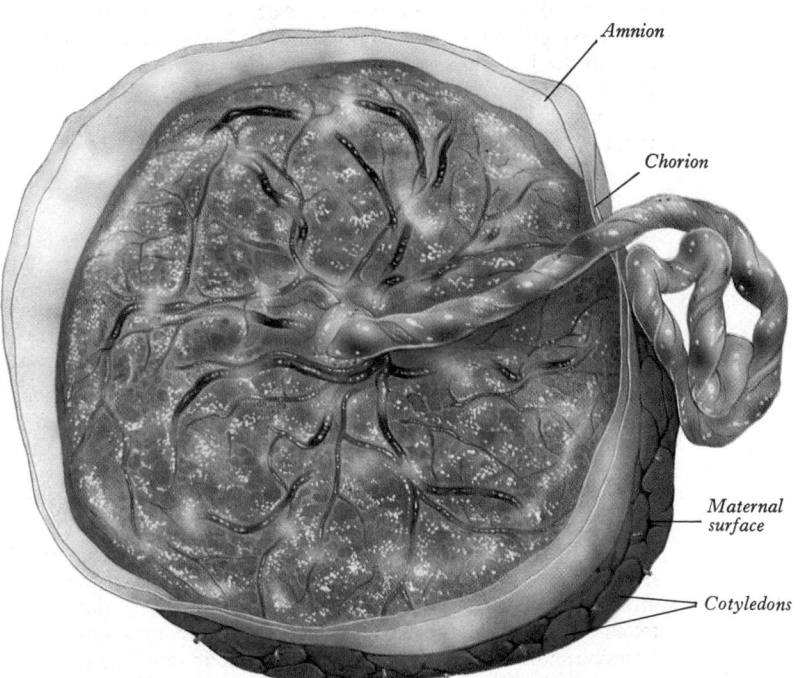

Amnion

Chorion

Maternal surface

Cotyledons

2.48 The fetal surface of a recently delivered placenta, drawn from a coloured photograph kindly provided by Mr. E. F. Gibberd. The maternal surface is exposed in the lower and right corner of the figure. Note the small branches of the uterine artery and the series of grooves. Note also the amnion and chorion which have been cut away near the placental margin.

Separation of the Placenta

After delivery of the fetus the placenta becomes separated from the uterine wall, and together with the so-called 'membranes' is expelled as the *afterbirth*. Separation takes place along the plane of the stratum spongiosum and extends beyond the placental area, detaching (1) almost the whole remaining thickness of the largely degenerate, confluent, decidua parietalis and decidua capsularis, (2) the chorion laeve, and (3) the amnion. These three layers are partially fused together and are continuous with the placenta at its margin; they constitute the *membranes* familiar in obstetrics. The process of separation requires rupture of many uterine vessels but their torn ends are closed by the firm contraction of the muscular wall of the uterus after delivery of the placenta and membranes, and thus, under normal circumstances, postpartum haemorrhage is limited in amount. When the placenta and membranes have been expelled, a thin layer of stratum spongiosum is left as a lining for the uterus, but it soon undergoes degeneration and is cast off in the early part of the puerperium. A new epithelial lining for the uterus is then regenerated from the remaining stratum basale.

Placental Lobes and Lobules

The placental *lobes* are demarcated by the grooves on its maternal surface, and they correspond in large measure to the major branches of distribution of the umbilical vessels, particularly well seen in specimens X-rayed after intravascular injection of radio-opaque media. However, the application of the term cotyledon to these major lobes does not correspond directly to its usage in comparative placentology.

Particularly in those forms with a diffuse kind of placentation, the whole chorion does not bear villous structures evenly, for these are restricted to a large number of scattered discontinuous patches (*fetal cotyledons*) with non-villous chorion between them. These fetal cotyledons come into apposition with similarly distributed discrete areas of endometrium (the *maternal cotyledons* or *uterine caruncles*); the combination of one fetal unit with its associated maternal unit is termed a *placentome*.

The fetal cotyledon of the human placenta evidently corresponds to a major villous stem and its branches; early in pregnancy, the chorion bears some 800–1,000 of such stems, but as pregnancy advances, with the formation of the chorion laeve and possibly some fusion between adjacent stems, the number is progressively reduced until only about 60 persist in the placental area in the last months of pregnancy. However, this number is distributed between the 15–30 lobes, i.e. each lobe (or 'cotyledon' of obstetrics) contains 2–4 major villous stems and is to be regarded as compounded of this number of *placentomes* or *lobules*.

THE PLACENTAL TISSUES

These are arranged as a chorionic plate, a basal plate, and between the two, the villous stems, their branches, and the intervillous space (**2.**41, 47).

The chorionic plate is covered on its fetal aspect by the amniotic epithelium, the cells of which have been briefly described elsewhere (p. 122), followed by a connective tissue layer carrying the main branches of the umbilical vessels, then a diminishing layer of cytotrophoblast, and finally, the inner syncytial wall of the intervillous space. The connective tissue layer is derived from fusion between the mesoderm-covered surfaces of the amnion and chorion, and is more fibrous and less cellular than the Wharton's jelly of the umbilical cord, except near the larger vessels. The latter radiate and branch from the cord attachment (with variations in the branching pattern), until they reach the bases of the trunks of the villous stems, which the branches enter and then arborize within their rami, ramuli and terminal villi. There is no cross anastomosis between the vascular trees of adjacent stems, but in contrast, the two umbilical arteries are normally joined by some form of substantial transverse (Hyrtl's) anastomosis at, or just before they enter, the chorionic plate.

The basal plate consists, from the fetal to the maternal aspect, of: (1) the syncytium (and perhaps in patches, maternal tissue) which forms the outer wall of the intervillous space, (2) Rohr's stria of fibrinoid, (3) what remains of the cytotrophoblastic shell, (4) Nitabuch's stria of fibrinoid, (5) the remains of the decidua.

The striae of fibrinoid are irregularly connected together, and strands pass from Nitabuch's stria into the adjacent decidua. The latter contains basal remnants of the endometrial glands, large and small decidual cells scattered in a connective tissue framework which also supports an extensive venous plexus. These deeper layers of the basal plate also contain a variety of *giant cells* which may possess either multilobate nuclei or are in fact multinucleate. Their origin has been disputed, but the majority probably have a fetal origin, either arising by the isolation of masses of peripheral syncytium, or by growth from the cells of the cytotrophoblastic shell. For some, however, a maternal origin cannot be excluded. Their role is uncertain but they may be one source of placental hormones. In this regard, the possession by the cytotrophoblastic cells in this region, of prominent nucleoli, a particularly basophilic cytoplasm rich in RNA, and varieties of cytoplasmic inclusions (Dallenbach-Hellweg and Nette 1964), both acidophil and basophil, may also be related to an endocrine secretory role.

Throughout the second half of pregnancy the basal plate is thinned and becomes progressively modified, with a relative diminution of the decidual elements, and with an increasing deposition of fibrinoid and admixture of fetal and maternal derivatives. It is, as we have seen, through the deeper decidual

layers of the plate that separation of the placenta occurs at parturition, and through its various layers the maternal blood vessels approach and reach the intervillous space. In the earlier stages of development the spiral arteries of the endometrium do not open directly into the intervillous space until the fifth or sixth week, after which they pass through the plate and open through gaps in the cytotrophoblastic shell and peripheral syncytium. The terminal parts of the vessels are dilated, lose their muscular walls by degeneration, and present a hypertrophy of their lining endothelium. Proliferations of cytotrophoblast often grow back for some distance along their lumina (Hamilton and Boyd 1960).

The veins which drain the blood away from the intervillous space pierce the basal plate and join tributaries of the uterine veins. The presence of a marginal venous sinus, which has hitherto been described as a constant feature, occupying the peripheral margin of the placenta and communicating freely with the intervillous space, has not been confirmed.

In the macaque monkey, radio-opaque material injected into the aorta passes in spurts or jets to the intervillous space and at sufficient pressure to drive it towards the chorion, thus preventing a short circuit of arterial blood into the venous openings. The openings of the coiled arteries show intermittent activity, probably due to alternating constriction and relaxation of the arteries themselves. Myometrial contractions alter the pressure in the intervillous space and promote placental venous drainage (Ramsey *et. al.* 1963; Martin 1965).

The Structure of a Villus

The terminal villi are the essential structures involved in exchanges between the mother and fetus, and, accordingly, the tissues which separate the fetal and maternal blood in this site are of considerable functional importance, and numerous investigations have been carried out on them.

Each villus is composed of a core of connective tissue bearing fetal capillaries and separated by a basement membrane from the ensheathing cyto- and syncytial trophoblast which is bathed by the maternal blood of the intervillous space (**2.**40, 41, 49). Cohesion between the cells of the cytotrophoblast and also between this layer and the syncytium is provided by numerous desmosomes between their apposed plasma membranes.

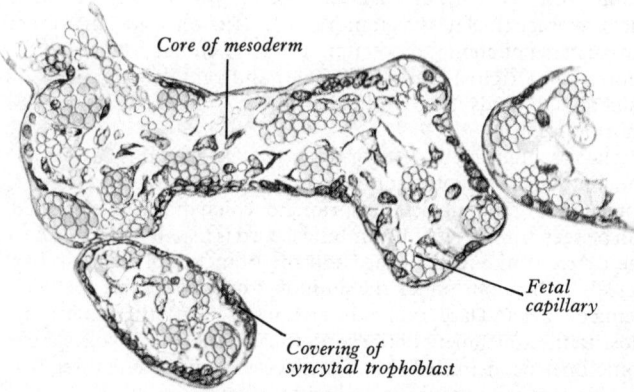

Core of mesoderm

Fetal capillary

Covering of syncytial trophoblast

2.49 Part of a section of a branching terminal villus in a mature placenta (ninth month). Note the close approach of many of the fetal capillaries to the subepithelial basement membrane; see text for details.

In the early stages, the cytotrophoblast forms an almost continuous layer on the basement membrane, but after the fourth month it gradually expends itself producing syncytium (Midgley *et al.* 1963). As it becomes reduced the syncytium is related to the basement membrane over an increasingly large area, and at the same time itself becomes progressively thinner, but cytotrophoblastic cells, usually disposed singly, can, however, be identified until term.

The cells of the villous cytotrophoblast (*Langhans' cells*) are pale-staining with only a slight basophilia. Ultrastructurally, they show a rather electron-translucent cytoplasm, bearing relatively few free polysomes, little granular endoplasmic reticulum, but a number of large mitochondria, a fairly extensive Golgi apparatus, microfilaments particularly in association with the desmosomes, and occasional vacuoles and dense inclusions. Between the desmosomes, the cell membranes of adjacent cells may be smooth and straight, or they may undulate, and are separated by a featureless gap of some 20 nm. Sometimes the intercellular gap widens to accommodate microvillous projections from the cell surfaces; in addition the gap occasionally contains patches of fibrinoid.

In contrast, the syncytial cytoplasm is more strongly basophilic and possesses many ultrastructural features which distinguish it from the cytoplasm of the Langhans' cells. Where the plasma membrane adjoins basement membrane it is often complexly infolded into the cytoplasm, whereas the surface bordering the intervillous space is beset with numerous long microvilli, the cores of which show linear densities. These microvilli are responsible for the appearance of a brush border seen with the light microscope.

The syncytial cytoplasm is exceedingly complex and more electron-dense than that of Langhans' cells. It contains a wealth of free ribosomes, cisternae of granular endoplasmic reticulum, scattered representations of the Golgi complex, a cytoskeleton of microfilaments, and a profusion of vesicles and vacuoles, some smooth and some coated, of a wide size range, numerous lysosomes, phagosomes and other electron-dense inclusions. It is evidently not, as was earlier claimed, a simple homogeneous layer functioning as a semipermeable membrane, but an intensely active tissue layer across which all transplacental transport must occur.

Glycogen is held to be present in both layers of the trophoblast at all stages but it is not always possible to demonstrate it by histochemical means. Lipid droplets are also present in both layers and free in the core of the villus. In the trophoblast they are found principally within the cytoplasm but also occur extracellularly between cytotrophoblast and syncytium, or between the individual cells of the cytotrophoblast, and also in the basement membrane. These droplets may represent fat in transit from mother to fetus, their numbers diminishing with advancing age. Membrane-bound granular bodies of moderate electron-density also occur in the cytoplasm, particularly in the syncytium; some of these are probably secretion granules of hormones or hormone precursors and are believed to arise in the Golgi apparatus. The lysosomes and phagosomes are evidently concerned in the degradation of materials engulfed from the intervillous space.

In the core of the villus there are fibroblasts, large phagocytic cells termed *Hofbauer cells* which are more numerous in early pregnancy, many plasma cells and free lipid granules. Electron microscopy shows the presence of fine collagen fibres which are thicker and collected into bundles in the villous stems. The fetal vessels include arterioles and capillaries. Their endothelial cells contain fine filaments in the cytoplasm, and they may present bulbous projections into the lumen (Rhodin and Terzakis 1962; Terzakis 1963; Iklé 1961, 1964). They are surrounded externally by the usual periendothelial basement membrane (p. 627).

The characteristics of both syncytium and cytotrophoblast, however, vary considerably from the brief account just given in sites other than a villus and at different stages of development, and for such details original sources should be consulted. There are, however, certain specialized regions of the villous syncytium which should be mentioned.

Many villi show localized ingrowths of syncytium (with some associated cytotrophoblast) into their mesodermal cores; these are termed *stromal trophoblastic buds*. They may remain attached by stems to the overlying trophoblast, or become detached to lie free, and they may even approach and project into the lumina of the fetal blood vessels. In the latter situation they may be a source of the syncytial masses which have been described in the blood of the umbilical vein. The frequency and fate of such fetal cellular emboli is uncertain. On the free surface of the villus two other types of specialization occur. The first, termed syncytial *knots* or *clumps*, are localized thickenings in which there is a close aggregation of a number of nuclei. Their significance has until recently remained obscure but in some cases appears to be related

to the maintenance of a neighbouring delicate, non-nucleated, *epithelial plate* of syncytium which comes into particularly close apposition with a fetal capillary. A recent ultrastructural study of syncytial knots (Jones and Fox 1977) has shown that their nuclear aggregates exhibit marked degenerative changes, and it is suggested that they represent a sequestration phenomenon involving removal of senescent nuclear material from adjacent metabolically active areas of syncytium. Fusion of knots from adjacent villi appears to be the basis of *syncytial bridge* formation, and for the latter a mechanical role as an internal placental strut system has been advanced. The second specialization is termed a *syncytial sprout* which continues to be formed throughout pregnancy. Some sprouts are unquestionably initial stages in the development of a new branch villus. Many others, however, project into the maternal bloodstream and become detached, forming syncytial emboli which pass to the maternal lungs. It has been computed that there is a passage of some 100,000 of such sprouts daily into the maternal circulation. In the lungs they provoke little local reaction and apparently disappear by lysis, but they may, on occasion, form a locus for neoplastic growth.

MATURATION AND FUNCTIONS OF THE PLACENTA

In the early stages of placental development the blood in the fetal vessels is separated from the maternal blood in the intervillous space by: the fetal vascular endothelial cells and their basement membranes, the relatively abundant connective tissue of the villus, the subepithelial basement membrane, and its covering of cyto- and syncytial trophoblast. These layers constitute a *placental barrier* which is interposed between the two bloodstreams, but it is a permeable barrier and allows water, oxygen and other nutritive substances and hormones, to pass from the mother to the fetus, and some of the products of excretion to pass from the fetus to the mother.

During the first half of pregnancy the placenta not only increases its surface area, but reaches its maximum thickness. The growth in depth results from increases in the size and length of the villous stems, and is not accompanied by any further invasion of the uterine wall. In the latter half of pregnancy the placenta further increases its surface area, doubling its diameter, but it does not increase in thickness (Stieve 1948; Hamilton and Boyd 1951). (Placental weight has also been correlated with parity—Chakraborty et al. 1975.) It therefore tends to fail to keep pace with the functional demands of the growing fetus, but, in compensation, the placental barrier becomes reduced in thickness. After the fourth month the trophoblastic covering of the villi becomes reduced; as it ages the syncytium becomes directly related to the subepithelial basement membrane over an increasing area, and it also becomes thinner. The fetal capillaries approach the surface of the villus but are always separated from the basement membrane of the trophoblast by some delicate connective tissue.

The mechanism of transfer of substances across the placental barrier is complex. The surface area of the villi has been variously estimated; it may, however, be as large as 14 square metres (Wilkin and Bursztein 1958). The volume of maternal blood circulating through the intervillous space has been assessed at 500 ml per minute (Assali et al. 1960). Simple diffusion suffices to explain gaseous exchange and a transfer of many dissolved substances of low molecular weight (e.g. sodium, potassium, chloride, iodide, phosphate). In later pregnancy, water is interchanged between fetus and mother (in both directions) at about 3·5 litres per hour. The transfer of substances of high molecular weight such as proteins, lipids and antibodies is not so readily understood. Energy-dependent selective transport mechanisms, including micropinocytosis, are probably involved.

Lipids may be transported unchanged through and between the cells of the trophoblast to the core of the villus. The passage of maternal antibodies across the placental barrier confers some degree of passive immunity on the fetus. The whole problem of investigations into transplacental mechanisms is complicated by the fact that the trophoblast itself is the site of synthesis and storage of certain substances, e.g. glycogen. For a comprehensive review of placental transfer mechanisms consult Boime and Boguslawski (1974).

The placenta is an important endocrine organ; the steroid hormones, various oestrogens and progesterone, and the proteinaceous placental lactogens, are synthesized by the syncytium, whilst chorionic gonadotrophin is probably secreted by both the syncytium and the cytotrophoblastic cells. It is of interest that steroid hormones are formed in isolated perfused placentae, and also by placental explants in tissue culture. The trophoblast is rich in birefringent lipids, and cytochemical methods show that it also contains enzyme systems which are associated with the synthesis of steroid hormones.

Leucocytes are more numerous in the blood of the umbilical vein than in that of the umbilical artery, suggesting that they may migrate from the maternal blood, through the placental barrier, into the fetal capillaries. It has also been shown that some fetal and maternal red blood cells may cross the barrier (Dancis 1959). The former may have important consequences, e.g. in Rhesus incompatibility (p. 68). The vitamins pass the placental barrier with different degrees of facility; vitamins B, C and D pass readily, whilst the remainder experience greater difficulty.

The majority of drugs pass the barrier and many are apparently tolerated by the fetus, but some may exert grave teratogenic effects on the developing embryo (e.g. thalidomide, *vide infra*).

Finally, a wide variety of bacteria, spirochaetes, protozoa and viruses are known to pass the placental barrier from mother to fetus, although the mechanism of transfer is uncertain. The presence of maternal rubella in the early months of pregnancy is of especial importance in relation to the production of congenital anomalies (*vide infra*).

Placental Variations

As a rule the placenta is attached to the posterior wall of the uterus near the fundus, with its centre in or near the median plane. The site of attachment is determined by the point where the blastocyst becomes embedded but the factors on which this depends are not yet fully understood. The placenta, however, may be attached at any point on the uterine wall, but these variations offer no complications to a normal labour unless it is attached so low down that it overlies the internal os uteri, when it may give rise to serious antepartum haemorrhage, especially if it is nearly central in position. This condition, which occurs in about 0·5 per cent of pregnancies, is known as *placenta praevia*. (*Extrauterine* sites of implantation are discussed on p. 110).

The umbilical cord, although usually attached near the centre of the organ, may reach it at any point between its centre and margin. In the latter event it is known as a *battledore* placenta. Occasionally the cord fails to reach the placenta itself and ends in the membranes in its vicinity. This is termed a *velamentous insertion* of the cord, and the larger branches of the umbilical vessels traverse the membranes before they reach and ramify on the surface of the placenta. A small accessory or *succenturiate* placental lobe is occasionally present connected to the main organ by membranes and blood vessels; it may be retained in the uterine cavity after delivery of the main placental mass and prolong postpartum haemorrhage. Occasionally other types of division of the placenta occur (*bipartite* or *tripartite* placentae). Other placental variations include *placenta membranacea*, in which villous stems and their branches persist over the whole chorion, and *placenta circumvallata*, in which its margin is undercut by a deep groove. Pathological forms of adherence or penetration include—*placenta accreta*, with exceptional adherence to the decidua basalis, *placenta incerta*, in which the myometrium is invaded, and *placenta perceta*, when the invasion by placental tissue has passed completely through the uterine wall.

At birth, when ligature of the umbilical cord is delayed, the blood volume of the child is, on the average, appreciably greater than it is when the ligature is applied at the earliest possible moment (de Marsh *et al.* 1942). It appears that in the former case most of the blood in the fetal placental vessels is transferred from the placenta to the fetus. The meaning of the phenomenon is far from clear, for in the first few days of life after late ligature of the cord the newly born suffers a loss both of plasma volume and of haemoglobin (Gotsev 1939).

Congenital Abnormalities—Teratology

The term teratology was at one time reserved for grosser examples of congenital abnormality—for the study of 'monsters'. Cleft-palate or thalassaemia could scarcely be regarded as monstrosities—something to be pointed at with surprise or aversion. But the Greek τέραξ and the Latin *monstrum* both imply pointing of a different kind—a pointing to the future, a warning or portent. In primitive cultures the advent of abnormal offspring or prodigy, whether human or otherwise, was usually accepted as a portent of ill-omen. Hence records of human congenital malformations, in cave-paintings, sculptures, and ultimately in writings, extend backwards into prehistory, as do also theories and beliefs in regard to their causes. Talipes, achondroplasia, and conjoined twins were portrayed in most ancient times, together with centaurs, sirens, mermaids and other fanciful creatures (Barrow 1971). The Hippocratic School identified hydrocephalus. Aristotle described many major and minor malformations with accuracy, but unlike his contemporaries, and his successors through many subsequent centuries, he denied the magical and divinatory aspects of teratology, ascribing 'monsters' to natural causes, a view which was not reawakened until the times of Harvey, Wolff, von Haller, and the Hunters. They and their contemporaries of the seventeenth and eighteenth centuries, stimulated by the growing knowledge of embryology, initiated the *theory of embryonic arrest* to explain malformations. Saint-Hilaire experimented on developing chick embryos in the early nineteenth century, and with this approach the study of teratology was consolidated as a science. In parallel, however, superstition and prejudice persisted, producing for example the 'hybrid theory' which attributed monstrous births to unnatural hybridization, a groundless hypothesis which cost many innocent men and women their lives. Another old theory, which is scarcely dead even today, was the concept that the experiences of the pregnant mother, usually visual, could influence her unborn offspring in an adverse manner. That factors in the *maternal environment* may influence the embryo has, of course, proved true, but in a different sense, for Watson, as long ago as 1749, suggested that fetal disease, contracted by a transplacental route, might be a cause of congenital abnormality, citing variola as an example. This theory slowly dwindled, and was almost extinguished a century later by the authority of Virchow and His, whose views dominated teratological opinion through most of the second half of the nineteenth century. A contemporary, working in obscurity, was Mendel; and as his work became known and genetics flowered into the twentieth century, attention inevitably turned to the hereditary aspect of congenital defects (already foreshadowed by Paré and John Hunter, centuries earlier). In 1941 Gregg's observation that congenital cataract is associated with the infection of pregnant mothers by rubella revived interest in environmental factors. Subsequent to this, teratology has been largely concerned with *genetic* and *environmental factors*, particularly in attempts to explain the mechanisms of abnormal development.

The expansion of experimental embryology has revealed a wide array of *environmental agents* capable of affecting normal development, including temperature variations, mechanical insult, variation in substances such as lithium and magnesium in culture media, irradiation exposure, hypoxia, hypo- and hypervitaminosis, hormonal effects (especially with oestrogens and androgens), nutritional defects, and exposure to various drugs and other chemicals. These experiments have stimulated and illumined research into teratogenic agents. Though much of such work has been pursued on the embryos of fish (minnow), amphibians (frog), and birds (chick), experiments on mammals in producing cleft palate, anophthalmia and other abnormalities suggest that similar mechanisms must operate in human maldevelopment. The identification of human teratogens may be said to have commenced with Gregg's demonstration of the effects of rubella virus. Other viral and bacterial maternal infections have since been implicated. But with the multiplication of drugs, for such purposes as abortion (aminopterin) and sedation (thalidomide), some of which have proved tragically teratogenic, teratological research has been much concentrated upon drugs. In parallel with this, however, has been the recognition of the genetic conditioning of a growing list of abnormalities (McKusick 1975), including metabolic aberrations (Garrod 1963). These remain the two major fields of current teratology, pharmacological and genetic.

While the great volume of recorded research has led to the identification of a large number of causal agents, the detailed mechanisms by which these produce abnormalities are for the most part obscure. In some cases it is clear that normal embryonic processes are delayed, arrested, or accelerated, leading to agenesis or to varying degrees of hypoplasia or hyperplasia. The recognition that cell death plays a normal role in prenatal growth and differentiation, leading to the disappearance, modelling, or reduction of structures (p. 79), has been extended to explanations of congenital defects. The available evidence (Henkes et al. 1970) suggests that cell death is a major factor in developmental aberrations associated with viruses, irradiation, nutritional defects and hypervitaminoses. Local hyper- or hypo-plasia and disturbed morphogenetic movements of cells may obviously produce distortions of normal development. It cannot be doubted that such factors as these are concerned, but in most cases the explanations are partial and missing links in the causal chain persist (Saxén 1970; Johnston and Platt 1975). This is not surprising, since the precise march of events in much of normal development is still imperfectly known.

A very large number of cytotoxic agents are now known, and many have been used as experimental teratogens, mostly on rodents. Some act as antimetabolites, amino-acid antagonists, anti-purines or spindle toxins, and most are highly selective in their effects, which are produced only by *controlled dosage* at *specific periods* during development. Connors (1975) has collated two decades of reports in this field, emphasizing the complex array of variable parameters which help to determine whether an agent acts as a teratogen or not. These include embryonic age, the amount, route and mode of administration of the agent, placental and embryonic permeability, maternal or embryonic ability to inactivate the agent, the state of differentiation of target cells and their ability to recover. Sullivan (1975) has reviewed the literature concerning teratogenic drugs taken by pregnant women, classifying them on their sites of action, whether directly on the embryo (thalidomide, tetracycline antibiotics), on embryonic endocrine balance (oestrogens and androgens), on the placenta, or on maternal tissues. These considerations are, of course, of intense clinical importance, but their contribution to *explanations* of teratogenic *mechanisms* is limited, except in so far as some drugs are known to be teratogenic at specific stages of development.

Genetic conditioning of congenital abnormalities, whether structural, metabolic, or behavioural, has been the subject of very widespread research and observation. A long list of inheritable defects has accumulated (McKusick 1975). According to Polani (1973) about a third of recognized defects in the newborn are due to single gene abnormalities, a further twelfth to chromosomal errors, and a further substantial, but unknown, fraction to polygenic interaction. About 2,000 birth defects occur in mankind which are either known to have, or are presumed to have, a genetic background. Of these 1,700 are probably autosomal and 150 sex-linked (McKusick 1975). Although much information is thus available which may prove of great value in genetic counselling, complete mechanisms from disturbed DNA coding to actual phenotypic expression of particular deformities or other aberrations are not likely to be definable. Such explanations are dependent upon a more complete knowledge of embryonic processes than is yet available. Presumably, different genetic errors may operate at different stages in the procession of events through multiplication, determination, aggregation, morphogenetic movement, differentiation, localized proliferation and cell death, each of which is itself a complex of biochemical activities. In some examples, however, it is possible to demonstrate or to hypothecate such events in detail, normal and abnormal metabolic processes in bacteria providing the most successful field. One of the earliest, if not the first of human abnormalities to be identified as genetic was alkaptonuria, now known to be due to deficiency of a single enzyme (homogentisic acid oxidase) and hence, presumably, to a single gene disturbance. Although the locus of the latter is not established as yet, the biochemical sequence during which 'alkapton' is produced is fully clarified. In sickle-cell anaemia, which is inherited as a Mendelian recessive characteristic, there is demonstrably a defect in the production of the protein moiety of the haemoglobin molecule (in fact an amino-acid substitution). The actual 'sickling' of erythrocytes only occurs in conditions of local oxygen concentration, an interaction between genetic effect and environmental factor which appears to be involved in the development of many basically genetic abnormalities. In the case of more elaborate structural deformities it is obvious that complete explanations are dependent upon adequate clarification of embryogenetic processes as noted above.

A relatively common example of an autosomal genetic abnormality often cited is the complex termed Down's Syndrome (mongolism), which occurs in 1 in 600 live births. It is due to a trisomy or tripling of chromosome No. 21. The incidence of Down's Syndrome increases with maternal age, suggesting that the aetiology is complicated and may not be simply genetic. The discovery that sex chromosomes can be identified has expedited the recognition of several sex-linked conditions, although, of course, familial history alone had indicated the sex-linked nature of haemophilia at a much earlier date. In Klinefelter's Syndrome, which is characterized by testicular atrophy and hence sterility and eunuchoidism, associated with a variable depression of intelligence, there is a condition of X-polyploidy, such as XXY, XXXY and even higher states of X. Development of the gonad into a testis is ensured by the Y chromosome, and the additional X chromosomes presumably retard differentiation. In another congenital deviation, Turner's Syndrome, which features dwarfism, rudimentary ovaries, amenorrhoea and failure of the appearance of secondary sexual characteristics, the Y chromosome is absent (XO) or imperfect. A similar sex inversion (XX) occurs in males. As in the case of autosomal abnormalities, sex-linked chromosomal defects are common in spontaneously aborted individuals. (XO is said to be the commonest, at about 1 per cent of all conceptions.)

It is clear from the foregoing that although the exact causology of the great majority of human congenital malformations is uncertain, the role of genetic and environmental factors (including teratogenic drugs) cannot be doubted. Wherever adequate familial studies can be established there is frequently an indication of genetic mechanism. However, an increased frequency of certain abnormalities may be equally associated with depressed socio-economic status, climate, or geographical regions, and maternal age.

Epidemiological studies of congenital aberrations are of considerable interest and value, not only in genetic counselling, but also in attempts to unravel the complex problems of causation. The incidence of congenital abnormality may be as high as 6 per cent in surviving infants. With the decline in incidence of, and death from, infectious and nutritional diseases, congenital abnormality is assuming a proportionately greater importance in the medical care of many populations. Several large epidemiological surveys, carried out in many countries, are now available (Kennedy 1967; Stevenson 1966; WHO 1972). These show not only the overall incidence of abnormality, but also the comparative frequency of various malformations, and the regional and racial variations which some display. For example, neurulation defects such as anencephaly and spina bifida, while being relatively frequent abnormalities, also show much variation in frequency (particularly high incidences have been recorded in Belfast and Bombay). Galactokinase deficiency was identified about 15 times more often in a Canadian series (Manitoba) than in the U.S.A. (Massachusetts). Such variations are at present largely inexplicable; but the well-known coincidence of sickle-cell gene for haemoglobin S, and of the thalassaemia trait, with the distribution of malaria are interesting exceptions.

The prenatal identification of certain malformations and aberrations must be mentioned. Aminocentesis and examination of the amniotic fluid and its suspended cells permits the diagnosis of some metabolic disorders, and potentially all chromosomal aberrations. It is of particular value in conditions of high incidence and early lethal effect such as Tay-Sachs disease, which occurs about 100 times more often in Ashkenazi Jews than in other races. Radiology alone may reveal anencephalic or hydrocephalic fetuses (Russell 1969); but the hazards of this technique have led to the substitution of ultrasonography, which is presumed to be safer and can discover similar conditions (Santos and Duenhoelter 1975; Taylor 1978). Amniography consists in radiological examination after the injection of a water-soluble radio-opaque substance into the amniotic fluid; fetography is a variant of this technique which employs an oily radio-opaque injectant (Wiesenhaan 1972). By absorption of the radio-opaque substance in the vernix caseosa, the latter technique is claimed to yield clearer delineation of surface features (For literature consult Persaud 1977).

Twinning

Twinning occurs once in about every eighty births. The twins may be *dizygotic* (binovular or fraternal) or *monozygotic* (uniovular or identical), of which the former occurs more frequently. Twin boys are most common, less common are a boy and a girl and least common two girls.

Dizygotic twins result from the discharge and fertilization of two ova, either from two separate or a single ovarian follicle. Each embryo develops in its own chorionic sac, but occasionally synchorial fusion occurs. The twins are of different genetic constitution and may be of the same or different sex. They bear no greater resemblance to one another than do other siblings of the same family. There is evidence that dizygotic twinning may be hereditary. The development of ovarian follicles, and the release of ova at ovulation, is under the hormonal control of circulating *gonadotrophins* produced by the β cells of the anterior lobe of the hypophysis cerebri. The control mechanisms are far from clear, but it has been demonstrated in a variety of mammals that raising the circulating gonadotrophic level by injections increases the number of ova discharged (Hammond 1961). Similar considerations apply to women who have had such treatment for long-standing amenorrhea (Gemzell and Roos 1966). In one series of about 100 cases so treated there were 43 pregnancies

including 20 singletons, 14 twins, 2 triplets, 3 quadruplets, 1 quintuplet, 2 sextuplets and 1 septuplet.

Monozygotic twins arise from a single ovum fertilized by a single sperm. At some stage up to the establishment of the axis of the embryonic area and the development of the primitive streak the formative material separates into two parts, each of which gives rise to a complete embryo. The twins may share one or have independent chorionic sacs and possess an individual or a common amniotic sac. The twins are of the same sex, have the same blood groups and tissues of identical antigenic potencies. They resemble each other closely. Monozygotic twinning is a hereditary character.

Interestingly, a number of workers have suggested that in addition to the widely recognized two foregoing types, a *third type* of twin (with several variants) may exist. Essentially this type is *uniovular* but *dispermatic* following an irregular meiotic cycle during oögenesis. It is considered that either the ovum, or the secondary oöcyte, or the primary oöcyte may undergo an equal cytoplasmic division (instead of throwing off a diminutive polar cell). Evidently each of the three variants would have different degrees of genetic dissimilarity depending upon the stage at which such an equal division occurred.

Multiple births greater than twinning, such as triplets or quadruplets, can arise from multiple ovulations or a single ovum, the formative material of which later separates into several parts, or from a combination of these mechanisms.

Conjoined twins or double monsters arise from the incomplete division of the formative material of a single early ovum. There is, however, no evidence that the condition is hereditary or that it occurs more frequently in families given to uniovular twinning. The conjoined twins are usually united by corresponding regions and the degree of duplication of parts varies.

For an excellent review of the types, frequency, inheritance and embryology of twinning, the diagnosis of zygosity, the course and outcome of twin pregnancies, their possible evolutionary significance, and some post-natal characteristics of twins, the interested reader should consult Bulmer 1970.

DEVELOPMENT OF INDIVIDUAL SYSTEMS

Development of the embryo has so far been considered as a whole, but, as the definition of its structures proceeds, overall description becomes so complicated as to be an actual impediment to clarity of appreciation of the events occurring. It is hence customary and convenient to limit attention to individual systems in their further development; but it must never be overlooked that the analysis of a whole organism into such divisions—however attractive on morphological and functional grounds—is largely a product of the sequential nature of human perception. Not only do the several systems into which we divide the organism develop simultaneously, they also interact and modify each other. This necessary interdependence is not only supported by the evidence of experimental embryology, but is also emphatically demonstrated by the phenomena of growth anomalies, which cut across the artificial boundaries of systems in most instances. For these reasons it is most desirable that the development of any one individual system should be frequently related to others, especially those most closely associated with it.

So far the development of the embryo has been taken to an age of between 3 and 4 weeks, the stage of early somite formation, equivalent to Horizons X or XI (*see* p. 222) on the scale established by the studies of Streeter and others (Streeter 1942–51). It is only partly constricted from the yolk sac, but the head and tail folds are well formed, with enclosure of the foregut and hindgut (proenteron and metenteron). The forebrain projection dominates the cranial end of the embryo, the oropharyngeal membrane and cardiac prominence being caudal and ventral to this. The intraembryonic mesoderm has begun to differentiate, and its paraxial region is undergoing segmentation into somites. The neural groove is in process of closure and is separated from the dorsal aspect of the gut by the notochord. The

earliest blood vessels have appeared and a primitive tubular heart occupies the pericardium. The chorionic circulation is about to be established, after which event the embryo will be completely dependent for its requirements upon the maternal bloodstream. The intraembryonic part of the coelom consists of the pericardial cavity, leading dorsally into the caudally directed right and left pericardioperitoneal canals, and the peritoneal cavity, into which the two canals open caudal to the septum transversum, thus establishing free communication with the extraembryonic coelom (**2.31**).

Development of the Skeletal System

The skeleton is a derivative of mesoderm (and in some parts neural crest), including not only its axial and appendicular divisions but also all accessory ossicles such as sesamoid bones and the osseous and cartilaginous elements of the branchial arches. Most of these parts pass through a first, *blastemal stage* of mesenchymal condensation, and a second, *cartilaginous stage* before becoming ossified. In some bones, however, ossification follows immediately upon the blastemal stage, the intermediate phase of chondrification being omitted (p. 260). (For a review of embryonic cartilage *see* Glenister 1976. Noback and Böving, 1962 and Tanaka, 1976, have recorded extensive observations of the times of appearance of centres of chondrification in human embryos.)

THE SKELETAL AXIS

Before reaching its final condition the skeletal axis passes through three preliminary states (Präder 1947; Sensenig 1949; Peacock

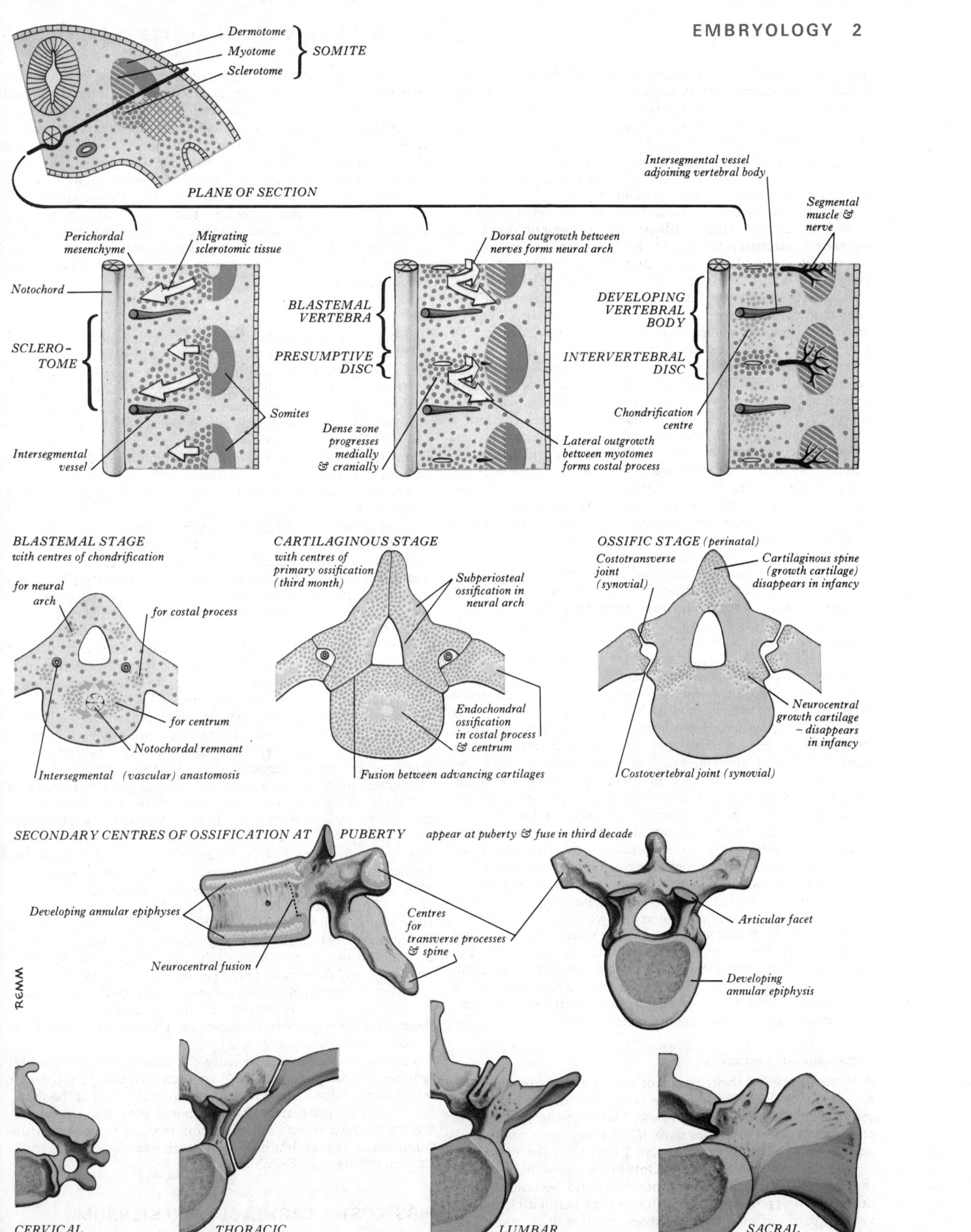

SOMITE
Dermotome
Myotome
Sclerotome

PLANE OF SECTION

Perichordal mesenchyme
Migrating sclerotomic tissue
Notochord
SCLERO-TOME
Intersegmental vessel
Somites

Dorsal outgrowth between nerves forms neural arch
BLASTEMAL VERTEBRA
PRESUMPTIVE DISC
Dense zone progresses medially & cranially
Lateral outgrowth between myotomes forms costal process

Intersegmental vessel adjoining vertebral body
Segmental muscle & nerve
DEVELOPING VERTEBRAL BODY
INTERVERTEBRAL DISC
Chondrification centre

BLASTEMAL STAGE
with centres of chondrification
for neural arch
for costal process
for centrum
Notochordal remnant
Intersegmental (vascular) anastomosis

CARTILAGINOUS STAGE
with centres of primary ossification (third month)
Subperiosteal ossification in neural arch
Endochondral ossification in costal process & centrum
Fusion between advancing cartilages

OSSIFIC STAGE (perinatal)
Costotransverse joint (synovial)
Cartilaginous spine (growth cartilage) disappears in infancy
Neurocentral growth cartilage – disappears in infancy
Costovertebral joint (synovial)

SECONDARY CENTRES OF OSSIFICATION AT PUBERTY   appear at puberty & fuse in third decade
Developing annular epiphyses
Neurocentral fusion
Centres for transverse processes & spine
Articular facet
Developing annular epiphysis

REMM

CERVICAL   THORACIC   LUMBAR   SACRAL

Parts of adult vertebrae derived from: CENTRA [ ] , NEURAL ARCHES [ ] , COSTAL PROCESSES [ ] of embryonic vertebrae

2.50   Sequential diagrams of vertebral development from early somitic and perichordal stages through blastemal, cartilaginous, and pre- and post-natal ossificatory stages. For views concerning segmentation see text. Bottom row indicates principal morphological parts of adult vertebrae.

1951–2; Walmsley 1953; Tondury 1958; **2**.50). First it is formed by the **non-segmental notochord**, a flexible rod of cells enclosed by a thick, membranous sheath. The notochord is not, however, limited to the region in which the vertebrae will replace it, for it extends into the head as far as the caudal limit of the hypophysis cerebri, this cranial extension being subsequently incorporated into the basilar part of the occipital bone and the dorsal part of the body of the sphenoid (**2**.51).

In the second stage of the skeletal axis, the notochord serves as a framework around which a **blastemal** or **mesenchymatous vertebral column** is formed. On each side the fusiform cells of the sclerotomes multiply rapidly and migrate ventromedially to enclose the notochord in a mesenchymal sheath (**2**.29), which at first retains a semblance of its segmental origin. Sclerotomic cells later migrate dorsally around the spinal cord, passing between and lateral to the spinal ganglion rudiments, and ventrolaterally into the intervals between myotomes.

Around the notochord each sclerotomic segment is divided into equal cranial and caudal parts by a transitory transverse split or loosening between the two groups of cells to form a *sclerotomic fissure* (**2**.50). The adjoining mesoderm condenses to form a transverse plate, a *perichordal disc*, and the less dense, caudal part of one segment fuses with the slightly smaller and less dense cranial part of the adjoining segment. This fusion defines the centrum of a vertebra. In the dorsal extensions of the sclerotomes segmental condensations between the spinal ganglia define the neural arch and, later, its processes, while their ventrolateral extensions outline the costal processes, which are continuous medially with the perichordal disc.

In the third stage the vertebral components become more clearly defined, and chondrification of these mesenchymatous models produces a **cartilaginous vertebral column**. Each centrum is chondrified from a pair of centres which appear during the sixth week and quickly coalesce. Each half of a neural arch is chondrified from a centre starting in its base and extending ventrally into the pedicles, to meet and blend with the centrum, and dorsally into the laminae; but the latter do not meet in the midline until the fourth month. The transverse and articular processes are chondrified in continuity with the neural arches; intervening zones of mesenchyme which do not become cartilage mark the sites of intervertebral and costovertebral joints, and synovial cavities appear later in these. The costal processes chondrify separately, and in the thoracic region they extend ventrally, the more cranial members curving round in the body wall to reach the developing sternal plates. They are separated from the developing transverse processes by non-chondrified mesenchyme in which the costotransverse joints will appear. At other than thoracic levels the developing costal process (or pleurapophysis) becomes incorporated into the 'transverse process' of descriptive adult anatomy (**2**.50).

In both the blastemal and chondrifying stages, the cranial and caudal aspects of each perichordal disc proliferate, contributing to the growth of adjoining centra and eventually merging with them. The main, intermediate mass of each perichordal disc (which contains the sclerotomic fissure), together with the enclosed part of the notochord, becomes an *intervertebral disc*.

Hypochordal Elements

A structure which, though present during the blastemal stage, only becomes clearly recognizable as an entity during chondrification is the *hypochordal arch* or bow. This connects the vertebral ends of two costal processes to each other and spans the ventral surface of a centrum; it is associated only with the first three or four cervical vertebrae in man. Only in the case of the *atlas* does the hypochordal arch persist, chondrify, and become ossified during the first post-natal year; it forms the anterior arch and the ventral parts of the lateral masses of this vertebra. (The hypochordal arch may correspond to the paired intercentra, ventral elements in the development of vertebrae in many other animals. They may persist at any level, forming such arrangements as haemal or intercentral arches, which enclose the caudal vessels.) The homologies of the various elements which fuse to form vertebrae are still subject to disagreements (Jollie 1962). It does, however, now appear certain that the original concept of

Gadow (1933), that all tetrapod vertebrae can be derived from four paired elements (basidorsals, basiventrals, interdorsals and interventrals) must be abandoned. Subsequent authorities, provided with much more extensive evidence, especially embryological data, have not substantiated Gadow's somewhat theoretical views (Mookerjee 1936; Devillers 1954; Williams 1959; Jenkins 1969). Another widely accepted view—that the centrum of the atlas is united with that of the axis as its dens—has been thrown into doubt by observations on developmental stages in mammals (Jenkins 1969). In the case of the axis and succeeding vertebrae the hypochordal arches degenerate, either disappearing entirely or becoming incorporated into the ventral parts of the adjoining centra.

Notochordal Vestiges

The notochord can still be identified for some time traversing the cartilaginous centra, but these parts of it ultimately atrophy and vanish. Between the developing vertebrae it expands as localizing aggregations of cells with intervening mucoid matrix to form the *nucleus pulposus* of the intervertebral disc. This nucleus is surrounded by the intermediate part of each perichordal disc to form the *annulus fibrosus*, which differentiates into an external laminated fibrous zone and an internal cuff around the nucleus pulposus. This inner zone contributes to the growth of the outer, and near the end of the second month of embryonic life it begins to merge with the notochordal tissue, being ultimately converted into fibrocartilage. After the sixth month of fetal life the notochordal cells in the nucleus pulposus commence to degenerate, being replaced by cells from the internal zone of the annulus fibrosus. This degeneration continues until the second decade of life, by which time all the notochordal cells have disappeared (p. 445). Thus, in the adult, notochordal vestiges are limited, at the most, to non-cellular matrix.

Each vertebra is thus usually described as formed from parts of two adjacent sclerotomes. The rearrangement of segmentation is said to result in an alteration in position of vertebrae with respect to myotomes, bringing the intersegmental vessels into line with the vertebral bodies, a relationship which persists in the lumbar and lower thoracic regions. However, an extensive study of the behaviour of the perichordal and sclerotomic mesenchyme in the embryos of sheep suggests that, in this species at least, 'resegmentation' of this kind does not occur, the primordia of vertebrae arising in their definitive positions, without any movement into new sites relative to surrounding structures (Verbout 1976). The essence of this view is that the vertebral *processes* and *centra* have different mesenchymal origins, the former from cells closely associated with myotomes (*sclerotomic tissue*), the latter from the non-segmented *perichordal mesenchyme*. The processes develop in a caudal condensation within their segment and are caudal to the numerically corresponding spinal nerve. It is suggested that a *gradient of segmentation* spreads medially from the myotomes to the precursors of the vertebral processes, and from these to the perichordal tissue (**2**.50).

Towards the end of the second month ossification commences in the cartilaginous vertebrae, and the column then enters the fourth and last stage in its development. The further details of this are described elsewhere (p. 282).

Applied Anatomy. Occasionally the coalescence of vertebral laminae is incomplete, a cleft of variable width being left through which dura and arachnoid mater may protrude. Part of the spinal cord, with its pia mater, also commonly projects, a condition known as *spina bifida*. The malformation is commonest in the lumbosacral region, but may occur at thoracic or cervical levels or even throughout the column.

RIBS, COSTAL CARTILAGES AND STERNUM

The ribs develop from the costal processes of the primitive vertebral arches, extending between the myotomic muscle plates. In the thoracic region (**2**.50c) of the vertebral column these processes grow laterally to form a series of *precartilaginous ribs*. The transverse processes grow laterally behind the vertebral ends of the costal processes, at first connected by mesenchyme, which later becomes differentiated into the ligaments and other tissues of the costotransverse joints. The costocentral joints are similarly formed from mesenchyme between the proximal end of the costal processes and centra. In *cervical vertebra* (**2**.50) the transverse process is dorsal to the foramen transversarium, while the costal process, corresponding to the head and neck of a rib, limits the foramen ventrolaterally. The distal parts of these cervical costal processes do not develop, but occasionally they do so in the case of the seventh cervical vertebra, even developing costovertebral joints. Such *cervical ribs* may even reach the sternum (p. 283). In *lumbar vertebrae* (**2**.50) the costal processes do not develop distally, but their proximal parts become the 'transverse processes' of these vertebrae, whose morphologically *true* transverse processes may be represented by their accessory processes (p. 279). Occasionally movable ribs may develop in association with the first lumbar vertebra. Only the upper two or three *sacral costal processes* usually develop (**2**.50). They fuse into the lateral mass of the sacrum, forming its ventral part. The *coccygeal* vertebrae are apparently devoid of costal processes.

The sternum is formed from bilateral mesenchymatous condensations, sternal plates, which begin in the dorsolateral region of the body wall. They are immediately ventral to the rudiments of the clavicles and ribs, but are independent of them in their formation, in so far as experiments on their primordia in the mouse can testify to the human condition (Chen 1952). These plates chondrify, move ventrally towards each other from both sides as the costal processes lengthen, and they eventually fuse together across the midline in a craniocaudal direction. This forms a longitudinal cartilaginous bar, with which the clavicles and upper seven pairs of costal cartilages establish contact. The xiphoid process develops as a caudal extension of the sternal bar. Hypertrophy of the cartilage cells as a preliminary to ossification occurs opposite future intercostal spaces, as the first indication of segmentation of the sternum into manubrium, four sternebrae for the 'body', and xiphoid process. The ossification and further growth of the sternum and ribs is described later (pp. 287, 290).

DEVELOPMENT OF THE CRANIUM (2.51)

The bones of the skull are developed in the mesenchyme which surrounds the cerebral vesicles; but, before the osseous state is reached, the cranium passes through blastemal and cartilaginous stages, like other parts of the skeleton. However, not all parts pass through a phase of chondrification; and hence the *chondro-cranium*, which includes all those bones doing so, is incomplete. Most of the cranial vault and parts of its base are not preformed in cartilage. Though the mesenchymatous (membranous) and cartilaginous parts of the skull will be considered in sequence, they develop together and complement each other in forming the complete cranium, some of whose bones are composite structures derived from both sources. All elements, of course, pass first through a mesenchymatous phase.

The Desmocranium

The blastemal skull (desmocranium) begins to appear at the end of the first month as a condensation and thickening of the mesenchyme which surrounds the developing brain, forming localized masses which are the earliest distinguishable cranial elements. The first masses evident are in the occipital region, outlining the basilar part of the occipital bone. These form an *occipital plate*, from which two extensions on each side grow laterally to complete a foramen around each hypoglossal nerve. At the same time the mesenchymal condensation extends forwards, dorsal to the pharynx, to reach the primordium of the hypophysis, thus establishing the clivus of the cranial base and the dorsum sellae of the future sphenoid bone. Early in the second month it surrounds the developing stalk of the hypophysis and extends ventrally between the two halves of the nasal cavity, where it forms the anlage of the ethmoid bone and of the nasal septum. The notochord traverses the occipital plate obliquely, being at first near its dorsal surface and then lying ventrally, where it comes into close relationship to the epithelium of the dorsal wall of the pharynx, being for a time fused with it. It then re-enters the cranial base and runs ventrally to end just caudal to the hypophysis (**2**.51).

During the fifth week the two *otocysts* (auditory vesicles) become enclosed in their mesenchymal *otic capsules*, which are soon differentiated into dorsolateral *vestibular* and ventromedial *cochlear* parts, enveloping the primordia of the semicircular canals and the cochlea. Between these two regions the facial nerve lies in a deep groove. The otocysts fuse with the lateral processes of the occipital plate, leaving a wide hiatus through which the internal jugular vein and the glossopharyngeal, vagus, and accessory nerves pass. At this stage the mesenchyme around the developing hypophysial stalk, which is forming the rudiment of the postsphenoid part of the sphenoid bone, spreads out laterally to form the future greater wings of this element. Smaller processes ventral to this indicate the sites of the lesser wings of the sphenoid, while other condensations reach the sides of the nasal cavity and also blend with the still mesenchymatous septum.

The first signs of the vault or neurocranial part of the skull appear about the thirtieth day; they consist of curved plates of mesenchyme at the sides of the skull and gradually extend cranially to blend with each other; they also extend towards and reach the base of the skull, which will become part of the chondrocranium.

The Chondrocranium

Chondrocranium (**2**.51) is a term applied to the parts of any vertebrate skull which pass through, and sometimes remain in, a cartilaginous stage. (In *Chondrichthyes*, the cartilaginous fishes, such as sharks, all the cranium chondrifies and persists in this state—*see* p. 295.) The crania of all land animals, the tetrapods, contain variable regions which ossify directly from mesenchyme. Such *dermal* (membranous) elements form much of the cranial vault, and in mammals chondrification is limited to the basal regions of the skull (**2**.51). In mammals this change occurs primarily in three regions: (1) dorsally, in relation to the notochord; (2) intermediately, in relation to the hypophysis; and (3) ventrally, between the orbits and the nasal cavity. These may be named *parachordal, hypophysial,* and *interorbitonasal* regions. The parachordal cartilage is developed from the mesenchyme related to the cranial end of the notochord; caudally it exhibits traces of four primitive segments separated by the roots of the hypoglossal nerves. The hypophysial cartilage ossifies to form the postsphenoid part of the sphenoid bone, but its morphological status is uncertain. The interorbitonasal cartilage is perhaps to be equated with the trabeculae cranii of lower vertebrates and is usually known as the *trabecular cartilage*, which is a bilateral structure, developing from two centres of chondrification The trabeculae cranii may largely be derived from branchial arch (neural crest) mesoderm, i.e. from the **viscerocranium**, having been adapted into the cartilaginous or basal part of the **neurocranium**, or 'brainbox'. From the evidence in embryos of earlier vertebrates, most of the chondrocranium and the majority of the viscerocranium are derived from neural crest tissue, including almost all of the branchial skeleton. Only the caudal parts of the trabeculae, the parachordal parts of the trabeculae, the parachordal bars, the otic capsules and the second basibranchial appear to be derived from general head mesenchyme. For these and other details of development and morphology of the mammalian chondrocranium, see references (Fawcett 1911, 1917, 1918; de Beer 1937; Stone 1926; Hörstadius 1950; Stark 1965; Balinsky 1975).

In the human embryo chondrification of the cranium begins in the second month, the first cartilaginous foci appearing in the occipital plate, one on each side of the notochord (parachordal cartilages); these later fuse—at about the end of the seventh week—around the notochord, whose oblique transit through the region has already been mentioned above (**2**.51). The cartilage of the posterior part of the sphenoid is formed from two hypophysial centres, which flank the stalk of the hypophysis, uniting at first behind it and then in front, thus enclosing a *craniopharyngeal canal* containing the hypophysial diverticulum. The canal is usually obliterated by the third month; its association with the derivation of the anterior lobe of the hypophysis from the pharyngeal diverticulum of Rathke has been denied (p. 199).

The otic capsules, the presphenoid, the bases of the greater wings and the lesser wings of the sphenoid, and finally the nasal

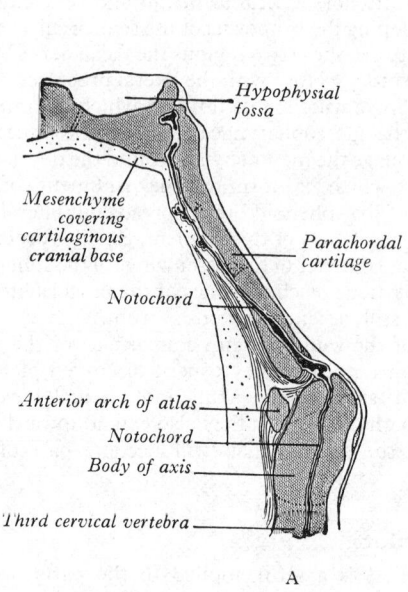

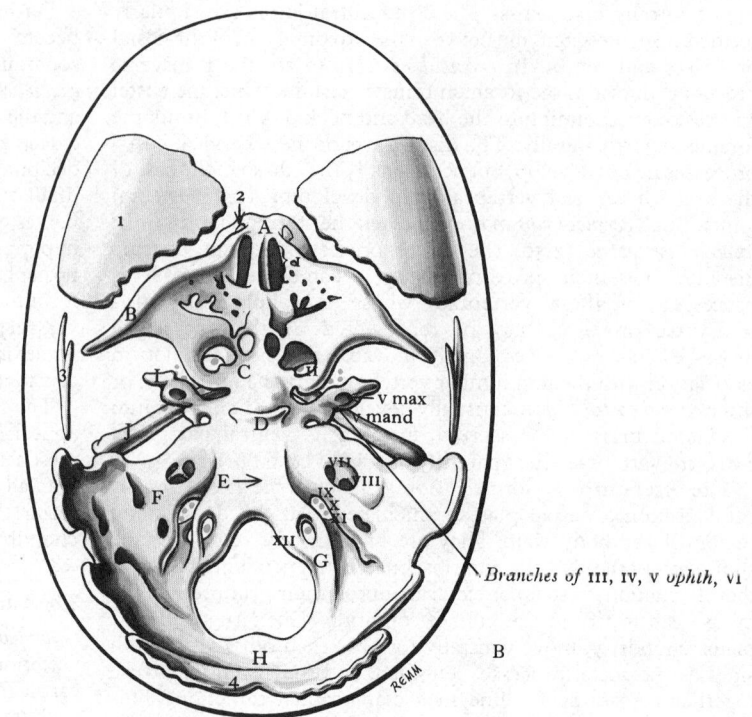

Key to chondral elements:

A   Nasal capsule
B   Orbitosphenoid
C   Presphenoid
D   Postsphenoid
E   Basi-occipital
F   Otic capsule
G   Exoccipital
H   Supra-occipital
I   Alisphenoid
J   Meckel's mandibular cartilage
K   Cartilage of malleus
L   Styloid cartilage
M   Hyoid cartilage
N   Thyroid cartilage
O   Cricoid cartilage
P   Arytenoid cartilage

Key to dermal (membrane) elements:

I   Frontal bone
2   Nasal bone
3   Squama of temporal bone
4   Squama of occipital bone
     (interparietal)
5   Palatine bone
6   Maxilla
7   Lacrimal bone
8   Zygomatic bone
9   Palatine bone
10   Vomer
11   Medial pterygoid plate
12   Tympanic ring
13   Mandible

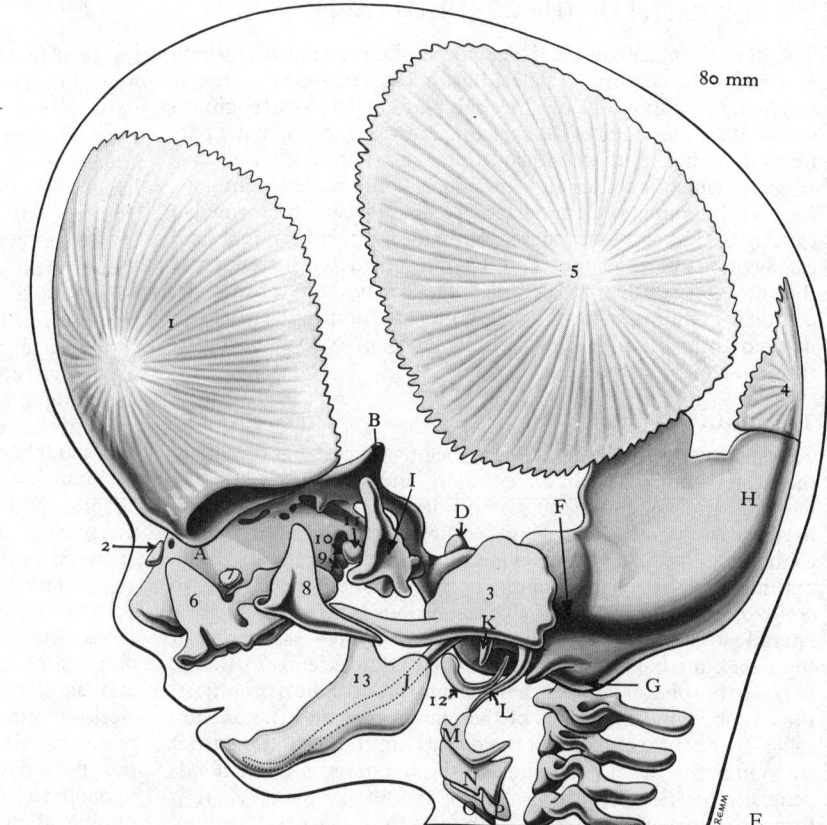

2.51 A–F   Representative stages in the development of the cranium. In all the diagrams the *chondrocranium* and cartilaginous stages of vertebrae are shown in blue, except where ossification is occurring and here the colour is green. The *desmocranium*, consisting of elements ossifying directly in mesenchyme is shown in yellow. Cranial nerves are indicated by the appropriate roman numeral.

A A sagittal section through the cranial end of the developing axial skeleton in an early human embryo of about 10 mm, showing the extent of the notochord. B Key for diagram C. C Superior aspect of cranium of human embryo at 40 mm. D Lateral aspect of C. E Key for diagram F. F Lateral aspect of cranium of human embryo at 80 mm.

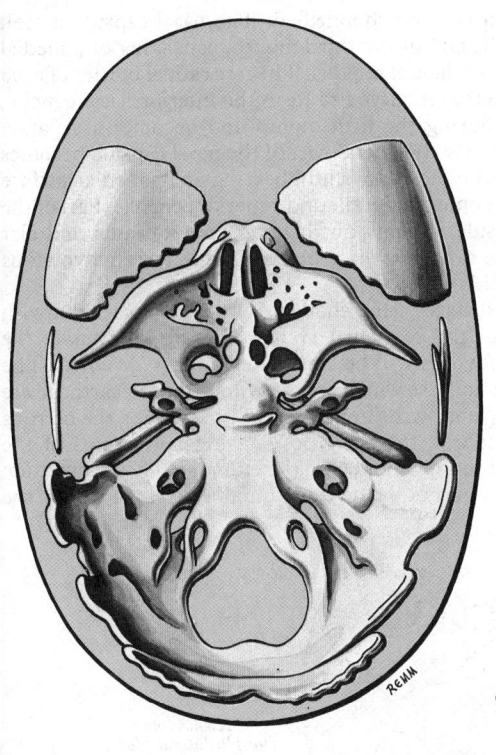

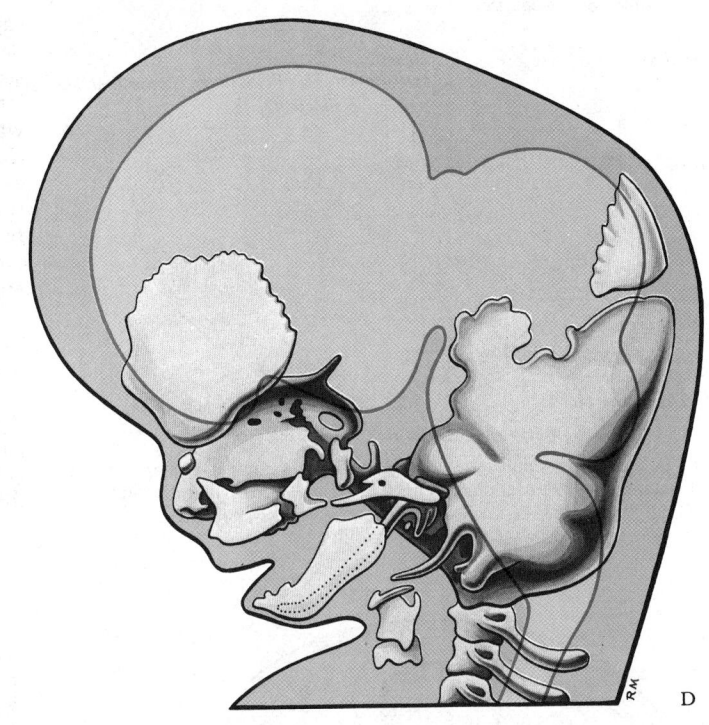

C

D

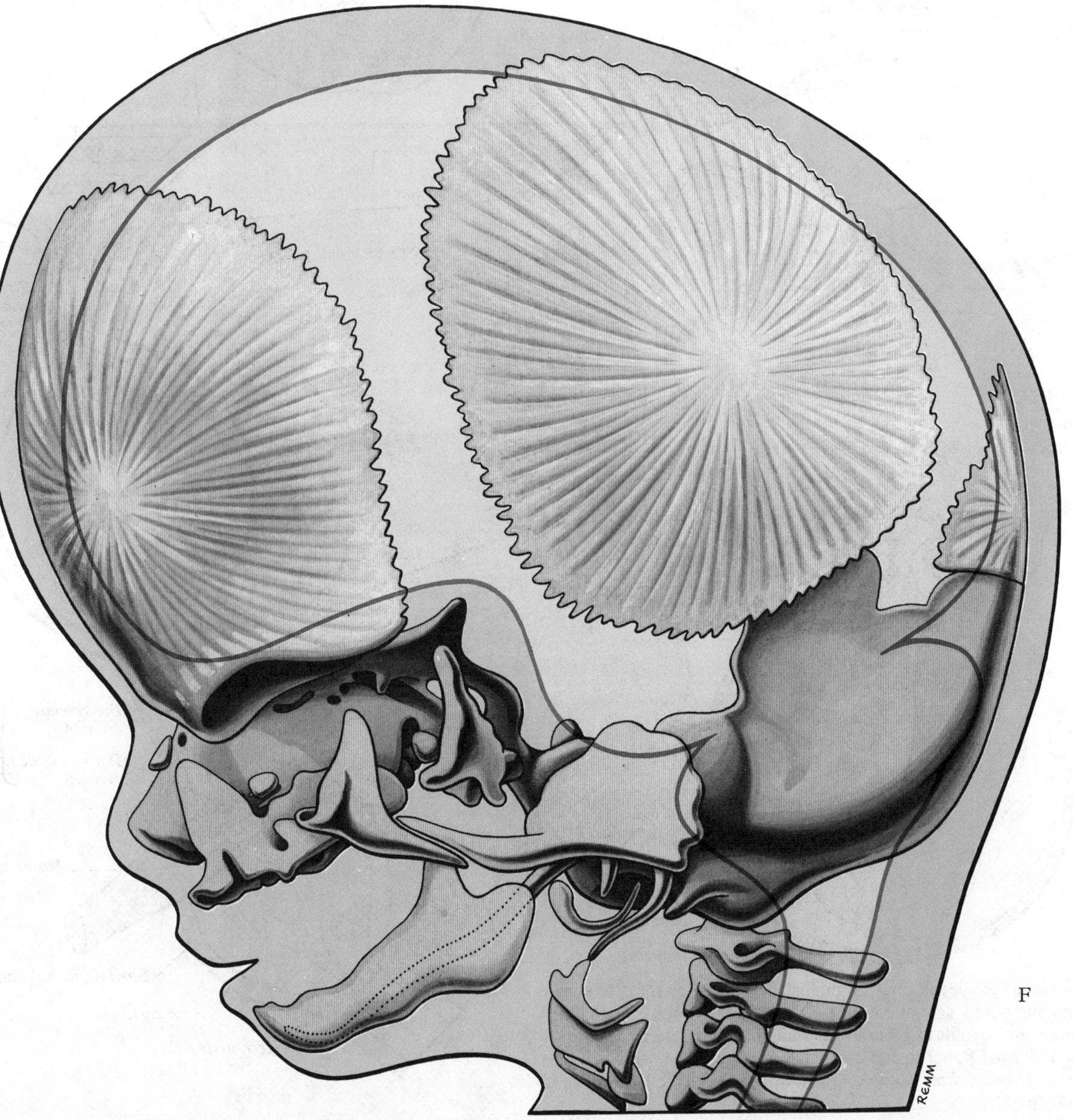

F

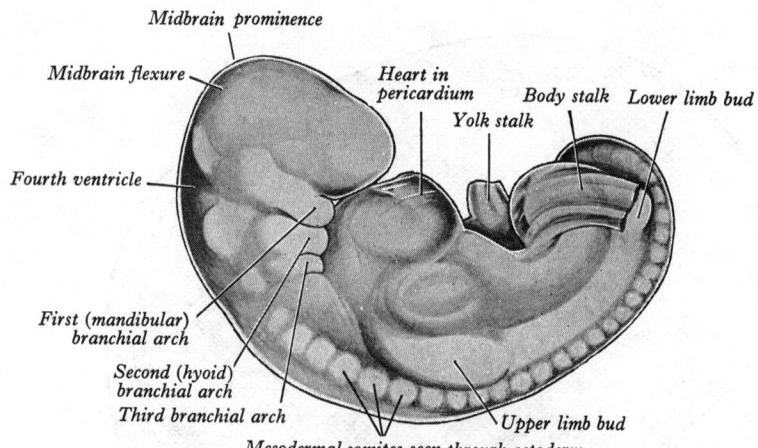

Midbrain prominence

Midbrain flexure

Heart in
pericardium

Yolk stalk

Body stalk   Lower limb bud

Fourth ventricle

First (mandibular)
branchial arch

Second (hyoid)
branchial arch

Third branchial arch

Mesodermal somites seen through ectoderm

Upper limb bud

2.52 A 26½-day macaque embryo, showing an early stage in the development of the branchial arches and the limb buds. Magnification × 12. (C. H. Heuser and G. L. Streeter, *Contr. Embryol.*, **29**, 1941.)

capsule, in turn become chondrified. The nasal capsule is well developed by the end of the third month, consisting of a medial septal part and two lateral regions. The free caudal border of each lateral part becomes incurved to form the inferior nasal concha, which ossifies during the fifth month and becomes a separate element. Posteriorly the lateral part of the nasal capsule becomes ossified as the ethmoidal labyrinth, bearing on its medial surface ridges which become the middle and superior conchae. Part of the rest of the capsule remains cartilaginous as the septal and alar cartilages of the nose, part is replaced by the mesenchymatous vomer and nasal bones.

The ventral surface of the chondrocranium is associated with the cartilages of the branchial arches, the *viscerocranium*, the development of which will be considered later (*vide infra*). The bones of the cranial base which are thus preformed in cartilage are the occipital (excepting the upper part of its squama), the petrous part of the temporal, the body, lesser wings, and roots of the greater wings of the sphenoid, and the ethmoid. These constitute the cartilaginous part of the neurocranium, the rest of which, the

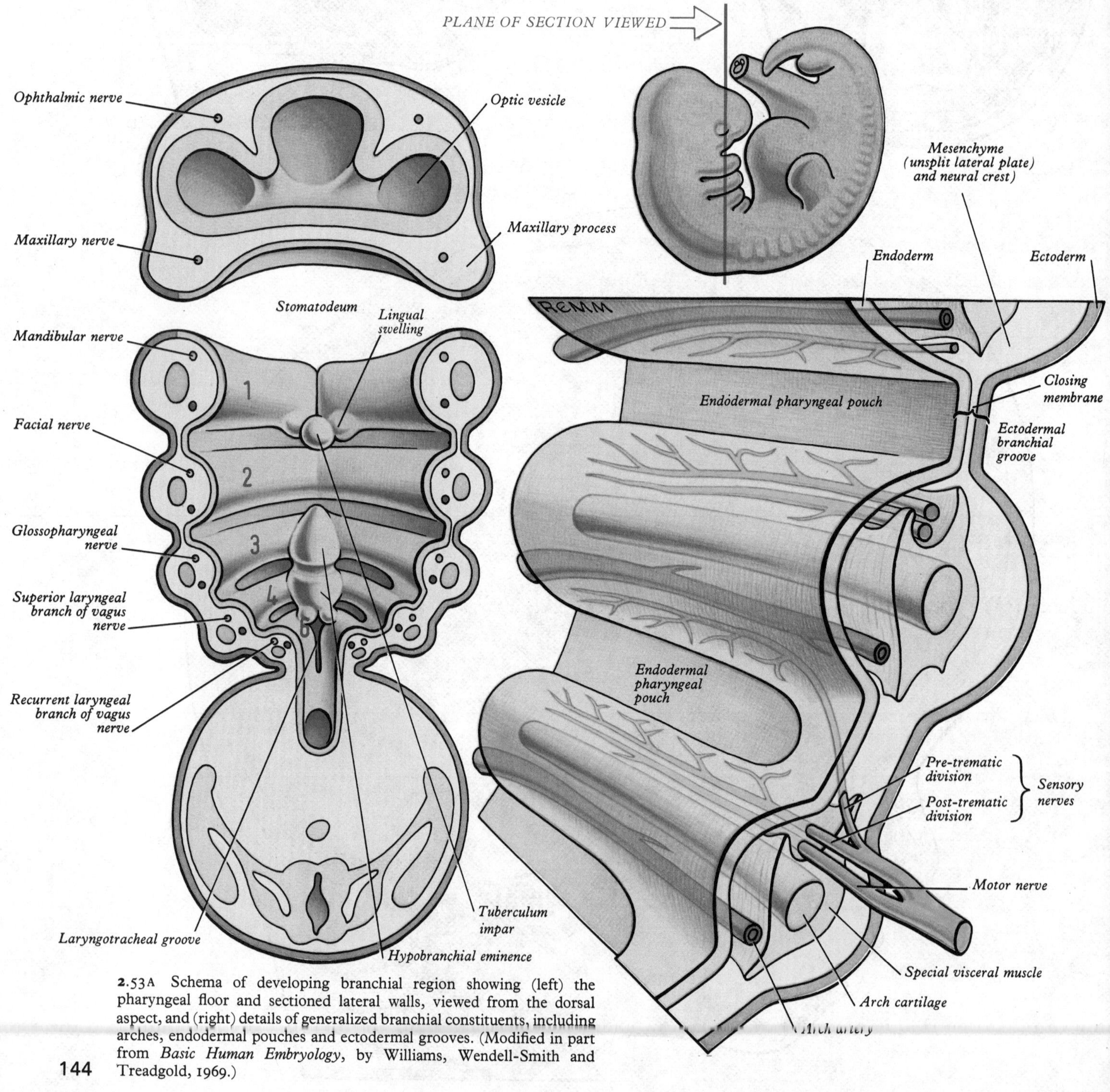

PLANE OF SECTION VIEWED ⟹

Ophthalmic nerve

Optic vesicle

Maxillary nerve

Maxillary process

Mesenchyme
(unsplit lateral plate)
and neural crest)

Endoderm

Ectoderm

Stomatodeum

Lingual
swelling

Mandibular nerve

Facial nerve

Glossopharyngeal
nerve

Superior laryngeal
branch of vagus
nerve

Recurrent laryngeal
branch of vagus
nerve

Laryngotracheal groove

Tuberculum
impar

Hypobranchial eminence

Endodermal pharyngeal pouch

Closing
membrane

Ectodermal
branchial
groove

Endodermal
pharyngeal
pouch

Pre-trematic
division

Post-trematic
division

Sensory
nerves

Motor nerve

Special visceral muscle

Arch cartilage

Arch artery

2.53A Schema of developing branchial region showing (left) the pharyngeal floor and sectioned lateral walls, viewed from the dorsal aspect, and (right) details of generalized branchial constituents, including arches, endodermal pouches and ectodermal grooves. (Modified in part from *Basic Human Embryology*, by Williams, Wendell-Smith and Treadgold, 1969.)

mesenchymatous (membranous) neurocranium, corresponding to the cranial vault, is not preformed in cartilage. Its elements, frequently described as *dermal* bones because of their probable origin (p. 295), are the frontal bones, the parietals, the squamous parts of the temporal bones, and the upper (interparietal) part of the occipital squama. To summarize, therefore, the base of the skull—except for the orbital plates of the frontal and the lateral parts of the greater sphenoidal wings—is preformed in cartilage, while the whole of the vault is ossified directly in mesenchyme (2.51 B–F).

Ossification commences before the chondrocranium has fully developed, and as this change extends, bone overtakes cartilage until little of the chondrocranium remains. However, parts of it still exist at birth, and small regions remain cartilaginous in the adult skull. At birth unossified chondrocranium still persists at the following points: (1) the alae and septum of the nose; (2) in the sphenoid bone (p. 326); (3) the speno-occipital and spheno-petrous junctions (p. 330), (4) the apex of the petrous bone (foramen lacerum); and (5) the occipital bone. The further development of these areas and of the cranial bones in general is included with their individual descriptions (*see* Section 3, Osteology, p. 293).

The Viscerocranium

Certain of the cranial components are derived from the branchial or visceral arches, the *viscerocranium*, and therefore this part of skeletal development will be considered next, before that of the appendicular skeleton (p. 153). It will also be convenient to include a general consideration of the branchial apparatus at this point.

The Branchial Apparatus

After the head fold has formed, the *stomatodeum* or the primitive mouth, is bounded cranially by the projection of the forebrain and caudally by the cardiac prominence (2.52). The mandibular region and the whole of the neck, which will subsequently intervene between the mouth and developing thorax, are as yet absent, but will be formed by the appearance and modification of six paired **branchial arches**, which develop on the lateral aspects of the head in the vicinity of the hindbrain (2.53). In the earliest vertebrates (*Agnatha*), which were jawless, the arches were a uniform series of bars between the gill clefts; but long before the evolution of the terrestrial vertebrates, remarkable adaptations had occurred in them. What are commonly regarded as the first pair of arches became the jaws, upper and lower, of the *Gnathostomata*, the jaw-bearing vertebrates, including most fish; they are, therefore, usually named the *mandibular arches*. It should, however, be noted that since this early identification, strong evidence has accumulated that a pair of *pre-mandibular arches* existed and have become adapted as the *trabeculae cranii* of subsequent vertebrate embryos. These are probably represented by the interorbitonasal cartilage of the human embryo, as mentioned above (p. 141), forming the visceral, or branchial element in the chondrocranium. The next (post-mandibular) arch in the series is the hyoid arch, whose skeletal derivatives form the varied hyoid elements present in all vertebrates with jaws. The most dorsal of these elements, the *hyomandibula*, is already present in cartilaginous fish as a strut between the skull and the primitive jaw joint, thereby reducing the cleft between the mandibular and hyoid arches to a small opening, the *spiracle*. The interesting further evolution of this region in land animals in connection with the auditory apparatus will be considered later (p. 178). The hyoid arch also contributes to the formation of a gill cover, or operculum in bony fish, and the remaining arches persist as the supports of the gill apparatus.

The mesenchyme of the branchial region (p. 115) at first exists as a thin sheet between the ectoderm and endoderm; but, as growth proceeds, it proliferates to form a series of cylindrical processes which constitute the mesodermal core of the branchial arches, covered, of course, by the two epithelia mentioned. At first, the arches produce rounded ridge-like prominences both of the overlying ectoderm and of the endodermal floor of the pharynx. In the furrows between these prominences the ectoderm

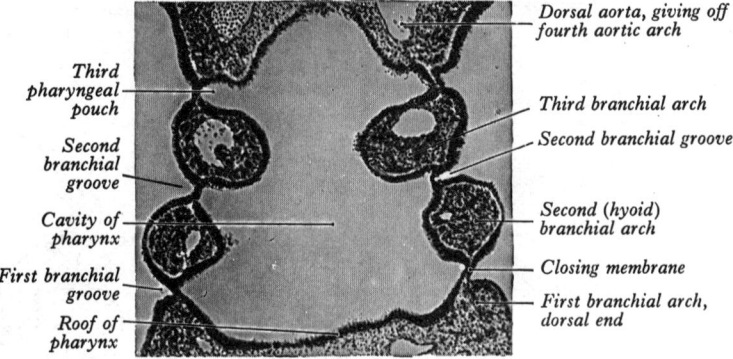

2.53B A transverse section through the pharynx of a human embryo. C.R. length—2 mm. Magnification× *c.* 50. (E. H. Norris, *Contr. Embryol.*, **27**, 1938.)

and endoderm are in virtual contact. The thin membranes so formed break down in gill-breathing vertebrates but persist in the tetrapods, in which true clefts are not formed. However, the external *branchial grooves* which correspond to them are often, less appropriately, called *branchial clefts*, and their internal counterparts are the *pharyngeal sacs* or *pouches*. At this stage in the human embryo the pharynx is transversely wide at its cranial end but rapidly narrows in a caudal direction; it is also dorsoventrally compressed, so that there is little true lateral wall (2.54).

In gill-breathing vertebrates the exchange of respiratory gases is directly from solution in water to solution in blood. From the cranial (arterial) end of the heart emerge two *ventral aortae* which traverse the ventral aspect of the pharynx, sending branches dorsally into the branchial arches, where they feed blood into capillary plexuses. These are drained by corresponding arteries, which join two *dorsal aortae* supplying the general circulation. As water is taken in through the mouth and passed back through the gill clefts, its dissolved oxygen diffuses through the pharyngeal endoderm of the gills to reach the blood, carbon dioxide diffusing out of the latter into the water. The intimate relationship between the developing mouth, branchial apparatus, and heart in water-breathing vertebrates is repeated in the embryos of their tetrapod descendants, but with many modifications necessary to changed respiratory function.

The human mandibular arch (2.55, 56) grows ventro-medially in the floor of the pharynx to meet its fellow in the midline, being situated between the primitive mouth and the cardiac (pericardial) swelling. The *hyoid arches* are caudal to the

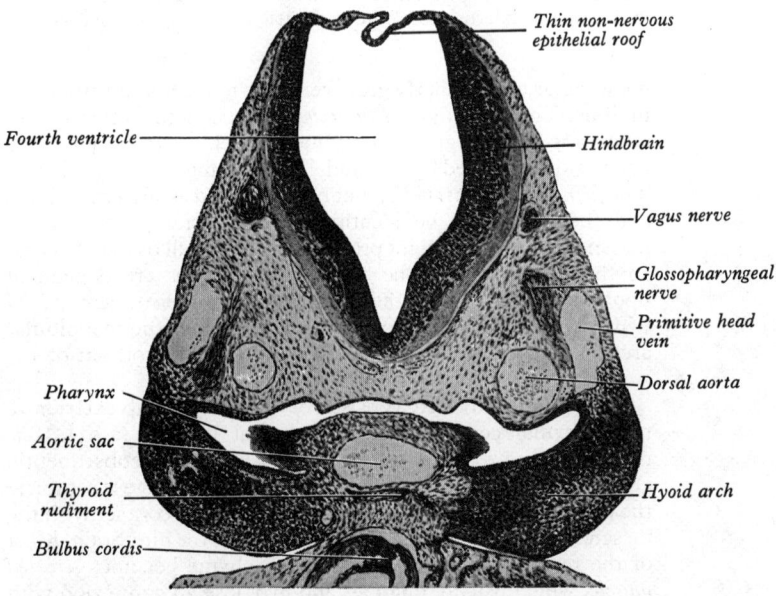

2.54 Coronal section through the head of a mole embryo, 4·5 mm long. The section passes through the hindbrain, the pharynx, the second (hyoid) and a part of the third branchial arches.

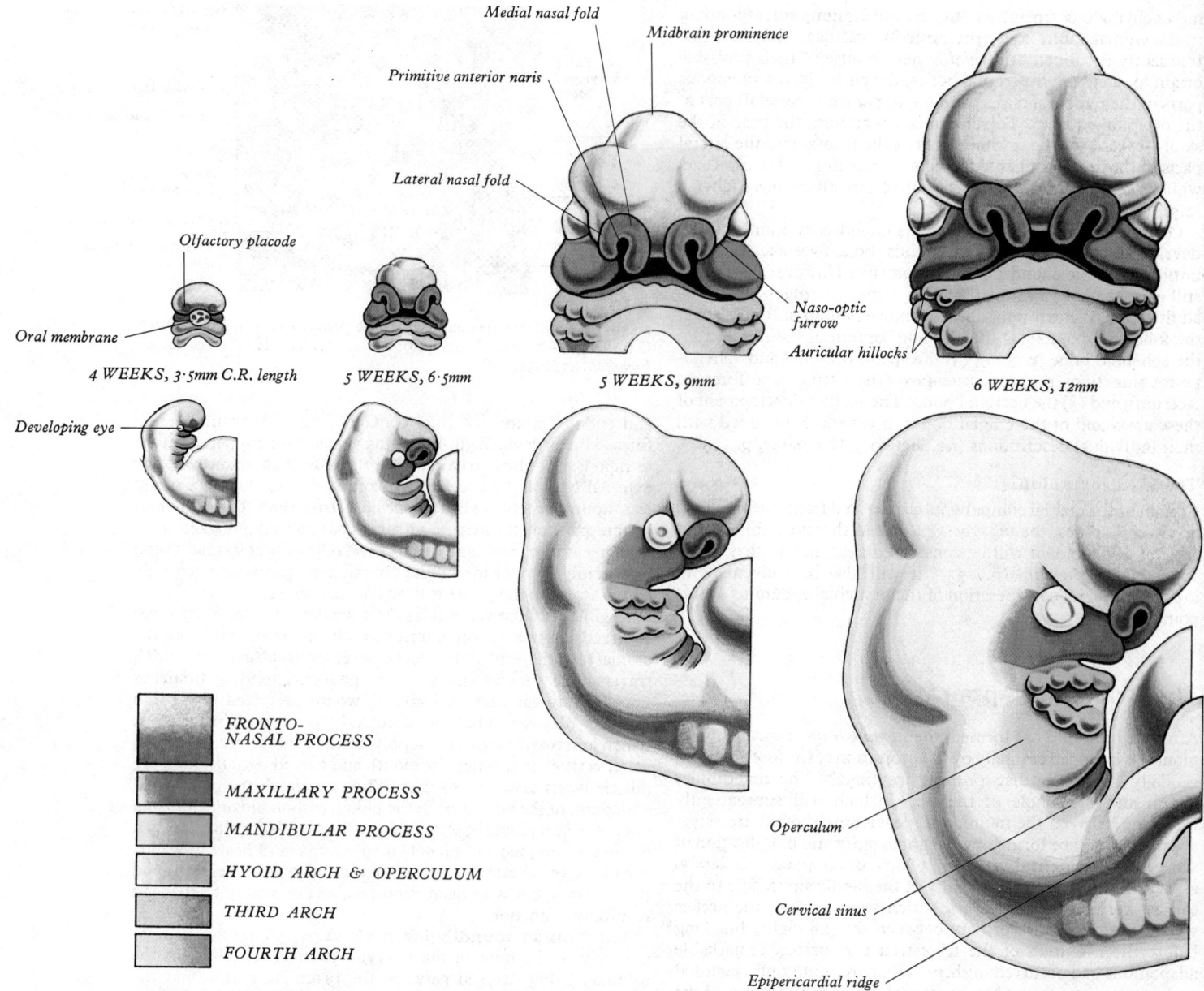

Medial nasal fold

Midbrain prominence

Primitive anterior naris

Lateral nasal fold

Olfactory placode

Oral membrane

4 WEEKS, 3·5mm C.R. length

5 WEEKS, 6·5mm

5 WEEKS, 9mm

Naso-optic furrow

Auricular hillocks

6 WEEKS, 12mm

Developing eye

FRONTO-NASAL PROCESS

MAXILLARY PROCESS

MANDIBULAR PROCESS

HYOID ARCH & OPERCULUM

THIRD ARCH

FOURTH ARCH

Operculum

Cervical sinus

Epipericardial ridge

2.55 A sequence of diagrams showing the superficial contributions of the facial processes and branchial elements to the development of the face, including the external nose, circumorbital structures, external acoustic meatus and pinna, and neck. All diagrams are drawn to scale. Note changes in general proportions and relative positions.

mandibular and similarly grow ventrally to meet and fuse in the midline. The *third* and *fourth arches*, especially the latter, do not attain any great degree of prominence, being largely sunk in a depression produced by the caudal overlapping of the hyoid arch. The *fifth* and *sixth arches* cannot be recognized as such externally, and they can only be identified by the arrangement of the mesenchyme and by slight projections in the wall of the pharynx. In their development the branchial structures are dependent upon the proximity of the pharyngeal endoderm, removal of which aborts branchial development. However, the mandibular arch is in part, at least, dependent upon the ectoderm of the stomatodeum.

Each branchial arch consists of an ectodermal exterior, a mesenchymal core, and an endodermal interior (**2.53**). The mesenchyme produces a *skeletal element*, which subsequently chondrifies either wholly or in part; and if this change is complete the element extends dorsally until it comes into contact with the mesenchymatous cranial base in the region of the hindbrain. Most of the remainder of the core of mesenchyme becomes *striated muscle*, which usually migrates and may lose all connexion with the skeletal elements in arches which cease to carry out their original respiratory function. The identities of these muscle masses, where they assume new functions, can nevertheless be traced by reference to their nerve supply. Motor nerves from the adjacent hindbrain (**2.54**) pass directly into the arches, which are ventral to it. However far muscle masses migrate from their sites of development, their original innervation almost always persists. The mandibular division of the trigeminal nerve innervates the musculature of the mandibular arch, the facial nerve supplies the hyoid arch, the glossopharyngeal the third arch, and the vagus and accessory nerves the rest of the arches. The recurrent laryngeal may be the nerve of the sixth arch and the superior laryngeal that of the fourth. The nerve of the fifth arch, itself difficult to identify in the human embryo, is uncertain.

The branchial arches play a large part in the formation of the face, oral cavity, neck, pharynx and larynx, but before the development and adaptations of each are detailed it is convenient to describe the events which lead to the construction of the face and nasal cavity.

THE FACE, NASAL CAVITY AND PALATE

While the **mandibular process** is invading the floor of the pharynx, the mesenchyme between the central aspect of the forebrain and the epithelial roof of the mouth proliferates and bulges under the ectoderm to form the *frontonasal elevation* or

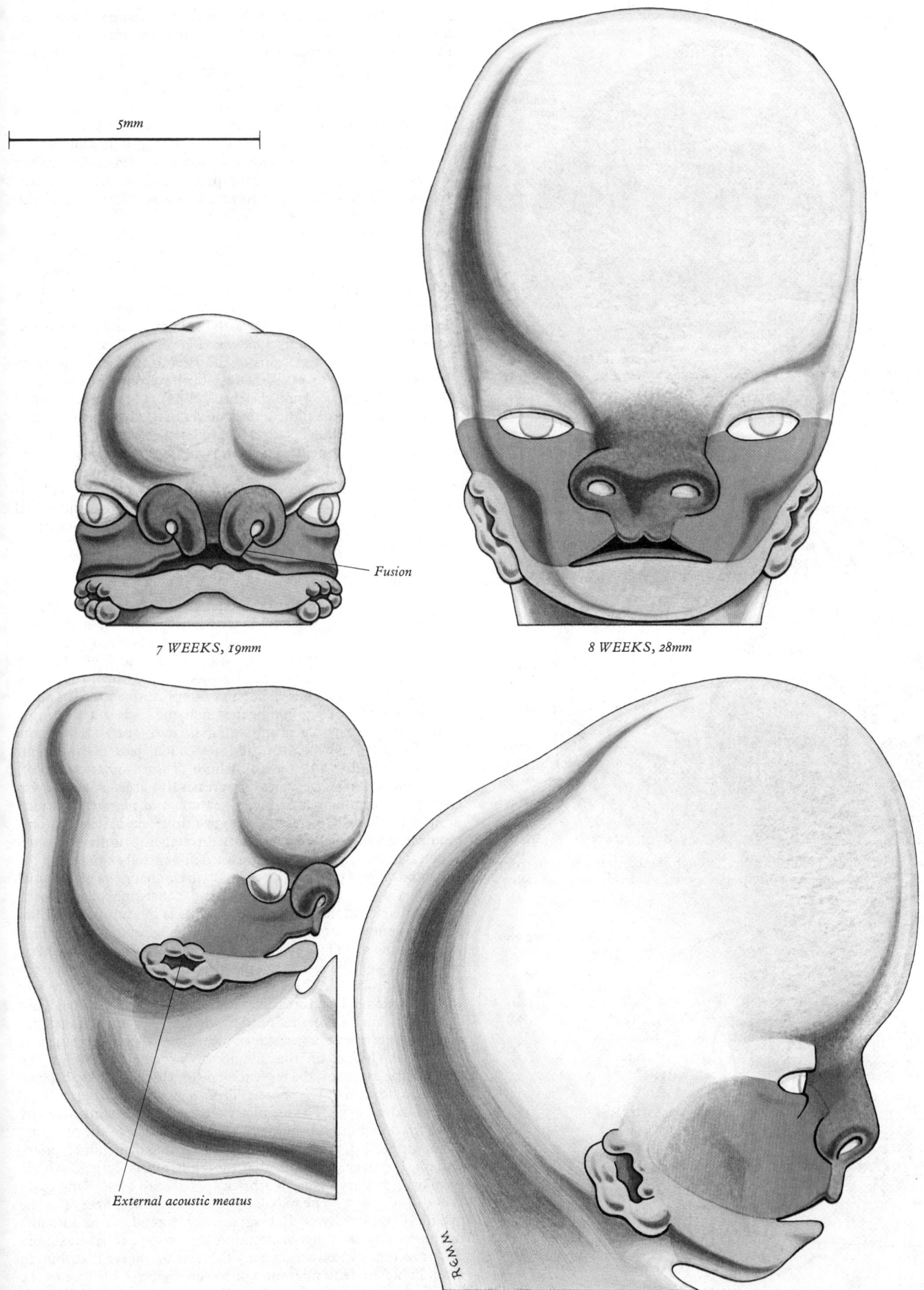

5mm

Fusion

7 WEEKS, *19mm*

8 WEEKS, *28mm*

External acoustic meatus

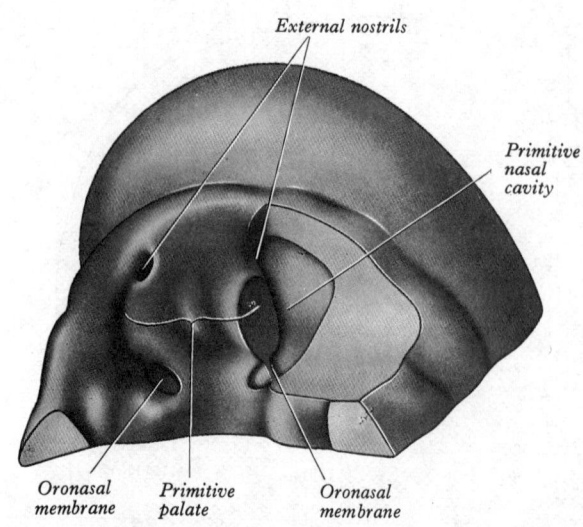

External nostrils

Primitive
nasal
cavity

Oronasal
membrane

Primitive
palate

Oronasal
membrane

2.56A The primitive palate of a human embryo in the seventh week. (From a model by K. Peter.) The figure shows the anterior part of the roof of the mouth; large parts of the left lateral nasal elevation and the left maxillary process have been removed to expose the left primitive nasal cavity.

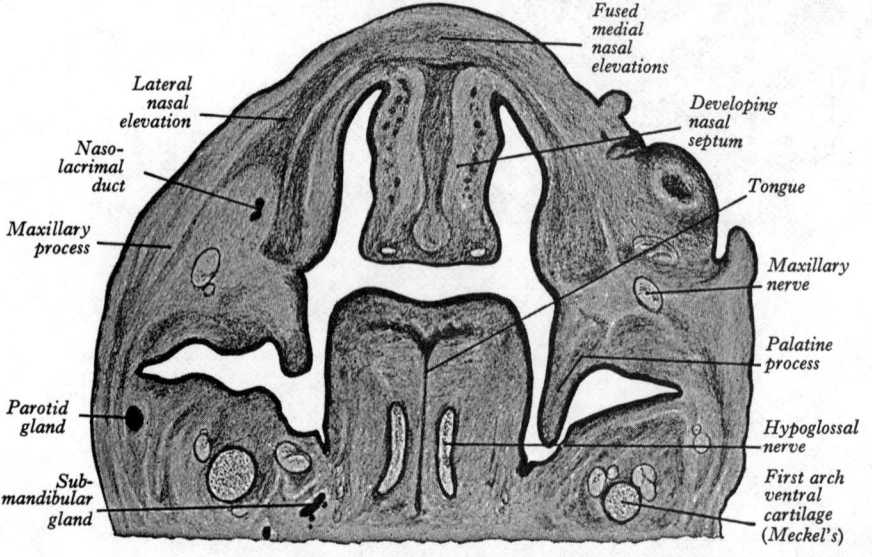

Fused
medial
nasal
elevations

Lateral
nasal
elevation

Developing
nasal
septum

Naso-
lacrimal
duct

Tongue

Maxillary
process

Maxillary
nerve

Palatine
process

Parotid
gland

Hypoglossal
nerve

Sub-
mandibular
gland

First arch
ventral
cartilage
(Meckel's)

2.56B Oblique coronal section through the head of a human embryo 23 mm long. The nasal cavities communicate freely with the cavity of the mouth.

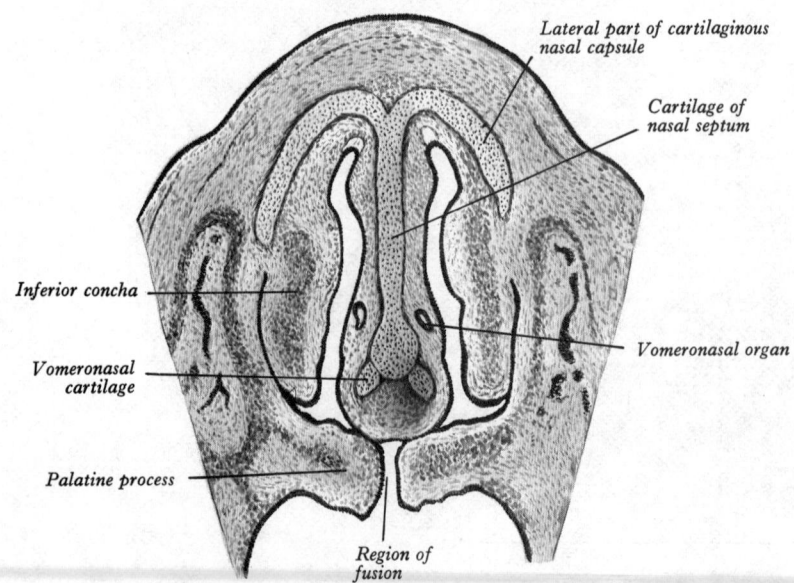

Lateral part of cartilaginous
nasal capsule

Cartilage of
nasal septum

Inferior concha

Vomeronasal
cartilage

Vomeronasal organ

Palatine process

Region of
fusion

2.56C A coronal section through the nasal cavity of a human embryo 28 mm long. (After Kollmann.)

process. During the fifth week a thickened plaque of ectoderm develops on each side ventrolateral to the frontonasal elevation, dividing this into *medial* and *lateral nasal elevations* or folds; these thickenings are the *olfactory* or *nasal placodes*. They are at first widely separated, and, as the elevations develop, they soon become depressed to form the *olfactory pits* (nasal pits). The lateral nasal elevations are the more prominent (**2.**55, 56B), but the medial nasal elevations, separated by the remainder of the frontonasal field, project caudally beyond the former. Extensions of mesenchyme from the medial processes into the roof of the stomatodeum proliferate to form the *premaxillary* or globular fields—the globular processes of His.

While these changes are progressing a somewhat triangular elevation swells ventrally from the cranial aspect of the dorsal region of the mandibular arch. This is the **maxillary process**, and like the frontonasal elevation it consists of proliferating mesenchyme covered by ectoderm. The maxillary process grows in a ventral direction and fuses with the lateral nasal fold, the two being at first separated by a *nasomaxillary groove* (**2.**55, Streeter 1948). The opposed margins of the lateral nasal and maxillary elevations growing together thus establish continuity between the side of the future nose and the cheek (**2.**55). The ectoderm along the boundary between them does not entirely disappear; it gives rise to a solid cellular rod, which at first develops as a linear surface elevation, the *nasolacrimal ridge*, and then sinks into the mesenchyme (Polityer 1952). Its caudal end proliferates to connect with the caudal part of the lateral nasal wall, while its cranial extremity later connects with the developing conjunctival sac. The solid rod becomes canalized to form the *nasolacrimal duct* (**2.**56B).

The Nasal Cavity

The rounded apex of the triangular maxillary process extends beyond the lateral nasal fold, crossing the caudal end of the olfactory pit to meet and fuse with the *premaxillary elevation* developing at the extremity of the frontonasal field. This closes off the lower or caudal edge of the olfactory pit, the upper part of the opening of which is thus defined as the primitive *external naris*. The growth of the surrounding mesenchyme leads to a deepening of the pit to become a primitive nasal cavity, or *nasal sac*, the epithelial wall of which, in the dorsocaudal part of its extent, retains contiguity with the epithelium of the stomatodeal roof. This contact area becomes progressively greater as growth continues, forming a thin layer, the *oronasal membrane*, which disappears later. Thereafter the primitive nasal cavity communicates with the stomatodeum through a primitive *choana*, which is at this stage still well forward or ventrally situated in the stomatodeal roof (Warbrick 1960). By these changes a new cranial boundary is set for the oral opening, consisting of the fused premaxillary and maxillary regions. This is the future upper lip, but it has not yet become separated from the deeper tissues which will form the maxillary alveolus (**2.**59). At the same time the nasal cavity acquires a floor through the fusion of the nasal folds and the maxillary process. At this stage the two external nares are still widely separated by an area derived from the frontonasal field, but this separation becomes reduced by the fusion of the premaxillary mesenchyme from the two sides. According to some investigators the mesenchyme of the maxillary processes invades the premaxillary regions, the mesenchyme of which is said to become buried, to form later the premaxilla or os incisivum (p. 341) (Boyd 1933; Baxter 1953). The maxillary mesenchyme is considered by some to contribute substantially to the formation of the philtrum of the upper lip, thus accounting for its maxillary innervation. Others, however, maintain that the philtrum is derived wholly from premaxillary tissue (Warbrick 1960; Keith 1948; King 1954; Wood *et al*. 1967). The maxillary nerve primarily innervates the maxillary mesenchyme but apparently extends later into the territory of the frontonasal elevation. (See also p. 1084.) It should be added that some workers deny that sensory nerve distribution is a reliable guide to migration of mesenchyme in the case of the maxillary process.

The Palate

Once the primitive nasal cavities are defined the ventral part of the roof of the oral cavity can be regarded as the *primitive palate* (**2**.56). It is formed by the premaxillary regions and maxillary processes, which become confluent and establish continuity with the thick median *primitive nasal septum*. As the head grows in size the region of mesenchyme between the forebrain and oral cavity increases greatly by proliferation, and the nasal cavities deepen, extending towards the forebrain. Simultaneously they also extend dorsally from the primitive choanae as two narrow and deep grooves in the oral roof (**2**.56) which are separated by a partition. The grooves and the partition deepen together, and the latter becomes the *nasal septum*, continuous ventrally with the *primitive* nasal septum (**2**.56B). The broad dorsocaudal border of the nasal septum is at first in contact with the dorsum of the developing tongue (**2**.56B), the right and left nasal cavities still communicating

former. The nasal cavities are thus extended dorsally, and the choanae reach their final position, leaving the caudal edge of the nasal septum free in about its dorsal quarter as the partition between them. Slightly later the dorsal extremities of the palatine processes, which extend dorsally beyond the choanae, fuse together to form the soft palate (**2**.57). This fusion, as is the case with most of the processes and elevations which form these regions, is more apparent than real, the epithelial cleft between the two processes being pushed to the surface by mesenchymal growth (Burdi and Faist 1967). There is later an upgrowth of mesenchyme from the third branchial arches into the palate and around the caudal margins of the auditory tubes, along a line corresponding in the final state to the palatopharyngeal arches (Baxter 1953).

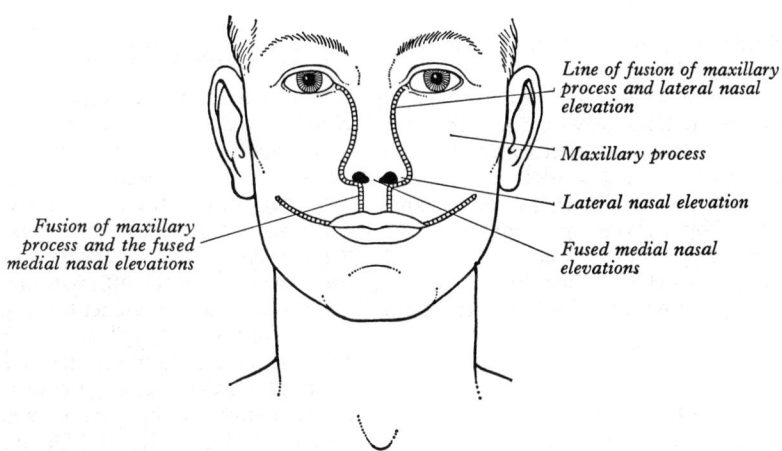

2.57 A diagram to show the parts of the adult face which are derived from the nasal elevations, and the maxillary and mandibular processes.

freely with the mouth except where the nasal floor is already established ventrally by the primitive palate.

During the sixth week the internal aspects of the maxillary processes produce *palatine processes*, which grow towards the midline but are for some time separated from each other by the tongue. At this stage the roof of the oral cavity projects ventrally beyond its floor, and the tip of the developing tongue actually lies in contact with the cranial (superior) surface of the primitive palate. A coronal section dorsal to this shows the maxillary palatine processes contiguous with the sides of the tongue and bent into a vertical position on each side of it (**2**.56B). With further growth, the mandibular region and the tongue are carried forwards (ventrally), and the lingual tip passes round to the caudal surface of the primitive palate. By some mechanism, still subject to much controversy (Kraus *et al.* 1966), the palatine processes assume a horizontal position which allows them to grow towards each other and thus to fuse (**2**.56C). The change of position occurs very rapidly in some experimental animals and may be due to a sudden increase in turgor in the processes (Ferguson 1977). Other views, such as active contraction of neighbouring striated and non-striated muscle, have been advanced (e.g. Wee, Wolfson and Zimmerman 1976). The change is also said to be rapid in the case of the human palate. These events obviously may have a direct bearing on maldevelopment of the palate (*vide infra*). Palatal elevation in mankind occurs during the eighth week; the development of the neck allowing some descent of the tongue and floor of the mouth, is then also occurring, and may be an additional factor in the elevation. This permits the palatine processes to grow medially along the inferior borders of the primitive choanae, uniting with them, except over a small area in the midline where a *nasopalatine canal* maintains connexion between the nasal and oral cavities for some time and marks the position of the incisive fossa. (The plates which form the primitive palate are sometimes known as *median* palatine processes, the maxillary contributions being then named the *lateral* palatine processes.) As the medial borders of the maxillary palatine processes fuse together, fusing also with the cranial free border of the nasal septum (**2**.62), the nasal and oral cavities are progressively separated and the tongue is excluded from the

On each side of the nasal septum, in a ventral or anterior position just above the primitive palate, the ectoderm is invaginated to form a pair of small diverticula, which extend dorsally and cranially into the septum. These are vestiges of the *vomeronasal organs*, whose openings are close to the junction between the two premaxillae and the maxillae; they are always rudimentary in mankind, but are well-developed auxiliary olfactory organs in many vertebrates (pp. 1055, 1142, 1144). For recent bibliographies in the field of facial development consult Sperber (1976) and Latham (1973).

Anomalies

Congenital malformations consequent upon arrest of development and failure of fusion of components in the formation of the face and palate are not uncommon. At the simplest, one maxillary process may fail completely to fuse with the corresponding premaxillary region (globular process), leading to a persistent fissure between the philtrum and lateral part of the upper lip on that side—*cleft* or 'hare' *lip*. A similar but rare malformation follows failure of fusion between the maxillary process and the lateral nasal elevations—*facial cleft*, in which the nasolacrimal duct is an open furrow, a condition usually associated with a cleft lip on the same side. The palatine processes may fail to fuse with each other and the nasal septum to variable degrees. In its severest form fusion is wholly lacking, leaving a wide fissure between the palatine processes through which the nasal septum is visible. On each side the premaxillary (primitive) parts of the palate are separated from the maxillary palatine processes by clefts which are continuous ventrally with bilateral clefts in the upper lip. In such cases the philtrum is a separate entity, continuous cranially and dorsally with the nasal septum. The floor of the nasal cavity is deficient throughout its extent and the choanae are not completed. Many varieties of milder degrees of cleft palate have been observed; the commonest type is unilateral, only one side of the nasal cavity being in communication with the mouth, the extent of the cleft being variable. In the mildest forms only the soft palate is cleft, or even merely the uvula. Such examples of arrested development may be associated with disturbances in embryonic nutrition during the second and the third months of gestation

(p. 136), and the grosser varieties are usually coupled with malformations in other regions of the body. In such cases the premaxillary region protrudes, with associated extension forwards of the nasal septum. For a recent discussion of this aspect of cleft palate consult Latham (1973).

The further growth of the face during the fetal period has received little attention, although this period is by no means characterized entirely by incremental growth. It is during fetal life that human facial proportions develop. The facial and cranial parts display different patterns of growth, though each influences the other. For an interesting analysis of the data observed from 280 fetuses consult Lavelle (1974).

Further Branchial Development

As has already been described to some extent, the branchial arches contribute extensively to the growth of the face, neck, nasal cavity and mouth, as they also do in the case of the larynx and pharynx. Although there are never complete clefts between the arches, the external ectodermal *branchial grooves* and internal endodermal *pharyngeal pouches* make various contributions. The first branchial groove does to some extent persist as the external acoustic meatus, but it is closed by the tympanic membrane. The remaining grooves disappear, except insofar as such derivatives of their corresponding pharyngeal pouches as the tympanic cavity, auditory tube, tonsil, thymus, parathyroid and thyroid glands can be counted.

ECTODERMAL DERIVATIVES

The ectoderm over the mandibular arches becomes the skin of the mandibular region of the face (2.55), and it also takes part in forming the tragus of the auricle (p. 1191). The ectoderm on the cranial aspects of these arches thickens along a curve which later becomes the *labiogingival* (vestibular) *groove*. This epithelial proliferation invades the subjacent mesenchyme and subsequently breaks down to separate the lower lip from the developing gum. Before this, however, an epithelial lamina develops from its internal surface (8.63) from which the enamel organs of the teeth will eventually be derived (p. 1293).

The first branchial groove is obliterated ventrally, as in all but the most ancient vertebrates. In man its dorsal end deepens to form the epithelium of the external acoustic meatus and the external surface of the tympanic membrane.

At the dorsal ends of the first, second and fourth grooves thickened patches of ectoderm appear, called the *epibranchial placodes*. These are closely related to the developing ganglia of the facial glossopharyngeal, and vagus nerves, to which they contribute (p. 174): other placodal cells also contribute to the trigeminal and vestibulocochlear ganglia.

At the end of the fifth week the third and fourth arches are sunk in a retrohyoid depression, the *cervical sinus*. Cranially the sinus is bounded by the hyoid arch, dorsally by a ridge produced by ventral extensions from the occipital myotomes and by mesenchyme developing into sternocleidomastoid and trapezius. Caudally, the smaller *epipericardial ridge* separates the sinus from the pericardium and curves cranially near the midline, and then with its fellow reaches the mandibular arch. The muscle cells which are often held to migrate from the occipital myotomes to the tongue follow the epipericardial ridge together with the hypoglossal nerve (but *see* p. 198). The established view that the sinus is obliterated by caudal growth of the hyoid arches to fuse with the cardiac elevation, excluding the succeeding arches from any part in the formation of the skin of the neck, has been criticized; an alternative view is that the sinus is reduced by gradual approximation of its walls from within outwards. During these changes the epibranchial placodes sink inwards as small vesicles which ultimately lose connexion with the surface ectoderm, and are then more closely associated with the glossopharyngeal and vagus nerves. The vesicle derived from the fourth groove placode is also for some time associated with the rudiments of the thymus and parathyroid arising from the third endodermal pharyngeal pouch (pp. 197–98 (Garrett 1948)), but

there is no evidence that the human thymus receives any contribution from this vesicle.

Experiment clearly shows that in lower vertebrates the facial, glossopharyngeal and vagus nerves derive elements from the overlying epibranchial placodes, or are dependent upon the associated placode for their full development. In man, the placodal vesicles associated with the glossopharyngeal and vagus nerves are believed to regress and disappear, abnormal persistence giving rise to *branchial cysts*, which are lined by stratified squamous epithelium and are situated between the carotid sheath and sternocleidomastoid (Wilson 1955). It is claimed that branchial cysts, lined by columnar epithelium, may be derived from persistent remnants of pharyngeal pouches, and that *branchial fistulae* may result from the second branchial cleft or possibly the cervical sinus (Martins 1961).

MESODERMAL DERIVATIVES

The mesoderm of the branchial arches (of mixed *unsplit lateral plate*, *neural crest*, and possibly *placodal* origin) is still 'pluripotential', as is demonstrated by the variety of its derivatives. These include the following: (1) An endothelial tube, an *aortic arch*, develops in the substance of each arch, connnecting the aortic sac, ventral to the pharynx (p. 182), to the dorsal aorta, which is dorsal to it. There are thus six pairs of aortic arches, but the appearance of the fifth pair in man is questioned; its position and connexions are subject to disagreement (Shaner 1921; Barry 1951). For further details *see* p. 187. (2) Some of the mesodermal cells condense to form a skeletal element in each arch; and this is typically a bar of cartilage connected by its dorsal end to the caudal region of the chondrocranium and meeting its fellow ventrally in the midline. (3) Other mesodermal cells differentiate into 'special visceral' striated muscle tissue; and partly associated with this developing *branchial musculature* is to be noted the invasion of the mesoderm by a nerve supply. The proportion of branchial mesoderm from each of its various sources noted above that contributes to the different types of derivative is unknown.

Typically each arch is invaded by two nerves, derived from the hindbrain. One runs along the rostral border of the arch and is hence described as *post-trematic*, because it is behind or caudal to the cleft or *trema* rostral to the arch. Similarly the other nerve, which is close to the caudal border, is *pre-trematic*—with respect to the cleft caudal to it. In the human embryo the pre- and post-trematic nerves can be identified only with certainty in the first, mandibular arch (*vide infra*).

The mesoderm of the branchial region also contributes to the fibro-areolar and other connective tissues of the lower face, tongue and neck. In the third arch it gives rise to some cellular elements of the *carotid glomus* or body (*see* p. 1462), which may also receive neuroblasts from the glossopharyngeal nerve and the developing superior cervical sympathetic ganglion (Boyd 1937; Adams 1958; Rogers 1965). Similarly *para-aortic* bodies are developed in association with the fourth and sixth arches, receiving accessions of neuroblasts from the ganglia of the vagus nerve (Hammond 1941). Kando (1975) has investigated the development of the carotid glomus in mice and has discussed the difficulties in identifying the sources of its neural elements (see also p. 1462).

Derivatives of the Branchial Skeleton (2.58A)

The first arch contains on each side a dorsal and ventral element. The former represents the *palatopterygoquadrate bar*, a prominent element in earlier vertebrates, forming part of the upper jaw, but much reduced in mammals. In human embryos it is transient in appearance, and its contribution to permanent cranial structures, such as the maxilla, is uncertain, (but *vide infra*). The *ventral cartilage* (of Meckel, 2.58A) extends from the developing otic capsule into the mandibular arch, meeting its fellow at its ventral end. The dorsal end of Meckel's cartilage becomes separated, and is often held to form the rudiments of both *malleus* and *incus*. However, there is strong paleontological (Romer 1970) and comparative anatomical (Shute 1956) evidence that the incus is to be regarded as a homologue of the *quadrate bone* of reptiles, and it is therefore probably more correctly regarded as a derivative of the palatopterygoquadrate cartilage. This cartilage may also contribute to the ala major of the sphenoid bone. A more precise developmental analysis of this region is needed. Beyond the rudiment of the malleus, the intermediate part of Meckel's cartilage disappears, but its sheath persists as the *anterior malleolar* and *sphenomandibular ligaments*. The ventral part, much the largest, is enveloped by the developing mesenchymatous mandible (p. 148); a small fraction of this, extending from the mental foramen almost to the site of the future symphysis, probably becomes ossified from invading mandibular tissue, into which it is incorporated, while the remainder of the cartilage is ultimately absorbed.

The cartilaginous element of the second arch (Reichert's cartilage) also extends from the otic capsule to the midline. Its dorsal end also separates and becomes enclosed in the developing tympanic cavity as the *stapes*. Thereafter the cartilage gives rise to the *styloid process*, *stylohyoid ligament*, the *lesser cornu* and probably the *cranial rim* of the body of the *hyoid bone* (**2**.58A).

Chondrification does not occur in the dorsal parts of the skeletal elements of the third to sixth arches. The ventral cartilage of the third arch becomes the *greater cornu* of the *hyoid bone* and the *caudal part* of its *body* (the whole of the body may be formed from the third arch cartilage). Alternatively, the hyoid body may be derived from cartilage formed in the base of the hypobranchial eminence (p. 197) and thus from third arch tissue alone (Frayer 1926), acquiring its connexion with the second arch cartilage secondarily.

The final adaptations of the cartilages of the skeletal elements in the fourth, fifth, and sixth arches are a source of much disagreement, but the following represents the general view. The thyroid cartilage develops from the fourth and fifth arches, which may also give rise to the arytenoid, corniculate, and cuneiform cartilages. The cricoid cartilage may be derived from the sixth arch cartilage, or it may be a modified tracheal ring. The epiglottis is developed in the substance of the hypobranchial eminence and probably not from 'true' branchial cartilage.

Derivatives of Branchial Musculature (2.58B)

The muscle mass of the mandibular arch forms the *tensor tympani*, *tensor veli palatini*, and the *masticatory muscles*, including *mylohyoid* and the *anterior belly* of *digastric*, all being supplied by the mandibular nerve, the post-trematic nerve of the arch. The tensor tympani retains its connexion with the skeletal element of the arch through its attachments to the malleus, and the tensor veli palatini to the base of the medial pterygoid process, which may be derived from the dorsal cartilage cartilage of the first arch, but the masticatory muscles transfer to the mandible, a dermal bone. The muscles derived from the hyoid arch mesoderm for the most part migrate widely, but retain their original nerve supply from the facial. The *stapedius, stylohyoid,* and *posterior belly of digastric* remain attached to the hyoid skeleton, but the *facial musculature, platysma, auricular muscles,* and *epicranius* all lose connexion with it. Their migration is facilitated by the early obliteration of most of the first cleft and pouch (p. 178). (This cleft, the spiracle in fishes, is already much reduced in all but the earliest vertebrates.)

The muscle masses from the remaining arches are adapted to form the musculature of the *pharynx, larynx* and *soft palate*. The stylopharyngeus can be attributed to the third and the cricothyroid to the fourth arch; the rest of the laryngeal muscles are derived from the sixth arch, but the precise origin of the remaining palatal muscles and the pharyngeal constrictors is uncertain in man. A mixed origin, partly from branchial mesoderm, and partly from adjacent myotomes, has been attributed to sternocleidomastoideus and trapezius (McKenzie 1955).

Neural Elements of the Branchial Arches

The nerves of the branchial arches, derived from the hindbrain, immediately enter the dorsal ends of them (**2**.54). They are typically mixed, their motor component supplying the muscles of the arch and their sensory fibres innervating the skin and mucous membrane derived from the region. In fish the trunks of the branchial nerves and their ganglia are close to the dorsal ends of

the true clefts existing in these forms, each sending a post-trematic branch into its own arch and a pretrematic branch into the arch cranial to this. In mammals both types of branch can be identified in the first arch, but only a single nerve can be identified with certainty in the second to sixth arches, with the exception of the fifth, the nerve of which is unknown and may have disappeared.

The trigeminal mandibular division is the post-trematic nerve of the first arch, the chorda tympani being generally regarded as its pretrematic nerve derived from the facial. The latter supplies the second arch, the glossopharyngeal the third, the superior laryngeal branch of the vagus the fourth, and the latter's recurrent laryngeal branch the sixth. In lower vertebrates the fifth arch is also supplied by a vagal branch.

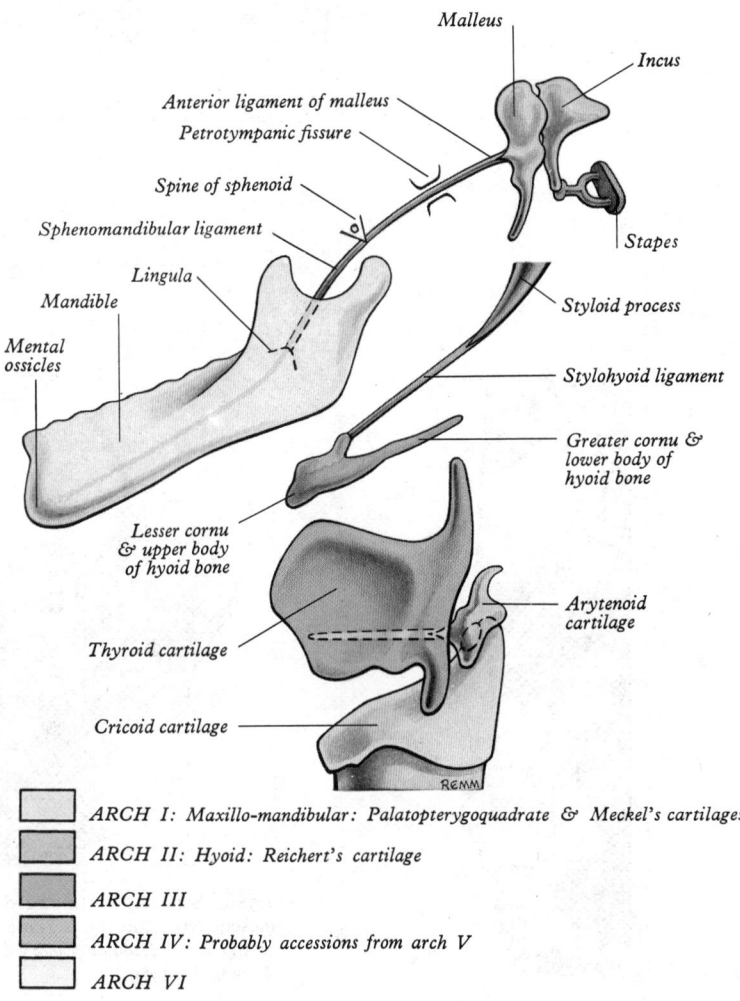

ARCH I: *Maxillo-mandibular: Palatopterygoquadrate & Meckel's cartilages*

ARCH II: *Hyoid: Reichert's cartilage*

ARCH III

ARCH IV: *Probably accessions from arch V*

ARCH VI

2.58A Schema illustrating the skeletal derivatives (osseous and cartilaginous) of the branchial arches (viscerocranium). (Modified from *Basic Human Embryology*, by Williams, Wendell-Smith and Treadgold, 1969.)

The difference in the courses of the recurrent laryngeal nerves can be explained by the development of the aortic arches. The nerve enters its sixth arch caudal to the aortic arch, retaining this position on the left side and hence being caudal to and looping round the ligamentum arteriosum in its final disposition. However, on the right, owing to the disappearance of the dorsal part of the sixth aortic arch and the whole of the fifth, the nerve loops round the caudal aspect of the *fourth* aortic arch, i.e. the subclavian artery.

DERIVATIVES OF THE PHARYNGEAL POUCHES

The first four pharyngeal pouches appear in sequence cranio-caudally, and their endoderm approaches the ectoderm of the overlying branchial grooves to form thin *closing membranes* (**2**.53,

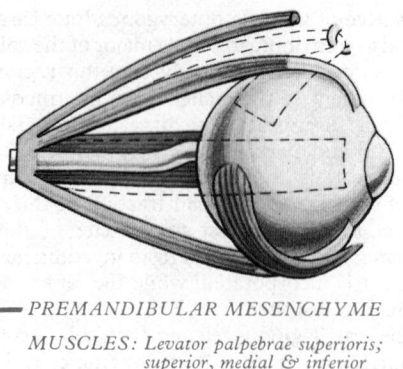

PRO-OTIC 'SOMITES'

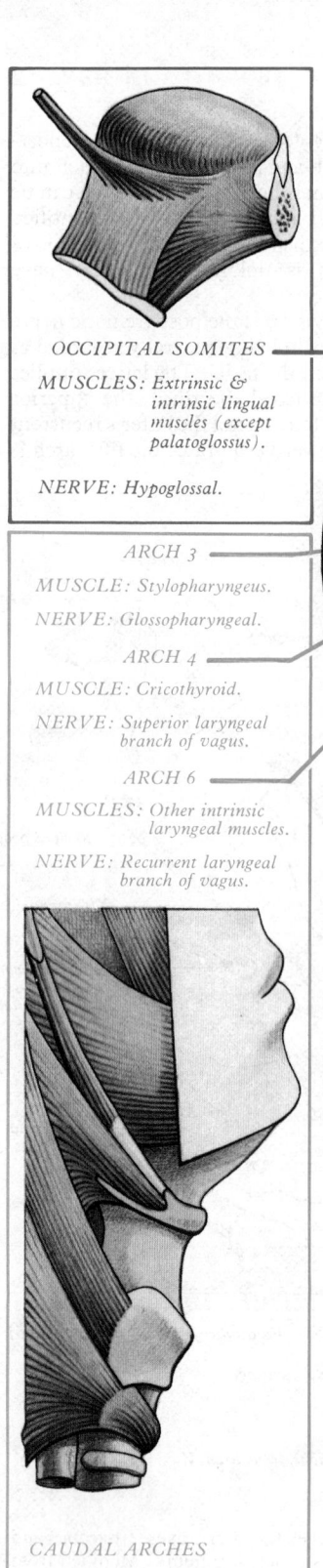

OCCIPITAL SOMITES

MUSCLES: *Extrinsic &*
intrinsic lingual
muscles (except
palatoglossus).

NERVE: *Hypoglossal.*

ARCH 3

MUSCLE: *Stylopharyngeus.*

NERVE: *Glossopharyngeal.*

ARCH 4

MUSCLE: *Cricothyroid.*

NERVE: *Superior laryngeal*
branch of vagus.

ARCH 6

MUSCLES: *Other intrinsic*
laryngeal muscles.

NERVE: *Recurrent laryngeal*
branch of vagus.

PREMANDIBULAR MESENCHYME

MUSCLES: *Levator palpebrae superioris;*
superior, medial & inferior
recti; inferior oblique.

NERVE: *Oculomotor.*

MAXILLOMANDIBULAR MESENCHYME

MUSCLES: *Superior oblique*
& lateral rectus.

NERVES: *Trochlear & Abducent.*

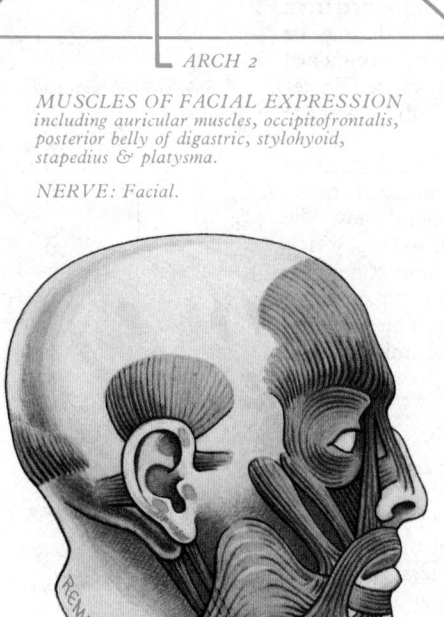

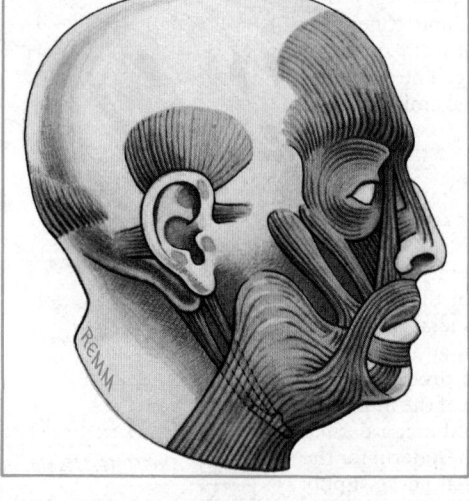

ARCH 2

MUSCLES OF FACIAL EXPRESSION
including auricular muscles, occipitofrontalis,
posterior belly of digastric, stylohyoid,
stapedius & platysma.

NERVE: *Facial.*

ARCH 1

MUSCLES OF MASTICATION
Temporalis, masseter, pterygoids,
mylohyoid, anterior belly of
digastric, tensor veli palatini
& tensor tympani.

NERVE: *Mandibular.*

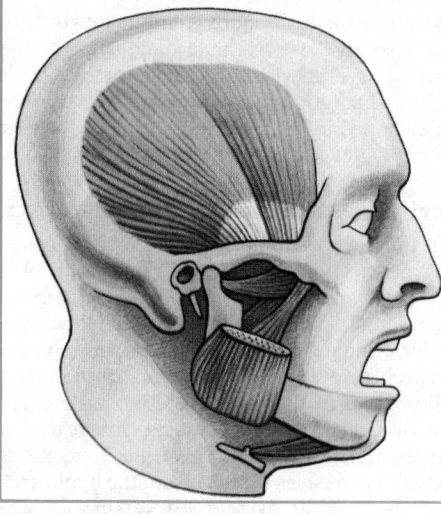

2.58B Schema illustrating the muscular derivatives of the branchial mesenchyme and the pre-otic and post-otic cranial 'somites'. Modified from *Basic Human Embryology*, by Williams, Wendell-Smith and Treadgold, 1969.)

CAUDAL ARCHES

Remaining palatine muscles
& constrictors – precise
sources uncertain.

NERVE: *Cranial accessory*
via branches of vagus.

2.58A, B). The blind recesses of the second, third and fourth pouches are prolonged dorsally and ventrally as angular, wing-like diverticula. From the fourth a diverticulum grows caudo-ventrally and is at first demarcated from the pouch by a groove in which may occur a transient fifth aortic arch artery. From this diverticulum a fifth pouch may develop and establish a connexion with ectoderm. The remainder of this diverticulum is the *ultimobranchial body*. This, together with the fourth pouch and the transitory fifth, when present, constitute the *caudal pharyngeal complex*. Its communication with the cavity of the pharynx is the *common pharyngobranchial duct*. The ultimobranchial body is almost a constant feature of vertebrate development

(Watzka 1955). Its form in the *human* embryo, however, has been a matter of controversy. Apparently it is incorporated into the rest of the caudal pharyngeal complex and contributes to the development of the lateral thyroid rudiment (p. 198). Ultimobranchial bodies exist in the adults of many lower vertebrates, and *calcitonin* has been isolated from such tissue (Copp *et al.* 1967). There is thus a strong presumption that the parafollicular cells of the human thyroid gland which are a source of calcitonin, are derived from ultimobranchial tissue. (See Proceedings of a symposium on: *Thyrocalcitonin and the C cells*, Taylor 1968; and Bussolati and Pearse 1967: also pp. 262, 1449).

The further development of the endodermal derivatives of the pharyngeal pouches is intimately associated with that of the mouth, pharynx and larynx, and will hence be considered with them later (pp. 197–200).

Development of Locomotor Structures

EARLY DEVELOPMENT OF THE LIMBS

Towards the end of the fourth week the limbs begin to appear as small elevations, the **limb buds**, from a slight lateral ridge extending along each side of the trunk (2.52, 59). The forelimb bud shows first, at the level of the more caudal cervical segments, the hind-limb bud being level with the lumbar and upper sacral segments. Each bud contains a small proliferating mass of mesenchyme derived from the somatopleure, covered by ectoderm. At this stage the limb bud is not more than a slight accentuation of the lateral ridge, or *crest of Wolff*, which—though not identified in all embryos—is considered to correspond to the *finfold of Gegenbaur*, the hypothetical origin of vertebrate limbs. From their initial appearance, the limb buds display an ectodermal crest at their developing apices, the *apical ectodermal ridge* or *cap*; and although there are some disagreements over details, there is general concurrence that the *interaction* between the mesenchymal core and the apical ectoderm is essential to the development of the limb. Most of the experimental evidence admittedly applies only to amphibian (Tschumi 1957), reptilian (Milaire 1957) and avian (Saunders 1948; Amprino and Camosso 1955; Amprino 1968; MacCabe *et al.* 1974) embryos, but some confirmation of similar behaviour in mammalian limb buds has been recorded, and observations on human embryos also (Horizons XII to XVII), have indicated the occurrence of ectodermal ridges (O'Rahilly *et al.* 1956). This evidence indicates that the ectoderm of the apical region of the limb bud, though itself relatively undifferentiated, is necessary to the development of successive accretions of mesenchyme as the bud grows. The potentiality of the initial mass of somatopleuric mesoderm is limited to the formation of the most proximal limb elements, i.e. the girdle region. The mesodermal development of the successive segments of the limb is induced, in a proximo-distal sequence by the ectodermal ridge, the cells of which apparently do not proliferate to add to the covering of the growing bud. The necessary increase in ectoderm for this appears to be derived by interstitial growth and from the body wall ectoderm. The apical ridge also appears to establish the polarity of the limb with respect to its pre- and post-axial borders and the arrangement of its basic vascular pattern. The evidence for these views has been surveyed by Zwilling (1961). Specific histochemical changes associated with morphogenetic events in the developing limb bud have been reviewed by Milaire (1962). Ultrastructural studies of the zone of epithelio-mesenchymal interaction at the apex of the bud have been reported (Kelly 1973). The evidence suggests that the sculpturing of the interdigital clefts is a process involving mesenchymal cell death with phagocytic removal (p. 79) accompanied by active epithelial invagination.

The mesenchyme of the limb buds is invaded by the ventral rami of adjoining spinal nerves—fourth cervical to second thoracic for the fore-limb, twelfth thoracic to fourth sacral for the hind-limb. The axial region of the mesenchyme proliferates and condenses to produce the skeleton of the limb, chondrification and ossification following as the limb grows. The musculature of the limbs is developed *in situ* from the mesenchyme surrounding the skeletal elements. It is important to emphasize that there is no migration of myotomic mesoderm into the limb bud, the so-called plurisegmental nature of the muscles in the limbs referring solely to their nerve supply. Such muscles as latissimus dorsi, with extensive attachments to the axial skeleton when fully developed, reach them secondarily by active migration.

The early stages of development of fore- and hind-limb buds are alike, except that the former precedes the latter in time by a few days in the human embryo (2.112). Differentiation in general proceeds in a proximo-distal direction. Flexion creases appear indicating the sites of elbow, wrist, knee and ankle; and the hands and feet are at first mere plate-like expansions (2.59, 2.111).

Towards the end of the sixth week there is still little difference in shape and posture between the arm and leg, the long axis of each being approximately orthogonal to the trunk with the prominences of elbow and knee both directed laterally (2.59, 2.111). At this time the pre- and post-axial borders of the limbs are respectively cranial and caudal in orientation, as their names indicate.

Some preliminary observations on human limb buds, removed from embryos of 4 to 6 weeks and cultured for 4 to 18 days *in vitro*, have been reported (Yasuda 1973). There was a general retardation of development *in vitro*, but the effect was more marked in fore- than in hind-limb buds, and amongst tissues, cartilage was the least retarded.

Limb Rotation

During the seventh and eighth weeks differential growth rates are exhibited by the limbs which bring them into a position of adduction towards the ventral aspect of the trunk, and flexion at elbow and knee is increased. By the end of the eighth week the limbs have attained the *fetal position*, with the conspicuous differences that the elbow points caudally and the knee in a rostral direction. This divergence of final positioning is of the utmost importance in the future functioning of the limbs (p. 350); it is achieved by a kind of rotation in each limb around its axis, presumably in part due to growth changes, for it occurs before any joints are fully defined. The upper limb rotates laterally, so that its pre-axial border becomes lateral and its ventral surface rostral or anterior. The changes are the reverse of this in the hind-limb; medial rotation carries the pre-axial border into a medial position, turning the dorsal surface to face anteriorly. The same re-orientation is illustrated by the radius and tibia, both pre-axial primitively in their limbs, the former becoming the lateral fore-limb bone, the tibia taking up a medial position in the foreleg. Inevitably this affects the hand and foot, whose pre-axial digits, thumb and great toe, take up lateral and medial positions. It should be noted that reverse *forearm* rotation at a later stage in most mammals, and the retention of supination-pronation activity in the primate forearm, modify this arrangement (p. 350). The nerve supply follows these movements, and this explains the cutaneous innervation of the 'lateral' aspect (pre-axial border) of the human upper limb from the more cranial nerves of the brachial plexus (7.200), and the 'medial' border by its more caudal nerves. In the hind-limb the pattern is similar but less clear; but as a generalization, the upper spinal nerve elements in the lumbosacral plexus innervate the medial border, the lower ones the lateral (7.209, 213). (See also p. 1117.)

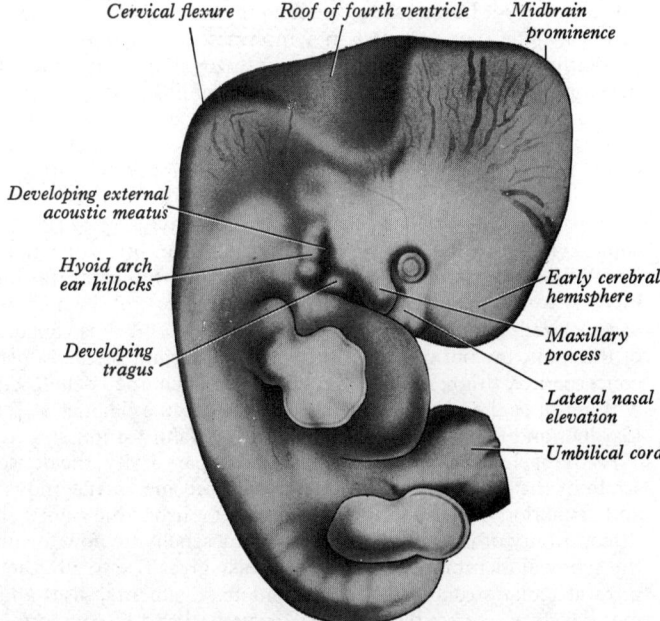

Cervical flexure — *Roof of fourth ventricle* — *Midbrain prominence*

Developing external acoustic meatus

Hyoid arch ear hillocks

Developing tragus

Early cerebral hemisphere

Maxillary process

Lateral nasal elevation

Umbilical cord

2.59 A human embryo, 15·5 mm long. Compare the upper and lower limb buds, concerning their stage of differentiation and degree of rotation. (G. L. Streeter, *Contr. Embryol.*, **32**, 1948.)

The foot and hand resemble each other closely at first, as flattened expansions at the terminations of the limb buds. The mesenchymal tissue in the periphery of these plates condenses to outline the pattern of the digits, and the thinner intervening regions break down from the circumference inwards. If this process is incomplete or becomes arrested varying degrees of webbing or *syndactyly* result. For a recent extended study of the phylogeny and development of the skeletal and muscular elements of hand and foot, with a full bibliography, consult Čihák (1972).

The further development and ossification of the skeletal elements of the limbs is described with the individual bones (*see* Section 3, Osteology). For a synopsis and bibliography concerning the chronology of appearance of ossific centres in human embryos consult O'Rahilly and Gardner (1972).

DEVELOPMENT OF JOINTS

Except in the vertebral column and the caudal part of the chondrocranium, which are derived from sclerotomes (pp. 118, 140), the mesenchyme from which other skeletal elements are derived at first shows no differentiation into primordia of individual bones. The skeletal mesenchyme appears as a continuous condensed mass which is not initially demarcated clearly from the surrounding myogenic tissue. Centres of chondrification and ossification develop in this mesenchymal core and rapidly extend to delineate individual skeletal elements, each of which contains its own centre or focus of change, the process of bone or cartilage formation spreading in an orderly and characteristic manner from it (Streeter 1949). Each element becomes limited by a compact layer of undifferentiated cells from surrounding tissues; this layer proliferates to produce cartilage cells and osteoblasts, which contribute to growth by surface accretion. This lamina gradually becomes more clearly demarcated from the underlying cartilage or bone as *perichondrium* or *periosteum*, continuing to generate chondroblasts and osteoblasts (p. 215). It appears to be an important factor in the determination of bone growth in the formation of individual elements (Fell 1939; Barnett *et al.* 1969). (For further details of development, *see* Section 3, Osteology.)

Connecting adjacent skeletal elements as they become defined from continuous mesenchyme masses are regions which do not undergo any change into cartilage or bone, but persist as plates of *interzonal mesenchyme*. These are the sites of future joints, and their development varies according to the type of joint formed. In fibrous joints the interzonal mesenchyme is converted into white fibrous tissue, as the definitive connecting medium between the bones involved. In the case of synchondroses it becomes cartilage of the hyaline type, whereas in symphyses the tissue formed is predominantly fibrocartilage. The interzonal mesenchyme of developing synovial joints becomes trilaminar, due to the appearance of a more tenuous intermediate zone between two dense strata next to the cartilaginous ends of the skeletal elements of the region. These latter are continuous peripherally with the adjoining perichondrium and, like it, are chondrogenic and thus concerned with growth of cartilaginous epiphyses (p. 241). In some synovial joints, however, this trilaminar interzone may be modified (Gardner and Gray 1953). The intermediate stratum merges with the general mesenchyme of the limb, which is vascularized. From this, a cuff condenses as the fibrous capsule of the joint, in continuity with the perichondrium of the bones concerned. A thinner layer of vascular mesenchyme is enclosed within this as the precursor of the synovial capsule (Haines 1947; Gardner and Gray 1950–1; Gardner and O'Rahilly 1968).

As the skeletal elements chondrify and in part ossify, the dense strata of the interzonal mesenchyme also become cartilaginous, and cavitation of the intermediate zone establishes the cavity or discontinuity of the joint. The synovial mesenchyme now forms the synovial membrane, and probably also gives rise to all other intra-articular structures, such as tendons, ligaments, discs and menisci. In joints containing discs or menisci and in compound articulations more than one cavity may appear initially, sometimes merging later into a single one. As development proceeds thickenings in the fibrous capsule can be recognized as the specializations peculiar to a particular joint. In some instances, however, such accessions to the fibrous capsule are derived from neighbouring tendons, muscles, or cartilaginous elements.

For a recent review of the literature concerning the chronology of developmental events in human embryonic limbs, consult O'Rahilly and Gardner (1975).

DEVELOPMENT OF SKELETAL MUSCULATURE

With the exception of certain muscles of the head and neck, which are developed from branchial mesenchyme (p. 152), and the limb muscles, which develop *in situ* from the mesenchyme of the limb buds (p. 153), all the somatic (striated or voluntary) muscles are derived from myotomes (p. 118). Typically, each myotome divides into dorsal (*epaxial*) and ventral (*hypaxial*) regions, the former sited dorsolateral to the vertebral column and innervated by the dorsal ramus of the corresponding spinal nerve, the latter migrating ventrally in the body wall or somatopleure and innervated by the corresponding ventral ramus. Primitively, in fishes, these two primary divisions of the myotomes are separated by vertebral transverse processes, and a fibrous septum extends from these to the lateral body line. The myogenic masses derived from the myotomes may divide longitudinally or tangentially, and the resultant parts may remain separate, e.g. intercostal muscles, or they may fuse with corresponding parts of adjacent myotomes to form sheets of muscle, such as the abdominal oblique and transverse layers. In mammals the derivatives of the ventral regions of the myotomes may subsequently migrate over the derivatives of their dorsal parts. Consequently, such muscles as the serrati posteriores, which reach the vertebral spines superficial to the erectores spinae, are nevertheless supplied by ventral rami of spinal nerves, indicating their origin from ventral myotomic mesoderm. It should be added that from experiments on chick embryos it is apparent that some parts of the intercostal and anterolateral abdominal musculature are not derived from myotomes but from the lateral plate of mesoderm (Straus and Rawles 1953). However, observations on amphibian (Detwiler 1955) and mammalian (Theiler 1957) embryos support their myotomic origin. Some muscles may also *migrate* in a rostral or caudal direction; the latter as in the case of latissimus dorsi, which is derived from lower cervical myotomes, and is innervated accordingly.

Although it is usually stated that the lingual muscles are derived from three or four occipital myotomes (Bates 1948; Deuchar 1958), with which the morphology of their supply from the hypoglossal nerve accords, the evidence for this in the human embryo is negligible, and earlier views that they arise *in situ* from mesenchyme associated with the developing tongue have not been controverted (Frazer 1926). The hypoglossal nerve is composite, representing the ventral roots of an uncertain number of segmental nerves cranial to the cervical series. It is considered to have acquired cranial status in the earliest land tetrapods, its exclusion from the modern amphibians being a secondary adaptation.

In lower vertebrates the extrinsic ocular muscles are developed from three paired head cavities which are considered to be three persistent pro-otic somites (Neal 1918). There is no direct evidence of this origin in man, although it is claimed that the most rostral of these cavities, in the *premandibular*, is derived from mesoderm originating in the prechordal plate, *see* p. 115 and Gilbert (1952). This condensation is held to give rise to the muscles innervated by the oculomotor nerve. The remaining extraocular muscles are considered to develop from a mass of mesenchymal cells dorsal to the region of conjunction of the mandibular arch and maxillary processes (which may correspond to the mandibular and hyoid head cavities identifiable in sharks, and which has been termed *maxillo-mandibular mesenchyme* by some authors). The dual origin of this mass is at least in accord with the innervation of these muscles by the trochlear and abducent nerves. The same mesenchymal mass is said to contribute to the formation of the scleral tunic of the eye (Gilbert 1957).

In the differentiation of striated muscle the primitive muscle cells, *myoblasts*, may arise from somite or non-somite (branchial or limb bud) mesenchyme. The myoblasts multiply by mitosis, but further recruitment from neighbouring mesenchyme also occurs. The cells elongate to form multinucleated *myocytes*, in which the nuclei move to the periphery as a myofibrillar structure and cross-striation begins to appear. Their maturation is described in more detail on p. 513. New fibres are said to form until mid-fetal stages of development, sometimes by splitting of pre-existing fibres. Thereafter growth of muscles is accomplished by enlargement of individual fibres (Cuajunco 1942). Muscle spindles can be identified as early as the twelfth week (Cuajunco 1940). The tendons of muscles develop independently in mesenchyme and their connexion with their muscles is secondary.

In man and other higher vertebrates some derivatives of myotomes degenerate and some disappear entirely. Others may persist as fibrous tissue vestiges; in some instances regions formed from myotomes which are muscular in ancestral forms may become aponeurotic or ligamentous, examples being the aponeuroses of the abdominal muscles and the sacrotuberous ligament (p. 474).

Non-striated and cardiac muscle are not obviously segmental in origin but arise from the mesoderm of the splanchnopleure and develop *in situ* (pp. 506, 515, 518).

Development of Skin and Appendages

The epidermis and its specialized appendages—hairs, nails, sudoriferous and sebaceous glands—are derivatives of the ectoderm, while the *dermis* (corium) is formed from the somatic layer of the mesoderm with some contribution from dorsolateral aspects of the mesodermal somites (p. 118). The ectoderm at first consists of a single stratum of cuboidal cells, but by the sixth week these have proliferated to form a double layer—a superficial *periderm*, or *epitrichium*, of flattened cells, and a subjacent *stratum germinativum*. The cells of the latter are at first cuboidal but become columnar. By multiplication and differentiation of these cells the characteristic layers of the epidermis are developed. The periderm is merely a temporary covering stratum. By the sixth month most of the periderm, which becomes keratinized, has disappeared, and the strata granulosum, lucidum, and corneum of the epidermis are established. The superficial cornified cells, together with sebaceous secretion and the remains of the periderm, form a cheesy or caseous material, the *vernix caseosa*, which may exercise the function of protecting the underlying epidermis from maceration by amniotic fluid. Vernix caseosa is characteristic of the surface of the fetus during its final three months of intrauterine existence, finally desquamating soon after birth. The developing epidermis has a high content of glycogen which, apart from its metabolic role, may aid in repelling the amniotic fluid, which also has a high concentration of glycogen.

Ultrastructural studies of the human periderm in recent years have shown that its cells possess microvilli, which are in contact, of course, with the amniotic fluid (Breathmach and Wylie 1965; Whittaker and Adams 1971). The functional significance of this transient surface layer remains nevertheless obscure.

At an early, but still undetermined age, neural crest cells invade the developing skin, and these elements differentiate into both epithelial and connective tissue pigment cells, such as *melano-blasts*. These assume a branching form and are hence named *dendritic cells*. In negro fetuses these cells become truly melanoblastic by developing granules of melanin (Rawles 1948; Billingham and Silvers 1960; Boyd 1960), which can be transferred to other cells in post-natal life. In less pigmented races similar cells exist in the epidermis and deeper tissues which, though normally containing little or no pigment, retain a melanogenic potentiality by which, given appropriate stimulation, such as intense sunlight, they are able to form melanin granules.

Ectodermal areas in the head and branchial region develop special propensities, producing such structures as the ocular lens and the ganglia of certain cranial nerves. These *ectodermal placodes* are described elsewhere (pp. 148–150). For the olfactory and otic placodes refer to pp. 174, 178, 225.

The dermis is a mesodermal derivative, and it is usually stated that some, at least, of its cells develop from the lateral wall of the somites which are hence named *dermatomes*. It is also held that dermal tissue is partly derived from mesenchyme associated with the somatopleuric sheet of lateral plate mesoderm and also the neural crest. Perhaps all such regions contribute; at present available evidence is not sufficient to support positive quantitative statements.

Towards the end of the third month the mesoderm next to the epidermis begins to condense and define the dermis, deep to which areolar connective tissue appears. A month later dermal papillae can be identified, and the characteristic patterns of ridges on the ventral hairless skin of the extremities, best seen and most familiar as fingerprints, are quickly established and remain substantially unchanged in the individual apart from growth in size (Cummins and Midlo 1961; Hale 1952). The histological elements of the dermis differentiate at various times; thus, collagenous fibres begin to appear and to aggregate during the third month, whereas elastic fibres become identifiable two or three months later.

Hairs begin to appear in the third month as solid cylinders of epidermis which grow obliquely into the dermis (**2.60**), each *hair bud* having an external layer of columnar cells and a core of polygonal elements (Hale 1952). The deepest part of the bud expands into a *hair bulb*, which becomes invaginated by a mesodermal *hair papilla*. The central cells become keratinized to form the substance of the hair itself; and as further cells are added from the inner layer of the bulb, these also contribute to the shaft of hair, which lengthens and begins to project above the skin surface. Further growth is due to continued proliferation of the epidermal cells around the mesodermal papilla (see also p. 1212). Hairs appear first on the face, and by the end of the third month there is a general covering over the body of somewhat scattered fine hairs, the *lanugo*. These are shed before birth and are replaced by thicker hairs arising in a new set of follicles. Melanoblasts become associated with developing hairs at an early stage, and these produce melanin in dark-haired individuals during the later months of gestation (Boyd 1950).

Sudoriferous glands first begin to develop in the skin of the palms and soles during the fourth month, appearing like hairs as solid downgrowths of epidermis into the dermis. These elongate and their deeper parts coil to form the main secreting part of the sweat glands. A lumen develops and opens superficially at about the seventh month. Both the duct and the coiled 'body' of each gland is double-layered, the inner epithelium being surrounded by a layer of myoepithelial cells (pp. 39, 521), outside which is a

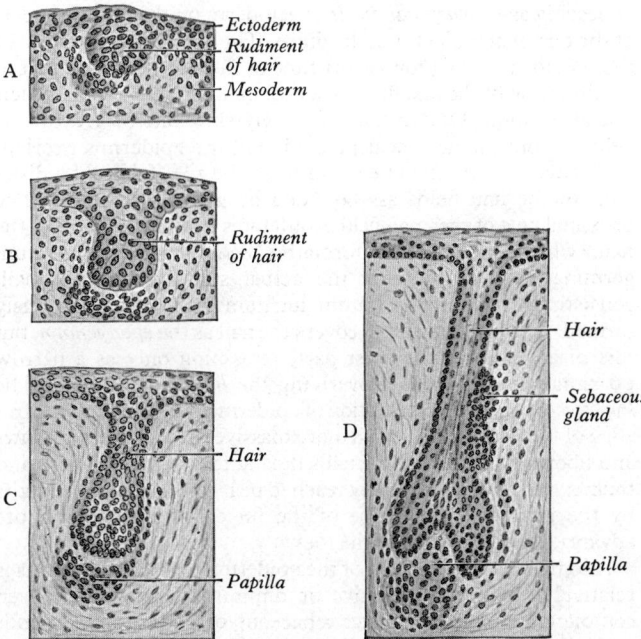

2.60 Successive stages in the development of hair.

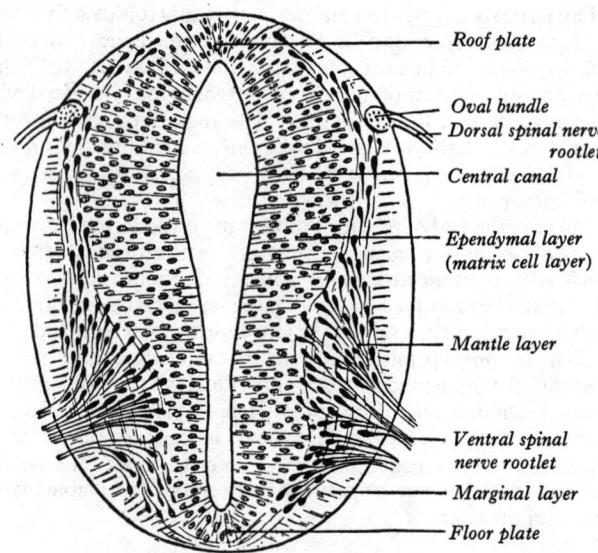

Roof plate

Oval bundle
Dorsal spinal nerve
rootlet

Central canal

Ependymal layer
(matrix cell layer)

Mantle layer

Ventral spinal
nerve rootlet

Marginal layer

Floor plate

2.61A A transverse section through the developing spinal cord of a human embryo four weeks old. (After His.)

basement membrane; the myoepithelial cells are contractile. Some glands arise from the superficial parts of hair follicles, acquiring a surface opening later. Specialized sudoriferous glands are formed in some sites, such as in the skin of the axillae, eyelids and external acoustic meatuses (Montagna and Ellis 1962).

Sebaceous glands develop as solid epidermal buds from the cuboidal cells of hair follicles during the fifth month. The buds extend into the mesenchyme, branching into several oval alveoli, the lining cells of which are derived from the stratum germinativum. These proliferate, the older cells being pushed centrally, where they degenerate to form sebum, this being extruded into the hair follicle. Sebaceous glands develop independently of such follicles in the nostrils, eyelids (where they form tarsal glands), and in the anal region. The arrectores pilorum develop near the sebaceous glands and hair follicles, but at some little distance from them, arising independently from the mesenchyme; their attachment to follicles is a secondary development (Pinkus 1958). Some of the independent sebaceous glands do not develop fully until after birth.

Nails are demonstrable as rudiments in the third month, appearing as *primary nail fields* of ectoderm on the dorsal aspects of the terminal segments of the digits. The thickened ectoderm of these fields is in fact for a short time at the *tips* of the digits; but proliferation of the nail fields in a proximal direction defines their dorsal position. Due to their relatively slow rate of growth the fields become somewhat depressed, and the epidermis overlaps their sides and proximal ends to form the *nail folds*. The distal ends of the nail fields are bounded by a shallow groove. The proximal part of each nail field proliferates to form the roots of the nail, which becomes the formative zone. From the stratum germinativum of the root the actual substance of the nail, consisting of modified stratum lucidum cells, is continuously formed. The stratum at first covers the nail as the *eponychium*, but this disappears for the most part, remaining only as a narrow proximal crescentric fold overlying the *lunule* of the nail. The *hyponychium* is an accumulation of epidermal cells beneath the free edge of the nail. (It is a much more massive development in claws and hooves.) Growth of the nails during fetal life is gradual, and their extremities have merely reached near to the tips of the digits by the time of birth, those of the fingers being rather more advanced than the nails of the toes.

Anomalous development of the epidermis and its derivatives is relatively common. Excessive or diminished growth, or even complete absence may affect sebaceous or sudoriferous glands and hair, either locally or generally. Similarly, the epidermis may be excessively pigmented (*melanism*), or lack melanocytes

(*albinism*). Excessive keratinization leads to *ichthyosis*. A *naevus* or 'mole' is a locus of excessive pigmentation.

Mammary glands are considered to be much modified sudoriferous glands, and as such they are basically ingrowths from the ectoderm, which forms their ducts and alveoli, supported by vascularized connective tissue derived from the mesenchyme. In embryos at about the fifth or sixth week two ventral bands of somewhat thickened ectoderm, the *mammary ridges*, extend from axilla to inguinal region, and in many mammals paired mammary glands develop at intervals along these ridges. In the human embryo the ridges are not prominent features, and a single pair only develops in the pectoral region. The ridges disappear later in embryonic life, but before this the cranial third of each begins to show proliferation to form the two rudiments of the glands. Supernumerary rudiments may form anywhere along the path of the mammary ridges, and these may develop into actual mammae or merely accessory or supernumerary nipples.

As the mammary primordium develops, its ectodermal ingrowth branches into fifteen to twenty solid buds of ectoderm which will become the lactiferous ducts and their associated lobes of alveoli in the fully formed gland. These are surrounded by mesenchyme which forms the connective tissue and fat. By proliferation and elongation and further branching the alveoli are formed and the duct system defined. During the last two months of gestation the ducts become canalized, and the epidermis at the point of original development of the gland forms a small *mammary pit*, into which the lactiferous tubules open. Around the time of birth, or slightly after it, the nipple is formed by mesenchymal proliferation. Should this fail the ducts open into shallow pits, a malformation known as inverted nipple. At birth the mammary glands are alike in their stage of development in both sexes, and in both some transient secretory activity may be observed, due presumably to circulating prolactin in the mother (Smith 1959). In males thereafter the mammary glands remain undeveloped, but in females at puberty, in late pregnancy, and during the period of lactation they undergo further developmental changes (p. 1436). For reviews of the prenatal histogenesis and ultrastructural appearances of mammary tissues consult—Hughes (1950); Raynaud (1971); Tobon and Salazar (1974).

The enamel of the teeth is also an ectodermal derivative (*see* p. 1296).

The Nervous System and Special Sense Organs

The gross initial appearance of a neural plate, and the development within it of a median neural groove bounded on each side by neural folds, which meet dorsally and fuse during the fourth week to form a neural tube, have already been outlined (p. 111).

Reference has also been made to the wealth of comparative experimental data, largely from submammalian forms, which indicate the fundamental role of *primary embryonic induction* in these early stages of development of the nervous system (p. 111).

Briefly, during gastrulation, the chorda-mesoderm, which is invaginated through the dorsal lip of the blastopore, forms a midline strip in the roof of the archenteron which is continuous cranially with the endoderm of the prechordal plate (p. 112), and becomes closely applied to the superjacent ectoderm. For a period the latter is competent to react to the inductive influence of the chorda-mesoderm, and in this manner, neural plate formation is initiated. The early midline neural tube is thus co-extensive with the notochord, stretching from the dorsal lip of the blastopore (primitive knot) to the oropharyngeal membrane. Further analyses also demonstrated a regional specificity in the inductive capacity of the various rostro-caudal regions of the archenteric roof. The region first invaginated, i.e. the endoderm of the prechordal plate and rostral tip of the notochord, induces the formation of forebrain and eyes, etc. (the so-called *archencephalon*); the intermediate part of the chorda-mesoderm induces hindbrain and associated structures (the *deuterencephalon*), whilst the remaining (caudal) roof of the archenteron operates as a *spino-*

caudal inducer promoting the formation of spinal cord and neighbouring muscle masses.

Although the search for the chemical identity of inducing substances continues, as pointed out on pp. 83, 85, there has, in recent years, been a great change of emphasis on the part of developmental biologists. Nevertheless, despite the profound developmental modifications introduced by the secondary reduction of yolk in mammalian ova associated with viviparity (p. 74), it is widely assumed that similar primary mechanisms operate throughout the *Chordata*, including mankind. Thus, from the outset, the development of the nervous system is heavily dependent upon morphogenetic influences from neighbouring non-nervous structures.

(For excellent reviews of early neurogenesis consult Hughes 1968; Jacobson 1970; Gaze 1970; Bellairs 1971, 1974; Gottlieb 1974.)

THE SPINAL CORD

When the neural tube is closing, its walls consist of a single layer of columnar *neural epithelial cells*, the extremities of which abut on internal and external limiting membranes. The mechanism of rounding up of the neural plate into a neural tube has been studied particularly closely in amphibia (Burnside 1971). The columnar cells increase in length and develop numerous longitudinally disposed microtubules, whilst the borders of their luminal ends are firmly attached to adjacent cells by junctional complexes, the cytoplasmic aspect of the complexes being associated with a dense paraluminal web of microfilaments (*see* p. 36; **1**.37; **2**.5, also Watterson 1965). It is proposed that this disposition of organelles imparts a slight wedge conformation on the cells resulting in neural groove, and eventually, neural tube formation. As the cells elongate their nuclei become clustered at varying depths in the deeper part of the wall (i.e. nearer the internal limiting membrane) and, for a period, the epithelium is *pseudostratified* with an inner nucleated zone, and an outer zone composed of the peripheral cytoplasmic processes of the cells. Soon, however, some of the peripheral cytoplasmic processes become detached from the (basal) external limiting membrane and rounded cells appear close to the inner membrane which by their repeated mitotic division, form descendants, which *migrate* outwards to take up an intermediate position in the wall of the tube. Histologically at this stage therefore, the wall of the tube presents three zones or layers (**2**.61–63). The internal **ventricular zone** (variously termed the *germinal, primitive ependymal* or *matrix layer*), consists of the nucleated parts of the columnar cells, and the round cells undergoing mitosis. The **intermediate zone** or *mantle layer* consists of the migrant cells from the divisions occurring in the deeper layer, just described. The outer **marginal zone** for a period consists of the external cytoplasmic processes of some of the original columnar cells, but they are soon invaded by tracts of axonal processes which grow from neuroblasts developing in the intermediate zone, together with other varieties of non-nervous cells (glioblasts, and later, vascular endothelium and perivascular mesenchyme).

HISTOGENESIS OF THE NEURAL TUBE

The classical view of neural tube histogenesis was first propounded by Wilhelm His in 1890 and soon, sometimes with minor modifications, gained quite wide acceptance (e.g. Kolliker 1896; Ramón y Cajal 1909). His proposed that almost from the first the wall of the tube was *stratified* and contained a *variety* of distinct cell types—primitive *spongioblasts*, *neuroblasts* and *germinal cells*. (Some contemporary cytologists even believed that the early neuroepithelium was syncytial, but this gained little acceptance.)

The primitve spongioblasts, originally elongate cells attached to both limiting membranes, with their nuclei in the ventricular zone, were considered to differentiate into a number of *sustentacular cell* varieties. By losing contact with one or both limiting membranes, they may differentiate into either astroblasts and astrocytes, oligodendroblasts and oligodendrocytes, or, retaining an internal attachment, into the definitive ependymal

cells which line the central canal. The *neuroblasts*, one of the cell types found in the intermediate zone, differentiated into the wide array of neurons. The round deeply placed *germinal cells* or *medulloblasts* (Glees 1963) he regarded as undifferentiated stem cells, which by repeated division gave further generations of both spongioblasts (*glioblasts*) and neuroblasts. On this view, therefore, the early neural epithelium was regarded as a *heterogeneous* grouping of cells and their various derivatives developed *simultaneously*.

The meticulous cytological studies of Sauer (1935 a, b; 1936) led him to oppose the classical view, and he maintained that the early neural epithelium, including the deeply placed ventricular mitotic zone, consisted of a *homogeneous* population of pluripotent cells, the varying appearances merely reflecting different phases in a *proliferative cycle*, the sequence being termed by Sauer *inter-kinetic migration*. In the intervening years much experimental evidence based upon colchicine studies (e.g. Watterson *et al.* 1956), spectrophotometric nuclear analysis (e.g. Sauer and Chittenden 1959), and upon autoradiographic studies following the distribution of cells at varying times after nuclear labelling with tritiated thymidine (e.g. Sidman *et al.* 1959, 1970; Fujita 1963; Fujita and Fujita 1963), and more recently ultrastructural studies (e.g. Hinds 1971) has supported the latter proposition. The scheme amplified by Fujita in a series of subsequent publications is illustrated in **2**.62A. The ependymal layer of earlier workers (termed by Fujita the *matrix layer*) is considered to be populated by a single basic type of *matrix cell*, and to exhibit three '*zones*' (the M or mitotic, the I or intermediate, and the S or synthetic zones). As they pass through a complete inter-mitotic and mitotic cycle, the matrix cells show an 'elevator movement' progressively approaching and then receding from the internal limiting membrane. DNA replication occurs whilst the cells are extended and their nuclei occupy the S zone; they then enter a pre-mitotic resting period whilst the cells shorten and their nuclei pass through the I zone. The cells now become rounded close to the internal limiting membrane (in the M zone) and undergo mitosis; thereafter they elongate again, their nuclei passing through the I zone during the post-mitotic resting period, finally to enter the synthetic zone once again. The daughter cells may then start another *proliferative cycle*, but others, presumably after a 'critical' or 'quantal' mitosis (*see* p. 84), rapidly migrate outwards (i.e. radially) and differentiate into *neuroblasts* as they approach and enter the adjacent stratum; possibly this differentiation is initiated as they pass through the I zone during the post-mitotic resting

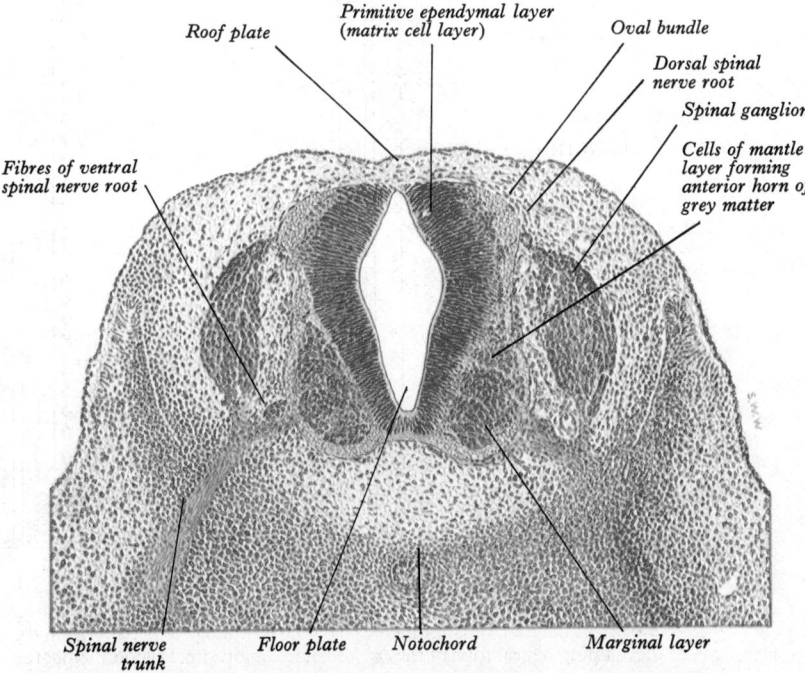

Roof plate

Primitive ependymal layer (matrix cell layer)

Oval bundle

Dorsal spinal nerve root

Spinal ganglion

Cells of mantle layer forming anterior horn of grey matter

Fibres of ventral spinal nerve root

Spinal nerve trunk

Floor plate

Notochord

Marginal layer

2.61B A transverse section of the developing spinal cord in the cervical region of a human embryo early in the sixth week. C. R. length=8 mm.

period. The *proliferative cycle*, with the production of clones of vast numbers of neuroblasts, continues for a time; but eventually neuroblast formation wanes, and the mitotic activity of the remaining matrix cells then begins to decline as their progeny differentiate into ependymal cells and the various *macroglial cell* varieties.

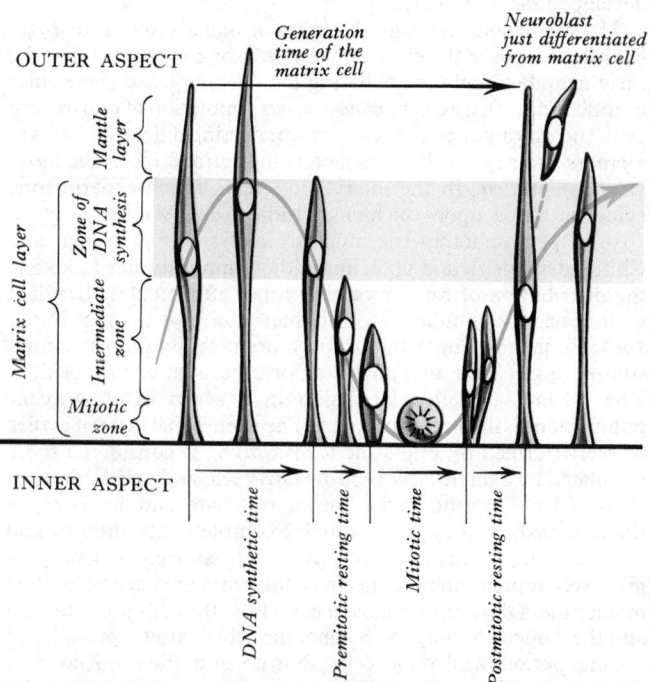

2.62A   A diagram showing the cytogenetic cycle in the matrix cell layer and mantle layer in the wall of the developing neural tube. Note the various zones and their associated time scales, through which the matrix cell nuclei pass during their postulated 'elevator movement' or 'interkinetic migration' which is described in greater detail in the text. (From S. Fujita, with the kind permission of the author and the *Journal of Comparative Neurology*.)

It should be emphasized here that differences of opinion persist concerning the precise details of the proliferative cycles, and differentiating mitoses in different species, sites in the central nervous system, and the possible mechanisms involved. Further, considerable confusion has obtained between different neuroembryologists in relation to the *terminology* to be adopted for the various 'layers' or 'zones' at different times and places in the developing neural tube. An international group of neurocytologists (The Boulder Committee: *see Anatomical Record*, **166**, 1970) proposed an unambiguous nomenclature which is increasingly adopted. They termed the early pseudostratified neuroepithelium, in which the 'elevator movement' occurs, the *ventricular zone* (its further development is summarized below, on pp. 160, 161, and in **2.63A**, **B**). For an excellent review of the biological and terminological problems posed, and an extensive bibliography consult Berry (1974), and for further comments on the natural history of neurons—p. 860.)

Illustration **2.63A** summarizes the main stages of development of the neural tube and the nomenclature proposed by the Boulder Committee—the complete series of stages A–E applying to the cerebral neocortex: only stages A–C are relevant in the context of the spinal cord. As detailed in the caption, it will be seen that the early pseudostratified *ventricular* zone (V) is followed by the sequential appearance of *marginal* (M), *intermediate* (I), *subventricular* (S) zones and, concurrently with the latter, the early *cortical plate* (CP)—*see* p. 173. It is unfortunate that this use of the term *zone* for these *major ontogenetic strata* does *not* correspond in any way to the M, I and S zones of Fujita detailed above (and in **2.62A**), the latter applying to subdivisions of the early neural tube only.

Illustration **2.63B** (which is based on the work of Berry and Rogers 1965, and Berry 1974) is included here for comparison with **2.63A**; it stems from labelling studies on the *patterns of migration* of neuroblasts in the developing neocortex of the rat and will be briefly mentioned later on p. 173.

In the future spinal cord the *roof plate* and *floor plate* of the neural tube do not participate in the cellular proliferation affecting the lateral walls and hence remain thin. Their cells contribute largely to the formation of ependyma.

The neuroblasts of the lateral walls of the tube are large and at first round or oval. Soon they develop processes at opposite poles

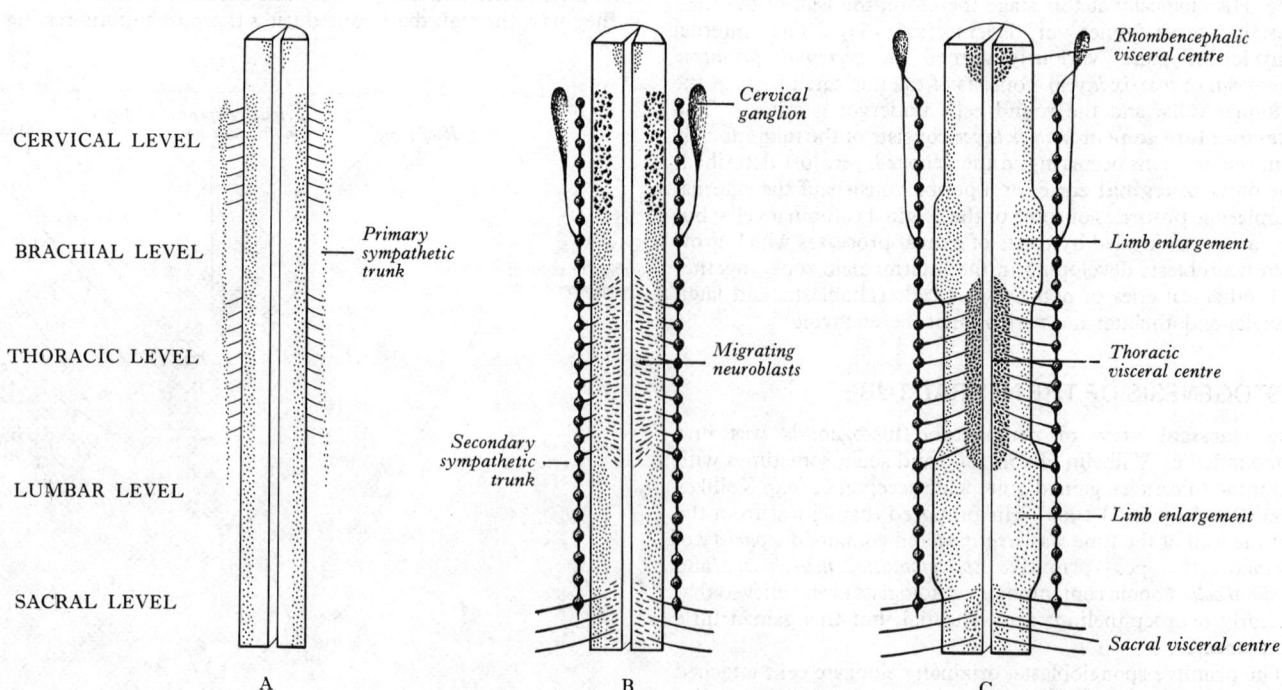

2.62B   A diagram illustrating the changing morphogenetic patterns in the developing neural tube of the *chick* seen in longitudinal section. Note the progression from A to C as the initial simple homogeneous columnar organization (A) is transformed into the definitive pattern (C). These changes follow variations in the turnover rate and differentiation rate of

neuroblasts in different regions of the tube, combined with migration of neuroblasts in some regions. For further description see the text. (From V. Hamburger, with the kind permission of the author and the New York Academy of Science.)

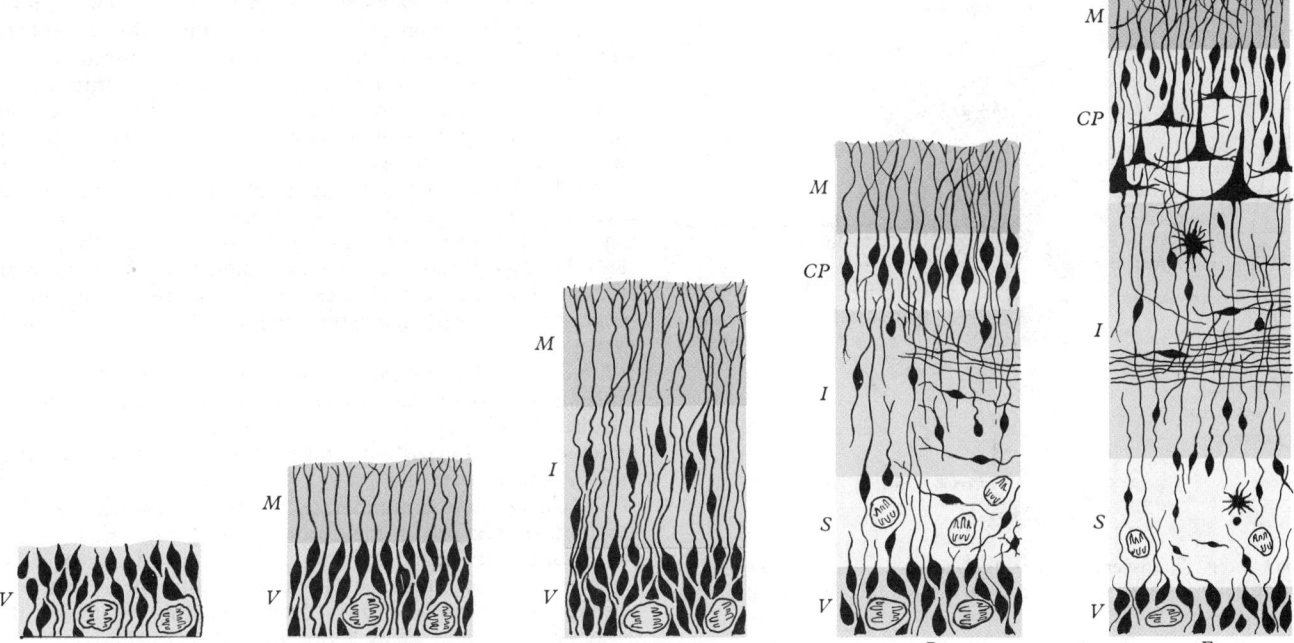

2.63A Schema of five sequential stages (*A* to *E*) in the development of part of the neural tube's wall in section to form cerebral cortex. (Modified from: Boulder Committee recommendations, *Anat. Rec.*, **166**, 1970.)

Abbreviations: V—ventricular zone; M—marginal zone; I—intermediate zone: S—subventricular zone; CP—cortical plate. See accompanying text and also p. 860 for details.

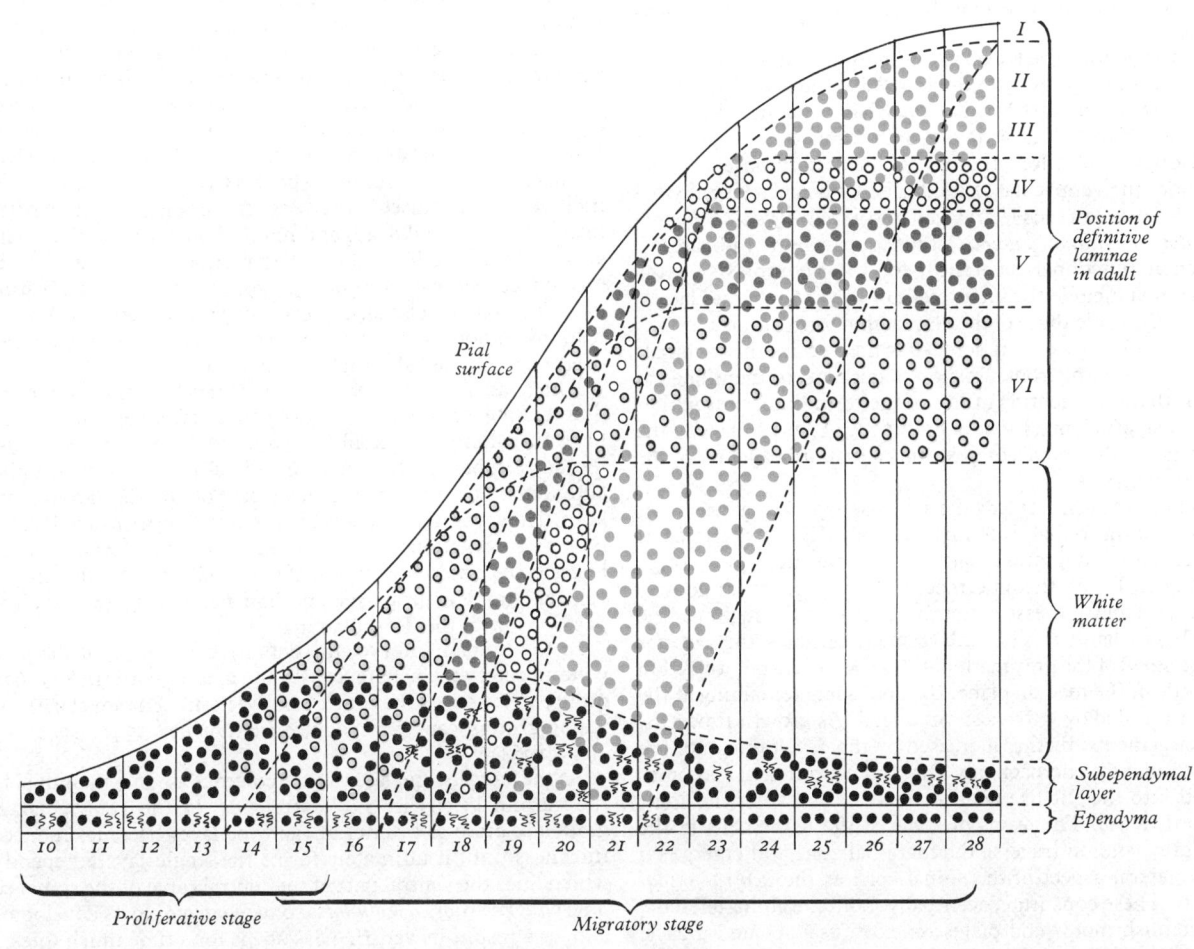

2.63B Schematic representation of the dynamics of neuroblast migrations during transformation of the early cranial neural tube to form the cerebral neocortex of the rat through days 10 to 28. Note the successive waves of migration. Coding: symbolic metaphase chromosomes—mitotic cells; *full black* discs—ventricular and subventricular zone neuroblasts; *full yellow discs*—infragranular neuroblasts destined for lamina VI; full *magenta discs*—infragranular neuroblasts destined for lamina V; *open black circles*—granular neuroblasts destined for lamina IV; *full blue discs*—supragranular neuroblasts destined for laminae III and II. Redrawn and colour coded by kind permission from data provided by Dr. M. Berry (1974), of the Anatomy Department, Birmingham University.

159

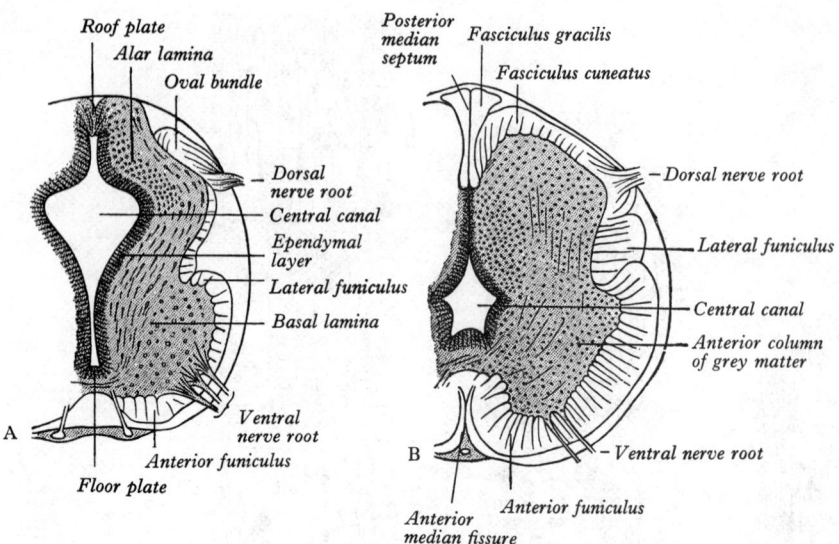

2.64A and B Transverse sections through the developing spinal cord of human embryos. A About six weeks old. B About three months old. (After His.)

becoming *bipolar neuroblasts*. One process is, however, withdrawn and the neuroblast becomes *unipolar* although this is not invariably so in the case of the spinal cord. Further differentiation leads to the development of dendritic processes and they become typical *multipolar* nerve cells. In the developing cord the cells occur in small clusters which represent clones of neuroblasts.

At first the neural tube is oval in outline and its lumen is narrow and slit-like (**2.61A**). As the lateral walls thicken, the lumen widens in its dorsal part and is somewhat diamond-shaped on cross-section (**2.61 B**). The widening of the canal is associated with the development of a longitudinal *sulcus limitans* on each side. This divides the ventricular and intermediate zones (ependymal and mantle layers) in each lateral wall into a *ventral (basal) lamina* and a *dorsal (alar) lamina*. This separation indicates a fundamental functional difference, for the neuroblasts in the ventral lamina include the motor cells of the anterior and lateral grey columns, while those of the alar lamina form 'interneurons', (both short and long axoned), some of which receive the terminals of primary sensory neurons. (It should, perhaps, be mentioned at this point that the scientific utility of the term 'interneuron' has been seriously challenged — *see* Gray 1974.) At its caudal end the central canal of the spinal cord exhibits a fusiform dilatation, the *terminal ventricle*.

The cells of the ventricular zone are closely packed at this stage and arranged in radial columns (**2.61B**). (For experimental studies of radial migration patterns consult Bergquist 1932, 1968.) The cells of the intermediate zone are more loosely scattered, and they increase in number at first in the region of the ventral (basal) lamina. This enlargement outlines the *anterior (ventral) column* of the grey matter and causes a ventral projection on each side of the median plane, the floor plate remaining at the bottom of the shallow groove so produced. As growth proceeds, these enlargements, further increased by the development of the anterior funiculi, encroach on the groove until it becomes converted into the slit-like anterior median fissure of the adult spinal cord (**2.64**). The axons of some of the nerve cells in the anterior grey column traverse the marginal zone and emerge on the anterolateral aspect of the spinal cord as the *ventral spinal nerve roots*. These constitute, eventually, both the alpha-efferents which establish motor end plates on extrafusal striated muscle fibres, and the gamma-efferents which innervate the contractile polar regions of the intrafusal muscle fibres of the muscle spindles (p. 847). (The histogenesis of beta-efferents is, of course, completely uncharted.)

In the thoracic and upper lumbar regions the intermediate zone neuroblasts in the dorsal part of the ventral lamina outline a

lateral column. Their axons join the emerging ventral nerve roots and pass as preganglionic fibres, to the ganglia of the sympathetic trunk, or related ganglia, the majority eventually myelinating to form the *white rami communicantes*. The fibres constituting the rami establish synapses with the autonomic ganglionic neurons (pp. 174–75, and the axons of some of the latter proceed as postganglionic fibres to innervate smooth muscle cells, adipose tissue, or glandular cells. Some of the preganglionic sympathetic efferent axons pass to the cells of the suprarenal medulla (p. 1458). The innervation of other 'chromaffin' tissues is less certain (but *see* the carotid body—p. 1462). Similarly an autonomic lateral column is also laid down in the mid-sacral region, and gives origin to the preganglionic parasympathetic fibres of the pelvic splanchnic nerves (p. 1132).

A number of investigations have been directed towards an understanding of the cell dynamics involved in the elaboration of the distinctive groupings of cells seen in the mature spinal cord of various species (Levi-Montalcini 1950; Hamburger 1952; Harris 1965; Romanes 1951, 1964).

The changes in cell number, position and density as seen in a longitudinal section of the chick spinal cord, based on the investigations of Hamburger (1952), are illustrated in **2.62B**. The anterior region of each ventral (basal) lamina forms at first a continuous column of cells throughout the length of the developing cord. In many forms this soon develops into two columns (on each side)—a medially placed one concerned with the innervation of axial musculature, and a laterally placed one which innervates the limbs. At limb levels the latter enlarges enormously but retrogresses at other levels.

Involved in transformations such as the foregoing, there is an interplay between a number of fundamental processes which vary in their prominence at different times and at different levels—cell *proliferation, migration,* followed either by progressive cell *growth and differentiation* or, in complete contrast, by cell *degeneration and death.* In the example quoted, cell proliferation persists as a prominent feature at the levels concerned with limb innervation, whilst at thoracic levels a dorsomedial migration of neuroblasts occurs to lay the foundation of the visceral efferent column. Further, save at limb levels, massive cell degenerations occur in the lateral 'motor' columns, whereas the medial columns, which innervate axial musculature, persist throughout the cord. The phenomenon of cell death and removal on a large scale, balanced against local proliferation and migration rates, has only been recognized relatively recently (p. 79) as a fundamental feature in many morphogenetic situations. For an excellent review of this field in relation to the developing nervous system, the interested reader should consult Hughes (1968; 1974).

Thus, as we have seen, some ventral laminal neuroblasts differentiate into the ventral horn neurons from which both alpha, beta and gamma efferent fibres arise, and these are accompanied at thoracic, upper lumbar, and mid-sacral levels, by preganglionic autonomic efferents from neuroblasts of the developing lateral horn. However, additionally, numerous interneurons develop in both these situations (including the well-studied Renshaw cells), but it is uncertain how many of these differentiate directly from ventral lamina neuroblasts, and how many migrate to their final positions from the dorsal lamina.

In the human embryo, the definitive grouping of the ventral column cells which characterizes the mature cord (p. 869) occurs quite early, and by the fourteenth week (the 80 mm stage) all the major groups can be recognized (Romanes 1941, 1942, 1946, 1955, 1964).

As the anterior and lateral grey columns assume their final form the germinal cells in the ventral part of the ventricular zone gradually cease to proliferate, and the layer becomes reduced in thickness until it ultimately forms the single-layered ependyma which lines the ventral part of the central canal of the spinal cord. The *posterior (dorsal) column* is somewhat later in its development, and, as a result, its ventricular zone is for a time much thicker in the dorsal lamina than it is in the ventral (basal) lamina (**2.61B**).

While the columns of grey matter are being defined, the dorsal region of the central canal becomes narrow and slit-like, and its walls come into apposition and fuse with each other (**2.64**). In this way the central canal becomes, relatively, reduced in size and somewhat triangular in outline.

About the end of the fourth week advancing axonal sprouts invade the marginal zone. The first to develop are those destined to become short *intersegmental* fibres from the neuroblasts in the intermediate (mantle) zone, and also fibres of *dorsal roots* of spinal nerves which pass into the spinal cord from neuroblasts of the early spinal ganglia. The earlier dorsal root fibres invade the dorsal marginal zone stem from *small* dorsal root ganglionic neuroblasts. By the sixth week the latter form a well-defined *oval bundle* near the peripheral part of the dorsal lamina (**2.61** B); this bundle increases in size and, spreading towards the median plane, forms the *primitive posterior funiculus*; its constituent fibres are destined to be of fine calibre. Later, fibres derived from new populations of *large* dorsal root ganglionic neuroblasts join the dorsal root and invade the cord nearer the medial plane; they are destined to become fibres of much larger calibre. As the posterior funiculi increase in thickness, their medial surfaces come into contact separated only by the *posterior median septum*, which is ependymal in origin, neuroglial in nature. (A more detailed analysis of the temporal sequence of modifications of the dorsal root plate, posterior median septum, and the lateral displacement of the *primitive posterior funiculus* with the later development of the fasciculus cuneatus followed by the fasciculus gracilis, based on a study of sectioned human embryos of 6–10 weeks, and dissections of fetuses up to the end of the fourth month has been provided by Hughes (1976). He proposed that the displaced primitive posterior funiculus may form the basis of the dorsolateral tract or fasciculus (of Lissauer), and also correlated the sequence, siting and calibre of the entrant dorsal root fibres, with the changing size-distribution of the somata of the dorsal root ganglionic neuroblasts.)

Long intersegmental fibres begin to appear about the third month and *corticospinal* fibres about the fifth month. All nerve fibres are at first without myelin sheaths, and different groups commence to develop sheaths at different times, e.g. the ventral and dorsal nerve roots about the fifth month, the corticospinal fibres after the ninth month. In peripheral nerves the myelin is formed by Schwann cells; in the central nervous system oligodendrocytes are believed by the majority of workers to be concerned (p. 845). Myelination, of course, persists until overall growth of the central and peripheral nervous system has ceased.

The cervical and lumbar enlargements first appear simultaneously with the development of their respective limb buds.

In early embryonic life the spinal cord occupies the *entire length* of the vertebral canal, and the spinal nerves pass outwards at right angles. After the embryo has attained a length of 30 mm the vertebral column begins to grow more rapidly than the spinal cord, the caudal end of which gradually becomes more cranial in the vertebral canal. Most of this relative rostral migration occurs during the *first half of* intrauterine life. By the twenty-fifth week the terminal ventricle of the spinal cord (p. 869) has altered in level from the second coccygeal vertebra to the third lumbar, a distance of nine segments, and there remain but two segments before the adult position is reached (Streeter 1919). As the change in level begins rostrally, the caudal end of the terminal ventricle, which has become adherent to the overlying ectoderm, remains *in situ*, and the walls of the intermediate part of the ventricle and its covering pia mater become drawn out to form a delicate filament, the *filum terminale*. The separated portion of the terminal ventricle persists for a time but it disappears before birth as a rule. It does, however, occasionally give rise to congenital cysts in the neighbourhood of the coccyx.

THE SPINAL NERVES AND NEURAL CREST

Each spinal nerve is connected to the spinal cord by a ventral root and a dorsal root. The fibres of the ventral roots grow out from neuroblasts in the anterior and lateral parts of the intermediate zone; these pass through the overlying marginal zone and external limiting membrane, to enter the *myotomes* of the mesodermal somites and ultimately form the alpha-, beta- and gamma-efferents. At appropriate levels these are accompanied by the outgrowing axons of preganglionic sympathetic neuroblasts (segments T1–L2), or preganglionic parasympathetic neuroblasts (S2–4).

The fibres of the dorsal roots are developed from the cells of the spinal ganglia. Before the neural groove is closed to form the neural tube (p. 112) a ridge of neurectodermal cells, termed the *neural crest (ganglion ridge)*, appears along the prominent margin of each neural fold (**2.29**). When the folds meet in the median plane the two neural crests fuse into a wedge-shaped mass along the line of closure of the tube. Opposite each primitive mesodermal segment the neural crest cells proliferate rapidly to form a bilateral series of oval-shaped *primordial spinal ganglia*, and these migrate for a short distance in a lateral and ventral direction. From the ventral *region* of each a small part separates to form *sympatho-chromaffin* cells (pp. 175, 1454), while the remainder becomes a *definitive* spinal ganglion. The spinal ganglia are arranged symmetrically at the sides of the neural tube and, except in the caudal region, are equal in number to the primitive segments. The cells of the ganglia, like the cells of the intermediate zone of the early neural tube, are glioblasts and neuroblasts. The glioblasts develop into the satellite cells, which become closely applied to the ganglionic nerve cell somata (perikarya), into Schwann cells, and possibly other cells. The neuroblasts, at first round or oval, soon become fusiform, with extremities gradually elongating into central and peripheral processes. The central processes grow into the neural tube, as the fibres of dorsal nerve roots, while the peripheral processes grow ventrolaterally to mingle with the fibres of the ventral root thus forming a mixed spinal nerve. As development proceeds the original bipolar form of the cells in the spinal ganglia changes; the two processes become approximated until they ultimately arise from a single stem in a T-shaped manner, to form a unipolar cell (sometimes, less appropriately, termed pseudo-unipolar). The biopolar form is, however, retained in the retina and in the ganglia of the vestibulocochlear nerve. Some observers hold that the T-form is derived from the branching of a single process which grows out from the cell.

It should be noted that the position of the early neural crest as a wedge-shaped mass along the line of tube closure noted above, and the identification of ganglionic cells in various positions in the wall of the early neural tube, and even within the central canal (Humphrey 1944; 1947), is strongly reminiscent of the developmental history of the *Rohon-Beard cells* in fish and amphibia (Rohon 1884; Beard 1896), which are thought to be important in the emergence of primitive locomotor patterns (Hughes 1968). In this regard, other investigators have claimed that in primitive chordates (Cyclostomes and Euselachians) the neural crest develops as an *evagination* of the dorsal region of the alar lamina of the late neural folds and early neural tube (Conel 1942). (For a review of the origin, widespread migration and differentiation of neural crest cells *see*, for example Weston 1970, Bellairs 1971, Leikola 1976, and p. 115.)

For more than a century there have been many attempts to plan investigations which would aid an understanding of the *causal morphogenetic mechanisms* which operate during the development of the nervous system, and to correlate the stage of development with emergent *behavioural patterns*. Whilst much has been established, the answers to many fundamental questions still remain in obscurity. A review of this important and interesting area of biology lies outside the scope of the present volume and the reader should consult the pioneering works which have appeared over the intervening years. Consult references (Bidder and Kupffer 1857; His 1879, 1883, 1887, 1890; Ramón y Cajal 1928, 1960; Waldeyer 1891; Harrison 1906, 1907, 1910, 1914, 1924; Coghill 1929; Detwiler 1936; Spemann 1938; Weiss 1950; Villier *et al.* 1955; Hughes 1968; Gottlieb 1974).

Briefly, the nervous system is closely interlocked, in terms of morphogenesis, with the 'periphery', i.e. surrounding non-nervous structures, and each is dependent upon the other for its effective structural and functional maturation.

Reference has already been made to the widely assumed initiating importance of a triggering system whereby the prechordal plate and chorda-mesoderm of the archenteron roof operate as a coordinated series of regional neural plate inducers—archencephalic, deuterencephalic, and spinocaudal. (However, *see* discussions on pp. 110, 112, 156 and consult Bellairs 1971.) Thereafter follows a period during which the main cell masses and subdivisions of the neural tube are established, and transplantation experiments have indicated that some of the earlier patterning is an expression of self-differentiation of the tube. (For a descriptive and experimental review of pattern formation in these comparatively early stages of neurogenesis in the chick consult Watterson 1965.) However, the polarization of ventral lamina neuroblasts, and those of the neural crest, and the formation of segmentally arranged ventral nerve roots and spinal ganglion primordia, is closely related to, and dependent upon, the orderly segmentation of the neighbouring mesodermal somites. Experimental removal or intercalation of somites results in corresponding disturbances of pattern in these nervous structures.

There are many other examples of the influence of peripheral structures on the development of the nervous system. For example, if the normal peripheral innervation of the tetrapod limb is prevented experimentally, there follows a gross disturbance in the balance between neuroblast proliferation, degeneration, migration and maturation in the relevant part of the neural tube and primitive spinal ganglia. Conversely, the denervated limb rudiment itself fails to complete its development in the absence of nerve-mediated influences. Other extensively investigated situations include the inductive influence of the optic vesicle on adjacent ectoderm with the formation of a lens vesicle (p. 176), and the reciprocal influences of the developing lens and peri-optic mesenchyme on the differentiation of the optic cup (p. 176); the profound influence of peripheral innervation on the regeneration of limbs in amphibia; the 'trophic' maintenance of striated muscle structure by peripheral nerves (p. 862); the dependency of regenerating peripheral nerves on the establishment of normal peripheral connexions before a restoration of their structural and functional parameters ensues. Thus, whilst there are numerous experimentally demonstrable examples of the *interdependence* of the developing nervous system and the periphery, the precise causal mechanisms at work remain obscure. In this regard, however, mention should be made of a fairly recent proposal (Prestige 1965, 1967) concerning the operation of *maintenance factors* thought to be produced by the tissues of developing tetrapod limbs, and which may enter the nervous system, perhaps to be stored in the cytoplasm of neurons, where their concentration influences the turnover of neuroblasts, i.e. the

balance between the rate of proliferation and degeneration. Similarly, other workers (Levi-Montalcini 1942, 1943, 1951, 1952, 1956, 1960, 1963; Cohen 1958, 1959, 1960; Cohen and Levi-Montalcini 1956, 1957) have isolated various proteinaceous *nerve growth factors* from tissue and tumour extracts, and snake venom, which influence the *in vitro* form and extent of nerve cell growth. Although the operation of such factors in normal ontogeny has not been claimed by these workers, others have demonstrated the presence of similar substances in homogenates of the axial structures of chick embryos during the period of early neurogenesis.

One problem which has attracted neuroembryologists for many years concerns the manner in which growing nerve fibres reach and make functional contact with their appropriate end organs (neuromuscular endings, secreto-motor terminals, or synapses with other neurons), during which they often pursue complex courses between the parent cell body and the sites of termination. If the cells concerned had been in contact from the first, the problem would have reduced to one concerning mechanisms of relative displacement and migration of both, with simple elongation of the nerve cell process between them. However, during the outgrowth of axonal processes from, for example, neuroblasts within the basal lamina to reach presumptive myoblasts in the limb buds, the earliest nerve fibres are known to cross appreciable distances occupied solely by loose general mesenchyme.

The growing tips of the neurites have been studied in tissue culture (Harrison 1910; Lewis 1945; Spiedel 1932, 1933, 1935; Nakai 1960; Nakai and Kawasaki 1959; Pomerat *et al.* 1967; and *vide infra*) and in the tail of larval amphibia (Spiedel 1932, 1933, 1935). Classically, the tip has been described as expanded into a *growth cone* which is constantly active, changing shape, extending and withdrawing processes which apparently 'explore' the local environment for a suitable surface along which extension may occur (*see* 'cell motility' p. 35). One process then enlarges at the expense of others and expands into a new growth cone, the exploratory behaviour recommencing. However, it has been suggested that the very rapid surface activity of growth cones is associated more with the micropinocytotic imbibition of water and solutes (perhaps in relation to a peripheral undulating cytoplasmic veil, or ruffled border, as in macrophages), forming cytoplasmic vacuoles which migrate centrally along the neurite. On this view the tip of the main body of the neurite moves in a series of discontinuous directional thrusts. Earlier ultrastructural studies also emphasized the presence within growth cones of microvesicles in the general cytoplasm (Estable *et al.* 1957) and vacuoles within the cisternae of the endoplasmic reticulum (Bellairs 1959), (but see also below). In this context it is also widely held that products important in the growth mechanisms, and synthesized within the cell bodies, are passed outwards by some form of proximodistal *axoplasmic flow* along the neurites. Bulk axoplasmic flow was first postulated following the experimental constriction of nerves in 1948 (Weiss and Hiscoe 1948), and since that time many intricate analyses of rapid and slow components of *bidirectional flow systems* within axons have been made (Lubinska 1964; and *see: Neurosciences Research Programme Bulletin 4*, vol 5, 1968; also p. 826).

However, in a series of ultrastructural studies on the *in vitro* outgrowth of neurites from explants of chick *spinal cord*, a number of established views concerning their form, structure and surface contacts have been challenged (Grainger *et al.* 1968; James and Tresman 1969; Grainger and James 1970). In addition to describing the formation of neuromuscular junctions and varieties of synapse under these experimental conditions, the authors have shown that the neurites usually grow out in bundles rather than as single processes. The neurites contain arrays of longitudinal microtubules which are probably interconnected, microfilaments, mitochondria, linear densities in the cytoplasm, and populations of vesicles. The expanded tip of the neurite bundle is not an aggregation of simple 'growth cones', but usually the neurite terminals end in association with a 'glial' cell. The latter has a relatively dense cytoplasm containing an array of organelles enabling it to be distinguished from the neighbouring neurites. Further, the glial cell is complex in form. It sends a major cytoplasmic process back towards the parent explant; it presents

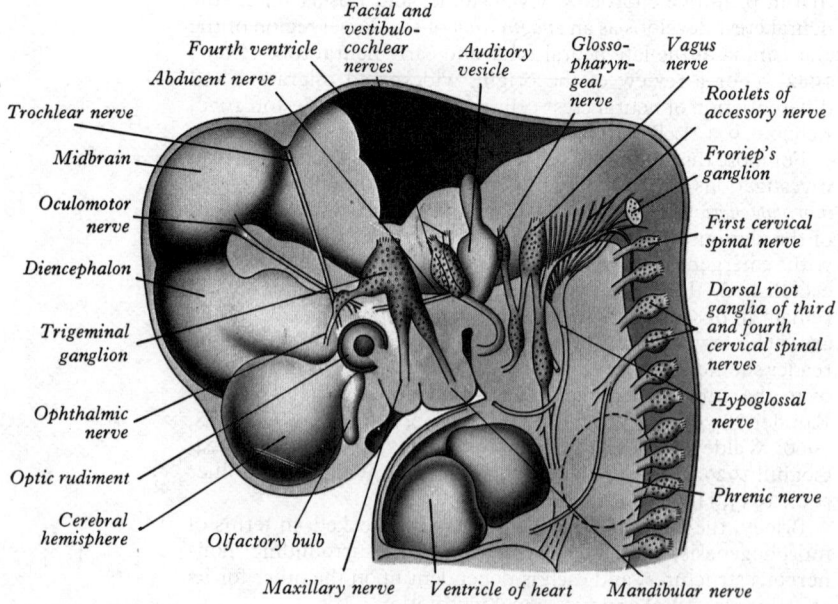

2.65 The brain and cranial nerves of a human embryo, 10·2 mm long. note the midbrain, cervical and pontine flexures, and the expanding fourth ventricle. Note also the ganglia (stippled) associated with the trigeminal, facial, vestibulocochlear, glossopharyngeal, vagus and spinal accessory nerves. Froriep's ganglion, an occipital dorsal root ganglion, is inconstant and soon disappears. (After His.)

surface projections which partly enfold the terminal neurites, and distally, beyond the neurites, cytoplasmic sheets or finger-like processes extend to come into intimate relationship with the culture substrate. Thus, the essential contacts appear to be between glial cell membrane and substrate, and between neurite membrane and glial cell membrane, posing the question whether the glial cell, rather than the tip of the neurite, is basically involved in determining the direction of outgrowth. Further investigations will be necessary to determine how far these appearances are reflected in the growth of both central and peripheral neurites in normal ontogeny, although some support-ing evidence has been adduced concerning the growth of *central* neuroblasts (Rakic 1971). Further details cannot be pursued here, but recently an excellent review has appeared concerning the morphology, motility and directional growth of axons and their growth cones, including the *in vitro* establishment of synaptic contacts—for this consult James (1974). It may be pointed out that some authors (e.g. Rakic 1971) have suggested that within the central nervous system, the direction followed by advancing neurites may stem from contact guidance (*see* pp. 91, 862, and below) with previously established glioblast processes. On the other hand, in recent studies of the growth cones of neurites from explanted dorsal root ganglia of the rat (including their endocytotic uptake and transport of horseradish peroxidase), using light, transmission and scanning electron microscopy, many surface features and ultrastructural details of the cones were provided; no 'leading' glial or Schwann cells were identified. (For further discussion and bibliography consult Fischer 1977; and for a literature review concerning *dendritic* growth—Berry 1974.)

Over the years two principal theories emerged concerning the directional growth of nerve fibres—the *neurotropism* of Ramón y Cajal (1919), and the principle of *contact-guidance* of Weiss (1941). The former, based upon observations on the innervation of epithelia, proposed that growing fibres were guided by some form of attraction, presumably chemical, which emanated from the area to be innervated. The second view denied the existence of such attractive forces and, based upon many series of tissue culture experiments, held that pioneer neurites were guided to their destination by preferential growth along pathways provided by an orientation of the micellar substructure of the intercellular matrix in the intervening spaces. Despite the forceful arguments brought to bear on this question, it should be noted that in a more recent critical review of this field (Hughes 1968), it was stated—'Both contact guidance and neurotropism, however associated or combined, are inadequate to explain the pattern of the peripheral nervous system and, *a fortiori*, of the enormously more intricate plan of the fully differentiated neural tube. No imaginable hypothesis is at present adequate to explain the phenomena of the regeneration of the optic nerve, which is the only central pathway which has so far been studied by experimental means. At present the only possible way of regarding the whole question is in terms of a succession of influences which operate in turn upon a growing nerve fibre, from its first emergence to its final link with an end organ. Such a sequence of controlling factors was postulated as far back as 1911.' Nevertheless, whilst few would disagree with the general tenets of these remarks, hypotheses abound, and there is currently a great upsurge of interest in neuroembryology. As an introduction to the extensive literature, the interested reader should consult, for example, Sperry (1958, 1963, 1965); Schmitt (1970); Jacobson (1970); Gaze (1970); Gottlieb (1973, 1974); Mark (1974). In addition to questions concerning the growth, guidance, and patterning of neurites and somata, the following kinds of problem confront neuroembryologists. What are the basic ontogenetic factors providing positional information (p. 87, controlling proliferative cycles, differentiating mitoses, and the initial polarization and migration of neuroblasts? How much of the initial patterning is preprogrammed in the genome, and how much environmentally and functionally dependent? To what extent do regeneration experiments reflect ontogeny? Do neurons bear individually unique chemical markers? What are the structural bases of neuronal plasticity, and the behavioural phenomena of instinct, learning, and memory? Answers still elude us and we remain at the stage of attempting to plan meaningful experiments.

Development of the Brain

Prior to the closure of the neural tube the neural folds become expanded considerably in the head region as a first indication of a brain. Subsequent to the closure of the anterior neuropore (p. 112) these expansions form the three *primary cerebral vesicles* (2.65, 68). The customary use of the term 'vesicle' though followed here, suggests an exaggerated view of these localized accelerations of growth in the wall of the brain (O'Rahilly and Gardner 1971). In 'higher' vertebrates, and especially in primates, including mankind, the bulging is not initially marked, nor are the constrictions between them particularly narrow. The vesicles are more like gently fusiform tubes. The three regions are named the *rhombencephalon* or *hindbrain*, the *mesencephalon* or *midbrain*, and the *prosencephalon* or *forebrain*—the first being continuous with the spinal cord. As the result of unequal growth of its different parts three flexures appear in the brain; two of these are concave ventrally and there are corresponding flexures of the head. The first is associated with the formation of the head fold and is a *midbrain flexure*; as a result the forebrain bends in a ventral direction around the cephalic end of the notochord and foregut and thus progresses until its floor lies almost parallel with that of the hindbrain (2.65, 68). The midbrain is thus, for a time, the most prominent part of the brain, and underlies the marked *midbrain prominence* of the overlying ectoderm, seen to best advantage when the embryo is viewed from the side. The second bend appears at the junction of the hindbrain and spinal cord and is termed the *cervical* or *neck flexure* (2.65, 68). This increases from the fifth to the end of the seventh week, by which the hindbrain forms nearly a right angle with the spinal cord; after the seventh week, however, extension of the head takes place and the cervical flexure diminishes and eventually disappears. These bends are important factors in determining the shape of the cranial end of the embryo. The third bend is named the *pontine flexure*, because it is at the level of the future pons. It differs from the other two in that its convexity is directed ventrally, and it does not substantially affect the outline of the head. (For more recent accounts of the early development of the brain consult Bartlemez and Dekaban (1962); Jacobson (1970); Gaze (1970); Eccles (1972); Gottlieb (1972, 1974); Mark (1974).)

The lateral walls of the hindbrain and midbrain, like those of the spinal cord, are divided into *dorsal* (*alar*) and *ventral* or (*basal*) *laminae* by the upward continuation of the limiting sulci of the spinal cord.

THE RHOMBENCEPHALON

By the time the midbrain flexure appears, the hindbrain exceeds in length the combined extent of the other two brain vesicles. Cranially it exhibits a constriction, the *isthmus rhombencephali* (2.68B), best viewed from the dorsal aspect. Ventrally the hindbrain is separated from the dorsal wall of the primitive pharynx only by the notochord, the two dorsal aortae, and a small amount of mesenchyme; on each side it is closely related to the dorsal ends of the visceral arches (2.54).

The formation of the pontine flexure appears to 'stretch' the thin, epithelial roof plate, which becomes widened, the greatest increase in width corresponding to the region of maximum convexity, so that the outline of the roof plate becomes rhomboidal. By the same change the lateral walls become separated, particularly dorsally, and the cavity of the hindbrain, subsequently the fourth ventricle, becomes flattened and somewhat triangular on cross-section. The pontine flexure becomes increasingly acute until, at the end of the second month, the laminae of its cranial (metencephalic) and caudal (myelen-cephalic) slopes are opposed to each other (2.67C) and, at the same time, the lateral angles of the cavity extend to form the lateral recesses of the fourth ventricle.

About the end of the fourth week, when the pontine flexure is first discernible, a series of six transverse *rhombic grooves* appears in the ventral lamina of the hindbrain. Between the grooves, the intervening masses of neural tissue have, at this stage, been termed *rhombomeres* (*neuromeres*). These are closely associated with the pattern of the underlying motor nuclei of certain of the

Roof plate

Special somatic
afferent column

General somatic
afferent column

Special visceral
afferent column

General visceral
afferent column

General visceral
efferent column

Branchial
efferent column

Somatic efferent
column

Floor plate

Branchial
striated muscle

Somatic striated
muscle

Non-striated muscle

Otocyst

Skin

Taste bud

Visceral epithelium

2.66 Diagram of a transverse section through the developing hindbrain of a human embryo, c. 10·5 mm long, to show the relative positions of the columns of grey matter from which the nuclei associated with the different varieties of nerve components are derived. Note the postganglionic neurons associated with the general visceral efferent column, the bipolar neurons associated with the otocyst and the unipolar afferent neurons associated with the other alar lamina columns.

cranial nerves. The first two overlie the trigeminal nucleus, the third the facial nucleus, the fourth that of the abducent nerve, the fifth that of the glossopharyngeal, and the sixth that of the vagus. These grooves, though transient, are constant in appearance, but their significance is uncertain. Some authorities regard them as evidence in support of a primitive segmental (neuromeric) origin for the brain and spinal cord (Berquist 1952).

The differentiation of the lateral walls of the hindbrain into ventral and dorsal laminae has a similar significance to the corresponding differentiation in the lateral wall of the spinal cord (p. 160), and ventricular, intermediate and marginal zones are formed in the same way.

The cells of the ventral lamina are often, in elementary accounts, simply termed 'motor' (but *vide infra*), and they form three elongated, but interrupted, columns. The most ventral column is continuous with the anterior grey column of the spinal cord and will supply muscles considered 'myotomic' in origin. It is represented in the caudal part of the hindbrain, as the hypoglossal nucleus, and it reappears at a higher level as the nuclei of the abducent, trochlear and oculomotor nerves, which are *somatic efferent nuclei*. The intermediate column, which is only represented in the upper part of the spinal cord, is for the supply of branchial musculature. It is interrupted also, but the caudal part, which gives fibres to the ninth, tenth and eleventh cranial nerves, forms the elongated *nucleus ambiguus*. At higher levels this column gives origin to the motor nuclei of the facial and trigeminal nerves. These three nuclei are termed *branchial (special visceral) efferent nuclei*. The most dorsal column of the ventral lamina (represented in the spinal cord by the lateral grey column) innervates viscera. It is interrupted also, its large caudal part forming a portion of the *dorsal nucleus of the vagus* and its cranial part the *salivatory nucleus*. These are termed *general visceral (general splanchnic) efferent nuclei* and their neurons give rise to preganglionic, parasympathetic nerve fibres.

It should be noted here that the neuroblasts of the ventral (basal) lamina and their three columnar derivatives are only 'motor' in the sense that *some* of their number, form either α, β or γ motor neurons, or preganglionic parasympathetic neurons. The remainder, which almost certainly greatly outnumber the former differentiate into functionally related interneurons.

The cell columns of the dorsal lamina are also interrupted and give rise to *general visceral (general splanchnic) afferent*, *special visceral (special splanchnic) afferent*, *general somatic afferent* and *special somatic afferent* nuclei (their relative positions are shown in (**2.66**). The general visceral afferent column is represented by a portion of the dorsal nucleus of the vagus (see also p. 1076), the special visceral afferent column by the nucleus of the tractus solitarius, the general somatic afferent column by the afferent nuclei of the trigeminal nerve (Brown 1974), and the special somatic afferent column by the nuclei of the vestibulocochlear nerve (Verbitskaja 1973). (Again it should be noted here that the relatively simple functional independence of these afferent columns implied by the foregoing classification is, in the main, an aid to elementary learning. The emergent neurobiological mechanisms are in fact much more complex, and less well understood—*see* p. 1084.) Although they tend to retain their primitive positions some of these nuclei are later displaced by differential growth patterns and by the appearance and growth of neighbouring fibre tracts, and possibly by active migration. It has been suggested that a nerve cell tends to remain as near as possible to its predominant source of stimulation and that when the danger of separation arises, owing to the development of neighbouring structures, it will migrate in the direction from which the greatest density of stimuli come. This phenomenon was termed neurobiotaxis (Kappers 1921, 1934). Cells can migrate in this way only by lengthening of their axons, which therefore trace the route taken by the cells on their transit. The curious courses of the fibres arising from the facial nucleus (p. 910) and nucleus ambiguus (pp. 900, 905) have been held to illustrate this. In the 10 mm embryo the facial nucleus lies in the floor of the fourth ventricle, occupying the position of the special visceral efferent column and at this stage it is placed at a higher level than the abducent nucleus. As growth proceeds the facial nucleus migrates at first caudally and dorsal to the sixth nerve nucleus and then ventrally to reach its adult position. As it migrates, the axons to which its cells give rise elongate and their subsequent course maps out the pathway along which the facial nucleus has travelled. Similarly the nucleus ambiguus arises initially, immediately deep to the ventricular floor but in the adult it is more deeply placed and its efferent fibres first pass dorsally and medially before curving laterally to emerge at the surface of the medulla oblongata. Attractive as the concept of neurobiotaxis may have seemed, it must be added that it is supported by little direct evidence.

The Myelencephalon

The caudal slope of the hindbrain constitutes the *myelencephalon*, which develops into the medulla oblongata. The nuclei of the ninth, tenth, eleventh and twelfth cranial nerves develop in the situations already indicated, and afferent fibres from the ganglia of the ninth and tenth nerves form an oval marginal bundle in the region overlying the dorsal (alar) lamina. The dorsal edge of this lamina throughout the rhombencephalon gives attachment to the thin expanded roof plate and is termed the *inferior rhombic lip*. As the walls of the rhombencephalon spreads outwards, the rhombic lip protrudes as a lateral edge which becomes folded over the adjoining area. According to this widely accepted view it later becomes adherent to this area, and the cells of the rhombic lip migrate actively into the marginal zone of the ventral *lamina*. In this way the oval bundle which forms the *tractus solitarius* becomes buried beneath the surface. Dorsal lamina cells which migrate from the rhombic lip are believed to give origin to the olivary and arcuate nuclei and the scattered grey matter of the nuclei pontis. While this migration is in progress the thin floor plate is invaded by fibres which cross the median plane (accompanied by neuroblasts which cluster in and near this plane), and it becomes thickened to form the *median raphe*. Some of the migrating cells from the rhombic lip in this region fail to reach the ventral lamina and form an oblique ridge across the dorsolateral aspect of the inferior cerebellar peduncle; this ridge constitutes the *corpus pontobulbare (nucleus of the circumolivary bundle)*.

The lower part of the myelencephalon takes no part in the formation of the fourth ventricle, and, in its development, it

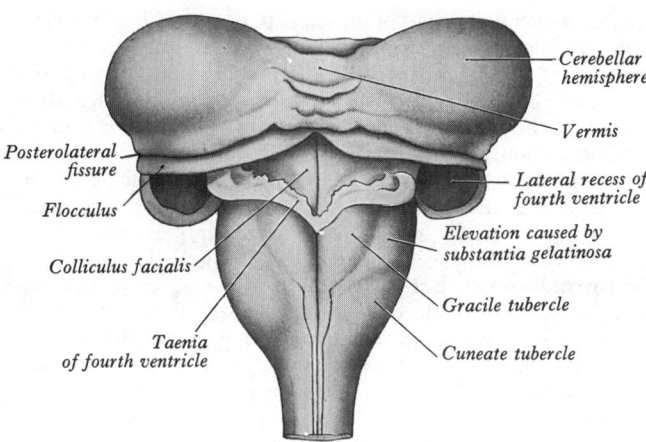

2.67A The cerebellum of a fetus in the fifth month. (After Kollmann.)

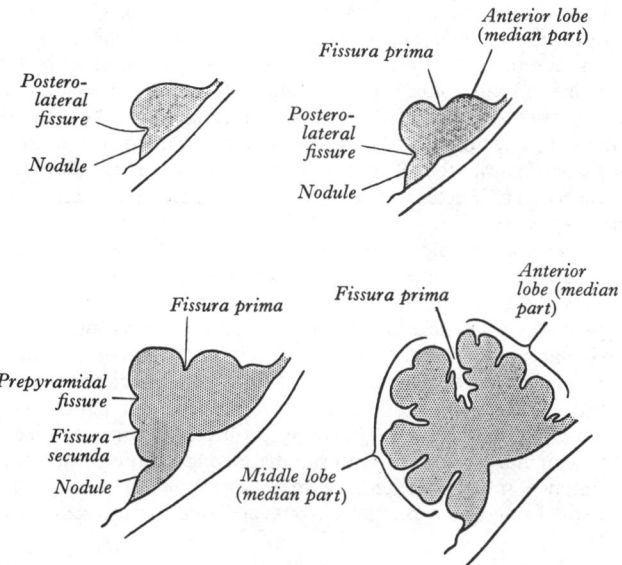

2.67B Median sagittal sections through the developing cerebellum, showing four different stages.

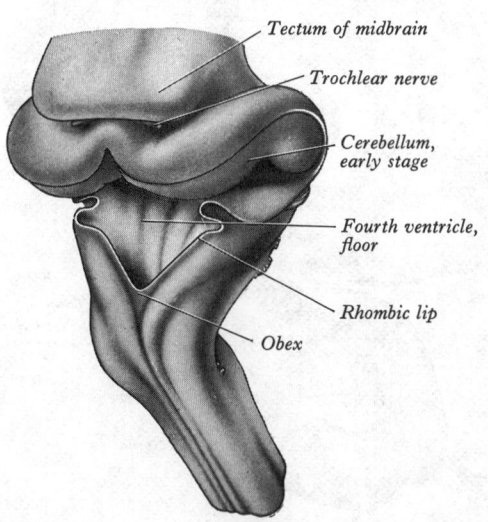

2.67C The dorsal aspect of the hindbrain of a human fetus about three months old. Viewed from behind and partly from the right side. (From a model by His.)

closely resembles the spinal cord. The large nuclei, gracilis and cuneatus, are derived from the dorsal lamina, and their efferent arcuate fibres play a large part in the formation of the median raphe.

About the fourth month the descending *corticospinal* fibres invade the ventral part of the medulla oblongata to form the pyramids, whilst dorsally, ascending fibres from the spinal cord, with olivocerebellar and parolivocerebellar fibres, external arcuate fibres, together with two-way reticulocerebellar and vestibulocerebellar interconnexions, form the inferior cerebellar peduncle. The reticular nuclei of the lower medulla probably have a dual origin from both laminae.

The Metencephalon

The rostral slope of the hindbrain is the *metencephalon*, and from it both cerebellum and pons are developed. Prior to the formation of the pontine flexure the dorsal laminae of the metencephalon are parallel with one another. Subsequent to its formation the roof plate of the hindbrain becomes rhomboidal and the dorsal laminae of the metencephalon lie obliquely, being close together at the cranial end of the fourth ventricle, but widely separated in the region of its lateral angles. The accentuation of the pontine flexure approximates the cranial angle of the ventricle to the caudal, and the dorsal laminae of the metencephalon now lie almost horizontally.

It should be noted that caudal to the developing cerebellum (*vide infra*), the roof of the fourth ventricle remains epithelial, covering an approximately triangular zone, from the lateral angles of the rhomboid fossa to the median obex (p. 933). Over this region the nervous tissue fails to develop and the vascular pia mater is closely applied to the subjacent ependyma. At each lateral angle, and in the midline caudally the membranes break through forming the lateral and median apertures of the roof of the fourth ventricle. Subsequently, these are the principal routes by which cerebrospinal fluid, produced in the ventricles, escapes into the subarachnoid space. The vascular pia mater, in an inverted V formation, cranial to the apertures invaginates the ependyma to form vascular fringes—the vertical and horizontal parts of the choroid plexuses of the fourth ventricle (p. 933).

The Cerebellum

While these changes are occurring the cells in the *superior rhombic lip* and adjacent dorsal part of the dorsal lamina of the metencephalon proliferate to form the *rudiment of the cerebellum* Two rounded swellings are formed which, at first, project partly into the ventricle (**2.67B, C**), and they form the rudimentary cerebellar hemispheres. The most cranial part of the roof of the metencephalon originally separates the two swellings, but it becomes invaded by cells, which form the rudiment of the vermis. These cells were regarded as derivatives of both ventral and dorsal laminae (Baxter 1953). At a later stage, *extroversion of the cerebellum* occurs, with reduction of its intraventricular projection and increasing prominence of a dorsal extraventricular projection. The cerebellum now consists of a bilobar (dumb-bell shaped) swelling stretched across the rostral part of the fourth ventricle (**2.67A**), continuous rostrally with the anterior medullary velum, which has formed from the isthmus, and caudally with the epithelial roof of the myelencephalon. As growth proceeds a number of transverse grooves appear on the dorsal aspects of the cerebellar rudiment, as the precursors of the numerous fissures which characterize the surface of the mature cerebellum **2.67B**, also p. 912).

The *posterolateral fissure*, in its lateral parts, appears first, demarcating the most caudal part from the rest of the cerebellar rudiment, enabling the *flocculi* to be identified. The lateral parts of this fissure extend medially and meet in the median plane, where they demarcate the nodule. The *flocculonodular lobes* can now be recognized and constitute the most caudal part of the cerebellum at this stage, but, owing to the growth of the adjoining areas, they progressively come to occupy the *anterior part* of the inferior surface in the adult. They are formed in close proximity to the line of attachment of the epithelial roof, i.e. to the rhombic lip (p. 933 and **2.67C**).

At the end of the third month a transverse sulcus appears on the

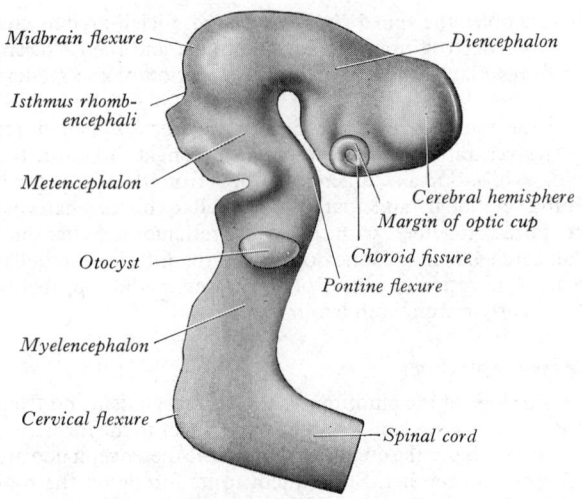

Midbrain flexure
Diencephalon
Isthmus rhomb-encephali
Metencephalon
Cerebral hemisphere
Margin of optic cup
Otocyst
Choroid fissure
Pontine flexure
Myelencephalon
Cervical flexure
Spinal cord

2.68A The right side of the brain of a human embryo, 9 mm long. (Drawn from a model by His.)

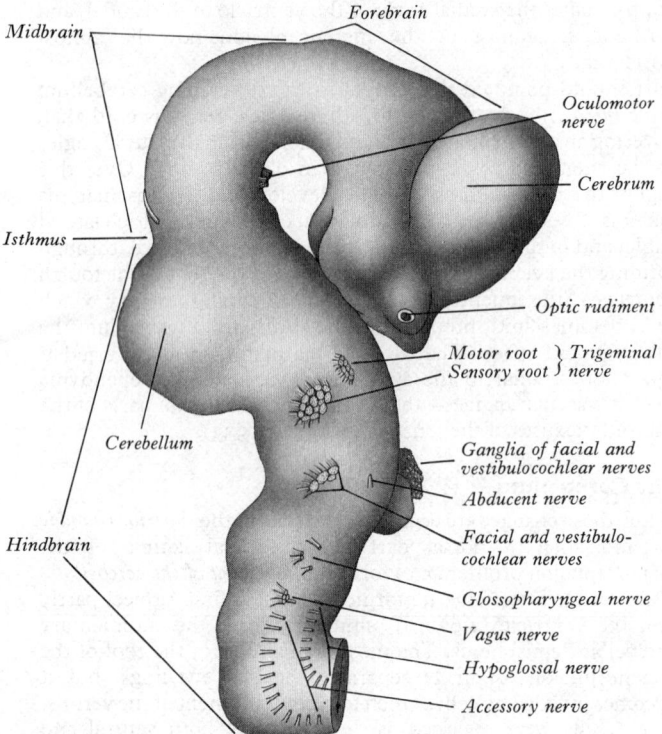

Forebrain
Midbrain
Oculomotor nerve
Cerebrum
Isthmus
Optic rudiment
Motor root } Trigeminal
Sensory root } nerve
Cerebellum
Ganglia of facial and vestibulocochlear nerves
Abducent nerve
Facial and vestibulo-cochlear nerves
Hindbrain
Glossopharyngeal nerve
Vagus nerve
Hypoglossal nerve
Accessory nerve

2.68B The brain of a human embryo about 10·2 mm long. Right lateral surface. (From a model by His.)

rostral slope of the cerebellar rudiment, and deepens to form the *fissura prima*, which cuts into the vermis and both hemispheres, separating off the most cranial region of the rudiment to form the anterior lobe.

About the same period two short transverse grooves appear on the inferior vermis; the first of these is the *fissura secunda*, which demarcates the uvula, and the second is the *prepyramidal fissure*, which demarcates the pyramid (their positions are indicated in 2.67B). The whole cerebellum now grows in a dorsal direction, and the caudal, or inferior, aspects of the hemispheres undergo much greater enlargement than the inferior vermis, which therefore becomes buried at the bottom of a deep hollow—the *vallecula*. While these changes are taking place numerous additional fissures develop, which are approximately parallel to, and intervene between, the foregoing. They are an expression of the relatively vast increase in surface area of the cerebellum which is occurring but their precise positions and systematic names have limited functional or morphological significance. The most extensive of these develops into the *horizontal fissure*.

In many mammals a part of the hemisphere immediately rostral to the floccular fissure becomes defined as an entity; in some it forms a very prominent part of the cerebellum; it is termed the *paraflocculus*, but the relationship is purely topographical, and, in contrast to the flocculus, the paraflocculus derives its afferent connexions mainly, but not entirely, from the cerebral cortex. It is uncertain whether any homologue of the paraflocculus exists in the human cerebellum, or whether it is represented by some small patches of grey matter which are found not infrequently on the inferior surface of the middle cerebellar peduncle (Larsell 1947).

Mammalian cerebellar histogenesis has been more completely described, both during normal ontogeny, and after experimental intervention, than in any other part of the nervous system, due in large measure to its precisely ordered geometry, and highly distinctive cell types. The connectivity, synaptology, and electrophysiology of the latter have also been intensively studied, and some knowledge of these aspects are a necessary prerequisite of any account of histogenesis (for this consult pp. 157–161). Particularly valuable in histogenetic investigations have been nuclear labelling with tritiated thymidine, the analysis of genetic variants, and the changes following surgical deafferentation, X-irradiation, and virus diseases. Most studies have been confined to mice and rats, and whilst it seems probable that *qualitatively* similar cell migrations and contacts occur, *quantitative* findings and ontogenetic *timings* cannot, of course, be extrapolated to the human cerebellum. Only the briefest introduction can be entertained in this volume, and the interested reader should enter the literature by consulting such key references as Miale and Sidman (1961), Fujita (1963), Fujita *et al* (1966), Kornguth *et al.* (1967), Mugnaini (1970), Eccles (1970), Hamori (1972), Altman (1972a, b and c), Eccles (1973), Swarz (1976), and Swarz and Del Cevro (1975, 1977).

The early cerebellar rudiment consists of a pseudostratified epithelium showing interkinetic migration (p. 157) which soon develops the three basic zones—ventricular, intermediate, and marginal—as elsewhere in the neural tube. Originating in the ventricular zone, *proliferation* of as yet *uncommitted germinal* (matrix) *cells* proceeds, however, in *two* quite distinct sites, which become increasingly separated as the rudiment expands and thickens. Germinal cells continue proliferating in the deep (subventricular) part of the intermediate zone forming what, for convenience, may be termed the *internal germinal layer*. Slightly later, spreading latero-medially from each side to meet centrally

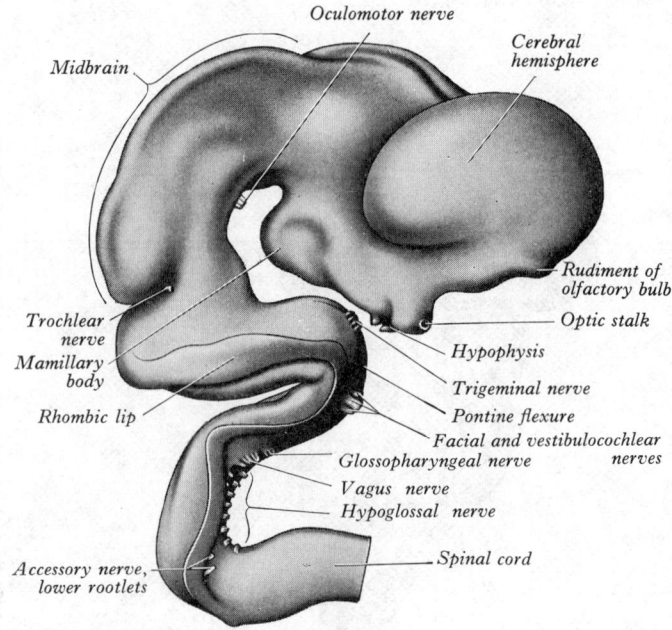

Oculomotor nerve
Cerebral hemisphere
Midbrain
Rudiment of olfactory bulb
Trochlear nerve
Optic stalk
Mamillary body
Hypophysis
Rhombic lip
Trigeminal nerve
Pontine flexure
Facial and vestibulocochlear nerves
Glossopharyngeal nerve
Vagus nerve
Hypoglossal nerve
Accessory nerve, lower rootlets
Spinal cord

2.68C The right side of the brain of a human embryo, 13·6 mm long. The roof of the hindbrain has been removed. Compare with 2.69B. (From a model by His.)

in a subpial position (i.e. the most superficial part of the marginal zone), similarly uncommitted germinal cells proliferate forming an *external germinal layer* (sometimes, less appropriately, called the external granular layer). Both germinal layers continue *proliferative mitoses*, giving clonal progeny, until they are several cells thick. This *proliferative phase* gradually diminishes with the onset of the *phase of neurogenesis*, and this, in its turn, is gradually followed by a *phase of gliogenesis*. Following a precisely ordered time sequence, sets of *critical (quantal) mitoses* occur (p. 84), the progeny becoming *committed neuroblasts* which do not divide again. Each set is destined to become one of the specific neuronal varieties that characterize the deep cerebellar nuclei and cortex, and they *migrate* to their definitive positions, develop their neurites and synaptic contacts, and mature, again in finely patterned spatio-temporal *sequences*, the latter often exhibiting some degree of overlap. Similarly as neurogenesis wanes sets of quantal mitoses of some remaining germinal cells continue to give generations of *committed glioblasts* which, although less well documented than the neuroblasts, also migrate to, and mature in their definitive positions. The fate of each germinal layer will now briefly be reviewed.

Quantal mitoses of the uncommitted cells of the *internal germinal layer* result in the emergence of two types, followed somewhat later by a third type, of definitive neuroblasts. Initially, primitive Purkinje neuroblasts and primitive nuclear neuroblasts in approximately equal numbers, but whether this follows a series of symmetrical or asymmetrical mitoses (p. 84) is unknown. The *nuclear neuroblasts* remain embedded in the developing future white matter adjacent to the roof of the rostral part of the fourth ventricle. The main mass of nuclear neuroblasts then slowly subdivides into the primordial fastigial, emboliform, globose and dentate deep cerebellar nuclei, and the individual neuroblasts differentiate into either small intranuclear interneurons, or into the larger projection neurons. The axons of the latter invade the early cerebellar peduncles and pursue complex paths to their multiple destinations (*see* pp. 916–917). The *Purkinje neuroblasts*, in contrast, *migrate* superficially towards their definitive position in the expanding cortex, where they slowly mature into their highly characteristic form of somata and dendritic trees. As they migrate, the terminals of one neurite—the future axon—remain adjacent to, and ultimately in synaptic contact with, the nuclear neuroblasts, the remainder of the axon elongating as it 'trails' behind the advancing soma. The mature and developing Purkinje cell has been a favourite object for *quantitative* cytological and ultrastructural studies, both during normal ontogeny, and after such experimental manipulations as suppression of granule cell (and therefore parallel fibre) development, or after prevention of climbing fibre growth (*see*, for example, Hamori 1972, and p. 924). Excellent illustrative examples of the maturation of normal rat cerebellar cortex are to be found in the writings of Altman (1972b). When the formation of nuclear and Purkinje neuroblasts has proceeded for some time, a number of the remaining internal germinal cells give rise to generations of *Golgi neuroblasts* which also migrate superficially to gradually occupy, and mature in, their definitive position and morphology.

The first set of quantal mitoses in the *external germinal layer* gives generations of *basket neuroblasts* which migrate deeply to meet, and ultimately synapse with, the somata and preaxons of the ascending Purkinje neuroblasts, their axons passing transversely in the primordial cerebellar folia. Secondly, following intense proliferation of the external germinal cells, a second set of quantal mitoses results in vast numbers of *granule cell neuroblasts*. A widely held view is that these microneurons are at first bipolar, each presumptive axon growing in opposite directions along the long axis of the folium: bundles of such axons form the parallel fibres of the mature cortex (p. 922), and the more recently formed granule cell neuroblasts (with their parallel fibre axons) are placed in successive layers approaching the pial surface. The soma of the granule cell does not, however, remain bipolar and external to the Purkinje neuroblast layer. A neurite develops and grows deeply *through* the Purkinje layer, its tip ultimately dividing, meeting, and synapsing with the ingrowing mossy afferent terminals (p. 924), thus forming the primordial cerebellar glomeruli. As this centripetal neurite develops, the granule cell nucleus migrates along it to reach, and helps to form, the definitive granular layer

of somata of the mature cortex. Rakic (1971a and b), however, reached a different conclusion from studies of migrating granule cell neuroblasts in the rhesus monkey, holding that they migrated deeply by contact guidance along the surface of the long radial processes of the Bergmann glial cells (Mugnaini and Forstrønen 1967). He considered that the migrating soma 'trailed' behind it an elongating neurite which subsequently bifurcated to form the parallel fibre axons, whilst dividing neurites passing in advance of the soma met the mossy afferents to form cerebellar glomeruli.

The final generations of neuroblasts from the external germinal layer merely migrate locally and differentiate into outer stellate cells.

The origin of the cerebellar glial cells remains much more problematical and has been the subject of dispute for almost a century. The view advanced by Obersteiner (1883) and Schaper (1897) was that the various macroglial cell varieties (Bergmann cells, astrocytes and oligodendrocytes) stemmed from final generations of germinal cells of *both* the internal and external germinal layers; this has received the more recent experimental support of such authorities as Fujita *et al.* (1966), Fujita (1967), Meller and Glees (1969) and Privat (1975). The second view proposed by Athias (1897) and Cajal (1911) held that the macroglia were formed exclusively from the internal germinal layer, and that the progeny of the external layer were solely the three varieties of neuroblast: this suggestion has been supported by the labelling studies of Swarz and Del Cerro (1977). Clearly such diametrically opposed views must await further critical evaluation using alternative methods, sites and species. All authorities seem agreed that the microglial elements are exogenous, invading the cerebellar rudiment from the surrounding mesenchyme.

The remainder of the metencephalon becomes the pons, but little is known of the individual stages in the transformation. Ventricular, intermediate and marginal zones are formed in the usual way, and the nuclei of the trigeminal, abducent and facial nerves develop in the mantle layer. It is probable that the grey matter of the formatio reticularis is derived from the ventral lamina and that of the nuclei pontis from the dorsal lamina by the active migration of cells from the rhombic lip. About the fourth month the pons is invaded by corticopontine and corticospinal fibres, becomes proportionately thicker, and takes on its adult appearance.

The region of the *isthmus rhombencephali* undergoes a series of changes which are difficult to interpret. As a result, the greater part of the region apparently becomes absorbed into the caudal end of the midbrain, only the roof plate, in which the anterior medullary velum is formed, and the dorsal parts of the dorsal laminae, which become invaded by the fibres of the superior cerebellar peduncles, remaining as recognizable derivatives in the adult. It should be noted that originally the decussation of the trochlear nerves is caudal to the isthmus, but as the growth changes occur it is displaced in a cranial direction until it reaches its adult position. These changes are also responsible (1) for the movement in the same direction of the trochlear nucleus, whereby it comes to lie in the midbrain, and (2) for the position of the mesencephalic nucleus of the trigeminal nerve, which is also a derivative of the isthmus rhombencephali (Frazer 1928).

THE MESENCEPHALON

The mesencephalon or midbrain is derived from the intermediate primary cerebral vesicle, which persists for a time as a thin-walled tube enclosing a cavity of some size, separated from that of the prosencephalon by a slight constriction and from the rhombencephalon by the isthmus rhombencephali. Later, its cavity becomes relatively reduced in diameter, and in the adult brain it forms the *cerebral aqueduct*. The ventral laminae of the midbrain increase in thickness to form the *cerebral peduncles*, which are at first of small size, but enlarge rapidly after the fourth month, when their fibre tracts begin to appear in the marginal zone. The neuroblasts of the ventral lamina give origin to the nuclei of the oculomotor nerve and grey matter of the tegmentum, while the nucleus of the trochlear nerve, and also the mesencephalic nucleus of the trigeminal nerve, migrate cranially into the midbrain owing to the developmental changes which occur in the isthmus

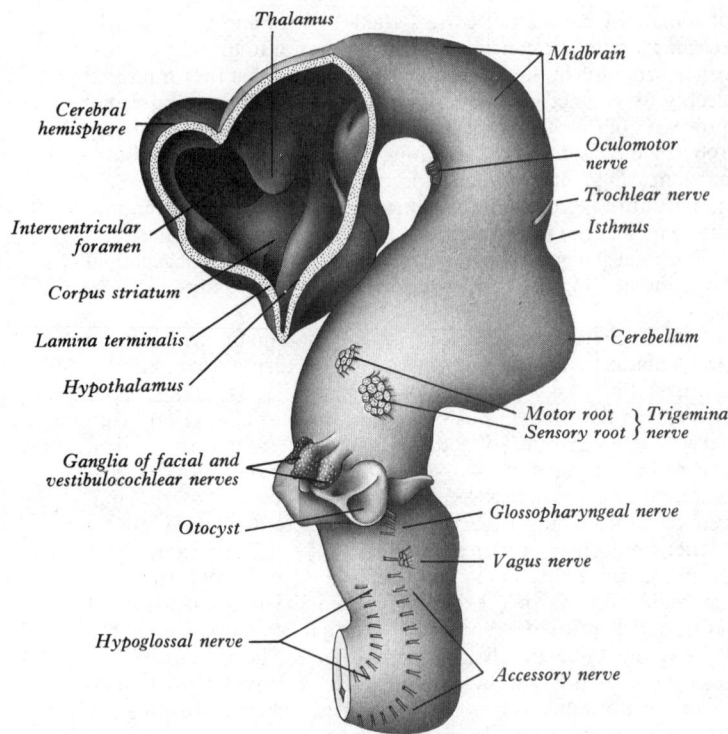

2.69A The brain of a human embryo, about 10·2 mm long. (From a model by His.)

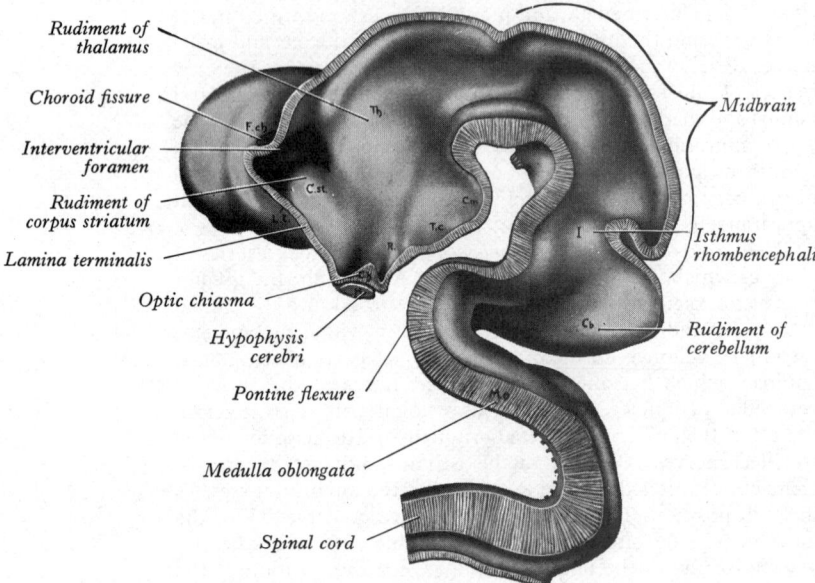

2.69B The brain of a human embryo, 13·6 mm long. Medial surface of right half. The roof of the hindbrain has been removed. (From a model by His.)

the principles outlined for the cerebellar cortex (p. 166) and the paleopallium and neopallium (p. 173) also apply to this region. It may be noted, however, that there exists a high degree of geometric order in the developing retinotectal projection (p. 941), and also a precise somatotopy in tectospinal projection. These facts coupled with the ability of the piscine and amphibian central nervous tracts to *regenerate* after severance, have led the retinotectal pathways to become classical sites for experimentation. A review of this field is beyond the scope of the present volume, but typical experiments included the electrophysiological and neuroanatomical mapping of the patterns of retinotectal connexions that were established after, for example, mere cutting of an optic nerve, or following nerve severance, the rotation of one or both eyes through 180°. Other experiments included bilateral optic nerve section together with median section of the optic chiasma; additionally the regeneration pattern was followed after deletion of different retinal sectors, or following various recombinations of half eye-cups. In general the results of such experiments led to the elaboration of a powerful theory of the *chemospecificity of neuroblasts*, the main proposition being that every neuroblast possessed a unique surface molecular configuration. It was further postulated that the growth pattern of a neurite followed a chemically coded path which led to a precise group of target cells bearing complementary surface configurations (*see* Sperry, 1951a and b, 1958, 1963, 1965, 1971). While the *overall* results have received much subsequent experimental support, more detailed analyses of the re-established connectivity patterns, and in particular the *changing* patterns that occur with tectal growth, and the modulation that occurs with altered environments, behaviour, experience and the phenomena of learning and memory, has cast doubt on the validity of a *rigid* theory of chemospecificity applying to *all* neuroblasts and neurons (e.g. *see* Székely 1974; Hunt and Jacobson 1974).

It may be pertinent at this point to mention the interesting suggestion put forward by Jacobson (1969, 1970a and b, 1974) that there exist two principal classes of neuron (an elegant modern refinement of the old, Golgi type I and type II varieties—*see* p. 821). Jacobson's class I neurons are phylogenetically ancient, develop early in ontogeny, possess prominent somata, long axons, specific target cells, and have invariant and unmodifiable connexions and functions from their first establishment. It is

rhombencephali. It has been claimed (Chu-wu and Wen-kuei 1965) that some of the migrating neuroblasts of the mesencephalic nucleus reach the posterior commissure and its nucleus, and also the interstitial nucleus. The cells of the dorsal part of the dorsal laminae proliferate and invade the roof plate, which therefore becomes thickened and is later divided into corpora bigemina by a median groove. At its caudal end this groove becomes a median ridge, which persists in the adult as the frenulum veli. The corpora bigemina are later subdivided into the *superior* and *inferior colliculi* by a transverse furrow. The *red nucleus* is clearly defined at the end of the third month. Its origin, whether from neuroblasts of the ventral or dorsal lamina, is uncertain.

The detailed histogenesis of the tectum, and its main derivatives, the colliculi, will not be followed here, but in general

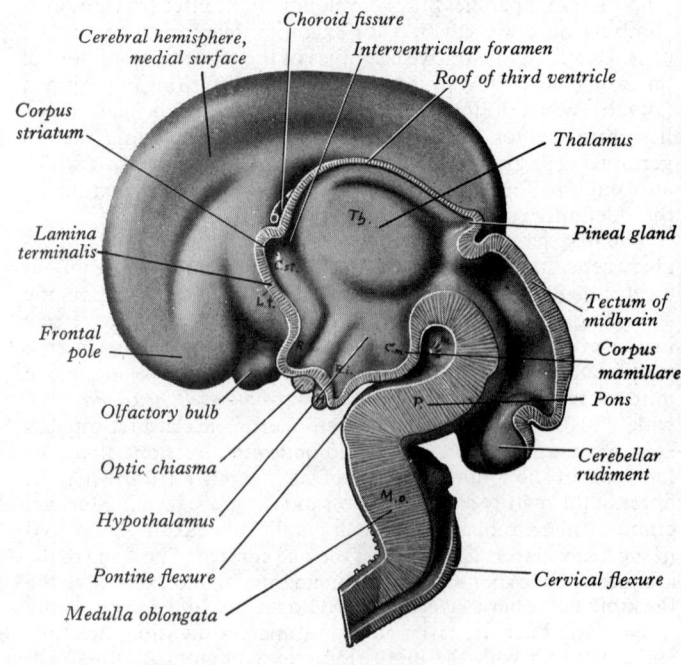

2.69C Medial surface of the right half of the brain of a human fetus, about three months old.

presumed that their final form and activities are almost wholly encoded in the genome. In contrast, class II neurons are small, possessing a short (or no) axon (microneurons); they are phylogenetically recent and develop relatively late in ontogeny. They are considered to possess, initially, *multiple* functional potentialities that may persist for considerable intervals, and only become progressively restricted during subsequent fetal or post-natal growth, the microneuron *ultimately* becoming 'unipotential and unmodifiable'. It is also proposed that this restriction of functional potential occurs at varying times and rates in different locations in the developing nervous system. Such microneurons are, on this view, therefore regarded as one of the main structural bases of *plasticity* in nervous systems whereby structural/functional modifications result from environmental changes, and the assessment of results of exploratory behaviour; they may thus be one of the sites where the elusive structural correlates of memory patterns should be sought (*see* essays by Young 1964; Gray 1974; Jacobson 1974).

THE PROSENCEPHALON

At an early stage, a transverse section through the forebrain shows the same parts as are displayed in similar sections of the spinal cord and medulla oblongata—thick lateral walls connected by thin floor and roof plates. Moreover, each lateral wall is divided into a dorsal area and a ventral area separated internally by the *hypothalamic sulcus*. This sulcus ends anteriorly at the medial end of the optic stalk, and in the fully developed brain, persists as a slight groove extending backwards from the interventricular foramen to the cerebral aqueduct. It is analogous to, if not the homologue of, the sulcus limitans.

At a very early period, before the closure of the anterior neuropore (p. 112), two lateral diverticula, the *optic vesicles*, appear, one on each side of the forebrain; for a time they communicate with its cavity by relatively wide openings. The distal parts of the optic vesicles expand, while the proximal parts become the tubular *optic stalks*; their further development is given on pp. 176–78. The forebrain next grows ventrally, and two diverticula rapidly expand from it to form two large pouches, one on each side. These diverticula subsequently form the *cerebral hemispheres*, and their cavities are the rudiments of the lateral ventricles; they communicate with the median part of the forebrain cavity by relatively wide openings which ultimately become the interventricular foramina. The anterior part of the roof plate of the forebrain consists of a thin sheet, the *lamina terminalis* (**2.69**A, B, C), which stretches from the interventricular foramina to the recess at the base of the optic stalks. The anterior part of the forebrain, including the rudiments of the cerebral hemispheres, is the *telencephalon*, and the posterior part the *diencephalon*; both contribute to the formation of the third ventricle, although the latter plays the predominant part.

The diencephalon develops into the (dorsal) thalamus and metathalamus along the dorsal area of its lateral wall. The *thalamus* (**2.69–70**) begins as a thickening which involves the anterior part of the dorsal area (Cooper 1950). Caudal to the thalamus the lateral and medial geniculate bodies, which constitute the *metathalamus*, are recognizable at first as surface depressions on the internal aspect and as elevations on the external aspect of the lateral wall (Cooper 1945). As the thalami enlarge, they gradually narrow the wide interval between them into a vertically compressed cavity which forms the greater part of the third ventricle. After a time these medial surfaces may come into contact and become adherent to each other over a variable area, the connexion (which may be single or multiple) constituting the *interthalamic adhesion*. The caudal growth of the thalamus excludes the geniculate bodies from the lateral wall of the third ventricle.

At first the lateral aspect of the developing thalamus is separated from the medial aspect of the cerebral hemisphere by a cleft, but as growth proceeds the cleft becomes obliterated (**2.70**) as the thalamus fuses with the part of the hemisphere in which the corpus striatum is developing. Later, with the development of the projection fibres of the neopallium (p. 173), the thalamus becomes

related to the internal capsule, which intervenes between it and the lateral part of the corpus striatum (lentiform nucleus). Ventral to the hypothalamic sulcus the lateral wall of the diencephalon forms a large part of the hypothalamus.

The *epithalamus*, which includes the pineal gland, the posterior commissure and the trigonum habenulae, develops in association with the caudal part of the roof plate and the adjoining portions of the lateral walls of the diencephalon. The *pineal gland* arises as a hollow outgrowth from the roof plate, immediately adjoining the mesencephalon. Its distal portion becomes solid by cellular proliferation, but its proximal stalk remains hollow, containing the pineal recess of the third ventricle. In many reptiles the pineal outgrowth is double. The anterior outgrowth (*parapineal organ*) develops into the pineal or parietal eye (p. 1445) while the posterior outgrowth is glandular in character. It is the *posterior* outgrowth which is homologous with the pineal gland in man. The anterior outgrowth also develops in the human embryo, but soon disappears entirely.

The *posterior commissure* is formed by fibres which invade the caudal wall of the pineal recess from both sides.

The *nucleus habenulae*, which is the most important constituent of the *trigonum habenulae*, is developed in the lateral wall of the diencephalon and is at first in close relationship with the geniculate bodies, from which it becomes separated by the dorsal growth of the thalamus. The habenular commissure develops in the cranial wall of the pineal recess.

The roof plate of the diencephalon, rostral to the pineal gland, remains thin and epithelial in character, and is subsequently invaginated by the choroid plexuses of the third ventricle. Before the development of the corpus callosum and the fornix it lies at the bottom of the longitudinal fissure, between the two cerebral hemispheres. Here, and elsewhere, choroid plexuses develop by the close apposition of vascular pia mater and ependyma with no nervous tissue intervening. With development, the vascular layer is infolded into the ventricular cavity and develops a series of small villous (finger-like) projections, each covered by a cuboidal epithelium derived from the ependyma. The cuboidal cells carry numerous microvilli on their ventricular surfaces whilst basally their plasma membrane becomes complexly folded into the cell. The *early* choroid plexuses secrete a protein-rich cerebrospinal fluid into the ventricular system which may provide a nutritive medium for the primitive epithelial neural tissues. With increasing vascularity of the latter, however, the histochemical reactions of the cuboidal cells and the character of the fluid change to the adult type (Klosovskii 1963). It should also be noted that in addition to choroid plexus formation, the remaining lining of the third ventricle does not simply form generalized ependymal cells.

Many regions become highly specialized, developing concentrations of tanycytes, or other modified cells such as those of the *subfornical organ*, the *organum vasculosum* (*intercolumnar tubercle*) of the lamina terminalis, the *subcommissural organ*, and those lining the *pineal, suprapineal,* and *infundibular recesses* (*see* Knigge *et al.* (eds) 1975)—collectively the *circumventricular organs*.

The floor of the diencephalon takes part in the formation of the *hypothalamus*, including the mamillary bodies, the tuber cinereum and infundibulum of the hypophysis.

The *mamillary bodies* arise as a single thickening, which becomes divided by a median furrow during the third month. Anterior to them the *tuber cinereum* develops as a cellular proliferation which extends forwards as far as the infundibulum. In front of the tuber cinereum the floor of the diencephalon gives origin to a wide-mouthed diverticulum, which grows towards the stomatodeal roof and comes into contact with the posterior aspect of a dorsally directed ingrowth from the stomatodeum (Rathke's pouch, p. 199). These two diverticula, the one derived from the floor of the neural tube and the other from the roof of the stomatodeum, together form the *hypophysis cerebri* (**2.94**). In the base of the neural outgrowth an extension of the third ventricle persists as the infundibular recess.

The optic vesicles, which are described with the development of the eye (p. 176), are derived from the lateral wall of the prosencephalon before the telencephalon can be identified. They are usually regarded as derivatives of the diencephalon, and the optic chiasma indicates the boundary between the diencephalon and the telencephalon.

The telencephalon consists of two lateral diverticula connected by a median region (the *telencephalon impar*). From the latter develops the anterior part of the cavity of the third ventricle, closed below and in front by the *lamina terminalis*. The lateral diverticula are outward pouchings of the lateral walls of the prosencephalon, which may correspond to the alar laminae, although this is uncertain; the cavities are the future lateral ventricles, and their walls the nervous tissue of the cerebral hemispheres. The roof plate of the median part of the telencephalon remains thin, and is continuous behind with the roof plate of the diencephalon. In the floor plate and lateral walls of the prosencephalon, ventral to the primitive interventricular foramina, the anterior parts of the *hypothalamus* are developed; these include the optic chiasma and the optic recess. The optic chiasma is formed by the meeting and partial decussation of the optic nerves in the ventral part of the lamina terminalis, and from it the optic tracts subsequently grow backwards to end in the diencephalon and midbrain.

The cerebral hemispheres arise as diverticula of the lateral walls of the telencephalon, with which they remain in continuity around the margins of the large, interventricular foramina except caudally, where they are continuous with the anterior part of the lateral wall of the diencephalon (2.69A, B); as growth proceeds the hemisphere enlarges forwards, upwards and backwards and acquires an oval outline, with medial and superolateral walls and a floor. As a result the medial surfaces are separated from each other by a cleft, the *longitudinal fissure*. At this stage the floor of the cleft is the epithelial roof plate of the telencephalon, which is directly continuous caudally with the epithelial roof plate of the diencephalon (2.69C), as already stated above.

The rostral end of the oval hemisphere becomes the *frontal pole*, but, as the hemisphere expands, its original posterior pole moves relatively in a caudoventral direction in association with the growth of the caudate nucleus, to form the *temporal pole*, and a new posterior part becomes defined, which persists as the *occipital pole* of the mature brain. The great expansion of the cerebral hemispheres is characteristic of mammals and especially of man, and in their subsequent growth they overlap, successively, the diencephalon, the mesencephalon and the cerebellum, and the temporal lobes grow round the flanks of the brainstem.

About the fifth week a longitudinal groove appears in the anteromedial part of the floor of each ventricle. This groove deepens and forms a hollow diverticulum connected to the

hemisphere by a short stalk. The diverticulum becomes connected on its ventral or inferior surface to a ganglionic mass, the cells of which receive the afferent axons of the sensory cells of the olfactory plate. As the head increases in size the diverticulum grows forwards and, subsequently losing its cavity, becomes converted into the solid *olfactory bulb*. The forward growth of the bulb is accompanied by elongation of its stalk, which forms the *olfactory tract*, and the portion of the floor of the hemisphere to which the tract is attached constitutes the *piriform area*.

The pia mater which covers the epithelial roof of the third ventricle at this stage is itself covered with loosely arranged mesenchyme. In the meshes of this tissue numerous blood vessels develop and, as we have seen (p. 169), on each side of the median plane these vessels subsequently invaginate the roof of the ventricle to form its *choroid plexuses* (Kappers 1966). The lower part of the medial wall of the hemisphere, which immediately adjoins the epithelial roof of the interventricular foramen and the anterior extremity of the diencephalon, also remains epithelial, consisting of ependyma and pia mater, while elsewhere the walls of the hemisphere are thickening to form the *pallium*. This thin part of the medial wall of the hemisphere is invaginated by vascular tissue, continuous in front with the choroid plexus of the third ventricle and constituting the *choroid plexus* of the *lateral ventricle*. This invagination occurs along a line which arches upwards and backwards, parallel with the anterior and upper boundaries of the interventricular foramen, and the curved indentation of the ventricular wall where no nervous tissue develops between ependyma and pia mater, is termed the *choroid fissure* (2.69C; 2.70A, B.)

At first growth proceeds more actively in the floor and the adjoining part of the lateral wall of the developing hemisphere, and elevations formed by the rudimentary *corpus striatum* (2.69A) encroach on the cavity of the lateral ventricle (Cooper 1946). The head of the *caudate nucleus* appears as three successive parts, medial, lateral and intermediate, which produce elevations in the floor of the lateral ventricle. Caudally these merge to form the tail of the caudate and the *amygdaloid complex*. From the beginning they are close to the temporal pole of the hemisphere and, when the occipital pole grows backwards and the general enlargement of the hemisphere carries the temporal pole downwards and forwards, the tail is continued from the floor of the central part of the ventricle into the roof of its temporal extension as the future *inferior horn*. Anteriorly the head of the caudate nucleus extends forwards to the floor of the interventricular foramen, where it is separated from the developing anterior end of the thalamus by a groove (2.69B). The lentiform nucleus is developed from two laminae of cells, medial and lateral, which are continuous with both the medial and lateral parts of the caudate nucleus. The internal capsule appears first in the medial lamina and extends laterally through the outer lamina to the cortex. It divides the laminae into two, the internal parts joining the caudate nucleus and the external parts forming the lentiform nucleus. In the latter, the remaining medial lamina cells give rise predominantly to the globus pallidus and the lateral to the putamen. Subsequently the putamen expands concurrently with the intermediate part of the caudate nucleus (Hewitt 1958, 1961).

As the hemisphere enlarges, the caudal part of its medial surface overlaps and hides the lateral surface of the diencephalon (thalamic part), being separated from it by a narrow cleft occupied by vascular connective tissue. At this stage (about the end of the second month) a transverse section made caudal to the interventricular foramen passes successively through (1) the developing thalamus, (2) the narrow cleft just mentioned, (3) the thin medial wall of the hemisphere, and (4) the cavity of the lateral ventricle, with the corpus striatum in its floor and lateral wall (2.70A). As the thalamus increases in extent it acquires a superior in addition to medial and lateral surfaces, and the lateral part of its superior surface fuses with the thin medial wall of the hemisphere, so that, finally, this part of the thalamus is covered with the ependyma of the lateral ventricle immediately ventral to the choroid fissure (2.70B). As a result the corpus striatum is approximated to the thalamus and separated from it only by a deep groove, which becomes obliterated by increased growth along the line of contact. The lateral aspect of the thalamus is now in continuity with the medial aspect of the corpus striatum, so that

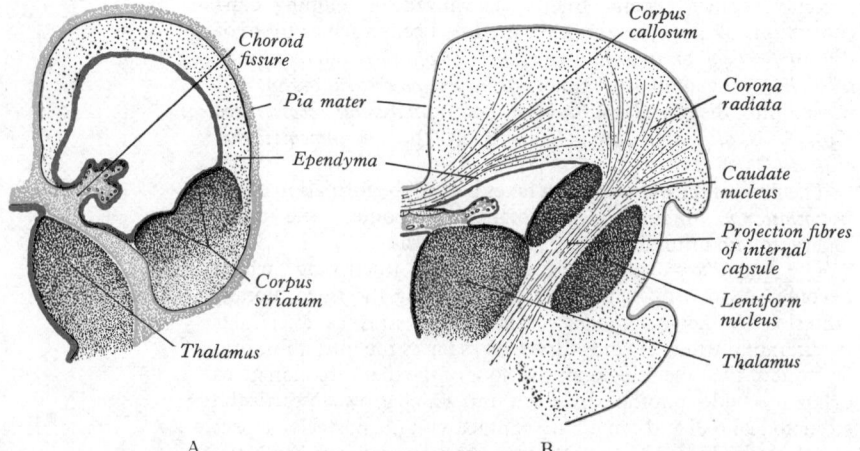

2.70A Diagrams illustrating transverse sections across the developing thalamus and cerebral hemisphere. Note that at the choroid fissure the vascular pia mater (blue) meets the ependyma (red) to form a choroid plexus. In A the lateral aspect of the thalamus is separated from the medial aspect of the hemisphere by an interval containing vascular mesenchyme. In B this interval has disappeared; the expanded upper surface of the thalamus is covered by the ependyma of the lateral ventricle, and the approximation of the thalamus and the corpus striatum has provided a pathway for the projection fibres of the internal capsule.

Labels in figure:
Choroid fissure
Pia mater
Ependyma
Corpus striatum
Thalamus
Corpus callosum
Corona radiata
Caudate nucleus
Projection fibres of internal capsule
Lentiform nucleus
Thalamus
A
B

a secondary union between the diencephalon and the telencephalon is effected over a wide area, providing a route for the subsequent passage of projection fibres to and from the cortex.

The *pallial* or *cortical* area which borders the interventricular foramen and lies outside the choroid fissure constitutes the *paleopallium*. It is the first part of the cortex to undergo differentiation (*vide infra*), and at first it forms a continuous, almost circular strip on the medial and inferior aspects of the hemisphere. Below and in front, where the stalk of the olfactory tract is attached, it constitutes a part of the *piriform area*. The portion outside the curve of the choroid fissure (**2.70B**) constitutes the *hippocampal formation*. In this region the neuroblasts of the *pallium* (mantle) or developing cortex, proliferate, and the wall of the hemisphere thickens and produces an elevation which projects into the medial side of the ventricle. This elevation is the *hippocampus* (Humphrey 1964, 1967). It appears first on the medial wall of the hemisphere above the area in front of the lamina terminalis (*paraterminal area*) and gradually extends backwards into the region of the temporal pole, where it adjoins the piriform area. The marginal zone in the region of the hippocampus becomes invaded by nerve cells which form the *dentate gyrus*. This structure is practically co-extensive with the hippocampus and, like it, extends from the paraterminal area backwards above the choroid fissure and follows its curve downwards and forwards towards the temporal pole, where it runs into the piriform area. A shallow surface depression (which has been termed the *hippocampal sulcus*) grooves the medial surface of the hemisphere in the region of the hippocampal formation, but it is not responsible for the elevation which the hippocampus forms in the interior of the ventricle.

The efferent fibres from the cells of the hippocampus collect along its medial edge and run forwards immediately above the choroid fissure. Anteriorly they turn ventrally and enter the lateral part of the lamina terminalis to gain the hypothalamus, where they end in and around the mamillary body. These efferent hippocampal fibres form the *fimbria hippocampi* and the *fornix*.

The development of the commissures effects a very profound alteration on the medial wall of the hemisphere. At the time of their appearance the two hemispheres are connected to each other by the median part of the telencephalon. The roof plate of this area remains epithelial, whilst its floor becomes invaded by the decussating fibres of the optic nerves. These two routes are thus not available for the passage of commissural fibres passing from hemisphere to hemisphere across the median plane, and these fibres therefore pass through the anterior wall of the interventricular foramen, i.e. the *lamina terminalis*. The first commissures to develop are those associated with the *olfactory areas* and *paleopallial cortex*. Fibres of the olfactory tracts cross in the ventral or lower part of the lamina terminalis and together with fibres from the piriform, and prepiriform areas, and the amygdaloid bodies, form the *anterior part* of the *anterior commissure*. In addition the two hippocampi become interconnected by transverse fibres which cross from fornix to fornix in the upper part of the lamina terminalis as the *commissure of the fornix*. Various other decussating fibre bundles (known as the *supra-optic commissures*, although they are not true commissures) develop in the lamina terminalis immediately dorsal to the optic chiasma, between it and the anterior commissure.

The commissures of the neopallium develop later and follow the pathways already established by the commissures of the limbic system. Fibres coming from the tentorial surface of the hemisphere join the *anterior commissure* and constitute its larger *posterior part*. All the other commissural fibres of the neopallium associate themselves closely with the commissure of the fornix and lie on its dorsal surface. These fibres increase enormously in number, and the bundle rapidly outgrows its neighbours to form the corpus callosum (**2.70B**).

The *corpus callosum* commences in a thick mass, the *precommissural area*, connecting the two cerebral hemispheres around and above the anterior commissure. The upper end of this area extends backwards to form the trunk of the corpus callosum. The *rostrum* of the corpus callosum develops later and separates the front end of the precommissural area from the remainder of the cerebral hemisphere. Further backward growth of the trunk of the corpus callosum then results in the entrapped precom-

missural area becoming stretched out to form the bilateral septum pellucidum (Hewitt 1962). As the corpus callosum grows backwards it extends above the choroid fissure, carrying the commissure of the fornix on its under surface. In this way a new floor is formed for the longitudinal fissure, and additional structures come to lie above the epithelial roof of the third ventricle. In its backward growth the corpus callosum invades the area hitherto occupied by the *upper* part of the hippocampal formation and the corresponding parts of the dentate gyrus (**2.70B**) and hippocampus are reduced to mere vestiges—the *indusium griseum* and the *longitudinal striae*.

The *inferior* regions of both the dentate gyrus and hippocampus persist and enlarge because, with the forward growth of the temporal lobes, the brainstem presents a complete barrier to further extension of the corpus callosum.

The growth of the *neopallium*, or *neocortex*, and its enormous expansion are associated with the appearance of projection fibres during the latter part of the third month. These fibres follow the pathway provided by the apposition of the lateral aspect of the thalamus with the medial aspect of the corpus striatum, and, as they do so, they divide the latter, almost completely, into a lateral part, the lentiform nucleus, and a medial part, the caudate nucleus, these two nuclei remaining confluent only in their antero-inferior regions. The corticospinal tracts begin to develop in the ninth week of fetal life and have reached their caudal limits by the twenty-ninth week. The fibres destined for the cervical and upper thoracic regions and implicated in the innervation of the upper limb are in advance of those concerned with the lower limbs, which in turn are in advance of those concerned with the face. The appearance of reflexes in these three parts of the body shows a comparable sequence (Humphrey 1960).

At the end of the third month the superolateral surface of the cerebral hemisphere shows a slight depression anterosuperior to the temporal pole. This corresponds to the site of the corpus striatum in the floor and lateral wall of the ventricle, and its presence is due to the more rapid growth of the adjoining cortical regions. This *lateral cerebral fossa* gradually becomes overlapped and submerged, and is converted into the *lateral cerebral sulcus*; its floor becomes the *insula* (**2.71 A–G**). The process, however, is not completed in its most anterior part until after birth.

The growth changes in the temporal lobe which help to submerge the insula produce important changes in the olfactory areas. The olfactory tract is continuous, on the one hand, with the *medial olfactory gyrus*, which turns upwards in front of the lamina terminalis, and, on the other hand, with the *lateral olfactory gyrus*, which is directly continuous with the piriform area. The forward growth of the temporal pole and the general expansion of the neopallium cause the lateral olfactory gyrus to bend laterally, the summit of the convexity lying at the antero-inferior corner of the

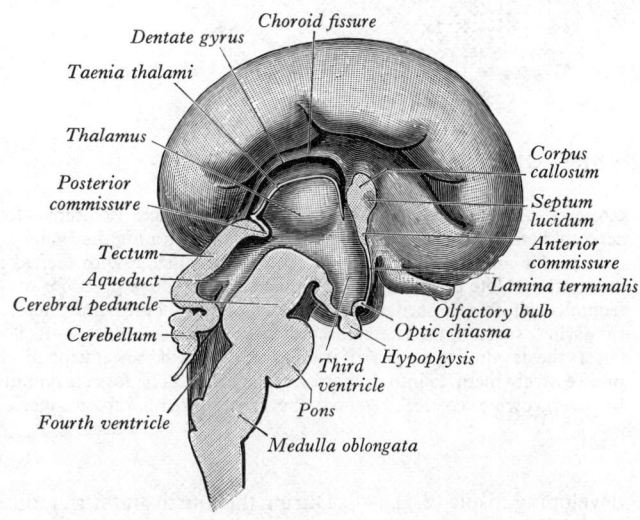

Choroid fissure
Dentate gyrus
Taenia thalami
Thalamus
Posterior commissure
Tectum
Aqueduct
Cerebral peduncle
Cerebellum
Corpus callosum
Septum lucidum
Anterior commissure
Lamina terminalis
Olfactory bulb
Optic chiasma
Hypophysis
Third ventricle
Pons
Fourth ventricle
Medulla oblongata

2.70B The brain of a human fetus, four months old. Medial aspect of left half.

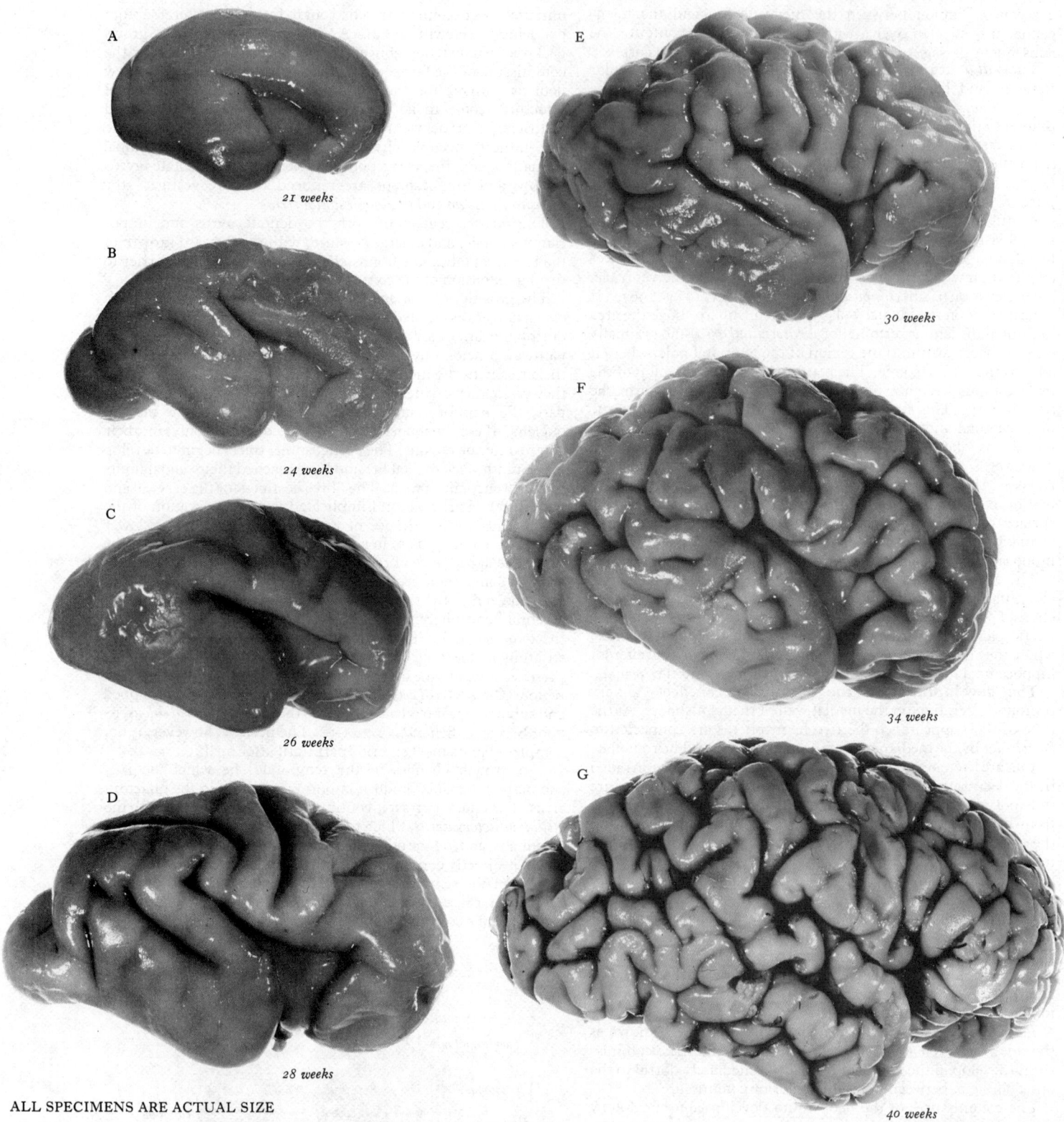

A
21 weeks

B
24 weeks

C
26 weeks

D
28 weeks

E
30 weeks

F
34 weeks

G
40 weeks

ALL SPECIMENS ARE ACTUAL SIZE

2.71A–G A series showing the superolateral surfaces of human fetal cerebral hemispheres at the ages indicated, demonstrating the changes in size, profile and the emerging pattern of cerebral sulci with increasing maturation. Note the changing prominence and relative positions of the frontal, occipital and particularly the temporal pole of the hemisphere. At the earliest stage (A) the lateral cerebral fossa is already obvious—its floor covers the developing corpus striatum in the depths of the hemisphere and progressively matures into the cortex of the insula. The fossa is bounded by overgrowing cortical regions, the frontal, temporal and parietal

opercula, which gradually converge to bury the insula; their approximation forms the lateral cerebral sulcus. By the sixth month the central, pre- and post-central, superior temporal, intraparietal and parieto-occipital sulci are all clearly visible. In the subsequent stages shown all the remaining principal and subsidiary sulci rapidly appear and by 40 weeks all the features which characterize the adult hemisphere in terms of surface topography are already present in miniature.

The photographs were kindly supplied by Dr. Sabina Strick of the Maudsley Hospital, London.

developing insula (**2.71**A–G). During the fourth and fifth months the *piriform area* becomes submerged by the adjoining neopallium, and in the adult only a part of it remains visible on the inferior aspect of the cerebrum.

Apart from the shallow hippocampal sulcus and the lateral cerebral fossa the surfaces of the hemisphere remain smooth and uninterrupted until early in the fourth month (**2.71**A–G). The parieto-occipital sulcus also appears about that time on the *medial* aspect of the hemisphere, and its appearance is associated with the increase in the splenial fibres of the corpus callosum. At about the same period the posterior part of the *calcarine sulcus* appears as a shallow groove extending forwards from the region of the occipital pole. It is a true infolding of the cortex and forms in the long axis of the *striate area*, producing at the same time an elevation, named the *calcar avis*, on the medial wall of the posterior horn of the ventricle.

During the fifth month the *sulcus cinguli* appears on the medial aspect of the hemisphere, but it is not until the sixth month that sulci appear on the inferior and superolateral aspects. The *central, precentral* and *postcentral sulci* appear each in two parts, upper and lower, which usually coalesce shortly afterwards, although they may remain discontinuous. The *superior* and *inferior frontal*, the *intraparietal, occipital, temporal* and *collateral sulci* make their appearance during the same period, and by the end of the eighth month all the important sulci can be recognized (**2**.71A–G).

Concerning the **histogenesis** of the pallial wall of the cerebral hemisphere, an impressive literature has accumulated since the early 1930s. Nevertheless, because of the immense complexity, multiplicity of cell types, and structural heterogeneity in different locations, descriptive and experimental analyses are less well understood and documented than those appertaining to the cerebellum, with its regular, geometrically ordered microstructure (p. 166). Only the briefest review of some basic principles, together with a few introductory key references, can be encompassed in this volume, and the interested reader will find it apposite to constantly cross-refer to the sections devoted to mature neuronal and cortical architecture (pp. 802–810, 1002–1009).

The wall of the earliest cerebral hemisphere, as elsewhere in the neural tube, consists of a pseudostratified epithelium, its cells exhibiting interkinetic migration (p. 157) as they proliferate to form clones of, as yet uncommitted, germinal cells. The columnar cells elongate and (following the nomenclature of the Boulder Committee, *Anat. Rec.*, 1970 and **2**.63A) their non-nucleated peripheral processes now constitute a *marginal zone*, whilst their nucleated, paraluminal and mitosing regions constitute the *ventricular zone*. Some of the mitotic progeny now leave the ventricular zone and continue to proliferate in an *intermediate zone*. This *proliferative phase* continues for a considerable period of fetal (and in some species post-natal) life, but, as in the case of the cerebellar cortex, after a period, groups of germinal cells undergo quantal mitoses forming initially generations of *definitive neuroblasts* and later, *definitive glioblasts*, which migrate to, and mature in, their final positions. It must be appreciated, however, that these phases of proliferation, differentiation, migration and maturation, are not absolutely sequential for each cell variety, but overlap each other in space and time.

The earliest migration of definitive neuroblasts from the ventricular and intermediate zones occurs radially until they approach, but do not reach, the pial surface, their somata becoming arranged as a transient *cortical plate*. Subsequently, proliferation wanes in the ventricular zone, but for considerable periods persists in the immediately adjacent *subventricular zone*. From the pial surface, inwards, therefore, there may now be defined the following zones: marginal, cortical plate, intermediate, subventricular, and ventricular. Briefly, whilst the foregoing *terminology* is relatively recent, it has, for long, been accepted that the *marginal zone* forms the outermost layer of the cerebral cortex, the neuroblasts of the *cortical plate* form the neurons of the remaining cortical laminae (the complexity, of course, varying in different locations, and with further additions of neuroblasts from the deeper zones), whilst the intermediate zone gradually transforms into the white matter of the hemisphere. Meanwhile other deep germinal cells have been producing generations of glioblasts which also migrate into the more superficial layers. As proliferation wanes and finally ceases in the ventricular and subventricular zones their remaining cells differentiate into general or specialized ependymal cells, tanycytes, or subependymal glial cells.

Although the short account just given of the *general* histogenetic history of the cortex gained, and maintains, wide acceptance, it is of the greatest interest that, with the advent of modern nuclear labelling techniques, there has been a fundamen-

tal reappraisal of the *sequences* involved in the migratory patterns of the neuroblasts, with consequently considerable ontogenetic and phylogenetic implications.

Pioneering studies into neuroblast migration in the developing mammalian *neocortex* were made by Tilney (1933), the technique available to him at this time being analysis of sections of Nissl stained tissue. Whilst it was clear that the subpial (marginal) zone formed the plexiform lamina (I), he considered that the remaining laminae stemmed from *three* quite distinct and separate migrations of neuroblasts up to the cortical plate. The first migration he considered to differentiate into the external granular lamina (II) and the pyramidal lamina (III); the second migration forming the internal granular lamina IV; the third migration he held to form the ganglionic lamina V and the multiform lamina VI. On this view, therefore, the *outermost* layers were the *earliest* to be formed, and progressively deeper layers at successively later times. (But see Rumyantsev 1979.)

The possibility that precisely the *reverse* sequence, progressing from *deep to superficial*, was first implied by the results of X-irradiation studies by Hicks *et al.* (1959). Further irradiation studies (Berry and Eayrs 1960), and autoradiographic nuclear labelling studies (Berry and Rogers 1965) supported this contention. These seminal investigations have, in the subsequent years, been amply confirmed in the rat, mouse, opossum and golden hamster. (The individual references are too numerous to quote in this volume, but for an excellent overview of this, and many other problems appertaining to cortical histogenesis, the reader should refer to Berry 1974.) Thus, in these various mammals, apart from the pre-existing anlage of Lamina I, the first laminae to be populated are VI and V, followed sequentially by laminae IV to II. Clearly, whilst the ontogenetic timings of migrations, from these experimental sources, are not appropriate to a volume on *human* anatomy, it is assumed from comparison of purely descriptive material, that similar *patterns* of migration and elaboration occur in the human cortex. It should also be noted that, as yet only in the human cortex, a thin subpial lamina of densely staining cells, of unknown origin, or destination, has been identified (Rabinowicz 1964, 1967; Brun 1965): they are not a prominent feature of cortical histogenesis, but an analogy with the external germinal layer of the cerebellum has been suggested.

No attempt will be made here to discuss neuroblast and glioblast differentiation, migration, and maturation with the establishment of intercellular contacts, for these the reader is referred to pp. 157–160, 837, 860 (see also Molliver *et al.* 1973 and Kostović and Krmpotić 1976). However, some general hypotheses may be mentioned, involving a comparison of ontogeny and phylogeny. Firstly, all parts of the neural tube, from the presumptive spinal cord to presumptive neocortex, pass through the stage of a pseudostratified epithelium and a proliferative phase, followed by differentiation sets, with the emergence of ventricular, intermediate, and marginal zones. Neuroblast migration, target cell contact and maturation (or, in some locations, *degeneration* and *cell death*), whilst *still confined to* the deeper reaches of the intermediate zone, are the principal events in spinal cord development. Throughout the encephalon, however, the primary difference is the *continued migration* of neuroblasts, and their ultimate maturation, far *beyond* the confines of the intermediate zone, forming either nuclear masses, variously displaced from the ventricular and aqueductual channels, or, in the 'roof-brain' regions, reaching the subpial marginal zones forming, initially, a simple cortical plate of neuroblasts. The latter then differentiates into subzones, showing a tangential *laminar* organization, whilst, in some locations, there emerges a well-defined *columnar* radial organization. Such cortical dispositions are evident in the hemispheric forebrain, tectal midbrain, and cerebellar hindbrain. In the pallial walls of the mammalian cerebral hemisphere, the phylogenetically *oldest regions*, and the *first* to differentiate during ontogeny, are those that *border* the interventricular foramen, and its extension the choroidal fissure. There exists an increasingly complex level of organization, from three to six tangential laminae, passing from the dentate gyrus and cornu ammonis, through the subiculum, until the general neocortex is reached. The *deepest*, and phylogenetically oldest *tangential laminae* are the first to be populated by migrating neuroblasts, more superficial layers being added in sequence, their neuroblasts migrating through the older layers; the number of superadded laminae depending upon the location with respect to the choroidal fissure. These broad patterns have been demonstrated by nuclear labelling studies, not only in the neocortex, but also in the dentate gyrus and hippocampus (*see*, for example, Angerine 1975; Altman and Bayer 1975).

The state of differentiation at birth and at various post-natal stages, as seen in Golgi (metal imprégnation) preparations, has been described in considerable detail elsewhere (Conel—a series of publications 1939–59). However, the *mechanisms* whereby variation in cell type, number, dendritic patterns and connectivity arise, are completely unknown. Nevertheless, gross nutritional deficiencies, endocrine imbalances, sensory deprivation, neurotropic viruses, vascular abnormalities and perinatal anoxia may all disturb the normal pattern.

At birth the volume of the brain is approximately *25 per cent* of its volume in adult life. The greater part of the increase occurs during the *first year*, at the end of which the volume of the brain has increased to *75 per cent* of its adult volume. The growth can be accounted for partly by increase in the size of nerve cell somata, the profusion and dimensions of their dendritic trees, axons and their collaterals, and by growth of the neuroglial cells and cerebral blood vessels, but it is the acquisition of myelin sheaths by the axons which is principally responsible for it. The great sensory pathways, visual, auditory and somatic, become myelinated first, and the motor fibres later. During the second and subsequent years, growth proceeds much more slowly, and the brain attains its adult size by the seventeenth or eighteenth year. This continued growth is connected with the continued myelination of various groups of nerve fibres.

A summary of the parts derived from the cerebral vesicles is as follows:

Rhombencephalon (or hindbrain)	1. Myelencephalon	Medulla oblongata Caudal part of the 4th ventricle Inferior cerebellar peduncles
	2. Metencephalon	Pons Cerebellum Middle part of the 4th ventricle Middle cerebellar peduncles
	3. Isthmus rhombencephali	Anterior medullary velum Superior cerebellar punduncles Rostral part of the 4th ventricle
Mesencephalon (or midbrain)		Cerebral peduncles Tegmentum Tectum Aqueduct
Prosencephalon (or forebrain)	1. Diencephalon	Thalamus Metathalamus Subthalamus Epithalamus Caudal part of the hypothalamus Caudal part of the 3rd ventricle
	2. Telencephalon	Rostral part of the hypothalamus Rostral part of the 3rd ventricle Cerebral hemispheres Lateral ventricles Pallium Corpus striatum

THE CRANIAL NERVES AND THE NEURAL CREST

With the exception of the olfactory and optic nerves, which will be considered separately, the cranial nerves are developed in a similar manner to the spinal nerves.

The motor fibres of the cranial nerves to striated muscle are the axons of cells in the ventral lamina of the midbrain and hindbrain which grow outwards to their muscle fibres of distribution, but whereas the motor fibres of the spinal nerves form one series, those of the cranial nerves form two, which are derived from the medial and lateral parts of the ventral lamina respectively. The first series (*somatic efferent*) comprises the oculomotor, trochlear, abducent and hypoglossal nerves, supplying somite-derived muscles or their homologues; the second (*branchial efferent*) comprises the accessory nerve, and the motor parts of the trigeminal, facial, glossopharyngeal and vagus nerves, all of which supply the *striated* muscles derived from the branchial arches, or from post-branchial 'visceral' mesoderm. (It should be recalled, however, that some authorities regard the mesoderm of the branchial arches and immediate post-branchial region as neither somatic nor visceral, but *mixed*, being *unsplit lateral plate* since it is devoid of a coelomic cavity.)

As the lips of the neural groove fuse with each other in the region of the hindbrain and midbrain, a *neural crest* is formed which is homologous with the neural crest of the trunk, flanking the spinal cord (p. 112). (For a review of the origin, migration and differentiation of neural crest cells, consult Weston 1970; Bellairs 1971; Leikola 1976; Fujita 1976; Pearse 1977.) The ganglia of the vagus, glossopharyngeal, vestibulocochlear (in part), facial and trigeminal nerves are derived from the neural crest, but they migrate ventrally and soon come to lie on the ventrolateral aspect of the hindbrain. The vestibulocochlear nerve ganglion is also believed to receive contributions from the wall of the otocyst. There is also descriptive and experimental evidence, mainly from lower vertebrates, that overlying ectodermal thickenings or *placodes* contribute to the ganglia of the trigeminal, facial, vestibulocochlear, vagus and glossopharyngeal nerves. It is claimed that these contributions give rise to the *special somatic afferent* (acoustic and lateral line) and the *special visceral afferent* (chiefly gustatory) components of these nerves. The development of these cranial nerve ganglia and the role of the ectodermal placodes in the human embryo, however, still need clarifying (*see* p. 116). Caudal to the ganglion of the vagus nerve the occipital region of the neural crest is concerned with the ganglia of the accessory and hypoglossal nerves. Rudimentary ganglion cells may occur along the hypoglossal nerve in the human embryo; they undergo regression later. Ganglion cells are also found on the developing spinal root of the accessory nerve and these are believed to persist in the adult. The central processes of the cells of these various ganglia, where they persist, form the sensory roots of the cranial nerves and enter the dorsal lamina of the hindbrain; their peripheral processes join the efferent components of the nerve to be distributed to the various tissues innervated. Some incoming fibres from the facial, glossopharyngeal and vagus nerves collect to form an oval bundle, termed the *tractus solitarius* (p. 903), on the lateral aspect of the myelencephalon. This bundle is the homologue of the oval bundle of the spinal cord, but in the hindbrain it becomes more deeply placed by the overgrowth, folding and subsequent fusion of tissue, derived from the rhombic lip, on the external aspect of the bundle.

The Autonomic Nervous System

The ganglion cells of the *sympathetic system* are derived from the neural crest through the medium of the *primitive spinal ganglia* (p. 160). Certain of the cells in the ventral parts of the latter migrate towards the sides of the aorta, where they subsequently form the *ganglia of the sympathetic trunks*, and certain other associated cells (*vide infra*). Others migrate still further and eventually form the subsidiary sympathetic ganglia such as the coeliac and renal (**2.72**). The original migration is limited to the thoracic and upper lumbar regions. Thereafter the chain grows headwards and tailwards until the whole trunk is laid down. The view has also been advanced that the sympathetic ganglion cells are, at least in part derived from cells which migrate from the *ventral lamina* along the ventral nerve roots. The results of destruction of the neural crest in chick embryos and of excision of the neural crest and portions of the neural tube in frog embryos have been somewhat contradictory in the hands of different investigators. On the whole the balance of the evidence favours the earlier view.

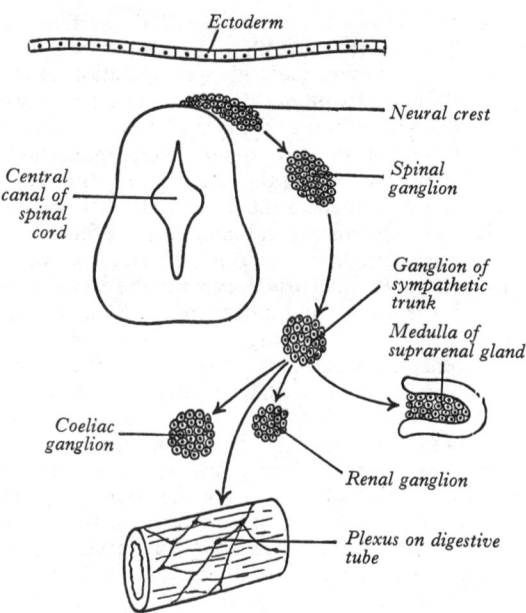

2.72 Diagram showing some derivatives of the neural crest.

The ganglion cells of the cranial part of the *parasympathetic system* are probably derived from the neural crest through the medium of the primitive ganglia of the oculomotor, trigeminal, facial, glossopharyngeal and vagus nerves. The *ciliary ganglion* is formed by cells which migrate from the trigeminal ganglion along the ophthalmic nerve, but it is almost certainly reinforced by cells migrating from the nucleus of the oculomotor nerve along which a few scattered cells are always demonstrable in postnatal life. The *pterygopalatine ganglion* receives contributions from the ganglia of the trigeminal and facial nerves, the *otic* from that of the glossopharyngeal, and the *submandibular* from that of the facial cranial nerve.

The origin of the enteric ganglia in the walls of the gastro-intestinal tract has been disputed, but a critical review of the available evidence (Andrews 1971), suggested that both the vagal neural crest, and that of the trunk may provide sources, but further evidence was awaited. Recently ('Le Douarin and Teillet 1974) studying explants of quail neural crest into the chick confirmed that avian enteric neuroblasts have both a vagal and lumbo-sacral neural crest origin.

The Neuroglia

The glial cells are partly neurectodermal and partly mesodermal in origin. The ventricular zone lining the early central canal of the spinal cord and the cavities of the brain give rise, as we have seen (p. 157), to two types of cell in the intermediate zone—initially neuroblasts which differentiate into neurons, and later, glioblasts which differentiate first into *astroblasts* and *oligodendroblasts*, which mature into astrocytes and oligodendrocytes (for their mature morphology and hypothesized functional roles, *see* p. 834). In the early stages the glioblasts have processes which extend outwards to form the outer limiting membrane deep to the pia mater and inwards to form the inner limiting membrane around the central cavity. As the glioblasts differentiate into primitive neuroglial cells some lose their connexions with both inner and outer limiting membranes, others gain an attachment to the pia by small pial feet and still others remain lining the central canal and cavities of the brain as generalized or specialized ependymal cells, including tanycytes, but lose their peripheral attachments. In some situations, as in the anterior median fissure of the spinal cord, the ependymal cells retain their attachments to both the inner and outer limiting membranes. (For further comments *see* pp. 157, 837.)

In contrast to astrocytes and oligodendrocytes which are ectodermal in origin, the microglia are mesodermal derivatives. They appear in the central nervous system after this has been penetrated by blood vessels and invade it in large numbers from certain restricted regions, whence they spread in what has picturesquely been called 'fountains of microglia', to extend deeply amongst the nervous elements. Their origin has been variously ascribed to perivascular mesenchyme cells, vascular endothelial cells, and even to blood-borne leucocytes (p. 837).

The Meninges

The meninges may be divided in development into the *pachymeninx* (dura mater) and *leptomeninges* (arachnoid and pia mater). Experimental work in lower vertebrates indicates that the dura mater is mesodermal (derived, at least in part, from the sclerotomes) in origin, whilst the arachnoid and pia mater are closely associated in their development, being ectodermal in origin and largely derived from neural crest cells (Harvey and Burr 1926, and for a recent review *see* Gil and Ratto 1973). However, morphological studies on the spinal meninges of human embryos suggest that all three membranes are derived from the loose mesenchyme of sclerotomic origin surrounding the spinal cord, hence termed the *meninx primitiva*. Neural crest cells mingle with this mesenchyme and, as mentioned above, appear to be predominantly involved in the formation of the pia mater (Sensenig 1951, Weston 1970, Gil and Ratto 1973, Leikola 1976). In the head, the mesoderm from which the cranial dura mater develops is closely associated at first with the mesenchyme which is chondrified and ossified to form the skull, and these layers are only clearly differentiated as the venous sinuses develop. For an interesting study of pre- and post-natal growth of the tentorium cerebelli, with a mathematical analysis *see* Klintworth (1967).

The Chromaffin Organs

The tissue from which the sympathetic ganglia are formed is, at first, a mass of relatively undifferentiated *sympatho-chromaffin cells* but later, a number of cell varieties which may be roughly grouped into *small* and *intermediate-sized neuroblasts* (*sympathoblasts*), and *large*, initially rounded *phaeochromocytoblasts*. The intermediate-sized neuroblasts differentiate into the typical multipolar postganglionic sympathetic neurons (which secrete noradrenalin at their terminals) of classical autonomic neuro-anatomy. The small 'neuroblasts' have only come into pro-minence in recent years: they are currently the subject of intense scrutiny, and they will unquestionably cause a radical reappraisal of autonomic terminology, and views concerning ganglionic transmission. Ultrastructural, and specialized light-microscopic fluorescence techniques led them to be termed firstly, small granulated cells, and latterly, *small intensely fluorescent* (SIF) cells, types I and II. Both have been shown (at least in some species and sites) to be dopamine-storing and secreting cells. It is postulated that type I function as true interneurons, synapsing with the principal post-ganglionic neurons. Type II are thought to operate as *local neuroendocrine cells*, secreting dopamine into the ganglionic microcirculation. Both types of SIF cells probably modulate the principal preganglionic/postganglionic synaptic transmission (for a discussion and bibliography *see* Chiba *et al.* 1977). The *large* cells differentiate into masses of columnar or polyhedral *phaeochromocytes* ('classical' chromaffin cells) which secrete either adrenalin or noradrenalin. These cell masses are termed *paraganglia* and may be situated near, on the surface of, or embedded in the capsules of the ganglia of the sympathetic chain, or in some of the large autonomic plexuses (*see* p. 1454 and **8**.204). The largest members of the latter are the *para-aortic bodies* which lie along the sides of the abdominal aorta in relation to the inferior mesenteric artery. During childhood the para-aortic bodies and the paraganglia of the sympathetic chain partly degenerate and can no longer be isolated by gross dissection, but even in the adult body chromaffin tissue can still be recognized microscopically in these various sites (p. 1454). It may be noted here that both the phaeochromocytes and the SIF cells, using a wider and more recent classification, are regarded as chromaffin; they belong to the APUD series of cells, and are paraneuronal in nature (*see* further discussions on pp. 115, 1455).

The Suprarenal Glands

Each suprarenal gland consists of a cortex of mesodermal origin and a medulla of ectodermal origin. The cortex develops during the second month as a proliferation of coelomic mesothelium into the underlying mesenchyme between the root of the dorsal mesogastrium and the mesonephros (Keene and Hewer 1927; Crowder 1957). (An older view that two separate proliferations form at first a fetal cortex before the definitive cortex, has not been corroborated.) The proliferating tissue extends from the level of the sixth to the twelfth thoracic segments. It is soon disorganized dorsomedially by invasion into it of sympathochromaffin tissue from adjacent sympathetic ganglionic masses to form the medulla and also by the development of venous sinuses. The latter are joined by capillaries which arise from adjacent mesonephric arteries and penetrate the cortex in a radial manner. When the proliferation of the coelomic epithelium ceases the cortex becomes enveloped ventrally, and later dorsally, by a mesodermal capsule which is probably derived from the mesonephros. The nests of cortical cells under the capsule are the rudiment of the *zona glomerulosa* of the suprarenal. These nests proliferate cords of cells which pass deeply between the capillaries and sinusoids. The cells in these cords degenerate in a somewhat erratic fashion as they pass towards the medulla, becoming granular, eosinophilic and ultimately being autolysed. These cords of degenerating cells constitute the *fetal cortex*, which undergoes a rapid degeneration during the first two weeks after birth resulting in a marked shrinkage of the gland. The *fascicular* and *reticular* zones of the adult cortex are proliferated from the glomerular zone after birth and are fully differentiated by about the twelfth year. Their cells mature more slowly than those of the fetal cortex.

Pigment cells occurring in connective and epithelial tissues are believed to be derived solely from neural crest cells. Such *melanoblasts* migrate to almost all parts of the body and synthesize melanin; when fully loaded they are termed *melanocytes* (Wilde 1961; see also p. 43).

Development of Special Sense Organs

THE NOSE

The development of the external nose and nasal cavities have already been considered (pp. 148–49).

The *olfactory nerves* are developed from the placodal cells which line the olfactory pits; these cells proliferate and give rise to *olfactory cells*. Their central processes are usually described as growing into the overlying olfactory bulb and thus forming the olfactory nerves. It has been claimed that the olfactory cells are from the first connected with the overlying brain by bridges of cytoplasm, within which the olfactory nerve fibres are developed (Smith 1908; Ballantyne 1925). More recent accounts, however, suggest that the earliest pioneer neurites are naked cytoplasmic processes which cross a mesenchyme-filled gap between the placode and the superjacent brain. Later, these, and subsequent generations of neurites become enclothed in Schwann cell processes, presumably derived from the rostral neural crest (Pearson 1941; Van Campenhout 1956; Dejean, Hervouët and Leplat 1958).

THE EYES

The rudiments of the eyeballs appear as two hollow diverticula from the lateral aspects of the forebrain (Mann 1964; Duke-Elder 1963; O'Rahilly 1966, 1975). These rudiments are visible some time before the closure of the rostral neuropore; after its closure they are known as the *optic vesicles*. Their formation is dependent on the organizing influence of the mesoderm of prechordal plate origin which underlies this part of the neural tube. They project towards the sides of the head, and the distal part of each expands while the proximal part remains narrow as the *optic stalk* (2.73A, B). Under the inductive influence of the subjacent optic vesicle the small area of ectoderm overlying it becomes thickened and depressed in its centre. (The classical experiments of Spemann in this field were largely responsible for initiation of the concept of embryonic induction—*see* p. 83.) This *lens pit* deepens and its edges come together and fuse to enclose a hollow *lens vesicle*

(2.73A) which soon loses its connexion with the surface ectoderm and is the rudiment of the lens. The outer wall of the optic vesicle increases in thickness and undergoes invagination to form the *optic cup*, consisting of two strata of cells (2.73A). These two strata are continuous with each other at the cup's margin, which grows forwards at the end of the third month, overlapping the front of the lens, and converge to form the edge of the future pupil. The invagination is not limited to the outer wall of the vesicle, but involves also its caudal surface and extends in the form of a groove for some distance along the optic stalk. Thus, for a time, a wide hiatus, the *optic* or *choroidal fissure*, exists in the caudal part of the cup. Through the groove and fissure mesenchyme extends into the optic stalk and cup, carrying the *hyaloid artery* with it; as growth proceeds, the edges of the fissure become approximated and they close during the seventh week, including the artery in the distal part of the stalk. Failure of the optic fissure to close is a very rare anomaly and there is always a corresponding deficiency in the choroid and the iris (*congenital coloboma*). It must be noted, however, that localized deficiencies of the choroid and iris may occur quite independently, and in such cases the retina is normal (Lopashov and Stroeva 1961). Although the cavity of the optic vesicle is largely obliterated by the invagination to form a cup, the space does persist, in a sense, at the microscopic level as the interval between rod and cone processes and the ciliated apices of the pigmented cells of the most external layer of the retina (p. 1163). The 'space' is of course originally continuous with that of the cerebral vesicle; it is also the site of pathological detachment of the retina.

The retina develops from the optic cup (2.73B, C). Its two layers are at first equipotential and mutually interchangeable. Depending on contact with surrounding mesoderm, however, the external stratum of the cup remains a single layer of cells, which assume a columnar shape, acquire pigment, and form the *pigmented epithelium of the retina*, the pigment first appearing in the cells near the edge of the cup. Under the influence of the lens the cells of the internal stratum proliferate and form a layer of considerable thickness from which are developed the nervous elements and the sustentacular fibres (cells) of the retina, together with a portion of the vitreous body. In the region of the cup which overlaps the lens the inner stratum is not differentiated into nervous elements, but persists as a layer of columnar cells which, together with the corresponding part of the pigmented layer, form the double epithelium of the *ciliary* and *iridial parts* of the retina.

The cells of the inner layer of the cup proliferate and form an outer *nuclear zone* and an inner *marginal zone*, devoid of nuclei. At 12 mm the cells of the nuclear zone invade the marginal zone, and at 17 mm the nervous stratum of the retina consists of inner and outer *neuroblastic layers*. The inner neuroblastic layer gives origin to the ganglion cells, the amacrine cells and the somata of the sustentacular fibres (of Müller); the outer neuroblastic layer is the source of the horizontal and rod- and cone-bipolar neurons and probably the rod and cone cells, which first appear in the central part of the retina. By the eighth month all the layers of the retina can be identified. For recent bibliographies on retinal development, including ultrastructural studies, consult Spira and Hollenberg (1973); Fisher and Linberg (1975); Warwick (1976).

The deepest part of the optic fissure is at the centre of the floor of the optic cup. In this situation, which later is the site of the *optic disc*, the inner (neural) cell layer of the cup is continuous with the corresponding invaginated cell layer of the optic stalk and, as a result, the developing nerve fibres of the ganglion cells can pass directly into the wall of the stalk to convert it into the *optic nerve*. The fibres of the optic nerve begin to acquire their myelin sheaths shortly before birth, but the process is not completed until some time after birth. The *optic chiasma* is formed by the meeting and partial decussation of the fibres of the two optic nerves, and it marks the junction of the telencephalon with the diencephalon in the floor of the third ventricle. Beyond the chiasma the fibres are continued backwards as the optic tracts principally to the lateral geniculate bodies.

The lens is developed from the lens vesicle (2.73A), which is overlapped by the margin of the cup and becomes separated from the overlying ectoderm by mesenchyme. The cells forming the posterior wall of the vesicle lengthen and are converted into the lens fibres, which grow into and fill its cavity (2.73B, C). The cells

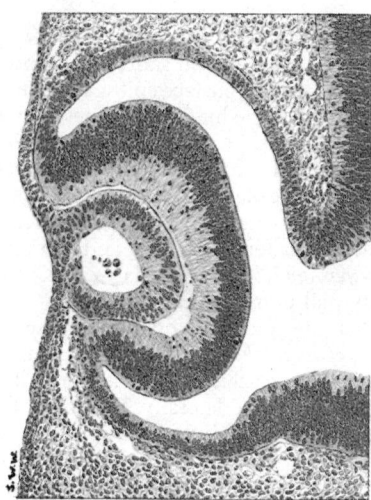

2.73A Section through the developing eye of a human embryo. 8 mm C.R. length. The thick nervous and the thinner pigmented layers of the retina and the developing lens are shown. Stained with haematoxylin and eosin. Magnification × c. 114. (From material loaned by Professor R. J. Harrison.)

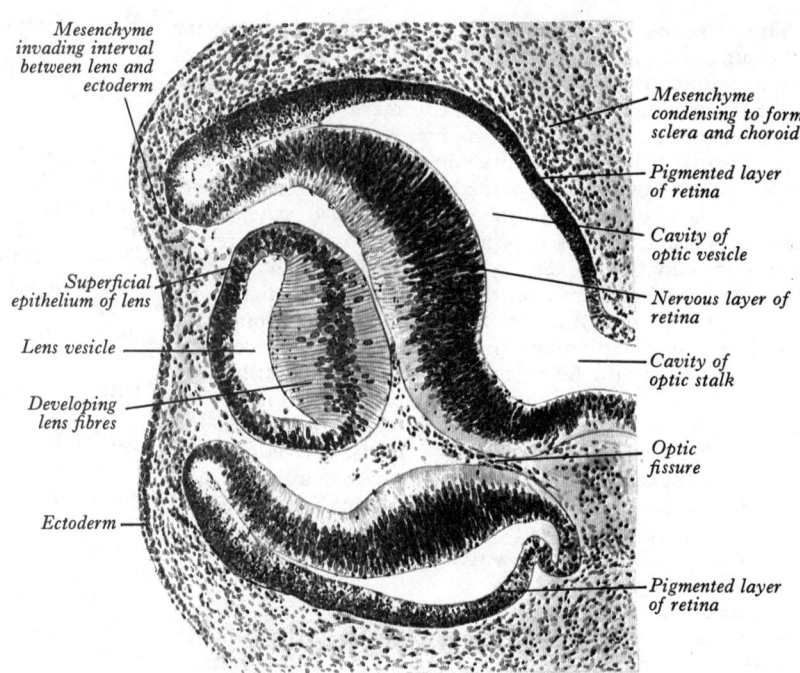

Mesenchyme invading interval between lens and ectoderm

Mesenchyme condensing to form sclera and choroid

Pigmented layer of retina

Cavity of optic vesicle

Superficial epithelium of lens

Nervous layer of retina

Lens vesicle

Cavity of optic stalk

Developing lens fibres

Optic fissure

Ectoderm

Pigmented layer of retina

2.73B A section through the developing eye of a human embryo, 13·2 mm long. (G. L. Street, *Contr. Embryol.*, **32**, 1948.)

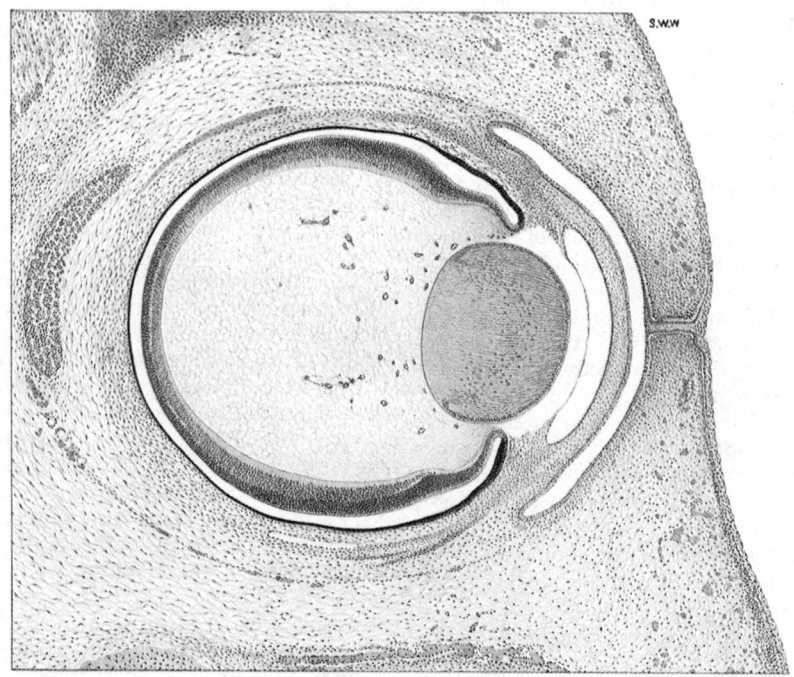

2.73C Section through the eye of a human embryo. 40 mm C.R. length. Note the layers of the retina, developing lens, pupillary membrane, cornea, conjunctival sac, anterior and posterior aqueous chambers, the developing vitreous body, and condensing circumoptic mesenchyme, and the fused eyelids. Stained with haematoxylin and eosin. Magnification × c. 62.

forming the anterior wall retain their cellular character, and form the epithelium on the anterior surface of the fully developed lens; at the equator of the lens the gradual transition of the cells into lens fibres can be observed; lens fibre differentiation and growth continues throughout life. Characteristic ultrastructural changes have been described (Wulle and Lerche 1967). By the second month the lens is invested by a vascular mesenchymal condensation termed the *vascular capsule* of the lens, the ventral part of which, covering the lens, is named the *pupillary membrane*; the blood vessels supplying the dorsal part of this capsule are derived from the *hyaloid artery*, those for the ventral part from the anterior ciliary arteries. By the sixth month all the vessels of the capsule are atrophied except the hyaloid artery, which becomes occluded during the eighth month of intrauterine life. Prior to this, during the fourth month, the hyaloid artery gives off retinal branches, and its proximal part persists in the adult as the *central artery of the retina*. The *hyaloid canal*, which carries the artery through the vitreous, persists after the vessel has become occluded. In the newly born child it extends more or less horizontally from the optic disc to the posterior aspect of the lens but, when the adult eye is examined with a slit-lamp, it can be seen to follow a wavy, curved course, sagging downwards as it passes forwards to the lens (Mann 1927). With the loss of its blood vessels the vascular capsule of the lens disappears, but sometimes the pupillary membrane persists at birth, giving rise to the condition termed *congenital atresia of the pupil*.

The vitreous body is developed between the lens and the optic cup. The lens rudiment and the optic vesicle are at first in contact with each other, but after the closure of the lens vesicle and the formation of the optic cup the former is withdrawn from the retinal layer of the cup; the two, however, remain connected by a network of delicate cytoplasmic processes. This network, derived partly from the cells of the lens and partly from those of the retinal layer of the cup, is the *primitive vitreous body*. At first these cytoplasmic processes spring from the whole of the retinal layer of the cup, but later are limited to the ciliary region, where by a process of condensation they appear to form the *ciliary zonule*. The mesenchyme which enters the cup through the choroidal fissure and around the equator of the lens becomes intimately united with this reticular tissue, and also contributes to the formation of the vitreous body, which is therefore derived partly from the ectoderm and partly from the mesoderm. Considerable controversy regarding the precise derivation of the vitreous still persists. There are probably three sources: the local *mesenchyme*, the lens (*ectoderm*) and the retina (*neurectoderm*). The mesodermal component is probably represented only by the juxtazonular vitreous and the vitreous immediately adjacent to the hyaloid vestiges.

The aqueous chamber of the eye appears as a cleft in that part of the mesenchyme which intervenes between the lens and the ectoderm. The mesenchyme superficial to the cleft forms the *substantia propria* of the cornea that deep to the cleft the mesenchymal stroma of the *iris* and the *pupillary membrane*. The cornea is induced by the lens and optic cup. The corneal epithelium is formed from the surface ectoderm and the endothelium of the anterior chamber from mesenchyme (O'Rahilly and Meyer 1959; Coulombre 1964). When the pupillary membrane disappears the anterior and posterior chambers of the eye communicate with each other.

The sclera and choroid are derived from the mesenchyme surrounding the optic cup, and the anterior part of the choroid is modified to form the *ciliary body* and *ciliary processes*. The fibres of the *ciliary muscle* are derived from the mesoderm, but those of the *sphincter* and *dilatator pupillae* are of neurectodermal origin, being developed from the cells of the pupillary part of the optic cup, as is the double epithelium on the posterior (lenticular) aspect of the iris.

The *eyelids* are formed as small cutaneous folds (**2.**73C). About the middle of the third month their edges come together and unite over the cornea; they are usually said to remain united until about the end of the sixth month. In the most recent study, however, Jain *et al.* (1972) concluded that the separation is slow but completed a month earlier (5·1 months, 155 mm stage). The same observers, from examination of 20 human embryos and fetuses, also stated that the tarsal plates and glands begin to develop respectively at 55 mm and 75 mm stages.

For the chronology of development of the human eye, at embryonic and fetal stages consult O'Rahilly (1966, 1975).

The Lacrimal Apparatus

The epithelium of the alveoli and ducts of the *lacrimal gland* arise as a series of tubular buds from the ectoderm of the superior conjunctival fornix; these buds are arranged in two groups, one forming the gland proper, and the other its palpebral process. The *lacrimal sac* and *nasolactimal duct* are considered to be derived from the ectoderm in the nasomaxillary groove between the lateral nasal elevation and the maxillary process (p. 149). This thickens to form a solid cord of cells which sinks into the mesenchyme; during the third month the central cells of the cord break down, and a lumen is acquired. In this way the *nasolacrimal duct* is established. The lacrimal canaliculi arise as buds from the upper part of the cord of cells and secondarily establish openings (*puncta lacrimalia*) on the margins of the lids; the inferior canaliculus cuts off a small part of the lower eyelid to form the *lacrimal caruncle*. The epithelium of the cornea and conjunctiva is of ectodermal origin, as are also the eyelashes and the lining cells of the tarsal and other glands which open on the margins of the eyelids.

For general accounts of ocular developmental abnormalities consult Dejean, Hervouet and Leplat (1958), and Mann (1964).

THE EARS

The rudiments of the **internal ears** appear shortly after those of the eyes, as two patches of thickened, surface epithelium, the *otic placodes*, situated in the region of the hindbrain. These patches, though surrounded by general skin ectoderm, are probably neurectodermal in character. Each placode is invaginated, becoming an *otic pit* (**2.**112A). The mouth of the pit is then closed, and a vesicle, termed the *otocyst* (auditory or otic vesicle), is formed (**2.**74A); it is initially piriform in shape, and from it the epithelial lining of the *membranous labyrinth* is derived (**2.**74A–F). A vertical infolding of its wall progressively marks off a tubular diverticulum on the medial side. This diverticulum differentiates into the *ductus* and *saccus endolymphaticus*, and it communicates with the remainder of the vesicle, which is termed the *utriculosaccular chamber* and is placed laterally. From the dorsal part of this chamber three compressed diverticula appear as disc-like evaginations; the central parts of the walls of the discs coalesce and disappear, while the peripheral portions of the discs persist to form the *semicircular ducts*; the anterior duct is the first, and the lateral the last, to be completed. From the ventral part of the utriculosaccular chamber arises a medially directed evagination which becomes coiled and forms the *cochlear duct*, the proximal extremity of which is subsequently constricted to form the *ductus reuniens*. The central part of the chamber now represents the membranous vestibule and is subdivided into a smaller ventral part, the *saccule*, and a larger dorsal part, the *utricle*, mainly by a horizontal infolding which extends deeply from the lateral wall, towards the opening of the ductus endolymphaticus, leaving only a narrow communication, the *utriculosaccular duct*, between the subdivisions. This duct becomes acutely bent on itself. During this period the membranous labyrinth undergoes a rotation so that the long axis, originally vertical, becomes more or less horizontal (Bast and Anson, 1949). Subsequently, otocyst derived cells (having contributed placodal cells to vestibulo-cochlear ganglion) differentiate into the specialized paraneuronal hair cells of the utricle, saccule, ampullae of the semicircular ducts, and organ of Corti; they also differentiate into various specialized sustentacular cells, and the unique epithelia of the stria vascularis and endolymphatic sac. The remainder form the general epithelial lining of the rest of the membranous labyrinth.

The mesenchyme surrounding the various parts of the epithelial labyrinth is converted into a cartilaginous *otic capsule*, and this is finally ossified to form the *bony labyrinth* of the internal ear. For a time the cartilaginous capsule is incomplete, and the cochlear, vestibular, and facial ganglia are situated in the gap between its canalicular and cochlear parts. These ganglia are soon covered by an outgrowth of cartilage, and at the same time the facial nerve is covered in by a growth of cartilage from the cochlear to the canalicular part of the capsule. In the embryonic connective tissue between the cartilaginous capsule and the epithelial wall of the labyrinth the perilymphatic spaces are developed. The rudiment of the *periotic cistern* or vestibular perilymphatic space can be seen in an embryo of from 30 to 40 mm in length, in the reticulum between the saccule and the fenestra vestibuli (Streeter 1917). The scala tympani is next developed, and begins opposite the fenestra cochleae; the scala vestibuli is the last to appear. The two scalae gradually extend along each side of the ductus cochlearis, and when they reach the tip of the ductus an opening, the helicotrema, is developed between them. The modiolus and the osseous spiral lamina of the cochlea are not preformed in cartilage but are ossified directly from connective tissue.

The Middle Ear Cleft

The auditory tube and tympanic cavity are developed from a hollow, termed the *tubotympanic recess* (Frazer 1914), between the first and third branchial arches, the floor of the recess consisting of the second arch and its limiting grooves. By the forward growth of the third arch the inner part of the recess is narrowed to form the tubal region, and the inner part of the second arch is excluded from this portion of the floor. The more lateral part of the recess subsequently develops into the *tympanic cavity*, and the floor of

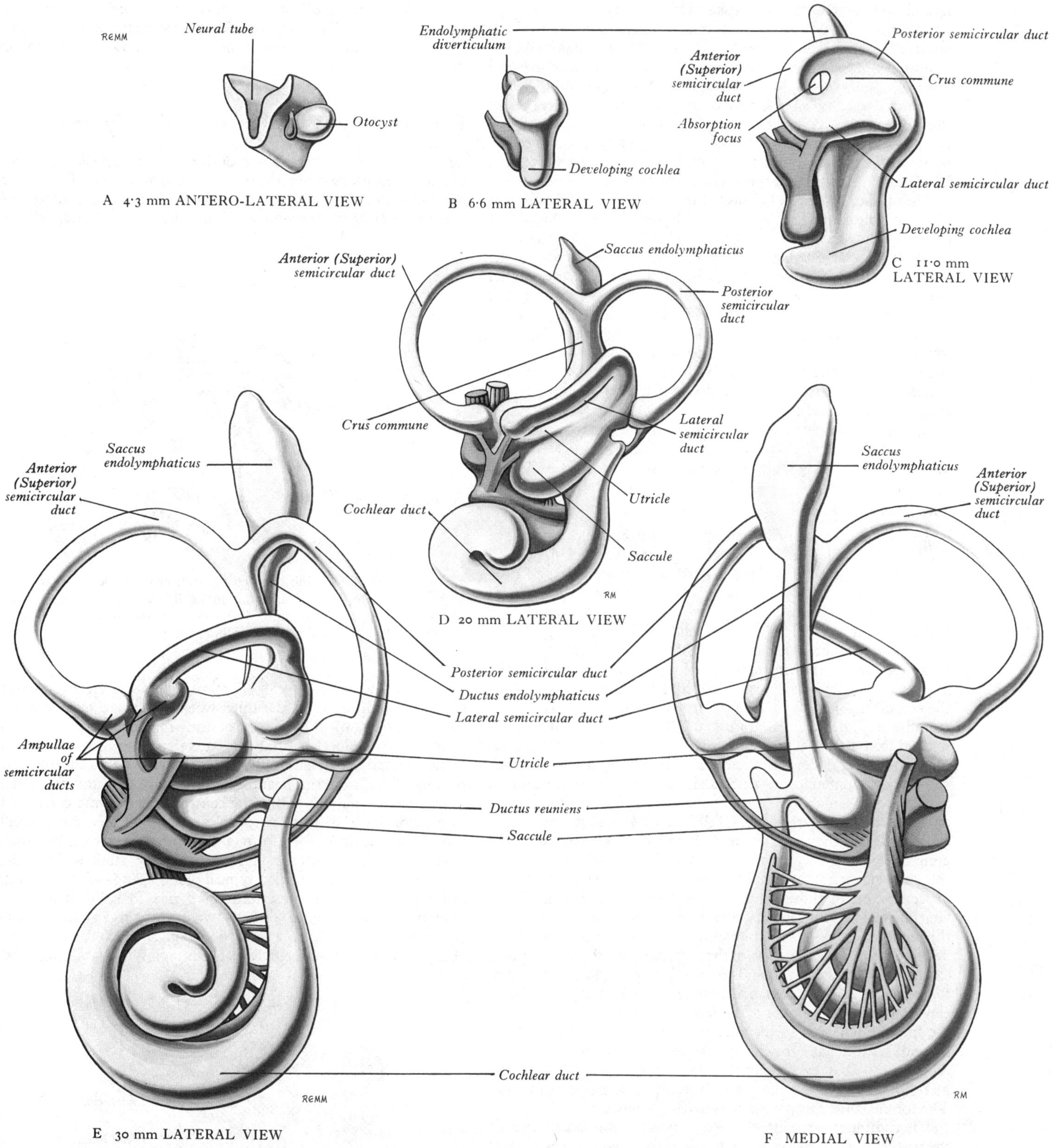

REMM

Neural tube

Otocyst

A 4·3 mm ANTERO-LATERAL VIEW

Endolymphatic diverticulum

Developing cochlea

B 6·6 mm LATERAL VIEW

Posterior semicircular duct

Anterior (Superior) semicircular duct

Crus commune

Absorption focus

Lateral semicircular duct

Developing cochlea

C 11·0 mm LATERAL VIEW

Anterior (Superior) semicircular duct

Saccus endolymphaticus

Posterior semicircular duct

Crus commune

Lateral semicircular duct

Cochlear duct

Utricle

Saccule

RM

D 20 mm LATERAL VIEW

Saccus endolymphaticus

Anterior (Superior) semicircular duct

Ampullae of semicircular ducts

Posterior semicircular duct

Ductus endolymphaticus

Lateral semicircular duct

Utricle

Ductus reuniens

Saccule

Cochlear duct

Saccus endolymphaticus

Anterior (Superior) semicircular duct

REMM

E 30 mm LATERAL VIEW

RM

F MEDIAL VIEW

2.74A–F A series of diagrams showing the stages in the development of the membranous labyrinth from the otocyst, at the embryonic stages and viewed from the aspects indicated. Note also the relationship of the vestibular (orange) and cochlear (yellow) parts of the vestibulocochlear nerve. (From a series of models prepared by His.)

this part forms the lateral wall of the tympanic cavity up to about the level of the chorda tympani nerve. From this it will be seen that the lateral wall of the tympanic cavity contains first and second arch elements, the first arch being limited to the part in front of the anterior process of the malleus. The second arch forms the outer wall behind this, and turns on to the back wall to take in the tympanohyal region. Recent observations, however, indicate that the tympanic cavity is derived wholly from the first pouch (Kanagasuntheram 1967). The tubotympanic recess is at first on the inferolateral aspect of the cartilaginous otic capsule, but as the latter enlarges the relations become altered and the tympanic cavity comes to lie anterolateral to the capsule. A cartilaginous process grows out from the lateral part of the capsule to form the tegmen tympani; and it curves caudally to form the

179

lateral wall of the auditory tube. In this way, subsequent to ossification, the tympanic cavity and the proximal part of the auditory tube become included in the petrous region of the temporal bone. During the sixth or seventh month the mastoid antrum appears as a dorsal expansion of the tympanic cavity. Much of the cavity's basic development thus occurs during *fetal* life (Bok 1966). A recent study of the posterior part of the tympanic cavity, and in particular of the sinus tympani, a recess between the promontory and pyramid (p. 1195), emphasizes the late fetal development of this region (Bollobás and Hajdu 1975).

The opinion generally held as to the development of the ossicles of the middle ear is that the *malleus* is derived from the dorsal end of the ventral mandibular (Meckel's) cartilage (2.58), the *incus* from the dorsal cartilage, which probably corresponds to the quadrate bone of birds and reptiles. The *stapes* is developed mainly from the dorsal end of the cartilage of the second or hyoid arch, and first appears as a ring (*annulus stapedis*) encircling the small stapedial artery (p. 187). The primordium of the stapedius muscle appears close to the artery and facial nerve at the end of the second month, and at almost the same time the tensor tympani begins to appear near the extremity of the tubotympanic recess (Candiollo and Levi 1969).

At first the ossicles are embedded in the mesenchymal roof of the tympanic cavity, and their extraneous origin is indicated in the adult by the covering which they receive from its mucous lining.

The External Ear

The external acoustic meatus is developed from the dorsal end of the hyomandibular or first branchial groove. Close to its dorsal extremity this groove extends inwards as a funnel-shaped tube (*primary meatus*) from which the cartilaginous portion and a small part of the roof of the osseous portion of the meatus are developed. From this funnel-shaped tube a solid epidermal plug extends inwards along the floor of the tubotympanic recess; by the breaking down of the central cells of this plug the inner part of the meatus (*secondary meatus*) is produced, while the deepest cells of the ectodermal plug form the epidermal stratum of the tympanic membrane. The fibrous stratum of the *tympanic membrane* is formed from the mesenchyme which extends between the meatal plate and the endoderm of the floor of the tubotympanic recess.

The development of the **auricle** is initiated by the appearance of six hillocks which form round the margins of the dorsal portion of the hyomandibular groove. Of the six, three are situated on the caudal edge of the mandibular arch and three on the adjoining cranial edge of the hyoid arch (2.111F). These hillocks appear at the 4 mm stage, but they tend to become obscured as development proceeds and of those on the mandibular arch only the most ventral, which subsequently forms the *tragus*, can be identified throughout (2.55). The remainder of the auricle owes its development to proliferation of the mesenchyme of the hyoid arch (Streeter 1922), which extends forwards round the dorsal end of the remains of the hyomandibular groove, forming a keel-like elevation—the forerunner of the *helix*. The contribution made by the mandibular arch to the auricle is greatest at the end of the second month; as growth continues, it becomes relatively reduced, so that eventually the area of skin supplied by the mandibular nerve extends very little above the tragus (7.280). The lobule is the last part of the auricle to develop.

The rudiment of the eighth nerve appears in the fourth week as the *vestibulocochlear ganglion*, which lies between the otocyst and the wall of the hindbrain. At first it is fused with the ganglion of the facial nerve (*acousticofacial ganglion*) but later the two separate. The cells of the ganglion are mainly derived from the neural crest, but probably some cells come from the neurectoderm of the otocyst. The auditory ganglion divides into vestibular and cochlear parts, each associated with the corresponding division of the eighth nerve. The cells of these ganglia remain bipolar throughout life, each sending a proximal fibre into the brainstem, and a peripheral fibre to the internal ear. These neurons are also unusual in that many of their *somata* become enveloped in thin *myelin sheaths*.

The ganglionic fibres just described provide, of course, the afferent, sensory innervation of the labyrinthine hair cells. The latter soon become associated with the outgrowing axons from cells of the superior olivary complexes of the pons which provide an efferent innervation termed the olivocochlear bundle (pp. 904, 909). Developmental details of the latter are, however, lacking in mankind.

Development of the Vascular System

The earliest blood vessels are derived from *angioblastic tissue*, which differentiates from the mesenchyme in three regions: (*a*) in the surface of the yolk sac, (*b*) in the body stalk, and (*c*) in the chorion (Hertig 1935; Bloom and Bartelmez 1940). In the yolk sac

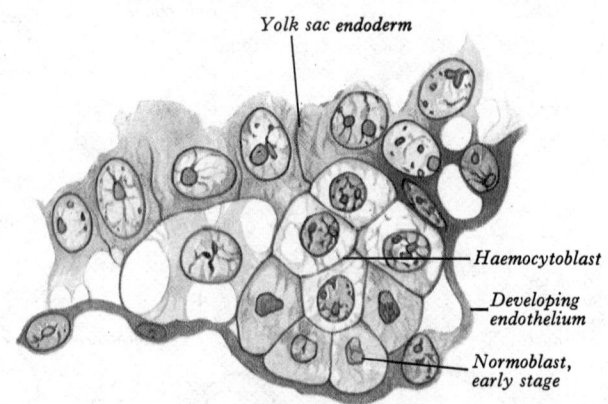

Yolk sac endoderm

Haemocytoblast

Developing endothelium

Normoblast, early stage

2.75A   A part of a section through the wall of the yolk sac of an early human embryo to show an early stage in the differentiation of angioblastic tissue. (Hamilton, Boyd and Mossman's *Human Embryology*, 1945. Reproduced by permission of the authors and publishers.)

and base of the body stalk small, more or less spherical groups of cells are found early in the third week. They are termed *blood islands* (2.75 A, B). The stages of the transformation of blood islands into blood-containing vessels have not yet been demonstrated in detail, but it is generally believed that the peripheral cells of the islands become flattened and form the vascular endothelium, while the central cells become converted into primitive red blood corpuscles (2.75B). Later these small blood-containing spaces form a continuous network of small vessels. In the chorionic end of the body stalk and in the mesoderm lining the chorion typical blood islands are not found, but the cells of the mesoderm give rise to solid strands of angioblasts. Each strand contains two or three rod-shaped nuclei arranged in a single row and soon comes to contain a space occupied by one or more nucleated haemoglobin-containing cells.

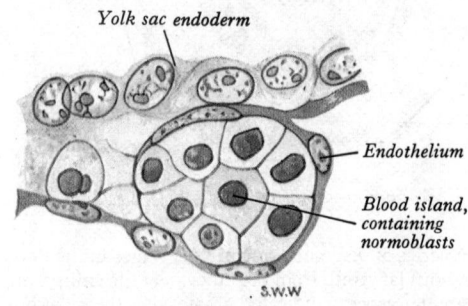

Yolk sac endoderm

Endothelium

Blood island, containing normoblasts

S.W.W.

2.75B   A part of a section through the wall of the yolk sac of an early human embryo, to show a developing blood vessel including a blood island. (Hamilton, Boyd and Mossman's *Human Embryology*, 1945. Reproduced by permission of the authors and publishers.)

These spaces coalesce to form blood vessels, which are therefore lined by derivatives of the mesoderm: the precise source of their contained blood cells is uncertain. The earliest blood vessels, therefore, are formed at several separate centres; from the walls of these vessels buds grow out, become canalized, converted into new vessels, and join with those of neighbouring areas to form a close meshwork. The heart and the blood vessels of the embryo arise from angioblastic tissue differentiated from the intra-embryonic mesoderm. Prior to the establishment of the circulation (p. 120), new vessels develop *in situ*, but thereafter they all take origin as outgrowths from pre-existing vessels.

The subsequent development of the blood corpuscles has already been described (p. 121).

Development of the Heart

The precocious development of the heart has already been mentioned (p. 120). In placental vertebrates it is the earliest major organ to function; for obvious nutritive reasons it must not only accommodate a stream of blood but also begin to propel it (p. 183). These early *functional* demands on the heart represent an important factor in the dynamics of its development (Sabin 1924; Johnstone 1925).

The early appearance of cardiac activity in the tubular hearts of chick (Sabin 1924; Johnstone 1925; Patten and Kramer 1933) and rat (Goss 1952) embryos was noted many years ago. This is first manifested by arrhythmic and sporadic ventricular contractions, which are rapidly superseded by regular peristaltic activity propagated along the atrioventricular tube. Earlier observers also described the formation of functional valves by the accumulation, subjacent to the endothelium, of *cardiac jelly* or *subendocardial reticulum*. This occupies a large space between the primitive endocardium and myoepicardial mantle (*vide infra* and **2.**76C, D). These rapid and critical early events in cardiac morphogenesis have been re-examined more recently with improved techniques, including electron microscopy, especially in the mouse (Challice and Viragh 1973). As the simple heart tube elongates and develops an asymmetrical twist, a succession of cavities, joined by more constricted regions, begin to define the sinus venosus, atrium, ventricle and bulbus cordis. At first these are somewhat cylindrical, but they rapidly become more spherical (within 24 hours in the mouse). Coincidentally, myoblastic strands invade the subendocardial reticulum from the cells of the myoepicardial mantle, and these form a complex network of intercommunicating *trabeculae* external to, but ultimately indenting and becoming clothed by endocardium. In many vertebrate groups the myocardium remains predominantly trabecular in arrangement, but in birds and mammals a compact layer of cardiac muscle develops external to the trabeculae. The latter also persist but constitute a lesser volume of the propulsive tissue.

These early events in cardiac development provide an approximate parallel between phylogeny and ontogeny. The

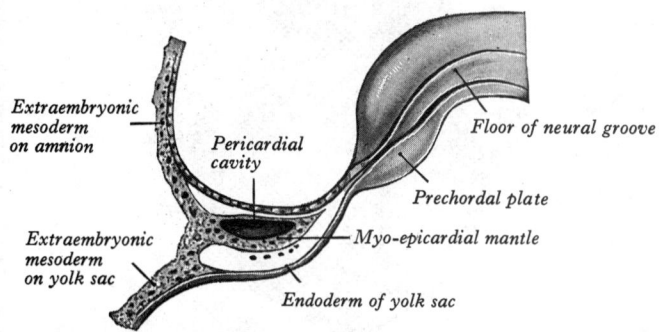

2.76A Median section through the cranial end of an early human embryo to show the position of the pericardium before the formation of the head fold. A few scattered angioblasts are seen between the cardiogenic plate and the yolk sac; they will ultimately form the endothelial heart tubes. (After C. L. Davis.)

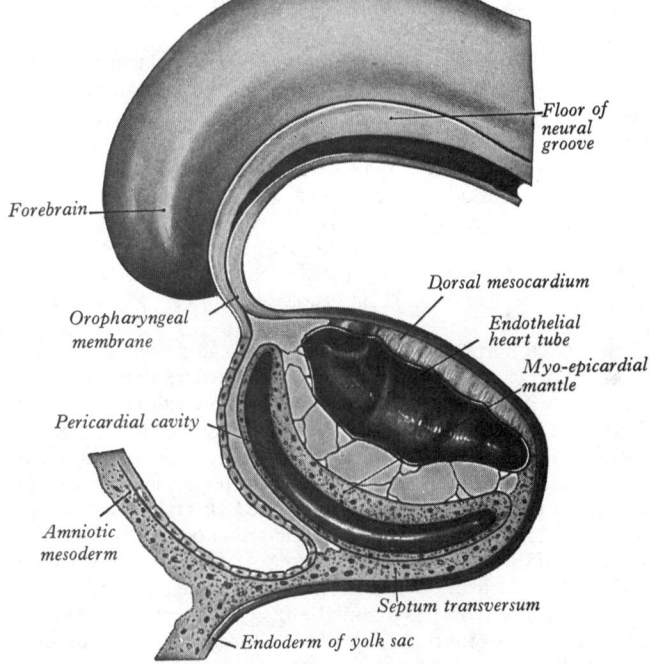

2.76C Median section through the cranial end of a young human embryo, after completion of the head fold and reversal of the pericardium.

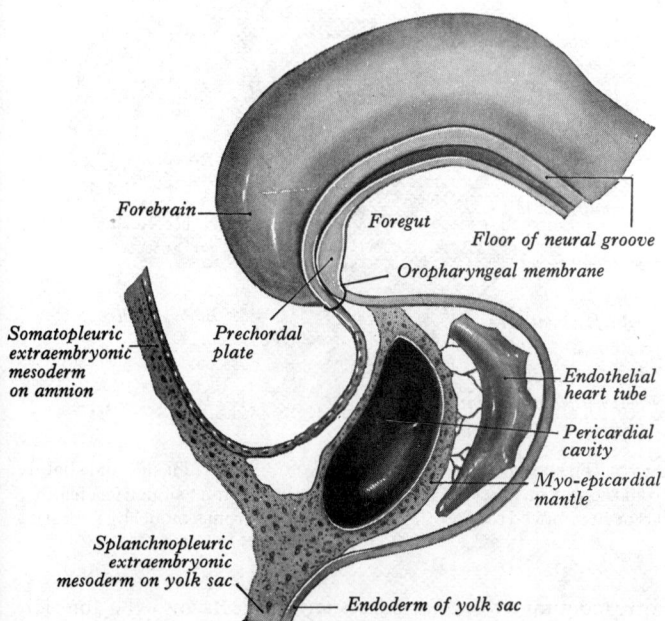

2.76B Median section through the cranial end of a young human embryo, showing the head fold in process of formation and its effect on the position of the pericardium.

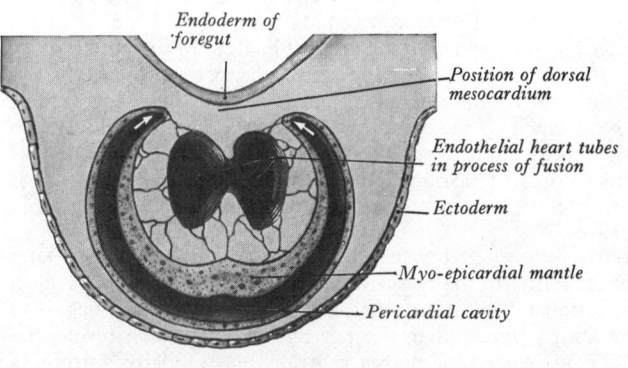

2.76D A horizontal section through the pericardium and developing heart of the embryo shown in **2.**76C. The arrows indicate the directions in which the dorsolateral recesses of the pericardium deepen so as to define the transient dorsal mesocardium. (After C. L. Davis.)

evolution from a 'trabecular' heart to an organ with rounder chambers and a largely compacted myocardium is likely to be an expression of increased efficiency. An attempt to demonstrate this by mathematical analysis (Challice and Viragh 1974) provides some corroboration and also an explanation for the persistence of the internal trabeculation in the mammalian heart (p. 638). In the account which follows, attention is necessarily concentrated upon the complex changes which transform a tube into a chambered septate human heart. It is also necessary to keep in mind that at every stage the heart must also function as a circulatory pump.

The human heart, as in all vertebrates, is formed by the fusion of two symmetrically developing tubes (p. 121), but the fusion is gradual, commencing at the bulbar or arterial end and extending to the venous end (Tandler 1912; Davis 1927).

The pericardial cavity can be identified before the head fold is formed or while it is in process of formation, at a stage when the embryo possesses only two somites. The heart is then represented by groups of angioblasts which lie between the pericardium and the endoderm of the yolk sac (2.76A). At this stage the ventral (or yolk sac) wall of the pericardium, which will form both the epicardium and the myocardium, is thicker than the dorsal wall and is termed the *myo-epicardial mantle*. When the head fold is formed, the mantle becomes the dorsal wall of the pericardium and lies ventral to the foregut. While this reversal of the pericardium is taking place (2.76B, C), the *cardiogenic mesoderm* gives rise to two paramedian endothelial tubes which rapidly fuse to form a tubular heart. Except at its venous end the tubular endothelial heart is separated from the myo-epicardial mantle by an interval occupied by the *cardiac jelly*.

The proximity of the presumptive cardiac mesoderm and yolk sac endoderm at this stage of development is suggestive, and in chick embryos, at least, experimental studies indicate that movements, determination of polarity and regional differentiation of the cardiac mesoderm are in some way dependent upon the adjacent endoderm, and perhaps also ectoderm. For discussions concerning such experimental studies and related questions such as the importance of haemodynamic influences consult Orts Llorca *et al.* (1964, 1967a and b), De Haan (1968), Stalsberg and De Haan (1968), Bellairs (1971), Balinsky (1975).

The dorsal aortae arise *in situ* as paired endothelial vessels. They extend caudally into the body stalk, where they establish continuity with the umbilical arteries, which precede them in time of appearance. At their cranial ends the dorsal aortae curve ventrally round the sides of the foregut to reach the pericardium and become continuous with the cranial end of the endothelial heart tube, thus forming the first pair of aortic arches (2.77A). In lower vertebrates the heart and aortae are laid down before the formation of the head fold and the arteries communicate with the caudal end of the heart. When the head fold forms, the ends of the heart are reversed and the cranial ends of the dorsal aortae are curved forwards round the sides of the foregut to form the first aortic arches.

A transverse groove appears on the surface of the heart tube about its middle and indicates the junction of the *bulbus cordis* with the *ventricle*. The bulbus is situated cranial to the groove and is continuous with the first pair of aortic arches. The ventricle shows a second groove at its caudal end where it opens into a *common atrium*, which lies at first in the floor of the pericardium (the future *septum transversum*) and is disposed transversely. On each side the common atrium is joined caudally by a short venous trunk, formed by the union of the corresponding umbilical vein with veins issuing from the *vitelline* (yolk sac) *plexus*. These trunks represent the right and left *horns* of the *sinus venosus* so that the common atrium may justifiably be termed a *common sinuatrial chamber*.

Early in the fourth week the heart tube undergoes a striking change. Hitherto the pericardium has been increasing in length proportionally with the heart, but now the heart tube begins to grow more rapidly than the pericardium and, as a result, the bulboventricular tube bulges ventrally and caudally, forming a U-shaped loop of which the bulb forms the right limb and the ventricle the left. On account of this loop—which is a conspicuous feature throughout the fourth and fifth weeks—a deep *bulbenventricular sulcus* is apparent on the outside of the heart (2.77A) and a

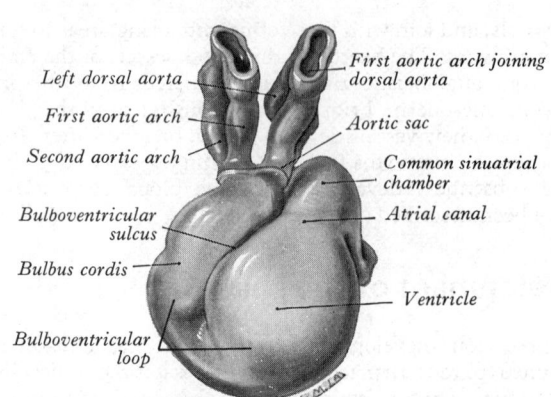

2.77A   The heart of a 0·95 mm rabbit embryo. Viewed from the ventral side. (Drawn from a model by G. Born.)

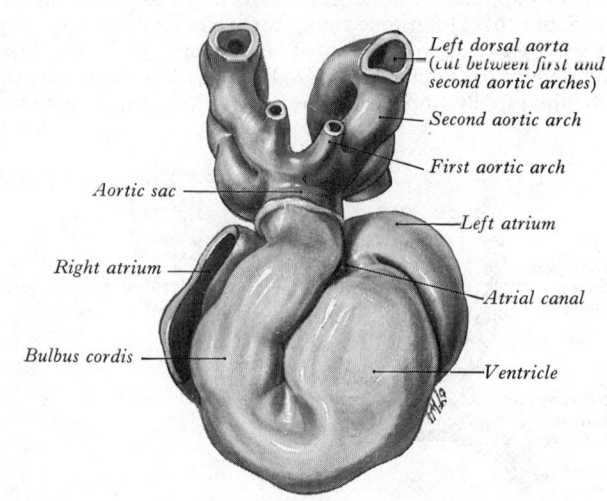

2.77B   The heart of a 1·7 mm rabbit embryo. Viewed from the ventral side. (Drawn from a model by G. Born.)

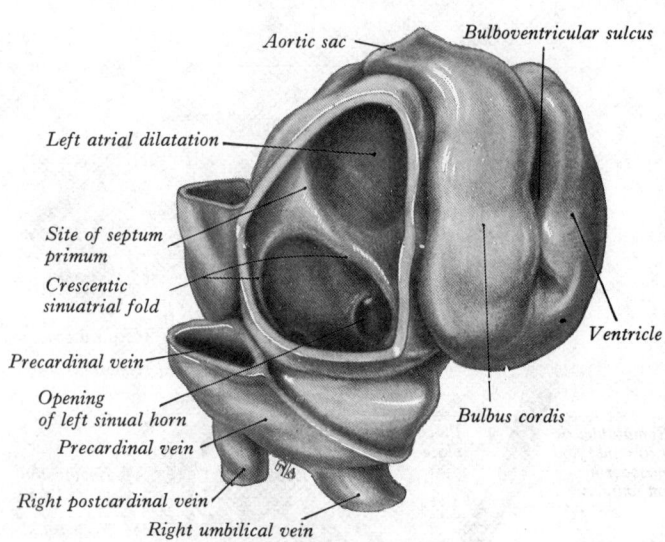

2.77C   The heart shown in **2.77B**, viewed from the right side and slightly from the ventral aspect. The right wall of the common sinuatrial chamber has been removed to show the interior. (Drawn from a model by G. Born.)

corresponding *bulboventricular ridge* projects into the interior. The dorsolateral recesses of the pericardium deepen and approach one another (2.76D) and their opposed walls fuse, completing the myo-epicardial covering of the heart and converting its hitherto broad dorsal attachment into a dorsal cardiac mesentery. This *dorsal mesocardium* is transient and when it breaks down early in the fourth week a passage is established across the pericardial cavity from side to side dorsal to the heart. This persists as the *transverse sinus* of the pericardium. While these changes are occurring in the bulboventricular region, the atrial part is not unaffected, for the atrioventricular opening moves cranially and to the left, and both parts of the common atrial or sinuatrial chamber grow cranially into the pericardial cavity dorsal to the ventricle. Owing to these changes the atrioventricular canal for a time connects the *left* atrium to the ventricle and venous blood from the right side has to pass through both atria.

At this stage (**2.**80A), reached about the middle of the fourth week, the bulbus cordis communicates with the dorsal aortae through the first pair of aortic arches, and these are connected with the capillary plexus associated with the developing cerebral vesicles. From this plexus the primitive head vein passes caudally, but it ends blindly before it reaches the heart. The intersegmental arteries are beginning to grow out from the dorsal aorta on each side but they have not yet established connexions with the corresponding veins; and the postcardinal veins, which later drain the body wall caudal to the heart, are only in process of development. The umbilical arteries and veins are defined and, early in the fourth week, their terminals and radicles, respectively, link up with the capillaries which have developed in the chorionic villous stems, establishing the chorionic part of the circulation. Despite the fact that the channels for the remainder of the circulation are only partially established there is good ground for assuming, on the analogy of observations made on the living chick embryo, that the heart begins to contract about this time. Under the conditions which exist the result can only be of an 'ebb and flow' nature, but this serves to effect some movement in the nutritive fluid which fills the pericardial cavity, the coelomic ducts and the exocoelom (p. 116) and on which the embryo still depends to a considerable extent:

Towards the end of the fourth week the connexion between the bulbus cordis and the first pair of aortic arches lengthens to form the *truncus arteriosus*, and the cranial end of this vessel becomes connected to the dorsal aortae by the remaining five pairs of aortic arches. By this time the venous drainage of the body wall and neural tube has been established. On each side a *precardinal vein*, from the cranial end of the embryo, unites with a *postcardinal vein* from the caudal region, to form the *common cardinal vein* (*duct of Cuvier*); the latter vessel opens close to the umbilical and vitelline veins into the dorsocaudal part of the common sinuatrial chamber.

As the chorionic circulation already exists, the embryo can now exchange material with the maternal blood in the intervillous space. This is not effected suddenly, for at first the blood cells in the heart and vessels of the embryo are not numerous enough to enable it to take full advantage of this new source of nourishment, and it would appear that, until sufficient are available, the embryo continues to draw upon the coelomic fluid.

The separation of the *sinus venosus* from the *atrium* completes the definition of the primitive chambers of the heart. A crescentic groove appears on the left wall of the sinuatrial chamber and rapidly deepens to the right. Hence the left horn of the sinus venosus loses its connexion with the left atrium and becomes linked to the right horn by separation of the caudal part of the sinuatrial chamber, which can be regarded as the *body* of the sinus venosus. At the same time the right horn becomes more clearly demarcated from the right atrium and its wide connexion with the atrium (**2.**77C) becomes relatively smaller (Foxon 1955). The right and left parts of the atrium grow cranially to occupy the dorsal part of the pericardial cavity, and later they bulge forwards at the sides of the bulb (**2.**78A).

The embryo has now attained a length of nearly 4 mm (**2.**80B). It possesses 28 somites and has almost completed the fourth week of development. From this stage onwards it is more convenient to deal with the individual chambers than with the development of the heart as a whole.

It must be noted at this point that the above account of early cardiogenesis, though widely subscribed to, is not without its critics. In considering the factors govering cardiac development—phylogenetic, ontogenetic, and physiological—

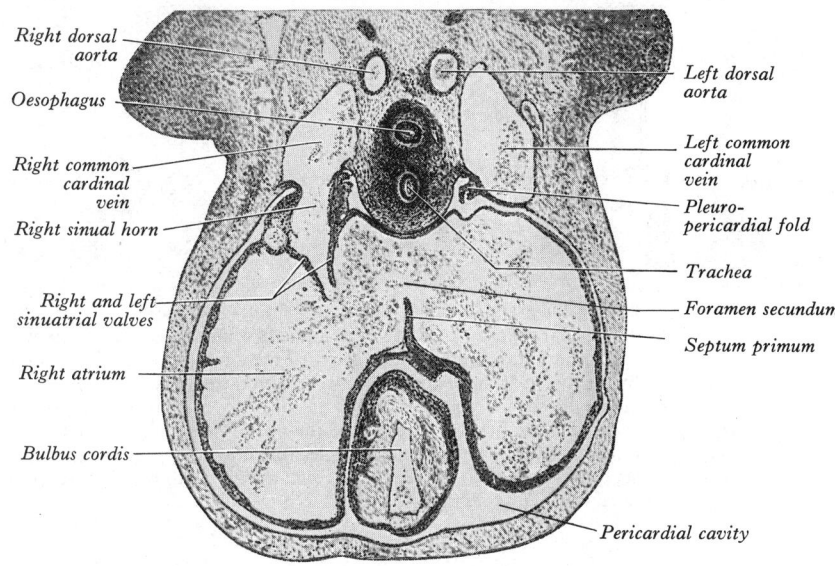

2.78A Transverse section of a human embryo, 8 mm long. Observe how the atria bulge forwards on each side of the bulbus cordis. The septum primum has broken down in its dorsal region and the two atria communicate through the ostium secundum or foramen ovale.

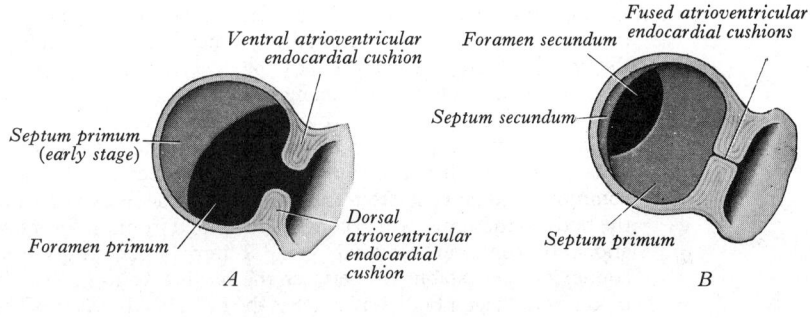

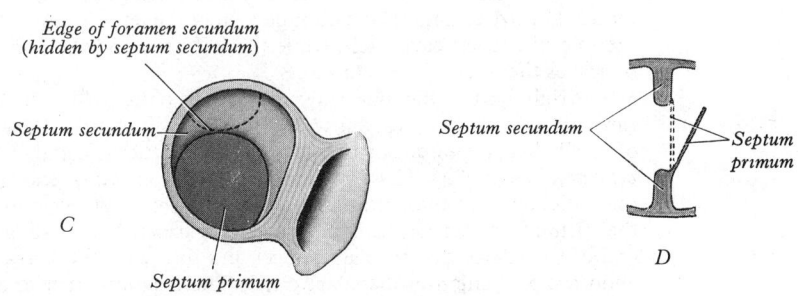

2.78B Diagrams representing three stages in the development of the atrial septum, viewed from the right side. The heart has been divided in its long axis to the right of its median plane and only the atria and the adjoining part of the ventricular cavity are depicted.

A The septum primum has not yet obliterated the original communication between the two atria and the atrioventricular endocardial cushions have not yet fused.

B The atrioventricular endocardial cushions have fused with each other and with the septum primum, which has broken down in its dorsal part. The foramen secundum, thus formed, subsequently moves to the position shown in C.

C The septum secundum has formed and hides the foramen secundum, the margins of which are indicated by the curved, dotted line.

D A section to show the valve-like character of the foramen secundum. When the pressure in the right atrium exceeds that in the left atrium, blood passes from the right to the left side of the heart, but when the two pressures are equal the septum primum assumes the position indicated by the dotted outline.

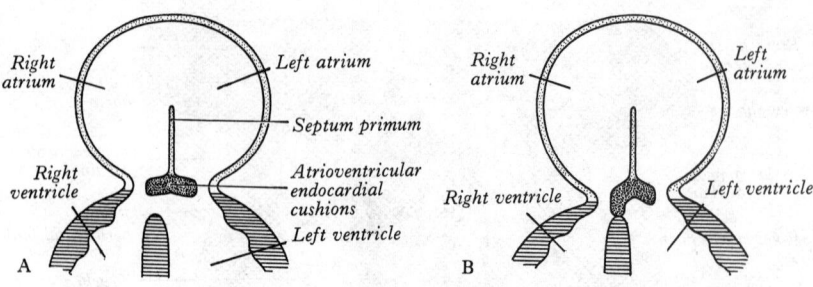

2.78C Diagram to show two stages in the formation of the adult ventricular septum. In A the right and left ventricles communicate with each other, but in B the interventricular communication has been closed by the fusion of the ventricular septum with the enlarged right extremity of the fused a.v. cushions. Note the position of the septum primum relative to the fused a.v. cushions, and observe that in B cushion tissue intervenes between the two ventricles (membranous part of ventricular septum), and also between the *right* atrium and the *left* ventricle (atrioventricular septum).

the last is usually underestimated and the first is perhaps stated too dogmatically (Foxon 1955). Ontogenetic mechanisms must conform to the early demand for a functioning heart, and cardiogenesis is not necessarily a mere repetition of phylogenetic steps, which are themselves uncertain, however plausible they may seem (De Vries and Saunders 1962).

THE SINUS VENOSUS

The right horn of the sinus venosus increases rapidly in size at the expense of the left, owing to the changes already outlined and to those brought about in the originally symmetrical arrangement of the umbilical and vitelline veins by the development of the liver (p. 193). As a result of these changes the vitello-umbilical blood flow enters the right horn through a wide but short vessel, the *common hepatic vein*, which becomes the cranial end of the inferior vena cava. In addition to this the right horn receives the right common cardinal vein (from the right side of the body wall) and the body of the sinus, which conveys the blood from the left horn and left common cardinal vein. Later, when transverse connexions are established between the cardinal veins (2.88), the blood from the left body wall reaches the heart via the veins of the right side. The left common cardinal vein then becomes much reduced in size and forms the oblique vein of the left atrium and the fold of the left caval vein, while the left horn and the body persist as the *coronary sinus* (2.88).

The right horn of the sinus venosus opens into the right atrium through its dorsal and caudal walls. The orifice, elongated and often slit-like, is guarded by two muscular folds, the *right* and *left venous valves* (2.78A). These two valves meet at the cranial end of the orifice and become continuous with a fold which projects into the atrium from its roof, termed the *septum spurium*. At the caudal end of the orifice the two valves meet and fuse with the dorsal endocardial cushion of the atrial canal. The cranial part of the right venous valve disappears, but its position is indicated in the adult heart by the crista terminalis of the right atrium; its caudal part forms the valve of the coronary sinus and most of the valve of the inferior vena cava. The medial (or left) end of the valve of the inferior vena cava is formed by a small fold which projects from the dorsal wall of the sinus venosus, the *sinus septum*. In the embryo the sinus septum intervenes between the orifice of the common hepatic vein and the opening of the body of the sinus.

The left venous valve blends with the right side of the atrial septum and usually no trace of it can be seen in the adult heart.

As the venous valves undergo these changes the right horn of the sinus venosus becomes incorporated in the right atrium and forms its dorsal wall, medial to the crista terminalis. This part of the adult atrium is termed the *sinus venarum*.

THE RIGHT AND LEFT ATRIA

As already stated, the common atrium is derived from the cranial part of the sinuatrial chamber. It receives the opening of the sinus

venosus in its dorsocaudal part and to the right of the median plane, while it communicates ventrally with the ventricle through the atrioventricular canal, which has resumed its median position by the middle of the fifth week and then permits both right and left atria to communicate with the common ventricular cavity. Swellings appear in the ventral and dorsal walls of the atrioventricular canal in the interval between the endothelial tube and the myo-epicardial mantle. These are termed the *atrioventricular endocardial cushions* and consist of a core of mesenchyme. They encroach on the canal and eventually fuse with each other, leaving a relatively small orifice on each side. The fused cushions constitute the *septum intermedium* (of His), and the two small orifices become the right and left atrioventricular openings.

The internal separation of the two atria is effected by the growth of two septa. The first to appear is the *septum primum*, and it grows from the dorsal and cranial wall as a sickle-shaped fold (2.78B, A), which is separated from the left venous valve by the *interseptovalvular space*. The ventral horn of the sickle reaches the ventral atrioventricular cushion, and the dorsal horn reaches the dorsal cushion. The terms ventral and dorsal refer to the positions of the cushions after the atrium has come to lie dorsal to the bulbus cordis. Actually the ventral cushion is ventrocranial and the dorsal cushion dorsocaudal in position, but the terms ventral and dorsal are in general use and will be retained in the subsequent description. Ventral and caudal to the advancing edge of the septum the two atria communicate through the *foramen primum* (2.78B, *A*). Free passage of blood from the right atrium to the left is essential throughout fetal life, as the oxygenated blood from the placenta reaches the heart via the inferior vena cava (p. 665); therefore, as the foramen primum becomes reduced in size, the dorsal part of the septum primum breaks down and a new communication, the *foramen secundum*, is formed between the two atria before the end of the fifth week. The foramen primum becomes occluded completely by the fusion of the edge of the septum primum with the fused atrioventricular cushions, in the median plane. The foramen secundum enlarges sufficiently to ensure the free passage of blood from the right atrium to the left (2.78B), and it persists throughout intrauterine life as the *foramen ovale*. At first this opening is situated in the cranial and dorsal portion of the septum primum but its position becomes modified until it is cranioventral. Towards the end of the second month the muscular wall of the atrium becomes invaginated to form a crescentic septum on the right side of the septum primum (2.78B, B, C). This is the *septum secundum*, and it involves more than the whole width of the interseptovalvular space, so that the dorsal attachments of the septum primum and the left venous valve are carried into the interior of the atrium on its left and right surfaces respectively.

The superior and inferior horns of the septum secundum at first grow ventrally; but the superior horn grows much more rapidly than the inferior and fuses first with the septum intermedium; it is then continued to form the *sinus septum* (*vide supra*). Thus the free edge of the septum secundum is at first directed caudoventrally and later caudally alone; it overlaps the foramen ovale (2.78B, C, D) so that the septum primum can act as a flap valve. Since the blood pressure is greater in the right atrium than it is in the left, the blood flows from right to left, but not in an opposite direction. After birth the intra-atrial pressures are equalized, and the free edge of the septum primum is therefore kept in contact with the left side of the septum secundum and fusion occurs. Not infrequently the fusion is incomplete, but the opening left is usually small and valvular and has no functional significance. The free margin of the septum secundum forms the *limbus fossae ovalis* and the septum primum the floor of the *fossa ovalis* of the adult heart. An alternative derivation of the septum secundum from a ridge developing to the right of the line of fusion between the confluent endocardial cushions and the septum primum has been advanced (Odgers 1934). The dorsal horn of the septum secundum is said to incorporate part of the left venous valve. Another view embodies the above suggestion, but regards the valve as of minor importance in this connexion, describing, however, another ridge—the *septum accessorium*, in the lower part of the dorsal border of the limbus (Christie 1963).

At an early stage in the development of the septum primum a

single, common pulmonary vein, of still unexplained origin, can be identified opening into the caudal part of the dorsal wall of the left atrium close to the septum. It is formed by the union of a right and a left pulmonary vein, and each of these is formed, in turn, by two small veins issuing from each developing lung bud. Subsequently the common trunk and the two veins forming it expand and are incorporated in the left atrium to form the greater part of its cavity. This expansion reaches as far as the orifices of the four veins, which thus open separately into the left atrium.

During the second month the two atria bulge ventrally one on each side of the bulbus cordis, which lies in a groove on their ventral surface (2.78A). These projecting parts of the atria form the auricles of the adult heart, while their dorsal regions expand to form the atria proper.

THE VENTRICLES BULBUS CORDIS AND THE TRUNCUS ARTERIOSUS

The process of separation of the ventricles is intimately related to that of the aortic and pulmonary orifices at the distal end of the bulbus, (2.79A) and to the division of the truncus arteriosus (p. 183) into pulmonary and aortic channels. These processes are so closely interdependent that the history of the truncus arteriosus will be dealt with in this section, although it takes no part in the formation of the heart itself, and both will be considered before the separation of the two ventricles is described.

Four endocardial cushions—ventral, dorsal, right and left—form in the distal part of the bulbus, and the right and left cushions fuse to constitute the *distal bulbar septum*. This septum separates a ventral, pulmonary orifice from a dorsal, aortic orifice, and later the cushions become modified to form the *semilunar valves*.

The separation of the pulmonary trunk from the aorta is a more complicated process. Two ridge-like thickenings project into the interior of the truncus arteriosus. Proximally, the ridges project from the *lateral* walls of the vessel, but further away from the heart, the right ridge passes obliquely on to the *ventral* and then the *left* wall, while the left ridge, extends on to the *dorsal* wall and then the *right* wall (2.79B). The ridges are therefore spiral, and their fusion forms the *spiral aorticopulmonary septum*. At its proximal end this septum meets and fuses with the distal bulbar septum, and on account of its spiral character the pulmonary trunk, which lies ventral to the aorta at its orifice, curves round to its left side as it ascends, and finally lies dorsal to it (2.79B). At its distal end the aorticopulmonary septum meets the dorsal wall of the aortic sac (*see* p. 187) cranial to the point where it is joined by the sixth pair of aortic arches, and as a result these arches become branches of the pulmonary trunk, while the other aortic arches are left in communication with the aorta (2.79B).

The separation of the two ventricles from each other leaves the right ventricle in communication with the right atrium and with the pulmonary artery, and the left ventricle in communication with the left atrium and the aorta. This involves a series of complex changes in which three distinct factors contribute to the formation of the adult ventricular septum—(*a*) the *ventricular septum*, (*b*) the *proximal bulbar septum*, and (*c*) the *atrioventricular endocardial cushions*.

(*a*) During the fifth week the right and left definitive ventricles are indicated as slight projections on the external surface of the primitive common ventricle. It is uncertain whether the right ventricle is a derivative of the common ventricle, or of the caudal end of the primitive bulbus, or of both. In either event, the appearance of a crescentic ridge in the inside of the heart indicates the separation between the two ventricles and, as the heart enlarges, this ridge deepens to form the *ventricular septum*. The dorsal and ventral horns of the septum meet and fuse with the corresponding endocardial cushions of the atrioventricular canal near their *right* extremities (2.79A). The septum has a free sickle-shaped margin, which bounds a circular *interventricular foramen* (2.79B).

At first the bulboventricular junction is marked by a distinct notch on the outside of the heart (2.79A), and on the inside of the heart there is a corresponding *bulboventricular ridge*. This ridge is between the atrioventricular orifice and the caudal part of the bulb

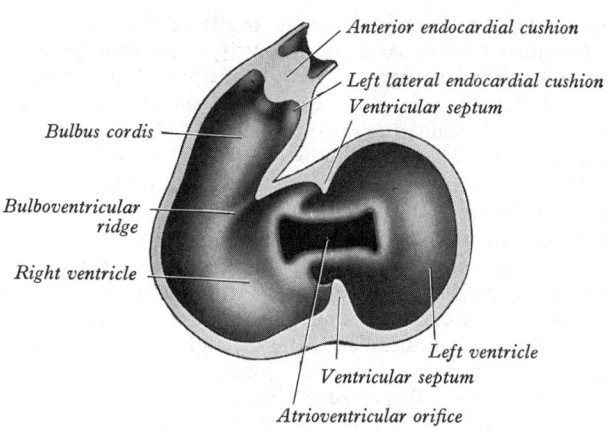

2.79A A diagram showing an early stage of the relations between the atrioventricular opening and ventricles, the cavity of the bulbus cordis, and the bulboventricular ridge. The endocardial cushions at the distal end of the bulb are shown in a more differentiated state than they really exhibit at this stage. (After J. E. Frazer.)

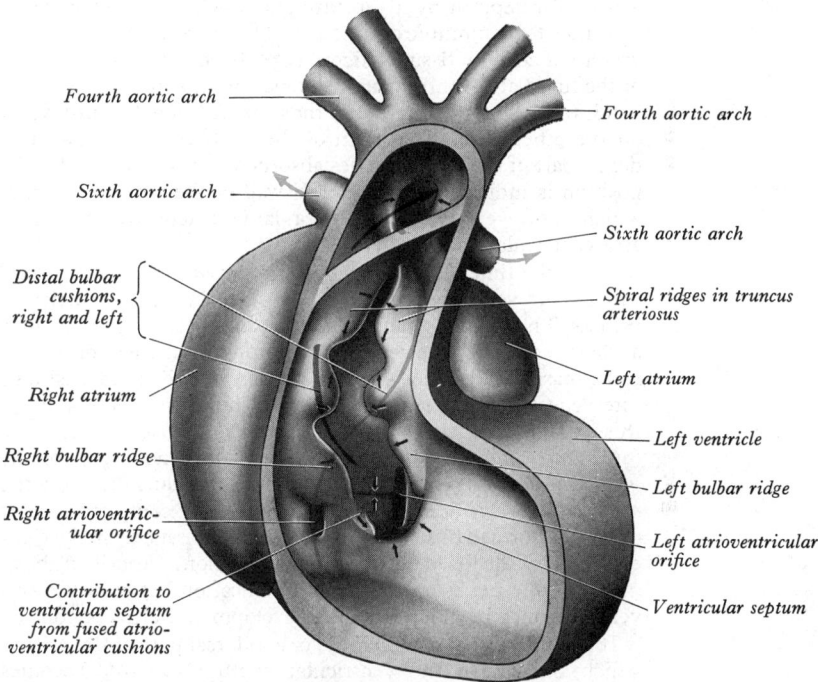

2.79B A diagram to show the mode of formation of the septa which separate the aortic and pulmonary channels in the embryonic heart. The red arrow indicates the aortic channel and the blue arrow the pulmonary. The small black arrows indicate the direction of growth. (From a model by James Whillis.)

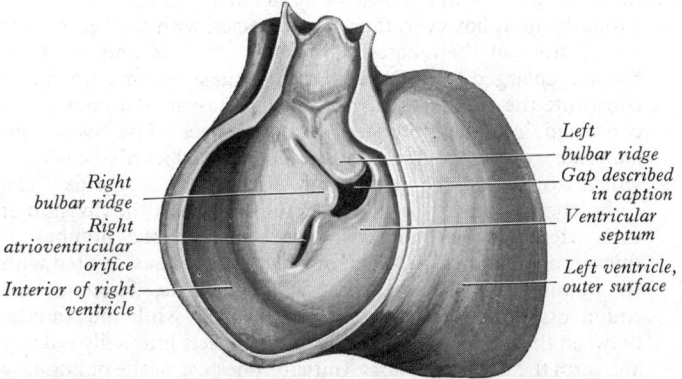

2.79C A diagram to show the part played by the fusion of the right and left bulbar ridges in the separation of the aortic and pulmonary channels. The darkly shaded area indicates the gap filled by the proliferation of cushion tissue (from the right extremity of the fused atrioventricular cushions) which establishes continuity between the proximal bulbar septum and the ventricular septum. Compare with 2.79B.

(2.79A) and its absorption is essential for the development of the four-chambered heart. As a result, partly of the absorption of the bulboventricular ridge and partly of the growth of the atrio-ventricular region, the right extremity of the atrioventricular canal comes to lie caudal to the orifice of the bulb (2.79B). This alteration in the relative positions of the structures concerned occurs while the ventricular septum is forming, and paves the way for the completion of the process of ventricular partition.

(b) The *proximal bulbar septum* separates the bulbus cordis into pulmonary and aortic channels, and is formed by the *right* and *left bulbar ridges*, which are in continuity with the corresponding bulbar endocardial cushions. The right bulbar ridge grows across on to the dorsal wall of the bulb and the right extremity of the fused atrioventricular endocardial cushions to reach the dorsal end of the free, sickle-shaped edge of the ventricular septum, and in doing so obliterates the ventral or cranial part of the right atrioventricular orifice (2.79B). The left bulbar ridge grows across on to the ventral wall of the bulb to reach the ventral or cranial end of the ventricular septum. Although the two bulbar ridges fuse with each other and so separate the conus arteriosus of the right ventricle from the aortic vestibule, the caudal edge of the septum so formed is separated from the free sickle-shaped edge of the ventricular septum by a gap through which the two ventricles continue to communicate (2.79C). This gap is closed by the growth of cushion tissue (Odgers 1938) from the right extremity of the fused atrioventricular cushions, and this fuses, on the one hand, with the caudal border of the proximal bulbar septum and on the other with the margin of the ventricular septum. The dorsal part of the bulb becomes absorbed almost entirely, but its position is indicated by the dorsal wall of the aortic vestibule, which, however, is formed to a large extent by the fused atrioventricular endocardial cushions.

(c) At the time of their fusion the *atrioventricular endocardial cushions* are very large relative to the size of the atrioventricular orifices. The atrial septum meets the atrial surface of the cushions at their centre, but the ventricular septum meets them near their right margins. It follows that a portion of the fused cushions intervenes between the *right* atrium and the *left* ventricle, and it is this part which forms the right wall of the aortic vestibule (*atrioventricular septum*, p. 643). The membranous part of the *interventricular septum*, which is continuous dorsally with the membranous *atrioventricular septum* in the completed heart (2.78C), is formed by proliferation of cushion tissue from the right extremity of the fused atrioventricular cushions, described above (2.78C). The persistence of a communication between the two ventricles may be due to arrested development in this region.

It should be observed that the craniodorsal part of the orifice, which lies above the ventricular septum (2.79B), becomes incorporated in the aortic vestibule and thereafter serves to connect the left ventricle with the aortic channel.

The valves of the heart are primarily derived from the endocardium.

The *atrioventricular valves* develop as endothelial projections directed towards the ventricles at the atrioventricular orifices. From the first, however, they are connected with the ventricular musculature at their bases; as growth proceeds and the flaps become enlarged, these muscular trabeculae become freed and constitute the papillary muscles, their juxta-valvular ends being converted into the fibrous *chordae tendineae*. The *aortic* and *pulmonary valves* are formed from the four endocardial cushions which appear at the distal end of the bulbus cordis. The completion of the distal bulbar septum results in the division of each lateral cushion into two parts, so that the number of thickenings is increased to six, of which three are associated with the pulmonary orifice and three with the aortic. These are the rudiments of the aortic and pulmonary valves, while the pouches between the valves and the walls of the vessels gradually enlarge and form their related *sinuses*. Initially, one cusp of the pulmonary valve lies anteriorly and the other two posterolaterally, whereas one cusp of the aortic valve lies posteriorly and the other two anterolaterally. However, a rotation of the heart to the left before birth changes the orientation of the cusps of the pulmonary and aortic valves and this is reflected in the various schemes for the designation of these cusps in the mature heart (*see* p. 651).

The development of the chambers of the heart has now been traced to a stage at which the main features of the adult heart are established. It is to be noted that the pattern has developed in such a way as to provide for the sudden establishment of the pulmonary circulation at birth, although it is adapted to the persistence of the placental circulation for the remainder of fetal life. The presence of the ductus venosus ensures that the oxygenated blood gains the right atrium with the minimum loss of oxygen to the liver. It has been claimed that relatively little admixture of oxygenated and deoxygenated blood occurs in the *right* atrium (Barclay *et al.* 1939) and that nearly all the oxygenated blood passes through the foramen ovale into the left atrium, so gaining the *left* ventricle, aorta, and systemic circulation. However, there is evidence for the opposing view that there is considerable mixing of the superior and inferior vena caval streams in the right atrium (Born *et al.* 1954; Lind and Wegelius 1954). Because the transition from a placental to a pulmonary circulation occurs suddenly at birth, the right ventricle and the pulmonary trunk of the fetus are relatively large, although only a small amount of blood passes through the lungs. In fact most of the blood expelled by the right ventricle into the pulmonary trunk passes through the ductus arteriosus into the descending aorta, and in order that this can occur it must be under higher pressure than the blood in the aorta. On this account the muscular wall of the right ventricle is thicker than the wall of the left ventricle, a condition which persists throughout fetal life but is progressively reversed after birth. The origin of the carotid and subclavian arteries from the aorta above the point at which it is joined by the ductus arteriosus may be correlated with the relatively rapid growth of the brain, demanding a copious blood supply, and with the more advanced development of the upper limbs, relative to the lower, at birth.

In the early stages of development the arteries of the embryo are disproportionately large and their walls consist of little more than a single layer of endothelium. At this time the cardiac orifices are also relatively large and the force of the cardiac contraction is weak. As a result, despite the rapid rate of contraction, the circulation is sluggish, but this is compensated for by the fact that the tissues are able to draw nourishment, not only from the capillaries, but also from the large arteries. As the heart muscle strengthens, the cardiac orifices become both relatively and absolutely reduced in size, the large arteries acquire their muscular walls and they too undergo a relative reduction in size. From this time onwards the embryo is dependent for its nourishment on the expanding capillary beds and henceforth the larger arteries function merely as distribution channels to keep them constantly supplied.

The conducting system of the heart is first identifiable towards the middle of the sixth week of embryonic life. Some investigators claim that the first part to appear is the *atrioventricular node* which develops as an outgrowth from the dorsal part of the muscular ring surrounding the atrioventricular canal. The *atrioventricular bundle* is an extension of this which passes distally behind the dorsal endocardial cushion to the free edge of the ventricular septum, where it gives rise first to the left and then to the right limbs of the bundle (Shaner 1929; Walls 1947). Others maintain that the stem of the atrioventricular bundle appears first as a derivative of the musculature of the dorsal wall of the atrioventricular canal. From this, differentiation spreads proximally to form the atrioventricular node and distally to form the right and left limbs of the bundle (Field 1951; Muir 1954). From its earliest appearance the whole complex is continuous with the muscular tissue of the atrium and ventricular wall, and this condition persists to adult life.

The *sinuatrial node* is the last part of the conducting system to appear, and its site is indicated by an aggregation of nerve cells and fibres in the wall of the superior vena cava just above its junction with the atrium, before nodal tissue itself can be recognized in the latter part of the third month. A narrow band of muscle in the ventrolateral surface of the superior vena cava is modified into the characteristic adult histological structure (p. 661). Although at first situated in the superior vena cava, in late fetal life it becomes incorporated into the atrial wall at the upper end of the sulcus terminalis. Its cranial edge, however, remains in continuity with the muscular coat of the vessel, although elsewhere it is continuous with the muscular wall of the atrium. The node does not appear until the right horn of the sinus venosus has been incorporated in the right atrium. Sinuatrial nodal tissue appears at an earlier relative stage in some rodents, and recently a *left* sinuatrial node has been described in the murine embryo (Heintberger 1974) as a transient structure. The occurrence of such a bilateral symmetry has been hypothecated in mankind but never substantiated (Patten 1956).

A recent study of conducting tissue development in human embryos emphasizes the basic difficulty of identifying such tissue, particularly in its earlier stages of differentiation (Wenink 1976). The previous views, stated above, which derived elements of the conducting system from part of the muscular rings between successive chambers of the primitive heart tube, were largely confirmed. However, it was claimed that *all* the rings were so involved, and that after complex repositioning during folding of the heart tube, the entire system is derived from them.

The *intrinsic nerves* of the heart, consisting of groups of ganglion cells associated with the great arteries, atria, and atrioventricular grooves, together with a developing epicardial plexus, are observable from the 15 mm stage (Navaratnam 1965; Smith 1970). For a highly detailed description of the state of development of the *extrinsic nerves* of the heart in a human embryo of 8 weeks consult Smith (1971), and Gardner and O'Rahilly (1976).

It will be noted that the heart commences to beat (p. 183) prior to the development of the conducting system, and that the circulation is established before a competent valvular mechanism has been laid down. The latter fact has been associated (Streeter 1945) with the incompressible but displaceable character of the cardiac jelly occupying the myoendocardial interval (p. 182).

A brief summary of the common congenital abnormalities of the heart will be found on p. 666.

FURTHER DEVELOPMENT OF THE ARTERIES

Apart from the aortae none of the main vessels of the adult arise as single trunks in the embryo. Along the course of each vessel a capillary network is first laid down, and by the selection and enlargement of definite paths in this network the larger arteries and veins are defined. The branches of the main arteries are not always simple modifications of the vessels of the capillary network, but may arise as new outgrowths from the enlarged stem.

As stated above, subsequent to the formation of the head fold each primitive aorta consists of ventral and dorsal parts which are continuous through the first aortic arch. The dorsal aortae run caudally, one on each side of the notochord, but in the fourth week they fuse from about the level of the fourth thoracic to that of the fourth lumbar segment to form a single *descending* aorta. Although in many animals paired ventral aortae arise from the truncus arteriosus and course headwards on the ventral surface of the pharynx, in the human embryo the ventral aortae are fused and form a dilated *aortic sac* (see **2.80**A and consult Congdon 1922). The first aortic arches run through the mandibular arches, and caudal to them five additional pairs are developed within the corresponding branchial arches; so that, in all, six pairs of aortic arches are formed (**2.81**A). The fifth arches are atypical and probably transient, at most, in man.

In fishes the aortic arches persist and give off branches to the gills, in which the blood is oxygenated. In mammals some of the arteries remain as permanent structures, while others disappear or are obliterated (**2.81**A–C).

The aortic sac represents fused, paired ventral aortae (**2.80**A, B). As the embryo grows and the aorticopulmonary septum is formed, part of the caudal end of the sac is incorporated in the pulmonary trunk. The cranial end of the sac becomes drawn out into *right* and *left* limbs as the neck lengthens. The right limb becomes the brachiocephalic trunk and the left limb forms that part of the arch of the aorta which lies between the origin of the brachiocephalic trunk and the left common carotid artery. The remainder of the sac contributes to the formation of the arch of the aorta.

The aortic arches (**2.81**A–C), with the exception of the fifth,

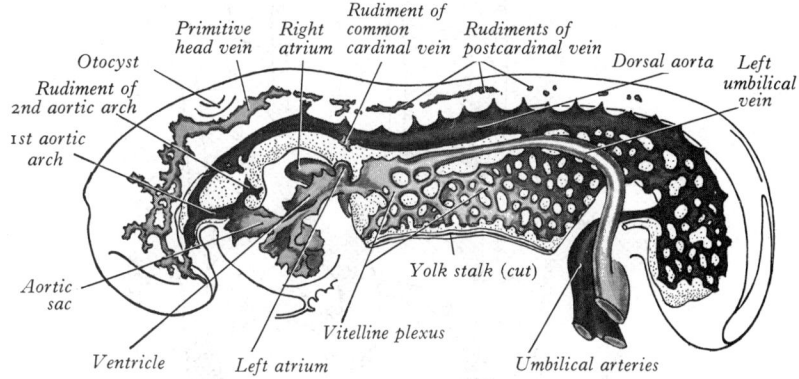

2.80A   The blood vascular system of a human embryo with 14 paired somites. Estimated age, 23½ days. C.R. length 2·4 mm. Magnification × c. 45. The arteries and veins are only in process of development, so that no true circulation is possible at this stage. Only the endothelial lining of the heart tube is shown. (G. L. Streeter, *Cont. Embryol.*, **30**, 1942.)

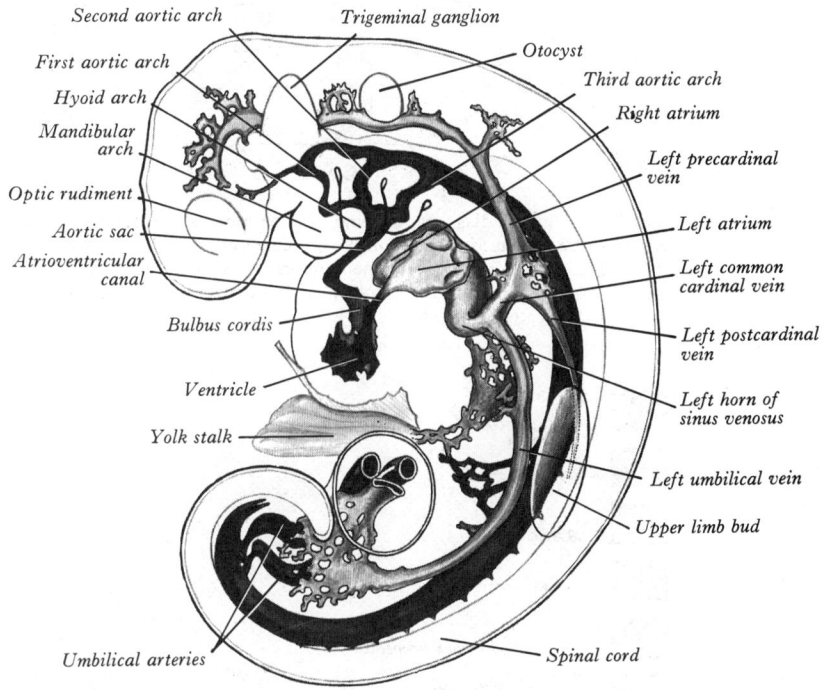

2.80B   Profile reconstruction of the blood vascular system of a human embryo, having 28 somites. C.R. length, 4 mm. Estimated age, 26 days. *Note.* Only the endothelial lining of the heart chambers is shown and, as the muscular wall has been omitted, the pericardial cavity appears much larger than the contained heart. Observe that the atrioventricular canal still connects the left atrium with the single ventricle. (G. L. Streeter, *Contr. Embryol.*, **30**, 1942.)

are developed in a craniocaudal sequence, but the more cranial are in process of disappearing before the caudal ones are completed. The *first* and *second aortic arches* are already dwindling by the time the third is established. The first disappears entirely. The dorsal end of the second arch or *hyoid* artery remains as the stem of the *stapedial artery*, while the remainder of this arch artery also disappears (**2.82**). The external carotid artery first appears as a sprout which grows headward from the aortic sac close to the ventral end of the third arch artery. The common carotid arises from an elongation of the adjacent part of the aortic sac, and the third arch artery becomes the proximal part of the internal carotid artery. (Evidence against this view, however, has also been recorded: *see*, for example. Moffat 1959; Adams 1957.) The *fourth aortic arch* on the right side forms the proximal part of the right subclavian artery, whilst the corresponding vessel on the left side is believed to constitute the arch of the aorta between the origins of the left common carotid and left subclavian arteries. It is

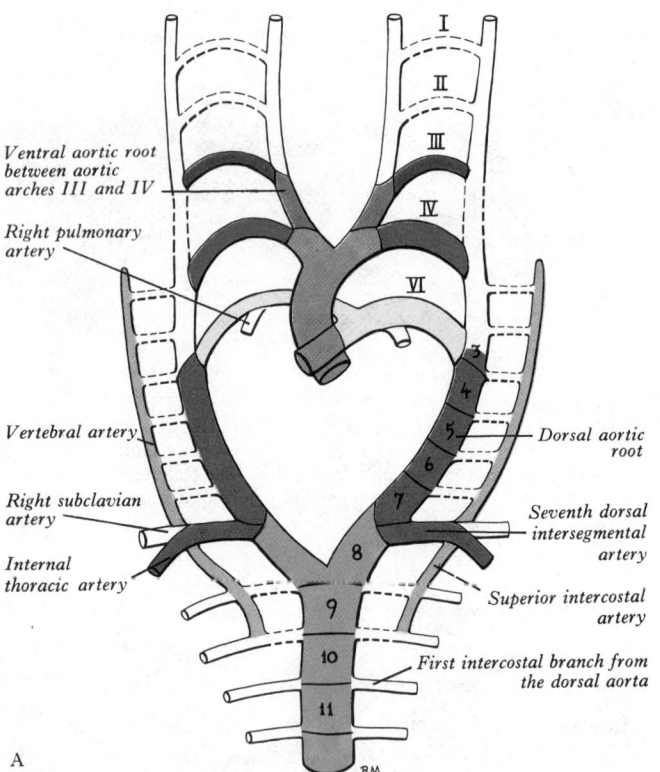

2.81A Schematic diagram showing the various components of the embryonic aortic arch complex in the human embryo. Structures which do not persist in normal development are indicated by interrupted lines. Roman numerals refer to the branchial arches concerned, and the arabic numerals indicate the metameric body segments.

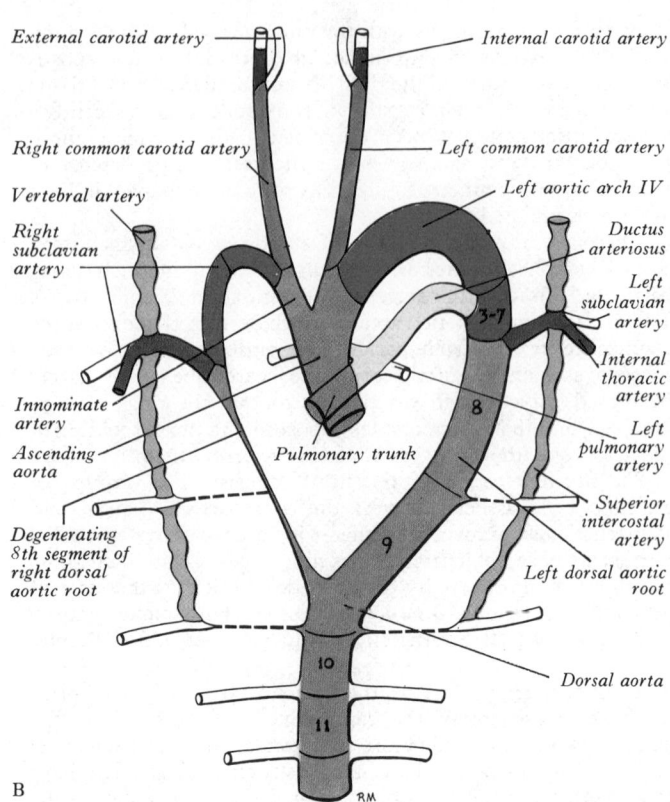

2.81B A diagrammatic ventral view of the aortic arch complex of a human embryo of 15 mm crown-rump length. Note the asymmetry in the pattern that has developed by this stage Compare with **2.81A** and C.

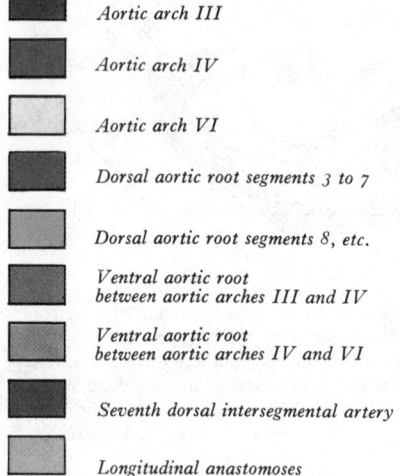

- Aortic arch III
- Aortic arch IV
- Aortic arch VI
- Dorsal aortic root segments 3 to 7
- Dorsal aortic root segments 8, etc.
- Ventral aortic root between aortic arches III and IV
- Ventral aortic root between aortic arches IV and VI
- Seventh dorsal intersegmental artery
- Longitudinal anastomoses

difficult to assess accurately the contributions of the *fourth aortic arches* and it has been variously claimed that the left fourth aortic arch is subsequently drawn into the descending or ascending limbs of the *definitive aortic arch* and that the corresponding vessel on the right contributes to the formation of the brachiocephalic artery. The identity and status of the *fifth aortic arch* is uncertain; it is incomplete and usually connects the fourth aortic arch or subjacent aortic sac with the dorsal end of the sixth aortic arch, whereas the other aortic arches pass between the aortic sac and dorsal aorta. The fifth aortic arch eventually disappears on both sides. From its inception the *sixth aortic arch* is associated with the developing lung buds. These are at first supplied by a capillary plexus from the aortic sac. Later this plexus becomes connected to the dorsal aorta, and the sixth aortic arch is constituted from the resulting vascular connexion between the aortic sac and the dorsal aorta, and continues to supply the developing lung buds. When the *truncus arteriosus* becomes

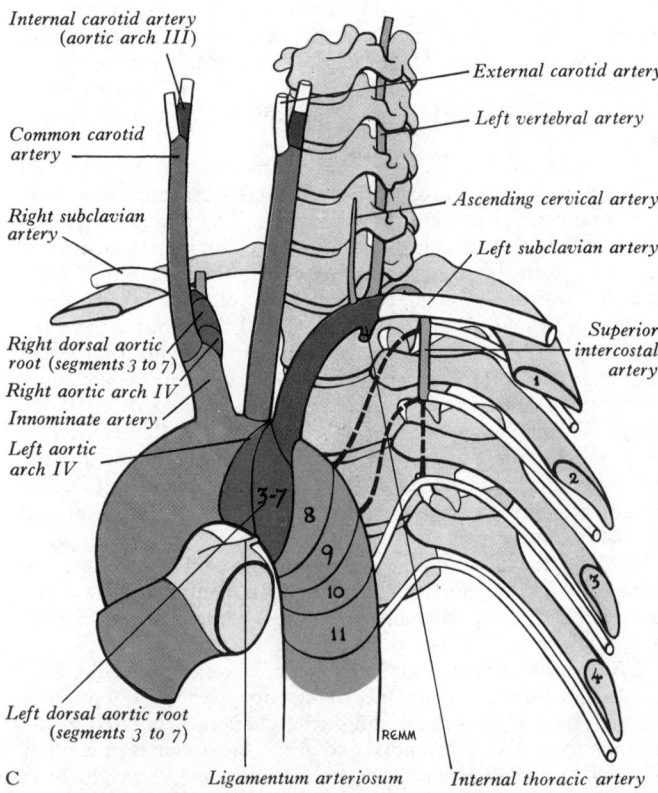

2.81C A diagram of the adult human aorta and its branches viewed from the left ventrolateral aspect, showing the position and relative sizes of the definitive contributions from the various embryonic components shown in **281A** and B.

(**2.81A–C** are based on the work of A. Barry, with the kind permission of the author and the *Anatomical Record*, **111**, 1951.)

divided into the pulmonary trunk and ascending aorta the sixth aortic arches remain continuous with the former. On the right side the ventral part of the sixth aortic arch persists as the proximal part of the right pulmonary artery, but its dorsal portion disappears, possibly due to a decreased blood flow resulting from the partitioning of the aortic and pulmonary bloodstreams (Navaratnam 1963). On the left side the ventral part of the sixth aortic arch is absorbed into the pulmonary trunk, while its dorsal portion persists as the *ductus arteriosus*, which functions during intrauterine life but becomes obliterated after birth and forms the fibrous *ligamentum arteriosum* (p. 665).

The transformation of the aortic arches described above is conditioned by environmental changes, and results largely from changes in the pharynx and from the descent of the heart. The whole period of transformation can be divided into two phases, the branchial and postbranchial. In the *branchial phase*, which lasts until about the 12 mm C.R. length stage, the arrangement of the aortic arches resembles that in lower vertebrates. In this phase the course of the blood from the heart to the dorsal aorta follows a succession of different pathways—first aortic arch, first and second aortic arches, second and third aortic arches, third and fourth aortic arches and finally third, fourth and sixth aortic arches. In the *postbranchial phase*, which extends onwards into intrauterine life the human pattern and disposition of the vessels is established.

The cranial arteries (2.82) develop as follows (Padgett 1948). The *internal carotid artery* is formed from the third arch artery, the dorsal aorta cranial to this and a forward continuation which differentiates, at the time of regression of the first and second aortic arches, from the capillary plexus extending to the walls of the forebrain and midbrain. At its anterior extremity the *primitive internal carotid artery* divides into cranial and caudal divisions, the former terminating as the *primitive olfactory artery*, which supplies the developing olfactory region, and the latter sweeping caudally to reach the ventral aspect of the midbrain. At the same time bilateral longitudinal channels differentiate along the ventral surface of the hindbrain from a plexus which is fed by intersegmental and transitory presegmental branches of the dorsal aorta and its forward continuation. The most important of the presegmental branches is closely related to the fifth nerve and is known as the *primitive trigeminal artery*. Otic and hypoglossal presegmental arteries also occur (Padgett 1948). Persistence of the latter has been surveyed by Nakayama *et al.* (1970). The longitudinal channels later connect, cranially, with the caudal divisions of the internal carotid arteries, each of which gives rise to an *anterior choroidal artery* and supplies branches to the diencephalon and the midbrain, and caudally with the vertebral arteries through the first cervical intersegmental arteries. Fusion of the longitudinal channels results in the formation of the *basilar artery*, whilst the caudal division of the internal carotid artery becomes the *posterior communicating artery* and the stem of the *posterior cerebral artery*. The remainder of the last-named vessel develops at a comparatively late stage, probably from the stem of the posterior choroidal artery which is annexed by the caudally expanding cerebral hemisphere, its distal portion becoming a choroidal branch of the posterior cerebral artery. In the rat, where the vascular pattern is essentially similar to that in man, this artery is derived from the posterior communicating artery, the common stem of origin of the posterior choroidal, mesencephalic and diencephalic arteries, together with a new channel formed in the plexus on the medial wall of the cerebral hemisphere initially supplied by the anterior choroidal artery (Moffat 1961a). The cranial division of the internal carotid artery gives rise to the *anterior choroidal, middle cerebral* and *anterior cerebral arteries*, the stem of the primitive olfactory artery remaining as a small medial striate branch of the anterior cerebral artery. In the rat the primitive olfactory artery and its recurrent branch form the anterior cerebral artery, the territory of which is initially supplied by the primitive maxillary and the cranial ramus of the internal carotid arteries (Moffat 1961b). The *cerebellar arteries*, of which the superior is the first to differentiate, emerge from the capillary plexus on the wall of the rhombencephalon.

The source of the blood supply to the territory of the fifth cranial nerve varies at different stages in development. When the

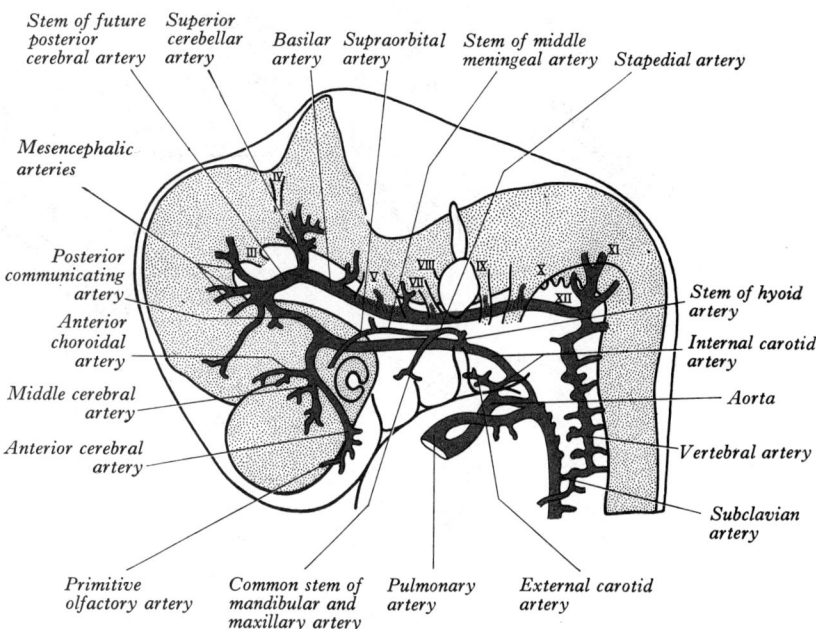

2.82 Diagram to show the origins of the main cranial arteries. (After D. H. Padgett, *Contr. Embryol.*, **32**, 1948.)

first and second aortic arch arteries begin to regress the blood supply to the corresponding arches is derived from a transient vessel termed the *ventral pharyngeal artery*, which grows into the region from the aortic sac. It terminates by dividing into *mandibular* and *maxillary branches*. Later a vessel, the *stapedial artery*, develops from the dorsal stem of the second arch artery, and passes through the ring of the stapes to anastomose with the cranial end of the ventral pharyngeal artery and thereby to annex its terminal distribution. The fully developed stapedial artery possesses three branches, the *mandibular, maxillary* and *supraorbital*, which follow the divisions of the fifth cranial nerve (2.82). The mandibular and maxillary branches arise by a common stem. When the external carotid artery emerges from the base of the third arch it incorporates the stem of the ventral pharyngeal artery, and its maxillary branch communicates with the common trunk of origin of the maxillary and mandibular branches of the stapedial artery and annexes these vessels. The proximal part of the common trunk forms the stem of the *middle meningeal artery*. More distally the meningeal artery is represented by the proximal part of the supraorbital artery. The maxillary artery becomes the infraorbital vessel and the mandibular branch forms the inferior alveolar artery.

When the definitive *ophthalmic artery* differentiates as a branch of the terminal part of the internal carotid artery, it communicates with the supraorbital branch of the stapedial artery, the distal

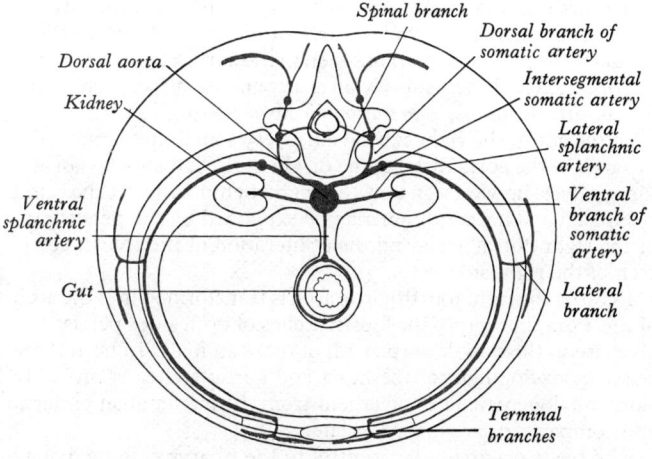

2.83 Diagram of the segmental and intersegmental arteries. Note the positions of the longitudinal anastomoses.

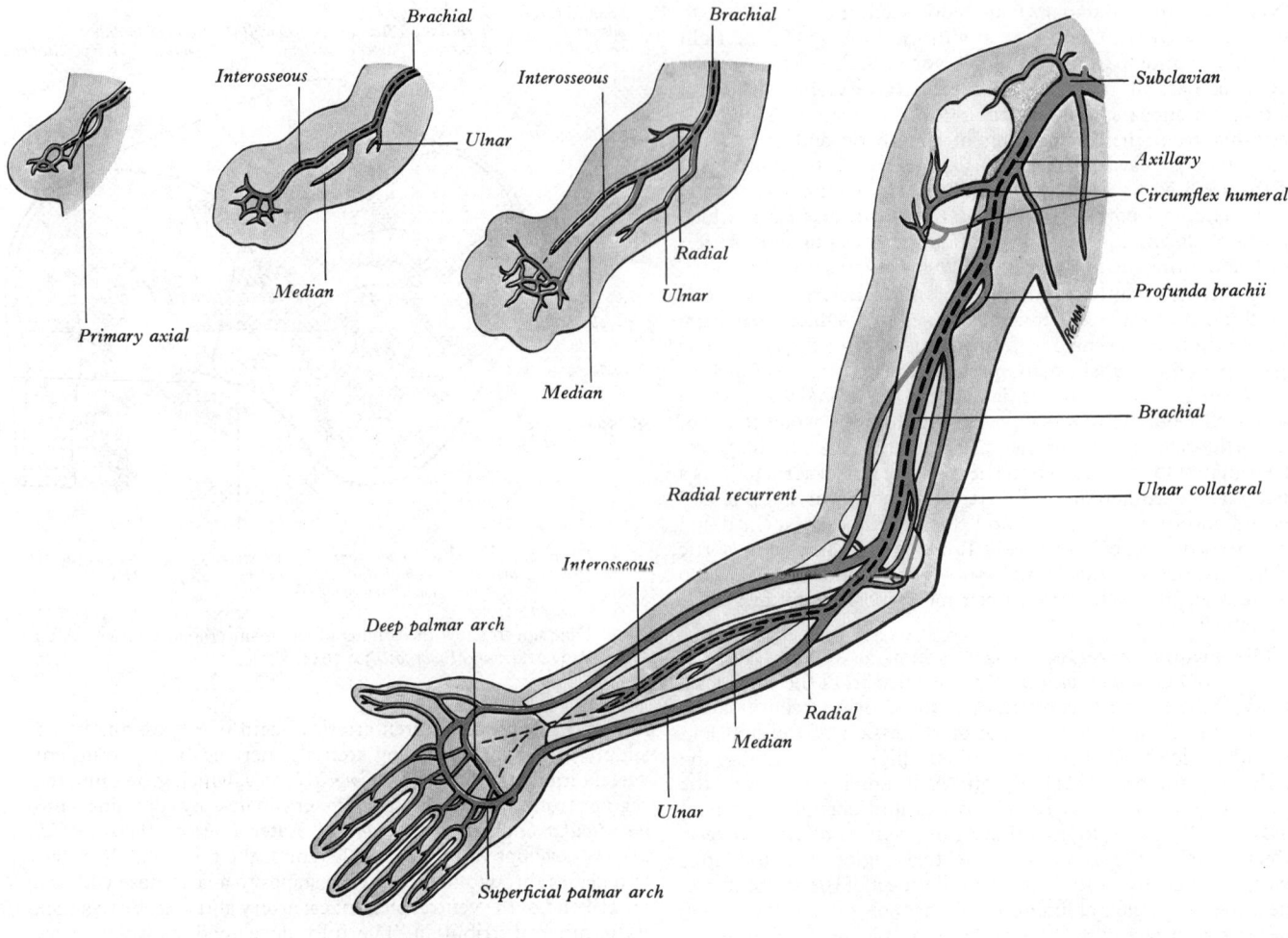

2.84 Stages in the development of the arteries of the arm. The original path of the axis artery is indicated by an interrupted line (After Patten.)

part of which becomes the *lacrimal artery*. This retains an anastomotic connexion with the middle meningeal artery. The dorsal stem of the original second arch artery remains as a *caroticotympanic branch* of the internal carotid artery.

The dorsal aortae (**2.**81A–C) persist on the cranial side of the third aortic arches and form the continuations of the internal carotid arteries. The part of the dorsal aorta between the third and fourth aortic arches is known as the *ductus caroticus* and it disappears; but from the fourth arch to the point of origin of the seventh intersegmental artery the dorsal aorta becomes part of the right subclavian artery (**2.**81A–C). Caudal to the seventh intersegmental artery the right dorsal aorta disappears as far as the point where the two dorsal aortae fuse to form the descending thoracic aorta. The part of the left dorsal aorta between the third and fourth arches disappears, while the remainder persists to form the descending part of the arch of the aorta. A constriction, the *aortic isthmus*, is sometimes seen in the aorta between the final site of origin of the left subclavian artery and the attachment of the ductus arteriosus (*see coarctation* of the aorta, p. 710).

In the adult, the right subclavian artery sometimes arises from the arch of the aorta distal to the origin of the left subclavian and then passes upwards and to the right behind the trachea and oesophagus. This condition may be explained by the persistence of the right dorsal aorta and the obliteration of the fourth aortic arch of the right side.

In birds the right fourth aortic arch is transformed into the arch of the aorta; in reptiles the fourth arches of both sides persist and give rise to the double aortic arch of these animals. In both these classes, development of the heart and aortic arches is probably along phylogenetic lines divergent from the mammalian pattern, and comparisons are inappropriate.

The heart originally lies ventral to the pharynx, immediately caudal to the stomatodeum (**2.**80A); with the elongation of the neck and the development of the lungs it recedes within the thorax, and, as a consequence, the vessels are drawn out, and the original position of the fourth and sixth aortic arches is greatly modified. Thus, on the right side the fourth aortic arch recedes to the thoracic inlet, while on the left side it descends into the thorax. The recurrent laryngeal nerves originally pass to the larynx caudal to the sixth pair of aortic arches, and are therefore affected by the descent of these structures, so that in the adult the left nerve hooks round the ligamentum arteriosum within the thorax; owing to the disappearance of the fifth and the dorsal part of the sixth aortic arch on the right side, the right recurrent laryngeal nerve hooks round the fourth aortic arch, i.e. the commencement of the right subclavian artery.

At first the aortae are the only longitudinal vessels present, for their branches all run at right angles to the long axis of the embryo. Later these transverse arteries become connected in certain situations by longitudinal anastomosing channels, which in part persist, forming such arteries as the internal thoracic, the inferior epigastric, the gastro-epiploic, etc. Each dorsal aorta gives off *segmental* branches to the digestive tube (*ventral splanchnic arteries*) and to the mesonephric ridge (*lateral splanchnic arteries*), and *intersegmental* branches to the body wall (*somatic arteries*).

The ventral splanchnic arteries are originally paired vessels which are distributed to the wall of the yolk sac, but after fusion of the dorsal aortae they appear as unpaired trunks and are distributed to the primitive digestive tube. Longitudinal anastomotic channels connect these branches along the dorsal and ventral aspects of the tube, forming dorsal and ventral splanchnic anastomoses (**2.**83) (Ennabli and Niveiro 1967). These longitudinal vessels obviate the need for so many 'subdiaphragmatic' ventral splanchnic arteries, and these are reduced to three— the coeliac trunk and superior and inferior mesenteric arteries. As the viscera supplied descend into the abdomen their origins migrate caudally, by differential growth; thus the origin of the

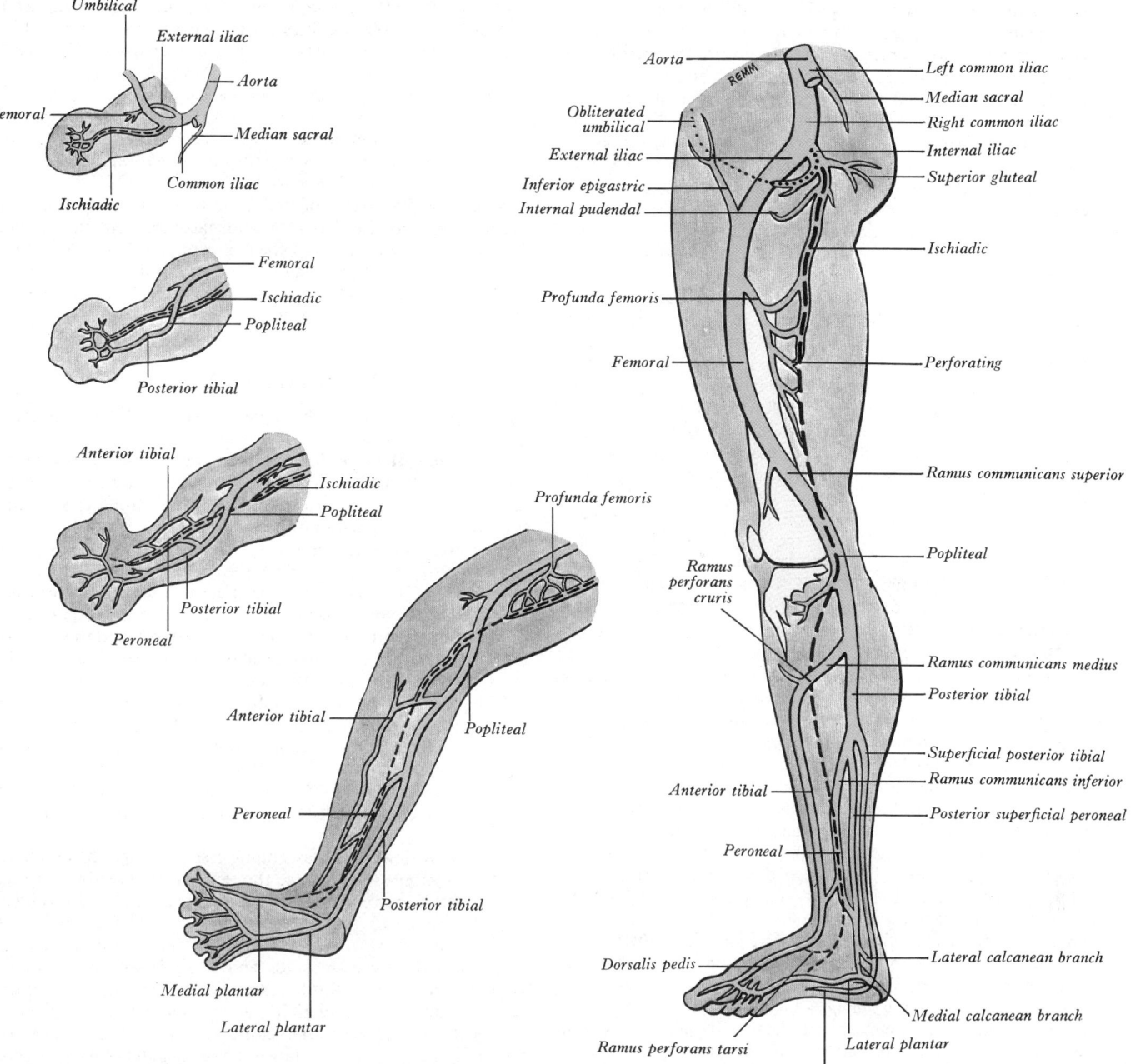

2.85 Stages in the development of the arteries of the leg. The original path of the axis is indicated by a dashed line. (After Senior.)

coelic artery is transferred from the level of the seventh cervical segment to the level of the twelfth thoracic, the superior mesenteric from the second thoracic to the first lumbar, and the inferior mesenteric from the twelfth thoracic to the third lumbar. (Above the diaphragm, however, a variable number of ventral splanchnic arteries persist, usually four or five, supplying the thoracic oesophagus.) The dorsal splanchnic anastomosis persists in the gastro-epiploic, pancreaticoduodenal, and the primary branches of the colic arteries, while the ventral splanchnic anastomosis forms the right and left gastric and the hepatic arteries. These arterial rearrangements have been investigated recently by angiography, and explanatory haemodynamic hypotheses have been advanced (Barth *et al.* 1975).

The lateral splanchnic arteries supply, on each side, the mesonephros, the testis or ovary, and the suprarenal gland; all these structures develop, in whole or in part, from the mesoderm of the mesonephric ridge (p. 213). One testicular or ovarian artery and three suprarenal arteries persist on each side. The phrenic artery branches from the most cranial suprarenal artery, and the renal artery arises from the most caudal. Additional renal arteries are frequently present and may be looked on as branches of persistent lateral splanchnic arteries.

The somatic arteries are intersegmental in position, and they persist, almost unchanged, in the thoracic and lumbar regions, as the posterior intercostal, subcostal and lumbar arteries. Each gives off a dorsal ramus which passes backwards in the intersegmental interval and divides into medial and lateral branches to supply the muscles and superficial tissues of the back (2.83). It also gives off a spinal branch, which enters the vertebral canal and divides into a series of branches to the walls of the osteoligamentous canal and neural branches to the spinal cord. Having produced its dorsal branch the intersegmental artery runs ventrally in the body wall, gives off a lateral branch and terminates in muscular and cutaneous rami. Before their division, the stems of the somatic arteries, at thoracic and lumbar levels, provide small rami which enter the developing vertebral bodies. For a study of these vessels in human fetuses consult Somogyi (1962).

Numerous longitudinal anastomoses link up the intersegmental arteries and their branches with one another (2.83). On both sides there develops a *postcostal anastomosis* which connects their dorsal branches in the intervals between the necks of the ribs and the vertebral transverse processes. This persists in the cervical region where it forms the greater part of the vertebral artery. A *post-*

transverse anastomosis also connects the dorsal branches and is responsible for the greater part of the deep cervical artery. A *precostal anastomosis* connects the intersegmental arteries beyond the origins of their dorsal branches. The ascending cervical and the superior intercostal arteries represent persistent portions of this vessel. Lastly, near the anterior median line the intersegmental arteries become linked up by a *ventral somatic anastomosis*. This bilateral vessel persists to a large extent in the adult and is represented by the internal thoracic, the superior epigastric and the inferior epigastric arteries.

The *umbilical arteries* at first appear to be the direct continuation of the primitive dorsal aortae and they are present in the body stalk before any vitelline (yolk sac) or visceral branches can be seen—an indication of the importance of the chorionic as compared with the vitelline circulation in the human embryo. After the fusion of the dorsal aortae the umbilical arteries arise from their ventrolateral aspect and pass medial to the primary excretory duct (Wolffian duct) on their way to the umbilicus. Later the proximal part of the umbilical artery is joined by a vessel which leaves the aorta at its termination and passes lateral to the primary excretory duct. This new vessel, which possibly may represent the fifth lumbar intersegmental artery, constitutes the dorsal root of the umbilical artery, the original stem being the ventral root. The dorsal root gives off the axial artery of the lower limb, branches to the pelvic viscera, and, at a more proximal point, the external iliac artery. The ventral root disappears entirely, and the umbilical artery now arises from that part of its dorsal root which lies distal to the external iliac artery, i.e. from the internal iliac artery.

The arteries of the limbs arise as a number of vessels contributing to a primitive capillary plexus, but in the upper limb bud, eventually only one trunk—the subclavian—persists, and it has the position and relations of the *seventh intersegmental artery* and probably represents its lateral branch. The main trunk to the upper limb (**2.84**), which later forms the *axillary* and *brachial arteries*, is continued into the forearm deep to the developing flexor muscle mass and terminates in a plexiform manner in the developing hand. This vessel ultimately persists as the *anterior interosseous artery* and the deep palmar arch. A branch from the main trunk passes dorsally between the developing radius and ulna and constitutes the *posterior interosseous artery*, while a second accompanies the median nerve into the hand, where it ends in a superficial capillary plexus. The *radial* and *ulnar arteries* are the latest arteries to appear in the forearm; at first the radial artery arises at a higher level than the ulnar and crosses in front of the median nerve, giving branches to the biceps. Later, the radial artery establishes a connexion with the main trunk at or near the site of origin of the ulnar artery and the upper portion of its original stem disappears to a large extent (see also p. 703). On reaching the hand the ulnar artery becomes linked up with the superficial palmar plexus, from which the superficial palmar arch is derived, while the median artery loses its distal connexions and becomes reduced to a very small vessel. The radial artery passes to the dorsal surface of the hand but, after giving off dorsal digital branches, it traverses the first intermetacarpal space and links up with the deep palmar arch.

The primary arterial trunk (**2.85**) or *axis artery* of the lower limb arises from the dorsal root of the umbilical artery, and courses along the *dorsal* surface of the thigh, knee and leg; below the knee it lies between the tibia and the popliteus and in the leg between the crural interosseous membrane and the tibialis posterior. It ends distally in a plantar network, and gives off a perforating artery which traverses the sinus tarsi and forms a dorsal network. The *femoral artery* passes along the ventral surface of the thigh, and opens up a new channel to the lower limb. It arises from a capillary plexus, connected proximally with the femoral branches of the external iliac artery and distally with the axis artery. At the proximal margin of the popliteus the axis artery gives off a *primitive posterior tibial* and a *primitive peroneal branch*, which run distally on the dorsal surface of that muscle and on the tibialis posterior to gain the sole of the foot. At the distal border of the popliteus the axis artery gives off a *perforating branch*, which passes ventrally between the tibia and the fibula and then runs downwards to the dorsum of the foot, forming the *anterior tibial*

artery and the *arteria dorsalis pedis*. The primitive peroneal artery establishes one communication with the axis artery at the distal border of the popliteus and another in its course in the leg (Senior 1919, 1920).

The femoral artery gradually increases in size, and coincidentally with this increase almost the whole of the axis artery disappears; proximal to its communication with the femoral the root of the axis artery, however, persists as the *inferior gluteal artery* and the *arteria comitans nervi ischiadici*.

The proximal parts of the primitive posterior tibial and peroneal arteries fuse, but their distal parts remain separate. Ultimately large portions of the axis artery and of the primitive peroneal artery disappear, although a part of the former vessel is incorporated in the permanent peroneal artery. The changes are shown in greater detail in **2.85**.

FURTHER DEVELOPMENT OF THE VEINS

The principal veins of the embryo may be divided into two groups, visceral and parietal.

The visceral veins include two *vitelline veins* bringing the blood from the yolk sac, and two *umbilical veins* returning the blood from the placenta; these four veins run through the septum transversum, and open into the sinus venosus.

The vitelline veins (**2.86**A–C) ascend, at first anterior to and subsequently one on each side of the digestive tube. A transverse anastomosis connects the two veins across the ventral surface of the tube, and beyond this they are connected to each other by two further anastomotic channels, one on the dorsal and the other on the ventral surface of the duodenal portion of the intestine, which is thus encircled by two venous rings forming a figure 8. The lengths of the vitelline veins within the septum transversum become surrounded by the trabeculae of the developing liver. The *liver sinusoids* develop as a plexus within the septum transversum and communicate freely with the vitelline and later the umbilical veins. The sections of the vitelline veins within the liver eventually lose their identity within the plexus of sinusoids.

The umbilical veins, ascending cranially from the umbilicus in the somatopleure, traverse the septum transversum on their way to the sinus venosus. After a time the right umbilical vein entirely disappears, but the left, which retains for a period its direct connexion with the left horn of the sinus venosus, pours its blood into the liver sinusoids and so communicates with the vitelline circulation (**2.86**B). In association with the establishment of the pulmonary circulation, venous channels develop which convey the blood from the left side of the liver to the right horn of the sinus venosus, where a large *common hepatic vein* now opens. The sinusoidal condition of the liver becomes condensed into a series of afferent vessels (*venae advehentes*) and a series of efferent vessels (*venae revehentes*) now leading to the common hepatic vein. In the process an oblique channel is differentiated from the liver sinusoids and conveys most of the blood brought by the umbilical vein direct to the common hepatic vein. This is the *ductus venosus* (**2.86**C) and it plays an essential part in the fetal circulation.

The superior mesenteric vein joins the left vitelline near the left extremity of the dorsal anastomosis. Later it is joined by the *splenic vein*, and the *portal vein* is thus established. The vitelline veins partly disappear (**2.86**C) and the portal vein is continued through the dorsal anastomosis and the cranial part of the right vitelline vein to the liver. The intrahepatic part of the right vitelline vein becomes the right branch of the portal vein, while the cranial ventral anastomosis and the intrahepatic part of the left vitelline vein form its left branch. The left umbilical vein joins the left branch of the portal vein, to which the ductus venosus is also connected. Some of the blood conveyed to the liver by the umbilical vein passes through the left venae advehentes, but the great majority of it finds its way through the ductus venosus to the *common hepatic vein* (which later forms the upper end of the inferior vena cava) and so to the right horn of the sinus venosus.

An alternative analysis (Dickson 1957) has shown that certain of the foregoing details require some modification. The ductus venosus is primarily a *median* channel linking the most proximal of the anastomoses, the *subhepatic anastomosis* between the vitelline veins, with a further anastomosis which connects these vessels in the cranial part of the liver, termed the *subdiaphragmatic anastomosis*. The left half of the subhepatic anastomosis is later incorporated in the left umbilical vein whilst the subdiaphragmatic anastomosis contributes to the formation of the left hepatic veins and the common hepatic vein, prior to the appearance of the prerenal part of the inferior vena cava (p. 195).

The parietal or somatic veins (2.87, 88) are at first represented by two large vessels on each side, the *precardinal* and *postcardinal* veins; the former drain the rostral part of the embryo, and the latter the caudal part. The two veins on each side unite to form a short vessel, the *common cardinal vein*, which passes ventrally, lateral to the pleuropericardial canal (p. 209), to open into the corresponding horn of the sinus venosus (2.80B).

Owing to the rapid development of the head and brain the precardinal veins become enlarged. They are further augmented by the *subclavian* veins from the upper limb buds, and so come to be the chief tributaries of the common cardinal veins; these gradually assume an almost vertical position in association with the descent of the heart into the thorax. The right and left common cardinal veins are originally of the same diameter. By the development of a transverse connexion, the *left brachiocephalic vein*, between the two precardinal veins, the blood is carried across from the left to the right precardinal (2.88). The part of the right precardinal vein between the left brachiocephalic and azygos veins forms the upper part of the superior vena cava; the caudal part of the latter vessel (below the entrance of the azygos vein) is formed by the right common cardinal. Caudal to the branching of the left brachiocephalic vein the left precardinal and left common cardinal veins atrophy, the former constituting the terminal part of the left superior intercostal vein, while the latter is represented by the ligament of the left vena cava and the oblique vein of the left atrium (2.88). The remainder of the left superior intercostal vein is developed from the cranial end of the postcardinal vein and drains the second, third and, frequently, the fourth intercostal veins. The oblique vein passes downwards across the back of the left atrium to open into the coronary sinus, which, as already indicated, represents the persistent left horn of the sinus venosus. Right and left superior venae cavae are present in some animals, and occasionally persist in the human body.

The inferior vena cava (2.87, 88) of the adult is a composite vessel, and the precise mode of development of its postrenal segment (caudal to the renal vein) is still somewhat uncertain. Its function is at first carried out by the right and left *postcardinal* veins, which receive the venous drainage of the lower limb buds and the pelvis and run in the dorsal part of the mesonephric ridges, receiving tributaries from the body wall (intersegmental veins) and from the mesonephroi.

A second pair of longitudinal veins, termed the *subcardinal veins*, form in the ventromedial part of the mesonephric ridges and become connected to the postcardinal veins by a number of vessels which traverse the medial part of the ridges. The subcardinal veins intercommunicate by a *pre-aortic anastomosis*, which later constitutes the part of the *left renal vein* crossing in front of the abdominal aorta.

The formation of an oblique transverse anastomosis between the iliac veins—which itself becomes the major part of the definitive *left common iliac vein*—diverts an increasing volume of blood into the right longitudinal veins, accounting for the ultimate disappearance of those on the left.

At its cranial end the subcardinal vein receives the *suprarenal vein* on each side, but on the right side it comes into intimate relationship with the liver. An extension of the vessel takes place in a cranial direction and meets and establishes continuity with a corresponding new formation which is growing caudally from the common hepatic vein. In this way on the right side a more direct route is established to the heart, and the *prerenal* (cranial) segment of the inferior vena cava is defined.

The enlargement of the metanephros diverts the postcardinal vein from its course, and the venous drainage of the mesonephric ridge is taken over by the subcardinal vein. At the same time new

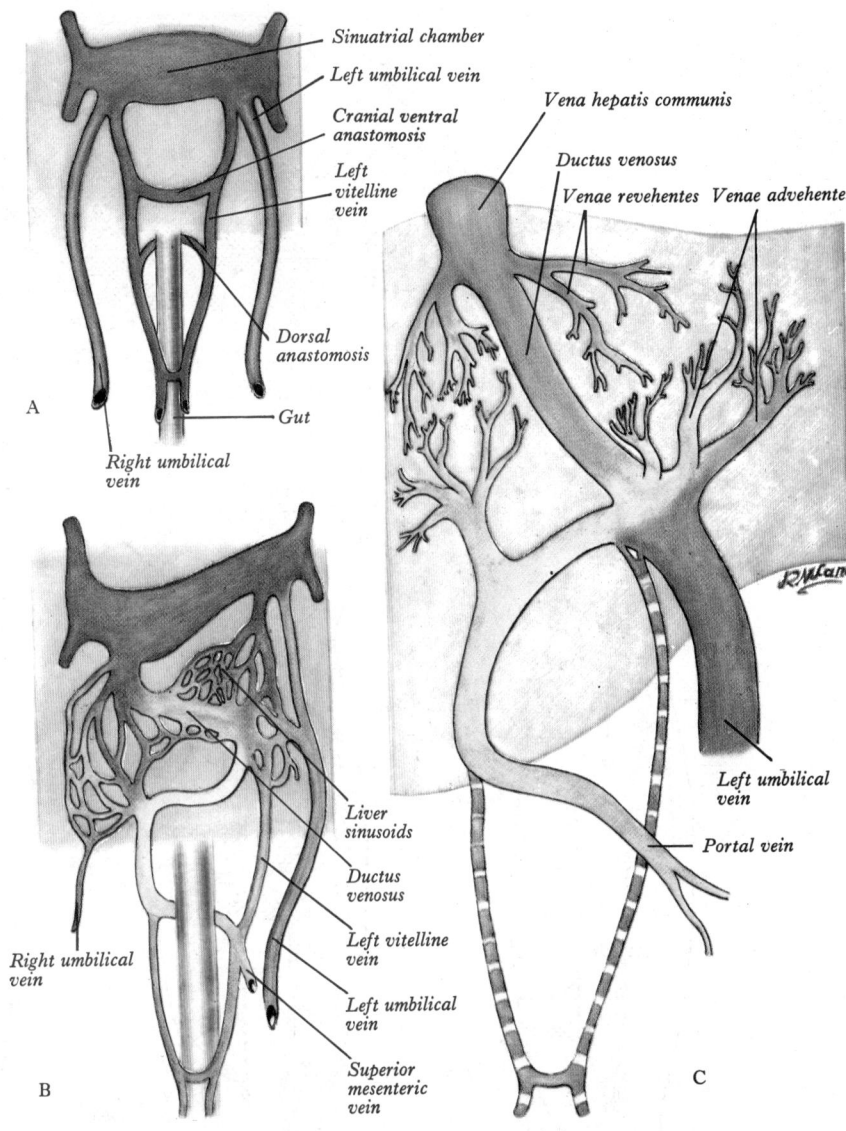

Sinuatrial chamber
Left umbilical vein
Cranial ventral anastomosis
Left vitelline vein
Dorsal anastomosis
Gut
Right umbilical vein

Vena hepatis communis
Ductus venosus
Venae revehentes Venae advehentes

Left umbilical vein
Portal vein

A

Liver sinusoids
Ductus venosus
Left vitelline vein
Left umbilical vein
Superior mesenteric vein
Right umbilical vein

B

C

R. McLane

2.86A–C Three stages in the development of the veins of the liver. Note, however, the results of a more detailed analysis mentioned in the text and in Dickson (1957), which differs in some respects from the views on which this illustration is based.

A Pink—umbilical veins; mauve—vitelline veins.

B Pink—umbilical veins and ductus venosus; mauve— vitelline veins and sinusoids of liver; blue—portions of vitelline veins later incorporated in the portal vein.

C Blue—portal vein; interrupted mauve—portions of vitelline veins which disappear completely; pink—left umbilical vein and ductus venosus.

longitudinal channels appear and take over intersegmental venous drainage and the whole of the postcardinal vein disappears with the exception of its extreme cranial and caudal ends. There are four such channels on each side, but, so far as is known at the present, only two of them persist as large vessels in the adult: (1) A bilateral longitudinal channel forms dorsolateral to the aorta and *lateral* to the sympathetic trunk and takes over the intersegmental venous drainage from the posterior cardinal vein. This is the *supracardinal* or *thoracolumbar 'complex'*. (2) A second channel is formed on each side, also dorsolateral to the aorta but *medial* to the sympathetic trunk. This is the *azygos line vein* and, gradually it takes over the intersegmental venous drainage from the thoracolumbar line. The intersegmental veins now reach their longitudinal channel by passing *medial* to the sympathetic trunk, the relationship which the lumbar and intercostal veins exhibit thenceforward. Cranially the azygos lines join the persistent cranial parts of the posterior cardinal veins. (3) Two *subcentral veins* are laid down directly dorsal to the aorta in the interval

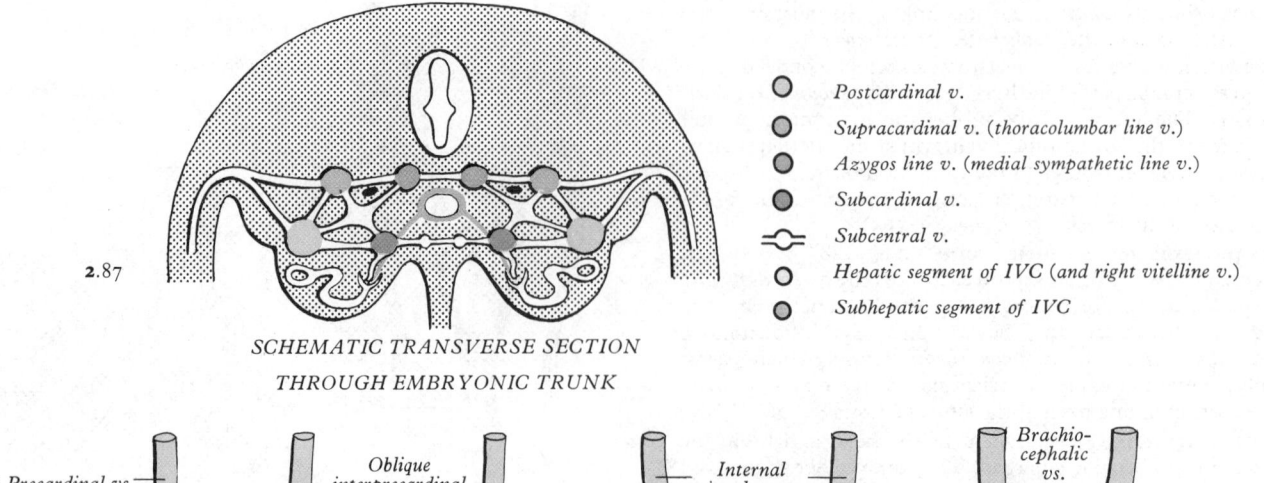

2.87

Postcardinal v.

Supracardinal v. (thoracolumbar line v.)

Azygos line v. (medial sympathetic line v.)

Subcardinal v.

Subcentral v.

Hepatic segment of IVC (and right vitelline v.)

Subhepatic segment of IVC

SCHEMATIC TRANSVERSE SECTION
THROUGH EMBRYONIC TRUNK

2.88

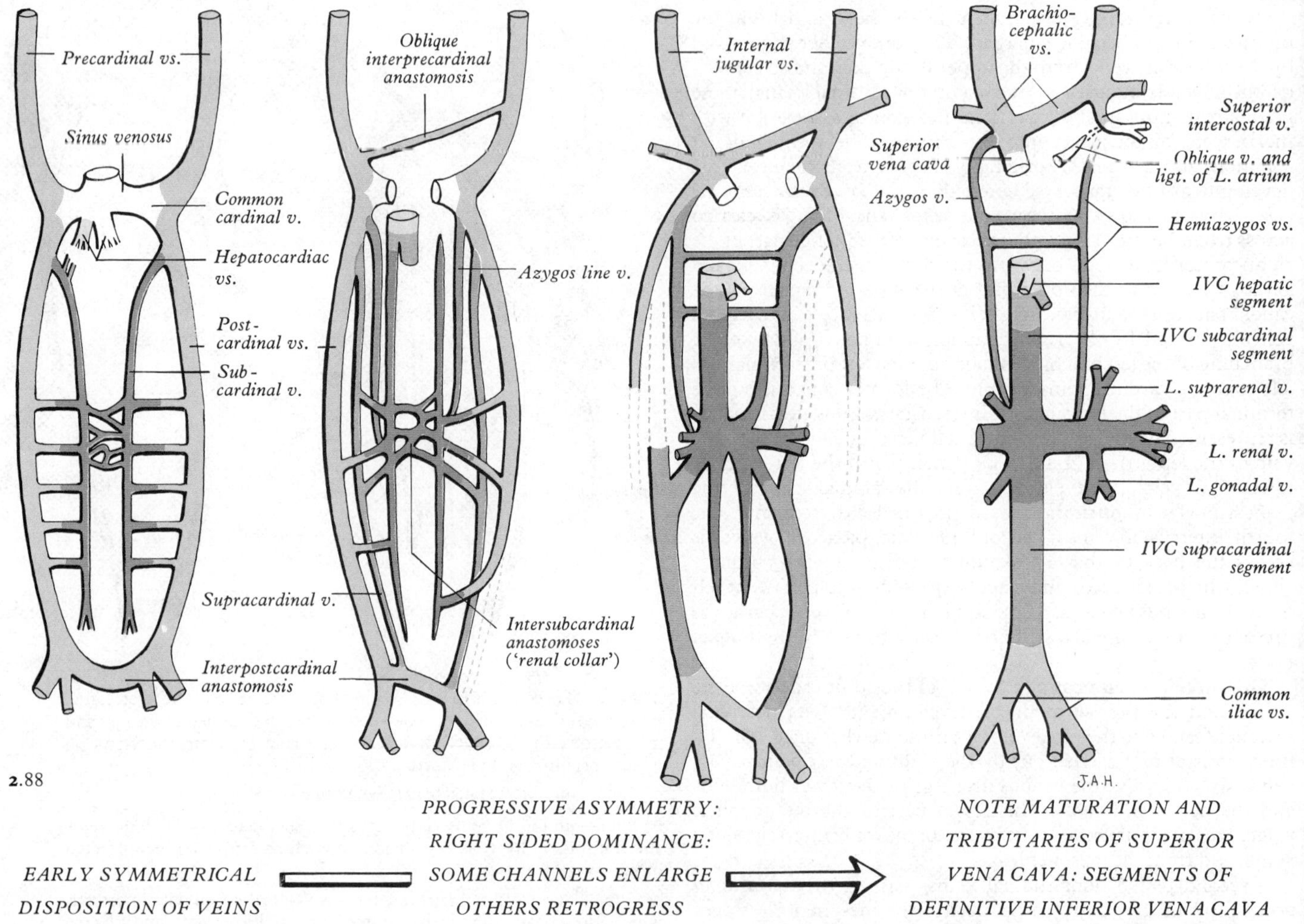

EARLY SYMMETRICAL
DISPOSITION OF VEINS

PROGRESSIVE ASYMMETRY:
RIGHT SIDED DOMINANCE:
SOME CHANNELS ENLARGE
OTHERS RETROGRESS

NOTE MATURATION AND
TRIBUTARIES OF SUPERIOR
VENA CAVA: SEGMENTS OF
DEFINITIVE INFERIOR VENA CAVA

2.87 Somatic venous development: schematic section through the embryonic trunk. Principal longitudinal veins are colour-coded. Interconnexions and intersegmental veins remain uncoloured. (Modified from *Basic Human Embryology*, by Williams, Wendell-Smith and Treadgold, 1969.)

2.88 Plan of development of principal somatic veins from the early symmetrical state, through states of increasing asymmetry, to the definitive arrangement. (Modified from *Basic Human Embryology*, by Williams, Wendell-Smith and Treadgold, 1969.)

between the origins of the paired intersegmental arteries. These veins communicate freely with each other and with the azygos line veins, and these connexions ultimately form the retro-aortic parts of the left lumbar veins and of the hemiazygos veins. The thoracolumbar lines or *supracardinal veins*, are laid down lateral to the aorta and sympathetic trunks, which therefore intervene between them and the azygos lines. These veins communicate caudally with the iliac veins and cranially with the subcardinal veins in the neighbourhood of the pre-aortic intersubcardinal anastomosis. In addition, the supracardinal veins communicate freely with each other through the medium of the azygos lines and the subcentral veins. The most cranial of these connexions,

together with the supracardinal-subcardinal and the intersubcardinal anastomoses, complete a venous ring around the aorta below the origin of the superior mesenteric artery, termed the *'renal collar'* (Huntington 1920).

The right supracardinal vein persists and forms the greater part of the postrenal segment of the inferior vena cava, the continuity of the vessel being maintained by the persistence of the anastomosis between the right supracardinal and the right subcardinal in the 'renal collar'. The left supracardinal disappears, but the portion of the 'renal collar' formed by the left supracardinal-subcardinal anastomosis in part persists in the left renal vein. It must be added that much confusion and disagreement exists with regard to the complicated array of longitudinal veins described above.

The *inferior vena cava* (**2**.88) is therefore formed, from below upwards, by (1) part of the right supracardinal vein, (2) an anastomosis between the right supracardinal and subcardinal veins, (3) the right subcardinal vein, (4) a new formation which connects the right subcardinal and the common hepatic veins, and (5) the common hepatic vein (Gladstone 1929). It should be noted that only the supracardinal part of the inferior vena cava receives the intersegmental venous drainage, and that the postrenal segment of the inferior vena cava is on a plane which lies dorsal to the plane of the prerenal segment. On this account the right phrenic, suprarenal and renal arteries, which represent persistent mesonephric arteries, pass behind the inferior vena cava, while the testicular or ovarian, which has a similar developmental origin, passes anterior to it.

In some animals the right postcardinal vein constitutes a large part of the postrenal segment of the inferior vena cava. In these cases the right ureter, on leaving the kidney, passes medially dorsal to the vessel and then, curving round its medial side, crosses its ventral aspect. Rarely, a similar condition is found in the human subject, and indicates the persistence of the right postcardinal vein and failure of the right supracardinal to play its normal part in the development of the vessel.

The ultimate arrangement of the embryonic abdominal and thoracic longitudinal somatic veins may be summarized as follows:

(1) The terminal part of the postcardinal vein on the left side forms the distal part of the left superior intercostal vein; on the right side its cranial part persists as the terminal portion of the vena azygos.

(2) The caudal part of the subcardinal vein is partly incorporated in the testicular or ovarian vein (McClure and Butler 1925) and partly disappears. The cranial end of the right subcardinal vein is incorporated into the inferior vena cava and also forms the right suprarenal vein. The left subcardinal vein, cranial to the intersubcardinal anastomosis, is incorporated into the left suprarenal vein. The renal and testicular or ovarian veins on both sides join the supracardinal-subcardinal anastomosis. On the left side this is connected directly to the part of the inferior vena cava which is of subcardinal status through an inter-subcardinal anastomosis.

(3) The right supracardinal vein forms the postrenal segment of the inferior vena cava. The left supracardinal vein disappears entirely.

(4) The right azygos line persists in its thoracic part to form all but the terminal part of the vena azygos. Its lumbar part can usually be identified as a small vessel which leaves the vena azygos on the body of the twelfth thoracic vertebra and descends on the vertebral column, deep to the right crus of the diaphragm, to join the posterior aspect of the inferior vena cava at the upper end of its postrenal segment. The left azygos line forms the hemiazygos veins.

(5) The subcentral veins give rise to the retro-aortic parts of the left lumbar veins and of the hemiazygos veins.

(6) The veins described by some observers as the thoracolumbar lines, and probably identical with the supracardinal veins, are, in any case, of uncertain fate in later development.

The veins of the head have a complicated developmental history (**2**.89A, B; and consult—Markoski 1911; Streeter 1918; Padgett 1957).

The primary vessels consist of a close-meshed capillary plexus drained on each side by the precardinal vein, which is at first continuous cranially with a transitory channel, the *primordial hindbrain channel*, lying on the neural tube medial to the cranial nerve roots. This is soon replaced by the *primary head vein* which runs caudally from the medial side of the trigeminal ganglion, lateral to the facial and vestibulocochlear nerves and the otocyst and then medial to the vagus nerve, to become continuous with the precardinal vein. A lateral anastomosis subsequently brings it lateral to the vagus nerve. The cranial part of the precardinal vein forms the internal jugular vein; its caudal moiety has already been described on p. 193.

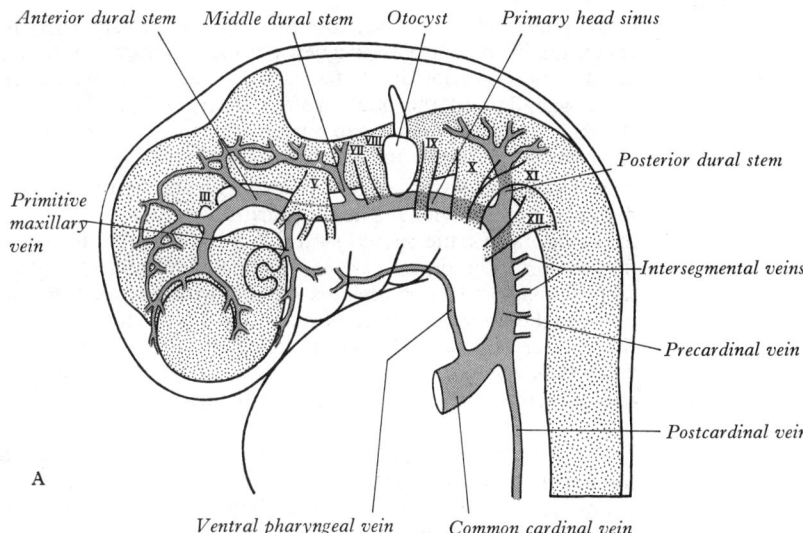

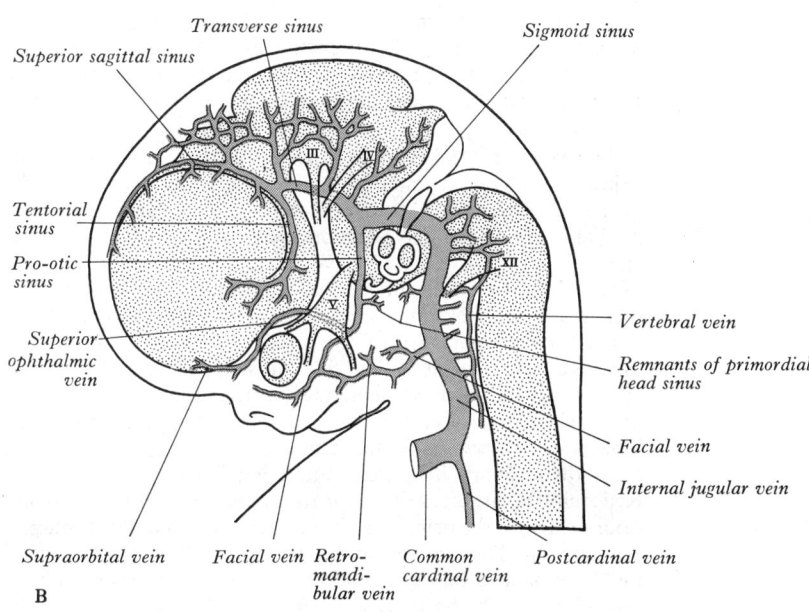

2.89A and B   Diagrams illustrating successive stages in the development of the veins of the head and neck, (A) at approximately 8 mm and (B) at approximately 24 mm C.R. length.

The capillary plexus of the head becomes separated into three more or less distinct strata by the differentiation of the skull and meninges. The superficial vessels, draining the integument and underlying soft parts, eventually drain in large part into the external jugular system. They retain some connexions with the deeper veins through the so-called emissary veins. Deep to this is the venous plexus of the dura mater, from which the dural venous sinuses differentiate. This plexus is drained on each side by *anterior, middle* and *posterior dural stems*. The anterior stem drains the prosencephalon and mesencephalon and enters the primary head vein in front of the trigeminal ganglion. The middle stem drains the metencephalon and empties into the primary head vein behind the trigeminal ganglion, while the posterior stem drains the myelencephalon into the commencement of the precardinal vein (**2**.89). The deepest capillary stratum is the plexus of the pial vessels from which the veins of the brain differentiate; it drains at the dorsolateral aspect of the neural tube into the adjacent dural venous plexus. In addition the primary head vein also receives, at its cranial end, the *primitive maxillary vein*, draining the maxillary process and region of the optic vesicle.

The vessels of the dural plexus undergo profound changes, largely resulting from the growth of the cartilaginous capsule of the labyrinth and expansion of the cerebral hemispheres. Owing to the growth of the cartilaginous otic capsule the primary head vein is gradually reduced and a new channel joining the anterior, middle and posterior dural stem appears dorsal to the cranial nerve ganglia and the otic capsule (Butler 1967). This new channel joining the middle and posterior stems together with the posterior dural stem forms the adult *sigmoid sinus*.

Between the growing cerebral hemispheres and along the dorsal margins of the anterior and middle plexuses there forms a curtain of capillary veins, the *sagittal plexus*, in the position of the future falx cerebri. The dorsal part of this plexus forms the *superior sagittal sinus* and is continuous behind with the anastomosis between the anterior and middle dural stems which forms most of the *transverse sinus*. The ventral part of the sagittal plexus differentiates into the *inferior sagittal* and *straight sinuses* and the *great cerebral vein*, and drains eventually into the left transverse sinus.

The vessels along the ventrolateral edge of the developing cerebral hemisphere form the transitory *tentorial sinus*, which drains the convex surface of the cerebral hemisphere and basal ganglia and the ventral aspect of the diencephalon to the transverse sinus. With expansion of the cerebral hemispheres, and in particular the emergence of the temporal lobe, the tentorial sinus becomes elongated, attenuated and eventually disappears, and its territory of supply is drained by enlarging anastomoses of pial vessels which become the *basal veins* and drain into the great cerebral vein.

The anterior dural stem disappears and the caudal part of the primary head vein dwindles; it is represented in the adult by the *inferior petrosal sinus*. The cranial part of the primary head vein, medial to the trigeminal ganglion, persists and still receives the stem of the primitive maxillary vein. The latter has now lost most of its tributaries to the anterior facial vein, its stem becoming the main trunk of the *primitive supra-orbital vein*, which will form the *superior ophthalmic vein* of the adult. Thus the main venous drainage of the orbit and its contents is now carried via the augmented middle dural stem, now termed the *pro-otic sinus*, into the transverse sinus and at a later stage into the cavernous sinus. The *cavernous sinus* is formed from a secondary plexus, derived from the primary head vein and lying between the otic and basioccipital cartilages. This forms the *inferior petrosal sinus* and drains through the primordial hindbrain channel into the internal jugular vein. The *superior petrosal sinus* arises later from a ventral metencephalic tributary of the pro-otic sinus; it communicates secondarily with the cavernous sinus (Butler 1957, 1967). The pro-otic sinus meanwhile has developed a new and more caudally situated stem draining into the sigmoid sinus; this new stem is the *petrosquamosal sinus* (p. 746) and the pro-otic sinus becomes, with progressive ossification of the skull, diploic in position. The development of the venous drainage and portal system of the hypophysis cerebri is closely associated with that of the venous sinuses (Wislocki 1937; Niemineva 1950).

The venous drainage of the face, scalp and neck becomes established after the development of the skull. The first identifiable vessel is the *ventral pharyngeal vein*, draining the massive mandibular and hyoid arches into the common cardinal vein. With the elongation of the neck its termination is transferred to the cranial part of the precardinal vein which later becomes the internal jugular. The ventral pharyngeal vein, receiving tributaries from the face and tongue, becomes the *linguofacial* vein. With development of the facial region the primitive maxillary vein extends its drainage into the territories of supply of the ophthalmic and mandibular division of the fifth nerve, including the pterygoid and temporal muscles. Over the lower jaw it anastomoses with the linguofacial vein. This anastomosis becomes the *facial vein*; it receives a strong tributary from the temporal region, the *retromandibular* vein, and drains through the linguofacial vein into the internal jugular. The stem of the linguofacial vein is now identifiable as the lower part of the facial vein, whilst the dwindling connexion of the facial with the primitive maxillary vein becomes the deep facial vein. The *external jugular vein* is developed from a tributary of the cephalic vein from the tissues of the neck and anastomoses secondarily with the anterior facial vein. At this stage the *cephalic vein* forms a *venous ring* around the clavicle from which it is connected with

the caudal part of the precardinal. The deep segment of the venous ring forms the *subclavian vein* and receives the definitive external jugular vein. The superficial segment of the venous ring dwindles, but may persist in adult life (Padgett 1957).

THE LYMPHATIC SYSTEM

Two different views are current as to the initial stages in the development of the lymphatic system (Rusznyák *et al.* 1960). According to the first view (Huntington 1908; McClure and Bulter 1925) lymphatic spaces commence as clefts in the mesenchyme, and their lining cells take on the characters of endothelium (Kampmeier 1969). These spaces form capillary plexuses from which certain *lymph sacs*, to be noted later, are derived. The connexions of the lymphatic and venous systems are regarded as entirely secondary. In contrast, however, according to Sabin (1902), the earliest lymph vessels arise as capillary offshoots from the endothelium of the veins, which form capillary plexuses. These plexuses lose their connexions with the venous system and become confluent to form lymph sacs. The balance of the evidence suggests that all but the earliest channels of the lymphatic system originate independently of the venous system and only acquire connexions with it at a later stage (Kampmeier 1969).

In the human embryo the lymph sacs from which the lymph vessels are derived are six in number; two paired (the *jugular* and the *posterior lymph sacs*) and two unpaired (the *retroperitoneal* and the *cisterna chyli*). In lower mammals an additional pair (the *subclavian*) is present, but in the human embryo these are merely extensions of the jugular sacs.

The position of the sacs is as follows (**2**.90): (1) the *jugular*, the first to appear, at the junction of the subclavian vein with the precardinal; (2) the *posterior*, at the junction of the iliac vein with the postcardinal; (3) the *retroperitoneal*, in the root of the mesentery near the suprarenal glands; (4) the *cisterna chyli*, opposite the third and fourth lumbar vertebrae. From the lymph sacs the lymph vessels bud out along lines corresponding more or less closely with the course of embryonic blood vessels, but many arise *de novo* in the mesenchyme and establish connexions with existing vessels. In the body wall and in the wall of the intestine (Heuer 1909), the deeper plexuses are the first to be developed; by continued growth of these the vessels in the superficial layers are gradually formed. The thoracic duct is, phylogenetically, a bilateral structure. In man it comprises the caudal part of the right vessel, a transverse anastomosis and the cranial part of the left vessel. According to the second view cited above it is formed from anastomosing outgrowths from the jugular sac and cisterna chyli. At its connexion with the cisterna chyli it is at first double, but the

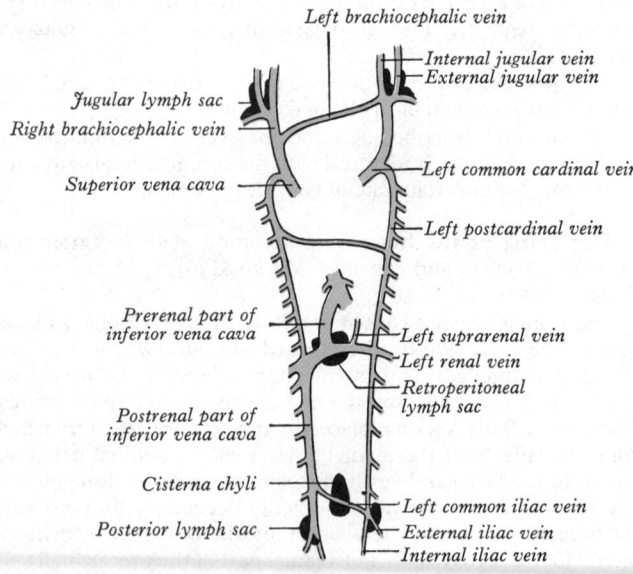

2.90 A scheme showing the relative positions of the primary lymph sacs. (After F. R. Sabin.)

vessels soon join. Numerous valves are laid down in the duct during the fifth month, but many of them disappear prior to birth. Those which persist are formed in situations where the duct may be subjected to pressure, e.g. where it is crossed by the oesophagus and the aortic arch.

All the lymph sacs except the cisterna chyli are, at a later stage, divided up by a number of slender connective tissue bridges. Later they are invaded by lymphocytes and transformed into groups of lymph nodes, the lymph sinuses representing portions of the original cavity of the sac. The lower portion of the cisterna chyli is similarly converted, but its upper portion remains as the adult cisterna; in many cases the cisterna chyli is plexiform (p. 785). The siting of the major groups of lymph nodes follows a similar basic pattern amongst the mammals (Spira 1962).

Haemal lymph nodes are said to develop as mesenchymal condensations in close relation to blood vessels rather than lymphatics (Meyer 1917).

The Development of the Alimentary and Respiratory Apparatus

THE BUCCAL CAVITY

The mouth is developed partly from the stomatodeum, and partly from the floor of the cranial portion of the foregut. By the growth of the head end of the embryo and the formation of the head fold, the pericardial area and the oropharyngeal membrane come to lie on the ventral surface of the embryo (p. 120). With the further expansion of the brain, and the bulging of the pericardium, the membrane is soon depressed between the two prominences. This depression constitutes the *stomatodeum* or *primitive mouth* (2.33). It is lined with ectoderm, and is separated from the cranial end of the foregut by the oropharyngeal membrane, which is formed by the apposition of the stomatodeal ectoderm with the foregut endoderm; at the end of the fourth week the membrane disappears and a communication is established between the stomatodeum and the cranial end of the foregut or future pharynx. No vestige of the membrane is found in the adult, and the communication just mentioned must not be confused with the permanent oropharyngeal isthmus. The epithelium of the lips and gums and the enamel of the teeth are ectodermal in origin, being formed from the walls of the stomatodeum, but the epithelium of the tongue, which is developed in the floor of the mouth and pharynx, is derived from the endoderm. The development of the teeth and gums is described on pp. 1293–1300.

The branchial arches grow in a ventral direction and come to lie

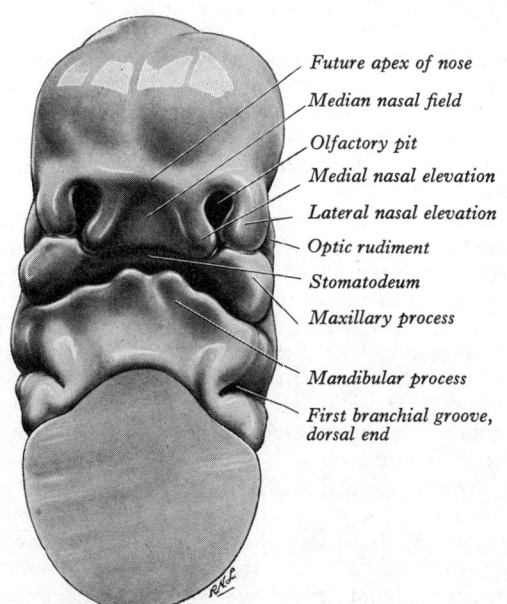

Future apex of nose

Median nasal field

Olfactory pit

Medial nasal elevation

Lateral nasal elevation

Optic rudiment

Stomatodeum

Maxillary process

Mandibular process

First branchial groove, dorsal end

2.91 The head of a human embryo in the sixth week. Ventral aspect. (From a model by K. Peter.)

between the stomatodeum and the pericardium; with the completion of the mandibular arch and the development of the maxillary processes (p. 148), the opening of the stomatodeum assumes a pentagonal form, bounded cranially by the frontonasal process, caudally by the mandibular arches, and laterally by the maxillary processes (2.91). With the inward growth and fusion of the palatine processes (2.56B, C), the stomatodeum is divided into a nasal and a buccal part. Along the free margins of the processes bounding the mouth cavity appears a shallow groove, and the ectoderm in the floor of this groove thickens and invades the underlying mesoderm; it divides into a medial *dental lamina* and a lateral *vestibular lamina*. The central cells of the latter degenerate and break down so that the furrow becomes deepened. It may now be termed the *labiogingival groove* or *sulcus*; its inner wall contributes to the formation of the alveolar processes of the maxillae and the mandible, while its outer wall forms the lips and cheeks.

The salivary glands arise from the epithelial lining of the mouth. The *parotid gland* can be recognized in human embryos 8 mm long as an elongated furrow running dorsally from the angle of the mouth between the mandibular arch and the maxillary process. The groove, which is converted into a tube, loses its connexion with the epithelium of the mouth, except at its ventral end, and grows dorsally into the substance of the cheek. The tube persists as the *parotid duct* and its blind end proliferates to form the gland. Subsequently the size of the oral fissure is reduced by partial fusion between the maxillary and mandibular processes and the duct opens thereafter on the inside of the cheek at some distance from the angle of the mouth. The *submandibular gland* is identifiable in human embryos 13 mm long as an epithelial outgrowth from the floor of the *linguogingival groove*. It increases rapidly in size by giving off numerous branching processes which later acquire lumina. At first the connexion of the submandibular outgrowth with the floor of the mouth lies at the side of the tongue, but the edges of the groove in which it opens come together, from behind forwards, and form the tubular part of the *submandibular duct*. As a result the orifice of the duct is shifted forwards till it is below the tip of the tongue, close to the median plane. The *sublingual gland* arises in embryos about 20 mm long as a number of small epithelial thickenings in the linguogingival groove and on the lateral side of the groove, which later closes to form the submandibular duct.

The tongue appears as a small median elevation, named the *median tongue bud* (*tuberculum impar*), in the endodermal floor of the pharynx, before the branchial arches meet ventrally; it subsequently becomes incorporated in the anterior part of the tongue. A little later two oval *distal tongue buds* (*lingual swellings*) appear on the endodermal aspect of the mandibular processes. They meet each other in front, and caudally they converge on the median tongue bud, with which they fuse (2.92A, B). A sulcus forms along the ventral and lateral margins of this elevation and deepens to form the linguogingival groove, while the elevation constitutes the anterior or buccal (presulcal) part of the tongue.

Caudal to the median tongue bud, a second median elevation, termed the *hypobranchial eminence* (copula of His), forms in the floor of the pharynx, and the ventral ends of the fourth, the third and, later, the second visceral arches converge into it. A transverse groove separates its caudal part to form the epiglottis, while ventrally it approaches the tongue rudiment, spreading ventrally in the form of a V, and forming the posterior or pharyngeal part of the tongue. In the process the third arch elements grow over and bury the elements of the second arch, excluding it from the tongue. As a result the mucous membrane of the pharyngeal part of the tongue receives its sensory supply from the glossopharyngeal, the nerve of the third arch. In the adult the union of the anterior and posterior parts of the tongue is marked by the angulated *sulcus terminalis*, its apex at the *foramen caecum*, a blind depression produced at the time of fusion of the constituent parts of the tongue, but also marking the site of the ingrowth of the median rudiment of the thyroid gland.

At first the tongue consists of a mass of mesoderm covered on its surface by endoderm. During the second month occipital myotomes have often been stated to migrate from the lateral aspects of the myelencephalon, invading the tongue to form its

197

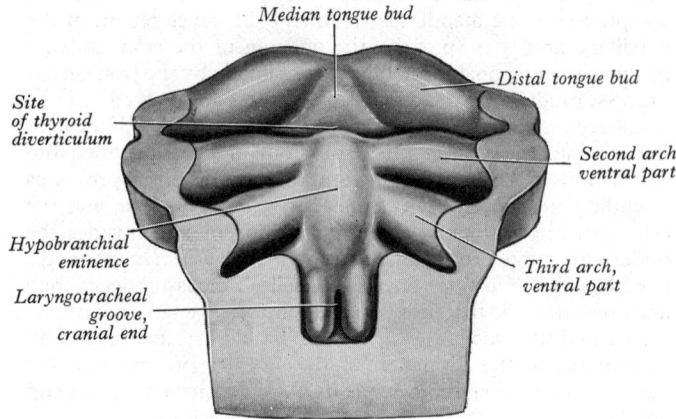

2.92A The floor of the pharynx of a human embryo at the beginning of the sixth week. (From a model by K. Peter.)

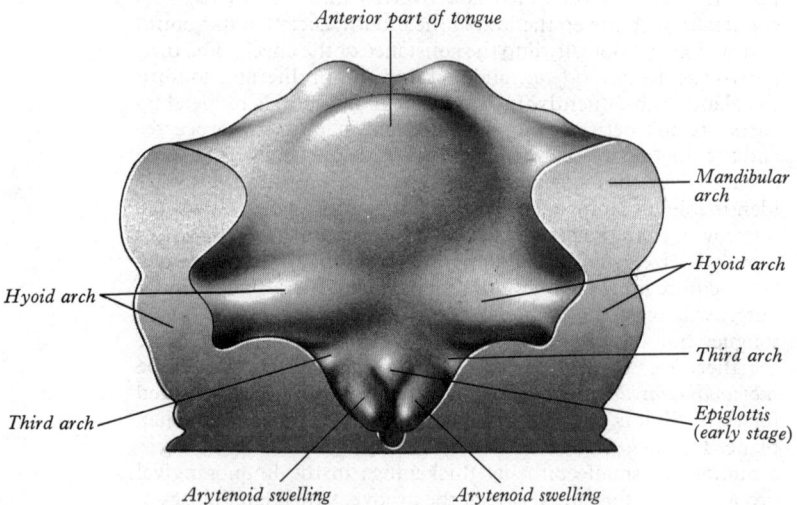

2.92B The floor of the pharynx of a human embryo, about six weeks old. (From a model by K. Peter.)

musculature. They are held to pass ventrally round the pharynx to reach its floor accompanied, necessarily, by their nerve (the hypoglossal), which therefore crosses superficial to both the internal and external carotid arteries; however, *see* p. 154.

The composite character of the tongue is indicated by its innervation. Impulses from the anterior, buccal part are mediated by (*a*) the lingual nerve, derived from the post-trematic nerve of the first arch (mandibular nerve), (*b*) the chorda tympani, the pretrematic nerve to the first arch. The posterior, pharyngeal part of the tongue is innervated by the glossopharyngeal, the nerve of the third arch. The muscles of the tongue being 'myotomic' in origin, receive their nerve supply from the hypoglossal nerve, which is serially homologous with spinal ventral nerve roots.

The sulcus terminalis cannot be distinguished earlier than the 52 mm stage according to some observers (Vij and Kanagasuntheram 1972). The vallate papillae appear at about the same time, increasing in number until the 170 mm stage. Serial reconstructions also suggest that the territory of the glossopharyngeal nerve extends considerably beyond these papillae (Watanbe *et al.* 1970).

The thyroid gland is first identifiable in embryos of about 20 somites, as a median thickening of endoderm in the floor of the pharynx between the first and second pharyngeal pouches and immediately dorsal to the aortic sac (Davis 1923). This area is later invaginated to form a median diverticulum, which appears in the latter half of the fourth week in the furrow immediately caudal to the tuberculum impar (2.92A). It grows caudally as a tubular duct which bifurcates and subsequently divides into a series of double cellular plates, from which the isthmus and the lateral lobes of the thyroid gland are developed. The *primary thyroid*

follicles differentiate by reorganization and proliferation of the cells of these plates. *Secondary follicles* subsequently arise by budding and subdivision (Norris 1916). The claim that the fourth pharyngeal pouches contribute thyroid tissue to the lateral lobes of the gland was long disputed and seemed perhaps unlikely on the grounds of comparative embryology. (But see also below—the *ultimobranchial body*.)

The connexion of the median diverticulum with the pharynx is termed the *thyroglossal duct*. The site of its connexion with the epithelial floor of the mouth is marked by the foramen caecum. From here it extends caudally in the middle line and ventral to the primordium of the hyoid bone, behind which it may later form a recurrent loop. The distal part of the duct usually differentiates to a variable extent to form the pyramidal lobe of the thyroid. The remainder fragments and disappears, but the lingual part is often identifiable until late in fetal life and may branch and give rise to miniature salivary glands (Boyd 1963). Occasionally parts of the midline thyroglossal duct persist (occurring in lingual, suprahyoid retrohyoid, or infrahyoid positions). They may give rise to the formation of aberrant masses of thyroid tissue, cysts, fistulae or sinuses, usually in the midline (*see* p. 1449). A *lingual thyroid* situated at the junction of the buccal and pharyngeal parts of the tongue is not uncommon, but nodules of glandular tissue may also be found other than in the midline, e.g. laterally placed posterior to sternocleidomastoid, and even below the level of the thyroid isthmus (*see* p. 1449).

The tonsils are developed from the parts of the second pharyngeal pouches which lie between the tongue and the soft palate. The endoderm lining these pouches grows into the surrounding mesenchyme in the form of a number of solid buds. These buds are excavated by the degeneration and shedding of their central cells, and by this means the tonsillar *fossulae* and *crypts* are formed. Lymphoid cells accumulate around the crypts, and become grouped as lymphoid follicles. A slit-like *intra-tonsillar cleft* (8.82 extends into the upper part of the tonsil and is a remnant of the second pharyngeal pouch.

The thymus is derived from the endoderm of the ventral part of the third pharyngeal pouch on each side (2.93). It cannot be recognized prior to the differentiation of the inferior parathyroid glands (*vide infra*), which occurs when the embryo is 10–12 mm long, but thereafter it is represented by two elongated diverticula which soon become solid cellular masses and grow caudally into the surrounding mesenchyme. Ventral to the aortic sac (p. 187) the two thymic rudiments meet and are subsequently united by connective tissue only; the rudiments themselves never fuse. The connexion with the third pouch is soon lost, but the stalk may persist for some time as a solid, cellular cord.

The development of thymic tissue from the ventral recess of the fourth pharyngeal pouch probably occurs in a proportion of embryos (Van Dyke 1941), although this has been denied by some authorities (Weller 1933; Norris 1938). Thymic tissue developing from this site is usually found outside the thyroid gland in close association with the superior parathyroid gland. An ectodermal contribution to the thymus, probably of placodal origin, occurs in some mammals, but a similar contribution in man cannot be regarded as proved (Garrett 1948).

Vascularized mesenchyme, including lymphoid stem cells, invades the cellular mass of the endodermal thymus and becomes partially lobulated. The cells of the cytoreticulum and the concentric corpuscles of the thymus are endodermal in origin. The epithelial character of these cells is more obvious in fetal life; some are even ciliated (Sebuwufu 1968). Lymphoid cells enter and colonize the thymus from the haemopoietic tissue stem cells during the third month, according to observations from *in vitro* and animal experiments (Averback 1960, 1961; Taylor 1965; Moore and Owen 1967).

At birth the thymus is large relative to total body weight. Its absolute weight increases in the first two years after birth, but its relative weight decreases. There is little change thereafter until about the seventh year, when rapid growth again occurs to reach a maximum at around eleven years. After this it begins to decline to an adult weight which is very variable but averages 12–15 gm. In old age the gland shrinks still further, especially after wasting diseases. For this and other reasons it is rarely identifiable in the preserved cadaver of the aged (Keynes 1954; Lasi 1959; see also p. 780).

The parathyroid glands are also derivatives of the endoderm. Prior to the appearance of the thymic rudiment from the third pharyngeal pouch, the epithelium on the dorsal aspect of the pouch and in the region of its duct-like connexion with the cavity of the pharynx becomes differentiated as the primordium of the *inferior parathyroid gland*, recognizable by its cells, which stain more lightly than the other endodermal cells lining the pouch. Although the connexion between the pouch and the pharynx is soon lost, the connexion between the thymic and parathyroid rudiments persists for some time, and the latter passes caudally with the developing thymus. The *superior parathyroid glands* develop in a similar manner from the dorsal recess of the fourth pharyngeal pouches. They come into relation with, and are anchored by, the lateral lobes of the thyroid gland and thus remain cranial to the parathyroid glands derived from the third pouch (*see* p. 1453).

The ultimobranchial body has already been noted as an endodermal diverticular part of the *caudal pharyngeal complex* (**2.**93). This separates from the ectoderm of the fourth branchial cleft and loses its connexion with the pharynx by attenuation and rupture of the common pharyngobranchial duct. It becomes closely associated with the expanding lateral lobe of the thyroid gland, with the superior parathyroid (parathyroid IV) component of the complex lying dorsally and outside the thyroid gland. The remainder of the complex, which includes the ultimobranchial body and possibly some vestiges of the ventral recess of the fourth pharyngeal pouch and of the transitory fifth pharyngeal pouch, is enveloped by the thyroid gland. Although some controversy persists, it is probable that the cells of the ultimobranchial body give rise to the 'C' or parafollicular cells producing calcitonin in the thyroid gland (p. 1449) of many if not all mammals. Calcitonin has been isolated from ultimobranchial tissue in vertebrates other than mammals (Copp *et al.* 1967; Taylor 1968). The derivation of thyroid parafollicular cells has now been clearly demonstrated in embryonic sheep (Jordan *et al.* 1973).

The hypophysis cerebri has anterior and posterior lobes; the former is derived from the ectoderm of the stomatodeum, the latter from the neurectoderm of the floor of the forebrain. Previous to the rupture of the oropharyngeal membrane a saccular recess is present in the ectodermal lining of the roof of the stomatodeum. This hypophysial recess (*pouch of Rathke*) (**2.**94A) is the rudiment of the anterior lobe of the hypophysis, lying immediately ventral to the cranial border of the membrane, extending in front of the cranial end of the notochord, and in contact with the ventral surface of the forebrain. It is constricted off by the surrounding mesenchyme to form a closed vesicle, but remains for a time connected to the ectoderm of the stomatodeum by a solid cord of cells, which can be traced down the posterior edge of the nasal septum. Masses of epithelial cells form on each side and in the ventral wall of the vesicle, and the development of the anterior lobe of the hypophysis is completed by the ingrowth of a mesenchymal stroma. Differentiation of epithelial cells into

stem cells and three differentiating types is said to be apparent during the early months of fetal development (Dubois 1967). Recent work also suggests that different types of cells arise in succession, and that they may be derived in differing proportions from different parts of the hypophysial recess (Conklin 1968). A *craniopharyngeal canal*, which sometimes runs from the anterior part of the hypophysial fossa of the sphenoid bone to the exterior of the skull, is often said to mark the original position of the hypophysial recess (of Rathke); traces of the stomatodeal end of the recess are occasionally present at the junction of the septum of the nose with the palate. It has been claimed, however, that the craniopharyngeal canal is a secondary formation caused by the growth of blood vessels, and is quite unconnected with the stalk of the anterior lobe (Arey 1949). Just behind the hypophysial recess a hollow diverticulum grows towards the mouth from the floor of the diencephalon (**2.**94B). This neural outgrowth forms an infundibular sac, the walls of which increase in thickness until the contained cavity is obliterated except at its upper end, where it persists as the *infundibular recess* of the third ventricle. Formed in this way the *posterior lobe* of the hypophysis becomes invested by the anterior, which extends dorsally on each side of it. In addition, the anterior lobe gives off two processes from its ventral wall which grow along the infundibulum and fuse to surround it, coming into relation with the tuber cinereum and constituting the *tuberal portion* of the hypophysis. The original cavity of the stomatodeal diverticulum remains first as a cleft, and later scattered vesicles, and can be identified readily in sagittal sections through the fully formed gland. The dorsal wall of the stomatodeal part, which remains thin, fuses with the adjoining part of the posterior lobe as the *pars intermedia*.

A small endodermal diverticulum, named *Seessel's pouch*, projects towards the brain from the cranial end of the foregut, close to the oropharyngeal membrane. In some marsupials this pouch forms a part of the hypophysis, but, in man, it apparently disappears entirely; but before doing so it may contribute some cells to the general head mesenchyme of mixed origin, and it possibly functions as an anterior head organizer (with the prechordal plate of endoderm).

The Pharynx

The Pharynx is the cranial end of the foregut, and the branchial arches and pharyngeal pouches play an important part in its development (**2.**95A, B). The endodermal aspect of the mandibular arch in its dorsal part contributes to the formation of the lateral wall of the nasopharynx in front of the orifice of the auditory tube. The ventral end of the first pouch becomes obliterated, but its dorsal end persists and deepens as the head enlarges. It remains close to the ectoderm of the dorsal end of the first cleft (p. 145) and, together with the adjoining lateral part of the pharynx and dorsal part of the second pharyngeal pouch, constitutes the *tubotympanic recess*, which forms the tympanic cavity and the auditory tube (p. 178). The site of the second arch is partly indicated by the *palatoglossal arch*, but its dorsal end is separated from its ventral end by the forward growth of the third arch, which obliterates the intermediate part. It is believed that the site of the second pharyngeal pouch is represented by the *supratonsillar cleft*, around which the tonsil is developed. The third arch forms the *lateral glosso-epiglottic fold*, and its dorsal end takes part in the formation of the floor of the auditory tube. The ventral ends of the fourth arches fuse with the posterior part of the hypobranchial eminence and so contribute to the formation of the epiglottis (p. 197). The adjoining portion becomes connected to the *arytenoid swelling* and may be identified in the *aryepiglottic fold*.

After the caudal portion of the hypobranchial eminence has separated from the dorsal part of the tongue (p. 197), it is in continuity with two linear ridges which appear in the ventral wall of the pharynx, the whole forming an inverted U, sometimes regarded as an independent formation, the *furcula* (of His). These vertical ridges have been identified as the sixth arches, placed very obliquely owing to the shortness of the pharyngeal floor compared with the greater extent of the roof. The contained groove of the furcula is carried downwards on the ventral wall of the foregut as the *laryngotracheal groove*, from which the lower part of the larynx, the trachea, bronchi and lungs are developed (p. 207). At

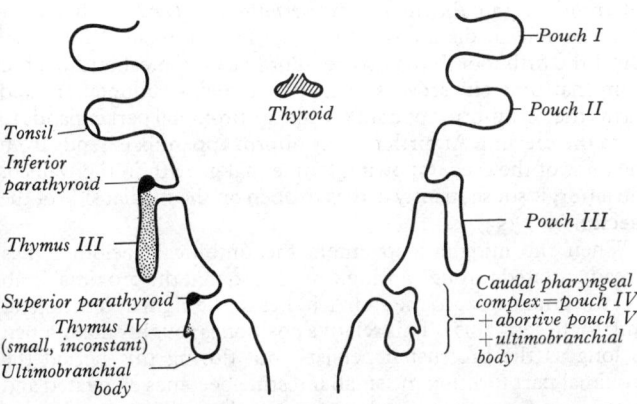

Tonsil
Inferior parathyroid
Thymus III

Superior parathyroid
Thymus IV (small, inconstant)
Ultimobranchial body

Thyroid

Pouch I
Pouch II
Pouch III
Caudal pharyngeal complex = pouch IV + abortive pouch V + ultimobranchial body

2.93   A scheme showing the development of the branchial epithelial bodies. The numbered sacs are the edodermal pharyngeal pouches.

2.94A and **B** Schematic sagittal sections of heads of early embryos to show first stages in the development of the hypophysis.

the cranial end of the groove paired arytenoid swellings arise which convert the slit-like upper aperture of the respiratory system into a T-shaped opening. The aryepiglottic folds can be recognized at this stage extending from the arytenoid swellings to the epiglottis.

Development of the Post-Pharyngeal Parts of the Alimentary Tube

The part of the **foregut** caudal to the pharynx remains as a splanchnopleuric tube which elongates to form the *oesophagus*. At five weeks its epithelium is two cells thick, the luminal columnar cells developing cilia at approximately 10 weeks, and it does not differentiate into stratified squamous epithelium until the fifth month. Non-striated circular and longitudinal layers of muscle are identifiable at about the ninth week. Striated fibres begin to appear at the tenth week (Jit 1974). As the neck and thorax develop the oesophagus lengthens rapidly. At the end of the fourth or the beginning of the fifth week the *stomach* can be recognized as a fusiform dilation (**2.**95A), and beyond this the gut opens into the yolk sac; this opening is at first wide (**2.**80A), but by the fifth week it has become narrowed into a tubular *yolk stalk* (containing the endodermal *yolk duct*) (**2.**95A, B), which soon loses its connexion with the digestive tube (**2.**96). At this stage the stomach is median in position and separated from the pericardium by the septum transversum (p.121), which extends ventrally on to the cranial side of the yolk duct. Dorsally, the stomach is related to the aorta and, owing to the presence of the pleuroperitoneal canals on each side, it is connected to the body wall by a short dorsal mesentery, the *dorsal mesogastrium* (**2.**97). This mesentery is directly continuous with the dorsal mesentery of the intestine. The liver develops as a hollow outgrowth from the ventral aspect of the foregut and grows cranially into the substance of the septum transversum (**2.**95A), this part of the septum now being termed the *ventral mesogastrium*.

In the human embryo, at the 10 mm stage, the characteristic curvatures of the stomach are already discernible. Growth proceeds more actively along the dorsal border of the viscus; its convexity is notably increased and the rudiment of the fundus appears. As a result of the more rapid growth of the dorsal border, the pyloric end of the stomach turns ventrally and the concavity of the lesser curvature becomes apparent (**2.**96). The stomach is now displaced to the left of the median plane and rotated so that its originally right surface becomes dorsal and its left ventral. As a result the right vagus nerve is distributed mainly to the dorsal and the left mainly to the ventral surface of the organ. The dorsal mesogastrium increases in depth and becomes folded on itself; the lesser omentum becomes more transverse than anteroposterior. The pancreatico-enteric recess (p. 205), hitherto a simple depression on the right side of the dorsal mesogastrium, becomes dorsal to the stomach and excavates downwards and to the left between the layers of the folded dorsal mesogastrium. It may now be termed

the inferior recess of the *bursa omentalis*. The displacement and rotation of the stomach has been attributed variously to its own growth changes, extension of the pancreatico-enteric recess and pressure by the rapidly growing liver (Kanagasuntheram 1957).

While these changes are occurring in the stomach, the **midgut** increases in length more rapidly than the vertebral column, and forms a loop, which acquires a dorsal mesentery as it lengthens, and projects into the coelomic cavity (**2.**97). The rapidly growing liver and the developing mesonephroi encroach on the available space in the coelom so much that the loop is extruded into the *umbilical coelom*, i.e., that part of the extra-embryonic coelom which lies in the proximal end of the umbilical cord (p. 123). This umbilical protrusion (often, quite inappropriately, called a 'physiological hernia') is a *normal* condition in human embryos between the 10 mm (end of fifth week) and the 40 mm (third month) stages; under abnormal conditions it may persist and be present at birth.

The rotation of the stomach reacts on the position of the duodenum, which prior to this stage is a ventrally directed loop, but is now carried dorsally and to the right. At this stage the duodenum possesses a thick mesentery which is continuous with the dorsal mesogastrium, on the one hand, and the mesentery of the intestinal midgut loop, on the other (**2.**97). Later, the approximation of the duodenum to the dorsal abdominal wall leads first to the adhesion of the right side of its mesentery to the parietal peritoneum, and later to the absorption of both layers. In this way the duodenum comes to be sessile or 'retroperitoneal'. The lining epithelium of the duodenum proliferates and almost occludes the lumen by the sixth week; the channel is re-established by the third month when the epithelium reverts to a simple columnar form.

Early in the sixth week a small diverticulum appears on the caudal limb of the midgut loop (**2.**96, 97), and this later differentiates into the *caecum* and *vermiform appendix*. Thereafter it is possible to distinguish the large from the small intestine. Until the fifth month the diverticulum has a conical outline, but from that time onwards its distal part remains rudimentary and forms the vermiform appendix, while its proximal part expands to form the caecum. At birth the vermiform appendix extends from the *apex* of the caecum; owing to unequal growth in the walls of the latter, it subsequently comes to open on the medial side of the caecum (p. 1353).

When the midgut loop enters the umbilical coelom it has already rotated through an angle of 90°, so that the proximal limb (i.e. the limb nearer to the stomach) lies to the right and the distal limb to the left (**2.**97). This relative position is roughly maintained so long as the protrusion persists, but during this period the proximal part forming the small intestine becomes elongated and coiled, and the mesentery adapts itself to these changes in the gut. The colic part of the loop elongates less rapidly and has no tendency to become coiled. By the time the fetus has attained a

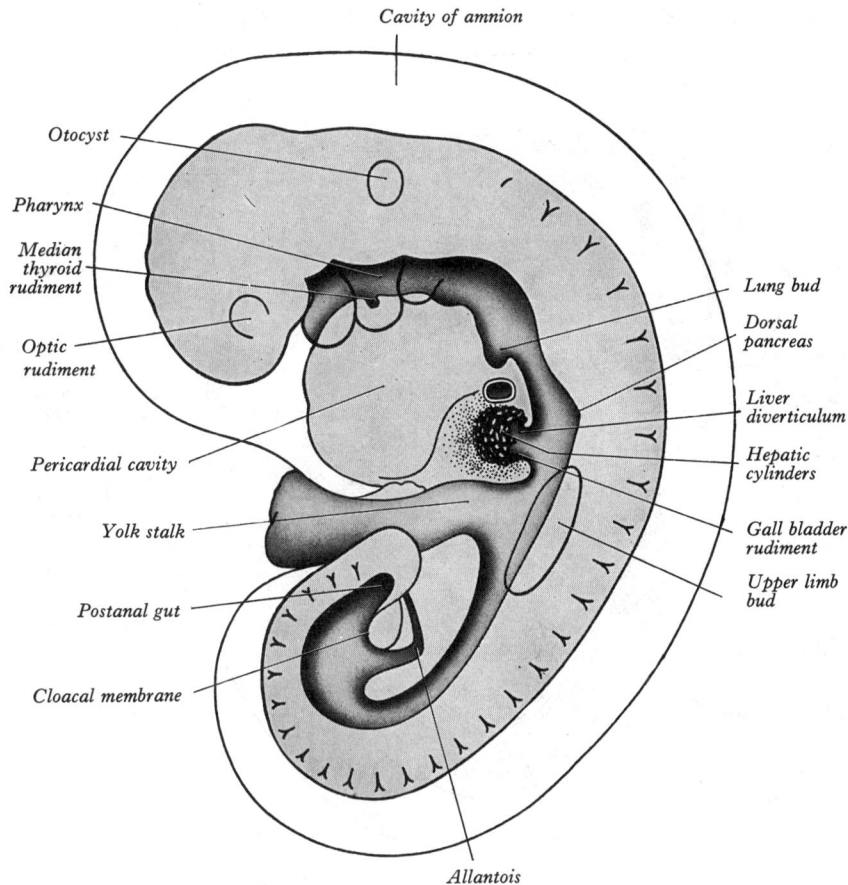

2.95A The digestive tube of a human embryo with 29 paired somites, a C.R. length of 3·4 mm and an estimated age of 27 days. Note branchial development. Magnification × *c.* 25. (G. L. Streeter, *Contr. Embryol.*, **30**, 1942.)

length of 40 mm (middle of third month), the peritoneal cavity has enlarged sufficiently to allow re-entry of the midgut, and this occurs fairly rapidly, but in a definite sequence. The manner of this is important, for at this stage the gut continues the process of *rotation*, resulting ultimately, after differential growth, in the establishment of the definitive relationships which the large intestine shows in the adult, including the relation of the transverse colon to the duodenum. The process has been analysed as follows: (Fraser and Robbins 1915, may be consulted for the classical account of gut rotation: the present account, however, differs substantially from previous accounts in accord with the data presented by Harris and Jones 1976, from a radiological study of 64 fresh fetuses of 11 to 20 weeks of gestation. It should be noted, however, that interpretation of observations of the latter authors has not been followed in its entirety, and some common misconceptions stemming from the classical account will be pointed out.)

While the gut is still in the umbilical coelom the dorsal mesentery forms a median partition from the dorsal wall to the umbilicus. As the gut re-enters the abdominal cavity, the coils of the small intestine, which return first, necessarily enter to the *right* of· this partition, thrusting it over to the *left* and thus determining the position of the descending colon. They pass dorsal to the superior mesenteric artery and determine its adult relationship to the horizontal part of the duodenum. This disposition of the duodenum is probably an important factor in initiating midgut rotation. The caecum is the last part to re-enter the abdomen, and it is at first on the surface of coils of ileum. Subsequently, with growth and some continued rotation the caecum is carried *dorsocaudally* and to the *right* where it comes to lie *in*, or even a little *caudal* to the diminutive *right iliac fossa*. It must be appreciated that at this stage the fetal liver is still, relatively, grossly enlarged and the caecum can be termed *subhepatic* in position, but that the term does *not* imply (as it is held to do in most textbook accounts) that the caecum returns to the right *upper* quadrant of the abdominal cavity. Harris and

Jones named the midgut-derived colon *presplenic* and the hind-gut-derived colon *postsplenic*, the transitional zone, of course, corresponding to the presumptive splenic flexure. At an early stage, after the completion of gut rotation, the proximal part of the presplenic colon runs largely *transversely* across the *lower* abdominal cavity, its left extremity curving cranially to reach the *primitive splenic flexure*, from which point the postsplenic colon first descends vertically, and then curves to the right, finally descending in the midline—the primitive rectum.

The longitudinal muscle coat differentiates earlier in the small than in the large intestine, and it has been suggested that this may play a role in the return of the midgut into the abdomen (Kanagasuntheram 1960). As the coils of small intestine re-enter the abdominal cavity, the mesentery of the *descending* post-splenic colon is thrust against the left dorsal abdominal wall and the opposed peritoneal surfaces become adherent and are gradually absorbed. The descending colon thus loses its mesentery and becomes sessile. Since this change takes place towards the end of the third month, the left colic vessels, whose position on the posterior abdominal wall is secondary, are ventral to such structures as the left ureter and gonadal vessels, which are primarily associated with the posterior wall. The pelvic part of the postsplenic colon, however, does not become sessile but retains its dorsal mesentery; nevertheless, the original midline parietal attachment of the latter, by *partial* adherence and fusion, gradually assumes the inverted V shape characteristic of the adult.

As the visceral surface of the enlarged liver gradually recedes towards the costal margin, the presplenic colon changes its early disposition, becoming firstly oblique, and then progressively developing its *ascending* and *transverse* parts. The former becomes sessile in the same manner as the descending colon, whilst the proximal part of the transverse colon passes ventral to the descending (second) part of the duodenum, to which it adheres. the remainder of the transverse colon retains its dorsal mesentery—the transverse mesocolon—which now has a *horizontal* parietal attachment to the head, neck and body of the

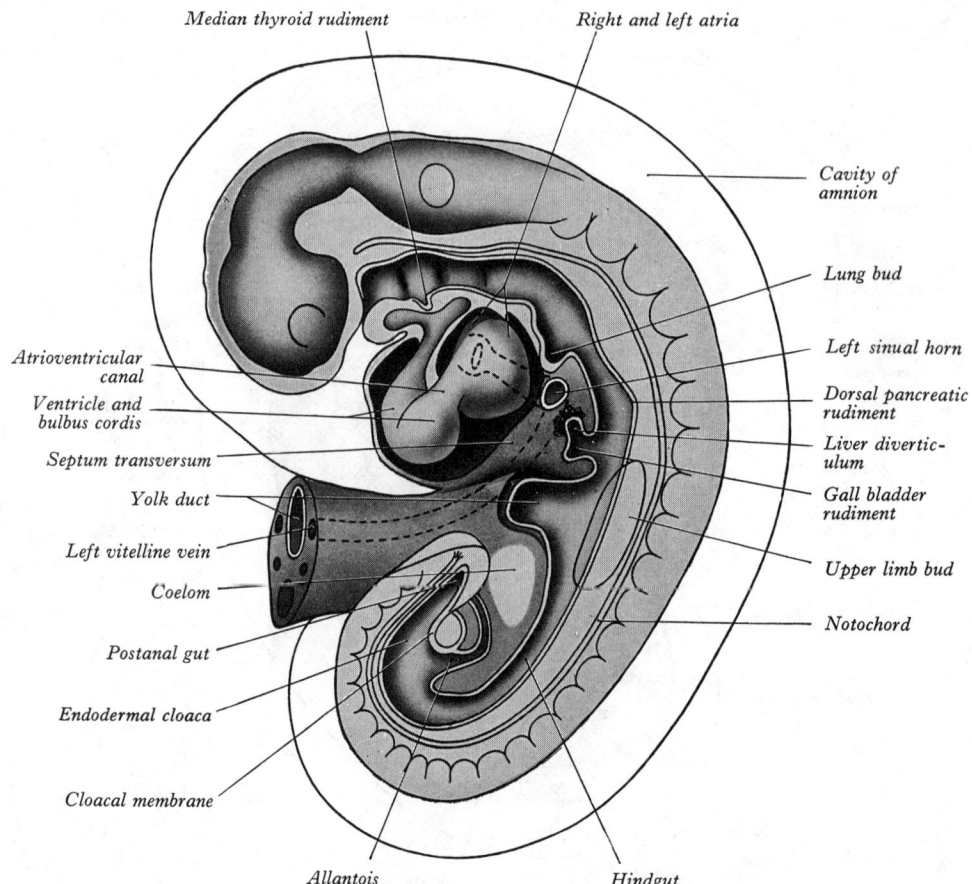

Median thyroid rudiment

Right and left atria

Cavity of amnion

Lung bud

Left sinual horn

Dorsal pancreatic rudiment

Liver diverticulum

Gall bladder rudiment

Upper limb bud

Notochord

Atrioventricular canal

Ventricle and bulbus cordis

Septum transversum

Yolk duct

Left vitelline vein

Coelom

Postanal gut

Endodermal cloaca

Cloacal membrane

Allantois

Hindgut

2.95B A composite diagram of a graphic reconstruction of a human embryo at the end of the fourth week. The alimentary canal and its outgrowths are shown in median section. The brain is shown in outline, but the spinal cord is omitted. The heart is shown in perspective, the left horn of the sinus venosus having been divided. The somites are indicated in outline. (After G. L. Streeter.)

developing pancreas. Whilst this is occurring the omental bursa is expanding and the transverse colon is carried dorsal to the bursa; a variable degree of fusion follows between the transverse mesocolon and the dorsal fold of that part of the dorsal

Facial and vestibulocochlear nerves

Otocyst

Glossopharyngeal nerve

Vagus nerve

Hyoid arch

Fourth ventricle

Third aortic arch

Hypoglossal nerve

Vertebral artery

First cervical spinal nerve

Midbrain flexure

Fourth aortic arch

Sixth aortic arch

Oesophagus

Left atrium

Trachea

Left ventricle

Upper limb bud

Dorsal root ganglion

Liver

Trigeminal ganglion

Mandibular process

Umbilical cord

Maxillary process

Olfactory placode

Allantois

Endodermal cloaca

Ureter

Umbilical artery

Mesonephric duct

Caecum

Roots of dorsolateral intersegmental arteries

Notochord

2.96 A human embryo of 7 mm greatest length. Fifth week. Left lateral aspect. (After Thompson.)

202

mesogastrium destined to become the greater omentum (**2.97**). Meanwhile, as the hepatic flexure has gradually become defined, there has been occurring a slow, short *ascent* of the ileo-caecal junction to the upper part of the right iliac fossa, and a much more rapid ascent of the splenic flexure to its definitive position. (It should be noted that the foregoing account is in complete contrast to earlier descriptions, which maintained that the early caecum gradually *descended* by slow growth from the subhepatic right upper abdominal quadrant caudally to the right iliac fossa.)

The formation of the rectum and anal canal is associated with the growth of the tail fold, described in an earlier section (p. 120). Further, their development is intimately associated with that of the bladder and other elements of the urogenital system, hence many of the relevant illustrations appear later (*see* p. 210 and **2.**107A–F). As growth proceeds, the gut lengthens, at first *pari passu* with the embryo, and a new section of the **hindgut** can be recognized between the caudal aspect of the yolk stalk and the origin of the allantois (**2.95B**). The part of the hindgut caudal to the latter point dilates to form a pouch, termed the *endodermal cloaca*, and in its ventral wall the cloacal membrane (p. 120) can be identified in the midline (**2.**107A). By growth of the surrounding mesoderm the cloacal membrane comes to lie at the bottom of a shallow depression, the *ectodermal cloaca* (**2.**107A). The hindgut and the allantois open into the endodermal cloaca from the time of its first appearance and in the fifth week the mesonephric ducts pierce its wall. By this time the ventral part of the cloaca is wider than its dorsal part, which remains very narrow, and it is into the dorsolateral corner (lateral horn) of the ventral portion that the mesonephric duct opens (**2.**107B, C).

The mesenchyme outside the line of union of these two parts of the cloaca grows rapidly and thrusts the endodermal epithelium inwards. As a result the two walls come into apposition and fuse. This process commences opposite the connexion of the allantoic canal with the cloaca and is continued caudally to form a septum, termed the *urorectal septum*, which separates the dorsal segment or rectum from the ventral segment, which forms the urinary bladder and the urogenital sinus (2.107B). At its caudal end the urorectal septum reaches the cloacal membrane and divides it into *anal* and *urogenital membranes*. For a time a *cloacal duct* connects the two parts of the cloaca proximal to the growing urorectal septum (2.107C), and occasionally persists as a passage between the rectum and the bladder or urethra. Anal tubercles form round the margin of the anal part of the cloacal membrane, which thus comes to lie at the bottom of a depression, the *proctodeum* (Tench 1936). With the absorption and disappearance of the anal membrane the rectum communicates with the exterior (2.107E). The lower part of the anal canal is formed from the proctodeum, but its upper part is endodermal in origin and is derived from the caudal end of the dorsal subdivision of the cloaca; the line of union corresponds with the edges of the anal valves in the adult (p. 1358). In the fourth and fifth weeks a small part of the hindgut, named the *postanal gut* (2.107A), projects caudally beyond the anal membrane; it usually disappears before the end of the fifth week.

The development of the *nerve supply* of the gut has received relatively little attention in human material. A migration of neuroblasts to establish a vagal innervation has been described in human embryos between the eighth and twelfth weeks (Okamoto and Ueda 1967), which is distributed in a craniocaudal direction to establish the myenteric plexus; the vagal ramifications in the oesophageal wall have been identified at the 15 mm stage (Lecco and Balli 1968). Ultrastructural and histochemical observations suggest that in the fetal small intestine and colon, the mucous membranes have already developed a potential for absorption (Kelly 1973; Lev and Orlic 1974).

Applied Anatomy. Abnormalities in the development of the digestive tube may lead to various disturbances which become apparent at birth or shortly after. Of these the following are the most common.

Closure of the laryngotracheal groove (p. 207) may be effected in such a way that the oesophagus is divided into two portions, a cranial, which communicates with the mouth above and ends blindly below in the neighbourhood of the tracheal bifurcation, and a caudal, which communicates below with the stomach and above with the trachea.

Congenital stricture of the small intestine is usually due to exuberant overgrowth of the lining epithelium and the formation of adhesions. This overgrowth is a normal occurrence at one stage in the development of the oesophagus and duodenum, but the lumen is soon restored. Some cases of atresia may result from interference with the blood supply (Louw 1959).

The umbilical herniation present between the 10 mm and 40 mm stages may fail to return into the abdominal cavity (p. 200) and may be present at birth.

The yolk duct (p. 200) may remain patent as a constituent of the umbilical cord, or its proximal part may persist as an *ileal diverticulum* (of Meckel), which may or may not be anchored to the umbilicus by a fibrous band. In its simplest form this diverticulum is a short, saccular protrusion from the anti-mesenteric border of the ileum about 3 ft above the ileocaecal valve (p. 1345). It may be lined by patches of heterotopic gastric, jejunal, or colonic mucosa, and on occasion it may contain pancreatic tissue. Most commonly, however, it is lined by mucosa similar to that of the surrounding ileum. Postnatal persistence of the yolk duct results in an *ileo-umbilical fistula*; persistence of its distal part only giving an *umbilical* sinus; finally, persistence of its intermediate part may result in a fluid-filled *enterocystoma* suspended from the ileum and umbilicus by fibrous cords.

Rotation of the gut (p. 200) may fail to occur. In these cases the colon occupies the left lower portion of the abdominal cavity and has no connexion with the greater omentum: the duodenum may be spirally coiled and the superior mesenteric vessels pass either behind it or to its left side.

The caecum may retain its fetal form with an apical vermiform appendix. It may fail to complete its rotation, and then lies ventral to the right kidney in close relation to the visceral surface of the liver. Occasionally it is in the transverse mesocolon. (For an analysis of the normal and abnormal development of the vermiform appendix consult Balthazar and Gade 1976.)

The separation of the endodermal cloaca into ventral and dorsal parts may be incomplete. The rectum then opens into the bladder, urethra or vagina and the anus is imperforate, a condition which may occur without other abnormalities. In some cases it is due to persistence of the anal membrane; in others, the colon may end blindly, considerably above the level of the pelvic floor. The proctodeum may or may not be present.

Because of its mode of development the anal canal is lined in its lower part by modified skin, and in its upper part by mucous membrane, which is endodermal in origin. Abrasions or tears of the wall of the lower part of the anal canal, such as occur in the condition known as anal fissure, are exceedingly sensitive and examination of them is acutely painful. On the other hand, lesions of the upper part of the anal canal are not associated with marked pain or tenderness (p. 1364).

THE PERITONEUM AND THE OMENTAL BURSA

Before the stomach rotates, two peritoneal pockets, the *right* and *left pneumato-enteric recesses*, appear in each side of the dorsal mesogastrium. The left recess is transitory and soon disappears. The right recess communicates with the peritoneal cavity by a small opening to the right and extends cranially along the right side of the oesophagus (2.98A) towards the root of the right lung bud. The right wall of the recess is formed by a mesenteric fold termed the *caval fold* (Kanagasuntheram 1957) which may be derived from the septum transversum. Just below the right pneumato-enteric recess, and sharing with it a common aperture of communication with the peritoneal cavity, is a second recess which extends to the left in the dorsal mesogastrium (2.98B). This is the *pancreatico-enteric recess* (formerly called the *omental bursa*). Its mouth extends from the oesophageal end of the stomach to the duodenum; the cranial and caudal margins of this opening occupy the positions of the gastro-pancreatic folds of the adult and contain the left gastric and hepatic arteries. The pancreatico-enteric recess does not, at first, involve the mesoduodenum. The rotation of the stomach deepens this recess, which lies dorsal to it extending further to the left in the dorsal mesogastrium (2.97). At the caudal end of the pneumato-enteric recess another peritoneal recess extends ventrally and progressively separates the stomach and liver. This is termed the *hepato-enteric recess*. This comes to lie behind the ventral mesogastrium and is carried caudally behind the mesoduodenum. When the caudate lobe of the liver develops it invades the caval fold and then projects from it into the hepato-enteric recess.

The cranial part of the pneumato-enteric recess is cut off by the development of the diaphragm and forms a small serous sac, the *infracardiac bursa*, within the right pulmonary ligament between the oesophagus and diaphragm. The remainder of this complicated system of peritoneal recesses together constitutes the adult *omental bursa* and the common communication with the peritoneal cavity is the *epiploic foramen*. The caudal part of the pneumato-enteric recess with the hepato-enteric recess form the superior part of the omental bursa in the adult, and the pancreatico-enteric recess gives rise to the inferior part of the omental bursa and its extensions.

As the stomach enlarges, the pancreatico-enteric recess keeps pace with it, and when the intestines return to the abdominal cavity they come to lie caudal and dorsal to its caudal part. In its cranial part the dorsal wall of the recess becomes pressed against the dorsal abdominal wall and the opposed peritoneal layers fuse (2.98D). Up to this time the dorsal mesogastrium is attached in the median plane, but as a result of the fusion the root of the mesogastrium acquires a new, curved attachment to the dorsal wall. From the oesophagus it passes caudally and to the left, becoming the *gastrophrenic* and *lienorenal (spleno-renal) ligaments* of the adult, and then turns to the right and runs somewhat caudally along the line of the pancreas (2.97).

The development of the spleen in the cranial part of the dorsal mesogastrium (p. 207) divides that region of it into *lienorenal* and *gastrosplenic ligaments*. The caudal part of the dorsal wall of the recess remains free and grows caudally, overlapping the transverse colon and the underlying coils of small intestine, forming the greater omentum. Later the two layers of the transverse mesocolon become adherent to the overhanging dorsal

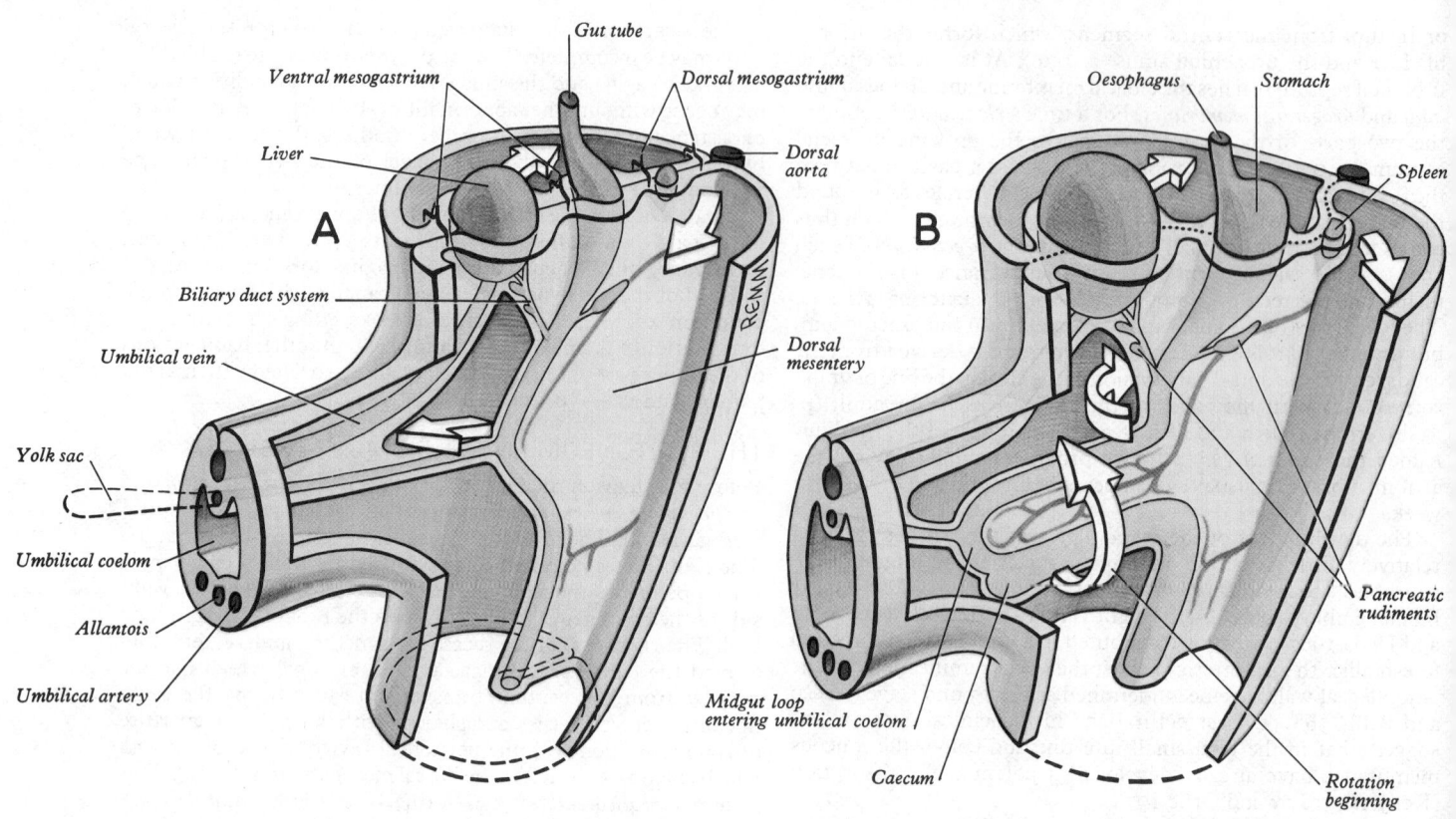

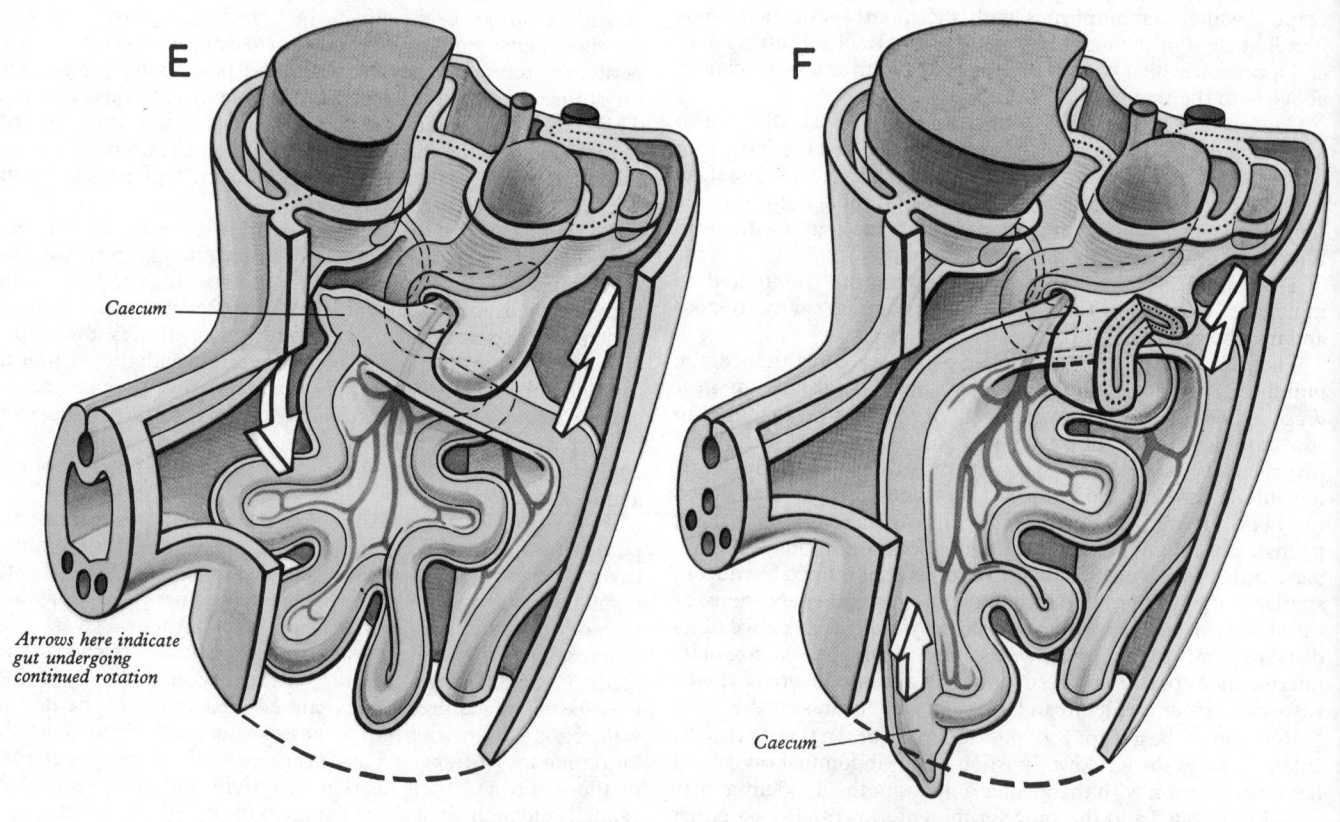

2.97   A three-dimensional schematization of the major developmental sequences of the subdiaphragmatic embryonic and fetal gut, together with its associated major glands, peritoneum and mesenteries, viewed from the left anterolateral aspect. The developmental sequence A–F spans 1½ months to the perinatal period. H denotes the general disposition of the

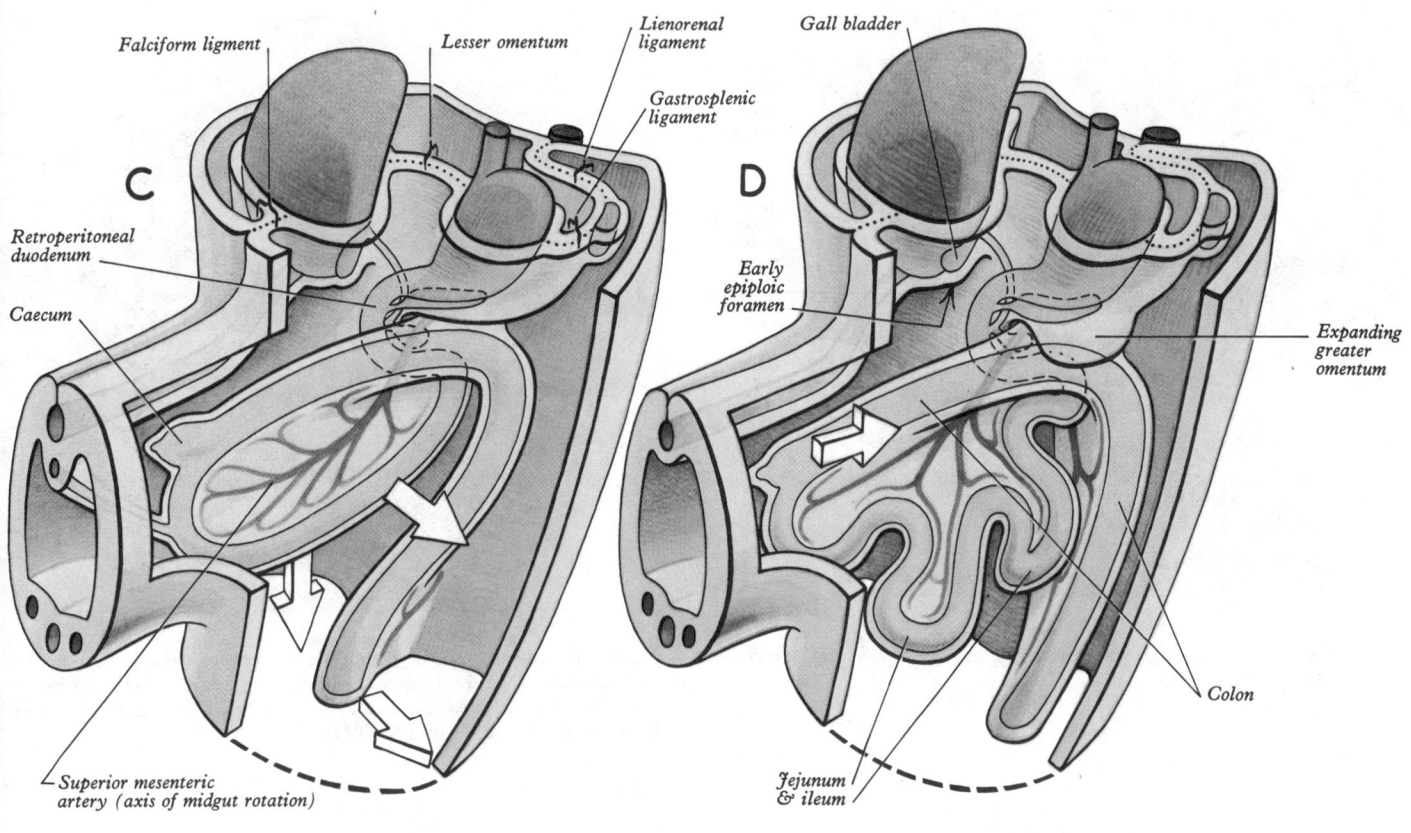

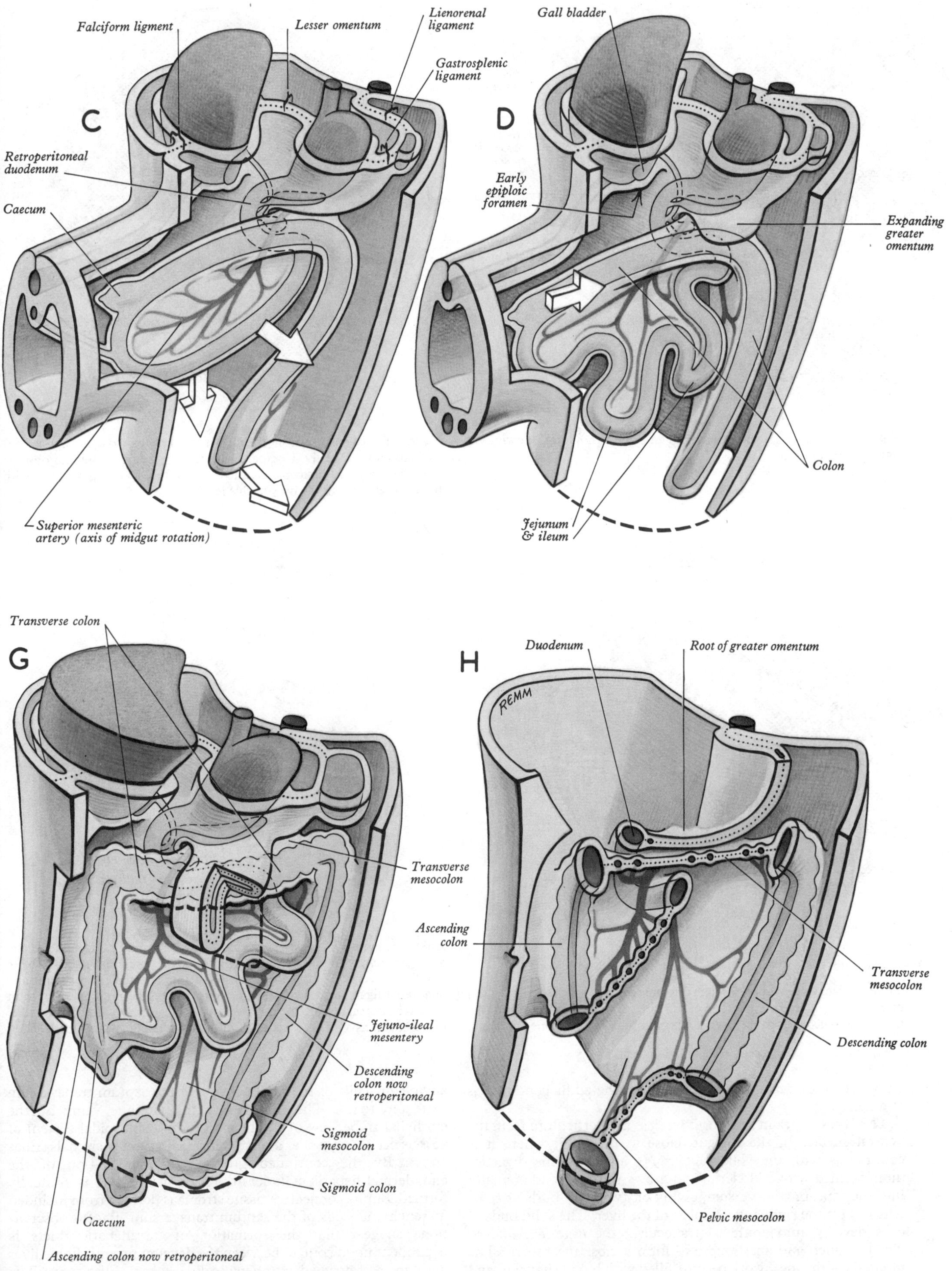

C

Falciform ligament

Lesser omentum

Lienorenal ligament

Gastrosplenic ligament

Retroperitoneal duodenum

Caecum

Superior mesenteric artery (axis of midgut rotation)

D

Gall bladder

Early epiploic foramen

Expanding greater omentum

Colon

Jejunum & ileum

G

Transverse colon

Transverse mesocolon

Jejuno-ileal mesentery

Descending colon now retroperitoneal

Sigmoid mesocolon

Sigmoid colon

Caecum

Ascending colon now retroperitoneal

H

Duodenum

Root of greater omentum

Ascending colon

Transverse mesocolon

Descending colon

Pelvic mesocolon

REMM

remaining viscera, mesenteric roots with their lines of attachment, and principal contained vessels, that approximate to the adult state for comparison. Some features of the sequence presented do not correspond to traditional descriptions—see text for further comment.

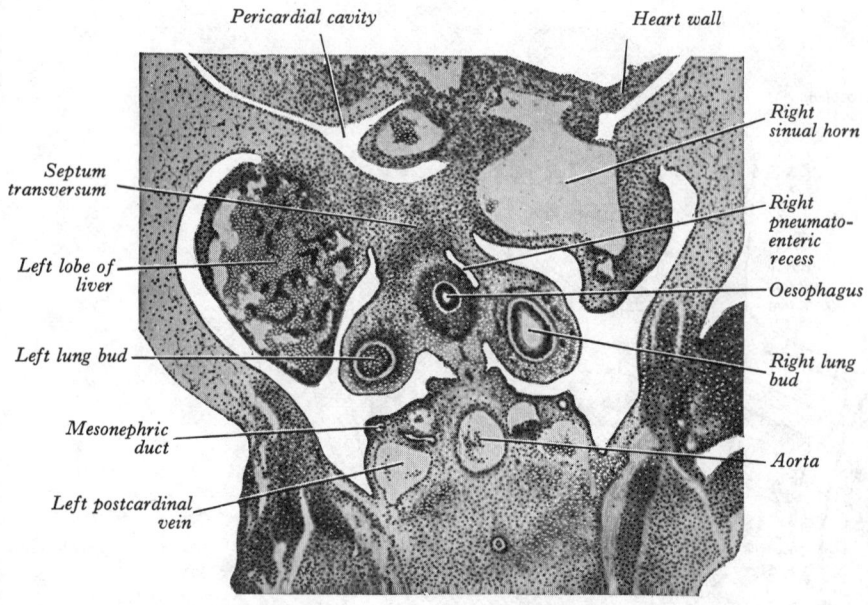

2.98A Transverse section of a human embryo, 8 mm long, showing the right pneumato-enteric recess.

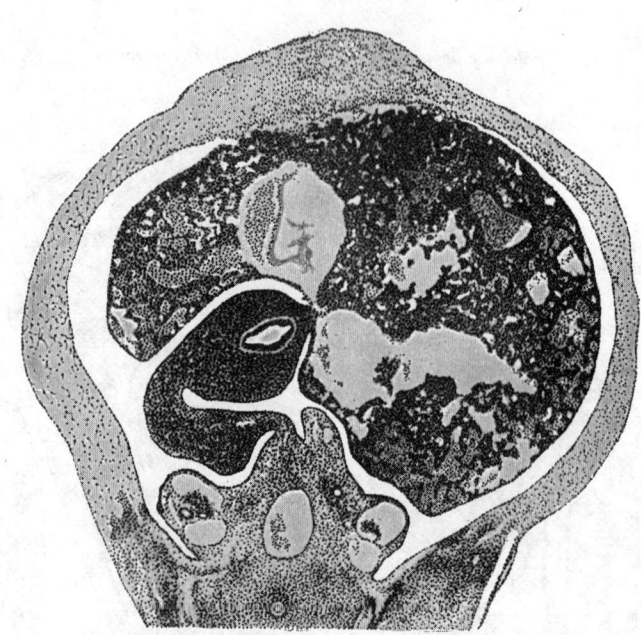

2.98C Transverse section through the same embryo as 2.98B, but 150 μm more caudally. Compare with the preceding figure and observe that the omental bursa (pancreatico-enteric recess) communicates with the general peritoneal cavity at this level.

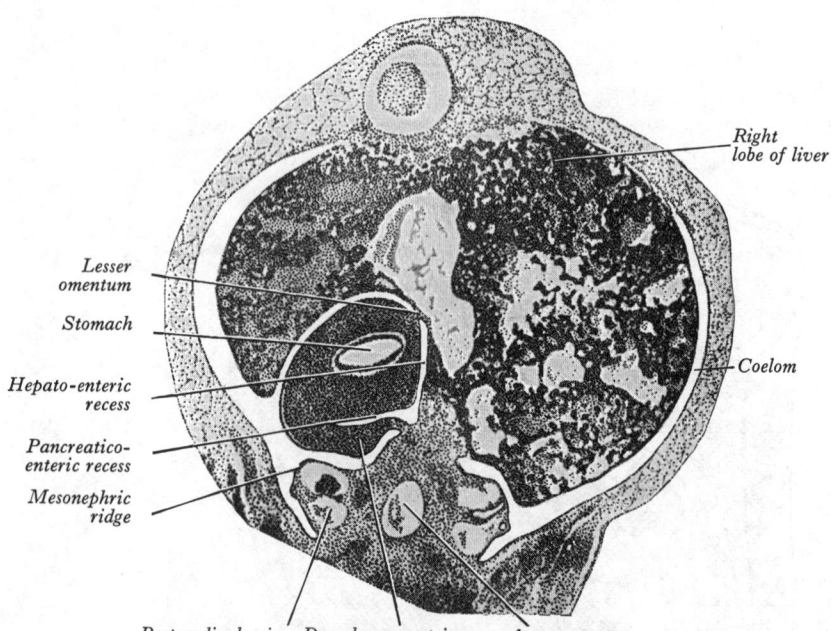

2.98B Transverse section through the same embryo as 2.98A but 530 μm more caudally. Note that rotation of the stomach has taken place and that the sinusoidal spaces in the liver communicate freely with one another.

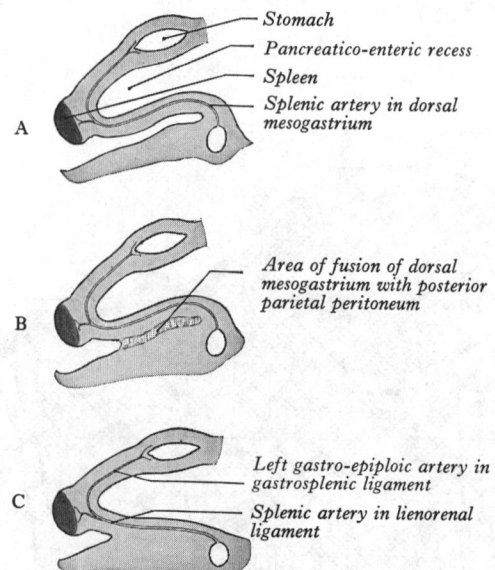

2.98D Diagram to show the fusion of the proximal part of the dorsal mesogastrium with the peritoneum on the posterior abdominal wall. Note also the conversion of the dorsal mesogastrium into the gastrosplenic and lienorenal ligaments. A represents a transverse section of an embryo in which the dorsal mesogastrium is still at the stage shown in 2.98B. B and C represent transverse sections of older embryos made at the same level.

surface of the greater omentum, so that the adult condition is attained (2.97).

The liver arises in the fourth week as a diverticulum from the ventral surface of the foregut, close to the point where it is continuous with the yolk stalk (2.95A and B). This diverticulum is lined with endoderm, and grows ventrally and cranially into the septum transversum, giving off two solid buds of cells, which form the right and left lobes of the liver. The solid buds of cells develop into epithelial trabeculae, the *hepatic cylinders*, which branch and anastomose to form a close meshwork. The intervals of the meshwork become filled with blood *sinusoids*, and on section the organ has the appearance of a vascular sponge (2.98C). These vessels arise *in situ* as the result of the influence exercised by the endodermal cells of the liver on the potentially

angiogenic cells of the mesenchyme of the septum transversum (Streeter 1942). The invasion of the vitelline veins by the epithelial trabeculae of the liver to form a sinusoidal system of vessels occurs only over a restricted area in a few mammals (Elias 1955). By the continued growth and ramification of the endodermal hepatic cylinders the mass of the liver is gradually formed, but its connective tissue stroma is derived from included mesenchymal cells of the septum transversum. Recent observations suggest that the formation of intrahepatic ducts is dependent upon contact between the developing embryonic liver mass and a preformed extrahepatic duct system (Elias 1967). The original diverticulum from the duodenum forms the *bile duct*, and from its distal part the cystic duct and gall bladder arise as an outgrowth, solid at first but later canalized. The bile duct first opens into the ventral wall of the duodenum; later, it migrates to the left across the dorsal (originally right) surface of the duodenum to the position which it occupies in the adult on the medial (or mesenteric) border. The migration is ascribed to differing rates of growth in the duodenal walls and is aided by proliferation of the lining epithelium of the duodenum, which is most marked in this position (Kanagasuntheram 1960).

As the liver enlarges, it projects more and more into the abdominal cavity from the caudal surface of the septum transversum. In the process, mesenchyme of the septum becomes drawn out ventral to the liver to form the falciform ligament, and craniodorsally to form the coronary and triangular ligaments, and the lesser omentum. These peritoneal folds are collectively the *ventral mesogastrium*. At three months the liver almost fills the abdominal cavity, and its left lobe is nearly as large as its right. Later the relative development of the liver is less active, more especially that of the left lobe, which actually undergoes some degeneration and becomes smaller than the right; but until birth the liver remains relatively larger than in the adult.

The pancreas (2.99A, B) is developed in *dorsal* and *ventral* parts. The former arises in the latter half of the fourth week as a diverticulum from the dorsal wall of the duodenum a short distance cranial to the hepatic diverticulum; growing craniodorsally in the mesoduodenum it enters that part of the dorsal mesogastrium which is forming the dorsal wall of the bursa omentalis. It forms the whole of the neck, body and tail of the pancreas and a part of the head. The *ventral* part grows out from the primitive bile duct at the point where the latter opens into the duodenum. This outgrowth is at first double, but the two soon fuse, and the resulting single mass grows round the gut into the mesoduodenum, where it enlarges to form the remainder of the head of the gland (Odgers 1930). The duct of the dorsal part (the *accessory pancreatic duct*) therefore opens directly into the duodenum, while that of the ventral part (the *main pancreatic duct*) opens with the bile duct. Early in the seventh week the two parts of the pancreas meet and fuse, and a communication is established between their ducts (2.99B). After this has occurred the terminal part of the accessory duct, i.e. the part between the duodenum and the point of meeting of the two ducts, undergoes little or no enlargement, while the duct of the ventral part increases in size and forms the terminal part of the main duct of the gland. The opening of the accessory duct into the duodenum is sometimes obliterated, and, even when it remains patent, it is probable that almost the whole of the pancreatic secretion is conveyed through the main duct.

At first the body of the pancreas grows craniodorsally between the layers of the dorsal mesogastrium, in the dorsal wall of the bursa omentalis. When this wall fuses with the dorsal parietal peritoneum, the process extends tailwards as far as the caudal (inferior) border of the pancreas, and thus, finally, the gland becomes sessile on the dorsal wall of the abdomen.

For details of cytogenesis, exocrine and endocrine, of the human fetal pancreas consult Conklin (1962); Laitio *et al.* (1974).

The spleen (2.97, 98D) is not a constituent part of the digestive system, but it is convenient to refer to its development here. It appears about the sixth week as a localized thickening of the coelomic epithelium of the dorsal mesogastrium near its cranial end, and the proliferating cells invade the underlying mesenchyme, which becomes condensed and vascularized. The process occurs simultaneously in several adjoining areas which soon fuse to form a lobulated spleen, derived in part from the coelomic epithelium and in part from the mesenchyme in the dorsal mesogastrium. As the organ enlarges it projects to the left so that its surfaces are covered by the peritoneum of the mesogastrium on its left aspect, thus forming a boundary of the greater sac. When fusion occurs between the dorsal wall of the lesser sac and the dorsal parietal peritoneum, the process does not extend so far to the left as the spleen (2.98D), which remains connected to the dorsal abdominal wall by a short lienorenal ligament, while its primitive connexion with the stomach persists as the gastrosplenic ligament. The earlier lobulated character of the spleen disappears, but is indicated by the presence of notches on its upper border in the adult.

The histogenesis of the spleen has attracted relatively little attention. For earlier accounts consult references (Sabin 1912; Holyoke 1936). The vascular reticulum is well developed at 8 to 9 weeks, with immature reticulocytes and numerous closely spaced thin-walled vascular loops. Differentiation of blood cells, macrophages, and of arteries, veins, capillaries and sinusoids has occurred by the eleventh to twelfth week. The capsule consists at first of cuboidal cells bearing cilia and microvilli (Weiss 1973). Later changes have not yet been described in detail.

The spleen is subject to various anomalies of development, including complete agenesis, multiple spleens or polysplenia, isolated small additional spleniculi and persistent lobulation. Recently attention has been directed to association of cardiac, pulmonary and other abnormalities with asplenia or polysplenia (Rose *et al.* 1975).

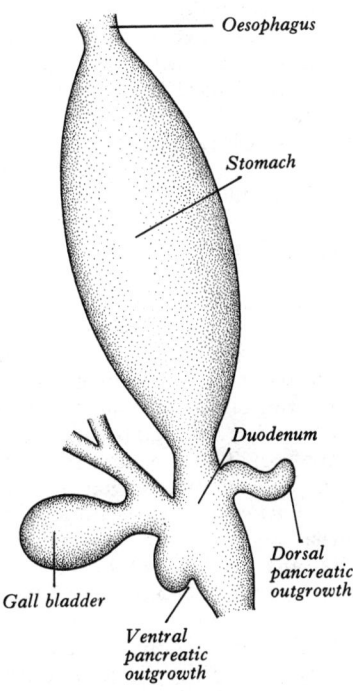

2.99A   Diagram of an early stage in the development of the pancreas in a human embryo, 7·5 mm long. (After Streeter.)

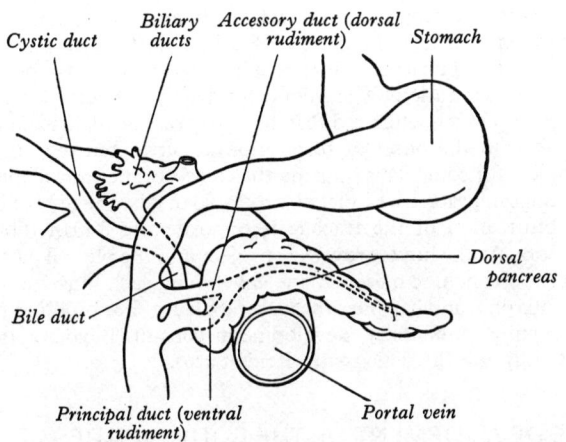

2.99B   A later stage in the development of the pancreas in a human embryo, 14·5 mm long. (After Streeter.)

DEVELOPMENT OF LARYNX TRACHEA AND LUNGS

The rudiment of the respiratory tree appears in the fourth week as a median *laryngotracheal groove* in the ventral wall of the pharynx. The groove deepens and its lips fuse to form a septum, converting the groove into a splanchnopleuric *laryngotracheal tube* (2.100A). The process of fusion commences at the caudal end of the groove

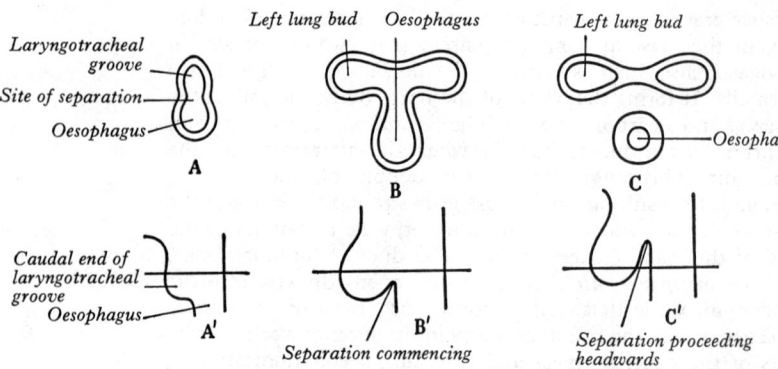

2.100A Diagrams to show the closure of the laryngotracheal groove and its separation from the oesophagus in the latter part of the fourth week. (After Streeter.) A, B and C represent transverse sections at the levels shown in A¹, B¹ and C¹, which are outline drawings of the oesophageal region in three closely following stages.

In A and A the laryngotracheal groove communicates freely with the oesophagus.

In B¹ the lower end of the laryngotracheal groove has begun to close and to form right and left evaginations, which represent the earliest rudiments of the lung buds, seen in B.

In C′ the separation of the laryngotracheal groove from the oesophagus has proceeded further in a headward direction, and in C the primitive lung buds are now freed from the oesophagus.

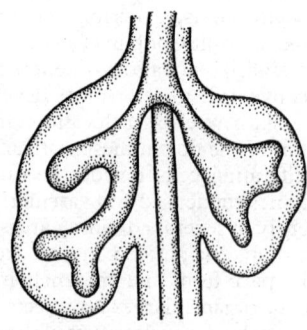

2.100B The lung buds from a human embryo, 11·8 mm long, showing commencing lobulation. (After Streeter, *Contr. Embryol.*, **32**, 1948.)

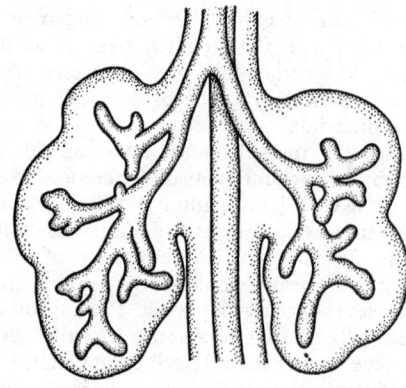

2.100C The lungs of a human embryo (14·2 mm long) in the early part of the sixth week. (After Streeter.)

and extends cranially, but it does not involve the cranial end of the groove, where the lips remain separate, bounding a slit-like aperture, through which the tube opens into the pharynx. The tube is lined with endoderm, and from this the epithelial lining of the respiratory tract is developed. The cranial end of the tube forms the larynx, and its succeeding part the trachea, while from its caudal end two lateral outgrowths arise and form the stem bronchi and the right and left *lung buds*. These grow into the pleural coeloms and are therefore covered with splanchnic mesenchyme (**2.98**A), from which the connective tissue, cartilage,

non-striated muscle and vasculature of the bronchi and lungs are developed.

The first rudiment of the **larynx** is the cranial end of the laryngotracheal groove, bounded ventrally by the caudal part of the hypobranchial eminence (p. 197) and on each side by the ventral ends of the sixth arches. Two *arytenoid* swellings appear, one on each side of the groove (**2.92**A, B), and, as they enlarge, they become approximated to each other, and to the caudal part of the hypobranchial eminence (**2.92**A, B), from which the *epiglottis* is developed. The *opening* into the larynx is at first a vertical slit, which is converted into a T-shaped cleft by the enlargement of the *arytenoid swellings*; the vertical limb of the T lies between the two arytenoid swellings and its horizontal limb between them and the epiglottis. Soon after its appearance the epithelial walls of the cleft adhere to each other, and the aperture of the larynx remains occluded until the third month, when its lumen is regained. However, with the upgrowth of the arytenoid swellings and the deepening of the *primitive aryepiglottic folds* forming the walls of the *vestibule*, a new definitive aperture is formed above the level of the primitive aperture, and the latter now corresponds to the level of the *glottis*. The arytenoid swellings differentiate into the *arytenoid* and *corniculate cartilages*, and the folds joining them to the epiglottis become the *aryepiglottic folds*, in which the *cuneiform cartilages* are developed as derivatives of the epiglottis. The *thyroid cartilage* is developed from the ventral ends of the cartilages of the fourth, or fourth and fifth branchial arches; it appears as two lateral plates, each chondrified from two centres and united in the mid-ventral line by a fibrous membrane in which an additional centre of chondrification develops. The *cricoid cartilage* arises from two cartilaginous centres, which soon unite ventrally, and gradually extend and ultimately fuse on the dorsal surface of the tube (see also p. 151). For a recent compilation of literature on the early development of the larynx, consult O'Rahilly and Tucker (1973).

The right and left *lung buds* make their appearance before the laryngotracheal groove is converted into a tube. (An analysis of early pulmonary rudiment development in various mammals is provided by Hjortsjo 1950.) They grow out into the pleural passages caudal to the common cardinal veins and divide into lobules, three appearing on the right, and two in the left lung bud; these divisions are the first indications of the corresponding lobes of the lungs (**2.100**B, C). The buds undergo further subdivision and ramification, and ultimately end in minute expanded extremities—the *infundibula*. After the fifth to sixth month the *air sacs* begin to make their appearance as minute pouches of the infundibula. (For the detailed chronology of the branching, subdivision, phases, and differentiation of the bronchial tree, consult Reid (1976).

Three phases of differentiation—glandular, canalicular and alveolar—are described. In the *glandular phase*, the bronchial divisions are differentiated; their epithelial cells have a prominent Golgi apparatus, scattered RNA granules and glycogen, but few organelles (Leeson and Leeson 1964). In the *canalicular phase* the respiratory parts are delineated and establish an intimate relation with the expanding blood vascular system. The *alveolar phase* extends from 6 months, but new bronchi and alveoli continue to be formed after birth (Wilson 1928). The mechanism of alveolar distension is uncertain; fetal respiratory movements, possibly involving aspiration of amniotic fluid into the lungs, may be involved (Duenhoelter and Pritchard 1973). The main expansion occurs with the onset of respiration at birth, but may not be complete for some days. During the course of their development the lungs migrate in a caudal direction, so that by the time of birth the bifurcation of the trachea is opposite the fourth thoracic vertebra. As the lungs grow they project into the pleural passages and the splanchnic mesenchyme enveloping each lung rudiment expands on it and becomes the visceral pleura. For bibliographies concerning pulmonary development consult Boyden (1972); O'Rahilly and Boyden (1973); Reid (1976).

THE DEVELOPMENT OF THE BODY CAVITIES

The formation of the intra-embryonic coelom and its mode of communication with the extra-embryonic coelom have already been described (p. 118). After the formation of the head fold the *pericardium* connects dorsally with the coelomic ducts, which open caudally into the peritoneal cavity. When the lung buds develop they project into the coelomic ducts, which may now be termed the *pleural coeloms*, and their communications with the pericardial and peritoneal coeloms, become the *pleuropericardial* and *pleuroperitoneal canals* respectively (2.98A, 100D).

A ridge of tissue, the *pulmonary ridge*, develops on the lateral side of the pleural coelom and partly encircles the pleuropericardial canal. The ridge is continuous dorsally with the septum transversum. The developing lung bud abuts on the ridge, which gives rise to two diverging membranes meeting at the septum transversum. One of these is cranially placed and is termed the *pleuropericardial membrane*; within its substance is the common cardinal vein and the phrenic nerve, which reach the septum transversum by this route. The other membrane is caudally placed and is termed the *pleuroperitoneal membrane*. As the apical part of the lung forms it invades the body wall and extends cranially on the *lateral aspect* of the common cardinal vein, carrying with it an extrusion of the pleural passage to form the definitive pleural sac. In this way the common cardinal vein and the phrenic nerve come to lie medially in the mediastinum. The pleuropericardial canal, which lies medial to the vessel, is gradually narrowed to a slit, which is soon obliterated by the apposition and fusion of its margins (2.101A). Its closure occurs early and is mainly effected by the growth and expansion of the surrounding viscera, the heart, lungs, trachea and oesophagus, and not by active growth of the pleuropericardial membrane across the opening to the root of the lung (2.101A).

In addition to its extension in a cranial direction the lung enlarges ventrally and medially, and the pleura is therefore carried into the body wall over the surface of the pericardium, thus separating it from the lateral thoracic walls (2.101B).

The separation of the pleural and peritoneal cavities from each other is effected by the development of the diaphragm. The septum transversum at first forms a sheet of mesoderm, caudal to the pericardial cavity and extending from the ventral and lateral regions of the body wall to the foregut. Dorsal to it on each side is the pleuroperitoneal canal. The liver grows into the septum transversum, which now can be seen to consist of two parts. One, the *pars diaphragmatica*, is disposed in the transverse plane and lies over the convex cranial surface of the liver. The other, the *pars mesenterica*, lies in the median sagittal plane and is expanded by the developing liver. At this stage the liver is widely attached to the pars diaphragmatica and to the ventral abdominal wall. These attachments are the forerunners of the coronary and triangular ligaments and of the falciform ligament respectively. Medial to the pleuroperitoneal canals, are the oesophagus and stomach, with the dorsal mesentery, and at the root of the latter the dorsal aorta. Dorsolateral to the canals are the pleuroperitoneal membranes, which remain small; dorsally are the mesonephric ridges, suprarenals and gonads. Just as the enlargement of the pleural cavity cranially and ventrally is effected by a process of burrowing into the body wall, so its caudal enlargement is effected in the same way. The expanding pleural cavities extend into the mesoderm dorsal to the suprarenal glands, gonads and mesonephric ridges. This mesoderm is peeled off the dorsal body wall to form a substantial portion of the lumbar part of the diaphragm. The pleuroperitoneal canal is closed by the fusion of its edges, which are carried towards one another by the growth of the organs surrounding it and in particular of the suprarenal, which carries the dorsal margin ventrally to meet the pars diaphragmatica of the septum transversum (Wells 1954). The right pleuroperitoneal canal closes earlier than the left. It is therefore on the left that an abnormal communication between the pleural and peritoneal cavities more frequently occurs. The further development of the peritoneal cavity has already been described (p. 205).

While these changes are in progress, the septum transversum undergoes a progressive alteration in position. In a 2 mm human embryo, the dorsal border of the septum transversum lies opposite the second cervical segment but as the embryo grows, and the heart enlarges, it migrates in a caudal direction. At first the ventral border moves more rapidly than the dorsal, but after the embryo has attained a length of 5 mm it is the dorsal border which

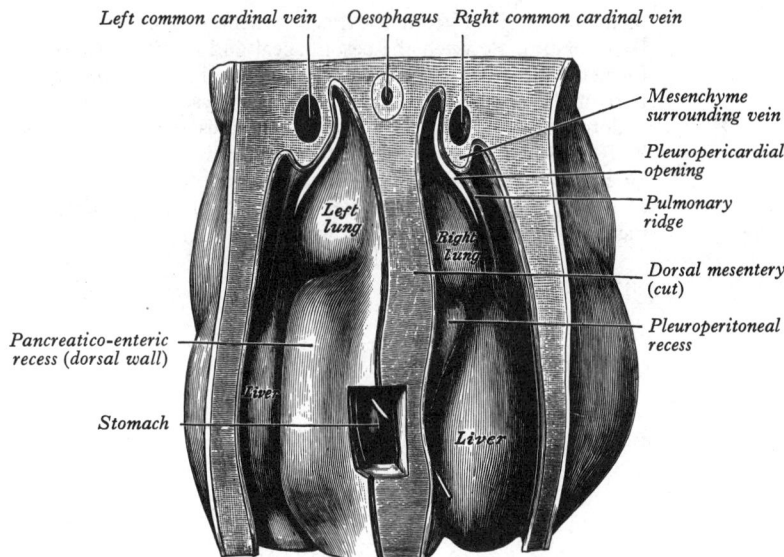

Left common cardinal vein　*Oesophagus*　*Right common cardinal vein*

Mesenchyme surrounding vein

Pleuropericardial opening

Pulmonary ridge

Dorsal mesentery (cut)

Pleuroperitoneal recess

Pancreatico-enteric recess (dorsal wall)

Stomach

Left lung　*Right Lung*　*Liver*

2.101A　A view, from the dorsal aspect, into the thoraco-abdominal part of the coelom of a human embryo 6·8 mm long, in the fifth week. Note that the dorsal body wall, including the spinal cord, developing vertebral column, the dorsal aorta and the mesonephroi, has been removed. A window has been made in the dorsal wall of the pancreatico-enteric recess to expose the posterior surface of the stomach and a wire has been passed through the epiploic foramen. (After Piper.)

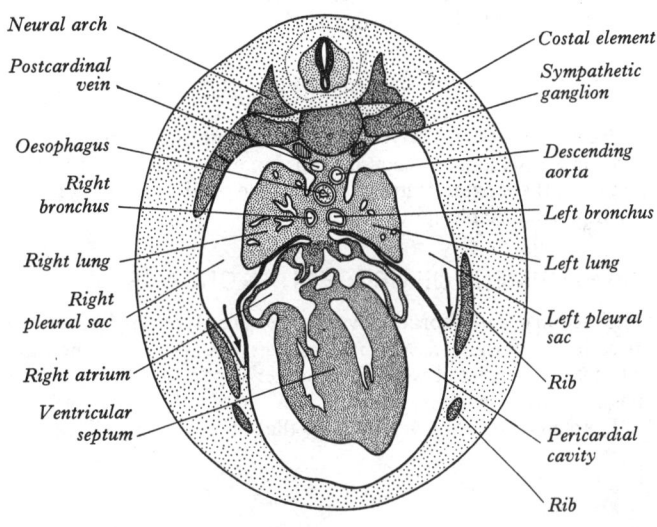

Neural arch

Postcardinal vein

Oesophagus

Right bronchus

Right lung

Right pleural sac

Right atrium

Ventricular septum

Costal element

Sympathetic ganglion

Descending aorta

Left bronchus

Left lung

Left pleural sac

Rib

Pericardial cavity

Rib

2.101B　Transverse section of a 21 mm human embryo, showing how the pleural sacs extend ventrally on each side of the pericardium and split the body wall. The arrows indicate the directions of growth of the two pleural sacs.

migrates more rapidly (2.102A). When the dorsal border of the septum transversum lies opposite the fourth cervical segment, the phrenic nerve (C 3, 4 and 5) and portions of the corresponding myotomes grow into it and accompany it in its later migrations. It is not until the end of the second month that the dorsal border of the septum transversum is opposite the last thoracic and first lumbar, segments, the final position occupied by the dorsal attachment of its derivative, the diaphragm.

The Development of the Diaphragm

The closure of the pleuroperitoneal openings completes a mesodermal partition which separates the thoracic from the abdominal viscera and which occupies the position of the future diaphragm. This partition has a composite origin. The sternal and costal parts are derived almost exclusively from the pars diaphragmatica of the septum transversum, with a very small dorsolateral contribution from the pleuroperitoneal membranes and from the thoracic wall (costal portion). Anterior to the oesophageal hiatus is a small contribution from the gastrohepatic

209

ligament, derived from the pars mesenterica of the septum transversum. Between the oesophageal and aortic hiatuses it is formed by the dorsal mesentery. The remainder of the lumbar part of the diaphragm is formed in the mesoderm around the abdominal aorta and more laterally in the mesoderm of the dorsal body wall behind the suprarenal, mesonephric ridge and gonad (Wells 1954). Some authorities consider that much greater areas of the adult diaphragm are derived from the pleuroperitoneal membranes and from the chest wall.

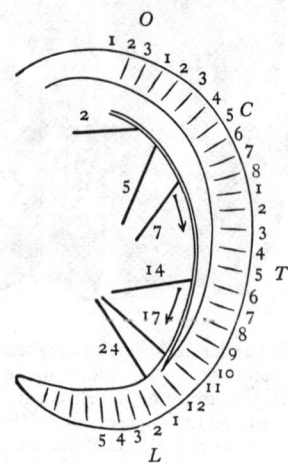

2.102A Schema showing stages in the descent of the septum transversum. The numerals on the heavy lines indicate the length of the embryo in mm, and the position of the occipital, cervical, thoracic and lumbar segments is also shown. (After Mall.)

Gaps between the lumbar and costal portions of the diaphragm are usually due to under-development of the latter.

Premuscle tissue, derived principally from the fourth cervical myotomes, invades the septum transversum as already described and from there extends into the rest of the partition, giving rise to the muscular diaphragm (2.102B).

DEVELOPMENT OF UROGENITAL ORGANS

The urinary and reproductive organs are developed from the mesoderm of the *intermediate cell mass* (p. 117) and they are intimately associated with one another especially in the earlier stages of their development.

Typically, in lower vertebrates, the intermediate cell mass is divided into a series of segments, the *nephrotomes* (2.103A), each of which develops a cavity, a *nephrocoele*, which communicates with the coelom through a *peritoneal funnel*. In each segment the dorsal wall of the nephrotome is evaginated to form a *nephric tubule* communicating with the nephrocoele by means of a *nephrostome*. The outer ends of the earlier developed nephric tubules bend caudally and fuse to form a longitudinal *primary excretory duct*, which grows first caudally and then ventrally to open into the cloaca. The more caudally placed and later developed tubules open secondarily into this duct or into tubular outgrowths from it. The *glomeruli* arise from the ventral wall of the nephrocoele (*internal glomeruli*) or in the roof of the coelom adjacent to the peritoneal funnels (*coelomic* or *external glomeruli*) or in both situations (2.103A). These external glomeruli are usually transitory.

It has been customary to regard the renal excretory system as consisting of three organs, the *pronephros*, *mesonephros* and *metanephros*, which succeed one another in time and space. The last of these to develop is retained as the permanent kidney. It must be noted that it is impossible to provide reliable criteria that distinguish them as individual organs or to define their precise limits in the embryos of all animals. A pronephros cannot be distinguished as a separate organ in man; the earliest and most cranially situated nephric tubules are rudimentary and transient and are generally regarded as marking the pronephric region. This region merges caudally without a clear line of demarcation into the mesonephros.

In the human embryo the development of nephrotomes and their cavitation to form nephrocoeles which communicate with the coelom is confined to the levels of the most cranial somites. Caudal to about the level of the eighth somite the mesoderm of the intermediate cell mass is fused into an unsegmented *nephrogenic cord*. This is connected at irregular intervals with the coelomic epithelium. No nephric tubules are developed at the most cranial levels.

The *primary excretory duct* begins to develop in embryos of about 14 somites (23 to 24 days) as a solid rod of cells in the dorsal part of the nephrogenic cord. At this stage its cranial end is placed at about the level of the ninth somite and its caudal end merges with the undifferentiated mesoderm of the nephrogenic cord. It begins to differentiate before any nephric tubules and is at first unconnected with them. In older embryos it has lengthened and its caudal end becomes detached from the nephrogenic cord to lie immediately beneath the ectoderm. From this level it grows caudally, independently of the nephrogenic cord, and then curves ventrally to reach the wall of the cloaca. It becomes canalized progressively from its cranial end to form a

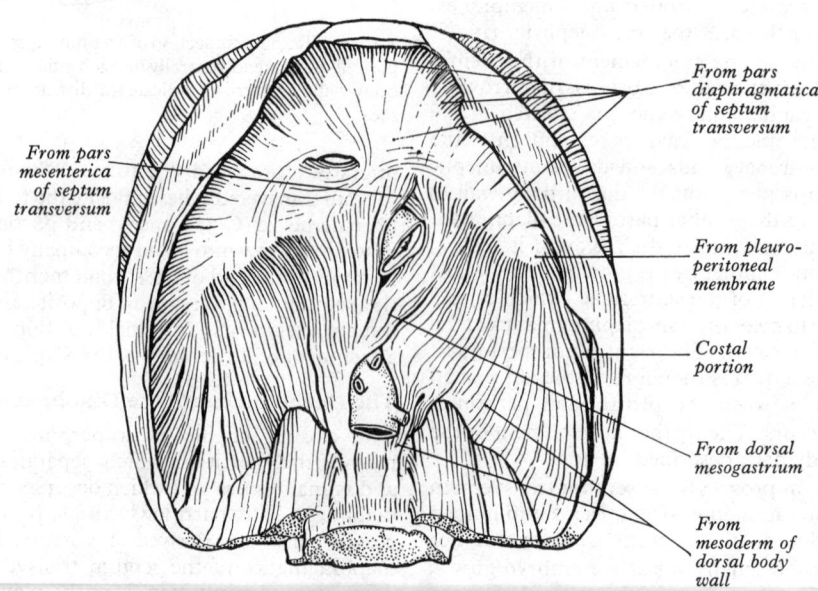

2.102B Inferior surface of the diaphragm showing the derivation of the different parts. (After L. J. Wells, *Contr. Embryol.*, **35**, 1954.)

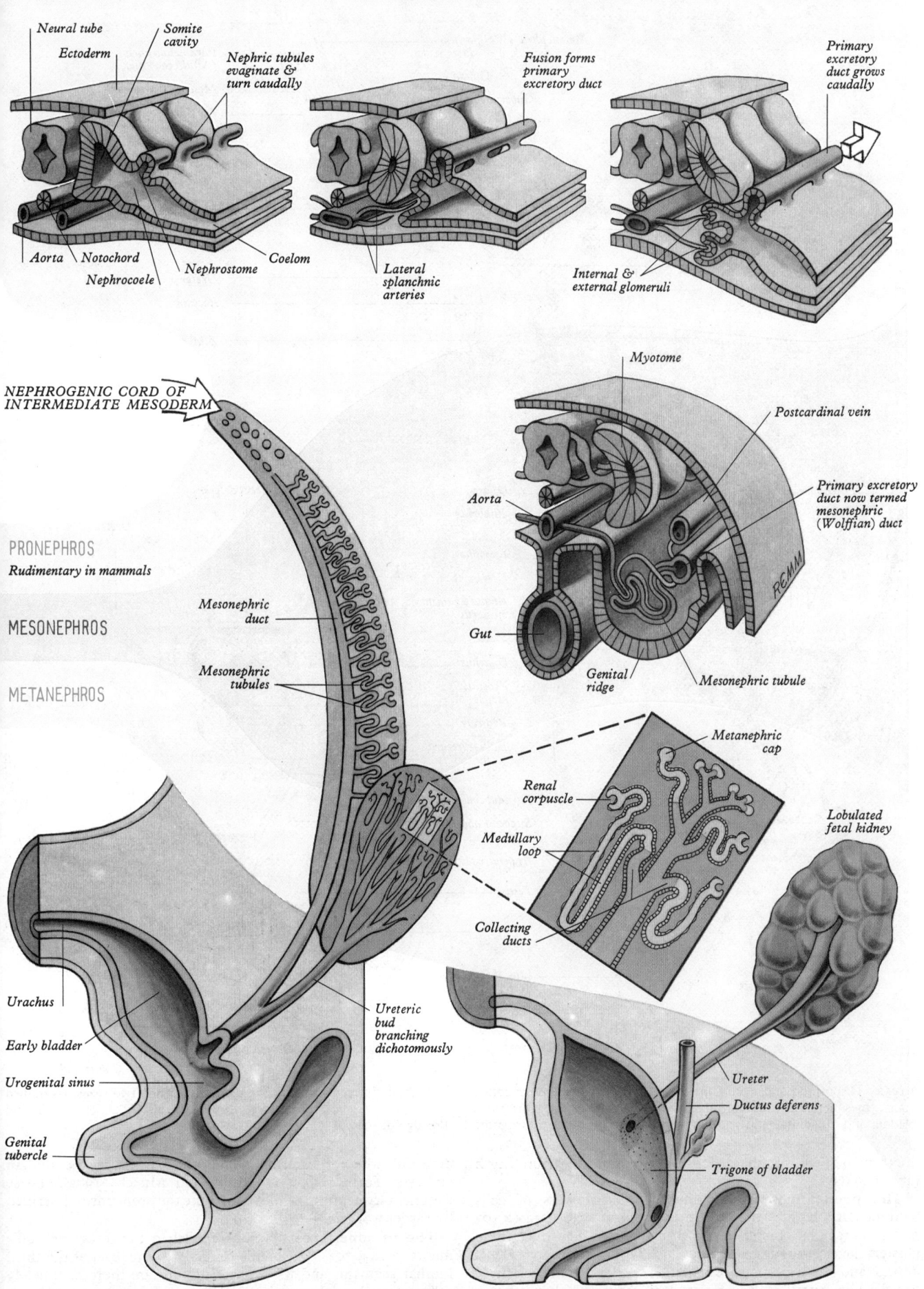

2.103A Principal features of the primitive vertebrate nephric system for comparison with the development of the human nephric system. It should be appreciated that a considerable period of embryonic and fetal life has been necessarily compressed into a single, static diagram. (Modified from *Basic Human Embryology*, by Williams, Wendell-Smith and Treadgold, 1969.)

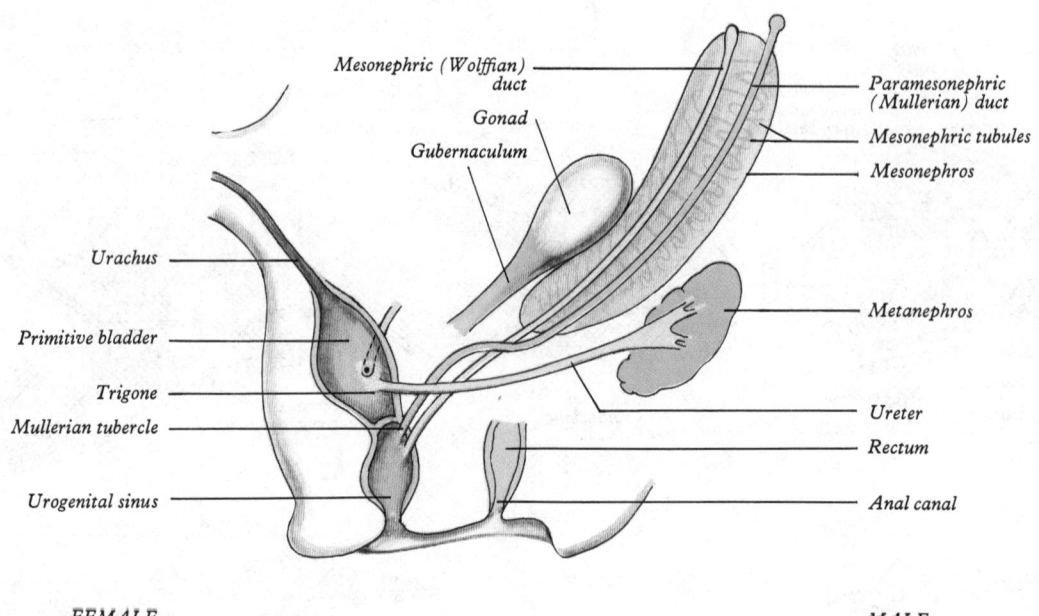

Mesonephric (Wolffian) duct
Gonad
Gubernaculum
Paramesonephric (Mullerian) duct
Mesonephric tubules
Mesonephros
Urachus
Primitive bladder
Metanephros
Trigone
Mullerian tubercle
Ureter
Rectum
Urogenital sinus
Anal canal

FEMALE

MALE

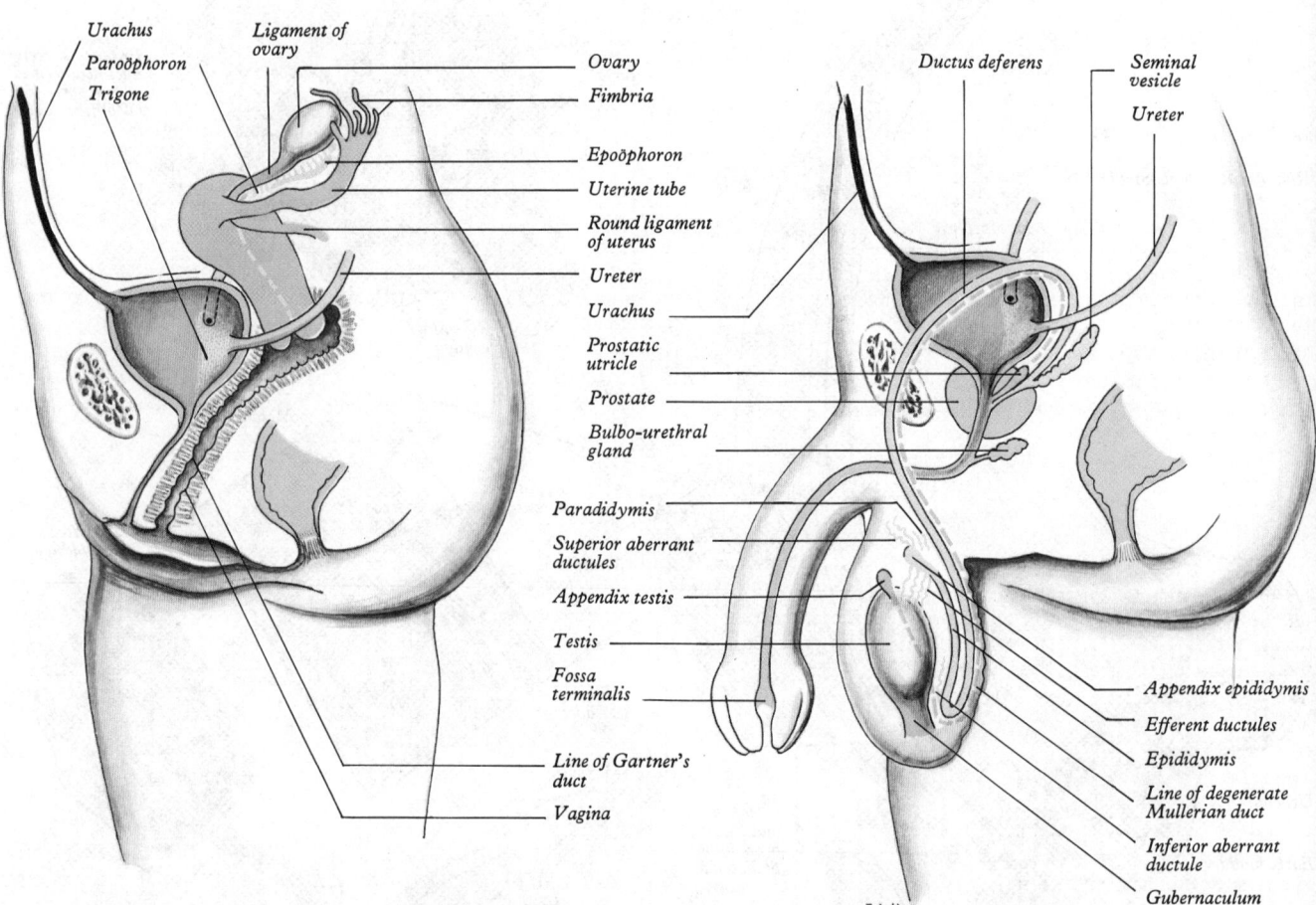

Urachus
Paroöphoron
Trigone
Ligament of ovary
Ovary
Fimbria
Epoöphoron
Uterine tube
Round ligament of uterus
Ureter
Urachus
Paradidymis
Superior aberrant ductules
Appendix testis
Testis
Fossa terminalis
Line of Gartner's duct
Vagina

Ductus deferens
Seminal vesicle
Ureter
Prostatic utricle
Prostate
Bulbo-urethral gland
Appendix epididymis
Efferent ductules
Epididymis
Line of degenerate Mullerian duct
Inferior aberrant ductule
Gubernaculum

J.A.H.

2.103B Disposition and fate of the mesonephric and paramesonephric ducts, mesonephric tubules, gonads and gubernaculum, primitive bladder and urogenital sinus, and ureters, as seen in the transformation from the indifferent stage to the condition in the two sexes. (Modified from *Basic Human Embryology*, by Williams, Wendell-Smith and Treadgold, 1969.)

duct which opens into the cloaca in embryos of 4 to 5 mm length (about 28 days).

The pronephros is first indicated as clusters of cells (rudimentary nephric tubules) in the nephrogenic cord (**2.**103 A, B). In regions cranial to the primary excretory duct these clusters develop no further. More caudally similar groups of cells appear and these become vesicular. The dorsal ends of these vesicles join the primary excretory duct, their central ends being connected with the coelomic epithelium by cellular strands probably representing rudimentary peritoneal funnels. Glomeruli are not developed in association with these cranially situated nephric tubules, which ultimately disappear. It is

doubtful whether external glomeruli develop in the human embryo (Torrey 1954; O'Rahilly and Muecke 1972). These rudimentary nephric tubules constitute the pronephros described by earlier workers.

The mesonephros develops caudal to the foregoing rudiments (**2.**103, 104). It extends caudally to the level of the third lumbar segment, and its nephric tubules are more completely differentiated and distinct internal glomeruli are formed. As in the more cranial region, each tubule first appears as a mass of cells which becomes hollowed in the centre. One end of the vesicle grows towards and opens into the primary excretory duct, which may now be termed the *mesonephric duct*, whilst the other forms the internal glomerulus. This becomes dilated and invaginated to form the *glomerular capsule*, the invaginated segment proliferating the tissue from which the glomerular capillaries are developed. These are supplied with blood through a lateral branch of the aorta (Streeter 1945). In all, it is estimated that about 70 to 80 tubules and a corresponding number of glomeruli develop, but these are not segmental in arrangement. All these tubules, however, are not present at the same time and it is rare to find more than 30 to 40 in any individual embryo, for the cranial tubules and glomeruli atrophy and disappear before the development of those which are situated more caudally. By the end of the sixth week the mesonephros forms an elongated spindle-shaped organ which projects into the coelomic cavity on each side of the dorsal mesentery from the septum transversum to the third lumbar segment. This projection is the *mesonephric ridge* (Wolffian body) and the genital gland is developed on its medial surface.

There are striking similarities in structure between the mesonephros and the permanent kidney or metanephros, but the former's nephrons lack a segment corresponding to the descending limb of the loop of Henle (Leeson 1957; Davies and Routh 1957).

In both sexes the cranial end of the mesonephros atrophies and disappears; in embryos of 20 mm length the organ is found only in the first three lumbar segments, although it may still possess as many as 26 tubules. (This massive degeneration of the more cranial tubules appears to have no parallel in other mammals.) The most cranial one or two tubules persist as the *rostral aberrant ductules*; the succeeding five or six tubules develop into the *efferent ductules of the testis* and the *lobules of the head of the epididymis* in the male, and the tubules of the *epoöphoron* in the female; the caudal tubules form the *caudal aberrant ductules* and the *paradidymis* in the male, and the *paroöphoron* in the female.

The *mesonephric duct* runs caudally in the lateral part of the mesonephric ridge, at the caudal end of which it is projected into the cavity of the coelom in the substance of a mesodermal fold (2.105A, C, D; 107A–F). As the ducts approach the urogenital sinus the two folds fuse with each other, between the bladder ventrally and the rectum dorsally, forming across the cavity of the pelvis a transverse partition which is termed the *genital cord* (2.105C). In the male the peritoneal fossa between the bladder and the genital cord becomes obliterated, but it persists in the female as the uterovesical pouch. The mesonephric duct itself becomes the canal of the *epididymis*, *ductus deferens* and *ejaculatory duct*. The *seminal vesicle* and the ampulla of the ductus deferens appear as a common swelling at the termination of the mesonephric duct around the end of the third and the start of the fourth month. This combined rudiment coincides in appearance with degeneration of the paramesonephric ducts, though no causal relation between the two processes is apparent. Separation into two rudiments occurs at about 125 mm crown–heel length. The seminal vesicle elongates, its duct becomes demarcated, and hollow diverticula bud from its wall. At about the sixth month (300 mm crown–heel length) the growth rate of both the vesicle and the ampulla is greatly increased, the cause of this being unknown. The acceleration may result from increased secretion of prolactin by the fetal or maternal hypophysis, or from the effects of placental hormones. The tubules of the prostate show a similar increase of growth rate at about the same time (Nilsson 1962). In the female the mesonephric duct is vestigial, becoming the longitudinal duct of the epoöphoron.

The metanephros, or permanent kidney, has a double origin (2.103). At about the 5 mm stage an outgrowth forms from the dorsomedial aspect of the mesonephric duct, near the point at which it opens into the cloaca. This is the *metanephric (ureteric) diverticulum*, and it grows dorsally at first and then cranially. Its blind extremity, which grows into the caudal end of the nephrogenic cord, becomes expanded and the adjoining mesoderm condenses around it to form the *metanephrogenic mass* or *cap* (2.103, 106). The presence of the actively growing extremity of the diverticulum may be regarded as an important factor in determining the differentiation of the mesoderm, and it may be noted that in cases where the ureter fails to develop, no metanephrogenic cap is formed. The stalk of the diverticulum

becomes the *ureter*, and its expanded cranial end gives origin not only to the *pelvis* of the kidney and its *calices* but also to the *collecting tubules* of the kidney. The secreting and convoluted tubules and the renal corpuscles are all derived from the metanephrogenic cap in a manner similar to the development of the mesonephric tubules and glomeruli. The blood vessels of the glomeruli develop *in situ* from mesenchyme in the concavity of the glomerular capsule (Lewis 1958) and extend by repeated splitting of the lumina of the first-formed capillaries (Zamboni and de Martino 1968). The secretory tubules must establish communication with the ends of the collecting tubules; should they fail to do so, congenital cysts of the kidney will be formed. (For details of the tubulo-vascular relationships in the developing kidney consult Speller and Moffat 1977.)

From the expanded extremity of the ureter arise four collecting tubules of the first order and it itself forms the primitive renal pelvis. Each of the tubules ends in an ampullated extremity, from which collecting tubules of the second order bud out. These, in turn, give rise to collecting tubules of the third order, and so on. In some animals the four collecting tubules of the first order are absorbed into the renal pelvis, which then presents a single renal papilla and no subdivision into calices. In man, however, the collecting tubules of the first and second order persist and constitute, respectively, the major and minor calices, while the tubules of the third and fourth orders are taken into the minor calices which, therefore, directly receive the openings of the collecting tubules of the fifth order.

When it first appears, the renal rudiment is in the pelvic region, but, as the ureteric outgrowth lengthens, it grows cranially, and, by the time the embryo has attained a length of 13 mm, its expanded pelvis lies on a level with the second lumbar vertebra. During this period the developing kidney receives its blood supply from arteries in its immediate neighbourhood, the middle sacral and common iliac arteries, and the definitive *renal artery* is not recognizable until the beginning of the third month. It arises from the most caudal of the three suprarenal arteries, all of which represent persistent mesonephric or lateral splanchnic arteries (p. 191). Additional renal arteries are by no means uncommon. They may enter at the hilum or at the upper or lower pole of the gland, and they also represent persistent mesonephric arteries.

At an early stage the kidney is lobulated, a condition which persists through fetal life but disappears during the first year after birth.

Further development of the ureter has attracted little attention, but Ruano-Gil *et al.* (1975) examined it in 45 human embryos (5–55 mm), and they state that it is permeable at first (5 mm). Its lumen then becomes obliterated (13–22 mm), to be subsequently recanalized. Both processes were observed to begin at intermediate levels of the ureter and to proceed in both directions. The recanalization was not associated with metanephric function, but possibly with the rapid elongation of the ureter in conformity with embryonic growth. Two fusiform enlargements appear subsequently, affecting its lumbar and pelvic levels. The lumbar enlargement appears during the fifth month, the pelvic not until the ninth month, the latter being inconstant. As a result the ureter shows a constriction at its upper end and another as it crosses the pelvic brim. A third narrowing is always present at its lower end, and is caused by the growth of the bladder wall.

At first the ureter is connected to the dorsomedial aspect of the mesonephric duct but, owing to differences in rates of growth, the connexion becomes lateral to the duct. Thereafter the caudal end of the duct becomes incorporated in the developing bladder, and the orifice of the ureter opens into the bladder on the lateral side of the opening of the duct. Later the two orifices become separated still further and, although the ureter retains its point of entry into the bladder, the mesonephric duct opens into that part of the urogenital sinus which subsequently becomes the prostatic urethra.

The female ducts (paramesonephric or Müllerian ducts), play an important part in the development of the female reproductive system, but do not begin to develop until the embryo reaches a length of 10–12 mm (beginning of sixth week). Each commences as a groove-like invagination of the coelomic epithelium (the *female duct groove*) on the lateral aspect of the   213

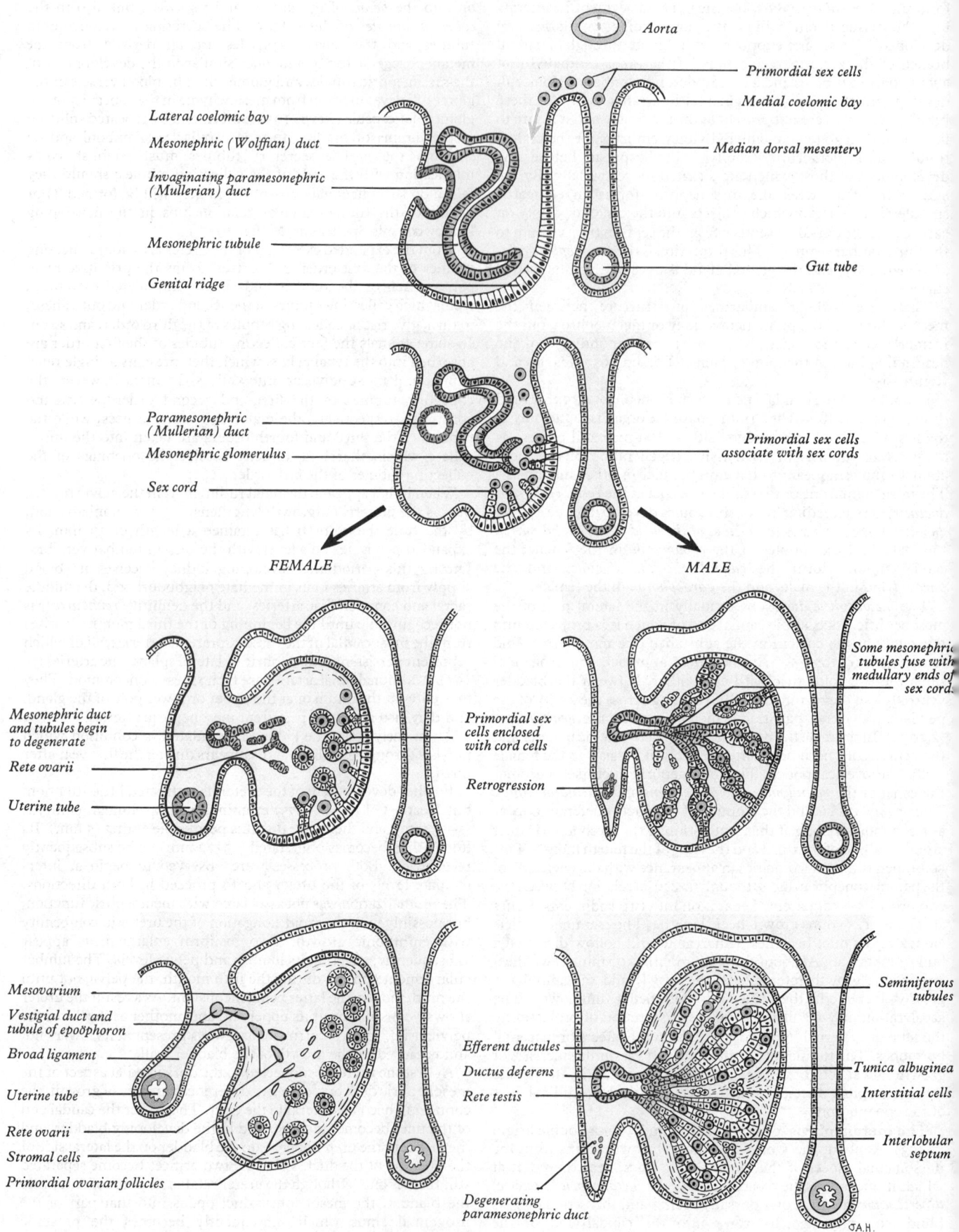

INDIFFERENT STAGES

Aorta

Primordial sex cells

Medial coelomic bay

Lateral coelomic bay

Mesonephric (Wolffian) duct

Median dorsal mesentery

Invaginating paramesonephric (Mullerian) duct

Mesonephric tubule

Gut tube

Genital ridge

Paramesonephric (Mullerian) duct

Mesonephric glomerulus

Sex cord

Primordial sex cells associate with sex cords

FEMALE   MALE

Mesonephric duct and tubules begin to degenerate

Rete ovarii

Uterine tube

Primordial sex cells enclosed with cord cells

Retrogression

Some mesonephric tubules fuse with medullary ends of sex cords

Mesovarium

Vestigial duct and tubule of epoöphoron

Broad ligament

Uterine tube

Rete ovarii

Stromal cells

Primordial ovarian follicles

Efferent ductules

Ductus deferens

Rete testis

Degenerating paramesonephric duct

Seminiferous tubules

Tunica albuginea

Interstitial cells

Interlobular septum

J.A.H.

2.104 Schema of the development of the gonads and associated ducts as seen in transverse section. Note fate of primordial sex cells, mesonephric duct and tubules and paramesonephric duct in the two sexes.

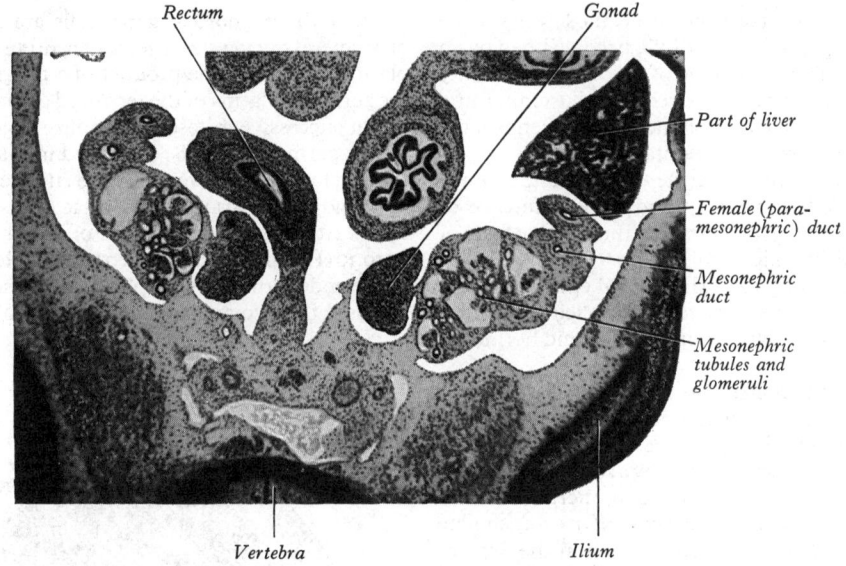

2.105A Transverse section through the lower part of the abdomen of a human fetus, nine weeks old, showing the connexions and relative positions of the structures derived from the mesonephric ridge.

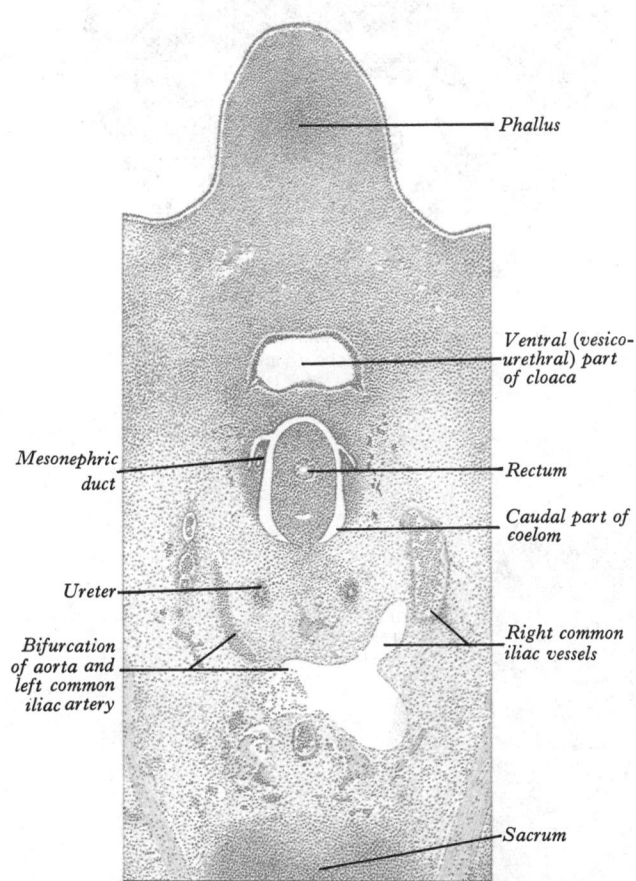

2.105B A transverse section of the tail end of a human embryo in the eighth week. C.R. length—22 mm. The projection from the dorsolateral aspect of the ventral portion of the cloaca marks the entry of the mesonephric duct. Stained with haematoxylin and eosin. Magnification × c. 20.

mesonephric ridge near its cranial end, and its blind end grows caudally in the ridge as a solid rod of cells which acquires a lumen as it lengthens. Throughout the extent of the mesonephros it is lateral to the mesonephric duct. At the caudal end of the mesonephros, which it reaches in the eighth week, the paramesonephric duct turns medially (2.103B) and crosses ventral

to the mesonephric duct to enter the genital cord (2.105C, D), where it bends caudally in close apposition with its fellow of the opposite side. The two ducts reach the dorsal wall of the urogenital sinus during the third month, and their blind ends produce an elevation on it termed the *sinus* or *Müllerian tubercle* (2.103B). Each duct consists of vertical cranial and caudal parts with an intermediate horizontal region. The cranial part forms the uterine tube, and the original coelomic invagination remains as the pelvic opening of the tube, the fimbriae becoming defined as the cranial end of the mesonephros degenerates. The caudal vertical parts of the two ducts fuse with each other (2.107F) to form the *uterovaginal primordium*. This gives rise to the lower part of the uterus, and as it enlarges it takes in the horizontal parts to form the fundus and most of the body of the adult uterus. The stroma of the endometrium and the uterine musculature, the myometrium, are developed from the surrounding mesoderm of the genital cord.

At about the 60 mm crown–rump length stage an epithelial proliferation (the *sinuvaginal bulb*) arises from the dorsal wall of the urogenital sinus in the region of the sinus tubercle, and its origin marks the site of the future hymenal orifice. Whether the epithelium involved in the proliferation is from the sinus (Bulmer 1957), or epithelium of the mesonephric duct which has extended over the Müllerian tubercle (Vilas 1932; Meyer 1938; Forsberg 1963) is uncertain. The proliferation gradually extends cranially as a solid, anteroposteriorly flattened plate, inside the tubular mesodermal condensation of the uterovaginal primordium which will eventually become the fibromuscular vaginal wall. The caudal tip of the female duct epithelium recedes until, at about the 140 mm stage, its junction with the sinus proliferation lies in the cervical canal.

Commencing from the lower end, and gradually extending cranially through its whole extent, the solid plate formed by the sinus proliferation enlarges into a cylindrical structure and the central cells desquamate to establish the vaginal lumen. According to one view, the paramesonephric ducts do not directly contribute to the formation of the vagina (Frutiger 1969), but it has also been suggested that the mesonephric and paramesonephric ducts are both concerned (Linkevich 1969). As the upper end of the vaginal plate enlarges it grows up into the mesoderm around the cervix to produce the vaginal fornices. As the lower end enlarges it invaginates the dorsal wall of the urogenital sinus around the hymeneal orifice, trapping a thin layer of mesoderm around the orifice to form the hymen (2.108). The hymen is lined on its upper surface by vaginal epithelium, which, despite its origin, is clearly distinct from the sinus epithelium on

215

the under surface. In the later stages, however, the lower surface of the hymen and a large part of the vestibule become lined by an epithelium indistinguishable from that of the vagina (Bulmer 1959). The urogenital sinus undergoes relative shortening in a craniocaudal direction to form the vestibule, which opens on the surface through the cleft between the genital folds.

During the later months of fetal life the vaginal epithelium is enormously hypertrophied, apparently under the influence of maternal hormones, but after birth it assumes the inactive form of childhood (Fraenkel and Papanicolaou 1938).

The differing embryonic origins of the vaginal epithelium and uterine epithelium have been correlated with their dissimilar responses in adult life to stimulation with oestrogenic hormones (Zuckerman 1940).

In the male the paramesonephric (female) duct mostly atrophies, but a vestige of its cranial end persists as the *appendix testis* (p. 1410). The fused terminal portions of the two ducts are connected to the wall of the urogenital sinus by a solid cord of cells, the *utricular cord*. In this position it soon merges with a proliferation of sinus epithelium, the *sinu-utricular cord*, similar to, but less extensive than, the sinus proliferation in the female. This proliferating epithelium is claimed to be an intermingling of the endoderm of the urogenital sinus with the lining epithelia of the mesonephric and paramesonephric ducts, which have extended on to the surface of the sinus tubercle. As the sinu-utricular cord grows, so the utricular cord recedes from the tubercle. In the second half of fetal life the composite cord acquires a lumen and dilates to form the *prostatic utricle*, the lining of which consists of hyperplastic stratified squamous epithelium. The sinus tubercle becomes the *colliculus seminalis* (Vilas 1933; Glenister 1962).

The primordial germ cells, from which the definitive germ cells are derived, are segregated very early in embryonic development. This has been demonstrated as early as the 2-somite stage in the chick (Clawson and Domm 1969). They are large cells, in comparison with most somatic cells, being from 12 to 20 μm in diameter, and they are characterized by vesicular nuclei with well-defined nuclear membranes, and by a tendency to retain yolk inclusions long after these have disappeared from somatic cells. (For the ultrastructure of human primordial germ cells consult Fukuda 1976, Fujimoto *et al.* 1977.) It is not yet established

whether the primordial germ cells are derived from particular blastomeres during cleavage, or constitute a clonal line from a single blastomere, or are the product of a progressive concentration of the germinal plasma of the fertilized ovum by unequal partition of this at successive mitoses (Buonoure 1939).

The germ cells can be identified in human embryos at the 13-somite stage (fourth week), in the endoderm at its caudal end and in the adjacent part of the yolk sac (Witschi 1948, and Fujimoto 1978). At this stage the number of cells is probably not more than 20 to 30 (Hardisty 1967). When the tail fold has formed they appear in the endoderm and the splanchnic mesoderm of the hindgut as well as in the adjoining region of the wall of the yolk sac. By amoeb-

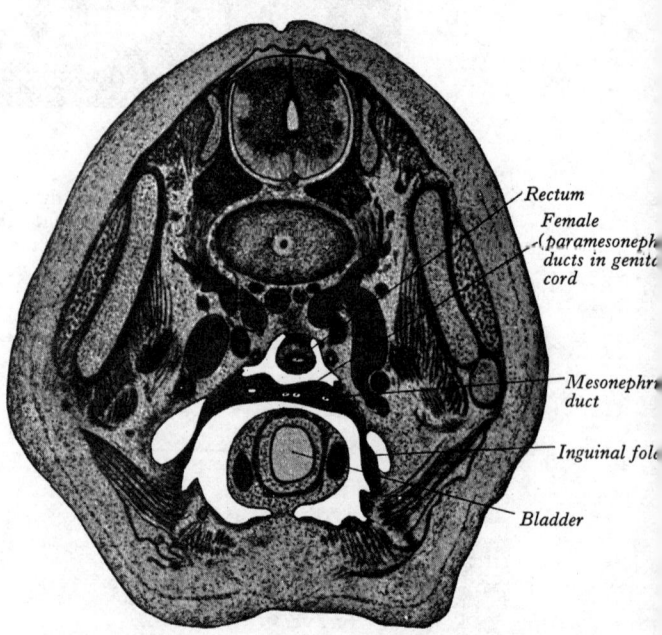

Rectum

Female (paramesonephric) ducts in genital cord

Mesonephric duct

Inguinal fold

Bladder

2.105C Transverse section through the pelvic part of a human fetus to show the formation of the genital cord and the inguinal folds.

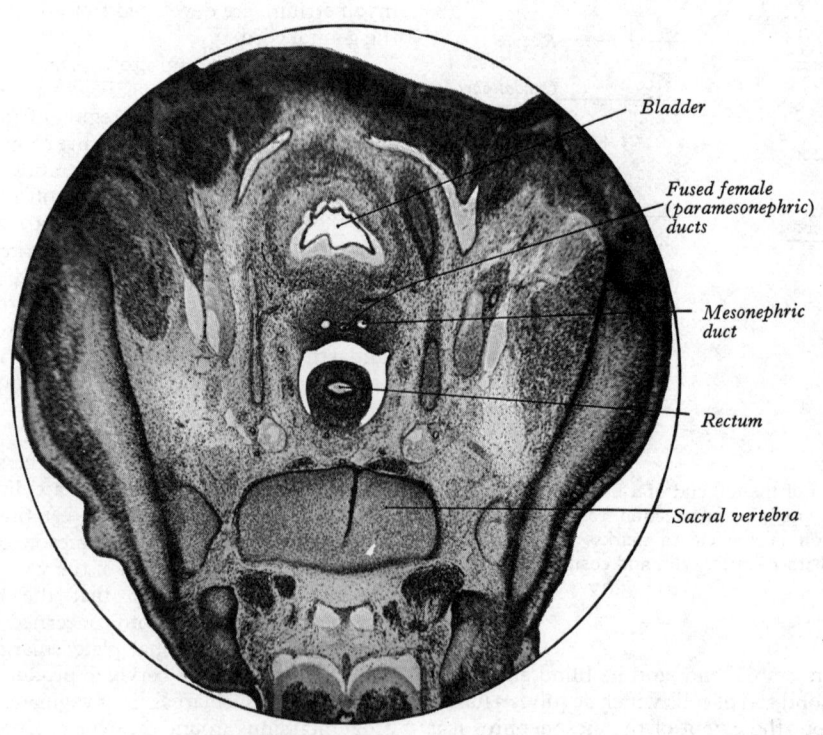

Bladder

Fused female (paramesonephric) ducts

Mesonephric duct

Rectum

Sacral vertebra

2.105D Transverse section through the pelvis of a nine-week-old male human embryo, showing the approximation of the genital cord to the dorsal wall of the urogenital sinus.

oid movements and by growth displacement they migrate dorso-cranially in the mesentery, passing around the dorsal angles of the coelom to reach the medial sides of the mesonephric ridges—the developmental sites of the gonads. In most vertebrates mitosis in the germ cells is arrested after their early segregation, to be renewed only when they reach the genital primordia. However, in mammals there is no such arrest, and the cells proliferate both during and after migration to the mesonephric ridges; cells which do not complete this migration degenerate. After segregation the primordial germ cells are often termed *primary gonocytes*, which in turn divide to form *secondary gonocytes*. The distinction between the two generations is clear in most vertebrates, but the absence of mitotic arrest in mammals leads to a merging of the two stages.

The early association of primordial germ cells with the endoderm suggests that they are in fact derived from primitive endodermal cells or perhaps from common stem cells (Franchi *et al.* 1961). Alternatively, the endoderm may exert an inductive effect on their development and differentiation without being their actual source. For a discussion of this problem and citation of pertinent references consult Carlon and Stahl (1973).

While sexual differences in germ cell numbers occur in some vertebrate species, it is uncertain whether this represents an original difference at segregation or results from earlier and more rapid proliferation in one sex or the other. No connexion between numbers of primordial germ cells and fertility has been detected, but there is evidence that gross deficiency in number may affect individual fecundity (Hardisty 1967).

The formation of the gonads is first indicated by the appearance of an area of thickened epithelium on the medial side of the mesonephric ridge in the fifth week (2.104). Elsewhere on the surface of the ridge the coelomic epithelium is one or two cells thick, but over this genital area it is many-layered. The thickening rapidly extends in a longitudinal direction until it covers nearly the whole of the medial surface of the ridge. The thickened epithelium continues to proliferate, displacing the renal corpuscles of the mesonephros in a dorsolateral direction, and forming a projection into the coelomic cavity, the *gonadal ridge*. Surface depressions form along the limits of the ridge which is thus connected to the mesonephros by an originally broad mesentery, the *mesogenitale*. In this way the mesonephric ridge becomes subdivided into a lateral part containing the mesonephric and paramesonephric ducts, which may be termed the *tubal fold*, and a medial part, termed the *gonadal fold*. The tubal fold contains the nephric tubules and glomeruli at its base.

Up to the seventh week the gonad possesses no differentiating feature. The proliferating epithelium now forms a number of cellular *gonadal cords*, separated by mesenchyme. These cords remain at the periphery of the primordium to form a cortex; more centrally a proliferation of the mesenchyme of the mesonephros constitutes a medulla. In the male all the progenitors of the definitive gonocytes become incorporated in the cords, but in the female a large number remain behind under the surface epithelium. At this stage in the male, an extension of the mesenchyme cuts off the gonadal cords from the surface and rapidly thickens to form the *tunica albuginea*. This also develops in the female but to a lesser extent and at a later stage.

The testis (2.104). The cellular cords lengthen partly by additions from the coelomic epithelium and encroach on the medulla, where they unite with the network derived from the mesenchyme which ultimately becomes the *testicular rete*. The primordial germ cells are incorporated in the cords, which later become enlarged and canalized to form the *seminiferous tubules*. (*See* Fukuda and Hedinger 1975.) The cells derived from the surface of the gonad form the *supporting cells* (of Sertoli). The *interstitial cells* of the testis are derived from the mesenchyme and possibly also from coelomic epithelial cells which do not become incorporated into the tubules. The cords of the testicular rete, which canalize later, become connected to the glomerular capsules in the cranial end of the persisting part of the mesonephros, and the glomerular tufts concerned become atrophied. The rete cords thus become connected to the mesonephric duct by the five to twelve most cranial persisting tubules and these become exceedingly convoluted and form the

lobules of the head of the epididymis. The mesonephric duct, which was the primitive 'ureter' of the mesonephros, becomes the canal of the *epididymis* and the *ductus deferens* of the testis. The seminiferous tubules do not acquire lumina until the seventh month, but the tubules of the testicular rete do so somewhat earlier.

The ovary (2.104). In its earliest stages, the ovary closely resembles the testis, although it is slower to differentiate its characteristically female features. Few, if any, of the gonadal cords invade the medulla, the majority remaining in the cortex, where they may be joined by a second proliferation from the epithelium covering the gonad. In sections of the ovary in the third and subsequent months the cords appear as clusters of cells which may or may not contain primitive germ cells. These clusters are separated by fine septa of undifferentiated mesenchyme. An *ovarian rete* arises in the medulla and some of its cords may form a junction with mesonephric glomeruli. The medulla regresses and connective tissue and blood vessels from this region invade the cortex to form the stroma of the ovary. As this invasion proceeds the cell clusters break up into individual groups which surround the *primordial ova*, now *primary oöcytes*, which have entered the prophase of the first meiotic division. These cells have been derived from a mitotic division of the primordial germ cells (*naked oögonia*). Their epithelial capsules (*pregranulosa cells*) consist of flattened cells derived from the proliferations of coelomic epithelium (Gillman 1948). The ovary now has its full complement of primary oöcytes. The majority undergo atresia at various stages during the course of their development, but the remainder resume development by completing the first meiotic division shortly before ovulation (Zuckerman 1951, 1956). The capsular cells at the same time enlarge and multiply to form the stratum granulosum, and as they do so they become surrounded by thecal cells which differentiate from the stroma.

Organ culture of human fetal ovaries confirms the above, but it was considered that the primary oöcytes had progressed to the leptotene stage. Strands of epithelial cells containing oögonia developed during culture (Baker and Neal 1974).

Only the middle part of the gonadal ridge produces the ovary. Its cranial part is sterile and becomes the *suspensory ligament of the ovary* (infundibulopelvic fold of peritoneum). Its caudal region, also sterile, is incorporated in the *ovarian ligament*.

The descent of the testis is not merely a simple migration. At first it lies on the dorsal abdominal wall, but, as it enlarges, its cranial end degenerates, and the organ therefore assumes a more caudal position. It is attached to the mesonephric fold by a peritoneal fold, the *mesorchium* (2.105A), the *mesogenitale* of the undifferentiated gonad, which contains the testicular vessels and nerves and a quantity of undifferentiated mesenchyme. In addition, it acquires a secondary attachment to the ventral abdominal wall, which has a considerable influence on its subsequent movements. At the point where the mesonephric fold

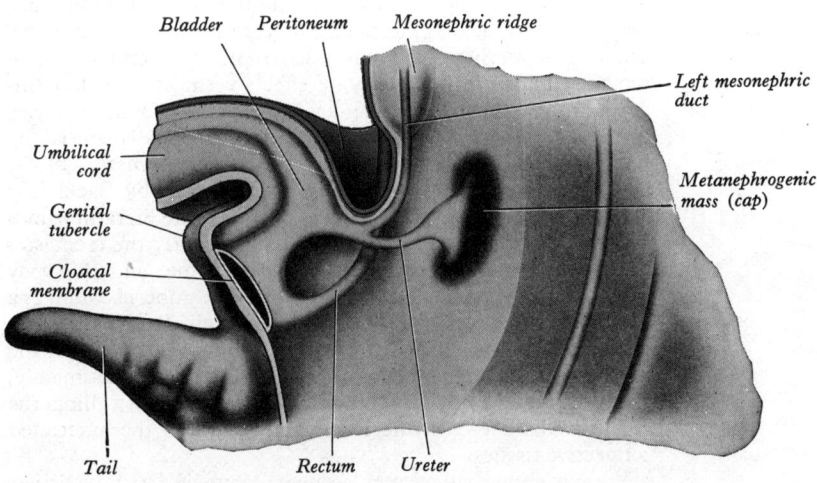

Bladder Peritoneum Mesonephric ridge
Left mesonephric duct
Umbilical cord
Genital tubercle
Cloacal membrane
Metanephrogenic mass (cap)
Tail   Rectum   Ureter

2.106 Schema, based on 2.107D, to show the formation of the pelvis of the kidney and the metanephrogenic mass or cap.

bends medially to form the genital cord (p. 213), it becomes connected to the lower part of the ventral abdominal wall by an *inguinal fold* of peritoneum (**2**.105C). The mesenchymal cells included in the inguinal fold form a cord, extending from the skin which will later form the scrotum through the inguinal fold and the mesorchium to the caudal pole of the testis. It traverses the site of the future inguinal canal, which is formed around it by the muscle of the abdominal wall as they become differentiated. At the end of the second month the caudal part of the ventral abdominal wall is horizontal but, after the return of the intestines to the peritoneal cavity (p. 201), it grows in length and becomes vertical. As a result the umbilical artery pulls up a falciform peritoneal fold, as it runs ventrally from the dorsal to the ventral wall, and this forms the medial boundary of a peritoneal fossa into which the testis projects. This fossa is termed the *saccus vaginalis* or *lateral inguinal fossa* (p. 1410) and its lower end protrudes down the inguinal canal along the gubernaculum, as the *processus vaginalis*. The caudal pole of the testis is retained in apposition with the deep inguinal ring by the gubernaculum until the seventh month, when it suddenly and rapidly passes through the inguinal canal and gains the scrotum. As it descends it is necessarily accompanied by its peritoneal covering, and the adjoining peritoneum from the iliac fossa is drawn down into the processus vaginalis. The distal end of the processes vaginalis, into which the testis projects, forms the *tunica vaginalis testis*, but the portion associated with the spermatic cord in the scrotum and in the inguinal canal normally becomes obliterated. The fascial coverings of the testis and spermatic cord, including the cremaster, are developed from the gubernaculum testis (Backhouse and Butler 1958).

The actual mechanism of the descent of the testis is still uncertain. It has been ascribed, by different investigators, to shortening and active contraction of the gubernaculum, to increased intra-abdominal pressure, to a simple growth process, and to the effect on the convex surface of the gland of the active contraction of the lower fibres of the internal oblique muscle, squeezing it through the canal. None of these explanations is entirely convincing. Whatever contribution the gubernaculum makes to the process, it cannot be by muscular contraction, because it contains no muscle, remaining a soft, mesenchymatous mass through its existence (Forssner 1928; Backhouse and Butler 1960). It precedes the testis both spatially and in rate of growth, forming a tapering column of soft tissue, with the diminutive testis at its cranial pole. It continues to grow until the seventh month, by which time its caudal part has filled the future inguinal canal and has begun to expand the developing scrotum. In this it also precedes the processus vaginalis, but does not develop attachments to skin, nor is there any evidence that it produces the radiating extensions into the suprapubic, perineal and femoral sites, which are often used as explanations for the various forms of ectopia testis. By its soft consistency the gubernacular tissue may offer a route of low resistance to the descending testis, and the cessation of its growth in the last two months of gestation, coupled with an accelerating rate of growth in the testis and epididymis, may also be a factor in testicular descent as far as the inguinal canal. The mechanism of the final, rapid intrascrotal descent remains unidentified. Endocrine effects seem certain, but this does not explain the actual agency. This account is based principally upon events as observed in porcine material by Backhouse and Butler (1960). A more recent discussion of the problems of testicular descent and maldescent by Backhouse (1964) should be consulted. He has reviewed the literature since Hunter's original description (1762). Apparently the cremaster muscle develops in gubernacular mesenchyme, and this may explain the development of the concept of the gubernaculum as a 'fibromuscular' ligament. The processus vaginalis also develops in close association with the soft, mucoid tissue of the gubernaculum and follows it into the genital swelling. Ultimately, as the testis and epidymis fill the scrotum, or genital swelling, the gubernacular mesenchyme is incorporated into the associated connective tissues.

Various abnormalities may occur in connexion with testicular descent and obliteration of the processus vaginalis (p. 1416). The testis may remain in the abdomen, or it may fail to reach the scrotum and may then lie in any of the following situations: (1) in the perineum, (2) at the root of the penis, (3) at the superficial inguinal ring (p. 1367), (4) in the upper part of the thigh. These malpositions have been traditionally associated with certain additional extensions of gubernacular tissue. The largest extension normally passes to the scrotum while lesser extensions have been described as gaining attachment to the perineum, the root of the penis, the pubis, the inguinal ligament, and the neighbourhood of the saphenous opening. The testis must follow the processus vaginalis, and should the latter for any reason follow any but the scrotal extension of the gubernaculum, malposition of the testis will result. It should be appreciated, however, that considerable doubt has now been expressed concerning these lesser expansions (previously the so-called 'tails of Lockwood'): possibly they reflect premature and abnormal fibrous partitioning of the gubernacular mesenchyme.

The processus vaginalis may remain completely patent, or its obliteration may be incomplete. When it retains a connexion with the general peritoneal cavity it provides a preformed sac for a potential oblique inguinal hernia. It may be occluded at its upper end and may be shut off from the tunica vaginalis and yet remain patent in the intervening section. The patent portion may become distended with fluid, constituting an encysted hydrocoele of the spermatic cord.

The descent of the ovary is less extensive. Like the testis, the ovary ultimately reaches a lower level than it occupies in the early months of fetal life, but it does not leave the pelvis to enter the inguinal canal save under abnormal conditions. Connected to the medial portion of the mesonephric fold by the *mesovarium*, which is homologous with the mesorchium, the ovary is also attached to the ventral abdominal wall through the medium of the inguinal fold. In this fold the mesenchymatous gubernaculum develops and, as it traverses the mesonephric fold, it acquires an additional attachment to the lateral margin of the uterus near the entrance of the uterine tube; its lower part, caudal to this attachment, becomes the *round ligament of the uterus* and the part cranial to this *the ovarian ligament*, these two structures together being homologous with the gubernaculum testis in the male. This new attachment restricts the extent of ovarian descent. At first the ovary is attached to the medial side of the mesonephric fold, but owing to the way in which the two mesonephric folds unite to form the genital cord (p. 213) its connexion is finally to the posterior layer of the broad ligament of the uterus. The gubernaculum thus persists in the female, unlike the male, as two fibrous bands or ligaments.

The *saccus vaginalis* also appears in the female; its prolongation into the inguinal canal (sometimes termed the canal of Nuck in the female) normally undergoes complete obliteration, but may remain patent and form the sac of an oblique inguinal hernia (p. 1367). At birth the ovary and the lateral end of the corresponding uterine tube lie above the pelvic brim, and they do not sink into the lesser pelvis until it enlarges sufficiently to contain both of them and the other pelvic viscera, including the bladder.

The urinary bladder (**2**.107A–E) is derived partly from the endodermal cloaca and partly from the caudal ends of the mesonephric ducts (**2**.103, 107, 109). After the separation of the rectum from the cloaca (p. 202), the ventral part of the latter becomes divided into three regions: (1) a cranial *vesico-urethral canal*, continuous with the allantoic duct—into this part the mesonephric ducts open; (2) a middle, narrow channel, the *pelvic portion*; and (3) a caudal, deep, *phallic* section, closed externally by the *urogenital membrane* (**2**.107D, E). The second and third parts together constitute the *urogenital sinus*. The ureter and the mesonephric duct come to open separately into the vesico-urethral part (p. 213). The termination of the mesonephric duct then moves caudally to open into that part which will form the prostatic urethra. This occurs by the formation of a caudally directed loop of the duct behind the urogenital sinus, followed by absorption of the apposed walls. In this way the mesonephric duct contributes to the trigone of the bladder and dorsal wall of the prostatic urethra. The remainder of the vesico-urethral part forms the body of the bladder and part of the prostatic urethra; its apex is prolonged to the umbilicus as a narrow canal, termed the *urachus*. In post-natal life the urachus is drawn downwards as the

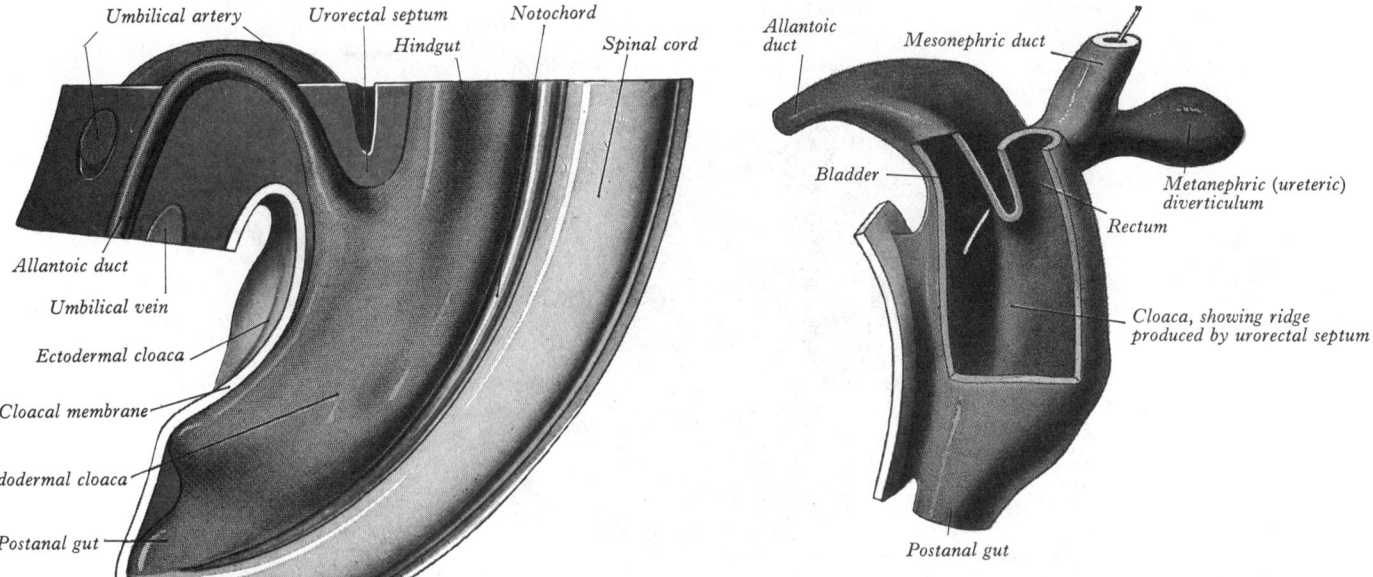

2.107A The tail end of a human embryo, about four weeks old. The model has been dissected to show the left lateral aspects of the spinal cord, notochord and endodermal cloaca. (After Keibel.)

2.107B The endodermal cloaca of a human embryo, near the end of the fifth week. A wire has been passed along the right mesonephric duct into the cloaca, and a part of the left wall of the cloaca, including the left mesonephric duct, has been removed, together with the adjoining portions of the walls of the developing bladder and rectum. A piece of the ectoderm around the cloacal membrane has been left *in situ* and is uncoloured. (After Keibel.)

bladder descends, but its blind upper end remains connected to one or both of the obliterated umbilical arteries. Its lumen persists throughout life, and its lower end frequently communicates with the bladder near its apex (Begg 1930).

The prostate arises during the third month from the proximal part of the urethra. The earlier outgrowths, some fourteen to twenty in number, arise from the endoderm around the whole circumference of the tube, but mainly on the lateral aspect and excluding the dorsal wall above the utricular plate. These outgrowths give rise to the outer glandular zone of the prostate (p. 1420). Later outgrowths from the dorsal wall above the mesonephric ducts arise from the epithelium of mixed urogenital, mesonephric and possibly paramesonephric origin covering the cranial end of the sinus tubercle. These produce the internal zone of glandular tissue. The outgrowths, which are at first solid, branch, become tubular and invade the surrounding mesenchyme, which is being differentiated into muscular tissue.

Similar outgrowths occur in the female, but remain rudimentary. The urethral glands correspond to the mucosal glands around the upper part of the prostatic urethra and the paraurethral glands to the true prostatic glands of the external zone (Glenister 1962).

The *bulbo-urethral glands* in the male, and *greater vestibular glands* in the female, arise as diverticula from the epithelial lining of the urogenital sinus.

The external genital organs, like the gonads, pass through an undifferentiated state before distinguishing sexual characters appear (2.110). A surface elevation, the *genital tubercle*, appears at the cranial end of the cloacal membrane and lengthens to form the *phallus* (2.105B). Within it is a longitudinal endodermal mass, the *urethral plate* (2.109, 110), which grows forwards from the walls of the cloaca and urogenital sinus towards the tip of the organ. The lower aspect of the plate is in contact with ectoderm lining a median groove, the *primary urethral groove*, which has meanwhile developed along the caudal surface of the phallus. The raised margins of the groove are the *genital folds* (2.110). Proximally they surround the urogenital membrane and terminate in a transverse ridge immediately ventral to the anus. The rupture of the urogenital membrane provides a common perineal orifice for the genital and urinary organs at the base of the phallus, bounded at the sides by the genital folds. Meanwhile disintegration of the cells of the urethral plate and contiguous ectoderm occurs, commencing at the base of the phallus and resulting in a

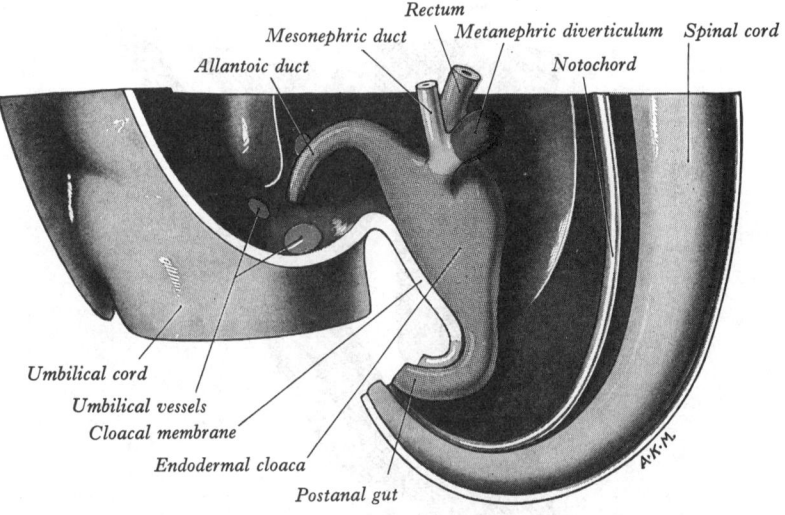

2.107C The caudal end of a human embryo, about five weeks old. (Drawn from a model by Keibel.)

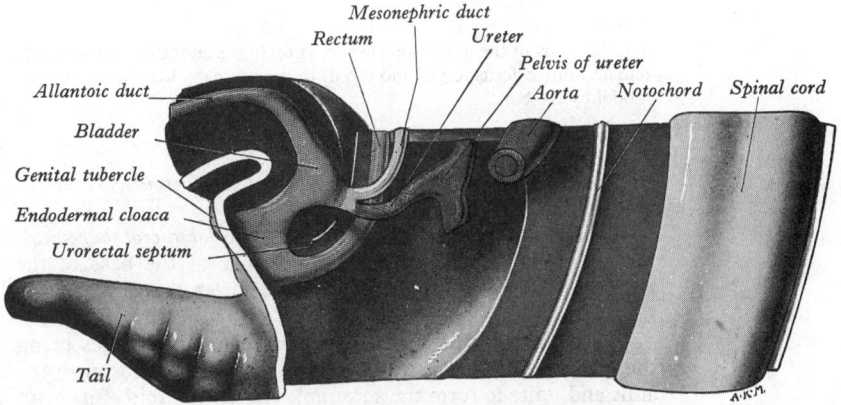

2.107D Part of the caudal end of a human embryo, about six weeks old. (Drawn from a model by Keibel. The model, which has been partly dissected, is seen from the left side.)

219

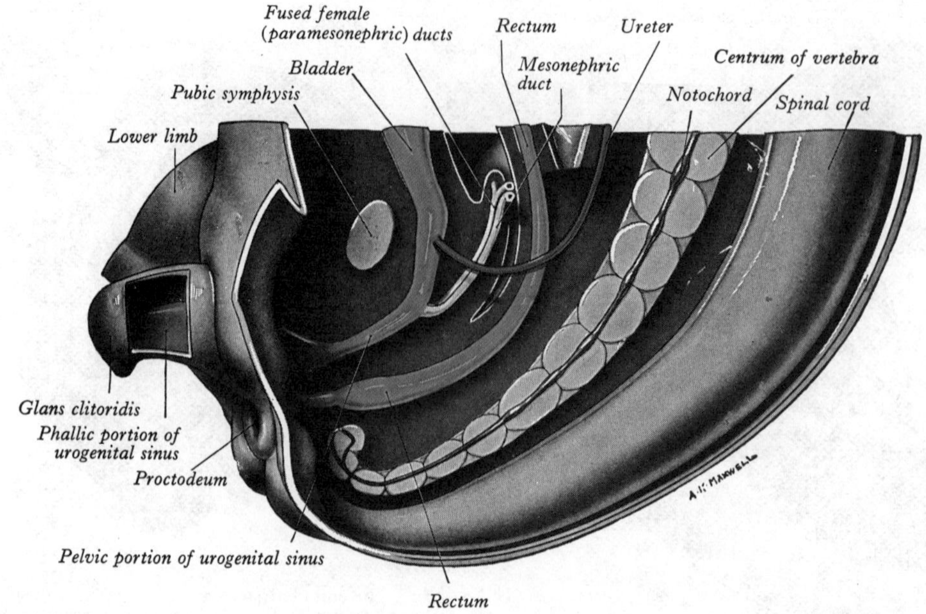

Fused female
(paramesonephric) ducts
Rectum   Ureter
Bladder
Mesonephric
duct
Centrum of vertebra
Pubic symphysis
Notochord   Spinal cord
Lower limb

Glans clitoridis
Phallic portion of
urogenital sinus
Proctodeum

Pelvic portion of urogenital sinus

Rectum

2.107E The tail end of a female human fetus, eight and a half to nine weeks old. The model has been dissected from the left side to show the structures in and near the median plane. Note that the cloaca has now been separated into urogenital and intestinal segments. (After Keibel.)

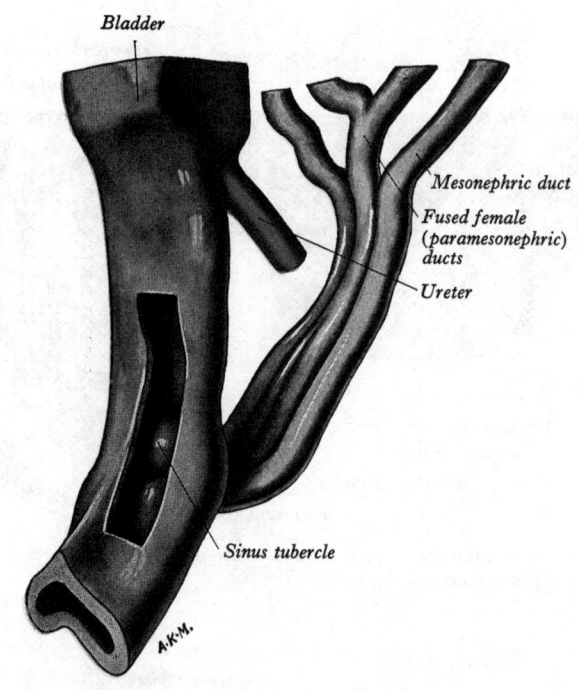

Bladder

Mesonephric duct

Fused female
(paramesonephric)
ducts

Ureter

Sinus tubercle

2.107F Part of the vesico-urethral portion of the endodermal cloaca of a female human fetus, eight and a half to nine weeks old. (Drawn from a model by Keibel.)

deepening of what can now be termed the *definitive urethral groove*.

While these changes are in progress two *labioscrotal (genital) swellings* have appeared, one on each side of the base of the phallus; these extend caudally, separated from the genital golds by distinct grooves (2.110).

The male phallus enlarges to form the penis, its apex being the glans. The genital swellings meet each other ventral to the anus and unite to form the scrotum. The genital folds fuse with each other from behind forwards enclosing the phallic part of the urogenital sinus behind to form the bulb of the urethra; similarly, the folds close the definitive urethral groove in front to form the

greater part of the spongiose urethra. Fusion of the folds results in the formation of a median raphe, and occurs in such a way that the lining of the urethra is mainly, if not wholly, endodermal in origin (Glenister 1954). Thus, as the phallus lengthens, the urogenital orifice is carried onwards until it reaches the glans. At the tip of the glans an ingrowth of surface epithelium has occurred to meet the anterior extremity of the urethral plate, and the disintegration of this gives rise to a groove which is entirely ectodermal in origin; closure of this creates the terminal part of the urethra contained within the glans.

The glans and shaft of the penis are hence recognizable by the third month. The prepuce begins to develop in the third month, when the urethra still exhibits its primary orifice at the base of the glans. A ridge consisting of a mesodermal core covered by epithelium appears proximal to the neck of the penis and extends forwards over the glans. Deep to this ridge is a solid lamella of epithelium which extends backwards to the base of the glans. The ventral extremities of the ridge curve backwards to become continuous with the genital folds at the margins of the urethral orifice. As the urethral folds meet to form the terminal part of the urethra the ventral horns of the ridge fuse to form the frenulum. Over the dorsum and sides of the glans, the epithelial lamella breaks down to form the preputial sac and thus free the prepuce from the surface of the glans. Thereafter the prepuce grows as a free fold of skin covering the terminal part of the glans. The preputial sac may not be complete until 6 to 12 months or more after birth and, even then, the presence of some connecting strands may still interfere with the retractability of the prepuce.

The mesodermal core of the phallus is comparatively undifferentiated in the first two months, but during the third month the blastemata of the corpora cavernosa become defined. Nerves are present in the differentiating mesenchyme from the seventh week (Dail and Evan 1974).

The female phallus, which exceeds the male in length in the early stages, becomes the clitoris. The genital swellings remain separate as the labia majora and the genital folds also remain separate, forming the labia minora. The perineal orifice of the urogenital sinus is retained as the cleft between the labia minora, above which the urethra and vagina open. The prepuce of the clitoris develops in the same way as the homologous structure in the male.

The urethra, in the *female* is derived entirely from the vesico-urethral region of the cloaca (p. 218). It is homologous with the

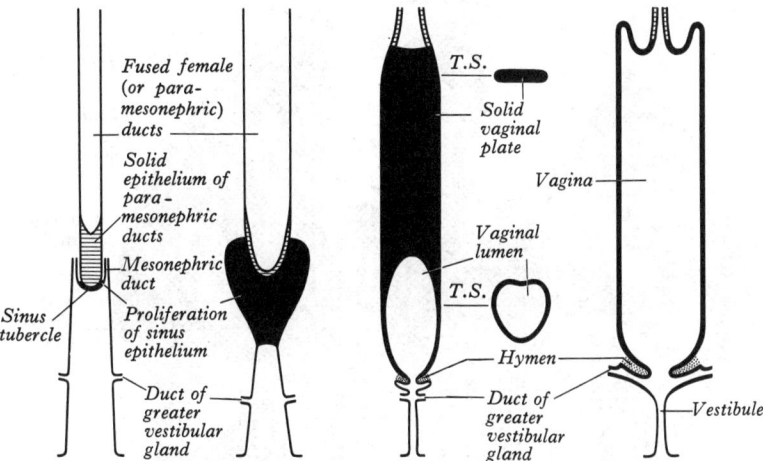

2.108   Diagrams to show the successive stages in the development of the uterus and vagina.

part of the prostatic urethra proximal to the orifices of the prostatic utricle and the ejaculatory ducts.

In the *male*, the prostatic urethra proximal to the orifice of the prostatic utricle is derived from the vesico-urethral part of the cloaca and the incorporated caudal ends of the mesonephric ducts (Bengmark 1958). The remainder of the prostatic part, the membranous part and probably the part within the bulb, are all derived from the urogenital sinus (p. 213). The succeeding section, as far as the glans, is formed by the fusion of the genital folds, while the section within the glans is formed in the way described above.

The homologies of the various parts of the urogenital system in the male and the female (**2**.103B, 110) can be summarized as follows:

Undifferentiated	Male	Female
Gonad	Testis	Ovary
Gubernacular cord	Gubernaculum testis	Ovarian and Round ligaments
Mesonephros (Wolffian body)	Appendix of epididymis(?) Efferent ductules Lobules of epididymis Paradidymis Aberrant ductules	Appendices vesiculosae (?) Epoöphoron Paroöphoron
Mesonephric duct (Wolffian duct)	Duct of epididymis Ductus deferens Ejaculatory duct Part of bladder and prostatic urethra	Duct of epoöphoron Part of bladder and urethra
Female (paramesonephric or Müllerian) duct	Appendix of testis Prostatic utricle	Uterine tube Uterus Vagina(?)
Allantoic duct	Urachus	Urachus
Cloaca: dorsal part	Rectum and upper part of anal canal	Rectum and upper part of anal canal
ventral part	Most of bladder, part of prostatic urethra	Most of bladder and the urethra
urogenital sinus	Prostatic urethra distal to utricle	
	Bulbo-urethral glands Rest of urethra	Greater vestibular glands Vestibule
Genital folds	Ventral penis	Labia minora
Genital tubercle	Penis	Clitoris

Congenital anomalies Defects of the urethra, due to arrests of development, are not uncommon in the male. The urethra may open on the ventral aspect of the penis at the base of the glans, and the part of the urethra which is normally within the

glans is absent. This constitutes the simplest form of *hypospadias*. In more severe cases the genital folds fail to fuse, and the urethra opens on the ventral aspect of a malformed penis just in front of the scrotum. A still greater degree of this malformation is accompanied by failure of the genital swellings to unite with each other. In these cases the scrotum is divided and, since the testes are also frequently undescended, the resemblance to the labia majora is very striking. Male children suffering from this deformity are often mistaken for girls.

Maldevelopment of the cloacal or urogenital membranes is a less common condition, but two varieties can be distinguished. (1) In *extroversion of the bladder* (ectopia vesicae) the lower part of the anterior abdominal wall is occupied by an irregularly oval area, covered with mucous membrane, on which the two ureters open (Wyburn 1937). Around its periphery this extroverted area, covered by transitional epithelium, becomes continuous with the skin. (2) In *extroversion of the cloaca* the condition is very similar, but is complicated by the presence of intestinal openings in the median plane. In (2) the cloacal membrane is probably abnormally elongated and ruptures prematurely and throughout its whole extent, prior to the formation of the urorectal septum. In (1) the maldevelopment occurs after the separation of the ventral from the dorsal part of the cloaca. The urogenital membrane extends further cranially than it does in normal cases and the genital tubercle forms at its *caudal* limit. Rupture of the membrane throws the bladder open to the exterior.

In *epispadias* the urethra opens on the dorsal aspect of the penis at its junction with the anterior abdominal wall. No entirely satisfactory explanation has yet been suggested for this anomaly. For a genetic and epidemiological study of urinary tract malformations consult Bois *et al.* (1975).

Differentiation of the genital organs (Burns 1955; Price and Jost 1957; Moore 1947) is primarily directed by genetic factors (p. 100) which may mediate their effects through endocrine controls, notably the cortex of the suprarenal gland. The differentiation of the gonad may be modified by several factors including chromosomal constitution (p. 102), degree of ripeness of the ovum at fertilization, the number and distribution of the primordial germ cells, and endocrine influences. A reduction in the number of primordial germ cells, believed to be associated with over-ripeness of the ovum at fertilization, exercises a masculinizing effect on genetic females. It is suggested that this is due to a failure of normal development of the superficial or cortical zone of the ovary, thus allowing the deep or medullary zone, which is the essentially male element, to dominate subsequent differentiation. The zones are not well marked in man but it is considered that they play an important role in the differentiation not only of the gonad but also of the genital ducts, mesonephric and paramesonephric, cloaca, external genitalia, and eventually of the secondary sexual characters. During their development the accessory sexual organs also pass through an indifferent or potentially bisexual stage.

221

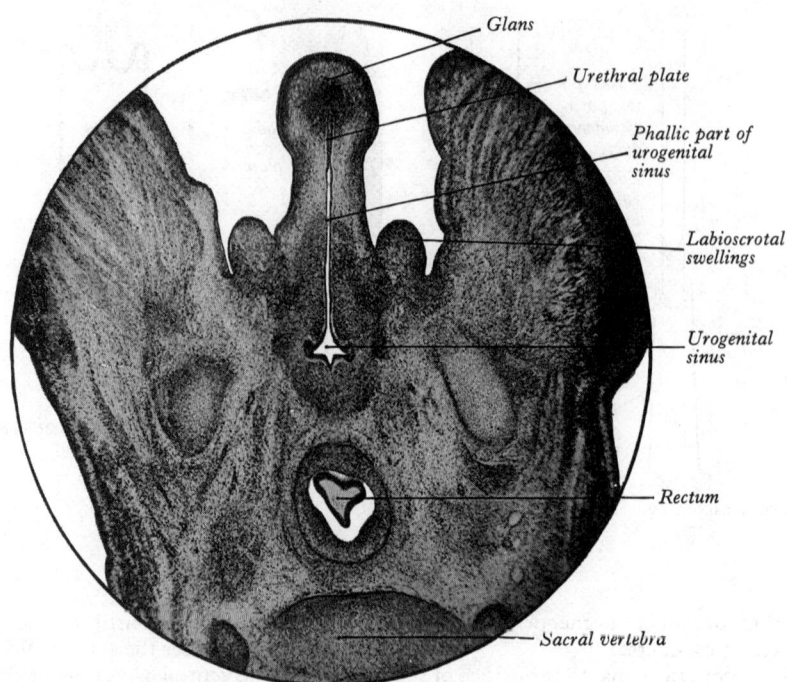

Glans

Urethral plate

Phallic part of urogenital sinus

Labioscrotal swellings

Urogenital sinus

Rectum

Sacral vertebra

2.109 Transverse section through the lower part of the pelvis of a nine-week-old human fetus.

The part played by hormones in the differentiation of the reproductive organs is not clear, but both androgens and oestrogens may profoundly modify this process, as instanced by work on the freemartin in cattle (Jost 1961). In twin pregnancy, where the fetuses are of opposite sex and their placental circulations anastomose, the male often exerts a masculinizing effect on its female twin. This has been attributed to the action of hormones elaborated in the interstitial tissue of the medullary zone in the testes of the male fetus. The gonad in turn affects the differentiation of the genital ducts, cloaca and external genitalia. It seems that the fate of the accessory organs, and particularly of the ducts, is determined largely by the presence or absence of a male factor, possibly hormonal, elaborated by the testes. The different parts of the accessory genital organs are most susceptible to this masculinizing effect over limited but differing periods.

Prenatal Growth in Form and Size

The absolute size of neither embryo nor fetus affords a reliable indication of either its true age or stage of structural organization, even though graphs based on large numbers of observations have been constructed to provide averages. All such data suffer from the difficulty of equating dimensions and degree of differentiation with the actual time of conception, which can rarely, if ever, be established with complete exactness. The life of the individual really commences with fertilization, but the date of this cannot be exactly determined in mankind. It has long been customary to compute the age, whether in a normal birth or an abortion, from the last menstrual period of the mother, but since ovulation usually occurs near the fourteenth day of a period, this 'menstrual age' is about two weeks too much. Where a single coitus can be held to be responsible for conception, a 'coital age' can be established, and the 'fertilization age' cannot be much less than this, because of the limited viability of both gametes, though it is usually held that the difference may be several days—a highly significant interval in the earlier stages of embryonic development. Even if the time of ovulation and coitus were known in instances of spontaneous abortion, not only would some uncertainty still persist with regard to the time of fertilization, but also there would remain an indefinable period between the cessation of development and the actual recovery of the conceptus. With the legalization of abortion in some countries the

latter source of inaccuracy may be expected to become less important.

To overcome these difficulties early embryos have been graded or classified, on the basis of both internal and external features, into developmental stages or 'horizons' (**2.**111 A–H). Classical contributions in this field have been made by Lillie (1917), Streeter (1942–51), Hamilton (1944), Hertig *et al.* (1956), Heuser and Corner (1957), and O'Rahilly (1973). Although it has become customary to describe these developmental levels as *stages*, the earlier descriptions are still accepted. However, Stages 1 to 9, covering the first three weeks of development, have been given a more reliable basis by O'Rahilly (1973), who has also gathered together the pertinent literature.

Stages 1 to 3 occupy the first 4 to 5 days after fertilization of the oöcyte in the ampulla of the uterine tube. The initial 24 hours (Stage 1) are occupied by fertilization, the dominant feature of which is the fusion of the male and female pronuclei. This is followed by the first mitotic division, which is the onset of segmentation, or cleavage, and is arbitrarily regarded as the transition from Stage 1 to Stage 2. Stage 2 is characterized by the continuation of cleavage, starting with two blastomeres and ending with about twelve. During this stage the developing morula moves along the uterine tube, by mechanisms still not wholly understood, a journey occupying about 4 days. During the fourth day a segmentation cavity appears within the cell mass, and this is taken as initiating Stage 3, which thus corresponds to the establishment of a free blastocyst.

Although comparatively few human embryos representing Stages 1 to 3 have been recovered (*see* pp. 103–5 and consult O'Rahilly 1973), a considerable degree of agreement exists and, moreover, the recent observations of *in vitro* fertilization between human spermatozoa and ova (**2.**18 A–H) support the earlier descriptions.

Stages 4 to 6 are concerned with the endometrial attachment of the blastocyst, trophoblastic development, implantation, further development of the blastocyst and the appearance of the primitive streak. Stage 4 corresponds to the fifth and sixth post-fertilization days. The blastocyst, now in the uterine cavity, loses its zona pellucida (**2.**19, 20) and it begins the rapid but complex activities of orientation in respect of the endometrium—adhesion, penetration and the cellular proliferation of trophoblastic growth. This establishment of dependence on the maternal circulation for nutritional requirements is far more rapid in primates than in

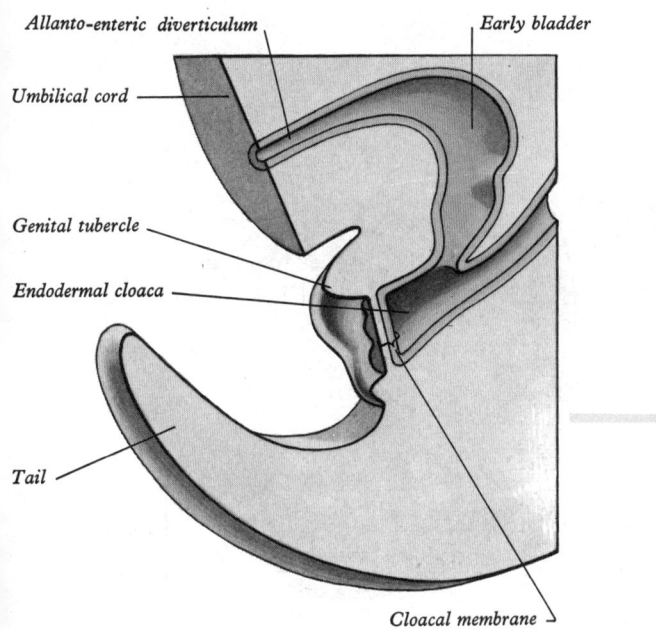

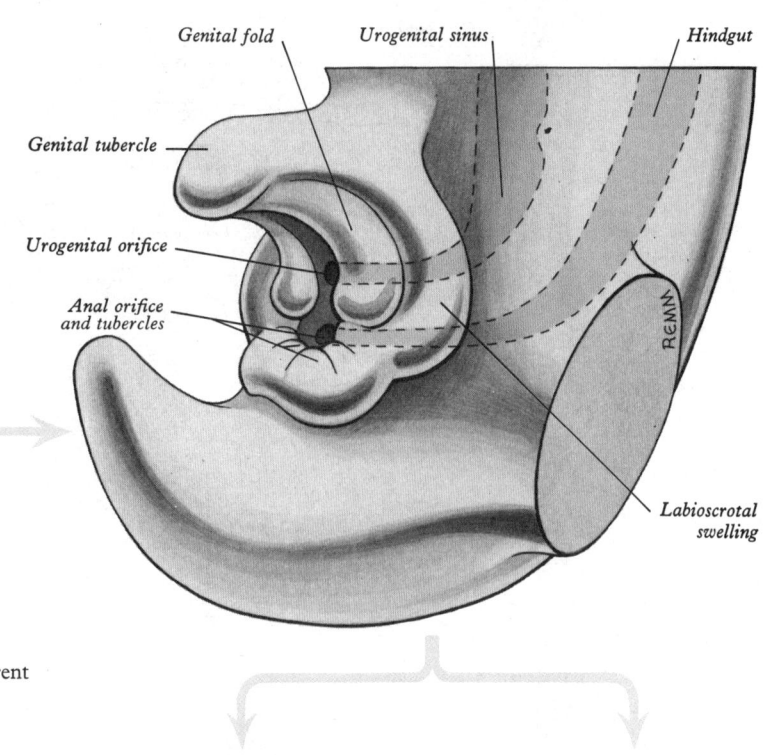

2.110 The development of the external genitalia from the indifferent stage to the definitive male and female conditions.

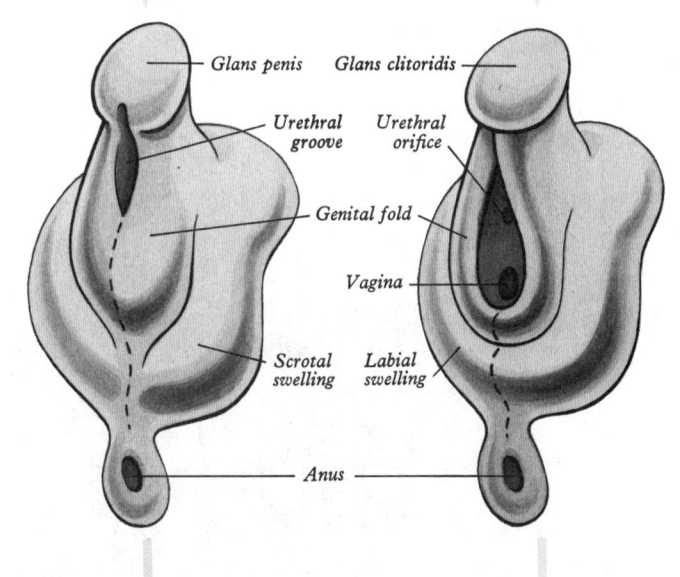

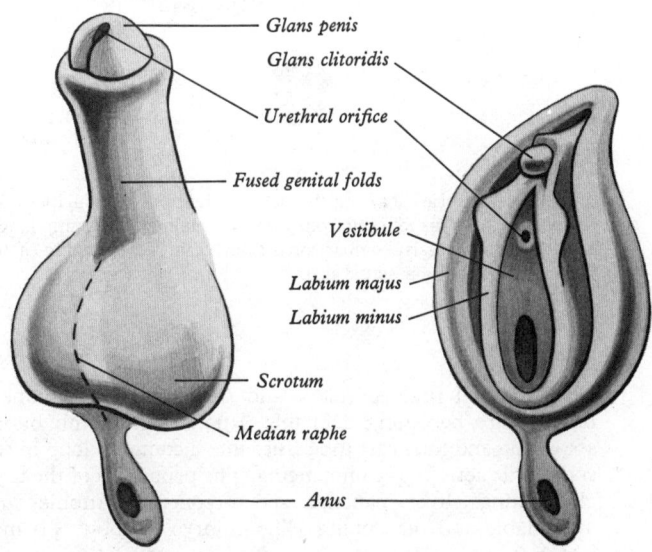

other mammals. Stage 5 is reached when implantation has occurred, occupying the seventh to twelfth days; syncytiotrophoblastic and cytotrophoblastic strata have differentiated, the proamniotic cavity has appeared, and a labyrinthine system of intercommunicating trophoblastic lacunae through which the maternal blood ebbs and flows, has developed. A little later in Stage 5 the exocoelomic membrane has been identified. In Stage 6 chorionic villous stems become defined and begin to develop side branches almost at once, producing an increasingly complex intervillous space. A little later the primitive streak becomes apparent and differentiation of the embryonic area has commenced, and may now be distinguished from the various extra-embryonic tissues.

Because of the complexity of the events in Stages 5 and 6, both have been subdivided by some authorities. A much greater number of human embryos have been recovered which represent Stages 4 to 6.

Stages 7 to 9 are characterized by basic embryogenic changes. During Stage 7 (sixteenth to eighteenth days) the primitive streak develops further and the notochordal ('head') process appears, together with the other mesodermal strata. The chorion and amnion continue to develop, villous stems being generally distributed over the former, but are more pronounced at the embryonic pole. Haemopoetic foci appear in the wall of the definitive yolk sac, and the cloacal membrane and allanto-enteric diverticulum are defined. Associated with the latter promordial germ cells have been noted. In Stage 8 (eighteenth to twentieth days) the prechordal plate, the primitive pit, the neural groove and the notochordal and neurenteric canals are all definable. By Stage 9 (twentieth to twenty-second days) the neural groove is deepening and the first somites begin to appear about midway along it. The cranial half of the groove, representing developing brain, is beginning to develop a cephalic flexure, optic primordia become visible, and the appearance of head and tail folds has commenced, as Stage 10 is approached. The foregut is becoming defined, and indications of early pharyngeal pouches may be identifiable. The embryo is now 1·5 to 2·0 mm in length.

Stages 10 to 12, occupying days 22 to 26, feature continued formation of somites, and during this, the fourth week, the head and tail folds are completed, the neural groove closes, and the primary cerebral vesicles appear. The cervical flexure can now be recognized, the optic vesicles form and the lens and otic placodes appear and become vesicular. The branchial arches are appearing,

223

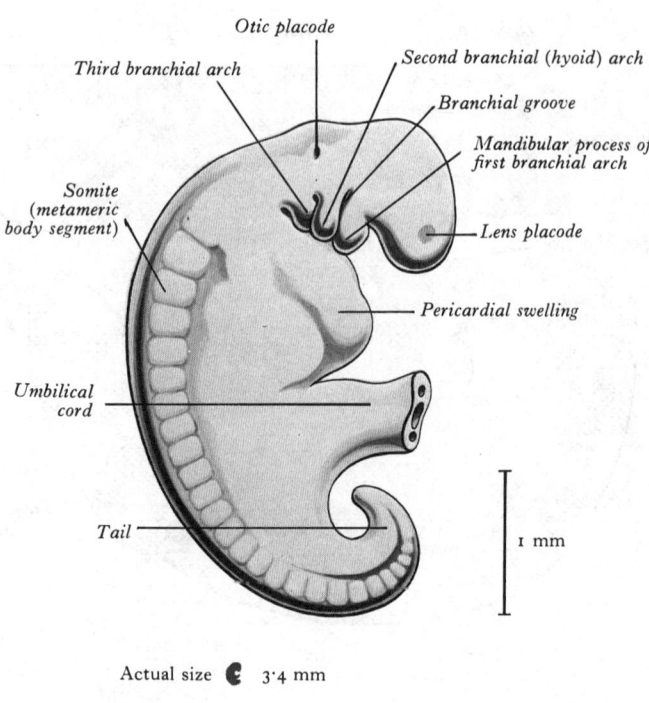

Otic placode

Third branchial arch

Second branchial (hyoid) arch

Branchial groove

Mandibular process of first branchial arch

Somite (metameric body segment)

Lens placode

Pericardial swelling

Umbilical cord

Tail

1 mm

Actual size   3·4 mm

A

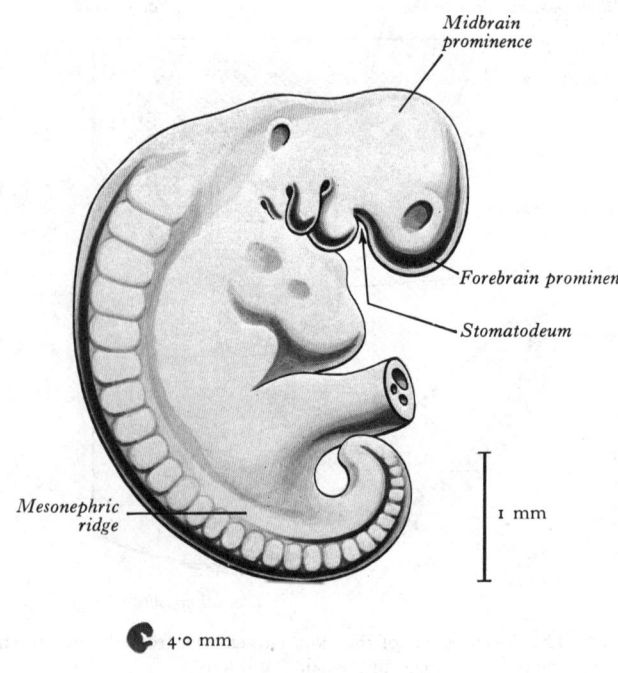

Midbrain prominence

Forebrain prominence

Stomatodeum

Mesonephric ridge

1 mm

4·0 mm

B

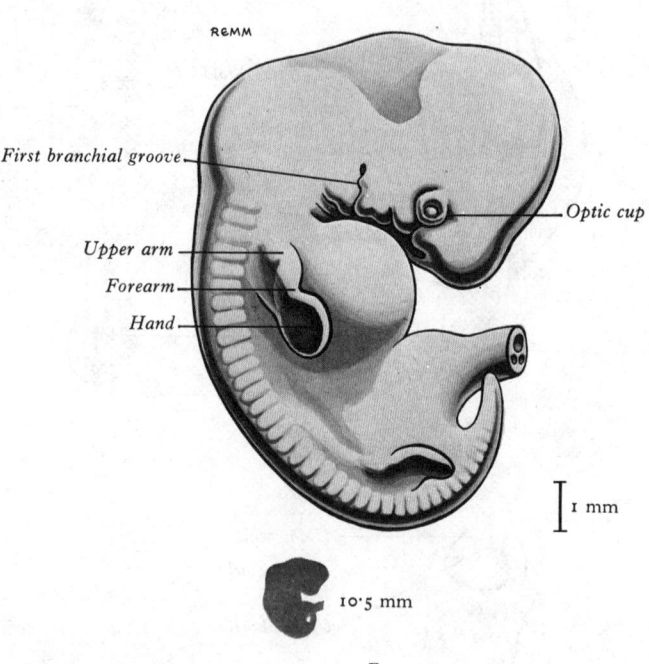

R&MM

First branchial groove

Optic cup

Upper arm

Forearm

Hand

1 mm

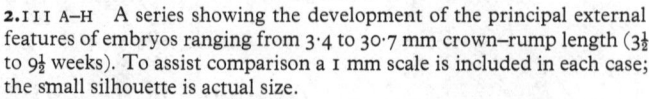

10·5 mm

E

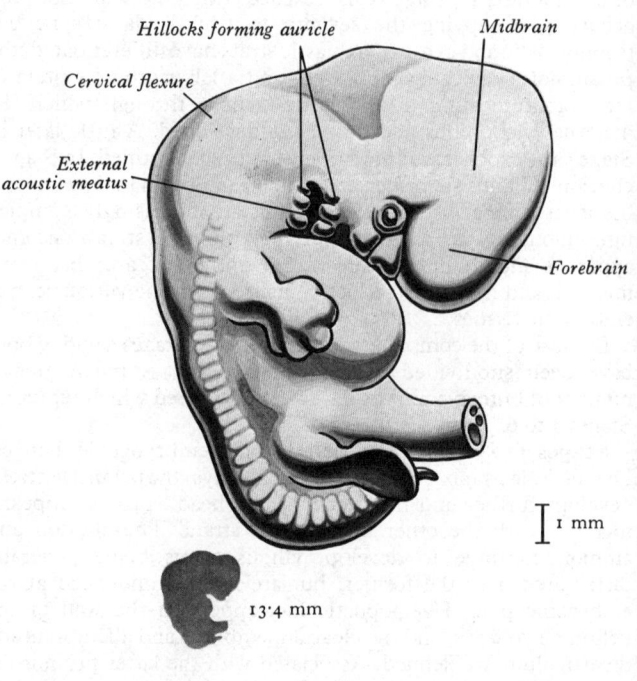

Hillocks forming auricle

Midbrain

Cervical flexure

External acoustic meatus

Forebrain

1 mm

13·4 mm

F

2.111 A–H A series showing the development of the principal external features of embryos ranging from 3·4 to 30·7 mm crown–rump length (3½ to 9½ weeks). To assist comparison a 1 mm scale is included in each case; the small silhouette is actual size.

head and tail folds develop, and the cloacal membrane and hindgut are becoming definable. Rudimentary limb buds are appearing and the heart tubes fuse into a common loop in which contractile activity is commencing. The primordia of the thyroid gland, lungs, liver, pancreas and mesonephric tubules are all identifiable and developing. The embryo is about 4·0 mm in length.

Stages 13 to 16 correspond approximately to the fifth week of

development, during which the embryo grows from about 4·0 to 8·0 mm. It becomes markedly curved and its junction with the yolk sac relatively constricted. The cervical flexure is increased and the mesencephalic flexure is appearing. The dorsolateral (alar) and ventrolateral (basal) laminae are differentiating, and the sites of the corpus striatum and thalamus are marked by proliferating cells. The cranial and spinal nerves are developing, together with their ganglia, from the associated placodes and

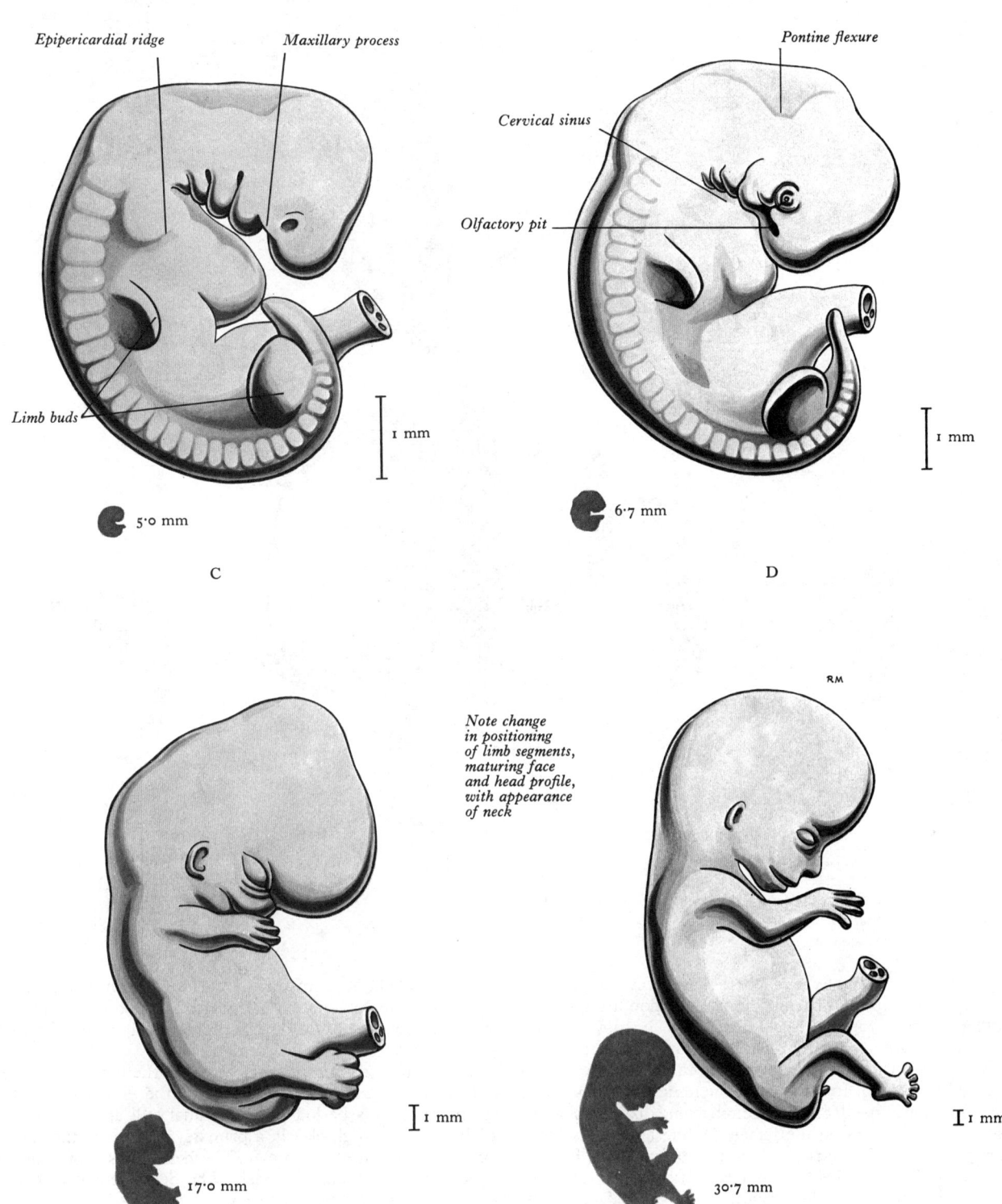

Epipericardial ridge
Maxillary process
Limb buds
5·0 mm
C

Pontine flexure
Cervical sinus
Olfactory pit
6·7 mm
D

Note change
in positioning
of limb segments,
maturing face
and head profile,
with appearance
of neck

17·0 mm
G

30·7 mm
H

neural crest elements. The limb buds are elongating, displaying joint flexures, and the hands and feet are differentiating. The olfactory placodes, maxillary processes, frontonasal elevations, tongue primordia, and the hypophysial pouch (of Rathke) are all appearing. The tubotympanic recesses are defined, and the primordia of the thymus and parathyroid glands can be identified. In the developing heart the septum primum appears, and the foramen primum follows. The mesonephric ducts reach the

cloaca, and subsequently the metanephric (ureteric) buds appear and extend to the metanephrogenic masses of mesoderm. The gonadal ridges, urorectal septum and genital tubercle are also developing.

Stages 16 to 20 are roughly equivalent to the sixth week, by the end of which the embryo has a length of about 13 to 15 mm. Its curvature has further increased, and its head and relatively long tail are in contact with the developing umbilical stump. The

225

ACTUAL SIZE

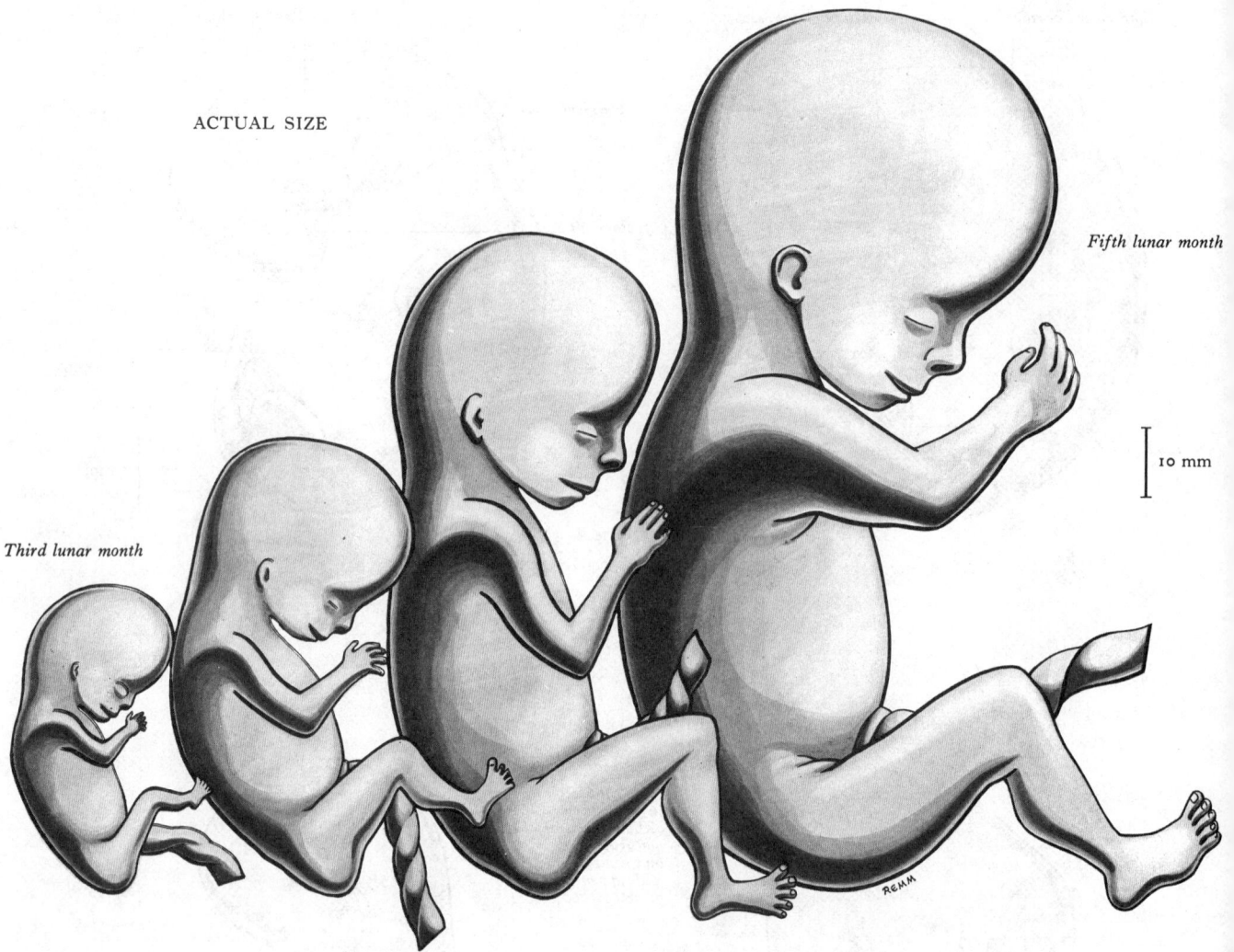

Third lunar month

Fifth lunar month

10 mm

2.112   The progressive changes in fetal size and proportions during the third, fourth and fifth months.

pontine flexure, cerebral hemispheres and cerebellum are developing. The upper limbs and the facial region are growing and differentiating rapidly; the palatal and primitive nasal processes are apparent, and the oronasal membrane ruptures. The liver produces a surface prominence between the cardiac region and the umbilical cord. Into the latter the midgut herniates as a loop in which the appendix and caecum become distinguishable. The spleen develops, as do the female (Müllerian) ducts. The foramen primum in the heart closes and septation of the bulbus cordis occurs. Cardiac muscle is differentiating. Haemopoiesis commences in the liver. Chondrification of skeletal elements begins and ossification commences in mesenchymatous bones, the mandible and clavicles.

Stages 21 to 23 complete the series, extending to about day 47. This represents the major part of the seventh week, and with the eighth week, the *formative* or *embryonic period* is regarded as coming to an end. This is a period during which *patterned differentiation* with consequent *organogenesis* tends to overshadow the growth which accompanies these events. Very considerable increase in size has, however, occurred; from a single cell about 0·14 mm in diameter the embryo has become a most complex and functioning creature, consisting of millions of cells and with a length of about 30 mm or more, and it has increased in weight many thousands of times. During the *fetal period*, which occupies the third to tenth lunar months (**2.112, 113**), the accent is upon growth rather than differentiation, but, of course, the latter continues through this period and to a lesser degree after birth (and in some tissues, throughout life). During this period the rate of growth in length is greater, but not markedly so; from the fourth to sixth weeks the rate is about 1 mm per day, with a

maximum of about 2 mm, during the fourth month. The increase in length in the fetal period is from 30/40 to about 500 mm, and the increase in weight from perhaps 2 or 3 gm to more than 3,000 gm.

During the *seventh* and *eighth weeks* there is a remarkable change in the external appearance of the embryo, for at the beginning of this period the individual still appears markedly 'embryonic', though clearly a primate, whereas at the end, the form is most definitely human. The head is less flexed and the neck longer and clearly defined. The development of the face proceeds much further, with completion of the upper lip and nostrils, though the latter are plugged and the palate still incomplete. Enamel organs are developed from the dental laminae. The external ears and the eyelids are developing, and the limbs elongate considerably, approaching much nearer to their ultimate proportions and displaying well-formed hands and feet with separated digits. Early in this period the interventricular septum is completed. Skeletal and visceral muscle tissues begin to differentiate about this time, and generalized ossification occurs in enchondral bones. In the metanephrogenic mass vesicles appear and the remainder of the nephrons and the collecting tubules of the kidneys are defined. The ovaries or testes are distinguishable and the female (Müllerian) ducts are fusing to form the primordia of uterus and vagina. The external genitalia are further advanced and show sexual differentiation by the beginning of the eighth week. The cloacal membrane becomes perforate, and the tail is retrogressing. By the end of the period the embryo possesses almost all the structural features, internal and external, characteristic of the human mammal, and it now passes into the fetal period. During this there is much growth in the dimensions of the established organs, extensive changes in proportions brought about by variations of growth rate, and widespread maturation in differentiating cells.

During the *third month* head flexion decreases further and the neck becomes proportionately longer. The eyelids meet and fuse and will remain temporarily united until the sixth lunar month. Nails appear on the digits and the limbs in general are comparatively accelerated in development. The umbilical protrusion of the gut is reduced by a proportionate augmentation of abdominal volume.

During the *fourth month* the primary covering of hair appears—the *lanugo* (p. 1223). As in the previous month, the head and upper limbs are still disproportionately large, and although the trunk and lower limbs begin to catch up by increased rates of growth during the rest of uterine life, the same disproportion is present after birth and to a diminishing degree throughout childhood into the years of puberty (p. 236). By the end of the fourth month the eyes have moved even further into an anteriorly directed position, but they are still relatively wide apart. The external ear is beginning to attain its characteristic form and it is nearer its ultimate position, at the side of the head and no longer in the upper part of the neck. The total fetal length, including the lower limb, is now of the order of 230 mm. Its weight at the end of the *fifth month* is about 300 gm, which will be increased more than tenfold during the second half of intrauterine life. Towards the end of this period sebaceous glands become active, and the sebum secreted blends with desquamated epidermal cells to form a cheesy covering to the skin, the *vernix caseosa*, which is usually considered to protect the former from maceration by the amniotic fluid. During this month the mother becomes conscious of fetal movements, the so-called 'quickening'.

The *sixth month* witnesses a further general change of bodily proportions and facial appearance towards those of the infant at birth. The lanugo darkens and the skin becomes markedly wrinkled, presumably through a disparity in the growth rates of cutaneous and subcutaneous tissues. The eyelids and eyebrows are now well developed. The vernix caseosa is more abundant. The length of the fetus is about 300 mm by the end of this month.

During the *seventh month* the hair of the scalp is lengthening and the eyebrow hairs and the eyelashes are well-developed. The eyelids themselves separate and the pupillary membrane disappears. The body becomes more plump and rounded in its contours and the skin loses its wrinkled appearance due to increased deposition of subcutaneous fat. Towards the end of this month the fetus is viable and may in fact be successfully raised if born prematurely. Its length has increased to about 350 mm and it weights about 1·5 kg.

Throughout the remaining three lunar months of normal gestation the covering of vernix caseosa is prominent. There is a progressive loss of lanugo, except for the hairs on the eyelids, eyebrows and scalp. The bodily shape is becoming more infantile, but despite some acceleration in its growth the leg has not quite

equalled the arm in length even at the time of birth. The thorax broadens relative to the head, and the infra-umbilical area of the abdominal wall shows a relative areal increase, so that the umbilicus gradually becomes more centrally situated. Average lengths and weights for the eighth, ninth and tenth months are 40, 45 and 50 cm and 2, 2·5 and 3 to 3·5 kg.

At the end of the *tenth lunar month*, just before birth, the lanugo has almost disappeared, the umbilicus is central, and the testes, which begin to descend with the vaginal process of peritoneum during the seventh month and are approaching the scrotum in the ninth month, are usually scrotal in position. The ovaries are not

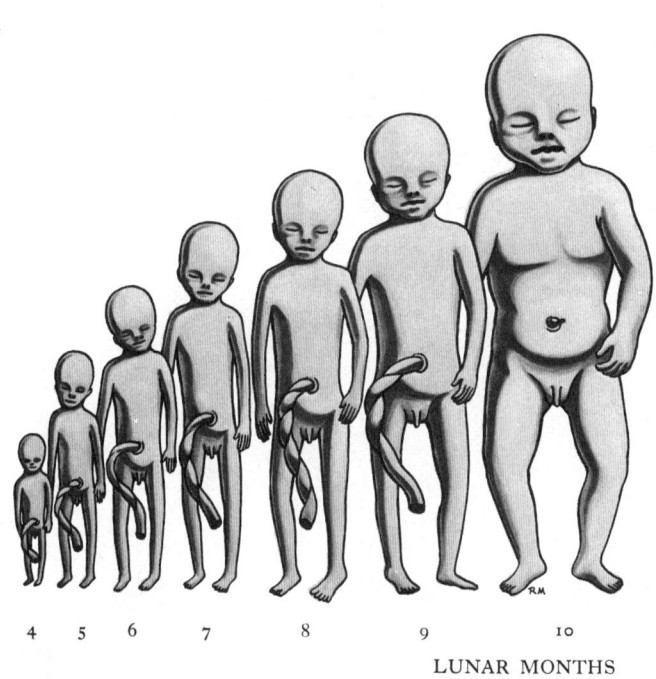

4   5   6   7   8   9   10

LUNAR MONTHS

2.113   Changes in *relative* size and bodily proportions at the fetal ages indicated.

yet in their final position at birth; though they have attained their final relationship to the uterine folds, they are still above the level of the pelvic brim.

The length of the *period of gestation* is regarded as nine calendar months in obstetric practice—approximately 270 days. It is usually about 266 days—ten lunar months less 14 days. Legally, the infant is regarded as viable if born after the end of seven calendar months, but considerably more premature infants have been successfully reared.

3

OSTEOLOGY

MORPHOLOGY OF THE HUMAN SKELETON

Introduction

The general shape of the human body accords with the bilaterally symmetrical pattern of its skeleton (**3**.IA, B). As in any typical vertebrate this consists of an axis, divided into segments to allow the flexibility necessary for movement, and of two paired appendages, or limbs, pectoral and pelvic in position, which are also divided into jointed parts for their basic activities in

locomotion, grasping, etc. To these must be added the expanded and highly modified cranial end of the axis, the skull, and also a variable series of osseo-cartilaginous nodules, the sesamoid bones, which develop in some tendons and ligaments. All these elements are implied in the collective term 'skeleton'.

The human skeleton, as in other vertebrate animals, is internal to the muscles with which it has evolved in intimate functional integration. It is an *endoskeleton*, in sharp contrast with the *exoskeleton* characteristic of many invertebrates, such as the *Insecta*, whose muscles are attached to the internal aspects of jointed elements of chitin, a rigid material which also offers considerable protection from external mechanical forces. The same defensive quality is usually ascribed to many features of the human endoskeleton, and often with an excess of emphasis sufficient to obscure the primary association of skeletal elements, osseous or cartilaginous, with muscle in the more effective production of movement. Perhaps only in the case of the vault of the skull may the protective role be said to be uppermost, and even here muscle attachments obtrude upon the scene. The superficial situation of these particular bones is in line with their evolutionary origin, as is generally accepted, from the dermal bony armour of many groups of earlier vertebrates, including fish, amphibians and reptiles, both extinct and extant. Of the latter the ossified dermal carapace of the tortoise is a familiar example. Not only the cranial vault but also the maxilla, mandible, clavicle, as well as nails and the dentine of teeth, are dermal derivatives; collectively these structures are the vestiges in mankind of the much more extensive systems of dermal bones from which they have been modified. They constitute the human 'exoskeleton'.

The skeletal elements of the branchial apparatus of fish also contribute a *visceral* component to the skeleton of higher vertebrates, including man. Much modified, these elements now appear as ear ossicles, perhaps a part of the mandible, and as the styloid process and hyoid bone. The more caudal branchial arches have persisted as the cartilaginous skeleton of the larynx, but it is not customary to include these as members of the human skeleton; much less has this concept been extended to the supportive visceral cartilages of the trachea and bronchi, though these, as stiffeners to maintain the respiratory lumen, subserve much the same function as the bones and cartilages of the nasal cavities.

In some ungulates bone develops in the connective tissue of the cardiac 'skeleton' as an *os cordis*, and an *os penis* occurs in many mammals. Even without these, the human skeleton is clearly a complex of components, derived from original endoskeleton, from exoskeletal dermal elements, from modified branchial arches, and extraskeletal ossification in structures such as tendons.

The Skeleton in Life

During life the skeleton may also be regarded as including other, non-osseous, but intimately related structures which are removed in the process of *maceration* by which bones are prepared for study and preservation. Not only the attached muscles, ligaments and periosteum, but also cartilages are stripped away. Costal cartilages, somewhat arbitrarily, are grouped with the bony skeleton without question; but articular cartilages, an equally integral part of most living bones, are excluded, as are ligaments and all other joint cartilages, including intervertebral discs. After maceration, therefore, a skeleton is completely disjointed into separated elements which, while convenient for individual examination, have lost much of their functional implications. The drying out of macerated bones also entails disappearance of the contained marrow, and a slowly progressive loss of the mechanical properties, such as elasticity, proper to living bone (Smith and Walmsley 1959).

The morphological properties of living bone have been studied extensively (Thompson 1942; Bell 1956), at macroscopic, microscopic and ultrastructural levels, in relation to the

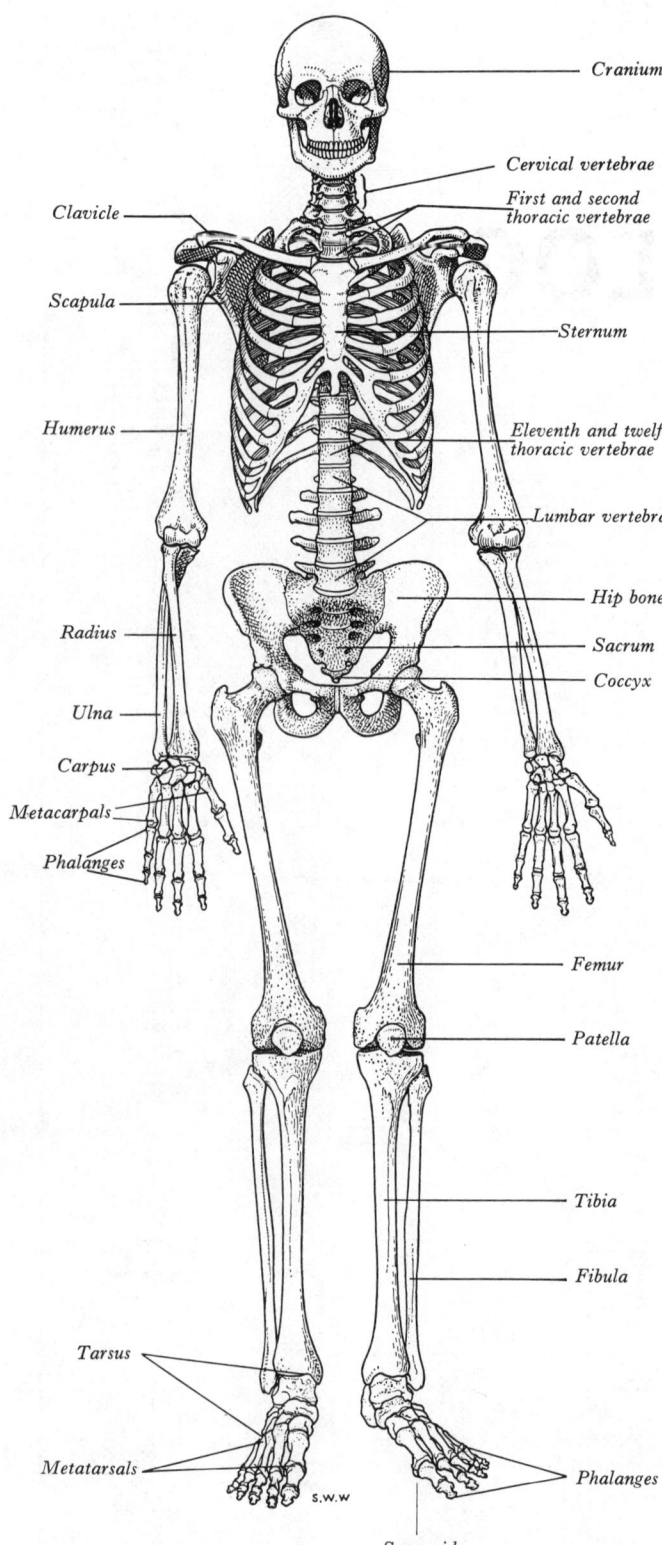

3.IA Anterior aspect of the male skeleton. The right hand is in the prone, the left in the supine position.

Cranium
Cervical vertebrae
First and second thoracic vertebrae
Clavicle
Scapula
Sternum
Humerus
Eleventh and twelfth thoracic vertebrae
Lumbar vertebrae
Hip bone
Radius
Sacrum
Coccyx
Ulna
Carpus
Metacarpals
Phalanges
Femur
Patella
Tibia
Fibula
Tarsus
Metatarsals
Phalanges
Sesamoid

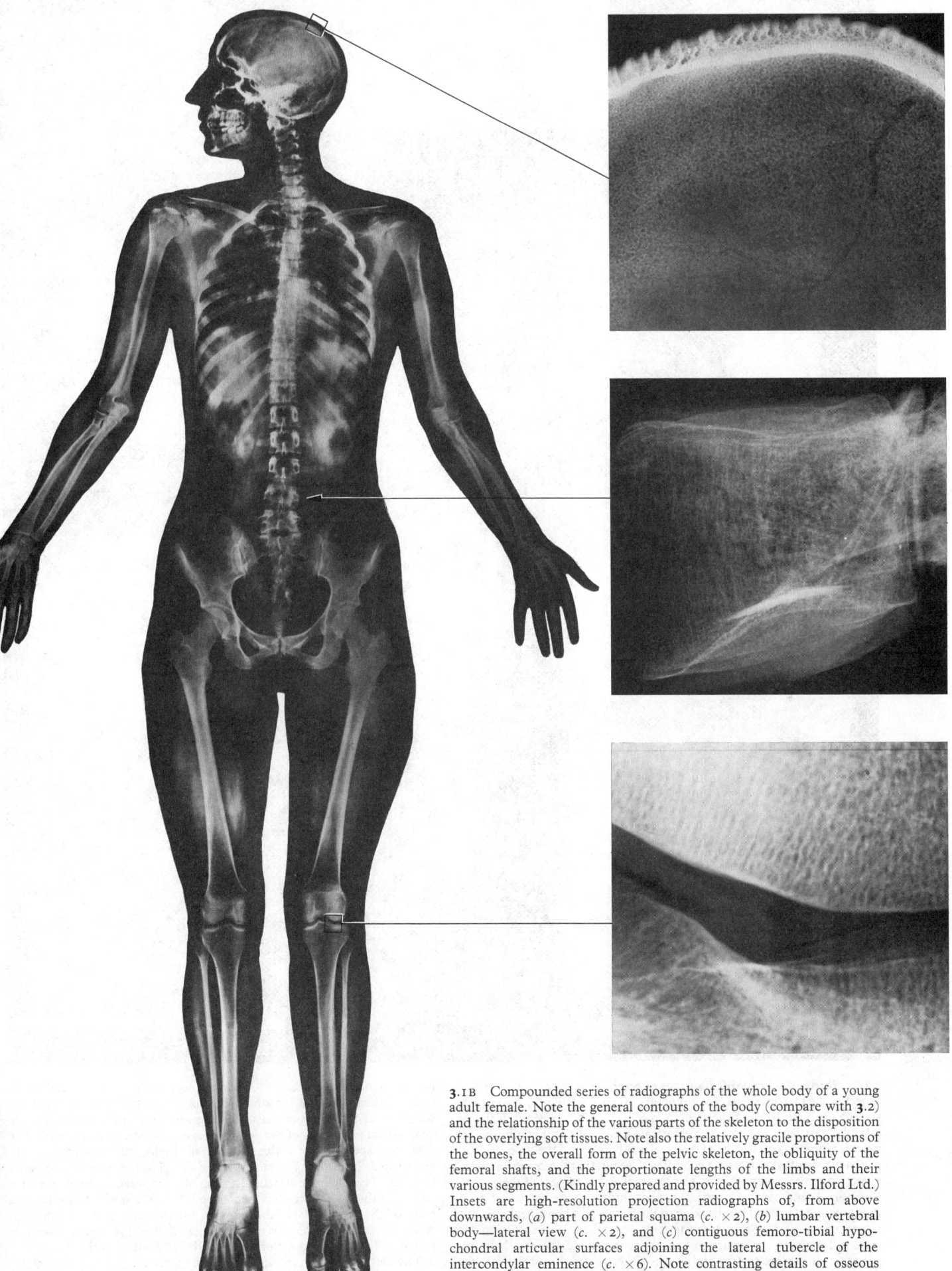

3.1B Compounded series of radiographs of the whole body of a young adult female. Note the general contours of the body (compare with **3**.2) and the relationship of the various parts of the skeleton to the disposition of the overlying soft tissues. Note also the relatively gracile proportions of the bones, the overall form of the pelvic skeleton, the obliquity of the femoral shafts, and the proportionate lengths of the limbs and their various segments. (Kindly prepared and provided by Messrs. Ilford Ltd.) Insets are high-resolution projection radiographs of, from above downwards, (*a*) part of parietal squama (*c*. ×2), (*b*) lumbar vertebral body—lateral view (*c*. ×2), and (*c*) contiguous femoro-tibial hypochondral articular surfaces adjoining the lateral tubercle of the intercondylar eminence (*c*. ×6). Note contrasting details of osseous architecture. (Projection radiographs kindly supplied by Dr. C. Buckland-Wright, Department of Anatomy, Guy's Hospital Medical School, London.)

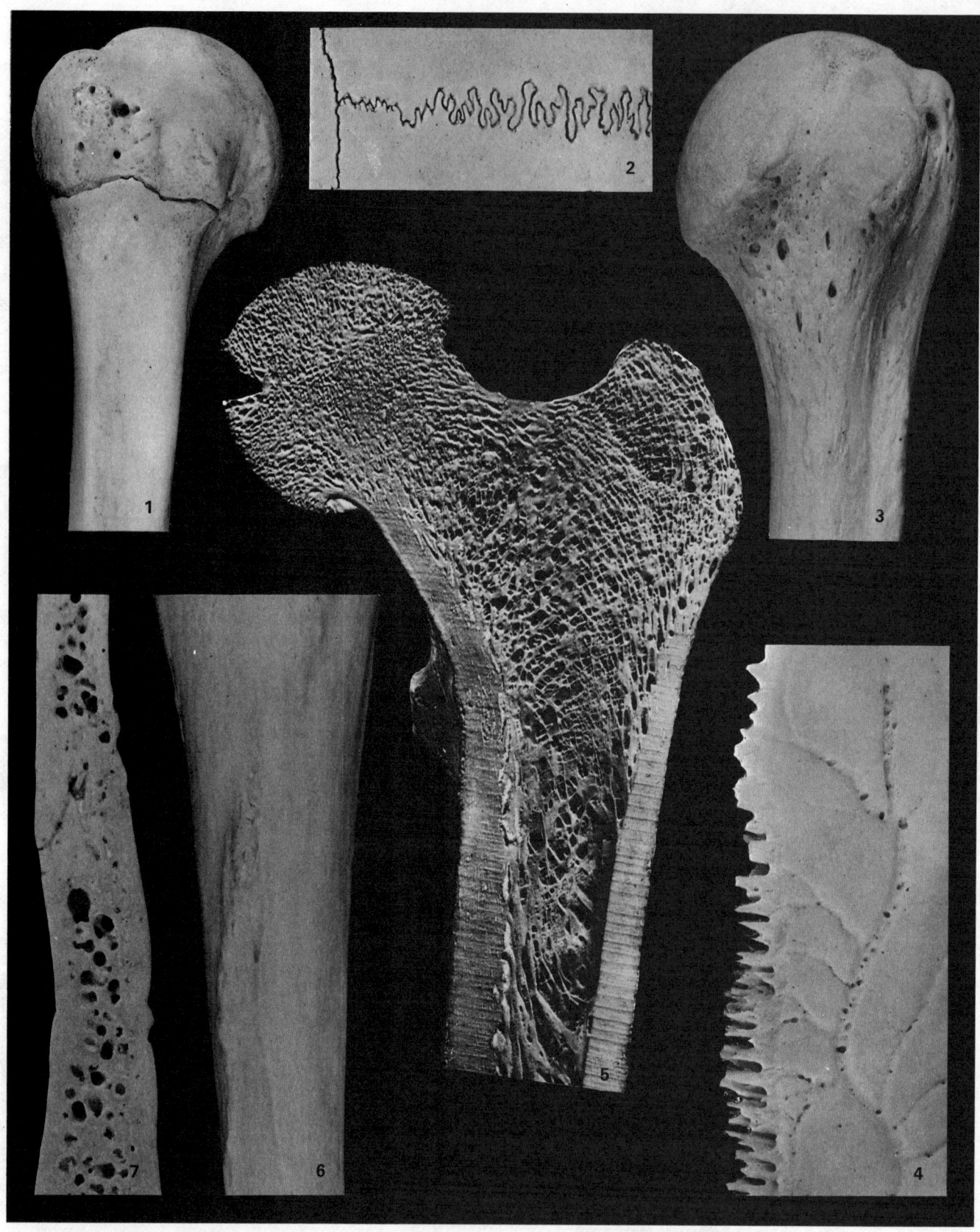

3.1 C Some macroscopic features of bone structure. 1. The ventral aspect of the proximal end of an immature humerus. Note the contrasting surface characteristics of the smooth articular surface covered in life by articular cartilage, the smooth periosteal surface of the metaphysis, the vascular foramina at the base of the greater tubercle, and the groove across which osseous fusion has not yet occurred, separating the compound epiphysis from the metaphysis. 2. The external aspect of part of the sagittal (horizontal) and coronal (vertical) cranial sutures. Note the variation in the form and degree of interlocking of the bones. 3. The ventral aspect of the proximal end of a mature humerus; complete osseous fusion has occurred between the compound epiphysis and the diaphysis. Note the variations in surface texture and the distribution and size of the vascular foramina. 4. The endocranial aspect of the sagittal margin of a parietal bone. Note the highly complex nature of the disarticulated sutural surface, the presence of vascular grooves and multiple vascular foramina

on the endocranial surface. 5. A coronal section through the head, neck, greater trochanter and proximal shaft of an adult femur, showing the variation in thickness of the shell of compact cortical bone, and the geometrical patterning of the deeper intersecting bony trabeculae. 6. The posterior aspect of part of the shaft of a tibia. Note the relatively smooth areas which bear 'fleshy' attachments of muscles, the ridge where dense lamellae of collagen are attached, and an oblique vascular foramen which transmits the main nutrient vascular bundle to the shaft. 7. A sectioned bone of the cranial vault showing the internal and external tables of compact bone separated by the trabecular diploë, the diploic spaces of which, in life, are filled with haemopoietic red bone marrow.

The sectioned femur in No. 5 has been reproduced using a specialized photographic technique which produces a 'bas-relief' effect. Photographs by Mr. Kevin Fitzpatrick, Department of Anatomy, Guy's Hospital Medical School, London.

mechanical factors in the environment. The intimate blending of hard inorganic and resilient organic components in bone tissue affords almost equal resistence to compression and tension, unlike most structural materials used by man, which are usually better in one respect than the other. In tensile strength bone is comparable to cast iron, with only a third of the weight, the breaking stress being 15·5 and 18 metric tons per square inch respectively (Bell *et al.* 1941)—for additional and more recent data consult tables on p. 241. In its superior flexibility bone resembles steel more closely than iron, having about half the strength of the former. Compression tests show that bone has very large margins of safety to weight-bearing (Koch 1917) and to impact forces in jumping. It should be remembered in this connexion that contracting muscles are responsible for much the larger component of pressure, even at weight-bearing joints, especially in active movement (Bell *et al.* 1941; Williams and Svensson 1968). In the hip joint only a fraction of the pressure loading is due to body weight (*vide infra*).

A tubular structure, as is typical of the shafts of many limb bones, is the strongest and lightest—and hence the most economical—arrangement which can be designed. This particular adaptation of form to habitual stresses may perhaps also be evident in the approximately longitudinal direction of many osteons in elongated bones. In such long limb bones the thickness of the compact cortical shell is greatest at midshaft, where torsional and bending stresses are most severe, and where internal trabecular structure is absent or comparatively trivial in development. At their articular extremities these same bones must withstand chiefly compression forces, and these may be surprisingly large. In standing, the hip joint takes half of the body weight, and the pull of muscles multiplies this by a factor of about six; in walking or running, the full body weight (less the weight-bearing leg) is carried alternately at each hip joint. Conservatively, in an average adult male, this means a total load of 600 lb. (*c.* 270 kg.), and there is evidence that this can easily be doubled in powerful exertion. At such articular regions bone structure differs considerably from the arrangements in shafts, the whole of the interior being occupied by trabecular bone, which supports a comparatively thin shell of compact bone tissue. The same mode of structure occurs in smaller bones, such as those of the carpus and tarsus and in vertebral bodies, all of which are probably subjected more to compression forces than to any other form of stress. It seems likely that in joints where weight-bearing is comparatively trivial, marked pressure may nevertheless exist in working conditions, due to the action of muscles.

It has long been recognized that the trabeculae of cancellous bone, though individually small, may collectively provide powerful underpinning to relatively thin shells of compact bone. This form of construction is indeed widespread throughout the human skeleton, and it is perhaps only modified where bending, twisting and tensile forces demand more considerable aggregations of compact bone. Numerous instances of both arrangements will be encountered below in the descriptions of individual bones; in most of the larger elements of the skeleton both forms of construction, and intermediate arrangements, will be seen to occur in adaptation to the local mechanical problems. Mechanical forces directly influence the growth and forms of bones; and it can be safely assumed that in the protracted permutations possible during evolutionary time, a most apt biological solution to the mechanical demands impressed has been arrived at, in interaction with the limiting factors of nutritional resources, available muscle power, and the best compromise between size and weight (see also p. 238).

Sections through cancellous bone (**3.**1C, 3A, 10B) show trabecular patterns which resemble to some extent the crisscrossed arrangements of girderwork. Attempts to equate this pattern in individual bones, such as the femur and calcaneus, with analyses of the lines of force corresponding to habitual stresses, were made at an early date (Ward 1838; Meyer 1867). Such architectural explanations of internal bone structure have attracted some marked criticism, (e.g. Murray 1936). Nevertheless, numbers of studies of individual bones, and even comprehensive works covering the entire skeleton (Hall 1966), continue to augment the evidence of a most intimate correlation

between stress and structure in bone, not only in the disposition of trabeculae but equally in the pattern of cortical bone. Wherever localized stresses are applied to bone, as in the attachment of tendons and ligaments, sections show that external features, such as ridges or facets, are not the only local change in osseous architecture. Thickening of compact cortical bone may also occur, and a subjacent condensation of trabeculae is usually evident and often pronounced. It is impossible to avoid the conclusion that patterns of stress and trabeculation are closely related, whatever the precise terms of the equation (see also p. 238). Many features of the *surface* architecture of bones remain unsuspected if they are merely examined superficially with the naked eye. Much additional information may be gleaned using a loupe, macrophotography, incident light microscopy, and recently, scanning electron microscopy (*see* the enormous variations in surface texture and incidence of foramina in selected regions of a normal human femur shown in **3.**1D).

The Shape and Proportions of Bones

Individual bones vary in shape, but some shared features prompted a categorization into long, short, flat and irregular bones. *Long* bones are typical of the limbs, their length in primate animals, such as man, reflecting the degree of speed and power in movement. They display elongate tubular shafts, containing a central medullary cavity, and expanded articular ends, which are epiphyses, having separate centres of ossification, often multiple (e.g. humerus and femur). Smaller examples of this shape are metacarpals, metatarsals and phalanges, although these bones are proportionately greater in diameter of shaft. The so-called *short* bones are exemplified in the carpus and tarsus, the elements of which are often approximately cuboid, cuneiform, trapezoid, or scaphoid in shape. Being subject to pressures rather than other forms of stress, they consist typically of a comparatively thin cortex of compact bone, supported by an interior completely occupied by cancellous bone. The *flat* bones include the curved laminar members of the cranial vault, in which an internal layer of trabecular bone, the diploë, variable in thickness, is enclosed between inner and outer laminae, or *tables*, of compact structure (**3.**1C). The scapulae, despite their irregular form, are described as flat. The class of *irregular bones* comprises any element not easily assigned to the foregoing groups. It is obvious that this classification, though time-honoured, has no great merit. Bones are better studied individually, and in relation to the functional demands made upon them.

In their overall shape and details of structure, bones are affected by a number of factors, which may be summarized as genetic, metabolic and mechanical. In all but its finer features and proportions, every bone is the product of a long functional history extended over countless successive generations. The basically genetic nature of the determination of the primary shape of bones has been demonstrated in organ culture, and by experimental transplantation of embryonic skeletal tissues (Murray and Huxley 1924; Fell and Canti 1934; Willis 1936). All the major characteristics of a particular bone appear to be self-differentiated. This is perhaps to be expected. Mechanical influences, such as moulding by the proximity and activity of associated muscles, often regarded as of first importance, can scarcely be operative at the early stage of development when the primary form of a bone is being established. As muscles become active during prenatal life, they may perhaps exert some influence upon the bone growth, though this is difficult to estimate. After birth, and up to adolescence, before all epiphyses are fused, increased use of muscles appears to augment growth of bones, both in length and girth. The reduction in osseous development, in limbs affected by paralysing diseases, such as poliomyelitis, implies that the play of active muscles is necessary to normal skeletal growth. Experimental studies (Appleton 1935; Washburn 1947; Wolffson 1950), usually consisting of removal of muscles, have confirmed, though in the same negative manner, the role of activity in bone growth. Again, comparison of the rate of increase in strength and stature shows some correlation in timing in adolescents (Jones 1949).

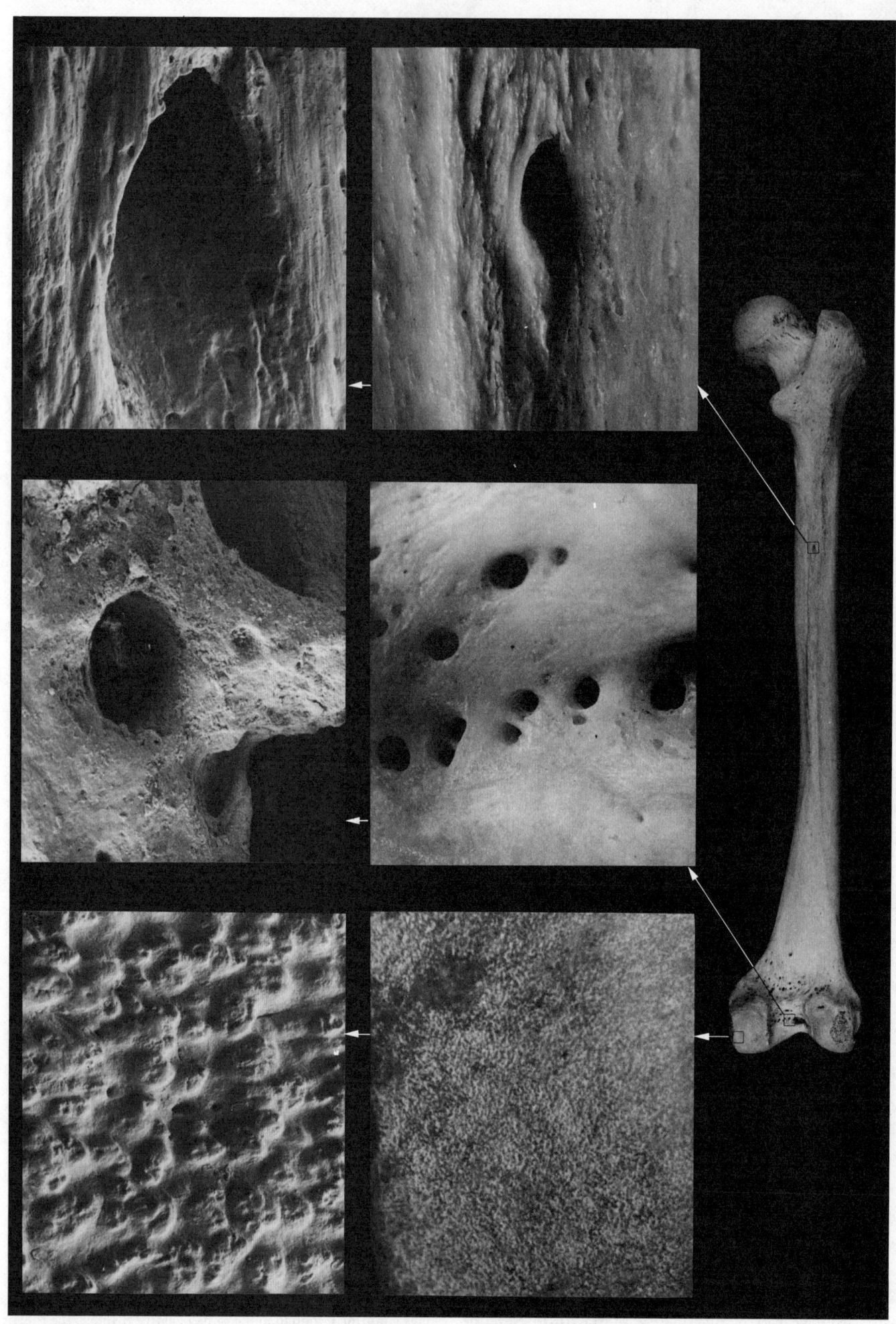

3.1D Surface details of a normal adult human femur photographed at the sites indicated, by macrophotography and scanning electron microscopy. The macrophotographs which are immediately adjacent to the whole bone photographs are at a magnification of ×5, except for that showing the diaphysial nutrient foramen which is ×4. All the scanning electron micrographs are at a magnification of ×80, with the exception of that

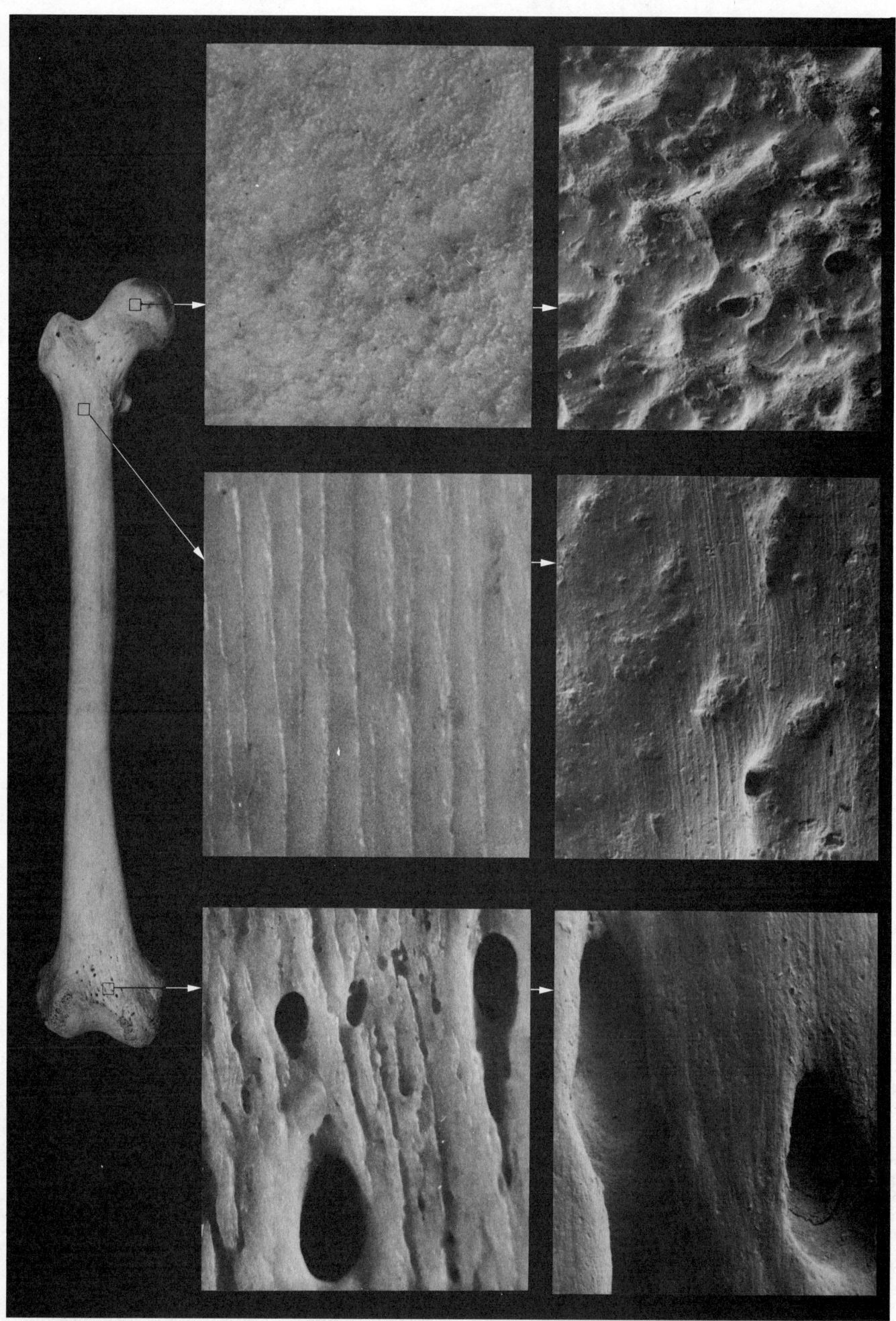

showing the diaphysial nutrient foramen, which is ×15. (Prepared and
provided by Kevin Fitzpatrick, Derrick Lovell, and Michael Crowder,
Department of Anatomy, Guy's Hospital Medical School, London.)

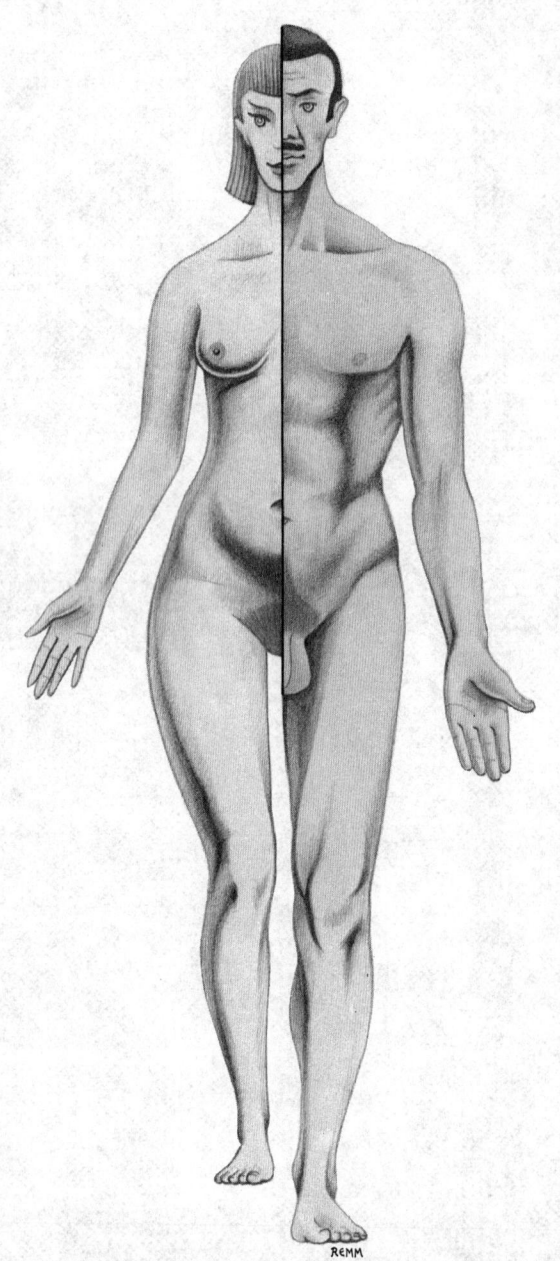

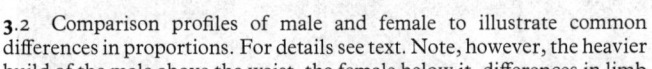

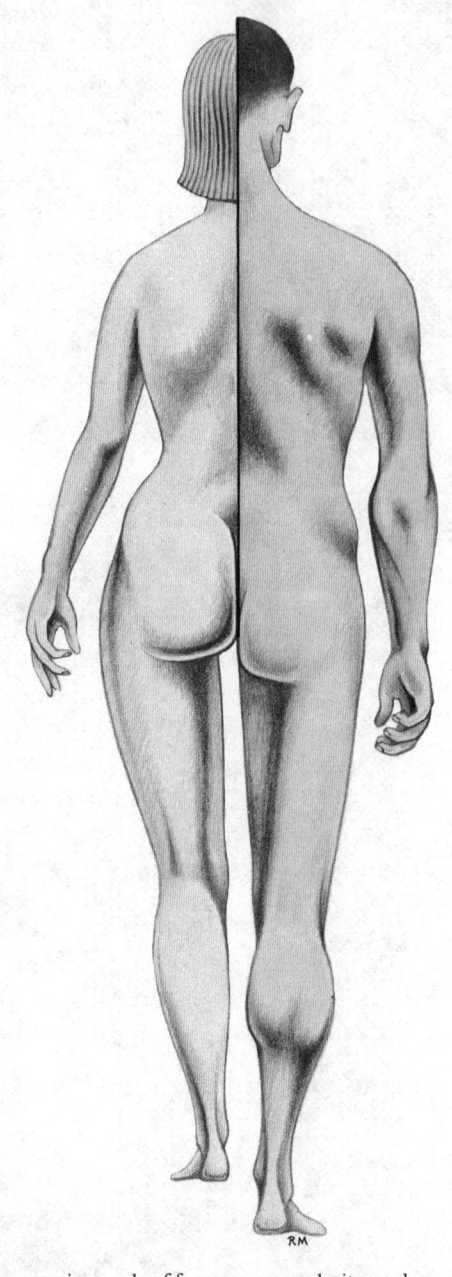

3.2 Comparison profiles of male and female to illustrate common differences in proportions. For details see text. Note, however, the heavier build of the male above the waist, the female below it, differences in limb proportions, carrying angle of forearm, muscularity, and apparent length of neck.

Metabolic influences on bone growth may operate at all stages of their development. The availability of calcium, phosphorus, vitamins A, C, and D, and the secretions of the hypophysis, the thyroid, parathyroid and adrenal glands, and of the gonads, all exercise essential roles in the growth of bone tissue (*vide infra*), and hence in the macroscopic features and dimensions of the skeleton. The effects of gross disturbance in many of these factors are well recognized pathological states; but in some of these it is impossible to define the limits between what is frankly pathological and merely variation from the norm. Body height provides an example; between the extremes of dwarfism and gigantism, demonstrably due to hormonal dysfunction, great variation from mean height can occur. Variations in stature and other dimensions associated with age, sex and race, are clearly to some degree genetically conditioned; but nutritional levels also have profound effects, especially in regard to race (Greulich 1951; Acheson 1960; Tanner 1962).

Bodily proportions, as well as absolute dimensions, exhibit a considerable range of variation in respect of age and sex (**3.**2) within a single racial group, and also in different races. While this

is in part dependent upon the varying degrees of development of muscles and adiposity, such differences are chiefly due to skeletal variation. *Anthropometry* is the study of such variations (Martin 1928; Hrdlička 1939). The data of anthropometry may be non-metrical, such as the presence or absence of a feature (e.g. the sagittal crest of the Eskimo skull, the pre-auricular sulcus of the female innominate bone), or the persistence of a characteristic which usually disappears (e.g. the interfrontal suture), or the degree of development of some detail (e.g. frontal ridges, projection of chin). However, the great majority of anthropometric observations are measurements, carried out by internationally accepted techniques, upon the living body or upon skeletal material. These measurements may involve the whole body, e.g. stature, or sections of it, such as the limbs; or they may be carried out on individual bones. The cranium early attracted special attention in such studies, but methods of recording the principal dimensions are established for all skeletal elements. Proportions are expressed by indices; for example, the breadth of a skull may be reckoned as a percentage of its length, this being the cranial index of that individual. Such indices may have an ethnic significance; Mongoloid people display a range of cranial indices at a higher level than some other races—they are 'broader' in the head, relatively and, in this instance, absolutely. A Mongolian child, however, might display a cranial breadth metrically less than that of a Negro adult, and yet be proportionately broader in this dimension. Ratios or indices between the length of the limbs and the 'sitting height', or between the arm and leg, the upperarm and forearm, the thigh and foreleg, are all used, and they show characteristic differences according to age, sex and race. Some details of such measurements and indices will be found in the appropriate sections on individual parts of the skeleton.

Any observations and measurements which aid in establishing the age, sex, size and race of an individual skeleton, or parts of it, are not only of interest in anthropology and archaeology (Brothwell 1968; Warwick 1968), but also of great value in forensic medicine (Glaister and Brash 1937; Boyd and Trevor 1953; Harrison 1953; Stewart 1970).

Estimation of skeletal age depends upon a number of criteria, which differ in usefulness at various stages of chronological age. Up to the age of about 25 years, including fetal life, the states of dentition and of ossification provide numerous data upon which an assessment can be made. The accuracy of this as an indication of actual, or chronological age is dependent upon the precision of the observations, the existence of adequate statistics for the sex and racial affinities of the individual under examination, and upon the nutritional history and endocrine activities of the individual. The latter data are not always forthcoming, and it is to be noted that available tables of appearance of ossific centres and fusion of epiphyses are usually based upon studies of well nourished children and adolescents. Moreover, these groups are usually derived from white communities in Europe or America. A few excellent studies of other racial groups exist (Todd, 1931; Modi 1957; Chang 1969). Racial variations in the order of the events of ossification would necessarily be genetic, and there is no clear evidence that such variation exists (Krogman 1962). Variations in timing of ossification are affected by wide divergence in standards of nutrition between, and even within, the racial groups so far studied. In all groups the data for females show an earlier appearance of centres of ossification and of fusion of epiphyses, a difference which is presumably genetic. Despite these complications, it is usually possible to assess the age of a complete skeleton within a year or so each way up to the age of 25, and with greater accuracy in the earlier years, especially if dental observations are also available. (For details see Teeth and sections on individual bones, and consult McKern 1970.)

From about 25 years onwards, skeletal age can be estimated to within ±5 years by the state of the cranial sutures and of the bony surfaces of the symphysis pubis. From the mid-twenties onwards into old age, sutures exhibit a progressive closure (see p.344 and Todd and Lyon 1924–5). Unfortunately, closure begins from within, and unless crania can be opened for internal inspection, observations are likely to be misleading. Subsequent studies have thrown doubt upon the value of this particular technique (Singer 1953; Genovese and Messmacher 1959), and possible complications due to racial variation have been recorded (Abbie 1950). A series of progressive changes in the borders and surface configuration of the articular aspects of the pubic bones at their symphysis have been shown to occur. Appearances characteristic of ages from late teens to the fifties and beyond have been established (Todd 1920–30; McKern and Stewart 1957; McKern 1968). Most authorities now regard this as the best method available for estimation of skeletal age in the decades of maturity. Sequences of age changes have also been described for other skeletal elements (see scapula, sternum and costal cartilages), but these provide less accurate estimates. Lipping of the borders of vertebral bodies and around the margins of articular surfaces, exaggerated secondary markings on bones, and ossification into tendons and ligaments, are all manifestations of advancing years, but they provide no more than vague indications of actual age.

Estimation of sex in the human skeleton, when this is complete, can be effected with little error, and without resort to measurement, provided the individual under examination has passed the age of puberty. Sexual differences are particularly marked in the pelvis (p. 387) and skull (p. 346), though not to the same degree in different populations. The 'sexing' of a collection of whole skeletons from the same racial group may be almost free from error, whereas assessment of a single individual, of unknown extraction, may be much less certain. Inspection of other postcranial bones than the pelvis, especially the larger limb bones, may provide clear evidence of sex, particularly if other individuals of the same race and of both sexes are available for comparison. Female bones are not only, on average, appreciably smaller than their male equivalents, they are also more slender. That is to say, they are of lesser diameter in the shaft relative to length. This is also reflected in their comparative *weights*. In a study of the femora of Hindus, for example, mean weights were approximately 385 gm. in males, and 279 in females (Singh and Singh 1974).

It has long been customary among anatomists, anthropologists, and forensic experts, to judge the sex of skeletal material by such non-metrical observations. Latterly, sexual divergence has been based upon actual measurements in many different bones; current texts should be consulted for details (Montagu 1960; Krogman 1962; Giles 1970). Of some recent discriminant analyses, one example may be quoted: Rother *et al.* (1977) examined 70 humeri by analysis of capital dimensions, and they concluded that the sex of unknown humeri was not easy to establish (whereas approximate age *could* be assessed). Perhaps the most useful aspect of such studies has been recognition of the need for appropriate standards of sexual dimorphism in different populations. The pelvis remains, however, by far the most reliable skeletal region in establishing sex, displaying differences not only before puberty, but even in infancy (Reynolds 1947) and fetal life (Boucher 1957).

Estimation of size, and particularly of height, from measurements of individual limb bones, has long been a matter of calculation (Rollet 1899), and with an increasing accuracy as successive studies have refined the formulae used (Trotter and Glaser 1958; Trotter 1970). The rationale of all such formulation is the observation—long familiar to artists (cf. Leonardo da Vinci)—that the major parts of the body, the trunk and limbs, exhibit characteristic ratios among themselves and in comparison with total height. These proportions are linked to age, sex and race. In infants the relatively huge head, long trunk, short arms and even shorter legs present a familiar picture of proportions which would appear grotesque even in an older child, and monstrous in an adult. As each infant grows in size it also changes in its proportions, gradually approaching adult shape, with a divergence towards one sex or the other around the time of puberty. The limbs become relatively longer in adult males, the shoulders relatively broader and pelvis narrower, and there are other differences (**3.2**). Between major races, at least, and to some extent between ethnic groups within each race, there are characteristic variations in bodily proportions. Negroes, for example, are comparatively long in leg and arm; moreover, the foreleg and forearm are long relative to the upperarm and thigh. In view of these variations it is clear that formulae designed to estimate height from long bones in one population cannot be applied indiscriminately to another. Alternative formulae for the sexes must be used, and immature bones must be recognized as such and subjected to suitable corrections.

The length of the femur has sometimes been used alone for estimates of stature, by application of a simple multiplier derived from the comparison of femoral length with the known height in a large number of individuals. However, from the earliest studies onwards, the humerus, radius, ulna, tibia and fibula have all been utilized; and more complex formulae, based upon the combined lengths of several long bones, have been shown to yield more accurate estimates. Whatever the methods used, it should be remembered that estimates are mean values with appropriate standard deviations. Assessment of the stature of unidentified skeletal remains, however carefully carried out, may show errors of several centimetres.

Since estimates of stature (and other assessments) must sometimes be made from *fragments* of bones, attempts to establish reliable formulae have been made (consult Steele 1970).

Estimation of race from skeletal observations has provided a central theme in anthropology since its emergence as a discipline. The skull in particular has attracted great attention in this regard. A large array of *non-metrical* features (Wood Jones 1931) and of cranial, facial and mandibular indices have been widely used and continue to be (Martin and Saller 1958–61; Berry and Berry 1967; Berry 1975), and elaborate statistical analysis of *metrical* data has been introduced (Giles and Elliot 1960). Skulls from the three major racial groups, Caucasian, Mongoloid and Negro, can be identified as such with considerable success (perhaps in 85–90 per cent of instances), and without elaborate measurement. However, with further division into racial groupings the percentage of errors rises steeply. The worldwide intermingling of races entails that 'pure' races are difficult to indicate. This has been recognized particularly in America, where the existence of large collections of skeletons of both Whites and Negroes, both well documented as to their derivation, has stimulated extensive comparative studies. The results were surveyed in detailed criticism (Krogman 1962), with the conclusion that only the skull (but not the mandible) and the pelvis (Todd and Lindàla 1928) have any value in estimating race. (For some details of *cranial* features of racial significance *see* the section on the skull, p. 346.) However, metrical peculiarities in limb bones of Japanese have been claimed to exist by Takahashi (1975, 1976). In addition, the racial significance of a number of *postcranial* non-metrical skeletal characteristics has recently been re-examined, using rigorous statistical techniques, by Finnegan (1978), and the same worker has summarized the findings of other observers. The value of non-metrical features in evaluating populations from skeletal data is currently receiving renewed and intense interest. Consult Finnegan and Faust (1974) for a bibliography.

Functions of Bone and Skeleton

Bone tissue and the skeletal struts and levers formed of it, are evidently adapted with exquisite precision to resist all forms of stress, and with a suitable degree of resilience, in the varying play of musculature. The skeleton is sometimes said to give shape and support to the body; but the shape is itself largely an expression of the characteristic motor activities of an animal, and this simply returns us to the role of bones in movement. In some lower vertebrates, of course, a cartilaginous skeleton, calcified to a varying degree, provides the levers of the locomotor system, and it is a matter of controversy whether cartilage was ancestral to bone in this respect (Romer 1942, 1970; Bassett 1962; Krompecher 1967). However that may be, a bony skeleton is not merely an advantage in the locomotor sense. Bone differs from cartilage in being much more intensely vascularized (p. 257) and for this reason it might be expected that the calcium salts with which it is hardened could be as easily removed as deposited there. As will be considered elsewhere (p. 261), this is indeed the case. At least some part of the calcium in bone is freely labile, and can be mobilized into the general circulation with but little delay. The well established effects of depressed calcium levels in rickets and osteomalacia indicate that calcium in bone and in the body at large are in a constant state of balanced interchange. This is confirmed by much evidence from more recent and precise observations (McLean and Urist 1968). The skeleton thus plays an integral part with other tissues in the general metabolism of calcium, providing the major store of calcium salts (97 per cent of the total bodily content).

Some skeletal elements obviously afford protection from extraneous forces; but the exoskeleton which is so clearly protective in earlier vertebrates persists in the human body only in a highly reduced and modified form. Even in the case of the skull, much of the cranial cavity of which is enclosed by dermal bones, primitively protective in nature, this function is rather more complex than the mere provision of a barrier to *external* assault. Powerful muscles, both masticatory and postural, have invaded the cranial walls for attachment, a situation which demands isolation of the brain and its circulation from these far more frequent *internal* sources of disturbance. Again, ribs are commonly described as protective, and they certainly reduce the risks of impact, even with pointed objects, but such dangers are sporadic, whereas respiratory movements occur many times a minute throughout life.

As already described, all bones are, to some extent, internally trabecular in arrangement. This pattern is not only concerned with mechanical stresses, but also with the siting of the bone marrow. The innumerable minute spaces between the trabeculae, together with the much larger cavities in the shafts of the longer limb bones, are occupied in life by marrow, whether haemopoietic or adipose (p. 63). In some cranial bones air-filled cavities develop and these are then termed *pneumatic*. Pneumatization of bone may take the form of single large hollows, such as the maxillary *sinus*, or of multiple, small, intercommunicating *air cells*, as in the mastoid process of the temporal bone (*see* **3.**95 A–D). In the human skull, the saving in weight thus effected can scarcely be significant, as it is in the extensively pneumatized cranial vault of the elephant. Nevertheless, the cancellous architecture of most bones is generally regarded as conducive to lightness without loss of strength, though perhaps also associated with economy in the use of materials. Some sinuses, particularly the smaller air cells, in fact appear to develop from the *diploë*, or cancellous tissue, of certain cranial bones, and might therefore be an expression of the above principle. The larger sinuses are said to affect the timbre of the voice, but they may be no more than the result of unusual growth patterns of the bones in which they occur.

The Mechanical Properties of Bone

The resemblance of many bones to man-made levers, supporting columns, arches, struts and girders, inevitably prompts the application of mechanical or engineering concepts to functional analysis of them. Not only are bones obviously analogous to such structures in external configuration and internal architecture, but also in their uses. In fact, the apparently fragile but collectively strong lattices of struts and trusses of trabecular bone, and the occurrence of skeletal forms such as tubes, H-girders, ridges and other thickenings of cortical bone, pre-dates by aeons of time the mechanical inventiveness of mankind. Once human technology began to overtake natural biomechanics, then comparisons became inevitable. Galileo (1638) recognized the possible significance of trabeculation and he also asserted that hollow cylinders are, weight for weight, stronger than solid rods. Havers described his 'systems' and their axial orientation in 1692; Ward (1838) and perhaps others likened trabecular patterns, with compressive and tensile elements, to supporting brackets and similar structures. However, Meyer (1867), with his mathematician collaborator, Culman, first clearly enunciated what is now known as the *trajectorial theory*; and this, despite much subsequent argument, still survives. No more satisfactory hypothesis has replaced it. The theory basically states that the trabecular pattern coincides with routes of stress; and this has not yet been convincingly demonstrated, nor disproved. It is clear that trabeculae must be concerned in the support of the shells of compact bone which adjoin them, for these cortical strata are usually thin; but the relations between stress and strain and trabecular pattern are presumably more complex than the simplified mathematical analyses which have been frequently applied to sections of bones or models of them (Kummer 1972).

In contrast to the largely theoretical approach to the intimate adaptation of bone tissue to skeletal functions is the direct estimation of the strength of bone as a material by the methods of engineering practice, to assess the response of either bone samples or entire skeletal elements. Wertheim (1847), anticipating Meyer's observations, applied physical tests to bone tissue, and many subsequent workers, particularly in recent times, have expanded such techniques. Developing interest in the mechanics of locomotion has added a stimulus to investigation of bone, and there are to be found, scattered through a large number of papers and monographs, a somewhat bewildering variety of numerical values for the various physical parameters of the tissue. The subject is, of course, fraught with difficulties and fallacies. Obviously differences are likely between bone material from different species, and sex and age are also likely to be factors. The nature of the specimen tested is also a variable; whole bones and

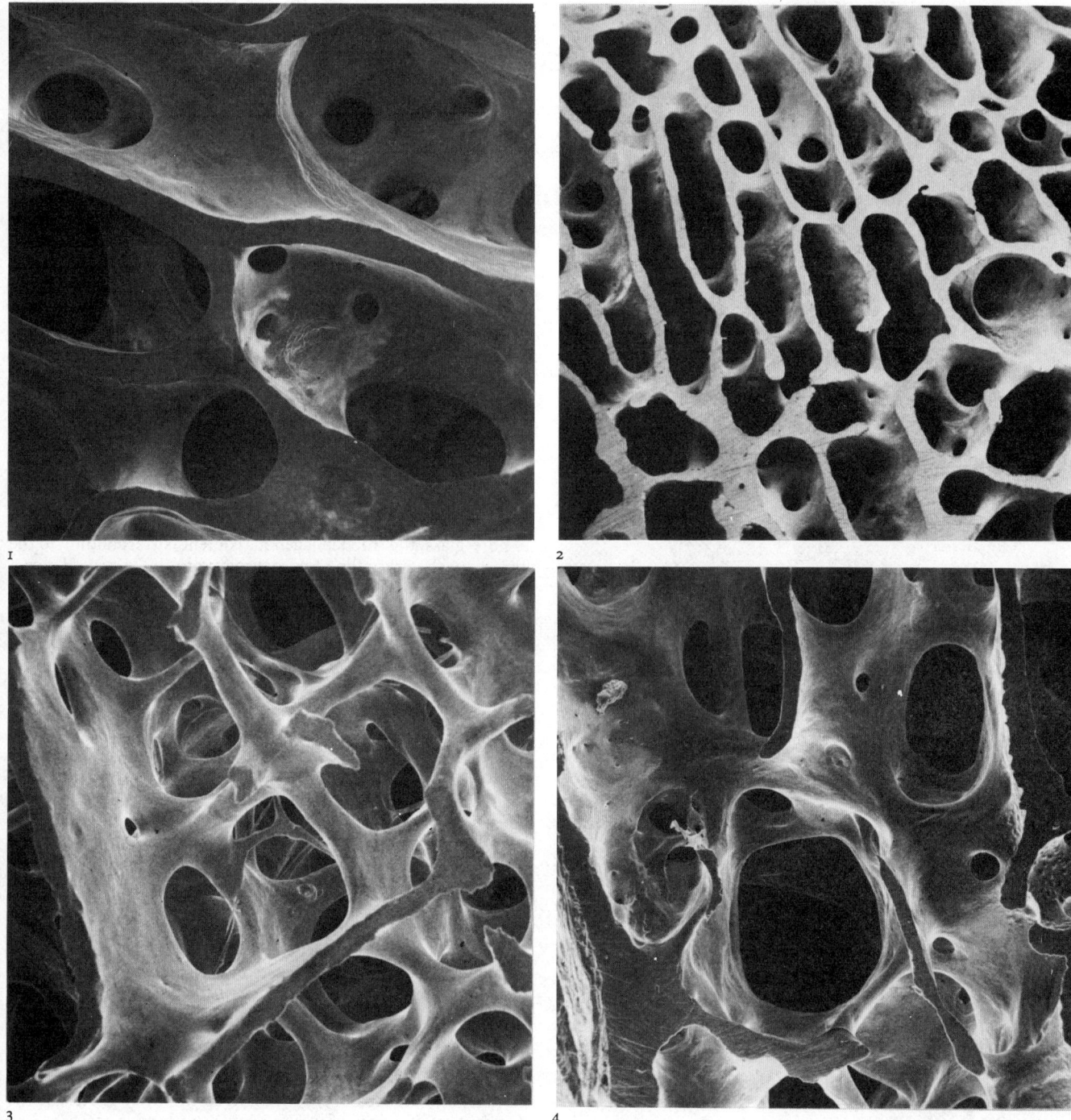

3.3A Scanning electron micrographs of trabecular bone at different sites in the proximal part of the same human femur. All fields are shown at the same magnification ($\times 20$). (1) is an area in the subcapital part of the neck, (2) and (3) are in the greater trochanter, and (4) in the rim of the articular surface of the head. Note wide variation in the thickness, orientation and spacing of the trabeculae. (Original photographs from Whitehouse and Dyson 1974, by kind permission of the authors, the *Journal of Anatomy*, and Cambridge University Press.)

blocks of cortical, trabecular, or a mixture of these behave differently. The immediate source of the specimen is thus important to define, if results are to be comparable. It is now established that bones from preserved cadavers yield misleading values, especially as regards plastic deformation (Reilly and Burstein 1974), but also in its elasticity, hardness, and compressive and tensile properties (Evans 1973). Dead bone, however, provided that it is kept wet, does not differ in its characteristics from living bone, tested *in vivo*, whereas, dried bone is harder but less deformable. Selecting a sample area from a given bone presents problems, because the difference in distribution of Haversian systems or trabeculae may affect parameters and vitiate comparisons. Testing of such specimens and intact bones, especially to the point of breakage, is rendered

very difficult by problems of support or attachment of tension devices, which may incur undesigned collapse at points of support or attachment. This is particularly true of the testing of axial tension.

Because of these and other difficulties, the experimental data, which have been recorded in profusion, present a highly complex picture. It is clear, for example, that it is impossible to give even useful mathematical means for the physical characteristics of bone as a material, because the resistance to various forms of stress and elasticity vary in different regions of the same bone. In view of the structural complexity of a complete skeletal element, in which occur well-recognized variations in thickness, density and modelling of the cortical shell, coupled with marked variation in the internal trabecular pattern (**3.3A**), it is not surprising that some authorities emphasize the improbability of ever establishing more than an approximate picture of the mechanical behaviour of an individual bone. The observations illustrated refer to the proximal end of the human femur (Whitehouse and Dyson 1974). A similar regional variability has been demonstrated in the distal end of the bone and in the proximal extremity of the tibia (Behrens et al. 1974). In the latter study, strength was not directly correlated with trabecular pattern or total specimen density, indicating a multi-factorial relationship. For quantitative data on the sternum see Whitehouse (1975), and, for the iliac squama, Whitehouse (1977).

A considerable variety of techniques, and combinations of them, have been employed to assess isolated physical properties of samples of bone tissue and whole bones, and also to assess the distribution of strains within the latter in response to various forms of stressing. The results of a very large number of tests, using engineering techniques, have been recorded and are summarized in the writings of Ascenzi and Bell (1972), Kummer (1972), and Evans (1973). Although these values vary amongst different workers, much greater disagreement characterizes the problems of how forces are transmitted in bones. The problem has been investigated in actual bones, commonly the femur, and also in plastic models. The use of such models is based on a false assumption, that bone is an isotropic, homogeneous material. In fact, bone tissue is a visco-elastic, biphasic substance, analogous to fibre-glass, the tropocollagenous content corresponding to the fibres and the crystalline hydroxyapatite to the 'glass'. Hence a plastic model can merely indicate the surface behaviour of stresses in the bone it is supposed to imitate. In both, the stress pattern can be studied by inspection of the cracks produced in a surface covering of colophonium resin (Kuntscher 1934) or special lacquers (Gurdjian and Lissner 1945), but there are grave doubts as to how far these superficial indications record more than a crude picture of the strains in bone. Hallermann (1934) introduced a photo-elastic technique, which reveals stress patterns in bones and plastic models using polarized light. Pauwels (1965) and Kummer (1966) have exploited this technique, but Brekelmans et al. (1972) have emphasized the limitations of such methods, not only in the use of models constructed of isotropic materials, but also in the over-simplification inherent in mathematical analyses of a two-dimensional nature.

Benninghoff (1925) used the split-line phenomenon in decalcified bones, claiming that the surface pattern of split-lines or cracks, induced by puncturing with a round-bodied needle or awl, followed the distribution of osteons orientated in the axes of compression or tension. He accepted the trajectorial theory of trabecular bone and considered that this was directly related to the behaviour of osteons in cortical bone. Tappen (1954) has made similar claims, but such views have been attacked; Tappen (1970) himself admits that the patterns may follow 'immature' osteons, and Isotupa (1972) considered that they represent the distribution of points of weakness due to vascular spaces. Most recently, Buckland-Wright (1977) has attributed the cracks to the existence of weaker zones of bone (which incidentally may be concerned in the genesis of fractures), and included cement-lines, interlamellar interfaces, osteocyte lacunae, and vascular canals. He considered the split-line technique an unreliable method for structural and force transmission analysis in bone.

Strain gauges have been used to record the results of stresses on whole bones, both as dead, isolated elements and during in vivo experiments in intact animals, and also on bone samples. Unfortunately they cannot be bonded on to bone in large numbers, and the data obtained are limited. Nevertheless, interesting in vivo results have been obtained by Lanyon (1973) on the vertebrae, tibiae, and calcanei of sheep, recording a deformation cycle during locomotion. Lanyon himself was critical of the method, which has, however, confirmed the elasticity of bone and shown that the elastic modulus is an inconstant quantity, varying with speed of movement.

Interference holography has been applied to the study of surface strains in the human mandible (Gupta and Knoell 1973), but the technique has proved difficult and has apparently not yet been adapted to experiments on the living.

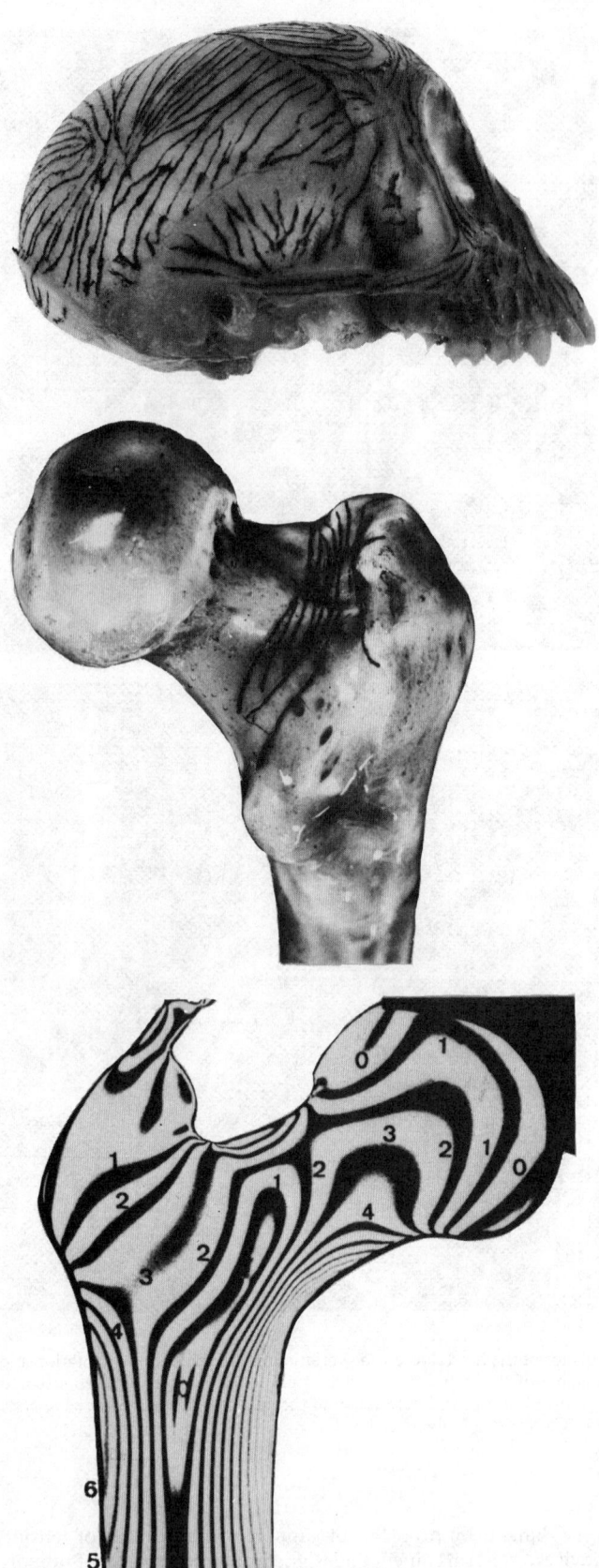

3.3B Photographs illustrating three methods used to investigate possible interrelations between structural organization and patterns of force transmission in bone. *Above*, split-line patterns produced in a decalcified skull (*Macaca mulatta*) by repeated puncture with an ink-tipped needle. *Centre*, pattern of cracks in colophonium resin coating of a femur subjected to a compressive stress of the head, vertically applied. *Below*, photo-elastic pattern produced in a flat plastic sheet modelled upon the proximal end of the femur. A compressive force has been applied to the femoral 'head', whilst a tensile force has been applied to its 'greater trochanter'. (The lowest photograph is from von Knief 1967, the others kindly contributed by Dr. C. Buckland-Wright, Department of Anatomy, Guy's Hospital Medical School, London.)

Mathematical analysis has been widely developed in this field, but since the basic data employed are usually limited, or are based upon the over-simplification inherent in the use of models, the usefulness of this approach is restricted. (For examples of such work consult the papers of Koch 1917; Kummer 1966; Rybicki et al. 1972.)

A combination of techniques appears to offer the most fruitful approach to the problems of bone mechanics, and it is therefore appropriate to end this brief review of a most complex and difficult field of research by reference to a recent study by Buckland-Wright (1978) of the patterns of transmission of strains in the feline skull during biting. This study combined the techniques of radiology and projection microradiography with the use of strain gauges, colophonium resin experiments, and histology, the limitations of each technique being offset by the others. The projection microradiographs were assessed by a stereoscopic technique, permitting a three-dimensional interpretation of the relation between bone structure and impressed stresses. On the basis of these observations a modified version of the trajectorial theory was propounded. Further work by such combined methods must be awaited with interest.

As already stated above there has been a profusion of estimates of the physical parameters of bone tissue recorded. The following results are cited as interesting representatives of such estimates, including a comparison with other materials.

Table of the approximate tensile breaking strengths of bone and a few other materials *(From Gordon 1968)*

MATERIAL	TONS/INCH	MN/m²
Traditional cast iron	5–10	75–150
Copper	10	150
Bone	10	150
Tendon	7	105
Wood, spruce (along grain)	7	105
Cotton	25	375

Growth of Individual Bones

As in any other mammal the bones of the human skeleton are preformed in hyaline cartilage or, in a minority of elements, in condensed mesenchyme. Thus, a soft tissue model at first appears, and this is gradually converted into hard osseous tissue by the development of osteogenesis, frequently in a central position, from which the process of transformation spreads, until the whole skeletal element is ossified. The appearance of such *centres of ossification* is spread over a long period of time. A large number are first detectable in embryonic life (3.4), some do not appear until much later in prenatal life, and others do not start until well into the growing period after birth. The process is complex, presenting phenomena which must be studied at different levels of organization. At the microscopic, ultrastructural and molecular levels are the basic problems of histogenesis, growth and transformation of skeletal tissues, including primitive mesenchyme and its differentiation into the various forms of connective tissue, cartilage and bone. All these matters will be considered in the next section which is concerned with the fine structure, histogenesis, growth, remodelling and regeneration of all the skeletal connective tissues (*vide infra* p. 245 *et seq.*).

At first, microscopic in size, ossific centres soon become macroscopic and their subsequent behaviour can be followed by the unaided eye, whether by plain inspection, dissection, or by radiological technique. It is this scale of events which will be considered here.

Many bones are ossified from a single centre; these include those of the carpus and tarsus, the lacrimal, nasal, and zygomatic bones, the inferior nasal concha, and the auditory ossicles. Even in this limited group the appearance of their centres extends from the eighth week of intrauterine development to the tenth year of childhood, thus providing in themselves a wide sequence of events in studying skeletal growth or in estimating age. However, most bones are ossified from several separate foci, one of which appears during late embryonic or early fetal life (seventh week to fourth month) near the middle of the future bone (3.4). This centre is concerned with progressive ossification towards the bone ends. In all such bones their ends are cartilaginous at birth (3.4), though displaying their typical shape and articular congruities. These terminal regions are ossified by separate centres,

Table of Young's modulus of wet bone with the direction of load parallel to the long axis of the bone *(From Reilly and Burnstein 1974)*

BONE SPECIES AND AUTHORS	TYPE OF LOADING	YOUNG'S MODULUS ($\times 10^9$ N/m²)	COMMENTS
HUMAN Dempster & Liddicoat (1952)	Tension, low strain rate	14·1	Dry, rewetted femur, tibia, humerus, mixed data, extensometer used.
Burnstein *et al.* (1972)	Tension, strain rate: 0·1 sec⁻¹	14·1	Femur; extensometer used for strain.
BOVINE Burnstein *et al.* (1972)	Tension, strain rate: 0·1 sec⁻¹	24·5 ± 5·10	Femur; extensometer used.
Simkin & Robin (1973)	Tension, low strain rate	23·8 ± 2·21	Tibia; unreported histology; extensometer used.
OTHER MATERIALS Methylmethacrylate: Plexiglass		8·6	
Wood: Douglas fir		13·4	(68 per cent moisture.)
Steel		210·0	

Note: MN = 10⁶ N   1 N = 1 kg × 1 m/s² = 10⁵ dyn.

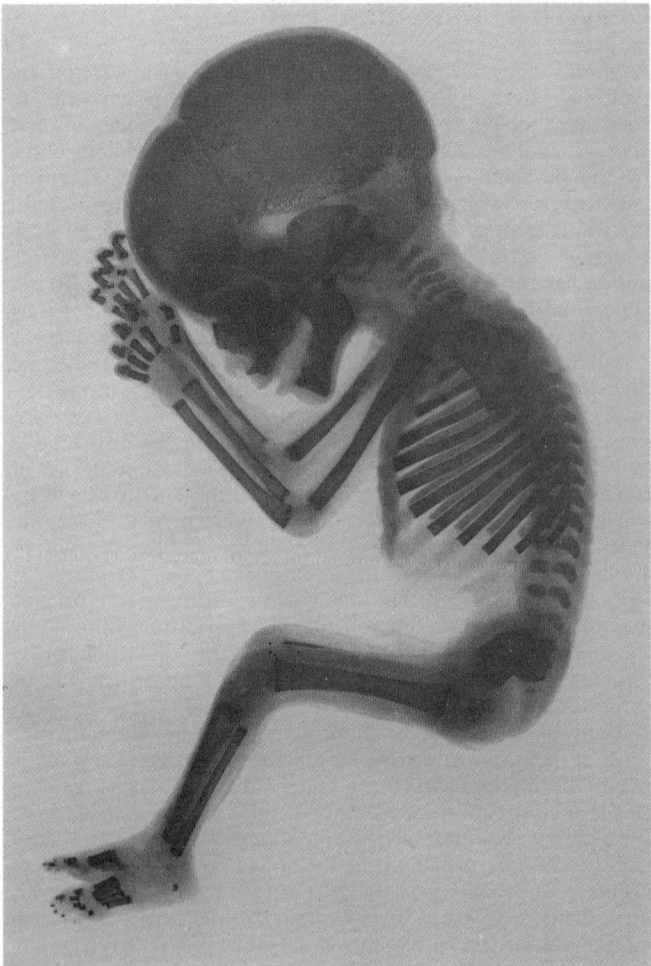

3.4A An alizarin stained and cleared human fetus of about 14 weeks *in utero*. Note the degree of progression of ossification from primary centres, which is endochondral in the appendicular and axial skeletons, excepting the clavicles, and the intramembranous centres for the majority of the cranial bones which are visible here. The carpus and tarsus are wholly cartilaginous except the primary centre for the calcaneus, as are, of course, the epiphyses of all the long bones. The centra and neural arches of the vertebrae are separate. The sternum is still unossified. The membranous anterolateral and posterolateral fontanelles are particularly obvious. (Photographed by Kevin Fitzpatrick, Department of Anatomy, Guy's Hospital Medical School, from a specimen prepared by Roslyn Holthouse, formerly of the same Department.)

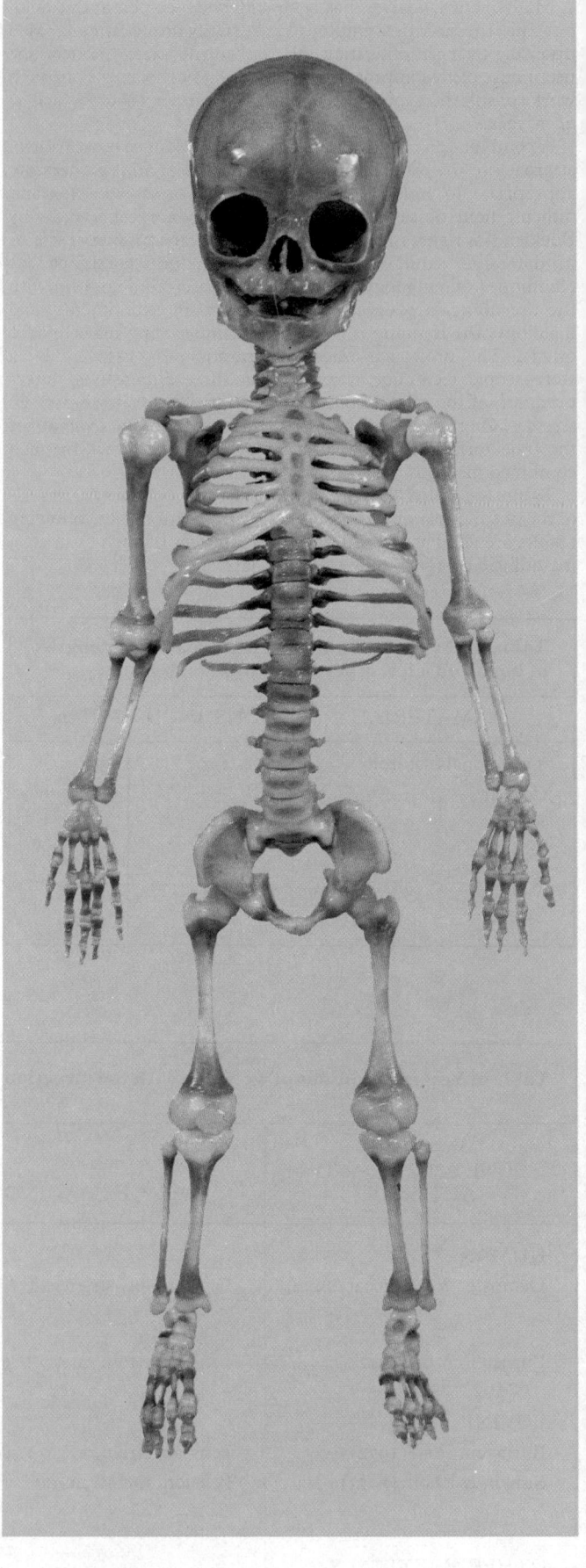

sometimes multiple, which develop from about the time of birth onwards into the late teens; they are said to be *secondary* to the earlier *primary* centre from which the majority of a bone is ossified. This is the mode of ossification in the long bones of the limbs and their shorter elements, metacarpals, metatarsals and phalanges; it is also characteristic of ribs and clavicles.

Typically, therefore, a bone such as the tibia has become, at birth, ossified throughout its shaft, or *diaphysis*, by the activity of a primary centre which appeared in the seventh week of embryonic existence; whereas its two extremities, its *epiphyses*, are subsequently transformed into osseous tissue by secondary centres. As the latter enlarge, replacing the surrounding cartilage, they come to occupy almost the whole of the articular ends of the bone. A layer of specialized hyaline cartilage persists over the actual joint surface (p. 249) and a thicker zone remains between the diaphysis and epiphysis. As described in detail below (p. 243) it is by the persistence of this growing *epiphysial plate* or *disc* of cartilage (synonymous terms are *growth plate* or *growth cartilage*), that the bone is able to increase in length until its characteristic dimensions are attained, at which stage the mechanism is abolished by ossification of the disc. The bone can then be said to have reached maturity. The process of coalescence of epiphysis and diaphysis is called *fusion*, final fusion usually representing the

3.4B The skeleton of a neonatal infant, with all cartilages preserved. Note particularly proportions of neurocranium, orbital cavities, and face, interfrontal (metopic) suture, sternebral ossification, extensive costal cartilages, large cartilaginous epiphyses, and metaphysial flaring of long bones, especially the humeri, femora, and tibiae. (Photographed by Kevin Fitzpatrick from a preparation by Dr. Michael C. E. Hutchinson, of the Department of Anatomy, Guy's Hospital Medical School, London.)

amalgamation into one osseous element of a number of osseous units, developed separately in the same cartilaginous or mesenchymal model, which has itself been growing throughout the process. Fusion also brings to an end the period of active gross dimensional growth in a bone.

Many bones show epiphysial arrangements at both extremities, others at one end only. The long limb bones illustrate the former mode of ossification, while metacarpals, metatarsals, phalanges, clavicles and ribs possess an epiphysis at one end only, although it might be suggested that costal cartilages may represent rib epiphyses which do not normally develop ossific centres. (For an interesting discussion of the *pseudo-epiphysis* of the distal end, or head, of the first metacarpal bone consult Haines 1974.) In some bones the epiphysial centres at one end or both ends are more complex. For example, in the proximal end of the humerus, which is wholly cartilaginous at birth, three separate centres appear during childhood. They soon coalesce to form a single epiphysial mass, which later fuses to the diaphysis. In this instance only one of the centres is involved in forming the articular surface, the others being concerned in the growth of the greater and lesser tubercles, projections to which muscles are attached. Similar composite epiphyses occur at the distal end of the humerus and in the femur, ribs and vertebrae. Because some centres ossify parts of bones exposed to pressure at joints, and others to regions subjected to tension by the pull of muscles, a classification into *pressure, traction* and *atavistic* has been proposed (Parsons 1903–5). Atavistic epiphyses are formed by centres of ossification which are considered to represent the skeletal elements which were separate in some earlier evolutionary stage. The evidence of comparative morphology suggests, strongly in some cases, that mammalian bones are composites of several separate reptilian or amphibian elements. The skull, clavicle, scapula and innominate bones afford examples. For instance, the small centre which regularly occurs in the coracoid process of the human scapula and the epiphysis at the medial end of the clavicle may be vestiges of separate skeletal elements present in earlier vertebrates, and repeated as transient features in the development of mammalian bones. Incidentally, the medial end of the clavicle could equally be regarded as a pressure epiphysis.

Many cranial bones develop from multiple centres of ossification, and the morphological evidence that there is an atavistic element in this is considerable. That there has been a marked reduction in the number of cranial bones in mammals, when compared with reptiles and other lower vertebrate groups, is certain. The sphenoid, temporal and occipital bones are examples of skeletal units which almost certainly are composites derived from a number of previously separate cranial elements. It is not difficult to equate at least some of their centres with these ancestral elements; and they present additional evidence of fusion in the existence of dermal (membranous) and cartilaginous derivatives, at first separate but united during growth to form a complex whole. Details of these derivations will be considered in the appropriate sections (*see* p. 315). It must be added that these events in cranial bone development are not precisely equivalent to the atavistic epiphyses in post-cranial bones, where such epiphyses are always *secondary* in nature. In the composite cranial bones, on the contrary, the human condition is the result of the coalescence of a number of elements, each with a *primary* centre of ossification.

The classification of epiphyses has perhaps little significance *per se*; it does, however, direct attention to matters of great interest concerning the evolution of the mammalian skeleton. Epiphysial centres of ossification do not occur in other vertebrates, with the exception of sporadic appearances in reptiles and birds (Haines 1937). When they do appear, they are pressure epiphyses, associated with articular surfaces. Traction epiphyses are peculiar to mammals, but may be regarded as genetically well established, for their appearance is not arrested by division of the structure attached to them (Appleton 1922; Barnett and Lewis 1958). There is, however, a general lack of such experimental evidence in connection with epiphysial development.

The rate of growth of ossific centres is more rapid than that of the cartilaginous mass in which they occur, though it must be emphasized that the latter is itself growing in concert with the general increase in size and change in proportions of the skeletal element to which it contributes. As a result the epiphysial region, at first wholly cartilaginous, is gradually converted into bone, with the exceptions noted above. The whole process starts at a time, and continues at a rate, which is characteristic for each bone. Individual, sexual and racial variations naturally occur, and must be taken into account, as far as established data permit, when assessing the skeletal age of an individual.

The rate of growth in bones varies both in those with epiphysial plates, and also those without epiphyses. Were this not so, and the rate uniform in all bones, their centres of ossification would appear in a strict descending order of size. That this is not so is illustrated by the primary centres for such widely different masses as the shaft of a terminal phalanx and a femur, separated by a week of embryonic life at the most. In the case of the primary centres for the carpal and tarsal bones there is some association between the size of each element and the order of ossification from the largest (the calcaneus in the fifth fetal month) to the smallest (the pisiform in the ninth to twelfth year of postnatal life). In individual bones, such as the humerus and the femur, the succession of ossific centres is approximately in the order of volume of bone which each produces. The largest epiphyses, the lower end of the femur and the upper end of the tibia, do ossify earliest (immediately before or after birth—points of forensic interest), but when comparisons between other bones are attempted, inconsistencies are numerous. Such dissimilar bone masses as a phalangeal epiphysis and the lesser tuberosity of the humerus begin to ossify about the same time, and both long before the massive greater trochanter of the femur, again emphasizing widely differing rates of bone development.

Experimental evidence in mammals indicates that at the epiphysial plate growth is initially equal in rate at both ends of bones possessing two epiphyses, but after birth one cartilage grows faster then the other (Brookes 1963). Since this end also usually fuses later with the diaphysis, the total contribution to the length of the bone is considerably greater at this extremity. Although the faster rate can only be presumed in man, the later fusion is a fact of radiological observation, and the curious recurrent direction of the nutrient arteries as they enter the femur, radius and ulna accords with the above facts (cf. p. 257).

The more active extremity of a long limb bone in respect of growth in length is often referred to as the *growing end* but this does not imply that growth is limited to this end. The variable rate of growth at epiphyses in general is evidenced by changes in the rate of increase of height characteristic of human development (Krogman 1941; Tanner 1962). The rate is characteristically rapid in infancy and again at puberty, but usually slower in between. The existence of a spurt at puberty, or slightly before, which decreases as epiphyses fuse during the later teens, has been much studied of late (Sinclair 1969).

Growth cartilages do not necessarily grow at the same rate at all points throughout their substance, and this accounts for changes in the relation between diaphysis and epiphysis—e.g. the alteration in angle between the shaft and neck of the humerus. On their diaphysial faces, epiphyses do not display a uniformly flat junction with the growth cartilage, nor indeed does the latter at its junction with the diaphysis. By differential rates of growth the two bony surfaces usually become reciprocally curved, commonly in such a way that the epiphysis fits like a shallow cup over the convex end of the shaft (but with cartilage intervening temporarily). This form of junction appears adapted to resist shearing forces at this relatively weak point in the growing bone. The reciprocity of the bone surfaces is usually carried further in the development of small nodules and ridges, which are highly characteristic of these surfaces when denuded of cartilage. Such adaptations emphasize the fact that all immature bones formed from more than one focus of ossification are complexes of several bones held together by epiphysial cartilages. When it is remembered that this is the case in the majority of the elements of the human skeleton, not only through the highly active years of childhood, but also during the even more vigorous period of adolescence, the strength of the bonding of bone to bone through cartilage is seen in true perspective.

The forces exerted at growth cartilages are largely compressive,

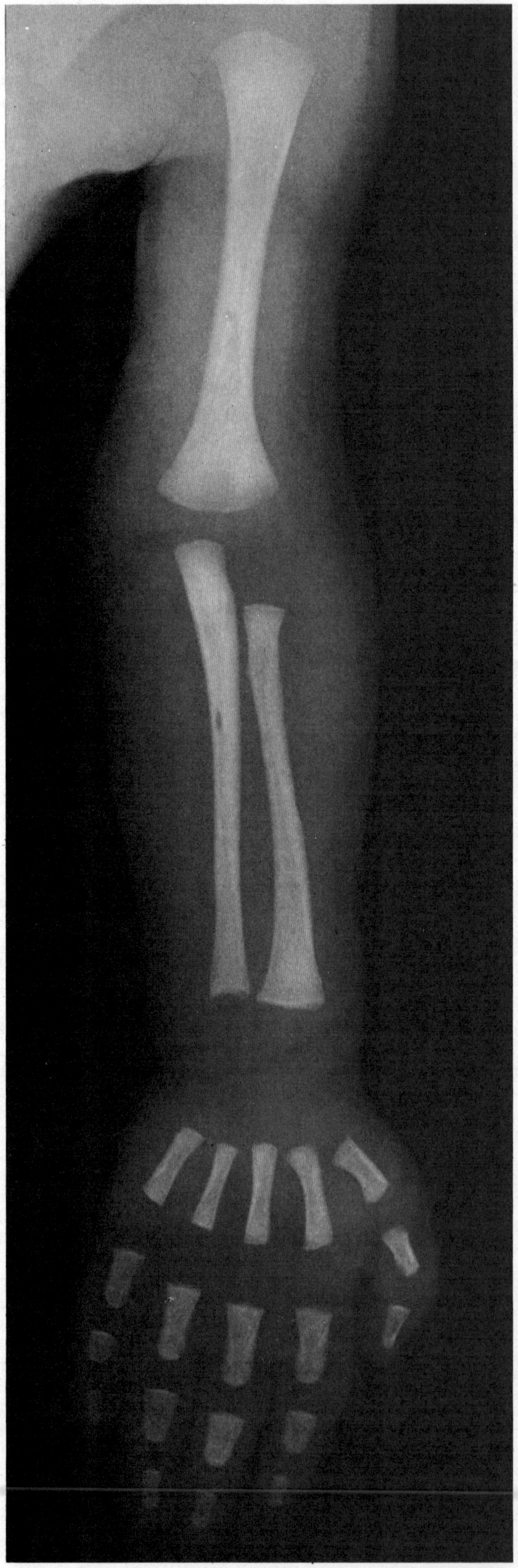

3·4 C

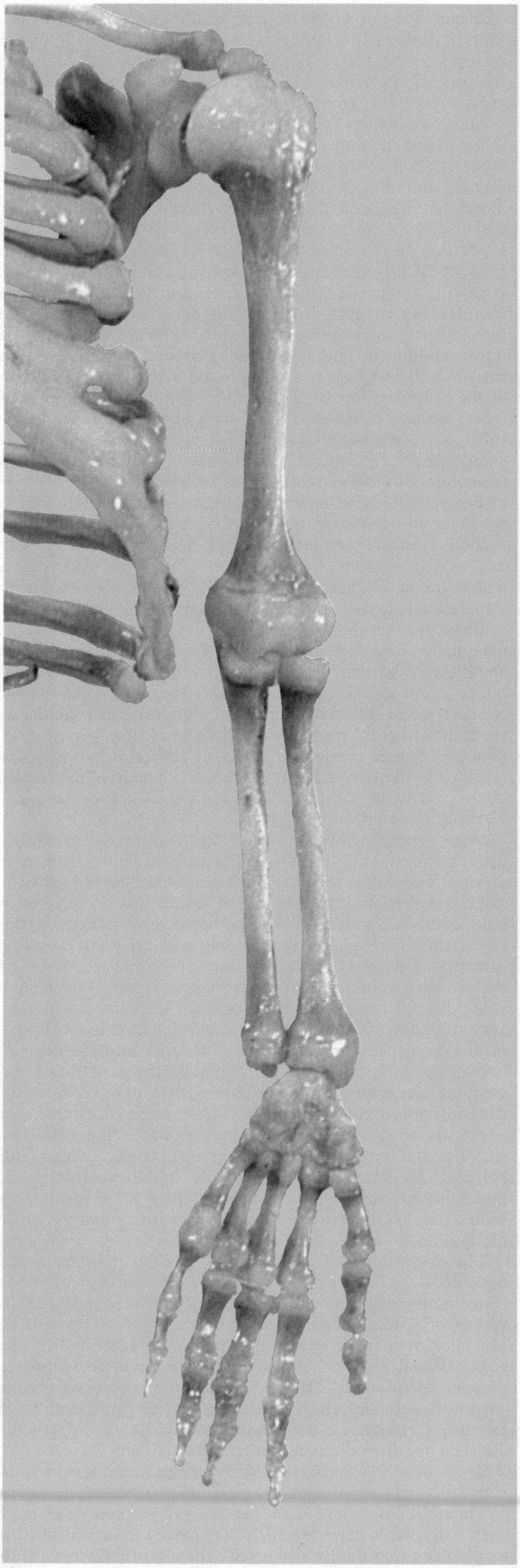

3·4 D

3.4C Radiograph of neonatal arm. Ossification from primary centres is well advanced in all bones except the carpals, which are still wholly cartilaginous. The gaps by which individual elements appear to be separated are, of course, filled by the radiolucent hyaline cartilage, in which epiphysial or carpal ossifications will subsequently occur. In the long bones note the flaring contours, with narrow midshaft and relatively expanded metaphyses. Note also the proportions of the limb segments characteristic of this age, and in particular the relatively large hand.

3.4D Photograph of a preparation of a neonatal left arm (from the specimen shown in **3.4B**). Compare the radiolucent areas in the radiograph (**3.4C**) with the preserved cartilaginous epiphyses and carpal elements in this specimen. (For acknowledgments see **3.4B**.)

but with also a considerable element of shear. Where traction epiphyses are attached by cartilage, prior to fusion, the pull of attached structures, such as muscles, creates tension stresses, and this has been shown to be associated with the development of abundant collagen fibres oriented along the supposed lines of stress (Smith 1962). Damage to epiphysial growth due to violence does occur, but disturbance due to severe constitutional disease, such as the fevers common in childhood, is more frequent. The latter may produce visible change in the trabecular pattern of the bone formed at that period, discernible in radiographs as a dense transverse so-called *line of arrested growth* (Harris 1933). Several such lines may be apparent in the limb bones of children who have suffered a succession of severe illnesses.

The actual process of fusion of the epiphyses with the diaphysis is described later (p. 266). The actively growing part of the diaphysis adjacent to the epiphysial cartilage, the *metaphysis*, appears to overtake the cartilage, but there is also macroscopic evidence of a contribution by the epiphysis to this process in the formation of a denticulated edge between the two bone elements as they bridge across the cartilage-filled gap. These appearances have been extensively described and are of importance in establishing criteria for estimations of the time of fusion, whether by inspection or radiography (Stevenson 1924). Knowledge of the sites, times of appearance, rates of growth and fusion times of epiphyses is important in clinical medicine, forensic practice and anthropology. Dependable chronological data can only be derived from adequate numbers of observations made with standardized techniques. The wide variation in published figures is considerably due to failure in these respects. Radiographs need to be taken at more frequent intervals than is usually the case, and the angle of view must be carefully controlled. Routine positions may be inadequate; for example, the separate centre for the lesser tubercle of the humerus has often been denied by radiologists, probably because of the superimposition of its image upon that of another centre.

Nevertheless, considerable variation in the chronology of skeletal development does occur on the basis of individuality, sex, and possibly race. It is generally agreed that the *sequence* of events shows little or no variation and that this lies chiefly in the *timing* of the appearance and fusion of ossification centres. In both regards, the *female antedates the male* in all groups studied so far, and the difference, trivial and perhaps insignificant before birth, becomes progressively greater thereafter, rising to as much as two years or so in the later fusions of adolescence. Estimates have been most thoroughly established for white races, but even here few parts of the skeleton have been studied in sufficient detail, numbers, or over adequate time periods. Perhaps the best studies available at present are those for the hand (Todd 1937; Greulich and Pyle 1959; Mathiasen 1973), the knee (Pyle and Hoerr 1955), and the ankle and foot (Hoerr *et al.* 1962).

THE SKELETAL CONNECTIVE TISSUES

The skeletal tissues, cartilage and bone, are essentially specialized connective tissues and consist of the same three tissue elements—cells embedded in an intercellular matrix, permeated by a system of fibres. Physically, however, the sclerous tissues differ from the soft, pliant, generalized connective tissues, for their matrix is solidified. Cartilage and bone are, nevertheless, quite distinct in their structure, physical properties, vascularization, and in their patterns of growth and regeneration.

Structure of Cartilage

This is a phylogenetically ancient tissue and is widespread as a permanent or temporary skeletal tissue, throughout the vertebrates. As we have seen, in man during early fetal life, the greater part of the skeleton is cartilaginous, but eventually much of this is replaced by bone. In adult life, cartilage persists as the articulating surfaces of synovial joints, in the walls of the thorax, larynx, trachea, bronchi, nose and ears, and as isolated small masses in the skull base. However, in many cases it is not a simple partial replacement of a temporary cartilaginous skeleton by bone, for the cartilaginous plates between the ossifying epiphyses are diaphyses of long bones and in other situations are essentially *growth plates* which continue to proliferate, allowing length increases of the bones concerned. They are only replaced by bone when growth ceases. (For a review of embryonic cartilage *see* Glenister 1976; visceral cartilage was also reviewed by Reid 1976; the structure and biochemistry of cartilage is analysed in depth by Serafini-Fracassini and Smith 1974; for an alternative view of cartilaginous growth plate biology *see* Lutfi 1974.)

The distinctive properties of cartilage as a tissue are, therefore, its low metabolic rate and vascularity, its capacity for continued and often rapid growth, both interstitial and appositional (*vide infra*), combined with rigidity, a high tensile strength and resistance to compressive and shearing forces, whilst retaining some resilience and elasticity. The surfaces of cartilage masses are usually clothed with a fibrous *perichondrium* or abut against bone, but where they form the articulating surfaces of a synovial joint they are modified as wear-resistant, smooth, polished surfaces which are bathed by lubricant synovial fluid and have an extremely low coefficient of friction.

The *matrix* in which the cartilage cells (*chondrocytes*) are embedded varies in its appearance when freshly cut, in its detailed composition, and also in the nature of its contained fibres. Accordingly, the tissue is often classified, for convenience, as either *hyaline cartilage* (from *hyalos* meaning glass), *white fibrocartilage* (containing much collagen), and *yellow elastic fibrocartilage* (containing a rich elastin network). A densely *cellular cartilage*, with only fine septa of matrix separating the cells, is a normal stage in the development of cartilage during early embryonic life and is also found as a permanent tissue in the external ears of many mammals. The reader should appreciate, however, that whilst white fibrocartilage and yellow elastic fibrocartilage are specific, distinct, relatively unvarying tissues, the term hyaline cartilage embraces virtually all remaining types of cartilage which show *great variation* in composition and properties according to species, age, and location.

The cartilage cells occupy small spaces or *lacunae* in the matrix which conform to the cell shape. Young cells (*chondroblasts*) are relatively small, often flattened, with an irregular contour and many small surface projections (filopodia) which are received into complementary recesses in the matrix (**3.5A**). (Early post-mitotic chondroblasts often show specialized intercellular contacts which are necessarily transient, disappearing as matrix synthesis proceeds.) Mature cells (*chondrocytes*) increase in size with age and, although more rounded, still present a few surface projections (p. 264), and bear an occasional isolated cilium. Cartilage cell nuclei are round or oval with a variable number of nucleoli; multinucleate cells are not uncommon. The cytoplasm contains a paranuclear Golgi apparatus and diplosome, scattered mitochondria, a concentration of granular endoplasmic reticulum and often a few fat globules, glycogen deposits and pigment

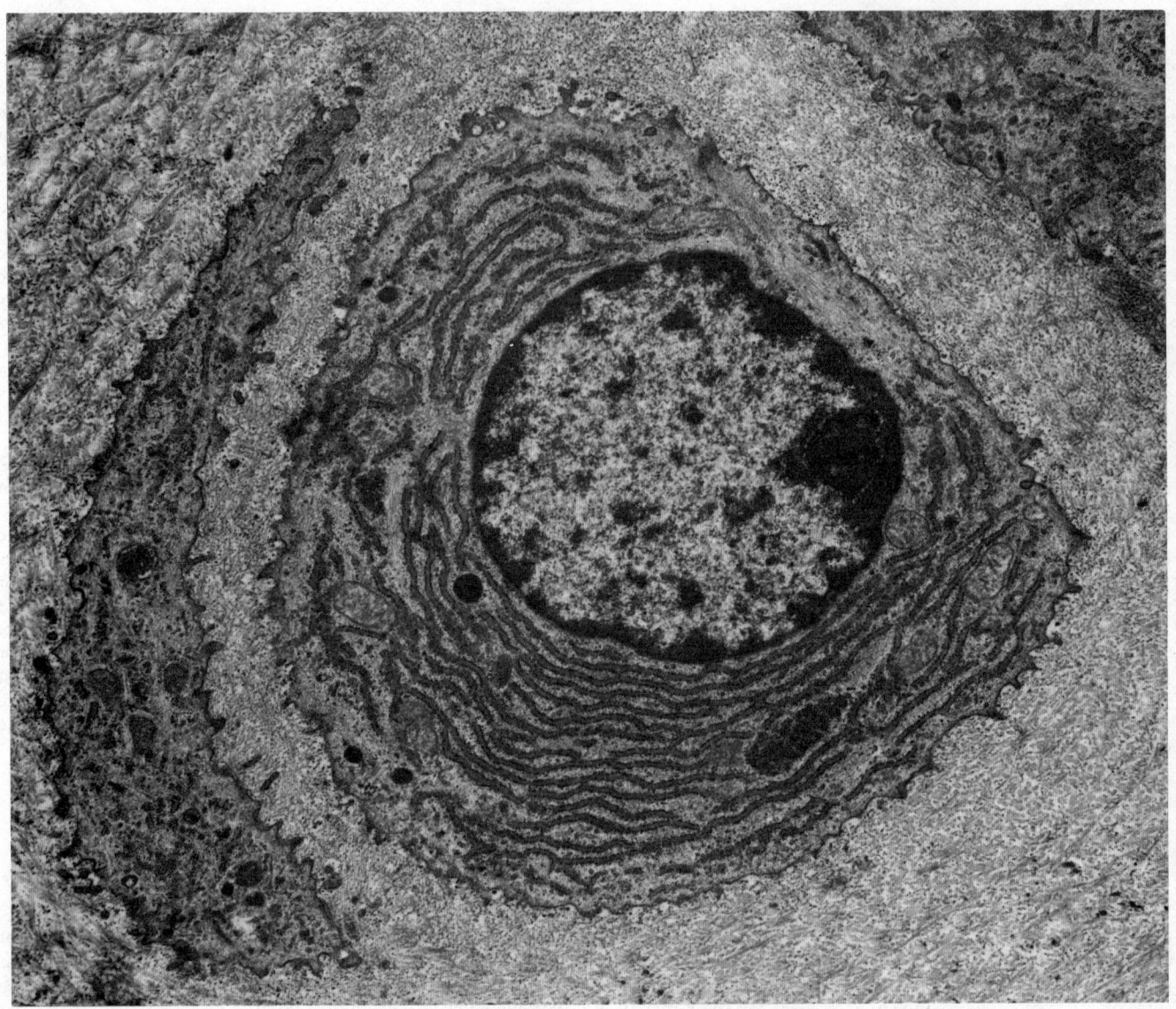

3.5A Transmission electron micrograph of an ultra-thin section of a perfusion-fixed specimen of a rabbit's femoral condylar cartilage. The centrally placed chondroblast contains an active euchromatic nucleus with a prominent nucleolus. Its cytoplasm contains a rich concentration of roughly parallel concentric flattened cisternae of rough endoplasmic reticulum, scattered mitochondria, lysosomes and aggregations of glycogen. Its plasma membrane bears numerous short filopodia which project into complementary recesses in the surrounding matrix. The latter shows a delicate fibrillary feltwork with a finely dispersed granular interfibrillary substance. No pericellular 'lacuna' is present; the matrix separates the central chondroblast from the sectioned cytoplasmic periphery of two adjacent chondroblasts. The left profile, crescentic in outline, is characteristic of chondroblasts with an almost squamous form, that are sited near the articular surface. (Preparation by Mrs. Susan Smith of the Department of Anatomy, Guy's Hospital Medical School, London.) Magnification × 14,500.

granules. In the chondroblasts of young and actively growing cartilage (3.5A) the Golgi apparatus and granular reticulum are particularly prominent but are much less so in mature, relatively quiescent cartilage (Silberberg *et al.* 1964).

The matrix of cartilage consists essentially of water, varieties of proteoglycan, some lipid, collagen, non-collagenous protein, and electrolytes: on superficial examination it presents as an amorphous ground substance surrounding a feltwork of collagen fibres which vary in their type, density and disposition in different regions. The *ground substance* is a firm gel which stains positively with the periodic acid-Schiff reaction, metachromatically with toluidine blue, and is basophilic in ordinary histological preparations. The *proteoglycans* consist of a series of copolymers of *core protein* (mucoprotein) conjugated to long unbranched side chains of the *glycosaminoglycans* chondroitin 4-sulphate, chondroitin 6-sulphate, and keratan sulphate, the proportions of which vary with age—keratan sulphate increasing with maturity (Kaplan and Meyer 1959, Chvapil, 1967). Little is known of the detailed molecular architecture of the core protein, but analysis of the polysaccharides is now well advanced. Each consists almost exclusively of repeating disaccharide units, one unit of which is invariably a substituted hexosamine, and the other an esterified hexuronic acid or a substituted hexose. Thus chondroitin 4-sulphate and chondroitin 6-sulphate consist of repeating N-acetylgalactosamine and glucuronyl sulphate disaccharides, differing only in the position of the sulphate ester linkage: keratan sulphate consisting of N-acetylgalactosamine and D-galactose together with a sulphate ester linkage which is variable in position. The acidic nature of these sulphated polysaccharides accounts for the basophilia, metachromasia, water and ion-binding power, selective permeability, and most of the visco-elastic properties of the matrix. The *collagen* present in cartilage differs in its composition from that found in other vertebrate tissues. The tropocollagen molecule of the latter consists of two $a1$ and one $a2$ polypeptide chains, whereas the tropocollagen molecule of cartilage consists of three identical $a1$ chains. Subsequent analysis has shown that two varieties of $a1$ polypeptide chains with differing amino acid sequences are present in different tissues, and these have been designated $a1(I)$ and $a1(II)$—*see* Miller (1971a & b). Using this classification the most widespread type of vertebrate collagen is $[a1(I)]_2 a2$; in contrast the collagen characteristic of cartilage is $[a1(II)]_3$.

It has long been held that the compression-resisting properties and visco-elasticity of cartilage are expressions of its glycosaminoglycan content; similarly its high tensile strength has been considered mainly a property of its contained collagen. These propositions, whilst undoubtedly partly true, are an oversimplification. Many investigations on a wide variety of connective tissues, including collagen, have demonstrated intimate and spatially precise relationships between the repeating regional substructure of the collagen filaments and the distribution of the proteoglycans (for an excellent review see Serafini-Fracassini and Smith 1974). Thus it has been proposed that the core protein part of the copolymer is applied tangentially to, or encircles, the collagen filaments at specific points (the a and b_1 loci) with some form of linkage between them. The sulphated and acetylated polysaccharide tails are thought to project laterally, radiating in all directions from the collagen filament, their terminals forming a variety of cross-links with either the polysaccharide or protein moiety of the proteoglycans of neighbouring filaments. The whole thus forms a complex *three-dimensional meshwork*, the polysaccharide elements of which are polyanionic bearing a fixed negative charge. The interstices of the meshwork are occupied by relatively large volumes of bound water, cations, and non-collagenous non-core protein, the structure of the latter being little understood. The reaction of cartilage to different forms of mechanical stress is accordingly a reflection of interacting contributions from *all* these structural elements: similarly, the properties of the 'walls' and 'contents' of the macromolecular labyrinth that constitutes cartilage affects the diffusion rates of both nutritive and degraded materials through the tissue.

For many years it was assumed, without direct evidence, that the chondroblasts and chondrocytes were somehow involved in the elaboration of *both* collagen and ground substance. In recent years histochemical, autoradiographic, and cell fractionation studies have confirmed and amplified this view (Davies and Young 1954; Bélanger 1956; Dziewiatkowski 1962; and see **3.5B**). Labelled collagen precursors were shown to accumulate sequentially over the ribosome bearing granular endoplasmic reticulum, the Golgi apparatus, and finally over the vesicular cell periphery and extracellular material (Revel and Hay 1963)—a common sequence in cells synthesizing exportable protein. However, evidence has also been presented consistent with the simpler view that, following initial synthesis at the ribosome, the elementary collagen is extruded directly from the ground cytoplasm of the chondroblast to the exterior, without involving endoplasmic reticulum cisterns of Golgi membranes (Cooper and Prockop 1968). The *details* of collagen synthesis in cartilage, however, appear similar to those obtaining in other tissues (p. 49).

Non-collagenous protein (mucoprotein) is also synthesized at the granular reticulum, and earlier work suggested that it was transferred to the Golgi apparatus, where sulphated polysaccharide synthesis was presumed to occur. More recently, however, the detailed structure of the chondroitin sulphate-protein linkage has been elucidated (Lindahl and Rodén 1966) and more detailed studies (Horowitz and Dorfman 1968) have now shown that core mucoprotein synthesis occurs at the ribosome; the linking sugar molecules are then added and chain initiation occurs in the rough endoplasmic reticulum, whilst chain completion occurs in the Golgi apparatus, and the final product is exported to the cell exterior by reversed pinocytosis. (For a

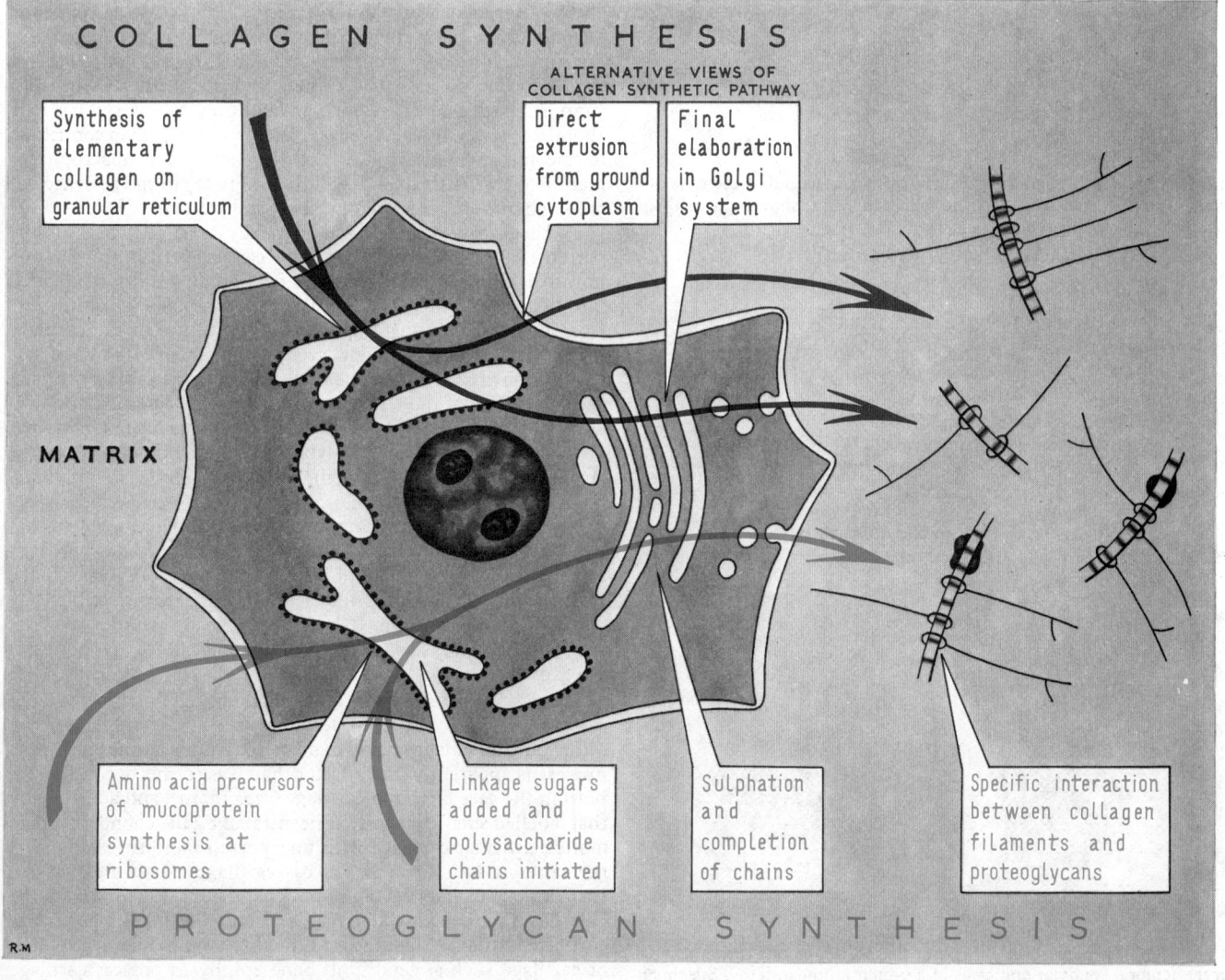

3.5B A summary of some of the important biosynthetic pathways of the chondroblast.

comprehensive review of cartilage biosynthesis *see* Serafini-Fracassini and Smith 1974.)

Such studies not only help to elucidate cellular synthetic pathways; but the speed of appearance of labelled precursors over various cartilage cell organelles (sometimes only a few minutes after injection, in the case of young growing cartilage), emphasizes the ready permeability of the cartilage matrix.

NUTRITION OF CARTILAGE

Cartilage is often described as a totally avascular tissue. While this is by no means true, the majority of cartilage cells are, compared with other tissues, an unusually long distance away from the nearest exchange vessels. In general, these lie in the perichondrial membrane on the surface of the blocks of cartilage. Between the perichondrial capillary network and the chondrocytes, nutrient substances and metabolites pass by diffusion down concentration gradients across the intervening cartilage matrix.

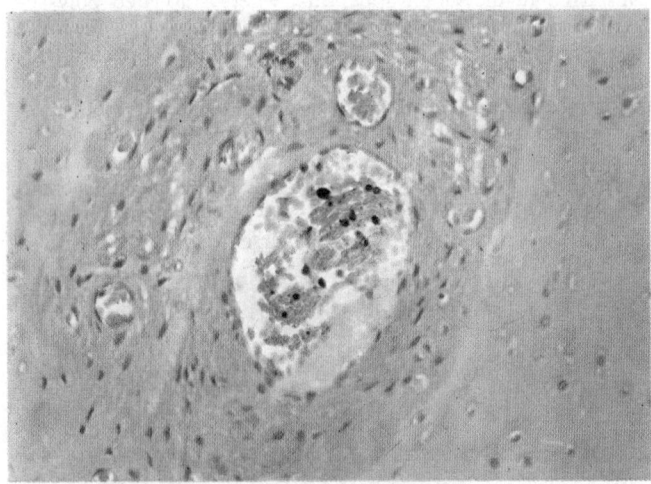

3.6A A section through the proximal cartilaginous end of the tibia of a young child stained with haematoxylin and eosin, showing a cartilage canal and its contained blood vessels.

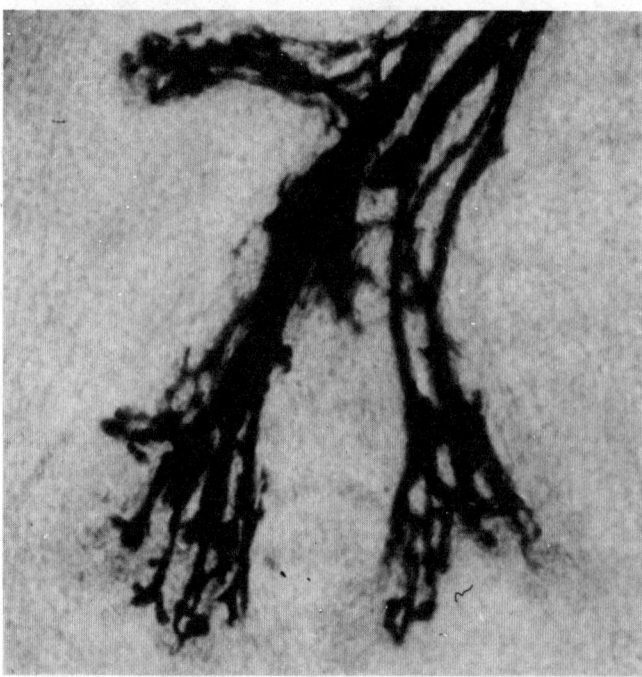

3.6B Capillary plexuses in the terminal arborization of a cartilage canal in the femoral condylar epiphysis of a 7-month human fetus. Specimen injected with Indian ink. (Kindly provided by Dr. Murray Brookes of the Department of Anatomy, Guy's Hospital Medical School, London.)

Cartilage Canals (Haines 1933–4; Hurrell 1934–5; Levene 1964; Blackwood 1965; Lufti 1970; Brookes 1971).

Nutrition is, however, augmented at least in cartilaginous epiphyses of large mammals (and analogous cartilaginous masses of many birds and reptiles) by an arborization of *cartilage canals* (Brookes 1971). Each canal contains a leash of small vessels which stem from centripetal offshoots from a perichondrial artery and vein. Perfusion techniques (Brookes 1958) and histological examination (Wilsman and Van Sickle 1972) suggest that the leash is composed of a central small artery or arteriole surrounded by numerous venules and perivascular capillaries (3.6A, B). The central vessel terminates in the blind end of the canal forming several capillaries from which venules pass back alongside the arterial twig to the parent perichondrial vein. Capillaries are also formed along the entire length of the canal, and these, together with the terminal capillaries, potentiate metabolic exchange with the cartilage mass (Wilsman and Van Sickle 1972). The blood vessels of cartilage canals are packed in loose connective tissues, rich in fibroblasts and macrophages, and continuous with the perichondrium.

How cartilage canals develop and grow is problematical (as are the detailed growth patterns of a cartilaginous epiphysis as a whole). If the latter, in general, grows mainly by surface accretion, then the canals may be generated by the *passive inclusion* of perichondrial vessels of the advancing surface of the cartilage mass. Harris (1933) adduced evidence that cartilaginous epiphyses grow interstitially from a zone of chondrocyte mitosis below the surface, and this would preclude the passive inclusion hypothesis. Hence several investigators have suggested that the formation of the canals is an *invasive* phenomenon, involving chondrolysis. The vascular content of the canals was held to be lytic by Eckert-Möbius (1924) and Hurrell (1934). Waterman (1961) proposed that chondrocytes lysed themselves 'free' in advance of an ingrowing vessel, which thereby, in certain regions, finds its canal already prepared for it. Brookes (1971) emphasized the chondrolytic activity of the loose connective tissue cells supporting the canal vessels, in addition to matrix lysis by chondrocytes in the canal wall. These 'freed' cells then contribute to the mesenchymal content of the canal. The macrophages identified by Wilsman and Van Sickle (1972) in the perivascular mesenchyme are, presumably, also chondrolytic.

Cartilage canals are not haphazardly arranged, but show a pattern which is specific for each cartilage mass. In general, however, the cartilage canals of a fetal epiphysis can be divided into two groups. The canals of the more conspicuous group pass from non-articular surfaces towards the centre of the cartilage mass. The less prominent group pass from vessels in the ossification groove (of Ranvier—p. 259), and the capillary tufts in the blind ends of the latter are most complex and glomerular in appearance, similar to the vascular tufts that characterize the post-natal subchondral circulation (Trueta and Morgan 1960). According to Brookes (1971) the pattern of arteries in individual adult bony epiphyses is closely similar to the vascular canal pattern in the corresponding fetal epiphyses. He concluded that the vessels in the major group of cartilage canals are the forerunners of the main epiphysial vessels to be found later in the mature bone; also that the vessels in the minor group of cartilage canals which originate in Ranvier's groove are the precursors of the subchondral vascular network found on the epiphysial side of the growth cartilage in post-natal life.

Cartilage canals are believed to augment the nutrition of large cartilage blocks, which otherwise would depend solely on diffusion gradients from and to the perichondrial capillaries. This process is apparently much faster than was formerly considered, both in the case of young cartilage (*vide supra*), and additionally that labelled small ions were concentrated by the chondrocytes of mature articular cartilage with unexpected rapidity (Hall 1965); the latter tissue does not possess cartilage canals and its free surface is, of course, devoid of perichondrium. Furthermore, the half-life of the protein-polysaccharide complex in rabbit articular cartilage is 4 days (Mankin and Lipiello 1969). While large blocks of cartilage such as fetal limb bone epiphyses possess cartilage canals, it is perhaps surprising that they are also found in the cartilaginous carpal elements and even diminutive sesamoids of the hand in early fetal life. They are present, for example, in the epiphyses of human fetal phalanges at 11 cm crown-rump length (Gray *et al.* 1957).

Hence, while not denying the canals a nutritional function, nearly all workers have opined that cartilage canals are in some way involved in the process of formation of secondary centres of ossification (and primary centres in some small bones without epiphyses). However, there are some puzzling anomalies. Distal phalangeal and proximal metacarpal cartilaginous extremities do develop elaborate arborizations of cartilage canals but, however, they lack separate centres of ossification. Nor is there any secure relationship between the onset of chondrification, the timing of the first appearance of the canals, and the onset of ossification. All these events occur at specific times peculiar to each skeletal element (Brookes 1971). The contents of the canals, blood vessels and loose connective tissue, appear to furnish the osteogenic cells and capillaries characteristic of ossification when, and if, a separate ossific centre appears. The control of the timing of these events remains a complete enigma.

In the condylar cartilage of the mandible canals continue to form until the second year, when they disappear. Laryngeal and nasal cartilage canals first form in the seventh month *in utero* and persist until old age. They form in costal cartilages in the first year and reach the centre of the shaft about the tenth year. In long-lived canals, marrow elements may begin to form after the twentieth year; this may persist until about the sixties when atrophy of the marrow supervenes and the canals remain filled with a mucinous material.

VARIETIES OF CARTILAGE

These include hyaline, white and yellow fibrocartilage and cellular cartilage.

Hyaline cartilage (**3**.7) has a pearly, bluish, translucent, homogeneous appearance and consists of a gristly mass of firm consistency and considerable elasticity. Costal, nasal, some laryngeal, tracheo-bronchial, all temporary cartilages and most articular cartilages are of the hyaline variety, but it must be emphasized that they present differences in the size, shape and arrangement of both cells and fibres, and composition of matrix, in different locations and at different ages. With the light microscope the cells in hyaline cartilage appear rounded or bluntly angular, but this appearance is belied by ultrastructural studies (*vide supra*). They are frequently arranged in groups of two or more cells which are the offspring of a common parent cell. These small clusters are known as *cell nests* or *isogenous cell groups* and where the cells are closely apposed they have a straight outline, but are rounded in the rest of their circumference. The matrix is typically basophilic and shows a metachromatic reaction, both of which are more pronounced where recently formed matrix forms the boundary of a cell-containing lacuna. This distinctive zone is sometimes called the *territorial matrix* or *lacunar capsule*, in contrast with the remaining pale-staining *interterritorial matrix*.

When viewed with transmitted light the matrix is either transparent and apparently structureless or it presents a dimly granular appearance like ground glass. However, when examined in polarized light or with electron microscopy it is shown to be permeated by a system of fine fibrils and fibres. These are 10–20 nm in diameter and show the 64 nm periodic banding which is characteristic of collagen elsewhere. In the lacunar capsules the fibres have been described as especially fine. However, a detailed comparison of transmission electron micrographs with scanning electron micrographs (Clarke 1974) has thrown some doubt on the latter. It has been widely held that, following the preparative techniques necessary for electron microscopy, each chondroblast is surrounded by four zones. Immediately surrounding the irregular cell surface is a clear apparently 'empty' zone, followed externally by an amorphous or fine fibrillar zone with a radial organization (these two zones correspond to the 'lacuna' of customary cytology). The fine radial fibrils are surrounded by a delicate, territorial, circumferential network of small diameter collagen fibres and associated ground substance, the whole being encompassed by the coarse interterritorial network. Scanning

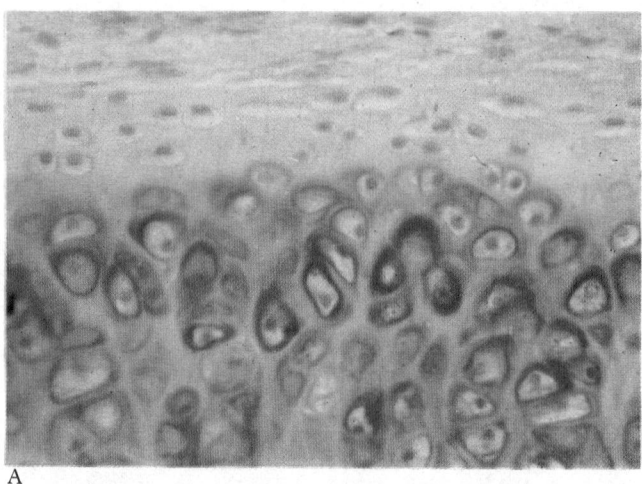

A

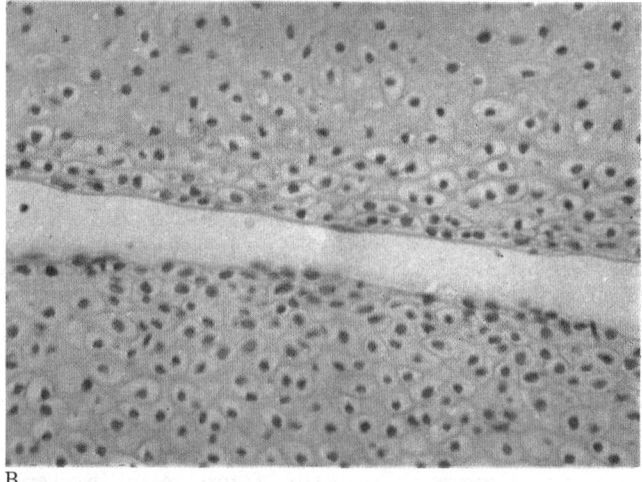

B

3.7   Hyaline cartilage.

A   Mature cartilage in a transverse section of rat trachea stained with haematoxylin and eosin. Note the gradation in cell types with increasing depth from the perichondrium (top) and the pronounced basophilia of the territorial matrix. See text for further details.

B   Young hyaline cartilage bordering a fetal joint cavity.

electron microscopy by Clarke failed to confirm this disposition of territorial collagen. However, in a similar, more recent study of articular cartilaginous blocks the perilacunar collagen fibres were found to be of a finer diameter and more closely packed than elsewhere (Minns and Steven 1977).

Articular hyaline cartilage (**3**.8). This covers the articulating ends of bones in synovial joints where it provides an extremely smooth, wear-resistant surface bathed by synovial fluid. This allows almost friction-free movement (coefficient of friction 0·01–0·02, decreasing as joint loading increases). Its elasticity, together with that of the periarticular structures, dissipates the effect of concussions and provides the whole articular mechanism with some degree of flexibility particularly near the end of its range of movement in any direction. The structure of articular cartilage is admirably adapted to resist the relatively enormous compressive forces generated during weight transmission, but especially during rapid changes of momentum and muscle action (p. 233).

It shows no tendency to ossification and its thickness varies from 1 to 7 mm from the smaller to the larger joints. Its deep surface is moulded to the shape of the underlying bone, but its free surface smooths and often accentuates and modifies the general surface geometry of the bone. On convex surfaces it is thickest at the centre, the reverse being the case on concave surfaces, whilst its thickness decreases progressively from maturity to old age. Its free surface is not covered by perichondrium, but the synovial membrane can be traced over a small part of its circumference, where its cell types grade insensibly into those of the surface layers of the cartilage.

Articular cartilage shows a gradual zonal variation in structure

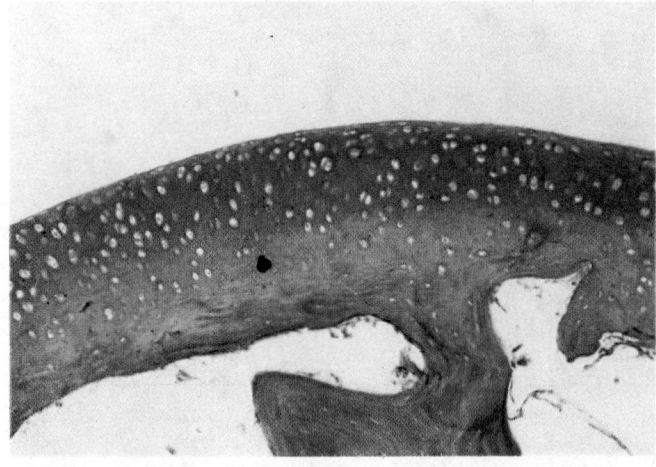

3.8 Articular cartilage. A section perpendicular to the joint surface of the head of a simian metacarpal bone, stained with Masson's trichrome method. Note the changing character of the cartilage (blue) with increasing depth from the surface, the zone of calcified cartilage (faint red) and the underlying epiphysial bone (red). Consult text for further details.

with increasing depth from the surface (Barnett *et al*. 1961; Davis *et al*. 1962; Palfrey and Davies 1966; Ghadially and Roy 1969). Except, possibly, for a thin surface lamina, the matrix is everywhere pervaded by collagen fibres. Those near the surface and surrounding the lacunae are fine, plexiform in arrangement and either faintly or not obviously banded, whilst elsewhere they are coarser and show the 64 nm banding typical of collagen (or some sub-unit of this periodicity). The arrangement of the fibres has been variously described as plexiform, coiled helices, or a series of arcades extending radially from the deepest zone towards the surface, where they pursue a short horizontal (tangential) course before returning vertically. (Most recent workers subscribe to the latter view.) It has long been known (Hultkrantz 1898; Amprino 1948) that if the surface of articular cartilage is pierced by a round-bodied pin, after withdrawal of the pin a longitudinal 'split-line' remains, and that on any particular joint surface, the patterned arrangement of the split-lines is constant and distinctive (cf. the cleavage lines of the skin). A more recent analysis (Bullough and Goodfellow 1968), using polarized light and electron microscopy, has shown that the splits follow the *predominant* direction of the collagen bundles in the *tangential* zone of the cartilage at any point (*vide infra*). It has also been proposed that they portray the 'tension trajectories' set up throughout the surrounding cartilage by compression at certain points during common activities. A series of subsequent investigations have largely confirmed the structural proposals of Bullough and Goodfellow—for example, detailed split-line analysis with transmission electron microscopy (Meachin *et al*. 1974), polarization microscopy (Ortmann 1975), and scanning microscopy (Minns and Steven 1977). Proposals concerning split-line disposition when superficial tangential fibre bundles are not parallel have been considered by Ortmann (1975), and theoretical models for the dynamics of articular cartilage have been put forward by Kempson *et al*. (1968) and Mow *et al*. (1974). It is perhaps pertinent at this point to recall that some of the earlier workers considered that split-line (and hence predominant collagen) orientation reflected the *direction* of the most *habitual movements*, the relationship of this to tension trajectories being unknown. Yet others (e.g. McCall 1969) deny any preferred collagen orientation in articular cartilage. Finally, Serafini-Fracassini and Smith (1974) propose that the most important function of articular cartilage collagen is not the direct resistance of tensile stresses, but to provide *fixation anchors* for the visco-elastic *domains* of the proteoglycan molecules when the latter are subjected to deforming and displacing stresses. Clearly, no theoretical model will prove adequate until it embraces the dyamics of *all* the *interacting* constituents of cartilage. (For an extensive review of the mechanical and biological properties of adult articular cartilage, *see* Freeman 1973.) (See also **4**.11.)

In the *superficial* or *tangential stratum* (*zone 1*) of articular cartilage the cells which are small, oval or elongated, flattened and disposed parallel to the surface, are surrounded by fine fibres arranged tangentially. These cells show a few short surface projections mainly from their lateral borders and deep surface, they possess scattered mitochondria and small cisternae of rough endoplasmic reticulum, and their Golgi apparatus is not prominent. Their cytoplasm is free of filaments, in contrast to the cells of deeper zones. The constitution of the *articulating surface* of articular cartilage—the thin superficial layer of zone 1—(often known as the *lamina splendens*), has long been uncertain. (Because of its wide usage the term lamina splendens needs clarification. It was introduced by MacConaill for the brilliantly illuminated surface zone, which contrasted with the dark main mass of cartilage, when *oblique* sections were studied using negative phase-contrast—*see* MacConaill 1951. The dimensions given are therefore not comparable with those obtained by other techniques, and the general utility of the term is questionable.) Recently, it has been claimed to be a cell-free layer about 3 μm in thickness containing tightly packed, non-banded, randomly arranged, fine filaments some 4 to 12 nm in diameter (Weiss *et al*. 1968). These filaments appeared to branch frequently and their associated proteoglycan ground substance was scanty. Deep to the lamina, typical banded collagen fibres are encountered; they are tangentially arranged, with directional regularity, and increase in diameter with depth. The authors point out that the flattened oval cells found in this deeper layer of zone 1 (*vide supra*) closely resemble quiescent mature fibrocytes. They propose that the articulating surfaces of all synovial joints should perhaps be regarded as formed almost exclusively by fine modified collagen filaments, with their characteristic cells of origin more deeply placed. Other authors (*see* Mow *et al*. 1974) designated the surface layer of zone 1 as the *superficial tangential zone*, considering it to consist of sheets of tightly *woven* fibrils that lie parallel to the free surface: they regard the fibrils as collagenous in nature, from 5 to 20 nm in diameter and forming a layer which varies from 1 to 200 μm in thickness in different locations. It has also been claimed that the surface zone although poor in glycosaminoglycans, is rich in hyaluronate.

The deeper cells of the *transitional* or *intermediate stratum* (*zone 2*) are larger, more rounded, and either single or in isogenous groups. They are, in the main, typical active chondrocytes in their ultrastructural appearances (*vide supra*) and around them the collagen fibres pursue oblique courses.

More deeply still in the *radiate stratum* (*zone 3*) the cells are large, rounded and often arranged in columns perpendicular to the surface, with intervening radially disposed collagen fibres.

The deepest part of the cartilage (*zone 4*) forms a *calcified stratum* which abuts on the *hypochondral osseous lamina* of the underlying epiphysis. These adjoining surfaces show a reciprocal fine ridging, grooving and interdigitation which, together with the confluence of their fibrous architectures, resists the shearing stresses generated during postural changes and muscle action.

The concentration of glycosaminoglycans varies with increasing depth from the surface: the chondroitin sulphate increases some fivefold in concentration in the intermediate zones, reducing to threefold in zone 4. In contrast keratan sulphate increases linearly with depth, and is found mainly in the interterritorial matrix, whereas the chondroitin sulphates are largely circumlacunar.

It should be noted that the sequence of structural changes, with increasing depth, from normal hyaline cartilage, through a columnar arrangement of cells, a zone of calcified cartilage and eventually epiphysial bone, is typical of cartilaginous growth plates elsewhere (p. 263). It reflects the radial growth of the epiphysis by extension of the process of endochondral ossification into the overlying calcified cartilage. The process becomes quiescent in maturity, but the sequence of zones persists throughout life. Such a terminal growth mechanism is also characteristic of bones which possess no epiphysis at one or both ends.

Division of articular cartilage cells is by mitosis, but mitotic figures are infrequently seen except in the young skeleton. There is no evidence that the superficial cells of young and normal joint surfaces are progressively worn away by joint movement, and equally there is no evidence for a continual replacement of surface cells from the deeper layers. However, degenerating cells may be found in any of the four zones and probably this accounts for the progressive reduction in cellularity of the cartilage with advancing age, particularly in the superficial layers (Stockwell 1967). Such cell degenerations probably contribute to the variable lipid content of the matrix.

Articular cartilage is believed to derive its nutriment by diffusion from three sources: the vessels of the synovial membrane, the synovial fluid and the hypochondral blood vessels of the underlying marrow cavity. Many capillaries derived from the latter penetrate into and even through the calcified layer of cartilage (Holmdahl and Ingelmark 1950; Barnett et al. 1961), and it has been estimated that they contact 1–7 per cent of the osseous surface of the cartilage. Opinions are divided concerning the *relative* contribution from these three sources, and as yet conclusive evidence is lacking. There have, however, in recent years been a considerable number of accurate determinations of the permeability both of articular cartilage (for detailed results and bibliography consult Marondas et al. 1975). Briefly, small molecules pass with ease through articular cartilage, with diffusion coefficients about half those in aqueous solution. Larger molecules pass less readily, their diffusion coefficients being inversely related to molecular size. For example, glucose, inulin and haemoglobin with molecular weights of 180, 5000 and 68000 respectively have molal distribution coefficients in the ratio 85:11:1. The permeability of cartilage to large molecules is extremely sensitive to variations in the glycosaminoglycan content; for example, a threefold increase in the latter is accompanied by a hundredfold increase in the partition coefficient.

In costal cartilage the cells and nuclei are large, and the matrix, which is usually homogeneous and transparent, has a tendency to fibrous striation, especially in old age. In the thickest parts of these cartilages a few large vascular channels and sometimes marrow elements may be detected (p. 248). The xiphoid process of the sternum and the cartilages of the nose, larynx and trachea (except the epiglottis and corniculate cartilages of the larynx which are composed of elastic fibrocartilage) resemble the costal cartilages in microscopic structure. The arytenoid cartilage of the larynx shows a transition from hyaline cartilage at its base to the elastic variety at the apex. Hyaline cartilages, especially in adult and advanced life, are prone to calcify and this occurs frequently in the costal and laryngeal cartilages. Quantitative studies have also demonstrated a reduction in cellularity throughout the thickness of costal cartilages with advancing age (Stockwell 1967).

White fibrocartilage consists of dense white fibrous tissue arranged in bundles, with attendant fibroblasts, and with small scattered groups of chondrocytes between the bundles; the cells are roughly ovoid in shape, and are surrounded by concentrically striated areas of cartilage matrix. When present in bulk, as in the intervertebral discs, it provides a tissue of great tensile strength combined with an appreciable degree of elasticity. When present in lesser amount, as in the articular discs (p. 423), the glenoidal and acetabular labra (pp. 458 and 477), the cartilaginous lining of bony grooves which lodge tendons, and certain articular cartilages (*vide infra*), it constitutes a tissue of considerable toughness and sufficient elasticity to enable it to resist the long-term effects of pressure and friction. The articular surfaces of the bones which ossify in mesenchyme are covered with white fibrocartilage (the squamous temporal, mandible and clavicle). The ultrastructure of such articular cartilages has been studied by Silva and Hart (1967). The *deep layers*, adjacent to the hypochondral bone resemble the calcified, degenerative, and hypertrophic regions of the radial zone of *hyaline* articular cartilage. The *superficial zone* consists of parallel bundles of closely packed, large diameter collagen fibres, interspersed with typical dense connective tissue fibroblasts, and scanty ground substance. Adjacent layers alternate in the direction of their fibre bundles, as in the cornea. An intermediate *transitional zone* with irregular bundles of coarse collagen, and fibroblasts with prominent Golgi vesicles separates the two. The latter are probably involved in the elaboration and

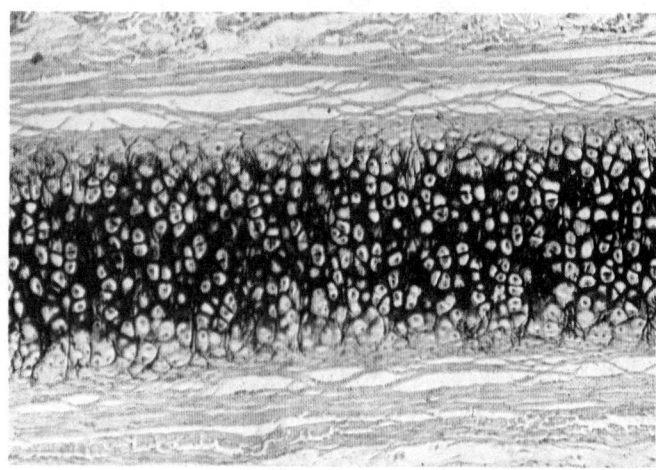

3.9 Elastic cartilage. A section of a rabbit's pinna stained with Verhoeff's method for elastin (black).

export of proteoglycans. Others consider that the transitional cells area germinal zone for the deeper cartilage.

For estimates of the permeability of the white fibrocartilage of intervertebral discs consult Marondas et al. (1975).

Yellow elastic fibrocartilage (**3**.9) is found in the external ears, the corniculate cartilages of the larynx, the epiglottis, and the apices of the arytenoids. It contains typical chondrocytes but its matrix is pervaded by a network of yellow elastic fibres except where it borders the lacunae; here it approaches typical hyaline matrix in its appearance. The fibres resemble those of yellow elastic tissue in other situations, in being unaffected by acetic acid, in their affinity for orcein, and in the absence of any periodicity under the electron microscope (Sheldon and Robinson 1958). For a recent review of the fine structure, histogenesis and growth of the chondroblasts and matrix of yellow elastic fibrocartilage (of the rabbit's pinna) consult Cox and Peacock (1977). As mentioned elsewhere, the histogenesis of the elastic fibres of cartilage (and probably other tissues) is a two-stage process. An initial glycoprotein microfibrillar framework of oxytalan is first laid down, and subsequently this is impregnated by elastin (p. 51).

It is of interest that all the sites in which elastic fibrocartilage is found have *vibrational* qualities as important functional characteristics (laryngeal sound-wave production in phonation, and their collection and transmission in the auditory apparatus).

Histogenesis, Growth and Regeneration of Cartilage

In normal development, cartilage is formed initially by the transformation of embryonic mesenchyme (Glenister 1976; Serafini-Fracassini and Smith 1974; Cox and Peacock 1977). The mesenchyme condenses, i.e. the cells proliferate and become more tightly packed, the shape of the condensation foreshadowing that of the subsequent mass of cartilage. The condensing mesenchyme cells lose their irregular surface projections and become rounded, with prominent round or oval nuclei, but with indistinct cell boundaries. Soon each cell becomes surrounded by a fine basophilic halo composed of a delicate three-dimensional feltwork of fine collagen filaments and associated proteoglycans, both secreted by the cells themselves. The cells have now differentiated into chondroblasts and each is situated within a primitive cartilage lacuna. Continued secretion of fibrils and matrix causes the cells to move further apart until typical hyaline cartilage is recognizable. In other situations collagen synthesis predominates, many of the cells differentiating into fibroblasts, whilst chondroblastic activity is seen only in isolated groups or rows of cells which soon become surrounded by dense bundles of collagen fibres to form white fibrocartilage. Similarly, in other situations the matrix of the early cellular cartilage becomes permeated first by anastomosing oxytalan, and later, elastin fibres. In each of the foregoing cases the mass of developing cartilage is surrounded by a zone of condensed mesenchyme which differentiates into a two-layered perichondrium. The outer cells form fibroblasts which secrete a feltwork of collagen fibres surrounded externally by vascular mesenchyme; the inner cells remain relatively undifferentiated, but are potentially chondrogenic.

Growth of a mass of cartilage is both *interstitial* and *appositional*.

Interstitial growth follows the continued mitotic division of the early chondroblasts throughout the thickness of the tissue mass. When such a cell divides, its descendants temporarily occupy the same lacuna but are soon separated by a fine matrix partition, which subsequently thickens with increasing separation of the cells. Further subdivision of these cells leads to an isogenous group. Interstitial growth is prominent only in young cartilage where it is assumed there is sufficient plasticity of the matrix to allow continued expansion from within. However, the mechanism of such expansions and internal reorganizations of the matrix and the factors determining the overall shape of a cartilaginous mass are little understood.

Appositional growth, as the name implies, follows proliferation of cells of the inner chondrogenic layer of perichondrium. Some of the resultant cells, applied to the outer surface of the cartilage, differentiate into chondroblasts and secrete a layer of matrix around themselves and thus become enclosed in a superficial lacuna. A continuation of this process leads to further surface increments and once enclosed in lacunae, the cells added by appositional growth now participate in interstitial growth. Appositional growth is often stated to be especially prevalent in the more mature cartilages, but interstitial growth of cartilage remains of fundamental importance for long periods, in the growth plate of cartilage between the epiphyses and diaphyses of long bones (pp. 243, 264).

In mammals the ability to regenerate a mass of cartilage which has been lost is quite low. Usually such a defect is slowly filled with vascularized connective (granulation) tissue which may become less vascular and persist for long periods as fibrous tissue. Occasionally, some cells in the centre of such a mass may transform to chondroblasts and secrete thin capsules of matrix around themselves.

Normal cartilage growth is dependent upon an adequate nutritional and hormonal background, and disturbances of these will be considered with the process of endochondral ossification (p. 263). With advancing years the matrix of many of the permanent cartilages of the body becomes *calcified*, and such a process is also an essential prelude to endochondral ossification.

An interesting feature of the biology of cartilage concerns the low antigenicity of its matrix, its relatively low vascularity and the isolation of its chondrocytes in lacunae. All these are features which permit successful homotransplantation of the tissue without evoking a marked cellular or humoral immune response.

Bone as a Tissue

Bone is essentially a highly vascular, living, constantly changing, mineralized connective tissue. It is remarkable for its hardness, resilience, characteristic growth mechanisms and its regenerative capacity. Whilst all bone consists of cells embedded in an amorphous and fibrous organic matrix permeated by inorganic bone salts, its fine structure varies widely with the age, site and natural history of the tissue. Thus, bone may develop either by the direct transformation of condensed mesenchyme or be preceded by a cartilaginous model which is later replaced by bone. Its collagen framework varies from an almost random network of coarse bundles to a highly organized system of parallel-fibred sheets or helical bundles. The inorganic matrix may exist as irregular dense masses with scattered bone cells or it may be arranged as a series of thin sheets (lamellae) in a variety of more or less precise patterns, with intervening rows of bone cells. Both lamellar and non-lamellar bone often develops as minute roughly cylindrical masses or *osteons*, each with a central vascular canal. Again, the various longitudinal and radial zones of a long bone have distinctive developmental histories and structures, and have been given regional names.

For all these reasons, a large and initially often rather confusing terminology has been developed to describe bony tissue. For clarification, the commonly used terms have been grouped as a table and they will be treated more fully in the following pages.

Classification of Bone Tissues

1. **Macroscopic appearance of cut surface** (p. 232).
   (a) *Compact bone*—the ivory-like surface layers of mature bone.
   (b) *Trabecular bone*—the interior of mature bones (also termed *cancellous* or *spongy* bone). Early embryonic bone is also spongy—the *primary spongiosa* (*Os spongiosum primum*).

2. **Development origin**
   (a) *Intramembranous* (*mesenchymal* or *dermal* bone)—from direct transformation of condensed mesenchyme.
   (b) *Intracartilaginous* (*cartilage* or *endochondral* bone)—replacing a preformed cartilage model.

3. **Regions of long bones**
   (a) *Diaphysis*—central region of shaft.
   (b) *Metaphysis*—more recently developed ends of shaft.
   (c) *Epiphysis*—the bone ends with a separate centre of ossification.

4. **Organization of collagen fibres**
   (a) *Woven-fibred* bone (*coarse-bundled* bone), with an irregular collagen network: includes embryonic bone; isolated patches in adult; formed during fracture repair.
   (b) *Parallel-fibred* bone—this includes all forms of lamellar bone and non-lamellar primary osteons.

5. **General microstructure**
   (a) *Non-lamellar* bone—includes early woven-fibred bone and primary osteons.
   (b) *Lamellar* bone—almost all mature bone.

6. **Disposition of lamellae**
   (a) *Circumferential* lamellae (*primary* lamellae)—parallel to both periosteal and endosteal bone surfaces.
   (b) *Osteonic* lamellae (*secondary* lamellae)—concentric lamellae around the vascular canals of mature bone.
   (c) *Interstitial* lamellae—in the crevices between osteons.

7. **Types of osteon**
   (a) *Primary osteons* (*atypical Haversian systems*)—the first formed, non-lamellar osteons.
   (b) *Secondary osteons* (*typical Haversian systems*)—concentric lamellae around the vascular canals of mature bone.

8. **General terms**
   (a) *Surface bone*—usually circumferential lamellae but may include woven-fibred areas.
   (b) *Interstitial bone*—found between osteons; often the lamellar remnants of secondary osteons but may include woven-fibred or primary osteon fragments (*vide infra*).

The Structure of Mature Bone

As we have seen, a mature bone is composed of two kinds of tissue, one of which is dense in texture like ivory, and termed *compact bone*; the other, consisting of a meshwork of *trabeculae* within which are easily visible intercommunicating spaces (or *cancelli*), is called *trabecular* (*spongy* or *cancellous*) bone. The compact bone is always on the exterior surrounding the spongy bone and their relative quantities and architecture vary characteristically for each bone, in a manner which reflects its overall shape, position and functional roles (p. 233). In long bones the origin and significance of their diaphysial, metaphysial, epiphysial regions and associated growth cartilages have already been considered in general (p. 241) and will be further considered with intracartilaginous ossification (p. 263). The thick cylinder of compact bone forming the shaft, presents a few trabeculae and spicules of bone on its inner surface; it encloses a large central *medullary* or *marrow cavity* which communicates freely with the intertrabecular spaces of the expanded bone ends. These various spaces are lined by a highly vascular condensation of areolar tissue, the *endosteum*, and are filled with bone marrow, either haemopoietic or adipose, its character varying with the age and region of the bone concerned. The marrow cavity begins to appear in the shaft during fetal life, and slowly extends in both directions through the subsequent years, until it almost reaches the epiphysial lines which mark the zone of fusion between epiphysis and diaphysis occurring when growth in length has ceased (p. 245).

Close examination of the compact bone shows it to be extremely porous, so that the difference between it and spongy bone depends largely upon the relative amount of solid matter and the size and number of spaces in each; in compact bone, spaces are small and the solid matter abundant, whilst the reverse is the case in spongy bone. During life, bone is permeated by many blood vessels and, except for its cartilage-covered articular surfaces, it is everywhere clothed by a membranous *periosteum*. When strong tendons or ligaments are attached to a bone, the periosteum blends with them. It consists of two layers closely united together, the outer formed chiefly of white fibrous tissue containing a few fat cells; the inner of fine elastic fibres forming dense membranous networks, which again can often be separated into several sub-layers. In young bones the periosteum is thick and very vascular, and is separated from the bone by a layer of soft *osteogenetic tissue*, containing a number of granular cells, the *osteoblasts*, through the agency of which surface ossification proceeds. Later in life the periosteum is thinner and less vascular, and the deeper layer consists of flattened, relatively quiescent cells which are, however, still potentially osteogenic. The periosteal blood vessels communicate with those of the underlying bone (p. 257), and a bone surface denuded of periosteum shows an increased susceptibility to exfoliation or necrosis.

Microstructure (3.10, 11, 12, 13, 14). In the adult, compact bone consists mainly of a number of irregularly cylindrical units, termed *typical Haversian systems* or *secondary osteons*, each consisting of a central *Haversian canal* which contains a neurovascular bundle and is surrounded by concentric lamellae of bony tissue. Between these lamellae are a number of small spaces termed *lacunae* which are connected with each other and with the central Haversian canal by many fine radiating channels called *canaliculi*. The lacunae and their canaliculi contain, respectively, the cell bodies and fine cytoplasmic extensions of the bone cells or *osteocytes*. The angular intervals between the secondary osteons are occupied by *interstitial bone* (most commonly in the form of interstitial lamellae with their lacunae and canaliculi running in various directions, but *vide infra*). Again, other lamellae are found encircling the inner and outer surfaces of the bone; and they are termed *circumferential* or *primary lamellae*.

The secondary osteons have a predominantly longitudinal orientation (Cohen and Harris 1958), but they frequently spiral, branch or intercommunicate and some terminate blindly. They may appear round, oval or ellipsoidal in transverse section and they vary considerably in diameter, containing on average six concentric lamellae, but these may be as many as fifteen in the largest osteons. Similarly, the contained Haversian canals vary in diameter but average about 50 μm, whilst those close to the marrow cavity are somewhat larger (Cohen and Harris 1958). Within each canal are one or two capillaries, and usually some non-myelinated and occasional myelinated nerve fibres (5–9 μm in diameter) together with their associated Schwann cells (Cooper 1969; Cooper *et al.* 1966).

The endothelial cells of the capillaries are fenestrated (cf. p. 628) and their cytoplasm contains occasional caveoli and electron-dense filaments about 10 nm in diameter. They are surrounded by a basal lamina which splits to enclose typical pericytes.

The perivascular cell population varies in different regions, even of a single osteon, depending on whether the bone is being actively deposited, absorbed or is relatively quiescent, and accordingly osteoblasts, osteoclasts or undifferentiated mesenchymal cells may predominate. Lining the wall of the canal are collagen fibres showing the usual 64 nm periodicity and of about 60 nm diameter.

The Haversian canals have long been claimed to communicate with the medullary cavity, with the spaces in spongy bone, and with the surfaces of bone by oblique or transverse channels termed *Volkmann's canals*, which lodge blood vessels and nerves, and are said not to be surrounded by concentric lamellae of bone.

It has been shown, however (Cohen and Harris 1958), that the majority of the canals so described are in fact simple interosteonic anastomotic canals, or radial canals passing through the circumferential lamellae, and that the frequency and calibre of periosteal vascular connexions in the diaphysis may have been over-emphasized (*vide infra*).

The trabeculae of the spongy bone consist of fragmentary systems of superimposed lamellae with numerous intervening cement lines (*vide infra*), and in places they may contain small islands of calcified cartilage. The trabeculae receive their nutriment from the blood vessels in the marrow around them and except for occasional large trabeculae which contain small Haversian systems, they are not penetrated by blood vessels.

Because of its regular microstructure, normal adult bone is termed *lamellar bone* and each lamella is a thin plate of bony tissue varying from 5 to 7 μm in thickness. The bony tissue consists of a ground substance or *matrix* in which are embedded fibres and which is impregnated with bone salts. In mature bone about one-fifth of the weight of the matrix is water; organic material forms 30–40 per cent and mineral salts 60–70 per cent of the dry weight of bone. The main organic components are collagen (90–95 per cent), mucopolysaccharide in combination with a non-collagenous protein (1 per cent) and a resistant protein (5 per cent), e.g. *see* Eastoe and Eastoe (1954). Organic matrix which is as yet uncalcified, is known as *osteoid matrix*. In normal bone its amount is very small, owing to the rapidity with which calcification follows. The amount of osteoid matrix is, however, greatly increased in certain disease states, such as rickets.

All Haversian and most interstitial systems of lamellae in adult bone are demarcated from neighbouring systems by a *cement line* which is strongly basophilic; it has little or no collagen and a high content of inorganic matrix. A cement line is formed when a cavity which has been formed by bone *resorption* is subsequently filled by the successive *deposition* of bony lamellae on its inner surfaces. The cement line indicates the limit of the erosive process and the commencement of the process of deposition and is, therefore, often called a *reversal line*.

The normal variation in the composition of bone from various sites and at different ages is not yet fully elucidated. It is, however, of considerable importance when attempting to assess (usually radiologically) the degree of rarefaction of bone (osteoporosis), which may occur in disease or as a result of disuse. Whether this arises from an actual reduction in the amount of bony tissue, rather than an overall reduction in its degree of mineralization, is difficult to determine experimentally because of the problems attendant on the separation of bone from other tissues. Further difficulties concern the quantitative determination of the related sizes of the vascular, lacunar and marrow spaces. Some have tried to minimize these difficulties by making estimates on cortical bone, it having been suggested that trabecular bone is more labile than cortical bone. It is maintained, however, that *normally* there is little difference in the composition of cancellous and compact bone. Overall, the water content decreases and the mineralization increases with age up to the sixth or seventh decade, whilst the organic content remains unchanged. Thereafter the degree of mineralization declines (Mueller *et al.* 1966).

Microradiographic and interferometric investigations on compact bone, however, show an uneven distribution of mineral salt. The inner and outer circumferential lamellae and the interstitial lamellae are in general evenly and highly mineralized. The secondary osteons exhibit varying degrees of mineralization, and in a single osteon there are differences in concentration in different areas. Young osteons have a low concentration which increases with age and initially is highest close to the central canal and decreases towards the periphery. This gradient in mineral content becomes less pronounced as the osteon grows older, and in very old highly mineralized osteons there is an even distribution of mineral salt. Bone resorption may be found both in areas of high and low concentration (*vide infra*). In developing fetal bone, mineralization up to 70–80 per cent is rapid (occurring in about 3 weeks), but thereafter, up to 100 per cent, it is a much slower process. Such sequences may also hold for the later developing osteons.

The lamellation of bone (3.10, 11, 12, 13, 14). There have

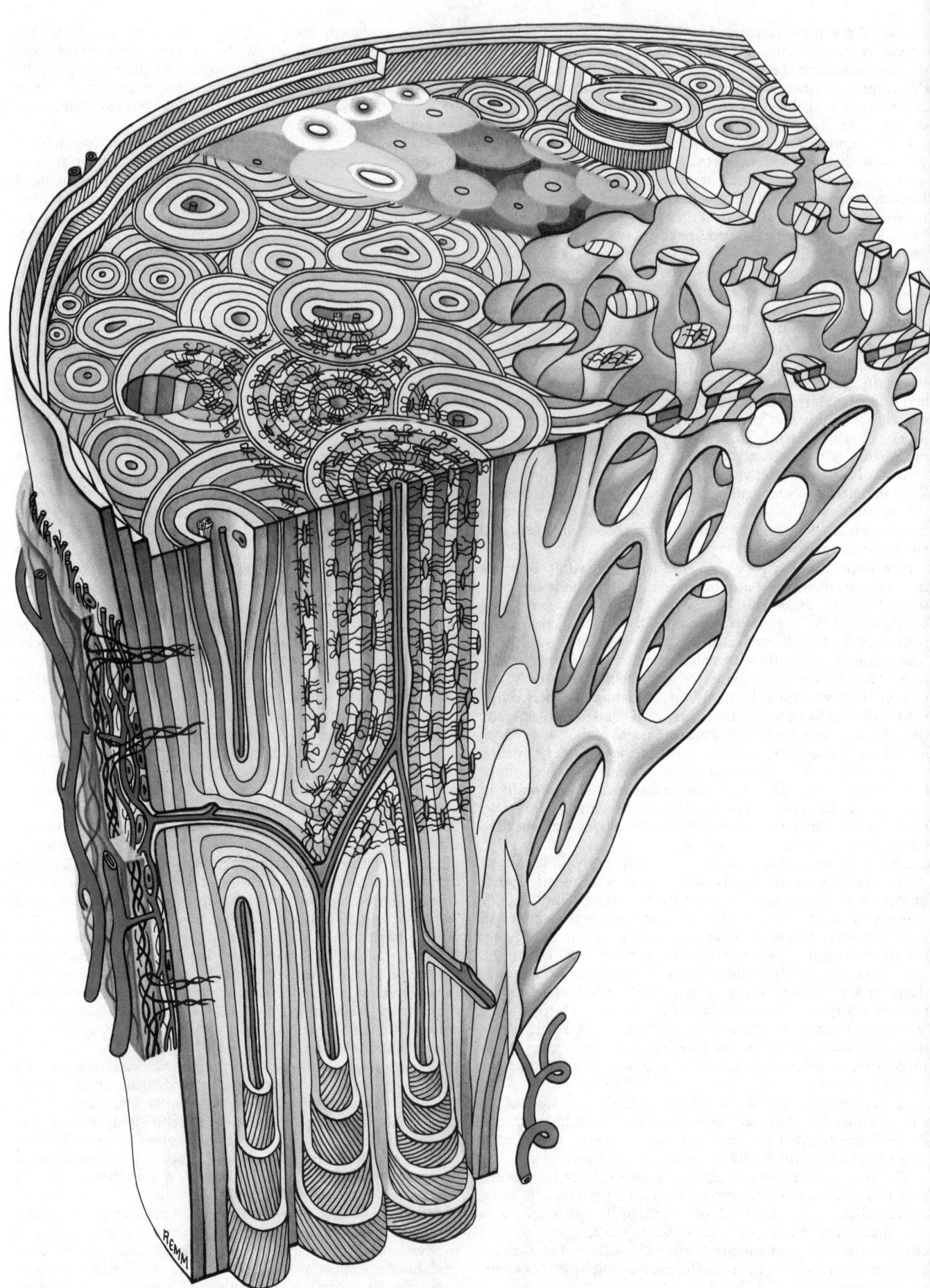

3.10A A schematic diagram of some of the main features of the microstructure of mature bone seen in both transverse section (top) and longitudinal section. Areas of compact and cancellous bone are included. The central area in transverse section simulates a microradiograph, the densities reflecting variations in mineralization. Note the general construction of the osteons, the distribution of the osteocyte lacunae, the Haversian canals and their contents, resorption spaces, and the different views of the structural basis of bone lamellation. See text for more detail.

been various explanations of the lamellar appearance of mature bone, but it seems largely due to the relative orientation of the anisotropic structures found in each layer and also the interposition of a thin interlamellar zone between successive layers. The anisotropic components are collagen fibre bundles and hydroxyapatite crystals, and these all lie parallel within a single lamella. In successive lamellae, however, the orientation of the fibre bundles (together with their associated crystallites) may change through an angle of 0–90°. Stratification is most apparent when this angle is large. Fibre bundles frequently leave one lamella and pass into the next through discontinuities in the interlamellar cementing zone. In the latter, the organic matrix is composed almost completely of irregularly disposed collagen, heavily flanked by very large concentrations of crystallites.

Earlier authorities (Gebhardt 1905) suggested that the collagen fibres are mainly longitudinal and circumferential in adjacent lamellae or that they have a spiral course, the pitch of the spiral differing in successive lamellae. The latter view has received recent support from a series of interesting investigations into the structure and compressive properties of isolated single osteons (Ascenzi and Bonucci 1968).

Other investigations, however, suggest that the fibre arrangement varies (Smith 1960). In some secondary osteons the fibres are predominantly longitudinal and circumferential in successive lamellae but in approximately equal concentration. In others, lamellae with dense, closely packed longitudinal fibres alternate with lamellae containing fewer dispersed circumferential fibres. In a third type of osteon, there is a predominance of the dense longitudinal fibre bundles with here and there very thinly dispersed circumferential fibres forming incomplete arcs, and in these cases the lamellation is indistinct. These different fibre patterns have been described as coexisting in different parts of the same osteon and may be associated with differences in mineral salt distribution. Since the mineral content is closely associated with the collagen fibres, it is believed to be in high concentration in those lamellae with a high fibre content and this may be reflected by variations in the X-ray absorption of different lamellae. Furthermore, it has been suggested that with ageing there is an increase in the number of longitudinal fibres and a simultaneous decrease in circumferential fibres with a loss of distinct lamellation (Smith 1960).

As mentioned above, the investigators analysing the mechanical properties of isolated osteons subscribe to a helical conformation of the lamellar fibrous architecture, with again, three main types of osteon (Ascenzi and Bonucci 1968). In the first type all lamellae have a predominantly transverse, spiral course of their collagen fibres; in the second type, alternate lamellae are longitudinal and transversal in their main helical fibre direction; type three osteons possess lamellae all showing a largely longitudinal orientation of fibres. The compressive strength and modulus of elasticity was greatest for osteons with transverse bundles and lowest in those with mainly longitudinal ones.

The circumferential lamellae are pierced by tapering bundles of collagen fibres which spring from the periosteum and run obliquely through them, apparently pinning them together, and are termed the *perforating* fibres of Sharpey.

The bone 'salts'. (For details consult Robinson 1952; Bourne 1956; McClean and Urist 1969.) The organic matrix of bone is impregnated by inorganic material, sometimes termed *bone salts*, and these confer on bone its hardness and rigidity. Their most important ionic constituents are calcium (Ca^{2+}), magnesium (Mg^{2+}), phosphate (PO_4^{3-}), carbonate (CO_3^{2-}), hydroxyl (OH^-), chloride (Cl^-), fluoride (F^-), and citrate ($C_6H_5O_7^{3-}$). The mineral substances of bone may be obtained by calcination, which destroys the organic matter, following which bone retains its original form but is white and brittle, loses about one-third of its original weight and crumbles under the slightest force. On the other hand, the mineral constituents of bone are mostly soluble in mineral acids and may be removed by prolonged immersion in dilute solutions. The bone then retains its shape but is highly flexible, so that a long thin bone (such as the fibula) can easily be tied into a knot. In a transverse section of such a softened bone the arrangement of the Haversian canals, lamellae, lacunae and canaliculi can still be recognized.

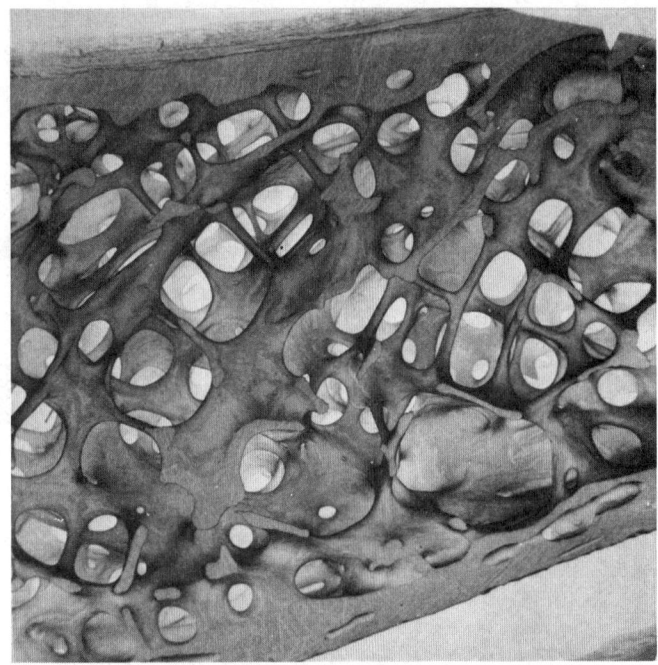

3.10B Low power scanning electron micrograph of a segment of a human rib. Note the cortical shell of compact bone and the delicate lattice of trabeculae, separated by labyrinthine marrow spaces. (Reproduced by kind permission of Drs. W. J. Whitehouse, E. D. Dyson, and C. K. Jackson, of the Medical Research Council Radiology Unit, Harwell. Such specimens are used currently for quantitative studies of bone architecture.) Magnification × 7.

The mineral salts consist of crystalline and amorphous components, and X-ray diffraction and ultrastructural studies have shown that the principal crystalline component is composed of needle-shaped crystals of hydroxyapatite ($Ca_{10}(PO_4)_6(OH)_2$) which are about 20–40 nm in length and 3–6 nm in breadth. In general the needles are arranged with their long axes parallel to those of the collagen fibres and they lie partly within the fibres, but their precise relationship to the periodic substructure of the collagen is still uncertain. Citrate ions are probably arranged on the surface of the crystallites, and carbonate ions may also be located on the surface or they may replace phosphate ions within the crystal. Substitution of various chemical groups for those normally present in the crystallites is of frequent occurrence. Thus, fluoride commonly substitutes for hydroxyl, and a series of *bone-seeking cations*, e.g. the cations of radium (Ra^{2+}), strontium (Sr^{2+}) and lead (Pb^{2+}), all readily substitute for calcium (Ca^{2+}). The accumulation of an isotope of strontium (^{90}Sr) may cause widespread radiation damage to the cells of its contained bone marrow.

The bone lacunae appear under the light microscope as small oblong or ovoid spaces situated along the margins, or in the middle, of the compact lamellae. At high magnification they are seen to possess rough walls and to be irregularly stellate spaces and the pointed extremities are continuous with the surrounding canaliculi. With few exceptions each lacuna is occupied in life by a branched bone cell or osteocyte, the processes of which extend into the canaliculi.

The bone canaliculi are minute channels, crossing the lamellae and connecting the lacunae of a Haversian system with one another and with the central Haversian canal. The peripheral canaliculi do not as a rule communicate with those of neighbouring systems, but form loops and return to their own lacunae. The average life of an osteocyte has been estimated at 25 years. Old osteocytes may retract their processes and their associated canaliculi may be empty. When osteocytes die, their lacunae and canaliculi may become plugged with cell débris and minerals and this may interfere with diffusion through the bone. Dead osteocytes are more numerous in interstitial bone than in

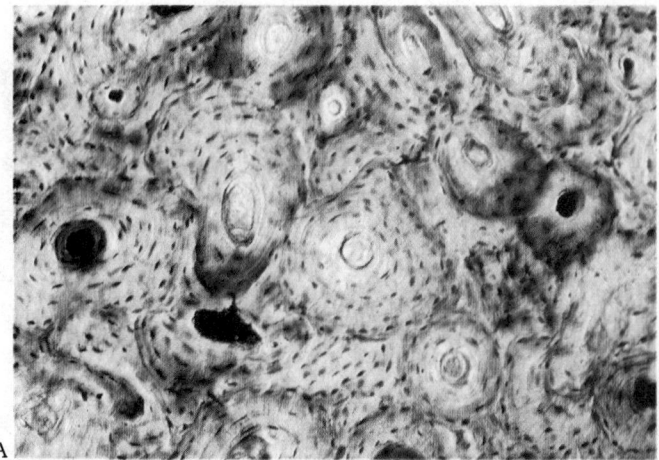

3.11A, B Transverse (A) and longitudinal (B) ground sections of compact bone from human femoral shaft. Note variation in shape and size of osteons and their canals and in distribution of lacunae in A. Haematoxylin staining in B has emphasized the osteons, predominantly longitudinal but showing intercommunications. (Material for **3.**11 to 14 prepared by Mr. David Ristow of Guy's Hospital Medical School.)

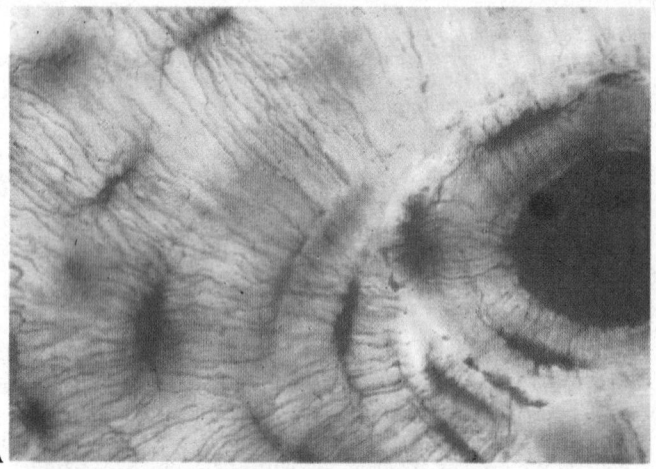

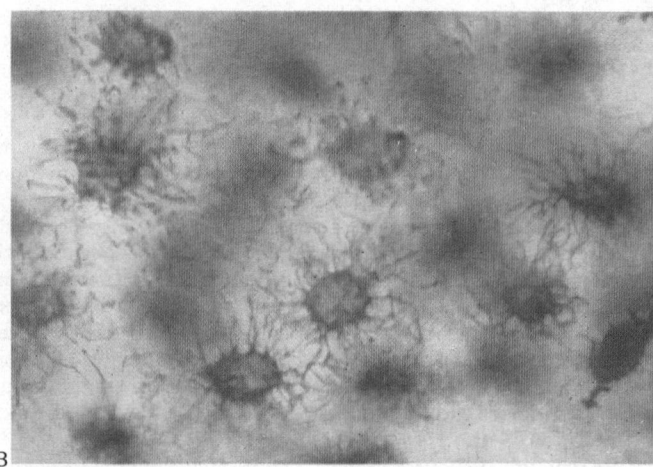

3.12A A high-power view of part of an osteon in transverse section seen with transmitted light. Note the relation of the osteocyte lacunae and their canaliculi to each other, and to the central Haversian canal (black).

3.12B A tangential section of osteocyte lacunae and their associated canaliculi—contrast with their appearance in A.

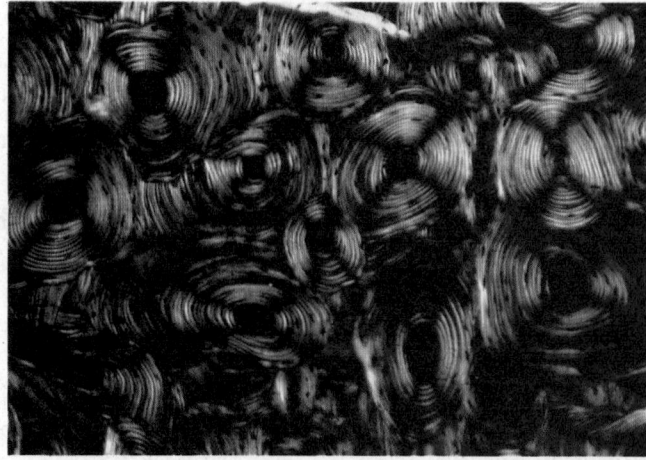

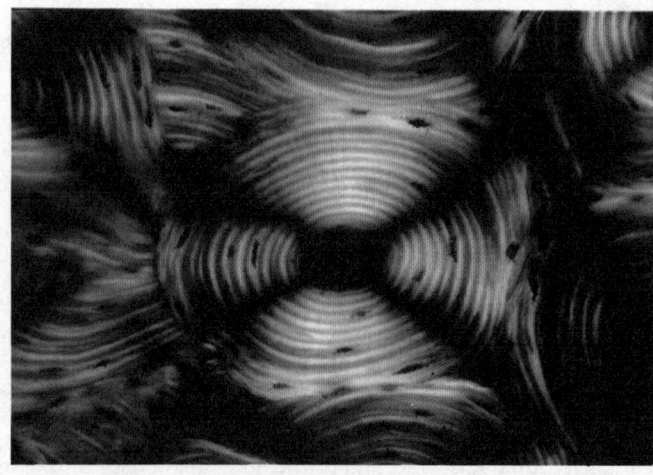

3.13 A ground transverse section of the compactum of the adult human femur photographed using polarization microscopy—compare with **3.**11.

3.14 A single secondary osteon viewed with high-power polarization optics to illustrate its lamellar architecture.

secondary osteons and they become particularly noticeable in the second and third decades.

The osteocytes vary in number, spatial arrangement and appearance. They are most numerous in young bone and they conform to the general shape of the lacunae in which they lie. Their processes, which have a more or less uniform diameter of 80–100 nm, pass into the canaliculi; they may be of considerable length and may also branch. Sometimes two processes lie alongside one another in a single canaliculus. The processes are not continuous but terminate in end-to-end or side-to-side juxtaposition without any obvious junctional specializations. It is not known whether some simply terminate in blind-ending canaliculi. The cytoplasm of the cell processes contains no distinctive organelles other than a few smooth-walled vesicles and electron-dense granules.

The cell bodies of mature relatively quiescent osteocytes contain an oval nucleus surrounded by a fairly small volume of faintly basophilic cytoplasm. There is a correspondingly small concentration of rough endoplasmic reticulum, some free ribosomes, centrioles and a small juxtanuclear Golgi apparatus. Like the osteoblasts from which they are derived, newly incorporated active osteocytes have a more basophilic cytoplasm corresponding to a greater concentration of rough endoplasmic reticulum, and the Golgi apparatus is more prominent. On the walls of the lacunae surrounding the osteocyte cell body, there is a small amount of non-mineralized or lightly mineralized collagen, but this is not seen in the canaliculi where the walls are heavily mineralized.

Neither the cell bodies of the osteocytes nor their processes occupy fully the lacunae and canaliculi in which they lie, and a space, presumably fluid-filled, surrounds them. There is thus a continuous extracellular space from the Haversian canal through its canaliculi and lacunae along which the cells of the osteon are nourished and through which the ions for the mineralization of the bone are transported. It has been estimated that the crystal surface exposed to extracellular fluid in the walls of this system in a 70 kilogram man is between 1,500 and 5,000 square metres, thus providing an enormous surface area across which exchanges of mineral ions can occur (Robinson 1964). In addition there is a further surface of some 8 square metres associated with the periosteal and endosteal membranes.

The osteocyte and its processes, deeply entrapped within the bone substance, probably maintain the architecture of the lacunae and canaliculi (i.e. the diffusion pathways), and they are possibly directly concerned in cellular transport of materials along these channels. There is increasing evidence that they are actively involved in modifying the neighbouring bone matrix, thereby facilitating the return of various ions, particularly calcium, from the bone salts (osteolysis) back into the bloodstream (Heller-Steinberg 1951; Bélanger 1963).

The Blood Vessels and Nerves of Bone

The osseous circulation supplies the bony tissue and cells, the bone marrow, the perichondrium, the epiphysial cartilage in young bones, and in part, the articular cartilages. Modern researches (Brookes 1964, 1967, 1971; Kelly 1968; Rhinelander 1968) have emphasized the essentially *centrifugal* flow of blood through the cortical bone of the shaft of long bones, revising the earlier concept of a substantial *centripetal* flow of arterial blood into the bone cortex from overlying periosteal vessels. In summary, the vascular supply of a long bone consists of several discrete points of arterial inflow, which feed complicated and regionally variable sinusoidal networks within the bone; in turn these drain into venous channels leaving the bone through all its surfaces which are not covered by articular cartilage. The principal features of these vascular patterns are summarized in **3.15A** and B.

One or two principal *diaphysial nutrient arteries* pierce the shaft of a long bone obliquely through, usually one, occasionally two, *nutrient foramina* leading into *nutrient canals*. These have long been recognized (Havers 1961; Bernard 1835). Their site of entry and angulation are fairly constant and, characteristically, the vessels point *away* from the dominant growing end of the bone (p. 259). This constitutes the widely accepted *growing-end hypothesis* concerning the mechanism leading to nutrient foramen and canal disposition (Clark 1965). Simple growth considerations, however, do not account for the exceptions to this pattern which some authors have described in various species and sites (Hughes 1952). However, in a study of human long bones, Mysorekar (1967) found anomalously directed canals only in the fibula, and

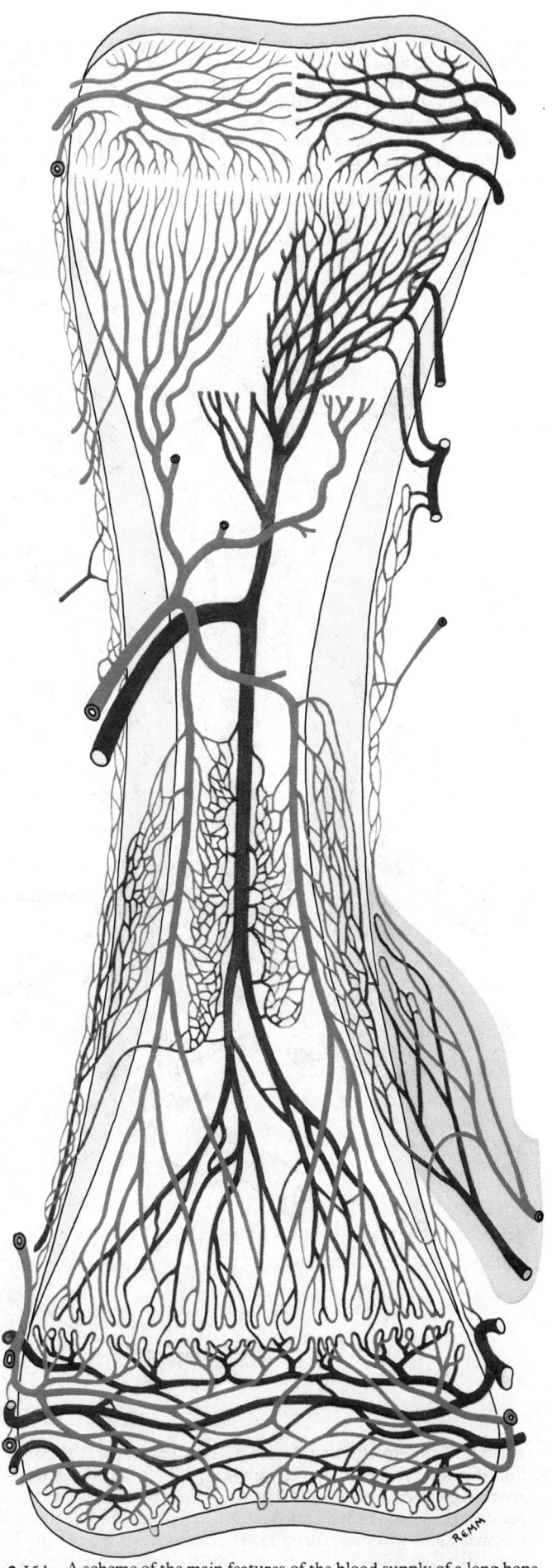

3.15A   A scheme of the main features of the blood supply of a long bone based upon descriptions by Dr. Murray Brookes (Guy's Hospital Medical School). Note the contrasting supplies of the diaphysis, metaphysis and epiphysis, and their connexions with periosteal, endosteal, muscular and periarticular vessels. Consult the text for a more extended description.

3.15B Diagram of the circulatory arrangements in part of the diaphysis of a typical long bone. Note, in the marrow cavity, the large central venous sinus, the dense network of medullary sinusoids, longitudinal medullary arteries and their circumferential rami. From the latter, longitudinally oblique, transcortical capillaries emerge through minute 'comet-shaped' foramina (**3.15C**) to become confluent with the periosteal capillaries and venules. Not to scale; obliquity of cortical capillaries is emphasized for clarity. (Constructed with the kind collaboration of Dr. Murray Brookes, Department of Anatomy, Guy's Hospital Medical School, London.)

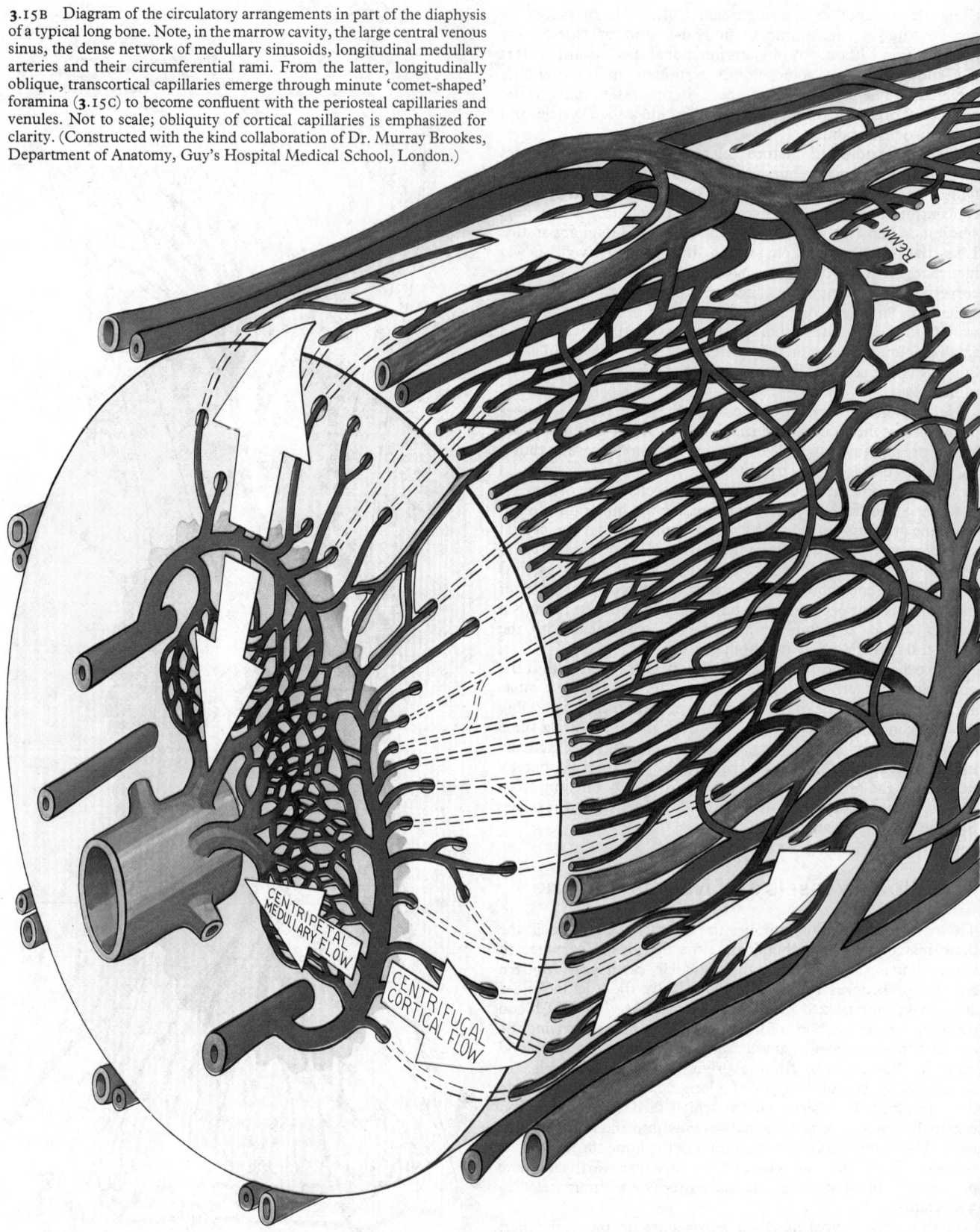

attributed this to the unique ossification pattern in that bone. Subsequently in a study of 848 metacarpals and 811 metatarsals, apart from a few bones with absent or double foramina, over 90 per cent possessed a single nutrient foramen in the middle third of the shaft, and without exception, all foramina, single or double, were directed away from the epiphysis, thus supporting the growing-end hypothesis. For details of the latter investigation, a bibliography, and an introduction to alternative hypotheses such as the '*periosteal slip*' theory of Schwalbe (1876), the 'vascular theory' of Hughes (1952), and the '*asymmetrical muscular*

development theory' of Lacroix (1951), consult Patake and Mysorekar (1977). Finally, for an interesting comparative study of the contrasting growth patterns of foramen and canal in the rat femur and tibia consult Henderson (1978). In its cortical canal, a nutrient artery does not branch, but once within the medullary cavity it divides into ascending and descending branches. These pass towards the bone ends and divide repeatedly into smaller branches which pursue a helical course in the juxta-endosteal zone of the marrow. Near the bone ends they are joined by the terminals of numerous *metaphysial* and *epiphysial arteries*. The

3.15C A scanning electron micrograph of the periosteal surface of the diaphysis of an adolescent tibia. Note the oblique 'comet-shaped' groove and foramen which in life transmitted a minute neurovascular bundle; numerous foramina of this general type are scattered over the surface of the diaphysis, but their pattern, direction, and degree of obliquity, vary with the overall subperiosteal deposition of bone that characterizes the various growth zones of the diaphysis. Magnification × 80. (Preparation by Mr. Michael Crowder, Department of Anatomy, Guy's Hospital Medical School.)

former pass directly into the bone from neighbouring systemic vessels, whereas the latter are derived from periarticular vascular arcades which are formed on the non-articular bone surfaces (p. 257). Numerous vascular foramina penetrate the bone surfaces, often at fairly specific points (see **3**.1D), near the bone ends and some are occupied by these arteries, but the majority contain thin-walled veins. Once within the bone these various arteries are distinctive in that they consist of a single endothelial layer surrounded by a thin layer of supportive connective tissue (Yoffey 1962). The epiphysial and metaphysial nutrient arteries are quantitatively much more important than the diaphysial supply, which they can effectively replace, when the latter has been obliterated experimentally.

The medullary arteries of the shaft (**3**.15A and B) give off (a) centrally directed branches which contribute to a hexagonal mesh of marrow sinusoids draining into an extremely wide, thin-walled central venous sinus; (b) a series of cortical branches which pass through endosteal canals to feed the fenestrated capillaries of the Haversian systems. The central venous sinus drains into veins which retrace the path of the nutrient arteries, or sometimes pierce the shaft elsewhere as independent emissary veins.

The cortical capillaries obviously conform to the pattern of the Haversian canals, often described as predominantly longitudinal with oblique interosteonic connexions. However, an oblique, radial pattern has been demonstrated in which the vessels largely radiate from the site of the primary ossification centre (Brookes 1964; 1967; 1971). At the bone surface, the cortical capillaries form capillary and venular connexions with the overlying periosteal plexuses (**3**.15A, B and C). The latter are formed by arteries from neighbouring muscles which form vascular arcades with longitudinal interconnexions on the fibrous periosteum. This external plexus supplies a capillary plexus which permeates the deeper osteogenetic layer of the periosteum. At points of direct muscular attachment to bone, the periosteal and muscular vascular plexuses become confluent and the cortical capillaries of the underlying bone drain into the interfascicular venules of the muscle.

The preceding account of an almost exclusively centrifugal supply to the cortex of the shaft of long bones is receiving increasing support, but some authorities (de Marneffe 1951; Morgan 1959) maintained that there exists an appreciable centripetal flow of arterial blood to the outer zones of the cortex from the overlying periosteal vessels.

The large nutrient arteries of the epiphysis form a free series of intraosseous anastomoses, and branches pass towards the articular surface in the trabecular spaces of the bone. Near the articular cartilage their branches form a series of anastomotic arcades (e.g. 3 or 4 in the femoral head) before giving off a series of end-arterial loops which often pierce the thin shell of hypochondral compact bone (**3**.1D) to enter, and sometimes perforate, the calcified zone of the articular cartilage before returning to the venous sinusoids of the epiphysis.

In immature long bones similar sources of supply exist but the epiphysis forms a discrete vascular zone; distinct epiphysial and metaphysial arteries can be distinguished, entering on either side of the growth plate of cartilage, and anastomoses between these vessels are either infrequent or absent. The growth cartilage probably receives its nutritional supply from both these sources and also from a collar of anastomotic vessels in the adjoining periosteum, although the relative contribution from each is not certain (Brookes 1964, 1967, 1971; Trueta and Morgan 1960; Rang 1969). Occasionally a cartilage canal becomes incorporated in the growth plate (p. 266). The metaphysial bone is nourished by the conjoined terminal branches of the metaphysial arteries and those of the primary nutrient artery of the shaft. Their terminals form sinusoidal loops in the zone of advancing ossification and the dilated bend of each loop is separated by a small gap from the nearest transverse partition of calcified cartilage (p. 264). This gap is often packed with red blood corpuscles which has been described as evidence of a 'microrupture' of the vessel at this point (Trueta and Morgan 1960). Other authors, however, have emphasized the intimate contact of the sinusoidal endothelium and the calcified cartilage, and in addition to providing a nutritive source for the growth cartilage, the vascular loops have been ascribed a chrondrolytic role (Cameron 1961; Brookes and Landon 1963). The periosteum of young bone is more vascular than that of the adult; its vessels communicate more freely with those of the shaft and also give rise to many metaphysial vessels.

Large irregular bones like the scapula and innominate bone receive a superficial blood supply from the periosteum and are frequently provided with large nutrient arteries which penetrate directly into the cancellous bone. The two systems anastomose freely.

Short bones receive numerous fine blood vessels from the periosteum. They enter at the non-articular surfaces and supply both compact and cancellous bone and bone marrow. In the vertebrae, arteries enter around the circumference but particularly close to the base of the transverse process. The marrow is drained by two voluminous basivertebral veins which converge at a large foramen on the posterior surface of the body.

The flat bones of the skull are supplied by numerous vessels from the covering periosteal or mucoperiosteal layers. The veins are large and run in tortuous canals in the diploë (cancellous bone). They have very thin walls so that when the bone is divided they remain open.

Lymphatic vessels are to be found accompanying the periosteal vascular plexuses but their presence within the bone substance has never been convincingly demonstrated.

Nerves are most numerous in the articular extremities of the long bones, in the vertebrae and in the larger flat bones. They are distributed freely within the layers of the periosteum, and fine myelinated and non-myelinated nerve fibres accompany the nutrient vessels into the interior of the bone and even into the perivascular spaces of the Haversian canals (p. 253).

The Histogenesis of Bone

As we have seen, some bones such as those in the roof and sides of the skull are preceded by a fibro-cellular membrane, whereas the majority of bones are preceded by shaped rods or masses of cartilage. Hence, two kinds of ossification are described; intramembranous and intracartilaginous, but there is no essential difference between these two methods of bone formation.

Intramembranous (Mesenchymal) Ossification

This process, which is essentially the direct mineralization of a highly vascular connective tissue, commences at certain constant points known as *centres of ossification*. At such a centre, the mesenchyme cells (*osteoprogenitor cells*) proliferate and condense around a profuse capillary network. Between the cells and around the vessels, is a fine meshwork of collagen fibres and an associated amorphous ground substance. The most central cells enlarge, and simultaneously fine strips of an eosinophilic matrix (the earliest bone) appears between the blood vessels. These strips rapidly extend and fuse to form a delicate labyrinth, the spaces of which enclose the blood vessels and transforming mesenchyme cells. The latter enlarge, becoming polygonal, roughly cuboidal, or low columnar, and arranged as an incomplete layer in close contact with the surfaces of the primitive, eosinophilic bony matrix. (It is now firmly established that the earliest hydroxyapatite crystallites appear inside minute extracellular *matrix vesicles* which are products of the differentiating osteoprogenitor cells, or osteoblasts —*see* p. 257 and **3.**16A, B).

The cells retain contact with each other via short side branches. Their cytoplasm becomes intensely basophilic and the nucleus often becomes eccentric with a pale area of cytoplasm on its inner side. The cells are now termed *osteoblasts* and ultrastructurally are seen to possess an open-faced nucleus with obvious nucleoli, a prominent paranuclear Golgi apparatus with centrioles, a profuse concentration of granular endoplasmic reticulum and some free ribosomes. Lipid droplets, scattered mitochondria and lysosomes are found throughout the cytoplasm which, using appropriate techniques, can be shown to contain periodic acid-Schiff positive granules and vesicles and to give a strong positive reaction for alkaline phosphatase. These cells are similar in some respects to the chondroblasts of developing cartilage (p. 245), and similar synthetic pathways are involved during their rapid production of collagen and its associated proteoglycans. These are extruded into the extracellular space, where they form the so-called osteoid matrix which rapidly becomes calcified with the appearance of minute deposits of calcium phosphate (*vide infra*). These are later transformed into typical needle-shaped hydroxyapatite crystallites intimately related first to the cores of matrix vesicles, and subsequently to the collagen fibres of the matrix. These collagen fibres form increasingly coarse, interweaving networks in the walls of this early sponge-work or labyrinth of *woven-fibred bone*—the so-called *primary spongiosa*. As further layers of calcifying matrix are added to the delicate early trabeculae, some of the osteoblasts become surrounded by matrix and are now included in primitive lacunae within the developing bone. The surface projections of the entrapped cells, however, retain contact with those of neighbouring cells, and as these projections elongate, the matrix condenses around them to form canaliculi. Further generations of osteoblasts form on the trabecular surfaces by transformation of adjacent vascular mesenchyme cells.

As the sequence of matrix secretion, calcification and entrapping of osteoblasts proceeds, the trabeculae gradually thicken and the intervening vascular spaces become progressively narrowed. Where the bone persists as cancellous bone, however, the process slows and the spaces later become occupied by haemopoietic tissue. In regions where compact bone is formed, the process of trabecular thickening and vascular space narrowing continues. As this occurs, the collagen fibres of the matrix which are secreted on to the narrowing walls of the spaces become rather more highly organized as parallel, longitudinal or spiralized bundles and the entrapped cells come to occupy roughly concentric sequential rows. These irregular, anastomosing, sometimes roughly cylindrical, masses of compacted, parallel-fibred bone containing a central vascular canal are termed *primary osteons* or *atypical Haversian systems*. These, and the intervening areas of woven-fibred bone are subsequently largely eroded and replaced by generations of lamellar *secondary osteons* (see also pp. 252, 253, 266).

As these changes are proceeding in the ossification centre, the surrounding mesenchyme condenses as a fibro-vascular periosteum around its edges and surfaces. Extension of the ossification process occurs through the agency of stem cells derived from the deeper layers of the periosteum. The cells are potentially osteoprogenitor and maintain themselves by mitotic division. Some of the derived cells progressively secrete new matrix and collagen fibres which radiate from the ossification centre, and enmesh further periosteal blood vessels and again become encrusted with ossifying bone matrix, eventually enclosing the young osteocytes within lacunae. Further growth of a bone involves a continuation of these processes together with much remodelling with varying rates of bone deposition in some areas and balanced resorption of bone in others. Obviously, these overall patterns vary with the topography and fine architecture of the particular bone in question and have been studied using a variety of techniques over many years. Examples of this will be considered elsewhere (p. 266) whilst the more intimate processes of calcification and bone resorption common to all forms of bony tissue, follow briefly.

Calcification and the Osteoblast

Whilst the essential crystallite unit in bone is in the nature of a complex hydroxyapatite with a number of other associated ions (p. 255), the necessary ions involved in the initial stages of calcification are only calcium (Ca^{2+}) and phosphate (PO_4^{3-}), and the problem, although unexplained in detail, was treated for many years as one simply involving precipitation dynamics. It was assumed that the product of the ion concentrations $[Ca^{2+}] \times [PO_4^{3-}]$ in the general body fluids was insufficient for precipitation to occur, whereas in osteoid matrix special conditions obtained and precipitation followed. General support for this view followed the observation of calcification deficiencies in subjects with a low intake of these ions and conversely, the phenomenon of calcification in sites other than the skeleton (e.g. blood vessel walls) with dietary excesses in experimental animals, or in clinical conditions involving massive doses of vitamin D or in hyperparathyroidism. The special conditions thought to occur in the osteoid matrix were sought in the alkaline phosphatase positivity of the osteoblasts (thereby raising the local concentration of phosphate) or in some mechanism whereby the local *p*H was raised to alkaline levels (with a consequent reduction in the solubility of the calcium salt). However, it is now apparent that simple consideration of the solubility product of the ions involved is inadequate, that their relevant *activity coefficients* must be considered, and that these are difficult to extrapolate from *in vitro* studies to the microenvironment of the osteoblast; as yet, it has proved impossible to measure them directly. For reviews of these various theories consult Urist (1966); Neuman and Neuman (1958).

There soon followed a change of viewpoint with the stating of the theory of *epitactic nucleation* (Neuman and Neuman 1953, 1958). The latter is based upon the concept of *seeding* or *epitaxy* in which a nucleus is somehow formed, with a crystalline structure which closely mimics that of hydroxyapatite and is effective in causing the initial aggregation of calcium and phosphate ions. Thereafter the hydroxyapatite crystal grows spontaneously by addition from the sufficiently saturated surrounding fluids.

With the wide acceptance of the general aspects of the theory of seeding, many investigators attempted to determine the nature and distribution of the nucleation sites involved. These were variously claimed to be points in the periodic substructure of collagen, the ground substance links between the collagen fibrils, or structural aspects of the proteoglycans (consult Bernard and Pease 1969 for a review).

The latter authors, however (foreshadowed by Bonucci 1967), proposed an interesting role for the osteoblast and claimed that the initial nucleation site was provided by cellular 'buds' or extrusions which appeared to have a polysaccharide core and which, possibly in association with protein, formed an 'image' of the crystal salt. Further aggregation of hydroxyapatite crystals around this initial locus was considered to lead to the formation of spherulitic bone nodules which eventually coalesced to form seams of bone, the association with neighbouring collagen fibres being a secondary phenomenon. On this view the 'cellular seeds'

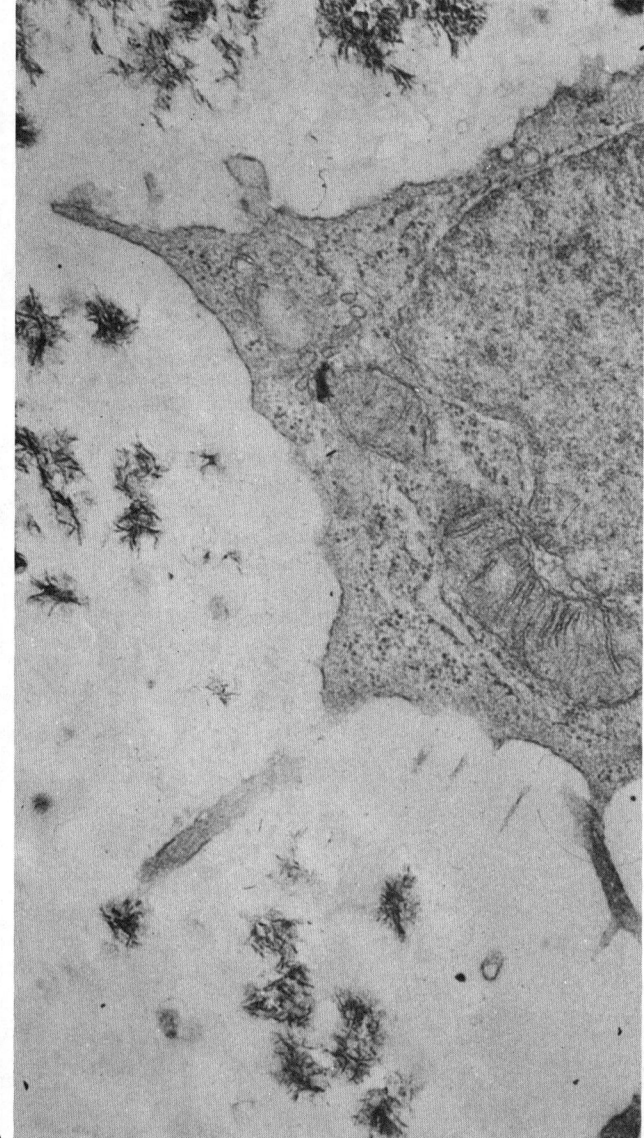

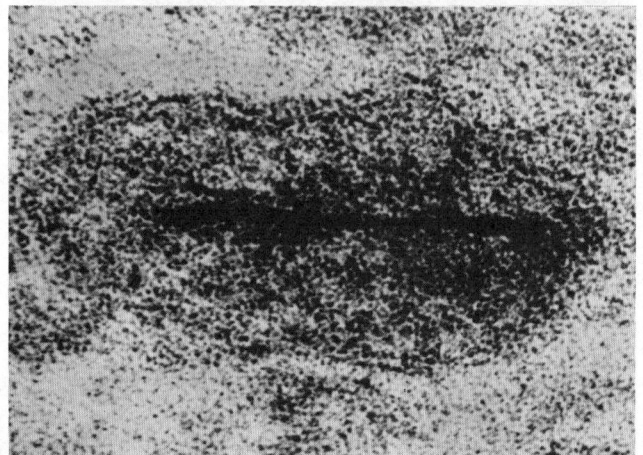

3.16A Transmission electron micrograph of part of an osteoblast, its surface bearing long filopodia, which project into the surrounding intercellular matrix. Within the latter are clusters of 'matrix vesicles' which provide the initial nucleation sites for the formation of hydroxyapatite crystallites in the early mineralization of bone. The specimen is from the subperiosteal ossifying front of a chick's tibia in tissue culture.

B A high power transmission electron micrograph of a single 'matrix vesicle' from the specimen shown in **3.16A**. The vesicle is membrane-bound, has a granular content and centrally, along its long axis, is a dense needle-shaped crystallite of hydroxyapatite. (Specimens kindly prepared and provided by Dr. J. J. Reynolds, Strangeways Laboratory, Cambridge.)

which initiate calcification were specific cell products of the osteoblasts only, and were absent in generalized connective tissues.

Although the theory of epitactic nucleation failed to obtain confirmatory and continuing support, the essential and intimate role of cellular products extruded into the surrounding matrix, as the site of initial mineralization in all vertebrate hard tissues, has aroused intense interest and numerous investigations. The cell products are minute membrane-bound spheres, with a characteristic granular, electron-dense core, and about $0.1 \, \mu$m in diameter; they have been named *matrix vesicles* (**3.16**). The vesicles have now been convincingly demonstrated as the loci within which the earliest needle-like hydroxyapatite crystallites appear. They have been identified in calcifying cartilage (Anderson 1967, 1969; Bonucci 1967, 1970), woven bone (Bernard and Pease 1969), dentine (Sisca and Provenza 1972; Katchburian 1973), and subperiosteal bone (Anderson and Reynolds 1973; Reynolds 1976). Thus it appears that matrix vesicles with similar properties can be produced by chondroblasts, odontoblasts and osteoblasts. The mechanism of initiation of calcification, however, remains a matter of some speculation. Being membrane-bound, the vesicles provide a partition between internal and external environments. They have been shown to carry a localized extracellular concentration of alkaline phosphatase, ATPase, inorganic pyrophosphatase and, in some sites, a relatively high concentration of a variety of lipids. It appears a sound working hypothesis that the matrix vesicles provide the necessary enzymes and environment to sufficiently concentrate calcium and phosphate so that crystallization is initiated, and thereafter proceeds spontaneously beyond the confines of the vesicle. Other hypotheses include the proposal that the membrane of the vesicle incorporates an inwardly-directed calcium pump, or alternatively, that there occurs calcium binding by the vesicle lipids.

It may also be noted here that there has been considerable controversy concerning the shape and size of the initial hydroxyapatite crystals. Some authorities hold that they are plate-like in form, whilst others claim a rod-like or needle-like conformation. Most agree that the smallest dimension of the crystals is about 5 nm, but disagreement persists concerning the largest (c-axial) length, estimates varying from 20 to 35 nm to values of some three times larger than this. These differing estimates reflect the variety of techniques used: thus polarized light microscopy, transmission electron microscopy, and X-ray diffraction all show that the long crystal axis lies *parallel* to adjacent collagen fibrils. However, X-ray diffraction and dark-field electron microscopy give a comparably short value for the c-axial length, whereas transmission electron microscopy of ion-beam thinned sections gives a much larger value (*see* Boyde 1976; Jackson *et al.* 1976).

Calcium Balance, Bone Resorption and the Osteoclast

A central theme of physiology concerns the dependency of many cellular activities upon the relative constancy of their micro-environment (in terms of osmolality, ionic species and concentrations, pH, etc.), and many delicately balanced feedback control systems operate to ensure such *homeostasis* (*see* e.g. Reynolds 1974). To this, the circulating level of calcium ions is no exception, and it is preserved despite wide variations in diet, rates of bone growth and remodelling. Involved in the mechanism, but varying quantitatively at different times, are both the labile and stable areas of bone salts (*vide infra*), the interrelated varieties of bone cells—osteoblasts, osteocytes and osteoclasts—and two endocrine secretions, parathyroid hormone and thyrocalcitonin.

Calcium in the blood and tissue fluids is in constant interchange with calcium salts in bone, and it has been shown that approximately one-quarter of the ionic blood calcium is so interchanged each minute. The blood calcium is in equilibrium with the tissue fluid calcium, especially the perivascular fluids which perfuse the Haversian canals and the systems of bone lacunae and canaliculi, by which means vast areas of bone salt are exposed for such physico-chemical exchanges (p. 255). The calcium content of the most recently developed bony tissue is more labile than that of older regions and the process of replacement of mature osteons by new ones, which continues gradually throughout life, provides a reservoir of readily available calcium ions. It has been proposed that bone salt release is in some way dependent upon active modification of the matrix by neighbouring osteocyte processes (Bélanger 1963), but it is uncertain to what quantitative extent such *active cellular osteolysis* contributes to the rapid continuous turnover of calcium and how much is simply a direct physico-chemical exchange.

A depression of the circulating ionic calcium level results in an increased secretion of parathyroid hormone, which in turn causes a rise in calcium level by a number of distinct mechanisms; firstly, by a direct action on bone, secondly, by increasing renal tubular reabsorption of calcium and thirdly, a minor effect by increasing absorption of calcium from the gut. The action of parathyroid hormone on bone is complex and imperfectly understood. It may have an effect by modifying the surface shells of ions which are associated with the hydroxyapatite crystals, or alternatively, by changing the structure of the collagen-associated mucopolysaccharides of the organic matrix. Finally, it has been proposed that it increases osteocyte activity and bone resorption by osteoclast action (*vide infra*). Conversely, a raised circulating calcium level results in an increased secretion of calcitonin (thyrocalcitonin) by the parafollicular cells of the thyroid gland (p. 1451 and **8**.203) which in turn causes a fall in circulating calcium levels by an ill-understood direct action upon bone. It has been suggested that calcium release from bone is inhibited by stabilization of the collagen of its organic matrix, and possibly by an increased cellular deposition of bone by osteoblasts.

simply degenerating cells released from bone during the erosive process, and otherwise unconnected with the erosion. It has also been proposed that they arise by fusion of osteoblasts which have undergone modulation and assumed a bone-erosive function (Tonna and Cronkite 1962). However, recent studies using parabiotic systems suggest that the osteoblast and osteoclast are derived from *different* sources. The evidence indicates a local connective tissue progenitor cell for the osteoblast, whilst the osteoclast originates from a circulating blood cell—either a monocyte or some other variety of phagocyte (Owen 1976).

Osteoclasts contain a variable number of oval, closely packed nuclei, often 15–20, which closely resemble those of osteoblasts. Their cytoplasm contains many mitochondria, a low concentration of granular endoplasmic reticulum, a small paranuclear Golgi apparatus, and is filled with characteristic vacuoles. Some of the latter show a positive reaction for acid phosphatase, as do the more typical lysosomes which are plentiful in the deeper parts of the cell. Where the cell surface abuts against bone, light microscopists have often identified a finely striated appearance which superficially resembled the brush border seen in many other cells (p. 18). In tissue culture this border is seen to be in a state of constant vigorous movement and to be the site of active pinocytosis. Ultrastructurally, however, the striated appearance is seen to be due to both the character of the cell surface and that of the adjacent bone undergoing resorption. The cell surface does not possess an array of microvilli (as in other brush borders) but is highly infolded with deep extracellular clefts extending between

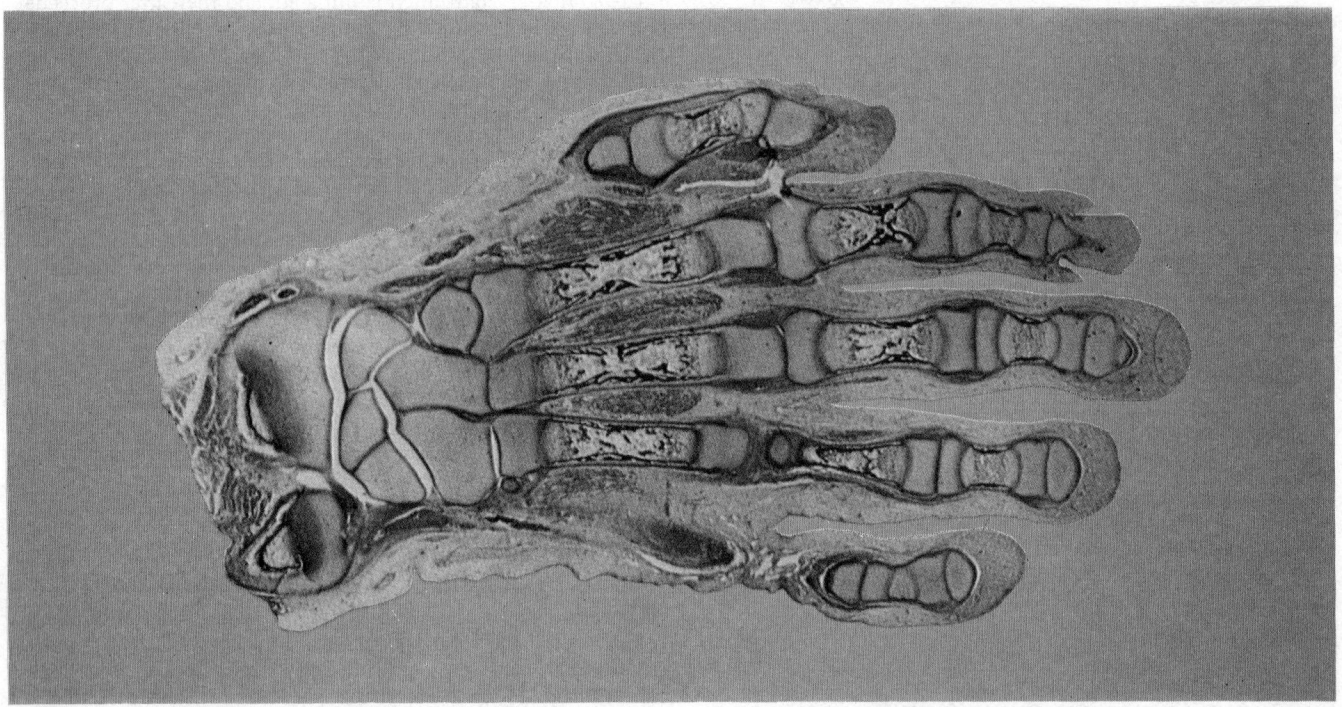

3.17A A survey photograph of a section of a fetal hand showing cartilaginous models of the carpal bones and various stages of development of primary ossification centres in the metacarpals and phalanges. Note that none of the carpal elements show any evidence of ossification.

When active bone growth is occurring, much remodelling takes place, involving deposition of bone on some surfaces and its removal from others. In mature bone also, continued remodelling of the osteon systems is preceded by bone removal. Surfaces undergoing massive bone resorption (in contrast to the less dramatic ionic interchanges referred to above) are associated with large, often multinucleate cells, with a foamy, lightly basophilic cytoplasm, termed *osteoclasts*. They are often found in close contact with the walls of localized erosions of the bone surface, and such pits are called the *resorption lacunae of Howship*. Their origin and significance has long been debated and many authors have held them to arise by endomitosis of osteoblasts or to be

irregular cell processes or lobopodia. The neighbouring bone surface is often partly demineralized and shows a frayed series of collagen strands extending towards the cell (and probably contributing to the striated appearance of the junctional zone). Minute crystals of bone mineral and collagen fragments have been observed deep within the clefts and also in cytoplasmic vesicles assumed to form from the depths of the clefts (Dudley and Spiro 1961; Hancox and Boothroyd 1963).

Whilst such appearances are suggestive of bone removal, the mechanisms of osteoclastic action are still unclear. The presence of lysosomes and a high level of proteolytic enzyme activity in the cytoplasm may be relevant in this context, whilst others have

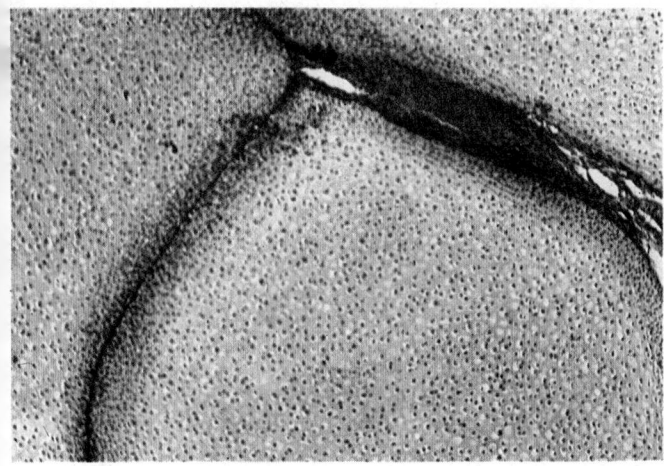

3.17B   Cartilaginous models of carpal bones and the connective tissue junctional zones between them.

suggested that they secrete chelating agents which form soluble, weakly ionized complexes with bone salt constituents. When the erosive process is complete, the osteoclasts disappear either by degeneration or by reversion to their parent cell type, but objective evidence concerning their fate is lacking.

Intracartilaginous (Endochondral) Ossification

As we have seen, most human bones are preformed in cartilage and each long bone is represented in early fetal life by a rod of hyaline cartilage (**3.17A**) which replaced a rod of condensed mesenchyme, its shape foreshadowing that of the early bone. (The smaller, e.g. carpal, bones are preceded by appropriately shaped cartilaginous 'models'—*see* **3.17A** and B.) Within and upon this cartilaginous model, orderly sequences of changes occur with the appearance of *centres of ossification*. The general significance of primary and secondary (epiphysial) ossification centres and their associated growth plates of cartilage have been discussed elsewhere (p. 241) and the microscopic changes follow here.

The cartilaginous model is surrounded by a highly vascular condensed mesenchyme or perichondrium, similar in every way to that which precedes and surrounds intramembranous ossification centres (p. 260), and again, its deeper layers contain osteoprogenitor cells.

The first indication of the appearance of a primary ossification centre involves the cartilage cells deeply placed in the centre of the primitive shaft (**3.18, 19**). These become greatly enlarged, their cytoplasm becomes vacuolated and develops accumulations of glycogen, whilst the intervening matrix becomes reduced to thin and often perforated partitions. As the process of lacunar enlargement continues, the enlarging cells, possibly due to auto-intoxication by chondrolysing matrix products, undergo progressive degenerative changes and ultimately die, leaving their enlarged and sometimes confluent lacunae as the *primary areolae*. With the death of the chondrocytes the thin walls of the primary areolae become calcified. Simultaneously, the subperichondrial cells transform to osteoblasts and lay down a peripheral layer of fenestrated young bone. This *'periosteal collar'* of bone is formed by the same sequence of cell changes described under intramembranous ossification (p. 260) and is at first a thin-walled tube enclosing the central region of the shaft. However, it progressively increases in girth and the process continues towards the extremities of the shaft (*vide infra*), whilst a series of further important changes are occurring deep within the cartilaginous model. (It should be noted that the calcifying cartilaginous areolar walls and the extracellular ground-substance of the periosteal collar are both sites containing *matrix vesicles—see* p. 261.)

Near the shaft centre where the periosteal collar overlies the calcified walls of the primary areolae, the collar is invaded by sprouts from the deeper layers of the periosteum. These sprouts consist of vascular channels and their accompanying cells which continually divide and transform into osteoblasts and osteoclasts. The latter excavate passages through the newly formed bone to

pass into the underlying calcified cartilage where they continue to erode some of the walls of the primary areolae, with a fusion of the original cavities to form an irregular system of larger, intercommunicating *secondary areolae* or *medullary spaces*. These spaces become filled with *embryonic bone marrow* (vascular mesenchyme, osteoblasts and osteoclasts), and their delicate residual walls of calcified cartilage become covered by a layer of osteoblasts. The latter lay down osteoid which soon transforms to patches of bony tissue which coalesce to form a complete lining for the spaces, and for a period, the process continues with the addition of further layers of bone, the entrapping of young osteocytes within lacunae and a narrowing of the vascular spaces. With continued deposition of subperiosteal bone, however, the formation of bone on the centrally placed areolar walls ceases, and a process of erosion begins with removal of these early spicules of bone and enlargement and confluence of the marrow spaces until a primitive marrow cavity forms. While these changes are proceeding in the centre of the shaft, the adjoining cartilaginous regions undergo a similar series of changes (**3.18, 19**). Since the changes are most advanced centrally and the ends of the bone are still cartilaginous, the various intervening zones show an orderly progression of change when the young bone is examined in longitudinal section.

The cartilaginous extremity (where an epiphysis usually forms) continues to grow in pace with the rest of the bone by appositional and interstitial mechanisms (p. 267). Where the extremity continues into the cartilaginous shaft, however, a particularly

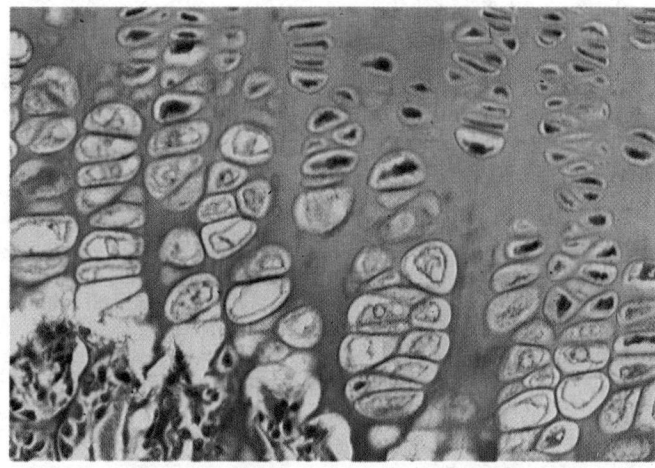

3.18   A section showing the transformation of cartilage cells and their lacunae as the ossifying front of an early primary centre of ossification is approached (below). Note the cell hypertrophy, lacunar enlargement with matrix partition reduction, and increased density of the partitions following calcification.

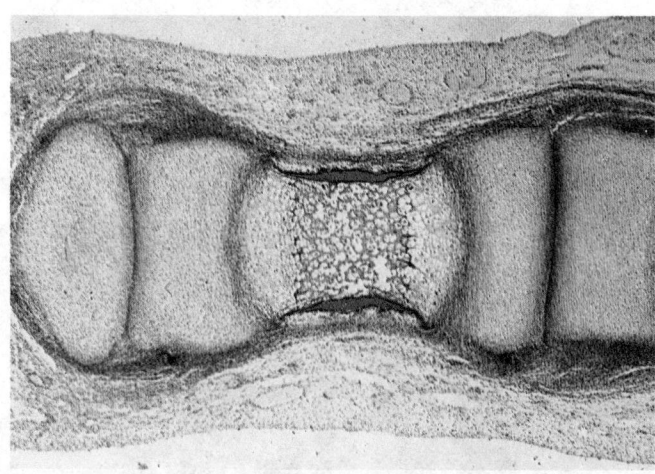

3.19   A longitudinal section of a phalanx (from the hand in **3.17A**) showing an early primary ossification centre. The cartilage cells in the shaft centre have hypertrophied and this region is surrounded by a delicate tube or collar of subperiosteal bone (red).

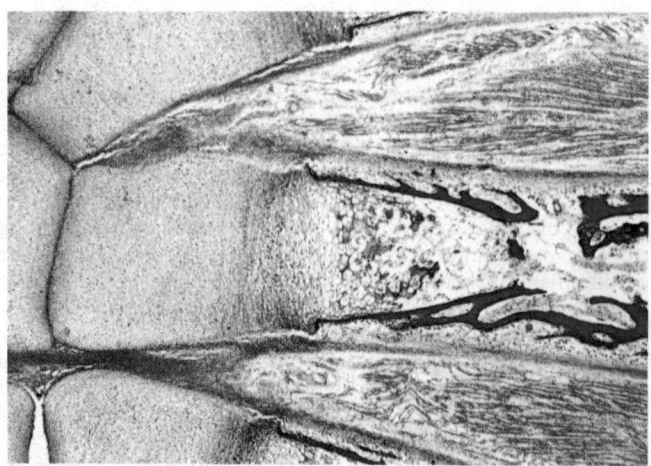

3.20 A longitudinal section of the proximal half of a fetal metacarpal bone (from **3.17A**) at a more advanced stage than the phalanx in **3.19**. The periosteal collar of woven bone is thicker, contains radially disposed vascular spaces, and vascular invasion of the shaft centre has occurred and is proceeding towards its extremities. See text for a detailed description.

rapid and geometrically organized region of cartilage growth becomes prominent, i.e. region of the future *growth plate* between epiphysis and diaphysis (*see* e.g. Siffert 1956; Rang 1969). This region grows both transversely and longitudinally (**3.18, 20, 21, 22**). Transverse or latitudinal growth follows the occasional mitoses of cells in a plane transverse to the long axis of the bone and also by appositional growth from cells derived from the encircling *perichondrial collar* (or *ring*) at this level. By this means the diameter of the future growth plate region increases in concert with the expanding shaft and adjacent future epiphysis. More frequent divisions in the long axis of the bone soon result in the formation of numerous longitudinal columns (palisades) of disc-shaped or wedge-shaped chondrocytes, each in its flattened lacuna. This process of proliferation and longitudinal cell column formation occupies the so-called *zone of cartilage growth* and it is the continued interstitial growth of this zone which is the essential mechanism whereby the whole bone increases in length. As a cell

column is traced towards the shaft centre, its component cells are increasing in maturity and their characteristics change. They gradually increase in size and accumulate glycogen in their cytoplasm which also begins to show positive histochemical reactions for oxidative enzymes. The younger chondrocytes of the columns possess a few surface projections which pass into reciprocal recesses in the lacunar walls, but as the cells enlarge and enter the *zone of cartilage transformation*, the number of surface projections increases greatly—possibly associated with removal of the surrounding cartilage matrix and a progressive enlargement of the lacunae across the zone. The largest cells finally withdraw their surface projections, expand to fill their large lacunae and then soon show degenerative changes leading to cell death. As this occurs, the walls of the empty lacunae (now arranged as a series of transverse and longitudinal partitions), become impregnated with hydroxyapatite crystals (sometimes called the zone of calcified cartilage). These calcified partitions now enter the *zone of bone formation* where they are invaded by vascular mesenchyme and its associated osteoblasts and osteoclasts from the adjacent primary ossification centre (**3.23**). The partitions, especially the transverse ones, are now partially eroded, and then on the remaining surfaces of the thinning longitudinal partitions occurs the same processes of osteoid deposition, bone formation, and osteocyte trapping, etc. as described above. (It should be appreciated that the calcifying walls of the enlarging lacunae and the osteoid matrix and sites of bone deposition, have all been shown to exhibit populations of *matrix vesicles*—(*see* Thyberg and Friberg 1970; Anderson and Reynolds 1973; and p. 261.) The erosion (chondrolysis) of the calcified partitions has been variously ascribed to osteoclast (chondroclast) action or to the possible chondrolytic action of the endothelium of the terminal loops of the vascular sinusoids which come to occupy each incomplete, columnar trabecular framework (**3.22**), and *see* Irving (1964). Opinion is divided whether these loops have an intact endothelium or whether they are temporarily fenestrated and the site of a microrupture (Brookes and Landon 1963; Trueta and Morgan 1960; Anderson and Parker 1966). Further, it has been claimed that longitudinal and transverse partitions of matrix are different, both in their structure and chondrolytic mechanisms. Transverse ones are lightly calcified, with sparse collagen, distinctive proteoglycans, and are susceptible to lysosome action, whilst the more resistant longitudinal ones only succumb to osteoclast action (Anderson 1962; Dingle 1962).

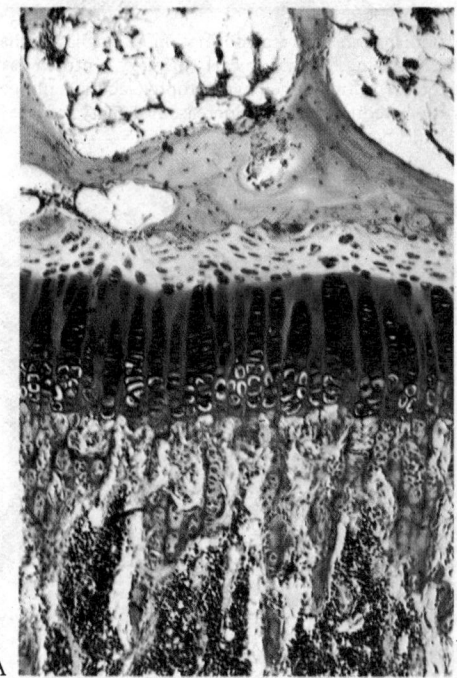

A

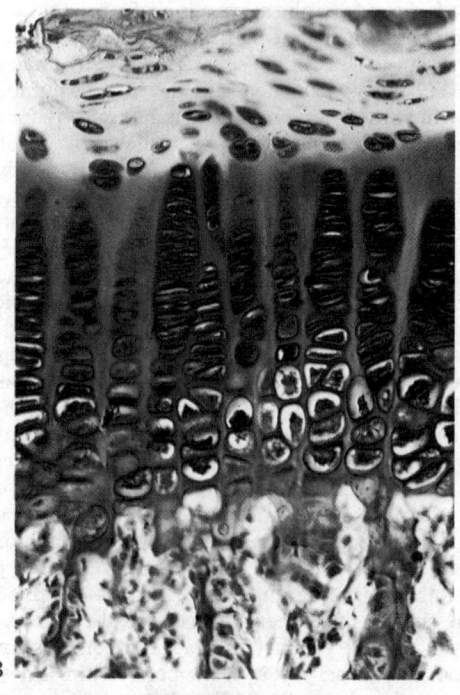

B

3.21A and B Low and intermediate magnifications of sections through the cartilaginous growth plate between the epiphysis and metaphysis (below) at the proximal end of a human tibia. Note the transition from hyaline cartilage, through zones of cell multiplication, hypertrophy, column formation, matrix calcification, and partial chondrolysis to the ossifying front. Compare with **3.22** and consult text for further details.

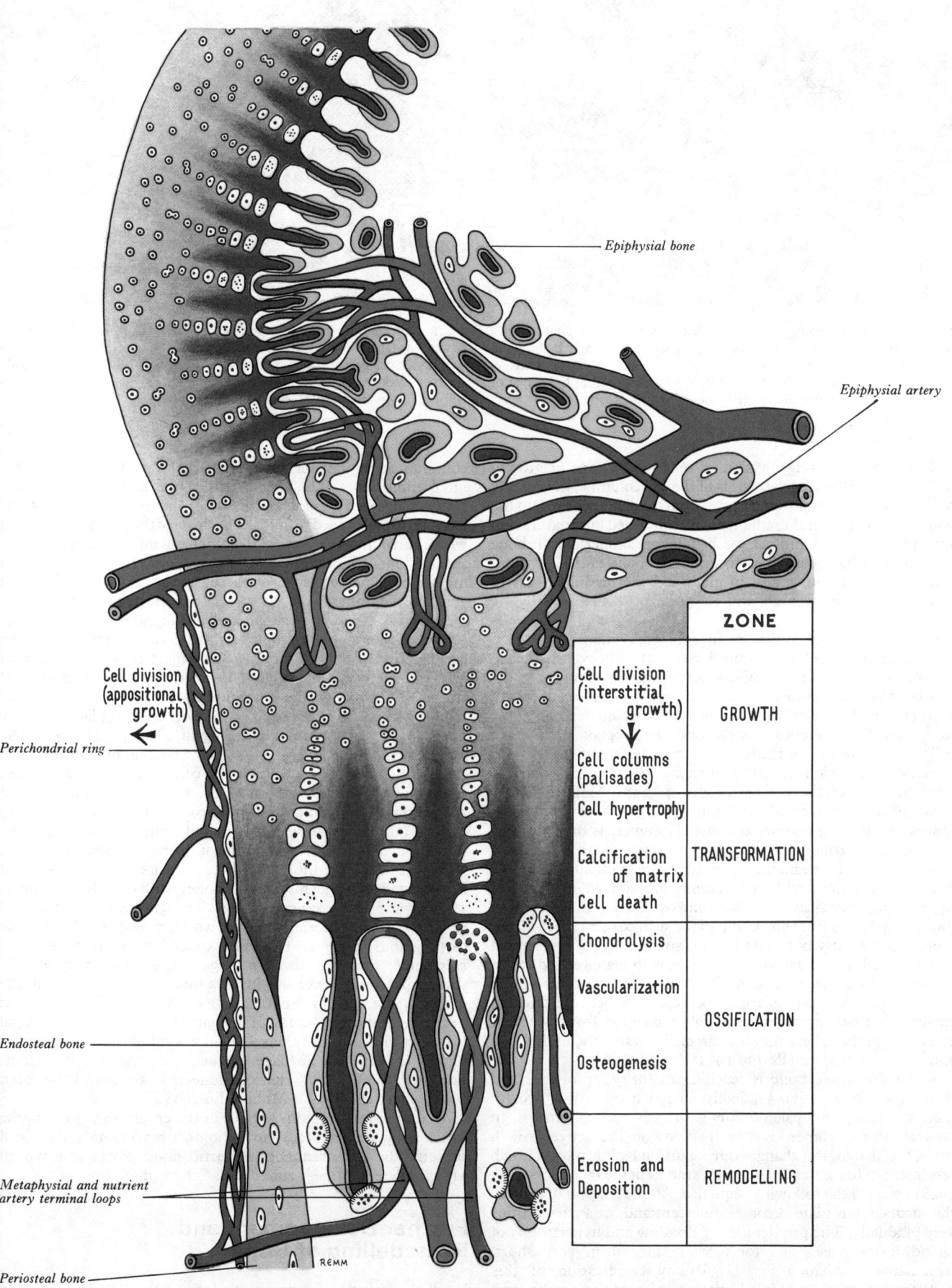

Epiphysial bone

Epiphysial artery

Cell division
(appositional
growth)

Perichondrial ring

Endosteal bone

Metaphysial and nutrient
artery terminal loops

Periosteal bone

REMM

	ZONE
Cell division (interstitial growth) ↓ Cell columns (palisades)	GROWTH
Cell hypertrophy Calcification of matrix Cell death	TRANSFORMATION
Chondrolysis Vascularization Osteogenesis	OSSIFICATION
Erosion and Deposition	REMODELLING

3.22 A scheme of the main features of an active growth cartilage and the
adjacent metaphysis and epiphysis. The different views concerning the
mechanism of chondrolysis described in the text are indicated.

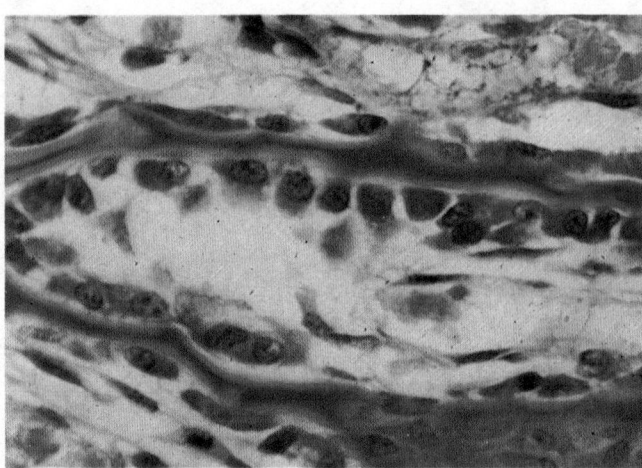

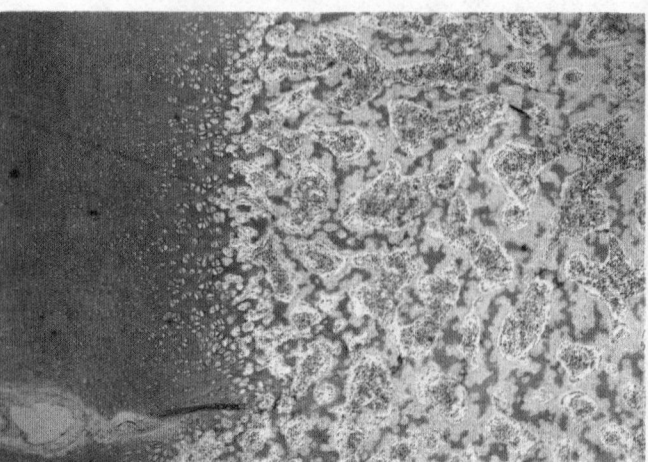

3.23A A section through the metaphysis of a fetal bone stained with haematoxylin and eosin showing developing spicules of early bone. Each spicule contains a deeply stained basophilic core of calcified cartilage, which is covered on both aspects by a lightly stained eosinophilic layer of young bone along which are ranged rows of active osteoblasts.

3.23B A transverse section through an epiphysis of a long bone at a more advanced stage of ossification. Note the gradation of cartilage cell lacunae (left) as the ossifying front is approached. The bony trabeculae (light pink) contain dark patches of basophilic calcified cartilage and the inter-trabecular spaces are filled with red bone marrow.

With the continuing division of cells in the zone of growth which adds to the epiphysial ends of the cell columns, the orderly sequence of changes proceeds away from the shaft centre, and the bone grows in length. Simultaneously, continued internal erosion and remodelling combines with further subperiosteal bone deposition, which continues towards the bone ends. By these means the bone grows also in girth and the medullary cavity enlarges both transversely and towards the extremities of the bone.

Growth continues in this manner for many months or years, depending upon the bone in question (p. 241), but eventually one or more secondary centres of ossification usually appears in the cartilaginous extremities. Such epiphysial centres (or ossifying bone ends without epiphyses) are not at first characterized by the formation of cell columns. Instead, scattered isogenous groups of cells hypertrophy and die, with matrix calcification and subsequent invasion of this matrix by vascular mesenchyme (sometimes of cartilage canal origin—p. 248). Many irregular bone spicules are formed, with intervening embryonic marrow spaces, in the same manner as above. However, as the epiphysis enlarges, its cartilaginous periphery forms a roughly spherical zone of proliferation and the whole structure gradually assumes a radial organization, with cell column formation and zones of hypertrophy, calcification, erosion and ossification to be found at increasing depths from any point on the surface (3.22). In a sense, therefore, the early bony epiphysis is surrounded by a spherical growth cartilage. Soon, however, the growth plate related to the metaphysis becomes a particularly prominent part of this sphere and in this region the enlarging cell columns become directed exclusively towards the metaphysial ossification front, whereas elsewhere in the sphere they are directed towards the underlying epiphysial ossification (Payton 1934; Rang 1969).

When the whole bone is reaching maturity, epiphysial and metaphysial ossification gradually encroach upon this growth plate and final bony fusion occurs with a cessation of growth. In contrast to the numerous investigations on the active growth plate, the histological changes during fusion have attracted much less interest (Rang 1969). However, studies in the rat (Becks *et al.* 1948) revealed the following sequence. After cessation of growth the cartilaginous plate becomes quiescent and for a period thins very gradually. The proliferation, palisading and hypertrophy of chondrocytes ceases and for a while they form very short, irregular cone-shaped masses. Patchy calcification of this cartilage is then accompanied by absorption of the patches and some of the adjacent metaphysial bone. These erosion channels are invaded by vascular mesenchyme and eventually some of the vascular sprouts pass completely through the thin plate of cartilage and at these points metaphysial and epiphysial vessels unite. Final bony fusion occurs by ossification in the walls of these

communicating channels, which then spread to involve the intervening zones. The bone formed is particularly dense and is easily recognized on radiographs as a so-called *epiphysial line* (the latter term is also used to indicate the level of the perichondrial ring which surrounds the growth cartilage of an immature bone, or the surface line of junction between epiphysis and metaphysis in a mature bone). More recently, there have appeared some detailed analyses of epiphysio-diaphysial union in the dog and man: for the histological stages, and an extended bibliography consult Haines (1975). The latter author points out that as union approaches new cartilage is added to the epiphysial aspect of the cartilaginous growth plate which initially retains its characteristic histological zoning. Then follows the formation of epiphysial and diaphysial bony plates on the two aspects of the cartilaginous plate: the bony plates being partly formed by calcifying cartilage (termed by the author *metaplastic* bone) and associated patches of lamellar bone. The preliminary union follows the coalescence of one or more columns of mineralized cartilage, accompanied by confluence of the epiphysial and diaphysial marrow spaces. In the smaller earlier-uniting epiphyses there is only one, usually eccentric, initial area of fusion, with thinning of the remaining cartilaginous plate. Subsequently the original site of fusion, now consisting of mineralized cartilage (metaplastic bone), with lamellar bone on its surface and containing marrow, are gradually eroded and replaced by new bone and marrow which extends until the whole cartilaginous plate is replaced, union is complete and no *epiphysial 'scars'* persist. In larger, later-uniting epiphyses, similar processes occur but involve multiple perforations in the growth plate, and in this case *islands* of epiphysial bone often persist as epiphysial scars. (It may also be noted that the special histological features of the so-called *metaplastic bone* that occurs underlying articular cartilage, and at the attachments of tendons, ligaments, and other dense connective tissues, have been reviewed by Haines and Mohuiddin 1968.)

When the cartilaginous surface of the epiphysis is to form the articulating surface of a synovial joint it remains unossified and presents the same sequence of cartilaginous zones as a typical growth plate, but these zones persist throughout life.

Further Development and Remodelling of Bone

In all the situations so far studied, the bone first laid down is spongy in texture and contains a continuous labyrinth of large vascular spaces; the lacunae are irregularly scattered, there is no lamellation and the collagen fibre bundles in the matrix are fairly randomly arranged and form a network. This type of bone is usually termed *woven-fibred*. It is best seen in young fetal bones but occurs in some situations in the adult (e.g. lining the tooth sockets) and is seen during the repair of fractures. Later, roughly concentric tracts of non-lamellated, parallel-fibred bone, termed *atypical Haversian systems* or *primary osteons* are deposited on the walls of the vascular spaces, which are thereby narrowed and often subdivided by osseous bridges. The collagen fibre bundles in the matrix are arranged in parallel longitudinally orientated spirals. No erosion precedes the deposition of primary osteons, and consequently no cement lines intervene between them and the surrounding woven bone. The circumferential bone which is deposited on both periosteal and endosteal surfaces, is generally composed of parallel-fibred lamelle with the fibres arranged longitudinally and circumferentially in successive lamellae, and where the latter are thin the lamellation is indistinct. Circumferential bone may, however, be non-lamellated or woven-fibred, particularly in some species, and the term *surface bone* is often preferable in these circumstances.

As the bone matures, *typical Haversian systems* (*secondary osteons*) gradually replace the primary osteons and woven-fibred bone. Their formation is always preceded by erosion, usually eccentric, of the walls of the vascular channels, which are composed of primary osteons and woven bone. This is followed by concentric deposition of lamellae of parallel-fibred bone on the

growth have, with increasing refinement, spanned the last two centuries. The earlier anatomists used metal markers in the form of wires encircling the shaft, or pellets embedded in the shaft, of growing bones. The pioneering observation, that when growing pigs were fed with the root of the madder plant, the bone newly formed during the period of feeding was identifiably pink, led to an elegant series of experiments on bone growth by alternating madder feeding with madder-free periods (Brash 1934). More recently the intraperitoneal injection of a solution of alizarin red has been used with greater precision for the same purpose (Hoyte 1960). These studies amply confirmed the accretionary nature of bone growth, and modern techniques have simply increased the resolution of the methods, enabling quantitative histological studies to be made. The latter include the administration of bone-seeking isotopes followed by autoradiography of sections (Leblond 1950); the exhibition of doses of a tetracycline at intervals followed by fluorescence microscopy and finally, historadiography of sections (Amprino and Engström 1952; Amprino 1968).

Relatively gross remodelling, with changes in the general shape of the whole, affects all bones during their growth, but particularly well studied examples are the bones of the skull cap and long bones with expanded extremities.

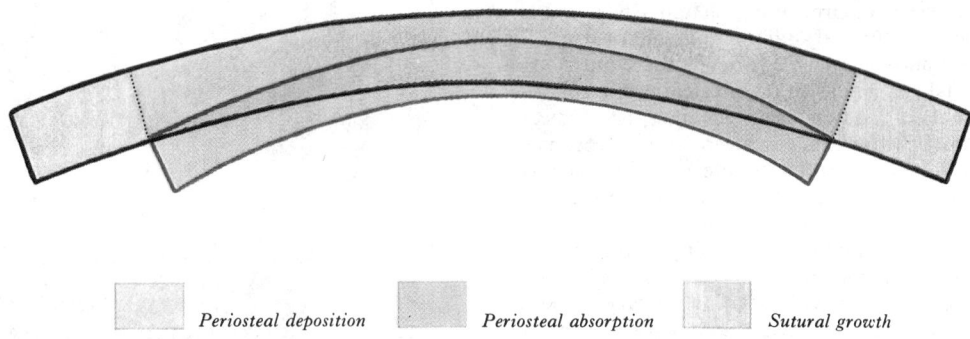

Periosteal deposition   Periosteal absorption   Sutural growth

3.24   A scheme illustrating the gross patterns of growth and remodelling of a bone of the cranial vault, in section. Note the changes in surface area, thickness and curvature. See text for further details.

walls of the *resorption cavity* and around the vascular channels which are thereby narrowed. A cement line is always deposited at such a site of reversal from erosion to deposition. The formation of secondary osteons does not end with the cessation of growth but continues at varying rates throughout life, and some of the first deposited secondary osteons are eventually involved in the erosive process. The remnants of woven bone, primary osteons, circumferential lamellae and eventually secondary osteons constitute the *interstitial bone* which occupies the crevices between the later formed osteons. (For some quantitative histological data on the ageing of human bone *see* Courpron *et al.* 1973, and for morphometric analyses of ageing effects of osteon remodelling in the human tibia *see* Ortner 1975.)

Implicit throughout the foregoing account of bone development is the fundamental feature of bone growth, that it proceeds exclusively by *appositional mechanisms* in which new layers of bony tissue are simply added sequentially to pre-existing surfaces; and, during the process, bone cells become immovably trapped in lacunae within its substance. Such cells, once within a lacuna, do not divide and, further, the rigidity which follows mineralization of the bone matrix prevents any expansion of the tissue from within. Thus interstitial growth mechanisms, which are so characteristic of many other tissues, including cartilage, are entirely absent in bone. They are, however, of supreme importance in the growth cartilages discussed above.

All remodelling of bone, therefore, whether involving major readjustments of the overall shape of a bone, or the less obvious internal remodelling which occurs throughout life, is dependent upon a delicate and geometrically ordered balance between bone deposition on some surfaces and bone removal from others. The experiments which demonstrated these main features of bone

A bone such as the parietal not only increases in thickness and surface area during growth, but its curvature also decreases (**3**.24). Whilst bone growth continues at its margins in association with interstitial growth of periosteal cells at the sutures, periosteal bone is also added on its outer surface and eroded internally. The latter processes, however, do not occur uniformly over the whole bone, but at a rate which increases with the radial distance from the bone centre (**3**.24).

As we have seen, increase in length of a long bone is due mainly to extension of the endochondral ossification process into the calcified zone of adjacent growth cartilages which are continually replaced by longitudinal interstitial growth of their proliferative zones, together with minor additions due to radial growth of the epiphyses. In company with these length increases there occur diametric increases of the growth cartilages and girth increments in the shaft. The latter occur in general by continuing subperiosteal deposition and endosteal erosion of bone. In many bones, however, such processes occur eccentrically, at different rates or even being reversed, at different points on the circumference. By such means a bone initially tubular may be transformed into one which is roughly triangular in cross-section such as the tibia. Meanwhile, the delicate waisted contours of the metaphysis are preserved by a continuous process of periosteal erosion and endosteal deposition (*see* **3**.25).

The tetracycline marker technique has emphasized the dynamic nature of even mature compact bone where at any time in life a single cross-section includes mature relatively quiescent osteons, recently formed osteons and osteons in process of formation, or of resorption. The rate of lamellar bone formation by apposition varies but is on average about 1–2 μm per day, and it has been shown that a resorption canal takes 1–3 months to form and a new osteon a similar period to regrow (Lee 1964). Such internal remodelling provides a continuous pool of young osteons with their labile calcium reserves (p. 261) and also provides a flexibility of bone architecture which may change in response to altered patterns of mechanical stress (p. 233). Whilst the morphogenetic control systems responsible for overall bone shape are not understood, interesting suggestions have been put forward concerning its responsiveness to altered conditions of stress. In common with many other crystalline substances, bone is piezo-electric; that is, it generates small electric currents when deformed, and the distribution of the potential difference variations, reflects the form of the distortion. It has been proposed that such currents could in some way polarize the cells responsible for osteolysis and bone deposition, so that the bone structure is reorganized in a manner which resists the predominating mechanical stresses (Bassett 1965; Cochran et al. 1968). Much work is now in progress concerning the sources, intensities and distributions of the bioelectric phenomena which accompany stressed connective tissues in general, and whilst further results are awaited with great interest, it would be premature to assume that firm causal relationships between such events and the morphogenesis of the tissues have yet been established.

Metabolic and Endocrine Effects on Bone

The role of the bone salts, parathyroid hormone and calcitonin in the homeostatic regulation of circulating calcium levels has been considered elsewhere (p. 261). In addition to these, to ensure the normal development and maintenance of bone there must be an adequate dietary intake and absorption of calcium, phosphorus, vitamins A, C and D and a balanced interplay between somatotrophic hormone, thyroxin, oestrogens and androgens.

Prolonged dietary deficiency of calcium eventually leads to a generalized loss of bone mineral or *osteoporosis*, and consequently an increased fragility of the bones. Vitamin D influences the intestinal absorption of calcium (and indirectly of phosphorus) and therefore circulating calcium levels. Again, prolonged vitamin D deficiency (with or without a low dietary intake of calcium) in the adult, leads to *osteomalacia* or *adult rickets*. The bones are poorly mineralized and often contain large areas of soft, deformable, uncalcified osteoid. Similar deficiencies, before growth has ceased, lead to classical *rickets* with serious disturbances of the growth cartilages and the process of ossification. The regular columnar organization of the growth plate is partially lost, there is a failure of calcification of the cartilage but the chondrocytes continue to proliferate and the cartilaginous plate becomes much thicker and less regular than in the normal bone. In the metaphysial region the cartilage is partially eroded, and osteoblasts secrete thick layers of osteoid which fail to ossify on the uncalcified or poorly calcified cartilage trabeculae. If the condition persists, weight-bearing produces gross deformity of such softened bones. (For a review of the role of vitamin D in bone metabolism, bone remodelling, and calcium homeostasis, with an extensive bibliography, *see* Reynolds 1974.)

Deficiency of vitamin C precipitates *scurvy*, with widespread changes in all the connective tissues of the body, and these are particularly noticeable in the growth cartilages and metaphyses of long bones. Vitamin C appears to be essential for the adequate synthesis of both collagen and mucopolysaccharides of the matrix of connective tissues including bone. The growth plate thins, bone formation almost ceases and metaphysial trabeculae and cortical bone are both excessively narrow, with fragility and delayed healing of fractures.

An adequate intake of vitamin A is also necessary for normal bone growth and is essential for the balanced activities of bone deposition and removal. Deficiency leads to growth retardation with a failure of internal erosion and remodelling, particularly of the skull base. Foramina become narrowed and this may proceed to a pressure degeneration of their contained nerves, while the cranial cavity and spinal canal may fail to expand in step with the central nervous system, leading to gross impairment of nervous function. Conversely, excessive vitamin A intake leads to a more rapid vascular erosion of the growth cartilages, which become thin or totally lost with a complete cessation of growth.

Balanced endocrine activities are essential to normal bone

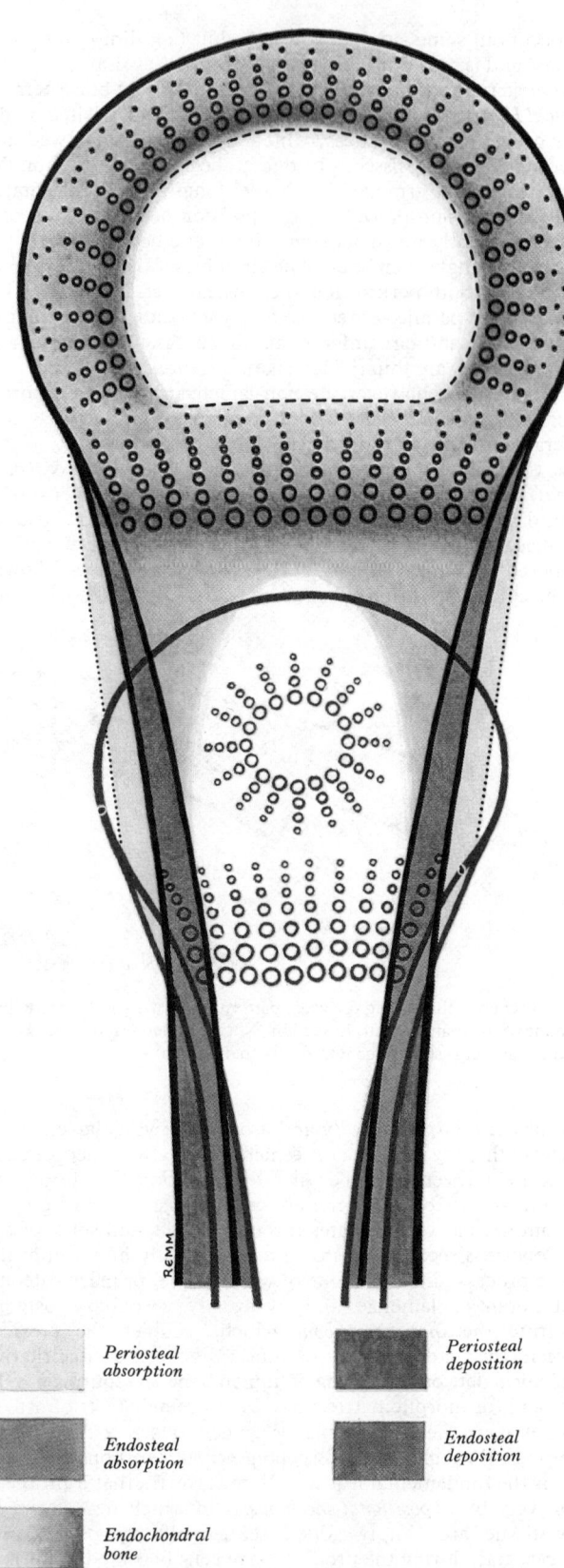

Periosteal absorption

Endosteal absorption

Endochondral bone

Periosteal deposition

Endosteal deposition

3.25 A scheme of the patterns of remodelling which occur during the growth of a long bone. Note the changing shape of the epiphysial ossification centre, the altered organization of the cartilaginous growth plates and the varying zones of bone deposition and absorption. See text for further details.

maturation, and endocrine disturbances may have profound effects. In addition to its normal role in calcium metabolism, parathyroid hormone in excess leads to osteocyte and osteoclast stimulation with widespread osteolysis and bone erosion leaving a demineralized framework—a condition known as *osteitis fibrosa*.

The eosinophil cells of the anterior part of the hypophysis cerebri secrete *growth hormone* and this is necessary for the continuance of normal proliferative interstitial growth of the growth cartilages and hence normal increase in stature. The mechanism whereby normal growth terminates is imperfectly understood but may result from a fall in hormone production or a lowered sensitivity of the cartilage cells to hormone action. Experimental hypophysectomy or, clinically, a reduction of growth hormone levels in the young, is accompanied by quiescence and thinning of the growth plates resulting in *pituitary dwarfism*. Conversely, continued hypersecretion of growth hormone in the same age group leads to *giantism*, but in the mature individual it results in an overgrowth and thickening of bones by subperiosteal deposition. Characteristically, the mandible and the skeleton of the hands and feet are particularly affected in this condition, known as *acromegaly*.

Whilst continued *growth* of the bones through the agency of their cartilages is dependent upon circulating growth hormone levels, effective remodelling and the assumption of a mature shape depends upon the concomitant action of thyroxin. Finally, the *rate* of growth and overall state of *maturity* is intimately related to the endocrine activities of the ovaries, testes and adrenal glands. High oestrogen levels are associated with an increased deposition of endosteal and trabecular bone, whereas the osteoporosis often seen in women of advanced years may reflect reduced ovarian function. The maturation level and growth rate which vary from phases of rapid change to relatively quiescent periods (p. 243) reflect the circulating levels of adrenal and testicular androgens. With hypogonadism, maturation including growth plate obliteration occurs late and the limbs become excessively long, whereas in hypergonadism, premature synostosis between epiphyses and diaphyses occurs with consequently a diminished stature.

Currently there is much interest and investigation into the natural history and pathology of growth cartilages and a large literature has accumulated. An exhaustive review of this field lies outside the scope of the present work, but several excellent accounts have appeared (Siffert 1956; Rang 1969; Serafini-Fracassini and Smith 1974).

General Features of Bones

The different bones of the skeleton vary amongst themselves not only in their *primary* or basic shape but also in their lesser details of surface structure, the *secondary* markings, which develop for the most part in postnatal life. Nevertheless, certain features, such as elevations and depressions, smooth areas and roughened ridges, appear in many different bones. It is convenient to use a repertoire of common terms for these common features; it is even more important to recognize that the same form of marking or the same surface texture usually has the same functional significance wherever it is encountered. For example, most bones display *articular surfaces* where they form joints with their neighbours; if small, these areas are termed *facets*, or occasionally *foveae*. Knuckle-shaped articular surfaces are called *condyles*, and a *trochlea* is a joint surface grooved like a pulley. Such areas are naturally adapted in shape to the characteristic movements of the joints in which they serve, and they are relatively smooth, though it must be remembered that in life they are covered by articular cartilages and that it is these which are primarily responsible for the smooth surfaces of synovial joints. The texture of articular surfaces is in fact due to another feature, frequently overlooked; they appear almost devoid of the small vascular foramina typical of many bone surfaces, articular cartilage itself being a poorly vascularized tissue. This does not, however, imply that osseous articular surfaces are completely impermeable, although the zone of cartilage contiguous with the subjacent bone is usually calcified, which makes any substantial interchange less probable (as described elsewhere, the nutritive avenue to articular cartilage may be largely via synovial fluid). It should be noted here, however, that with the advent of the scanning electron microscope it is now clear that *all* bony surfaces, including those that appear smooth when examined with the naked eye or hand lens, possess large numbers of minute foramina. What proportion

of these are occupied by vascular, nervous, or collagenous elements, is uncertain. (For surface architecture consult **3**.1D and **3**.150C.)

Substantial tendons (e.g. adductor magnus, subscapularis) are attached to bony facets which, though lacking the smooth planes or curves of articular surfaces, are similar to them in texture, and for the same reason, tendons being poorly vascularized. Such tendon facets are sometimes slightly depressed (the adductor tubercle of the femur is quite often replaced by a small pit), or they surmount considerable elevations, for example, the tubercles (tuberosities) of the humerus.

Depressions and elevations, varying in size and shape, frequently interrupt the otherwise featureless surfaces of bones. A depression is termed a *fossa*, and some articular surfaces are in fact fossae (cf. the temporomandibular joint). Elongated depressions are called grooves or *sulci* (e.g. the bicipital sulcus of the humerus); a notch-like depression is an *incisura*, and an actual gap is termed a *hiatus*. A projection of considerable size is usually referred to as a *process*; if particularly elongated, slender or pointed it may be termed a *spine*. A curved or hook-like process is named a *hamulus* or *cornu* (cf. the pterygoid hamuli of the sphenoid bone and the cornua of the hyoid bone). Any localized and roughly rounded projection is styled a *tuberosity* or *tubercle*, occasionally a *trochanter*. Elongated elevations are dubbed *crests*, or *lines* if less developed features; crests are usually wider and therefore may present boundary edges, or *lips*; however, these terms are not carefully differentiated, and one of the most substantial of crests is called the *linea aspera* (cf. femur). An *epicondyle* is a relatively small projection adjacent to a condyle and usually concerned with the attachment of collateral ligaments of the joint in which the condyle is involved (cf. humerus). The terms protuberance, prominence, eminence and torus·also have a more limited application to certain bony projections. The expanded proximal ends of many elongated bones are frequently spoken of as the '*head*' or *caput* (cf. humerus, femur, radius, etc.).

A hole through bone is a *foramen*; foramina are called *canals* when considerable in length. Larger holes are sometimes called *apertures*. Clefts in or between bones are usually known as *fissures*. A *lamina* is a thin plate of bone; larger laminae may be called *squamae* (e.g. the squama of the temporal bone).

Large areas on many bones are comparatively featureless, and indeed are often smoother to the touch than articular surfaces. They differ from the latter in being liberally speckled with small vascular foramina. Such a surface texture is observed wherever muscle tissue is directly attached, the foramina transmitting numerous small blood vessels which pass uninterruptedly from bone to muscle and perhaps vice versa. The same obtains in areas covered only by periosteum, but the vessels are much less numerous.

Tendons, as already noted, are usually associated at their point of attachment with some degree of roughening of the otherwise smooth bone surface. In the same way, wherever any considerable aggregation of collagen in a muscle reaches bone, the surface displays irregularities corresponding closely in form and extent to the pattern of such 'tendinous fibres', as they are often called. Such markings are almost always elevated to some degree from the general level of the surface, as if ossification were proceeding into the collagen bundles from the periosteal bone. It is, however, not yet certain what is the precise mechanism of production of such secondary markings. The evidence suggests that their prominence may be associated with the power of the muscles involved, and they certainly increase with advancing years, as if the pull of muscles, and indeed ligaments, may exercise an accumulative effect. However this may be, the surface markings on bone faithfully delineate the shape of attached connective tissue structures, whether it be obvious tendon, intramuscular tendon or septum, or aponeurosis, or merely tendinous fibres scattered through an otherwise direct muscular attachment. Hence the markings may be facets, ridges, nodules, roughened areas, or complex mixtures of these. Wherever they occur, they afford most accurate data of the junction of bone with muscle, tendon, ligament or articular capsule. Numerous examples of this will be encountered when the structures are individually considered.

It is important to add that even in a so-called fleshy or direct attachment of muscle to bone, the actual muscle fibres are not themselves adherent to periosteum or bone. The precise mechanism by which tension is transmitted from contracting muscle fibres to bone is, in some details, still uncertain, but it must be effected somehow through the connective tissue which pervades all muscles in the form of perimysium and endomysium. It is not clear how muscle fibres are attached, if indeed they are, to such collagenous elements in their structure; but it is certain that whether the attachment to bone is by tendon or is direct, the actual agent of attachment is connective tissue. (*See*, however, comments on the *myotendinal junction*—p. 515.) The difference between these two extreme forms of attachment—a considerable spectrum of intermediate admixtures exists—is in the amount of collagen fibres intermediating between muscle and bone, and their arrangement. Where collagen is concentrated in visible masses, visible markings appear on the bone surface at localized sites. In contrast, the multitude of microscopic connective tissue ties of a direct attachment of muscle, necessarily over a larger area, do not mark it appreciably. Hence the bone here appears smooth—at least, to unaided vision and to touch. (*See* 3.1D for magnified views of osseous surfaces.)

THE AXIAL SKELETON

Introduction

Division of the skeleton into axial and appendicular sections is not a mere arbitrary convenience, for the axial structures, the cranium and vertebral column and their associates, the ribs and sternum, constitute the primary skeleton; the skeletal elements of the appendages, in fins, limbs, or wings were subsequent though early additions, becoming of increasing importance with the elaboration of vertebrate locomotor habits. An axial endoskeleton, first as a notochord and then a vertebral column, provides the basic distinguishing feature of the phylum *Chordata* and its subphylum, the *Vertebrata*, to which, of course, mankind belong. A stiff but flexible axis, in bilaterally symmetrical animals with an early tendency to elongate shape, is an obvious locomotor advantage. It prevents telescoping of the body during the waves of contraction which pass through successive segmental muscles to produce the sinuous side-to-side flapping movements, especially in the tail, which are the basic mode of locomotion in aquatic vertebrates. A chain of ossified elements, flexibly connected by intervening discs of deformable substance, soon replaced the notochord by actually growing around it. It is due to this that notochordal vestiges occur in the vertebrae of many fish, amphibians and reptiles, and in the central part of the intervertebral discs of mammals. This replacement is repeated in every vertebrate embryo. The vertebral elements which grow round the notochord to replace it are complex and variable in pattern in earlier vertebrates, but in the reptiles onwards the most important is the *centrum*, which forms most of the vertebral body, ventral to the spinal cord (or spinal medulla). The plan of a typical vertebra also includes a *neural arch*, encircling the spinal cord, which fuses ventrally with the back of the centrum. The arch usually bears a midline dorsal projection, a *spinous process* or spine, and paired lateral *transverse processes* near the neurocentral junctions. This enclosure of the spinal cord isolates it from the activities of neighbouring axial musculature and affords protection from external forces.

The centrum and each half of the neural arch ossify from separate centres; and when these extend through their cartilaginous precursors to meet and fuse, the dorsolateral parts of the *body* of the fully developed vertebra are in fact formed from the ventral extremities of the neural arch. The terms centrum and body are therefore not synonymous, nor is the *vertebral arch* of a fully ossified vertebra exactly equal to a neural arch. A centrum is somewhat less than a vertebral body, a neural somewhat more than a vertebral arch (p. 282).

The segmental muscles which flex the vertebral axis are only in part attached directly to vertebrae; in the connective tissue septa (*myocommata*) between adjacent myotomes develop ribs as levers for such attachments. Such costal struts evolved in a dorsal position in the axial musculature and extended ventrally into the body wall in early vertebrates, and in fish ventral ribs also appeared, which in the tail enclosed the caudal vessels. It is generally agreed that the ribs of land tetrapods correspond to the dorsal series in fish (Romer 1970).

Ribs are intersegmental in position, and since the basic action of segmental muscles derived from myotomes is to bend the vertebral column, vertebrae also become intersegmental, though their embryonic development (p. 138) shows primarily a segmental pattern. In lower vertebrates there are dorsal ribs associated with most of the vertebrae, and the latter themselves show little regional adaptation except in the caudal region, posterior to the anal opening. With the appearance of land vertebrates, and the modification and elaboration of their appendages to various forms of locomotion, the vertebral column was exposed to new patterns of force in the distribution of weight and muscular tensions.

In the case of the hind limbs the pelvic girdle formed an articulation with several vertebrae, which fused with the ribs, or costal elements, of the region to form a sacrum. Such vertebrae lose their individual movement upon each other, and in mammals there are three to five *immovable sacral vertebrae*; distal to these are a variable number of *caudal vertebrae*, reduced to four degenerate elements which become fused into a common mass, the coccyx, in mankind during adult life.

The presacral vertebrae show considerable variation in mammals, in numbers and differentiation; but there is always a distinction into *cervical* (neck) vertebrae, *thoracic* vertebrae where ribs persist, and *lumbar* vertebrae devoid of mobile ribs. In the neck, ribs are small or 'absent' (but *vide infra*); their disappearance in the cervical and lumbar regions is in both cases linked to the change from gill to lung breathing and to the development of a neck for rapid and independent movements of the head upon the trunk. There is no true neck in fish, the immediate postcranial region being occupied by the branchial apparatus. The caudal shift of the respiratory apparatus in land vertebrates was a necessary precursor to the development of a neck, whose vertebrae became early stabilized to a series of seven in the mammals (except in tree sloths and manatees), despite such extremes of external form as whales and giraffes, which both conform. With the adaptation of ribs to respiratory function in land animals, and the development of a diaphragm in mammals, ribs as well-formed functional entities, articulating with but separate from vertebrae, become limited to the region thus defined as thoracic. It is not, however, true to state that ribs disappear completely in the *cervical* and post-thoracic, *lumbar* regions, because small vestiges, called *costal elements*, can be identified in association with the transverse processes of all these vertebrae, as will be apparent when they are individually considered (see also 2.50, and 3.29).

In primate mammals there is a general reduction in the total number of vertebrae, excluding the tail, from the lemuroid and tarsioid to the anthropoid primates. The latter suborder, including monkeys, apes and men (extinct and extant species), shows some uniformity of arrangement. The cervical, thoracic, lumbar and sacral vertebrae number respectively 7, 11 to 15, 4 to 7, and 3 to 6, the figures for mankind being 7, 12, 5 and 5.

With these brief preliminary remarks on the origin of the human vertebral column it is now appropriate to consider the individual elements in detail, beginning with the form and structure of a typical vertebra. Some reference to the phylogeny of the other components of the axial skeleton, the ribs and sternum, will be found in the sections describing them.

General Vertebral Characteristics

A typical vertebra (**3**.26) consists basically of an anterior or ventral part, the *body*, and a posterior or dorsal *vertebral (neural) arch* which is extended by lever-like processes and encloses a *vertebral foramen*, occupied in life by the spinal cord, the meninges, and associated vessels. The opposed surfaces of adjacent vertebral bodies are strongly bound to each other by *intervertebral discs* of fibrocatilage. In the complete column the bodies and discs form a continuous, flexible pillar which is the central axis of the body and supports, in a biped such as man, the weight of the head and trunk. It also transmits considerably greater forces due to the muscles attached to it directly and otherwise. The foramina afford a *vertebral canal* in which is the spinal cord. Between the arches of adjacent vertebrae, on each side, near their junction with vertebral bodies, are *intervertebral foramina* which transmit the spinal nerves and vessels.

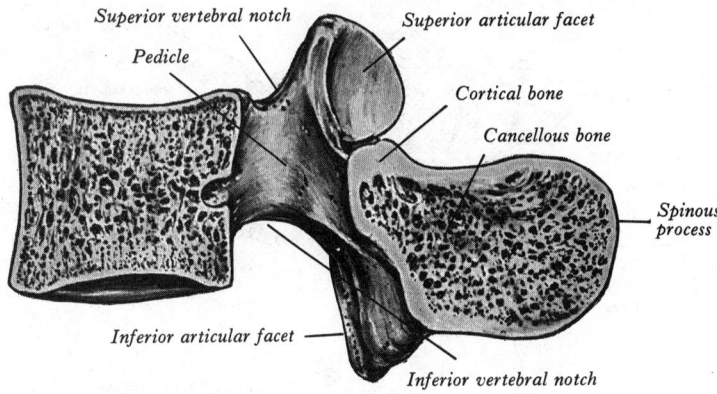

3.27 A median sagittal section through a lumbar vertebra.

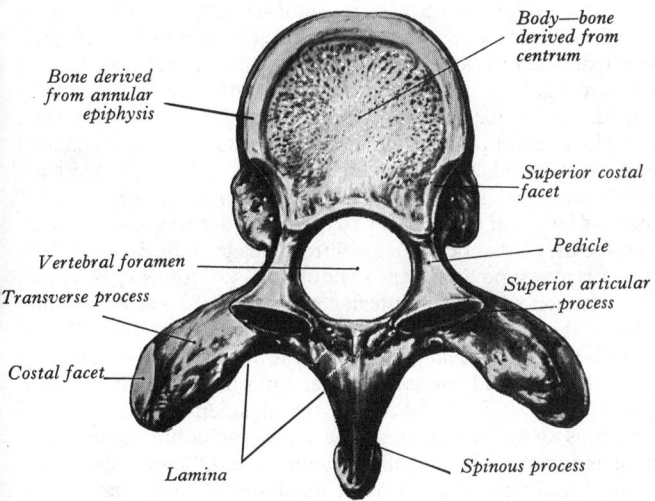

3.26 A typical thoracic vertebra. Superior aspect.

The vertebral body is roughly cylindrical, but varies widely in size, shape and proportions in different animals and in the different regions of the column of the same species. Its two surfaces are flattened and display a peripheral smooth zone formed from the annular epiphysial disc (p. 283), within which the surface is roughened. These differences in surface texture correspond to variations in the early structure of the intervertebral discs (p. 140). In the transverse plane the body is convex except where its dorsal aspect completes the vertebral foramen; here it is concave. Vertically it is concave except on its dorsal aspect, where it is flat. Small venous foramina show on the front and sides of the body; posteriorly there are a number of small foramina for arteries (Willis 1949) and a large irregular orifice (sometimes double) for the exit of basivertebral veins (**3**.27). It should be noted that the adult vertebral *body* is *not* coextensive with the developmental *centrum* (p. 140) but in addition, posterolaterally, includes parts derived from the neural arch.

The vertebral arch has on each side a vertically narrower anterior part, the *pedicle* and behind this the broader *lamina*. Projecting from it are seven processes, paired transverse and superior and inferior articular processes, and a single spinous process.

The two *pedicles* are short, thick and rounded bars projecting back from the body at the junction between its lateral and dorsal surfaces, and nearer the superior surface, so that the concavity above the pedicle is shallower than the one below it (**3**.27). These concavities are the *vertebral notches*; when vertebrae are articulated together by their discs, adjacent notches enclose an *intervertebral foramen*.

The two *laminae* are vertical, broad and plate-like, and are, of course, directly continuous with the pedicles. They curve

backwards and medially to fuse in the midline with the spinous process, thus completing the vertebral foramen.

The *spinous process* or spine projects dorsally and often downwards from the junction of the laminae. Spines vary much in size, shape and direction. They are levers for muscles which extend the vertebral column or, to a lesser extent, rotate it.

The *articular processes* (*zygapophyses*), paired superior and inferior, project from the vertebral arch at the junctions of pedicles and laminae. The superior articular processes jut upwards and bear surfaces which face dorsally and often laterally; the inferior processes bulge downwards and their articular facets are directed forwards and often somewhat medially. Thus, the articular processes of adjoining vertebrae meet to form small synovial joints (p. 445) which, while permitting a limited degree of movement, are particularly concerned in guiding and restricting the range of movement between vertebrae.

The *transverse processes* project laterally from the junctions of pedicles and laminae, and they serve as levers for the muscles and ligaments concerned in rotation and lateral bending of the spinal column. In the thoracic region they also articulate with ribs. In other parts of the vertebral column, the transverse process of mature vertebrae is a compound structure composed of the 'true' transverse process (*diapophysis*) and the incorporated costal element.

Costal elements (or *pleurapophyses*) develop as basic constituent parts of each vertebral arch in mammalian embryos, but they become independent units, ribs, only in the thoracic region. Elsewhere they remain less prominently developed and unrecognizable as ribs, fusing with their vertebrae, as part of the 'transverse process' of descriptive adult anatomy (*see*, for example, **3**.29).

Internally, vertebrae are composed throughout of trabecular bone (*see* **3**.1B), covered by an external shell of compact bone, which is traversed by numerous vascular foramina. The cortical bone is relatively thin on the upper and lower surfaces of the body but rather thicker in the arch and its processes. The trabecular interior of the body is traversed antero-posteriorly by one or two large canals containing the commencement of basivertebral veins.

The Cervical Vertebrae

The seven cervical vertebrae (**3**.28, 29, 34, 35) are the smallest movable vertebrae and identifiable by their transverse processes, which are perforated by a foramen. The first, second, and seventh vertebrae present special features and will be considered separately; the rest conform to a common type.

A typical cervical vertebra has a relatively small and transversely broad body; the *vertebral foramen* is comparatively large and triangular rather than round. In conformity with this the *pedicles* project somewhat laterally (**3**.28A) as well as backwards, and the *laminae* are angled markedly from them in a medial direction. Superior and inferior vertebral notches are almost equal in depth, the pedicles being attached about midway

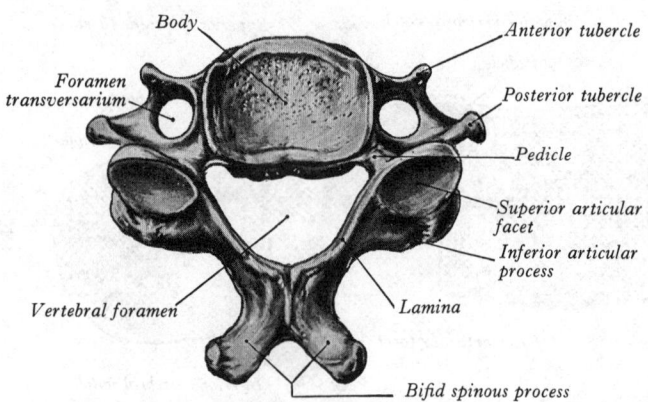

3.28A A typical cervical vertebra. Superior aspect.

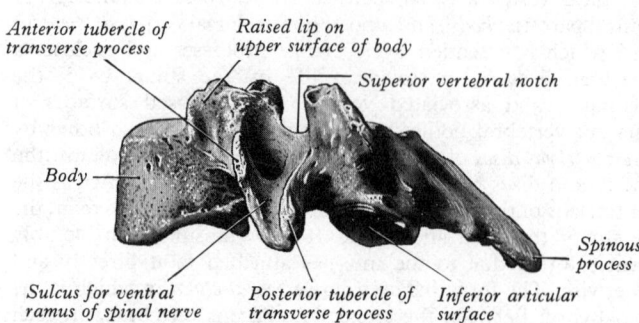

3.28B A typical cervical vertebra. Left lateral aspect.

between upper and lower borders of the body. The *laminae* are relatively long and narrow, with a thin upper border. The *spinous process* is short and bifid, with terminal tubercles often unequal in size. The *superior* and *inferior articular processes* form an *articular pillar*, which bulges laterally at the junction of pedicle and lamina. The '*transverse process*' of adult anatomy is morphologically a compound structure containing the *foramen transversarium*. It displays anterior and posterior *roots* or *bars* which terminate laterally as anterior and posterior *tubercles*. The roots are connected lateral to the foramen by an *intertubercular lamella* of bone (often, quite incorrectly, called the costotransverse bar). Of these separately named parts of the transverse process, all except the medial portion of the posterior root constitute the homologue of a *rib*. The medial moiety of the posterior root is homologous with a true transverse process (*diapophysis*) such as exists in a thoracic vertebra (**3.26**). Cave (1975) has examined the morphology of the mammalian, including the human, cervical 'costal element' or *pleurapophysis*, and has defined its extent more precisely, especially in relation to cervical ribs (*see* **3.29**, and p. 140). It should also be noted that the foramen transversarium may be divided into two in any of the cervical vertebrae.

The anterior surface of the body, convex transversely, is marked at its upper and lower margins by the attached fibres of the anterior longitudinal ligament, and on each side of the midline a slight depression shows the point of union of the vertical part of the longus colli. The flattened posterior surface displays centrally two or more large foramina for basivertebral veins. To its upper and lower borders the posterior longitudinal ligament is united in

life. The superior surface is transversely concave with a prominent lip on each side, and its anterior rim may be slightly bevelled. The inferior surface is reciprocally convex transversely and also slightly concave anteroposteriorly, i.e. it is saddle-shaped. Its bevelled lateral edge is separated in the living state from the lateral lip of the subjacent vertebra by a small synovial joint, oblique and cleft-like, which develops at about 9 to 10 years. It is limited medially by the intervertebral disc, laterally by a capsular ligament (Cave *et al.* 1955). Some authorities, however, describe the cleft as being in the fibrocartilage of the disc and deny its status as a synovial joint (Tondury 1943; Orofins *et al.* 1960). The anterior rim of the inferior surface of the vertebral body projects downwards in front of the intervertebral disc. To the upper borders of the laminae and the lower part of their anterior surfaces are bound the paired ligamenta flava. To the cervical spinous processes are attached the ligamentum nuchae and numerous deep extensors of the neck, including semispinalis thoracis and cervicis, multifidus, spinalis and the interspinales.

The dorsal rami of the cervical spinal nerves turn back close to the anterolateral aspects of the articular pillars and actually groove those of the third and fourth vertebrae. Their superior facets are oval and flat, and are directed backwards and upwards; in conformity the inferior facets, similar in shape, are turned forwards and downwards. In all but the seventh cervical vertebra the transverse foramen transmits the vertebral artery and its accompanying venous and sympathetic plexuses. The anterior root of the transverse process ends in a roughened tubercle to which are attached tendinous slips of scalenus anterior, longus

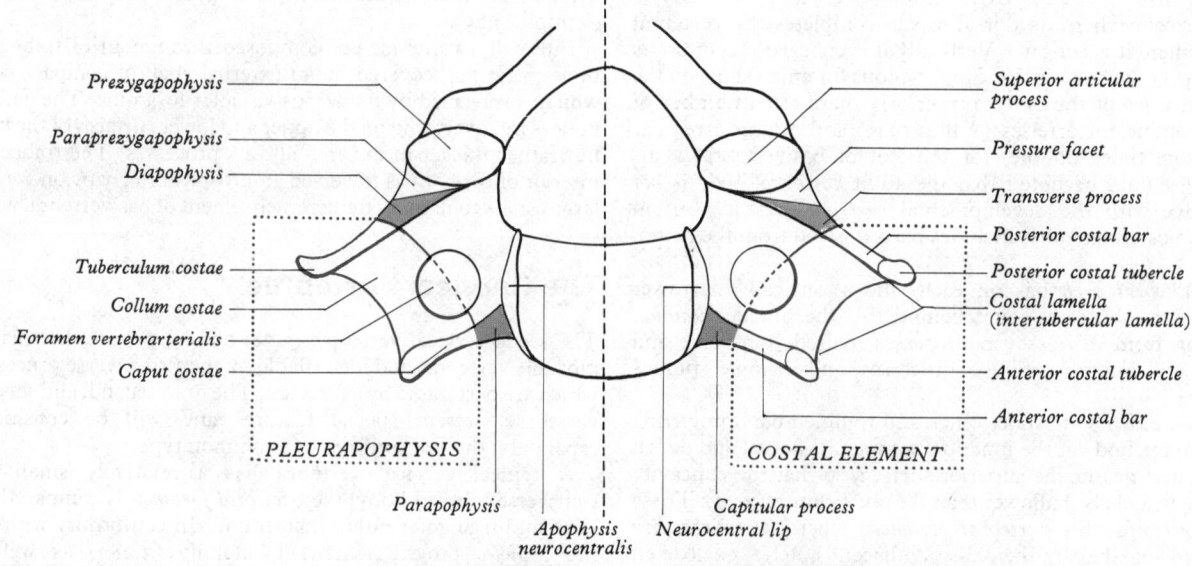

3.29 The morphology of a generalized cervical vertebra, with particular reference to the pleurapophyses. On the left the terms are zoological, on the right are alternatives for human anatomy suggested by Professor

A. J. E. Cave. (Reproduced by kind permission of the author, the *Journal of Zoology* and Cambridge University Press.)

capitis, and longus colli (5.36). The anterior tubercle of the sixth vertebra is particularly large, and the common carotid artery, being immediately anterior to it, can be effectively compressed here. It is hence called the *carotid tubercle*. The intertubercular lamella is oblique in the third and fourth cervical vertebrae, passing downwards, backwards and laterally; it is increasingly grooved from the fourth to the sixth by the emerging anterior rami of the spinal nerves. The groove across the intertubercular lamella of the sixth vertebra is conspicuously wide and shallow. The posterior root of the transverse process also ends in a rounded tubercle which is lateral to the anterior tubercle and at a slightly lower level in all except the sixth vertebra, in which the two tubercles are approximately level. Attached to the posterior tubercles are splenius, longissimus and iliocostalis cervicis, levator scapulae, and scalenus posterior and medius (the latter extending forwards along the intertubercular lamella).

The atlas, the first cervical vertebra (3.30), supports the 'globe' of the head and hence its name. It differs from all other vertebrae in lacking a body; its topographic position is occupied by the *dens* of the axis which forms a projecting pivot around which the atlas turns (p. 447). Moreover, it has no true spine, consisting of two bulky *lateral masses*, connected in front by a short *anterior arch* and posteriorly by a longer curved *posterior arch*. It is thus somewhat annular. When the atlas and axis are articulated together, the upward projection of the latter, the *dens*, is behind the anterior arch and is retained there in life by a *transverse ligament* (4.28). Some comparative morphological and developmental studies have thrown much doubt on the traditional view that the centrum of the atlas is secondarily fused with the axis as its dens (for discussion and bibliography *see* Jenkins 1969).

The anterior arch is slightly curved, its forward convexity being accentuated by a roughened, medially placed *anterior tubercle*. Attached to this is the anterior longitudinal ligament, and on each side of this the superior oblique part of the longus colli. The upper and lower borders of the arch provide attachment respectively to the anterior atlanto-occipital membrane and the lateral parts of the anterior longitudinal ligament.

The *lateral masses* converge, their long axes running forwards and medially. Each bears on its upper surface an elongated, somewhat reniform, concave facet jointing with the corresponding occipital condyle. The facet is usually constricted near its middle and may be divided into two. It faces upwards and medially and is adapted to nodding and lateral movements of the head which occur at these two atlanto-occipital joints. The inferior surface of each lateral mass presents an almost circular facet, flat or slightly concave, which articulates with a similar superior articular facet of the axis. The inferior facets of the atlas face down, medially, and slightly backwards. Attached to the edges of superior and inferior facets are the corresponding articular capsules. The medial aspect of each lateral mass has a small, rough tubercle for the transverse ligament of the atlas (3.30), which passes behind the dens and retains it in place. This ligament divides the cavity of the atlas into a smaller, anterior part, containing the dens, and a larger, posterior region, occupied by the spinal cord and meninges. The anterior aspect of the lateral mass has the rectus capitis anterior attached to it.

The *posterior arch* accounts for about two-fifths of the ring of the atlas. Its upper surface shows a wide groove behind each lateral mass, behind the overhang of which the vertebral artery winds (3.30). Quite frequently (about 37 per cent of 66 macerated specimens) the groove is partly or wholly converted into a foramen by a spicule of bone arching backwards from the upper surface of the lateral mass (for details *see* Lamberty and Živanović 1973). The first cervical spinal nerve also lies in the groove, between the artery and bone. Behind the two grooves the posterior atlanto-occipital membrane joins the posterior arch, and to its lower margin are attached the highest pair of ligamenta flava. The posterior tubercle of the arch represents a spinous process; to its roughened surface are attached the ligamentum nuchae and flanking this the two recti capitis posteriores minores.

The *transverse processes* are unusually long (3.34), and hence the width of the atlas exceeds that of all cervical vertebrae save the seventh. The length and strength of these processes makes them adequate levers for the muscles which aid in rotation of the head.

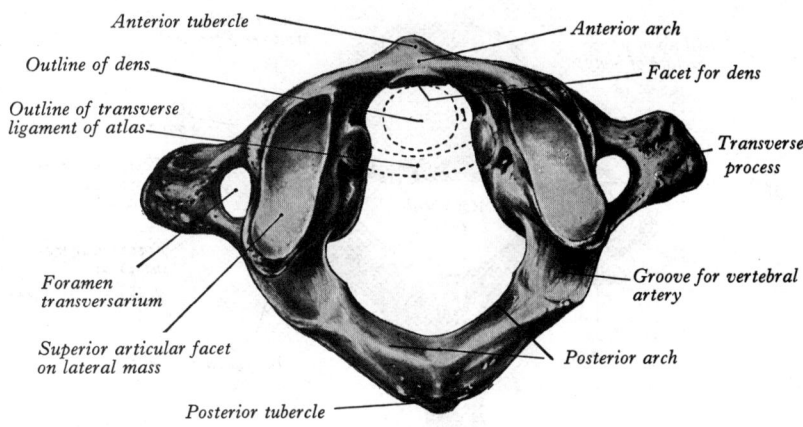

3.30 The first cervical vertebra, or atlas. Superior aspect.

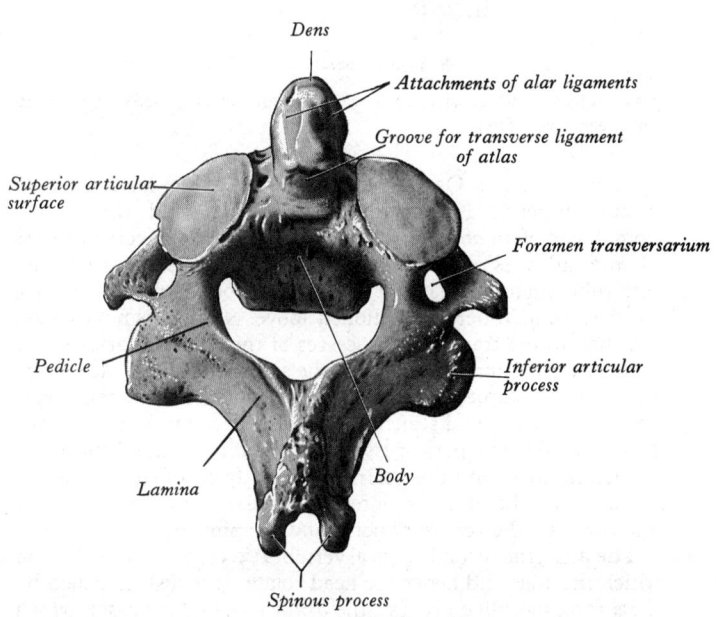

3.31 The second cervical vertebra, or axis. Posterosuperior aspect.

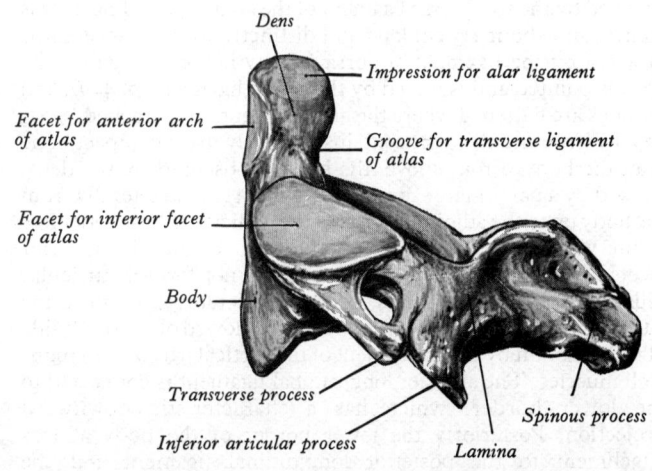

3.32 The second cervical vertebra, or axis. Left lateral aspect.

(The maximal width of the atlas shows a range of 74 to 90 mm in the male and 65 to 76 mm in the female of European and American whites, providing a useful criterion of sex in identifying human remains.) The apex of the transverse process is homologous with the *posterior* tubercle in a typical cervical vertebra; it can be felt through the overlying tissues between the mastoid process and

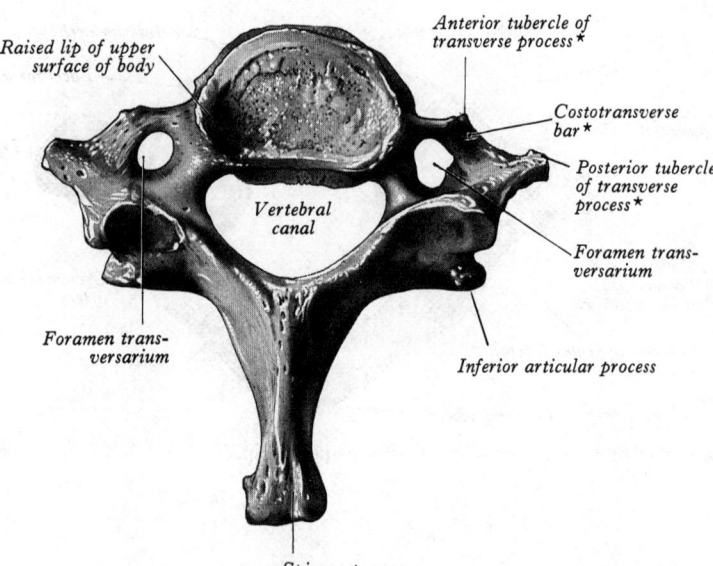

3.33 The seventh cervical vertebra. Superior aspect. (*See text and **3.29** for alternative terms.)

mandibular angle. On the anterior aspect of the lateral mass a minute tubercle is sometimes perceptible, and this is the homologue of an *anterior* tubercle. Hence the transverse process of the atlas is the equivalent of the posterior root and intertubercular lamella only, the anterior root being absent, save for the minute tubercle mentioned above. Numerous muscles are attached to the transverse processes of the atlas; superiorly, the rectus capitis lateralis in front of the superior oblique, at the tip, the inferior oblique, laterally and below, slips of levator scapulae, splenius cervicis, and scalenus medius. The ventral ramus of the first cervical spinal nerve passes forwards on the lateral surface of the lateral mass and then turns down in front of the transverse process and behind the internal jugular vein, crossed here anteriorly by the accessory nerve and occipital artery.

The axis, the second cervical vertebra (**3.31, 32**), is the pivot on which the atlas and hence the head rotate. It is distinguished by the strong toothlike process, the *dens* (odontoid process), which juts vertically upwards from the body. On the front of the dens is a small oval articular facet which forms a joint with the similar facet on the back of the anterior arch of the atlas; posteriorly the dens is grooved by the transverse ligament of the atlas (**3.31**). The dens is conical and about 1·5 cm long and distinctly waisted, where it is grooved by the ligament, a bursa usually being interposed. Its apex is pointed and is joined by the apical ligament (p. 449), and its sides are flattened where the alar ligaments are attached (**3.31, 32**). In structure the dens contains relatively more compact bone than the body. From above, the body is obscured by the dens, flanked by a pair of large, oval facets, which extend laterally from the body on to the adjoining parts of the pedicles. These articulate in life with the inferior facets of the atlas. Unlike the superior facets of other cervical vertebrae, they do not form an articular pillar with the inferior facets, being considerably anterior to the latter. The anterior aspect of the body is hollowed out on each side of the midpoint by the attachment of the vertical part of the longus colli muscles. The anterior longitudinal ligament is connected to the lower border, which has a characteristic downward projection. Posteriorly the lower border of the body affords attachment to the posterior longitudinal ligament and the membrana tectoria (p. 449). The pedicles are stout, the inferior vertebral notches deep, and the superior ones scarcely discernible. The laminae are thicker than in any other cervical vertebra and provide attachment to ligamenta flava. They fuse behind with a large and powerful spinous process which takes the pull of muscles which extend, retract, and rotate the head. The inferior oblique arises from a rough impression on each side of the spine, and the rectus capitis posterior major a little posterior to this. The ligamentum nuchae is attached to the broad gap at the apex of the spine, together with slips of the semispinalis and spinalis cervicis, interspinalis, and multifidus.

The transverse processes of the axis are small; their blunt tips present a single tubercle, a homologue of the posterior tubercle at other levels. The anterior tubercle is at or near the junction of the anterior root of the transverse process with the body, as in the atlas. To the tip of the process is attached levator scapulae, between scalenus medius in front and splenius cervicis behind. To its upper and lower surfaces are attached the corresponding intertransverse muscles (p. 545). Each transverse foramen is directed superolaterally, because the vertebral arteries must deviate laterally as they pass up to the more widely separated foramina in the atlas. The inferior articular facets are carried at the junction of pedicles and laminae and face downwards and forwards, as in a typical cervical vertebra. The vertebral foramen is relatively large.

In a study of 60 cervical vertebral series (Krmpotić-Nemanić and Keros 1973) the longitudinal axis of the dens relative to the body of the vertebra varied according to the cervical vertebral curvature, and also showed changes with age.

For a discussion concerning the homologies of the transverse processes of the atlas and axis, and of the atlanto-occipital and atlanto-axial joints and their neuromuscular relationships, consult Cave (1975).

The seventh cervical vertebra (**3.33**) is sometimes called the *vertebra prominens*; its long spinous process is visible through the skin at the lower end of the nuchal furrow. (In fact, the spine of the first thoracic vertebra is usually just as prominent and sometimes more so.) The seventh cervical spine is thick, almost horizontal, and ends in a single tubercle, to which the lower end of the ligamentum nuchae is tied. Also attached near the tip of the spine are trapezius, rhomboideus minor, serratus posterior superior, splenius capitis, spinalis cervicis, semispinalis thoracis, multifidus, and interspinales. The transverse processes are large, particularly their prominent posterior parts. Anterior to the transverse foramina, which are relatively small and may be reduplicated, the anterior part of the process is slender and shorter. It is the costal element and may be a separate bone, a cervical rib. The intertubercular lamella anterolateral to the foramen (**3.33**), is grooved superiorly by the ventral ramus of the seventh cervical nerve; it is often partly deficient. The prominent posterior tubercle of the transverse process provides attachment

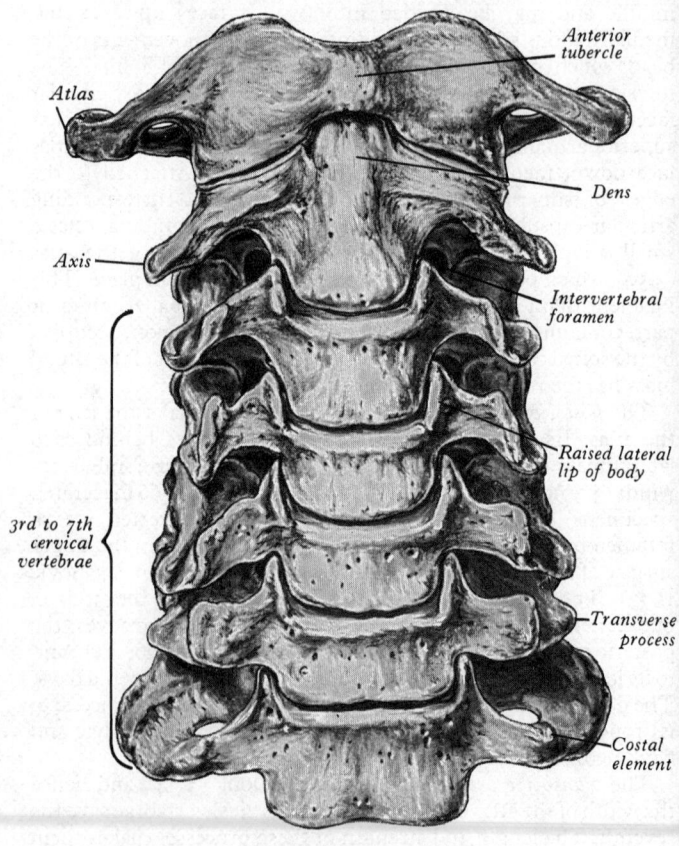

3.34 The cervical vertebrae. Anterior aspect.

r scalenus minimus, when present, and also to an aponeurotic yer which covers the cervical dome of the pleura, the prapleural membrane. The first pair of levatores costarum are nnected to the lower edge of the transverse processes.

he Thoracic Vertebrae

he twelve thoracic vertebrae (**3**.26, 36, 37, 38) show a gradual crease in size downwards, like other vertebrae, an expression of e increasing load carried from head to sacrum. All are stinguished by costal facets on the sides of their bodies, and all ut the last two or three by facets on their transverse processes. hese articulate respectively with the heads and tubercles of ribs. he first and the ninth to twelfth thoracic vertebrae possess rtain peculiarities and will be considered separately; the rest, espite minor differences, may be regarded as 'typical'.

The body of a typical thoracic vertebra is a waisted cylinder xcept where the vertebral foramen encroaches upon it behind; its ansverse and anteroposterior dimensions are therefore almost qual. On each side it bears two *costal facets*; the superior pair are sually larger and are at the upper border in front of the root of the edicle; the inferior facets are at the lower border immediately in ont of the vertebral notch (**3**.36). The *vertebral foramen* is latively small, and its circular outline can be linked with the fact at the *pedicles* do not diverge as in cervical vertebrae and that the noracic part of the spinal cord is smaller and more nearly circular. he *laminae* are hence also short, and they are thick, broad, and verlap each other from above downwards. The typical *spinous rocess* slants back and downwards. The superior *articular rocesses*, thin plates of bone, project upwards at the junctions of edicles and laminae; they are almost flat and face back, and a little terally and upwards. The *inferior processes* are downward rojections of the laminae, their facets directed forwards and a ttle medially and upwards. The *transverse processes* are ubstantial and club-shaped projections from the vertebral arch at nctions of pedicles and laminae. They point laterally and omewhat backwards; they bear, anteriorly near their tips, oval cets which articulate with the tubercles of the numerically orresponding ribs.

The first thoracic vertebra (**3**.37) is distinguished by the pper costal facets on its body, for these are circular, articulating ith the whole of the facet on the head of each first rib, hereas the inferior facets are smaller and semilunar; they rticulate, as is usual through the thoracic series, with a demifacet n the head of a rib. The spine is thick, long and horizontal, and is ommonly as prominent as that of the seventh cervical vertebra.

The ninth thoracic vertebra (**3**.37), though otherwise ypical, often fails to form joints with the tenth ribs, and the nferior demifacets on the body are then absent.

The tenth thoracic vertebra (**3**.37) articulates only with enth pair of ribs, taking no part in the costovertebral joints of the leventh pair. Superior facets only are therefore present on the ody, and these are usually large and semilunar, but they are oval hen the tenth ribs fail to reach the ninth vertebra and ntervening disc. The transverse process may or may not present a cet for the tubercle of the tenth rib.

The eleventh thoracic vertebra (**3**.37) articulates only with ne eleventh ribs by their heads. The circular costal facets are lose to the upper border of the body and extend slightly on to the edicles. The transverse processes are small and lack articular acets.

The twelfth thoracic vertebra (**3**.37) forms joints with the eads of the twelfth ribs by circular facets somewhat below the pper border of the body and spreading on to the pedicles. The ody is large and approximates to the lumbar type. The ransverse processes are insignificant and present lumbar features *vide infra*).

The bodies of the upper thoracic vertebrae show a gradual ransition from cervical to thoracic type, while the bodies of the ower thoracic vertebrae show a similar transition from thoracic to umbar characteristics. The body of the first is cervical in form, its ansverse is nearly twice as great as its anteroposterior easurement. The body of the second retains the cervical shape,

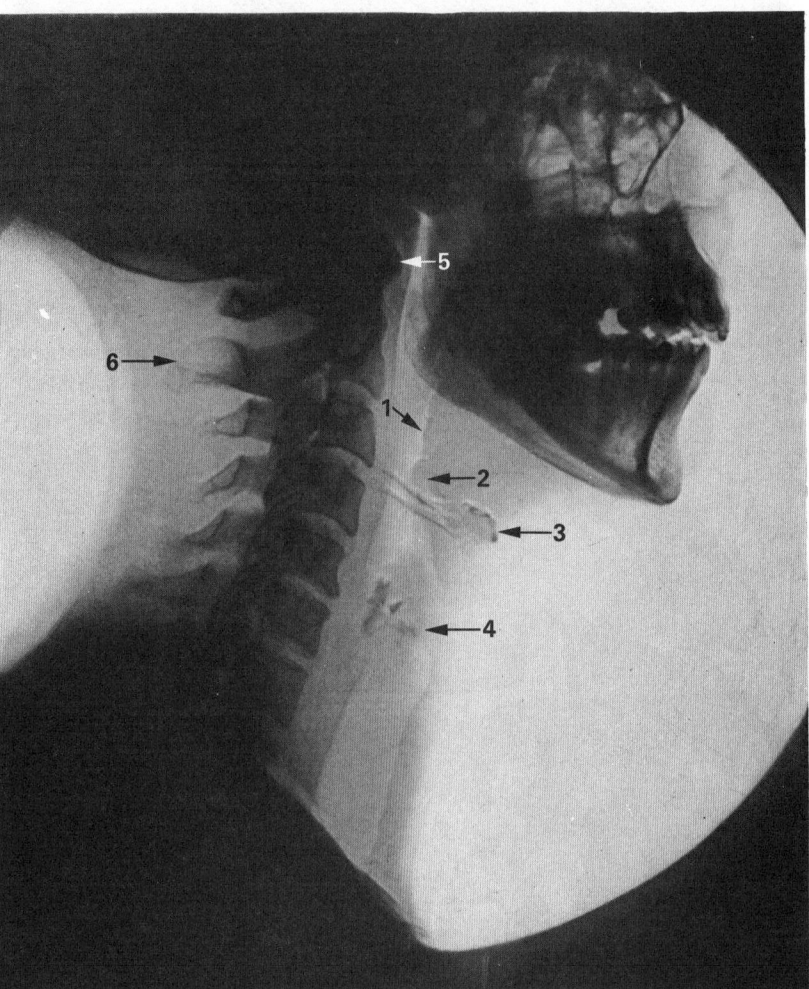

3.35   Lateral radiograph of the neck. Positive print. The cervical curve of the vertebral column is well shown. The arrows point (1) to the pharyngeal part of the tongue; (2) to the epiglottis; (3) to the body of the hyoid bone; (4) to the thyroid cartilage, which is undergoing calcification; (5) to the anterior tubercle of the atlas; (6) to the spinous process of the axis.

In front of the body of the first thoracic vertebra, a small part of the apex of the lung shows as a narrow, clear area which crosses the opacity caused by the oesophagus and encroaches very slightly on the broad clear area caused by the presence of air in the trachea.

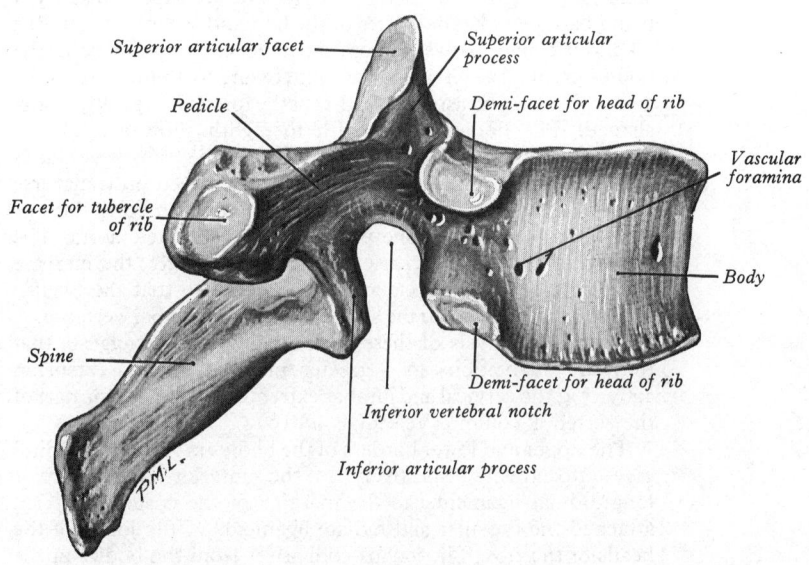

3.36   A typical thoracic vertebra. Right lateral aspect.

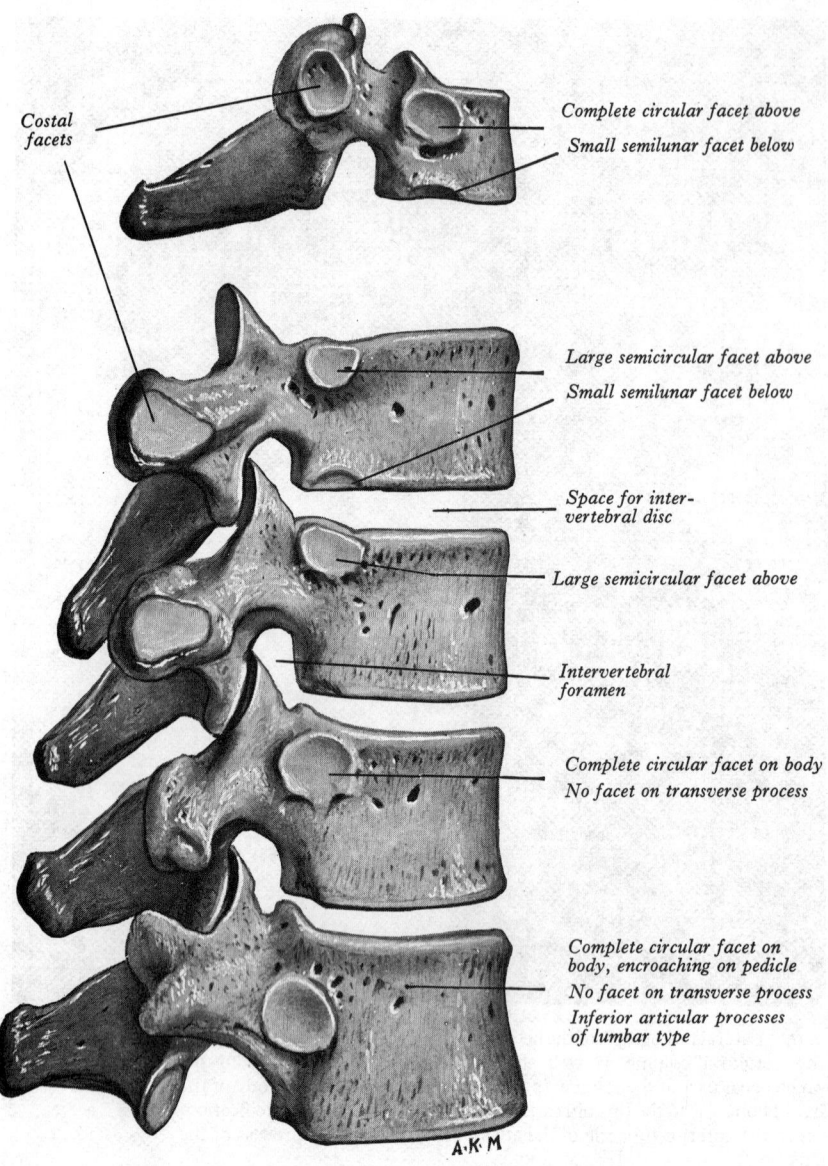

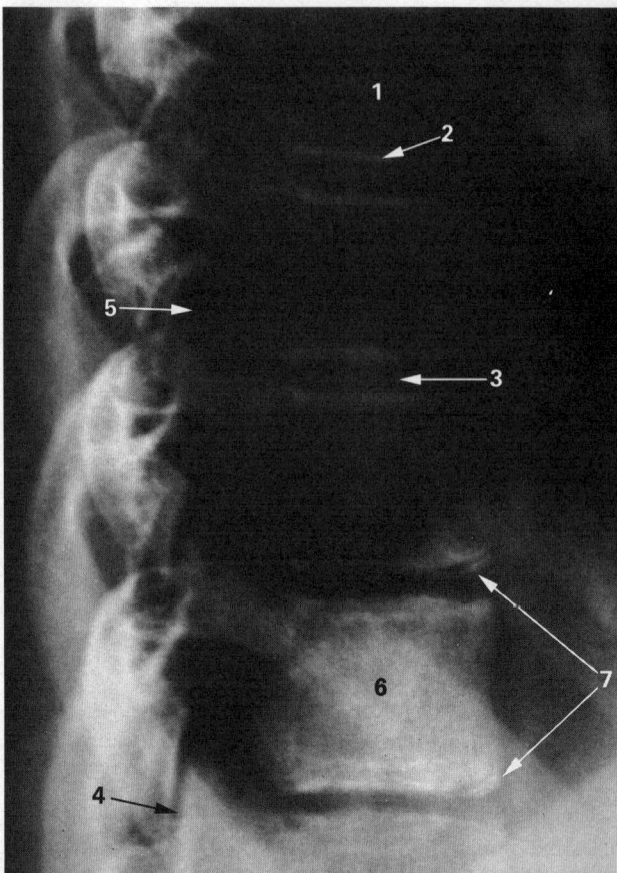

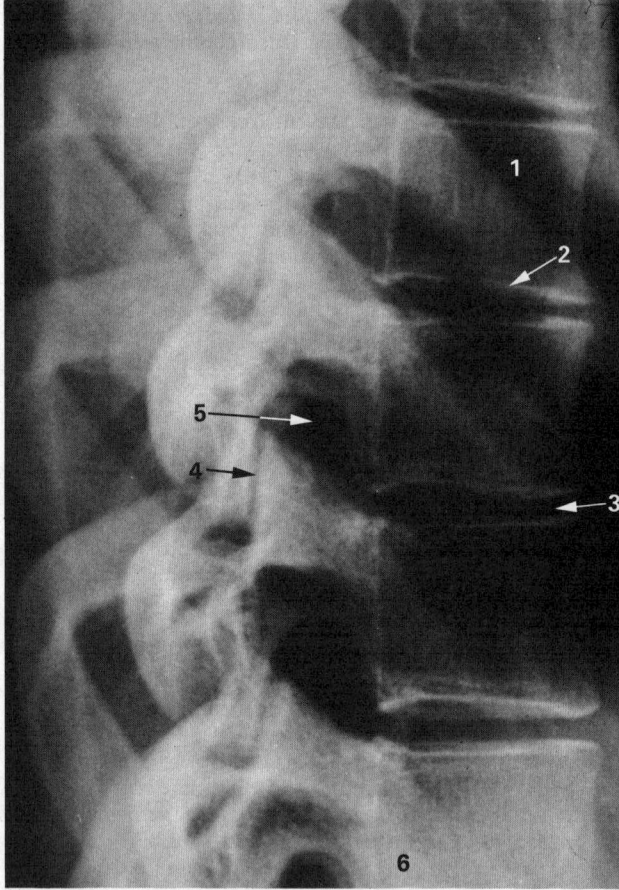

Costal facets

Complete circular facet above

Small semilunar facet below

Large semicircular facet above

Small semilunar facet below

Space for inter-vertebral disc

Large semicircular facet above

Intervertebral foramen

Complete circular facet on body

No facet on transverse process

Complete circular facet on body, encroaching on pedicle

No facet on transverse process

Inferior articular processes of lumbar type

A·K·M

3.37   The first, ninth, tenth, eleventh and twelfth thoracic vertebrae. Right lateral aspect.

but its breadth is less and the disproportion between its two measurements is diminished. The body of the third is actually the smallest of the thoracic bodies, but its anterior aspect, instead of being flattened like the bodies of the first and second, is rounded off and is convex forwards from side to side. From this point the bodies gradually increase in size and, owing to an increase in the anteroposterior measurement, that of the fourth is typically heart-shaped. The bodies of the fifth to eighth show a gradually increasing anteroposterior dimension while the transverse shows little alteration. These four vertebrae, when seen on transverse section, are asymmetrical, for the left side of each body shows a flattening produced by the pressure of the thoracic aorta. The remaining vertebrae increase in size more rapidly, the increase affecting all the measurements of the body, so that the twelfth approximates closely to the shape of a typical lumbar vertebra. A geometrical analysis of these regional differences suggests that they are adaptations to a greater range of flexion-extension activity at the cervical and lumbar extremes of the thoracic part of the vertebral column (Veleanu *et al.* 1972).

The upper and lower borders of the bodies in front and behind give attachment respectively to the anterior and posterior longitudinal ligaments; to the margins of the costal facets are attached the capsular and radiate ligaments of the joints of the heads of the ribs. The longus colli arises from the bodies of the first three thoracic vertebrae, lateral to the anterior longitudinal

3.38A and B   Lateral radiographs of mid-thoracic vertebral column in girl of 12 years (A) and an adult female of 22 years (B). 1. Cancellous bo of vertebral body. 2. Shell of compact bone. 3. Site of intervertebral dis 4. Synovial joint between articular processes. 5. Intervertebral forame 6. Superimposed shadow of rib. 7. Ossification occurring in unfuse annular epiphyses of vertebral bodies.

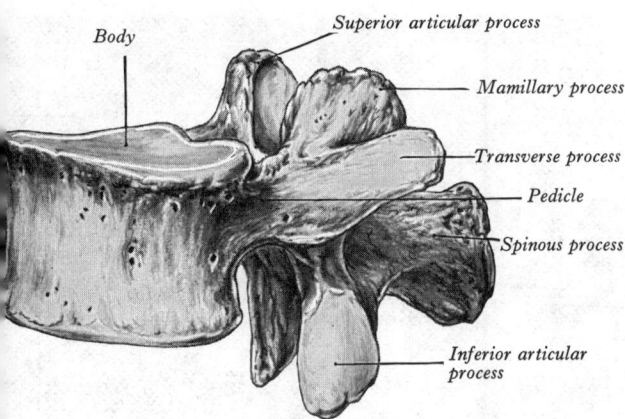

3.39   A lumbar vertebra. Left lateral aspect.

Usually these two vertebrae can be distinguished by the size and character of the transverse process and the distance between the costal facet and the upper border of the vertebra (p. 275).

The change in the character of the articular processes from the thoracic to the lumbar type occurs suddenly, usually in the eleventh thoracic vertebra, but sometimes in the twelfth or tenth, at one of which the articular processes commonly interlock in tenon and mortice fashion. In the transitional vertebra the superior articular processes are thoracic in type; their articular facets are directed backwards and a little laterally and upwards, while the inferior articular processes are turned laterally and their facets are slightly convex from side to side, facing laterally and forwards. This vertebra marks the site of a sudden change in function (see p. 446, and Davis 1955).

The Lumbar Vertebrae

The five lumbar vertebrae (3.39, 40, 41) differ from other vertebrae in their greater size and absence of costal facets and transverse foramina.

The *body* is large, wider from side to side and a little deeper in

gament. The psoas major and minor arise from the lateral aspect f the body of the twelfth near its lower border.

The pedicles increase in thickness from above downwards. The uperior vertebral notch is scarcely recognizable except in the first horacic vertebra, but the inferior notch is deep and conspicuous. The upper borders of the laminae and the lower parts of their nterior surfaces serve for attachment of the ligamenta flava; their orsal aspects give insertion to the rotatores muscles.

The transverse processes gradually diminish in length from bove downwards. In the upper six (sometimes five) the costal acets are concave and face forwards and laterally; in the others the acets are flattened and face upwards, laterally, and slightly orwards. The anterior surface of the process medial to the facet ;ives attachment to the costotransverse ligament, its tuberculated xtremity to the lateral costotransverse ligament and its lower order to the superior costotransverse ligament. In addition, the upper and lower borders of the transverse process provide ttachment for intertransverse muscles or their fibrous vestiges, nd the posterior surface for the deep muscles of the back, the evator costae arising from the dorsal aspect of the tuberculated xtremity under cover of the longer muscles.

The spines overlap from the fifth to the eighth, which is the ongest and most nearly vertical of the thoracic spines. Above and elow they are less oblique in direction. (In quadrupeds the najority of the spines of the thoracic vertebrae project dorsally nd caudally, while those in the lumbar region are directed lorsally and cranially. The change in inclination is effected in one of the lower thoracic vertebrae, the spine of which points almost traight dorsally. This vertebra is known as the *anticlinal*, and in nan its representative is the eleventh thoracic.) In man the upraspinous and interspinous ligaments, the trapezius, rhomb- oideus major and minor, latissimus dorsi, the serratus posterior uperior and inferior, and many of the deep muscles of the back re attached to the thoracic spines.

The first thoracic resembles a cervical vertebra in the shape of ts body. In addition, the posterolateral parts of its upper border re raised, as they are in the cervical region, and this projection orms the anterior border of the superior vertebral notch, which is a distinctive feature of this vertebra. The upper facet on the side of the body is not always complete, as the head of the first rib often rticulates with the intervertebral disc between the seventh cervical and the first thoracic vertebra. Immediately below the facet there is frequently a small, deep depression in the bone.

In the eleventh and twelfth thoracic vertebrae the spinous processes are characteristically triangular, with blunted apices. In each case, the lower border is horizontal, or nearly so, and the upper border is oblique. In the region of the transverse process of the twelfth thoracic vertebra three small tubercles can be distinguished. Of these the superior is the largest and juts upwards. It corresponds to the mamillary process of a lumbar vertebra, but it is not so closely connected with the superior articular process. The lateral tubercle is small and corresponds to the true transverse process. The inferior tubercle, directed downwards, corresponds to the accessory process of a lumbar vertebra.

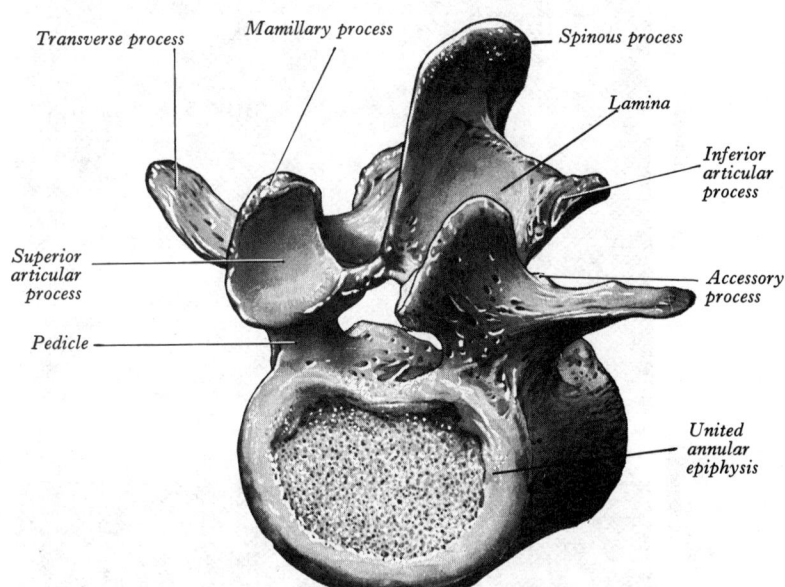

3.40   A lumbar vertebra. Posterosuperior aspect, viewed obliquely from the left side.

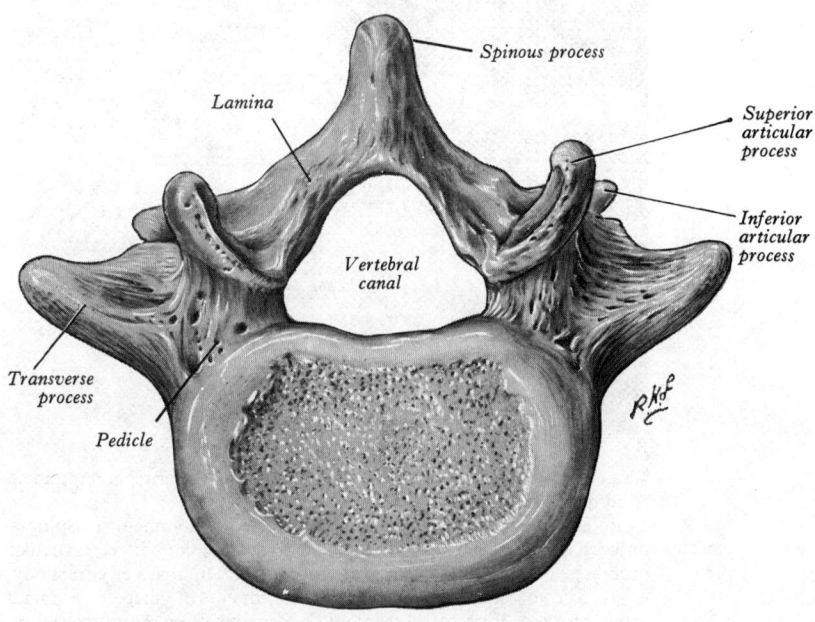

3.41   The fifth lumbar vertebra. Superior aspect.

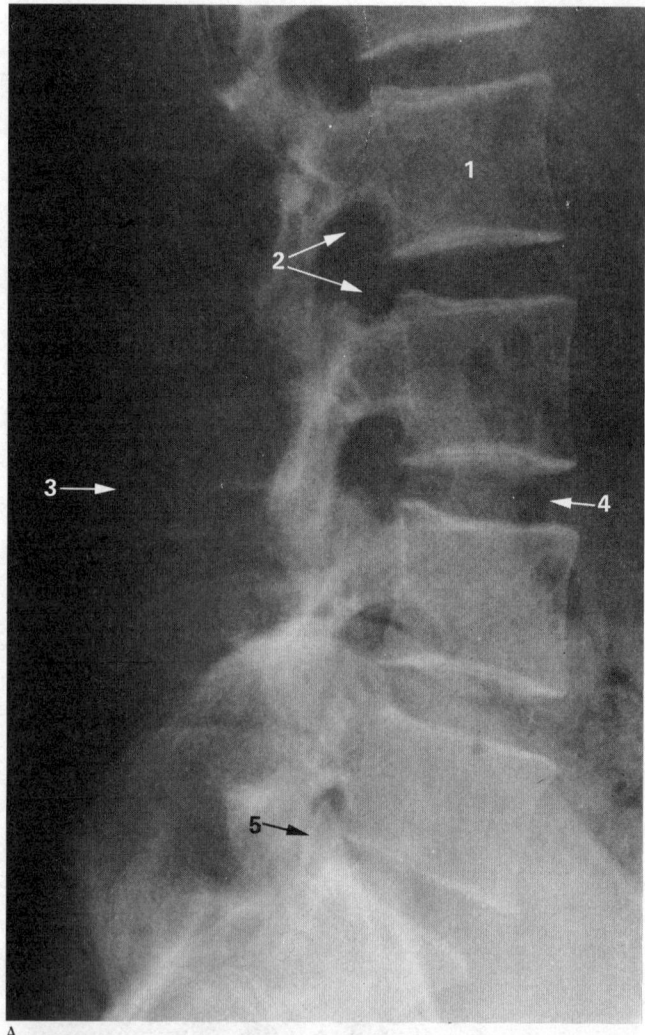

A

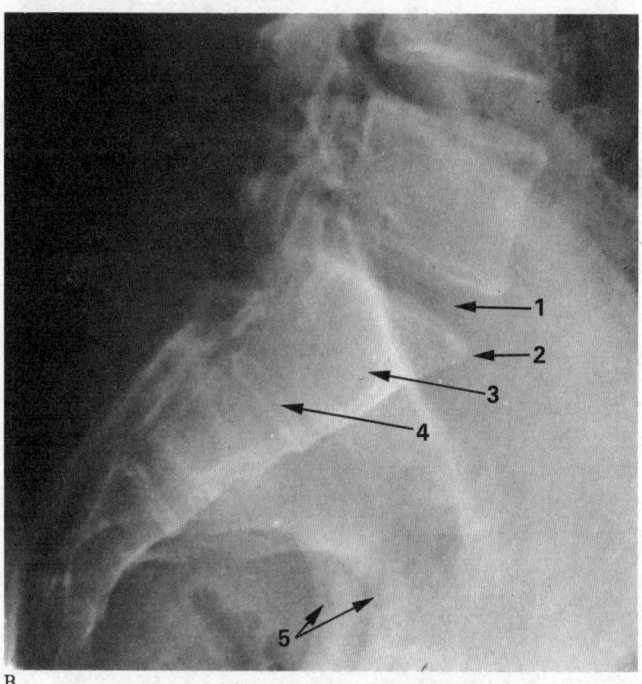

B

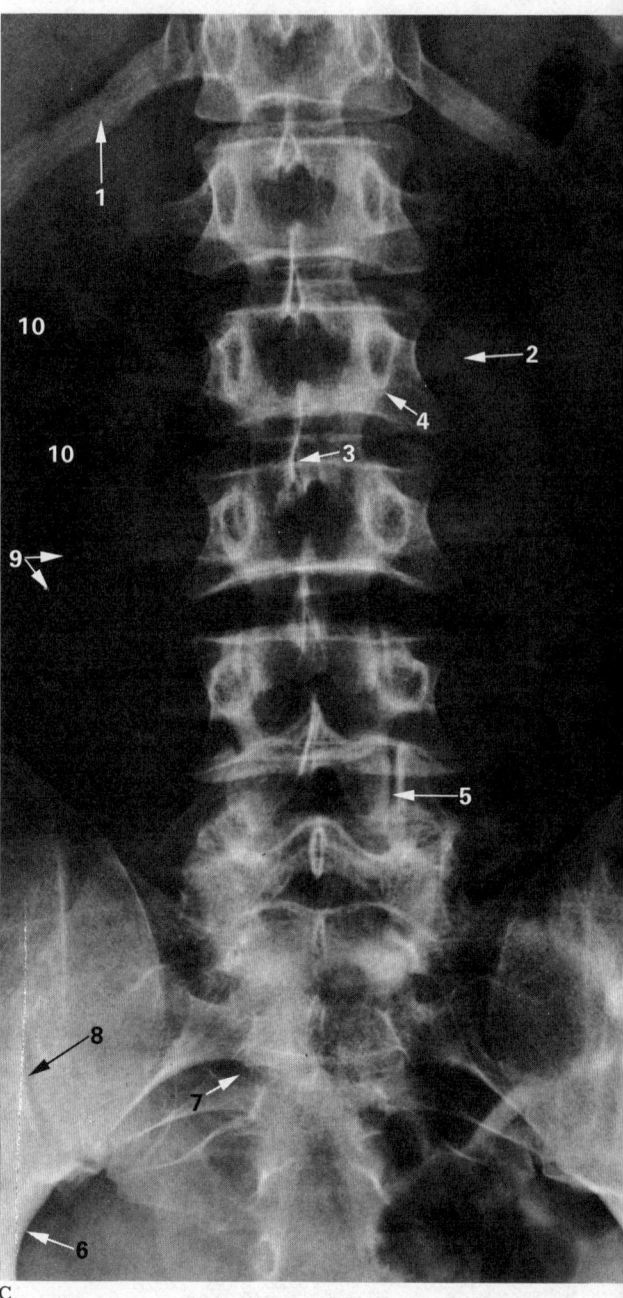

C

3.42C Anteroposterior radiograph of lumbosacral vertebral column in young adult male aged 22 years. 1. Twelfth rib. 2. Transverse process. 3. Spinous process (2nd lumbar). 4. Compact bony shell of pedicle (2n lumbar). 5. Joint between articular processes (4th and 5th lumbar). Pelvic brim. 7. Anterior sacral foramen. 8. Sacro-iliac joint. 9. Latera border of psoas major. 10. Gas in colon.

3.42A and B Lateral radiographs of lumbosacral vertebral column in an adult male aged 26 years.

A 1. Lumbar vertebral body. 2. Intervertebral foramen. 3. Spinous process. 4. Site of intervertebral disc. 5. Synovial joint between articular processes. Note slightly cuneiform profile of fifth lumbar vertebral body.
B 1. Site of lumbosacral disc. Note cuneiform shape. 2. Sacral promontory. 3. First sacral segment. 4. Remains of sacral intervertebral disc. 5. Profiles of greater sciatic notches.

front. The *vertebral foramen* is triangular, larger than in th thoracic but smaller than in the cervical region. The *pedicles* ar short. The *spinous process* is almost horizontal, quadrangular, an thickened along its posterior and inferior borders. The *superio articular processes* bear slightly concave vertical articular facet which face medially and backwards. The posterior border of eac has a rough elevation, the *mamillary process*. The *inferior articula processes* have slightly convex vertical articular facets which fac laterally and forwards. The *transverse processes* are thin and long except those of the fifth vertebra, which are more substantial. small, rough *accessory process* marks the postero-inferior aspect o the root of each transverse process. (Measurement of the third an fourth lumbar vertebrae in 338 skeletons, male and female, age 20–90 years demonstrates that the *breadth* of the body increase with age. In males, the posterior height shows a relative decrease and in both sexes the anterior height of the vertebral bod decreases relative to breadth—Ericksen 1976.)

The fifth lumbar vertebra (**3**.41) is distinguished by its massive *transverse process;* it is connected to the *whole of the lateral suface of the pedicle and encroaches on the side of the body.* Its body is usually the largest of the lumbar vertebrae and it is markedly deeper in fronter than behind, in association with the prominence of the sacrovertebral angle.

The upper and lower borders of lumbar *bodies* in front and behind give attachment to the anterior and posterior longitudinal ligaments. Lateral to the anterior longitudinal ligament, the bodies of the upper lumbar vertebrae (three on the right side, two on the left) give origin to the crura of the diaphragm. Behind this the psoas major arises from the upper and lower margins of the bodies of all the lumbar vertebrae; tendinous arches carry its attachments across the concave sides of the bodies. The vertebral foramen of the first lumbar vertebra contains the lower part of the spinal cord—the conus medullaris; the lower foramina contain the cauda equina and the spinal meninges. The pedicle is strong, and springs from the posterolateral aspect of the body just below its upper border. The superior vertebral notch is shallow; the inferior notch is deep. The laminae are broad, short and strong, but do not overlap one another to the same extent as in the thoracic region. They give attachment to the ligamenta flava. The spinous processes provide attachment for the posterior lamella of the thoracolumbar fascia, the erector spinae, spinalis thoracis, multifidus, interspinal muscles and ligaments and supraspinous ligaments. The spinous process of the fifth lumbar vertebra is the least substantial and its extremity is often rounded and down-turned. The superior articular processes are wider apart than the inferior in the upper lumbar region, but the difference is very slight in the fourth, and in the fifth the two measurements are approximately equal. The articular facets are reciprocally concave (superior) and convex (inferior), so that they permit some rotation as well as flexion and extension. The transverse processes, with the exception of the fifth, are compressed from before backwards and pass laterally and slightly backwards. In the fifth lumbar vertebra the lower border of the transverse process is angulated and passes at first laterally and then upwards and laterally to a blunt tip, giving the whole process the appearance of a greater upward inclination than those of the vertebrae above. The blunt angle on the inferior border is said to represent the tip of the costal element and the outer extremity, the tip of the true transverse process. The transverse processes increase in length from the first to the third—which are the longest of all transverse processes—and then become shorter. The fifth incline *upwards,* passing laterally and slightly backwards. A faint, vertical ridge marks the anterior surface of the transverse process nearer the tip than the root. It gives attachment to the anterior layer of the thoracolumbar fascia and separates the surface into a medial area for the psoas major and a lateral area for the quadratus lumborum. To the tip of the process is attached the middle layer of the thoracolumbar fascia, but, in addition, the tip of the first gives attachment to the medial and lateral arcuate ligaments and the tip of the fifth to the iliolumbar ligament. The posterior surfaces of the transverse processes are covered by the deep muscles of the back and give origin to fibres of the longissimus thoracis. The upper and lower borders of the process give attachment to lateral intertransverse muscles. The mamillary process is homologous with the superior tubercle in the twelfth thoracic vertebra. It gives attachment to the multifidus and to the medial intertransverse muscle. The accessory process varies in prominence and may be difficult to identify. It gives attachment to the medial intertransverse muscle. The costal element is incorporated in the transverse process. (For a discussion concerning the possible homologies of the accessory, transverse and mamillary processes consult Wood Jones 1912.)

The Sacrum

The sacrum (**3**.43–7) is large, triangular and formed by fusion of the five sacral vertebrae. It is situated at the upper and posterior part of the pelvic cavity, inserted like a wedge between the two innominate bones. Its narrow, blunted *apex* is at the inferior end of the bone and articulates with the coccyx. At the opposite end the wide *base* articulates with the fifth lumbar vertebra, with

which it forms the *sacrovertebral angle.* In the erect position the bone is very oblique and is also curved longitudinally so that its dorsal surface is convex and its pelvic surface is concave (**3**.45). This ventral concavity increases the capacity of the true pelvis. In addition to a base and an apex the sacrum possesses *dorsal, pelvic,* and *lateral surfaces* and encloses the *sacral canal.*

In childhood the individual sacral vertebrae are connected by cartilage and can be separated by maceration. The adult bone shows many signs of its vertebral constitution, especially on its basal aspect.

The base (**3**.46) is formed by the upper surface of the *first sacral vertebra* and presents all the features of a typical vertebra in a slightly modified form. The *body* is large and much wider transversely. Its anterior projecting edge is the *sacral promontory.* The vertebral foramen is triangular, because the *pedicles* are short, widely separated, and are directed backwards and laterally. The *laminae* are very oblique and incline downwards, medially and backwards. Where they meet, the spinous process is represented by a *spinous tubercle.* The *superior articular processes* project upwards, with concave articular facets directed dorsomedially to articulate with the inferior articular processes of the fifth lumbar vertebra. The posterior part of each process projects backwards, and its lateral aspect bears a roughened area which corresponds to the mamillary process of a lumbar vertebra. The region of the *transverse process* shows marked modifications. A broad, sloping mass of bone projects from the lateral side of the body, pedicle and superior articular process (**3**.46)—a feature which is not found in any other vertebra, although foreshadowed in the fifth lumbar. It consists of the transverse process and the costal element fused to each other and to the rest of the vertebra, and forms the upper surface of the *lateral part* or *ala* of the sacrum.

The pelvic surface (**3**.43) faces downwards and forwards. Overall it is concave from above down and from side to side, but occasionally the body of the second sacral vertebra protrudes to produce a convexity in this region. It displays four pairs of *pelvic sacral foramina,* which communicate through intervertebral foramina with the sacral canal. They transmit the ventral rami of

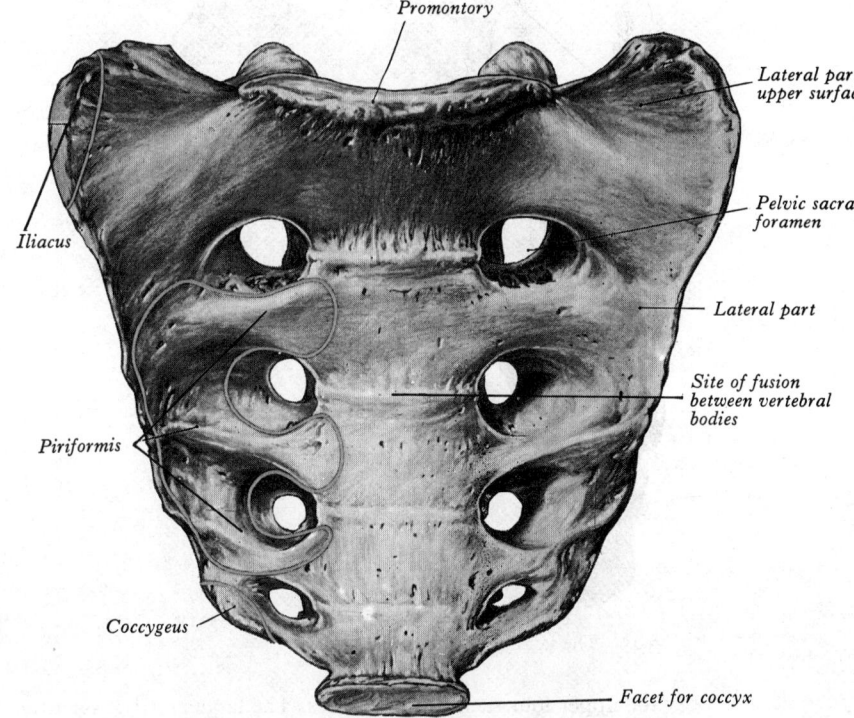

Promontory

Lateral part, upper surface

Pelvic sacral foramen

Lateral part

Site of fusion between vertebral bodies

Facet for coccyx

Iliacus

Piriformis

Coccygeus

3.43   The sacrum. Pelvic surface.

(The reader should appreciate that the old terms 'origin' and 'insertion' are functionally obsolete in terms of modern myokinetics. Accordingly they have been abandoned in the present volume: likewise the need for distinguishing colours has also disappeared. Thus, here and elsewhere in this edition, *all muscle attachments* are outlined in a single colour—a full blue line.)

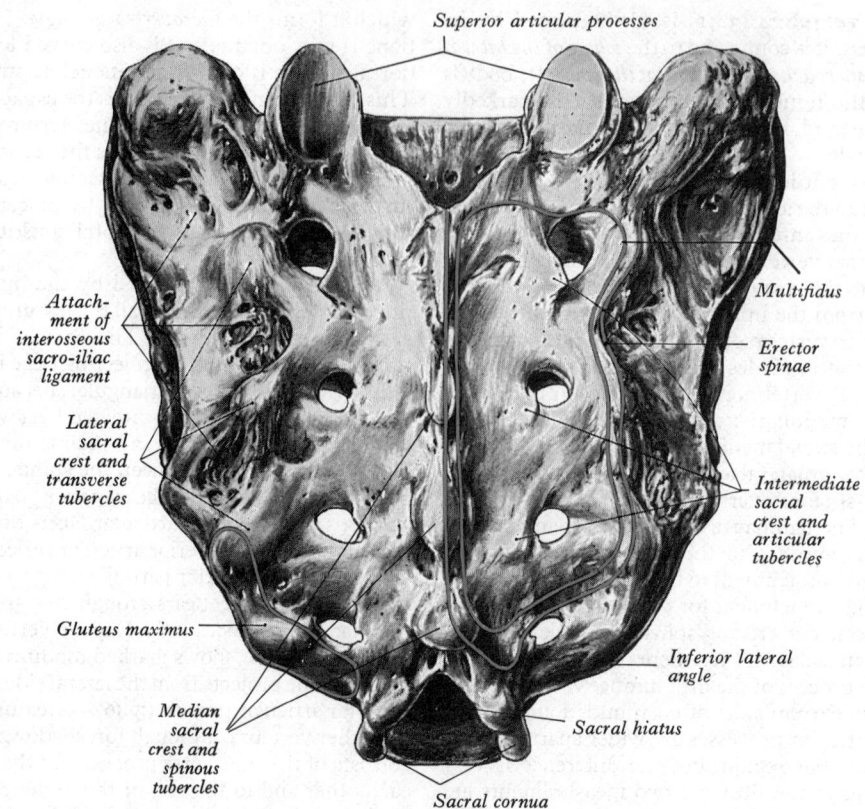

3.44   The sacrum. Dorsal surface.

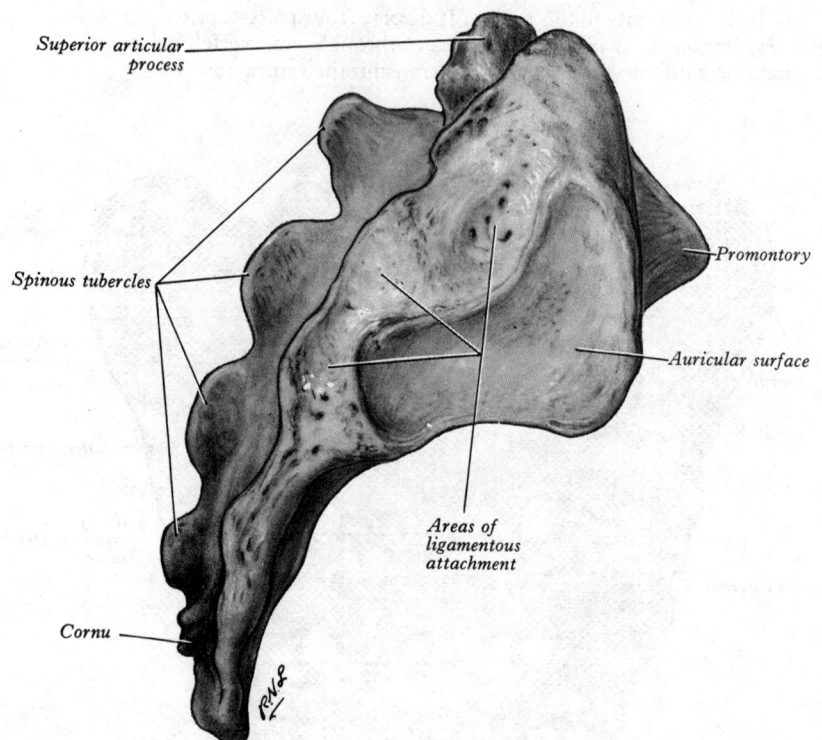

3.45   The sacrum. Right lateral aspect.

the upper four sacral spinal nerves. The large area between the foramina of the right and left sides is formed by the flattened pelvic surfaces of the bodies of the sacral vertebrae, and their lines of fusion are clearly visible as four raised *transverse ridges*. The bars of bone between the foramina on each side are the *costal elements*, which are fused to the vertebrae. Lateral to the foramina the costal elements unite with one another and, posteriorly, with the transverse processes to form the *lateral part* of the sacrum.

The dorsal surface (**3**.44) is convex backwards and upwards.

In the median plane is a raised interrupted *median sacral crest*, which bears four (sometimes only three) *spinous tubercles*. The crest represents the fused spines of the sacral vertebrae. Below the fourth, or sometimes the third, there is a ∩-shaped gap in the posterior wall of the sacral canal, the *sacral hiatus*. This is due to the failure of the laminae of the fifth sacral vertebra to meet in the median plane; as a result, the posterior surface of its body is exposed on the dorsal aspect of the sacrum. Lateral to the median crest the posterior surface is formed by the fused laminae. Lateral to this area on each side are four *dorsal sacral foramina*. Like the pelvic foramina they communicate with the sacral canal through the intervertebral foramina; each transmits the dorsal ramus of a sacral spinal nerve. Medial to the foramina, and vertically below each superior articular process of the first sacral vertebra, is a row of four small tubercles, which together constitute the *intermediate sacral crest*; the tubercles are sometimes termed *articular* since they represent contiguous articular processes fused together. The inferior articular processes of the fifth sacral vertebra project downwards at the sides of the sacral hiatus. They are the *sacral cornua* and are connected to the cornua of the coccyx by the intercornual ligaments. The interrupted crest lateral to the dorsal sacral foramina is the *lateral sacral* crest; it is formed by the fused transverse processes, the tips of which appear as a row of tubercles sometimes called the *transverse tubercles*.

The lateral surface (**3**.45) of the sacrum is formed by fused transverse processes and costal elements. It is wide above but rapidly diminishes in breadth in its lower part. The broad, upper part bears an ear-shaped surface, termed the *auricular surface*, for articulation with the ilium, and the area behind it is rough and deeply pitted by the attachment of ligaments. The auricular surface is borne by the costal elements and is shaped like an inverted letter L. The cephalic limb is shorter and restricted to the first sacral vertebra; the caudal limb extends downwards to the middle of the third sacral vertebra. The lower part of the lateral surface is not articular and consequently reduced in breadth. At its lower end it curves medially to reach the side of the body of the fifth sacral vertebra. The point at which the change of direction occurs is the *inferior lateral angle*. Below the angle the lateral surface becomes a thin border. An *accessory* sacral articular facet, of variable disposition, sometimes occurs.

The apex of the sacrum is formed by the inferior surface of the body of the fifth sacral vertebra and bears an oval facet for articulation with the coccyx.

The sacral canal (**3**.47) is formed by the vertebral foramina of the sacral vertebrae and is triangular on transverse section. Its upper opening, seen on the basal surface, is set obliquely but, owing to the inclination of the sacrum, it is directed upwards in the erect position. The lateral wall of the canal presents four intervertebral foramina (**3**.47), through which the canal is connected with both the pelvic and the dorsal sacral foramina. The lower median opening is the *sacral hiatus*.

To the ventral and dorsal surfaces of the body of the first sacral vertebra are attached the lowest fibres of the anterior and posterior longitudinal ligaments. The upper borders of the laminae of the first sacral vertebra give attachment to the lowest pair of ligamenta flava. The upper surface of the lateral part is smooth and slightly concave in its medial part but is irregularly roughened laterally. It is covered almost entirely by the psoas major. The smooth area is marked by an oblique, shallow groove which lodges the lumbosacral trunk. The rough area gives attachment to the lower band of the iliolumbar ligament (p. 447) which lies lateral to the fifth lumbar spinal nerve, and to the ventral ligament of the sacro-iliac joint. The anterolateral part of the area affords attachment to a portion of the iliacus (**3**.43).

The pelvic surface of the sacrum gives origin on each side to the piriformis (**3**.43). On emerging from the pelvic sacral foramina the ventral rami of the first three sacral nerves pass at once on to the anterior surface of the muscle. Along the medial margins of the foramina, on each side, the sympathetic trunk descends in contact with bone, and in the median plane the median sacral vessels are a direct relation. Lateral to the foramina the lateral sacral vessels bear a variable relation to the bone. The ventral surfaces of the bodies of the first and second and part of the third sacral vertebra are covered with parietal peritoneum and crossed obliquely, to the left of the median plane, by the root of the sigmoid mesocolon. The rectum lies in contact with the pelvic surfaces of the bodies of the third, fourth and fifth sacral vertebrae, but the bifurcation of the superior rectal artery into right and left branches intervenes between it and the third sacral vertebra.

The dorsal surface of the sacrum is rough and irregular. The erector spinae arises by an elongated U-shaped origin from the spinous and transverse tubercles, and covers the multifidus, which arises from the intervening area (**3**.44). The dorsal rami of the upper three sacral spinal nerves pierce these muscles after they emerge from the dorsal sacral foramina.

The auricular surface is covered with hyaline cartilage in the recent state and is formed entirely by costal elements. It shows cranial and caudal elevations and an intervening central depression; in the elderly a third elevation is found dorsal to this depression. The rough area behind the auricular surface shows two or three well-marked depressions and gives attachment to the strong interosseous sacro-iliac ligaments. Below the auricular surface to the lateral aspect of the sacrum are attached the gluteus maximus, the sacrotuberous and sacrospinous ligaments and the coccygeus, the structures being enumerated from behind forwards.

The sacral canal (**3**.47) contains the cauda equina (including the filum terminale) and the spinal meninges. Near the mid level of the sacrum the subarachnoid and subdural spaces become closed, and the lower sacral nerve roots and the filum terminale pierce the arachnoid and dura mater at that level. The filum terminale emerges below at the sacral hiatus and passes downwards across the dorsal surface of the fifth sacral vertebra and the sacrococcygeal joint to reach the coccyx. The fifth sacral spinal nerves also emerge through the sacral hiatus close to the medial side of the sacral cornua and groove the lateral aspects of the fifth sacral vertebra.

No difficulty will be experienced in distinguishing typical male or female sacra, but, as the characters are not always pronounced, there are many cases in which this is by no means easy. In the female the sacrum is shorter and wider than in the male, producing the wider pelvic cavity. Sacral width expressed as a percentage of vertical length is known as the *sacral index* (see

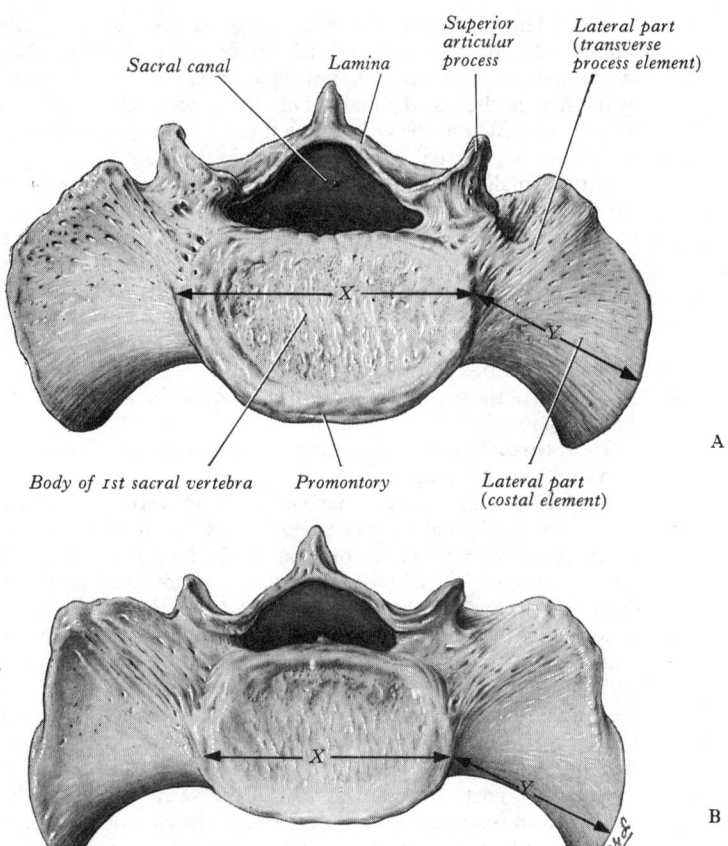

3.46 The base of the sacrum in the male (A) and in the female (B). The body of the first segment (X) forms a larger part of the breadth of the base in the male than it does in the female. In the latter the body is relatively smaller and the lateral, costal part (Y) is relatively broader.

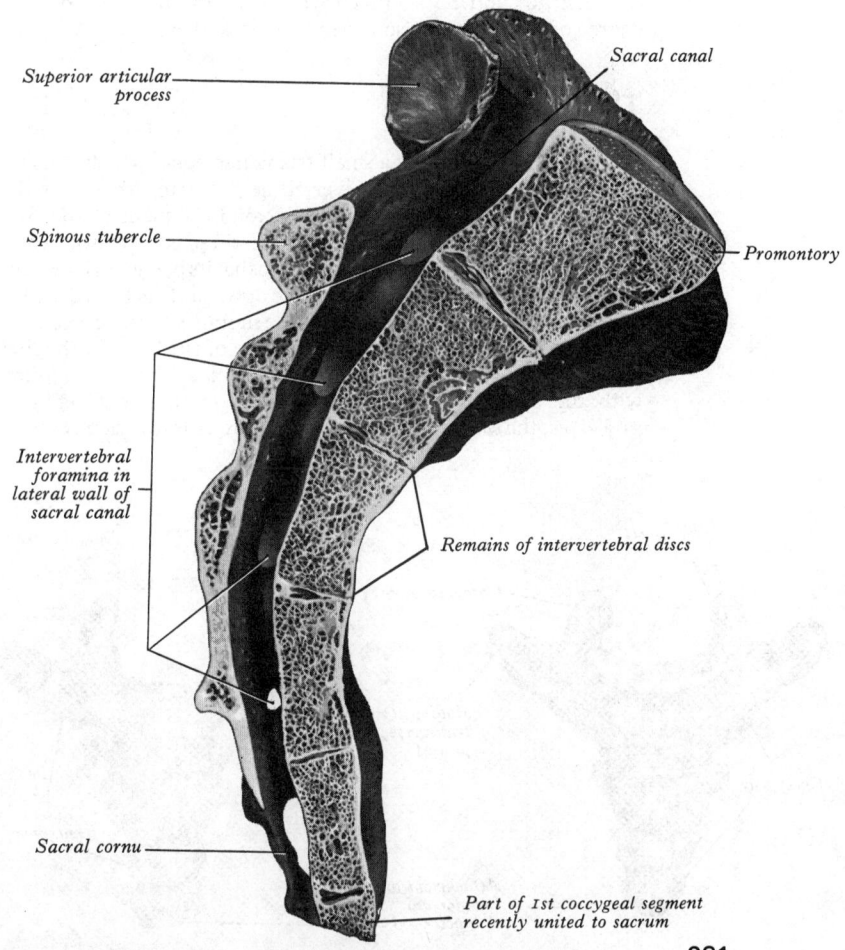

3.47 A median sagittal section through the sacrum.

p. 389). The ventral concavity of the sacrum is deeper in the female and its deepest point is usually placed at a higher level than in the male; the curvature is somewhat greater above this point in the female than in the male. The dorsal convexity produced by the protrusion of the second sacral vertebra (p. 389) occurs in both sexes but is usually less prominent in the male.

In the female the pelvic surface of the bone faces downwards more than in the male; this increases the size of the pelvic cavity and renders the sacrovertebral angle more prominent. In the female the auricular surface for articulation with the ilium is shorter than that in the male but in both sexes it usually extends along the sides of the first three sacral vertebrae. Owing to the great size of the body of the fifth lumbar vertebra the upper surface of the body of the first sacral vertebra occupies a larger proportion of the base of the sacrum in the male than it does in the female (3.46).

Structure. The sacrum consists of trabecular bone enveloped by a thin layer of compact bone.

Variations. The sacrum may consist of six vertebrae. There may be an additional sacral element or the last lumbar or first coccygeal vertebra may be incorporated. The inclusion of the fifth lumbar vertebra (*sacralization*) is usually incomplete and it may be limited to one or other side. In the most minor degree of the abnormality the transverse process of the fifth lumbar vertebra is unusually large, and articulates, sometimes by a synovial joint, with the sacrum at the dorsolateral angle of the base. Reduction of the number of the constituents of the sacrum is less common, but *lumbarization* of the first sacral vertebra does occur; it remains partially or completely separate from the rest of the sacrum. A considerable part of the dorsal wall of the sacral canal may be deficient, in consequence of imperfect development of the laminae and spines. The orientation of the superior sacral articular facets (and hence the relation between the planes of the two zygapophysial joints made with the fifth lumbar vertebra) displays wide variation according to Cihák (1970). The chords of the concave sacral facets formed an angle with the sagittal plane which varied from 20° to 90° in 132 sacra, the majority lying between 40° and 60°. The curvatures of the facets, and their degree of asymmetry, also varied to a marked extent.

The Coccyx

The coccyx (3.48A, B) is a small triangular bone, which consists usually of four rudimentary vertebrae fused together, but the number varies from five to three. Not infrequently, the first coccygeal vertebra is separate. The bone is directed downwards and ventrally from the apex of the sacrum, so that its pelvic surface faces up and forwards and its dorsal surface down and backwards. This orientation is, of course, varied by the mobility of the coccyx.

The *base*, formed by the upper surface of the body of the *first coccygeal vertebra*, presents an oval, articular facet which joints with the apex of the sacrum. Dorsolateral to the facet, two processes, the *coccygeal cornua*, project upwards to articulate with

the sacral cornua; they are the homologues of the pedicles and superior articular processes of fully formed vertebrae. A rudimentary *transverse process* projects laterally and slightly upwards from each side of the body of the first coccygeal vertebra and may ascend to articulate or fuse with the inferior lateral angle of the sacrum. In that event five pairs of foramina exist in the sacrum.

The *second*, *third* and *fourth coccygeal vertebrae* diminish successively in size and are usually fused together. They are mere nodules of bone, which represent rudimentary vertebral bodies, although the second may show traces of transverse processes and pedicles.

The lateral parts of the *pelvic surface*, including the rudimentary transverse process, have attached to them the levatores ani and the coccygei. The ventral sacrococcygeal ligament is attached to the front of the body of the first, and may extend downwards to reach the second coccygeal vertebra (4.56). To the *cornua* are connected the intercornual ligaments. The interval between the body of the fifth sacral vertebra and the articulating sacral and the coccygeal cornua on each side represents the intervertebral foramen between the fifth sacral and the first coccygeal vertebrae, and transmits the fifth sacral nerve. The dorsal ramus of that nerve descends behind the rudimentary transverse process, but its ventral ramus passes forwards through a foramen between the transverse process and the sacrum, bounded laterally by the lateral sacrococcygeal ligament, which connects the process to the inferior lateral angle of the sacrum. The *dorsal surface* of the coccyx gives origin, on each side, to the gluteus maximus and, at its tip, to the sphincter ani externus. The median area gives attachment to the deep and superficial dorsal sacrococcygeal ligaments. The latter extends downwards from the margins of the sacral hiatus and may close the lower end of the sacral canal. The filum terminale, which is situated between the two ligaments, blends with them on the dorsal surface of the first coccygeal vertebra.

Ossification of the Vertebral Column

For the earlier stages of vertebral development consult p. 138 and illustration 2.50.

Each typical vertebra is ossified from three primary centres (3.49A), one in each half of the vertebral arch and one in the body. The primary centres in the vertebral arch appear at the roots of the transverse processes; ossification spread backwards into the laminae and spine, forwards into the pedicles and posterolateral portions of the body, laterally into the articular processes and upwards and downwards into the articular processes. The classical view has been that the centres in the *vertebral arches* appear first in the upper cervical vertebrae at the ninth to tenth week of intrauterine life, and subsequently in successively lower vertebrae, reaching the lower lumbar about the end of the third month. However, in a radiographic study of 195 unsexed human fetuses, of which 33 proved to lie in the age range appropriate to the present study (Bagnall *et al.* 1977), a pattern was found which differed substantially from the simple cranio-caudal sequence described above. A regular progression was not present in the cervical region. A group of centres first appeared in the lower cervical/upper thoracic region, quickly followed by a second group in the upper cervical region. After a short interval a third group appeared in lower thoracic/upper lumbar region, and the remaining centres then appeared, spreading regularly and rapidly in both cranial and caudal directions. The remaining, major portion of the body constitutes the *centrum* and is ossified from a primary centre which appears dorsal to the notochord. (The centrum is occasionally ossified from bilateral centres which sometimes fail to unite. The suppression of one of these centres leads to the formation of a wedge-shaped vertebra, and is well-recognized cause of lateral curvature of the vertebral column. The condition is frequently multiple.) This centre is first seen in the lower thoracic region at the ninth to tenth week of intrauterine life and then develops at successively higher and lower levels, reaching the second cervical vertebra by the end of the fourth month. In the study referred to above (Bagnall *et al.* 1977), the

A *Pelvic aspect*
B *Dorsal aspect*
Coccygeal cornua
Rudimentary transverse process
Coccygeus
Gluteus maximus
Attachment area of—Levator ani Sphincter ani

3.48A and B The coccyx. A Pelvic aspect. B Dorsal aspect.

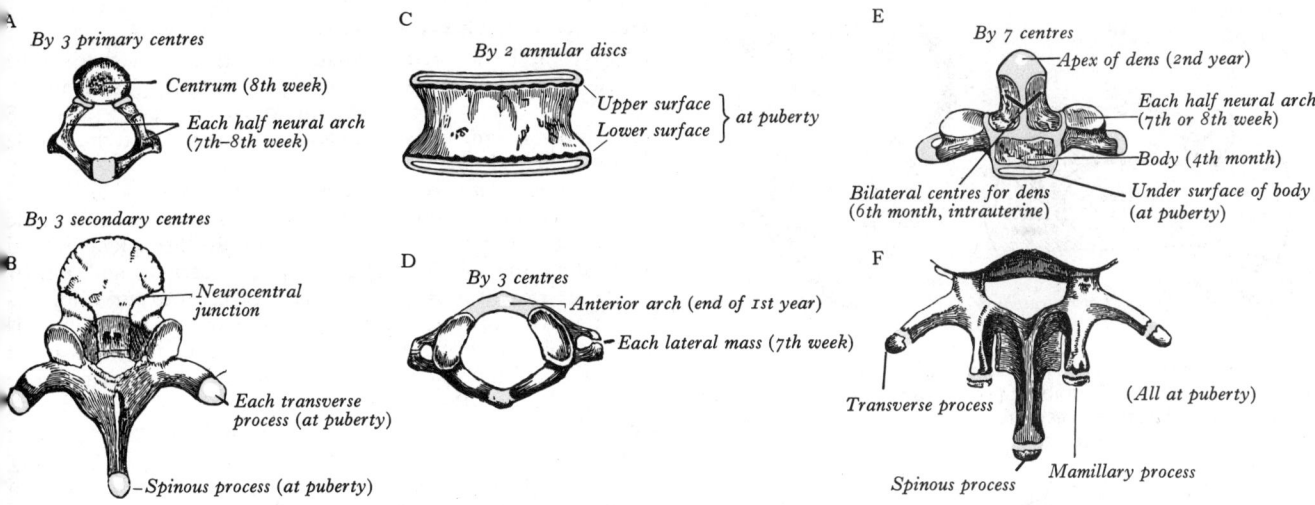

3.49A–F Ossification of the vertebral column. A A typical vertebra. B A typical vertebra at puberty. C Body of a typical vertebra at puberty. D The atlas. E The axis. F A lumbar vertebra.

account just given was largely confirmed, but the earliest centres were found in the lower thoracic *and* upper lumbar regions; cranial progression was rather faster than caudal progression, the last centre to appear being for the centrum of the fifth sacral vertebra. (The latter, of course, ignores the much later and erratic ossification of the coccyx—*vide infra.*) During the first few years of life the centrum is connected to each half of the vertebral arch by a synchondrosis, termed the *neurocentral joint*. In the thoracic region the costal facets on the bodies lie behind the neurocentral joints. At birth a vertebra consists of three ossified elements, the centrum and the halves of the vertebral arch, united, of course, by cartilage. During the first year the halves of the arch unite behind, first in the lumbar region and then upwards through the thoracic and cervical regions. In the upper cervical vertebrae centra unite with arches about the third year, but in the lower lumbar vertebrae union is not completed until the sixth year. Until puberty the upper and lower surfaces of the bodies and the ends of the transverse and spinous processes are cartilaginous, but about this time five secondary centres appear, one for the tip of each transverse process, one for the end of the spinous process, and two annular epiphysial rings for the circumferential parts of the upper and lower surfaces of the body (**3.38**, 49B and C). The costal articular facets are extensions of the annular epiphysial rings (Dixon 1920). These secondary centres fuse with the rest of the bone about the age of 25 years. In the bifid spinous processes of the cervical vertebrae there are two secondary centres. It is doubtful whether the annular 'epiphyses' of vertebrae can be equated with the epiphyses of long bones. In most mammals the vertebral epiphyses are complete osseous discs. For discussion of these arrangements consult François and Dhem (1974).

Exceptions to this mode of ossification occur in the first, second and seventh cervical, and in the lumbar vertebrae.

The atlas is usually ossified from three centres (**3.49D**). One appears in each lateral mass about the seventh week of intrauterine life, and gradually extends into the posterior arch, where the two unite between the third and fourth years, either directly or through the medium of a separate centre. At birth, the anterior arch consists of fibrocartilage; in this a separate centre appears about the end of the first year, and unites with the lateral masses between the sixth and eighth years—the lines of union extending across the anterior portions of the superior articular facets. Occasionally the anterior arch is formed by the forward extension and ultimate union of the centres for the lateral masses; sometimes it is ossified from two laterally placed centres.

The axis is ossified from five primary and two secondary centres (**3.49E**). The vertebral arch is ossified from two primary centres, and the centrum from one, as in a typical vertebra; the centres for the arch appear about the seventh or eighth week of intrauterine life, the centre for the centrum about the fourth or

fifth month. The dens is ossified almost entirely from two laterally placed centres; these appear about the sixth month of intrauterine life, and join before birth to form a conical bony mass, deeply cleft by cartilage above. A wedge-shaped piece of cartilage fills the cleft and forms the summit of the process: in this cartilage a centre appears about the second year and unites with the main mass of the process about the twelfth year. It is regarded as representing a part of the cranial sclerotome half of the first cervical segment or the *pro-atlas* (Gadow and Green 1933). The base of the process is separated from the body of the axis by a cartilaginous disc, the circumference of which ossifies, but the centre remains cartilaginous until advanced age; in this cartilaginous disc what are possibly rudiments of the lower epiphysial lamella of the atlas and the upper epiphysial lamella of the axis may sometimes be found. In addition a thin epiphysial plate is formed on the under surface of the body of the bone about the time of puberty. It has long been held, almost universally, that the dens represents the centrum of the atlas, secondarily fused with cranial aspect of the centrum of the axis. Considerable doubt has now been cast on this traditional view following comparative morphological and developmental studies on a variety of mammals by Jenkins (1969), who regards the dens as a new formation. On the other hand, Ganguly and Singh-Roy (1965) consider that the apical centre for the dens is derived from the *pro-atlas*, and the latter may also contribute to the lateral masses of the atlas. Torklus and Gehle (1968) consider that the dens is the centrum of the pro-atlas.

The seventh cervical vertebra shows an additional feature in that the costal processes are usually ossified from separate centres, which appear about the sixth month of intrauterine life, and join the body and transverse processes between the fifth and sixth years. As already stated (p. 274), the costal processes may remain separate, and grow laterally and forwards, as cervical ribs.

Separate ossific centres have also been found in the costal processes of the fourth, fifth and sixth cervical vertebrae. (See also p. 292.)

The lumbar vertebrae (**3.49F**) have two additional centres, one for each mamillary process. In the fifth lumbar vertebra there is also a pair of scale-like epiphyses on the tips of the costal elements.

The sacrum (**3.50A–E**) resembles a typical vertebra in the ossification of each of its vertebral segments. Primary centres for the centrum and each half of the vertebral arch appear between the tenth and twentieth weeks. Primary centres for the costal elements of the upper three or more sacral vertebrae appear, above and lateral to the pelvic sacral foramina, between the sixth and eighth months of intrauterine life. The costal element unites with its half of the vertebral arch between the second and fifth years, and the conjoined element so formed unites ventrally with the centrum and dorsally with its fellow of the opposite side about

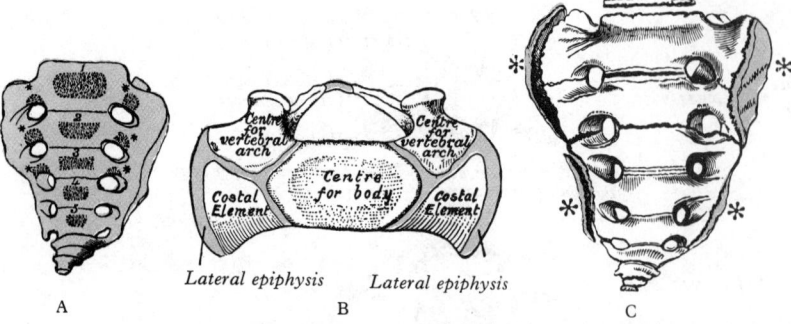

3.50A–C The ossification of the sacrum and coccyx. A, at birth. B, the base of the sacrum of a child of about 4 years of age. C, at twenty-fifth year. In C the epiphysial plates for each lateral surface are marked by asterisks.

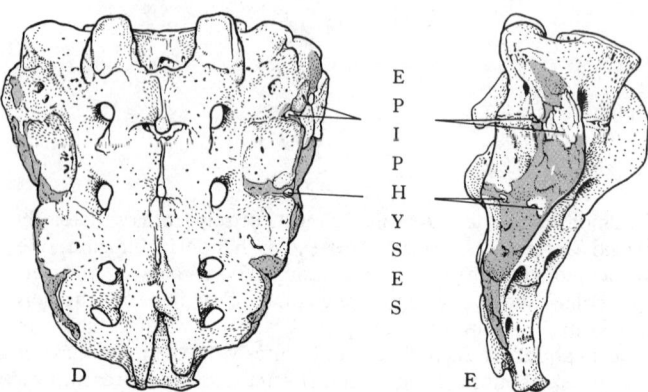

3.50D and E The epiphyses of the costal and transverse process of the sacrum at the eighteenth year.

the eighth year. When this has occurred the upper and lower surfaces of the body of each sacral vertebra are covered by an epiphysial plate of cartilage which is separated from its neighbour by a plate of fibrocartilage representing an intervertebral disc. Laterally the successive conjoined vertebral arches and costal elements are separated by hyaline cartilage and there is a cartilaginous epiphysis, sometimes divided into upper and lower parts, on the auricular surface and lower part of the lateral surface of the bone. Soon after puberty the conjoined vertebral arches and costal elements of adjacent vertebrae begin to fuse with one another from below upwards. At the same time epiphysial centres develop (1) for the upper and lower surfaces of the bodies, (2) for the spinous tubercles, (3) fo r the transverse tubercles and (4) for the costal elements. The costal epiphysial centres appear at the lateral extremities of the hyaline cartilages which intervene between adjacent costal elements. Two such centres, placed anteriorly and posteriorly, appear in each of the intervals between the first and second and the second and third sacral vertebrae. From these ossification spreads into the epiphysial plate covering the auricular surface. One costal epiphysial centre, placed anteriorly, appears in each of the remaining intervals and from them ossification spreads to the epiphysial plate covering the lower part of the lateral surface of the sacrum.

The bodies of the sacral vertebrae unite with one another at their adjacent margins after the twentieth year but the central area and greater part of each intervertebral disc remains unossified up or or even after middle life; available information is based on relatively few specimens (Fawcett 1907; McKern and Stewart 1957).

The coccyx. Each segment of the coccyx is ossified from one primary centre but the incidence and times of ossification are imperfectly known. The centre for the first segment appears about birth and its cornua may ossify from separate centres appearing soon after birth. The remaining segments ossify at widely separated intervals up to the twentieth year or later. As age advances, the segments unite with one another, the union between the first and second being frequently delayed until the age of 30 years. At a late period of life, especially in females, the coccyx often fuses with the sacrum.

The Vertebral Curvatures

The vertebral column is median and dorsal in the body—the common vertebrate plan. In the human male its length is about 70 cm, its cervical part measuring approximately 12 cm, thoracic 28 cm, lumbar 18 cm, and sacrum and coccyx 12 cm. In the female the total length is about 60 cm.

The curves of the vertebral column viewed from the side (**3.51B**), are cervical, thoracic, lumbar and pelvic. The thoracic and pelvic curves are *primary*; they are concave ventrally during fetal life and retain the same curvature after birth. The cervical and lumbar curves are *secondary* or compensatory; the cervical curve appears in intrauterine life and is further accentuated when the child is able to hold up its head (at three or four months), and to sit upright (about nine months); the lumbar curve appears at twelve to eighteen months, when the child begins to walk; its development is neccessary to bring the centre of gravity of the trunk over the legs. This change in the mechanics of the lumbar region appears to be a potent factor in the necessary alterations in the proportions of the vertebral bodies and the intervertebral discs, especially in the fifth lumbar vertebra and the disc below it (Taylor 1975). It should be noted that although it has long been held that the secondary cervical curve first appears either late in intrauterine life, or early in postnatal life, this view has recently been challenged (Bagnall *et al.* 1977). In a careful radiographic study of 195 human fetuses ranging in conceptual age from 8 to 23 weeks, it was shown that 83 per cent already possessed a cervical curve as soon as the ossification centres became sufficiently radio-opaque—namely 9½ weeks. Further, these authors emphasized the importance of the early onset of *fetal movements*, in relation to the emergence of an identifiable cervical curve.

In the adult, the *cervical* curve is convex forwards, and is the least marked; it begins at the atlas, and ends at the second thoracic vertebra. The *thoracic* curve is concave forwards, and reaches from the second to the twelfth thoracic vertebra; it is caused by the greater depth of the posterior parts of the vertebral bodies. The *lumbar* curve is convex forwards and is more pronounced in the female than in the male; it reaches from the last thoracic vertebra to the lumbosacral angle, and the convexity of the lower three segments is greater than that of the upper two; it is caused mainly by the greater depth of the anterior parts of the intervertebral discs, but the shape of the vertebral bodies also helps to produce it. Sex differences in the lumbar curvature, and also in the cervical region, are confirmed by Knussman and Finke (1977). The *pelvic* curve extends from the lumbosacral joint to the apex of the coccyx; its concavity faces downwards and forwards.

In the upper part of the thoracic region of the vertebral column there is often a slight *lateral* curvature, with its convexity directed towards the right side in right-handed persons, and to the opposite side in the left-handed.

When viewed from the front (**3.51A**), the width of the vertebral bodies is seen to increase from the second cervical to the third lumbar vertebra. This change is associated with the increase in weight bearing from above downwards. There is some variation in the size of the bodies of the last two lumbar vertebrae, but thereafter the width diminishes rapidly to the apex of the coccyx. In the last two lumbar vertebrae there is an inverse relation between the areas of the upper and lower surfaces of the bodies and the size of the pedicles and transverse processes which suggests that the latter transmit some of the vertebral compressive forces from the spine to the pelvis (Davis 1961). In a detailed dimensional study of primate lumbar vertebrae, human and otherwise, Rose (1975) considered that the size of their bodies is as much concerned with bending as compressive stresses. He also indicated that in man only the *fifth* body *regularly* displays a greater depth ventrally, concluding that other factors must also contribute to the lumbar curvature: as noted above this is principally the conformation of the intervertebral discs.

The posterior surface of the column presents the spinous processes of the vertebrae in the median plane (**3.51C**). In the cervical region (with the exception of the second and seventh vertebrae) these are short and nearly horizontal, with bifid ends. In the upper part of the thoracic region they are directed obliquely downwards; in the middle part they are long and almost vertical; in the lower part of the thoracic region and in the lumbar region

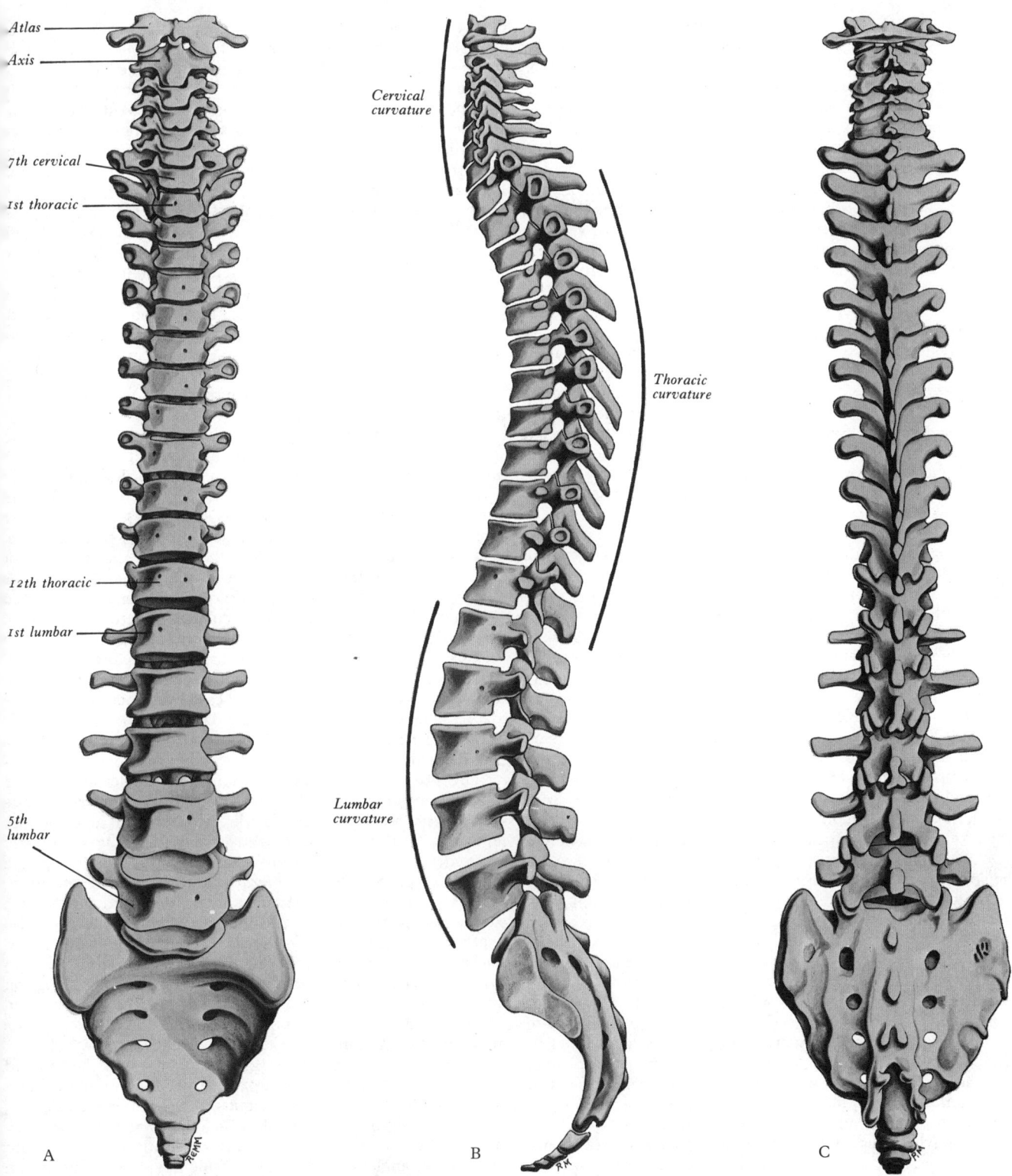

Atlas

Axis

7th cervical

1st thoracic

Cervical
curvature

Thoracic
curvature

12th thoracic

1st lumbar

Lumbar
curvature

5th
lumbar

A   B   C

3.51 A–C   The vertebral column. A Anterior aspect. B Lateral aspect. (Note
curvatures.) C Dorsal aspect. (Note slight thoracic curvature to left.)

they are nearly horizontal. They are separated by considerable
intervals in the cervical and lumbar regions, but are closely
approximated in the middle of the thoracic region. Occasionally a
spinous process may deviate from the median plane—a fact to be
remembered in practice, as irregularities of this sort occur also in
fractures or dislocations of the vertebral column. The seventh
cervical can be felt at the lower end of the nuchal furrow, and
below it the first and the second thoracic are prominent. The third
thoracic lies opposite the root of the scapular spine and the
seventh lies opposite the inferior angle of the scapula, provided
that the arm is at rest by the side. The spine of the fourth lumbar
vertebra lies on a level with the highest parts of the iliac crests and

that of the second sacral vertebra with the posterior superior iliac
spines. At the sides of the spinous processes the *vertebral grooves*
contain the deep muscles of the back. In the cervical and lumbar
regions these grooves are shallow and are formed by laminae; in
the thoracic region they are deep and wide, and are formed by
laminae and transverse processes. Lateral to the laminae are the
articular processes, and still more lateral the transverse processes.
In the thoracic region the transverse processes lie on a plane
considerably behind that of the same processes in the cervical and
lumbar regions. In the cervical region the transverse processes are
placed in front of the articular processes, lateral to the pedicles,
and between the intervertebral foramina. In the thoracic region

they are behind the pedicles, intervertebral foramina and articular processes. In the lumbar region they are in front of the articular processes, but behind the intervertebral foramina. The size of the transverse processes of the atlas has already been emphasized (p. 273), and the breadth from the tip of one transverse process to the other has been contrasted with the same measurement in the axis. Breadth varies little from the second to the sixth cervical vertebra, but in the seventh it shows a substantial increase. In the thoracic region it is greatest in the first and then gradually diminishes, being least in the twelfth, where the transverse process elements are usually reduced to mere vestiges. In the first lumbar vertebra the measurement is greater; in the second it is further increased; while in the third it is greater than it is in any of the other vertebrae. In the fourth and fifth it diminishes.

The *lateral aspects* of the vertebral column are separated from the posterior surface by the articular processes in the cervical and lumbar regions, and by the transverse processes in the thoracic region. The anterior part of the lateral surface of the column is formed by the sides of the bodies of the vertebrae, marked in the thoracic region by the facets for articulation with the heads of the ribs. The intervertebral foramina are behind the bodies and between the pedicles; they are oval in shape, smallest in the cervical and upper part of the thoracic regions, and gradually increasing in size to the last lumbar; they contain the spinal nerves and vessels.

The *vertebral canal* follows the curves of the column; it is large and triangular in the cervical and lumbar regions, where movement is free, but small and circular in the thoracic region, where motion is more limited. These differences may also be linked with the variations in diameter of the spinal cord.

Although the range of movement between adjacent vertebrae is small, these small ranges add up to considerable degrees of bending or rotation over the vertebral column as a whole (p. 445). The intervertebral discs not only tie the vertebrae together but are the sites of all movement between them. By their elastic deformability they not only permit slight tilting and torsion between adjoining vertebral bodies, but they also provide a very considerable element of compressibility in the spinal column. This ability to absorb and neutralize compression stresses is augmented by the sinuous curvature of the column itself; forces which would be transmitted from end to end of a straight column are expended completely against the slight 'give' and recovery of the spinal curves. The effects of body weight, muscle pull and thrust due to impact of the feet, whether trivial as in walking, or of greater magnitude, in running and jumping, are largely smoothed out by the discs and curvatures.

Applied Anatomy. Fractures or fracture dislocations of the vertebral column are the result of (1) forced flexion, e.g. by a violent blow on the back from a large object, and (2) violence transmitted along the long axis of the column, e.g. by falling on to the feet from a height, or by diving on to the head. In the first group the injury commonly occurs at the level of the fifth or sixth thoracic. In the second group, owing to the normal curvature of the vertebral column, the injury is also a flexion fracture and its site is between the ninth thoracic and second lumbar.

In spondylolisthesis, which is present in 5 per cent of skeletons, the spine, laminae and inferior articular processes of the fifth (sometimes the fourth) lumbar vertebra are united to one another but are separate from the rest of the bone. The condition may give rise to low back pain and is usually regarded as due to a congenital defect. It has been suggested that each part of the vertebral arch ossifies from two primary centres which fail to fuse but there is no evidence for this.

The Sternum

The sternum is probably confined to land vertebrates; it is absent in fish. The elongate form seen in the human skeleton is typical of mammals, the sternum being more of a complex plate in amphibians and reptiles. It has been associated with both the shoulder girdles and the ribs from its earliest appearance, but is usually considered to be axial in nature. The details of its embryonic development, including its ossification, indicate a separate origin from the ribs.

The human sternum, as is usual in mammals, consists of three parts (3.52, 53): a cranial element, the *manubrium*; a middle part the *body* or *mesosternum*; and a caudal element, the *xiphoid process*. Its total length in males is about 17 cm and considerably less in females. The ratio between manubrial and mesosternal length differs in the sexes, but clear racial characteristics have not been established. Such estimates are complicated by the probability that the sternum continues to grow beyond the third decade and perhaps throughout life (Ashley 1956, Rother *et al.* 1976).

In addition to the three major sections noted above, the sternum shows further division in its body, which in early life consists of four segments or *sternebrae*; from their relationship to the ribs sternebrae appear to be intersegmental in character.

In its natural position the sternum is inclined downwards and a little forwards. It is slightly convex in front, and concave behind it is broadest at the level of the first costal cartilages, narrow at the junction of the manubrium with the body, below which it gradually widens as far as the level of the articulations of the cartilages of the fifth ribs, and then narrows quickly to its lower end.

The manubrium sterni is somewhat triangular, broad and thick above, narrow below at its junction with the body. Its *anterior surface* is smooth, convex from side to side and concave from above downwards. Its *posterior surface* is concave and featureless. The *superior border* is thick, and presents at its centre the *jugular (suprasternal) notch*; on each side of this there is an oval articular surface, directed upwards, backwards and laterally, for articulation with the sternal end of the clavicle, and termed the *clavicular notch*. The *inferior border*, oval and rough, is covered with a thin layer of cartilage, for articulation with the upper end of the body. The *lateral borders* are each marked above by a depression for the first costal cartilage, and below by a small articular facet, which, with a similar one on the upper angle of the body, forms a joint with the second costal cartilage. Between the depression for the first costal cartilage and the facet for the second, the narrow curved edge slopes from above downwards and medially.

The mesosternum (body) is longer, narrower and thinner than the manubrium, and is broadest close to its lower end. Its *anterior surface*, nearly flat, is directed forwards and slightly upwards, and is marked by three transverse ridges, variably defined, marking the lines of fusion of four originally separate *sternebrae* (p. 288). A *sternal foramen*, of varying size and form, occurs occasionally at the junction of the third and fourth sternebrae (p. 288). The *posterior surface*, slightly concave, is also marked by three transverse lines, less distinct, however, than those on the anterior surface. The *upper end* is oval and articulates with the manubrium at the *sternal angle*, which forms a ridge on the anterior surface that can usually be felt through the skin without difficulty. On the posterior surface the position of the manubriosternal joint is marked by a transverse groove (3.53A). The *lower end* is narrow, and is continuous with the xiphoid process. Each *lateral border* (3.53B), at its superior angle, has a small notch, which, with a similar one on the manubrium, receives the second costal cartilage; below this, four *costal notches* articulate with the third, fourth, fifth and sixth costal cartilages; the inferior angle has a small facet, which, with a similar one on the xiphoid process, forms a notch for the reception of the seventh costal cartilage. These articular depressions are separated by a series of curved edges, which diminish in length from above downwards, and correspond to the anterior ends of the intercostal spaces.

The xiphoid process is the smallest and most variable part of the sternum. It may be broad and thin, pointed, bifid, perforated, curved or deflected to one side or the other. It is cartilaginous in youth, but, at its upper part, more or less ossified in the adult. Above, it is continuous with the lower end of the body of the bone, and on the front of each superior angle there is a facet for a part of the cartilage of the seventh rib (3.53).

The manubrium is level with the third and fourth thoracic vertebrae. Its *anterior surface*, on each side, gives attachment to the sternal origins of the pectoralis major and sternocleidomastoid. To its *posterior surface* are attached the sternothyroid, opposite the first costal cartilage; above this level the most medial bres of the sternohyoid usually arise from the bone. This surface forms the anterior boundary of the superior mediastinum; its lower part is related to the arch of the aorta, and its upper part to the left brachiocephalic vein and the brachiocephalic, left common carotoid and left subclavian arteries. Its lateral portions are related to the lungs and pleurae. The *jugular notch* gives attachment to some of the fibres of the interclavicular ligament. On the *lateral border* the junction between the manubrium and the first costal cartilage is a synchondrosis in contrast to the other sternocostal joints.

The sternal body lies opposite the fifth to ninth thoracic vertebrae. Its *anterior surface* gives attachment, on each side, to the articular capsules of the sternocostal joints and to the sternal origin of the pectoralis major. Its *posterior surface* affords origin inferiorly to the transversus thoracis (sternocostalis), and has numerous important relationships. On the right of the median plane it is related to the right pleura and the thin, anterior border of the right lung, which intervene between it and the pericardium. To the left the upper two sternebrae are related to the left pleura and lung, but the lower two are directly related to the pericardium. The *borders* give attachment to the external intercostal membranes in the intervals between the costal notches. With the exceptions of the first and the sixth, the cartilages of the true ribs articulate with the sternum at the lines of junction of its primitive component segments.

The xiphoid process lies in the epigastric region. To its *anterior surface* are attached the most medial fibres of the rectus abdominis and the aponeuroses of the external and internal obliques. Its *lower end* gives attachment to the linea alba, and its *borders* to the aponeuroses of the internal oblique and transversus abdominis. Its *posterior aspect* gives origin, on each side, to some of the fibres of the diaphragm, and is related to the anterior surface of the liver.

The sternum is composed of highly vascular trabecular bone covered by a compact layer which is thickest in the manubrium between the clavicular notches. The spaces in the spongy substance contain *red* marrow. The internal trabecular pattern of the manubrium and the first and second sternebrae has been the subject of an intensive examination by scanning electron microscopy and computer analysis of its metrical parameters (Whitehouse 1975). Centrally the bone is lightly constructed, trabeculae being thicker and more widely spaced in the lateral regions of the sternum. No mechanical or other explanation of these variations, or of the noted differences in the patterns already studied in the ribs and lumbar vertebrae, has yet been formulated.

Pismenov and Zapetski (1977) have described the vascular supply of the sternum in detail.

Ossification (3.54A, B). The sternum is formed by the fusion of two cartilaginous *sternal plates* (p. 141), one on each side of the median plane. The number and arrangement of the centres of ossification are variable and are related to the completeness and time of fusion of the sternal plates and to the width of the adult bone (Ashley 1956). Incomplete fusion results in the presence of a

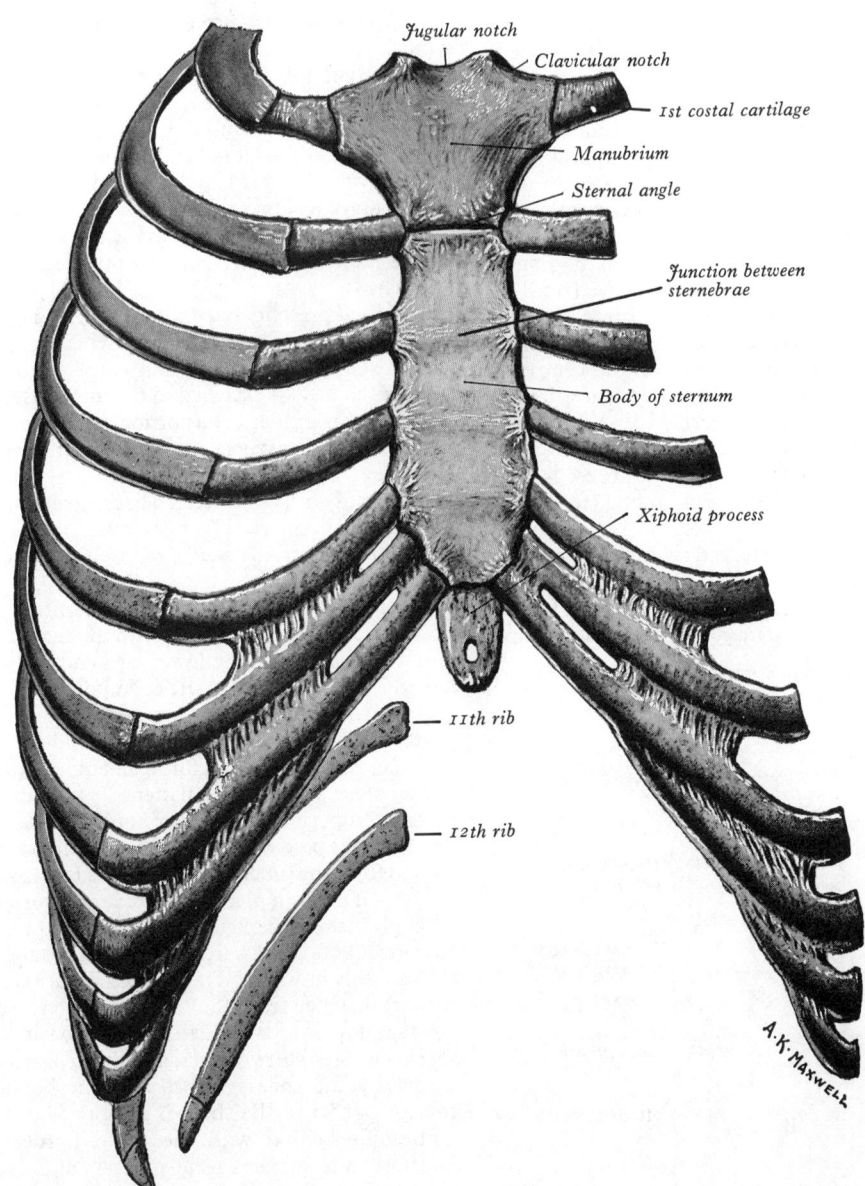

Jugular notch
Clavicular notch
1st costal cartilage
Manubrium
Sternal angle
Junction between sternebrae
Body of sternum
Xiphoid process
11th rib
12th rib

A.K. MAXWELL

3.52   The sternum and costal cartilages. Anterior aspect.

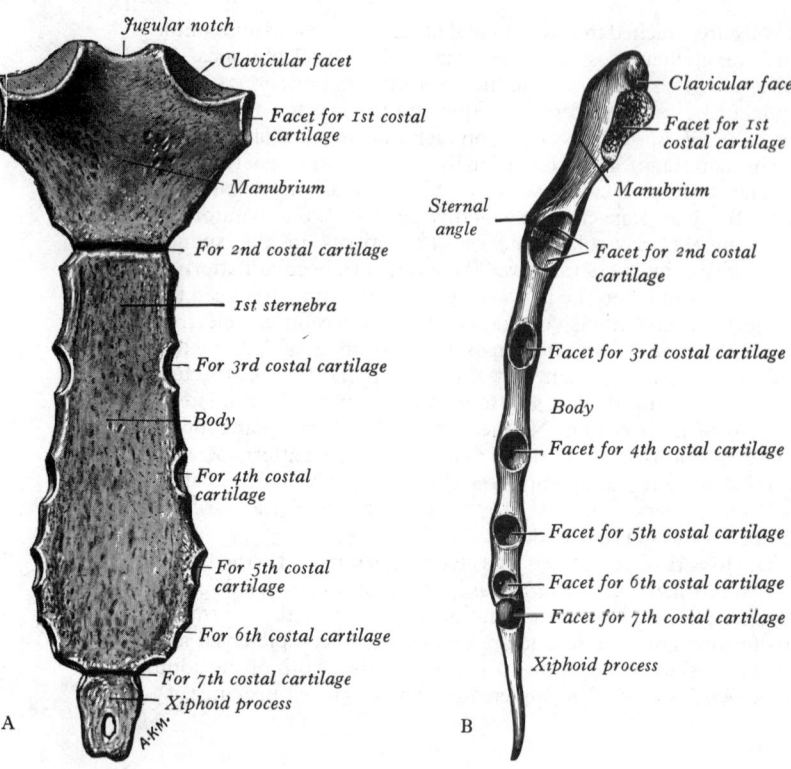

3.53 The sternum. A Posterior aspect. B Lateral aspect.

sternal foramen in the adult bone (p. 286). The manubrium is ossified from one, two or even three centres which appear in the fifth month of fetal life. The first and second sternebrae usually ossify from single centres appearing in the fifth month. The centres of ossification in the third and fourth sternebrae are generally paired and appear in the fifth and sixth months respectively, but one member of the pair may be delayed until the seventh or even eighth month, and the fourth sternebral centre may be absent (Paterson 1904). The xiphoid process begins to ossify in the third year or much later. In some sterna all the centres of ossification are single and median; in others the centre for the manubrium is single whilst those for the sternebrae are all paired, and may be symmetrical or asymmetrical. Union between the centres for the body begins about puberty, and proceeds from below upwards; by the age of 25 they are all united (see also p. 452).

Suprasternal ossicles, which may be paired or single, are present in about 7 per cent of sterna. They may be fused to the manubrium or articulate with it at the lateral border of the jugular notch posteriorly. When well formed they are pyramidal in shape,

the base of the pyramid forming the articular surface. Cartilaginous at birth, they become ossified during adolescence.

The Ribs

The ribs are elastic arches of bone, which are connected behind with the vertebral column, and form a large part of the skeleton of the thorax. There are twelve pairs, but this number may be increased by the development of a cervical or a lumbar rib, or may be reduced to eleven by the absence of the twelfth pair. The first seven pairs are connected in front, through the costal cartilages, to the sternum (3.52); they are called *true* ribs. The remaining five are so-called *false* ribs; of these the cartilages of the eighth, ninth and tenth are joined to the cartilage of the rib immediately above; the eleventh and twelfth are free at their anterior ends, and are *floating* ribs. (The latter statement requires qualification: the tenth rib is also *usually* 'floating' in Japanese, a condition also recorded in other races: *see* p. 291.)

The ribs are situated one below the other in such a manner that *intercostal spaces* intervene. These are deeper in front than behind, and deeper between the upper than between the lower ribs. The ribs vary in direction, the upper ones being less oblique than the lower; the obliquity reaches its maximum at the ninth rib, and gradually decreases to the twelfth. The ribs increase in length from the first to the seventh, and then diminish to the twelfth. In breadth they decrease successively from above downwards; in the upper ten the greatest breadth is at the anterior extremity. The first two and the last three ribs present special features, but the remaining seven conform to a common plan and are referred to as typical.

A typical rib has a posterior and an anterior end, and an intervening portion—the shaft (3.55, 56).

The *anterior end* can be distinguished by a small cup-shaped depression, for the lateral end of the costal cartilage. The shaft is curved with the convexity outwards, and is grooved along the lower part of its inner surface so that the lower border of the shaft is thin and sharp in contrast to the thick, rounded upper border.

The *posterior or vertebral end* possesses a head, a neck and a tubercle.

The *head* presents two facets, separated by a transverse ridge, named the *crest*. The lower facet, which is the larger, articulates with the body of the numerically corresponding vertebra, and the crest of the head is attached to the intervertebral disc.

The *neck* is the flattened portion which succeeds the head; it lies in front of the transverse process of the corresponding vertebra. It is oblique, so that its anterior surface faces forwards and upwards. Its posterior surface is directed backwards and downwards, and is roughened and pierced by numerous foramina. Its upper border is sharp and forms the *crest of the neck of the rib*: its lower border is rounded.

The *tubercle* is on the outer surface of the posterior part of the rib, at the junction of the neck with the shaft; it is more prominent in the upper than in the lower ribs and is divided into a medial articular and a lateral non-articular portion. The articular portion bears a small, oval facet for articulation with the transverse process of the numerically corresponding vertebra; the non-articular portion is rough for ligamentous attachments.

The *shaft* is thin and flattened, with external and internal surfaces, superior and inferior borders. It is not only curved but also bent, the *posterior angle* (often simplified to 'the angle'), being 5–6 cm from the tubercle. The shaft is twisted in its long axis, as is apparent if the rib is placed on a horizontal surface. The part behind the angle inclines medially and upwards, and its outer surface faces downwards and backwards; in front of the angle the outer surface faces slightly upwards. The *external surface* is convex and smooth. A short distance from the tubercle it is crossed by a rough line, directed downwards and laterally, the position of the (posterior) angle. The *internal surface* is smooth and is marked along its lower border by the *costal groove*, which is bounded below by the inferior border of the shaft. The upper border of the groove is continuous behind with the lower border of the neck, but anteriorly it terminates at the junction of the middle and anterior thirds of the shaft; in front of this point, the groove is absent. (*Vide infra* concerning the *anterior angle*.)

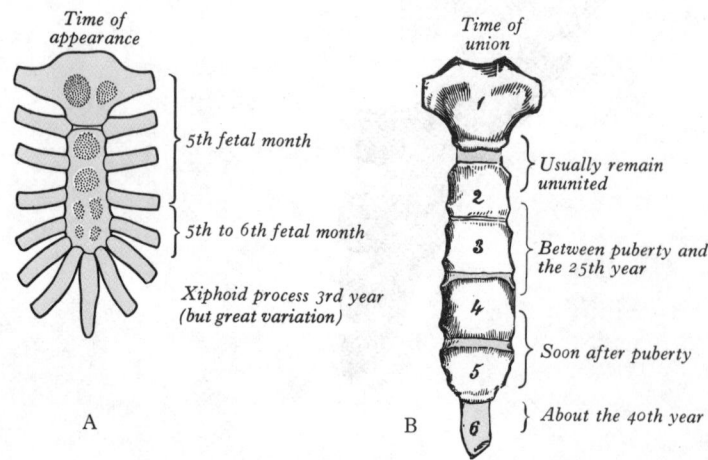

Time of appearance

5th fetal month

5th to 6th fetal month

Xiphoid process 3rd year (but great variation)

A

Time of union

1

Usually remain ununited

2

3
Between puberty and the 25th year

4

5
Soon after puberty

6
About the 40th year

B

3.54A and B The ossification of the sternum. A Before birth. B At puberty.

The first rib (3.57) is the most curved and usually the shortest of the ribs; it is broad and flat, its surfaces facing 'upwards' and 'downwards' and its borders inwards and outwards. It is placed very obliquely in the body, sloping downwards and forwards from its vertebral to its sternal end. The *head* is small and rounded, and bears a single, nearly circular, articular facet, which articulates with the body of the first thoracic vertebra. The *neck* is rounded, and directed upwards, backwards and laterally. The *tubercle*, wide and prominent, is directed upwards and backwards; medially an oval facet articulates with the transverse process of the first thoracic vertebra. At the tubercle the rib is bent, so that the head of the bone is directed slightly downwards; the angle and the tubercle therefore coincide. The *upper surface* of the shaft is crossed obliquely by two shallow grooves, separated from each other by a slight ridge, which ends at the inner border of the rib in a small projection, the *scalene tubercle*. The *under surface* is smooth and has no costal groove. The *outer border* is convex, thick behind, but thin in front. The *inner border* is concave and thin, and marked near its centre by the scalene tubercle. The *anterior end* is larger and thicker than in any other rib.

The second rib (3.58) is about twice the length of the first, but has a similar curvature. The non-articular portion of the *tubercle* is often small. The *angle* is slight, and close to the tubercle. The *shaft* is not twisted, so that both ends touch any plane surface upon which the rib may be laid; but at the tubercle there is an upward convexity, similar to, but smaller than, that in the first rib. The *external surface* of the shaft is convex, and looks upwards and a little outwards; near its middle it is marked by a rough, muscular impression. The *internal surface,* smooth and concave, is directed downwards and a little inwards; on its posterior part there is a short costal groove.

The tenth rib has a single articular facet on its head, which may articulate with the intervertebral disc above as well as with the upper border of the tenth thoracic vertebra close to its pedicle.

The eleventh and twelfth ribs (3.59) each have a single, relatively large, articular facet on the head; they have no necks or tubercles; their anterior ends are pointed and tipped with cartilage. The eleventh has a slight angle and a shallow costal groove. The twelfth has neither; it is much shorter than the eleventh, and its vertebral end is directed slightly upwards. The inner surfaces of both ribs look *upwards* as well as inwards, the upward inclination being more marked in the twelfth.

The *head of a typical rib* gives attachment along its anterior border to the radiate ligament, and on its crest to the intra-articular ligament. The anterior surface of the head is related to the costal pleura, and, in the lower ribs, to the sympathetic trunk. The *anterior surface* of the neck is divided into an upper and a lower area by a faint ridge, which affords attachment to the internal intercostal membrane and is continuous with the inner of the two lips of the superior border of the shaft. The upper area, of varying size and more or less triangular in shape, is separated from the internal intercostal membrane by some fatty tissue; the lower area is smooth and covered with the costal pleura. The *posterior surface* of the neck gives attachment to the costotransverse ligament and is pierced by numerous vascular foramina. The *crest of the neck* is rough for the attachment of the superior costotransverse ligament, and it can be traced laterally into the outer lip of the superior border of the shaft. The *inferior border* of the neck is rounded and can be traced laterally into the upper border of the costal groove; it gives attachment to the internal intercostal membrane. The *articular part of the tubercle* in the upper six ribs is convex and faces backwards and medially; in the succeeding three or four ribs it is flattened and faces downwards, backwards and slightly medially. The *non-articular part of the tubercle* gives attachment to the lateral costotransverse ligament.

The ridge which marks the (posterior) *angle* on the *external surface* of the *shaft* of a typical rib gives attachment to the upward continuation of the thoracolumbar fascia and the most lateral fibres of the iliocostalis thoracis. From the second to the tenth ribs the distance between the angle and the tubercle becomes progressively greater. Medial to the angle, the external surface gives attachment to the corresponding levator costae and is covered by the erector spinae. Near the sternal end of this surface an indistinct, oblique line (which marks the *anterior* 'angle')

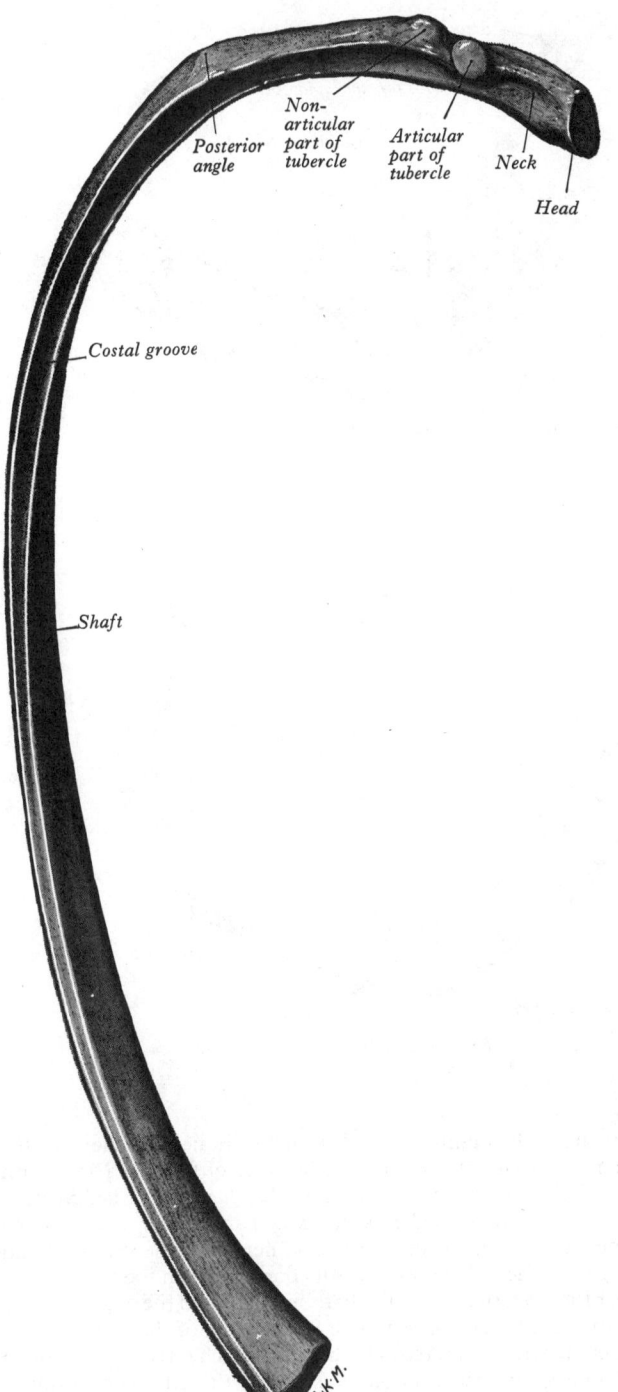

3.55 A typical rib of the left side. Inferior aspect.

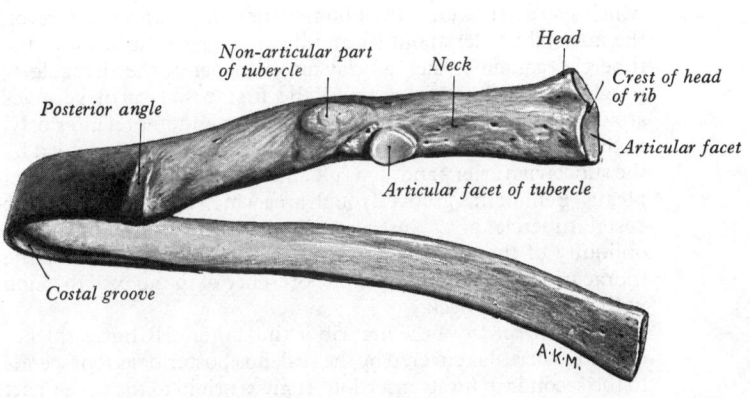

3.56 A typical rib of the left side. Posterior aspect.

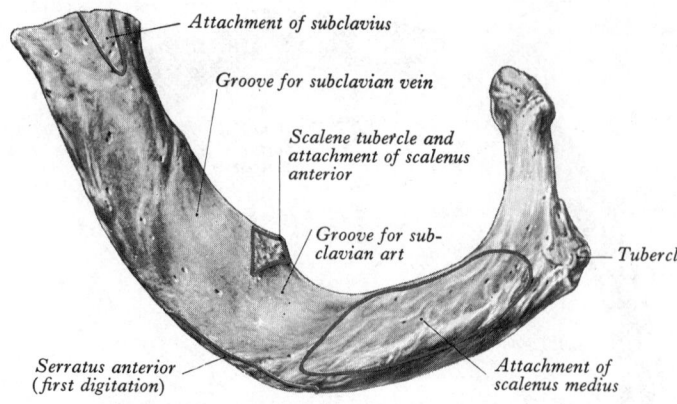

3.57 The first rib of the left side. Superior aspect.

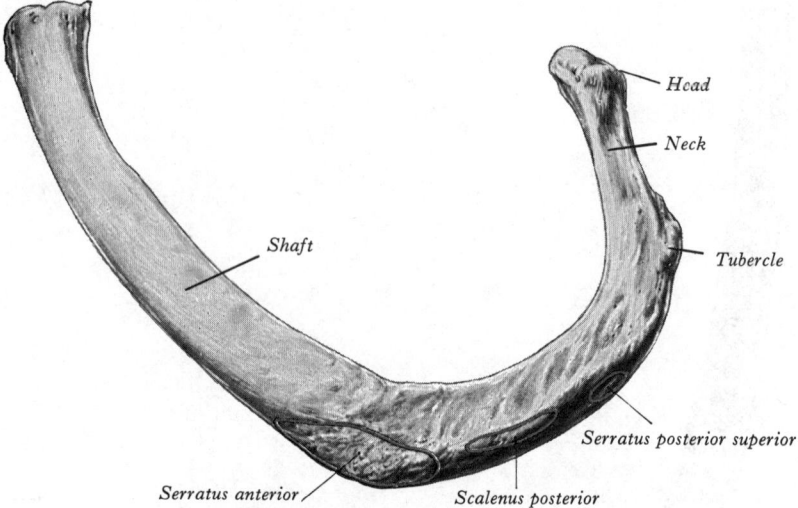

3.58 The second rib of the left side. Superior aspect.

separates the origins of the external oblique and the serratus anterior (or latissimus dorsi, in the cases of the ninth and tenth ribs). The *costal groove* on the *internal surface* gives attachment to the internal intercostal muscle, which intervenes between the bone and the intercostal vessels and nerve. At the vertebral end the groove faces downwards, as its borders lie on the same plane. Near the posterior angle the shaft broadens and the groove passes on to the internal surface. The upper edge of the groove gives attachment to the intercostalis intimus, which rarely extends on to the anterior fourth of the rib. Posteriorly this edge is continuous with the lower border of the neck. The sharp lower border of the rib gives origin to the external intercostal muscle. Its upper border is marked, posteriorly, by inner and outer lips: the inner gives attachment to both the intercostalis internus and intimus; the outer only to the intercostalis externus.

The first rib (**3.57**) presents the *tubercle for the scalenus anterior,* which also extends on to the adjoining part of the *upper surface,* on the internal border about its middle. The groove in front of the tubercle contains the subclavian vein, and the irregularly roughened area between it and the first costal cartilage gives attachment to the costoclavicular ligament and, more anteriorly, to the subclavius. The groove behind the tubercle is occupied by the subclavian artery and, as a rule, the lower trunk of the brachial plexus. Behind this groove a rough area which extends as far as the costal tubercle gives insertion to the scalenus medius. The obliquity of first rib is responsible for the obliquity of the thoracic inlet and accounts for the presence of the apex of the lung in the root of the neck.

The outer border of the first rib is thin anteriorly but is thicker behind, where it is covered by the scalenus posterior as it descends to the second rib for its insertion. It gives origin to the upper part of the first digitation of the serratus anterior, behind and opposite

to the groove for the subclavian artery. The *inner border* gives attachment to the suprapleural membrane, which covers the cervical dome of the pleura.

The *second rib* (**3.58**) bears a rough *tubercle for the serratus anterior* on its external surface just behind its midpoint; to this tubercle the lower part of the first and the whole of the second digitation are attached. The *costal groove* is very poorly marked on the *internal surface* and is restricted to its posterior part. The second intercostal nerve lies between the second rib and the pleura in most of its course. The inner and outer lips of the upper border are distinct and are widely separated behind. Immediately in front of the poorly marked angle the outer lip is roughened above for the insertion of the scalenus posterior and, below this, of the serratus posterior superior.

The *twelfth rib* (**3.59**), although short, has attached to it numerous muscles and ligaments. The lower part of its *anterior surface,* in its medial half to two-thirds, gives insertion to the quadratus lumborum and its covering fascia. Superior to these attachments the surface is related to the costodiaphragmatic recess of the pleura. At or close to the upper border are the attachments of the internal intercostal medially and the diaphragm laterally. The *lower border* affords attachment to the middle lamella of the thoracolumbar fascia and, at the lateral border of the quadratus lumborum, to the lateral arcuate ligament (p. 549). Posteriorly, close to the head, is attached the lumbocostal ligament (p. 451), by which it is connected to the transverse process of the first lumbar vertebra. The *external surface* has attached to it the lowest levator costae, the longissimus thoracis, and the iliocostalis in its medial half; more laterally, it gives insertion to the serratus posterior inferior, and origin to the latissimus dorsi and the external oblique muscle of the abdomen. Along the *upper border* is the insertion of the external intercostal. These muscle attachments show considerable variation. The insertions of the internal intercostal, levator costae and erector spinae merge and are difficult to define. The attachments of the latissimus dorsi, diaphragm and external oblique may extend on to the costal cartilage. The lower limit of the pleural sac crosses the anterior aspect of the twelfth rib, approximately at the point where the rib is crossed by the lateral border of the iliocostalis. The lateral extremity of the rib usually lies below the line of pleural reflection and is therefore not covered with pleura.

The ribs consist of the highly vascular trabecular bone enclosed in a thin layer of compact bone and containing a large proportion of red marrow (*see* **3.10B**).

Ossification. Each rib, with the exception of the first and the last two, is ossified from four centres—a primary centre for the shaft, and three secondary centres, one for the head and one each for the articular and non-articular parts of the tubercle (Fawcett 1911). No epiphysis is usually present on the non-articular part of the tubercle below the sixth or seventh rib. The primary centre appears near the angle, towards the end of the second month of intrauterine life, first in the sixth and seventh ribs. The secondary centres for the head and tubercle appear about puberty, and unite with the shaft soon after the twentieth year. The first rib has three centres, viz.: a primary one for the shaft, a secondary centre for the head, and one for the tubercle. The eleventh and twelfth ribs, being destitute of tubercles, have each only two centres.

The Costal Cartilages

The costal cartilages (**3.60**) are flattened bars of hyaline cartilage which extend forwards from the anterior ends of the ribs and contribute materially to the mobility and elasticity of the walls of the thorax. The first seven pairs are connected with the sternum; the eighth, ninth and tenth articulate with the lower border of the cartilage immediately above; the lowest two are pointed, and end in the muscular wall of the abdomen. The cartilages increase in length from the first to the seventh, and then gradually decrease to the twelfth. They diminish in breadth from the first to the last, like the intervals between them. They are broad at their attachments to the ribs, and taper towards their medial extremities, with the exception of the first and second which are of the same breadth throughout, and the sixth, seventh, and eighth which are enlarged where their margins are in contact. The first cartilage descends a little, the second is horizontal, the third

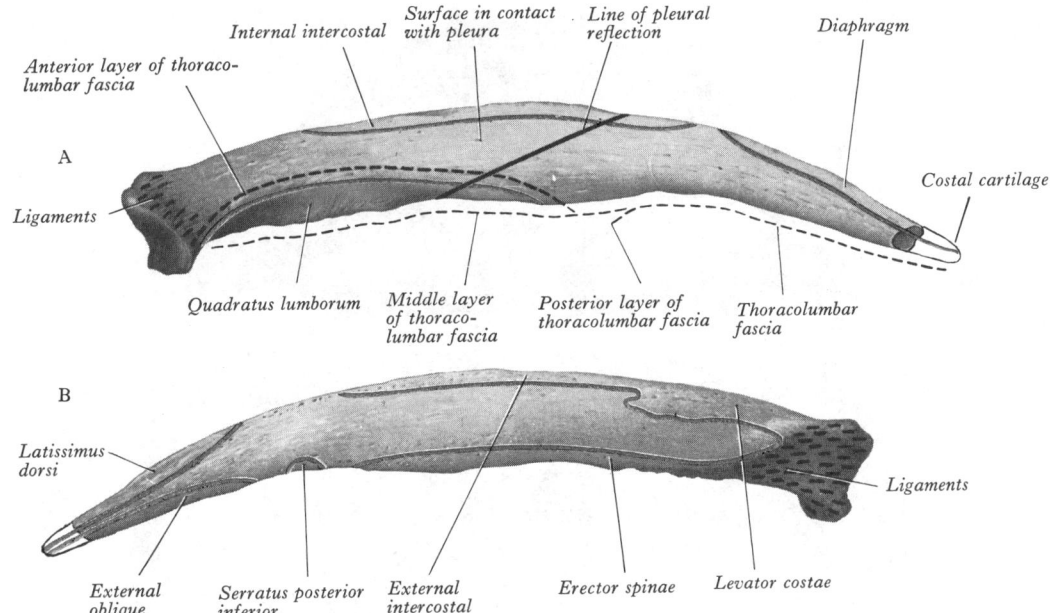

Labels for figure:
Anterior layer of thoraco-lumbar fascia
Internal intercostal
Surface in contact with pleura
Line of pleural reflection
Diaphragm
A
Ligaments
Costal cartilage
Quadratus lumborum
Middle layer of thoraco-lumbar fascia
Posterior layer of thoracolumbar fascia
Thoracolumbar fascia
B
Latissimus dorsi
Ligaments
External oblique
Serratus posterior inferior
External intercostal
Erector spinae
Levator costae

3.59   The twelfth rib of the left side. A anterior aspect and B, posterior aspect.

ascends slightly, while the others are angular, continuing the course of the ribs for a short distance, and then inclining upwards to the sternum or cartilage above.

Each costal cartilage has two surfaces, two borders, and two ends. The *anterior surface* is convex, and faces forwards and upwards; that of the first gives attachment to the sternoclavicular articular disc, the costoclavicular ligament and subclavius; those of the first six or seven at their medial ends, to the pectoralis major. The others are covered by, and give partial attachment to, some of the flat muscles of the anterior abdominal wall. The *posterior surface* is concave, and directed backwards and downwards; that of the first gives attachment to the sterno-thyroid, those of the second to the sixth inclusive to the transversus thoracis, and the six lower ones to the transversus abdominis and the diaphragm. The *superior border* is concave, and the *inferior* convex; they afford attachment to the internal intercostal muscles, and the external intercostal membranes. The inferior borders of the sixth, seventh, eighth and ninth cartilages present heel-like projections at the points of greatest convexity; a similar projection is frequently present on the lower border of the fifth cartilage. On these projections are oblong facets which articulate respectively with facets on slight projections from the superior borders of the sixth, seventh, eighth, ninth and tenth cartilages. The *lateral end* of each cartilage is continuous with the osseous tissue of the corresponding rib. The *medial end* of the first is continuous with the sternum; the medial ends of the six succeeding cartilages are rounded and articulate with the shallow costal notches on the lateral margins of the sternum. The medial ends of the eighth, ninth and tenth costal cartilages are pointed, and each is connected with the cartilage immediately above. Those of the eleventh and twelfth are pointed and free. With the exception of the junction of the first rib with the sternum, all these articulations are synovial (p. 451).

The tenth rib is usually described as united, at its ventral extremity, with the ninth by a fibrous joint (p. 452); but it is frequently not so connected, presenting a free pointed extremity like the eleventh and twelfth ribs. The incidence of a 'floating' tenth rib varies from 35 to 70 per cent in different racial studies (Shimaguchi 1974).

In old age the costal cartilages are prone to undergo superficial ossification, losing their pliability and becoming brittle. The histology (Amprino and Bairati, 1933), histochemistry (Quintarelli and Delaro, 1966) and ultrastructure of senile changes in costal cartilage have been reviewed by Dearden *et al.* (1974). An increase in keratan sulphate is noted, maximally in the subperichondral zone.

The Thorax

The skeleton of the thorax or chest (**3**.60), is an osseocartilaginous framework within which are the principal organs of respiration and circulation. It is conical in shape, narrow above and broad below, flattened from before backwards, and longer behind than in front. It is reniform on horizontal section on account of the forward projection of the vertebral bodies.

Posteriorly the thoracic skeleton includes the twelve thoracic vertebrae and the posterior parts of the ribs. At each side of the vertebral column there is a wide and deep groove in consequence of the lateral and backward direction which the ribs follow from their vertebral extremities to their angles. *Anteriorly* are the sternum, the anterior ends of the ribs and the costal cartilages, and this aspect is flattened or slightly convex. *Laterally* the thorax is convex, and is formed by the ribs. The ribs and costal cartilages are separated from each other by the intercostal spaces, eleven in number, which are occupied by the intercostal muscles and membranes, and neurovascular bundles and lymphatics.

The *inlet* of the thorax is reniform in shape; its anteroposterior diameter is about 5 cm, its transverse about 10 cm. Its plane slopes downwards and forwards, and it is bounded by the first thoracic vertebra behind, the superior border of the manubrium sterni in front, and the first rib on each side. The *outlet* is bounded by the twelfth thoracic vertebra behind, by the twelfth and eleventh ribs at the sides, and in front by the cartilages of the tenth, ninth, eighth and seventh ribs, which ascend on each side and form an angle, termed the *infrasternal angle*. The outlet is wider transversely than from before backwards, and slopes obliquely downwards and backwards; it is closed by the diaphragm, which forms the floor of the thoracic cavity.

Like any other part of the skeleton, the thorax displays variation in dimensions and proportions which are partly individual and also linked to age, sex and race. In the newborn the transverse diameter is relatively less, but a change to adult proportions occurs when the child begins to walk. In the female the capacity is less, absolutely and proportionately; the sternum is relatively shorter and the thoracic inlet more oblique, the suprasternal notch being level with the third thoracic vertebra, rather than the second, as in the male. The upper ribs are also more movable in females, permitting comparatively greater expansion of the upper thorax. In tall and thin individuals the thoracic skeleton usually shows similar proportions, as also in the short and broad. Racial variations are likewise chiefly linked to stature and general proportions. In a race which is characteristically short and broad, the thorax shares in these proportions.

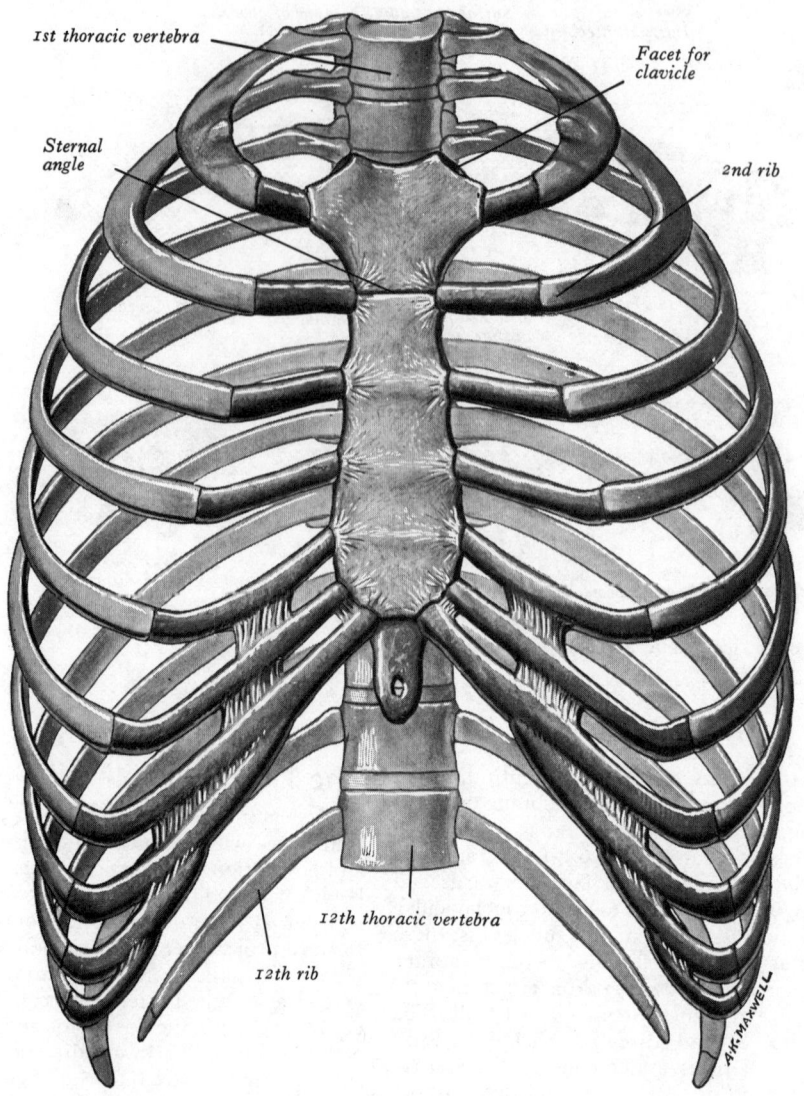

1st thoracic vertebra

Facet for clavicle

Sternal angle

2nd rib

12th thoracic vertebra

12th rib

3.60   The skeleton of the thorax. Anterior aspect.

The outstanding functional significance of the thorax is in respiration (p. 453). The obvious protection afforded is perhaps best regarded as fortuitous, but a large number of muscles are attached to the thoracic skeleton, not all of which are primarily concerned with respiratory movements, though they may also share in these. The muscles of the arm, especially those acting on its girdle, those of the abdominal wall, and also those of the spinal column have widespread thoracic attachments.

Applied Anatomy. The elastic recoil of the ribs, which suspend the sternum, may account for the comparative rarity of sternal fractures. Despite their ability to bend under stress, ribs are much more frequently broken. The middle ribs are most liable; and since the stress is usually of a bending nature, due to compression of the thorax, the weakest point, which is just in front of the angle, is the most frequent site of fracture. Direct impact may fracture a rib anywhere and in this case in particular the broken ends of the bone may be driven inwards, with the possible complication of injury to thoracic or upper abdominal viscera.

A *cervical rib*, derived from the costal element of the seventh cervical vertebra, may be a mere epiphysis articulating only with its transverse process; more commonly it has a definite head, neck

and tubercle, with or without a shaft. According to the length of the shaft it extends laterally and forwards into the posterior triangle of the neck, where it may have a free end or join the first rib or costal cartilage, or even the sternum. It may be in part a fibrous cord, but the effects of its presence are not necessarily related to the size and development of its osseous part. If it reaches far enough forwards its relations are like those of a first thoracic rib: part of the brachial plexus (usually the lower trunk) and the subclavian vessels are *superior* to it, and they are apt to suffer some degree of compression here, being confined in a narrow angle between the cervical rib and the scalenus anterior. As a result, cervical ribs may first be discovered after the onset of nervous and vascular symptoms. In particular, pressure on the fibres of the eighth cervical and first thoracic spinal nerves may lead to motor and sensory effects in the structures which they supply.

According to Cave (1975) a cervical rib or pleurapophysis may show a variety of forms of junction—synostosis or diarthrosis—with either the anterior (parapophysial) or posterior (diapophysial) 'roots' of the so-called transverse process of a seventh cervical vertebra, or, more usually, with both of them. (*See* **3**.29 for a recent morphological analysis of the cervical pleurapophysis.)

THE SKULL

Introduction

The vertebrate skull is the most highly modified region in the axial skeleton. Whatever truth may reside in the hypothesis that the skull is partly derived from metameric elements, in the form of modified vertebrae, the evidence of palaeontology and comparative anatomy clearly indicates that from the earliest times it has been a complex of skeletal elements, adapted to support and contain the brain and special senses and to secure and devour food. The siting of special receptors, especially those of the chemical senses and of vision, and the feeding orifice, at the leading or head end of an elongate creature such as a vertebrate, offers advantages so fundamental that they are often overlooked. Associated with this concentration of functions is the enlargement of the fore end of the central nervous system into a brain, whose relative size and dominance have increased throughout the vertebrate series. The

circulation. Moreover, the reputed buffering mechanism, due to the meninges, subarachnoid space and its contained fluid, could only work effectively within such a rigid container.

The extreme dependence of the brain upon an uninterrupted flow of blood is well known; the independence of the cerebral arterial pressure from extracranial variations in vascular pressures, due to some form of auto-regulation as yet unidentified, is also well established. It appears likely that the situation of the brain in a rigidly contained space is a factor of some importance in these arrangements, though it appears to have evaded any exact assessment. Of course, the cranial cavity is not completely closed; cerebrospinal fluid passes freely through the foramen magnum and can be displaced in either direction. The brainstem itself can also be displaced downwards to some extent through the same orifice under the influence of raised intracranial pressure. Variability in the volume of fluid in the cerebral

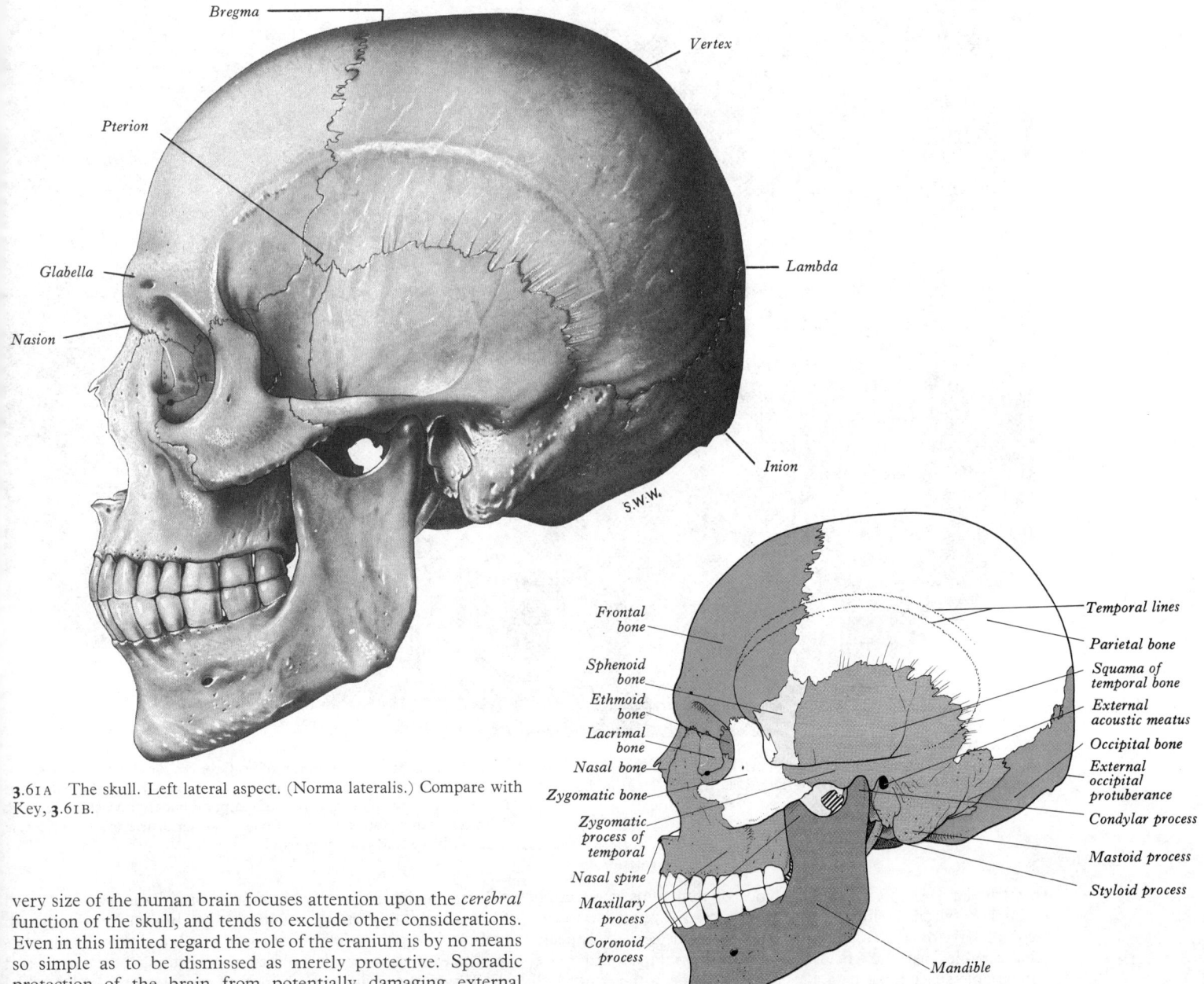

3.61A   The skull. Left lateral aspect. (Norma lateralis.) Compare with Key, **3**.61B.

very size of the human brain focuses attention upon the *cerebral* function of the skull, and tends to exclude other considerations. Even in this limited regard the role of the cranium is by no means so simple as to be dismissed as merely protective. Sporadic protection of the brain from potentially damaging external impacts is of undoubted value; the need of a barrier against the stresses and pressures due to the play of powerful masticatory and axial musculature is perhaps less obvious, but is habitual and frequent. Quite apart from these extraneous forces, the rigid walls of the cranium afford a continuous isolation to the cerebral

3.61B   The skull. (Norma lateralis.) Key to **3**.61A. *Note.—Blue*=frontal and occipital; *brown*=temporal, lacrimal and nasal; *magenta*=mandible and ethmoid; *green*=maxilla; *yellow*=sphenoid; *white*=zygomatic and parietal.

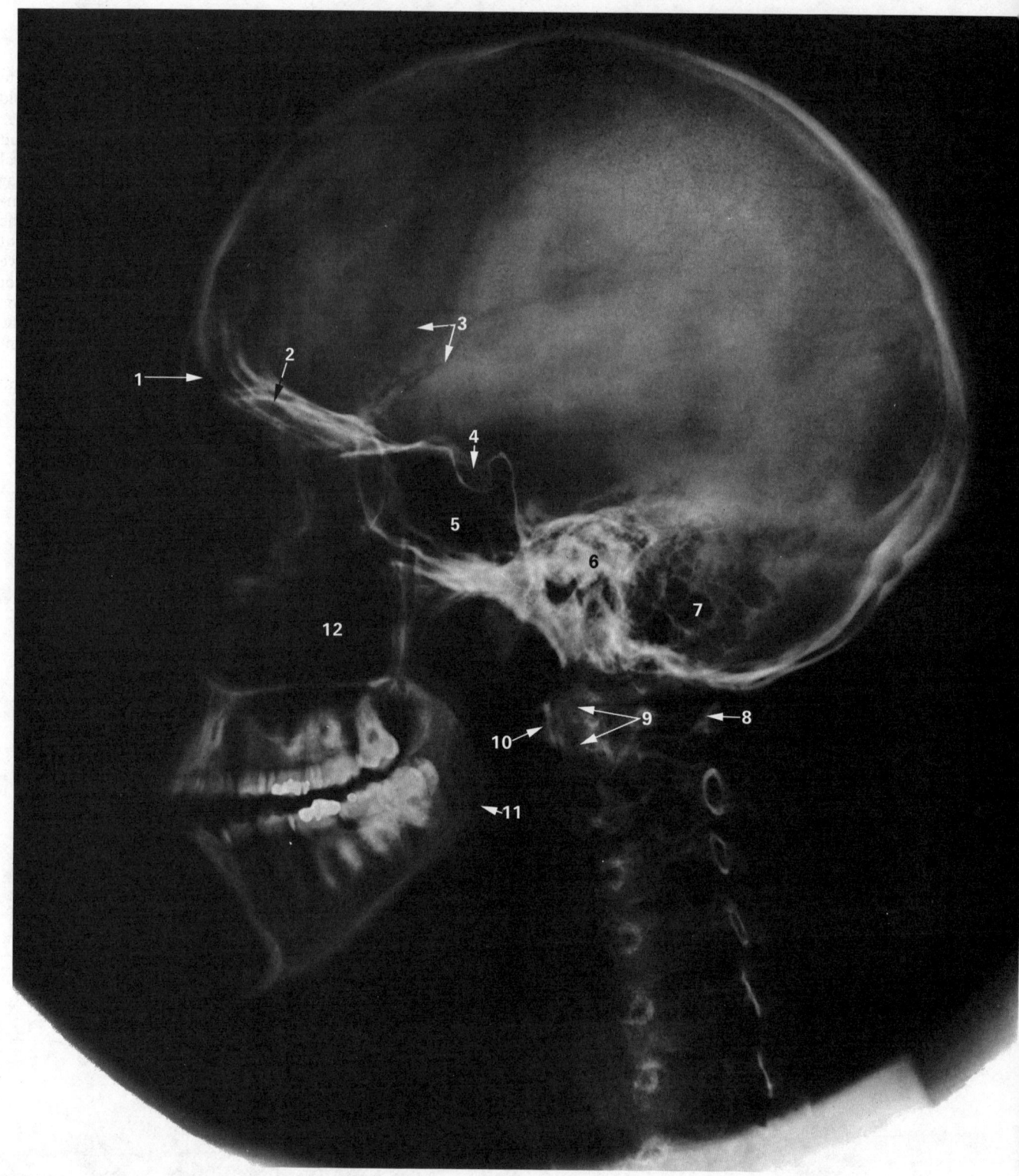

3.62A and B Lateral (A) and anteroposterior (B) radiographs of adult female skull. (Kindly provided by Dr. R. D. Hoare, Neuroradiologist, Guy's Hospital.)

A 1. Frontal sinus. 2. Orbital roof. 3. Grooves for anterior branches of middle meningeal vessels. 4. Hypophysial fossa. 5. Sphenoidal sinus. 6. Dense shadow of petrous part of temporal bone. 7. Mastoid air cells. 8. Posterior arch of atlas. 9. Dens of axis. 10. Anterior arch of atlas. 11. Angle of mandible. 12. Maxillary sinus.

ventricular system and the existence of numerous connecting vessels between the intra- and the extra-cranial veins add to the complexity of the total situation. Nevertheless, it appears inescapable that the enclosure of the brain in an otherwise invariable space must be a considerable factor in the control of cerebral circulation. This peculiar, perhaps unique, location of the brain entails some penalties, however well adapted to ordinary circumstances. Lesions which occupy space can raise pressure within the cranium far more easily than elsewhere, and with far more devastating effects.

Even in the earlier vertebrates the *neurocranium*, developed from a series of cartilages ventral to the brain as in the human embryo (p. 141), has been joined by special cartilaginous supports for the external nares and olfactory receptors, the eyeballs, and the labyrinths. With the addition of jaw elements of branchial origin, and of dermal or 'membrane' bones, arising in the mesenchyme underlying the skin over the dome of the head, over the jaws and in the roof of the buccal cavity, the vertebrate skull becomes defined in all its complexity. It must be emphasized that all the cartilaginous elements usually become ossified, except in

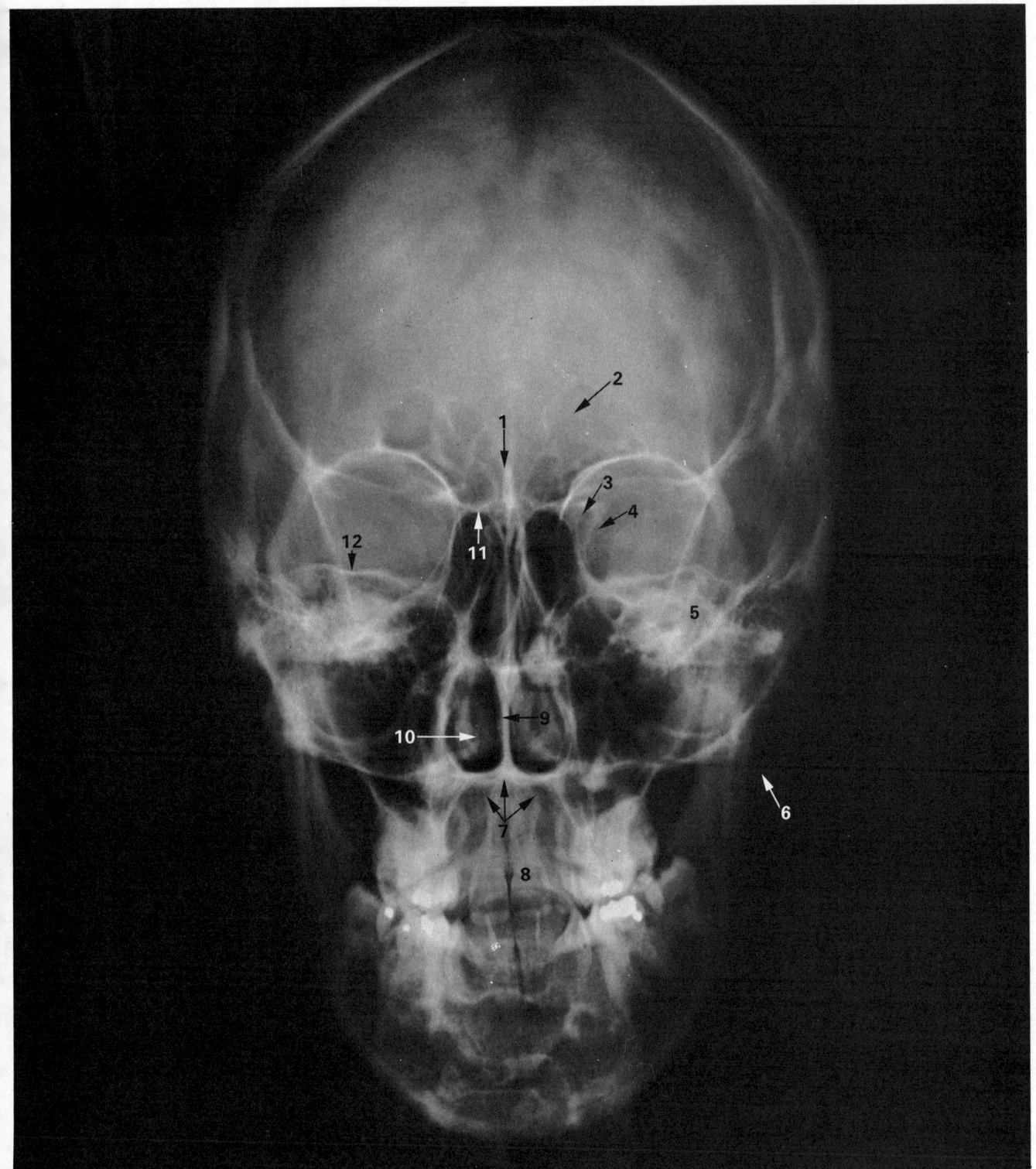

B   1. Crista galli. 2. Frontal sinus. 3. Optic canal. 4. Superior orbital
fissure. 5. Dense shadow of petrous part of temporal bone. 6. Apex of
mastoid process. 7. Dens of axis. 8. Upper central incisor tooth. 9. Nasal
septum. 10. Inferior concha. 11. Cribriform plate. 12. Superior border of
petrous part of temporal bone.

minor groups, such as the sharks, where the persistence of
cartilage is considered to be degenerate rather than primitive.
Moreover, some of the dermal elements appear to be as ancient—
judged by the fossil record—as 'cartilage' bones, and may even
have preceded them in evolution. It may therefore be misleading
to regard cartilage as the primeval substance in this respect. Bone
may arise by two major routes; either by direct ossification in
mesenchymal tissues, or indirectly in cartilages, themselves
derived from such tissue. The roof and sides of the neurocranium,
or 'brainbox', are developed in the former manner from sheets, or

membranes, of mesenchyme in or subjacent to the dermis; hence
the terms 'membrane' or 'dermal' bone, the latter being perhaps
preferable as it is more informative. These dermal members of the
cranial box are currently considered to be at least as primitive in
origin as its chondrocranial base. In the case of the jaws the march
of events appears more certain; the primary cartilaginous
elements, derived from the branchial apparatus, have been
replaced by overlying dermal bone.

The 'capsules', enclosing the special sense organs, which have
become integral parts of the vertebrate skull, obviously afford a

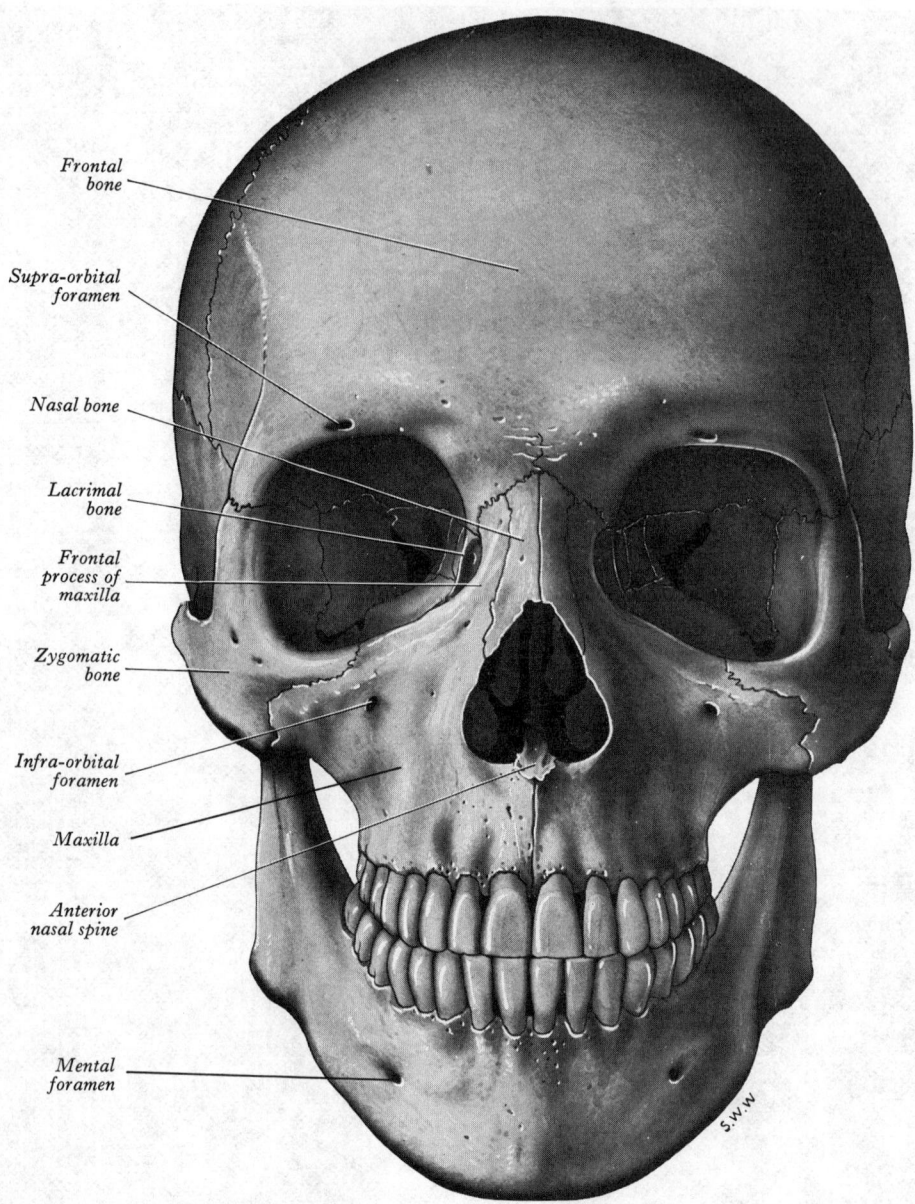

Frontal bone

Supra-orbital foramen

Nasal bone

Lacrimal bone

Frontal process of maxilla

Zygomatic bone

Infra-orbital foramen

Maxilla

Anterior nasal spine

Mental foramen

3.63A   The skull. Anterior aspect. (Norma frontalis.) Compare with the Key, **3**.63B.

measure of protection; but they also confer advantages which, though less obvious, may be of greater significance in terms of survival. The primitive nostrils of lower vertebrates are concerned in olfaction; and even when they lead backwards into nasal cavities with the development of a secondary palate in the majority of land vertebrates, the olfactory function persists side by side with the new respiratory arrangement. In both functions the maintenance of open airways, whatever sphincteric mechanisms are added, is a clear necessity. The skeletal elements surrounding the eyes have from the beginning provided sockets, more or less complete, containing not only the eyeballs but also their muscles. The latter, in their attachments to the walls of the optic 'capsule', have varied only in minor details of arrangement from their appearance in the earliest vertebrates, judging from their extant forms. With little exception their effects are rotatory, and it is difficult to envisage how this mode of action could be effective in the absence of a socket. The bony orbit not only aids the stationing of the eyeball during its rotations, but also ensures that the distance between the two eyes remains equal, a necessary prelude to binocular vision.

The deep situation of the labyrinths in cartilage or bone entails a fixed relationship between the three semicircular canals on each side. The fusion of the otic capsules into the base of the cranium determines a strict orientation of the six canals not only with

respect to each other but also in relation to the head itself. Without such an arrangement no orderly correlation between these peripheral receptors and their central nervous connexions could be developed.

Axial muscles have extended from the vertebral column to the caudal aspects of the skull from the earliest differentiation of these regions, and they become further elaborated and more massive in land animals with the development of a neck and the increased problems of suspending the head in quadrupeds. In all but the jawless fishes (Agnatha), such as the lamprey, a numerically insignificant class of vertebrates, the primitive skull is invaded by the ligamentous and muscular apparatus of the jaws, the maxillae becoming integrated with the rest of the skull at a very early stage. In both cases the large muscle masses, whether axial or mandibular, create stresses transmitted to the skull which are at least considerable and sometimes very great, and which are associated with extensive cranial modifications to resist and absorb them. Even in human beings, whose masticatory and neck musculatures are of modest development, the whole body weight can be suspended from the bite of the teeth. In other mammals, especially amongst quadrupeds, the far greater development of jaws and greater weight of head accentuate these problems and lead to the development of large plates, bars and buttresses of bone which owe nothing in their origin to the protection which they undoubtedly increase. A perfect example of this is to be seen in the high cranial dome of the elephant, which is associated with the production of a large area of attachment for the massive extensor neck muscles necessary to the suspension of so weighty a head. However, enough has perhaps now been said to emphasize the diversity of functions in which the skull is involved beyond the truism of its protectiveness.

General Cranial Features

When the mandible is left out of account the remainder of the skull, strictly speaking, constitutes the *cranium*, but the term skull is widely used with the same significance. The upper part of the cranium forms a box to enclose and protect the brain, and is often termed the *calvaria*. The remainder of the skull forms the *facial skeleton*, of which the upper part is immovably fixed to the calvaria and the lower part is the freely movable mandible.

The skull, considered as a whole, is of greater practical importance than the individual bones of which it is made up. Nevertheless, the position of its individual constituents must be familiar before the student can follow more detailed description (**3**.61 A, B; 62 A, B).

The skull may be viewed from above (*norma verticalis*), from below (*norma basalis*), from behind (*norma occipitalis*), from in front (*norma frontalis*) and from the side (*norma lateralis*). The roof of the calvaria, or *calva* (skull cap), may be removed and the interior may be examined. In the erect attitude the lower margins of the orbital openings and the upper margins of the external acoustic meatuses lie on the same horizontal (the Frankfurt plane), and it is important that the student should bear this in mind when examining the various aspects of the skull.

The region of the forehead is formed by the *frontal bone* (**3**.61 A, B), which passes backwards in the vault of the skull as far as the *coronal suture*, where it meets the anterior borders of the right and left *parietal bones*. These two bones together form the greater part of the cranial vault, and they articulate with each other at the serrated *sagittal suture*. Posteriorly they extend backwards to meet the *occipital bone*, which forms the back of the head. The suture between the two parietals and the occipital bone is called *lambdoid*, after the Greek letter lambda, **Λ**, which it resembles in shape. Each parietal bone curves downwards on the side of the vault until it meets the upper limit of the *greater wing* of the *sphenoid bone* in front, and the *squamous part* of the *temporal bone* behind. When the *calva*, or *skull cap* is removed, the section passes through the frontal bone and usually cuts across the lower part of the parietal bone, but it may involve the squamous part of the temporal bone. Posteriorly the section cuts the occipital bone. Consequently, the calva consists of (1) a large part of the frontal bone; (2) most of the two parietal bones; (3) possibly, small parts of the squamae of the temporal bones; and (4) a small part of the occipital bone.

When the calva is removed, the floor of the calvaria, the *base of the skull*, is exposed. It shows a natural subdivision into three regions, which are named the anterior, middle and posterior cranial fossae (**3**.75 A, B).

The anterior cranial fossa forms less than the anterior third of the base and is limited behind by a sharp edge on each side of the median plane. It is important to observe that the floor of the anterior cranial fossa constitutes the roof of the orbit, on each side, and the roof of the nasal cavity, in the median area. On each side of the median plane an *orbital part* projects backwards from the *frontal bone* and constitutes most of the roof of the orbit. These two plates are separated by a relatively narrow interval, which is occupied by a perforated strip of bone. This is termed the *cribriform plate* of the *ethmoid bone*; it forms a large part of the roof of the nose, while the rest of the bone to which it belongs participates in the side walls and the septum of the nasal cavity. In the median plane the cribriform plate bears a crestlike elevation on its upper surface, the *crista galli*. The posterior part of the floor of the anterior cranial fossa is formed by the *sphenoid bone*. In the median area the front of the *body of the sphenoid* meets the cribriform plate of the ethmoid. On each side a narrow *lesser wing* projects laterally from the body and meets the posterior margin of the orbital part of the frontal bone. It is the sharp posterior border

of the lesser wing of the sphenoid bone which forms the posterior limit of the floor of the anterior cranial fossa on each side of the midline. This border is related to the lateral cerebral fissure (p. 984).

The middle cranial fossa (**3**.75 A, B), which lies immediately behind the anterior fossa, is of small extent in the median region but is expanded, in a backward and lateral direction, on each side. The narrow median portion of the floor is formed by the body of the sphenoid, the upper surface of which is hollowed out to contain the hypophysis cerebri (pituitary gland). The floor of the lateral part of the fossa is formed by the *greater wing* of the sphenoid in front, and by the *petrous part* of the *temporal bone* behind. The greater wing extends laterally from the side of the body of the sphenoid and curves upwards in the side of the skull to reach the antero-inferior part of the parietal bone. Behind it the floor of the fossa is formed by the anterior surface of the petrous part of the temporal bone, which is continuous laterally with its squamous region.

The posterior cranial fossa (**3**.75 A, B) is almost circular in outline and occupies roughly two-fifths of the base of the skull. It is formed to a large extent by the *occipital bone*. The large opening in its floor, termed the *foramen magnum*, is entirely within the bone and through it the brain becomes continuous with the spinal cord. The anterior part of the fossa is formed by the *basilar part* of the occipital bone, which is fused in front with the posterior part of the sphenoid. On each side the lateral wall of the fossa is formed by the posterior surface of the petrous part of the temporal bone above, and by the *lateral (condylar) part* of the occipital bone, below. The *mastoid part* of the temporal bone, which lies immediately posterolateral to the petrous part, joins the *squamous part* of the occipital bone to complete the fossa.

When the skull is viewed from in front (*norma frontalis*, **3**.63)

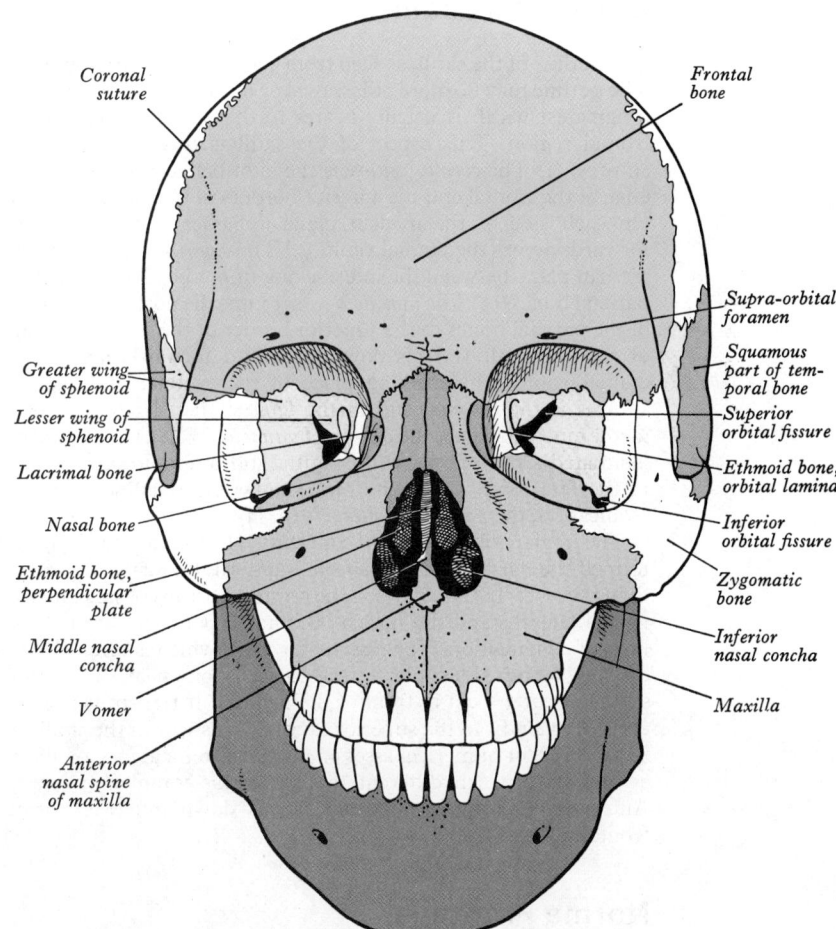

3.63B The skull. Anterior aspect. (Norma frontalis.) Key to **3**.63A. *Note.*—Blue=frontal bone; *yellow*=sphenoid bone; *green*=maxillae; *brown*=lacrimal, nasal, temporal bones and vomer; *magenta*=mandible; *uncoloured*=parietal, zygomatic and ethmoid bones.

the *orbits*, which contain the eyeballs, and the anterior aperture of the *nasal cavity*, are apparent. The part below the mouth is formed entirely by the *body* of the *mandible*; the part above the mouth is formed almost entirely by the *maxillae*, or upper jaws. The latter form the upper boundary of the mouth, and the lower and lateral boundaries of the anterior nasal aperture. In addition, on each side the maxilla forms the medial part of the lower margin of the orbit, which it helps the zygomatic bone to complete, while its frontal process ascends in the medial margin of the orbit to reach the *frontal*. The frontal processes of the two maxillae are separated from each other by the two *nasal bones*, which are the upper boundary of the anterior nasal aperture.

When the skull is viewed from the side (*norma lateralis*, **3**.61) the *ramus* of the *mandible*, which passes from the posterior end of the body of the bone upwards and slightly backwards to reach the cranium, is visible. The *head* of the mandible, which lies at the upper end of the posterior border of the ramus, is adapted to the *articular fossa* on the under surface of the squamous part of the temporal bone. The back of the mandibular head is separated from the ear passage, the *external acoustic meatus*, by the *tympanic part* of the temporal bone. Above and in front of the meatus the *zygomatic process* of the temporal bone passes forwards to meet the zygomatic or cheek bone, and the two form the *zygomatic arch*, or *zygoma*, which is separated widely from the rest of the side of the skull. The *zygomatic bone* is responsible for the prominence of the upper and anterior part of the cheek. It forms the lateral part of the lower margin of the orbital opening, as already stated, and ascends in the lateral margin to meet the frontal bone.

When the mandible is removed (**3**.70) a process of bone can be seen immediately behind the maxilla and above the level of the maxillary teeth. This is the *pterygoid process*, which project downwards from the sphenoid along the line of union of it greater wing with its body. It consists of a large lateral plate with a smaller one medial to it.

The inferior aspect of the cranium (*norma basalis*, **3**.71 A, B) i the external aspect of the base of the skull. Posteriorly is th *occipital bone*, with the foramen magnum. Lateral to the foramer magnum the occipital bone articulates with the *mastoid part o* the temporal bone. Anterolaterally it articulates with the *petrou part*, which extends forwards almost to the root of the pterygoic process. In the anterior part of the inferior aspect of the cranium the *bony palate*, which lies in the roof of the mouth, can be seer within the arch of the teeth of the maxilla. Four bones contribute to its formation, viz., the two maxillae and the two palatine bones The anterior three-fourths of the bony palate are formed by the *palatine processes* of the maxillae, which meet each other in the median plane; the posterior fourth is formed by the *horizonta plates* of the *palatine bones*; their *perpendicular plates* are hidden a they ascend, on each side, from the lateral border of the horizonta plate to form the posterior part of the lateral wall of the nose.

The *lacrimal bone*, which lies in the anterior part of the media wall of the orbit, the *vomer*, which forms a large part of the nasa septum (**3**.77), and the *inferior concha*, which lies in the lateral wal of the nasal cavity, can be seen only when the orbits and the nose are examined (pp. 300, and 313). With these exceptions all the bones of the skull have now been identified, and a more detailed study can be undertaken.

THE EXTERIOR OF THE SKULL

Norma Verticalis

The outline of the skull, as seen from above (**3**.64), varies greatly. The outline may be more or less oval or more nearly circular, but its greatest width is usually nearer to the occipital than to the frontal region. This aspect of the skull is traversed by three sutures. (1) The *coronal suture* is the joint between the posterior edge of the frontal and the anterior borders of the parietal bones. On each side of the median plane it passes downwards and forwards across the cranial vault. (2) The *sagittal suture* is in the median plane between the interlocking upper borders of the two parietal bones. (3) The *lambdoid suture* joins the posterior borders of the parietal bones to the superior border of the squamous part of the occipital. It runs downwards and forwards across the cranial vault. The meeting place of the coronal and sagittal sutures is the *bregma*, and in the fetal skull it is the site of a membrane-filled gap, the *anterior fontanelle* (p. 345). The *lambda* is situated at the junction of the sagittal and lambdoid sutures, and in the fetal skull is the site of a similar but smaller defect in ossification, the *posterior fontanelle* (p. 345).

The region of maximum convexity of the parietal bone is termed the *parietal tuber* (*eminence*) and can be easily felt in the living subject. In this situation the norma verticalis passes into the norma lateralis and the norma occipitalis, but there are no real lines of demarcation. The *parietal foramen*, which is often absent on one or both sides, pierces the parietal bone near the sagittal suture about 3·5 cm in front of the lambda. It transmits a small emissary vein from the superior sagittal sinus within the skull (p. 740). A recent numerical assessment gives an incidence of about 40–60 per cent (in different races) for this foramen (p. 340). Anteriorly the norma verticalis slopes down into the norma frontalis.

Norma Frontalis

Viewed from the front (**3**.63), the skull exhibits a more or less oval outline, wider above than below. Its upper part is formed by the frontal bone and is smooth and convex. Its lower part, the face, is very irregular and is interrupted by the orbits and the anterior

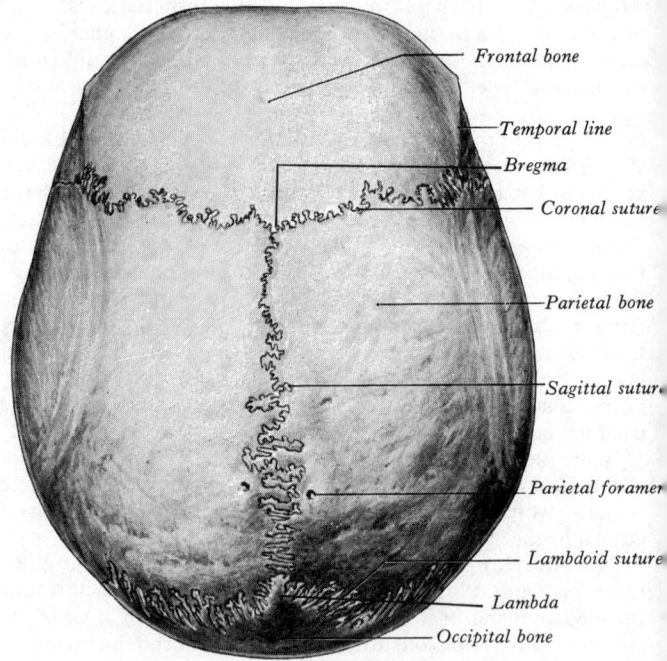

3.64   The skull, superior aspect. (Norma verticalis.)

Frontal bone

Temporal line

Bregma

Coronal suture

Parietal bone

Sagittal suture

Parietal foramen

Lambdoid suture

Lambda

Occipital bone

bony aperture of the nasal cavities. Immediately above the medial part of each orbit the *superciliary arch* forms a rounded elevation, better marked in the male than in the female skull, and these two arches are connected by a median elevation, the *glabella*. Below the glabella the skull recedes to the point where the nasal bones meet the frontal, forming the floor of a depression at the root of the nose. The point where the internasal and frontonasal sutures meet is named the *nasion*. Above the superciliary arch on each side there is a slight rounded elevation termed the *frontal tuber*, or tuberosity. These bony landmarks can be felt in the living subject, and provide reference points for the surgeon and the anthropologist.

The orbital opening is more or less quadrangular in shape. Its *supra-orbital margin* is formed entirely by the frontal bone; at the junction of its sharp lateral two-thirds with its rounded and medial third, is the supra-orbital notch (or foramen, as the case may be), which transmits the supra-orbital vessels and nerve. The *lateral margin* is formed almost entirely by the frontal process of the zygomatic bone, but it is completed above by the zygomatic process of the frontal bone, and the suture which connects these two bones can be felt in the living subject as a slight depression. The zygomatic bone laterally and the maxilla medially share in the formation of the *infra-orbital margin*. Both these margins are sharp and can be felt easily through the skin. The *medial margin* is not so clear-cut; formed above by the frontal bone and below by the lacrimal crest of the frontal process of the maxilla, which is sharp and distinct in its lower half only.

The anterior nasal aperture is piriform (pear-shaped), wider below than above and bounded by the nasal bones and the maxillae. The two nasal bones articulate with each other in the median plane and both articulate with the frontal bone above. On each side the nasal bone articulates behind with the frontal process of the maxilla, but its lower border, to which the lateral nasal cartilage is attached, is free and forms the upper boundary of the piriform aperture (**3**.63).

The maxillae predominate in the skeleton of the face, and it is their growth which is responsible for the elongation of the face between the ages of 6 and 12 years. Only the anterior surface of the maxilla is visible in the norma frontalis. Medially this surface presents the well-marked *nasal notch*, which forms the lower border and part of the lateral border of the piriform aperture. A prominent, sharp projection marks the meeting of the two maxillae in the lower boundary of the aperture and is termed the *anterior nasal spine*. It is palpable in the lower border of the free part of the nasal septum. About 1 cm below the infra-orbital margin the maxilla is perforated by the *infra-orbital foramen*, which transmits the infra-orbital vessels and nerve; it lies on, or just lateral to, a vertical line passing through the supra-orbital notch. The *alveolar process* of the maxilla, which contains the sockets for the maxillary teeth, can be examined most satisfactorily in the basal view (p. 304). The *zygomatic process* of the maxilla is a short but stout projection from the upper and lateral part of the anterior surface of the bone. Its upper surface is oblique and articulates with the zygomatic bone at the zygomaticomaxillary suture. Inferiorly it presents a free lower border, which meets the body of the bone above the first molar tooth, and can be palpated through the skin of the cheek or through the mucous membrane of the vestibule of the mouth. The *frontal process* of the maxilla ascends behind the nasal bone, forming the lower part of the medial margin of the orbital opening, and reaches the frontal bone. The *prominence of the cheek* below and lateral to the orbit is produced by the zygomatic bone. It is the convex lateral surface of the bone which can be examined both in the norma frontalis and in the norma lateralis. The mandible is described on p. 315.

The *glabella* may show the remains of the frontal suture, which in about 9 per cent of skulls extends upwards to the coronal suture. It indicates that the adult frontal bone is formed by the fusion of right and left halves, which ossify independently of each other. The medial part of the *superciliary arch* gives origin to the corrugator supercilii. The nasal part of the frontal bone and the frontal process of the maxilla give origin to the orbital part of the orbicularis oculi. Between these two areas the medial palpebral ligament is attached to the frontal process of the maxilla (**3**.63A). The procerus arises from the nasal bone near the median plane. The lower margin of the nasal bone usually presents a small notch, converted into a foramen by the lateral cartilage of the nose. It transmits the external nasal nerve. In front of the orbicularis oculi the levator labii superioris alaeque nasi takes origin from the frontal process of the maxilla. More laterally, the levator labii superioris arises from the maxilla in the interval between the infra-orbital margin and the foramen of the same name.

The stout root of the canine tooth produces an elevation termed

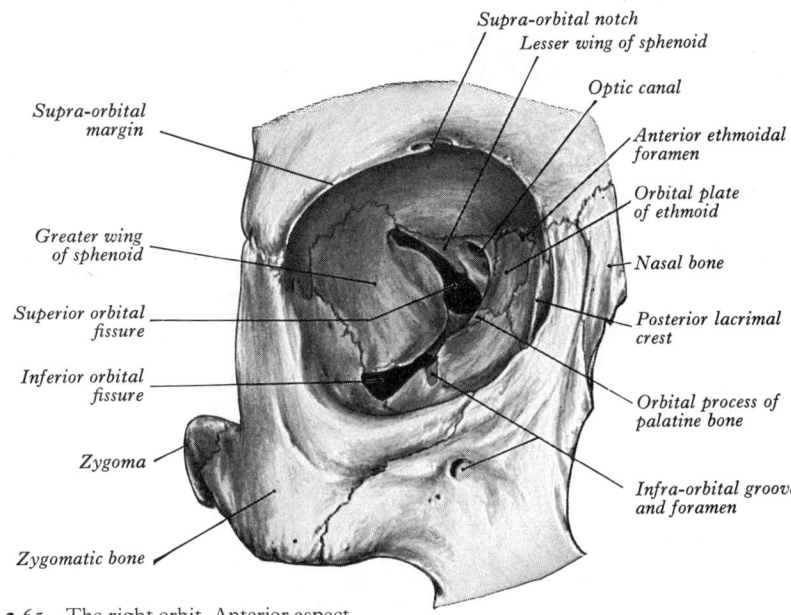

3.65 The right orbit. Anterior aspect.

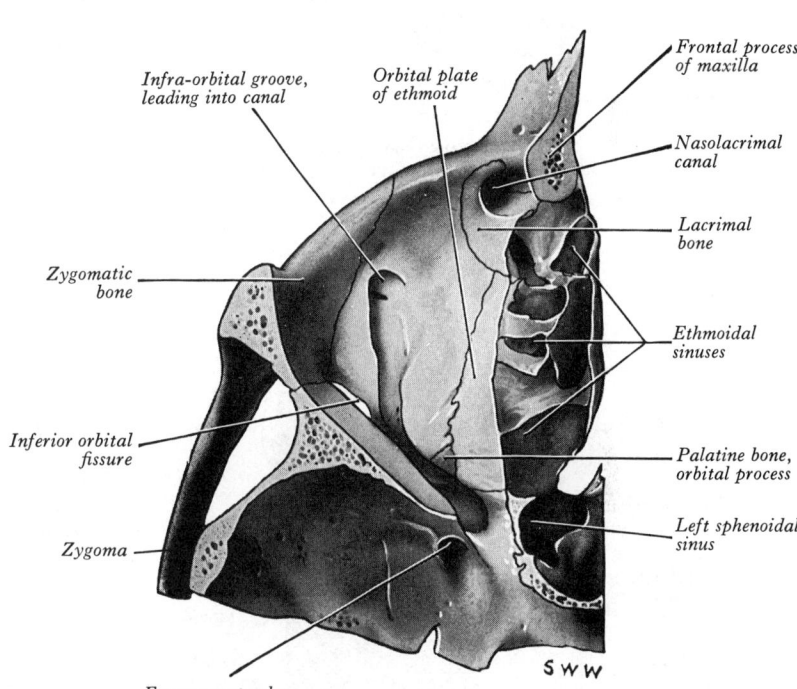

3.66 A horizontal section through the left orbit and nasal cavity viewed from above.

the *canine eminence*, which separates the *canine fossa* on its lateral side from the *incisive fossa* on its medial side. The levator anguli oris arises from the canine fossa. The surface bordering the nasal notch gives origins to the nasalis and depressor septi, below which is the origin of the incisive muscle.

The zygomatic bone is marked opposite the junction of the infra-orbital and lateral margins of the orbit by the small *zygomaticofacial foramen* (**3**.63) which transmits the nerve of the same name and a minute artery. The foramen, which is sometimes duplicated, opens laterally and downwards. Below the foramen the zygomatic bone gives origin to the zygomaticus minor, and more laterally to the zygomaticus major. The zygomaticofacial foramen is often absent; it was missing in 12 to 30 per cent of different populations in a series of 580 crania (*see* p. 348).

The Orbital Cavity

The orbits (**3**.65, 66, 67) are two recesses which contain the eyeballs, their associated muscles, vessels, nerves, etc., and most of the lacrimal apparatus, together with a variable amount of soft 299

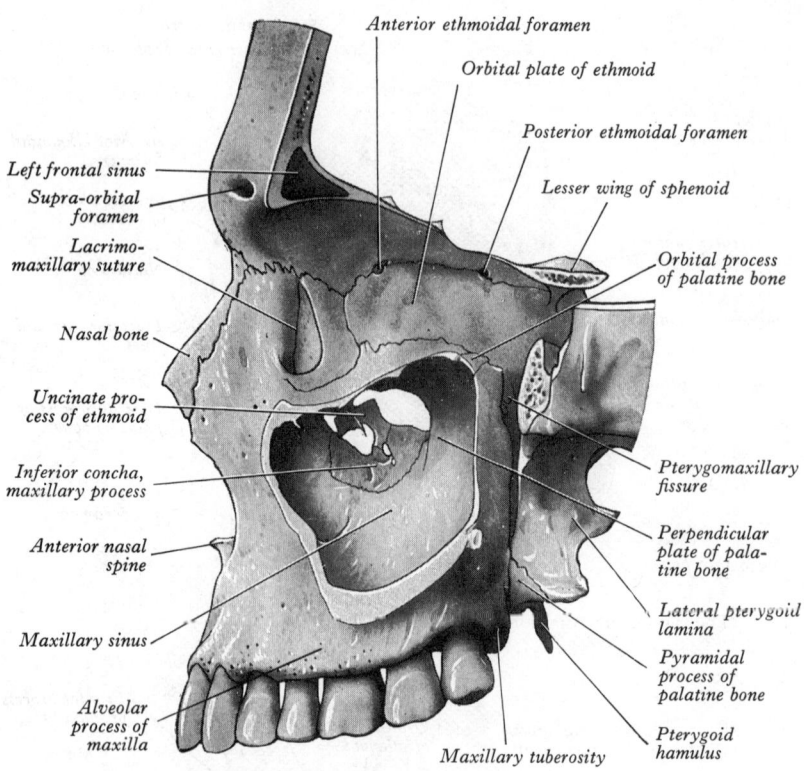

3.67A   Oblique parasagittal section through the anterior part of the skull, showing the medial wall of the left orbit and the medial wall of the left maxillary sinus.

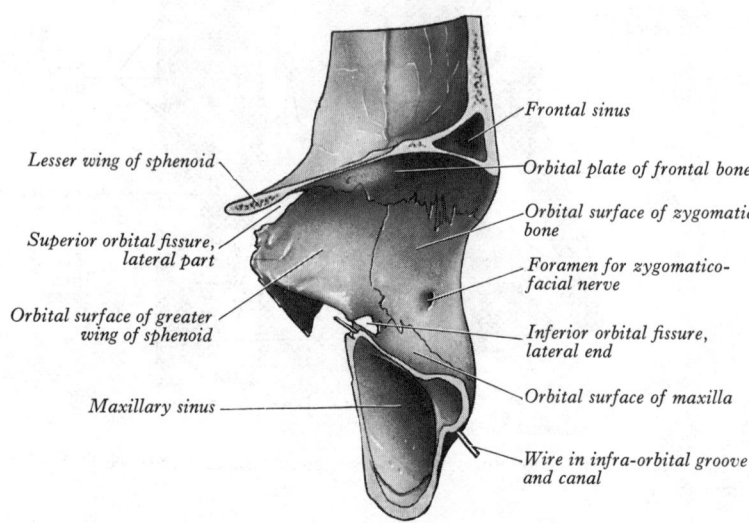

3.67B   The lateral wall of the left orbit, viewed from the medial side. Compare with 3.67A, which represents the opposite part of the same section of the skull.

fat. The orbital cavity is pyramidal; its base is the orbital opening, and its long axis is directed backwards and medially. Each orbit presents a roof, floor, medial and lateral walls, a base or orbital opening and an apex.

The superior wall or roof is a thin, gently concave plate which intervenes, throughout most of its extent, between the orbit and the brain in the anterior cranial fossa. Anteromedially it is separated into two laminae by the frontal sinus, an air space in the bone communicating with the nasal cavity. Anterolaterally is a deep hollow, the *lacrimal fossa*, in which is the orbital part of the lacrimal gland. At the posterior end of the junction of the roof with the medial wall the *optic canal or foramen* establishes communication between the orbit and the middle cranial fossa. It contains the optic nerve and the ophthalmic artery. Close to the superior, medial and lower margins of the opening of the canal

into the orbit a common tendinous ring is attached to the orbital walls for the origin of certain muscles of the eyeball (7.269).

The medial wall (3.67A) is exceedingly thin, except at its most posterior part, and slopes gently downwards and laterally into the floor. Anteriorly is the *lacrimal groove*, for the lacrimal sac. The groove communicates below with the nasal cavity through the *nasolacrimal canal*, which is little more than 1 cm long and contains the nasolacrimal duct. The floor of the groove separates the orbital from the nasal cavity, but behind it air-containing ethmoidal sinuses intervene between them. Posteriorly, the medial wall is related to the anterior part of the sphenoidal sinus and forms its lateral wall.

The inferior wall or floor of the orbit (3.66) is relatively thin and in most of its extent forms the roof of the maxillary sinus (3.67A). It is not quite horizontal, but faces upwards and slightly laterally. In front it is continuous with the lateral wall, but posteriorly the two walls are separated by the *inferior orbital fissure*. This leads into the orbit from the pterygopalatine fossa posteriorly, and anteriorly from the infratemporal fossa. The maxillary nerve is the most important structure traversing the fissure. The medial lip of the fissure is notched by the *infra-orbital groove*, which passes forwards in the floor, sinking into it anteriorly and becoming the *infra-orbital canal* whose anterior opening is the *infra-orbital foramen*. The groove, canal and foramen transmit the infra-orbital nerve, the continuation of the maxillary nerve. Through the anterior part of the inferior orbital fissure a vein passes to connect the inferior ophthalmic with the pterygoid plexus in the infratemporal fossa. The infra-orbital foramen is sometimes double or even multiple (Harris 1933), the supernumerary foramina being usually smaller and hence named *accessory*. Such accessory infra-orbital foramina have been recorded at incidences of 2 to 18 per cent in various populations (*see* p. 348).

The lateral wall (3.67B) is the thickest, especially where it separates the orbit posteriorly from the middle cranial fossa. In front it is between the orbit and the temporal fossa. The lateral wall and the roof are continuous anteriorly, but they are separated posteriorly by the *superior orbital fissure*. This is noticeably widened at its medial end (3.65), and its long axis is directed medially, backwards and slightly downwards. It communicates with the middle cranial fossa and transmits the oculomotor, trochlear and abducent nerves and the terminal branches of the ophthalmic nerve, and the ophthalmic veins. Where the fissure begins to widen its lower border is marked by a bony projection, often sharp in character, which gives attachment to the lateral part of the common tendinous ring (7.269). Royle (1973) has drawn attention to an 'infra-orbital' *sulcus*, present in 22 of 64 orbits, which extends from the superolateral end of the superior orbital fissure towards the floor of the orbit. It may be associated with an anastomotic vessel between the middle meningeal and infra-orbital arteries.

The boundaries of the *orbital opening* have already been described (p. 299). The *apex* of the orbit is at the medial end of the superior orbital fissure.

Further details of the orbital walls are given in the following paragraphs.

The roof of the orbit is formed almost entirely by the orbital plate of the frontal, but the under surface of the lesser wing of the sphenoid forms its posterior part. The suture between the two bones is almost horizontal. The *optic canal* lies between the two roots of the lesser wing and is bounded medially by the body of the sphenoid bone. Near the junction of the roof and the medial wall, and close to the orbital opening, the small *trochlear fossa* (occasionally replaced by a trochlear spine) marks attachment of the fibrous loop for the tendon of the superior oblique muscle of the eyeball.

The medial wall (3.67A) is limited in front by the anterior lacrimal crest of the frontal process of the maxilla, which gives attachment to the orbicularis oculi and to the lacrimal fascia. Behind this crest the maxilla and the lacrimal bone join to form the *lacrimal groove*, the suture between them being in its floor. The upper opening of the *nasolacrimal canal* lies at the lower end of the groove, and its lateral boundary is formed by the tiny *hamulus of the lacrimal bone*, which curves forwards and medially to meet the lower part of the anterior lacrimal crest. The posterior border of the groove is formed for the most part by the *crest of the lacrimal bone*, which gives origin to the lacrimal part of the orbicularis oculi (p. 531) and to the lacrimal fascia, which bridges over the groove. The posterior part of the orbital surface of the lacrimal bone is flattened, and articulates behind, by an almost vertical suture, with the orbital plate of the labyrinth of the ethmoid bone. The frontolacrimal and the lacrimomaxillary sutures indicate the other limits of the orbital aspect of the lacrimal bone. The *orbital plate of the ethmoid* makes the largest contribution to the medial wall. Almost rectangular, it consists of very thin bone, forming the lateral walls of the ethmoidal sinuses. Above, it articulates with the medial edge of the orbital plate of the frontal, and the line of this suture is interrupted by the *anterior* and *posterior ethmoidal foramina*, which lead into minute bony canals. These transmit vessels and nerves of the same names—the posterior ethmoidal nerve, however, is often absent—and lead into the anterior cranial fossa, where they open at the lateral edge of the cribriform plate. Below, the orbital plate articulates with the medial edge of the orbital surface of the maxilla and, at its most posterior part, with the orbital process of the palatine bone. Posteriorly, the orbital plate of the ethmoid articulates with the body of the sphenoid, which forms the most posterior part of the medial wall of the orbit, separated from the roof by the optic canal.

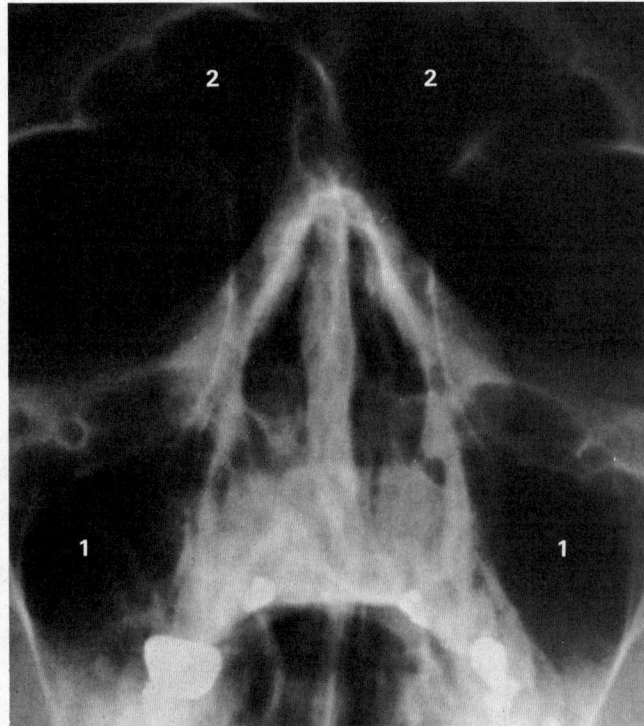

3.68   Radiograph of adult skull. Frontal view. 1. Maxillary sinus. 2. Frontal sinus. Both infra-orbital canals are shown. Compare with the radiographs in illustrations **3**.62A, B.

Racial and sexual variation in the position and incidence of the ethmoidal canals was analysed in a series of 580 crania from several populations (*see* p. 348). The anterior foramen was observed to lie outside the fronto-ethmoidal suture in about 10–20 per cent of several modern races, and in as much as 62 per cent of 53 Peruvian crania.

The inferior wall or floor of the orbit (**3**.66) is mostly formed by the orbital surface of the maxilla and, anterolaterally, by the zygomatic bone. At its posteromedial corner, where the floor meets the medial wall, a small triangular area is formed by the orbital process of the palatine bone. In addition to the maxillary nerve the *inferior orbital fissure* transmits the infra-orbital vessels, the zygomatic nerve, and a few minute twigs from the pterygopalatine ganglion. The fissure is bounded above by the greater wing of the sphenoid, below by the maxilla and the orbital

process of the palatine bone, and laterally by the zygomatic bone or the zygomaticomaxillary suture. In 35 to 40 per cent of skulls the maxilla and the sphenoid bone articulate with each other at the anterior end of the fissure and exclude the zygomatic bone from it. In the anteromedial part of the floor, just lateral to the hamulus of the lacrimal bone, a small depression may mark the origin of the inferior oblique muscle from the maxilla.

The lateral wall of the orbit (**3**.67B) is formed by the orbital surface of the greater wing of the sphenoid behind, and by the orbital surface of the frontal process of the zygomatic bone in front. These two bones meet at the sphenozygomatic suture. This aspect of the zygomatic bone presents the openings of minute canals for the zygomaticofacial and the zygomaticotemporal nerves. The former lies near the junction of the floor of the orbit with its lateral wall, the latter at a slightly higher level, sometimes close to the sphenozygomatic suture. The *superior orbital fissure* is bounded above by the lesser and below by the greater wing of the sphenoid and medially by its body. The lacrimal and frontal nerves traverse the narrow, lateral part of the fissure, which transmits also the meningeal branch of the lacrimal artery and the occasional orbital branch of the middle meningeal artery. The trochlear nerve is situated more medially and lies just outside the common tendinous ring of the recti (p. 1178). The two divisions of the oculomotor nerve, the nasociliary and the abducent nerves, pass within the tendinous ring, and therefore traverse the wider, medial part of the fissure. They may be accompanied by the superior and inferior ophthalmic veins, but the superior ophthalmic vein may accompany the trochlear nerve whilst the inferior vein may pass through the medial end of the fissure below the ring (**7**.269).

Norma Occipitalis

The outline of the skull, viewed from behind, is like a broad arch, convex above and on each side, and flattened below. The *lambdoid suture* can now be seen through its entire length. Its serrations are deep and prominent above and behind, but become much less conspicuous as the suture is traced downwards and forwards. Inferiorly it meets the *occipitomastoid* and *parietomastoid sutures* at the postero-inferior angle of the parietal bone (**3**.61). The posterior portions of the parietal bones, parietal tuberosities and foramina, visible in the norma occipitalis, have already been viewed from above. Sutural bones are relatively common at the lambda (meeting of lambdoid and sagittal sutures) and along the lambdoid suture (*see* pp. 322 and 338).

The most outstanding feature of the norma occipitalis is the *external occipital protuberance* (**3**.61) and the ridges which lead away from it. The protuberance is in the lower part of the field in the median plane and may be overhanging. It can readily be identified in the living at the upper end of the median furrow at the back of the neck. The *superior nuchal lines* are the ridges, often sharp in character, which pass laterally from the protuberance. They are the boundary between the scalp and the neck, and the portions of the occipital bone below them, now seen fore-shortened in perspective, will be better seen in the norma basalis. The *highest nuchal lines*, when present, are curved ridges, about 1 cm above the superior nuchal lines. Commencing at the upper part of the protuberance, they are more arched than the superior nuchal lines. In various ethnic groups the incidence of this feature varied from 3·6 to 40 per cent (*see* p. 348).

The mastoid process and the mastoid part of the temporal bone are inferolateral in this aspect of the skull, and can be examined more satisfactorily in the norma lateralis.

The *inion* is the most salient point on the external occipital protuberance in the median plane. To the lower part of the protuberance are attached the upper end of the ligamentum nuchae and, to its upper part, fibres of trapezius, which arises also from the adjoining part of the *superior nuchal line*. The lateral part of the line (**3**.86) gives attachment to the posterior fibres of sternocleidomastoid and, under cover of that muscle, to fibres of splenius capitis. The *highest nuchal line* has attached medially the galea aponeurotica and laterally the occipital belly of the occipitofrontalis.

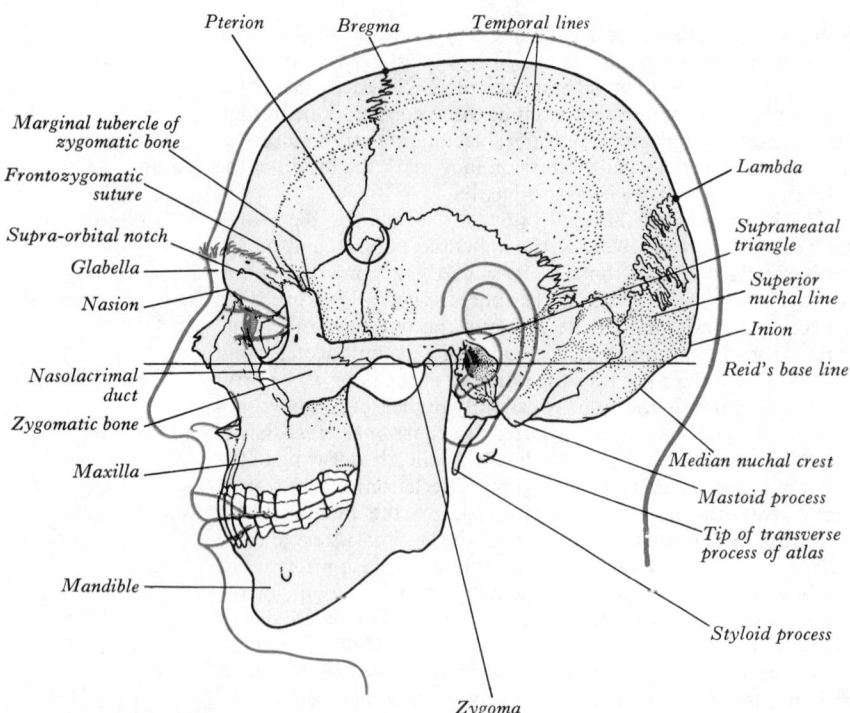

Pterion *Bregma* *Temporal lines*

Marginal tubercle of zygomatic bone

Frontozygomatic suture

Supra-orbital notch

Glabella

Nasion

Nasolacrimal duct

Zygomatic bone

Maxilla

Mandible

Zygoma

Lambda

Suprameatal triangle

Superior nuchal line

Inion

Reid's base line

Median nuchal crest

Mastoid process

Tip of transverse process of atlas

Styloid process

3.69   Left lateral aspect of the head, showing the surface relations of the bones.

Norma Lateralis

Much of the side of the skull (**3**.61) has already been seen and described when it was examined from other aspects, but in its central area many features have not yet been considered. This region is limited above by the *temporal line*, which arches upwards and backwards from the zygomatic process of the frontal across the coronal suture to the parietal bone. Salient at first, and felt easily through the skin, it arches across the parietal bone and becomes much less distinct and is usually represented by *two curved ridges*, which enclose between them a smooth and often polished strip of bone. Posteriorly the upper of these two lines fades away, but the lower becomes more prominent again as it curves downwards and forwards across the squamous part of the temporal bone, just above the base of the mastoid process. This part of the line, which is often termed the *supramastoid crest*, becomes continuous with the posterior root of the zygomatic process. Throughout its extent the temporal line marks the periphery of the temporalis and its covering fascia. On the parietal bone the muscle arises from the lower ridge, while the fascia is attached to the upper ridge and to the bone immediately below.

The temporal fossa is the region bounded by the zygomatic arch, the temporal line and the frontal process of the zygomatic bone, and to its floor is attached the temporalis. An irregularly H-shaped arrangement of sutures can be seen in the anterior part of the fossa, the horizontal limb of the H being formed by the suture between the antero-inferior angle of the parietal and the upper border of the greater wing of the sphenoid bone. In this situation the frontal, the sphenoid, the parietal and the squamous part of the temporal bone closely adjoin one another (**3**.69), and a small circular area can be outlined so as to include portions of all four. This area is termed the **pterion**, and its centre—an important landmark for the surgeon—lies 4·0 cm above the zygomatic arch and 3·5 cm behind the frontozygomatic suture (**3**.61). (The anterior branch of the middle meningeal artery is marked by the pterion. It also gives an approximate indication of the position of the more deeply lying lesser wing of the sphenoid which, in life, is in the stem of the Sylvian (lateral) cerebral fissure. Hence, centrally, the pterion is also sometimes termed the *Sylvian point*.) The anterior wall of the fossa is formed by the temporal surface of the zygomatic bone, the adjoining part of the greater wing of the sphenoid and a small portion of the frontal bone. It is between the

fossa and the orbit. Inferiorly the fossa communicates freely with the infratemporal fossa through the gap which separates the zygomatic arch from the side of the skull. In this situation the tendon and some fleshy fibres of the temporalis descend to reach their insertion into the mandible (p. 316).

The zygomatic arch, formed by the temporal process of the zygomatic bone and the zygomatic process of the temporal bone, is easily palpable where the cheek and temple meet each other. Its sharp, upper border is obscured by the attachment of the temporal fascia, and its lower border by the origin of masseter, which arises also from its deep surface. The arch stands away from the rest of the skull, separated from it by a gap which is deeper in front than behind. Anteriorly, the arch is crossed obliquely downwards and backwards by the zygomaticotemporal suture.

The *zygomatic process of the temporal bone* widens posteriorly as it approaches the squamous part, and divides into anterior and posterior roots. The *anterior root* passes medially in front of the *mandibular fossa* into the smooth *articular tubercle*, the anterior boundary of the fossa. The *posterior root* passes backwards, lateral to the fossa, and its upper border becomes continuous with the supramastoid crest of the temporal bone.

The external acoustic meatus opens immediately below the posterior part of the posterior root of the zygoma. Its margins are roughened, especially below and in front, for the attachment of the cartilaginous segment of the meatus. The upper margin and the upper part of the posterior margin are formed by the squamous part of the temporal bone; the rest is formed by the *tympanic part* of the temporal bone. The squamotympanic suture is antero-superior to the opening, but the suture on the posterior wall is usually obliterated in the adult. Below the meatus the tympanic plate is drawn downwards, forming a somewhat triangular roughened area. Immediately above and behind the meatus there is frequently a small depression with a bony spicule (*suprameatal spine*) in its anterior margin. This lies within the area of the *suprameatal triangle*, which is bounded above by the supramastoid crest, in front by the posterosuperior margin of the orifice of the meatus and behind a vertical line, drawn as a tangent to the curve of the posterior margin of the meatal orifice. This triangle forms the lateral wall of the mastoid (tympanic) antrum (**3**.69 and p. 329), which is an air space of importance to the surgeon and is contained in the petrous part of the temporal bone.

The mastoid part of the temporal bone appears in the norma lateralis behind the meatus. Above, it is continuous with the squamous portion in front, but, behind, it possesses a free upper border, which articulates with the postero-inferior part of the parietal bone at the horizontal *parietomastoid suture*. Its posterior border is free and articulates with the squamous part of the occipital bone at the *occipitomastoid suture*. These two sutures meet each other at the lateral extremity of the lambdoid suture. The meeting point of the three sutures is the *asterion*. The *mastoid process* (**3**.61) is a strong, nipple-shaped projection from the lower part of the mastoid portion of the temporal bone. It lies immediately behind and below the external acoustic meatus and can be felt through the skin under cover of the lobule of the auricle. The *mastoid foramen* pierces the bone above the base of the mastoid process and near, or on, the occipitomastoid suture; it transmits an emissary vein from the sigmoid sinus. A sutural ossicle may occur in the parietomastoid suture, usually at the asterion, in the position of the posterolateral fontanelle (Le Double 1903), but sometimes elsewhere along the suture.

The styloid process (**3**.61) is a slender, elongated projection which, although attached to the base of the skull, is better seen in the norma lateralis. It lies a short distance in front of the mastoid process but on a deeper plane, its base being partly ensheathed by the lower margin of the tympanic plate. Directed downwards, forwards and slightly medially, its tip is usually hidden by the posterior margin of the ramus of the mandible, when that bone is in place. From its extremity the stylohyoid ligament passes downwards and forwards to the lesser cornu of the hyoid bone (p. 319), which is therefore suspended from the skull.

The infratemporal fossa (**3**.70) is an irregular space behind the maxilla. It communicates with the temporal fossa through the gap between the zygomatic arch and the side of the skull. Medial o this gap, the roof is formed by the infratemporal surface of the greater wing of the sphenoid, and a small part of the squamous temporal. In this situation the greater wing is pierced by the foramen ovale and the foramen spinosum. Medially is the lateral pterygoid plate. These walls are considered in detail in the norma basalis (p. 305). Behind, below and on the lateral side the fossa is freely open. The anterior and medial walls meet below but they are separated above by the *pterygomaxillary fissure*, through which the infratemporal fossa communicates with the pterygopalatine fossa. The upper end of the pterygomaxillary fissure is continuous with the posterior end of the *inferior orbital fissure*, which the infratemporal fossa communicates with the pterygopalatine fossa. The upper end of the pterygomaxillary fissure is connects the infratemporal fossa with the orbit (p. 300).

The pterygopalatine fossa is a small pyramidal space situated below the apex of the orbit. It communicates with the infratemporal fossa through the pterygomaxillary fissure, with the nasal cavity through the sphenopalatine foramen, and with the orbit through the medial end of the inferior orbital fissure. The foramen rotundum opens on its posterior wall, and the maxillary nerve, which runs forwards and laterally from the foramen across the upper part of the fossa, is the most important of its contents.

The floor of the *temporal fossa* is marked by a few small vascular furrows, of which the most constant are above the external acoustic meatus; they are produced by the middle temporal vessels. In the anterior wall of the fossa the *zygomaticotemporal foramen* pierces the posterior, temporal surface of the zygomatic bone in an upward and backward direction. It transmits the zygomaticotemporal nerve and a minute artery. In addition to the tendon of the temporalis, the deep temporal vessels and nerves traverse the gap between the zygomatic arch and the side of the skull and ascend into the temporal fossa.

As the anterior root springs from the zygomatic process, it is marked by a small tubercle often termed the *tubercle of the root of the zygoma*, to which are attached fibres of the lateral ligament (4.18) of the temporomandibular joint. It is palpable immediately in front of the head of the mandible. Behind the mandibular fossa a small downward projection from the posterior root of the zygoma, *the postglenoid tubercle*, meets the tympanic plate at the anterosuperior part of the orifice of the external acoustic meatus (3.93), and its anterior aspect takes a small part in the formation of the mandibular fossa.

The posterior part of the lateral surface of the *mastoid process* and its rounded apex are roughened by the insertions of the sternocleidomastoid, splenius capitis and longissimus capitis, from before backwards. In front of and parallel to this roughened area, the partially obliterated remains of the squamomastoid suture may be visible. From this position of the suture it will be obvious that the floor of the suprameatal triangle, and therefore the lateral wall of the mastoid antrum, is formed by the squamous part of the temporal bone. The tympanomastoid fissure is placed on the anterior aspect of the base of the process. The outer opening of the mastoid canaliculus (3.96), which transmits the auricular branch of the vagus nerve, is between the lips of the fissure.

The *styloid process* is covered on its lateral aspect by the parotid gland and intervenes between it and the internal jugular vein. It gives origin to three muscles. The stylohyoid arises by a slender tendon from its posterior aspect, nearer to the base than the tip, the styloglossus from the tip and the adjacent part of the anterior aspect, and the stylopharyngeus from the medial aspect of the base. The stylomandibular ligament is attached to the lateral aspect of the process near to the stylohyoid ligament at its tip. Behind the base of the process the facial nerve emerges from the stylomastoid foramen and passes forwards lateral to the process in the substance of the parotid gland.

The *infratemporal fossa* (3.70) contains the lower part of the temporalis as it passes to the coronoid process. The maxillary artery and its branches and the pterygoid plexus of veins are medial to the temporalis and usually on the lateral surface of the lower head of the lateral pterygoid muscle. The deepest part of the fossa is occupied by the medial pterygoid, the mandibular nerve and the chorda tympani. The mandibular nerve enters the fossa through the foramen ovale in its roof and breaks up into terminal

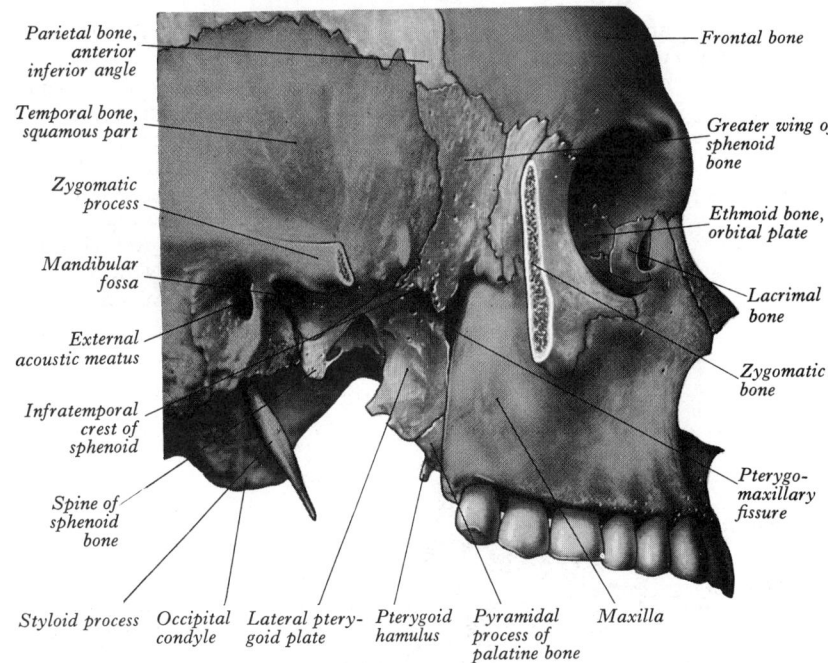

Parietal bone, anterior inferior angle	Frontal bone
Temporal bone, squamous part	Greater wing of sphenoid bone
Zygomatic process	Ethmoid bone, orbital plate
Mandibular fossa	Lacrimal bone
External acoustic meatus	Zygomatic bone
Infratemporal crest of sphenoid	Pterygo-maxillary fissure
Spine of sphenoid bone	

Styloid process   Occipital condyle   Lateral ptery-goid plate   Pterygoid hamulus   Pyramidal process of palatine bone   Maxilla

3.70 The right infratemporal fossa. *Note.—Blue*=frontal bone; *yellow*=sphenoid and lacrimal bones; *brown*=temporal and nasal bones; *green*=maxilla. The parts shown of the parietal, zygomatic, ethmoid and palatine bones are uncoloured.

branches medial to the lateral pterygoid muscle. Its branches traverse the fossa and most of them leave it to gain other regions. The chorda tympani enters the fossa on the medial side of the spine of the sphenoid and runs downwards and forwards to join the lingual nerve. The maxillary nerve appears at the upper part of the fossa as it passes between the upper end of the pterygopalatine fossa and the inferior orbital fissure. The anterior wall of the fossa is pierced by two or three small foramina for the posterior superior alveolar vessels and nerves. It is limited below by the alveolar part of the maxilla in the region of the molar teeth, and in this situation a horizontal strip of the bone is closely covered with the mucous membrane of the gum. Immediately above this strip are attached the upper fibres of buccinator, which extends backwards behind the last molar tooth on to the tuberosity of the maxilla. The medial wall of the fossa, formed by the lateral pterygoid plate, is completed below and in front by the pyramidal process (tubercle) of the palatine bone which is wedged in between the tuberosity of the maxilla and the lateral pterygoid plate. The superficial head of the medial pterygoid muscle arises from this surface of the pyramidal process and the adjoining part of the maxillary tuberosity.

The *pterygomaxillary fissure* is a triangular interval formed by the divergence of the maxilla from the pterygoid process of the sphenoid bone. It admits the terminal part of the maxillary artery to the pterygopalatine fossa, and its uppermost part gives passage to the maxillary nerve, which appears for a brief part of its course in the upper part of the infratemporal fossa before it enters the inferior orbital fissure to gain the orbit (p. 300).

The *pterygopalatine fossa* (3.70) communicates *above* with the orbit through the medial—or posterior—part of the inferior orbital fissure, and it is closed *inferiorly* where the lower part of the lateral surface of the perpendicular plate of the palatine bone meets the postero-inferior part of the medial surface of the maxilla. It is bounded *behind* by the root of the pterygoid process and the adjoining part of the anterior surface of the greater wing of the sphenoid, *medially*, by the upper part of the perpendicular plate of the palatine bone with its orbital and sphenoidal processes, *in front*, by the medial portion of the upper part of the posterior surface of the maxilla. *Laterally* it communicates with the infratemporal fossa through the pterygomaxillary fissure. The most important contents of the fossa are the maxillary nerve, the pterygopalatine ganglion and the terminal part of the maxillary artery. The maxillary nerve enters the fossa through the foramen rotundum and passes forwards and laterally to the posterior end of the infra-orbital groove in the floor of the orbit. Below and medial to the foramen rotundum the pterygoid canal transmits the nerve (and artery) of the same name from the lower part of the anterior wall of the foramen lacerum to the pterygopalatine ganglion, and inferomedially the palatovaginal canal transmits the pharyngeal nerve (and artery) from the ganglion to the roof of the pharynx. A fourth foramen, on the medial wall, is the *sphenopalatine foramen* (**3**.78). It is bounded above by the body of the sphenoid, in front by the orbital process of the palatine bone,

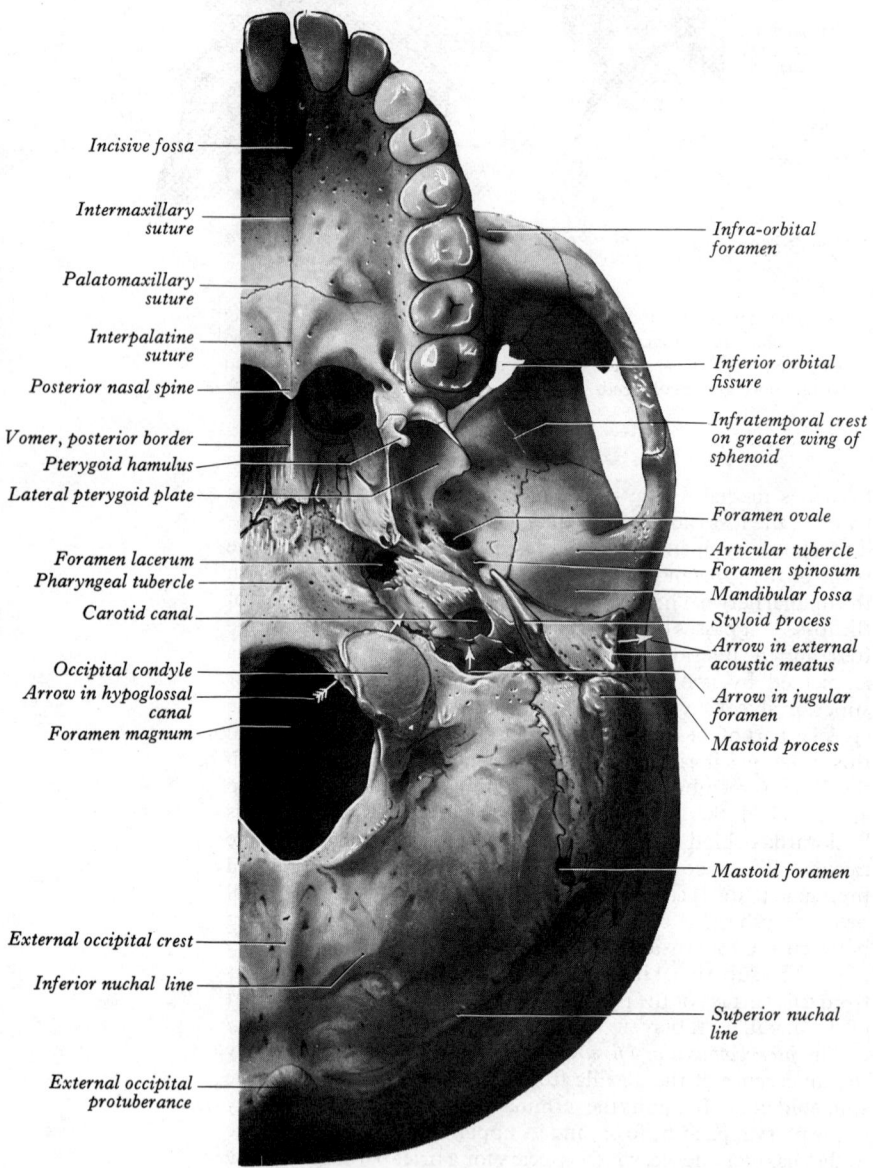

Incisive fossa

Intermaxillary suture

Palatomaxillary suture

Interpalatine suture

Posterior nasal spine

Vomer, posterior border
Pterygoid hamulus
Lateral pterygoid plate

Foramen lacerum
Pharyngeal tubercle
Carotid canal

Occipital condyle
Arrow in hypoglossal canal
Foramen magnum

External occipital crest
Inferior nuchal line

External occipital protuberance

Infra-orbital foramen

Inferior orbital fissure

Infratemporal crest on greater wing of sphenoid

Foramen ovale

Articular tubercle
Foramen spinosum
Mandibular fossa
Styloid process
Arrow in external acoustic meatus

Arrow in jugular foramen

Mastoid process

Mastoid foramen

Superior nuchal line

3.71A The inferior surface of the left half of the base of the skull. (Norma basalis.) Compare with Key, **3**.71B.

behind by the sphenoidal process and below by the upper border of the perpendicular plate. It carries into the nasal cavity the nasopalatine nerve and accompanying vessels. The fifth foramen is placed inferiorly at the junction of the anterior and posterior walls, and leads into the *greater palatine canal*. Bounded anterolaterally by the maxilla and posteromedially by the perpendicular plate of the palatine bone, this canal transmits the anterior, middle and posterior palatine nerves and the greater and lesser palatine vessels, which emerge through foramina on the bony palate (*vide infra*).

Norma Basalis

The inferior surface of the base of the skull is very irregular (**3**.71, 72, 73). It extends from the incisor teeth of the maxillae, back to the superior nuchal lines of the occipital bone. Laterally are the other teeth, the zygomatic arch and its posterior root, and the mastoid process. It is useful to divide the area into anterior, middle and posterior parts. The anterior part, formed by the hard palate and the alveolar arches, is on a lower level than the rest, which is divided, in an arbitrary manner, into middle and posterior parts by a transverse line through the anterior margin of the foramen magnum.

The Anterior Part of the Norma Basalis

The *bony palate* (**3**.72) lies within the upper dental arch, and is formed by the palatine processes of the maxillae and the horizontal plates of the palatine bones, separated by a *cruciform suture*, made up of the intermaxillary, interpalatine and palatomaxillary sutures. Owing partly to the downward projection of the alveolar arches, the palate is arched both from before backwards and from side to side. The depth and the breadth of the palatine vault are subject to considerable variation but are always greatest in the region of the molar teeth. The *incisive fossa* lies anteriorly in the median plane. The *lateral incisive foramina*, which lead into the incisive canals and so to the floor of the nasal cavity (p. 315), are situated in its lateral walls; the *median incisive foramina*, which are present in some skulls, open on its anterior and posterior walls. The *greater palatine foramen*, which is the lower orifice of the canal of the same name, opens close to the lateral border of the palate immediately behind the palatomaxillary suture (**3**.72). A vascular groove, deep behind and becoming shallower in front, leads forwards from the foramen. The *lesser palatine foramina*, usually two on each side, are situated behind the greater foramina. They pierce the *pyramidal process of the palatine bone*, which projects backwards and laterally from the posterolateral corner of the bony palate and becomes wedged into the notch between the lower ends of the two pterygoid plates. (When more than one lesser palatine foramen is present, the supernumerary foramina are sometimes described as *accessory*. The incidence of such accessory foramina is high and shows some racial and sexual variation from 30 to 70 per cent—*see* p. 348.) The vault of the bony palate is pierced by numerous small foramina and marked by depressions for the palatine glands. Near the posterior border it presents a slightly curved ridge of variable prominence, termed the *palatine crest*, which commences behind the greater palatine foramen and runs medially. The free posterior border of the bony palate projects backwards in the median plane as the *posterior nasal spine*.

The *alveolar arch* provides sixteen sockets, or alveoli, for the teeth. These vary in size and depth and are single or subdivided by septa according to the roots of the teeth which they contain.

The lateral incisive foramen transmits the terminal branches of the greater palatine vessels and the nasopalatine nerve. When median incisive foramina are present the left nasopalatine nerve passes through the anterior and the right through the posterior foramen. The lateral foramina are sometimes held to be in the line of fusion of the os incisivum (premaxilla) with the maxilla proper, and represent a primitive communication between the mouth and the nose. In young skulls what is possibly a suture line between the os incisivum and the maxilla may be visible, extending from the posterior part of the incisive fossa to the septum between the sockets of the lateral incisor and canine teeth (but *see* p. 340).

The greater palatine foramen transmits the greater palatine nerve and vessels, and the vessels groove the lateral part of the palate as they run forwards to the incisive fossa. The lesser palatine foramina, usually two, sometimes one and occasionally three in number, perforate the inferior and medial aspects of the pyramidal process of the palatine bone; they contain the lesser palatine nerves and vessels. The palatine crest, which commences on the tubercle and extends on to the horizontal plate of the palatine bone, gives attachment to part of the tendon of the tensor veli palatini muscle. To the free posterior border of the palate is attached the palatine aponeurosis and to the posterior nasal spine the musculus uvulae. These collagenous elements are continuous.

The margins of the median palatal intermaxillary suture are sometimes raised forming a longitudinal midline ridge, the *palatine torus*, the surface of which may be variously smooth, pitted or irregularly roughened. Similarly a longitudinal ridge, the *maxillary torus* is occasionally present on the alveolar process spanning the palatal aspect of the subcervical roots of the upper molar teeth.

The Middle Part of the Norma Basalis

The middle part of the external surface of the base of the skull (**3**.71, 73) extends from the posterior border of the bony palate to an arbitrary line drawn transversely through the anterior margin of the foramen magnum. In the median plane anteriorly the *posterior border of the vomer* separates the two *posterior nasal apertures*. Immediately behind the vomer the posterior part of the inferior surface of the body of the sphenoid is directly continuous with the inferior surface of the *basilar part of the occipital bone*, which forms a broad bar of bone extending backwards and downwards to the foramen magnum. It is convex from side to side and wider behind than in front. A short distance in front of the foramen magnum is a small midline elevation, the *pharyngeal tubercle* the attachment of the highest fibres of the superior constrictor muscle of the pharynx.

The pterygoid process descends behind the third molar tooth from the junction of the greater wing and body of the sphenoid. It has two laminae, the medial and lateral pterygoid plates, separated from each other by a cuneiform interval, directed backwards and somewhat laterally and named the *pterygoid fossa*. Anteriorly the pterygoid plates are fused except inferiorly, where they are separated by a narrow gap occupied by the *pyramidal process* of the palatine bone; the suture lines can usually be identified. Medially they articulate with the posterior border of the perpendicular plate of the palatine bone in front, and form with it the flattened area of bone which lies in the lateral wall of the posterior nasal aperture and nasopharynx. On the lateral side the fused laminae are separated from the posterior surface of the maxilla in front by the pterygomaxillary fissure (**3**.70). The *medial pterygoid plate* is the narrower and projects directly backwards. Its medial surface is covered in life with mucous membrane and forms the lateral boundary of the posterior nasal aperture and part of the lateral wall of the nasopharynx. The posterior border of the medial pterygoid plate is sharp, and presents a small projection about its midpoint. Above this projection the border is curved and is attached to the pharyngeal end of the auditory tube. At its upper end the border divides to enclose the shallow, *scaphoid fossa* (**3**.73); below, it projects beyond the rest of the plate as the slender *pterygoid hamulus*. This process curves downwards and laterally and is groved anteriorly at its root by the tendon of the tensor veli palatini. The *lateral pterygoid plate* projects backwards and laterally and its lateral surface forms the medial wall of the infratemporal fossa. Superiorly it is continuous with the *infratemporal surface of the greater wing of the sphenoid*, which forms the anterior part of the roof of the infratemporal fossa. This surface of the greater wing is directed downwards and, sometimes, slightly to the lateral side. It is roughly pentagonal; its anterior margin is the posterolateral border of the inferior orbital fissure; anterolaterally it is limited by the infratemporal crest. Laterally it articulates with the squama of the temporal bone; medially it is continuous with the root of the pterygoid process and the side of the body of the sphenoid; posteromedially it articulates with the petrous part of the temporal bone.

The foramina ovale et spinosum are important openings on the infratemporal surface of the greater wing of the sphenoid. The *foramen ovale*, irregularly oval, lies close to the upper end of the posterior margin of the lateral pterygoid plate. It transmits the mandibular division of the trigeminal nerve. Posterior and slightly lateral to the foramen ovale the *foramen spinosum* pierces the greater wing and transmits the middle meningeal artery to the middle cranial fossa. It is much smaller than the foramen ovale and is circular. Immediately posterolateral to the foramen spinosum this surface of the greater wing has an irregular downward projection, the *spine of the sphenoid*. The medial surface of the spine is flattened, and together with the adjoining

part of the posterior border of the greater wing forms the anterolateral border of a groove which is completed posteromedially by the petrous part of the temporal bone. This groove contains the cartilaginous part of the *auditory tube*, and leads backwards into the canal for the tube in the petrous part of the temporal bone and forwards to the upper part of the posterior border of the medial pterygoid plate. In the roof of the groove the posterior border of the greater wing and the anterior border of the petrous temporal are apposed at the petrosphenoidal suture.

Occasionally the foramen ovale is deficient posteriorly, being then confluent with the foramen spinosum, the frequency varying

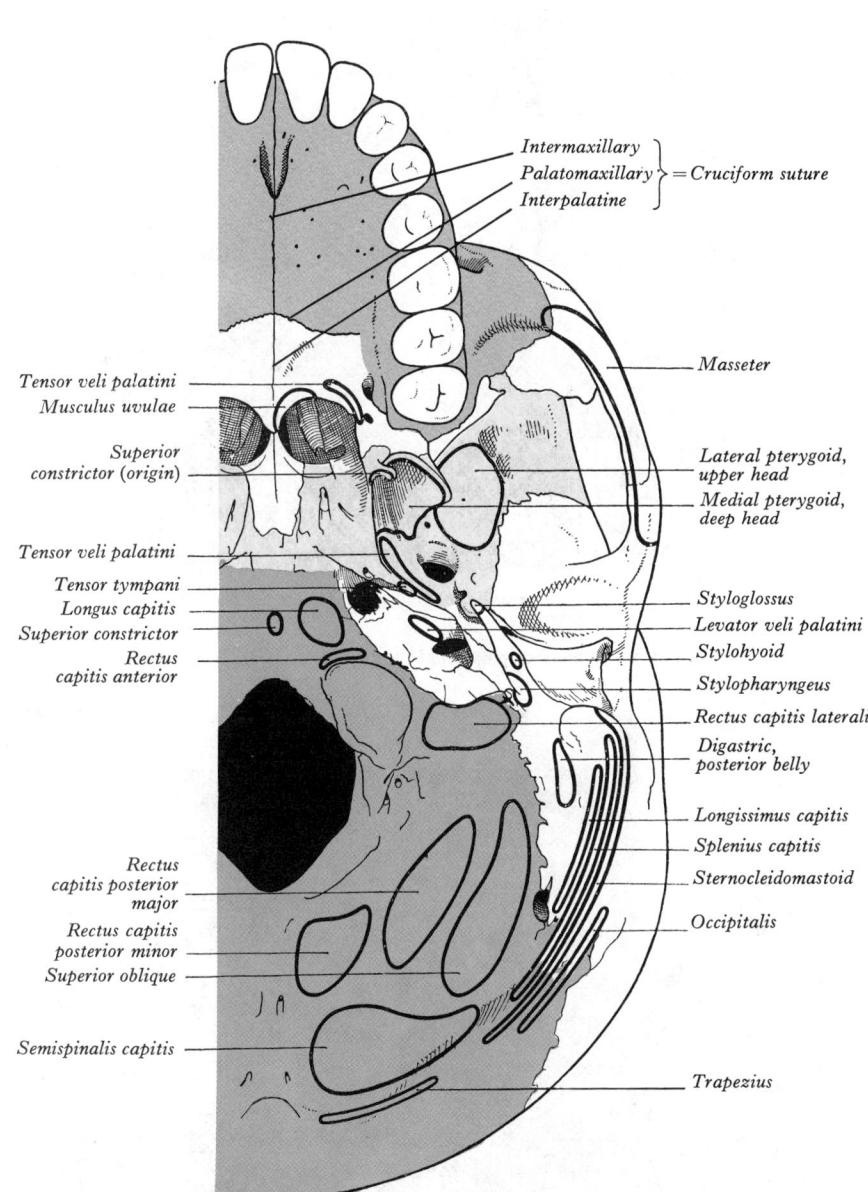

3.71B Outline drawing of norma basalis showing the attachments of muscles. Key to **3**.71A. *Blue*=occipital bone; *yellow*=sphenoid bone; *green*=maxilla.

from 0·7 to 10·4 per cent in modern populations. Similarly, the posterior circumference of the foramen spinosum may be defective with an incidence of 2 to 17 per cent (*see* p. 348).

Behind and medial to the groove for the tube the *inferior surface of the petrous temporal* occupies the interval between the greater wing of the sphenoid and the basilar part of the occipital bone. The anterior part of this surface is rough and uneven, and its apex is separated from the posterolateral part of the body of the sphenoid by an irregular bony canal, termed the *foramen lacerum*. Behind this rough area a large and approximately circular foramen leads upwards into the bone, the inferior opening of the *carotid canal*, occupied by the internal carotid artery. The canal turns forwards and medially and opens on the posterior wall of the foramen lacerum. Emerging from the canal the artery turns upwards to gain the cranial cavity. The lower part of the foramen

lacerum is occupied by fibrocartilage, and no large structure enters or leaves the skull through this structure.

From the base of the spine of the sphenoid the *squamotympanic fissure* runs laterally and slightly backwards between the upper part of the tympanic plate and the floor of the mandibular fossa. The fissure can usually be traced to the upper part of the anterior margin of the orifice of the external acoustic meatus, but it is sometimes obliterated near its lateral end. The *mandibular fossa* is deeply concave from before backwards and gently concave from side to side, and is wider at its lateral end. It contains the head of the mandible when the mouth is closed. Anteriorly the articular surface passes on to a transverse rounded elevation, the *articular tubercle*, continuous laterally with the anterior root of the zygoma. In front is the part of the squamous temporal which lies in the roof of the infratemporal fossa. Behind the squamotympanic fissure the *tympanic part of the temporal bone* separates the mandibular fossa from the external acoustic meatus. This part of the temporal bone is roughly triangular, the apex being at the medial end of the squamotympanic fissure close to the root of the spine of the sphenoid. Its lower border is free and skirts the anterolateral margin of the lower opening of the carotid canal, extending backwards and laterally to reach the root of the styloid process. There it forms the *sheath of the styloid process*, which is longer and more apparent on the lateral than on the medial side. At its lateral margin the tympanic part is fused with the rest of the temporal bone below and behind and is free above, where it forms the anterior border of the orifice of the external acoustic meatus.

The upper border of the *vomer*, which is applied to the inferior surface of the body of the sphenoid, is expanded into an *ala* on each side (**3.113**), and the groove between the alae fits the *rostrum of the sphenoid*. The lateral border of each ala reaches to a thin lamella which projects medially from the root of the medial pterygoid plate, and is termed the *vaginal process*. The two may come merely into contact or the edge of the ala may extend into the narrow interval between the body of the sphenoid above and the vaginal process below. The inferior surface of the vaginal process is marked by an anteroposterior groove, which is converted into a canal anteriorly by the upper surface of the sphenoidal process of the palatine bone. This is the *palatovaginal canal*, opening anteriorly through the posterior wall of the pterygopalatine fossa. It transmits the pharyngeal branch of the pterygopalatine ganglion and a minute pharyngeal branch from the third part of the maxillary artery. A second canal may be present in this situation on the medial side of the palatovaginal canal. It lies between the ala of the vomer and the upper surface of the vaginal process and is termed the *vomerovaginal canal*. When present it leads forwards into the anterior end of the palatovaginal canal.

In front of the pharyngeal tubercle the inferior surface of the basilar part of the occipital bone is intimately related to the roof of the nasal pharynx and the pharyngeal tonsil. Lateral to the tubercle is attached the longus capitis, and the area extends forwards on each side beyond the tubercle. Behind the longus capitis the rectus capitis anterior is inserted, in front of the occipital condyle and medial to the hypoglossal canal.

At the upper part of the posterior border of **the medial pterygoid plate** the *scaphoid fossa* gives origin to the anterior fibres of the tensor veli palatini, which descends along the lateral surface and posterior border of the plate to reach the *hamulus*. The tendon of the muscle twists medially round the lateral and anterior aspects of the process to gain the soft palate. The posterior border of the medial pterygoid plate, notched above by the auditory tube (p. 325), gives attachment to the pharyngobasilar fascia. Its lower part and the posterior aspect of the hamulus give origin to the highest fibres of the superior constrictor muscle of the pharynx, which curve upwards and medially to be inserted into the pharyngeal tubercle. The tip of the hamulus gives attachment to the pterygomandibular raphe. At its upper end the posterior border of the medial pterygoid plate is marked by a small tubercle, which lies on the medial side of the scaphoid fossa. This tubercle projects backwards below the posterior opening of the *pterygoid canal*, which leads forwards to open on the posterior wall of the pterygopalatine fossa. It transmits the nerve and vessels of the pterygoid canal and lies in the line of fusion of the pterygoid process and greater wing with the sphenoidal body.

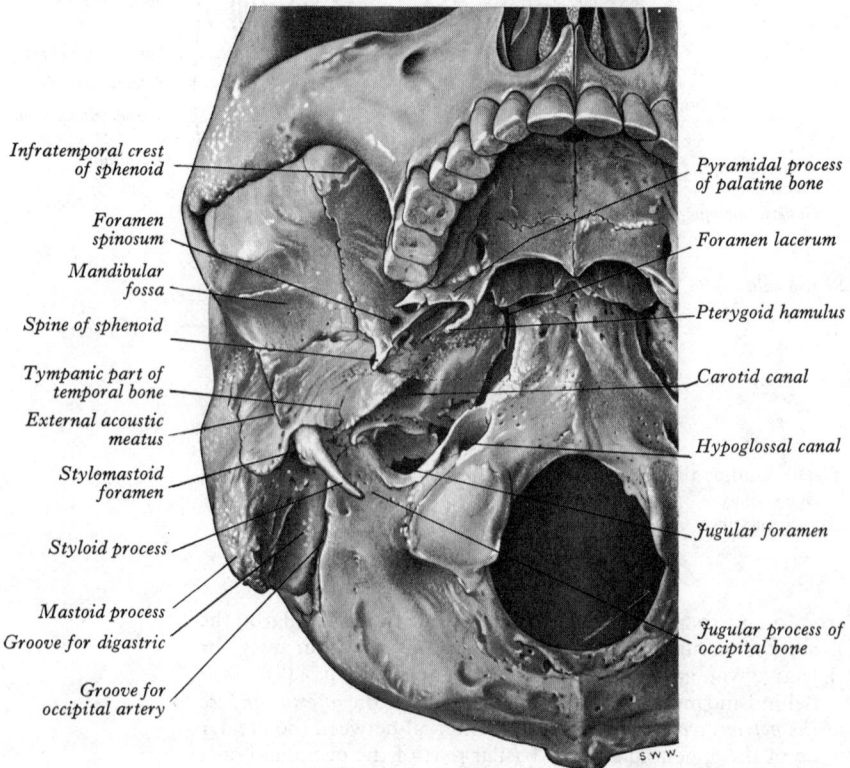

Intermaxillary suture

Incisive fossa

Palatine process of maxilla

Groove for greater palatine vessels

Greater palatine foramen

Lesser palatine foramina

Interpalatine suture

A.K.MAXWELL

Palatomaxillary suture

Horizontal plate of palatine bone

3.72 The bony palate and the alveolar arch. Inferior aspect.

Infratemporal crest of sphenoid

Foramen spinosum

Mandibular fossa

Spine of sphenoid

Tympanic part of temporal bone

External acoustic meatus

Stylomastoid foramen

Styloid process

Mastoid process

Groove for digastric

Groove for occipital artery

Pyramidal process of palatine bone

Foramen lacerum

Pterygoid hamulus

Carotid canal

Hypoglossal canal

Jugular foramen

Jugular process of occipital bone

S.W.W.

3.73 The right and part of the left side of the norma basalis of the skull. To show some features to better advantage, the anterior end of the skull has been elevated so that the Frankfurt plane is tilted to an angle of about 45° to the horizontal.

The *pterygoid fossa*, between the opposed surfaces of the two pterygoid plates, is completed below and in front by the pyramidal process of the palatine bone.

The lateral pterygoid plate (3.70) is a wider lamina than the medial plate, and its irregular posterior border may present a backward projection, termed the *pterygospinous process*, which is connected by a ligament (sometimes ossified) to the spine of the sphenoid. The lateral surface, which is the rougher of the two, gives origin to the lower head of the lateral pterygoid muscle; the medial surface gives origin to the greater part of the medial pterygoid. The lateral aspect of the *pyramidal process of the palatine bone*, which appears between the tuberosity of the maxilla and the lower part of the lateral pterygoid plate, gives origin to some fibres of the superficial slip of the medial pterygoid.

The infratemporal surface of the greater wing of the sphenoid affords origin to the upper head of the lateral pterygoid muscle, and is crossed by the deep temporal and masseteric nerves, which run between the muscle and bone. The *foramen ovale*, in addition to the mandibular nerve, contains the accessory meningeal artery. Its sharp posterior border provides attachment for the fibres of the tensor veli palatini, which intervenes between the mandibular nerve and auditory tube. **The foramen spinosum** transmits the small meningeal branch of the mandibular nerve in addition to the middle meningeal artery. In the interval between the foramen ovale and the scaphoid fossa the bone sometimes presents a small foramen, the *sphenoidal emissary foramen*, which transmits an emissary vein from the cavernous sinus. The *spine of the sphenoid*, which varies greatly in size, and may be sharply pointed or blunt, has attached to it the sphenomandibular ligament. It is related laterally to the auriculotemporal nerve and medially to the chorda tympani, by which its medial aspect is sometimes grooved, and to the auditory tube. Anteriorly the most posterior fibres of the tensor veli palatini are attached to it. The groove for the tube varies in width and depth and its roof is occasionally completed by fibrous tissue. The lateral or sphenoidal wall of the groove gives origin posteriorly to fibres of the tensor tympani.

To the lateral part of the rough inferior surface of the petrous part of the temporal bone the levator veli palatini is attached. The **foramen lacerum** is bounded in front by the posterolateral part of the body of the sphenoid and the adjoining roots of the pterygoid process and greater wing, behind and laterally by the apex of the petrous part of the temporal, and medially by the basilar part of the occipital. It is a canal nearly 1 cm long, but no large structure passes completely through it. The anterior orifice of the carotid canal opens on its posterior wall, and the vessel with its venous and sympathetic plexuses, ascends through the upper end of the foramen. In the foramen the deep petrosal nerve from the carotid sympathetic plexus is joined by the greater petrosal nerve to form the nerve of the pterygoid canal, which opens on the lower part of the anterior wall. Meningeal branches of the ascending pharyngeal artery and emissary veins from the cavernous sinus pass right through the foramen. The cartilage which fills its lower part is a remnant of the primitive chondrocranium.

The floor of the *mandibular fossa* is very thin and corresponds to the most lateral part of the floor of the middle cranial fossa. It is covered in the recent state by white fibrocartilage (p. 440). The *tubercle of the root of the zygoma* gives attachment to the lateral ligament of the temporomandibular joint. A thin edge of bone may be visible in the depths of the medial end of the *squamotympanic fissure*. It is the lower border of the downturned lateral portion of the tegmen tympani and therefore a part of the petrous temporal. It divides the upper part of the squamotympanic fissure into *petrotympanic* and *petrosquamous* fissures. Through the petrotympanic fissure the chorda tympani travels in its anterior canaliculus, as it passes downwards and forwards from the tympanic cavity. The anterior tympanic branch of the maxillary artery also traverses the petrotympanic fissure.

The tympanic part of the temporal bone (3.73) is separated from the capsule of the temporomandibular joint by a portion of the parotid gland, which usually contains the auriculotemporal nerve. It is thinnest near the centre of this surface and is occasionally deficient (p. 327). Its grooved upper aspect forms the anterior wall, the floor and the lower part of the posterior wall of the external acoustic meatus. Except where it ensheathes the styloid process its posterior surface is fused with the petromastoid part of the bone.

The Posterior Part of the Norma Basalis

The median region of the posterior part of the cranial base (3.71) features anteriorly **the foramen magnum**, which leads into the posterior cranial fossa. The foramen is oval, wider behind than in front, and its anteroposterior measurement exceeds its transverse. It transmits a number of structures, of which the most important is the lower end of the medulla oblongata. Anteriorly the margin of the foramen magnum is overlapped slightly on each side by an *occipital condyle*, which projects downwards to articulate with the superior articular facet on the lateral mass of the atlas. Oval in outline, the condyle is placed obliquely so that its anterior end is nearer the median plane. It shows a pronounced convexity from before backwards and is less convex from side to side. The medial aspect is roughened for ligamentous attachments. Above the anterior part of the condyle the occipital bone is pierced by the *hypoglossal (anterior condylar) canal*, which runs laterally and slightly forwards from the posterior cranial fossa and contains the hypoglossal nerve.

A depression of variable depth occurs behind the condyle. It is the *condylar fossa*, and may be pierced by a *condylar canal* which, when present, transmits an emissary vein from the sigmoid sinus. (The incidence of condylar canals shows racial variation—from 13·3 per cent in modern Palestinians to 70 per cent in Peruvian crania. It also illustrates sexual variation, being present, for example, in 58 per cent of male Burmese, but only 31 per cent of females, *see* p. 348.) Lateral to the condyle the *jugular process of the occipital bone* articulates with the petrous temporal. The anterior border of the process is free and forms the posterior boundary of the *jugular foramen*.

The jugular foramen is a large irregular hiatus between the occipital bone and the jugular fossa of the petrous temporal, and is set at the posterior end of the petro-occipital suture. In front it is separated from the lower orifice of the carotid canal by a raised ridge of bone, and on its lateral side it is related to the medial aspect of the sheath of the styloid process. Medially it is separated from the hypoglossal canal by a thin bar of bone. The foramen is usually larger on the right side of the skull, and its long axis is directed forwards and medially. The anterior part of the foramen transmits the inferior petrosal sinus; its intermediate part, the glossopharyngeal, vagus and accessory nerves; its posterior part the internal jugular vein. When the superior bulb of the internal jugular vein is well developed the jugular fossa of the temporal bone is recessed out in an upward and lateral direction in company.

Posterior to the root of the styloid process the *stylomastoid foramen* transmits the facial nerve. Posterolateral to the foramen the tip of the *mastoid process* projects downwards and forwards, and forms the lateral wall of the *mastoid notch*, from which the posterior belly of the digastric takes origin. Medial to the notch this part of the temporal bone may be grooved by the occipital artery (3.73).

In the median plane behind the foramen magnum the squamous part of the occipital bone presents the *external occipital crest*, to which is attached the upper end of the ligamentum nuchae. It terminates behind at the external occipital protuberance. Near its midpoint the *inferior nuchal line* begins and curves backwards and laterally. It is nearly parallel to the *superior nuchal line*, which extends in the same direction from the external occipital protuberance and may be raised into a distinct crest in its medial part.

The foramen magnum provides a wide communication between the posterior cranial fossa and the vertebral canal. Anteriorly the apical ligament of the dens and the membrana tectoria pass through it, both being attached to the upper surface of the basilar part of the occipital bone. Its wider, posterior part transmits the lower end of the medulla oblongata and the meninges. In the subarachnoid space the spinal roots of the accessory nerves, and the vertebral arteries, with their sympathetic plexuses, ascend into the cranium, and the posterior spinal arteries descend, one on each posterolateral aspect of the brainstem, as does the anterior spinal artery on the front of the brainstem in the median plane. In addition, the lower part of the tonsils of the cerebellum may project into the foramen on each side of the medulla oblongata. The *anterior margin* of the foramen gives attachment to the anterior atlanto-occipital membrane, which is continuous on each side with the capsular ligament of the atlanto-occipital joint. To the *posterior margin* is attached the posterior atlanto-occipital membrane, and to the roughened medial aspect of the condyle, the alar ligament.

In addition to the hypoglossal nerve the *hypoglossal canal* contains a meningeal branch of the ascending pharyngeal artery and a small emissary vein from the basilar plexus. Not uncommonly the canal is divided into two by a spicule of bone (p. 322). The inferior surface of the jugular process of the occipital bone provides attachment for the rectus capitis lateralis.

The jugular foramen (3.71) is directed upwards, medially and backwards, and on the external surface of the skull its apparent size is increased owing to the presence of the jugular fossa of the temporal bone on its lateral side. The floor of the fossa separates the superior bulb of the internal jugular vein from the tympanic cavity, and its lateral wall is pierced by a minute canal, the *mastoid canaliculus*, which transmits the auricular branch of the vagus nerve. Passing laterally through the bone, this nerve is very near to the facial canal, and finally emerges in the tympanomastoid suture. It is extracranial at birth but becomes surrounded by bone as the tympanic plate and the mastoid process develop. On or near the ridge between the jugular fossa and the orifice of the carotid canal, the *canaliculus for the tympanic*

nerve pierces the bone to transmit to the middle ear the tympanic branch from the glossopharyngeal nerve. On the upper boundary of the jugular foramen, near its medial end, there is a small notch—more easily identified on the internal surface—which contains the inferior ganglion of the glossopharyngeal nerve. The orifice of the *cochlear canaliculus* (p. 329) lies at the apex of the notch, the projecting edges of which may reach the occipital bone and divide the foramen into three parts.

The stylomastoid foramen lies behind the root of the styloid process and at the anterior end of the mastoid notch. As the facial nerve emerges from the foramen it is close to the posterior belly of the digastric, which it supplies before entering the parotid gland. In addition to the facial nerve the foramen carries the stylomastoid branch of the posterior auricular artery. A vascular groove crosses the inferior aspect of the posterior part of the temporal bone medial to the mastoid notch. It is occupied by the occipital artery, and its absence indicates that the vessel lies at a lower level than usual and between the splenius and longissimus capitis instead of deep to both.

The area below the *inferior nuchal line* gives insertion medially to the rectus capitis posterior minor and laterally to the rectus capitis posterior major (3.71B). The interval between the inferior and the superior nuchal lines provides insertion medially for the semispinalis capitis and laterally for the obliquus superior. In its medial part the *superior nuchal line* gives origin to the highest fibres of the trapezius; laterally, fibres of the sternocleidomastoid and, more anteriorly, splenius capitis are attached to it.

THE INTERIOR OF THE CRANIUM

The cranial cavity contains the brain and its meninges, and their blood vessels. Its walls are formed by the frontal, parietal, sphenoid, temporal and occipital bones and, to a small extent by the ethmoid bone. They are lined with a fibrous membrane, the *endocranium*, which is the outer zone of the dura mater. It passes through the various foramina which lead to the exterior, and becomes continuous with the periosteum on the outer surface of the skull, the *pericranium*. Both these fibrous membranes are continuous with the sutural ligaments, or cartilages, which occupy the narrow interosseous intervals at the sutures.

The walls of the cranial cavity vary in thickness in different skulls and in different parts of the same skull; but they tend to be thinner in situations where they are well covered with muscles externally, e.g. the temporal and posterior cranial fossae. Most of the cranial bones display *outer* and *inner tables*, formed of compact bone and separated from each other by the *diploë*, cancellous bone containing red bone marrow in its interstices. The inner table is thinner and more brittle than the outer table, which is generally very resilient. Many of the bones are so thin that the two tables are continuous, e.g. the vomer, pterygoid plates.

The skull is thicker in some races and some individuals, but no relationship exists between thicknesses of skull and cranial capacity, and in all races, the bones of the skull are thinner in women and children than in men.

The interior of the skull is described in two parts: the internal surface of the skull cap, the calva, and the internal surface of the base of the skull.

The Internal Surface of the Cranial Vault

The *calva* or *skull cap* (3.74) includes most of the frontal and parietal bones and the upper part of the squama of the occipital bone. It is marked, therefore, by the coronal, sagittal and lambdoid sutures, which may or may not be visible because the cranial sutures tend to become obliterated in old age and the process commences on the cerebral surface. The cranial vault is deeply concave in all directions and presents numerous vascular furrows and other markings.

Anteriorly, in the median plane, the *frontal crest* projects backwards. It gives attachment to the falx cerebri and is grooved

by the commencement of the *sagittal sulcus*. This accommodates the superior sagittal sinus, and widens progressively as it runs backwards in the median plane below the sagittal suture. On each side of the sagittal sulcus the bone presents a number of irregular depressions, the *granular foveolae*. They become more numerous and obvious as skulls age and are adapted to the arachnoid granulations.

The frontal branch of the middle meningeal vein, and sometimes its accompanying artery, groove the bone deeply about 1 cm behind the coronal suture, corresponding more or less to the precentral sulcus of the cerebrum. The rami of these vessels and

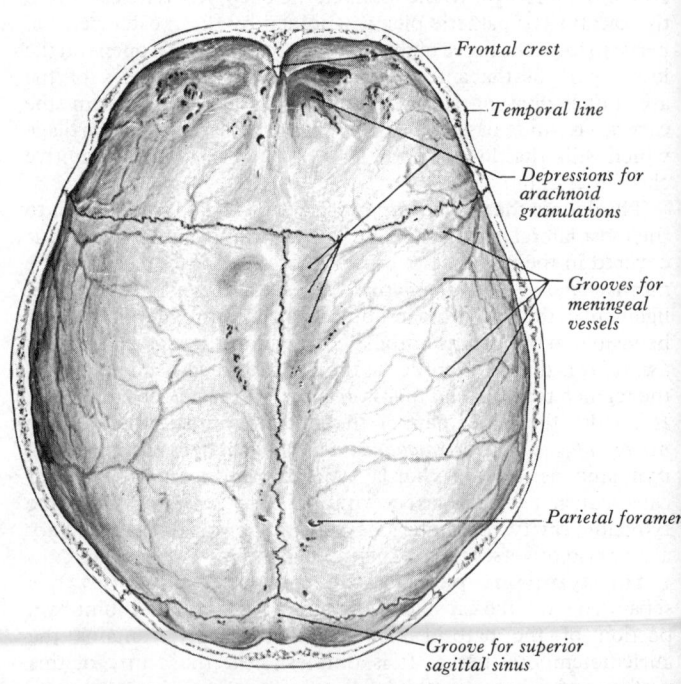

Frontal crest

Temporal line

Depressions for arachnoid granulations

Grooves for meningeal vessels

Parietal foramen

Groove for superior sagittal sinus

3.74 The internal (endocranial) surface of the skull cap or calva.

those of their parietal branches course upwards and backwards in grooves on the inner surface of the parietal bone. Smaller grooves produced by meningeal vessels may be present on the inner surfaces of the frontal and occipital bones. When present, the *parietal foramina* open on this surface near the sagittal groove about 3·5 cm in front of the lambdoid suture. Each transmits an emissary vein from the superior sagittal sinus.

The *impressions for the cerebral gyri* are less distinct on the vault than on the base of the skull and are seen best towards the latter.

The Internal Surface of the Cranial Base

The internal surface of the base of the skull (**3**.75) shows a natural division into anterior, middle and posterior cranial fossae. It is very irregular owing, partly, to the impressions for the cerebral gyri, which are especially conspicuous in the anterior and middle fossae, where they reflect accurately the pattern of the surface of the corresponding parts of the cerebrum. The dura mater is firmly adherent to the whole area, and through the numerous foramina and fissures its outer layer, the endocranium, is continuous with the periosteum on the exterior of the skull.

The Anterior Cranial Fossa

The anterior cranial fossa (**3**.75) is limited in front and on each side by the frontal bone. Its floor is formed by the orbital parts of the frontal bone, the cribriform plate of the ethmoid and the lesser wings and anterior part of the body of the sphenoid.

The cribriform plate of the ethmoid, which stretches across the median plane, lies between the two orbital parts of the frontal bone, and is depressed below the level of the rest of the floor. It separates the fossa from the nasal cavity, the roof of which it helps to form (**3**.78). Anteriorly it presents a median crest-like elevation, termed the *crista galli*, which projects upwards between the two cerebral hemispheres. A depression intervenes between the front of the crista galli and the *crest of the frontal bone*, the floor of which is crossed by the fronto-ethmoidal suture and marked by the presence of the *foramen caecum*. On each side the crista galli is separated from the orbital part of the frontal bone by a narrow interval. The numerous small foramina which perforate the cribriform plate transmit the minute olfactory nerves from the nasal mucosa to the olfactory bulb. Posteriorly the cribriform plate articulates with the body of the sphenoid at the spheno-ethmoidal suture.

The orbital part of the frontal bone forms the greater part of the floor of the fossa on each side of the median plane and separates the orbit and its contents from the inferior surface of the frontal lobe of the brain. Its cranial surface is convex and marked by impressions for the cerebral gyri and by one or two small grooves for meningeal vessels. In its anteromedial part it is split into two laminae to contain part of an air space, termed the *frontal sinus*. The medial part of the orbital plate covers the ethmoidal labyrinth and shuts it out from the floor of the anterior cranial fossa. Posteriorly it articulates with the anterior border of the *lesser wing of the sphenoid bone*. In the median plane the cerebral surface of the frontal bone is marked by the *frontal crest*, which projects backwards between the cerebral hemispheres and extends upwards on the interior of the skull cap.

The sphenoid bone completes the posterior region of the floor of the fossa. Centrally is the anterior part of the upper surface of its body, termed the *jugum sphenoidale*; this separates the fossa from bilateral air spaces in the body of the sphenoid and named the *sphenoidal sinuses* (**3**.77). Anteriorly the jugum articulates with the posterior margin of the cribriform plate; posteriorly it is limited by the anterior border of a groove, termed the *sulcus chiasmatis*, which crosses the body of the sphenoid in the forepart of the middle cranial fossa and leads from one optic canal to the other. Lateral to the jugum the floor of the anterior fossa is formed by the *lesser wing of the sphenoid*. The posterior margin of the lesser wing, which curves medially and backwards, is free and overhangs the middle cranial fossa. Laterally the lesser wing tapers to a point and meets the suture between the frontal bone and the greater wing at or near the lateral end of the superior orbital fissure. The medial extremity of its posterior border forms a projection termed the *anterior clinoid process*. Medially the lesser

wing is connected to the body of the sphenoid by two roots, separated from each other by the *optic canal*. The anterior root, broad and flat, is continuous with the jugum sphenoidale; the posterior root, smaller and thicker, is connected to the body of the sphenoid opposite the posterior border of the chiasmal sulcus.

The crista galli and the frontal crest afford attachment to the falx cerebri. The foramen caecum between them usually ends blindly, but on rare occasions it is patent and accomodates a vein from the nasal mucosa to the superior sagittal sinus. The narrow groove on the lateral side of the crista galli is related to the gyrus rectus, and the olfactory bulb lies on the medial edge of the orbital part of the frontal bone. The *anterior ethmoidal canal* opens on the line of the suture between the orbital part of the frontal bone and cribriform plate (**3**.67, 104). It is placed behind the crista galli and is difficult to identify, for it is directed medially and is overlapped above by the medial edge of the orbital plate. It transmits the anterior ethmoidal nerve and vessels, which run forward under the dura mater and gain the nasal cavity by passing downwards through a slit-like foramen at the side of the crista galli. The *posterior ethmoidal canal* opens at the posterolateral corner of the cribriform plate and is overhung by the anterior border of the sphenoid. It transmits the posterior ethmoidal vessels.

The posterior border of the lesser wing of the sphenoid, fits into the stem of the lateral cerebral sulcus and may be grooved by the sphenoparietal sinus. Above, the lesser wing is related to the posterior part of the inferior surface of the frontal lobe and medially to the anterior perforated substance. Inferiorly it forms the upper boundary of the superior orbital fissure and helps to complete the roof of the orbit. The anterior clinoid process has attached to it the free border of the tentorium cerebelli and is grooved on its medial aspect by the internal carotid artery as it pierces the roof of the cavernous sinus. Not infrequently the anterior clinoid process is connected to the middle clinoid process by a thin bar of bone, which completes a caroticoclinoid foramen around the internal carotid artery. The jugum sphenoidale underlies posterior ends of the gyri recti and olfactory tracts.

The Middle Cranial Fossa

The middle cranial fossa (**3**.75A and B), is deeper than the anterior; it is more extensive on each side than in the midline. In front it is bounded by the posterior borders of the lesser wings of the sphenoid, the anterior clinoid processes and the anterior margin of the sulcus chiasmatis, behind by the superior borders of the petrous parts of the temporal bones and the dorsum sellae of the sphenoid bone, laterally by the temporal squamae, the frontal angles of the parietal bones and the greater wings of the sphenoid.

Centrally the floor is formed by the body of the sphenoid. In front, the chiasmal sulcus leads on each side into the optic canal. The sulcus is rarely in contact with the optic chiasma, which is usually above and behind it. The **optic canal** is between the two roots of the lesser wing with, medially, the body of the sphenoid. It extends forwards, laterally and somewhat downwards, containing the optic nerve, ophthalmic artery and meninges. Immediately behind the sulcus the upper surface of the body of the sphenoid is shaped like a Turkish saddle and is hence termed the *sella turcica*. Its anterior slope bears a median elevation, the *tuberculum sellae*, and behind that the surface is hollowed out as the *hypophysial fossa* (**3**.75), which contains the hypophysis cerebri. The floor of the hypophysial fossa forms part of the roof of the sphenoidal sinuses (**3**.62A). Posterior to the fossa a plate of bone projects upwards and forwards to form the *dorsum sellae* (**3**.62A). Each superolateral angle of the dorsum sellae is expanded into a *posterior clinoid process*. Lateral to the sella turcica the body of the sphenoid presents a shallow *groove for the internal carotid artery*, as it runs forwards from the foramen lacerum. A small elevation marks the anterior part of the medial edge of the carotid groove and is termed the *middle clinoid process*; it may be joined to the anterior clinoid process by a thin bar of bone to make a foramen around the internal carotid artery. Posteriorly the lateral edge of the carotid groove may be deepened by a small projection termed the *lingula*.

Laterally the middle cranial fossa is deep and supports the temporal lobe of the cerebrum. It is formed in front by the cerebral surface of the greater wing of the sphenoid, behind by the anterior surface of the petrous part of the temporal bone, and laterally the cerebral surface of the temporal squama occupies the interval between these two bones. It is related in front to the apical region of the orbit, laterally to the temporal fossa, below to the infratemporal fossa. Anteriorly it communicates with the orbit through **the superior orbital fissure**, which is bounded above

the foramen ovale, and transmits the middle meningeal artery. The artery, with its accompanying veins, runs laterally to gain the temporal squama, on which it runs upwards, forwards and laterally. Crossing the sphenosquamosal suture for a second time it ascends on the greater wing and divides into frontal and parietal branches. The frontal branch proceeds upwards across the cerebral surface of the pterion (p. 302) and gains the anterior part of the parietal bone. In the region of the pterion the artery is often

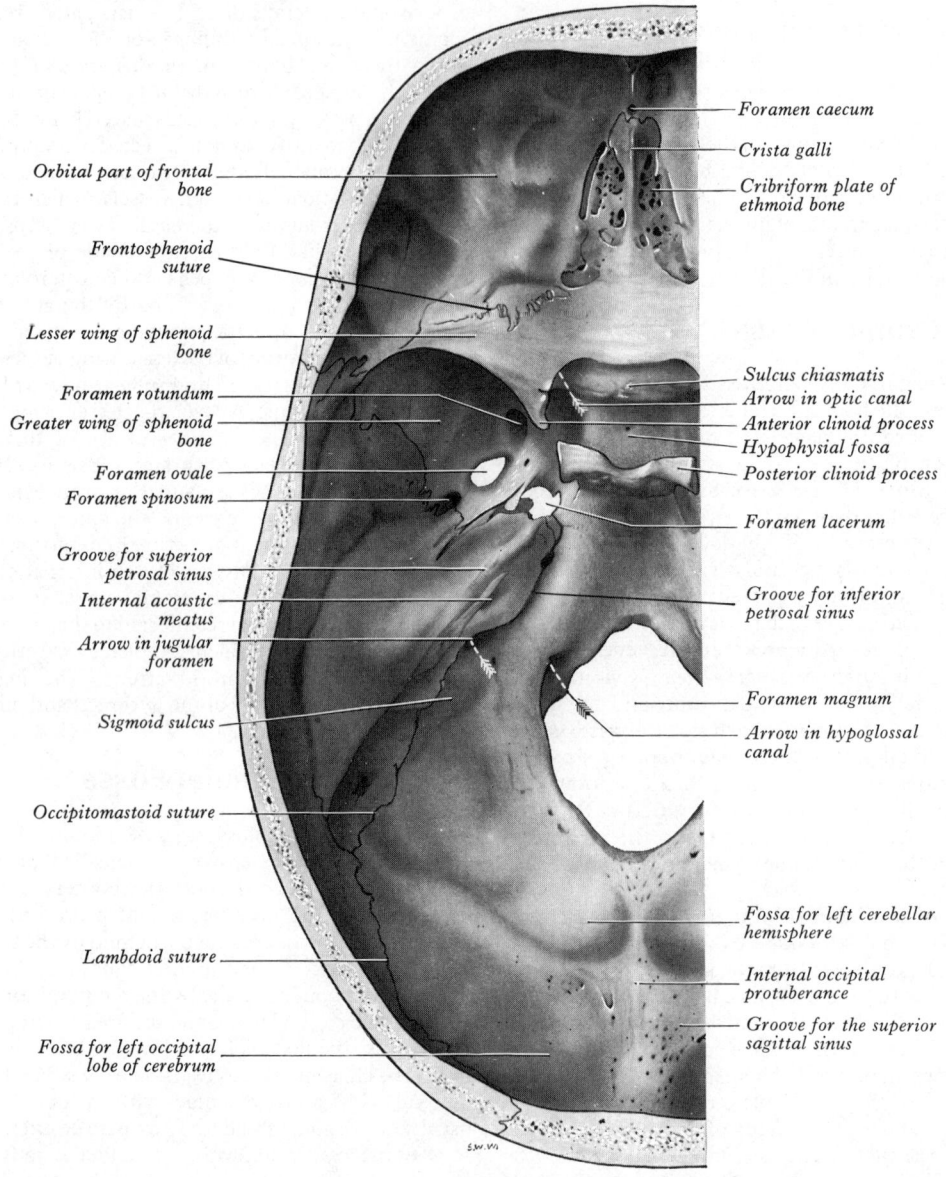

Foramen caecum

Crista galli

Cribriform plate of ethmoid bone

Orbital part of frontal bone

Frontosphenoid suture

Lesser wing of sphenoid bone

Foramen rotundum

Greater wing of sphenoid bone

Foramen ovale

Foramen spinosum

Groove for superior petrosal sinus

Internal acoustic meatus

Arrow in jugular foramen

Sigmoid sulcus

Occipitomastoid suture

Lambdoid suture

Fossa for left occipital lobe of cerebrum

Sulcus chiasmatis

Arrow in optic canal

Anterior clinoid process

Hypophysial fossa

Posterior clinoid process

Foramen lacerum

Groove for inferior petrosal sinus

Foramen magnum

Arrow in hypoglossal canal

Fossa for left cerebellar hemisphere

Internal occipital protuberance

Groove for the superior sagittal sinus

3.75A   The internal (endocranial) surface of the left half of the base of the skull. Compare with Key, **3.**75B

by the lesser wing, below by the greater wing, and medially by the side of the body of the sphenoid. The fissure is wider medially and its long axis is directed upwards, laterally and forwards. It transmits the terminal branches of the ophthalmic nerve, the ophthalmic veins, the oculomotor, trochlear and abducent nerves, and some smaller vessels. (*Vide infra.*)

The **foramen rotundum** pierces the greater wing of the sphenoid immediately below and a little behind the medial end of the superior orbital fissure. It leads forwards into the pterygopalatine fossa, to which it conducts the maxillary nerve. (*Vide infra.*)

The **foramen ovale** passes through the greater wing of the sphenoid posterior to the foramen rotundum and lateral to the lingula and the posterior end of the carotid groove. It leads downwards into the infratemporal fossa and transmits the mandibular nerve to that region.

The **foramen spinosum** is near the posterolateral margin of

enclosed in a bony canal. The parietal branch runs backwards and upwards on to the upper part of the temporal squama and crosses the squamosal suture to gain the posterior part of the parietal bone. These arteries and (laterally) their accompanying veins lie in conspicuous grooves in the floor and lateral wall of the middle cranial fossa. (*Vide infra.*)

At the posterior end of the carotid groove and posteromedial to the foramen ovale is **the foramen lacerum**. It is bounded behind by the apex of the petrous temporal and in front by the body of the sphenoid and the posterior border of its greater wing. This end of the foramen lacerum contains the internal carotid artery and its accompanying sympathetic and venous plexuses. (*Vide infra.*)

Behind the foramen lacerum the anterior surface of the petrous temporal presents a shallow depression adjoining the apex of the bone, occupied by the trigeminal ganglion, the *trigeminal impression.* Posterolateral to this the surface presents a shallow depression, limited posteriorly by a transversely rounded elevation, the *arcuate eminence.* This elevation is produced by the anterior semicircular canal, which is closely related here to the floor of the middle cranial fossa.

Lateral to the trigeminal impression the anterior surface of the petrous temporal displays a narrow groove passing backwards and laterally into the *hiatus for the greater petrosal nerve.* Lateral to this

internal carotid artery and its plexus of sympathetic nerves, the sinus contains the oculomotor, trochlear, abducent and ophthalmic nerves, but these structures are not in contact with bone. An anterior intercavernous sinus, which crosses the tuberculum sellae, and a posterior intercavernous sinus, which crosses the front of the dorsum sellae, connect the two cavernous sinuses to each other. The diaphragma sellae, which surrounds the infundibulum, is connected to the tuberculum in front and to the

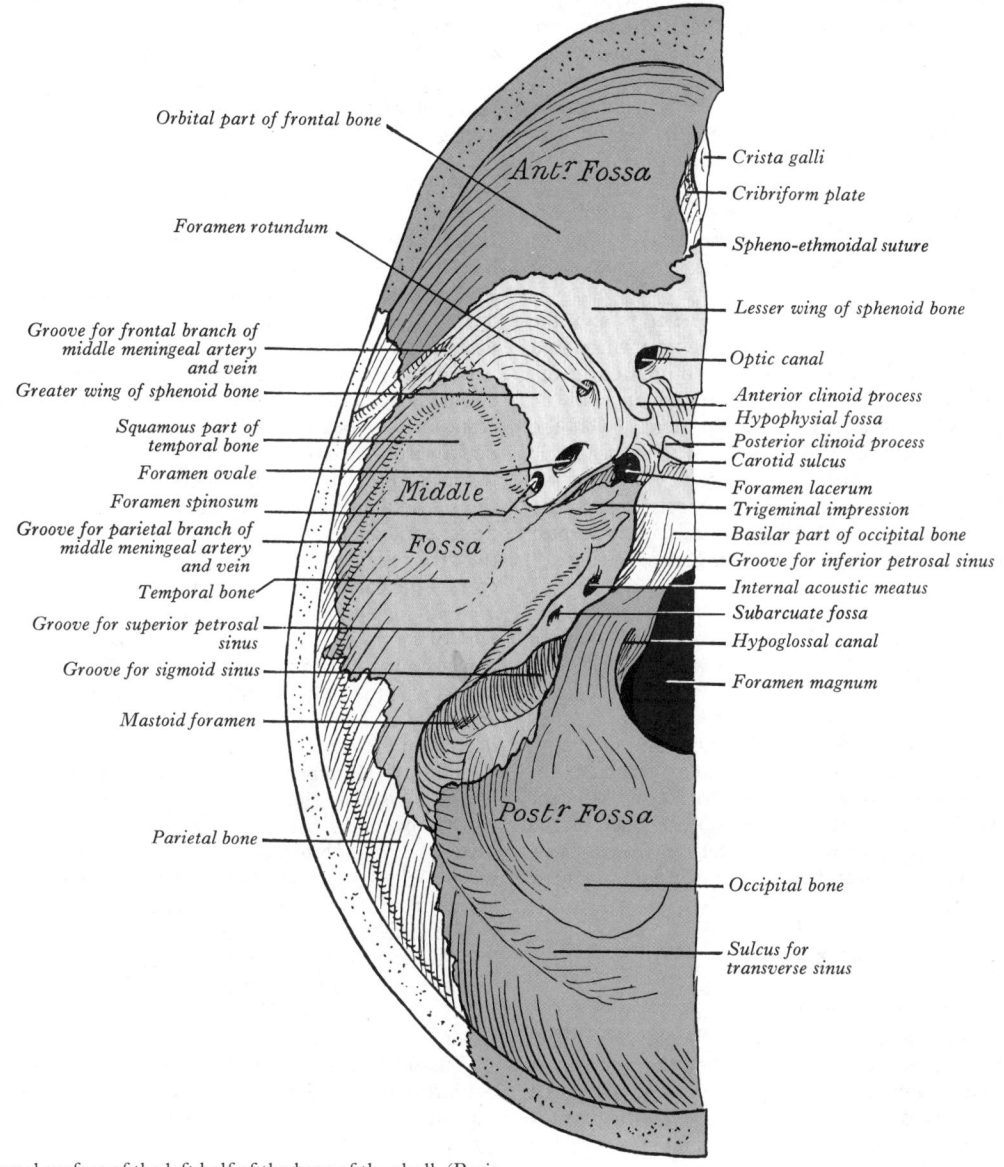

Orbital part of frontal bone

Foramen rotundum

Groove for frontal branch of middle meningeal artery and vein

Greater wing of sphenoid bone

Squamous part of temporal bone

Foramen ovale

Foramen spinosum

Groove for parietal branch of middle meningeal artery and vein

Temporal bone

Groove for superior petrosal sinus

Groove for sigmoid sinus

Mastoid foramen

Parietal bone

Ant.ʳ Fossa

Middle

Fossa

Post.ʳ Fossa

Crista galli

Cribriform plate

Spheno-ethmoidal suture

Lesser wing of sphenoid bone

Optic canal

Anterior clinoid process

Hypophysial fossa

Posterior clinoid process

Carotid sulcus

Foramen lacerum

Trigeminal impression

Basilar part of occipital bone

Groove for inferior petrosal sinus

Internal acoustic meatus

Subarcuate fossa

Hypoglossal canal

Foramen magnum

Occipital bone

Sulcus for transverse sinus

3.75B   The internal surface of the left half of the base of the skull. (Basis cranii interna.) Key to **3.**75A. *Blue*=frontal and occipital bones; *yellow*=sphenoid bone; *brown*=temporal bone.

is the *hiatus for the lesser petrosal nerve.* Anterolateral to the arcuate eminence the anterior surface of the petrous temporal is formed by the *tegmen tympani,* a thin osseous lamella which forms the roof of the tympanic cavity, and extends forwards and medially above the auditory tube. Lateral to the arcuate eminence the posterior part of the tegmen tympani forms the roof of the mastoid antrum, an air space in the bone continuous in front with the tympanic cavity.

The superior border of the petrous temporal separates the middle cranial fossa from the posterior cranial fossa. Behind the trigeminal impression it is grooved by the superior petrosal sinus, which connects the posterior end of the cavernous sinus to the upper end of the sigmoid sinus.

On each side of the body of the sphenoid the cavernous sinus extends from the medial end of the superior orbital fissure to the apex of the petrous part of the temporal bone. In addition to the

dorsum sellae behind. The posterior clinoid process gives attachment to the anterior extremity of the attached margin of the tentorium cerebelli and to the petrosphenoidal ligament (p. 1069).

The superior orbital fissure (3.65) opens into the orbit between its roof and its lateral wall. Its lower border is marked by a small projection which gives attachment to the lateral part of the common tendinous ring. At the lateral extremity of the fissure the greater wing articulates with the orbital part of the frontal bone (See also p. 300.)

The foramen rotundum, like the medial end of the superior orbital fissure, is intimately related to the lateral wall of the sphenoidal sinus. Originally a part of the fissure, it becomes separated secondarily. A small foramen may be present at the root of the greater wing medial to the foramen ovale; it transmits an emissary vein from the cavernous sinus and is often termed the *emissary sphenoidal foramen.* In addition to the mandibular nerve the foramen ovale transmits the accessory meningeal artery, and, sometimes, the lesser petrosal nerve. Together with the middle meningeal artery the foramen spinosum transmits the meningeal branch of the mandibular nerve. Both these foramina are represented at first by notches on the margin of the greater wing, which subsequently become converted into foramina.

The foramen lacerum, as described above, is a short bony canal, traversed in its whole extent only by minute meningeal branches from the ascending pharyngeal artery and a few small veins. The internal carotid artery pierces its posterior wall and ascends through its upper opening. The greater petrosal nerve emerges from its hiatus and runs forwards in the groove which marks the anterior surface of the petrous part of the temporal bone. It turns downwards through the foramen lacerum on the lateral side of the internal carotid artery and is joined by the deep petrosal nerve to form the nerve of the pterygoid canal. This nerve leaves the foramen lacerum above its lower opening by traversing the pterygoid canal, which opens on its anterior wall. The lesser petrosal nerve lies to the lateral side of the greater as it emerges from its hiatus on the anterior surface of the petrous part of the temporal bone and may occupy a small groove.

In a young skull the suture between the petrous and the squamous parts of the temporal bone may be visible at the lateral limit of the tegmen tympani, but it is usually obliterated in the adult skull. In this situation anteriorly, the lateral margin of the tegmen tympani turns downwards, forming the lateral wall of the bony part of the auditory tube, and its lower border may be visible in the floor of the squamotympanic fissure (p. 306). Lateral to the anterior part of the tegmen tympani the temporal squama is thin and translucent over a small area. This corresponds to the deepest part of the mandibular fossa on the external surface of the base of the skull.

In front of the commencement of the groove for the superior petrosal sinus the upper border of the petrous temporal shows a shallow, smooth notch, often termed the trigeminal notch, which leads into the trigeminal impression. In this situation the roots of the trigeminal nerve intervene between the superior petrosal sinus and bone. A tiny spicule, directed forwards and medially, marks the anterior extremity of the notch and gives attachment to the lower end of the petrosphenoidal ligament. The abducent nerve bends forwards sharply across the upper border of the petrous part of the temporal bone immediately in front of this bony spicule, and so lies between the petrosphenoidal ligament and the side of the dorsum sellae.

The Posterior Cranial Fossa

The posterior fossa (3.75A and B) is the largest and deepest of the three cranial fossae. It is bounded in front by the dorsum sellae, the posterior part of the body of the sphenoid and the basilar part of the occipital bone; behind, by the lower portion of the squamous part of the occipital bone; on each side, by the petrous and mastoid parts of the temporal bone, the lateral part of the occipital and, above and behind, by a small part of the mastoid angle of the parietal bone. It contains the cerebellum behind and the pons and medulla oblongata in front.

The foramen magnum (p. 307) is in the floor of the fossa and surrounded by the parts of the occipital bone; by the basilar part in front, the lateral part on each side, and a small portion of the squama behind. Just in front of its transverse diameter it is encroached on by the medial aspects of the occipital condyles, so that it is somewhat ovoid in shape and wider behind. Its narrower, anterior part lies above the dens of the axis vertebra; its wider, posterior part communicates below with the vertebral canal, and through it the medulla oblongata becomes continuous with the spinal cord.

In front of the foramen magnum the basilar part of the occipital bone, the posterior part of the sphenoidal body and the dorsum sellae form a sloping surface, the clivus, gently concave from side to side, antero-inferior to the pons and medulla oblongata. On each side this area is separated from the petrous part of the temporal bone by the petro-occipital fissure, occupied in life by a thin plate of cartilage. The fissure is limited behind by the jugular foramen, and its margins are grooved by the inferior petrosal sinus.

The jugular foramen (p. 307) lies at the posterior end of the petro-occipital fissure, and leads forwards, downwards and laterally to the exterior. Its upper border is sharp and irregular and presents a notch for the glossopharyngeal nerve. Its lower border is smooth and regular. The posterior part of the foramen contains the sigmoid sinus, continuous below with the internal jugular vein. In front of the vein the accessory, vagus and glossopharyngeal nerves, in that order from behind forwards, traverse the foramen.

Medial to the lower border of the jugular foramen a rounded elevation, termed the jugular tubercle, marks the lateral part of the occipital bone. It lies above and somewhat in front of the inner opening of the hypoglossal canal, which pierces the bone at the junction of the basilar with the lateral part and contains the hypoglossal nerve.

The posterior surface of the petrous part of the temporal bone forms a large portion of the lateral (or anterolateral) wall of the posterior fossa. Above the anterior part of the jugular foramen the internal acoustic meatus (3.75) runs transversely in a lateral direction. It is a short passage, about 1 cm long, closed laterally by a perforated plate of bone which separates it from the internal ear. Through it pass the facial and vestibulocochlear nerves, the nervus intermedius and labyrinthine vessels.

Behind the petrous temporal the lateral wall of the posterior cranial fossa is formed by the mastoid part of the temporal bone. Anteriorly it bears a wide groove, which runs forwards and downwards, then downwards and medially, and finally forwards to the posterior limit of the jugular foramen. This groove contains the sigmoid sinus and is termed the sigmoid sulcus (3.75). At its upper end, where it touches the mastoid angle of the parietal bone, the groove is continuous with that for the transverse sinus and crosses the parietomastoid suture. As it descends, it lies behind the mastoid antrum, a most important relation of that cavity. In this part of its course the mastoid foramen opens near its posterior margin and transmits an emissary vein from the sinus. In its lowest part the sigmoid sulcus crosses the occipitomastoid suture and grooves the jugular process of the occipital bone. It is usually deeper on the right than on the left side.

Behind the foramen magnum the squamous part of the occipital bone shows near the midline the internal occipital crest, which ends above and behind in an irregular elevation named the internal occipital protuberance. On each side of the protuberance a wide shallow groove curves laterally with a light upward convexity to the mastoid angle of the parietal bone. It is produced by the transverse sinus, is usually deeper on the right, and at its lateral extremity is continuous with the sigmoid sulcus. Below the transverse sulcus the internal occipital crest divides the bone into two gently hollowed fossae, adapted to the cerebellar hemispheres.

When the condylar canal is present (3.86), its inner orifice lies behind and lateral to the orifice of the hypoglossal canal. It transmits an emissary vein from the lower end of the sigmoid sinus.

The anterior wall of the posterior fossa (the clivus) is related to the plexus of basilar sinuses which connects the two inferior petrosal sinuses and communicates below with the internal

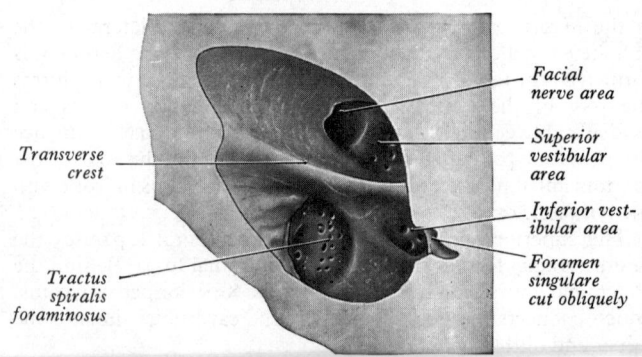

Facial nerve area

Superior vestibular area

Inferior vestibular area

Foramen singulare cut obliquely

Transverse crest

Tractus spiralis foraminosus

3.76 The fundus of the right internal acoustic meatus, exposed by a section through the petrous part of the right temporal bone nearly parallel to the line of its superior border.

vertebral venous plexus. A little in front of the foramen magnum the membrana tectoria is attached to the basilar part of the occipital bone (4.29), covering the attachment of the apical ligament of the dens. The jugular tubercle is often grooved by the glossopharyngeal, vagus and accessory nerves, as they pass to the jugular foramen. In addition to the hypoglossal nerve **the hypoglossal canal**, which is often subdivided (p. 322), transmits a meningeal branch of the ascending pharyngeal artery. To the roughened medial aspect of the occipital condyle (3.86) the alar ligament is attached.

The lower and posterior borders of **the jugular foramen** (pp. 307 and 329) are smooth and regular, but its upper border is sharp and interrupted by a notch, the ends of which may divide the foramen into two or sometimes three compartments. At its deepest part the notch is pierced by the *cochlear canaliculus*, which contains the perilymphatic 'duct' (*see* p. 1262).

The internal acoustic meatus is separated at its lateral end or *fundus* from the internal ear by a vertical plate which is divided into two unequal portions by a *transverse crest* (3.76). Above the crest anteriorly the bone is pierced by **the facial canal**, which conducts the facial nerve through the petrous temporal to the stylomastoid foramen. Behind the opening of the facial canal there is a small depression, termed the *superior vestibular area*, which presents a number of openings for the passage of the nerves to the utricle and the anterior and lateral semicircular ducts. Below the transverse crest anteriorly lies the *cochlear area*, in which a number of small, spirally arranged openings encircle the central canal of the cochlea and constitute the *tractus spiralis foraminosus*. Behind the cochlear area the *inferior vestibular area* presents several openings for the nerves to the saccule. Below and behind the inferior vestibular area the *foramen singulare* gives passage to the nerve to the posterior semicircular duct.

Behind the orifice of the internal acoustic meatus a thin plate of bone with an irregularly curved margin projects backwards, and the slit which it bounds contains the external opening of the *aqueduct of the vestibule* (3.94). Within the aqueduct the saccus and ductus endolymphaticus (p. 1203) are situated together with a small artery and vein. In the area between the internal acoustic meatus and the external opening of the aqueduct of the vestibule there is a small depressed area, termed the *subarcuate fossa* (3.75B), which lodges a small process of the dura mater. It lies nearer to the upper border of the bone (3.94) and is pierced by a small vein. In the infant the fossa is relatively large and extends as a short blind tunnel under the anterior semicircular canal; it corresponds to the floccular fossa in some animals.

In addition to an emissary vein, the *mastoid foramen* transmits a meningeal branch of the occipital artery, which is sometimes large enough to produce a groove on the squamous part of the occipital bone.

The internal occipital crest, to which the falx cerebelli is attached, may be grooved by the occipital sinus. Its lower end is related to the inferior vermis of the cerebellum. The internal occipital protuberance is related to the confluence of sinuses and is grooved on each side by the commencement of the transverse sinus. The margins of the groove for the transverse sinus give attachment to the two layers of the tentorium cerebelli. On each side of the internal occipital crest the bone is thin and translucent, in marked contrast to the regions of the crest and of the internal occipital protuberance.

The Nasal Cavity

The nasal cavity, the first of the respiratory passages, is an irregularly shaped space which extends from the roof of the mouth upwards to the base of the skull. It is divided into right and left halves by a *septum* (3.77), which is approximately median. In the dried skull the septum is deficient anteriorly, and as a result a single *anterior nasal aperture* presents in the norma frontalis. The septum, however, reaches the posterior limit of the cavity, which communicates with the nasal part of the pharynx through a pair of *posterior nasal apertures*, immediately above the posterior border of the bony palate. The cavity is wider below than above and is widest and vertically deepest in its central region. It com-

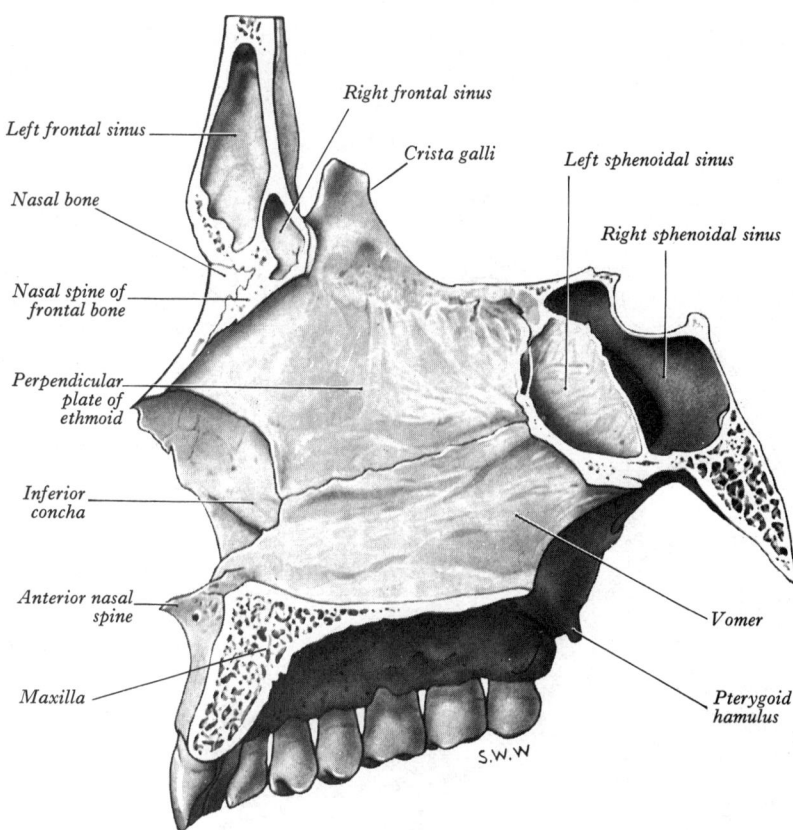

3.77 The bony nasal septum. Left side. An enlarged part of **3.**80.

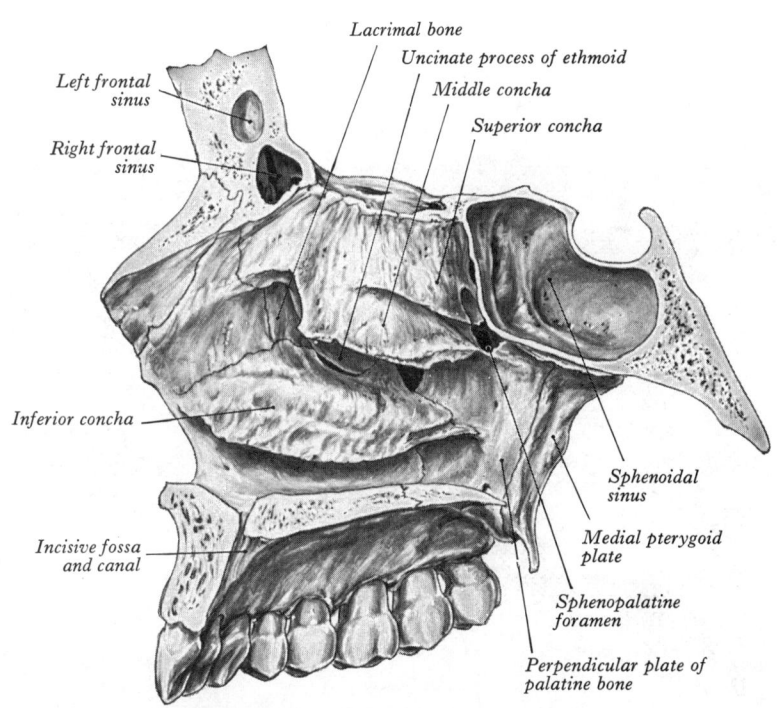

3.78 The roof, floor and right lateral wall of the right nasal cavity.

municates with the frontal, ethmoidal, maxillary and sphenoidal sinuses.

Each half of the cavity has a roof, a floor, lateral and medial walls, the medial wall being formed by the nasal septum.

The roof (3.77, 78) is horizontal in its middle part but slopes downwards in front and behind. The anterior sloping part is formed by the nasal spine of the frontal and the nasal bones and contributes to the formation of the external nose. The horizontal part is formed by the cribriform plate of the ethmoid bone and 313

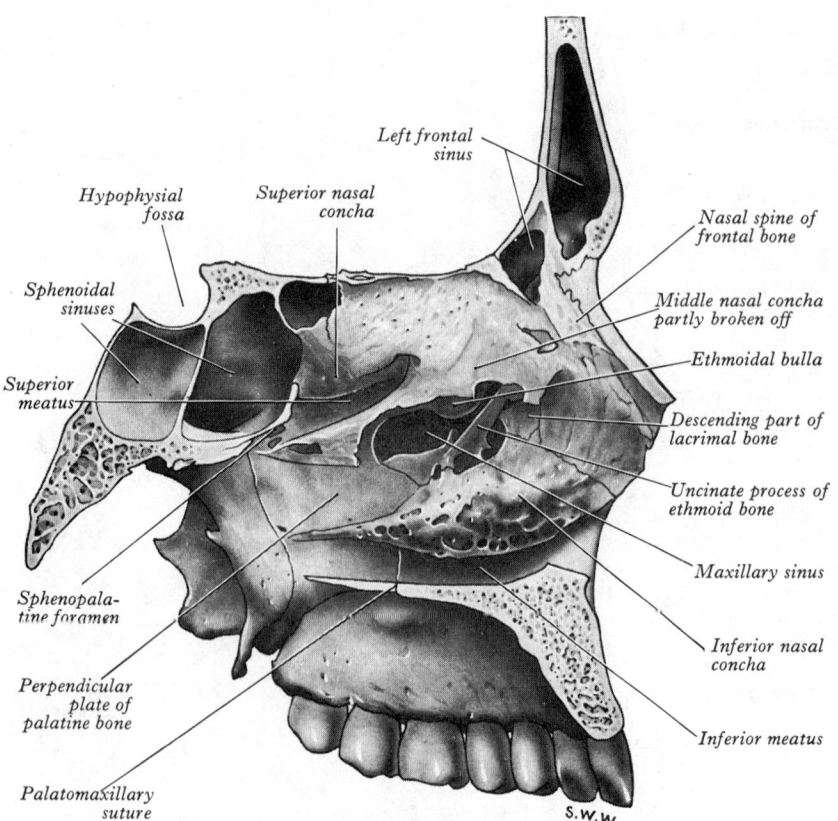

3.79 The lateral wall of the left nasal cavity, with an irregularly shaped portion removed from the lower part of the middle concha.

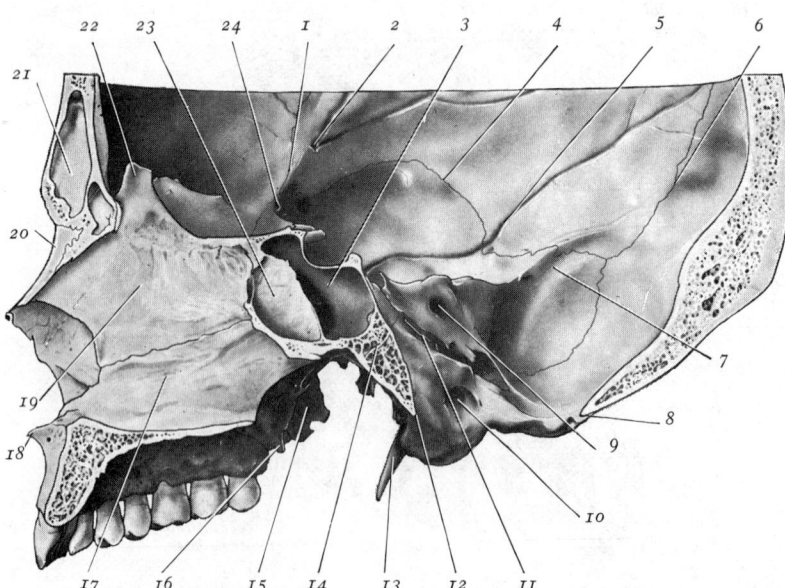

3.80 A sagittal section through the lower part of the skull, slightly to the left of the median plane. 1. Frontosphenoidal suture. 2. Bony canal for middle meningeal vessels, frontal branches, upper orifice. 3. Right sphenoidal sinus. 4. Squamosal suture. 5. Groove for parietal branches of middle meningeal vessels. 6. Lambdoid suture. 7. Groove for transverse sinus. 8. Posterior margin of foramen magnum. 9. Internal acoustic meatus. 10. Hypoglossal canal. 11. Petro-occipital suture in floor of groove for inferior petrosal sinus. 12. Anterior margin of foramen magnum. 13. Styloid process. 14. Line of occipitosphenoidal junction. 15. Lateral pterygoid plate. 16. Pterygoid hamulus. 17. Vomer. 18. Anterior nasal spine. 19. Perpendicular plate of ethmoid. 20. Nasal bone. 21. Frontal sinus. 22. Crista galli. 23. Left sphenoidal sinus. 24. Bony canal for frontal divisions of middle meningeal vessels, lower orifice.

separates the nasal cavity from the medial part of the floor of the anterior cranial fossa. It presents numerous small openings for the passage of the olfactory nerves. The posterior sloping part is formed by the body of the sphenoid and is interrupted, on each side, by the rounded orifice of a sphenoidal sinus.

The floor is smooth, gently concave from side to side, and slopes upwards a little as it passes backwards from the anterior aperture to the posterior aperture. It is formed by the upper surface of the bony palate and therefore intervenes between the nasal and oral cavities. Anteriorly the palatine processes of the two maxillae meet in the median plane, and behind them the horizontal plates of the palatine bones articulate with each other in the midline and with the palatine processes of the maxillae in front. In its anterior part the floor close to the septum presents a small funnel-shaped opening which leads into the *incisive canals* (p. 341).

The medial wall is formed by the bony *nasal septum* (3.77), which extends between the roof and the floor. It is a thin sheet of bone and presents a wide deficiency in front, occupied in the recent state by the septal cartilage. It is formed almost entirely by the vomer and the perpendicular plate of the ethmoid. The *vomer* extends from the inferior surface of the sphenoid body to the bony palate and forms the lower and posterior part of the septum, including its posterior border. It is marked by small furrows for vessels and nerves. The *perpendicular plate of the ethmoid* forms the upper and anterior part of the septum (3.77) and is continuous above with the cribriform plate (p. 334). The septum is often deflected to one side and the deviation occurs most commonly in the line of the vomero-ethmoidal suture.

The lateral wall (3.78, 79) is very irregular, owing to the presence of three bony projections termed the inferior, middle and superior nasal *conchae*. It is formed, for the most part, by the nasal surface of the maxilla below and in front, by the perpendicular plate of the palatine bone posteriorly, and above by the nasal surface of the ethmoidal labyrinth, which intervenes between the nasal cavity and the orbit. The three conchae project downwards and slightly medially, and each forms the roof of a groove which communicates freely with the nasal cavity. These grooves are termed the meatuses of the nose.

The *inferior concha* is a thin curved lamina and is an independent bone. It articulates with the nasal surface of the maxilla and the perpendicular plate of the palatine bone, and possesses a free lower border, which is gently curved. The *inferior meatus* is under the inferior concha and extends downwards to the floor of the nasal cavity. It is the largest of the three meatuses and extends almost the entire length of the lateral wall of the nose. The inferior meatus is deepest at the junction of its anterior and middle thirds, and at this point is the lower orifice of the nasolacrimal canal.

The *middle* and *superior conchae* are projections from the medial surface of the ethmoidal labyrinth. The *middle concha* is much the larger and extends backwards to articulate with the perpendicular plate of the palatine bone. The *middle meatus* is between the middle and inferior conchae. Its lateral wall displays several important features which can be examined only after the removal of the middle concha (3.79). Its upper part is occupied by a rounded elevation, the *ethmoidal bulla*, which contains the middle ethmoidal air cells. Below and in front of the bulla a thin, curved lamina of bone, named the *uncinate process of the ethmoid*, passes downwards and backwards, crossing the large bony orifice of the maxillary sinus. The curved gap (3.79) which intervenes between this process and the ethmoidal bulla is the *hiatus semilunaris*. At its upper end it becomes continuous with the *ethmoidal infundibulum*, a short, curved canal which receives the openings of the anterior ethmoidal air cells and then leads upwards through the labyrinth into the frontal sinus. In about 50 per cent of skulls, however, the infundibulum ends blindly and the frontal sinus then opens directly into the upper and anterior part of the middle meatus. The middle ethmoidal air cells open above, or near, the bulla.

The *superior concha* is a small curved lamina which lies above and behind the middle concha. It roofs in the *superior meatus*, which is much the shortest and shallowest of the three meatuses; it receives the opening of the posterior ethmoidal air cells. Immediately behind the superior meatus the sphenopalatine foramen, which opens into the pterygopalatine fossa, pierces the lateral wall of the nasal cavity. A narrow interval, the *spheno-ethmoidal recess*, separates the superior concha from the anterior surface of the body of the sphenoid, through which the sphenoidal sinus opens into the nasal cavity.

The posterior nasal apertures, or *choanae*, are separated from each other by the posterior border of the vomer. They are bounded below, on each side, by the posterior border of the horizontal plate of the palatine bone, above by the sphenoid bone, and laterally, on each side, by its medial pterygoid plate.

The anterior nasal aperture has been described on p. 299. In addition to the numerous small foramina for the transmission of the olfactory nerves, the horizontal part of the roof presents a separate foramen, situated anteriorly, for passage of the anterior ethmoidal nerve and vessels. The posterior sloping part of the roof is formed above by the anterior aspect of the body of the sphenoid, with which the sphenoidal concha (p. 323) is fused, and below by the ala of the vomer and the sphenoidal process of the palatine bone.

The floor (3.78) is crossed at the junction of its middle and posterior thirds by the palatomaxillary suture. Close to the median plane anteriorly it is pierced by the incisive canal (p. 341). Both incisive canals open into the incisive fossa on the bony palate and they possibly traverse the line of union of the os incisivum (premaxilla) with the maxilla; they represent a primitive communication between the mouth and the nose (but *see* p. 341).

At the upper and lower borders of **the medial wall** (3.77) other bones, in addition to the vomer and the perpendicular plate of the ethmoid, make minor contributions to the septum. Above and in front, the nasal bones and the nasal spine of the frontal bone, above and behind, the rostrum and crest of the sphenoid, and below, the nasal crests of the maxillae and palatine bones all take small parts in its formation. The vomer is grooved by the nasopalatine nerves, as they run downwards and forwards to reach the incisive canal.

The lateral wall (3.78, 79) is formed anteriorly and above by the nasal bone and the frontal process of the maxilla. Behind the frontal process of the maxilla, and articulating with its posterior border, the lacrimal bone lies in the lateral wall of the middle meatus and articulates below with the lacrimal process of the inferior concha. These two bones form the medial wall of the

nasolacrimal canal (3.79), which conveys the nasolacrimal duct to the inferior meatus. Posteriorly the lacrimal bone articulates with the ethmoidal labyrinth and helps to close some of the ethmoidal air cells. The *uncinate process of the ethmoid* springs from this part of the labyrinth and curves downwards and backwards in the lateral wall of the middle meatus. It is a very thin and fragile process, about 3 mm wide, which curves across the maxillary hiatus and articulates near its extremity with the ethmoidal process of the inferior concha. The concave, posterior border of the process is free and forms the medial edge of the hiatus semilunaris; the convex anterior border is free in its upper part only. Owing to its position relative to the maxillary hiatus the uncinate process helps to form the medial wall of the maxillary sinus. The *maxillary hiatus*, which forms such a wide opening on the nasal surface of the maxilla (3.116), is greatly reduced in size by the neighbouring bones. Its lower part is covered by the inferior concha and its maxillary process; above the inferior concha the uncinate process of the ethmoid, as already stated, encroaches on the gap. Posteriorly the anterior part of the perpendicular plate of the palatine bone closes it in still further, and above and in front small portions of the ethmoidal labyrinth and the lacrimal bone overlap its margins (3.79). As a result, the maxillary hiatus is reduced sometimes to a single orifice in the floor of the posterior part of the hiatus semilunaris, although as a rule additional openings exist behind the uncinate process, and between its lower border and the upper border of the inferior concha. The *ethmoidal bulla* is very variable in its size and shape and may be fused with the upper part of the uncinate process. In that event the duct of the frontal sinus opens into the upper part of the middle meatus medial to the blind end of the infundibulum. A third concha (*concha suprema*) is often present on the medial surface of the ethmoidal labyrinth above and behind the posterior end of the superior concha; it is little more than a slight ridge, separated from the superior concha by a shallow depression. The *sphenopalatine foramen* (3.78) is posterior to the superior meatus. It transmits the sphenopalatine artery and the nasopalatine and superior nasal nerves from the pterygopalatine fossa. It is bounded above by the body of the sphenoid and sphenoidal concha, below by the notched upper border of the perpendicular plate of the palatine bone, and in front and behind by its orbital and sphenoidal processes.

THE INDIVIDUAL CRANIAL BONES

The Mandible

The mandible (3.81), which is the largest and strongest bone of the face, has a curved, horizontal *body*, which is convex forwards, and two broad *rami*, which project upwards from the posterior ends of the body.

The body of the mandible is curved like a horseshoe, and possess external and internal surfaces, separated by upper and lower borders. The *external surface* is marked in the upper part of the median plane by a faint ridge, often indistinguishable, which indicates the line of fusion of the two halves of the fetal bone (*symphysis menti*). Inferiorly the ridge divides to enclose a triangular raised area, termed the *mental protuberance*, the base of which is depressed in the centre but raised on each side to form the *mental tubercle*. Below the interval between the two premolar teeth, or below the second premolar, the *mental foramen*, from which emerge the mental nerve and vessels, opens on the surface. The posterior border of the foramen is smoothed out by the direction of the emerging mental nerve, largely backwards but also somewhat upwards (Warwick 1950). A faint ridge, termed the *oblique line*, runs upwards and backwards from the mental tubercle, to become pronounced behind, where it is continuous with the anterior border of the ramus.

The lower border of the body is termed the *base of the mandible*. It extends backwards and laterally from the mental symphysis, and becomes continuous with the lower border of the ramus behind the third molar tooth. Near the median plane it presents a

small, roughened depression, the *digastric fossa*. Behind the digastric fossa the base is thick and rounded and presents a slight downwards convexity.

The upper border of the body is the *alveolar part*, which is hollowed by sixteen *sockets* for the roots of the teeth. These sockets vary in size and depth, and are single or subdivided by septa according to the teeth which they contain.

The *internal surface* is divided into two areas by an oblique ridge, termed the *mylohyoid line*. Sharp and distinct in the region of the molar teeth, it becomes almost indiscernible in front. It commences behind the third molar tooth, almost a centimetre from the upper border, and runs forwards and downwards to reach the mental symphysis in the interval between the two digastric fossae. Below the mylohyoid line the surface is slightly concave, the hollow being the *submandibular fossa* for the submandibular salivary gland. The area above the mylohyoid line widens as it is traced forwards and presents in front a triangular depression, the *sublingual fossa*, for the sublingual salivary gland. Above the sublingual fossa and extending backwards to the third molar tooth a strip of the bone is closely covered with the mucous membrane of the mouth. Above the anterior ends of the mylohyoid lines the posterior surface of the symphysis menti is marked by a small irregular elevation, which may be divisible into upper and lower parts, termed the *mental spines* (genial tubercles). Posteriorly, the *mylohyoid groove* extends downwards and forwards on to the body from the ramus and passes below the posterior end of the mylohyoid line. Immediately superior to the mental spines, the majority of mandibles present a midline pit which continues into a bony canal (Ingram—personal communication). As yet its development and contents are uncertain, but clinically it has proved a useful radiological landmark (*see* p. 349): the name *genial foramen* has been proposed. (See also *accessory mandibular foramina*, below.)

Above the mylohyoid line, the bone medial to the roots of the molar teeth is sometimes developed into a rounded ridge or *torus mandibularis*. For the incidence of this feature consult Mayhall *et al.* (1970) and Berry (1975).

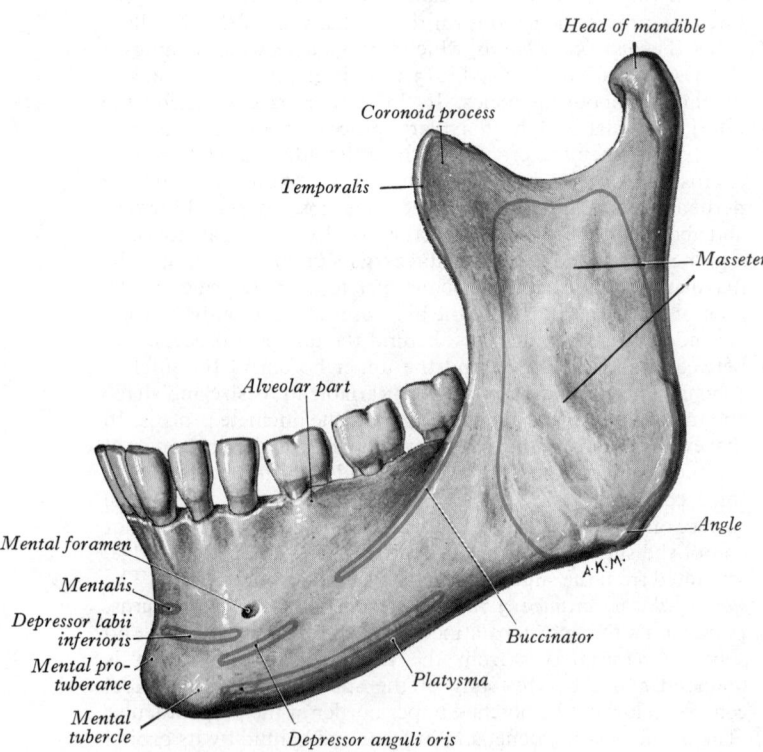

3.81A The left half of the mandible. Lateral (external) aspect.

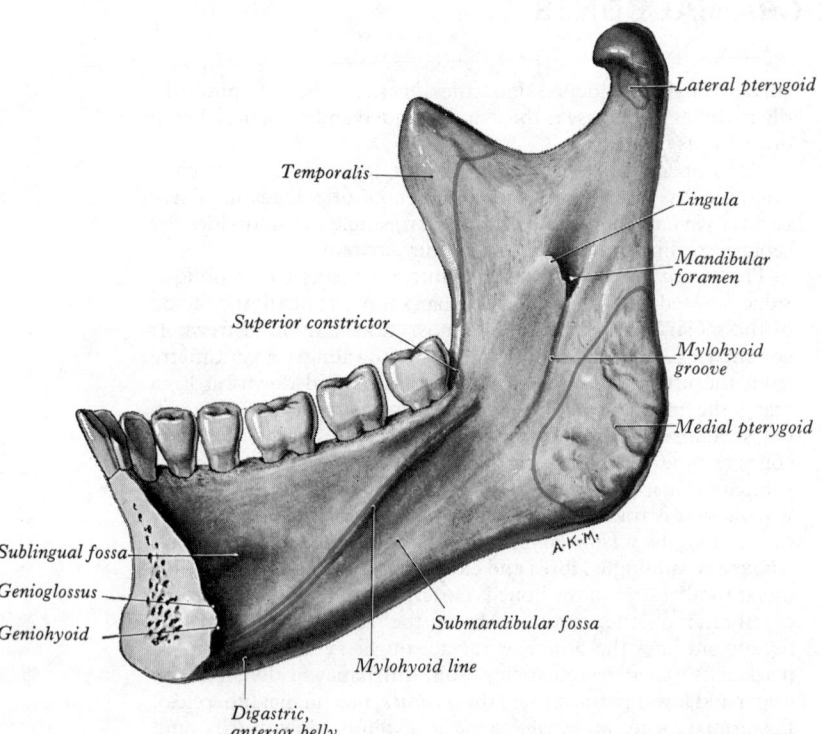

3.81B The right half of the mandible. Medial (internal) aspect.

The ramus of the mandible (3.81) is quadrilateral and has two surfaces, four borders and two prominent processes. The *lateral surface* is flat and marked by oblique ridges in its lower part. The *medial surface* presents, a little above its centre, an irregular opening, the *mandibular foramen*. This opening leads into the *mandibular canal*, which curves downwards and forwards into the body of the bone to open on the external surface at the mental foramen. In front and on the medial side the foramen is obscured by a thin triangular process termed the *lingula*. The *mylohyoid groove* commences behind the lingula and runs downwards and forwards to reach the internal surface of the body. The part of the medial surface which lies behind the groove is marked by a number of short ridges. The *inferior border* of the ramus is continuous in front with the base of the mandible; behind, it meets the posterior border at the *angle of the mandible*. Eversion of the angle is characteristic of the male mandible; in the female it is frequently incurved. The *upper border* is thin and bounds a wide notch, the *mandibular incisure*. It is surmounted in front by a triangular, flattened projection, termed the coronoid process, and behind by a strong, articular process named the condylar process. The *posterior border*, thick and rounded, extends from the back of the condylar process to the angle of the mandible. It is gently curved, being convex backwards above and concave below, and is intimately related to the parotid salivary gland. The *anterior border* is thin above, where it is continuous with the anterior border of the coronoid process, and thicker below, where it is continuous with the oblique line.

The coronoid process is a flattened triangular projection, directed upwards and slightly forwards in the living subject. Its posterior border bounds the mandibular incisure; its anterior border is continuous with the anterior border of the ramus. Its margins and medial surface provide insertion for most of the fibres of the temporalis.

The condylar process is expanded above to form the *head* of the mandible, which is covered with fibrocartilage. It articulates with the mandibular fossa of the temporal bone—an articular disc intervening. It is knuckle-shaped, being convex in all directions, and its transverse measurement is greater than its anteroposterior. The lateral aspect of the head has a blunt point which projects beyond the lateral surface of the rest of the ramus and can be felt in the living just in front of the tragus of the auricle. When the mouth is opened the head passed downwards and forwards, and the examining finger sinks into a small depression. The constricted portion below the head is termed the *neck of the mandible*. It is slightly flattened from before backwards, and its anterior aspect is limited on the lateral side by the backward continuation of the margin of the mandibular incisure. Medial to this ridge the anterior surface of the neck presents a rough muscular depression, the *pterygoid fovea*.

The mandibular canal runs from the mandibular foramen obliquely downwards and forwards in the ramus, and then horizontally forwards in the body below the sockets of the teeth, with which it communicates by small canals. It contains the inferior alveolar nerve and vessels, from which branches enter the roots of the teeth. Between the roots of the first and second premolars, or below the root of the second premolar tooth, the mandibular canal divides into *mental* and *incisive canals*; the mental canal swerves upwards, backwards and laterally to reach the mental foramen; the incisive canal is continued below the incisor teeth. (But *see* p. 1287 for canal variations.)

More detailed inspection of **the body of the mandible** shows a small shallow fossa, sometimes termed the incisive fossa, below the incisor teeth and providing attachment for mentalis and a part of orbicularis oris. The anterior end of the oblique lines gives origin to the depressor labii inferioris and the depressor anguli oris. Platysma is inserted into the bone below these muscles and extends backwards beyond them. The lower margin of the *mental foramen* is sharp and the mental nerve is directed upwards and backwards as it emerges from the bone. Adjoining the alveolar border, the bone is closely covered with the mucous membrane of the mouth. Immediately below this area, in the region of the molar teeth, the buccinator has a linear attachment which extends medially behind the last molar tooth to the pterygomandibular raphe.

The mylohyoid line gives origin to the mylohyoid muscle. Above its posterior end the bone gives origin to fibres of the superior constrictor muscle of the pharynx, and the *pterygomandibular raphe* is attached immediately behind the third molar tooth. The lingual nerve gains the tongue by passing above the posterior end of the mylohyoid line and here is closely related to the inner surface of the mandible (*see* p. 1067). The upper mental spine gives origin to the genioglossus and the lower to the geniohyoid. The *submandibular fossa* lodges some of the submandibular lymph nodes in addition to the submandibular salivary gland, and the facial artery may come into contact with this region as it descends to curl round the base of the mandible, where it sometimes produces a shallow groove. The *digastric fossa* gives attachment to the anterior belly of the digastric and lies below the anterior end of the mylohyoid line.

The ramus and its processes provide insertion for all the principal muscles of mastication. Its *lateral surface* gives insertion to the masseter, except at its upper and posterior part, where it is covered by the parotid gland.

The *medial surface* gives insertion to the medial pterygoid at the roughened area which lies behind and below the mylohyoid groove. To the *lingula*, the medial border of the mandibular foramen, the sphenomandibular ligament is attached. Posterior to the lingula the mylohyoid nerve and vessels enter the *mylohyoid groove*, which reaches the body of the mandible below the posterior end of the mylohyoid line, and the nerve and vessels then pass on to the superficial aspect of the mylohyoid. In front of the mylohyoid groove and below the lingula the medial surface of the ramus is related to the medial pterygoid, but the lingual nerve intervenes between the muscle and the bone. The lowest fibres of insertion of the temporalis descend beyond the coronoid process and are attached to the anterior border of the ramus and the adjoining part of the medial surface. The area above and behind the mandibular foramen is related to the maxillary artery and its inferior alveolar branch, and next to the part adjoining the mandibular incisure is the lateral pterygoid. The *mandibular notch* transmits the masseteric nerve and vessels from the infratemporal fossa.

The coronoid process is covered on its lateral aspect by the anterior fibres of the masseter as they pass downwards and backwards to be inserted into the ramus. If the finger is pressed into the yielding part of the cheek below the zygomatic bone, the anterior border of the coronoid process can be identified in the living when the mouth is opened.

The condylar process is expanded and projects beyond the surfaces of the ramus, but more so on the medial than on the lateral side. The *articular surface* of the mandible extends only for a short distance down the anterior surface of the process, but it covers the whole of its superior aspect and descends for 5 mm or more on its posterior aspect. Its superior aspect slopes medially and slightly downwards and backwards. Its projecting lateral part is separated from the cartilaginous part of the external acoustic meatus by a portion of the parotid gland. The smooth lateral aspect of the *neck of the mandible* gives attachment to the lateral ligament of the temporomandibular joint (**4**.18) and is covered by the parotid gland. The pterygoid fovea on the front of the neck receives the insertion of the lateral pterygoid muscle. The medial surface of the neck is related to the auriculotemporal nerve above and to the maxillary artery below.

The relation of the *parotid gland* to the mandible requires special mention. It occupies the interval below the external acoustic meatus, bounded in front by the posterior border of the ramus, behind by the mastoid process and medially by the styloid process; but it extends forwards beyond this area and covers the lateral aspect of the temporomandibular joint and the part of the lateral surface of the ramus behind the masseter. In addition it curls round the posterior border and comes into contact with the medial aspect of the ramus just above the insertion of the medial pterygoid.

Accessory foramina of the mandible, as in the case of other bones, are commonly present, usually unnamed, and infrequently described. Yet a recent study of 300 mandibles yielded a count of 2,449 accessory foramina (Sutton 1974). Since there is evidence that many of these transmit auxiliary nerve supplies to the teeth

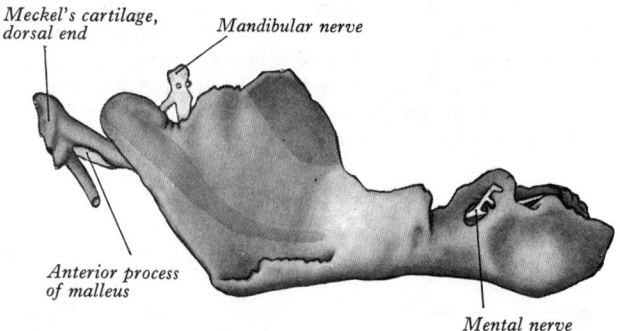

3.82A The right half of the mandible of a human embryo. 95 mm long. Lateral aspect. (Reconstruction model by Prof. A. Low.) *Note.—Blue* = cartilage; *yellow* = bone.

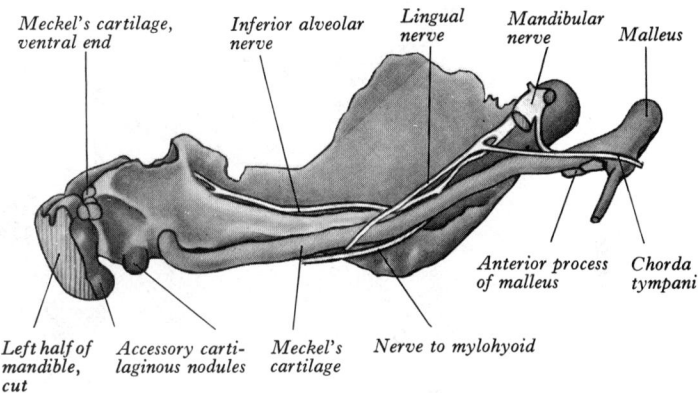

3.82B The right half of the mandible of a human embryo. 95 mm long. Medial aspect. The developing bone is shown in yellow, and the cartilage in blue. (Reconstruction by Prof. A. Low.)

and their sockets (from such nerves as the facial, mylohyoid, buccal, transverse cervical cutaneous, and others), their occurrence is of importance in dental nerve block techniques.

Further mandibular variants include *lingual depressions* molar or canine in position; *variable position* of the mental foramen; *multiple mental foramina* either double or triple; *lingual fenestrations* of molar sockets; *retromolar foramina*; and *condylar defects*. For incidences in 125 mandibles *see* Azaz and Lustmann (1973).

Aspects of mandibular structure. In addition to variations in the form and course of the mandibular canal (e.g. Fawcett 1895; Barros 1959; Carter and Keen 1971) and extensions from accessory mandibular foramina (Sutton 1974), there have been numerous attempts to analyse the fine architecture of the surface tables and buttresses of compact bone and the geometry of the internal trabeculae, and to relate these to the habitual functional stresses imposed on the bone. These are too numerous to review in detail here, but as an introduction to the literature the reader should consult Beltrami (1945); Seipel (1948); Dal Pont (1960); Weiss (1965); Scott and Symons (1970); and Mercier *et al.* (1970). An interesting preliminary report of the use of holographic interferometry in the study of surface strains induced by orthodontic forces has recently appeared (Hewitt 1976, *et seq.*).

Ossification. The mandible is ossified in dense fibromembranous tissue which lies lateral to the inferior alveolar nerve and its incisive branch and the lower portion of Meckel's cartilage (**3**.82). Each half is ossified from one centre which appears near the mental foramen about the sixth week of fetal life, i.e. just after the appearance of the primary centre for the clavicle (p. 358). From this centre ossification spreads to form the body and ramus; it extends medially, first below, then around the inferior alveolar nerve and its incisive branch and upwards to form a trough and later crypts for the developing teeth. By the tenth week the part of Meckel's cartilage below the incisor teeth is surrounded and invaded by bone (p. 150). Somewhat later secondary or accessory cartilages appear, which have no connexion with Meckel's cartilage (3.82). A cone-shaped mass of cartilage, the *condylar cartilage*, extends from the head of the mandible downwards and forwards through the ramus; this contributes to the growth in height of the ramus and though it is largely invaded and replaced

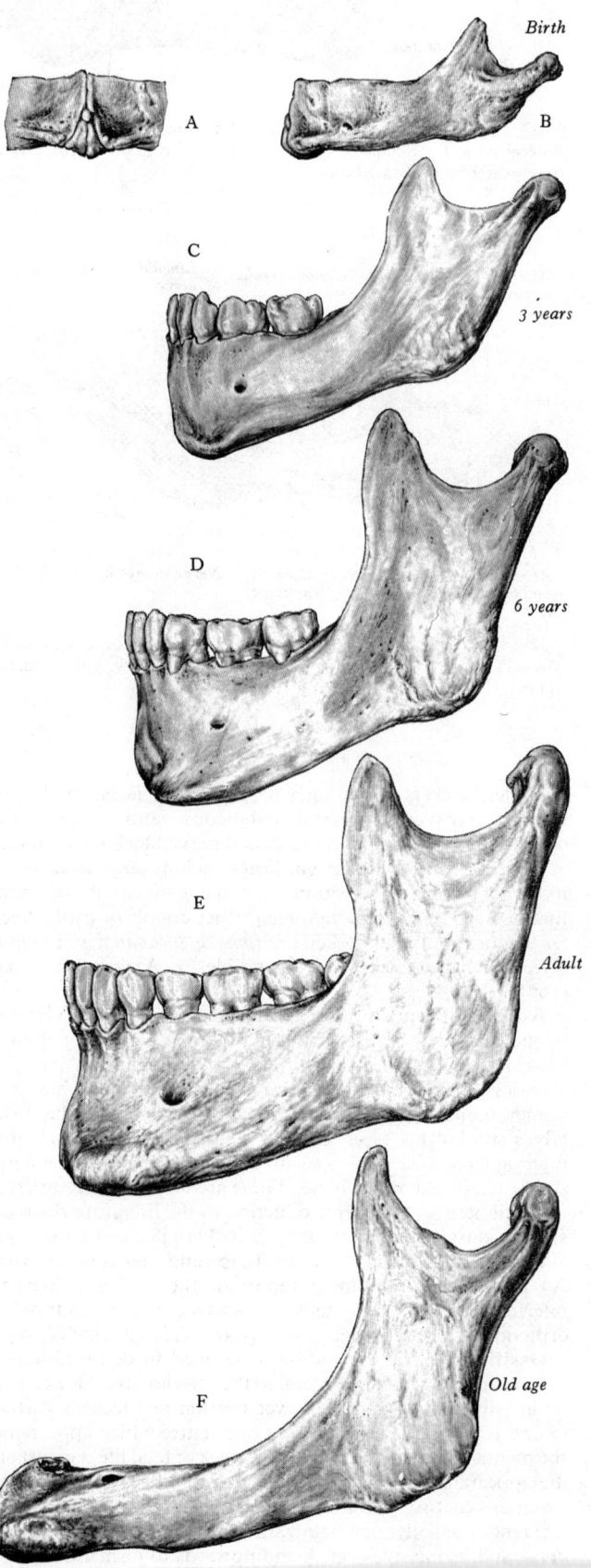

Birth

3 years

6 years

Adult

Old age

3.83A–F   The mandible at different periods of life.

by bone by the middle of fetal life, its upper end persists as a zone of proliferating cartilage beneath the fibrous articular surface of the head until the third decade. Another secondary cartilage, into which ossification soon extends, occurs along the anterior border of the coronoid process; it has disappeared before birth. One or two cartilaginous nodules also appear on each side at the symphysis menti. About the seventh month of fetal life these may ossify to form a variable number of small ossicles, the *mental ossicles*, in the fibrous tissue of the symphysis. These ossicles unite with the remainder of the bone before the end of the first year of life (*see* Lebourg and Champagne 1951; Sicker 1962; Scott and Symons 1970; and consult 3.83A).

It may be mentioned here that ^{3}H-thymidine labelling in the *rat* showed no interstitial growth in the condylar cartilage; it expanded by surface accretion only (Kvinnsland and Kvinnsland 1975).

Age Changes in the Mandible

At birth (3.83A and B) the mandible is in two separate halves, united in the median plane by fibrous tissue. This union is usually termed the *symphysis menti*. The anterior extremities of both mandibular rudiments are covered by cartilages, which are separated only by the fibrous tissue of the symphysis. Until fusion occurs across the midline new cells are added to each cartilage from the fibrous tissue on its symphysial aspect, while ossification proceeds in its mandibular side towards the midline. When the latter process overtakes the former, and extends into the median zone of fibrous tissue, fusion of the two halves of the mandible occurs. This description accords with recent observations, but many uncertainties still remain. The body of the bone is a mere shell, enclosing the sockets of the deciduous teeth, imperfectly partitioned off from one another. The mandibular canal runs near the lower border of the bone, and the mental foramen opens below the socket of the first deciduous molar tooth, and its direction is forwards (Warwick 1950). The coronoid process is relatively large and projects above the level of the condyle.

After birth (3.83C), in the first year, the two halves of the bone become joined at the symphysis from below upwards; but a trace of separation near the alveolar margin may still be visible at the beginning of the second year. The body elongates, more especially behind the mental foramen, to provide space for the three additional teeth developed in this part (*see* pp. 1300–1301). During the first and second years, as the prominence of the chin develops, the mental foramen alters in direction from forwards to upwards and backwards, and then almost horizontally backwards, as in the adult (Warwick 1950). This change accompanies the altering direction of the emerging mental nerve. The persisting upper end of the condylar cartilage fulfils the role of the epiphysial plate of a long bone. Its proliferation contributes to the vertical height of the ramus and to the downward and forward growth of the mandible as a whole. It also increases the distance between the condyles as the base of the skull widens. At this stage the cartilage is covered on its articular aspect by a self-perpetuating fibrous layer, deep to which is the proliferating intermediate zone which is responsible for the growth of the ramus. Beneath this is a layer of hypertrophic cartilage cells and then bone (Blackwood 1959). The depth of the body increases; growth of the alveolar part of the bone affords room for the roots of the teeth, and the subalveolar portion becomes thicker and deeper. After the second dentition the mandibular canal is situated a little above the level of the mylohyoid line, and the mental foramen occupies the position usual to it in the adult. As the mandible increases in size, bone is laid down along the posterior borders of the ramus and the coronoid process, while at the same time absorption of bone is occurring along their anterior borders. This process of remodelling goes on continuously until the bone has reached its adult size, and it enables the alveolar part to lengthen sufficiently to provide the necessary space for the permanent molar teeth (Enlow and Harris 1964). Organ culture of the condylar cartilage in rats showed no growth potential, but *in vivo* all the mandibular growth cartilages respond to mechanical forces (Petrovic 1972).

In the adult (3.83E) the alveolar and subalveolar portions of the body are of about equal depth. The mental foramen opens midway between the upper and lower borders of the bone, and the mandibular canal runs nearly parallel with the mylohyoid line. The *angle* subtended by the lower border of the body of the mandible to a plane surface which touches the posterior surface of the condyle above and the posterior border of the ramus below, necessarily diminishes as the height of the ramus increases with age, but X-ray photographs of the same child at different ages (Brodie 1941) show that the *contour* of the angle of the mandible remains unaltered.

In old age (**3**.83F) the bone is reduced in size, as the teeth are lost. Following the loss of teeth the alveolar part is absorbed, and consequently the mandibular canal and the mental foramen are close to the alveolar border. The foramen and part of the canal may even disappear, thus exposing part of the inferior alveolar nerve (Gabriel 1958). The ramus is oblique in direction, the angle measures about 140°, and the neck of the mandible is more or less bent backwards (Fawcett *et al.* 1924). The process of absorption affects chiefly the thinner of the two alveolar walls and, after its completion, a linear *alveolar ridge* is found on the alveolar border of the bone. In the mandible the labial wall is the thinner in the incisor and canine regions, but it is the lingual wall which is the weaker in the molar region. The alveolar ridge lies therefore within the line of the teeth in the incisor region but lies outside that line in the molar region, forming a curve which is wider than the curve of the line of the teeth and intersects it on each side in the premolar region. In the maxilla, however, the labial wall is everywhere the thinner and after absorption, the alveolar ridge lies wholly within the curve of the line of the teeth.

The Hyoid Bone

The hyoid bone (**3**.84) is U-shaped and is suspended from the tips of the styloid processes of the temporal bones by the stylohyoid ligaments. It has a body, two greater and two lesser cornua.

The body of the hyoid bone is an irregular, elongated quadrilateral in form. Its *anterior surface* is convex and directed forwards and upwards. Its upper part is crossed by a well-marked ridge, which has a slight downward convexity, and in many cases a vertical median ridge divides the body into lateral halves. The portion of the vertical ridge above the transverse line is present in the majority of specimens, but that below the transverse line is rarely seen. The *posterior surface* is smooth, concave, directed backwards and downwards, and separated from the epiglottis by the thyrohyoid membrane and a quantity of loose areolar tissue; a bursa intervenes between the bone and the membrane. In early life the lateral extremities of the body are connected to the greater cornua by cartilage, but after middle life they are usually united by bone.

The greater cornua of the hyoid bone project backwards from the lateral limits of its body; they are flattened from above downwards and diminish in size from before backwards. Each cornu ends posteriorly in a tubercle. When the throat is gripped between finger and thumb just above the thyroid cartilage, the greater cornua can be felt in the living subject and the bone can be moved from side to side.

The lesser cornua of the hyoid bone are two small, conical eminences attached by their bases at the angle of junction of the body and greater cornua. They are connected to the body of the bone by fibrous tissue and occasionally to the greater cornua by synovial joints, which usually persist throughout life, but occasionally become ankylosed (*vide infra*).

Attached to the *anterior surface of the body* is the geniohyoid in the greater part of its extent both above and below the transverse ridge; a portion of the origin of the hyoglossus invades the lateral margin of the geniohyoid area (**3**.84B). The lower part of this surface gives insertion to the mylohyoid, and below that to the sternohyoid medially and the omohyoid laterally. The *superior border* of the body is rounded and gives attachment to the lowest fibres of the genioglossi, to the hyo-epiglottic ligament and to the thyrohyoid membrane. The *inferior border* gives insertion to the sternohyoid medially and the omohyoid laterally, and sometimes to the medial fibres of the thyrohyoid. It gives attachment also to the levator glandulae thyroideae, when present.

The *upper surface of the greater cornu* gives origin to the middle

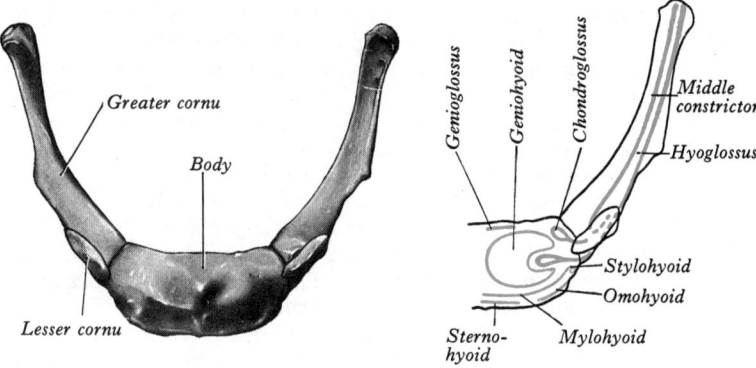

3.84A   The hyoid bone. Anterosuperior aspect.

3.84B   A drawing of the left half of the hyoid bone to show the muscular attachments. Superior aspect.

constrictor of the pharynx and, more laterally, to the hyoglossus, both of which extend throughout its whole length. Near the junction of cornu with body the stylohyoid is inserted lateral to the hyoglossus, and a little posterior to this insertion the fibrous loop through which the tendon of the digastric runs is attached to the bone. The *medial border* gives attachment to the thyrohyoid membrane, the *lateral border* receives, anteriorly, the insertion of the thyrohyoid. The *inferior surface*, which is oblique, is separated from the thyrohyoid membrane by some fibro-areolar tissue.

The *posterior* and *lateral aspects of the lesser cornu* give origin to fibres of the middle constrictor of the pharynx. To its apex is attached the stylohyoid ligament, which is often in part ossified. The medial aspect of its base gives origin to the chondroglossus.

Ossification. The hyoid bone is developed from the cartilages of the second and third visceral arches—the lesser cornua from the second, the greater cornua from the third, and the body from the fused ventral ends of both (p. 150). Chondrification begins in the fifth week in these elements, and is completed in the third and fourth months. It is ossified from six centres; a pair for the body, and one for each cornu. Ossification commences in the greater cornu towards the end of intrauterine life, in the body before or shortly after birth, and in the lesser cornua around puberty. The centre in the body may be paired; those for the lesser cornua are very variable. The apices of the greater cornua remain cartilaginous until the third decade, and epiphyses may occur here. They fuse with the body. The synovial joints between the lesser cornua and the rest of the bone may be obliterated by ossification in the later decades.

The Occipital Bone

The occipital bone (**3**.85, 86, 87), forming much of the back and base of the cranium, is trapezoid in shape and, naturally, internally concave. It encloses basally a large oval opening, termed the *foramen magnum*, through which the cranial cavity communicates with the vertebral canal. The expanded plate above and behind this foramen is named the *squamous part*; the thick, somewhat quadrilateral piece in front of it is the *basilar part*; on each side of the foramen is a *lateral part*.

The squamous part of the occipital bone, above and behind the foramen magnum, is curved from above downwards and from side to side.

The *external surface* is convex and presents, midway between the summit of the bone and the foramen magnum, a prominence termed the *external occipital protuberance*. On each side two curved lines, one above the other, extend laterally from this bony prominence. The upper line, faintly marked and often almost imperceptible, is the *highest nuchal line*, and to it the galea aponeurotica is attached. The lower line is the *superior nuchal line*. The external surface above the highest nuchal lines is smooth, and covered with the occipital belly of the occipitofrontalis. The part below the highest nuchal lines is rough and irregular for the attachment of several muscles. From the external occipital protuberance the *external occipital crest*, often faintly marked, descends to the foramen magnum, and affords attachment to the ligamentum nuchae; on each side the *inferior nuchal line* runs laterally from the midpoint of the crest. The areas of muscular attachments are shown in **3**.86. The posterior atlanto-occipital membrane is attached around the posterolateral part of the foramen magnum immediately outside the margin.

The *internal surface* of the squamous part is deeply hollowed out, and divided into four fossae by an irregular elevation, the *internal occipital protuberance*. The upper two fossae are triangular, and adapted to the poles of the occipital lobes of the cerebrum; the lower two are quadrilateral and shaped to the hemispheres of the cerebellum. A wide groove, with raised edges, extends upwards from the protuberance to the superior angle of the bone; this is the *sulcus of the superior sagittal sinus*. The posterior part of the falx cerebri is attached to its margins. A prominent ridge, the *internal occipital crest*, runs downwards and forwards from the protuberance; it gives attachment to the falx cerebelli, and bifurcates near the foramen magnum; the occipital sinus, which is sometimes duplicated, lies in the attached margin of the falx. At the lower part of the internal occipital crest a small depression is sometimes distinguishable; it is often termed the *vermian fossa*, since it is occupied by part of the inferior vermis of the cerebellum. On each side a wide *sulcus of the transverse sinus* extends laterally from the internal occipital protuberance; to the margins of these sulci the tentorium cerebelli is attached. The right sulcus is usually larger and continuous with the sulcus of the superior sagittal sinus; but the left may be larger than the right, or the two may be almost equal in size. The position of the *confluence of the sinuses* is indicated by a depression on one or other side of the protuberance.

The *superior angle* of the squamous part articulates with the occipital angles of the parietal bones, and corresponds in position with the *posterior fontanelle* of the fetal skull. The *lateral angles* are at the ends of the sulci of the transverse sinus; each projects into the interval between the parietal bone and the mastoid part of the temporal bone. The *lambdoid borders* extend from the superior to the lateral angles; they are serrated for articulation with the

occipital borders of the parietal bones to form the lambdoid suture. The *mastoid borders* extend from the lateral angles to the jugular processes; each articulates with the mastoid portion of the corresponding temporal bone. A variety of partially or wholly independent ossicles may occur at or near the lambda (*vide infra*, p. 338 and consult Strivastava 1977). Amongst others these include the 'interparietal' (Inca bone, or ossicle of Goethe).

The basilar part of the occipital bone extends forwards and upwards from the foramen magnum, and presents *anteriorly* a cut surface, more or less quadrilateral in shape, since it is fused with the sphenoid in adults. In the young skull this surface is rough and uneven and is joined to the body of the sphenoid bone by a growth plate of cartilage. By the twenty-fifth year this plate of cartilage has undergone ossification and the occipital and sphenoid bones are fused.

On the *inferior surface* of the basilar part, about 1 cm in front of the foramen magnum, a small elevation, the *pharyngeal tubercle*, gives attachment to the fibrous raphe of the pharynx. The longus capitis is inserted into the bone lateral to the pharyngeal tubercle, and the rectus capitis anterior, into a small depression in front of the occipital condyle. (The latter depression may, instead, be represented by a small *precondylar tubercle*. Its incidence is low in most crania but relatively frequent in Mexican and Burmese crania; *see* p. 348.) The anterior margin of the foramen magnum gives attachment to the anterior atlanto-occipital membrane.

The *superior surface* of the basilar part consists of a broad, shallow groove, the *clivus*, which inclines upwards and forwards from the anterior border of the foramen magnum; it supports the medulla oblongata and the lower part of the pons, and near the margin of the foramen gives attachment to the membrana tectoria and the apical ligament. On the lateral margins of this surface are the *sulci of the inferior petrosal sinuses*, and below each of these sulci the lateral margin of the basilar part is rough for articulation with the petrous part of the temporal bone.

The lateral (condylar) parts of the occipital bone are situated at the sides of the foramen magnum; on their *inferior surfaces* are the *occipital condyles*, two processes for articulation with the superior facets of the atlas. They are oval or reniform in

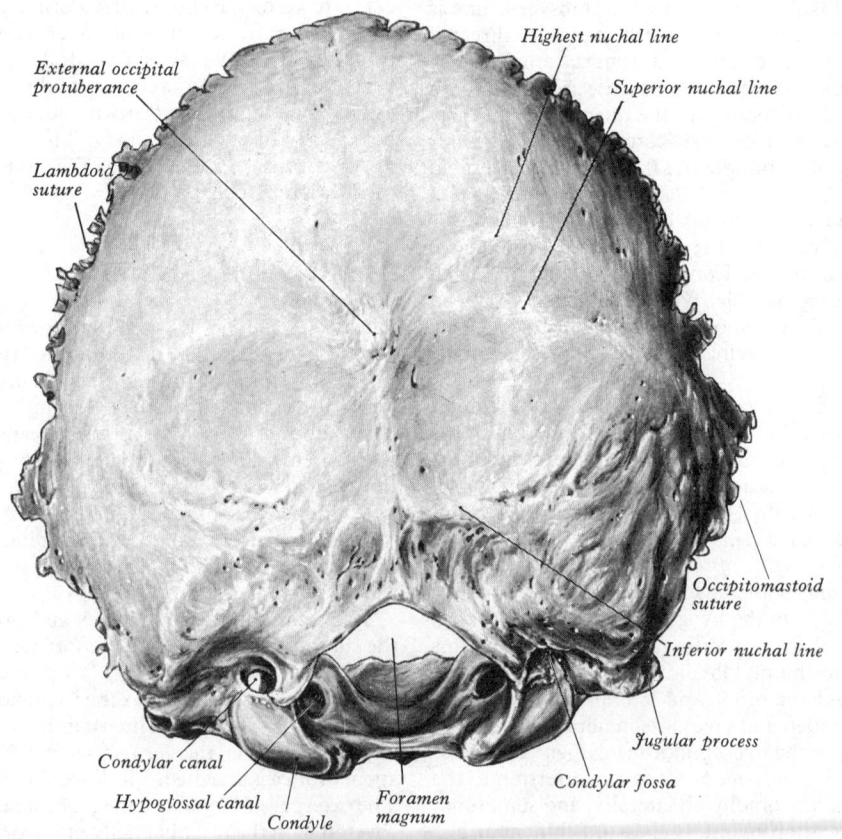

External occipital protuberance

Lambdoid suture

Highest nuchal line

Superior nuchal line

Occipitomastoid suture

Inferior nuchal line

Jugular process

Condylar fossa

Condylar canal

Hypoglossal canal

Condyle

Foramen magnum

3.85 The occipital bone. Posterior aspect. The condylar canal was present on the left side only in this specimen.

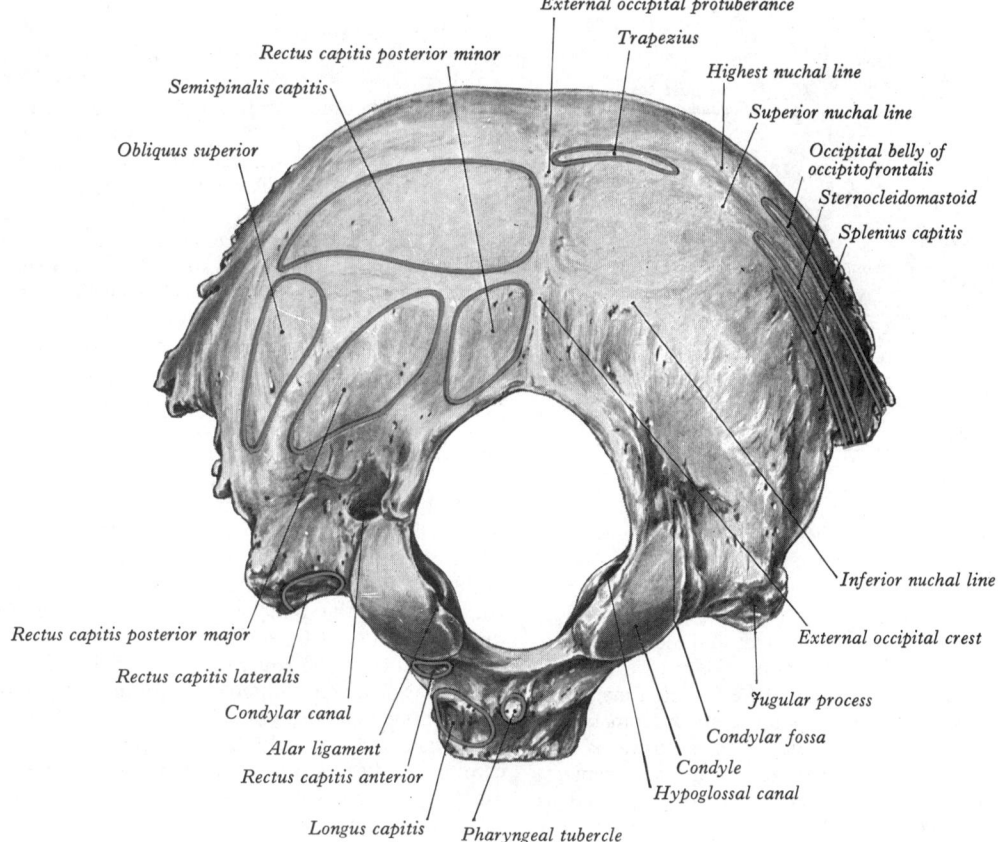

External occipital protuberance

Rectus capitis posterior minor

Trapezius

Semispinalis capitis

Highest nuchal line

Superior nuchal line

Obliquus superior

Occipital belly of occipitofrontalis

Sternocleidomastoid

Splenius capitis

Inferior nuchal line

External occipital crest

Rectus capitis posterior major

Jugular process

Rectus capitis lateralis

Condylar fossa

Condylar canal

Condyle

Alar ligament

Hypoglossal canal

Rectus capitis anterior

Longus capitis

Pharyngeal tubercle

3.86   The occipital bone. Inferior aspect. Drawn from the same specimen as **3**.85.

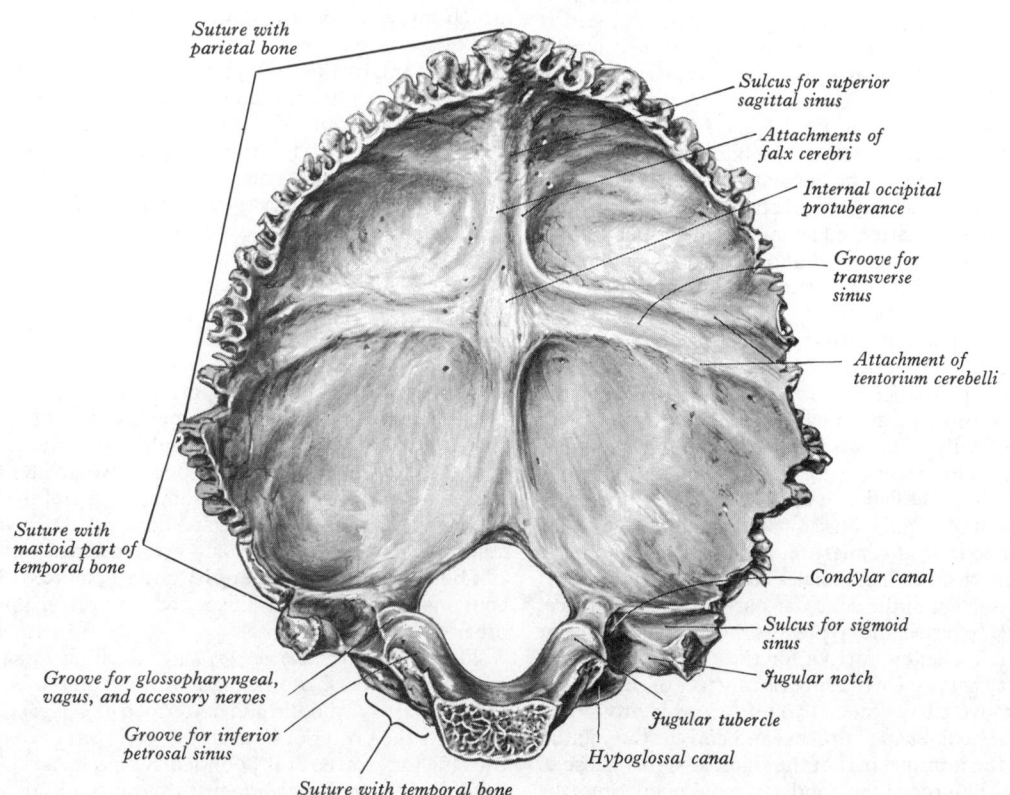

Suture with parietal bone

Sulcus for superior sagittal sinus

Attachments of falx cerebri

Internal occipital protuberance

Groove for transverse sinus

Attachment of tentorium cerebelli

Suture with mastoid part of temporal bone

Condylar canal

Sulcus for sigmoid sinus

Jugular notch

Groove for glossopharyngeal, vagus, and accessory nerves

Groove for inferior petrosal sinus

Jugular tubercle

Hypoglossal canal

Suture with temporal bone

3.87   The occipital bone. Internal aspect.

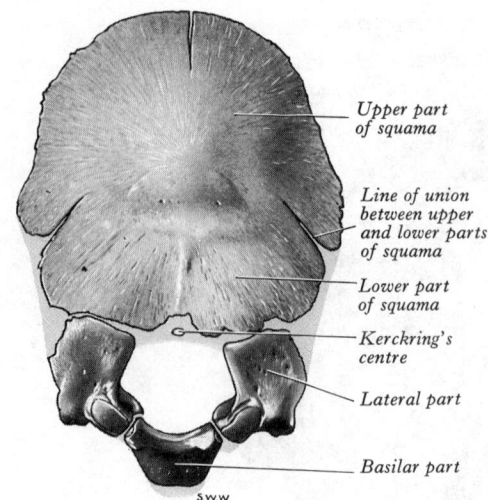

Upper part
of squama

Line of union
between upper
and lower parts
of squama

Lower part
of squama

Kerckring's
centre

Lateral part

Basilar part

S.W.W

3.88 The occipital bone of a newborn child. External surface. Parts of the chondrocranium still unossified are shown in *blue*.

shape, with their long axes running forwards and medially; their anterior ends encroach on the basilar portion, their posterior ends extend back to the level of the middle of the foramen magnum. The articular surfaces of the condyles are convex from before backwards and from side to side; they face downwards and laterally, and are occasionally constricted near their centres. Rarely a condyle may be in two parts, usually described as 'double' (*see* p. 348): It may be recalled that the superior articular facets of the atlas are likewise frequently constricted and occasionally 'double' (Singh 1965). On the medial side of each a rough impression or tubercle gives attachment to the alar ligament. Above the anterior part of each condyle the bone presents the *hypoglossal canal*, which begins on the cranial surface of the bone a short distance above the anterior part of the foramen magnum, and is directed laterally and forwards. It may be partially or completely divided into two by a spicule of bone; it transmits the hypoglossal nerve, and a meningeal branch of the ascending pharyngeal artery. A depression, the *condylar fossa*, lies behind the condyle and approximates to the posterior margin of the corresponding superior facet of the atlas when the head is extended; the floor of this fossa is sometimes perforated by the *condylar canal*, through which an emissary vein passes from the sigmoid sinus. The *jugular process* extends laterally from the posterior half of the condyle. It is a quadrilateral plate of bone, indented in front by the *jugular* notch, which forms the posterior part of the jugular foramen. The jugular notch is sometimes partly divided into two by a bony spicule named the *intrajugular process*, which projects forwards and laterally. The under surface of the jugular process is rough and affords attachment to the rectus capitis lateralis; from this surface an eminence, the *paramastoid process*, sometimes projects downwards, and may be of sufficient length to articulate with the transverse process of the atlas. Laterally the jugular process presents a rough quadrilateral or triangular area which is joined to the jugular surface of the temporal bone by a growth plate of cartilage; about the age of 25 this plate begins to ossify. For a study of variations of the jugular foramen, processes, and neighbouring features, in relation to age and sex, consult Solter and Paljan (1973). While it is apparently rare or absent in many populations, a paramastoid process was recorded in 44 per cent of 149 American Indian males, and in 31 per cent of 137 American Indian females (Finnegan 1972).

On the *superior surface* of the condylar part an oval eminence, the *jugular tubercle*, overlies the hypoglossal canal; its posterior part often presents a shallow furrow for the glossopharyngeal, vagus and accessory nerves. On the superior surface of the jugular process a deep groove curves medially and forwards around an upwardly directed, hook-shaped process and ends at the jugular notch; it contains the terminal part of the sigmoid sinus. Close to the medial end of the groove the condylar canal opens into the posterior cranial fossa.

322

The *foramen magnum* is considered on p. 307.

Structure. The occipital, like the other cranial bones, consists of two compact lamellae, called the *outer* and *inner plates*, between which there is spongy bone or *diploë*; the bone is thick at ridges, protuberances and condyles, and at the anterior portion of the basilar part; in the lower parts of the fossae for the cerebellar hemisphere it is thin, semi-transparent, and devoid of diploë.

Ossification (**3**.88). A commonly stated, but almost certainly *oversimplified* account of the ossification of the occipital bone (*see* further below) is that above the highest nuchal line the squamous part is developed in a membrane of fibrous tissue, and is ossified from two centres, one appearing on each side of the median plane about the second month of intrauterine life; this part may remain separate throughout life, and is then known as the *interparietal bone*. The rest of the occipital bone is preformed in cartilage. Below the highest nuchal line, the squamous part is ossified from two centres, which appear about the seventh week of intrauterine life and soon unite to form a single piece. The upper and lower portions of the squamous part unite in the third month of intrauterine life, but their line of union can be recognized in the bone at birth (**3**.88).

An occasional centre appears in the posterior margin of the foramen magnum about the sixteenth week (Kerckring); it unites with the rest of the squamous parts before birth. Having surveyed the previous literature, and examined a series of 620 human skulls for anomalies of the squamous part of the occipital, Strivastava (1977) has proposed a much more complex developmental history of the bone. He regards the *membranous* (*dermal*) part of the occipital squama, which lies above the nuchal lines as compounded of an *interparietal* and a *pre-interparietal* part: the interparietal consisting of *two lateral plates* and a *centre piece*. The intramembranous centres of ossification for these he proposes as—a pair of centres for each lateral plate and two centres for the central part of the interparietal, plus a pair of centres for the pre-interparietal. There may be partial or complete failure of fusion between any or all of these subregions. The *cartilaginous supra-occipital* he regards as presenting five endochondral centres of ossification, a pair for each of its right and left *lateral segments* and a single ossific centre for the *central segment*, (the latter may correspond to Kerckring's centre mentioned above). Each of the lateral (condylar, or exoccipital) parts of the occipital bone ossifies from a single centre, which appears during the eighth week of intrauterine life. The basilar portion is ossified from one centre which appears about the sixth week of intrauterine life. About the end of the second year the squamous part unites with the lateral portions, and by the sixth year the bone consists of a single piece. Between the eighteenth and twenty-fifth years the occipital and sphenoid bones unite to form a single bone. Metrical study of the occipital bone (Olivier 1975) suggests that the squamous and basilar parts of the complex evince independent parameters of growth. The same observations indicated that sexual differences are chiefly evident in the condylar regions of the bone.

The Sphenoid Bone

The sphenoid bone (**3**.89, 90, 91) is in the base of the skull, in front of the temporal bones and the basilar part of the occipital. In shape it resembles a bat or a bird with wings outstretched. It consists of a central portion or body, two greater and two lesser wings, which spread laterally from the sides of the body, and two pterygoid processes, which are directed downwards from the adjoining parts of the body and greater wings.

The body of the sphenoid bone is more or less cubical; it contains two large air sinuses, which are separated from each other by a septum.

The *cerebral* or *superior surface* of the body (**3**.89) articulates in front with the cribriform plate of the ethmoid bone. Anteriorly this surface is smooth and is termed the *jugum sphenoidale*; it supports the posterior ends of the gyri recti of the cerebrum and the olfactory tracts. It is bounded behind by a ridge, the anterior border of the *sulcus chiasmatis*; this sulcus leads laterally to the *optic canal* on each side. Posterior to the groove there is an elevation, the *tuberculum sellae*; and behind this a hollow, the *sella turcica*, the deepest part of which, the *hypophysial fossa*, lodges the hypophysis cerebri. The anterior boundary of the sella turcica is completed laterally by two small eminences, the *middle clinoid processes*, whilst posteriorly is a square plate of bone, the *dorsum sellae*. The superior angles of this plate ends in two tubercles of varying size, the *posterior clinoid processes*, which give attachment to the fixed margin of the tentorium cerebelli. On each side of the body below the dorsum sellae a small projection articulates with the apex of the petrous portion of the temporal bone and is often termed the *petrosal process*. (For variations in the sella turcica *see* Kinnman 1977, and Lang 1977.)

The sloping area behind the dorsum sellae is the *clivus*, and is uninterruptedly continuous with the clivus of the occipital bone in the adult skull; it supports the upper part of the pons.

The *lateral surfaces* of the body are united with the greater wings and with the medial pterygoid plates. Above the attachment of each wing a broad groove, the *carotid sulcus*, forms a curve somewhat like the italic letter *f*; it lodges the internal carotid artery and the cavernous sinus. The carotid sulcus is deepest at its posterior end, where it is overhung medially by the petrosal process, and is limited laterally by a sharp margin called the *lingula*; the latter is continued backwards to overlie the posterior opening of the pterygoid canal.

The *anterior surface* of the body (**3**.91) presents, in the median plane, a triangular crest, which forms a small part of the septum of the nose and is termed the *sphenoidal crest*. The anterior border of this crest articulates with the perpendicular plate of the ethmoid bone. On each side of the crest a rounded opening leads into the corresponding *sphenoidal sinus*. The sphenoidal sinuses (p. 1149) are two large, irregular cavities in the body of the bone, separated from each other by a bony septum which is commonly deflected to one side or the other. They vary in form and size, are seldom symmetrical, and are often partially divided by bony laminae. A lateral recess may extend from one or other sinus into the greater wing and lingula (Cope 1917); the sinuses occasionally reach into the basilar part of the occipital bone nearly as far as the foramen magnum. In relation to the trans-sphenoidal surgical approach to the hypophysis cerebri the sphenoidal sinuses have been divided into three types: *conchal*—a small sinus separated from the sella turcica by cancellous bone some 10 mm thick; *presellar*—in which the sinus does not extend dorsal to the tuberculum sellae; and *sellar*—in which the sinus extends, as mentioned above, very variable distances beyond the tuberculum (Hammer and Rådberg 1961; for further details *see* Kinnman 1977). In the articulated skull they are closed in front and below by the *sphenoidal conchae* (p. 325), but a round opening is left in the anterior wall of each sinus, by which it communicates with the spheno-ethmoidal recess of the nasal cavity. Each half of the anterior surface of the body of the sphenoid bone consists of two parts: (*a*) an upper and lateral depressed area, which articulates with the labyrinth of the ethmoid bone and completes the posterior ethmoidal sinuses; its lateral margin articulates with the orbital plate of the ethmoid bone above, and with the orbital process of the palatine bone below; (*b*) a lower and medial, smooth, triangular area, which forms the posterior part of the roof of the nose; near its superior angle is the orifice of the sphenoidal sinus.

The *inferior surface* of the body presents in the median plane a triangular spine, the *sphenoidal rostrum* (**3**.91), which projects into a deep fissure between the anterior parts of the alae of the vomer. The posterior, triangular parts of the sphenoidal conchae extend backwards on the sides of the rostrum, and articulate with the alae of the vomer. On each side of the posterior part of the rostrum, and immediately behind the apex of the sphenoidal concha, a thin lamina, the *vaginal process*, projects medially from the base of the medial pterygoid plate (**3**.91).

The greater wings are two strong processes which curve laterally and upwards from the sides of the body. The posterior part of each is triangular and fits into the angle between the petrous and squamous parts of the temporal bone (**3**.71) forming the sphenosquamosal suture.

The *cerebral surface* of the greater wing (**3**.89) forms part of the floor of the middle cranial fossa; it is deeply concave and presents a pattern of elevations and depressions corresponding with the

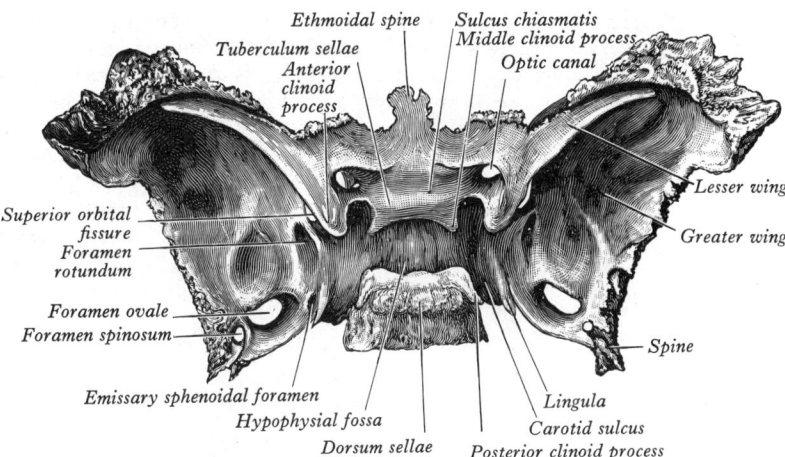

3.89   The sphenoid bone. Superior aspect.

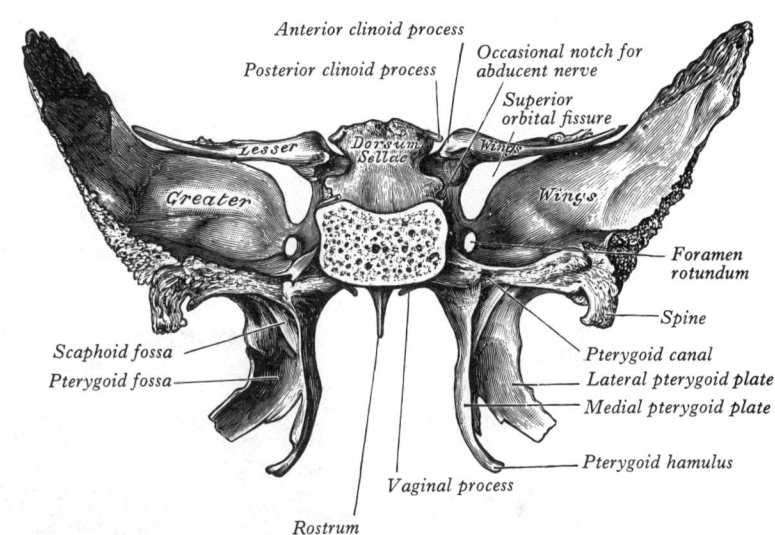

3.90   The sphenoid bone. Posterior aspect.

convolutions of the anterior part of the temporal lobe. At its anteromedial part the *foramen rotundum* gives passage to the maxillary nerve. Behind and lateral to this the *foramen ovale* transmits the mandibular nerve, the accessory meningeal artery, and sometimes the lesser petrosal nerve which may, however, pass through a special canal (*canaliculus innominatus*) on the medial side of the foramen spinosum. Medial to the foramen ovale, a small aperture, termed the *emissary sphenoidal foramen*, is present on one or both sides in nearly 40 per cent of skulls; it opens below at the lateral side of the scaphoid fossa and transmits a small vein from the cavernous sinus. In the posterior angle, anteromedial to the spine, there is a short canal, termed the *foramen spinosum*, which transmits the middle meningeal artery and the meningeal branch of the mandibular nerve.

The *lateral surface* of the greater wing (**3**.70) is convex from above downwards, and is divided by a transverse ridge, termed the *infratemporal crest*, into an upper or temporal and a lower or infratemporal surface. The *temporal surface*, concave from before backwards, forms a portion of the temporal fossa and gives origin to a part of the temporalis. The *infratemporal surface* is concave and directed downwards; together with the infratemporal crest, it gives origin to the upper head of the lateral pterygoid. It is pierced by the *foramen ovale* and *foramen spinosum*, and its posterior part bears the *spine of the sphenoid* bone (p. 305), which forms a small, sometimes pointed process, projecting downwards. Its medial side may be marked by a faint groove, directed downwards and forwards, for the chorda tympani nerve, and helps to form the lateral wall of the sulcus for the auditory tube (p. 305). Its tip gives attachment to the sphenomandibular ligament. Medial to the anterior extremity of the infratemporal crest a triangular process serves to increase the attachment of the lateral pterygoid. A ridge runs downwards and medially from this triangular process to the front of the lateral pterygoid plate; it forms the posterior boundary of the pterygomaxillary fissure.

The *orbital surface* of the greater wing (**3.91**A), quadrilateral in shape, is directed forwards and medially and forms the posterior part of the lateral wall of the orbit. Its upper, serrated edge articulates with the orbital plate of the frontal bone, its lateral,

serrated margin with the zygomatic bone. Its inferior smooth border forms the posterolateral boundary of the inferior orbital fissure. Its medial sharp margin constitutes the lower, or lateral, boundary of the superior orbital fissure, and from it a small tubercle affords attachment to part of the common tendinous ring, from which the rectus muscles of the eyeball take origin. Below the medial end of the superior orbital fissure there is a grooved area which forms the posterior wall of the pterygopalatine fossa and is pierced by the foramen rotundum.

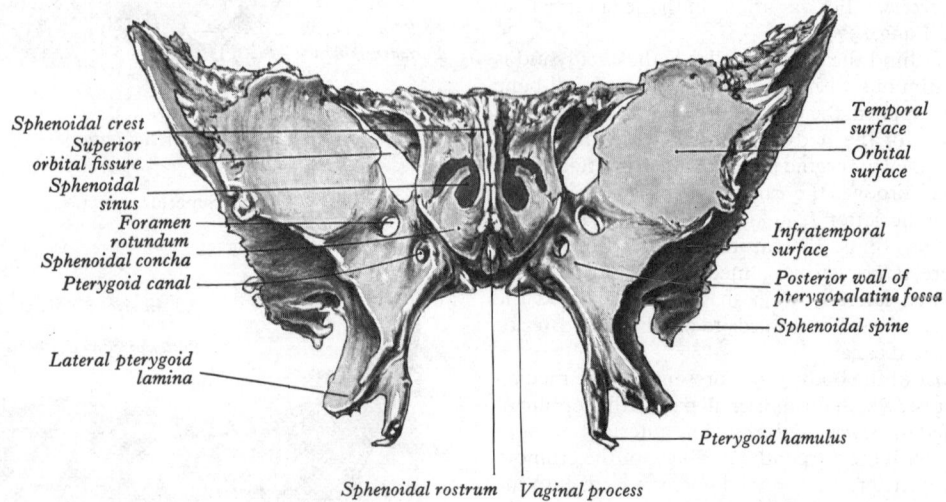

Sphenoidal crest
Superior orbital fissure
Sphenoidal sinus
Foramen rotundum
Sphenoidal concha
Pterygoid canal

Lateral pterygoid lamina

Temporal surface
Orbital surface

Infratemporal surface

Posterior wall of pterygopalatine fossa

Sphenoidal spine

Pterygoid hamulus

Sphenoidal rostrum   Vaginal process

3.91A   The sphenoid bone of an eight-year-old child. Anterior aspect.

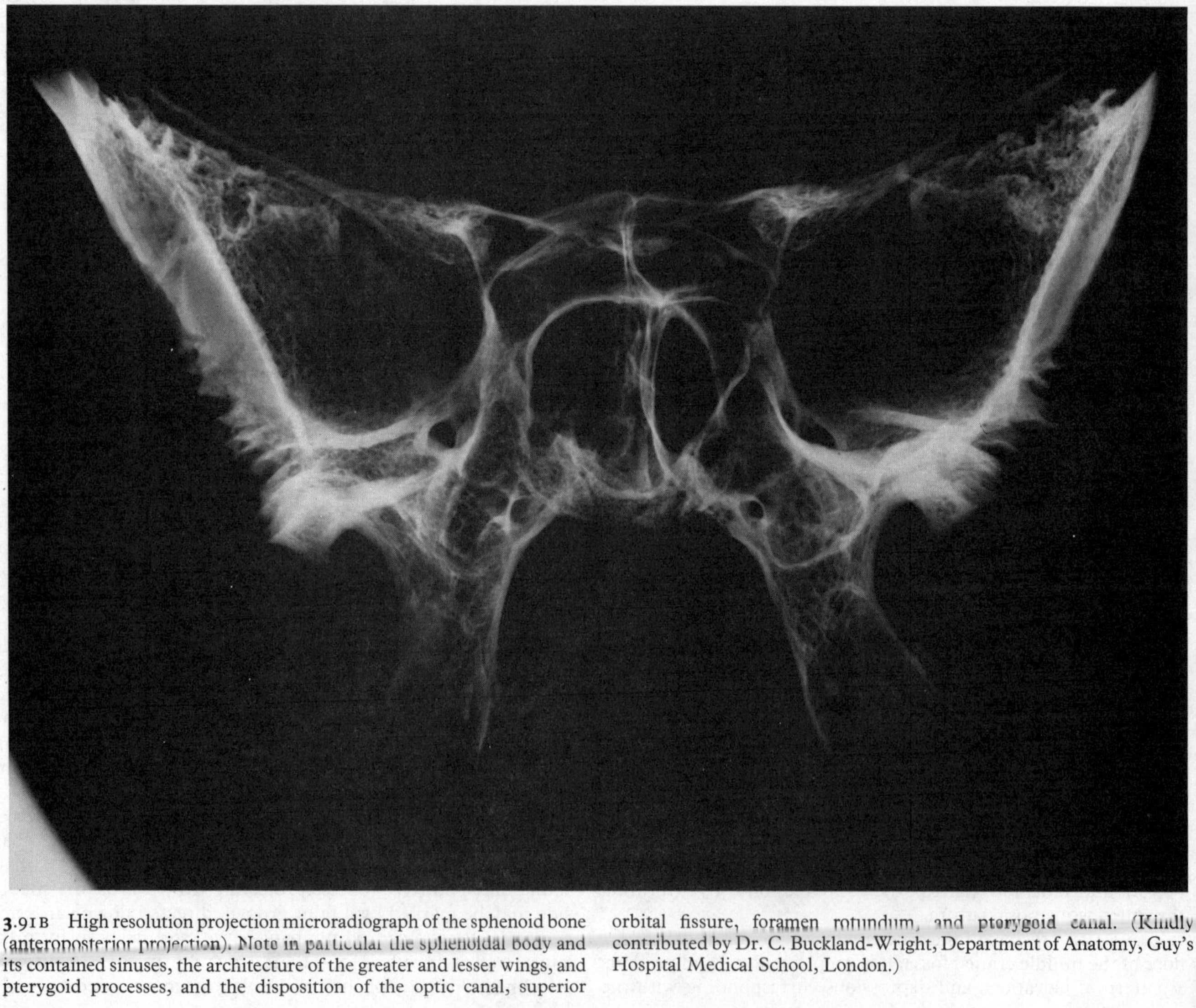

3.91B   High resolution projection microradiograph of the sphenoid bone (anteroposterior projection). Note in particular the sphenoidal body and its contained sinuses, the architecture of the greater and lesser wings, and pterygoid processes, and the disposition of the optic canal, superior orbital fissure, foramen rotundum, and pterygoid canal. (Kindly contributed by Dr. C. Buckland-Wright, Department of Anatomy, Guy's Hospital Medical School, London.)

The *margin of the greater wing* (3.90), where it extends from the body to the spine of the sphenoid, is irregular. Its medial half forms the anterior boundary of the *foramen lacerum* and presents the posterior aperture of the pterygoid canal for the passage of the corresponding nerve and artery. Its lateral half articulates with the petrous portion of the temporal bone, by means of a cartilaginous joint, the *sphenopetrosal synchrondrosis*. Between the two bones, on the under surface of the skull, there is a furrow, the *sulcus tubae*, which lodges the cartilaginous part of the auditory tube. Extending forwards from the sphenoidal spine the *squamosal margin* forms a concave, serrated edge, bevelled at the expense of the inner surface below and of the outer surface above, for articulation with the squamous part of the temporal bone. The tip of the greater wing is bevelled at the expense of the inner surface and articulates with the sphenoidal angle of the parietal bone at the *pterion*. Medial to this, there is a triangular rough area, for articulation with the frontal bone; the medial angle of this area is continuous with the sharp edge which forms the lower boundary of the superior orbital fissure, and the anterior angle with the serrated margin for articulation with the zygomatic bone.

The lesser wings of the sphenoid bone are two triangular plates which project laterally from the upper and anterior parts of the body and end in sharp points (3.89, 90). The *superior surface* of each is smooth, and supports a small part of the frontal lobe of the cerebrum. The *inferior surface* forms the posterior part of the roof of the orbit and the upper boundary of the *superior orbital fissure*; it overhangs the anterior part of the middle cranial fossa. The *anterior border* of the lesser wing is serrated for articulation with the posterior edge of the orbital plate of the frontal bone. The *posterior border* is smooth and projects into the lateral cerebral fissure; the medial end of this border forms the *anterior clinoid process*, which gives attachment to the anterior end of the free border of the tentorium cerebelli. The anterior and middle clinoid processes are sometimes united by a spicule of bone, and when this occurs the end of the groove for the internal carotid artery is converted into a *caroticoclinoid foramen*. The lesser wing is connected to the body by two roots, the anterior thin and flat, the posterior thick and triangular; the *optic canal*, which lies between them, transmits the optic nerve and ophthalmic artery. The variations in the growth of the posterior root are closely associated with variations in the optic canal itself (Kier 1966). Its cranial opening may be duplicated (Warwick 1951). More commonly the division of this part of the canal is incomplete. (See also Lang, 1977.)

The superior orbital fissure is triangular in shape and leads from the cranial cavity into the orbit; it is bounded medially by the body of the sphenoid bone; above, by the lesser wing, below, by the medial margin of the orbital surface of the greater wing; it is completed laterally, between the greater and lesser wings, by the frontal bone. Through it the oculomotor, trochlear and abducent nerves, the three branches of the ophthalmic division of the trigeminal nerve, the orbital branch of the middle meningeal artery, and some filaments from the internal carotid plexus of the sympathetic enter the orbit; and the recurrent meningeal branch of the lacrimal artery, and the ophthalmic veins pass backwards into the cranial cavity.

The pterygoid processes of the sphenoid (3.90, 91), one on each side, descend perpendicularly from the regions where the greater wings unite with the body. Each process consists of a medial and a lateral plate, the upper parts of which are fused anteriorly. The plates are separated below by an angular cleft, termed the *pterygoid fissure*, the rough margins of which articulate with the pyramidal process of the palatine bone. The two plates diverge behind, and the wedge-shaped *pterygoid fossa* between them contains the medial pterygoid and tensor veli palatini muscles. Above this fossa there is a small, oval, shallow depression, named the *scaphoid fossa*, which is formed by the division of the upper part of the posterior border of the medial pterygoid plate; it gives origin to part of the tensor veli palatini. The anterior surface of the pterygoid process is broad and triangular near its root, where it forms the posterior wall of the pterygopalatine fossa; it is pierced by the anterior orifice of the *pterygoid canal*.

The lateral pterygoid plate is broad, thin and everted; its *lateral surface* forms part of the medial wall of the infratemporal

fossa and gives origin to the lower head of the lateral pterygoid; its *medial surface* forms the lateral wall of the pterygoid fossa and gives origin to the greater part of the medial pterygoid. The upper part of the *anterior border* forms the posterior boundary of the pterygomaxillary fissure; the lower part articulates with the palatine bone; its *posterior border* is free.

The medial pterygoid plate is narrower and longer than the lateral; its lower end curves laterally into a hook-like process, the *pterygoid hamulus*, around which the tendon of the tensor veli palatini is deflected, and to which the pterygomandibular raphe is attached. The *lateral surface* of this plate forms the medial wall of the pterygoid fossa, and the tensor veli palatini lies against it; the *medial surface* is the lateral boundary of the corresponding

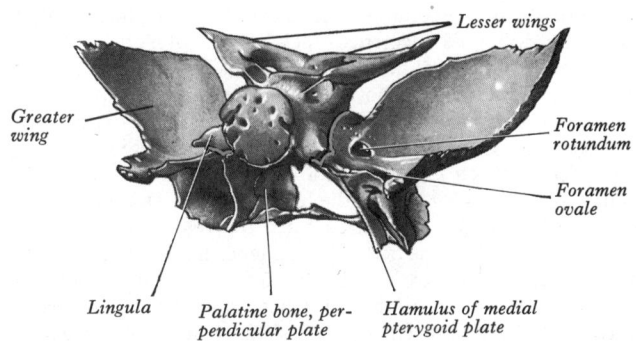

3.92   The sphenoid bone at birth. Viewed from behind and from the right side. The blue strip indicates the cartilage between the central and the lateral parts on the right side; this is only partly visible on the left. *Note.*—The two palatine bones are *in situ*.

posterior nasal aperture. Superiorly the medial pterygoid plate is prolonged on to the under surface of the body as a thin lamina, the *vaginal process*, which articulates anteriorly with the sphenoidal process of the palatine bone and medially with the ala of the vomer. Inferiorly it bears a furrow, the anterior part of which is converted into a canal by the sphenoidal process of the palatine bone; this is the *palatovaginal canal*, which transmits the pharyngeal branch of the maxillary artery and the pharyngeal branch of the pterygopalatine ganglion. To the whole of the posterior margin of the medial pterygoid plate is attached the pharyngobasilar fascia, and from its lower end the superior constrictor muscle of the pharynx takes origin. The upper end of this margin has a small projection, the *pterygoid tubercle*, which lies immediately below the posterior opening of the pterygoid canal. Projecting backwards from near the middle of the margin is an angular process, the *processus tubarius*, which supports the pharyngeal end of the auditory tube. The lower part of the anterior margin of the medial pterygoid plate articulates with the posterior border of the perpendicular plate of the palatine bone.

The sphenoidal conchae (3.91) are two thin, curved platelets, at the anterior and lower parts of the body of the sphenoid bone; the upper, concave surface of each forms the anterior wall and a part of the floor of the corresponding sphenoidal sinus. The sphenoidal conchae are usually more or less destroyed in the process of disarticulating the skull, but, when seen *in situ* each consists of an anterior, vertical, quadrilateral part and a posterior, horizontal, triangular part. The anterior part consists of (*a*) an upper and lateral depressed area, which completes the posterior ethmoidal sinuses and articulates below with the orbital process of the palatine bone, and (*b*) a lower and medial area, smooth and triangular, which forms part of the roof of the nasal cavity, and is perforated above by a round opening through which the sphenoidal sinus communicates with the spheno-ethmoidal recess of the nasal cavity. The anterior vertical parts of the two bones meet in the median plane and are protruded forwards as the sphenoidal crest. The horizontal triangular portion of the concha forms a part of the roof of the nasal cavity and completes the sphenopalatine foramen; its medial edge articulates with the rostrum of the sphenoid and the ala of the vomer; its apex, directed backwards, lies medial to and above the vaginal process of the medial pterygoid plate, and articulates with the posterior part of the ala of the vomer. A small piece of the sphenoidal concha sometimes appears in the medial wall of the orbit, between the orbital plate of the ethmoid in front, the orbital process of the palatine bone below, and the frontal bone above.

Ossification. Until the seventh or eighth month of intrauterine life the body of the sphenoid consists of two parts—viz.: one in front of the tuberculum sellae, the *presphenoidal part*, with which the lesser wings are continuous; the other comprising the sella turcica and dorsum sellae, forming the *postsphenoidal part*, with which the greater wings and pterygoid processes are associated. A considerable part of the bone is preformed in cartilage. There are six centres for the presphenoidal and eight for the postsphenoidal part. This multiplicity of ossific centres and of parts in the development of the mammalian sphenoid bone corresponds with its evolution in land vertebrates from a number of separate sphenoid elements, such as the presphenoid and basisphenoid in the midline and homologous with the parts of the human sphenoid defined above. The lesser wings are primarily separate as the orbitosphenoids, but show a tendency to fusion in mammals.

Presphenoidal part. About the ninth week a centre of ossification appears for each of the lesser wings, just lateral to the optic canal; shortly afterwards two centres appear in the presphenoidal part of the body. The sphenoidal conchae are each developed from a centre which appears in the fifth month of intrauterine life in the upper and posterior part of the nasal capsule. As the centre enlarges it partially surrounds a small backward expansion of the upper and posterior part of the nasal cavity, which subsequently becomes the sphenoidal sinus. The posterior wall of the concha becomes absorbed, allowing the sinus to invade the presphenoid. In the fourth year the concha fuses with the ethmoidal labyrinth, and before puberty it fuses with the sphenoid and palatine bones. The deficiency in its anterior wall persists at the opening of the sphenoidal sinus.

Postsphenoidal part. The first centres of ossification are those for the greater wings. About the eighth week one appears below the foramen rotundum in the cartilage which forms the base of each wing. This centre forms only the root of the greater wing in the neighbourhood of the foramen rotundum and the pterygoid canal. The whole of the rest of the greater wing is ossified in membrane and the process extends downwards into the lateral pterygoid plate. About the fourth month, two centres appear in the postsphenoidal part of the body, one on each side of the sella turcica, and soon fuse. Each medial pterygoid plate is ossified in membrane, and its centre probably appears about the ninth or tenth week; the hamulus is *chondrified* during the third month and almost at once begins to ossify (Fawcett 1905). The medial and lateral pterygoid plates join about the sixth month. About the fourth month a centre appears for each lingula and speedily joins the rest of the bone.

The presphenoidal and the postsphenoidal parts of the body fuse about the eighth month of intrauterine life, but a wedge-shaped piece of cartilage persists for some time after birth in the lower part of the line of fusion. At birth the bone is in three pieces (**3**.92): a central, consisting of the body and lesser wings, and two lateral, each comprising a greater wing and pterygoid process. In the first year after birth the greater wings and body unite around the margins of the pterygoid canal, and the lesser wings extend medially above the anterior part of the body and meet to form an elevated smooth surface, termed the *jugum sphenoidale*. By the twenty-fifth year the sphenoid and occipital bones are completely fused. In the *anterior part* of the hypophysial fossa there is occasionally a vascular foramen, which is usually, but erroneously, termed the *craniopharyngeal canal* (p. 199). (For an analysis and bibliography concerning the cartilage canals in the fetal spheno-occipital synchondrosis consult Moss-Salentÿn 1975.)

The sphenoidal sinus is present before birth as an extension of the nasal cavity into the sphenoidal concha. In the second or third year it extends backwards into the presphenoid and subsequently invades the postsphenoid, reaching its full, normal size in adolescence. As age advances it frequently undergoes a further enlargement, associated with absorption of its bony walls.

Certain parts of the sphenoid bone are connected by ligaments which occasionally ossify. The more important of these ligaments are: the *pterygospinous*, stretching between the spine of the sphenoid and the upper part of the lateral pterygoid plate (p. 307); the *interclinoid*, joining the anterior to the posterior clinoid process; and the *caroticoclinoid*, connecting the anterior to the middle clinoid process. For details of the fetal and perinatal development of the optic foramen and canal consult Kier (1966). Lang (1977) has surveyed the literature of ossificatory variations of the sella turcica.

Premature ossification or synostosis of the suture between the pre- and post-sphenoidal parts and of the spheno-occipital suture produces a characteristic physiognomy. This is best seen in profile, and consists of an abnormal depression of the bridge of the nose; it is a feature often observed in achondroplasia. Anomalous development of the presphenoidal elements may lead to excessive separation between the orbits and an abnormally broad nasal bridge (*hypertelorism*).

The Temporal Bones

The temporal bones (**3**.93–96) take part in the formation of the sides and base of the skull. Each consists of four parts, viz.: the *squamous, petromastoid,* and *tympanic parts*, and the *styloid process.* These are morphologically distinct elements which have become fused with one another during the evolution of higher vertebrates. The squamous part is a dermal bone developed to assist in enclosure of the cerebrum. The petromastoid portion is preformed in cartilage and it stabilizes and preserves a precise orientation for the membranous labyrinth. The tympanic part, formed in membrane, is homologous with the angular bone, which constitutes a part of the composite lower jaw of many reptilians and bony fishes; it has become incorporated in the skull and adapted to form part of the walls of the tympanic cavity and external acoustic meatus and to support the tympanic membrane—all concerned with sound wave transmission. The styloid process represents the dorsal end of the skeletal element of the hyoid arch. The fusion of these distinct elements to form the temporal bone and the inclusion of the tympanic cavity and auditory ossicles within the bone during the process have been considered on p. 180.

The squamous part of the temporal forms the anterior and upper part of the bone, and is scale-like, thin and partly translucent. Its *temporal surface* (**3**.93) is smooth and slightly convex; it forms part of the temporal fossa and gives origin to the temporal muscle; above the opening of the external acoustic meatus it is marked by a vertical groove for the middle temporal artery. A curved line, often termed the *supramastoid crest*, extends backwards and upwards across its posterior part; it marks the attachment of the temporal fascia and limits the origin of the temporal muscle. The boundary between the squamous and mastoid portions of the bone lies about 1·5 cm below the supramastoid crest and is frequently indicated by traces of the original squamomastoid suture. Between the anterior end of the supramastoid crest and the posterosuperior sector of the opening of the external acoustic meatus there is a depression, the *suprameatal triangle*; this triangle marks the mastoid antrum, which lies medial to it, at a depth of about 1·25 cm (p. 331). The anterior part of the depression is usually marked by a small projection, the *suprameatal spine*.

The *zygomatic process*, part of the **zygoma**, projects anteriorly from the lower region of the temporal surface. Its posterior part is triangular in shape and springs from a broad base; it is directed laterally, and its surfaces are superior and inferior. The process is then twisted forwards and medially, and the surfaces of its anterior portion are therefore medial and lateral. The superior surface of the posterior part is concave, and continuous with the temporal surface of the squamous part; the inferior surface is bounded by two roots, *posterior* and *anterior*, which converge as they approach the anterior part of the process. At the meeting point of the two roots the *tubercle of the root of the zygoma* gives attachment to the lateral ligament of the temporomandibular joint. The posterior root is prolonged forwards from the surface of the squamous part immediately above the opening of the external acoustic meatus; its upper border is continuous behind with the supramastoid crest. The anterior root juts almost horizontally from the side of the squamous temporal; its inferior surface, with an anteroposterior convexity covered by cartilage, is in contact with the articular disc of the joint, and the whole root presents the form of a short semicylindrical bar, the *articular tubercle*, which forms the anterior boundary of the mandibular fossa. Very rarely the squamous part is perforated just above the posterior root of the zygoma. When present, this *squamosal foramen* transmits the petrosquamous sinus (p. 746).

The anterior part of the temporal zygomatic process is thin and flat. To the superior border, which is long and thin, the temporal fascia is attached; the inferior border, which is short and arched, gives origin to some fibres of the masseter. The lateral surface is convex and subcutaneous; the medial is concave and gives origin to part of the masseter. The anterior end is deeply serrated and cut obliquely at the expense of the lower border, it articulates with the temporal process of the zygomatic bone. In front of the articular tubercle a small triangular area forms a part of the roof of the infratemporal fossa and is separated from the temporal surface of the squamous part by a ridge; this ridge is continuous behind with the anterior root of the zygomatic process, and in front, in the articulated skull, with the infratemporal crest on the greater wing of the sphenoid bone.

The mandibular fossa is bounded in front by the articular tubercle; it consists of an anterior, articular portion, formed by the squamous part of the temporal bone, and a posterior, non-articular part, formed by the tympanic element. The *articular surface*, smooth, oval and deeply concave, articulates with the articular disc of the temporomandibular joint; the non-articular portion sometimes lodges a small part of the parotid gland. A small, conical eminence, the *postglenoid tubercle*, separates the lateral region of the articular surface from the anterior margin of the tympanic part of the bone, and is the representative of a prominent tubercle which, in some mammals, descends behind the condyle of the mandible and prevents its backwards displacement; the postglenoid tubercle is sometimes described as the third root of the zygomatic process. (For a multivariate analysis of the conformation of this articular surface in modern and fossil mankind and the great apes, and an evaluation of the validity of functional deductions from such data, consult Ashton *et al.* 1976.) The medial part of the articular fossa is separated from the tympanic part of the bone by the *squamo-tympanic fissure*, into which projects the downturned anterolateral edge of the tegmen tympani of the petrous bone; the *petrotymp-*

anic fissure is situated between this plate and the tympanic part. This fissure leads into the tympanic cavity; it contains the anterior ligament of the malleus and the anterior tympanic branch of the maxillary artery. The medial end of the fissure presents the anterior opening of the *anterior canaliculus for the chorda tympani*. Very rarely a *postglenoid foramen* is present just in front of the external acoustic meatus and in the line of fusion of the squamous and tympanic portions of the bone. It replaces the squamosal foramen already mentioned, and transmits the petrosquamous sinus (p. 746).

The *cerebral surface* of the squama (**3**.94) is concave; its depressions correspond to the convolutions of the temporal lobe

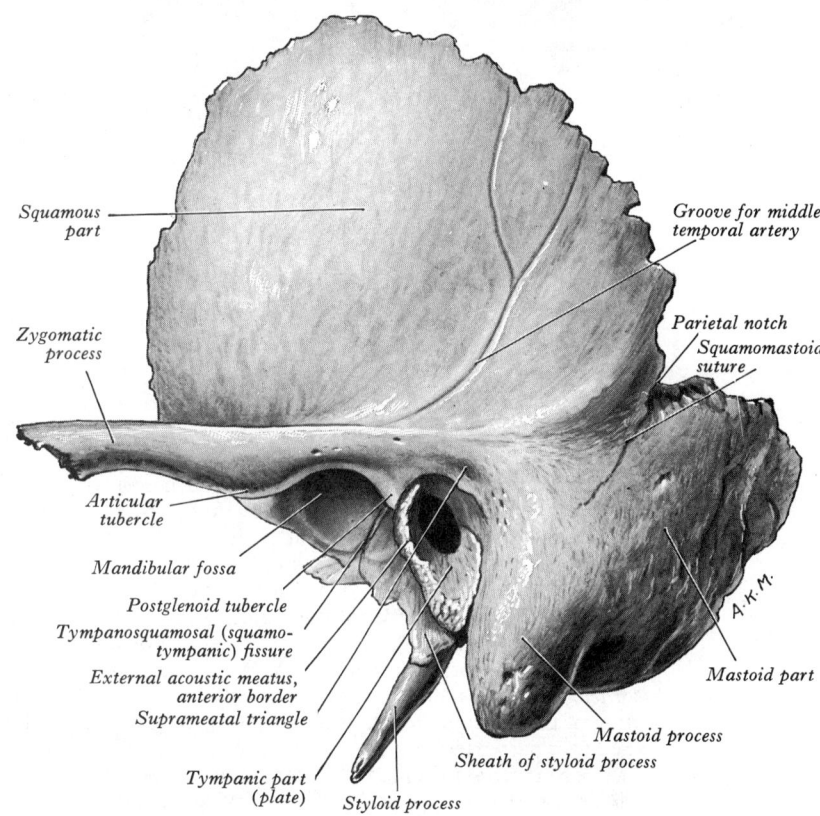

Squamous part
Zygomatic process
Articular tubercle
Mandibular fossa
Postglenoid tubercle
Tympanosquamosal (squamo-tympanic) fissure
External acoustic meatus, anterior border
Suprameatal triangle
Tympanic part (plate)
Styloid process
Groove for middle temporal artery
Parietal notch
Squamomastoid suture
Mastoid part
Mastoid process
Sheath of styloid process

3.93   The left temporal bone. External aspect.

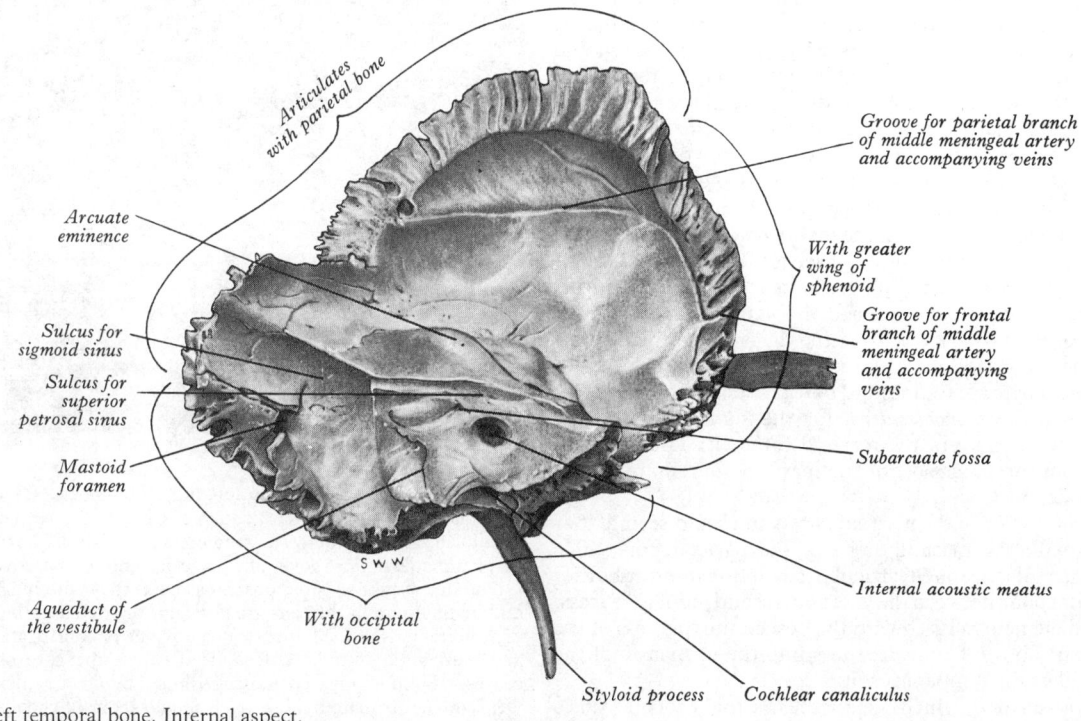

Articulates with parietal bone
Arcuate eminence
Sulcus for sigmoid sinus
Sulcus for superior petrosal sinus
Mastoid foramen
Aqueduct of the vestibule
With occipital bone
Styloid process
Cochlear canaliculus
Groove for parietal branch of middle meningeal artery and accompanying veins
With greater wing of sphenoid
Groove for frontal branch of middle meningeal artery and accompanying veins
Subarcuate fossa
Internal acoustic meatus

3.94   The left temporal bone. Internal aspect.

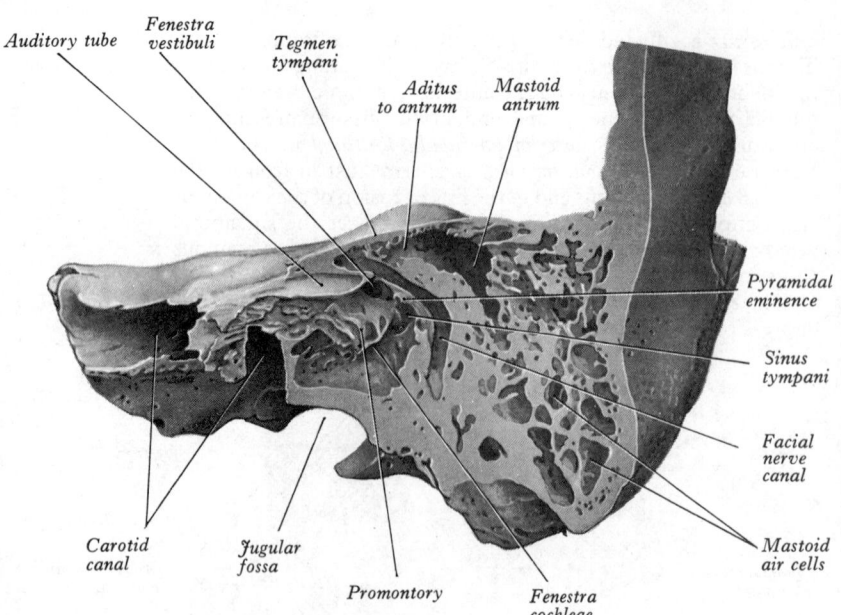

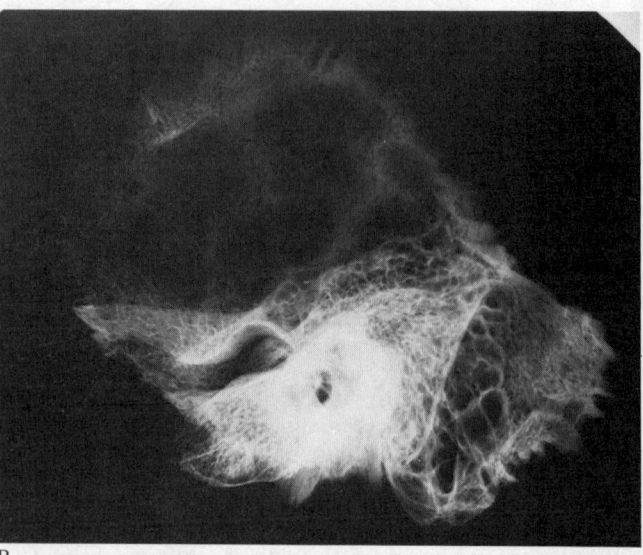

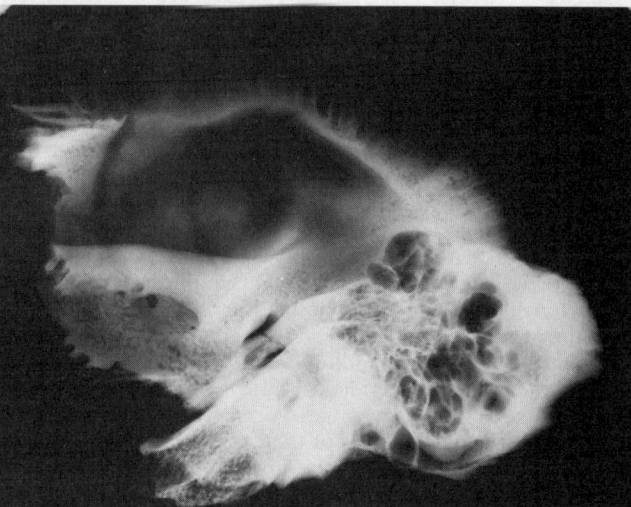

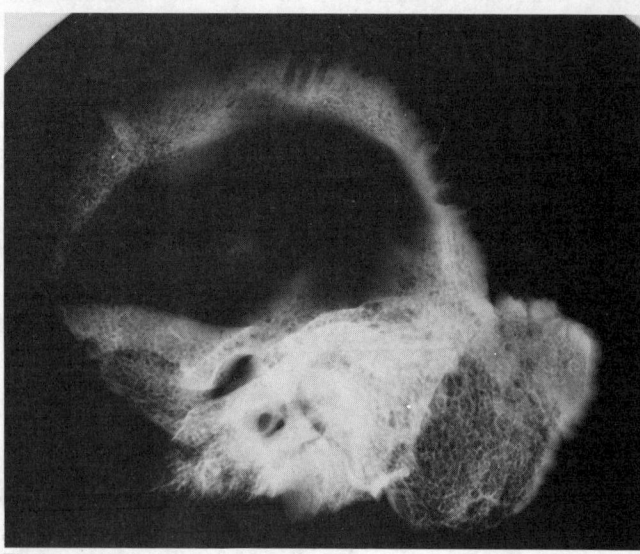

3.95A A section through the left temporal bone in the long axis of the tympanic cavity. The lateral surface of the medial half of the bone is shown.

of the cerebrum, and the grooves to the branches of the middle meningeal vessels; its lower border is united to the anterior surface of the petrous portion but traces of a petrosquamosal suture are frequently seen in the adult bone.

The *superior border* is thin, bevelled on its cerebral aspect, and overlaps the inferior border of the parietal bone forming with it the squamosal suture. Posteriorly the superior border forms an angle with the mastoid portion of the bone. The *antero-inferior border*, thin above and thick below, articulates with the greater wing of the sphenoid bone; its upper part is bevelled internally, the lower part externally.

The petromastoid part of the temporal bone is morphologically a single element (p. 326) but it is convenient to describe it in two parts.

The mastoid part forms the posterior part of the temporal bone. Its *outer surface* (**3**.93) is roughened by the attachments of the occipital belly of the occipitofrontalis, and the auricularis posterior. It is frequently perforated near its posterior border by the *mastoid foramen*, which is traversed by a vein from the sigmoid sinus and transmits a small branch of the occipital artery to the dura mater; the position and size of this foramen are variable; it may be situated in the occipital bone, or in the suture between the temporal and the occipital bones. (This foramen is parasutural in 40–50 per cent of crania—*see* p. 348; it may also be absent.) The mastoid portion is continued below into a conical projection, **the mastoid process**; it is larger in the male than in the female. The lateral surface of this process gives insertion to the sternocleidomastoid, splenius capitis, and longissimus capitis; on its medial side there is a deep groove, termed the *mastoid notch*, for the attachment of the posterior belly of the digastric; medial to this notch the shallow *occipital groove* lodges the occipital artery.

On the *inner surface* of the mastoid portion (**3**.94) there is a deep, curved groove, the *sigmoid sulcus*, for the sigmoid sinus; and posteriorly the opening of the mastoid foramen. The sulcus is separated from the innermost of the mastoid air cells by a thin lamina of bone.

The *superior border* of the mastoid part is thick and serrated for articulation with the mastoid angle of the parietal bone. The *posterior border*, also serrated, articulates with the inferior border of the occipital bone between the lateral angle and jugular process. The mastoid element is fused with the descending process of the squamous part above: below, it enters into the formation of the posterior wall of the tympanic cavity.

The *mastoid air cells*, which occupy the mastoid process (**3**.95),

3.95B, C and D High resolution projection microradiographs of the petromastoid, tympanic, and squamous parts of three temporal bones (mediolateral projection), showing marked variation in extent and mode of pneumatization. Above, the air-cells are comparatively large and extend beyond the mastoid process as far as the mandibular fossa. In the middle photograph the cells are large but restricted to the post-otic and mastoid regions of the bone. Below, obvious pneumatization is absent; a fine meshwork of trabecular bone occupies the core of the mastoid process. (Kindly prepared and contributed by Dr. C. Buckland-Wright, Department of Anatomy, Guy's Hospital Medical School, London.)

and the *mastoid antrum*, by which they communicate with the tympanic cavity, are described on p. 1196.

The petrous part of the temporal bone is wedged between the sphenoid and occipital bones at the base of the skull (**3**.75 A and B). It is directed medially, forwards, and a little upwards; it has a base, an apex, three surfaces, and three margins. The labyrinth lies within it.

The *base* is a somewhat artificial concept; it corresponds to the suture between the petrous and squamous elements of the temporal bone until this disappears by fusion soon after birth. Since the mastoid process is a post-natal development of the petrous element, the base is here an arbitrary division, but is indicated by the partial separation of the two by the mastoid antrum.

The *apex*, blunt and irregular, projects into the angular interval between the posterior border of the greater wing of the sphenoid bone and the basilar part of the occipital bone; in it is the anterior orifice of the carotid canal and it forms the posterolateral boundary of the foramen lacerum.

The *anterior surface* helps to form the floor of the middle cranial fossa and is continuous with the cerebral surface of the squamous part, although remains of the petrosquamosal suture are often distinct even at a late period of life.

The whole surface is marked by impressions for the gyri of the inferior surface of the temporal lobe. Immediately behind the apex is a slight hollow for the trigeminal ganglion, the *trigeminal impression*. The bone, anterior and slightly lateral to the impression, forms the roof of the anterior part of the carotid canal; it is often deficient here. An irregular ridge separates the trigeminal impression posteriorly from a second hollow, which forms part of the roof of the internal acoustic meatus and covers the cochlea. This concavity is limited behind by an elevation, the *arcuate eminence* (**3**.94), raised by the anterior semicircular canal. In its lateral part it roofs in the vestibule and the beginning of the facial canal. Between the squamous part on the lateral side, and the arcuate eminence and the hollows just described on the medial side, the surface is formed by the *tegmen tympani*. This thin plate of bone forms the roof of the mastoid antrum behind and extends forwards above the tympanic cavity (**3**.95) and the canal for the tensor tympani. Its lateral margin meets the squama at the site of the petrosquamosal suture and turns downwards in front to form the lateral wall of the canal for the tensor tympani and the bony part of the auditory tube; the lower edge of this downturned portion has already been observed in the floor of the squamotympanic fissure (p. 307). Anteriorly the tegmen tympani presents a narrow groove, which runs backwards and laterally and enters the bone through an opening placed in front of the lateral part of the arcuate eminence. This hiatus transmits the greater petrosal nerve, which runs forwards to the foramen lacerum. A second, smaller hiatus, lying lateral to the one just described, transmits the lesser petrosal nerve from the tympanic plexus; the bone in front of the opening may be grooved. The posterior slope of the arcuate eminence covers the posterior and lateral semicircular canals, and lateral to it the posterior part of the tegmen tympani roofs in the mastoid antrum.

The *posterior surface* (**3**.94) forms the anterior part of the posterior cranial fossa and is continuous with the inner surface of the mastoid portion. Near the centre of the posterior surface is the orifice of **the internal acoustic meatus** (p. 313). Behind this orifice there is a small slit almost hidden by a thin plate of bone; it leads to a canal, the *aqueduct of the vestibule*, which contains the saccus and ductus endolymphaticus together with a small artery and vein. The terminal half or more of the saccus endolymphaticus protrudes through the narrow orifice and lies between the periosteum and the covering dura mater. Above and between these two openings the *subarcuate fossa* (p. 313) forms an irregular depression.

The *inferior surface* (**3**.96), rough and irregular, is part of the external surface of the base of the skull. Near the apex a quadrilateral rough area serves partly for the attachment of the levator veli palatini and the cartilaginous portion of the auditory tube, and partly for connexion with the basilar part of the occipital bone, some dense fibrocartilage intervening. Behind this a large, nearly circular aperture leads into **the carotid canal** (p. 306).

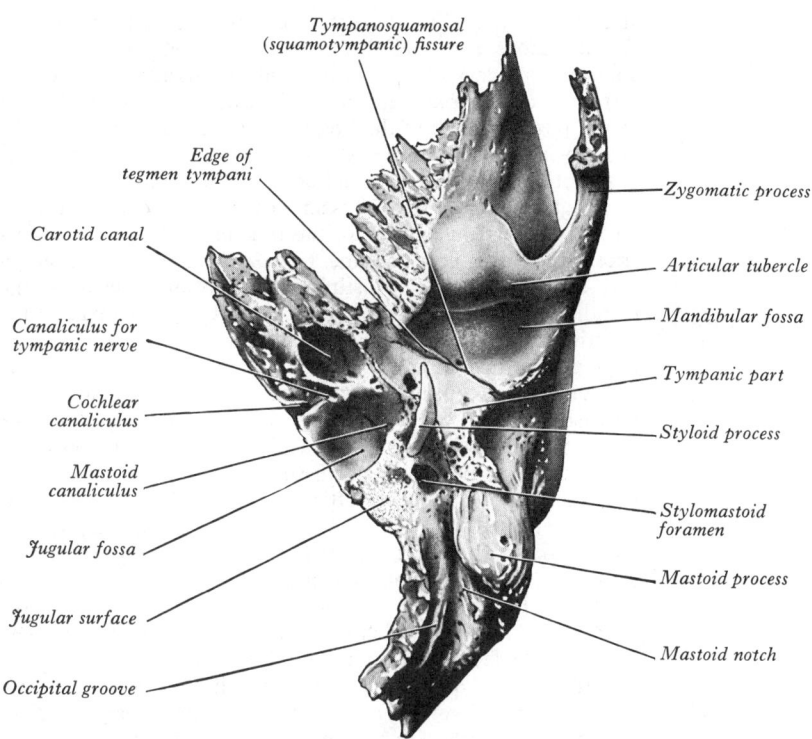

3.96 The left temporal bone. Inferior aspect.

Behind this opening there is a deep depression, termed **the jugular fossa**, of variable depth and size in different skulls; it lodges the superior bulb of the internal jugular vein. In front of the medial part of the jugular fossa and directly below the internal acoustic meatus, the bone is marked by a triangular depression, which lodges the inferior ganglion of the glossopharyngeal nerve. At the apex of this notch a small opening leads into the *cochlear canaliculus*, which is occupied by the perilymphatic duct and a tubular prolongation of the dura mater, and transmits a vein from the cochlea to join the internal jugular vein. On the bony ridge dividing the carotid canal from the jugular fossa there is a small *canaliculus for the tympanic nerve*, which is derived from the glossopharyngeal nerve (p. 308). In the lateral part of the jugular fossa the bone is pierced by the *mastoid canaliculus* for the entrance of the auricular branch of the vagus nerve. Behind the jugular fossa is the *jugular surface*, a rough quadrilateral area covered with cartilage and articulating with the jugular process of the occipital bone.

The *superior border*, the longest, is grooved for the superior petrosal sinus, the dura mater of the tentorium cerebelli being attached to the sharp edges of the groove, except at its medial extremity, where it is crossed by the roots of the trigeminal nerve. The *posterior border* is intermediate in length between the superior and the anterior. Its medial part is marked by a sulcus, which forms, with a corresponding sulcus on the occipital bone, the channel for the inferior petrosal sinus. Behind this there is **the jugular fossa**, which, with the jugular notch on the occipital bone, forms the jugular foramen; it presents a notch for the glossopharyngeal nerve, and bone on either or both sides of the notch may meet the occipital bone and divide the foramen into two parts, sometimes three. The *anterior border* is divided into two parts: a lateral, joined to the squamous part at the *petrosquamosal suture*; a medial, free, for articulation with the greater wing of the sphenoid bone.

At the angle of junction of the petrous and squamous parts two canals are placed one above the other, and separated by a thin plate of bone. Both canals lead into the tympanic cavity; the upper contains the tensor tympani; the lower is the osseous part of the canal of the auditory tube.

The tympanic part of the temporal bone (**3**.96) is a curved plate lying below the squamous part and in front of the mastoid process. Internally, it is fused with the petrous part, and appears in the angle between it and the squamous part, where it lies below

329

Labels on figure:
Tympanosquamosal (squamotympanic) fissure
Edge of tegmen tympani
Carotid canal
Canaliculus for tympanic nerve
Cochlear canaliculus
Mastoid canaliculus
Jugular fossa
Jugular surface
Occipital groove
Zygomatic process
Articular tubercle
Mandibular fossa
Tympanic part
Styloid process
Stylomastoid foramen
Mastoid process
Mastoid notch

and lateral to the orifice of the auditory tube. Behind, it fuses with the squamous part and the mastoid process, and forms the anterior boundary of the tympanomastoid fissure. Its *posterior surface* is concave and forms the anterior wall, the floor, and a part of the posterior wall of the bony external acoustic meatus (the postero-inferior wall of the external meatus occasionally presents a smooth to roughened longitudinal elevation—the *auditory torus*); at the medial end of this surface there is a narrow furrow, termed the *tympanic sulcus*, for the attachment of the circumference of the tympanic membrane. Its *anterior surface*, quadrilateral and slightly concave, constitutes the posterior wall of the mandibular fossa and is sometimes in contact with a part of the parotid gland. Its *lateral border* is free and roughened; it forms a large part of the margin of the opening of the external acoustic meatus and is continuous with the cartilaginous part of the meatus. The lateral part of the *upper border* is fused with the back of the postglenoid tubercle; its medial part forms the posterior boundary of the petrotympanic fissure. The *lower border* is sharp; its lateral part splits to enclose the root of the styloid process and is therefore named the *sheath of the styloid process* (vaginal process). The central portion of the tympanic part of the temporal bone is thin, and in a considerable percentage of skulls is perforated by a foramen. Between the styloid and mastoid processes the *stylomastoid foramen*, which is the lower end of the facial canal, transmits the facial nerve and the stylomastoid artery.

The external acoustic meatus, which is about 16 mm long, is directed inwards and slightly forwards and downwards; its floor is convex upwards. On sagittal section the meatus is oval or elliptical in shape with the long axis directed downwards and slightly backwards. Its anterior wall, its floor and the lower part of its posterior wall are formed by the tympanic part of the bone, its roof and the upper part of its posterior wall by the squamous part. Its inner end is closed by the tympanic membrane; its outer end is bounded above by the posterior root of the zygomatic process, below which the small *suprameatal spine* is sometimes seen at the upper and posterior part of the orifice.

The styloid process of the temporal bone, slender, pointed, and averaging about 2·5 cm in length, projects downwards and forwards, from the under surface of the bone. Its proximal part (the *tympanohyal*) is surrounded by a bony sheath, derived from the tympanic plate and best marked on its anterolateral aspect, while its distal part (the *stylohyal*) gives attachment to certain muscles and ligaments (p. 302). The process is covered laterally by the parotid gland, the facial nerve crosses its base, and the external carotid artery its tip, as they lie within the gland. On its deep surface the process is separated from the commencement of the internal jugular vein by the origin of the stylopharyngeus.

The structure of the squamous part is like that of the other cranial bones: the mastoid portion is trabecular and variably pneumatized, and the petrous portion dense and hard.

Ossification. The four elements of the temporal bone are ossified independently (**3**.97). The *squamous part* is ossified in membranous condensed mesenchyme from a single centre which appears in the region of the roots of the zygomatic process about the seventh or eighth week of intrauterine life. The *petromastoid part* is ossified from several centres which appear in the cartilaginous *otic capsule* (p. 178) during the fifth month of intrauterine life. As many as fourteen centres have been described and their order of appearance is somewhat variable. Several of these centres are small and inconstant and soon fuse with adjacent centres. This multiplicity of centres accords with numerous otic bones existing in lower forms. The whole otic capsule is more or less completely ossified by the end of the sixth month of intrauterine life. The *tympanic part* is ossified in collagenous fibrous tissue (membrane) from a centre which appears about the third month of intrauterine life. At birth it is represented by an incomplete ring, the *tympanic ring*, which is deficient above (**3**.97, 98A). Its concavity is grooved by the tympanic sulcus for the attachment of the circumference of the tympanic membrane. Running obliquely downwards and forwards across the inner aspect of the anterior part of the ring is the *malleolar sulcus* (**3**.97), which lodges the anterior process of the malleus, the chorda tympani nerve and the anterior tympanic artery. The *styloid process* is developed from the cranial end of the cartilage of the

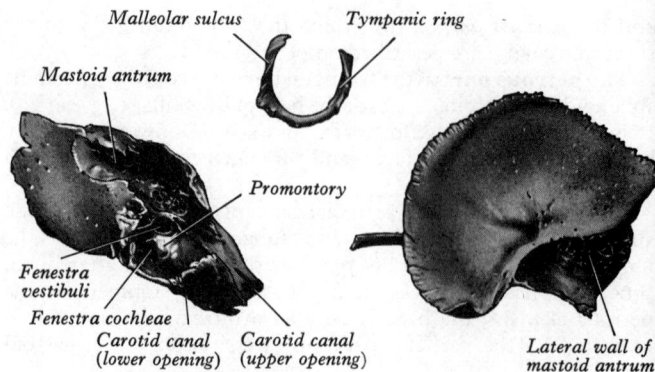

3.97 The three principal parts of the right temporal bone at birth. From left to right: lateral aspect of petromastoid part; medial aspect of tympanic ring; medial aspect of squama.

second visceral or hyoid arch (p. 151) by two centres: one for the proximal part of the process, termed the tympanohyal, appears before birth; the other, for the distal part of the process, termed the stylohyal, does not appear until after birth. The tympanic ring unites with the squamous part shortly before birth, the petromastoid fuses with the squamous part during the first year, and with the tympanohyal portion of the styloid process about the same time. The stylohyal does not unite with the rest of the bone until after puberty, and in some skulls never.

During ossification, the tympanic cavity, the mastoid antrum and the posterior end of the auditory tube are all enclosed within the bone. The petrous part forms the roof, floor and medial wall of the cavity, while the squamous and tympanic parts, together with the membrana tympani, form the lateral wall. *At birth*, the tympanic cavity, mastoid antrum and tympanic membrane are all of approximately *adult size*; so also are the auditory ossicles, but the anterior process of the malleus does not join the rest of the

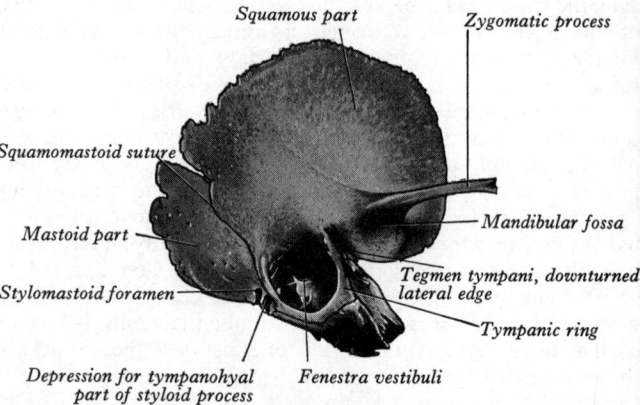

3.98A The right temporal bone at birth. Lateral aspect. *Note.*—The petromastoid part has not been coloured: the tympanic part is coloured *yellow* and the squamous part is coloured *brown*. The whole of the styloid process has been removed.

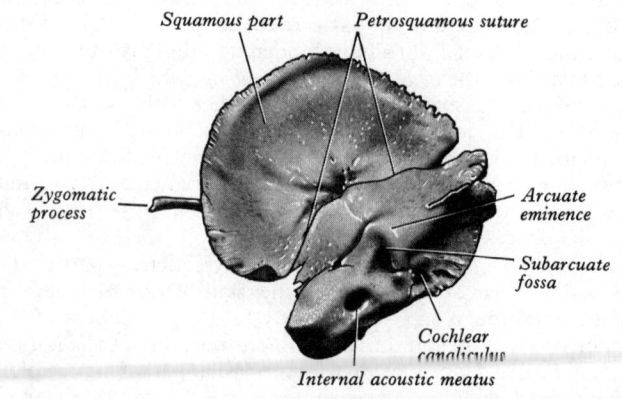

3.98B The right temporal bone at birth. Medial aspect.

bone until six months later. The internal acoustic meatus is about 6 mm in horizontal diameter, 4 vertically, and 7 in length at birth, adult values being 7·7 and 11 mm (Bergström 1973).

After birth the chief changes in the temporal bone, apart from the increase in size, are: (1) The tympanic ring extends laterally and backwards to form a more cylindrical structure, the tympanic part of the bone. It grows into a fibrocartilaginous plate, sometimes termed the *fibrous tympanic plate*, which forms the adjacent part of the external acoustic meatus at this stage. This growth does not, however, take place at an equal rate all round the ring, but occurs most rapidly on its anterior and posterior portions, and these outgrowths meet and blend, and thus, for a time, there exists in the floor of the meatus an opening (the *foramen of Huschke*); this deficiency is usually closed about the fifth year, but may persist through life. Its persistence has been noted in 5 to 46 per cent of adult crania from several ancient and modern populations (p. 348). In Burmese crania a marked sexual difference was noted: 25 per cent in males, 40 per cent in females. In its backward extension the tympanic element ensheathes the styloid process, and also extends medially over the petrous part as far as the carotid canal. (2) The mandibular fossa is at first extremely shallow and looks more laterally than downwards; it becomes deeper and is ultimately directed downwards. The postero-inferior portion of the squamous part grows downwards behind the tympanic ring and forms the lateral bony wall of the mastoid antrum. (3) The mastoid part is at first flat, and the stylomastoid foramen and rudimentary styloid process lie immediately behind the tympanic ring. With the development of the mastoid air cells the lateral part of the mastoid portion grows downwards and forwards to form the mastoid process, and the styloid process and stylomastoid foramen come to lie on the under surface of the bone. The descent of the stylomastoid foramen is necessarily accompanied by a corresponding increase in the length of the canal for the facial nerve. It is not until the latter part of the second year that the mastoid process forms a definite

elevation on the surface of the skull. (4) The subarcuate fossa on the posterior surface of the petrous portion is gradually filled and almost obliterated.

The external acoustic meatus is relatively as long in the child as in the adult, but in the child the canal is fibrocartilaginous, whereas in the adult the inner two-thirds of it are osseous. Surgical access to the tympanic cavity is through the mastoid antrum. In the child only a thin scale of bone requires to be removed from the suprameatal triangle to open into the antrum (p. 1196).

The Parietal Bones

The parietal bones (3.99, 100) form the sides and the roof of the cranium. Each is irregularly quadrilateral, and hence has two surfaces, four borders, and four angles.

The *external surface* (3.99) is convex, smooth, and marked near the centre by a slight elevation, the *parietal tuber (tuberosity)*. Two curved lines, the *superior and inferior temporal lines*, cross the middle of the surface, forming an arch which is convex upwards and backwards; to the former is attached the temporal fascia; the latter marks the upper limit of origin of the temporal muscle. The bone above these lines is covered by the *galea aponeurotica* (epicranial aponeurosis); that below the lines forms a part of the temporal fossa. Posteriorly, and close to the upper or sagittal border, the *parietal foramen* transmits a vein from the superior sagittal border, the *parietal foramen* transmits a vein from the superior sagittal sinus and, sometimes, a small branch of the occipital artery; the foramen is not always present.

The *internal surface* (3.100) is concave; it presents impressions for the cerebral gyri, and numerous furrows for the ramifications of the middle meningeal vessels; these furrows run upwards and backwards from the *sphenoidal (antero-inferior) angle*, and from the middle and posterior part of the inferior border. Along the sagittal border there is the shallow *groove for the superior sagittal*

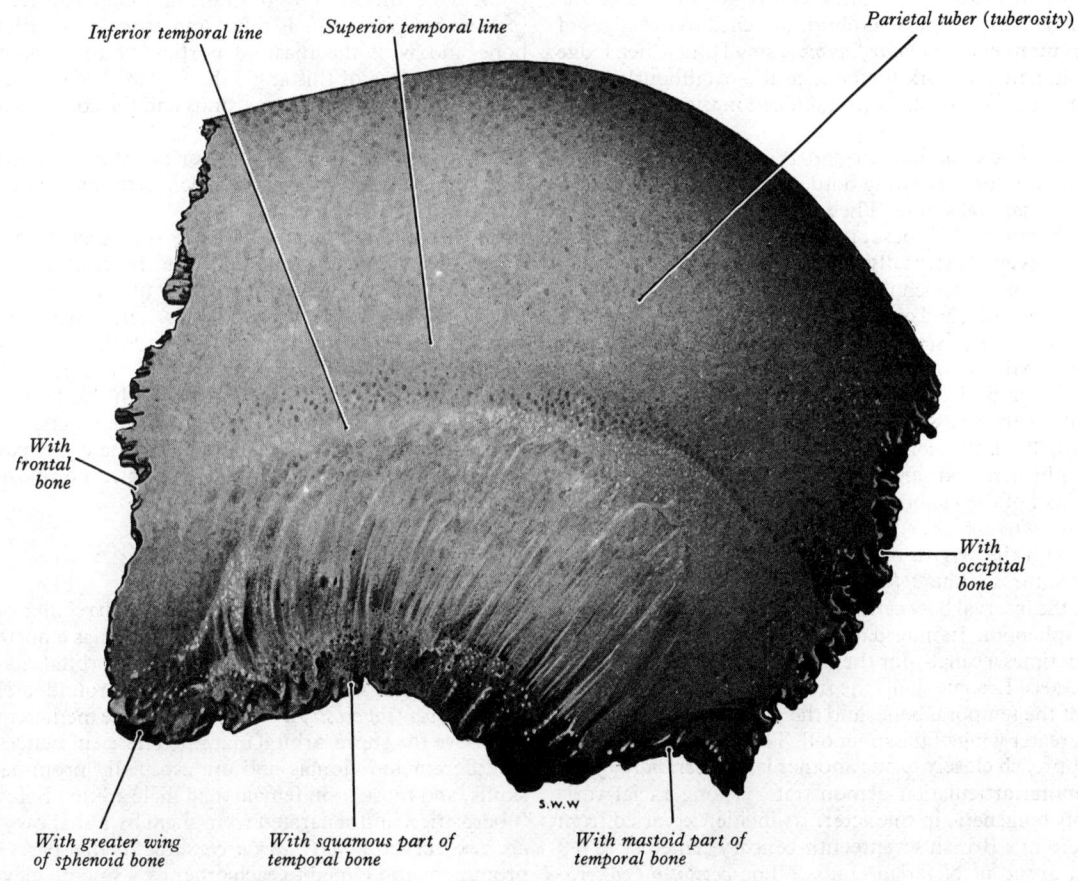

Inferior temporal line   Superior temporal line   Parietal tuber (tuberosity)

With frontal bone

With occipital bone

S.W.W

With greater wing of sphenoid bone   With squamous part of temporal bone   With mastoid part of temporal bone

3.99 The left parietal bone. External surface.

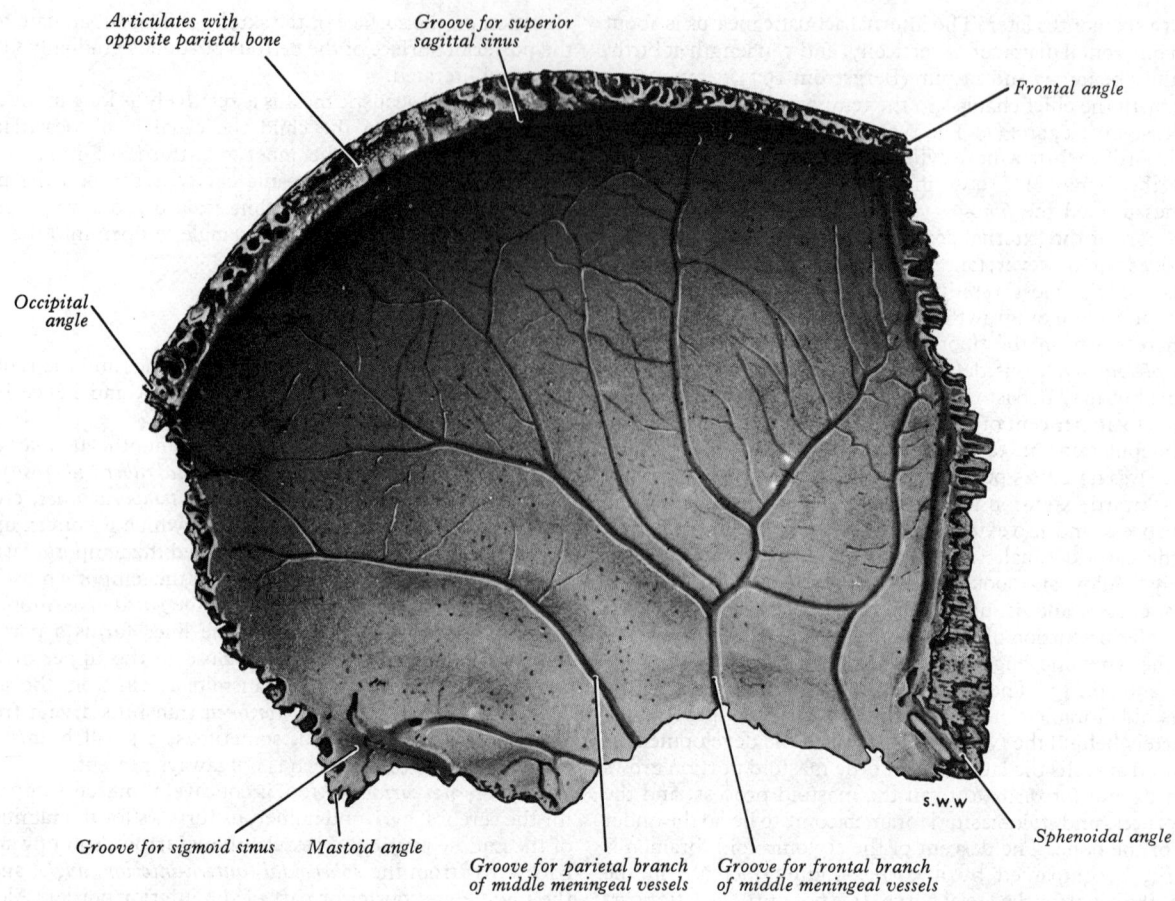

Articulates with opposite parietal bone

Groove for superior sagittal sinus

Frontal angle

Occipital angle

Groove for sigmoid sinus *Mastoid angle*

Groove for parietal branch of middle meningeal vessels

Groove for frontal branch of middle meningeal vessels

Sphenoidal angle

S.W.W

3.100 The left parietal bone. Internal aspect.

sinus, which, in the articulated skull, is completed by that on the opposite parietal bone; the falx cerebri is attached to the edges of this groove. A number of *granular foveolae*, small pits which lodge arachnoid granulations, mark the bone in the neighbourhood of the sagittal sulcus; they are most pronounced in the skulls of old persons.

The *sagittal border*, the longest and thickest, is dentated; it articulates with the corresponding border of the opposite parietal bone to form the sagittal suture. The *squamosal (inferior) border* is divided into three parts: of these, the anterior is short, thin and truncated; it is bevelled externally, and is overlapped by the tip of the greater wing of the sphenoid; the middle portion is arched, also bevelled externally, and overlapped by the squamous part of the temporal bone; the posterior part is short, thick and serrated, and articulates with the mastoid portion of the temporal bone. The *frontal border* is deeply serrated, and bevelled externally above and internally below; it articulates with the frontal bone, forming one-half of the coronal suture. The *occipital border*, which is deeply dentated, articulates with the occipital bone, forming one-half of the lambdoid suture.

The *frontal (anterosuperior) angle* is almost a right angle, and corresponds with the *bregma* or point of meeting of the sagittal and coronal sutures. The *sphenoidal (antero-inferior) angle* is received into the interval between the frontal bone and the greater wing of the sphenoid. Its internal surface is marked by a deep groove—sometimes a canal—for the frontal branch of the middle meningeal vessels. In some skulls the frontal bone articulates with the squama of the temporal bone, and the parietal bone then fails to reach the greater wing of the sphenoid. The region where these four bones approach closely to one another is the *pterion* (p. 302). (Fronto temporal articulation demonstrates strong racial variation probably epigenetic in character. Its incidence varied from practically zero in a British seventeenth-century cemetery to 9·8 per cent in a group of Nigerian crania.) The *occipital (postero-superior) angle* is rounded and corresponds with the *lambda* or

point of meeting of the sagittal and lambdoid sutures. The *mastoid (postero-inferior) angle* is blunt and articulates with the occipital bone and with the mastoid portion of the temporal. On the internal surface of this angle there is the broad, shallow groove, for the end of the transverse sinus and the commencement of the sigmoid sinus.

At birth there are unossified or membranous intervals in the skull at the angles of the parietal bones; they are named *fontanelles* and are described on p. 345.

Ossification. The parietal bone is ossified from two centres, which appear one above the other at the parietal tuber about the seventh week of intrauterine life in fibrous, condensed mesenchyme. These centres unite early, and ossification gradually extends in a centrifugal manner towards the margins of the bone; the angles are consequently the parts last to be ossified; and it is here that the fontanelles exist. At birth the temporal lines are situated low down; they reach their permanent position only after the eruption of the molar teeth. Occasionally the parietal bone is divided into upper and lower parts by an anteroposterior suture.

The Frontal Bone

The frontal bone is shaped like a shallow, irregular cap and forms the region of the forehead; on each side it has a horizontal *orbital part*, which forms most of the roof of the orbital cavity.

The *external surface* (**3**.101) presents a rounded elevation, the *frontal tuber (tuberosity)*, on each side of the median plane, about 3 cm above the supra-orbital margin. These eminences vary in size in different individuals and are especially prominent in young skulls, and more so in female than male adults. Below the frontal tuberosities, and separated from them by a shallow groove, there are two curved *superciliary arches*, the medial parts of which are prominent and joined to each other by a smooth elevation named the *glabella*. These arches are more prominent in the male, to a

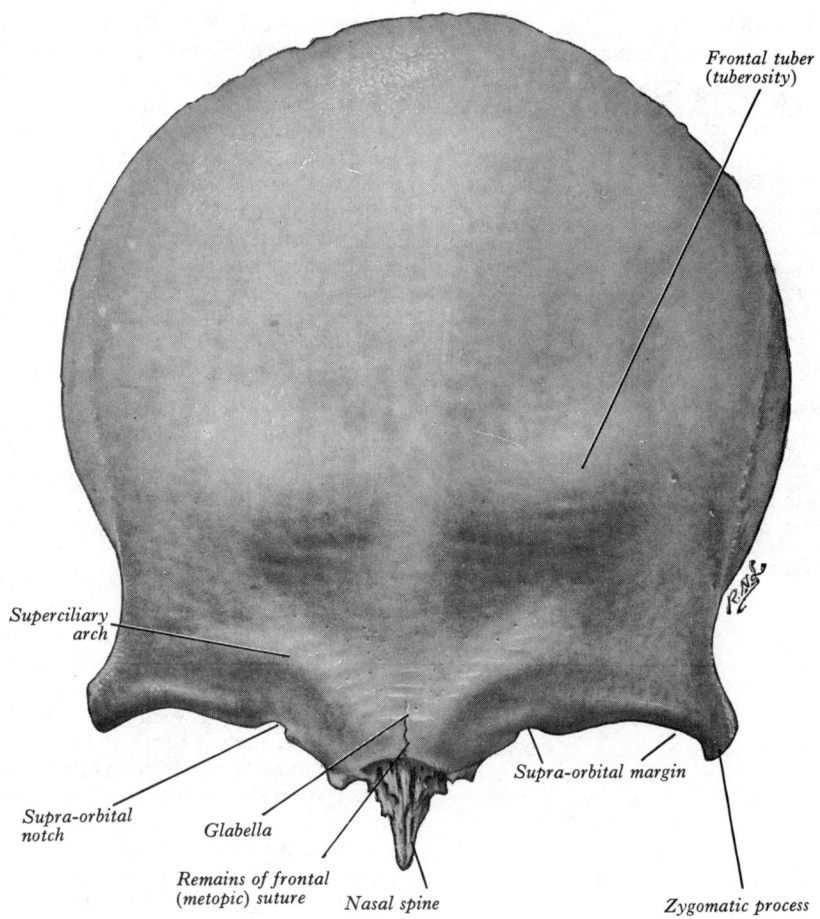

3.101   The frontal bone. External aspect.

degree dependent to some extent on the size of the frontal sinuses; prominent superciliary arches are, however, occasionally associated with small sinuses. Beneath the superciliary arches the curved *supra-orbital margins* form the upper borders of the orbital openings. The lateral two-thirds of each supra-orbital margin are sharp; the medial one-third is rounded. At the junction of these two parts is the *supra-orbital notch*, which may, on occasion, be a *foramen*, and contains the supra-orbital vessels and nerve. Medial to this notch, the small *frontal notch* or *foramen* is present in about 50 per cent of skulls. Berry (1975) has recorded the incidences of frontal and supra-orbital *foramina* and *notches* (incisures). While usually a notch, the supra-orbital feature is as frequently a foramen in some populations (e.g. 51 per cent of Mexican crania). The frontal foramen also shows great racial variation from 15 to 87 per cent in different ethnic groups. Both features show a sexual dimorphism. For further information on racial variation consult Kimura (1977).

The supra-orbital margin ends laterally in the *zygomatic process*, which is strong and prominent and articulates with the zygomatic bone. From this process a line curves upwards and backwards and soon divides into the *superior* and *inferior temporal lines*.

The portion of the bone which projects downwards between the supra-orbital margins is named the *nasal part*. It presents a rough, serrated area, sometimes termed the *nasal notch*, which articulates on each side of the median plane with the nasal bone, and lateral to this with the frontal process of the maxilla and with the lacrimal bone. From the centre of the notch posteriorly, the nasal part projects downwards and forwards behind the nasal bones (**3**.77) and frontal processes of the maxillae, and supports the bridge of the nose. The nasal part ends below in a sharp *nasal spine*, and on each side of this there is a small grooved surface which forms a part of the roof of the corresponding nasal cavity. The nasal spine forms a very small part of the septum of the nose; in front it articulates with the crest of the nasal bones, behind with

the perpendicular plate of the ethmoid bone (**3**.77).

The *temporal surface*, below and behind the temporal lines, forms the anterior part of the temporal fossa and gives origin to a part of the temporal muscle (**3**.61).

The *internal surface* (**3**.102) of the frontal bone is concave. In the upper part of the median plane it is marked by a vertical groove, the sulcus for the sagittal sinus, the edges of which unite below to form the *frontal crest*; the sulcus lodges the anterior part of the superior sagittal sinus, while to its margins and to the frontal crest the anterior part of the falx cerebri is attached. The crest ends below in a small notch, which is converted into the *foramen caecum* (p. 309) by articulation with the ethmoid bone. On each side of the median plane the surface is marked by impressions for the cerebral gyri, and minute furrows for meningeal vessels. Several small, irregular pits, *granular foveolae*, may be seen on each side of the sulcus for the sagittal sinus, for the reception of arachnoid granulations.

The *parietal (posterior) margin* is thick, strongly serrated and bevelled at the expense of the internal surface above and the temporal surface on each side; it is continued below into a triangular, rough surface, for articulation with the greater wing of the sphenoid bone.

The orbital parts of the frontal bone consist of two thin triangular laminae, which form the major part of the roofs of the orbits, and are separated from each other by a wide gap named the *ethmoidal notch*.

The *orbital surface* (**3**.102) of each orbital plate is smooth and concave, and presents, in its anterolateral part, a shallow depression, the *fossa for the lacrimal gland*. Below and behind the medial end of the supra-orbital margin, about midway between the supra-orbital notch and the frontolacrimal suture, there is a small depression, sometimes a tiny spine, the *trochlear fovea* or *spine*, for the attachment of the fibrocartilaginous trochlea of the superior oblique muscle. The *internal surface* is convex, and marked by impressions for the gyri on the inferior surface of the

333

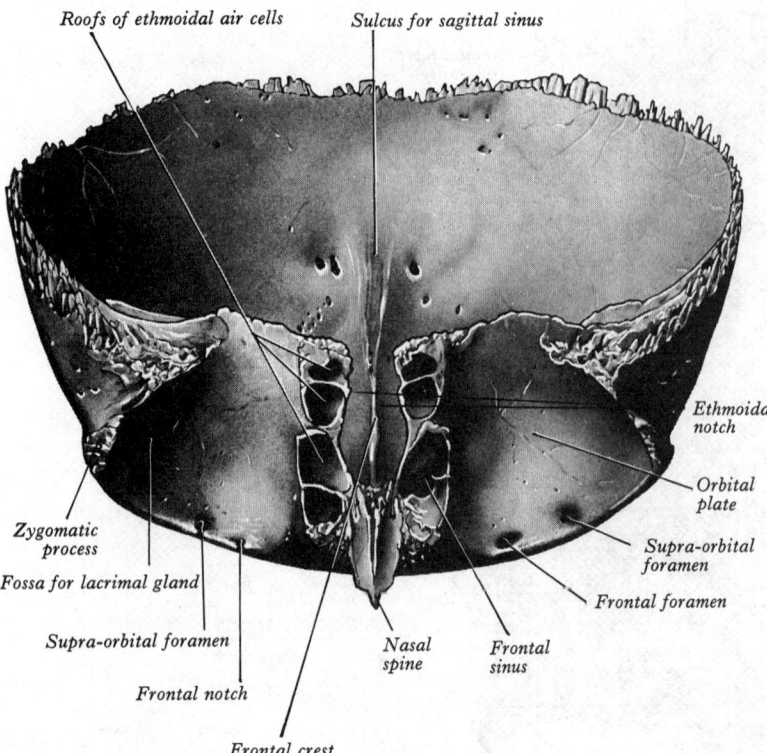

Roofs of ethmoidal air cells   Sulcus for sagittal sinus

Ethmoidal notch

Orbital plate

Supra-orbital foramen

Frontal foramen

Zygomatic process

Fossa for lacrimal gland

Supra-orbital foramen

Nasal spine

Frontal sinus

Frontal notch

Frontal crest

3.102   The frontal bone. Inferior aspect.

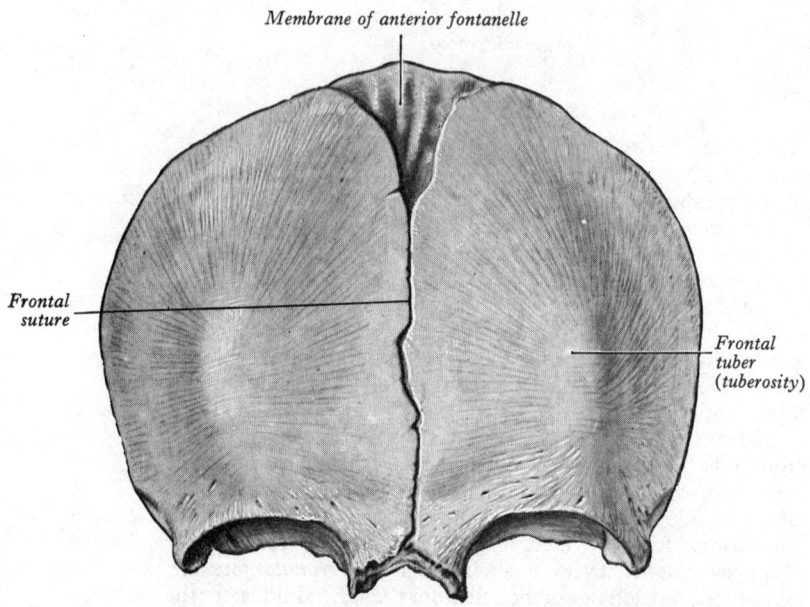

Membrane of anterior fontanelle

Frontal suture

Frontal tuber (tuberosity)

3.103   The frontal bone at birth. Anterior aspect. Note that at this stage the bone consists of right and left halves connected by the frontal suture.

frontal lobe of the brain, and by faint grooves for the meningeal branches of the ethmoidal vessels.

The *ethmoidal notch* (**3**.102) is quadrilateral and is occupied, in the articulated skull, by the cribriform plate of the ethmoid bone. On the under surfaces of the lateral margins of the notch parts of several air cells are present: they complete the ethmoidal air cells when the ethmoid bone is in position. Two transverse grooves cross each margin of the notch; they are converted into the *anterior* and *posterior ethmoidal canals* by the ethmoid bone, and open on the medial wall of the orbit; they transmit the anterior and posterior ethmoidal nerves and vessels.

The openings of the *frontal sinuses* (**3**.102) are situated in front of the ethmoidal notch, and lateral to the nasal spine. These

sinuses are two irregular cavities, which extend backwards, upwards, and laterally for a variable distance between the laminae of the frontal bone; they are separated from each other by a thin bony septum, which is often deflected to one or other side of the median plane, with the result that the sinuses are seldom symmetrical. Rudimentary at birth, the frontal sinuses are usually fairly well developed by the seventh and eighth years, but reach their full size only after puberty. They vary in size in different persons and are larger in males. Each communicates with the middle meatus of the corresponding half of the nasal cavity by means of a passage called the *frontonasal canal*. The development of the frontal sinuses is associated with the appearance of the superciliary arches, and the latter owe their origin to the mechanical stresses to which the frontal bone is subjected by the masticatory apparatus (Weinmann and Sicher 1955). The sinuses undergo a primary expansion with the eruption of the first deciduous molars, and a secondary expansion when the permanent molars begin to appear at the sixth year. With advancing age absorption of bone from the inner walls of the sinuses may occur as an atrophic change leading to further enlargement.

The *posterior borders of the orbital plates* are thin and serrated, and articulate with the lesser wings of the sphenoid; the lateral part of each usually appears in the middle cranial fossa between the greater and lesser wings of the sphenoid bone.

Structure. The frontal bone is thick and consists of cancellous tissue contained between two compact laminae; the former is absent in the regions occupied by the frontal sinuses. The orbital part, composed entirely of compact bone, is thin and translucent in its posterior two-thirds. It may be absorbed in patches in old age.

Ossification (**3**.103). The frontal bone is ossified in fibrous tissue (membrane) from two primary centres which appear in the eighth week of intrauterine life, one in the region of each frontal tuber. From each of these centres ossification extends upwards to form the corresponding half of the bone, backwards to form the orbital part and downwards to form the nasal part of the bone. No secondary centres of ossification occur in connexion with any part of the bone, with the possible exception of the nasal spine, for which two secondary centres have been described as appearing about the tenth year (Inman and Saunders 1937). At birth the bone consists of two halves separated by the *frontal* or *metopic suture*, but union begins in the second year, and the suture is usually obliterated by the eighth year. In a percentage of persons this shows some racial variation: the two halves of the frontal bone remain separate, and the metopic suture persists (Ashley Montagu 1951; Tongerson 1951; Linc and Fleischmann 1968). Recently, metopism has been assessed at 0–7·4 per cent of individuals in various ethnic groups (Berry 1975).

The Ethmoid Bone

The ethmoid bone is cuboidal, and exceedingly light in build; it is situated at the anterior part of the base of the cranium, and assists in forming the medial walls of the orbits, the septum of the nose, and the roof and lateral walls of the nasal cavity. It consists of four parts: a horizontal, perforated lamina named the cribriform plate, a perpendicular plate and two lateral masses, named labyrinths.

The cribriform plate (**3**.104) occupies the ethmoidal notch of the frontal bone, and forms a part of the roof of the nasal cavity. A thick, smooth, triangular process, called the *crista galli* from its resemblance to a cock's comb, projects upwards from this lamina in the median plane. Its posterior border, long, thin and curved, gives attachment to the falx cerebri. Its anterior border, short and thick, articulates with the frontal bone by two small projecting *alae* which complete the foramen caecum (p. 309). Its sides are smooth but sometimes bulge owing to the presence of a small air sinus in the interior. On each side of the crista galli the cribriform plate is narrow and depressed; the gyrus rectus and the olfactory bulb are immediately above it, and it presents numerous foramina for the passage of the olfactory nerves. In the front part of the cribriform plate, on each side of the crista galli, there is a small slit-like fissure, which is occupied by a process of dura mater. The

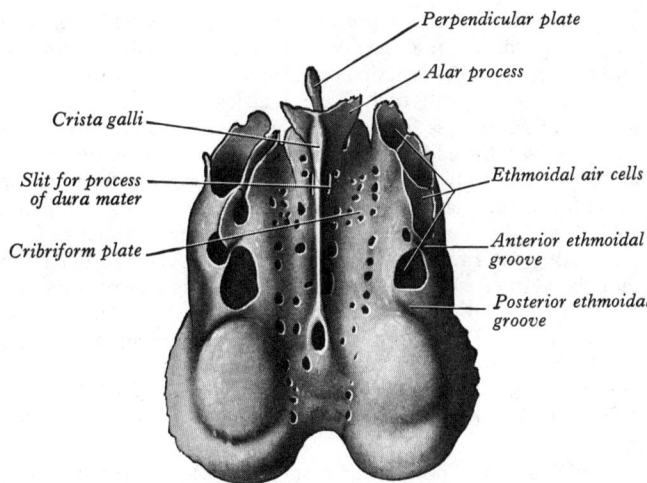

3.104 The ethmoid bone. Superior aspect.

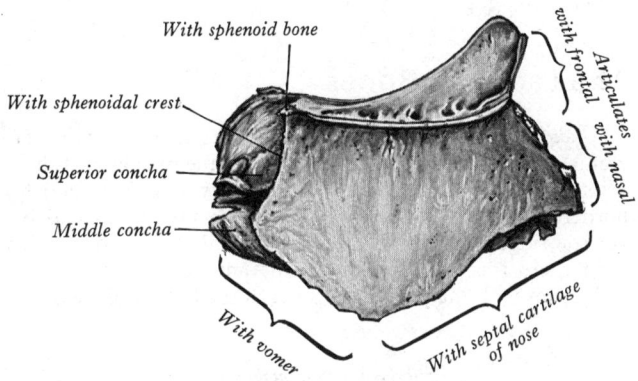

3.105 The perpendicular plate of the ethmoid bone. Right lateral aspect. Shown after removal of the right ethmoidal labyrinth.

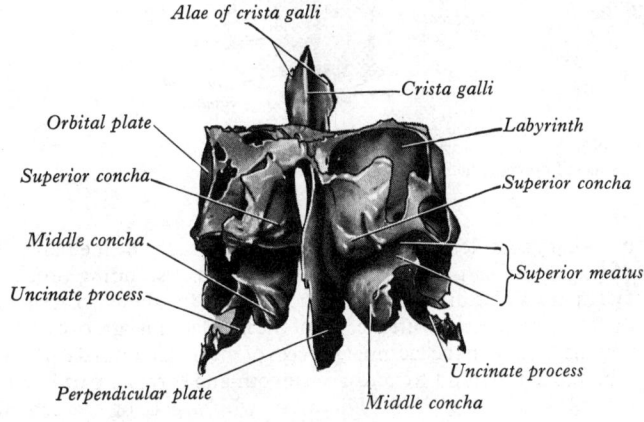

3.106 The ethmoid bone. Posterior aspect.

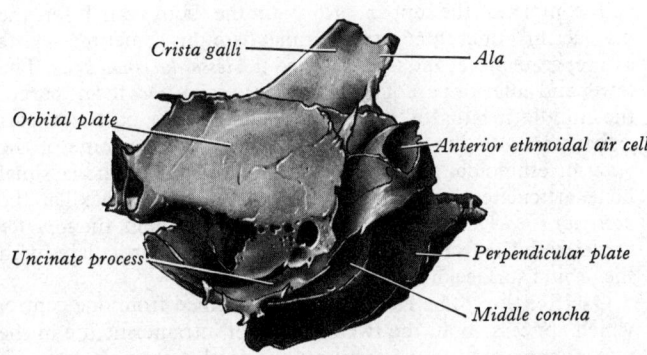

3.107 The ethmoid bone. Right lateral aspect.

foramen which transmits the anterior ethmoidal nerve to the nasal cavity is placed lateral to the anterior end of the fissure, and to it a groove runs forwards from the orifice of the anterior ethmoidal canal.

The perpendicular plate (3.105, 106), is thin, flat and quadrilateral. It descends from the cribriform plate and forms the upper part of the nasal septum; it is generally deflected a little to one or other side. Its *anterior border* articulates with the nasal spine of the frontal bone and the crest of the nasal bones. The *posterior border* articulates with the sphenoidal crest above and with the vomer below. The *superior border* is attached to the cribriform plate. The *inferior border* is thick, and serves for the attachment of the septal cartilage of the nose. The surfaces of the lamina are smooth, except above, where numerous grooves and canals are seen; these lead to the medial foramina in the cribriform plate and contain filaments of the olfactory nerves.

The ethmoidal labyrinths consist of a number of thin-walled *ethmoidal air cells*, arranged in three groups—*anterior, middle* and *posterior*—and interposed between two vertical plates of bone (p. 1149); the lateral or *orbital plate* forms part of the medial wall of the orbit, the medial plate, part of the lateral wall of the nasal cavity. In the disarticulated bone many of these ethmoidal air cells are opened, but in the articulated skull they are everywhere closed, except at their apertures of communication with the nasal cavity. The *upper surface* of the labyrinth (3.104) presents a number of open air cells, the walls of which are completed, in the articulated skull, by the edges of the ethmoidal notch of the frontal bone (3.102). This surface is crossed by two grooves which are converted into the *anterior* and *posterior ethmoidal canals* by articulation with the frontal bone. On the *posterior surface* of each labyrinth (3.106) large air cells are visible, and their walls are completed by the sphenoidal concha and the orbital process of the palatine bone. The *lateral surface* (3.107) consists of a thin, smooth, oblong plate, named the *orbital plate*, which covers the middle and posterior ethmoidal sinuses and forms a large part of the medial wall of the orbit; it articulates superiorly with the orbital part of the frontal bone, inferiorly with the maxilla and the orbital process of the palatine bone, anteriorly with the lacrimal bone, and posteriorly with the sphenoid bone (3.67).

A few air cells lie in front of the orbital plate and their walls are completed by the lacrimal bone and the frontal process of the maxilla. A thin, curved bar of bone, termed the *uncinate process*, subject to considerable variation in size, projects downwards and backwards from this part of the labyrinth; it can be seen in the medial wall of the maxillary sinus (3.67) as it crosses the anterior part of the hiatus maxillaris to reach the ethmoidal process of the inferior nasal concha, with which it articulates. The upper edge of this process is free and forms the medial boundary of the hiatus semilunaris in the middle meatus of the nose.

The *medial surface* of the labyrinth (3.108) forms a part of the lateral wall of the corresponding half of the nasal cavity; it consists of a thin lamella, which descends from the inferior surface of the cribriform plate and ends in a free, convoluted portion, named the *middle nasal concha*. The upper part of the medial surface is marked by numerous grooves, directed nearly vertically downwards; they lodge branches of the olfactory nerves. The posterior part of the medial surface is subdivided by a narrow, oblique fissure, termed the *superior meatus* of the nose, which is bounded above by a thin, curved plate, named the *superior nasal concha*; the posterior ethmoidal air cells open into this meatus. Below and in front of the superior meatus the convex surface of the middle nasal concha extends along the whole length of the medial surface of the labyrinth. Its lower margin is free and thick, while its lateral surface is concave and assists in forming the *middle meatus* of the nose. The middle ethmoidal air cells produce a rounded swelling, named the *bulla ethmoidalis*, on the lateral wall of the middle meatus (3.79); on the bulla, or immediately above it, these cells open into the meatus. A curved passage, named the *infundibulum*, extends upwards and forwards from the middle meatus; it communicates with the anterior ethmoidal sinuses, and in rather more than 50 per cent of crania is continued superiorly as the frontonasal duct into the frontal sinus.

Ossification. The ethmoid bone is ossified in the cartilaginous *nasal capsule* from three centres: one for the perpendicular plate, and one for each labyrinth.

The centre for each labyrinth appears in the region of the orbital plate between the fourth and fifth months of intrauterine life, and extends into the conchae. At birth, the two labyrinths, which are small and ill developed, are partially ossified, but the rest of the bone is cartilaginous. During the first year after birth, the perpendicular plate and crista galli begin to ossify from a single centre, and they fuse with the labyrinths about the beginning of the second year. The cribriform plate is ossified partly from the perpendicular plate and partly from the labyrinth. The ethmoidal air cells begin to develop during intrauterine life, and in the newborn infant have the form of narrow pouches.

The Inferior Nasal Conchae

The inferior nasal conchae are curved laminae which lie horizontally in the lateral walls of the nasal cavity (3.108). Each bone has two surfaces, two borders, and two ends.

The *medial surface* (3.109A) is convex, perforated by numerous apertures, and traversed by longitudinal grooves for the lodgement of vessels. The *lateral surface* is concave (3.109B), and forms part of the inferior meatus of the nasal cavity. The *superior border* is thin and irregular, and may be divided into three parts, of which the anterior articulates with the conchal crest of the maxilla, and the posterior with the conchal crest of the palatine

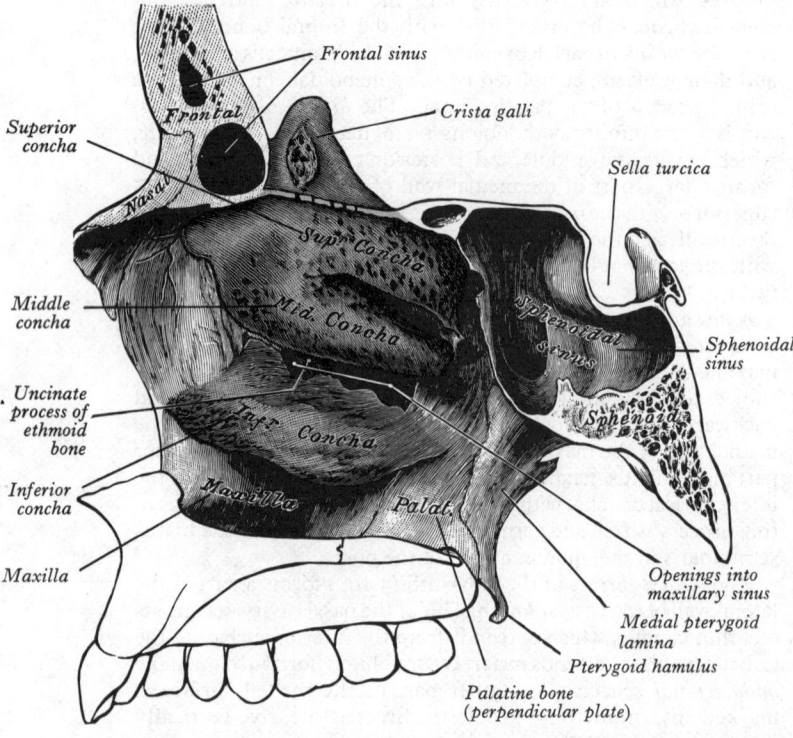

3.108 The lateral wall of the right half of the nasal cavity, showing the ethmoid bone (coloured *brown*) and the inferior nasal concha (coloured *blue*) in position.

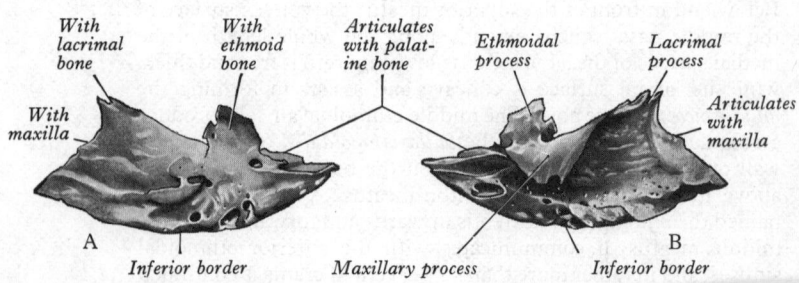

3.109A and B The right inferior nasal concha. (A) Medial aspect. (B) Lateral aspect.

bone. The middle region presents *three processes*, which vary in size and form. Of these, the *lacrimal process* is small and pointed and is situated at the junction of the anterior one-fourth with the posterior three-fourths of the bone. It articulates, by its apex, with a descending process from the lacrimal bone (3.110) and, by its margins, with the edges of the nasolacrimal groove on the medial surface of the body of the maxilla; it thus assists in forming the canal for the nasolacrimal duct. Behind this process a thin plate, the *ethmoidal process*, ascends to join the uncinate process of the ethmoid (3.79). From the middle part of the superior border a thin lamella, termed the *maxillary process*, curves downwards and laterally; it articulates with the maxilla and the maxillary process of the palatine bone, forming a part of the medial wall of the maxillary sinus (p. 340). The *inferior border* is free, thick and spongy in structure, more especially in the middle of the bone. Both *ends* are more or less pointed, the posterior being the more tapered.

Ossification. The inferior nasal concha is ossified from one centre; this appears about the fifth month of intrauterine life in the incurved lower border of the lateral wall of the cartilaginous nasal capsule. It loses continuity with the nasal capsule during ossification.

The Lacrimal Bones

The lacrimal bones, which are the smallest and most fragile of the cranial bones, are situated at the front of the medial walls of the orbits (3.67A). Each lacrimal bone has two surfaces and four borders. The *lateral* or *orbital surface* (3.110) is divided by a vertical ridge, termed the *posterior lacrimal crest*. In front of this crest there is a vertical groove, the anterior border of which articulates with the posterior border of the frontal process of the

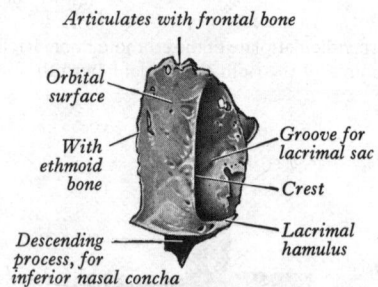

3.110 The right lacrimal bone. Lateral aspect.

maxilla to complete the *fossa for the lacrimal sac*. The medial wall of the groove is prolonged downwards as a descending process (3.111) to assist in forming the bony canal for the nasolacrimal duct by articulating with the lips of the nasolacrimal groove of the maxilla, and with the lacrimal process of the inferior nasal concha. The portion behind the crest is smooth and forms a part of the medial wall of the orbit. The crest, with a part of the orbital surface immediately posterior to it, gives origin to the lacrimal part of the orbicularis oculi. The crest ends below in a small hook, termed the *lacrimal hamulus*, which articulates with the maxilla and completes the upper orifice of the bony canal for the nasolacrimal duct (3.66); the lacrimal hamulus sometimes exists as a separate piece, and is then called the *lesser lacrimal bone*. The lower and anterior part of the *medial* or *nasal surface* forms part of the middle meatus of the nose; its upper and posterior part articulates with the ethmoid bone and completes some of the anterior ethmoidal air cells. The *anterior border* of the lacrimal bone articulates with the frontal process of the maxilla, the *posterior border* with the orbital plate of the ethmoid, the *superior border* with the frontal bone. The *inferior border* articulates with the orbital surface of the maxilla (3.66).

Ossification. The lacrimal bone is ossified from one centre, which appears about the twelfth week of intrauterine life in the mesenchyme around the cartilaginous nasal capsule. In later life the lacrimal bone is subject to patchy erosion.

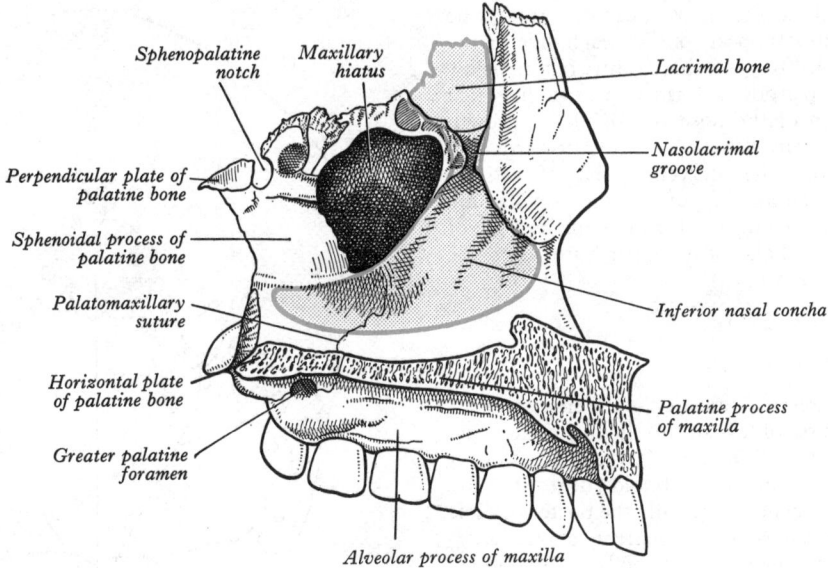

3.111 A drawing to show how the medial wall of the nasolacrimal canal is formed by the articulation of the descending process of the lacrimal bone with the lacrimal process of the inferior nasal concha.

The Nasal Bones

The nasal bones are two small oblong bones, varying in size and form in different individuals; they are placed side by side between the frontal processes of the maxillae, and form, by their junction, the bridge of the nose (**3.**63A and B, 79).

Each nasal bone has two surfaces and four borders. The *external surface* (**3.**112A), is concavoconvex from above downwards, and convex from side to side; it is covered by the procerus and nasalis muscles, and is perforated near its centre by a foramen for the transmission of a small vein. The *internal surface* (**3.**112B) is concave from side to side, and is traversed from above downwards by a groove for the anterior ethmoidal nerve. The *superior border*, thick and serrated, articulates with the nasal part of the frontal bone. The *inferior border*, thin and notched, is continuous with the lateral cartilage of the nose. The *lateral border* articulates with the frontal process of the maxilla. The *medial border*, thicker above than below, articulates with the opposite nasal bone and is prolonged behind into a vertical crest, which forms a small part of the septum of the nose, and articulates from above downwards with the nasal spine of the frontal, the perpendicular plate of the ethmoid, and the cartilage of the nasal septum.

Ossification. The nasal bone is ossified from one centre, which appears at the beginning of the third month of intrauterine life in the membrane overlying the anterior part of the cartilaginous nasal capsule.

The Vomer

The vomer is a thin, flat bone, almost trapezoid in shape; it forms the postero-inferior part of the septum of the nose (**3.**77) and has two surfaces and four borders. Each *surface* (**3.**113A) is marked by small furrows for blood vessels, and is traversed by a groove which runs obliquely downwards and forwards, and lodges the nasopalatine nerve and vessels. The *superior border* is the thickest, and presents a deep furrow, bounded on each side by a projecting *ala*: the furrow fits over the rostrum of the sphenoid; the alae articulate with the sphenoidal conchae, the sphenoidal processes of the palatine bones and the vaginal process of the medial pterygoid plates. When the edge of the ala occupies the interval between the body of the sphenoid and the vaginal process, its lower surface takes part in the formation of the vomerovaginal canal (p. 306). The *inferior border* articulates with the nasal crest formed by the maxillae and palatine bones. The *anterior border* is

the longest; its upper half articulates with the perpendicular plate of the ethmoid, its lower is cleft for the reception of the inferior margin of the cartilage of the septum of the nose. The *posterior border* is free, concave, and separates the posterior nasal apertures; it is thick and bifid above, thin below. The anterior end of the vomer articulates with the posterior margin of the incisor crest of the maxillae and projects downwards between the incisive canals.

Ossification. At an early period the septum of the nose consists of a plate of cartilage. The superior part of this cartilage is ossified to form the perpendicular plate of the ethmoid; its anteroinferior portion persists as the septal cartilage, whilst the vomer is ossified in the strata of connective tissue covering each aspect of

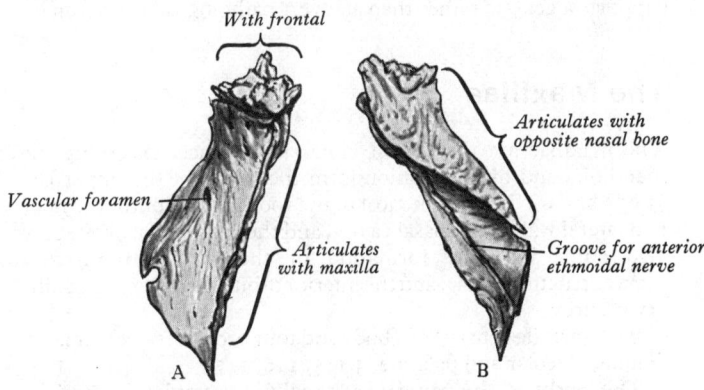

3.112A and B   The left nasal bone. (A) External aspect. (B) Internal aspect.

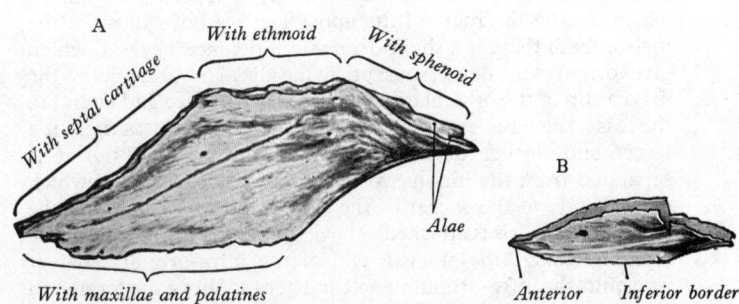

3.113A and B   The vomer. (A) Left lateral aspect. (B) The vomer at birth.   337

its postero-inferior part. About the eighth week of intrauterine life two centres of ossification appear, one on each side of the median plane. About the twelfth week these centres unite below the cartilage, and thus a deep groove is formed (3.113B) in which the cartilage of the septum of the nose is lodged. As growth proceeds the union of the bony lamellae extends upwards and forwards, and at the same time the intervening plate of cartilage undergoes absorption. By the age of puberty the lamellae are almost completely united, but evidence of the bilaminar origin of the bone is seen in the everted alae of its upper border and the groove on its anterior margin (3.113A).

The Sutural Bones

In addition to the usual centres of ossification of the cranial bones, others may occur in the course of the sutures, giving rise to what are often irregular, isolated, *sutural bones* (3.114). They occur most frequently in the course of the lambdoid suture, but are occasionally seen at the fontanelles, especially the posterior. The latter may represent a *pre-interparietal* element, a true *interparietal* element or some partially or wholly compounded variation of the two. An isolated bone at the lambda is sometimes termed the *Inca bone* or *Goethe's ossicle* (p. 322). One or more *pterion ossicles* or *epipteric bones*, sometimes exist between the sphenoidal angle of the parietal bone and the greater wing of the sphenoid bone. These bones vary much in size, but have a tendency to be more or less symmetrical on the two sides of the skull. Often, a sutural bone or bones appear to have little morphological significance, but there are notable exceptions (*vide infra* and p. 322). Their number is generally limited to two or three; but they may be present in great numbers in the skulls of hydrocephalic subjects. Sutural bones have therefore been associated with rapid cranial expansion but no causal relationships have been established For a detailed analysis of these, and many other *epigenetic variations* in the human cranium, derived from a study of 585 adult crania, consult Berry and Berry (1967). These authors discuss the anthropological and anatomical implications of being able to characterize populations genetically by such criteria; subsequently many quantitative data have been provided (Berry 1975). A recent report of the occurrence of sutural bones in a small series of fetal skulls (El-Najjar and Davson 1977) also supports a genetic rather than an overt pathological causation.

The Maxillae

The maxillae are the largest bones in the face, excepting the mandible, and by their union form the whole of the upper jaw (3.63A and B). They form most of the roof of the mouth, the floor and lateral wall of the nasal cavity, and the floor of the orbit; they also enter into the formation of the infratemporal and pterygopalatine fossae, and the inferior orbital and pterygomaxillary fissures.

Each maxilla consists of a body and four processes—zygomatic, frontal, alveolar and palatine (3.115, 116, 117).

The body of the maxilla is roughly pyramidal. It has four surfaces—anterior, infratemporal (or posterior), orbital and nasal—which enclose a large cavity, the maxillary sinus.

The *anterior surface* (3.115A and B) is directed forwards and laterally. Its lower part displays a number of slight elevations, which overlie the roots of the upper teeth. Above those of the incisor teeth there is a slight depression, the *incisive fossa*, which gives origin to the depressor septi; to the alveolar border below the fossa a slip of the orbicularis oris is attached; above and lateral to the fossa, the nasalis arises. Lateral to the incisive fossa there is a larger and deeper depression, named the *canine fossa*; it is separated from the incisive fossa by the *canine eminence*, which corresponds to the socket of the canine tooth; in the fossa the levator anguli oris is attached. Above the canine fossa is the *infra-orbital foramen*, the anterior end of the infra-orbital canal; it transmits the infra-orbital vessels and nerve. Above the foramen a sharp border marks the junction of the anterior and orbital surfaces. This border forms a small part of the circumference of

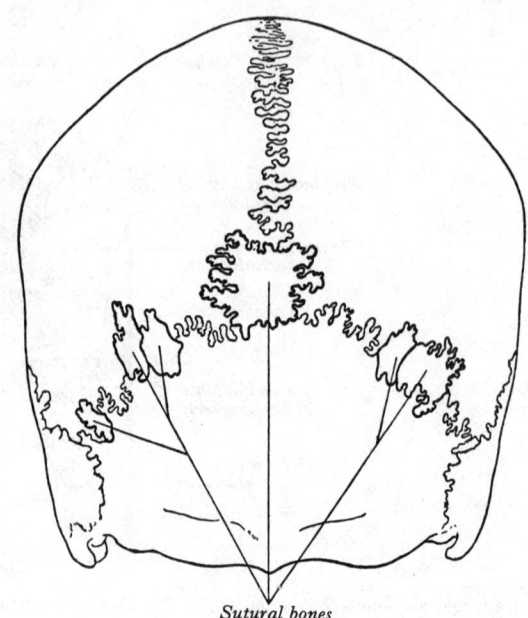

Sutural bones

3.114   Sutural bones in the lambdoid and sagittal sutures.

the orbital opening, and attached to it is the levator labii superioris. Medially the anterior surface terminates at a deeply concave border, the *nasal notch*, ending below in a pointed process which, with the corresponding process of the opposite maxilla, forms the *anterior nasal spine*. The anterior surface bordering the notch gives origin to the nasalis and depressor septi.

The *infratemporal surface* (3.115A) is convex, directed backwards and laterally, and is the anterior wall of the infratemporal fossa. It is separated from the anterior surface by the zygomatic process and by a ridge which runs upwards to that process from the socket of the first molar tooth. It presents near its centre the apertures of two or three *alveolar canals*, which contain the posterior superior alveolar vessels and nerves. At the lower and posterior part of this surface there is a round eminence, the *maxillary tuberosity*, which is roughened on the upper part of its medial aspect where it articulates with the pyramidal process of the palatine bone (3.115, 117); it gives origin to a few fibres of the medial pterygoid muscle and, in some cases, articulates with the lateral pterygoid plate of the sphenoid bone. Above this a smooth surface forms the anterior boundary of the pterygopalatine fossa and is grooved by the maxillary nerve; the groove for this nerve is directed laterally and slightly upwards and is continuous with the infra-orbital groove on the orbital surface.

The *orbital surface* (3.115A) is smooth and triangular, and forms the greater part of the floor of the orbit. Its *medial border* presents a depression anteriorly, the *lacrimal notch*, behind which it articulates from before backwards with the lacrimal bone, the orbital plate of the ethmoid, and the orbital process of the palatine bone (3.117). Its *posterior border* is smooth and rounded; it forms most of the anterior margin of the inferior orbital fissure, and its middle part is notched by the commencement of the infra-orbital groove. The *anterior border* forms a small part of the circumference of the orbital opening and is continuous medially with the lacrimal crest on the frontal process (p. 340). The *infra-orbital groove*, for the passage of the infra-orbital vessels and nerve, begins at the middle of the posterior border, where it is continuous with the groove near the upper edge of the posterior surface; it passes forwards and ends in the *infra-orbital canal*, which opens on the anterior surface of the bone just below the infra-orbital margin. Near its midpoint the canal gives off a small branch from its lateral side for the passage of the anterior superior alveolar nerve and vessels. This small canal, sometimes called the *canalis sinuosus* (Wood Jones 1939), passes forwards and downwards in the floor of the orbit lateral to the infra-orbital canal and then curves medially in the anterior wall of the maxillary sinus, passing below the infra-orbital foramen.

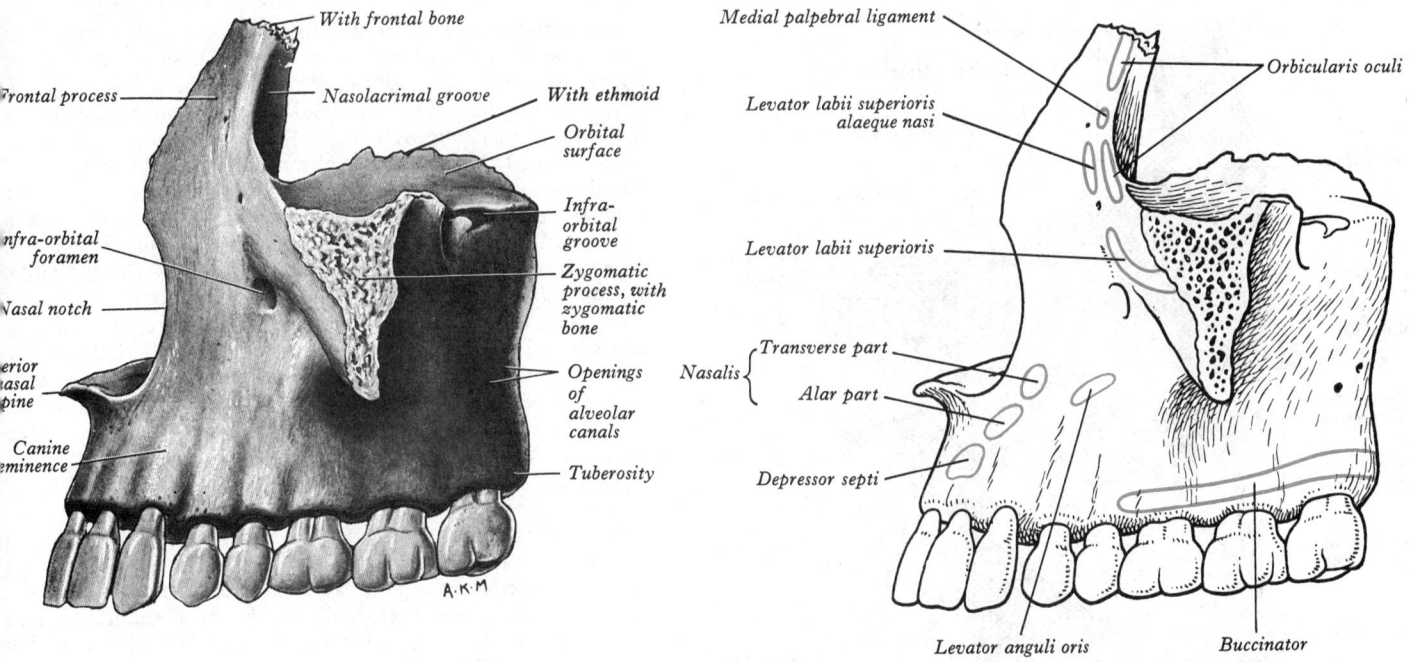

3.115A   The left maxilla. Lateral aspect.

3.115B   Outline of left maxilla, showing muscular attachments.

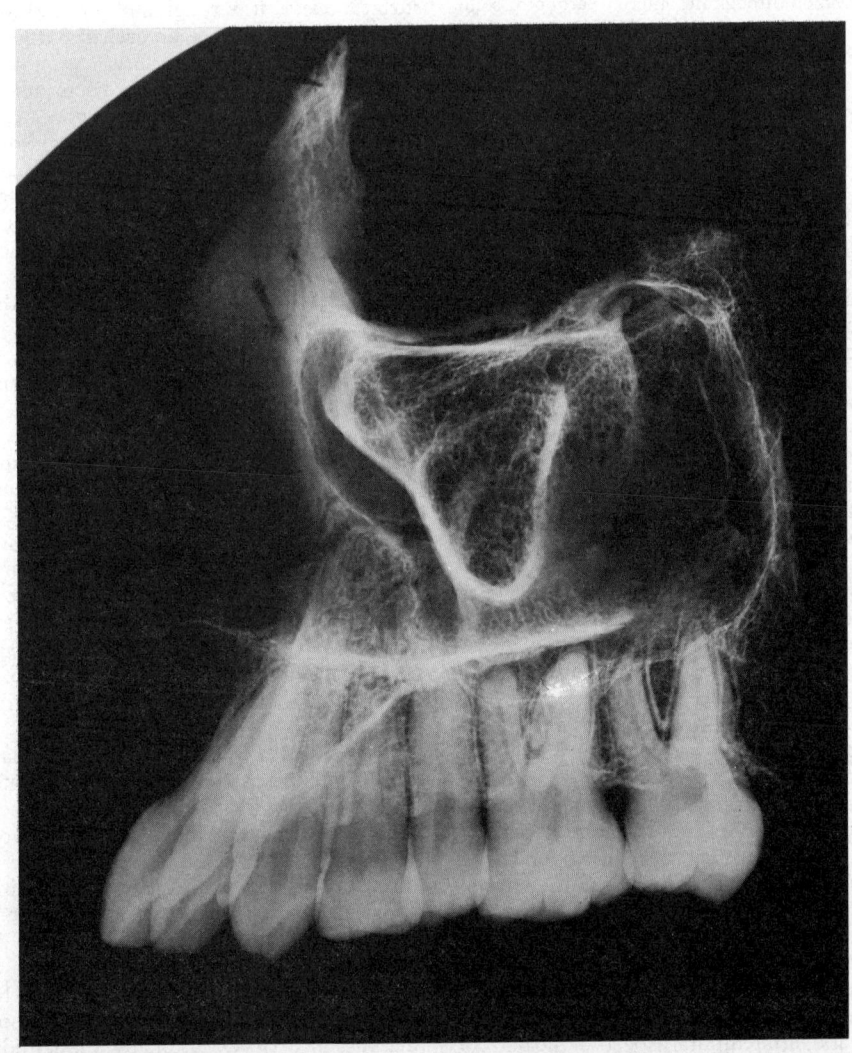

3.116 High resolution projection microradiograph of a maxilla (mediolateral view). Note particularly the distribution of compact and trabecular bone, the extent of the profile of the maxillary sinus and its relation to the dental alveoli, the compact bone of the *laminae durae* of the latter, the hard tissues, pulp cavities, root canals and foramina of the teeth.

Contrast the dense shells of compact bone in the frontal, zygomatic and palatine processes, with the fine filigree of trabecular bone in much of the alveolar process, interior of the zygomatic process, and the posterior wall of the sinus. (Kindly contributed by Dr. C. Buckland-Wright, Department of Anatomy, Guy's Hospital Medical School, London.)

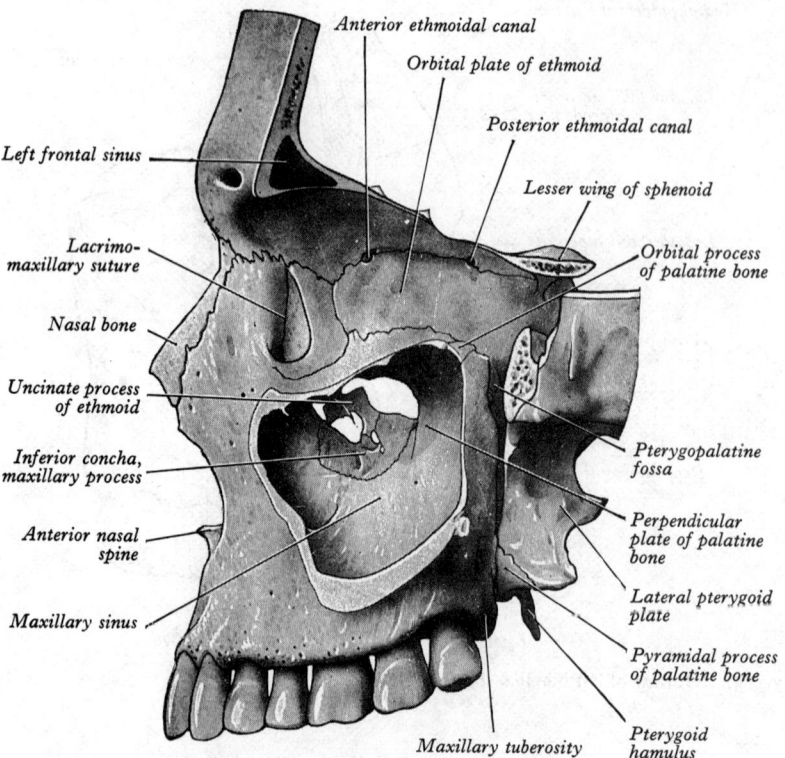

Left frontal sinus

Lacrimo-
maxillary suture

Nasal bone

Uncinate process
of ethmoid

Inferior concha,
maxillary process

Anterior nasal
spine

Maxillary sinus

Anterior ethmoidal canal

Orbital plate of ethmoid

Posterior ethmoidal canal

Lesser wing of sphenoid

Orbital process
of palatine bone

Pterygopalatine
fossa

Perpendicular
plate of palatine
bone

Lateral pterygoid
plate

Pyramidal process
of palatine bone

Maxillary tuberosity

Pterygoid
hamulus

3.117 An oblique sagittal section through the anterior part of the skull, showing the medial wall of the left orbit and the medial wall of the left maxillary sinus. The inferior concha is shown in *yellow* and the perpendicular plate of the palatine bone in *blue*.

Reaching the margin of the anterior nasal aperture just in front of the anterior end of the inferior concha, it then follows the lower margin of the aperture and opens at the side of the nasal septum in front of the incisive canal. At the medial and front part of the orbital surface, and lateral to the lacrimal groove, the origin of the inferior oblique muscle of the eyeball may be marked by a small depression.

The *nasal surface* (3.116) displays posterosuperiorly a large, irregular opening, the *maxillary hiatus*, which leads into the maxillary sinus. At the upper border of this aperture there are parts of air sinuses, which are completed by the ethmoid and lacrimal bones. Below the maxillary hiatus a smooth concave surface forms part of the inferior nasal meatus, and behind it there is a rough surface for articulation with the perpendicular plate of the palatine bone; this surface is traversed by a groove, which begins near the middle of the posterior border, runs obliquely downwards and forwards, and is converted into the *greater palatine canal* by the perpendicular plate of the palatine bone. In front of the maxillary hiatus a deep groove, continuous above with the groove on the lacrimal bone (p. 338), constitutes about two-thirds of the circumference of the nasolacrimal canal, the remaining one-third being formed by the descending part of the lacrimal bone and the lacrimal process of the inferior nasal concha (3.111); this canal opens into the inferior meatus (3.79) and contains the nasolacrimal duct. More anteriorly the bone has an oblique ridge, the *conchal crest*, for articulation with the inferior nasal concha. The shallow concavity below this ridge forms part of the inferior meatus, and the surface above the ridge, part of the atrium of the middle meatus.

The maxillary sinus (3.116, 117) is a large pyramidal cavity in the body of the maxilla. Its walls are thin and correspond to the orbital, alveolar, anterior and infratemporal aspects of the body of the bone. Its *apex*, directed laterally, extends into the zygomatic process and may reach the zygomatic bone itself; its *base*, which faces medially, is the lateral wall of the nasal cavity and presents the maxillary hiatus in the disarticulated bone. In the articulated skull this aperture is much reduced in size by the uncinate process

of the ethmoid and the descending part of the lacrimal bone above, the maxillary process of the inferior nasal concha below, and the perpendicular plate of the palatine bone behind (3.79, 117). The maxillary sinus communicates with the middle meatus of the nose, generally by two small apertures, one of which is usually closed by mucous membrane in life. Its *posterior wall* contains the *alveolar canals*, which transmit the posterior superior alveolar vessels and nerves to the molar teeth; these canals occasionally project as ridges into the maxillary sinus. The *floor* is formed by the alveolar process of the maxilla, and its lowest part is usually about 1·25 cm below the level of the floor of the nasal cavity. In a large proportion of bones, radiating septa of varying sizes spring from the floor of the sinus in the intervals between the adjacent teeth; in some cases the floor is perforated by the roots of the molar teeth (p. 1149). The infra-orbital canal usually projects into the sinus, a well-marked ridge extending from the *roof* to the *anterior wall*. The size of the cavity varies in different skulls, and even on the two sides of the same skull (p. 1149).

Applied Anatomy. Because of the extreme thinness of the walls of the sinus a tumour growing in it and encroaching upon the adjacent parts may push up the floor of the orbit and displace the eyeball, project into the nasal cavity, protrude forwards on to the cheek, or make its way backwards into the infratemporal fossa or downwards into the mouth. Extraction of molar teeth may damage the floor of the sinus.

The zygomatic process is an irregular, pyramidal projection, where the anterior, infratemporal and orbital surfaces converge. *In front* it forms part of the anterior surface of the body of the bone; *behind*, it is concave, and continuous with the infratemporal surface; *above*, it is rough and serrated for articulation with the zygomatic bone; *below*, an arched border separates the anterior from the infratemporal surface.

The frontal process projects upwards and backwards between the nasal and lacrimal bones (3.61, 118A). Its *lateral surface* (3.115A) is divided by a vertical ridge, the *anterior lacrimal crest*, which gives attachment to the medial palpebral ligament and is continuous below with the infra-orbital margin. At the junction of the crest with the orbital surface there is a small tubercle, a guide to the position of the lacrimal sac. The area in front of the lacrimal crest is smooth and merges below with the anterior surface of the body; a portion of the orbicularis oculi and the levator labii superioris alaeque nasi are attached here. Behind the anterior lacrimal crest there is a vertical groove which combines with a similar groove on the lacrimal bone to complete a fossa for the lacrimal sac.

The *medial surface* of the frontal process (3.116) is part of the lateral wall of the nasal cavity. A rough, uneven area at its upper part articulates with the ethmoid bone and closes the anterior ethmoidal air cells. Below this there is an oblique ridge, the *ethmoidal crest*, the posterior part of which articulates with the middle nasal concha, while the anterior underlies the *agger nasi*, a ridge anterior to the middle concha, on the lateral wall of the nasal cavity; the ethmoidal crest forms the upper limit of the atrium of the middle meatus of the nose. The *upper end* of the frontal process articulates with the nasal part of the frontal bone, the *anterior border* with the nasal bone, and the *posterior border* with the lacrimal bone.

The alveolar process of the maxilla is thick and arched, broader behind than in front, and excavated to form sockets for the reception of the roots of the teeth. These cavities are eight in number and vary in size and depth according to the teeth they contain. That for the canine tooth is the deepest; those for the molars are the widest, and are subdivided into three minor sockets by septa; those for the incisors and the second premolar are single; that for the first premolar is sometimes divided into two. The buccinator arises from the outer surface of this process, as far forwards as the first molar tooth. When the maxillae are articulated with each other, their alveolar processes together form the *alveolar arch*. Occasionally a longitudinal ridge, of variable prominence and roughness, the *maxillary torus*, is present on the palatal aspect of the alveolar process opposite the subcervical parts of the upper molar tooth sockets.

The palatine process of the maxilla, which is thick and strong, is horizontal and projects medially from the lowest part of ne nasal surface of the bone. It forms a considerable part of the loor of the nasal cavity and the roof of the mouth, and is much hicker in front than behind. Its *inferior surface* (**3**.72) is concave, ough and uneven, and forms, with the palatine process of the opposite bone, the anterior three-fourths of the bony palate. It is marked by numerous vascular foramina and displays depressions or the palatine glands; it is channelled at the posterior part of its ateral border by two grooves, which contain the greater palatine essels and nerve. When the two maxillae are articulated, a unnel-shaped depression, the *incisive fossa*, is seen in the median lane, immediately behind the incisor teeth. Caudal to the ncisive fossa, the midline palatal intermaxillary suture on its oral spect, although a little uneven, is usually relatively flat. However, in some cases its bony margins are raised, forming a ongitudinal prominent ridge, the *palatine torus* (see also p. 305).

bone appears above the canine fossa at about the sixth week of intrauterine life; two others have been ascribed to the anterior or *premaxillary* part of the bone, the so-called *os incisivum*. This corresponds, in position at least, with the premaxilla, a distinct and separate entity in most mammalian upper jaws, including those of primates.

Of the two 'premaxillary' centres, reputed to occur in the human embryonic upper jaw (Woo 1949; Noback and Moss 1953), the main one is said to appear above the incisor tooth germs in the seventh week. The second centre, sometimes called *paraseptal* or *prevomerine*, is considered to commence ossifying on the medial wall of the paraseptal cartilage, at the ventral margin of the nasal septum; it fuses almost at once with the palatal process of the maxilla. Classically, the bone formed by the main pre-maxillary centre is considered to be overgrown superficially by

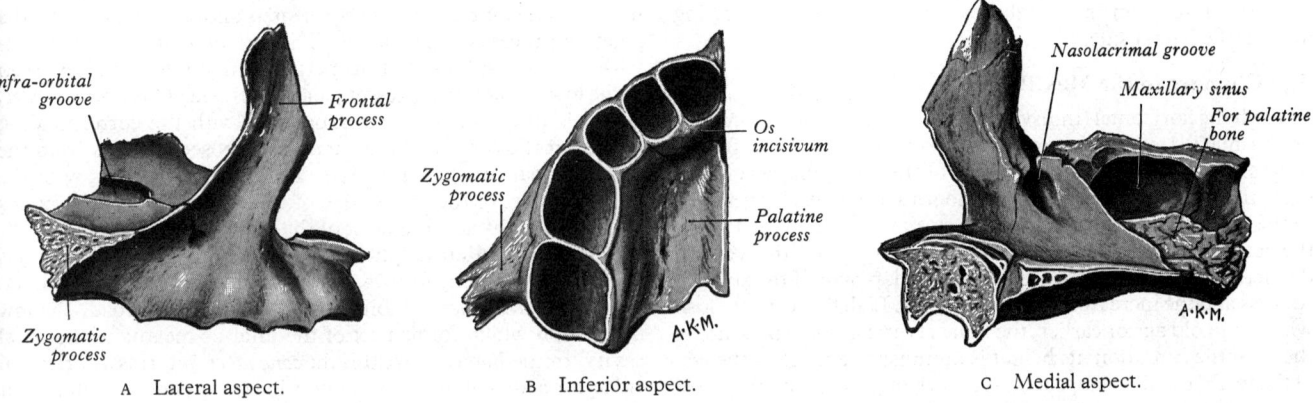

nfra-orbital groove — *Frontal process* — *Zygomatic process* — *Os incisivum* — *Palatine process* — *Nasolacrimal groove* — *Maxillary sinus* — *For palatine bone*

Zygomatic process

A·K·M.   A·K·M.

A   Lateral aspect.   B   Inferior aspect.   C   Medial aspect.

**.118   The right maxilla at birth. Note alveoli for deciduous teeth in B nd rough articular area for opposite maxilla in C.

Its incidence shows racial variation in modern and ancient populations (p. 348). It is rare in Burmese, for example, but is frequent in English crania (29 per cent in males, 48 per cent in females of a group of nineteenth-century Londoners). In the depths of the *incisive fossa* the orifices of two lateral canals are visible: they are named the *incisive canals*; each leads upwards into the corresponding half of the nasal cavity and through it pass the terminal branch of the greater palatine artery and the nasopalatine nerve. Occasionally there are two additional apertures in the median plane: the *anterior* and *posterior incisive foramina*; when present, these transmit the nasopalatine nerves, the left passing through the anterior, and the right through the posterior foramen. On the inferior surface of the palatine process, a delicate groove, sometimes termed the *incisive suture*, prominent in young skulls, may sometimes also be noticed in adults extending laterally and forwards from the incisive fossa to the interval between the lateral incisor and the canine teeth. The small volume of bone anterior to this supposed suture is the *os incisivum*, and has been considered for a long time to represent the premaxilla, which is a separate bone in most vertebrates, including primates. This classical view has been seriously challenged (*vide infra*). The *superior surface* of the palatine process is concave from side to side, smooth, and forms the greater part of the floor of the nasal cavity; close to the anterior part of its median margin the bone is traversed by its incisive canal. The *lateral border* of the process is fused with the rest of the bone. The *medial border*, thicker in front than behind, is raised into a ridge, the *nasal crest*, which, with the same ridge on the opposite maxilla, forms a groove for the reception of the vomer. The front of this ridge rises to a considerable height and is sometimes termed the *incisor crest* (**3**.116); it is prolonged forwards into a sharp process, which forms with the similar process of the opposite bone, the *anterior nasal spine*. The posterior border is serrated for junction with the horizontal plate of the palatine bone.

Ossification. The maxilla develops by ossification in a sheet of mesenchymal tissue superficial to the nasal capsule. Three centres of ossification have been described; one for the main mass of the

bone from the main mass of the maxilla, fusing along its anterior limit with the maxilla's alveolar process (this junction may remain discernible for a time as the *interalveolar suture of Jarmer*). This explains the fact that after the third month of intrauterine life there is no indication on the facial aspect of the upper jaw of a premaxilla (os incisivum). However, a suture, or what appears to be a suture, may be observed somewhat rarely on the floor of the nasal cavity, a little behind its anterior margin (Ashley Montagu 1953). This groove, when identifiable in the skull post-natally, passes medially on each side to the incisive canal. Furthermore, a suture line, or cleft, can always be seen at birth in the anterior region of the palate, diverging on each side from the incisive fossa. It runs into the septum between the lateral incisor and canine teeth (rarely between the canine and first premolar). This *palatal* indication of separation between the os incisivum and the rest of the maxilla may persist until the middle decades. Considered with the inconstant grooves in the floor of the nasal cavity mentioned above, these palatal features apparently delineate a separate element, anterior to the incisive fossa and canals, and forming parts of the alveoli of the incisor teeth. Being fused anteriorly with an overlap of bone derived from the maxillary centre, this os incisivum has no facial representation.

The regular occurrence of a premaxilla in other primates, as an undeniable separate osseous element, has naturally prompted attempts to identify its homologue in man. Despite early disagreements (Fawcett 1911), the above description of the development of the upper jaw has become the standard account, conveniently equating the os incisivum with the mammalian premaxilla. This argument depends primarily upon the supposed centres of ossification in the os incisivum. In this respect, a recent study of serial sections of the region in human embryos provides a denial, the evidence strongly supporting the view that ossification in the os incisivum is merely an extension of the maxillary centre (Wood *et al.* 1969). The lamina of bone developing from this displays, as in the case of the mandible, a complex shape, such that it appears in more than one place in some sections, suggesting multiple centres. It is claimed that the essential continuity of these can be clearly established by serial sectioning. This provides an agreeable simplification for the complexities of maxillary development. The extension of bone from the supposed initial 'premaxillary' centre to form frontal and palatine processes, can thus be ascribed to the main maxillary centre. Nevertheless, this simplification of the ossific process does not explain the various sutures, clefts and grooves which are held to delineate the os incisivum in the human upper jaw. Whether the pattern of pre-ossification mesenchymal condensations in the region has any relevant phylogenetic implication remains uncertain; the status of the premaxilla in man thus becomes problematical. There is strong evidence, at least, that it is not distinguished by any specific centre of ossification.

The maxillary sinus appears as a shallow groove (3.118c) on the nasal aspect of the bone about the fourth month of intrauterine life. The infra-orbital vessels and nerve are for a time in an open groove in the floor of the orbit, the anterior part of which is converted into the infra-orbital canal by a lamina of bone growing in from the lateral side.

Age Changes in the Maxilla

At birth the horizontal (transverse and anteroposterior) diameters of the maxilla are both greater than the vertical. The frontal process is well marked, but the body of the bone consists of little more than the alveolar process, the tooth sockets reaching almost to the floor of the orbit. The maxillary sinus is a mere furrow on the lateral wall of the nasal cavity. In the adult the vertical diameter is the greatest, owing to the development of the alveolar process and the increase in size of the sinus. If all the teeth are lost, whether in old age or earlier, the bone reverts in some measure to the infantile condition: its height is diminished and, after the loss of the teeth, the alveolar process is absorbed (p. 346) and the lower part of the bone contracted and reduced in thickness at the expense of the labial wall (p. 319). The differences in the way in which the alveolar processes are absorbed in the maxilla and in the mandible are of considerable practical importance in the provision of dentures.

The Palatine Bones

The palatine bones are in the posterior part of the nasal cavity, between the maxillae and the pterygoid processes of the sphenoid bone (3.79). Each assists in forming the floor and lateral wall of the nasal cavity, the roof of the mouth, and the floor of the orbit, parts of the pterygopalatine and pterygoid fossae, and the inferior orbital fissure.

The palatine bone bears some resemblance to the letter L,

consisting of horizontal and perpendicular plates, and three processes, *pyramidal* directed backwards, laterally and downwards from the junction of the horizontal and perpendicular plates, *orbital* and *sphenoidal*, which surmount the perpendicular plate and are separated by the deep sphenopalatine notch.

The horizontal plate of the palatine bone (3.72, 121) is quadrilateral, with two surfaces and four borders. The *nasal surface*, concave from side to side, forms the posterior part of the floor of the nasal cavity. The *palatine surface* forms, with the corresponding surface of the opposite bone, the posterior quarter of the bony palate; near its posterior margin there is often a curved ridge, the *palatine crest*. The *posterior border* is thin and concave; to it, and to the palatine surface as far forwards as the palatine crest, the expanded tendon of the tensor veli palatini is attached. The medial end of the posterior border is pointed and, when united with that of the opposite bone, forms a projecting process, the *posterior nasal spine*, which gives attachment to the musculus uvulae. The *anterior border* is serrated and articulates with the palatine process of the maxilla. The *lateral border* is continuous with the inferior border of the perpendicular plate and is marked at the lower end by the greater palatine groove. The *medial border*, which is thick and serrated, articulates with the corresponding border of the opposite bone, and the opposed borders from the posterior part of the *nasal crest*. This crest articulates with the posterior part of the lower edge of the vomer and is continuous anteriorly with the nasal crest of the maxillae.

The perpendicular plate of the palatine bone (3.120, 121) is thin and of an oblong form, and has two surfaces and four borders.

The *nasal surface* exhibits at its lower part a broad, shallow depression which forms part of the inferior meatus of the nasal cavity. Immediately above this the *conchal crest* forms a horizontal ridge for articulation with the inferior nasal concha; still higher there is a second broad, shallow depression, which forms part of the middle meatus and is limited above by the *ethmoidal crest* for articulation with the middle nasal concha. Above the ethmoidal crest there is a narrow, horizontal groove, which forms part of the superior nasal meatus.

The *maxillary surface* is rough and irregular throughout the greater part of its extent, for articulation with the nasal surface of the maxilla; its upper and posterior part is smooth and forms the medial wall of the pterygopalatine fossa; its front portion, which is also smooth, projects beyond the posterior border of the maxillary hiatus and forms the posterior part of the medial wall of the maxillary sinus (3.117). On the posterior part of the maxillary surface there is a deep vertical groove, named the *greater palatine groove*, which in the articulated skull is converted into the *greater palatine canal* by the maxilla; this canal transmits the greater palatine vessels and nerve.

The *anterior border* is thin and irregular; at the level of the conchal crest a pointed lamina projects forwards below and behind the maxillary process of the inferior nasal concha. It articulates with the latter and assists in forming the medial wall of the maxillary sinus (3.117). The *posterior border* (3.121) is serrated for articulation with the medial pterygoid plate of the sphenoid bone. This border is continuous above with the sphenoidal process; it expands below into the pyramidal process of the palatine bone. The *superior border* supports the orbital process in front and the sphenoidal process behind. These processes are separated by the *sphenopalatine notch*, which is converted into the *sphenopalatine foramen* by the inferior surface of the body of the sphenoid. In the articulated skull this foramen leads from the pterygopalatine fossa into the posterior part of the superior meatus of the nose, and transmits the sphenopalatine vessels and posterior superior nasal nerves. The *inferior border* is continuous with the lateral border of the horizontal plate and, in front of the pyramidal process, is marked by the lower end of the greater palatine groove.

The pyramidal process of the palatine bone projects backwards, laterally, and downwards from the junction of the horizontal and perpendicular plates of the bone, and fits into the angular interval between the lower ends of the pterygoid plates. On its *posterior surface* there is a smooth, grooved triangular area, limited on each side by a rough articular furrow. The furrows articulate with the pterygoid plates, while the grooved triangular

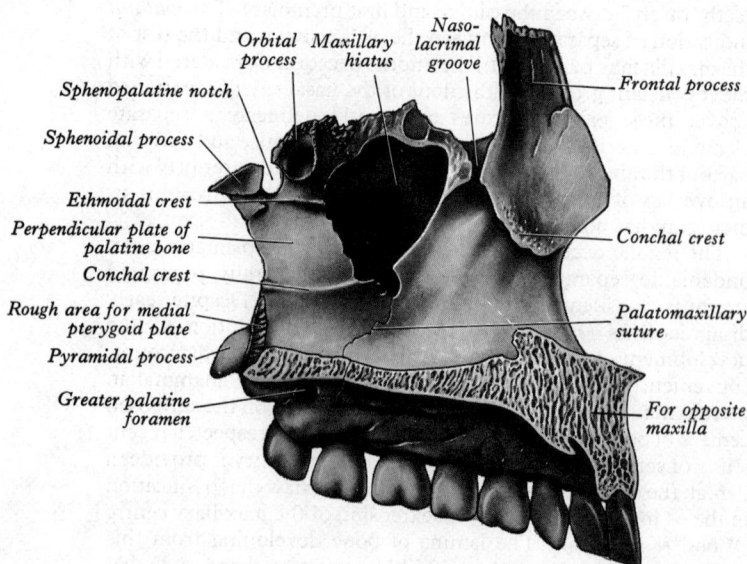

Orbital process Maxillary hiatus Naso-lacrimal groove

Sphenopalatine notch

Sphenoidal process

Ethmoidal crest

Perpendicular plate of palatine bone

Conchal crest

Rough area for medial pterygoid plate

Pyramidal process

Greater palatine foramen

Frontal process

Conchal crest

Palatomaxillary suture

For opposite maxilla

3.119 The left palatine bone in articulation with the left maxilla. Medial aspect.

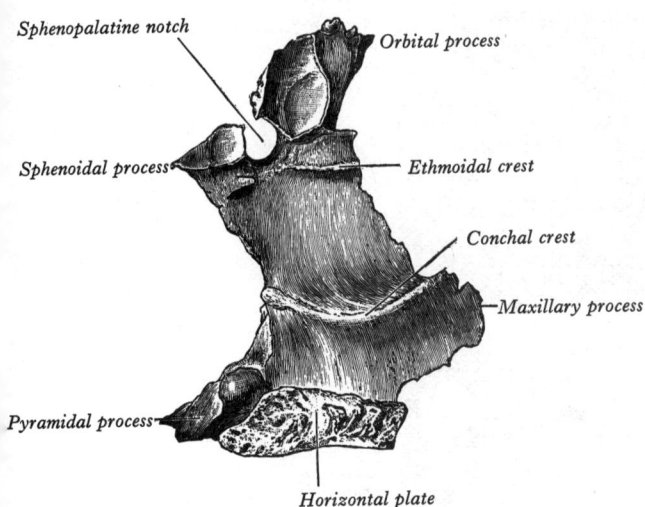

3.120 The left palatine bone. Medial aspect. (Enlarged.)

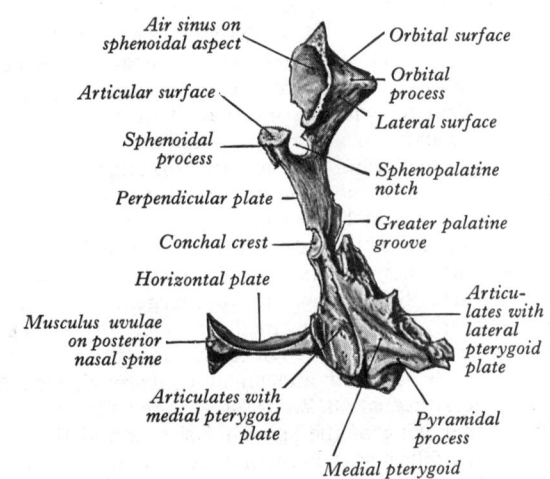

3.121 The right palatine bone. Posterior aspect.

area completes the lower part of the pterygoid fossa for attachment of some fibres of the medial pterygoid. The anterior part of the *lateral surface* is rough for articulation with the maxillary tuberosity; the posterior part consists of a smooth triangular area, which appears at the lower part of the infratemporal fossa between the maxillary tuberosity and the lateral pterygoid plate (3.70). The *inferior surface* of the pyramidal process, close to its union with the horizontal plate of the bone, presents the *lesser palatine foramina* for the transmission of the lesser palatine nerves and arteries (3.72).

The orbital process (3.120, 121) is directed upwards and laterally from the front of the perpendicular plate, to which it is joined by a constricted neck. It encloses an air sinus, and presents three articular and two non-articular surfaces. Of the articular surfaces, (1) the *anterior* or *maxillary*, of an oblong form, is directed forwards, laterally, and downwards, and articulates with the maxilla; (2) the *posterior* or *sphenoidal*, directed backwards, upwards, and medially, presents the opening of the air sinus, which usually communicates with the sphenoidal sinus; the margins of the opening articulate with the sphenoidal concha; (3) the *medial* or *ethmoidal* is directed medially and forwards, and articulates with the labyrinth of the ethmoid bone. In some cases, the air sinus opens on this surface and then communicates with the posterior ethmoidal air cells: more rarely it opens on the ethmoidal *and* sphenoidal surfaces, and them communicates with the posterior ethmoidal air cells and the sphenoidal sinus. Of the non-articular surfaces, (1) the *superior* or *orbital*, triangular in shape, is directed upwards and laterally, and forms the posterior part of the floor of the orbit (3.66); (2) the *lateral*, of an oblong

form, is directed towards the pterygopalatine fossa and is separated from the orbital surface by a rounded border, which forms the medial part of the lower margin of the inferior orbital fissure; the lower part of this surface may present a groove, directed laterally and upwards, which lodges the maxillary nerve and is continuous with the groove on the upper part of the posterior surface of the maxilla (p. 338). The border between the lateral and posterior surfaces is prolonged downwards as the anterior boundary of the sphenopalatine notch.

The sphenoidal process (3.120, 121) is a thin, compressed plate, smaller and on a lower level than the orbital process; it is directed upwards and medially. Its *superior surface* articulates with the under surface of the sphenoidal concha and the root of the medial pterygoid plate; it presents a groove which contributes to the formation of the palatovaginal canal. The *inferomedial surface* is concave and forms a small part of the roof and lateral wall of the nasal cavity. The posterior part of the *lateral surface* articulates with the medial pterygoid plate; the anterior part is smooth and forms a portion of the medial wall of the pterygopalatine fossa. The *posterior border* is rough and articulates with the vaginal process of the medial pterygoid plate. The *anterior border* forms the posterior boundary of the sphenopalatine notch. The *medial border* articulates with the ala of the vomer.

The orbital and sphenoidal processes are separated from each other by the *sphenopalatine notch*, which is converted into the sphenopalatine foramen by the under surface of the body of the sphenoid; sometimes the two processes are united by bone to form a foramen.

Ossification. The palatine bone is ossified in a membrane of connective tissue from one centre, which appears during the eighth week of intrauterine life in the perpendicular plate. From this point ossification spreads upwards into the orbital and sphenoidal processes, medially into the horizontal plate, and downwards into the pyramidal process. At the time of birth the height of the perpendicular plate is about equal to the transverse width of the horizontal plate, whereas in the adult it measures nearly twice as much, a change in proportions which accords with the growth changes in the adjoining maxilla.

The Zygomatic Bones

The zygomatic bones are superolateral elements in the facial skeleton. Each forms the prominence of the cheek, and contributes to the formation of the lateral wall and floor of the orbit and to the walls of the temporal and infratemporal fossae (3.122).

The zygomatic bone is roughly quadrangular in shape, with a flange-like anteromedial projection, the frontal process. It has three surfaces, five borders and two processes.

The *lateral surface* (3.122, 123A), directed laterally and forwards, is convex and pierced near its orbital border by the *zygomaticofacial foramen* (often double), for the passage of the zygomaticofacial nerve and vessels; below this foramen a slight elevation provides attachment for the zygomaticus minor and more posteriorly the zygomaticus major takes origin. (The zygomaticofacial foramen is also absent in a proportion of cases, *see* p. 299.) The *temporal surface* (3.123B), which is directed medially and backwards, is concave. It presents anteriorly a roughened area for articulation with the maxilla, and posteriorly a smooth, concave area which extends upwards on the posterior aspect of the frontal process as the anterior boundary of the temporal fossa. It also extends backwards on the medial aspect of temporal process to form an incomplete lateral wall for the infratemporal fossa. The *zygomaticotemporal foramen*, for the nerve of the same name, pierces this surface near the base of the frontal process. The *orbital surface* (3.123B), smooth and concave, forms the anterolateral part of the floor and the adjoining part of the lateral wall of the orbit, extending upwards on to the medial aspect of the frontal process. It usually presents the orifices of two canals, the *zygomatico-orbital foramina*, one of which leads to the zygomaticofacial and the other to the zygomaticotemporal foramen.

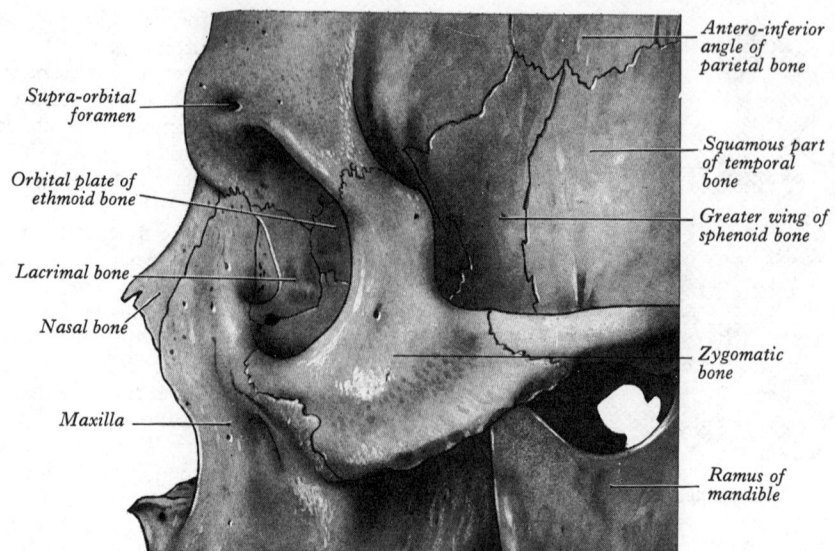

Supra-orbital foramen

Orbital plate of ethmoid bone

Lacrimal bone

Nasal bone

Maxilla

Antero-inferior angle of parietal bone

Squamous part of temporal bone

Greater wing of sphenoid bone

Zygomatic bone

Ramus of mandible

3.122   The left zygomatic bone *in situ*.

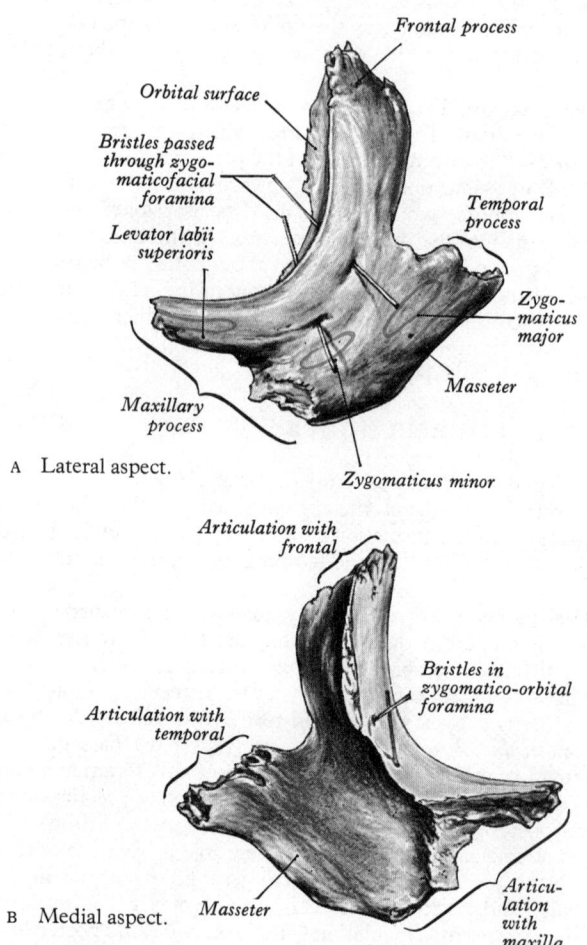

Frontal process

Orbital surface

Bristles passed through zygo-maticofacial foramina

Levator labii superioris

Maxillary process

Temporal process

Zygo-maticus major

Masseter

A   Lateral aspect.

Zygomaticus minor

Articulation with frontal

Articulation with temporal

Bristles in zygomatico-orbital foramina

Masseter

Articulation with maxilla

B   Medial aspect.

3.123   The left zygomatic bone, showing muscular attachments.

The *anterosuperior* or *orbital border* is smooth and concave, and forms a considerable part of the circumference of the orbital opening, below and on the lateral side. It separates the orbital from the lateral surface. The *antero-inferior* or *maxillary border* is rough and articulates with the maxilla; its medial extremity is pointed and lies above the infra-orbital foramen; near the orbital margin a part of the levator labii superioris is attached to it. The *posterosuperior* or *temporal border* is curved, being convex in its upper and concave in its lower part; it is continuous above with the posterior border of the frontal process and below with the upper border of the zygomatic arch; it gives attachment to the temporal fascia. A little below the frontozygomatic suture this border frequently presents a small, rounded projection, the *marginal tubercle*, which can be felt easily through the skin. The *postero-inferior border* affords attachment by its rough edge to the masseter. The *posteromedial border* is serrated for articulation with the greater wing of the sphenoid above, and the orbital surface of the maxilla below. Between these two serrated portions there is usually a short, concave, non-articular part, which forms the lateral boundary of the inferior orbital fissure. This non-articular part is sometimes absent, and the fissure is then completed by the junction of the maxilla and the sphenoid, or by the interposition of a small sutural bone in the angular interval between them.

The *frontal process* is thick and serrated; it articulates above with the zygomatic process of the frontal bone and behind with the greater wing of the sphenoid. On its orbital aspect, just within the orbital opening and about 1 cm below the frontozygomatic suture, there is a tubercle of varying size and form in 95 per cent of skulls (Whitnall 1921). Attached to this are the lateral check ligament of the rectus lateralis, part of the aponeurosis of the levator palpebrae superioris, the suspensory ligament of the eyeball and the lateral palpebral ligament (p. 1178). The *temporal process* is directed backwards and ends in an oblique, serrated margin which articulates with the zygomatic process of the temporal bone, and helps to form the zygomatic arch.

Ossification. The zygomatic bone is ossified from one centre, which appears in a fibrous tissue precursor about the eighth week of intrauterine life. The bone is sometimes divided by a horizontal suture into an upper larger, and a lower smaller division.

CRANIAL CHARACTERISTICS AT DIFFERENT AGES

The skull at birth is large in proportion to other parts of the skeleton; but the facial part of the cranium is relatively small, forming but one-eighth of the neonatal cranium, compared with a half in adult life. The smallness of the face at birth is due to the rudimentary condition of the mandible and maxillae, the non-eruption of the teeth and the small size of the maxillary sinus and the nasal cavity. The latter lies almost entirely between the orbits, the lower border of the piriform aperture of the nose being only a little below the orbital floor. On the other hand, the large size of the calvaria—especially the cranial vault—is related to the recocious growth of the brain. At this stage the base of the skull is elatively short and narrow, and though the middle and internal ear re almost adult in size, the petrous part of the temporal bone is not et approaching its adult length as measured from its apex to lateral urface.

The bones of the cranial vault are smooth and unilaminar and without diploë. The frontal and parietal tuberosities are rominent, and from above the greatest width of the skull is seen o be between the parietal tuberosities (p. 348). The glabella, uperciliary arches and mastoid processes are not developed.

Ossification of the skull bones is incomplete, many consisting of everal bony elements united by fibrous tissue or cartilage. The os incisivum' is continuous with the maxilla (p. 341) and the pre- nd post-sphenoids have united just before birth, but the two halves of the frontal bone and mandible are separate, as are the quamous, lateral and basilar parts of the occipital bone. The econd centre for the styloid process (stylohyal) has not appeared nd the bony elements of the temporal bone are separate except or the commencing union of the tympanic with the petrous and quamous parts. Areas of the fibrous tissue membrane, which orms the primitive cranial vault before ossification begins, are till unossified at the angles of the parietal bone. These *fonticuli* or *fontanelles* are six in number; two, the *anterior* and *posterior*, are in he median plane, and two pairs, the *sphenoidal* and *mastoid*, appear on each side.

The *anterior fontanelle* (**3**.124A), the largest, is at the junction of he sagittal, coronal and frontal sutures; it is diamond-shaped and measures about 4 cm in anteroposterior and 2·5 cm in transverse diameters. The *posterior fontanelle* (**3**.124A) is triangular and ituated at the junction of the sagittal and lambdoid sutures. The *sphenoidal* (anterolateral) and *mastoid* (posterolateral) *fontanelles* (**3**.124B) are small and irregular in shape, and lie respectively at the sphenoidal and mastoid angles of the parietal bones.

At birth the orbits are large and the germs of the developing teeth lie close to their orbital floors. The temporal bone shows marked differences from that of the adult. While the internal ear, tympanic cavity, auditory ossicles and mastoid antrum are almost adult in size, the tympanic plate is represented by an incomplete ring, and the mastoid process has not commenced to develop. The external acoustic meatus is short and straight and its bony part unossified and represented by a fibrocartilaginous plate. The external aspect of the tympanic membrane faces caudally rather than laterally, coinciding with the contour of the skull base at this point. Owing to the absence of a mastoid process, the stylomastoid foramen is exposed on the lateral surface of the skull. The styloid process has not fused with the temporal bone and the mandibular fossa is flat and more laterally placed; the articular tubercle has not developed. The paranasal sinuses are rudimen- tary or absent, only the early maxillary sinuses being usually identifiable.

During birth the skull is altered in shape or moulded by slow compression and the part of the scalp which lies more centrally in the birth canal is often temporarily swollen and oedematous due to interference with the venous return. The swollen area is called the *caput succedaneum*. Owing to the presence of the fontanelles and the width and lack of interlocking of sutures, the bones of the skull cranial vault are able to overlap to a small extent. The skull is compressed in one plane and shows compensatory elongation at right angles to this. These effects disappear within a week of birth.

Postnatal growth of the skull. Though co-ordinated, the postnatal growth of the calvaria and facial skeleton proceed at different rates and over different periods, that of the cranial cavity being related to the growth of the brain and that of the facial skeleton to the development of dentition, muscles of mastication and tongue. Growth of the base of the skull does not proceed at the same rate as that of the vault; in the following account the three parts, vault, base and facial skeleton, will be considered separately. The anterior part of the cranial base is thus a zone of interaction between facial and cranial growth (Brash 1924; Scott 1967).

The growth of the vault proceeds rapidly during the first year and thereafter more slowly to the seventh year, by which time it has almost reached adult dimensions. Over most of this period its expansion is, broadly speaking, concentric and its form is

determined in the early part of the first year, after which time it remains largely unaltered (Weinmann and Sicher 1955; Scott and Symon 1971; Brodie 1941; Hoyte 1966). That the *shape* of the vault is not directly related to the growth of the brain but to independent genetic factors is to some extent supported by the great range of cranial indices and the association of certain forms of head shape with racial groups. During the first and early part of the second years the growth of the vault is achieved principally by ossification at the apposed margins of the bones, which are covered by an osteogenic layer (p. 420). This expansion is necessarily accompanied by some accretion and absorption of bone at the surfaces in order to adapt them to the continually altering curvatures of the vault. Growth in breadth occurs at the sagittal suture, the sutures bordering the greater wing of the sphenoid, the occipitomastoid sutures and the petro-occipital cartilaginous joints in the base of the skull. Growth in height occurs at the frontozygomatic suture, the pterion (p. 302), the squamosal suture and the asterion (p. 347). During this period the fontanelles are closed by growth and expansion of the bones which surround them, but sometimes they are the sites of separate

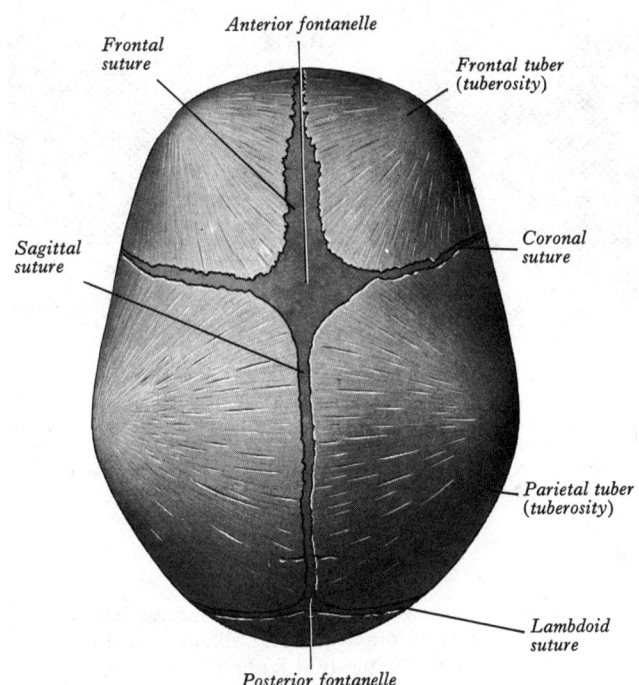

A   Superior aspect.

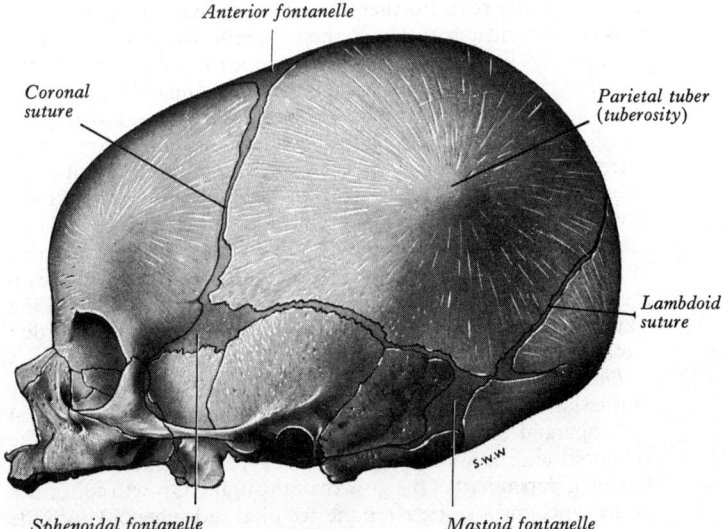

B   Lateral aspect.

3.124   The skull of a newborn infant.

345

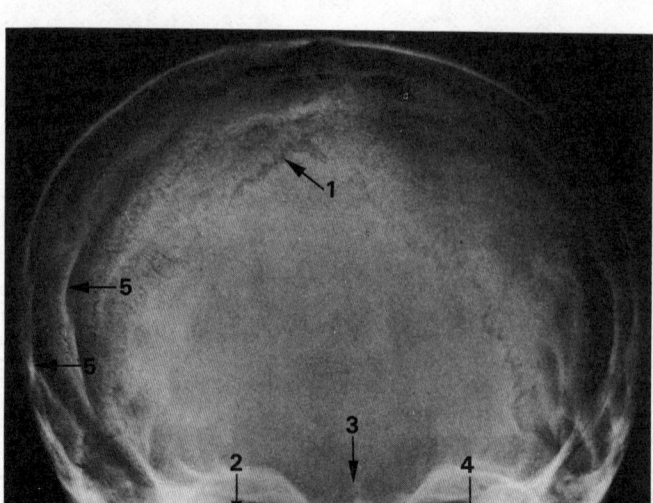

3.125 Radiograph of child's skull, aged 7. Occipitofrontal view. 1. Lambdoid suture. 2. Petrous portion of right temporal bone, seen through the cavity of the orbit. 3. Crista galli of the ethmoid bone. 4. Fracture through petrous portion of left temporal bone, seen through the cavity of the orbit. 5. Impressions for cerebral gyri. 6. Second permanent molar tooth, not yet erupted.

centres of ossification which develop into sutural bones. The sphenoidal and posterior fontanelles are obliterated within two or three months of birth; the mastoid fontanelle is usually closed about the end of the first year and the anterior fontanelle about the middle of the second year. By the second year the bones of the vault have interlocked at the sutures, a process which commences early in the first year. Further expansion of the vault is principally by accretion and absorption on their superficial and deep surfaces respectively (Ford 1956). At the same time the bones thicken, but not uniformly. At birth the cranial vault is unilaminar, the two tables and the intervening diploë appearing about the fourth year, the differentiation reaching its maxim about thirty-five years, when diploic veins form a characteristic and prominent feature in radiographs. The thickness of the bones of the vault and the development of muscular markings on the outer table are related to the development of the muscles of mastication and the posterior muscles of the neck. The mastoid process forms a visible bulge on the surface in the second year and is invaded by air cells in the sixth year. Galli, Ottaviani, and Galli (1976) have recorded metrical studies of the process.

The growth of the base is responsible for much of the lengthening of the skull and mostly occurs at the cartilaginous joints between the sphenoid and ethmoid on the one hand and between the sphenoid and occipital on the other, especially the latter. It is largely independent of the growth of the brain. Growth continues at the synchondrosis between the occipital and sphenoid until the eighteenth to twenty-fifth year; this prolonged period is related to the continued expansion of the jaws to accommodate the erupting teeth, provision of room for the muscles of mastication and

growth of the nasopharynx. Doubts have been expresse regarding this dating; there is some evidence that growth ma cease at about 15 years (Latham 1966). A pubertal growth spu has been ascribed to both sexes, occurring about two years earlie in females. The same study indicated considerable post-puberta growth, up to 17·5 years in males (Roche and Lewis 1974). For recent review of the literature on basal growth consult Hoyt (1975), and for labelling studies of growth in craniofacia cartilages (in rats) *see* Kvinnsland and Kvinnsland (1975).

The growth of the face occupies a period long in comparison wit that of the calvaria. Much of our knowledge has been obtaine from studies of serial lateral radiographs (Salymann 1961).

The ethmoid bone, orbital cavities and upper part of the nasa cavities have almost completed their growth by the seventh year Growth of the orbits and upper nasal region is achieved by sutura growth in their walls with deposition of bone on the facial aspect of the orbital margins. The maxilla is carried downwards an forwards by the expansion of the orbits, growth of the nasa septum and sutural growth, especially at the fontanelles and at th zygomaticomaxillary and pterygomaxillary sutures. In the firs year, growth in width occurs at the midpalatal sutures and th symphysis menti in association with growth at the internasal an frontal sutures, but growth at these sutures is considerabl diminished or even terminated when the symphysis menti an frontal sutures are closed during the first few years of life, eve though the midpalatal suture persists until mature years. Muc facial growth occurs during this period; it continues, however, t puberty and later, in association with the eruption of th permanent dentition. After the termination of sutural growth around the end of the second year, expansion of the facial skeleto occurs by surface accretion on the face, alveolar processes an palate associated with resorption in the walls of the maxillar sinus, upper surface of the hard palate and inner surface of th alveolar process. Associated growth and divergence of th pterygoid processes occurs due to deposition and resorption o bone of their surfaces. The growth of the mandible is correlate with that of the facial skeleton and is described on pp. 317–319.

Obliteration of the sutures of the vault of the skull takes place as age advances. It may commence between the ages of 30 and 40 on the inner surface, and about ten years later on the outer surface of the skull, but the times at which the sutures close are subject to great variations (Todd and Lyon 1924, 1925; Abbie 1950; Singer 1953). Obliteration usually occurs first at the bregma, extending into the sagittal, coronal and lambdoid sutures, in that order.

In old age the skull generally becomes thinner and lighter, but in a small proportion of cases it increases in thickness and weight. The most striking feature of the senile skull is the diminution in the size of the mandible and maxillae consequent on the loss of the teeth and the absorption of the alveolar processes. This is associated with a marked reduction in the vertical measurement of the face and with an alteration in the angles of the mandible (**3.**83) already described (pp. 317–319).

Sexual and Racial Differences in the Cranium

Until the age of puberty there is little difference between the skulls of the two sexes. The skull of an adult female is as a rule a little lighter and smaller, and its capacity is about 10 per cent less than that of the male. Its walls are thinner and its muscular ridges less marked; the glabella, superciliary arches, and mastoid processes are less prominent, and the corresponding air sinuses are smaller. The tympanic part of the temporal bone is smaller and its margins less roughened than in the male. The upper margin of the orbit is sharp, the forehead vertical, the frontal and parietal tuberosities are prominent, and the vault somewhat flattened. The contour of the face is rounder, the facial bones are smoother, and the mandible and maxillae and their contained teeth smaller. Speaking generally, more of the childhood characteristics are retained in the skull of the adult female. A well-marked male or female skull can easily be recognized as such, but in some skulls the characteristics are so indistinct that the determination of the sex may be difficult or impossible.

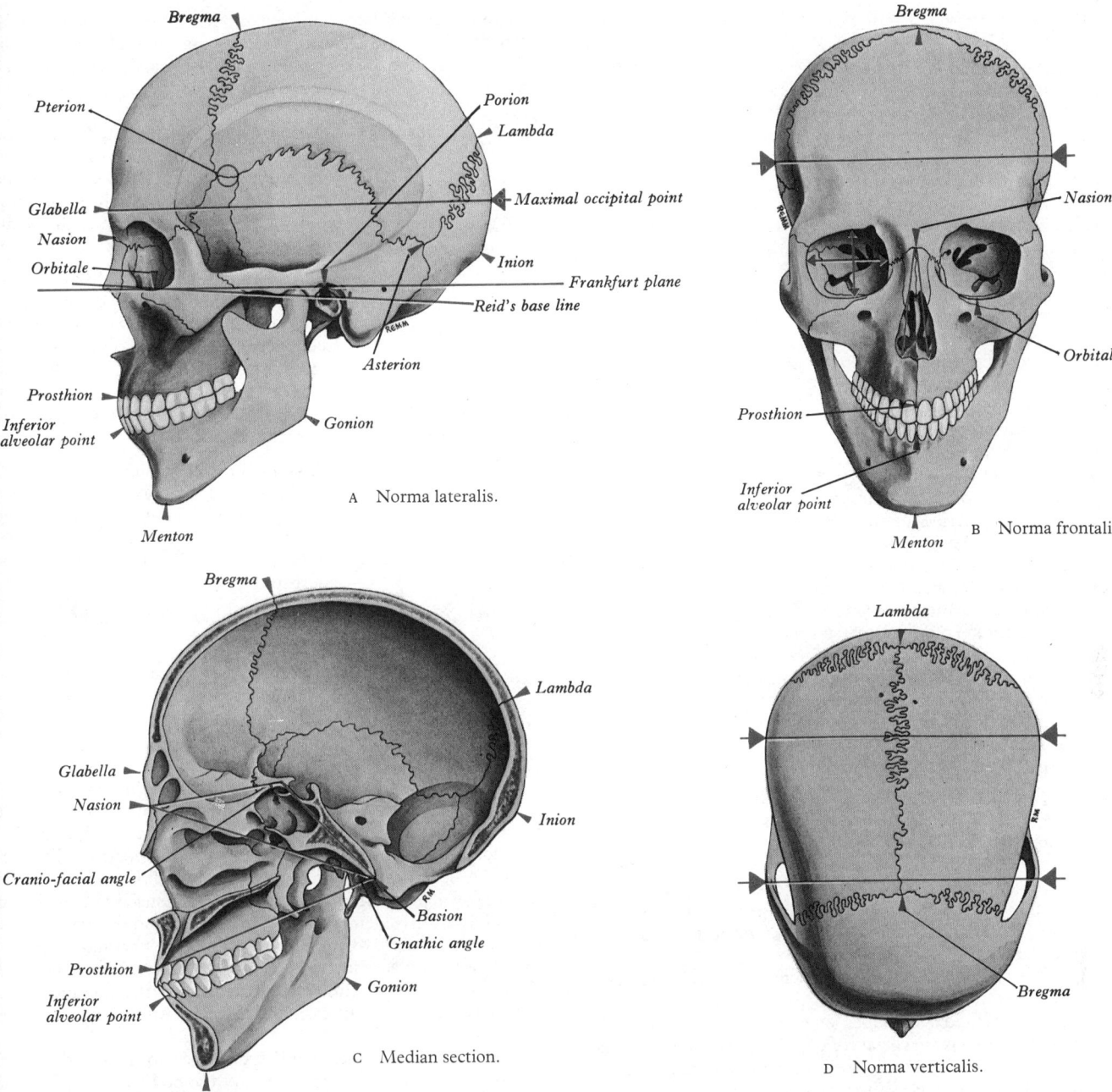

A   Norma lateralis.

B   Norma frontalis.

C   Median section.

D   Norma verticalis.

3.126A These diagrams illustrate the cranial points used, by international agreement, in making linear and certain angular measurements in anthropometry. In all four views the skull is in the standard orientation, that is, with the Frankfurt plane as a horizontal. The point at which the cranio-facial angle is measured (**3.**126C) is not named; it corresponds to the midpoint of the chiasmal groove.

Sexual dimorphism is generally less marked in mankind than in some other primates. This is to be associated with the *paedomorphic* tendency of the human stock, not only in females but also males being less divergent in adult development, from their own juvenile form, than is the case among some other primates, especially in the males (Abbie 1952; Schultz 1956). Exaggerated sexual differences in the skull occur in most of the anthropoid apes and some monkeys, in the form of enlarged canines in the male, together with associated jaw development, accentuated muscular ridges, and in the total size of the skull. These excessive developments are typified in the gorilla, orang-utan, and in many species of baboon, in which the male also exceeds the female considerably in general stature and physique. Even in extinct species of mankind sexual dimorphism appears—on the slender fossil evidence available—to have been less than in other primates. In modern man the differences are further reduced, but the degree of divergence between the sexes in cranial dimensions and proportions varies in different racial groups. These sexual differences have attracted less exact evaluation than

have general studies of ethnic variation. In both cases, assessment depends upon observation of two kinds of feature: (*a*) those which cannot be measured, or have so far not been subjected to a satisfactory technique of quantitative study (e.g. size of mastoid process, prominence of chin), and (*b*) those which are customarily expressed as actual measurements or indices (e.g. cranial capacity, orbital index). Examples of the former category in distinguishing sex have been cited above and more exhaustive lists can be consulted in appropriate texts (Keen 1950). The use of such features in judging sex or race in the case of isolated crania is much dependent upon the observer's experience; but where a number of crania, known to be from a single ethnic group, are available, both assessments are more certain. Distinction of the three major races by non-metrical morphological traits can be effected with some confidence (Todd and Tracey 1930). The incidences of a number of variations have been observed in a number of racial groups; these include metopism and other sutural variants (Berry and Berry 1967), and are of considerable ethnological but lesser forensic interest. Berry (1975) has made a special study of *non-*

metrical human cranial characteristics; these include the presence or absence of the following:

1. Highest nuchal line (p. 361).
2. Os suturale at lambda (pp. 322, 328).
3. Ossa suturalia in lambdoid suture (pp. 332, 325).
4. Parietal foramen (p. 298).
5. Os suturale at bregma (p. 338).
6. Frontal suture (metopism) (p. 334).
7. Ossa suturalia in coronal suture (p. 332).
8. Epipteric os suturale (p.338).
9. Frontotemporal articulation (p. 302).
10. Parietomastoid os suturale (p. 302).
11. Os suturale at asterion (p. 302).
12. Tympanic foramen of Hüschke (p. 331).
13. Extrasutural mastoid foramen (p. 328).
14. Absence of mastoid foramen (p. 328).
15. Patent condylar canal (p. 332).
16. Double condylar facet (p. 322).
17. Precondylar tubercle (p. 332).
18. Double hypoglossal canal (p. 322).
19. Incomplete foramen ovale (p. 305).
20. Incomplete foramen spinosum (p. 307).
21. Accessory palatine foramina (p. 304).
22. Maxillary torus (p. 305).
23. Palatine torus (pp. 305, 341).
24. Absence of zygomaticofacial foramen (pp. 299, 343).
25. Supra-orbital foramen or incisure (p. 333).
26. Frontal foramen or incisure (p. 333).
27. Extrasutural anterior ethmoid foramen (p. 301).
28. Absence of posterior ethmoid foramen (p. 301).
29. Accessory infra-orbital foramen (p. 338).
30. Interparietal (Inca) bone (pp. 322, 338).
31. Paramastoid process (p. 332).
32. Auditory torus (p. 330).

Although many observers have studied these and other cranial variants (e.g. the mandibular torus, p. 316), few of these studies have been quantitative. The literature has been reviewed by Berry and Berry (1967) and Berry (1975), who also provide most extensive statistical data, and have assessed which cranial variations exhibit racial (and sometimes sexual) correlation. The interested reader should also consult Czarnetzki (1971), and Hjarno *et al.* (1974).

To achieve more objective judgement of the racial affinities of crania, metrical studies have long been practised, and internationally accepted techniques of *craniometry* have led to the accumulation of a large corpus of comparable data for the males, and to a lesser extent the females, of many ethnic groups. The number of standard measurements used is considerable, and only the major dimensions and examples of the indices derived from them can be included here (**3.**126). The calvarial part of the skull is characterized as follows:

A. *Maximal cranial length* from the glabella to the furthest point at the occiput.
B. *Maximal cranial breadth* greatest breadth, measured at right angles to the median plane.
C. *Cranial height* from the basion (median point on the anterior rim of the foramen magnum) to the bregma.

All measurements are made to the nearest millimetre. From these three dimensions, three indices are calculated: B/A, C/A, and C/B, and these are expressed as percentages.

The breadth/length ratio is the *cranial index* (or *cephalic index* if measured in the living subject). Its recorded range of variation is high, as with other skull or head indices. In all of these the range of values encountered is arbitrarily divided into several steps, usually covering 5 per cent sections of the total range, and to each of the steps an appropriate term is applied. For example:

(*Maximal breadth*/*Maximal length*) × 100 = Cranial index or Cephalic index

Up to 74·9	= Dolichocranic or Dolichocephalic
75·0–79·9	= Mesocranic or Mesocephalic
80·0–84·9	= Brachycranic or Brachycephalic

The distinction between measurement of dried crania and of living subjects, in the form of the two groups of terms, is not rigidly adhered to. The 'cephalic' terms are more commonly used, and often for both purposes. Other indices in common use are:

(*a*) *Total facial index* = (Nasion–Gnathion height/Bizygomatic breadth) × 100. The nasion is the point where the internasal suture meets the frontal bone. The gnathion is the mid-point of the lower border of the mandible.

(*b*) *Upper facial index* = (Nasion–Prosthion length/Bizygomatic breadth) × 100. The prosthion is the midpoint of the alveolar rim of the maxilla, between the two central incisors. The bizygomatic breadth is the greatest distance that can be found by trial between the two zygomatic arches on their external aspects.

(*c*) *Nasal index* = (Nasal breadth/Nasal height) × 100. The breadth is the maximum across the nasal aperture, and the height is from the nasion to the mean between the two lowest points on the lower border of the aperture.

(*d*) *Orbital index* = (Maximal orbital height/Maximal orbital breadth) × 100.

(*e*) *Palatal index* = (Maximal palatal breadth/Maximal palatal length) × 100.

(*f*) *Gnathic index* = (Basion–Prosthion/Basion–Nasion) × 100.

All these indices provide a system for *metrical* description of sizes and proportions of cranial features, in place of subjective impressions. If a given racial group is said to be dolichocephalic, this does not (or should not) imply a vague comparison, but that the index lies within certain numerical limits. It is possible to state that the orbital opening appears rounder in females than males, as it does, but this statement can be quantified and is then far more useful. Again, the jaws project forwards somewhat more in some races than others, a fact which can be detected by inspection alone, when the condition is pronounced; the gnathic index permits a numerical expression of this trait, and comparisons are then possible.

The following measurements are applied to the mandible: (*a*) height of the symphysis, (*b*) length of the body, (*c*) length of the ramus, (*d*) bigonial width (between the angles), (*e*) bicondylar width, and so on. The angle between body and ramus is also often estimated. Since the mandible is often missing from dried skulls and moreover is held to be a less reliable element in assessing racial affinity than the rest of the skull (Morant 1936), its measurements are not so widely used.

Other forms of cranial measurement are also employed. The horizontal circumference of the head and various arcs and contours are sometimes measured. Radiographic studies make it possible to extend the classical series of craniometric dimensions described above and to measure directly the angles between certain of them, for example, the *gnathic angle*, measured between the basion-nasion and basion-prosthion lines. The *craniofacial* or *cranial base angle* is of special interest, representing the degree of cranial flexure. The angle has a value of about 130°, and this is probably achieved by the time of birth, little further change occurring. Its high value in man is sometimes associated with the movement of the tongue towards the pharynx, a change sometimes regarded as essential to speech. (For a review of the literature *see* Schulter 1976.) A radiographic technique for estimation of endocranial volume, or cranial capacity, has also been evolved (Haas 1952). *Cranial capacity*, which is taken as an indication of brain volume, can be assessed directly by filling the cranial cavity with lead shot, millet seed, or other particulate materials suitable for volumetric measurement when poured out again. A number of formulae have also been calculated to yield cranial capacity from length, breadth and height of the cranium. Examples are:

Males: 0·000337 $(L-11)(B-11)(H-11)+406·01$ c.c.
Females: 0·000400 $(L-11)(B-11)(H-11)+206·60$ c.c.

In these formulae, L and B are length and breadth, and H is the *auricular height*, which is measured to the vertex from the external acoustic meatus. All measurements are expressed in millimetres. These methods are subject to considerable inaccuracy, and various corrections and allowances do not entirely remove this (Hrdlička 1947, Montagu 1960).

Attempts to establish reliable craniometric differentiation between races are as old as craniometry itself; and although the mandible and cranial capacity are in this connection less dependable, satisfactory characterization has been established for some racial groups and especially for Caucasians and Negroes. Where a number of measurements are treated by discriminant analysis, an accuracy of over 90 per cent can be expected (Giles and Elliot 1960). Multivariate analysis may also elucidate problems of cranial growth (Liebgott 1977). Craniometric methods have a special usefulness in forensic practice where cranial remains can be compared with existing photographic and radiographic records in making an identification (Glaister and Brash 1937). They also play a part in attempts to reconstruct the appearance in life of individuals represented only by skeletal remains (Stewart 1954). Cephalometry has a further application in clinical specialities, such as plastic and oral surgery, concerned with cranio-facial deformity. A recent development in this field is *morphanalysis*, an approach which uses gridded radiographs and photographs that are universally related in three dimensions (**3.126B**). Such standardized records enable diagnostic comparisons to be made between patients and the normal population (Rabey 1977).

Applied Anatomy. Features which tend to prevent fracture of the skull are its elasticity, its rounded shape, and its construction from a number of secondary elastic arches, each made up of a single bone. Moreover, in some places where the bone is thin, overlying muscles may offer some resistance in cushioning blows. Examples are the temporal squama and the inferior occipital

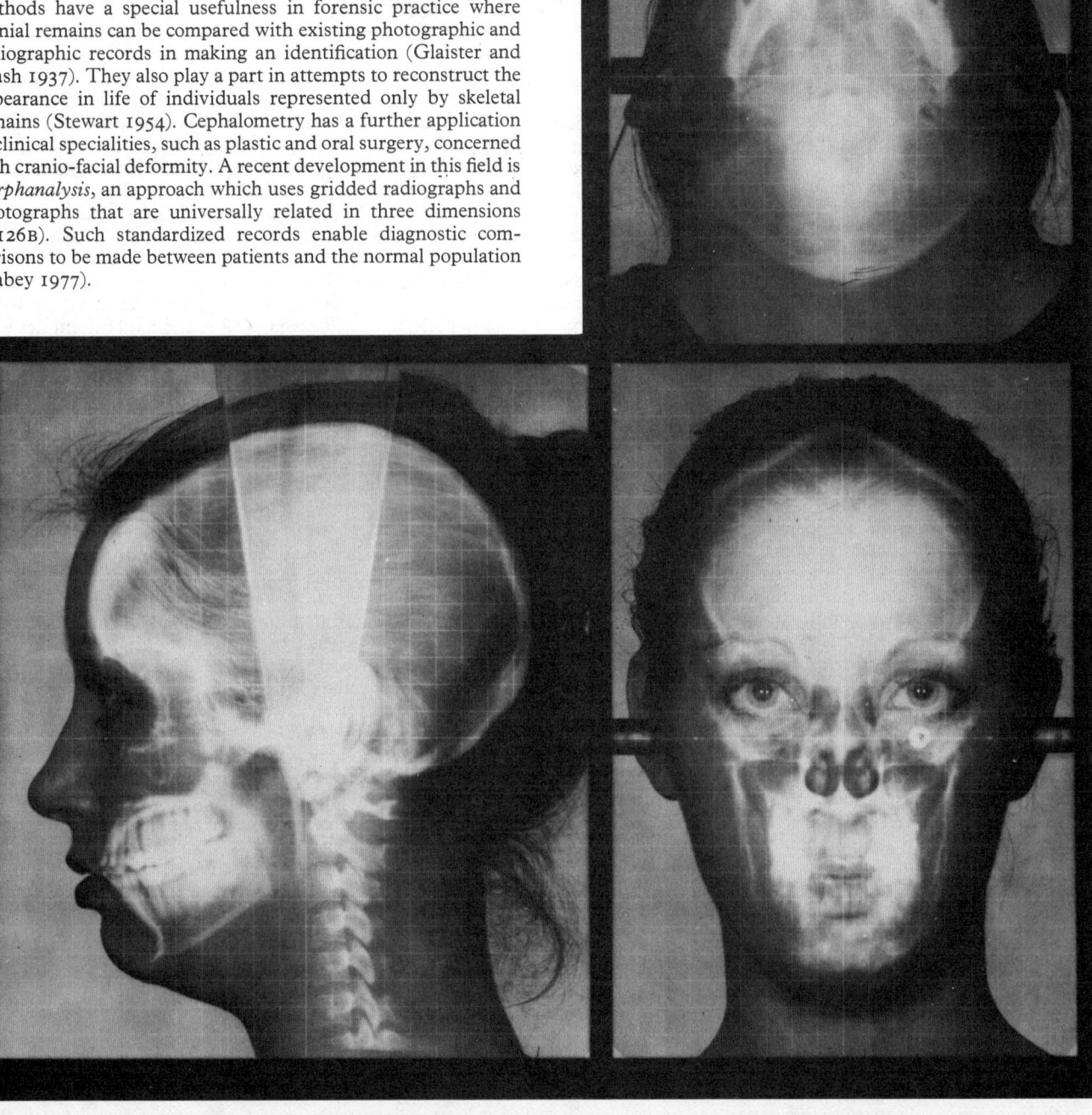

3.126B Combined, accurately superimposed, photographs and radiographs of the kind shown here are currently being used to produce a computerized analysis of metrical relations between a selection of datum points in the skeletal and soft tissue components of human heads and faces. The technique, known as morphanalysis, has been elaborated by Dr. G. P. Rabey, to whom we are indebted for these records. Both photographs and radiographs are produced in the three primary planes, and hence the morphometric analysis is three-dimensional. For further information, *see* Rabey (1968, 1971, 1980).

fossa. A metrical study of the thickness of the calvaria in 28 Caucasians (aged from 15 to 82 years) yielded an average thickness of 5·8 mm. In male specimens, thickness increased with age; in females, the reverse was observed (Lippert and Käfer 1974).

The most common place for fracture of the base to occur is through the middle fossa, and here the fissure usually takes a fairly definite course. Starting from the point struck, usually near the parietal tuber, it runs downwards through the parietal and the squamous temporal and across the petrous portion, frequently traversing and implicating the internal acoustic meatus, to the foramen lacerum. This course explains the symptoms which may arise: thus, if the fissure passes across the internal acoustic meatus, injury to the facial and vestibulocochlear nerves may result, with consequent facial paralysis and deafness; if the fissure extends through the semicircular ducts vertigo will ensue, or the tubular prolongation of the arachnoid around the nerves in the meatus may be torn and thus permit the escape of the cerebrospinal fluid, should there be a communication between the internal ear and the tympanic cavity together with rupture of the tympanic membrane, as is frequently the case.

The bones of the face are sometimes fractured by direct violence. Those most commonly broken are the nasal bones and the mandible; the latter is by far the most frequently fractured of all facial bones. Fracture of the nasal bone is usually transverse, about 1·25 cm from the free margin. The broken portion may be displaced backwards or more generally to one side by the force which produced the lesion. The most common situation for a fracture of the mandible is in the neighbourhood of the canine tooth, as at this spot the bone is weakened by the deep socket for its root; it is next most frequently fractured at the angle. Occasionally a double fracture may occur, one in each half of the bone. Such fractures are usually compound, from laceration of the mucous membrane covering the gums.

THE APPENDICULAR SKELETON

Phylogeny and Functions

The ancestral vertebrates, the *Agnatha*, were not only jawless but also without appendages, like their surviving representatives, such as the lampreys (Young 1962; Romer 1970). It has been hypothecated that two continuous finfolds, ventrolateral in position, were the forerunners of the separated pairs of fins which appear in the fossil remains of the earliest fish which possessed jaws, the *Placoderms* (Westoll 1958). These sometimes displayed more than two pairs of appendages, but in their descendants, the bony and cartilaginous classes of fish, the *Osteicthyes* and *Chondricthyes*, the characteristic vertebrate pattern of four appendages, arranged in pectoral and pelvic pairs, became stabilized—departed from only in such forms as the snakes, which have subsequently lost their limbs. The appendicular skeletons of fossil and extant fish show much variation; but there is a basic division into 'girdle' bones, embedded in the musculature of the body wall, and the rodlike elements radiating into the mobile fin, the two articulating at a joint of approximately ball and socket type. In many orders of bony fish the homologues of the components of the pectoral and pelvic girdles can already be identified, as well as those of the proximal two segments, arm and forearm or thigh and foreleg, of the terrestrial limbs (Bolk *et al.* 1931–9). Distal to this the equivalence is less clear; but in the multiplicity of its small bony units the fin bears at least some resemblance to the extremities, hand and foot, of the land tetrapods.

The terrestrial vertebrates are grouped as the *tetrapods* because their four limbs are an outstanding common feature, even though the pectoral pair may become extensively modified for flight. The tetrapods are widely considered to be derived from a primitive order of bony fish, the *Crossopterygii* ('fringe-finned'), related to another ancient group, the *Dipnoi* or lungfish, several species of which still survive today. The crossopterygians, in contrast, were known only as fossil remains until 1938, when a living representative of a suborder, *Coelacanthus*, was captured off the coast of South Africa; further living coelacanths have been secured subsequently near the Comoro Islands and elsewhere in the Indian Ocean. These forms possess lobed fins, with a narrowed attachment to the body, and it is from such appendages that tetrapod limbs can most plausibly be derived (Gregory and Raven 1944).

In the change to a terrestrial environment, during the evolution of primitive amphibians from crossopterygian fish, two major adaptions commenced. The first, already foreshadowed in their fins, was a change in the positioning of these limb-like paddles to bring them into more effective use for support and propulsion on land. The *fore-limbs* became rotated in a *lateral* direction at their joints with the pectoral girdles, so that the hinge of the elbow, at first pointing outwards away from the body, as it still does in many amphibians and reptiles, finally projected backwards. In contrast, the *hind-limbs* were rotated *medially* at their junction with the pelvic girdle into opposite positions, the knees projecting forwards. These opposite changes in the positioning of the limbs entailed that the *pre-axial* or leading *border*, corresponding with the 'first' digit, pollex or hallux, was rotated from an anterior to a lateral position in the pectoral limb, but from an anterior to a medial position in the pelvic limb. (As described later on p. 352 this change in orientation of the fore-limb necessitates a complementary *pronation* of its radio-ulnar segment to ensure a plantigrade position of the appendage. The compensation is unnecessary in the hind-limb.) Coupled with these growing structural distinctions between the fore and hind extremities was a second major modification—the articulation of the pelvic girdles with the spinal column, a condition not present in any fish, but uniform in tetrapods, except in some early fossil amphibians.

With the evolution of the amphibians, the general skeletal pattern of the tetrapod limb was defined, and apart from relatively minor modifications this has persisted throughout subsequent forms. It is often called the *primitive pentadactyl limb*, since it terminates typically in five digits. Among mammals especially, reduction in the number of these is frequent, but the primates, including mankind, have retained the full number and are in this respect unspecialized.

Of the *limb girdles*, the *pectoral* always consists of a dorsal element, the *scapula*, but displays considerable variation in its ventral components, which are commonly a mixture of endochondral and dermal bones. Of these the *coracoid* and *clavicle* persists in mammals, the former being reduced in size and fused to the scapula. The clavicle, the sole mammalian survivor of a considerable variety of dermal elements, such as the *cleithrum* and *interclavicle*, present in the pectoral girdle of lower vertebrates, is also reduced and frequently lost in mammals, especially in fast-running or bounding types of quadruped. Its persistence in the primates can be associated with the development of the fore-limb for climbing and prehensile activity; the clavicle acts as a mobile strut, at the lateral end of which the rest of the limb can be steadied in a variety of positions (Watson 1917). The *pelvic girdle*, in contrast to the pectoral, contains only endochondral elements; the dermal armour of many primitive fish extended far enough caudally to permit the incorporation of dermal elements into the pectoral but not the pelvic girdle. The latter developed a primitive ventral *symphysis* in fish, and in tetrapods this region is composed of distinguishable pubic and ischial elements. The dorsal component, the ilium, though present in some fish, is not well developed. In land animals the need of larger muscles to support the body's weight and to impart thrust to it through the hind-limbs, is associated with greater size of the ilium and of its articulation with the sacral part of the spinal column. Throughout the tetrapods these three pelvic entities, *pubis, ischium* and *ilium*, show a greater uniformity of arrangement than in fish. They also show an increasing tendency to fuse together into an *innominate* one, especially in adult mammals. The two innominate bones are ointed together ventrally at a symphysis pubis and each to the acrum dorsally at sacro-iliac articulations. Thus is established a tructural arrangement, the *pelvis*, the primary functions of which re locomotor, in conducting stresses between the axial skeleton nd that of the limbs. The term pelvis, meaning a basin, is nisleading, for it is more of an irregular ring of bone which, being n the body wall at the caudal end of the body cavity, inevitably urrounds certain viscera. This relationship may, in human .natomy, obscure the essentially appendicular origin and unction of the pelvis. Since the cloaca is immediately caudal to it, 1ot only the hindgut, but other tubes, urinary or genital, must >ass through the pelvis, even when they achieve separate external >penings in the mammals. During birth, offspring must therefore raverse the pelvis, and their comparatively large size in human emales entails obstetric considerations (p. 384), which again may >vershadow the somatic and locomotor nature of the hind-limb ;irdle.

In mammals the functional differences between the fore- and 1ind-limbs, which are accentuated in a primate such as man, are 1ssociated with marked structural divergence in their girdles. These major differences are as follows:

Pectoral Girdle	**Pelvic Girdle**
1. Dermal and endochondral.	Entirely endochondral.
2. Two components, clavicle and scapula, which remain separate.	Three components, pubis, ischium and ilium, which fuse into a single innominate bone.
3. No articulation with vertebral column.	Articulates with sacral vertebrae.
4. No direct ventral articulation (clavicles connected only by interclavicular ligament).	Direct ventral articulation at symphysis pubis.
5. Articulations of clavicles with axial skeleton (sternum) are relatively small, mobile, and ventral.	Articulations of innominate bones with axial skeleton (sacrum) are relatively large, capable of little movement, and dorsal.
6. Comparatively lightly built for mobility.	Massively constructed for resistance to stress rather than for mobility.
7. Resilient to thrust.	Transmits thrust between vertebral column and leg.
8. Shallow joint with limb, allowing wide range of movement.	Deep joint with limb, limiting range of movement.

Both girdles articulate with the distal, freely movable parts of the limbs by 'ball and socket' joints, the first segment of each limb containing a single element, *humerus* or *femur*. The shoulder and hip joints are essentially the same in mechanism, allowing some degree of movement in all directions, though usually to a more limited extent in the latter. Even in quadrupedal mammals, where the limbs tend to function as walking props, mainly adapted to fore and aft movement in the long axis of the body, greater mobility is usual at the shoulder, particularly as regards abduction, which permits outward splaying of the fore-limbs. Throughout the tetrapods the shoulder region shows a greater range of structural adaptation with changing function than the hip.

The orientation of the scapulae in relation to the thorax and the humerus varies far more than do their hind-limb equivalents, even in the mammals. This variation is perhaps most extreme in the bipedal mammals, the primates (Ashton *et al.* 1976). Their widespread adoption of an arboreal habitat, in which they climb, walk, run and swing, is linked with the development of a highly mobile and prehensile fore-limb, and with similar but much less marked changes in the hind-limb. Perhaps equally to be emphasized is the common habit of *sitting* upright, for in even those primates, such as some baboons, which have largely deserted trees and live a predominantly quadrupedal life on the

ground, the arms are released for prehensile and manipulative use, activities of the profoundest significance in the evolution of man.

In modern mammals, the *Eutheria*, the rounded articular 'head' of the humerus is usually directed dorsocaudally and the glenoid socket of the scapula in a reciprocal manner. While this favours the quadrupedal use of the fore-limb it prevents movement of the humerus upwards into the vertical position of the arm so easily and characteristically assumed by primates in reaching upwards above the head. This is associated with a re-orientation of the humeral head relative to its shaft, the head being directed *medially* rather than caudally. Coincident with this reorganization of the shoulder is a change in the maximum diameter of the thorax from dorsoventral to transverse. The former is typical of quadrupeds, the latter of man. This change is coupled with the retention of the clavicle in primate mammals, and with a changed orientation of the scapulae. The medial aspects of these come to face largely ventrally and their glenoid fossae more laterally. These osseous changes, and correlated muscular alterations, are clearly linked with a far freer range of movement at the shoulder joint.

The tubercles and bicipital groove of the humerus share in the migration of its articular head, though to a lesser extent, but the lower end of the bone does not change its transverse orientation (Martin 1932). This means that a radial axis through the centre of the articular head is approximately at right angles to the axis of the lower end in most modern mammals (*Eutheria*) because the head is directed caudally. As it becomes more medial this angle increases above 90°. It is still some 95° in carnivora, rises to about 100° in monkeys, 120° in apes, and ranges from 135° to 165° or more in man. This change in the primate humerus has been regarded as a 'torsion', the angle being termed the *angle of torsion*. The term is perhaps misleading and there is no evidence of any true 'twisting' of the humeral shaft; the spiralling of the radial nerve and its associated groove pre-dates this change. Nor is there any particularly human character in the torsion angle. It reaches 180° (where the axes are parallel) in birds and was apparently nearly as high in mammal-like reptiles, returning to about 90° in the *Metatheria* (marsupials). A few words must be added in regard to the method of recording humeral torsion angles. As defined here the angle is measured between the articular 'axes' of the shoulder and elbow joints. However, Krahl (1944, 1976) has contributed many papers on this subject, and these are sometimes quoted in textbooks. In his method the angle measured is that between the axis of the elbow joint and an orthogonal line relative to the axis of the shoulder joint; his values, therefore, differ by 90° from those cited here. Neither method is completely satisfactory in conceptual terms, but very similar figures are obtained. Krahl's work largely supports the view put forward here, that the change in orientation of the head of the humerus, relative to its lower end, has involved alterations at the proximal end of the bone, and has not been accomplished by an evolutionary twisting of the shaft. Racial and sexual differences have been recorded (Kate 1968).

In the lower extremity the state is like that of the mammal-like reptiles (and birds), in which the two axes are nearly parallel. (In man the femoral head and the long axis of its neck are anteverted about 16° relative to the transcondylar axis of the lower end of the femur.) It is said that this condition has not varied in the lower limb, unlike the upper. This lability of the 'torsion' angle of the humerus may be coupled with the greater variability in fore-limb muscular activity, especially in those groups which demonstrate bipedalism and hence release of the fore-limbs.

The intermediate segments of the two limbs, and their joints with the humerus and femur, show similarities and differences. The differences are greater in bipeds, as would be expected; in quadrupeds the similarities are more evident. Both segments consist of two parallel elements, radius and ulna, tibia and fibula. Primitively both articulate with the single proximal element in each limb, but in mammals in general, as in man, the fibula withdraws from the articulation of the knee. This leaves the *pre-axial* element, the tibia, as the major lever and prop. In the forearm the ulna, the *post-axial* bone, becomes the main force-transmitting strut at the elbow.

Both the elbow and knee are hinge joints in quadruped

mammals, and while this was the primitive tetrapod condition in the knee, the elbow also contained a potentiality for *rotation* from the beginning. The hind-limb turns medially bringing its extensor surface forwards, in which position the foot, hinged at the ankle, projects forwards with its plantar surface on the ground. This is the typical *plantigrade* habit of many quadrupeds—it brings the hind-limb well under the body as a series of hinged levers well adapted for propulsion and body support. It is to be noted that the tibia and fibula are parallel in this position. On the other hand, the *lateral* rotation of the fore-limb which turns the extensor aspect backwards, is accompanied by *pronation* at the elbow bringing the 'hand' forwards, palm downwards to achieve plantigrade locomotion. This brings the pre-axial bone, the radius forwards and medially across the ulna, and the two bones are stabilized in this crossed relation in a large majority of quadrupeds, including all four-*footed* mammals. It should be added that the forearm and foreleg bones show a general tendency to fuse together in such mammals. Such fusions are particularly well marked in those which show a reduction in the number of digits and are usually *digitigrade* (e.g. Ungulata), in which contact with the ground is restricted to the flexor aspects of the tips of their remaining digits.

The primitive rotatory potential at the elbow has been retained and carried further in the primates, in the movements of *pronation* and *supination*, which are essential to the development of manual skills. These movements are impossible in the leg. At the primate elbow, the ulna forms a massive hinge with the humerus but tapers to a relatively small distal end. The extremities of the radius are reciprocal in size and shape—its upper end is small and adapted for rotation, while its lower end is enlarged and carries the hand. Thus, when the radius revolves it takes the hand with it, turning round the distal end of the ulna.

Both wrist and ankle joints are hinges, but adduction and abduction are also possible at the wrist. The *carpus* and *tarsus* are groups of small bones intermediate between the forearm or foreleg and the skeletal components of the digits. They show considerable modification throughout the tetrapod series, and also some divergence from each other. Nevertheless, however modified, both groups are considered to be derived from the same primary arrangement in the primitive *pentadactyl limb* (Barde-

leben 1894). This consisted of a proximal row of three elements, a distal row of five, articulating with the bases of the digits, and a central or intermediate group, probably four in number, wedged between these two rows. By fusion, loss and modification in size, shape and articulation, the carpal and tarsal bones display much variation (particularly reduction). Tracing the homologies between the primitive elements and their subsequent transformations (structural and functional) has aroused much controversy (Broom 1901; Wood Jones 1949; Lewis 1964), but a considerable measure of agreement has now been reached. Authorities may disagree in minor details, but the generally accepted homologies are shown in the table below.

The main divergencies between the mammalian carpus and tarsus, apart from greater size of the latter, especially in primates, are three. Firstly, the carpus has usually remained more primitive in retaining a full proximal row, all articulating at the wrist joint; in the tarsus only one of these (the talus) retains this role in most mammals, and hence primates. Secondly, one of the proximal tarsal bones (calcaneus) projects backwards as a lever behind the tibia and fibula; this does not occur in the carpus. Thirdly, the tarsal bones, at their intermediate joints, permit some rotation— inversion and eversion—an adaptation to uneven surfaces. (Such contingencies are met in some fore-limbs by the movements of pronation and supination between the *forearm* bones and by some measure of abduction and adduction at the *wrist*.) In these three modifications the tarsus is less primitive. It is to be emphasized that the evolution of the calcanean lever system is a mammalian, not just a primitive or human modification. Its high development in the human foot, and the fact that other primate feet are more like hands, may give rise even to the superficial assumption that it is a *human* characteristic.

Because of the occasional occurrence of small supernumerary bones, pre- or post-axial in position, in association with the carpus and tarsus, it has been argued that these are evidence of additional digits (Wood Jones 1941), suggesting that the ancestral pattern comprises more than five digits. Known tetrapods appear to be content with the orthodox five, and often many less. The five metacarpal and metatarsal bones originally articulated each with a separate carpal or tarsal, but the latter are reduced to four by

HOMOLOGIZATION OF THE PRIMITIVE TETRAPOD AND HUMAN CARPUS AND TARSUS

	TETRAPOD CARPUS	HUMAN CARPUS		HUMAN TARSUS		TETRAPOD TARSUS	
PROXIMAL ROW	Os Radiale	Scaphoid	Scaphoid minus tubercle	Talus	Navicular minus tubercle	Os Tibiale	PRE-AXIAL
	Os Intermedium	Lunate			Talus	Os Intermedium	
	Os Ulnare	Triquetral		Calcaneus		Os Fibulare	POST-AXIAL
CENTRAL ELEMENTS	Os Centrale	Absent	Tubercle of Scaphoid	Navicular	Tubercle of Navicular	Os Centrale	
DISTAL ROW	Os Carpale 1	Trapezium		Medial Cuneiform		Os Tarsale 1	PRE-AXIAL
	Os Carpale 2	Trapezoid		Intermediate Cuneiform		Os Tarsale 2	
	Os Carpale 3	Capitate		Lateral Cuneiform		Os Tarsale 3	
	Ossa Carpalia 4 and 5	Hamate		Cuboid		Ossa Tarsalia 4 and 5	POST-AXIAL

N.B. Alternative views have been stated for the carpal scaphoid and the tarsal navicular. In each case the most widely accepted view is on the left. In the case of the talus the anomalous persistence of its posterior process as a separate element, an *os trigonum*, may have influenced the orthodox view of the talus as an amalgam of ossa tibiale et intermedium. If it becomes accepted that the talus is merely the os intermedium, the main residual controversies concern the scaphoid and navicular, centred on the possible derivation of each bone's tubercle from the corresponding os centrale. It should be added that the *centrale* element has been shown as singular in this table, though there were probably a variable number of centralia in the ancestral forms.

usion of the fourth and fifth even in a primate such as man. In mammals with reduced or lost digits the metacarpals and metatarsals are also involved, as well as the phalanges. A minor example of reduction is provided by the Carnivora, whose first digit, pollex or hallux, is degenerate and useless, although it often still carries a claw, an extreme example being the horse, which retains only the middle digit, with a greatly enlarged metapodial bone.

The *metapodial* (metacarpal or metatarsal) *formula* puts these bones in order of length and indicates their number. In man it is usually 2>3>4>5>1 in the hand, and 1>2>3>4>5 in the foot. The *phalangeal formula* likewise starts from the first digit—pollex or hallux—and details the number of phalanges in each digit. In many primitive reptiles, for example, it is 2.3.4.5.3. In primitive mammals, it is 2.3.3.3.3. for hand and foot, as in man. In the horse it is 0.0.3.0.0 A *digital formula* denoting the relative projection of the tips of the digits is sometimes employed, especially when comparing the extremities of primates.

Throughout the tetrapods the pollex and hallux evince a tendency to diverge from the other digits, perhaps as a prop. It is believed that some early mammals were arboreal, and could use the hand as primates do, with a grasp between the pre-axial digit and the remaining digits. Among the primates this grasping ability is developed in both hand and foot, although it is absent from the human foot and not always effective in the primate hand. In gibbons, for example, the hand is used rather as a hook in brachiating or swinging from branch to branch, and the thumb does not effectively *oppose* to the other digits; the foot, in contrast, has an excellent grasp. In most primates the distal end of the first metatarsal is not tied by ligaments to the second, which allows the hallux marked freedom of movement. In mankind, however, the first and second metatarsals are so connected, all five being tied by ligaments. The human thumb, on the contrary, is even better developed for *opposition* than in other primates. In this movement the joint between the thumb and its carpal bone—the *trapezium* (the *os carpale 1* of the primitive tetrapod wrist)—is adapted to permit the thumb to flex and rotate towards the fingers. This action can be used as a 'power' grip, when the thumb and fingers wrap around opposite sides of an object such as a branch, weapon, or tool. It can also be used in a 'precision' grip, in which the thumb is opposed to a single finger, often the index, in lighter, more accurate manipulations (Napier 1966).

The skeletal elements of the human limbs, especially when compared on the basis of their functional differences and similarities, provide beautiful examples of adaptation. In the leg, the massive, almost immobile pelvic girdle, directly articulated to the axial skeleton, and the limited ball and socket mechanism of the hip, adapted for fore and aft swinging, but allowing some abduction to permit a steady stance in a biped; the hinging joints at the knee and ankle for resilience and power in walking, running and springing, and the powerful lever of the foot, arched for resilience in the final thrust of takeoff and the impact of landing—all these characterize a limb highly developed for *locomotion*.

The arm presents a different system of adaptions, analysable into components, but integrated in action. The scapula floats in a muscular suspension, lightly strutted by a slender mobile clavicle. A shallow ball and socket joint allows a far wider range for the reaching, grasping, 'inquisitive' hand. Flexion, combined with rotation at the elbow, brings the hands within the arena of the eyes, maintaining a steady basis at the wrists for the fine movements of the opposable thumb and highly mobile fingers. Endlessly variable manipulations, involving most delicate muscular adjustments, can thus be developed with the constant feedback of stereoscopic vision.

We see the arm and hand of man at the end of several hundred million years of evolution. Perhaps it is not the 'end', but even were the human upper limb to evolve no further, it has, with an accompanying high development of cerebral control, accomplished such a mastery over man's environment as to deflect the whole course of primate evolution into a new channel (Huxley 1942).

THE SKELETON OF THE UPPER LIMB

The Scapula

The scapula (**3.**127, 128, 129) is a large, flattened, triangular bone on the posterolateral aspect of the thorax, overlapping parts of the second to the seventh ribs. It has costal and dorsal surfaces, upper, lateral and medial borders, inferior, superior and lateral angles, and three bony processes, the spine, its continuation the acromion, and the coracoid process. The lateral angle is truncated by the glenoid cavity for articulation with the head of the humerus. This part of the bone may be regarded as the head, and it is connected to the plate-like body by an inconspicuous neck. The long axis of the scapula is nearly vertical and the relatively featureless costal surface contrasts with the dorsal surface, which is interrupted by the spine (**3.**129). The bone is very much thickened in the immediate neighbourhood of the lateral border, which runs from the inferior angle below to the glenoid cavity above.

The costal surface (**3.**127), facing medially and forwards when the arm is by the side, is slightly hollowed out, especially above. Near the lateral border it presents a longitudinal, rounded ridge, prominent in the neighbourhood of the neck, but less so below, which is separated from the lateral border by a narrow, grooved area. The *dorsal surface* (**3.**128) is divided into a smaller, upper area and a larger, lower one by a shelf-like projection, the *spine of the scapula*. These two areas communicate with each other through the *spinoglenoid notch*, which lies between the free, lateral border of the spine and the dorsal aspect of the neck of the bone. A flattened strip, for muscular attachments, marks the dorsal surface along the lateral border.

The lateral border of the scapula is a clearly defined, sharp, roughened ridge, which runs sinuously from the inferior angle to the glenoid cavity. At its upper end it widens into a rough, somewhat triangular area, which is termed the *infraglenoid tubercle* (**3.**129). Throughout its whole length the lateral border is thickly covered with muscles and cannot be felt clearly in the living. The lateral border is sometimes described as thick, but the grooved lateral part of the costal surface and frequently the flattened strip along the lateral part of the dorsal surface are incorporated in it. The *medial border* of the scapula extends from the inferior to the superior angle. In its lower two-thirds this border can easily be felt through the skin, but its upper third is more deeply placed and cannot be palpated. The *superior border*, thin and sharp, is the shortest. At its anterolateral end it is separated from the root of the coracoid process by the *suprascapular notch*.

The inferior angle lies over the seventh rib, or seventh intercostal space. It can be felt through the skin and the muscles which cover it and, when the arm is raised above the head, it can be seen to pass forwards round the chest wall. The *superior angle* is placed at the junction of the superior and medial borders, and is obscured by the muscles which cover it. The *lateral angle* is truncated and broadened. It constitutes the head of the bone. On its free surface it bears the *glenoid cavity*, forming the shoulder joint with the head of the humerus. Being shallow, it forms a limited socket for the humeral head. It is narrow above and wider below, and is pear-shaped in outline. Immediately above the glenoid cavity a small roughened area encroaches on the root of the coracoid process, the *supraglenoid tubercle*. The *neck* of the scapula is the constriction immediately adjoining the head. It can be identified most easily on its inferior and dorsal aspects. Ventrally, it can be regarded as extending between the infraglenoid tubercle and the anterior margin of the suprascapular notch.

The spine of the scapula (**3.**128) projects on the upper part of

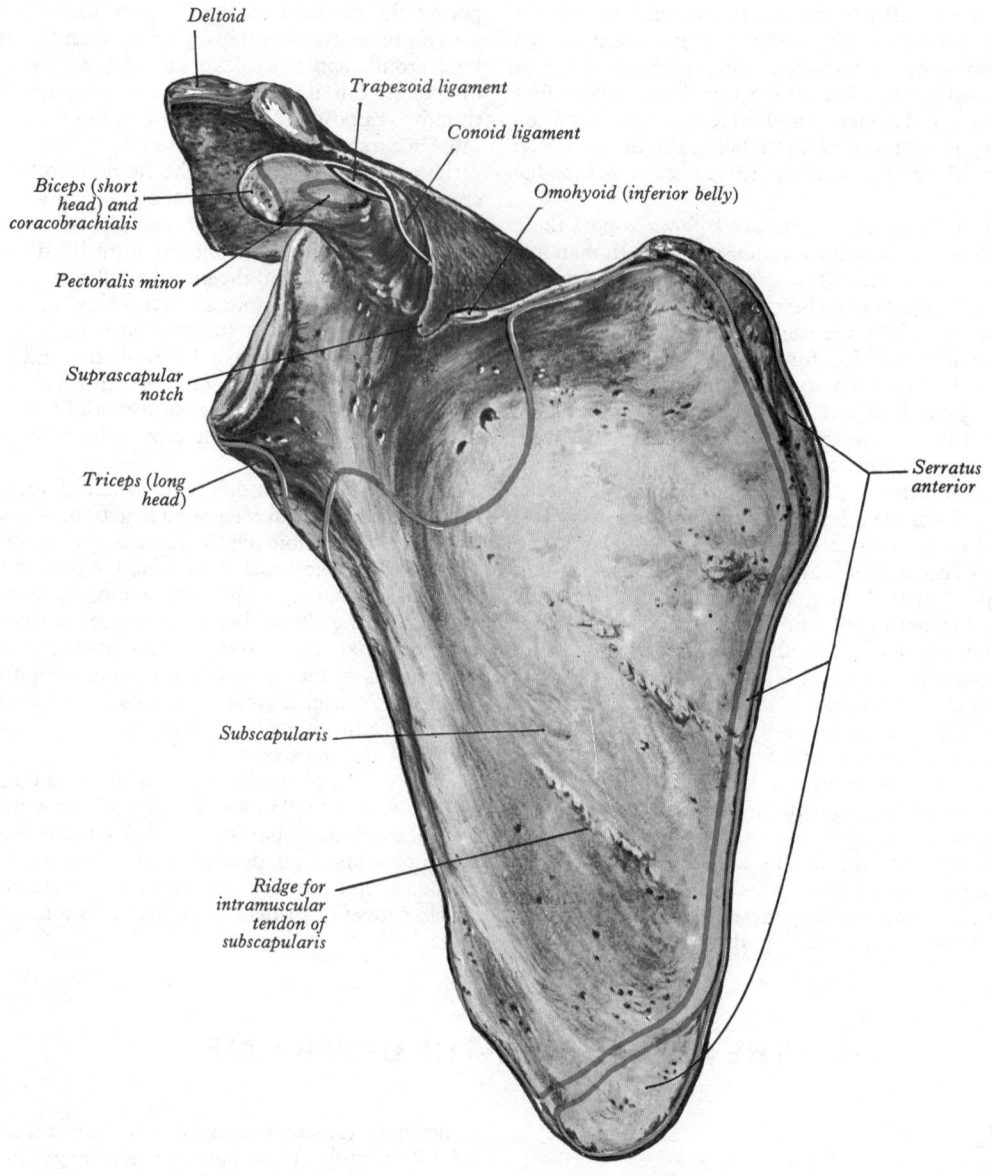

Deltoid

Trapezoid ligament

Conoid ligament

Biceps (short
head) and
coracobrachialis

Omohyoid (inferior belly)

Pectoralis minor

Suprascapular
notch

Triceps (long
head)

Serratus
anterior

Subscapularis

Ridge for
intramuscular
tendon of
subscapularis

3.127   The right scapula. Anterior (costal) aspect.

the dorsal surface of the bone, and is triangular in shape. Its lateral border is free, thick and rounded and helps to bound the *spinoglenoid notch*, which lies between it and the dorsal surface of the neck of the bone. Its anterior border joins the dorsal surface of the scapula along a line which runs laterally and slightly upwards from the junction of the upper and middle thirds of the medial border. It should be noted that the plate-like body of the bone is bent along this line, and this fact accounts for the more marked concavity of the upper part of the costal surface. The third border is termed the *crest of the spine*, and is subcutaneous throughout nearly its whole extent. At its medial end the crest expands into a smooth, triangular area. Elsewhere the upper and lower edges and the surface of the crest are roughened by muscular attachments. The upper surface of the spine widens as it is traced laterally, and is slightly concave. Together with the upper area of the dorsal surface of the bone, the upper surface of the spine forms the *supraspinous fossa*. The lower surface is overhung by the crest at its medial, narrow end, but is gently convex in its wider, later portion. Together with the lower area of the dorsal surface of the bone, the lower surface of the spine forms the *infraspinous fossa*, which continues into the supraspinous fossa through the spinoglenoid notch.

The acromion projects forwards, almost at right angles, from the lateral end of the spine, with which it is continuous. The lower border of the crest of the spine becomes continuous with the lateral border of the acromion at the *acromial angle*, which forms a

subcutaneous, bony landmark. The medial border of the acromion is short and is marked anteriorly by a small, oval facet, directed upwards and medially, for articulation with the lateral end of the clavicle. The lateral border, tip and upper surface of the acromion can all be felt through the skin without difficulty. An accessory articular facet may occur on the inferior surface of the acromion.

The coracoid process (3.127) springs from the upper border of the head of the scapula and is bent sharply so as to project forwards and slightly laterally. When the arm is by the side, the coracoid process points almost straight forwards and its slightly enlarged tip can be felt through the skin, although it is covered by the anterior fibres of the deltoid. It lies about 2·5 cm below the junction of the lateral fourth of the clavicle with the rest of the bone. The supraglenoid tubercle marks the root of the coracoid process where it adjoins the upper part of the glenoid cavity. Another impression is placed on the dorsal aspect of the coracoid process at the point where it changes direction. This gives attachment to the conoid part of the coracoclavicular ligament, which will be mentioned again in connection with the clavicle.

From the costal surface springs the subscapularis (3.127), which arises from nearly the whole of this aspect including the grooved area immediately adjoining the lateral border, but excluding the area adjoining the neck of the bone. Small intramuscular tendons are attached to the roughened ridges which subdivide this surface incompletely into a number of smooth areas. The anterior aspect

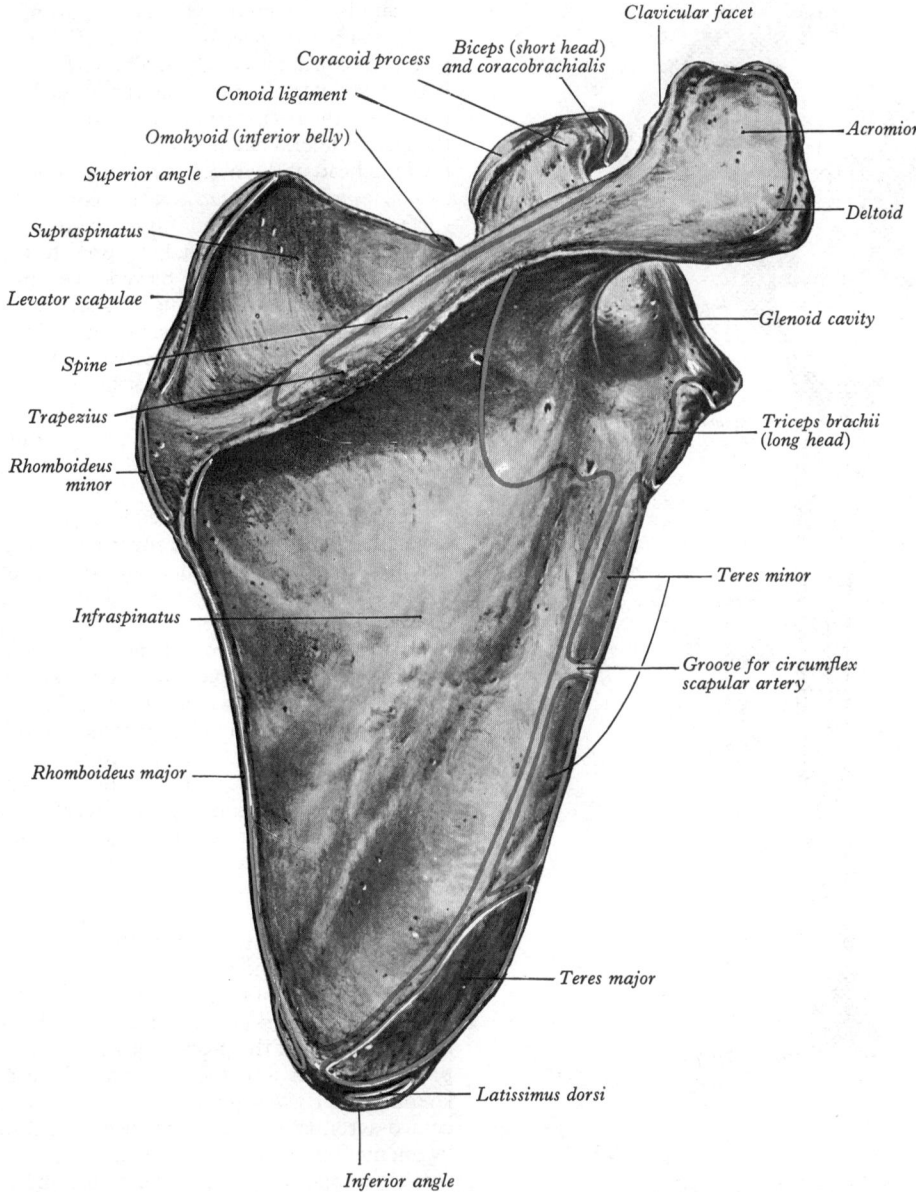

Clavicular facet

Coracoid process

*Biceps (short head)
and coracobrachialis*

Conoid ligament

Omohyoid (inferior belly)

Superior angle

Supraspinatus

Levator scapulae

Spine

Trapezius

*Rhomboideus
minor*

Infraspinatus

Rhomboideus major

Acromion

Deltoid

Glenoid cavity

*Triceps brachii
(long head)*

Teres minor

*Groove for circumflex
scapular artery*

Teres major

Latissimus dorsi

Inferior angle

3.128   The right scapula. Dorsal aspect.

of the neck is separated from the tendon of the subscapularis by a synovial protrusion of the shoulder joint in the form of a bursa. Near the inferior angle a somewhat oval area gives insertion to the lower five or six digitations of the serratus anterior (**3.**127). The remainder of the muscle is inserted into a narrow strip along the ventral aspect of the medial border, which is wider above, where it receives the large first digitation. The longitudinal thickening of the bone near the lateral border provides a lever capable of withstanding the pull of the serratus anterior at the inferior angle during rotation of the scapula forwards. In this movement the glenoid cavity faces more directly upwards and the upper limb is raised from the side and carried above the head against the action of gravity.

On the dorsal surface, the supraspinous fossa in its medial two-thirds is the origin of the supraspinatus, and its margins give attachment to the fascia which covers the muscle. The flattened strip adjoining the lateral border has attached to it, in its upper two-thirds, the teres minor and is grooved, near its midpoint, by the circumflex scapular vessels, which pass between the muscle and the bone as they enter the infraspinous fossa (**3.**128). The lower limit of the origin of the teres minor is indicated by an oblique ridge, which runs from the lateral border to the neighbourhood of the inferior angle and cuts off a somewhat oval area for the teres major. With the exception of an area near the neck of the bone, the rest of the infraspinous fossa, which is hollowed out laterally and convex medially, gives origin to the

infraspinatus. The strong infraspinatus fascia passes on to the teres minor and the teres major and sends fascial partitions between them to reach the bone along the ridges which mark the limits of their attachments.

The *lateral border* separates the attachment of the subscapularis from those of the teres minor and major. These muscles project laterally beyond the bone, and together with the latissimus dorsi cover it so completely that it cannot be felt through the skin. The infraglenoid tubercle gives origin to the long head of the triceps brachii. The *medial border* is thin and often angled opposite the root of the spine. A narrow strip, extending from the superior angle to the root of the spine, gives insertion to the levator scapulae. Below this, and opposite the root of the spine, the rhomboideus minor gains insertion. The remainder of the border is taken up by the rhomboideus major (p. 566).

The *upper border* of the scapula is thin and sharp. Near the suprascapular notch the inferior belly of the omohyoid is attached. Bridging the notch is the superior transverse ligament, attached laterally to the root of the coracoid process and medially to the limit of the notch, and the ligament is sometimes ossified. The foramen thus completed transmits the suprascapular nerve to the supraspinous fossa, whereas the suprascapular vessels pass backwards above the ligament.

The *inferior angle* of the scapula is covered dorsally by the upper border of latissimus dorsi, which frequently receives a small slip of origin from this part of the bone. The *superior angle* of

355

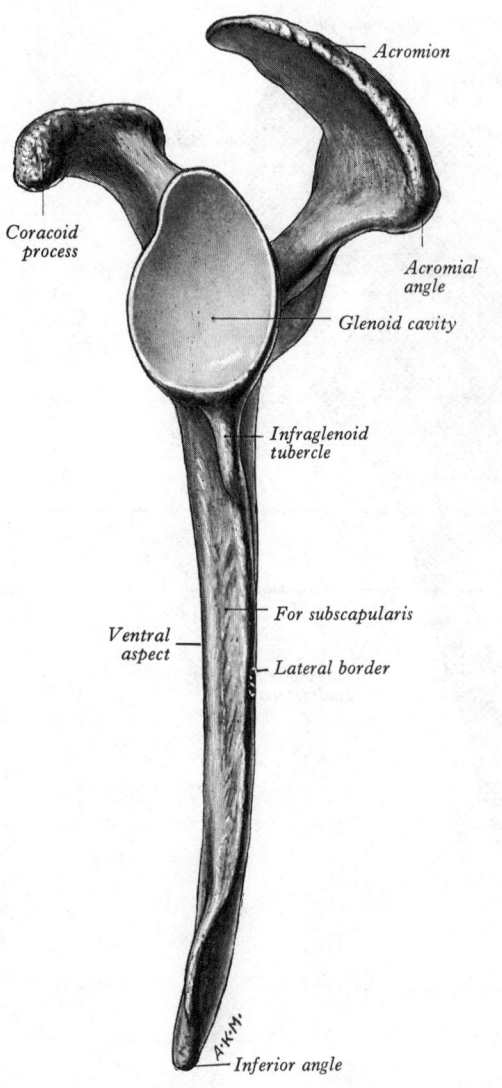

Coracoid process

Acromion

Acromial angle

Glenoid cavity

Infraglenoid tubercle

For subscapularis

Ventral aspect

Lateral border

Inferior angle

3.129   The left scapula. Lateral aspect.

the scapula is covered by the upper part of trapezius. The *lateral angle* bears the *glenoid cavity* to the margins of which the glenoidal labrum is attached. The surface of the cavity is covered with a layer of hyaline articular cartilage, thinnest at the centre and thickest at the periphery. Its anterior margin gives attachment to the glenohumeral ligaments (p. 457). To the *supraglenoid tubercle* the long head of the biceps brachii is attached.

The *spine of the scapula* gives attachment by its upper and lower surfaces to the supra- and infra-spinatus muscles. The flattened, triangular area at its root lies opposite the spine of the third thoracic vertebra and is played over by the tendon of the trapezius, a bursa intervening to facilitate this. The lower border of the crest is occupied by the posterior fibres of the deltoid. The upper border of the crest receives the insertion of the middle fibres of the trapezius. The lowest fibres of this muscle terminate in a flat triangular tendon which glides over the smooth area mentioned above and inserts into a roughened prominence, erroneously called the *deltoid tubercle*, on the dorsal aspect of the spine lateral to the area.

The *acromion* is subcutaneous over its dorsal surface, being covered only by the skin and superficial fascia. The lateral border, which is thick and irregular, and the tip of the process, as far round as the clavicular facet, give origin to the middle fibres of the deltoid. The medial aspect of the tip gives attachment, below the deltoid, to the lateral end of the coraco-acromial ligament. The articular capsule of the acromioclavicular joint is attached around the margins of the clavicular facet. Behind the facet, the medial border of the acromion gives insertion to the horizontal fibres of the trapezius. The inferior aspect of the acromion is relatively smooth, and together with the coraco-acromial ligament and the coracoid process forms an arch over the shoulder joint. The tendon of the supraspinatus passes below the overhanging acromion and is separated from it and from the deltoid by the subacromial bursae.

The *coracoid process* lies below the clavicle at the junction of the lateral fourth with the rest of the bone and is connected to its under surface by the coracoclavicular ligament. The attachment of the *conoid* part of the ligament has already been considered (p. 354); the *trapezoid* part is attached to the upper aspects of the horizontal part of the process (3.127). The superior aspect of the process receives also the insertion of the pectoralis minor. Its lateral border gives attachment to the wider, medial end of the coraco-acromial ligament and, below that, to the coracohumeral ligament. The enlarged tip of the process gives origin to the coracobrachialis, medially, and to the short head of the biceps,

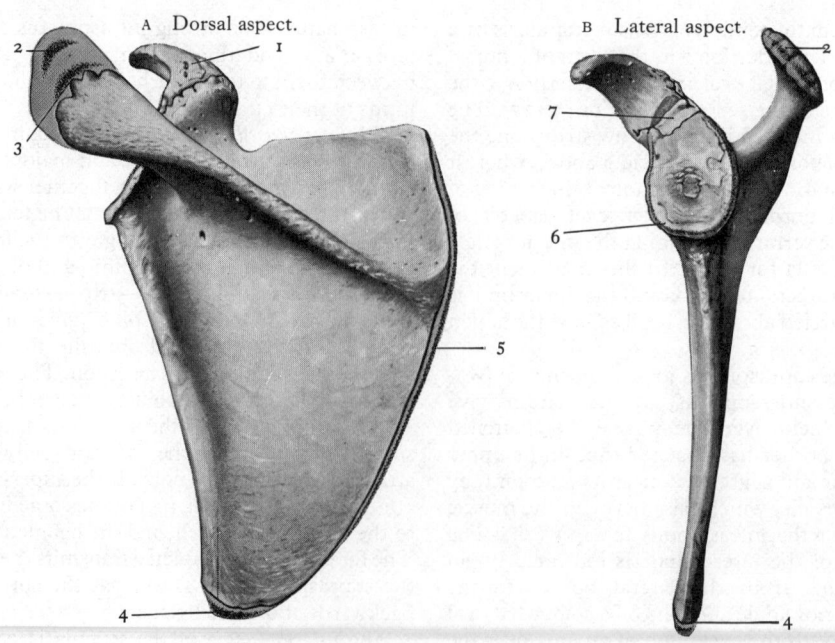

A   Dorsal aspect.

B   Lateral aspect.

3.130A and B   The ossification of the scapula. 1. Coracoid centre. 2. Distal acromial centre. 3. Proximal acromial centre. 4. Centre at inferior angle. 5. Centre for medial border. 6. Glenoid centre. 7. Subcoracoid centre.

laterally. It is covered by the anterior fibres of the deltoid and can be felt only on moderately deep pressure through the lateral border of the infraclavicular fossa.

It has recently been emphasized that the presence of a coraco-acromial ligament in man is a trait shared only with other hominoids amongst the anthropoid primates (Ciochou and Corruccini 1977). For comparative studies these authors devised a *coraco-acromial projection index*, which represents the *projection height* (i.e. vertical distance from the supraglenoid tubercle to a line connecting the most lateral points on the tips of the acromion and coracoid process, divided by the *height of the glenoid cavity*. They suggest that their observations reflect the specialized locomotor/feeding adaptations of *Hominoidea*.

The main processes, and thickened parts of the scapula contain cancellous bone; the rest consists of a thin layer of compact bone. The central part of the supraspinous fossa and the greater part of the infraspinous fossa are thin and even translucent; occasionally the bone is deficient in these situations, the gaps being filled by fibrous tissue.

Ossification (3.130A and B). The cartilaginous scapula is ossified from eight or more centres: one in the body, two in the coracoid process, two in the acromion, and one each in the medial border, the inferior angle and the lower part of the rim of the glenoid cavity.

The centre for the body appears in the eighth week of intrauterine life. Ossification begins in the middle of the coracoid process in the first year, or in a small proportion of individuals before birth, and the process joins the rest of the bone about the fifteenth year. At or soon after puberty centres of ossification occur in the rest of the coracoid process (subcoracoid centre), in the rim of the lower part of the glenoid cavity, frequently at the tip of the coracoid process, in the acromion, in the inferior angle and contiguous part of the medial border and in the medial border. A variable area of the part of the glenoid cavity, usually the upper third, is ossified from the subcoracoid centre; it unites with the rest of the bone in the fourteenth year in the female and the seventeeth year in the male. A horseshoe-shaped epiphysis appears for the rim of the lower part of the glenoid cavity; thicker at its peripheral than at its central margin, it converts the flat glenoid cavity of the child into the gently concave fossa of the adult. The base of the acromion is formed by an extension from the spine; the rest of the acromion is ossified from two centres which unite and then join the extension from the spine. The various epiphyses of the scapula have all joined the bone by about the twentieth year.

The Clavicle

The clavicle (3.131, 132) extends almost horizontally across the root of the neck laterally towards the point of the shoulder; it is subcutaneous throughout its whole extent. It acts as a prop which struts the shoulder and enables the limb to swing clear of the trunk; it transmits part of the weight of the limb to the axial skeleton. The *lateral* or *acromial end* of the bone is flattened and articulates with the medial side of the acromion, whereas the *medial* or *sternal end* is enlarged and articulates with the clavicular notch of the manubrium sterni. The *shaft* is gently curved and in shape resembles the italic letter *f*, being convex forwards in its medial two-thirds and concave forwards in its lateral third. The *inferior* aspect of the intermediate third is grooved in its long axis.

The lateral third of the clavicle is flattened and has superior and inferior surfaces, limited by anterior and posterior borders. The *anterior border* is concave, thin and roughened and may be marked by a small *deltoid tubercle*. The *posterior border*, also roughened for muscular attachments, is convex backwards. The *superior surface* is roughened near its margins but smooth centrally, where it can be felt through the skin. The *inferior surface* presents two obvious markings. Close to the posterior border, at the junction of the lateral fourth with the rest of the bone, a prominent tubercle gives attachment to the conoid part of the coracoclavicular ligament and is termed the *conoid tubercle*. From the lateral side of this tubercle a narrow, roughened strip runs forwards and laterally, reaching almost as far as the acromial end (3.132). It is the *trapezoid line* to which the trapezoid ligament is attached.

The medial two-thirds of the clavicle is cylindrical or prismatic in form and possesses four surfaces, but the inferior is often a mere ridge. The *anterior surface* is rough over most of its extent but smooth and rounded at its lateral end, where it is the upper boundary of the infraclavicular fossa (p. 567). The *superior surface*, also, is rough in its medial part and smooth laterally. The *posterior surface* is smooth, while the *inferior surface* is marked, near the sternal end, by a roughened oval impression, which is often depressed below the surface. Its margins give attachment to the costoclavicular ligament, which connects the clavicle to the upper surface of the first rib and its cartilage. Rarely, this area is smooth or raised to form an eminence which may articulate with the upper surface of the first rib by means of a synovial joint (Cave 1961). The lateral half of the inferior surface displays a groove in the long axis of the bone.

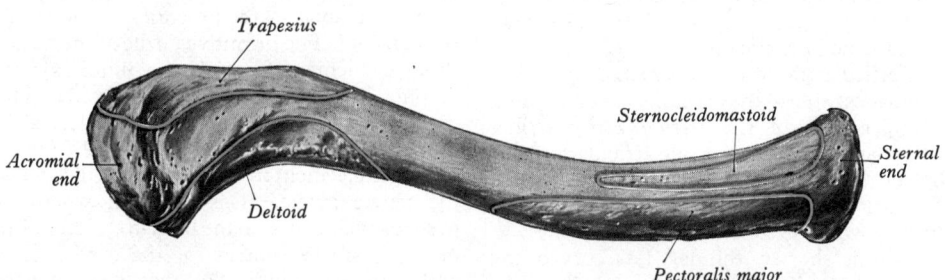

3.131   The right clavicle. Superior aspect.

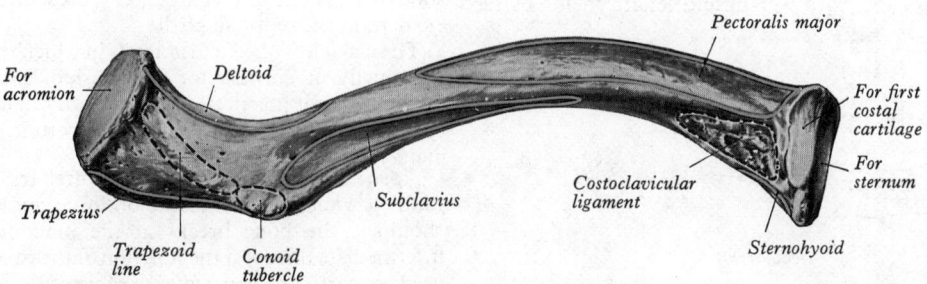

3.132   The right clavicle. Inferior aspect.

The acromial end of the clavicle is flattened and has a small oval articular facet, which articulates with the medial aspect of the acromion at the acromioclavicular joint. The facet faces laterally and slightly downwards.

The sternal end of the clavicle is directed medially, and a little downwards and forwards, to articulate with the clavicular notch of the manubrium sterni. The sternal surface is quadrangular (sometimes triangular) and its upper part is slightly roughened for ligamentous attachments. Elsewhere the surface is smooth and articular, extending on to the inferior surface for a short distance, where it articulates with the first costal cartilage. The sternal end of the clavicle projects upwards beyond the manubrium sterni and can be felt, and often seen, in the lateral wall of the jugular fossa.

The *lateral third* of the shaft has attached to it the deltoid at its anterior border and the trapezius on its posterior border. Both encroach on the upper surface. The coracoclavicular ligament, attached to the conoid tubercle and the trapezoid line (**3**.132), transmits the weight of the upper limb to the clavicle. This weight is counteracted by activity of the trapezius which supports the lateral part of the bone (p. 565). From the conoid tubercle the weight is transmitted through the medial two-thirds of the shaft to reach the axial skeleton. Fracture of the clavicle medial to the conoid tubercle interrupts the line of weight transmission, so that practically the whole weight of the limb has to be supported by trapezius. The muscle is unable to meet the demand and the limb therefore drops on the affected side.

The *medial two-thirds* provides attachment anteriorly over most of its extent to the clavicular head of the pectoralis major, and, as a rule, the area is clearly indicated on the bone. The clavicular head of the sternocleidomastoid arises from the medial half of the upper surface, but the marking on the bone is not conspicuous. The smooth, *posterior surface* is devoid of muscular attachments except at its lower part immediately adjoining the sternal end, where the lateral fibres of the sternohyoid arise. Medially, this surface is related to the lower end of the internal jugular vein (from which it is separated by the sternohyoid), the termination of the subclavian vein and the commencement of the brachiocephalic vein. More laterally, it arches in front of the trunks of the brachial plexus and the third part of the subclavian artery. The suprascapular vessels are related to the upper part of this surface. The *inferior surface* gives insertion to the subclavius in the subclavian groove (**3**.132), and to the edges of the groove the clavipectoral fascia, which encloses the muscle, is attached. The posterior lip of the groove runs into the conoid tubercle and carries the fascia into continuity with the conoid ligament. A nutrient foramen occurs at the lateral end of the groove, running in a lateral direction. The nutrient artery concerned is derived from the suprascapular artery. The impression for the costoclavicular ligament is very variable in character.

The margins of the articular facet at the *acromial end* afford attachment to the articular capsule of the acromioclavicular joint.

The roughened, upper part of the *sternal end* provides attachment for the interclavicular ligament, the articular capsule and the articular disc of the sternoclavicular joint. The sternal surface, denuded of its articular cartilage, is rarely smooth and is usually irregular and somewhat pitted.

In women, the clavicle is shorter, thinner, less curved and smoother, and the acromial end is a little below the level of the sternal end; in men it is on a level with, or slightly higher than, the sternal end, with the arm hanging by the side, of course. The midshaft circumference is the most reliable single indicator of sex; but a combination of this with weight and length yields better

results (Olivier 1951; Jit and Singh 1966). In those who perform hard manual labour the clavicle is thicker and more curved, and it ridges for muscular attachment are better marked. The clavicl consists of cancellous bone internally, enveloped by a layer c compact bone which is much thicker in the intermediate part tha at the ends. Unlike most other long bones it does not possess medullary cavity.

Ossification (**3**.133). The clavicle begins to ossify before an other bone in the body. The shaft is ossified from a pair of primar centres, medial and lateral, which appear in condense mesenchyme between the fifth and sixth weeks of intrauterine life and fuse about the forty-fifth day. A secondary centre for th sternal end appears in the late teens, or even early twenties usually about two years earlier in the female. Fusion probabl occurs quickly, but reliable figures are not available. A secondar centre sometimes develops at the acromial end at about 18 to 20 but the epiphysis so formed is always small and rudimentary an rapidly joins the rest of the bone (Todd and D'Erico 1928). In th most recent study of the sternal epiphysis, carried out on 68. Punjabees, ossification occurred from 11 to 19 years (both sexes and fusion at 18 to 25 years (Jit and Kulkaria 1976). Compariso with other studies suggests racial differences. Assessment b radiological appearances is, of course, not devoid of fallacy.

However, the full development of the clavicle is not exclusivel by a process of simple intramembranous ossification. In a 14 mn embryo the future clavicle is represented by a band of condense mesenchyme which extends from the acromion of the scapula t the tip of the first rib, and is continuous with the rudiment of th sternum. In this band medial and lateral zones of early cartilage transformation (so-called 'precartilage') occur, and in the mesenchyme between them two intramembranous centres of ossification for the body appear and soon fuse. The sternal an acromial zones soon transform into true cartilage, into which ossification of the body extends. Growth in length of the clavicle continues by longitudinal, interstitial growth of these terminal masses of cartilage, which develop zones of hypertrophy, calcification and an advancing front of endochondral ossification as in other growth cartilages. Girth increments follow apposi- tional subperichondrial growth of the cartilaginous extremities and continued subperiosteal deposition of bone throughout the shaft. The epiphyses are formed endochondrally and fusion probably occurs as in other long bones, which, in contrast to the clavicle, are *primarily* cartilaginous.

Imperfections of ossification in the clavicle and other primarily intramembranous skull bones, are occasionally encountered together, in the condition of cleido-cranial dysostosis.

The primitive reptilian pectoral girdle comprises a dorsal element—the scapula—and two ventral elements, of which the anterior (cranial) is the precoracoid and the posterior (caudal) is the coracoid. The primitive girdle of the hind-limb also possesses three elements, of which the ilium is homologous with the scapula, the pubis with the precoracoid and the ischium with the coracoid. The clavicle, which is a dermal bone in origin and therefore morphologically distinct from the others, is an additional element in the pectoral girdle and is not represented in the pelvic girdle. It is doubtful whether any trace of the precoracoid persists in the human skeleton, although the presence of two primary centres for the clavicle is regarded by many authorities as an indication that the human clavicle corresponds both to the precoracoid and to the clavicle in the reptilian shoulder girdle. Others believe that the first coracoid centre to appear in man represents the precoracoid element, while the subcoracoid centre is regarded as representing the caudal ventral element in the reptilian girdle.

The clavicle is absent in animals in which the forelimbs are used principally or entirely for progression, e.g. the ungulates and carnivores, but it is present and well developed in animals which use the limb for prehension, e.g. many rodents, the primates and man.

Applied Anatomy. The clavicle is often fractured, commonly by indirect violence, the result of force applied to the hand or shoulder. the bone breaks at the junction of its lateral and intermediate thirds, at the junction of the two curves, for this is its weakest part. The consequent deformity is caused by the weight of the arm, which acts on the lateral fragment through the

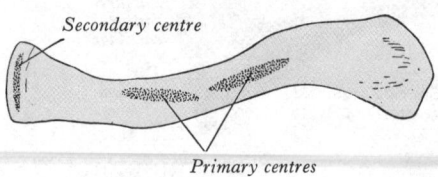

Secondary centre

Primary centres

358   **3**.133   Diagram showing the three centres of ossification of the clavicle.

racoclavicular ligament and draws it downwards. The medial agment, as a rule, is little displaced.

he Humerus

he humerus (**3**.134–139) is the longest and largest bone of the pper limb. It comprises expanded upper and lower extremities d a shaft. Proximally a round 'head' articulates with the glenoid vity of the scapula in a ball and socket joint (but *see* p. 456). The

lower extremity is loosely termed 'condylar' in form, adapted to the forearm bones at the hinge of the elbow joint.

The upper end of the humerus consists of the articular head, and the greater and lesser tubercles.

The head of the humerus (**3**.134, 135) forms rather less than half a spheroid, and its smooth surface is covered with hyaline articular cartilage. When the arm is by the side, the head is directed medially, backwards and upwards to articulate with the glenoid cavity of the scapula. The humeral articular surface is much more extensive than the glenoid cavity, and only a portion

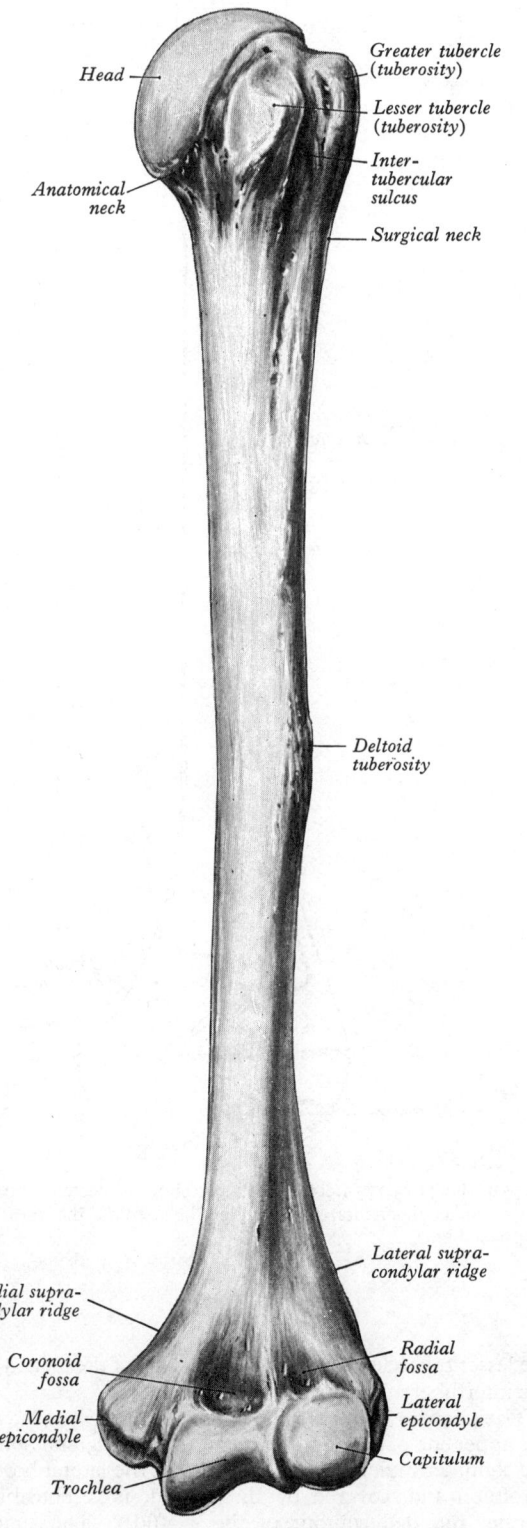

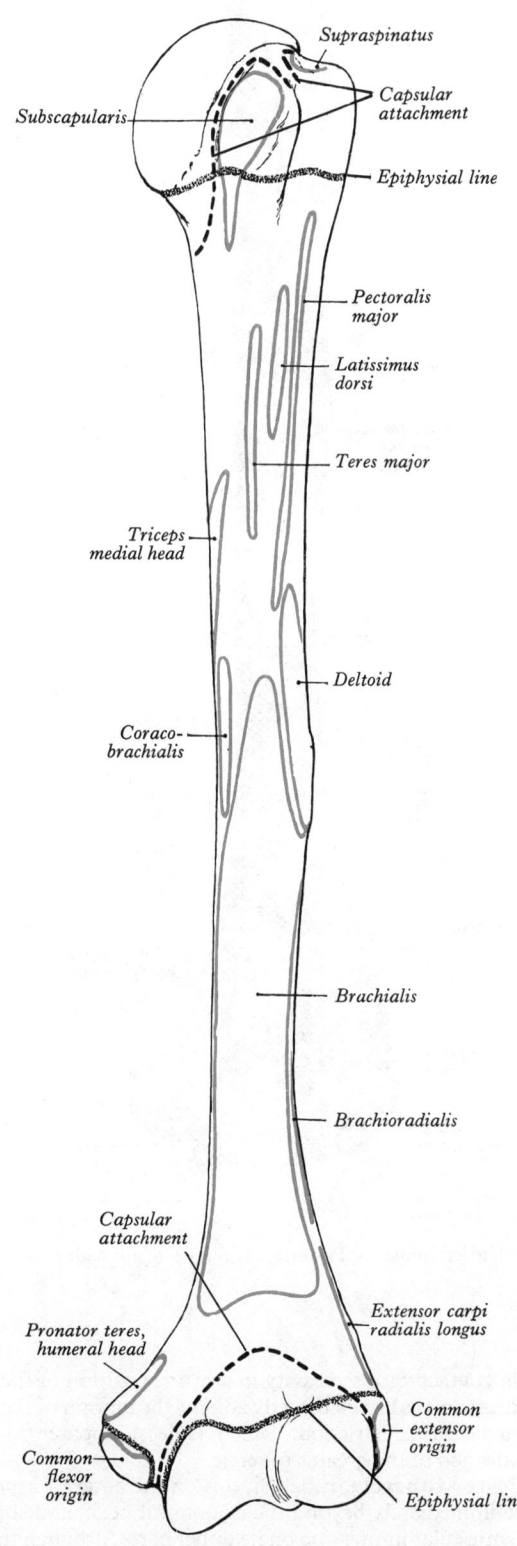

3.134A The left humerus. Anterior aspect. Compare with Key, **3**.134B.

3.134B Key to **3**.134A. The interrupted lines indicate the attachment of the capsular ligaments; the stippled lines mark the position of the epiphysial lines.

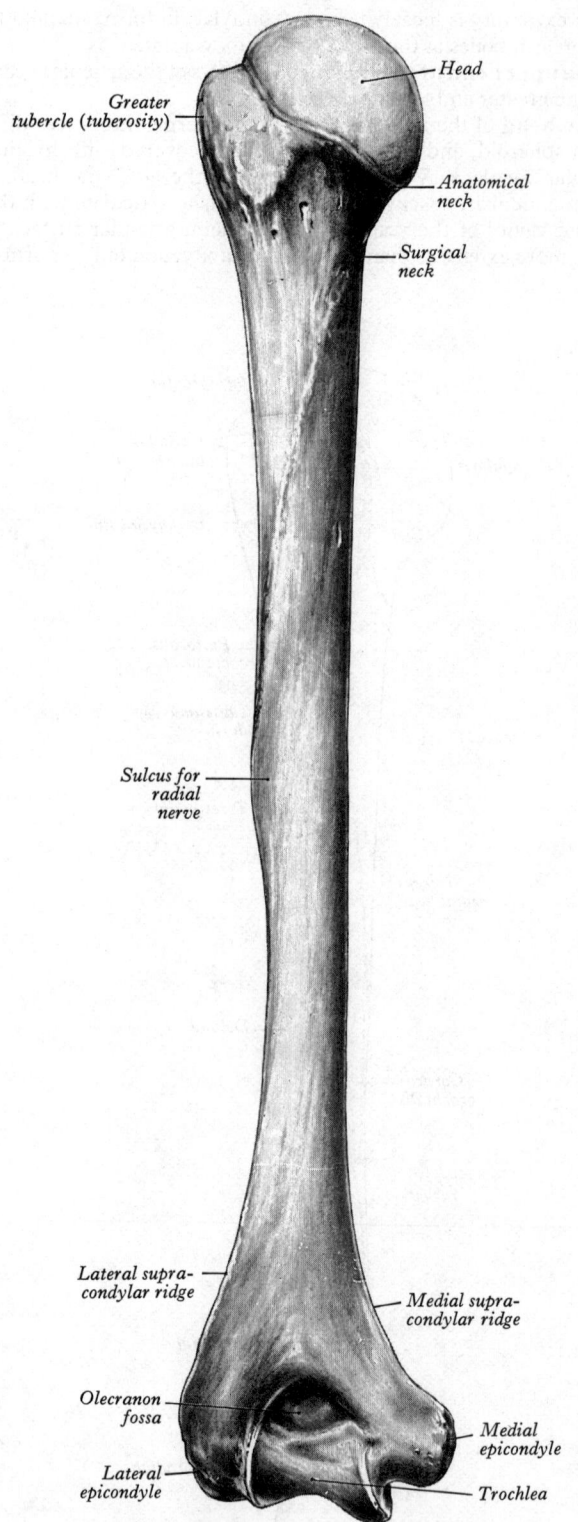

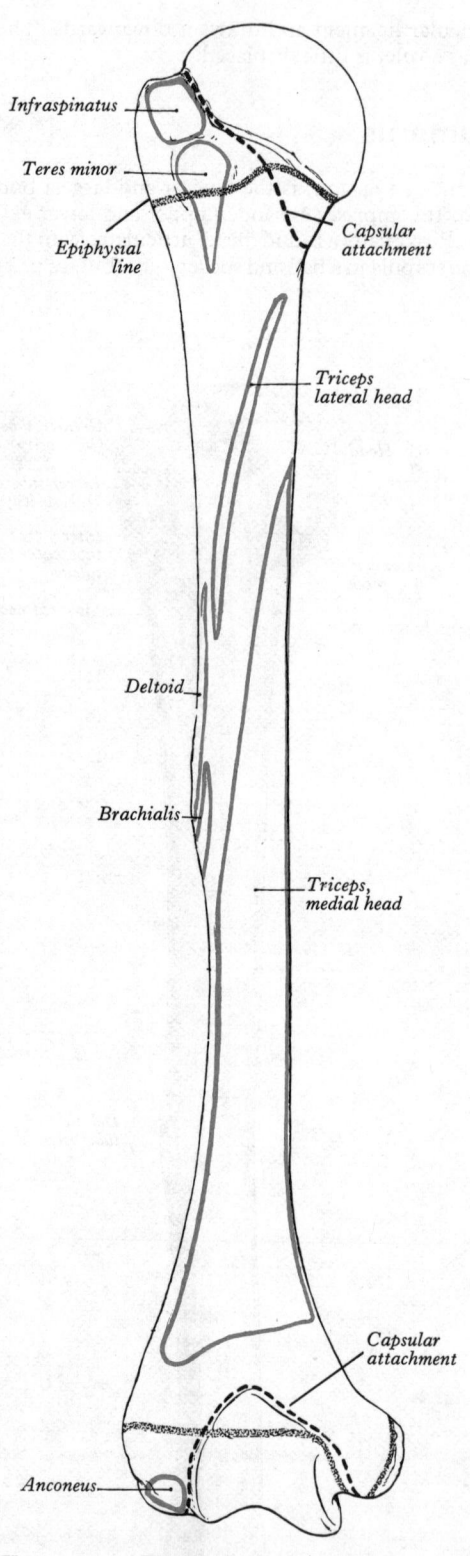

3.135A The left humerus. Posterior aspect. Compare with Key, 3.135B.

3.135B Key to 3.135A. The interrupted lines indicate the attachment o
the capsular ligaments; the stippled lines mark the position of th
epiphysial lines.

of it is in contact with the cavity in any one position of the arm.

The anatomical neck directly adjoins the margin of the head and is a slight constriction, which is least apparent in the neighbourhood of the greater tubercle.

The lesser tubercle (tuberosity) is on the anterior aspect of the bone immediately beyond the anatomical neck, and shows a smooth, muscular impression on its upper part. Although thickly covered by the deltoid it can be felt on deep pressure 3 cm below the tip of the acromion. The bony prominence moves under the examining finger when the humerus is rotated. The lateral edge of

the lesser tubercle is sharp and forms part of the medial border o
the intertubercular sulcus.

The greater tubercle (tuberosity) occupies the lateral part o
the upper end of the humerus and is the most lateral bony point in
the shoulder region. It projects beyond the lateral border of the
acromion and, covered by the deltoid, is responsible for the
normal rounded contour of the shoulder. The superior and
posterior aspect of the tubercle adjoining the anatomical neck
bears three flattened impressions for muscular attachments.

The two tubercles are separated by the **intertubercular**

ilcus. Where the upper extremity joins the shaft is a tapering region called the *surgical neck*. Close to this medially are the illary nerve and the posterior humeral circumflex artery .135).

The shaft of the humerus is almost cylindrical in its upper .lf but triangular on section below this, where it is compressed in a anteroposterior direction. It can be felt when the arm is asped firmly, but it is obscured by the strong muscles which othe it. It presents three surfaces and three borders—which are ot everywhere equally obvious.

The *anterior border* commences above on the front of the eater tubercle and runs downwards almost to the lower end of e bone. Its upper third forms the lateral edge of the tertubercular sulcus and is roughened for muscular attach-ents. The succeeding portion is also roughened and forms the terior limit of the deltoid tuberosity, but the lower half of the rder is smooth and rounded.

The *lateral border* is conspicuous inferiorly where its sharp ige is roughened along its anterior aspect. In its middle and pper thirds the border is barely discernible to the eye, but in a ell-marked bone it can be traced upwards to the posterior irface of the greater tubercle. About its middle the border is iterrupted by a wide, shallow groove which crosses the bone bliquely, passing downwards and forwards from its posterior to s anterior surface. It is termed the *sulcus for the radial nerve*.

The *medial border*, although rounded, is clear in the lower half f the shaft. A little below the middle of the bone it presents a oughened strip, and superiorly it becomes indistinct until it eappears as the medial lip of the intertubercular sulcus. In this tuation the border is again roughened and can be traced into the sser tubercle.

The *anterolateral surface* of the humerus lies between the nterior and the lateral borders. A little above its middle it is larked by a roughened area, tapering to a point below, which is ermed the *deltoid tuberosity*. Behind this the groove for the radial erve runs downwards and fades away on the lower part of the irface.

The *anteromedial surface* is bounded by the anterior and the nedial borders of the bone. Rather less than its upper third forms ne rough floor of the intertubercular sulcus, but the rest of the urface is smooth. A little below its middle the nutrient foramen, 'hich is directed downwards, opens close to the medial border.

The *posterior surface* lies between the medial and the lateral orders and is the most extensive surface of the three. Its upper nird is crossed by a faint ridge, sometimes roughened, which uns obliquely downwards and laterally. The middle third is rossed by the commencement of the groove for the radial nerve. tather more than the lower third forms an extensive, flattened urface, which widens considerably below.

The lower end of the humerus (3.134, 135, 136), basically a 1odified *condyle*, is expanded transversely, and presents articular nd non-articular parts.

The *articular part* adjoins the radius and the ulna in the ormation of the elbow joint. It is divided by a faint groove into a iteral, convex surface, the capitulum, and a medial, pulley-haped surface, the trochlea.

The capitulum is a rounded, convex projection, considerably ess than half a sphere, formed of the anterior and inferior surfaces f the lateral part of the condyle of the humerus but not extending n to its posterior surface. It articulates with the disc-like head of he radius, which lies in contact with its inferior surface in full xtension of the elbow but moves on to its anterior surface when he joint is flexed.

The trochlea is a grooved surface, like part of the ircumference of a pulley, which covers the anterior, inferior and)osterior surfaces of the condyle of the humerus. On its lateral ide it is separated from the capitulum by a faint groove, but its nedial margin is salient and projects downwards beyond the rest)f the bone. The trochlea articulates with the trochlear notch of he ulna. When the elbow is extended the inferior and posterior ispects of the trochlea are in contact with the ulna, but, as the joint s flexed, the trochlear notch slides forwards on to the anterior ispect and the posterior aspect is then left uncovered. The lownward projection of the medial edge of the trochlea is a major actor in determining the angulation which is present between the

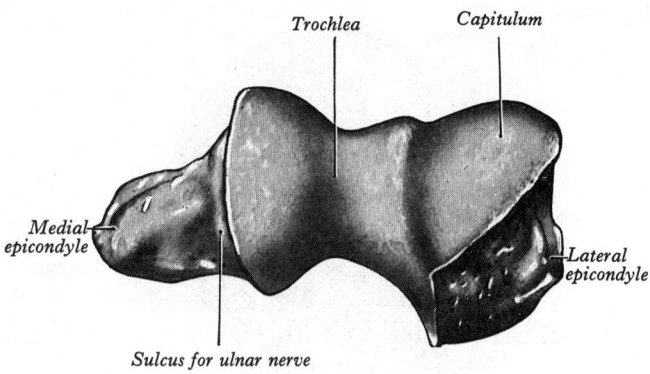

3.136   The lower end of the left humerus. Distal (inferior) aspect.

long axis of the humerus and the long axis of the ulna and supinated forearm when the elbow is extended (p. 461).

The *non-articular part* of the condyle of the humerus includes the medial and lateral epicondyles, together with the olecranon, coronoid and radial fossae.

The medial epicondyle is a conspicuous, blunt projection on the medial side of the condyle. It is subcutaneous and easily

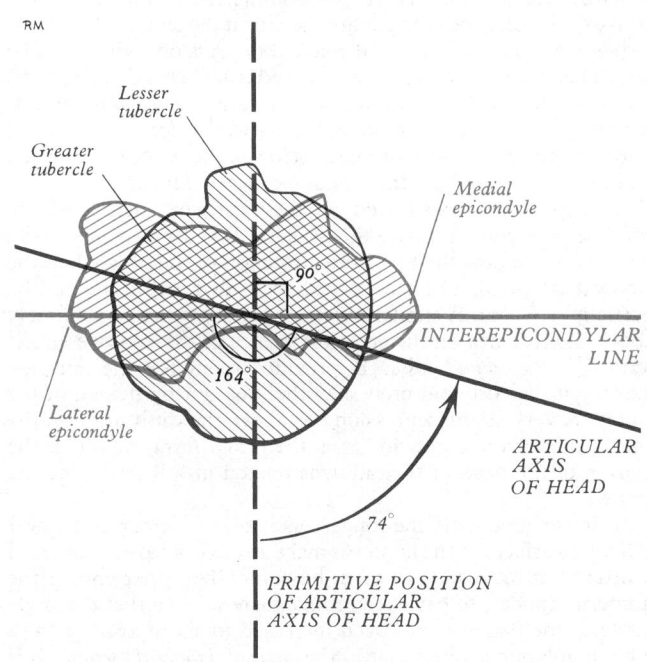

3.137   This diagram illustrates the concept of humeral 'torsion'. The left humerus is viewed distally along its length, with its proximal end (*magenta*) superimposed upon the distal (*blue*). The 'axes' of the two extremities are shown at right angles, as in early tetrapods and most quadrupedal mammals. In primates, including mankind, the axis of the head is directed medially, and the angle which this now makes with the 'primitive' anteroposterior position of the axis is the so-called angle of torsion. This is on average 74° in man: some authorities add to this the original angle of 90°, giving a total of 164°. See pp. 351 and 362 for more detailed explanation.

identified through the skin, especially in passive flexion of the elbow. Its posterior surface is smooth and crossed by the ulnar nerve, in a shallow sulcus, as it runs down into the forearm. In this situation the nerve can be felt and rolled against the bone. If the pressure exerted is sufficient, sensations are aroused identical with those produced when the nerve is jarred against the epicondyle. The lower part of the anterior surface of the medial epicondyle bears an impression for attachment of the superficial group of forearm flexor muscles.

The lateral epicondyle is the lateral part of the non-articular

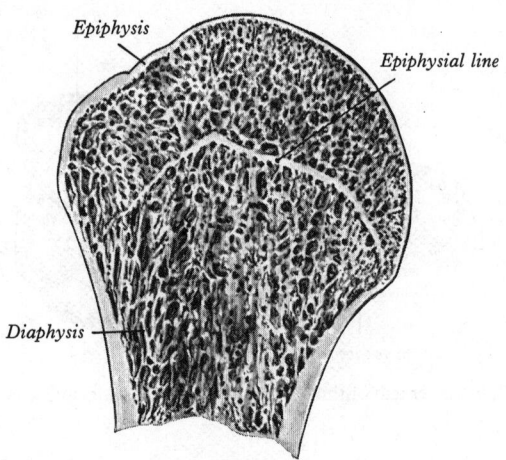

Epiphysis

Epiphysial line

Diaphysis

3.138 A longitudinal section through the head of the right humerus in adolescence.

portion of the condyle, but does not project beyond the lateral supracondylar ridge. Its lateral and anterior surfaces display a well-marked impression for the origin of the superficial group of forearm extensor muscles. Its posterior surface, which is slightly convex, can easily be felt through the skin at the back of the elbow. It lies at the bottom of a well-marked depression, which can be seen when the extended elbow is viewed from behind. The lateral border of the humerus terminates at the *lateral epicondyle*, and its lower portion is the *lateral supracondylar ridge*. The medial border of the humerus terminates below at the *medial epicondyle*, and its lowest portion is the *medial supracondylar ridge*.

A deep hollow is situated on the posterior surface of the condyle, immediately above the trochlea. It is the *olecranon fossa*, because it contains the tip of the olecranon of the ulna when the elbow is extended. The floor of the fossa is always thin and may be partially deficient. A similar but smaller hollow lies immediately above the trochlea on the anterior surface of the condyle and is termed the *coronoid fossa*. It provides room for the anterior margin of the coronoid process of the ulna during flexion of the elbow. A very slight depression lies above the capitulum on the lateral side of the coronoid fossa, the *radial fossa*, to which the margin of the head of the radius is related in full flexion of the elbow.

In lower mammals the longest axes of the upper and lower articular surfaces of the humerus make an angle with each other of a little more than 90°. In man, however, the upper end of the humerus appears to have been rotated laterally, so that the angle between the two axes has been increased to about 164° (**3.137**). This angulation is referred to as the *angle of 'humeral torsion'*. It is greater in men than women, in adults than children, and in man than in anthropoid apes. Krahl (1976) claims that the angle also increases with age, up to the time of fusion of the epiphysis (see also p. 351).

The *articular cartilage* which covers the *head* of the humerus is thickest at its centre and becomes thinner towards the circumference.

The *anatomical neck* gives attachment to the *capsular ligament* of the shoulder joint (**3.134, 135**), except at the upper end of the intertubercular sulcus, where the long tendon of the biceps emerges from the joint. On the medial side, however, the attachment extends downwards for 1 cm or more on to the shaft of the bone.

To the *lesser tubercle* the subscapularis is attached (**3.134B**) and to its sharp lateral margin the transverse ligament of the shoulder joint.

The *greater tubercle* shows three impressions for muscle insertions; the uppermost for supraspinatus, the middle for infraspinatus, the lowest, placed on the posterior surface of the tubercle, for the teres minor (**3.135B**). The projecting lateral surface of the tubercle presents numerous vascular foramina and is covered by the deltoid; a part of the subacromial bursa

may cover the upper part of this area and separate it from t muscle.

The *intertubercular sulcus* (bicipital groove) is occupied by t long tendon of the biceps, its accompanying synovial sheath, a an ascending branch from the anterior circumflex humeral arter The rough, lateral lip of the groove gives insertion to the tendon the pectoralis major, its floor, to the tendon of latissimus dor and its medial lip, to the tendon of the teres major. The insertic of the pectoralis major extends lower than the insertion of ter major, while the insertion of latissimus dorsi is the least extensi of the three. Below the intertubercular sulcus the anteromedi surface of the humerus is devoid of muscular attachment over small area, but to its lower half the medial portion of the brachial is attached (**3.134B**). The roughened strip on the middle of tl medial border of the bone gives insertion to the coracobrachiali Close to the lowest part of the medial supracondylar ridge narrow area provides attachment for the humeral head of tl pronator teres, and the ridge itself to the medial intermuscul: septum of the arm.

The oblique ridge which crosses the upper part of the *posteri surface* gives origin to the lateral head of the triceps. Abov this muscle the axillary (circumflex) nerve and the posteri circumflex humeral vessels wind round this aspect of the bor under deltoid. Below and medial to the origin of the lateral head triceps, a shallow groove, containing the radial nerve and tl profunda vessels, runs downwards and laterally to gain tl anterolateral surface of the shaft. The area for the fleshy medi head of the triceps includes a very large part of the posteri surface of the bone. It is an elongated triangular area, the apex which is on the medial part of the posterior surface above the lev of the lower limit of the insertion of teres major. The area wide below and covers the whole dorsal surface almost down to tl lower end of the bone (**3.135B**).

The *anterolateral surface* of the humerus is smooth in its upp part, which is covered by the deltoid. A little above the middle this surface the deltoid is attached to the deltoid tuberosity, an below this the surface is occupied by the lateral fibres of tl brachialis, which extend upwards into the floor of the lower end the radial groove (**3.135B**). The rough anterior aspect of the later supracondylar ridge gives origin by its upper two-thirds t brachioradialis and by its lower third to extensor carpi radial longus. Behind these muscles the ridge is joined by the later intermuscular septum of the arm.

The *articular portion of the condyle* of the humerus is curve forwards, so that its anterior and posterior surfaces lie in front the corresponding surfaces of the shaft. The groove of tl trochlea winds backwards and laterally as it is traced from tl anterior to the posterior surface of the bone, and it is wide deeper and more symmetrical posteriorly. Anteriorly, the media flange of the pulley is much longer than the lateral, and the surfac adjoining its projecting medial margin is convex to accommodat itself to the medial part of the upper surface of the coronoi process of the ulna. These asymmetries are, of course, associate with the varying angulation between the humeral and ulnar axes.

The *capsular ligament of the elbow joint* (**3.134B, 135B**), i attached anteriorly to the upper limits of the radial and coronoi fossae, so that both these bony depressions are intracapsular an therefore covered by synovial membrane. It is also attached to th medial non-articular aspect of the projecting lip of the trochle and to the root of the medial epicondyle. Posteriorly it ascends to or almost to, the upper margin of the olecranon fossa, which i therefore intracapsular and covered with synovial membrane Laterally it skirts the lateral borders of the trochlea an capitulum, lying medial to the lateral epicondyle.

The common origin of the superficial group of flexor muscle arises from the epiphysis for the medial epicondyle, entirel outside the articular capsule of the elbow joint. The commor origin of the superficial group of extensor muscles of the forearn is attached to the lateral epicondylar part of the lower humera epiphysis, and as in the case of the flexors, is outside the articula capsule. To a small area on the posterior surface of the lateral epicondyle the anconeus is attached (**3.135B**). The medial epicondyle is directed backwards a little at its extremity, where as the lateral epicondyle shows a slight trend in the opposite direction.

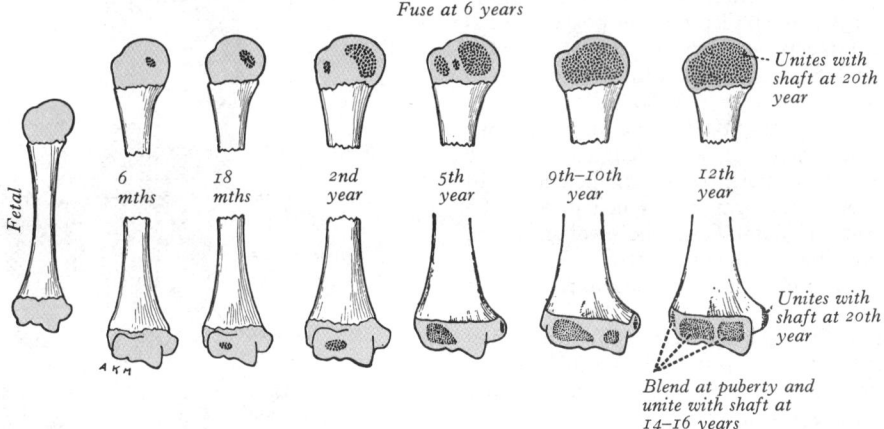

Fuse at 6 years

Unites with shaft at 20th year

Unites with shaft at 20th year

Blend at puberty and unite with shaft at 14-16 years

3.139   Stages in the ossification of the humerus.

It should be observed that when the humerus is at rest by the side, the medial epicondyle lies on a plane which is much posterior to the plane of the lateral epicondyle. In this position the head of the humerus is directed almost equally *backwards* and *medially*, and the posterior surface of the shaft looks laterally and also backwards. Since the glenoid cavity faces forwards, as well as laterally, the humerus is not, in fact, rotated medially *relative to the scapula*, in this position of rest; but it *is* so rotated relative to the position of the arm in the *conventional anatomical position* of descriptive anatomy. This position of the bone must be kept in mind when the movements of the arm and forearm are considered (pp. 458 and 465; see also *Introduction*).

A hook-shaped process of bone, the *supracondylar process*, varying from 2 mm to 20 mm in length, is occasionally found projecting from the anteromedial surface of the shaft of the humerus, about 5 cm above the medial epicondyle. It is curved downwards and forwards, and its pointed end is connected to the medial border, just above the epicondyle, by a fibrous band from which part of the pronator teres arises; the foramen completed by this fibrous band usually encloses the median nerve and brachial artery. Sometimes the nerve travels through it alone, or it may be accompanied by the ulnar artery in cases of high division of the brachial. A groove, occupied by the artery and nerve, usually appears behind the process. This foramen is the homologue of the *entepicondylar foramen* found in many animals, and probably serves in them to protect the nerve and artery from compression during the contraction of the muscles in this region. These skeletal features are regarded as valuable in assessing racial affinities (p. 238).

Ossification (3.139). The humerus is ossified from eight centres—one for each of the following parts: the shaft, the head, the greater tubercle, the lesser tubercle, the capitulum with the lateral part of the trochlea, the medial part of the trochlea, and one for each epicondyle. The centre for the shaft appears near its middle in the eighth week of intrauterine life, and gradually extends towards its ends. Ossification usually commences in the humeral head in the first six months after birth, but just before this in a fifth of individuals. The greater and lesser tubercles begin to ossify in the second and fifth years in males, and about a year earlier in females. By the sixth year the centres for the head and tubercles have joined to form a single large epiphysis, which is hollowed out on its inferior surface (3.138) to adapt it to the somewhat conical upper end of the diaphysis. It fuses with the shaft of the humerus about the twentieth year in males, about two years earlier in females. Some observers deny the existence of a centre for the lesser tubercle, perhaps because it is often obscured in the usual anteroposterior views of the shoulder (3.140). The lower end is ossified as follows. During the first year ossification begins in the capitulum and extends medially to form the chief part of the articular surface; the centre for the medial part of the trochlea appears in the ninth year in females and tenth year in

males. Ossification begins in the medial epicondyle in the fourth year in females and the sixth year in males and in the lateral epicondyle about the twelfth year. The centres for the lateral epicondyle, capitulum and trochlea fuse around puberty and the large epiphysis thus formed unites with the shaft in the fourteenth

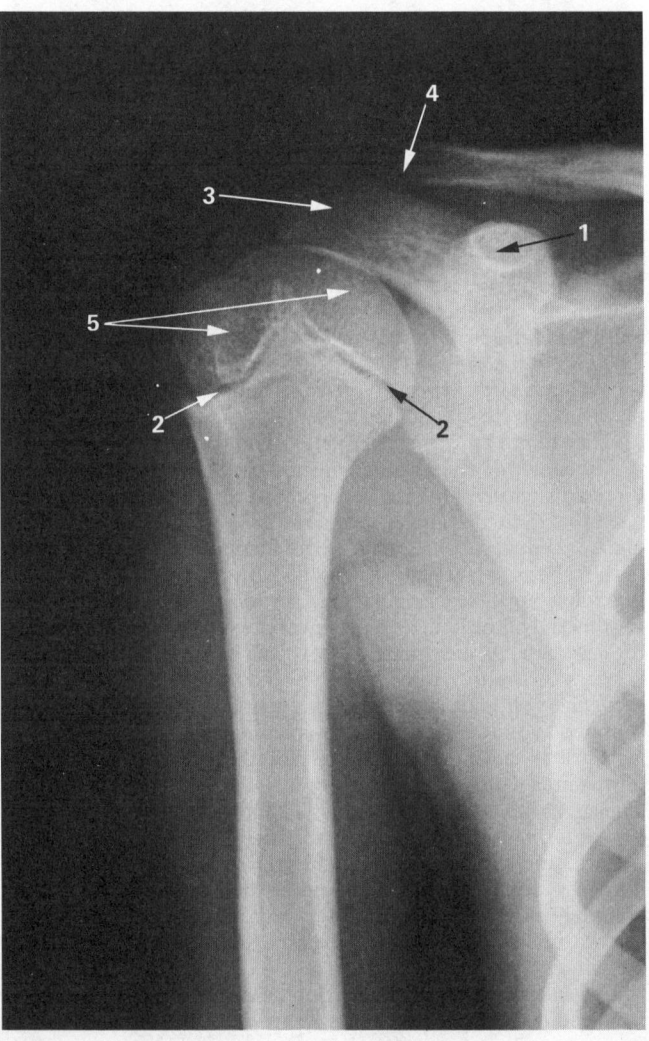

3.140   Anteroposterior radiograph of the right shoulder in a boy aged 11 years. 1. Coracoid process. 2. Growth plate of cartilage at upper end of humeral diaphysis. 3. Acromion. 4. Lateral end of clavicle, not yet completely ossified. 5. Proximal humeral epiphysis—note its conical junction with the diaphysis.

year in females and sixteenth year in males. The centre for the medial epicondyle forms a separate epiphysis, which is entirely extracapsular (3.134B, 3.135B) and is placed on the posteromedial aspect of the epicondyle. It is separated from the rest of the lower epiphysis by a downgrowth from the shaft, with which it unites about the twentieth year.

Applied Anatomy. The upper epiphysis of the humerus joins the shaft later than the lower, and the growth in length of the bone is due predominantly to the proximal growth cartilage. Hence, in cases of amputation through the arm in young subjects, the humerus continues to grow considerably, and the lower end of the bone, which immediately after the operation is covered with a thick cushion of soft tissue, begins to project, rendering the stump conical.

Fractures of the humerus are comparatively common and may occur at almost any level. It is fractured by muscular action probably more frequently than any other long bone; it is usually the shaft, just below the attachment of deltoid, which is thus broken. The radial nerve may be injured as it lies in its groove; rarely it is involved later in the growth of callus. Non-union is frequent and, indeed, commoner than in any other bone except the tibia. Fractures of the upper end of the bone may damage the axillary nerve, while fractures of the medial epicondyle may be complicated by damage to the ulnar nerve. The late fusion of the ossifying medial epicondyle sometimes leads to a fallacious diagnosis of fracture in this region.

The Radius

The radius (3.141–144) is the lateral bone of the forearm. Its upper and lower ends are expanded, but the lower end is much the wider of the two. The shaft increases in breadth rapidly towards the lower end, is convex laterally and is concave forwards in its lower part.

The upper end of the radius includes a head, neck and tuberosity. *The head* is disc-shaped and its upper surface is a shallow cup for articulation with the capitulum of the humerus. The *articular circumference* of the head is smooth and widest medially, where it articulates with the radial notch of the ulna. Its posterior surface can be felt through the skin as it lies at the bottom of a small depression which is visible in the living subject on the lateral side of the posterior surface of the extended elbow. The *neck* of the radius is the constricted part below the head, and is overhung by it, especially on the lateral side. *The tuberosity* is placed below the medial part of the neck. Its posterior part is roughened but anteriorly is usually smooth.

The shaft of the radius is gently curved with the convexity directed laterally. On transverse section it is triangular, but only one of its three margins is sharp, the *interosseous border*, a salient crest, except at its upper end, where it approaches the lower part of the tuberosity. At its lower end it forms the posterior margin of a small, elongated, triangular area, an additional, medial, surface for this end of the bone. To the lower three-fourths of this border the interosseous membrane is attached; it connects the opposed borders of the two forearm bones. The *anterior border* can be recognized without difficulty at its upper and lower ends, but it is rounded and indefinite in the intervening region. It commences just below the anterolateral part of the tuberosity and runs downwards with a lateral inclination. This part is described as the *anterior oblique line* of the radius. The lower part of the anterior border forms a sharp crest along the lateral margin of the anterior surface, and can even be felt through the skin. The *posterior* border is clearly defined in its middle third only. Above, it runs obliquely upwards and medially towards the postero-inferior part of the tuberosity. Below, it forms a rounded ridge which is difficult to trace.

The *anterior surface* lies between the anterior and interosseous borders. It is slightly concave from side to side and curves forwards at its lower end. A little above its middle is a nutrient foramen, directed upwards. The *posterior surface*, which is bounded by the interosseous and the posterior borders, is generally flat but may be slightly hollowed out in its upper part The *lateral surface* is gently convex in all directions. Above, owing

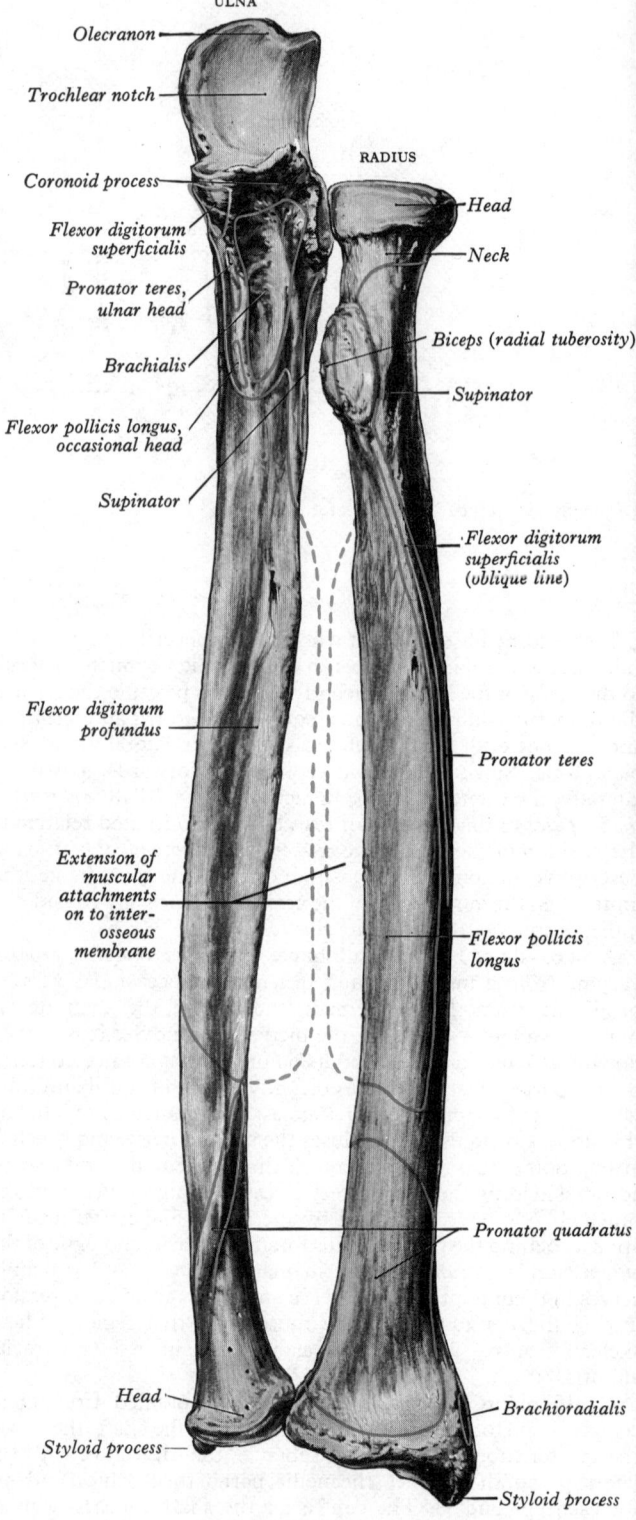

ULNA

Olecranon

Trochlear notch

RADIUS

Coronoid process

Head

Flexor digitorum superficialis

Neck

Pronator teres, ulnar head

Biceps (radial tuberosity)

Brachialis

Supinator

Flexor pollicis longus, occasional head

Supinator

Flexor digitorum superficialis (oblique line)

Flexor digitorum profundus

Pronator teres

Extension of muscular attachments on to interosseous membrane

Flexor pollicis longus

Pronator quadratus

Head

Brachioradialis

Styloid process

Styloid process

3.141 The bones of the left forearm. Anterior aspect.

to the obliquity of the upper parts of the anterior and posterior borders, it encroaches on the anterior and posterior aspects of the bone, and this widened portion is usually slightly roughened. A finely irregular, rough surface occupies an oval area near the middle of the shaft, but below this the surface of the bone is smooth and featureless.

The lower end of the radius, the widest part, is four-sided on transverse section. Its *lateral surface* is slightly rough and projects downwards beyond the rest of the bone to form *the styloid process.* This projection can be felt through the skin, when the tendons which conceal it in the living body are relaxed. The inferior

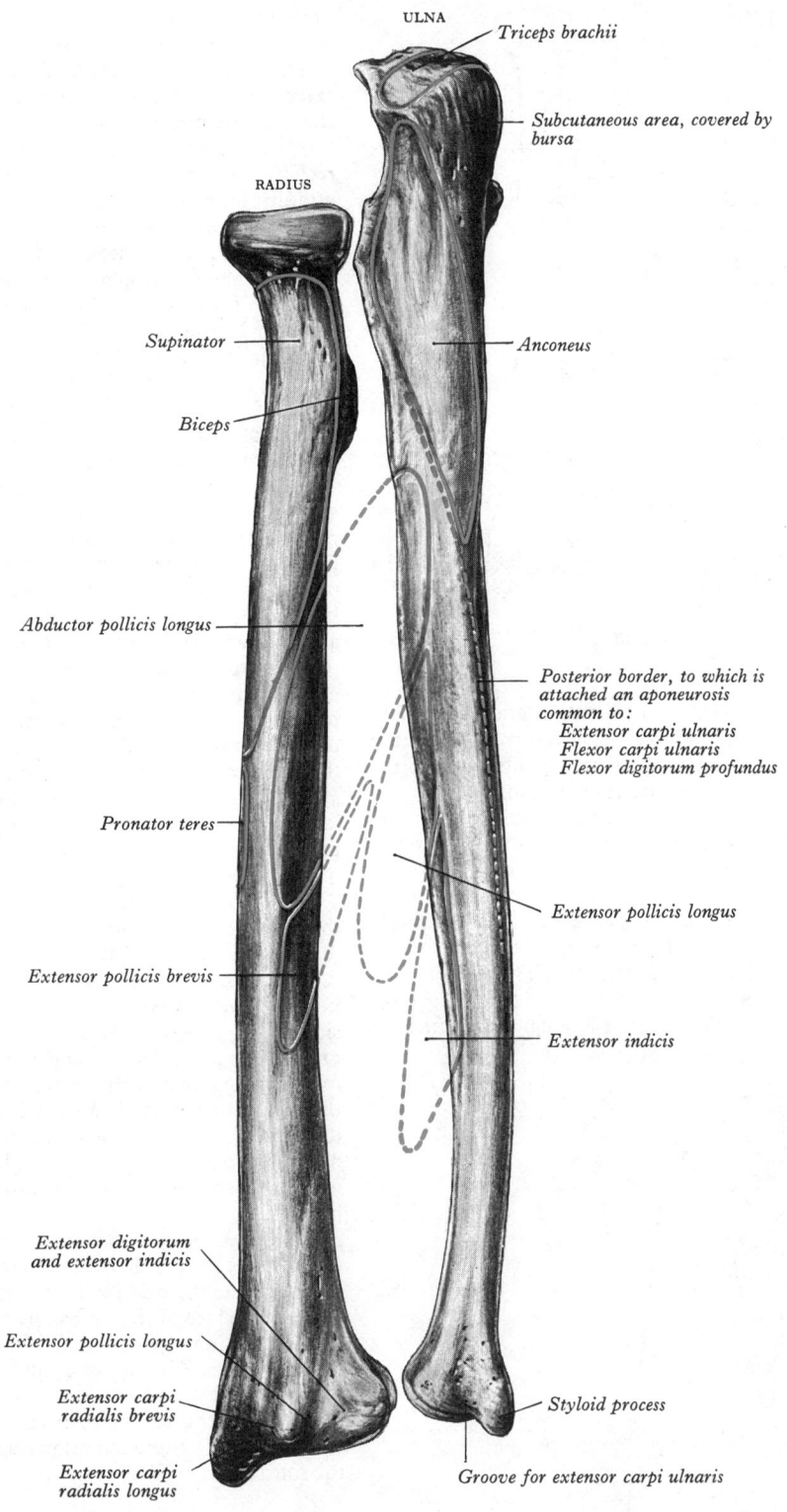

ULNA — *Triceps brachii*

— *Subcutaneous area, covered by bursa*

RADIUS

Supinator —

— *Anconeus*

Biceps —

Abductor pollicis longus —

Posterior border, to which is attached an aponeurosis common to:
Extensor carpi ulnaris
Flexor carpi ulnaris
Flexor digitorum profundus

Pronator teres —

Extensor pollicis longus

Extensor pollicis brevis —

Extensor indicis

Extensor digitorum and extensor indicis —

Extensor pollicis longus —

Extensor carpi radialis brevis —

Styloid process

Extensor carpi radialis longus —

Groove for extensor carpi ulnaris

3.142   The bones of the left forearm. Posterior aspect.

surface of the lower end (**3.147**) is smooth and takes part in the formation of the wrist joint. This *carpal articular surface* is divided by a faint ridge into medial and lateral areas. The medial area is quadrangular, the lateral is triangular and covers the medial side of the styloid process. The *anterior surface* is a thick, prominent ridge, which can be palpated in the living subject, despite the overlying tendons. It is nearly 2 cm above the base of the thenar eminence. The *medial surface* is the *ulnar notch*, a smooth strip, concave from before backwards, for articulation with the head of the ulna in the inferior radio-ulnar joint. The *posterior surface* is marked by the *dorsal tubercle*, which is limited medially by a

narrow, oblique groove. It lies in line with the cleft between the index and middle fingers and can readily be felt through the skin. A wide, shallow groove lies lateral to the tubercle and is divided into two parts by a very faint vertical ridge. A similar but undivided groove marks the medial part of the posterior surface.

The upper surface of the *head* and its *articular circumference* are everywhere covered with hyaline cartilage. The upper margin of the head fits into the groove between the capitulum and the trochlea, and, when the forearm is flexed, it invades the radial fossa. The articular circumference joints with the radial notch of the ulna, and in the rest of its extent is surrounded by the annular

365

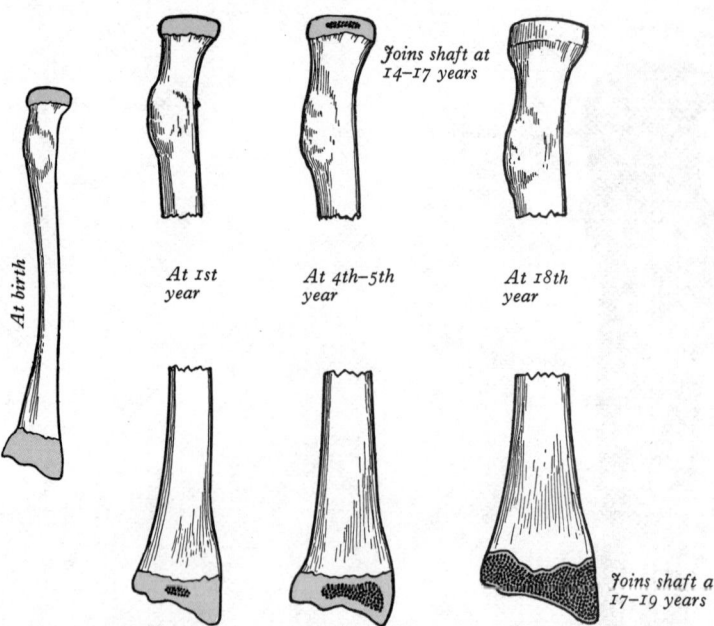

3.143 Stages in the ossification of the radius.

At birth

At 1st year

At 4th–5th year

At 18th year

Joins shaft at 14–17 years

Joins shaft at 17–19 years

ligament, within which it rotates in pronation and supination. The *neck* of the bone is surrounded by the narrower, lower part of the ligament, but is separated from it by a protrusion of the synovial membrane of the superior radio-ulnar joint.

To the rough, posterior part of the *tuberosity* the biceps tendon is attached, but its smooth, anterior part is separated from the tendon by a bursa. A little below the tuberosity the oblique cord is attached to the radius.

The upper, oblique part of the *anterior border* and a variable portion of the border below provide origin for the thin, expanded radial head of the flexor digitorum superficialis. The conspicuous lower part of the anterior border has attached to it the lateral edge of the extensor retinaculum. The small, triangular area in front of

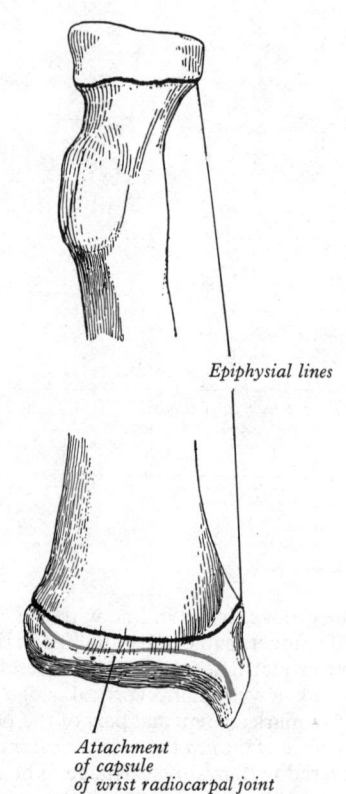

Epiphysial lines

Attachment of capsule of wrist radiocarpal joint

3.144 The epiphysial lines of the left radius in adolescence. Anterior aspect. The line of attachment of the articular capsule of the wrist joint is in blue.

the lower end of the interosseous border and above the ulnar notch is for the deepest fibres of the pronator quadratus.

The upper two-thirds of the *anterior surface* provides an extensive area for the origin of flexor pollicis longus, which conceals the nutrient foramen. The lower fourth of this surface together with the triangular area on the medial side of the bone receives the pronator quadratus. The roughened area at the middle of the *lateral surface* is situated at the region of maximum curvature for the attachment of the pronator teres. Above, the surface widens at the expense of the anterior and posterior surfaces, and the elongated V-shaped area (**3.141, 142**) provides insertion for the supinator. Below the insertion of the pronator teres the lateral surface is smooth and covered by the tendons of the radial extensors of the wrist. The upper part of the *posterior surface* gives origin to the abductor pollicis longus, above, and the extensor pollicis brevis, below. The lower part of this surface is devoid of muscular attachments but is closely covered by the extensor pollicis brevis and longus muscles.

The *styloid process* of the radius projects below the ulnar styloid, and its tip is concealed by the tendons of the abductor pollicis longus and extensor pollicis brevis. It gives attachment by its tip to the lateral ligament of the wrist joint. The lateral surface of the radius a little above the styloid process receives the insertion of the brachioradialis and is crossed obliquely, from above downwards and forwards, by the tendons of the abductor pollicis longus and extensor pollicis brevis. The ridge-like *anterior surface* of the lower end gives attachment to the palmar radiocarpal ligament of the wrist joint. The *ulnar notch* is limited below by a smooth ridge to which the base of the articular disc of the inferior radio-ulnar joint is attached. A small protrusion of the synovial membrane of the joint extends upwards in front of the lower end of the interosseous membrane. The lateral, triangular part of the *carpal articular surface* joints with the scaphoid and the medial, quadrangular part with the lateral part of the lunate bone. When the hand is adducted fully the whole of the upper surface of the lunate bone comes into contact with the radius.

The *dorsal tubercle* gives attachment to a slip from the extensor retinaculum and is grooved on its medial side by the extensor pollicis longus tendon. The wide, shallow groove to the lateral side of the tubercle contains the tendons of the extensor carpi radialis longus, laterally, and the extensor carpi radialis brevis, medially, together with the synovial sheaths. The medial part of the posterior surface is grooved by the tendons of the extensor digitorum, but the extensor indicis and the posterior interosseous nerve intervene between them and the bone. To the lower margin of the posterior surface is attached the dorsal radiocarpal ligament.

Ossification (**3.143, 144**). The radius is ossified from three centres: one for the shaft, and one for each end. That for the shaft appears near the middle in the eighth week of intrauterine life. About the close of the first year, ossification begins in the lower end. At the upper end it begins at the fourth year in females and fifth in males. The upper epiphysis fuses with the shaft at the fourteenth year in females and seventeenth in males; the lower epiphysis at the seventeenth and nineteenth years respectively. An additional centre sometimes appears in the tuberosity about the fourteenth or fifteenth year.

The Ulna

The ulna (**3.141, 142, 145**) is the medial bone of the forearm and is parallel with the radius when the forearm is supine. The upper end is thick, strong and hook-like (**3.145**), the concavity of the hook being directed forwards. The lateral border of the shaft is a thin, sharp crest. The bone diminishes in size from its upper to its lower end, which bears a small, rounded enlargement termed the head of the ulna. The shaft is triangular on section.

The bone is not perfectly straight, and shows a slight but appreciable double curve. Throughout its whole length it forms a gentle curve, the convexity of which is directed backwards. In addition, the upper half or more shows a slight curvature to the lateral side, and the lower half or less a similar curvature in the opposite direction.

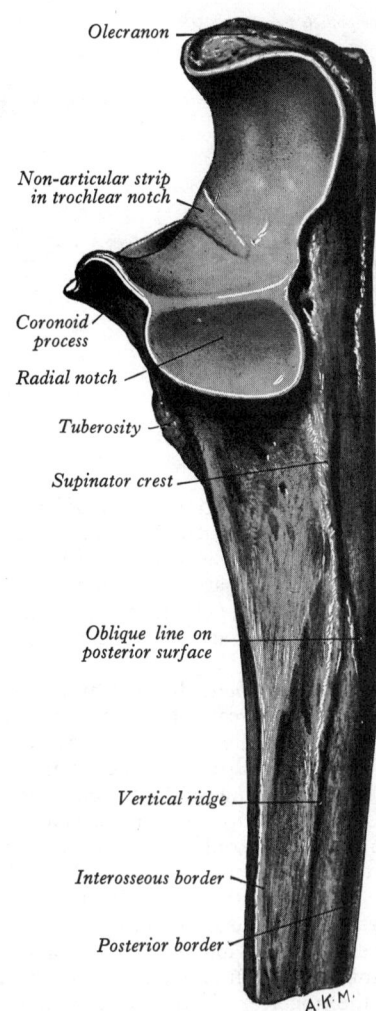

Olecranon

Non-articular strip
in trochlear notch

Coronoid
process

Radial notch

Tuberosity

Supinator crest

Oblique line on
posterior surface

Vertical ridge

Interosseous border

Posterior border

A·K·M·

3.145 The proximal part of the left ulna. Lateral aspect.

The upper end of the ulna (3.145) displays two substantial processes, the olecranon and coronoid, and two articular areas, termed the trochlear and radial notches, which articulate, respectively, with the humerus and the radius.

The *olecranon* is the uppermost part of the bone. It is bent forwards at its summit to form a prominent beak, which projects into the olecranon fossa of the humerus when the forearm is extended. Its posterior surface, which is smooth and triangular in outline, can easily be felt through the skin, and its upper border forms the point of the elbow. When the elbow is extended, this border can be felt on, or just above, the line joining the two epicondyles of the humerus, but when the elbow is flexed, it descends, and the three bony points then form the angles of an isosceles triangle. The anterior surface of the olecranon is smooth and articular, and forms the upper part of the trochlear notch. The base of the olecranon is constricted where it joins the shaft, and this is the narrowest part of the upper end of the ulna.

The *coronoid process* is a bracket-like projection from the front of the bone immediately below the olecranon. Its upper surface forms the lower part of the trochlear notch and is therefore smooth and articular. The upper part of the lateral surface presents the shallow *radial notch* for articulation with the side of the head of the radius, and the bone below it is hollowed out to make room for the tuberosity of the radius during the movements of pronation and supination. The anterior surface of the process is triangular in shape and bears on its lower part the rough *tuberosity of the ulna*. The medial border of the process is sharp and bears a small, but prominent tubercle at its upper end.

The *trochlear notch* articulates with the trochlea of the humerus. It is formed by the anterior surface of the olecranon and the superior surface of the coronoid process. The bone is constricted at the junction between these two areas, and they may be separated completely by a narrow, roughened strip. A smooth ridge, which corresponds to the groove of the trochlea, divides the notch into a larger, medial, and a smaller, lateral part. The medial part conforms to the large flange of the trochlea of the humerus.

The *radial notch* (3.145) is an oblong, articular depression on the upper part of the lateral aspect of the coronoid process. It forms a joint with the articular circumference of the head of the radius, and is separated from the lateral part of the trochlear notch by a smooth ridge.

The lower end of the ulna is slightly expanded and comprises the rounded head and the styloid process. The *head* forms a surface elevation on the medial part of the posterior aspect of the wrist when the hand is pronated, and it can be gripped between the finger and thumb when the supinated hand is passively flexed. It presents a convex articular surface on its lateral side for articulation with the ulnar notch of the radius. Its inferior surface (3.146) is smooth and is separated from the carpus by the articular disc of the inferior radio-ulnar joint, which is attached by its apex to the small rough area between the articular surface and the styloid process. The *styloid process* is a short, rounded projection which springs from the posteromedial aspect of the lower end of the ulna. Its tip can be felt through the skin on the posteromedial aspect of the wrist, where it lies about 1 cm above the tip of the styloid process of the radius. On the dorsal surface is a shallow groove between the head and styloid process.

The shaft of the ulna is triangular in section (3.146) in its upper three-fourths, but is almost cylindrical in its lower quarter. The surfaces of the shaft are anterior, posterior and medial; the borders, interosseous, posterior and anterior.

The *interosseous border* is the lateral margin of the bone and forms a conspicuous crest in its middle two-fourths. The upper part is continuous with the posterior border of the depression below the radial notch and is here termed the *supinator crest*; its lower part fades away on the cylindrical, lower portion of the shaft. The *anterior border* is thick and rounded. It commences above medial to the ulnar tuberosity and inclines backwards below, where it can usually be traced to the base of the styloid process. The *posterior border*, also thick and rounded, commences at the apex of the posterior aspect of the olecranon and curves laterally as it decends. Below, very indistinct, it descends to the styloid process. Throughout its whole length this border can easily be felt through the skin. It lies in a longitudinal furrow, which is seen best when the elbow is in full flexion.

The *anterior surface* of the ulna (3.141), between the interosseous and anterior borders, is grooved, sometimes deeply, in its long axis. A little above the middle of the bone on this surface is the nutrient foramen (3.141), which is directed upwards and transmits a branch of the anterior interosseous artery. Its inferior part is crossed obliquely by a rough strip, of variable prominence, which runs downwards from the interosseous border to the anterior border. The *medial surface* is bounded by the anterior and posterior borders. Convex from side to side, it is smooth and featureless. The *posterior surface* (3.142) lies between the posterior and interosseous borders. It is divided into three areas, of which the uppermost is limited by an oblique line—not always easily discernible—which runs upwards and laterally from the junction of the middle and upper thirds of the posterior border to the posterior end of the radial notch (3.145). The region below this line is divided into a larger medial and a narrower lateral strip by a vertical ridge, usually distinct in its upper three-fourths but less clear below.

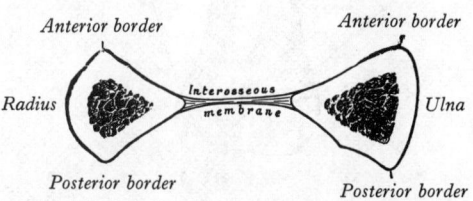

Anterior border   Anterior border

Radius   Interosseous membrane   Ulna

Posterior border   Posterior border

3.146 A transverse section through the left radius and ulna, showing the attachment of the antebrachial interosseous membrane. Superior aspect.

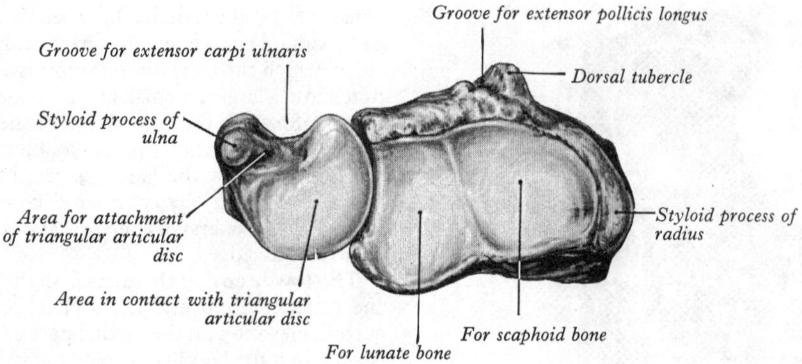

Groove for extensor carpi ulnaris

Groove for extensor pollicis longus

Dorsal tubercle

Styloid process of ulna

Area for attachment of triangular articular disc

Styloid process of radius

Area in contact with triangular articular disc

For scaphoid bone

For lunate bone

3.147   The distal ends of the right radius and ulna. Distal (inferior) aspect.

The upper surface of the *olecranon* gives attachment, in front, to the capsular ligament of the elbow joint, in its roughened posterior two-thirds, to the tendon of the triceps. Occasionally these two areas are separated by a smooth bursal area. Its medial surface is marked in its upper part by a rough elevation, the site of attachment of the posterior and oblique bands of the ulnar collateral ligament of the elbow joint and the ulnar head of flexor carpi ulnaris. Its lower part is smooth and gives origin to the uppermost fibres of the flexor digitorum profundus. The lateral surface of the process, and the adjoining part of the posterior surface of the shaft down to the oblique line already mentioned (**3.**145), affords attachment to the anconeus. Its posterior surface is separated from the skin by a bursa.

To the anterior surface of the *coronoid process*, including the tuberosity of the ulna, is attached the brachialis. Its medial border is sharp, and a small, rounded tubercle is situated at its upper end to which are attached the oblique and anterior bands of the ulnar collateral ligament of the elbow joint and the lowest part of the humero-ulnar head of the flexor digitorum superficialis. Below the tubercle the margin gives origin to the ulnar head of the pronator teres. An ulnar head of the flexor pollicis longus may arise from the lateral or, more rarely, the medial border of the coronoid process (Martin 1958). The medial surface of the process is concave and has attached to it fibres of the flexor

digitorum profundus. The anterior border of the radial notch provides attachment to the annular ligament of the radius, the posterior attachment of which is usually marked by a rough ridge at or just behind the posterior margin of the notch. The depressed area below the notch is limited behind by the *supinator crest*; the supinator arises from the crest and from much of the depression anterior to it.

The part of the *trochlear notch* formed by the olecranon is, typically, divided into three areas. Of these the most medial faces forwards and slightly medially and is hollowed out to fit the medial flange of the humeral trochlea: the intermediate area is flattened and fits the lateral flange of the trochlea; the most lateral area, which forms a narrow strip directed to the radial side, comes into contact with the trochlea only when the elbow is extended. The constriction of the articular surface is more pronounced than the constriction of the base of the olecranon. The resulting small non-articular parts of the anterior surface of the olecranon are covered in the fresh specimen and in life by tag-like processes of the synovial membrane which contain a little fat (**4.**44). The coronoid part of the trochlear notch is divided into medial and lateral parts, which correspond, respectively, to the medial and intermediate parts of the olecranon area. Of these the medial is hollowed out much more than the lateral, in conformity with the convexity of the medial flange of the trochlea (*see* carrying angle, p. 462). The medial and anterior borders of this area give attachment to the medial and anterior portions of the capsular ligament of the elbow joint.

The subcutaneous *posterior border* of the ulna gives attachment to the deep fascia of the forearm, which acts, in its upper three-fifths, as an additional origin for the flexor carpi ulnaris, and in its middle third as an additional origin for the extensor carpi ulnaris. Both these muscles are therefore connected to the posterior border. The *interosseous border*, except at its upper end, gives attachment to the interosseous membrane. The rounded *anterior border* is covered in its upper three-fourths by the flexor digitorum profundus, which is attached to it.

The *anterior surface* gives origin in its upper three-fourths to the flexor digitorum profundus. In the same extent the muscle arises also from the anterior border and the *medial surface*, extending upwards on to the medial sides of the coronoid process and the olecranon. The rough strip which crosses the lower fourth of the anterior surface marks the attachment of the pronator quadratus. The anconeus is inserted into the *posterior surface* above the oblique line already mentioned, and extends upwards on to the lateral aspect of the olecranon. The narrow strip between the interosseous border and the vertical ridge affords partial origin to three of the deep muscles of the forearm. The abductor pollicis longus arises from its upper fourth, and a ridge may separate this area from the succeeding fourth, which gives attachment to the extensor pollicis longus. The extensor indicis is

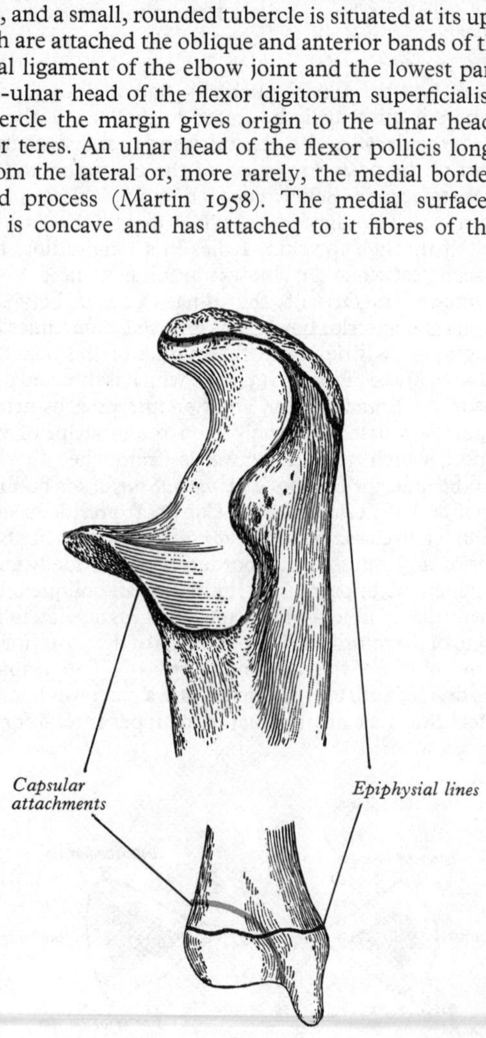

Capsular attachments

Epiphysial lines

3.149   Anteroposterior radiograph of the forearm of a girl aged 11 years. 1. Proximal radial epiphysis. 2. Conjoined epiphyses of capitulum and lateral epicondyle. 3. Epiphysis of medial epicondyle. 4. Diaphysial bone. 5. Trochlear epiphysis. 6. Cartilaginous growth plates. 7. Distal ulnar epiphysis. 8. Distal radial epiphysis.

3.148   The epiphysial lines of the left ulna in adolescence. Lateral aspect. The lines of attachment of the articular capsules are in blue.

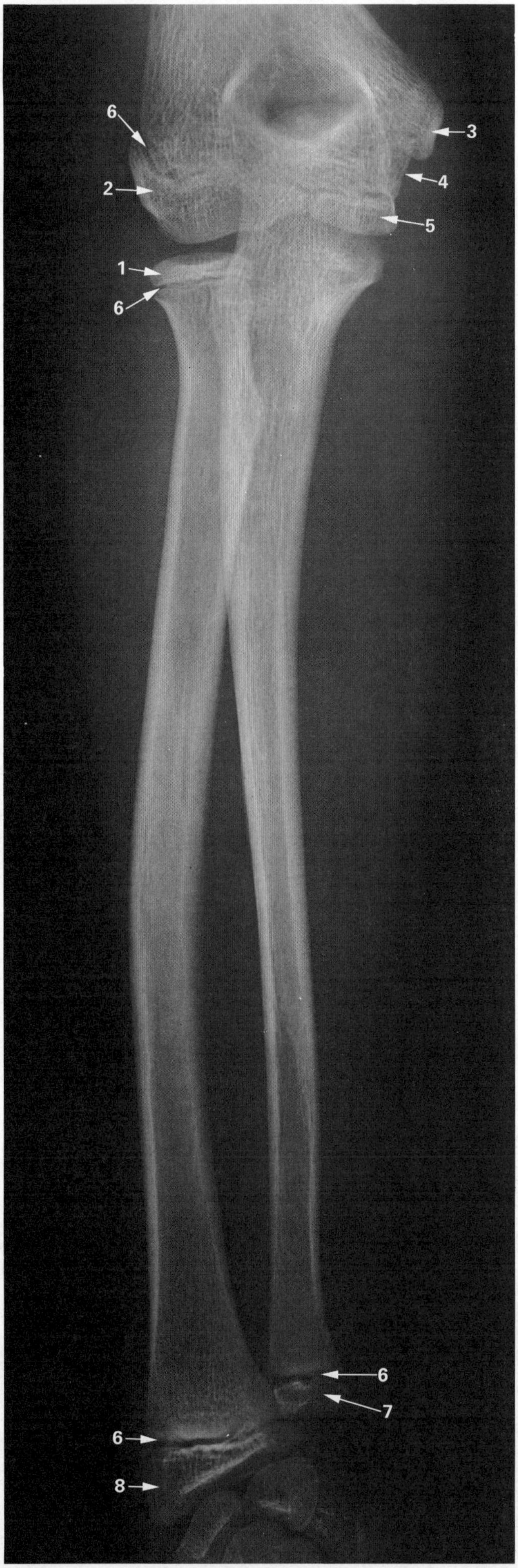

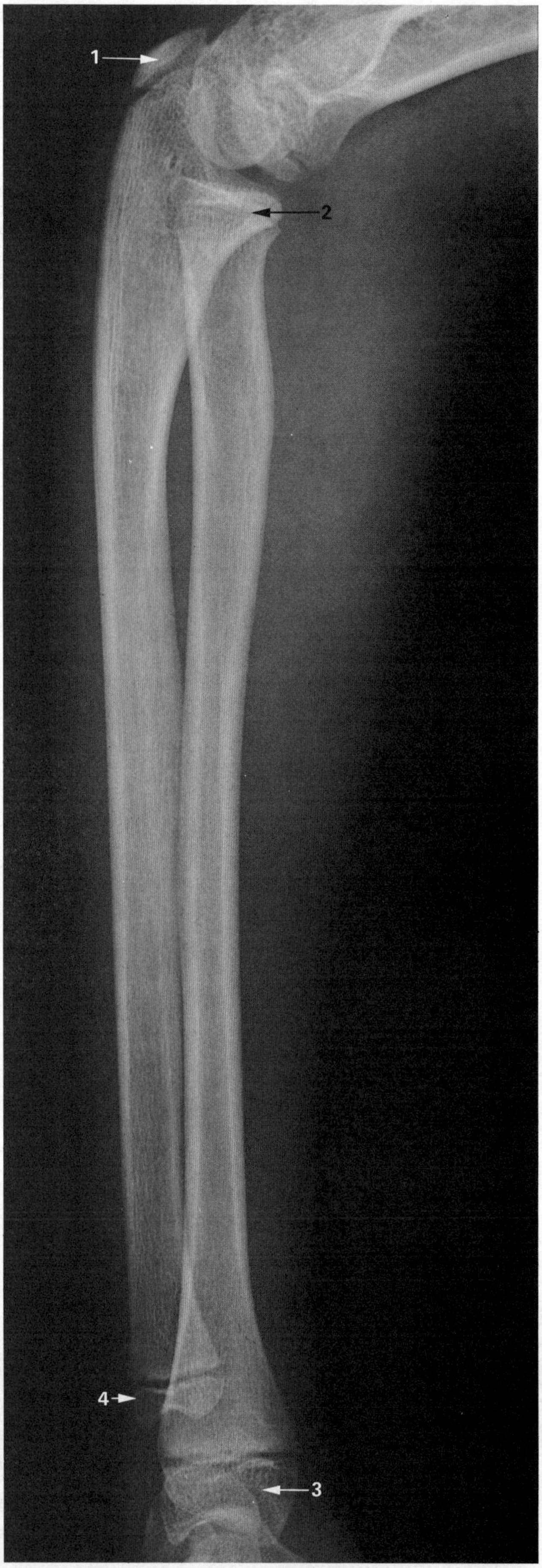

3.150 Lateral radiograph of forearm of a girl of 11 years, semiflexed at the elbow. Note following epiphyses and adjacent radio-translucent growth cartilages—1. Olecranon. 2. Proximal radial. 3. Distal radial. 4. Distal ulnar.

attached to the third quarter of this area. The broad strip to the medial side of the vertical ridge is devoid of muscular attachments but is covered by the extensor carpi ulnaris, the tendon of which occupies the groove on the posterior aspect of the lower end of the bone. The ulnar collateral ligament of the wrist joint is attached to the tip of the *styloid process*. The articular disc separates the *head* of the ulna from the medial part of the lunate bone and, in adduction of the hand, from the triquetral bone.

For an account of the vascularization of the ulna consult Fischer *et al.* (1974).

Ossification (**3.**148, 151). The ulna is ossified from three centres: one each for the shaft, the lower end, and the proximal part of the olecranon. Ossification begins near the middle of the shaft, about the eighth week of intrauterine life, and soon extends through its greater part. In the fifth year in females and sixth in males the centre for the lower end appears in the middle of the head, and extends into the styloid process. The lower part of the olecranon is ossified by an upward extension of the shaft. The remainder is ossified from two centres; the lower forms the upper part of the trochlear surface; the upper is a thin scale-like epiphysis on top of the process (Porteous 1960). The latter appears in the ninth year in females and eleventh in males, and the whole of the upper epiphysis has joined with the shaft by the fourteenth year in females and sixteenth in males. The lower epiphysis of the ulna unites with the shaft in the seventeenth year in females and the eighteenth in males. All these dates are, of course, subject to some variation.

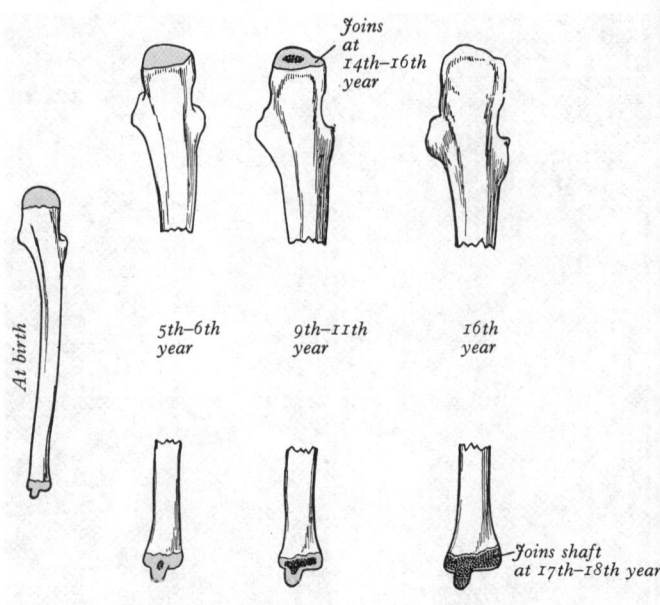

3.151   Stages in the ossification of the ulna. The diagram is simplified in relation to the epiphysial ossification of the olecranon, which is held to have two centres (*see* Porteous 1960 and the description in the text).

THE SKELETON OF THE HAND

The skeleton of the hand has three segments: (1) the carpal or wrist bones, (2) the metacarpal bones of the palm, and (3) the phalanges or bones of the digits. In the description which follows, the terms *proximal* and *distal* will be used to the exclusion of the terms *superior* and *inferior* hitherto employed, and in addition the terms *palmar* and *dorsal*, which are self-explanatory, will be utilized instead of *anterior* and *posterior*.

The Carpus

General features. The carpus (**3.**152, 153) is composed of eight bones, which are arranged in proximal and distal rows, each containing four elements. The bones of the proximal row, from lateral to medial side, are the *scaphoid, lunate, triquetral* and *pisiform*; those of the distal row, the *trapezium, trapezoid, capitate*

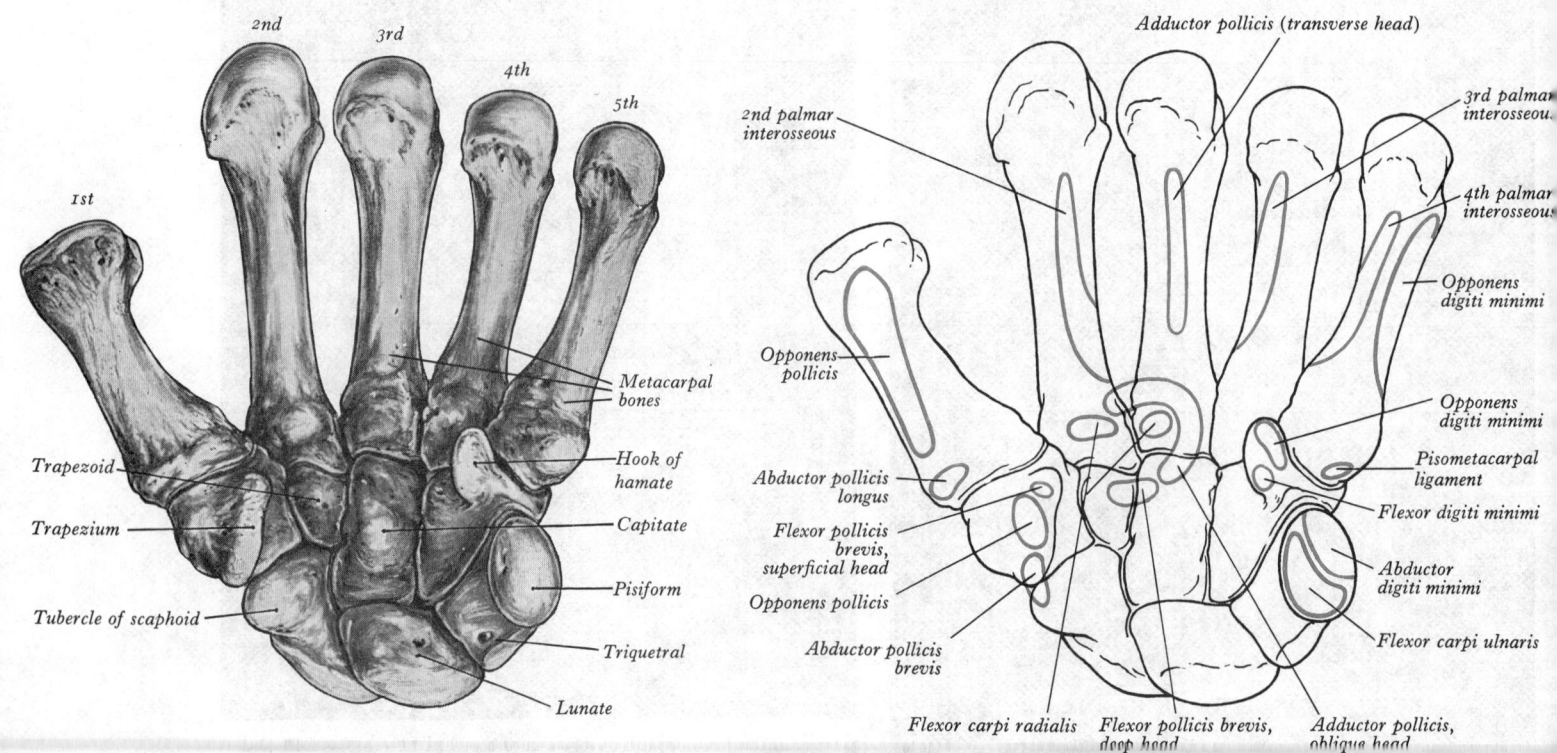

370   **3.**152A   The carpal and metacarpal bones of the left hand. Palmar aspect.

3.152B   Muscle attachments to the carpal and metacarpal bones of the left hand. Palmar aspect. Dorsal interossei not shown.

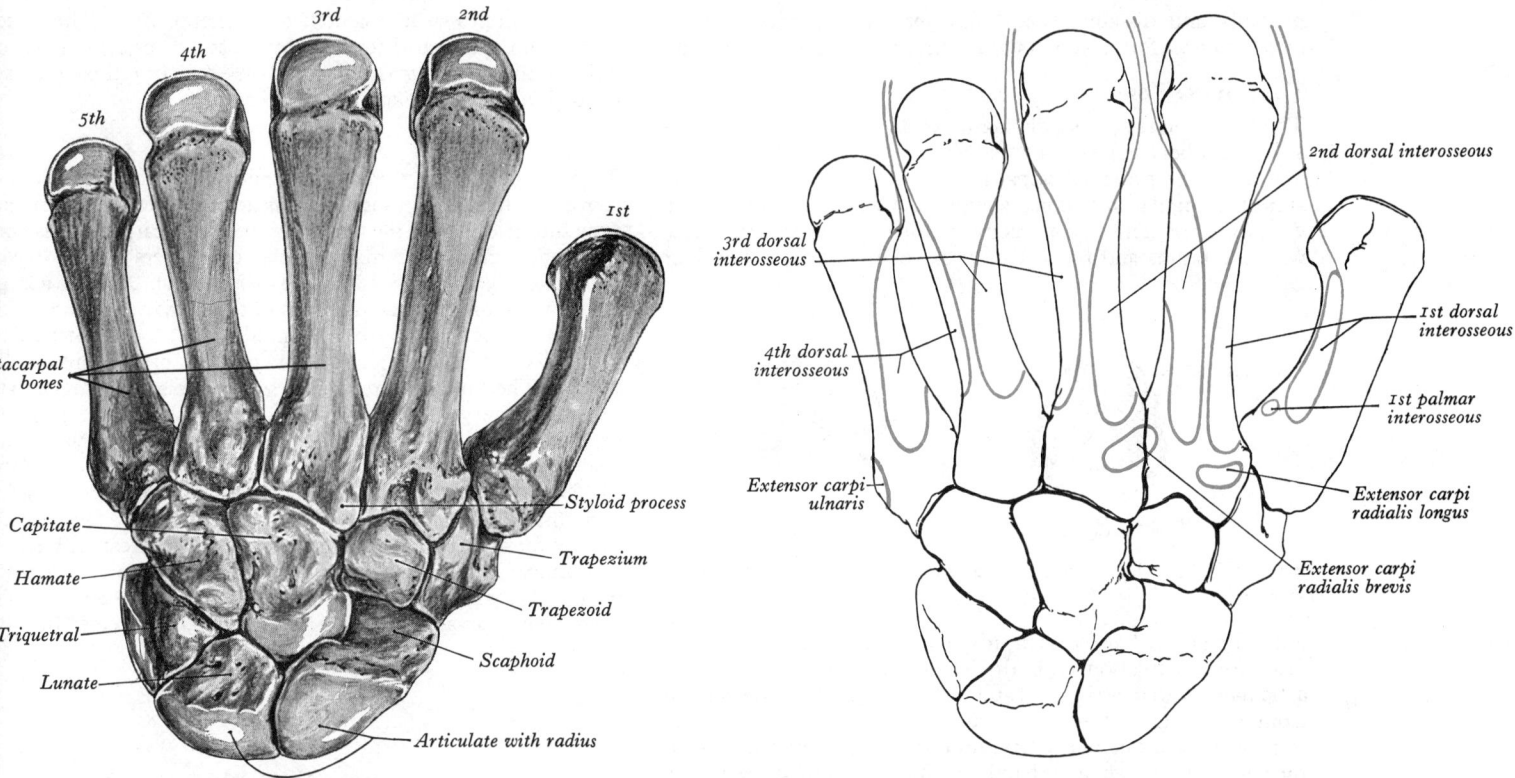

Metacarpal bones

Capitate
Hamate
Triquetral
Lunate

Styloid process
Trapezium
Trapezoid
Scaphoid
Articulate with radius

3.153A   The carpal and metacarpal bones of the left hand. Dorsal aspect.

2nd dorsal interosseous

3rd dorsal interosseous
4th dorsal interosseous
Extensor carpi ulnaris

1st dorsal interosseous
1st palmar interosseous
Extensor carpi radialis longus
Extensor carpi radialis brevis

3.153B   Muscle attachments to the carpal and metacarpal bones of the left hand. Dorsal aspect.

and *hamate*. The pisiform is on the palmar surface of the triquetral and is separated from the other carpal bones, all of which articulate with their immediate neighbours. The other bones of the proximal row form an arch convex proximally, which articulates with the radius and the articular disc of the inferior radio-ulnar joint. The concavity of the arch is directed distally and forms a recess into which fit the proximally projecting parts of the capitate and hamate bones. In this way the two rows are adapted to each other strongly and yet without sacrifice of all movement.

The dorsal surface of the carpus is gently convex from side to side, but the palmar surface is a deep concavity, the *carpal groove*, partly owing to the presence of certain ventral projections on its lateral and medial borders. The medial border of the concavity is formed by the *pisiform bone* and the *hamulus*, a hook-like process on the palmar surface of the *hamate* bone. The pisiform lies in the medial part of the proximal border of the muscular hypothenar eminence which forms the medial part of the palm, and its position in front of the triquetral makes it easy to feel. The hook of the hamate is concave on its lateral side; its tip can be identified in the living subject 2·5 cm distal to the pisiform and in line with the radial border of the ring finger. In this situation the superficial division of the ulnar nerve can be rolled from side to side over this bony point. The projecting lateral border of the carpal groove is formed by the *tubercle of the scaphoid* and the *tubercle of the trapezium*. The former is on the distal part of the anterior surface of the scaphoid and can be felt—and sometimes seen—as a small, rounded knob, in the medial part of the proximal border of the muscular thenar eminence ('ball' of the thumb), which forms the lateral part of the palm. The tubercle of the trapezium is a rounded ridge which runs vertically across the anterior surface of the bone, being slightly hollowed out on its medial side. It lies immediately distal and slightly lateral to the tubercle of the scaphoid, and can be felt only on deep pressure. The carpal groove is converted into an osseofibrous *carpal tunnel* by a strong fibrous retinaculum attached to the bony margins. the tunnel transmits the flexor tendons and median nerve to the hand (*see* p. 1102). The fibrous retinaculum increases the strength of the carpus and the efficiency of the flexor muscles. The palmar and dorsal

surfaces of the carpal bones, apart from the triquetral and pisiform, are rough for the attachment of ligaments (radiocarpal, intercarpal and carpometacarpal).

The Individual Carpal Bones

The Scaphoid Bone

The scaphoid bone (**3.**154) is the largest element in the proximal row and lies with its long axis directed distally, laterally and slightly forwards. Its tubercle is a rounded elevation on the distal part of the *palmar surface* and is directed anterolaterally. It gives attachment to the flexor retinaculum and a few fibres of the abductor pollicis brevis, and is crossed by the tendon of the flexor carpi radialis. The *dorsal surface* is rough and slightly grooved, and is narrower than the palmar surface. It is pierced by a number of small nutrient foramina and in a small proportion of cases (13 per cent) these are restricted to the distal half of the bone (Obletz and Halbstein 1938). The *lateral surface*, also narrow and roughened, has attached to it the radial collateral ligament of the wrist joint. The remaining surfaces of the bone are articular. The *radial surface* is convex and is directed proximally and laterally. The *lunate surface* is a flattened, narrow semilune, directed medially. The *capitate surface*, large and concave, is directed

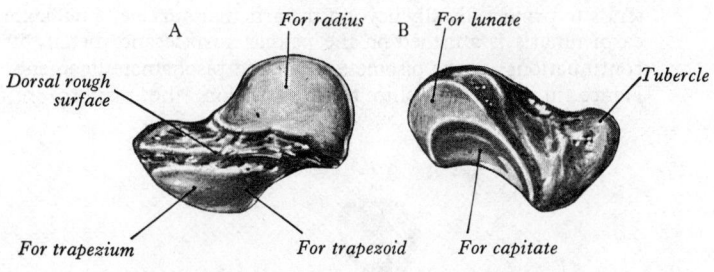

A   For radius   B   For lunate

Dorsal rough surface   Tubercle

For trapezium   For trapezoid   For capitate

3.154   The left scaphoid bone. A—dorsal and B—palmar aspects.

medially and distally. The *surface for the trapezium and the trapezoid bone* forms a continuous convex area, directed distally.

The Lunate Bone

The lunate bone (**3**.155), distinguished by its crescentic outline, articulates between the scaphoid and the triquetal bones in the middle of the proximal carpal row. The rough *palmar surface*, almost triangular in outline, is larger and wider than the rough, *dorsal surface*. The smooth, convex, *proximal surface* articulates with the radius and the articular disc of the distal radio-ulnar

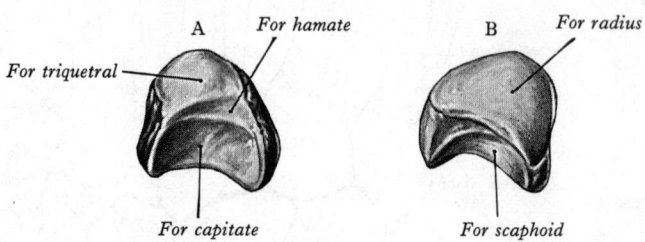

3.155 The left lunate bone. A distomedial and B—proximolateral aspects.

joint. The *lateral surface* is narrow and presents a flat, semilunar facet for articulation with the scaphoid. The *medial surface* articulates with the triquetral and is almost square. It is separated from the distal surface by a curved ridge, which is usually somewhat hollowed out (**3**.155) for articulation with the edge of the hamate bone, when the hand is adducted. The *distal surface* is deeply concave to accommodate the medial part of the head of the capitate bone.

The Triquetral Bone

The triquetral bone (**3**.156), somewhat pyramidal in shape, is distinguished by an oval, isolated, smooth facet for articulation with the pisiform, which marks the distal part of its rough, *palmar* surface. The *medial* and *dorsal surfaces* are confluent. Rough distally for the attachment of the ulnar collateral ligament of the

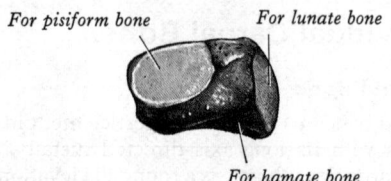

3.156 The left triquetral bone. Palmar aspect.

wrist joint, this aspect is smooth in its proximal part, which articulates with the articular disc of the distal radio-ulnar joint in full adduction of the hand. The *hamate surface*, directed laterally and distally, is a concavoconvex area, broad proximally and narrow distally. The *lunate surface*, almost square, is directed proximally and laterally.

The Pisiform Bone

The pisiform bone (**3**.157) is shaped like a pea with one flattened surface, which is its only articular facet. This is on the *dorsal surface* of the bone to articulate with the triquetral, and its long axis runs distally and laterally. The non-articular part of the bone tends to project distally beyond the articular surface. The flexor carpi ulnaris is attached on the palmar surface and distally its continuations are the pisometacarpal and pisohamate ligaments. Hence in its relationship to this tendon, the pisiform has

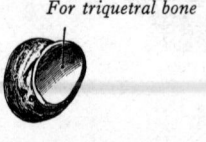

3.157 The left pisiform bone. Dorsal aspect.

all the attributes of a sesamoid bone (*see* p. 417). The flexor retinaculum is attached to the palmar part of the lateral aspect while the abductor digiti minimi and the extensor retinaculum are attached to the medial and distal aspects.

The Os Trapezium

The os trapezium (**3**.158) is characterized by the tubercle and groove which mark its rough, *palmar surface*. The groove, medial to the tubercle, lodges the tendon of the flexor carpi radialis, and its margins give attachment to the two layers of the flexor retinaculum (**5**.76 and p. 583). The *tubercle* is obscured to a large extent by the origin of the muscles of the thenar eminence. The opponens pollicis arises from its middle part, the flexor pollicis brevis distally and the abductor pollicis brevis proximally (**3**.152B). The elongated, rough, *dorsal surface* is closely related to the radial artery, before it passes forwards into the palm to become the deep palmar arch. The *lateral surface* also is large and rough for the attachment of the radial collateral ligament of the wrist joint and the capsular ligament of the carpometacarpal joint of the thumb. A large *saddle-shaped (sellar) surface* faces disto-laterally for articulation with the base of the metacarpal bone of the thumb. The most distal part of the bone projects slightly between the bases of the first and second metacarpal bones and is covered with a small, quadrilateral facet which is directed distally

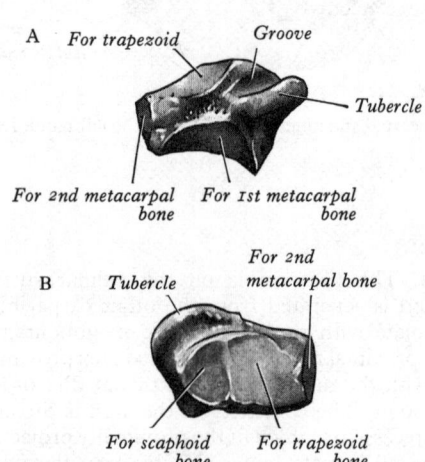

3.158 The left trapezium. A—palmar and B—proximomedial aspects.

and medially to articulate with the base of the second metacarpal bone. The medial surface is a large, gently concave facet for articulation with the trapezoid. The proximal surface is occupied by a small, slightly hollowed-out facet, which articulates with the scaphoid. In view of its great contribution to the mobility of the pollex, the distal, metacarpal articular surface of the trapezium has attracted particular attention. The ridge or 'summit' which fits the concavity of the first metacarpal base, extends in a palmar direction but also somewhat laterally, making an angle of about 60° with the plane of the second and third metacarpals (Kuczynski 1974; see also pp. 471 and 593). The movements of abduction and adduction occur in the plane of the axis of this ridge. The trapezial ridge is shorter than the corresponding metacarpal groove, and the contours of both vary reciprocally along their axes, being more curved near to the base of the second metacarpal and assuming a longer radius of curvature in a direction away from this. The two surfaces are not completely congruent, and the area of close contact probably moves towards the palm in adduction and reverses in abduction (*see* p. 471).

The Trapezoid Bone

The trapezoid bone (**3**.159) is small and irregular in shape. The *palmar surface* is rough, narrow and considerably smaller than the rough, *dorsal surface*; it is continued for a short distance on to the lateral aspect. The *distal surface* articulates with the grooved base of the second metacarpal bone. Triangular in outline, it is convex from side to side and concave from before backwards. The *medial*

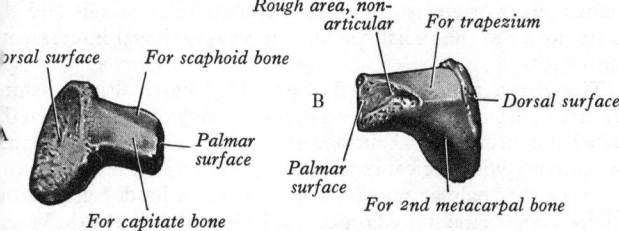

159 The left trapezoid bone. A—proximomedial and B—distolateral aspects.

...rface articulates with the distal part of the capitate bone, by means of a slightly concave facet. The *lateral surface* articulates with the trapezium and the *proximal surface* with the scaphoid bone.

The Capitate Bone

The capitate bone (3.160), the largest of the carpal bones, articulates with the base of the third metacarpal bone; it is therefore central in position. The *distal surface*, roughly triangular, forms a concavoconvex facet for this articulation. Its lateral border carries a concave strip which articulates with the medial side of the base of the second metacarpal bone, and its dorsomedial angle usually bears a small facet for the fourth metacarpal bone. The convex *head* projects into the concavity

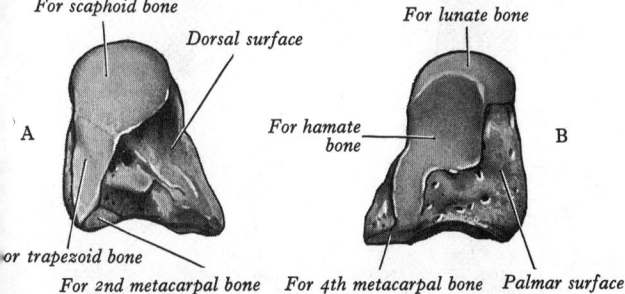

160 The left capitate bone. A—lateral and B—medial aspects.

formed by the lunate and scaphoid bones. Its *proximal surface* articulates with the lunate and its *lateral surface* with the scaphoid. The facet for the scaphoid is usually continuous with the facet for the trapezoid on the distal part of the lateral surface of the bone, but the two may be separated by a rough interval. The *medial surface* presents a large facet for the hamate bone, deeper proximally than it is distally where a part of the surface is non-articular. The *palmar* and *dorsal surfaces* are roughened for carpal ligaments; the dorsal is the larger.

The Hamate Bone

The hamate bone (3.161) is cuneiform and has a hook-like process, the *hamulus, palmar surface*. The concavity of the hamulus is directed to the lateral side, and takes part in the formation of the carpal tunnel. To its tip is attached the flexor retinaculum. The distal aspect of the base of the hamulus occasionally shows a slight transverse groove in contact with the deep, terminal branch of the ulnar nerve. The rest of the *palmar surface*, like the *dorsal surface*, is

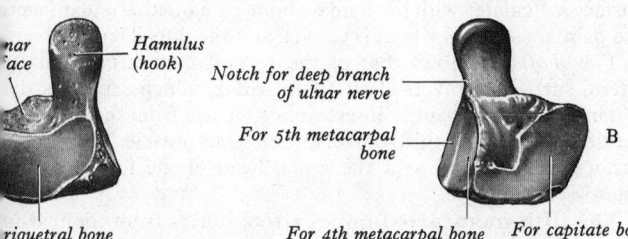

161 The left hamate bone. A—medial and B—lateral aspects.

rough for the attachment of ligaments. The *distal surface* is divided into two articular surfaces by a faint ridge: of these, the smaller, lateral facet articulates with the base of the fourth, and the larger, medial facet with the base of the fifth metacarpal bone. The *proximal surface* is the thin margin of the wedge and usually bears a narrow facet which comes into contact with the lunate bone where the hand is adducted. The *medial surface* is covered by a broad articular strip, convex proximally and concave distally, for the triquetral bone. The *lateral surface* articulates with the capitate bone by means of a facet which covers all but the distal and palmar angle of the surface.

The Metacarpus

The metacarpus (Singh 1959) consists of five metacarpal bones, conventionally numbered from lateral to medial side. The metacarpals are miniature long bones, each possessing a rounded head, a shaft and an expanded base. The *heads* are distal ends and articulate with the proximal phalanges. Their oblong, articular surfaces are convex, the degree of convexity being less in the transverse than in the anteroposterior direction, and extend farther proximally on the palmar surface, than on the dorsal, especially at the margins. The prominence of the knuckles is produced by the metacarpal heads. The *bases* of the metacarpals are their expanded proximal ends, which articulate with the distal row of the carpus and with one another, save that the first metacarpal bone is isolated and does *not* articulate with the second metacarpal bone. The *shafts* are concave longitudinally on their palmar surfaces, forming a hollow containing certain palmar muscles. The dorsal surface of each presents a flattened triangular area in its distal part, continued proximally as a rounded ridge. These flattened areas can be felt on the back of the hand immediately proximal to the knuckles.

The medial four metacarpal bones lie side by side, but the first metacarpal bone lies on a more anterior plane and is rotated medially round its long axis through an angle of 90°. Hence its morphologically *dorsal* surface is directed to the *lateral* side, its radial border forwards, its palmar surface medially, and its ulnar border dorsally. By virtue of this position the thumb moves medially in front of the palm when it is flexed and it can be rotated into opposition with each of the fingers in turn. This ability to oppose the thumb to the fingers is rendered possible by the rotation of the bone medially. It is the most important factor in rendering the hand an efficient instrument for prehension, for, when an object is grasped in the hand, the fingers encircle it from one side and the thumb from the other, and the power of the grip is increased very greatly thereby. Lewis (1977) has recently contributed an extended anatomical, functional and morphological study of the human metacarpals and their articulations with the carpus and phalanges, in explanation of the evolution of human mammalian skills.

The Individual Metacarpal Bones

The first metacarpal bone (3.162) is shorter and stouter than the others. Its dorsal surface can be felt to face laterally; its long axis passes distally and laterally, diverging from its neighbour. The *shaft* is flattened and its dorsal surface is uniformly broad and convex from side to side. The palmar surface is concave in its long axis and is subdivided by a rounded ridge into a larger lateral and a smaller medial part. The opponens pollicis is attached to the radial border and adjoining palmar surface; the radial head of the first dorsal interosseous muscle arises from the ulnar border and the adjoining palmar surface. The *base* presents a concavoconvex surface for articulation with the trapezium (p. 471). On its lateral side the abductor pollicis longus is attached to it; its ulnar side and adjoining shaft gives origin to the first palmar interosseous muscle (p. 590). The *head* is less convex than the heads of the other metacarpal bones, and also in contrast to them is broader from side to side than from palmar to dorsal surface. On its palmar surface the ulnar and radial angles are enlarged to form two articular eminences, on each of which a sesamoid bone glides.

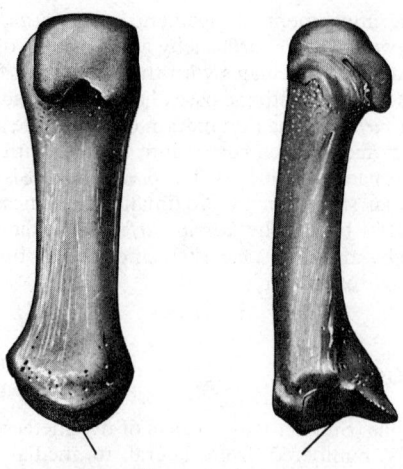

3.162 The first right metacarpal bone. Palmar and lateral aspects.

The second metacarpal bone (3.163) is the longest and its base is the largest. The *base* is characteristically grooved in a dorso-palmar direction for articulation with the trapezoid; the groove is bounded medially by a deep ridge, the edge of which articulates with the capitate bone. The lateral side of the base, nearer the dorsal than the palmar surface, bears a small, quadrilateral facet for the trapezium. Immediately behind this, i.e. on the lateral part of the dorsal surface of the base, a small rough impression marks the attachment of extensor carpi radialis longus. On the palmar surface is a small tubercle or ridge for one insertion of the flexor carpi radialis. The medial side of the base articulates with the lateral side of the base of the third metacarpal bone by a strip-like facet, constricted at its middle.

The *shaft* is prismoid in form and curved so as to be convex dorsally in its long axis and concave towards the palm. It has medial, lateral and dorsal surfaces. The dorsal surface is broad near the head but narrows into a ridge as it approaches the base. This surface is covered by the extensor tendons of the index finger; its converging borders begin at two little tubercles, one on each side of its head; collateral ligaments are attached to these (p. 472). The *lateral surface* inclines dorsally at its proximal end: it gives origin to the ulnar head of the first dorsal interosseous muscle. The *medial surface* inclines similarly and is divided into two nearly parallel strips by a faint ridge. Of these, the more

palmar gives origin to the second palmar interosseous and the more dorsal to the radial head of the second dorsal interosseou muscle.

The third metacarpal bone (3.164) has a distinguishin feature in the short *styloid process*, projecting proximally from th radial side of the dorsal surface of its base. The *base* articulate proximally, with the capitate bone by means of a facet which convex in its palmar portion but concave in its dorsal portion where it covers the styloid process. The lateral aspect of the base marked by a strip-like facet, constricted at its middle, fc articulation with the metacarpal bone of the index. On its medi side it articulates with the base of the fourth metacarpal bone b means of two small, discrete oval facets. Sometimes the mor palmar facet is absent, and less frequently the two may b

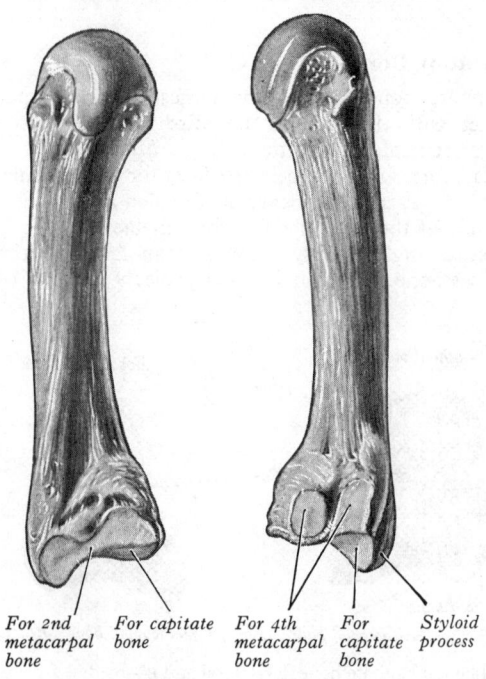

3.164 The left third metacarpal bone. Lateral and medial aspects.

connected by a narrow bridge along the medial border of the base The palmar surface of the base receives a slip from the flexor carp radialis tendon, while its dorsal surface, immediately beyond th styloid process, gives insertion to the extensor carpi radiali brevis.

The *shaft* resembles that of the index metacarpal. Its latera surface gives origin to the ulnar head of the second dorsa interosseous muscle and its medial surface to the radial head of th third dorsal interosseous muscle. The palmar ridge whicl separates these two surfaces gives origin, in its distal two-thirds to the transverse head of the adductor pollicis. Its dorsal surface i covered by the extensor tendon of the middle finger.

The fourth metacarpal bone (3.165) is shorter and thinne than the second and third. It can be identified by its *base*. Th lateral aspect bears two oval facets for articulation with the third metacarpal bone (*vide supra*). Of these the more dorsal is usuall (but not always) the larger, and its proximal part comes into contact with the capitate. The medial aspect is marked by a single elongated facet for the fifth metacarpal bone. The proximal surface articulates with the hamate bone by a quadrangular facet the palmar aspect of which is convex and the dorsal concave.

The *shaft* resembles that of the second metacarpal, but its lateral surface is traversed by a faint ridge, which separates the origin of the third palmar interosseous muscle from the origin o the ulnar head of the third dorsal interosseous muscle. The media surface has attached to it the radial head of the fourth dorsal interosseous muscle.

The fifth metacarpal bone (3.166) differs from the rest in that the medial surface of its *base* is non-articular and presents a tubercle for the insertion of the extensor carpi ulnaris. The

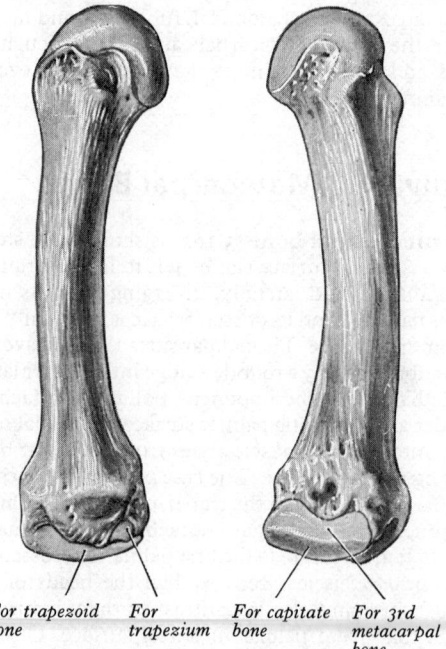

3.163 The left second metacarpal bone. Dorsolateral and medial aspects.

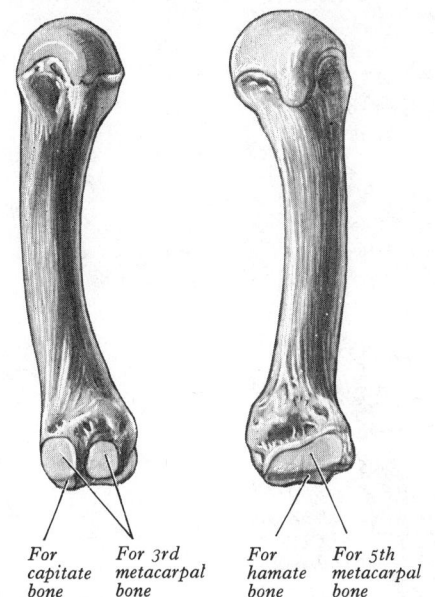

3.165 The left fourth metacarpal bone. Lateral and medial aspects.

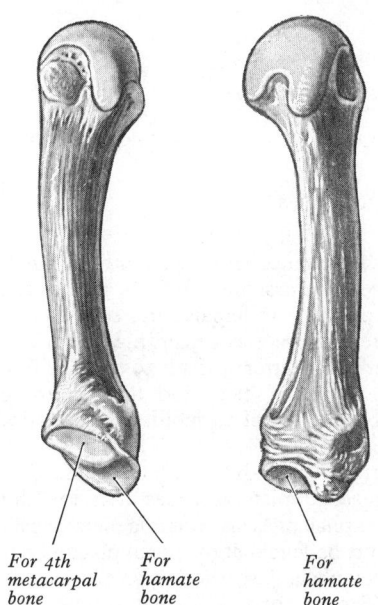

3.166 The left fifth metacarpal bone. Lateral and medial aspects.

proximal surface of the base is a facet, concave from side to side and convex from its palmar to its dorsal aspect, for articulation with the hamate. Its lateral aspect presents an elongated strip-like facet for the fourth metacarpal bone.

The *shaft* is characterized by the fact that the triangular area on its dorsal surface reaches almost to the base and only the lateral surface inclines dorsally at its proximal end. The medial surface gives insertion to the opponens digiti minimi; the lateral surface is divided by a longitudinal ridge, sometimes quite sharp and distinct, into a palmar strip for the origin of the fourth palmar interosseous muscle, and a dorsal strip for the origin of the ulnar head of the fourth dorsal interosseous muscle.

The Phalanges of the Hand

There are fourteen phalanges, three in each finger and two in the thumb. Each has a head, shaft and base or proximal end. In each the *shaft* tapers to its distal end and its dorsal surface is convex from side to side. The palmar surface is flattened from side to side, but is gently concave forwards in its long axis. The *bases* of the

proximal phalanges are marked by concave, oval facets for articulation with the heads of the metacarpal bones. Their *heads* are smoothly grooved like pulleys and encroach farther on the palmar than dorsal surfaces. To conform to the shape of the head of the proximal phalanx, the *base* of the *middle phalanx* is marked by two small, concave facets separated by a smooth ridge. The head of the middle phalanx also is pulley-shaped and the *base* of the *distal phalanx* conforms to it. The head of the distal phalanx is non-articular, but is marked on its palmar surface by a rough, horseshoe-shaped *tuberosity*, to which connective tissue strands attach the soft tissues (pulp) of the finger tip.

In addition to providing attachment for the ligaments of the joints in which they participate, the phalanges afford attachment to numerous muscles. The base of the distal phalanx gives attachment on its palmar surface to the corresponding tendon of the flexor digitorum profundus and on its dorsal surface to the extensor digitorum. The sides of the middle phalanx receive the insertion of the flexor digitorum superficialis tendon (p. 575) and the fibrous flexor sheath. Its base gives attachment on its dorsal surface to a part of the extensor digitorum tendon. To the sides of the proximal phalanx the fibrous flexor sheath is connected. Its base receives, laterally, part of the insertion of the corresponding dorsal interosseous, and medially another dorsal interosseous (p. 589).

The phalanges of the little finger and the thumb differ in certain respects from the other three. The medial side of the base of the proximal phalanx of the little finger receives the insertion of the abductor digiti minimi and the flexor digiti minimi. The base of the proximal phalanx of the thumb receives on its dorsal surface the tendon of the extensor pollicis brevis, on its lateral side the abductor pollicis brevis, the flexor pollicis brevis and the lateral part of the oblique head of the adductor pollicis, and, on its medial side, the transverse and the remainder of the oblique head of the adductor pollicis; the latter is sometimes conjoined with the first palmar interosseous muscle (p. 589). The margins of the proximal phalanx of the thumb are not sharp like those of the other digits, as the fibrous sheath is not so strongly developed as it is in the other digits.

Ossification of the Bones of the Hand (3.167, 168, 169, 170)

The carpal bones are usually cartilaginous at birth but centres of ossification may be present in the capitate and hamate. Each bone is usually ossified from one centre; the capitate first and the pisiform last, but the order in which the other carpal bones ossify is subject to considerable variation. The capitate begins to ossify in the second month, the hamate at the end of the third, the triquetral in the third *year*, the lunate during the fourth and the scaphoid, trapezium and trapezoid in the fourth year in females and fifth in males; the pisiform commences to ossify in the ninth or tenth year in females and the twelfth in males. Though the foregoing is the most common order of ossification there are considerable variations according to sex (Garu and Rohmann 1960), nutrition and, possibly, race. (For literature in respect of race consult Shakir and Zaini 1974, and Wingerd *et al.* 1974.)

Occasionally an additional bone, named the *os centrale* (p. 352), is found between the scaphoid, trapezoid and capitate bones. During the second month of intrauterine life it is represented by a small cartilaginous nodule which usually fuses with the cartilaginous scaphoid. Sometimes the styloid process of the third metacarpal bone is detached and forms an additional ossicle. Occasionally, the lunate and triquetral elements may fuse; a considerable range of other fusions and accessory ossicles have been described (O'Rahilly 1953, 1956).

Each **metacarpal bone** is ossified from two centres, a primary centre for the shaft and a secondary centre for the base or proximal end of the first and for the head or distal end of each of the other four. Ossification begins in the middle of the shaft about the ninth week of intrauterine life. The centres for the heads of the second, third, fourth and fifth metacarpals appear in that order during the second year in females and between one and a half and two and a half years in males. They unite with the shafts about the fifteenth or

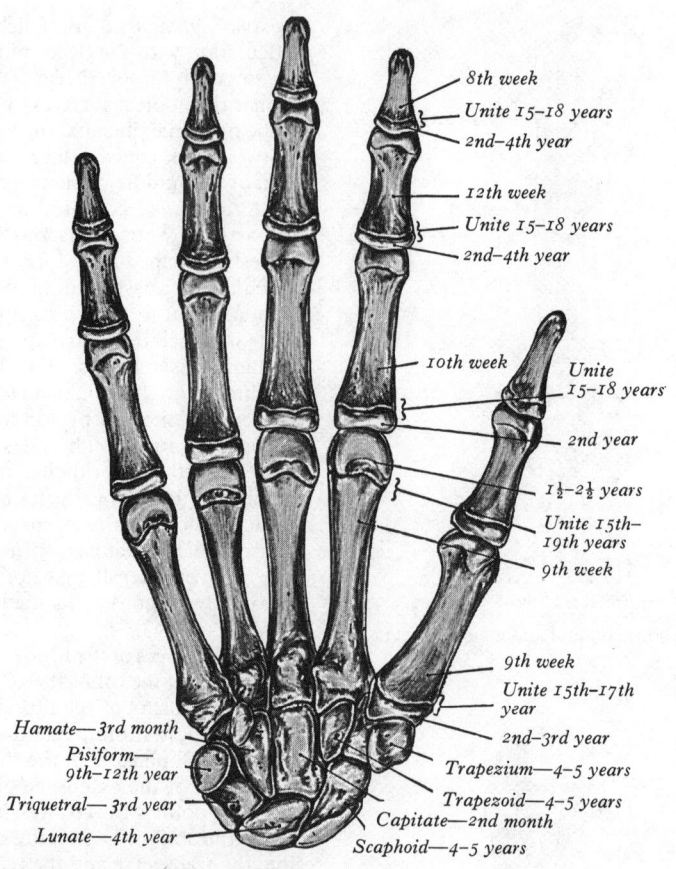

8th week
Unite 15–18 years
2nd–4th year

12th week
Unite 15–18 years
2nd–4th year

10th week
Unite 15–18 years
2nd year
1½–2½ years
Unite 15th–19th years
9th week

9th week
Unite 15th–17th year
2nd–3rd year
Trapezium—4–5 years
Trapezoid—4–5 years
Capitate—2nd month
Scaphoid—4–5 years

Hamate—3rd month
Pisiform—9th–12th year
Triquetral—3rd year
Lunate—4th year

3.167   The bones of the hand of a child, indicating the general plan of ossification.

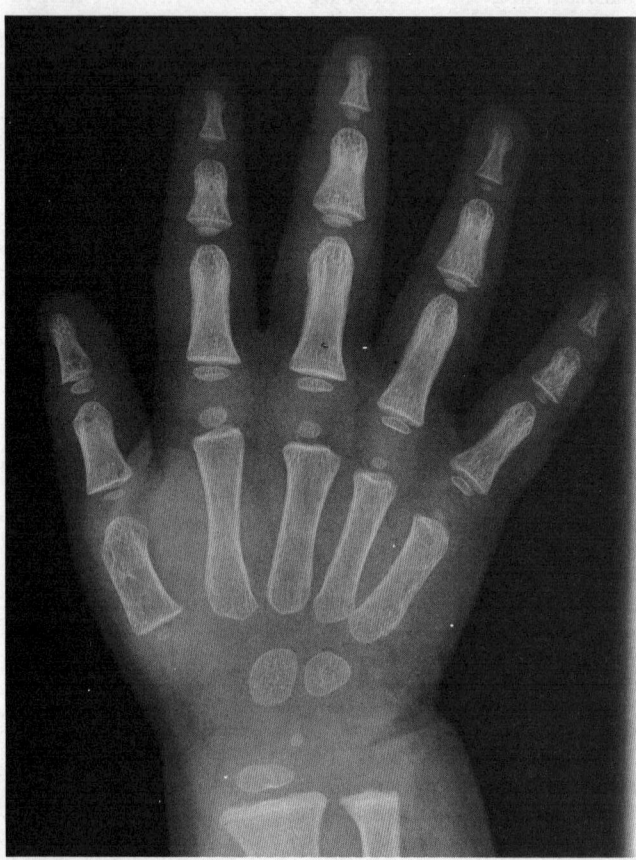

3.168   Radiograph of hand at 2½ years (male). Note early stages of ossification in epiphyses at proximal ends of phalanges and first metacarpal, at distal ends of remaining metacarpals and radius, and in the capitate, hamate, and lunate bones. The last is more usually preceded by the centre for the triquetral. Compare with **3.**169 and **3.**170.

sixteenth year in females and eighteenth or nineteenth year in males. The base of the first metacarpal begins to ossify in the latter part of the second year in females and early in the third year in males. It unites before the fifteenth year in the former and seventeeth year in the latter (Joseph 1951). The foregoing figures represent, of course, a select and limited group. For wider assessments of the range of variability consult Modi (1957) and Krogman (1962).

The metacarpal bone of the thumb is ossified like a phalanx and therefore some anatomists consider that the thumb skeleton consists of three phalanges and not of a metacarpal bone and two phalanges; others believe that the distal phalanx represents fused middle and distal phalanges, a condition occasionally observed in the little toe (Broom 1930). (When the thumb possesses *three* phalanges, the metacarpal bone has a distal, as well as a proximal epiphysis. The metacarpal occasionally bifurcates distally; in these circumstances the medial branch, which has no distal epiphysis, bears *two* phalanges, whereas the lateral branch possesses a distal epiphysis, and *three* phalanges—Nicholson 1937.) The presence of an epiphysis only at the distal end of a typical metacarpal may be associated with the greater range of movement which the metacarpophalangeal joint enjoys. In the thumb, on the other hand, it is the carpometacarpal joint which possesses the wider range of movement, and the presence of a basal epiphysis in the first metacarpal bone may be attributable to this fact. However, a distal epiphysis also is present occasionally in the first, and a basal or proximal epiphysis sometimes occurs in the second metacarpal.

Growth studies of the phalanges of 1,700 children, aged 1 to 18 years, suggested that the hypothesis noted above, that the pollicial metacarpal is a proximal phalanx (as first suggested by Vesalius in 1543), may still be a valid possibility. The latter postulated view, however, does not take account of earlier observations that the so-called distal epiphysis of the first metacarpal bone is usually (if not always) a prolongation of the diaphysis into the terminal unossified cartilage, rather than a separate centre of ossification. A recent review of the earlier literature concerning such 'pseudo-epiphyses' has corroborated this description (Haines 1974).

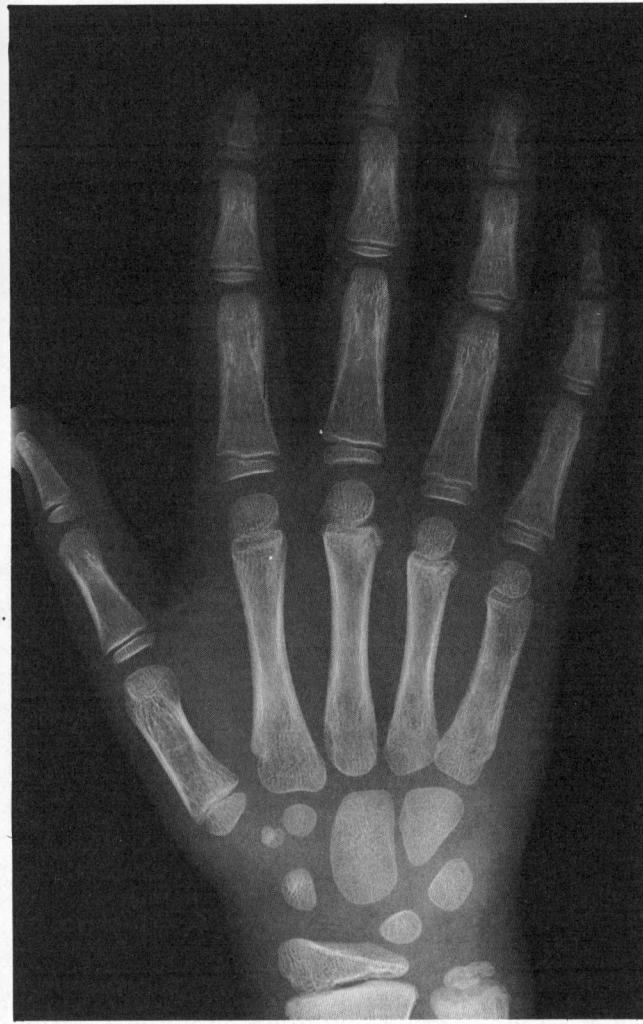

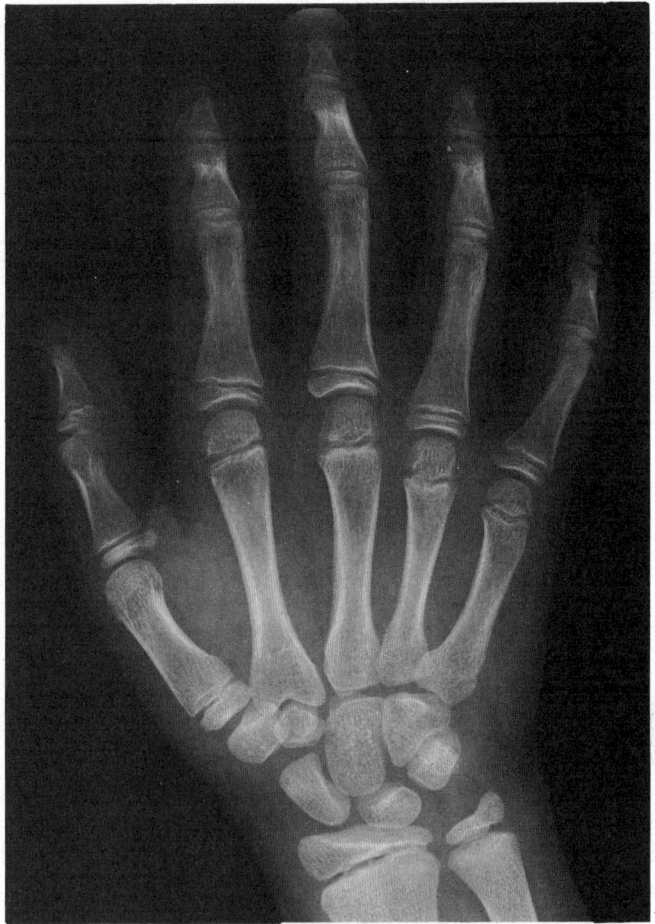

3.169   Radiograph of hand at 6½ years (male). Note the more advanced state of the centres of ossification already visible in **3**.168, and additional centres in the distal ulnar epiphysis and in the triquetral, scaphoid, trapezium and trapezoid carpal bones.

The phalanges are each ossified from two centres: a primary centre for the shaft, and a secondary or epiphysial centre for the proximal extremity. Ossification of the shafts occurs as follows: distal phalanges in the eighth or ninth week, proximal phalanges in the tenth and middle phalanges in the eleventh or later week of intrauterine life. Epiphysial centres appear in the proximal phalanges early in the second year in females and in the latter half of the second year in males, and in the middle and distal phalanges in the second year in females and in the third or fourth year in males. All these epiphyses unite with the shafts about the fifteenth to sixteenth year in females and the seventeenth to eighteenth year in males.

Applied Anatomy. The scaphoid is the most frequently fractured of the carpal bones. As a rule, the fracture runs at right angles to the long axis of the bone. Following fracture through its proximal part or through the 'waist' of the bone, non-union is by no means uncommon and may, in some cases, be due to the fact that the proximal fragment, which is devoid of nutrient foramina in 13 per cent of cases (p. 371), has been cut off from its blood supply.

Dislocation forwards of the lunate bone is not uncommon and it is often associated with fracture of the scaphoid. The displaced lunate bone may compress the median nerve against the flexor retinaculum.

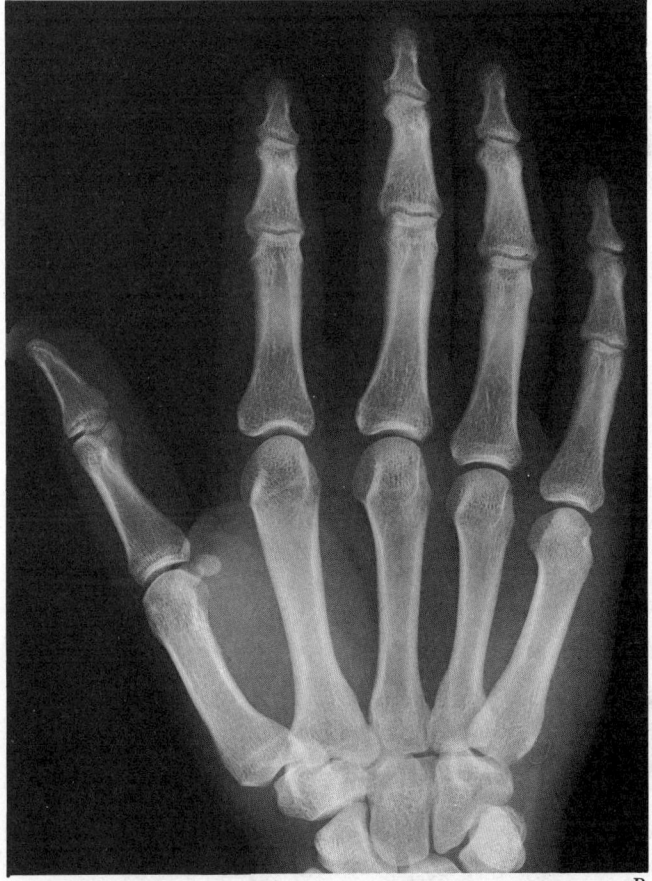

3.170A and B.
A   Radiograph of hand at 11 years (female). Note the maturing shapes of all the ossifications previously seen in **3**.168 and **3**.169, with the addition of the pisiform bone.
B   Radiograph of adult hand for comparison (male of 19 years). Note additional ossification in the sesamoid bones of the thumb.

377

THE SKELETON OF THE LOWER LIMB

The Innominate Bone

The innominate or hip bone (**3**.171, 172) is large, irregularly shaped, constricted in the middle and expanded above and below. Its *lateral surface* near its middle bears a deep, cup-shaped hollow, termed the *acetabulum*, which articulates with the rounded head of the femur. Below and in front of the acetabulum the bone presents a large, oval, or triangular gap, the *obturator foramen*. Above the acetabulum the bone forms a wide, flattened plate, with a long, curved upper border termed the *iliac crest*.

The hip bone articulates in front with the corresponding bone of the opposite side and the two bones form the pelvic girdle of the lower limbs (p. 351). Each consists of three parts, named the **ilium**, the **ischium** and the **pubis**, which are connected by cartilage in the young subject but are united by bone in the adult; the union of the three parts takes place in the walls of the acetabulum. The lines of fusion are shown as stippled bands in illustrations **3**.171B, 172B. The *ilium* includes the upper part of the acetabulum and the expanded, flattened area of bone above it; the *ischium* includes the lower part of the acetabulum and the bone below and behind; the *pubis* forms the anterior part of the acetabulum and separates the ilium from the ischium in this situation; in addition, it forms the anterior part of the lower portion of the hip bone and meets the pubis of the opposite side in the median plane.

The Ilium

The ilium, so named because it supports the *flank*, possesses two extremities and three surfaces. The lower extremity is the smaller and forms rather less than the upper two-fifths of the articular surface of the acetabulum; the upper extremity is greatly expanded to form the *iliac crest*. The surfaces are the gluteal, sacropelvic, and the iliac fossa. The *gluteal surface* is directed backwards and laterally and is an extensive rough area; the *iliac fossa* is smooth and gently hollowed out and occupies the anterior and upper part of the medial aspect of the ilium; the *sacropelvic surface* also is placed on the medial aspect and lies behind and below the iliac fossa, separated from it by a ridge, termed the *medial border*.

The iliac crest, the upper border of the ilium, is convex upwards, but sinuously curved, being concave inwards in front and concave outwards behind. Its anterior and posterior extremities project a little beyond the bone below and are termed respectively the anterior and the posterior superior iliac spines. The *anterior superior spine* lies at the lateral end of the fold of the groin and is easily felt in the living; the *posterior superior spine* cannot be felt, but its position may be indicated by a dimple, about 4 cm lateral to the second spinous tubercle of the sacrum above the medial part of the buttock. Morphologically the crest consists of a long ventral and a shorter dorsal segment. The ventral segment forms rather more than the anterior two-thirds of the crest and is associated with alterations in the form of the ilium which were necessitated by the adoption of the erect attitude; the dorsal segment forms rather less than the posterior third of the crest and can be identified in all land vertebrates. The ventral segment of the crest is bounded by *outer* and *inner* lips, enclosing a rough *intermediate zone*, which is narrowest at its middle and becomes wider both in front and behind. The *tubercle of the crest* (**3**.171A) is a prominent projection on the outer lip about 5 cm or more behind and above the anterior superior spine. The dorsal segment presents two sloping surfaces separated by a well-marked ridge, which terminates in the posterior superior spine. The highest point of the crest, which is a little behind its midpoint, is on the same level as the interval between the spines of the third and fourth lumbar vertebrae.

The *lower end* of the ilium will be described with the acetabulum (p. 383).

The *anterior border* of the ilium descends to the acetabulum from the anterior superior spine. Its upper part is rounded and concave forwards; its lower part presents a roughened projection, the *anterior inferior iliac spine*, which lies immediately above the anterior part of the acetabulum.

The *posterior border* is irregularly curved (**3**.172B). It commences at the posterior superior spine and runs at first downwards and forwards, with a backward concavity, forming a small notch. At the lower end of the notch the bone presents a wide, low projection, the *posterior inferior iliac spine*, where the posterior border makes a sharp bend. It then runs almost horizontally forwards for about 3 cm and finally turns downwards and backwards to become continuous with the posterior border of the ischium. As a result the posterior border shows a deep *greater sciatic notch*, which is bounded above by the ilium and below by the ilium and ischium (**3**.172B).

The *medial border* separates the iliac fossa from the sacropelvic surface. Indistinct near the crest, it is roughened in its upper part, then sharp and clear-cut where it bounds the articular surface for the sacrum, and finally smooth and rounded. The latter portion forms the *arcuate line*; at its inferior end it reaches the posterior part of the *iliopubic (iliopectineal) eminence*, which marks the union of ilium and pubis.

The gluteal surface (**3**.171A) faces backwards and laterally in its posterior part, and laterally and slightly downwards in front. It is bounded above by the iliac crest, below by the upper border of the acetabulum, and in front and behind by the anterior and posterior borders. The surface is rough and curved, being convex in front and concave behind, and marked by three uneven ridges,

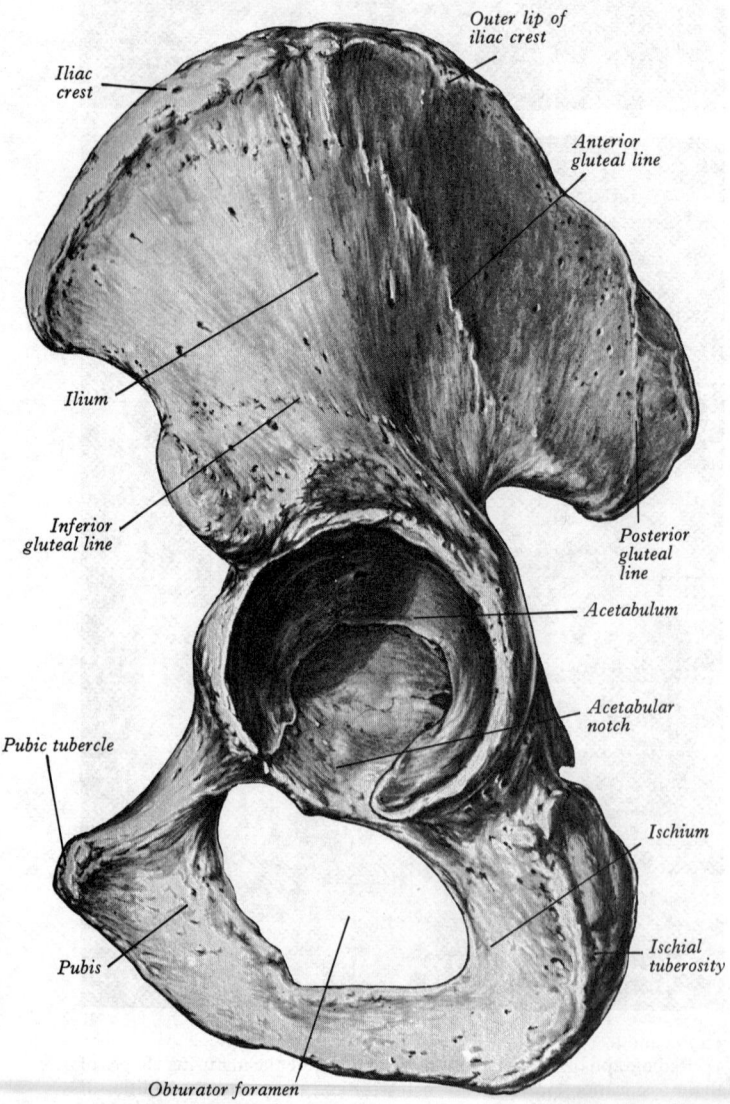

Iliac crest

Outer lip of iliac crest

Anterior gluteal line

Ilium

Inferior gluteal line

Posterior gluteal line

Acetabulum

Acetabular notch

Pubic tubercle

Ischium

Pubis

Ischial tuberosity

Obturator foramen

3.171A The left innominate bone. Lateral (external) aspect. Compare with Key, **3**.171B.

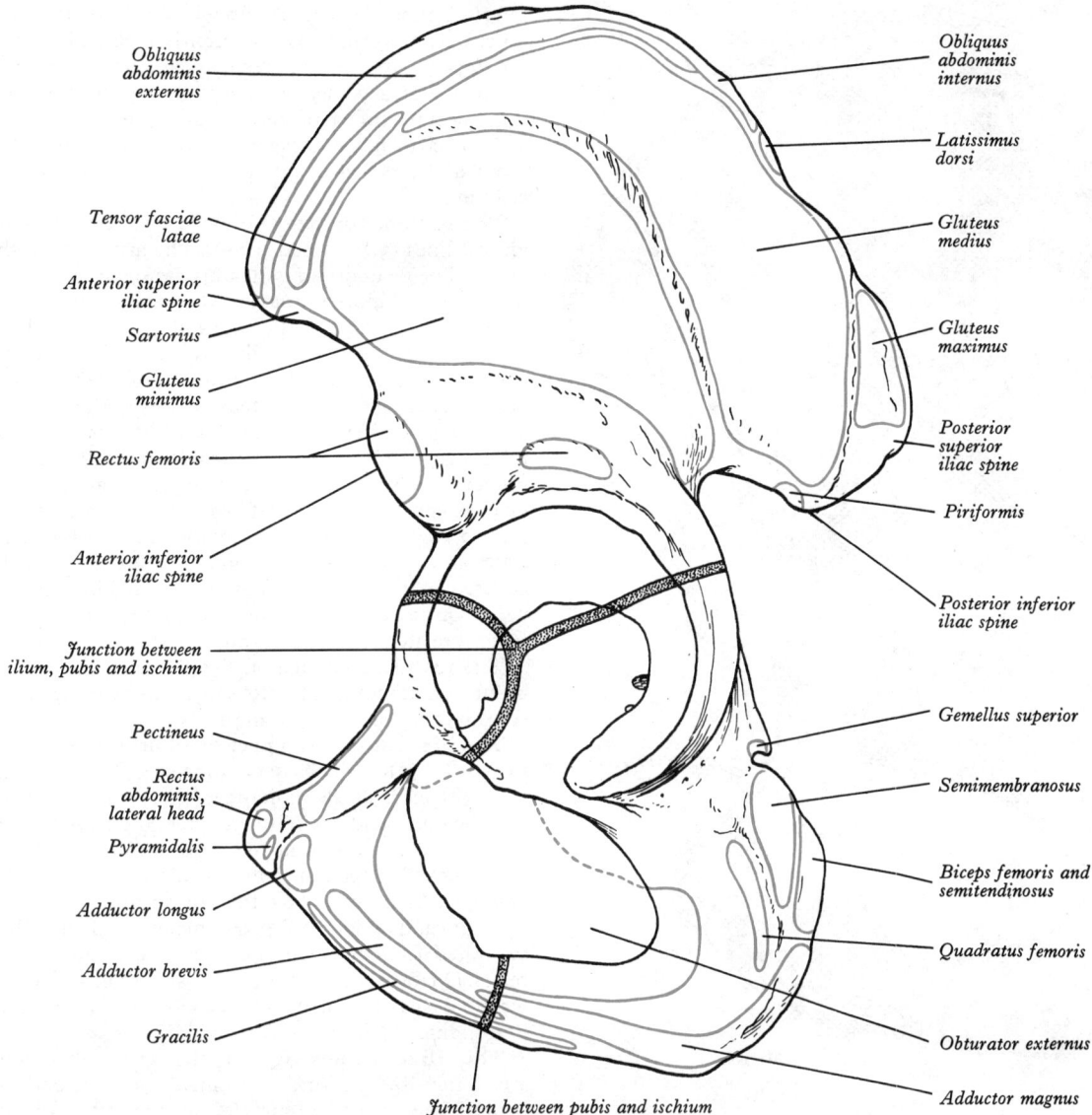

Obliquus
abdominis
externus

Tensor fasciae
latae

Anterior superior
iliac spine

Sartorius

Gluteus
minimus

Rectus femoris

Anterior inferior
iliac spine

Junction between
ilium, pubis and ischium

Pectineus

Rectus
abdominis,
lateral head

Pyramidalis

Adductor longus

Adductor brevis

Gracilis

Obliquus
abdominis
internus

Latissimus
dorsi

Gluteus
medius

Gluteus
maximus

Posterior
superior
iliac spine

Piriformis

Posterior inferior
iliac spine

Gemellus superior

Semimembranosus

Biceps femoris and
semitendinosus

Quadratus femoris

Obturator externus

Adductor magnus

Junction between pubis and ischium

3.171B   Key to **3.**171A the stippled bands indicate the limits of the iliac, pubic and ischial parts of the bone.

the posterior, anterior and inferior gluteal lines. The *posterior gluteal line*, which is the shortest, begins above on the external lip of the crest about 5 cm in front of its posterior extremity and ends below, a short distance in front of the posterior inferior spine. Its upper part is usually distinct, but inferiorly it is ill-defined and frequently absent. The *anterior gluteal line*, the longest of the three, begins about the middle of the upper margin of the greater sciatic notch and runs upwards and forwards to become confluent with the outer lip of the crest a little in front of the tubercle. The *inferior gluteal line*, which is rarely a well-marked feature, begins a little above and behind the anterior inferior spine and curves backwards and downwards to end near the apex of the greater sciatic notch. Between the inferior gluteal line and the margin of the acetabulum there is a rough, shallow groove on the bone. Behind the acetabulum the lower part of the gluteal surface becomes continuous with the posterior surface of the ischium. The site of the union of these two elements is marked by a low elevation.

The iliac fossa is the hollow of the anterior and upper part of the internal aspect of the ilium. It is limited above by the iliac crest, in front by the anterior border, and behind by the medial border, which separates it from the sacropelvic surface. The surface is smooth and gently concave and forms the posterolateral wall of the greater pelvis. Below, it is continuous with a shallow groove (3.172A) bounded laterally by the anterior inferior spine and medially by the iliopubic eminence.

The sacropelvic surface (3.172A) is the posterior and lower

part of the medial aspect of the ilium. It is bounded behind and below by the posterior border, in front and above by the medial border, and above and behind by the iliac crest. It is subdivided into three areas, viz., the iliac tuberosity, the auricular surface and the pelvic surface. The *iliac tuberosity* is an extensive, roughened area which lies immediately below the dorsal segment of the iliac crest. It bears cranial and caudal depressions separated by an oblique ridge and is connected to the sacrum by the interosseous sacroiliac ligament. The *auricular surface* (3.172A) is placed immediately below and in front of the tuberosity, and articulates with the lateral mass of the sacrum. It is shaped like an ear, the wide expanded portion lying above and in front, and the lobule below and behind, covering the medial aspect of the posterior inferior spine. Its edges are clearly defined, but the surface, although articular, is finely roughened and irregular. The *pelvic surface* lies below and in front of the auricular surface and helps to form the wall of the lesser pelvis. It comprises an upper and a lower portion. Its upper part faces downwards and lies between the margin of the auricular surface and the upper border of the greater sciatic notch; its lower region faces inwards and is separated from the iliac fossa by the arcuate line. The line of union of the ilium with the ischium is completely obliterated on this surface.

The iliac crest corresponds to the lower limit of the waist and provides attachment for the lateral muscles of the abdominal wall, fasciae and muscles of the lower limb, and muscles and fasciae of the back (3.171B, 172B). The *outer lip* of the ventral segment

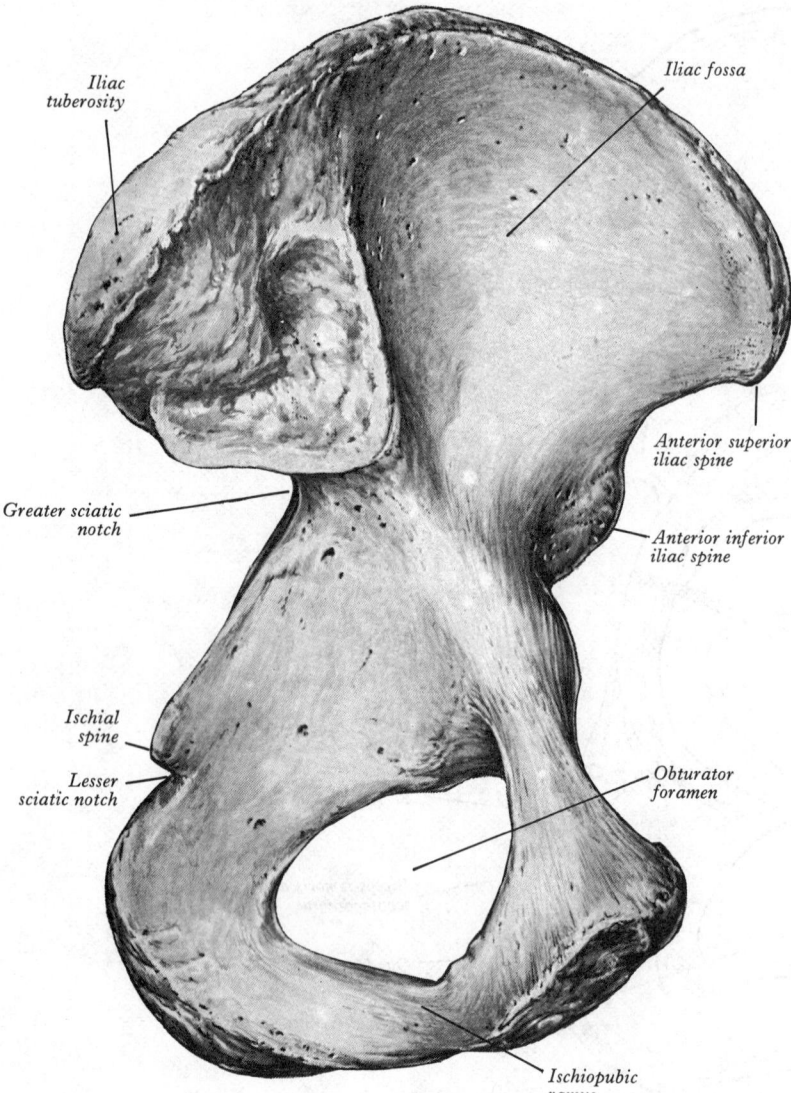

Iliac tuberosity

Iliac fossa

Anterior superior iliac spine

Greater sciatic notch

Anterior inferior iliac spine

Ischial spine

Lesser sciatic notch

Obturator foramen

Ischiopubic ramus

3.172A   The left innominate bone. Medial (internal) surface. Compare with Key, **3**.172B.

(p. 378) gives attachment to the fascia lata, including the iliotibial tract; in front of the tubercle of the crest the tensor fasciae latae arises: in its anterior two-thirds it provides insertion for the lower fibres of the external oblique; just behind its highest point the lowest fibres of the latissimus dorsi arise. An interval of variable size intervenes between the posterior limit of the attachment of the external oblique and the anterior limit of the latissimus dorsi; in this situation the crest forms the base of the lumbar triangle. The *intermediate line* gives origin to the internal oblique. The *inner lip* in its anterior two-thirds provides attachment for the transversus abdominis, and behind that to the lumbodorsal fascia and the quadratus lumborum. The dorsal segment of the iliac crest (p. 378) gives origin on its lateral slope to the highest fibres of the gluteus maximus, on its medial slope to the erector spinae and, along its medial margin, to the interosseous and dorsal sacroiliac ligaments.

The **anterior superior spine** provides attachment for the lateral end of the inguinal ligament and below that sartorius, which extends downwards for a short distance on the *anterior border*. The **anterior inferior spine** is divided indistinctly into two areas. To the upper the straight head of the rectus femoris is attached. The lower covers the inferior part of the spine and extends in a lateral direction along the upper margin of the acetabulum; it is a rough triangular impression for the iliofemoral ligament.

To the upper part of the **posterior border** the upper fibres of the sacrotuberous ligament are attached. In front of the posterior

inferior spine (i.e. on the upper border of the greater sciatic notch), it gives origin to fibres of piriformis and, in front of that on the upper border of the greater sciatic notch, it is related to the superior gluteal vessels and nerve as they emerge from the pelvis. The lower part of the posterior border (i.e. the lower margin of the greater sciatic notch) is covered by the piriformis and is related to the sciatic nerve, although the nerve lies for the most part on the ischium.

The gluteal surface is divided into four areas by the three gluteal lines (**3**.171A and B). (*a*) The area behind the posterior gluteal line gives origin in its upper roughened part to the upper fibres of the gluteus maximus; its lower, smooth part gives attachment to some of the fibres of the sacrotuberous ligament and the iliac head of the piriformis. (*b*) The area between the posterior and anterior gluteal lines, bounded above by the iliac crest, gives origin to the gluteus medius. (*c*) The area between the anterior and inferior gluteal lines gives origin to the gluteus minimus. (*d*) The area below the inferior gluteal line is marked by numerous vascular foramina. The groove above the acetabulum gives origin to the reflected head of the rectus femoris, and the area adjoining the rim of the acetabulum affords attachment to the articular capsule of the hip joint. The greater part of this area is covered by the gluteus minimus, but behind and below in the neighbourhood of the site of union of the ilium and ischium, the bone is related to the piriformis.

The vascular foramina on the gluteal aspect of the ala of the ilium sometimes lead into large vascular canals within the substance of the bone (Sirang 1973).

The iliac fossa in its upper two-thirds provides origin for the iliacus (**3**.172B), which covers the lower third but is not attached to it. Branches of the iliolumbar artery run between the muscle and the bone, and one of them enters the large nutrient foramen which is often present at the postero-inferior part of the fossa. The groove between the anterior inferior spine and the iliopubic eminence is occupied by the converging fibres of the iliacus laterally and tendon of psoas major medially; the tendon is separated from the bone near the acetabulum by its synovial bursa. On the right side the iliac fossa contains the caecum and the terminal part of the ileum, on the left side the terminal part of the descending colon.

The iliac tuberosity of the sacropelvic surface gives attachment to the dorsal sacroiliac ligaments and, immediately behind the auricular surface, to the interosseous sacroiliac ligament. The upper and anterior part of the tuberosity gives attachment to the iliolumbar ligament, and this area lies immediately below the medial part of the origin of the quadratus lumborum from the iliac crest. **The auricular surface** articulates with the auricular surface of the sacrum and is reciprocally curved (p. 473). Its anterior and inferior borders are sharp and give attachment to the ventral sacroiliac ligament. The lower part of the **pelvic surface** between the inferior margin of the auricular surface and the upper margin of the greater sciatic notch is often marked by a roughened groove, which is termed the *pre-auricular sulcus*; it gives attachment to the lower fibres of the ventral sacroiliac ligament. This sulcus is more often well marked in the female than the male. However, the unreliability of the presence of this sulcus in determination of sex is shown by a study of 237 Indian pelves, as in other races (Jit and Gandhi 1966). For further pertinent analyses, however, consult Finnegan and Faust (1974), and Finnegan (1978). Lateral to the sulcus the bone sometimes gives origin to fibres of the piriformis. The rest of the pelvic surface gives origin to the upper half or less of the obturator internus.

The Pubis

The pubis forms the ventral part of the innominate bone and meets the opposite pubis in the median plane to form a cartilaginous joint, the pubic symphysis. It possesses a body, which lies anteriorly, a superior ramus, which passes upwards and backwards to the acetabulum, and an inferior ramus, which passes backwards, downwards and laterally to unite with the ramus of the ischium on the medial side of the obturator foramen.

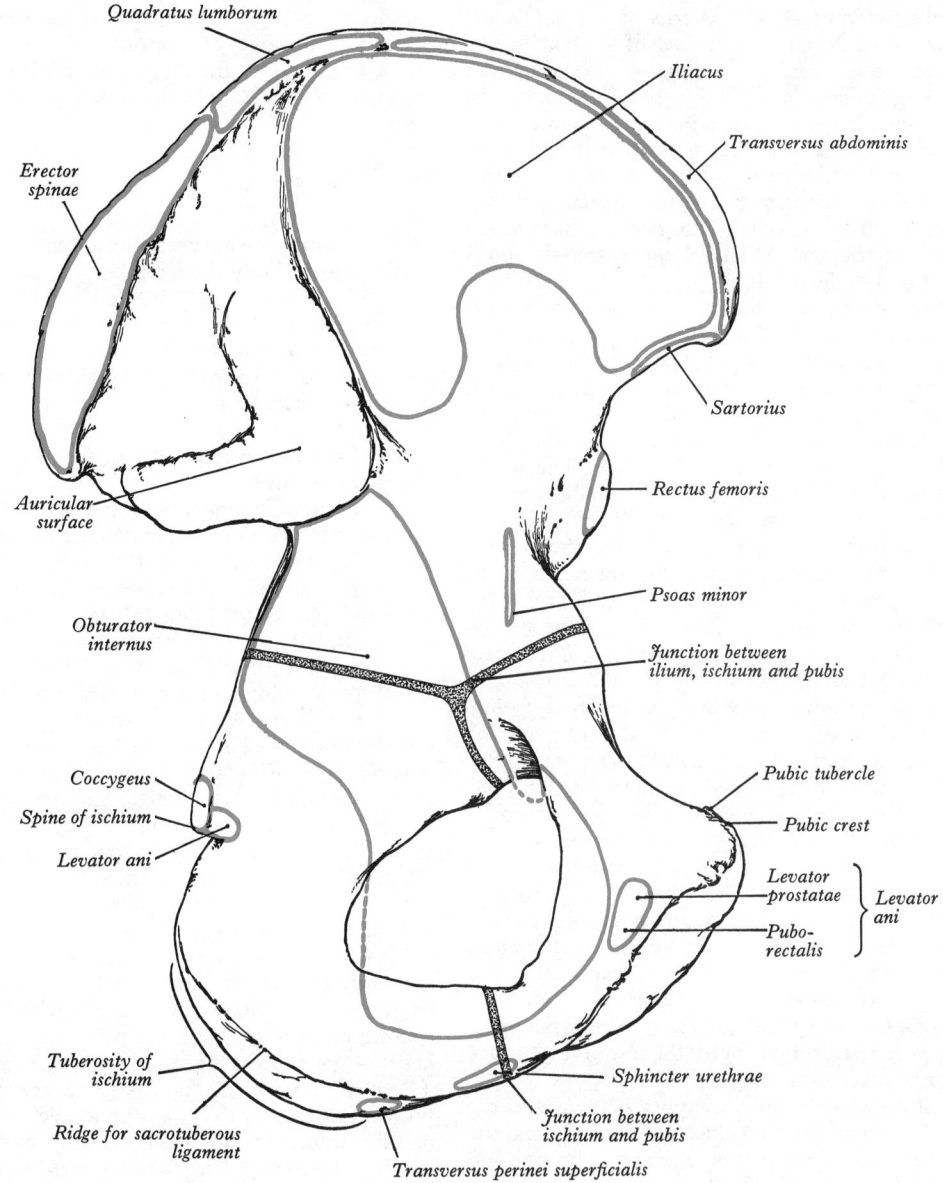

Quadratus lumborum

Iliacus

Transversus abdominis

Erector spinae

Sartorius

Rectus femoris

Auricular surface

Psoas minor

Obturator internus

Junction between ilium, ischium and pubis

Pubic tubercle

Coccygeus

Pubic crest

Spine of ischium

Levator prostatae

Levator ani

Levator ani

Pubo-rectalis

Tuberosity of ischium

Sphincter urethrae

Ridge for sacrotuberous ligament

Junction between ischium and pubis

Transversus perinei superficialis

3.172B Key to **3.172A**. The stippled bands indicate the limits of the iliac, pubic and ischial parts of the bone.

The body is compressed from before backwards and presents anterior, posterior and symphysial (or medial) surfaces and a free upper border termed the *pubic crest*. The *anterior surface* faces downwards, forwards and slightly laterally in the upright posture; rough in its upper and medial parts, it presents a smooth surface elsewhere. It is directed towards the lower limb and affords attachment for the medial group of muscles of the thigh. The *posterior surface* is smooth and faces upwards and backwards, forming the anterior wall of the pelvis minor; it is related to the urinary bladder. The *symphysial surface* is an elongated oval area, covered with cartilage in the recent state and articulating with the opposite pubis at the pubic symphysis. When denuded of cartilage it presents, in the elderly, an irregular surface, marked by a number of small ridges and furrows or by small nodular elevations; but these appearances show wide variations at different ages (Todd 1920, 1921; Brooks 1955; Suchey 1979). The forensic application of such observations is obvious. The *pubic crest* is the rounded upper border of the body. It is projected forwards and overhangs the upper part of the anterior surface (**3.171A**). Its lateral extremity forms a rounded projection termed the *pubic tubercle*. Both the crest and the tubercle can be felt through the skin, but the latter is obscured in the male by the spermatic cord, which crosses its upper aspect as passes upwards from the scrotum to pierce the abdominal wall.

The superior ramus of the pubis springs from the upper and lateral part of the body, and passes backwards, upwards and laterally above the obturator foramen to reach the acetabulum. It is triangular on section and has three surfaces and three borders. The *pectineal surface* is directed forwards and slightly upwards. Triangular in outline, it extends from the pubic tubercle to the iliopubic eminence (**3.171A**). It is bounded in front by a rounded ridge termed the *obturator crest*, and behind by a sharp edge termed the *pecten pubis* (*pectineal line*), which together with the pubic crest constitutes the pubic part of the *linea terminalis* (i.e. the anterior part of the pelvic brim). The *pelvic surface*, which is directed upwards, backwards and medially, is smooth and featureless; it is narrow at its lateral extremity where it is continuous with the posterior surface of the body. It is bounded above by the pecten pubis and below by a sharp edge which forms the *inferior border*. The *obturator surface* is directed downwards and backwards, and is crossed from behind forwards and downwards by a groove termed the *obturator groove*. It is bounded in front by the obturator crest and behind by the inferior border.

The inferior ramus springs from the lower and lateral part of the body and passes backwards, downwards and laterally to unite with the ramus of the ischium on the medial side of the obturator foramen. The site of union may be marked by a localized thickening, often difficult to identify in the adult bone. The ramus has two surfaces and two borders. The *anterior* or *external surface* is continuous above with the anterior surface of the body of the pubis; it faces the thigh and is roughened for muscular attachments. It is bounded laterally by the margin of the obturator foramen and medially by a rough anterior border. The *posterior* or *internal surface* is continuous above with the posterior surface of the body, and is transversely convex from side to side. Its medial part is often prominently everted in male subjects (**3.**180) and is connected to the crus of the penis. It is directed medially towards the perineum. Its lateral part is smooth and is directed upwards towards the pelvis.

The pubic tubercle provides attachment for the medial end of the inguinal ligament; it lies in the floor of the superficial inguinal ring and is crossed by the spermatic cord. The ascending limbs of the loops of the cremaster are attached to the tubercle and to the anterior wall of the sheath of the rectus abdominis. From the lateral part of the **pubic crest** arises the lateral head of the rectus abdominis, and from the bone below the pyramidalis. The medial part of the crest is crossed by the medial head of the rectus abdominis, which arises from an interlacement of fibres in front of the upper part of the pubis and pubic symphysis. The *anterior surface* of the body is directed towards the adductor region of the thigh. A roughened strip, usually wider in the female, marks the medial part of the surface and gives attachment to the ventral pubic ligament. In the angle between the upper end of this strip and the pubic crest the rounded tendon of the adductor longus takes origin. At a slightly lower level the gracilis arises from a linear origin close to the medial border of the body and extending downwards on to the inferior ramus. Lateral to the gracilis the adductor brevis arises from the body and the inferior ramus. The lateral part of the anterior surface, and the adjoining portions of both rami provide origin for the obturator externus (**3.**171B).

The *posterior surface* of the body is separated from the urinary bladder by the retropubic pad of fat. About its middle it provides origin for the anterior fibres of the levator ani, and more laterally the obturator internus arises from this surface and extends on to both rami. Medial ·to the origin of the levator ani, the puboprostatic ligaments are attached to the bone.

The *pectineal surface of the superior ramus* gives origin, along its upper part, to the pectineus, which covers the rest of the surface (**3.**171B) but is not attached to it.

The pecten pubis, which forms the upper boundary of the pectineal surface, is a salient, sharp ridge. At its medial end the conjoint tendon and the lacunar ligament are attached, and throughout the rest of its extent it affords attachment to a strong fibrous band often termed the *pectineal ligament* (p. 552). About its middle it receives the insertion of the psoas minor when present. The *pelvic surface* is smooth and is not covered with muscle or fascia. It is separated from the parietal peritoneum only by the intervening subperitoneal tissue, in which the lateral umbilical ligament runs downwards and forwards across the ramus and, near its lateral end, the ductus deferens passes backwards. The *obturator groove* on the obturator surface is converted into a canal by the upper borders of the obturator membrane, the obturator internus and the obturator externus. It transmits the obturator vessels and nerve from the pelvis to the thigh, where they emerge under cover of the pectineus. The *obturator crest* (**3.**171B) at its lateral end gives attachment to some of the fibres of the pubofemoral ligament.

To the *outer surface of the inferior ramus* are attached the gracilis, the adductor brevis and the obturator externus, named from the medial to the lateral side. In addition, the origin of the adductor magnus usually extends from the ramus of the ischium on to the lower part of the inferior ramus of the pubis in the interval between the adductor brevis and the obturator externus. The *inner surface* is divided into a medial, an intermediate and a lateral area, but they are not separated from one another by clear-cut markings on the bone. The medial area faces downwards and medially and is in direct contact with the crus penis; it is limited above and behind by an indistinct ridge which gives attachment to the inferior fascia of the urogenital diaphragm (p. 562). The intermediate area is related to the dorsal nerve of the penis and the internal pudendal vessels and the fascial sheath in which they are enclosed, and it may give origin to some of the inner fibres of the

sphincter urethrae. From the lateral area arise fibres of the obturator internus. The *medial margin* of the ramus is strongly everted in the male and gives attachment to the fascia lata and to the membranous layer of the superficial fascia of the perineum.

The Ischium

The ischium is the lower, posterior part of the innominate bone. It comprises a body and a ramus. The body has upper and lower extremities, and femoral, dorsal and pelvic surfaces. The *upper extremity* of the body forms the lower and posterior part of the acetabulum, and its *lower extremity* gives off the *ramus*, which runs upwards, forwards and medially at an acute angle to fuse with the descending ramus of the pubis, so completing the obturator foramen.

The *femoral surface* of the body faces downwards, forwards and laterally towards the thigh. It is bounded in front by the margin of the obturator foramen, and laterally by the lateral border, indistinct above but clearly defined below, where it forms the lateral border of the ischial tuberosity. The *dorsal surface* faces backwards, laterally and upwards. Above, it is continuous with the lower part of the gluteal surface of the ilium, and where the two elements meet, the bone presents a low convexity, which corresponds to the curvature of the posterior part of the acetabulum. Below, the surface is marked by the upper part of the ischial tuberosity. Above the tuberosity the bone presents a wide and shallow groove both on the lateral and on the medial side. The *ischial tuberosity* is a large, roughened impression which marks the lower part of the dorsal surface and the inferior extremity of the body of the ischium. Although obscured by the gluteus maximus when the hip joint is extended, it can be identified without difficulty when the joint is flexed. It lies 5 cm from the median plane and about the same distance above the gluteal fold (p. 600). It is an elongated area, widest near its upper end and tapering inferiorly, and provides attachment for the posterior femoral (hamstring) muscles. The dorsal surface is placed between the lateral and the posterior borders of the body. The *posterior border* is continuous above with the posterior border of the ilium and helps it to complete the lower margin of the *greater sciatic notch*. The posterior end of that margin is marked by a conspicuous projection, termed the *ischial spine*. Below the spine the border becomes rounded and indefinite, forming the floor of a rounded notch, termed the *lesser sciatic notch*, which lies between the ischial spine and the tuberosity. The *pelvic surface* is smooth and relatively featureless and is directed towards the pelvic cavity; its lower portion forms part of the lateral wall of the ischiorectal fossa in the perineum.

The ramus of the ischium presents anterior and posterior surfaces, continuous with the corresponding surfaces of the inferior pubic ramus. The *anterior surface* is directed forwards and downwards towards the thigh and is rough for the attachment of the medial femoral muscles. The *posterior surface* is smooth, and partly subdivided into a perineal and a pelvic area, like the inferior ramus of the pubis. The *upper border* helps to complete the margin of the obturator foramen; the *lower border* is roughened and free, and together with the medial border of the inferior ramus of the pubis forms the lateral boundary of the subpubic angle and part of the pubic arch.

Attached to the *femoral surface* of the **body of the ischium** below is a part of obturator externus (**3.**171B), and along the lateral border of the upper part of the ischial tuberosity the quadratus femoris. Just below the acetabulum the lateral border gives attachment to the ischiofemoral ligament.

Immediately above the ischial tuberosity the *dorsal surface* is crossed by the tendon of the obturator internus and the gemelli; the nerve to the quadratus femoris runs downwards between these structures and the bone. At a higher level, the bone is covered by the piriformis, which is partially separated from it by the sciatic nerve and the nerve to the quadratus femoris. The **ischial tuberosity** is divided by a nearly transverse ridge into an upper and a lower area (**3.**173). The upper area is associated with the hamstring muscles; it is divided by an oblique line into an upper and lateral part from which arise the semimembranosus, and a lower and medial part from which the long head of the biceps femoris arises in common with the semitendinosus. The lower portion of the tuberosity narrows as it passes forwards on to the lower end of the ischium. It is divided by an irregular vertical ridge into a lateral and a medial area; the lateral area is the larger and affords origin to part of the adductor magnus, the medial is covered by fibro-adipose tissue, usually containing the ischial bursa of the gluteus maximus. It is the medial area on the lower part of the tuberosity which supports the body in the sitting posture. On its medial side the tuberosity is limited by a curved ridge which extends forwards on to the ramus of the ischium and to which the sacrotuberous ligament and its falciform process are attached (3.172B). Many of the fibres of the biceps femoris can be traced into the sacrotuberous ligament, and this intimate relationship is noteworthy, for the sacrum and the posterior part of the ilium constitute the primitive mammalian origin of the biceps femoris. The origin of the muscle from the tuberosity in man is secondary and the sacrotuberous ligament represents the remains of its primitive tendon of origin.

Above and medial to the tuberosity the posterior surface presents a wide, shallow groove. Here the bone is usually covered with a thin layer of hyaline cartilage and a bursa is interposed between it and the tendon of the obturator internus, which lies in the groove. To the lower margin of the groove, close to the tuberosity, the inferior gemellus is attached; the upper margin of the groove close to the ischial spine, gives origin to the superior gemellus.

The ischial spine projects downwards and slightly medially. To its margins is attached the sacrospinous ligament, which separates the greater sciatic foramen from the lesser (4.56). Its dorsal surface is crossed by the internal pudendal vessels and the nerve to obturator internus, as they pass through the gluteal region. The pelvic surface of the spine provides an origin for coccygeus and the most posterior fibres of levator ani. The structures passing through the greater and lesser sciatic foramina are detailed on p. 475.

The *pelvic surface* of the body of the ischium is smooth. To its upper part is attached the obturator internus, the fibres of which converge on the lesser sciatic notch and cover the remainder of this surface, with the exception of the pelvic aspect of the ischial spine. The muscle and its covering fascia separate the bone from the ischiorectal fossa.

The *anterior surface* of the *ramus of the ischium* faces the adductor region of the thigh. It gives origin to obturator externus above, the anterior fibres of the adductor magnus and, near the lower border, to the gracilis. Between the adductor magnus and the gracilis the origin of the adductor brevis may extend downwards from the inferior ramus of the pubis for a short distance. The *posterior surface* is divided into pelvic and perineal areas. The pelvic area is directed upwards and backwards and part of the obturator internus is attached to it. The perineal area is directed medially; its upper part is related to the crus of the penis or clitoris and gives origin to the sphincter urethrae; its lower part has the ischiocavernosus and the superficial transversus perinei attached to it. The inferior fascia of the urogenital diaphragm is attached to the ridge which separates the perineal from the pelvic area below, and the area for the crus from the origin of the sphincter urethrae above. The *lower border* of the ramus provides attachment for the fascia lata of the thigh and the membranous layer of the superficial fascia of the perineum.

The acetabulum (3.171A) is an approximately hemispherical cavity on the lateral aspect of the innominate bone about its centre, and is directed laterally, downwards and forwards. It is surrounded by an irregular projecting margin which is deficient *inferiorly*; this gap is the *acetabular notch*. The floor of the cavity is roughened and non-articular and is termed the *acetabular fossa*. The sides of the cup present an articular, *lunate surface*, which is widest superiorly; in this situation the weight of the trunk is transmitted to the femur in the erect attitude. This horseshoe-shaped strip is covered with articular cartilage and provides the surface on which the head of the femur moves within the hip joint. All three elements of the hip bone contribute to the formation of the acetabulum in man, but not in equal proportions. The pubis forms the upper and anterior fifth of the articular surface, the

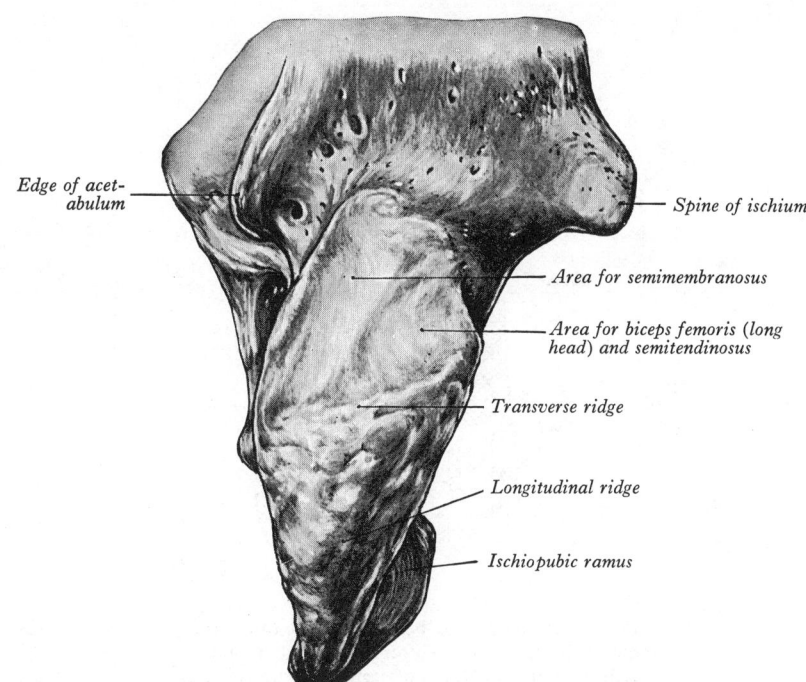

Edge of acet-abulum

Spine of ischium

Area for semimembranosus

Area for biceps femoris (long head) and semitendinosus

Transverse ridge

Longitudinal ridge

Ischiopubic ramus

3.173 The left ischial tuberosity. Posterior aspect. The *transverse ridge* forms the lower boundary of the area for the hamstring muscles and separates it from the lower half of the tuberosity, which is divided into lateral and medial areas by the *longitudinal ridge*. To the lateral area is attached the adductor magnus; the medial area is covered with fibro-adipose tissue and supports the body in the sitting posture.

ischium, the floor of the acetabular fossa and rather more than the lower and posterior two-fifths of the articular surface, the ilium forms the remainder of the articular surface. A roughly linear defect occasionally crosses the acetabular surface from its *superior* border to the acetabular fossa; it does *not*, however, correspond to the junctions between ilium and ischium or ilium and pubis.

The obturator foramen is a large gap in the hip bone, below and slightly in front of the acetabulum, between the pubis and ischium. It is bounded above by the grooved obturator surface of the superior pubic ramus, medially by the body and inferior ramus of the pubis, below by the ramus of the ischium, and laterally by the anterior border of the body of the ischium, including the margin of the acetabular notch. The foramen is occupied in the recent state by a fibrous sheet, the *obturator membrane*, which is attached to its margins, except above, where a communication is left between the pelvis and the thigh. The free upper edge of the membrane is attached in front to the *anterior obturator tubercle*, which marks the anterior end of the inferior border of the superior ramus of the pubis, and behind to the *posterior obturator tubercle*, which is placed on the anterior border of the acetabular notch. These tubercles are not always easy to identify. Since the obturator groove cuts across the upper border of the foramen, the total border is in the form of a spiral. The foramen is large and oval in the male, but is smaller and nearly triangular in the female.

Structure. The thicker parts of the hip bone consist internally of cancellous bone, enclosed between two layers of compact bone; the thinner parts, at the bottom of the acetabulum and centre of the iliac fossa, are usually translucent and composed entirely of compact bone. At the upper part of the acetabulum and along the arcuate line, i.e. along the line of weight transmission from the sacrum to the head of the femur, the amount of compact bone shows a considerable increase. In this situation the underlying spongy bone displays two sets of pressure lamellae. The first arise near the upper part of the auricular surface and diverge to impinge on two stout buttresses formed by the compact bone. From there two similar sets of lamellar arches start and converge on the acetabulum (Wakeley 1929). Because of its frequent use as a site for biopsy puncture for bone marrow examination, the iliac crest,

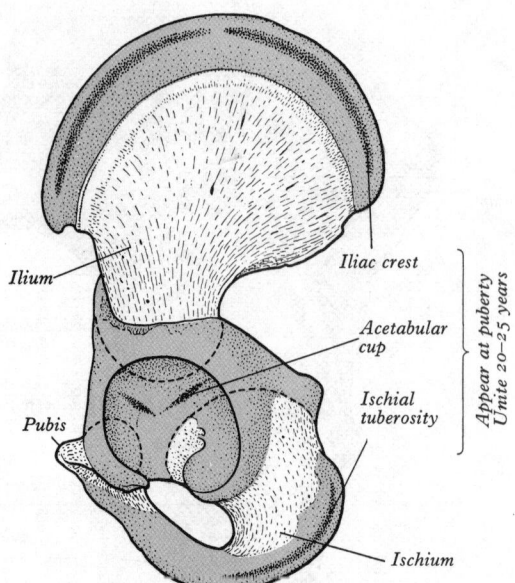

Ilium

Iliac crest

Acetabular cup

Ischial tuberosity

Pubis

Ischium

Appear at puberty
Unite 20–25 years

3.174 The innominate bone at birth. More heavily stippled areas indicate the secondary centres of ossification and the Y-shaped medial extension of the cartilaginous acetabular cup is indicated by interrupted lines. The ossified part of the ischium is shown in the floor of the acetabulum.

especially in its ventral region, has been frequently studied in regard to the distribution of cortical and trabecular bone. Whitehouse (1977) has surveyed these studies; his own observations, based on scanning electron micrography, indicate that the cortical bone of the crest is peculiarly porous, being only 75 per cent bone, and as low as 35 per cent near the anterior superior iliac spine. Denser cortical bone was observed to commence at the margins of the crest, and to increase rapidly in thickness below it, on both aspects of the iliac squama.

Ossification. The hip bone is ossified from three primary centres, one each for the ilium, ischium and pubis: the centre for the ilium appears immediately above the greater sciatic notch about the eighth week of intrauterine life, the one for the ischium appears in the body in the fourth month, and that for the pubis in the superior ramus between the fourth and fifth months. At birth three parts of the hip bone are still cartilaginous; these are the whole of the iliac crest, the floor of the acetabulum and a strip along the inferior margin of the bone (3.174). At this stage the acetabulum is a cartilaginous cup with a triradiate stem extending medially from its deep aspect to appear on the pelvic surface as a Y-shaped epiphysial plate between the ilium, ischium and pubis (Harrison 1957). This cartilage also includes the anterior inferior iliac spine. The strip of cartilage along the inferior margin of the hip bone covers the surface of the ischial tuberosity, forms (temporarily) the full conjoined rami of the ischium and pubis and then continues to the symphysial surface of the pubis and then along the pubic crest to the pubic tubercle.

The ossifying ischium and pubis fuse to form the conjoined bony ramus at the seventh or eighth year. The secondary centres appear about puberty and join the rest of the bone between the fifteenth and twenty-fifth years. There are usually two secondary centres for the iliac crest but they fuse rapidly with one another. Ossification of the acetabular cartilage begins as two separate centres, one or other of which is often termed the *os acetabuli*, between the ilium and pubis and between the ilium and ischium; these two centres become fused as ossification spreads to include the whole stem of the cartilage. This centre forms a substantial part of the articular surface of the adult bone along the junction of the triradiate stem and cup of cartilage and around the periphery of the cup. The anterior inferior iliac spine may be ossified as an extension from this centre or from a separate centre. The cartilage around the inferior margin of the bone begins to ossify over the ischial tuberosity and ossification spreads forwards. The part

which ossifies first fuses almost at once, before the more anterior parts have ossified and the fusion also extends forwards. The pubic tubercle and crest and the symphysial surfaces of the pubic bone may have separate centres.

The Skeletal Pelvis

The term *pelvis*, meaning a basin, is indiscriminately applied to the skeletal ring formed by the two innominate bones and the sacrum, the cavity of this arrangement, and also, by extension, to the entire region where trunk and lower limbs meet. It is used here in the skeletal sense, as the somewhat unsuitable name for the irregular bony girdle interposed between the femoral heads and the fifth lumbar vertebra. It is massively constructed in conformity with its primary, commonplace function of withstanding the compression and other stresses due to body weight and powerful musculature. Its mechanisms in this regard will be considered elsewhere (p. 475). Here, we shall be concerned with metrical and other characteristics, particularly in respect to sexual differences, which are of obstetric, forensic and anthropological application.

The pelvis can be regarded as separable into greater and lesser segments, sometimes called the true and false pelves, which are arbitrarily divided by an oblique plane. This passes through the sacral promontory behind and the *lineae terminales* at the sides and in front. Each linea terminalis includes the arcuate line of the ilium (p. 378) and the pecten (iliopectineal line) and crest of the pubis. The greater and lesser pelves are, however, structurally continuous, and the parts of the body cavity which they enclose are also continuous through the superior pelvic aperture, or pelvic inlet (3.176).

The greater pelvis consists of the flanged parts of the iliac bones above the lineae terminales on each side and the base of the sacrum posteriorly. The bone structure along this junctional zone is particularly massive, forming the main pathway on each side from the acetabular fossae to the vertebral column around the visceral cavity. The cavity of the greater pelvis is, of course, part of the abdomen; and because of the inclination of the pelvis as a whole (*vide infra*) the cavity has little anterior wall, as far as bone is concerned.

The lesser pelvis encloses a true basin when the soft tissues of the pelvic floor are in place. From the skeletal point of view it is a narrowed continuation of the greater pelvis, with irregular but more complete walls, bounding the pelvic cavity or canal. This cavity, which is naturally of the greatest obstetric importance, has an axis which is curved in the median plane. It is limited above and below by superior and inferior openings, the superior being occupied in life by viscera traversing it, the inferior being largely closed by the pelvic floor and its sphincteric mechanisms.

The superior pelvic aperture (pelvic inlet) is variable in contour, being rounded or oval, but encroached upon behind by the sacral promontory. The boundaries of the aperture, as described above, are conveniently referred to as the *pelvic brim* (3.176), which has particular obstetric significance (*vide infra*). It has long been subject to measurement for anthropological and obstetric reasons—as has the pelvic cavity in general—and a very considerable array of values is available, especially for the female. Naturally, the means recorded by different observers show some variation, being founded on groups differing greatly in racial and economic background. The figures cited here are merely approximate samples of the values to be expected in Europeans (Martin 1928). The following three dimensions have become conventional.

A. The *anteroposterior diameter* (true conjugate) is measured between the midpoints of the sacral promontory and the upper border of the symphysis pubis (male 100 mm, female 112 mm).

B. The *transverse diameter* is the maximum distance that can be found between similar points (as far as they can be assessed by eye) on opposite sides of the pelvic brim (male 125 mm, female 131 mm).

C. The *oblique diameter* is measured from one iliopubic

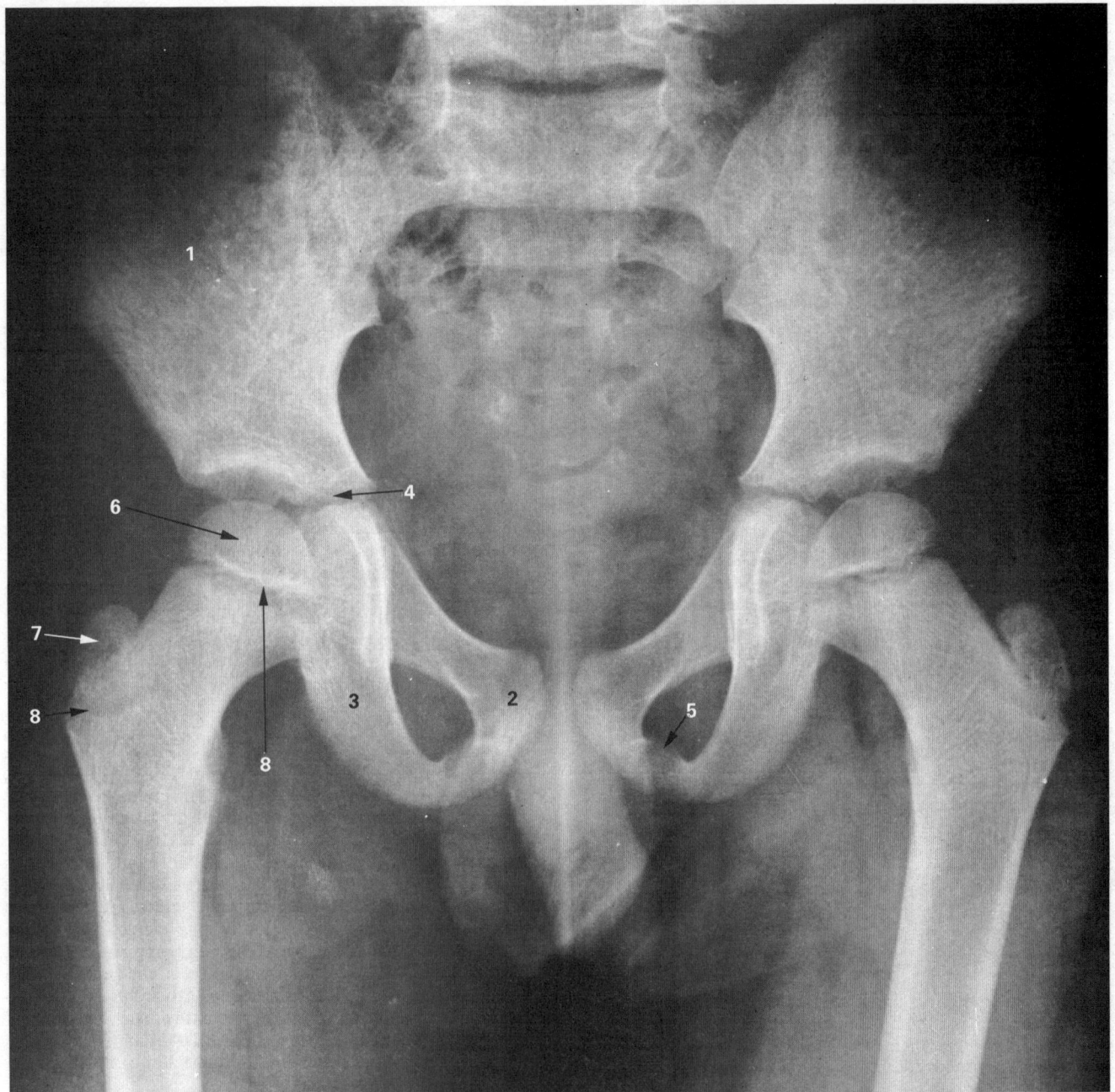

3.175　Anteroposterior radiograph of the pelvis of a boy aged 7 years. 1. Ilium. 2. Pubis. 3. Ischium. 4. Part of triradiate growth cartilage. 5. Cartilage between pubic and ischial rami. 6. Superior femoral epiphysis. 7. Ossifying greater trochanter. 8. Cartilaginous growth plates.

eminence to the opposite sacroiliac joint (male 120 mm, female 125 mm).

The cavity of the lesser pelvis is short and curved, being markedly longer in its posterior wall than in its anterior. In front and below, it is bounded by the body of the pubis and its rami and symphysis, limited behind by the concave anterior surface of the sacrum and by the coccyx, and on each side is a smooth quadrangular area formed by the pelvic aspect of the fused ilium and ischium. The region thus enclosed is the pelvic cavity proper, through which pass, in all land vertebrates, the end of the alimentary canal and the urogenital ducts. It thus contains, in human beings, the rectum, bladder and parts of the reproductive organs. The rectum is most posterior, the bladder anterior, and the uterus intermediate in position.

The diameters of the pelvic cavity are also sometimes estimated, and while this could be done theoretically at a large number of levels, the measurements are usually carried out at approximately mid-level.

A. The *anteroposterior diameter* is measured between the midpoints of the third sacral segment and the posterior surface of the symphysis pubis (male 105 mm, female 130 mm).

B. The *transverse diameter* is the widest transverse distance that can be found between the bony side walls of the cavity. It is in fact often the greatest transverse dimension in the whole extent of the cavity (male 120 mm, female 125 mm).

C. The *oblique diameter* is defined as the distance from the lowest point of one sacroiliac joint to the midpoint of the opposite obturator membrane (male 110 mm, female 131 mm).

The inferior pelvic aperture (pelvic outlet) (**3**.177) is much less simple in outline than the superior aperture, being indented behind by the coccyx and sacrum and on each side by the ischial tuberosities. The perimeter of the opening thus consists of three wide notches, or arcs, the anterior being the *pubic arch*, between the converging ischiopubic rami. Between the sacrum and coccyx

385

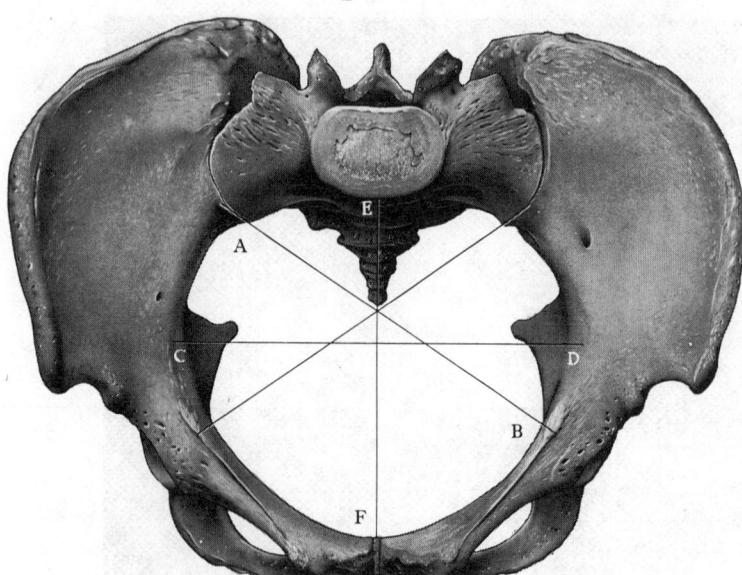

3.176 The diameters of the superior aperture of the lesser pelvis (female). A=sacroiliac joint; B=iliopubic eminence; C and D=middle of pelvic brim; E=sacral promontory; F=pubic symphysis.

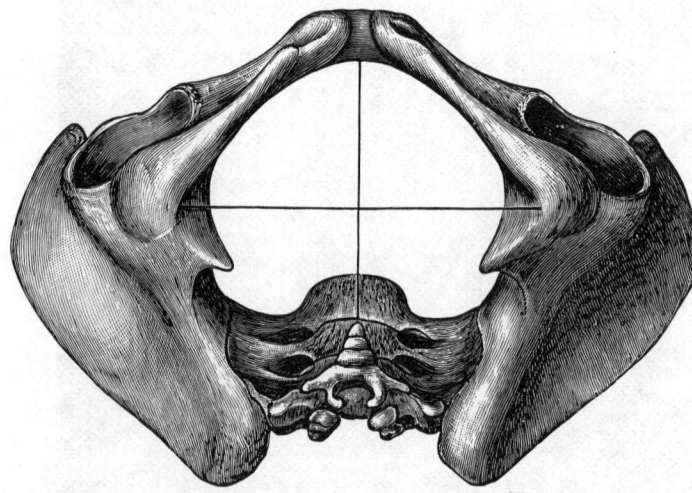

3.177 The diameters of the inferior aperture of the lesser pelvis (female). (The oblique diameter is not shown.)

behind and the ischial tuberosities at the sides, are two large *sciatic notches*, divided by the sacrotuberous and sacrospinous ligaments into the *greater* and *lesser sciatic foramina* (p. 475). Taking into account these ligaments, the inferior aperture is rhomboidal or diamond-shaped, its anterior limbs being the ischiopubic rami (joined together by the inferior pubic ligament in the midline), the posterior sides being the sacrotuberous ligaments, with the coccyx in the midline. The pelvic outlet is thus not entirely rigid in its posterior half, being limited only by the ligaments and the coccyx, which is mobile on the sacrum. Even when the sacrum is taken as the posterior midline limit of the aperture, as is more satisfactory for measurement, it is to be remembered that the sacrum also is slightly mobile at the sacroiliac joints. It is also to be noted that the plane of the inferior aperture is theoretical rather than real. The anterior, ischiopubic part of the opening has a plane which inclines downwards and backwards to a transverse line joining the lower limits of the ischial tuberosities. The posterior half of the aperture has a plane corresponding roughly to the sacrotuberous ligaments and sloping downwards and forwards to the same transverse line. The dimensions of the inferior aperture are measured as follows:

A. The *anteroposterior diameter* is usually measured from the apex of the coccyx, despite the bone's mobility, to the midpoint of the lower rim of the symphysis. Sometimes the lowest part of the sacrum is used for the posterior point (male 80 mm, female 125 mm).

B. The *transverse diameter* is measured between the ischial tuberosities, from the lower border of their medial surfaces. It is consequently termed the *bi-tuberous diameter* (male 85 mm, female 118 mm).

C. The *oblique diameter* extends from the midpoint of the sacrotuberous ligament on one side to the junction of the ischial and pubic rami on the other (male 100 mm, female 118 mm).

Apart from these three pelvic planes of measurement, which by agreement form the basis of pelvic osteometry, certain other planes and measurements are also used, especially in obstetric practice. The *plane of greatest pelvic dimensions* is an obstetrical concept, representing the level in the pelvic cavity where it is most capacious. It is between the pelvic brim and the midlevel plane, corresponding with the latter in its anterior level (middle of the symphysis) but inclining slightly above it to reach the sacrum between its second and third segments. The *plane of least pelvic dimensions* is said to be at about the midlevel of the cavity, but its transverse diameter is the shortest distance between the apices of the ischial spines. It is at this level that most difficulty with labour is encountered. The *posterior median diameter* is a useful concept in obstetric application. It corresponds to the part of any anteroposterior diameter behind its intersection with the transverse diameter at that level. It can thus be considered at any of the conventional levels or planes, and is an indicator of the capacity of the posterior segment of the pelvic cavity. Because of the impossibility of direct measurement of the osseous dimensions of the pelvic cavity in living patients, except by radiological methods, some indirect measurements have long been practised.

The *diagonal* (or *oblique*) *conjugate* is the distance between the sacral promontory and the lower border of the symphysis, measured *per vaginam*, and is taken as a guide to the 'true' conjugate. The latter, and, of course, all the diameters usually described anatomically in the pelvic cavity, disregard the existence of soft tissues. The *intercristal* and *interspinous diameters*, which are respectively the greatest widths between the iliac crests and the anterior superior iliac spines, are sometimes estimated and compared in females. Average values are 250 mm and 275 mm, the difference between them being regarded as an indication of the width of the pelvic cavity, on the generalization that the iliac crests do not turn medially at their anterior ends to the same extent in the female as they do in the male. The value of this method is dubious, because correlation between sexual characteristics in the pelvis is still the subject of controversy.

Morphological Classification of the Pelvis

The importance of the dimensions detailed above is mainly obstetric, but also, less frequently, forensic. All diameters and other measurements of the pelvis display, as elsewhere in the skeleton, considerable variation in individuals, and the figures quoted are merely means taken from limited surveys. Naturally, sexual differences also occur, and there are differing racial ranges in values. The range in any group, both for males and females, is found to be wide wherever it has been adequately assessed. Measurements have necessarily been carried out on series of skeletonized pelves in the most extensive studies; the data from these are not only of anthropological interest but have also formed the basis of attempts to classify variant forms of pelvis which could be clinically identified, without resorting to the full range of basic measurements. These cannot all be measured in the living, but radiological pelvimetry has become a highly refined technique (Borell and Fernström 1960), although radiation risks impose a limiting factor. In the female patient it is obviously her individual measurements which are significant, and not available average values, and it is the comparison between these dimensions and those of the particular fetal head involved which is of principal concern. These are all dimensions which can only be satisfactorily measured by radiographic techniques, and hence modern pelvimetry and fetal cephalometry have, to a considerable extent, displaced older caliper measurement in obstetric practice, whenever the requisite expertise is available (Clyne 1963; Lewis 1964).

The basic pelvic measurements, however obtained, have been analysed by a succession of anatomists, anthropologists, obstetricians and radiologists in an attempt to classify the human pelvis, more especially in the female. The *pelvic brim index*— (ant.-post./trans. diameters) × 100—was one of the earliest attempts to characterize the shape of the pelvic cavity (Turner 1886). Like the cephalic index (p. 348) its range is divided into steps, on which basis pelves are described as *platypellic* (transversely flattened), *mesatipellic* (intermediate form) and *dolichopellic* (long anteroposteriorly). This classification has been carried further, on anatomical and radiological data, to add a *brachypellic* form, also known as *android* and responsible for most cases of severe difficulty in childbirth (Greulich and Thoms 1938, 1939; Thoms 1940). It should be emphasized that these methods of classification depend upon exact measurements and are equally applicable to males and females. For example, in one large series, children and males were found to be predominantly dolichopellic, females mostly mesati- and brachy-pellic.

Platypellic pelves are comparatively rare in all series. Another method of classification depends jointly upon basic anatomical data and analyses of radiological data, the latter including inspectional as well as metrical observations (Caldwell and Moloy 1933; Caldwell *et al.* 1940). This approach to the problem involves dividing the superior pelvic aperture by its transverse diameter into an anterior segment, the *forepelvis* and a posterior *hindpelvis*, the latter being more variable in shape and capacity. The slope of the pelvic walls, whether straight, convergent, or divergent (as seen in radiographs), the pelvic depth and the shape and size of the greater sciatic notch (lateral radiographs), and the shape of the subpubic arch are all taken into consideration. This study led to the wide acceptance of the classification into four basic pelvic types shown in illustration **3**.181. It should be emphasized that the classification depends fundamentally upon measurement of skeletal material, though it is frequently applied in practice in a somewhat subjective manner. Though the existence of intermediate forms of pelvis is not denied, the classification is claimed to embrace all but a very small percentage. The basic skeletal data of the original study were as follows:

PELVIC TYPE	CON-JUGATE DIA. (mm)	TRANS-VERSE DIA. (mm)	WHITE FEMALES (%)	NEGRO FEMALES (%)
Gynaecoid (=mesatipellic)	108·5	137·6	41·4	42·1
Android (=brachypellic)	105·9	135·6	32·5	15·7
Anthropoid (=dolichopellic)	117·5	129·4	23·5	40·5
Platypelloid (=platypellic)	85·5	144·5	2·6	1·7

Apart from racial differences (e.g. the high rate of anthropoid pelves in Negro women), these observations emphasize the low incidence of the highly 'flattened' platypelloid form. It is interesting to note that children (under 9 years) and adult males also show a high incidence of dolichopelly; perhaps this is a paedomorphic trait in the male. The platypelloid pelvis has been described as 'ultra-human'.

Correlation of pelvic type with general bodily physique is still a matter of controversy (Smout *et al.* 1969). It has been claimed, for example, that women with the dangerous android form of pelvis are thick-set, short-necked, and broad-shouldered (Kenny 1944), but such views have been regarded as unreliable by other observers (Ince and Young 1940).

Pelvic Axes and Inclination

The axes of the superior pelvic aperture passes through the centre of and at right angles to the opening's plane. It is directed downwards and backwards, and if prolonged it passes through the umbilicus and the middle of the coccyx. The axis is similarly established for the inferior aperture, and when projected upwards

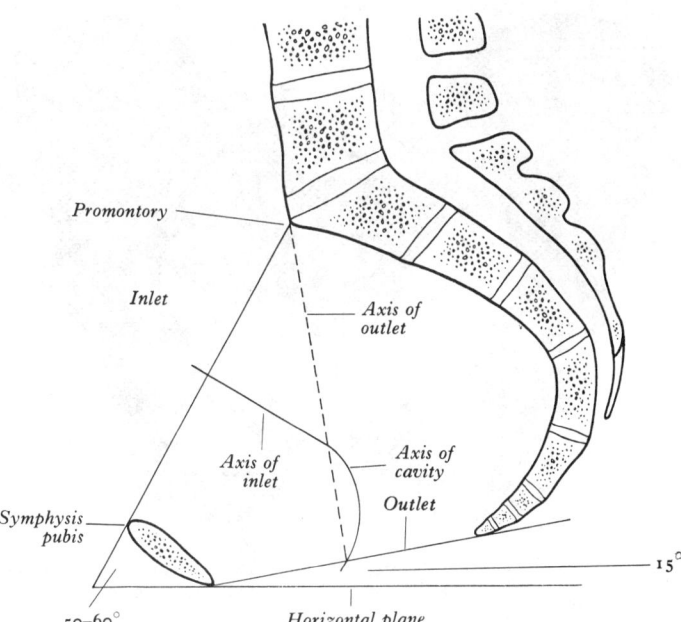

3.178   A median sagittal section through the female pelvis to show the planes of inlet and outlet and their relation to each other and the horizontal plane. The curved axis of the pelvic cavity is also shown. It should be observed that, as depicted in this section, the curve of the sacrum affects the lower part of the third and the upper part of the fourth sacral vertebrae. In many cases the curve is restricted to the fourth vertebra only.

it impinges on the sacral promontory. An axis could be likewise constructed for any intermediate plane, and an axis for the whole cavity can be regarded as the concatenation of an infinite series of such lines. It follows the curvature of the cavity, indicated by the profile of the sacrum and coccyx in lateral views (**3**.178). The form of the *pelvic axis* and the disparity in depth between the anterior and posterior contours of the cavity are both important factors in the mechanism of passage of the fetus through the pelvic canal.

In the standing position the pelvic canal curves obliquely backwards relative to the trunk and the abdominal cavity (**3**.178). The whole pelvis is tilted forwards so that the plane of the pelvic brim makes an angle of 50 to 60° with the horizontal. The plane of the inferior aperture is tilted in the same sense to about 15°, the dorsal parts of the perimeters of both planes being above the ventral. Strictly speaking, the pelvic outlet has two planes, an anterior passing backwards from the symphysis and a posterior passing forwards from the apex of the coccyx, the two sloping down to meet at the intertuberous line. With the pelvis orientated as in standing, the pelvic aspect of the symphysis pubis faces as much upwards as backwards, and the concavity of the sacrum is directed antero-inferiorly. The front of the symphysis and the anterior superior iliac spines are in the same vertical plane. In the sitting position the body weight is transmitted through the inferomedial parts of the ischial tuberosities, a variable amount of soft tissues intervening. In this posture the same spines are in a vertical plane through the centres of the acetabular fossae. Thus the whole pelvis is tilted backwards, and the lumbosacral angle is somewhat diminished at the promontory.

Sex Differences in the Pelvis

As is to be expected, the pelvis provides the most marked and typical skeletal differences between the male and female (**3**.179, 180, 181). Perhaps less to be expected is the distinction that can be made between the sexes in this respect even during fetal life, particularly in the subpubic angle or arch (Boucher 1957). Radiographic studies of the pelvis in American children during the first postnatal year have shown that in general male infants exceed females in dimensions of the whole pelvis, whereas the females exceed males in measurements of the pelvic cavity (Reynolds 1945). In the prepubertal period of 2 to 9 years similar tendencies are apparent; but the maximal difference is said to

387

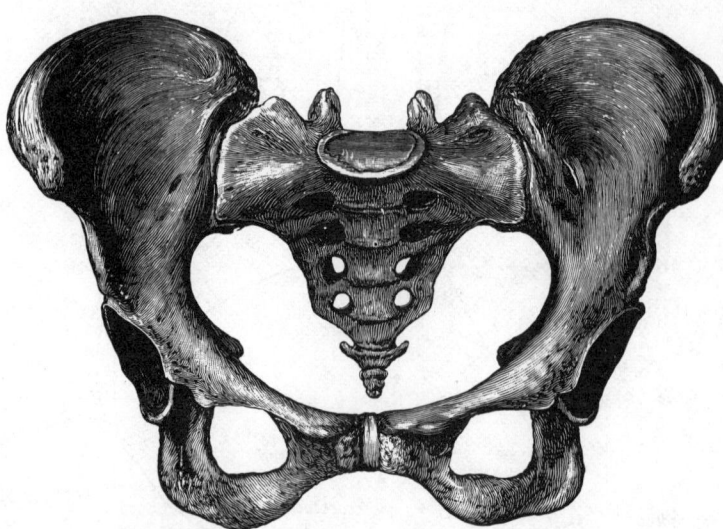

3.179A The female pelvis. Anterior aspect. (From a specimen in the museum of the Royal College of Surgeons of England).

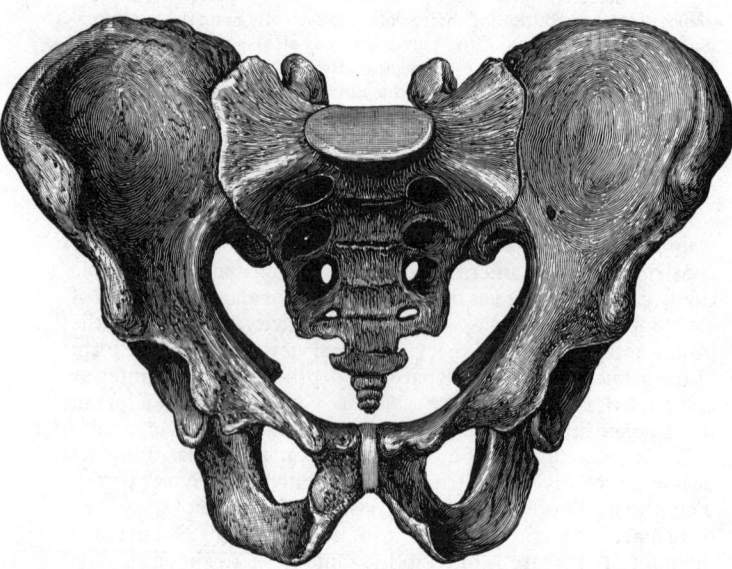

3.179B The male pelvis. Anterior aspect.

the pelvis than the female, and this produces characteristic differences at the lumbosacral and hip joints. The articular facet on the base of the sacrum for the fifth lumbar vertebra and intervening disc occupies distinctly more than a third of the width of the base, but less than a third in the female, whose sacrum is also relatively broader, accentuating this difference. Similarly at the hip joint, the acetabulum is absolutely larger in the male, and its diameter is about equal to the distance of its anterior rim from the symphysis pubis. In the female, however, the acetabular diameter is usually distinctly less than this distance, not only because it is absolutely smaller than in the male, but also because the whole of the anterolateral wall of the pelvic cavity is comparatively wider in the female. The symphysis and the adjoining parts of the pubis and ischium, forming the anterior pelvic wall, are also absolutely less in height than in the male, and this is commonly associated with a somewhat triangular shape in the obturator foramen, which is more ovoid in males. The different mode of growth in the pubic region in the sexes is also expressed in the subpubic arch or angle, situated below the symphysis and between the inferior pubic rami. It is more angular in the male, being of the order of 50 to 60°. In the female it is less easy to measure because of the rounded nature of the angle, but it is usually 80 to 85°. Also associated with the pubic width of the female is greater separation between the pubic tubercles. The ischiopubic rami display some differences. They are much more lightly built in the female and are narrowed near the symphysis. In the male they bear a distinctly roughened and everted area for the attachment of the crus penis, the corresponding attachment for the clitoris being poorly developed. The ischial spines are closer together in the male, being characteristically inturned. The greater sciatic notch is usually much wider in the female; comparisons have been made of its width and angle, mean angular values for males and females being 50·4° and 74·4° (Hanna and Washburn 1953). The greater value of this angle in the female, and the greater width of the notch, are associated with the increased backward tilt of the sacrum, and generally greater anteroposterior diameter of the female pelvic cavity, especially at lower levels. For a method of comparing the *depth* of the notch in the sexes, consult Jovanović and Živanović (1965). More recently,

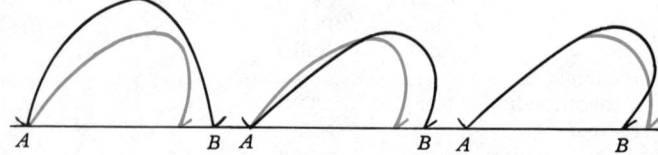

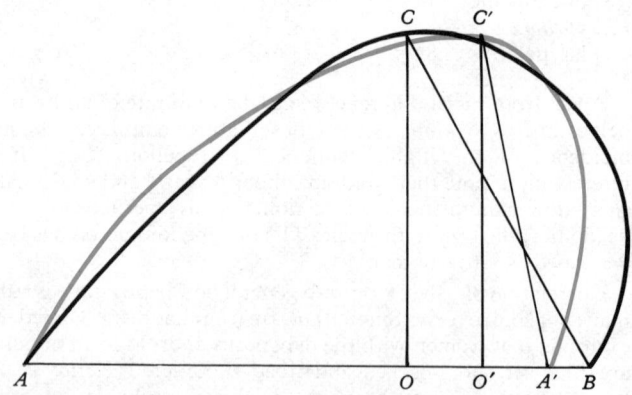

3.180 *Above*: three common profiles of the greater sciatic notch in males (blue) and females (black). *Below*: points, axes and angles utilized in mensuration of the greater sciatic notch. AB equals *maximal width*, i.e. distance between the tip of the ischial spine and a tubercle marking attachment of the piriformis muscle. OC equals *maximal depth* (perpendicular to AB), OB equals *posterior segment* of width. Index I = OC × 100/AB. Index II = OB × 100/AB. ACB = *total angle*. BCO = *posterior angle*. (Modified, with kind permission, from Singh and Potturi 1978).

occur at 22 months, the sexual divergence decreasing in the later years of childhood (Reynolds 1947).

Sexual differences in the adult pelvis have been studied and measured extensively (Genovese 1959). They can be divided into metrical and non-metrical characteristics, the range of most showing some overlap between the sexes.

The differences are inevitably associated with function, and while the primary function of the pelvis in *both* sexes is locomotor, it is specially adapted to the needs of childbirth in the female. This reproductive adaptation affects particularly the lesser pelvis; but the changes in this affect the proportions and dimensions of the greater pelvis to a variable degree in the female. Since the male is distinctly more muscular, and therefore more heavily built, the overall dimensions of the pelvis, such as intercristal measurement, are greater in males, markings for muscles and ligaments are more pronounced, and the general architecture is relatively stouter. The iliac crest is more rugged and curves medially at its anterior end more acutely than in the female, whose crests are less curved in every particular. The iliac blades are more vertical in the female, but do not extend so far upwards; the iliac fossae are therefore shallower, and the iliopectineal line, or pecten pubis, is more vertical. These iliac peculiarities probably account for the greater prominence of hip in the female.

The male is relatively and absolutely more heavily built above

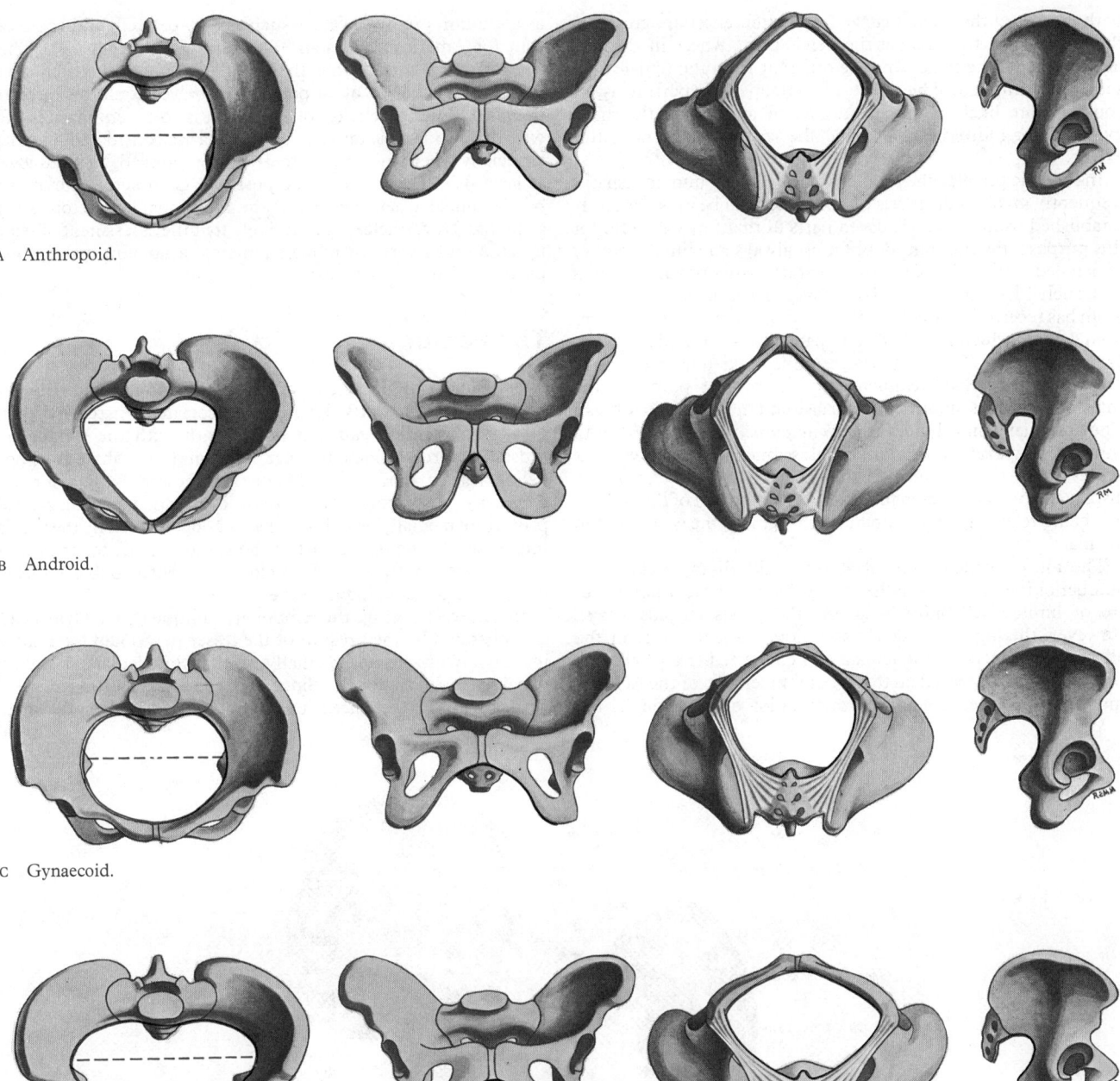

A   Anthropoid.

B   Android.

C   Gynaecoid.

D   Platypelloid.

3.181   The major differences between the four types of pelvis in the most widely accepted classification. Types shown in A and B are commonest in males, B and C in females, while D is rare, even in females. Note the variations in superior and inferior apertures, greater sciatic notch, and subpubic arch. Note varying proportion between fore- and hind-pelvis, anterior and posterior to transverse diameter of the inlet. (Caldwell and Moloy's Classification, after Clyne 1963.)

a detailed analysis of various dimensions and derived indices in relation to sex determination and the greater sciatic notch has appeared (Singh and Potturi 1978). Having reviewed the literature, the latter studied 200 adult hip bones (120 males and 80 females) of known sex. Using defined demarking points (Jit and Singh 1966) they measured maximal width, maximal depth, posterior segment of width, total angle, posterior angle, and indices I and II—see **3**.180 and original paper for details. The width and depth of the notch *per se* were found to be valueless as criteria for determining sex: the posterior angle was the best single parameter, whilst the length of the posterior segment and index II were also highly effective, especially in females.

The sacrum presents a considerable number of sexual differences, in addition to the differences in orientation just mentioned. The female sacrum is less curved, its curvature being most marked between the first and second segments and between the third and fifth, with an intervening flattened region. The male sacrum is more evenly curved, relatively long and narrow, and more often consists of over five segments (due to the addition of a lumbar or coccygeal vertebra). A *sacral index* can be assessed by comparing sacral breadth (measured between the most anterior points of the two auricular surfaces) with length (measured between the midpoints on the anterior margins of the promontory and apex); average values for male and female are 105 and 115 per cent. The auricular surfaces differ in some respects, being relatively smaller and more obliquely set in the female; contrary to a commonly made statement, they extend on to the upper three sacral vertebrae in *both* sexes. The concavity of the dorsal border of the auricular surface is more marked in females (Weisl 1954).

Many of these differences have been summarized in the generalization that the pelvic cavity is longer and more conical in the male, shorter and more cylindrical in the female, though in

389

both the axis of the canal is curved. The differences are greater at the inferior aperture than at the pelvic brim, where in absolute measurements the male is not as different from the female as is sometimes implied. The shape of the superior aperture is, of course, more likely to be anthropoid or android in the male, gynaecoid or android in the female, the sexes overlapping to this extent.

In forensic practice the problem of identifying human remains frequently involves diagnosis of sex, which can be most certainly established from the pelvis. Even parts of this may be useful for this purpose, though some doubt must always remain. A number of detailed studies of the metrical characteristics of various parts of the pelvis have been carried out, involving various indices. The ilium has received particular attention. For example, an index has been devised which essentially compares the pelvic and sacroiliac parts of the bone (Derry 1923/24). A line is extended backwards from the iliopectineal eminence to the nearest point on the anterior margin of the auricular surface and then to the iliac crest. The auricular point divides this *chilotic line*, as it is called, into anterior (pelvic) and posterior (sacral) segments, each of which is expressed as a percentage of the other. These *chilotic indices* display reciprocal values in the sexes, the pelvic part of the chilotic line being relatively predominant in the female, the sacral part in the male.

The most detailed metrical study of the ilium, involving a number of indices, leads to the conclusion that its characteristics are of limited reliability in sexing the pelvis (Straus 1927). However, the higher incidence and more marked nature of the preauricular sulcus in the female was confirmed (*see* p. 380). A *pubo-ischial index*, based on the maximum lengths of the ischium and pubis, as measured from their point of junction in the acetabulum, produces values such as 83·7 per cent and 100·0 per cent for American males and females (Washburn 1949). When this index was correlated with the angle of the sciatic notch it was claimed that the sex of 98 per cent of pelves could be correctly deduced. The correlation of such data is to be emphasized, for when a considerable range of pelvic data are considered together, especially where they can be metrically expressed, an accuracy of at least 95 per cent should be possible. Complete accuracy has been claimed when the rest of the individual's skeleton is also available. Nevertheless, it is unlikely that the assessment of sex in isolated and restricted human remains, a not unusual forensic problem, can always be absolutely certain.

The Femur

The femur or thigh bone is, of course, the longest and strongest bone (**3**.182–189). Its length is associated, naturally, with mankind's striding gait and its strength with the weight and muscular forces which it must withstand. Its shaft is almost cylindrical in most of its length and bowed with a forward convexity. Its upper extremity is a rounded articular head, projecting medially on a short neck of bone, formed by the medial inclination of the upper part of the shaft. The distal or inferior extremity is more massive, being in the form of a double 'knuckle' or condyle, articulating with the tibia.

In the erect posture the femora are oblique (**3**.1). Their heads are separated by the breadth of the lesser pelvis and their shafts incline downwards and medially, so that the medial sides of the two knees almost touch. Since the bones of the legs descend vertically from the knees, the obliquity of the femoral shafts

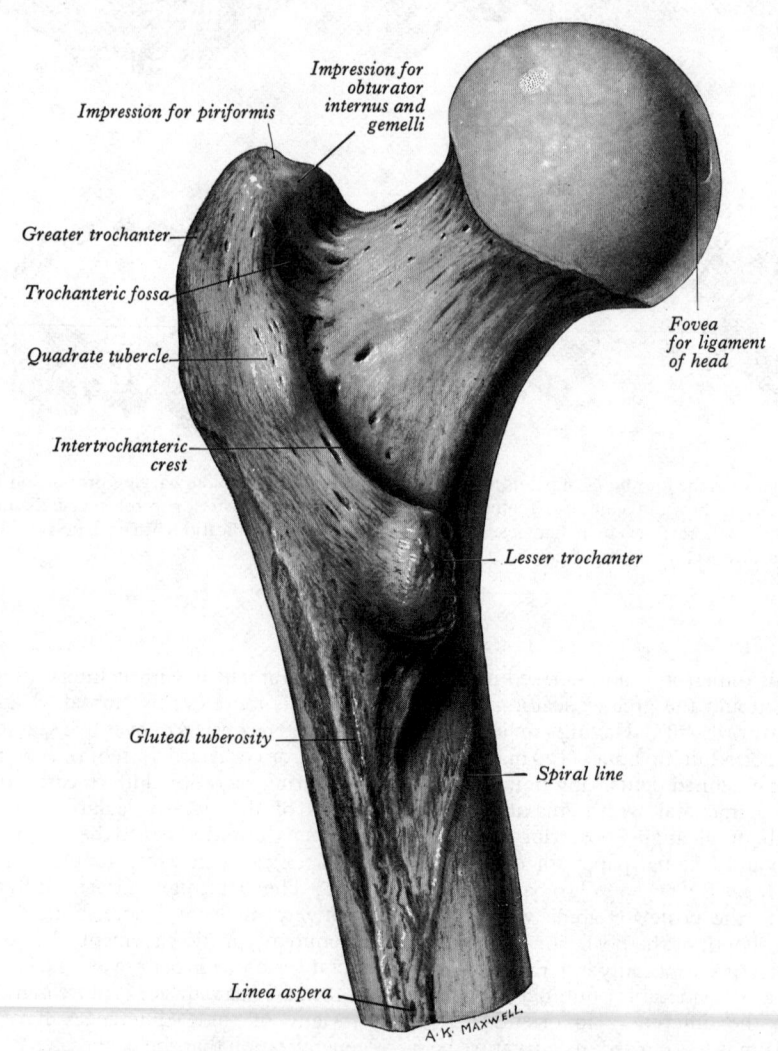

Impression for obturator internus and gemelli

Impression for piriformis

Greater trochanter

Trochanteric fossa

Quadrate tubercle

Intertrochanteric crest

Fovea for ligament of head

Lesser trochanter

Gluteal tuberosity

Spiral line

Linea aspera

A·K·MAXWELL.

results in the approximation of the feet in the erect attitude, bringing them directly under the line of body weight. The narrowness of the base detracts from the stability of the body but greatly facilitates forward movement and increases the speed and smoothness with which it can be executed. The degree of obliquity of the shafts varies in different individuals, but is greater in women on account of the greater breadth of the pelvis and relative shortness of the femora.

The upper end of the femur (3.182) comprises a head, a neck, a greater and a lesser trochanter.

The head of the femur is rather more than half a 'sphere' (*see* pp. 477, 432); it is directed upwards, medially and slightly forwards, to articulate with the acetabulum. Its surface is smooth, but a little below and behind its centre is a small roughened pit or *fovea*.

The neck of the femur, which is about 5 cm long, connects the head and the shaft, with which it forms an angle of about 125°. This arrangement facilitates the movement of the hip joint and enables the lower limb to swing clear of the pelvis. The neck is narrowest at its middle and is wider at its lateral than at its medial end. Its two borders are rounded. The upper border is roughly horizontal and is gently concave upwards. The lower border is straighter but oblique, and directed downwards, laterally and backwards to meet the shaft near the lesser trochanter. The anterior surface of the neck is flattened and its junction with the shaft is marked by a prominent rough ridge, termed the *intertrochanteric line*. The posterior surface is convex backwards and upwards in its transverse axis, and concave in its long axis, and its junction with the shaft is marked by a rounded ridge, termed the *intertrochanteric crest*.

The greater trochanter is a large, quadrangular projection at the upper part of the junction of the neck with the shaft. Its posterosuperior portion projects upwards and medially (3.185) so as to overhang the adjoining part of the posterior surface of the neck; in this situation its medial surface presents a roughened, depressed area, the *trochanteric fossa*. The upper border of the trochanter lies one hand's breadth below the tubercle on the iliac crest, and is on a level with the centre of the head of the femur. The anterior surface of the trochanter bears a roughened impression; its lateral surface is divided into two areas by an oblique, flattened strip, wider above than below, which runs downwards and forwards across it. The lateral surface of the trochanter can be palpated in the living (3.183), and, when the adjoining muscles are relaxed, the trochanter can be gripped. The trochanteric fossa occasionally presents a tubercle or exostosis, which has been regarded as a useful non-metrical racial characteristic (*see* p. 238).

The lesser trochanter (3.182) is a conical eminence, which projects medially and backwards from the shaft at its junction with the lower and posterior part of the neck. Its summit and anterior surface are roughened, but its posterior surface, which lies at the lower end of the intertrochanteric crest, is smooth. It is placed too deeply to be felt in the living.

The intertrochanteric line marks the junction of the anterior surface of the neck with the shaft of the femur (3.184). It is a prominent roughened ridge, which commences in a tubercle at the upper and medial part of the anterior surface of the greater trochanter and runs downwards and medially. It reaches the lower border of the neck on a level with the lesser trochanter, but in front of it. It often presents a second tubercle near its lower end. Below, it is continuous with the *spiral line* (p. 392).

The intertrochanteric crest (3.182) marks the junction of the posterior surface of the neck with the shaft of the femur. It is a smooth rounded ridge, which commences at the posterosuperior angle of the greater trochanter and runs downwards and medially to terminate at the lesser trochanter. A little above its middle it presents a low rounded elevation, the *quadrate tubercle*.

The head of the femur is entirely intracapsular and is encircled immediately lateral to its greatest diameter by the acetabular labrum. Its circumference is sharply defined, except anteriorly, where the cartilage-covered surface extends on to the front of the neck. The fovea which marks the head below and behind its centre (3.182) affords attachment to the ligament of the head of the femur. The inferomedial part of the anterior surface of the head is related to the femoral artery, from which it is separated by the tendon of the psoas major and the articular capsule. The blood supply of the head is of surgical importance. Around the neck, at the attachment of the fibrous capsule of the hip joint, there is a vascular ring supplied by branches of the medial and lateral circumflex arteries. From this arteries arise which pierce the capsule and pass under the zona orbicularis (p. 477) to ascend along the neck beneath the reflection of synovial membrane. These arteries divide into metaphysial branches which enter the neck and epiphysial branches which pass into the peripheral non-articular portion of the head to supply the epiphysis. During growth the territories of the two groups of branches are separated by the epiphysial plate; after osseous union of the head with the neck they anastomose freely. A small supply to the head of the femur is carried by the vessels in the ligament of the head (pp. 397 and 728) which are acetabular branches of the obturator and medial femoral circumflex arteries; these anastomose with the epiphysial vessels (Crock 1965). Observations on the developmental patterns of this blood supply in late fetal and early post-natal periods do not entirely corroborate the above description (Ogden 1974). Although the medial and lateral circumflex arteries contribute equally to the vascular network initially, the final pattern consists of two major branches of the medial circumflex, both related to the posterior aspect of the femoral neck. The supply from the lateral circumflex diminishes and the arterial 'ring' becomes incomplete. As the femoral neck elongates, the extracapsular ring becomes more and more distant from the epiphysial head of the femur.

The neck of the femur is marked by numerous vascular foramina, especially on its anterior surface and on the upper part of its posterior surface. The angle which it makes with the shaft is widest at birth and diminishes steadily until the adult condition is reached. It is less in the female than in the male, owing to the increased breadth of the lesser pelvis and the greater obliquity of the shaft of the femur. The *anterior surface* of the neck is entirely intracapsular and on this surface the capsular ligament extends laterally to the intertrochanteric line. Facets, often covered by an extension of the articular cartilage of the head, and various imprints frequently occur on the anterior surface of the neck. Although attributed to squatting habits, they are also seen in non-

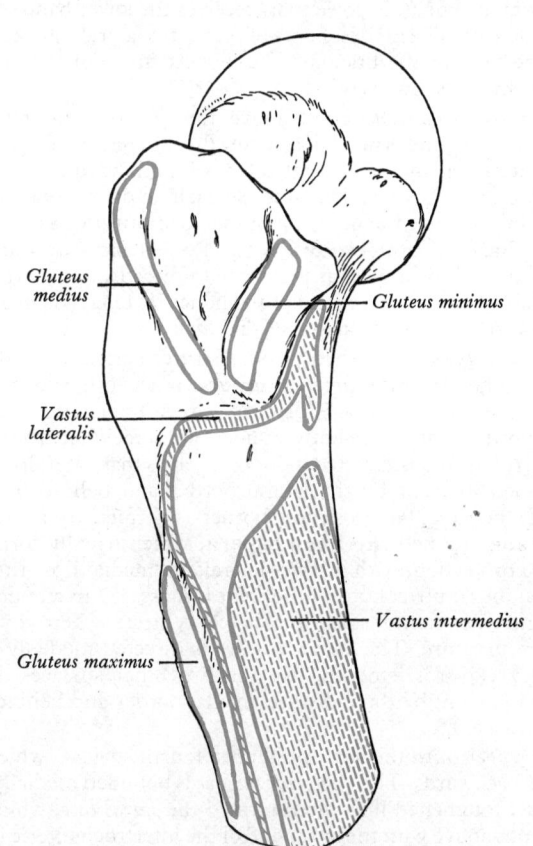

Gluteus medius

Gluteus minimus

Vastus lateralis

Vastus intermedius

Gluteus maximus

3.183   The proximal part of the right femur. Lateral aspect.

squatters and there is no agreement about their causation (Kostick 1963; Kate 1963). One such feature, the *cervical fossa* (of Allen) may be of racial significance (*see* p. 238). On the *posterior surface* the capsular ligament does not reach the intertrochanteric crest (**3**.182), and only a little more than the medial half of the neck lies within the capsule. The part of the anterior surface adjoining the head covered with cartilage is related to the iliofemoral ligament in the erect posture. A faint groove crosses the posterior surface in an upward and lateral direction; it is produced by the obturator externus tendon as it passes to the trochanteric fossa. The neck of the femur does not lie in the same plane as the shaft, but is carried forwards as it passes upwards and medially. On this account the transverse axis of the head of the femur makes an angle with the transverse axis of the lower end of the bone, and this angle is known as the *angle of femoral torsion* (approximately 15°).

The greater trochanter (**3**.183) provides insertion for most of the muscles of the gluteal region. The gluteus minimus is inserted into the rough impression on its anterior surface. The gluteus medius is inserted into the oblique, flattened strip, which runs downwards and forwards across its lateral surface. The area in front of this insertion is separated from the tendon by the trochanteric bursa of the gluteus medius; the area behind the insertion is covered by the deep fibres of the gluteus maximus, and part of the trochanteric bursa of that muscle may be interposed. The upper border of the trochanter gives insertion to the piriformis, and its medial surface to the common tendon of the obturator internus and the two gemelli. At their insertions these two tendons are frequently blended with each other to a variable extent. The trochanteric fossa receives the insertion of the obturator externus.

The lesser trochanter has attached to it psoas major on its summit and on the medial part of its anterior surface. The base of the trochanter is expanded and its medial or anterior surface has the iliacus attached to it extending downwards for a short distance behind the spiral line. The upper fibres of the adductor magnus play over the posterior surface of the lesser trochanter and a bursa is sometimes interposed between them.

The *intertrochanteric line* marks the lateral limit of the capsular ligament of the hip joint. Its upper part, including the tubercle already noticed, receives the attachment of the upper band of the iliofemoral ligament; its lower part receives the lower band of the same ligament. The highest fibres of the vastus lateralis arise from the upper end of the line, and the highest fibres of the vastus medialis from its lower end.

The *intertrochanteric crest*, above the *quadrate tubercle*, is covered by the gluteus maximus; below the tubercle it is separated from that muscle by the quadratus femoris and the upper border of the adductor magnus. The tubercle itself, and a portion of the bone below, receive the insertion of the quadratus (**3**.185B).

The shaft of the femur (**3**.184, 185) is narrowest in its middle: it expands a little as it is traced upwards, but it widens appreciably near the lower end of the bone. Its long axis inclines about 10° from the vertical axis of the tibia.

In its *middle third* the shaft possesses three surfaces and three borders. The extensive *anterior surface*, is smooth and gently convex in all directions. It is placed between the lateral and the medial borders, which are both rounded and somewhat arbitrary. The *lateral surface* is directed more backwards than laterally, and is bounded in front by the lateral border and behind by the posterior border. The posterior border is formed by a broad, rough ridge, termed **the linea aspera**, which usually forms a crestlike projection, with distinct lateral and medial lips. In this situation the compact bone of the shaft is increased in amount to withstand the compression forces concentrated here by its anterior curvature. The *medial surface* is directed medially and slightly backwards; smooth, like the two other surfaces, it is bounded in front by the indistinct medial border and behind by the linea aspera.

In its *upper third* the shaft presents a fourth surface, which is directed backwards. This *posterior surface* is bounded medially by a narrow, roughened line, often termed the *spiral line*, which is continuous above with the lower end of the intertrochanteric line, and below with the medial lip of the linea aspera. On the lateral side the surface is bounded by a broad, roughened vertical ridge

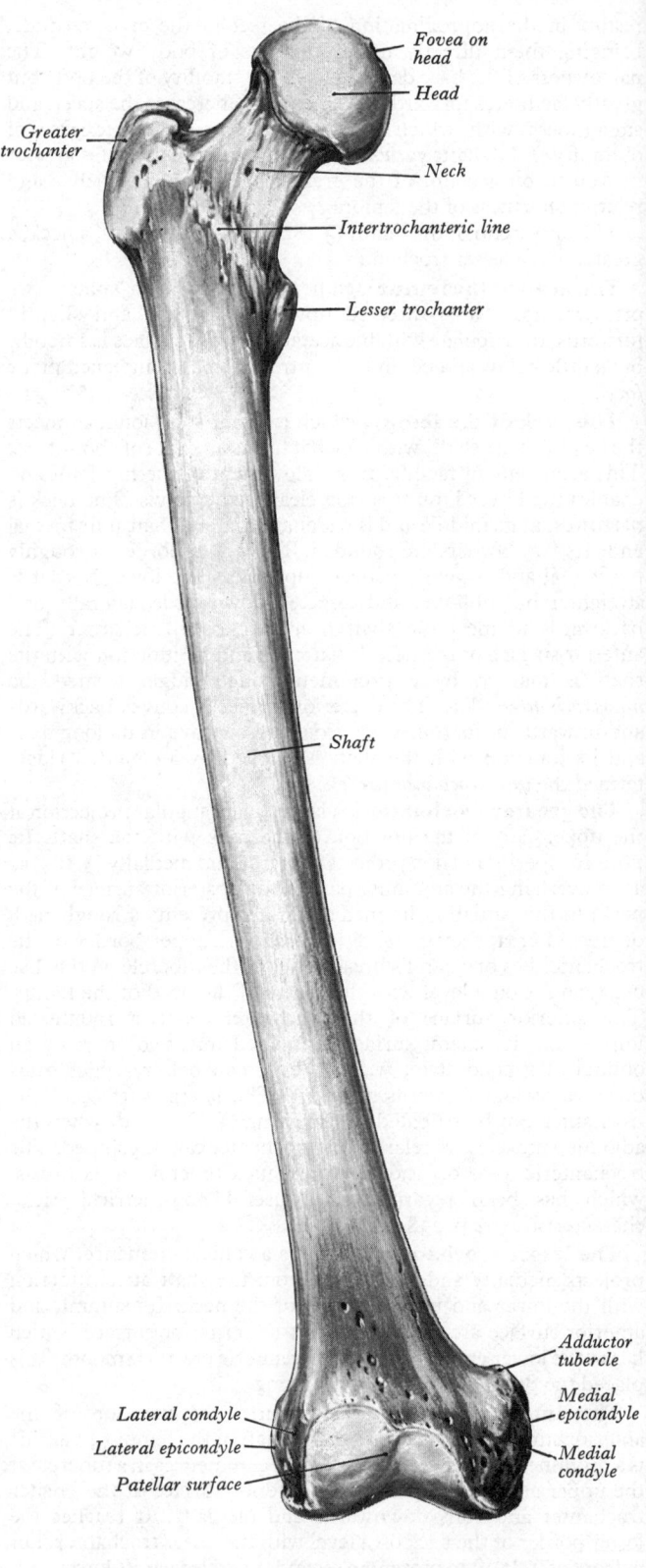

3.184A The right femur. Anterior aspect. Compare with Key, 3.184B.

termed the *gluteal tuberosity*, which extends upwards and laterally to the root of the greater trochanter and is continuous below with the lateral lip of the linea aspera. This posterior surface is therefore triangular.

In its *lower third* the shaft also possesses a fourth, *posterior surface*. This is placed between the *medial* and *lateral supracondylar lines*, which are continuous above with the corresponding lips of the linea aspera. These lines form definite but not conspicuous ridges, of which the lateral is the more distinct. Near

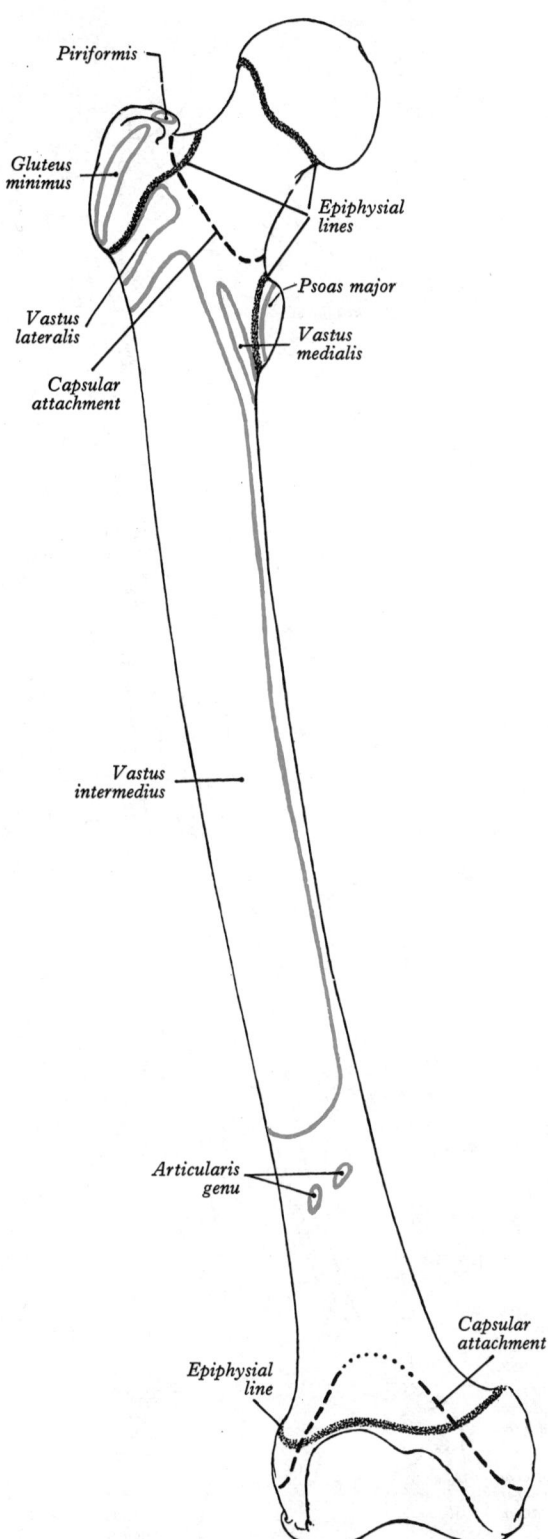

Piriformis

Gluteus
minimus

Epiphysial
lines

Psoas major

Vastus
lateralis

Vastus
medialis

Capsular
attachment

Vastus
intermedius

Articularis
genu

Capsular
attachment

Epiphysial
line

3.184B Key to **3.184A**. The epiphysial lines are stippled. The interrupted lines correspond to the attachments of the capsular ligaments. The dotted part of the lower line indicates the site of communication between the cavity of the knee joint and the suprapatellar synovial bursa.

its upper end the medial supracondylar line is in part obliterated; here the femoral artery lies in close relation with the bone as it passes from the thigh to the popliteal fossa. The posterior surface of the lower third is a flattened, triangular area, the *popliteal surface* of the femur (**3.185**); in its lower and medial part it presents a rough and slightly elevated area.

The shaft (**3.184B, 185B**) is thickly surrounded with muscles and cannot be felt through the skin. Its *anterior* and *lateral surfaces* provide attachment in their upper three-fourths for the

vastus intermedius; below that muscle the articularis genu arises by several small slips from the front of the bone. The lower portion of the anterior surface for 5 or 6 cm above the patellar articular surface is covered by the suprapatellar bursa, which intervenes between the bone and the muscles mentioned. The lower portion of the lateral surface is covered by vastus intermedius. The *medial surface* is devoid of muscular attachments and is covered by the vastus medialis.

The vastus lateralis has a linear origin which commences in front at the root of the greater trochanter and follows it to the upper end of the gluteal tuberosity. It then descends along the lateral margin of the tuberosity to the lateral lip of the linea aspera, from the upper half of which it takes origin. The vastus medialis also has a linear origin. It commences at the lower end of the intertrochanteric line and follows the spiral line to the medial lip of the linea aspera. At the lower end of the linea aspera it follows the upper part of the medial supracondylar line; below this the origin may be partly or wholly from the aponeurotic insertion of the adductor magnus.

The *gluteal tuberosity* may take the form of an elongated, roughened depression or it may project as a ridge. Occasionally a part of it is sufficiently prominent to merit the name of a *third trochanter*. It receives the deeper fibres of the lower half of the gluteus maximus. The medial edge of the tuberosity provides insertion for the pubic fibres of the adductor magnus; the succeeding fibres of that muscle are inserted into the linea aspera and the upper part of the medial supracondylar line; the remaining fibres form a stout tendon which is inserted into the adductor tubercle (p. 395) and sends a membranous expansion to the lower part of the medial supracondylar line.

Between the gluteal tuberosity and the spiral line the posterior surface receives the insertions of the pectineus and the adductor brevis. The pectineus is inserted into a line, sometimes slightly roughened, which descends from the root of the lesser trochanter to the upper end of the linea aspera. The adductor brevis is attached lateral to the pectineus and extends downwards to the upper part of the linea aspera, where it is attached medial to the adductor magnus.

In addition to the attachments already described, the **linea aspera** receives the adductor longus, the intermuscular septa and the short head of biceps femoris. The structures connected to the linea aspera are inseparably blended at their bony attachments. The perforating arteries cross the linea aspera from medial to lateral sides, under tendinous arches in the adductor magnus and the short head of biceps femoris. The foramina for the nutrient arteries are situated close to the linea aspera. They vary in number and position. One is usually near the upper end of the linea aspera, and a second, which is not always present, near its lower end. The foramina are directed upwards through the compact bone.

The popliteal surface of the shaft is the floor of the upper part of the popliteal fossa. It is covered by a variable amount of fat, which separates the popliteal artery from the bone. The superior medial genicular artery arises from the popliteal artery as it lies in the intercondylar notch. It arches medially above the medial condyle, but is separated from the bone by the medial head of gastrocnemius, attached to the rough elevation a little above the medial condyle. Below this there may be a smooth facet which underlies the bursa under the medial head of the gastrocnemius. More medially, above the articular surface, there is often an imprint which, in acute flexion, receives the rough tubercle on the medial tibial condyle for the attachment of semimembranosus (Kostick 1963). The superior lateral genicular artery arches upwards and laterally above the lateral condyle, but is separated from the bone by the attachment of plantaris to the lower part of the lateral supracondylar line.

The lateral supracondylar line is most distinct in its upper two-thirds, to which the short head of the biceps femoris and the lateral intermuscular septum are attached. Its lower part is marked by a small roughened area which gives origin to the plantaris and often encroaches on to the popliteal surface. The **medial supracondylar line** is feebly marked in its upper two-thirds, where it gives origin to the vastus medialis. Near its upper end it is crossed by the femoral vessels as they enter the fossa from the adductor canal. It is often sharp and prominent for 3 or 4 cm

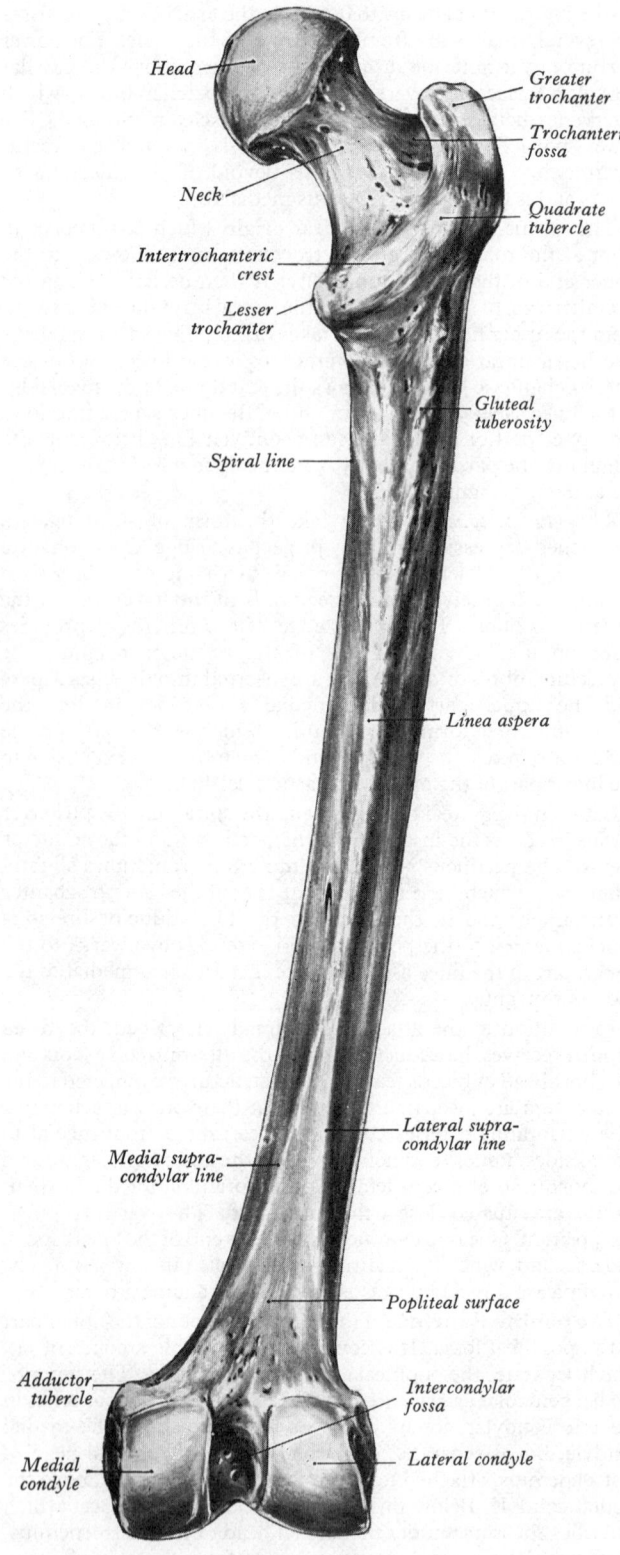

3.185A   The right femur. Posterior aspect. Compare with Key, **3.**185B.

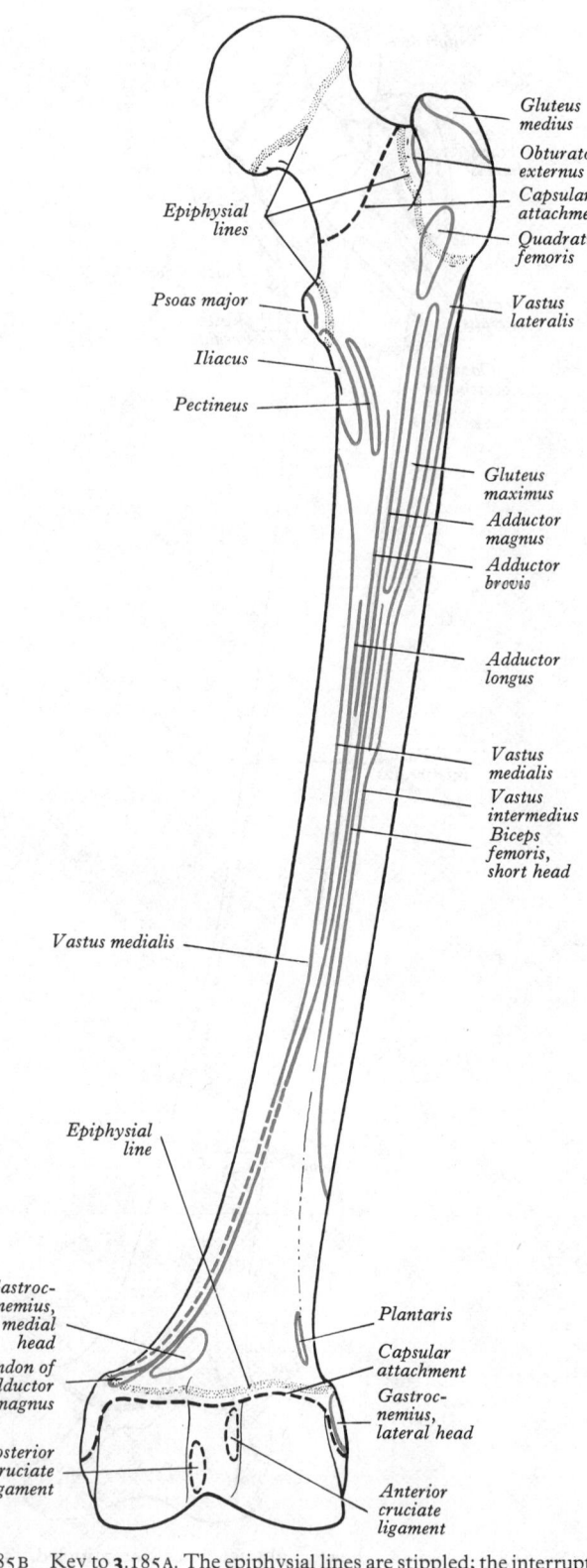

3.185B   Key to **3.**185A. The epiphysial lines are stippled; the interrupted lines indicate the attachments of the capsular ligaments.

above the adductor tubercle and in this situation it gives attachment to a membranous expansion from the tendon of the adductor magnus.

The lower end of the femur is widely expanded and thus provides a good bearing surface for the transmission of the weight of the body to the top of the tibia. It consists of two prominent masses of bone, the *condyles*, which are partially covered by a large *articular surface*. Anteriorly the two condyles are united and are continuous with the front of the shaft; posteriorly they are separated by a deep gap, the *intercondylar fossa* (intercondylar notch), and they project backwards considerably beyond the

plane of the popliteal surface.

The *articular surface* forms a broad ∩-shaped area for articulation with the patella above and tibia below (**3.**186). The *patellar surface* extends over the anterior surfaces of both condyles, but much the larger part of it is on the lateral condyle. It is concave from side to side, being grooved proximo-distally to accommodate the posterior surface of the patella. The *tibial surface* is divided into medial and lateral parts by the intercondylar fossa, but anteriorly each part is directly continuous with the patellar surface. The medial part forms a broad strip which covers the convex inferior and posterior surfaces of the medial condyle, and is gently curved with the convexity of the curve directed medially. The lateral tibial surface covers the same aspects of the lateral condyle but is broader and passes straight backwards. Heiple and Lovejoy (1971) suggested that a deep patellar groove is a hominid feature, concerned with bipedal gait. Wanner (1978) has studied its variations and disagrees with this view. The role of the groove, in preventing lateral subluxation of the patella therefore remains *sub judice*.

The lateral condyle (3.187) is flattened on its lateral surface and is not so prominent as the medial condyle, but it is stouter and stronger, for it is placed more directly in line with the shaft and probably takes a greater share in the transmission of the weight to the tibia. The most prominent point on its lateral aspect is termed the *lateral epicondyle*, and the whole of this surface can be felt through the skin. A short groove, deeper in front than behind, separates the lateral epicondyle from the articular margin below and behind. The medial surface of the condyle forms the lateral wall of the intercondylar fossa.

The medial condyle possesses a bulging, convex medial aspect, which can be palpated without difficulty. Its uppermost part is marked by a small projection, termed the *adductor tubercle* (3.185A) because it gives insertion to the tendon of the adductor magnus. The tubercle is an important bony landmark for the surgeon, and can be identified most readily when it is approached from above. It is, however, often a facet, rather than a tubercle. The most prominent point on the medial surface of the condyle is below and a little in front of the adductor tubercle and is termed the *medial epicondyle*. The lateral surface of the condyle is the roughened medial wall of the intercondylar fossa.

The intercondylar fossa (intercondylar notch) separates the two condyles below and behind. In front it is limited by the lower border of the patellar surface, and behind by the *intercondylar line*, which separates it from the popliteal surface. It lies within the capsular ligament of the knee joint, but is covered with synovial membrane only over a very limited area.

The patellar surface extends higher on the lateral than on the medial side; its upper border therefore is oblique and runs downwards and medially (3.186). It is separated from the tibial surfaces by two faint grooves; which cross the condyles obliquely. The lateral groove is the better marked (3.186); it runs laterally and slightly forwards from the front part of the intercondylar fossa and expands to form a faint triangular depression, which rests on the anterior edge of the lateral meniscus when the knee joint is fully extended. The medial groove is restricted to the medial part of the medial condyle and likewise rests on the anterior edge of the medial meniscus in full extension of the knee. Where this groove ceases the patellar surface is continued backwards on to the lateral part of the medial condyle as a semilunar area adjoining the anterior part of the intercondylar fossa. This area articulates with the medial vertical facet of the patella in full flexion of the knee joint; it is not distinctly outlined in most femora. In habitual squatters the articular cartilage may extend on to the lateral aspect of the lateral condyle where it underlies the vastus lateralis. **The tibial surfaces** are convex from side to side and from before backwards. The anteroposterior curvature of the two surfaces is not uniform, being much sharper in both posteriorly than it is in front. The medial tibial surface is longer than the lateral anteroposteriorly, and shows a slight curve, which is concave to the lateral side; these differences are important determinants of the rotatory (spin) movements, both adjunct and conjunct (p. 436), which are possible at the knee joint (p. 488).

The medial condyle projects medially and downwards to such an extent that, despite the obliquity of the shaft, the lower surface of the lower end of the bone is practically horizontal. A curved strip, about 1 cm wide, adjoining the medial margin of the articular surface, is covered with synovial membrane and lies within the capsule of the knee joint. The medial epicondyle, which lies above this area, gives attachment to the tibial collateral ligament of the knee joint.

The lateral condyle is less prominent and its lateral surface projects but little beyond the lateral surface of the shaft. The lateral epicondyle has attached to it the fibular collateral ligament of the knee joint, and above and behind it the bone bears an

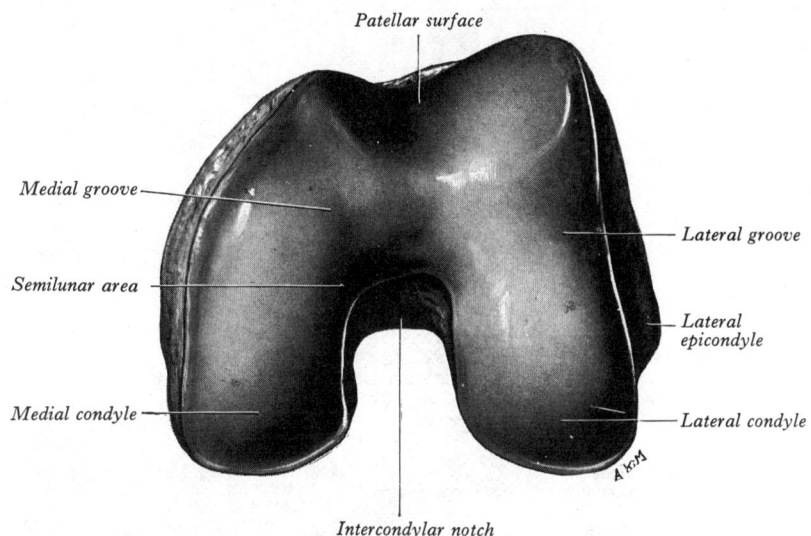

3.186 The distal end of the left femur. Inferior aspect.

impression which is the origin of those fibres of the lateral head of the gastrocnemius that do not arise from the capsular ligament. The deepened, anterior end of the groove which lies between the epicondyle and the articular margin is the attachment of popliteus (3.187); the posterior (or upper) end of the groove contains the tendon of the muscle only in full flexion of the knee. In extension the tendon passes across the margin of the articular surface below, and sometimes grooves it. Immediately adjoining the articular margin a strip of the lateral condyle, 1 cm broad, is intracapsular, and is covered with synovial membrane, with the exception of the attachment of popliteus.

The intercondylar fossa separates the projecting portions of the two condyles, and is intracapsular but, to a large extent, extrasynovial. Its lateral wall, formed by the medial surface of the lateral condyle, bears a flattened impression which occupies its upper and posterior part and extends on to the floor of the fossa close to the intercondylar line. This is for the attachment of the upper end of the anterior cruciate ligament. The medial wall of the fossa, formed by the lateral surface of the medial condyle, bears a similar but rather larger impression for the upper end of the posterior cruciate ligament; it is placed anteriorly and extends on to the anterior part of the floor of the fossa. These two impressions look relatively smooth, being largely devoid of vascular foramina, like most impressions for tendons or ligaments; but the rest of the fossa is rough and pitted by vascular

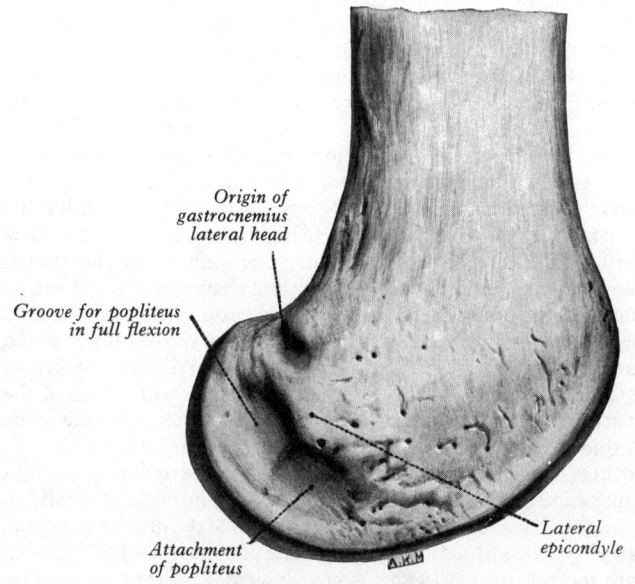

3.187 The distal end of the right femur. Lateral aspect.

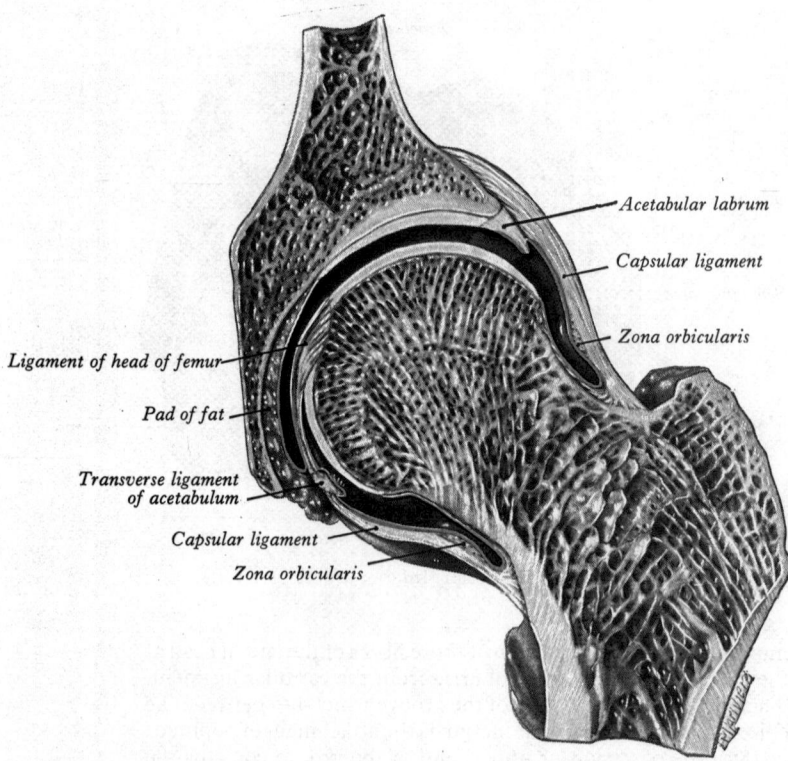

3.188 A section through the hip joint.

Acetabular labrum

Capsular ligament

Zona orbicularis

Ligament of head of femur

Pad of fat

Transverse ligament of acetabulum

Capsular ligament

Zona orbicularis

foramina, although occasionally the bursal recess between the two ligaments extends upwards to reach it. The *intercondylar line* gives attachment to the capsular ligament and, laterally, to the oblique popliteal ligament of the knee joint. To the anterior border of the fossa the infrapatellar synovial fold is attached (p. 485).

Structure. The shaft of the femur is a cylinder of compact bone, with a large medullary cavity. The wall of the cylinder is thick in the middle third of the shaft, where the bone is narrowest and the medullary cavity most capacious; but above and below this the wall becomes thinner, while the medullary cavity is gradually filled up with trabecular bone; the upper and lower ends of the shaft—and the articular extremities more especially—consist of trabecular bone, invested by a thin compact layer.

The trabeculae in the ends of the femur are approximately disposed along the lines of greatest compression and stress (p. 478). In the upper end (**3.**188, 189) the chief trabeculae are arranged as a series of bony plates at right angles to the articular surface of the head which converge to a central dense wedge; this wedge is supported by strong trabeculae which extend to the sides of the neck and are specially marked along its upper and lower borders. Any force therefore applied to the head of the femur is transmitted directly to the central wedge and thence to the junction of the neck with the shaft. This junction is in turn strengthened by a series of dense trabeculae which extend from the lesser trochanter to the lateral end of the superior border of the neck; this arrangement will obviously offer considerable resistance to either tensile or shearing forces applied to the neck through the head. A smaller bar stretching across the junction of the greater trochanter with the neck and shaft resists the shearing force of the muscles attached to this prominence. These two bars—one at the junction of shaft and neck, the other at the junction of shaft and greater trochanter—form the upper layers of a series of arches which extend across between the sides of the shaft and transmit to it forces applied to the upper end of the bone. A thin vertical plate of bone, named the *calcar femorale* (**3.**189), springs from the compact wall of the shaft in the region of the linea aspera and extends into the spongy bone of the neck. Medially, it joins the inner surface of the posterior wall of the neck of the bone; laterally, it continues the plane of the posterior wall of the neck into the greater trochanter, where it shades off into the general spongy bone. It is thus situated in a plane anterior to the

trochanteric crest and to the base of the lesser trochanter. Intensive study of the proximal end of the femur by scanning electron microscopy, admittedly on a small series of specimens, had largely confirmed the above views (Whitehouse and Dyson 1974). A surprisingly wide variation in the metrical parameters of the trabecular pattern in different regions of the femoral head and neck was nevertheless observed.

Tensile and pressure tests of 40 human femora also indicated that the axial trabecular tissue of the femoral head withstands such stresses at much higher levels than peripheral trabeculae (Kaessmann *et al.* 1972).

In the lower end, the trabeculae spring on all sides from the inner surface of the cylinder, and descend in a direction perpendicular to the articular surface—the trabeculae above the condyles being the strongest and having a more accurately perpendicular course. In addition to this, there are horizontal planes of spongy bone, arranged like crossed girders forming a series of cubical compartments.

Ossification (**3.**184B, 185B, 190). The femur is ossified from five centres; one each for the shaft, head, greater trochanter, lesser trochanter and lower end. Except for the clavicle, it is the first long bone to show traces of ossification. Ossification begins in the middle of the shaft in the seventh week of intrauterine life, and extends upwards and downwards so that the miniature shaft is largely ossified at birth. The secondary centres appear as follows: in the lower end, during the ninth month of intrauterine life (from this centre the condyles and epicondyles are formed); in the head during the first six months after birth; in the greater trochanter during the fourth year: and in the lesser trochanter between the twelfth and fourteenth years. The manner in which the epiphysis for the head develops is noteworthy. The ossific centre appears in the upper part of the cartilaginous head and until the age of 10 is restricted to that part of the bone, so that the epiphysial line, as seen in X-ray photographs (**3.**175), is horizontal and the lower and medial part of the articular surface is borne on the neck (**3.**190).

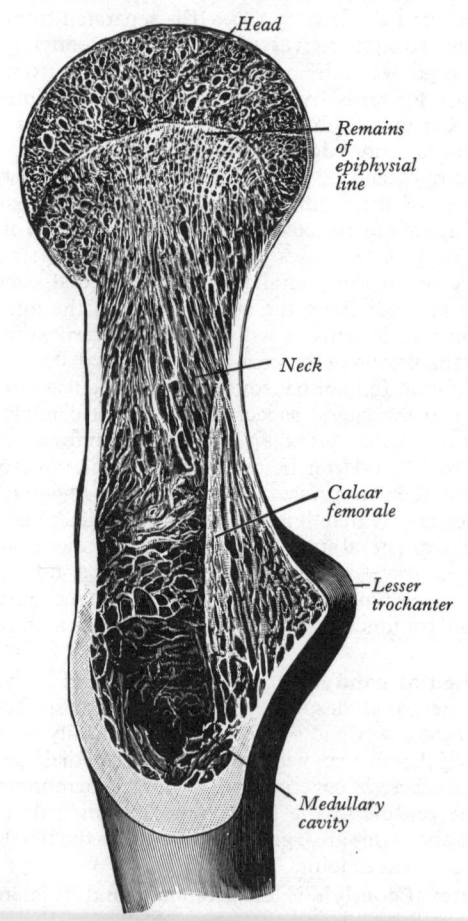

Head

Remains of epiphysial line

Neck

Calcar femorale

Lesser trochanter

Medullary cavity

3.189 An oblique section through the proximal end of the left femur showing the *calcar femorale*. Compare with **3.**1C and **3.**188.

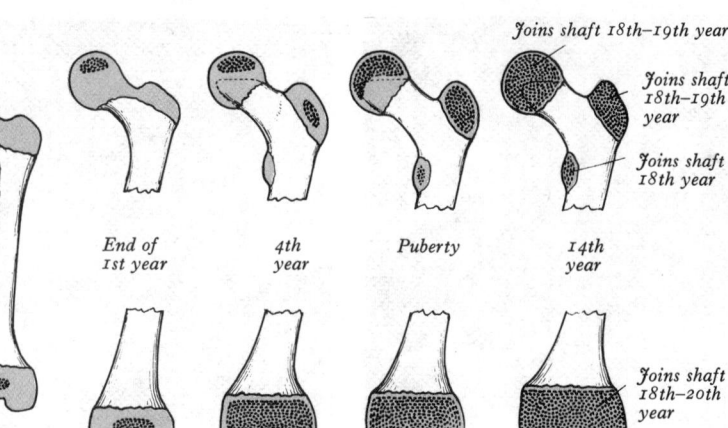

3.190 Stages in the ossification of the femur. Note how the neck, which is ossified as an extension from the shaft, invades the cartilaginous head.

Thereafter the medial margin of the epiphysis grows over the lower and medial part of the articular surface and covers it completely. As a result, when fully developed, the epiphysis forms a gently hollowed-out cup on the summit of the neck. The epiphysial line at the upper end follows the articular margin except superiorly where it is separated from the articular surface by a non-articular area through which blood vessels enter the head (Trueta 1957; Smith 1962). The epiphyses fuse independently with the shaft, the lesser trochanter soon after puberty, then the greater trochanter, then the head at the fourteenth year in females and seventeenth in males and finally the lower end at the sixteenth year in females and eighteenth in males. It should be noted that the lower epiphysial plate passes through the adductor tubercle (**3**.184B).

Applied Anatomy. The distal extremity of the femur is the only epiphysis in which ossification constantly starts just before birth. The presence of this centre of ossification is always relied upon in medico-legal investigations as proof that a newborn child found dead was viable. The position of the epiphysial plate should be carefully noted. It is on a level with the adductor tubercle, and the epiphysis does not, therefore, form the whole of the synovial covered portion of the lower end of the bone. It is essential to bear this point in mind when operations are performed on the lower end of the femur, since growth in length of the bones takes place chiefly from the lower epiphysial cartilage, and any interference with it in a young child may involve marked shortening of the limb.

Fractures of the neck of the femur are due to indirect violence and usually occur through tripping over some minor obstruction. The trunk continues to move forwards and, overbalancing, falls to the same side, imposing full medial rotation on the thigh and leg. Below the age of 16 the usual injury suffered from such an accident is a spiral fracture of the shaft of the femur. In contrast between the ages of 16 and 40 the result is often a 'bucket-handle' tear of the medial meniscus of the knee joint. In older patients, between 40 and 60, a common result is a Potts' fracture of the leg, but, in patients over 60, a fracture of the neck of the femur is commonplace owing to the senile osteoporotic changes which have occurred in the bone. Women are more liable to this injury than men, for their bones are more lightly built in the first instance.

In the normal, when the lower limb is aligned with the trunk, the tip of the greater trochanter is on the line joining the anterior superior iliac spine and the most prominent part of the ischial tuberosity (Nelaton's line). In displacements due to fracture of the neck and other conditions the tip of the greater trochanter may lie above this line.

The Patella

The patella (**3**.193), the largest of the sesamoid bones, is in the front of the knee joint embedded in the back of the tendon of the quadriceps femoris. It is flattened, triangular below, curved above, and has anterior and posterior surfaces, three borders, and an apex. In the living subject in the erect attitude, its lower limit lies more than 1 cm above the line of the knee joint.

The *anterior surface*, readily palpable, is convex, perforated by nutrient vessels, and marked by numerous rough, longitudinal striae. It is separated from the skin by a bursa and covered by an expansion from the tendon of the quadriceps femoris; this expansion is continuous below with the superficial fibres of the so-called patellar *ligament*, which is, of course, the *tendon* of the quadriceps. The *posterior surface* presents in its upper part a smooth, oval, articular area, divided into two facets by a smooth vertical ridge, which corresponds to the groove on the patellar surface of the femur, and the facets to the medial and lateral parts of the same surface; the lateral facet is the broader and deeper. The ridge is also covered by articular cartilage, and there is hence no true division of the articular surface, which really consists of two regions angulated with respect to each other by the 'ridge'. A narrow strip, broader above than below and often inconspicuous in the macerated specimen, is marked off from the medial part of the medial facet. This strip comes into contact with the medial condyle of the femur in extreme flexion of the knee joint. Below the articular surface the *apex*, which points downwards, is roughened in its lower part for the attachment of the ligamentum patellae; its upper part is covered by the infrapatellar pad of fat.

The *superior border* is thick, and slopes from behind, downwards and forwards: except near its posterior margin, it has attached to it the part of the quadriceps femoris which is derived from the rectus femoris and vastus intermedius. The *medial* and *lateral borders* are thinner and they converge below: they give attachment to those portions of the quadriceps femoris which are derived from the vasti medialis et lateralis. Near the junction of the base and lateral border there is a small, shallow, circular depression into which a part of the tendon of the vastus lateralis is inserted.

Structure. The patella consists of a nearly uniform dense cancellous bone, covered by a thin compact lamina. The trabeculae immediately beneath the anterior surface are arranged parallel with it. In the rest of the bone they radiate from the articular surface towards the other parts of the bone.

Ossification. The patella begins to ossify from several centres which appear in the third to sixth years and quickly coalesce. Accessory marginal centres appear later and fuse with the central mass (Hellmer 1935). (See also Prakash *et al.* 1979, for bibliography.)

The Tibia

The tibia (**3**.195, 196) is the medial and much the stronger of the two bones of the leg; excepting the femur it is the longest bone of the skeleton. It is prismoid in section in its shaft and has expanded extremities. Its lower end is smaller than the upper, and on its

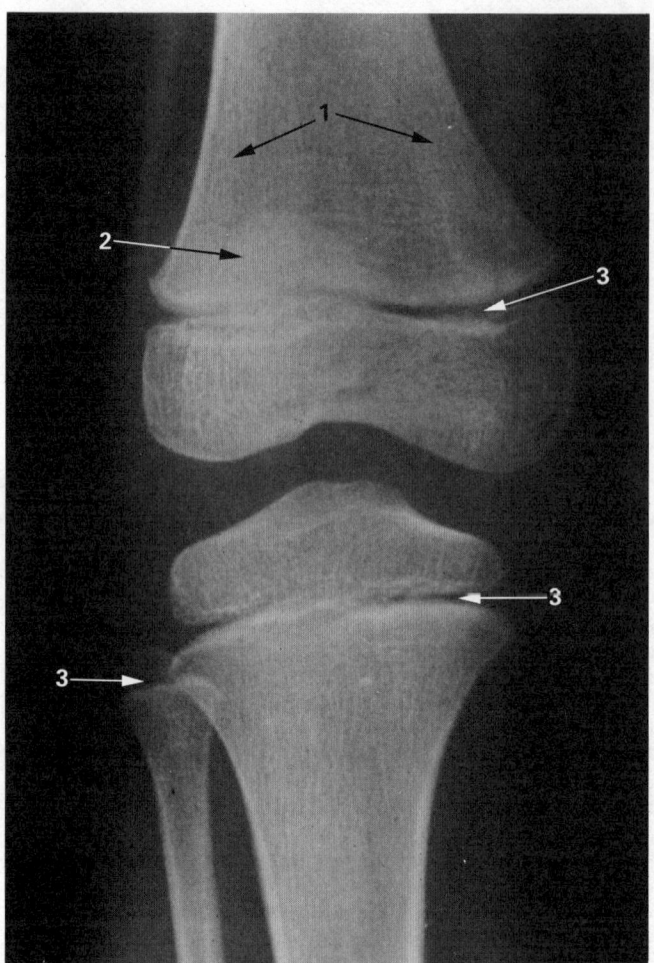

A

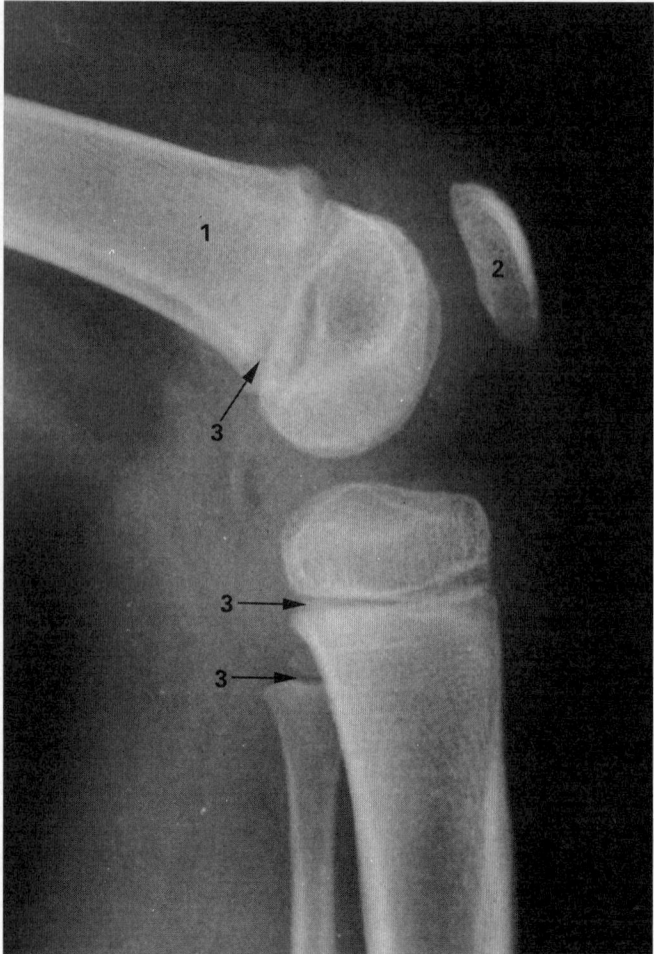

B

3.191 A and B Anteroposterior (A) and lateral (B) radiographs of the knee in a girl aged 6 years. 1. Flared femoral metaphysis. 2. Patella. 3. Carti-laginous growth plates with adjacent epiphyses. Note early stage of ossification in fibular epiphysis.

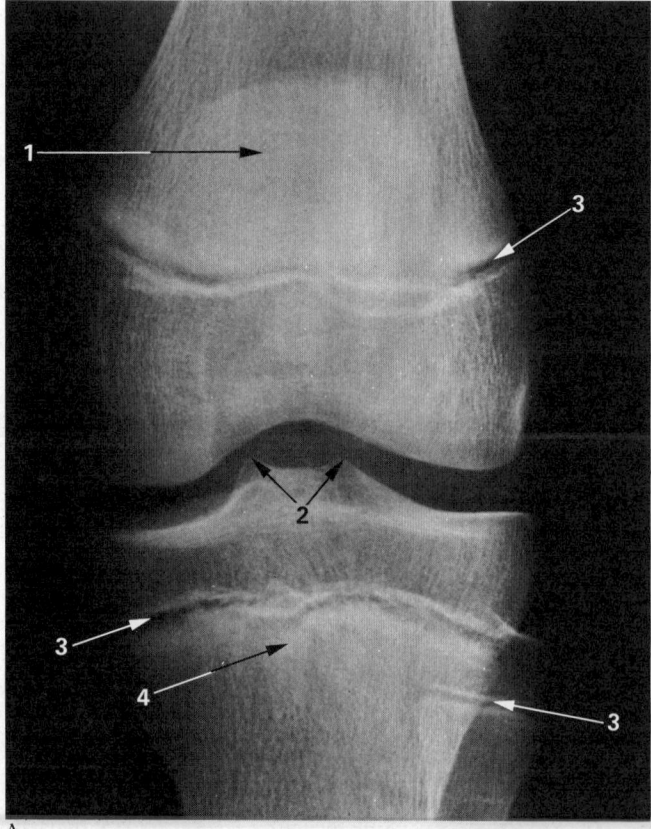

A

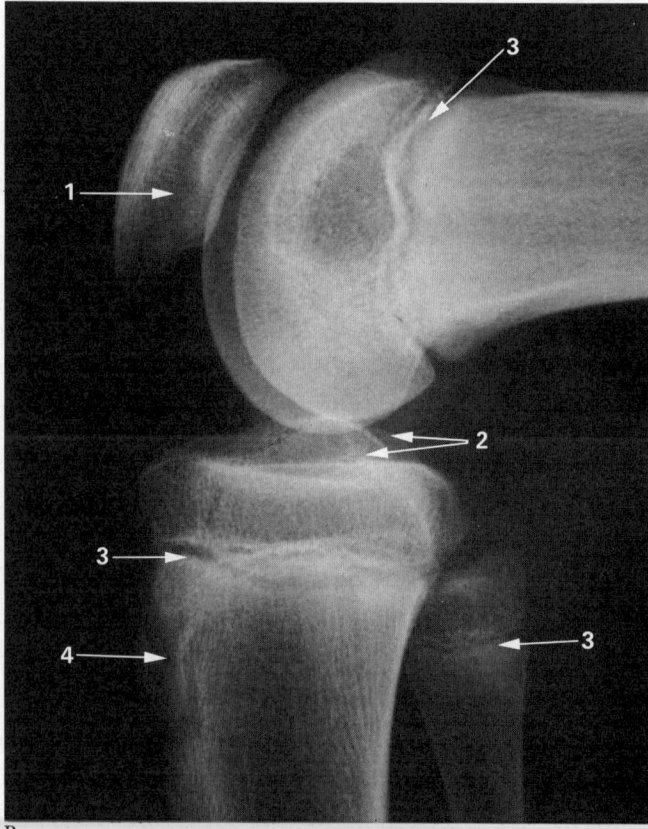

B

3.192 A and B Anteroposterior (A) and lateral (B) radiographs of the knee in a boy aged 14 years. 1. Patella. 2. Intercondylar eminences. 3. Carti-laginous growth plates with adjacent epiphyses. 4. Prolongation of proximal tibial epiphysis and growth plate forming the tibial tuberosity.

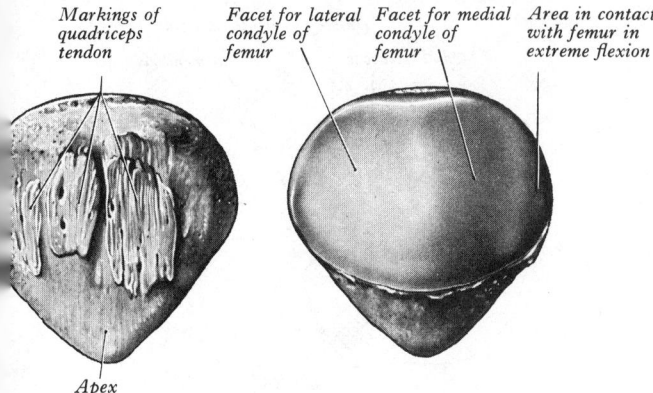

Markings of quadriceps tendon — Facet for lateral condyle of femur — Facet for medial condyle of femur — Area in contact with femur in extreme flexion

Apex

3.193   The left patella. Anterior and posterior aspects. Drawn from a fresh, unmacerated specimen, with the articular cartilage preserved.

medial side a stout process, termed the medial malleolus, projects downwards beyond the rest of the bone. The anterior border of the shaft is a conspicuous, sharp crest, which curves medially at the lower end towards the medial malleolus; it is the most prominent of the three borders.

The upper end of the tibia is expanded, especially in its transverse axis, providing an adequate bearing surface for the body weight transmitted through the lower end of the femur. It comprises two prominent masses, the *medial* and *lateral condyles*, and a smaller projection, the *tuberosity of the tibia*. The condyles project backwards a little, so as to overhang the upper part of the posterior surface of the shaft. Superiorly each is covered with an articular surface, the two being separated by an irregularly roughened *intercondylar area*. They form visible and palpable landmarks at the sides of the ligamentum patellae, the lateral condyle being the more prominent. When the knee is flexed passively, the anterior margins of the tibial condyles can be felt readily, and each forms the lower border of a depression at each side of the patellar ligament.

The medial condyle is the larger but does not overhang so much as the lateral condyle. Its upper articular surface (**3**.194A), oval in outline, is concave in all diameters, and its lateral border projects upwards, deepening the concavity and covering an elevation, the *medial intercondylar tubercle*. The posterior surface of the condyle is marked, immediately below the articular margin, by a horizontal, roughened groove. Its medial and anterior surfaces form a rough strip, separated from the medial surface of the shaft by an inconspicuous ridge.

The lateral condyle overhangs the shaft, especially at its posterolateral part, which bears on its inferior surface a small circular facet for articulation with the upper end of the fibula. The upper surface (**3**.194A) is covered with an articular surface for the lateral condyle of the femur. Nearly circular in outline, it is slightly hollowed in its central part, and its medial border extends upwards to cover an elevation, termed the *lateral intercondylar tubercle*. The posterior, lateral and anterior surfaces of the condyle are rough.

The anterior surfaces of the two condyles became continuous in front with a large triangular area, the apex of which is directed downwards and is formed by the tuberosity of the tibia. The lateral edge of this area forms a sharp ridge which separates the lateral condyle from the lateral surface of the shaft.

The intercondylar area (**3**.194A) is a roughened area on the superior surface, which intervenes between the articular surfaces of the two condyles. It is narrowest at its middle, where it is elevated into the *intercondylar eminence*. The lateral and medial parts of the eminence project slightly upwards, and constitute the *lateral* and *medial intercondylar tubercles*. Both behind and in front of the eminence the intercondylar area becomes wider, as the curved margins of the articular surfaces recede from each other.

The tuberosity of the tibia is at the upper end of the anterior border of the shaft, and is the truncated apex of the triangular area on the front of the bone where the anterior surfaces of the two condyles become continuous. It is a low eminence, divided into a lower roughened and an upper smooth region. The lower can be

felt through the skin, from which it is separated only by a bursa, the subcutaneous infrapatellar bursa; to the upper part is attached the ligamentum patellae.

The upper surface is tilted backwards relative to the long axis of the shaft. The tilt is maximal in the newborn and decreases with age; it is also more marked in habitual squatters (Kate and Robert 1965). The *articular surface of the medial condyle* is oval in shape, with its long axis anteroposterior and is perceptibly the longer of the two, in conformity with the differences which exist between the tibial surfaces of the two femoral condyles (p. 482). It is related around its anterior, medial and posterior margins to the medial meniscus, and the area of contact is flattened. The imprint of the cartilage, which is widest behind and narrower at the medial side and in front, can often be recognized on the bone. The rest of the surface is concave, and its raised lateral margin covers the medial intercondylar tubercle. The *articular surface of the lateral condyle* is more nearly circular in shape. Like the medial articular surface it is related to the corresponding meniscus, and bears its flattened imprint. Elsewhere the surface is very slightly concave to adapt it to the surface of the corresponding femoral condyle, and its raised medial margin is continued on the lateral intercondylar tubercle. The edges of the two articular surfaces are sharp except at the posterior part of the lateral surface, where the margin is smooth and rounded; in this situation the tendon of the popliteus is intimately related to the bone.

The anterior intercondylar area (**3**.194B) is widest anteriorly. In its anteromedial part, just in front of the medial articular surface, it bears a slight depression to which is attached the anterior horn of the medial meniscus. Behind that depression a relatively smooth area affords attachment to the lower end of the anterior cruciate ligament. The anterior horn of the lateral meniscus is attached in front of the intercondylar eminence and lies lateral to the anterior cruciate ligament. The *intercondylar eminence* with its medial and lateral tubercles, occupies the narrow, middle part of the area. The posterior slope of the eminence gives attachment to the posterior horn of the lateral meniscus, and behind that the **posterior intercondylar area** inclines downwards and backwards. A depression behind the base of the medial intercondylar tubercle gives attachment to the posterior horn of the medial meniscus. The rest of the area is smooth and affords attachment to the lower end of the posterior cruciate ligament, as far back as the ridge to which the capsular ligament is attached.

To the superior lip of the groove on the posterior surface, the capsular ligament and the posterior part of the tibial collateral ligament of the knee joint are attached; its inferior lip receives the insertion of semimembranosus. At the lateral limit of the groove is a rough tubercle which provides the principal attachment of the semimembranosus tendon (Cave and Porteous 1958). The medial and anterior surfaces of the condyle, which are marked by numerous vascular foramina, give attachment to the medial patellar retinaculum.

In a recent study of 13 healthy knee joints, which were used for measurements (together with observations on 75 macerated specimens), a more detailed plan of ligamentous and other attachments was evolved (*see* Jacobsen 1974; and **3**.194C). Considerable variation in the areas of these attachments was noted. Dense fibrous attachments produced facets, separated by areas of a more porous appearance (p. 400).

The tibia has been investigated extensively by Ljungren (1976) using stereometric techniques; analysis has shown functional adaptation of the curvatures and inclinations of the tibial condylar facets and the tibial tuberosity, particularly in regard to habits of locomotion in different racial groups.

The fibular facet on the *lateral condyle* is directed downwards and slightly backwards and laterally. Above and to its medial side the posterior surface of the condyle is grooved by the tendon of the popliteus, but a synovial recess intervenes between the tendon and the bone. The lateral and anterior surfaces of the condyle are separated from the lateral surface of the shaft by a sharp margin which gives attachment to the deep fascia of the leg. An impression on the anterior surface, often well marked though flattened, affords attachment for the iliotibial tract. Small slips from the tendon of the biceps femoris are inserted above and in

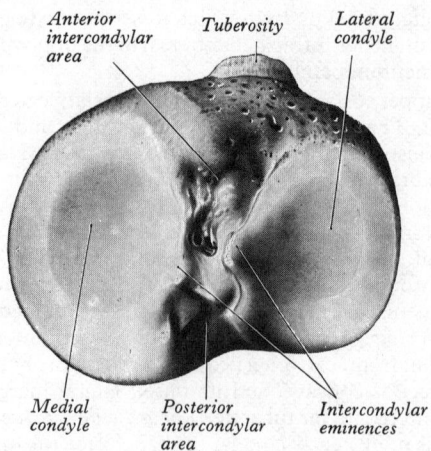

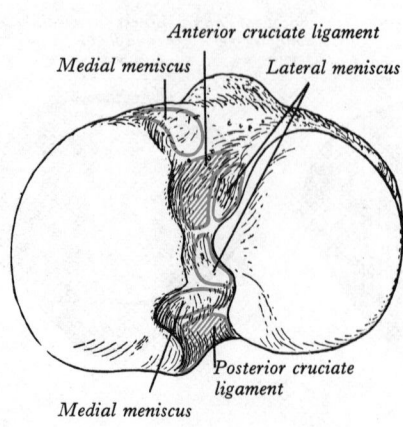

3.194A The proximal articular surface of the right tibia. The imprints of the menisci were very conspicuous in this specimen.

3.194B An outline of A showing the attachments of the menisci an cruciate ligaments.

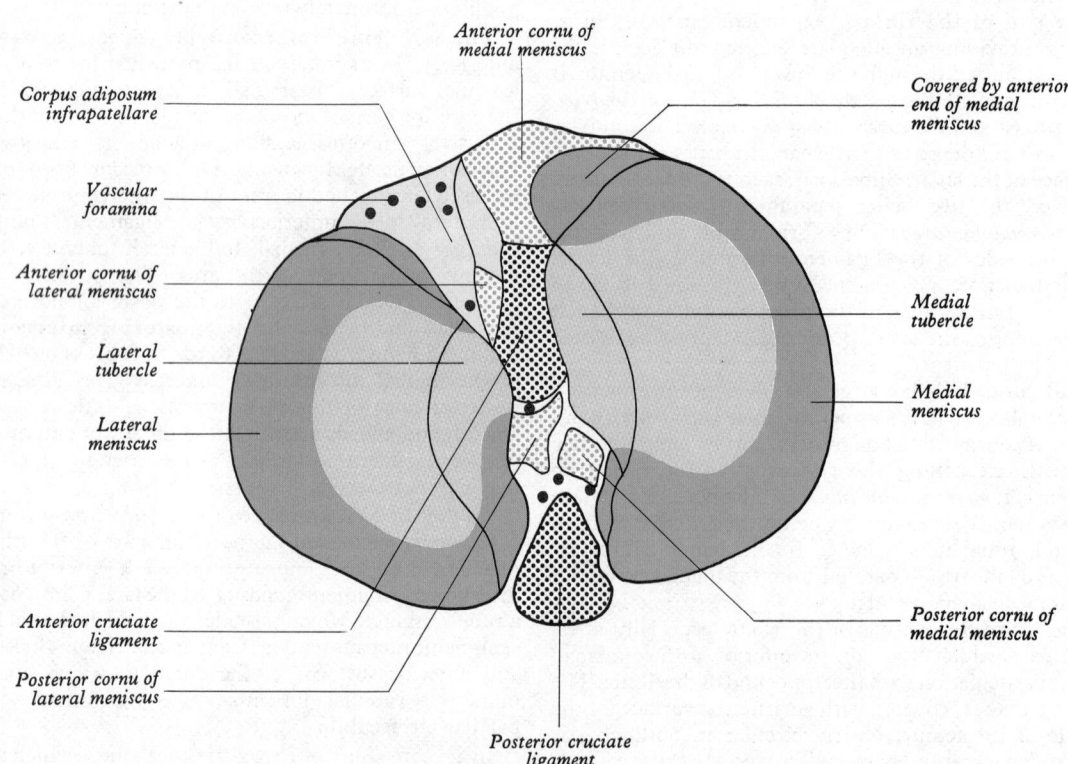

3.194C A detailed analysis of the surface features of the proximal aspect of the human tibia (left). The condylar areas are shown in full blue, the parts in contact with the menisci being in deep blue, while the remaining condylar regions are in light blue. The attachments of the meniscal cornua are in blue stipple, and those of the cruciate ligaments are in black stipple. The yellow area indicates the extent of contact with the corpus adiposum and the red dots signify vascular foramina. (Adapted by kind permissio of Dr. Klaus Jacobsen, Department of Orthopaedic Surgery, Th Gentofte Hospital, Copenhagen, *Journal of Anatomy* and Cambridg University Press.)

front of the fibular facet; below these the uppermost fibres of the extensor digitorum longus and occasionally those of the peroneus longus arise from the lateral surface.

The line which divides the **tibial tuberosity** into upper and lower part marks the position of the epiphysial line between the upper epiphysis and shaft (p. 404). The ligamentum patellae is attached to the smooth portion; its superficial fibres may extend to the crest and lower rough portion where the lower limits of the attachment may be marked by a slight ridge (Lewis 1958). The lower part of the tuberosity is subcutaneous. Above the tuberosity the bone is related to the deep surface of the ligament, but the deep infrapatellar bursa and fibro-adipose tissue

intervene. In habitual squatters there is usually a vertical groov on the front of the lateral condyle, which is occupied by th margin of the ligamentum patellae in flexion.

The shaft of the tibia (3.195, 196), being triangular i section, possesses medial, lateral and posterior surfaces, separate by anterior, interosseous and medial borders. It is thinnest at th junction of its middle and lower thirds, but expands considerabl towards its upper and lower ends.

The *anterior border* commences at the tuberosity of the tibia an runs downwards to the medial malleolus. It is subcutaneou throughout its length, and, except in its lower fourth, where it i rounded and indistinct, forms a sharp crest (the 'shin'). It i

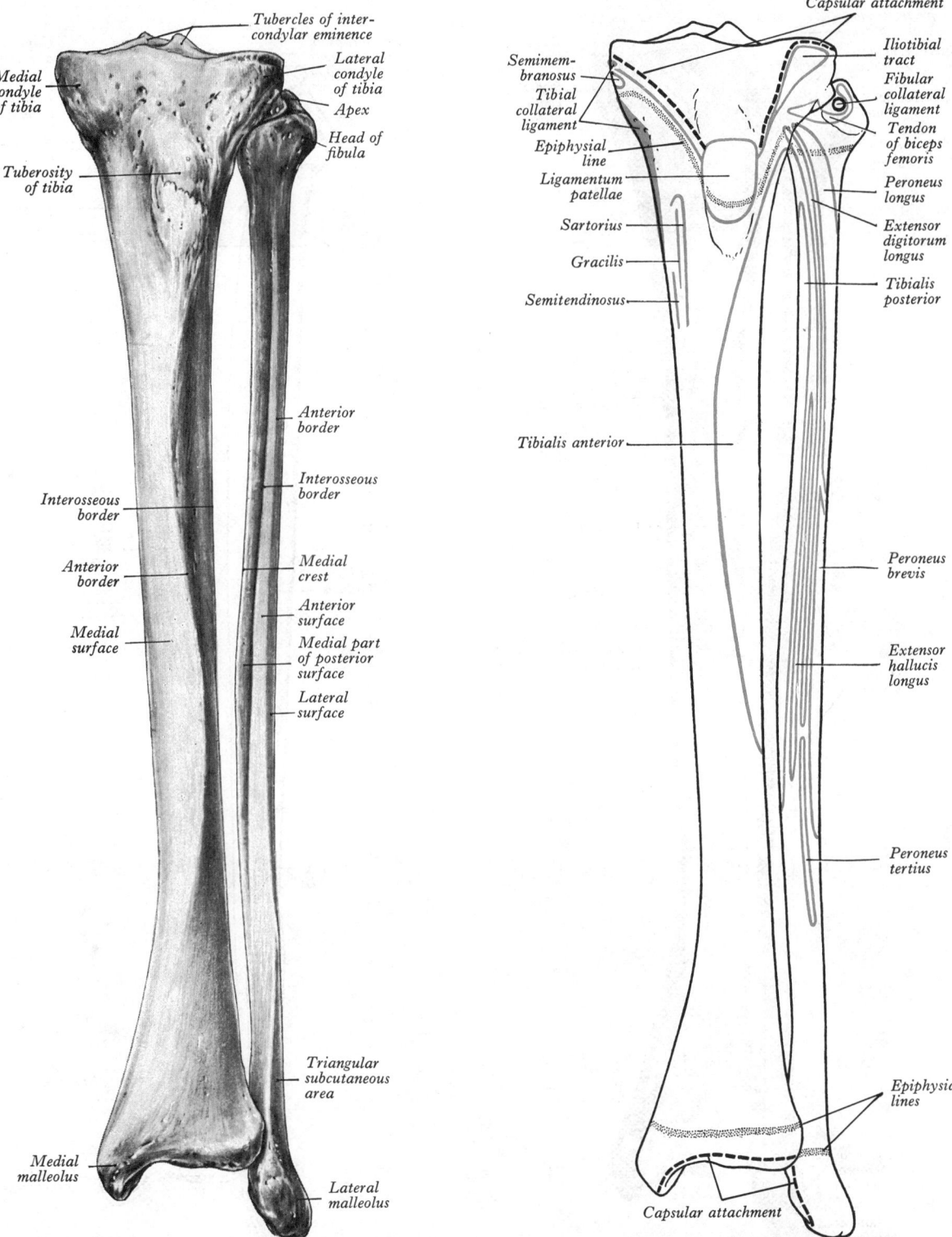

3.195A Left tibia and fibula. Anterior aspect. Compare with key, shown in **3.195B**.

3.195B Key to **3.195A**. The epiphysial lines are stippled; the interrupted lines correspond to the attachments of the capsular ligaments.

slightly sinuous and its lower fourth diverges towards the medial side. The *interosseous border* commences below and a little in front of the fibular facet on the lateral condyle and descends to reach the anterior border of the fibular notch, which marks the lateral aspect of the lower end of the tibia. In nearly the whole of its length it affords attachment to the interosseous membrane which connects the tibia to the fibula. As a rule it is poorly defined at its upper end, but is obvious below this. The *medial border* commences below the anterior end of the groove on the medial condyle and runs downwards to the posterior margin of the medial malleolus. Its upper and lower fourths are rounded and ill-defined, but its middle region is sharp and distinct.

The *medial surface* is bounded in front by the anterior border, and behind by the medial border. It is broad and smooth, and is subcutaneous throughout practically its whole extent. The *lateral surface*, also broad and smooth, is placed between the anterior and the interosseous borders. In its upper three-fourths it is directed laterally and is slightly concave from before backwards. Its lower fourth is carried round on to the front of the bone, owing to the deviation of the anterior border to the medial side and the forward inclination of the lower part of the interosseous border. This part of the surface is somewhat convex forwards. The *posterior surface* is bounded by the interosseous and the medial borders, and is widest at its upper end, where it is crossed from above downwards

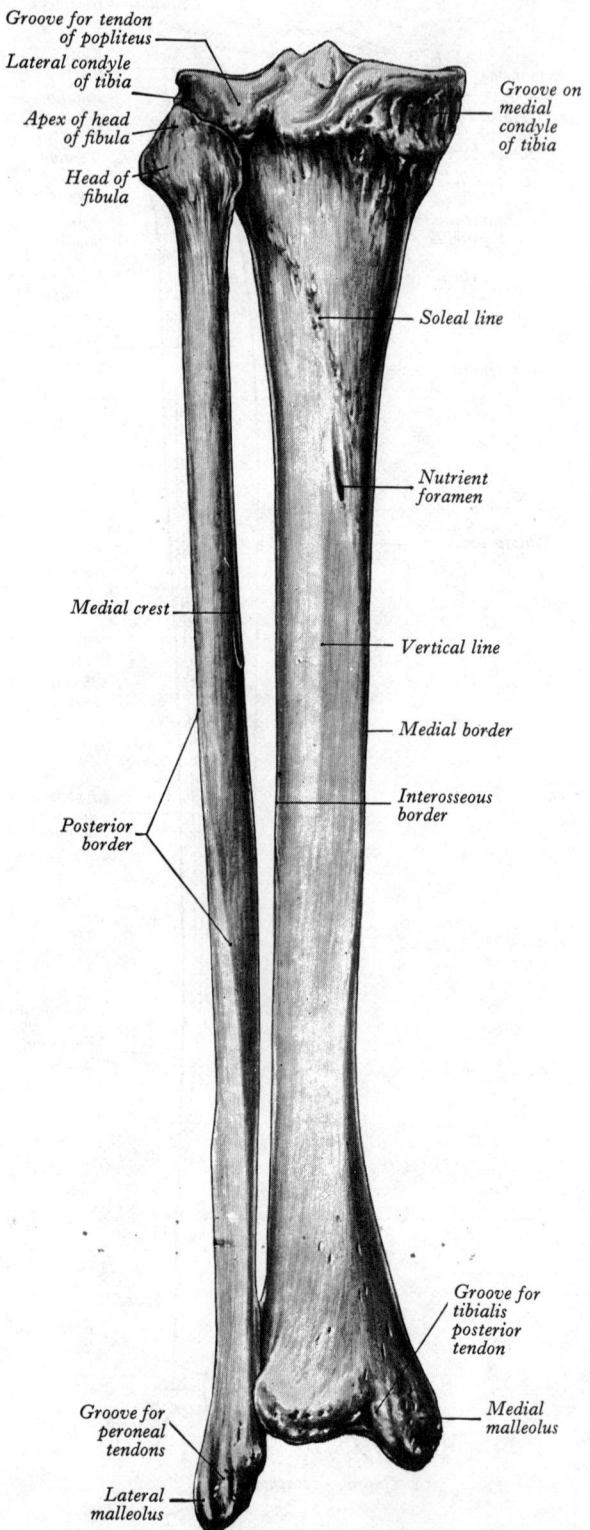

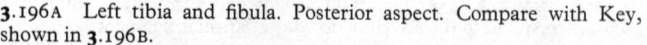

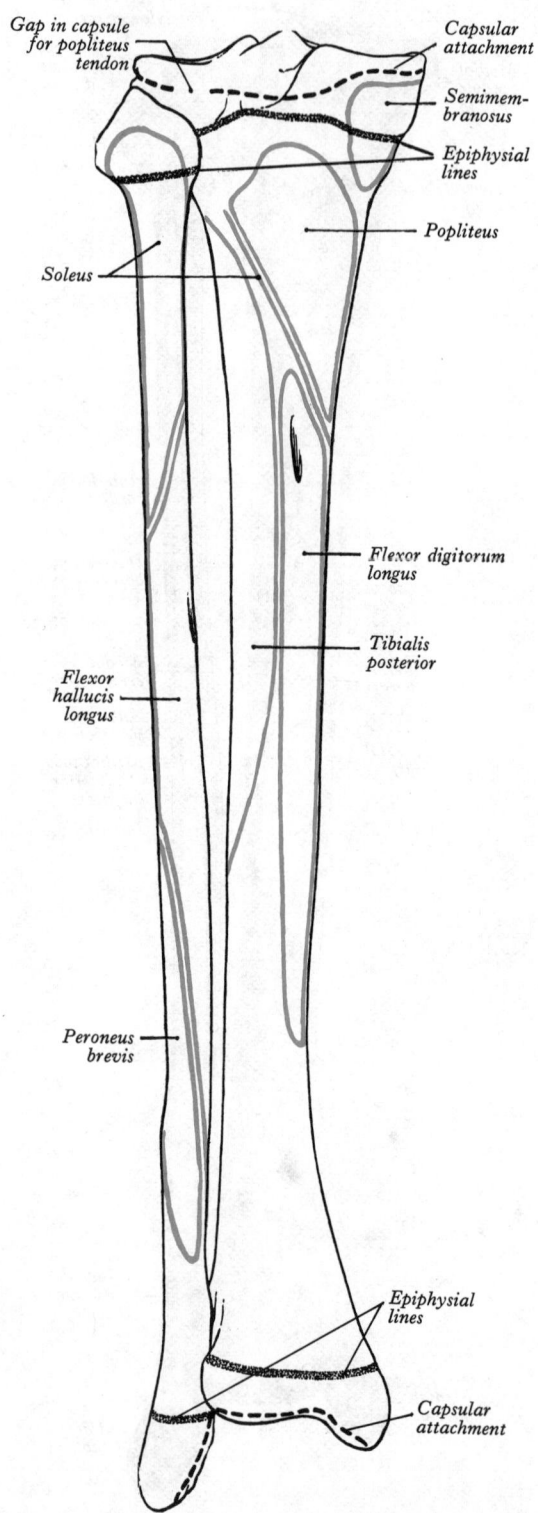

3.196A Left tibia and fibula. Posterior aspect. Compare with Key, shown in **3.196B**.

3.196B Key to **3.196A**. The epiphysial lines are stippled; the interrupted lines correspond to the attachments of the capsular ligaments.

and medially by an oblique, roughened ridge, termed the *soleal line*. The area below this line is subdivided by a faint *vertical line*, which begins at or just below the middle of the soleal line and soon fades away. A prominent vascular groove marks the bone near the upper end of the vertical line and descends to a large nutrient foramen; it may be situated lateral or medial to the vertical line.

The *anterior border* provides attachment for the deep fascia of the leg. A little above the medial malleolus it receives the medial end of the superior extensor retinaculum. Above the soleal line the *medial border* gives attachment to the fascia covering the popliteus, to the posterior fibres of the tibial collateral ligament of the knee joint and to slips of the semimembranosus; below the

soleal line it gives origin for a short distance to fibres of the soleus and attachment to the fascia which covers the deep muscles of the leg. At its lower end it becomes continuous with the medial border of the groove which lodges the tendon of the tibialis posterior. The *interosseous border* gives attachment to the interosseous membrane of the leg, except at its upper and lower ends. Its upper end is scarcely recognizable, and in this situation there is a large gap in the interosseous membrane for the passage of the anterior tibial vessels. Its lower end forms the anterior boundary of the fibular notch and gives attachment to the anterior tibiofibular ligament.

The *medial surface* is usually rough, close to the upper part of the medial border over an area nearly 5 cm long and 1 cm wide, for attachment of the anterior part of the tibial collateral ligament of the knee joint and behind this to the semimembranosus. In front of this roughened area the surface provides insertion for the tendons of sartorius, gracilis and semitendinosus, which rarely mark the bone. The gracilis, above, and semitendinosus, below and behind, are inserted immediately in front of the ligamentous area; the sartorius has a linear attachment which commences above and descends in front of the other two (**3.**195 B). The rest of the surface is covered only by superficial fascia and skin, but its lower part is crossed obliquely by the great saphenous vein as it ascends from in front of the medial malleolus.

The *lateral surface* in its upper two-thirds or less, provides attachment for tibialis anterior. Its lower part is devoid of muscular attachments, but is crossed by the tendon of the tibialis anterior (which lies along the lateral side of the anterior border), the extensor hallucis longus, the anterior tibial vessels and nerve,

The *anterior surface* is smooth and bulges forwards beyond the inferior surface, from which it is separated by a narrow groove. It is continuous above with the lateral surface of the shaft (p. 401). The *medial surface* also is smooth and is continuous above with the medial surface of the shaft and below with the medial surface of the malleolus. It is subcutaneous and can be seen and felt through the skin. The *posterior surface* is crossed at its medial end by a groove, usually conspicuous, which can be traced down on to the posterior surface of the medial malleolus. Elsewhere this surface of the lower end is smooth and is continuous above with the posterior surface of the shaft. The *lateral surface* is the triangular *fibular notch*, which is bound by ligaments to the lower end of the fibula. The anterior and posterior borders of the notch project and converge to meet above on the interosseous border of the bone. The floor of the notch is roughened in its upper part for the attachment of an interosseous ligament. Its lower part is smooth and is sometimes covered with articular cartilage. The *inferior*

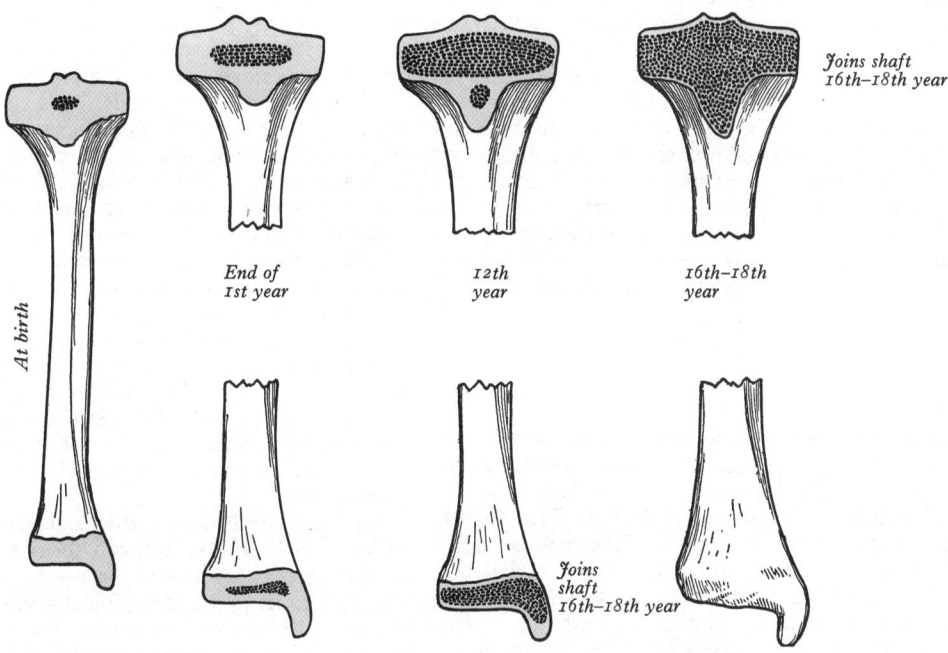

3.197 Stages in the ossification of the tibia.

the extensor digitorum longus and peroneus tertius, in that order from the medial to the lateral side.

The *posterior surface* gives insertion to the popliteus over the triangular area above the soleal line, with the exception of the area adjoining the fibular facet. To the *soleal line* are attached the strong fascia which covers the popliteus, and the soleus, its covering fascia, and the fascia covering the deep muscles of the leg. The upper end of the line does not reach the interosseous border and is marked by a tubercle, to which is attached the medial end of the tendinous arch in the soleus. Lateral to the tubercle the posterior tibial vessels and nerve descend on the surface of the tibialis posterior. Below the soleal line, the *vertical line* separates the attachments of flexor digitorum longus on the medial side from that of tibialis posterior (**3.**196 B). The lower quarter, or more, of the posterior surface is devoid of muscular attachments, but is intimately related to the tendon of the tibialis posterior as it runs downwards and medially to reach the groove on the back of the medial malleolus. The flexor digitorum longus lies on the posterior surface of the tibialis posterior crossing it obliquely from the medial to the lateral side, but the posterior tibial vessels and nerve and the flexor hallucis longus come into contact only with the lateral part of this surface for a short distance inferiorly.

The lower end of the tibia is slightly expanded, and projects medially and downwards as the medial malleolus. It possesses anterior, medial, posterior, lateral and inferior surfaces.

surface is smooth for articulation with the body of the talus. Wider in front than behind, it is concave from before backwards and slightly convex from side to side. Medially it is continuous with the articular surface of the medial malleolus. The tibial articular surface may extend into the narrow groove (*vide supra*) which separates this surface from the anterior aspect of the tibial shaft. Such extensions, which may be medial or lateral or both in situation, are described as *squatting facets*, and they articulate with corresponding facets on the neck of the talus (p. 407), when the foot is in extreme dorsi-flexion. Such features have been used in racial evaluation (p. 238).

The medial malleolus is a short but stout process. Its lateral surface is smooth and occupied by a comma-shaped articular facet, which articulates with the medial side of the talus. Its anterior surface is rough, and its posterior surface bears the lower end of the groove that marks the posterior surface of the lower end of the shaft. The lower border of the malleolus is pointed anteriorly and depressed posteriorly.

The *anterior surface* of the lower end is related to the tendons, vessels and nerve which lie on the lower part of the lateral surface of the shaft and have already been enumerated (p. 402). To the narrow groove adjoining the anterior border of the inferior surface is bound the anterior part of the articular capsule of the ankle joint. The groove on the *posterior surface* is adapted to the tendon of tibialis posterior, which at that level usually separates the tendon of the flexor digitorum longus from the bone. More laterally the posterior tibial vessels and nerve and the flexor hallucis longus tendon are in contact with this surface. The anterior and posterior borders of the *fibular notch* give attachment respectively to the anterior and posterior tibiofibular ligaments. The *medial malleolus* ends at a higher level than the lateral, which also lies on a more posterior plane. Its anterior surface has attached to it the anterior part of the articular capsule of the ankle joint. The groove for the posterior tibial muscle's tendon, on the posterior surface, has a prominent medial border, to which the flexor retinaculum is attached. The upper end of the deltoid ligament is attached to the lower border of the malleolus, both to the pointed anterior part and to the depression behind it.

Ossification. The tibia is ossified from three centres (3.195B, 196B, 197): one for the shaft and one for each end. Ossification begins in the middle of the shaft about the seventh week of intrauterine life. The centre for the upper end is usually present at birth; from it, at about 10 years of age, a thin tongue-shaped process extends downwards in front to form the smooth part of the tuberosity (3.197). An additional centre for the tongue-shaped process is not uncommon; it appears about the twelfth year and soon fuses with the upper epiphysis. In its lower two-thirds the epiphysial plate of the upper end consists of dense collagenous tissue, the fibres of which are aligned with those of the ligamentum patellae. This peculiar structure is attributed to the large tensile stresses to which this part is subjected through the ligament (Lewis 1958; Smith 1962). The centre for the lower end appears early in the first year. The lower end joins the shaft about the fifteenth year in females and seventeenth year in males, the upper in the sixteenth year in females and eighteenth year in males. The medial malleolus is usually formed by a downward extension from the lower epiphysis, commencing in the seventh year; it occasionally possesses a separate centre of ossification.

The Fibula

The fibula (3.195, 196), the lateral bone of the leg, is much more slender than the tibia, for it is not called upon to share in the transmission of body weight. It possesses an upper end or head, a shaft, and a lower end, the lateral malleolus. The slightly constricted part of the shaft adjoining the head is sometimes called the neck of the fibula. The shaft shows considerable variation in its form, for it is moulded by the muscles to which it gives attachment; and these variations may prove confusing.

The head of the fibula is slightly expanded in all its diameters, and projects beyond the shaft in front, behind and on the lateral side. On its upper surface is a nearly circular facet, which articulates with the inferior surface of the lateral condyle of the tibia; it faces upwards, and slightly forwards and medially. A blunt elevation, the *apex of the head* (*styloid process*), projects upwards from the lateral part of its posterior surface. The head of the fibula can be felt through the skin on the posterolateral aspect of the knee, nearly 2 cm below the level of the knee joint. Immediately below the head the common peroneal nerve crosses the posterolateral aspect of the neck and can be rolled against the bone in the living subject. If sufficient pressure is exerted, tingling sensations will be experienced on the dorsum of the foot, radiating to the toes, and especially to the medial side of the big toe.

The lower end or lateral malleolus projects downwards to a lower level than the tibia and lies on a more posterior plane. Its *lateral aspect* is subcutaneous and easily felt; its *posterior aspect* carries a broad groove with a prominent lateral border. Its *anterior aspect* is rough and rounded and is continuous below with the inferior border. The *medial surface* presents a triangular articular facet, with its apex pointing downwards (3.200). It articulates with the lateral surface of the talus in the ankle joint, and is convex from above downwards. Behind the articular facet the bone is marked by a roughened depression, the *malleolar fossa*, which readily admits the tip of a finger.

The shaft of the fibula (3.195, 196) possesses three borders and three surfaces, each of which is associated with a particular group of muscles. The borders are anterior, posterior and interosseous. At its lower end the *anterior border* commences at the apex of an elongated, triangular area which is continuous

below with the lateral surface of the lateral malleolus. Traced upwards, it ascends to reach the anterior aspect of the head. The *posterior border* is continuous with the medial margin of the groove on the back of the lateral malleolus. Usually sharp and distinct in its lower part, it is often rounded in the upper half of its extent. The *interosseous border* lies to the medial side of the anterior border and as a rule is on a more posterior plane (3.198), but in the upper two-thirds of the bone these two borders are very close to each other, and the intervening surface may be reduced to 1 mm or less in width.

The *lateral surface* is bounded by the anterior and posterior borders. It is associated with the peroneal muscles, and is directed laterally in its upper three-fourths. Its lower quarter inclines backwards and becomes continuous with the groove on the back of the lateral malleolus. The *medial surface* is bounded by the anterior and the interosseous borders. It is usually directed forwards and medially, but frequently faces directly forwards. Wider below, it becomes very narrow in its upper half, and may be reduced to little more than a rounded ridge. It is associated with the extensor muscles of the leg. The *posterior surface* is the largest of the three and is placed between the interosseous and the posterior borders. It is associated with the flexor muscles of the leg. In its upper two-thirds it is divided into two areas by a longitudinal ridge, termed the *medial crest*, which is separated from the interosseous border by a grooved surface, directed medially. The rest of the posterior surface faces backwards in its upper half or more, but its lower part curves round on to the medial aspect and faces medially. The lower part of this area fits into the fibular notch on the tibia and is roughened for the attachment of the interosseous tibiofibular ligament.

The elongated, triangular area immediately above the lateral surface of the lateral malleolus (3.195 A) is subcutaneous, but the rest of the shaft is covered with muscles and cannot be palpated.

The *head of the fibula* affords origin to fibres of the extensor digitorum longus in front, peroneus longus anterolaterally, and soleus behind. The fibular collateral ligament of the knee joint is attached just in front of the apex and is embraced by the principal insertion of the biceps femoris. The margins of the articular facet provide attachment for the capsular ligament of the tibiofibular articulation.

The *anterior border* of the fibula divides inferiorly into two ridges which enclose between them a subcutaneous triangular surface (3.195 A). The anterior intermuscular septum of the leg is attached to its upper three-fourths, and the lateral extremity of the superior extensor retinaculum to the lower part of the anterior border of the triangular area. The lower part of the posterior margin of the triangular area gives attachment to the lateral extremity of the superior peroneal retinaculum. The *interosseous border* terminates below at the upper extremity of the roughened area for the interosseous ligament. It provides attachment for the interosseous membrane and does not reach the head of the bone, on account of the gap in the upper part of the membrane for the transmission of the anterior tibial vessels. The *posterior border* is not always recognizable at its upper end; except at its lower end it gives attachment to the posterior intermuscular septum of the leg. The *medial crest* of the bone is intimately related to the peroneal

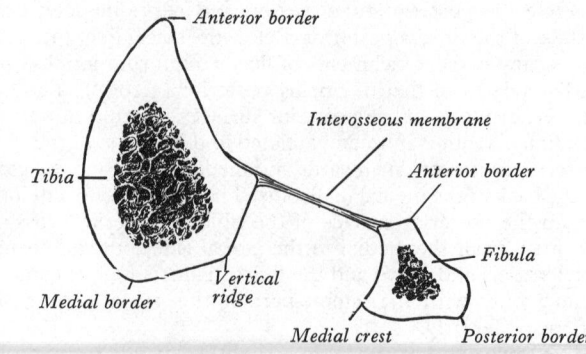

3.198 A tranverse section through the right tibia and fibula, showing the attachment of the crural interosseous membrane. Proximal aspect.

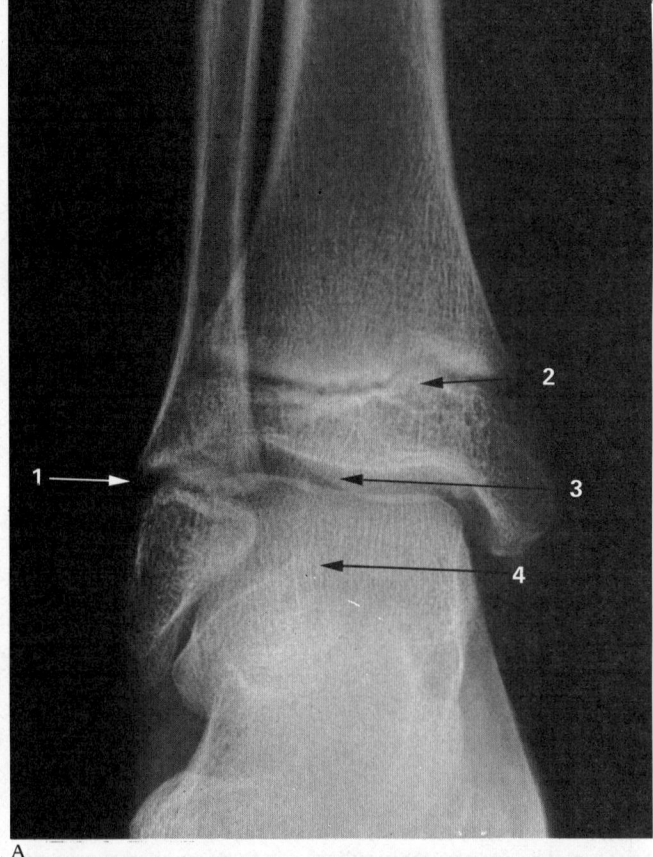

A

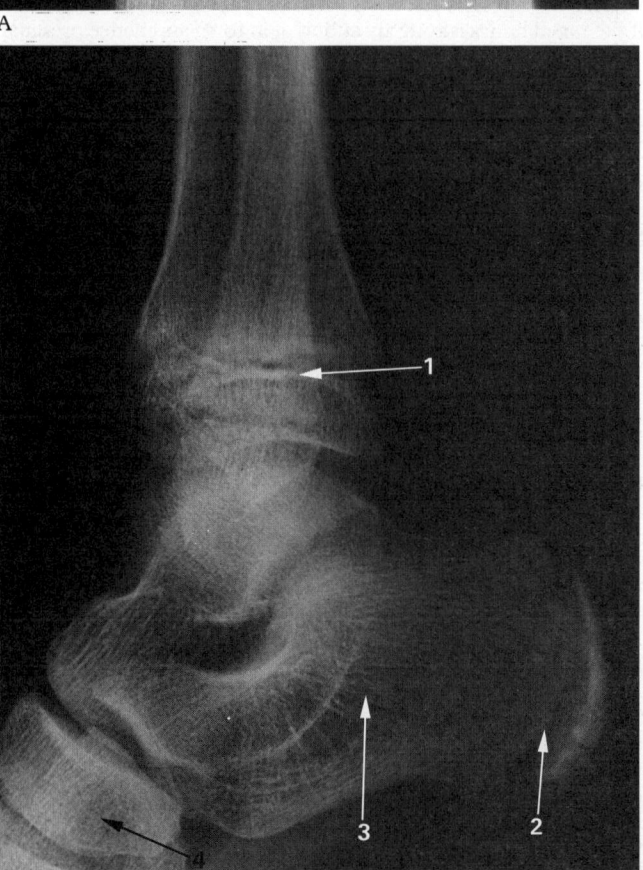

B

3.199A and B   Ankle region of a child of 10 years in plantar flexion. A. Obliquely anteroposterior. B. Lateral.
A   1. Inferior growth cartilage of fibula. 2. Inferior growth cartilage of tibia. 3. Talocrural joint. 4. Talus. Note that the fibular growth cartilage is approximately at the level of the talocrural joint.
B   1. Tibial growth cartilage. 2. Epiphysis of posterior surface of calcaneus. 3. Note trabecular pattern in calcaneus. 4. Shadow of navicular bone superimposed on that of cuboid.

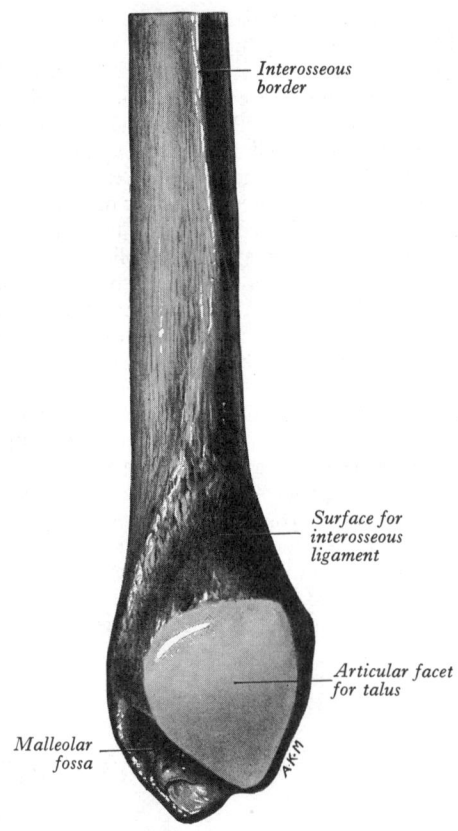

3.200   The lower end of the left fibula. Medial aspect.

artery, and the nutrient foramen of the fibula is situated either on the crest or in its immediate vicinity near the middle of the shaft. A layer of the deep fascia of the leg is attached, which separates the tibialis posterior from the flexor hallucis longus and the flexor digitorum longus.

The *medial surface* of the fibula is often termed the *extensor surface*, for attached to it are the extensor digitorum longus, the extensor hallucis longus and the peroneus tertius. The extensor digitorum longus arises from the whole breadth of the upper fourth of the surface and from the anterior part of the succeeding two-fourths; the extensor hallucis longus arises from its middle two-fourths behind the extensor digitorum longus; the peroneus tertius from its lower fourth or more, being directly continuous with the lower part of the extensor digitorum longus. The *lateral* or *peroneal surface* gives origin to the peroneus longus and the peroneus brevis. The peroneus longus arises from the whole

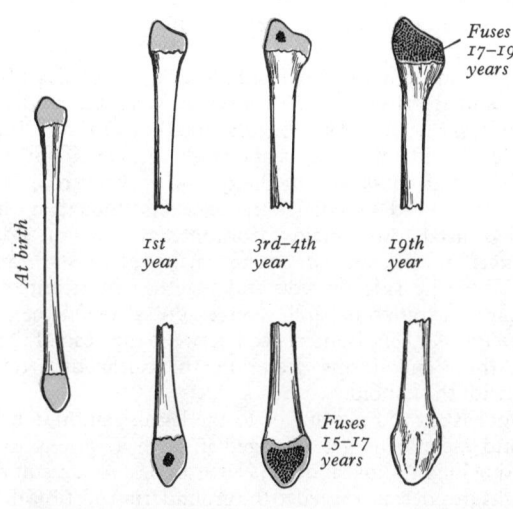

3.201   Stages in the ossification of the fibula.

extent of the upper third of the surface and from the posterior part of the middle third. The peroneus brevis arises in front of the lower half of the peroneus longus and extends downwards beyond it almost to the lower end of the bone. On account of the relative attachments of their fleshy bellies the tendon of the peroneus brevis is closely applied to the bone below and separates it from the tendon of the peroneus longus. The *posterior surface*, which is divided longitudinally into two parts by the medial crest, is often termed the *flexor surface*. The region which lies between the crest and the interosseous border is slightly hollowed out and gives origin to the tibialis posterior; it is often crossed by an oblique ridge which gives attachment to an intramuscular tendon. This part of the surface is usually confined to the upper three-fourths of the shaft, and at its lower end the medial crest becomes confluent with the interosseous border. The portion of the posterior surface which lies between the medial crest and the posterior border gives origin in its upper fourth to the soleus, which extends upwards on to the posterior aspect of the head; near the upper end of the medial part of this origin a roughened tubercle marks the lateral end of the tendinous arch which crosses the posterior tibial vessels and nerve as part of the attachment of soleus. Below the origin of the soleus the remainder of this surface gives origin to the flexor hallucis longus, almost to the lower end of the bone. A little above its middle, this surface is pierced by the nutrient foramen, which

directed downwards and transmits a branch from the peroneal artery.

The anterior surface of the *lateral malleolus* gives attachment to the anterior talofibular ligament. The lower border is marked in front by a slight notch and behind by a small projection which constitutes the apex of the malleolus. It is to the notch that the calcaneofibular ligament is attached. The tendons of the peroneus brevis and peroneus longus groove the posterior aspect; the latter is the more superficial and is closely covered by the superior peroneal retinaculum. The *malleolar fossa* (**3**.200) is pitted by numerous small vascular foramina; to its upper part the posterior tibiofibular ligament is attached, to its lower part the posterior talofibular ligament.

Ossification. The fibula is ossified from three centres (**3**.201): one for the shaft, and one for each end. Ossification begins in the shaft about the eighth week of intrauterine life, in the lower end in the first year, and in the upper about the third year in females and fourth year in males. The lower epiphysis—the first to ossify—unites with the shaft about the fifteenth year in females and seventeenth year in males, whereas the upper epiphysis, which does not begin to ossify until later, does not unite with the shaft until about the seventeenth year in females and nineteenth year in males. In this respect the fibula contrasts with the ossification pattern of other long bones.

THE SKELETON OF THE FOOT

Functionally the skeleton of the foot may be divided into segments: the tarsus, the metatarsus, and the phalanges or digital bones. In the description which follows use will be made of the terms *plantar* and *dorsal*, which are self-explanatory, and this will obviate, almost entirely, the employment of the terms anterior and posterior, which are not entirely apposite in the foot. The terms proximal and distal will also be employed, wherever reasonable, with the same significance with which they were used in the hand. Reference to the rotation which occurs in the early stages of the development of the limbs (p. 153) will explain how it comes about that, whereas the thumb is the most lateral of the digits in the hand (when in the 'anatomical position' but not, of course, in the much more 'functional position' of medial rotation of the shoulder joint combined with some degree of flexion at the elbow joint and a mid-prone position of the forearm bones and hand), the great toe is the most medial in the foot. (See also p. 350.)

The Tarsus

The seven bones of the tarsus make up the skeleton of the posterior half of the foot (**3**.202, 203). The tarsus is homologous with the carpus, but its constituent elements are larger and stronger, on account of the part they play in supporting and distributing the weight of the body. As in the carpus the tarsal bones are arranged in proximal and distal rows, but an additional element is interposed between the two rows on the medial side. The proximal row comprises the *talus* and the *calcaneus*. These do not lie side by side; the talus is above the calcaneus, but its long axis is directed forwards, medially and downwards, and its anterior end or head is medial to the calcaneus, though at a higher level. The distal row contains, from medial to lateral side, the *medial cuneiform, intermediate cuneiform, lateral cuneiform* and the *cuboid*. These lie side by side and together contribute to the formation of a transverse arch, convex dorsally. On the medial side the *navicular bone* is interposed between the head of the talus and the three cuneiforms. Laterally the calcaneus articulates directly with the cuboid.

The foot is set at right angles to the leg in standing, and the tarsus and metatarsus are arranged in such a way as to form intersecting longitudinal and transverse arches. As a result thrust and weight are not transmitted to the ground from the tibia directly through the tarsus, but are distributed through the tarsal and is

metatarsal bones to the extremities of the arches (*see* p. 617). Each of the tarsal bones is roughly cuboidal in form and hence presents six surfaces.

The Individual Tarsal Bones

The Talus

The talus (**3**.204 A–D) is the connecting link between the rest of the foot and the bones of the leg, through its part in the formation of the ankle joint. The rounded head at the distal end of the bone, the trochlear surface for the tibia on the dorsal surface, the large triangular facet for the lateral malleolus on the lateral surface, the neck, and the body are the main features of the bone.

The head of the talus, directed distally and slightly downwards and medially, has a *distal surface*, which is oval and convex, with its long axis also directed downwards and medially to articulate with the proximal surface of the navicular bone. Its *plantar surface* has three articular areas, separated by indistinct ridges. Of these the most posterior is largest; oval in outline, it is slightly convex and rests on the upper surface of a shelf-like projection from the medial side of the calcaneus, the *sustentaculum tali*. Anterior and lateral to this area, and usually continuous with it, a flattened articular facet rests on the anteromedial part of the upper surface of the calcaneus; it is continuous in front with the navicular surface. Medial to the two calcanean facets a part of the head of the talus is covered with articular cartilage and is continuous on the one hand with the calcanean areas and on the other with the navicular area (**3**.204 B). It is in contact with the *plantar calcaneonavicular ligament* (p. 497). When the foot is inverted passively, the dorsal and lateral part of the head can be both seen and felt; it lies 3 cm in front of the lower end of the tibia and becomes obscured by the extensor tendons when the toes are dorsiflexed.

The neck of the talus is the slightly constricted part which connects the head to the body. It is set very obliquely on the body and inclines medially. Its roughened surfaces give attachment to ligaments, and the medial part of its plantar surface exhibits a deep groove, the *sulcus tali*. When the talus and calcaneus are articulated together this groove forms the roof of a bony canal, the *sinus tarsi*, which is occupied by the interosseous talocalcanean and cervical ligaments.

The body of the talus is cuboidal in shape. Its dorsal surface

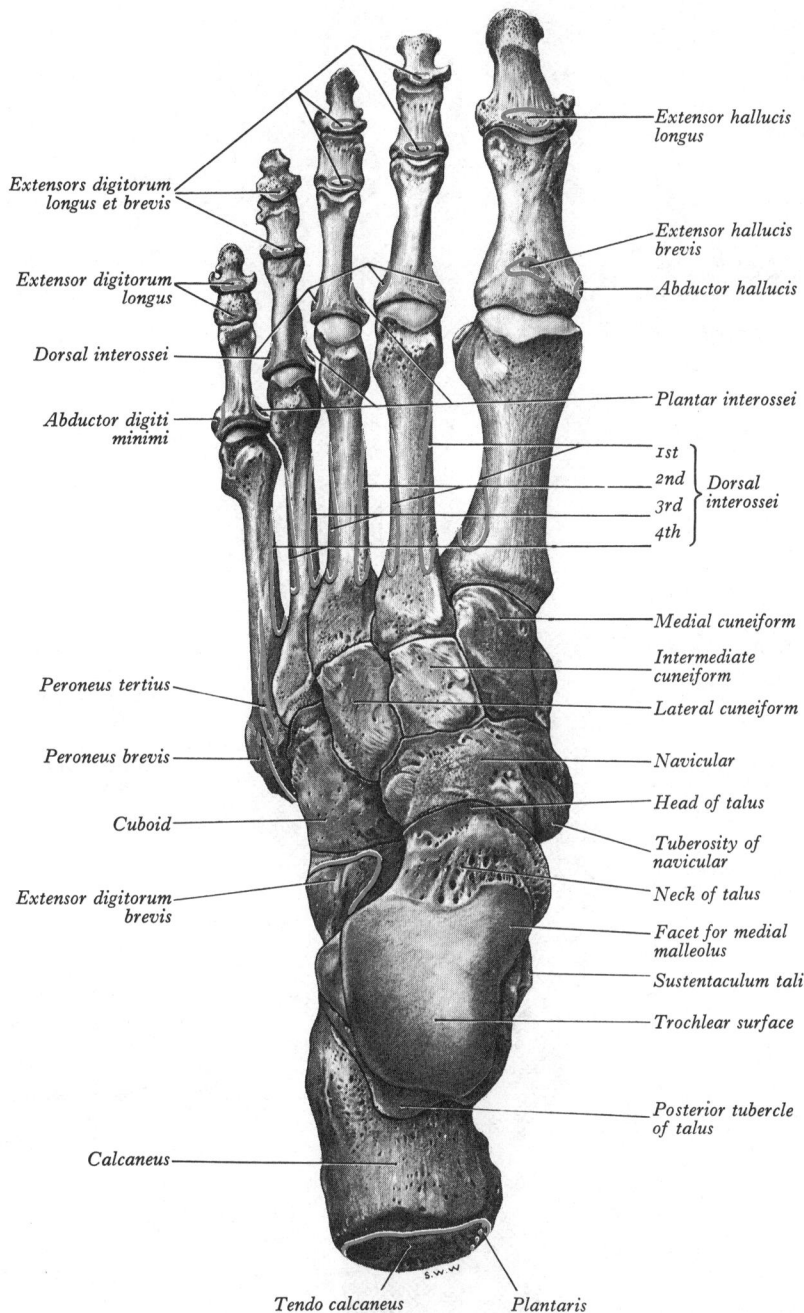

Extensor hallucis longus

Extensor hallucis brevis

Abductor hallucis

Extensors digitorum longus et brevis

Extensor digitorum longus

Dorsal interossei

Abductor digiti minimi

Plantar interossei

1st
2nd
3rd
4th
} Dorsal interossei

Medial cuneiform

Intermediate cuneiform

Lateral cuneiform

Navicular

Head of talus

Tuberosity of navicular

Neck of talus

Facet for medial malleolus

Sustentaculum tali

Trochlear surface

Posterior tubercle of talus

Peroneus tertius

Peroneus brevis

Cuboid

Extensor digitorum brevis

Calcaneus

Tendo calcaneus

Plantaris

3.202 The skeleton of the left foot. Dorsal aspect. Attachments of extrinsic and intrinsic muscles.

is covered by a *trochlear* articular surface, which articulates with the lower end of the tibia in the ankle joint. It is convex from before backwards and gently concave from side to side, and it is widest anteriorly. The *lateral surface*, triangular in outline, is smooth for articulation with the lateral malleolus, and is concave from above downwards. Superiorly it is continuous with the trochlea; inferiorly its apex forms the *lateral process* of the talus. The *medial surface* is covered in its upper part by a comma-shaped articular facet which is deeper in front than behind and articulates with the medial malleolus. Below this facet the surface is rough and is pitted by numerous vascular foramina. The *posterior surface* is rough, small in extent and projects to form the *posterior process* of the bone. It is marked by an oblique groove placed between two tubercles. The *lateral tubercle*, usually the larger, is on the lateral side of the groove; the *medial tubercle* is less prominent and lies immediately behind the sustentaculum tali of the calcaneus (**3.**203). The *plantar surface* rests on the dorsal surface of the calcaneus and is covered with an oval concave facet,

the long axis of which runs distally and laterally at an angle of about 45° with the median plane.

The talus is devoid of any muscular attachments, but it provides attachment for numerous ligaments (**4.**86, 87), for it takes part in the formation of the talocrural, subtalar and talocalcaneonavicular joints.

The long axis of the neck is inclined downwards, distally and medially, making an angle of about 150° with the long axis of the body. This angle is smaller (130°–140°) in the newly born (**3.**205), and helps to account for the inverted position of the foot in young children. The dorsal surface of the neck gives attachment distally to the dorsal talonavicular ligament and the articular capsule of the ankle joint; the proximal part of this surface therefore lies within the capsular ligament of the ankle joint. The medial articular facet and, less frequently, part of the trochlear articular surface may extend forwards on to the neck, especially in the child (**3.**205). A *squatting facet* is commonly found on the upper and lateral part of the neck in Indians but seldom in adult Europeans.

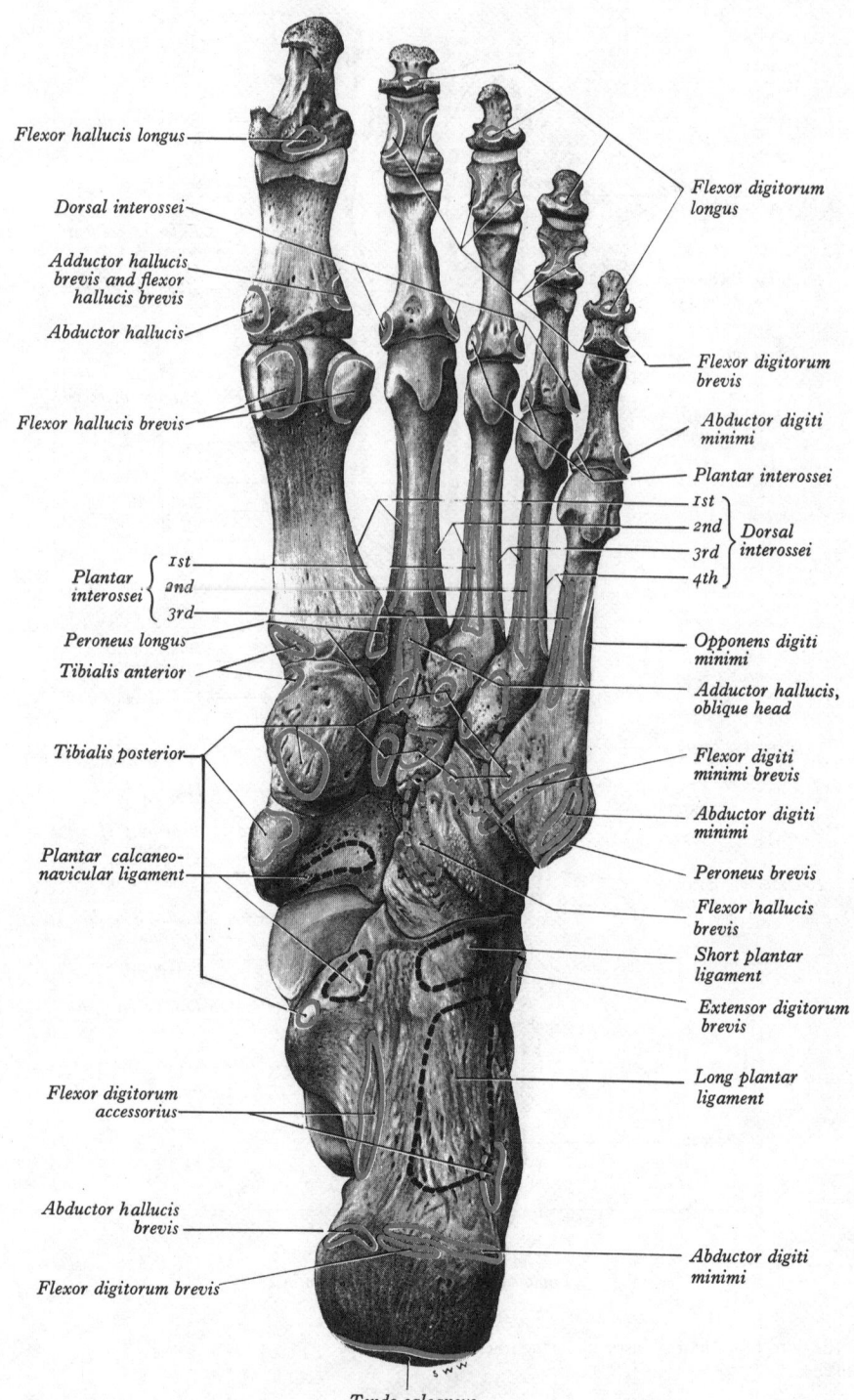

Flexor hallucis longus

Dorsal interossei

Adductor hallucis
brevis and flexor
hallucis brevis

Abductor hallucis

Flexor hallucis brevis

Plantar { 1st
interossei { 2nd
          { 3rd

Peroneus longus

Tibialis anterior

Tibialis posterior

Plantar calcaneo-
navicular ligament

Flexor digitorum
accessorius

Abductor hallucis
brevis

Flexor digitorum brevis

Flexor digitorum
longus

Flexor digitorum
brevis

Abductor digiti
minimi

Plantar interossei

1st
2nd } Dorsal
3rd } interossei
4th

Opponens digiti
minimi

Adductor hallucis,
oblique head

Flexor digiti
minimi brevis

Abductor digiti
minimi

Peroneus brevis

Flexor hallucis
brevis

Short plantar
ligament

Extensor digitorum
brevis

Long plantar
ligament

Abductor digiti
minimi

Tendo calcaneus

3.203   The skeleton of the left foot. Plantar aspect. The attachments of the tibialis posterior to the metatarsals vary; those to the third and fifth may be absent. The adductor hallucis (oblique head) and the flexor hallucis brevis arise in large part from the ligaments and tendinous extensions in the sole of the foot and not directly from bone; these attachments are shown by interrupted lines.

It articulates with the anterior margin of the tibia in extreme dorsiflexion (Barnett 1954; Singh 1959). The facet may be double, forming *medial* and *lateral talar facets*. The lateral part of the neck gives attachment to the anterior talofibular ligament, which extends downwards along the neighbouring anterior border of the lateral surface. The inferior surface of the neck gives attachment to the interosseous talocalcanean and cervical ligaments (p. 495).

The medial border of the *trochlear articular surface* is straight but its *lateral border* inclines medially at its posterior part, which is often broadened and flattened to form a small elongated triangular area. It is this part of the bone which comes into contact with the posterior tibiofibular ligament in dorsiflexion of the ankle joint.

The *posterior process* receives the attachment of the posterior

talofibular ligament, which extends upwards to the groove, or depression, between the process and the posterior border of the trochlea. Its plantar border gives attachment to the posterior talocalcanean ligament. The *groove* between the two tubercles of the process contains the tendon of the flexor hallucis longus and is continuous below and in front with the groove on the plantar surface of the sustentaculum tali. The *medial tubercle* gives attachment on its medial aspect to the medial talocalcanean ligament below, and to the most posterior of the superficial fibres of the deltoid ligament above.

To the roughened area below the comma-shaped articular facet on the medial surface the deep fibres of the deltoid ligament are attached.

Sex differences in the dry weight of the talus have been

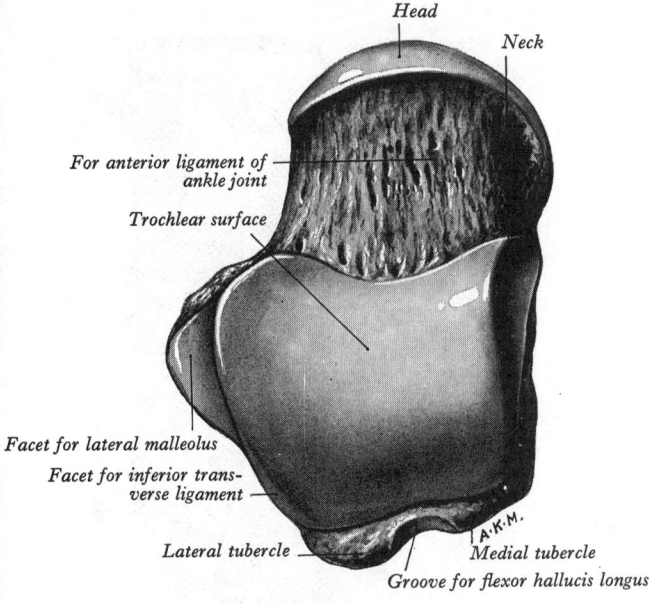

Head

Neck

For anterior ligament of
ankle joint

Trochlear surface

Facet for lateral malleolus

Facet for inferior trans-
verse ligament

Lateral tubercle

Medial tubercle

Groove for flexor hallucis longus

A   Dorsal (superior) aspect.

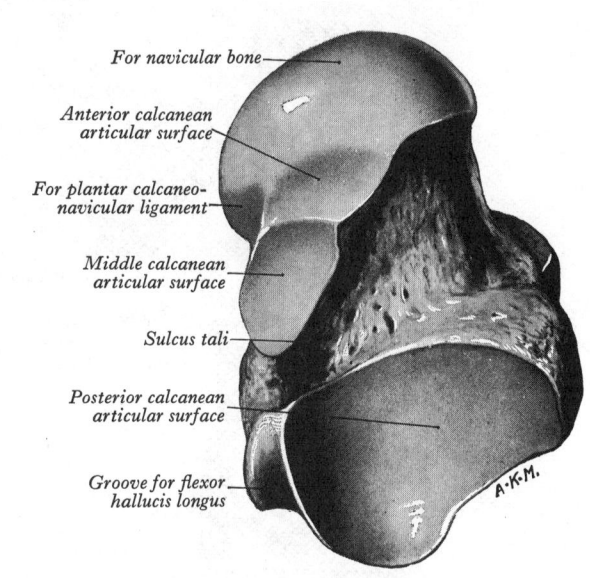

For navicular bone

Anterior calcanean
articular surface

For plantar calcaneo-
navicular ligament

Middle calcanean
articular surface

Sulcus tali

Posterior calcanean
articular surface

Groove for flexor
hallucis longus

B   Plantar (inferior) aspect.

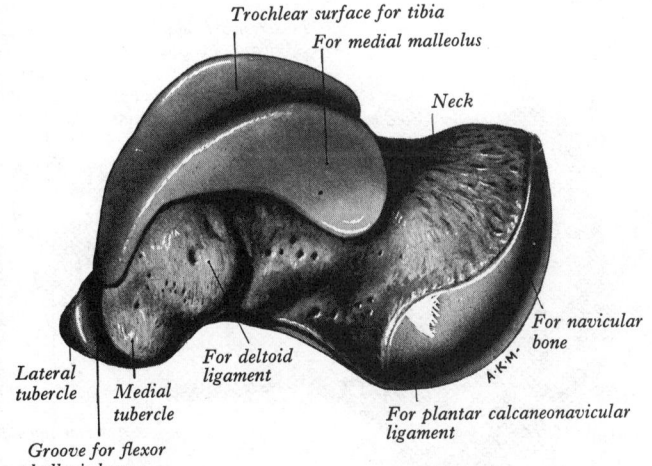

Trochlear surface for tibia

For medial malleolus

Neck

For navicular
bone

For deltoid
ligament

Lateral
tubercle

Medial
tubercle

For plantar calcaneonavicular
ligament

Groove for flexor
hallucis longus

C   Medial aspect.

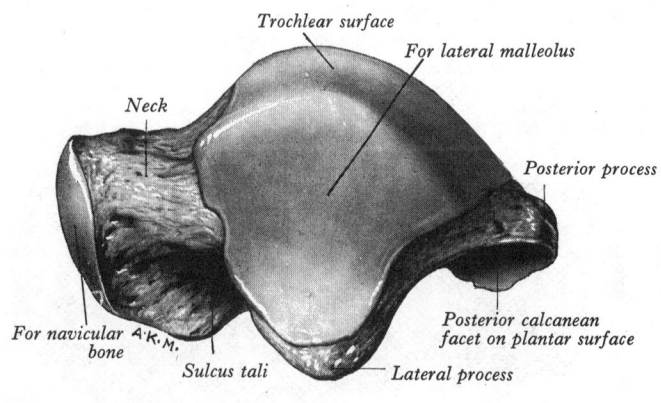

Trochlear surface

For lateral malleolus

Neck

Posterior process

For navicular
bone

Sulcus tali

Posterior calcanean
facet on plantar surface

Lateral process

D   Lateral aspect.

3.204A–D   THE LEFT TALUS

recorded (Singh and Singh 1975); the ranges in adult Indian males and females were 15·1 to 36·8 and 6·0 to 20·5 gm; the respective means being 23·50 and 15·32 gm. Estimation of sex on dimensional data has been claimed by Steele (1976).

The Calcaneus

The calcaneus (**3**.206A–D) is the largest of the tarsal bones; it projects backwards beyond the bones of the leg so as to provide a short lever for the muscles of the calf, which are inserted into its posterior surface. It is irregularly cuboidal in shape, and its long axis is directed forwards, upwards and somewhat laterally. The small articular anterior end contrasts with the larger, roughened, posterior end. The dorsal surface bears a large, articular facet about its middle which distinguishes it from the rough plantar surface. Finally the lateral surface is flattened, whereas the medial surface is hollowed out from above downwards and backwards.

   The dorsal or superior surface is divisible into three areas. The posterior third is roughened, convex from side to side and concave from behind forwards; it supports a mass of fibro-adipose

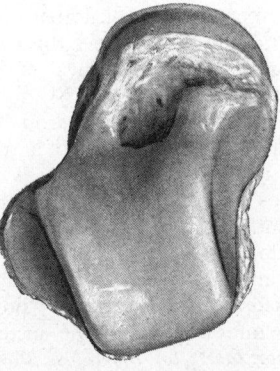

3.205   The left talus of a newborn infant. Superior aspect. Compare with **3**.204A, and note the angle which the axis of the neck makes with the long axis of the body of the bone.

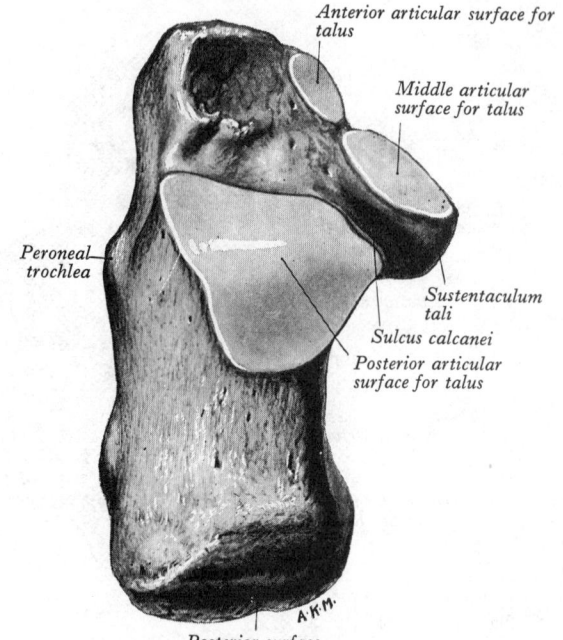

A   Dorsal aspect.

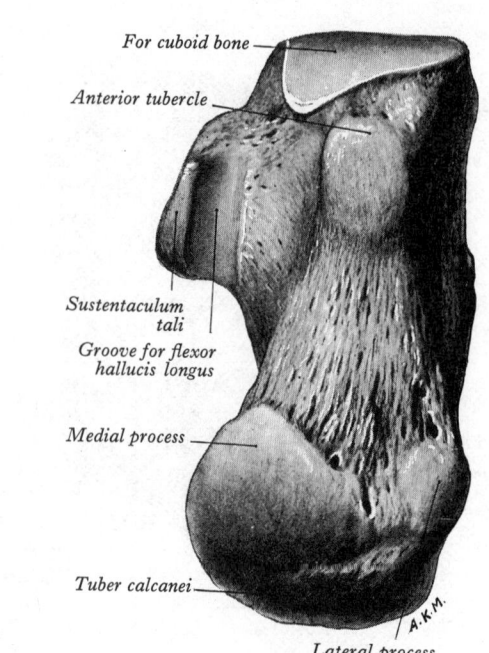

B   Plantar aspect.

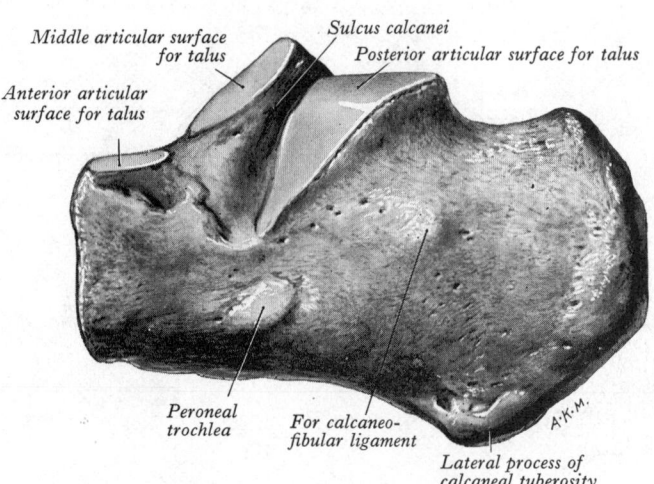

C   Lateral aspect.

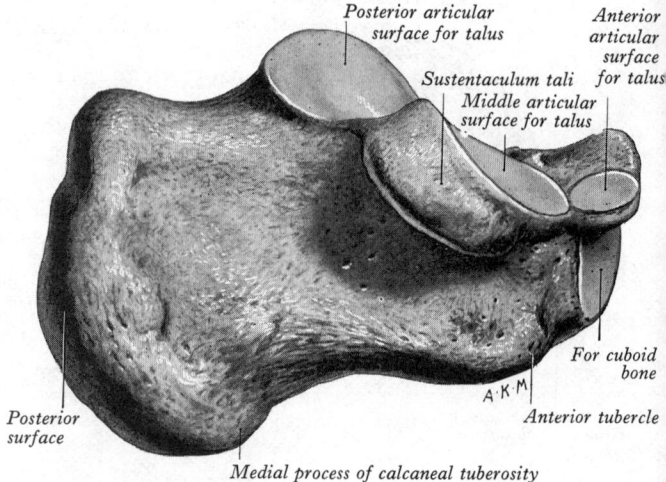

D   Medial aspect.

3.206A–D   THE LEFT CALCANEUS

tissue located between the tendo calcaneus and the back of the ankle joint. The middle third carries the *posterior facet for the talus*, which is oval in outline and convex anteroposteriorly. The anterior third is partly articular and partly non-articular. In front of the posterior articular facet there is a rough depression, which becomes narrower and takes the form of a groove on the medial side. This is termed the *sulcus calcanei*; it corresponds to the sulcus tali and helps it to complete the sinus tarsi in the articulated foot. In front of and medial to this groove an elongated articular area covers the dorsal surface of the sustentaculum tali, and extends forwards and laterally on to the body of the bone. This area is often divided into two by a narrow non-articular interval which marks the anterior limit of the sustentaculum tali, the *middle* and *anterior facets for the talus*. The incidence of this division varies with sex and race. Rarely, all three facets on the upper surface of the talus are fused into one large irregular area (Bunning and Barnett 1963, 1965). Recently, a detailed analysis of

the patterns of the *anterior* facets for articulation with the head of the talus has been made on a series of 401 normal Indian calcanei: four types were described. Type I (67 per cent) showed one continuous facet on the sustentaculum which extended to the anteromedial corner of the distal calcaneal end; Type II (26 per cent) presented two facets, one sustentacular and one distal calcaneal; Type III (5 per cent) possessed only a single sustentacular facet; and Type IV (2 per cent) was similar to Type I, but was also confluent with the posterior talar facet of the calcaneus.

The anterior surface is the smallest of the six surfaces, and is entirely covered by an obliquely set concavoconvex facet which articulates with the cuboid bone.

The posterior surface is divided into three areas. The uppermost is smooth and separated from the tendo calcaneus by a bursa and some fatty tissue. The middle area is the largest and is limited above by a groove and below by a rough ridge; it is for the tendo calcaneus. The lowest area slopes downwards and forwards and is vertically striated; it is the subcutaneous weight-bearing surface.

The plantar surface is rough, especially posteriorly, where it bears the *calcanean tuberosity*. The *lateral* and *medial processes* of the tuberosity extend distally for a short distance, separated by a notch. The medial process is the broader of the two (**3.206**B). More distally a tubercle marks the distal limit of the attachment of the long plantar ligament.

The lateral surface is almost flat and is deeper behind than in front; it can be palpated on the lateral aspect of the heel and can be traced forwards below the lateral malleolus. In its anterior part it presents a small elevation, termed the *peroneal trochlea* (*tubercle*) (**3.206**C), which is exceedingly variable in size and, when well developed, can be felt in the living subject 2 cm below the tip of the lateral malleolus. It exhibits an oblique groove on its postero-inferior aspect, for the peroneus longus tendon, and a shallower anterosuperior groove, for the peroneus brevis tendon. About 1 cm or more behind the peroneal trochlea a second elevation may mark the bone; it is for attachment of the calcaneofibular part of the lateral ligament of the ankle joint.

The medial surface is concave from above, downwards and backwards, and its concavity is accentuated by a shelf-like process, the *sustentaculum tali*, which projects medially from the anterior part of its upper border (**3.206**D). The superior surface of this process bears the middle facets for the talus, and its inferior surface is marked by a groove which is continuous with the groove on the posterior surface of the talus and is adapted to the flexor hallucis longus tendon (**3.206**B). The medial surface of the sustentaculum tali can be felt indistinctly through the skin immediately below the tip of the medial malleolus; occasionally it presents a groove for the flexor digitorum longus tendon.

The *sulcus of the calcaneus* gives attachment to the interosseous talocalcanean and cervical ligaments (p. 495) and to the medial root of the inferior extensor retinaculum. In addition, the non-articular area in front of the posterior facet for the talus is part of the attachment of extensor digitorum brevis and the attachment of the principal band of the inferior extensor retinaculum and the stem of the bifurcated ligament.

The *medial process of the tuberosity* affords attachment by its prominent medial margin to abductor hallucis and the superficial part of the flexor retinaculum, and in front to the plantar aponeurosis and the flexor digitorum brevis. To the *lateral process* is attached abductor digiti minimi, which also extends medially to the front of the medial process. The roughened region between the two processes, proximally, and the tubercle, distally, gives attachment to the long plantar ligament, while the short plantar ligament springs from the tubercle and the narrow rough area distal to it. The lateral tendinous head of the flexor digitorum accessorius arises in front of the lateral process close to the lateral margin of the long plantar ligament.

The *posterior surface* is wider below than above. Close to the medial side of the attachment of the tendo calcaneus the muscle plantaris terminates.

The anterior part of the *lateral surface* is crossed by the peroneal tendons, but in most of its extent is covered only by the skin and superficial fascia. The peroneus brevis tendon, after passing behind the lateral malleolus, runs forwards and slightly downwards above and in front of the *peroneal trochlea*; the peroneus longus tendon passes downwards and forwards below and behind the trochlea, which gives attachment to a slip from the inferior peroneal retinaculum (p. 611). The calcaneofibular ligament is attached to the bone about 1 cm behind the peroneal trochlea, and the site is usually indicated by a low rounded elevation.

The *sustentaculum tali* assists, by its dorsal surface, in the formation of the talocalcaneonavicular joint; its plantar surface is grooved by the flexor hallucis longus tendon, and the margins of the groove give attachment to the deep part of the flexor retinaculum. The medial margin of the process is narrow, roughened, and convex from before backwards. Anteriorly the plantar calcaneonavicular ligament is attached to it (p. 497), behind that a slip from the tibialis posterior tendon and some of the superficial fibres of the deltoid ligament, and posteriorly the

medial talocalcanean ligament. Below the attachment of the deltoid ligament the tendon of the flexor digitorum longus is related to this aspect of the process and sometimes its position is indicated by a groove. Below the groove for the flexor hallucis longus the medial surface gives origin to the large, fleshy, medial head of the flexor accessorius.

Singh and Singh (1975) have studied variations in the dry weight of the calcaneus in both sexes in an adult Indian population. The male range was 23·3 to 56·8 gm (mean=36·64), and 13·0 to 33·0 gm (mean=24·75) in females. Steele (1976) has advocated sex estimation on dimensional data.

The calcaneus has been the subject of an interesting comparison between the right and left bones (Webber and Garnett 1976); comparisons in 52 adults showed a significantly greater density of the calcaneus corresponding to the side of hand preference. It was concluded that the concordance between upper limb dominance and the increased density of the ipsilateral os calcis had been determined during fetal life.

The Navicular Bone

The navicular bone (**3.207**) is interposed between the head of the talus proximally and the cuneiform bones distally.

The *distal surface* is convex from side to side and is divided into three facets (of which the most medial is the largest) for articulation with the three cuneiform bones. The *proximal surface*, oval and concave, articulates with the head of the talus. The *dorsal surface* is roughened and convex from side to side. The *medial surface* also is rough and is continued downwards to form a prominent projection termed the *tuberosity*. It can be felt through the skin about 2·5 cm below and in front of the medial malleolus. The *plantar surface*, also rough, is concave, and separated from the tuberosity on the medial side by a groove. The *lateral surface* is rough and irregular, but frequently presents a facet for articulation with the cuboid.

The facet for the medial cuneiform is triangular, with its rounded apex on the medial side. The facets for the intermediate

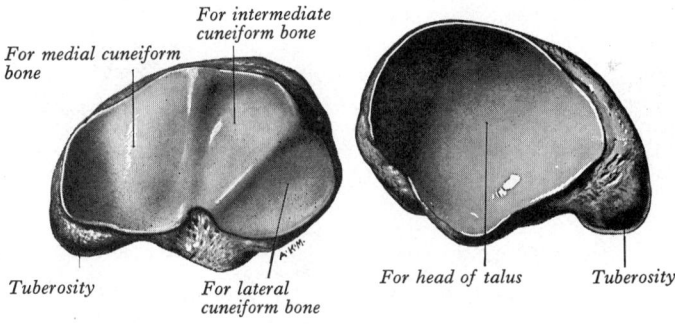

For intermediate cuneiform bone

For medial cuneiform bone

Tuberosity

For lateral cuneiform bone

For head of talus   Tuberosity

A   Distal aspect.   B   Proximal aspect.

3.207A, B   THE LEFT NAVICULAR BONE

and lateral cuneiform bones are also triangular, but their apices point towards the sole. To the dorsal surface are attached the dorsal talonavicular, cuneonavicular and cubonavicular ligaments. The **tuberosity** of the navicular bone provides the principal attachment for the tibialis posterior tendon, and the groove which lies on its lateral side transmits the part of this tendon which runs forward to reach the cuneiform bones and the bases of the middle three metatarsal bones. A slight projection marks the plantar surface of the bone on the lateral side of the groove. Together with the proximal part of this surface it provides attachment to the plantar calcaneonavicular ligament. To the roughened part of the lateral surface is attached the calcaneo-navicular part of the bifurcated ligament.

The Cuneiform Bones

The three cuneiform bones (**3.208, 209, 210**) are wedge-shaped and articulate with the navicular bone proximally and the bases of 411

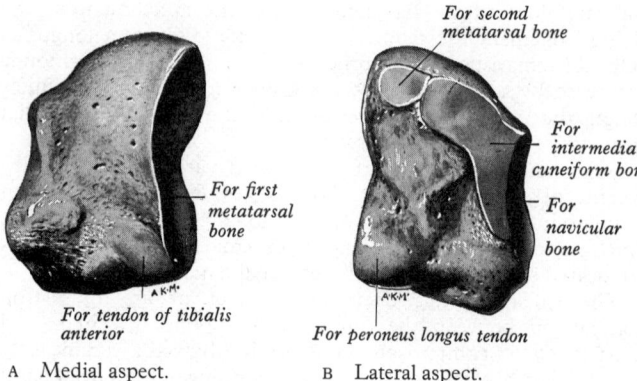

A Medial aspect.

B Lateral aspect.

For first
metatarsal
bone

For tendon of tibialis
anterior

For second
metatarsal bone

For
intermediate
cuneiform bone

For
navicular
bone

For peroneus longus tendon

3.208A, B   THE LEFT MEDIAL CUNEIFORM BONE

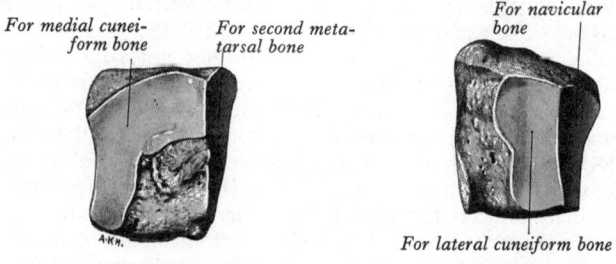

For medial cunei-
form bone

For second meta-
tarsal bone

For navicular
bone

For lateral cuneiform bone

A Distal and medial aspect.

B Proximal and lateral aspect.

3.209A, B   THE LEFT INTERMEDIATE CUNEIFORM BONE

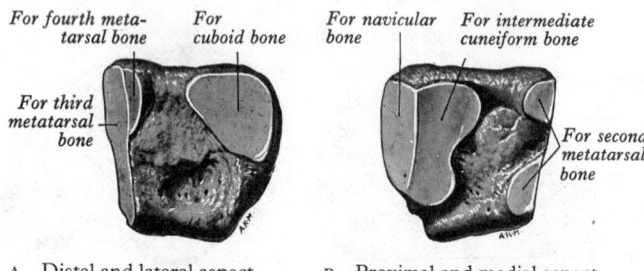

For fourth meta-
tarsal bone

For
cuboid bone

For third
metatarsal
bone

For navicular
bone

For intermediate
cuneiform bone

For second
metatarsal
bone

A Distal and lateral aspect.

B Proximal and medial aspect.

3.210A, B   THE LEFT LATERAL CUNEIFORM BONE

the first, second and third metatarsal bones distally. The medial cuneiform is the largest, the intermediate the smallest. In the intermediate and in the lateral cuneiform bones the dorsal surface is the base of the wedge and the plantar surface represents the edge, but in the medial cuneiform the wedge is reversed so that its edge is represented by the narrow, dorsal surface of the bone. This arrangement is an important factor in the construction of the transverse arch of the foot. The proximal surfaces of the three cuneiform bones form a slight concavity for the navicular, but the distal parts of the medial and lateral cuneiforms project distally beyond the intermediate cuneiform, and bound a deep recess into which the base of the second metatarsal bone projects.

The medial cuneiform bone (**3.**208) articulates with the navicular proximally and with the base of the first metatarsal distally. The *dorsal surface* is rough and narrow. The *plantar surface* receives a substantial slip from the tendon of tibialis posterior. The *distal surface* bears a large reniform facet for articulation with the base of the first metatarsal bone, the little notch representing the hilum being on its lateral margin. The *proximal surface* bears a piriform facet, which articulates with the navicular bone. Concave in its vertical axis, it is narrower

dorsally. The *medial surface* is rough and subcutaneous; it is slightly convex in its vertical axis and its distal-plantar angle is marked by a large, flattened impression for most of the fibres of the tibialis anterior tendon (**3.**208A). The *lateral surface* is partly articular and partly non-articular. Along its proximal and dorsal margins it is covered with a smooth Γ-shaped strip which articulates with the intermediate cuneiform bone. The dorsal and distal part of this strip is separated by a vertical ridge from a small, almost square, facet which articulates with the dorsal part of the medial surface of the base of the second metatarsal bone. Below this facet the medial cuneiform is attached to the medial side of the base of the second metatarsal bone by a strong interosseous ligament, and proximally the interosseous intercuneiform ligament connects it to the medial side of the intermediate cuneiform. The distal and plantar part of the lateral surface is roughened where it receives the insertion of part of the peroneus longus tendon (**3.**208B).

The intermediate cuneiform bone (**3.**209) articulates distally with the base of the second metatarsal and proximally with the navicular bone. It is of regular wedge-like form, the base of the wedge forming the roughened, *dorsal surface*, and the edge the narrow, *plantar surface*, which receives a slip from the tibialis posterior tendon. The *distal* and *proximal surfaces* are each covered with a triangular articular facet, for the base of the second metatarsal and the navicular bone respectively. The *medial surface* is partly articular and partly non-articular. Along its proximal and dorsal margins it bears a smooth, Γ-shaped strip, occasionally divided into two, articulating with the medial cuneiform. The *lateral surface* also is partly articular and partly non-articular. Along its proximal margin is a vertical strip, usually indented at its middle, for articulation with the lateral cuneiform. Strong interosseous ligaments connect the non-articular parts of the lateral and medial surfaces to the lateral and medial cuneiform bones, respectively.

The lateral cuneiform bone (**3.**210) is placed between the intermediate cuneiform and the cuboid. Distally it articulates with the base of the third metatarsal and proximally with the navicular bone. Like the intermediate cuneiform, its *dorsal surface*, rough and almost rectangular, represents the base of the wedge, and its *plantar surface*, narrow and rough, the edge. The latter receives a slip from the tibialis posterior tendon and may share in the origin of flexor hallucis brevis. The *distal surface* is completely covered with a triangular articular facet for the base of the third metatarsal bone. The *proximal surface* is rough in its plantar part but is smooth and articular in its dorsal two-thirds, which articulate with the navicular by means of a triangular facet. The *medial surface* is partly articular, partly non-articular. Along its proximal margin is a vertical strip, indented at its middle, for the intermediate cuneiform, and along its distal margin a narrower strip, often divided into two small facets, serves for articulation with the lateral aspect of the base of the second metatarsal bone. The *lateral surface* also is partly articular and partly non-articular. A large triangular or oval facet is situated proximally for articulation with the cuboid; a small facet, semi-oval in shape, is placed at the dorsal part of its distal margin for articulation with the dorsal part of the medial side of the base of the fourth metatarsal bone. The non-articular portions of the medial and lateral surfaces provide attachment for strong intercuneiform and cuneocuboid ligaments, which play an important part in maintaining the transverse arch of the foot.

The Cuboid Bone

The cuboid bone (**3.**211) is the most lateral one in the distal row of the tarsus, and is situated between the calcaneus proximally and the fourth and fifth metatarsal bones distally.

The *dorsal surface*, directed laterally as well as dorsally, is rough for ligamentous attachments. The distal part of the *plantar surface* is crossed by an oblique *groove for the peroneus longus tendon*, bounded proximally by a prominent ridge. This ridge ends laterally in an enlargement, the *tuberosity of the cuboid bone*. The lateral aspect of the tuberosity is faceted for the sesamoid bone or cartilage which is frequently found in the peroneus longus tendon. Proximal to the ridge of the plantar surface is rough and, owing to the obliquity of the calcaneocuboid joint, extends

Occasional facet for   *For lateral cunei-*
navicular bone   *form bone*

A   Medial aspect.

Groove for peroneus
longus tendon

Facet on tuberosity, for   *For*
sesamoid bone in peroneus   *calcaneus*
longus tendon

B   Proximal and lateral aspect.

3.211 A, B   THE LEFT CUBOID BONE

proximally and medially so that the medial border of this surface is much longer than the lateral border. The *lateral surface* is rough and exhibits a deep notch on its plantar edge which marks the commencement of the groove for the peroneus longus tendon. The *medial surface* is much more extensive and is partly articular and partly non-articular. It is marked near its middle by a smooth oval facet for articulation with the lateral cuneiform bone; proximal to this a small facet for the navicular bone is sometimes present. The two form a continuous articular surface but are separated by a vertical ridge. The *distal surface* is divided by a vertical ridge into two articular areas; the medial facet is quadrilateral and articulates with the base of the fourth metatarsal bone; the lateral facet, triangular in outline, with the apex on the lateral side, articulates with the base of the fifth metatarsal. The *proximal surface*—smooth, triangular and concavoconvex—articulates with the distal surface of the calcaneus; its medial plantar angle projects proximally as a process which helps to support the distal end of the calcaneus.

To the dorsal surface are attached the dorsal calcaneocuboid, cubonavicular, cuneocuboid and cubometatarsal ligaments. The ridge on the plantar surface gives attachment to the deep fibres of the long plantar ligament, which conceals the attachment of the short plantar ligament to the proximal border of this surface. The projecting proximomedial part of the plantar surface receives a slip from the tendon of the tibialis posterior and gives origin to the flexor hallucis brevis. The rough part of the medial surface of the cuboid affords attachment for the interosseous cuneocuboid and cubonavicular ligaments, and in its proximal part to the medial calcaneocuboid, which is the lateral limb of the bifurcated ligament.

The Metatarsus

The metatarsus consists of five metatarsal bones situated in the distal part of the foot, connecting the tarsus to the phalanges. They are numbered from the medial side. Like the metacarpals, the metatarsals are miniature long bones, and each possesses a shaft, a base or proximal end, and a head or distal end. With the exception of the first and, to a lesser degree, the fifth, the *shafts* are long and slender, and are slightly convex longitudinally on their dorsal aspects and concave on their plantar aspects. They are prismatic in section and taper from base to head. The *bases* articulate with the distal row of the tarsus and with one another. The line of each tarsometatarsal joint, excluding the first, passes proximally and laterally, and the bases of the metatarsals are therefore set somewhat obliquely relative to their shafts. The *heads* articulate with the proximal phalanges of their own digits, each by means of a convex articular surface which extends farther on the plantar than on the dorsal surface; the plantar extension ends on each side on the summit of a slight eminence. The sides of

the heads are flattened, and each shows a depression surmounted dorsally by a tubercle, for one of the collateral ligaments of the metatarsophalangeal joint.

The Individual Metatarsal Bones

The first metatarsal bone (**3.**212) is the shortest and thickest. Its *body* is strong, and of well-marked prismatic form. The *base* has no articular facets on its sides, but there is occasionally a facet or ill-defined smooth area on the lateral side caused by contact with the second metatarsal. Its proximal surface, large and usually indented on both its medial and lateral margins, articulates with the medial cuneiform. Its circumference is grooved for its tarsometatarsal ligaments, and medially provides attachment for a part of the tendon of the tibialis anterior; its plantar angle presents a rough, oval prominence on its lateral aspect for the tendon of

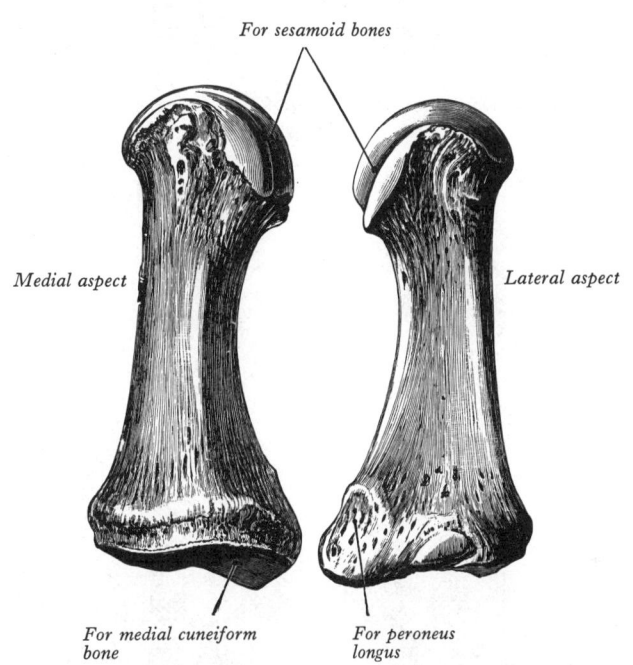

For sesamoid bones

Medial aspect   *Lateral aspect*

For medial cuneiform   *For peroneus*
bone   *longus*

3.212   The left first metatarsal bone. Medial and lateral aspects.

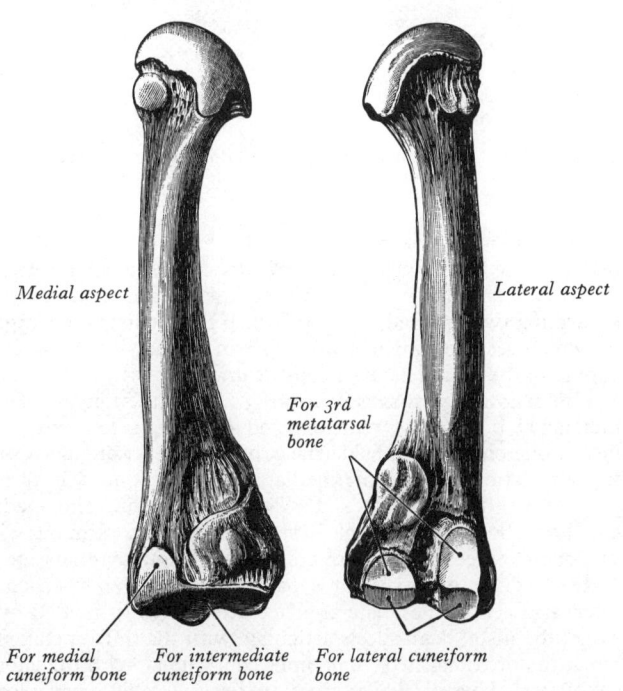

Medial aspect   *Lateral aspect*

For 3rd
metatarsal
bone

For medial   *For intermediate*   *For lateral cuneiform*
cuneiform bone   *cuneiform bone*   *bone*

3.213   The left second metatarsal bone. Medial and lateral aspects.

413

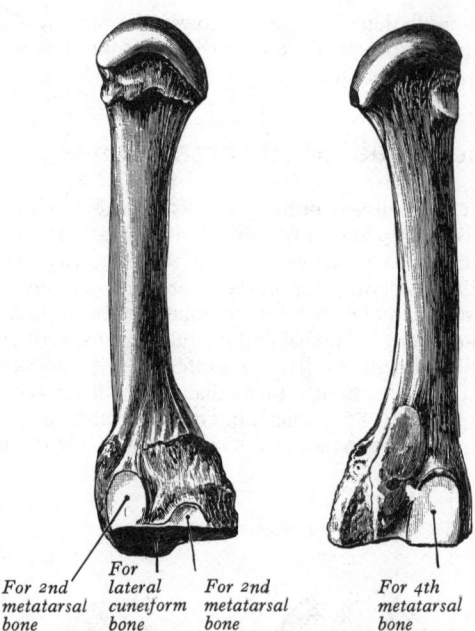

For 2nd   For   For 2nd   For 4th
metatarsal lateral   metatarsal   metatarsal
bone   cuneiform bone   bone
            bone

3.214   The left third metatarsal bone. Medial and lateral aspects.

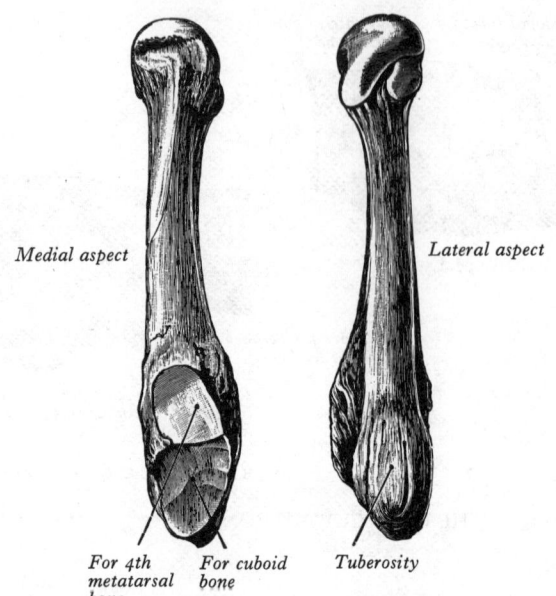

Medial aspect   *Lateral aspect*

For 4th   For cuboid   Tuberosity
metatarsal   bone
bone

3.216   The left fifth metatarsal bone. Medial and lateral aspects.

1960). A pressure facet is occasionally present due to contact with the first metatarsal bone; it is oval in shape, and is situated on the medial side of the base, plantar to the facet for the medial cuneiform. To the medial and lateral surfaces of the shaft are attached respectively the lateral head of the first dorsal interosseous muscle and the medial head of the second.

The third metatarsal bone (**3**.214) has a flat triangular *base*, which articulates proximally with the lateral cuneiform; medially it articulates by two facets, dorsal and plantar, with the second metatarsal bone, and laterally, by a single facet situated at the dorsal angle, with the fourth metatarsal. The plantar facet on the medial side of the base is frequently absent. From the medial surface of the shaft arises the lateral head of the second dorsal interosseous muscle and the first plantar; to the lateral surface gives origin to the medial head of the third dorsal interosseous muscle.

The fourth metatarsal bone (**3**.215) is smaller than the third. The proximal surface of its *base* bears an oblique quadrilateral facet for articulation with the cuboid, on its lateral side a single facet, for the fifth metatarsal bone, and on its medial side an oval facet for the third metatarsal bone, sometimes divided by a ridge, the proximal portion then constituting an articular surface for the lateral cuneiform (**3**.215). The medial surface of the shaft gives origin to the lateral head of the third dorsal and to the second plantar interosseous muscles; the lateral surface gives origin to the medial head of the fourth dorsal interosseous muscle.

The fifth metatarsal bone (**3**.216) has a rough eminence, its *tuberosity* (*styloid process*), on the lateral side of its base. The *base* articulates proximally with the cuboid by a triangular, obliquely cut surface and medially with the fourth metatarsal bone. The tendon of peroneus tertius is attached to the medial part of the dorsal surface, and along the medial border of the shaft, and that of the peroneus brevis on the dorsal surface of the tuberosity. A strong band of the plantar aponeurosis, sometimes containing muscle fibres, connects the projecting part of the tuberosity with the lateral process of the calcaneal tuberosity. The plantar surface of the base is grooved by the tendon of the abductor digiti minimi, and gives origin to the flexor digiti minimi brevis. To the medial side of the shaft are attached the lateral head of the fourth dorsal and the third plantar interosseous muscles. The tuberosity can be both seen and felt, half-way along the lateral border of the foot. In acute inversion of the foot it may be fractured.

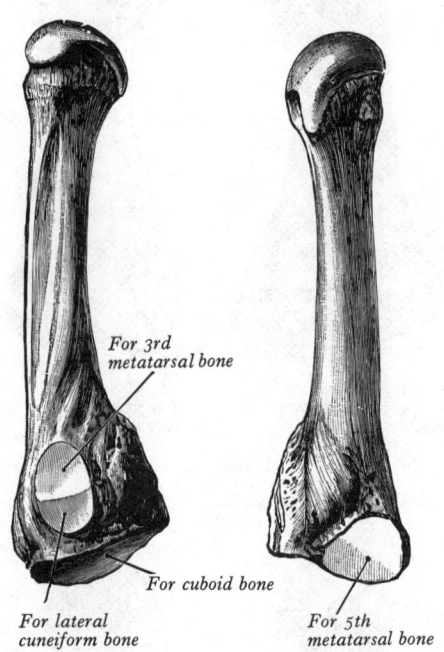

For 3rd
metatarsal bone

For cuboid bone

For lateral   For 5th
cuneiform bone   metatarsal bone

3.215   The left fourth metatarsal bone. Medial and lateral aspects.

peroneus longus. The lateral surface of the shaft is flat and gives origin to the medial head of the first dorsal interosseous muscle. The *head* is large; on its plantar surface there is a median elevation separating two grooved facets, of which the medial is the larger, on which sesamoid bones glide. For an account of the arterial supply of the first metatarsal consult Jaworek (1973).

The second metatarsal bone (**3**.213) is the longest of the metatarsal bones. Its wedge-shaped *base* bears four articular facets: one on its proximal surface, of concave, triangular form, for articulation with the intermediate cuneiform; one at the dorsal part of its medial surface, for articulation with the medial cuneiform bone, very variable in size and usually continuous with the articular area for the intermediate cuneiform; two on its lateral surface, a dorsal and a plantar, separated by a rough non-articular interval. Each of these lateral articular surfaces is divided by a ridge; the distal demi-facets articulate with the third metatarsal bone; the proximal two (sometimes continuous) with the lateral cuneiform. These articular areas are very variable, particularly those on the plantar facet; this facet may even be absent (Singh

The Phalanges of the Foot

The phalanges (**3**.202, 203) of the foot correspond in number and general arrangement with those of the hand; there are two in the llux, and three in.each of the other toes. They are, however, uch shorter, and their shafts, especially those of the bones of the rst row, are compressed from side to side. The proximal halanges closely resemble those of the hand. The *shaft* of each is ompressed from side to side, convex dorsally, concave on its lantar aspect. The *base* is concave for articulation with the head f the corresponding metatarsal bone, and the *head* possesses a ochlear surface for articulation with the middle phalanx. The iddle phalanges are remarkably small and short, but rather roader than the proximal phalanges. The distal phalanges esemble those of the fingers; but they are smaller, and are attened; each presents a broad base for articulation with the orresponding middle phalanx, and an expanded distal extremity. ach bears a roughened *tuberosity* on the plantar aspect of its istal end, which gives attachment to the pulp of the tip of the toe nd provides a wider area to take pressure.

The base of the distal phalanx of each of the lateral four toes ives attachment on its plantar aspect to a tendon of the flexor igitorum longus and on its dorsal aspect to the extensores igitorum. In the hallux the corresponding surfaces have ttached to them the flexor hallucis longus and the extensor allucis respectively. The base of the middle phalanx provides ttachment on each side of its plantar aspect to the flexor igitorum brevis and on its dorsal aspect to the extensores igitorum. The base of the proximal phalanx of each of the econd, third and fourth toes receives the attachment of a umbrical on the medial side and to an interosseous on each side Wood Jones 1949). The base of the proximal phalanx of the fifth oe receives the insertion of a plantar interosseous on the medial ide and of the flexor digiti minimi brevis and abductor digiti ninimi on the lateral side. In addition the margins of the proximal nd middle phalanges of the lateral four toes provide the

attachments for the fibrous flexor sheaths. The base of the proximal phalanx of the big toes receives, on its medial side, the insertion of the abductor hallucis and part of the flexor hallucis brevis and, on its lateral side, of the adductor hallucis and the remainder of the flexor hallucis brevis. The margins of the articular surfaces of all phalanges provide attachments for the capsules and ligaments of the corresponding metatarsophalangeal and interphalangeal joints. The terminal phalanx of the hallux normally shows a small degree of valgus (lateral) deviation; such deviation may also be seen in the proximal phalanx even in persons who have never worn shoes (Barnett 1962). A comparison of the angulation in 30 students (18–21 years) and 35 aged individuals showed no increase in the angle in the latter group. The deviation was also observed in fetal specimens (Wilkinson 1954).

Ossification of the Bones of the Foot

The tarsal bones are each ossified from a single centre, excepting the calcaneus, which has a scale-like epiphysis for its posterior part. The centres for the calcaneus and talus appear in intrauterine life, in the third month (Kleiger and Mankin 1961) and sixth month respectively. (A recent detailed study of early ossification in the calcaneus of 177 human fetuses between 49 and 150 mm crown–rump length showed about 16 per cent possessing a laterally placed perichondral mesenchymatous centre of ossification; some 11 per cent possessed a centrally placed endochondral centre, while only about 2 per cent possessed both—Meyer and O'Rahilly 1976.) The cuboid frequently commences to ossify before birth and its centre is generally present by six months after birth. The medial cuneiform, which may have two centres, begins to ossify in the second year, the intermediate cuneiform and navicular in the third and the lateral

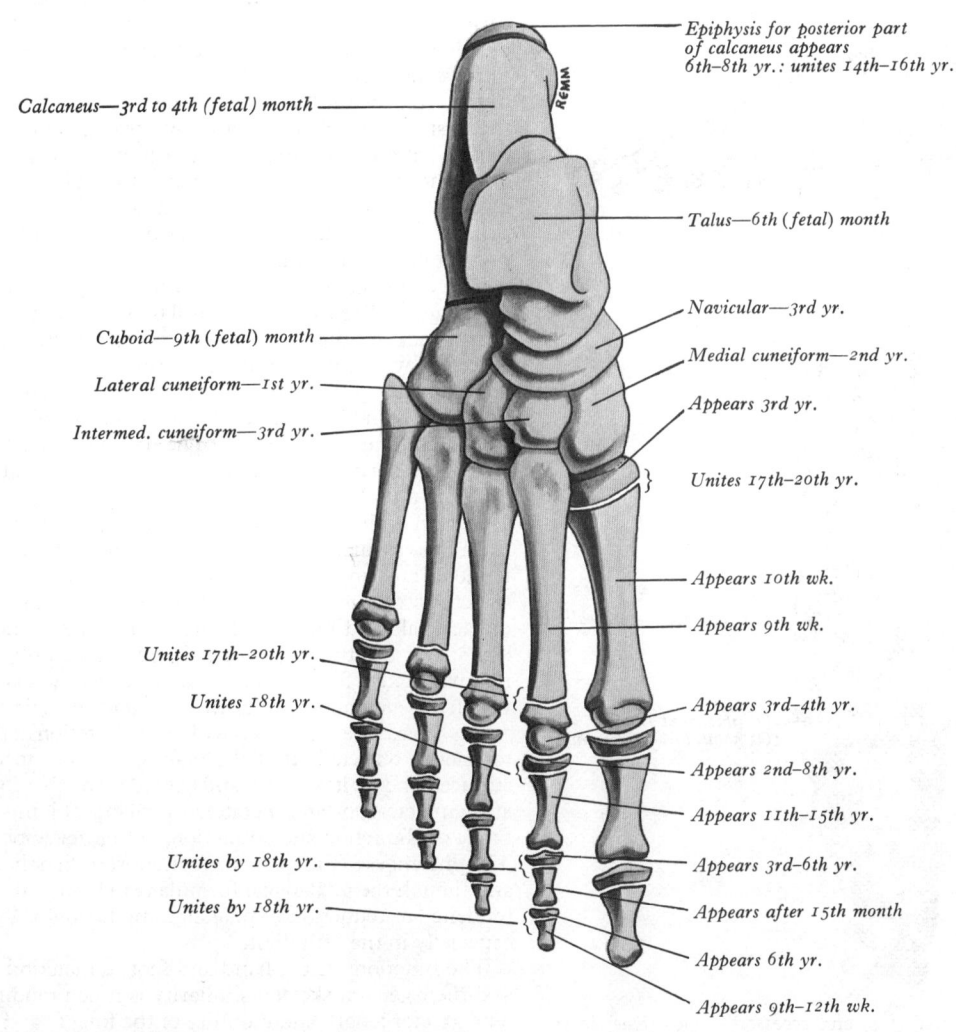

Epiphysis for posterior part of calcaneus appears 6th–8th yr.: unites 14th–16th yr.

Calcaneus—3rd to 4th (fetal) month

Talus—6th (fetal) month

Navicular—3rd yr.

Cuboid—9th (fetal) month

Medial cuneiform—2nd yr.

Lateral cuneiform—1st yr.

Appears 3rd yr.

Intermed. cuneiform—3rd yr.

Unites 17th–20th yr.

Appears 10th wk.

Appears 9th wk.

Unites 17th–20th yr.

Unites 18th yr.

Appears 3rd–4th yr.

Appears 2nd–8th yr.

Appears 11th–15th yr.

Appears 3rd–6th yr.

Unites by 18th yr.

Appears after 15th month

Unites by 18th yr.

Appears 6th yr.

Appears 9th–12th wk.

3.217   A plan of the ossification of the bones of the foot.

cuneiform in the first year. The scale-like epiphysis of the calcaneus covers the greater part of the posterior surface and extends slightly on to the plantar surface; it begins to ossify in the sixth year in females and eighth in males and unites in the fourteenth year in the former and sixteenth in the latter. The posterior process of the talus is sometimes ossified from an independent centre, and may then remain separate or be connected to the rest of the bone by cartilage. This additional ossicle is named the *os trigonum*. A number of other accessory bones occur in the foot and may lead to radiological misinterpretation (*see* Trolle 1948; and **3.218**).

The metatarsal bones are ossified each from two centres: a primary centre for the shaft, and a secondary or epiphysial centre for the base of the first, and for the head of the other four. In the

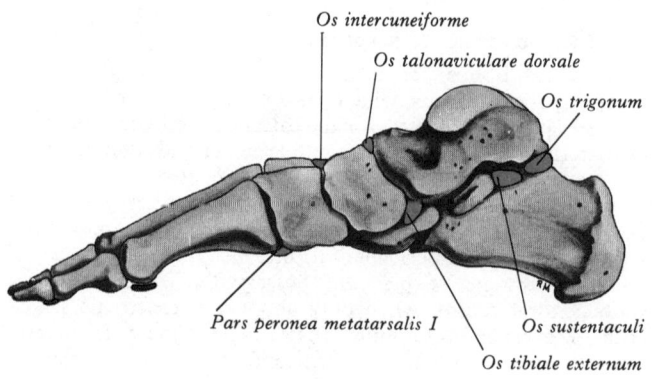

Os intercuneiforme

Os talonaviculare dorsale

Os trigonum

Pars peronea metatarsalis I

Os sustentaculi

Os tibiale externum

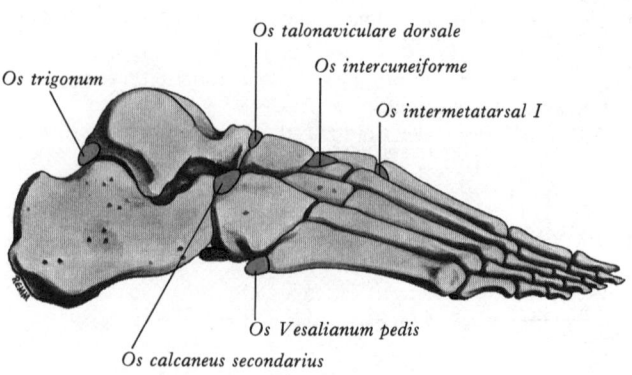

Os trigonum

Os talonaviculare dorsale

Os intercuneiforme

Os intermetatarsal I

Os Vesalianum pedis

Os calcaneus secondarius

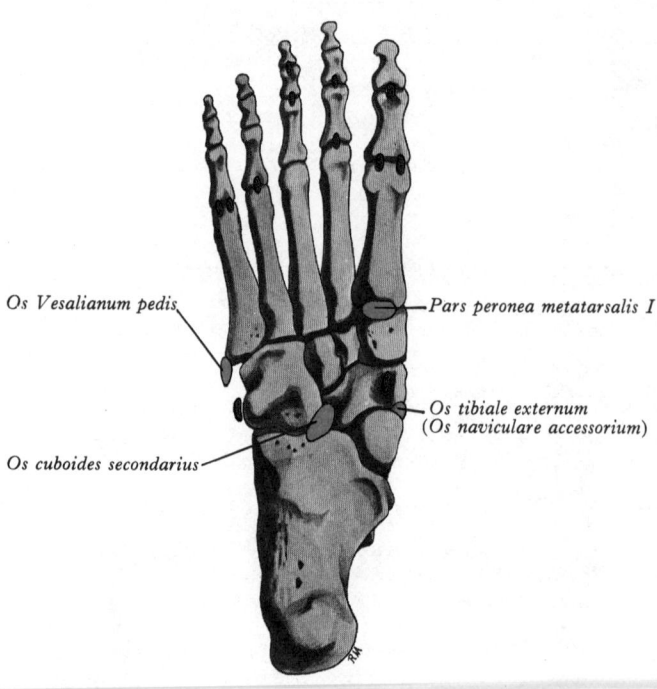

Os Vesalianum pedis

Pars peronea metatarsalis I

Os tibiale externum
(Os naviculare accessorium)

Os cuboides secondarius

3.218 The sites of sesamoid (*red*) and accessory bones (*blue*) in the human foot.

second, third and fourth metatarsals ossification begins in the middle of the shaft about the ninth week of intrauterine life and in the first and fifth about the tenth week. The epiphysis for the base of the first metatarsal appears about the third year; those for the heads of the other metatarsals between the third and fourth year; all unite with the shafts between the seventeenth and twentieth years. As in the first metacarpal bone (p. 376), so in the first metatarsal there is sometimes an epiphysis for the head as well as one for the base. An epiphysis is frequently present on the tubercle of the base of the fifth metatarsal bone.

The phalanges are each ossified from two centres: a primary one for the shaft and an epiphysis for the base. The primary centres for the distal phalanges appear between the ninth and twelfth weeks of intrauterine life or even later in the fifth toe; those for the proximal phalanges between the eleventh and fifteenth weeks and those for the intermediate phalanges after the fifteenth week, with a wide range of variation. The epiphysial centres appear between the second and eighth years (usually in the second or third year in the big toe) and unite with the shaft by about the eighteenth year (Venning 1956). The variations in the order of appearance of these centres is apparent in different reports, and racial differences probably exist (Kraus 1961).

Comparison of the Bones of the Hand and Foot

The origin of the skeletal elements of the hand and the foot from the terminal segment of the primitive pentadactyl limb has already been considered in general principles (p. 350), and it is now possible to examine their similarities and differences in greater detail. Both hand and foot consist of proximal, intermediate, and distal series of components, i.e. carpus or tarsus, metacarpus and metatarsus, and phalanges. The carpus and tarsus both contain seven elements (omitting the sesamoid pisiform), and in a general way these can be described as more or less cubical bones; but they vary much in shape and size. The carpal bones are smaller, being concerned with the transmission of smaller forces; and they have retained the primitive alignment in rows more clearly than have the tarsal bones. The articulation of the tarsus with the foreleg is reduced to one bone (talus), and the tarsus as a whole is divisible into two moieties, corresponding to the longitudinal arches of the foot, with a small degree of independence of movement which is absent from the carpus. The lever-like extension of the calcaneus also has no equivalent in the primate carpus, though carpal elements arranged similarly are present in some mammals.

The components of the digits, metacarpus or metatarsus and phalanges, display more similarity. The metacarpals and metatarsals are much alike in form, and both series are bound together by soft parts into a structural entity. Their mobility is very limited proximally at their bases, where they not only articulate with the distal carpals or tarsals but also to some extent with each other. Between them lie the interosseous muscles, and closely associated with the first and fifth members are short intrinsic muscles proper to these particular digits. On the flexor aspects in both hand and foot are large arterial anastomoses in the form of vascular arches, superficial and deep to long flexor tendons crossing to enter the free parts of the digits, whose skeletal structures are the phalanges. The distal end of the metacarpals and metatarsals are slightly more mobile than the bases; but the third metacarpal and the second metatarsal are relatively fixed, forming axes in their respective extremities about which the neighbouring elements exercise their limited movements. These are concerned with the adaptation of the hand's grip to uneven objects and of the plantigrade foot to inequalities of surface. The digits can flex and extend through a considerable arc at the metacarpo- and metatarso-phalangeal joints, with a lesser range of abduction and adduction, better developed in the hand. The phalanges of the foot are much shorter than those in the hand; and though the phalangeal formula is the same, there is a greater tendency to reduction of phalanges by fusion or loss in the foot, especially in the fifth digit.

The functions of the hand and foot in mankind differ greatly, and their general skeletal similarity is much modified in details. The greater length and mobility of the fingers and especially the preservation of the free-ranging opposable thumb are associated with the grasping habit of a prehensile limb. The divergent position of the metacarpal of the pollex and its specialized joint with the trapezium contrasts markedly with the arrangements in the hallux. The thumb is rotated about 90° around its own long axis relative to the fingers, so that it is already, in the resting position, halfway round to opposition with them. The great toe is in the same plane as the other toes, its flexor surface facing the ground like them. Moreover, it is tied by transverse metatarsal ligaments to the second metatarsal, thus forming an integral entity with all its fellows, adapted to the propulsive role of the foot. In association with this the digital formula is different in the foot; in the hand the middle digit projects furthest; in the foot it is the great toe, which takes the major stresses transmitted through the lever of the foot in locomotion. The dorsiflexion of the foot to a right angle with the leg is an obvious adaptation to plantigrade standing and progression. This position is sometimes said to be peculiar to man; but it is the primitive plantigrade adaptation of ancestral mammal-like reptiles; and while most become to varying degrees digitigrade, bringing the foot and hand into line with the rest of their limbs, most primates and some carnivores, such as bears display the plantigrade stance in the foot. The outstanding human peculiarity is the loss of opposition in the great toe, which distinguishes man even from other primates. The human foot is, indeed, in its general proportions far more specialized for bipedal habit than is the case in any other primate animal. The strong build of the great toe, both in its metatarsal and phalanges, and the elongation and stouter proportions of the tarsal bones, together constitute a powerful lever, on which the whole weight of the body can be elevated to add further impetus and spring in running and jumping. The arched form of the foot, and the development of a lateral arch to steady it on the ground and a medial to transmit the main force of thrust in propulsion, are specializations absent from the hand (pp. 350–353, **4**.92 and **4**.93). Finally, it is noteworthy that the rotational element in the foot, in the movements of inversion and eversion, is absent from the hand, which rotates with the forearm in supination and pronation.

The Sesamoid Bones

Sesamoid bones, like the seeds after which they are named, are usually more or less ovoid nodules a few millimetres in diameter, although their shape and size vary, some being quite large, for example, the patella (Bizarro 1921). They are not always completely ossified, and may consist of dense fibrous tissue, cartilage and bone, in varying proportions, but the majority are, to some extent at least, ossified. They are almost always embedded in tendons either in close relation to articular surfaces or in situations where tendons are sharply angled round a bony surface. In both cases, the surface of the sesamoid related to another bone is covered with articular cartilage and actually slides over the opposed bone, which is itself usually an extension of an articular surface, as in the metacarpophalangeal joint of the thumb. This arrangement entails that the tendons concerned are to some extent fused with the joint's articular capsule, as is most clearly seen in the relationship between the patella, quadriceps femoris tendon, and the knee joint (p. 483). In other situations, well exemplified by the sesamoid mechanism in the tendon of the peroneus longus (p. 606), the cartilaginous articular aspect of the sesamoid glides freely over a cartilage-covered surface of opposed bone, in this instance a facet on the inferolateral aspect of the tuberosity of the cuboid bone. Moreover, the arrangement is enclosed within a bursa, which could as well be termed a joint capsule, for the complete system has all the essentials of a synovial joint. Some authors consider that sesamoids are *primarily* articular in nature, i.e. embedded in articular capsules and that their association with tendons is a *secondary* phenomenon (Patterson 1946).

Despite their close association with articulations, the precise functions of sesamoids are regarded as still uncertain. It has been suggested that they may modify pressure, diminish friction, and sometimes alter the direction of pull of a muscle, the patella being commonly advanced as an example of the last of these suggestions. Where a tendon is acutely deflected in direct proximity to bone, the presence of a sesamoid, if ossified, may aid in maintaining the

local circulation, bone being more resistant to pressures which could compress the vessels in tendon. However, the latter are never numerous and pressure is not likely to be long maintained, except perhaps in the foot.

The fact that sesamoid elements appear during fetal life, and indeed in greater numbers than in the adult, demonstrates that they are phylogenetically integral parts of the skeleton, and not merely the result of local physical circumstances; but the latter possibility cannot be dismissed in every instance. They are much more numerous in the extremities of many other mammals. In reptiles, intratendinous ossification in the proximity of the ends of some long limb bones produces nodules closely resembling sesamoids, but later some of these fuse and may be equated with traction epiphyses (Haines 1942). The proximal epiphysis of the olecranon is an example. Experimental procedures involving the patella suggest that both phylogenetic and other factors are involved in the formation of sesamoid bones. Transplantation of limb bud fragments from chick embryos to the chorio-allantoic membrane has been followed by development of a patella, indicating a self-differentiating mechanism (Murray and Huxley 1924/5). Conversely, removal of the patella in young dogs has been followed by regeneration of the bone, provided that the normal motor activities of the quadriceps femoris were permitted (Carey *et al.* 1927), suggesting that local mechanical factors are perhaps also operative in the formation of sesamoids. This may indicate an interesting parallel between sesamoids and bursae, in that the latter also appear regularly before birth in certain sites, but can also develop in other positions, as adventitious structures, apparently in response to disturbed local mechanical conditions.

The incidence of sesamoid bones has received much attention, at least insofar as their siting is concerned, but numerical data in regard to individual sites are not always available, and in particular the onset and progress of ossification in these elements has been but little studied.

In the upper limb, sesamoid nodules associated with joints are limited to the palmar aspect of the hand. Almost constantly present in the tendons of the adductor pollicis and flexor pollicis brevis, are two relatively large sesamoid bones, the medial being the larger. They articulate with special facets on the palmar aspect of the joint surface on the head of the first metacarpal bone, and since they are firmly attached in the capsule of the articulation, as well as in the tendons involved, it is likely that they conduct some of the pull of the thumb musculature to the ligaments of the joint. A sesamoid is embedded in the tendons anterior to the metacarpophalangeal joint of the index figure (35 per cent of hands), and of the fifth digit (70 per cent), where it is occasionally doubled. Less frequently, similarly placed sesamoids are encountered in the third and fourth digits, and in a majority of hands (73 per cent) one is present in the palmar aspect of the interphalangeal joint of the thumb (Gray *et al.* 1957). Sesamoid elements in non-articular sites are seldom present in the upper limb, but one is occasionally encountered in the biceps tendon near its insertion into the radial tuberosity.

In the lower limb, the patella is the largest articular sesamoid bone; its relationship to the tendon of the quadriceps femoris and the capsule of the knee joint exemplify the arrangement of structures around all the smaller sesamoids which are integrated into synovial articulations. In the foot, their distribution is similar to that in the hand. Two, both of which may be duplicated (sometimes leading to fallacious diagnosis of fracture), are always present in the tendons of flexor hallucis brevis, on the plantar aspect of the metatarsophalangeal joint of the hallux. As in the case of the pollex, they are firmly tied in with the ligamentous structures around them, including the most medial part of the plantar aponeurosis. Single sesamoids occur not infrequently in the plantar aspect of the capsule of the same joint in all the other toes, and also in the interphalangeal joint of the great toe.

Sesamoid bones, or cartilages, which are not associated with a synovial joint, occur more frequently in the lower limb. The element in the tendon of peroneus longus has already been mentioned above. Another sesamoid, usually appearing late in life and therefore perhaps adventitious in character, occurs in the tendon of tibialis anterior where it is in contact with the distal part of the medial surface of the medial cuneiform bone. Similarly, the tendon of tibialis posterior may contain a sesamoid nodule where it glides over the medial side of the head of the talus. Sesamoids also occur occasionally in the lateral part of the gastrocnemius, behind the lateral femoral condyle, in the tendon of psoas major where it is in contact with the ilium, in the tendon of the gluteus maximus where it passes over the greater trochanter, and in tendons which are deflected round the medial and lateral malleoli. In some, if not all these sites, bursal arrangements occur, and the opposing bony surfaces involved are covered with articular cartilage, so that a true synovial articulation exists. The sesamoid which occurs in the gastrocnemius (lateral head) is sometimes called a *fabella*; when present, it articulates directly with the lateral femoral condyle, and this arrangement should therefore really be classified with the 'articular' sesamoids.

Sesamoid bones begin to ossify relatively late, apart from the patella (p. 397). Those in the hand are usually said to begin in the early teens. In a selected group of Caucasians, those associated with the thumb began to ossify in males between 12 and 15 years, and sesamoids in the other digits between 15 and 18, the dates for females being about three years earlier (Joseph 1951). No such data appear available for large mixed or other racial groups, and figures for sesamoid ossification in the lower limb are vague or lacking.

4

ARTHROLOGY

Introduction

Skeletal structures, both of vertebrate and invertebrate animals, are occasionally of a purely protective nature. Far more often, however, they provide for the attachment and leverage of muscles; they are essential to *motor* functions involving either the whole body, or the relative positioning of its members. Since *movement* of any kind, including locomotion, is almost always accomplished by the bending or straightening of appendages, or of the trunk itself, a skeleton of rigid bony members— in contrast to a pliant structure such as a notochord—must contain intervening arrangements permitting some degree of angulation, torsion, or displacement. This applies equally to exo- and endo-skeletons. The jointed exoskeleton of crustaceans and insects accounts for the name of the largest phylum, the *Arthropoda*. All vertebrates possess joints; even those in which the skeleton is wholly cartilaginous display junctional regions of greater pliancy between their individual components.

Not all junctions between bones, however, are primarily constructed to allow *movement*; thus, for example, mature sutures are usually so shaped as to *interlock*, despite the osseous discontinuities which they delineate. In bones with epiphyses, temporary zones of cartilage intervene between these regions of ossification and their ossifying diaphyses; these arrangements are clearly essential to the processes of *growth* rather than *mobility*. (It should perhaps be noted here, however, that the post-cranial cartilaginous masses, together with the cranial fibrous and cartilaginous sutures of the full-term fetus, possess a pliancy and flexibility that permits advantageous 'moulding' of the parts during childbirth: *see* illustration **2.**3B.) Nevertheless, to group these two forms of skeletal junction together with articulations whose function is locomotor may be convenient, but is otherwise uninformative; and the application to the various forms of juncture of terms which are largely morphological does little to improve matters. Moreover, joints at which motion *does* occur display a considerable variety of mechanism, and this again has led to a classification, based partly upon structural features and to some extent upon the kind of movement permitted, with an array of terms for different forms. This diverse assortment of osseous junctions is, indeed, held together by one criterion alone—that *joints* are merely the *meeting places of bones*.

Thus the study of arthrology brings together, in what is functionally a rather unsatisfactory association, skeletal dispositions concerned with *growth*, *rigidity* and *movement*. The undesirable nature of this custom is even more apparent in its **current textbook classification:**

Coaptations between skeletal components may be called *juncturae, articulationes, arthroses*, or by vernacular terms such as *joints* with little discrimination. They are habitually divided into three groups:

(1) **fibrous joints** (*'fixed' joints, articulationes fibrosae*, or *synarthroses*);
(2) **cartilaginous joints** (*'slightly movable' joints, articulationes cartilagineae*, or *amphiarthroses*);
(3) **synovial joints** (*'freely movable' joints, articulationes synoviales*, or *diarthroses*).

The inconsistencies of this classification can be illustrated at once by the fact that fibrous joints include not only *fixed* joints (*sutures*) but also *slightly movable* joints (*syndesmoses*, such as the one between the inferior ends of the tibia and fibula), and in addition the minimally resilient junctions between teeth and their sockets (gomphoses). Again, the cartilaginous joints comprise not only symphyses (*secondary cartilaginous joints*), which permit limited movement, but also *synchondroses* (*primary cartilaginous joints*), where movement is absent or minimal. (It may also be noted that whilst the majority of the 'sutures' of the facial skeleton and cranial vault vault are fibrous, those of the cranial base are cartilaginous—*see* p. 295.) The conventional classification also fails to emphasize the basic mechanical differénce between joints at which rigid skeletal elements are able to move relative to each other through the interposition of a deformable tissue (fibrous *and* cartilaginous joints) and those where movement is dependent upon opposed sliding surfaces (synovial joints). In engineering practice this distinction is obvious, and there seems no reason

against applying it as the primary feature separating biological joints into two, rather than three, major groups. There is, however, no overall term for fibrous and cartilaginous joints in current use, though Galen grouped them as *synarthroses*. In any case the histological distinction between them is also unsatisfactory, not only because it cuts across functional grouping as pointed out above, but also because many of these joints contain admixtures of fibrous tissue and cartilage, as will become apparent below. What might be termed the *non-synovial* joints could be separated more appropriately into growth mechanisms and true, mobile joints, but this also cuts right across established morphological custom.

These preliminary remarks are occasioned by the defects of current classifications of joints, and by the confusion which may be entailed by the vagaries in their terminology. It is not proposed, nevertheless, to add to this confusion by disregarding the customary scheme, but its deficiencies cannot pass without comments, which may, perhaps, also help to encourage a more rational and practical order.

Fibrous and Cartilaginous Joints

Since all bones ossify in pre-existing fibrous tissue or cartilage, those two tissues inevitably appear at junctions such as sutures and synchondroses, where separate skeletal elements are still in the process of growth. The actual osseous margins or surfaces gradually approach, and when growth ceases the junctional tissue is obliterated, leading to complete bony continuity. Such junctures are therefore temporary; and although the obliteration of sutures is retarded relative to the fusion of epiphyses, it is ultimately as complete in most instances.

Synchondroses include the numerous temporary cartilaginous junctions between diaphyses and epiphyses in the immature post-cranial skeleton (**3.**3B), and also the regions of unossified cartilage between skull components developing in the chondrocranium (e.g. between the sphenoid and occipital bones). The structure and growth phenomena of these junctions have been considered elsewhere (p. 251). The cartilages involved are hyaline, but it is apposite to note that special accumulations of collagen fibres, aligned in a regular manner, occasionally appear in growth cartilages (Smith 1962). The majority of the latter, extended across the axis of elongated bones near their extremities, are thus subjected to compression forces approximately at right angles to their transverse planes. In some situations, notably in man where the superior tibial epiphysis inclines downwards to form the tibial tuberosity, the direction of pull of a muscle (quadriceps femoris in this example) is oblique in respect to the growth cartilage, thus introducing shearing forces. In such sites the cartilage may be almost completely replaced by collagen fibres. The microscopic structure of bone trabeculae in the metaphysis adjacent to the fibrous and cartilaginous regions in such mixed growth 'cartilages' also evinces interesting differences, which may be equated with the impress of tensile or compressive forces. Trabeculae immediately adjoining a fibrous zone in the cartilage are long, slender, parallel and sparsely linked by cross-connexions, as if adapted to tension, whereas the trabeculae elsewhere (adjacent to cartilage proper) exhibit the usual irregular cancellous pattern associated with compression.

Sutures are limited to the skull, occurring wherever margins or broader surfaces of bones meet and articulate, separated only by a zone of connective tissue, the *sutural ligament* or membrane. This is the surviving unossified part of the mesenchymatous sheet in which dermal bones develop (p. 295). The sutural ligament is not simple, displaying regions of differentiation concerned in the growth and binding together of the apposed bone surfaces (Pritchard *et al.* 1956). Each bone is covered on its sutural aspect by a layer of somewhat flattened osteogenic cells (the 'cambial' layer of some observers), which is itself overlaid by a lamella of fibrous tissue (**4.**1A, B, C). These two layers together correspond to periosteum, and are continuous with the latter at the margins of the sutural surfaces, both inside and outside the skull. Between these two layers of sutural periosteum is a central stratum of loosely arranged fibrous connective tissue, the width of which

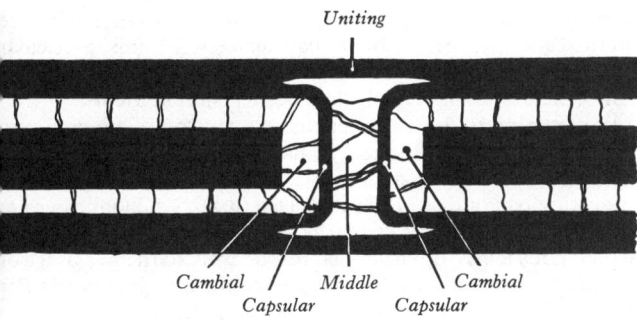

Uniting

Cambial / *Middle* / *Cambial*
Capsular / *Capsular*

4.1A The general structure of a suture. Note the five layers which intervene between the bone ends: cambial, capsular, middle, capsular and cambial. The cambial and capsular layers are continuous with similar layers on the inner and outer aspects of the bones, whilst the superficial fibrous strata of the periosteia cross the suture as uniting layers. (By courtesy of Professor J. J. Pritchard, Drs. J. H. Scott, F. G. Girgis and the *Journal of Anatomy*.)

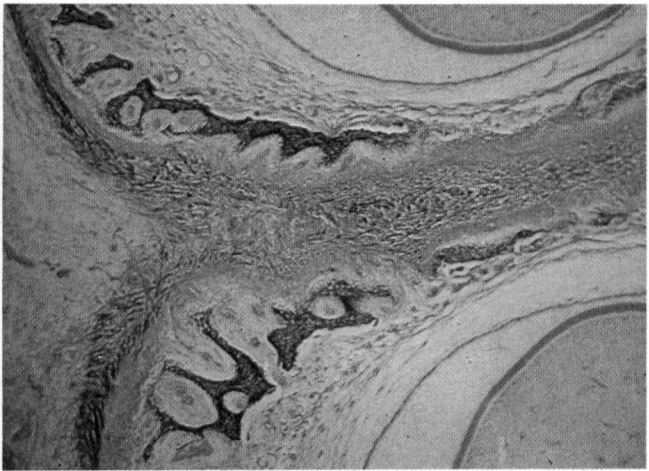

B

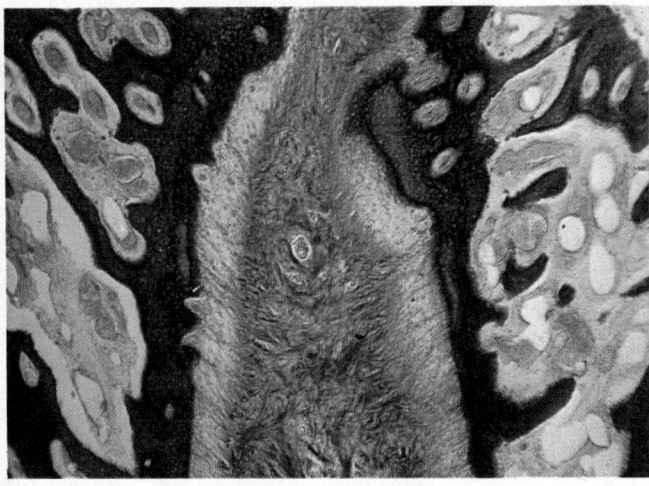

C

4.1B and C The mid-palatal sutures of human fetuses at 165 mm crown–rump length and near full term respectively; trichrome stain. The sagittal plane is approximately horizontal in B and vertical in C. Compare with **4.1A**. (Kindly provided by the late Dr. J. H. Scott.)

varies according to the interval between the bones involved. This central region of the sutural ligament contains thin-walled blood vessels, the veins of which communicate with diploic vessels and intracranial venous sinuses and with external veins such as those in the scalp. The fibrous element in the periosteum adherent to the surfaces of adjacent bones, connected by a suture, crosses the interval between them, thus closing in the sutural ligament and adding to its binding effect. There is some experimental evidence that during active growth the orientation of collagen fibres in sutural membranes is rapidly adaptable to a variety of factors, and

particularly to the direction of growth of minute bone spicules (Kośkinen *et al.* 1976).

In view of the occurrence of patches of fibrous tissue in synchondroses, as noted above, it is of considerable interest to find that, during the period of growth, areas of secondary cartilage formation are frequently observed in sutural ligaments, a further evidence of the close relationship between fibrous and cartilaginous joints.

When growth at sutures comes to an end, the osteogenic cells bring about complete transformation of the sutural ligament into bone, a process which is slow (p. 260), but leads finally to obliteration and, of course, rigid synostosis. Sutural fusion does not even commence until the late twenties, proceeding slowly thereafter; yet it is clearly necessary that sutures should cease to function as mobile joints as rapidly as possible after birth. Although sutural ligaments may be sufficient to effect an immovable bond between large areas of bone, especially where these show reciprocally adapted irregularities, even if these be fine as in the junction between the two maxillae, no such immobility is to be expected at the narrow articulations at the edges of the bones of the cranial vault. At sutures such as these, however, the bone margins become highly complex and irregular, developing spikes and recesses which interlock so intimately that bones may be difficult to separate even when denuded of all connective tissue. Where the edges are like a saw, or serrated, the junction is a *serrate suture*. A *denticulate suture*, as its name suggests, is characterized by small toothlike projections, which often widen towards their free ends, providing an even more effective interlocking than do serrate sutures. When such sutures are closely tied by the sutural ligament and periosteum complete immobility results. The sagittal suture is serrate, and much of the lambdoid is usually denticulate. Where one bone overlaps its neighbour, as at the suture between the temporal and parietal bones, the arrangement is a *squamous suture*. At sutures of this kind the bone surfaces are reciprocally bevelled, one internally, one externally. The bevelled surfaces may be mutually ridged or serrated, in which case the junction is sometimes termed a *limbous suture*. Where there is simple apposition of contiguous surfaces, usually roughened and irregular in a complementary manner, the junction is nevertheless usually named a *plane suture*. An instance of this has been cited above, and others are the sutures between the palatine and zygomatic bones and the maxillae. Although the linear demarcations visible on the surface between such bones show none of the intricate interlocking evident at serrate or denticulate sutures, the irregular surfaces of contact, held together by wide expanses of sutural ligament, provide a high degree of resistance to shearing or torsional forces; like other sutures they are, for all practical purposes, immovable.

A schindylesis is an articulation where a ridged bone fits into a groove on a neighbouring element, as in the junction between the vomer and the rostrum of the sphenoid bone.

A gomphosis, or peg-and-socket joint, is the specialized type of fibrous articulation restricted to the fixation of teeth in the mandible and maxillae. It is more appropriate to describe it with the teeth (pp. 1284, 1292).

A syndesmosis is a form of articulation in which closely apposed bony surfaces are bound together by an interosseous ligament, affording a small degree of movement between adjoining bones. It is considered by some to be a rarity in mammals, in the current usage of the term at least, having for long been confined in man to the inferior tibiofibular joint (but *vide infra*). However, although not usually described as such, the dorsal part of the junction between the ilium and sacrum, through the medium of the interosseous sacro-iliac ligament, closely resembles a syndesmosis. Incidentally, the sacro-iliac joints themselves, which are primarily synovial, are frequently invaded by fibrous tissue in the later decades of life, and may become totally converted into fibrous articulations, differing little from syndesmoses p. 473). It would be reasonable to regard the inferior tibiofibular syndesmosis as no more than an interosseous ligament, adjacent to the ankle joint or to a synovial extension of it when this exists, as it occasionally does in man and regularly in some other primates. If this be accepted there is little valid objection to extending the designation syndesmosis to numbers of

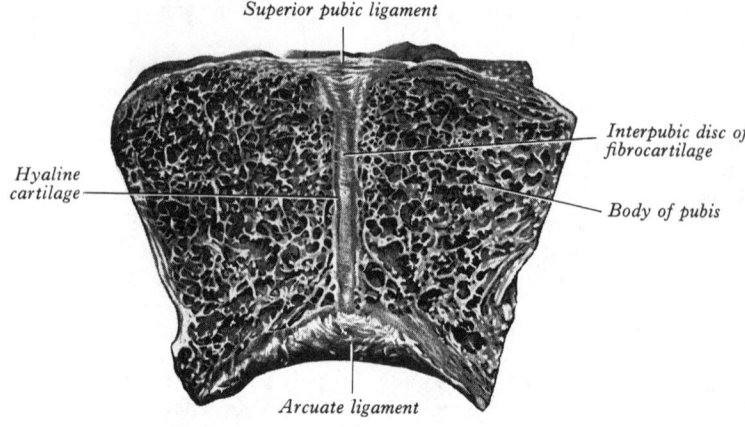

Superior pubic ligament

Interpubic disc of fibrocartilage

Hyaline cartilage

Body of pubis

Arcuate ligament

4.2 Section through the pubic region to show the structure of a symphysis.

other interosseous ligaments, such as those in the carpus and tarsus, or to including the interosseous membranes of the forearm and foreleg, especially since the latter are already described as intermediate joints in the radio-ulnar and tibiofibular series. In fact, since ligaments are almost all 'interosseous', it would become difficult to restrict the use of the term 'syndesmosis' at all, unless it be insisted that only short ligaments in close proximity to a synovial joint be so designated. These would appear to be the criteria which limited the term in orthodox nomenclature to a single joint in the human skeleton, and this restricted usage rendered all the more unsatisfactory the suggested substitution of the term 'syndesmology' for arthrology. The former is a word of unfamiliar derivation and usage and hence obscure in meaning, whereas the stem **arthron**, a joint, is widely employed and

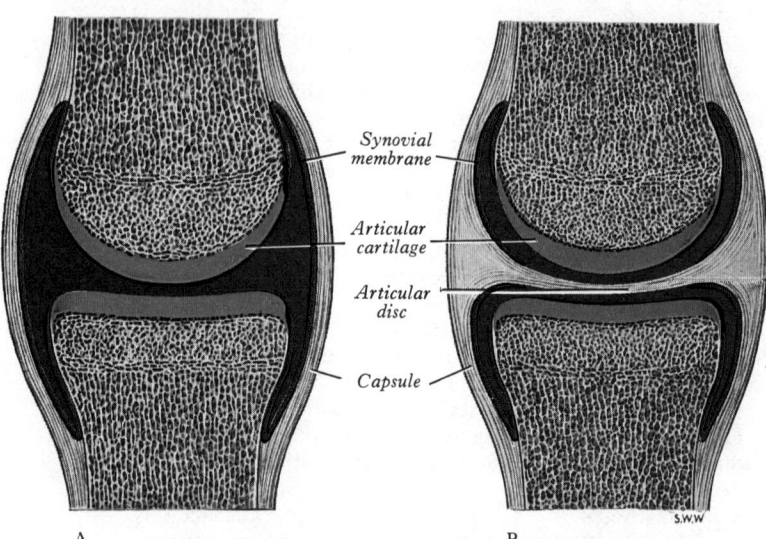

Synovial membrane

Articular cartilage

Articular disc

Capsule

A   B

4.3A A diagrammatic section through a simple synovial joint.

4.3B A diagrammatic section through a synovial joint with an articular disc. Articular surfaces are shown artificially separated for clarity.

familiar in both medical and biological applications – witness: arthritis, arthroplasty, and the name of the largest animal phylum, the arthropods. The re-adoption of the internationally accepted term **arthrology** is thus most welcome. The reader will, in addition, be interested to note that in the recently published *Nomina Anatomica* (1977) the term **syndesmosis** has now been extended to include the following ligaments: pterygospinous, stylohyoid, interspinous, supraspinous, intertransverse, the ligamenta flava, and the ligamentum nuchae. The *syndesmosis radio-ulnaris* includes the antebrachial interosseous membrane and oblique cord, while the *syndesmosis tibiofibularis* includes the

crural interosseous membrane and ligament, together with the anterior and posterior tibiofibular ligaments. This particular choice of ligaments does, however, seem somewhat arbitrary, for many others could lay equal claim to inclusion in the list.

A symphysis is a cartilaginous or, more precisely, a fibrocartilaginous articulation, of non-synovial form, at which a limited range of movement is enabled by deformation of a connecting pad or disc of fibrocartilage (**4.2**). Synchondroses are associated with symphyses by the terms *primary* and *secondary* cartilaginous joints, often applied respectively to them, a usage of dubious morphological value and also confusing, in view of their functional dissimilarity. However, a symphysis resembles a synchondrosis in one small histological feature – in containing some hyaline cartilage, arranged as a thin film covering the articular surfaces, to the growth of which it contributes. Nevertheless, the distinguishing feature of this kind of joint is the resilient bond provided by the fibrocartilage, whose thickness is a prime factor in the range of movement possible. It is to be emphasized that all strata of the joint – bone, hyaline and fibrocartilage – are intimately united, there being no internal surfaces between any elements in the articulation. However, in some symphyses, such as the pubic and manubriosternal joints, cavities filled with fluid do occur, but there is no development of sliding surfaces. While these arrangements lack the distinctive specializations of synovial joints, they may represent an intermediate phase in the evolution of the latter, as will be noted later (p. 424). Fibrous ligaments extend across symphyses from one bone to the other, fusing with the periphery of the fibrocartilage and thus providing an additional flexible linkage; but such structures do not form a complete capsule, as they do in synovial joints, although they are like the latter in containing terminal plexuses of afferent nerves, which also invade the marginal zone of the fibrocartilage.

Unlike synchondroses, which are temporary junctions disappearing after the period of growth, symphyses are permanent and concerned with movement. There are some exceptions to this: the manubriosternal symphysis may be obliterated by bony union (synostosis) in the later years of life. The fibrocartilages in different symphyses are not of uniform structure, and there are often regional variations in them, especially in regard to the distribution and alignment of collagen and elastic fibres. The latter features are most probably local adaptations to the stresses imposed by the specific activities of the joints in question (e.g. *see* the structural details of intervertebral discs, p. 444).

It is sometimes pointed out that all symphyses are in the median plane of the human body. Since they are confined to the axial skeleton, save for the symphysis pubis, this is to be expected. The articulation between the two halves of the mandible is usually described as a symphysis, and the symphysis menti does bear some resemblance in its histological features (p. 318). It is rapidly abolished during the first postnatal year, and since it is improbable that it is the site of any functionally significant movement while present, it is of doubtful propriety to assign it to this group of joints.

Synovial Joints

Articulations of the *synovial* type utilize an entirely different principle from the *non-synovial* fibrous and cartilaginous joints (**4.3**A, B; **4.4**A). Although the bones involved are linked together by a fibrous capsule, and frequently by accessory ligaments inside or outside of this, the major parts of the osseous surfaces concerned are in contact but *not* continuity. They are covered by a relatively thin stratum of specialized hyaline cartilage (occasionally of fibrocartilage), and the actual contact is between these cartilaginous surfaces, which are characterized by a very low coefficient of friction (Charnley 1959). This sliding contact is facilitated by a viscous *synovial fluid* (synovia), which acts like a lubricant in some respects, and is also concerned in the maintenance of living cells in the articular cartilages. The somewhat complex properties of this fluid are considered later (p. 428). The low level of friction in synovial joints has been picturesquely described as equivalent to 'ice on ice'.

The fibrous capsule encloses the joint completely, with certain exceptions, such as the hip joint (p. 477), where its continuity is interrupted by protrusions of the synovial stratum; these exceptions are described with the individual articulations concerned. The capsule is lined throughout by *synovial membrane*, which also extends on to all intra-articular surfaces except those actually involved in compression contact during the activities of the joint. This includes nonarticulating bone surfaces and also the tendons and ligaments which may be partly or wholly within the fibrous capsule, as at the shoulder and knee. Where a tendon is attached within a joint and issues from it, a prolongation of the synovial stratum commonly accompanies it for a short distance external to the capsule. In some situations tendons which are extracapsular may be separated from it by a synovial bursa (p. 458), continuous with the interior of the joint. Such details of synovial arrangements are important in that they offer avenues for the spread of infection from extra-articular sites into the joint itself.

A third type of intra-articular structure, which is *not* covered by synovial membrane, is an *articular disc* or *meniscus*. Such structures intervene between certain articular surfaces, where the degree of congruity is low, and they consist of fibrocartilage, the fibrous element usually predominating. If they extend wholly across a synovial joint, they effectively divide it into two parts, both structurally and functionally, i.e. there are two separate synovial cavities. Peripherally, articular discs are connected to the fibrous capsule, usually through the medium of vascularized connective tissue; but the union between the two is sometimes more intimate and strong, as in the knee and jaw joints. The peripheral zone of the disc is invaded by vessels and afferent and motor (sympathetic) nerves. The main bulk of the disc contains few cells, but its surface may display an incomplete stratum of

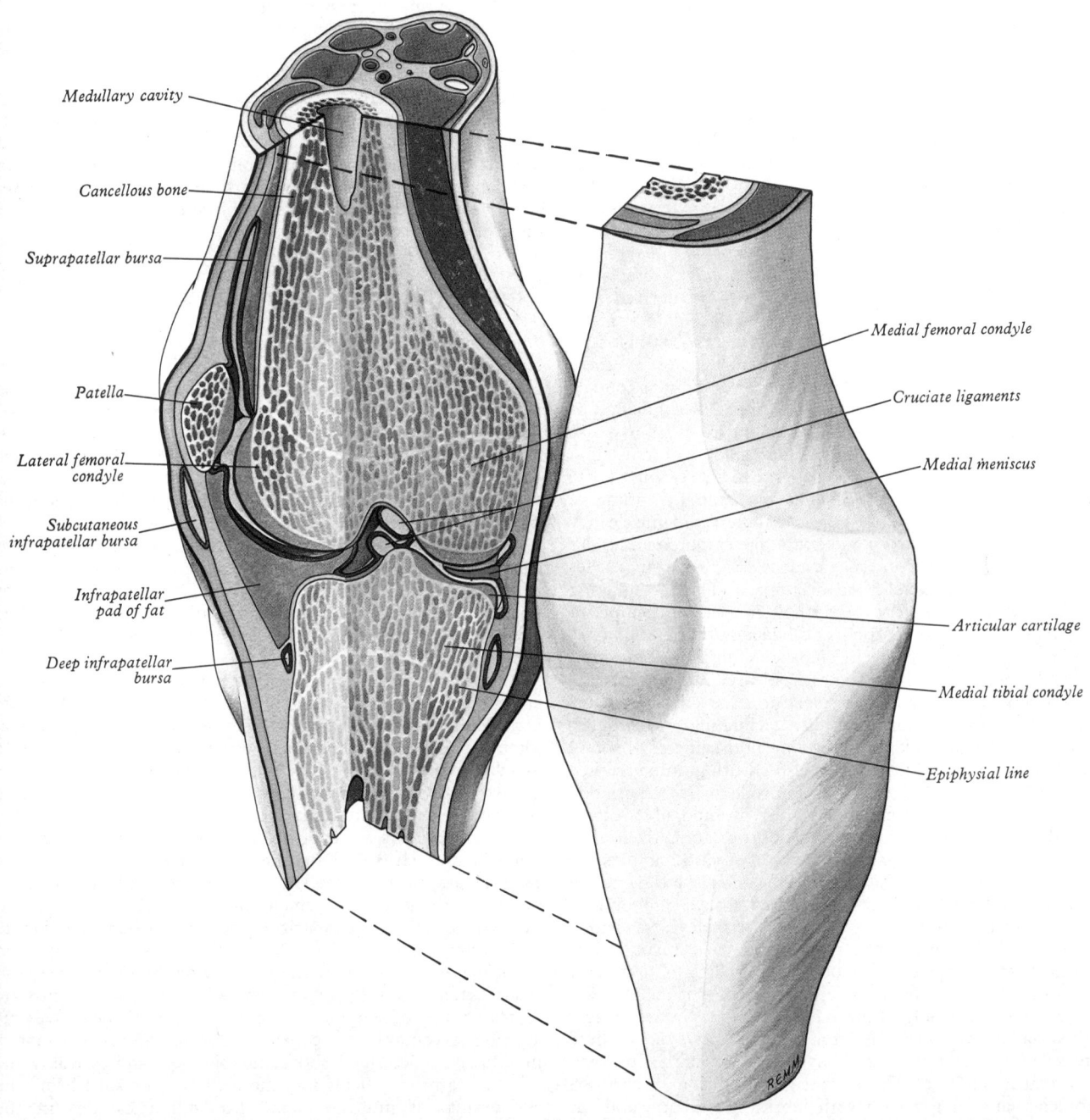

Medullary cavity

Cancellous bone

Suprapatellar bursa

Patella

Lateral femoral condyle

Subcutaneous infrapatellar bursa

Infrapatellar pad of fat

Deep infrapatellar bursa

Medial femoral condyle

Cruciate ligaments

Medial meniscus

Articular cartilage

Medial tibial condyle

Epiphysial line

4.4A The general organization of the knee joint, which is compounded of three pairs of articular surfaces and also possesses two menisci of fibrocartilage. Note the following bony parts and their covering of articular cartilage: (1) the patella, (2) the patellar surface of the femur, (3) the medial femoral condyle, (4) the medial tibial condyle. Note also, (5) the cut edge of the synovial membrane (*purple*), (6) menisci in section (*green*), (7) the cruciate ligaments, which are intra-articular in position, (8) the suprapatellar synovial bursa, continuous with the general joint cavity, (9) the deep infrapatellar bursa, (10) the infrapatellar pad of fat and its associated infrapatellar fold or synovial membrane, which passes back to the anterior wall of the intercondylar fossa of the femur, (11) sectional surfaces of patellar, femoral and tibial articular cartilages (*bright blue*), (12) the epiphysial line of dense bone indicating the zone of fusion between epiphysis and metaphysis.

flattened cells, continuous at the attached perimeter with the adjacent synovial membrane. There is, nevertheless, little structural similarity between the two groups of cells (p. 427). Strictly, the term meniscus should be reserved for incomplete discs, such as those in the knee joint and, occasionally, the acromioclavicular joint. Complete fibrocartilages occur in the sternoclavicular and inferior radio-ulnar joints, while the disc in the temporomandibular joint may be complete or incomplete, with about equal incidence. Complete discs often present small perforations and, where menisci are usual, complete discs may occur, or they may be merely slightly perforated. The functional role of intra-articular fibrocartilages remains unproven. An experimental approach to this problem is difficult to devise or to put into effect. Therefore the views advanced are largely deductions from structural or phylogenetic data, with the help of mechanical analogies. The plethora of suggestions is hence not surprising, and it includes such activities as shock absorption, improvement of fit between joint surfaces, facilitation of combined movements, limitation of translation movements at joints such as the knee, distribution of weight over a larger surface, protection of the edges of articular surfaces, facilitation of rolling movements, and spread of lubricant.

In studies of the evolution of articular discs attention has been focused more often upon equating them with vestigial skeletal elements, such as the quadrate bone (jaw joint) or the os intermedium carpi (inferior radio-ulnar joint), rather than upon comparative function. Such contentions are in any case usually based upon somewhat tenuous evidence, although nodules of bone, or lunules, do occur in some primate discs, but rarely in mankind (Lewis *et al.* 1970). Perhaps a more interesting observation from the functional point of view—also derived from comparative studies—is that those joints which contain articular discs usually also display translation movements, combined, of course, with others (*vide infra*, p. 430). For example, in the temporomandibular joint translation of the mandibular condyle in forward and backward directions accompanies angulation or hinge movements. However, the disc, though usually thin, is also present in carnivores, despite the negligible amount of translation in their hinge-like joints. It is also apposite to note (Moffet 1957) that during evolution of the mammalian from the primitive reptilian jaw joint, the tendon of the lateral pterygoid muscle may have become partially incorporated into the temporomandibular disc; but such a structure is nevertheless absent in earlier mammals, the monotremes and marsupials. These conflicting observations illustrate very clearly how indecisive attempts to elucidate function from morphological data must often remain.

Despite these phylogenetic discrepancies, it is a most plausible view that the pliant and sometimes elastic qualities of discs and menisci provide adaptable additional articular surfaces in the joints where they occur, not only facilitating synchronous movements of different kinds in the two compartments of a joint thus divided (p. 440), but perhaps also ensuring uninterrupted lubrication (MacConaill 1932). In joints in which translation movements can occur, complete congruity is impossible, except in the theoretical case of an entirely plane articulation, the nearest approximation to this being the joints of some sesamoids. In general, however, all articular surfaces exhibit some degree of curvature, and hence translation is almost always compounded with another form of motion (p. 434). Nevertheless, translation does occur, and intra-articular cartilages may aid in controlling the shift from one position to another, as in rolling motion in the knee joint (p. 437). It has also been suggested that the tapering profile of menisci enables them to fit easily between incongruent joint surfaces, thus obviating large collections of synovial fluid between them and preserving a filmy distribution of the fluid over the articular surfaces. The advantages of this in preventing turbulence and drag accord with lubrication theory, and the argument provides a most attractive hypothesis; unfortunately, it is difficult to envisage at present how it could be experimentally demonstrated.

It is important to appreciate that a complete articular disc creates in effect two joints in series, a fact which can be compared with the concatenation of multiple joints, each affording a relatively small range of movement, which nevertheless summate to produce a much larger motion. This kind of arrangement is present in the carpus and tarsus, but is perhaps more clearly appreciable in the multiplication of interphalangeal joints in the flippers of many extinct aquatic reptiles and certain extant aquatic mammals, such as some whales. It is even possible in this kind of analysis to liken the talus in some mammals to a meniscus, in terms of function, of course; it may also be equally tempting to equate some menisci with degenerate skeletal elements. It should be added that the various views which have been propounded in connexion with intra-articular discs are not reciprocally exclusive; such structures may have arisen by more than one morphological route, and likewise they may contribute in more than one way to the smooth and steady performance of synovial joints.

The functions of two other types of intra-articular structure are also uncertain, these being *labra* and *fat pads*. A labrum is a fibrocartilaginous annular *lip* (as the word implies), which is usually triangular in cross-section, like a meniscus, and attached to the margin of an articular surface. Typical examples are the glenoid and acetabular labra. These deepen their respective sockets and add to the area of contact between articulating bone surfaces. The latter fact prompts the supposition that they may act as lubricant spreaders and that, like menisci, they may cut down synovial fluid space to capillary dimensions, thus limiting drag. It is important to note that, unlike menisci, labra are not in a position to be compressed between articular surfaces. Fat pads are closely associated with synovial membrane, with which they are therefore described (p. 427).

The Evolution of Synovial Joints

Much speculation and a considerable volume of observation and experiment have been directed in recent years to the mechanics of synovial joints by workers in several disciplines, including orthopaedic specialists, physicists and engineers. The basic problem is to account convincingly in terms of movement, loading and lubrication for the unusual and perhaps unique efficiency with which biological joints preserve their smoothness of action, without any real tendency to jamming, even under the most variable conditions, except when diseased. It is at present impossible even to postulate satisfactorily, much less to construct, a single mechanical joint capable of meeting all the requirements of the biological arrangement. Accepted lubrication theory has not yet provided a connected explanation of the outstanding effectiveness of synovial joints, though considerable progress has been made (p. 428).

Since not all joints are synovial it is pertinent to inquire into the specific advantages of sliding articulations. That these are considerable is indicated by the general observation that synovial joints have become increasingly dominant as vertebrates have evolved to higher forms. The evolutionary history of joints has attracted little attention, and few joints or vertebrate groups have received systematic study. It is clearly established that synovial articulation occurs as far down the scale as the more primitive bony fishes, such as the lungfishes (Dipnoi), especially in the jaw joint (Haines 1942). Hence, the synovial type of joint is not, *per se*, a particularly novel arrangement, in its general features at least. Nevertheless, the great majority of the movable bony junctions in the piscine ancestors of the land vertebrates were simpler in construction, as is so in their surviving representatives and relatives today. These simpler forms of joint are likely to illustrate stages in the development of synovial joints themselves. It appears acceptable to consider fibrous and cartilaginous junctures, as described above, the simplest and probably the primary form of articulation, the next step forward being the appearance of multiple small fluid-filled cavities in the deformable tissue of the joint. A further advance is held to be the union of these spaces into a single joint cavity, centrally placed and still surrounded by a substantial cuff of the fibrous or fibrocartilaginous tissue uniting together the skeletal components involved. This pattern of organization resembles those symphyses in which cavitation occurs. The final step would be the complete dissolution of continuity between the two bones, eading to a markedly increased potential for amplitude of movement. With the subsequent development of a synovial stratum and external to this a fibrous capsule, the fluid-containing cavity of the joint becomes more strictly confined, the fluid component being reduced to a mere film between joint surfaces now approximated in smooth, sliding contact.

Examples of all the stages mentioned above can be demonstrated in living vertebrates; it is of some interest to note that in the more primitive land animals the synovial joints, which increasingly replace other arrangements in their limb articulations, are in some respects less well organized than the piscine jaw joint. It is not improbable that the jaw joint was the first vertebrate articulation to reach the refinement of synovial status.

The hypothetical stages described above are in general accord with the major events in the prenatal development of human synovial joints. Perhaps the most significant happening is the breakdown of the interzonal mesenchyme (p. 154) to form the precursor of the synovial cavity. This potentiality of mesoderm for cleavage (with the formation of either considerable cavities or spaces which are no more than clefts between surfaces in appositional contact and hence freely moving) is obviously most significant to the evolution of synovial joints. In addition to the regularly occurring mesodermal discontinuities, including the whole range of coelomic and synovial arrangements, fully differentiated connective tissue appears to retain this potentiality, which may be considered to reveal itself in the occurrence of such acquired dispositions as adventitious bursae. In view of this, it is reasonable to suppose that, once a region of pliant connective tissue is established between rigid skeletal elements, their ultimate separation by a synovial cavity appears to be an evolutional probability.

It is considered likely that synovial joints may have developed not only as discontinuities in previously uninterrupted skeletal bars, but also by the approximation of already separate elements. This may have occurred in the formation of the tibiofibular joints of mammals, the two bones being out of contact in reptiles, but the synovial arrangements in this instance are probably extensions of the knee and ankle joints, and not new formations. Perhaps a better example is the occasional synovial joint between the clavicle and the coracoid process (p. 455), with articular cartilages covering the opposed bone surfaces (Lewis 1959). Even more interesting is the observation that the joint may consist merely of a bursa or, more frequently than either of these arrangements, that it may be a fibrocartilaginous junction. This provides an excellent indication of the lability of joints both in occurrence and in structure. As we have seen, a synovial joint such as the sacro-iliac may exhibit retrogression to a simpler form, and this, and other joints such as the interphalangeal, carpal and tarsal articulations, may disappear entirely, either by synostosis in the individual or as an evolutionary phenomenon in a whole group of animals. The tibiotarsus and tarsometatarsus in the avian leg illustrate simplification by re-union of once separate bones, whereas the opposite tendency towards the multiplication of elements by new joint formation is apparent in the phalangeal pattern in the paddle-like extremities of some aquatic mammals (and extinct reptiles).

The further evolution of synovial articulations at the mammalian level shows two particular tendencies. Firstly, the proportion of such joints increases by the replacement of non-synovial arrangements; this change affects in particular the limbs, eventually reaching even their smallest terminal articulations. Secondly, synovial joints display an increasing specialization, especially towards limitation of their potentiality for movement to what is required at individual joints. A typical non-synovial joint, such as a symphysis, is essentially multi-axial, inasmuch as it permits movement in all directions, however limited in range. These movements are basically bending or angulation, and torsion or rotation ('swing' and 'spin'—see p. 434) and translation; all other movements are combinations of these basic possibilities. Judging from the restricted data available, it is probable that the earliest synovial joints are likely to have permitted the same repertoire of movement, with improvement in range and smoothness. This implies that the ancestral vertebrate limb contained joints roughly approximating to a 'ball-and-

socket' form of construction; paleontological data, as well as the status of synovial joints in extant primitive amphibians and reptiles, support this supposition. This viewpoint also entails that the proximal joints of the limbs have in this sense remained less specialized than the more distal articulations; they certainly evince a closer approximation to multi-axial activity. A multi-axial joint requires considerably more complex muscular control than a bi-axial one, in which more reliance can be placed upon ligaments, an advantage which is even more marked in uni-axial joints. This advantage obtains in both dynamic and static situations: for when a multi- or bi-axial articulation is involved in a substantially uni-axial form of movement, muscle effort must be utilized to some extent to prevent unwanted movement in other axes. Similarly, when a joint is to be maintained static in some phase of movement or stance, this can be effected more economically and efficiently if the joint surfaces themselves, together with the disposition of adjoining ligaments, impose limitations upon movement in some directions. Unless this interferes with the overall activity of the joint, the ultimate trend will be towards uni-axial functioning, and this appears to be the evolutionary tendency in the limbs. This is really tantamount to saying that joints are adapted to the range of motion required of them in the characteristic activities of any particular animal form. To take the argument further, a joint may even disappear if its activity ceases; examples of this can be indicated in many mammals, and in the human skeleton the process is apparent in synostosis of the segments of the sternum and sacrum, an abolition of joints present and functional in other mammals.

The refinement of a joint evident in the limitation of its direction and range of movement to whatever is functionally habitual in motor behaviour is conducive to the development of skilled control, and this is linked to the most advantageous distribution of available muscle power. There are practical limitations to the sheer bulk of muscle concentrated around a joint, and it is clearly an advance if restraining activities can be transferred from muscle to the influence of ligaments and to the shape of the articular surfaces. A marked instance of this is the arrest of joints in temporary but perhaps prolonged static positions, as is required of the human hip and knee in standing. If the joints can be maintained in a nearly close-packed position (p. 438) by the action of gravity, muscular effort can be reduced to a minimum, and this is, of course, the basis of the mechanism of standing in mankind (p. 617). This commonly cited example of joint control is associated with the concept of *stability* at joints, a now somewhat overworked term, whose *static* implications are obvious. Joints are, however, primarily dynamic, and during actual movements the qualities which render them stable in one position are equally important in controlling transit from one attitude to another. In the medical context it is perhaps natural to view these factors as preserving the integrity of the articulation, a somewhat negative approach, which appears to be more concentrated upon the features of joints likely to prevent dislocation than upon a functional interpretation of their structural details. The synovial joints of higher vertebrates such as man are superbly engineered to accomplish their customary movements with greatest effectiveness commensurate with the limited resources available. However, before taking analysis of these activities further, it is appropriate to consider in greater detail the intimate structure of the tissues involved in synovial joints.

The Structure of Synovial Joints

The articular surfaces of the majority of bones are formed of a special variety of hyaline cartilage which reflects their preformation as cartilaginous models in embryonic life (Barnett *et al.* 1961; Ghadially and Roy 1969). In contrast, the clavicular surface at the sternoclavicular joint and both surfaces at the temporomandibular joint are of dense fibrous tissue containing only isolated groups of chondrocytes with little surrounding matrix, again reflecting the initiation of these bones by mesenchymatous centres of ossification. Nevertheless, it is an oversimplification to regard the cartilaginous articular surfaces as unmodified hyaline

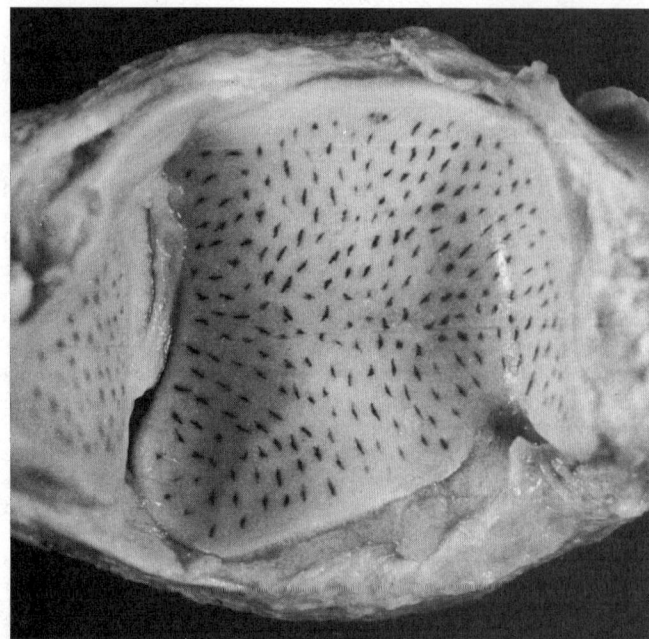

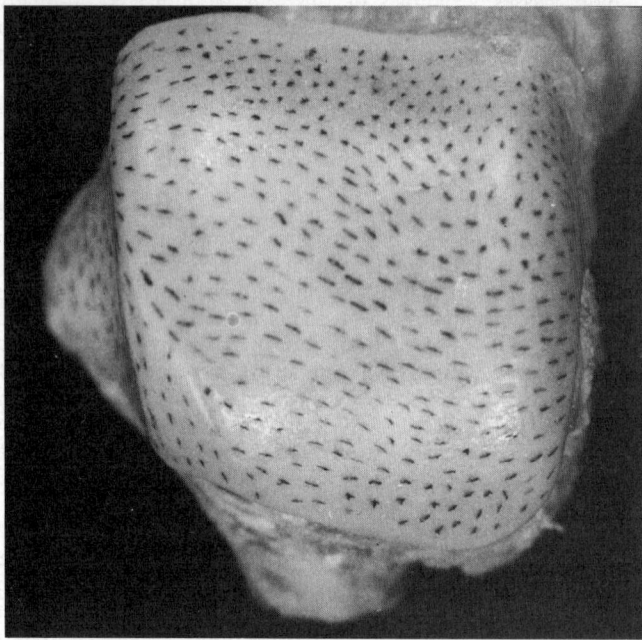

4.4B Articular surfaces of the talocrural joint (left), demonstrating the patterns of split-lines in the articular cartilages produced by multiple insertions of a round-bodied needle, previously immersed in Indian ink. *Above*, tibiofibular mortice; note interosseous ligament. *Below*, superior (tibial) and lateral (fibular) surfaces of the talus. For possible significances of split-line configuration, *see* p. 250.

cartilage and merely a surface sector of the primitive model which has grown, but remained unossified. In long bones it can be recognized as a specialized tissue long before the subjacent bone has ossified (Davies and Edwards 1948), whilst mature articular cartilage shows a highly distinctive radial organization with marked variations in cell type and arrangement, fibrous architecture and calcification level with increasing depth from the surface. Four strata have been described and these are considered in greater detail elsewhere (p. 349). The split-line phenomenon in articular cartilage, and its possible relationships to the architecture and mechanical properties of the cartilage are discussed on p. 250. Illustration 4.4B displays examples of the split-line patterns in a talocrural joint.

Essentially articular cartilage provides a wear-resistant, low-friction lubricated surface, both slightly compressible and elastic, which is ideally constructed for ease of movement over a similar

surface but able to accommodate the relatively enormous forces of compression and shear generated during weight-bearing and muscle action. The latter properties are of especial importance at one end of the range of the most habitual movement of the joint, when it enters the so-called close-packed position (p. 438).

The thickness of articular cartilage varies considerably and is often stated to range from 1 to 2 mm, but this is more typical of small bones in individuals of advanced age. In youth it may reach a maximum thickness of 5–7 mm in the larger joints, and such young cartilages are typically white, smooth and glistening to the naked eye, with well marked compressibility. In contrast, ageing cartilages are thinner, less cellular, firmer, more brittle, with a less regular surface, and they develop a yellowish opacity. Despite the 'glassy' appearance of moist, freshly exposed articular cartilage, and the earliest measurements which supported this impression of smoothness, later estimates using more refined techniques emphasized the relative 'roughness' of the surfaces (Dowson *et al.* 1969; Longfield *et al.* 1969). The 'centre-line average' of the undulations described by these authors varied between 30×10^{-6} inch and 200×10^{-6} inch, which compares very unfavourably with an engineering bearing (5×10^{-6} inch to 15×10^{-6} inch); but when lubricated with synovial fluid the surface exhibits an extremely low coefficient of friction (< 0.002). It was proposed that under conditions of loading the 'valleys' between the 'peaks' trap pools of synovial fluid. Others (McCutchen 1959) have discussed the porous nature of articular cartilage (a hypothetical pore size of 6 nm was estimated) and suggest that it may be regarded as a spongework, which under resting conditions is saturated with synovial fluid. With increasing load and compression of the cartilage it is presumed that synovial fluid 'weeps' from its porous surface. The advent of scanning electronmicroscopy has greatly renewed interest in articular surfaces. When *detached* blocks of articular cartilage were prepared for the scanning microscope many authors described, initially, patterns of ridges and undulations as a constant and characteristic feature of the surface (e.g. McCall 1968; Gardner and Woodward 1969; Walker *et al.* 1969; Redler and Zimney 1970; Gardner 1972; Mow *et al.* 1974; Redler 1974). However, this view has been strongly challenged, since the ridges and undulations were absent when the cartilage was *still attached* to subchondral bone; such appearances were therefore regarded as artefactual, atypical or pathological (Clarke 1973a; Ghadially *et al.* 1976). The only constant feature seen with the scanning microscope in what the authors regarded as 'normal', mature articular cartilage was the occurrence of numerous oval, fusiform or rounded *shallow pits*, thought to correspond to the sites of underlying chondrocytes (Clarke 1973a, b; Ghadially *et al.* 1974, 1975, 1976). However, in juvenile cartilage, instead of pits, these features were replaced by elevations or small 'humps' of similar dimensions. Doubt was therefore expressed concerning the 'enrichment theory' of joint lubrication (p. 428 and *see* Ghadially *et al.* 1977). It must be stressed, however, that despite the impressive micrographs produced, and the fact that the articular cartilage was still attached to subchondral bone (unquestionably of great technical importance), the fact remains that the cartilage was subjected to all the rigorous preparative procedures for scanning microscopy. To what extent the observed pits and humps reflect this is uncertain.

Recently, however, some of the problems associated with tissue preparation for scanning electron microscopy have been circumvented by studying fresh, moist, necropsy specimens of the weight-bearing area of the lateral femoral condyle using reflected light interference microscopy (Longmore 1976, 1978; Longmore and Gardner 1978). Curvatures, undulations and irregularities of the articular surface were classified into four groups of decreasing dimensions: (1) the *primary anatomical contours*, e.g. ovoid or sellar surfaces; (2) *secondary undulations* of 100–500 μm crest to crest; (3) *tertiary hollows* which were 20–50 μm in diameter and 0.5–2 μm deep. (4) *quaternary ridges* which were 1–4 μm in diameter and 130–275 μm deep. In their most recent publication (Longmore and Gardner 1978) the four classes were confirmed; particular emphasis was placed on the tertiary hollows, and their size, shape, and distribution were shown to be strongly age-dependent (a phase of *maturation* from 0–21 years eing followed by *ageing* from 22 to 50 years). In the youngest age roup the tertiary hollows were predominantly circular and ghtly packed, when viewed *en face*. With increasing age they radually separated, becoming circular to oval, sometimes ppearing in clusters or 'figure of eight' pairs, while the areas etween hollows although 'granular and fuzzy' were relatively mooth. From 21 years onwards, quaternary ridges, that had first ppeared sporadically between 10 and 20 years, were now present ore frequently in focal patches occupying some of the 'inter-ollow areas'. It was concluded that the tertiary hollows almost ertainly corresponded to underlying chondrocyte lacunae, and hat the quaternary ridges may represent incipient cartilage brillation.

Articular cartilages are closely moulded to the bone ends but heir zonal variations in thickness often accentuate the overall urface geometry of the underlying bone. Typically male (convex) urfaces are thickest centrally, gradually thinning peripherally, vhilst the converse holds for female (concave) surfaces. The recise configurations of these, their degree of congruence closeness of mutual fit) in various positions, together with the lisposition and mechanics of the surrounding fibrous capsule and gaments are, of course, intimately related to the types and ranges f movement permitted at any specific joint (*vide infra*). (For a implified classification based upon the approximate geometry of oint surfaces *see* p. 432, and for more detailed comments *see* p. 37.)

Articular cartilage contains neither nerves nor blood vessels except occasional looped vessels which penetrate as far as and ometimes even through the calcified zone (p. 251). Its nutrition is onsidered to be derived from three sources: the vascular net in he synovial membrane near its periphery (the *circulus vasculosus rticuli*), from the synovial fluid, and from the blood vessels in the underlying marrow spaces, but the relative importance of these ources is still disputed.

The free surface of articular cartilage has no covering of erichondrium and was for long regarded as a thin, fibre-free *amina splendens* (*vide infra*) but recently this has been shown to onsist of a feltwork of fine filaments—probably a form of ollagen with minimal interstitial substance (p. 250), and, more leeply placed, their attendant, flattened, elongate fibroblasts. This view, having received further support, implies that the oundary layer on the articular surface of *all* synovial joints is omposed almost exclusively of collagen. (The origin of the term amina splendens is explained on p. 250, and it is appropriate that ts systematic use should now be abandoned.)

With advancing age, undulations of the articular surfaces ncrease in prominence and they become studded with minute, rregular, shred-like projections, perhaps indicating the effects of rolonged wear and tear. Erosive effects undoubtedly occur in athologically 'dry' joints and also in those with an altered viscosity of the synovial fluid, but in substantially normal joints uch processes are extremely slow. There is no direct evidence for ny continual replacement of eroded surface layers by proli-eration of the deeper tissues—mitotic figures are absent in adult rticular cartilage and previous claims of amitotic cell division ave not been substantiated.

The fibrous capsule consists of parallel and interlacing undles of white connective tissue fibres. It forms a cuff or orassard, each end of which is attached to a continuous line round the articular end of one of the bones concerned, usually in he immediate neighbourhood of the periphery of the articular urface, but this arrangement is subject to considerable variation. t is perforated by the articular vessels and nerves, and may resent one or more apertures through which the synovial nembrane protrudes to form a pouch or sac. The fibrous capsule usually shows two or more localized thickenings in which the constituent fibre bundles are generally parallel to one another. These thickenings are the *capsular ligaments* of the joint and they re named according to their position or attachments. In some oints the fibrous capsule is reinforced or replaced by the tendons, or by expansions from the tendons of neighbouring muscles. Some joints possess *accessory ligaments* which stand clear of the ibrous capsule. Such accessory ligaments may be either extracapsular or intracapsular in position.

All ligaments are tough and unyielding, but at the same time flexible and pliant, so that they offer no resistance to normal movements. They are designed to prevent the occurrence of excessive or abnormal movements, and every ligament becomes taut at the normal limit of some particular movement (p. 439). They are elastic only within narrow limits, and are protected from excessive tension by reflex contraction of appropriate muscles (Smith 1954).

Synovial membrane (Barnett *et al.* 1961) is a characteristic tissue derived from embryonic mesenchyme which lines the nonarticular parts of synovial joints, synovial bursae and synovial tendon sheaths—all regions where the essential function of the tissues entails movement between contiguous planes. In each case, the apposed surfaces are lubricated by a fluid which superficially resembles egg-albumin (therefore named *synovia*), and which is secreted and absorbed by the synovial membrane.

In joints, the membrane lines the fibrous capsule and clothes any bony surfaces, ligaments and tendons which are intracapsular in position. It is absent, however, from the surfaces of intra-articular discs or menisci and ceases at the margins of the articular cartilages. The peripheral few millimetres of the latter are a *transitional zone*, with intermediate grades of structure between typical membrane and cartilage surface.

Pink, smooth and shiny throughout much of its extent, the inner surface of the synovial membrane bears occasional small finger-like projections, the *synovial villi* (Ghadially and Roy 1969). Elsewhere the membrane is thrown into folds and fringes which project into the joint cavity—some of these are sufficiently constant and prominent to be named, e.g. the alar folds and the ligamentum mucosum of the knee joint. Accumulations of adipose tissue are characteristic of the synovial membrane of many joints, and these *articular fat pads* were mistakenly identified as 'mucilaginous glands' by earlier anatomists. These folds, fringes and pads form flexible cushions which fill those potential spaces and irregularities in the joint 'cavity' which are not filled with the small volume of synovial fluid, and they accommodate to the changing shape and volume of these irregularities with joint movement. They effectively increase the surface area of the synovial membrane and they may promote an adequate distribution of lubricant synovial fluid over the articular surfaces (cf. intra-articular discs and menisci). The synovial villi are few in number in normal joints but are more numerous where the synovial membrane rests upon loose areolar tissue near the articular margins and over the surfaces of folds and fringes. They increase in number with age and become very prominent in various pathological states.

Structurally, synovial membrane varies considerably in different regions of a joint but essentially consists of a cellular *intima* which rests upon a vascular fibrous *subintimal lamina* (often less fittingly called the *subsynovial tissue*). The latter is often loose and areolar, but sometimes contains organized laminae of collagen and elastin fibres running parallel to the membrane surface, between which are scattered fibroblasts, macrophages, mast cells and fat cells (Davies 1950). The elastic component prevents the formation of excessively redundant folds during joint movement which would become compressed between the articular surfaces.

The subintimal adipose cells which accumulate as fat pads are unremarkable, but they are arranged in compact lobules surrounded by fibro-elastic interlobular septa. These septa are richly vascularized and they impart a firmness, deformability and elastic recoil, which again are important during joint movement.

In contrast, where synovial membrane lines prominent intrinsic ligaments or clothes intracapsular tendons the subintima is difficult to distinguish as a separate zone, being formed of fibrous tissue which merges with that of the adjacent capsule, ligament or tendon.

The synovial lamina intima (*lamina propria synovialis*) also shows regional variations but consists basically of one to four layers of *synovial cells* embedded in a granular, amorphous, fibre-free intercellular matrix. The cells vary from polyhedral with round or oval nuclei, to flattened, elongate cells with spindle-shaped nuclei. Many possess delicate branching processes which pass horizontally and intermingle with those of adjacent cells.

The free surface of some superficial cells carries a few filopodia, whilst others are uncomplicated or slightly wavy. In many areas, but particularly over areolar subintimal tissue, the cells are ill-opposed with intervening narrow gaps which sometimes contain extensions of the subintimal connective tissue (which therefore borders the synovial cavity at these points). Conversely, over a fibrous subintima, the cells are more flattened, approach each other more closely and may even form endothelioid sheets.

Ultrastructurally, two cell types (A and B cells) have been recognized (Barland et al. 1962), but cells with intermediate characteristics are common, and perhaps the described differences merely reflect quantitative variations in functional activity rather than distinct cell lineages.

Synovial cells of *type A* predominate and are characterized by surface filopodia, plasma membrane invaginations and associated micropinocytotic vesicles. Their cytoplasm contains numerous mitochondria, varieties of lysosome, a system of cytoplasmic filaments and a particularly prominent Golgi apparatus and associated smooth-walled vesicles, but profiles of endoplasmic reticulum are scanty. Neighbouring cells are often separated by distinct gaps, but where they approach more closely their surfaces may be complex and interdigitate. The latter is common in compact areas of rat synovial membrane, as are tight junctions and desmosomes between adjacent cells, but these have not been identified in human joints.

In *type B synovial cells* most of the above characteristics are poorly developed, but in contrast they contain a wealth of rough endoplasmic reticulum, varying from small round or oval profiles to large flattened, intercommunicating cisternae, together with scattered free cytoplasmic ribosomes.

Both cell types contain glycogen deposits, but lipid inclusions are rare, as are paranuclear centrioles. The latter correlates well with the low mitotic rate of intimal synovial cells, and any cell replacement of the surface layers probably follows transformation of primitive fibroblast-like subintimal cells.

The significance of the synovial cells is best considered in relation to synovial fluid production and absorption, and the removal of other substances from joint cavities.

Synovial fluid is found in the cavities of synovial joints, bursae and tendon sheaths, but its composition has not been extensively investigated in the latter two sites. That obtained from synovial joints is a clear or pale yellow, viscous (glairy) fluid, of slightly alkaline *p*H at rest (though diminishing on exercise), which carries a small mixed population of cells and some amorphous particles which exhibit metachromatic staining. Its viscosity, volume and colour vary widely between different joints and species, and it has proved difficult to correlate these variations with a particular joint or with the size, weight or condition of exercise of the animal studied. In man its volume is low, and usually less than 0·5 ml can be aspirated from a large joint such as the knee.

The physical properties of synovial fluid are particularly characteristic and it shows viscous, elastic and plastic components. Many of the wide divergencies and inconsistencies of earlier investigations stemmed from the relatively crude technical methods available and a failure to appreciate the non-Newtonian properties of the fluid, i.e. with low rates of shear, the fluid is highly viscous, but the measured viscosity drops dramatically with increased rates of shear. It was therefore proposed that with slow joint movement the weight-bearing capacity would be maximal and that with fast movement there would be reduced impedance by fluid drag. However, more recent workers have pointed out that the *product* of the viscosity and shear rate is approximately constant and therefore also the weight-bearing capacity. Additionally it has been shown that the viscosity is very sensitive to changes in dilution and that it falls with increasing temperature, and increasing *p*H. The elasticity of the synovial fluid is similarly affected by changes in dilution, *p*H and temperature, but it increases with increased rates of shear, in contrast to the viscosity changes under similar conditions. (For extended earlier reviews of the physical properties of synovial fluid and theories of lubrication consult Barnett 1958; McCutchen 1959; Barnett et al. 1961; Dintenfass 1963; Negami 1964; Davies and Palfrey 1965; MacConaill 1966; Dowson et al. 1969;

Longfield et al. 1969. For a more recent, excellent, collection of essays on the physical and biological properties of adult articular cartilage including nutrition and lubrication, *see* Freeman 1973; and for an analysis of the structure, properties and biochemistry of cartilage in general *see* Serafini-Fracassini and Smith 1974.)

The composition of synovial fluid is consistent with the view that it is a dialysate of blood plasma containing some protein (about 0·9 mgm/100 ml) and with added mucin. The latter is principally hyaluronate, a sulphate-free mucopolysaccharide containing equimolar concentrations of glucuronic acid and N-acetyl-glucosamine, and much evidence has accumulated to show that the visco-elastic and thixotropic (plastic) properties of the fluid are largely determined by its hyaluronate content. The origin of the synovial mucin proved elusive for many years and was variously ascribed to an attrition product of the matrix of articular cartilage, transformed products of synovial mast cell secretion, or products of the synovial cells themselves. Investigations by light microscopy, histochemistry and tissue culture have proved inconclusive, but recent studies using colloidal iron techniques combined with electronmicroscopy revealed a strong positive reaction for polysaccharides in relation to the plasma membrane, filopodia, cytoplasmic vesicles and Golgi apparatus of the A type of synovial cells. This evidence, together with comparable results in other cells synthesizing polysaccharides, is regarded as strong support for the view that A cells are the site of synovial hyaluronate synthesis.

Synovial fluid protein is partly free and partly bound to mucopolysaccharides including hyaluronate. Most workers agree that much of this protein derives from the blood plasma, but the possible addition of some of the hyaluronate-bound protein by secretion from the B synovial cells cannot be excluded.

The small content of cells in synovial fluid (about 60 per ml in resting human joints) consists of monocytes, lymphocytes, macrophages, free synovial cells and occasional polymorphonuclear leucocytes (Bauer et al. 1940). The higher cell counts characteristic of younger age groups, and many species other than man, probably reflect more active movements of the joints in the period immediately before sampling of the fluid. The amorphous, metachromatic particles, and the fragments of cells and fibrous tissue sometimes found in synovial fluid, are thought to be the result of slow wear and tear of the joint surfaces.

The functions of synovial fluid are considered to include the provision of a liquid environment, with a narrow *p*H range for the joint surfaces, a nutritive source for the articular cartilages, discs and menisci, and as a lubricant which increases joint efficiency and reduces erosion of surfaces. To what degree it operates as a source of nutrition compared with direct diffusion from neighbouring vascular plexuses, remains unclear. Similarly, whilst most are agreed concerning the general lubricant significance of the fluid, and in particular, its hyaluronate content, the detailed analysis of its action is still disputed.

Various models have been proposed for the lubrication phenomena in joints and, interestingly, these have largely paralleled the current advances in engineering physics. Initially, it was proposed that the principal mechanism involved was 'fluid film' or 'hydrodynamic' lubrication, familiar in engineering practice, and that the bearing surfaces are usually separated by a substantial layer of lubricant fluid, the effectiveness of which depends upon the rheological properties of the fluid when in bulk. A later refinement of this view included the important consequences of the elastic properties of the fluid—the so-called 'elastohydrodynamic model'. However, some investigators criticized the suitability of a joint environment for simple fluid-film lubrication on several counts, and the principles of 'boundary lubrication' were proposed, in which the properties of the solid surfaces, probably combined in some way with an extremely thin layer of lubricant molecules, were considered most important. Then followed the new concept of 'weeping lubrication' in which particular emphasis was placed on the porous, sponge-like, fluid-filled nature of the deformable articular surfaces. Under conditions of load, it was proposed that the surfaces were lubricated by a fine film of fluid expressed from the interstices of the sponge. Recently a theory of 'boosted lubrication' is receiving increasing attention, which suggests that as articular cartilages are compressed, fluid pools become trapped in the 'valleys' on their irregular surfaces. With increasing compression, the small-moleculed, mobile fraction of the synovial fluid passes *into* the cartilage over the contact area, and that the fluid remaining in the pools becomes increasingly enriched in hyaluronate content (and therefore of greater viscosity) and thus forms a progressively more effective coat which protects and lubricates the surfaces. However, as pointed out on p. 426, the presence of major ridges and valleys are almost certainly preparative artefacts, and the existence of small 'pits' or 'humps' *in vivo* cannot be regarded as proven.

When consideration is given to the wide variations in joint geometry, structure and activity, it seems quite possible that multiple mechanisms may operate under different conditions, and whilst many investigations are currently in progress, no statement of joint lubrication theory can yet be regarded as definitive (*see* references quoted above, and in particular Freeman 1973).

The synovial membrane is involved not only in the *production* of synovial fluid, but also in the *removal* of materials from the joint cavity. Small molecules of crystalloids and soluble dyes can cross the membrane and pass directly into subintimal capillaries and venules. The former are fenestrated according to Kos (1970). Particulate substances, however, pass preferentially into the subintimal lymphatic capillaries and are transported to the regional lymph nodes. Studies using thorotrast, colloidal carbon and ferritin as tracers, introduced into the joint cavity, followed by electronmicroscopy, have emphasized the marked phagocytic powers of the A-cells of the synovial intima, which rapidly enclose the particles within their micropinocytotic vesicles. The origin of the subintimal macrophages is not certain; some may enter the joint tissues from the bloodstream, but many are probably phagocytic A-cells of intimal origin which have passed into the deeper tissues (Ghadially and Roy 1969).

The Classification and Movements of Synovial Joints

Various criteria have been used as the bases for several classifications of synovial joints including their movements, and they differ considerably in their universality, scientific accuracy and practical utility. These criteria include the complexity of organization and number of articulating surfaces; the number and distribution of the principal axes about which movements occur; the approximate geometrical form of the articular surfaces and the gross movements permitted; finally, a much more precise attempt to define the geometry of the surfaces and their associated movements. The latter study forms the basis for part of a now rapidly advancing science of *human kinesiology*, and whilst this important subject can only be considered briefly in this volume (p. 431) several excellent extended treatments of the subject are available (Steindler 1955; Barnett *et al.* 1961; MacConaill and Basmajian 1969).

Complexity of organization—Many synovial joints possess only two articulating surfaces (*male* and *female*) and are termed *simple joints*. In some, one surface is convex in all directions (male) and is always greater in surface area than that of its adjoining concave (female) surface. In other simple joints, however, both surfaces are concavo-convex and in these cases, the one of greater area is considered male.

A joint possessing more than one pair of articulating surfaces is termed *compound* (e.g. at the elbow joint the distal end of the humerus presents two distinct male surfaces, the capitulum and the trochlea, which articulate with the female surfaces of the radius and ulna respectively, forming two mating pairs. In addition the convex, male articular circumference of the head of the radius articulates with the female radial notch of the ulna at the superior radio-ulnar joint. This makes a third mating pair all contained within a common articular capsule.) In all such compound joints the articulating territory of each pair remains distinct, the male surface of one pair never passing on to the female surface of an adjoining pair in any position of the joint.

Finally, where an intracapsular disc or meniscus of fibrocartilage is present, the joint is termed *complex*.

Degrees of freedom of joints. When the main positional changes of one member of a pair of articulating bones are analysed, it is often useful to consider them as rotations of the bone around one or more of three mutually perpendicular axes. The directions of the particular set of axes chosen may, for convenience, vary with the joint under consideration. In many cases they are referred to the principal planes of the body when in the anatomical position, i.e. a vertical, a transverse and an anteroposterior axis. Alternative directions may, however, be more appropriate; for example, many anatomists prefer to consider movements of the humerus on the scapula at the shoulder joint to occur around a vertical axis, an obliquely transverse axis in the plane of the body of the scapula, and a third axis at right angles to these (**4**.5).

When the movement of a bone at a joint is substantially limited to rotation about a single axis, the joint is termed *uni-axial* and it possesses *one degree of freedom*. Similarly, if *completely independent* movements can occur around two axes, the joint is classed as *bi-axial* and it possesses *two degrees of freedom*. Since there are three axes about which independent rotations may occur, a joint may exhibit one, two or *three degrees of freedom*, but no more.

When using this apparently straightforward classification, however, a number of additional points should constantly be borne in mind.

Firstly, a bone articulating at a so-called ball-and-socket joint can not only rotate about the three main axes, but also around many others intermediate in position. Such a *multi-axial* joint is still considered to possess only three degrees of freedom, since all these intermediate rotations can be resolved mathematically into components involving only three mutually orthogonal axes.

Secondly, if the movement of a bone referred to one plane is carefully examined throughout its full range, it is found that a relevant axis cannot be adequately described by a single line fixed in space, but rather by a succession of such lines which change continuously in position as the movement progresses. This positional change reflects the fact that articular surfaces are not simple geometric forms, but that their degree of curvature changes at each point across any profile (*see* p. 433). Such changes vary greatly in amount with different forms of joint (*vide infra*), but it should be fully realized that it is an approximate *mean position* of such a moving axis that is often referred to as *the axis* of conventional description. For many purposes, however, such compromise axes can be a most useful concept, particularly when a study of only the gross aspects of movement of a part are sufficient.

Thirdly, movements which superficially appear simple and limited to one plane only are, on closer examination, seen to be compound. For example, what initially appears as a simple change in angulation between two long bones is often seen to be accompanied by some degree of rotation of one of the bones about its long axis. This is an inevitable consequence of the curved nature of biologic joint surfaces (p. 432) and when, as is often the case, such a rotation cannot be carried out independently, it does not constitute an additional degree of freedom (e.g. *see* movements of the ulna, p. 465).

Finally, it should be noted that many movements at joints contain an element of *translation*, in which one articular surface slides bodily across its partner, and in which any element of rotation or angulation may be slight (but *see* plane joints, below). Translations are also a feature of many joints in association with large angulation changes (e.g. the shoulder joint). It will be appreciated that such movements of translation are not accommodated by the foregoing classification, which includes only three rotational degrees of freedom.

GROSS MORPHOLOGICAL CLASSIFICATION OF SYNOVIAL JOINTS

The approximate shape of synovial joints has formed the basis for a widely used classification and whilst seven varieties have been recognized officially, and may indeed have some practical utility, a closer examination shows that there is no exclusive basis for their separate classification. They are, as we shall see, merely variations (although sometimes extreme) of two basic geometrical forms (*see* p. 432).

Plane joints are formed by the apposition of fairly flat articular surfaces (e.g. the intermetatarsal joints and some of the intercarpal joints) but even here close examination shows them to possess some degree of curvature. For many purposes the latter are disregarded, and the movements are considered to be pure translations, or slides, of one bone on its neighbour. In more advanced treatments of joint dynamics, however, the consequences of the slight curvatures assume a greater importance (*vide infra*).

Hinge joints (ginglymi) roughly resemble the hinges of a door, where the surfaces are so moulded as to largely restrict the to-and-fro movement to one plane, i.e. they are uni-axial joints. The sides of the joint are typically provided with strong collateral ligaments, and the interphalangeal and humero-ulnar joints are good examples. However, the surfaces of a biological hinge depart from the regular cylinders of the engineer, and in section their profiles are not arcs of circles, but varieties of spiral. Consequently the main to-and-fro swing of the bone is inevitably accompanied by some rotation about its own long axis (*vide infra*).

Pivot (trochoid) joints are also uni-axial and comprise a central bony pivot surrounded by an osteoligamentous ring. Movement is restricted to a rotation around a longitudinal axis through the centre of the pivot. The pivot may habitually rotate within the ring, as in the case of the proximal radio-ulnar joint, in which the head of the radius rotates within a ring formed by the annular ligament and the radial notch of the ulna. Conversely, the ring may rotate around the pivot, as in the articulation between the dens of the axis and the ring formed by the anterior arch of the atlas and its transverse ligament.

Bicondylar joints are allied to the other uni-axial joints having a principal movement occurring largely in one plane, but in addition, a limited amount of rotation is possible about a second axis set at 90° to the first. These joints have male surfaces formed by two distinct, convex, knuckle-shaped *condyles* which articulate with two concave female surfaces (which in some cases are also, less appropriately, known as condyles). The condyles may be almost parallel as in the knee joint, where they are enclosed in a common fibrous capsule. Alternatively, the condyles may be in line and even enclosed in separate capsules. An example of this is the pair of temporomandibular joints which necessarily cooperate in all mandibular movements, and which *together* form a true condylar mechanism (and should always be studied as such).

Ellipsoid joints are bi-axial and formed by the reception of an oval, convex, male surface into an elliptical female concavity. Examples are the radiocarpal joint and the metacarpophalangeal joints. Primary movements are possible about two axes at right angles (e.g. flexion-extension, and abduction-adduction). Whilst these movements may be combined as a movement of circumduction (*vide infra*), there is little appreciable rotation around the third axis, as this is prevented by the overall shape of the articular surfaces.

Sellar (saddle) joints are also bi-axial and their apposing surfaces are concavoconvex. If either surface is examined it is found to have a particular direction in which it is maximally convex, whilst at *right angles* to this, it is maximally concave (p. 433). In the articulated joint, of course, the convexity of the larger (male) surface is apposed to the concavity of the female surface and vice versa. Primary movements occur in two planes at right angles, but because of the articular geometry these are accompanied by a degree of axial rotation of the moving bone. Such *conjunct rotations* cannot occur independently, but they must not be regarded as simply a by-product of 'imperfect' mechanics. As we shall see, they are of considerable functional significance during the habitual positioning of bones and in the limitation of joint movement (pp. 438, 439).

Perhaps the best known and most fully investigated saddle joint is the carpometacarpal joint of the thumb; others include the talocrural (ankle) joint and the calcaneocuboid joint.

Spheroidal joints (often, somewhat loosely, termed ball-and-socket joints) are formed by the reception of a male globular 'head' of one bone into a cup-like concavity—excellent examples

are the hip and shoulder joints. Such joints are, of course, multi-axial and possess three degrees of freedom. However, it important to realize that their articular surfaces, althoug superficially resembling parts of spheres, are in fact not spherica but slightly egg-shaped (*articulatio ovoidalis* has been proposed a an appropriate term for official recognition). Consequently, i most positions of the joint there is *not* a perfect fit (or congruence between the apposed surfaces. The latter occurs only in on position of the joint, at the end of the range of its commones movement (*vide infra* for further discussion).

The Movements and Mechanisms of Joints

In the following section the subject of joint movement will first b considered using the terms and concepts usually encountered i standard textbooks of anatomy. This is then followed by a brie review of the more important approaches and ideas which ar emerging in modern texts on kinesiology.

The movements permitted in joints are conventionall considered to be of four kinds: gliding and angular movement circumduction and rotation. Almost always these are combined t produce an infinite variety. Where movement is limited in exten the reciprocal articular surfaces approach each other in size, bu where movement is free the bone which is habitually more mobil possesses the larger surface.

Translation is the simplest kind of motion that can take plac in a joint, one surface sliding over another without an appreciable angular or rotatory movement. It occurs as a important element combined with other movements in man joints, but in some carpal and tarsal articulations it is ofte regarded as the only motion permitted. However, even here examination of cineradiographs reveals a considerable degree o rotation and angular change in the relative positions of the smal bones during most movements of the carpus and tarsus.

Angular movement implies diminution or increase in th angle between adjoining bones. Two types of angular movemen are so common, especially in the limbs, that attempts must b made to define them. They occur around axes set at right angles t each other and are named (1) *flexion* or bending, and its opposite *extension* or straightening, and (2) *abduction* and also its opposite *adduction*.

Flexion, although in wide biological use, is difficult to defin with precision. It often occurs around a transverse axis and result in the approximation of two morphologically ventral surfaces The thumb, however, lies in a plane set almost at right angles t that of the fingers. Its dorsal surface is directed laterally and accordingly, flexion and extension at each of its joints occu around an anteroposterior axis. At the shoulder joint flexion is more appropriately considered to occur around an obliquel transverse axis through the centre of the head of the humerus, and in the plane of the scapular body. Thus the arm is carried forwards and medially and is brought no nearer to the ventral aspect of the trunk. Again, at the hip joint, flexion which here occurs around a transverse axis, approximates a morphologically *dorsal* surface of the thigh, to the *ventral* aspect of the trunk—this reflects the characteristic rotation of the hind limb bud in early embryonic life (p. 153).

The status of the movements at the talocrural (ankle) joint is again complicated by the posture of the foot, which is set at a righ angle to the leg. Elevation of the foot produces a diminution of this angle and is sometimes termed flexion, but it results in the approach of two morphologically dorsal surfaces and might, with equal justification, be called extension.

Flexion has also been defined as the posture of the fetus 'in utero' and this implies that elevation of the foot at the talocrural joint is flexion. Such a view is also supported by an examination of withdrawal reflexes in which elevation of the foot is always associated with flexion at the knee and hip, whereas the converse holds in the crossed extensor reflexes. Thus the definitions of flexion based on morphological and physiological grounds are contradictory. To avoid confusion the self-explanatory terms

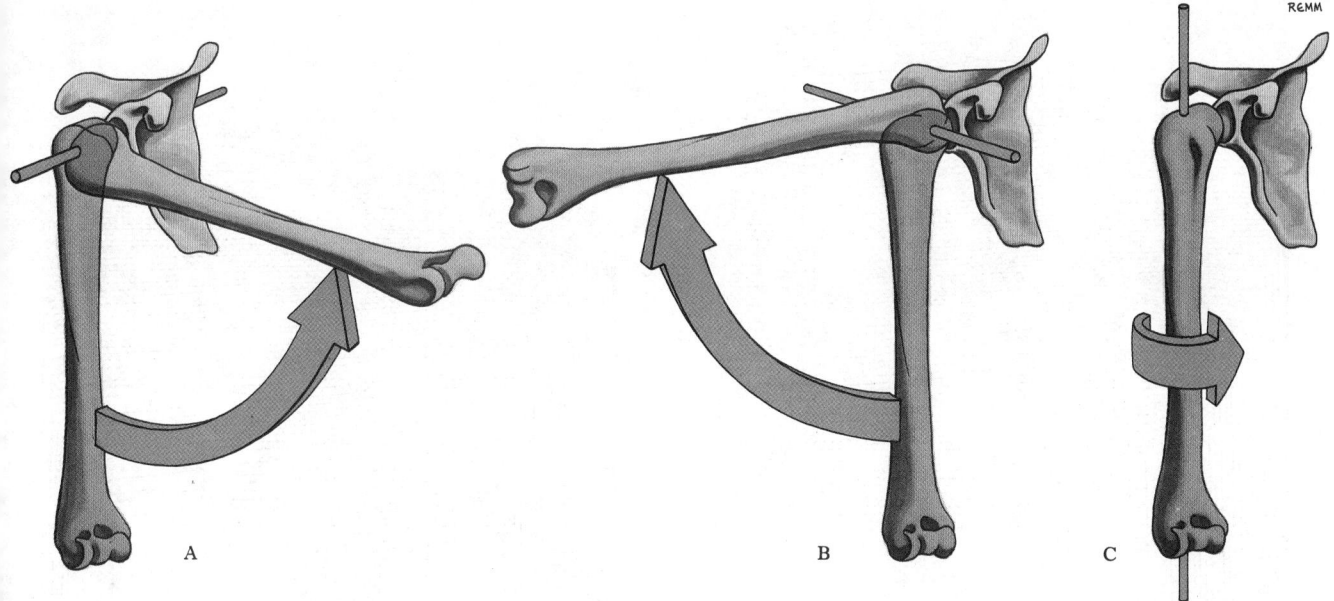

RGMM

A   B   C

4.5 The shoulder joint, which is polyaxial and possesses three degrees of freedom. In A, B and C the three mutually perpendicular axes around which the principal movements of flexion-extension (A), abduction-adduction (B), and medial and lateral rotation (C) occur are shown. Note that in the case of the shoulder the axes are referred to the plane of the scapula and not to the coronal and sagittal planes of the erect body as a whole. An infinite variety of additional movements may, of course, occur at such a joint, e.g. those involving intermediate planes, or there may be movement combinations or sequences. However, these can always be resolved mathematically into components related to the three axes which are illustrated.

dorsiflexion and *plantar flexion* will be used throughout this volume when talocrural movements are being considered.

Abduction and adduction occur around an anteroposterior axis except in the case of the carpometacarpal joint of the thumb, and the shoulder joint, for reasons already stated. The terms generally imply movement away from, and towards the midline of the body, except in the case of the digits, where the arbitrary planes of reference chosen are the midline of the middle digit in the hand, and the midline of the second digit in the foot, because these are least mobile in this respect. Abduction of the thumb occurs around a transverse axis and carries the thumb away from the palm, the movement occurring in a plane at right angles to that of the palm. Similarly, abduction of the humerus on the scapula at the shoulder joint occurs in the plane of the scapula around an oblique axis at right angles to this plane (**4.5**).

Circumduction occurs when a long bone circumscribes a conical space—the base of the cone is described by the distal end of the bone, whilst its apex is at the articular cavity; it is best seen in movements at the shoulder and hip joints. It is, of course, a derived movement in which elements of flexion, extension, abduction and adduction are compounded.

Rotation is another term which is widely used but imprecisely described in many textbooks of anatomy. *In a restricted sense*, it is often used to denote a form of movement in which a bone moves around some 'longitudinal' axis. This may be in a separate bone, as with the pivot formed by the dens of the second cervical ('axis') vertebra around which the (cranium and) atlas vertebra rotates. Alternatively, the axis may roughly coincide with the shaft of a long bone, as in medial and lateral rotation of the humerus (**4.5**). Again, the axis may not be quite parallel to the long axis of the bone, as in the movement of the radius on the ulna during pronation and supination, where it joins the centre of the head of the radius to the base of the ulnar styloid process. Another example is medial and lateral rotation of the femur, where the axis joins the centre of the femoral head and the lateral femoral condyle (p. 478).

In the examples given, the rotations can be carried out as independent movements; they are termed *adjunct rotations* and they provide an additional degree of freedom. They must be carefully distinguished from the *conjunct rotations* which occur at many joints as an inevitable accompaniment of some other main movement resulting from the geometry of the articular surfaces (*see* p. 437). As will be noted below, however, in some types of articulation, conjunct rotations may be combined with some additional degree of adjunct rotation, and in different circumstances, the latter may either increase or relatively nullify the effect of the former.

Further, it should be noted that with the exception of simple movements of translation, *all* other movements of bones are in fact rotations. For example (p. 459), as we have seen, medial and lateral rotations of the humerus occur around its own longitudinal axis (which is vertical, with the body in the anatomical position). However, the swinging 'angular' movements called flexion-extension and abduction-adduction are clearly rotations of the bone around the other two axes.

Kinesiology

The preceding account of joint movement embodies many of the more conventional views and terminologies built up over many years, often following simple inspection of articular anatomy and the more obvious positional changes of limb segments and the axial skeleton. For many purposes this will be found adequate.

In recent years, however, a closer examination of these matters has led to the rapid accumulation of a body of more objective observations, experimental results, and theory, in relation to the study of movement (the emerging science of kinesiology—a branch of biomechanics). This has necessitated a new terminology, a revision or expansion of many older concepts, and, to many, unfamiliar mathematical treatments. Only a few of these terms and ideas are considered briefly here and the interested reader should consult MacConaill (1946; 1950; 1953; 1964; 1966; 1969); Steindler (1955); Barnett *et al.* (1961), Kapandji (1970; 1974).

In the present context two fields of study are relevant, (*a*) *osteokinematics* which deals primarily with the overall movements of bones, with little reference to their related joints, and (*b*) *arthrokinematics*, concerned with the more intimate mechanics of joints.

It is appropriate to consider first certain generalizations about the shape of articular surfaces, and then the concept of replacing an irregularly shaped bone by a single straight line called the

431

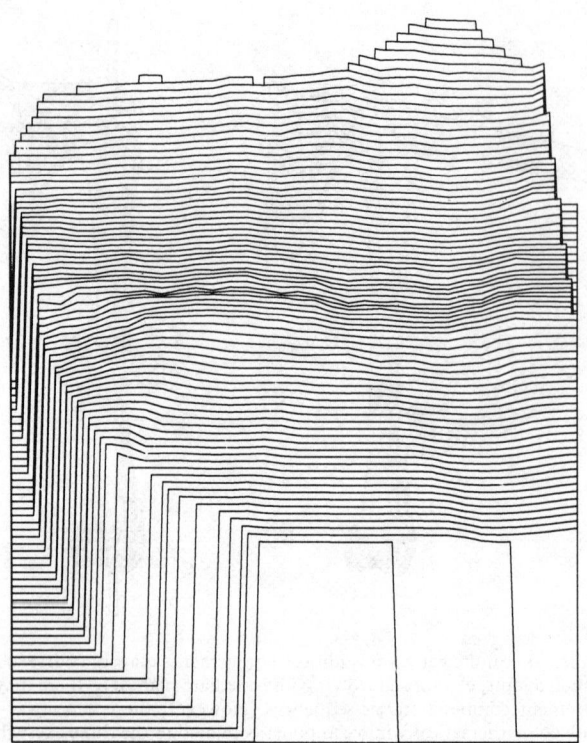

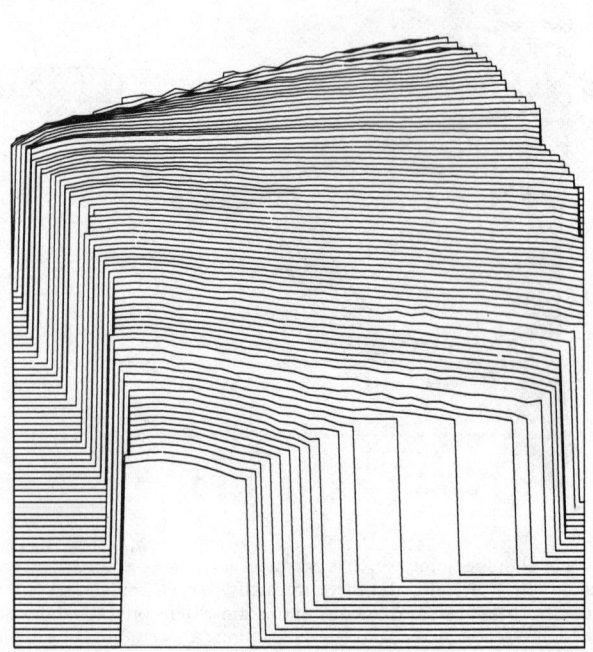

4.6A *Left:* A computer reconstruction of a sagittally sectioned cast of the trochlear surface of a fresh human tibia (left side, female, aged 63) viewed posteriorly. The medial malleolar surface is not included. It is a modified female sellar surface and reciprocally curved to its mating surface on the talus. Anteroposteriorly the surface is concave; lateromedially (left to right) there is a central convexity flanked by anterior and posterior concavities.

Right: A computer reconstruction of a coronally sectioned cast of the trochlear surface of a fresh human talus (left side, male, aged 62) viewed anteriorly. The triangular facet for the 'inferior transverse ligament' part of the posterior tibiofibular ligament has been excluded. It is a modified male sellar surface, which is convex anteroposteriorly; mediolaterally (left to right) it presents a medial convexity, a central concavity, and a rather flattened lateral convexity. Reconstructions kindly provided by Drs. A. Palfrey and Linda K. Ziemer of the Department of Anatomy, Charing Cross Hospital Medical School, London.

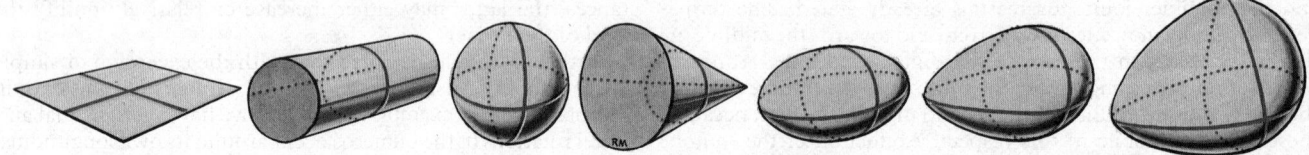

4.6B A variety of geometric figures to which reference is commonly made in simple classifications of synovial joints. However, such comparisons are rough approximations only—no synovial joint possesses articular surfaces which are truly plane, cylindrical, spherical, conical or ellipsoid. A simple and compounded ovoid which are commonplace are included for comparison.

mechanical axis. Using these, certain basic types of movement of bones and of joint surfaces may be contemplated. Finally, the significance of certain joint positions will be discussed.

The Shape of Articular Surfaces

Despite the official recognition of seven types of synovial joint based on their general form, this does *not* reflect a similar number of fundamentally different surface shapes. Close examination of articular surfaces shows that they are never perfectly flat, neither are they parts of spheres, cylinders, cones, or true ellipsoids (**4.**6). They approach much more nearly parts of the surface of *ovoids* (egg-shaped bodies, which are compared with some other solid forms in illustration **4.**6), or are compounded of more than one ovoid surface. When such surfaces are either convex in all directions or concave in all directions, they are termed male or female *ovoid articular surfaces* respectively. The other well-recognized type of surface is *sellar* or *saddle-shaped*, and it is convex in one plane and concave at right angles to this; even here the principal curvatures can be seen in section to be convex or concave ovoid profiles. Whilst, of course, there are great variations in the degree of curvature of different joint surfaces from 'nearly flat' to 'nearly part of a sphere' (spheroidal), evidence is increasing that they may *all* be classed as either **ovoid** or **sellar** (**4.**7). It should, however, be emphasized at this point that the detailed quantitative examination of the topology of conarticular surfaces has, until recently, been lacking, particularly in relation to human joints. It is of particular interest, therefore, that Palfrey and Ziemer (1978) have attempted such analyses on three fresh talocrural joints (amputation specimens). Physical damage and desiccation of the surfaces were avoided and multiple casts, using dental wax, were made of the trochlear surfaces of the talus and tibia; coronal and sagittal slices (3 mm or less in thickness) were prepared. Multiple points were then marked along the curved profiles of the articular cartilage and their spatial coordinates determined by projection and tracing upon grid paper. Similar data were then accumulated for the subchondral bone surfaces. The information obtained was subsequently subjected to computer analysis resulting in two-dimensional contour maps and three-dimensional reconstructions. In this manner a much more intimate knowledge of the geometry, congruence, thickness variations in cartilage, conarticular surface areas, and contrasting profiles of cartilage and subchondral bone was obtained.

It is convenient to consider next certain descriptive terms applied to ovoid surfaces and certain of their properties, because they are relevant both to the movements of articular surfaces and to those of the bone as a whole.

Two points on an ovoid surface may be joined by a curved line across the surface which is the shortest distance between the points (as when a thread is tightly stretched between them). Such a line has been termed a *chord*, and it is distinct from any other, longer, curved line joining the points, which is an *arc* (4.8)—*see* MacConaill and Basmajian (1977). Obviously, any point moving across an ovoid surface traces either a chordal or an arcuate path (or a succession of these)—these will be considered further below.

The terms chord and arc as used in the present context may be unfamiliar (more commonly they refer to a chord or an arc *of a circle* on a plane surface in Euclidean geometry). To assist understanding, one may imagine two points on a plane but deformable surface. The two points may be joined by the shortest route between them which is of course a straight line—in this application—a *chord* (or *geodesic*). In a general sense *any* longer route between them may be called an *arc* (or *non-geodesic*). The surface with the points and lines drawn upon it may now be deformed in any number of ways (e.g. into a sphere, cylinder, or ovoid or sellar surfaces) but the relationship of the shortest route between the points (chord or geodesic), and longer routes (arcs or non-geodesics) is of course preserved.

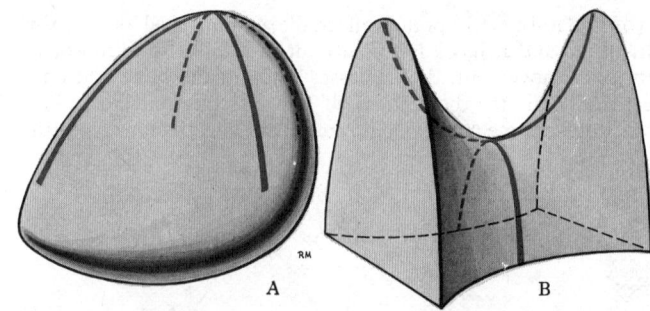

4.7 The two fundamental geometric types of articular surfaces. (A) Ovoid, which may be convex (male) or concave (female). Note that the solid body illustrated presents ovoid profiles in two planes at right angles and that the curvatures of the two may be different. (B) Sellar or saddle-shaped surfaces, which are concavo-convex. In practice both types of surface may vary from only slightly, to highly curved. Thus ovoid surfaces may be 'almost flat' or 'almost spherical', but the majority show intermediate grades of curvature, and much variation in change of radius from place to place.

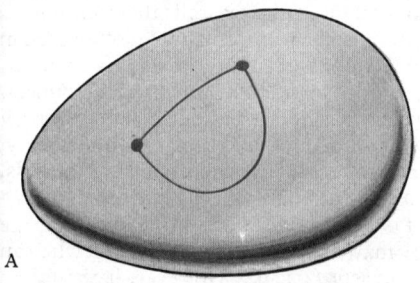

A

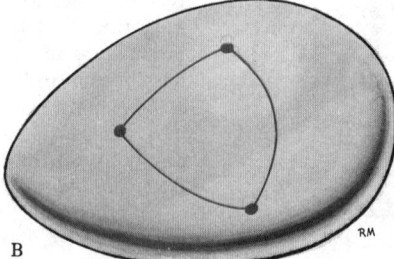

B

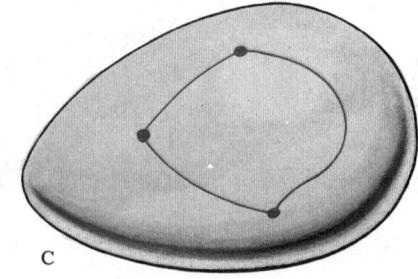

C

4.8 Geometric paths across ovoid surfaces, and three-sided figures enclosed by such paths.

A   A chord (the shortest path between two points), and one example of an arc (any longer path between the points), are illustrated.

B   A triangle—enclosed by three chords.

C   A trigone—a three-sided figure in which one or more sides is an arc.

The figure enclosed by three chords is termed a *triangle* (4.8), and the sum of its angles exceeds 180° on an ovoid surface, is less than 180° on a sellar surface, and is, of course, precisely 180° if the surface is flat. It is clear that the amount that this sum deviates from 180° depends upon the degree of curvature of the surface, and this is of importance in joint movements (*vide infra*). Any other three-sided figure in which at least one side is an arc, is called a *trigone*.

Examination of the profile of a longitudinal section through an ovoid surface (4.9A) reveals two properties of importance in joint mechanics. Firstly, the radius of curvature of the surface varies continuously across the profile, which may be considered as being formed of a series of short segments of circles of changing diameter; the line joining their centres is known as the *evolute* of the profile. Clearly, the rotation of a body sliding across such a profile cannot be referred to any single axis, but to successive points on the evolute. Secondly, if we consider a more extensive male ovoid surface, and a smaller segment of a similarly curved female surface slides across it (4.9B), it is evident that the two will only fit perfectly in one position (cf. the close-packed position of a joint, described below). In all other positions, the surfaces are not congruent, the area of contact between them is greatly reduced, and elsewhere, wedge-shaped intervals separate the two. (Theoretically, in a 'perfect', incompressible ovoid, the contact would be linear, but in practice it is an *area* because of the fine undulations of the surface—p. 426—and the compressibility of articular cartilage.)

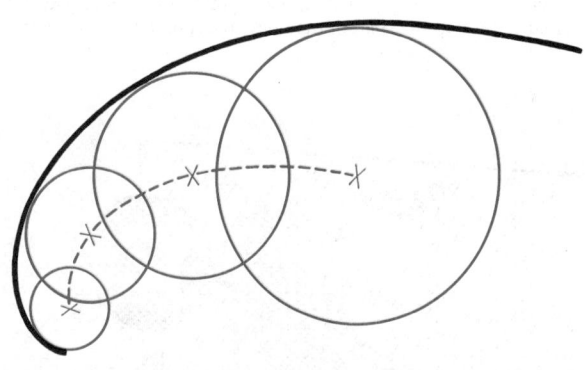

A

4.9A The profile of a section through an ovoid surface showing that it may be considered as a series of segments of circles of changing radius. The line joining the centres of the circles is the evolute of the profile.

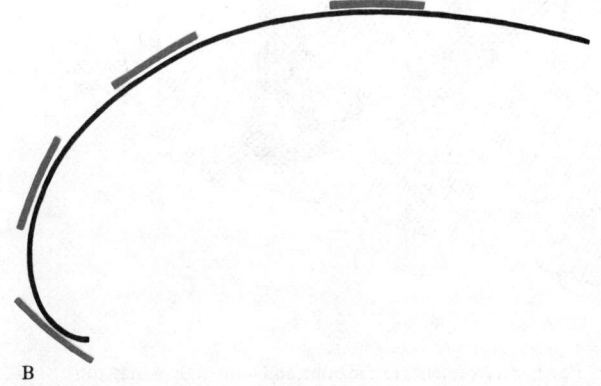

B

4.9B A small section of an ovoid profile in various positions, in relation to a more extensive profile—the two fit perfectly (i.e. are fully congruent) in only one position.

The Mechanical Axis of a Bone

Bones are irregular in shape and in general their articular surfaces bear no simple symmetrical relationship to the rest of the bone. Because of this, comparison of the movements of individual *bones* may be quite misleading in terms of the *joint mechanics* involved, and for this reason, to allow more effective comparisons, the concept of the *mechanical axis* has been introduced. This is perhaps understood more easily by reference to illustrations

4.10A and B. In A, a hypothetical simple, symmetrical long bone is illustrated and it has a terminal joint in the mid-position of its ranges of movement. A rod passes through the bone and ends perpendicular to the centre of its articular surface. This constitutes its mechanical axis, and it can effectively replace the bone when *movements* are being considered. This is particularly obvious, because in this special case of symmetry, it coincides with the long dimension of the bone. In B, however, the bone is not symmetrical, but the joint is again in its mid-range and the mechanical axis is again shown. It is clear that similar *articular movements* will result in similar displacements of the mechanical axis, but that the accompanying movements of the remainder of the bone may appear dissimilar. However, despite the usefulness of the concept of a mechanical axis in some contexts, in others it is less so, because of the somewhat arbitrary choice of a 'central point' on the articular surface which is difficult to define with precision. (The chosen axis would, perhaps, be better defined as that around which the *most habitual conjunct rotation* occurs, and the limiting position of which is the *close-packed* position of the joint under consideration—*vide infra.*) Again, in the case of spheroidal multi-axial joints, the definition of a *single* mechanical axis may be an insufficient reference system for the many movement combinations which are possible.

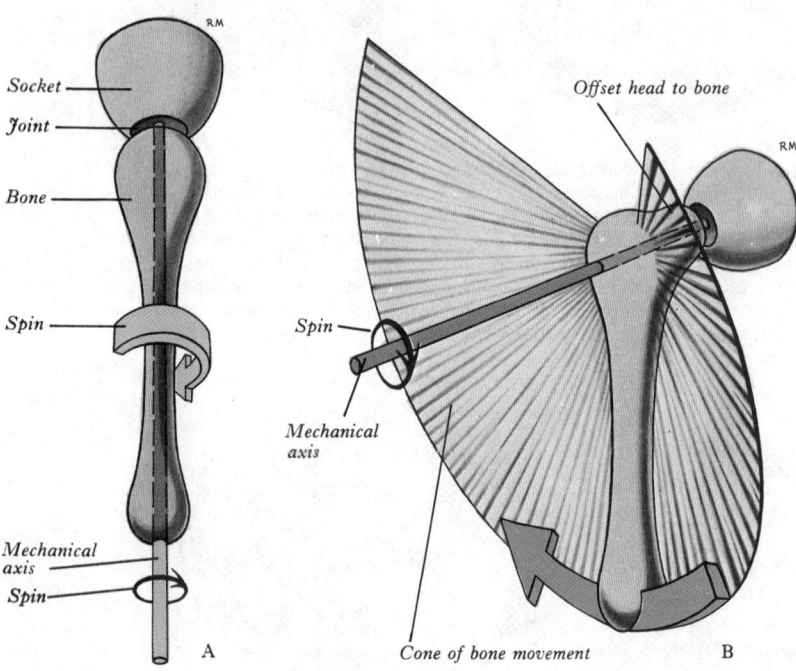

Movements of Bones

Apart from certain passive accessory movements of bones caused by the application of external forces (p. 437), all other movements are forms of rotation, and since the word 'rotation' is often used in a restricted sense in conventional anatomy (p. 431), new terms have been introduced in more advanced treatments of the subject.

Any bone which is simply rotated around its stationary mechanical axis has been said to undergo a pure *spin*, and clearly, any point on the bone or its articular surface, outside the axis, describes the arc of a circle, with the axis as its centre (4.10).

All other displacements of the bone and its mechanical axis are termed *swings*, and these may be *pure* (4.11A), or *impure*, when an element of spin is also present (4.11B). (However, it should be realized that although the mechanical axis is a useful concept, in a spheroidal joint, pure spin is theoretically possible around a large number of alternative 'mechanical axes'.)

Another useful concept when considering positional changes of a bone, is the so-called *ovoid of motion*. During any swing, a point on the mechanical axis of the bone some distance from its related joint, will describe a curved path in space, and if all such possible

4.10 The mechanical axis.

A In a hypothetically simple, symmetrical long bone, with a terminal joint. A 'spin' is occurring around this axis.

B In a bone with an offset head. A similar 'spin' is occurring between the articular surfaces, but the shaft of the bone traces part of the surface of a cone.

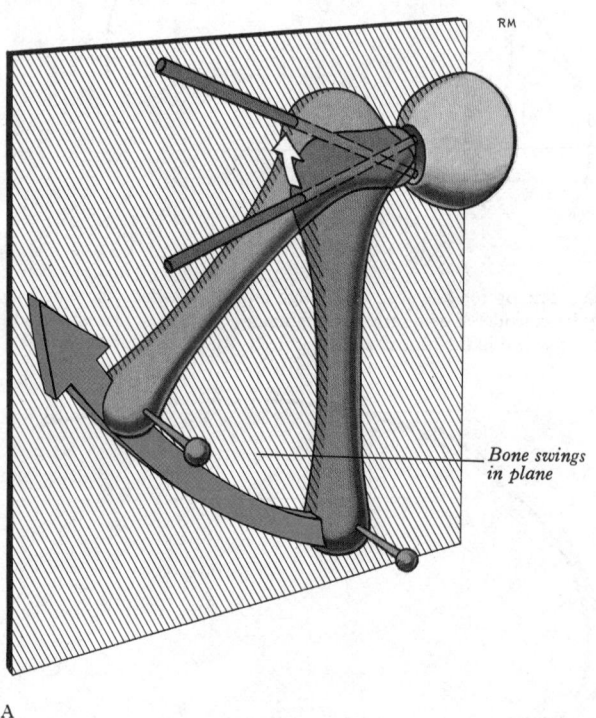

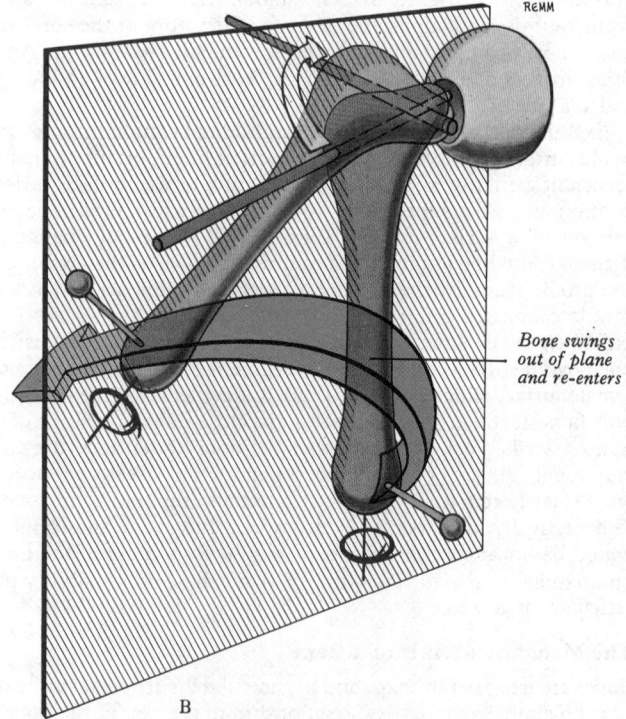

4.11 Further movements of the bone and joint shown in 4.10B.
A A cardinal swing. Note that the mechanical axis moves in one plane (its proximal end traces a chordal path at the joint, and its distal end a chordal path on the ovoid of motion—see 4.12). There is no spin.

B An arcuate swing. Note that the bone moves out of the plane illustrated and then returns to it. The ends of the mechanical axis trace arcuate paths at the joint and on the ovoid motion. There is an associated spin or conjunct rotation around the mechanical axis (and a rotation around the long axis of the bone shaft).

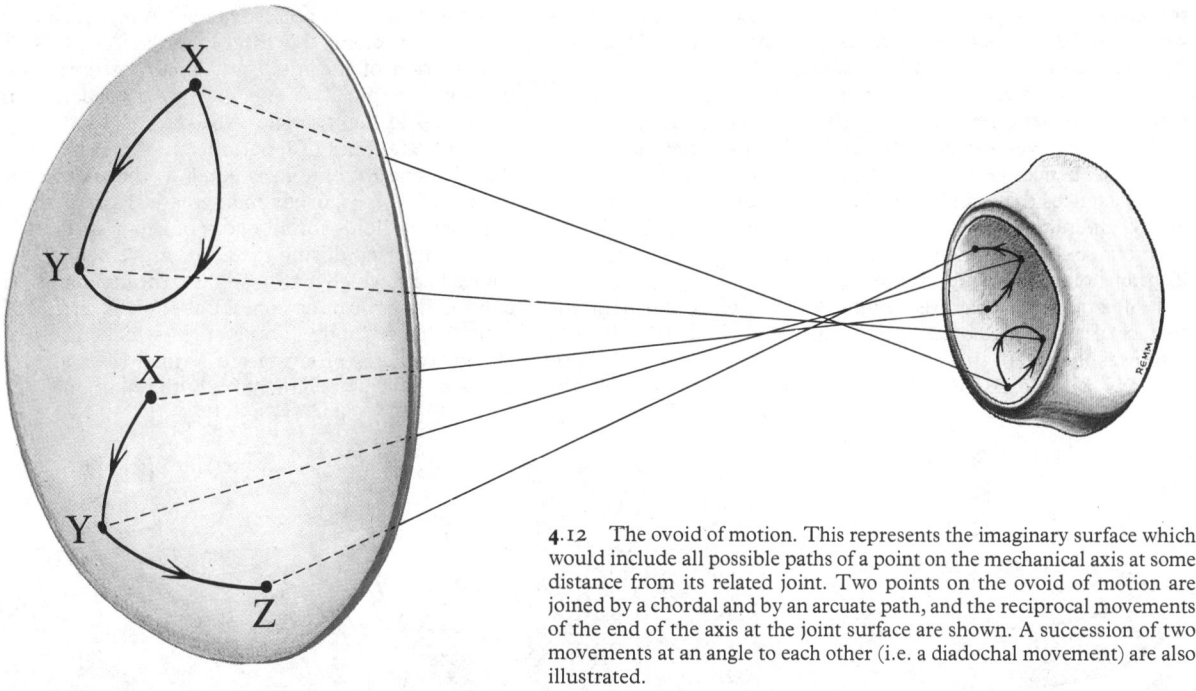

4.12 The ovoid of motion. This represents the imaginary surface which would include all possible paths of a point on the mechanical axis at some distance from its related joint. Two points on the ovoid of motion are joined by a chordal and by an arcuate path, and the reciprocal movements of the end of the axis at the joint surface are shown. A succession of two movements at an angle to each other (i.e. a diadochal movement) are also illustrated.

paths are considered, they are found to fall upon part of the surface of an ovoid (4.12). The area and shape of such ovoids of motion will of course vary greatly with individual bones and articular surfaces. During any particular swing, therefore, the point on the mechanical axis will move from position X to position Y on the ovoid of motion, either along a chord (i.e. the shortest distance, a so-called *cardinal swing*), or it may take a longer route from X to Y during an *arcuate swing*. At the same time, the

articular end of the axis is tracing a reciprocal chordal or arcuate path across its neighbouring joint surface. An important consequence of these distinct kinds of swing relates to the presence or absence of any associated spin of the bone. During a single cardinal swing along a chord (in practice an unusual kind of movement), there is *no associated spin* of the bone. In an arcuate swing, however, there is inevitably an associated spin of the bone, which is of functional importance (*vide infra*). Similarly, a spin is

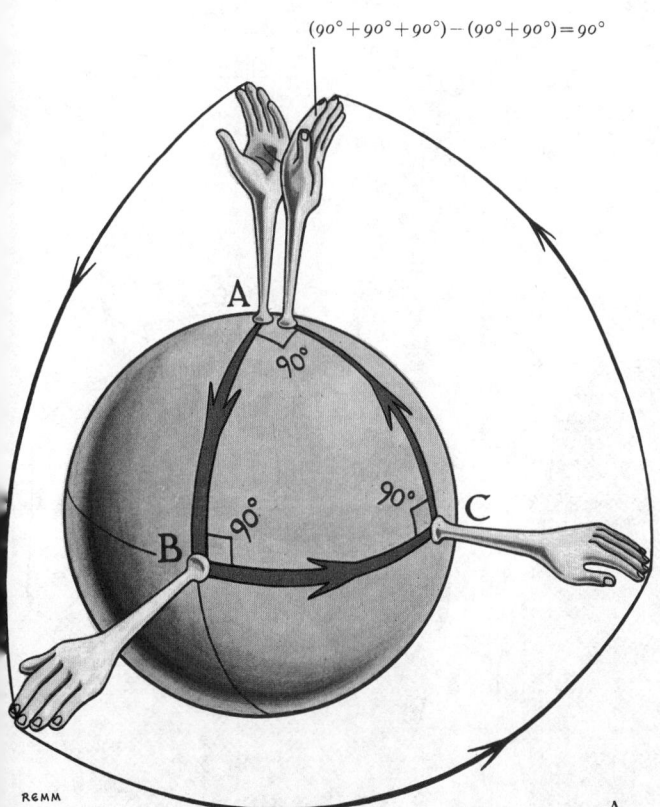

$(90° + 90° + 90°) - (90° + 90°) = 90°$

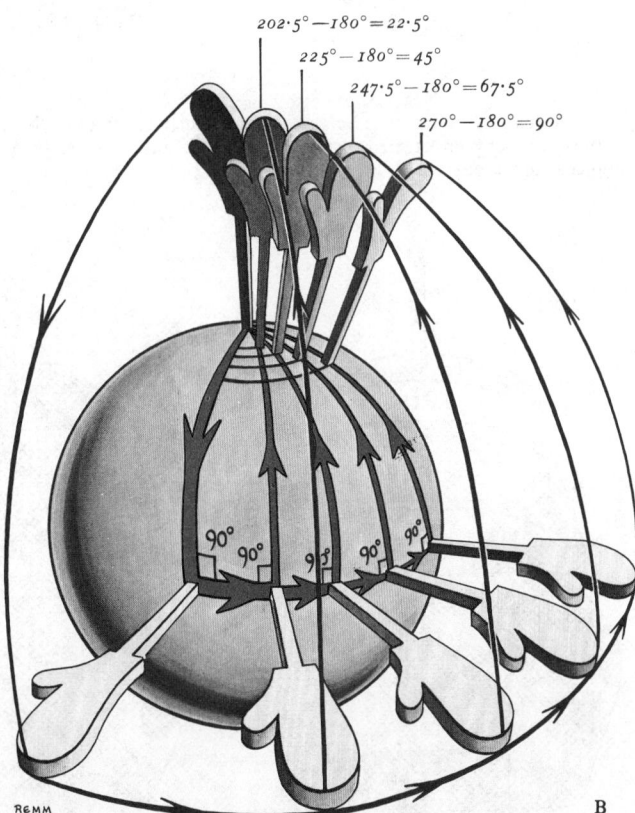

$202·5° - 180° = 22·5°$
$225° - 180° = 45°$
$247·5° - 180° = 67·5°$
$270° - 180° = 90°$

4.13 The 'spin' or conjunct rotation which accompanies a diadochal movement. A model hand and arm moves over the surface of a sphere, first along a line of longitude, then one of latitude to return along a line of longitude.

A The arm traverses three chordal paths which enclose a 'triangle' with

three right angles (i.e. the sum of the angles = 270°). Note that as it passes along the line of latitude from B to C, a spin of 90° is imparted.

B This illustrates how the amount of spin imparted during a diadochal movement varies with the length and angulation of the second stage of the movement.

present if the movement involves a succession of two chords which are at an angle to each other (one form of *diadochal movement*), e.g. X to Y, and Y to Z, in **4.12**.

In some apparently single swinging movements of bones, a spinning element remained unsuspected for many years. Its presence, however, can be predicted by a mathematical analysis of the consequences of the curved nature of joint surfaces, and in most instances they can be verified by careful simple inspection, or by cineradiography. Whilst a full review of the relevant geometry is beyond the scope of this volume, a series of basic diagrams may assist understanding.

In illustration **4.13**A three points, A, B and C are shown on the surface of a sphere, and the lines AB, BC, and CD are the shortest distances between the points (i.e. they are chords). In this instance AB and AC are 'lines of longitude' and they cut the

equator at 90° at the points B and C. A model hand and arm move successively along the three chords, A→B, B→C, C→A and examination of its initial and final postures shows that it ha undergone a spin of 90° (i.e. the number of degrees that the sum o the angles of the 'triangle' ABC *exceeds* 180°). Fig. **4.13**B show: intermediate 'lines of longitude' and in each case it will be seer that the amount of spin is equal to the sum of the three angle: minus 180°. A rigorous mathematical proof shows that simpl relationship holds for all ovoid or sellar surfaces. The spin is o course imparted during the B→C movement (i.e. along a secon chordal path which is at an angle to the first path). An arcuate path may be thought of as compounded of a series of chordal paths which change angle. These spins which inevitably accompany certain varieties of swing are termed *conjunct rotations* and are characteristic of movements at both sellar joints and the majority

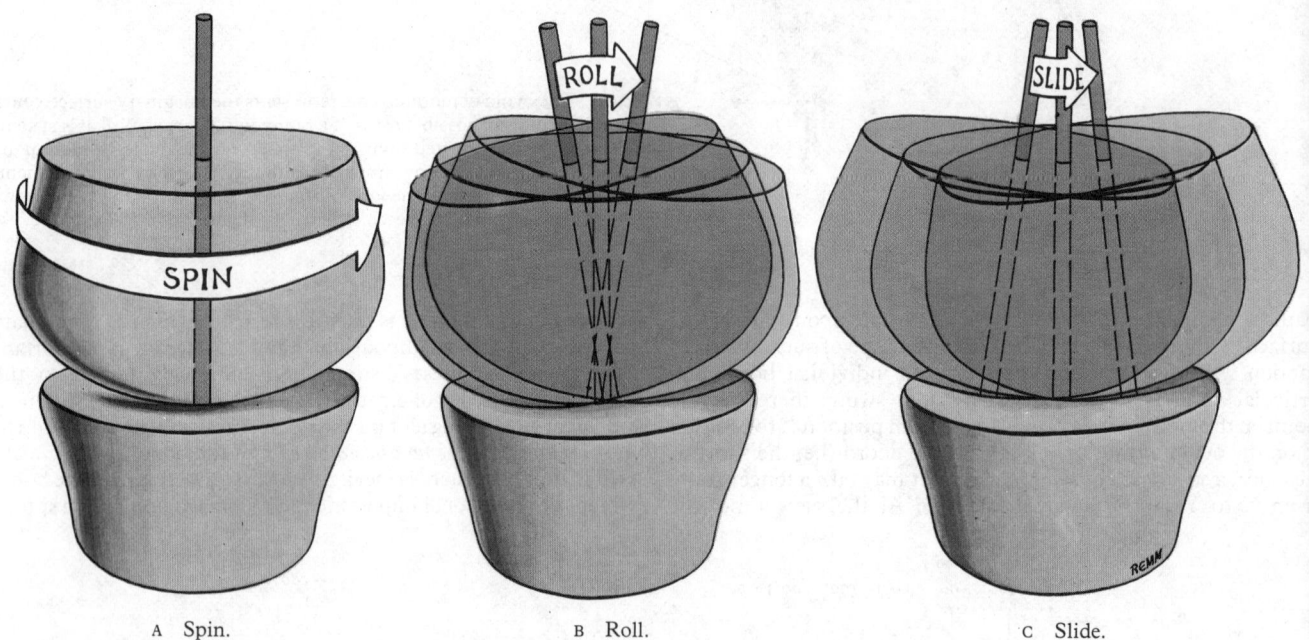

A Spin.   B Roll.   C Slide.

4.14 An analysis of the types of movement which occur (usually in combination) between articular surfaces when a male surface moves over a stationary female surface.

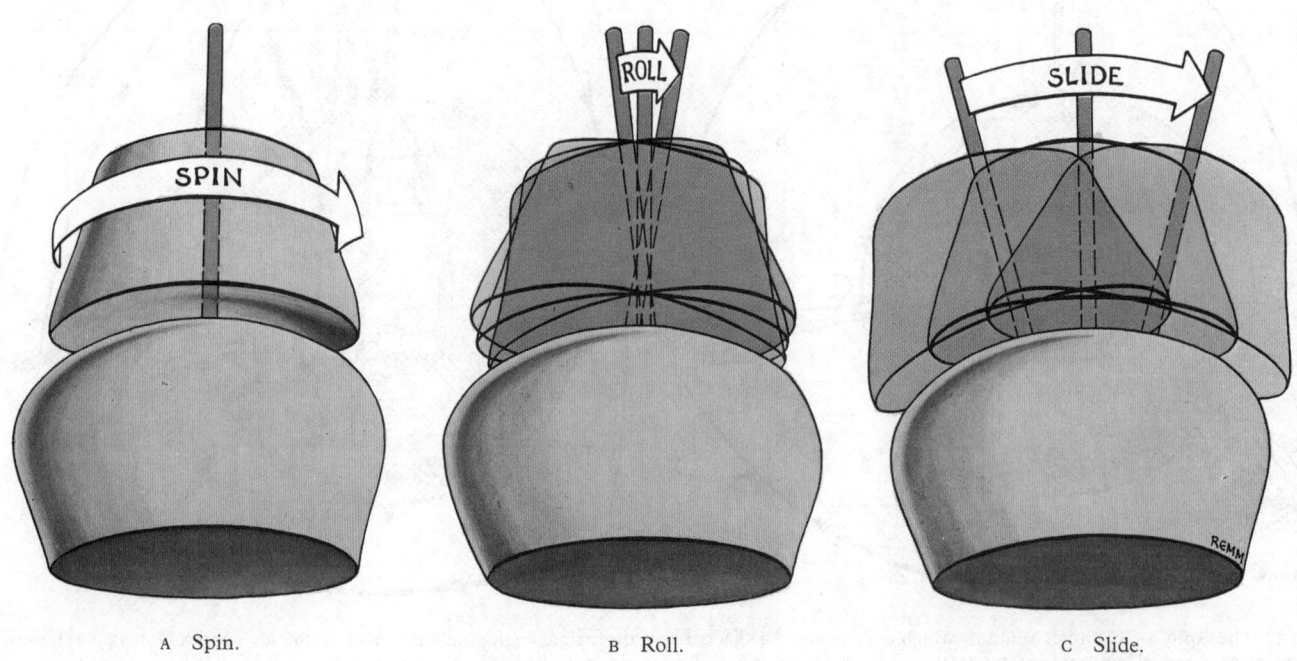

A Spin.   B Roll.   C Slide.

4.15 Articular movements—a female surface moves on a stationary male surface.

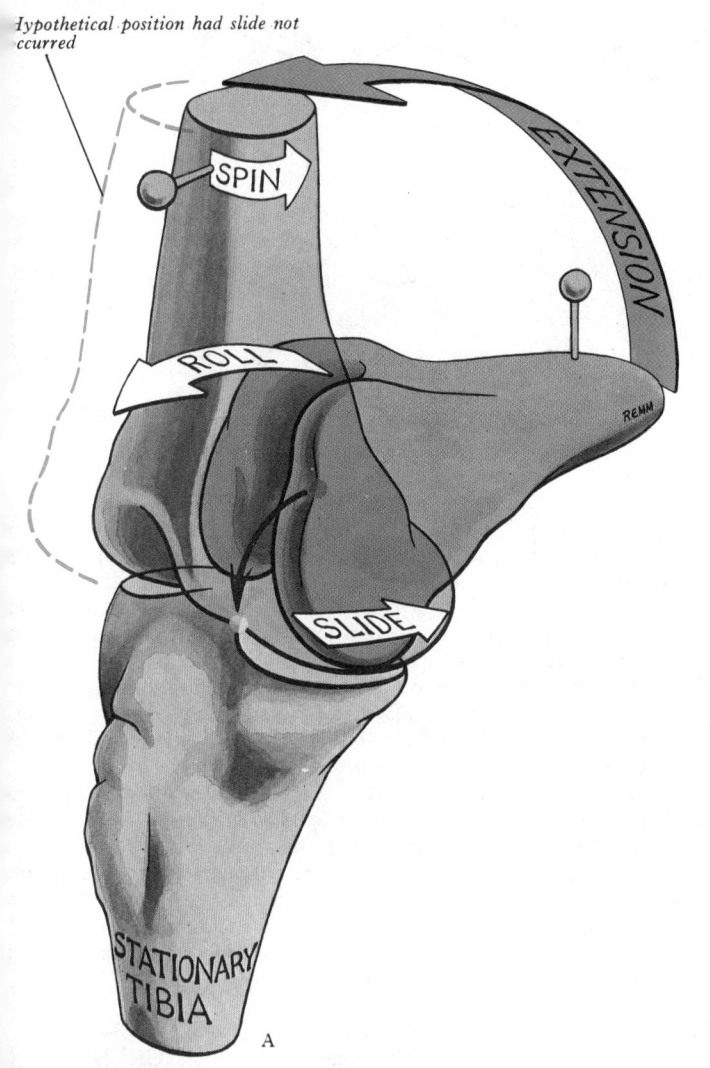

Hypothetical position had slide not occurred

A

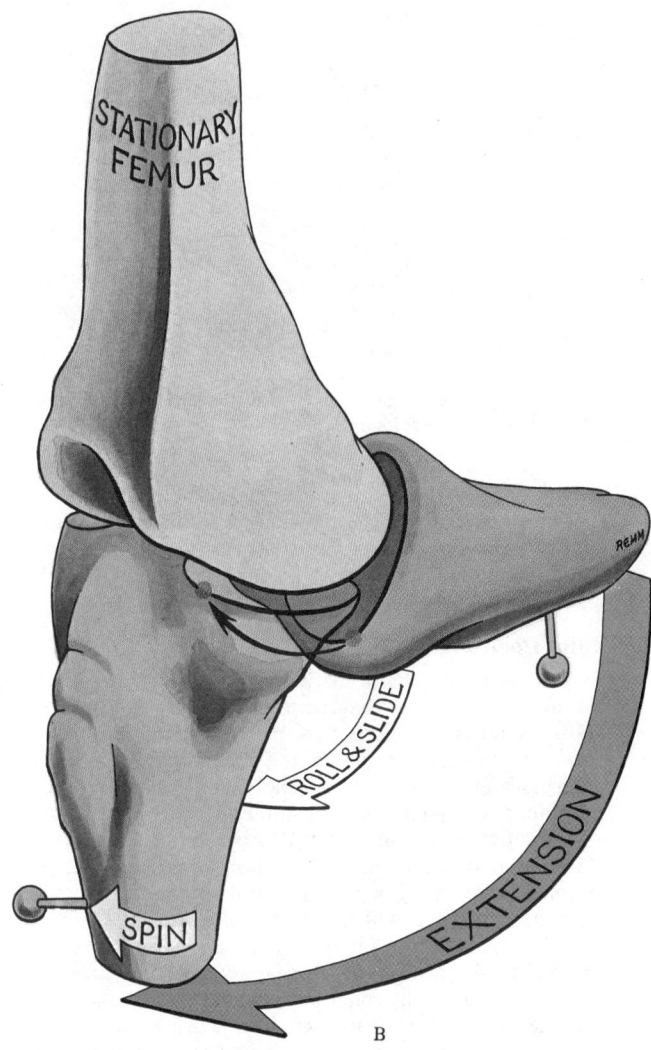

B

4.16 An analysis of the articular movements which are combined during extension at the knee joint.

A   With a stationary tibia, i.e. a moving male surface.

B   With a stationary femur, i.e. a moving female surface. Notice that in each case elements of slide, roll and spin occur together. In A the roll and slide are in opposite directions, whereas in B they are in the same direction. *See* text for further description.

of movements at ovoid joints. It has been shown that the *habitual* movements at *all* joints are always accompanied by some degree of conjunct rotation, and an important consequence of this can be best appreciated after the brief consideration of the related types of movement and 'fit' of the articular surfaces (*vide infra*).

The fundamental arthrokinematical considerations that explain concepts of *conjunct rotations* that *necessarily* accompany particular movements or successions of movements, at a joint has been noted in the foregoing paragraphs. Any other rotation of the bone (resulting from the interplay of gravity, additional externally applied forces, and muscle action) is defined as an *adjunct rotation*. Such adjunct rotations are restricted to joints with two or more degrees of freedom, and they may either involve a bone that is undergoing *no* simultaneous conjunct rotation, or one in which the latter *is* to be considered. In the first case, the factors causing the adjunct rotation may, of themselves, generate a *pure spin* of the bone. In the second case, the effects of the adjunct rotation may be of the *same sense* (e.g. both may be 'clockwise' when viewed from a particular aspect), i.e. the effect of the adjunct rotation additive, increasing the rotation, and it is thus termed a *cospin*. Alternatively, the adjunct rotation may be of the *opposite sense*, reducing or completely nullifying the effect of the conjunct rotation; such adjunct rotations are termed *antispins*. Thus a bone which is embarking upon one or a succession of arcuate swings which, if unmodified, would involve some degree of conjunct rotation, may also be subjected to a *nullifying antispin*: the latter may occur *gradually*, being applied throughout the evolution of

the movement, or *suddenly*, near its termination. It will be clear, therefore, that whilst the *path* of the bone is a more or less complex *arcuate* swing, it is so modified that it becomes *quasichordal* because there is no net accompanying spin.

Movements at Articular Surfaces

When considering some general properties of ovoid surfaces (p. 433) it was seen that in one position only do the surfaces fit together precisely. This occurs at the end of the most common excursion of the joint and is but one important feature of the close-packed position of a joint discussed below. In all other positions the surfaces are a poor fit and the joint is said to be loosely-packed. These changes in congruity are illustrated with reference to the shoulder joint in **4.17**.

Provided the surfaces are ill-adapted, any relative movement between two articular surfaces can be analysed into three basic components, namely, spin, slide and roll. It is convenient to consider these in two circumstances, firstly with a concave, female surface moving over a stationary male surface, and vice versa—reference to illustrations **4.14, 15**, should make the distinction between these components clear. However, in most actual movements these are combined in various ways (**4.16**A, B). It can be seen that with a moving male surface, slide and roll occur simultaneously in *opposite* directions, whereas with a moving female surface they occur in the *same* direction, and these combinations increase the amount of angulation possible without increasing the size of the articular surfaces.

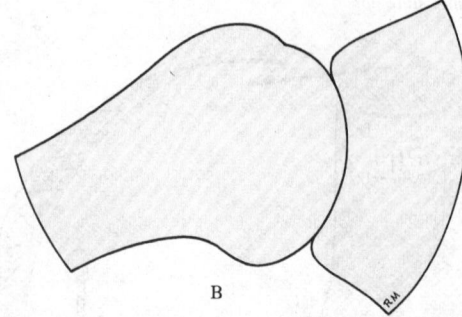

4.17 Congruence of articular surfaces.

A In loose-packed positions of a joint, the surfaces are not congruent (this has been over-emphasized for clarity).

B The close-packed position of the joint, with close-fitting, or fu[ll] congruence, of the surfaces.

The significance of these various types of articular movement and variations in congruity are best considered in relation to the two fundamental states which a joint may assume, namely 'loose-pack' and 'close-pack'.

Joint Positions

It has been seen that a mating pair of ovoid or of sellar joint surfaces are perfectly congruent in only one position of the joint. This occurs at one extreme of the *most habitual movement* of the joint (e.g. full extension in the case of the knee, wrist and interphalangeal joints, dorsiflexion at the ankle, and abduction combined with lateral rotation at the shoulder joint). As this state of full congruence is approached, other important changes occur. The points of attachment of the fibrous capsule and ligaments become increasingly separated and their tension increases. The conjunct rotation which occurs in all habitual movements increases this effect by imposing a spiral twist on these structures. In the final position of **close-packing** therefore, the joint surfaces become fully congruent, their area of contact is maximal, and they are tightly compressed, having in a sense been 'screwed-home', whilst the fibrous capsule and ligaments are maximally spiralized and tense, and no further movement is possible. In a close-packed joint the surfaces cannot be separated by distractive forces (as they often can in other positions) and the two articulating bones can be regarded as temporarily locked together, as if they had no joint between them. Such an extreme position is only assumed when a special effort is to be undertaken and, because of the rigidity of the position and the enormous stresses which may be generated, the articular structures are maximally liable to trauma.

Whilst the close-packed position is a final, limiting position of a joint and any force which tends to further change the position is actively resisted by reflex contraction of the appropriate musculature, the final stages of the movement *just short* of close-packing are physiologically of great importance. Ligaments and articular cartilage are, as we have seen, to a small degree elastically deformable tissues, and in these final stages the actual position taken up by a joint is an equilibrium in which the external torque applied to the joint (often the effect of gravity) is balanced by the resistance to further deformation of the joint by the stretched and twisted capsule and the compressed articular surfaces. Thus, in symmetrical easy-standing the knee and hip joints approach their close-packed positions sufficiently to maintain the erect posture with the minimal expenditure of muscular energy (Joseph 1960).

In all other positions of the joint, the articular surfaces are not congruent and some parts of the articular capsule are lax—the joint is said to be **loose-packed**. The laxity of the capsule is sufficient in many cases near the mid-range of joint movement to allow a separation of the articular surfaces by an externally applied distractive force. The ill-fitting of the surfaces (in which the male surface has a smaller radius of curvature than the female) is advantageous in a number of ways. Firstly it is necessary to allow the combined elements of spin, roll and slide, which characterize most movements at joints. Secondly, the contact area between the surfaces is greatly reduced, and frequently changing,

and this may well diminish the frictional and erosive effect[s]. Thirdly, the small wedge-shaped intervals which separate th[e] surfaces around the contact area are filled with a small volume o[f] synovial fluid, and their shape has been proposed as an importan[t] factor in maintaining efficient joint lubrication and nutrition o[f] the avascular articular cartilages. Finally, it should be noted tha[t] the combination of simultaneous sliding and rolling movement[s] of articular surfaces increases the effective range of movement at [a] joint. With a slide or a roll alone, such a range could only b[e] achieved by considerably more extensive articular surfaces. Reference to illustration **4.**16 will make it clear how the range o[f] angulation of a bone bearing a female surface is increased whe[n] the slide and roll occur in the *same* direction, whereas with [a] moving male surface the increased range follows sliding an[d] rolling in *opposite* directions.

According to one authority (MacConaill and Basmajian 1977[)] the close-packed and least-packed positions of some principa[l] joints are as shown in the following table:

JOINT	CLOSE-PACKED POSITION	LEAST-PACKED POSITION
Shoulder	Abduction + lateral rotation	Semiabduction
Ulnohumeral	Extension	Semiflexion
Radiohumeral	Semiflexion + semipronation	Extension + supination
Wrist	Dorsiflexion	Semiflexion
Metacarpophalangeal (2–5)	Full flexion	Semiflexion + ulnar deviation
Interphalangeal (fingers)	Extension	Semiflexion
First carpometacarpal	Full opposition	Neutral position of thumb
Hip	Extension + medial rotation	Semiflexion
Knee	Full extension	Semiflexion
Ankle	Dorsiflexion	Neutral position
Tarsal joints	Full supination	Semipronation
Metatarsophalangeal	Dorsiflexion	Neutral position
Interphalangeal (toes)	Dorsiflexion	Semiflexion
Vertebral	Dorsiflexion	Neutral position

Whilst general concepts such as the foregoing have greatly increased our knowledge of general joint mechanics, many aspects of joint lubrication, the maintenance of normal articular cartilage, and the detailed mechanics of many individual joints still await resolution (*see*, for example, Freeman 1973).

Accessory movements. The movements which can be performed actively at any joint do not necessarily include all the movements which the structure of the joint would permit. Certain movements which cannot be performed voluntarily can nevertheless be produced when resistance is encountered to active movements (accessory movements, first type), e.g. it is only when some solid object, such as a cricket ball, is grasped in the hand, that the fingers can be rotated at the metacarpophalangeal joints (see p. 472). Other movements can be produced only passively (accessory movements, second type) and their widest range is obtained when the muscles acting on the joint are fully relaxed, e.g. when the supported arm is partially abducted at the shoulder joint, a distractive force can draw the humerus away from the glenoid cavity, and this is characteristic of many joints when they are in a loose-packed position. Such movements are commonly termed 'passive movements', but as all movements, whether active or not, can be performed passively when the muscles concerned are relaxed, the term 'accessory movements' will be used to designate all movements which cannot be performed actively in the absence of resistance (Salter 1955).

Limitation of movements is effected by a number of different factors, of which the tension of ligaments is very important, as can be seen when attempts are made to produce hyperextension of the dissected knee or hip joints, and as we have seen (p. 438) increasing ligamentous tension is an integral factor in producing the close-packed position of a joint which limits its most habitual movement. In life, however, the tension of the muscles which are antagonistic to the movement is equally important as a limiting factor. The latter involves both the *passive elastic component* of the muscles (and the other soft tissues which cross the joint) and the *reflex contraction* of the appropriate musculature which occurs when the stimulation of mechanoreceptors in the articular and periarticular tissues reaches a critical level. The part played by muscles in limiting movement is well seen in flexion of the hip joint. When this movement is performed with the knee extended, it is much more limited in range than when it is performed with the knee flexed. In the latter case flexion of the knee relaxes the hamstring muscles, and this permits the thigh to be flexed until it comes into contact with the anterior abdominal wall. The movement is then limited by *the approximation of the soft parts concerned*—a third factor which is present in connexion with some other movements, e.g. flexion of the elbow and knee. Similarly, contact of the teeth obviously limits elevation of the mandible.

In synovial joints, where the bones concerned are connected by ligaments and muscles only, some part of their articular surfaces are in constant apposition in all positions of the joint. (Some authorities maintain that 'apposition' in this sense implies a fine film of synovial fluid between the adjacent surfaces of about 10 μm or less.) The maintenance of this apposition is assisted by atmospheric pressure and by cohesive forces between the articular surfaces, but these factors are merely subsidiary to the influence exerted by the balanced contraction of the various muscle groups which cross the joint. When such muscles contract the force generated may be resolved (vectorially) into various components (see p. 528). Some are directed towards maintaining or altering the three-dimensional positions of the bones ('swing' and 'spin' components), and oppose any internal and external resistances, including, on occasion, the effects of gravity. Another component of the force is termed *transarticular* (the 'shunt' component), and it increases the compression between the articular surfaces and helps to maintain their apposition in various postures, and during movement. (See also p. 528.) The reader should also note, however, that the effects of externally applied compressive or tensile forces, including gravity, will vary greatly according to the particular posture of the whole body and its various members in space, and the direction of the applied force. Thus, for example, gravity or load bearing may, on occasion, provide a *distractive force* tending to separate the conarticular surfaces, alternatively they may predominantly exert a *translatory/swing* force between the surfaces, and finally, they are often so directed that they, themselves, exert a considerable *compressive force* between the conarticular fellows.

The preceding pages have furnished an elementary introduction to some aspects of the theoretical biology currently engaging kinesiologists. Further brief references to myokinetics will be found on p. 525. For more thorough analyses, the references quoted should be consulted: for readers with a greater facility in mathematics and physics, it will be of considerable interest that attempts are now being made to present a 'generalized mechanics of articular swing', ranging from Aristotelian and Newtonian physics, to embrace relativity theory and quantum mechanics! (consult MacConaill 1978 a, b and c).

Blood supply and lymphatics. Joints receive blood vessels from periarticular arterial plexuses, from which numerous rami pierce the capsular ligament to form a vascular plexus in the deeper parts of the synovial membrane. The blood vessels of the synovial membrane terminate around the articular margins in a fringe of looped anastomoses termed the *circulus articularis vasculosus*. (See also, the blood supply of bones and their epiphyses, p. 257.)

The lymphatics form a plexus in the subintima of the synovial membrane and drain along the blood vessels to regional deep lymph nodes.

Nerve supply. Movable joints are innervated in general by the nerves of supply to the muscles which act on them, and it is probable that this arrangement establishes local reflex arcs involved both in active movements and postural maintenance. Though the branches concerned vary, each nerve innervates a specific region of the capsule but the territories of supply of adjacent nerves freely overlap. The part of the articular capsule which is rendered taut on the contraction of a given muscle or group of muscles is usually innervated by the nerve or nerves supplying their antagonists (Gardner 1948). For example, the inferior part of the articular capsule of the hip joint, which is put on the stretch in abduction, is supplied by the obturator nerve. Tension of this part of the capsule produces a reflex contraction of the adductor muscles which, usually, is successful in preventing over-stretching or tearing of the ligament. This concept does not hold good for the shoulder joint, where the axillary nerve is described as innervating the antero-inferior part of the articular capsule.

The myelinated fibres in articular nerves terminate in Ruffini endings, lamellated articular corpuscles and endings similar to the neurotendinous endings of Golgi. Free nerve endings are numerous at the attachments of the fibrous capsule and ligaments. They are the terminals of non-myelinated and finely myelinated fibres and are believed to mediate pain sensation (Gardner 1950). The Ruffini endings are orientated in various directions, and in the knee joint they are found principally on the flexor aspect of the fibrous capsule. They respond to stretch and are slowly adapting. The lamellated articular corpuscles are less numerous than the Ruffini endings, placed laterally and are quickly adapting. They respond to rapid movement and vibration. Together with the Ruffini type they register the speed and direction of movement. The Golgi type of end organs are connected to the largest myelinated nerve fibres (10–15 μm diameter), are similar to those at neuromuscular junctions and are slowly adapting (Boyd and Roberts 1953; Boyd 1954; Skoglund 1956). Such articular innervation mediates position sense (Stopford 1921; Mountcastle and Powell 1959; Gardner 1967), and is also concerned in stereognosis—the recognition of shape in objects held in the hand (Renfrew and Melville 1960). Many of the non-myelinated fibres are sympathetic in origin; they terminate in relation to the smooth muscle of blood vessels and are believed to be vasomotor or vasosensory in function but little evidence of this function is available.

In the synovial membrane no special end organs and few free nerve endings are present, apart from those which are associated with the blood vessels. The membrane is relatively insensitive to pain (Kellgren and Samuel 1950; Barnett *et al.* 1961).

(For further histological and functional details, and a proposed classification of nerve endings in joints, see p. 858.)

THE INDIVIDUAL ARTICULATIONS

The Temporomandibular Joints

Each temporomandibular joint involves the articular tubercle and the anterior region of the mandibular fossa of the temporal bone above and the condyle of the mandible below. The articular surfaces are covered with a variety of white fibrocartilage in which collagen fibres predominate and cartilage cells are few. An articular disc divides the joint into upper and lower parts. The joint is commonly described as 'condylar' but a preferable term is *ellipsoid*. The right and left joints constitute together a *bicondylar* articulation and they always work in concert.

The fibrous capsule is attached, above, to the articular tubercle in front, to the lips of the squamotympanic fissure behind, and between these two attachments to the circumference of the mandibular fossa; below, to the neck of the mandible. Above the articular disc, the capsule forms a loose envelope, but below the disc it is taut. The **synovial membrane** lines the fibrous capsule. Below it is reflected upwards on the neck of the mandible, and the lateral pterygoid tendon to the margin of the articular cartilage of the condyle.

The lateral temporomandibular ligament (4.18) is intimately related to the fibrous capsule. It is attached, above, to the tubercle on the root of the zygoma, below, to the lateral surface and posterior border of the neck of the mandible. Its fibres are directed obliquely downwards and backwards, covered superficially by the parotid gland.

The sphenomandibular ligament (4.19) is medial to the joint and clearly separated from the articular capsule. It is a flat, thin band attached above to the spine of the sphenoid; becoming broader as it descends, it is fixed to the lingula of the mandibular foramen. Laterally it is related, above, to the lateral pterygoid and the auriculotemporal nerve; lower down, it is separated from the neck of the mandible by the maxillary vessels; still lower, the inferior alveolar vessels and nerve and a lobule of the parotid gland lie between it and the ramus of the mandible, and it is pierced here by the vessels and nerve to the mylohyoid. Medially it is related, below, to the medial pterygoid muscle; above, it is separated from the pharynx by fat and pharyngeal veins and, near its upper end, it is crossed by the chorda tympani. Some of its fibres can be traced through the medial end of the petrotympanic fissure to its primary attachment, the anterior process of the

malleus, a vestige of the dorsal extremity of Meckel's cartilage. These fibres are a part of the anterior ligament of the malleus (p. 1198). The ligament is a vestige; its participation in the mechanic of mandibular movement is doubtful.

The articular disc (4.20, 21A, B, C) a roughly oval plat consisting of fibrous tissue, which has been likened in genera form to a schoolboy's peaked cap, and which completely divide the joint. Its upper surface is saddle-shaped or concavoconve from before backwards, accommodated to the form of th articular fossa and articular tubercle. Its inferior surface, i contact with the head of the mandible, is concave. It circumference is connected to the fibrous capsule and, in front, t the tendon of the lateral pterygoid (p. 535). Medially and laterally short but strong fibrous bands pass from the margin of the disc t the medial and lateral poles of the mandibular condyle. Thes ensure that the disc and condyle move together during protrac tion and retraction of the mandible. Posteriorly the disc contain a venous plexus and divides into upper and lower lamellae. Th upper lamella consists of fibro-elastic tissue and is attached to th posterior margin of the mandibular fossa. The lower lamella i white fibrous tissue and is attached to the back of the condyle; it i not elastic. It varies in thickness in its different parts and i thickest a little behind its centre, where it occupies the deepes part of the mandibular fossa. A more detailed analysis of th structure of the disc (Rees 1954), has revealed two relatively thicl regions (the anterior and posterior bands) with thinner zone intervening. These various subdivisions have been named from before backwards—the anterior extension, anterior band intermediate zone, posterior band and finally the bilaminar regio mentioned above; these are illustrated in lateral and superio view, and in section in illustration 4.21A, B, C. The postnata development of the disc, up to the age of 21, has been described b Wright and Moffet (1974); at no stage were chondrocyte observed in the disc, which is flat at birth and develops its sigmoic profile as the articular eminence enlarges. The disc shows, from the fifth decade of life onwards, macroscopic evidence o degeneration (from fraying and thinning to actual perforation) s frequently, that such changes can be regarded as 'normal' ageing Weisengreen (1975) found about 40 per cent of 183 individuals between the ages of 40 to 90 years, showing such frank degeneration. The disc is often incomplete, being perforated to a variable degree.

The stylomandibular ligament (4.18) is a specialized band of the deep cervical fascia (p. 537), which stretches from the apex of the styloid process to the angle and posterior border of the mandibular ramus. Although classed among the ligaments of the joint, it can only be considered as accessory and its functional status is uncertain.

The *nerves* of the joint are derived from the auriculotemporal and masseteric branches of the mandibular nerve, the *arteries*, from the superficial temporal and maxillary arteries.

Movements. The mandible may be depressed or elevated, protruded or retracted; a considerable amount of rotation also occurs. These activities involve gliding, spin, and angulation.

In the *position of rest* the teeth of the mandible and maxillae are not in contact but slightly separated. On closure of the jaws the teeth come into apposition in the *occlusal position*. When the mouth is *opened* the mandibular heads or 'condyles' rotate around a common horizontal axis and this movement is then combined with gliding forwards and downwards in contact with the lower surface of the articular discs. At the same time the discs slide forwards and downwards on the temporal bones. This movement results from the attachments of each disc to the medial and lateral poles of the head of the mandible and from the contraction of the lateral pterygoid which carries each head with its articular disc on to the articular tubercle. The forward sliding of a disc ceases when the fibro-elastic tissue attaching it to the temporal bone posteriorly has been stretched to its limit. Thereafter, there is some further hinging and gliding forwards of each head of the mandible until it articulates with the most anterior part of the disc

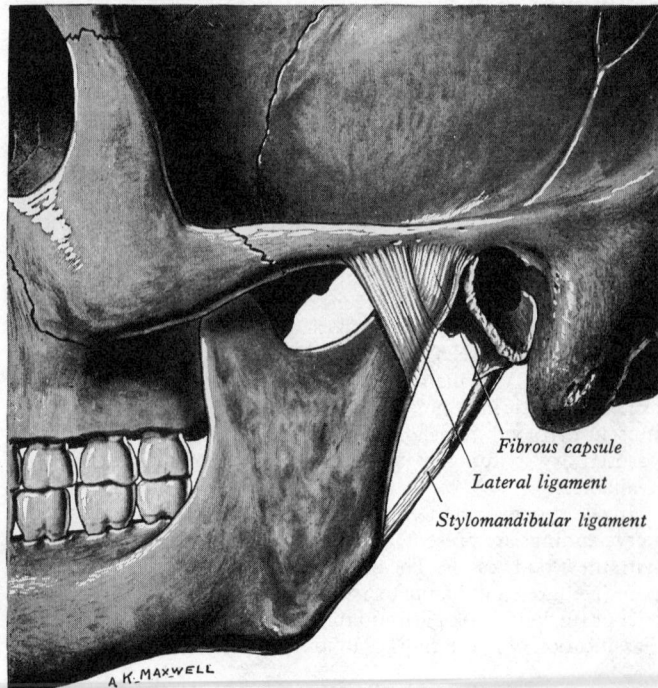

Fibrous capsule
Lateral ligament
Stylomandibular ligament

A.K. MAXWELL

4.18   The left temporomandibular joint. Lateral aspect.

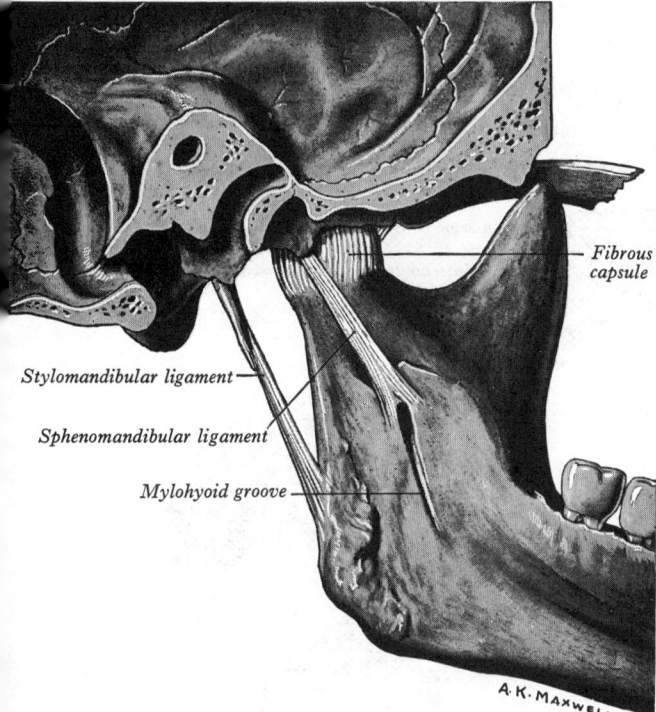

Fibrous capsule

Stylomandibular ligament

Sphenomandibular ligament

Mylohyoid groove

A·K·MAXWELL

4.19 The left temporomandibular joint. Medial aspect.

and the mouth is opened fully. In *closure* of the mouth the movements are reversed. In the first phase of this movement each mandibular head glides backwards and then hinges on its disc, which is held forwards by the lateral pterygoid; finally this relaxes gradually to allow the articular disc to glide backwards and upwards on the temporal bone. The sequential changes in the position of disc and condyle during a full cycle of opening and closing the mouth are illustrated in **4.22**, which is derived from observations by Rees (1954). In *protrusion* the teeth are retained parallel to the occlusal position, but variably separated, and the lower teeth are drawn forwards over the upper teeth by both lateral pterygoid muscles. In *retraction* the mandible is drawn backwards to the position of rest. In the rotatory movements of grinding or chewing the head of one side with its disc glides forwards, rotating around a vertical axis which passes immediately behind the head of the opposite side. It then glides backwards rotating in the opposite direction as the head of the opposite side comes forwards in its turn. These alternating movements swing the mandible from side to side (Sarnat 1951). It is important to note that *both* temporomandibular joints are necessarily involved in every jaw movement, forming together a *bicondylar* arrangement. Consult Hylander (1975) for recent literature and discussion.

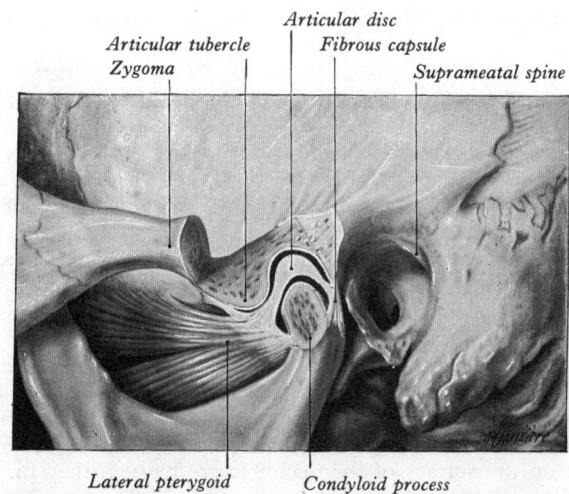

Articular tubercle
Articular disc
Zygoma
Fibrous capsule
Suprameatal spine

Lateral pterygoid   Condyloid process

4.20 A sagittal section through the left temporomandibular joint.

Anterior temporal attachment of meniscus

ANTERIOR

Temporomandibular ligament

Anterior extension

Anterior band

Intermediate zone

Posterior band

Lateral wall of capsule

Medial wall of capsule

Bilaminar region

Sphenomandibular ligament

Posterior temporal attachment

Posterior wall of capsule

A   Superior aspect.

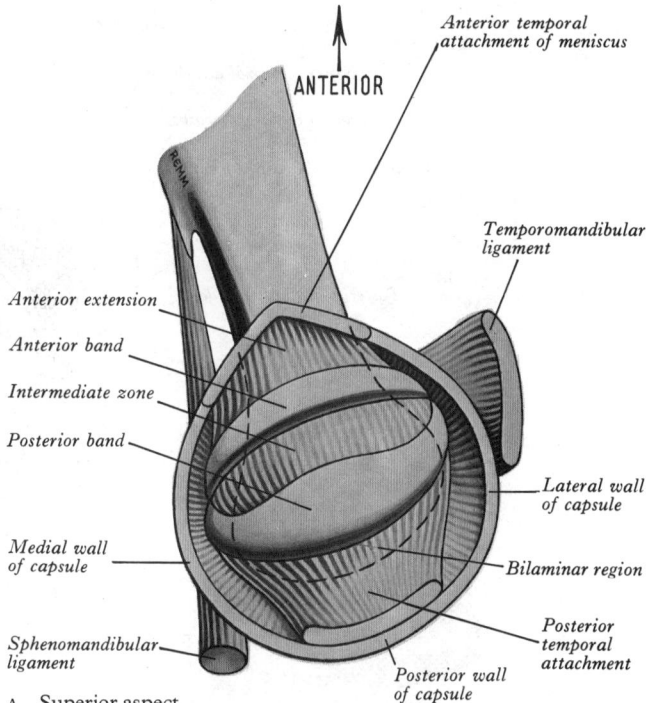

Bilaminar region   Posterior band

Posterior attachment   Intermediate zone

Anterior band

Anterior temporal attachment

Anterior extension

Posterior wall of capsule

Anterior mandibular attachment

Posterior mandibular attachment

ANTERIOR →

Medial wall of capsule (cut)

B   Lateral aspect.

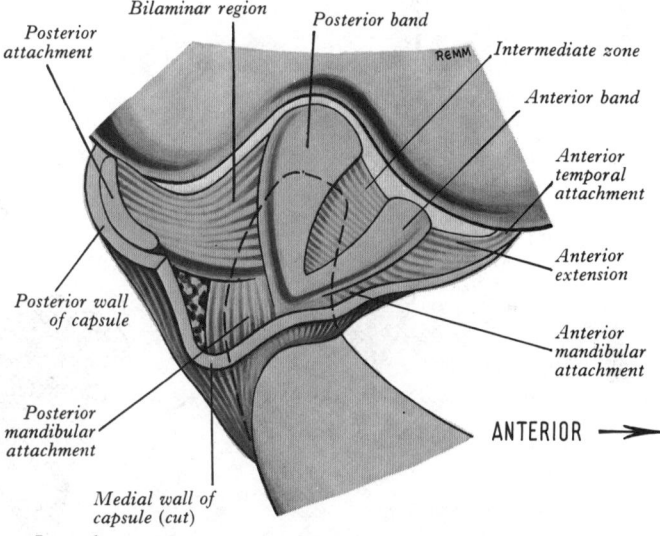

Roof of mandibular fossa   Upper joint compartment

Bilaminar region   Posterior band

Posterior temporal attachment   Intermediate zone

Anterior band

Anterior extension

Anterior temporal attachment

Posterior mandibular attachment

Lower joint compartment

Posterior wall of capsule

ANTERIOR →

Head of condyle

Anterior mandibular attachment

C   In sagittal section.

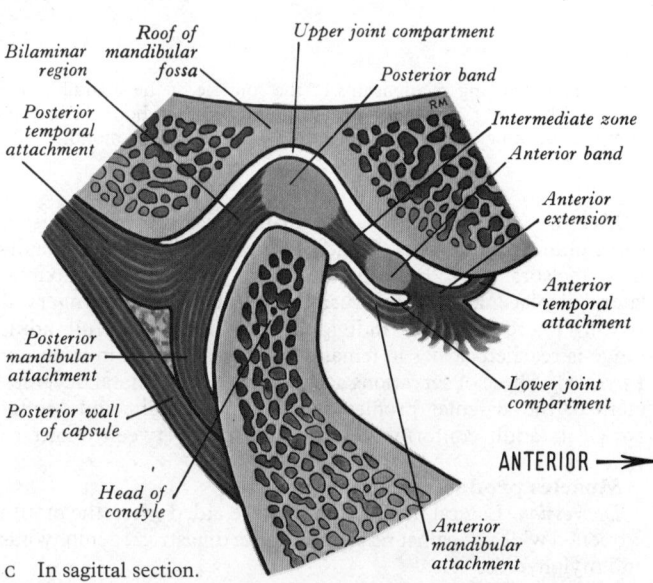

4.21 The form, subdivisions and thickness variations of the intra-articular disc in the temporomandibular joint. *See* text for details.

441

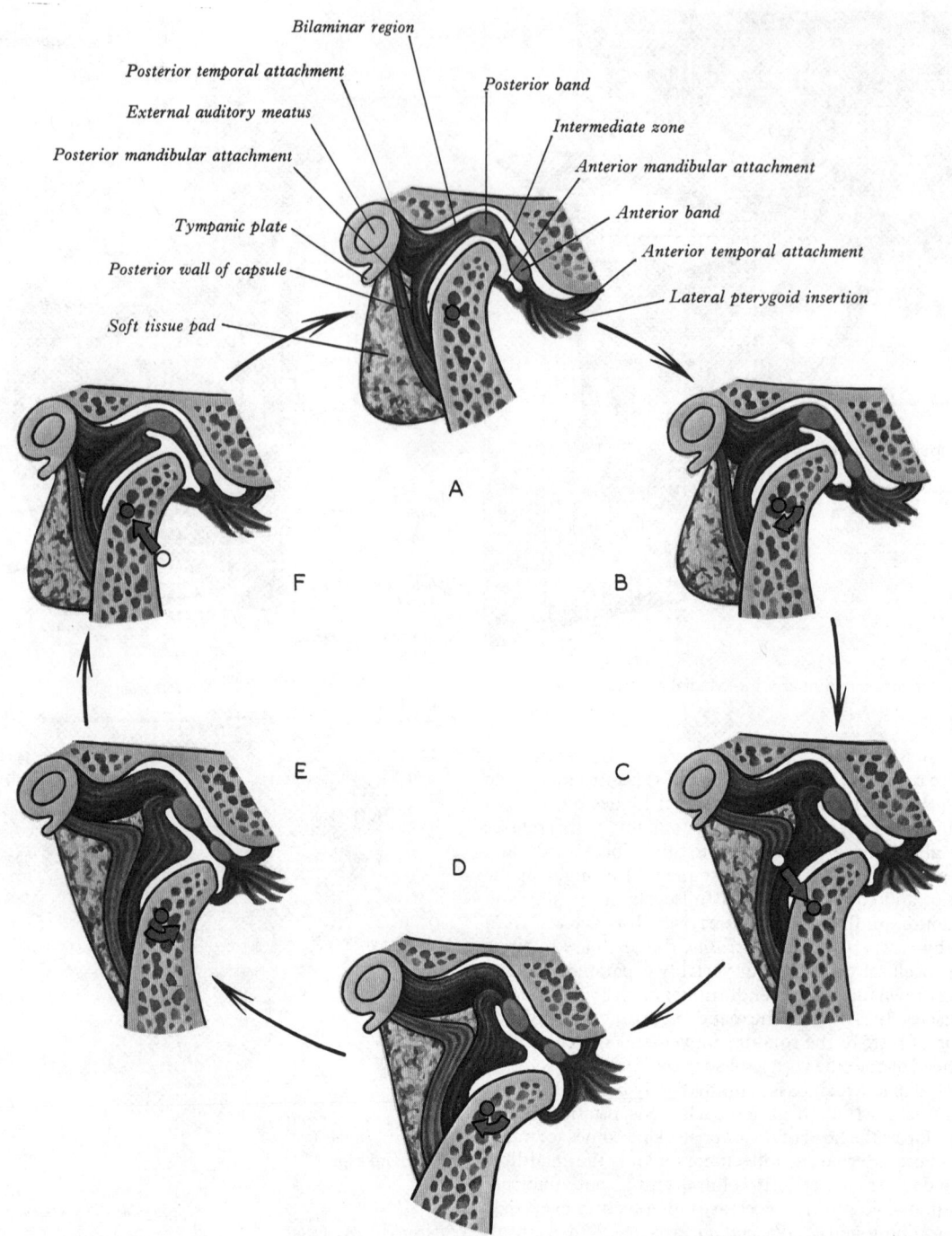

Bilaminar region

Posterior temporal attachment

External auditory meatus

Posterior mandibular attachment

Tympanic plate

Posterior wall of capsule

Soft tissue pad

Posterior band

Intermediate zone

Anterior mandibular attachment

Anterior band

Anterior temporal attachment

Lateral pterygoid insertion

A   B   C   D   E   F

4.22 The changing relationships of the condyle of the mandible, the articular disc and the articular surface of the temporal bone during one complete opening (A→D) and closing (D→A) cycle of the mouth.

4.21 and **4.22** are based upon Rees (1954) with kind permission of the late Professor J. J. Pritchard, on behalf of the late Leonard A. Rees, and the *British Dental Journal*.

Measurement of the range of mandibular movements is of some clinical importance. In adults the mandibular and maxillary incisors may be separated by 50–60 mm, maximal lateral displacement and protrusion being about 10 mm (Ingervall 1970). There is much individual variation. The full adult range is reached earlier in females (*c.* 10 years) than in males (*c.* 15 years). These observations accord with the postnatal development of the articular profiles of the joint (Wright and Moffet 1974); its adult conformation is established between 6 and 12 years.

Muscles producing the movements.

Depression. Lateral pterygoids; they are aided when the mouth is opened wide or against resistance by the digastrics, geniohyoids and mylohyoids.

Elevation. Temporalis, masseter, medial pterygoid, of both sides. In closing the mouth the head of the mandible is retracted by the posterior fibres of the temporalis and then the mandible is

elevated. The temporalis maintains the position of rest (Latif 1957).

Protrusion. Lateral and medial pterygoids.

Retraction. Temporales (posterior fibres). In forcible movements they may be assisted by the middle and deep parts of the masseters, digastrics and geniohyoids.

Lateral movements. Medial and lateral pterygoid (of each side, acting alternately).

A simple listing of muscles, such as the foregoing, fails to convey the fact that, in practice, mandibular movements are the result of complex integrated patterns of simultaneous contraction and lengthening of groups of muscles. Similarly, alternative methods of describing mandibular movements have been proposed (Kraus *et al.* 1969). If all the possible positions of a fixed point on the surface of the mandible are plotted, they form a complex three-dimensional shape (the so-called *envelope of movement*). *Border movements* of the mandible involve movement of the fixed point over the surface of the envelope; in *free movements* the point moves within the envelope and in *contact movements* the maxillary and mandibular teeth maintain apposition. For a more detailed analysis of mandibular positions and movements in relation to incision, chewing and deglutition the reader should consult Kraus *et al.* (1969).

Mention must be made of the unresolved controversy concerning the forces transmitted at the temporomandibular joints. A hypothesis that the joints carry no load or, at least, are subject to little transarticular compression during biting has been supported by Wilson (1920), Scott (1955) and others, while later views (e.g. Smith and Savage 1959; Turnbull 1970) are opposed to this. Most of the literature concerns vertebrates other than mammals, and no reliable experimental data are available for man (*see* Gingerich 1971). Biomechanical analysis favours a force-transmitting function (Barbenel 1972), which is supported by experimental observations (on the opossum) by Crompton and Hiiemäe (1969) and (on the cat) by Buckland-Wright (1978). In view of the complex sliding and rotatory activities at the temporomandibular joints, both the above views may be at times

correct. It is only concerning almost pure hinge movements of limited extent, as in biting, that any controversy need exist.

Applied Anatomy. The mandible can be dislocated only forwards. When the mouth is open, the condyles of the mandible are on the articular eminences, and any sudden violence, or even a sudden muscular spasm, as during a convulsive yawn, may displace one or both forwards into the infratemporal fossa. Reduction is accomplished by depressing the jaw with the thumbs placed on the last molar teeth, and at the same time elevating the chin. The downward pressure overcomes the spasm of the masseter, temporal, and pterygoid muscles, and elevation of the chin throws the head of the mandible backwards. Derangement of the articular disc may follow trauma, overclosure of the mouth with backward displacement of the head of the mandible or malocclusion, and results in clicking and pain on movement of the jaw. In operations on the joint care must be taken to preserve the branches of the facial nerve overlying it.

Changes in the occlusal relationships of the teeth lead to small degrees of remodelling of the articular surfaces of the temporomandibular joints (Moffet *et al.* 1964).

THE JOINTS OF THE VERTEBRAL COLUMN AND THORAX

The vertebrae from the second cervical to the first sacral inclusive are articulated to one another by a series of cartilaginous joints between the vertebral bodies, and a series of synovial joints between the vertebral arches (**3.**35, 38, 42).

The Joints of the Vertebral Bodies

The vertebral bodies are united by anterior and posterior longitudinal ligaments, and by intervertebral discs of fibrocartilage; they are classed as symphyses.

The anterior longitudinal ligament (4.23A) is a strong band which extends along the anterior surfaces of the vertebral bodies. It is broader below than above, thicker and narrower in the thoracic than in the cervical and lumbar regions, and somewhat thicker and narrower opposite the bodies of the vertebrae than opposite the intervertebral discs. It is attached,

above, to the basilar part of the occipital bone, from which it extends to the anterior tubercle of the atlas, then to the front of the body of the axis and is continued down as far as the upper part of the front of the sacrum. It consists of longitudinal fibres, firmly fixed to the intervertebral discs and to the margins of the vertebral bodies, but loosely attached at intermediate levels of the bodies. In the latter situation the ligament is thick and fills up the concavities on the anterior surfaces, and makes the profile of the vertebral column flatter (4.23A). It is composed of several layers of fibres, of which the most superficial are the longest and extend over three or four vertebrae. The intermediate fibres extend between two or three vertebrae, while the deepest reach from one vertebra to the next. At the sides of the bodies the ligament consists of a few short fibres which connect adjacent vertebrae.

The posterior longitudinal ligament (4.23A, 24A) is inside the vertebral canal on the posterior surfaces of the bodies of the vertebrae. Above, it is attached to the body of the axis, and is

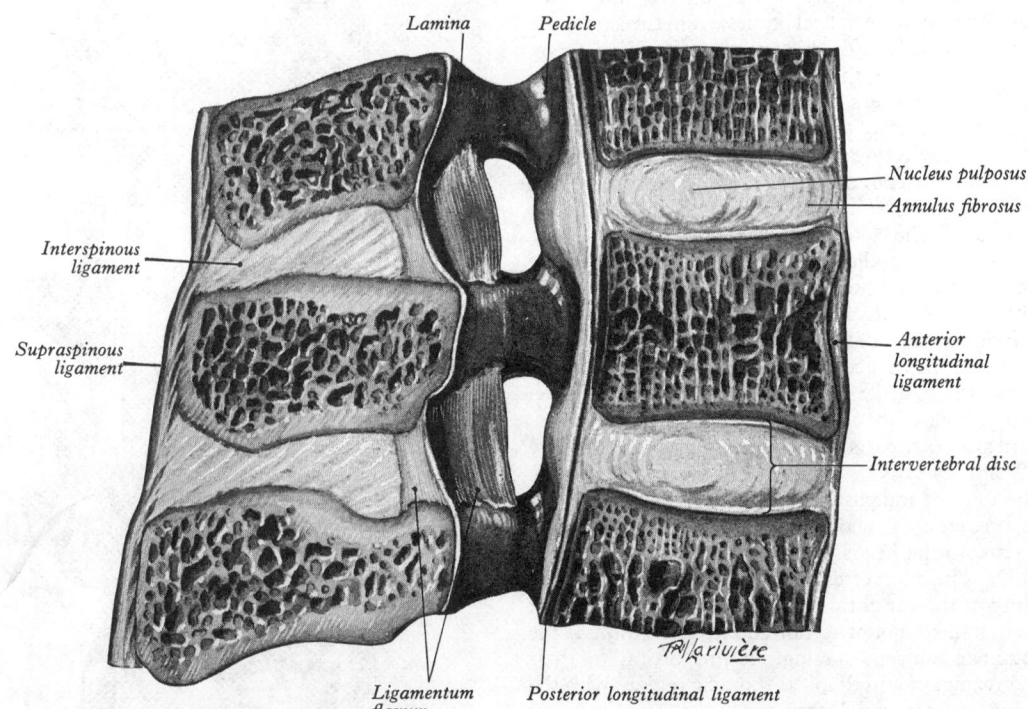

Lamina   *Pedicle*

Interspinous ligament

Supraspinous ligament

Nucleus pulposus

Annulus fibrosus

Anterior longitudinal ligament

Intervertebral disc

Ligamentum flavum   *Posterior longitudinal ligament*

4.23A A median sagittal section through part of the lumbar region of the vertebral column. For contrasting details concerning the direction of fibre bundles in the interspinous ligaments *see* text and Heylings (1978).

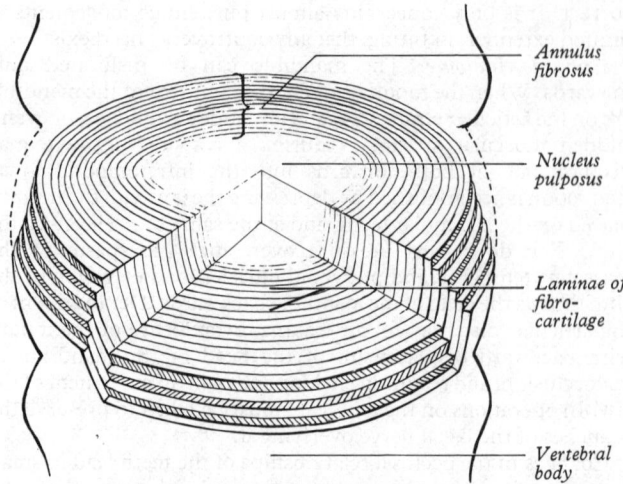

Annulus
fibrosus

Nucleus
pulposus

Laminae of
fibro-
cartilage

Vertebral
body

4.23B A simplified schema of the main structural features of an intervertebral disc. The fibrocellular structure of the nucleus pulposus is omitted. For clarity the number of fibrocartilaginous laminae has been greatly reduced, since they are in fact of microscopic dimensions. Note alternating obliquity of collagen fascicles in adjacent laminae. (Modified after Inoue 1973.)

thence continued downwards to the sacrum; its upper end is continuous with the membrana tectoria (p. 449). It consists of smooth, glistening fibres, attached to intervertebral discs and to the margins of vertebral bodies, but separated between these attachments by emerging basivertebral veins, and by veins which drain these into the anterior internal vertebral plexuses. At cervical and upper thoracic levels the ligament is broad and nearly uniform in width, but in the lower thoracic and lumbar regions (4.24A) it is denticulated, being narrow over the vertebral bodies and broad over the discs. It consists of superficial layers bridging the interval between three or four vertebrae, and deeper layers which extend between adjacent vertebrae. These deeper layers, sometimes termed *perivertebral ligaments* (François 1975), are in turn close to and, in adult life, fused with the annulus fibrosus of the adjoining intervertebral disc. The layers are more easily identifiable in immediate post-natal years.

The intervertebral discs (4.23A and B) are interposed between adjacent surfaces of vertebral bodies, from the axis to the sacrum, and are the chief bonds of connexion between them. Their shape corresponds with that of the bodies between which they are placed. Their thickness varies in different regions of the column, and in different parts of the same disc. They are thicker in front than behind in the cervical and lumbar regions, and thus contribute to the anterior convexities of these levels, while they are of nearly uniform thickness in the thoracic region, the anterior concavity of this part of the column being almost entirely due to the shape of the vertebral bodies. The discs are thinnest in the upper thoracic and thickest in the lumbar region. They are adherent to thin layers of hyaline cartilage which cover the superior and inferior surfaces of vertebral bodies. Except for their most peripheral parts, which receive a supply from adjacent blood vessels, the discs are avascular and are supported by diffusion through the spongy bone of the adjacent surfaces of the vertebrae. The vascular and avascular parts hence differ in reaction to injury (Smith and Walmsley 1951). The intervertebral discs are connected to the anterior and posterior longitudinal ligaments; in the thoracic region they are joined laterally, by means of the intra-articular ligaments, to the heads of those ribs, which articulate with two vertebrae. The intervertebral discs constitute about one-fifth of the length of the vertebral column, exclusive of the first two vertebrae; but this amount is not equally distributed, the cervical and lumbar portions having, in proportion to their length, a much greater amount than the thoracic region, with the result that they are more pliant (Harris 1939).

Structure of intervertebral discs. Each disc consists of an outer laminated periphery, the *annulus fibrosus* and an inner core, the *nucleus pulposus* (4.23A and B).

The *annulus fibrosus* consists of a narrower outer zone of collagenous fibres and a wider inner zone of fibrocartilage. Its laminae are convex from above downwards and form incomplete collars which are connected by strong fibrous bands and overlap or dovetail into one another. In the posterior region of the disc the laminae join with each other in a complex fashion. Within each lamina the majority of the fibres lie in parallel and run obliquely between two vertebrae; the fibres in contiguous laminae run in different directions (4.23B) and lie at an obtuse angle to each other, thus exercising control over the rotatory movements in different directions. A predominantly *vertical* direction of fibres in the posterior part of the annulus fibrosus has been described (Zaki 1973) with the suggestion that this predisposes to herniation. The obliquity of such fibres of the deeper zones varies in different concentric lamellae (Inone 1973).

The *nucleus pulposus* is better developed in the cervical and lumbar regions than in the thoracic part of the spine. It lies nearer the posterior than the anterior surface of the disc. At birth it is soft, gelatinous, relatively large and consists of mucoid material containing a few multinucleated notochordal cells, into the

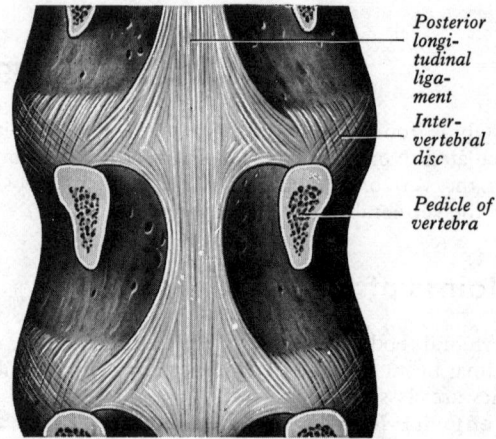

Posterior
longi-
tudinal
liga-
ment

Inter-
vertebral
disc

Pedicle of
vertebra

4.24A The lumbar region of the posterior longitudinal vertebral ligament.

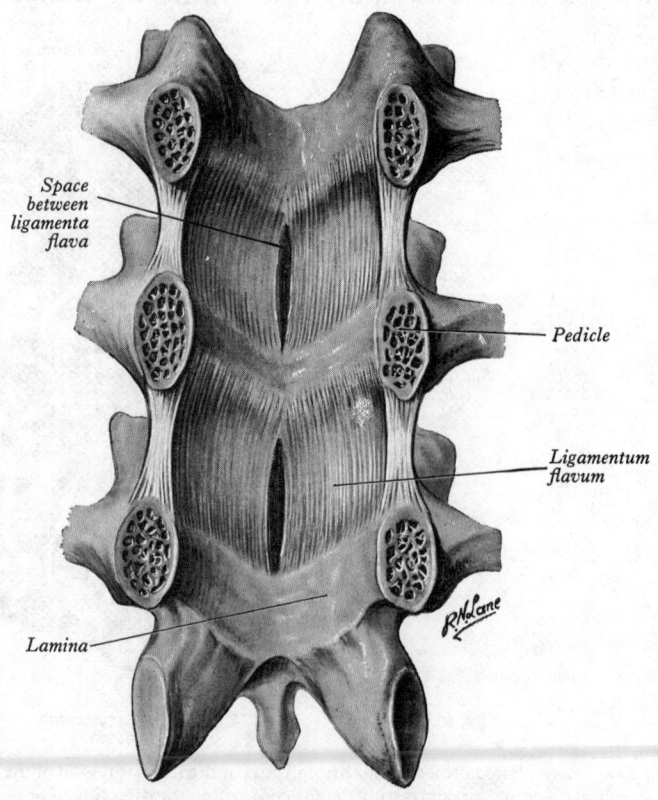

Space
between
ligamenta
flava

Pedicle

Ligamentum
flavum

Lamina

4.24B The ligamenta flava of the lumbar region. Anterior aspect.

periphery of which extend cells and fibres from the inner zone of the adjacent annulus fibrosus. The notochordal cells have disappeared by the end of the first decade, after which time there is a gradual replacement of the mucoid material by fibrocartilage (Sylvén 1951), derived mainly from the cells of the annulus fibrosus but also from the cartilaginous plates covering the upper and lower surfaces of the vertebrae. With this transformation the nucleus pulposus, hitherto distinct, becomes more and more difficult to differentiate from the remainder of the disc (Walmsley 1950; Peacock 1952; Tondury 1958). Recent observations on lumbar intervertebral discs showed that the cellularity of the structure (6000 cells/mm³ taken overall) is highest in the periphery of the annulus fibrosus and in the hyaline cartilage nearest to the vertebral bodies (Maroudas *et al.* 1975). The same study recorded a glucose diffusion coefficient for these structures of about 2·5 cm² per second, comparable to the values for cartilage elsewhere in the body. It was, however, concluded that nutritional conditions in the intervertebral discs might be more critical. This is especially likely in the large discs of the lumbar region.

With these structural changes the nucleus pulposus alters in appearance and becomes amorphous and sometimes discoloured. Its water-binding capacity diminishes and its elasticity is reduced (Püschel 1930), these physical properties of the nucleus pulposus being related in particular to its mucopolysaccharide and protein component (Inman and Saunders 1947; Hendry 1958). When the disc is not loaded, the pressure in the nucleus pulposus is low at all ages (Nachemson 1960).

Applied Anatomy. In the young adult the intervertebral discs are so strong that when violence is applied to the vertebral column the bones always give way first, provided that the discs are healthy. It is impossible to damage a healthy disc except through a fracture of a vertebral body passing through the upper region, where, as the result of forcible flexion, rupture of the disc may occur without fracture of a vertebral body.

After the second decade, however, degenerative changes are liable to occur in the discs, and these may result in necrosis, and sequestration of the nucleus pulposus, and in softening and weakening of the annulus fibrosus. Under these circumstances a comparatively minor strain may cause *either* an internal derangement of the joint tissues with eccentric displacement of the nucleus pulposus *or* an external derangement in which the nucleus pulposus bulges or actually bursts through the annulus fibrosus, usually in a posterolateral direction. In the former the unequal tension within the joint is responsible for muscle spasm and for the sudden violent pain of an acute attack of lumbago; in the latter the projecting nucleus may press upon and irritate the adjacent nerve roots with resulting referred pain, such as sciatica. Derangements of this kind occur most commonly in the lower lumbar region, especially at the lumbosacral joint, and not infrequently in the cervical region at the levels of C. 5–7, but they may occur at any level. Motor effects, with loss of power and reflexes, may also ensue.

The Joints of the Vertebral Arches

The joints between the articular processes of the vertebrae are synovial and vary in shape (p. 446); the laminae, spines and transverse processes are connected by ligamenta flava, interspinous, supraspinous and intertransverse ligaments, and the ligamentum nuchae, which can all be regarded as accessory ligaments of these joints. Each has also an articular capsule.

The articular capsules, thin and loose, are attached just peripheral to the margins of the articular facets of adjacent articular processes; they are longer and looser in the cervical than in the thoracic and lumbar regions.

The ligamenta flava (4.23, 25) connect the laminae of adjacent vertebrae, and are best seen from the interior of the vertebral canal. Their attachments extend from the articular capsules to the regions where the laminae fuse to form the spine; here their posterior margins come into contact and are to a certain extent united, small intervals being left for the passage of veins from the internal to the posterior external vertebral venous

plexuses. The predominant component of a ligamentum flavum is yellow elastic tissue, the fibres of which, almost perpendicular in direction, are attached to the lower part of the anterior surface of the lamina above and the posterior surface and upper margin of the lamina below. The ligaments are thin, but broad and long in the neck; thicker in the thoracic region, and thickest at lumbar levels. They permit separation of the laminae in flexion and at the same time brake the movement so that its limit is not reached abruptly. In this way they assist in restoring the vertebral column to the erect attitude after it has been flexed and may protect the discs from injury.

The supraspinous ligament (4.23) is a strong fibrous cord which connects together the apices of the spines from the seventh cervical vertebra to the sacrum. It is thicker and broader in the lumbar region than in the thoracic, and intimately blended in both situations with the neighbouring fascia. The most superficial fibres of this ligament extend over three or four vertebrae; those more deeply seated pass between two or three vertebrae; while the deepest connect the spines of neighbouring vertebrae and are continuous with the interspinous ligaments. Between the spine of the seventh cervical vertebra and the external occipital protuberance it is much expanded and called the ligamentum nuchae. Heylings (1978) has described the supraspinous ligaments as ceasing at the fifth lumbar spinous process.

The ligamentum nuchae is a fibro-elastic membrane or intermuscular septum, which, in the neck, is homologous with the supraspinous and interspinous ligaments of other levels. Its superficial part extends from the external protruberance and external occipital crest to the spine of the seventh cervical vertebra. From this a fibrous lamina is given off which is attached to the median part of the squamous occipital bone below the external occipital protuberance, to the posterior tubercle of the atlas and to the spines of the cervical vertebrae, forming a septum for attachment of muscles of the two sides of the neck. In bipedal man it is the reduced representative of a much thicker elastic ligament which, in quadrupedal mammals, is an important factor suspending the head and modifying its flexion, functioning in the same way as the ligamenta flava in man.

The interspinous ligaments (4.23), thin and almost membranous, connect adjoining spines, and their attachments extend from the root to the apex of each process. They meet the ligamenta flava in front and the supraspinous ligament behind. They are narrow and elongated in the thoracic region, broader, thicker and quadrilateral in form in the lumbar region, and only slightly developed in the neck. The direction of fibres in these ligaments has commonly been described as obliquely dorso-caudal. In a recent study of 28 cadavers Heylings (1978) found their direction to be obliquely dorsocranial at lumbar levels.

The intertransverse ligaments are between the transverse processes. In the cervical region they consist of a few, irregular, scattered fibres and are largely replaced by intertransverse muscles; in the thoracic region they are rounded cords intimately connected with the deep muscles of the back; in the lumbar region they are thin and membranous.

Movements of the Vertebral Column

The range of movement possible between any two adjoining vertebrae is restricted, due to the limited range of deformation of the intervertebral disc connecting the vertebral bodies. The greater thickness of the discs in the cervical and lumbar regions as compared with the thoracic region is associated with the greater individual ranges of movement occurring in those regions. But, although movement between adjoining vertebrae is small, the summation of these movements gives a relatively wide range to the vertebral column as a whole.

Flexion, or forward bending, extension, or backward bending, bending to one or other side (usually termed lateral flexion), rotation and circumduction are all possible in the vertebral column.

In *flexion* the anterior longitudinal ligament is relaxed and the anterior parts of the intervertebral discs are compressed; at the limit of the movement the posterior longitudinal ligament, the ligamenta flava, and the interspinous and supraspinous ligaments are stretched, as well as the posterior fibres of the intervertebral discs. The intervals between the laminae are widened, and the inferior articular processes glide upwards upon the superior articular processes of the subjacent vertebrae. However, tension of the extensor muscles of the back is an important factor in limiting further movement—e.g. when carrying a load on the shoulders. Flexion is most extensive in the cervical region.

In *extension* an opposite disposition of the parts takes place. This movement is limited by the tension of the anterior longitudinal ligament, and by the approximation of the spines. It is free in the cervical and lumbar regions, but is restricted in the thoracic region, partly because the discs are thinner, perhaps also due to the effects of the thoracic skeleton and its musculature. It has been suggested that in full extension the axis of movement is behind the articular processes and that it moves forwards as the vertebral column straightens and passes into flexion, reaching the centre of the vertebral bodies when flexion is full (Wiles 1935).

In *lateral flexion* the sides of the intervertebral discs are compressed, the extent of motion being limited by the resistance offered by the tension of the antagonist muscles and surrounding ligaments. It is always associated with some degree of rotation. Lateral movements may take place in any part of the column, but are most free in the cervical and lumbar regions.

Circumduction is limited, and is merely a succession of the preceding movements.

Rotation is produced by the twisting of the vertebrae relative to each other with a torsional deformation of the intervening intervertebral discs. Although only slight between any two vertebrae, this allows a considerable extent of movement when it takes place along the length of the column, the upper part being turned to one or other side. This movement occurs to a slight extent in the cervical region, is freer in the upper part of the thoracic region and is least in the lumbar region.

The extent and variety of vertebral movements are influenced by the shape and direction of the articular facets. Though often described as *plane*, these articulations are never truly flat, but are of ovoid type, the opposing surfaces being reciprocally concave and convex. In the *cervical* region the upward inclination of the superior articular facets allows free flexion and extension. Extension can usually be carried further than flexion; at the upper end of the region it is checked by the locking of the posterior edges of the superior atlantal facets in the condylar fossae of the occipital bone; at the lower end it is limited by a mechanism whereby the inferior articular processes of the seventh cervical vertebra slip into grooves behind and below the superior articular processes of the first thoracic vertebra. Flexion is arrested just beyond the point where the cervical convexity is straightened; the movement is checked by the apposition of the projecting lower lips of the bodies of the vertebrae with the shelving surfaces on the bodies of the subjacent vertebrae. Lateral flexion and rotation in the cervical region are always combined; the upward and medial inclinations of the superior articular facets impart a rotatory movement during attempts at lateral flexion. In the *thoracic* region, notably in its upper part, all the movements are limited, thus reducing interference with respiration to a minimum. The almost complete absence of an upward inclination of the superior articular facets prohibits any marked flexion, while extension is checked by the contact of the inferior articular margins with the laminae, and the contact of the spines with one another. Rotation is freer in the thoracic region. The axis of rotation lies within the vertebral bodies in the mid-thoracic region but in front of the bodies elsewhere, so that rotation is accompanied by some lateral displacement of them (Davis 1959, 1965). The direction of the articular facets would allow of free lateral flexion, but this movement is considerably limited in the upper part of the region by the resistance of the ribs and sternum. Rotation is usually associated with a slight degree of angulation or flexion towards the same side as the rotation. In the *lumbar* region extension is free and wider in range than flexion. A considerable amount of lateral flexion and a small amount of rotation can also occur. The range of rotation is in part limited by the fact that right and left articular facets do not have a common centre of curvature (Putz 1976). The functional transition between the thoracic and lumbar regions occurs most frequently between the eleventh and twelfth thoracic vertebrae (p. 277). At this transition the joints of the

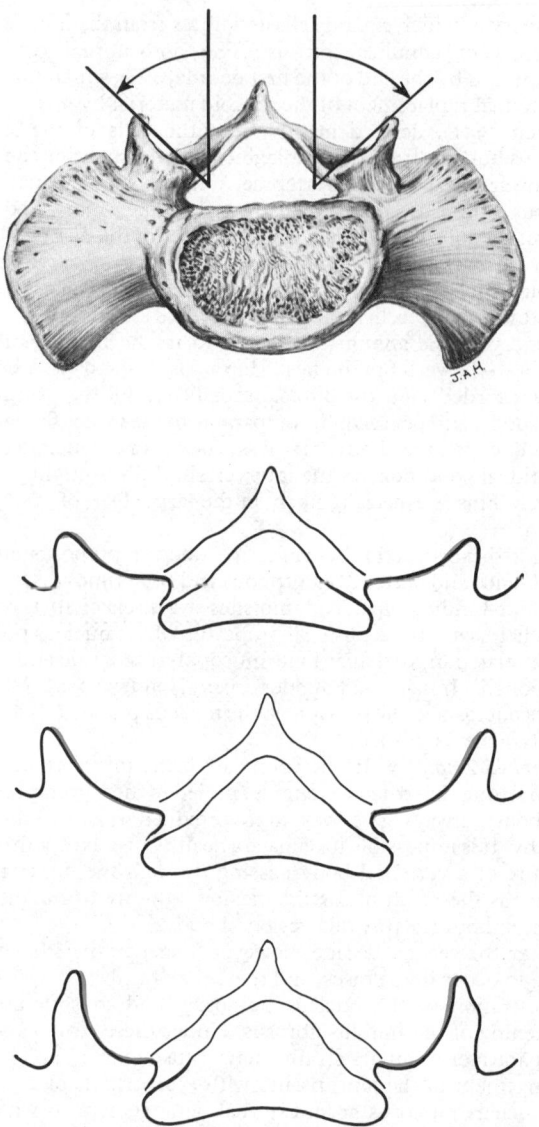

4.25 The method employed by Čihák (1970), to measure the inclinations of articular surfaces of the lumbosacral zygapophysial joints (above). The lower profiles depict three degrees of increasing curvature and inclination. For quantitative data derived from 132 human sacra consult the original paper.

vertebral arches usually fit very intimately, as in a carpenter's mortice and tenon joint, and a small amount of compression locks the joint, preventing all movement other than flexion.

Muscles producing vertebral movements. The spinal column may be moved by muscles attached to it and acting directly on it, and by muscles attached to other bones and acting indirectly on the column. Gravity also plays a part.

Flexion. Longus cervicis, scaleni, sternocleidomastoid and rectus abdominis of both sides.

Extension. Erector spinae, splenius and semispinalis capitis of both sides.

Lateral flexion. Longissimus and iliocostocervicalis components of erector spinae, the oblique muscles of abdominal wall, and the muscles concerned in *flexion* of the side towards which lateral flexion is occurring.

Rotation. Rotatores, multifidus and splenius cervicis.

Extension, principally of the lumbar spine, occurs when raising the body from a stooping position. In the initial phase, extension is mainly at the hip and knee joints. There is a delay in the onset of continuous lumbar extension during which there is little or no activity in the erector spinae. When lifting heavy weights in the initial phase there is considerable compression of the lumbar intervertebral discs with large rises in pressure in the thoracic and abdominal cavities which may resist flexion (Davis 1963; Davis *et al.* 1965).

HE LUMBOSACRAL JOINTS

The articulations between the fifth lumbar vertebra and the first segment of the sacrum resemble the joints between any two typical vertebrae. The bodies of the fifth lumbar and first sacral vertebra are united by a very large intervertebral disc which is thicker ventrally. Its anterior margin is at the sacrovertebral angle (p. 279) and the anterior and posterior longitudinal ligaments are adherent to it. The right and left zygapophyseal joints between the inferior articular processes of the fifth lumbar vertebra and the superior articular processes of the sacrum are separated by a wider interval than those of the vertebrae above. Associated with them are ligamenta flava, interspinous and supraspinous ligaments. In addition the fifth lumbar vertebra is attached to the ilium and sacrum by the iliolumbar ligament. Marked variation occurs in the geometry of these joints (see p. 279). Examples of the range of variation are shown in 4.25.

The iliolumbar ligament (4.56) is attached to the tip and to the lower and front part of the transverse process of the fifth lumbar vertebra, and occasionally has an additional, weak attachment to the transverse process of the fourth. It radiates as it passes laterally and is attached by two main bands to the pelvis. The lower band, which is often termed the *lumbosacral ligament*, runs from the inferior aspect of the fifth lumbar transverse process to the anterior part of the upper surface of the lateral part of the sacrum, blending with the ventral sacro-iliac ligament; the upper, which gives partial origin to the quadratus lumborum, is attached to the crest of the ilium immediately in front of the sacro-iliac joint and is continuous above with the thoracolumbar fascia.

THE SACROCOCCYGEAL JOINT

The sacrococcygeal joint is a symphysis between the apex of the sacrum and the base of the coccyx, the bones being united by a disc of fibrocartilage and by ventral, dorsal and lateral sacrococcygeal ligaments.

The fibrocartilage is a thin disc interposed between the contiguous surfaces of the sacrum and coccyx; it is somewhat thicker in front and behind than at the sides. Occasionally the coccyx is freely movable and articulates with the sacrum by a synovial joint.

The ventral sacrococcygeal ligament (4.56) consists of a few irregular fibres which descend along the pelvic surfaces of the sacrum and coccyx, resembling the anterior longitudinal ligament in its attachments.

The superficial dorsal sacrococcygeal ligament is a flat band which passes from the margin of the sacral hiatus and descends to the dorsal surface of the coccyx (4.57). This ligament completes the lower part of the sacral canal.

The deep dorsal sacrococcygeal ligament extends from the back of the fifth sacral vertebra to the back of the coccyx, corresponding to the posterior longitudinal ligament.

A lateral sacrococcygeal ligament exists on each side and, like an intertransverse ligament, connects the transverse process of the coccyx to the inferior lateral angle of the sacrum; it completes the foramen for the fifth sacral nerve.

The intercornual ligaments connect the cornua of the sacrum and coccyx on each side. A fasciculus also passes from the sacral cornu on each side to the transverse process of the coccyx.

THE INTERCOCCYGEAL JOINTS

These also are symphyses, thin discs of fibrocartilage being interposed between the segments in the young subject. The segments of the coccyx are further connected together by the extension downwards of the ventral and dorsal sacrococcygeal ligaments. In the adult male all the pieces become ossified together at a comparatively early period, but in the female this does not commonly occur until a later period of life. At a more advanced age the joint between the sacrum and coccyx is obliterated. Occasionally the joint between the first and second pieces of the coccyx is synovial. The tip of the last piece of the coccyx is connected to the overlying skin by a bundle of white fibrous tissue.

The Craniovertebral Joints

The articulation of the vertebral column with the cranium involves not only the paired atlanto-occipital joints but also ligaments connecting the axis and occipital bone. It is appropriate to include as well the joints between atlas and axis, since it is here that most rotation of the head on the neck occurs.

THE ATLANTO-AXIAL JOINTS

The articulation of atlas and axis comprises three synovial joints, one on each side between the inferior facet of the lateral mass of the atlas and the superior facet of the axis, the other median, between the dens and the anterior arch and transverse ligament of the atlas.

The lateral atlanto-axial joints are often classified as plane, but when covered with cartilage, the *articular surfaces* involved are, in fact, ovoid, the atlantal facets being slightly concave, those of the axis convex and reciprocal in form (pp. 273, 274). According to Kapandji (1974), however, both facets are transversely cylindrical, engaging like two wheels.

The fibrous capsules are thin and loose; they are attached at the articular margins and are, of course, lined with synovial membrane. Each is strengthened posteromedially by an *accessory ligament*, which is attached below to the body of the axis near the base of the dens, and above to the lateral mass of the atlas near the transverse ligament.

In front the two vertebrae are connected by a continuation of the anterior longitudinal ligament (4.26). In this position it is a strong and wide band, fixed above to the lower border of the anterior arch of the atlas, and below to the front of the body of the axis. It is thickened in the median plane into a rounded cord which connects the tubercle on the anterior arch of the atlas to the body of the axis.

Behind, the atlas and axis are joined by a broad, thin membrane (4.27) attached above to the lower border of the posterior arch of the atlas, below to the upper edges of the laminae of the axis; it is in series with the ligamenta flava and is pierced near its lateral extremity by the second cervical nerve.

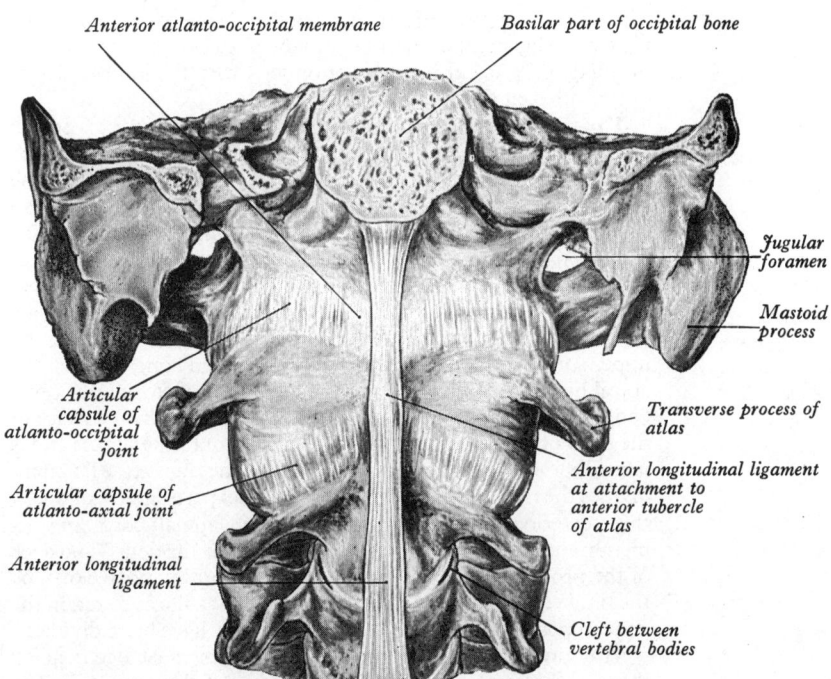

Anterior atlanto-occipital membrane

Basilar part of occipital bone

Jugular foramen

Mastoid process

Articular capsule of atlanto-occipital joint

Transverse process of atlas

Articular capsule of atlanto-axial joint

Anterior longitudinal ligament at attachment to anterior tubercle of atlas

Anterior longitudinal ligament

Cleft between vertebral bodies

4.26 The atlanto-occipital and atlanto-axial joints. Anterior aspect. On each side a small cleft has been opened between the lateral part of the upper surface of the body of the third cervical vertebra and the bevelled, inferior surface of the body of the axis. (See also p. 272.)

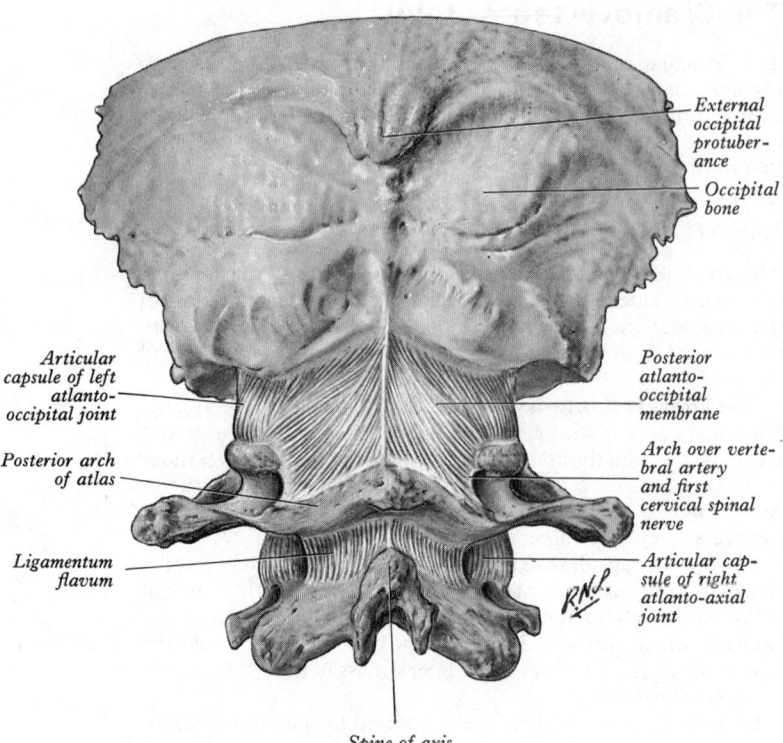

4.27 The atlanto-occipital and atlanto-axial joints. Posterior aspect.

Labels on figure 4.27:
- External occipital protuberance
- Occipital bone
- Articular capsule of left atlanto-occipital joint
- Posterior atlanto-occipital membrane
- Posterior arch of atlas
- Arch over vertebral artery and first cervical spinal nerve
- Ligamentum flavum
- Articular capsule of right atlanto-axial joint
- Spine of axis

The median atlanto-axial joint is a pivot joint between the dens of the axis and the ring formed by the anterior arch and the transverse ligament of the atlas; in a sense it is a double joint, as it consists of two separate 'articulations' and synovial cavities. The latter, clearly, can only act in concert.

Anteriorly, the facet on the anterior surface of the dens articulates with that on the posterior aspect of the anterior arch of the atlas; this articulation is surrounded by a fibrous capsule which is weak and loose and lined with synovial membrane. Posteriorly a second and larger synovial cavity (sometimes termed a bursa), lies between the cartilage-covered anterior surface of the transverse ligament of the atlas and the posterior grooved surface of the dens (4.30): it is often continuous with the joint cavity of one or other of the atlanto-occipital joints.

The transverse ligament of the atlas (4.28, 29, 30) is a thick strong band which arches across the ring of the atlas and retains the dens in contact with the anterior arch. It is broader in the middle than at the ends, and firmly attached on each side to a small tubercle on the medial surface of the lateral mass of the atlas. The median part of its anterior surface is covered by a thin layer of articular cartilage. As it crosses the dens, a small longitudinal band is prolonged upwards, and another downwards, from its superficial or posterior fibres. The upper band is attached to the upper surface of the basilar part of the occipital bone between the apical ligament of the dens and the membrana tectoria; the lower band, which may be absent, is attached to the posterior surface of the body of the axis; hence the whole ligament forms a cross and is named the *cruciform ligament of the atlas*. The transverse ligament divides the ring of the atlas into two unequal parts (4.28): of these, the posterior and larger surrounds the spinal cord and its membranes; the anterior and smaller contains the dens. The neck of the process is constricted where it is embraced posteriorly by the transverse ligament, so that this ligament suffices to retain the dens in position after all the other ligaments have been divided.

Movement at the atlanto-axial joints must occur at all three at the same time, consisting largely of the rotation of the atlas (and with it the skull) upon the axis, the extent being limited by the alar ligaments (p. 449).

The opposed articular facets of the atlas and axis are both slightly convex in their anteroposterior axes. When, therefore, the upper facet glides forwards or backwards on the lower it also

descends. The stretching of the fibres of the fibrous capsule that would result from this forward movement is diminished owing to the contemporaneous descent of their upper attachments, and in this way excessive laxity of the capsule is obviated. The lateral atlanto-axial joints support the weight of the head via the atlas; the pivot joint guides the rotatory movement. For the most elaborate analysis of these and other cervical movements consult Kapandji (1974), who emphasizes the slight helical component in movements of the lateral atlanto-axial joints, as described above.

Muscles producing the movements. Obliquus capitis inferior, rectus capitis posterior major and splenius capitis of one side, acting with the opposite sternocleidomastoid.

THE ATLANTO-OCCIPITAL JOINTS

Each of the paired atlanto-occipital joints involves a superior articular facet of the lateral mass of the atlas and a condyle of the occipital bone: it is ellipsoid in type. The *articular surfaces* are reciprocally curved. Each atlantal facet is concave and tilted somewhat medially. The shape of the facet varies but it is usually constricted about its middle; the articular surface is thus partially, sometimes completely, divided (Singh 1965). The bones are united by the articular capsules and the anterior and posterior atlanto-occipital membranes.

Fibrous capsules surround the condyles of the occipital bone and the superior articular facets of the atlas. They are thickened posteriorly and laterally, but are very thin and sometimes deficient medially, where the synovial cavities frequently communicate with the synovial bursa between the dens and the transverse ligament of the atlas (Cave 1934).

The anterior atlanto-occipital membrane (4.26) is broad and composed of densely woven fibres which pass between the anterior margin of the foramen magnum above, and the upper border of the anterior arch of the atlas below; laterally it is continuous with the capsular ligaments; in front, it is strengthened in the median plane by the continuation of the anterior longitudinal ligament, a strong, rounded cord which connects the basilar part of the occipital bone to the tubercle on the anterior arch of the atlas (4.26).

The posterior atlanto-occipital membrane (4.27), broad but thin, is connected above to the posterior margin of the foramen magnum, below, to the upper border of the posterior arch of the atlas. On each side it arches over the groove for the vertebral artery, and with this groove bounds an opening for the entrance of the artery and the exit of the first cervical spinal nerve. The free border of the membrane, arching over the artery and nerve, is sometimes ossified.

In movements at the atlanto-occipital joints their long axes are set obliquely and run from behind forwards and medially. As a result of this obliquity and of the curvature of the occipital condyle, the corresponding articular surfaces of the two sides are in reality portions of the surface of one ellipsoid, the long axis of

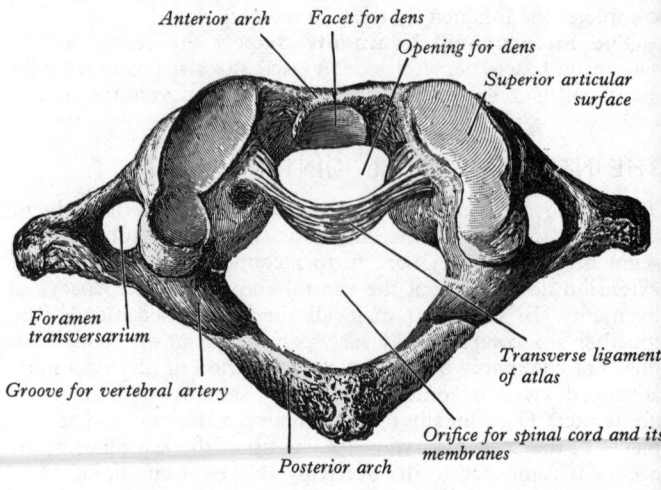

Labels on figure 4.28:
- Anterior arch
- Facet for dens
- Opening for dens
- Superior articular surface
- Foramen transversarium
- Transverse ligament of atlas
- Groove for vertebral artery
- Orifice for spinal cord and its membranes
- Posterior arch

4.28 The atlas vertebra, with the transverse ligament.

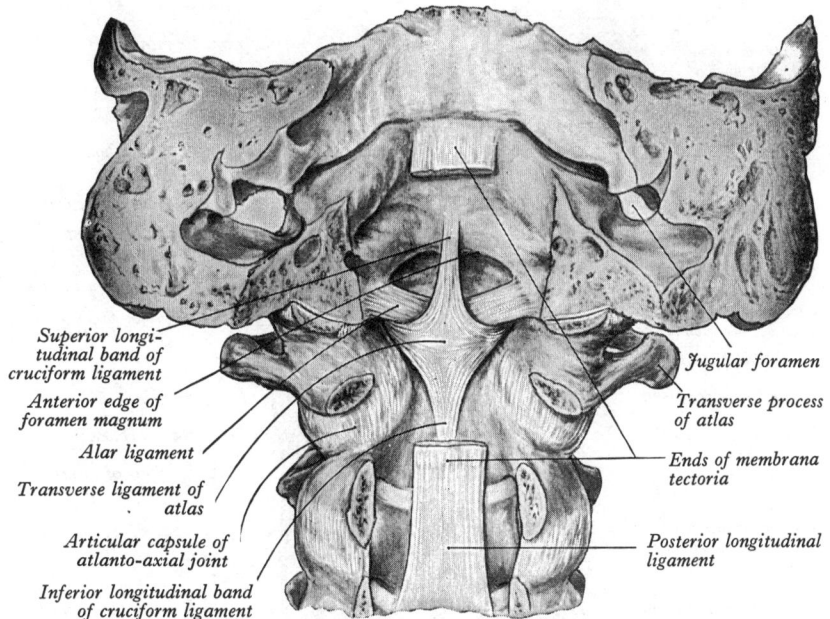

Superior longi-
tudinal band of
cruciform ligament

Anterior edge of
foramen magnum

Alar ligament

Transverse ligament of
atlas

Articular capsule of
atlanto-axial joint

Inferior longitudinal band
of cruciform ligament

Jugular foramen

Transverse process
of atlas

Ends of membrana
tectoria

Posterior longitudinal
ligament

4.29   Posterior aspect of the atlanto-occipital and atlanto-axial joints, after removal of the posterior part of the occipital bone and the laminae of the upper cervical vertebrae. The atlanto-occipital joint cavities have been opened.

which is set transversely. The two joints therefore act as one, and movement may occur around transverse and anteroposterior axes, but not round a vertical axis. The movements permitted therefore are (a) flexion and extension (nodding of the head), and (b) a slight lateral tilting motion to one or other side. (Consult Kapandji 1974.)

Muscles producing the movements.

Flexion. Longus capitis and rectus capitis anterior.

Extension. Recti capitis posteriores major et minor, obliquus capitis superior, semispinalis capitis, splenius capitis, and trapezius (upper part).

Lateral flexion. Rectus capitis lateralis, semispinalis capitis, splenius capitis, sternocleidomastoid and trapezius (upper part).

LIGAMENTS CONNECTING THE AXIS WITH THE OCCIPITAL BONE

The membrana tectoria, paired alar ligaments and the apical ligament extend between axis and occipital bone.

The membrana tectoria (4.29, 30) is situated within the vertebral canal. It is a broad, strong band, which covers the dens and its ligaments and appears to be a prolongation upwards of the posterior longitudinal ligament of the vertebral column. It consists of superficial and deep lamellae, both of which are attached below to the posterior surface of the body of the axis. The superficial lamella expands as it ascends and is attached above to the upper surface of the basilar part of the occipital bone, in front of the foramen magnum, blending with the cranial dura mater. The deep lamella consists of a median band which extends to the basilar part of the occipital bone and two lateral bands which ascend on the medial sides of the atlanto-occipital joints to the margins of the foramen magnum.

The alar ligaments (4.29), two strong, rounded cords, start on each side of the upper part of the dens, and, passing obliquely upwards and laterally, are attached to rough impressions on the medial sides of the condyles of the occipital bone. The alar ligaments are relaxed on extension of the head but become taut on flexion and help to limit the movement. They are so disposed that they would render free rotation of the head impossible, were it not that rotation is accompanied by a slight descent of the atlas. This descent causes sufficient relaxation of the alar ligaments to compensate for the tension brought about by rotation. Rotation to the right is eventually checked by the tension of those fibres of the right alar ligament which are attached to the dens in front of the axis of movement, and by tension of those fibres of the left alar

ligament which are attached to the process behind the axis of movement. Rotation to the left is checked by the opposite fibres on each side.

The apical ligament of the dens (4.30), which extends from the tip of the process to the anterior margin of the foramen magnum, lies between the alar ligaments, being intimately blended with the deep portion of the anterior atlanto-occipital membrane and with the upper longitudinal band of the cruciform ligament of the atlas. It is said to represent the core of the centrum of the pro-atlas vertebra and contains traces of the notochord in its substance (Ganguly and Roy 1964).

In addition to the ligaments which unite the atlas and axis to the skull, the ligamentum nuchae (p. 445) connects the cervical vertebrae with the cranium.

Applied Anatomy. Dislocation of the atlas from the axis, with rupture of the transverse ligament of the atlas and consequent injury to the spinal cord, is the mode by which death is produced in many cases of execution by hanging. Hanging may, however, produce a fracture through the axis, or a separation through the disc between the axis and the third cervical vertebra. Occasionally a pathological dislocation of the atlas on the axis may result from softening of the cruciform ligament.

The Costovertebral Joints

The articulations of ribs with vertebral column are divisible into two groups, one connecting the heads of ribs with the bodies of vertebrae, the other, the costotransverse joints, uniting the necks and tubercles of the ribs with the transverse processes.

THE JOINTS OF THE HEADS OF THE RIBS

These are formed by the articulation of the heads of the typical ribs with the facets on the margins of the bodies of adjacent thoracic vertebrae, and with the intervertebral discs between them (4.31). They are customarily classed as plane joints, though the surfaces of many are in fact curved. The first, tenth, eleventh, and twelfth ribs each articulate with a single vertebra; in each of the other joints, an intra-articular ligament divides the joint into two distinct parts. The ligaments of each joint are capsular, radiate, intra-articular.

The fibrous capsules connect the heads of the ribs with the circumference of the articular cavities formed by the intervertebral discs and the adjacent vertebrae. Some of their upper

449

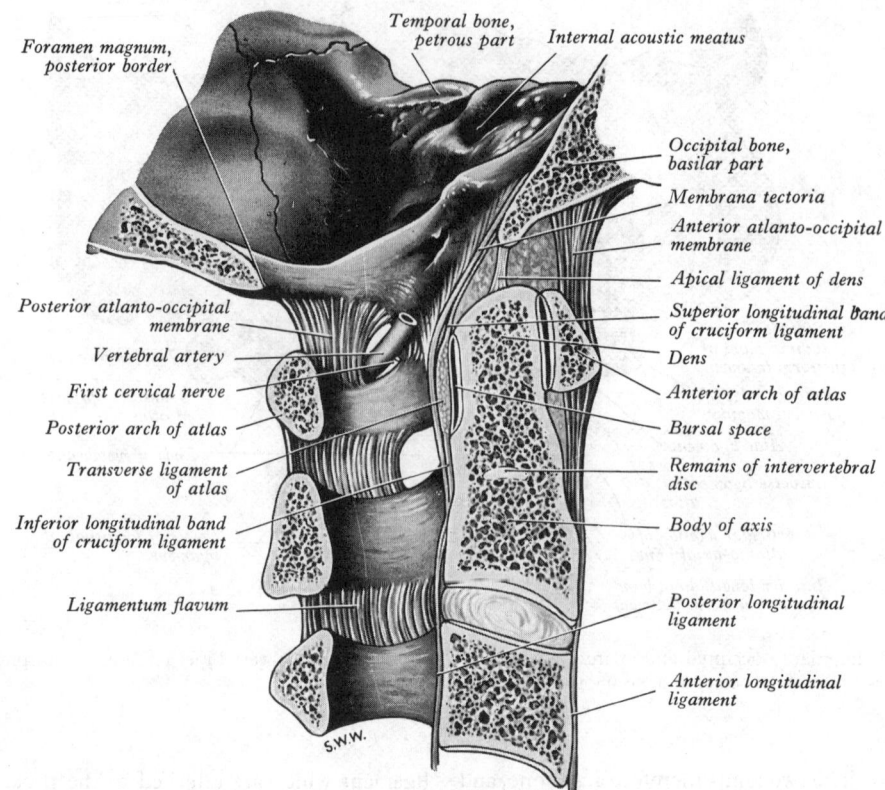

Foramen magnum, posterior border
Temporal bone, petrous part
Internal acoustic meatus
Occipital bone, basilar part
Membrana tectoria
Anterior atlanto-occipital membrane
Apical ligament of dens
Posterior atlanto-occipital membrane
Superior longitudinal band of cruciform ligament
Vertebral artery
Dens
First cervical nerve
Anterior arch of atlas
Posterior arch of atlas
Bursal space
Transverse ligament of atlas
Remains of intervertebral disc
Inferior longitudinal band of cruciform ligament
Body of axis
Ligamentum flavum
Posterior longitudinal ligament
Anterior longitudinal ligament

4.30 A median sagittal section through the occipital bone and the first to third cervical vertebrae.

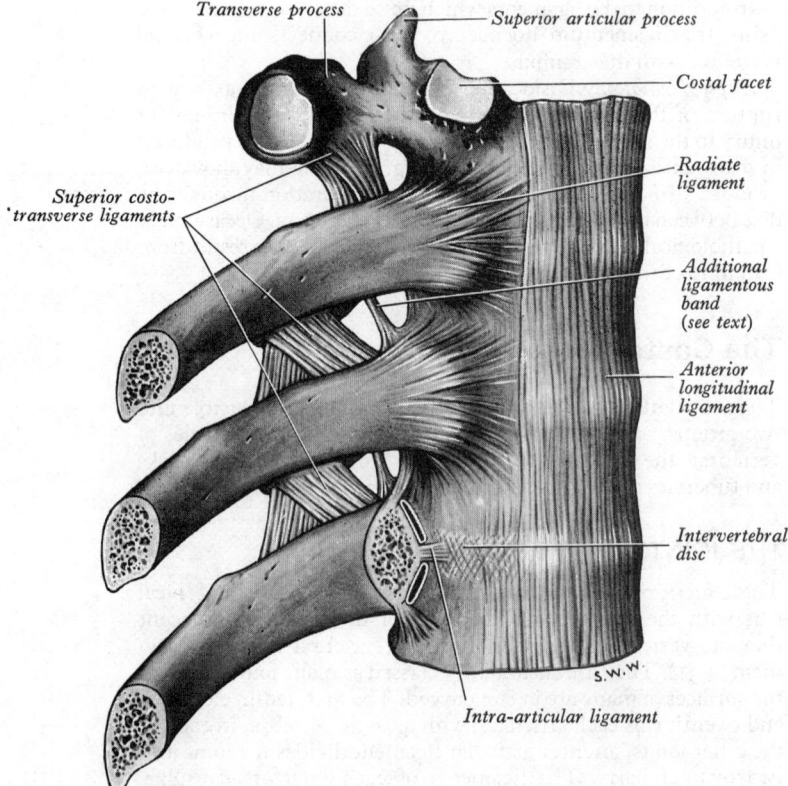

Transverse process
Superior articular process
Costal facet
Radiate ligament
Superior costo-transverse ligaments
Additional ligamentous band (see text)
Anterior longitudinal ligament
Intervertebral disc
Intra-articular ligament

4.31 The costovertebral joints. Right anterolateral aspect. In the lowest joint shown most of the radiate ligament and the anterior part of the head of the rib have been excised to show the two joint cavities and the intra-articular ligament between them.

fibres pass through the intervertebral foramen to the back of the intervertebral disc, while the posterior fibres are continuous with the costotransverse ligament.

The radiate ligament connects the anterior part of the head of each rib with the sides of the bodies of two vertebrae and the intervertebral disc between them. It is attached to the anterior part of the head of the rib, just beyond the articular surface. The superior fibres ascend and are connected with the body of the vertebra above; the inferior fibres descend to the body of the vertebra below; the middle fibres, the shortest and least distinct, are horizontal and attached to the intervertebral disc.

In the articulation of the first rib, the radiate ligament is attached to the body of the last cervical vertebra as well as the first thoracic. In the joints of the tenth, eleventh and twelfth ribs, each of which articulates with a single vertebra, the radiate ligament is connected to this vertebra and also to the one above.

The intra-articular ligament is a short band, flattened from above downwards, attached laterally to the crest separating the two articular facets on the head of the rib and medially to the intervertebral disc; it divides the joint completely. In the joints of the first, tenth, eleventh and twelfth ribs intra-articular ligaments do not exist; consequently these are single synovial joints.

THE COSTOTRANSVERSE JOINTS

The articular part of the tubercle of a rib articulates with a reciprocal facet on the transverse process of the vertebra to which it corresponds numerically (4.32). In the eleventh and twelfth ribs this articulation is absent. In the upper six (or five) joints the articular surfaces are reciprocally curved, but in the lower joints the surfaces are more flattened (4.33).

The ligaments of these joints are costotransverse, superior and lateral costotransverse, and the fibrous capsule.

The fibrous capsule is a thin membrane attached to the circumference of the articular surfaces, and lined with synovial membrane.

The superior costotransverse ligament has an anterior layer attached between the crest of the neck of a rib and the lower aspect of the transverse process above it (4.31); laterally it is continuous with the internal intercostal membrane and it is crossed by the corresponding intercostal vessels and nerve. The posterior layer is attached to the dorsal aspect of the neck of the rib and passes up medially behind to reach the transverse process above. Laterally it forms a continuous layer with the external intercostal muscle.

The first rib has no superior costotransverse ligament. The shaft of the twelfth rib, close to its head, is connected to the base of the transverse process of the first lumbar vertebra by a *lumbocostal ligament* in series with the superior costotransverse ligaments.

An accessory ligament is usually present medial to the superior costotransverse, and separated from it by the dorsal ramus of a thoracic spinal nerve and accompanying vessels (4.31). Somewhat variable in attachments, such bands commonly extend from a depression medial to the tubercle of a rib to the inferior articular process immediately above; some fibres may pass to the base of the transverse process.

The costotransverse ligament occupies the *costotransverse foramen*, the interval between the neck of the rib and the anterior surface of the corresponding transverse process. Its short but numerous fibres are attached in front to the rough surface on the back of the neck of the rib and behind to the anterior surface of the adjacent transverse process. In the case of the eleventh and twelfth ribs the ligament may be rudimentary or absent.

The lateral costotransverse ligament is a short, thick, strong fasciculus, passing obliquely from the apex of a transverse process to the rough nonarticular portion of the tubercle of the adjacent rib. The ligaments attached to the upper ribs ascend from the transverse processes; they are shorter and more oblique than those attached to the lower ribs, which descend slightly.

Movements at the costotransverse joints. The heads of ribs are so closely connected to the bodies of the vertebrae by the radiate and intra-articular ligaments that only slight gliding movements of the articular surfaces on one another can take place. Similarly, the strong ligaments binding the necks and tubercles of the ribs to the transverse processes limit the movements of the costotransverse joints to slight gliding, guided by the shape and direction of the articular surfaces (4.33). The articular surfaces on the tubercles of the upper six ribs are oval in shape and convex from above downwards; they fit into corresponding concavities on the anterior surfaces of the transverse processes, so that upward and downward movements of the tubercles are associated with rotation of the rib-neck on its long axis. On the seventh to tenth ribs the articular surfaces of the tubercles are almost flat, and face obliquely downwards, medially and backwards. The surfaces with which they articulate are on the upper aspects of the transverse processes; when, therefore, the tubercles are drawn up they are at the same time carried backwards and medially. The joints of the heads of the ribs and the costotransverse joints move simultaneously and in the same directions, the total effect being that the neck of the rib moves as if on a single joint, of which the two articulations form the ends. In the upper six ribs the neck of the rib moves but slightly upwards and downwards; its chief movement is one of rotation round its own long axis, rotation downwards of the front of the neck of the rib being associated with depression, rotation upwards with elevation of the anterior end of the rib and its costal cartilage. In the seventh to tenth ribs the neck of the rib moves upwards, backwards and medially, or downwards, forwards and laterally, with resultant increase or diminution of the infrasternal angle; a small degree of rotation accompanies these movements.

The muscles involved in the movements of respiration are considered on p. 550.

The Sternocostal Joints

The costal cartilages are received into small concavities on the lateral sternal borders (*chondrosternal articulations* (4.34)). The perichondrum and periosteum are continuous. The first costal cartilage is directly united to the sternum by a synchondrosis. The cartilages of the second to seventh ribs articulate with the lateral border of the sternum at synovial joints. The articular cavity is, however, frequently absent, particularly in the lower members of this series of joints. The articular surfaces are covered with *fibrocartilage* which also unites the costal cartilages to the sternum

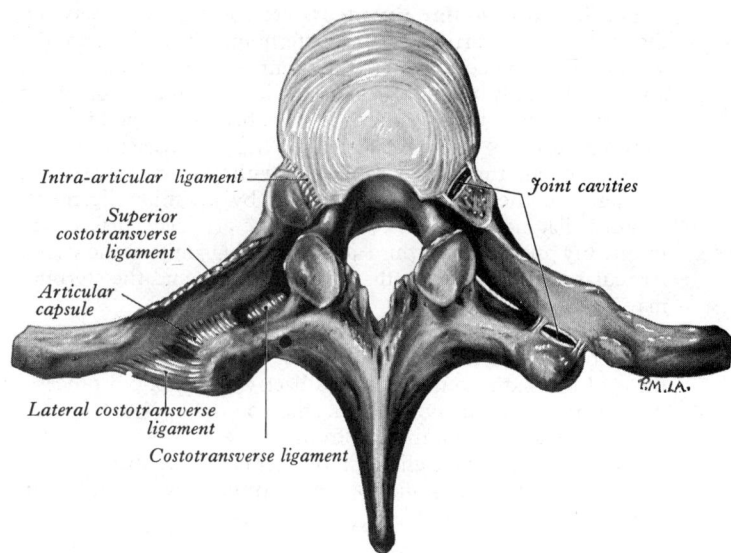

4.32 The costovertebral joints. Superior aspect.

Intra-articular ligament
Superior costotransverse ligament
Articular capsule
Lateral costotransverse ligament
Costotransverse ligament
Joint cavities

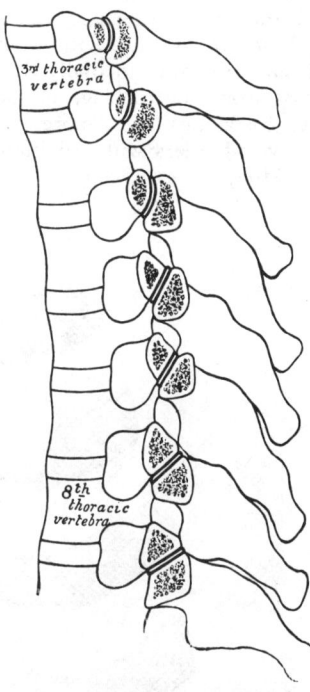

3rd thoracic vertebra

8th thoracic vertebra

4.33 A section through the costotransverse joints from the third to the ninth inclusive. Contrast the concave facets on the upper with more flattened facets on the inferior transverse processes.

when joint cavities are absent (Gray and Gardner 1943). The ligaments involved are capsular, radiate sternocostal, intra-articular and costoxiphoid (*see* Sick and Koritke 1976, for variations of the seventh costosternal joint, which may be synovial or 'symphysial').

The fibrous capsules surround the joints between the sternum and the second to the seventh costal cartilages. They are very thin, intimately blended with the sternocostal ligaments, and strengthened above and below by a few fibres which connect the cartilages to the side of the sternum.

The radiate sternocostal ligaments are broad, thin, membranous bands which radiate from the front and back of the sternal ends of the cartilages of the true ribs to the anterior and posterior surfaces of the sternum. Their superficial fibres intermingle with the fibres of the ligaments above and below them, with those of the opposite side, and on the front of the sternum with the tendinous fibres of origin of the pectoralis major, forming a thick fibrous membrane which envelops the bone, and is more distinct at its lower part than at its upper part.

The intra-articular ligaments are constant only between the second costal cartilages and the sternum. The cartilage of the second rib is connected with the sternum by means of a ligament, attached laterally to the cartilage of the rib, and medially to the fibrocartilage which unites the manubrium and body of the sternum, and is therefore intra-articular. Occasionally the cartilage of the third rib is connected with the first and second segments of the body of the sternum by a similar ligament. Fibrocartilaginous strands connecting the articular surfaces are frequently present in the third and succeeding joints of the series. Articular cavities between the costal cartilages and the sternum may be absent at all ages.

The costoxiphoid ligaments connect the anterior and posterior surfaces of the seventh costal cartilage, and sometimes those of the sixth, to the front and back of the xiphoid process. They vary in length and breadth, those on the back of the joint being less distinct than those in front.

Movements. Slight gliding movements are permitted in the sternocostal joints, sufficient to permit easy mobility in respiration.

THE INTERCHONDRAL JOINTS

The contiguous borders of the sixth and seventh, the seventh and eighth, and the eighth and ninth costal cartilages articulate with each other by small smooth, oblong facets. Each articulation is enclosed in a thin *fibrous capsule*, lined with synovial membrane and strengthened laterally and medially by *interchondral ligaments*, which pass from one cartilage to the other (**4.**34). Sometimes the fifth costal cartilages, more rarely the ninth, articulate by their lower borders with adjoining cartilages by small oval facets; more frequently this connexion is made by a few

ligamentous fibres. The articulation between the ninth and tenth cartilages is never synovial and sometimes it is absent (p. 290).

THE COSTOCHONDRAL JUNCTIONS

The costal cartilages are, of course, the persistent, unossified anterior parts of the cartilaginous models which precede the fully developed ribs. When artificially separated from its rib a costal cartilage displays a rounded end and the rib a depression. The junction is covered by periosteum, continuous with the adjoining perichondrium. No movement occurs at these junctions.

THE STERNAL JOINTS

The manubriosternal joint, between the manubrium and the body of the sternum, is usually a symphysis, the bony surfaces being coated with the hyaline cartilage and connected by a disc of fibrocartilage, which may become ossified in the aged. In rather more than 30 per cent of people the central part of the disc undergoes absorption and the joint *resembles* a synovial one; the same change may occur in the symphysis pubis. The two segments of the bone are also connected by the fibrous membrane which envelops the bone. In 10 per cent of all aged over 30 years the manubrium is joined to the body of the sternum by bone; when this occurs only the superficial part of the intervening cartilage may be ossified, and only in the aged is this complete. Early synostosis has been attributed to the persistence of a synchondrosis in place of a symphysis at this junction (Ashley 1954). In the newborn the union of the manubrium and body of the sternum is by collagenous and elastic fibres without obvious chondrocytes.

Movements. The manubriosternal symphysis permits a small

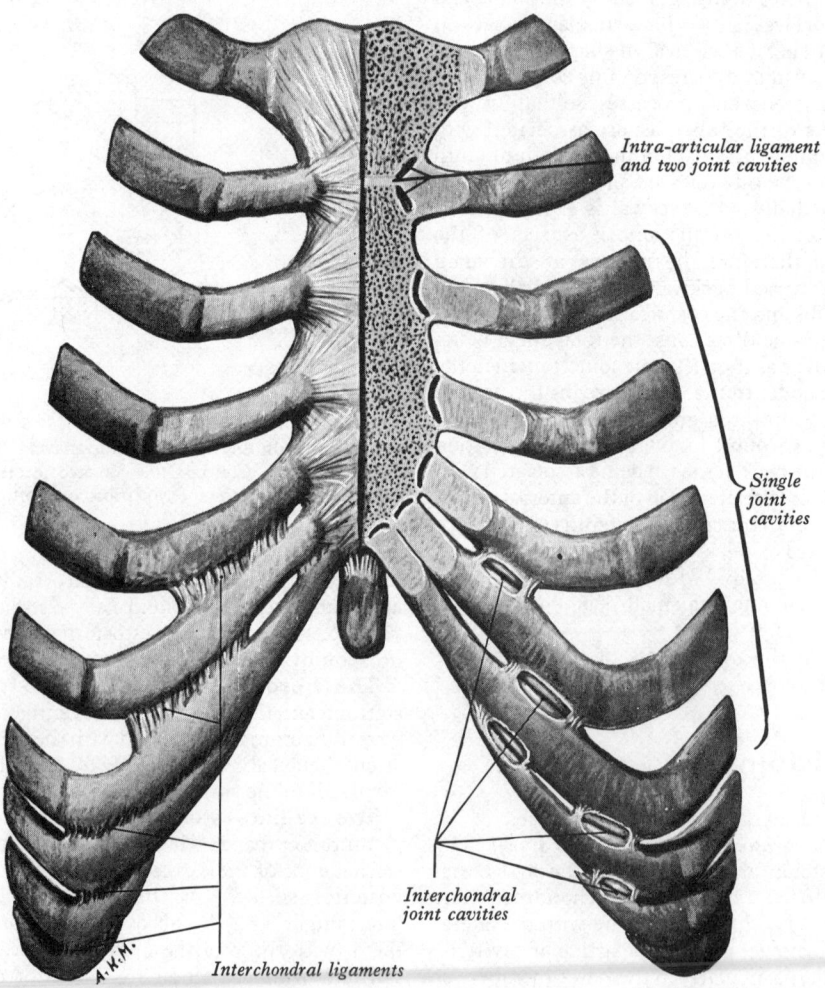

*Intra-articular ligament
and two joint cavities*

*Single
joint
cavities*

*Interchondral
joint cavities*

Interchondral ligaments

A·K·M·

4.34   The sternocostal and interchodral joints. Anterior aspect.

range of angular change between the longitudinal axes of the manubrium and corpus sterni, and also a limited degree of anteroposterior displacement between these two sternal elements. A detailed study (Constantinesco 1974) of these movements in 62 male athletes yielded (in the standing position) mean values of 162·7° (full inspiration) and 164·7° (full expiration) for the manubriosternal angle. The angular movement and the displacement both contribute to the ventral movement of the sternum in inspiration and its reversal in expiration (*vide infra*).

The xiphisternal joint is between the xiphoid process and the body of the sternum and is also a symphysis; it usually becomes a synostosis by the fortieth year, but the xiphoid process may remain separate even in old age.

The Mechanism of the Thorax

Each rib possesses its own range and direction of movement, contributing to the combined respiratory excursions of the thorax. Each rib may be regarded as a lever, the fulcrum of which is situated immediately lateral to its costotransverse articulation, so that when the shaft of the rib is elevated the neck is depressed, and vice versa; because of the large difference in length of the arms of this lever a slight movement at the vertebral end of the rib is much magnified at the anterior extremity.

The anterior ends of the ribs are lower than the posterior, and therefore, when the shaft of the rib is elevated, they rise in a forward direction. Again, the middle of the shaft of the rib lies in a plane below that passing through the two extremities, so that when the shaft is elevated relatively to its ends it is at the same time carried outwards from the median plane; further, each rib forms the segment of a curve which is greater than that of the rib immediately above. Therefore the elevation of a rib increases the transverse diameter of the thorax in the plane to which it is raised. The modifications of the rib movements at their vertebral ends have already been described (p. 451). Further modifications result from the attachments of their anterior extremities, and it is convenient therefore to consider separately the movements of the ribs in three groups, vertebrosternal, vertebrochondral and vertebral.

Vertebrosternal ribs (4.35A). The first rib moves little, except in deep respiration. Its movement occurs about an oblique axis through the neck; the shaft is displaced upwards and laterally in inspiration and the under surface comes to face more directly downwards. This movement is impossible if the costal cartilage is calcified, as it sometimes is; the first rib on each side and the manubrium sterni then move as a single unit about a transverse axis through the costotransverse joints. Movement of the second rib is slight also in quiet respiration. Elevation of the third to sixth ribs raises and thrusts forwards their anterior extremities, the greater part of the movement being effected by the rotation of their necks backwards. The thrust of the anterior extremities carries forwards and upwards the body of the sternum, which moves on the joint between it and the manubrium, and thus the anteroposterior thoracic diameter is increased. This movement, however, is soon arrested, and the elevating force is then expended in raising the middle part of the shaft of the rib and everting its lower border; at the same time the costochondral angle is opened out. By these latter movements a considerable increase in the transverse diameter of the thorax is effected. Measurements of sternal movements during respiration reveal extensive excursions, especially in physically fit adult males (Constantinescu 1974). Between full inspiration and expiration the suprasternal

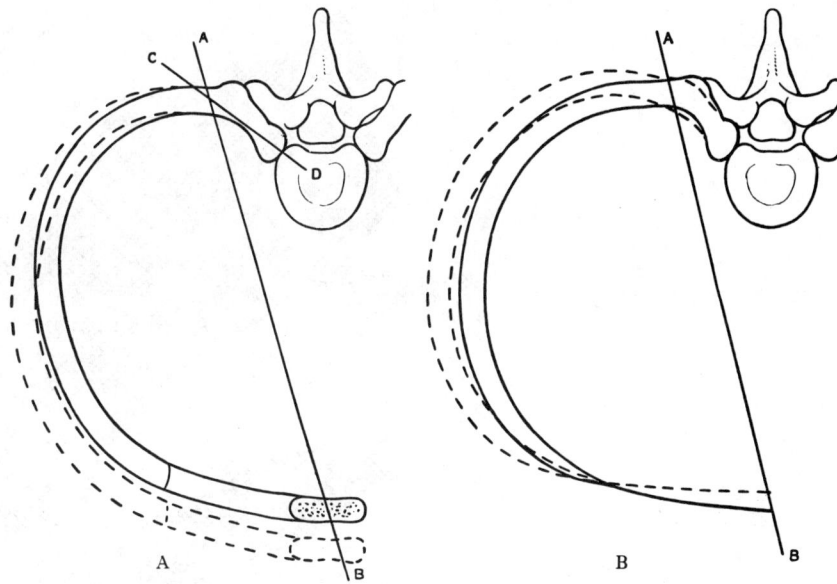

4.35A A diagram showing the axes of movement (AB and CD) of a vertebrosternal rib. The interrupted lines indicate the position of the rib in inspiration.

4.35B A diagram showing the axes of movement (AB) of a vertebrochondral rib. The interrupted lines indicate the position of the rib in inspiration.

notch may show an excursion of 31 mm; the excursions at the superior (34 mm) and the inferior (37 mm) extremities of the corpus sterni being greater, due to changes in the sternal angle.

Vertebrochondral ribs (4.35B). The seventh rib is included with this group, as it conforms more closely to their type of motion. While the movements of these ribs assist in enlarging the thorax for respiratory purposes, they are also concerned in increasing the upper abdominal space for viscera displaced by the action of the diaphragm, although relaxation of the abdominal wall can account at least in part for this. The costal cartilages articulate with one another, so that each pushes up that above it, the final thrust being directed to pushing forwards and upwards the lower end of the body of the sternum. The amount of elevation of the anterior extremities is limited on account of the very slight rotation of the necks of these ribs. Elevation of the shaft is accompanied by an outward and backward movement; the outward movement everts the anterior end of the rib and opens up the infrasternal angle, while the backward movement pulls back the anterior extremity and counteracts the forward thrust due to its elevation; this latter is most noticeable in the lower ribs, which are the shortest. The total result is a considerable increase in the transverse and a diminution in the median anteroposterior diameter of the upper part of the abdomen; at the same time, however, the lateral anteroposterior diameters of the abdomen are increased.

Vertebral ribs. Since these ribs have free anterior extremities and only costovertebral articulations with no intra-articular ligaments, they are capable of slight movements in all directions. When the other ribs are elevated these are depressed and fixed by the quadratus lumborum muscles to form fixed points of action for the diaphragm.

The muscles producing these movements are discussed on p. 550.

THE JOINTS OF THE UPPER LIMB

The Sternoclavicular Joint

Entering into the formation of the sternoclavicular joint are the sternal end of the clavicle, the clavicular notch of the manubrium sterni, and the cartilage of the first rib (4.36). The *articular surface*

of the clavicle is much larger than that of the sternum and is covered with a layer of *fibrocartilage*, which is considerably thicker than that on the sternum. It is convex vertically, and slightly concave anteroposteriorly and is therefore a type of sellar (saddle) joint; the clavicular notch of the sternum is reciprocally

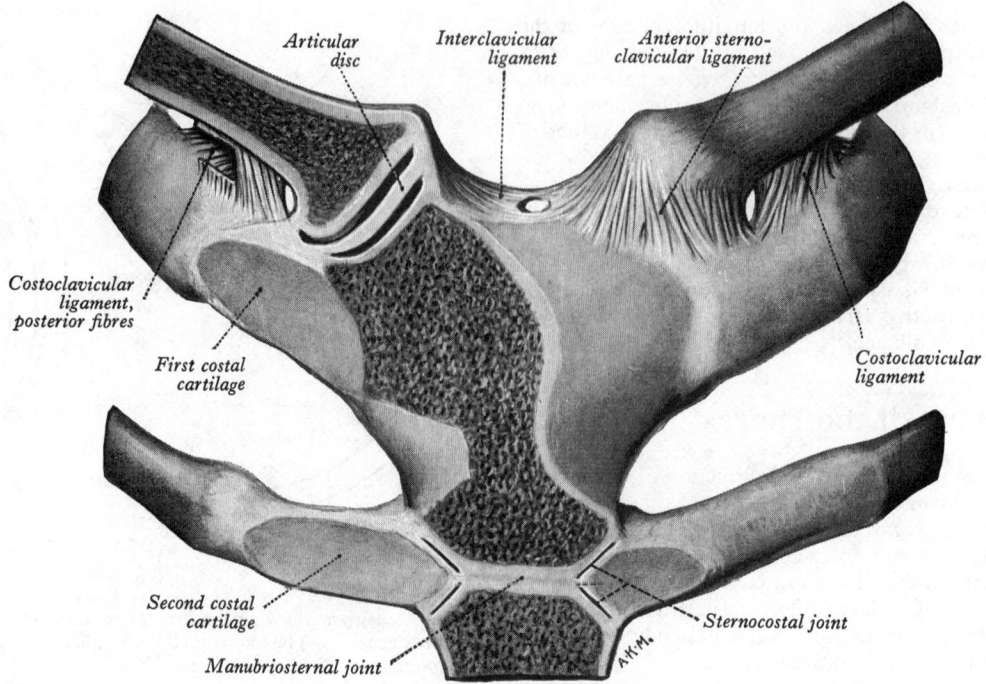

4.36   The sternoclavicular joints. Anterior aspect.

curved, but the two surfaces are not perfectly congruent. (These two joint surfaces show some variation, particularly in their peripheral shape but also in their degrees of curvature—*see* p. 358). The joint is completely divided by an articular disc. Its ligaments are capsular, anterior and posterior sternoclavicular, interclavicular and costoclavicular.

The fibrous capsule surrounds the articulation; in front and behind, it is thickened, but above, and especially below, it is thin and partakes more of the character of loose areolar tissue.

The anterior sternoclavicular ligament is a broad band, covering the anterior surface of the joint; it is attached above to the anterosuperior area of the sternal end of the clavicle, and, passing obliquely downwards and medially, is attached below to the front of the upper part of the manubrium sterni and more laterally to the first costal cartilage.

The posterior sternoclavicular ligament is a weaker band which covers the posterior aspect of the joint. It is attached to the posterior aspect of the sternal end of the clavicle and passes obliquely downwards and medially to be attached to the back of the upper part of the manubrium sterni.

The interclavicular ligament is continuous above with the deep cervical fascia; it unites the upper part of the sternal ends of both clavicles but some of its fibres are attached to the upper margin of the manubrium. When present, the *suprasternal ossicles* (p. 288) are in this ligament.

The costoclavicular ligament is short, shaped like an inverted cone and flattened anteroposteriorly. It consists of anterior and posterior laminae, which are attached below to the upper surface of the first rib and the adjacent part of its cartilage, and above to the margins of the impression on the inferior surface of the medial end of the clavicle. The fibres of the anterior lamina are directed upwards and laterally and those of the posterior lamina, which are shorter, upwards and medially (4.36). Laterally the laminae fuse with each other and medially they merge with the fibrous capsule of the sternoclavicular joint. Between the laminae is a bursa, an indication that they have discrete functions (Cave 1961). It seems probable that each of these two laminae is tensed towards the opposite extremes of clavicular axial rotation.

The articular disc, flat and nearly circular, is interposed between the articulating surfaces of the sternum and clavicle. It is attached, above, to the upper and posterior border of the articular surface of the clavicle, below, to the cartilage of the first rib, near its junction with the sternum, and by the rest of its circumference

to the fibrous capsule. It is thicker peripherally—especially at its upper and posterior part—than at its centre, and divides the joint into two parts, the larger above and laterally and the smaller below and medially. The capsule around the former is more lax, and movements between the clavicles and the discs are more extensive than those between the discs and the sternum and first costal cartilages. Movement at the joint is further considered below (p. 455), but it should be noted here that the sellar shape of the joint surfaces permits not only *angulation* in approximately anteroposterior and vertical planes but also a small range of rotation (about the clavicular long axis) or *spin* (30° according to Kapandji 1970). This spin is conjunct. The close-packed position of the joint probably coincides with maximum posterior spin associated with full rotation of the scapula (*vide infra*). Some anteroposterior gliding (translation) also occurs.

The *arteries* supplying the joint are branches from the internal thoracic and suprascapular arteries; the *nerves*, from the anterior supraclavicular nerve and the nerve to the subclavius. For details of vascularization consult Sick and Ring (1976).

Applied Anatomy. The strength of this joint depends upon its ligaments, and especially on the articular disc. Owing to this, and to the fact that the force of a blow is usually transmitted along the long axis of the clavicle, dislocation is rare, and the clavicle is usually broken rather than displaced.

The Acromioclavicular Joint

The acromioclavicular articulation (4.40, 42), between the acromial end of the clavicle and the medial margin of the acromion of the scapula, is of approximately plane type; but both surfaces are not quite flat, either of them being sometimes slightly convex, the other reciprocally concave. Both *articular surfaces* are covered with *fibrocartilage*; on the acromial end of the clavicle is a narrow, oval area, directed downwards and laterally to overlap a corresponding area on the medial border of the acromion. The long axis of the joint is anteroposterior. Its ligaments are capsular, acromioclavicular and coracoclavicular.

The fibrous capsule completely surrounds the articular margins, and is strengthened above by the acromioclavicular ligament.

The acromioclavicular ligament is quadrilateral and covers the superior part of the joint, extending between the upper part of the acromial end of the clavicle and the adjoining part of the upper surface of the acromion. It is composed of parallel fibres, which interlace with the aponeuroses of trapezius and deltoid.

An articular disc is often found in this joint; when present, it occupies the upper part of the articulation, and only partially separates the articular surfaces (de Palma 1957). More rarely, it divides the joint completely.

Movements at the acromioclavicular joint (see also below) are like those of the sternoclavicular articulation. The range of axial rotation of the clavicle here is said to be about 30° (Kapandji 1970). The two joints working together, therefore, permit some 60° rotation of the scapula. Angulation between scapula and clavicle can occur in any direction. Both spin and angulation tense the coracoclavicular ligament (*vide infra*) as extremes are reached, and spin tightens the capsular ligament by spiralization. According to MacConaill and Basmajian (1977) the close-packed position of the joint is achieved in forward rotation of the scapula (*vide infra*), when the angle between the superior scapula border and clavicular shaft becomes about 90°. This 'opening' or *ouverture* of the angle leads to tension of the conoid part of the coracoclavicular ligament (Kapandji 1970). When this position is reached, further rotation of the scapula is permitted by rotation at the sternoclavicular joint.

The coracoclavicular ligament (4.37) connects the clavicle with the coracoid process of the scapula. Though separated by an interval from the acromioclavicular joint it forms a most efficient means of preventing the clavicle from losing contact with the acromion. It consists of two parts, viz. *trapezoid* and *conoid*, which are usually separated by fat or, frequently, a bursa. This bursa thus intervenes between the medial end of the horizontal part of the coracoid process and the lateral end of the groove for the subclavius on the clavicle. These bony regions may be closely apposed and covered with cartilage to form a coracoclavicular joint (Lewis 1959)

The *trapezoid part*, the anterolateral fasciculus, is broad, thin and quadrilateral. It is attached below to the upper surface of the coracoid process and above to the trapezoid line on the inferior surface of the coracoid process and above to the trapezoid line on the inferior surface of the clavicle. The ligament is almost horizontal. Its anterior border is free, its posterior is joined with the conoid part, the two forming, by their junction, an angle projecting backwards.

The *conoid part*, the posteromedial fasciculus, is a dense band above, triangular in form, its base superior. Its apex is attached to the medial and posterior edge of the root of the coracoid process just in front of the scapular notch; its base is fixed to the conoid tubercle on the inferior surface of the clavicle, and to a line proceeding medially from it for a short distance.

The *arterial supply* to the acromioclavicular joint is derived from the suprascapular and thoraco-acromial arteries; the *nerve* supply from the suprascapular and lateral pectoral nerves.

Applied Anatomy. In dislocation of the acromioclavicular joint the coracoclavicular ligament is torn and the scapula falls away from the clavicle. Owing to the flatness and orientation of the joint surfaces, the dislocation readily recurs.

Movements of the shoulder girdle. The clavicular movements which occur at the sternoclavicular and acromioclavicular joints must always be associated with movement of the scapula, and movements of the scapula are usually accompanied by movements of the humerus at the shoulder joint. Therefore this account should be read in association with the description of movements of the shoulder joint (p. 458), and scapular musculature (p. 569).

The acromioclavicular joint allows the acromion, and hence the whole scapula, to glide forwards and backwards, and to rotate on the clavicle, but the range of scapular movement is greatly increased by associated movements at the sternoclavicular joint. The analysis of all these movements can be best made clear by a study of the movements of the scapula.

The movements of the scapula can be primarily analysed as (1) elevation and depression, (2) forward and backward movement round the chest wall, (3) rotation forwards or upwards, using the inferior angle as a reference point, and the reverse movement.

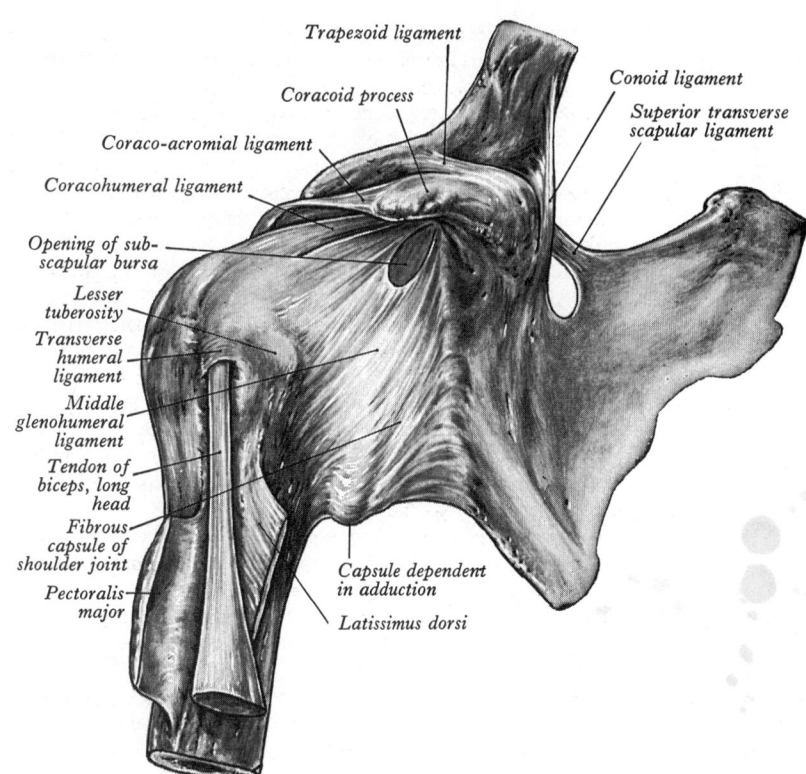

4.37 The right shoulder joint. Anterior aspect.

1. *Elevation and depression of the scapula*, picturesquely exemplified by shrugging of the shoulders, do not necessarily imply movement at the shoulder joint.

During elevation a slight degree of angulation or swing occurs at the acromioclavicular joint, but the sternal end of the clavicle, which rotates round an anteroposterior axis passing through the bone above the attachment of the medial end of the costoclavicular ligament, slides downwards over the surface of the articular disc (*translation*). This movement at the joint is checked by tension of the antagonist muscles, the costoclavicular ligament and the lower part of the capsule. It is brought about by the trapezius (upper part) and the levator scapulae, and, since these muscles tend to rotate the scapula in opposite directions, a purely upward movement can be effected.

In the reverse movement a little angular movement occurs at the acromioclavicular joint, but at the sternoclavicular joint the clavicle slides upwards on the disc, and this is checked by tension of the antagonist muscles, and by the interclavicular and sternoclavicular ligaments and the articular disc. This particular role of the interclavicular ligament in suspension of the clavicle in the depressed position has been confirmed by experiment (Bearn 1967). As a rule this movement is carried out with the help of gravity, but it can be performed actively by the serratus anterior (lower part) and the pectoralis minor.

2. *Forward movement of the scapula (protraction)* round the chest wall occurs in all forward-pushing, thrusting and especially reaching movements, and is usually accompanied by some degree of forward rotation. The acromion moves forwards over the clavicular facet to the limit of its range of movement, and at the same time the point of the shoulder is advanced further by a forward movement of the lateral end of the clavicle, associated with a backward glide of the sternal end of the bone, which moves backwards over the sternal facet, carrying the articular disc with it. Tension of antagonist muscles, of the anterior sternoclavicular ligament and posterior lamina of the costoclavicular ligament checks backward slide of the sternal end of the clavicle. The serratus anterior and the pectoralis minor are the prime movers, and this combination ensures the continuous apposition of the scapula, especially its medial border, in a smooth glide over the chest wall. In addition, the upper part of the latissimus dorsi acts

like a strap to keep the inferior angle in close contact with the thoracic wall, both in this movement and in forward rotation.

In *backward movement of the scapula (retraction)*, such as occurs when the shoulders are braced back, the reverse movements take place and are checked at the sternoclavicular joint by tension of the posterior sternoclavicular ligament and the anterior lamina of the costoclavicular ligament. Trapezius and the rhomboids are the prime movers, but gravity may also produce backward scapular movement, when the weight of the trunk is taken by the arms in leaning forwards; the degree of 'retraction' permitted is controlled through the protraction musculature.

It may be noted here that when force is applied at the extremity of the outstretched arm, e.g. by a fall on the hand, the pressure transmitted to the glenoid cavity tends to drive the sloping acromial facet below the acromial end of the clavicle, but at the same time it causes tension of the trapezoid ligament, which serves to resist the displacement. It must also be noted that, unless the fall is unexpected and sudden, even more extensive forces are available to prevent such dislocation.

3. *Forward rotation of the scapula* serves to increase the range of movement of the humerus, by turning the bone so that the glenoid cavity faces almost directly upwards, the position which it assumes when the arm is raised above the head. This movement is always associated with a degree of elevation of the humerus and accompanied by some forward movement of the scapula round the chest wall. Rotation of the scapula necessarily involves movements at the sternoclavicular and acromioclavicular joints. The sternoclavicular joint permits elevation of the lateral end of the clavicle; this movement is almost complete when the arm has been abducted to 90°. Movement at the acromioclavicular joint occurs in the first 30° of abduction at which point the conoid ligament becomes taut; thereafter abduction is accompanied by a rotation of the clavicle at the sternoclavicular joint around the longitudinal axis of the bone, accompanied by further depression, as its lateral end continues to rise. Some acromioclavicular movement occurs also in the final stages of abduction of the arm (Inman *et al.* 1944). In this important movement the upper part of the trapezius and the lower part of the serratus anterior are the prime movers.

The opposite or *return rotation* is usually effected mainly by gravity, the gradual lengthening of the trapezius and serratus anterior being sufficient to bring it about in a controlled manner. When force is necessary the levator scapulae, rhomboids, and, in the initial stages at least, pectoralis minor are the prime movers in rotating the scapula to its position of rest.

Attention should be drawn to the fact that muscles which are antagonists for one type of movement may nevertheless combine together and act as prime movers for another. Movements and not individual muscles are represented in the cerebral motor centres, and, more significantly, muscles are not associated, in their nervous control, in unalterable partnerships, but can be combined with great plasticity as the demands of movement dictate. Thus the serratus anterior and the trapezius are opposed in forward and backward movements of the scapula round the chest wall, but combine together as prime movers for forward rotation of the bone.

In all the movements of the scapula the subclavius probably serves to steady the clavicle by drawing it medially and downwards, though it is naturally difficult to be certain of its role, due to its relative inaccessibility.

The movements of the scapula on the thoracic wall are greatly facilitated by the loose connective tissue which intervenes between the subscapularis and serratus anterior and between the latter muscle and the costal structures. When the upper limb is pendant, the maintenance of the normal position of the shoulder girdle relative to the trunk involves a comparatively small degree of activity in the trapezius and serratus anterior (Basmajian 1962), which obviously must increase as the weight of the limb is increased by loading.

THE LIGAMENTS OF THE SCAPULA

The ligaments of the scapula (4.37) are the coraco-acromial and transverse scapular.

The coraco-acromial ligament is a strong triangular band, extending between the coracoid process and the acromion. Its apex is attached to the edge of the acromion just in front of the articular surface for the clavicle; and its base to the whole length of the lateral border of the coracoid process. This ligament, together with the coracoid process and the acromion, forms an arch above the head of the humerus. It sometimes consists of two strong marginal bands and a thinner intervening portion. When the pectoralis minor is inserted, as it is occasionally, into the capsule of the shoulder joint instead of into the coracoid process, the tendon of the muscles passes between the two bands of the coraco-acromial ligament (p. 568). The subacromial bursa (p. 458 and 4.39) facilitates movement between the coraco-acromial arch and the subjacent supraspinous muscle and shoulder joint, acting thus almost as a secondary articulation.

The superior transverse scapular (suprascapular) ligament converts the scapular notch into a foramen, and is sometimes ossified. It is a flat fasciculus, narrowing toward its extremities, which are attached to the base of the coracoid process and the medial end of the scapular notch. The suprascapular nerve runs through the foramen; the suprascapular vessels cross over the ligament.

A weak, membranous band, the *inferior transverse (spino-glenoid) ligament*, stretches from the lateral border of the spine of the scapula to the margin of the glenoid cavity. It forms an arch under which the suprascapular nerve and vessels enter the infraspinous fossa. It is frequently absent.

The Humeral (Shoulder) Joint

The shoulder joint is a multi-axial spheroidal joint (3.140, 4.37, 38A, B, C, 41, 42). The bones involved are the roughly hemi-spherical head of the humerus and the shallow glenoid cavity of the scapula, a construction which permits of very considerable movement but reduces the security of the joint. Skeletally the shoulder joint is weak, since, for such strength as it possesses, it is dependent on the support given by the muscles which surround it and not on its bony conformation or the presence of any strong ligaments. It is, however, covered above by an arch, formed by the coracoid process, the acromion and the coraco-acromial ligament.

The *articular surfaces* are reciprocally curved and are in fact not sections of true spheres but *ovoids* (*see* p. 432). Here, and in the hip joint, where the ovoid surfaces approximate to a spherical form, as indicated above, they are often classed as *spheroidal*. As the convexity of the humerus is much larger than the glenoid con-cavity (4.38A and B) only a minor part of it can be in contact with the cavity in any given position of the joint and the remainder of its articular surface is in contact with the inner aspect of the capsule. The glenoid cavity (4.38A, C, 39) is deepened somewhat by a fibrocartilaginous rim attached to its margins, the *glenoidal labrum*. Both articular surfaces are covered with a layer of hyaline cartilage; that on the head of the humerus is thickest at its centre and thinner peripherally, while the reverse is the case in the glenoid cavity. However, in most positions of the joint, the curvatures of the adjacent parts of the surfaces are not precisely the same, i.e. they are not congruent and the joint is loose-packed (Saha 1961). Full congruence and the close-packed position is reached when the humerus is abducted and laterally rotated (*see* pp. 438, 459; and 4.5, 17). When the arm is by the side the anterior edge of the glenoid cavity can be represented on the front of the shoulder by a line, 3 cm long, drawn downwards from a point just lateral to the tip of the coracoid process. This line, which should be very slightly concave laterally, lies over the lower half of the joint.

The ligaments of the articulation are the *glenoidal labrum, fibrous capsule, glenohumeral, coracohumeral,* and *transverse humeral*.

The fibrous capsule (4.37, 38) envelops the joint, and is attached, medially, to the circumference of the glenoid cavity beyond the glenoidal labrum; above, it encroaches on the root of the coracoid process so as to include the origin of the long head of biceps within the joint. Laterally it is attached to the anatomical neck of the humerus, that is to say close to the articular margin,

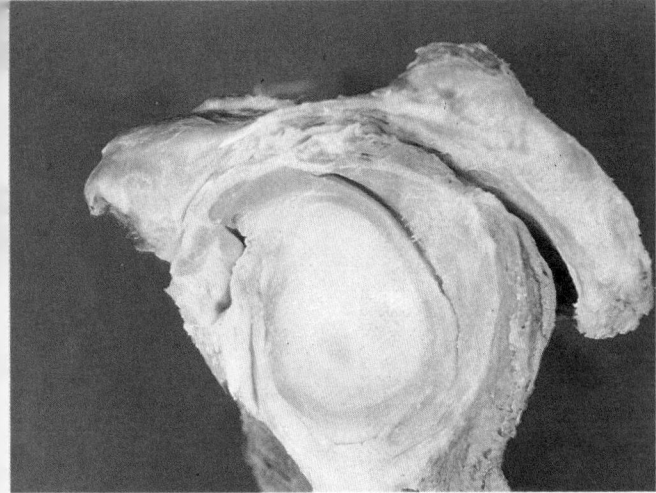 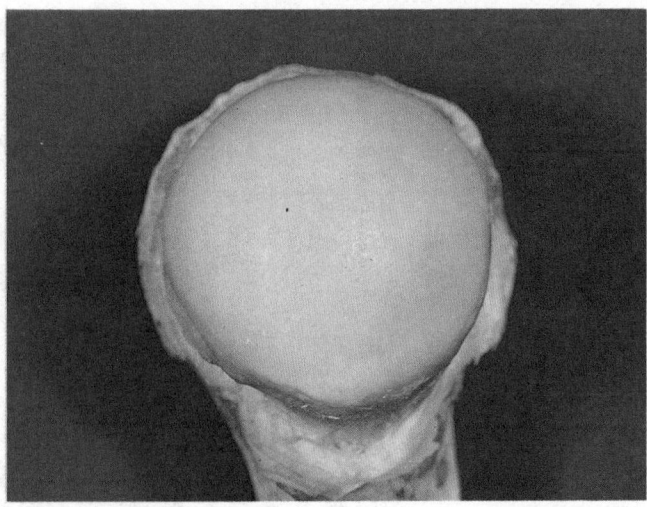

4.38A and B   Dissections from a preserved cadaver showing the articular surfaces and periarticular structures of the glenohumeral joint. A shows the glenoid cavity covered by articular cartilage and the glenoidal labrum; note the intracapsular part of the tendon of the long head of the biceps sectioned near the supraglenoid tubercle, the fibrous capsule intimately blended with 'rotator cuff' musculature, the subacromial bursa, and the

superjacent coraco-acromial arch complex. B displays the much more extensive spheroidal articular surface of the head of the humerus, and also the pendant inferior part of the fibrous articular capsule. (Preparation by Dr. M. C. E. Hutchinson and photography by Mr. Kevin Fitzpatrick, Guy's Hospital Medical School, London.)

except on the medial side where the attachment descends for rather more than 1 cm on the shaft of the bone. It is so remarkably loose and lax that the bones may even be separated from each other for a distance of 2 or 3 cm by a distractive force. This considerable laxity is obviously to be correlated with the great range of movement which is possible at the shoulder articulation. However, this degree of unphysiological separation can be affected only after the superior part of the ligament has been relaxed by some degree of abduction. The fibrous capsule is supported superiorly, by the supraspinatus; inferiorly, by the long head of the triceps; posteriorly, by the tendons of the infraspinatus and teres minor; anteriorly, by the tendon of the subscapularis. The tendons of the subscapularis, supraspinatus, infraspinatus, and teres minor are all intimately blended with the fibrous capsule forming a cuff, sometimes termed the *rotator cuff*; they reinforce the capsule and provide active support for the joint during movement, and indeed at all times unless fully relaxed. The relationship of the long head of the triceps is not so close, for it is separated from the inferior part of the capsule by the axillary nerve and the posterior circumflex humeral vessels as they pass backwards on leaving the axilla (4.41). It is the inferior part of the capsule, therefore, which is least supported, and it is just this part which is subjected to the greatest strain, because it is stretched tightly across the rounded head of the humerus when the arm is fully abducted.

There are usually two, occasionally three, openings in the capsule. One situated anteriorly, below the coracoid process, establishes a communication between the joint and a bursa behind the tendon of the subscapularis; another, placed between the tuberosities of the humerus, gives passage to the long tendon of the biceps and its synovial sheath; the third, which is not constant, is posterior and between the joint and a bursal sac under the tendon of infraspinatus.

Three thickenings, the *glenohumeral ligaments* (4.40), reinforce the capsule. These are best seen by opening the joint from behind and removing the head of the humerus. At their scapular ends they are all attached to the upper part of the medial margin of the glenoid cavity and are blended with the glenoidal labrum. The superior band passes along the medial edge of the tendon of biceps and is attached to a small depression above the lesser tubercle of the humerus; the middle band reaches to the lower part of the lesser tubercle; the inferior band extends to the lower part of the anatomical neck of the humerus. In addition to these, the capsule is strengthened in front by extensions from the tendons of pectoralis major and teres major.

The *synovial membrane* lines the inner surface of the fibrous capsule, and covers the lower part and sides of the anatomical neck

of the humerus as far as the articular cartilage on the head of the bone. The tendon of the long head of the biceps passes through the joint and is enclosed in a tubular sheath of synovial membrane, which is continued round the tendon into the intertubercular sulcus as far as the surgical neck of the humerus (4.37, 38).

The coracohumeral ligament (4.37), a broad band strengthening the upper part of the capsule, is attached to the lateral border of the root of the coracoid process, and passes obliquely downwards and laterally to the front of the greater tubercle of the humerus, blending with the tendon of supraspinatus. The posterior and lower border of the ligament is united to the fibrous capsule; its anterior and upper border is free, and overlaps it.

The transverse humeral ligament (4.37) is a broad band passing from the lesser to the greater tubercle of the humerus; it converts the intertubercular sulcus into a canal, and its

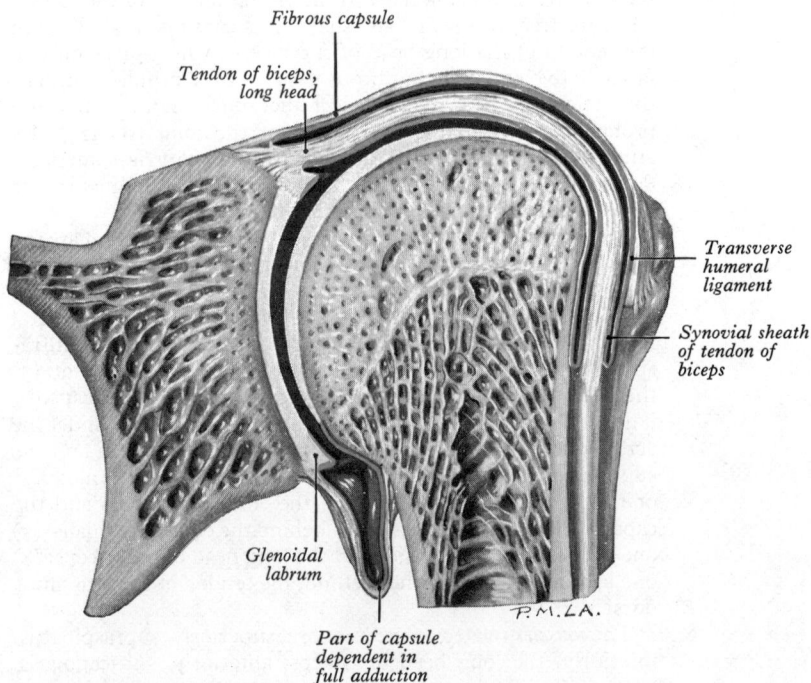

Fibrous capsule

Tendon of biceps, long head

Transverse humeral ligament

Synovial sheath of tendon of biceps

Glenoidal labrum

P.M.LA.

Part of capsule dependent in full adduction

4.38C   A section through the shoulder joint. The synovial membrane is in blue.

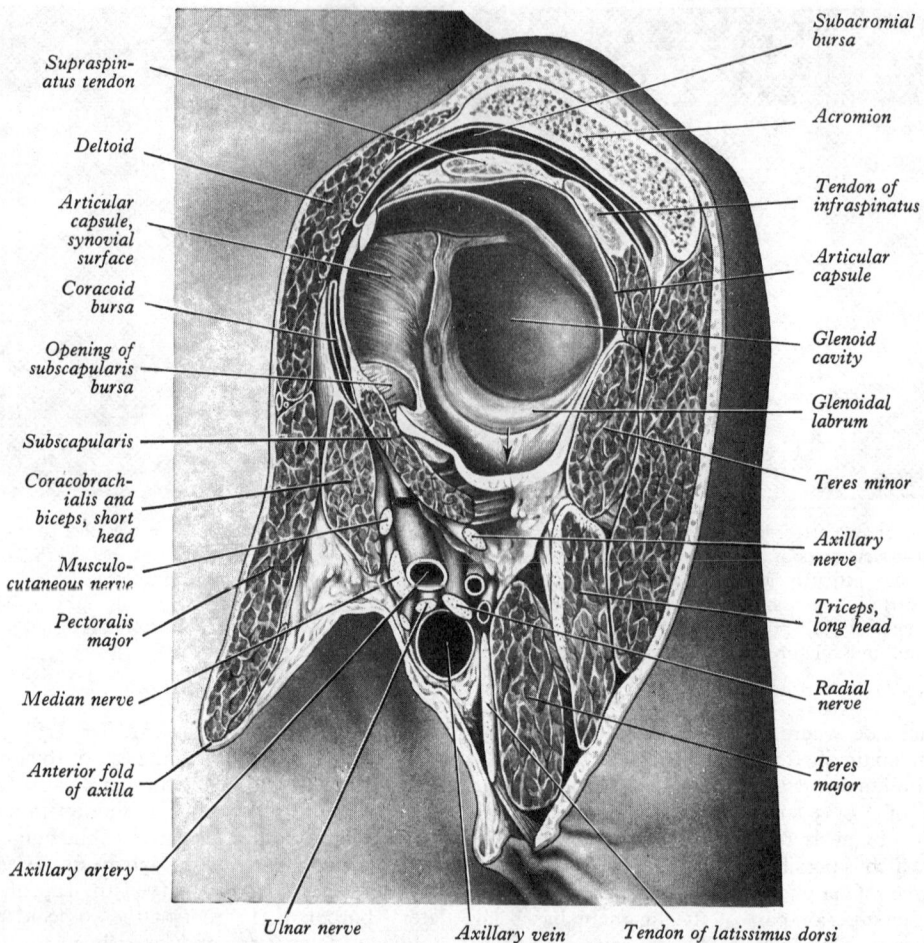

Supraspin-
atus tendon

Deltoid

Articular
capsule,
synovial
surface

Coracoid
bursa

Opening of
subscapularis
bursa

Subscapularis

Coracobrach-
ialis and
biceps, short
head

Musculo-
cutaneous nerve

Pectoralis
major

Median nerve

Anterior fold
of axilla

Axillary artery

Subacromial
bursa

Acromion

Tendon of
infraspinatus

Articular
capsule

Glenoid
cavity

Glenoidal
labrum

Teres minor

Axillary
nerve

Triceps,
long head

Radial
nerve

Teres
major

Ulnar nerve   Axillary vein   Tendon of latissimus dorsi

4.39 An obliquely coronal section through the left shoulder and shoulder joint, dissected after removal of the upper limb. The arrow points to the dependent part of the articular capsule. *Note*—The relations of the axillary vessels to each other and to the branches of the brachial plexus as displayed in this dissected section may be misleading, and the reader is referred to **6.**69 for the usual view.

attachment lies above the epiphysial line. The ligament functions simply as a retinaculum for the tendon of the long head of biceps.

The glenoidal labrum (**4.**39) is a fibrocartilaginous rim attached round the margin of the glenoid cavity. It is triangular on section, the base being fixed to the circumference of the cavity, while the free edge is thin and sharp. It is continuous above with the tendon of the long head of the biceps, which gives off two fasciculi to blend with the fibrous tissue of the labrum. It deepens the articular cavity, may protect the edges of the bone, and probably assists in the lubrication of the joint (p. 424). Its attachment to the margin of the glenoid cavity is sometimes deficient in parts, and a small fringe of the synovial membrane occasionally protrudes through such gaps.

The bursae in the neighbourhood of the shoulder joint are the following: (1) one between the tendon of the subscapularis and the joint capsule (**4.**40), communicating with the joint through an opening between the superior and the middle glenohumeral ligaments; (2) one sometimes separates the tendon of infraspinatus and the capsule; it occasionally opens into the joint; (3) the *subacromial bursa* (**4.**39) lies between deltoid and the capsule; it does not communicate with the joint, but is prolonged under the acromion and coraco-acromial ligament, between them and the supraspinatus; (4) one on the upper surface of the acromion; (5) one is frequently found between the coracoid process and the capsule; (6) one sometimes exists behind the coracobrachialis; (7) one between the teres major and the long head of the triceps; (8) one in front of, and another behind, the tendon of the latissimus dorsi.

The *muscles* related to the joint are, superiorly, supraspinatus; inferiorly, the long head of triceps; anteriorly, subscapularis; posteriorly, infraspinatus and teres minor; within, the tendon of the long head of biceps. Deltoid covers the joint in front, behind and laterally (**4.**39).

The *arteries* supplying the joint are derived from the anterior and posterior circumflex humeral and suprascapular arteries.

The *nerves* of the joint are derived mainly from the posterior cord and from the suprascapular, axillary and the lateral pectoral nerves. The suprascapular supplies the superior and posterior parts of the articular capsule, the axillary, the inferior and anterior parts, and the lateral pectoral, the anterior and superior parts (Gardner 1948).

Movements. The shoulder is a multi-axial spheroidal joint, and therefore is capable of an infinite variety of combinations of swinging and spinning movements (p. 435) but these may all be described as rotations of the moving bone about three mutually perpendicular axes, i.e. it possesses three degrees of freedom (p. 429). Classically the joint is considered to permit the movements of flexion-extension, abduction-adduction, circumduction, and medial and lateral rotation (p. 430). The laxity of its articular capsule and the large size of the head of the humerus compared with that of the shallow glenoid cavity give to the shoulder a wider range of movement than is possible at any other joint. Notwithstanding this, however, when the arm is dependent, even when it carries a moderate load, the supraspinatus, combined with tension in the upper part of the fibrous capsule, suffices to prevent downward displacement of the humerus in relation to the glenoid cavity (Basmajian 1962).

When the movements at the shoulder joint are being analysed, it is preferable and easier to consider the movements of the humerus in relation to the scapula rather than to the sagittal and coronal planes, and the relevant axes are illustrated in **4.**5A, B, C. When the arm is by the side in the resting position, the glenoid cavity faces almost equally forwards and laterally, and the position of the humerus corresponds to that of the scapula, although relative to the anatomical position it is rotated medially (p. 431). As a result *flexion* carries the arm forwards and medially across the front of the chest, and the movement takes place around an axis which passes through the head of the humerus at right angles to the plane of the glenoid cavity at (approximately) its centre. *Abduction and adduction* occur in a vertical plane at right angles to the plane of flexion and extension, and the axis passes horizontally through the head of the humerus parallel to the plane of the glenoid cavity (Flecker 1929). Pure abduction therefore carries the arm anterolaterally away from the trunk, and *the movement occurs in the plane of the body of the scapula*. If, however, the movements of the humerus are considered in relation to the *trunk* and not to the scapula, flexion and extension occur in the paramedian plane and abduction and adduction in the coronal plane. Then the deltoid and supraspinatus produce a combination of flexion and extension, and raising the arm towards the vertical from a position of flexion (in this sense) involves medial rotation of the humerus; raising it to the vertical from the position of abduction (in this sense) is accompanied by lateral rotation of the humerus. Whether 'scapular' or any other form of abduction is described, it must be emphasized that these are merely selected planes in an infinite series. In pure (scapular) abduction points on the humeral surface pursue vertical chords, whereas in rotation the chords are horizontal. In 'pure' flexion-extension, in a plane orthogonal to the scapular plane movement is around a notional mechanical axis projected from the centre of the glenoid cavity.

At the glenohumeral joint the limit of abduction is about 90° (Kapandji 1970), but figures as high as 120° have been stated (Inman *et al.* 1944). About 60° further abduction is contributed by the sterno- and acromio-clavicular articulations. Contralateral flexion of the vertebral column aids in bringing the arm to a vertical position. It must be emphasized, however, that during active elevation of the arm the movement at the shoulder joint and the separate movements of the scapula and clavicle occur simultaneously—except in the initial stages (25°–30°), when most and in some cases the whole of the movement takes place at the shoulder joint. For every 15° of elevation the glenohumeral joint is said to contribute 10° and the scapular movement 5°.

In flexion, however, the humerus moves in a plane at right angles to the plane of the body of the scapula, and no amount of rotation of the scapula can increase the degree of elevation (120°) obtained in full flexion. If the fully flexed humerus is gradually abducted, the degree of elevation increases *pro rata*, until, when the humerus comes to lie in the plane of the body of the scapula, i.e. when the position of true abduction is reached, the full 180° of elevation may be obtained. In *rotation*, which may be medial or lateral, the humerus revolves for about one-quarter of a circle about a vertical axis. The range of rotatory movement is greatest when the arm is by the side, and least when it is raised to the vertical. In practice, when assessing the range of medial and lateral rotation of the humerus at the glenohumeral joint, it is advisable to do so with the forearm bones flexed to a right angle at the elbow joint—this prevents misinterpretation due to super-added movements of either pronation or supination which may occur with the dependent limb. In *circumduction*, which results from a succession of the foregoing movements, the lower end of the humerus describes the base of a cone, the apex of which is at the head of the bone, but this movement at the shoulder joint can be increased very substantially by the movements of the scapula, and the combination is well exemplified in the arm movements of a fast bowler in cricket.

In the newer terminology, it will be realized that because of the offset position of the articular head of the humerus in relation to the long axis of its shaft, the flexion-extension movements of the bone are accompanied by almost pure 'spins' of the head in relation to the glenoid cavity (p. 434). The remaining movements possible at the joint are an infinite variety of cardinal or arcuate swings or successions of these—some of the general consequences of such movements are discussed on pp. 434, 436.

The peculiar relation of the tendon of the long head of the biceps to the shoulder joint appears to subserve various purposes. By its connexion with both the shoulder and elbow the muscle harmonizes the action of the two joints, and acts as an elastic ligament during all the movements which occur at these articulations. It strengthens the upper part of the shoulder joint, and helps to prevent the head of the humerus from being pressed

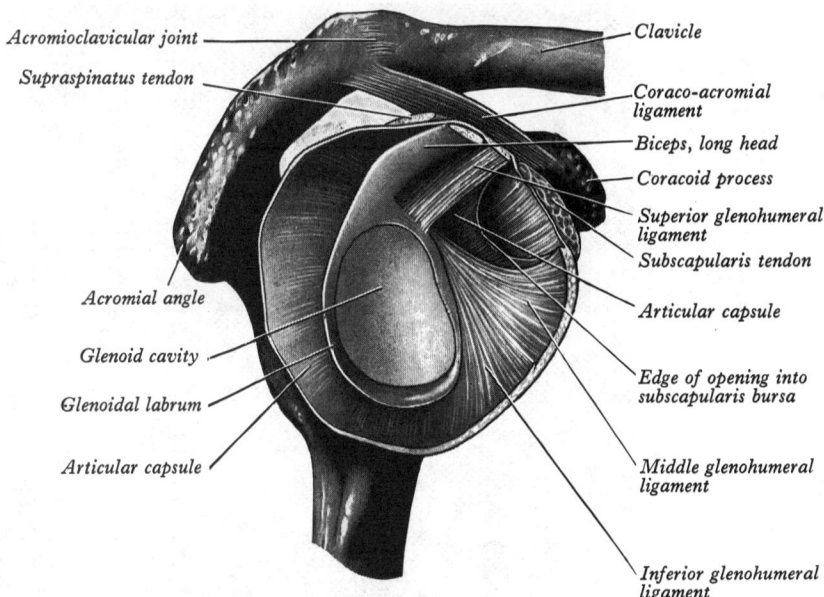

4.40   Interior of the right shoulder joint. Anterolateral aspect.

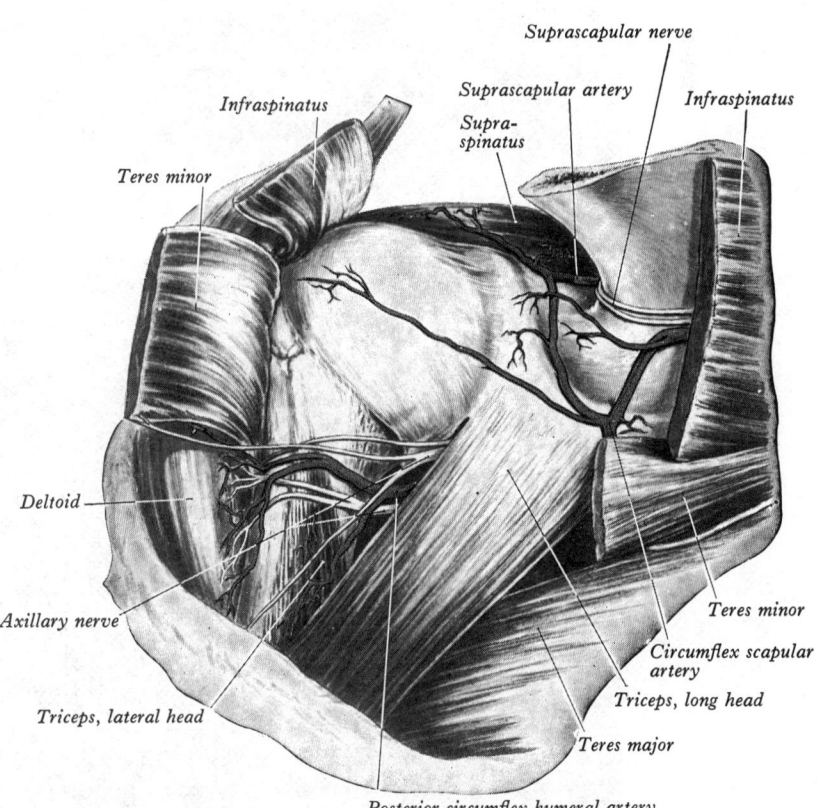

4.41   The posterior aspect of the left shoulder joint. An oblique section has been made through the spine of the scapula and the acromion has been removed. Parts of the infraspinatus and teres minor have been excised and their tendons have been turned forwards.

up against the acromion when the deltoid contracts (*vide infra*). By its passage along the intertubercular sulcus it assists in steadying the head of the humerus in the various movements of the arm.

Owing to the shallowness of the glenoid cavity and the laxity of the capsule, a wide range of *accessory movements* (p. 429) is possible at the shoulder joint. The head of the humerus can be moved backwards, forwards, upwards and downwards in relation to the glenoid cavity and, when the arm is abducted—the position in which the accessory movements are most free—it can be separated from the cavity by traction.

Muscle producing the movements. The muscles moving   459

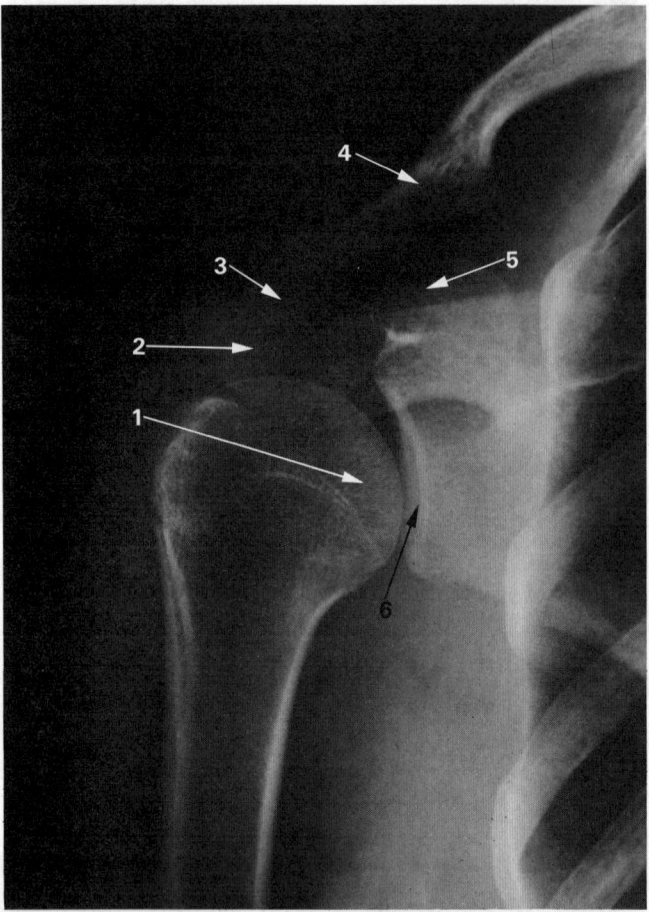

A

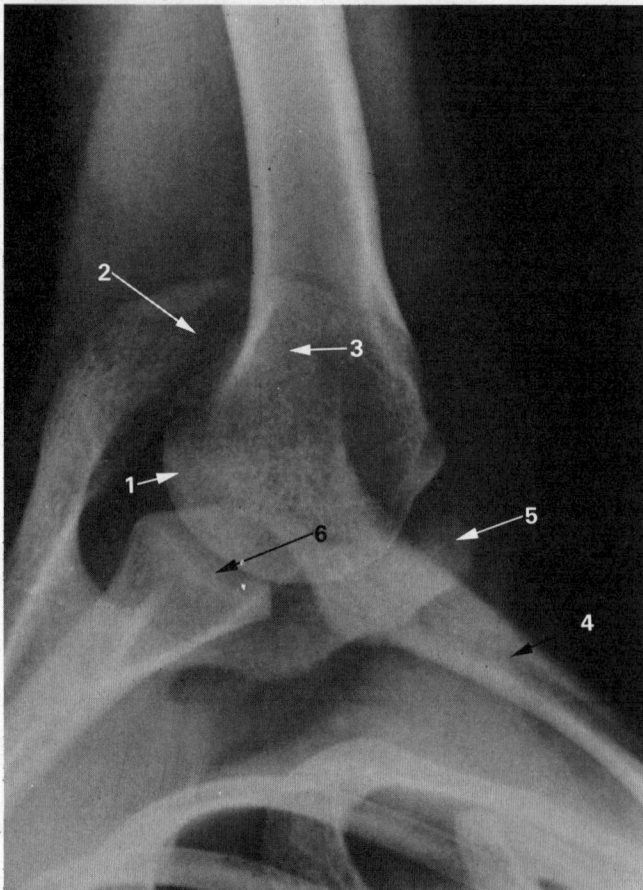

B

4.42A and B   Radiograph of shoulder in a young female of 18 years in anteroposterior view (A) and axillary view with the arm abducted (B). 1. Head of humerus. 2. Acromion. 3. Acromioclavicular joint. 4. Clavicle. 5. Coracoid process. 6. Glenoid articular surface.

the shoulder may be divided into: (a) those acting on the shoulder girdle itself and (b) those acting on the shoulder joint.

(a) **Muscles acting on the shoulder girdle**. These muscles have already been considered in connection with the movements of the girdle (p. 455).

(b) **Muscles acting on the shoulder joint**. The principal muscles concerned—the deltoid, pectoralis major, latissimus dorsi and teres major all converge on the humerus and act to considerable mechanical advantage on a joint which, owing to the shallowness of the glenoid cavity and the laxity of the articular capsule, is less stable than it would otherwise be. This is counteracted by the influence of the short muscles inserted close to the upper end of the humerus, viz. the subscapularis, supraspinatus, infraspinatus and teres minor, for they function frequently as postural muscles to retain the head of the humerus in its correct alignment relative to the glenoid cavity, and to resist the tendency to uncontrolled skid during active movements.

Flexion. Pectoralis major (clavicular part) (p. 567), deltoid (anterior fibres) and coracobrachialis, assisted by biceps. The sternocostal part of pectoralis major plays an important part when the arm is drawn forwards to the plane of the trunk from the fully extended position.

Extension. Deltoid (posterior fibres) and teres major, when the movement starts with the arm by the side. When the fully flexed arm is extended against resistance, the latissimus dorsi and the sternocostal part of the pectoralis major are powerful adjuvants until the arm reaches the plane of the body.

Abduction. Deltoid. It should be noted that, in the initial stages, its force is exerted mainly in an upward direction and, unless opposed, would result in upward displacement of the head of the humerus. The subscapularis, infraspinatus and teres minor provide the opposing force by exerting downward traction on the head of the bone and these three muscles and the deltoid constitute a 'couple', the effect of which is to produce true abduction, i.e. elevation of the humerus in the plane of the scapula. The supraspinatus assists in effecting and maintaining the movement, but its precise role is a matter of controversy.

Medial rotation.. Pectoralis major, deltoid (anterior fibres), latissimus dorsi, teres major and, when the arm is by the side, subscapularis.

Lateral rotation. Infraspinatus, deltoid (posterior fibres) and teres minor.

Applied Anatomy. Owing to the adaptation of the shoulder joint to provide a wide range of movement, it is more often dislocated than any other. Dislocation usually occurs when the arm is abducted, when the head of the humerus presses against the lower and front part of the capsule, which is the thinnest and least supported part of the ligament. The rent in the capsule is almost invariably here, and through it the head of the bone escapes, so that the dislocation in most instances is primarily subglenoid. If, after the dislocation has been reduced, abduction of the arm is prevented, the dislocation cannot recur.

When the shoulder joint is ankylosed, the loss of movement in the joint is partly compensated for by increased mobility of the scapula. In treating conditions of the shoulder joint likely to lead to ankylosis, the humerus should be kept in the position it assumes when the palm of the hand is placed on the back of the neck, i.e. abducted, so as to make full use of this compensating mobility of the scapula.

The Cubital (Elbow) Joint

The elbow joint (3.149, 150; 4.48), includes two articulations: (1) *humero-ulnar*, between the trochlea of the humerus and the trochlear notch of the ulna, and (2) *humeroradial*, between the capitulum of the humerus and the facet on the head of the radius. It is hence a *compound* synovial joint. Its complexity is increased further by its continuity with the superior radio-ulnar joint, and the complete articular complex is the *cubital articulation*. The superior radio-ulnar joint will, however, be considered separately.

The *articular surfaces* are the trochlea and capitulum of the humerus, on the one hand, and the trochlear notch of the ulna and the head of the radius, on the other. The trochlea is not part of a simple pulley, for its medial flange is much more extensive than its lateral flange and projects downwards to a lower level, so that the line of the joint, which is roughly 2 cm below the line joining the two epicondyles, passes downwards and medially. In addition, the trochlea is widest posteriorly and in this position its lateral edge is a sharp rim. The trochlear notch is by no means perfectly congruent with the trochlea. In full extension the medial part of its upper (olecranon) half is not in contact with the trochlea, and a corresponding strip on the lateral side loses contact on flexion. In addition, the description of the trochlea just given should not obscure the fact that it is an asymmetric type of sellar surface—largely concave transversely and convex anteroposteriorly—and that sections of these surfaces show them to be compounded spirals. Consequently the principal swing, or to-and-fro movement of the hinge, is accompanied (as in all hinge joints) by screwing and conjunct rotation (pp. 434, 436). The olecranon and coronoid parts of the trochlear notch are usually separated by a narrow, roughened strip of bone, devoid of articular cartilage and covered with a little fibro-adipose tissue covered by synovial membrane. The capitulum and the head of the radius are reciprocally curved, but the best contact is obtained when the semiflexed radius is in the midprone position. The rim of the head, which is more prominent medially, fits into the groove between the capitulum and the trochlea.

The humero-ulnar and humeroradial articulations together form a largely uni-axial joint, the ligaments of which are the articular capsule, and ulnar and radial collateral ligaments.

The articular capsule (4.44, 45). The anterior part of the *fibrous capsule* is a broad and thin layer. It is attached, above, to the front of the medial epicondyle and to the front of the humerus immediately above the coronoid and radial fossae below, to the anterior surface of the coronoid process of the ulna and to the annular ligament (p. 462), being continuous at the sides with the ulnar and radial collateral ligaments. Anteriorly the capsule is related to the brachialis and receives numerous fibres from its deep surface.

The posterior part of the fibrous capsule is thin. Above, it is attached to the humerus immediately behind the capitulum and close to the lateral margin of the trochlea, to all but the lower part of the rim of the olecranon fossa, and the back of the medial epicondyle some distance from the trochlea. Inferomedially it is fixed to the upper and lateral margins of the olecranon, and laterally it is continuous with the capsule of the superior radio-ulnar joint deep to the annular ligament (p. 462). It is related, behind, to the tendon of triceps and to anconeus.

The *synovial membrane* (4.43, 44) extends from the margin of the articular surface of the humerus, lines the coronoid, radial and olecranon fossae of that bone and covers the flattened medial surface of the trochlea (3.136); it is reflected over the deep surface of the fibrous capsule and lines the deep surface of the lower part of the annular ligament. Projecting into the joint between the radius and ulna from behind there is a crescentic fold of synovial membrane, partly dividing the joint into humeroradial and humero-ulnar parts. This fold is irregularly triangular and contains a variable quantity of extrasynovial fat (4.46).

Between the fibrous capsule and the synovial membrane there are three other pads of fat. The largest, over the olecranon fossa, is pressed into this by the triceps during flexion of the joint; the second, over the coronoid fossa, and the third, over the radial fossa, are pressed by the brachialis into their respective fossae during extension. They are, of course, displaced from these fossae during the opposite of these movements. In addition, smaller tags of fat covered with synovial membrane project into the joint cavity opposite to the constrictions on each side of the trochlear notch (p. 367), and cover the small nonarticular areas of the bone in these situations.

The ulnar collateral ligament of the elbow joint (4.45A) is a thick triangular band consisting of two parts, an anterior and a posterior, united by a thinner intermediate portion; it is sometimes called the medial ligament. The *anterior part* is attached above, by its apex, to the front of the medial epicondyle of the humerus, and below, by its broad base, to a tubercle on the upper part of the medial margin of the coronoid process. The

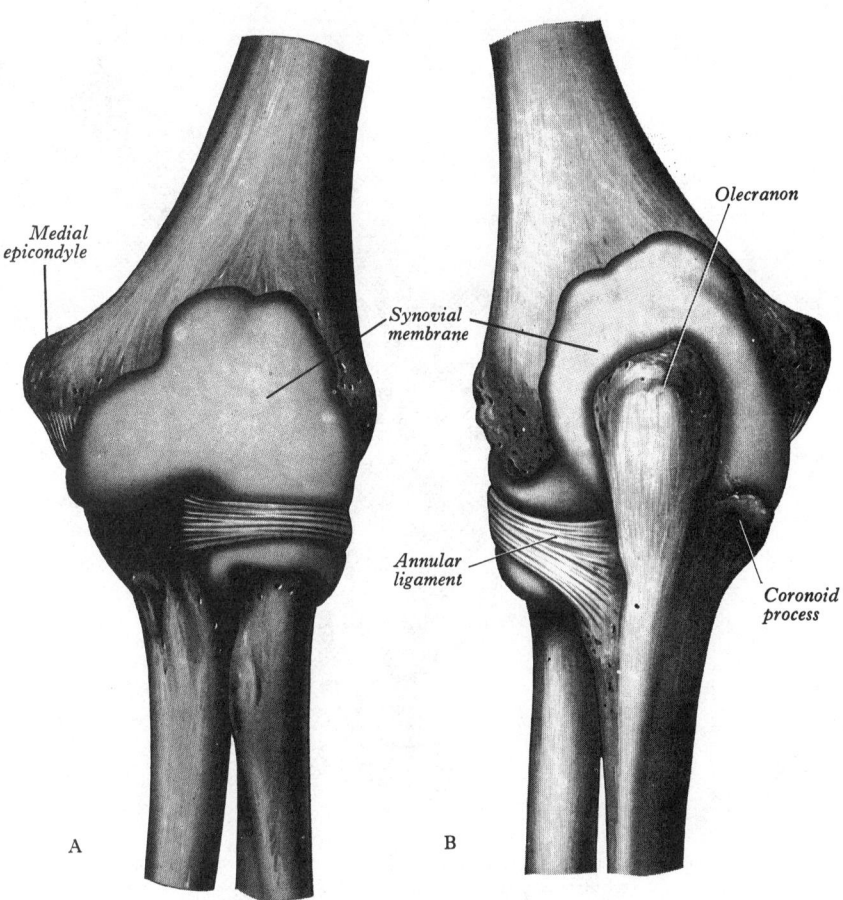

4.43A The synovial cavity of the left elbow joint, partially distended. Anterior aspect. (Originally drawn from a specimen prepared by J. C. B. Grant.) The fibrous capsule of the elbow joint has been removed but the annular ligament has been left *in situ*. Note that the synovial membrane descends below the lower border of the annular ligament.

4.43B The synovial cavity of the left elbow joint, partially distended. Posterior aspect of the specimen represented in 4.43A.

posterior part, also of triangular form, is attached, above, to the lower and back part of the medial epicondyle; below, to the medial margin of the olecranon. Between these two bands a few intermediate fibres descend from the medial epicondyle to an *oblique band*—often feebly developed—which stretches between the olecranon and coronoid processes. This converts the depression on the medial margin of the trochlear notch into a foramen, through which the intracapsular pad of fat is continuous with the extracapsular fat on the medial side of the joint. The ulnar collateral ligament is related to the triceps and flexor carpi ulnaris *and the ulnar nerve*. Along its anterior portion the origin of the flexor digitorum superficialis extends from the medial epicondyle of the humerus downwards and laterally to the medial border of the coronoid process of the ulna.

The radial collateral ligament of the elbow joint (4.45B) is attached to the lower part of the lateral epicondyle of the humerus, and below to the annular ligament, some of its most posterior fibres passing over it to the upper end of the supinator crest of the ulna. It is intimately blended with the origins of the supinator and the extensor carpi radialis brevis.

The *muscles* in relation with the joint are, in front, brachialis, behind, triceps and anconeus, laterally, supinator and the common tendon of origin of the extensor muscles, medially, the common tendon of origin of the flexor muscles, and the flexor carpi ulnaris.

The *articular arteries* are derived from the anastomotic network around the joint (6.74).

The *articular nerves* are mainly from the musculocutaneous and radial nerves, but the ulnar, median and, sometimes, the anterior interosseous nerves also contribute articular branches. The

461

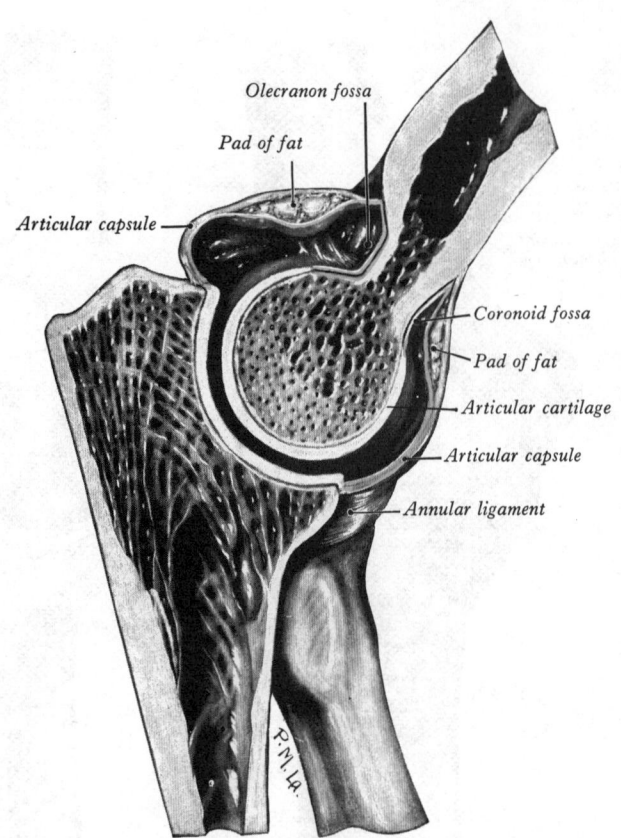

Olecranon fossa
Pad of fat
Articular capsule
Coronoid fossa
Pad of fat
Articular cartilage
Articular capsule
Annular ligament

4.44 A sagittal section through the left elbow joint. The synovial membrane is shown in blue. Medial aspect.

branch from the musculocutaneous arises from the nerve to brachialis and is distributed to the anterior part of the articular capsule. The articular branches of the radial nerve supply the posterior and anterolateral portions of the articular capsule and are derived both from the nerve to anconeus and from the ulnar collateral branch to the medial head of triceps. The ulnar nerve supplies twigs to the ulnar collateral ligament as it lies behind the medial epicondyle (Gardner 1948). These articular nerves are closely associated with the blood vessels supplying the synovial membrane, fat pads and epiphyses; presumably they contain some vasomotor fibres in addition to afferent fibres subserving the modalities of pain and proprioception.

Movements. The elbow is a uni-axial or hinge joint and its movements consist of flexion and extension, the ulna moving on the trochlea, and the head of the radius on the capitulum of the humerus. However, the flexion-extension movements of the ulna do not constitute a pure swing but are accompanied by a small degree of conjunct rotation—the ulna is slightly pronated during extension, and supinated during flexion. Since the capitulum of the humerus is smaller than the head of the radius, the posterior edge of the head of the radius can be felt projecting at the back of the joint when the forearm is fully extended. The movement of extension is limited by the tension of the fibrous capsule and muscles on the front of the joint (extension constituting the close-packed position of the joint), that of flexion chiefly by apposition of soft parts.

When the forearm is fully extended and the hand supinated, the upper arm and forearm are not in the same line; the forearm is directed somewhat laterally, and, in both sexes, forms with the upper arm the 'carrying angle' of about 163° open to the lateral side (Steel and Tomlinson 1958). Owing to the existence of this angulation the ulnar border of the supinated and extended forearm cannot be brought into contact with the lateral surface of the thigh when the arm is by the side. The 'carrying angle' is caused partly by the medial edge of the trochlea of the humerus, which projects about 6 mm below the lateral edge, and partly by

the obliquity of the superior articular surface of the coronoid process, which is not set at right angles to the shaft of the ulna. The angles which the articular surfaces of the humerus and the ulna make with the long axes of the bones are approximately equal, and as a result the carrying angle disappears on full flexion of the forearm and the two bones come to lie in the same plane. When this movement is carried out with the arm by the side, the ulnar border of the little finger lies over the clavicle on account of the position of the resting humerus (p. 459). If the humerus is rotated laterally during the movement, the hand is carried upwards in front of the shoulder. The carrying angle is masked also in pronation of the extended forearm, and this has the effect of bringing the upper arm, the semipronated forearm and the hand into the same straight line. This arrangement increases the precision with which the hand, and any instrument or weapon held in the hand, can be controlled in full extension of the elbow or while the elbow is being extended.

The *accessory movements* of the elbow joint are limited to slight screw action, abduction and adduction of the ulna, and forward and backward movement of the head of the radius on the capitulum of the humerus. In the latter movement, the head of the radius is moved on the radial notch of the ulna also, the annular ligament being slewed backwards and forwards at the same time. The extent of this movement is greater when the elbow joint is half-flexed.

Muscles producing the movements.

Flexion. Brachialis, biceps and brachioradialis. In slow flexion or in the maintenance of flexion against gravity, even when the forearm carries a moderate load, the brachialis and biceps are principally involved. With increasing speed of movement, activity in brachioradialis becomes increasingly prominent (MacConaill 1949; Basmajian 1959). Owing to the position of its attachments it acts most effectively when the forearm is in the mid-prone position. Against resistance pronator teres and flexor carpi radialis may act.

Extension. Triceps and anconeus. In rapid extension brachioradialis again acts as a shunt muscle (pp. 528, 578).

The Radio-ulnar Joints

The radius and the ulna are connected at their upper and lower extremities by synovial joints, the *proximal* and *distal radio-ulnar joints*. In addition, the shafts of the bones are connected by an interosseous membrane and a ligament, which together are commonly regarded as a non-synovial, *middle radio-ulnar joint*.

THE PROXIMAL RADIO-ULNAR JOINT

This articulation forms a uni-axial pivot between the circumference of the head of the radius and the osseofibrous ring formed by the radial notch of the ulna and the annular ligament.

The annular ligament (4.45, 46, 47, 49), a strong band, encircles the head of the radius, and retains it in contact with the radial notch of the ulna. It forms about four-fifths of the osseofibrous ring. In front it is attached to the anterior margin of the radial notch. Posteriorly it broadens and may divide into several bands to attach to a rough ridge on the ulna at or just behind the posterior margin of the radial notch; outlying bands may reach the lateral margin of the trochlear notch above and the upper end of the supinator crest below. The upper border of the ligament blends with the fibrous capsule of the elbow joint except posteriorly where the fibrous capsule passes as a separate entity deep to the ligament to be attached around the posterior and inferior margins of the radial notch. From the lower border a few fibres pass over the reflexion of the synovial membrane to be loosely attached to the neck of the radius. A thin fibrous layer, often termed the *quadrate ligament*, covers the synovial membrane which closes the distal aspect of the joint between the radius and ulna (Martin 1958). The *superficial surface* of the annular ligament blends with the radial collateral ligament of the elbow, and affords origin to part of the supinator. It is related posteriorly to the anconeus and the interosseous recurrent artery. On its *internal surface* the annular ligament has a thin coating of cartilage where it

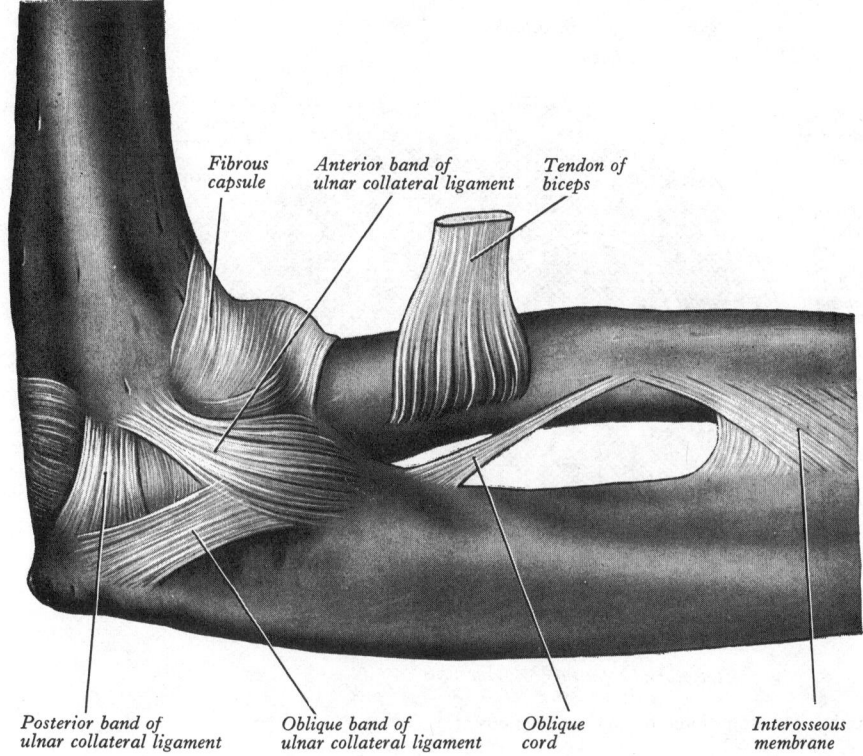

Fibrous capsule   *Anterior band of ulnar collateral ligament*   *Tendon of biceps*

Posterior band of ulnar collateral ligament   *Oblique band of ulnar collateral ligament*   *Oblique cord*   *Interosseous membrane*

4.45A   The left elbow joint. Medial aspect.

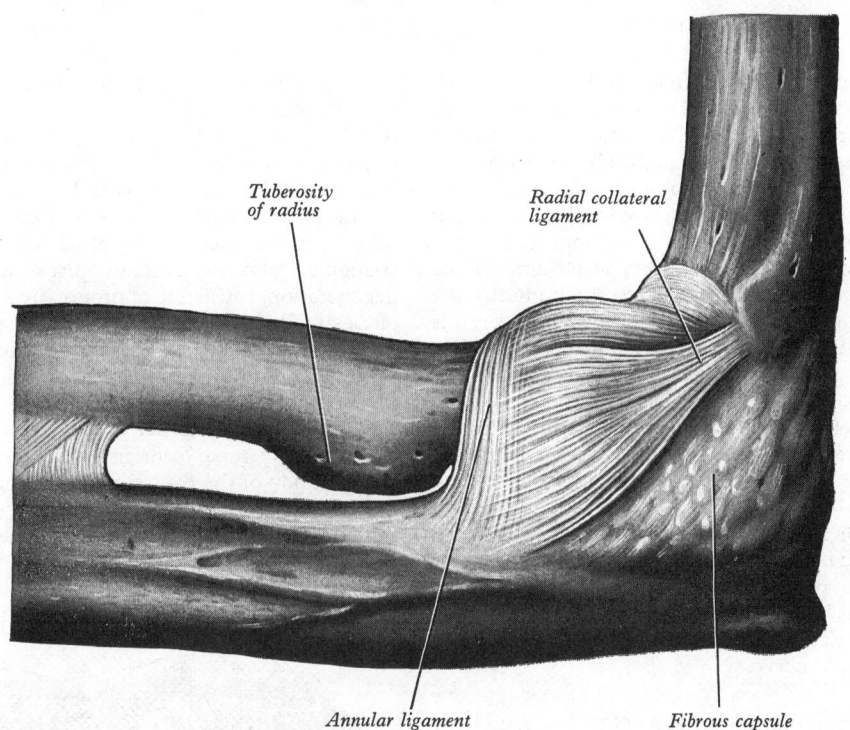

Tuberosity of radius   *Radial collateral ligament*

Annular ligament   *Fibrous capsule*

4.45B   The left elbow joint. Lateral aspect.

comes into contact with the circumference of the head of the radius; its lower part is covered with synovial membrane which is reflected upwards on to the neck of the bone.

THE MIDDLE RADIO-ULNAR UNION

The shafts of the radius and ulna are connected by the oblique cord and the interosseous membrane of the forearm.

The oblique cord (4.49) is a small, inconstant, flattened band or cord formed in the fascia overlying the deep head of supinator and extending from the lateral side of the tuberosity of the ulna to the radius a little below the radial tuberosity. Its fibres run at right

angles to those of the interosseous membrane. It is sometimes absent and can scarcely be considered of much functional significance.

The interosseous membrane of the forearm (4.49) is a broad and thin sheet, the fibres of which slant downwards and medially from the interosseous border of the radius to that of the ulna; the lower part of the membrane is attached to the posterior of the two lines into which the interosseous border of the radius divides. Two or three bands are occasionally found on the posterior surface of this membrane; their fibres descend obliquely from the ulna towards the radius, i.e. across the other fibres. The membrane is deficient above, commencing about 2 or 3 cm below

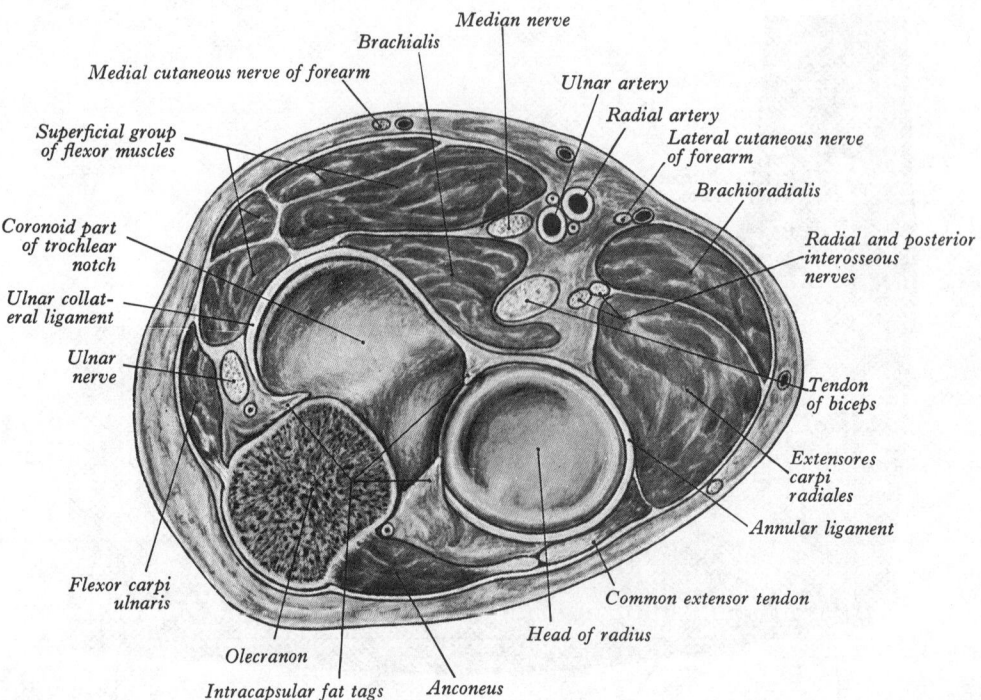

4.46 Transverse section of the right elbow joint to show the relations of the joint. Superior aspect.

the tuberosity of the radius, is broader in the middle than at each end, and presents an oval aperture a little above its lower margin, for the passage of the anterior interosseous vessels to the back of the forearm. Between its upper border and the oblique cord there is a gap, through which the posterior interosseous vessels pass. The membrane increases the extent of surface for the attachment of the deep muscles of the forearm and also connects the two bones. The arrangement of its fibres has often been claimed to be such as to transmit to the ulna and thence to the humerus any force acting upwards through the hand and radius. However, it is only tense when the hand is midway between the prone and supine positions and is somewhat relaxed in complete pronation and supination, and the hand is usually pronated when subjected to such forces. Further, it has been shown that the radius can itself transmit substantial forces directly to the humerus (Travill 1964). *In front* the membrane is related, in its upper three-fourths, to flexor pollicis longus on the radial side, flexor digitorum profundus on the ulnar side, and between these to the anterior interosseous vessels and nerve, in its lowest quarter with pronator quadratus; *behind* it is related to supinator, abductor pollicis longus, extensor pollicis brevis, extensor pollicis longus, extensor indicis and, near the wrist, with the anterior interosseous artery and posterior interosseous nerve.

THE DISTAL RADIO-ULNAR JOINT

This is a uni-axial pivot joint between the convex lower end (head) of the ulna and the concave ulnar notch of the lower end of the radius; the surfaces are enclosed in an articular capsule and are also held together by an articular disc.

The fibrous capsule is slightly thickened in front and behind; above, it is lax and lined with synovial membrane which projects upwards as a pouch (*recessus sacciformis*) in front of the lower part of the interosseous membrane.

The articular disc (4.52), triangular in shape, binds the lower ends of the ulna and radius together. Its periphery is thicker than its centre, which is occasionally perforated. It is attached by a broad apex to a depression between the styloid process and the inferior surface of the head of the ulna, and by its base, which is thin, to the prominent edge which separates the ulnar notch from the carpal articular surface of the radius. Its margins are united to

the ligaments of the wrist joint. Its *upper and lower surfaces* are both smooth and concave. The upper articulates with the head of the ulna, whilst the lower forms a part of the radiocarpal joint and articulates with the medial part of the lunate bone; when the hand is adducted, it articulates with the triquetral bone.

Mikić (1978) studied 180 wrist joints from 100 fresh cadavers ranging in age from fetuses to 94 years, and concluded that the triangular fibrocartilaginous disc showed an age-dependent degeneration, consisting of progressively reduced cellularity, loss of elastic fibres, mucoid degeneration of the ground substance, exposure of collagen fibres, fibrillation, ulceration, abnormal thinning, and ultimately disc perforation. No perforations were found in the first two decades, 7·6 per cent in the third, 18·1 per cent in the fourth, 40 per cent in the fifth, 42·8 per cent in the sixth, and 53·1 per cent in those over sixty. Associated with degeneration of the disc, comparable degenerative changes were frequently found in the 'discal' surface of the ulnar head, and the 'discal' area of the proximal articular facet of the lunate bone.

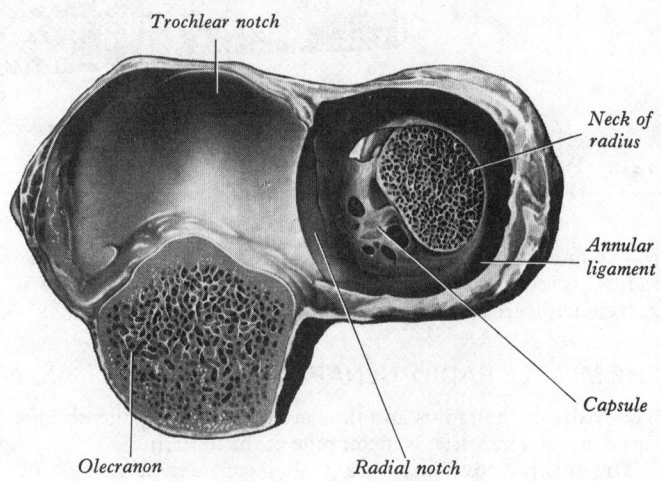

4.47 The annular ligament of the right radius. Superior aspect. The head of the radius has been sawn off and the bone dislodged from the ligament.

Movements. The movements which take place at the radio-ulnar joints result in pronation and supination of the hand. In *pronation* the radius, carrying the hand with it, is turned obliquely across the front of the ulna, its upper end remaining lateral and its lowest end becoming medial to that bone. In *supination* the movement is reversed and the radius lies lateral to and parallel with the ulna. When the movement is limited to the radio-ulnar joints, the hand can be turned through an angle of 140°–150°, but, provided that the elbow joint is extended, the apparent range can be increased to nearly 360° by rotation of the humerus accompanied by forward and backward movements of the shoulder girdle. The power of supination is greater than that of pronation. This point has been appreciated in the industrial design of nuts, bolts and screws, which are tightened or inserted

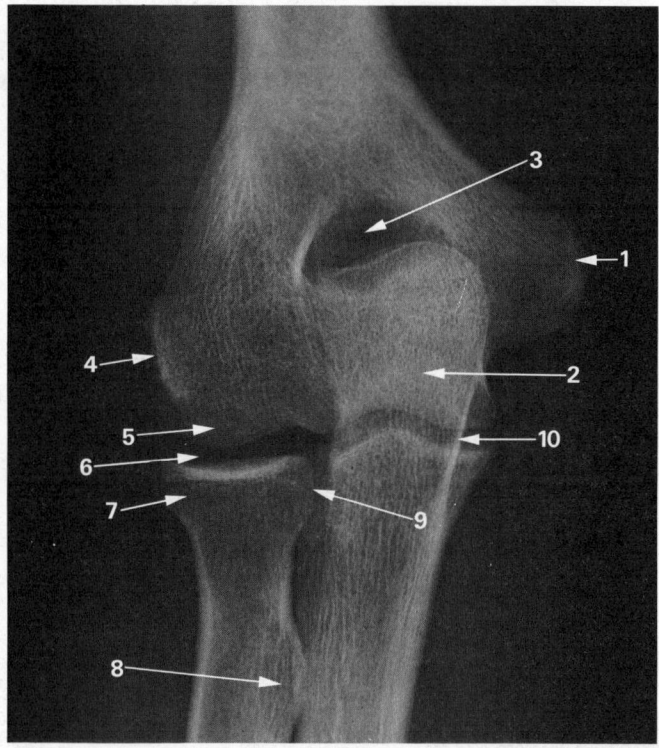

A

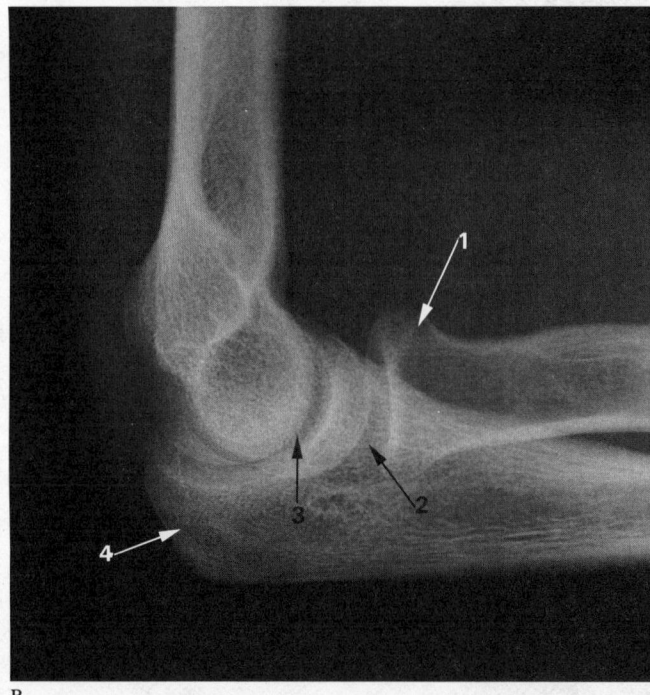

B

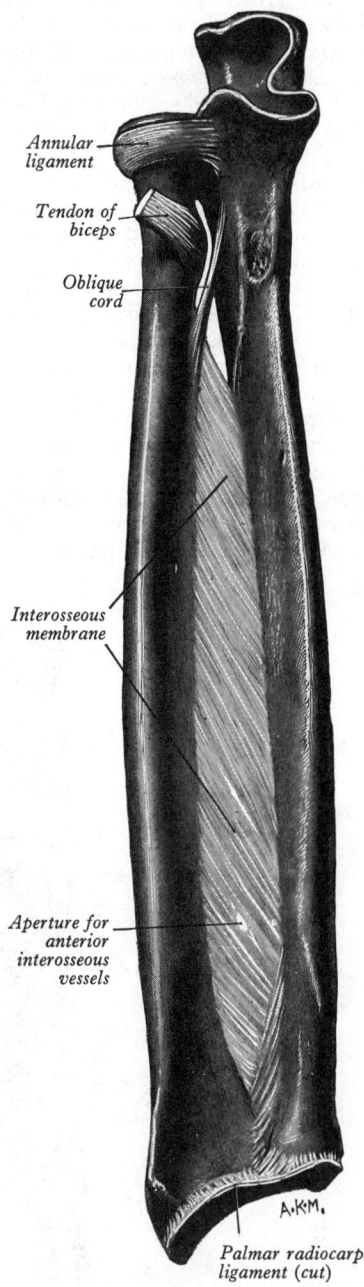

Annular ligament
Tendon of biceps
Oblique cord
Interosseous membrane
Aperture for anterior interosseous vessels
Palmar radiocarpal ligament (cut)

A·K·M.

4.49 The interosseous membrane of the forearm. Anterior aspect.

4.48A and B Anteroposterior (A) and lateral (B) radiographs of an adult elbow joint. The joint is semiflexed in B.

A 1. Medial humeral epicondyle. 2. Shadow of olecranon superimposed on trochlea. 3. Olecranon fossa. 4. Lateral epicondyle. 5. Capitulum. 6. Humeroradial joint. 7. Head of radius. 8. Radial tuberosity. 9. Radial notch of ulna. 10. Humero-ulnar joint.

B 1. Head of radius. 2. Profile of capitulum. 3. Profile of trochlea. 4. Olecranon.

by supinative movements in right-handed persons. Moreover, supination is an anti-gravity movement and in seizing objects for examination (or many other purposes) pronation is merely a preliminary activity, aided by gravity.

The axis about which the movements of pronation and supination occur is often somewhat inadequately represented by a line drawn through the centre of the head of the radius above, and through the ulnar attachment of the articular disc below. More precisely this is solely the axis of movement of *radius relative to ulna*, and, as we shall see, it *does not remain stationary*. The head of the radius rotates within the ring formed by the annular ligament and the radial notch of the ulna, while the lower end and the articular disc swing on the head of the ulna. During the rotation of the head of the radius within the osteoligamentous ring, its proximal surface is, of course, spinning in relation to the articular surface of the capitulum of the humerus. However, the lower end of the ulna *is not stationary* during these movements. It moves a variable amount, along a curved course, backwards and laterally during pronation, and forwards and medially during supination. The occurrence of this movement means that the axis, as defined above, becomes displaced laterally in pronation and medially in supination. The axis, around which *supination and pronation* of the *whole forearm and hand* occurs, may therefore be considered as passing *between* the forearm bones at both the superior and inferior radio-ulnar joints, when there is marked ulnar movement, and through the centre of the head of the radius and ulnar styloid when ulnar movement is minimal; below it may run towards any one of the digits depending on the amount of medial or lateral displacement of the lower end of the ulna (Ray *et al.* 1951). It is important to state that this lateral displacement is not a conjunct rotation, for it can be varied at will. For example, the index finger and thumb, in a precision grip (p. 591) can be made to describe an arc of varying radius according to prevailing functional demands. The greatest radius of arc involves minimal ulnar deviation, whereas rotation of the opposed digits around a virtually fixed point (often employed in the manipulation of tools and instruments) involves marked to maximal ulnar displacement. These ulnar movements are permitted by the incongruence of the trochlea and trochlear notch, and may therefore occur *without* any rotation of the humerus at the glenohumeral joint, whether the elbow joint be in a flexed, semi-flexed or extended position. The fact that the trochlear incongruence persists in all positions of the elbow joint perhaps indicates that the 'close-

packed position' of the humero-ulnar articulation in relation to terminal elbow extension is more dependent upon the disposition of its ligaments than upon the articular surface geometry.

Accessory movements. In addition to the backward and forward movement of the head of the radius on the radial notch of the ulna (p. 367), the head of the ulna can be moved backwards and forwards on the ulnar notch of the radius.

Muscles producing the movements.

Pronation. Pronator quadratus (p. 578), aided during rapid movement and movement against resistance by pronator teres (Basmajian and Travill 1961). Gravity also assists.

Supination. Supinator, in slow unresisted movement and when the elbow is extended. It is assisted by biceps in fast movements with the elbow flexed, especially when resistance is encountered.

Electromyographic studies have not confirmed activity in brachioradialis during pronation and supination.

Applied Anatomy. Dislocation backwards, and abduction dislocation are the commonest forms at the elbow joint. Owing to the shapes of the bones, dislocation backwards is often complicated by fracture of the coronoid process; owing to the strength of the collateral ligaments, the medial epicondyle is frequently torn away in abduction dislocations.

The elbow joint is occasionally the seat of acute synovitis. The joint cavity then becomes distended with fluid, the bulging showing itself principally around the olecranon, in consequence of the laxness of the articular capsule. Again, there is often some swelling, just above the head of the radius in the line of the humeroradial joint, or the whole elbow may assume a fusiform appearance.

Dislocation of the head of the radius alone is a not uncommon

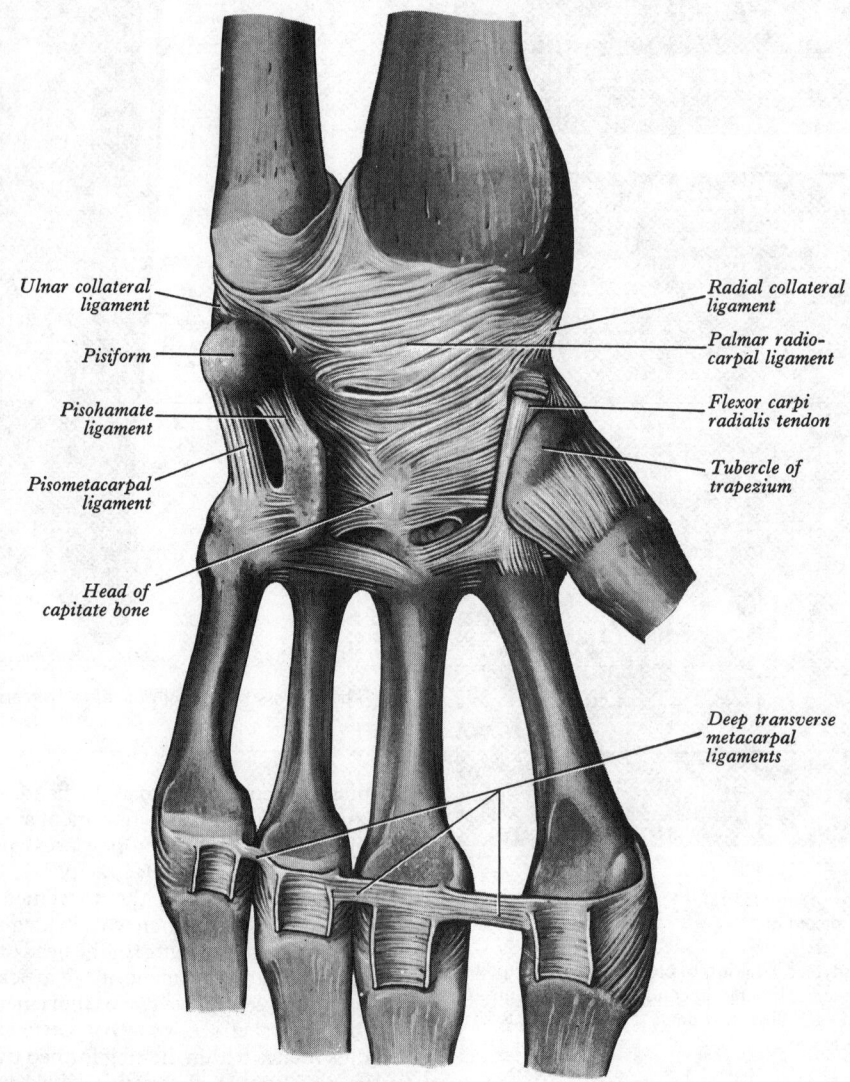

4.50   The ligaments of the left wrist and metacarpus. Palmar aspect.

accident, and occurs most frequently in young persons from falls on the hand when the forearm is extended and supinated, the head of the bone being displaced forward. It is attended by rupture of the annular ligament. Occasionally a peculiar injury, which is supposed to be a subluxation, occurs in young children. It is believed that the head of the radius is displaced downwards in the annular ligament, the upper border of which becomes folded over the head of the bone, between it and the capitulum of the humerus; the small size of the head of the radius in the child predisposes to this injury. The forearm becomes fixed in a position of semiflexion, midway between supination and pronation, and pain is complained of when any attempt is made to move the joint.

The Radiocarpal or Wrist Joint

The radiocarpal or wrist joint (4.50, 51, 52) is a bi-axial joint, commonly classed as ellipsoid (*vide infra*). The parts forming it are the distal end of the radius and lower surface of the articular disc, above, and the scaphoid, lunate and triquetral bones, below. The *articular surface* of the radius and the lower surface of the articular disc form together a roughly elliptical, concave surface with its long axis transversely disposed. However, the radial surface is divided into two by a low ridge and therefore bears two concavities. A similar ridge is usually distinguishable between the medial concavity on the radius and the concave distal surface of the articular disc. The proximal articular surfaces of the scaphoid, lunate and triquetral bones form a fairly smooth convex surface, which is received into the proximal concavity. The line of the joint corresponds to a line, convex upwards, joining the styloid process of the radius to that of the ulna (4.53, 54).

The **articular capsule** is lined by *synovial membrane* which is usually distinct from that of the inferior radio-ulnar joint and from that of the intercarpal joints; but a protrusion of it, ventral to the triangular articular disc, is closely related to the styloid process, the *prestyloid recess*, which is bounded distally by a small fibrocartilaginous meniscus, projecting inwards from the ulnar collateral ligament, between the styloid process and triquetral (4.52). Occasionally ossification occurs within this meniscus (Lewis *et al.* 1970). The fibrous capsule is strengthened by ligaments—palmar radiocarpal and ulnocarpal, dorsal radiocarpal, and radial and ulnar collateral.

The **palmar radiocarpal ligament** (4.50) is a broad membranous band, attached above to the anterior margin of the lower end of the radius and to its styloid process; its fibres pass downwards and medially to be attached to the anterior surfaces of the scaphoid, lunate and triquetral bones, some being continued to the capitate. It is partly intracapsular.

The **palmar ulnocarpal ligament** is a rounded fasciculus which runs from the base of the styloid process of the ulna and the anterior margin of the triangular articular disc of the distal radio-ulnar joint to the lunate and triquetral bones. Lewis, Hamshire and Buckwill (1970) and Mayfield, Johnson and Kilcoyne (1976) regard both palmar carpal ligaments as completely intracapsular. The latter workers also divide the radiocarpal into three parts and the ulnocarpal into two, named according to their attachments, e.g. radiocapitate, ulnolunate, etc. It is not entirely clear in these accounts whether the ligaments are totally intracapsular, i.e. completely separate from the overlying articular capsule. (See also pp. 423 and 468.)

The palmar ligaments of the wrist are perforated by apertures for the passage of vessels and are in relation, in front, with the tendons of the flexor digitorum profundus and flexor pollicis longus.

The **dorsal radiocarpal ligament** (4.51), thinner and weaker than the palmar, as attached, above, to the posterior border of the distal end of the radius; its fibres are directed obliquely downwards and medially, and are fixed, below, to the dorsal surfaces of the scaphoid, lunate and triquetral bones, being continuous with those of the dorsal intercarpal ligaments. It is related behind, with the extensor tendons of the wrist and fingers, their synovial sheaths and the posterior interosseous nerve; in front, it is blended with the articular disc of the inferior radio-ulnar articulation.

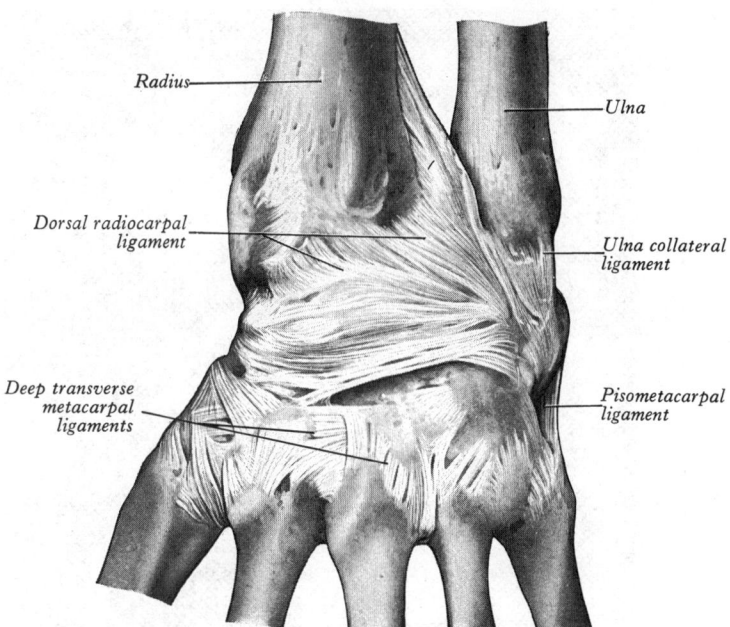

4.51   The ligaments of the left wrist. Dorsal aspect.

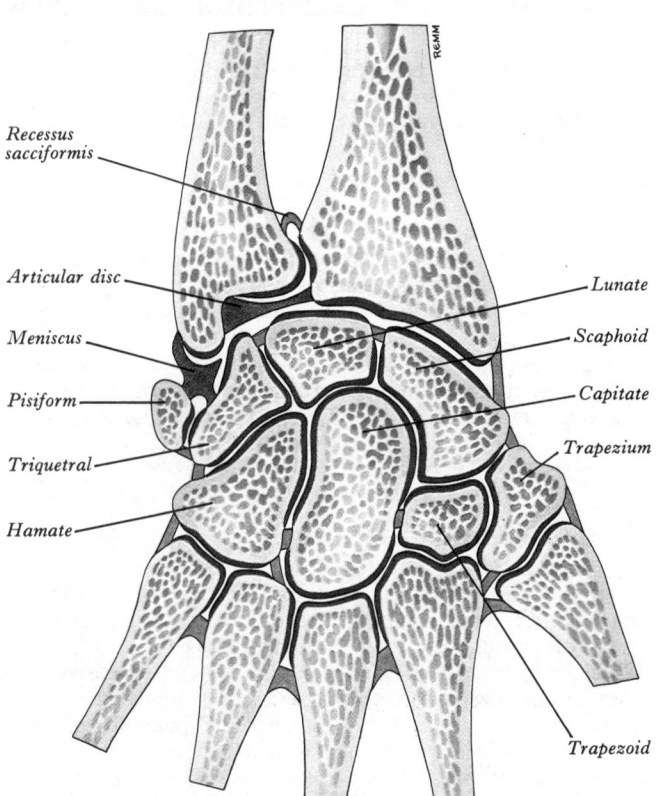

4.52   A coronal section through the distal ends of the radius and ulna, the carpus and the proximal ends of the metacarpals, showing the general form of the articular surfaces, synovial cavities, interosseous ligaments and fibrocartilages. Partly after Lewis *et al.* (1970).

The **ulnar collateral ligament** of the wrist joint (4.50, 51) is attached to the end of the styloid process of the ulna; it divides into two fasciculi, one of which is fixed to the medial side of the triquetral, the other to the pisiform bone.

The **radial collateral ligament** of the wrist joint (4.50, 51) extends from the tip of the styloid process of the radius to the radial side of the scaphoid bone, some of its fibres being prolonged to the trapezium. It is in relation with the radial artery as it winds round the lateral side of the wrist separating the ligament from the tendons of the abductor pollicis longus and extensor pollicis brevis. Both collateral ligaments are relatively poorly developed.   467

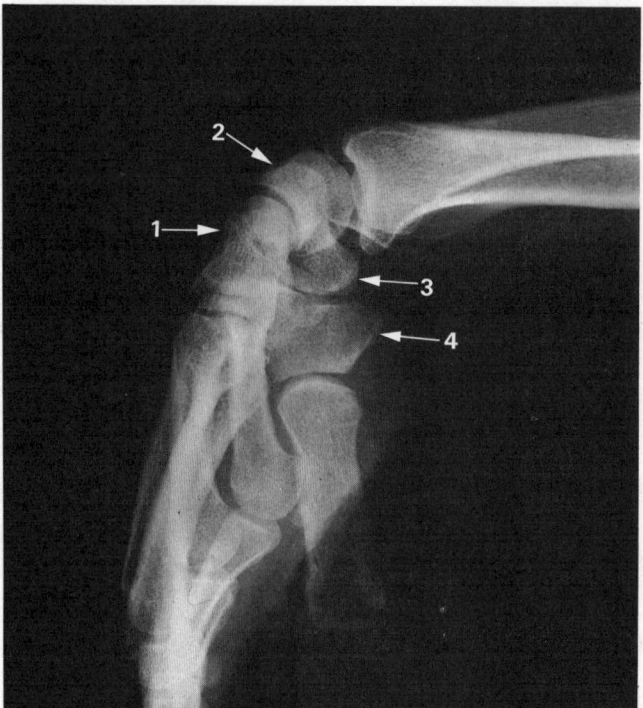

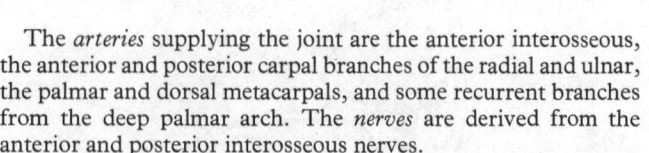

4.53A Radiograph of the hand and wrist in full flexion. Lateral aspect. The arrows point: (1) to the capitate bone; (2) to the lunate bone; (3) to the tubercle of the scaphoid bone; and, (4) to the tubercle of the trapezium. Compare with 4.53B and note the relative positions of the capitate and lunate, and the lunate and radius.

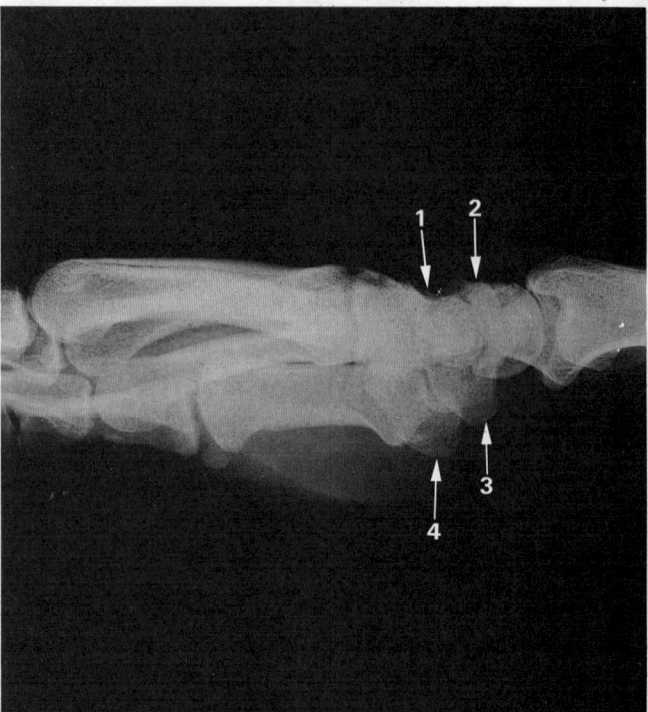

4.53B Radiograph of the hand and wrist. Lateral aspect. The long axes of the third metacarpal bone, the capitate and the lunate are, approximately, in line with the long axis of the radius. The arrows point to the same structures as in 4.53A. Note the relative positions of the capitate and lunate, and the lunate and radius.

The *arteries* supplying the joint are the anterior interosseous, the anterior and posterior carpal branches of the radial and ulnar, the palmar and dorsal metacarpals, and some recurrent branches from the deep palmar arch. The *nerves* are derived from the anterior and posterior interosseous nerves.

Movements. The movements at this joint are closely associated with those in the intercarpal and midcarpal joints. All these are described together (p. 469). The close-packed position of the radiocarpal joint is in full extension.

The Intercarpal Joints

The intercarpal joints connect the carpal bones to one another and may be subdivided into: (1) joints between the bones of the proximal row of the carpus; (2) joints between the bones of the distal row; (3) a somewhat complicated and extensive joint between the two rows, termed the midcarpal joint. The carpal elements are connected by a most extensive system of ligaments, not all of which are specifically named. In addition to the ligaments described below, the flexor retinaculum (p. 583) is an important accessory ligament of the intercarpal joints. The joint surfaces are either sellar, ellipsoid, or spheroidal.

THE JOINTS OF THE PROXIMAL ROW OF CARPAL BONES

(a) The scaphoid, lunate and triquetral bones are connected by dorsal, palmar and interosseous ligaments.

The dorsal and palmar ligaments are placed transversely between the bones of the first row; they connect the scaphoid to the lunate and the lunate to the triquetral. The palmar ligaments are weaker than the dorsal. Mayfield *et al.* (1976) described a distinct capito-triquetral ligament (in 28 dissections), extending from the palmar aspect of the capitate, across the hamate, to the palmar surface of the triquetrum.

The interosseous ligaments (4.52) are two narrow bundles, one connecting the lunate and scaphoid bones, the other the

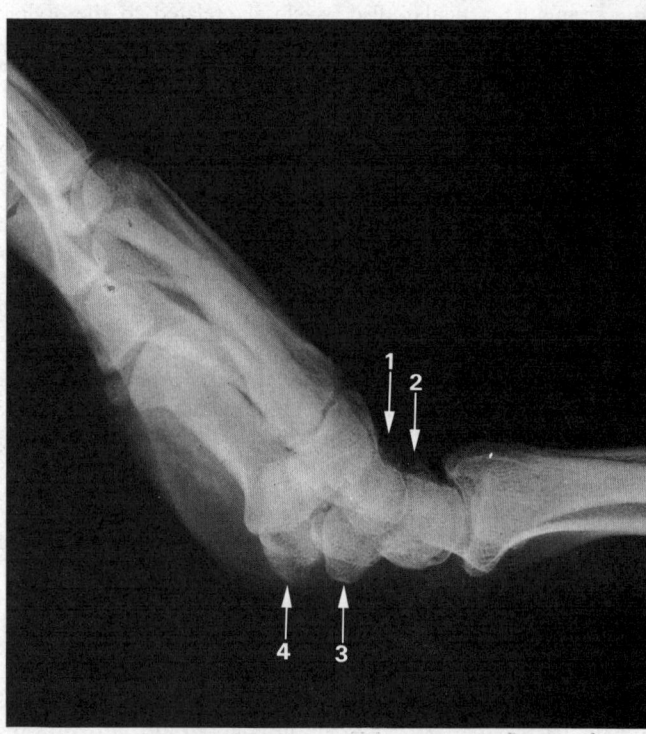

4.53C Radiograph of the hand and wrist in full extension. Lateral aspect. The arrows point to the same structures as in 4.53A. Compare with 4.53B and note the alterations in the relative positions of the capitate and lunate bones, and the lunate bone and the radius.

lunate and triquetral bones. They are on a level with the proximal surfaces of these bones, and form part of the convex articular surface of the radiocarpal joint.

(b) The pisiform articulates with the palmar surface of the triquetral bone and the following are ligaments associated with the joint: fibrous capsule, pisohamate and pisometacarpal.

The fibrous capsule is thin, and surrounds the joint; the synovial membrane·is usually distinct from that of the other carpal joints but may communicate with the radiocarpal.

The pisohamate and pisometacarpal ligaments connect the pisiform to the hook of the hamate, and to the base of the fifth metacarpal bone respectively (4.50). Both are really continuations of the tendon of flexor carpi ulnaris and are, as such, misnamed.

THE JOINTS OF THE DISTAL ROW OF CARPAL BONES

The bones of the distal row of the carpus are connected by dorsal, palmar and interosseous ligaments.

The dorsal and palmar ligaments extend transversely between the trapezium and trapezoid, the trapezoid and capitate, and the capitate and hamate bones.

The interosseous ligaments are much thicker than those of the proximal row; one unites the capitate and hamate, a second the capitate and trapezoid, and a third the trapezium and trapezoid. The first is the strongest and rarely missing, while the second and third are frequently absent.

THE MIDCARPAL JOINT

The joint between the scaphoid, lunate and triquetral bones on the one hand, and the second row of carpal bones on the other is a compound articulation, which may be divided for descriptive convenience into medial and lateral parts. On the medial side the head of the capitate and the hamate articulate with the concavity formed by the scaphoid, lunate and triquetral bones, and constitute a compound sellar joint; on the lateral side the trapezium and trapezoid articulate with the scaphoid bone and constitute a second compound articulation, often described as plane, but in fact another sellar joint. The ligaments are dorsal, palmar and collateral.

The dorsal and palmar ligaments consist of short, irregular bundles passing between the bones of the first and second rows. On the palmar surface the fascicles radiating from the head of the capitate to the surrounding bones are termed the *radiate carpal ligament*.

The collateral ligaments, radial and ulnar, are very short. The former, the stronger and more distinct, connects the scaphoid and trapezium, the latter the triquetral and hamate; they are continuous with the corresponding ligaments of the wrist joint. In addition to these ligaments, a slender, interosseous band sometimes connects the capitate and scaphoid bones, but it does not completely divide the midcarpal synovial cavity.

The synovial membrane of the carpus is most extensive (4.52), and bounds an articulation of very irregular shape. The proximal part of the cavity intervenes between the distal surfaces of the scaphoid, lunate and triquetral bones and the proximal surfaces of the bones of the second row. It sends two prolongations upwards—between the scaphoid and lunate bones, and between the lunate and triquetral bones—and three downwards between the four bones of the second row. The prolongation between the trapezium and the trapezoid, or that between the trapezoid and capitate bone, is, owing to the absence of the interosseous ligament, often continuous with the cavity of the carpometacarpal joints, sometimes of the second, third, fourth and fifth metacarpal bones, sometimes of the second and third only. In the latter condition the joint between the hamate and the fourth and fifth metacarpal bones has a separate synovial membrane and is separated from the others (4.52) by the carpometacarpal interosseous ligament (p. 471). The synovial cavities of the carpometacarpal joints are prolonged for a short distance between the bases of the metacarpal bones. There is usually a separate synovial joint between the pisiform and triquetral bones.

Movements. The movements which occur at the radiocarpal and intercarpal joints are considered together, for the joints concerned form parts of the same mechanism and are acted on by the same muscle groups. The active movements which can be carried out are flexion (*c.* 85°), extension (*c.* 85°), adduction (ulnar deviation), abduction (radial deviation) and circumduction.

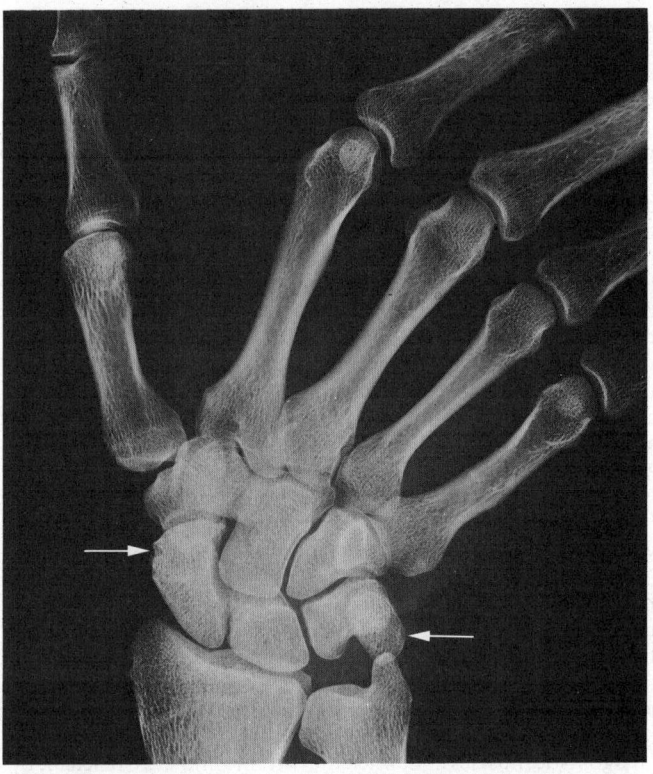

4.54A Radiograph of the hand in full adduction. The arrows point to the scaphoid bone on the lateral side and to the pisiform bone on the medial side. Note that the shadow of the pisiform bone overlaps the shadow of the tip of the styloid process of the ulna. Compare with 4.54B and observe that the movements occur at both the radiocarpal and intercarpal joints.

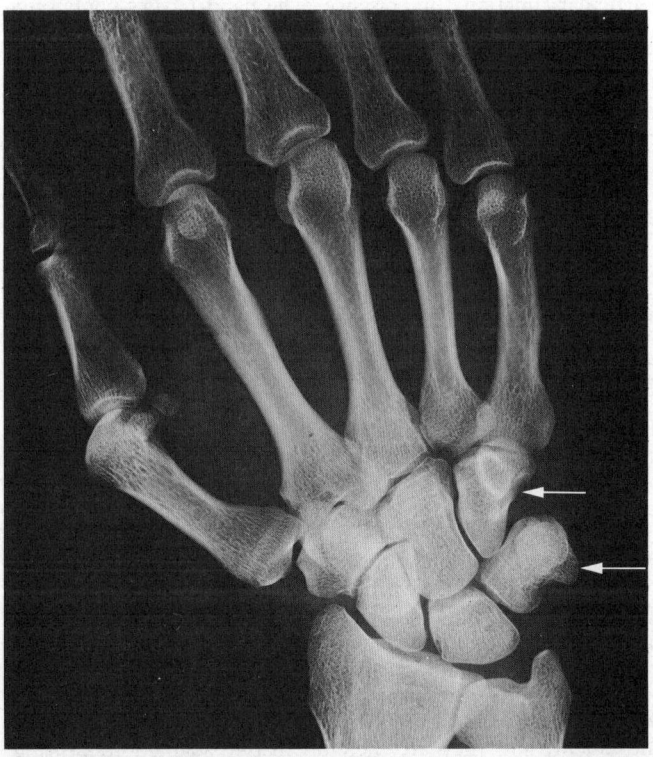

4.54B Radiograph of the same hand in full abduction. The arrows point to the hamate and pisiform bones. Compare with 4.54A and note: (1) that the scaphoid and lunate bones have passed medially so that the latter articulates to a large extent with the articular disc of the inferior radio-ulnar joint; (2) that the pisiform is now widely separated from the styloid process of the ulna; (3) that the scaphoid, having rotated round a transverse axis, is much foreshortened; (4) that the apex of the hamate bone has been thrust away from the lunate by the rotation of the capitate around an anteroposterior axis; (5) that a gap has opened up between the distal portions of the hamate and triquetral bones; and (6) that the long axes of the capitate and lunate bones are now almost in the same straight line.

When the wrist is *flexed*, both the radiocarpal and the midcarpal joints are implicated, but the range of movement is greater at the latter. In *extension* the reverse is the case and more of the movement takes place at the radiocarpal joint (4.53). In consequence, the proximal surfaces extend further on the posterior than on the anterior surfaces of the lunate and scaphoid bones. These movements are limited chiefly by the tension of the antagonistic muscles and, in this connexion, it may be noted that the range of flexion is perceptibly diminished when the fingers are flexed, owing to the increased tension of the extensor muscles. It is only when the joints are subjected to violence at the limits of flexion or extension that the dorsal or palmar ligaments, respectively, are fully stretched (but see also below).

The range of *adduction* of the hand (*c.* 45°) is considerably greater than the range of abduction (*c.* 15°), a fact which may be associated with the shortness of the styloid process of the ulna. In adduction, most of the movement occurs at the radiocarpal joint, and the lunate bone, which articulates both with the radius and with the articular disc when the hand is in line with the forearm (4.52), passes off the articular disc and comes to articulate only with the radius (4.54A).

In *abduction* the movement takes place almost entirely at the midcarpal joint and radiographs of the abducted hand show that the capitate bone rotates round an anteroposterior axis so that its head passes medially, and the hamate conforms to the movements of the capitate. As a result, the distance between the lunate bone and the apex of the wedge of the hamate is perceptibly increased (4.54B). The movements are limited by the tension of the antagonistic muscles and, when the extreme limits are reached, by the radial and ulnar collateral radiocarpal ligaments.

Circumduction of the hand does not result from axial rotation but from the movements of flexion, adduction, extension and abduction carried out in that order or in reverse.

It should be noted that abduction-adduction movements are of particular functional importance. The hand is very commonly used with the carpus slightly extended and the forearm in a position intermediate between full pronation and supination. Highly skilled abduction-adduction movements are then used to manipulate a large variety of precision tools, from fine needles to hammers.

The accessory movements are obtained for the most part at the radiocarpal joint. They are relatively free and can be demonstrated more easily in flexion than in extension of the wrist. The carpus can be moved bodily backwards and forwards on the radius and articular disc and *it can be rotated axially to a considerable extent*. A little side-to-side movement is also possible.

ADDITIONAL ANALYSES OF RADIOCARPAL AND CARPAL MOVEMENTS

It has been suggested that the mechanical equivalent of the radiocarpal and the medial part of the midcarpal joint is a 'link' joint, such as exists in its simplest form between the units of the chain of a bicycle (Gilford *et al.* 1943). This type of joint is stable only when it is under tension and 'on centre', i.e. when the links are in line, and unless it is strengthened by the addition of a 'stop' mechanism, it buckles when subjected to a compression force acting in its long axis, especially when it is 'off centre'. Certain advantages, however, are inherent in the 'link' joint. Since the range of movement at each of its constituent parts is appreciably less than the total range, the articular surfaces can be flatter than would be possible in a single joint giving the same total range of movement, and they are therefore better adapted to bearing pressure. Further, there is less tendency for the overlying tissues to be squeezed at the extremes of movement.

Other investigators, however, have made an analysis of carpal movement which contrasts sharply with the foregoing account (MacConaill 1941). Firstly it is pointed out that in most positions of the hand, the carpal bones are loosely packed and relatively mobile, and that they only become a rigid block in full extension which is the close-packed position for both radiocarpal and most carpal joints; further, close packing is achieved in *two* distinct stages (*vide infra*). The carpus is regarded as divisible into four functional parts: (1) trapezium; (2) scaphoid; (3) hamate, capitate

and trapezoid; and (4) triquetrum and lunate. When passing from full flexion to full extension the following stages are considered to occur. Initially, the distal row (3) moves on the proximal row (2 and 4) until the hand is approximately in line with the forearm, at which point the hamate, capitate, trapezoid and scaphoid come into a mutual close-pack and form a rigid mass (i.e. 2+3). During the second stage, the rigid mass moves as a whole upon the triquetrum and lunate which themselves move at the radiocarpal joint until full extension is reached, with close-packing of the radiocarpal joint and most of the carpal joints (i.e. except the articulations of pisiform and trapezium). In such a position, 'which is adopted only when a special effort is to be made, exceptionally large forces may be transmitted across the articular structures. A similar position is often assumed during a fall on the outstretched hand and commonly results in traumatic damage, e.g. a 'supination' (Colles') fracture of the radius, a fracture of the scaphoid, or dislocation of the lunate. Mayfield *et al.* (1976), in a study of the behaviour of carpal ligaments in experimental production of carpal fractures, have corroborated these views of the dorsal 'locking' of the wrist, and have analysed further the role of carpal ligaments in producing the progressive intercarpal slides which terminate in the close-packed position.

It may be noted that during the initial stages of the movement, the scaphoid bone functions as a part of the proximal row of carpals, but in the later stages joins the distal row in its movements. The joint between the scaphoid and trapezium is not considered to be involved in the general close-packing of the rest of the carpus and it is proposed that this is concerned with a continued functional independence of the thumb even when the remainder of the carpus has been converted into a rigid mass.

A more recent analysis of carpal movements (Kauer 1974) continues the concept of articular links or chains, but places particular emphasis upon rotations of the lunate and scaphoid relative to the radius and articular disc. During flexion-extension both rotate relative to each other and to the capitate bone. The curvatures of the proximal aspects of the lunate and scaphoid are different and this dictates some degree of independence of movements, which nevertheless are interdependent as a result of the strong interosseous ligament between the two bones.

Muscles producing the movements.

Flexion. Flexor carpi radialis, flexor carpi ulnaris and palmaris longus, assisted by flexores digitorum superficialis et profundus, flexor pollicis longus and abductor pollicis longus.

Extension. Extensores carpi radiales longus et brevis and extensor carpi ulnaris, assisted by extensor digitorum, extensor digiti minimi, extensor indicis and extensor pollicis longus.

Adduction. Flexor carpi ulnaris in association with extensor carpi ulnaris.

Abduction. Flexor carpi radialis, in association with extensores carpi radiales longus et brevis, abductor pollicis longus and extensor pollicis brevis.

As a result of these combinations of carpal flexors and extensors abduction or adduction can be carried out without simultaneous flexion or extension.

The Carpometacarpal Joints

CARPOMETACARPAL JOINT OF THE THUMB

This is a saddle-shaped (sellar) joint between the base of the first metacarpal bone and the trapezium, and it enjoys a great range of movement on account of the configuration of its articular surfaces. The joint is surrounded by a fibrous capsule, which is thick but loose, and passes from the circumference of the base of the metacarpal bone to the rough edge bounding the articular surface of the trapezium; it is thickest laterally and dorsally. The synovial membrane which lines the fibrous capsule is distinct from that of the other carpometacarpal joints (4.52).

The metacarpal bone of the thumb is connected to the trapezium by lateral, anterior and posterior ligaments in addition to the capsular ligament.

The *lateral ligament* is a relatively broad band, running from the lateral surface of the trapezium to the radial side of the base of the first metacarpal bone. The *palmar* and *dorsal ligaments* are oblique bands which converge on the ulnar side of the base of the metacarpal bone from the palmar and dorsal surfaces, respectively, of the trapezium. These two ligaments play an important part in connexion with the movements of the thumb (Haines 1944).

Relations of the joint. On the *palmar surface* the joint is covered by the muscles of the thenar eminence. On the *dorsal surface* are the long and short extensors of the thumb. *Medially* is the first dorsal interosseous muscle and the radial artery which is passing from the dorsal to the palmar surface of the hand through the first interosseous space. *Laterally* is the tendon of the abductor pollicis longus.

Movements. In this joint the *active* movements that can be carried out are flexion-extension, abduction-adduction, rotation and circumduction. Owing to the set of the first metacarpal bone in the position of rest (p. 592), flexion and extension take place in a plane parallel to the plane of the palm, while abduction and adduction occur in a plane at right angles to the flexion-extension plane. Except in its early stages, flexion of the metacarpal bone is always associated with medial rotation and, conversely, medial rotation cannot be performed actively apart from flexion. The intimate association of these two movements is largely dependent upon the geometry of the articular surfaces, which are sellar and impose some degree of conjunct rotation, and upon the obliquity of the fibres of the dorsal ligament, which when taut anchors the ulnar side of the base of the metacarpal bone, while its radial side is free to move. Under these conditions contraction of the flexor brevis muscle, assisted by the opponens pollicis, will produce medial rotation combined with flexion (Napier 1955). This combined movement, together with abduction at the carpometacarpal joint, enables the pulp of the thumb to be brought into contact with the palmar surfaces of the pulps of the slightly flexed fingers, a movement which is termed *opposition*.

Full extension of the metacarpal bone of the thumb is conversely associated with slight lateral rotation (Kuczynski 1974). This is attributable, again, to the sellar form of the joint surfaces, and to the fact that the ulnar side of the base of the metacarpal bone becomes anchored by tension of the palmar ligament while the radial side is still free to move, and continued action of the extensors of the thumb produces lateral rotation in addition to further extension.

As indicated, however, these various rotations are conjunct and the inevitable accompaniment of flexion-extension, of spiralization and tautening of the joint capsule, and of balanced conarticular compression, occurring at the extremes of the movements. The position of close-packing has been claimed to occur in powerful, complete opposition, when the greatest forces would be transmitted to the joint. (For interesting analyses of this articulation consult Kapandji 1963, 1970; Kuczynski 1974; MacConaill and Basmajian 1977.)

Accessory movements are limited to axial rotation in the position of rest, and distraction.

Muscles producing the movements.
Flexion. Flexor pollicis brevis and opponens pollicis, aided by flexor pollicis longus when the other joints of the thumb are flexed. Flexion is associated with medial rotation.

Extension. Abductor pollicis longus and extensores pollicis brevis et longus. In full extension the extensor pollicis longus, partly owing to the obliquity of its line of pull and partly also to the disposition of the palmar ligament of the joint, rotates the thumb laterally and draws it dorsally, i.e. slightly adducts it.

Abduction. Abductor pollicis brevis and abductor pollicis longus. When the movement reaches its maximum the digit and its metacarpal bone are not in line, for the thumb is then abducted both at its metacarpophalangeal and at its carpometacarpal joints.

Adduction. Adductor pollicis alone; the first palmar interosseous muscle acts on the metacarpophalangeal joint.

Opposition. Opponens pollicis and flexor pollicis brevis which simultaneously flex and medially rotate the abducted thumb. The pressure which the opposed thumb can exercise on the tips of the fingers is increased by the reinforcing action of the adductor pollicis and the flexor pollicis longus.

Circumduction. The above muscle groups acting consecutively, the extensors, the abductors, the flexors and the adductors following one another in that order.

THE SECOND TO FIFTH CARPOMETACARPAL JOINTS

The joints between the carpus and the second to fifth metacarpal bones, though often classed as plane, are in fact formed by curved articular surfaces often of complex sellar shape. (*See* the description of the relevant facets given on pp. 373–375.) The bones are united by articular capsules, strengthened by dorsal, palmar and interosseous ligaments.

The dorsal ligaments, which are the strongest and most distinct, connect the carpal and metacarpal bones on their dorsal surfaces. The second metacarpal bone receives two fasciculi, one each from the trapezium and trapezoid bone; the third metacarpal receives two, one each from the trapezoid and capitate bones; the fourth two, one each from the capitate and hamate bones; the fifth receives a single fasciculus from the hamate, and this is continuous with a similar ligament on the palmar surface, forming an incomplete fibrous capsule.

The palmar ligaments have a somewhat similar arrangement, with the exception of those of the third metacarpal bone, which are three in number: a lateral one from the trapezium, situated superficially to the sheath of the tendon of the flexor carpi radialis; an intermediate one from the capitate; and a medial one from the hamate bone.

The interosseous ligaments consist of short, thick fibres, and are limited to one part of the carpometacarpal articulation; they connect the contiguous distal margins of the capitate and hamate bones with the adjacent surfaces of the third and fourth metacarpal bones, and they may be united at their proximal ends.

The synovial membrane is often a continuation of that of the intercarpal joints. Occasionally, the joint between the hamate and the fourth and fifth metacarpal bones has a separate synovial membrane, and is then bounded on its lateral side by the more medial of the two interosseous ligaments just described, and by extensions from it to the palmar and dorsal parts of the capsule (4.52).

THE INTERMETACARPAL JOINTS

The bases of the second, third, fourth and fifth metacarpal bones articulate with one another by small surfaces covered with cartilage, and are connected together by dorsal, palmar and interosseous ligaments.

The dorsal and palmar ligaments pass transversely from one bone to another on the dorsal and palmar surfaces. **The interosseous ligaments** connect the contiguous surfaces of the bones, just distal to their collateral articular facets.

The synovial membrane of these joints is continuous with that of the carpometacarpal articulations.

Movements. The movements permitted in the carpometacarpal and intermetacarpal articulations of the fingers are limited to slight gliding of the articular surfaces upon each other, the extent of which varies in the different joints. They are partly accessory movements of the first type (p. 431) and only come into prominence when the palm of the hand becomes 'cupped' as a solid object is grasped. The metacarpal bone of the little finger is the most movable, then that of the ring finger; the metacarpal bones of the index and middle fingers are almost immovable. The close-packed position probably coincides with extension of carpal articulations as in gripping.

Further accessory movements are limited to spiral twisting of the metacarpus as a whole on the carpus.

The Metacarpophalangeal Joints

These articulations (4.55A, B) are usually classified as ellipsoid, but the heads of the metacarpals, fitting into shallow concavities on the bases of the proximal phalanges, are not regularly convex; their articular surfaces are partially divided on their palmar aspect and are almost bicondylar (3.162–167). Each joint has a palmar and two collateral ligaments.

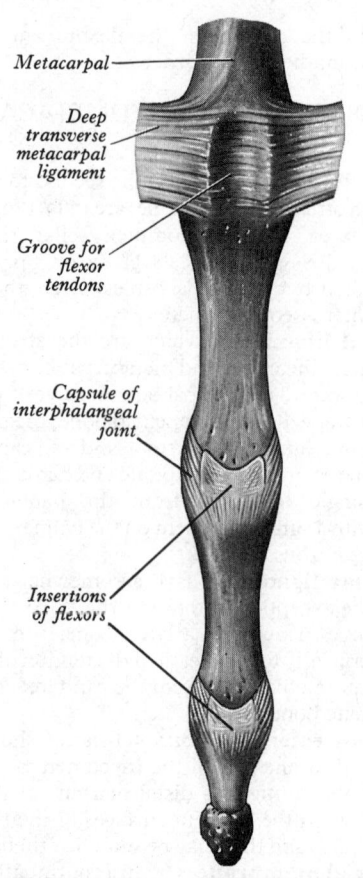

4.55A The metacarpophalangeal and digital joints of the middle finger. Palmar aspect.

The palmar ligaments are unusual in that they are thick, dense, fibrocartilaginous structures, placed upon the palmar surfaces of the joints in the intervals between the collateral ligaments, to which they are connected; they are loosely united to the metacarpal bones, but are very firmly attached to the bases of the proximal phalanges. Their palmar surfaces are intimately blended with the deep transverse ligaments of the palm, and are grooved for the flexor tendons, the fibrous sheaths of which are connected to the sides of the grooves. Their deep surfaces form parts of the articular areas for the heads of the metacarpal bones.

The deep transverse metacarpal ligaments consist of three short, wide, flattened bands which connect the palmar ligaments of the second, third, fourth and fifth metacarpophalangeal joints (**4.55A**). They are related, anteriorly, to the lumbricals and the digital vessels and nerves and, posteriorly, to the interossei. Offsets from the digital slips of the central portion of the palmar aponeurosis join the palmar surface (p. 585). On both sides of the metacarpophalangeal joint of the middle and ring fingers (but only on the ulnar side of the index and on the radial side of the little finger) the transverse band of the dorsal digital expansion (p. 580) joins the transverse metacarpal ligament; deep to it is the phalangeal attachment of the dorsal interosseous and superficial to it the remainder of this muscle, the palmar interosseous and lumbrical (p. 589 and **5**.70).

The collateral ligaments are strong, rounded cords, along the sides of the joints; each is attached to the posterior tubercle and adjacent depression on the side of the head of the metacarpal bone, and passes obliquely distally and forwards to reach the side of the ventral aspect of the base of the phalanx (**4.55B**).

On the dorsal surfaces of these joints the fibrous capsule is thin and is separated from the extensor tendon by a bursa (**5**.70).

Movements. The active movements are flexion, extension, adduction, abduction, circumduction, and some limited rotation. However, rotation cannot be instigated voluntarily as an isolated movement; it is accessory to flexion-extension and usually due to resistance of a grasped object.

Extension is limited to a few degrees, while flexion is possible to almost 90°; both are limited chiefly by tension of antagonistic muscles, but flexion is commonly limited by the resistance of a grasped tool or other object. Abduction and adduction are also limited to a few degrees, and in the flexed position of these joints the movements are neglible in range. The pollicial metacarpophalangeal joint has a flexion-extension range limited to about 60° (almost entirely flexion) or even less; other active movements are almost nil apart from slight conjunct rotation accompanying the former. Of the other metacarpophalangeal joints, that of the index is the most mobile in adduction/abduction (c. 30°), followed by the minimus, annularis, and with the medius least mobile.

The accessory movements comprise rotation (which in the case of the thumb may be considerable), gliding of the phalanx on the head of the metacarpal bone forwards, backwards and from side to side, and distraction.

Muscles producing the movements.

Flexion. Flexores digitorum superficialis et profundus, assisted by lumbricals and interossei (Long *et al.* 1961; and *see* p. 591) and, in the little finger, by flexor digiti minimi; in the thumb flexores pollicis longus et brevis and first palmar interosseous. Slight lateral rotation accompanies finger flexion.

Extension. In the middle and ring fingers, extensor digitorum, assisted, in the index and little fingers, by extensor indicis and extensor digiti minimi respectively; in the thumb extensores pollicis longus et brevis only.

Adduction. With extended fingers, palmar interossei. During flexion the long flexors of the fingers play the principal part. The slight degree of this movement in the thumb is attributable to the adductor pollicis and the first palmar interosseous.

Abduction. In the extended fingers, dorsal interossei, assisted by the long extensors except in the case of the middle finger. In the little finger, abductor digiti minimi. Abductor pollicis brevis produces the slight movement possible in the thumb which contributes towards opposition. When the fingers are in the flexed position, abduction cannot be performed actively, but, provided that the long flexors of the fingers are not in active contraction, it can easily be carried out passively. The inability to perform the movement actively in this position may be due to the fact that the dorsal interossei and the abductor digiti minimi are so shortened by flexion that they are unable to function, but the altered relation of the line of pull of the interossei to the axis of movement is probably the determining factor, for whereas in extension of the digits the axis of side to side movements is anteroposterior, in flexion it is proximodistal and the line of pull of the interossei is then nearly parallel to the axis; the inability is certainly not to be ascribed solely to tension of the collateral ligaments.

The Interphalangeal Joints

The interphalangeal articulations are uni-axial hinge joints (**4.55A, B**). In addition to a fibrous capsule, each has a palmar and two collateral ligaments. The arrangement of these ligaments is similar to those in the metacarpophalangeal joints (p. 471). The extensor tendons take the place of dorsal ligaments.

Movements. The only active movements which occur at the interphalangeal joints are flexion and extension; these movements are greater in range between the proximal and middle phalanges than between the middle and distal. The amount of flexion is very considerable, but extension is limited by tension of the digital flexors and, in violent movements, by the palmar ligaments. Full extension is the close-packed position of these joints, and it is utilized whenever the fingers are used as props to transmit the body weight, or powerful thrust, though the fully clenched position is also used under these circumstances.

Finally, careful inspection of a finger during flexion and extension reveals that these principal movements are accompanied by a small amount of conjunct rotation. During flexion, the rotation turns the pulp of the finger slightly laterally, i.e. it faces more fully the pulp of the opposed thumb. An opposite rotation occurs, of course, during finger extension. (*See* also above.)

The *accessory movements* comprise a limited range in each case of rotation, abduction, adduction and gliding forwards and

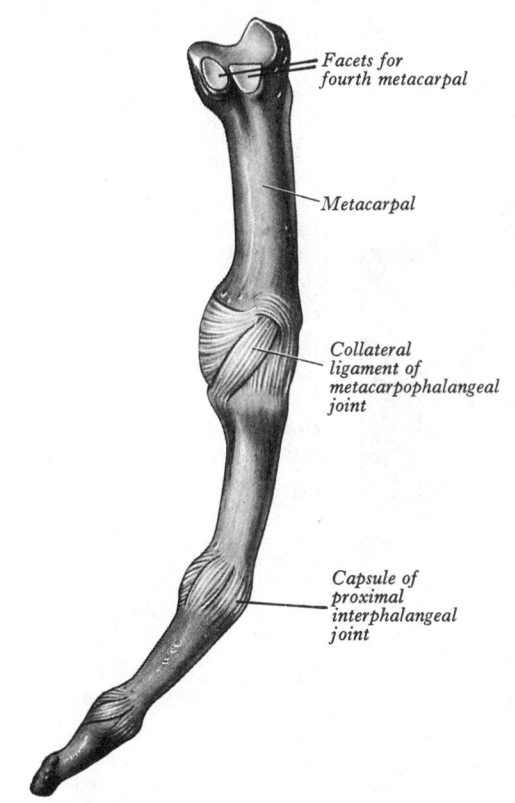

Facets for
fourth metacarpal

Metacarpal

Collateral
ligament of
metacarpophalangeal
joint

Capsule of
proximal
interphalangeal
joint

4.55B The metacarpophalangeal and digital joints of the right third finger. Medial aspect.

interossei are not only the active agents in producing flexion at metacarpophalangeal joints, but are also the active agents in extending the interphalangeal joints through their attachments to the dorsal digital expansions (p. 579). It has been urged that, when the lumbricals and interossei flex the metacarpophalangeal joints, the balance between the tone of the digital flexors and extensors is altered in favour of the extensors, and that this factor alone is responsible for the extension of the interphalangeal joints (Braithwaite *et al.* 1948). However, both the lumbricals and interossei alone can extend these joints (Sunderland 1945; *see* also p. 589).

The Flexure Lines of the Wrist and Hand

A number of more or less constant flexure lines or skin creases traverse the flexor surfaces of the wrist and hand (**6.**82). Though not necessarily overlying the bony joints, these result from anchorage of the skin to the underlying deep fascia and mark sites of folding of the skin during movements of the parts. (Here, and of course elsewhere, such flexures have often thus been termed 'skin joints'.) These lines are useful landmarks.

At the wrist there are usually three transverse lines; the proximal marks the upper limit of the synovial sheaths of the flexor tendons, the intermediate overlies the wrist joint, and the distal is placed at the upper margin of the flexor retinaculum.

In the palm a curved longitudinally directed flexure line, often termed the *radial longitudinal line*, encircles the thenar eminence and ends at the lateral margin of the hand. Medial and roughly parallel to it are one or two less constant longitudinal lines. Two transverse lines, proximal and distal, cross the palm. The proximal line begins at the distal end of the radial longitudinal line and runs to the hypothenar eminence; it is placed obliquely across the shafts of the metacarpal bones. The distal line begins in the cleft between the index and middle fingers and runs with a slight convexity upwards across the palm, over the heads of the second, third and fourth metacarpal bones, at a level near the commencement of the fibrous flexor sheaths.

In each of the medial four digits there are three sets of transverse lines, proximal, middle and distal. The proximal lines are placed at the roots of the fingers and lie about 2 cm distal to the metacarpophalangeal joints. The middle digital lines are typically double, the proximal member of the pair lying over the proximal interphalangeal joint. The distal flexure lines, usually single, lie proximal to the distal interphalangeal joint, the level of which may, however, be indicated by a second but fainter line nearer the tip of the finger. In the thumb, the base is partially encircled by a line which starts on the radial side and proceeds distally crossing the metacarpophalangeal joint obliquely to end in the cleft between the thumb and index finger on a level with the base of the proximal phalanx. A second shorter crease lies about 1 cm distal to it. At the interphalangeal joint of the thumb there are two lines which are comparable in position to the middle digital lines on the other fingers.

backwards. They permit the fingers to adapt themselves to the shape of any object gripped in the hand, and they provide against the stresses and strains which occur during the ordinary use of the hand.

Muscles producing the movements.

Flexion. At the proximal joint, flexor digitorum superficialis and flexor digitorum profundus, at the distal joint, flexor digitorum profundus, at the interphalangeal joint of the thumb, flexor pollicis longus.

Extension. Extensor digitorum and extensor pollicis longus. Extension occurs simultaneously in both joints as in opening the closed fist.

Attention should be drawn at this point to the combined movements of flexion at the metacarpophalangeal joint and extension at the interphalangeal joints, which can be carried out simultaneously and are of such importance in fine movements as executed in writing, drawing and threading a needle. For many years it has been widely taught that the lumbricals and the

THE JOINTS OF THE LOWER LIMB

The Sacro-iliac Joint

The sacro-iliac articulation is synovial and between the auricular surfaces of the sacrum and ilium (4.59). Although often described as plane, the *articular surfaces* are nearly flat only in the infant; in the adult they exhibit irregular elevations and depressions. These curvatures, more pronounced in the male, fit into one another, restrict movements, and contribute to the strength of the joint, which transmits weight from the vertebral column to the lower limb. According to Putschar (1931) and Schunke (1938) the sacral articular surface is covered by hyaline cartilage, that on the ilium by fibrocartilage, an unusual arrangement, which apparently has not been subsequently confirmed. Fibrous adhesions and gradual obliteration of the synovial cavity occurs in both sexes, earlier in

males, and after the menopause in females. A radiological study (Cohen *et al.* 1967) of 94 individuals without demonstrable joint disease, showed such changes in 6 per cent before 50 years and 24 per cent after this. (In old age the joint may become completely fibrosed and even ossified.) The *articular capsule* is attached close to the margins of the articular surfaces of the sacrum and ilium. The ligaments of the joints are ventral, interosseous, and dorsal sacro-iliac.

The ventral sacro-iliac ligament (4.56) is a thickening of the anterior and inferior parts of the fibrous capsule. It is particularly well developed at the level of the arcuate line and inferiorly at the level of the posterior inferior iliac spine, where it connects the third piece of the sacrum to the lateral margin of the pre-auricular sulcus, but it is thin elsewhere.

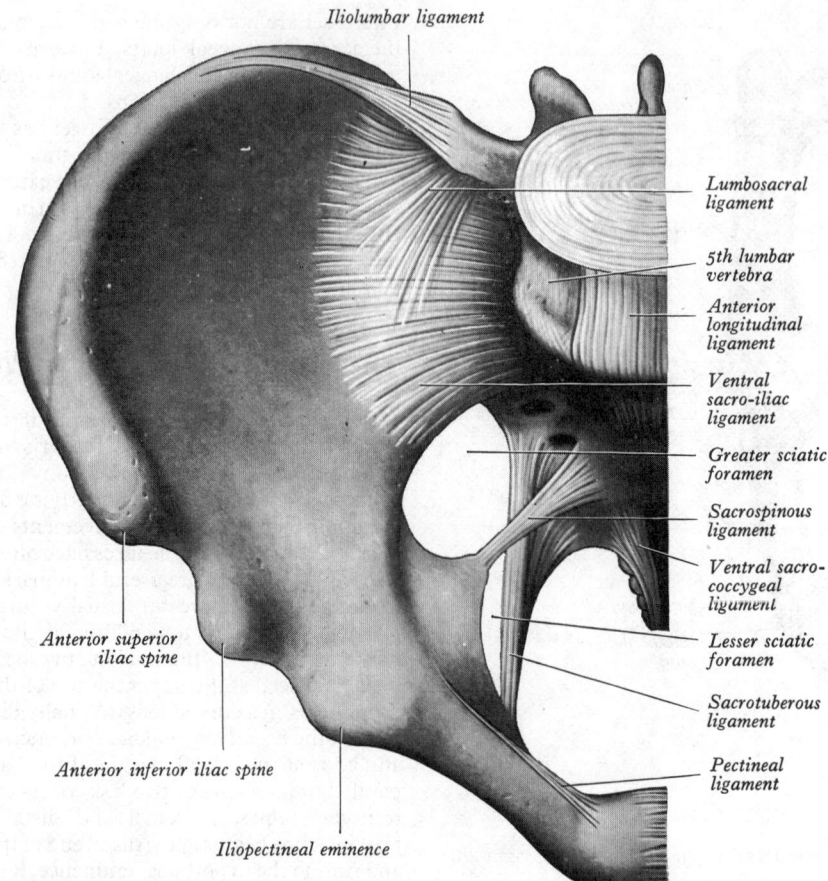

Iliolumbar ligament

Lumbosacral ligament

5th lumbar vertebra

Anterior longitudinal ligament

Ventral sacro-iliac ligament

Greater sciatic foramen

Sacrospinous ligament

Ventral sacro-coccygeal ligament

Lesser sciatic foramen

Sacrotuberous ligament

Pectineal ligament

Anterior superior iliac spine

Anterior inferior iliac spine

Iliopectineal eminence

4.56   The joints and ligaments of the right half of the pelvis. Ventral aspect.

The interosseous sacro-iliac ligament is massive and forms the chief bond between the two bones. It fills the irregular space immediately above and behind the joint (**4**.60) and is covered by the dorsal sacro-iliac ligament. It consists of deeper and more superficial parts. The deeper part has cranial and caudal bands which pass from the depressions behind the auricular surface of the sacrum to depressions on the iliac tuberosity. These bands are covered by and blend with the more superficial part which forms a fibrous sheet connecting the cranial and dorsal margins of the rough area behind the auricular surface on the sacrum to the corresponding margins of the iliac tuberosity. This sheet is often partially divided into cranial and caudal parts. The cranial one unites the superior articular process and lateral crest on the first two pieces of the sacrum to the ilium and is sometimes termed the *short posterior iliac ligament* (**4**.57).

The dorsal sacro-iliac ligament overlies the interosseous ligament, from which it is separated by the dorsal rami of the sacral spinal nerves and vessels. It consists of several weak fasciculi which arise above from the intermediate and below from the lateral crest of the sacrum and pass with varying degrees of obliquity to the posterior superior iliac spine and inner lip of the dorsal part of the iliac crest. The lower fibres passing from the third and fourth segments of the sacrum to the posterior superior iliac spine may form a separate fasciculus which is sometimes termed the *long posterior sacro-iliac ligament* (**4**.57) and is continuous laterally with some of the fibres of the sacrotuberous ligament and medially with the posterior layer of the thoracolumbar fascia (Weisl 1954).

It is not uncommon to find accessory synovial articulations between the lateral sacral crest and the posterior superior iliac spine and tuberosity of the ilium (Trotter 1937).

Movements. A small amount of anteroposterior rotatory movement occurs at the sacro-iliac joint around a transverse axis which is usually about 5–10 cm vertically below the promontory of the sacrum (Weisl 1955). These movements occur during flexion and extension of the trunk and the range is the same in the

male and non-pregnant female. The range is increased temporarily in pregnancy. The greatest change in position of the sacrum relative to the iliac bones occurs when rising from the recumbent to the standing position (**4**.61). The sacral promontory moves forwards as much as 5–6 mm as the body weight is taken upon the sacrum. Backward movement of the lower end of the sacrum is considerably less. The movement is not a simple rotation, the axis being dynamic (Weisl 1953); some translation or glide is thus associated with the rotation.

THE VERTEBROPELVIC LIGAMENTS

The ilium is connected to the fifth lumbar vertebra by the *iliolumbar ligament* (p. 272), and the sacrum to the ischium by the *sacrotuberous* and *sacrospinous ligaments*.

The sacrotuberous ligament (**4**.56, 57) is attached by a broad base to the posterior iliac spines (where it is partly blended with the dorsal sacro-iliac ligament), to the lower transverse tubercles of the sacrum, and to the lateral margin of the lower part of the sacrum and upper part of the coccyx. Its fibres run obliquely downwards and laterally, and converge to form a thick, narrow band; this widens below and is fixed to the medial margin of the ischial tuberosity, continuing along the ramus of the ischium under the name of the *falciform process*, the free concave edge of which blends with the fascial sheath of the internal pudendal vessels (p. 721) and pudendal nerve. To its posterior surface are attached the lowest fibres of the gluteus maximus, and some of the superficial fibres of its lower part are continued into the tendon of the long head of biceps femoris. The ligament is pierced by the coccygeal branches of the inferior gluteal artery, by the perforating cutaneous nerve, and by minute filaments of the coccygeal plexus.

The sacrospinous ligament (**4**.56) is thin and triangular in form; it attaches to the spine of the ischium and, medially, by its broad base to the lateral margins of sacrum and coccyx in front of the sacrotuberous ligament, with which its fibres are intermingled. It is related in front to the coccygeus, to which it is closely connected, and of which it is often held to represent a degenerated part.

The sacrotuberous and, to a lesser extent, the sacrospinous ligaments oppose upward tilting of the lower part of the sacrum under the downward thrust which is imparted to the upper end of the bone by the weight of the trunk (*vide infra*). These two ligaments convert the sciatic notches into foramina.

The greater sciatic foramen is bounded in front and above by the greater sciatic notch, behind by the sacrotuberous ligament, and below by the sacrospinous ligament and the spine of the ischium. It is partially filled by the piriformis, which emerges from the pelvis through it. Above this muscle the superior gluteal vessels and nerve pass out of the pelvis; below it, the inferior gluteal vessels and nerve, the internal pudendal vessels and pudendal nerve, the sciatic and the posterior femoral cutaneous nerves, and the nerves to the obturator internus and quadratus femoris make their exit from the pelvis.

The lesser sciatic foramen is bounded in front by the body of the ischium, above, by the spine of the ischium and the sacrospinous ligament, behind, by the sacrotuberous ligament. It transmits the tendon of the obturator internus, the nerve to this muscle, and the internal pudendal vessels and pudendal nerve.

The Pubic Symphysis

The pubic bones meet each other in the median plane, where they form a cartilaginous joint, the *pubis symphysis* (**4**.2, 58). They are connected by superior and arcuate pubic ligaments, and by an interpubic disc of fibrocartilage.

The superior pubic ligament connects the pubic bones superiorly, and extends as far as the pubic tubercles.

The arcuate pubic ligament is a thick arch of fibres, connecting the lower borders of the symphysial surfaces of the two pubic bones, and forming the upper boundary of the pubic arch. Above, it is blended with the interpubic disc, laterally, it is attached to the inferior rami of the pubic bones; its base is free, and is separated from the free ventral border of the urogenital diaphragm by an opening through which the deep dorsal vein of the penis (or clitoris) enters the pelvis.

The interpubic disc connects the adjacent surfaces of the pubic bones. Each of these surfaces is covered with a thin layer of hyaline cartilage firmly joined to the bone. This junction is not a flat zone (p. 381), being marked by ridges and papilliform elevations, with their reciprocal depressions. This arrangement would, theoretically at least, present a resistance to shearing forces. The opposed surfaces of hyaline cartilage are connected by a lamina of fibrocartilage, which varies in thickness in different subjects. It often contains a cavity, probably formed by softening and absorption since it is rarely seen before the tenth year of life and is not lined with synovial membrane. This cavity, which is better developed in the female, is usually limited to the upper and back part of the joint; it occasionally reaches the front, and may even occupy the whole of the sagittal length of the cartilage (**4**.58). In front the disc is strengthened by several superimposed layers of fibres, which pass obliquely from one bone to the other, decussating and forming an interlacement with the fibres of the external oblique aponeuroses and the medial tendons of origin of the recti abdominis.

Movements at this articulation have been the subject of little recorded observation. Obviously, as at any symphysis, small degrees of angulation, rotation and displacement are possible, and are likely to occur in activities at the sacro-iliac and hip joints. Some separation between the pubic bones is held to occur late in gestation and particularly during childbirth.

The Mechanism of the Pelvis

While the skeletal pelvis supports and protects the contained viscera, it is primarily part of the lower limb. It affords surfaces for the attachments of the muscles of the trunk and lower limb. Its

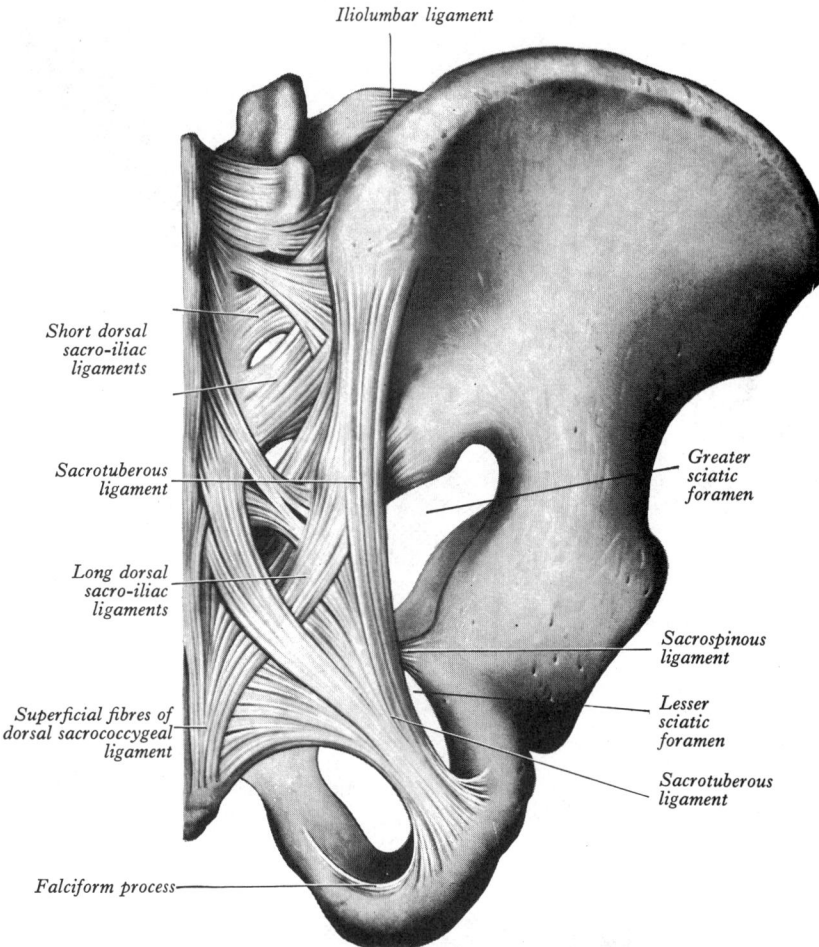

Iliolumbar ligament

Short dorsal sacro-iliac ligaments

Sacrotuberous ligament

Long dorsal sacro-iliac ligaments

Superficial fibres of dorsal sacrococcygeal ligament

Falciform process

Greater sciatic foramen

Sacrospinous ligament

Lesser sciatic foramen

Sacrotuberous ligament

4.57   The joints and ligaments of the right half of the pelvis and fifth lumbar vertebra. Posterior aspect.

most important mechanical function is to transmit the weight of the head, trunk and upper limbs to the lower extremities. It may be divided into two arches by a vertical plane passing through the acetabular cavities; the posterior of these arches is the one chiefly concerned in the function of transmitting the weight of the trunk. Its essential parts are the upper three sacral vertebrae and two strong pillars of bone running from the sacro-iliac joints to the acetabular fossae (p. 384). The anterior arch is formed by the pubic bones and their superior rami. It connects the bases of the lateral pillars of the posterior arch and so acts as a tie-beam to prevent their separation. The anterior arch also acts as a compression strut, resisting the medial thrust of the femoral heads. The sacrum is the summit of the posterior arch; transmitted weight falls on it at the lumbosacral joint and, theoretically, has a component in each of two directions. One component of the force drives the sacrum downwards and backwards between the iliac bones, while the other thrusts the upper end of the sacrum downwards and forwards towards the pelvic cavity.

The movements of the sacrum are regulated by its form and ligamentous attachments. Viewed as a whole, it presents the shape of a wedge with its base upwards and forwards. The first component of the force is therefore acting against the resistance of the wedge, and its tendency to separate the iliac bones is resisted by the sacro-iliac and iliolumbar ligaments and by the ligaments of the symphysis pubis.

If a series of coronal sections be made through the sacro-iliac joints, the articular portion of the sacrum may be divided into three segments: anterior, middle and posterior. In the *anterior segment*, which involves the first sacral vertebra, the articular surfaces show slight sinuosities and are almost parallel to one another. In the *middle segment* (**4**.60) the width between the dorsal margins of the sacral articular surfaces is greater than that between the ventral margins, and in the centre of each surface

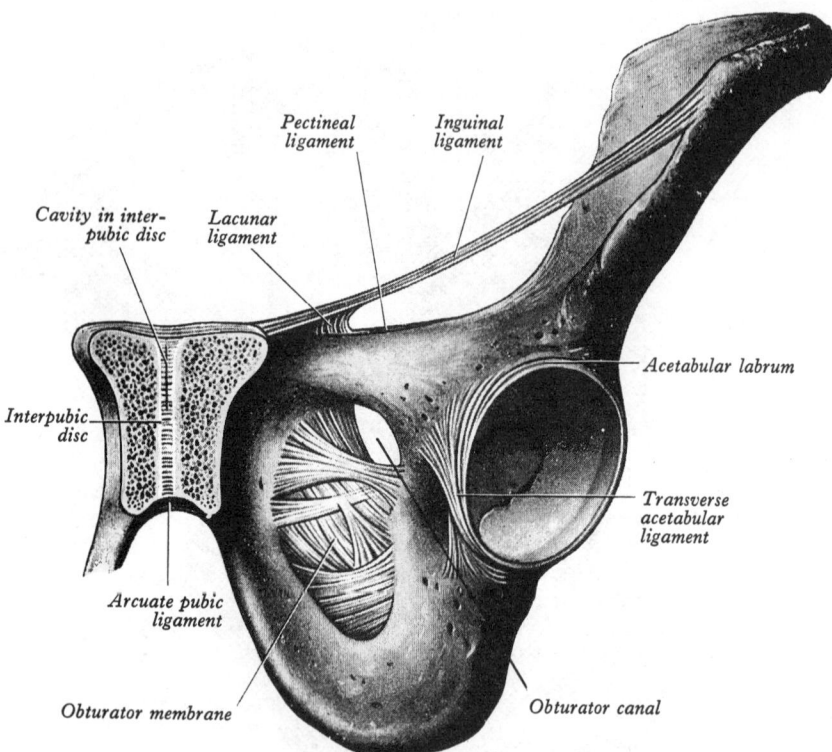

Pectineal ligament

Inguinal ligament

Cavity in inter-pubic disc

Lacunar ligament

Interpubic disc

Acetabular labrum

Transverse acetabular ligament

Arcuate pubic ligament

Obturator membrane

Obturator canal

4.58 An obliquely coronal section through the pubic symphysis. Antero-inferior aspect.

there is a concavity into which a corresponding convexity of the iliac surface fits. This forms an interlocking mechanism which relieves the strain of the body weight on the ligaments. In the *posterior segment* the ventral width of the sacrum is greater than the dorsal, and the articular surfaces are only slightly concave.

Dislocation downwards and forwards of the sacrum by the second component of the force applied to it is prevented therefore by the middle segment, which interposes the resistance of its wedge shape and that of the interlocking mechanism on its surfaces; a rotary movement, however, is produced, by which the anterior segment is tilted downwards and the posterior upwards (4.61). The movement of the anterior segment is slightly limited by its wedge form, but chiefly by the dorsal and interosseous sacro-iliac ligaments; that of the posterior segment is checked to a slight extent by its wedge form, but the chief limiting factors are the sacrotuberous and sacrospinous ligaments. In all these movements the effect of the sacro-iliac and iliolumbar ligaments and the ligaments of the symphysis pubis in resisting the separation of the iliac bones must be recognized.

Statistical analysis of the locomotor significance of structural data has been carried out more extensively on the primate, including human, pelvis than any other part of the skeleton, except perhaps the femur. As an example of the application of metrical analysis to the interactions of structure and function consult a paper by Zuckerman, Ashton *et al.* (1973). Muscular gravitational and inertial forces acting on the pelvis have been analysed by means of a mathematical model by Goel and Svensson (1977).

Applied Anatomy. During pregnancy the pelvic joints and ligaments are relaxed and capable of more extensive movements. This relaxation renders the locking mechanism of the sacro-iliac joint less restrictive and permits greater rotation. This change may allow alterations in the diameter of the pelvis at childbirth

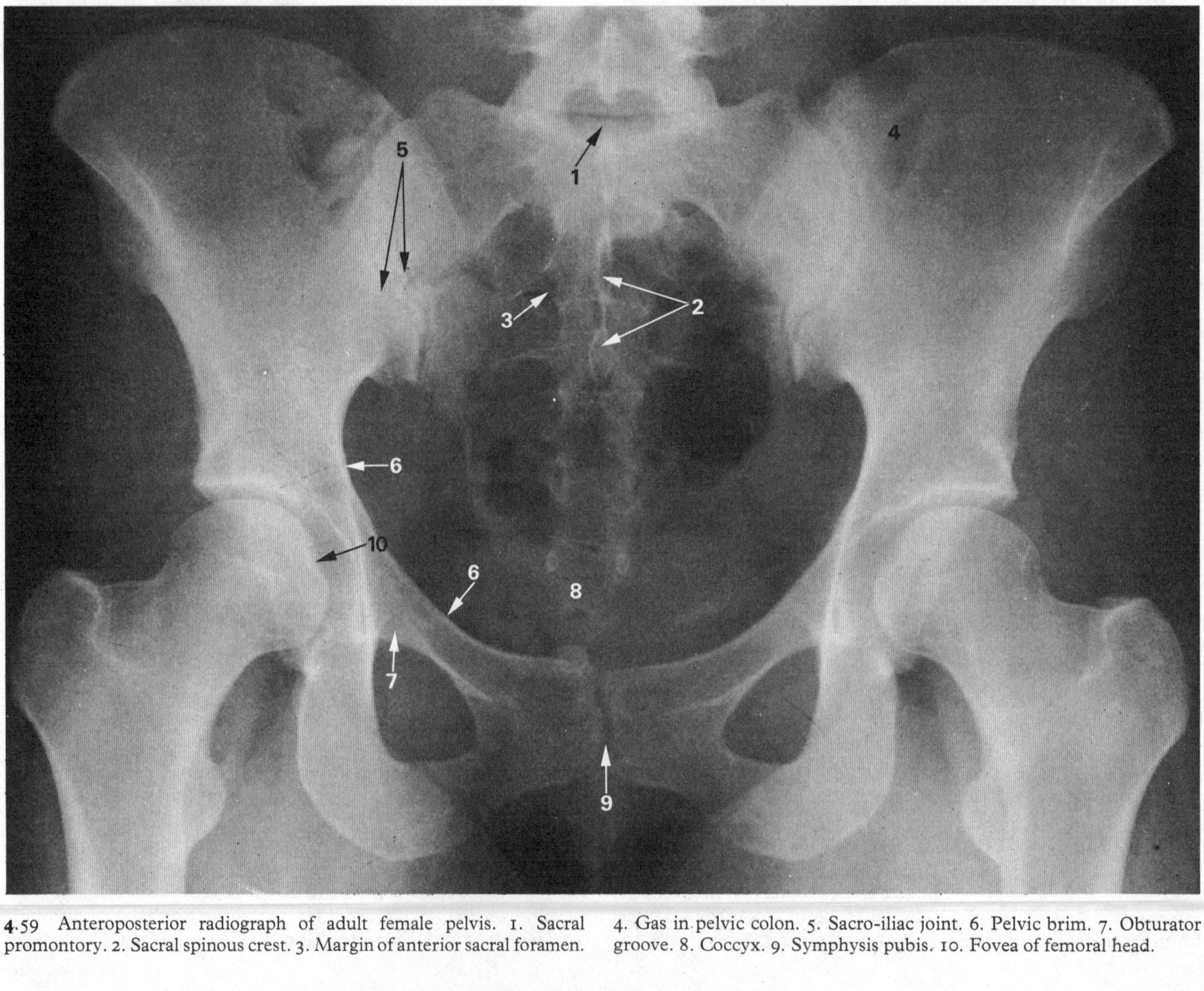

4.59 Anteroposterior radiograph of adult female pelvis. 1. Sacral promontory. 2. Sacral spinous crest. 3. Margin of anterior sacral foramen. 4. Gas in pelvic colon. 5. Sacro-iliac joint. 6. Pelvic brim. 7. Obturator groove. 8. Coccyx. 9. Symphysis pubis. 10. Fovea of femoral head.

but its effect is, at the best, small (Young 1940). The less the locking mechanism, the more the strain of weight-bearing falls on the ligaments, leading to the frequent occurrence of sacro-iliac strain after pregnancy. After childbirth the ligaments become tightened up again and the locking mechanism increases again, but, in some cases, the locking may occur in the position of rotation of the hip bones adopted during the pregnancy. This so-called subluxation of the sacro-iliac joint causes pain by the unusual tension which it imposes on the ligaments, and reduction by forcible manipulation may be attempted. The common position found in this condition is believed to be backward rotation of the innominate bone on the sacrum; it is usually unilateral.

The Hip (Coxal) Joint

The hip joint is multi-axial and of the ball-and-socket type. The head of the femur articulates with the cup-shaped fossa of the acetabulum. Its centre lies some 1·2 cm below the middle third of the inguinal ligament (4.59); this is, of course, an average estimate for an adult male. The articular surfaces are reciprocally curved, but are not co-extensive, nor are they completely congruent, the close-pack position being one of full extension, with some degree of abduction and medial rotation. As in the case of the shoulder joint, the articular surfaces are spheroid rather than spherical. The sphericality of these surfaces has been subject to controversy, and the literature should be consulted, e.g. Hammond and Charnley 1967; Bullough and Goodfellow 1968; Greenwald and Haynes 1972. The weight of evidence favours spheroid (slightly ovoid—p. 432) surfaces, which may become more nearly spherical with advancing age. The head of the femur is completely

structures of the joint are the fibrous capsule, the acetabular labrum, the ligament of the head of the femur, and the iliofemoral, ischiofemoral, pubofemoral and transverse acetabular ligaments.

The fibrous capsule (4.64, 65) is strong and dense. It is attached above to the margin of the acetabulum, 5 or 6 mm beyond the acetabular labrum, in front, to the outer margin of the labrum, and, opposite the acetabular notch, to the transverse acetabular ligament and the edge of the obturator foramen. It surrounds the neck of the femur, and is attached in front to the trochanteric line, above, to the base of the neck, behind, to the neck about 1 cm above the trochanteric crest, below, to the lower part of the neck close to the lesser trochanter (4.64). From its attachment to the front of the femoral neck many of the fibres are reflected upwards along the neck as longitudinal bands, termed *retinacula*; these contain blood vessels supplying the head and neck of the bone (p. 727). The fibrous capsule is much thicker at the upper and fore part of the joint, where the greatest amount of resistance is required, particularly in standing; behind and below, it is thin and only loosely connected to the bone. It consists of two sets of fibres, circular and longitudinal. The circular fibres (*zona orbicularis*) are the deeper (4.63A, B) and form a collar or ring round the neck of the femur; although partially blended with the pubofemoral and ischiofemoral ligaments, these fibres have no direct attachment to bone. The longitudinal fibres are greatest in number at the upper and front part of the capsule, where they are reinforced by the *iliofemoral ligament*. The articular capsule is also strengthened by the *pubofemoral* and the *ischiofemoral ligaments*. The external surface of the capsule is rough, covered by numerous muscles, and separated in front from the psoas major and iliacus by a bursa.

The synovial membrane is extensive. Commencing at the margin of the articular cartilage of the head of the femur, it covers

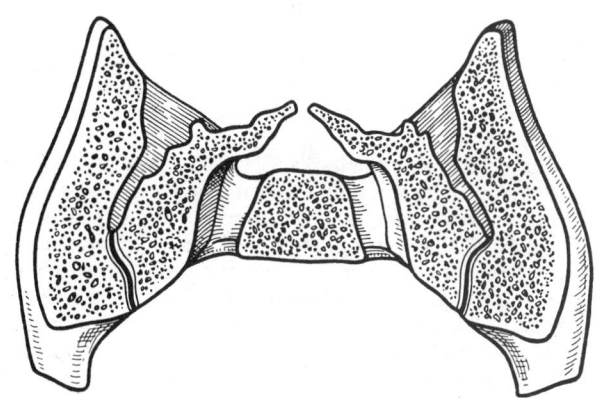

4.60 A coronal section passing through the middle of the sacro-iliac joints. The sacrum is in its normal position with the body in the erect attitude.

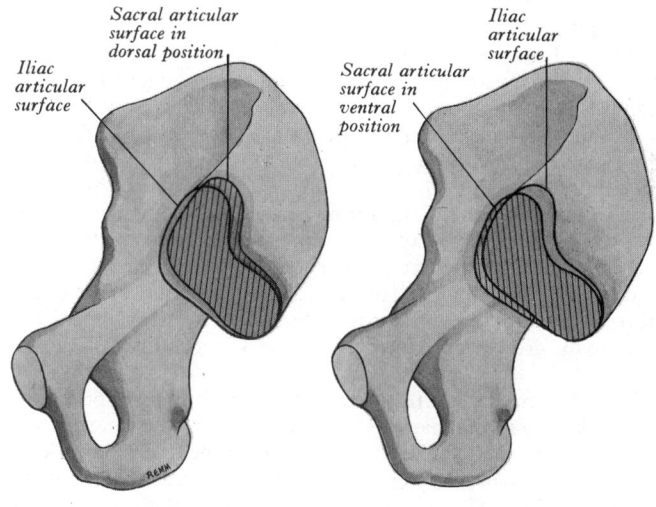

Recumbent   *Erect*

4.61 This illustrates the changing relation (rotation) of the auricular surface of the sacrum and that of the ilium when changing from a recumbent to an erect posture. (Based upon work by Dr. H. Weisl, 1953, with kind permission of the author.)

covered with articular cartilage, except over the small, roughened pit to which the ligament of the head is attached (4.62A). In front, the cartilage extends laterally to cover a small area on the adjoining part of the neck of the femur; it is thickest at the centre of the head and thinner towards its periphery. Kurrat and Oberländer (1978) have measured the thicknesses of the acetabular and femoral articular cartilages in 10 human joints. They find the maximal thickness in the acetabulum to be in its ventrocranial quadrant, whereas that on the femoral head is ventrolateral (4.62C, D). The articular surface of the acetabulum forms an incomplete ring, termed the *lunate surface*, broadest at its upper part where the pressure of the body weight falls in the erect attitude and narrowest where it covers the pubic constituent. It is deficient below opposite the acetabular notch (4.62B). It is covered with articular cartilage which is thickest where the lunate surface is broadest; the floor of the acetabular fossa within this surface is devoid of articular cartilage, lodging a fibroelastic mass of fat largely covered with synovial membrane. The depth of the acetabulum is appreciably increased by a fibrocartilaginous rim, the *acetabular labrum*. The ligamentous

the portion of the neck which is contained within the joint capsule; from the neck it is reflected on the internal surface of the capsule, covers both surfaces of the acetabular labrum, ensheathes the ligament of the head of the femur, and covers the mass of fat contained in the acetabular fossa. It is very thin over the deep surface of that part of the iliofemoral ligament which is compressed against the head of the femur in the erect attitude and has even been described as absent in this situation. The interior of the joint communicates sometimes with the subtendinous iliac bursa (p. 594) beneath the psoas major tendon, through a circular aperture between the pubofemoral ligament and the vertical band of the iliofemoral ligament.

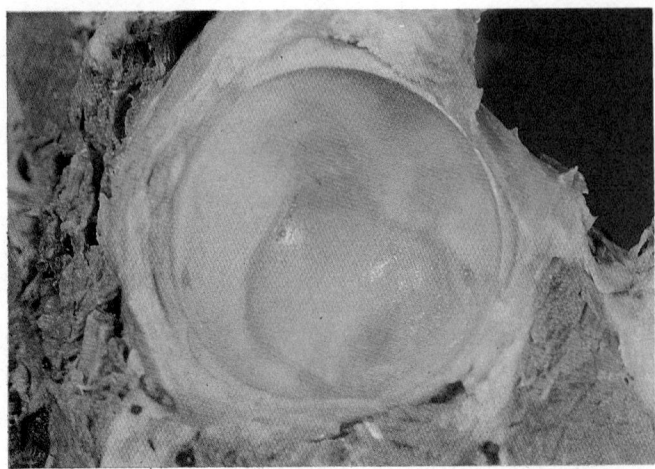 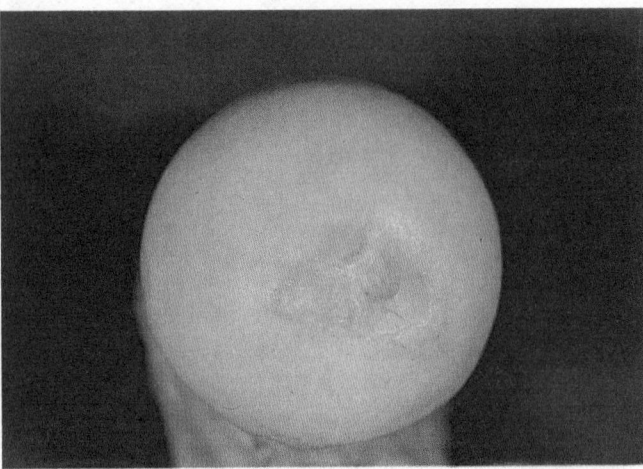

4.62A and B The articular surfaces of a cadaveric hip joint. A shows the acetabulum, comprising the articular cartilage-covered lunate surface, the acetabular fossa with its articular pad of fat, the acetabular labrum, the transverse acetabular ligament, and portions of the peri-articular musculature are included in the field of view. B displays the spheroidal

articular cartilage covering the femoral head; note the severed ligament of the head of the femur. (Preparation by Dr. M. C. E. Hutchinson and photography by Mr. Kevin Fitzpatrick, Guy's Hospital Medical School, London.)

The iliofemoral ligament (4.64), triangular in shape and of great strength, lies in front of the joint and is intimately blended with the capsule. Its apex is attached to the lower part of the anterior inferior iliac spine, its base to the trochanteric line of the femur. The medial and lateral parts of the ligament are strong bands, while the central part is relatively thin and weak; the medial band is vertical in direction and is fixed to the lower part of the trochanteric line; the lateral band is oblique and is attached to the tubercle at the upper part of the same line. The iliofemoral ligament is frequently called the Y-shaped ligament, from its likeness to the inverted letter Y. Its lateral band is sometimes termed the *iliotrochanteric ligament*.

The pubofemoral ligament (4.64) is also triangular, with its base at the innominate bone, where it is attached to the iliopectineal eminence, the superior ramus of the os pubis, the obturator crest and obturator membrane; below, it blends with the capsule and with the deep surface of the medial band of the iliofemoral ligament.

The ischiofemoral ligament (4.65) has a somewhat spiral disposition on the back of the joint. From its attachment to the ischium below and behind the acetabulum it is directed upwards and laterally over the back of the neck of the femur. Some of its fibres are continuous with those of the zona orbicularis, others are fixed to the greater trochanter anterior and deep to the iliofemoral ligament.

The ligament of the head of the femur (4.66) is a triangular, somewhat flattened band implanted by its apex on the anterosuperior part of the pit on the head of the femur; its base is attached by two bands, one into each side of the acetabular notch, and between these bony attachments it blends with the transverse ligament. It is ensheathed by synovial membrane, and varies greatly in strength in different subjects; occasionally only its synovial sheath exists, and in rare cases even this is absent. From these attachments it can be concluded that the ligament is tense when the thigh is semiflexed and then adducted and relaxed when the limb is abducted.

The acetabular labrum (4.58, 62) is a fibrocartilaginous rim attached to the margin of the acetabulum, the cavity of which it deepens. It bridges the acetabular notch as the *transverse ligament of the acetabulum*, and thus forms a complete circle. It is triangular on cross-section; the base is attached to the edge of the acetabulum, the apex being the free margin of the labrum. The rim of the acetabular cavity is constricted by the free edge of the labrum, which is turned in and embraces the head of the femur closely.

The transverse ligament of the acetabulum (4.58) is in reality a portion of the acetabular labrum, though differing from it in having no cartilage cells among its fibres. It consists of strong, flattened fibres, which cross the acetabular notch, and convert it into a foramen through which vessels and nerves enter the joint.

Relations of the hip joint. The capsule is surrounded by muscles on all sides (4.67). *Anteriorly*, the lateral fibres of the pectineus intervene between the most medial part of the capsule and the femoral vein. Lateral to the pectineus the tendon of the psoas major, with the iliacus on its lateral side, runs downwards across the front of the capsule, partly separated from it by a bursa. Here the femoral artery is on the psoas tendon and the femoral nerve lies deeply in the groove between the tendon and the iliacus. More laterally the straight head of the rectus femoris crosses the joint and, under its lateral border, the deep layer of the iliotibial tract blends with the fibrous capsule. *Superiorly*, the reflected head of the rectus femoris is in contact with the medial part of the capsule; the gluteus minimus covers the lateral part and is closely adherent to it. *Inferiorly*, the lateral fibres of the pectineus lie on the capsule as they incline backwards and, more posteriorly, the obturator externus crosses obliquely to gain the posterior aspect of the joint. *Posteriorly*, the lower part of the capsule is covered with the tendon of the obturator externus, which separates it from the quadratus femoris and is accompanied by the ascending branch of the medial circumflex femoral artery. Above that the tendon of the obturator internus and the two gemelli are in contact with the joint and intervene between it and the sciatic nerve. The nerve to the quadratus femoris lies deep to the obturator internus tendon and descends on the most medial part of the capsule. The uppermost part of the posterior surface of the articular capsule is crossed by the piriformis.

The *arteries* supplying the joint are derived from the obturator, medial circumflex femoral, and superior and inferior gluteal arteries.

The *nerves* are derived from the femoral, directly or through its muscular branches, the obturator, accessory obturator, nerve to quadratus femoris and superior gluteal (Gardner 1948).

Movements. The *active movements* of the hip joint can be categorized as flexion-extension and adduction-abduction, a combination of the foregoing—circumduction, and medial and lateral rotation. These various movements are conveniently considered as rotations occurring around three mutually perpendicular axes (p. 434), but when movements of the whole femur are being considered in relation to the accompanying movements of the articular surfaces, due regard must be paid to the effect of the length and angulation of the neck of the femur in relation to the long axis of its shaft. Thus, when the thigh is flexed or extended, the head of the femur rotates ('spins') within the acetabulum around an approximately transverse axis, and of course conversely, the acetabula rotate around similar axes during flexion and extension of the trunk and pelvis on relatively stationary femoral heads. Medial and lateral rotation of the femur occurs around a vertical axis which passes through the centre of the head of the femur and its lateral condyle, when the foot is *in firm contact with the ground*. Such rotations are an inevitable

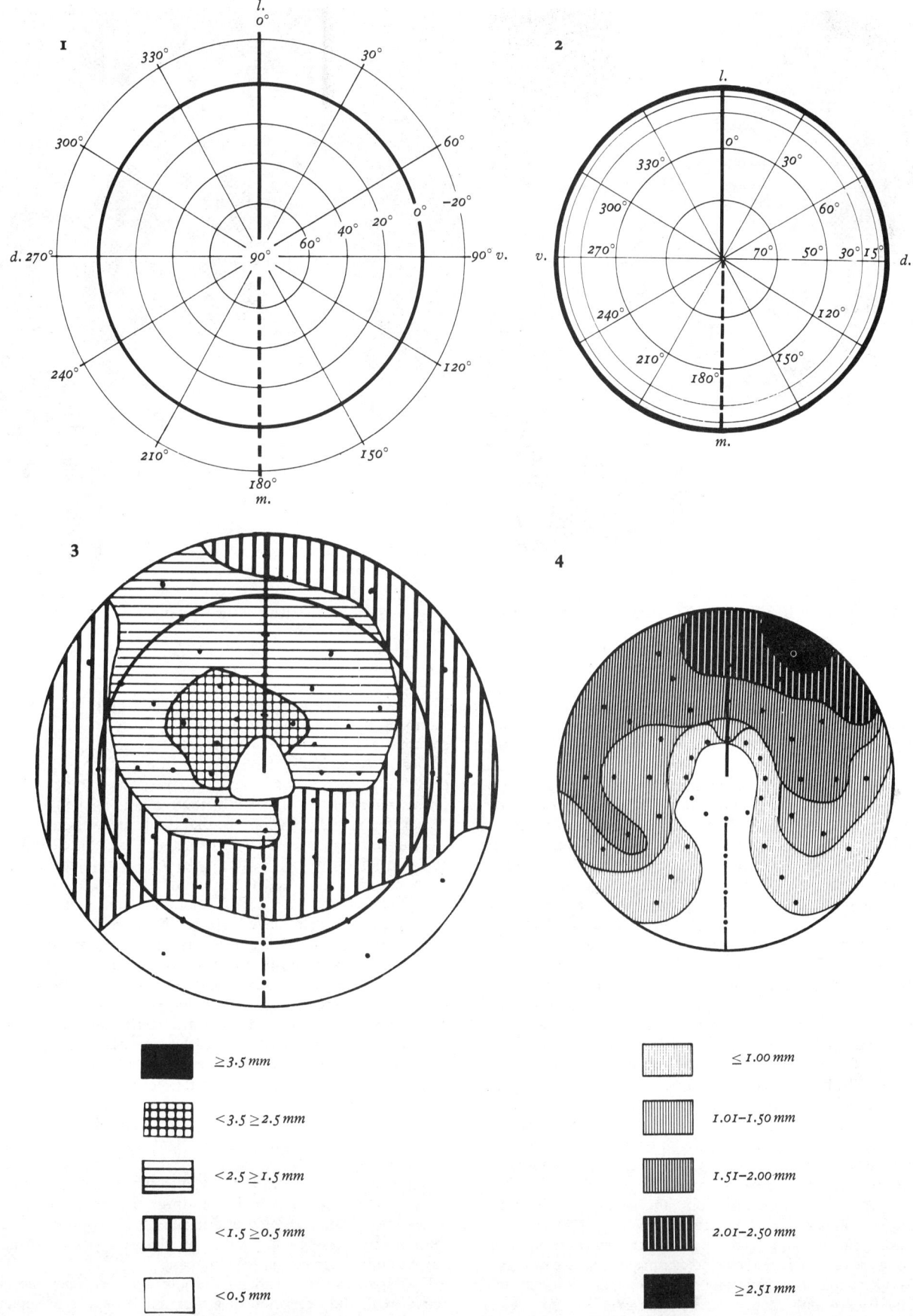

4.62c Diagrams showing variations in the thickness of the articular cartilages of the femoral head (1 and 3) and the lunate surface of the acetabulum (2 and 4). Diagrams 1 and 2 are reference grids by which the distance and angular direction of sampling points are measured from the centres of the femoral head (2) and of the acetabulum (4). Diagrams 3 and 4 show the average contours of the thickness ranges indicated in the shading codes included below. Dorsal and ventral aspects are denoted as 'd' and 'v'; 'l' and 'm' denote superolateral and inferomedial points on the circumference of the femoral articular surface. The black dots indicate the intersections of lines of 'longitude' and 'latitude' and also represent sampling points. It is immediately apparent that this method does not provide a detailed representation of the graded topology of the articular surfaces (*see* p. 432 for an alternative methodology, recently introduced). (Kindly provided from the paper by Drs. H. J. Kurrat and W. Oberländer, *Journal of Anatomy* 1978, by courtesy of Cambridge University Press.)

479

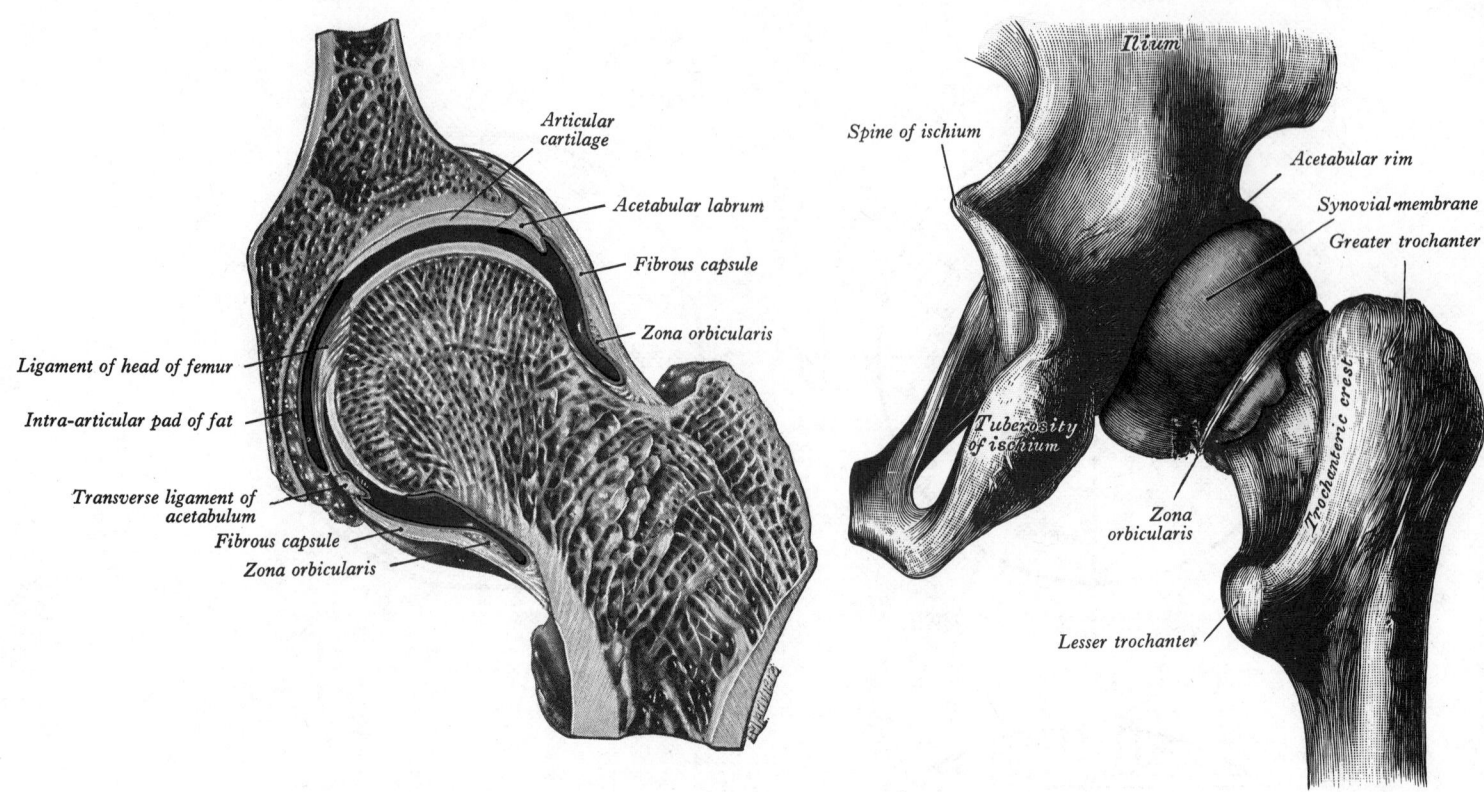

4.63A A section through the hip joint. The synovial membrane is shown in blue.

4.63B The synovial cavity of the right hip joint (distended). Posterior aspect.

accompaniment of the conjunct rotation which occurs during the terminal stages of extension, or initial stages of flexion, at the knee joint (pp. 434–437). It should be noted that because of the relation of the axis to the remainder of the femur (i.e. the angulated neck and oblique position of its shaft), during such a medial rotation, the medial condyle of the femur moves in an arc *backwards* in relation to the medial tibial condyle, whereas the greater trochanter is simultaneously moving *forwards* along an arc: obviously, converse movements occur during lateral rotation. With the foot in loose contact with, or free of the ground, however, medial and lateral rotation of the whole lower limb may occur around variable axes, all of which pass through the femoral head above but through any part of the foot below. Conversely, with one foot firmly on the ground, and the opposite foot free, the whole trunk and pelvis may rotate about a single femoral head— well instanced by the cross-kick of the footballer. The movements of abduction and adduction occur around an anteroposterior axis which passes through the centre of the head of the femur. As the head of the femur is not truly spherical none of these axes is stationary (Walmsley 1928). When considering hip joint movements, some kinesiologists prefer to consider them in relation to a *mechanical axis* coincident with the long axis of the femoral neck and impinging on the approximate centre of the articular surface of its head (p. 433). In such an analysis it is seen that the movements of extension and flexion of the thigh are relatively pure 'spins' at the hip joint and that they will be most effective in spiralizing and tautening, or in straightening and relaxing the capsule and ligaments, respectively (*vide infra*). All remaining movements are regarded as pure or impure swings (p. 434). While such a simplification permits an analysis much closer to actual functioning than a cardinal tri-axial system, related arbitrarily to the anatomical position, it must be understood that the mechanical axis itself is dynamic, and that 'spin' may occur in many positions of the joint. It makes an angle of about 125° with the long axis of the femoral shaft, the actual angle varying, of course, with age and sex.

Simple flexion of the hip joint is possible to a forward angulation of about 90 to 100° from the *vertical*; extension posteriorly from the *vertical* is highly limited (perhaps 10 to 20°).

Both movements are given much greater amplitude by adjustments of the spinal column and pelvis, by flexion of the knee and concomitant medial or lateral rotation at the hip. For example, flexion of the knee (by lessening tension in the posterior femoral muscles) increases coxal flexion to 120°, and the thigh can be drawn passively to the trunk, though this also involves some spinal flexion. Extension of the thigh in walking, running, dancing, etc., is increased by forward bending of the trunk, tilting and rotation of the pelvis, and by lateral rotation at the hip joint. (For analyses of these and other hip movements, consult Kapandji 1965, 1974; and Joseph 1975.) Abduction and, to a lesser extent, adduction can be similarly increased in range.

The hip joint presents a very striking contrast to the shoulder joint as regards the limitation of its movements. In the shoulder, as has been seen, the head of the humerus has an articular surface which much exceeds that of the glenoid cavity, and its ordinary movements are restrained but little by the capsule. In the hip joint, on the contrary, the head of the femur is fairly closely fitted to the acetabulum for an area extending over nearly half a sphere, and at the margin of the bony cup it is embraced still more closely by the acetabular labrum, so that the head of the femur is held in place by that ligament even when the fibres of the capsule have been divided. The iliofemoral ligament is the strongest of all the ligaments in the body, and is progressively spiralized and tautened, when the femur in extension reaches the line of the trunk as in symmetrical standing. In this position the pubofemoral and ischiofemoral ligaments also tighten up, and as the joint approaches its position of close-packing, resistance to any extending torque rapidly increases.

Owing to the structure of the joint, no *accessory movements* occur, with the exception of a very small degree of separation which can be effected by strong traction.

Muscles producing the movements.

Flexion. Psoas major and iliacus, assisted by pectineus, rectus femoris and sartorius. Adductors, particularly adductor longus, participate in the movement, especially in its early stages.

Extension. gluteus maximus and hamstring muscles. In the erect posture a vertical line passing through the centre of gravity of the trunk falls just behind the line joining the centres of the

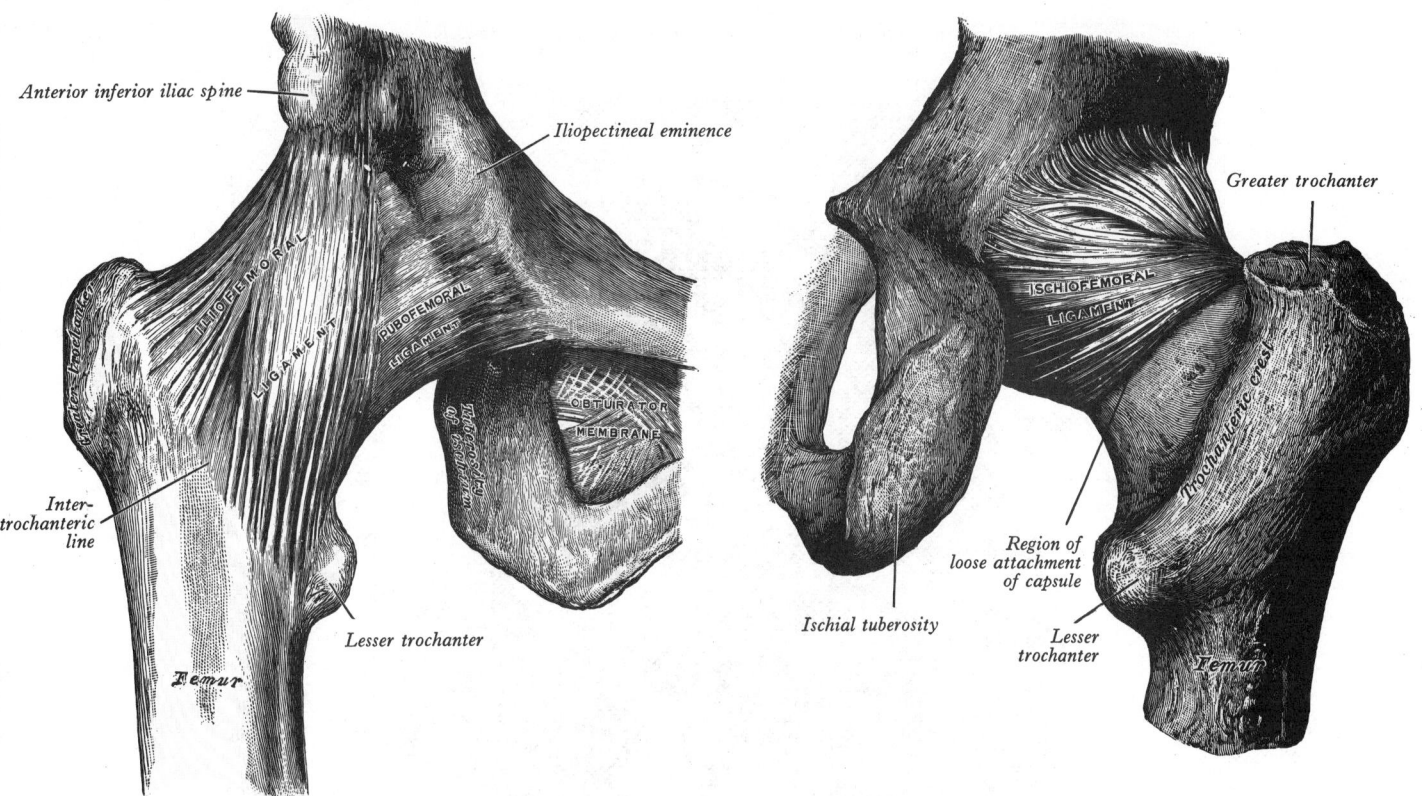

Anterior inferior iliac spine

Iliopectineal eminence

Greater trochanter

Inter-
trochanteric
line

Lesser trochanter

Region of
loose attachment
of capsule

Ischial tuberosity

Lesser
trochanter

4.64   The right hip joint. Anterior aspect.

4.65   The right hip joint. Posterior aspect.

femoral heads and therefore the pelvis (and trunk) tends to fall backwards, but this is counterbalanced by the ligamentous tension, congruence and compression of the articular surfaces which characterize the close-packed position of the joint. Under conditions of increased loading of the trunk or in backward swaying, these passive mechanisms are assisted by contraction of the joint flexors. During forward swaying, holding the arms in front of the trunk, or forward bending at the hip joints, the line of body weight is carried forwards in front of the transverse axis and the posture adopted, or rate of change of posture, is largely controlled by contraction of the hamstrings. The gluteus maximus is active more particularly when the thigh is extended against resistance, as in rising from a bending or sitting position, or when walking upstairs.

Abduction. Glutei medius et minimus (p. 601) assisted by tensor fasciae latae and sartorius. Abduction is a relatively free movement and it is limited by the tension of the adductor muscles, the pubofemoral ligament and the medial band of the iliofemoral ligament. These muscles are most constantly involved during walking or running (p. 617).

Adduction. Adductores longus, brevis et magnus, assisted by pectineus and gracilis. Adduction is limited by contact with the opposite limb, but a wider range of movement can be obtained when the thigh is flexed. Adduction of the flexed thigh is limited by the tension of the abductor muscles, the lateral band of the iliofemoral ligament and the ligament of the head of the femur.

Medial rotation. Tensor fasciae latae and anterior fibres of the glutei minimus and medius. The movement is relatively weak and is limited by the tension of the lateral rotator muscles, the ischiofemoral ligament and the posterior part of the fibrous capsule. Electromyographical data suggest that the adductors also assist in medial (rather than lateral) rotation, but this is dependent upon the primary position.

Lateral rotation. The obturator muscles, gemelli and quadratus femoris, assisted by piriformis, gluteus maximus and sartorius. The movement is powerful and is limited by the tension of the medial rotator muscles and by the lateral band of the iliofemoral ligament.

Applied Anatomy. The iliofemoral ligament is rarely torn in

dislocations of the hip, and this fact is taken advantage of by the surgeon in reducing these uncommon dislocations. It is made to act as the fulcrum to a lever, of which the long arm is the body of the femur, and the short arm the neck of the bone.

Congenital dislocation is met with more commonly in the hip joint than in any other articulation. The displacement usually

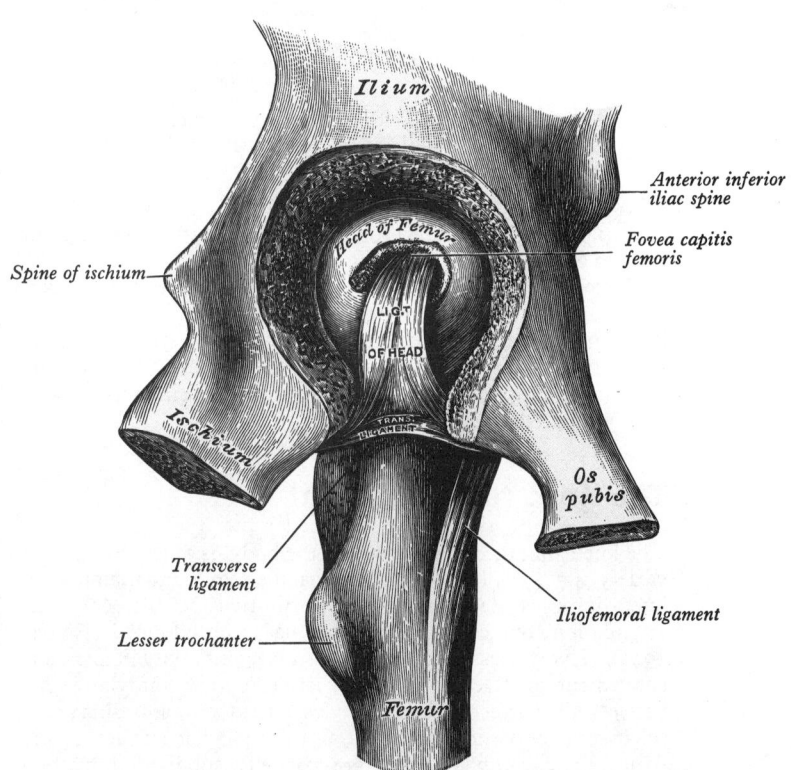

Ilium

Anterior inferior
iliac spine

Fovea capitis
femoris

Head of Femur

Spine of ischium

LIG.
OF HEAD

Ischium

TRANS.
LIGAMENT

Os
pubis

Transverse
ligament

Iliofemoral ligament

Lesser trochanter

Femur

4.66   The left hip joint, opened by the removal of the floor of the acetabulum from within the pelvis.

481

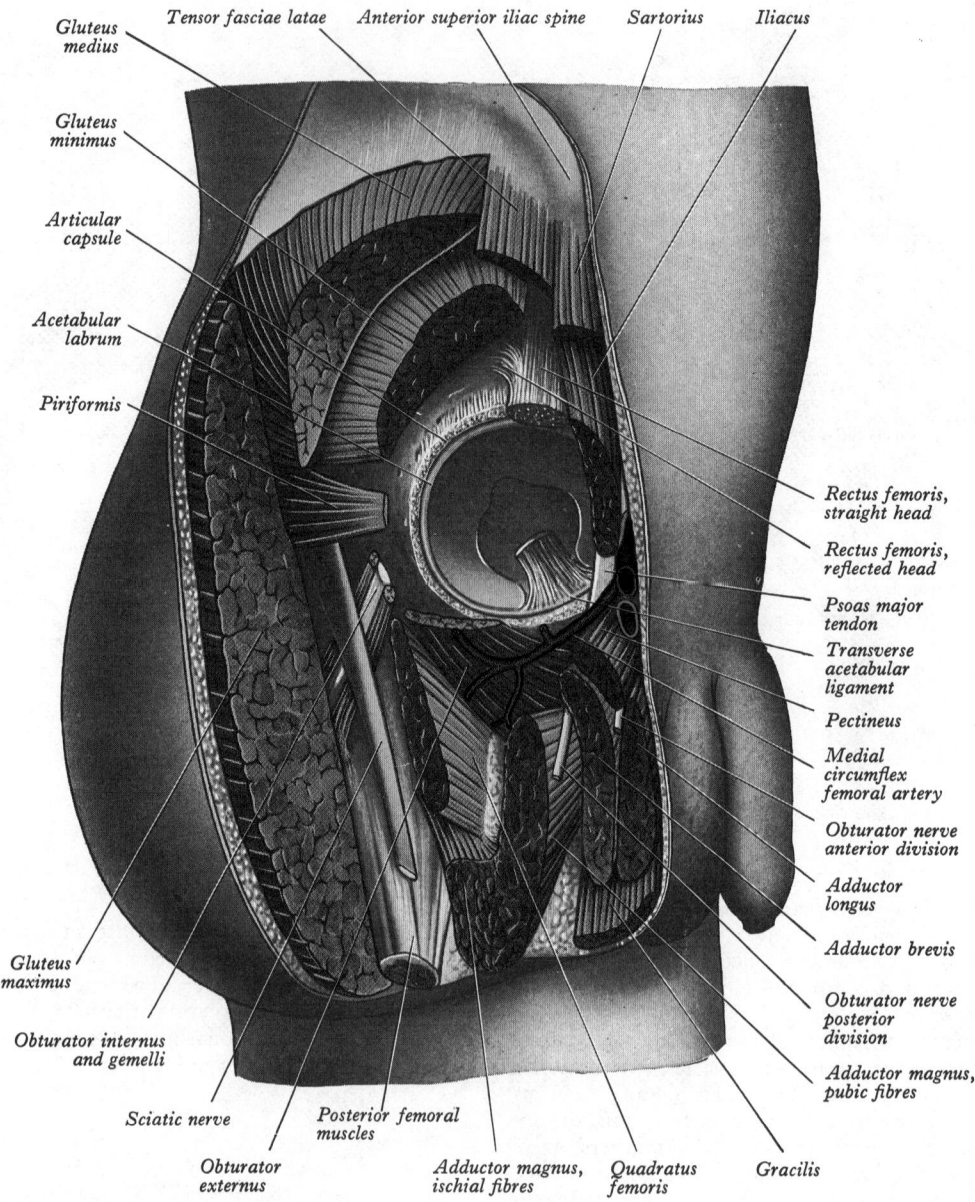

Gluteus medius *Tensor fasciae latae* *Anterior superior iliac spine* *Sartorius* *Iliacus*

Gluteus minimus

Articular capsule

Acetabular labrum

Piriformis

Rectus femoris, straight head

Rectus femoris, reflected head

Psoas major tendon

Transverse acetabular ligament

Pectineus

Medial circumflex femoral artery

Obturator nerve anterior division

Adductor longus

Adductor brevis

Obturator nerve posterior division

Adductor magnus, pubic fibres

Gluteus maximus

Obturator internus and gemelli

Sciatic nerve

Obturator externus

Posterior femoral muscles

Adductor magnus, ischial fibres

Quadratus femoris

Gracilis

4.67 A dissection to display the structures surrounding the right hip joint. The head of the femur has been disarticulated and removed.

takes place on to the gluteal surface of the ilium, the upper part of the rim of the acetabulum being deficient.

When manipulating the sacro-iliac joint the surgeon takes advantage of the fact that the iliofemoral and ischiofemoral ligaments are taut in extension of the hip joint. So strong are these ligaments that forcible attempts to produce hyperextension of the hip joint accompanied by forward pressure on the iliac crest result in movement at the sacro-iliac joint.

The Knee (Genual) Joint

The knee joint (**4.**4), the largest in the body, is of the compound variety (**4.**77, 78). It has been held that the compound mammalian knee joint is derived from a primitive double condylar articulation, but contrary evidence has been adduced (Haines 1942). Nevertheless, and despite its single joint cavity in man, it is convenient to describe it as consisting of two condylar joints between the corresponding condyles of the femur and tibia and a sellar joint between the patella and the patellar surface of the femur. The condylar joints are partially subdivided by two menisci between the corresponding articular surfaces. The level of the joint corresponds to the upper margins of the tibial condyles (**4.**77).

The articular surfaces (pp. 394, 397, 399) are obviously not congruent. The femoral condyles are convex from side to side and from before backwards; in profile they are both spiral in shape with the curvature greatly accentuated posteriorly (**3.**187), but the lateral condyle flattens from back to front more rapidly than the medial one. The tibial articular surfaces are the cartilage-covered areas of the condyles which are separated by the intercondylar area and each surface is gently hollowed out centrally and flattened peripherally where it is covered by the corresponding meniscus. The articular surface on the lateral tibial condyle is almost circular, and smaller than that on the medial condyle, which is oval, with its long axis lying anteroposteriorly. Both tibial articular surfaces are somewhat raised where they border the intercondylar area and pass on to the intercondylar eminences. Posteriorly the lateral tibial surface is prolonged over the back of the condyle in relation to the popliteus tendon, whilst anteriorly, the surface adjacent to the anterior horn of the lateral meniscus passes for a short distance on to the anterior sloping surface of the condyle (*vide infra*). The opposing femorotibial surfaces are adapted to one another more closely by the menisci, which are so shaped that they increase the concavity of the tibial surfaces, but the combined tibiomeniscal upper surface of the lateral condyle has a deeper concavity than the medial one. The articular surface of the lateral femoral condyle is marked in front

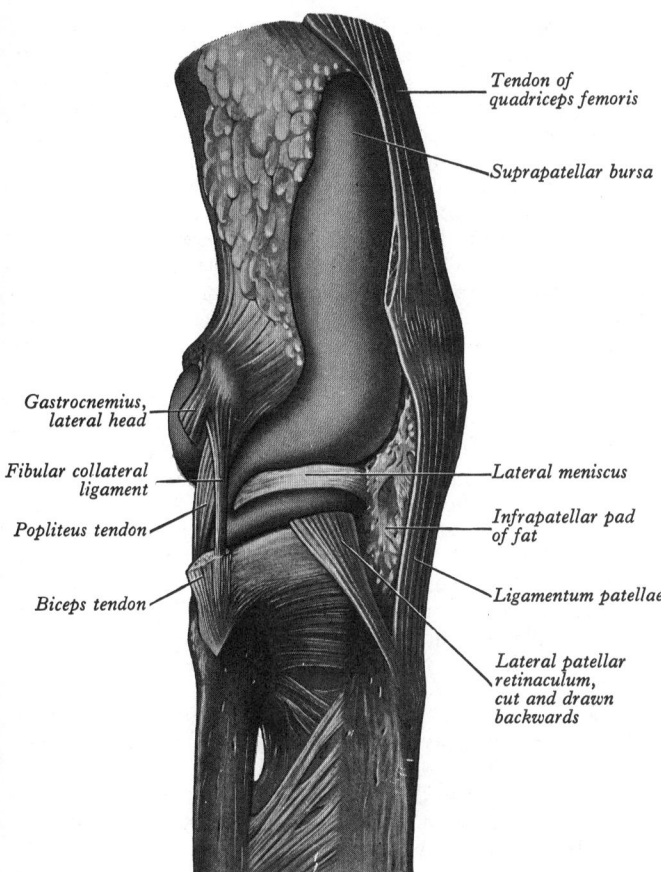

Tendon of
quadriceps femoris

Suprapatellar bursa

Gastrocnemius,
lateral head

Fibular collateral
ligament

Popliteus tendon

Biceps tendon

Lateral meniscus

Infrapatellar pad
of fat

Ligamentum patellae

Lateral patellar
retinaculum,
cut and drawn
backwards

4.68 A dissection of the right knee joint. Lateral aspect. The joint cavity
has been injected and the synovial membrane is coloured blue.

by a faint groove (**3**.186) which rests on the peripheral border of
the lateral meniscus in full extension of the joint. A similar groove
marks the articular surface of the medial condyle, but it does not
reach the lateral border of the condyle, where a narrow strip is
marked off which comes into contact with the medial part of the
patellar articular surface in full flexion of the knee. Thus the
grooves just described demarcate the patellar surface of the femur

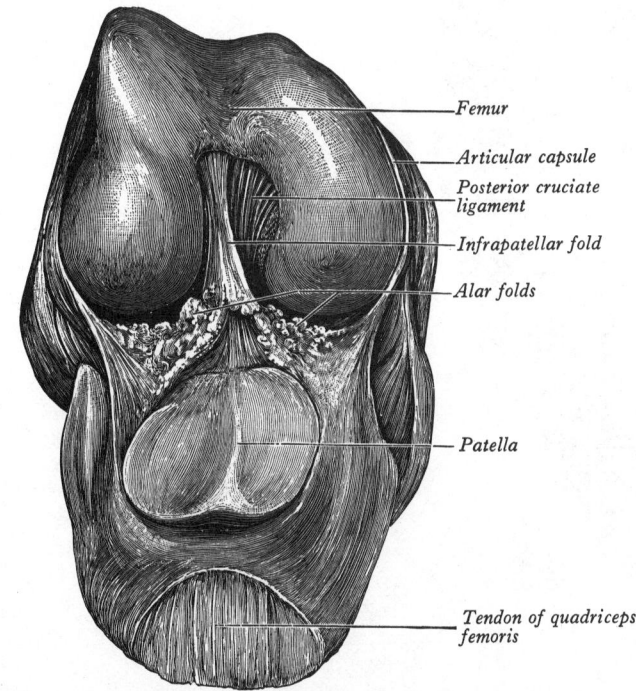

Femur

Articular capsule

Posterior cruciate
ligament

Infrapatellar fold

Alar folds

Patella

Tendon of quadriceps
femoris

4.69 The right knee joint. Anterior exposure.

from its condylar surfaces which articulate exclusively with the
tibial surfaces and menisci. When seen from below the *outline* of
the femoral (**4**.69) surfaces conform in a general way to the shape
of the tibial articular surfaces. That on the lateral femoral condyle
is almost circular in outline whereas the medial, larger, surface is
somewhat oval, elongated anteroposteriorly, and curved with its
concavity facing laterally. These differences in shape, of course,
correlate well with the movements of the joint (*vide infra*). The
joint surfaces approach full congruence in full extension, the
position of close-packing. This has been confirmed by casting
techniques, and by radiological studies of excised human knee-
joints. (*Vide infra* for a more detailed discussion of close-packing
in the knee joint. Consult Kettlekamp *et al.* 1972 for a relevant
bibliography.)

The articular surface of the patella is adapted in a general way to
the patellar surface of the femur (p. 397). The latter covers the
anterior surfaces of both condyles and its outline is like an
inverted U. An oblique groove, which passes downwards and a
little laterally, divides the patellar surface of the femur into a more
extensive lateral and a smaller medial area. The lateral part is
wider, passes more steeply on to the prominent anterior boss of
the lateral condyle, and ascends higher on its anterior surface.
Since the whole patellar surface of the femur is concave
transversely and convex in a parasagittal plane it can be regarded
as an asymmetric saddle-shaped or sellar surface (p. 432). The
rounded obliquely vertical ridge which divides the articular
surface of the *patella* into a larger, lateral part and a smaller,
medial part fits into the corresponding groove on the femur, but
the lateral and medial parts are only imperfectly congruent with
the corresponding parts of the femur. The articular surface of the
patella may be divided still further by two faint, horizontal ridges
which, with the vertical ridge, map out three pairs of facets. In
many patellae only one horizontal ridge can be made out. It is
better marked on the lateral area, and the upper and lateral facet
differs from the others in being more deeply hollowed out. On the
medial side a second vertical ridge cuts off a narrow, elongated,
semilunar strip from the medial border of the surface. This strip
comes into contact with the lateral part of the anterior end of the
medial femoral condyle in full flexion, and in that position of the
joint the uppermost lateral facet on the patella is in contact with
the anterior part of the lateral condyle. As the knee is extended the
middle facets of the patella make contact with the lower half of the
femoral patellar surface, and in full extension only the lowest
patellar facets are in contact with the femur (**4**.79). Seedham and
Tsubuku (1977) have elaborated a special technique for investigation of contact areas between visco-elastic bodies which they
have applied specifically to the patella-femoral joint. They
emphasize the temporal factor in such compressive contacts.

The ligaments of the joints are the fibrous capsule, ligamentum
(tendo) patellae, tibial and fibular collateral, oblique and arcuate
popliteal, anterior and posterior cruciate, and transverse
ligaments.

The fibrous capsule is a very complicated structure, for in
part it is deficient and in part it is augmented by strong expansions
from the tendons of the muscles which surround the joint.
Posteriorly, it consists of vertically running fibres which are
attached above to the margins of the femoral condyles and the
posterior margin of the intercondylar fossa, and below to the
posterior margins of the tibial condyles and the posterior border
of the intercondylar area. This part of the capsule is blended
above on each side with the origin of the corresponding head of
the gastrocnemius, and centrally it is strengthened by the oblique
popliteal ligament. On the medial side the fibres are attached to the
medial surfaces of the femoral and tibial condyles beyond the
articular margins. In this situation the fibrous capsule blends with
the posterior part of the tibial collateral ligament of the joint.
Between the medial epicondyle and the convex border of the
medial meniscus is a thickening of the capsule which can be
regarded as a deep component of the tibial collateral ligament
(Last 1948). On the lateral side the fibres are attached to the femur
above the origin of popliteus; they descend over the tendon to the
lateral condyle of the tibia and head of the fibula. The fibular
collateral ligament of the joint stands clear of the capsule and is
separated from it by a little fat and the inferior lateral genicular

483

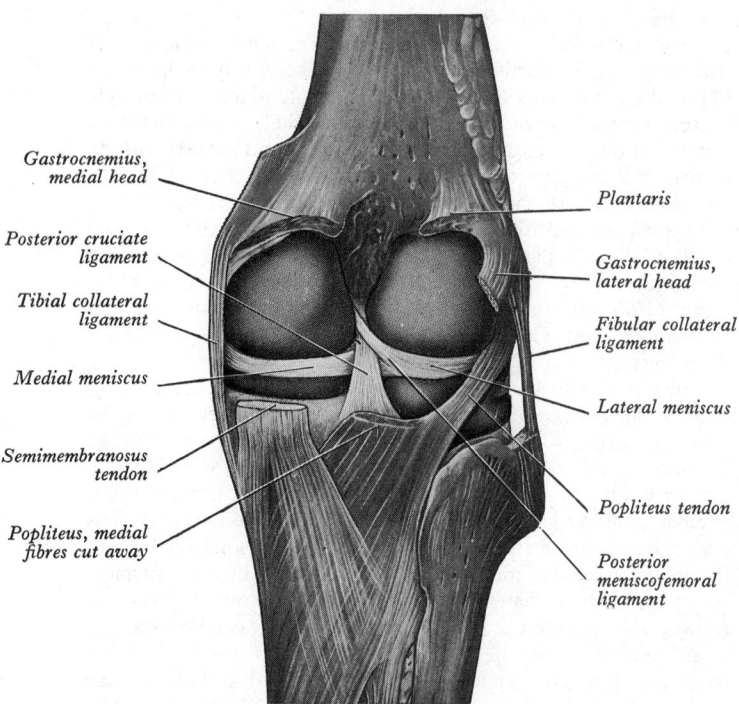

Gastrocnemius, medial head

Posterior cruciate ligament

Tibial collateral ligament

Medial meniscus

Semimembranosus tendon

Popliteus, medial fibres cut away

Plantaris

Gastrocnemius, lateral head

Fibular collateral ligament

Lateral meniscus

Popliteus tendon

Posterior meniscofemoral ligament

4.70 A posterior dissection of the right knee joint. The fibrous capsule has been removed, exposing the unopened synovial membrane, which is coloured blue. The cavity of the joint had been partially distended by injection.

vessels and nerve. Anteriorly, the fibrous capsule does not extend above the patella nor, of course, over the patellar area. Elsewhere it blends indistinguishably with expansions from the vastus medialis and vastus lateralis. The expansions are attached to the margins of the patella and ligamentum patellae and extend backwards on each side as far as the corresponding collateral ligament and downwards to the condyles of the tibia. They form the *medial* and *lateral patellar retinacula*, and the latter is further strengthened by the iliotibial tract. Above the patella the absence of the fibrous capsule allows the suprapatellar bursa (p. 598) to communicate freely with the cavity of the joint. Posteriorly, the attachment of the fibrous capsule to the posterior surface of the lateral tibial condyle is interrupted where the popliteus emerges from within the capsule (4.72). The oblique popliteal ligament, which is derived from the tendon of the semimembranosus, strengthens the posterior aspect of the capsule. Laterally, a prolongation from the iliotibial tract fills the interval between the oblique popliteal and the fibular collateral ligaments of the joint, and partly covers the latter; expansions from the tract also reach the lateral patellar retinaculum and patellar ligament. Medially, expansions from the sartorius and semimembranosus pass upwards to the tibial collateral ligament and strengthen the capsule. On its deep surface the fibrous capsule is attached to the periphery of each meniscus and connects it to the adjacent margin of the head of the tibia. This connexion is often termed the *coronary ligament*.

The synovial membrane of the knee joint is the most extensive and complex in the body. Commencing at the upper border of the patella, it forms a large pouch, termed the *suprapatellar bursa*, under cover of the quadriceps femoris on the lower part of the front of the femur (4.68, 75). This 'bursa' is preferably regarded, for practical purposes, as an extension of the joint cavity of the knee. The pouch is sustained during the movements of the knee, by a small muscle, the articularis genus,

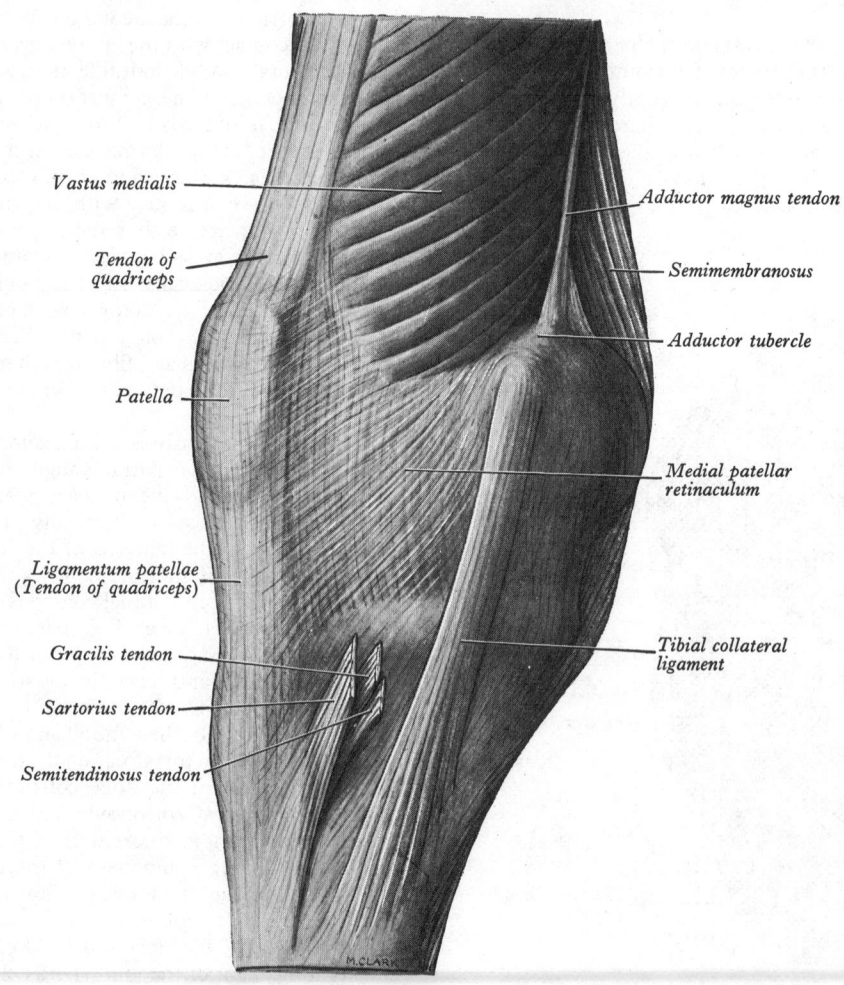

Vastus medialis

Tendon of quadriceps

Patella

Ligamentum patellae (Tendon of quadriceps)

Gracilis tendon

Sartorius tendon

Semitendinosus tendon

Adductor magnus tendon

Semimembranosus

Adductor tubercle

Medial patellar retinaculum

Tibial collateral ligament

4.71 The right knee joint. Anteromedial aspect.

which is inserted into it. On each side of the patella, the synovial membrane extends beneath the aponeuroses of the vasti, more extensively beneath that of the vastus medialis. Below the patella it is separated from the ligamentum patellae by the *infrapatellar pad of fat*. Opposite the medial and lateral borders of the lower part of the articular surface of the patella, the synovial membrane covering the infrapatellar pad is projected into the interior of the joint in the form of two fringe-like *alar folds*; behind, these folds converge and are continued as a single band, the *infrapatellar fold* (or *ligamentum mucosum*), to the front of the intercondylar fossa of the femur (4.69). This fold is often regarded as a vestige of the inferior boundary of an originally separate femoropatellar joint (p. 482).

At the sides of the joint the synovial membrane passes downwards from the femur, lining the fibrous capsule as far as its attachment to the menisci, the free surfaces of which possess no covering of synovial membrane. At the back of the lateral meniscus the synovial membrane forms a cul-de-sac, the *subpopliteal recess*, between the groove on the surface of the meniscus and the tendon of the popliteus (4.70). This may communicate with the superior tibiofibular joint (p. 490). The relation of the synovial membrane to the cruciate ligaments is described on p. 486.

The ligamentum patellae (4.71) is, despite its name, the central portion of the common tendon of the quadriceps femoris, which is continued from the patella to the tuberosity of the tibia. It is a strong, flat band, about 8 cm in length, attached, above, to the apex, adjoining margins and the rough depression on the lowest part of the posterior surface of the patella; and below, to the upper part of the tuberosity of the tibia. Its superficial fibres are continous over the front of the patella with those of the tendon of the quadriceps femoris. The medial and lateral portions of the tendon of the quadriceps pass down, on each side of the patella, to be inserted into the upper extremity of the tibia, on the sides of the tuberosity; they merge into the fibrous capsule, as stated above, forming the *medial* and *lateral patellar retinacula*. The posterior surface of the ligamentum patellae is separated from the synovial membrane by a large infrapatellar pad of fat, and from the tibia by a bursa (4.75).

The oblique popliteal ligament (4.75) is an expansion from the tendon of the semimembranosus close to its insertion into the tibia. It partially blends with the fibrous capsule passing upwards and laterally to be attached above to the lateral part of the intercondylar line and to the lateral condyle of the femur. It consists of fasciculi separated from one another by apertures for the passage of vessels and nerves; it forms part of the floor of the popliteal fossa and the popliteal artery is in contact with it.

The arcuate popliteal ligament (4.72) consists of a Y-shaped system of capsular fibres, the stem of which is attached to the head of the fibula. The posterior limb arches medially over the emerging tendon of popliteus to be attached to the posterior border of the intercondylar area of the tibia. The anterior limb, sometimes absent, extends to the lateral epicondyle of the femur, where it is connected with the lateral head of the gastrocnemius. This limb is often termed the *short lateral ligament*.

The tibial collateral ligament (4.71, 76), a broad, flat band, nearer to the back than the front of the joint, is attached, above, to the medial epicondyle of the femur immediately below the adductor tubercle and, below, to the medial condyle and medial surface of the shaft of the tibia. Its anterior part is a flattened band, about 10 cm long, easily distinguishable from the fibrous capsule deep to it. One or more bursae may separate it from the fibrous capsule and medial meniscus (Brantigan and Voshell 1943). It inclines forwards as it descends to the medial margin and the posterior part of the medial surface of the shaft of the tibia (3.195). It is crossed, below, by the tendons of sartorius, gracilis and semitendinosus, a bursa being interposed. Its deep surface covers the medial inferior genicular vessels and nerve, and the anterior part of the tendon of semimembranosus, to which it is connected to a slight extent. The posterior part of the ligament is fan-shaped and blends with the capsule posteriorly; it is short and inclines backwards and downwards to the medial condyle of the tibia above the groove for semimembranosus.

The fibular collateral ligament (4.73) is a strong, rounded

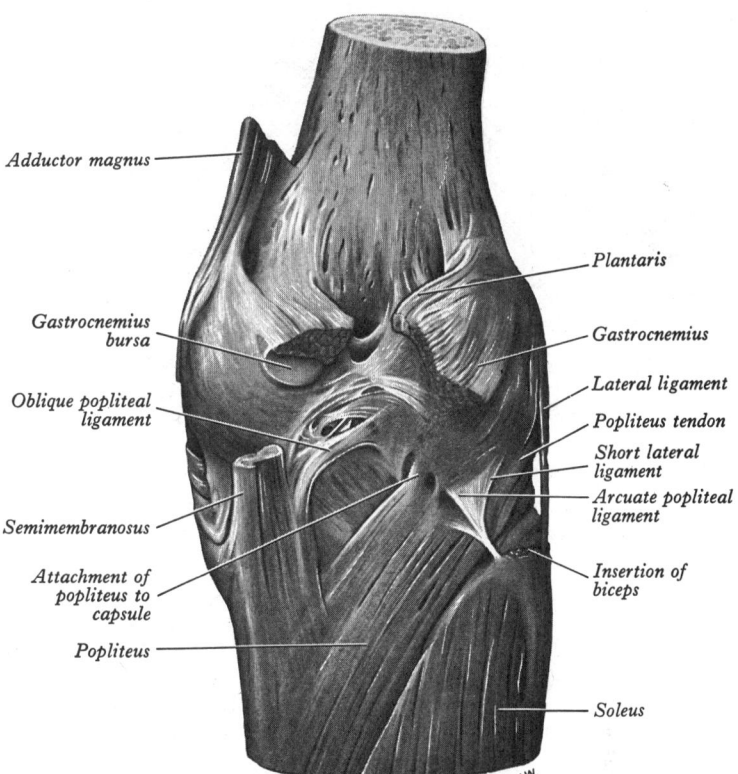

Adductor magnus

Gastrocnemius bursa

Oblique popliteal ligament

Semimembranosus

Attachment of popliteus to capsule

Popliteus

Plantaris

Gastrocnemius

Lateral ligament

Popliteus tendon

Short lateral ligament

Arcuate popliteal ligament

Insertion of biceps

Soleus

4.72 The right knee joint. Posterior aspect.

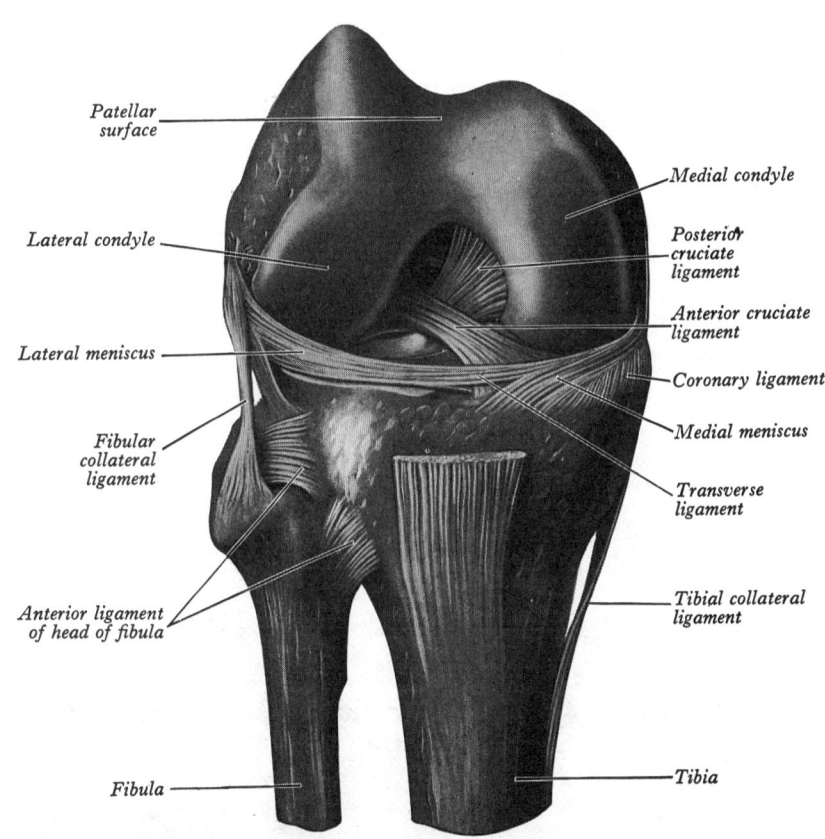

Patellar surface

Lateral condyle

Lateral meniscus

Fibular collateral ligament

Anterior ligament of head of fibula

Fibula

Medial condyle

Posterior cruciate ligament

Anterior cruciate ligament

Coronary ligament

Medial meniscus

Transverse ligament

Tibial collateral ligament

Tibia

4.73 The right knee joint in full flexion. Dissection from anterior aspect.

cord, attached above to the lateral epicondyle of the femur, immediately above the groove for the tendon of the popliteus and, below, to the head of the fibula, in front of its apex. It is largely hidden by the tendon of biceps femoris, which embraces and is partly attached to the ligament. Deep to the ligament are the tendon of popliteus and the inferior lateral genicular vessels and nerve. The ligament is *not* attached to the lateral meniscus.

485

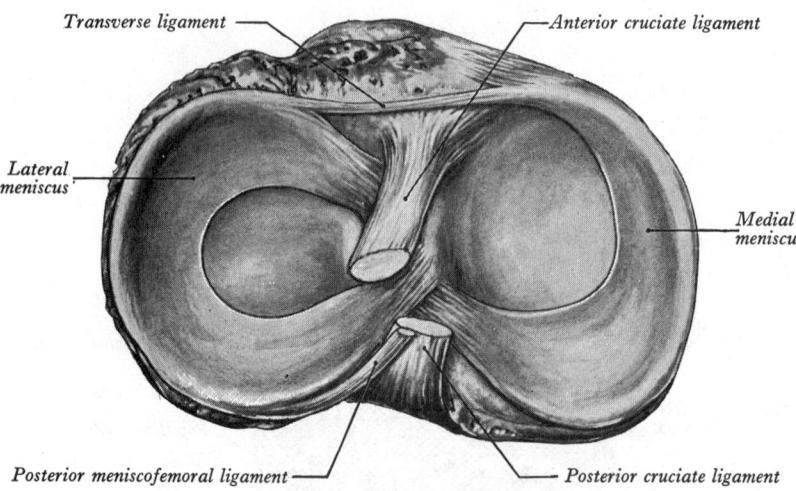

Transverse ligament — — Anterior cruciate ligament

Lateral meniscus —

Medial meniscus

Posterior meniscofemoral ligament — — Posterior cruciate ligament

4.74   The superior aspect of the left tibia, showing the menisci and the tibial attachments of the cruciate ligaments. See also **3**.194A.

The cruciate ligaments are of considerable strength, and are situated a little posterior to the centre of the joint. They are called *cruciate* because they cross each other, and *anterior* and *posterior* from the position of their attachments to the tibia. Their position within the joint suggests their identification as collateral ligaments of originally separate medial and lateral femorotibial joints (p. 482, 485).

The anterior cruciate ligament (4.74) is attached to the medial part of the anterior intercondylar area of the tibia, being partly blended with the anterior end of the lateral meniscus; it

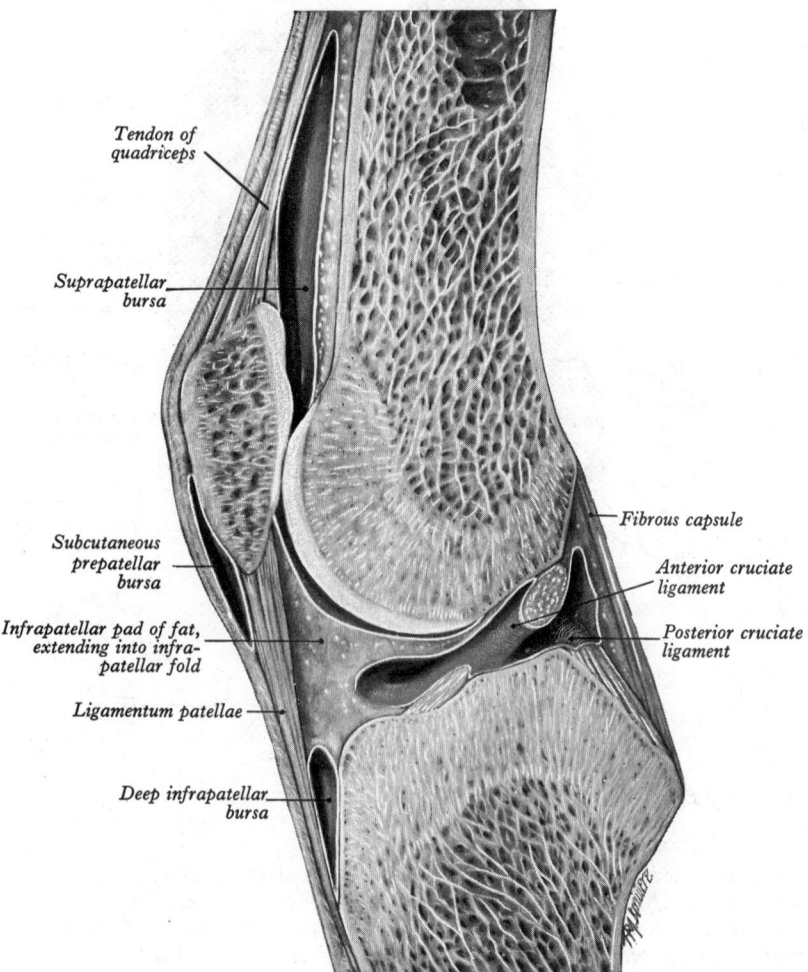

Tendon of quadriceps —

Suprapatellar bursa —

Subcutaneous prepatellar bursa —

Infrapatellar pad of fat, extending into infra- patellar fold —

Ligamentum patellae —

Deep infrapatellar bursa —

Fibrous capsule

Anterior cruciate ligament

Posterior cruciate ligament

4.75   A sagittal section through the left knee joint. Lateral aspect. The synovial membrane is shown in colour.

passes upwards, backwards and laterally, twisting on itself, and fans out to be attached to the posterior part of the medial surface of the lateral condyle of the femur (Last 1951). It lies anterolateral to the posterior cruciate ligament.

The posterior cruciate ligament (4.73, 74) is stronger, but shorter and less oblique in its direction than the anterior. It is attached to the posterior intercondylar area of the tibia, and to the posterior extremity of the lateral meniscus; it passes upwards, forwards and medially, broadening out to be attached to the lateral surface of the medial condyle of the femur.

In a recent study of the cruciate ligaments (Girgis *et al.* 1975) observations on 24 fresh knee joints suggested that each ligament has two parts—a main posterolateral band and a smaller anteromedial band, which behave differently in joint movements (*vide infra*). In each ligament some spiralization is apparent and becomes more pronounced in certain positions. Average dimensions were given: anterior cruciate length 38 mm, width 11 mm; posterior cruciate length 38 mm, width 13 mm.

Synovial membrane covers the cruciate ligaments anteriorly and on each side but posteriorly it is reflected from the sides of the posterior cruciate ligament on to the adjoining parts of the fibrous capsule. Thus the middle, or intercondylar, part of the posterior portion of the fibrous capsule is devoid of any synovial covering on its anterior surface. A bursal recess passes in between the two ligaments from their lateral aspect(4.75) and may reach the medial wall of the intercondylar fossa of the femur.

The menisci (semilunar cartilages) (4.74) are two crescentic lamellae which deepen the surfaces of the upper end of the tibia in articulation with the femoral condyles. The peripheral attached border of each meniscus is thick and convex; the free border is thin and concave. The thick peripheral zone is vascularized by capillary loops from the fibrous capsule and synovial membrane, while the inner region is avascular (Davies and Edwards 1948). The upper surfaces of the menisci are smooth and concave, and in contact with the condyles of the femur; their lower surfaces are smooth and flat, and rest upon the tibia. Each covers approximately the peripheral two-thirds of the corresponding articular surface of the tibia.

The medial meniscus is nearly semicircular in form, but is broader behind than in front (4. 74); its anterior end is attached to the anterior intercondylar area of the tibia, in front of the anterior cruciate ligament, its posterior fibres being continuous with the transverse ligament. This end or cornu of the meniscus lies in the floor of the depression on the medial side of the upper part of the ligamentum patellae. The posterior end of the meniscus is fixed to the posterior intercondylar area of the tibia, between the attachments of the lateral meniscus and the posterior cruciate ligament. Its peripheral border is attached to the fibrous capsule and is firmly adherent to the deep surface of the tibial collateral ligament (4.78).

The lateral meniscus forms about four-fifths of a complete ring (4.74) covering a larger area of articular surface than the medial. It is of the same breadth throughout its extent, and is grooved posterolaterally by the tendon of popliteus, which separates it from the fibular collateral ligament. Its anterior end is attached in front of the intercondylar eminence of the tibia, behind and lateral to the anterior cruciate ligament, with which it partly blends; the posterior end is attached behind the intercondylar eminence of the tibia, in front of the posterior end of the medial meniscus. The anterior attachment of the lateral meniscus is twisted so that its free margin looks backwards and upwards, its anterior end resting on a sloping shelf of bone on the front of the lateral intercondylar tubercle. Close to its posterior attachment it commonly sends off a strong fasciculus, the *posterior meniscofemoral ligament* (4.74), which passes upwards and medially behind the posterior cruciate ligament to the medial condyle of the femur. Another oblique band, the *anterior meniscofemoral ligament*, may connect the posterior part of the meniscus and pass to the medial condyle of the femur in front of the posterior cruciate ligament. The meniscofemoral ligaments are often the sole attachments of the posterior horn of the lateral meniscus (Hiller and Langman 1964). It is important to note that the tendon of the popliteus intervenes between the lateral meniscus and the fibular collateral ligament. The more medial part of the tendon is inserted into the lateral meniscus and thus the mobility of the posterior horn of the lateral meniscus may be controlled by meniscofemoral ligaments and popliteus (Last 1948). A distinct *meniscofibular ligament* has been described as present in about 80 per cent of knee joints (Živanović 1973).

That the menisci meet a functional need is demonstrated by the fact that they are re-formed following excision, provided that their whole breadth is removed, since regeneration can only occur from the vascular fibro-areolar tissue around their periphery. After complete excision of a meniscus and prior to re-formation, the knee joint can be used actively without any sign of instability; but if it is subjected to continued active and violent exercise, the subsequent clinical history indicates that the articular cartilage suffers permanent damage, and this has been attributed to inefficient lubrication during the period of regeneration.

The transverse ligament (4.74) connects the anterior convex margin of the lateral to the anterior end of the medial meniscus; its thickness varies considerably in different subjects, and it is sometimes absent. According to Živanović (1974) this anteriorly situated *menisco-meniscal* ligament varies much in development and is even absent in about 40 per cent of joints. A posterior menisco-meniscal ligament was observed in 20 per cent of 300 knee joints. The same authority described other rare ligamentous arrangements in this articulation.

Bursae associated with the knee joint are numerous. **Anteriorly** there are four; (1) a large *subcutaneous prepatellar bursa* is interposed between the lower part of the patella and the skin; (2) a small *deep infrapatellar bursa* between the upper part of the tibia and the ligamentum patellae; (3) a *subcutaneous infrapatellar bursa* between the lower part of the tuberosity of the tibia and the skin; and (4) one of large size, the *suprapatellar bursa* lies between the anterior surface of the lower part of the femur and the deep surface of quadriceps femoris (4.75). They develop as separate bursae in the fetus and later communicate with the joint, and better regarded as an extension of it. **Laterally** there are four bursae: (1) one (which sometimes communicates with the joint) *between* the *lateral head of gastrocnemius* and the capsule; (2) one *between* the *fibular collateral ligament* and the *tendon of biceps femoris*; (3) one *between the* same *ligament* and the *tendon of popliteus* (this is sometimes only an expansion from the next bursa); (4) one *between* the *tendon of popliteus* and the *lateral condyle* of the femur, usually an extension from the synovial membrane of the joint. **Medially**, bursae may be found in the following positions: (1) *between* the *medial head of gastrocnemius* and the capsule: this sends a prolongation between the tendon of the medial head of gastrocnemius and the tendon of semimembranosus and often communicates with the joint; (2) *superficial* to the *tibial collateral ligament*, between it and the *tendons of sartorius, gracilis and semitendinosus*; (3) *deep* to the *tibial collateral ligament* intervening in variable numbers and positions between this and the femur, capsule of the knee joint, medial meniscus, tibia or the tendon of semimembranosus; (4) *between* the *tendon of semimembranosus* and the *medial condyle* of the tibia and the medial head of gastrocnemius: this is termed the semimembranosus bursa and may communicate with (1) above; (5) occasionally between the tendons of semimembranosus and semitendinosus. **Posteriorly**, bursal extensions are variable.

Structures around the joint. *Anteriorly*, the quadriceps femoris covers the joint, and tendinous expansions from the vastus medialis and vastus lateralis extend backwards from its margins over the *anteromedial* and *anterolateral* aspects of the capsule respectively, forming the patellar retinacula. On the *posteromedial* side the sartorius, with the gracilis tendon lying along its posterior border, descends across the joint; on the *posterolateral* side the biceps tendon, with the common peroneal nerve on its medial side, is in contact with the fibrous capsule,

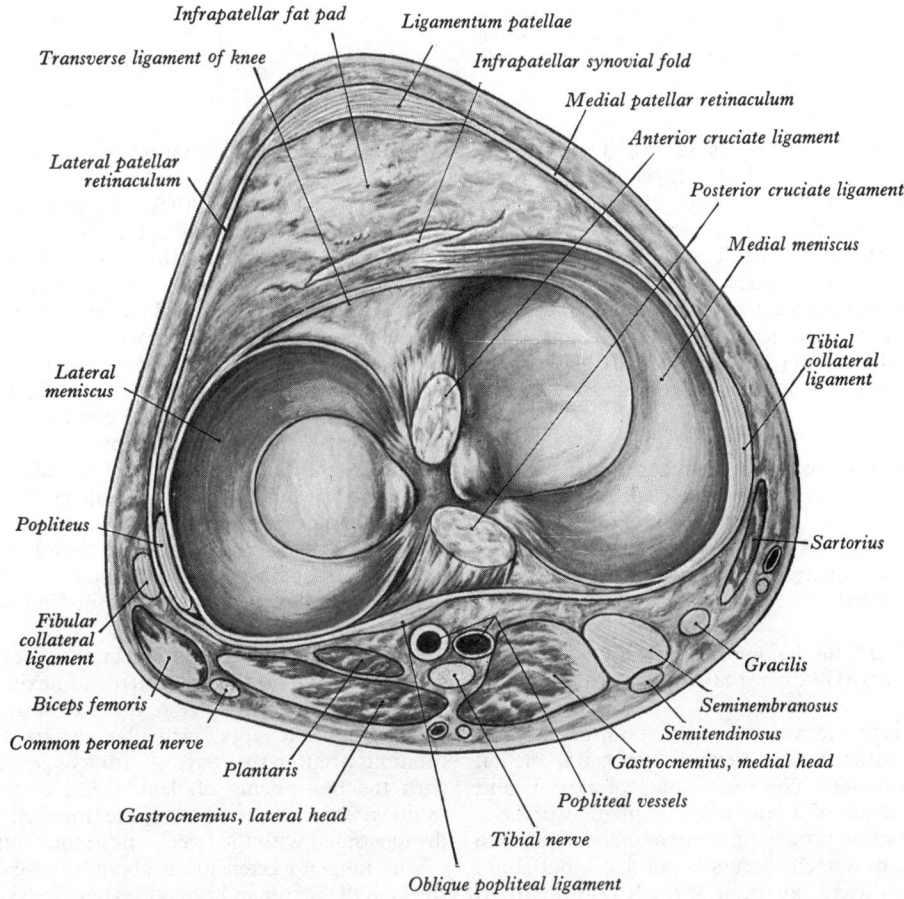

Infrapatellar fat pad
Ligamentum patellae
Transverse ligament of knee
Infrapatellar synovial fold
Medial patellar retinaculum
Lateral patellar retinaculum
Anterior cruciate ligament
Posterior cruciate ligament
Medial meniscus
Tibial collateral ligament
Lateral meniscus
Popliteus
Sartorius
Fibular collateral ligament
Gracilis
Biceps femoris
Seminembranosus
Common peroneal nerve
Semitendinosus
Gastrocnemius, medial head
Plantaris
Popliteal vessels
Gastrocnemius, lateral head
Tibial nerve
Oblique popliteal ligament

4.76 A transverse section of the left knee joint, superior aspect, to show the relations of the joint.

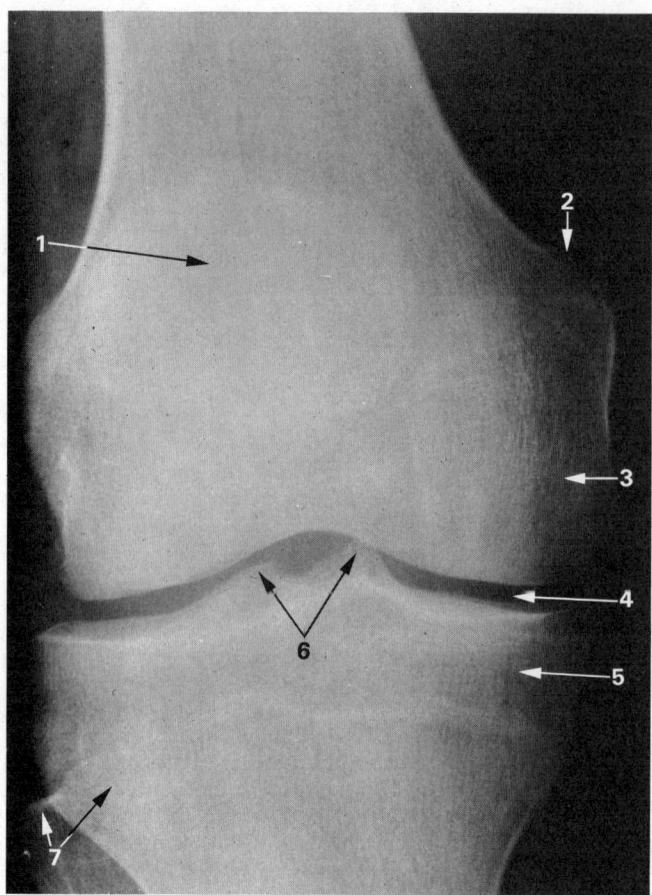
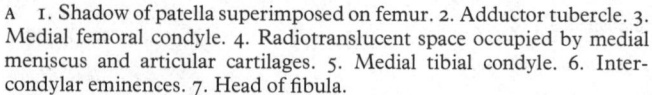
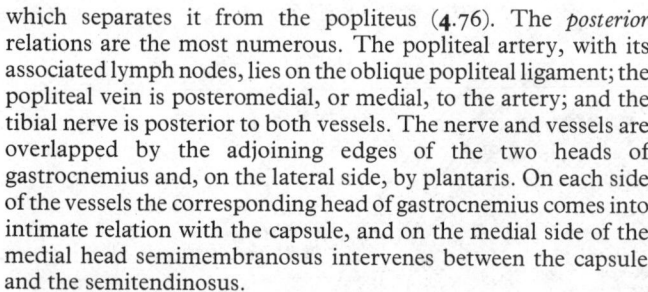

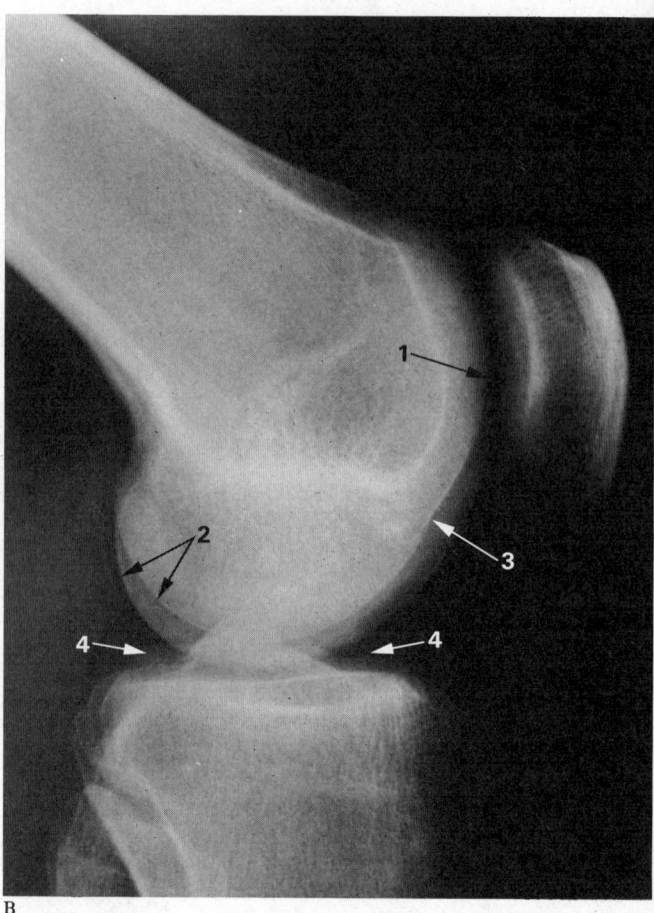

A

B

4.77A and B Anteroposterior (A) and lateral (B) radiographs of an adult knee (male of 22 years), partly flexed in B.

A 1. Shadow of patella superimposed on femur. 2. Adductor tubercle. 3. Medial femoral condyle. 4. Radiotranslucent space occupied by medial meniscus and articular cartilages. 5. Medial tibial condyle. 6. Intercondylar eminences. 7. Head of fibula.

B 1. Patellar surface of femur. 2. Spiral profiles of femoral condyles. 3. Groove impinging on anterior end of meniscus in full extension. 4. Note marked incongruity of femorotibial joint surfaces.

which separates it from the popliteus (**4.**76). The *posterior* relations are the most numerous. The popliteal artery, with its associated lymph nodes, lies on the oblique popliteal ligament; the popliteal vein is posteromedial, or medial, to the artery; and the tibial nerve is posterior to both vessels. The nerve and vessels are overlapped by the adjoining edges of the two heads of gastrocnemius and, on the lateral side, by plantaris. On each side of the vessels the corresponding head of gastrocnemius comes into intimate relation with the capsule, and on the medial side of the medial head semimembranosus intervenes between the capsule and the semitendinosus.

The *arteries* supplying the joint are the descending genicular, the genicular branches of the popliteal, the recurrent branches of the anterior tibial, and the descending branch from the lateral circumflex femoral branch of the arteria profunda femoris. For details consult Scapinelli (1968) and Wladmirow (1968); these studies indicate a more penetrative ligamentous supply than usually described, even parts of the cruciate ligaments receiving a few small vessels.

The *nerves* are derived from the obturator, femoral, tibial and common peroneal nerves (Gardner 1948; Freeman and Wyke 1967).

Movements. The active movements carried out at the knee joint are customarily described as flexion, extension, medial rotation and lateral rotation. The movements of flexion and extension differ from those of a true hinge joint in two ways. Firstly, because of the spiral profiles of the femoral condyles, the axis around which the movement occurs is not fixed, but shifts upwards and forwards during extension of the leg on the thigh and backwards and downwards during flexion. Secondly, with the foot fixed on the ground, the last 30° of extension is associated

with conjunct medial rotation of the femur, and the first few degrees of flexion with a corresponding degree of lateral rotation; being conjunct rotations they are a consequence of the geometry of the articular surfaces and the disposition of the ligamentous structures. Conversely, with the foot off the ground extension is associated with lateral rotation of the tibia and flexion with medial rotation of this bone. In full flexion the posterior parts of the tibial articular surfaces are in contact with the posterior parts of the articular surfaces of the femoral condyles. During extension the tibia and its menisci glide forwards on the femoral condyles, and the area of contact between the two bones increases and moves forwards also, carrying the menisci with it. As the movement progresses the flatter curve of the femoral condyles makes contact with the tibia, and the menisci are opened out, the net result being that their anterior ends move forwards, while their posterior ends suffer little change in position. In flexion the reverse movement occurs, so that the menisci, moving with the tibia, adapt their outline to the curve of the parts of the femoral condyles which are making contact with that bone.

The movements of rotation at the knee joint have a much smaller range than the movements of flexion and extension, and during their execution the menisci move with the femoral condyles on the upper articular surfaces of the tibia. These rotations occur in two ways — either as *conjunct rotations* integral with the movements of flexion and extension, or as *adjunct rotations* which can occur quite independently and are best demonstrated with the knee joint in the semiflexed position.

The range of extension is about 5–10° beyond a vertical axis through the femur and tibia. Flexion can be carried to about 120° with the hip joint extended, to some 140° when the latter is flexed, and to 160° when a passive element, such as sitting on the heels, is introduced. *Passive* rotation is about 60–70° in range; in contrast *conjunct* rotation between femur and tibia is limited to about 20°.

The movement of conjunct medial rotation of the femur on the tibia associated with the later stages of extension of the knee is a constituent part of a 'locking' mechanism, which is an asset when the knee, fully extended and under load, is subjected to strain. The position of full extension of the joint is of course a condition of close-packing, with maximum congruence, compression and contact area of the articular surfaces, and with maximum spiralization and tautening of the ligaments. The precise role of the various factors—geometry of articular surfaces, activity of musculature, disposition of ligaments—in the generation of the conjunct rotations has been the subject of considerable dispute (consult, for example, Barnett 1952; Kapandji 1974; Girgis *et al.* 1975), but the following points should be borne in mind. As we have seen, the lateral, upper, combined meniscotibial 'receiving surface' is smaller, more circular and more deeply concave than its medial counterpart. Additionally, the tibial surface of the lateral femoral condyle is also smaller, rounder and flattens more rapidly in an anterior direction, than the medial condyle. Consequently, the lateral femoral condyle approaches a fully congruent relationship with its apposed tibiomeniscal surface some 30° before full extension has been reached (and of course long before the medial condyle is in a similar state). A continuation of simple extension is therefore impossible, but medial rotation of the femur about a vertical axis (through its head and lateral condyle) is possible. As this occurs, the medial femoral condyle moves backwards in an arc, whilst the rotation of the lateral femoral condyle and meniscus causes the anterior horn of the meniscus to pass progressively on to the sloping anterior surface of the lateral tibial condyle. In this manner, full congruence of the lateral condyle is delayed (by deformation of its 'receiving surface') and continuing extension is possible. Of course in practice, these elements of rotation and extension continue simultaneously and smoothly until final close-packing of both condyles occurs together. At the beginning of flexion from the fully extended position (with the foot on the ground) lateral rotation of the femur occurs—often called 'unlocking' of the joint. While the conformation of the joint surfaces and disposition of ligaments are again inevitably involved in this movement, electromyographic

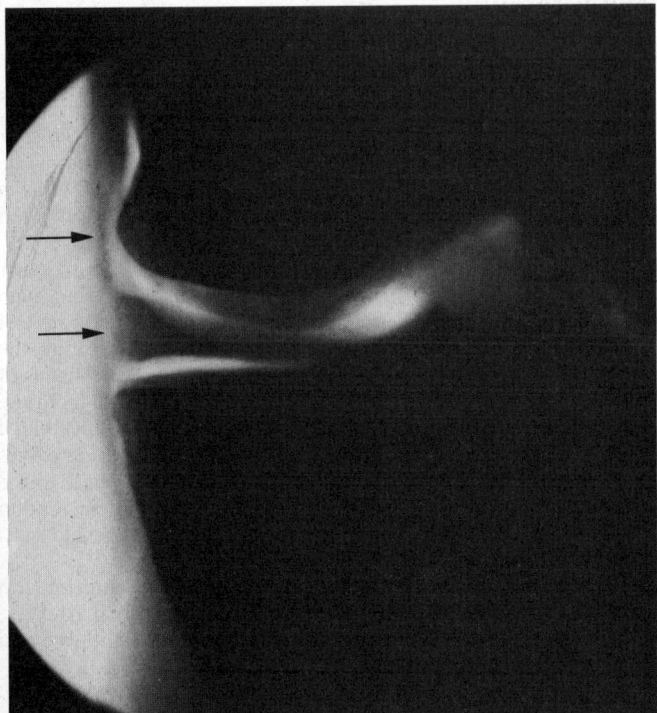

4.78   Radiograph taken after injection of air into the knee joint, showing the shadow thrown by the medial meniscus. The upper arrow points to the medial ligament and the lower to the medial meniscus. (Aerogram by Wing-Commander A. A. Butler, R.A.F.)

evidence indicates the importance of contraction of the popliteus muscle, which pulls downwards and backwards on the area of attachment of its tendon to the lateral condyle of the femur, which is of course lateral to the axis of femoral rotation. Furthermore, through its attachment to the lateral meniscus, its contraction displaces the posterior horn backwards during lateral rotation and continuing flexion of the knee joint, thereby reducing the risk of traumatic compression of the fibrocartilage (Last 1950).

It has been seen that during extension of the knee joint the first discernible rotation of the femur occurs some 30° before full extension; it progresses only slowly at first, but increases more rapidly in the last 5°. Over this terminal phase there is a progressive increase in the passive mechanisms which resist further extension, i.e. not only the spiralizing and tautening ligaments and increasing congruence and compression of articular surfaces, but gradually increasing tension in all the extra-articular tissues which cross the posterior aspect of the joint (Smith 1956). Accordingly, the actual position of extension assumed by the joint is the result of a balance between the forces (torque) acting to extend the joint and these passive mechanisms which resist this movement. A range of positions just short of the fully close-packed position is functionally of great importance. In the erect symmetrical standing posture, the line of body weight passes in front of the transverse axes of the knee joints but the passive mechanisms discussed above are sufficient to preserve the posture with a minimal expenditure of muscular energy (Joseph 1960). Active contraction of the joint extensors and the assumption of a fully extended, close-packed position of the knee joints only occurs in asymmetrical postures, or during forward swaying, heavy loading of the trunk, or when a powerful thrust is to be exerted.

Thus, in extension parts of both cruciate ligaments, the tibial and fibular collateral ligaments, the posterior aspect of the capsule and oblique posterior ligament, and the skin and fasciae are all taut, and there is passive and sometimes active tension in the hamstring and gastrocnemius muscles; the anterior parts of the menisci are compressed between the femoral condyles and the tibia. (Kaplan 1958 and Evans 1977, have pointed out the importance of the iliotibial tract as a strong tibial collateral ligament. *See* p. 596.) In the act of extending the knee the ligamentum patellae is tightened by the quadriceps femoris but in the erect attitude it is relaxed. When the knee is flexed the fibular collateral ligament and the posterior part of the tibial collateral ligament are relaxed, but the cruciate ligaments and anterior part of the tibial collateral ligament remain taut; the posterior parts of the menisci are compressed between the femoral condyles and the tibia. Flexion is normally checked by the tension of the quadriceps extensor and its tendon, and tension in the anterior parts of the capsule, posterior cruciate ligament, and compression of soft tissues behind the knee joint. In extreme *passive* flexion, contact with the flexor aspect of the thigh may become a limiting factor. Parts of both cruciate ligaments become tense in this extreme movement. In addition to the movements of conjunct rotation associated with the termination of extension or the commencement of flexion, because of the relaxation of the collateral ligaments, medial and lateral rotation of the leg can be effected independently (adjunct rotation) when the joint is flexed. One at least of the cruciate ligaments is taut in all positions and they act as a direct bond between the tibia and femur and to prevent the former bone being carried too far backwards or forwards. In extension they are assisted in this function by the collateral ligaments. Forward gliding of the tibia on the femur is prevented by the anterior cruciate ligament and backward gliding of the tibia on the femur is prevented by the posterior cruciate ligament (Voshell 1956).

During extension of the femur on a relatively fixed tibia, the movements of the femoral articular surfaces involve simultaneous forward rolling, backward sliding and a medial spin. However, the extension of a freely moving tibia upon a relatively fixed femur involves a simultaneous forward roll and slide, and a lateral spin of the tibial articular surfaces. This analysis of the arthrokinematics of the knee joint is illustrated in **4**.16A, B. For a method of geometrical analysis of the knee joint, allowing for individual variation, consult Low and Lewis (1977).

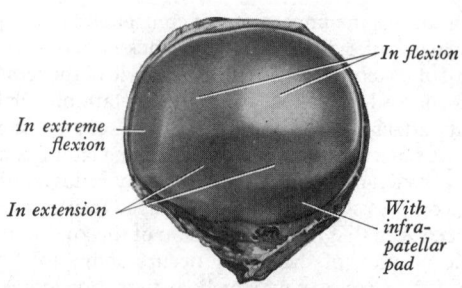

4.79 The posterior surface of the right patella, showing areas of contact with the femur and infrapatellar fat pad in different positions of the knee joint as indicated.

The menisci probably assist lubrication, facilitate combinations of sliding, rolling and spinning of the surfaces, and may also act as cushions at the extremes of flexion and extension (p. 424). Seedham and his collaborators (1974, 1979) have adduced considerable evidence of the role of menisci in accepting part of compression forces in the knee joint.

Accessory movements. A wider range of rotation can be obtained by passive movements than can be performed actively when the joint is semiflexed, and in this position the tibia can also be made to glide backwards and forwards on the femur. Abduction and adduction are prevented in the fully extended position by the taut collateral and cruciate ligaments. When the knee is slightly flexed a limited amount of adduction and abduction can be obtained, and it may be noted that these movements can be performed actively, provided that the foot is on the ground. A slight amount of separation of the femur and tibia can be obtained on strong traction.

Muscles producing the movements.

Flexion. biceps femoris, semitendinosus and semimembranosus, assisted by gracilis, sartorius and popliteus. When the foot is on the ground, gastrocnemius and plantaris are capable of participating in the movement.

Extension. Quadriceps femoris with some assistance from tensor fasciae latae.

Medial rotation of the flexed leg. Popliteus, semimembranosus and semitendinosus, assisted by sartorius and gracilis.

Lateral rotation of the flexed leg. Biceps femoris alone.

Applied Anatomy. It would appear that the knee is one of the least secure joints in the body. It is formed between the two longest bones, and therefore the amount of leverage which can be brought to bear upon it is considerable; the articular surfaces are but ill-adapted to each other, and the range of motion which it enjoys is great. Nevertheless, on account of the powerful ligaments which bind the bones together and the strength of the muscles concerned, the joint is one of the strongest in the body, and traumatic dislocation is a rare accident. Many of these muscles have a direct attachment to the fibrous capsule of the joint.

Injuries to one or other meniscus are of common occurrence and result from twisting strains applied to the knee when it is slightly flexed or when it is in full flexion. The damage may be a tear *in* the cartilage, or occasionally its periphery may be detached from the capsule over a variable extent. The torn or detached portion may become displaced towards the centre of the joint and jammed between the articular surfaces of the femur and tibia, arresting all movement. It should be noted that the menisci are largely avascular and that a tear cannot be expected to heal unless it has occurred close to the capsule (p. 487). The medial meniscus is much more commonly affected than the lateral, probably because it is more securely attached to neighbouring structures and is therefore less able to adapt itself to sudden changes of position. Further, during rotation of the flexed or partially flexed joint it moves through a greater interval than the lateral meniscus. On the other hand, it has been suggested that the medial fibres of the popliteus may draw the posterior part of the lateral meniscus backwards on to the groove on the back of the lateral tibial condyle and so prevent its being trapped between the articular surfaces.

Injuries to the cruciate ligaments are also common and range from simple sprain to complete rupture. The anterior ligament is more commonly affected. Sprain or rupture of the tibial collateral ligament is a less common injury, since excessive strain is likely to fall upon it only when the knee is in full extension.

Osteoligamentous preparations have aided the understanding of the excessive movements resulting from ligamentous injuries. These movements occur particularly in the flexed knee, the fibular collateral ligament and the posterior part of the tibial collateral ligament being relaxed in this position. If the anterior or posterior cruciate ligament is cut, gliding of the tibia forwards or backwards is increased. Similarly rotation is increased if either of the cruciate ligaments or the tibial collateral ligament is cut. Division of both collateral ligaments leads to excessive lateral rotation with untwisting of the cruciate ligaments, but medial rotation is unaffected. Abduction and adduction occur to an excessive degree if both cruciate ligaments are cut (Voshell 1956).

Acute synovitis, the result of injury, is of frequent occurrence in the knee joint. When the cavity is distended with fluid, the swelling shows itself above and at the sides of the patella, reaching 5 cm or more above the patellar surface of the femur, and extending a little higher under the vastus medialis than under the vastus lateralis. The lower level of the synovial membrane is just below the upper end of the tibia.

The bursae about the knee joint are sometimes distended. The subcutaneous prepatellar or infrapatellar bursae are frequently affected in those who are in the habit of kneeling. The bursa beneath the semimembranosus tendon also occasionally becomes enlarged, and forms a fluctuating swelling at the back of the knee. During extension the swelling is firm and tense; but during flexion it may become soft, since the bursa often communicates with the synovial cavity of the joint. The fluid it contains can then be made to disappear by pressure when the knee is flexed and hence loosely packed.

The Tibiofibular Articulations

The tibia and fibula are connected at both extremities, by a synovial articulation above and a fibrous joint below. In addition, the shafts of the bones are connected by the crural interosseous membrane.

THE SUPERIOR TIBIOFIBULAR JOINT

This articulation (4.80) is approximately a plane joint between the lateral condyle of the tibia and the head of the fibula. The *articular surfaces* of the bones are variable in size, form and inclination. The facet on the head of the fibula is generally elliptical or circular in shape and almost flat or slightly grooved. These facets are covered with cartilage, and by the anterior and posterior ligaments.

The fibrous capsule is attached to the margins of the articular facets on the tibia and fibula; it is much thicker in front than behind. In about 10 per cent of persons the synovial membrane of the joint is continuous with that of the knee joint through the subpopliteal recess (p. 485). **The anterior ligament** consists of two or three flat bands, which pass obliquely upwards from the front of the head of the fibula to the front of the lateral condyle of the tibia. **The posterior ligament** is a thick band, which passes obliquely upwards from the back of the head of the fibula to the back of the lateral condyle of the tibia. It is covered by the tendon of the popliteus. These ligaments are not entirely separable from the fibrous capsule.

The *arteries* to the joint are derived from the anterior and posterior tibial recurrent branches of the anterior tibial artery.

The *nerves* are derived from the common peroneal nerve and from the nerve to popliteus.

THE CRURAL INTEROSSEOUS MEMBRANE

The crural interosseous membrane (4.80) connects the interosseous borders of the tibia and fibula, and intervenes between the muscles on the front and those on the back of the leg, some of

distance of about 4 mm by an upward prolongation of the synovial membrane of the talocrural joint, and may be covered with articular cartilage in the lowest millimetre. Apart from this, the part of the joint which is of the fibrous type is usually described as a syndesmosis (p. 421).

The anterior tibiofibular ligament (4.86) is a flattened band which extends obliquely downwards and laterally between the adjacent margins of the tibia and fibula, on the front of the syndesmosis.

The posterior tibiofibular ligament (4.81), stronger than the preceding, is disposed in a similar manner on the posterior surface of the syndesmosis. Its lower and deep portion forms the *inferior transverse ligament*—a strong, thick band of yellowish fibres which passes transversely from the upper part of the lateral malleolar fossa of the fibula to the posterior border of the articular surface of the tibia, almost as far as the medial malleolus. The inferior transverse ligament projects below the margins of the bones, and forms part of the articulating surface for the talus. It contains a proportion of yellow elastic fibres.

The interosseous ligament is continuous, above, with the crural interosseous membrane and consists of numerous, short, strong bands which pass between the adjacent rough surfaces of the tibia and fibula, and constitute the largest bonds between the lower ends of the bones.

Movements. The joints between the tibia and fibula permit very slight movements. Owing to the varying inclination of the articular surface on the talus for the lateral malleolus, the fibula undergoes a small amount of lateral rotation during dorsiflexion of the ankle (Barnett and Napier 1952). The bones are also forced apart to a small extent in full dorsiflexion. Although it is possible that slight bending or torsion of the fibular shaft could permit the movements of the inferior tibiofibular joint (caused by activities of the talocrural joint) the existence of a proximal synovial tibiofibular articulation doubtless improves efficiency. (See also movements of talocrural articulation.)

The *arteries* to the joint are derived from the perforating branch of the peroneal artery and the medial malleolar branches of the anterior and posterior tibial arteries.

The *nerves* are derived from the deep peroneal, tibial and saphenous nerves.

The Talocrural Joint

The talocrural or ankle joint is of uni-axial type. The lower end of the tibia and its malleolus, the malleolus of the fibula, and the inferior transverse tibiofibular ligament together form a deep recess in which the body of the talus is embraced. The line of the joint can be gauged from the anterior margin of the lower end of the tibia, which can be felt through the skin in the living when the overlying tendons are relaxed. Although anatomically this joint appears to be a simple hinge, and is usually styled 'uniaxial', it must be emphasized that the axis of rotation is dynamic, taking up a series of different positions during dorsiflexion-plantar flexion changes. (*See* Sammarco 1977, for an analysis.)

The articular surfaces are covered with hyaline cartilage. The trochlear surface of the talus, which is convex from before backwards and gently concave from side to side, is wider in front than behind, and the inferior articular surface of the tibia is reciprocally shaped. The articular surface for the medial malleolus is restricted to the upper part of the medial surface of the talus. It is fairly flat and comma-shaped, being deeper in front than behind. The articular surface on the lateral side of the talus is triangular in outline and concave from above downwards; that on the lateral malleolus is reciprocally curved. Posteriorly, the edge between the trochlear and fibular articular surfaces of the talus is bevelled to form a flattened triangular area which articulates with the inferior transverse tibiofibular ligament (4.84, 85). It is to be emphasized that all these talar surfaces are continuous; separate description is a mere convenience.

The bones are connected by a fibrous capsule, and by deltoid, anterior and posterior talofibular, and calcaneofibular ligaments.

The fibrous capsule surrounds the joint; it is thin in front and behind and attached above to the borders of the articular surfaces

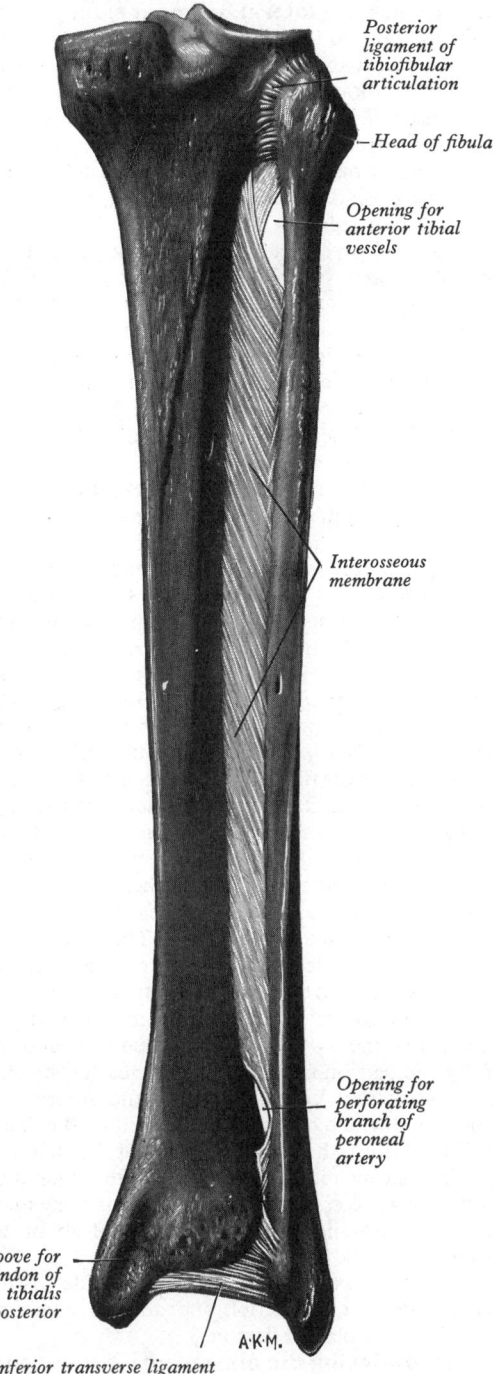

Posterior ligament of tibiofibular articulation

—Head of fibula

Opening for anterior tibial vessels

Interosseous membrane

Opening for perforating branch of peroneal artery

Groove for tendon of tibialis posterior

A·K·M.

Inferior transverse ligament

4.80   The crural interosseous membrane. Posterior aspect.

which are attached to it. The anterior tibial artery passes to the front of the leg through a large oval opening in the uppermost part of the membrane, and the perforating branch of the peroneal artery pierces it below. Its fibres, largely oblique, descend laterally for the most part, but a few descend medially, including a bundle which forms the upper border of the opening for the anterior tibial artery. The membrane is continuous below with the interosseous ligament of the inferior tibiofibular joint. It is related, in front, with tibialis anterior, extensor digitorum longus, extensor hallucis longus, peroneus tertius, and the anterior tibial vessels and deep peroneal nerve, behind, the tibialis posterior and flexor hallucis longus.

THE INFERIOR TIBIOFIBULAR JOINT

This joint is between the rough, *convex* surface on the medial side of the lower end of the fibula and the rough, *concave* surface of the fibular notch of the tibia. Below, these surfaces are separated for a

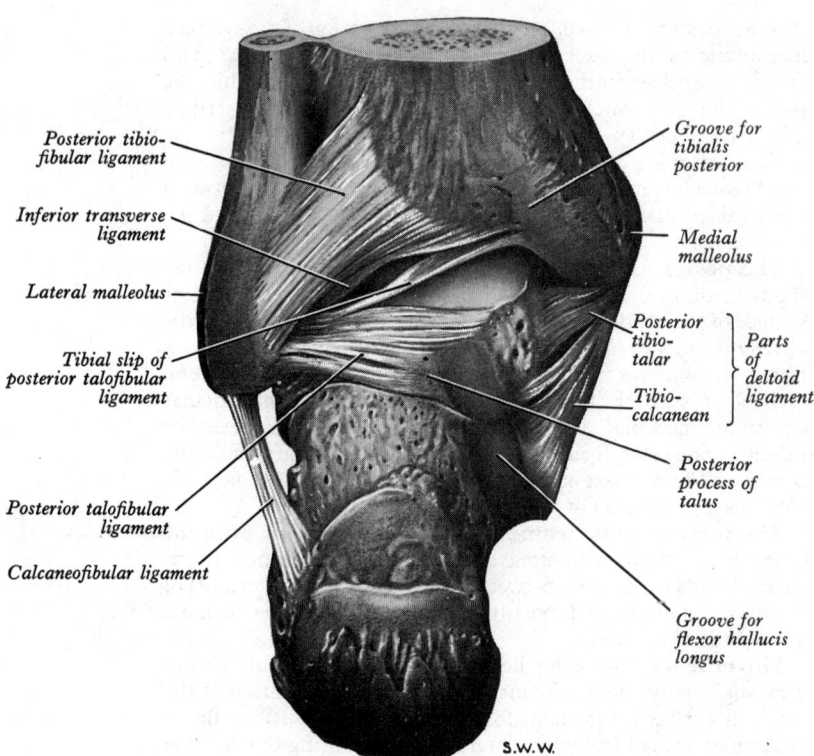

Labels on figure (left column):
Posterior tibio-
fibular ligament

Inferior transverse
ligament

Lateral malleolus

Tibial slip of
posterior talofibular
ligament

Posterior talofibular
ligament

Calcaneofibular ligament

Labels on figure (right column):
Groove for
tibialis
posterior

Medial
malleolus

Posterior
tibio-
talar

Tibio-
calcanean

Parts
of
deltoid
ligament

Posterior
process of
talus

Groove for
flexor hallucis
longus

S.W.W.

4.81 Left talocrural joint. Posterior aspect. (From a specimen in the Museum of the Royal College of Surgeons of England, by permission of the Council.)

of the tibia and malleoli, and below, to the talus close to the margins of the trochlear surface except in front where it is attached to the dorsum of the neck of the talus at some distance in front of its superior articular surface. It is supported on each side by strong collateral ligaments. The posterior part of the capsule consists principally of transverse fibres. It blends with the inferior transverse ligament and is somewhat thickened laterally where it reaches as far as the malleolar fossa of the fibula.

A synovial membrane lines the fibrous capsule, and the joint cavity ascends for a short distance between the tibia and fibula (4.87).

The medial ligament, or deltoid collateral ligament (4.82, 87) is a strong, triangular band, attached above to the apex and anterior and posterior borders of the medial malleolus. It consists of superficial and deep fibres. Of the superficial fibres the anterior (*tibionavicular*) pass forwards to the tuberosity of the navicular bone, and immediately behind this they blend with the medial margin of the plantar calcaneonavicular ligament; the middle fibres (*tibiocalcanean*) descend almost perpendicularly to the whole length of the sustentaculum tali of the calcaneus; the posterior fibres (*posterior tibiotalar*) pass backwards and laterally to the medial side of the talus, and to its medial tubercle. The deep fibres (*anterior tibiotalar*) are well developed and are fixed above to the tip of the medial malleolus and below to the non-articular part of the medial surface of the talus. The deltoid ligament is crossed by the tendons of tibialis posterior and flexor digitorum longus.

The anterior talofibular ligament (4.86) passes from the anterior margin of the fibular malleolus, forwards and medially, to the talus, where it is attached in front of the lateral articular facet and to the lateral aspect of the neck. **The posterior talofibular ligament** (4.81), strong and deeply seated, runs almost horizontally from the lower part of the lateral malleolar fossa to the lateral tubercle of the posterior process of the talus. A bundle of fibres (the 'tibial slip') leaves it to be attached to the medial malleolus. **The calcaneofibular ligament** (4.86) is a long rounded cord, running from the depression in front of the apex of the fibular malleolus downwards and backwards to a tubercle on the lateral surface of the calcaneus. It is crossed by the tendons of peroneus longus and brevis. The foregoing three ligaments

together constitute the **lateral ligament** of the talocrural joint.

Relations. The tendons, vessels and nerves in relation with the joint are: in front, from the medial side, the tibialis anterior, extensor hallucis longus, anterior tibial vessels, deep peroneal nerve, extensor digitorum longus and peroneus tertius; behind, from the medial side, the tibialis posterior, flexor digitorum longus, posterior tibial vessels, tibial nerve, flexor hallucis longus; in the groove behind the fibular malleolus, the tendons of the peroneus longus and brevis (4.83).

The *arteries* supplying the joint are derived from the malleolar branches of the anterior tibial and from the peroneal arteries.

The *nerves* are derived from the deep peroneal and tibial nerves.

Movements. When the body is in the erect position the foot is at right angles to the leg. The *active movements* of the talocrural joint are those of dorsiflexion (*c.* 10°) and plantar flexion (*c.* 20°); in dorsiflexion the angle between the front of the leg and the dorsum of the foot is diminished; in plantar flexion the angle is increased, the heel being raised and the toes pointed downwards. Dorsiflexion is the position of 'close-pack' with maximal congruence of the joint surfaces and ligamentous tension. This is the initial position from which all major thrusting movements are developed, whether in walking, jumping or running. The malleoli embrace the talus in the position of rest, and no appreciable degree of side-to-side movement can occur without stretching the ligaments of the tibiofibular syndesmosis, and slight bending of the fibula. The superior articular surface of the talus is broader in front than behind. In dorsiflexion, therefore, greater space is required between the two malleoli. This is obtained by a slight lateral rotation of the lower end of the fibula and is consequent on slight movement at the tibiofibular syndesmosis; this is facilitated by a minor degree of gliding at the superior tibiofibular joint. The deltoid ligament is exceedingly strong—so much so, that it usually resists a force which fractures the process of bone to which it is attached. Its middle portion, together with the calcaneofibular ligament, binds the bones of the leg firmly to the foot, and resists displacement in every direction. The posterior talofibular ligament assists the calcaneofibular in resisting displacement of the foot backwards, and deepens the cavity for the reception of the talus. The anterior talofibular ligament is a security against displacement of the foot forwards. Plantar flexion of the foot is limited by the tension of the opposing muscles, by the anterior fibres of the deltoid ligament and by the anterior talofibular ligament. Dorsiflexion of the foot is limited by the tension of the calcareal tendon, by the posterior fibres of the deltoid ligament and by the calcaneofibular ligament (4.86). Dorsiflexion and plantar flexion are commonly increased in range by movement at the intertarsal articulations, which can add about 10° to the former and 20° to the latter.

Accessory movements. Slight amounts of side-to-side gliding movement, rotation, abduction and adduction are permitted, when the foot is in plantar flexion.

Muscles producing the movements (see also pp. 604–611, 617–619).

Dorsiflexion. Tibialis anterior, assisted by extensor digitorum longus, extensor hallucis longus and peroneus tertius.

Plantar flexion. Gastrocnemius and soleus, assisted to a lesser degree by plantaris, tibialis posterior, flexor hallucis longus and flexor digitorum longus.

In the symmetrical erect posture, the line of the body weight falls in front of the ankle joint which is not in, or even near, its close-packed position. Stabilization of this posture requires persistent activity of the soleus muscles (and often intermittent activity of the gastrocnemius muscles). However, the force of contraction of these muscles undoubtedly fluctuates—increasing with forward sway and decreasing with backward sway. If backward sway is sufficient to carry the line of weight behind the transverse axis of the talocrural joint there is immediate relaxation of the plantar flexors and contraction of the dorsiflexors.

Applied Anatomy. Owing to the depth of the tibiofibular embrasure into which the talus projects, the talocrural joint is rarely dislocated unless one of the malleoli is fractured. So-called sprains of the talocrural joint are almost always abduction sprains of the subtalar joints, although some of the fibres of the deltoid ligament also are torn. True sprains of the talocrural joint are

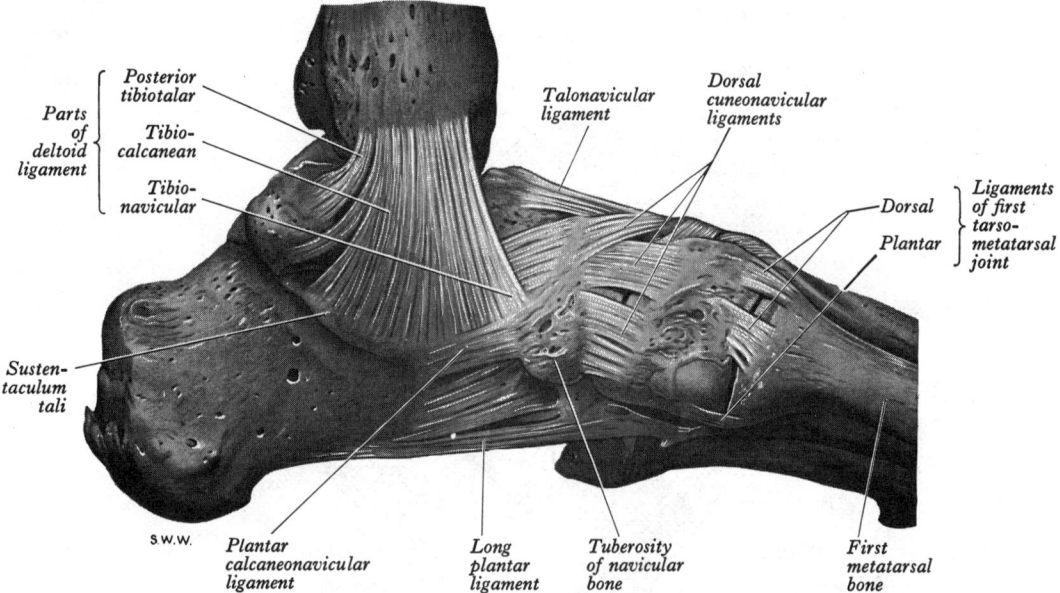

4.82 The ligaments of the left talocrural and tarsal joints. Medial aspect. (Drawn from a specimen in the Museum of the Royal College of Surgeons of England, by kind permission of the Council.)

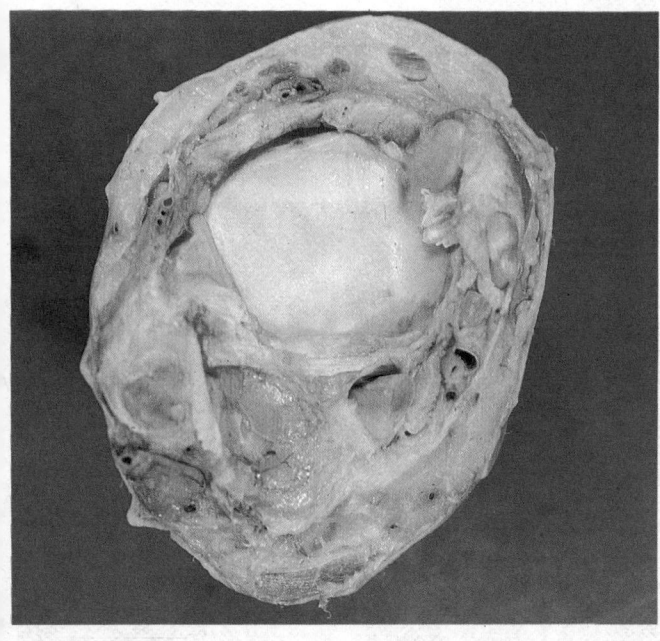

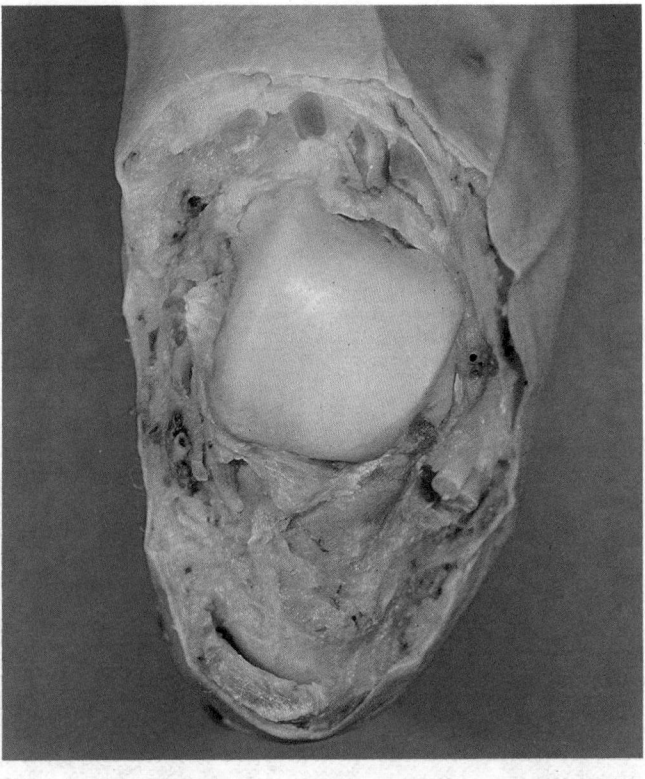

4.82A and B The articular surfaces of a human cadaveric talocrural joint. A displays the tibiofibular mortice, which includes the principal talar surface of the tibia with its medial malleolar facet, the interosseous tibiofibular ligament, and the articular surface of the fibular (lateral) malleolus. B shows the characteristically sellar articular surface of the talus; note the obliquity of its principal axes relative to the sagittal axis of the foot. (Preparation by Dr. M. C. E. Hutchinson and photography by Mr. Kevin Fitzpatrick, Guy's Hospital Medical School, London.)

usually caused by forced plantar flexion and they result in tearing of the capsular ligament on the front of the joint and bruising by impaction of the structures at the back of the joint.

When disease or injury of the talocrural joint is likely to lead to ankylosis, the optimal position to encourage is slight plantar flexion.

The Intertarsal Joints

THE SUBTALAR (TALOCALCANEAN) JOINT

There are two articulations between the calcaneus and talus, anterior and posterior. These form a single functional unit which is often termed the 'subtalar joint', but this term is used here for the posterior joint only, the anterior being described as part of the talocalcaneonavicular joint (*vide infra*).

The subtalar joint, as defined above, involves the concave posterior calcanean facet on the inferior surface of the talus and the convex posterior facet on the upper surface of the calcaneus. The joint is of modified multi-axial type, the more usually stationary surface (talus) being concave or 'female' (*see* p. 429). The two bones are connected by a fibrous capsule and by lateral, medial, interosseous talocalcanean and cervical ligaments.

The fibrous capsule (4.88) envelops the joint, and consists for the most part of short fibres; it is split into slips, and between these there is only a weak fibrous investment. It is lined with synovial membrane, and the joint cavity does not communicate with any other tarsal joint.

The lateral talocalcanean ligament is a short, flattened

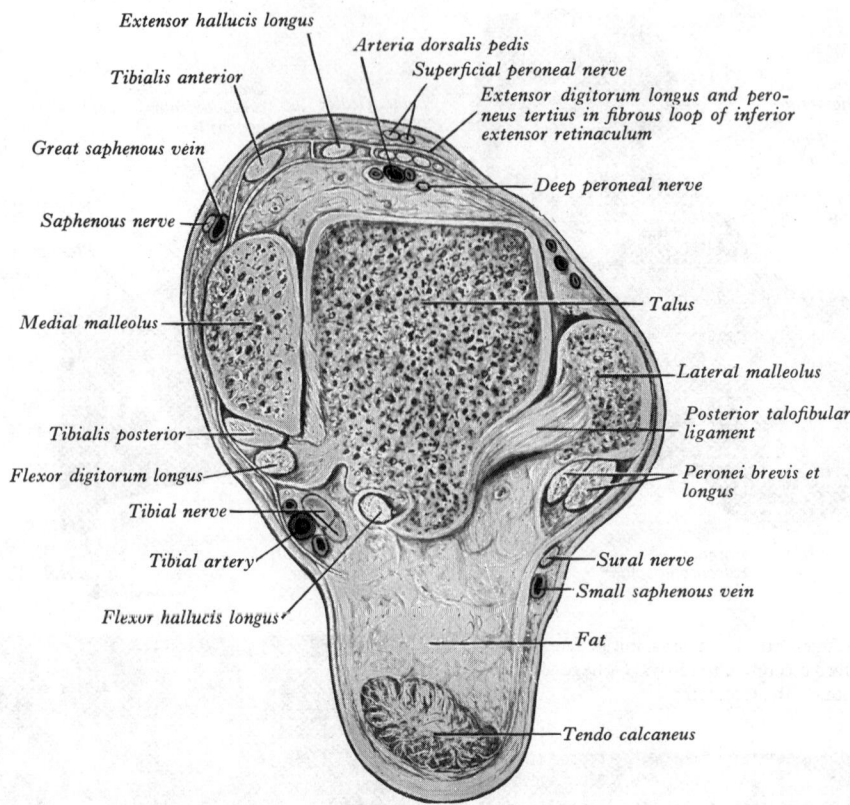

Extensor hallucis longus
Tibialis anterior
Great saphenous vein
Saphenous nerve
Medial malleolus
Tibialis posterior
Flexor digitorum longus
Tibial nerve
Tibial artery
Flexor hallucis longus
Arteria dorsalis pedis
Superficial peroneal nerve
Extensor digitorum longus and pero-
neus tertius in fibrous loop of inferior
extensor retinaculum
Deep peroneal nerve
Talus
Lateral malleolus
Posterior talofibular
ligament
Peronei brevis et
longus
Sural nerve
Small saphenous vein
Fat
Tendo calcaneus

4.83   Transverse section through the lower part of the right talocrural joint. Superior aspect.

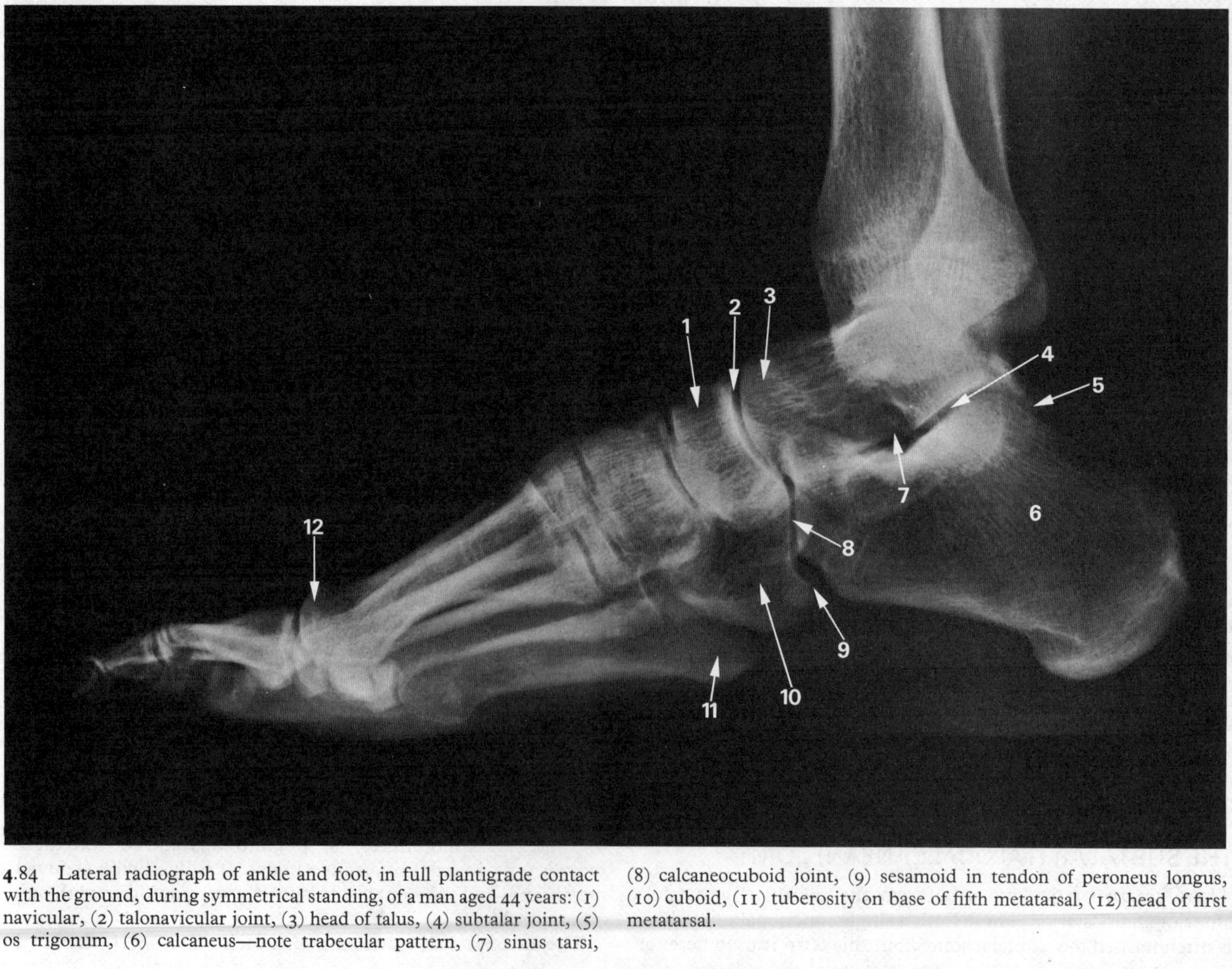

4.84   Lateral radiograph of ankle and foot, in full plantigrade contact with the ground, during symmetrical standing, of a man aged 44 years: (1) navicular, (2) talonavicular joint, (3) head of talus, (4) subtalar joint, (5) os trigonum, (6) calcaneus—note trabecular pattern, (7) sinus tarsi, (8) calcaneocuboid joint, (9) sesamoid in tendon of peroneus longus, (10) cuboid, (11) tuberosity on base of fifth metatarsal, (12) head of first metatarsal.

asciculus, which passes downwards and backwards from the ateral process of the talus to be attached to the lateral surface of he calcaneus, above and in front of the calcaneofibular ligament.

The **medial talocalcanean ligament** connects the medial uberercle of the talus with the back of the sustentaculum tali and he medial surface of the calcaneus immediately behind the latter. ts fibres blend with those of the deltoid ligament; the most posterior fibres line the groove for the flexor hallucis longus between the talus and the calcaneus.

The **interosseous talocalcanean ligament** (4.87, 88) is a broad, flattened band lying somewhat transversely in the sinus tarsi. It passes obliquely downwards and laterally from the sulcus tali to the sulcus calcanei. Its medial fibres are rendered taut in eversion.

The **cervical ligament** is placed just lateral to the sinus tarsi and is attached to the upper surface of the calcaneus, medial to the origin of extensor digitorum brevis. From here it passes upwards and medially to a tubercle on the inferior and lateral aspect of the neck of the talus (Barclay Smith 1896, Smith 1958). It is regarded as being taut when the foot is inverted and prevents the occurrence of too great a degree of this movement.

Movements. The movements permitted between the talus and calcaneus are closely associated with the movements at the talocalcaneonavicular joint and will be described with them.

THE TALOCALCANEONAVICULAR JOINT

This articulation is multi-axial; the compounded ovoid head of the talus fits the complex concavity formed by the posterior surface of the navicular, the middle and anterior facets for the talus on the calcaneus, and the upper surface of the plantar calcaneonavicular ligament. In contrast to the subtalar joint, the proximal and less mobile surface is convex or 'male' (*see* p. 429). The position of the joint can be gauged from the head of the talus, which can be both seen and felt 3 cm in front of the lower end of the tibia, when the foot is passively inverted. The bones forming the joint are connected by a fibrous capsule, and by the talonavicular, the plantar calcaneonavicular and bifurcated (calcaneonavicular part) ligaments.

The **fibrous capsule** is imperfectly developed except posteriorly, where it is considerably thickened and forms the anterior part of the interosseous ligament which fills the sinus tarsi.

The **talonavicular ligament** (4.82, 86) is a broad, thin band, connecting the neck of the talus to the dorsal surface of the navicular bone; above it are the extensor tendons. The *plantar calcaneonavicular ligament* (p. 496) is the plantar and the *calcaneonavicular part of the bifurcated ligament* (4.86) the lateral ligament for this joint.

Movements. A considerable range of gliding and rotation occurs at both the talocalcanean and the talocalcaneonavicular joints by which the calcaneus and the navicular, carrying the foot with them, can rotate medially on the talus. This results in elevation of the medial border and corresponding depression of the lateral border of the foot, so that the plantar aspect of the foot faces medially. This is the position of *inversion*. The greater part of the movement occurs at the subtalar joint and at the articulation between the head of the talus and the sustentaculum tali, front of the calcaneus, plantar calcaneonavicular ligament and the posterior surface of the navicular. The surfaces of these two joints are curved in opposite directions and the movements occurring at them have sometimes been likened to the movements between the radius and ulna at the superior and inferior radio-ulnar joints (but in the foot the terms supination and pronation are used in a different context—*vide infra*). The axis of the movements may be considered as a line joining the approximate centres of curvature of the two joints, and runs forwards, upwards and medially, from the back of the calcaneus through the sinus tarsi to emerge at the superior and medial aspect of the neck of the talus (Shepard 1951). The obliquity of this axis accounts in part for the adduction and slight plantar flexion of the forefoot which occur in inversion. The movement of the calcaneus around the talus is, however, accompanied by movement at the *transverse tarsal articulation* which comprises the talonavicular and the calcaneocuboid

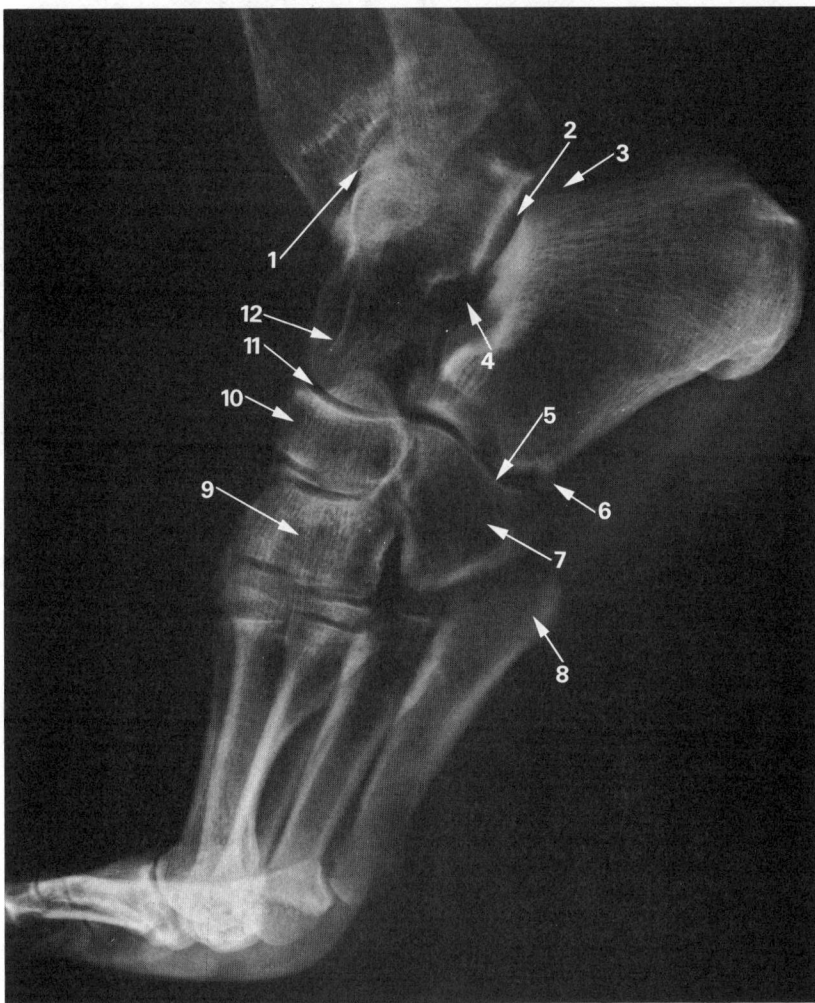

4.85   The same foot as in 4.84 with heel raised from the ground, as near the end of the 'stance phase' in walking. Note forward angulation of leg bones, plantar flexion at the talocrural joint, and the position of the metatarsals and phalanges. (1) talocrural joint, (2) subtalar joint, (3) os trigonum, (4) sinus tarsi, (5) calcaneocuboid joint, (6) sesamoid in peroneus longus tendon, (7) cuboid, (8) tuberosity on base of fifth metatarsal, (9) cuneiforms, (10) navicular, (11) talonavicular joint, (12) head of talus.

articulations and the range of movement is thereby increased. These lie almost in the same transverse plane and during inversion the navicular rotates on the head of the talus whilst the cuboid glides downwards and rotates on the front of the calcaneus at the sellar (or saddle-shaped) calcaneocuboid joint. The range of the movement of inversion is also further increased in plantar flexion of the foot, for in this position the narrow part of the trochlear surface of the talus occupies the tibiofibular mortice and a slight amount of movement of the talus in the mortice gives an increased range of adduction with inversion. The opposite movement, which is much more limited in range, is termed *eversion*.

The chief factor in the limitation of inversion is the tension of the peronei and the strong lateral part of the interosseous talocalcanean ligament (*vide supra*). The other tarsal interosseous ligaments and the calcaneofibular ligament are less powerful factors. Eversion is arrested by the tension of the tibialis anterior and tibialis posterior and the deltoid ligament.

The complex movements of inversion and eversion as described above refer to changes in position and form of the whole foot (with only minor movements of the talus), *when the foot is off the ground*. When the foot is fully inverted it is also plantar flexed, and conversely, eversion is associated with dorsiflexion.

When the foot is on the ground, transmitting weight and thrust, these movements are modified to maintain plantigrade contact. The distal part of the tarsus and metatarsus are *pronated* or *supinated*, relative to the calcaneus and talus, the former

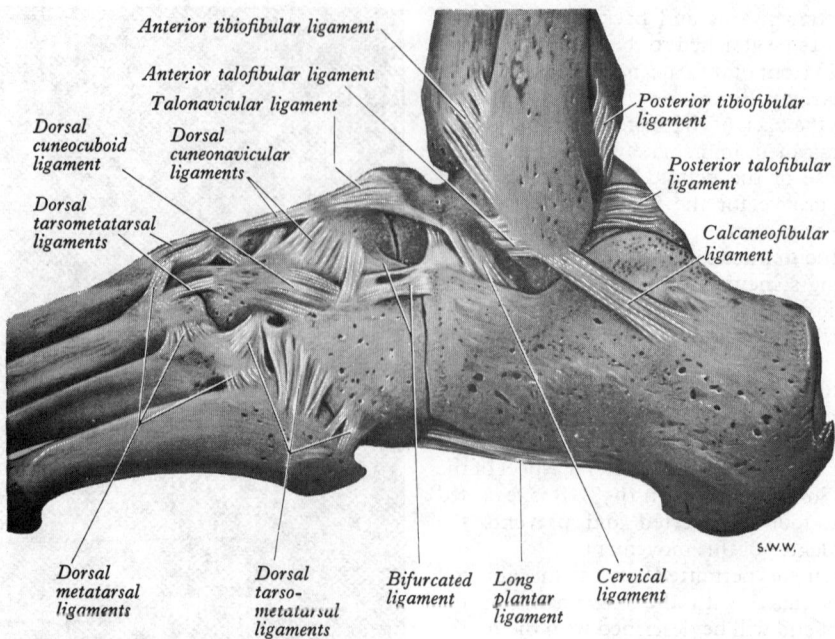

4.86 The ligaments of the left talocrural and tarsal joints. Lateral aspect. (Drawn from a specimen in the Museum of the Royal College of Surgeons of England, by kind permission of the Council.)

movement involving (by analogy with the hand) a downward rotation of the medial border and hallux towards the ground, supination being the reverse of this, to bring the lateral border into more direct plantigrade contact. Unfortunately, the terms inversion and eversion are sometimes equated with or confused with pronation and supination, whereas the latter are really components of the former, just as are the slight movements of adduction and abduction which may accompany inversion and eversion. Moreover, supination and pronation can be dissociated from inversion and eversion in adaptation of the forefoot to surfaces in plantigrade standing or progression (pp. 501–503, 617–619).

The terminology of movements of the foot is a matter of some confusion amongst orthopaedists, theoreticians and others. The

above account represents widely accepted usage. (Consult also MacConaill 1950; Kapandji 1977; MacConaill and Basmajian 1977; Keller 1977). According to Kapandji (1965, 1977) inversion is a combination of adduction and supination whereas eversion involves simultaneous abduction and pronation.

Muscles producing the movements.

Inversion. Tibialis anterior and posterior.

Eversion. Peroneus longus and brevis.

THE CALCANEOCUBOID JOINT

The articular surfaces of the calcaneocuboid joint, which lies 2 cm behind the tubercle on the base of the fifth metatarsal bone, are saddle-shaped. The ligaments of the joint are: the fibrous capsule, the calcaneocuboid portion of the bifurcated ligament, the long plantar and the plantar calcaneocuboid ligaments.

The fibrous capsule is slightly thickened over the dorsal surface of the joint, as the *dorsal calcaneocuboid ligament*. The synovial membrane is distinct from that of the other tarsal articulations (4.91).

The bifurcated ligament (4.86) is a strong band attached behind to the anterior part of the upper surface of the calcaneus and dividing in front into calcaneocuboid and calcaneonavicular parts. The *calcaneocuboid ligament* is fixed to the dorsal part of the medial side of the cuboid bone and forms one of the principal bonds between the first and second rows of the tarsal bones. The *calcaneonavicular ligament* is attached to the dorsolateral aspect of the navicular bone.

The long plantar ligament (4.89), the longest of the tarsal ligaments, is attached posteriorly to the plantar surface of the calcaneus in front of the medial and lateral processes of the tuberosity, and to the anterior tubercle, and anteriorly to the ridge and tuberosity on the plantar surface of the cuboid bone, to which the deep fibres are attached, the more superficial fibres being continued forwards to the bases of the second, third, fourth and, sometimes, the fifth metatarsal bones. This ligament converts the groove on the plantar surface of the cuboid bone into a tunnel for the tendon of the peroneus longus. It possesses great strength and is an important factor in limiting flattening of the lateral longitudinal arch of the foot (p. 501).

The plantar calcaneocuboid ligament (*short plantar ligament*) lies nearer to the bones than the preceding ligament, from which it is separated by a little areolar tissue. It is a short but wide band of great strength, and stretches from the anterior tubercle of the calcaneus and the depression in front of it, to the adjoining part of the plantar surface of the cuboid bone. Like the

Interosseous ligament of tibiofibular syndesmosis

Lateral malleolus

Posterior talofibular ligament

Interosseous talo-calcanean ligament

Body of calcaneus

Tendon of peroneus brevis

Tendon of peroneus longus

Medial malleolus

Body of talus

Deltoid ligament

Tendon of tibialis posterior

Sustentaculum tali

Tendon of flexor digit-orum longus

Tendon of flexor hallucis longus

4.87 A coronal section through the left talocrural and talocalcanean joints.

preceding ligament, it limits flattening of the lateral longitudinal arch of the foot.

Movements. The movements permitted between the calcaneus and the cuboid bone are limited to gliding with conjunct rotation of the bones upon each other during the complex movements of inversion and eversion of the free foot and during the pronative or supinative changes of relationship of the forefoot to the hindfoot (pp. 501–503).

THE LIGAMENTS CONNECTING THE CALCANEUS AND NAVICULAR BONE

Though the calcaneus and the navicular bone do not articulate directly, they are connected by two ligaments, the calcaneonavicular and the plantar calcaneonavicular.

The calcaneonavicular ligament has been described above; it forms the medial band of the bifurcated ligament.

The plantar calcaneonavicular (spring) ligament (4.82, 88, 90) is a broad, thick band connecting the anterior margin of the sustentaculum tali of the calcaneaus to the plantar surface of the navicular bone. This ligament ties the calcaneus to the navicular bone, and lies below the head of the talus, forming part of the articular cavity for its head; it limits flattening of the medial longitudinal arch of the foot (p. 501). The *dorsal surface* of the ligament presents a triangular fibrocartilaginous facet upon which a portion of the head of the talus rests (4.88). Its *plantar surface* is supported by the tendon of the tibialis posterior medially, and by the tendons of the flexor hallucis longus and the flexor digitorum longus, laterally; its *medial border* is blended with the anterior fibres of the superficial part of the deltoid ligament of the talocrural joint. (Despite its vernacular name there is no real evidence that the 'spring' ligament is peculiarly resilient.)

THE CUNEONAVICULAR JOINT

The navicular articulates in front with the three cuneiform bones. The joint is often described as of the plane variety, but in fact the distal surface of the navicular is convex from side to side and

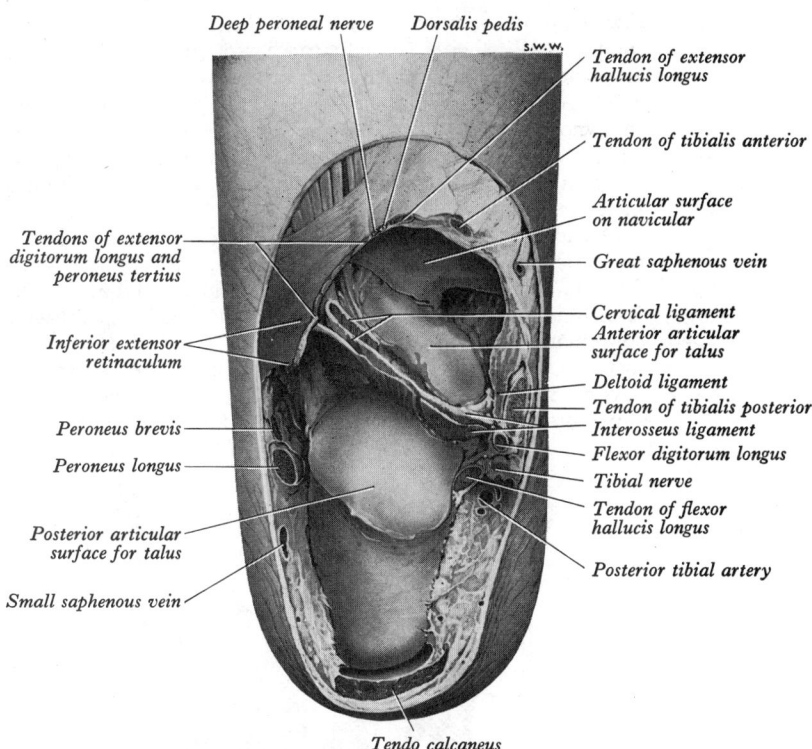

4.88A The left talocalcanean and talocalcaneonavicular joints. Exposed from above by removal of the talus.

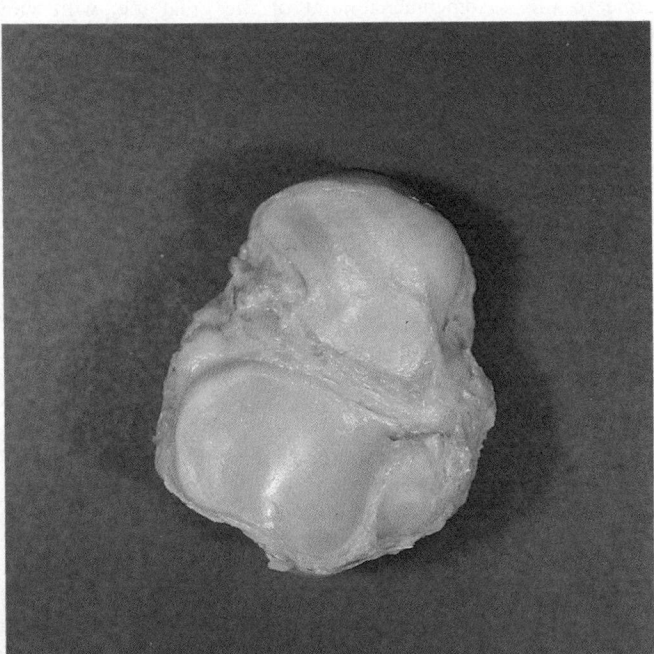

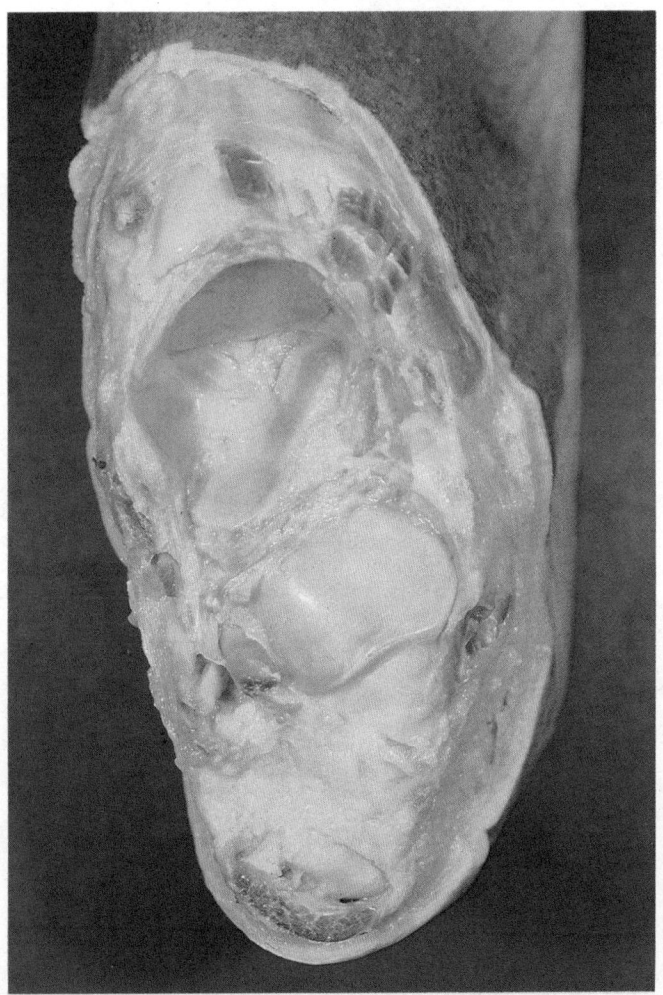

4.88B and C A dissection in which the talus has been removed from the subjacent osseous and ligamentous tarsal elements. Compare the reciprocal articular curvatures of the inferior surface of the talus (B) and those of the superior aspects of their conarticular tarsal elements (C). Note the ovoid surfaces of the (posterior) talocalcaneal articulation (the *subtalar* joint of some authorities). Anteriorly, the complex convexity of the talar head is received into the '*acetabulum pedis*', which is formed by the calcaneus, navicular, the plantar calcaneonavicular ligament, and part of the deltoid ligament. The detailed ligamentous architecture between these articulations is described in the text, and illustrated diagrammatically (A).

(Preparation by Dr. M. C. E. Hutchinson and photography by Mr. Kevin Fitzpatrick, Guy's Hospital Medical School, London.)

divided into three facets by low ridges, each of which is adapted to the proximal, slightly curved surface of a cuneiform bone. Its articular capsule is continuous with those of the intercuneiform and the cuneocuboid joints. Its *synovial cavity* is continuous with the synovial cavities of these joints and with those of the second and third cuneometatarsal joints and of the intermetatarsal joints between the bases of the second and third, and third and fourth metatarsal bones (p. 499).

Dorsal and plantar ligaments connect the navicular to each of the cuneiform bones. **The dorsal ligaments** are three small fasciculi, one attached to each of the cuneiform bones. The fasciculus connecting the navicular with the medial cuneiform is continuous round the medial side of the joint with the plantar ligament which unites these two bones. **The plantar ligaments** have a similar arrangement to the dorsal, and are strengthened by slips from the tendon of the tibialis posterior.

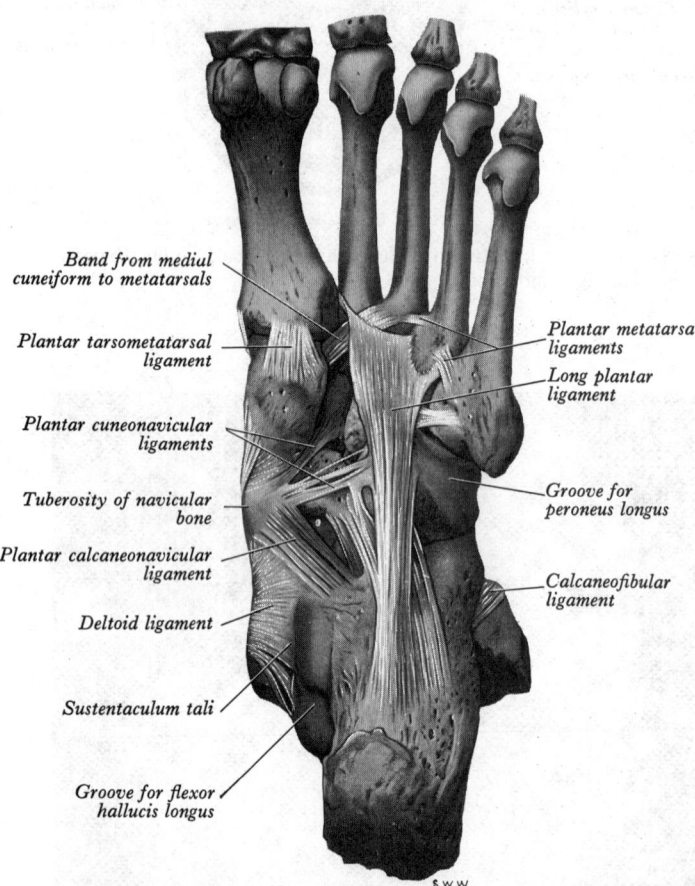

Band from medial
cuneiform to metatarsals

Plantar tarsometatarsal
ligament

Plantar cuneonavicular
ligaments

Tuberosity of navicular
bone

Plantar calcaneonavicular
ligament

Deltoid ligament

Sustentaculum tali

Groove for flexor
hallucis longus

Plantar metatarsal
ligaments

Long plantar
ligament

Groove for
peroneus longus

Calcaneofibular
ligament

S.W.W.

4.89 The ligaments of the plantar surface of the left foot. Some of the fibres of the long plantar ligament which arise in front of the medial tubercle of the calcaneus have been removed. (Drawn from a specimen in the Museum of the Royal College of Surgeons of England, by kind permission of the Council.)

THE CUBOIDEONAVICULAR JOINT

The cuboideonavicular joint is usually a fibrous joint, the two bones being connected by dorsal, plantar and interosseous ligaments. **The dorsal ligament** extends obliquely forwards and laterally, while the **plantar** passes nearly transversely from the cuboid to the navicular bone. **The interosseous ligament** consists of strong transverse fibres, and connects the rough nonarticular parts of the adjacent surfaces of the two bones (4.91).

Not infrequently the syndesmosis is replaced by a synovial joint. In that event the joint is of an almost plane variety and its articular capsule and synovial cavity are continuous with those of the cuneonavicular joint.

THE INTERCUNEIFORM AND CUNEOCUBOID JOINTS

The intercuneiform joints and the joint between the lateral cuneiform bone and the cuboid are all synovial in character and approximately plane or with only minor degrees of curvature. Their articular capsules and synovial cavities are continuous with those of the cuneonavicular joint.

The bones are connected together by dorsal, plantar and interosseous ligaments.

The dorsal and plantar ligaments each consist of three transverse bands: one connects the medial and intermediate cuneiform bones, another the intermediate and lateral cuneiform bones, and another the lateral cuneiform and cuboid bones. The plantar ligaments are strengthened by slips from the tendon of the tibialis posterior. **The interosseous ligaments** connect the rough nonarticular portions of the adjacent surfaces of the bones and possess considerable strength; they are a factor in the transverse arch of the foot (p. 502).

Movements. The movements permitted at the cuneonavicular, cuboideonavicular, intercuneiform and cuneocuboid joints are limited to a slight amount of gliding and rotation of the bones concerned on each other. These movements occur during pronation or supination of the foot (p. 499), i.e. during positional alterations of the loaded foot when in contact with the ground. For example, they greatly increase the suppleness of the foot when its forepart is stressed, as in the initial thrust of running and jumping.

The Tarsometatarsal Articulations

These are synovial joints of approximately plane variety. The first metatarsal articulates with the medial cuneiform; the second is recessed between the medial and lateral cuneiforms, articulating between them with the intermediate cuneiform; the third articulates with the lateral cuneiform; the fourth joints with the lateral cuneiform and cuboid, and the fifth, with the cuboid alone. These joints lie on a line joining the tubercle of the fifth metatarsal bone to the tarsometatarsal joint of the great toe, with the exception of the joint between the second metatarsal and the intermediate cuneiform bone, which lies 2 mm to 3 mm proximal to the line of the others (4.91). The first joint possesses an independent articular capsule and synovial cavity. The articular capsules and synovial cavities of the second and third joints are continuous with those of the intercuneiform and cuneonavicular joints, but are shut off from those of the fourth and fifth joints by an interosseous ligament which passes between the lateral cuneiform and the base of the fourth metatarsal bone. The bones are connected by dorsal and plantar tarsometatarsal and interosseous cuneometatarsal ligaments.

The dorsal ligaments are strong, flat bands. The first metatarsal is joined to the medial cuneiform bone by an articular capsule; the articular capsules of the remaining tarsometatarsal joints are largely blended with their dorsal and plantar ligaments. the second metatarsal receives three bands, one from each cuneiform; the third, one from the lateral cuneiform; the fourth one from the lateral cuneiform, and another from the cuboid and The fifth, one from the cuboid bone.

The plantar ligaments consist of longitudinal and oblique bands, disposed with less regularity than the dorsal ligaments. Those for the first and second metatarsal bones are the strongest; the second and third metatarsal bones are joined by oblique bands to the medial cuneiform; the fourth and fifth metatarsal bones are connected by a few fibres to the cuboid.

The interosseous cuneometatarsal ligaments are three in number. The first is the strongest; it passes from the lateral surface of the medial cuneiform to the adjacent angle of the second metatarsal bone and is constant (4.91). The second connects the lateral cuneiform with the adjacent angle of the second metatarsal bone; it does not completely divide the joint between the second metatarsal and the lateral cuneiform and is inconstant. The third connects the lateral angle of the lateral cuneiform with the adjacent side of the base of the fourth metatarsal bone.

Movements. The movements permitted between the tarsal and metatarsal bones are limited to gliding of the bones upon each other. This movement is very limited in range except in the case of the joint between the medial cuneiform and the first metatarsal bone, where an appreciable amount both of flexion and extension and rotation of the metatarsal bone can be obtained passively when the muscles concerned are relaxed. These movements are carried out actively in standing and walking and form part of the mechanism by which the foot is kept in adequate ground contact. Because of the mortice-like recession of the proximal end of the second metatarsal between the medial and lateral cuneiform bones, it is the least mobile of the metatarsals. The shape of the proximal articular surfaces of the latter is such that they become more closely pressed together in plantar flexion (the close-packed position of the tarsometatarsal joints), and they splay out in dorsiflexion.

It is the movements of **pronation** and **supination** which make it possible to keep the feet in plantigrade apposition to flat surfaces in a wide range of positions, from straddling to actual crossing of the feet. This is usually ascribed to inversion or eversion, but these affect the whole foot below the talus, and even the talus if the slight additional movements of adduction and abduction possible at the plantar-flexed ankle joint are added.

When the feet are planted far apart, the whole foot inverts, but there is a change between the relative positions of the forefoot and the hindfoot, to maintain full plantigrade contact. This is necessary because the talus is tilted medially with the tibiofibular mortice, and is followed by the calcaneus as soon as the limited degree of movement between them is exhausted. Were the rest of the foot to follow the same medial tilting (inversion) weight would be largely taken from the lateral border and concentrated on the medial. To correct this, and bring about a more even distribution of weight in full plantigrade contact with the ground, the forefoot rotates in the opposite sense. That is, it *supinates* or *untwists*. Much of this movement occurs at the transverse tarsal joint, but other tarsal and tarsometatarsal joints also contribute and in particular the joint between the medial cuneiform and first metatarsal. Conversely, when the feet are close together or even crossed, full ground contact can only be maintained if the forefoot is maximally *twisted* or *pronated* relative to the calcaneus and talus. (See also pp. 501–503.)

THE INTERMETATARSAL JOINTS

The *base* of the first metatarsal bone is not connected with that of the second by any ligaments; in this respect the great toe resembles the thumb. (It differs, of course, in being so connected at its distal end.) A small bursa often occurs between the lateral side of the base of the first metatarsal bone and the medial side of the shaft of the second (**4.91**).

The bases of the second, third, fourth and fifth metatarsal bones are connected by dorsal, plantar and interosseous ligaments.

The heads of all the metatarsal bones are connected indirectly by the deep transverse metatarsal ligaments (p. 500).

The dorsal and plantar ligaments pass transversely between the bases of the adjacent bones.

The interosseous ligaments consist of strong transverse fibres which connect the rough nonarticular portions of the adjacent surfaces (**4.91**).

Movements. The movements between the tarsal ends of the metatarsal bones are limited to a slight gliding of the articular surfaces one upon another when the anterior part of the foot is working under load (cf. movements of the intercuneiform joints, etc., p. 498).

THE SYNOVIAL ARRANGEMENT OF THE TARSUS AND METATARSUS

The synovial cavities (**4.91**) in the tarsus and metatarsus are six in number, one for the subtalar, a second for the talocalcaneo-navicular, a third for the calcaneocuboid, a fourth for the cuneonavicular, intercuneiform, and cuneocuboid articulations, the articulations of the intermediate and lateral cuneiform bones

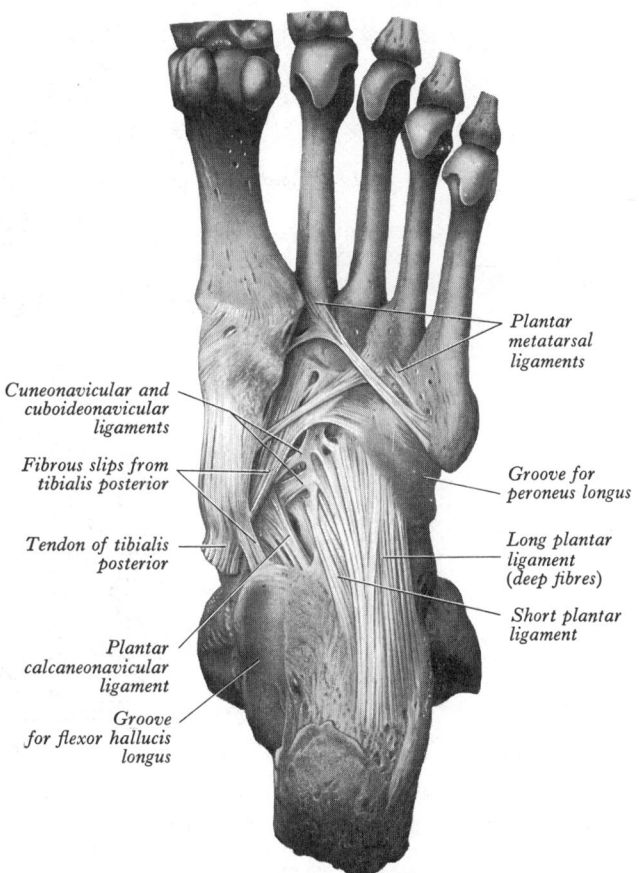

Plantar metatarsal ligaments

Cuneonavicular and cuboideonavicular ligaments

Fibrous slips from tibialis posterior

Tendon of tibialis posterior

Plantar calcaneonavicular ligament

Groove for flexor hallucis longus

Groove for peroneus longus

Long plantar ligament (deep fibres)

Short plantar ligament

S.W.W.

4.90 The ligaments on the plantar surface of the left foot. The long plantar ligament has been removed. (Drawn from a specimen in the Museum of the Royal College of Surgeons of England, by kind permission of the Council.)

with the bases of the second and third metatarsal bone, and the adjacent surfaces of the bases of the second, third and fourth metatarsal bones; there is a fifth for the medial cuneiform with the metatarsal bone of the great toe, and a sixth for the articulation of the cuboid with the fourth and fifth metatarsal bones. A small synovial cavity is sometimes found between the contiguous surfaces of the navicular and cuboid bones; it usually communicates with that between the cuboid and lateral cuneiform bones.

The Metatarsophalangeal Articulations

The metatarsophalangeal joints are ovoid or ellipsoid; they are formed by the reception of the rounded heads of the metatarsal bones in shallow cavities on the bases of the proximal phalanges. These are all ovoid surfaces. They lie 2·5 cm proximal to the webs of the toes.

The *articular surfaces* cover the distal and plantar surfaces of the heads of the metatarsal bones but do not extend on to their dorsal surfaces. The plantar part of the head of the first metatarsal presents two longitudinal grooves separated by an intervening ridge. Each groove articulates with a sesamoid bone embedded in the plantar part of the capsule of the joint, formed by the tendons of the intrinsic muscles of the hallux. The articular surface of the base of the proximal phalanx is concave in all diameters.

The ligaments of the joints are capsular, plantar, deep transverse metatarsal and collateral.

The fibrous capsules surround the joints and are attached to the margins of the articular surfaces. Dorsally, they are thin and may be separated from the tendons of the long extensors by small bursae: they are inseparable from the deep surfaces of the plantar and collateral ligaments.

The plantar ligaments are thick, dense, fibrous structures. 499

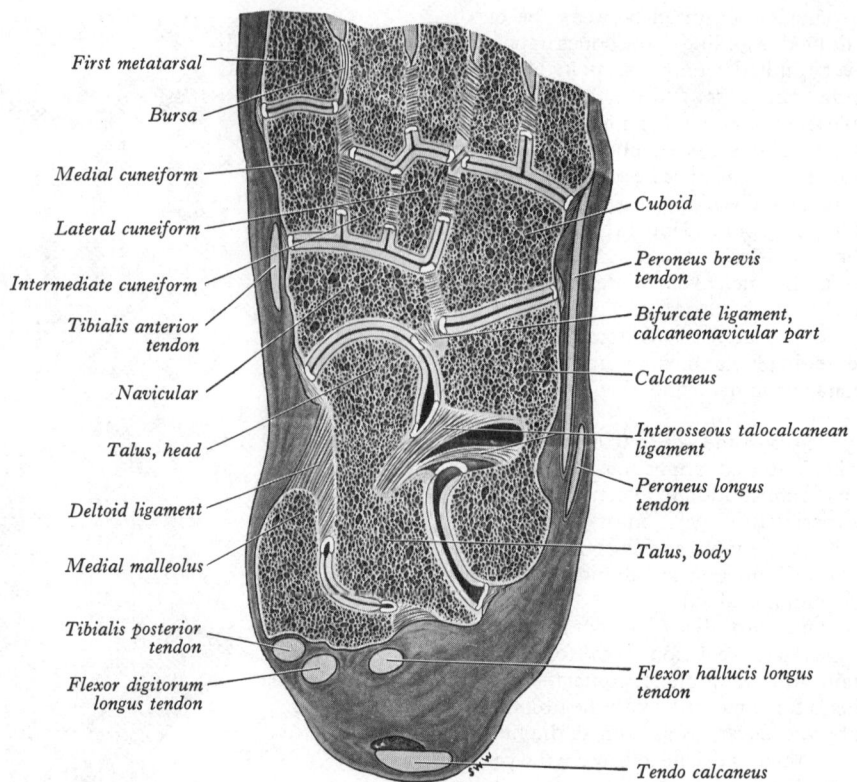

First metatarsal
Bursa
Medial cuneiform
Lateral cuneiform
Intermediate cuneiform
Tibialis anterior tendon
Navicular
Talus, head
Deltoid ligament
Medial malleolus
Tibialis posterior tendon
Flexor digitorum longus tendon

Cuboid
Peroneus brevis tendon
Bifurcate ligament, calcaneonavicular part
Calcaneus
Interosseous talocalcanean ligament
Peroneus longus tendon
Talus, body
Flexor hallucis longus tendon
Tendo calcaneus

4.91 An oblique section through the right foot, showing the synovial cavities of the intertarsal and tarsometatarsal joints. Superior aspect. *Note.* The section passed below the joint between the medial cuneiform and the base of the second metatarsal bone; no synovial joint was present between the navicular and cuboid bones.

They are placed on the plantar surfaces of the joints in the intervals between the collateral ligaments, to which they are connected; they are loosely united to the metatarsal bones, but are firmly fixed to the bases of the proximal phalanges. Their margins are continuous with the deep transverse metatarsal ligaments, and their plantar surfaces are grooved for the flexor tendons, the fibrous sheaths of which are connected to the sides of the grooves; the deep surfaces of the ligaments form parts of the articular facets for the heads of the metatarsal bones.

The deep transverse metatarsal ligaments consist of four short, wide, flattened bands which connect the plantar ligaments of adjoining metatarsophalangeal joints to one another. Their dorsal surfaces are related to the interossei and their plantar aspects to the lumbricals and the digital vessels and nerves. They closely resemble the deep transverse metacarpal ligaments (p. 472), but, in contrast, they are connected to the plantar ligament of the first metatarsophalangeal joint.

The collateral ligaments are two strong, rounded cords, on each side of each joints; each is attached by one end to the dorsal tubercle on the side of the head of the metatarsal bone, and runs obliquely forwards and downwards to reach the corresponding side of the base of the phalanx.

Movements. The active movements possible at the metatarsophalangeal joints are very similar to those permitted at the corresponding joints in the hand but differ in their range. In marked contrast to the condition in the hand, the range of active extension (50–60°) is greater than the range of flexion (30–40°), and this is associated with the requirements of walking. This is especially the case in the metatarsophalangeal joint of the great toe, where flexion is limited to a few degrees but extension may be possible up to 90°. In this connexion it should be remembered that when the foot is on the ground the metatarsophalangeal joints are already extended to rather more than 25° owing to the participation of the metatarsal bones in the longitudinal arches of the foot (4.92 A, B). The *passive* range of extension–flexion of these joints is increased to 90° (extension) and 45° (flexion), according to Kapandji (1974). Adduction is associated with flexion and abduction with extension, except that abduction of the little toe is

always associated with a slight degree of flexion. As in the hand, the *accessory movements* comprise gliding movements and rotation of the phalanges around their long axes.

Muscles producing the movements.

Flexion. Flexor digitorum brevis, lumbricals and interossei, assisted by flexor digitorum longus and accessorius. In the little toe the flexor digiti minimi brevis assists and in the great toe the flexores hallucis longus et brevis are the only muscles concerned.

Extension. Extensores digitorum longus et brevis and extensor hallucis longus.

Adduction. Of the great toe, adductor hallucis; of the third, fourth and fifth toes, the first, second and third plantar interossei respectively.

Abduction. Of the great toe, abductor hallucis; of the second toe, to the medial and lateral sides, the first and second dorsal interossei respectively; of the third and fourth toes, the corresponding dorsal interossei; of the little toe, abductor digiti minimi.

Note. The line of reference in adduction and abduction passes through the *second toe*, whose metatarsal is the least mobile.

THE INTERPHALANGEAL ARTICULATIONS

The interphalangeal joints are hinge joints in which the trochlear surface on the head of a phalanx articulates with a reciprocally curved surface on the base of the neighbouring phalanx. Each has an articular capsule and two collateral ligaments. The arrangement of these ligaments is similar to that in the metatarsophalangeal joints (p. 499). The plantar surface of the articular capsule is strengthened to form a fibrous plate, similar to the plantar metatarsophalangeal ligament. This is often termed the *plantar ligament.*

Movements. The only active movements possible in the joints of the digits are flexion and extension; these movements are of greater amplitude between the proximal and middle phalanges than between the middle and distal. The amount of flexion is very considerable, but extension is limited by the tension of the flexor muscles and by the plantar ligaments.

The *accessory movements* comprise abduction, adduction and rotation.

Muscles producing the movements.

Flexion. Flexores digitorum longus, brevis et accessorius, flexor hallucis longus.

Extension. Extensor hallucis longus, extensores digitorum, longus et brevis.

Movements of the Foot

Movements of the foot take place both when it is off the ground and free to move on the leg, and when it is on the ground and bearing weight or taking thrust. In the latter case the movements are more limited in range and are, to some extent, imposed on the foot by body weight acting through the femur, tibia and talus; but they are also the resultant of muscular contraction. (See also pp. 617–619.)

Active movements occur at the talocrural, talocalcaneonavicular and subtalar joints. At the talocrural joint the movements are almost restricted to dorsi- and plantar-flexion, though a small degree of rotation may occur in plantar-flexion (abduction and adduction). At the talocalcaneonavicular and subtalar joints the range of movement is greater, and it is here that the inversion and eversion largely occur.

When the foot is on the ground in the resting position, the body weight brings about some degree of supination with flattening of the longitudinal arches. About one-third of the weight borne by the fore part of the foot is carried through the head of the first metatarsal. When this position of rest, as in standing, is changed to the active position on commencing to walk, the foot becomes pronated by muscular effort; the head of the first metatarsal (and to a lesser extent the second) is depressed and the longitudinal arch is accentuated to its maximum height (Hicks 1953).

Similar movements can be imposed on the weight-bearing foot by active rotation of the femur, lateral rotation of which is transmitted through the tibia to the talus. This entails passive inversion of the foot at the subtalar joints. Medial rotation of the femur has the opposite effect.

It is important to note at this point that, when the foot is relatively immobilized in contact with the ground, muscles which move it when freely suspended may exert their effects upon the leg itself. For example, the dorsiflexors of the foot can pull the leg forwards at the ankle joint when the foot is prevented from moving by contact with the ground.

The foot has two equally important functions to carry out—to support the weight of the body in standing or progression, and to act as a lever to propel the body forwards in walking, running or jumping.

To fulfil the first function the foot must be an adequate platform on which to spread the stresses of standing and moving; it must also be pliable enough to do so on uneven and sloping surfaces. To fulfil the second function the foot must be capable of transformation into a strong and adjustable lever which will not collapse under body weight or powerful muscular thrust. A segmented lever such as the foot can best meet such stresses if it is built upon an arched form.

In the child, due to the fatty connective tissue in the sole, the foot may appear flattened; and soft tissues also modify external appearances to a variable extent at all ages. Nevertheless, the human foot, unlike that of other primates, normally presents an arched or curved form in its skeletal basis, and this is usually associated with a visible concavity in the sole.

It is worthy of note at this point that the word 'arch', as applied to the human foot, has perhaps been customarily interpreted in too architectural a sense, and this has led to some rigidity in classical descriptions of the curved form of the foot, and to differences of opinion based sometimes more upon linguistics than fact. In common usage the word has several meanings, and doubtless the expression, 'arches of the foot', has different implications. As a starting point, it is perhaps wisest to accept the simplest meaning, which in this case implies little more than a curved form, concave on the plantar aspect. Such an arch need not be confused with a piece of static masonry, with two extremities

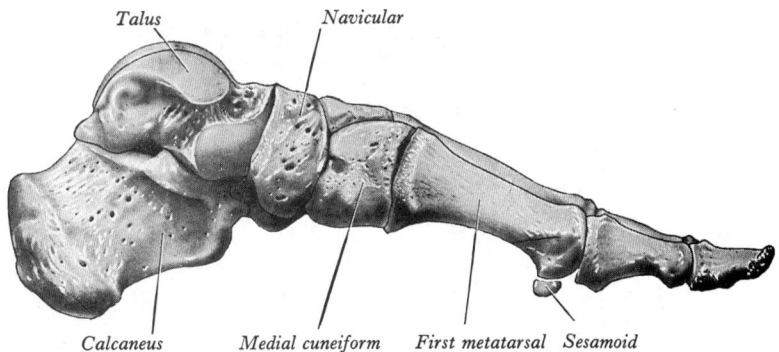

Talus *Navicular*

Calcaneus *Medial cuneiform* *First metatarsal* *Sesamoid*

4.92A The skeleton of the left foot. Medial aspect. Note the height of the medial part of the longitudinal arch and compare with **4.92B**.

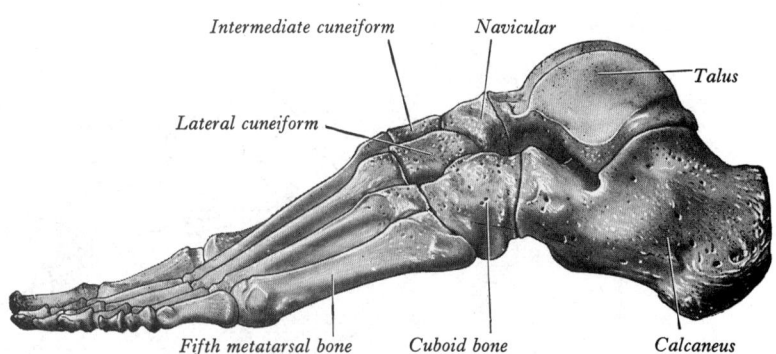

Intermediate cuneiform *Navicular*

Lateral cuneiform *Talus*

Fifth metatarsal bone *Cuboid bone* *Calcaneus*

4.92B The skeleton of the left foot. Lateral aspect.

set on terra firma and an intermediate structure of keystone pattern. The pedal arch is a dynamic arrangement, from which muscles and ligaments are functionally inseparable; it is, moreover, much of the time off the ground at its calcanean end. In the account which follows, therefore, the term 'arch' is used to denote no more than a curved form, just as the back is merely curved when it is 'arched'.

The curved or bowed form of the foot is customarily analysed into subsidiary components—longitudinal and transverse arches—and this has some value as an initial analysis. These arches, especially the longitudinal and particularly its medial part, vary in height in different individuals; even more importantly, being dynamic, they vary thus in the same individual in different phases of activity. The arched form is often said to be largely dependent on bony configurations and the effects of ligamentous ties, the muscles associated with the arches playing a secondary role (Jones 1941; Hicks 1955). On the other hand, clinical experience points to muscular insufficiency as the commonest cause of flat foot, in which ligaments elongate and bones ultimately alter their configuration. It is therefore unwise, perhaps, to pick one or other factor for particular emphasis, especially since all work together in the functioning of the living foot. Nevertheless, loading experiments on the knee, both in amputated legs and by electromyography in the living, strongly suggest that in *standing* ligaments play the major role. Immediately movement upon the foot occurs, however, the muscular factor comes into play.

The longitudinal curvature of the foot is usually accounted as consisting of medial and lateral arches or components, and this is to some degree justified by the arrangement of bones in the foot and by functional differences in its medial and lateral parts.

The medial arch comprises the calcaneus, talus, navicular, the cuneiforms and the medial three metatarsal bones. Its summit is the superior articular surface of the talus, taking the full thrust of the tibia; from this, stresses are passed backwards to the calcaneus, and forwards through the navicular, cuneiforms, to the medial three metatarsals. When the foot is on the ground, forces are transmitted to the ground through the heads of the three metatarsals and the plantar aspect of the calcaneus (more

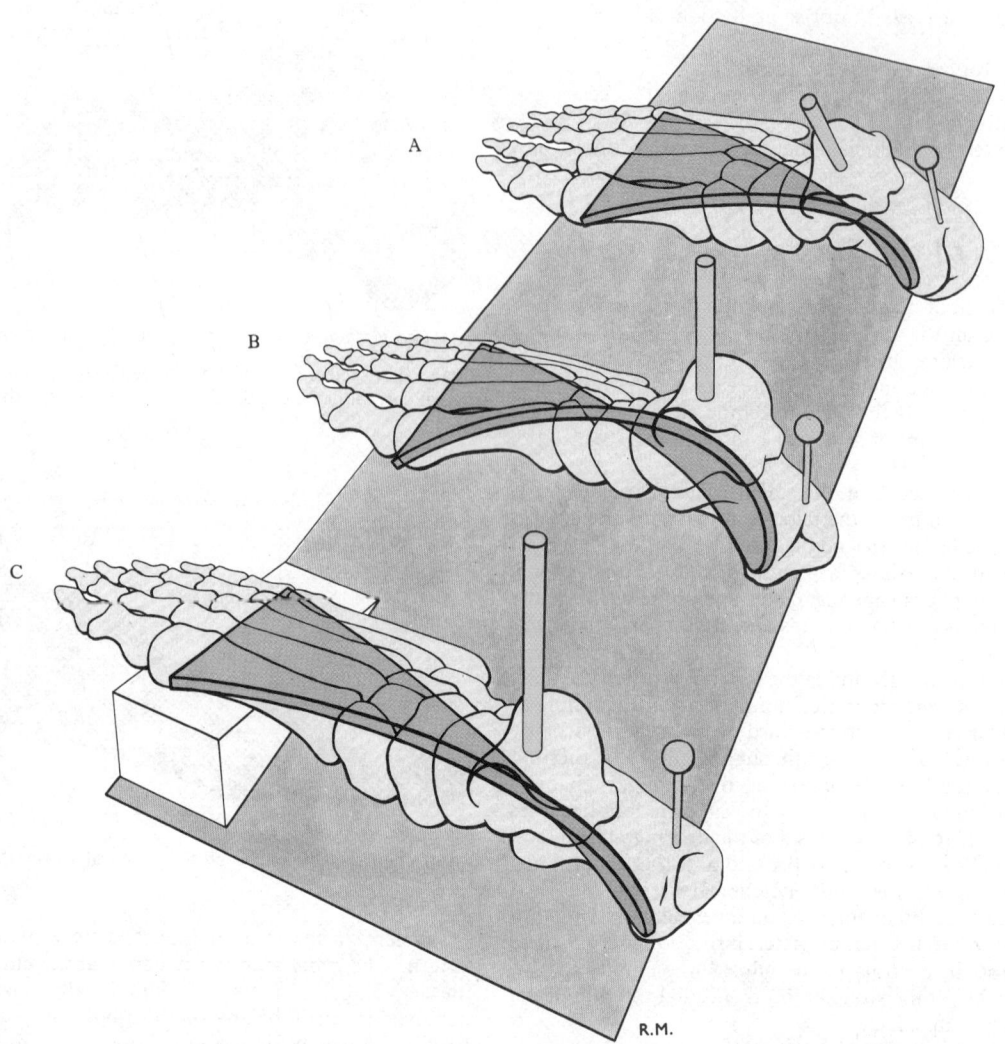

R.M.

4.93 The concept of the pedal skeleton as a twisted plate which may be untwisted (supination) or further twisted (pronation) during the maintenance of a plantigrade stance in various positions of the foot. (Based upon MacConaill 1945, 1950.) A The pedal skeleton in supination, as in standing with the feet widely separated. Note the marked medial tilting of the talus and, to a lesser degree, of the calcaneus, and the depression of the medial longitudinal arch. B Relative pronation of the foot, as in standing with the feet close together. C Supination of the foot when standing on an inclined surface; if the position of the wedge had been reversed, the pedal skeleton would, of course, approach maximal pronation.

especially its tuberosity). This arch is considerably higher, more mobile and resilient than the lateral. Flattening of the medial arch progressively tightens up the plantar ligaments of all the joints involved, including the plantar calcaneonavicular (p. 497), and the plantar fascia. The tibialis posterior, flexor digitorum longus, flexor hallucis longus and intrinsic mucles in the foot all aid in controlling the medial arch, with the advantage of being almost infinitely adjustable, unlike the ligaments. Nevertheless, they are often considered to play a secondary part, and it is claimed that they are not usually active in standing, but come into marked activity in any movement which involves elevation of the arch. This view, as regards static posture, does not accord with clinical views on the aetiology of flat foot.

The lateral arch is composed of the calcaneus, cuboid and lateral two metatarsal bones. Its summit is considered to be the subtalar articulation, and it is hence skeletally much lower than the medial arch. Its chief joint is the calcaneocuboid part of the 'transverse tarsal joint' (p. 495), with a markedly limited range of movement. The lateral arch is characteristically low, of limited mobility, and built to transmit weight and thrust to the ground, rather than to provide a mechanism for the absorption of such forces. As it flattens under stress, the long plantar and plantar calcaneocuboid ligaments are tightened; muscles of particular significance in the activities of the arch are the peroneus longus and the short muscles of the fifth toe. The lateral arch makes more extensive contact with the ground than the medial, under the stress of weight and thrust. With the foot flat on the ground its anterior and posterior extremities (heads of lateral two metatarsals and calcanean tuberosity) of course transmit these forces; but as the arch flattens, an increasing fraction of the load is transmitted through the soft tissues inferior to the whole of the arch. In fact, in the living part, the whole of the lateral border usually contacts the ground, whereas the medial border does not, exhibiting a concavity in most subjects, even when standing. This accounts for the usual outline of the human footprint, though this varies according to the positioning of the feet, whether apart or together (4.93) and according to the development of soft tissues within the arch. As soon as the heel is raised, preparatory to walking or standing on tip-toe, the toes become extended and the muscular structures (including the plantar aponeurosis) tighten up in the sole with a consequent accentuation of the longitudinal arches. It is deduced that in this phase of activity tension diminishes in the deeper plantar ligaments (Hicks 1955).

Since the sole is to some extent transversely concave, both in skeletal configuration and in external form, it is usual to describe a series of transverse arches, most developed inferior to the metatarsus and adjoining distal part of the tarsus. Apart from the region of the metatarsal heads, and to some degree along the lateral border of the foot, the transverse arch mechanism is not capable of transmitting forces to the ground though the subjacent soft tissues of the sole; on the medial side only the heads of the metatarsals can do so. For this reason the foot has been likened to a half-dome, with a concavity directed downwards and medially, which becomes a complete dome when the feet are placed together. The peroneus longus has a particular function in maintaining the transverse curvature.

The foregoing, largely structural analysis, is chiefly useful as a prelude to considerations of the integrated activities of the living foot, direct observation of which coupled with the findings of electromyography, kinesiology and clinical data are all necessary to a complete comprehension of the dynamic events in the foot during natural use. Unfortunately, disagreements persist, with the usual indication of uncertainties. The following remarks are hence not to be taken as dogmatic.

In standing, with body weight only to support, there is a tendency to relax the intrinsic and extrinsic muscles of the foot, and to rely upon the tension in plantar ligaments to tie the bones into their arched form. This they can do only if the longitudinal arches are allowed to sink by muscular relaxation. If the feet are close together, the medial arch is more elevated than when they are set well apart. That is to say, a degree of inversion with supination becomes increasingly apparent as the feet are separated. This sag of the medial arches can, of course, be taken up by voluntary contraction of muscles such as the anterior tibial.

Pronation and supination of the foot ensure that, in standing, whatever the position of the feet, a maximal weight-bearing area, from the metatarsal heads along the lateral border to the calcaneus, is close to the ground. The twist imparted to the foot by pronation, and to some extent undone in supination, prompted MacConaill (1945, 1950) to liken the foot to a twisted plate (4.93), able to untwist in a resilient manner. This mechanism not only ensures adequate ground contact whatever the state of angulation between the foot and leg, but also imparts an adaptable resilience to the foot in standing and progression. It is likely, too, that such a form adds strength to the foot in leverage, perhaps by spiral tautening of ligaments.

In walking, the suspended foot, with arches accentuated by muscular action, is swung forward until the heel meets the ground. As it plantar-flexes, the lateral border, or arch, and then the metatarsal heads roll into contact with the ground, the foot being supinated to bring about maximum contact. As the foot is now raised, heel first, pronation (with elevation of the medial arch) occurs, and thrust is largely transferred to the ball of the hallux. At this point, the ankle (in dorsiflexion) and the metatarsus (well elevated) are admirably adapted to meet the stresses of maximum thrust. As in the repetitive activities at the hip and knee joints, there is a cycle of events from a position of near close-pack, coupled with maximal effort, through a return to the mobile range of the joints which is merely a preparation for a re-entry into the phase of thrust. For a wide and detailed analysis of foot movements consult Kapandji (1970).

5

MYOLOGY

Introduction

One of the characteristics of living organisms is their ability to react to changes in their environment by an appropriate response, which may be chemical, electrical, photic or mechanical. All such responses involve the utilization of metabolic energy by some sort of effector system. Several types of mechanical effector system exist. One, common to probably all nucleated cells, is that which produces cytoplasmic streaming, exemplified particularly well in unicellular organisms such as *Amoeba* and in the motile cells of vertebrates typified by macrophages and leucocytes. Intracellular streaming also forms an important transport mechanism which facilitates diffusion; the axoplasmic flow in neurons is a well-studied example of this.

Other examples of mechanical effectors include the cilia and flagella of motile cells and of the respiratory and genital tracts, the mitotic apparatus of all dividing cells, and the relatively unspecialized contractile systems of endothelial and myo-epithelial cells. It is interesting that in all these examples the proteins *actin* and *myosin*, or proteins similar to them, have been demonstrated in the region of the mechanical responses (*see* p. 36). Muscles, also, contain high proportions of these proteins in their total mass, and can be viewed as cells which are specialized for the synthesis of the contractile apparatus, which is present in a rudimentary form in most cells; muscle cells also possess specially excitable plasma membranes capable of initiating cellular contractions. The organization of such cells into distinct groups, their association with surrounding tissues such as those of the skeleton, and their coordination by the nervous system, transform them into an effector system capable of a wide range of highly complex actions.

Muscle cells are mainly derived from mesenchymal cells at various sites throughout the body of the embryo, and may differentiate along one of three pathways to produce physiologically and anatomically distinct muscle types. Two of these categories are similar in that they possess a highly organized structure capable of relatively fast contractions, and constitute the types of **striated muscle**. The latter include firstly those muscles usually associated with the bony skeleton, composed of a tissue termed here **skeletal muscle**. It is also, less appropriately, sometimes known as voluntary muscle because it is, in some sites, under direct voluntary control. The second type is peculiar to the myocardium of the heart and is termed **cardiac muscle**. Both skeletal and cardiac muscle, when viewed microscopically, exhibit regularly spaced transverse bands and are therefore often loosely grouped as either **striated muscle**, or **striped muscle**. It should be added that the terminology chosen is not wholly satisfactory since not all 'skeletal' muscle is attached directly to the hard skeleton but may transmit forces to connective tissue sheets instead. It must also be pointed out that the terms 'striated' or 'striped' muscle are often used exclusively for skeletal muscle.

The third type of muscle has a relatively poorly organized contractile apparatus; it is able to make prolonged tonic contractions of considerable extent. This type is termed here **non-striated muscle**, but it is also known as smooth, plain or, alternatively, involuntary muscle, since it is rarely under direct volitional control. It is found in many regions of the body such as the walls of the alimentary tract, blood vessels, the urinary and genital tracts, and also as the arrector pili muscles of the skin, where slow but maintained responses are required. Its particular type of contraction makes it especially suitable for the regulation of the internal environment, and disturbances in its function often result in impaired homeostasis.

The various muscle types originate from distinct regions of the embryo (*see* p. 154). Skeletal muscles are largely derived from the myotomes of the paraxial mesodermal somites; others, however, appear to differentiate *in situ*, e.g. the extrinsic ocular muscles in condensations of premandibular and maxillo-mandibular mesenchyme near the developing optic cup (p. 154), the tongue muscles in pharyngeal floor mesenchyme (often considered a derivative of occipital myotomes—p. 197), the limb musculature in condensed limb-bud mesenchyme of somatopleuric origin (p. 153), and the various striated 'special visceral' muscles which develop in condensations of the unsplit lateral plate mesenchyme (of mixed origin) in the branchial arches and post-branchial region (p. 150). Cardiac muscle stems from splanchnopleuric mesenchyme of the primitive pericardium, whilst the non-striated muscle of the viscera also arises in splanchnopleuric cells elsewhere, or those derived from intermediate mesoderm. Vascular non-striated muscle may, however, develop at any point from unspecialized mesenchymal cells. The non-striated muscle of the iris is widely held to be derived from cells (of ectodermal origin) near the margin of the optic cup; similarly the various myoepitheliocytes (p. 39) are considered to have an ectodermal origin, whereas the non-striated ciliaris oculi and arrectores pili stem from local mesenchymal sources. In a number of situations fasciculi of non-striated muscle develop in close association with, or even intermingle with, skeletal or striated 'special visceral' muscle. Such sites include the anal and vesical sphincters, in relation to the tarsi of the upper and lower eyelids and the orbitalis muscle, the suspensory muscle of the duodenum (but *see* p. 1344), the intermediate zones of the oesophagus, and the many fasciculi which are admixed with the fasciae and ligaments on the pelvic aspect of the pelvic diaphragm (which some authors have called the *smooth muscle diaphragm*).

Skeletal Muscle

The units of skeletal muscle are the *muscle fibres*, each of which some regard as a 'single cell' provided with many hundreds of nuclei (i.e. each fibre is a *syncytial striated myocyte*). The fibres are arranged in bundles (*fasciculi* or *myonemes*) of various sizes and patterns within each muscle. Connective tissue fills the spaces between muscle fibres within a fasciculus, where it is known as the *endomysium*; each fasciculus is also surrounded by a stronger connective tissue sheath or *perimysium*; surrounding the whole muscle lies the more substantial *epimysium*, which is continuous on the one hand with the outer perimysial septa and externally with the connective tissues of surrounding structures.

With the light microscope skeletal muscle fibres appear as closely packed cylinders in teased preparations, as parallel ribbons in longitudinal section, and with either circular, elliptical, or polygonal profiles in cross-section, of $10-100\mu m$ in diameter in adult muscles (**5**.1, 2). Each fibre is elongated, and may stretch from one end of a muscle to the other, even achieving a length of over 30 cm in long muscles. Elsewhere they may traverse only part of the length of a muscle, ending in tendinous or other connective tissue intersections which penetrate the body of the muscle.

The flattened nuclei of muscle fibres lie peripherally in the zone immediately within the cell membrane or *sarcolemma*, whilst their cytoplasm, or *sarcoplasm*, is divided into longitudinal threads or *myofibrils* each about $1 \mu m$ in diameter (Gould 1973). In transverse sections prepared by older techniques myofibrils frequently appear aggregated in small groups, the *fields of Cohnheim*, but this is now widely regarded as artefactual and better methods of preservation show them to be evenly distributed.

In longitudinal section, or surface view, the myofibrils are seen to be traversed by striations apparently continuous right across the fibre (**5**.1), which vary in their staining reactions and optical properties. Some stain lightly with basic dyes such as haematoxylin, and only rotate the plane of polarization of light slightly, being termed therefore the *isotropic* or *I bands* (also sometimes called J bands). Others, alternating with the former, have the converse properties of staining deeply with haematoxylin and strongly rotating the plane of polarization of light (**5**.3) in a manner which indicates a highly ordered longitudinal structure—these are the *anisotropic* or *A bands* (also sometimes called Q bands). The I bands are bisected transversely by a thin line also stained by basic dyes, the *zwischenscheibe* or *Z band* (Krause's membrane). The A band is also bisected by a paler (Hensen's) line—the *H band*. This description applies to relaxed or minimally contracted muscle; with continued unopposed contraction the I and H bands narrow to extinction, but the A bands remain unaltered, and *contraction bands* appear on each side of the Z band.

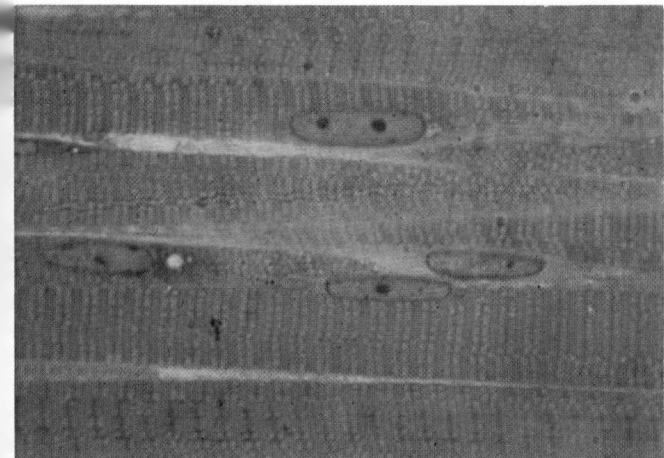

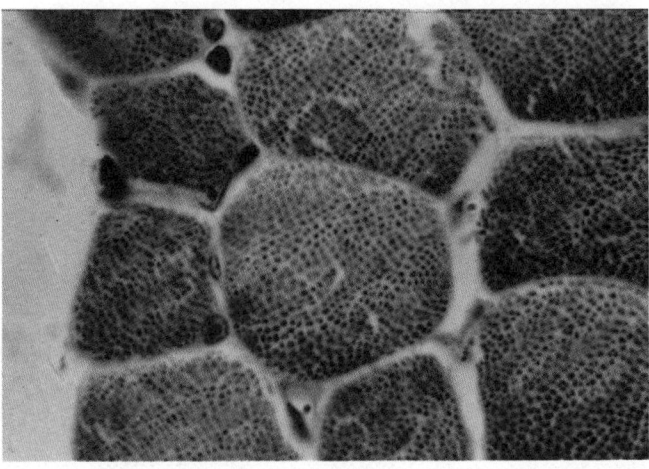

5.1 Longitudinal section of uncontracted human skeletal muscle showing characteristic banding pattern of large A bands (stained dark purple) and I bands (light pink). The Z bands are thin and transect the I bands. Araldite section stained with methyl green-P.A.S. (Material kindly supplied by Dr. R. O. Weller, Guy's Hospital Medical School.)

5.2 Transverse section of skeletal muscle showing muscle fibres containing myofibrils and muscle cell nuclei; endomysial sheaths lie between the muscle fibres. Silver stain. Magnification × 800.

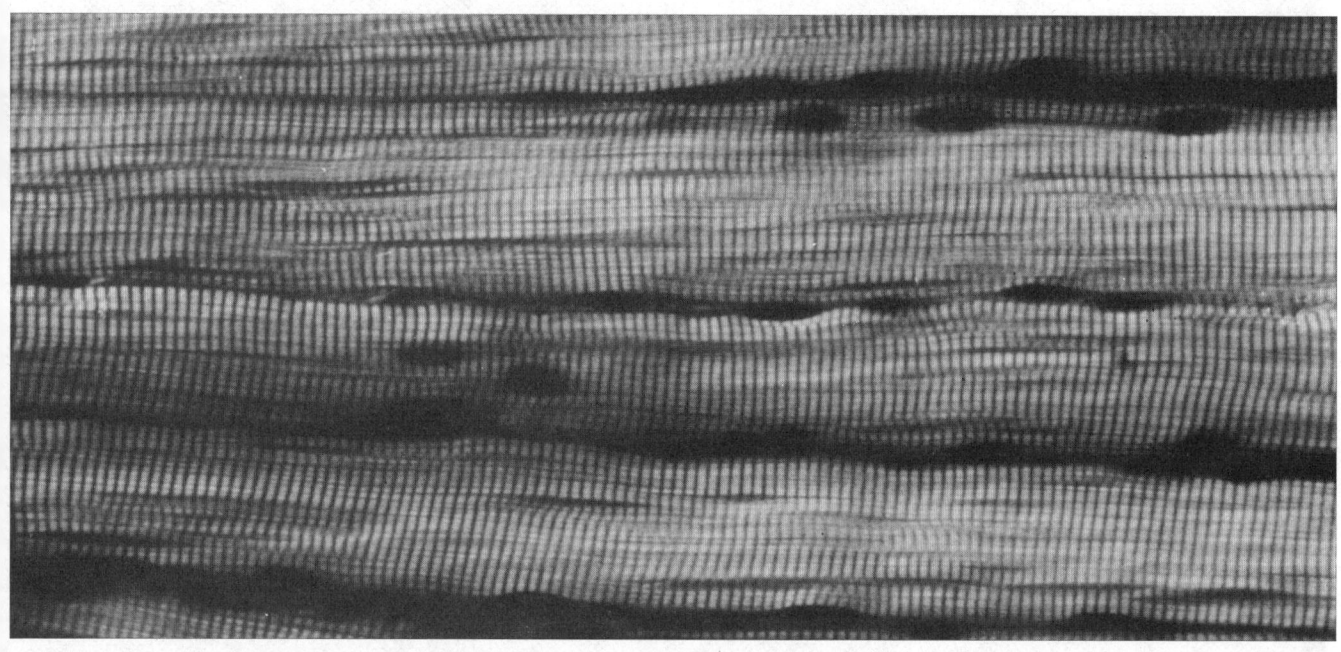

5.3 Longitudinal section of skeletal muscle viewed microscopically using crossed polaroid filters to show the birefringence of the A bands. Note also the dark muscle cell nuclei, and longitudinal striations corresponding to the myofibrils. (Photography by Mr. Kevin Fitzpatrick, Department of Anatomy, Guy's Hospital Medical School, London.) Magnification × 500.

The significance of these various bands and their behaviour during contraction were discovered independently by two groups of workers (A. F. Huxley and Neidegerke 1954; H. E. Huxley and Hanson 1954); to appreciate these concepts the ultrastructural appearance of skeletal muscle must first be described.

The Ultrastructure of Skeletal Muscle

Electron microscopy shows each myofibril to be composed of longitudinally disposed fine *myofilaments* (5.4–8). These are divided transversely by the Z bands into serially repeating regions termed *sarcomeres*, each about 2·5 μm long in resting muscle. Two types of myofilament are distinguishable in each sarcomere, fine ones about 5 nm in diameter and thicker ones about 12 nm across; these have been characterized chemically as actin and myosin respectively. The actin filaments are each attached at one end to a Z band and are free at the other to interdigitate with the myosin filaments. The latter are co-extensive with the A band of light microscopy. The I bands represent those regions of the actin filaments which do not overlap with the myosin. The H bands are the middle region of the A bands into which the actin filaments have not penetrated. Another line, the M band, lies transversely across the middle of the H band and close examination shows this to consist of fine strands interconnecting adjacent myosin filaments (*see* Page 1965, and 5.7). In contracted muscle the actin filaments have slid in relation to the myosin towards the centre of the sarcomere, so bringing the attached Z bands closer together, with shortening of the whole contractile unit. X-ray diffraction studies of unfixed muscle support this interpretation (Huxley and Brown 1969). The manner in which this sliding is brought about has been the subject of much study. The myosin filaments show lateral projections which extend towards the adjacent actin filaments (Huxley 1969, 1972), and the latter are arranged around the myosin filaments in a hexagonal pattern (5.5, 6, 7). It has been shown that, when relaxed, the lateral projections of the myosin lie close to their parent filament, whereas in contraction they project to contact the adjacent actin strands (5.8A and B). This and many other observations, indicate that muscle contraction may be caused by the successive making and breaking of cross-connexions between the thick myosin and thin actin filaments in a cyclical fashion, so pulling the thin ones between the thick ones towards the sarcomere centre. If contraction is continued to give a sarcomere length of less than 2 μm the actin filaments from each

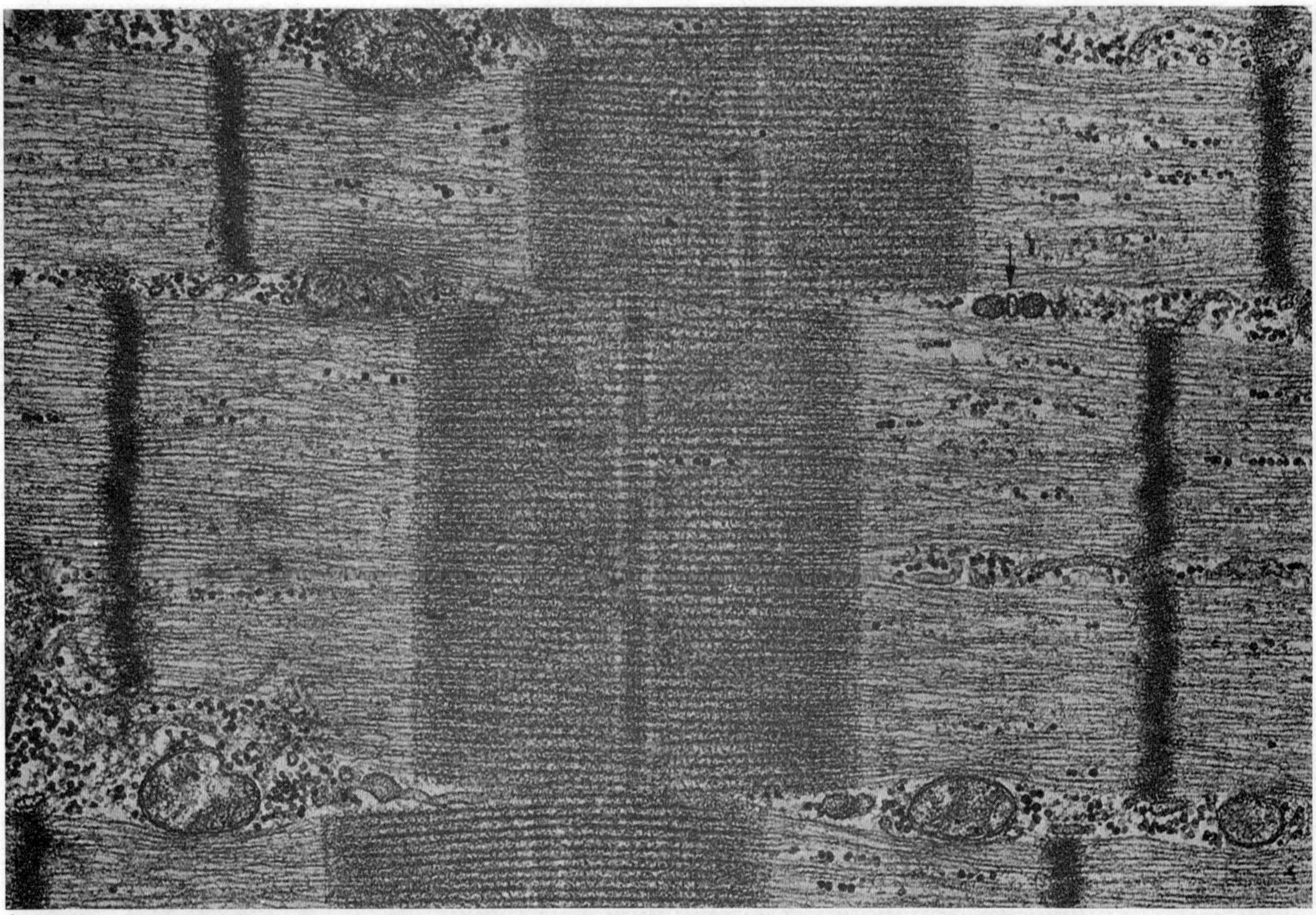

A

B

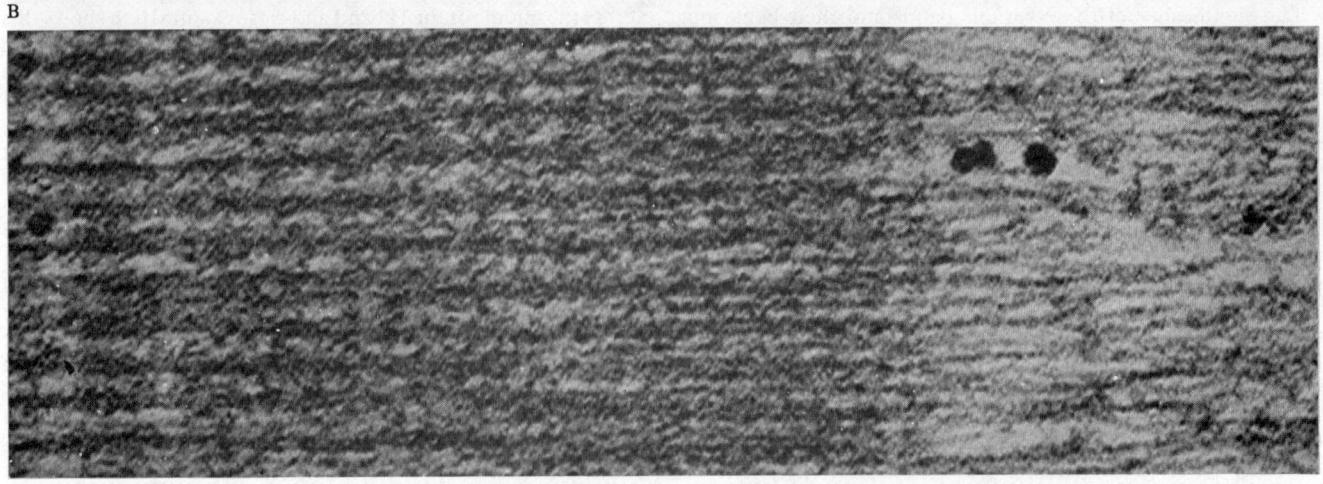

C

5.4A–C Electron micrographs of skeletal muscle in longitudinal section. A Uncontracted myofibrils. B Contraction has occurred shortening sarcomeres. C Detail of unshortened muscle showing cross bridges between actin and myosin filaments. Arrows indicate the position of the T-system components. Compare with 5.6, 7. (Material for A provided by Dr. R. O. Weller; *cf*. 5.1.) A and B × 15,000 and C × 100,000.

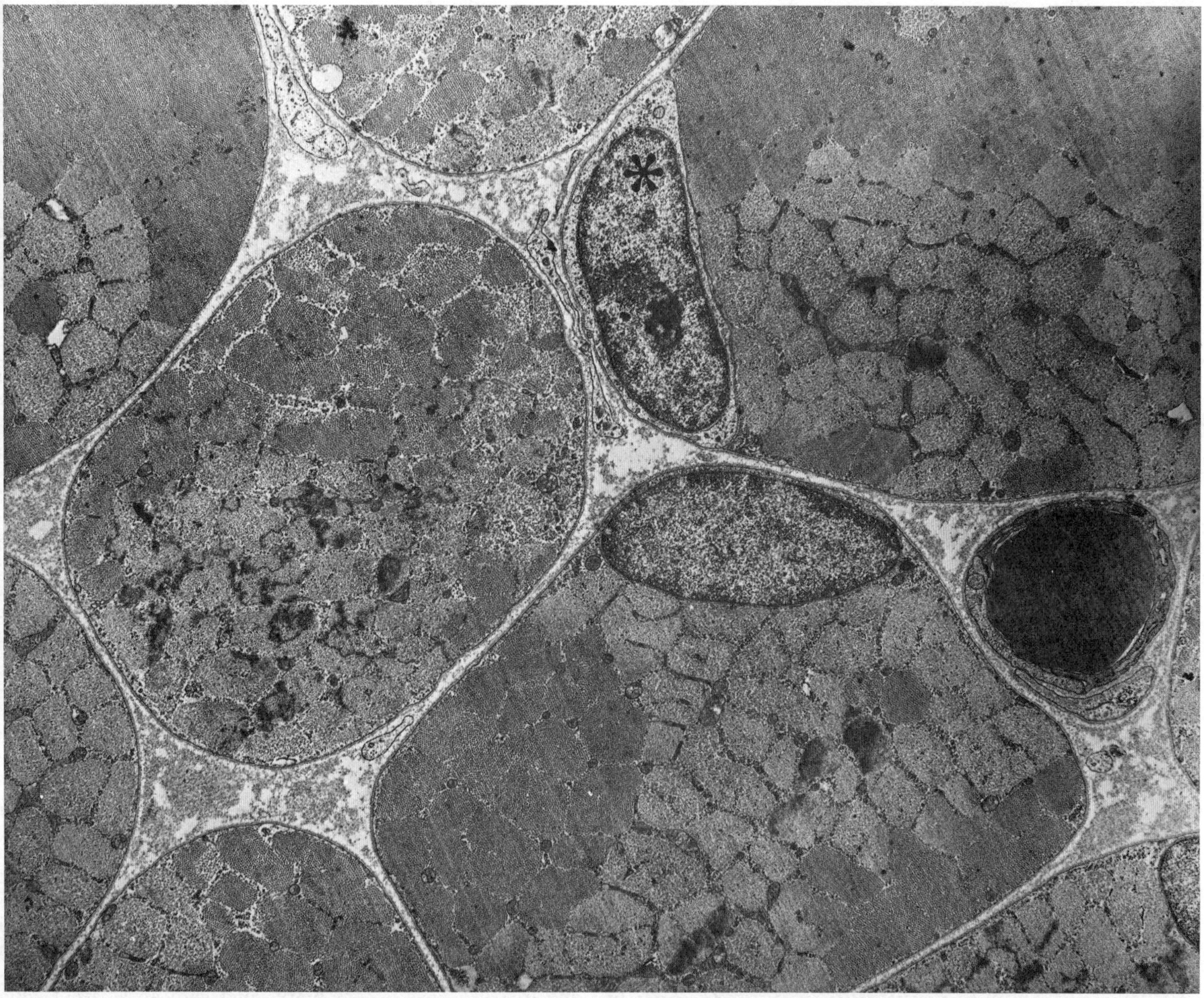

5.5A   Transverse section of skeletal muscle from a child showing muscle fibres, myofibrils and mitochondria. A satellite cell★ is also present, lying beneath the basement membrane. Immediately below this is a muscle fibre nucleus. Note the transversely cut endomysial capillary. × 4,000.

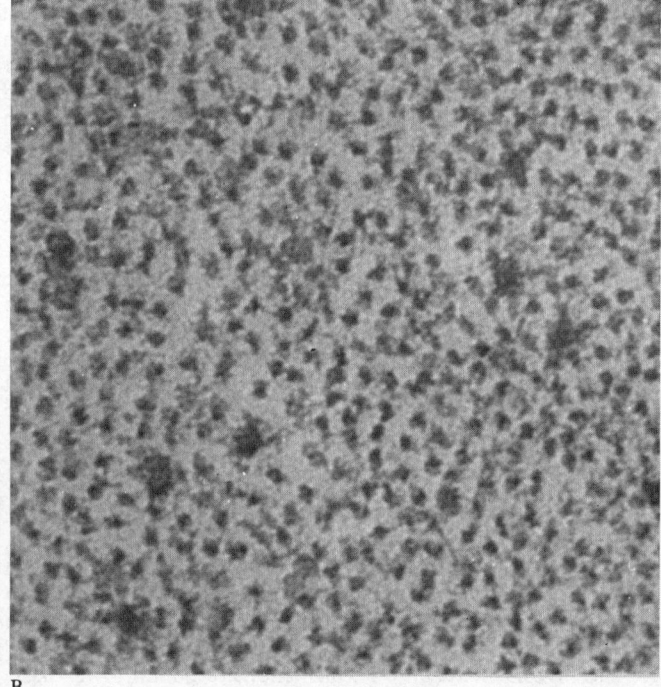

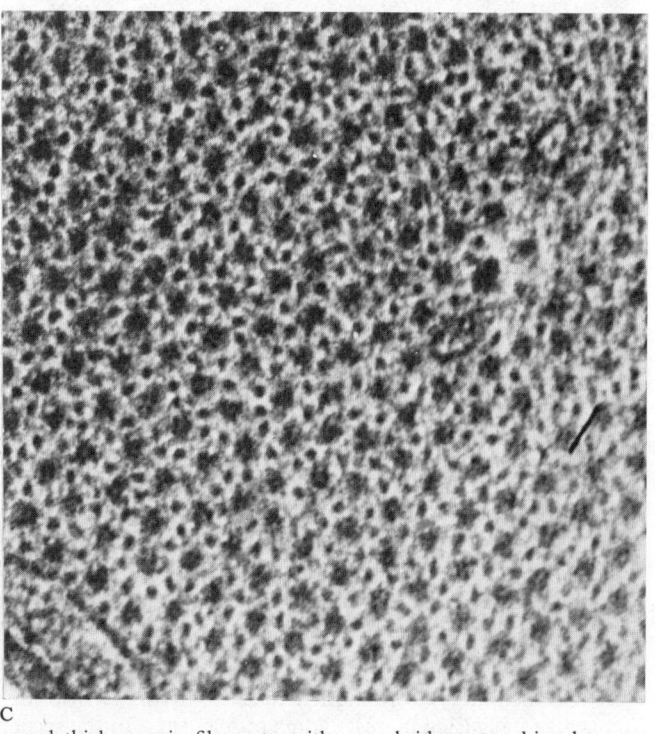

5.5B and C   Details of the configuration of actin and myosin filaments in transverse section. B is through the I band and shows only the actin filaments, C is through the A band and shows thin actin filaments arranged round thick myosin filaments, with cross bridges stretching between them. × 200,000.

end of the sarcomere come to overlap in the middle of the A band and their hexagonal arrangement is disrupted. On further contraction the Z bands meet the ends of the myosin filaments and distort them, forming the 'contraction band' of light microscopy.

As the length of a sarcomere changes, the amount of overlap between its actin and myosin filaments also changes proportionately; since the numbers of possible cross-links between the two depend upon the overlap between them, it might be expected that a muscle would generate different tensions if it were made to contract at different lengths without being allowed to shorten (isometric contraction). Measurements confirm this expectation: if a muscle is held experimentally in clamps so that its sarcomeres are in excess of $3 \cdot 5$ μm long, when little or no overlap of actin and myosin exists, practically no tension is generated. At shorter lengths of sarcomere, the observed tension rises to a

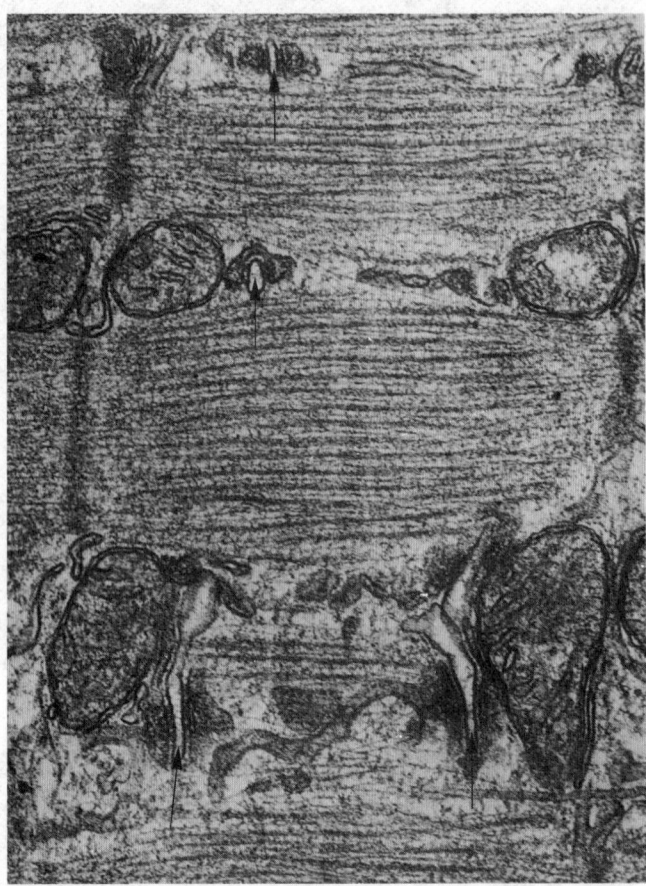

5.5D Longitudinal section through a group of myofibrils, showing the sarcoplasmic reticulum and associated transverse tubules (arrow). Note the darkly granular terminal cisternae of the sarcoplasmic reticulum where it lies adjacent to the transverse tubules. Magnification × 40,000.

maximum value at a sarcomere length of about $2 \cdot 0$ μm, when there is maximum overlap of filaments without distortion of their arrangement. At sarcomere lengths shorter than this the tension drops rapidly to zero at $1 \cdot 9$ μm which is coincident with the meeting of the myosin filaments with the Z bands.

It is useful when correlating the fine structure of muscles with their behaviour *in vivo* to consider briefly three ways in which they may contract. First, a muscle may shorten under constant load thus performing positive external work, a type of contraction termed *isotonic*; second, it may contract at a fixed length, performing no external work but creating tension in its attached structures—*isometric contraction*; third, a muscle may generate tension whilst it is being actively stretched, for example when decelerating a limb segment at the termination of a movement (p. 525). In the first type of contraction, actin-myosin cross bridges are active in causing mutual sliding of filaments. In the second, cross bridges are made and broken repetitively to maintain length under conditions of external loading. In the third case the precise

behaviour of the filaments has not been established, but it is probable that cross bridges interact in the same manner whilst the filaments are sliding apart.

In practice there are, of course, many combinations of the foregoing three 'types' of contraction, with variations in the conditions of external loading, initial length, etc. It should be appreciated, however, that in the body muscles are so arranged that their habitual contractions involve only limited length changes compared with the overall length of the muscle. Further, maximum overlap of the myosin and actin filaments occurs at a sarcomere length which is close to that of resting muscle, and it is probable that they normally operate at near maximum force per contracting unit, i.e. in the energetically most efficient range.

Detailed Structure of the Myofilaments (5.7, 8)

Each myosin filament can be separated by appropriate chemical treatment into its constituent myosin molecules of which there are about 180 members per filament. Each has a molecular weight of 500,000 daltons and consists of a long 'tail' and a 'head'; the latter, on close examination, is seen to be a double structure. On further treatment the molecule can be broken into two moieties, *light meromyosin* comprising most of the tail, and *heavy meromyosin* representing the head with part of the tail. The heavy meromyosin can be further cleaved enzymatically to release the head. These cleavage points are probably regions where the molecule can flex during muscle contraction. (Huxley 1972; Offer 1974). The myofilament is formed by the tails of the molecules, which lie parallel in a bundle, with their free ends directed towards the midpoint of the long axis. The heads project laterally from the filament in pairs or triplets, at 14·3 nm intervals. Each pair is rotated 120° with respect to its neighbours to form a spiral pattern along the filament. The precision of this arrangement probably has much to do with the rapidity with which contraction occurs in skeletal muscle. It has been demonstrated biochemically that each myosin head bears two sites for splitting adenosine triphosphate with the release of chemical energy. This enzymic action is stimulated by the presence of calcium ions, which are liberated during contraction from adjacent cytomembranes. The manner in which this energy is converted into a mechanical movement is still not clear, but may involve a change in the structure of the myosin molecule such that the head projects laterally at a different angle, and can also make a temporary bond with an adjacent actin filament.

The actin filaments (*f-actin*) are composed of globular sub-units of *g-actin*, each about 5·5 nm in diameter with a molecular weight of 42,000 daltons; these sub-units are attached end to end in two longitudinal filaments wound round each other in an extended helix with 13 sub-units to each complete turn. Another protein, *tropomyosin B* is associated with the actin filaments, and lies in the groove between the two strands of the helix; at 40 nm intervals yet another protein, *troponin*, is present, bound to the tropomyosin B. Tropomyosin B and troponin exert a regulatory effect on muscle contraction; in resting muscle, tropomyosin B prevents the myosin heads from attaching to actin, but when calcium ions are released before contraction, the ions cause troponin to push tropomyosin B away from the actin, allowing the myosin to bind. It has been proposed that the shearing movement of actin and myosin filaments may be caused by a change in the angulation of the myosin head (5.8B) after it has been bound, thus exerting a pull via the attached myosin tail to the whole filament (Huxley 1972). It is interesting that the actin filament helices on each side of a Z band spiral in opposite directions, and this polarity may underlie the directional sliding of the actin on the myosin during contraction.

At the Z band the actin filaments enter into a square lattice with a complex substructure (Landon 1970; Kelly and Cahill 1972). Many other proteins are also known to be associated with the contractile apparatus. Some of them are probably so soluble that they are beyond the reach of available structural methods, but recently, various proteins have been located in the neighbourhood of the myosin filaments to which they are bound at regularly spaced intervals. They include the so-called '*C' proteins* (Offer 1973) which may have enzymic activities, and also creatine phosphokinase (Turner *et al.* 1973).

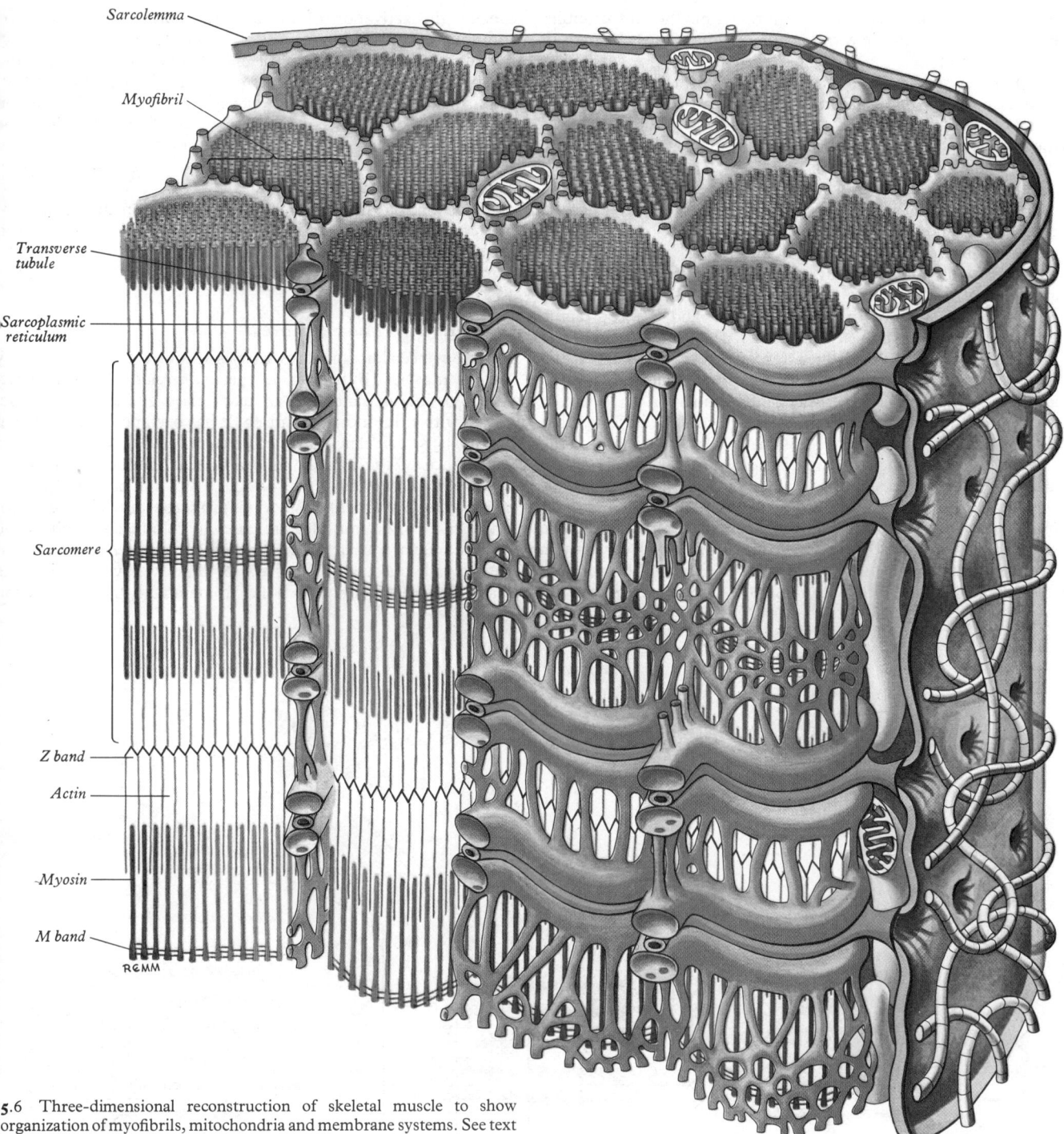

Sarcolemma

Myofibril

Transverse
tubule

Sarcoplasmic
reticulum

Sarcomere

Z band

Actin

Myosin

M band

REMM

5.6 Three-dimensional reconstruction of skeletal muscle to show
organization of myofibrils, mitochondria and membrane systems. See text
for further details. (Modified from a diagram by Professor D. Fawcett.)

The Sarcoplasm and Cytomembranes of Muscle Fibres

Ribosomes are present in the perinuclear cytoplasm and also in
that scattered between the myofibrils. The maintenance of a
protein-synthesizing machinery is obviously important for the
manufacture of enzymes and other cell components which have a
rapid rate of turnover. Perinuclear Golgi bodies, lysosomes and
lipid vascuoles also occur, while glycogen is distributed as clusters
of small granules between the myofibrils and amongst the actin
filaments of the I bands. Muscle glycogen plays an important role
in the economy of energy within the body; its immediate function
is to serve as a source of energy which can be mobilized rapidly at
the site of its utilization.

It has been found, by physiological, biochemical and
anatomical studies, that the membranes of striated muscle play a
vital role in the initiation of contraction (Taylor 1972).

The outer membrane of muscle fibres (*sarcolemma*) has the fine
structural appearance of typical plasma membranes elsewhere. It
is coated externally by a basement membrane to which adhere
reticulin and collagen fibres of the endomysium. Crushed fibres
viewed by light microscopy *in vivo* show a thin coherent sheath,
often interpreted as sarcolemma but in reality a complex of cell
membrane, basement membrane and extracellular fibres.

The sarcolemma also sends tubular invaginations, the
centrotubules, into the cell interior (forming the transverse or *T-
system* of membranes). These penetrate amongst the myofibrils,
running transversely at the level of the junctions between the A
and I bands, and they branch and anastomose in these planes
(Franzini-Armstrong 1973). There are, accordingly, two planes
of tubular structures in each sarcomere (**5**.6) and this characterizes the musculature of reptiles, birds and mammals; in contrast
other vertebrates (fish and amphibia) show centrotubules only at
the levels of the Z bands. The connexion between the lumina of
the centrotubules and the extracellular space was for some time in
dispute but has been convincingly demonstrated by infusion of
colloidal suspensions of heavy metals into the endomysium: such

511

materials penetrate deeply into the fibres within the centrotubules but do not communicate with the other membranous channels of the sarcoplasm.

The latter constitute the true intracellular membranes of the cytoplasm, or *sarcoplasmic reticulum*. This consists of a plexiform series of branching and anastomosing channels which fill much of the space between the myofibrils (*see* **5**.6). A typical feature is the presence of transverse channels, the *terminal cisternae* at the level of the A–I band junctions where they lie on both sides of, and in close apposition to, a transverse centrotubule, the complex of the three membranous channels being termed a *muscle triad*. The two membranous components show dense fibrous interconnexions, and are probably in close functional relationship. However, they are thought to play quite distinct roles in the coupling of the electrical excitation of the cell membrane and the contractile events consequent upon it. Electrical recordings show that the sarcolemma is depolarized when appropriately stimulated at the motor end plate, and that it propagates a wave of depolarization to all parts of the muscle fibre. Electrical stimulation of restricted areas of cell membrane with fine microelectrodes indicates that electrical changes are carried along the membranes of the centrotubular system, and so reach the depths of the fibre. It has also been shown biochemically that the sarcoplasmic reticulum is capable of binding calcium ions strongly to its constituent membranes and of releasing them under appropriate ionic stimuli. It is thought that when electrical disturbances occur in the T-system these cause the release of calcium and magnesium from the sarcoplasmic reticulum into the adjacent myofibrils, so

causing the activation of the myosin-bound ATPase and th consequent contraction of the muscle. As electrical stability returns to the sarcolemma at the end of contraction, calcium an magnesium are re-bound to the endoplasmic reticulum (a ste requiring the expenditure of energy), and the ATPase actio ceases, with the consequent relaxation of the muscle. Th significance of the T-system seems to be that all parts of the fibre can be stimulated to contract more or less simultaneously, t produce the characteristic 'twitch'.

Mitochondria (*sarcosomes*) are found between the myofibrils they are usually highly elongate structures with densely packed cristae, and their number varies according to the type of muscle being more plentiful in slow than in fast twitch muscles (Hes 1970).

Slow and Fast Muscle Fibres

It has long been known that two types of contraction are shown b skeletal muscles (Buller 1969), that is, a relatively slow 'tonic twitch lasting (in mammals) about 75 msec. and a faster 'phasic twitch with a duration of 25 msec. It has been possible in severa mammals to correlate the speed of contraction with the gross appearance of the muscle. 'Slow twitch' muscles such as the soleus of the cat are red because of the large amounts of the protein myoglobin (a pigment similar in properties to haemoglobin) while 'fast twitch' muscles are paler in colour, although in the extrinsic ocular muscles of mammals (Salpeter *et al.* 1974), and i the general musculature of some of the lower vertebrates this correlation does not always hold. In mammals, the fibres of the

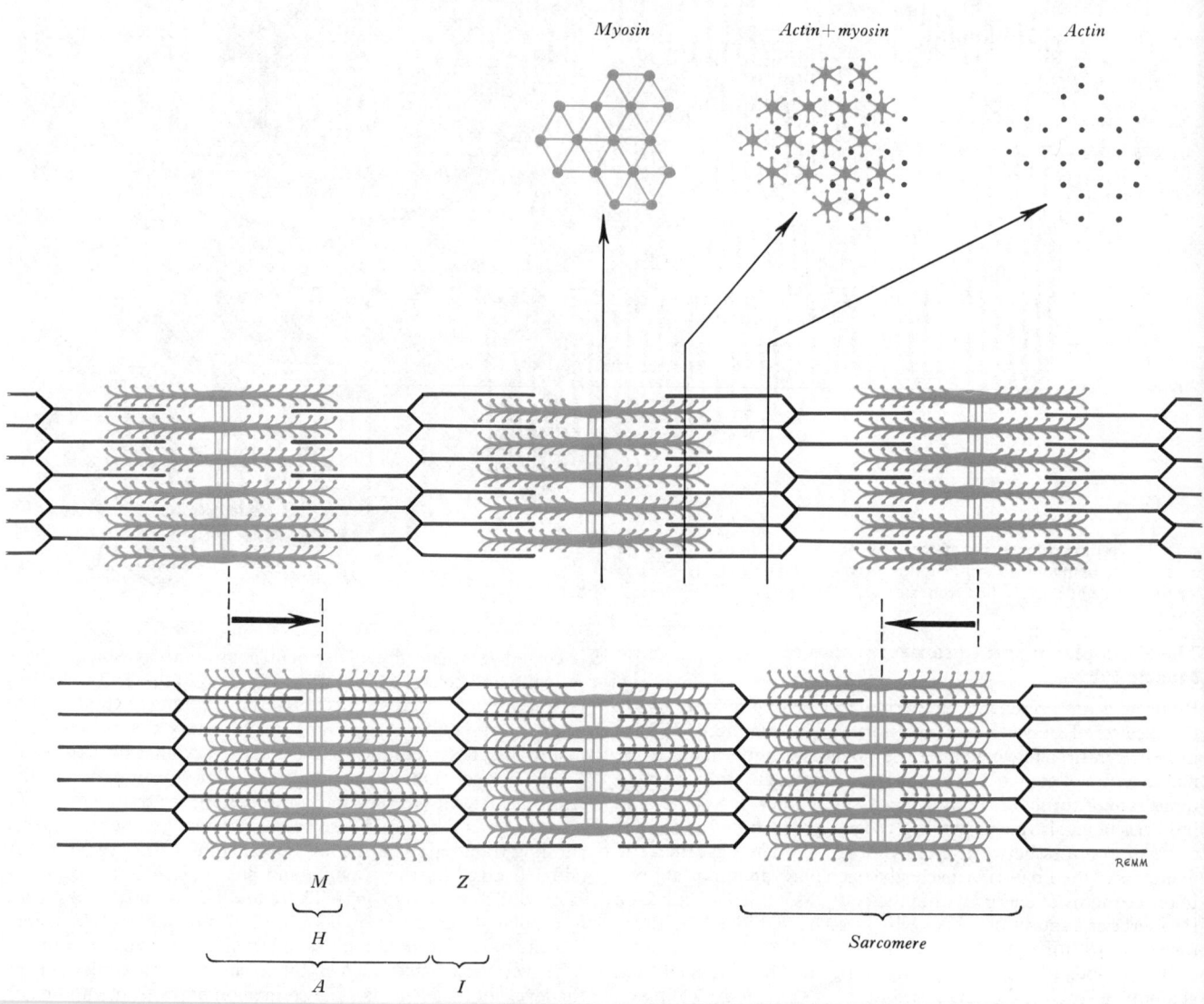

5.7 Diagram showing the organization of sarcomeres in skeletal and cardiac muscle and the changes occurring during shortening. Transverse sections are shown at various levels and indicate the packing of actin and myosin filaments. Compare with **5**.4 and see text for a detailed description.

slow muscles and those of fast muscles are different in their structure and histochemical properties. Most muscles show a mixture of two types of fibre, which are known as type I (slow) and type II (fast) fibres (Dubowitz 1969; Gauthier 1974). Slow fibres in mammals are narrower than fast fibres; they have poorly defined myofibrils, irregular in size, with thick Z bands (Rowe 1973; Eisenberg et al. 1974), and are rich in mitochondria and oxidative enzymes but poor in phosphorylases (Beckett and Bourne 1974). Fast fibres, in contrast, have fewer mitochondria, more extensive sarcoplasmic reticulum, thin Z bands, and are poor in oxidative enzymes but rich in phosphorylases; they also possess more glycogen. The significance of these differences lies partly in their respiratory metabolism; fast fibres obtain energy primarily by glycolytic respiration but are quite easily fatigued, whereas slow fibres also have a well-developed aerobic metabolism and are highly resistant to fatigue. Predominantly 'red' muscles also possess a richer circulation than 'white', relating to their respective oxygen demands (Carrow et al. 1967).

Type I fibres seem well suited to a relatively slow but repetitive type of contraction which generates the sustained 'tonic' forces characteristic of postural muscles. Type II fibres in contrast are adapted to produce more rapid phasic forces which operate in large-scale movements of body segments. Recent studies combining electrophysiological and histochemical approaches (Burke et al. 1973) have shown that type II ('fast') muscle fibres can be subdivided into those which fatigue easily, and those which are relatively resistant to fatigue, but not to the same extent as the type I fibres. Other histochemical investigations have also demonstrated two subdivisions of the type I fibres, although their functional significance is not yet clear (Askanas and Engel 1975).

The variations in enzyme content of muscle fibres appear to be, in part, dependent upon the activity of the fibres; this is shown in experiments where the nerves supplying 'fast' and 'slow' muscles have been crossed, thereby inducing a partial change in enzyme patterns corresponding to the type of nerve now serving the muscle (Buller 1970; Close 1972). It was first suggested that this alteration could be caused by the secretion of a chemical 'trophic factor' from the nerve, but subsequent work indicated that the type of muscle fibre is determined primarily by the frequency of neural activity in the axon serving it (Riley and Allin 1973). Indeed it has been suggested that three chief categories of muscle fibre type exist (types I, II, and intermediate), and that these correspond to the different types of motor unit commonly found in mammals (Burke et al. 1974).

The distribution of the various muscle fibre types has not been investigated extensively in man, but, where known, their proportions reflect the habitual activities of the muscle (Schmalbruch and Kamieniecka 1974; Thorstensson 1977).

Development, Growth and Regeneration of Skeletal Muscle

The precursors of muscle fibres are the embryonic mesenchyme cells of the myotomes or other regions of the body (p. 78, 114, 154). The various stages of differentiation and maturation of these cells to form mature muscle fibres have been the subject of some controversy, from which a number of different schemes of nomenclature have emerged (Boyd 1960; Fischman 1972; Murray 1972). However, it is generally agreed that specific mesenchyme cells (*promyoblasts*), grow into large fusiform elements, up to some 400 μm long (*myoblasts*), which cease dividing, and then fuse end to end, thus forming long multinucleate cylinders or *myotubes*, so named because, initially, the nuclei form a centrally placed row with a surrounding 'tube' of cytoplasm. The myotubes grow in length as further myoblasts fuse to form the fetal muscle fibres. Almost immediately after fusion the assembly of myofilaments begins, and these form myofibrils with highly organized sarcomeres. The myofibrils first appear near the centre of the cell, and further myofilaments are added to the outside of the existing ones, and in this manner the nuclei become displaced to the periphery of the fibre. Somewhat later, sarcoplasmic reticulum begins to form, followed by the ingrowth of the transverse tubule system.

Myotubes have been classed in three grades of increasing maturity, and it has been shown that for a period areas of the surface of early myotubes are enveloped by a characteristic multinucleate syncytium presenting numerous pseudopodial projections which invaginate the surface of the myotube especially opposite the myotube nuclei (Kelly and Zacks 1969; Landon 1970). This surface syncytium has been variously interpreted as a stage in fusion with the myotube, or as a close conjunction which may have a morphogenetic significance, but does not lead to fusion, the surface syncytium eventually disappearing. The final number of fibres within a muscle is reached some time before birth (Jaubert 1955). Subsequent longitudinal growth of the muscle is brought about by an increase in the number of sarcomeres which are added primarily at the ends of the muscle fibres where fusion of myoblasts with the muscle fibres is continuing. A further elongation is also achieved by a slight increase in the resting lengths of the sarcomeres themselves. Growth in diameter is by addition of myofilaments around the periphery of myofibrils which, after they have reached a critical size, appear to each split longitudinally into two myofibrils. These processes continue well into postnatal life, with myoblasts probably fusing with established muscle fibres to increase the number of nuclei in each fibre. Some of these myoblasts are thought to persist into adult life as *satellite cells* which lie close to muscle fibres beneath their basement membrane (Schmalbruch and Hellhammer 1976). The fibre diameter is subsequently much affected by use, reacting to exercise by hypertrophy and to disuse by atrophy (Rowe and Goldspink 1969). The postnatal growth of skeletal muscle is greatly affected by hormone levels, particularly the anabolic steroids such as testosterone, which is responsible for the greater development of muscle fibre size in males. When denervated the fibres become progressively reduced in diameter and finally degenerate, being replaced by connective tissue, which with time becomes increasingly fibrous and may undergo contracture.

Skeletal muscle is capable of limited regeneration in man; fibres on each side of a damaged zone break up into nucleated cylinders of cytoplasm, macrophages enter the necrotic area and engulf dead materials but leave the basement membrane intact. The muscle fibre cylinders now fuse and grow back inside the original basement membrane to form a myotube until eventually the two growing undamaged ends fuse and fill the gap completely (Carlson 1973). Varying degrees of maturation of this zone then follow. Satellite cells (*myosatellitocytes*) are thought to take part in this process, fusing at the ends of the existing cytoplasm to form part of the new fibre (Bischoff 1975; Hall-Craggs 1974). If large regions are damaged, however, regeneration may not occur and the missing muscle is replaced with connective tissue.

The Innervation of Skeletal Muscle

Each skeletal muscle receives one or more nerves of supply. In the limbs, face and neck, the supply is usually single, but where a muscle more obviously retains its segmental arrangement (e.g. the muscles of the abdominal wall), its nerve supply is multiple. Usually, after dividing into several small branches, the nerve of supply enters the deep surface nearer to its customarily less mobile attachment (origin) and, together with the principal blood vessels of the muscle, enters at a small elongate oval area, the *neurovascular hilum*, which is fairly constant in position for each muscle (Coërs and Woolf 1959; Brash 1955). Each nerve contains both motor and sensory fibres; the motor fibres comprise the large myelinated efferents of ventral grey column motor neurons (*alpha-efferents*) which supply extrafusal muscle fibres, the smaller myelinated *gamma-efferents* which run to the muscle spindles, and the fine nonmyelinated *autonomic efferents* which supply vascular smooth muscle. The sensory fibres comprise a range of myelinated fibre diameters distributed to the muscle spindles, neurotendinous sensory endings and terminals in the fasciae, and non-myelinated pain afferents of uncertain origin. Details of these various nerve terminals are considered further elsewhere (p. 847).

Once the nerve has entered the muscle it breaks up into a plexus which runs in the epi- and peri-mysial septa before passing into the endomysial spaces around the muscle fibres. The alpha-efferents then branch and finally lose their myelin sheaths as they terminate on a variable number of individual muscle fibres. These terminals are distributed within a fairly narrow transverse (motor) band near the centre of the muscle, whilst some of the afferents enter the muscle spindles often in much the same region. The autonomic fibres ramify in the endomysium throughout the whole muscle supplying its vasculature. (For the innervation of associated tendons *see* pp. 522, 855.)

The somatic motor axons break up into a number of branches each of which terminates on an individual muscle fibre in the form of a specialized structure, the *neuromuscular termination* or *motor end plate*. In most muscles this, as its name suggests, is a plate-like ('*en plaque*') terminal but in some slow muscle fibres, such as those of the extrinsic ocular muscles (Namba *et al.* 1968), the motor terminal is itself branched. The latter are named '*en grappe*' terminals because of a supposed resemblance to a bunch of grapes. This arrangement may be necessary because the sarcolemma of such slow muscle fibres does not show large propagated changes of potential when stimulated locally.

A motor unit is a functional division of a muscle and is defined as a single alpha motor neuron together with the muscle fibres which it innervates. (The term *myone* has been proposed for the group of *structural units* or muscle fibres that form a *functionally contractile motor unit*—*see* MacConaill and Basmajian 1977.) The size of motor units varies between muscles, smaller units occurring where precise control of muscular action is required. In the extrinsic ocular muscles each motor neuron innervates only 6–12 muscle fibres (Bors 1926), whereas in large proximal limb muscles the ratio may reach 1 to 2,000. Consequently, the force generated by each motor unit is inversely related to the precision of control.

Physiological investigations have shown that the muscle fibres innervated by one motor neuron are often widely spread within a muscle and do not necessarily correspond to its myonemes or fascicular divisions (Burke and Tsairis 1973). Thus, even when only a few motor units are active, the force is generated diffusely. It has also been shown that each axon terminates within a round or oval territory, with considerable overlap between adjacent zones of innervation (Sissons 1969). (It should also be noted here, that in the terminology proposed by MacConaill and Basmajian (1977) the muscle fascicle or *myoneme* is defined by them as the *kinematic unit* of muscle. A further classification of various structural/functional types of myoneme has also been advanced, and some of these will be mentioned briefly below.)

CONTROL OF MUSCLE CONTRACTION

In considering the contraction of a muscle as a whole it is important to understand how its individual elements operate. As seen above, each fibre is in receipt of a single terminal branch of a motor axon. When this is active, volleys of nerve impulses pass along its membrane, and, reaching the terminal, cause the liberation of a chemical transmitter, *acetylcholine*, close to the adjacent muscle fibre membrane. The transmitter is then rapidly broken down by enzymic action so that it operates only briefly. Electrical recordings show that the transmitter causes changes in the permeability of, and hence the electrical potential across, the sarcolemma: in resting fibres the so-called *resting potential* has a steady value of about 80 mV, the interior of the cell being negative with respect to the exterior. When stimulated by low concentrations of acetylcholine, a small transient localized depolarization of the membrane, the *end plate potential* occurs. This does not itself cause contraction, but with increased concentrations of transmitter it builds up until a critical level is reached, at which point the neighbouring membrane abruptly undergoes a massive rapid depolarization, continuing to reversal of the membrane potential so that the interior becomes 20 mV positive relative to the exterior. This is known as the *action potential*; the swing rapidly reverts to the resting potential but it initiates a wave of depolarization, potential reversal and return to resting levels, which sweeps at a speed of 5 metres per second over the whole surface of the muscle fibre, and this causes its momentary contraction. The mechanical response is a single twitch lasting 25–75 msec. Repeated action potentials can be elicited from a single muscle fibre if the activity of its motor axon is maintained, but there is a maximum limit to this since, after a

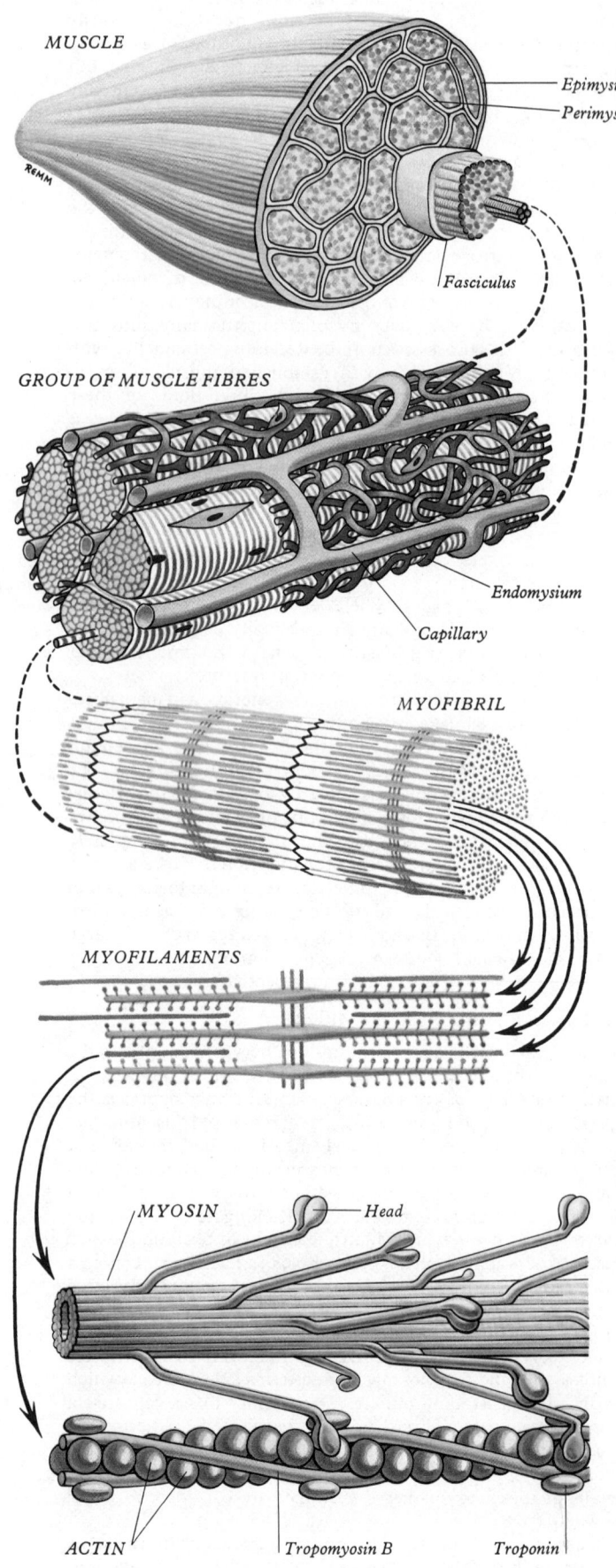

MUSCLE

Epimysium
Perimysium

Fasciculus

GROUP OF MUSCLE FIBRES

Endomysium

Capillary

MYOFIBRIL

MYOFILAMENTS

MYOSIN — Head

ACTIN | Tropomyosin B | Troponin

5.8A Diagram showing successive levels of organization within a skeletal muscle, from whole muscle, through fasciculi, fibres, myofibrils, myofilaments, down to molecular dimensions.

single muscle action potential has occurred, there is a *refractory period* of about 10 msec. until another can be elicited.

Since all action potentials are identical, as are all isolated twitches of a given muscle fibre, they are said to show an 'all or none' response, i.e. if they contract at all, they contract maximally within the limits imposed by their initial length and conditions of loading. Whole muscles, however, exhibit considerable gradation in their contraction and this is achieved by differential activity of the motor units (p. 526). Individual units can vary greatly in their twitch frequency, and the number of units that are active also fluctuates. In small contractions only a few units are operative, but with increasing contraction more are recruited until many or all are active. The sum of these activities results in a steady contraction of the whole muscle even though the individual units are twitching repetitively (but in an asynchronous manner). *Experimentally*, above a certain frequency of stimulation, its rapidity outstrips the contraction-relaxation time of the fibre so that relaxation cannot occur before the next contraction is initiated. This results in a partial or complete fusion of twitches, or *tetanus*, but this does not occur under physiological conditions. The frequency of stimulation which produces tetanus differs with muscle type—slow muscles at 30 impulses per second, whereas fast muscles may need 100 impulses per second.

It must be noted, however, that since all forms of muscle contraction are the summation of individual twitches, even those that appear 'steady' on superficial examination, in fact show a fine oscillation on closer examination with sensitive recording devices. Thus, in the same sense as one regards the energetic states of subatomic particles, or the release of a neurotransmitter at an interneuronal synapse or neuromuscular junction as *quantal*, muscle contraction may also be regarded as quantal, and for some purposes, therefore, quantum mechanical physics may prove the most appropriate.

The Blood Vessels of Skeletal Muscle

The blood supply of muscles is derived by muscular branches from the neighbouring arteries. In many the branches of the principal artery and nerve, as we have seen, enter together along a strip, often fairly constant in position, termed the neurovascular hilum. Subsidiary arteries are generally present and enter at the periphery or close to the ends of the muscle. These branch into smaller arteries and arterioles which ramify in the perimysial septa, and give off capillaries which serve the muscle fibres. These capillaries lie in the endomysium, mainly parallel with the muscle fibres, but they present frequent transverse anastomoses, forming a three-dimensional lattice. It has been shown in cats, that the capillary bed of predominantly red muscle is much denser than that of white muscle.

The supply and drainage of the capillary beds have been the subject of some controversy (*see* Zweifach and Metz 1955, and p. 625). Physiological studies with isotopic sodium as a tracer (Barlow *et al.* 1961), and microscopic examination (Grant and Payling Wright 1968), indicate that there are two distinct circulations in skeletal muscle. The *nutritive circulation* is derived from arteriolar branches of arteries entering by way of the

neurovascular hilum. These penetrate to the endomysium where all the blood passes through the capillary bed before collection into venules and veins to leave again through the hilum. Alternatively, some of the blood passes into the arterioles of the epi- and peri-mysium in which few capillaries are present. Arteriovenous anastomoses are abundant here, and most of the blood returns to the veins without passing through capillaries; this circuit therefore constitutes a *non-nutritive* pathway through which blood may pass when the flow in the endomysial capillary bed is impeded, e.g. during contraction. (See also p. 632.)

The lymphatic drainage commences in lymphatic capillaries which lie in the epi- and peri-mysial sheaths, apparently not penetrating the endomysium, and these converge into larger lymphatic vessels which accompany the veins leaving the muscles.

The Connective Tissues of Skeletal Muscle

The general arrangement of the connective tissue as various so-called 'sheaths' has already been mentioned (p. 506). In fact these perform important roles in the organization of the muscle fibres since they give a degree of mechanical coherence to the muscle, whilst allowing a measure of relative movement within it. These connective tissue planes are, of course, the route whereby nerves and blood vessels penetrate to all parts of the muscle. The endomysium is of particular importance since it forms the external environment of the muscle fibres, and mediates the exchange of metabolites between them and neighbouring capillaries. Its mucopolysaccharide matrix is also important in the flow of ions during excitation of the membrane of the muscle fibre.

The collagen, reticulin and elastin fibres, and their associated ground substances, play a large part in determining the mechanical characteristics of muscle during contraction and passive extension (Hill 1970).

The manner in which muscle fibres transmit their contractile force to adjacent structures depends upon the interaction between muscle fibres and the surrounding connective tissue sheaths. Important in this regard is the junctional region between the ends of the muscle fibres and the tendinous attachments, loosely called the *myotendinal junction* (Schippel and Reissig 1968; Hanak and Böck 1971). Here, the connective tissue of the endo-, peri and epi-mysia become strongly fibrous and thicken to continue as the fibrous bundles of the tendon. The muscle fibres at this point may taper, be flat ended, or may show a terminal expansion. Electron microscopy shows these ends to have a highly indented sarcolemma with a dense internal layer of cytoplasm into which are inserted the actin filaments of the adjacent sarcomeres. Externally, the basement membrane is prominent, and collagen and reticulin fibres lie in close contact. There are, however, no indications of desmosomal attachments, and it seems likely that here, as along the whole length of the fibre, the contractile forces are transmitted from the cell surface to the matrix of the surrounding connective tissues primarily by viscous adhesion with its fibres and ground substance.

Cardiac Muscle

Heart muscle has structural and physiological properties which distinguish it from both skeletal and smooth muscle. Histologically, it is made up of tracts of cells (**5.9**) each about 80 μm long and 15 μm in diameter in man (resting length), containing a single centrally placed nucleus. Each cell may be partially divided into two or more branches at its ends and these abut against other cells: the appearance by light microscopy is of a network of branching and anastomosing cylinders, and indeed for many years cardiac muscle was, quite incorrectly as we shall see below, thought to be a syncytium. In the spaces between the cells lies an endomysium comprised of fibroblasts, collagen and reticulin fibres, ground substance, capillaries and nerves. The larger bundles or tracts of cells, varying in size from a few hundred to many thousands of cells in diameter, are encompassed and separated by perimysium in much the same fashion as skeletal muscle.

Each cardiac muscle cell (*cardiac myocyte*) shows striations which are identical in organization with those of skeletal muscle, with A, I, Z and H bands. In histological preparations these bands

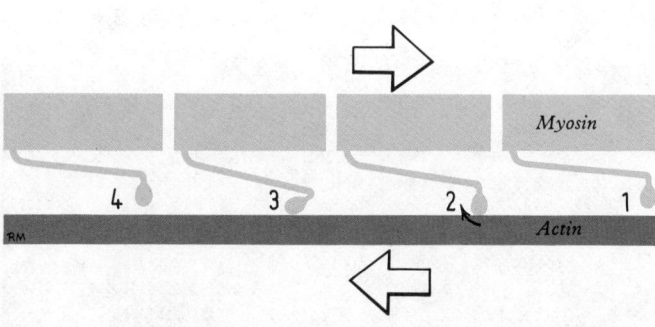

5.8B Diagram showing the changes in myosin-actin interactions during the generation of shearing forces between the two types of filament. The cycle of attachment, flexing, and detachment of the myosin heads is indicated in the sequence 1–4.

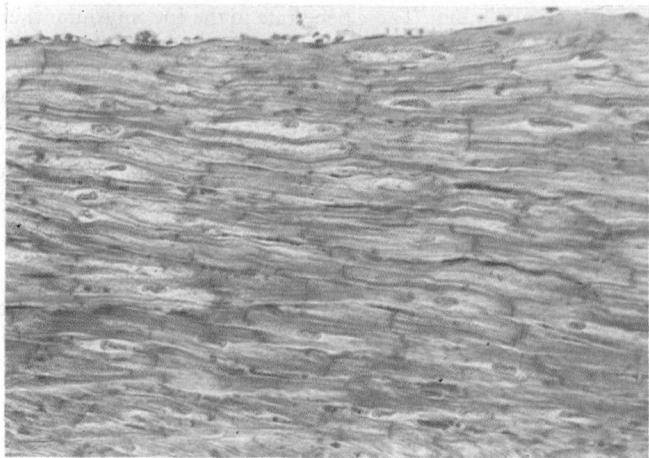

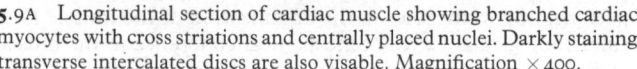

5.9A Longitudinal section of cardiac muscle showing branched cardiac myocytes with cross striations and centrally placed nuclei. Darkly staining transverse intercalated discs are also visable. Magnification ×400.

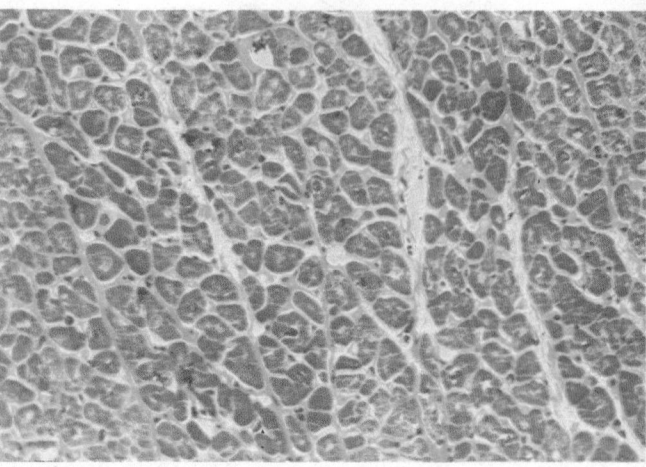

5.9B Transverse section of cardiac muscle showing cardiac myocytes with central nuclei and peripheral myofibrils. Magnification ×400.

are not as conspicuous as in skeletal muscle. At the ends of the cells are conspicuous cross striations, the *intercalated discs*. With the electron microscope the same pattern of actin and myosin filaments is seen, as in skeletal muscle (Fawcett and McNutt 1969; Page and Fozzard 1973). They are not, however, grouped into distinctive myofibrils, and the numerous mitochondria found in heart muscle cells, together with their various intracytoplasmic membranes, are closely surrounded on every side by myofilaments (**5.11**). T-systems are present (Forssmann and Girardier 1970), but they penetrate the cells at the level of the Z band rather than at the A–I band junction (**5.11**), possibly reflecting the relatively slower contraction time of cardiac muscle compared with skeletal muscle. The sarcoplasmic reticulum is present but not as abundant as in fast skeletal muscle. Glycogen and

myoglobin are also prominent features of cardiac muscle, and histochemical studies show the cells to be rich in oxidative enzymes. These characteristics, somewhat similar to those of mammalian slow skeletal muscle, reflect the high oxygen demand and the constant rhythmic expenditure of energy which typifies heart muscle. The density of the vascular bed and blood flow to the heart also emphasizes these considerations.

Each cardiac myocyte shows a spontaneous rhythm of contraction and relaxation, even when isolated. This *myogenic rhythm* is accompanied by a parallel oscillation in membrane potential, but it is not known how this is achieved. The rhythm is slower than that of the heart as a whole, but can be synchronized with that of the other cells by the action of the pacemaker and conducting system (*vide infra*, and also p. 659).

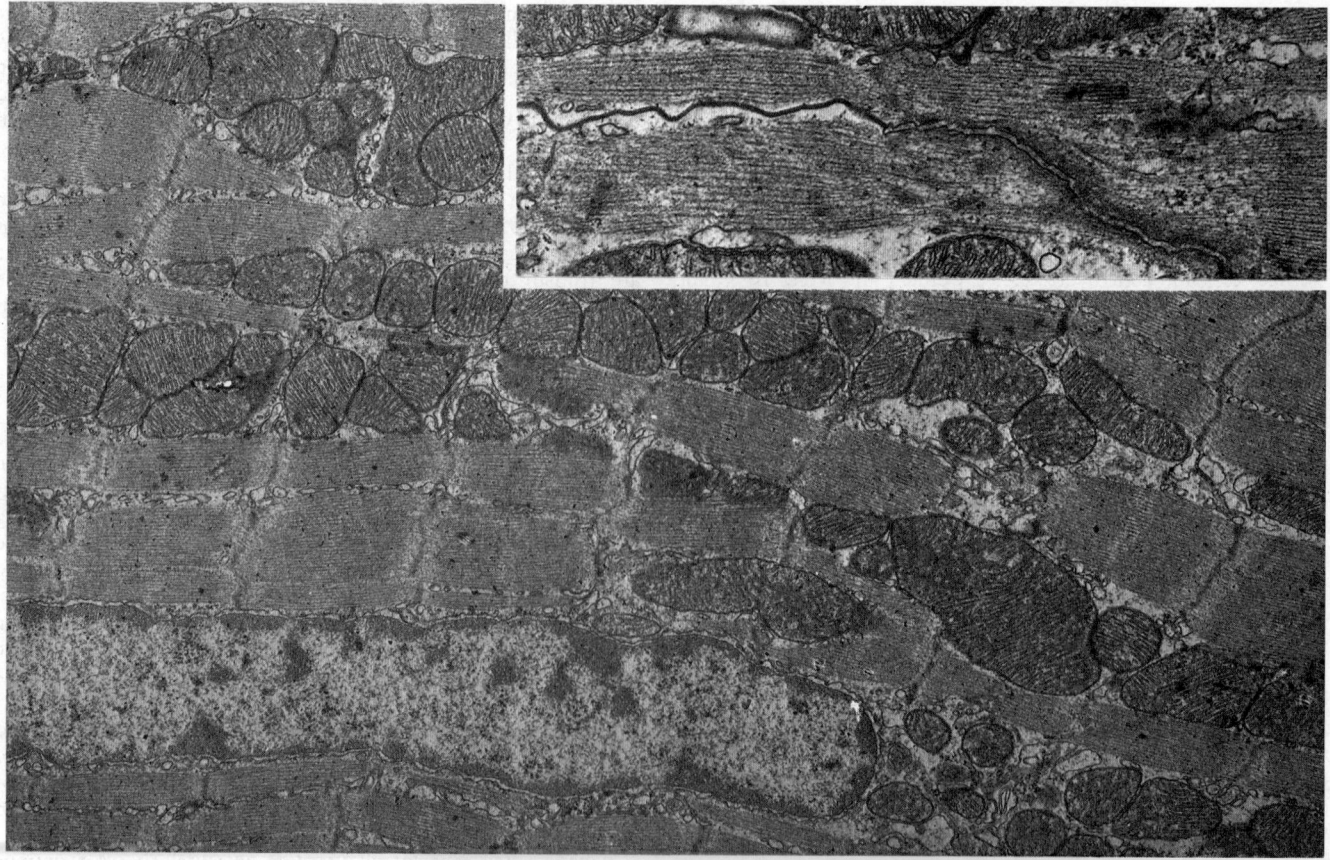

5.10 Electron micrograph of cardiac muscle in longitudinal section showing abundant mitochondria and intercalated discs. ×4,000. Detail shows part of intercalated disc with gap junction in left half of field. ×15,000.

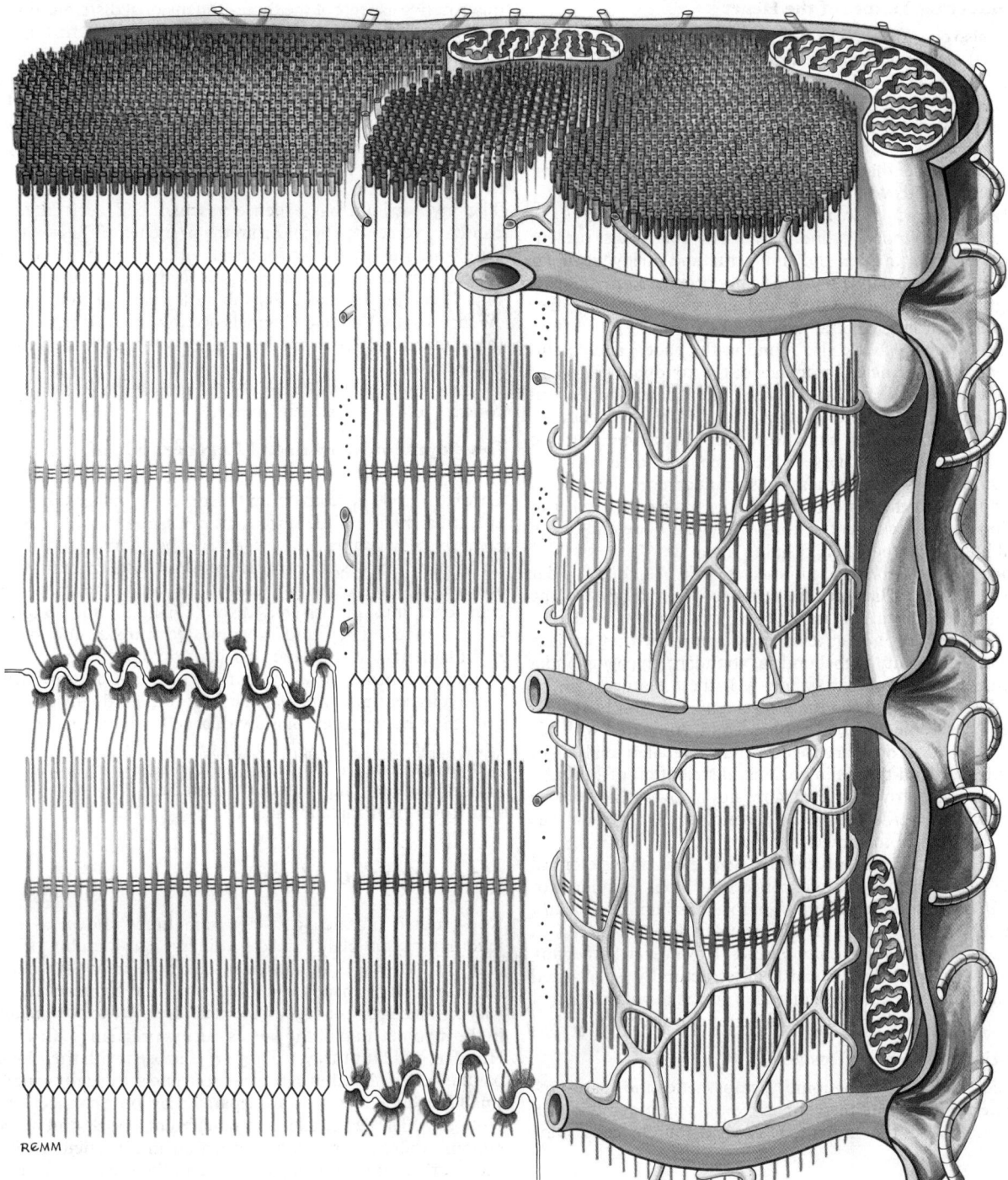

5.11 Three-dimensional reconstruction of cardiac muscle showing organization of myofibrils and membrane systems. As in previous diagrams, actin is shown *red* and myosin *green*; mitochondria are coloured *blue*; the sarcolemma and luminal surface of the invaginated transverse tubules are *purple*. The intracellular surface of the transverse tubule is *orange*, and the sarcoplasmic reticulum is coloured *yellow*. An intercalated disc with intermittent desmosomes and communicating junctions (nexuses) is depicted on the lower left of the picture. Compare with **5**.6, and see text for further details. (Modified from a diagram by Professor D. Fawcett.)

Intercalated Discs

With the electron microscope these have been shown to constitute specialized regions of cytoplasm and cell membrane which mark the junctions between the ends of the cells. The cell membrane in such regions runs transversely across the end of the cell but often forms a series of steps (**5**.9, 10). The cytoplasm adjacent to the membrane is particularly dense and appears to form an anchor for the actin filaments of the adjacent sarcomere. Numerous desmosomes are scattered between the adjacent ends of cardiac myocytes, and amongst the line of attachments there are special regions where the cell membranes come to lie in close apposition; these regions are similar in structure to the specialized contacts (gap junctions) of epithelial cells and to the 'nexuses' of smooth muscle cells (*vide infra*). It has been shown by electrical recording and stimulation, that such regions permit the relatively un-impeded electrotonic flow of ionic current between cells and are thus of great importance in the dissemination of contraction within the heart. Because of these electrical contacts between cells, excitation, and therefore contraction, in one part of a muscle bundle can spread to all the cells of that bundle. For this reason the heart behaves *electrically* as a syncytium (i.e. as though there were no cell membranes separating its cells), and, as intimated above, before the advent of electron microscopy, it was regarded as structurally syncytial.

The Conducting Tissues of the Heart

These are also considered with the heart (*see* p. 659). Three major types of cell are found in the conducting system, namely *nodal myocytes, transitional myocytes,* and *Purkinje myocytes,* each present in particular regions (Sommer and Johnson 1968; James and Sherf 1974). (It should perhaps be noted here, that rows of Purkinje myocytes constitute the Purkinje 'fibres', and it seems less than appropriate that the official term Purkinje *fibre* has been taken as synonymous with single Purkinje myocytes.)

Nodal myocytes, which are responsible for pacemaker activities, are uninucleate, rounded, cylindrical, or polygonal cells, present in clusters or rows in both the sinu-atrial and atrioventricular nodes. They possess only a few randomly orientated myofibrils, lack a regular sarcotubular system, and have a pale organelle-free zone around their large central nuclei.

Transitional myocytes are also found in the nodes, and extend into the stem and principal branches of the conducting system; they form connecting pathways between the nodal myocytes and the final elements of the system, the Purkinje myocytes. Transitional myocytes are narrower than the general cardiac myocytes, but otherwise possess the same type of contractile apparatus. It is possible that these cells conduct more slowly than the larger fibres of the system, and are the basis of the delay in conduction which occurs at the atrioventricular node and in the principal bundles immediately distal to the node.

Purkinje myocytes are large uninucleate cells which are wider and shorter (30 μm in diameter and some 20–50 μm long) than the surrounding general cardiac myocytes. They also possess larger nexuses both laterally and at their ends. Internally they contain fewer myofibrils, many mitochondria, abundant glycogen, and a well developed sarcotubular system. All these features can be correlated with the rapid conduction velocities of impulses along Purkinje fibres, some 2–3 metres per second, compared with 0.6 metre per second in ordinary myocardium. Detailed analysis of the organization and three-dimensional distribution of Purkinje myocyte nexuses terminating in apposition to normal cardiac myocytes within the depths of the ventricular myocardium has not been achieved, at the time of writing.

As will be mentioned elsewhere (p. 663) electrophysiological evidence has been presented concerning the existence of *preferential conduction paths* from right to left atrium (*interatrial*), between the sinu-atrial and atrioventricular nodes (*internodal*), and also *accessory atrioventricular bundles.* However, whether any or all of these pathways consist of structurally specialized *conducting tissue,* or reflect the elongation and axial polarization of tracts of *ordinary cardiac myocytes,* remains controversial.

Innervation of Cardiac Muscle

Non-myelinated, post-ganglionic, sympathetic and parasympathetic nerves permeate the connective tissue septa and end in close proximity to the individual cardiac muscle cells. The

quantitative aspects of the three-dimensional distribution of both autonomic efferent and afferent nerve terminals in the various localities of the myocardium (with its varieties of cardiac myocyte, including the conducting system), the valve annuli, cusps or leaflets, and chordae, the fibrous 'skeleton', endocardium, epicardium, and coronary vasculature is virtually an uncharted field.

Blood Vessels of Cardiac Muscle

The myocardium has a dense vascular bed derived from the coronary vessels (p. 667). From the branches of these, capillaries arise and run in the endomysial septa to form a network of branching and anastomosing vessels so arranged that each cardiac muscle cell lies not more than about 8 μm away from a capillary (Wearn 1941). Vascular channels occupy about 60 per cent of the total interstitial space in the rabbits' myocardium (Frank and Langer 1974).

The lymphatic capillaries also penetrate the endomysium, in contrast to those of skeletal muscle.

Development, Growth and Regeneration of Cardiac Muscle

Cardiac muscle develops from the mesenchymal cells of the myo-epicardial mantle of the embryonic heart (p. 181). This source has recently been confirmed, with details, by Morris (1976) working on human embryos. The cardiac promyoblasts are at first stellate, adhering to each other by desmosomal attachments which later transform into intercalated discs (Challice and Virágh 1973; Morris 1976). They undergo repeated mitoses and synthesize myofilaments to become cardiac myoblasts, which begin rhythmic contraction early in fetal life. The multiplication of these cells continues into postnatal life, but gradually gives way to growth by hypertrophy of individual cells which is to some extent dependent upon the work load of the muscle.

Regeneration of damaged muscle does not occur to any noticeable extent; repair of the myocardium is by means of fibrous scar tissue (Hudson and Field 1973).

Non-striated Muscle

Non-striated muscle (5.12–15), sometimes known as smooth, involuntary or plain muscle, differs considerably from the previous two types of muscle tissue both in its structure and behaviour. It is made up of mononucleate fusiform myocytes, varying in length from 15 μm in small arterioles to 500 μm or more in the myometrium of the uterus during pregnancy (Csapo 1962; Huddart and Hunt 1975). They are disposed with their long axes parallel to the direction of contraction, and are usually arranged in small fasciculi. The latter are separated by loose connective tissue septa in which lie the afferent and efferent vessels and nerves. In regions where a concerted contraction in a particular direction occurs, these fasciculi lie parallel, as in the separate layers of the muscularis externa of the intestines; where contraction effects a reduction in surface area in a particular plane, the fasciculi interweave freely in all directions in that plane, as in the muscularis mucosae of the intestines and in the bladder. In muscular arteries non-striated muscle is present in thick sheets into which capillaries do not penetrate. Within a fasciculus much of the surface of each cell is covered by a prominent basal lamina (basement membrane), and between and inside the laminae fine elastin, reticulin, and collagen fibres abound, forming complex networks around each cell, which are separated therefore, except at the special points of contact described below, by a space of 40–80 nm. There is usually a lack of fibroblasts or other connective tissue cells within the fasciculi, and the extracellular fibres may be a product of the myocytes themselves at some stage in their development. At the boundaries of the fasciculi these connective tissue fibres interweave with those of the interfascicular septa, and presumably provide the means whereby the forces generated by the contraction of individual cells are transmitted throughout the tissue.

Two types of non-striated muscle have been distinguished on the basis of their innervation and behaviour, *multi-unit* and

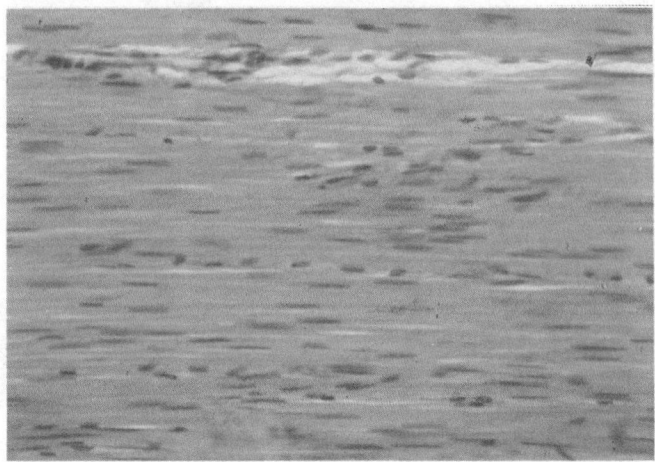

5.12 Longitudinal section of non-striated muscle from the ileum, showing groups of non-striated myocytes arranged in a fasciculus. Haematoxylin and eosin.

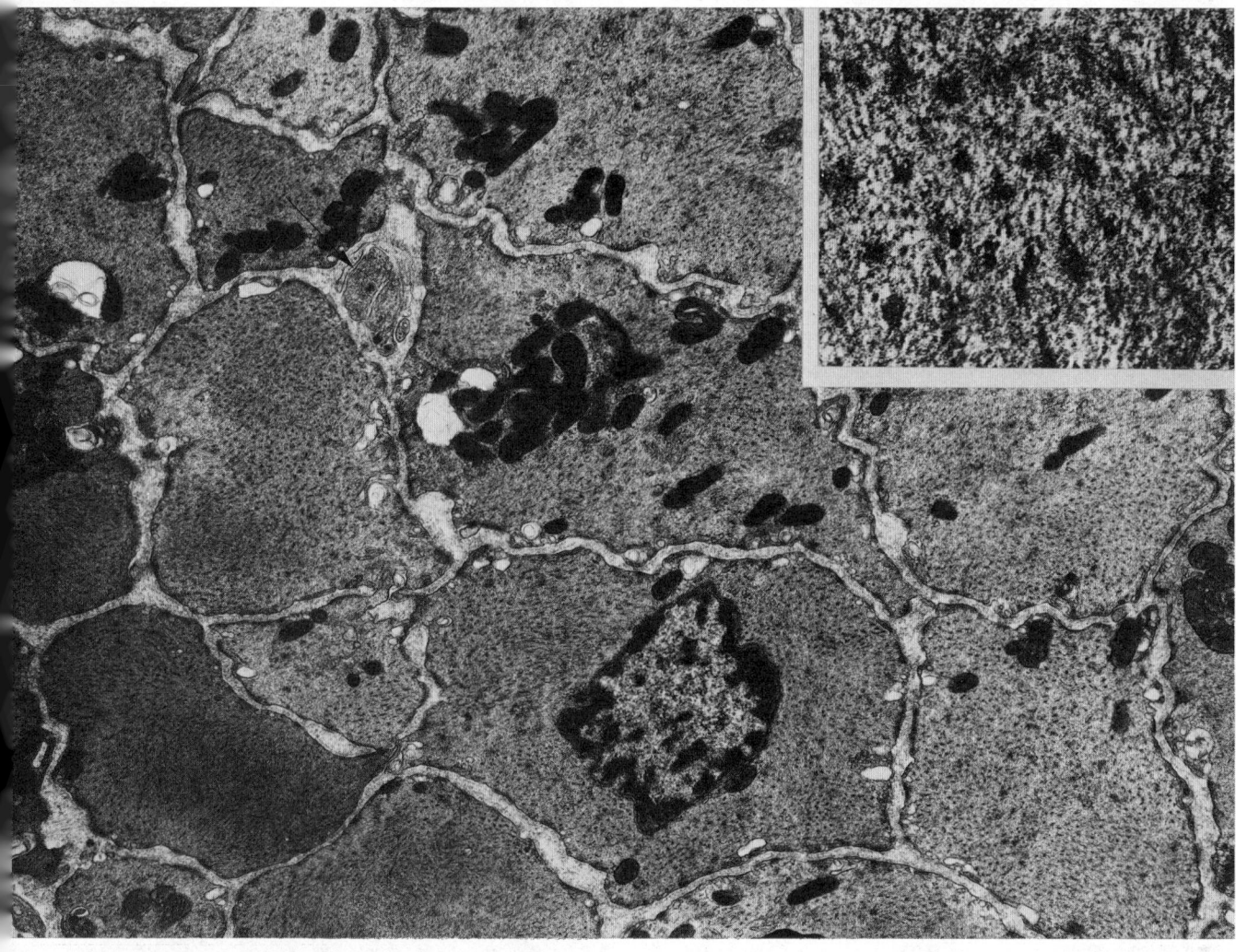

5.13 Low-power electron micrograph of a group of non-striated myocytes in transverse section, showing a centrally placed nucleus in one of the cells, and numerous darkly staining mitochondria. An autonomic nerve fibre (arrow) is visible among the myocytes. Specimen taken from the ileum. Magnification ×6,000. The inset shows a highly magnified sample of cytoplasm from one myocyte demonstrating thin actin and thick myosin filaments irregularly arranged. Magnification ×100,000.

unitary muscles (Bozler 1948; Burnstock 1970). In the former there are rich nerve plexuses so that many of the cells receive motor terminals; contraction is usually initiated by nervous action, it is *neurogenic*, rather than by stretching or arising spontaneously, and small groups of muscle cells may operate independently. The muscle of the iris and those of larger arteries and ductus deferens are of this type. Unitary muscles, in contrast, have relatively few motor nerves; their rhythmic contractions are spontaneous, i.e. *myogenic* in origin, and may be governed by *pacemaker regions* in the muscle, or may be elicited by previous stretching of the muscle. Typical·examples are the muscle layers of the stomach, intestines, uterus, ureter and some of the smaller blood vessels. The function of the nervous control in unitary muscles is to enhance or depress the rate and force of the pre-existing rhythmic contraction. Intermediate types of muscle also exist and the two divisions outlined here probably represent the extremes of a wide spectrum of types and activities.

Individual non-striated myocytes when seen with the light microscope show a centrally placed nucleus, which may be highly elongated in stretched muscle, or ovoid with a crenated surface in contracted muscle. The cells are weakly birefringent, indicating some degree of longitudinal orientation of the cell components.

Ultrastructurally the cytoplasm of the non-striated myocyte is seen to consist mainly of closely-packed fine filaments lying parallel to the long axis of the cell (**5**.13, 14, 15B). Most of the usual cellular organelles including Golgi complexes, agranular and granular endoplasmic reticulum, free ribosomes, and lyosomes are positioned near both ends of the nucleus (the nuclear poles), but clusters of mitochondria are also found elsewhere in the cytoplasm, either peripherally or in deeply placed scattered groups, the site varying with different types of non-striated muscle (Gabella 1973; Huddart and Hunt 1975). A branched, irregular sarcoplasmic reticulum, capable of sequestering calcium ions, also ramifies within the masses of filaments (Popescu and Diculescu 1975), and its terminal sacs lie immediately beneath the myocyte's plasma membrane. Numerous small endocytotic vesicles line the cell surface, either in rows or distributed randomly, and the cell surface is also characterized by occasional projections (*see* **5**.13) and various specialized junctions with other cells.

The myofilaments are responsible for cellular contraction, but until recently the detailed organization of actin and myosin was unknown—only actin was visible in the earlier micrographs. Improvements in fixation resulted in the demonstration of myosin in the form of thick ribbons, or rod-like aggregates, of myosin molecules which are up to 15 μm long and some 0.1 μm wide (Somlyo *et al.* 1973). It is probable that the myosin molecules are arranged in non-parallel directions on the two sides of such assemblies, presumably allowing them to generate shearing forces in relation to adjacent actin filaments, thus shortening the myocyte, without the necessity for the sarcomeric arrangement of striated muscle. Each myosin aggregate, together with up to 15 surrounding actin filaments, forms an obliquely orientated '*contractile unit*' which can be visualized in glyceri-nated preparations with the light microscope (Small 1974). Actin filaments are also inserted into cytoplasmic densities associated with the inner surface of the plasma membrane (Cooke 1976). In addition to the myofilaments, bundles of '*intermediate filaments*',

519

5.14A A low-power electron micrograph of a longitudinal section through two non-striated myocytes, one showing a nucleus. Note the irregular outlines of the cells, and the endoplasmic reticulum and mitochondria within. Fine collagen fibrils lie in the intercellular spaces. Specimen from ileal wall. Magnification × 8,000.

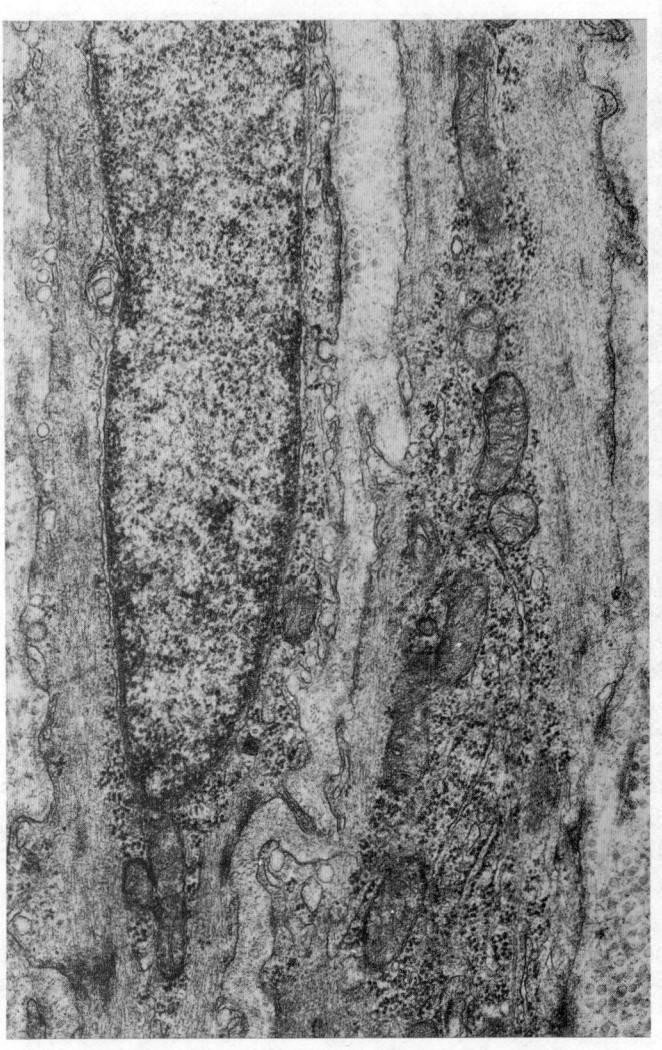

10 nm in diameter, traverse the cytoplasm becoming attached to the occasional dense bodies distributed throughout the cytoplasm. They resemble the tonofilaments of other cells, and may act as a cytoskeleton binding the various contractile units together.

Excitation of non-striated myocytes occurs when their plasma membranes are depolarized and calcium ions enter, thereby activating the actin-myosin systems. The myocytes can also excite each other directly by way of the nexuses (*maculae communicantes, communicating junctions, see* 5.15A, B) which are present as small irregular patches distributed over the cell surfaces (Cobb and Bennett 1969; Lowry and Small 1970; Watanabe and Yamamoto 1974). It has been estimated that in some locations a few score such junctions are present on each cell. Individual nexuses are associated with an underlying expanded region of the sarcoplasmic reticulum; the latter may therefore be directly affected by signals from other cells (Fry *et al.* 1977) perhaps causing the mobilization of calcium stores during excitation. Electrical coupling ensures that a group of myocytes act in concert, and also enables a single nerve fibre to control many myocytes without exciting each individually. In addition to nexuses, myocytes are also interlinked by desmosomes, and to the surrounding basal lamina by surface specializations similar to zonulae adherentes (p. 7).

5.14B An electron micrograph of the cytoplasm of a non-striated myocyte in longitudinal section showing the irregularly distributed thick myosin and thin filaments. Note also the numerous endocytic vesicles and sarcoplasmic reticulum at the cell boundaries. Specimen from ileal wall. Magnification × 20,000.

Other Activities of Non-striated Muscle

As already mentioned, extracellular fibres are always present surrounding the muscle fasciculi, and it has been shown *in vitro* that non-striated myocytes can secrete elastin and other fibres. This resemblance to fibroblasts has called into question the sometimes rather arbitrary division between those two cell types. Certainly in the aorta, for example, non-striated myocytes generate large quantities of extracellular material including the elastic laminae, and are presumably limited in their contractile powers. Conversely, in regenerating connective tissue there are contractile fibroblast-like cells which are responsible, at least in part, for the movement that occurs during scar contracture. It is probable that both non-striated myocytes and fibroblasts arise from similar types of mesenchyme cells, and that in certain circumstances they develop some common features (*see* review by Ross 1971).

Development of Non-striated Muscle

The origins of non-striated muscle have already been outlined (p. 506). The mesenchyme cells involved become fusiform as the contractile proteins are synthesized, to form first myoblasts, and then mature myocytes which may continue to grow in size. New myocytes continue to originate, even in adult life, from mesenchymal cells, as in regenerating blood vessels, and there is evidence that non-striated myocytes are able to multiply mitotically, for example, in the myometrium after treatment with progesterone (Crandall 1938), although to what extent this occurs physiologically is uncertain. (For more recent studies of hormonal modification of myometrial growth, consult Ross and Klebanoff 1967.) Non-striated muscle can regenerate to a limited extent if damaged, by the multiplication and differentiation of mesenchyme cells, and by division of pre-existing muscle cells.

Innervation and Blood Supply of Smooth Muscle

Autonomic motor and sensory nerves form nonmyelinated plexuses of fibres which send branches to the muscle fasciculi—the nature of these will be considered elsewhere (pp. 847, 1136).

Capillary networks run in the connective tissue sheaths of the smaller fasciculi in most smooth muscles, but the vascularization is not as dense as in skeletal muscle. In some situations, e.g. the tunica media of muscular arteries, the capillaries may not penetrate the muscle blocks at all, reflecting their low energy requirements.

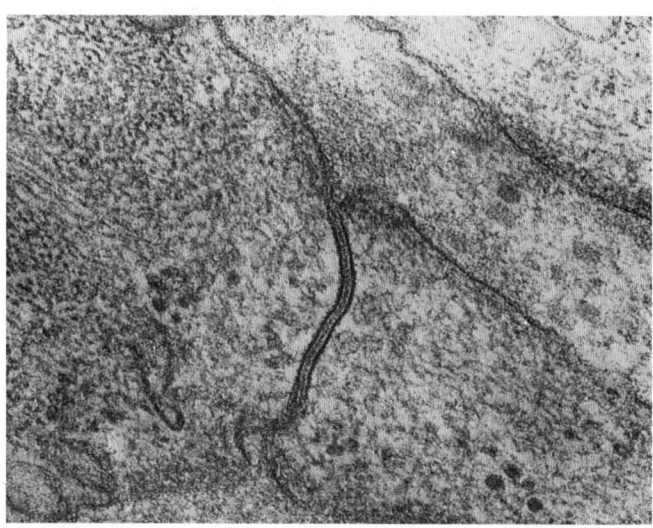

5.15A A gap junction (nexus) between two smooth muscle cells. × 70,000.

Myoepitheliocytes

These are contractile cells whose embryonic origin is believed to be ectodermal; they occur in association with several types of secretory tissue, either in close contact with the glandular part or the surrounding ducts. Myoepitheliocytes may be either *stellate*, with long dendritic extensions appearing to clasp an adjacent acinus, as in the submandibular gland, where they are known as basket cells, or *fusiform*, as in the intralobular ducts of salivary glands. They are also present around the acini of the mammary gland and the secretory portion of sweat glands. They are often under autonomic nervous control and their contraction initiates a rapid flow of secretion from the apical parts of the glands into the wider ducts.

With the electron microscope (Ellis 1965; Tandler *et al.* 1970), these cells are seen to have extensive desmosomal attachments with surrounding tissues, and internally there are large numbers of fine filaments, 5–10 nm in diameter, similar to the actin filaments of smooth muscle. Histochemically they are rich in alkaline phosphatase, ATPase, and adenyl cyclase (Han *et al.* 1976).

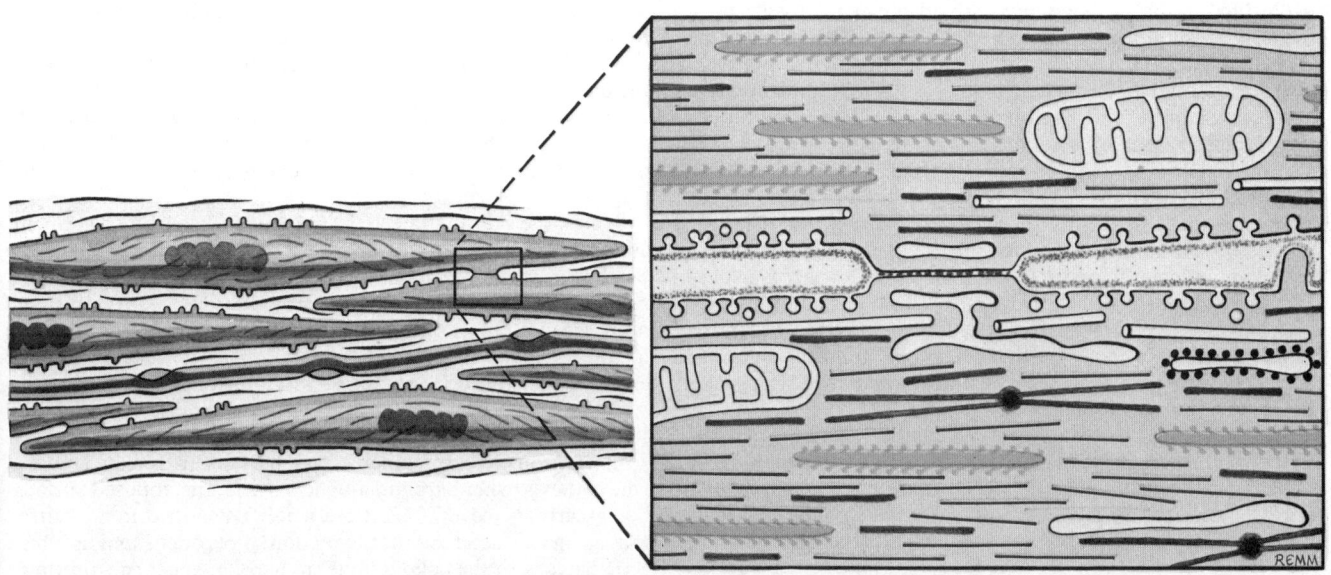

5.15B Diagram showing the principal features of organization of non-striated myocytes. On the left a group of myocytes shows the fusiform shape of the contracted cells, and the arrangement of the autonomic efferent innervation; on the right the main features of the cytoplasmic filaments and membrane systems is depicted. Thick myosin filaments (*green*) are surrounded by irregularly arranged actin filaments (*red*); microtubules, and intermediate filaments, some with associated dense bodies, are also illustrated. The elongated membranous cisternae correspond to the sarcoplasmic reticulum; endocytic vesicles, and a communicating junction between two adjacent myocytes are also visible.

ATTACHMENTS, GENERAL FORM AND ACTIONS OF SKELETAL MUSCLES

Tendons, Aponeuroses and Fasciae

Tendons are integral parts of muscles, forming a component which is for practical purposes often considered to be of unvarying length (but *vide infra*). Being largely composed of collagen fibres they are highly resistant to extension; but they are relatively flexible, and can therefore be angulated around bone surfaces or deflected beneath retinacula (*vide infra*) to change the final direction of pull. Since vascular networks are of low density in tendons, these appear white. Usually taking the form of cords or straps, they are round, oval or elongate in cross-sectional profile and consist of fascicles of collagen fibres, largely running parallel to the long axis but also partly interweaving to some degree. The fascicles may be large enough to give the tendon a longitudinally striated appearance visible to the naked eye, or even to form a duplicated or multiple structure. Some tendons are commonly doubled in this way, and some give rise to minor strands or slips which connect with adjacent tendons. In general the surface of tendons is smooth, but longitudinal ridging due to the arrangement of coarse fasciculi is quite common in large tendons, and may be marked, as on the osseous aspect of the angulated tendon of obturator internus (p. 602). The areolar connective tissue which permeates the tendon between its fascicles, providing a route for vessels and nerves, is condensed on its surface into a so-called sheath, or *epitendineum*; this usually contains elastic as well as irregularly arranged collagen fibres. Such a sheath does not provide a true free surface, being continuous with surrounding areolar tissue, the loose arrangement of which imposes minimal drag on the tendon in its movements. In many situations, however, tendons are separated completely or partially from their surroundings by synovial membrane (*vide infra*).

The tensile strength of tendon is similar to that of bone, i.e. about half that of steel. This is much in excess of ordinary demand. A tendon of cross-sectional area of 1 cm² will support 600–1000 kilograms. Although only slightly extensile, they *are* elastic, and considerable contractile energy is transferred to them when their muscles contract. Tendon tissue, in addition to its limited elasticity, also exhibits viscosity. Arnold (1974) has reviewed the literature of biomechanical and rheological properties of tendons, with special reference to the human foot. Certain avian wing tendons are highly elastic; the ultrastructural relations and interactions between collagen and elastin fibres in such structures have been investigated by Oakes and Bialkower (1977).

The blood supply of tendons is provided by a relatively sparse array of small arterioles which run longitudinally from the adjacent muscular tissue to ramify in the interfascicular intervals, where they intercommunicate freely and are accompanied by venae comitantes and lymphatic vessels. The longitudinal network in the tendon is augmented by small vessels from the surrounding areolar tissue or synovial sheath, where the latter is present (Edwards 1946). At their bony attachments the vessels appear to consist of no more than transversely arranged capillaries. There is no evidence of passage of vessels between bone and tendon at these sites, and the osseous surface is usually devoid of foramina (p. 297). The metabolism of tendon tissue is low, but must be increased in reaction to infection or injury. Repair is almost exclusively due to proliferation of connective tissue cells associated with collagen fibres, which take little part in the process (Potenza 1963).

The nerve supply of tendons appears to be largely, if not exclusively, afferent; no clear evidence of vasomotor control is available. Specialized afferent receptors, *neurotendinous endings* (p. 522), exist in tendons, especially near their junctions with muscle. The nerve fibres proceeding centrally from these receptors are carried partly in branches of the 'motor' nerves to the muscle, extending from these longitudinally into the tendon, and also in small branches which pass directly to neighbouring peripheral nerves (Stillwell 1957).

During postnatal growth tendons increase in length interstitially, but particularly at the musculotendinous junction, where there are concentrations of cells for rapid elaboration and deposition of collagen. From this point there is a decreasing growth gradient towards the osseous attachment.

Synovial sheaths and bursae occur in situations where structures which move relative to each other are in tight apposition. This applies particularly to sites where tendons are deflected around bones or under retinacula in the vicinity of joints; but the arrangement is seen at its simplest in a few locations such as the olecranon and the patella, where the skin must move freely over subcutaneous bony surfaces, usually under conditions of pressure. The *bursae* formed at such sites are simply flattened sacs of synovial membrane supported by dense irregular connective tissue; they are interposed in the loose areolar tissue (superficial fascia) between skin and bone, and because of their position are usually known as *subcutaneous bursae*. The degree of flattening of a bursa is obscured by the term 'sac'. Its opposed walls are separated merely by a film of fluid, and it is more of an enclosed cleft. These walls are to some extent tethered to periosteum and dermis, moving with these structures and hence sliding over each other. The whole arrangement is often described as a device to reduce friction, but its most fundamental characteristic is the creation of absolute discontinuity between tissues, yielding complete freedom of movement over a limited range. Each bursa contains a capillary film of synovial fluid acting as a lubricant and providing the cells of the synovial membrane with a wet environment on their free surfaces and a metabolic intermediary between tendons and their surroundings.

The majority of synovial bursae do in fact occur between tendons and bone, tendons and ligaments, or between one tendon and another, being then termed *subtendinous*. They may, however, be sited between muscle on the one hand and bone, tendon or ligament on the other—*submuscular synovial bursae*. They may also separate aponeurotic areas from bone (*subfascial bursae*), or are sometimes *interligamentous*. Not only are the majority of bursae near joints but may communicate with them, so that their synovial membranes are continuous. Such *communicating bursae* may develop separately and are only subsequently connected with the articular cavity; sometimes they ensheath a tendon for a short distance after it has emerged from an intracapsular attachment.

Tendon synovial sheaths occur where tendons pass under ligamentous bands, retinacula, through fascial slings or osseofibrous tunnels. They consist of two concentric layers, separated by a capillary film of synovial fluid and continuous at their extremities. The arrangement is thus a closed double-walled cylinder, the internal or *visceral layer* of which is attached to the tendon by loose areolar tissue, the external or *parietal layer* to neighbouring connective tissue structures or periosteum. Since the tendon is invaginated into the sheath, the visceral and parietal layers are often connected by an elongated mesentery or *mesotendon*. In some sheaths the mesotendons are reduced to localized cords, often multiple, such as the vincula tendinum of the digital synovial tendon sheaths (p. 577).

Where the skin is subjected to repetitive lateral displacement, associated with pressure, as in forearm or elbow in writing, or the buttock in certain sedentary occupations such as hand-weaving, *adventitious bursae* may appear, providing the skin with much more freedom of movement. This may throw some light on the factors involved in the evolution of bursae in general, but the established series develop during intrauterine life.

Some bursal arrangements are more elaborate and illustrate some of the characteristics of a synovial articulation (p. 424). Thus, osseous or cartilaginous sesamoids occur in some tendons, where they are bent around a bone surface, the apposed surfaces being cartilage covered, and the whole enveloped in a localized bursa, e.g. the sesamoid in the tendon of peroneus longus which articulates with the cuboid bone (p. 606), in effect constituting a synovial joint.

Aponeuroses are flat sheets of densely arranged collagen fibres, hence white and showing, like tendons, a surface iridescence, when exposed, due to the regular disposition of the fibres. They are frequently striated, the large fascicles of collagen being separated by loose interfascicular connective tissue. Aponeuroses usually consist of several layers, the fasciculi being parallel within a layer but inclined in different directions in adjacent layers. In earlier usage the term aponeurosis was applied to any sheet of connective tissue spreading from ('apo-') the edge of a tendon ('neuron', a term originally used indiscriminately for tendons and peripheral nerves, which are not dissimilar in macroscopic appearance), and providing an extension of its attachment. Such an arrangement is exemplified by the levator palpebrae superioris and by the retinacula of the quadriceps extensor tendon, but the term is more widely used for any broad sheet of dense connective tissue associated with the attachment of muscle. Sometimes the whole attachment may be aponeurotic (e.g. obliquus externus abdominis); elsewhere aponeuroses provide extensive auxiliary areas from which muscle fibres arise (e.g. supra- and infra-spinous aponeuroses or fasciae). In these cases the aponeuroses are themselves attached to bone, often providing an additional attachment for muscles partly attached directly to bone.

Fascia is a term so wide and elastic in usage that it signifies little more than a collection of connective tissue large enough to be described by the unaided eye. The advent of tissue fixatives, and especially formalin, which preserve and accentuate areolar tissue, was a great stimulus to the regional naming of fasciae; and the habit of attaching a specific local term to any aggregation of connective tissue, sizeable enough to dissect, is still current, though perhaps on the wane. During development a large number of mesodermal cells are differentiated into bone, muscle, vessels, renal and splenic tissue, etc., but large amounts of mesoderm persist as less specialized connective tissues permeating all regions of the body, not only as the microscopic areolar component between, for example, the fibres of muscle, nerves and tendons, but also in larger macroscopic accumulations between whole muscles, viscera and other large structures. The arrangement of such connective tissue is highly variable. As a result of dissection it appears as condensations on the surfaces of muscles, etc. and is hence spoken of as *investing* fascia, though its function in such a situation is by no means so simple. Between muscles which move extensively upon each other it is usually in the form of very loose areolar tissue, presumably to facilitate movement. The peripheral nerves, blood and lymph vessels travel in the loose 'fascia' between other structures, often bound together as *neurovascular bundles*; and sometimes, in the case of larger vessels, such as the common carotid and femoral arteries, much denser sheaths of connective tissue exist, their precise functions being a matter of conjecture. They may aid venous return by surrounding large veins in company with the pulsating artery, but they may be no more than incidental growth patterns which result from the pulsatile tensions generated by the arteries.

Fats may accumulate in the cells of fibro-areolar tissue wherever it occurs, forming adipose tissue, often well developed in the superficial region between muscles and skin. The dermis is, of course, compacted irregular connective tissue; and deep to this and continuous with it, loose areolar tissue, often adipose and of a variable thickness, is known as the **superficial fascia**. It allows the skin considerable freedom of movement and acts as a thermal insulator. Branches from the subcutaneous nerves, vessels and lymphatics travel in the superficial fascia, but their main trunks are in its deepest part where adipose tissue is scanty or absent. The superficial fascia also contains skin muscles, such as platysma, a representative of much more extensive sheets of skin-twitching musculature in other mammals. The quantity and distribution of its contained fat is different in the sexes. It is somewhat more abundant in females and more generally distributed; in males it diminishes from the trunk towards the extremities and this becomes more obvious in middle age when the total amount increases in both sexes. There is also an association with climate (rather than race), fat in the superficial fascia being more developed in the colder zones of the earth. Superficial fascia is most distinct over the lower part of the anterior abdominal wall, where it contains much elastic tissue and may present the appearance of several layers as it passes through the inguinal region into the thigh. It is also well differentiated in the limbs and perineum. It is thinnest over the dorsal aspects of

the hands and feet, the side of the neck, face, around the anus, and especially over the penis and scrotum. It is practically absent from the external ears, or auricles. It is particularly dense in the scalp, the palms and the soles where it is permeated by numerous strong bands of connective tissue binding the superficial fascia and the skin to underlying fibrous structures, which come under the general term, *deep fascia*, but are known regionally as the aponeuroses of the scalp, palm, and sole.

Deep fascia is also composed chiefly of collagen fibres, but these are here arranged more compactly and with a high degree of regularity in their directions, so that deep fascia is often indistinguishable from aponeurotic tissue. As in the latter, the parallel fibres of one layer are usually set at an angle to those of another. In the limbs, where the deep fascia is particularly well developed, many fibres are longitudinal in arrangement, while others encircle the limb, binding the longitudinal fibres together into a tough, inelastic sheath for the musculature. In both arm and leg some of the muscles arise from the internal aspect of the deep fascia, which at these sites performs the same function as an aponeurosis and has the same appearance. Wherever the deep fascia comes into direct contact with bone, or rather its periosteum, it fuses with this and is hence well able to take the pull of attached muscles. A particularly powerful band of deep fascia exists in the thigh to transmit the tension of the tensor fasciae latae, and most of the gluteus maximus, to the tibia (p. 596)—the iliotibial tract. The biceps brachii is partly attached to the deep fascia of the forearm, while the palmar aponeurosis, which not only takes the pull of palmaris longus but also has the small intrinsic muscles of the thumb and minimus partly attached to it, is the deep fascia of the region.

Both in the neck and the limbs, laminae, continuous with the deep fascia and of identical structure, pass inwards between the groups of muscles and connect extensively with bone. These are termed *intermuscular septa*, and they do incidentally separate muscles, or groups of them, sometimes with somewhat different actions, developmental histories, and innervations. Except as a phenomenon of growth, however, most of the smaller resulting 'compartments' are of little significance, for the septa often serve to connect rather than separate muscles, in fact arising from both aspects of a particular septum, which thus functions as an *intermuscular aponeurosis*.

In some situations, the wrist and ankle being prime examples, localized transverse thickenings occur in the deep fascia and are attached at both ends to local bony prominences. These are the *retinacula*, thus termed because they retain tendons passing deep to them which would otherwise be dragged or 'bowed' out of position by the activities of their muscles. In their osseofibrous channels the tendons are deflected to work effectively, smooth action being aided by synovial sheaths. Similar arrangements of deep fascia and bone form the fibrous sheaths or tunnels in which the flexor tendons of the digits, in both hand and foot, are retained—again surrounded by synovial sheaths.

Additionally, and especially in the lower limb, deep fascia contributes to the efficiency of venous return, its contained muscles exerting a pumping effect upon the deep veins during contractions. Regurgitation in the deep veins, or backflow into the superficial veins, is largely prevented by venous valves.

Regional details of the deep and superficial fasciae, retinacula, intermuscular septa, aponeuroses, bursae and tendon sheaths, are included at the appropriate places throughout this section.

The Form of Muscles

There is, of course, a wide variation in the size, shape, complexity, fascicular architecture and, as we have seen, the form of attachment of muscles. Each, however, is well adapted to provide an appropriate range, direction and force of contraction, to meet the habitual requirements at the articulations over which it passes. In this section our particular concern is the general shape of whole muscles, how their fibres vary in number, size and direction relative to the ultimate 'line of pull' at their attachments, and certain aspects of muscle action in which these structural variants are involved.

Individual muscle fibres are by no means uniform in diameter, ranging from 10 to 60 μm, and in length, from a few millimetres in the shortest muscles to many centimetres (15–30 cm and even longer have been quoted for the sartorius muscle). The diametric distribution of the fibres varies considerably with the particular muscle and species under consideration, as does the grouping of fibres into fasciculi by perimysial septa (p. 506). The general texture of the muscle naturally reflects the proportion and distribution of its connective tissue elements, muscles with fine fasciculi often being associated with precision movement patterns (e.g. the extrinsic ocular muscles), whereas coarse fasciculation characterizes many of the larger limb muscles (e.g. gluteus maximus). However, much remains to be investigated concerning the biomechanical significance, morphogenesis, and relation to

similar arrangement of fasciculi occurs in the 'bellies' of *fusiform* muscles, which may, however, be quite short, converging to a tendon, sometimes of considerable length at one or both ends. The tendon, of course, may concentrate the force of contraction on to a restricted area of bone surface, or alter the direction of pull, whilst the proportion of the total span of the muscle occupied respectively by belly and tendon is closely related to its maximal range of contraction, which is often quite limited.

The maximum *force* which a muscle can generate is finally dependent upon its effective *mass* of contractile tissue, which in turn reflects the number and dimensions of its contained fibres, whilst, as already implied, its maximum *range* of unrestrained contraction is a function of the *length* of its muscle fibres. Where the fasciculi are parallel to the line of pull at its attachments, full

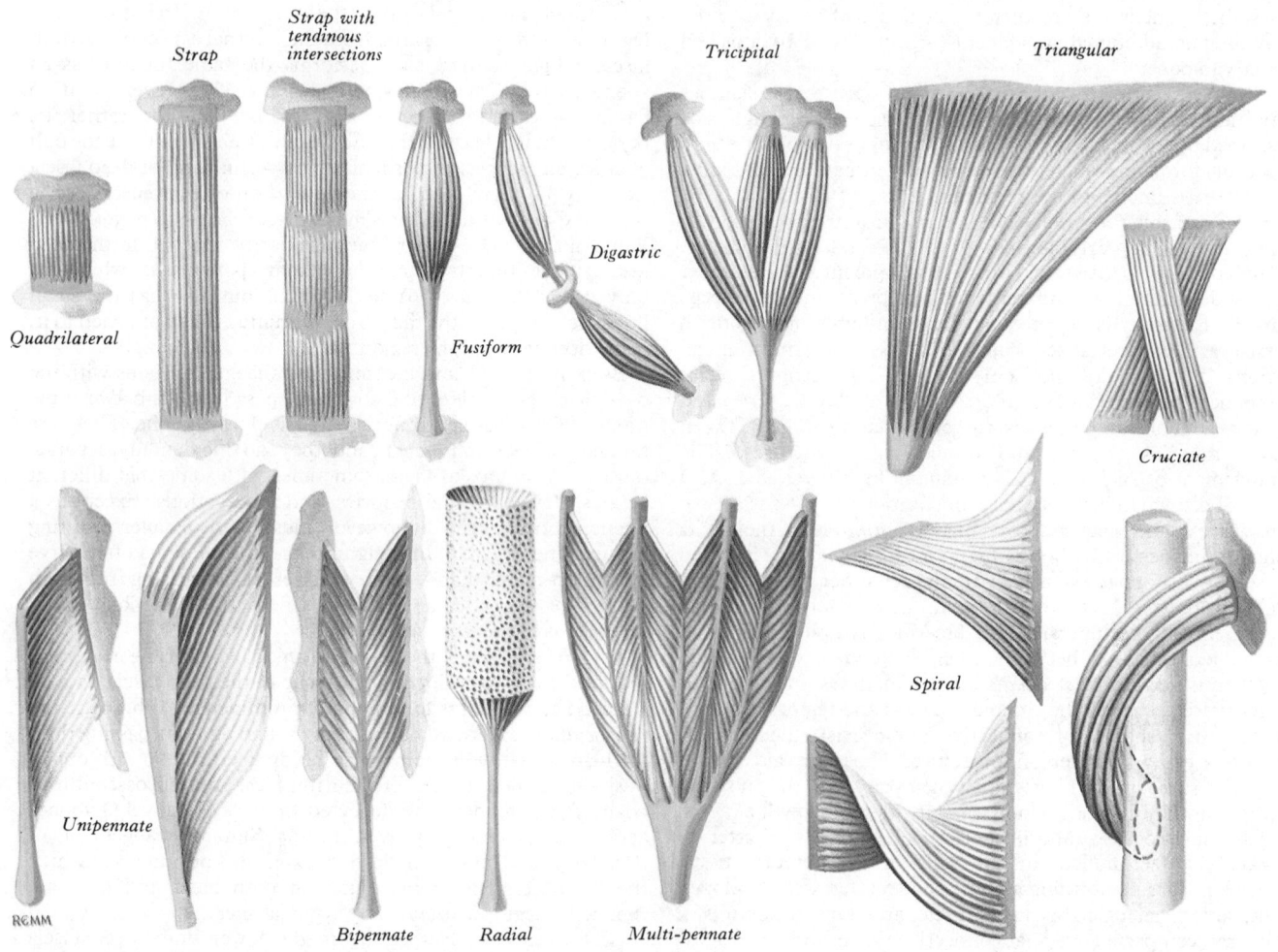

5.16 The morphological 'types' of muscle based upon their general form and fascicular architecture.

vascular and innervation patterns, of the fascicular units in muscles. They must also be carefully distinguished from the motor units which are the essential neuro-muscular functional divisions of a muscle, and which have a different distribution (p. 514), the interrelation of the two being at present obscure.

Broadly speaking, muscles may be grouped according to the orientation of their fasciculi, which may be *parallel*, *oblique*, or *spiralized* relative to the final direction of pull at their attachments (**5.16**).

Where the fasciculi are approximately parallel to the line of pull, they vary from flat, short and *quadrilateral* (e.g. thyrohyoid) to long and *straplike* (e.g. sternohyoid, sartorius), the individual muscle fibres often running almost the whole length of the muscle. Alternatively, the fibres may traverse shorter segments between irregular, transverse, tendinous intersections distributed at intervals along the muscle (e.g. rectus abdominis). A rather

advantage can be taken of both the available force and range (**5.17**A). However, in straplike muscles, the fibres, although long, are often relatively few in number, and such muscles are particularly effective in circumstances demanding a considerable range of action, coupled with modest power. The bellies of fusiform muscles, however, may be slender or massive, long or short, with similar variations in the length and strength of their associated tendons, and these structural variations show corresponding grades of potential power and range.

Where the fasciculi are *oblique* to the line of pull (**5.17**B), muscles are variously classed as *triangular* (e.g. temporalis, adductor longus) or *pennate* (feather-like) in their construction. The latter vary in complexity (see **5.16**). Examples are— *unipennate*, flexor pollicis longus; *bipennate*, rectus femoris and the dorsal interossei; *multipennate*, deltoid. A similar general architecture also obtains in certain other muscles, e.g. in the main part of soleus, where the fasciculi pass obliquely between deep and superficial aponeuroses and the muscle when seen in sagittal section is therefore somewhat 'unipennate' in form; in other situations muscle fibres arise from around the walls of osteofascial compartments and converge obliquely towards a centrally sited tendon—these are sometimes termed *circumpennate*, e.g. tibialis anterior. (The terms unipennate and bipennate are unfortunate; feathers or *pennae*, rarely, if ever, consist of a one-sided set of plumules attached to a midrib, like a 'unipennate' muscle. They are almost invariably symmetrical, corresponding to a 'bipennate' muscle. More logical terms would be semipennate and pennate.)

In all these muscle forms, it is evident that only a proportion of the force and range of action available are effective along the line of the tendon. It is seen in **5**.17B that the force generated by contraction of an oblique fibre can be resolved into two components, one acting along the tendon, and another at 90° to this; the latter is largely counterbalanced in bi- or multi-pennate forms by the opposite fibres, but it tends to deviate the line of the tendon in unipennate muscles. Similarly, the range is restricted (and is in fact proportional to the length of the muscle fibre × the cosine of the angle of its attachment to the tendon). In practice the loss of efficiency in force transmission to the tendon is greatly outweighed by the large number of short fibres accommodated in such muscles, and they are found typically where limited range coupled with considerable power are habitually required.

Yet other muscles are to some degree *spiralized* or 'twisted' in their general organization, the form and degree of the spiral varying with the muscle in question, its state of contraction and the posture adopted. For example, the large triangular trapezius

sometimes termed *cruciate* (**5**.16); examples are the sterno-cleidomastoid, masseter and adductor magnus, all of which are both partially spiral and cruciate in form.

Finally, it should be noted that many muscles exhibit regional variations in structure and are often compounded of more than one of the foregoing classical types, and this often reflects contrasting functional roles for the muscle. In this regard it should be emphasized that in some actions, only one region of a muscle may be active, and the remainder quiescent (e.g. see the action of trapezius in upward rotation of the scapula, p.565).

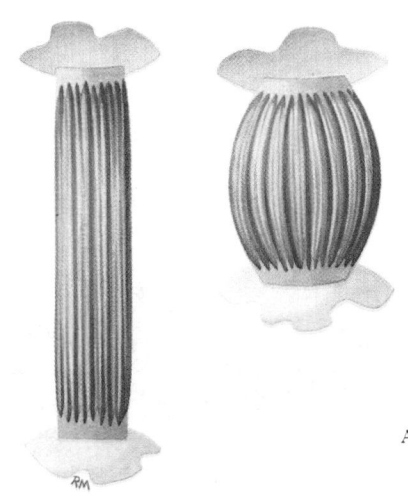

A

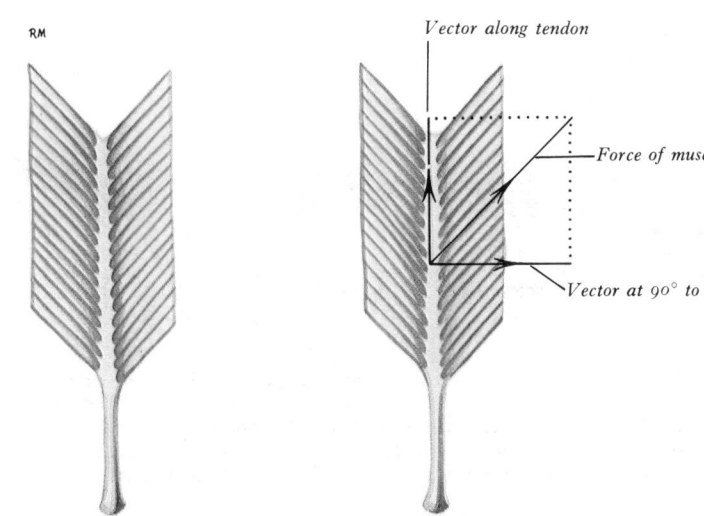

Vector along tendon

Force of muscle contraction

Vector at 90° to line of tendon

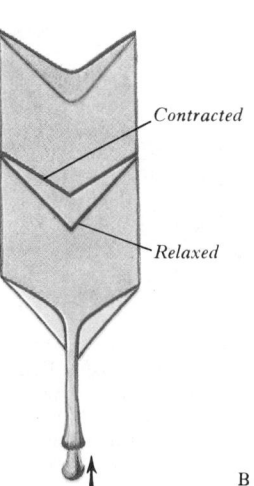

Contracted

Relaxed

B

5.17A and B   Simple mechanical considerations related to the actions of A strap, and B bipennate muscles. (For details of the actions of these different muscle types consult text.)

muscle has a long linear *vertical* midline attachment, and a *horizontal* curved lateral attachment, the muscle having a 90° twist between the two, with the body erect and the arms pendent. Similar considerations apply to the clavicular head of pectoralis major, whereas its sternocostal fibres, and those of latissimus dorsi, undergo a 180° twist between their midline and lateral attachments. When such muscles contract they not only act to approximate their attachments, but also to bring them into the same plane, and in so doing undergo some 'despiralization' (**5**.20). Elsewhere, muscles are spirally wrapped around a bone which forms one of their attachments, e.g. the supinator, which winds obliquely around the proximal radial shaft; their contraction again involves a reduction of the spiral and the imparting of a strong rotational force to the bone (*vide infra*).

Allied to spiral muscles are those which consist of two or more planes in which the fasciculi run in different directions and are

The Actions of Muscles

The analysis of sarcomere construction and its relationship to muscle contraction has made rapid progress up to the present time, and the views current at the time of writing have been set out elsewhere (p. 506). The nature of a simple muscle twitch, the partial or complete fusion of the mechanical responses into a tetanic contraction upon repetitive stimulation, and the concept of the functional neuromuscular territories or motor units, have also been discussed (pp. 514, 894). Ideas of isotonic and isometric forms of contraction, and the varying relationships between sarcomere length and the tension developed on contraction under isometric conditions, and during lengthening, have also received consideration (p. 510).

All these approaches are, of course, essential stages in the development of our understanding of the biology of muscle, but the distribution of such events in space and time within a whole intact muscle, whilst difficult to analyse experimentally, provides a much greater range and flexibility of response than might at first appear from the previous analysis.

Rapid, repetitive, synchronous stimulation of all, or the majority of the motor units does not occur *in vivo*, and habitual physiological demands are met by the *asynchronous* contraction of only a proportion of the units available. During normal activities, therefore, at any instant in time, the muscle consists of an admixture of motor units in different states, some quiescent, some undergoing stimulation, some contracting and some relaxing, the overall state of the muscle being the integral of these activities.

accompanied by slow 'background' activity rotating among a few motor units. On such accounts many misconceptions concerning the use of the term 'muscle tonus' were based. However, it must be realized that even a fully relaxed muscle is a highly complex three-dimensional structure endowed with *passive viscous, elastic* and other mechanical properties which provide some degree of resistance to deformation (e.g. by stretching). During activity also, such properties of the inactive motor units and their connective tissues provide an initial internal source of resistance, to be taken up and overcome by the active units, before external forces can be opposed and work performed. Such passive attributes of muscle, therefore, contribute greatly to the space-

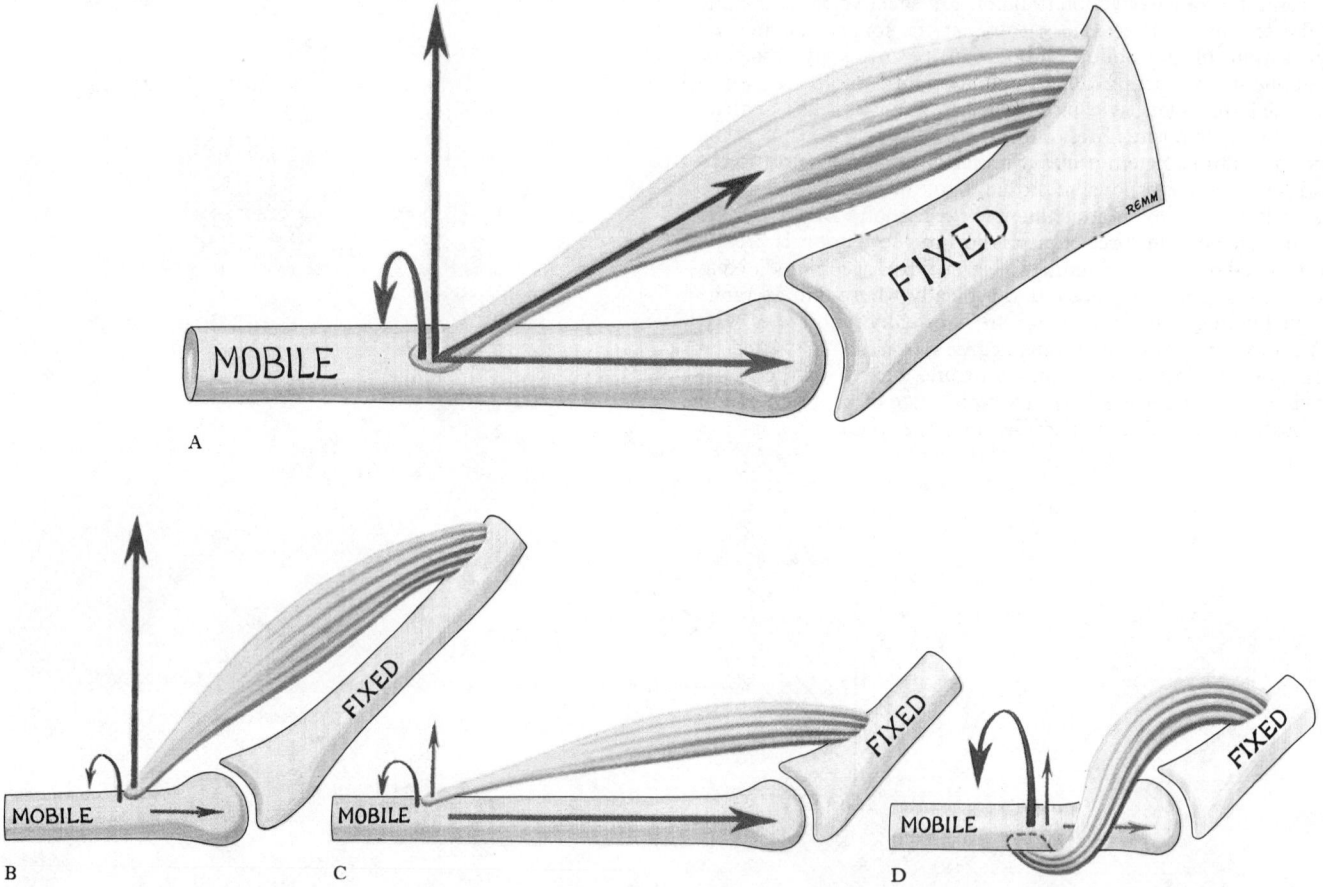

5.18A–D   A—Vectorial analysis of the *force* of contraction generated by a muscle which is attached to a fixed base, crosses a single multiaxial joint and is attached eccentrically to the shaft of a mobile bone. Colour code of arrows: *purple*—muscle force; *red*—transaxial 'swing' vector; *green*—paraxial 'shunt' vector; *blue*—tangential 'spin' vector. Similar analyses relating to muscles which are *predominantly*—B 'spurt', C 'shunt', and D

'spin' in their actions. It should be appreciated that these diagrams have been constructed to illustrate certain principles discussed in the text: they do *not* represent any specific human bone or muscle. It is also clear that, in B and C, if the fixity and mobility of the two bones were interchanged, B would now operate as a 'shunt' muscle, and C as a 'spurt' muscle.

With the passage of time there follows an ever-changing mosaic of active, relaxing and quiescent units. During such a rotation of states of activity between the available motor units, however, the *proportion* active may of course increase, decrease, or remain fairly constant with varying functional demands. Whether the muscle lengthens, shortens, or maintains its length, and what tension changes ensue depend upon many factors. These include the initial length of the muscle, the proportion of active motor units, or rate of change of this proportion, and the sum of the various forces both extrinsic and intrinsic to which the muscle is subjected, partly through its attachments. It must be emphasized that such a mechanism allows an infinite variety of possible quantitative responses of any particular muscle.

At one extreme, the muscle may be fully relaxed, and under such conditions much electromyographic evidence has accumulated to show that there is no motor-unit activity and the muscle is electrically silent (Joseph *et al.* 1955; Joseph 1960)—these views contrast with many earlier accounts of electrophysiologists who held that the nearest approach to full relaxation was still

time-force relationship of muscles during the initiation, increase, and maintenance of particular states of activity; they are, of course, of equal importance during diminution and/or cessation of activity.

When an individual muscle fibre or motor unit contracts, it of course tends to approximate its points of attachment, but whether it in fact does so depends upon the force it is capable of generating (a function of its geometry, mass and initial average sarcomere length), balanced against the *extrinsic* and *intrinsic* forces which are transmitted to it through its connective tissues, and which oppose its shortening.

For an active whole muscle, therefore, the following variations in activity are possible. Its original length may be *maintained* (with, of course, slight fluctuations) and the tension generated at its attachments may persist approximately unchanged, or it may increase or decrease, depending upon external circumstances and the number and state of its active motor units. Similarly, the muscle may *shorten* but at the same time, its tension may increase, persist or even decrease with varying conditions. Finally, a muscle may *lengthen*, and yet again it is particularly important to realize that its tension may decrease, persist or increase with varying demands. It should be clear from the foregoing that when considering whole muscles it is a naïve appraisal to simply correlate the term 'contraction' with only shortening or tension increases, and 'relaxation' with only lengthening or decreases of tension—many other subtle grades of combination are not only possible, but form the whole foundation for the infinitely varied patterns of activity in human behaviour.

When a muscle contracts to initiate a movement, for example a limb muscle across a single joint, the circumstances which oppose its shortening stem from many sources. These include the passive

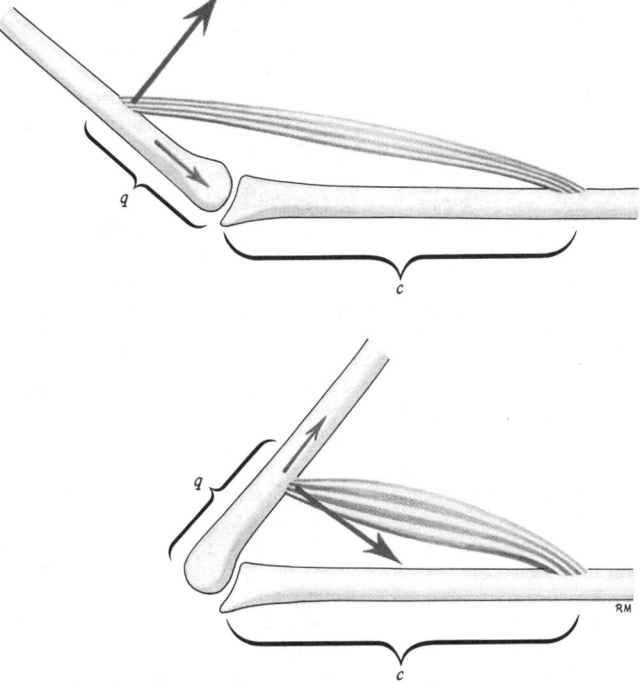

5.19A Some features of a 'spurt' muscle—note that it has a large 'swing' component of its force which acts to impart an angular acceleration to the mobile bone. Its small 'shunt' component acts towards the joint up to 90° of flexion, and thereafter away from the joint. *c*—cisarticular length; *q*—transarticular length.

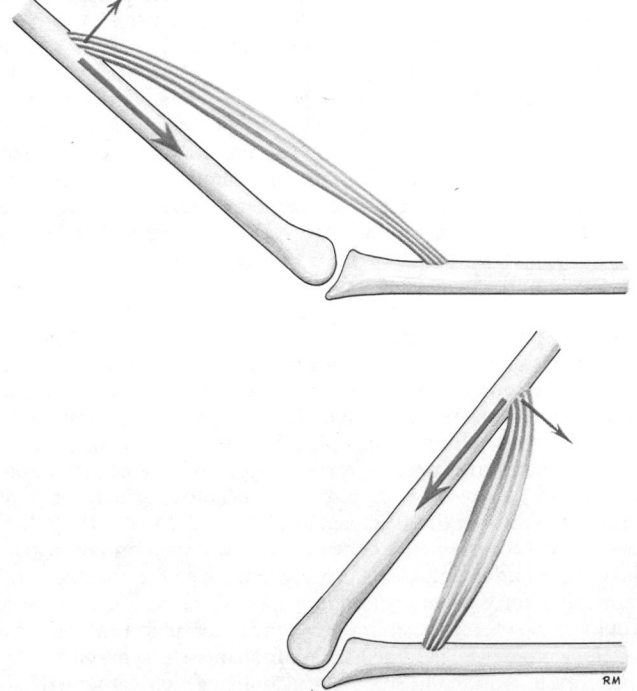

5.19B Some features of a 'shunt' muscle—the large shunt component of its force acts towards the joint in all positions. In figures **5.19**A and B no specific human muscle or bone is intended; in each case it is clear that the right bone is regarded as fixed whilst the left bone is mobile. Clearly their mobilities (and designations) could be interchanged.

mechanical properties of the muscle itself mentioned above, similar properties of the articular tissues, opposing musculature, and the other soft tissues of the limb, any contractile activity of the opposing musculature, the inertia of the segment to be moved, together with any additional burden carried by it, and finally, the inescapable effect of gravity upon all the foregoing. In general, sufficient motor units are thrown into activity to overcome these resistances and to impart an angular acceleration on the limb segment until a required velocity has been reached, after which only sufficient units continue contracting to maintain this throughout the remainder of the movement.

In practice, the various movements at different joints are

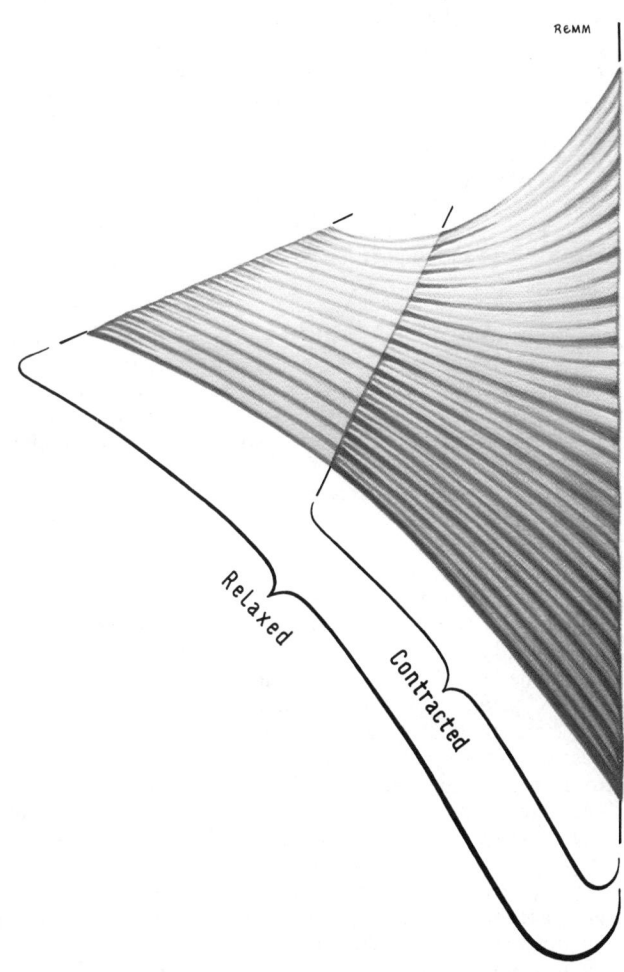

5.20 An illustration of the 'detorsion' or untwisting which follows the contraction of a spirally disposed muscle.

carried out by *ordered patterns of activity* in muscle groups and these have been classified according to their role.

Ignoring for a moment movements carried out with the assistance of gravity (*vide infra*), one or more muscles are constantly active in the initiation and maintenance of a particular movement, and these are classed as *prime movers* (e.g. brachialis in elbow flexion); and the muscle or muscles which can wholly oppose such a movement, or initiate and maintain its converse, are termed *antagonists* (e.g. triceps in the example quoted). However, views concerning the interrelation of the two have undergone extensive revision in recent years. For long it was widely held that at the outset of a particular movement, contraction of the prime movers was immediately accompanied by partial contraction of the antagonists, after which smooth continuation of the movement depended upon progressive contraction of the prime movers with simultaneous lengthening and relaxation of the antagonists, the latter finally increasing their contraction to decelerate and terminate the movement. The neural connexions presumed to be responsible for this coordinated activity were incorporated in a '*principle of reciprocal innervation*'. However, it has now been demonstrated that apart from a variable transient burst of activity in the antagonists at initiation of the movement, they then remain quiescent until the final brief deceleration. Nevertheless, the activity of the prime movers is not unrestrained, but is balanced against the various *passive, inertial* and *gravitational* effects mentioned above.

There are, however, occasions when prime movers and antagonists *contract together* as *fixators*, increasing the *transarticular compression* (*vide infra*), and *stabilizing* the position of a joint, often to provide an immobile base from which other prime movers can act. Sometimes this stability can be achieved by the effects of gravity alone, e.g. when the joint is in or near its close-packed position (cf. the knee and hip joints in the erect posture, pp. 438, 617), or by prime movers balancing the various passive forces (cf. the stabilization of the humeral head in the glenoid cavity by supraspinatus with a pendent upper limb, pp. 458, 569); when this is not the case or when powerful external forces are encountered, both groups of muscles contract together, holding the joint in any desired position.

It is important to emphasize that virtually all movements (except the theoretically possible, but highly improbable case of a uniaxial joint acting precisely in the horizontal plane), are either opposed or assisted by gravitational forces. This has a profound effect upon the play of musculature during various movements, and prime-mover activity is always replaced by gravity whenever this is appropriate—in such cases the conventionally described roles of muscles are often reversed. For example, the prime mover for extension of the forearm on the arm at the elbow joint is, of course, the triceps (e.g. when thrusting, or partially supporting body weight on the hands with semiflexed elbows). When a weight is lowered from face level to a table, however, the movement of extension is accompanied by active controlled lengthening of the joint flexors. Similar considerations apply in many other circumstances, for example, the increasing length and force of contraction in the hamstrings (extensors of the hip joint), during controlled forwards bending or flexion of the trunk at the hip joints.

In some cases the contraction of a prime mover, acting across a uniaxial joint, simply produces the desired movement; but elsewhere, particularly with multiaxial joints, or when muscles cross more than one joint, the unrestrained contraction of such prime movers may produce additional unwanted movements. The elimination of the latter is effected by the contraction of *synergic* muscles which are, in a sense, partial antagonists to the prime mover. For example, the prime movers of powerful flexion of the fingers at the interphalangeal and metacarpophalangeal joints are the long flexors, superficial and deep, which also cross the intercarpal and radiocarpal joints and which, with unrestrained contraction, would also flex with a loss of efficiency. However, simultaneous contraction of the synergic carpal extensors not only eliminates this, but often produces some degree of extension at the wrist, with further increase of efficiency. (For further examples of synergy, *vide infra*.) Electromyography is now showing that, wherever feasible, movements are effected without using energy-consuming, synergist-dependent muscles. However, the majority of complex movements are the outcome of a finely graded interplay between external forces including gravity, inertia of the masses involved, passive mechanical properties of the various tissues, and the integrated patterns of tension-length variations in prime movers, antagonists, synergists and fixators. These patterns, and the positions of the joints between the body segments are constantly monitored by afferent nerve terminals in the various connective, periarticular and muscular tissues, and are in turn controlled and welded into functionally organized sequences by mechanisms which involve the cooperation of all levels in the central nervous system.

In addition to the foregoing account of muscle action some further concepts and an accompanying terminology have emerged from parallel kinesiological studies of joint and muscle mechanics (MacConaill and Basmajian 1977). These are only touched on briefly here and for a more comprehensive treatment, the references quoted should be consulted.

Illustration 5.18A shows two bones, one for our present purposes is considered mobile and one fixed in position, articulating at a joint. A muscle which is attached slightly eccentrically to the mobile bone is contracting and exerting a force along the line of its tendon. It can be seen that this *force* can, in relation to the mobile bone, be resolved vectorially into three components. One is *transaxial* acting at right angles to the bone, tending to change its angulation at the joint (i.e. a *swing component*), another is *paraxial* and acts along the shaft of the bone towards the joint and tends to compress the articular surfaces (i.e. a *transarticular* or *shunt component*), and the third acts tangentially around the long axis of the bone and tends to rotate it about this axis (i.e. a *spin component*). The terms spurt and shunt may be unfamiliar to the reader: they originated from terminology used by nineteenth-century British engineers—*see* MacConaill (1949).

The actual movement produced will of course depend upon any constraints built into the shape and ligaments of the articulation, adjacent musculature, and extrinsic and intrinsic forces opposing the movement. Nevertheless, each of the three components has its own functional importance and is enhanced in certain muscles, and less prominent in others (5.18B, C, D). For example, when a muscle is markedly eccentric in its attachment to, or even wrapped around the mobile bone, it has a prominent spin component. Thus, the transversely wrapped pronator quadratus is almost a pure muscle of spin, whereas the longer oblique pronator teres has spin, transarticular and swing components. Finally, in contrast, the hamstring muscles exert a powerful swing force (flexion) on the knee joint and a much weaker spin force in medial and lateral rotation of the semiflexed leg. Such spin components of the force may be used as prime movers, e.g. in pronation, and supination of the forearm, medial and lateral rotation at the hip joint, etc., or they may be used synergically balancing an opposite spin force, so allowing pure swing to occur (*see* comments on *antispin* p. 437).

Ignoring any spin component, the relative values of the transarticular and swing components will vary according to the distances of the fixed and mobile attachments of the muscle from the joint (5.19). These distances are termed the *cisarticular* (c) and *transarticular* (q) *lengths* respectively, and c/q denotes the *partition ratio* (p). When the fixed attachment is distant from the joint and the mobile attachment close to it (5.19A), i.e. $p > 1$, there is a powerful swing component and only a modest transarticular (shunt) component, and such a muscle has been designated a *spurt* muscle (and is constructed of spurt myonemes). A typical example is the brachialis muscle acting on the elbow joint from a fixed humerus. Such muscles acting alone are the most effective in initiating and continuing relatively slow movements of swing whether these are weak or powerful, and in carrying out postural readjustments which demand graded alterations of tension and length. In these circumstances the shunt component of the muscle involved is often sufficient to maintain contact between the articular surfaces at the joint. However, it is to be noted that when the angulation of the joint passes 90° the component of force acting along the bone is now directed away from the joint.

In contrast, a *shunt* muscle (composed of shunt myonemes) arising from a fixed base close to a joint and having a mobile attachment at some more distant point, i.e. $p < 1$, shows a large persistent (shunt) force vector acting along the bone towards the joint in all its positions, but only has a trivial swing component (5.19B). An often quoted example is brachioradialis acting on the elbow joint from a fixed humerus. Shunt myonemes or whole muscles are necessary in any circumstances which operate strongly to distract the articular surfaces. Such forces may follow loading, e.g. of the pendent arm, but it was also proposed that during a *rapid* swing of a limb segment, e.g. the forearm, the centrifugal force generated tending to separate the joint surfaces, would be too great to be opposed by spurt musculature alone. The latter was therefore assumed to be supplemented synergically by appropriate shunt muscle contractions which provided an opposing centripetal force towards the joint. Evidently, the generation of a centrifugal force, and the need for its opposition, would be independent of the direction of the swing, and shunt activity of the brachioradialis has been demonstrated electromyographically in both rapid flexion and rapid extension of the forearm (however, *vide infra*). In the foregoing brief discussion of spurt and shunt myonemes and whole muscles it will have become clear to the reader that in any particular instance under consideration, one of the attachments is regarded as fixed (i.e. the 'functional origin') while the other (the 'functional insertion') is attached to a mobile unit. In practice, of course, the relative fixity and mobility of the two attachments are often reversed: in such circumstances, for example, a spurt myoneme would now operate as a shunt myoneme (and *vice versa*). For such reasons the terms origin and insertion have been abandoned in the following pages devoted to descriptions of individual muscles, the more general term attachment being preferred.

Since its first proposal, subsequent refinement, and explicit mathematical statement (for literature *see* MacConaill and Basmajian 1977; and MacConaill 1978a, b, c), the anatomico-kinetic classification of myonemes and whole muscles into spurt and shunt varieties has prompted much discussion and, in some cases, adverse criticism (*see*, for example, the writings of Stern 1971; Rozendal and Molen 1972; Joseph 1973; Jackson *et al.* 1977; and Stanier 1977). In his most recent publications, however, the proposer has stoutly maintained the validity of his *classification* (*see* references above), but has withdrawn certain *hypotheses* based upon this classification. In an attempt to provide a comprehensive 'generalized mechanics of articular swing', the interesting proposal has been made that the centripetal force acting along a bone moving at *any* speed is a *consequence* of its curvilinear motion; not a *cause* of it.

Another absorbing problem which is currently engaging both thoretical and experimental kinesiologists is the analysis of the *patterns* of sequential muscle contractions that occur when two or more muscles are capable of producing the same movement at a particular joint (consult Joseph 1973; and Jackson *et al.* 1977). The latter contend that there is a 'primary' muscle which always initiates the movement, and that 'secondary' muscles are only recruited when a critical level of activity has been reached in the primary muscle. These findings are held to apply to both isotonic contraction and isometric contraction, and the authors emphasize the importance of variations in force and velocity as determinants of sequential patterning. However, the values of the various parameters which, in combination, constitute a critical level of 'activity' have not yet been determined in a wide range of human muscles. It is also of interest that few authors have discussed explicitly the phenomenon of *increasing length* accompanied by *increasing tension* of a myoneme, or a whole muscle—a combination which is commonplace in habitual movements, and clearly should *not* be termed 'relaxation'.

Finally, it must be emphasized that these more advanced considerations of the activities of living myonemes, muscles, and the interplay of groups of muscles, coordinated by a most complex hierarchy of central nervous control systems, render much of the customary myological nomenclature as sometimes redundant, often inaccurate, and frequently frankly misleading.

THE FASCIAE AND MUSCLES OF THE HEAD

These are habitually divided into the *facial muscles* which are in large measure related to the orbital margins plus eyelids, the external nose and nostrils, the lips, cheeks and mouth, the pinna, scalp and cervical skin (all, not entirely appropriately, often grouped as 'the muscles of facial expression'); and the *masticatory muscles* primarily concerned with movements of the temporo-mandibular joint. This dual division of head musculature reflects the different embryonic origins and innervations of the two groups, but it must be emphasized that in many functional activities, such as mastication, deglutition, vocalization, communicative and emotional expression, respiration, ocular, aural and nasal action, the two groups are closely cooperative and interdependent.

CRANIOFACIAL MUSCLES

(A) Epicranial Musculature

The epicranius consists of two main parts: M. occipitofrontalis and M. temporoparietalis.

The superficial fascia in the scalp is firm and fibro-adipose, adherent to skin and to the underlying epicranius and its aponeurosis, the *galea aponeurotica* (epicranial aponeurosis); behind, it is continuous with the superficial fascia of the back of the neck; laterally, it is prolonged into the temporal region, where it is looser in texture.

The occipitofrontalis, a broad, musculofibrous layer, covers the dome of the skull, from the nuchal lines to the eyebrows. It consists of four parts—two occipital and two frontal—connected by the galea aponeurotica. (The osseous attachment necessarily interrupts the direct continuity of the galea aponeurotica and the cervical fascia, although, of course, the two are in a sense confluent, and *adherent to bone*, through the medium of the occipital periosteum.)

Each *occipital* part, thin and quadrilateral, arises by tendinous fibres from the lateral two-thirds of the highest nuchal line of the occipital bone and the mastoid part of the temporal bone. It ends in the galea aponeurotica.

Each *frontal* part (**5.**21) is thin, quadrilateral, and adherent to the superficial fascia. It is broader than the occipital part and its fibres are longer and paler. It has no bony attachments. Its medial fibres are continuous with those of the procerus; its intermediate fibres blend with the corrugator supercilii and orbicularis oculi;

its lateral fibres are also blended with the latter muscle over the zygomatic process of the frontal bone. From these attachments the fibres are directed upwards to join the galea aponeurotica in front of the coronal suture. The medial margins of the frontal slips are joined together for some distance above the root of the nose; but between the occipital bellies there is a considerable, though variable interval, occupied by an extension of the galea aponeurotica.

The temporoparietalis is a variably developed sheet of

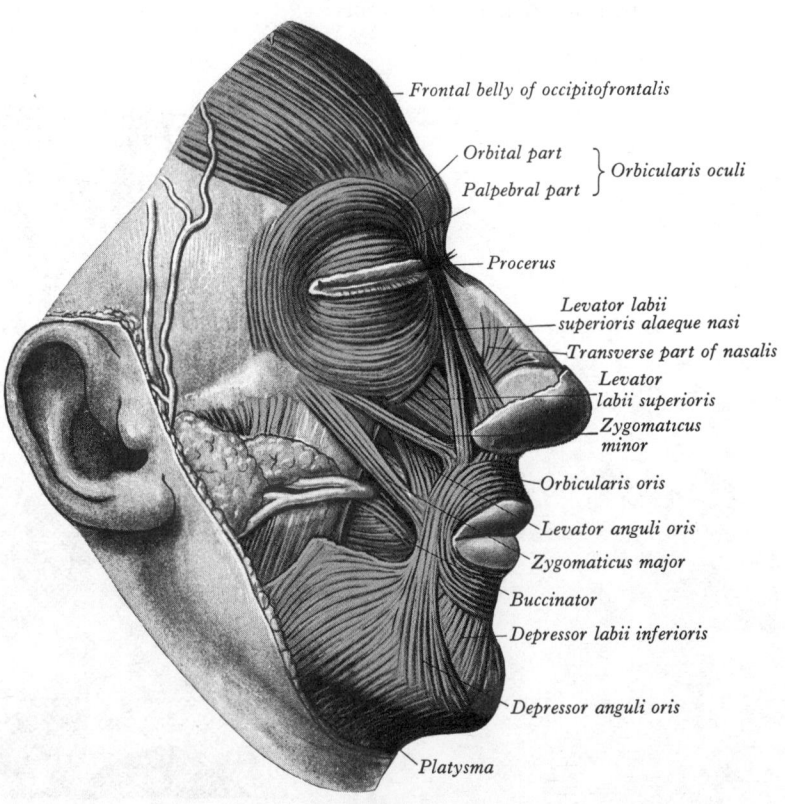

Frontal belly of occipitofrontalis

Orbital part
Palpebral part $\Big\}$ Orbicularis oculi

Procerus

Levator labii
superioris alaeque nasi

Transverse part of nasalis

Levator
labii superioris

Zygomaticus
minor

Orbicularis oris

Levator anguli oris

Zygomaticus major

Buccinator

Depressor labii inferioris

Depressor anguli oris

Platysma

5.21 Muscles of the scalp and face. Right lateral aspect.

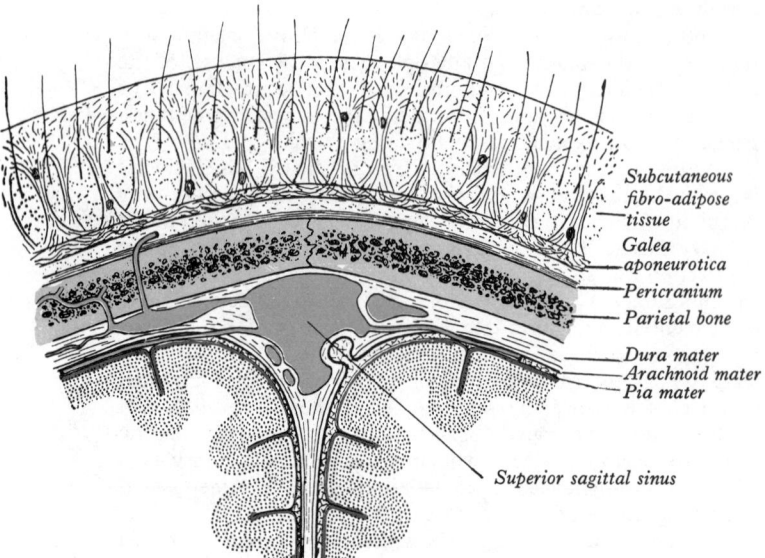

5.22 A coronal section through the scalp, skull and brain.

Labels on figure:
Subcutaneous fibro-adipose tissue
Galea aponeurotica
Pericranium
Parietal bone
Dura mater
Arachnoid mater
Pia mater
Superior sagittal sinus

Nerve supply. The occipital part is supplied by the posterior auricular branch, and the frontal part by the temporal branches, of the facial nerve.

Actions. The occipital slips draw the scalp backwards; the frontal slips acting from above raise the eyebrows and the skin over the root of the nose; acting from below they draw the scalp forwards, throwing the integument of the forehead into transverse wrinkles. The occipital and frontal parts, acting alternately, move the entire scalp backwards and forwards. In the ordinary action of the frontal parts the eyebrows are elevated, a common accompaniment of glancing upwards. The action is also a part of expressions of surprise, horror, or fright, etc.

A thin muscular slip, termed the *transversus nuchae*, is present in about 25 per cent of people; it arises from the external occipital protuberance or from the superior nuchal line, either superficial or deep to the trapezius; it is frequently inserted with the auricularis posterior, but may join the posterior edge of the sternocleidomastoid.

Applied Anatomy. The scalp consists of five layers—skin, subcutaneous tissue, epicranius (and its aponeurosis), sub-aponeurotic areolar tissue, and pericranium (**5**.22). But it is better to regard the first three as a single layer, since when torn off in accidents, or turned down surgically, they remain firmly connected to each other. Because of the denseness of the subcutaneous tissue, any inflammatory swelling is slight and a wound which does not involve epicranius or the galea aponeurotica does not gape. Contraction and retraction of arteries is impeded by the dense nature of this tissue and therefore haemorrhage from scalp wounds is often considerable.

The subaponeurotic areolar tissue is surgically important. It is loose and lax, and is easily torn through; hence, it is this tissue which is torn when the scalp is avulsed. The vessels are in the avulsed tissue, and since they anastomose freely, necrosis is unusual.

muscle which lies between the frontal part of the occipitofrontalis and the anterior and superior auricular muscles.

The galea aponeurotica (epicranial aponeurosis) covers the upper part of the cranium and forms with the epicranius a continuous fibromuscular sheet extending from the occiput to the eyebrows. Behind, in the interval between the occipital parts of the occipitofrontalis, it is attached to the external protuberance and highest nuchal line of the occipital bone. In front it splits to enclose the frontal parts and sends a short narrow prolongation between them. On each side the auriculares anterior et superior are attached to it; in this situation it becomes thinner, and is continued over the temporal fascia to the zygomatic arch. It is united to the skin by the firm, fibrous superficial fascia; it is connected to the pericranium by loose areolar tissue which allows it free movement, the latter carrying with it the skin of the scalp (**5**.22). Subdivision of this 'loose' subaponeurotic tissue into three layers has been described (Chayen and Nathan 1974), the intermediate of which is dense and resembles the galea aponeurotica in its attachments.

(B) Circumorbital and Palpebral Musculature

Of the muscles to be considered in this connexion—orbicularis oculi, corrugator supercilii, and levator palpebrae superioris—the last is described with the adnexa of the eye, including the extra-ocular muscles (p. 1178).

The orbicularis oculi (**5**.21, 23) is a broad, flat elliptical muscle which occupies the eyelids, surrounds the circumference

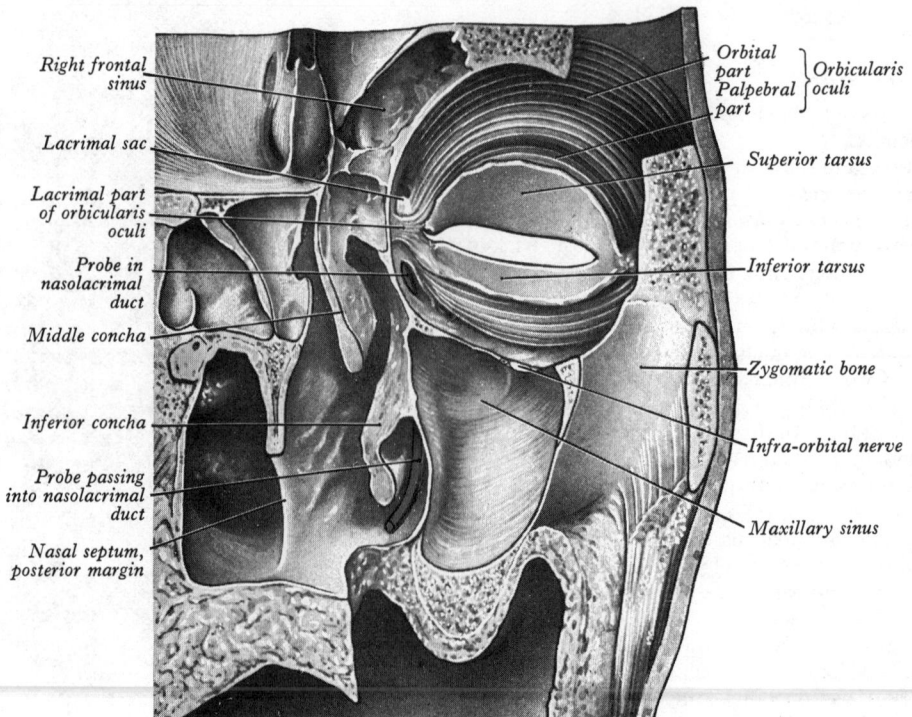

Labels on figure:
Right frontal sinus
Lacrimal sac
Lacrimal part of orbicularis oculi
Probe in nasolacrimal duct
Middle concha
Inferior concha
Probe passing into nasolacrimal duct
Nasal septum, posterior margin
Orbital part / Palpebral part } Orbicularis oculi
Superior tarsus
Inferior tarsus
Zygomatic bone
Infra-orbital nerve
Maxillary sinus

5.23 A dissection to expose the right orbicularis oculi from behind.

of the orbit, and spreads on the temporal region and cheek. It consists of orbital, palpebral and lacrimal parts.

The *orbital* part, reddish and thicker than the palpebral fasciculi, arises from the nasal part of the frontal bone, from the frontal process of the maxilla (**5.24**), and from the medial palpebral ligament, which interrupts the bony attachment. Its fibres form complete ellipses without interruption on the lateral side, the upper ones blending with the frontal part of occipitofrontalis and the corrugator. Many of the upper orbital fibres are inserted into the skin and subcutaneous tissue of the eyebrow. These constitute the *depressor supercilii*.

The *palpebral* part is thin and pale; it arises from the medial palpebral ligament, chiefly from its superficial but also from its deep surface, though not from its lower margin; it arises also from the bone immediately above and below the ligament. The fibres sweep across the eyelids in front of the orbital septum (p. 1185) and at the lateral commissure interlace to form the lateral palpebral raphe. A small group of fine fibres lies close to the margin of each eyelid, behind the eyelashes; it is named the *ciliary bundle*.

The *lacrimal* part lies behind the lacrimal sac, but is separated from it by the lacrimal fascia. It is attached to the lacrimal fascia, to the upper part of the crest of the lacrimal bone, and adjacent part of the lateral surface of the lacrimal bone (**5.24**). Passing laterally behind the lacrimal sac the muscle divides into upper and lower slips; some fibres are inserted into the tarsi of the eyelids close to the lacrimal canaliculi, but most continue across in front of the tarsi and interlace in the lateral palpebral raphe.

The *medial palpebral ligament*, about 4 mm in length and 2 mm in breadth, is attached to the frontal process of the maxilla in front of the nasolacrimal groove. Crossing the lacrimal sac, it divides into upper and lower parts, each attached to the medial end of the corresponding tarsus. It is separated from the lacrimal sac by the lacrimal fascia.

The *lateral palpebral raphe* is a much weaker structure, formed by the interlacing lateral ends of the palpebral fibres of the orbicularis oculi, strengthened on its deep surface by the orbital septum. A few lobules of the lacrimal gland or, more frequently, a small lobule of fat may lie between it and the more deeply placed lateral palpebral ligament.

Nerve supply. Temporal and zygomatic branches of the facial nerve.

Actions. The orbicularis oculi is the sphincter muscle of the eyelids. The palpebral portion can act under voluntary control, or reflexly, closing the lids gently, as in sleep or in blinking; the orbital portion is more frequently under voluntary control. It should be appreciated that during eye closure, although the major factor is lowering of the upper eyelid, there is also considerable elevation of the lower lid. Thus the palpebral part has upper *depressor* and lower *elevator* fascicles. When the entire muscle contracts, the skin of the forehead, temple and cheek is drawn *towards* the *medial* angle of the orbit, and the eyelids are not only firmly closed, but displaced *in toto* a little medially. The skin is thrown into folds, radiating from the lateral angle of the eyelids; these often become permanent—the so-called 'crow's feet'. The levator palpebrae superioris is an antagonist of the superior palpebral depressor fibres of this muscle since the former raises the upper eyelid and exposes the front of the bulb of the eye. The lacrimal part of the orbicularis oculi draws the eyelids and the lacrimal papillae medially; at the same time it exerts traction on the lacrimal fascia and is said to dilate the lacrimal sac (see also p. 1188). Thus the orbicularis oculi is importantly involved in tear transport. It may affect the disposition and pressure relationships of the palpebral part of the lacrimal gland, its ducts and their orifices; it assists in establishing the sinuous trajectories of the medially coursing tear film across the cornea; it adjusts the deep positioning of the puncta lacrimalia in the lacus lacrimalis, it may reduce pressure in the lacrimal sac, and possibly affects the operation of the ciliary and tarsal glands. The muscle is also an important element in facial expression and ocular reflexes. Partial closure of the palpebral fissure together with bunching and protrusion of the eyebrows diminishes the entry of light, into the eyes. The latter action of the upper orbital fibres, and their peripheral extensions, cause vertical furrowing above the bridge

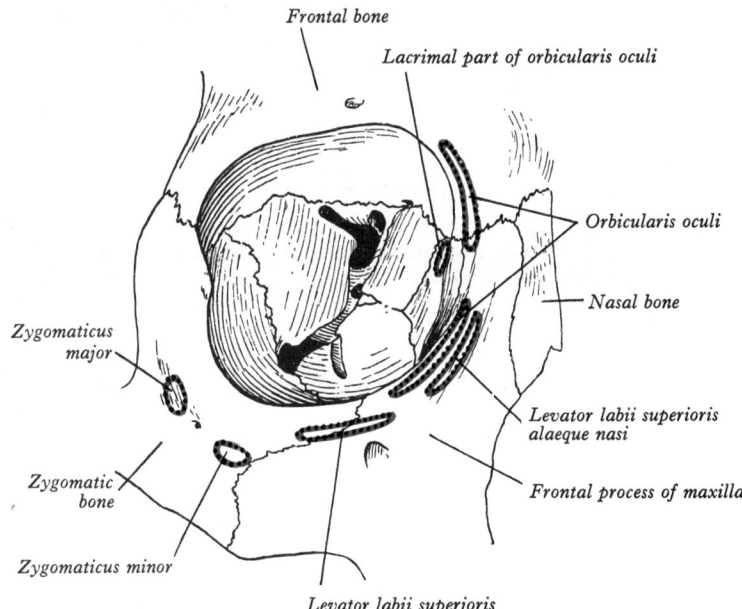

5.24   Attachments of muscles around the right orbital opening.

of the nose—features particularly well developed in inhabitants of the tropics. The protective value of the blink reflex is obvious.

The corrugator supercilii is a small pyramidal muscle, at the medial end of the eyebrow, deep to the frontal part of occipitofrontalis and the orbicularis oculi. From the medial end of the superciliary arch its fibres pass laterally and slightly upwards to the deep surface of the skin above the middle of the supra-orbital margin.

Nerve supply. The temporal branches of the facial nerve.

Actions. The muscle draws the eyebrow medially and downwards, producing, together with orbicularis oculi, the *vertical* wrinkles of the forehead. It assists in drawing the eyebrows downwards in bright sunlight, and is also involved in frowning.

(C) The Nasal Musculature

This group comprises the procerus, nasalis and depressor septi (**5.21**).

The procerus is a small pyramidal slip continuous with the medial side of the frontal part of the occipitofrontalis. It arises from the fascia covering the lower part of the nasal bone and the upper part of the lateral nasal cartilage; it is inserted into the skin over the lower part of the forehead between the eyebrows.

Actions. The procerus draws down the medial angle of the eyebrow and incidentally produces the transverse wrinkles over the bridge of the nose. It is active in frowning and 'concentration'. It also aids in reducing the glare of bright sunlight.

The nasalis consists of transverse and alar parts which may be continuous at their origins. The *transverse part* (compressor naris) arises from the maxilla just lateral to the nasal notch; its fibres proceed upwards and medially and expand into a thin aponeurosis, which is continuous on the bridge of the nose with that of the muscle of the opposite side, and with the aponeurosis of the procerus. The *alar part* (dilatator naris) arises from the maxilla, below and medial to the transverse part. It is attached to the cartilaginous ala nasi.

Actions. The transverse part of the nasalis compresses the nasal aperture at the junction of the vestibule with the nasal cavity. The alar part draws the ala downwards and laterally and so assists in widening the anterior nasal aperture. These actions are visible in deep respiration, especially its inspiratory phase, and they also accompany certain emotional states.

The depressor septi, often regarded as part of the dilatator naris, is attached to the maxilla above the central incisor tooth; it ascends to the mobile part of the nasal septum. It is immediately deep to the mucous membrane of the upper lip.

Actions. The depressor septi assists the alar part of the nasalis in widening the nasal aperture in deep inspiration.

Nerve supply. All the muscles of this group are supplied by the superior buccal branches of the facial nerve.

(D) The Buccolabial Musculature

The shape of the buccal orifice and the posture of the lips are controlled by a complex assembly of muscular slips (5.21). These include elevators and retractors of the upper lip (levator labii superioris alaeque nasi, levator labii superioris, zygomaticus major and minor, levator anguli oris, and risorius), depressors and retractors of the lower lip (depressor labii inferioris, depressor anguli oris, and mentalis), and a sphincter (orbicularis oris and buccinator).

The levator labii superioris alaeque nasi arises from the upper part of the frontal process of the maxilla and, passing obliquely downwards and laterally, divides into medial and lateral slips. The medial slip is inserted into the greater alar cartilage and

skin of the ala of the nose; the lateral slip is prolonged into the lateral part of the upper lip, and blends with the levator labii superioris and orbicularis oris.

Actions. The lateral slip raises and everts the upper lip; the medial slip acts as a dilatator of the nostril.

The levator labii superioris starts from the lower margin of the orbital opening immediately above the infra-orbital foramen, arising from the maxilla and the zygomatic bone. Its fibres converge into the muscular substance of the upper lip between the lateral slip of the levator labii superioris alaeque nasi and levator anguli oris.

Actions. The levator labii superioris raises and everts the upper lip. Together with the zygomaticus minor it forms the nasolabial furrow, from the side of the nose to the upper lip. The furrow is often deepened in the expression of sadness or seriousness.

The zygomaticus minor arises from the lateral surface of the zygomatic bone immediately behind the zygomaticomaxillary suture, and passes downwards and medially into the muscular substance of the upper lip. It is separated from the levator labii superioris by a narrow interval (5.21).

Actions. The zygomaticus minor elevates the upper lip and also produces the nasolabial furrow. When the levator labii superioris alaeque nasi, the levator labii superioris and the zygomaticus minor are in action together, they express contempt and disdain.

The levator anguli oris arises from the canine fossa, just below the infra-orbital foramen, and is inserted into the angle of the mouth, intermingling with the fibres of zygomaticus major, depressor anguli oris, and orbicularis oris. Between the levator anguli oris and the levator labii superioris are the infra-orbital vessels and plexus of nerves (5.26).

Actions. The levator anguli oris raises the angle of the mouth and is instrumental in producing the nasolabial furrow.

The zygomaticus major extends from the zygomatic bone, in front of the zygomaticotemporal suture, to the angle of the mouth, where it blends with the fibres of the levator anguli oris, orbicularis oris, and depressor anguli oris. The zygomaticus major and minor and the levator labii superioris are sometimes more or less concealed by a thin sheet of muscle, named the *musculus malaris* and continuous with the orbicularis oculi (Lightoller 1925).

Actions. The zygomaticus major draws the angle of the mouth upwards and laterally as in laughing.

Nerve supply. All the five preceding muscles are supplied by the buccal branches of the facial nerve.

The mentalis is a conical fasciculus, at the side of the frenulum of the lower lip. It arises from the incisive fossa of the mandible and descends to be attached to the skin of the chin.

Actions. The mentalis raises and protrudes the lower lip, and at the same time wrinkles the skin of the chin, in drinking and also in expressing doubt or disdain. Electromyography shows, it is claimed, fairly continuous activity in mentalis, even to some extent in sleep. The finding is unexplained.

The depressor labii inferioris is quadrilateral and arises from the oblique line of the mandible, between the symphysis menti and the mental foramen, passing upwards and medially into the skin of the lower lip, blending with its fellow and orbicularis oris. At its origin it is continuous with platysma. Much fat is intermingled with the superficial fibres of this muscle.

Actions. The depressor labii inferioris draws the lower lip downwards and a little laterally in masticatory activity; it also contributes to the expression of irony.

The depressor anguli oris arises from the oblique line of the mandible, below and lateral to the depressor labii inferioris; it converges into a narrow fasciculus blending with other muscles at the angle of the mouth. At its origin it is continuous with

Maxillary artery
Lateral pterygoid plate, partly excised
Maxilla
Tensor veli palatini
Tuberosity of maxilla
Mandibular nerve
Middle meningeal artery
Spine of sphenoid
Levator veli palatini
Pterygoid hamulus
Superior pharyngeal constrictor
Buccinator
Stylopharyngeus
Parotid duct
Glossopharyngeal nerve
Pterygo-mandibular raphe
Styloglossus
Middle pharyngeal constrictor
Hygoglossus
Stylohyoid ligament
Mylohyoid
Greater cornu of hyoid bone
Geniohyoid
Lateral thyrohyoid ligament
Lesser cornu of hyoid bone
Thyrohyoid membrane
Internal laryngeal nerve
Inferior pharyngeal constrictor
Superior laryngeal vessels
Cricothyroid ligament
Recurrent laryngeal nerve
Cricothyroid
Oesophagus

5.25 Buccinator and the muscles of the pharynx. The zygomatic arch, the masseter, the ramus of the mandible, the temporalis, and a large part of the lateral pterygoid plate and the pterygoids have all been removed. In addition, the upper parts of the stylopharyngeus and styloglossus have been excised, together with the postero-inferior part of the hyglossus and all the infrahyoid muscles.

platysma, and at its insertion with orbicularis oris and risorius; some of its fibres are directly continuous with those of levator anguli oris, and others occasionally cross to the muscle of the other side; these latter fibres constitute the *transversus menti*.

Actions. The depressor anguli oris draws the angle of the mouth downwards and laterally in opening the mouth and in the expression of sadness.

Nerve supply. The mentalis, depressor labii inferioris and the depressor anguli oris are all supplied by the mandibular marginal branch of the facial nerve.

The buccinator (**5**.25) is a thin quadrilateral muscle, occupying the interval between the maxilla and the mandible, in the cheek. It is attached to the outer surfaces of the alveolar processes of the maxilla and mandible, opposite to the three molar teeth and, behind, to the anterior border of the pterygomandibular raphe, which separates the muscle from the superior constrictor of the pharynx. Between the tuberosity of the maxilla and the upper end of the pterygomandibular raphe a few fibres spring from a fine tendinous band which bridges the interval between the maxilla and the pterygoid hamulus. The tendon of the tensor veli palatini on its way to the soft palate pierces the pharyngeal wall in the small gap which lies behind this tendinous band (**5**.25). The fibres of the buccinator converge towards the angle of the mouth, where the central fibres intersect each other, those from below being continuous with the upper segment of the orbicularis oris, and those from above with the lower segment; the highest and lowest fibres are continued forward into the corresponding lip without decussation.

Relations. The buccinator is on the same plane as the superior pharyngeal constrictor and is covered by the buccopharyngeal fascia. *Superficially*, posteriorly, a large mass of fat separates it from the ramus of the mandible, masseter and a small portion of temporalis; this fat was originally named the suctorial pad, but its association with the act of sucking is not obvious. Anteriorly, the superficial surface of buccinator is related to zygomaticus major, risorius, levator and depressor anguli oris, and the parotid duct, which pierces it opposite the third upper molar tooth; the facial artery and facial vein cross it; it is also crossed by branches of the facial and buccal nerves. The *deep surface* is related to the buccal glands and mucous membrane of the mouth.

Nerve supply. The buccinator is supplied by the lower buccal branches of the facial nerve.

Actions. The buccinators compress the cheeks against the teeth, so that during the process of mastication the food is passed between them. Also when the cheeks have been previously distended with air, the buccinators expel it between the lips, as in any act of blowing; hence the name *buccinator*=a trumpeter.

The pterygomandibular raphe is an interlacing of tendinous fibres stretching from the hamulus of the medial pterygoid plate down to the posterior end of the mylohyoid line of the mandible. *Medially* it is covered by the mucous membrane of the mouth. *Laterally* it is separated from the ramus of the mandible by a quantity of adipose tissue. *Posteriorly* it gives attachment to the superior constrictor of the pharynx, and *anteriorly* to a part of the buccinator (**5**.25).

The orbicularis oris (**5**.21, 25) is made up of several strata which surround the orifice of the mouth but have different directions. It consists partly of fibres derived from the other facial muscles which pass into the lips, and partly of fibres proper to them. Of the former, a considerable number are derived from the buccinator, and form the deeper stratum. Some of the buccinator fibres—namely, those near the middle of the muscle—decussate at the angle of the mouth; the uppermost and lowermost fibres pass across the lips from side to side without decussation. Superficial to this is a second stratum, formed by the levator and depressor anguli oris, which cross each other at the angle of the mouth; the fibres from the levator pass to the lower lip, and those from the depressor to the upper lip, along which they run to the skin near the anterior median line. Fibres are also derived from the levator labii superioris, the zygomaticus major and minor, and the depressor labii inferioris; these intermingle with the transverse fibres described above, and have principally an oblique

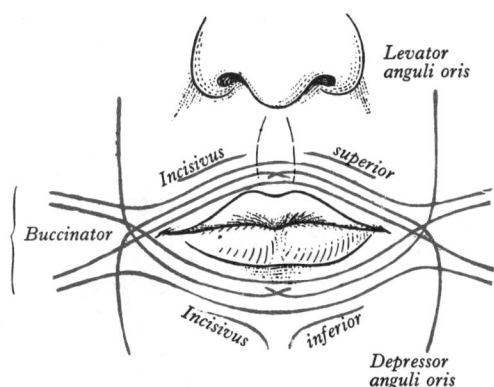

5.26   Arrangement of the fibres of the orbicularis oris.

direction. Some eight or nine muscles thus converge on each of the two angles of the mouth and interlace here at a palpable nodular mass, the *modiolus*. This can be fixed in a given position by the combined action of the zygomaticus major, levator anguli oris and the depressor anguli oris. These thus serve to fix the attachments of the orbicularis oris and the buccinator in this area. Within the lips the fibres of the orbicularis oris are divisible into two fasciculi, the *marginal* and *peripheral*. These combine laterally to form the *labial bands* which are traceable to the modiolus (Lightoller 1925; Burkitt and Lightoller 1926, 1927). The proper fibres of the lips are oblique, and pass from the deep surface of the skin to the mucous membrane, through the thickness of the lip. Finally, there are fibres by which the muscle is connected with the maxillae above, and with the mandible below. In the upper lip these constitute the *incisivus labii superioris*, which arises from the alveolar border of the maxilla, opposite the lateral incisor tooth, and arching laterally is continuous with the other muscles at the angle of the mouth. The additional fibres for the lower lip constitute a slip, *incisivus labii inferioris*, on each side; this slip arises from the mandible, lateral to the mentalis, and mingles with the other muscles at the angle of the mouth. In a study of a small number of fetal lips (14–25 weeks) Latham and Deaton (1976) concluded that orbicular fibres interlace and cross the midline in their cutaneous insertions, thus creating the ridges of the philtrum of the upper lip.

Nerve supply. The orbicularis oris is supplied by the lower buccal and the mandibular marginal branches of the facial nerve.

Actions. The orbicularis oris in its ordinary action effects the direct closure of the lips; by its deep, and oblique, fibres it compresses the lips against the teeth. The superficial part, consisting principally of the decussating fibres, brings the lips together and protrudes them. The orbicularis oris and other muscles of the lip play an important part in articulation, as well as in mastication (Duckworth 1947).

The risorius arises from the parotid fascia and is inserted into the skin at the angle of the mouth. It is a narrow bundle of fibres, broadest at its origin, but varying much in its size and form. The muscle may be absent or reduplicated; it may arise from the zygomatic arch, external ear, or the fascia over the mastoid process.

Nerve supply. The risorius is supplied by the buccal branches of the facial nerve.

Actions. The risorius retracts the angle of the mouth, and produces a sardonic expression.

Revelation of emotional state, intentions, and so forth by facial expression serves obvious social functions; but the facial musculature is even more frequently and usefully concerned in speech. There can be no doubt which is the more important form of communication. The niceties of speech immeasurably outstrip the repertoire of even the most expressive face. Nine muscles converge upon the mouth from each side to control the intricate labial movements of speech; occasionally they may curl the lips in irony or a sardonic grin. It is also important not to overlook the contribution of the facial musculature in feeding and drinking, in

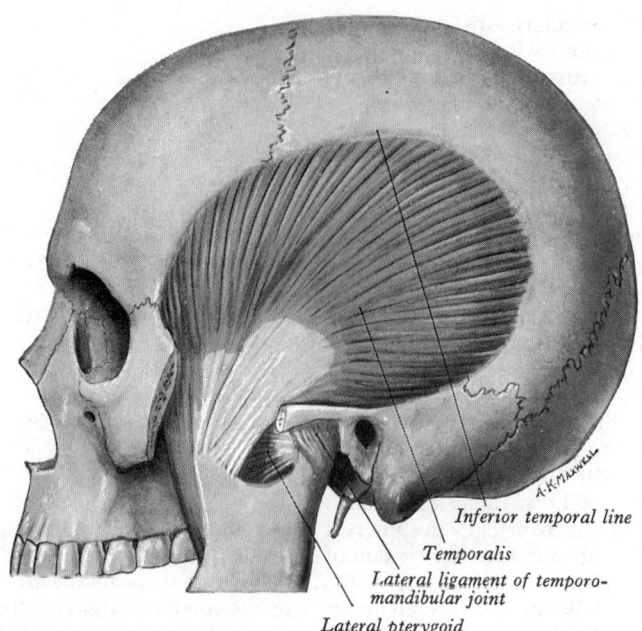

Inferior temporal line

Temporalis

*Lateral ligament of temporo-
mandibular joint*

Lateral pterygoid

5.27 Left temporalis. The zygomatic arch and masseter have been removed.

all their phases. Unfortunately, the individual activity of facial muscles has been little studied by experimental techniques and their participation in different movements is largely a matter of simple observation. Their role in mastication has been the subject of electromyographic study (Vitti *et al.* 1975).

MASTICATORY MUSCLES

The muscles immediately concerned with the movements of the mandible in mastication (and speech), are the masseter, temporalis and pterygoid muscles.

A strong layer of fascia, derived from the deep cervical fascia and named the **parotid fascia**, covers the masseter and is firmly connected with it. It is attached to the lower border of the zygomatic arch, and invests the parotid gland (p. 1272).

The masseter (**5**.34) is quadrilateral, and consists of three superimposed layers blending anteriorly. The *superficial layer*, the largest, arises by a thick aponeurosis from the zygomatic process of the maxilla, and from the anterior two-thirds of the lower border of the zygomatic arch; its fibres pass downwards and backwards, to be inserted into the angle and lower half of the lateral surface of the ramus of the mandible. Intramuscular tendinous septa in this layer are responsible for the ridges on the bone. The *middle layer* arises from the deep surface of the anterior two-thirds of the zygomatic arch and from the lower border of the posterior third and is inserted into the middle of the ramus of the mandible. The *deep layer* arises from the deep surface of the zygomatic arch and is inserted into the upper part of the ramus of the mandible and into the coronoid process. The middle and deep layers together constitute the deep part of the masseter of the Nomina Anatomica (MacDougall 1955); these two layers form a *cruciate* muscle (p. 525). On account of its proximity to the skin, the masseter can be palpated when it is thrown into contraction vigorously, as in clenching the teeth. It has been stated (McConaill 1975) that the most superficial fibres of masseter are continued through their attachment at the lower border of the mandible, into the attachment of the medial pterygoid muscle. For more complex accounts of the lamination of the masseter consult Schumacher (1961), and Yoshikawa and Suzuki (1969).

Relations. *Superficial* to the muscle are the integument, platysma, risorius, zygomaticus major, and parotid gland; the parotid duct, branches of the facial nerve, and the transverse facial vessels cross the muscle. The *deep surface* overlies the insertion of temporalis and the ramus of the mandible; a mass of fat separates it in front from buccinator and the buccal nerve. The masseteric

nerve and artery reach the deep surface of the muscle by passing through the dorsal part of the mandibular incisure. The *posterior margin* is overlapped by the parotid gland; the *anterior margin* projects over the buccinator and is crossed below by the facial vein.

Nerve supply. The masseter is supplied by a branch of the anterior trunk of the mandibular nerve. (See also medial pterygoid p. 536.)

Actions. The masseter elevates the mandible to occlude the teeth in mastication. Its activity in the resting position of the mandible is minimal. It has a small effect in side-to-side movements, protraction or retraction. An analysis of muscle fibre numbers and their diameter distribution, together with attempts to equate these data with dental occlusive forces, has been made by Rinqvist (1974).

The temporal fascia covers the temporalis. It is a strong, fibrous investment covered, laterally, by the auriculares anterior et superior, the galea aponeurotica and part of orbicularis oculi. The superficial temporal vessels and the auriculotemporal nerve ascend over it. Above, it is a single layer, attached to the whole of the superior temporal line; below, it has two layers, one attached to the lateral and the other to the medial margin of the upper border of the zygomatic arch. A small quantity of fat, the zygomatic branch of the superficial temporal artery, and the zygomaticotemporal branch of the maxillary nerve are between these layers. The deep surface of the fascia affords attachment to the superficial fibres of the temporalis.

The temporalis (**5**.27) is fan-shaped and arises from the whole of the temporal fossa (except the part formed by the zygomatic bone) and from the deep surface of the temporal fascia. Its fibres converge and descend into a tendon which passes through the gap between the zygomatic arch and the side of the skull, to be attached to the medial surface, apex, anterior and posterior borders of the coronoid process, and the anterior border of the ramus of the mandible nearly as far as the last molar tooth.

Relations. *Superficial* to the muscle are the skin, the auriculares anterior et superior, the temporal fascia, the superficial temporal vessels, the auriculotemporal nerve, the temporal branches of the facial nerve, the zygomaticotemporal nerve, the galea aponeurotica, the zygomatic arch, and the masseter. The *deep surface* is in relation with the temporal fossa, the lateral pterygoid, the superficial head of the medial pterygoid and a small part of the buccinator, the maxillary artery and its deep temporal branches, the deep temporal nerves, and the buccal vessels and nerve. *Behind* the tendon of the muscle the vessels and masseteric nerve traverse the mandibular incisure. The *anterior border* is separated from the zygomatic bone by a mass of fat.

Nerve supply. The temporalis is supplied by the deep temporal branches of the anterior trunk of the mandibular nerve. (See also medial pterygoid p. 536.)

Actions. The temporalis elevates the mandible and so closes the mouth and approximates the teeth. This movement requires both the upward pull of the anterior fibres and the backward pull of the posterior fibres, because the head of the mandible rests on the articular eminence when the mouth is open. The posterior fibres draw the mandible backwards after it has been protruded. It is also a contributor to side-to-side grinding movements. Despite numerous electromyographic studies little can be added in further analysis of these activities, there being considerable unresolved disagreement between workers. The most recent study (Vitti and Basmajian 1977), however, suggests that the temporalis is active in forcible elevation, but not involved in slow elevation without occlusion: otherwise this study confirmed the above description. Owing to the strength of the temporal fascia, the muscle is not easy to palpate, but its contraction is easy to feel. Its upper limit can be made out along the inferior temporal line when the teeth are firmly clenched.

The lateral pterygoid (5.28), is a short, thick muscle with two parts or heads: an *upper* from the infratemporal surface and infratemporal crest of the greater wing of the sphenoid bone, and a *lower* from the lateral surface of the lateral pterygoid plate. Its fibres pass backwards and laterally, to be inserted into a depression on the front of the neck of the mandible, and into the articular capsule and disc of the temporomandibular articulation.

Early in the third month of intrauterine life the lateral pterygoid is inserted into the mesenchyme condensed around the developing condyle of the mandible, but a part of its tendon sweeps backwards above the condyle and gains insertion into that portion of Meckel's cartilage which later forms the head of the malleus (Harpman and Woollard 1938). This part of the tendon becomes incorporated into the articular disc of the temporomandibular joint, although its attachment to the malleus does not persist (Rees 1954).

Relations. Its *superficial surface* is related to the ramus of the mandible, the maxillary artery, which crosses it, either deep or superficial to it, the tendon of the temporalis and the masseter. Its *deep surface* rests against the upper part of the medial pterygoid, the sphenomandibular ligament, the middle meningeal artery, and the mandibular nerve; its *upper border* is in relation with the temporal and masseteric branches of the mandibular nerve, its *lower border* with the lingual and inferior alveolar nerves. The buccal nerve and the maxillary artery pass between the parts of the muscle (5.29).

Nerve supply. The lateral pterygoid is supplied by a branch from the anterior trunk of the mandibular nerve. (See also nerve supply of medial pterygoid p. 536).

Action. The lateral pterygoid assists in opening the mouth by pulling forward the condylar process of the mandible and the articular disc, while the head of the mandible rotates on the articular disc (Posselt 1952). In the reverse movement of closure the backward gliding of the articular disc and condyle of the mandible is controlled by the slow elongation of the lateral pterygoid, while the masseter and temporalis restore the jaw to the occlusal position. Acting with the medial pterygoid of the same side the lateral pterygoid advances the condyle of that side so that the jaw rotates about a vertical axis through the opposite condyle. When the medial and lateral pterygoids of the two sides act together they protrude the mandible so that the lower incisors project in front of the upper. Some authorities have ascribed different actions to the two parts of the lateral pterygoid muscle, the upper (superior) head being involved in chewing, the inferior in protrusion. Electromyographic records in rhesus monkeys (Macnamara 1972) favour this view. For a review and a mechanical assessment, consult Grant (1973).

The medial pterygoid (5.28), a thick, quadrilateral muscle, is attached to the medial surface of the lateral pterygoid plate and the grooved surface of the pyramidal process of the palatine bone; it also has a more superficial slip from the lateral surfaces of the pyramidal process of the palatine bone and tuberosity of the maxilla, and lies at first on the surface of the lower part of the lower head of the lateral pterygoid. Its fibres pass downwards, laterally, and backwards, and are attached, by a strong tendinous

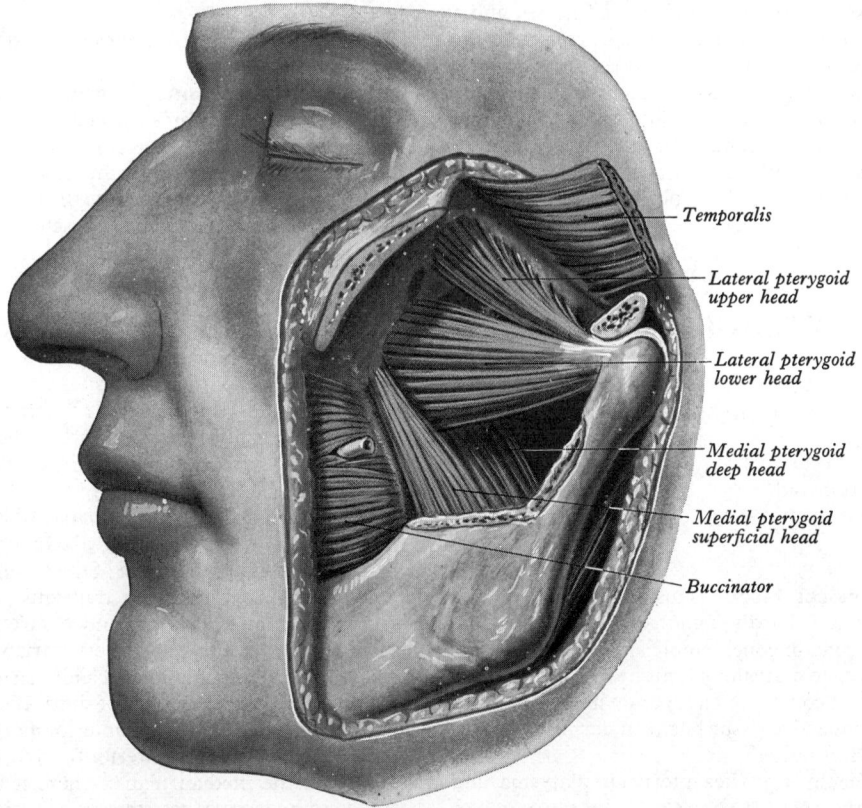

Temporalis

Lateral pterygoid upper head

Lateral pterygoid lower head

Medial pterygoid deep head

Medial pterygoid superficial head

Buccinator

5.28 The left pterygoid muscles. The zygomatic arch and part of the ramus of the mandible have been removed.

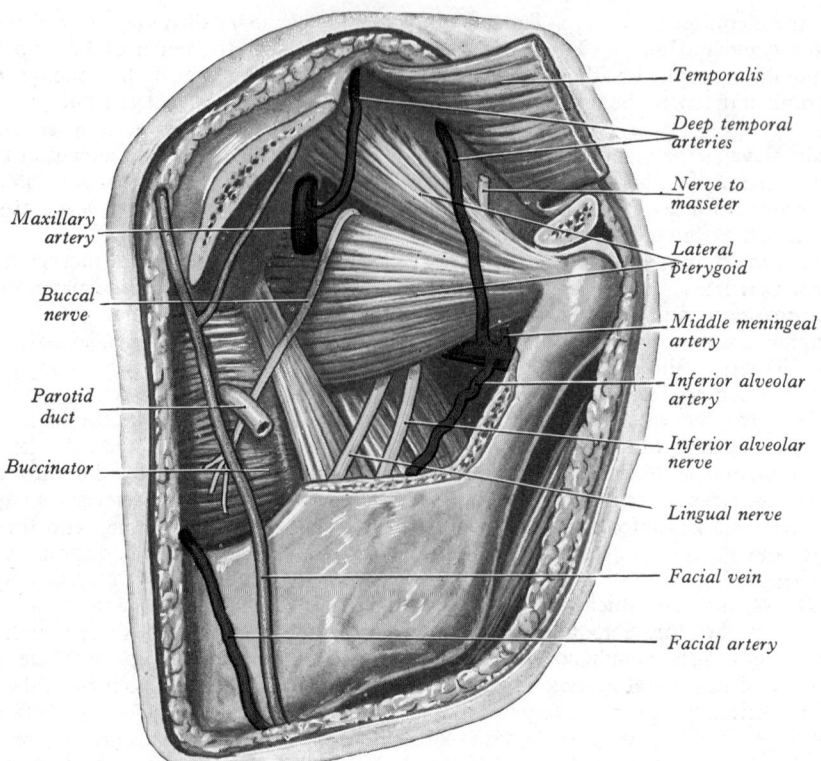

Temporalis

Deep temporal arteries

Nerve to masseter

Lateral pterygoid

Middle meningeal artery

Inferior alveolar artery

Inferior alveolar nerve

Lingual nerve

Facial vein

Facial artery

Maxillary artery

Buccal nerve

Parotid duct

Buccinator

5.29   The structures related to the left pterygoid muscles.

lamina, to the postero-inferior part of the medial surfaces of the ramus and angle of the mandible, as high as the mandibular foramen and nearly as far forwards as the mylohyoid groove (**3**.81). This area is often rugged, due to the tendinous fasciculi in this attachment.

Relations. The *lateral surface* is related to the ramus of the mandible, from which it is separated, above, by the lateral pterygoid, the sphenomandibular ligament, the maxillary artery, the inferior alveolar vessels and nerve, the lingual nerve, and a process of the parotid gland. The *medial surface* is related to the tensor veli palatini, and is separated from the superior constrictor by the styloglossus, the stylopharyngeus and some areolar tissue.

Nerve supply. The medial pterygoid is supplied by a branch from the mandibular nerve. Schumacher and his colleagues (1976) have investigated the ramifications of the muscular nerves of the masticatory muscles in detail. They observed a very similar mode of branching in all these muscles.

Actions. The medial pterygoids assist in elevating the mandible. Acting with the lateral pterygoids they protrude it. When the two pterygoid muscles of one side are in action, the corresponding side of the mandible is swung forwards and to the opposite side, while the head of the mandible on the other side undergoes a slight degree of rotation (p. 440); by an alternating action of the muscles of the two sides, the side-to-side movements, which take place during trituration of the food, are effected.

For a further analysis of these activities *see* pp. 440–443 and references therein.

The pterygospinous ligament, which is occasionally replaced by muscle fibres, stretches between the spine of the sphenoid bone and the posterior border of the lateral pterygoid plate near its upper end. It is sometimes ossified and then helps to bound a foramen which transmits the branches of the mandibular nerve going to the temporalis, masseter and lateral pterygoid.

THE ANTEROLATERAL MUSCLES AND FASCIAE OF THE NECK

The anterolateral muscles of the neck will be described in the following groups:

I. Superficial and lateral cervical.
II. Suprahyoid and infrahyoid.
III. Anterior vertebral.
IV. Lateral vertebral.

The superficial cervical fascia is usually a thin lamina investing the platysma and is hardly demonstrable as a separate layer. It may, however, contain considerable amounts of adipose tissue, and usually does so to a greater extent in the female. Like all superficial fascia it is, of course, not a separate membrane, but merely a zone of loose connective tissue between dermis and deep fascia, to both of which it is united.

The deep cervical fascia (**5**.30) lies internal to platysma, and invests the muscles of the neck. It consists of the fibro-areolar tissue which exists between the muscles, viscera, vessels, etc., of the neck. In certain situations it forms well-defined fibrous sheets;

elsewhere it is loosely arranged. It becomes condensed around the blood vessels, as fibrous sheaths which here, as elsewhere in the body, bind the arteries and their accompanying veins closely together.

The *superficial (investing) lamina* of deep cervical fascia is continuous behind with the ligamentum nuchae and the periosteum covering the spine of the seventh cervical vertebra. It forms a thin investment for trapezius, and from the anterior border of this muscle is continued forwards, as a rather loose areolar layer covering the posterior triangle of the neck, to the posterior border of the sternocleidomastoid, where it becomes denser. Along the hinder edge of sternocleidomastoid it divides to enclose the muscle, and at the anterior margin again forms a single lamina, which covers the anterior triangle of the neck and reaches forwards to the median plane, where it is continuous with the corresponding lamina from the opposite side. In the median plane it is adherent to the symphysis menti and the body of the hyoid bone.

Above, the fascia is fused with periosteum along the superior nuchal line of the occipital bone, over the mastoid process, and along the whole of the base of the mandible. Between the mandibular angle and the anterior edge of the sternocleidomastoid it is particularly strong. Between the mandible and the mastoid process it ensheathes the parotid gland; the layer which covers the gland extends upwards under the name of the *parotid fascia* and is fixed to the zygomatic arch. From the part which passes deep to the parotid gland a strong band ascends to the styloid process, as the *stylomandibular ligament* (p. 440).

Below, the fascia is attached to the acromion, the clavicle and the manubrium sterni, fusing with their periostea. Some little distance above the last, it splits into superficial and deep layers. The former is attached to the anterior border of the manubrium, the latter to its posterior border and to the interclavicular ligament. Between these two layers there is a slit-like interval, the *suprasternal space*; it contains a small quantity of areolar tissue, the lower portions of the anterior jugular veins and the jugular arch, the sternal heads of the sternocleidomastoid muscles, and sometimes a lymph node. Over the lower part of the posterior triangle, between trapezius and sternocleidomastoid, the superficial lamina of deep fascia is also divided into superficial and deep layers. The superficial layer is attached below to the upper border of the clavicle. The deep layer surrounds the inferior belly of omohyoid and, deep to sternocleidomastoid, its intermediate tendon. Below, this deep layer blends with the fascia around subclavius and is attached to the back of the clavicle and anterior end of the first rib.

The *carotid sheath* is a condensation of the cervical fascia in which the common and internal carotid arteries, the internal jugular vein, the vagus nerve and the constituents of the ansa cervicalis are embedded. It is thicker around the arteries than the vein, and peripherally it is connected to the neighbouring layers by loose areolar tissue (5.31).

The *prevertebral lamina* of the cervical fascia covers the prevertebral muscles and extends laterally on the scalenus anterior, scalenus medius and levator scapulae, i.e. it forms a fascial floor for the posterior triangle of the neck. As the subclavian artery and the brachial nerves emerge from behind the scalenus anterior they carry the prevertebral fascia downwards and laterally behind the clavicle to form the *axillary sheath*. Traced laterally, the prevertebral fascia becomes thinner and areolar in character and is lost as a definite fibrous layer under cover of trapezius. Superiorly it is attached to the base of the skull, and inferiorly continues downwards in front of the longus colli muscles into the superior mediastinum, where it blends with the anterior longitudinal ligament. Anteriorly the prevertebral lamina is separated from the pharynx and its covering buccopharyngeal fascia by a loose cellular interval which is termed the *retropharyngeal space*. Further from the median plane, the same loose areolar tissue connects the prevertebral lamina to the carotid sheath and the fascia on the deep surface of the sternocleidomastoid. It should be observed that all the ventral rami of the cervical nerves lie at first behind the prevertebral lamina, and the nerve to the rhomboids, the nerve to serratus anterior and the phrenic nerve, retain this position throughout their course in the neck. However, the accessory nerve lies superficial to the prevertebral fascia (p. 1082).

The *pretracheal lamina* of the cervical fascia is very thin, but it provides a fine fascial sheath for the thyroid gland. Above, it is attached to the arch of the cricoid cartilage, and, below, it is continued into the superior mediastinum with the inferior thyroid veins.

Applied Anatomy. The superficial lamina of deep cervical fascia opposes the extension of abscesses towards the surface, and pus beneath it has a tendency to extend laterally. If the pus is in the anterior triangle, it may find its way into the mediastinum, in front of the pretracheal lamina of fascia; but owing to the thinness of the fascia in this situation it more frequently approaches the surface and points above the sternum. Pus behind the prevertebral lamina may extend towards the lateral part of the neck and point in the posterior triangle, or may perforate this layer of fascia and the buccopharyngeal fascia to bulge into the pharynx as a retropharyngeal abscess.

I. THE SUPERFICIAL AND LATERAL CERVICAL MUSCLES

These include the platysma, trapezius, and sternocleidomastoid muscles.

The platysma (5.21) is a broad sheet arising from the fascia covering the upper parts of the pectoralis major and deltoid; its fibres cross the clavicle, and proceed obliquely upwards and medially in the side of the neck. The anterior fibres interlace, below and behind the symphysis menti, with the fibres of the

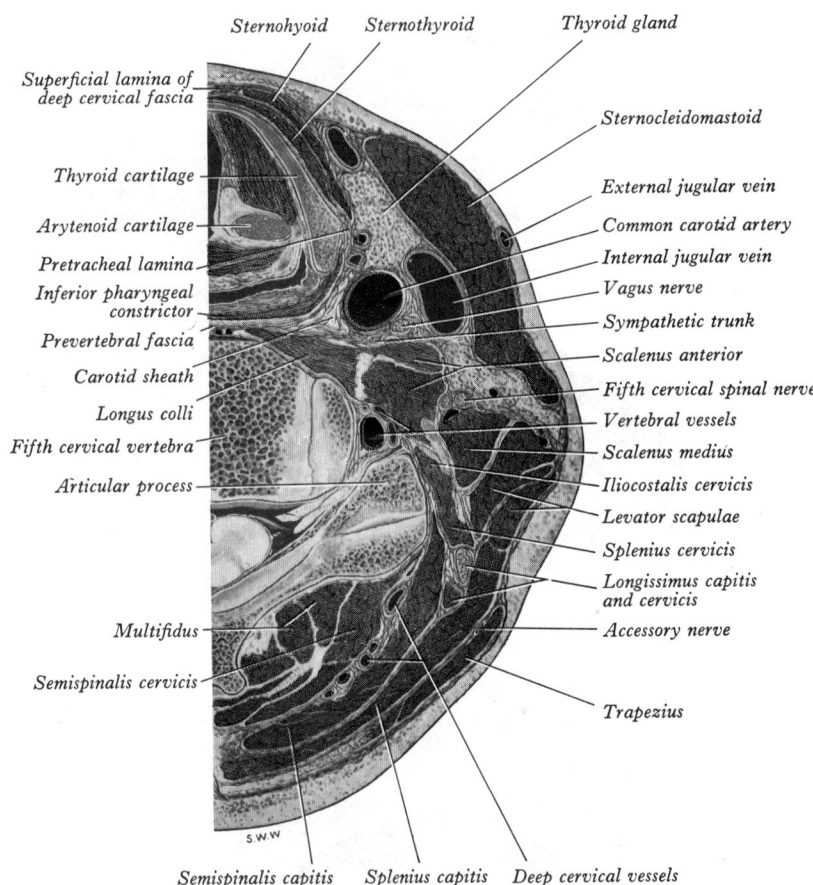

5.30 A transverse section through the left half of the neck to show arrangement of the deep cervical fascia. (Specimen provided by Professor R. E. M. Bowden, Royal Free Hospital School of Medicine.)

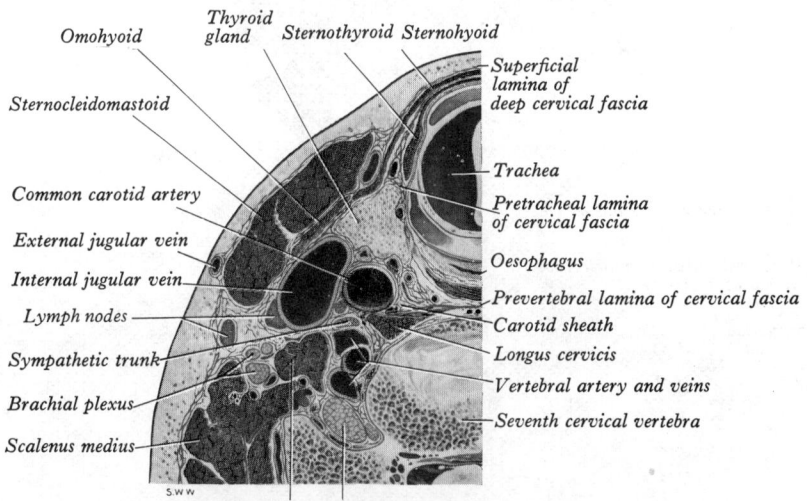

5.31 Part of a transverse section through the lower part of the neck at the level of the seventh cervical vertebra to show the arrangement of the deep cervical fascia. (Specimen provided by Professor R. E. M. Bowden.)

opposite muscle; the succeeding fibres are inserted into the lower border of the body of the mandible, while the posterior fibres cross the mandible and the lower, anterior part of masseter to be attached to the skin and subcutaneous tissue of the lower part of the face, many blending with the muscles about the angle and lower part of the mouth. Under platysma the external jugular vein descends from the angle of the mandible to the middle of the clavicle. The platysma varies considerably in extent, and may

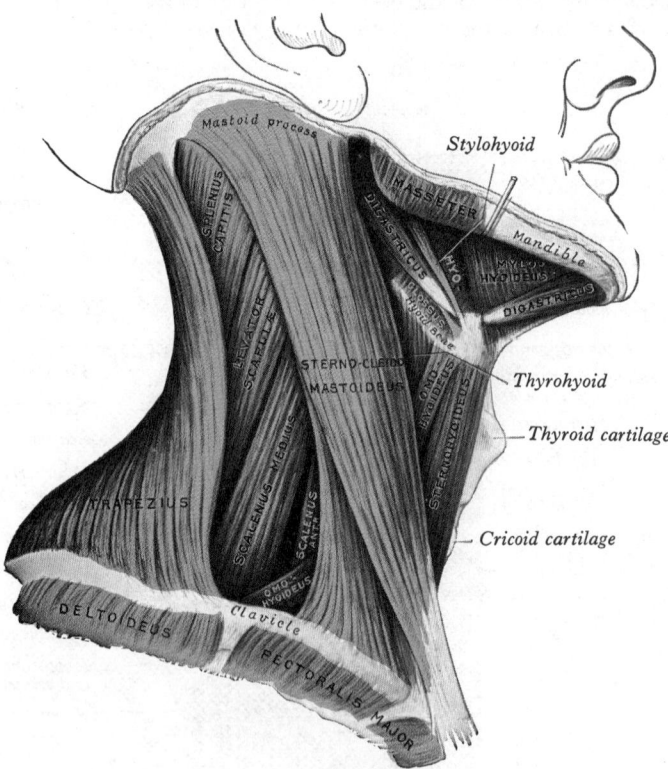

5.32 Muscles of the neck. Right lateral aspect.

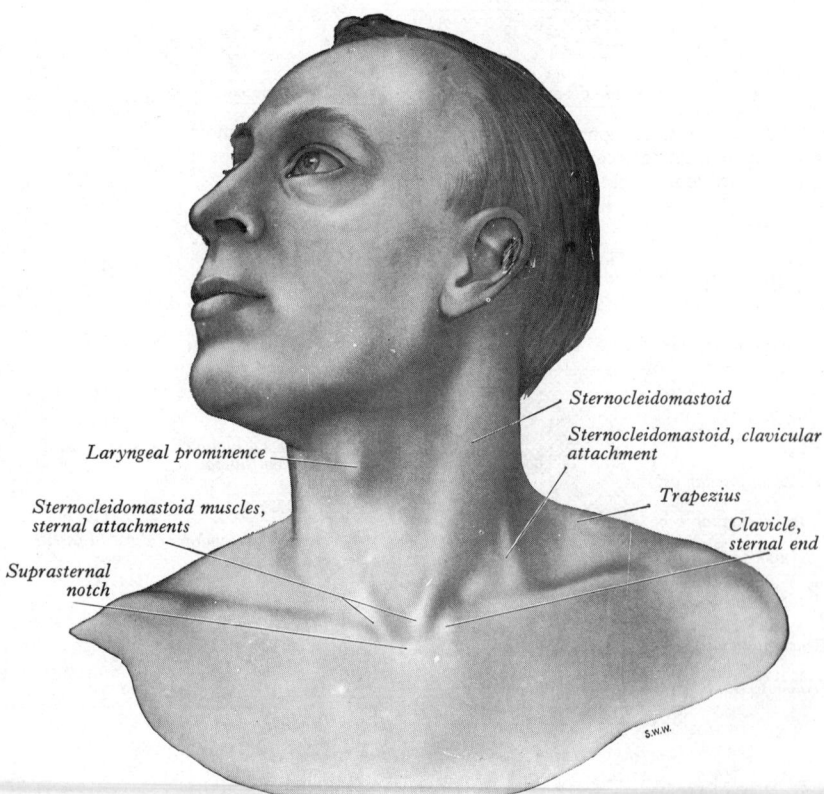

Laryngeal prominence

Sternocleidomastoid muscles, sternal attachments

Suprasternal notch

Sternocleidomastoid

Sternocleidomastoid, clavicular attachment

Trapezius

Clavicle, sternal end

5.33 Surface landmarks of the neck.

even be absent on one side or both. It may be joined by slips from the mastoid process, the occipital bone or from the fascia over the upper part of trapezius (occipitalis minor).

Nerve supply. The platysma is supplied by the cervical branch of the facial nerve.

Actions. When the entire platysma is in action it wrinkles the surface of the skin of the neck in an oblique direction, and tends to diminish the concavity between the jaw and the side of the neck. Its anterior portion, which is the thickest part of the muscle, may assist in depressing the mandible; it also serves to draw down the lower lip and angle of the mouth (as in the expression of horror or surprise).

Electromyographic studies show that the muscle is active in sudden deep inspiration (de Sousa 1964). Its action in widening the mouth and pulling down the angle is easily confirmed by mere palpation. The usefulness of platysma's activities are not clear. It is often very much contracted in situations of sudden and violent effort.

The trapezius is described on p. 564 with the scapular musculature.

The sternocleidomastoid (5.32) passes obliquely down across the side of the neck and forms a prominent landmark (5.33) especially when contracted. It is thick and narrow centrally, but broader and thinner at each end. *Below* it has two heads. The *medial* or *sternal head* is a rounded tendinous fasciculus, attached to the upper part of the anterior surface of the manubrium sterni, and passes upwards, laterally, and backwards. It can be seen and felt in the root of the neck. The *lateral* or *clavicular head*, which varies in width and is composed of muscular and fibrous fasciculi, passes almost vertically upwards from the upper surface of the medial third of the clavicle. The two heads are separated at their attachments by a triangular interval; but, as they ascend, the clavicular head spirals behind the sternal head and blends with its deep surface below the middle of the neck, forming a thick, rounded belly. *Above*, the muscle is inserted by a strong tendon into the lateral surface of the mastoid process, from its apex to its superior border, and by a thin aponeurosis into the lateral half of the superior nuchal line. The clavicular fibres are chiefly attached to the mastoid process; the sternal fibres are more oblique, superficial, and extend on to the occiput. The direction of pull of the two heads is therefore different, and the muscle may be classed as 'cruciate' and slightly 'spiralized' (*see* p. 525).

This muscle divides the side of the neck into two triangles, anterior and posterior. The boundaries of the *anterior triangle* are, in front, the median line of the neck; above, the base of the mandible, and a line continuing this from the angle of the mandible to the sternocleidomastoid; behind, the anterior border of the sternocleidomastoid. The apex of the triangle is at the upper border of the sternum. The boundaries of the *posterior triangle* are, in front, the posterior border of the sternocleidomastoid; below, the middle third of the clavicle; behind, the anterior margin of the trapezius. The apex corresponds with the meeting of the sternocleidomastoid and trapezius on the occipital bone. The subdivisions and contents of these triangles are given on pp. 685–686.

Relations. *Superficial* to the sternocleidomastoid are skin and platysma; it is separated from platysma by the external jugular vein, the great auricular and transverse cervical nerves, and the superficial lamina of the deep cervical fascia. Near its insertion the muscle is overlapped by a small portion of the parotid gland. The *deep surface* of the muscle is related at its origin to the sternoclavicular joint; it lies upon the sternohyoid, sternothyroid and the omohyoid, while the anterior jugular vein crosses deep to it, but superficial to the infrahyoid muscles, immediately above the clavicle. The carotid sheath and the subclavian artery are internal to these muscles. Between the omohyoid and the posterior belly of digastric the anterior part of sternocleidomastoid is superficial to the common, internal and external carotid arteries, internal jugular, facial and lingual veins, the deep cervical lymph nodes, the vagus nerve, and the rami of the ansa cervicalis. The sternocleidomastoid branch of the superior thyroid artery crosses deep to the muscle at the upper border of

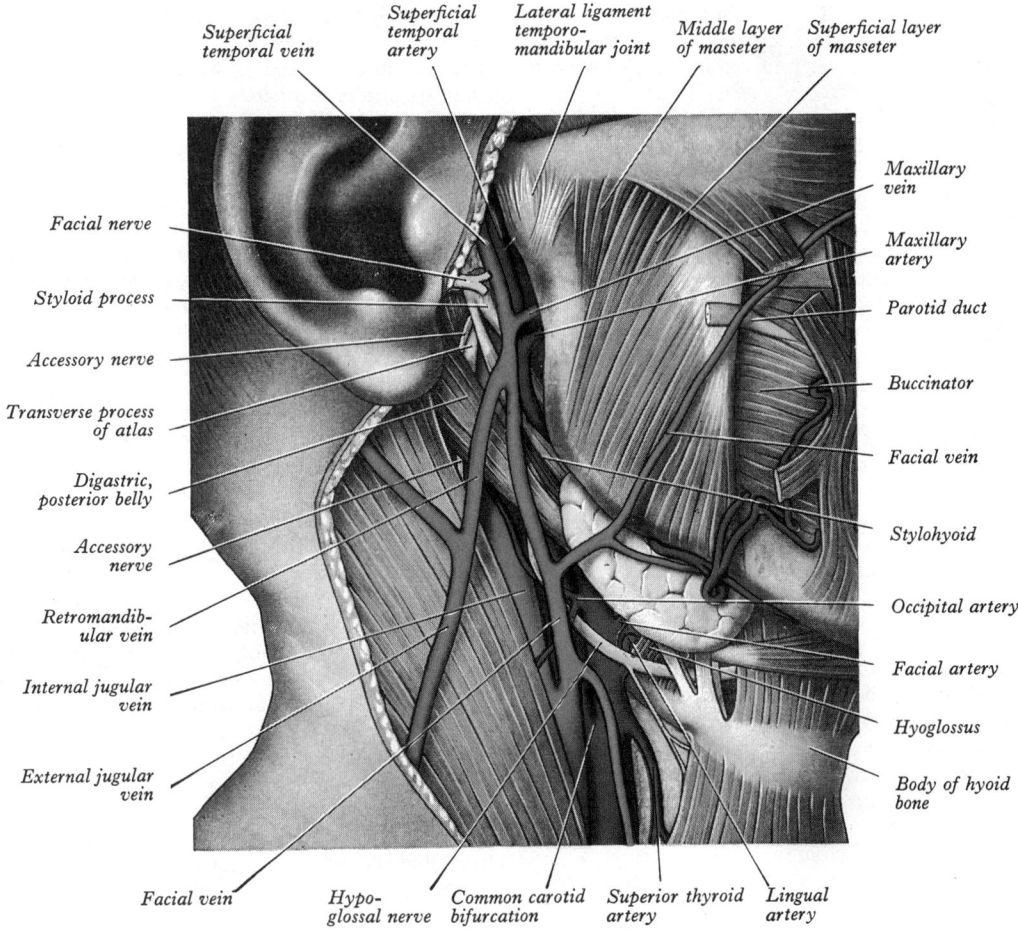

Superficial temporal vein
Superficial temporal artery
Lateral ligament temporo-mandibular joint
Middle layer of masseter
Superficial layer of masseter
Maxillary vein
Maxillary artery
Parotid duct
Buccinator
Facial vein
Stylohyoid
Occipital artery
Facial artery
Hyoglossus
Body of hyoid bone

Facial nerve
Styloid process
Accessory nerve
Transverse process of atlas
Digastric, posterior belly
Accessory nerve
Retromandib-ular vein
Internal jugular vein
External jugular vein

Facial vein
Hypo-glossal nerve
Common carotid bifurcation
Superior thyroid artery
Lingual artery

5.34 Relations of the posterior belly of the digastric, exposed by the removal of skin, fasciae, parotid gland, and cutaneous branches of the cervical plexus.

the omohyoid. The posterior part of the muscle is related on its internal aspect to the splenius, levator scapulae and scaleni, the cervical plexus, the upper part of the brachial plexus, the phrenic nerve, and the transverse cervical and suprascapular arteries. The occipital artery crosses deep to the muscle at, or under cover of, the lower border of the digastric, where the accessory nerve, which pierces the muscle, runs downwards and laterally deep to it. At its insertion the muscle lies superficial to the mastoid process, and to the splenius, longissimus capitis, and the posterior belly of the digastric.

Nerve supply. The sternocleidomastoid is supplied by the accessory nerve, which usually traverses it, and by branches from the ventral rami of the second, third and sometimes the fourth cervical spinal nerves. Although long believed to be solely proprioceptive, clinical evidence now suggests that a proportion of these cervical fibres are motor.

Actions. When one sternocleidomastoid acts, it tilts the head towards the shoulder of the same side; it also rotates the head so as to carry the face towards the opposite side. The movement occurs in looking to the side and upwards, but more or less level rotation from side to side is a far more common visual movement and probably the most frequent in which the sternocleidomastoids (with, of course, other muscles) are involved. Acting together from below the sternocleidomastoid muscles draw the head forwards and so help the longi colli to flex the cervical part of the vertebral column. This is a common movement in feeding. The two muscles are also active when the head is raised while the body is supine. If the head is fixed, they assist in elevating the thorax in forced inspiration. Electromyographic observations (e.g. de Sousa 1973) suggest that the sternal fibres are more active in contralateral rotation, but that both parts of the muscle are involved to some extent in all the above movements. This study

also indicated that the muscle is involved in cervical extension as well as flexion.

Applied Anatomy. The deformity termed *torticollis* is a contracture of the sternocleidomastoid. There is also a condition coming on in adult life (spasmodic torticollis) which begins with tonic or clonic spasm of one sternocleidomastoid muscle, soon followed by a spasm of the trapezius, particularly its clavicular portion. Such abnormal conditions illustrate the isolated muscle's action. It must be emphasized, however, that this is a caricature of its ordinary activities, which are invariably modified by synergists and antagonists, such as splenius capitis (p. 543), and many others.

II (A). THE SUPRAHYOID MUSCLES

The *suprahyoid muscles* include the digastric, stylohyoid, mylohyoid and geniohyoid.

The digastric (5.34) has two bellies united by an intermediate rounded tendon. It lies below the body of the mandible, and extends, in an angled form, from the mastoid process to the chin. The *posterior belly*, longer than the anterior, is attached in the mastoid notch of the temporal bone and passes downwards and forwards. The *anterior belly* is attached to the digastric fossa on the base of the mandible close to the median plane, and passes downwards and backwards. The two bellies meet in an intermediate tendon, which perforates the stylohyoid. It is held to the side of the body and the greater cornu of the hyoid by a fibrous loop, sometimes lined by a synovial sheath. An aponeurotic layer is given off from the tendon of the digastric to the body and greater cornu of the hyoid bone. The digastric may lack its intermediate tendon and is then attached midway along the body of the mandible. The posterior belly may be augmented by a slip from the styloid process or arise wholly from it. The anterior belly may cross the midline in part, and is not uncommonly more or less fused with mylohyoid. (*See* Lennartsson 1979 for proprioceptors.)

Relations. Its *superficial surface* is related to the platysma, sternocleidomastoid, part of splenius, longissimus capitis, mastoid process, stylohyoid, retromandibular vein and parotid and submandibular salivary glands. The *medial surface* of the anterior belly adjoins the mylohyoid, that of the posterior belly the superior oblique, rectus capitis lateralis, the transverse process of the atlas vertebra, the accessory nerve, internal jugular vein, occipital artery, hypoglossal nerve, the internal and external carotid, facial and lingual arteries and the hyoglossus (**5.**34).

Nerve supply. The anterior belly of digastric is supplied by the mylohyoid branch of the inferior alveolar nerve, the posterior belly by the facial nerve. These different supplies are, of course, associated with the separate derivation of the two parts of the muscle from first and second branchial arch mesenchyme (p. 150).

Actions. The digastric depresses the mandible and can elevate the hyoid bone. Electromyography indicates that the digastric muscles always act together and that they are secondary to the lateral pterygoids in mandibular depression, coming into play especially in maximal depression (Moyers 1950). The posterior belly is especially active in swallowing and chewing.

The stylohyoid (**5.**34, 35) arises by a small tendon from the posterior surface of the styloid process, near its base and, passing downwards and forwards, is inserted into the body of the hyoid bone, at its junction with the greater cornu, and just above the omohyoid. It is perforated, near its insertion, by the tendon of the digastric. Occasionally the muscle may be absent or double. It may be medial to the external carotid artery. It may end in the digastric or the supra- or infra-hyoid muscles.

Nerve supply. The stylohyoid is supplied by the facial nerve.

Actions. The stylohyoid elevates and draws back the hyoid bone. This elongates the floor of the mouth. With the other supra- and infra-hyoid muscles it can fix the hyoid bone as a basis for the action of tongue muscles attached to the bone. Its precise roles in masticatory and vocal movements have not been satisfactorily analysed (but *see* p. 1237). The roles of the stylohyoid muscle and the hyoid apparatus in speech and swallowing have, however, been investigated radiographically in mankind by Delmas and Senecail (1977).

The stylohyoid ligament is a fibrous cord, attached to the tip of the styloid process and the lesser cornu of the hyoid bone. It gives origin to the highest fibres of the middle constrictor of the pharynx and is intimately related to the lateral wall of the oral pharynx (**5.**25). Below, it is overlapped by the hyoglossus. It is derived from the cartilage of the second branchial arch (p. 150), and may be partially ossified; in many mammals it forms a distinct bone, the *epihyal*.

The mylohyoid (**5.**32, 35) is superior or deep to the anterior belly of digastric, and forms, with its fellow, a muscular floor for the oral cavity. It is a flat, triangular sheet attached to the whole length of the mylohyoid line of the mandible. The posterior fibres pass medially and slightly downwards, to the front of the body of the hyoid bone near its lower border. The middle and anterior fibres from each side intersect in a median fibrous raphe which stretches from the symphysis menti to the hyoid bone. This raphe is sometimes wanting; if so, the two muscles form a continuous sheet. Sometimes it is fused with the anterior belly of digastric.

Relations. Its *inferior* or *superficial surface* is related to the platysma, the anterior belly of digastric, the superficial part of the submandibular gland, the facial and submental vessels, and the mylohyoid vessels and nerve. Its *superior* or *internal surface* is in relation with the geniohyoid, part of the hyoglossus, and the styloglossus, the hypoglossal and lingual nerves, the submandibular ganglion, sublingual gland, deep part of the submandibular gland and the submandibular duct, the lingual and sublingual vessels, and, posteriorly, with the mucous membrane of the mouth. In about one-third of subjects there is a hiatus in the muscle through which a process of the sublingual gland protrudes (Gaughran 1963).

Nerve supply. The mylohyoid is supplied by the mylohyoid branch of the inferior alveolar nerve.

Actions. The mylohyoid elevates the floor of the mouth in the first stage of deglutition (Whillis 1946). It also elevates the hyoid bone or depresses the mandible.

The geniohyoid (**5.**35) is a narrow muscle, above the medial part of mylohyoid. From the inferior mental spine on the back of the symphysis menti it runs backwards and slightly downwards to the anterior surface of the body of the hyoid bone; it is in contact with its fellow and may occasionally be fused with it, or with the genioglossus.

Nerve supply. The first cervical spinal nerve through the hypoglossal nerve (p. 1083).

Actions. the geniohyoid elevates the hyoid bone and draws it forwards, thus acting as a partial antagonist to stylohyoid. When the hyoid is fixed, it depresses the mandible.

II (B). THE INFRAHYOID MUSCLES

The *infrahyoid muscles* are the sternohyoid, sternothyroid, thyrohyoid and omohyoid. As a group they are antagonists to the suprahyoid group, in depressing the hyoid bone; but they may act as fixators of the hyoid, or cooperate in cyclic hyoid movements.

The sternohyoid (**5.**32, 35), a thin, narrow strap muscle, arises from the posterior surface of the medial end of the clavicle, the posterior sternoclavicular ligament, and the upper and posterior part of the manubrium sterni. Passing upwards and medially, it is attached to the inferior border of the body of the hyoid bone. Below, it is separated from its fellow by a considerable interval; but the two muscles usually come into contact in the middle of their course, and are contiguous above. The muscle may be absent or double, or augmented by a clavicular slip (*cleidohyoid*), or interrupted by a tendinous intersection.

Nerve supply. Branches from the ansa cervicalis (c. 1, 2, 3).

Action. the sternohyoid depresses the hyoid bone after it has been elevated in deglutition. It also doubtless plays a part in speech and mastication.

The sternothyroid (**5.**32, 35) is shorter and wider than sternohyoid, and lies internal to it. It passes from the posterior surface of the manubrium sterni below the origin of the

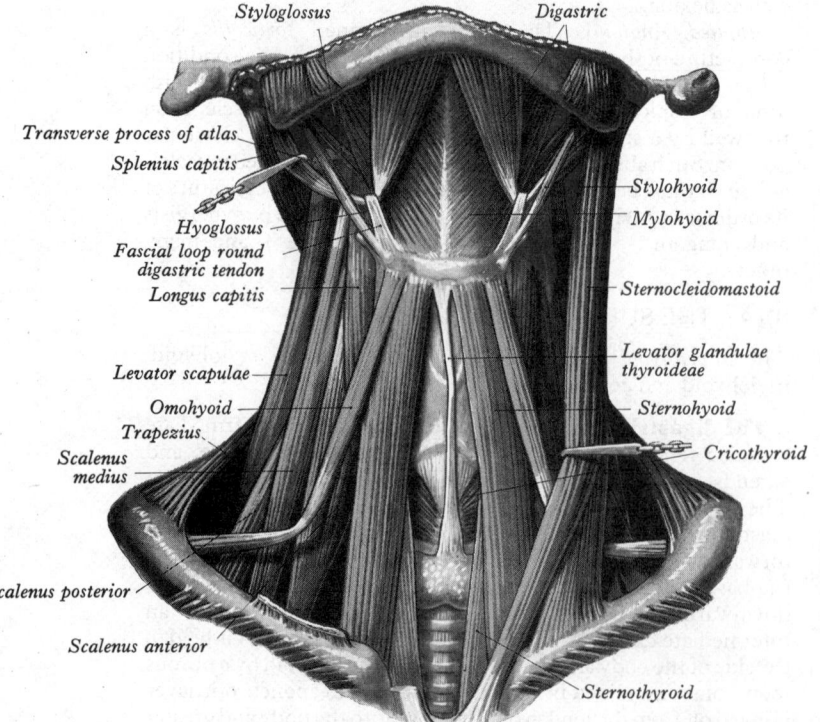

Styloglossus   *Digastric*

Transverse process of atlas
Splenius capitis
Hyoglossus
Fascial loop round digastric tendon
Longus capitis
Levator scapulae
Omohyoid
Trapezius
Scalenus medius
Scalenus posterior
Scalenus anterior

Stylohyoid
Mylohyoid
Sternocleidomastoid
Levator glandulae thyroideae
Sternohyoid
Cricothyroid
Sternothyroid

5.35 Muscles of the front of the neck. On the right side the sternocleidomastoid has been removed. In this subject, the origin of the scalenus medius extended up to the transverse process of the atlas.

sternohyoid and the edge of the cartilage of the first rib to the oblique line on the lamina of the thyroid cartilage. In the lower part of the neck this muscle is in contact with its fellow, but it diverges as it ascends. It is applied to the anterolateral surface of the lobe of the thyroid gland.

Nerve supply. Branches from the ansa cervicalis (c. 1, 2, 3).

Action. The sternothyroid draws the larynx downwards after it has been elevated, as in swallowing or in vocal movements.

The thyrohyoid, a small, quadrilateral muscle, may be looked upon as an upward continuation of the sternothyroid. From the oblique line on the lamina of the thyroid cartilage it passes up to be attached to the lower border of the greater cornu and adjacent part of the body of the hyoid bone.

Nerve supply. The thyrohyoid is supplied by a branch from the hypoglossal nerve. Like the nerve to the geniohyoid, this branch contains fibres of the first cervical spinal nerve which supply the muscle.

Actions. The thyrohyoid depresses the hyoid bone, or raises the larynx.

The omohyoid (5.32, 35) consists of two fleshy bellies united at an angle by an intermediate tendon. It arises from the upper border of the scapula near the scapular notch, and occasionally from the superior transverse scapular ligament (p. 436). The *inferior belly* is a flat, narrow band, which inclines forwards and slightly upwards across the lower part of the neck; it then passes behind the sternocleidomastoid and there ends in the intermediate tendon..The inferior belly divides the posterior triangle of the neck into upper, *occipital* and lower, supraclavicular triangles (p. 686). The *superior belly* passes almost vertically upwards from this tendon, close to the lateral border of the sternohyoid, and is attached to the lower border of the body of the hyoid bone, lateral to the insertion of the sternohyoid. The intermediate tendon, which varies in length and form, usually lies adjacent to the internal jugular vein, opposite the arch of the cricoid cartilage. It is held by a band of the deep cervical fascia which ensheathes it and is attached below to the clavicle and the first rib; it is by this fascial process that the angular form of the muscle is maintained. A variable amount of striated muscle may occur in this fascial band. Either belly may be absent or double; the inferior may be attached directly to the clavicle, the superior is sometimes fused with sternohyoid. These variations have been confirmed by Buntine (1970).

Nerve supply. The superior and inferior bellies of the omohyoid are supplied respectively by branches from the ramus superior of the ansa cervicalis (c. 1), and from the ansa itself (c. 2, 3).

Actions. The omohyoid depresses the hyoid bone after it has been elevated. It has been suggested also in prolonged inspiratory efforts; by rendering tense the lower part of the deep cervical fascia they may lessen the inward suction of the soft parts, which would otherwise compress the great vessels and the apices of the lungs. Such interesting views are, of course, speculations. So far, sparse data in regard to the hyoid musculature as a whole have been contributed by electromyographers. However, a combined cineradiographic and electromyographic study of mastication (in the opossum, *Didelphys virginiana*) strongly suggests that the omohyoid muscles are involved in initiating hyoid movements in this activity (Crompton *et al.* 1975); see also p. 1313.

III. THE ANTERIOR VERTEBRAL MUSCLES

This group of muscles includes the longi colli et capitis and the recti capitis anterior et lateralis, all to some extent flexors of the head and neck (**5**.36).

The longus colli is applied to the anterior surface of the vertebral column, between the atlas and the third thoracic vertebra. It is divisible into three parts, inferior oblique, superior oblique, and vertical; its origin and insertion consist of tendinous slips. The *inferior oblique part*, which is the smallest, runs upwards and laterally from the front of the bodies of the first two or three thoracic vertebrae to the anterior tubercles of the transverse processes of the fifth and sixth cervical vertebrae. The

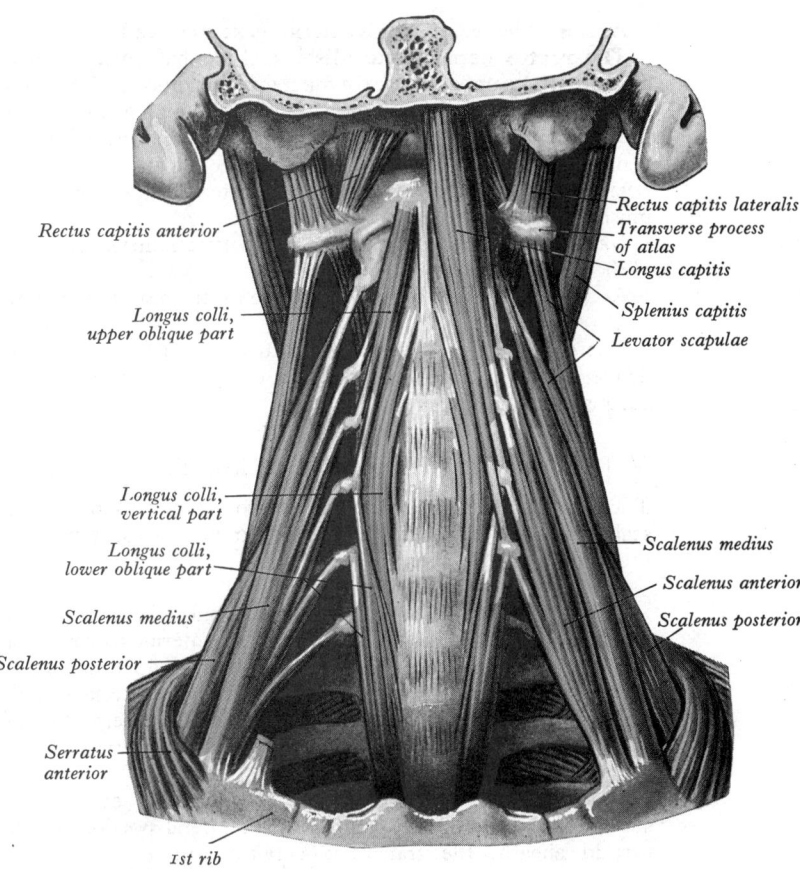

5.36 The anterior and lateral vertebral muscles. On the right side the scalenus anterior and the longus capitis have been removed.

superior oblique part passes from the anterior tubercles of the transverse processes of the third, fourth and fifth cervical vertebrae upwards and medially, to be attached by a narrow tendon into the anterolateral surface of the tubercle on the anterior arch of the atlas. The *vertical* intermediate part ascends from the front of the bodies of the upper three thoracic and lower three cervical vertebrae to the front of the bodies of the second, third and fourth cervical vertebrae.

Nerve supply. The muscle receives branches from the ventral rami of the second, third, fourth, fifth and sixth cervical spinal nerves.

Actions. The longus colli bends the neck forwards; in addition, the oblique portions flex it laterally, and the inferior oblique portion rotates it to the opposite side. Despite its deep situation the muscle has been studied electromyographically (Fountain *et al.* 1966) largely in confirmation of the above, except that it is claimed that longus colli has little effect in lateral flexion. Its main antagonist is the longissimus cervicis.

The longus capitis, broad and thick above, narrow below, is attached below by tendinous slips to the anterior tubercles of the transverse processes of the third, fourth, fifth and sixth cervical vertebrae, and above to the inferior surface of the basilar part of the occipital bone.

Nerve supply. Branches from the ventral rami of the first, second and third cervical spinal nerves.

Action. The longus capitis flexes the head.

The foregoing anterior vertebral muscles vary chiefly in the number of their vertebral slips.

The rectus capitis anterior is a short, flat muscle situated behind the upper part of the longus capitis. From the anterior surface of the lateral mass of the atlas and the root of its transverse process, it ascends almost vertically to the inferior surface of the basilar part of the occipital bone in front of the occipital condyle.

Nerve supply. Branches from the loop between the ventral rami of the first and second cervical spinal nerves.

Action. The rectus capitis anterior flexes the head.

The rectus capitis lateralis is a short, flat muscle which arises from the upper surface of the transverse process of the atlas and is inserted into the inferior surface of the jugular process of the occipital bone. In view of its attachments and its relation to the ventral ramus of the first spinal nerve, the rectus capitis lateralis is regarded as homologous with the posterior intertransverse muscles.

Nerve supply. Branches from the loop between the ventral rami of the first and second cervical spinal nerves.

Action. The rectus capitis lateralis bends the head to the same side.

The actions of the above deep-seated muscles can only be deduced, and their precise roles in everyday activities await adequate technical appraisal.

IV. THE LATERAL VERTEBRAL MUSCLES

The scaleni, anterior, medius and posterior, extend obliquely like scaling ladders from the upper two ribs to the cervical transverse processes (5.36).

The scalenus anterior lies deeply at the side of the neck behind the sternocleidomastoid. From the anterior tubercles of the transverse processes of the third, fourth, fifth and sixth cervical vertebrae, it descends almost vertically, to be attached by a narrow, flat tendon to the scalene tubercle on the inner border of the first rib, and to the ridge on the upper surface of the rib in front of the groove for the subclavian artery.

Relations. In *front* of it are the clavicle, subclavius, sternocleidomastoid, and omohyoid, the lateral portion of the carotid sheath, the transverse cervical, suprascapular and ascending cervical arteries, the subclavian vein, the prevertebral fascia and the phrenic nerve. Its *posterior surface* is in relation with the suprapleural membrane (p. 1249), the pleura, the nerves forming the brachial plexus and the subclavian artery; the latter two separate it from the scalenus medius. *Below*, it is separated from the longus colli by an angular interval (5.36), in which the vertebral artery, with its companion vein on its lateral side, ascends to reach the foramen transversarium of the sixth cervical vertebra. The inferior thyroid artery crosses the interval from the lateral to the medial side near its apex. The sympathetic trunk and its cervicothoracic ganglion are closely related to the postero-medial side of this part of the vertebral artery (6.63). On the *left side* the thoracic duct crosses this interval at the level of the seventh cervical vertebra and usually comes into contact with the medial edge of the muscle. Above, it is separated from the longus capitis by the ascending cervical branch of the inferior thyroid artery.

Nerve supply. Scalenus anterior is supplied by branches from the ventral rami of the fourth, fifth and sixth cervical spinal nerves.

Actions. Acting from below, the scalenus anterior bends the cervical portion of the vertebral column forwards and laterally and rotates it towards the opposite side. When the muscle acts from above it assists in elevating the first rib.

The scalenus medius, the largest and longest of the scaleni, is attached above to the transverse process of the axis and the front of the posterior tubercles of the transverse processes of the lower five cervical vertebrae, and frequently extends upwards to the transverse process of the atlas (5.35); below it is attached to the upper surface of the first rib, between the tubercle of the rib and the groove for the subclavian artery.

Relations. Its *anterolateral surface* is related to the sterno-cleidomastoid; it is crossed by the clavicle and omohyoid: *anteriorly,* it is separated from the scalenus anterior by the subclavian artery and the cervical nerves. The levator scapulae and the scalenus posterior are posterolateral to it. The upper two roots of the nerve to serratus anterior and the dorsal scapular nerve (to the rhomboids) pierce the substance of the muscle and appear on its lateral surface.

Nerve supply. Branches from the ventral rami of the third to eighth cervical spinal nerves.

Actions. The scalenus medius, acting from below, bends the cervical part of the vertebral column to the same side; acting from above it helps to raise the first rib. The scalene muscles, in particular the scalenus medius, are active during inspiration, even during quiet breathing in the erect attitude (Campbell 1955).

The scalenus posterior, the smallest and most deeply situated of the scaleni, passes from the posterior tubercles of the transverse processes of the fourth, fifth, and sixth cervical vertebrae, to be attached by a thin tendon to the outer surface of the second rib, behind the tubercle for serratus anterior. It is occasionally blended with the scalenus medius. The scalene muscles vary a little in the number of vertebrae to which they are attached, in their degree of separation, and segmental innervation.

Nerve supply. Branches from the ventral rami of the lower three cervical spinal nerves.

Actions. The scalenus posterior bends the lower end of the cervical part of the vertebral column to the same side, when the second rib is fixed; when its upper attachment is fixed it helps to elevate the second rib.

The scalenus minimus (pluralis) is associated with suprapleural membrane, and cervical pleura, and is considered with them (p. 1249).

THE FASCIAE AND MUSCLES OF THE TRUNK

The muscles of the trunk may be arranged for convenience of description in six groups:

I. Deep muscles of the back.
II. Suboccipital muscles.
III. Muscles of the thorax.
IV. Muscles of the abdomen.
V. Muscles of the pelvis.
VI. Muscles of the perineum.

I. The Deep Muscles of the Back

The deep or intrinsic muscles of the back consist of a complex group of muscles extending from the pelvis to the skull. These include extensors and rotators of the head and neck (splenius capitis and cervicis), short segmental muscles (interspinales and intertransversarii), and the complex extensor and rotator masses (erector spinae and transversospinalis, the latter including the

semispinales, rotatores, and multifidus). Collectively these muscles control the vertebral column (5.37).

The *superficial* and the *deep fasciae* of the back of the neck are described on p. 564.

The thoracolumbar (lumbar) fascia covers the deep muscles of the back of the trunk. Above, it passes anterior to the serratus posterior superior and is continuous with the superficial lamina of the deep cervical fascia on the back of the neck.

In the thoracic region the thoracolumbar fascia is a thin fibrous lamina covering the extensor muscles of the vertebral column and separating them from the muscles connecting the vertebral column to the upper extremity. It is attached, *medially,* to the spines of the thoracic vertebrae; *laterally,* to the angles of the ribs.

In the lumbar region the thoracolumbar fascia is in three layers (5.38). The posterior layer is attached to the spines of the lumbar and sacral vertebrae and to the supraspinous ligaments; the middle layer is attached, *medially,* to the tips of the transverse

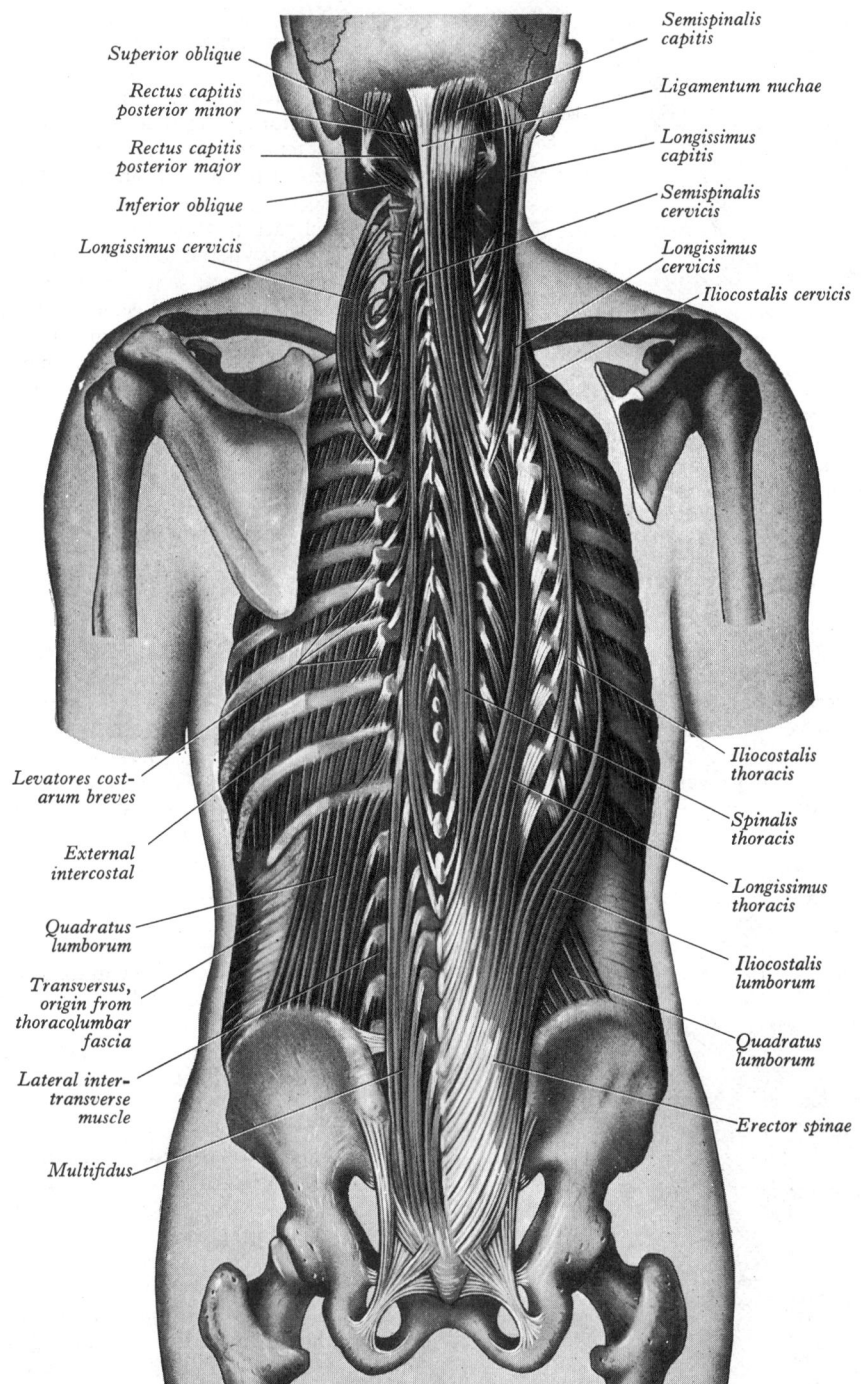

5.37 The deep muscles of the back. On the left side the erector spinae and its upward continuations (with the exception of the longissimus cervicis, which has been displaced laterally) and the semispinalis capitis have been removed.

processes of the lumbar vertebrae and the intertransverse ligaments, *below*, to the iliac crest, and *above*, to the lower border of the twelfth rib and the lumbocostal ligament (p. 451). The anterior layer covers the quadratus lumborum and is attached *medially* to the anterior surfaces of the transverse processes of the lumbar vertebrae behind the lateral part of the psoas major. *Below*, it is attached to the iliolumbar ligament and the adjoining part of the iliac crest; *above*, it forms the lateral arcuate ligament (p. 549). The posterior and middle layers unite at the lateral margin of the erector spinae, and at the lateral border of the quadratus lumborum they are joined by the anterior layer to form the aponeurotic origin of the transverse abdominis.

The splenius capitis (5.58) is attached below to the lower half of the ligamentum nuchae, the spine of the seventh cervical vertebra, and the spines of the upper three or four thoracic vertebrae, deep to the rhomboids and trapezius. The muscle passes upwards and laterally under cover of sternocleidomastoid, to the mastoid process of the temporal bone, and the rough surface on the occipital bone just below the lateral third of the superior nuchal line. The muscle forms a part of the floor of the posterior triangle of the neck, above and behind the levator scapulae.

Nerve supply. Lateral branches of the dorsal rami of the middle cervical spinal nerves.

The splenius cervicis ascends from the spines of the third to the sixth thoracic vertebrae to the posterior tubercles of the transverse processes of the upper two or three cervical vertebrae immediately anterior to the attachment of levator scapulae. The splenii may be absent or variable in their vertebral attachments. Accessory slips also occur.

Nerve supply. Lateral branches of the dorsal rami of the lower cervical spinal nerves.

Actions. The splenii of the two sides, acting together, draw the

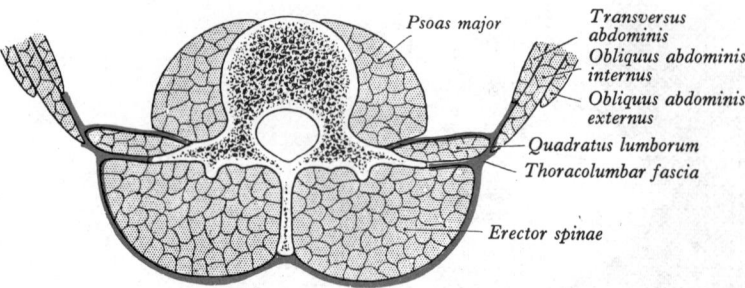

5.38 Transverse section through the posterior abdominal wall, showing disposition of the thoracolumbar fascia. Note that all other connective tissue strata have been omitted.

head directly backwards; acting separately, they draw the head to one side, and slightly rotate it, turning the face to the same side. Thus they act as synergists with the opposite sternocleidomastoid.

THE ERECTOR SPINAE (SACROSPINALIS)

This muscle and its prolongations in the thoracic and cervical regions lie in the groove on the side of the vertebral column (5.37), covered in the lumbar and thoracic regions by the thoracolumbar fascia, the serratus posterior inferior below, and the rhomboid and splenius muscles above. It forms a large muscular and tendinous mass, which varies in size and composition at different levels of the vertebral column. In the sacral region it is narrow and pointed, and at its attachment chiefly tendinous. In the lumbar

region it forms a thick fleshy mass which can readily be felt in the living subject. Its lateral border is flanked by a visible groove (5.39), and, traced upwards over the back of the thorax, it crosses the ribs at their angles, inclining medially as it ascends.

The erector spinae arises from the anterior surface of a broad and thick tendon, which is attached to the median sacral crest, to the spines of the lumbar and the eleventh and twelfth thoracic vertebrae, to their supraspinous ligaments, to the medial aspect of the dorsal part (p. 380) of the iliac crest, and to the lateral sacral crest (p. 280), where it blends with the sacrotuberous and dorsal sacroiliac ligaments; some of its fibres are continuous with the gluteus maximus. The muscular fibres form a large fleshy mass which splits in the upper lumbar region into three columns: lateral, the *iliocostocervicalis*, intermediate, the *longissimus*, and medial, the *spinalis*. Each of these consists, from below upwards, of three parts as follows:

Lateral **Iliocostalis.**	*Intermediate* **Longissimus.**	*Medial* **Spinalis.**
(a) I. lumborum.	(a) L. thoracis.	(a) S. thoracis.
(b) I. thoracis.	(b) L. cervicis.	(b) S. cervicis.
(c) I. cervicis.	(c) L. capitis.	(c) S. capitis.

Iliocostocervicalis

The iliocostalis lumborum is attached, by flattened tendons, to the inferior borders of the angles of the lower six or seven ribs.

The iliocostalis thoracis starts from the upper borders of the angles of the lower six ribs *medial* to the tendons of insertion of the iliocostalis lumborum; it ascends to the upper borders of the angles of the upper six ribs and the back of the transverse process of the seventh cervical vertebra.

The iliocostalis cervicis passes from the angles of the third, fourth, fifth and sixth ribs *medial* to the tendons of insertion of the iliocostalis thoracis to the posterior tubercles of the transverse processes of the fourth, fifth and sixth cervical vertebrae.

Nerve supply. These three muscles are supplied by the dorsal rami of the lower cervical, thoracic and upper lumbar spinal nerves.

Actions. These muscles are extensors of the vertebral column; they are also lateral flexors (p. 445).

Longissimus

The longissimus thoracis is the intermediate and largest of the continuations of the erector spinae. In the lumbar region, where it is as yet blended with the iliocostalis lumborum, some of its fibres are attached to the whole length of the posterior surfaces of the transverse processes and the accessory processes of the lumbar vertebrae, and to the middle layer of the thoracolumbar fascia. In the thoracic region it is attached, by rounded tendons, to the tips of the transverse processes of all the thoracic vertebrae, and by fleshy processes to the lower nine or ten ribs between their tubercles and angles.

The longissimus cervicis, situated medial to the longissimus thoracis, is attached by long thin tendons to the transverse processes of the upper four or five thoracic vertebrae, and also by tendons to the posterior tubercles of the transverse processes of the cervical vertebrae from the second to the sixth.

The longissimus capitis lies between the longissimus cervicis and the semispinalis capitis. It arises by tendons from the transverse processes of the upper four or five thoracic vertebrae, and the articular processes of the lower three or four cervical vertebrae, and is attached to the posterior margin of the mastoid process, deep to the splenius capitis and sternocleidomastoid. It is usually crossed by a tendinous intersection near its insertion.

Nerve supply. The longissimi are supplied by the dorsal rami of the lower cervical, the thoracic and the lumbar spinal nerves.

Actions. The longissimi thoracis et cervicis bend the vertebral column backwards and laterally; the longissimus capitis extends the head, and turns the face towards the same side.

Spinalis

The spinalis thoracis, the medial continuation of the erector spinae, is scarcely separable as a distinct muscle. It is situated at

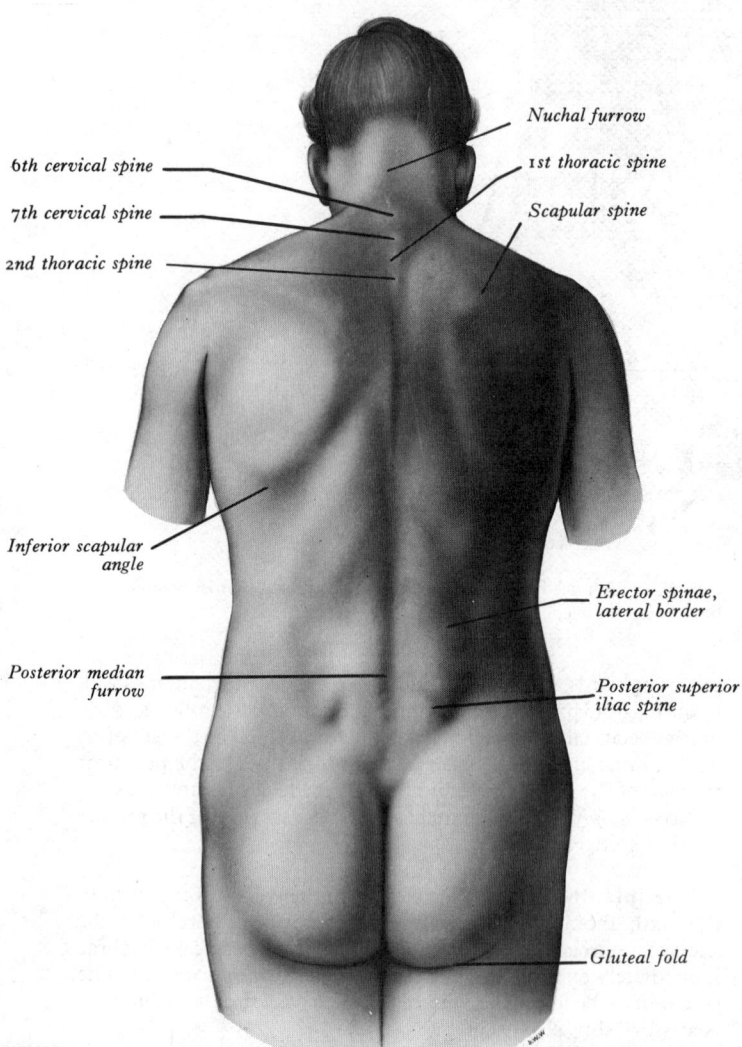

6th cervical spine

7th cervical spine

2nd thoracic spine

Nuchal furrow

1st thoracic spine

Scapular spine

Inferior scapular angle

Erector spinae, lateral border

Posterior median furrow

Posterior superior iliac spine

Gluteal fold

5.39 Dorsal aspect of the trunk, showing principal surface landmarks.

the medial side of the longissimus thoracis, and is intimately blended with it; it is attached below by three or four tendons to the spines of the eleventh and twelfth thoracic, and first and second lumbar vertebrae; these, uniting, form a small muscle which is attached above by separate tendons to the spines of the upper thoracic vertebrae, the number varying from four to eight. It is intimately united with the semispinalis thoracis, which lies anterior to it.

The spinalis cervicis is an inconstant muscle, which ascends from the lower part of the ligamentum nuchae, the spine of the seventh cervical, and sometimes from the spines of the first and second thoracic vertebrae to the spine of the axis, and occasionally the spines of the two vertebrae immediately below it. This muscle is often absent.

The spinalis capitis is usually more or less blended with the semispinalis capitis.

Nerve supply. The spinales are supplied by the dorsal rami of the lower cervical and thoracic spinal nerves.

Actions. The spinales are extensors of the vertebral column.

TRANSVERSOSPINALIS

This consists of the the following muscles:

Semispinalis thoracis.   Multifidus.   Rotatores thoracis.
Semispinalis cervicis.   Rotatores cervicis.
Semispinalis capitis.   Rotatores lumborum.

These muscles run obliquely upwards and medially from the transverse processes to the spines of the vertebrae.

The semispinalis thoracis consists of thin, fleshy fasciculi, interposed between tendons of considerable length. It arises by a series of tendons from the transverse processes of the thoracic vertebrae from the sixth to the tenth inclusive, and is inserted, by tendons, into the spines of the upper four thoracic and lower two cervical vertebrae.

The semispinalis cervicis, thicker than the preceding, arises by a series of tendinous and fleshy fibres from the transverse processes of the upper five or six thoracic vertebrae, and is inserted into the cervical spines, from the axis to the fifth inclusive. The fasciculus connected with the axis is the largest, and is chiefly muscular in structure.

The semispinalis capitis (5.37) is situated at the back of the neck, under cover of the splenius, and medial to the longissimi cervicis et capitis. It arises by a series of tendons from the tips of the transverse processes of the upper six or seven thoracic and the seventh cervical vertebrae, from the articular processes of the fourth, fifth, and sixth cervical vertebrae and, occasionally, from the spine of the seventh cervical or first thoracic vertebra. The tendons are succeeded by a broad muscle, which passes upwards to the medial part of the area between the superior and inferior nuchal lines of the occipital bone. The medial part, usually more or less distinct from the rest of the muscle, is named the *spinalis capitis*; it is sometimes called the *biventer cervicis*, since it is transversed by an imperfect tendinous intersection.

Nerve supply. The semispinales are supplied by the dorsal rami of the cervical and thoracic spinal nerves.

Actions. The semispinales thoracis et cervicis extend the thoracic and cervical regions of the vertebral column, and rotate them towards the opposite side; the semispinalis capitis extends the head, and turns the face slightly towards the opposite side.

The multifidus consists of a number of fleshy and tendinous fasciculi, which lie deep to the foregoing muscles and fill the groove at the side of the spines of the vertebrae from the sacrum to the axis. Most caudally, its fasciculi arise from the back of the sacrum as low as the fourth sacrum foramen, from the aponeurosis of the erector spinae, posterior superior iliac spine and dorsal sacro-iliac ligaments; in the lumbar region, from all the mamillary processes; in the thoracic region, from all the transverse processes; in the cervical region, from the articular processes of the lower four vertebrae, Each fasciculus passes obliquely upwards and medially, and is attached to the whole length of the spine of one of the vertebrae above. The fasciculi vary in length; the most superficial pass from one vertebra to the third or fourth

above; those next in depth run from one vertebra to the second or third above; while the deepest connect contiguous vertebrae.

Nerve supply. The dorsal rami of the spinal nerves.

Deep to the multifidus are the *rotatores* and these are best developed in the thoracic region.

The rotatores thoracis are eleven in number on each side, and are small and somewhat quadrilateral in form. Each connects the upper and posterior part of the transverse process of one vertebra to the lower border and lateral surface of the lamina of the vertebra next above, the fibres extending as far as the root of the spine. The first is found between the first and second thoracic vertebrae; the last, between the eleventh and twelfth. Sometimes the number of these muscles is diminished by the absence of one or more from the upper or lower end of the series.

The rotatores cervicis et lumborum are represented only by irregular and variable muscle bundles with similar attachments to those of the rotatores thoracis.

Nerve supply. The dorsal rami of the spinal nerves.

The interspinales are short paired muscular fasciculi, between the spines of contiguous vertebrae, one on each side of the interspinous ligament. In the *cervical region* they are most distinct, and consists of six pairs; the first is situated between the axis and third vertebra, and the last between the seventh cervical and the first thoracic vertebrae. They are small narrow bundles, attached, above and below, to the apices of the spines. In the *thoracic region* they occur between the first and second vertebrae, and sometimes between the second and third, and the eleventh and twelfth vertebrae. In the *lumbar region* there are four pairs in the intervals between the five lumbar vertebrae. A pair is occasionally found between the last thoracic and first lumbar vertebrae, and another between the fifth lumbar vertebra and the sacrum. Sometimes the cervical interspinales span more than two vertebrae.

Nerve supply. The dorsal rami of the spinal nerves.

The intertransversarii are small muscles placed between the transverse processes of the vertebrae. They are best developed in the *cervical region* where they consist of anterior and posterior slips, which are separated by the ventral rami of the spinal nerves. The *posterior intertransverse muscles* are divisible into medial and lateral slips, which are supplied by the dorsal and ventral rami of the spinal nerves respectively. The *medial*, which is the proper intertransverse muscle, is often further subdivided into medial and lateral parts by the passage through it of the dorsal ramus of the spinal nerve. The *anterior intertransverse muscles* and the lateral parts of the posterior muscles connect the costal processes of contiguous vertebrae, and the medial parts of the posterior muscles connect the true transverse processes. There are seven pairs of these muscles, the highest between the atlas and axis, and the lowest between the seventh cervical vertebra and the first thoracic, but the anterior muscles between the atlas and axis is often absent. In the *thoracic region* they consist of single muscles, which are present between the transverse processes of the last three thoracic vertebrae only, and between the transverse processes of the last thoracic vertebra and the first lumbar. In the *lumbar region* they again consist of two sets of muscles, one named the *intertransversarii mediales*, connecting the accessory process of one vertebra with the mamillary process of the next, and the other named the *intertransversarii laterales*, which are really divisible into ventral and dorsal parts (Cave 1937; Morrison 1954). The ventral parts connect the transverse processes (costal elements) of the lumbar vertebrae, and each dorsal part connects the accessory process to the transverse process of the succeeding vertebra. The thoracic intertransverse muscles and ligaments are homologous with the medial slips of the proper posterior intertransverse muscles of the cervical region, and the *levatores costarum* are homologous with their lateral moieties. The lateral branch of the dorsal ramus of the spinal nerve separates the thoracic intertransverse from the levator costae muscle. In the lumbar region the levatores costarum are represented by the medial intertransverse muscles, while the lateral intertransverse are homologous with the intercostal muscles. For later views on the homologies and classification of transversospinal musculature consult Sato (1973).

Nerve supply. The intertransversarii mediales lumborum, the thoracic intertransversarii and the medial parts of the posterior intertransversarii of the cervical region are supplied by the dorsal rami of the spinal nerves; the others are supplied by the ventral rami.

Actions. The short muscles of the back probably function, for the most part, as postural muscles. The vertebral column consists, in effect, of a series of short levers jointed together. A mechanical arrangement of this nature is unstable to compression forces and it will tend to buckle unless there is some mechanism by which the movements of the individual joints relative to one another can be controlled. This mechanism is provided by the short muscles of the back, which steady adjoining vertebrae and control their movements during motion of the vertebral column as a whole. In this way they ensure the efficient action of the long muscles.

Theoretically, these muscles are capable of producing extension (multifidi, spinales), lateral flexion (multifidus, intertransversarii), and rotation (multifidus and rotatores), but their detailed patterns of activity are not known.

The deep muscles of the back are certainly involved in control of posture. They show intermittent contractions during the swaying movements which occur in the upright posture. Contraction of the erectores spinae extends the trunk, but the continued control of extension is to a large extent dependent on the activity of the rectus abdominis muscles. Conversely during flexion of the trunk, movement is initiated by the flexor muscles such as the recti abdominis, but as the centre of gravity moves forward control is by the erectores spinae. In lateral flexion the control is by the contralateral erector spinae. In the position of full flexion of the trunk the erectores spinae, however, are relaxed and electromyographically quiet. It is believed that this position is maintained by tension in the ligaments of the spine and by passive resistance to further deformation of the intervertebral disc (Floyd and Silver 1955; Joseph 1960). Jonsson (1974) has demonstrated that electromyographic activity in the erector spinae group is in most subjects greater in working postures assumed when they are working on lower surfaces from a standing position. This would be expected from the views expressed above.

II. The Suboccipital Muscles

This group of small muscles (5.40) is concerned in extension of the head at the atlanto-occipital joints (recti capitis posteriores major and minor), and rotation of the head and atlas on the axis (obliqui capitis superior and inferior).

The rectus capitis posterior major starts by a pointed tendon from the spine of the axis, and, becoming broader as it ascends, is attached to the lateral part of the inferior nuchal line of the occipital bone and the bone immediately below the line. As the muscles of the two sides pass upwards and laterally, they leave between them a triangular space, in which parts of the recti capitis posteriores minores are visible.

Actions. The rectus capitis posterior major extends the head, and turns the face towards the same side.

The rectus capitis posterior minor arises by a narrow pointed tendon from the tubercle on the posterior arch of the atlas, and, widening as it ascends, is attached to the medial part of the inferior nuchal line of the occipital bone and also the bone between the line and the foramen magnum (p. 307). Either of the rectus capitis posterior muscles may be doubled longitudinally.

Action. The rectus capitis posterior minor extends the head.

The obliquus capitis inferior, the larger of the two oblique muscles, passes laterally and slightly upwards from the lateral surface of the spine and adjacent part of the upper part of the lamina of the axis to the lower and back part of the transverse process of the atlas.

Action. The obliquus capitis inferior turns the face towards the same side. Owing to the length of the transverse process of the atlas the muscle has considerable mechanical advantage.

The obliquus capitis superior, narrow below, wide and expanded above, arises by tendinous fibres from the upper surface of the transverse process of the atlas. It passes upwards and backwards, and is attached to the occipital bone, between the superior and inferior nuchal lines, lateral to the semispinalis capitis and overlapping the insertion of the rectus capitis posterior major.

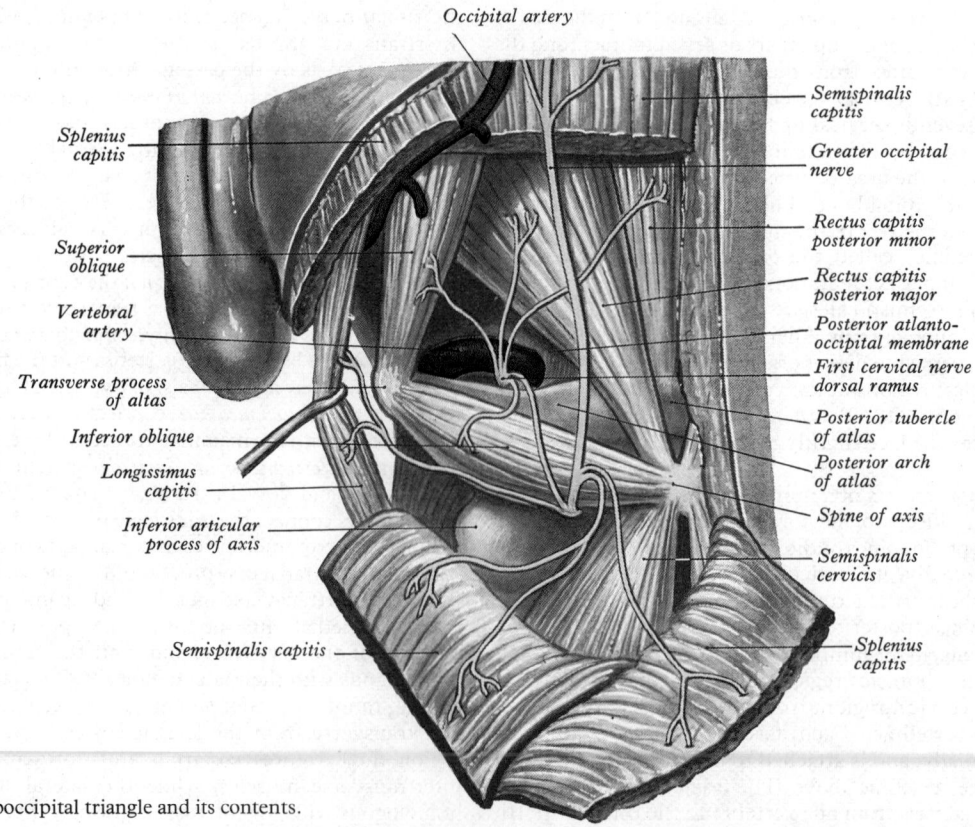

Occipital artery

Splenius capitis

Superior oblique

Vertebral artery

Transverse process of atlas

Inferior oblique

Longissimus capitis

Inferior articular process of axis

Semispinalis capitis

Semispinalis capitis

Greater occipital nerve

Rectus capitis posterior minor

Rectus capitis posterior major

Posterior atlanto-occipital membrane

First cervical nerve dorsal ramus

Posterior tubercle of atlas

Posterior arch of atlas

Spine of axis

Semispinalis cervicis

Splenius capitis

5.40 The left suboccipital triangle and its contents.

Actions. The obliquus capitis superior bends the head backwards and to the same side. This muscle and the two recti are probably employed more frequently as postural muscles than as prime movers, but direct observation is difficult.

Nerve supply. All the suboccipital muscles are supplied by the dorsal ramus of the first cervical spinal nerve.

The Suboccipital Triangle

This triangle is a region bounded, above and medially, by the rectus capitis posterior major, above and laterally, by the obliquus capitis superior, below and laterally, by the obliquus capitis inferior. Medially, it is covered by a layer of dense adipose tissue, situated deep to the semispinalis capitis, and, laterally, by the longissimus capitis and sometimes by the splenius capitis, which overlap the obliquus capitis superior. The 'floor' of the triangle is formed by the posterior atlanto-occipital membrane and the posterior arch of the atlas; the vertebral artery and the dorsal ramus of the first cervical nerve (5.40) lie in a groove on the upper surface of the posterior arch of the atlas.

III. The Muscles of the Thorax

This group (5.41, 42, 43) consists of muscles which connect adjoining ribs (intercostales—externi, interni, et intimi), span several ribs between attachments (subcostales), connect ribs to sternum (transversus thoracis), or ribs to vertebrae (levatores costarum, serratus posterior superior et inferior) and the diaphragm. They are all concerned in the movements of ribs and hence, potentially, with respiration.

The intercostales (5.42) are thin superimposed layers of muscle and tendinous fibres occupying the intercostal spaces. They are named from their spatial relations—the intercostales externi are most superficial and the intercostales intimi are the innermost.

The intercostales externi are eleven in number on each side. Their attachments extend from the tubercles of the ribs, where they are blended with the posterior fibres of the superior costotransverse ligaments, almost to the cartilage of the ribs in front, where each is replaced by an aponeurotic layer named the *external intercostal membrane*, which is continued forwards to the sternum. Each muscle passes from the lower border of one rib to the upper border of the rib below. In the lower two spaces they extend to the ends of the costal cartilages, and in the upper two or three spaces they do not quite reach the ends of the ribs. They are thicker than the intercostales interni and their fibres are directed obliquely downwards and laterally at the back of the thorax, and downwards, forwards and medially at the front.

The intercostales interni are also eleven in number on each side. Their attachments commence anteriorly at the sternum, in the interspaces between the cartilages of the true ribs, and at the anterior extremities of the cartilages of the false ribs, and extend backwards as far as the posterior costal angles, where each is replaced by an aponeurotic layer named the *internal intercostal membrane*, which is continuous posteriorly with the anterior fibres of the superior costotransverse ligament and anteriorly with the fascia intervening between the internal and external intercostal muscles. Each muscle descends from the floor of the costal groove and the corresponding costal cartilage, and is inserted into the upper border of the rib below. Their fibres are also directed obliquely and nearly at right angles to those of the external intercostal muscles.

Actions. Opinions concerning the actions of the intercostals are not unanimous and at various times the following views have been maintained: that both the internal and external groups act as elevators of the ribs, that the external intercostals are elevators and the internal intercostals depressors of the ribs, that the intercartilaginous parts of the internal intercostals act with the external intercostals in inspiration and that both intercostals form strong elastic supports which prevent the contents of the intercostal spaces being drawn in or bulged out during

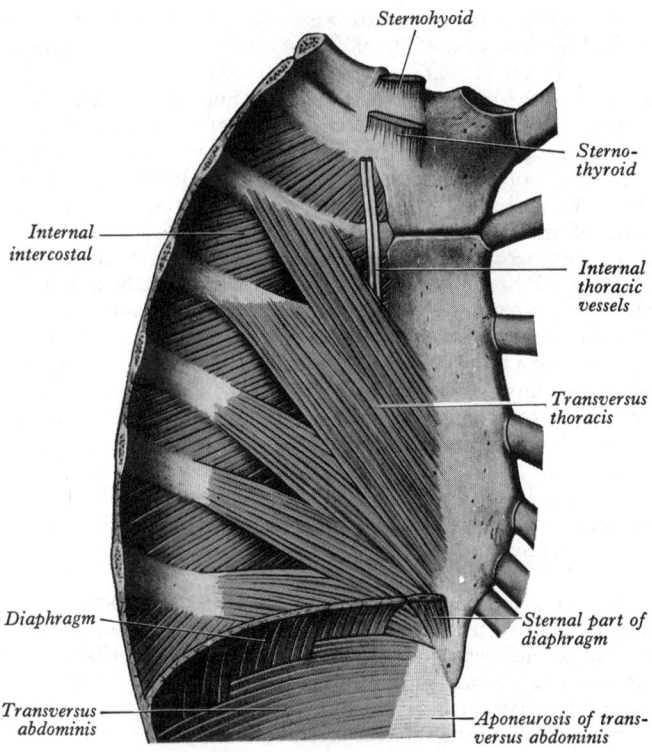

5.41 The left transversus thoracis, exposed and viewed from its posterior aspect. Note that, in the interval between the sternal and the costal origins of the diaphragm, the lower border of the transversus thoracis is in contact with the upper border of the transversus abdominis.

respiration. Experiment has shown that impulses pass along the nerves to the external intercostals of the lower intercostal spaces (5–9) in normal inspiration and that this activity is continued into the early part of expiration (Bronk and Ferguson 1935; Draper *et al.* 1960). With forced inspiration the contraction of these muscles is greater and more sustained. Electromyography has yielded some apparently divergent results. It has been claimed that the intercostals of the 6th, 7th and 8th spaces contract in the inspiratory phase of quiet breathing and are relaxed in expiration (Campbell 1958) and that in quiet breathing the intercostals of the upper four intercostal spaces show sustained and not rhythmic activity. In forced inspiration this activity is increased and continues at a diminished level in forced expiration (Jones *et al.* 1953). It is possible that the type and extent of activity in the intercostals varies with the depth of respiration (see also p. 550). These controversies have not yet been resolved (Hoshiko 1962; Basmajian 1967). The intercostals show constant activity during speech.

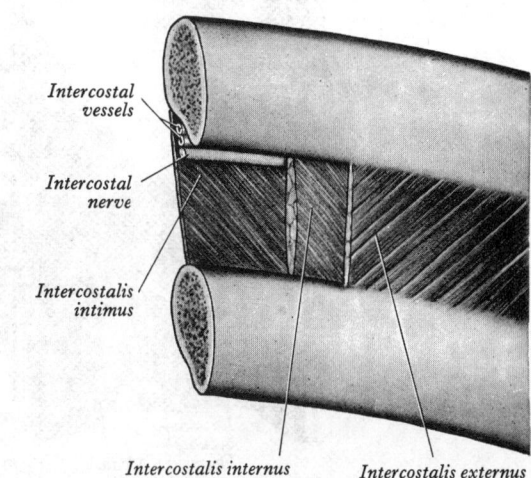

5.42 A dissection of a part of the thoracic wall, showing the position of the intercostal vessels and nerve relative to the intercostal muscles.

The intercostales intimi were once regarded as merely internal laminae of the internal intercostal muscles, with which their fibres do in fact coincide in direction. Each muscle is attached to the internal aspects of two adjoining ribs. They are insignificant and sometimes absent at highest thoracic levels, but become progressively more extensive below this, extending through about the middle two quarters of lower intercostal spaces. Walmsley (1915) considered them to correspond with the plane of the transversus abdominis and suggested the term musculi intracostales; his description was confirmed by Davies *et al.* (1931). Posteriorly the intercostales intimi, in those spaces where they are well developed, may become contiguous with the corresponding subcostales, which Davies *et al.* regarded as representing a fourth layer. The consensus of opinion, however, favours Walmsley's description of a three-layered arrangement. The intercostales intimi are related internally to the endothoracic fascia and parietal pleura, and externally to the intercostal nerves and vessels.

Actions. In the absence of specific information it can only be presumed that these muscles act in conjunction with the internal intercostal muscles (*vide supra*).

The subcostales consist of muscular and aponeurotic fasciculi, and are usually well developed only in the lower part of the thorax; each descends from the inner surface of one rib near its angle to the internal surface of the second or third rib below. Their fibres run in the same direction as those of the internal intercostals. Like the intercostales intimi they are between the intercostal vessels and nerves and the pleura. The incidence and distribution of these muscles have been recorded in detail by Eister (1912), and by Satoh (1974).

Actions. The subcostales probably depress the ribs.

The transversus thoracis (sternocostalis) is situated upon the internal surface of the anterior wall of the thorax (5.41). It arises from the lower third of the posterior surfaces of the body of the sternum, the xiphoid process, and the costal cartilages of the lower three or four true ribs near their sternal ends. Its fibres diverge as they pass upwards and laterally, to be attached by slips into the lower borders and inner surfaces of the costal cartilages of the second, third, fourth, fifth and sixth ribs. The lowest fibres of this muscle are horizontal, and are contiguous with the highest fibres of the transversus abdominis; the intermediate fibres are oblique, while the highest are almost vertical. This muscle varies in its attachments not only in different subjects but on opposite sides of the same subject. Like the intercostales intimi and subcostales, the transversus thoracis separates the intercostal nerves from the pleura.

Actions. The transversus thoracis draws down the costal cartilages to which it is attached.

Nerve supply. All the above muscles are supplied by the adjacent intercostal nerves.

The levatores costarum (5.37), twelve in number on each side, are strong bundles which arise from the ends of the transverse processes of the seventh cervical and 1st to 11th thoracic vertebrae; they pass obliquely downwards and laterally, parallel with the posterior borders of the external intercostals, and each is attached to the upper edge and external surface of the rib immediately below the vertebra from which it takes origin, between the tubercle and the angle (*levatores costarum breves*). Each of the four lower muscles divides into two fasciculi, one of which is attached as described above; the other passes down to the *second* rib below its origin (*levatores costarum longi*).

Nerve supply. The lateral branches of the dorsal rami of the corresponding thoracic nerves. (Consult Morrison 1954.)

Actions. The levatores costarum elevate the ribs but their importance in respiration is disputed (Primrose 1952); they are also said to act from their rib attachments as rotators and lateral flexors of the vertebral column.

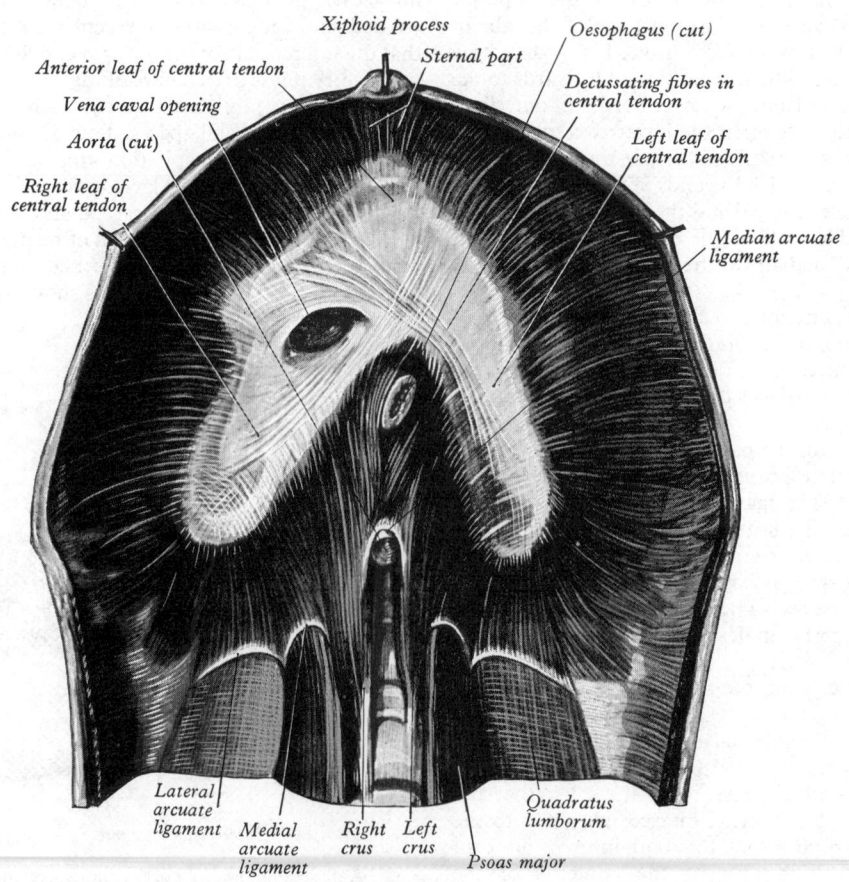

Xiphoid process

Anterior leaf of central tendon

Vena caval opening

Aorta (cut)

Right leaf of central tendon

Sternal part

Oesophagus (cut)

Decussating fibres in central tendon

Left leaf of central tendon

Median arcuate ligament

Lateral arcuate ligament

Medial arcuate ligament

Right crus

Left crus

Quadratus lumborum

Psoas major

5.43   The abdominal aspect of the diaphragm.

The serratus posterior superior is a thin, quadrilateral muscle, external to the upper and posterior part of the thorax. It arises by a thin aponeurosis from the lower part of the ligamentum nuchae, from the spines of the seventh cervical and upper two or three thoracic vertebrae and from their supraspinous ligaments. Inclining downwards and laterally it is attached by four fleshy digitations, into the upper borders and external surfaces of the second, third fourth and fifth ribs, a little lateral to their angles. It lies superficial to the thoracic part of the thoracolumbar fascia and anterior to the rhomboids. The number of slips may vary from three to six, and the muscle may even be absent.

Nerve supply. The second, third, fourth and fifth intercostal nerves.

Actions. By their attachments it is clear that this muscle could elevate ribs, but although experiments in dogs indicate that it is not a respiratory muscle (Ogawa *et al.* 1960), its role in man is not clear.

The serratus posterior inferior (5.58) is a thin irregularly quadrilateral muscle, situated at the junction of the thoracic and lumbar regions. It arises by a thin aponeurosis from the spines of the lower two thoracic and upper two or three lumbar vertebrae and from the supraspinous ligament; this aponeurosis is blended with the lumbar part of the thoracolumbar fascia. Passing obliquely upwards and laterally, it becomes fleshy, and is attached by four digitations into the inferior borders and outer surfaces of the lower four ribs, a little lateral to their angles. The number of slips may be reduced, and rarely the whole muscle is absent.

Nerve supply. The ventral rami of the ninth, tenth, eleventh and twelfth thoracic spinal nerves.

Actions. The serratus posterior inferior draws the lower ribs downwards and backwards, but perhaps not in respiration.

The Diaphragm

The diaphragm (5.43) is a dome-shaped, musculofibrous sheet which separates the thoracic from the abdominal cavity, its convex upper aspect facing the former, and its concave inferior surface directed towards the latter. Its periphery consists of muscular fibres attached to the circumference of the thoracic outlet and converging into a central tendon. The muscular fibres may be grouped into three parts—sternal, costal and lumbar.

The *sternal part* arises by two fleshy slips from the back of the xiphoid process, the *costal part* from the internal surfaces of the cartilages and adjacent parts of the lower six ribs on each side, interdigitating with the transversus abdominus (5.41), the *lumbar part* from two aponeurotic arches, the medial and lateral arcuate ligaments (sometimes termed lumbocostal arches) and from the lumbar vertebrae by two pillars or *crura*. The xiphoid slip is sometimes absent.

The *lateral arcuate ligament,* which is a thickened band in the fascia covering the quadratus lumborum, arches across the upper part of that muscle, and is attached, medially, to the front of the transverse process of the first lumbar vertebrae, and, laterally, to the lower margin of the twelfth rib near its mid point.

The *medial arcuate ligament* is a tendinous arch in the fascia covering the upper part of the psoas major; medially, it is continuous with the lateral tendinous margin of the corresponding crus, and is attached to the side of the body of the first or second lumbar vertebra; laterally, it is fixed to the front of the transverse process of the first lumbar vertebra.

The *crura* are tendinous in structure at their attachments, and blend with the anterior longitudinal ligament of the vertebral column. The *right crus*, broader and longer than the left, arises from the anterolateral surfaces of the bodies and intervertebral discs of the upper *three* lumbar vertebrae, while the *left crus* arises from the corresponding parts of the upper *two*. The medial tendinous margins of the crura meet in the median plane to form an arch across the front of the aorta, the *median arcuate ligament*; it is often poorly defined.

From this circumferential attachment the fibres of the diaphragm converge into a central tendon. The fibres from the xiphoid process are short, and occasionally aponeurotic; those from the medial and lateral arcuate ligaments, and more especially those from the ribs and their cartilages, are longer, and describe

marked curves as they ascend and converge to their central attachment. The fibres from the crura diverge as they ascend, the most lateral being directed upwards and laterally to the central tendon. The medial fibres of the right crus ascend on the *left* side of the oesophageal opening, and occasionally a fleshy fasciculus from the medial side of the left crus crosses the aorta and runs obliquely through the fibres of the right crus towards the vena caval opening, but this fasciculus is not continued upwards around the oesophageal passage on the right side (Low 1907).

The *central tendon* of the diaphragm is a thin but strong aponeurosis of closely interwoven fibres situated near the centre of the dome formed by the muscle, but closer to the front of the thorax, so that the posterior muscular fibres are longer. It is immediately below the pericardium, with which it is partially blended. It is trifoliate in shape, consisting of three folia separated by slight indentations. The middle leaf has the form of an equilateral triangle, the apex directed towards the xiphoid process. The right and left folia are tongue-shaped and curve laterally and backwards, the left being a little narrower. The central area of the tendon consists of four well-marked *diagonal bands* radiating from a thick central point and expanding in a fan-shaped manner; the central point of decussation is a thick *node* of compressed tendinous strands situated in front of the oesophageal aperture, and to the left of the vena caval opening.

The Diaphragmatic Apertures

The diaphragm is pierced for the passage of structures between the thorax and abdomen: three large openings—the aortic, the oesophageal and the vena caval (5.43)—and a number of smaller ones exist.

The *aortic aperture* is the lowest and most posterior of the large openings; it is at the level of the lower border of the twelfth thoracic vertebra and the thoraco-lumbar intervertebral disc, slightly to the left of the median plane. Strictly speaking, it is an osseo-aponeurotic opening lying between the diaphragmatic crura laterally, the vertebral column posteriorly and the diaphragm anteriorly. It therefore lies *behind* the diaphragm and, more specifically, its median arcuate ligament when present. Occasionally some tendinous fibres from the medial parts of the crura also pass *behind* the aorta, and convert the opening into a fibrous ring. The aortic opening transmits the aorta and the thoracic duct and *occasionally* the azygos and hemiazygos veins (p. 754); some lymphatic trunks also *descend* through it from the lower posterior thoracic wall.

The *oesophageal aperture* is in the muscular part of the diaphragm at the level of the tenth thoracic vertebra; it is elliptical in shape, and is formed by the splitting of the medial fibres of the right crus (Low 1907). It is above, in front, and a little to the left of the aortic opening, and transmits the oesophagus, the complex vago-sympathetic gastric nerves, oesophageal branches of the left gastric vessels, and some lymphatic vessels. There is no direct continuity between the oesophageal wall and the muscle around the oesophageal opening. The fascia on the inferior surface of the diaphragm, which is continuous with the transversalis fascia (p. 559), and is rich in elastic fibres, extends upwards into the opening in a conical fashion to be attached to the wall of the oesophagus about 2 cm above the gastro-oesophageal junction. Some of its elastic fibres penetrate to the submucosa of the oesophagus. This fascial expansion is sometimes termed the *phreno-oesophageal ligament*; it constitutes a flexible connexion between the oesophagus and diaphragm, permitting some freedom during swallowing and respiration and at the same time limiting upward displacement of the oesophagus (Allison 1951; Hayward 1961).

The *vena caval aperture*, the highest of the three large openings, is at about the level of the disc between the eighth and ninth thoracic vertebrae. It is quadrilateral, and at the junction of the right leaf with the central area of the tendon so that its margins are aponeurotic. It is traversed by the inferior vena cava, which is adherent to the margin of the opening, and some branches of the right phrenic nerve.

There are two *lesser apertures* in each crus; one transmits the greater and the other the lesser splanchnic nerve. The ganglionated trunks of the sympathetic usually enter the abdominal cavity behind the diaphragm, deep to the medial arcuate ligament. Openings for minute veins frequently occur in the central tendon.

On each side there are two small areas where the muscular fibres of the diaphragm are deficient and are replaced by areolar tissue. One, between the sternal and costal parts, contains the superior epigastric branch of the internal thoracic artery and some lymph vessels from the abdominal wall and convex surface of the liver. The other, between the costal part and the fibres springing from the lateral arcuate ligament, is less constant; when it exists, the posterosuperior surface of the kidney is separated from the pleura only by areolar tissue.

Relations. The upper surface of the diaphragm is in relation with three serous membranes, viz. on each side with the pleura which separates it from the base of the corresponding lung, and over the middle folium of the central tendon with the pericardium, which intervenes between it and the heart. The central area lies on a slightly *lower* level than the summits of the lateral regions, which are usually termed the *cupolae*. The greater part of the inferior surface is covered by peritoneum. The right side is accurately moulded over the convex surface of the right lobe of the liver, the right kidney, and right suprarenal gland; the left over the left lobe of the liver, the fundus of the stomach, the spleen, the left kidney, and the left suprarenal gland.

Nerve supply. The diaphragm is supplied with motor fibres by the phrenic nerves. The lower six or seven intercostal nerves distribute sensory fibres to the peripheral part of the muscle. There has been some controversy concerning the motor supply to the crural fibres stemming from a suggestion that these are supplied by intercostal nerves. However, Collis *et al.* (1954) and Bottia (1957) have described a supply by the phrenic nerves, and this has been confirmed by Shehata (1966). All three observers state that the crural fibres (irrespective of their origin) on the right and left aspects of the oesophageal hiatus are supplied by the ipsilateral phrenic nerve.

Actions. The diaphragm is the essential muscle of inspiration. Its dome presents a concave surface towards the abdomen. The central part of the dome is tendinous, with the pericardium attached to its upper surface; the circumferential part muscular. During inspiration the lowest ribs are fixed, and from these and the crura the muscular fibres contract and draw the central tendon downwards and forwards with the attached pericardium. In this movement the curvature of the diaphragm is scarcely altered, the dome moving downwards nearly parallel to its original position and pushing before it the abdominal viscera. The descent of the abdominal viscera is permitted by the extensibility of the abdominal wall, but the limit of this is soon reached. The central tendon, applied to the abdominal viscera, then becomes a fixed point for the action of the diaphragm, the effect of which is to elevate the lower ribs and through them to push forwards the body of the sternum and the upper ribs. The right cupola of the diaphragm, lying on the liver, has a greater resistance to overcome than the left, which lies over the stomach, but to compensate for this the right crus and the fibres of the right side are more substantial than those of the left. The balance between descent of the diaphragm and protrusion of the abdominal wall ('abdominal' breathing) and elevation of the ribs ('thoracic' breathing) varies in different individuals and with the depth of respiration. The thoracic element is usually more marked in females and in both sexes in deep inspiration.

In all expulsive acts the diaphragm is called into action to give additional power to each effort. Thus, before sneezing, coughing, laughing, crying, or vomiting, and previous to the expulsion of urine, or faeces, or of the fetus from the uterus, a deep inspiration takes place. A relatively deep inspiration, followed by closure of the glottis, is a common preliminary to powerful action of the trunk muscles, as in lifting heavy weights.

It has been suggested that the fibres of the right crus exert a sphincteric action on the lower end of the oesophagus in man. The act of expiration, which immediately succeeds the act of swallowing, relaxes these fibres and allows the contents of the oesophagus to pass into the stomach (*see* Whillis 1931; and p. 1318). In recent years less emphasis has been placed on this action of the right crus, and the intrinsic muscle of the lower

2 cm of the oesophagus below the attachment of the phreno-oesophageal ligament is considered to have some effective sphincteric action. (See also p. 1336.)

The level of the diaphragm varies continuously during respiration; it is also affected by the degree of distension of the stomach and intestines and the size of the liver. After a forced expiration the right cupola is level in front with the fourth costal cartilage, at the side with the fifth, sixth and seventh ribs, and behind with the eighth; the left cupola is a little lower. The excursion of the diaphragm is about 1·5 cm during quiet breathing; in deep respiration the maximum movement ranges from 6 to 10 cm (Campbell 1958).

Radiographs show that the level of the diaphragm in the thorax also varies considerably with posture. It is highest when the body is supine, and in this position it performs the largest respiratory excursions with normal breathing. When the body is erect the dome of the diaphragm is lower, and its respiratory movements become smaller. The dome is still lower when the sitting posture is assumed, and respiratory excursions are smallest. When the body is horizontal and on one side, the two halves of the diaphragm do not behave alike. The uppermost half sinks to a level lower even than in sitting, and moves little with respiration; the lower half rises higher in the thorax than it does in the supine position, and its respiratory excursions are much increased.

It appears that the position of the diaphragm in the thorax depends upon three main factors: (1) the elastic retraction of the lung tissue, tending to pull it upwards; (2) the pressure exerted from below by the viscera: this naturally tends to be a negative pressure, or downward suction, when the patient sits or stands, and a positive, or upward pressure, when he is lying down; (3) the intra-abdominal pressure due to the abdominal muscles. These muscles are in a state of contraction in the standing position but not in the sitting; hence the diaphragm is pushed up higher in the former position.

Applied Anatomy. Abdominal organs, usually the stomach, occasionally herniate through the diaphragm into the thorax, either alongside the oesophagus or between the fibres arising from the lateral arcuate ligament and the costal fibres.

The Movements of Respiration

The rhythmic movements of respiration produce alterations in the capacity of the thoracic cavity which in turn bring about the inflow and expulsion of air in the lungs to ventilate the alveoli, where the gaseous exchanges between the blood and alveolar air occur. The capacity of the thoracic cavity can be increased, either independently or in combination, in the transverse, antero-posterior and vertical dimensions.

The transverse and anteroposterior dimensions are increased by movements of the ribs. When elevated, the anterior ends of the vertebrosternal ribs are thrust forwards and upwards with some slight movement at the sternal angle, thereby increasing the anteroposterior dimension of the cavity. At the same time these same ribs, rotating around an anteroposterior axis through their ends (p. 453), are everted, thereby increasing the transverse diameter. Elevation of the vertebrochondral ribs (4.35) results in an outward and backward movement to produce an increase in the transverse diameter. This is associated with some widening of the infrasternal angle and an increase in the capacity of the upper part of the abdominal cavity. The vertical dimension of the thoracic cavity is increased by contraction of the diaphragm. This presses on the abdominal viscera, which are permitted to descend by relaxation of the abdominal wall, during inspiration. When the limit of this descent is reached the abdominal viscera provide resistance for the central tendon of the diaphragm from which its muscle fibres elevate the lower ribs.

Each respiratory cycle consists of an inspiratory and an expiratory phase. When the body is at rest the former is of about 1 second and the latter about 3 seconds duration. The increased thoracic capacity in inspiration reduces the intrapleural pressure with consequent expansion of the lung, decreasing intra-pulmonary pressure and entry of air through the respiratory passages. The expiratory phase is largely passive, the recoil of the chest wall and lungs raising the intrathoracic pressure and expelling air.

In *quiet respiration* about 500-750 ml of *tidal air* are inspired or expired in each respiratory cycle. About one-third of this occupies the '*anatomical dead space*'. This space extends from the nostrils to the terminal bronchioles, its epithelium is not respiratory, and practically no gaseous exchanges with the blood can occur. The remainder of the inspired air is available for ventilation of the alveoli. The inspired air is warmed, humidified and filtered in the nasal cavity.

In *quiet inspiration* the principal and often the sole muscle concerned is the diaphragm. In an average adult the right cupola descends to the level of the disc between the tenth and eleventh thoracic vertebrae and the left cupola descends to the level of the disc between the eleventh and twelfth. The dome descends about 1·5 cm. This diaphragmatic movement is responsible for the greater part and perhaps sometimes the whole of the tidal volume of about 500 ml of air inspired. There is little or no alteration in the position of the bony boundaries of the thoracic inlet. Electromyography, however, shows rhythmic activity of the scalene muscles (anterior and medius). They serve to fix the upper ribs, while the intercostal muscles of the upper six intercostal spaces may show continuous activity, thereby preventing the sucking in or blowing out of the tissues of these spaces with changes in the intrapleural pressure. At the same time in quiet respiration the intercostals of the lower spaces may undergo rhythmic contraction, thereby contributing to the increase of the transverse diameter at this level in the thoracic cavity. These muscles may also resist pressure changes.

In inspiration the intrapleural and the intrapulmonary pressure fall below atmospheric. At the end of inspiration the intrapulmonary pressure is equal to the atmospheric. During inspiration the air which flows into the lung is not evenly distributed. Regional differences occur in both the ventilation and blood flow. The ventilation is greater in the lower than in the upper lobes of the lung. This accords with the fact that the movement of the upper chest in quiet respiration is inconspicuous, whereas that of the lower chest is greater, affecting principally the transverse diameter. These regional differences may also be related to the varying elasticity of different parts of the lungs and the dimensions of the air passages leading to them. It has been suggested that the alveoli which expand earliest in inspiration receive more of the air lying in the dead space.

In *quiet expiration* the elastic recoil of the lungs and chest wall is graduated by the slow relaxation of the inspiratory muscles. Electromyography demonstrates that the diaphragm and intercostals continue contracting but with progressively decreasing intensity well into the expiratory phase. Another component believed to be important in the recoil of the lungs is the thin film of surface-active material at the interphase between the alveolar air and the surface of the alveolar epithelium (*see* p. 1258). Provided the ventilation rate is below 40 litres/min there is no active participation of the intercostal or abdominal muscles in expiration.

In *deep inspiration* the movements already described are increased and additional muscles are brought into action. The muscles contract with greater force, the number of motor units called into action increases, the discharges along their nerves are of longer duration and the contribution of the different motor units is better synchronized.

The intercostal muscles become active in a progressively increasing number of intercostal spaces to produce movements of the upper ribs and sternum. The first rib is elevated by the scaleni, anterior and medius, and by the sternocleidomastoid indirectly through the clavicle and costoclavicular ligament and directly through the manubrium sterni. For each 1 cm increase in the circumference of the chest the capacity increases by about 200 ml.

The twelfth rib is fixed by the quadratus lumborum so that the diaphragm is able to exert a more powerful downward thrust on the abdominal viscera (Boyd *et al.* 1965). It has also been suggested that this action might provide a fixed basis for controlled relaxation of the diaphragm in the nice adjustment of expiration necessary to speech and singing (Taylor 1960). In addition, the erectores spinae come into action and the concavity

of the thoracic part of the vertebral column is diminished, a movement which results in a slight increase in the width of the intercostal spaces and allows the ribs a greater range of movement. The intrapulmonary pressure at the beginning of *deep expiration* may rise to 30 mm of mercury above the atmospheric pressure.

In *forced respiration*, as in breathing against resistance, further muscles are called into action. In forced *inspiration* the diaphragm contracts maximally, the right cupola descends to about the level of the eleventh thoracic vertebra, while the left cupola may reach the level of the body of the twelfth. Such maximal excursions of the diaphragm correspond to a maximal change in level of 9 to 10 cm. The action of the erectores spinae is appreciably augmented and muscles connecting the upper limb to the trunk may show some activity. In forced *expiration* additional expulsive factors are provided by the strong contraction of the muscles of the abdominal wall, particularly the oblique and transverse muscles, and by both the latissimi dorsi, which contract suddenly and energetically with such efforts as coughing and sneezing. The muscles of the abdominal wall raise the intra-abdominal pressure, forcing the relaxing diaphragm upwards and drawing the lower ribs downwards and medially.

The range and character of the movements of the thoracic parietes exhibit very striking individual variations, which may be dependent on the conformation of the thoracic skeleton, on habit or on other factors, and this extreme variability must be borne in mind when the movements are being analysed in any particular subject.

In all these phases of respiration it must be remembered that there are changes in the nostrils, larynx, trachea and bronchial tree. The actions of the dilators of the nostrils become evident in forced respiration. During quiet respiration the rima glottidis (p. 1237) is widened by lateral rotation of the vocal processes abducting the vocal cords and this is widened even further in forced inspiration by increasing abduction of the vocal cords and by abduction of the arytenoid cartilages. The calibre of the trachea, bronchi and bronchioles can be changed passively by distension or actively by alteration of the tone of the smooth muscle in their walls mediated through their parasympathetic (vagus nerve) and sympathetic nerve supplies, so that they dilate in inspiration and contract during expiration. The precise role of these antagonistic influences is not yet certain.

Applied Anatomy. The changes in the height of the diaphragm during alterations in posture explain why patients suffering from severe dyspnoea are most comfortable and least short of breath when they sit up. In unilateral disease of the pleura or lungs interference with the position or movement of the diaphragm can generally be observed on X-ray examination.

IV. The Muscles of the Abdomen

The muscles of the abdomen may be conveniently divided into anterolateral and posterior groups.

THE ANTEROLATERAL MUSCLES OF THE ABDOMEN

This group consists of four large flat muscular sheets forming the abdominal wall (obliqui externus and internus, transversus and rectus abdominis), and two smaller elements (cremaster and pyramidalis) concerned with suspension of the testis and tensing of the midline tendinous raphe of the abdominal wall.

The superficial fascia of the abdominal wall consists, over the greater part of it, of a single layer containing a variable amount of fat; but over the lower part, particularly in obese individuals, the fascia differentiates into layers. The extent and significance of these layers is disputed (Tobin and Benjamin 1949), but in practice the following obtains. The superficial fascia is divisible into two layers, between which are superficial vessels, nerves and the superficial inguinal lymph nodes.

The superficial layer (or layers) of the fascia is thick, areolar in texture, and contains in its meshes a varying quantity of fat. Below, it passes over the inguinal ligament, and is continuous with the superficial fascia of the thigh. In the male this layer is

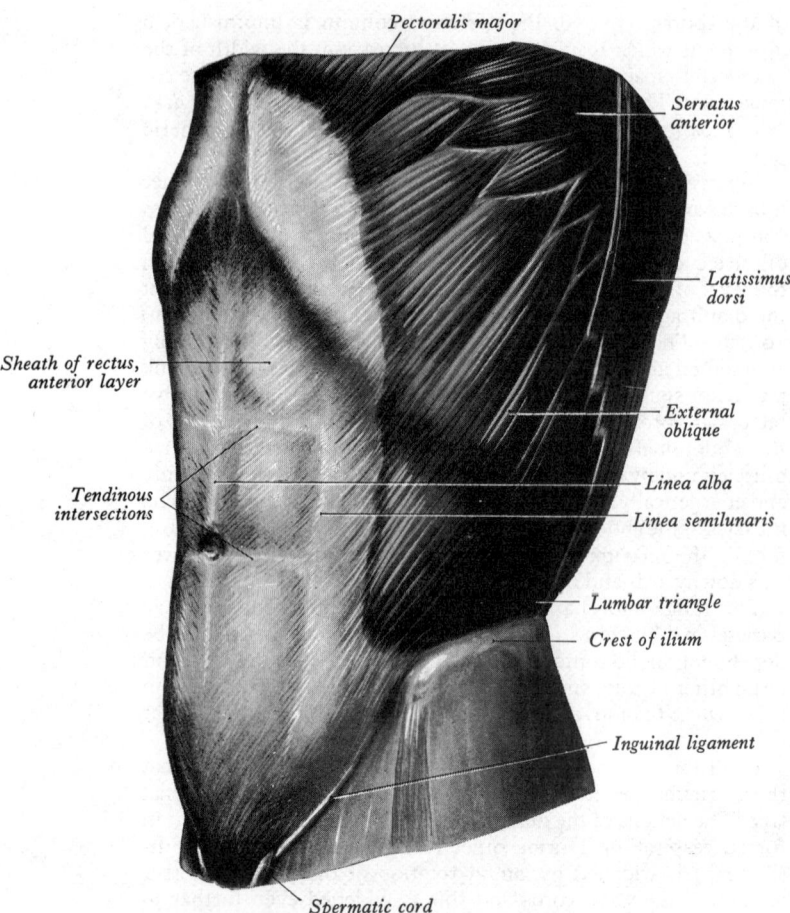

Pectoralis major

Serratus anterior

Latissimus dorsi

Sheath of rectus, anterior layer

External oblique

Linea alba

Linea semilunaris

Tendinous intersections

Lumbar triangle

Crest of ilium

Inguinal ligament

Spermatic cord

5.44 The left obliquus externus abdominis.

continued over the penis and outer surface of the spermatic cord into the scrotum, where it changes its character, becoming thin, destitute of adipose tissue, and of a pale reddish colour. In the scrotum it also contains non-striated muscular fibres, forming the *dartos* muscle, which is supplied by the genitofemoral nerve. From the scrotum it may be traced backwards into continuity with the superficial fascia of the perineum. In the female it is similarly continued from the abdomen into the labia majora and perineum.

The deep layer (or layers) of the fascia, more membranous than the superficial, contains elastic fibres. It is loosely connected by areolar tissue to the aponeurosis of the external oblique; but in the median plane it is intimately adherent to the linea alba and symphysis pubis, and is prolonged on the dorsum of the penis, contributing to the *fundiform ligament* (**5**.45). Above, it is continuous with the superficial fascia over the rest of the trunk; below and laterally, it passes over the inguinal ligament and blends with the overlying superficial layer and the underlying fascia lata in the flexure line or skin crease of the thigh (**5**.45); below and medially, it is continued over the penis and spermatic cord to the scrotum, and from there may be traced backwards into continuity with the membranous layer of the superficial fascia of the perineum (p. 561). In the female it is continued into the labia majora and thence to the fascia of the perineum.

In the child the testis can frequently be retracted out of the scrotum into the interval occupied by loose areolar tissue between the external oblique and the deep layer of superficial fascia over the inguinal canal. This interval is sometimes called the *superficial inguinal pouch* (Browne 1938).

The obliquus externus abdominus (**5**.44), curved round the lateral and anterior parts of the abdomen, is the largest and the most superficial of the three flat muscles in this region. It arises, by eight fleshy slips, from the external surfaces and inferior borders of the lower eight ribs; these slips interdigitate with the digitations of serratus anterior and latissimus dorsi, and are arranged in an oblique line which extends downwards and backwards, the upper ones being attached close to the cartilages of

the corresponding ribs, the lowest to the apex of the cartilage of the last rib, the middle ones to the ribs at some distance from their cartilages. From these attachments the fibres diverge as they pass to their insertions. Those from the lower two ribs pass nearly vertically downwards, and are attached to the anterior half or more of the outer lip of the ventral segment of the iliac crest (p. 379); the middle and upper fibres, directed downwards and forwards, end in an aponeurosis, opposite a line drawn vertically from the ninth costal cartilage to a little below the level of the umbilicus, and then inclining laterally to the anterior superior iliac spine. The muscle fibres rarely descend beyond a line from the anterior superior iliac spine to the umbilicus. The posterior border of the muscle is free (**5**.58).

The *aponeurosis* of the external oblique is a strong tendinous sheet, the fibres of which are directed downwards and medially. In the median plane its fibres end in the *linea alba* (**5**.44), a tendinous raphe which stretches from the xiphoid process to the symphysis pubis. At the raphe it is continuous with the aponeurosis of the opposite muscle and the two together cover the front of the abdomen. Below and medially the aponeurosis is attached to the upper border of the pubic symphysis and the pubic crest as far as the pubic tubercle. The margin of the part of the aponeurosis between the anterior superior iliac spine and the pubic tubercle is a thick band, folded internally upon itself to present a grooved upper surface; this is the *inguinal ligament*. Expansions, continuous with the medial end of the inguinal ligament are attached to the pecten pubis and constitute the *lacunar ligament complex*. From this end of the inguinal ligament fibres also pass upwards and medially to join the rectus sheath and the linea alba; these constitute the *reflected part* of the inguinal ligament (**5**.50).

The muscular and aponeurotic parts of the external oblique are invested by external and internal layers of fascia of which the former is better developed. The upper and lower digitations may be absent; digitations or even the whole muscle may be reduplicated. Digitations may also be continuous with pectoralis major or serratus anterior.

Nerve supply. The ventral rami of the lower six thoracic spinal nerves.

The inguinal ligament (**5**.46), as noted above, is the thick folded lower border of the aponeurosis of the external oblique, presenting a grooved, superior, abdominal surface (the 'floor' of the inguinal canal), and which stretches from the anterior superior iliac spine to the pubic tubercle. (It has variously been called the *crural arch*, the *superficial crural arch*, and *Poupart's ligament*.) The inguinal ligament is convex downwards towards the thigh and continuous with the fascia lata. In the adult it is some 12–14 cm in length and is inclined 40 to 35° to the horizonal. Its lateral half is rounded and more oblique; its medial half gradually widens towards its attachment to the pubis, is more horizontal and supports the spermatic cord. The fibres of the external oblique aponeurosis are *not* initially parallel to the long axis of the inquinal ligament: they approach the latter obliquely at an angle of 10–20°, and then, on reaching the ligament, each fibre turns *medially*, the majority now running in the long axis of the ligament to reach the pubic tubercle. The deeper fibres, however, splay out posteromedially to reach the pecten pubis (*see* Lytle 1974).

The lacunar ligament (*pectineal part of the inguinal ligament*) (**4**.58) has been described as composite in nature (Lytle 1974). Its abdominal, 'deep' part (i.e. the lacunar ligament of classical anatomy) consists of the extension of the aponeurosis of the external oblique, which continues backwards and laterally from the medial part of the inguinal ligament to the medial end of the pecten pubis. It is triangular, and almost horizontal when the body is upright; it is a little larger in the male, and measures about 2 cm from base to apex. Its base, directed laterally, is concave and thin, and forms the medial boundary of the femoral ring; its apex is attached to the pubic tubercle. Its posterior margin is attached to the pecten pubis, and is continuous with the *pectineal fascia*; its anterior margin is continued into the inguinal ligament. It has superior and inferior surfaces. A strong fibrous band, the *pectineal ligament* (of *Astley Cooper*) extends laterally from its base (**4**.58)

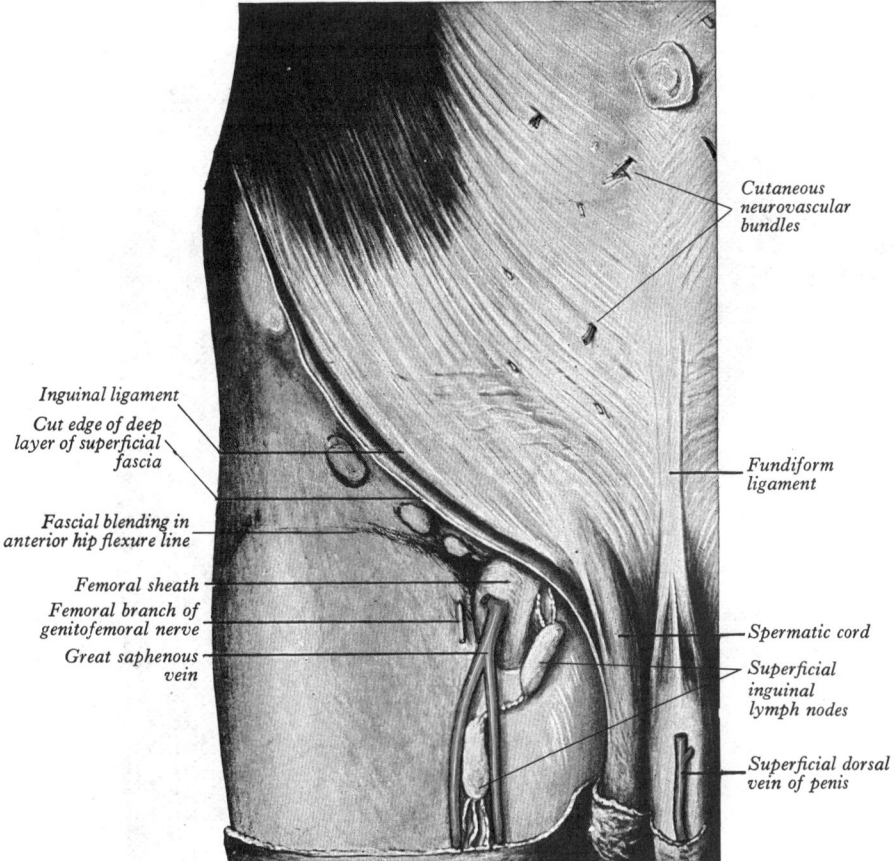

5.45   Superficial structures of the inguinal region and the lower part of the anterior abdominal wall. Right side.

along the pecten pubis to which it is attached. It is augmented by the pectineal fascia and by a lateral expansion from the lower end of the linea alba (*adminiculum lineae albae*, p. 559) It should be noted that the lacunar formation just described is best examined during dissection, and at operation, from its abdominal aspect (in the recent account by Lytle, this part is named as a subregion of the inguinal ligament). When approached from the thigh, however, in the living, a second, distinct lacunar fibrous sheet is distinguishable. This is derived from an inflexion of the *fascia lata* which, becoming confluent with the posterior border of the inguinal ligament, and receiving a reinforcement from the transversalis fascia, then fuses with pectineal fascia for a distance of about 1 cm, before proceeding upwards to reach and fuse with the thickened periosteum of the pecten pubis. This *fascial lacunar ligament* also presents a thickened curved lateral border which fits closely around the medial wall of the femoral sheath, and because of its pectineal fascial attachment is sited about 1 cm *below* and *anterior* to the pecten pubis, and some 3 cm lateral to the pubic tubercle. It should be noted that the traditional account of the boundaries of a femoral hernia, may need considerable revision. Further details will not be pursued here, expansion and confirmation of these new findings being awaited. (The interested reader should consult McVay and Anson 1940; Madden *et al.* 1971; Lytle 1957, 1974.)

The reflected ligament (part of the inguinal ligament) (**5**.46, 50) is an expansion from the lateral crus of the superficial inguinal ring. It passes upwards and medially behind the medial end of the superficial inguinal ring behind the external oblique and in front of the falx inguinalis; the fibres of the right and left ligaments interlace at the linea alba.

The superficial inguinal ring (**5**.45, 46) is a hiatus in the aponeurosis, just above and lateral to the crest of the pubis. The aperture is somewhat triangular, its long axis corresponding with the oblique course of the fibres of the aponeurosis. Its size is variable but usually it does not extend laterally beyond the medial third of the inguinal ligament. It is smaller in the female, the

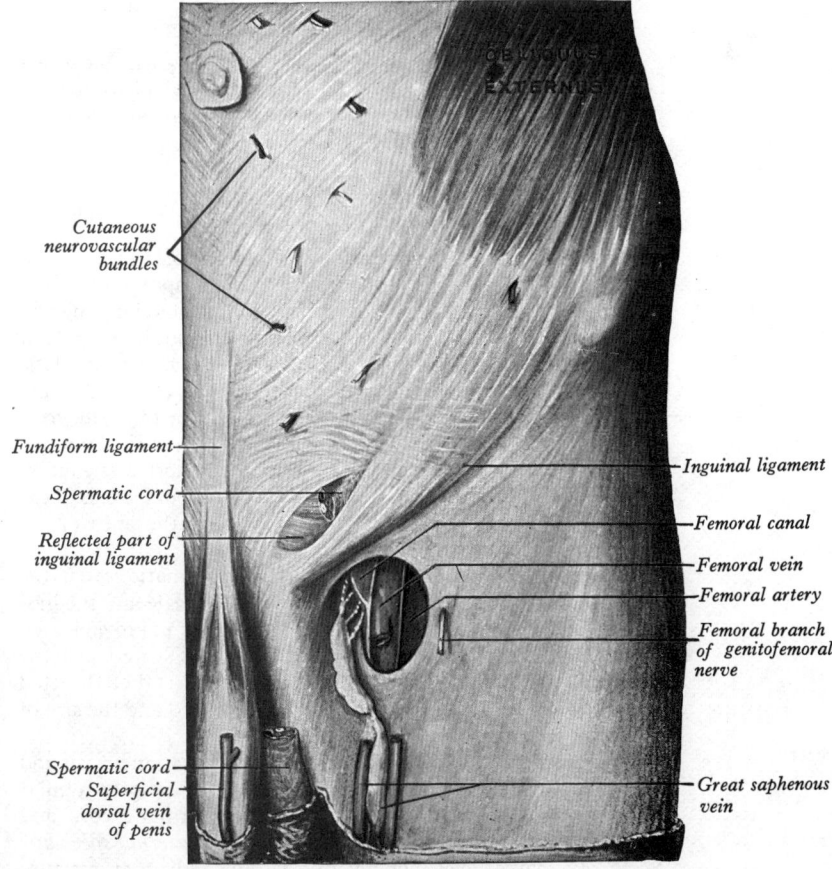

5.46   Superficial structures of the inguinal region and the lower part of the anterior abdominal wall. Left side.

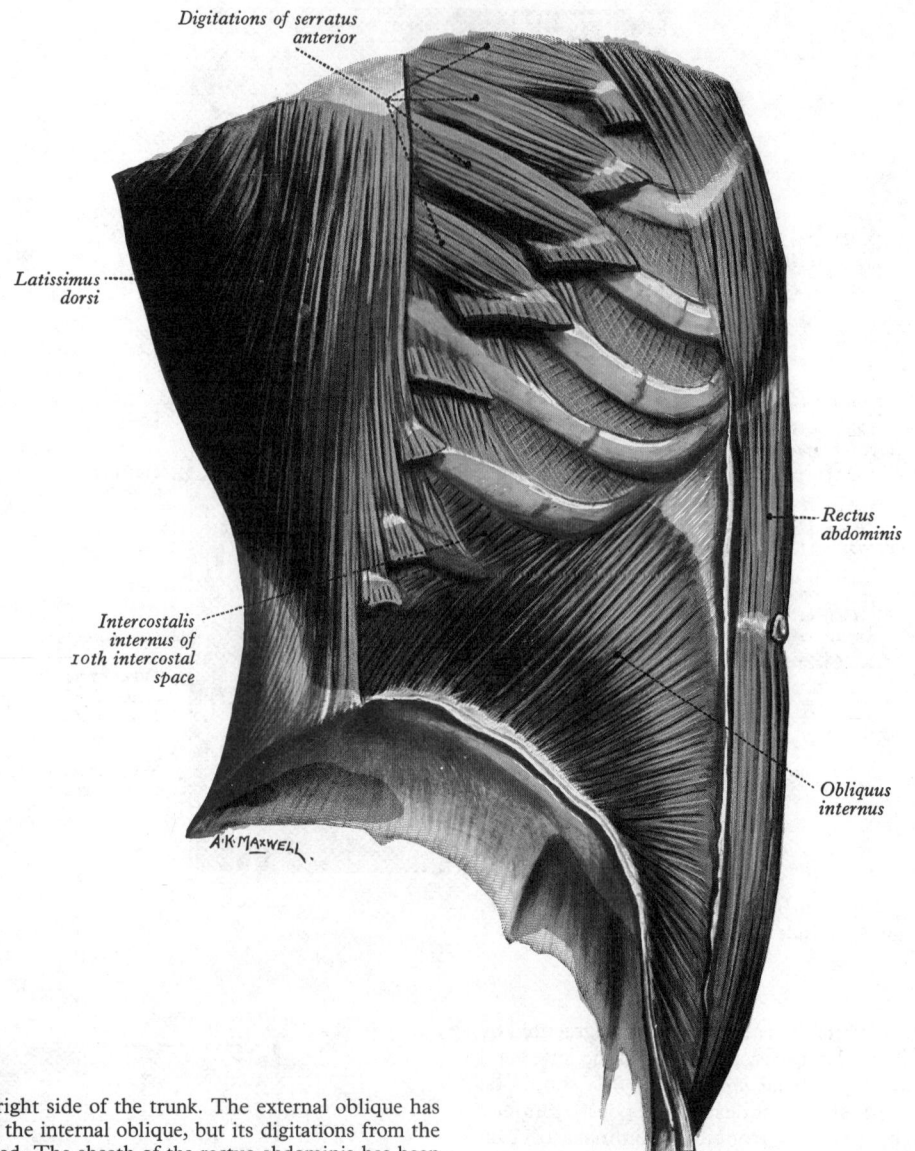

Digitations of serratus anterior

Latissimus dorsi

Intercostalis internus of 10th intercostal space

A·K·MAXWELL.

Rectus abdominis

Obliquus internus

5.47   Muscles of the right side of the trunk. The external oblique has been removed to show the internal oblique, but its digitations from the ribs have been preserved. The sheath of the rectus abdominis has been opened and its anterior lamina removed.

spermatic cord being absent. Its base is the crest of the pubis, and its sides are the margins of the opening in the aponeurosis, the *crura* of the ring. The lateral crus is the stronger, and is formed by fibres of the inguinal ligament inserted into the pubic tubercle; it is curved so as to form a kind of groove, in which the spermatic cord rests. The medial crus is a thin, flat band, attached to the front of the symphysis pubis, interlacing with the opposite crus. In the external layer of the investing fascia of the external oblique are bands of fibres of variable development and distribution which course at right angles to the fibres of the aponeurosis of the muscle. Some of these bands may arch above the apex of the superficial inguinal ring as *intercrural fibres*.

The superficial inguinal ring contains the spermatic cord in the male, the round ligament of the uterus in the female and the ilio-inguinal nerve in both. If the skin of the scrotum is invaginated upwards and laterally the spermatic cord can be followed up to the superficial inguinal ring. If the examining finger is then directed *backwards*, the crura of the ring can be recognized and the size of the ring appreciated.

At the margins of the ring the external oblique aponeurosis and overlying fascia are continued downwards as a delicate tubular prolongation of fibrous tissue around the spermatic cord and testis, as the outermost of their coverings, the *external spermatic fascia*. The superficial inguinal ring is only a distinct aperture after the continuity of this fascia and the aponeurosis has been severed (Anson *et al.* 1960).

The obliquus internus abdominis (5.47), internal to the external oblique, is thinner and less bulky. It arises, by muscular fibres, from the lateral two-thirds of the grooved upper surface of the inguinal ligament, from the anterior two-thirds of the intermediate line of the ventral segment of the iliac crest, and from the thoracolumbar fascia (5.38). It has been described as attached to the iliac fascia and not directly to the inguinal ligament (McVay and Anson 1940); fascia and ligament are adherent at this point (p. 564). The posterior fibres pass upwards and laterally to the inferior borders of the lower three or four ribs, and are there continuous with the internal intercostals. The uppermost fibres form a short, free superomedial border. The fibres from the inguinal ligament, paler in colour, arch downwards and medially across the spermatic cord in the male and the round ligament of the uterus in the female. Becoming tendinous, they are attached with the corresponding part of the aponeurosis of the transversus abdominis to the crest and the medial part of the pecten pubis, forming the *falx inguinalis* (*conjoint tendon*). The rest of the fibres of the internal oblique diverge and end in an aponeurosis which gradually broadens from below upwards. In its upper two-thirds this aponeurosis splits at the lateral border of the rectus abdominis into two laminae which pass around it and reunite in the linea alba, which they help to form. The anterior layer blends with the aponeurosis of the external oblique, the posterior with the aponeurosis of the transversus abdominis, and its upper part is attached to the cartilages of the seventh, eighth and ninth ribs. In the lower part of the abdominal wall the whole aponeurosis passes with that of the transversus in front of the rectus to the linea alba (p. 558).

Nerve supply. The internal oblique is supplied by the ventral rami of the lower six thoracic and the first lumbar spinal nerves.

The cremaster (5.49) consists of a number of loosely arranged muscle fasciculi lying along the spermatic cord. They are united by areolar tissue to form the sac-like *cremasteric fascia* around the cord and testis within the external spermatic fascia. The lateral part of the muscle, arising from the inguinal ligament, has been variously described as in continuity with the medial edge of the internal oblique, deep to the internal oblique, extending as far as the anterior superior iliac spine and in continuity with either the internal oblique or transversus, or as a pointed tendon from the middle of the inguinal ligament, piercing the internal oblique near its medial margin. Of these possibilities one extensive study revealed that in sixty subjects, all exhibited both a tendinous attachment and cremasteric fibres derived from the obliquus internus and transversus (Blunt 1951). The fibres pass along the lateral aspect of the spermatic cord through the superficial inguinal ring and then spread out into fasciculi in loops of increasing length along its anterolateral aspect. The shortest and most superior fasciculi turn inwards in front of the cord to join the medial part, while the longer fasciculi gain attachment to the fascia over the cord and upper part of the tunica vaginalis. The medial part of the muscle is variably developed and may be absent. It arises from the pubic tubercle and possibly from the pubic crest, falx inguinalis and lower border of transversus. Its fasciculi loop on the posteromedial aspect of the cord, interlacing with those of the lateral part. The whole muscle may be described as forming continuous loops from the middle of the inguinal

ligament as far as the tunica vaginalis and then returning to be attached to the pubic tubercle. In the female a few fibres descending on the round ligament of the uterus represent the lateral part of cremaster.

Nerve supply. the genital branch of the genitofemoral nerve, derived from the first and second lumbar spinal nerves.

Action. The cremaster pulls up the testis towards the superficial inguinal ring. Although its fibres are striated, it is not usually under voluntary control. Stroking of the medial side of the thigh evokes a reflex contraction of the muscle and this *cremasteric reflex* is much more active in children. Doubtless this may be a protective response, but the cremaster also appears to play an essential role in testicular thermoregulation. Shafik (1977) considers that the position of the testis is adjusted by conjoint action of the cremaster (which he divides into two distinct parts— *cremaster internus* and *externus*, separated by the internal spermatic fascia) and the dartos, the latter being attached to the testis by a 'scrotal ligament' (p. 1416).

The transversus abdominis (5.48), so-called from the direction of its fibres, is the innermost of the flat muscles of the abdominal wall, being internal to the internal oblique. Its muscle fibres arise from the lateral third of the inguinal ligament, the anterior two-thirds of the inner lip of the ventral segment of the iliac crest, the thoracolumbar fascia between the iliac crest and the twelfth rib, and the internal aspects of the lower costal cartilages (usually six), where it interdigitates with the diaphragm (5.41). The precise origin of the fibres, whether from the inguinal ligament direct, or from adjacent iliac fascia has been the subject of dispute (McVay and Anson 1940). The muscle ends in an aponeurosis of variable extent, the lower fibres of which curve downwards and medially together with those of the aponeurosis

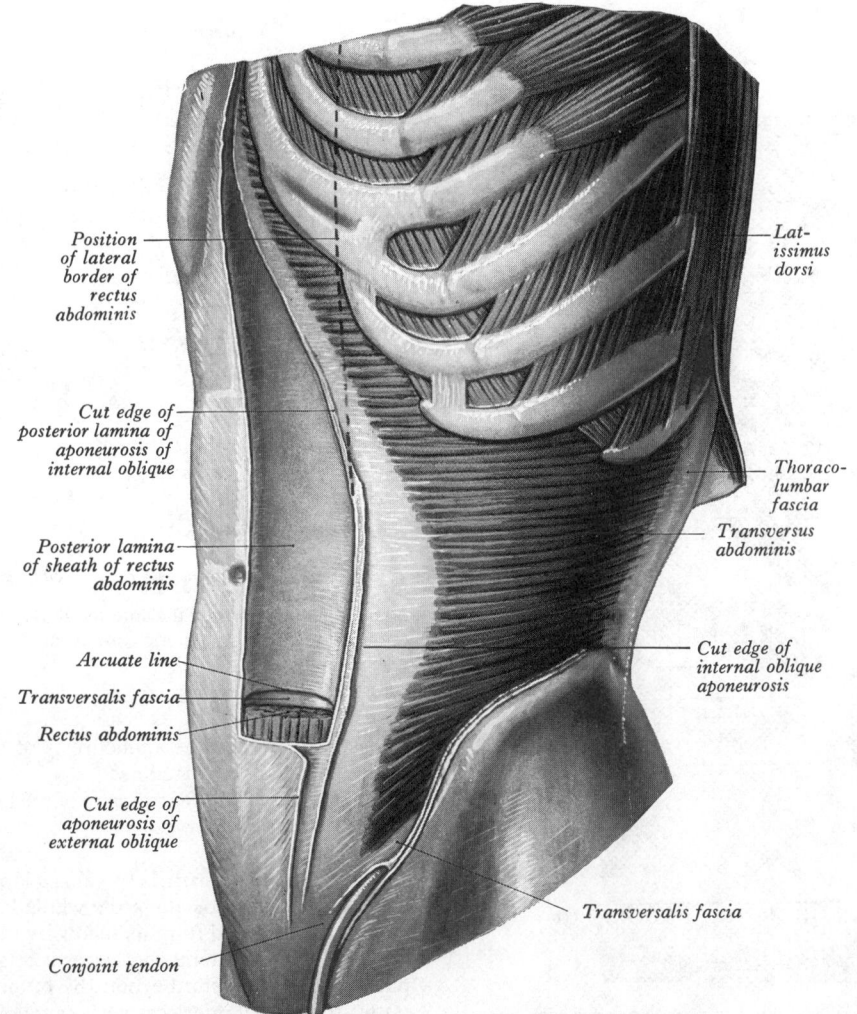

Position of lateral border of rectus abdominis

Cut edge of posterior lamina of aponeurosis of internal oblique

Posterior lamina of sheath of rectus abdominis

Arcuate line

Transversalis fascia

Rectus abdominis

Cut edge of aponeurosis of external oblique

Conjoint tendon

Lat- issimus dorsi

Thoraco- lumbar fascia

Transversus abdominis

Cut edge of internal oblique aponeurosis

Transversalis fascia

5.48   The left transversus abdominis.

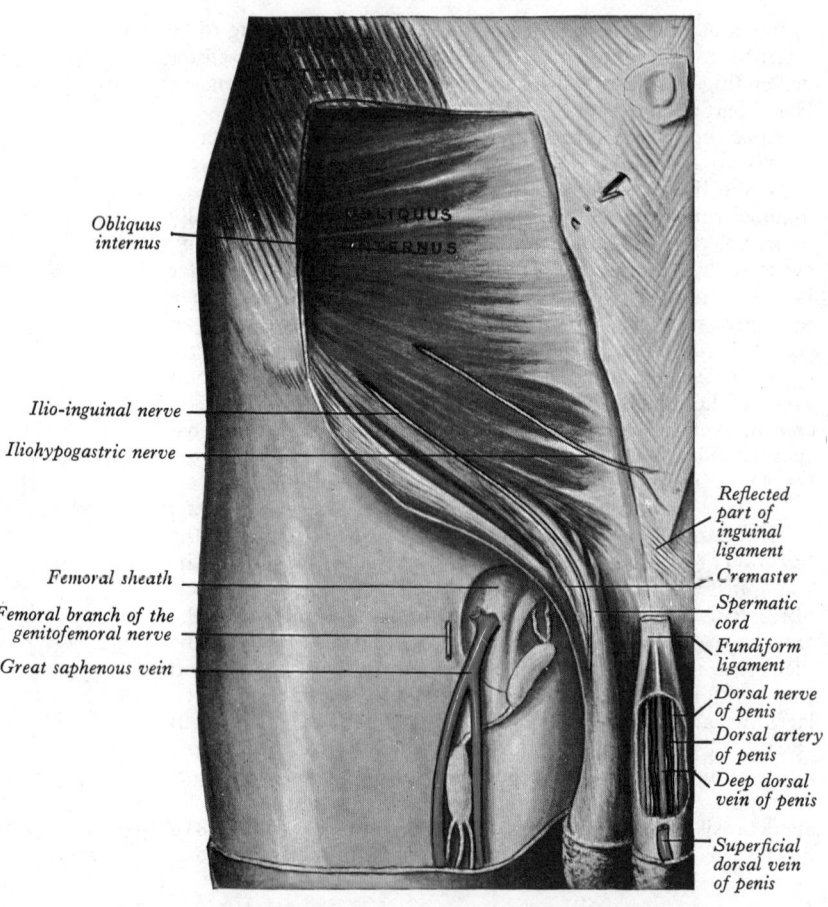

Obliquus internus

Ilio-inguinal nerve

Iliohypogastric nerve

Femoral sheath

Femoral branch of the genitofemoral nerve

Great saphenous vein

Reflected part of inguinal ligament

Cremaster

Spermatic cord

Fundiform ligament

Dorsal nerve of penis

Dorsal artery of penis

Deep dorsal vein of penis

Superficial dorsal vein of penis

5.49 A dissection of the regions shown in 5.45, with part of the obliquus externus removed.

Obliquus internus

Fascia transversalis

Spermatic cord

Falx inguinalis

Reflected part of inguinal ligament

Fundiform ligament

Dorsal nerve of penis

Dorsal artery of penis

Deep dorsal vein of penis

Spermatic cord

Superficial dorsal vein of penis

Ascending branch of deep circumflex iliac artery

Femoral canal

Femoral vein

Femoral artery

Femoral branch of genitofemoral nerve

Great saphenous vein

5.50 A dissection of the regions shown in 5.46, with parts of the external and internal oblique muscles removed.

of the internal oblique to the crest and pecten of the pubis, forming the *falx inguinalis*. The rest of the aponeurosis passes horizontally to the median plane, and blends with the linea alba; its upper three-fourths lie behind the rectus abdominis and blend with the posterior lamina of the aponeurosis of the internal oblique; its lower fourth is in front of the rectus. The upper muscular fibres of the transversus abdominis are continued medially behind the rectus (5.48) and the posterior lamina of the aponeurosis of the internal oblique, sometimes being continuous across the midline with the opposite transversus. Near the xiphoid process they reach to within 2 or 3 cm of the linea alba. The muscular fibres of the transversus run into the aponeurosis along a line which is concave medially (5.48), the aponeurosis being widest opposite the origin of the muscle from the thoracolumbar fascia.

Fusiform defects filled with fascia occur in the lower muscular and the aponeurotic parts of both the internal oblique and the transversus abdominis. The two muscles are sometimes fused, or the transversus may be absent.

The falx inguinalis (conjoint tendon) of the internal oblique and transversus (5.50, 51) is mainly formed by the lower part of the aponeurosis of the transversus, and is inserted into the crest and pecten of the pubis; it descends behind the superficial inguinal ring, thus serving to protect from behind what would otherwise be a weak point in the abdominal wall. The attachment to the pecten pubis is frequently absent. Medially the falx inguinalis is directly continuous with the anterior wall of the sheath of the rectus abdominis. Laterally, it may be continuous with an inconstant ligamentous band, named the *interfoveolar ligament* (5.51), which sometimes connects the lower margin of the transversus to the superior ramus of the pubis; it occasionally contains a few muscular fibres. Muscular fasciculi, attached to the pecten pubis behind the falx inguinalis, may reach the

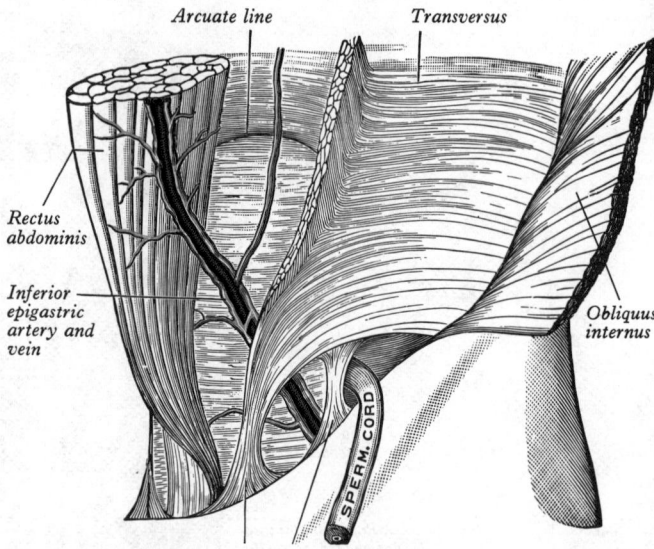

Arcuate line

Transversus

Rectus abdominis

Inferior epigastric artery and vein

Obliquus internus

Falx inguinalis

Interfoveolar ligament

5.51 The lower part of the anterior abdominal wall (left side) showing the relations of the spermatic cord at the deep inguinal ring. (Modified from Braune.)

transversalis fascia, the aponeurosis of the muscle, or even the lateral end of the arcuate line.

Nerve supply. The ventral rami of the lower six thoracic and the first lumbar spinal nerves.

The rectus abdominis (5.52) is a long strap muscle, broader above, which extends along the whole length of the front of the abdomen, separated from its fellow by the linea alba. It arises by two tendons; the lateral and larger is attached to the crest of the pubis and may extend beyond the pubic tubercle to the pecten pubis; the medial interlaces with its fellow and is connected with the ligamentous fibres covering the front of the symphysis pubis.

Some fibres may spring from the lower part of the linea alba. The muscle is attached by three slips of unequal size into the fifth, sixth and seventh costal cartilages; the most lateral fibres are usually attached to the anterior end of the fifth rib but this slip may be absent, although conversely, the muscle may even reach the fourth and third; the most medial fibres are occasionally connected with the costoxiphoid ligaments and the side of the xiphoid process.

The muscle fibres of the rectus are interrupted by three fibrous bands, named *tendinous intersections*; one is usually situated opposite the umbilicus, another opposite the free end of the xiphoid process, and a third about midway between the xiphoid process and the umbilicus. These intersections pass transversely or obliquely across the muscle in a zigzag course; they rarely extend completely through its substance and may pass only halfway across it; they are intimately adherent to the anterior lamina of the sheath of the muscle. Sometimes one or two incomplete intersections are present below the umbilicus. The intersections may be secondary occurrences and opinions vary as to whether they represent the myosepta delineating the myotomes forming the muscle.

The rectus abdominis is enclosed between the aponeuroses of the obliqui and transversus, forming the so-called *rectus sheath* (5.44, 48, 53A). At the lateral margin of the rectus, the aponeurosis of the internal oblique divides into two, one lamina passing anterior to the rectus, to blend with the aponeurosis of the external oblique, the other behind it to blend with the aponeurosis of the transversus. These join again at the medial border of the rectus, to reach and help to form the linea alba. (It should be noted, however, that some authors do not regard the foregoing account as the common disposition—see e.g. Walmsley 1937; McVay and Anson 1940). This arrangement is, nevertheless, widely held to exist from the costal margin to a variable level, usually midway between the umbilicus and symphysis pubis, where the posterior wall of the sheath ends in a curved margin, named the *arcuate line*, its concavity directed downwards or downwards and laterally. As already stated (p. 566) the muscular fibres of the upper part of the transversus abdominis are continued behind the corresponding part of the rectus abdominis to within 2 or 3 cm of the linea alba (5.48, 53A), so that the posterior layer of the sheath is here muscular to a varying degree. Below the level of the arcuate line the aponeuroses of all three muscles pass in front of the rectus; those of the transversus and internal oblique are intimately fused together, but the aponeurosis of the external oblique is bound to them merely by loose connective tissue except in and near the median plane; this part of the rectus is separated from the peritoneum by the transversalis fascia (5.53B). Since the aponeuroses of the internal oblique and

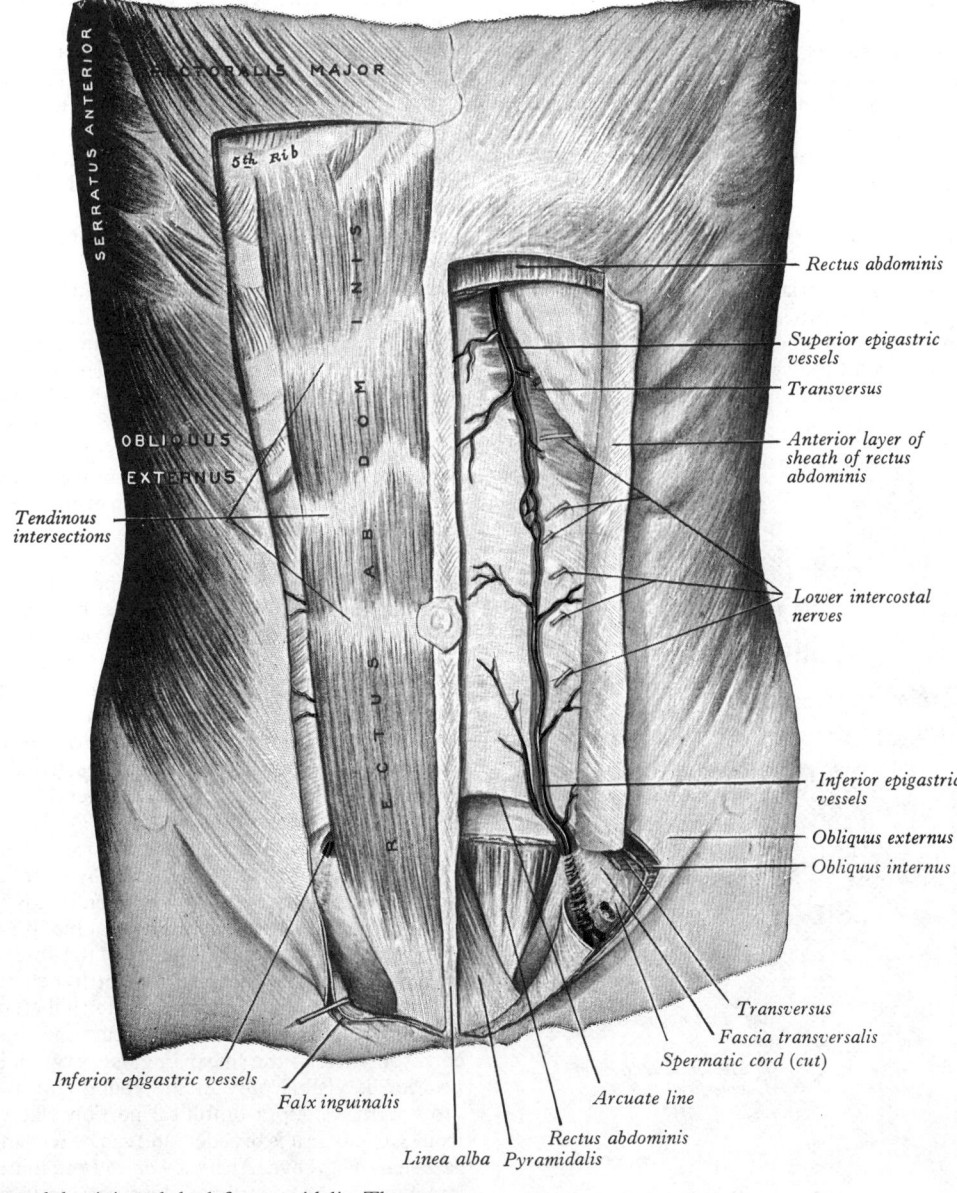

5.52   The right rectus abdominis and the left pyramidalis. The greater part of the left rectus abdominis has been removed to show the superior and inferior epigastric vessels.

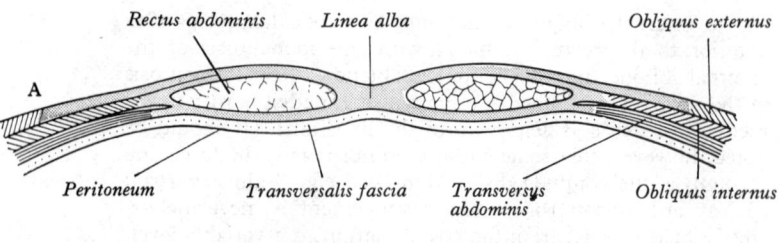

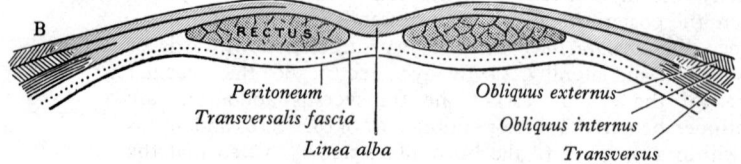

5.53A and B   Transverse sections through the anterior abdominal wall. A Immediately above the umbilicus. B Below the arcuate line. Note the extent to which the external oblique aponeurosis remains as a separate entity, passing medially, ventral to rectus, before blending with the other aponeuroses: these have already fused, lateral to rectus.

transversus reach only as high as the costal margin it follows that above that level the muscle rests directly on the cartilages of the ribs; the front of this part of the rectus is overlapped merely by the aponeurosis of the external oblique.

The medial border of the muscle is closely related to the linea alba: its lateral border is marked on the surface of the anterior abdominal wall by a curved groove, termed the *linea semilunaris*, which extends from the tip of the ninth costal cartilage to the pubic tubercle. It is readily seen in a muscular subject even when the muscle is not actively contracting, but may be completely obscured in the corpulent.

Nerve supply. The rectus abdominis is supplied by the ventral rami of the lower six or seven thoracic spinal nerves.

The pyramidalis (5.52) is a triangular muscle in front of the lower part of the rectus abdominis and within its sheath. It is attached by tendinous fibres to the front of the pubis and the ligamentous fibres in front of the symphysis; the muscle passes

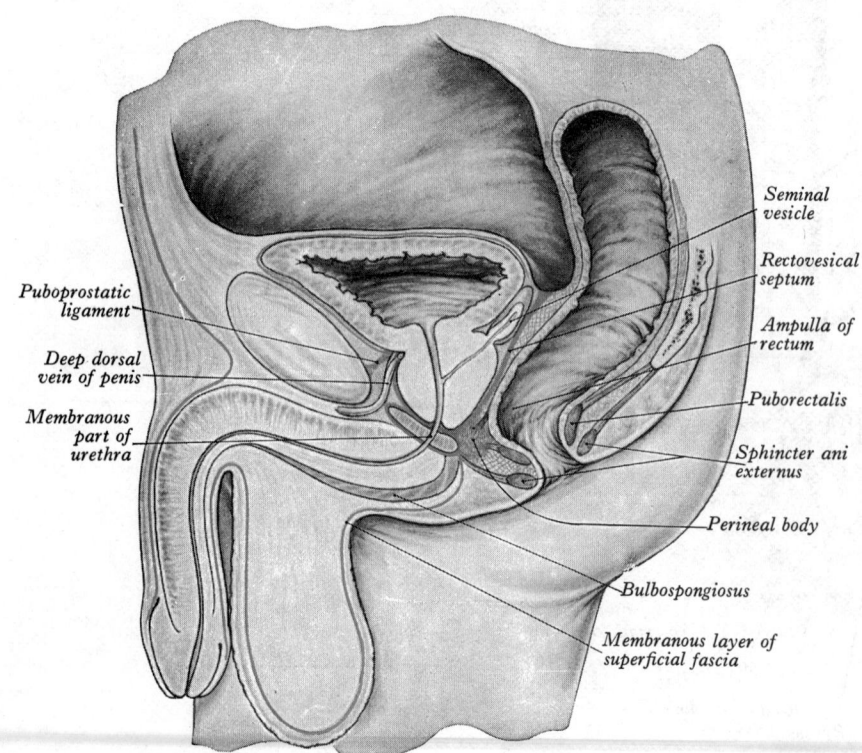

5.54   Median sagittal section through the pelvis, showing the arrangement of the fasciae (*blue*).

Puboprostatic ligament

Deep dorsal vein of penis

Membranous part of urethra

Seminal vesicle

Rectovesical septum

Ampulla of rectum

Puborectalis

Sphincter ani externus

Perineal body

Bulbospongiosus

Membranous layer of superficial fascia

558

upwards, diminishing in size as it ascends, and ends in a pointed extremity which is attached to the linea alba midway between the umbilicus and pubis, but may extend to a higher level. The muscle varies much in size and may be larger on one side than on the other, or may be absent on one or both sides. It is occasionally double. Anson *et al.* (1938) found the muscle absent in 76 of 430 cadavers (17.7 per cent).

Besides the rectus and pyramidalis, the sheath of the rectus contains the superior and inferior epigastric vessels, and the terminal portions of the lower intercostal nerves.

Nerve supply. The pyramidalis is supplied by the subcostal nerve, which is the ventral ramus of the twelfth thoracic spinal nerve.

Actions. The anterolateral group of abdominal muscles provide a firm but elastic wall to retain the abdominal viscera in position and to oppose the action of gravity on them in the erect and sitting postures. This function is principally dependent on the normal contraction of the oblique muscles, especially the internal oblique. However, their most frequent involvement is in respiratory movements.

When the thorax and pelvis are fixed, the active contraction of these muscles exercises a compressing force on the abdominal viscera. In this way they play an important part in expiration, and they assist in expelling faeces from the rectum, the fetus from the uterus, the urine from the bladder and the contents of the stomach in vomiting. This action is mainly due to the oblique muscles. They tense the rectus sheath, whilst the rectus abdominis itself plays a minor role. If the pelvis and vertebral column are fixed, the external oblique muscles aid further in expiration by depressing and compressing the lower part of the thorax.

When the pelvis is fixed, the recti and to a lesser extent the obliqui of the two sides, acting together, bend the trunk forwards and flex the lumbar part of the vertebral column; when the thorax is fixed, they draw the front of the pelvis upwards and have the same effect as before on the vertebral column (Floyd and Silver 1950). If the muscles of only one side act, the trunk is bent towards that side. In addition, the external oblique tends to turn the front of the abdomen towards the opposite side, and the internal oblique turns it to the same side. Electromyographic studies suggest that in most movements of the trunk, whether in the sitting or the standing position, the abdominal musculature is little involved, unless considerable resistance is applied. Extension of the trunk raises activity in these muscles, flexion does not. All activity ceases in the supine position, but the recti in particular at once spring into action even when merely the head is raised. Further flexion brings the obliques into action, but less forcibly. In general the obliques appear to be largely concerned in compressive and forcible twisting movements, the recti in flexion when resistance has to be overcome. In normal standing there is little postural activity in the entire musculature, except for the lower part of the internal oblique, where it is accessible to electromyographic recording through the aponeurosis of the external oblique (*see* inguinal canal, p. 559).

The transversus probably acts only on the abdominal contents and has no appreciable effect on the vertebral column. It has not proved possible to carry out effective electromyographic investigations on this muscle.

The pyramidalis is a tensor of the linea alba, but the advantage of, or the necessity for this action is by no means clear.

The linea alba (5.44, 53) is a tendinous raphe between the xiphoid process and the symphysis pubis. It is between the recti, and is formed by the interlacement of the fibres of the aponeuroses of the obliqui and transversi. A little below its midpoint is a fibrous cicatrix, covered by a puckered adherent area of skin—the *umbilicus*. In its infra-umbilical part the linea alba is narrow, corresponding to the linear interval between the recti; in life its position is visible only in the young and muscular as a slight groove. In its supra-umbilical portion where the recti diverge from each other it is broader, and can be recognized on the surface as a shallow groove. At its lower end the linea alba has a double attachment—its superficial fibres passing in front of the medial heads of the recti to the front of the symphysis pubis, while its deeper fibres form a triangular lamella, attached behind the recti o the posterior surface of the crest of the pubis, and named the *adminiculum lineae albae*. The linea alba is traversed by a few minute vessels; in the fetus the umbilicus transmits the umbilical vessels, urachus and, up to the third month, the vitelline or yolk stalk. The umbilicus is closed a few days after birth, but the vestiges of the vessels and urachus remain attached to its deep surface (i.e. the *ligamentum teres of the liver*—remnant of the fetal left umbilical vein; the *medial umbilical ligaments*, enclosed in peritoneal folds bearing the same names, and being the obliterated umbilical arteries; and the *median umbilical ligament*—the partially obliterated remains of the urachus).

The transversalis fascia is a thin connective tissue stratum between the internal surface of transversus and the extraperitoneal fat. It is part of the general layer of fascia between the peritoneum and the abdominal walls, and is continuous with the iliac and pelvic fasciae. In the inguinal region it is thick and dense, and augmented by the aponeurosis of the transversus, but it becomes thin as it ascends to the diaphragm, and blends with the fascial covering of its inferior surface. Behind, it fuses with the anterior lamina of the thoracolumbar fascia. Below, it has the following attachments: posteriorly, to the whole length of the iliac crest, between the origins of the transversus and iliacus; between the anterior superior iliac spine and the femoral vessels, and is connected to the posterior margin of the inguinal ligament, and is there continuous with the iliac fascia. Medial to the femoral vessels it is thin and is fixed to the pecten pubis, behind the falx inguinalis, with which it is united; it descends in front of the femoral vessels to form the anterior part of the femoral sheath (p. 725). In front of the femoral vessels the transversalis fascia is strengthened by transversely arched fibres which spread laterally towards the anterior superior iliac spine and medially diverge behind the rectus abdominis, while some descend to the pecten pubis behind the falx inguinalis. These arched fibres constitute the *deep crural arch*. The spermatic cord or the round ligament of the uterus pass through the transversalis fascia at the *deep inguinal ring*. This opening is not visible externally since the transversalis fascia is prolonged on these structures as the *internal spermatic fascia*. Around the testis this blends with the areolar tissue on the parietal layer of the tunica vaginalis (p. 1418) and may contain smooth muscle fibres (Barrett 1951). The curved fibres of the deep crural arch thicken the inferomedial part of the rim of the deep inguinal ring.

The deep inguinal ring is situated in the transversalis fascia, midway between the anterior superior iliac spine and the symphysis pubis, and about 1·25 cm above the inguinal ligament. It is oval, the long axis being vertical; it varies in size, and is much larger in the male. It is related above to the arched lower margin of transversus abdominis, and medially to the inferior epigastric vessels and the interfoveolar ligament, when present. The drag on the fascial ring exerted by the internal oblique muscle constitutes a valve-like safety mechanism when intra-abdominal pressure is raised (Lytle 1970).

The inguinal canal contains the spermatic cord or the round ligament of the uterus, and in both sexes the ilio-inguinal nerve. It is oblique and about 4 cm long, slanting downwards and medially, parallel with, and a little above, the inguinal ligament; it extends from the deep to the superficial inguinal ring. It is bounded *in front* throughout by the skin, superficial fascia, and aponeurosis of external oblique, and in its lateral one-third by the muscular fibres of the internal oblique; *behind*, by the reflected inguinal ligament, the falx inguinalis, and the transversalis fascia, which separate it from extraperitoneal connective tissue and peritoneum; *above*, by the arched fibres of the internal oblique and transversus abdominis; *below*, by the union of the transversalis fascia with the inguinal ligament, and at its medial end by the lacunar ligament.

The presence of the canal would appear to weaken the lower part of the anterior abdominal wall, but this is compensated partly by the obliquity of the canal and partly by the arrangement of the structures in its walls. Owing to the oblique direction of the canal the two inguinal rings do not coincide and increases in intra-abdominal pressure exercise their effect not only at the deep inguinal ring but also on the posterior wall of the canal thus approximating it to the anterior. The posterior wall is strengthened by the falx inguinalis and the reflected inguinal ligament posterior to the superficial inguinal ring, and the internal oblique takes part in the formation of the anterior wall, where it lies opposite to the deep inguinal ring (*see* p. 554). This part of the internal oblique, and that of the transversus which also arises from the inguinal ligament (i.e. the parts of those muscles which 'arch' over the oblique canal), are constantly active in standing; any increase in intra-abdominal pressure (as in coughing or straining) accentuates contraction in the internal oblique (and probably the transversus).

The extraperitoneal connective tissue. Between the peritoneum and the inner surface of the general layer of the fascia which lines the interior of the abdominal and pelvic cavities, there is a considerable amount of areolar connective tissue. This extraperitoneal tissue varies in quantity in different situations. It is especially abundant on the posterior wall of the abdomen, and particularly around the kidneys, where it contains much fat. It is scanty on the anterolateral wall, except in the pubic region and above the iliac crest; there is a considerable amount in the pelvis.

It is perhaps worthy of note that such fasciae are no more than the connective tissue usually present between differentiated structures. The extraperitoneal tissue is, of course, in continuity with the epimysium of the muscles of the abdominal wall and through this with the internal connective tissue of these structures.

THE POSTERIOR MUSCLES OF THE ABDOMEN

The psoas major and minor, and the iliacus, with the fasciae covering them, though forming part of the abdominal parietes, are muscles of the lower limb and are described with them. Only the quadratus lumborum will be considered here.

The *fascia* covering the quadratus lumborum is the *anterior layer of the thoracolumbar fascia* (p. 542). It is attached, medially, to the anterior surfaces of the transverse processes of the lumbar vertebrae; below, to the iliolumbar ligament; above, to the apex and lower border of the last rib. Its upper margin, which extends from the transverse process of the first lumbar vertebra to the apex and lower border of the last rib, is the lateral arcuate ligament (p. 543). Laterally, the fascia blends with the fused posterior and middle layers of the thoracolumbar fascia (**5.38**).

The quadratus lumborum (**5.37, 43**) is irregularly quadrilateral, being broader inferiorly. It is attached below by aponeurotic fibres to the iliolumbar ligament and the adjacent portion of the iliac crest for about 5 cm, and above to the medial half of the lower border of the last rib, and by four small tendons to the apices of the transverse processes of the upper four lumbar vertebrae, and sometimes to the transverse process or body of the twelfth thoracic. Occasionally a second layer of this muscle is found in front of the preceding; it passes from the upper borders of the lumbar transverse processes of the lower three or four lumbar vertebrae to the lower margin and the lower part of the anterior surface of the last rib.

Anterior to the quadratus lumborum are the colon, kidney, psoas major et minor, and diaphragm; the subcostal, iliohypogastric, and ilio-inguinal nerves are anterior to the fascia over the muscle, but are bound down to it by the medial continuation of the transversalis fascia.

Nerve supply. The ventral rami of the twelfth thoracic and upper three or four lumbar spinal nerves.

Actions. The quadratus lumborum fixes the last rib, and acts as a muscle of inspiration by helping to steady the origin of the diaphragm. If the pelvis is fixed, it may act upon the vertebral column, flexing it to the same side; and when both muscles act together they probably help to extend the lumbar part of the vertebral column.

V. The Pelvic Muscles and Fasciae

The muscles within the pelvis may be divided into two groups: (1) the piriformis and obturator internus, which are described with the muscles of the lower limb (pp. 601, 602); (2) the levator ani and coccygeus, which, with the corresponding muscles of the

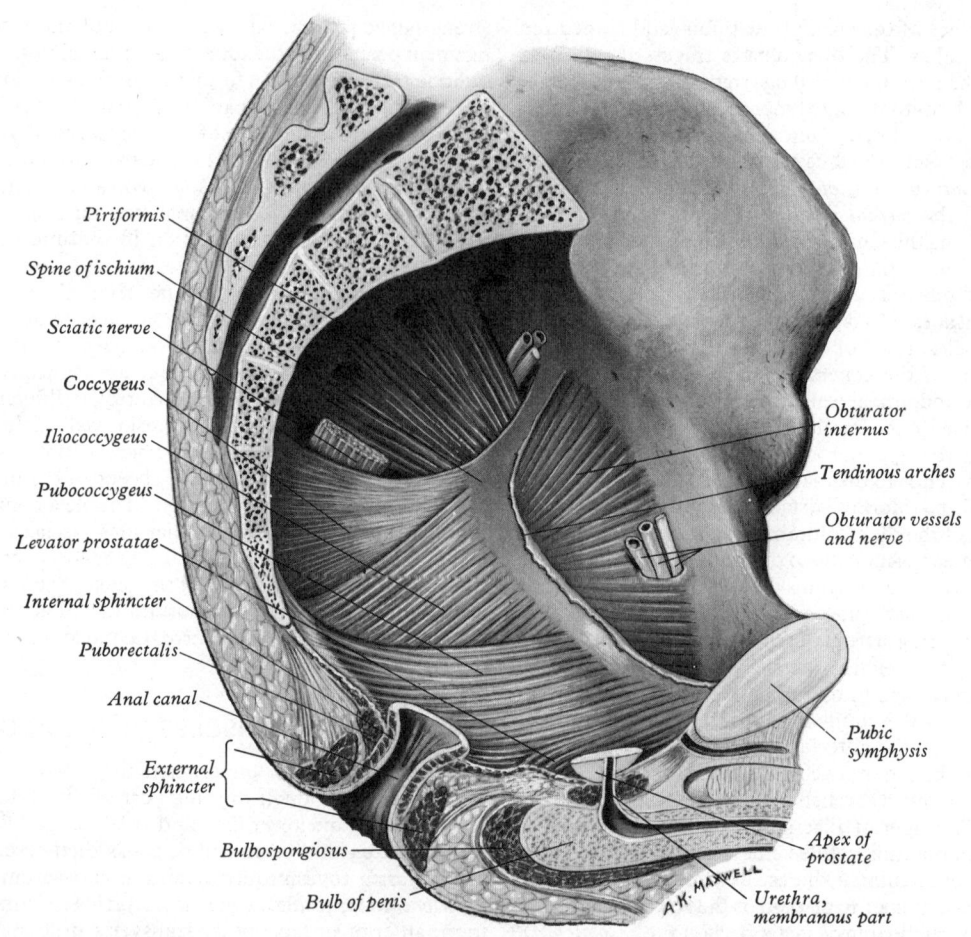

Piriformis

Spine of ischium

Sciatic nerve

Coccygeus

Iliococcygeus

Pubococcygeus

Levator prostatae

Internal sphincter

Puborectalis

Anal canal

External sphincter

Bulbospongiosus

Bulb of penis

Obturator internus

Tendinous arches

Obturator vessels and nerve

Pubic symphysis

Apex of prostate

Urethra, membranous part

A.K. MAXWELL

5.55 Pelvic aspect of the left levator ani and coccygeus. The superior gluteal vessels and nerve have been cut close to the upper border of the piriformis; the anal canal has been divided below the anorectal flexure and the greater part of the prostate has been removed. The constituent parts of the levator ani are shown.

opposite side, form the **pelvic diaphragm**. The fasciae investing the muscles form a connective tissue continuum (**5.54**), which connects with the fascial coverings of the pelvic viscera above and with the fasciae of the perineum below. (For the muscles and fasciae of the **urogenital diaphragm** see pp. 562–564.)

The pelvic fascia. The fascia of the pelvis may be resolved for description into: (1) The *parietal pelvic fascia*, the fascial sheaths of the pelvic muscles; (2) the *visceral pelvic fascia*, the fascial sheaths of the pelvic viscera and of their blood vessels and nerves (*see* individual organs).

The parietal pelvic fascia covering the pelvic surface of obturator internus is well differentiated as the *obturator fascia*. Above, it is connected to the posterior part of the arcuate line of the hip bone, and is there continuous with the iliac fascia. In front of this, as it follows the line of origin of the obturator internus, it gradually separates from the iliac fascia, and the continuity between the two is retained only through the periosteum. It arches below the obturator vessels and nerve, completing the obturator canal, and at the front of the pelvis is attached to the back of the body of the pubis. Above the *tendinous arch of the levator ani* (*vide infra*) the obturator fascia is markedly aponeurotic and is concerned in the attachment of levator ani (p. 561); below that level it is thin and membranous and forms the lateral wall of the ischiorectal fossa (p. 561). Here it blends with the fascial sheath of the internal pudendal vessels. Behind, it is indirectly continuous with the fascia on piriformis.

The *fascia of the piriformis* is very thin and fuses with periosteum on the front of the sacrum around the margins of the anterior sacral foramina. At its sacral attachment it ensheathes the nerves emerging from these foramina; hence the sacral anterior primary rami are frequently described as lying behind the fascia. The internal iliac vessels and their branches, on the other hand,

are in the extraperitoneal tissue in front of the fascia, and branches of these vessels carry sheaths of this tissue into the gluteal region, above and below piriformis.

The *fascia of the pelvic diaphragm* extends over both surfaces of the levator ani. On the inferior surface of the muscle it is thin, and is the so-called, *inferior fascia of the pelvic diaphragm* (anal fascia); it covers the medial wall of the ischiorectal fossa, and above is continuous with the fascia of the pudendal canal and with the obturator fascia along the line of attachment of levator ani; it is continuous below with the fasciae on the sphincter urethrae and the sphincter ani externus. The fascia on the muscle's superior surface is often named the *superior fascia of the pelvic diaphragm*. Laterally it follows the line of attachment of the muscle, and is therefore somewhat variable. In front it is attached to the back of the symphysis pubis about 2 cm above its lower border, and can be traced laterally across the back of the superior ramus of the pubis for a short distance to the obturator fascia. It blends with this along a somewhat irregular line to the spine of the ischium. This irregularity is explained by the fact that while in lower mammals the levator ani arises posteriorly from the pelvic brim, in man its attachment is at a lower level, leaving its aponeurosis as the thickened upperpart of the obturator fascia. In some cases tendinous fibres of attachment extend up towards, and may reach, the pelvic brim. Medially, the superior fascia of the pelvic diaphragm blends with the visceral pelvic fascia. The fascia covering the part of obturator internus above the attachment of levator ani is therefore a composite structure and includes, (*a*) the obturator fascia, (*b*) the fascia of levator ani, and (*c*) the degenerated aponeurosis of the levator. A thickening in the obturator fascia along the line of fusion of these structures is the *tendinous arch of the levator ani*.

To be distinguished from the latter is the *tendinous arch of the pelvic fascia* which is a thickened whitish band in the superior fascia of the pelvic diaphragm along a line from the lower part of the symphysis pubis to the spine of the ischium. It marks the line of attachment of the lateral 'true' ligament of the urinary bladder. Anteriorly the fascia forms two thickened bands, the *puboprostatic (pubovesical) ligaments*, one on each side of the median plane. Both tendinous arches may be more or less well defined.

The levator ani (5.55), a broad, thin muscle, is attached to the inner surface of the side of the true pelvis, and unites with the opposite muscle to form the greater part of the floor of the pelvic cavity. It is attached, in front, to the pelvic surface of the body of the pubis lateral to the symphysis, behind, to the medial surface of the spine of the ischium, and between these two points, to the obturator fascia. Posteriorly this attachment to the obturator fascia corresponds, more or less closely, with the tendinous arch of the levator ani, but in front, the muscle arises from the fascia at a varying distance above the arch, in some cases reaching nearly as high as the canal for the obturator vessels and nerve. The muscle's fibres pass towards the median plane with varying degrees of obliquity. The most anterior sweep backwards and downwards across the side of the prostate to end in the perineal body (p. 562). They constitute the *levator prostatae*, but in the female they cross the sides of the vagina to reach their attachment, and so constitute an additional and important sphincter for it, the *pubovaginalis*. Intermediate fibres pass backwards and downwards across the side of the prostate and then turn medially to become continuous with those of the opposite side and form a muscular sling around the anorectal flexure. This relatively thick part of the muscle is termed the *puborectalis*. Posteriorly its fibres are inextricably mingled with those of the deep part of the external anal sphincter. Other intermediate fibres blend with the longitudinal muscle coat of the rectum and descend within the puborectal sling as the conjoined longitudinal coat of the anal canal (p. 1359). More posterior fibres are attached to the last two coccygeal segments and to the anococcygeal ligament or raphe (p. 1359).

Morphologically, the levator ani may be divided into *pubococcygeus* and *iliococcygeus*. Although this is a useful approximation for convenience of description, it must be added that the homologies of these muscles with arrangements in other mammals are still matters of controversy (Wendell-Smith 1967).

The *pubococcygeus* arises from the back of the pubis and from the anterior part of the obturator fascia, and is directed backwards almost horizontally along the side of the anal canal. It is attached to the front of the coccyx by a tendinous plate which is continuous with the ventral sacrococcygeal ligament. The *iliococcygeus* arises from the ischial spine and from the posterior part of the tendinous arch of the levator ani. Its fibres attach to the sides of the coccyx and to those of the opposite side in a median raphe on the undersurface of the tendinous plate of pubococcygeus, contributing to the anococcygeal ligament. Iliococcygeus is usually thin and may be mostly aponeurotic. An accessory slip at its posterior part is sometimes named *iliosacralis*. In lower mammals both muscles are inserted only into the caudal vertebrae. The iliococcygeus is then responsible for side to side movements of the tail, and the pubococcygeus draws it downwards and forwards between the hind limbs. The gradual disappearance of the tail sets free these muscles to meet the demands for a more complete pelvic floor, partially necessary to adoption of the erect attitude.

The levator and depressor caudae of tailed mammals are represented by rudimentary *ventral* and *dorsal sacrococcygeal* muscles in man. These run from the sacrum to the coccyx anterior and posterior to the sacrococcygeal joint.

Relations. The *superior* or *pelvic surface* of the levator ani is separated by its covering fascia from the bladder, prostate, rectum and peritoneum. Its *inferior* or *perineal surface* forms the medial boundary of the ischiorectal fossa, and is covered by the inferior fascia of the pelvic diaphragm. Its *posterior border* is free and separated from the coccygeus by areolar tissue. The *medial borders* of the two muscles are separated by the *visceral outlet*, an interval through which the urethra, vagina, and the ano-rectum pass from the pelvis.

Nerve supply. The levator ani is supplied by a branch from the fourth sacral spinal nerve and by a branch which arises either from the inferior rectal nerve (p. 1116) or from the perineal division of the pudendal nerve.

Actions. The levatores ani constrict the lower end of the rectum and vagina and probably fix the perineal body. Together with the coccygei they form a muscular diaphragm which supports the pelvic viscera and opposes itself to the downward thrust produced by any increase in the intra-abdominal pressure. (These muscles are, however, not accessible for direct or electromyographic examination, and dogmatic views are to be avoided.)

The coccygeus (5.55) is posterosuperior to, but in the same tissue plane, as the levator ani. It is a triangular sheet of muscular and tendinous fibres, arising by its apex from the pelvic surface of the spine of the ischium and the sacrospinous ligament, and attached at its base to the margin of the coccyx and the side of the fifth segment of the sacrum. The coccygeus is occasionally absent, but more often varies in the proportions of muscle and fibrous tissue in it. It lies on the pelvic aspect of the sacrospinous ligament, and the latter is commonly regarded as a degenerate part of the muscle, or as an aponeurosis of it. The muscle is well developed, and the ligament often absent, in mammals with a mobile tail. Morphologically, the coccygeus (*ischiococcygeus*) is probably to be associated with the iliococcygeal moiety of the levator ani.

Nerve supply. A branch from the fourth and fifth sacral spinal nerves.

Actions. The coccygei may pull forward and support the coccyx, after it has been pressed backwards during defecation or parturition. With the levatores ani and piriformes, they close the posterior part of the pelvic outlet.

Applied Anatomy. Injury to the muscles forming the pelvic floor occurs not infrequently during parturition. Rupture of the perineal body (p. 562), which includes fibres of the pubovaginalis, permits divarication of the levatores and may contribute to uterovaginal prolapse at a later date.

VI. The Perineal Muscles and Fasciae

The perineum overlies the **inferior pelvic aperture** (pelvic outlet). Its deep boundaries are—in front, the pubic arch and the arcuate pubic ligament; behind, the tip of the coccyx; on each side, the inferior ramus of the pubis and the ramus of the ischium, the ischial tuberosity and the sacrotuberous ligament. The space within these boundaries is somewhat trapezoid. On the surface of the body the perineum in the male is limited by the scrotum in front, the buttocks behind, and the medial sides of the thighs laterally. For descriptive purposes a line drawn transversely in front of the ischial tuberosities divides the region into two triangular parts. The posterior contains the termination of the anal canal, and is known as the **anal region** (*anal triangle*); the anterior contains the external urogenital organs, and is termed the **urogenital region** (*urogenital triangle*).

The muscles and fasciae of the perineum may therefore be divided into two groups—anal, and urogenital (male and female), but it should be remembered that the two groups meet in the perineal body and constitute a single morphological unit derived from the *sphincter cloacae*.

THE MUSCULATURE AND FASCIAE OF THE ANAL REGION

The superficial fascia of the region is very thick, areolar in texture, and contains many fat cells in its meshes. On each side a pad of fatty tissue extends deeply into the vertical, cuneiform cleft between the levator ani and obturator internus called the *ischiorectal fossa*.

The deep fascia lines the ischiorectal fossa; it comprises the inferior fascia of the pelvic diaphragm and the part of the obturator fascia below the attachment of levator ani.

The ischiorectal fossa, as noted above, is somewhat wedge-shaped, with its base directed to the surface of the perineum and its thin edge at the line of meeting of the obturator internus and

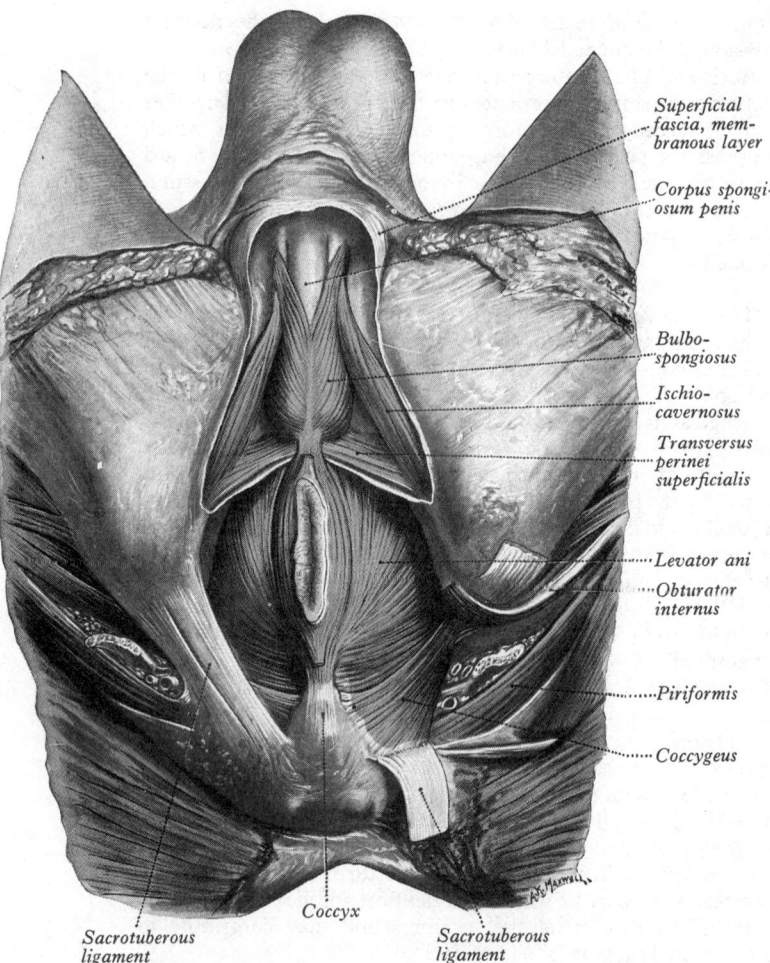

Superficial fascia, membranous layer

Corpus spongiosum penis

Bulbospongiosus

Ischiocavernosus

Transversus perinei superficialis

Levator ani

Obturator internus

Piriformis

Coccygeus

Sacrotuberous ligament

Coccyx

Sacrotuberous ligament

5.56   The muscles of the male perineum. (From Quain's *Anatomy*, 11th edition.)

levator ani, covered by the obturator fascia and inferior fascia of the pelvic diaphragm. It is bounded medially by the sphincter ani externus and the inferior fascia of the pelvic diaphragm and laterally, by the tuberosity of the ischium and the obturator fascia. Posteriorly, the fossa is partly limited by the lower border of the gluteus maximus but extends deep to it as far as the sacrotuberous ligament. Anteriorly, the fossa is partly limited by the posterior region of the urogenital diaphragm, but it is prolonged as a narrow recess above it, reaching almost to the posterior surface of the pubis. This *anterior recess* has the same relations to the obturator internus and levator, but inferiorly it is separated from the surface of the perineum by the muscles and fascia of the urogenital diaphragm. The anterior extent of the recess is variable and limited by fusion between the inferior fascia of the pelvic diaphragm with the superior fascia of the urogenital diaphragm. It may, however, reach the retropubic space (p. 1405).

The internal pudendal vessels and their accompanying nerves are in the lateral wall of the ischiorectal fossa, and are enclosed in a special sheath of fascia forming the *pudendal canal*. This sheath is fused with the lower part of the obturator fascia, extending upwards to blend with the inferior fascia of the pelvic diaphragm and downwards to become continuous with the falciform process of the sacrotuberous ligament (p. 474). It is sometimes termed the *lunate fascia* (**8.**124B), and regarded as forming the roof and lateral wall of the fossa. In front it passes above the urogenital diaphragm, blending with its lateral margin at its attachment to the inner surface of the inferior ramus of the pubis.

The sphincter ani externus (**5.**56, **8.**124, p. 1359) surrounds the lowest part of the anal canal. Below, it is intimately adherent to the skin; above, it overlaps the sphincter ani internus. Details of its attachments, subdivisions, nerve supply and actions are given

on p. 1359. Suffice it here to say that it has deep, superficial, and subcutaneous parts which relate to other muscles of the visceral outlet. The deep part exchanges fibres not only with the puborectal sling above it (p. 561) but also via the perineal body with the muscles of the deep perineal pouch which lie anterior to it in the same curved plane. Similarly, the superficial part exchanges fibres with the superficial urogenital muscles, with which they are coplanar. The superficial part, and to a lesser extent the other parts, gains a substantial attachment to the coccyx and contributes to the anococcygeal ligament. Above this attachment there is a *retrosphinteric space* connecting the ischiorectal fossae of the two sides (Courtney 1949).

The anococcygeal ligament is a layered musculotendinous structure occupying the midline between the anorectum and the coccyx. With the overlying presacral fascia it has been termed the *postanal plate* (Wendell-Smith and Wilson 1977). From above downwards it comprises the tendinous plate of pubococcygeus, the muscular raphe of iliococcygeus and the posterior attachments of the external anal sphincter.

Nerve supply. The perineal branch of the fourth sacral spinal nerve and by twigs from the inferior rectal branch of the pudendal nerve (second and third sacral spinal nerves).

MUSCLES AND FASCIAE OF THE MALE UROGENITAL REGION

The musculature of the urogenital region consists in both sexes (**5.**56, 57) of the bulbospongiosus, ischiocavernosus, transversi perinei superficialis et profundus, and sphincter urethrae, concerned in the functions of urination, copulation, and general support of the pelvic contents. The muscles may be grouped into superficial and deep strata: (1) the *superficial urogenital muscles*, namely, the median bulbospongiosus, and the right and left ischiocavernosus and transversus perinei superficialis; these occupy the *superficial perineal space* which is detailed below. (2) Occupying the *deep perineal space* (also detailed below) are the sphincter urethrae and the right and left transversus perinei profundus; collectively this deep muscular stratum is called **the urogenital diaphragm** (and this accords with the modern terminology applied to the deep fasciae of the region).

The superficial fascia of this region consists of a superficial, fatty, and a deeper, membranous or fibrous layer.

The *fatty layer* is thick, loose, areolar in texture, and contains a variable accumulation of fat cells in its meshes. In front, it is continuous with the dartos muscle of the scrotum, behind, with the subcutaneous areolar tissue surrounding the anus, and, on each side, with the same fascia on the medial sides of the thighs. In the median plane, it is adherent to the skin and to the membranous layer of the superficial fascia.

The *membranous layer* (**5.**54) is thin, aponeurotic in structure, and of considerable strength, perhaps serving to bind down the muscles of the root of the penis. It is continuous, in front, with the dartos muscle, the fascia penis, and the membranous layer of the superficial fascia upon the anterior wall of the abdomen; on each side it is attached to the margins of the rami of the pubis and ischium, lateral to the crus penis and as far back as the tuberosity of the ischium; posteriorly it curves round the transversi perinei superficiales to join the posterior margins of the fasciae of the urogenital diaphragm and the perineal body. Between the membranous layer of superficial fascia and the inferior fascia of the urogenital diaphragm is the **superficial perineal space** or *pouch*.

The perineal body (the official term, centrum tendineum, is unsuitable; the body is not an insubstantial centre, nor is it tendinous) is a pyramidal fibromuscular node in the median plane occupying the angle between the anal canal and the urogenital apparatus with the *rectovesical (rectovaginal) septum* at its apex. Below this, muscles and their fasciae converge and interlace. From above downwards these are the two levatores prostatae (pubovaginales), the deep and some of the superficial perineal muscles. The deep muscles comprise the sphincter urethrae, the two deep transversi perinei, and the deep part of the sphincter ani externus. The superficial muscles comprise the bulbospongiosus, the two superficial transversi perinei, and the superficial part of the sphincter ani externus. The attachments of these muscles to the pubes, ischia, and coccyx, determine the position of the perineal body and thus of the visceral canals in all three dimensions.

The transversus perinei superficialis is a narrow muscular strip which passes more or less transversely across the superficial space in front of the anus. It is very variable and often feebly developed, sometimes being absent. It arises by tendinous fibres from the medial and anterior part of the tuberosity of the ischium, and, running medially, it ends in the perineal body, joining in this situation with the muscle of the opposite side, the superficial part of the sphincter ani externus behind, and bulbospongiosus in front. In some cases, the fibres of the deeper layer of the sphincter ani externus decussate in front of the anus, become more superficial, and are continued into this muscle. Occasionally it gives off fibres which join the bulbospongious of the same side or the sphincter ani externus.

Action. The simultaneous contraction of the two transversi perinei superficiales probably helps to fix the perineal body, but the precise value of this is not entirely clear.

The bulbospongiosus (bulbocavernosus) is in the median line of the perineum, in front of the anus, and consists of two symmetrical parts, united by a median tendinous raphe. It arises from this median raphe and from the perineal body. Its fibres diverge like two halves of a feather; the most posterior form a thin layer, which peters out on the inferior fascia of the urogenital diaphragm; the middle fibres encircle the bulb and the adjacent part of the corpus spongiosum penis, and are attached to a strong aponeurosis on their upper surfaces; the anterior fibres spread out over the side of the corpus cavernosum penis, ending partly in the corpus, anterior to the ischiocavernosus, and partly in a tendinous expansion which covers the dorsal vessels of the penis.

Actions. The bulbospongiosus aids emptying of the urethra, after the bladder has expelled its contents; during the greater part of the act of micturition its fibres are relaxed, and they only come into action at the end of the process, and can be used to arrest it. The middle fibres assist in the erection of the corpus spongiosum penis, probably by compressing the erectile tissue of the bulb. The anterior fibres also contribute to the erection of the penis by compressing the deep dorsal vein of the penis, as their tendinous expansion is inserted into, and is continuous with, the deep fascia covering the deep dorsal vessels of the penis (see also p. 1419). This, and the other perineal muscles have been little investigated by electromyography, but this technique suggests that the bulbospongious (and the sphincter urethrae) contract repeatedly in ejaculation, as simple palpation in part reveals.

The ischiocavernosus covers the crus penis. It is attached by tendinous and fleshy fibres to the inner surface the ischial tuberosity behind the crus penis and to the ramus of the ischium on both sides of the crus. The muscular fibres end in an aponeurosis attached to the sides and under surface of the crus penis.

Action. The ischiocavernosus compresses the crus penis, and so may play a part in maintaining erection of the penis.

Between the above muscles a triangular space exists, bounded medially by bulbospongiosus, laterally by ischiocavernosus, and behind by transversus perinei superficialis; its floor is the inferior fascia of the urogenital diaphragm. The posterior scrotal vessels and nerves, and the perineal branch of the posterior femoral cutaneous nerve traverse the space from behind forwards; the transverse perineal artery courses along its posterior boundary on the transversus perinei superficialis.

The deep stratum of urogenital muscles, together with their fasciae, i.e. transversus perinei profundus and sphincter urethrae, as noted above, constitute the **urogenital diaphragm**. Superficial to the transversus perinei profundus and the sphincter urethrae is the *inferior fascia of the urogenital diaphragm* (also widely termed the *perineal membrane*). This strong layer stretches almost horizontally across the pubic arch. Its base, directed backwards, is connected to the perineal body and continuous with

the superior fascia of the urogenital diaphragm, behind the transversus perinei superficialis, and with the membranous layer of the superficial fascia. Its lateral margins are attached to the inferior ramus of the pubis and the ramus of the ischium, above the crus penis. Its apex, directed forwards, is thickened to form the *transverse perineal ligament*; between this ligament and the arcuate pubic ligament the deep dorsal vein of the penis (or clitoris) enters the pelvis and the dorsal nerve of the penis passes forwards. The diaphragm is perforated by the following: from 2 to 3 cm behind the inferior edge of the symphysis the urethra, the aperture for which is circular and about 6 mm in diameter; the arteries and nerves to the bulb and, close to the urethra, the ducts of the bulbo-urethral glands; the deep arteries of the penis, one on each side close to the pubic arch and about halfway along its attached margin; the dorsal arteries of the penis near its apex. Its base is also perforated by the posterior scrotal vessels and nerves.

Superior (deep) to the transversus perinei profundus and the sphincter urethrae is a less definite layer of fascia, the *superior fascia of the urogenital diaphragm*. It stretches across the pubic arch and is continuous laterally with the obturator fascia. Behind, it blends with the inferior fascia of the diaphragm, the perineal body and the membranous layer of the superficial fascia; above, it is pierced by the urethra and is continuous at this point with the prostatic fascia.

Between the superior and inferior fasciae of the urogenital diaphragm is the **deep perineal space** or *pouch*. In it are the membranous section of the urethra, the two transversi perinei profundi and the sphincter urethrae, the bulbo-urethral glands and the proximal parts of their ducts, the pudendal vessels and dorsal nerves of the penis in the forward continuation of the pudendal canal, the arteries and nerves of the bulb of the penis, and a plexus of veins.

The transversus perinei profundus starts from the inner surface of the ramus of the ischium and runs to the median plane, where it ends in the perineal body, joining in this situation with its fellow of the opposite side, the deep part of the sphincter ani externus behind and sphincter urethrae in front, with any of which it may exchange fibres.

Action. The transversus perinei profundus can be said to help to steady the perineal body. In this way it is likely to contribute to the general supportive function of the region, but clear functional observations are lacking.

The sphincter urethrae surrounds the membranous part of the urethra, as far as the inferior fascia of the urogenital diaphragm. Its *superficial* or *inferior fibres* arise in front from the transverse perineal ligament and from the neighbouring fasciae. They pass backwards on each side of the urethra and converge on the perineal body. Its *deep fibres*, some of which arise from the inner surface of the ramus of the pubis and pass medially, form a continuous circular investment for the membranous urethra.

Actions. The muscles of both sides act together as a sphincter, compressing the membranous region of the urethra, particularly if the bladder contains fluid (Basmajian and Spring 1955). During micturition, like the bulbospongiosus, they are relaxed, and only come into action at the end of the process to eject the last drops of urine; they are also concerned in ejaculation.

Nerve supply. All the muscles of the urogenital region are supplied by the perineal branch of the pudendal nerve (second, third and fourth sacral spinal nerves).

MUSCLES AND FASCIAE OF THE FEMALE UROGENITAL REGION

This group, in the female, includes the same five paired muscles as in the male, with some differences in size and disposition due to the presence of the vagina and female external genitalia (5.57). They are similarly grouped into superficial and deep strata, the latter constituting the *urogenital diaphragm*.

The transversus perinei superficialis in the female is a narrow muscular slip, which differs little from the corresponding muscle in the male (*vide supra*).

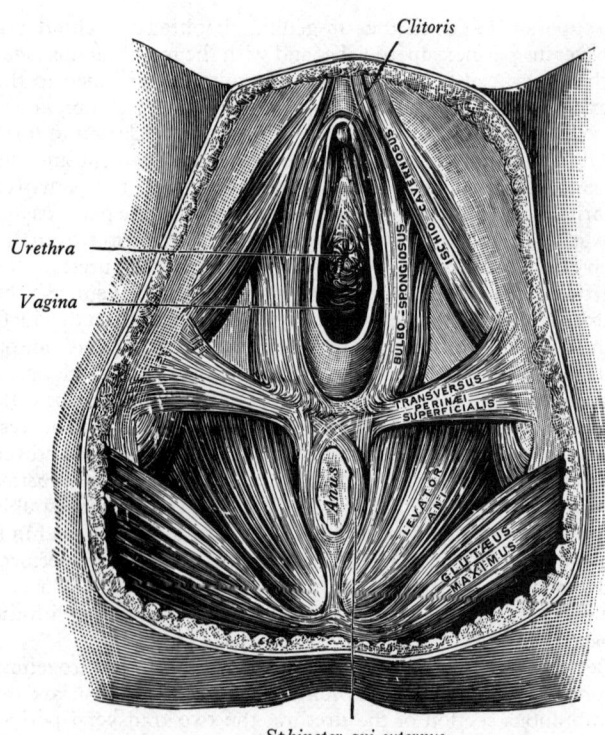

5.57 The muscles of the female perineum.

The bulbospongiosus surrounds the orifice of the vagina. It covers the lateral parts of the vestibular bulbs, and is continuous posteriorly with the perineal body, where it blends with the sphincter ani externus. Its fibres pass forwards on each side of the vagina, to be attached to the corpora cavernosa clitoridis; a fasciculus crosses over the body of the clitoris as if to compress the deep dorsal vein.

Actions. The bulbospongiosus diminishes the orifice of the vagina. The anterior fibres probably contribute to the erection of the clitoris by the compression of its deep dorsal vein.

The ischiocavernosus, smaller than it is in the male, covers the unattached surface of the crus clitoridis. It arises by the tendinous and fleshy fibres from the inner surface of the tuberosity of the ischium, behind the crus clitoridis and from the adjacent surface of the ramus of the ischium. The muscular fibres end in an aponeurosis which is attached to the sides and under surface of the crus clitoridis.

Actions. The ischiocavernosus may compress the crus clitoridis and thus retard the return of blood through the veins, thus serving to erect the clitoris.

The *inferior fascia of the urogenital diaphragm* in the female is less extensive than it is in the male, and is pierced by the aperture of the vagina, with the external coat of which it blends, as well as by the urethra and vessels and nerves corresponding to those already enumerated for the male (p. 562). It covers the following structures: parts of the urethra and the vagina, the transversus perinei profundus and sphincter urethrae muscles, the internal pudendal vessels, the dorsal nerves of the clitoris, the arteries and nerves of the vestibular bulbs, and a plexus of veins.

The transversus perinei profundus extends from the inner surface of the ramus of the ischium and runs across behind the vagina to meet the muscle of the opposite side and other structures in the perineal body. The more anterior fibres sink into the vaginal wall.

Action. The transversus perinei profundus probably aids in fixing the perineal body.

The sphincter urethrae, like the corresponding muscle in the male, consists of inferior and superior fibres.

The *inferior fibres* arise on each side from the transverse perineal ligament and sweep backwards on each side of the urethra. Some of the fibres interlace with those of the opposite side between the urethra and the vagina, while others can be traced into the vaginal wall. The *superior fibres* encircle the lower end of the urethra.

Actions. The muscles of the two sides act as a constrictor of the urethra.

THE FASCIAE AND MUSCLES OF THE UPPER LIMB

The muscles of the upper limb may be considered in the following groups:

  I. Muscles connecting the upper limb with the vertebral column.
  II. Muscles connecting the upper limb with the thoracic wall.
  III. Muscles of the scapula.
  IV. Muscles of the upper arm.
  V. Muscles of the forearm.
  VI. Muscles of the hand.

In the remainder of this section the use of brackets around numerals indicates that it is doubtful whether the spinal nerve in question actually contributes to the motor innervation of a muscle. On the other hand, the use of heavy type indicates that the nerve or nerves indicated are the predominant source of motor supply. (See also p. 1120.)

I. The Muscles connecting the Upper Limb and Vertebral Column

The trapezius, rhomboid major and minor, and levator scapulae, all extend from the cervico-dorsal part of the vertebral column to the pectoral girdle; the latissimus dorsi reaches beyond this to the humerus.

The superficial fascia of the cervico-dorsal region is of considerable thickness and strength, and contains much fat. It is continuous with the general superficial fascia. In the upper part of the neck it forms a thick, tough layer, characterized by the presence of numerous white connective tissue fibres, by which it is firmly attached to the overlying skin.

The deep fascia of the region is a dense fibrous layer where it is attached above to the superior nuchal line of the occipital bone, and in the median plane to the ligamentum nuchae and supraspinous ligament, and to the spines of all the vertebrae below the seventh cervical. Elsewhere it is, for the most part, a thin fibrous membrane. Laterally, in the neck, it is attached to the spine and acromion of the scapula, and is continued downwards over the deltoid to the arm; on the thorax it is continuous with the deep fascia of the axilla and chest, and on the abdomen with the fascia covering the abdominal muscles; below, it is attached to the crest of the ilium.

The trapezius (5.58) is a flat, triangular muscle, extending over the back of the neck and upper thorax. It is attached to the medial one-third of the superior nuchal line of the occipital bone, the external occipital protuberance, the ligamentum nuchae, the seventh cervical and all the thoracic vertebral spinous processes, and the corresponding supraspinous ligaments. The superior fibres proceed downwards, the inferior upwards, and the middle horizontally, all converging laterally upon the shoulder; the superior fibres are there attached to the posterior border of the

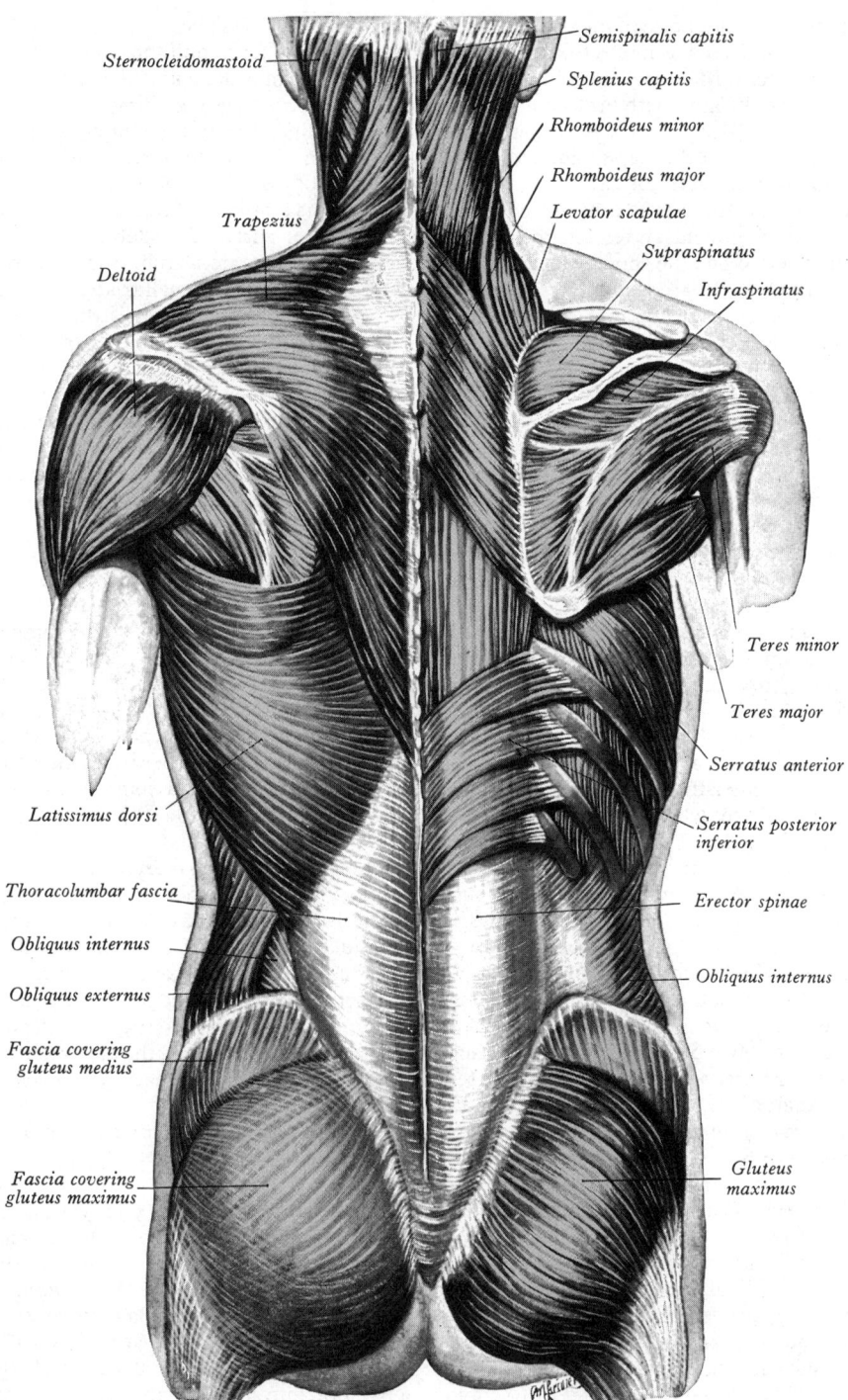

5.58 Superficial muscles of the back of the neck and trunk. On the left only the skin, superficial and deep fasciae have been removed; on the right, the sternocleidomastoid, trapezius, latissimus dorsi, deltoid and obliquus externus abdominis have been dissected away.

lateral third of the clavicle; the middle fibres to the medial margin of the acromion and superior lip of the crest of the spine of the scapula; the inferior fibres pass into an aponeurosis, which glides over the smooth triangular surface at the medial end of the spine of the scapula and is attached to a tubercle at the apex of this surface. The upper part of the midline attachment of the muscle is connected to the occipital bone by a thin fibrous lamina, firmly adherent to the skin; the middle part consists of a broad triangular aponeurosis, which reaches from the sixth cervical to the third thoracic vertebra; the lower part contains short tendinous fibres. the two muscles together resemble a trapezoid; the lateral angles correspond to the shoulders, the superior to the occipital protuberance and the inferior to the spine of the twelfth thoracic vertebra.

The clavicular attachment of this muscle varies in extent; it sometimes reaches as far as the middle of the clavicle, and occasionally blends with the posterior edge of the sternocleidomastoid. The vertebral attachment is also variable and may not descend beyond the eighth thoracic spine. The occipital attachment may be absent. The cervical and dorsal parts of the muscle are occasionally separate.

Nerve supply. The trapezius is supplied by the accessory nerve and by branches, generally believed to be entirely sensory (proprioceptive), from the ventral rami of the third and fourth cervical spinal nerves.

Actions. Acting with other muscles attached to the scapula, the trapezius steadies that bone and controls its position and movements during active use of the upper limb. In this way it is concerned in maintaining the level and poise of the shoulder; but it should be noted that in the 'unloaded' arm electromyographic activity is often low or even absent. Moreover, many people can suspend considerable loads with the arm, with little contribution from the upper part of trapezius (Bearn 1960). Acting with the levator scapulae, its upper fibres elevate the scapula and with it the point of the shoulder; acting with the serratus anterior, it rotates the scapula in a forward direction so that the arm can be raised above the head; acting with the rhomboids, it retracts the scapula and so braces back the shoulder. When the shoulder is fixed, the trapezius may draw the head backwards and laterally. The trapezius, levator scapulae, rhomboids and serratus anterior are a concerted group in producing varieties of scapular rotation (*see* p. 456).

The latissimus dorsi (5.58) is a large, triangular, flat muscle, extended over the lumbar region and the lower half of the thorax, from which its fibres converge to a narrow tendon. It arises by tendinous fibres from the spines of the lower six thoracic vertebrae anterior to trapezius, and from the posterior layer of the thoracolumbar fascia (p. 542), by which it is attached to the spines of the lumbar and sacral vertebrae, to the supraspinous ligaments, and to the posterior part of the crest of the ilium. In addition, it arises by muscular fibres from the posterior part of the outer lip of the crest of the ilium, lateral to the margin of the erector spinae, and by fleshy slips from the three or four lower ribs; the latter interdigitate with the lower slips of the obliquus abdominis externus (5.44). From this extensive attachment the fibres pass laterally with varying degrees of obliquity, the upper ones horizontally, the middle obliquely upwards, and the lower almost vertically upwards, to converge into a thick fasciculus, the upper part of which crosses, and usually receives a few fibres from, the inferior angle of the scapula. The muscle is wrapped round the lower border of the teres major, curving round to its anterior surface. Here it ends in a quadrilateral tendon, about 7 cm long, which passes in front of the tendon of the teres major, and is attached to the bottom of the intertubercular sulcus of the humerus, giving an expansion which blends with the deep fascia of the upper arm; it extends higher on the humerus than the teres major. The lower border of the tendon is united with the tendon of teres major, the surfaces of the two being separated near their attachment by a bursa; another bursa sometimes occurs between the muscle and the inferior angle of the scapula. On account of the way in which the muscle curves round the lower border of the teres major, its constituent fibres are spiralized, consequently the fibres which were *lowest* at the midline attachment are attached highest on the humerus, while the *highest* midline fibres pass into the lower part of its tendon.

The latissimus dorsi, and intimately related teres major, together produce the posterior fold of the axilla. When attempts are made to adduct the arm against resistance, this fold is accentuated and the inferolateral border of the latissimus dorsi can be traced downwards to its attachment to the iliac crest.

A muscular slip, the *axillary arch*, varying from 7 to 10 cm in length, and from 5 to 15 mm in breadth, occasionally springs from the edge of the latissimus dorsi about the middle of the posterior fold of the axilla, and crosses the axilla in front of the axillary vessels and nerves, to join the under surface of the tendon of the pectoralis major, the coracobrachialis, or the fascia over the biceps. It is present in about 7 per cent of subjects and may be multiple. (For a bibliography and views on the nerve supply of such arches consult Kasai and Chiba 1977.) The vertebral and costal attachments of the latissimus dorsi may be reduced or, more rarely, increased.

A fibrous slip usually passes from the lower border of the tendon of the latissimus dorsi, near its insertion, to the long head of the triceps. This is occasionally muscular, and is the homologue of the *dorso-epitrochlearis brachii* of apes.

Nerve supply. The thoracodorsal nerve from the posterior cord of the brachial plexus, C. **6, 7** and 8.

Actions. The latissimus dorsi is active in the movements of adduction, extension and especially medial rotation of the humerus. Further, it acts with the sternocostal part of the pectoralis major and teres major to depress the raised arm against resistance. It is active also in backward swinging of the arm. When the arms are raised above the head and fixed, as in climbing, the same muscles act to pull the trunk upwards and forwards.

The latissimus dorsi is said to take part in all violent expiratory movements, such as coughing or sneezing, and palpation largely confirms this. Electromyography suggests that it aids in deep inspiration. When the fibres of the muscle are stretched, as in elevation of the arm, sufficient pressure is exerted on the inferior angle of the scapula to keep it in contact with the chest wall.

The lower part of the lateral margin of the latissimus dorsi is commonly separated from the posterior free border of the external oblique by a small triangular interval, named the *lumbar triangle*, the base of which is the iliac crest, and the floor the internal oblique (5.58). Another triangle, sometimes termed the *triangle of auscultation*, is situated medial to the scapula. It is bounded above by trapezius, below by latissimus dorsi, and laterally by the medial border of the scapula; part of rhomboideus major is exposed in the triangle. If the scapula is drawn forwards by folding the arms across the chest, and the trunk bent forwards, parts of the sixth and seventh ribs and the interspace between them become subcutaneous in this situation.

The rhomboideus major (5.58) arises by tendinous fibres from the spines of the second, third, fourth and fifth thoracic vertebrae and the supraspinous ligaments. The muscle passes downwards and laterally to be attached to the medial border of the scapula between the triangular surface of the root of the spine and the inferior angle. Usually the muscular fibres end in a tendinous band which is fixed at its ends to the two points mentioned and is joined to the medial border by a thin membrane; occasionally the arch is incomplete, and some of the muscular fibres are then inserted directly into the scapula.

The rhomboideus minor (5.58) arises from the lower part of the ligamentum nuchae and from the spines of the seventh cervical and first thoracic vertebrae; it is attached to the base of the triangular smooth surface at the medial end of the spine of the scapula. It is usually separated from the rhomboideus major by a slight interval, but the edges of the two muscles are occasionally united. The vertebral and scapular attachments of the rhomboids may vary in extent. Along the upper border of the minor a slender strip of muscle may reach the occipital bone (*rhomboideus occipitalis*).

Nerve supply. The nerve to the rhomboids is the dorsal scapular, C. 4, **5**.

The levator scapulae (5.36, 58) is attached by tendinous slips to the transverse processes of the atlas and axis and the posterior tubercles of the transverse processes of the third and fourth cervical vertebrae. It descends diagonally to be attached to the medial border of the scapula between the superior angle and the triangular smooth surface at the medial end of the spine. This muscle is very variable in its vertebral attachments, its degree of separation into slips, and the occurrence of accessory attachments to the mastoid process, occipital bone, first or second rib, scalene, trapezius and serrate muscles.

Nerve supply. The levator scapulae is supplied directly by the third and fourth cervical nerves, and from the fifth cervical through the dorsal scapular nerve.

The blood supply of this muscle is chiefly from the transverse cervical and ascending cervical arteries; its vertebral extremity is, however, supplied by rami of the vertebral artery, an important consideration in transposition of the muscle (Smith *et al.* 1974).

Actions. In association with the other muscles inserted into the scapula, the rhomboids and levator scapulae help to control its position and movements during active use of the upper limb.

Acting with the trapezius, the rhomboids retract the scapula and brace back the shoulder; acting with the levator scapulae and pectoralis minor, they rotate the scapula so as to depress the point of the shoulder.

When the cervical part of the vertebral column is fixed, the levator scapulae acts with the trapezius to elevate the scapula, or to sustain a weight carried on the shoulder. If the shoulder is fixed, the muscle inclines the neck to the same side.

II. The Muscles connecting the Upper Limb and Thoracic Wall

The group includes the serratus anterior and pectoralis minor, which extend from costal attachments to the scapula, the pectoralis major, which acts from the upper thorax to move the humerus, and the subclavius, a much smaller muscle, which is connected only to the first rib and clavicle.

The superficial fascia of the anterior thoracic region is, of course, continuous with that of the neck and upper limb above, and of the abdomen below. It contains the mammary gland and gives off numerous septa which pass between its lobes. From the fascia superficial to the gland, fibrous processes, termed the *mammary suspensory ligaments*, extend to the skin and nipple.

The pectoral fascia is a thin lamina, covering the surface of pectoralis major and sending numerous prolongations between its fasciculi. It is attached in the median plane to the front of the sternum, above, to the clavicle; laterally and below, it is continuous with the fascia of the shoulder, axilla and thorax. It is very thin over the upper part of the pectoralis major, but thicker in the interval between it and latissimus dorsi, which it crosses as the *axillary fascia*; this divides at the lateral margin of the latissimus dorsi into two layers which ensheathe the muscle and are attached behind to the spines of the thoracic vertebrae. As the fascia leaves the lower edge of the pectoralis major to cross the concavity of the axilla, it sends a layer upwards under cover of the muscle; this lamina splits to envelop the pectoralis minor, and at the upper edge of this muscle is continuous with the clavipectoral fascia (p. 568). The hollow of the armpit, seen when the arm is abducted, is produced mainly by the tethering effect of this fascia between the skin and structures of the axillary floor, and hence the lamina is sometimes named the *suspensory ligament of the axilla*. The axillary fascia is pierced by the axillary tail of the mammary gland (p. 1435). At the lower part of the thoracic region the deep fascia is well developed, and is continuous with the fibrous sheath of the rectus abdominis.

The pectoralis major (5.59) is a thick, triangular muscle arising from the anterior surface of the sternal half of the clavicle, from half the breadth of the anterior surface of the sternum, as low down as the attachment of the cartilage of the sixth or seventh rib, from the cartilages of all the true ribs, with the exception, frequently, of the first, or seventh, or both, from the ventral extremity of the sixth rib, and from the aponeurosis of the obliquus externus abdominis. The clavicular fibres are usually separated from the sternal fibres by a slight interval. The muscle is attached by a flat tendon, about 5 cm broad, into the lateral lip of the intertubercular sulcus of the humerus. This tendon has two laminae, anterior and posterior, usually blended together below. The *anterior lamina* is thicker and formed by fibres arising from the manubrium joined, superficially, by clavicular fibres, and, deeply, by the fibres from the margin of the sternum and from the second to the fifth costal cartilages. The clavicular fibres may be prolonged downwards into the tendon of the deltoid. The *posterior lamina* receives the fibres from the sixth and frequently the seventh costal cartilages and ribs, from the front of the sternum and from the aponeurosis of the obliquus abdominis externus. The costal fibres join the lamina directly without twisting. The fibres from the sternum and from the aponeurosis curve round the lower border of the rest of the muscle, the lower turning successively behind those cranial to them so that this part of the muscle is twisted on itself and the abdominal fibres rise highest upon the tendon (Ashley 1952). The posterior lamina of the tendon reaches higher on the humerus than the anterior, and gives off an expansion which covers the intertubercular sulcus and blends with the capsular ligament of the shoulder joint. From the deepest fibres of this lamina, at its insertion, an expansion is given off which lines the intertubercular sulcus, while from the lower border of the tendon a third expansion passes downwards to the fascia of the upper arm.

The number of costal attachments of the muscle varies, as does the degree of separation of its clavicular and costal parts. The muscles may decussate across the sternum. Superficial to the pectoralis a vertical slip, or slips, may pass from the lower costal

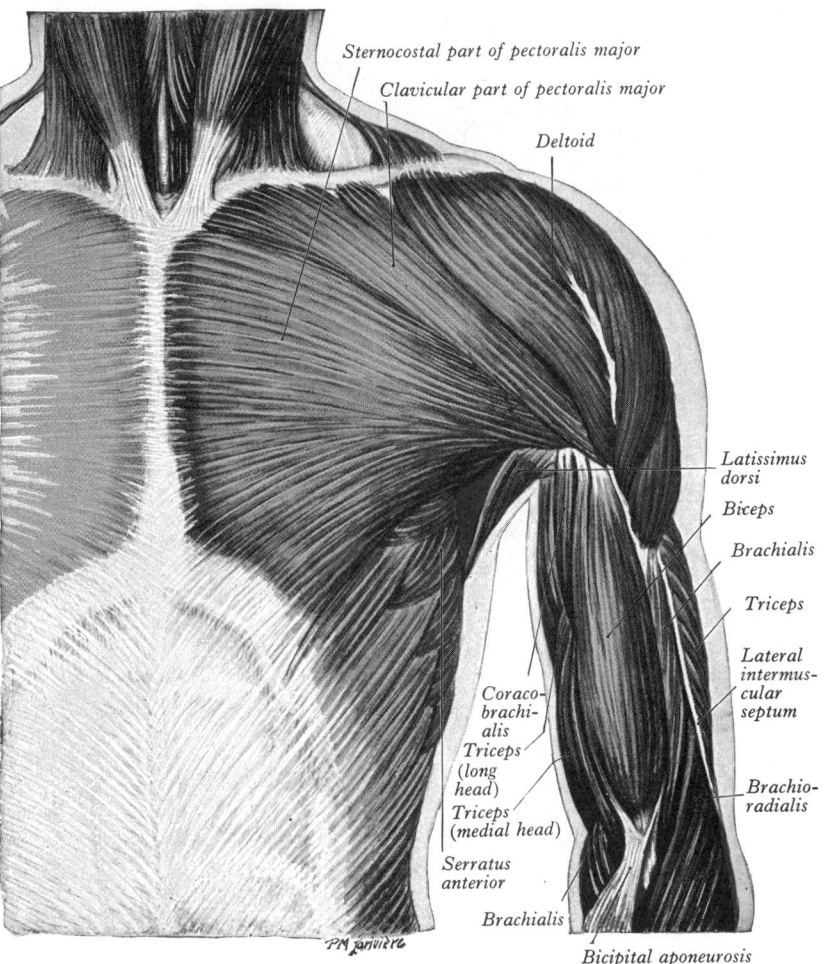

5.59 Superficial muscles of the front of the chest and upper arm. Left side.

Labels on figure:
Sternocostal part of pectoralis major
Clavicular part of pectoralis major
Deltoid
Latissimus dorsi
Biceps
Brachialis
Triceps
Lateral intermuscular septum
Brachioradialis
Coracobrachialis
Triceps (long head)
Triceps (medial head)
Serratus anterior
Brachialis
Bicipital aponeurosis

cartilages and rectus sheath to blend above with the sternocleidomastoid, or to be attached to the upper sternum or costal cartilages. This constitutes the *sternalis* muscle (termed by some the *rectus sternalis*). For other variations, including agenesis, consult Čihák and Popelka (1961) and Kasai and Chiba (1977).

The rounded lower border of the muscle forms the anterior axillary fold and becomes conspicuous in the living subject when the abducted arm is adducted against resistance. When the arm is swung forwards to a right angle, the clavicular head is thrown into contraction, while the sternocostal head is relaxed; but when the flexed arm is swung backwards against resistance, the clavicular head becomes relaxed while the sternocostal head stands out in bold relief.

Relations. *In front*, the pectoralis major is related to skin, superficial fascia, platysma, anterior and middle supraclavicular nerves, mammary gland, and deep fascia; its *posterior surface* is in contact with the sternum, ribs and costal cartilages, clavipectoral fascia, subclavius, pectoralis minor, serratus anterior and intercostal muscles; it forms the superficial stratum of the anterior wall of the axilla, and is hence anterior to the axillary vessels and nerves and the upper parts of the biceps and coracobrachialis. Immediately below the clavicle its *upper border* is separated from the deltoid by the *infraclavicular fossa*, in which are the cephalic vein and deltoid branch of the thoraco-acromial artery. Its *lower border* forms the anterior fold of the axilla; it is separated from the latissimus dorsi by a considerable interval at the medial wall of the axilla, but the two muscles gradually converge towards the lateral wall of the axilla (the floor of the intertubercular sulcus of the humerus between the attachments of these muscles).

Nerve supply. The pectoralis major is supplied by the lateral and medial pectoral nerves; the fibres for the clavicular part of the muscle are derived from C. 5 and **6**, and those for the sternocostal part arise from C. **7, 8,** and T. **1**.

567

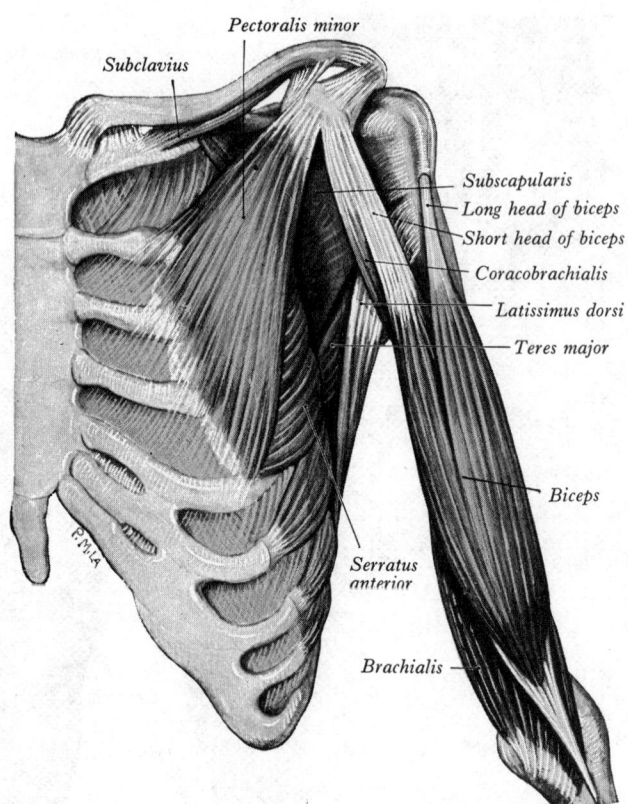

Pectoralis minor

Subclavius

Subscapularis

Long head of biceps

Short head of biceps

Coracobrachialis

Latissimus dorsi

Teres major

Biceps

Serratus anterior

Brachialis

5.60 The deep muscles of the front of the chest and arm. Left side.

The pectoralis minor (5.60), thin and triangular, is posterior or deep to the pectoralis major. From the upper margins and outer surfaces of the third, fourth and fifth ribs (frequently the second, third and fourth ribs) near their cartilages, and from the fasciae covering the external intercostals its fibres pass up and laterally, converging into a flat tendon, attached to the medial border and upper surface of the coracoid process of the scapula. Sometimes part or the whole of the tendon passes over the coracoid process into and even through the coraco-acromial ligament; when the latter occurs the tendon blends with the coracohumeral ligament and is thus attached to the humerus. The slips of this muscle are sometimes fully separated, and they vary in number and level. A slip passes from the first rib to the coracoid (*pectoralis minimus*) on rare occasions. Anson *et al.* (1953, 1963) recorded the costal attachments of 1,000 muscles as follows: 2nd to 5th ribs—337, 3rd to 5th—344, 2nd to 4th—193, 3rd to 4th—67. In this series the muscle was never absent.

Relations. The muscle's *anterior surface* is contiguous with the pectoralis major, the lateral pectoral nerve and the pectoral branches of the thoraco-acromial artery, its *posterior surface*, with the ribs, external intercostals, serratus anterior, the axilla, the axillary vessels and brachial plexus of nerves. Its *upper border* is separated from the clavicle by a narrow triangular interval occupied by the clavipectoral fascia, posterior to which are the axillary vessels and nerves. Parallel with the *lower border* is the lateral thoracic artery; piercing and partly supplying the muscle is the medial pectoral nerve.

Nerve supply. Both the pectoral nerves, C. 6, **7** and 8.

Actions. The pectoralis minor assists the serratus anterior in drawing the scapula forwards round the chest wall. Acting with the levator scapulae and the rhomboids, it rotates the scapula so as to depress the point of the shoulder. Both pectoral muscles are electromyographically 'quiet' in inspiration, unless this is very forcible.

The subclavius (5.60) is a small, triangular muscle, between the clavicle and first rib. It arises from the junction of the first rib and its costal cartilage, in front of the costoclavicular ligament, by a thick tendon which is prolonged along the inferior margin of the muscle (Cave and Brown 1952). It proceeds obliquely upwards and laterally to the groove on the under surface of the middle third of the clavicle where it is attached by muscular fibres. Occasionally, the latter attachment may be to the coracoid process in addition to, or instead of, the clavicle. It may also reach the upper border of the scapula.

Relations. Its *posterior surface* is separated from the first rib by the subclavian vessels and brachial plexus of nerves, its *anterior surface* from the pectoralis major by the clavipectoral fascia.

Nerve supply. A branch of the brachial plexus which derives its fibres from C. **5** and 6 (p. 1095).

Action. The subclavius probably pulls the point of the shoulder downwards and forwards and steadies the clavicle, during movements of the shoulder, by bracing it against the articular disc of the sternoclavicular joint. It is, however, inaccessible to palpation and difficult to investigate by electromyography.

The serratus anterior (5.60) is a muscular sheet which passes backwards around the thorax from an extensive costal attachment to a more limited attachment on the scapula. Its muscular digitations arise from the outer surfaces and superior borders of the upper eight, nine or even ten ribs, and from the fasciae covering the intervening intercostals. The first springs from both the first and second ribs, and from the fascia covering the first intercostal space, the others from a single numerically corresponding rib. The lower four interdigitate with the upper five slips of obliquus externus abdominis. Closely applied to the chest wall, and ventral to the scapula, the muscle reaches its medial border in the following manner. The first digitation is attached to a triangular area on the costal surface of the superior angle; the next two or three spread out to form a thin, triangular sheet, the base of which is directed backwards to nearly the whole length of the costal surface of the medial border. The lower four or five digitations converge like a fan to be attached by muscular and

Actions. The two parts of the pectoralis major are capable of acting in combination or independently of each other. As a whole the muscle takes an active part in the movements of adduction and medial rotation of the humerus, but the activity is only marked if resistance is to be overcome; when the arm has been drawn backwards and laterally, i.e. extended, the pectoralis major draws it forwards and medially. When the arm is swung forwards and medially, the sternocostal fibres take no part in the movement, which is carried out by the clavicular fibres (*portio attollens*) acting with the anterior fibres of deltoid and coracobrachialis. When the opposite movement, usually carried out with the assistance of the force of gravitation, is resisted, the sternocostal part (*portio deprimens*) helps latissimus dorsi and teres major. When the raised arms are fixed, e.g. by gripping a branch of a tree, the same combination of muscles operates to draw the trunk upwards and forwards in climbing. Pectoralis major is also active when inspiration is deep and forcible. Electromyography (de Sousa *et al.* 1969) suggests that only the clavicular part of pectoralis major is active in medial rotation, with or without resistance.

The clavipectoral fascia is a strong fibrous sheet posterior to the clavicular part of pectoralis major. It occupies the interval between the pectoralis minor and subclavius, and covers the axillary vessels and nerves. Traced upwards, it splits to enclose subclavius, and is attached to the clavicle, anterior and posterior to the muscle: the posterior layer fuses with the deep cervical fascia connecting the omohyoid to the clavicle (*see* p. 541) and with the sheath of the axillary vessels. Medially, the clavipectoral fascia blends with that covering the first two intercostal spaces, and is attached also to the first rib medial to subclavius. Laterally, it is thick and dense and attached to the coracoid process, blending with the coracoclavicular ligament. The part between the first rib and coracoid process is often a thickened band, the *costocoracoid ligament*. Below this, the fascia is thin; it splits around pectoralis minor and from its lower border passes downwards to the axillary fascia and laterally to the fascia covering the short head of biceps. The clavipectoral fascia is pierced by the cephalic vein, thoraco-acromial artery and vein, and lateral pectoral nerve.

tendinous fibres to a triangular impression on the costal surface of the inferior angle. Digitations may be lacking, particularly the first and the eighth; the middle part of the muscle is also sometimes absent. The muscle may be partly fused with the levator scapulae, adjacent external intercostals, or the external oblique.

Nerve supply. The long thoracic nerve, C. 5, **6** and **7**, which descends on the external surface of the muscle.

Actions. The serratus anterior, with the pectoralis minor, draws the scapula forwards, and is the chief muscle concerned in all reaching and pushing movements. The upper part of serratus anterior, together with levator scapulae and the upper fibres of trapezius, provide a muscular suspension for the scapula, for which a mild grade of activity is sufficient to support the unloaded arm. The lower and stronger fibres of serratus anterior draw the lower angle of the scapula forwards round the chest wall and assist the trapezius in upward rotation of the bone. They thus play an important part in the movement of raising the arm above the head (p. 455). In the initial stages of abduction the serratus anterior acts with other scapular muscles to steady it so that deltoid is able to exert its action on the humerus and not on the scapula. While deltoid is raising the arm to a right angle with the scapula, the serratus anterior and the trapezius are rotating the scapula, and the arm can be raised even to the vertical as the result of this combination. During such an upward rotation of the scapula, the muscles cooperate in the following manner: the forward pull exerted on the inferior angle of the scapula by the lower digitations of serratus anterior is coupled with an upward and medially directed pull on the lateral end of the clavicle and acromion by the trapezius (upper fibres) and a downward and medial pull on the base of the scapular spine by the lower fibres of trapezius. Conversely, a slow downward rotation of the scapula, assisted by gravity, follows controlled lengthening of the foregoing muscles, whereas a more powerful downward rotation follows balanced contraction of the upper fibres of serratus anterior, the levator scapulae, the rhomboids, the pectoralis minor and the middle part of the trapezius. When weights are carried in front of the body, activity of the serratus anterior increases to prevent backward rotation of the scapula. It has been claimed that electromyography has finally disproved the popular view that the serratus anterior is an accessory inspiratory muscle; but of the references cited Jefferson *et al.* 1960 refer to work on dogs, whereas an electromyographic study on man (Catton and Gray 1951) ignores the effect of fixing the scapula by holding on to, say, a bedrail, railing, and so on as asthmatics and athletes may be seen to do! This problem is clearly not yet resolved.

Applied Anatomy. When the muscle is paralysed, the medial border and especially the lower angle of the scapula stand out prominently, giving a peculiar 'winged' appearance to the back. The patient is unable to raise the arm fully or to carry out pushing movements, and attempts to do so are followed by a further projection of the lower angle of the scapula from the back of the thorax.

III. The Scapular Muscles

The shoulder joint is surrounded by a group of six muscles (deltoid, supra- and infra-spinatus, subscapularis, and teres major and minor) which all extend from scapula to humerus. By their general proximity of attachment to the joint, which they can move in any direction, they also maintain safe contact of the articular surfaces, in both static and dynamic conditions, in a particularly shallow multiaxial joint (p. 430).

The deep fascia over the deltoid sends numerous septa between its fasciculi. In front, it is continuous with the pectoral fascia, behind, where it is thick and strong, with the infraspinous fascia; above, it is attached to the clavicle, acromion, and crest of the spine of the scapula; below, it is continuous with the brachial fascia.

The deltoid (5.59), a thick, triangular muscle, like the Greek letter delta inverted, is attached at its base above to the anterior border and superior surface of the lateral third of the clavicle, the lateral margin and superior surface of the acromion, and the lower edge of the crest of the spine of the scapula, as far as the smooth triangular surface at its medial end. The muscle converges into a short and substantial tendon attached to the deltoid tuberosity on the lateral aspect of the humeral shaft. The anterior and posterior fibres descend obliquely and without interruption into the tendon, whereas the intermediate part of the muscle is multipennate; four intramuscular tendinous septa descend from the acromion to interdigitate with three which ascend from the deltoid tuberosity, and short fibres pass between them, providing a short but powerful pull. The fasciculi of the muscle are comparatively large, giving it a coarsely striped appearance. The deltoid is wrapped around the humeral articulation on all sides except below and medially, and its bulk determines the rounded profile of the shoulder. When contracted its borders are easily seen and palpated. The tendon gives off an expansion into the brachial deep fascia, and it may on rare occasions reach the forearm. Commoner variations are partition into the three parts described above, union with the pectoralis major, or additional slips from trapezius, the infraspinous fascia, or from the axillary border of the scapula.

Relations. Superficially are skin, superficial and deep fasciae, platysma, lateral supraclavicular and upper lateral brachial cutaneous nerves. Deep to the muscle are the coracoid process, coraco-acromial ligament, subacromial bursa, the tendon of insertion of the pectoralis minor, the tendons of origin of the coracobrachialis and both heads of the biceps, the tendon of the pectoralis major, the insertions of the subscapularis, supra-spinatus, infraspinatus, and teres minor, the long and lateral heads of the triceps, the circumflex humeral vessels, the axillary nerve and the surgical neck and upper part of the shaft of the humerus, including both tuberosities. Its *anterior border* is separated above from the pectoralis major by the infra-clavicular fossa, in which the cephalic vein and deltoid branch of the thoraco-acromial artery lie; lower down the two muscles are in contact and their tendons usually unite. Its *posterior border* overlies infraspinatus and triceps.

Nerve supply. The deltoid is supplied by the axillary nerve, C. **5** and 6.

Actions. The muscle is capable of acting in part or as a whole. The anterior fibres cooperate with pectoralis major in drawing the arm forwards, and they are potential medial rotators of the humerus. Conversely, the posterior fibres act with latissimus dorsi and teres major in drawing the arm backwards and they can act as lateral rotators of the humerus. The multipennate, acromial part is the strongest part of the muscle. Aided by the supraspinatus it raises the arm from the side until the inferior part of the capsule of the shoulder joint is put on the stretch, the movement being in the same plane as the body of the scapula. This point is of importance because only thus can scapular rotation have its full effect in raising the arm above the head. When the arm is actively maintained in this position of true abduction (4.5), the acromial fibres are strongly contracted, but the clavicular and posterior fibres are put on the stretch steadying the limb and preventing side-sway. While the deltoid is effecting this movement of the humerus on the scapula the latter is itself rotated (p. 456). In the early stages of abduction the traction exerted by the deltoid is in an upward direction, but the head of the humerus is prevented from gliding upwards on the glenoid cavity by the downward pull of the subscapularis, infraspinatus and teres minor, which come into action synergically with the deltoid. As a result deltoid and supraspinatus, on the one hand, and the subscapularis, infraspinatus and teres minor, on the other, constitute a mechanical couple which ensures the perfect performance of abduction. Electromyography suggests that deltoid contributes little to medial and lateral rotation, but otherwise the muscle is to some degree active in most movements of the shoulder. Deltoid may aid the supraspinatus in resisting downward drag of the loaded arm (*vide infra*). Perhaps the commonest action of the deltoid is arm-swinging in walking. An interesting ergonomic study of the clavicular part of the muscle by Jonsson and Hagberg (1974) demonstrates the importance of these fibres in adjusting the hand to various heights in different manual tasks.

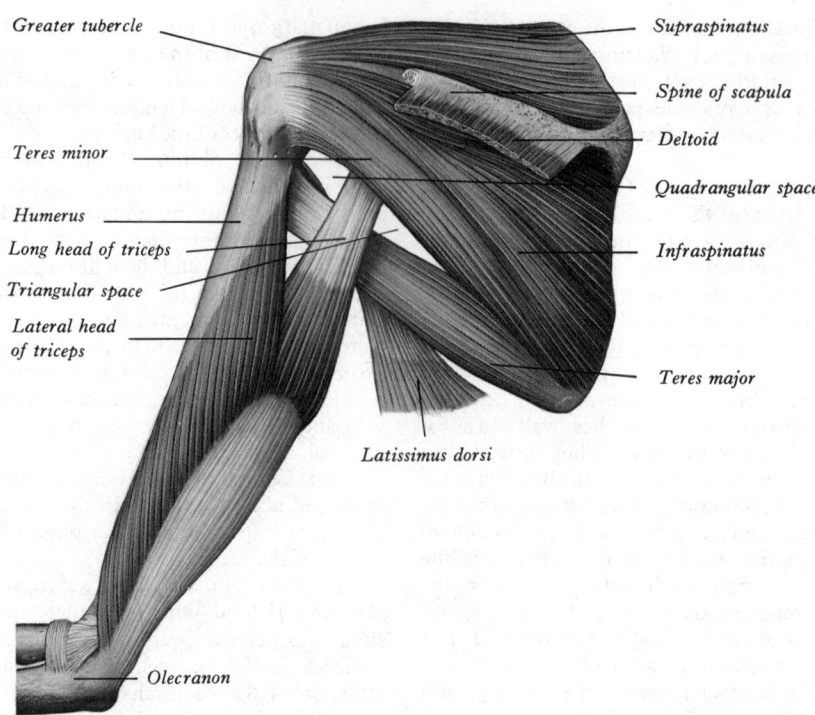

Greater tubercle

Teres minor

Humerus

Long head of triceps

Triangular space

Lateral head
of triceps

Olecranon

Supraspinatus

Spine of scapula

Deltoid

Quadrangular space

Infraspinatus

Teres major

Latissimus dorsi

5.61   The dorsal scapular muscles and triceps. Left side. The spine of the
scapula has been divided near its lateral end and the acromion has been
removed together with a large part of the deltoid.

Applied Anatomy. The muscle atrophies after division of the
axillary nerve, and in this condition dislocation of the shoulder
joint is simulated by flattening of the shoulder and apparent
prominence of the acromion; the distance between the acromion
and the humeral head is increased also, and the tips of the fingers
can be inserted between them.

The subscapular fascia is a thin aponeurosis attached to the
entire circumference of the subscapular fossa, and from its deep
surface some of the fibres of the subscapularis arise.

The subscapularis (5.60) is a large, triangular muscle, which
fills the subscapular fossa and arises from its medial two-thirds,
including the grooved area adjoining the lateral border of the
scapula. Some fibres arise from tendinous laminae which intersect
the muscle and are attached to ridges on the bone, others from
aponeuroses which cover the muscle and separate it from teres
major and the long head of the triceps. The fibres pass laterally,
gradually converging into a tendon attached to the lesser tubercle
of the humerus and the front of the capsule of the shoulder joint.
The tendon is separated from the neck of the scapula by a large
bursa, which communicates with the shoulder joint through an
aperture in its fibrous capsule and is hence really a protrusion of
the joint cavity. Variation in this muscle is unusual; a separate slip
may pass from the upper end of the medial scapular border to the
capsule of the shoulder joint or medial to the intertubercular
sulcus of the humerus.

Relations. The muscle forms a considerable part of the
posterior wall of the axilla. The lower and medial two-thirds of its
anterior surface are apposed to the serratus anterior, and its upper
and lateral third is related to coracobrachialis and biceps, the
axillary vessels and brachial plexus of nerves, and the subscapular
vessels and nerves. Its *posterior surface* is of course attached to the
scapula and anterior to the capsule of the shoulder joint. Its *lower
border* is in contact with the teres major and latissimus dorsi.

Nerve supply. The upper and lower subscapular nerves, C. 5, **6**
and (7).

The supraspinous fascia completes the osseofibrous
compartment in which the supraspinatus arises, and its deep
surface affording attachment to part of the muscle. It is thick

medially, but thinner laterally under the coraco-acromial
ligament.

The supraspinatus (**5**.61) arises from the medial two-thirds
of the supraspinous fossa and from the supraspinous fascia. The
muscle passes under the acromion, and converges into a tendon
which crosses above the shoulder joint to become attached to
the highest of three facets sited on the greater tubercle of the
humerus; the tendon is intimately adherent to the capsule of the
shoulder joint. A slip may pass from it to the tendon of pectoralis
major.

Nerve supply. The suprascapular nerve, C. 4, **5** and 6.

Applied Anatomy. The tendon of supraspinatus is separated
from the coraco-acromial ligament, acromion and deltoid by the
subacromial bursa; when this is inflamed, abduction of the
shoulder joint is painful. The tendon is the most frequently
partially or completely ruptured element of the musculotendinous cuff which surrounds the shoulder joint.

The infraspinous fascia covers the infraspinatus, and is fixed
to the borders of the infraspinous fossa; from its deep surface
some fibres of the muscle arise. It is continuous with the deltoid
fascia along the overlapping posterior border of the deltoid.

The infraspinatus (**5**.61) is a thick triangular muscle
occupying most of the infraspinous fossa; it arises by muscular
fibres from the medial two-thirds of the fossa, and by tendinous
fibres from the ridges on its surface it also arises from the
infraspinous fascia, which covers it and separates it from the teres
major and minor. The fibres converge to a tendon, which glides
over the lateral border of the spine of the scapula, and, passing
across the posterior part of the capsule of the shoulder joint, is
attached to the middle facet on the greater tubercle of the
humerus. The tendon is sometimes separated from the capsule of
the shoulder joint by a bursa, which may communicate with the
joint cavity. The muscle is sometimes fused with teres minor.

Nerve supply. The suprascapular nerve, C. (4), **5** and 6.

The teres minor (**5**.61) is a narrow, elongate muscle, which
arises from the upper two-thirds of a flattened strip on the lateral
part of the dorsal surface of the scapula, immediately adjoining
the lateral border, and from two aponeurotic laminae, one

separating it from infraspinatus, and the other from teres major. It runs obliquely upwards and laterally; its upper fibres end in a tendon attached to the lowest of the three facets on the greater tubercle of the humerus; the lower fibres are inserted directly into the humerus immediately below this impression and just above the origin of the lateral head of the triceps. The tendon passes across, and is united with, the lower part of the posterior surface of the capsule of the shoulder joint. The muscle may be fused with infraspinatus.

Nerve supply. The axillary nerve, C. (4), **5** and 6.

Actions. The subscapularis, supraspinatus, infraspinatus and teres minor together have an essential steadying effect on the head of the humerus, maintaining it in stable apposition to the glenoid cavity and controlling its tendency to excessive sliding (p. 458). (As we have seen—p. 459—swings of the humerus at the glenohumeral joint are accompanied by complex movements between the articular surfaces which contain elements of rolling, sliding and often spinning. In the absence of a closely integrated muscular control, excessive degrees of sliding can easily occur at such a shallow articulation.) During the initial stages of abduction the subscapularis, infraspinatus and teres minor counteract the strong upward component of the pull of the deltoid and thus, aided by the supraspinatus, enable the force component which is *perpendicular* to the humeral shaft to draw the arm away from the body. Infraspinatus and teres minor, in association with the posterior fibres of the deltoid, act as lateral rotators of the humerus, and the subscapularis assists in medial rotation when the arm is by the side. The above four muscles are widely termed the '*rotator cuff*' of the shoulder joint, for obvious reasons. When the loaded, or unloaded, upper limb is hanging by the side, the tendency for downward translation of the humeral head in relation to the glenoid cavity is resisted by supraspinatus rather than deltoid, according to electromyographic data; the lack of tension in deltoid can also be appreciated by palpation alone. On the contrary, anterior and posterior drag on the humerus are strongly resisted by the corresponding parts of deltoid.

The teres major (**5**.61) is a thick, somewhat flattened muscle which arises from the oval area on the dorsal surface of the inferior angle of the scapula, and from the fibrous septa interposed between the muscle and teres minor and infraspinatus; the fibres are directed upwards and laterally and end in a flat tendon, about 5 cm long, which is attached to the medial lip of the inter-tubercular sulcus of the humerus. At its insertion the tendon lies behind the tendon of the latissimus dorsi, from which it is separated by a bursa, the two tendons being, however, united along their lower borders for a short distance. The muscle may be united with the scapular part of latissimus dorsi, and a slip of it may join the long head of triceps or the brachial fascia.

Nerve supply. The lower subscapular nerve, C. **6** and 7.

Actions. The teres major draws the humerus medially and backwards, and rotates it medially. Electromyographic studies, however, evince some disagreement upon the muscle's role during even violent active movements, but its activity during the maintenance of static postures and during arm-swinging appears well established. Note that, despite their common name, the teres major and teres minor have different nerve supplies and belong to different functional groups.

IV. The Muscles of the Upper Arm

This group comprises the coracobrachialis, which acts only on the shoulder joint; the biceps and triceps which cross both the shoulder and elbow joints; and brachialis which alone acts entirely upon the latter joint.

The brachial fascia, or deep fascia of the upper arm, is continuous with that covering the deltoid and pectoralis major; it forms a thin, loose sheath for the muscles of the upper arm, and sends septa between them. It is thin over biceps, but thicker where it covers triceps, and over the epicondyles of the humerus; it is strengthened by fibrous aponeuroses derived from pectoralis major and latissimus dorsi medially, and from deltoid laterally.

On each side it gives off a strong intermuscular septum, which is attached to the corresponding supracondylar ridge and epicon-dyle of the humerus.

The *lateral intermuscular septum* extends from the lower part of the lateral lip of the intertubercular sulcus, along the lateral supracondylar ridge, to the lateral epicondyle; it is blended with the tendon of deltoid, gives attachment to triceps behind, to brachialis, brachioradialis and extensor carpi radialis longus in front, and is perforated at the junction of its upper and middle thirds by the radial nerve and the radial collateral branch of the arteria profunda brachii. The *medial intermuscular septum*, thicker than the preceding, extends from the lower part of the medial lip of the intertubercular sulcus below teres major, along the medial supracondylar ridge to the medial epicondyle; it is blended with the tendon of coracobrachialis, and affords attachment to triceps behind and brachialis in front. It is perforated by the ulnar nerve, the superior ulnar collateral artery, and the posterior branch of the inferior ulnar collateral artery.

At the elbow, the brachial fascia is attached to the epicondyles of the humerus and the olecranon of the ulna, and is continuous with the antebrachial fascia. Just below the middle of the medial side of the upper arm, the basilic vein and some lymphatic vessels pass through it, and at various points, branches of the brachial cutaneous nerves.

The coracobrachialis (**5**.60, 62) arises from the apex of the coracoid process, in common with the tendon of the short head of biceps, and by muscular fibres from the upper 10 cm of this tendon; it is attached to an impression, from 3 to 5 cm in length, midway along the medial border of the shaft of the humerus between the attachments of triceps and brachialis. The muscle forms an inconspicuous rounded ridge on the upper part of the medial side of the arm, and the pulsations of the brachial artery can be felt, and often seen, in the depression behind it. Accessory parts of this muscle may be attached, above, to the lesser tubercle, and below, to the medial epicondyle or medial intermuscular septum.

Relations. It is perforated by the musculocutaneous nerve, and is related, in *front*, to pectoralis major above, and at its humeral attachment with the brachial vessels and median nerve, which cross it; *behind* it are the tendons of the subscapularis, latissimus dorsi, and teres major, the medial head of triceps, the humerus and the anterior circumflex humeral vessels; *medially*, are the third part of the axillary artery, the upper part of the brachial artery, the median and musculocutaneous nerves; *laterally*, the biceps and brachialis.

Nerve supply. The musculocutaneous nerve, C. 5, **6** and 7.

Actions. The coracobrachialis draws the arm forwards and medially, especially from a position of extension. When the arm is raised from the side, it acts with the anterior fibres of the deltoid to prevent side-sway.

The biceps brachii (**5**.60, 62, 63), a large, fusiform muscle in the flexor compartment of the upper arm, has received its name from the fact that it has two heads or attachments proximally. The *short head* arises by a thick flattened tendon from the apex of the coracoid process, in common with coracobrachialis. The *long head* starts within the fibrous capsule of the shoulder joint. It arises by a long narrow tendon from the supraglenoid tubercle at the apex of the glenoid cavity, and is continuous with the glenoidal labrum (p. 458). The tendon of the long head, enclosed in a sheath of the synovial membrane of the shoulder joint, arches over the head of the humerus, emerges from the joint by passing behind the transverse humeral ligament, and then descends in the inter-tubercular sulcus; it is retained in the groove by the transverse humeral ligament and by a fibrous expansion from the tendon of pectoralis major. Each tendon is succeeded by an elongated muscular belly, and the two bellies, although closely applied to each other, can be readily separated until within about 7 cm of the elbow joint. Here they end in a flattened tendon which is attached to the rough posterior area of the radial tuberosity, a bursa being between the tendon and the anterior area of the tuberosity. As the tendon of the muscle approaches the radius it is twisted, its anterior surface becoming lateral and then applied to the

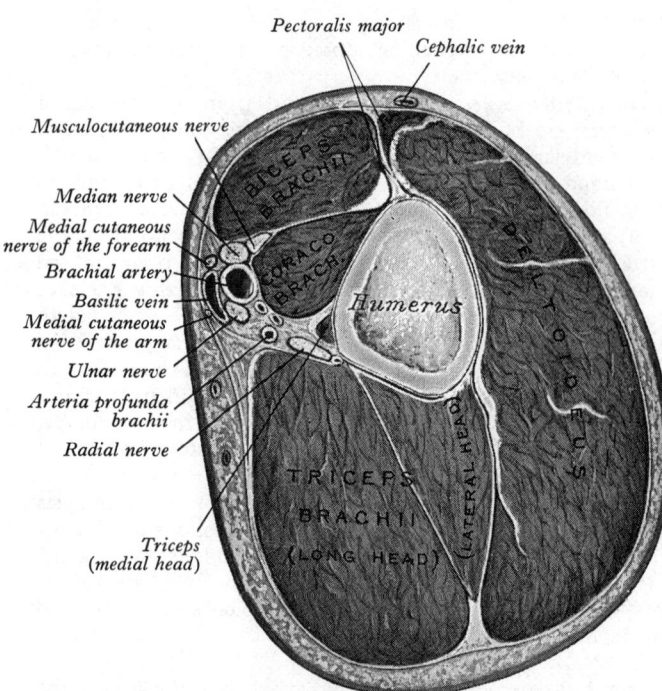

5.62 Transverse section through the right arm at the junction of the proximal and middle thirds of the humerus. Superior (proximal) aspect.

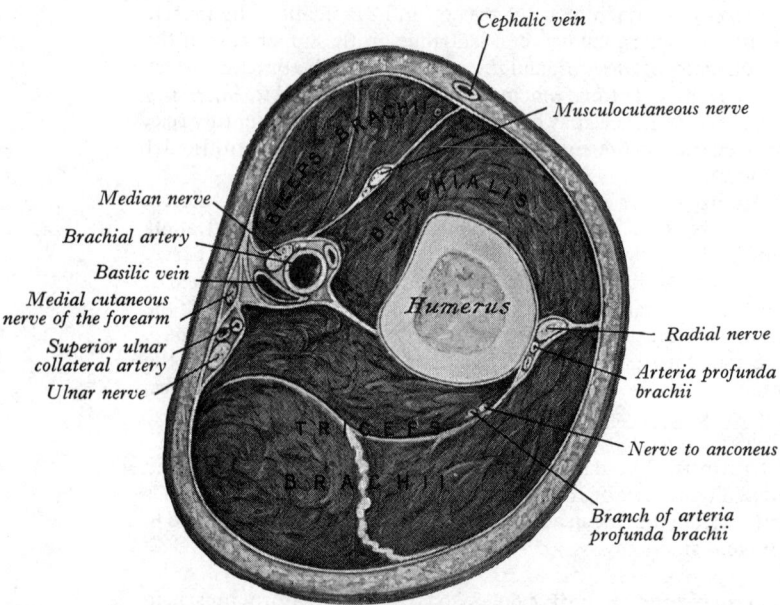

5.63 Transverse section through the right arm, a little below the middle of the shaft of the humerus. Superior (proximal) aspect.

tuberosity of the radius. Opposite the bend of the elbow the tendon produces, from its medial side, a broad band, the *bicipital aponeurosis*, which passes obliquely downwards and medially across the brachial artery, and fuses with the deep fascia covering the origins of the flexor muscles of the forearm (**5.65** and **6.**70). With little force the tendon of insertion can be split distally as far as the radial tuberosity, when it can be verified that its anterior part receives the fibres of the short head, and the posterior those of the long head.

A third head to the biceps often occurs (10 per cent), arising at the upper and medial part of the brachialis, with which it is blended, and attached to the bicipital aponeurosis and medial side of the tendon of the muscle; in most cases this additional slip lies behind the brachial artery. In some instances the third head consists of two slips, which pass down, one in front of, the other behind the artery. Less often, a further part may spring from the

lateral aspect of the humerus, or the intertubercular sulcus. Other additional 'heads' may occur.

Relations. The biceps is overlapped above by pectoralis major and deltoid; in the rest of its extent it is superficial, being covered only by the fasciae and skin, and it forms a conspicuous elevation on the front of the arm. Its long head passes through the shoulder joint, and its short head is anterior to the joint and to the upper part of the humerus; below, it is anterior to brachialis, the musculocutaneous nerve, and supinator. Its *medial border* is contiguous with coracobrachialis, and overlaps the brachial vessels and median nerve; its *lateral border* is related to the deltoid and brachioradialis.

Nerve supply. The musculocutaneous nerve, C. 5 and **6**, which gives a branch to both bellies.

Actions. The biceps is a powerful supinator, employed when the movement is rapid or carried out against resistance; it also flexes the elbow joint, most effectively with the forearm in supination, and is also to a slight extent a flexor of the shoulder joint. Through the bicipital aponeurosis, it is also attached to the posterior border of the ulna, the distal end of which is drawn medially in supination (p. 462). The long head helps to prevent the head of the humerus from gliding upwards during contraction of the deltoid. When the elbow is flexed against resistance, the tendon of insertion of the biceps can be grasped between the finger and thumb and the bicipital aponeurosis can be traced down from its medial edge to the deep fascia covering the origins of the flexor muscles of the forearm.

It should be noted that many habitual movements involve *increasing tension* of contraction, with *increasing length*. For example, if the hand holding an object is lowered to a table under the influence of gravity, the extension of the elbow joint involved is controlled by such activities in the joint *flexors*. As the loaded hand descends and the elbow extends, the load is carried further and further from the fulcrum of movement at the joint and the effective load increases.

Applied Anatomy. The long tendon of the biceps is sometimes dislocated from its groove on the humerus. When this occurs, the arm is fixed in a position of abduction, but the head of the humerus can be felt in its proper position. The tendon can generally be replaced by flexing the forearm on the arm and rotating the limb.

The brachialis (**5.**60, 63, 64) arises from the lower half of the front of the humerus, commencing above at the insertion of deltoid, which it embraces by two pointed processes, and extending below to within 2·5 cm of the margin of the articular surface. It also arises from the intermuscular septa, but more extensively from the medial than from the lateral; it is separated from the lower part of the lateral intermuscular septum by brachioradialis and extensor carpi radialis longus. Its fibres converge to a thick and broad tendon which is attached to the tuberosity of the ulna and the rough impression on the anterior surface of the coronoid process. The brachialis may be divided into two or more parts. It may be fused with the brachioradialis, pronator teres, or biceps. It may send a tendinous slip to the radius or bicipital aponeurosis.

Relations. Biceps, the brachial vessels, musculocutaneous and median nerves are *anterior* to it; *posteriorly* are the humerus and capsule of the elbow joint, *medially*, pronator teres, and also the medial intermuscular septum, which separates it from triceps and the ulnar nerve; *laterally*, are the radial nerve, radial recurrent and radial collateral arteries, the brachioradialis and extensor carpi radialis longus.

Nerve supply. The musculocutaneous nerve, C. 5 and **6**, and the radial nerve (C. 7) to a small lateral part of the muscle (Ip and Chang 1968).

Action. The brachialis is a flexor of the elbow joint with the forearm either prone or supine, with or without resistance to the movement.

The triceps (**5.**61, 64), in the extensor compartment of the upper arm, is of large size, and arises by three heads (long, lateral and medial), hence its name.

The *long head* arises by a flattened tendon from the infraglenoid

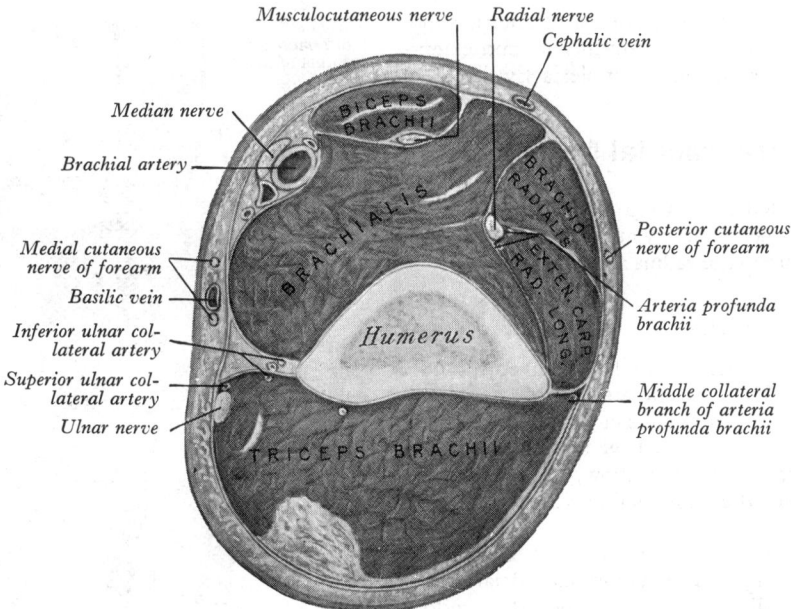

5.64 Transverse section through the right arm, a little proximal to the medial epicondyle of the humerus. Superior (proximal) aspect.

tubercle of the scapula, blended above with the fibrous capsule of the shoulder joint; the muscular fibres pass downwards medial to the lateral head and superficial to the medial head, and join them in a common tendon of insertion.

The *lateral head* arises by a flattened tendon from a narrow ridge on the posterior surface of the shaft of the humerus, from the lateral border of the humerus and the lateral intermuscular septum. The bony origin extends upwards with varying obliquity from the lateral border of the humerus above the radial groove, behind the posterior limb of the deltoid tuberosity, to the surgical neck medial to the insertion of teres minor. These fibres converge towards the common tendon.

The *medial head,* which is overlapped posteriorly by the lateral and the long heads, arises from the posterior surface of the shaft of the humerus, below the radial groove; it is narrow and pointed above, and extends from the insertion of teres major to within 2·5 cm of the trochlea of the humerus; it also arises from the medial border of the bone and from the back of the whole length of the medial and lower part of the lateral intermuscular septa. Some of the muscle fibres are directed downwards to the olecranon, while others converge to the common tendon.

The tendon of triceps begins about the middle of the muscle. It consists of two aponeurotic laminae, one of which is superficial in the lower half of the muscle; the other is in the substance of the muscle. After receiving the muscular fibres, the two layers unite above the elbow, and are attached, for the most part, posteriorly on the upper surface of the olecranon; on the lateral side a band of fibres is continued downwards, over the anconeus, to blend with the antebrachial fascia.

The long head descends between teres minor and teres major, dividing the triangular interval between these two muscles and the humerus into two smaller parts, one triangular, the other quadrangular (**5**.61). The triangular space contains the circumflex scapular vessels; it is bounded by teres minor above, teres major below, and the long head of triceps laterally. The quadrangular space transmits the posterior circumflex humeral vessels and the axillary nerve; it is bounded by the subscapularis, teres minor and capsule of the shoulder joint above, the teres major below, the long head of triceps medially, and the humerus laterally.

The lateral head of triceps forms an elevation, parallel with and medial to the posterior border of the deltoid; it stands out prominently when the elbow is extended actively. The mass which lies to its medial side, and disappears under cover of deltoid, is produced by the long head.

The articularis cubiti (subanconeus) is the name given to a few fibres which spring from the deep surface of the lower part of the triceps, and blend with the posterior part of the fibrous capsule of the elbow joint.

Nerve supply. The radial nerve, C. 6, **7** and **8** with separate branches supplying the three heads (p. 1101).

Actions. The triceps is the great muscle of extension of the forearm on the arm at the elbow joint. The medial head is active in all forms of extension. The actions of the lateral and long heads, however, are minimal, except when the forearm is acting against resistance (Basmajian 1967). Examples of the latter are during forward thrusting or pushing, or when the body weight is partly supported on the hands, with semiflexed elbow joints. When the flexed arm is extended at the shoulder joint, the long head of the muscle may assist in drawing the humerus backwards and in adducting it to the thorax. The long head supports the lower part of the capsule of the shoulder joint, when the arm is raised from the side. The articularis cubiti draws up the posterior part of capsule of the elbow joint during extension of the forearm.

During forceful supination of the semiflexed forearm, involving active contraction of both supinator and biceps brachii, the triceps contracts synergically to fix the position of semiflexion at the elbow joint.

The human triceps brachii has been the subject of an interesting comparison of its geometric and dynamic parameters by Cnockaert and Pertuzon (1974).

V. The Antebrachial Muscles

The antebrachial fascia (the deep fascia of the forearm), continuous above with the brachial fascia, is a dense general sheath for the muscles in this region; it is attached to the olecranon and posterior border of the ulna, and from its deep surface many intermuscular septa pass inwards for muscle attachment, some reaching bone. Muscles also arise from its internal aspect especially in the upper forearm, and it also ensheathes the different muscles. Transverse septa occur in the anterior and posterior regions of the forearm, separating deep from superficial muscles. It is much thicker posteriorly and in the lower part of the forearm. It is strengthened above by tendinous fibres derived from the biceps in front and triceps behind. Two localized thickenings, the *flexor* (p. 583) and *extensor retinacula* (p. 583), occur in it near the wrist. These retain the digital tendons in position and so increase their efficiency. Apertures exist in the fascia for the passage of vessels and nerves; one of these, of large size and anterior to the elbow, transmits a venous communication between superficial and deep veins.

The antebrachial or forearm muscles consist of anterior and posterior groups which are morphologically flexors and extensors, though some are modified for more complex activities.

1. The Anterior Antebrachial Muscles

These muscles can be divided into two groups, superficial and deep, the superficial flexors being largely attached proximally to the humerus, the deep group to the radius and ulna.

SUPERFICIAL FLEXOR GROUP

The muscles of this group take origin from the medial epicondyle of the humerus by a common tendon, and are the pronator teres, the palmaris longus, flexores carpi radialis et ulnaris, and flexor digitorum superficialis (5.65, 66). They have additional attachments to the antebrachial fascia near the elbow, and to the septa which pass from this between the individual muscles.

The pronator teres (5.65, 66) has humeral and ulnar attachments. The *humeral head*, larger and more superficial, arises immediately above the medial epicondyle, from the tendon common to the origin of the flexor muscles, from the intermuscular septum between it and flexor carpi radialis and from the antebrachial fascia. The much smaller *ulnar head* arises from the medial side of the coronoid process of the ulna below the attachment of flexor digitorum superficialis, and joins the humeral head at an acute angle. The median nerve usually (83 per cent of cases, *see* p. 1098) enters the forearm between the parts of the muscle, and is separated from the ulnar artery by the ulnar head. The muscle passes obliquely, across the forearm and ends in a flat tendon, attached to a rough area midway along the lateral surface of the radial shaft. The lateral border of the muscle is the medial boundary of the triangular hollow in front of the elbow joint termed the *cubital fossa*. The coronoid attachment of the muscle may be absent. Additional slips may be derived from a supracondylar process of the humerus, if present, from biceps, or brachialis, or the medial intermuscular septum.

Nerve supply. The median nerve, C. 6 and 7.

Actions. The pronator teres rotates the radius upon the ulna, turning the palm of the hand backwards, i.e. it pronates the forearm and hand; it is also a weak flexor of the elbow joint. It acts with pronator quadratus which is always active during any act of pronation. Pronator teres reinforces the quadratus only during rapid or forcible pronation, but its activity is always less than that of quadratus (Basmajian and Travill 1961).

The flexor carpi radialis (5.65, 66, 69) is medial to pronator teres and arises from the medial epicondyle by the common flexor tendon, from the antebrachial fascia, and from the intermuscular septa between it and the adjacent muscles. Its belly is fusiform and, rather more than halfway down the forearm, ends in a long tendon, which passes through a canal in the lateral part of the flexor retinaculum and occupies a groove on the trapezium which is lined by a synovial sheath to the palmar surface of the base of the second metacarpal bone, sending a slip to the base of the third metacarpal bone. These attachments are hidden by the oblique head of the adductor pollicis (5.80). In the lower part of the forearm the radial artery is between the tendon of this muscle and that of brachioradialis.

The muscle may be absent. Proximally it may have additional slips from the biceps tendon, bicipital aponeurosis, coronoid process or radius. Distally it may be additionally attached to the flexor retinaculum, trapezium, or fourth metacarpal.

Nerve supply. The median nerve, C. 6 and 7.

Actions. Acting with the flexor carpi ulnaris, and sometimes the flexor digitorum superficialis, the flexor carpi radialis flexes the wrist; acting with the radial extensors of the wrist, it helps to abduct the hand.

The palmaris longus (5.65, 66, 79) is slender, fusiform and medial to flexor carpi radialis. It arises from the medial epicondyle of the humerus by the common tendon, from intermuscular septa between it and adjacent muscles, and from the antebrachial fascia.

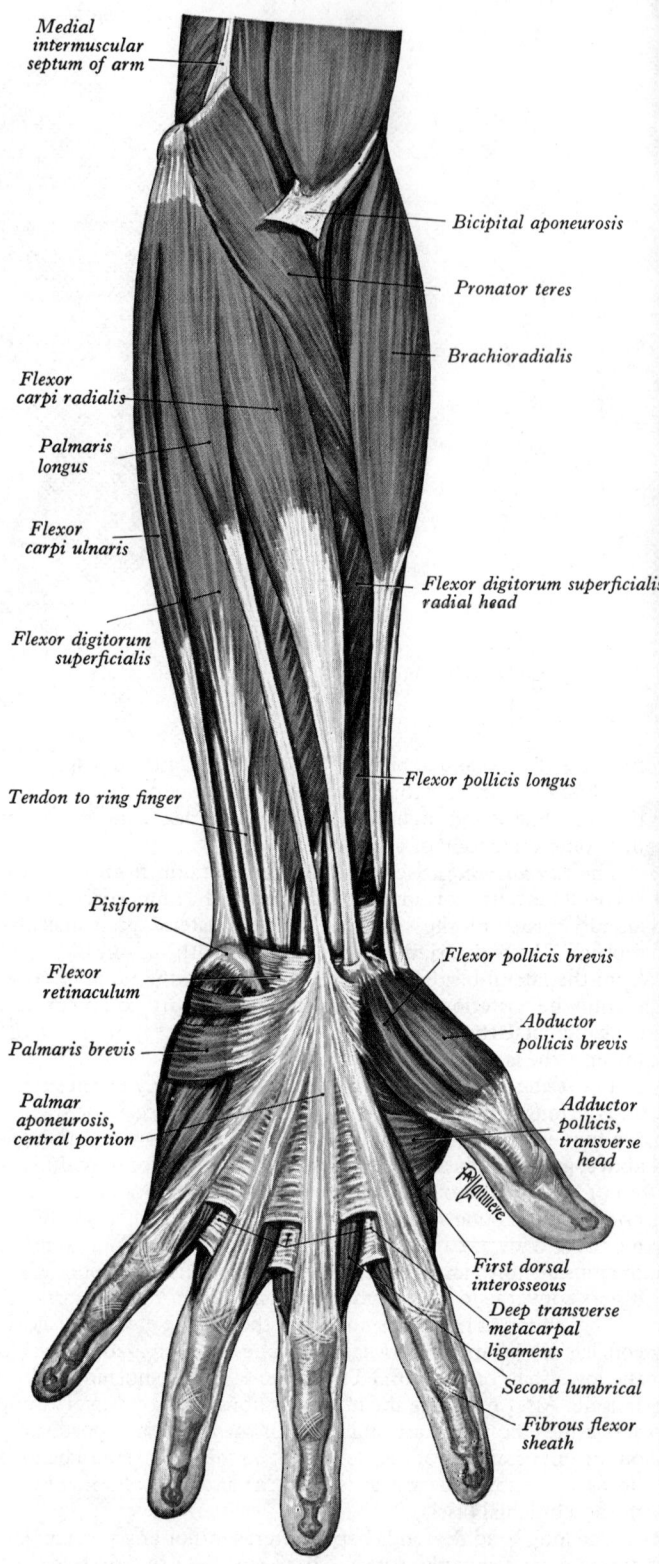

5.65 The superficial flexor muscles of the left forearm, the palmar aponeurosis, and the digital fibrous flexor sheaths.

It ends in a long slender tendon, which passes in front of the flexor retinaculum to be attached to the anterior surface of its distal half and the central part of the palmar aponeurosis, frequently sending a tendinous slip to the thenar muscles. Just proximal to the wrist the median nerve lies deep to the tendon, and projects a little beyond its lateral edge. This muscle is often absent on one or both sides, and is much subject to variation (Machado and DiDio 1967). It may have a proximal tendon, or be reduced to a tendinous strand. It may be digastric or reduplicated. It may end in the antebrachial fascia, tendon of flexor carpi ulnaris, the

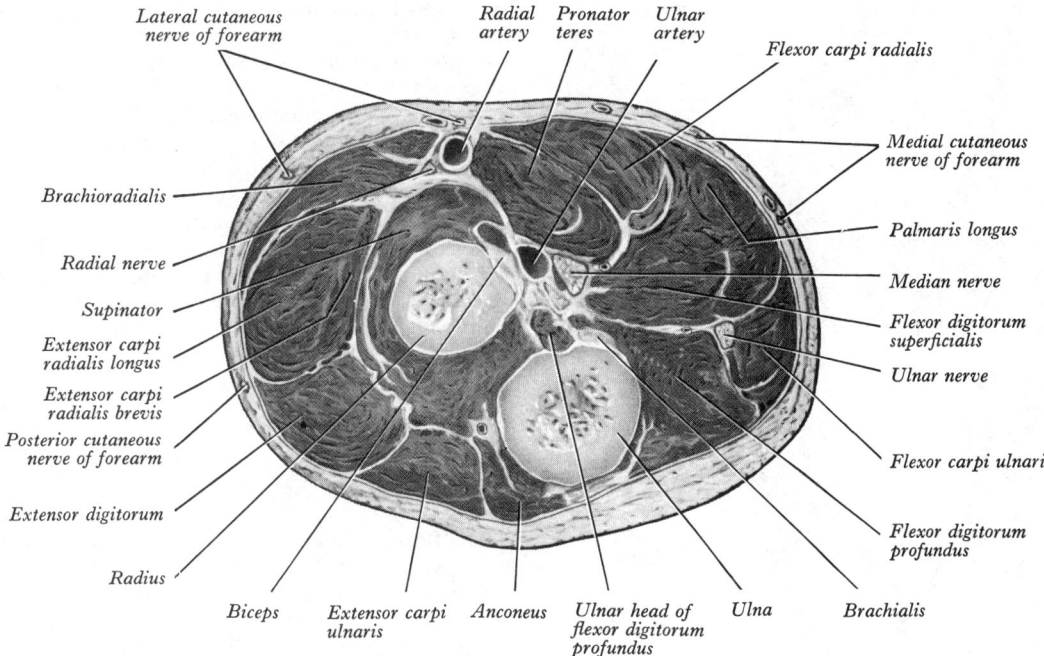

Lateral cutaneous nerve of forearm — Radial artery — Pronator teres — Ulnar artery — Flexor carpi radialis

Brachioradialis — Medial cutaneous nerve of forearm

Radial nerve — Palmaris longus

Supinator — Median nerve

Extensor carpi radialis longus — Flexor digitorum superficialis

Extensor carpi radialis brevis — Ulnar nerve

Posterior cutaneous nerve of forearm

Extensor digitorum — Flexor carpi ulnaris

Radius — Flexor digitorum profundus

Biceps — Extensor carpi ulnaris — Anconeus — Ulnar head of flexor digitorum profundus — Ulna — Brachialis

.66   Transverse section through the forearm at the level of the radial tuberosity.

pisiform, scaphoid, etc. Reimann *et al.* (1944) found the muscle absent in 281 of 2205 specimens (12·7 per cent); the muscle displayed accessory slips in 15 instances and was double in 4 forearms.

Nerve supply. The median nerve, C. 7 and 8.

Actions. The palmaris longus flexes the wrist, and may act as a tensor of the palmar fascia.

The flexor carpi ulnaris (5.65, 66, 69) is the most medial of the superficial forearm flexors. It arises by two heads, humeral and ulnar, connected by a tendinous arch, beneath which the ulnar nerve passes distally and the posterior ulnar recurrent artery proximally. The *humeral head* is small and arises from the medial epicondyle of the humerus by the common tendon, the *ulnar head* from the medial margin of the olecranon, and the upper two-thirds of the posterior border of the ulna by an aponeurosis common to it and extensor carpi ulnaris and flexor digitorum profundus, and from the intermuscular septum between it and the flexor digitorum superficialis. A tendon is formed along the anterolateral border of the muscle in its distal half and is attached to the pisiform bone, whence it is prolonged to the hamate and fifth metacarpal bones by the so-called pisohamate and pisometacarpal ligaments (4.50); it is also attached by a few fibres to the flexor retinaculum. A coronoid slip is sometimes present, and distally a more substantial attachment to the flexor retinaculum, or to the fourth or fifth metacarpal bones may be observed. The ulnar vessels and nerve lie on the lateral side of its tendon of insertion.

Nerve supply. The ulnar nerve, C. 7 and **8**.

Actions. With flexor carpi radialis, palmaris longus when present, and flexor digitorum superficialis, the flexor carpi ulnaris flexes the wrist; with the extensor carpi ulnaris, it adducts the hand. Both the flexor and extensor carpi ulnaris act as synergists to prevent abduction of the hand when the thumb is actively extended at its carpometacarpal joint, as can easily be confirmed by palpation. Flexor carpi ulnaris also fixes the pisiform bone during abduction and flexion of the little finger.

When the wrist is flexed against resistance, two tendons are prominent; the lateral is the flexor carpi radialis and medial is the palmaris longus, approximately in the middle line of the limb. Flexor carpi ulnaris can also be identified proximal to the pisiform bone.

The flexor digitorum superficialis (sublimis) (5.65, 66, 69) is deep to the preceding muscles; it is the largest of the muscles of the superficial flexor group, and arises by two heads, humero-ulnar and radial. The *humero-ulnar head* arises from the medial epicondyle of the humerus by the common tendon, from the anterior band of the ulnar collateral ligament of the elbow joint, from the intermuscular septa between it and the preceding muscles, and from the medial side of the coronoid process, above the ulnar origin of pronator teres. The *radial head* which is a thin sheet of muscle, arises from the anterior border of the radius, extending from the radial tuberosity to the insertion of pronator teres. The median nerve and the ulnar artery pass downwards through the gap which intervenes between these two heads. The muscle usually separates into two strata of muscular fibres, superficial and deep; the superficial, joined laterally by the radial head, divides into two tendons for the middle and ring fingers; the deep stratum gives off a muscular slip to join the part of the superficial stratum associated with the tendon of the ring finger, and then ends in two tendons for the index and little fingers. As the four tendons pass behind the flexor retinaculum, they are arranged in pairs, the superficial pair going to the middle and ring, the deep pair to the index and little fingers. The tendons diverge in the palm, and opposite the bases of the proximal phalanges each divides, to allow the passage of a tendon of the flexor digitorum profundus; the surfaces of the two slips so formed become reversed and they then reunite, partially decussate, and form a grooved channel for the profundus tendon. Finally the superficialis tendon divides again and ends on the sides of the shaft of the middle phalanx. The dimensions of these tendons have been described in detail by Shrewsbury and Kuczynski (1974), who have also estimated the tendon excursions during activity, and the bowing ranges of the tendons. They have, in addition, attempted to correlate these data with an analysis of their fibrous architecture, in particular the significance of the *scissural bands* and *chiasmata*, during digital flexor activity.

The flexor digitorum superficialis is subject to several variations. The radial head may be absent. The slip from the deep stratum may provide most or all of the muscular fibres acting on the index. The part of the muscle for the fifth digit may be absent and replaced by a separate slip arising from the ulna, flexor retinaculum, or palmar fascia. The number and formation of the tendons may also vary. Dylevský (1968) has advanced developmental

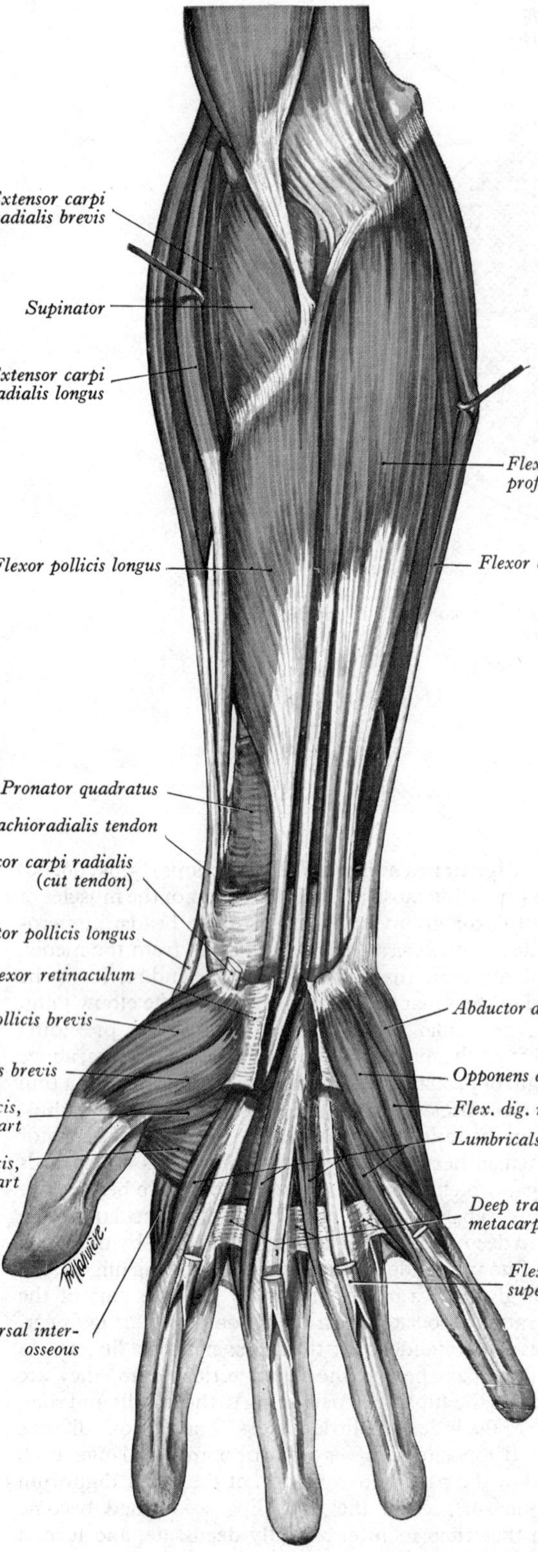

Extensor carpi
radialis brevis

Supinator

Extensor carpi
radialis longus

Flexor pollicis longus

Flexor digitorum
profundus

Flexor carpi ulnaris

Pronator quadratus

Brachioradialis tendon

Flexor carpi radialis
(cut tendon)

Abductor pollicis longus

Flexor retinaculum

Abductor pollicis brevis

Flexor pollicis brevis

Adductor pollicis,
oblique part

Adductor pollicis,
transverse part

First dorsal inter-
osseous

Abductor digiti minimi

Opponens digiti minimi

Flex. dig. min. brevis

Lumbricals

Deep transverse
metacarpal ligaments

Flexor digitorum
superficialis tendon

5.67 The deep flexor muscles of the right forearm.

tissue) and flexor pollicis longus. *At the wrist*, the tendons of flexor digitorum superficialis pass behind the flexor retinaculum, lying in front of the tendons of flexor digitorum profundus in a common synovial sheath (5.76); flexor pollicis longus tendon and the median nerve lie to their lateral side. *In the hand*, the tendons lie behind the palmar aponeurosis, superficial palmar arch and digital branches of the median and ulnar nerves, but in front of the tendons of flexor digitorum profundus and the lumbricals.

Nerve supply. The median nerve, C. 7 and **8** and T. **1**.

Actions. The flexor digitorum superficialis flexes first the middle and then the proximal phalanges. It is also a flexor of the wrist. It is particularly involved in rapid, forceful flexion of the digits in grasping movements, being electromyographically silent in gentle unresisted flexion.

DEEP FLEXOR GROUP

The group consists of the flexor digitorum profundus, flexor pollicis longus, and the pronator quadratus.

The flexor digitorum profundus (5.66, 67, 69) arises deep to the superficial flexors from the upper three-fourths of the anterior and medial surfaces of the ulna, embracing the insertion of brachialis above, and extending to within a short distance of pronator quadratus below. It also arises from a depression on the medial side of the coronoid process, and from the upper three-fourths of the posterior border of the ulna by an aponeurosis, in common with flexor and extensor carpi ulnaris; it also springs from the anterior surface of the ulnar half of the interosseous membrane. The muscle ends in four tendons, which run posterior to the tendons of flexor digitorum superficialis (and hence posterior to the flexor retinaculum). The part of the muscle acting on the index finger is usually distinct throughout, but the tendons for the middle, ring and little fingers are connected together by areolar tissue and tendinous slips, as far as the palm. Anterior to the proximal phalanges the tendons pass through openings in the tendons of flexor digitorum superficialis (p. 575), to their attachments on the palmar surfaces of the bases of the distal phalanges. The tendons of the profundus undergo changes in shape with rearrangement of their constituent fibre bundles (Wilkinson 1953; Martin 1958) as they pass between the diverging halves of the superficialis tendons.

Flexor digitorum profundus forms most of the muscular elevation which can be seen and felt on the dorsum of the forearm medial to the subcutaneous posterior border of the ulna.

The muscle may be joined by accessory slips from the radius (acting on the index), and from the flexor superficialis, flexor pollicis longus, the medial epicondyle, or the coronoid process.

Nerve supply. The medial part of the muscle is supplied by the ulnar and the lateral part by the anterior interosseous branch of the median nerve, C. **8** and T. **1**.

Actions. The flexor digitorum profundus flexes the distal phalanges, after the flexor digitorum superficialis has bent the middle phalanges; it can also assist in flexing the wrist. When all the fingers are flexed, the digital extensors relax, but when individual digits are flexed, activity is sustained in the antagonist extensors (Person and Roschina 1958). More complex movements involve the synergic action of the interossei and lumbricals (p. 589). Electromyographic observations (Long 1964) suggest that the flexor digitorum profundus is alone active in gentle, unresisted flexion of the fingers, the superficialis coming into play when greater force and/or velocity are required.

The lumbricals are attached to the tendons of the flexor digitorum profundus in the palm, and are described with the muscles of the hand (p. 589).

Fibrous Sheaths of the Flexor Tendons

In the digits the tendons of the flexores digitorum superficialis et profundus lie in osseo-aponeurotic canals (5.77, 79), formed behind by the phalanges, and by fibrous bands, the *digital fibrous sheaths*, which arch across the tendons anteriorly, and are attached to the margins of the phalanges and to the palmar ligaments of the interphalangeal joints. Anterior to the middle of the proximal and middle phalanges the bands are very strong, and the fibres are

explanations of such variations in the deep and superficial digital flexors, which he regards as derived from separate strata, and not from a single blastema. However, Chaplin and Greenlee (1975), following development of digital tendons in human embryos and fetuses confirmed the usual description—an initial single blastemal arrangement of the digital flexor tendons.

Relations. *In the forearm*, the flexor digitorum superficialis is deep to palmaris longus, flexor carpi radialis, pronator teres, brachioradialis, the radial artery and the radial nerve. It is in front of flexor digitorum profundus, the upper part of the ulnar artery, the median nerve (which is closely bound to it by fibroareolar

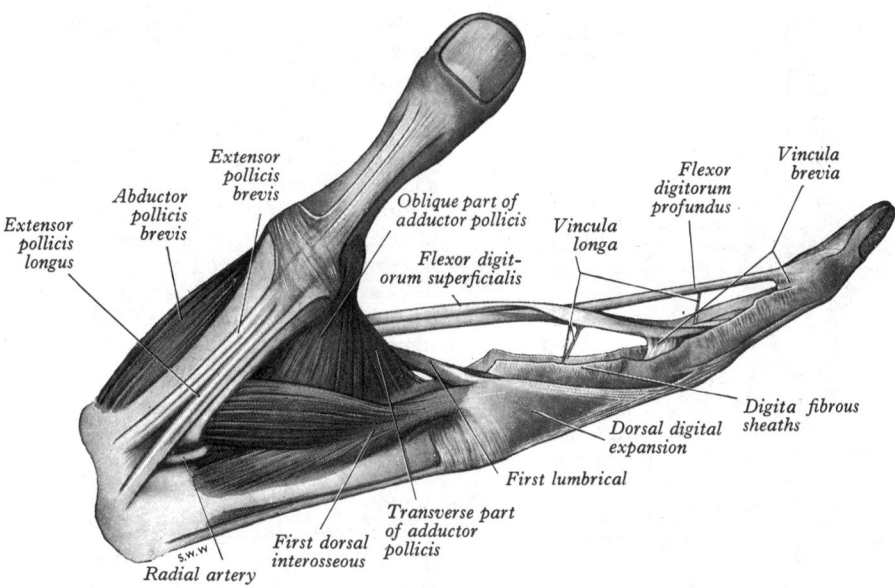

5.68   Lateral part of the right hand showing the tendons and vincula tendinum of the index finger and the muscles in the first intermetacarpal space.

transverse (*annular part*); but opposite the joints they are much thinner, and consist of oblique fibres (*cruciform part*). Each canal is lined by a synovial sheath, which is reflected on to the contained tendons. The synovial membrane is directly adherent to the fibrous sheath.

As the flexor tendons approach their attachments they are connected to the dorsal parts of the enclosing synovial sheaths by triangular or thread-like bands of synovial membrane termed *vincula tendinum* (**5**.68), which convey minute vessels to them. They are of two kinds, (*a*) vincula brevia and (*b*) vincula longa.

The *vincula brevia*, two in number in each finger, are triangular bands attached to the deep surfaces of the tendons close to their attachments; one connects the tendon of the flexor digitorum superficialis to the front of the proximal interphalangeal joint and adjacent part of the proximal phalanx, and the other the tendon of the flexor digitorum profundus to the front of the distal interphalangeal joint and adjacent part of the middle phalanx. The *vincula longa* are thread-like slips, of which two are usually

attached to each tendon of the flexor digitorum superficialis, and one to each tendon of the flexor digitorum profundus. Those of the flexor digitorum superficialis are connected to the slips of that tendon where these fold over the tendon of the flexor digitorum profundus, and, passing one on each side of the latter tendon, are attached to the sheath at the lateral margins of the proximal end of the proximal phalanx. That of the flexor digitorum profundus is fixed to its tendon shortly after the latter has pierced the tendon of the flexor digitorum superficialis. It runs upwards and backwards, perforates one of the two slips of the latter tendon, or passes between the two slips; thereafter it blends with the vinculum breve of the flexor digitorum superficialis and is attached to the dorsal wall of the synovial sheath at the distal end of the proximal phalanx.

The flexor pollicis longus (**5**.67, 76) is lateral to the flexor digitorum profundus. It arises from the grooved anterior surface of the radius, extending from immediately below its tuberosity to within a short distance of pronator quadratus. It arises also from

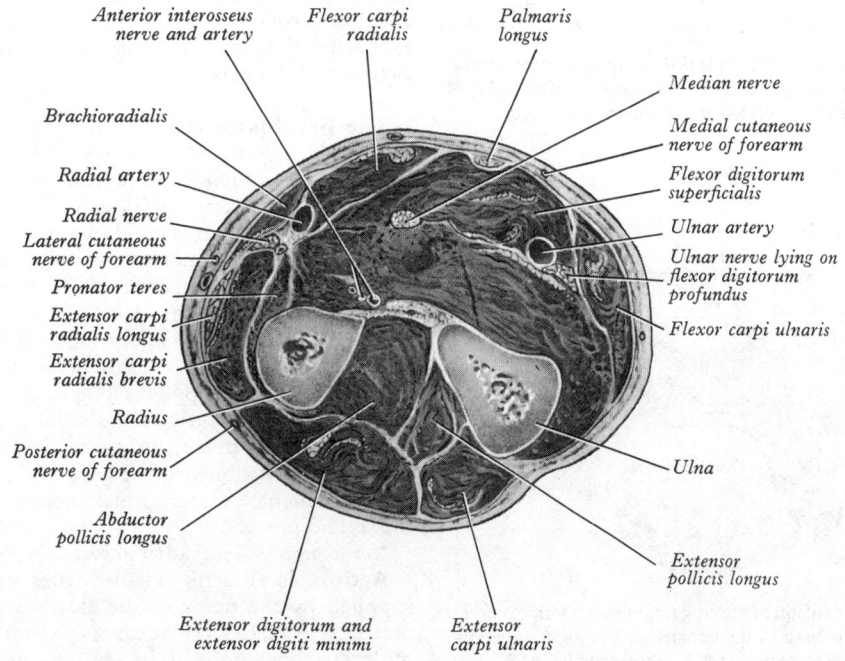

5.69   A transverse section through the middle of the left forearm.

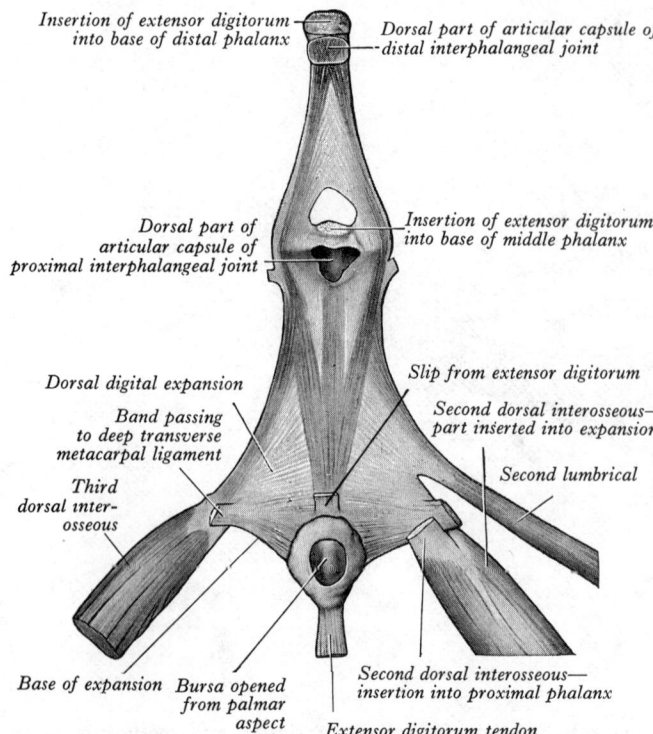

5.70A The dorsal digital expansion of the middle finger, showing some of its principal connexions. Viewed from the palmar surface, the basal angles having been drawn out and the whole expansion kept taut and flattened. Compare with 5.70B and C. Immediately distal to the bursa which overlies the metacarpophalangeal joint, a small slip from the extensor digitorum tendon is inserted into the base of the proximal phalanx. This slip is not always present.

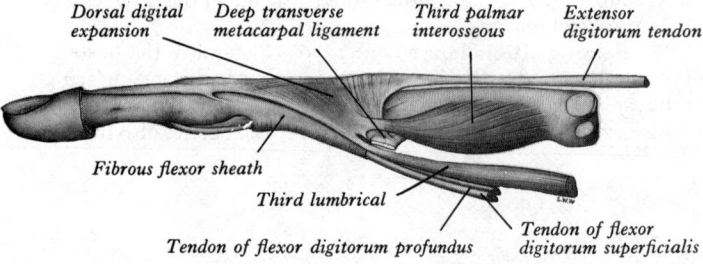

5.70B The ring finger dissected to show the dorsal digital expansion and its principal connexions. Lateral aspect. Note the position of the base of the expansion when the finger is extended and compare with 5.70C.

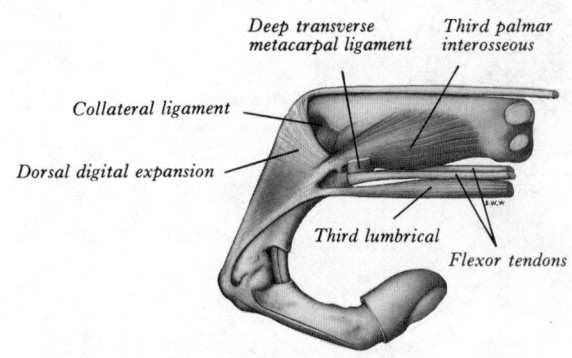

5.70C The dorsal digital expansion of the ring finger in flexion. Compare with 5.70B, and note that the base of the expansion has moved distally. Compare 5.70A, B and C which were prepared by Professor James Whillis. Additional details and nomenclature are shown in 5.71A, B and C.

578

the adjacent part of the interosseous membrane, and frequently by a variable slip from the lateral or, more rarely, the medial border of the coronoid process, or from the medial epicondyle of the humerus (Martin 1958). The fibres end in a flattened tendon which passes behind the flexor retinaculum, between the opponens pollicis and the oblique head of the adductor pollicis and enters an osseo-aponeurotic canal similar to those for the flexor tendons of the fingers, to be attached to the palmar surface of the base of the distal phalanx of the thumb. The anterior interosseous nerve and vessels descend on the front of the interosseous membrane between the flexor pollicis longus and flexor digitorum profundus. The long flexor of the thumb is also sometimes connected by slips to the flexor digitorum superficialis, or profundus, or pronator teres. The interosseous attachment may be absent, as may the whole muscle.

Nerve supply. The anterior interosseous branch of the median nerve, C. **8** and T. **1**.

Action. The flexor pollicis longus flexes the phalanges of the thumb.

The pronator quadratus (5.67) is a flat, quadrilateral muscle, extending across the front of the lower parts of the radius and ulna. It arises from the oblique ridge on the lower part of the anterior surface of the shaft of the ulna (3.141), from the medial part of the anterior surface of the lower fourth of the ulna and from a strong aponeurosis which covers the medial third of the muscle. The fibres pass laterally and slightly downwards, to be attached to the lower fourth of the anterior border and surface of the shaft of the radius; the deeper fibres are inserted into the triangular area above the ulnar notch of the radius.

Nerve supply. The anterior interosseous branch of the median nerve, C. **8** and T. **1**.

Action. This muscle is the principal pronator of the forearm, and its activity is only reinforced by pronator teres during rapid or forceful pronation. The deeper fibres of pronator quadratus oppose separation of the lower ends of the radius and ulna when upward thrusts are transmitted through the carpus.

2. The Posterior Antebrachial Muscles

These muscles are divided for convenience of description into two groups, superficial and deep.

SUPERFICIAL EXTENSOR GROUP

All the muscles in this group (5.72, 75), except brachioradialis and anconeus, which act on the elbow alone, act on a series of joints as extensors. The extensores carpi radiales longus et brevis and the extensor carpi ulnaris act on the radiocarpal and carpal joints, the extensores digitorum et digiti minimi extend the digits and may also extend the wrist.

The brachioradialis (5.65, 66, 69) is the most superficial muscle on the radial side of the forearm and forms the lateral border of the cubital fossa. It arises from the upper two-thirds of the lateral supracondylar ridge of the humerus, and from the anterior surface of the lateral intermuscular septum. The radial nerve and the anastomosis between the arteria profunda brachii and the radial recurrent artery are between it and brachialis. The fibres end above the mid level of the forearm in a flat tendon attached to the lateral side of the lower end of the radius, just above the styloid process. The tendon is crossed at its termination by those of abductor pollicis longus and extensor pollicis brevis; the radial artery is on its ulnar side. The muscle is often fused proximally with brachialis. Its tendon may divide into two or three separately attached slips. Rarely it is double or absent. Its radial attachment may be much more proximal than the base of the styloid process.

Nerve supply. The radial nerve, C. 5, **6** and (7).

Action. The brachioradialis is a flexor of the elbow joint, but is supplied by the nerve of the extensor muscles, i.e. the radial nerve. It acts to best advantage when the forearm is in the midprone position and stands out prominently in the living subject when the semi-pronated forearm is flexed forcibly; it shows little activity in slow, easy flexion movements, or when the forearm is supine. It is active in *rapid* movements of *both flexion and extension* of the elbow, when it is considered to function as a 'shunt' muscle (p. 528), providing a centripetal force towards the elbow joint which balances the centrifugal force generated by a rapid swing of the forearm in either direction (MacConaill 1949; Basmajian 1959). (But see also p.529.)

The extensor carpi radialis longus (**5**.69, 72) is partly overlapped by the brachioradialis. It arises mainly from the lower one-third of the lateral supracondylar ridge of the humerus and the front of the lateral intermuscular septum, but a few fibres come from the common tendon of origin of the forearm extensors. The muscle ends at the junction of the upper and middle thirds of the forearm in a flat tendon, which runs along the lateral surface of the radius, deep to abductor pollicis longus and extensor pollicis brevis; it then passes under the extensor retinaculum, where it lies on the back of the radius in a groove immediately behind the styloid process. It is attached to the radial side of the dorsal surface of the base of the second metacarpal bone. It may also send slips to the first or third metacarpal bones and it contributes to inter-metacarpal ligaments in the region (F. Bojsen-Møller 1978).
Nerve supply. The radial nerve, C. 6 and 7.

The extensor carpi radialis brevis (**5**.69, 72, 76) is shorter than the preceding muscle and is covered by it. It arises from the lateral epicondyle of the humerus, by a tendon common to it and other forearm extensors (*vide infra*), from the radial collateral ligament of the elbow joint, from a strong aponeurosis which covers its surface, and from the intermuscular septa between it and adjacent muscles. Its belly ends about the middle of the forearm in a flat tendon which closely accompanies that of the preceding muscle to the wrist; it passes deep to abductor pollicis longus and extensor pollicis brevis, then under the extensor retinaculum. It is attached to the dorsal surface of the base of the third metacarpal bone, on its radial side and distal to its styloid process, and into the adjoining part of the base of the second metacarpal bone. Under the extensor retinaculum the tendon is on the back of the radius in a shallow groove, on the ulnar side of the groove occupied by the tendon of extensor carpi radialis longus, and separated from it by a variably marked ridge.
Nerve supply. The extensor carpi radialis brevis is supplied by the posterior interosseous nerve, C. **7** and 8.

The tendons of the two preceding muscles pass through the same compartment of the extensor retinaculum in a single synovial sheath. They can be identified on the back of the carpus in the living subject, when the fist is alternately clenched and relaxed.

Both the radial carpal extensors may display splitting of their tendons into two or three slips, variably attached to the second and third metacarpal bones. They may be united or exchange muscular slips.

Actions. Both the preceding muscles act synergically with the flexors of the fingers, and this action can readily be appreciated when the fist is clenched. Working with the extensor carpi ulnaris, they extend the wrist; working with the flexor carpi radialis, they abduct the hand. It should be noted that they are more often called into action as synergic muscles than as prime movers. This applies particularly to the extensor carpi radialis longus which is much less active than the brevis in pure wrist extension (Tournay and Paillard 1953), even when this is rapid. In contrast the longus is synergically much more active than the brevis in grasping or clenching.

The extensor digitorum (**5**.69, 72, 76) arises from the lateral epicondyle of the humerus by the common extensor tendon, from the intermuscular septa between it and adjacent muscles, and from the antebrachial fascia. It divides below into four tendons, which pass, together with that of the extensor indicis, through a separate tunnel under the extensor retinaculum, with a common synovial sheath. The tendons then diverge on the dorsum of the hand, one to each finger. The tendon to the index is accompanied by extensor indicis, which lies medial to it. On the dorsum adjacent tendons are variably tied by three *intertendinous connexions*, directed obliquely downwards and laterally. The most

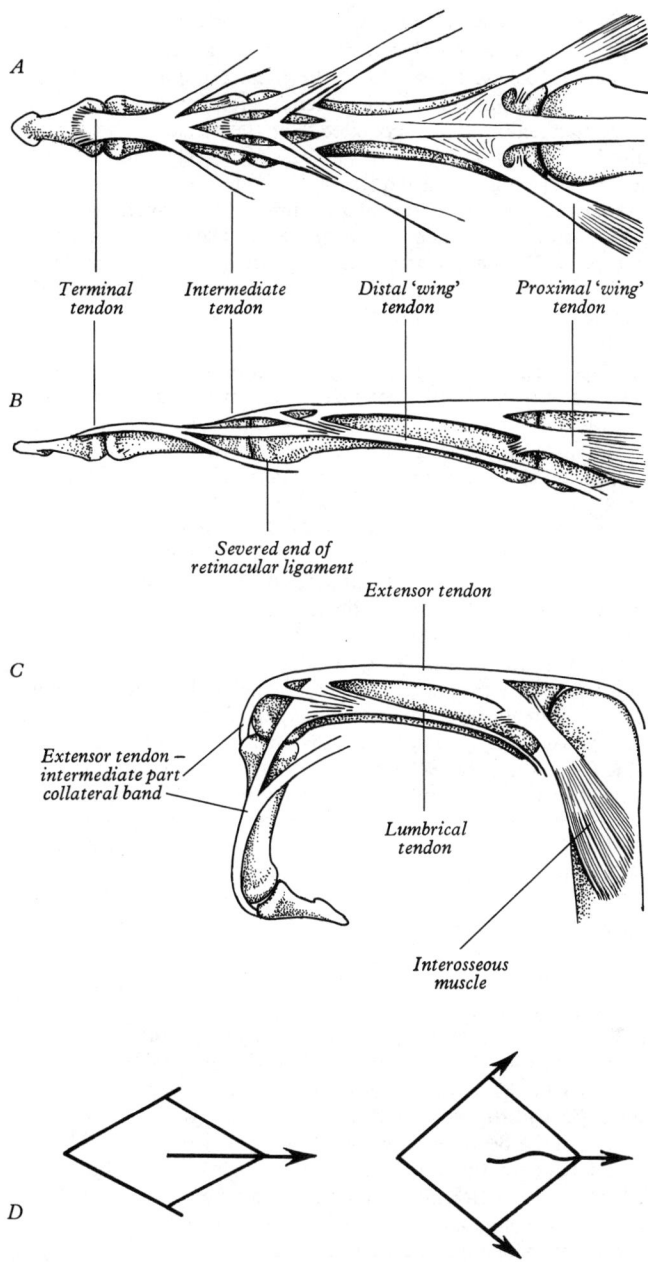

Terminal tendon   Intermediate tendon   Distal 'wing' tendon   Proximal 'wing' tendon

Severed end of retinacular ligament

Extensor tendon

Extensor tendon – intermediate part collateral band

Lumbrical tendon

Interosseous muscle

Distal 'wings' relaxed   Distal 'wings' contracted

5.71A–C Detailed diagrams to illustrate the digital extensor mechanisms. A—viewed from dorsal aspect with digit in full extension. B—viewed from the side, also in full extension. C—viewed from the side, in full flexion. Note particularly the terminal and intermediate parts of the long extensor tendon, the proximal and distal 'wing' tendons, the dorsal extensor expansion, the diamond-shaped complex of tendons dorsal to the proximal interphalangeal joint, and the retinacular ligaments. Note also the changing disposition of the tendinous and ligamentous elements in relation to the digital joints, as the digit passes from extension (B) to flexion (C). D is a diagram showing changes in the 'diamond' complex with contraction of the lumbrical musculature. *See* text for discussion. (Modified from diagrams by Mr. Graham Stack, 1962, and 1963.)

medial is strong and draws the tendon of the fifth digit towards that of the fourth. The intertendinous connexion between the tendons to the third and second digits is weak and may be absent (Leslie 1954). The effect or function of these bands is not clear; they *may* affect independent extension of the digits, and it is a tenet of apocryphal anatomy that they have been divided to benefit pianists.

The digital attachments are complicated by a fibrous expansion dorsal to the proximal phalanx in which the lumbrical, interosseous and digital extensor tendons all participate.
The dorsal digital expansion is a small aponeurosis which

covers the dorsum of the proximal phalanx and the sides of its base. It is triangular and its base, which is proximal, enwraps the dorsal and collateral aspects of the metacarpophalangeal joint. The extensor digitorum tendon blends with the expansion along its median region, and is separated from the metacarpophalangeal joint by a small bursa (5.70A). The base of the expansion connects the extensor digitorum tendon on each side with the adjoining interosseous muscles and contains numerous transverse fibres. A short band links each basal angle to the deep transverse metacarpal ligament; it separates the *phalangeal* attachment of the dorsal interosseous from the rest of the muscle and from the palmar interosseous and lumbrical.

The margins of the expansions are thickened laterally by the tendons of lumbrical and interosseous muscles and medially by the tendon of an interosseous alone, or, in the case of 'the fifth digit, the abductor digiti minimi. These points of attachment of the intrinsic digital muscles are often termed '*wing tendons*' in clinical practice. Distinct proximal and distal 'wings' can be identified in the fingers (5. 71) usually on both sides of each dorsal extensor expansion. The attachments of the interossei to these has led to the classification of these muscles as 'distal' and 'proximal' rather than the more usual palmar and dorsal (p. 589). Between the margins and the extensor digitorum tendon the expansion is a thin and translucent sheet (5.70A). As it approaches the proximal interphalangeal joint the extensor digitorum tendon divides into an intermediate part and two collateral slips. The former, which receives some fibres from the lumbrical and interosseous tendons (Landsmeer 1949), passes over the dorsal aspect of the joint to be attached to the base of the middle phalanx. Each collateral slip is joined by the corresponding thickened border of the dorsal digital expansion, and the two then converge and unite to be attached to the dorsal aspect of the base of the distal phalanx. The distal 'wing tendon' is just proximal to this attachment.

The expansion forms a movable hood (Bunnell 1949), which passes distally when the metacarpophalangeal joint is flexed, and proximally when it is extended, in which position it is most closely applied around the metacarpophalangeal joint. Landsmeer (1949) and Haines (1951) have described *retinacular ligaments* ('*link*' *ligaments*) which correlate movements at the interphalangeal joints (5.71A–C). From a proximal attachment to the side of the proximal phalanx where the fibrous sheath reaches bone, and from the sheath itself, they extend distally to blend with the margin of the dorsal extensor expansion, which is formed by the lateral band of the extensor tendon 5. 71C). They thus reach the base of the terminal phalanx. There are, of course, two retinacular ligaments in each digit. (See also p. 591.)

The extensor digitorum tendons may be to a variable degree deficient, but more often they are doubled or even tripled in the case of one or more digits, more often the index or middle fingers. A slip of tendon to the thumb occasionally exists.

Nerve supply. The posterior interosseous nerve, C. **7** and 8.

Actions. The extensor digitorum extends the fingers at the metacarpophalangeal and interphalangeal joints in opening the hand to relax the grip or in preparation for grasping. It is also an equal prime move with the carpal extensors in extending the wrist (McFarland *et al.* 1962). It tends to abduct the index, ring and little fingers as it extends them, but it has no corresponding action on the middle finger. The attachment of the lateral bands of the extensor tendons to the deep transverse metacarpal ligaments (as noted above) greatly limits the extensor effect of these tendons at the interphalangeal joints, an action which can only be completed by virtue of the attachments of intrinsic muscles, especially the lumbricals, to the lateral extensor bands *distal* to the metacarpophalangeal joints.

The movement of *flexion* of the metacarpophalangeal joint combined with *extension* of the interphalangeal joints, and its converse, enter into most of the fine movements of the digits, especially in the activities of the index finger and thumb. They can be illustrated well by attempts to thread a needle; but the fine adjustments effected, in approach (flexion at the metacarpophalangeal and extension at interphalangeal joints) and withdrawal (extension at metacarpophalangeal and flexion at interphalangeal joints) of the thread, are, of course, used in many other and commoner activities. The metacarpophalangeal flexion, for which

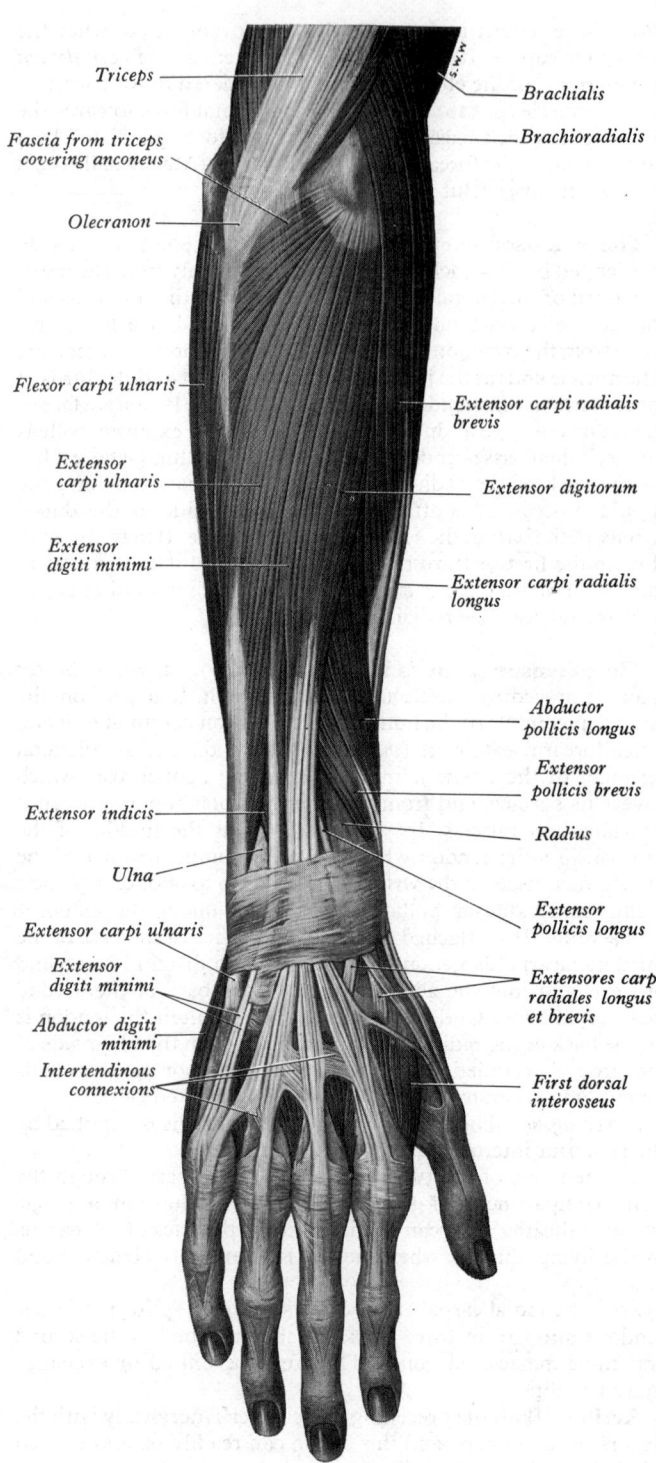

5.72 Muscles of the extensor aspect of the right forearm, superficial layer.

lumbricals and interossei are responsible, allows sufficient relaxation of the digital flexors to enable the extensor digitorum to extend the interphalangeal joints (Braithwaite *et al.* 1948). It should be noted, however, that the interossei, working over the distal border of the deep transverse metacarpal ligaments, are not disadvantageously placed to produce extension of the interphalangeal joints through the medium of the dorsal digital expansion, and can in fact be trained to do so in cases of radial nerve palsy (Sunderland 1945).

The extensor digiti minimi (5.72) is a slender muscle medial to, and usually connected with, extensor digitorum. It arises from the common extensor tendon by a thin tendinous slip and from intermuscular septa shared with adjacent muscles. Its tendon runs through a separate compartment of the extensor retinaculum behind the inferior radio-ulnar joint, then divides into two, the lateral slip being joined by the tendon of extensor digitorum (**5**.72). All are attached to the dorsal digital expansion of the fifth digit, which conforms with that of the other digits. This muscle usually arises in addition from the antebrachial fascia. Its tendon does not always divide and it may send a slip of tendon to the fourth digit. It is rarely absent, but sometimes fused with the extensor digitorum.

Nerve supply. The posterior interosseous nerve, C. **7** and 8.

Actions. The extensor digiti minimi extends the little finger and the wrist working in conjunction with extensor digitorum.

The extensor carpi ulnaris (**5**.69, 72) arises from the lateral epicondyle by the common extensor tendon, from the posterior border of the ulna by an aponeurosis shared with flexor carpi ulnaris and flexor digitorum profundus, and from the antebrachial fascia. It can be identified immediately lateral to the groove which overlies the posterior border of the ulna. It ends in a tendon which slides in the groove between the head and the styloid process of the ulna, in a separate compartment of the extensor retinaculum, and is attached to the tubercle on the medial side of the base of the fifth metacarpal bone.

Nerve supply. The posterior interosseous nerve, C. 7 and **8**.

Actions. In association with the extensores carpi radiales, longus et brevis, the extensor carpi ulnaris acts synergically with the flexors of the fingers in order to extend and fix the wrist when objects are being gripped or when the fist is clenched. Simple observation shows that it is impossible to grip strongly unless the wrist is extended. Acting with the extensores carpi radiales, it extends the wrist; acting with the flexor carpi ulnaris, it adducts the hand.

The anconeus (**5**.72, 73) is a small, triangular muscle at the back of the elbow joint, and is partially blended with triceps, of which it is an integral part in many other primates. It arises by a separate tendon from the posterior surface of the lateral epicondyle of the humerus; its fibres diverge medially towards the ulna, covering the posterior aspect of the annular ligament. They are attached to the lateral side of the olecranon and upper fourth of the posterior surface of the shaft of the ulna.

The muscle varies in its degree of fusion with triceps or extensor carpi ulnaris.

Nerve supply. The radial nerve, C. 7, 8 and (T. **1**).

Action. Anconeus assists triceps in extending the elbow joint and is often held to be responsible for the movement of the ulna in pronation (p. 465). Electromyographic studies, however, contradict this opinion (Basmajian and Griffith 1972). The abduction of the ulna in pronation, which is necessary if the forearm is to turn over the hand without translating it medially, can be voluntarily inhibited. A tool in the hand can hence be revolved 'on the spot', or it can be swept through an arc.

DEEP EXTENSOR GROUP

The deep forearm extensor muscles (**5**.73, 75) consist of three acting on the thumb (abductor pollicis longus and extensores pollicis longus et brevis), the extensor indicis, and also the supinator. Apart from the latter, all are attached proximally to the forearm bones alone.

The supinator (**5**.67, 73, 74) surrounds the upper third of the radius, and consists of superficial and deep parts, between which the posterior interosseous nerve passes (**5**.73). The two arise in common—the superficial by tendinous and the deep by muscular fibres—from the lateral epicondyle of the humerus, from the radial collateral ligament of the elbow joint and the annular ligament of the superior radio-ulnar joint, from the supinator crest of the ulna and from the posterior part of the triangular depression in front of it, and from an aponeurosis covering the muscle. The muscle is attached distally to the lateral surface of the proximal third of the radius, reaching as far as the distal attachment of pronator teres. The radial attachment extends on to its anterior and posterior aspects, for the anterior and posterior

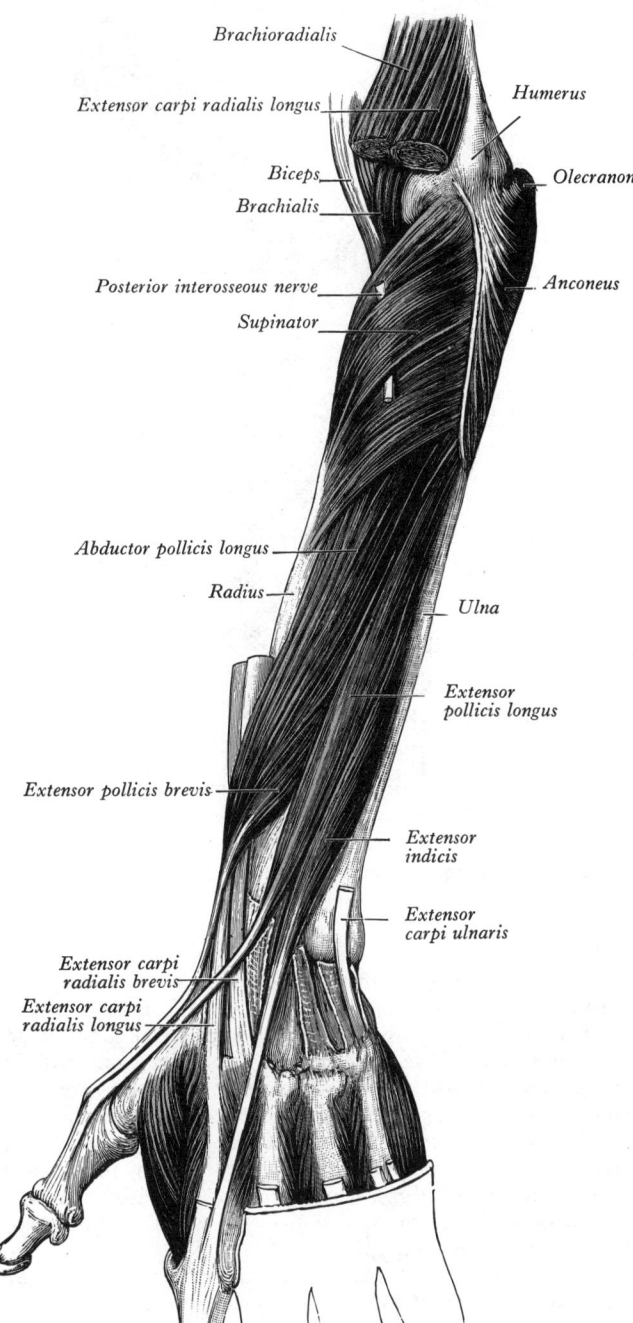

5.73 The deep extensor muscles of the left forearm.

borders of the bone incline medially at their proximal ends (p. 364, **3**.142).

Nerve supply. The posterior interosseous nerve, C. 5 and **6**.

Action. The supinator rotates the radius so as to turn the palm of the hand forwards, and acts alone in slow, unopposed supination, and even in fast supination if the forearm is extended. The muscle is joined by biceps brachii in forceful supination particularly when the forearm is flexed. It is worthy of note that objects, which may be heavy, are picked up frequently with the forearm pronated. The more powerful supinators lift it against gravity, and the rotation is often combined with increasing elbow flexion to bring the hand's burden up towards the face.

The abductor pollicis longus (**5**.69, 72, 73) is distal to supinator and closely related to extensor pollicis brevis. It arises from the lateral part of the posterior surface of the shaft of the ulna below the anconeus, from the interosseous membrane, and from the middle third of the posterior surface of the radius immediately distal to the attachment of supinator. It passes downwards and

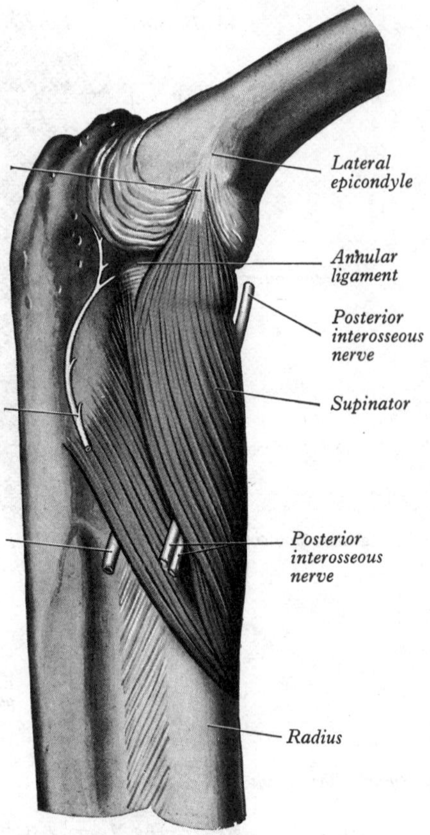

Radial collateral ligament of elbow joint
Lateral epicondyle
Annular ligament
Posterior interosseous nerve
Supinator
Interosseous recurrent artery
Posterior interosseous artery
Posterior interosseous nerve
Radius

5·74 The right supinator muscle. Posterolateral aspect.

laterally, becoming superficial in the distal quarter of the forearm, visible there as an oblique elevation (5·72). It ends in a tendon which becomes free of muscle fibres just proximal to the wrist. It then runs in a groove on the lateral side of the lower end of the radius accompanied by the tendon of extensor pollicis brevis. The tendon is usually split into two components, one attached to the radial side of the base of the first metacarpal bone, the other to the trapezium. Further fasciculi of the tendon may become continuous with the opponens pollicis or the fleshy belly of the abductor pollicis brevis. Occasionally the whole muscle is divided into two, or cleavage may be partial.

Nerve supply. The posterior interosseous nerve, C. 7 and **8**.

Actions. The abductor pollicis longus, acting with the abductor pollicis brevis, abducts the thumb; with the extensores pollicis, it extends the thumb at the carpometacarpal joint (p. 471).

The extensor pollicis brevis (5·72, 73) is medial to, and closely connected with, the abductor pollicis longus. It arises from the posterior surface of the radius below the abductor, and from the interosseous membrane. Its direction is similar to that of the abductor pollicis longus, its tendon passing through the same groove on the lateral side of the lower end of the radius, to be attached to the dorsal surface of the base of the proximal phalanx of the thumb. There is often an additional attachment to the base of the distal phalanx usually through a fasciculus which joins the tendon of extensor pollicis longus (Muller 1959). The muscle may be absent or completely fused with the abductor longus. Its tendon sometimes fuses completely with that of the long extensor.

In the lower third of the forearm the abductor pollicis longus and extensor pollicis brevis become superficial by emerging between the extensor carpi radialis brevis and extensor digitorum. Thence obliquely crossing the tendons of the radial extensors they cover the termination of brachioradialis, and, passing through the most lateral compartment of the extensor retinaculum in a single synovial sheath, cross superficial to the styloid process of the radius and the radial artery.

Nerve supply. The posterior interosseous nerve, C. 7 and **8**.

Actions. The extensor pollicis brevis extends the proximal phalanx of the thumb and, in continued action, helps to extend the metacarpal bone.

The extensor pollicis longus (5·72, 73) is larger than the extensor pollicis brevis, the proximal attachment of which it partly covers. It arises from the lateral part of the middle third of the posterior surface of the shaft of the ulna below abductor pollicis longus, and from the interosseous membrane. It ends in a tendon which passes through a separate compartment of the extensor retinaculum, in a narrow, oblique groove on the back of the distal end of the radius. It then crosses obliquely the tendons of the extensores carpi radiales longus et brevis (5·73), and is separated from the extensor pollicis brevis, when the thumb is fully extended, by a triangular depression, jocularly termed the 'anatomical snuff-box', in which the radial artery can be felt. Finally, the tendon reaches its attachment to the base of the distal phalanx of the thumb. On the dorsum of the proximal phalanx the lateral and medial sides of the tendon are joined by expansions from the tendons of the abductor pollicis brevis, and of the first palmar interosseous and adductor pollicis respectively.

Nerve supply. The posterior interosseous nerve, C. 7 and **8**.

Actions. The muscle extends the distal phalanx of the thumb and, acting in association with the extensor pollicis brevis and abductor pollicis longus, it extends the proximal phalanx and the metacarpal. In continued action, owing to the obliquity of the course of its tendon, the extensor pollicis longus adducts the extended thumb and rotates it laterally (p. 471).

Applied Anatomy. Avascular necrosis is probably a causative factor in rupture of the tendon of the extensor pollicis longus which may follow Colles' fracture or occur spontaneously. Above

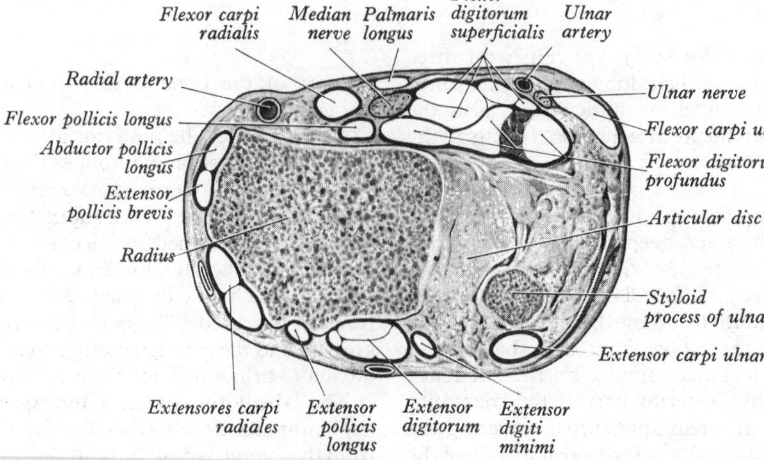

Flexor carpi radialis
Median nerve
Palmaris longus
Flexor digitorum superficialis
Ulnar artery
Radial artery
Ulnar nerve
Flexor pollicis longus
Flexor carpi ulnaris
Abductor pollicis longus
Flexor digitorum profundus
Extensor pollicis brevis
Articular disc
Radius
Styloid process of ulna
Extensor carpi ulnaris
Extensores carpi radiales
Extensor pollicis longus
Extensor digitorum
Extensor digiti minimi

5·75 Transverse section, passing through the distal end of the right radius and the styloid process of the right ulna, made with the hand and forearm in full supination. Distal (inferior) aspect.

the wrist the blood supply is derived mainly from the anterior interosseous artery and below this joint from the radial artery and blood vessels of the thumb (Davies 1951).

The extensor indicis (5.73) is a narrow, elongated muscle, medial to, and parallel with, the preceding. It arises from the posterior surface of the ulna below the extensor pollicis longus, and from the interosseous membrane. Its tendon passes under the extensor retinaculum in the compartment containing the tendons of extensor digitorum; opposite the head of the second metacarpal bone it joins the ulnar side of the tendon of the extensor digitorum for the index finger. The muscle occasionally sends accessory slips to the extensor tendons of other digits. Rarely its tendon may be interrupted on the dorsum of the hand by an additional muscle belly ('*extensor indicis brevis manus*').

Nerve supply. The posterior interosseous nerve, C. 7 and **8**. The distribution of muscle spindles in this muscle has been studied by Gorp and Kennedy (1974). The greatest density is concentrated near the point of entry of its nerve supply.

Actions. The muscle helps to extend the index finger and can assist in extending the wrist.

The Hand

Human hands, acting with the eyes in a complex synergy mediated by central nervous connexions of apparently unlimited variability, have been a paramount factor in the evolution of human skills; and, despite the production of an astonishing multitude of tools and techniques to extend these skills, one may presume that the hands will continue in this role, in further refinement, development and invention. Being put to such multiple and frequent uses they are subject to much injury; and for both reasons a large literature, impossible to quote even in skeleton here, has accumulated on the structure, activities, injuries and reparative surgery of the hand. For guidance to this literature the works of Wood Jones (1941), Napier (1965), Kaplan (1966), Stack (1973) and Landsmeer (1976) are particularly valuable, especially since each has contributed notably to knowledge in this field.

The Retinacula, Fasciae, and Synovial Sheaths of the Wrist and Hand

Before proceeding to the muscles of the hand, it is appropriate to consider first the disposition of the retaining bands and deep fascia, and the synovial tendon sheaths of the region.

The flexor retinaculum (5.76, 77), a strong, fibrous band, crosses the front of the carpus and converts its anterior concavity into the *carpal tunnel*, through which pass the flexor tendons of the digits and the median nerve (p. 1102). The retinaculum is transversely short, measuring a mere 2·5 to 3 cm, with a similar breadth. It is attached, medially, to the pisiform and the hook of the hamate; laterally, it splits into two laminae, a superficial attached to the tubercles of the scaphoid and trapezium, and a deep, to the medial lip of the groove on the latter bone (**5**.76). With the groove on the trapezium the two laminae form a tunnel, lined by a synovial sheath containing the tendon of flexor carpi radialis. The retinaculum is continuous, proximally, with the fascia covering flexor digitorum superficialis and the antebrachial fascia. It is these two layers which separate on reaching the trapezium. The retinaculum is crossed superficially by the ulnar vessels and nerve, and the palmar cutaneous branches of the median and ulnar nerves. To its anterior surface the tendons of palmaris longus and flexor carpi ulnaris are partly attached; distally some of the short muscles of the thumb and little finger are attached to it, and it is continuous with the deep fascia of the palm, including its central longitudinal or aponeurotic fibres, which are associated with palmaris longus.

A localized thickening in the general investing layer of the antebrachial fascia which extends laterally from the pisiform is termed the *superficial part of the flexor retinaculum*. It crosses superficial to the ulnar vessels and nerves and blends with the rest of the retinaculum lateral to them.

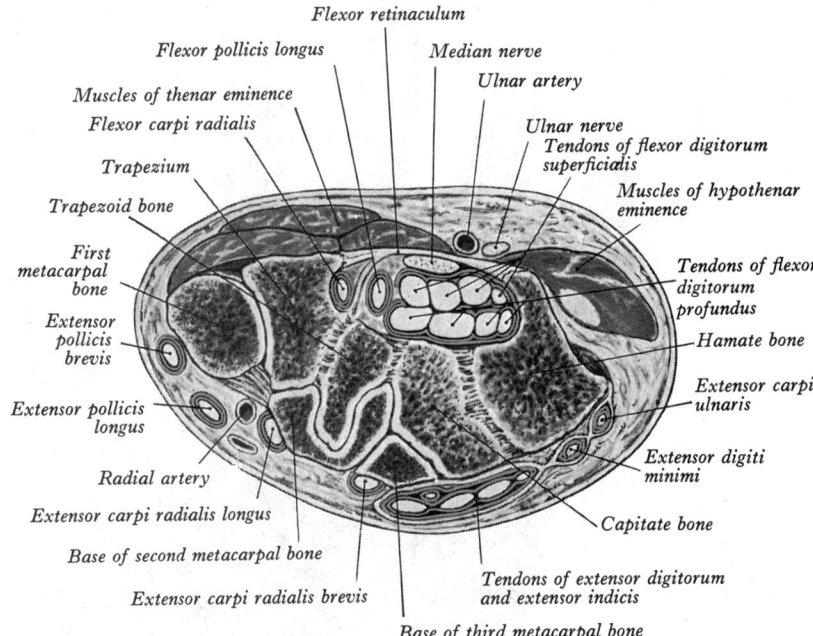

5.76 Transverse section through the left wrist, showing the tendons and their synovial sheaths. The section is slightly oblique and divides the distal row of the carpus, and the bases of the first, second and third metacarpal bones. The arrangement of the tendons of the flexors of the fingers shown in the figure is not diagrammatic but represents the actual condition at this level. Observe that the carpometacarpal joint of the thumb is separate from the joint between the trapezium and the base of the second metacarpal bone.

The extensor retinaculum (5.78), a strong, fibrous band, extends obliquely across the back of the wrist, consisting of part of the antebrachial fascia, strengthened by the addition of some obliquely transverse fibres. It is attached laterally to the anterior border of the radius, medially to the triquetral and pisiform bones, and, in its passage across the wrist, to the ridges on the posterior aspect of the distal end of the radius.

The Synovial Sheaths of the Carpal Flexor Tendons

Two synovial sheaths envelop the flexor tendons as they traverse the carpal tunnel, one for the flexores digitorum superficialis et profundus, the other for flexor pollicis longus (**5**.76). These sheaths extend into the forearm for about 2·5 cm proximal to the flexor retinaculum, and occasionally communicate with each other deep to it. The sheath of the flexores digitorum tendons reaches about halfway along the metacarpal bones, where it ends in blind diverticula around the tendons to the index, middle and ring fingers (**5**.77). It is prolonged around the tendons to the little finger and is usually continuous with their digital synovial sheath. A transverse section through the carpus (**5**.76) shows that the tendons are invaginated into the sheath from the lateral side. The parietal layer lines the flexor retinaculum and the floor of the carpal tunnel and is reflected, laterally, on to the tendons of flexor digitorum superficialis from in front, and on to those of flexor digitorum profundus from behind. Medially a recess formed by the inner layer of the sheath is insinuated between the two groups of tendons and passes laterally for a variable distance. The sheath of the flexor pollicis longus, which is usually separate but occasionally communicates with the common flexor sheath behind the flexor retinaculum, is continued along the thumb as far as the insertion of the tendon. The fibrous sheaths enveloping the terminal parts of the tendons of the flexores digitorum have already been described (p. 576).

The Synovial Sheaths of the Carpal Extensor Tendons

Deep to the extensor retinaculum there are six tunnels for the passage of the extensor tendons, each containing a synovial sheath (**5**.78). They are as follows: 1, on the lateral side of the styloid

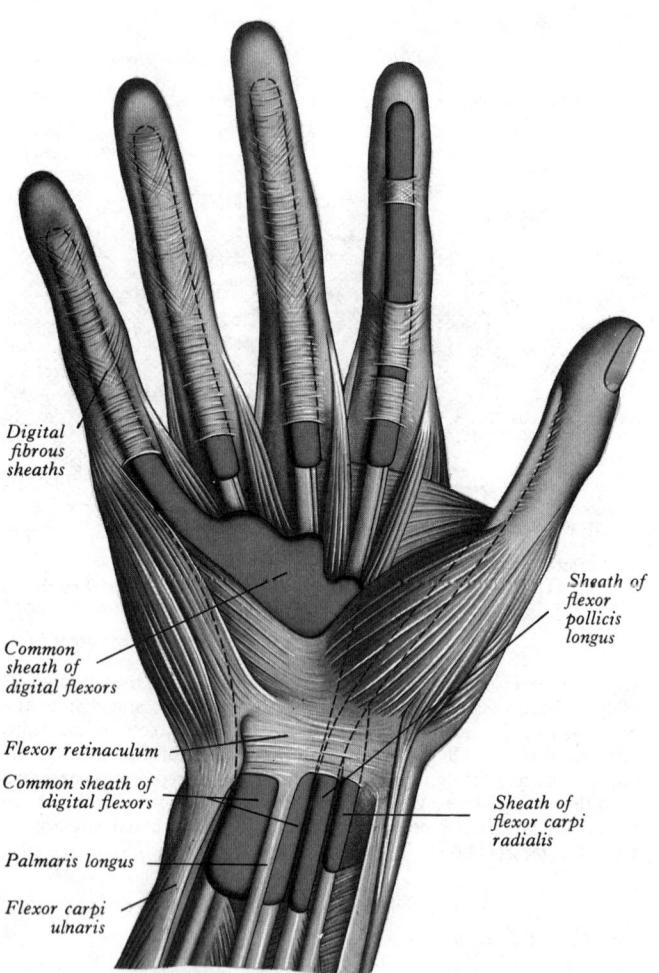

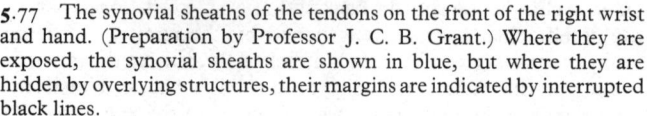

5.77 The synovial sheaths of the tendons on the front of the right wrist and hand. (Preparation by Professor J. C. B. Grant.) Where they are exposed, the synovial sheaths are shown in blue, but where they are hidden by overlying structures, their margins are indicated by interrupted black lines.

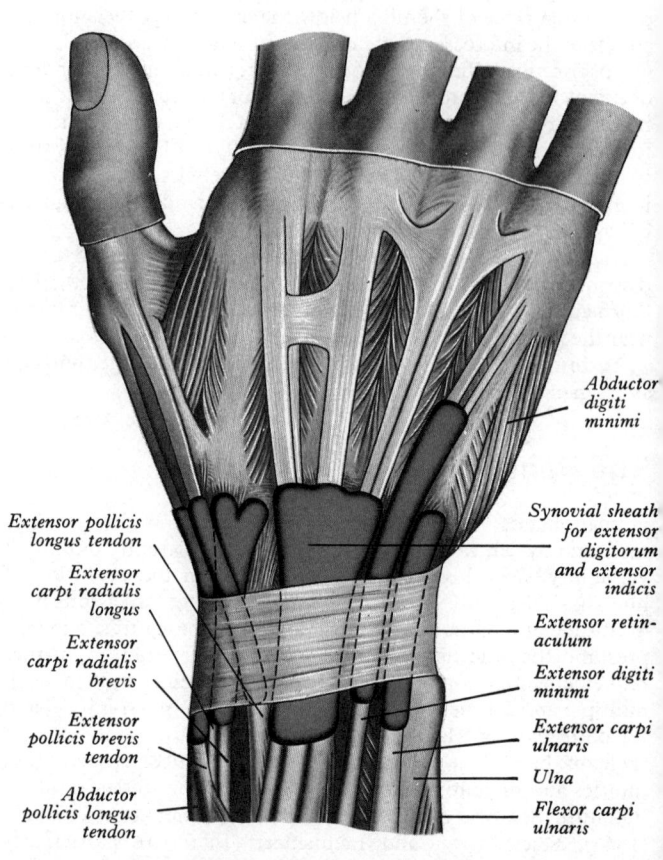

5.78 The synovial sheaths of the tendons on the back of the right wrist. (Preparation by Professor J. C. B. Grant.) The synovial sheaths are shown in blue, but they have not been coloured where they lie deep to the extensor retinaculum. In this situation, and where one sheath lies deep to another, the margins of the sheaths are indicated by the broken lines.

process of the radius, for the tendons of abductor pollicis longus and extensor pollicis brevis; 2, behind the styloid process, for the tendons of the extensores carpi radiales longus et brevis; 3, on the medial side of the dorsal tubercle of the radius, for the tendon of the extensor pollicis longus; 4, on the medial side of the latter, for the tendons of the extensor digitorum and extensor indicis; 5, opposite the interval between the radius and ulna, for extensor digiti minimi; 6, between the head and the styloid process of the ulna, for the tendon of extensor carpi ulnaris. The sheaths of the tendons of the abductor pollicis longus, extensores pollicis brevis et longus, extensores carpi radiales, and extensor carpi ulnaris stop immediately proximal to the bases of the metacarpal bones, while those of the extensor digitorum, extensor indicis and extensor digiti minimi are sometimes prolonged to the junction of the proximal with the intermediate third of the metacarpus.

Fascial Arrangements in the Hand

No region of the body has provided so fertile an arena for dissection of connective tissue as the hand, with the consequent identification and naming of numerous sheets, septa, bands and ligaments. From a great number of studies, spread over more than two centuries, the larger features have received general accord, and these include the fibrous flexor sheaths, dorsal digital expansions ('wing tendons') of the digital extensor tendons, the transverse metacarpal ligaments, and the palmar fascia or aponeurosis. As Wood Jones (1941) emphasized, every structure—muscle, tendon, nerve and vessel—is surrounded by

connective tissue separating it from adjacent structures to a degree dependent upon the local need of individual mobility. Recognized thickenings of such 'sheaths' also serve to restrain or bind together some structures; and in the hand (as in the foot) the subcutaneous tissue is traversed by numerous localized dense bands or 'skin ligaments', which tie the skin to the deeper fasciae and hence to bone, for obvious functional reasons. These are little developed on the dorsum of the hand but marked in the palm and flexor aspects of the phalanges, especially in the pulps of the terminal digital segments. The deeper connective tissue is densely compacted and in places even aponeurotic, and it blends with periosteum whenever nothing else intervenes, extending as strong septa between certain groups of muscles.

Successive studies have added many details to this description. Since in such finer points authorities differ, individual variation doubtless occurs, and functional comprehension is not always advanced by them, such complexities must be here reduced to simpler concepts. At the simplest it is clear that the deep (muscular) fasciae of the hand are extensively connected on one side to skin and on the other to bone, including not only the carpals, metacarpals, and phalanges, but also proximally the radius and ulna. At well-recognized places, and especially where cutaneous flexion furrows are apparent, skin fixation is most marked. Between these regions of maximum stability large numbers of smaller fixation bands exist, but these are interspersed by cushioning pads of adipose tissue. These, though largely restrained by the connective tissue bands, permit a small range of tangential movement; and, since some displacement is possible the skin and subcutaneous tissue are also easily indented, with an immediate resilient recovery when the distorting external force is removed. It is obvious that these arrangements have a protective value; the fat pads which exist in the interdigital webs are said to separate and 'protect' the digital vessels and nerves from adjacent tendons, and the thenar and hypothenar pads are clearly useful cushions when these parts of the hand are used to apply force to various tools. Less obvious, but equally necessary is the pliant conformation of the grasping surfaces of the hand to the contours of the object grasped, however irregular and potentially damaging. The hand can thus evaluate the shape, weight and consistency of an infinite number of external objects (see p. 583), putting them to a limitless variety of uses, often involving large forces, the skilful and safe transmission of which is as much due to the skin and fasciae of the hand as to muscles and tendons. Like the latter, the skin and fasciae respond to use by hypertrophy, as may the bones while still capable of growth. The skin commonly shows not only a general responsive thickening but also various localized callosities (over the metacarpal heads, for example), sometimes characteristic of particular occupations. The palmar fascia, including its central aponeurosis sometimes responds to an excessive degree, with resultant disabling contractures. For further details regarding the fasciae of the hand, including their diseases and injuries consult Kaplan (1966), Stack (1973), Bojsen-Møller and Schmidt (1974), and Landsmeer (1976).

In addition to the functions noted above, the fasciae of the hand also help to maintain the concavity of the palm, to restrain the flexor tendons, and perhaps to aid in venous return.

The palmar aponeurosis (5.79) is in direct relation with the muscles of the palm, and accordingly consists of central, lateral and medial parts.

The *central part* is triangular, and of marked strength and thickness. Its apex is continuous with the distal margin of the flexor retinaculum, and the expanded tendon of palmaris longus is attached to it. Its base divides into four slips, extending to each finger. Superficial fibres of them reach the skin of the palm and fingers; those to the palm join the dermis at the furrow corresponding to the metacarpophalangeal joints, those to the fingers pass into it at the transverse folds at the roots of the fingers. The deeper part of each slip divides into two processes which are continuous with the fibrous sheaths of the flexor tendons; from the sides of these processes offsets are attached to the deep transverse metacarpal ligaments, the capsule of the metacarpophalangeal joint and the sides of the base of the proximal phalanx. By this arrangement short channels are formed anterior to the heads of the metacarpal bones, through which the flexor tendons pass. The intervals between the four slips are traversed by the digital vessels and nerves, and the tendons of the lumbricals. At the divisions into slips, numerous strong, transverse fibres bind them together. The central part of the palmar aponeurosis is strongly bound to the skin by dense fibroareolar tissue, and to the proximal part of its medial margin the palmaris brevis is attached. It covers the superficial palmar arch, the tendons of the flexores digitorum, the terminal part of the median nerve, and the superficial branch of the ulnar nerve.

The *lateral* and *medial parts* of the palmar aponeurosis are thin, fibrous coverings of the thenar and hypothenar muscles; they are continuous with its central region and with the fascia on the dorsum of the hand.

A septum passes dorsally from each border of the central portion of the palmar aponeurosis. The *medial palmar septum* lies close to the lateral side of the opponens digiti minimi and reaches the palmar surface of the fifth metacarpal bone. Distally, it is continuous with the slip of the palmar aponeurosis to the lateral side of the fibrous sheath of the little finger; proximally it reaches the hook of the hamate bone and the pisohamate ligament and is pierced by the deep branches of the ulnar nerve and artery. The *lateral palmar septum* passes dorsally to reach the palmar surface of the first metacarpal bone. It skirts the medial side of the flexor brevis and opponens pollicis, between them and the flexor pollicis longus tendon and its synovial sheath. Proximally it reaches the tubercle of the trapezium and is pierced by the branch from the median nerve to the muscles of the thenar eminence.

The fascial spaces of the palm. The central part of the palm, which lies behind the central region of the palmar aponeurosis

and between the lateral and medial palmar septa, is divided into medial and lateral parts by a thin intermediate palmar septum. The medial area has been termed the *middle palmar space* and the lateral area the *thenar space*.

The intermediate palmar septum is between the flexor tendons of the index finger laterally and the second lumbrical muscle medially. Distally, it is continuous with the slip given by the palmar aponeurosis to the medial side of the fibrous sheath of the index finger and the adjoining deep transverse metacarpal ligament. Dorsally, it blends with the fasciae covering the distal part of the second palmar interosseous and the transverse head of the adductor pollicis, and it can be traced medially on the latter muscle to the third metacarpal bone. Anteriorly, it is attached to the deep surface of the palmar aponeurosis, but at its proximal end it meets the common flexor synovial sheath and blends with the connective tissue on its posterior surface. However, as an anatomical entity, the intermediate palmar septum is poorly developed. Nevertheless, when pus accumulates on either side of it, the adjoining areolar tissue is compressed against it forming an effective septum, capable of determining the direction of spread.

The *middle palmar space* lies between the medial and intermediate septa. Its dorsal wall is formed by the third, fourth and fifth metacarpal bones, by the fascia covering the interosseous muscles in the third and fourth spaces and the medial part of the transverse head of the adductor pollicis. Anteriorly are the central zone of the palmar aponeurosis and, proximally, the common flexor synovial sheath. The space contains the flexor tendons of the fifth, fourth and third fingers, the fourth, third and second lumbrical muscles, the superficial palmar arch and the digital vessels and nerves for the fifth, fourth, third and ulnar side of the index finger. Distally, the space communicates with the subcutaneous tissues at the webs between the fingers; proximally, it may extend dorsal to the common flexor synovial sheath.

Bojsen-Møller and Schmidt (1974) regard the distal part of the middle palmar 'space' as divided into numerous independent channels not only for the lumbrical muscles but also for the common digital vessels and nerves (see 5.80D).

The *thenar space* is flanked by the lateral and the intermediate

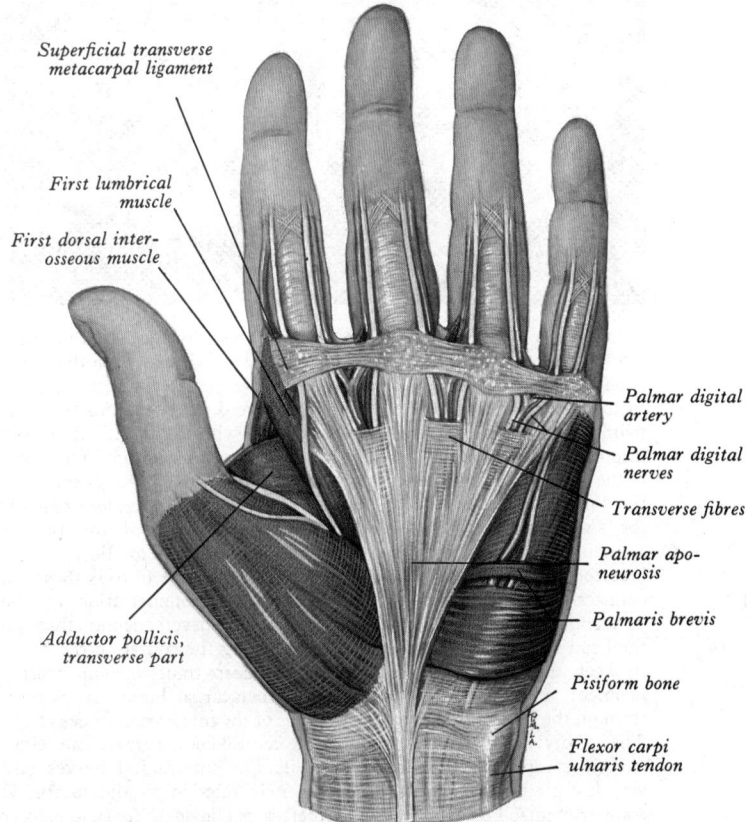

Superficial transverse metacarpal ligament

First lumbrical muscle

First dorsal inter-osseous muscle

Adductor pollicis, transverse part

Palmar digital artery

Palmar digital nerves

Transverse fibres

Palmar apo-neurosis

Palmaris brevis

Pisiform bone

Flexor carpi ulnaris tendon

5.79 The left palmar aponeurosis.

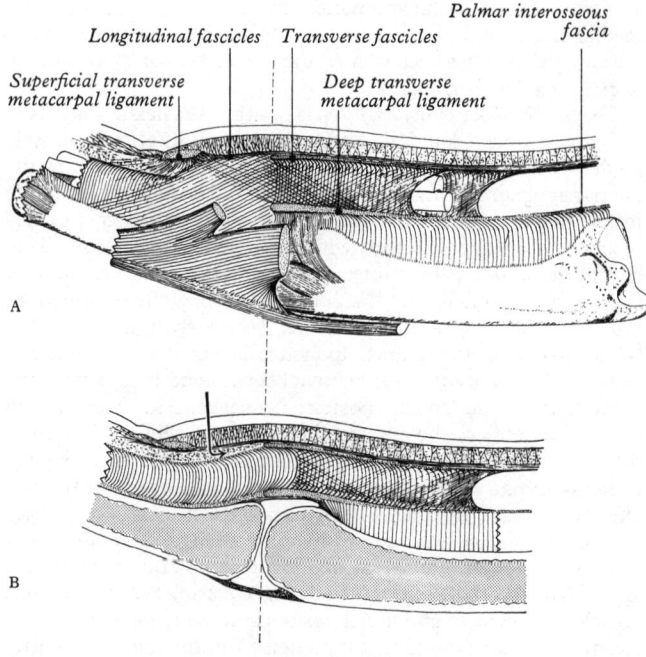

Superficial transverse metacarpal ligament | Longitudinal fascicles | Transverse fascicles | Palmar interosseous fascia | Deep transverse metacarpal ligament

A

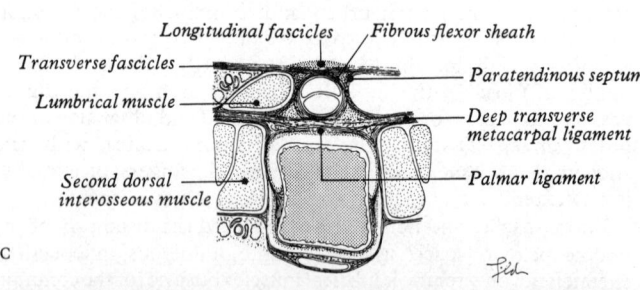

B

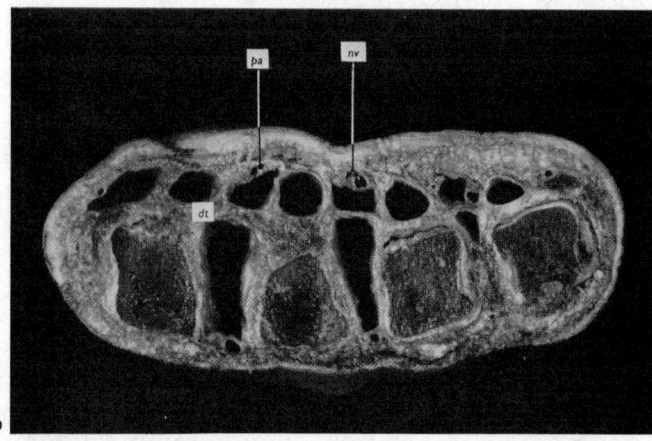

Transverse fascicles — Longitudinal fascicles — Fibrous flexor sheath — Paratendinous septum

Lumbrical muscle

Second dorsal interosseous muscle — Deep transverse metacarpal ligament — Palmar ligament

C

5.80A–D (A) Drawing of the course of the fibres of the fasciae around the left third metacarpal bone and proximal phalanx viewed from the radial side. The deep transverse metacarpal ligament connects both with the metacarpal bone and with the longitudinal and transverse fibres of the palmar aponeurosis and, via their retinacula, with the skin. The palmar interosseous fascia is connected by means of a sagittal septum to the shaft of the third metacarpal bone. (B) sagittal section through the centre of the third metacarpal bone. The fibrous sheath of the flexor tendons extends for some distance between the tranverse fascicles of the palmar aponeurosis and the deep transverse metacarpal ligament. Between the longitudinal fibres in the palmar aponeurosis, and the fibrous sheath, a connective tissue space affords cleavage for demonstration of the paratendinous septa by dissection (arrow). (C) Transverse section through the head of the third metacarpal bone showing the course of the fibres between the palmar aponeurosis and the deep transverse metacarpal ligament, and their attachments to the metacarpal bone. (D) Section through the hand at the level of the heads of the metacarpal bones (2–5). The subdivision of the distal part of the central compartment into eight narrow compartments is demonstrated. The interdigital nerves and vessels (nv) have been left to show their relation to the lumbrical compartments. The deep transverse metacarpal ligament (dt) and palmar aponeurosis (pa) are also shown. (From F. Bojsen-Møller and L. Schmidt 1974, by courtesy of the authors and the Cambridge University Press.)

palmar septa. Dorsally, are the fasciae covering the transverse head of the adductor pollicis and, beyond its lower border, the first dorsal interosseous muscle. Anteriorly, is the palmar aponeurosis. It contains the flexor pollicis longus tendon and its synovial sheath, the flexor tendons of the index finger and the first lumbrical muscle, and the palmar digital vessels and nerves of the thumb and radial side of the index finger. Distally the space communicates with the subcutaneous tissues of the web of the thumb; proximally it extends behind the flexor retinaculum.

All the 'spaces' described above are, of course, in fact *regions* or *localities*, in which relatively loose connective tissue, whilst allowing some mobility during normal functioning, also offers little opposition to the spread of infection, whereas their boundaries, being of denser structure, may contain an infection, temporarily at least.

The superficial transverse metacarpal ligament is a thin band (5.79) which stretches across the roots of the fingers in the superficial fascia, and is attached to the skin of the clefts, and medially to the fifth metacarpal bone, forming a kind of rudimentary web. The digital vessels and nerves pass posterior to these fasciculi. Since this ligament adjoins the bases of the proximal phalanges, rather than the metacarpal heads, it would be more appropriately named the *palmar digital ligament*, according to Bojsen-Møller and Schmidt (1974).

Applied Anatomy. The palmar aponeurosis is liable to undergo contraction, producing a disabling deformity known as 'Dupuytren's contracture' (*see* Stack 1973).

Owing to their constant exposure to injury and sources of contamination, the fingers are very liable to infection. In the *pulp* over the distal three-fourths of the terminal phalanx, pus is confined under tension by strong fibrous septa which radiate between the skin and periosteum. This may lead to necrosis of the diaphysis of the terminal phalanx from obstruction of its blood supply from the digital arteries which pass through the pulp. The epiphysis of the terminal phalanx may escape because its vessels of supply leave the digital arteries before the latter enter the compartments of the pulp. Inflammation may involve and spread along the sheath of the flexor tendons. Infection of the synovial sheaths of the pollex or minimus may prove more serious than in the other digits, because their sheaths may communicate with the principal sheath of the flexor tendons, and the infective process may extend into the palm and behind the flexor retinaculum into the forearm.

Chronic inflammation of the common flexor sheath is occasionally met with, a condition known as *compound palmar ganglion*; it presents a bilobar outline, with a swelling in front of the wrist and another in the palm of the hand, and a constriction, corresponding to the flexor retinaculum, between the two. The fluid contents can be forced from the one swelling to the other behind the retinaculum.

Although the applied, medical significance of the fasciae and ligaments of palm and digits must receive due emphasis, their basic function is, of course, to transmit tangential forces from the skin to the skeleton of the hand, Contraction of the palmaris longus and extension at the wrist and metacarpophalangeal joints both assist in tensing the palmar aponeurosis.

VI. The Muscles of the Hand

The intrinsic muscles of the hand can be considered conveniently in three groups: 1, three muscles of the thumb, which produce the *thenar eminence*; 2, those of the fifth digit, producing the *hypothenar eminence*; 3, the central palmar and interosseous muscles, together with the adductor pollicis.

1. THENAR MUSCLES, ADDUCTOR POLLICIS

The thumb has intrinsic muscles which flex, abduct, oppose and adduct it. The three former occupy the thenar eminence.

The abductor pollicis brevis is a thin, subcutaneous muscle in the radial part of the thenar eminence; its chief origin is from the flexor retinaculum, but a few fibres spring from the tubercles of the scaphoid bone and trapezium, and some arise from the

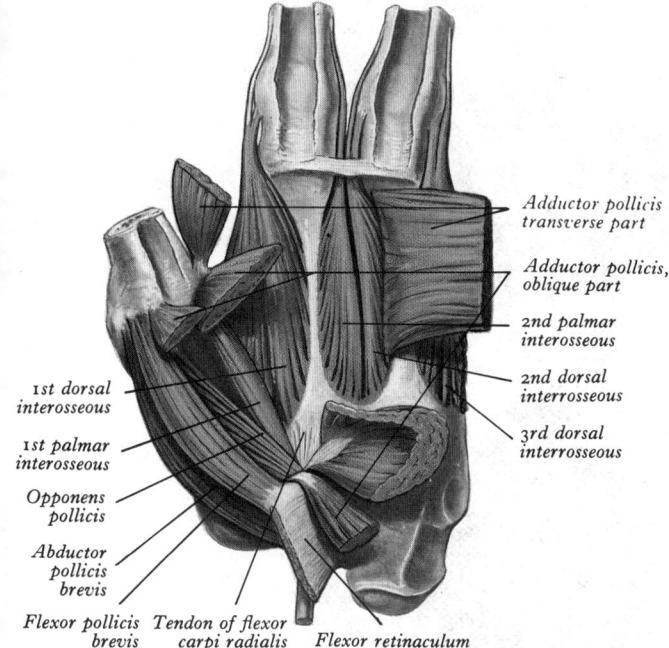

1st dorsal interosseous

1st palmar interosseous

Opponens pollicis

Abductor pollicis brevis

Flexor pollicis brevis

Tendon of flexor carpi radialis

Flexor retinaculum

Adductor pollicis transverse part

Adductor pollicis, oblique part

2nd palmar interosseous

2nd dorsal interosseous

3rd dorsal interosseous

5.81A Dissection of the left hand to show the first palmar interosseous muscle and the deep head of the flexor pollicis brevis.

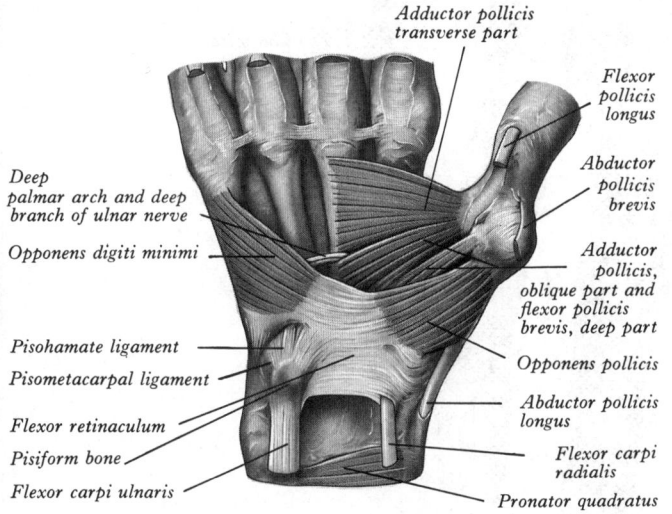

Adductor pollicis transverse part

Flexor pollicis longus

Abductor pollicis brevis

Adductor pollicis, oblique part and flexor pollicis brevis, deep part

Opponens pollicis

Abductor pollicis longus

Flexor carpi radialis

Pronator quadratus

Deep palmar arch and deep branch of ulnar nerve

Opponens digiti minimi

Pisohamate ligament

Pisometacarpal ligament

Flexor retinaculum

Pisiform bone

Flexor carpi ulnaris

5.81B A dissection of the right palm, showing the opponens pollicis and opponens digiti minimi and the two parts of the adductor pollicis.

tendon of abductor pollicis longus (p. 581). Its medial fibres are attached by a thin, flat tendon to the radial side of the base of the proximal phalanx of the thumb; its lateral fibres join the dorsal digital expansion of the thumb. The muscle may also receive accessory slips from the long and short extensors of the thumb, opponens pollicis or from the styloid process of the radius.

Nerve supply. The lateral terminal branch of the median nerve, C. **8** and T. **1**.

Actions. The abductor pollicis brevis draws the thumb forwards in a plane at right angles to the palm of the hand and rotates it medially. The movement occurs both at the metacarpophalangeal joint and at the carpometacarpal joint, so that in full abduction of the thumb the digit is not in line with its metacarpal bone. 'Pure' abduction cannot occur because there is some degree of conjunct medial rotation accompanying the movement at both joints.

The opponens pollicis (5.80, 81) is deep to the abductor pollicis brevis. It arises from the tubercle of the trapezium and the flexor retinaculum, and is attached to the whole length of the lateral border, and the lateral half of the palmar surface, of the metacarpal bone of the thumb.

Nerve supply. The lateral terminal branch of the median nerve,

C. **8** and T. **1** and commonly a twig from the deep terminal branch of the ulnar nerve. Dissections by Day and Napier (1961), Forrest (1967), and electromyographic tests by Harness *et al.* (1974), indicate a double innervation in 92 out of 120 hands, and this must, therefore, be regarded as the usual arrangement.

Actions. The opponens pollicis flexes the metacarpal bone of the thumb, i.e. bends it medially across the palm of the hand, and rotates it medially (p. 471). By this combination, which is termed *opposition*, the palmar ('pulpal') surface of the terminal segment of the thumb can be brought into contact with corresponding parts of any of the fingers. It should be noted that the movements of flexion and medial rotation are not sequential but occur simultaneously, and the earlier phases of the displacement are also accompanied by some degree of abduction. The continuous smooth movement of opposition involves both carpometacarpal and metacarpophalangeal joints.

During opposition of the thumb, the fingers are flexed at their metacarpophalangeal joints, and to a variable degree of their interphalangeal joints. Thus the thumb may contact any part of the palmar surface of the finger from base to tip. Flexion of the fingers is also accompanied by some degree of lateral rotation (thus partially 'meeting' the pulp of the opposing thumb *en face*). This lateral rotation is least marked in the index finger and greatest in the minimus. In the latter there are also displacement at the carpometacarpal joint (*vide infra*—opponens digiti minimi).

The flexor pollicis brevis (5.82) is medial to abductor pollicis brevis. It arises by superficial and deep parts. The superficial head arises from the distal border of the flexor retinaculum and the lower part of the tubercle of the trapezium; it passes along the radial side of the tendon of the flexor pollicis longus, and is attached, by a tendon containing a sesamoid bone, to the radial side of the base of the proximal phalanx of the thumb. It is frequently more or less blended with the medial border of the opponens pollicis. The deep part arises from the trapezoid and capitate bones and from the palmar ligaments of the distal row of carpal bones. It passes deep to the tendon of the flexor pollicis longus and unites with the superficial head on the sesamoid bone and base of the first phalanx. The deep head varies much in size and may even be absent.

Nerve supply. The superficial head is usually supplied by the lateral terminal branch of the median nerve and the deep head by the deep branch of the ulnar nerve, C. **8** and T. **1**. Some variation of this arrangement is not uncommon (Day and Napier 1961).

Actions. The flexor pollicis brevis flexes the proximal phalanx of the thumb and, continuing to act, flexes the metacarpal bone and rotates it medially (p. 593). In the latter movement it cooperates with opponens pollicis.

The adductor pollicis (5.81) arises by oblique and transverse heads. The *oblique head* is attached to the capitate bone, the bases of the second and third metacarpal bones, the palmar ligaments of the carpus and the sheath of the tendon of the flexor carpi radialis. Most of its fibres converge into a tendon, which contains a sesamoid bone, unites with the tendon of the transverse head and is attached to the ulnar side of the base of the proximal phalanx of the thumb. Some of its deepest fibres may pass into the medial side of the dorsal digital expansion of the thumb. On the lateral side of the oblique head a considerable fasciculus, which passes deep to the tendon of flexor pollicis longus to join flexor pollicis brevis, has been described as the deep head of flexor pollicis brevis (Day and Napier 1961). The *transverse head* (5.81) is the deepest of the pollicial muscles. It is triangular, and arises from the distal two-thirds of the palmar surface of the third metacarpal; the fibres converge to be attached, with the oblique head of the muscle and the first palmar interosseous muscle, to the base of the proximal phalanx of the thumb. The two parts of the adductor vary in relative size and degree of connexion.

Nerve supply. The deep branch of the ulnar nerve, C. **8** and T. **1**.

Action. The adductor pollicis approximates the thumb to the palm of the hand and it acts to greatest advantage when the abducted and flexed thumb is opposed to the fingers in gripping.

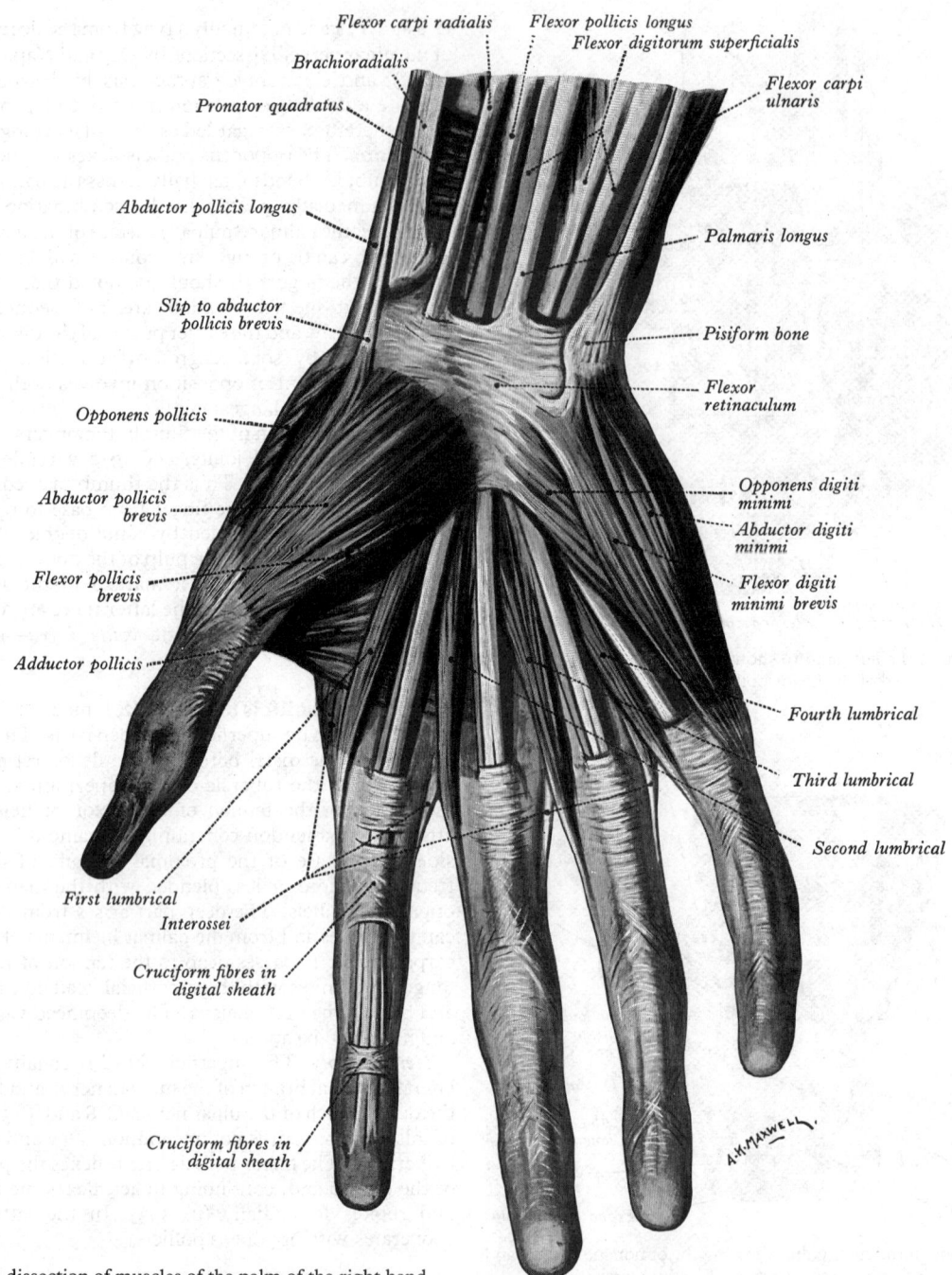

Flexor carpi radialis

Brachioradialis

Pronator quadratus

Abductor pollicis longus

Slip to abductor pollicis brevis

Opponens pollicis

Abductor pollicis brevis

Flexor pollicis brevis

Adductor pollicis

First lumbrical

Interossei

Cruciform fibres in digital sheath

Cruciform fibres in digital sheath

Flexor pollicis longus

Flexor digitorum superficialis

Flexor carpi ulnaris

Palmaris longus

Pisiform bone

Flexor retinaculum

Opponens digiti minimi

Abductor digiti minimi

Flexor digiti minimi brevis

Fourth lumbrical

Third lumbrical

Second lumbrical

A.K.MAXWELL.

5.82 Superficial dissection of muscles of the palm of the right hand.

2. THE HYPOTHENAR MUSCLES

Apart from palmaris brevis, a superficial muscle, this group consists of muscles which flex, abduct and oppose the fifth digit (**5**.79, 81, 82).

The palmaris brevis (**5**.79) is a thin, quadrilateral muscle, beneath the skin of the ulnar side of the palm. It arises from the flexor retinaculum and the medial border of the central part of the palmar aponeurosis; it is attached to the dermis on the ulnar border of the hand. It is superficial to the ulnar artery and the superficial terminal branch of the ulnar nerve.

Nerve supply. The superficial branch of the ulnar nerve, C. 8 and T. **1**.

Action. The palmaris brevis wrinkles the skin on the ulnar side of the palm of the hand and deepens the hollow of the palm by accentuating the prominence of the hypothenar eminence. In this way it may contribute to the security of the palmar grip.

The abductor digiti minimi (**5**.82) arises from the pisiform bone, the tendon of flexor carpi ulnaris and the pisohamate ligament. It ends in a flat tendon which divides into two slips, one

attached to the ulnar side of the base of the proximal phalanx of the little finger, the other to the ulnar border of the dorsal digital expansion of extensor digiti minimi. This muscle may be arranged in two or three slips; it may be fused with the flexor brevis. An additional slip may arise from the flexor retinaculum, antebrachial fascia, or tendons of palmaris longus or flexor carpi ulnaris. The muscle may be partly attached to the fifth metacarpal by a slip which sometimes arises from the pisiform.

Action. The abductor digiti minimi abducts the little finger away from the fourth. It thus takes part in the habitual spreading of the digits when they are extended.

The flexor digiti minimi brevis (**5**.82) lies on the radial side of the preceding muscle. It arises from the convex surface of the hook of the hamate bone and the palmar surface of the flexor retinaculum, and is inserted into the ulnar side of the base of the proximal phalanx of the little finger with the abductor digiti minimi. Its origin is separated from that of the abductor by the deep branches of the ulnar artery and nerve. This muscle may be missing, or may be fused with the abductor. It may have a slip attached to the distal end of the fifth metacarpal.

Action. The flexor digiti minimi brevis flexes the little finger at its metacarpophalangeal joint.

The opponens digiti minimi (5.81) is of a triangular form, and placed under cover of the flexor and abductor. It arises from the convexity of the hook of the hamate bone, and contiguous portion of the flexor retinaculum; it is inserted into the whole length of the ulnar margin of the fifth metacarpal bone. It is often divided into two lamellae by the deep branches of the ulnar artery and nerve. The muscle is to a variable degree blended with its neighbours.

Action. The opponens digiti minimi draws the fifth metacarpal bone forwards and rotates it laterally at the carpometacarpal joint, thus deepening the hollow of the palm and, together with flexion and some lateral rotation at the metacarpophalangeal and interphalangeal joints, bringing the digit into opposition with the thumb.

Nerve supply. All the muscles of the minimus are supplied by the deep branch of the ulnar nerve, C. 8 and T. **1**.

3. THE LUMBRICAL AND INTEROSSEOUS MUSCLES

The lumbricals (5.82) are four small fasciculi which arise from the tendons of flexor digitorum profundus—the first and second from the radial sides and palmar surfaces of the tendons of the index and middle fingers respectively, the third, from the adjacent sides of the tendons of the middle and ring fingers, and the fourth, from the adjoining sides of the tendons of the ring and little fingers. Each passes to the radial side of the corresponding finger, and is attached to the lateral margin of the dorsal digital expansion of extensor digitorum covering the dorsal surface of the finger (5.71). Variations in the attachments of the lumbricals are common. Any of them may be unipennate or bipennate, arising in the latter case from adjoining tendons of flexor digitorum profundus, and also from that of flexor pollicis longus when the first is bipennate (Goldberg 1970). Accessory lumbrical slips may be attached to an adjacent tendon of flexor digitorum superficialis.

Nerve supply. The first and second lumbricals are supplied by the median nerve, C. 8 and T. **1**; the third and fourth lumbricals by the deep terminal branch of the ulnar nerve, C. 8 and T. **1**. The third lumbrical frequently receives a supply from the median nerve. This arrangement is remarkably constant but the manner in which the branches to the lumbricals arise varies. The first and second are also occasionally supplied by the deep terminal branch of the ulnar nerve (Mehta and Gardner 1961).

Actions. Since the lumbrical muscles link one tendon (flexor) to another (extensor), a unique arrangement in the human body, apart from the foot, their actions are likely to be complex (p. 591), depending not only upon their own length and tension, but also on the state of these parameters in the long flexors and extensors. Descriptions of lumbrical activities often disregard these facts, being based merely upon reference to a proximal attachment (considered as fixed), an insertion into the dorsal extensor apparatus, and a direction of pull slightly ventral or palmar to the axes of the metacarpophalangeal joints. Regarded because of the last feature as flexors, the lumbricals are probably only weakly so; since the axes of these joints appear to move ventrally during flexion, the lumbrical flexor effect does not necessarily increase with increasing flexion. Through their attachments to the dorsal expansion the lumbricals would be expected to extend the interphalangeal joints, and there is direct electromyographic evidence for this. As long ago as 1954 Backhouse and Catton confirmed this action by inserting needle electrodes into a number of 2nd lumbrical muscles, and Long and Brown (1964) have made similar observations. The well-known 'claw' deformity following lesions of the ulnar nerve provides further evidence.

When the middle, ring or little finger is flexed fully at the metacarpophalangeal and the proximal interphalangeal joints, its distal phalanx can neither be flexed nor extended by voluntary effort so long as the other fingers are kept extended. The inability to extend the distal phalanx under these conditions is due to the fact that, owing to the way in which the extensor digitorum tendons are inserted, the terminal phalanx can be extended only

when the middle phalanx also is extended. For a further analysis of the flexor-extensor apparatus of the digits *see* p. 591.

The *interossei* occupy the intervals between the metacarpal bones, and are divided into a dorsal and a palmar set.

The dorsal interossei (5.83A), are four bipennate muscles, each arising from the adjacent sides of two metacarpal bones, but more extensively from the metacarpal bone of the finger into which the muscle passes. They are attached to the bases of the proximal phalanges and the dorsal digital expansions (p. 579). Between the double origin of each of these muscles there is a narrow triangular interval; through the first of these intervals the radial artery passes, and through each of the others a perforating branch from the deep palmar arch. The *first*, largest muscle is sometimes named the *abductor indicis*; it is attached to the radial side of the proximal phalanx of the index finger and to the capsule of the adjoining metacarpophalangeal joint. The *second* and *third* are attached to the middle finger, the former to its radial, the latter to its ulnar side. Whereas the second generally reaches the digital expansion and the proximal phalanx, the third usually extends only to the digital expansion (5.71). The *fourth*, too, may be wholly attached to the digital expansion but it often sends an additional slip to the proximal phalanx.

The palmar interossei (5.83B) are smaller than the dorsal and are upon the palmar surfaces of the metacarpal bones, rather than between them. With the exception of the first, each of the four arises from the entire length of the metacarpal bone of one

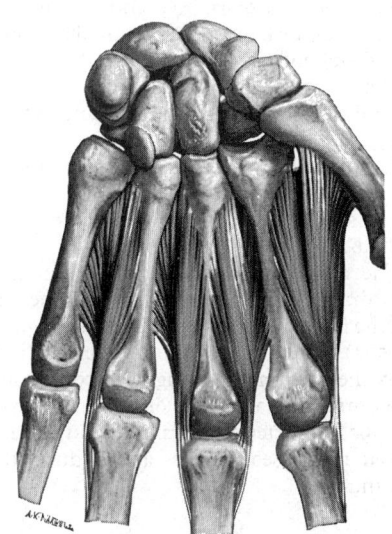

5.83A   The dorsal interosseous muscles of the left hand. Palmar aspect.

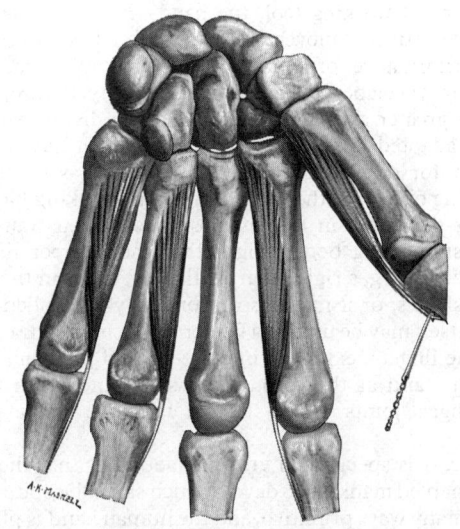

5.83B   The palmar interosseous muscles of the left hand. Palmar aspect.   589

finger, and passes to the appropriate side of the dorsal digital expansion (Salsbury 1937; Landsmeer 1949).

The *first* arises from the ulnar side of the palmar surface of the base of the first metacarpal bone and is inserted into the sesamoid bone on the ulnar side of the proximal phalanx and usually also into the dorsal digital expansion (Lewis 1965). It lies in front of the lateral head of the first dorsal interosseous muscle, and is overlapped anteriorly by the oblique head of adductor pollicis (5.81A). (Some authorities have preferred to regard the first palmar interosseous muscle as a deep stratum of the flexor pollicis brevis; but this view is now largely disregarded.) The *second* arises from the ulnar side of the second metacarpal bone, and is inserted into the same side of the digital expansion of the index finger. The *third* arises from the radial side of the fourth metacarpal bone, and is inserted in common with the third lumbrical (5.71). The *fourth* arises from the radial side of the fifth metacarpal bone, and is attached in common with the fourth lumbrical with an additional attachment to the base of the proximal phalanx. The attachment of these muscles to the base of each dorsal digital expansion (5.71) ensures the stability of the extensor tendon on the convexity of the head of the corresponding metacarpal bone during the movements of flexion and extension at the metacarpophalangeal joint. The interossei are not subject to gross variation; they are occasionally reduplicated, a condition perhaps associated with the origin of palmar interossei from paired short flexors (Lewis 1965).

Stack (1962) and others prefer to classify the interosseous muscles as 'proximal' and 'distal'. The former, chiefly the dorsal interossei, are attached to the proximal part of the extensor expansion, the 'hood' or 'dorsière', and to the base of the phalangeal tubercle at the base of the proximal phalanx (5.70, 71). The palmar or 'distal' interossei, apart from the first, are attached only to the extensor expansion, distal to the 'hood' and usually opposite to a lumbrical, both attachments being to the distal wing tendons (p. 579).

Nerve supply. All the interossei are supplied by the deep branch of the ulnar nerve, C. 8 and T. **1**. Rarely, the first dorsal interosseous is supplied exclusively by the median nerve (Sunderland 1946).

Actions. The dorsal interossei abduct the fingers from an imaginary axis drawn longitudinally through the centre of the middle finger; the palmar interossei adduct the fingers to this axis (see also p. 592). The interossei, in conjunction with the lumbricals, flex the proximal phalanges; in consequence of their insertions into dorsal digital expansions they are able, in certain conditions (p. 592), to extend the middle and distal phalanges. The first palmar interosseous flexes and adducts the proximal phalanx of the thumb.

THE NATURE OF MANUAL DEXTERITY

The human upper limb—with its mobile base, the shoulder girdle, its extensible and folding member, the arm and forearm, and its terminal working tool, the hand—has an extraordinary range and versatility of movement. One has but to see the slow and limited performance of even the most complex mechanical manipulators yet elaborated by human ingenuity, to recognize the immensely greater skill and repertoire of the living hand. Preeminently adapted for reaching and grasping, it has also much potentiality for non-prehensile activities, such as pushing and manipulating objects without grasping them, striking blows, and supporting the trunk in leaning positions. When transmitting thrust or supporting body weight the whole upper limb may become, like the leg, a rigid pillar, with most joints in their close-packed positions; or it may absorb forces by controlled flexion. The hand itself may be used as a fist, or forces may be transmitted through the fingers, extended in close-packed positions at most digital joints and at the carpus, but sometimes with the first interphalangeal joints slightly flexed, as in the toes at take-off (p. 617).

However, it is specially in grasping activities that the human hand has enabled mankind to develop such skills. Despite this, the hand is in many ways primitive, and the human hand is physically capable of only a limited range of movements, though with much

versatility within this range. It is important to recognize that human hands are not specialized in their movements beyond those of some other primates. Some, of course, like the gibbon, have specialized hands and have lost the power of opposition, but

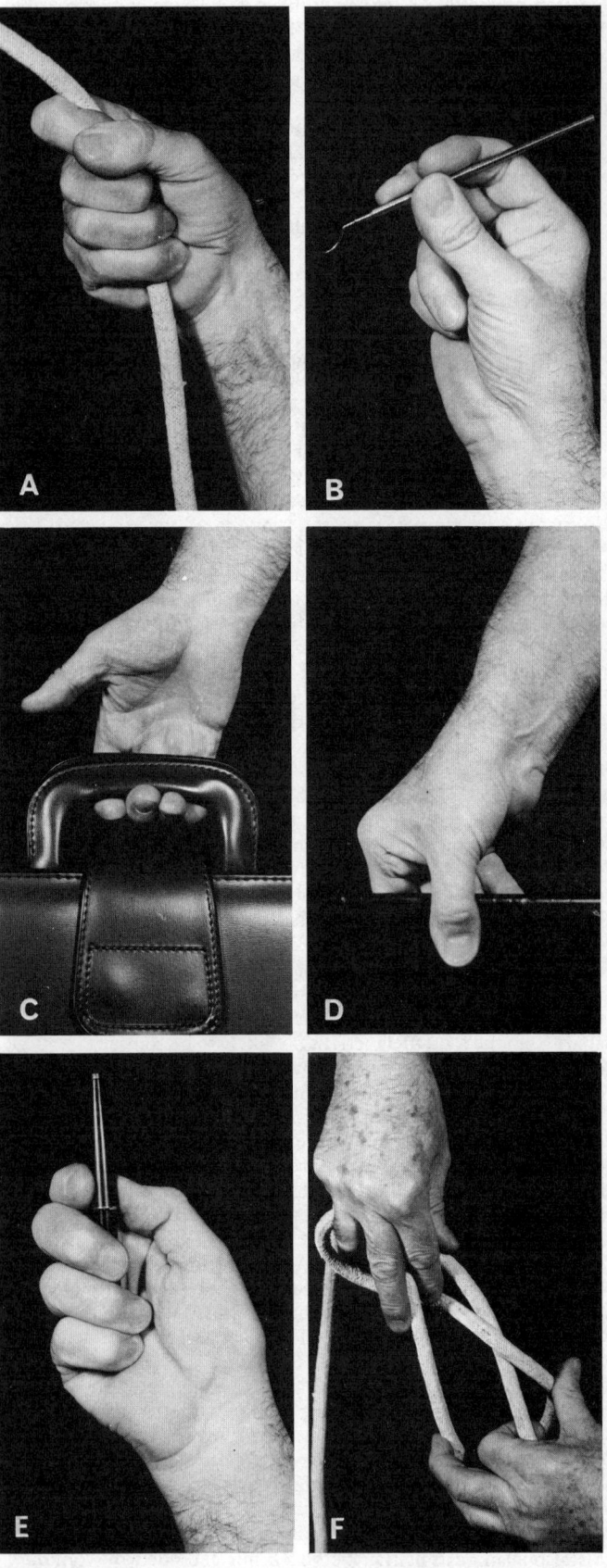

5.84 A few of the many varieties of functional postures which may be adopted by the human hand, A, power grip, B, precision grip, C, hook grip, D, pincers grip, E, simultaneous power and precision grip, F, complex manipulation.

the chimpanzee, gorilla and many monkeys have the same repertoire of hand movements as man. The difference lies in the range of uses to which they are put (Jones 1941). Mankind goes far beyond any other creature in his endlessly variable applications of manipulative skills, the hands being the responsive tools of his unrivalled mental development. A note of caution must be added on use of the terms 'primitive' and 'specialized', which may lead to a semantic morass. Biologically such terms are usually employed by morphologists, with little reference to function. In this sense the human hand is obviously more generalized in structure than most mammalian extremities, though it is in many minor structural details more highly adapted to its intricate functions, and is from this point of view extremely 'specialized' (Bishop 1964).

Prehensile activities have been variously analysed—as, for example, cylinder, ball, ring, pliers and pincer grips (Griffiths 1943), but a simpler view is possible, all such grips being regarded as variants of a *power grip* and a *precision grip* (Napier 1956). A combination of the two occurs in some activities, and the *hook grip* is in some features distinct from the other forms.

In a **power grip** (5.84) the fingers are flexed around an object, with counter pressure from the thumb, which is positioned to bring either its pad or its medial border firmly against the object held. The purpose is to hold an object firmly, either to be wielded as a tool or weapon by the whole hand, or to be worked upon by the other hand. Once the grip is made, the hand is held still or moved as a whole by the rest of the limb. Any skill in movements is due to the limb, including the carpus, and relative movements of the thumb and fingers are not involved in imparting skill.

In **the precision grip** (5.84) on the contrary, not only is the method of holding more precise, but small movements of the digits are usually, but not always, the main source of the adroit activities carried out. The object is gripped between the tips of the fingers and thumb, sometimes by all the former, more often by the thumb and index, with the middle finger frequently involved as a third partner, as in the grasping of a pen, pencil or the handle of any small tool. The positioning of the instrument may be carried out at the wrist, or it may even involve supination, pronation and small adjustments at the elbow and shoulder. In the finest work, however, the most characteristic actions are those in which the lumbricals and interossei act in associating metacarpophalangeal flexion with interphalangeal extension, and vice versa. These are utilized in the motions of advancing and withdrawing any small objects to or from another (e.g. a thread to the eye of a needle), one being held still in a power grip, and the other manipulated to and fro using a precision grip. Such activities are commonplace in picking up any small object for inspection, use, or manipulation. The flexor-extensor musculature of the digits is highly organized for such activities, and their individual contributions have been described and investigated in considerable detail (Long 1968; MacConaill and Basmajian 1969; Kapandji 1970).

Relatively small objects may be held steady in a power grip by flexion of the medial three digits against the palm, while the index and thumb carry out manipulations upon them by a precision grip. For example, a fountain pen may be thus held while its cap is screwed off or on. Thus a single hand may be used simultaneously for holding and manipulating.

The hook grip (5.84) is used to suspend or pull upon objects; it may also be used like the power grip, to suspend or elevate the body in climbing. The fingers are looped around such things as handles, straps, cords, branches, rocks, etc. and flexed towards the palm, to a degree depending upon the dimensions of whatever is grasped. The thumb may or may not be involved. It is a grip for the transmission of forces and not for skilful manipulation.

Thus, by flexion and extension of the wrist and digits, adduction and abduction of the digits, and opposition of the thumb, a relatively small repertoire of manual actions is possible. Accessory to the above, it is to be noted, are minor degrees of rotation in the fingers and a freer range of this in the thumb. Though the repertoire is essentially limited, the scope of the basic movements and the nicety of control with which they can be exercised, especially with practice, are unrivalled perquisites of man. This manual skill, guided by discriminative vision, and directed by a highly imaginative and inquisitive mentality into a

seemingly endless range of activities, has enabled mankind to master much of the natural environment, and to create around himself a culture of art, science and technology. The ever-increasing complexity of this artificial environment has carried us far beyond any other form of life of which we are aware.

THE MECHANICAL ANALYSIS OF MOVEMENT IN THE HAND

The mechanics of movements of the hand involve many factors, and not least amongst these are the effects of external resistance from the objects manipulated. These initiate a continuous feedback of stimuli, to which the motor apparatus, the extrinsic and intrinsic muscles, respond in intricate nuances of adaptive adjustment at a complex chain of articulations. Ignoring the carpus, each digit has four such joints (three in the thumb), whose individual changes must be concerted; and each digit is actuated by not less than six muscles (medius and annularis), and as many as nine (minimus), a total of thirty-six. It is therefore scarcely practicable to deal exhaustively with the myokinetics of manual movements, even if such analyses were fully supported by experiment or direct observation. In fact, the most detailed accounts (Wood Jones 1941; Napier 1965; Kaplan 1966; Stack 1973; and Landsmeer 1976) are a mixture of mechanical theorization and factual data. Such theoretical interpretations of the anatomical facts are, however, highly plausible, and the experimental data available are largely confirmatory. The following account is necessarily based upon this mixture of direct observation and derived interpretation.

The hand is clearly a flexor-extensor mechanism, complicated by adduction-abduction and slight rotational movements, the latter being usually conjunct in nature. To be added to this is opposition. The working, flexor aspect of the hand, the palm, is necessarily hollowed, the dynamic arching of the palm being carried proximally into the rigid concavity of the carpus. Analogies between this arched form and that of the foot have been indicated; but the hand is only occasionally weight-bearing, and this comparison will not be pursued here. In a prehensile apparatus it is to be expected that the flexors will exceed the extensors in available power. Electromyographic data show that in slow and less forceful flexion of the digits only the flexor digitorum profundus is active, the flexor digitorum superficialis adding its effect in more powerful and rapid gripping. Extension, usually accompanied by digital abduction, is of course usually a preparation for grasping. But digital manipulation demands more than simple extension and flexion; as already emphasized above, it is the dissocation of events at the metacarpophalangeal and interphalangeal joints which converts the hand from relatively crude grasping organ to an adroit manipulator. It is conceivable that separate flexors and extensors, operating in opponent pairs at each joint in the digital articular chain, might serve to produce independent flexor-extensor adjustments, as they do in the thumb. But the latter has only two phalanges; the triple phalangeal series of the other digits, with two intervening articulations, is difficult and perhaps even impossible to control completely with even such a multiplicity of long tendons from extrinsic muscles. However that may be, only one long extensor tendon is available; and this, through its attachment to the base of the proximal phalanx and to the deep transverse metacarpal ligament (p. 479 and 5.70B), is limited to extension at the metacarpophalangeal joint, at which point retraction of the extensor 'hood' becomes arrested. This entails that the part of the extensor tendon distal to this attachment must be of sufficient length to allow full interphalangeal flexion, even when the metacarpophalangeal joint is fully extended. In full flexion of an average digit the distance along its extensor aspect increases by 2 to 3 cm. Therefore there must exist some means by which the digital part of a long extensor tendon can be varied in its effective length. Mere elasticity might provide this; but elongation, to this extent, of structures which are largely collagenous is clearly impossible. In fact, the extensor apparatus (it is far more than a simple tendon) is so arranged as to exploit this apparent difficulty (5.71). The lateral bands of each extensor tendon, which end at the base of a terminal phalanx, can be tensed and drawn proximally by the intrinsic muscles attached to them, and can also be drawn apart by the same forces. This shortens the distal part of the extensor tendon. The arrangement resembles a diamond (5.71C), the forces being transmitted at its four angles. An increase in the transverse width of the diamond, by traction at its lateral angles (by the intrinsic muscles), shortens the total length of the extensor apparatus. This theoretically, would permit flexion or extension at the interphalangeal joints whatever the state of the metacarpophalangeal, and this is precisely what occurs in digital movements. Moreover the existence of proximal and distal attachments of the intrinsic muscles provides a potentiality for separate extension at the two interphalangeal joints.

The interosseous and lumbrical muscles thus supply the third and balancing force between the long or extrinsic digital flexors and extensors. Without this the fine and independent control of movements at three, serially arranged digital articulations would be impossible. This can be stated with confidence; the effects on finger movements of paralysis of the intrinsic muscles are well known, and the data of electromyography are convincing, even though limited.

Something more must be added in regard to the role of the lumbrical muscles. Like the interossei, their tractive effect is slightly ventral to the metacarpophalangeal joints, and they can therefore act here as somewhat inefficient flexors. Again like the interossei, they exert traction upon the extensor expansions, at their distal wings and in their case exclusively so. In contrast to the interossei they do not act from a static proximal attachment, but from the adjustable base of the deep flexor tendons. It is difficult to envisage the precise changes in the lengths of these muscles which must occur in different states of flexor-extensor balance, nor are any direct observations available. However, Stack (1962) has contributed a detailed analysis of such changes in length of the lumbrical (and interosseous) muscles in different positions of the digits, basing his estimates of the proximal positioning of the lumbricals upon excursions of the deep flexor tendons. These excursions have been assessed at 5 to 20 mm by Bunnell (1948), Kaplan (1953), and Landsmeer (1955). As will be appreciated from the diagram in 5.71, the lumbrical muscles are shown as most contracted and most distal in position in full digital extension, and on the contrary they are most elongated and proximal in full flexion. Stack interprets these reconstructions (which cannot be far from reality) as indicating that the intrinsic muscles, and especially the lumbricals, act as servo-mechanisms, the main power of extension being derived from the long digital extensor tendons, while the precise points of application of power at the interphalangeal joints is determined and transmitted by the intrinsic muscles. (It is perhaps reasonable, therefore, to enquire whether the *extensor* forces cannot also be exerted by the deep digital *flexor* via the lumbricals.) Whatever the real activities of the lumbrical muscles, it is undeniable that they perform a key role in digital movements. It has also been suggested that the stretch receptors in these muscles are of particular importance. This appears to be a reasonable assumption, although it is perhaps unwise to select for special emphasis any one such factor in the complex apparatus of digital movement.

The contribution of the *digital retinacular ligaments* (p. 580) to movements of the fingers is held to be an integration between the two interphalangeal joints. Flexion at the distal joints appears to tauten the ligaments on each side of the digit and to pull the proximal joint into flexion. Certainly it can be observed that when flexion at the proximal joint is prevented, flexion at the distal joint is limited by some resistance, and this seems most likely to be due to the retinacular ligaments. Similarly, the two joints are linked together in extension movements. Passive flexion of the distal joint automatically produces a like movement at the proximal joint. Stack (1962) has constructed an interesting model to illustrate these and other activities of the digits.

Nothing has yet been said of the slight rotations which occur at the digital joints. The interphalangeal joints are usually regarded as pure hinges or uni-axial articulations, but slight conjunct rotation also occurs (p. 472), more obvious rotation is present at the metacarpophalangeal joints, and rotational movements of the metacarpals at the carpometacarpal junctions also make a contribution to digital movements. When the fingers and thumb

are opposed together in precise manipulations all the digits except the medius undergo slight metacarpal rotation; in the case of the fingers these rotations are considered to be effected by the appropriate interossei (Braithwaite *et al.* 1948). This is necessary to bring together the tips of the fingers and thumb in various forms of precision grip (*vide supra*), which can involve more than the pollex and index. These two provide a '*pliers*' or '*pincer*' grip; any or all of the remaining digits may be added to produce varieties of '*chuck*' grip (5.84). In these and other flexor actions the metacarpal movements, though small (except in the thumb), are easily observed. In gripping any object powerfully with the whole hand the more mobile metacarpals (5th, 4th, 2nd) are flexed to the relative extents indicated at their carpometacarpal joints, and the 5th, at least, may also rotate a little. These movements are, however, dependent upon the shape, dimensions, and malleability of the object grasped. For example, in grasping a long cylinder across the whole palm metacarpal rotation cannot occur, unless the object is of small diameter or yielding consistency, such as a thin rope. In the latter case the flexion and rotation of the fifth metacarpal in particular contributes to the well-known effectiveness of the minimus in gripping small, elongated objects. In all such grasping movements the ultimate events at the articulations involved depend upon the size, contours and resistances presented by the object gripped as much as upon the deployment of muscular forces. For these reciprocally adaptive adjustments the multiplicity of skeletal components, and hence articulations, in the hand appears to be peculiarly suitable.

The thumb, in the greater accessibility to direct observation of its muscles and joints, presents less problems in descriptions of its activities than the other digits. But there is much more controversy in the definition and naming of its *cardinal* movements (compare Haines 1944, Napier 1955, de la Caffiniere 1970, and Pieron 1973), and especially in how these movements contribute to the *customary* pollicial activities. These latter are not, of course, the selected actions—flexion, extension, abduction, adduction, and circumduction—of formal analysis. These actions have dictated the names of muscles involved, despite the fact that few of them have the 'pure' or cardinal effects indicated by these names. For example, the abductor pollicis brevis is not merely an abductor of the thumb, for it also assists in medial rotation towards the palm, a combination of effects which is clearly adapted to the initial stages of opposition, 'pure' abduction being a much less frequently required movement.

It is important to clarify at this point the cardinal movements alluded to. Since the carpometacarpal joint of the thumb is sellar the planes of the two sets of cardinal 'swings'—flexion-extension and adduction-abduction—must be approximately at right angles to each other; but concerning the orientation of these planes with respect to the rest of the hand there is considerable variation in different accounts, and this perhaps is the source of controversies in naming the movements. Commonly the plane of adduction-abduction is regarded as being approximately at right angles to the plane of the palm, and this entails that flexion and extension are carried out parallel to the latter plane. Inspection clearly shows, however, that major axes of the sellar carpometacarpal joint of the thumb are not so orientated, the ridge on the articular surface of the trapezium (along which the metacarpal base slides in adduction and abduction) being inclined at an angle nearer to 60° than 90° with respect to the 'palmar' plane. This entails that abduction of the thumb is not only away from the palm but also somewhat laterally. Doubtless the plane of this movement varies individually, and even to regard it as a plane is probably a simplification. The sellar surfaces at this joint are not perfectly symmetrical; Kuczynski (1974) has examined a short series (15 hands) in great detail, and has shown that the contours of the surfaces are such that pure swings are unlikely, being accompanied in fact by slight rotation. He also considers that the rotations are greater in flexion and extension. This is perhaps to be expected in a digit which is constantly moving from positions of opposition to an open grasp, with the thumb somewhere between extension and abduction. As both Kuczynski and Kapandji (1970) emphasize, movements of the thumb should be regarded in the light of what is occurring at all its joints simultaneously. The medial metacarpal rotation, which occurs with flexion and adduction in the action of opposition, is accompanied by a similar rotation at the metacarpophalangeal joint. Medial rotation at the latter joint is usually compounded with some abduction of the proximal phalanx, as in the act of opposing the tips of the thumb and fifth digit. The movements occurring at both ends of the pollicial metacarpal could be integrated by the thenar muscles, all of which, except the opponens, act on the carpometacarpal and metacarpophalangeal joints. The muscles of the thumb, both intrinsic (thenar) and extrinsic, have been likened to a series of stays or guy-ropes, which can be altered in length to adjust either of the above-mentioned to a new position, from which the rest of the digit can then be manipulated. The movements of the terminal phalanx at the interphalangeal joint are controlled by the long flexor and extensor. Having only two phalanges the thumb does not present the same extensor problems as do the other digits, and no complex extensor expansion is required here.

In the position of rest the thumb is in contact with or, perhaps more often, close to the lateral border of the palm and the index finger. The rest of the digits are loosely flexed and to a degree increasing from the second to the fifth. The metacarpal heads form an arch, corresponding in its palmar concavity to that of the carpus, at which the hand is about half extended. The whole apparatus, though 'open', is semi-flexed or 'cupped' and ready for any grasping or oppositional effort required of it. For larger objects an initial extensor 'spread' may be necessary, but this is only a preliminary to active progress towards some form of power or precision grip. If the latter involves all the digits, as in holding a ball between their tips, the variable mobility of the metacarpals is

particularly noticeable; this allows the fingertips to turn laterally towards the thumb and to space themselves at different angles around such an object. The opponens minimi digiti enables the most mobile of the metacarpals to assist in bringing the fifth digit in complete opposition to the thumb across the diameter of larger spherical objects. The influence of ligaments, especially at the carpometacarpal joint, in pollicial movements has received little attention. The intermetacarpal ligament, between the first and second metacarpal bases (described by Haines 1944), has been re-examined by Bojsen-Møller, who regards it as particularly significant in such movements.

All of the movements described above are concerned with prehension; but the hand can also be used in transmitting force to objects, as in pushing or pressing down upon surfaces. The forces to be transmitted may then be very great, amounting even to the weight of the whole body. In dynamic situations, where the hand is used to fend off moving bodies or to avoid collision with static obstructions, the forces transmitted through the outstretched and extended hand may be even greater. Most of the joints in the hand and wrist are then in the close-packed position, and structures on the flexor aspect, tendons, ligaments, and fasciae are tensed to produce maximum resistance to further extension.

The great variety and skill of manual movements is sustained by a most complex motor equipment and associated sensory apparatus, both cutaneous and neuromuscular. It is therefore not surprising that the hand monopolizes a greater area of motor and sensory cortex (7.131 A) than the rest of the arm or the whole of the leg. The importance of the thumb is epitomized in the allocation to it of about half of the cortical 'areas' for the hand.

THE FASCIAE AND MUSCLES OF THE LOWER LIMB

The muscles of the lower limb may be conveniently considered in the following groups:

I. Muscles of the iliac region.
II. Muscles of the femoral and gluteal regions.
III. Muscles of the leg.
IV. Muscles of the foot.

I. The Muscles of the Iliac Region

The three muscles of this group arise from the lumbar vertebral column (psoas major et minor) and the iliac bone (iliacus), and two are attached together on the femur as flexors. The psoas minor falls short of this and acts on the spine and sacro-iliac joint.

The iliac fascia covers the psoas and iliacus. It is thin above, but gradually thickens towards the inguinal ligament.

The *part covering the psoas* is thickened above as the medial arcuate ligament (p. 549). Medially, the fascia over the psoas is attached by a series of arched processes to the intervertebral discs, margins of vertebral bodies, and the upper part of the sacrum. Laterally, above the iliac crest, it blends with the fascia anterior to quadratus lumborum (p. 559), below the crest, with the fascia covering the iliacus.

The *iliac part* is connected, laterally, to the whole of the inner lip of the iliac crest, and medially, to the pelvic brim, where it blends with the periosteum. It is attached to the iliopectineal eminence and there receives a slip from the tendon of the psoas minor, when present. The external iliac vessels are anterior to the fascia but the branches of the lumbar plexus are posterior to it; it is separated from the peritoneum by loose extraperitoneal tissue.

Lateral to the femoral vessels, the iliac fascia is continuous with the posterior margin of the inguinal ligament and the transversalis fascia. Medially it passes behind the femoral vessels to become the pectineal fascia and attached to the pecten pubis. At the junction of its lateral and medial parts it is attached to the iliopectineal eminence and the capsule of the hip joint. It thus forms a septum between the inguinal ligament and the hip bone dividing the space here into a lateral part, the *lacuna musculorum*, containing psoas

major, iliacus and the femoral nerve, and a medial part, the *lacuna vasorum*, transmitting the femoral vessels. The downward continuation of the iliac fascia forms the posterior wall of the femoral sheath (p. 725).

It should be noted that the segmental values of the nerves to the muscles of the lower limb are those given by Sharrard (1955).

The psoas major (5.85) is a long fusiform muscle lateral to the lumbar region of the vertebral column and the pelvic brim. It arises from—(1) the anterior surfaces and lower borders of the transverse processes of all the lumbar vertebrae, (2) by five digitations, each from the bodies of two vertebrae and their intervertebral disc, the highest slip arising from the lower margin of the body of the twelfth thoracic vertebra, the upper margin of the body of the first lumbar vertebra and the interposed disc, the lowest from the adjacent margins of the bodies of the fourth and fifth lumbar vertebrae and the interposed disc; and (3) from a series of tendinous arches extending across the constricted parts of the bodies of the lumbar vertebrae between the preceding slips. The lumbar arteries and veins, and filaments from the sympathetic trunk, pass medial to these arches. The muscle descends along the pelvic brim, continues posterior to the inguinal ligament and anterior to the capsule of the hip joint, and finally converges into a tendon which, having received on its lateral side nearly the whole of the fibres of the iliacus, becomes attached to the lesser trochanter of the femur. The large subtendinous iliac bursa, which occasionally communicates with the cavity of the hip joint, separates the tendon from the pubis and the capsule of the joint. The somewhat complex vertebral attachments of the psoas major sometimes display minor numerical variations.

Relations. The upper limit of psoas major is posterior to the diaphragm in the lowest part of the posterior mediastinum. It may be in contact with the posterior extremity of the pleural sac. In the abdomen its *anterolateral surface* is related to the medial arcuate ligament, its own fascia, extraperitoneal tissue and peritoneum, the kidney, psoas minor, renal vessels, ureter, testicular (or ovarian) vessels, and genitofemoral nerve. In front the right psoas

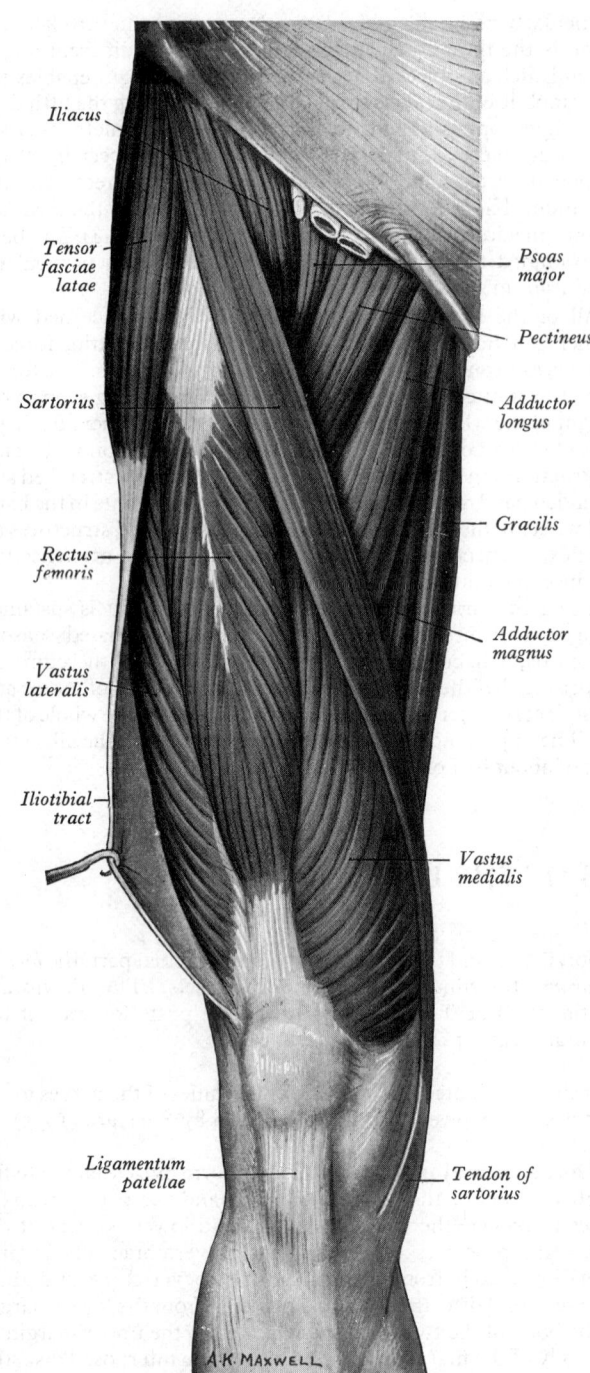

5.85 Superficial muscles of the thigh. Extensor aspect. (From Quain's *Anatomy*, 11th edition.)

is overlapped by the inferior vena cava and crossed by the end of the ileum, and the left is crossed by the colon. Its *posterior surface* is in relation with the transverse processes of the lumbar vertebrae, and the medial edge of the quadratus lumborum. The lumbar plexus is situated posteriorly in the substance of the muscle (p. 1106). *Medially* the muscle is related to the bodies of the lumbar vertebrae and lumbar vessels. Along its anteromedial margin the muscle is in contact with the sympathetic trunk, aortic lymph nodes and, along the pelvic brim, with the external iliac artery. This margin is covered by the inferior vena cava on the right side, and on the left side lies posterior and lateral to the abdominal aorta. In the thigh it is related to the fascia lata and the femoral artery *in front*, *behind*, to the capsule of the hip joint, from which it is separated by a bursa, at its *medial border*, to pectineus and medial circumflex femoral artery, and also to the femoral vein, which may overlap it slightly, at its *lateral border*, to the femoral nerve and the iliacus. The femoral nerve descends at first through

the fibres of psoas major, and then in the furrow between it and iliacus.

Nerve supply. From the ventral rami of the lumbar nerves, L. **1**, **2** and 3.

Actions. The psoas major acts conjointly with the iliacus (*vide infra*).

The psoas minor is anterior to psoas major within the abdomen. It arises from the sides of the bodies of the twelfth thoracic and first lumbar vertebrae and from the disc between them. It ends in a long, flat tendon which is attached to the pecten pubis and iliopectineal eminence and, laterally, to the iliac fascia. This muscle is absent in about 40 per cent of subjects.

Nerve supply. A branch from the first lumbar nerve.

Action. The psoas minor is a weak flexor of the trunk.

The iliacus (**5**.85) is a triangular sheet of muscle which arises from the superior two-thirds of the concavity of the iliac fossa, from the inner lip of the iliac crest, from the ventral sacro-iliac and iliolumbar ligaments, and from the upper surface of the lateral part of the sacrum (**3**.43); in front, it reaches as far as the anterior superior and anterior inferior iliac spines, and receives a few fibres from the upper part of the capsule of the hip joint. Most of its fibres converge into the lateral side of the tendon of psoas major, but some of them are attached directly to the femur for about 2·5 cm below and in front of the lesser trochanter.

Relations. In the abdomen, the *anterior surface* of iliacus is related to its fascia, which separates the muscle from extraperitoneal tissue and peritoneum, to the lateral femoral cutaneous nerve, on the right side, the caecum, on the left side, the iliac part of the descending colon; its *posterior surface* is related to the iliac fossa, its *medial border*, to the psoas major and femoral nerve. In the thigh, its *anterior surface* is in contact with the fascia lata, rectus femoris, sartorius and arteria profunda femoris, its *posterior surface* with the capsule of the hip joint, a bursa common to it and psoas major being interposed.

Nerve supply. Branches of the femoral nerve, L. **2** and 3.

Actions. The psoas major, acting from above, flexes the thigh upon the pelvis, being assisted by iliacus. Electromyographic studies do not support the commonly held view that the psoas major acts as a medial rotator of the hip joint, but activity has been described in *lateral* rotation, particularly in the young. This view has received support from dissection studies on infants (McKibbin 1968). However, because of the variety of conflicting views over a number of years, further confirmation of these results must be awaited. When the psoas major and iliacus of both sides act from below, they contract powerfully to bend the trunk and pelvis forwards against resistance, as in raising the trunk from the recumbent to the sitting posture.

Geometrical reasoning suggests that when the body is erect and the lower limb fixed in position, contraction of one psoas major might flex the vertebral column (and trunk) forwards and laterally; electromyography, however, again does not support such a prediction, indicating maximum activity when the lumbar curvature is increased. However, other investigators, who carried out direct electromyography of the muscle during the operation of sympathectomy in the lumbar region, suggest that in addition to its role as a hip flexor, it has important activities in balancing the trunk while sitting (Keagy *et al.* 1966).

In symmetrical standing, electromyographic recordings from iliopsoas have varied from electrical silence to *slight* intermittent or continuous activity. These results probably reflect differences between individuals and between techniques of investigation. Nevertheless, most are agreed that the low level of activity correlates well with the fact that the line of body weight falls behind the transverse axis of the hip joints, and that the latter are near their close-packed positions, with spiralization and tautening of the ligaments (especially the iliofemoral) and marked compression and congruence of the articular surfaces, these passive mechanisms largely counterbalancing the extending torque of the weight of the trunk.

Applied Anatomy. When the neck of the femur is fractured the psoas major acts as a lateral rotator of the femur, thus accounting for the characteristic posture of the lower limb. Occasionally pus arising from disease of the thoracic and lumbar vertebrae tracks into the thigh within the fascial sheath of the psoas major.

II. The Muscles of the Thigh and Gluteal Region

The musculature of the thigh and hip can be divided on functional and morphological grounds into flexors and extensors; but in consequence of the rotation of the limb into a new working position in land vertebrates (p. 350), the primitive dorsal or extensor surface of the thigh has become anterior, and the ventral aspect containing the flexor muscles is now posterior. This change has not affected the 'girdle' muscles; iliopsoas, a flexor, already described above, is still ventral in position, and the gluteal muscles, which represent primitive extensors, have preserved a dorsal situation. Moreover, the evolution of multiaxial function at the hip joint has been coupled with the development of adductor and abductor groups of muscles, so that the joint is surrounded on all sides, not merely ventrally and dorsally, by muscle forces collectively capable of moving the femur in any direction, or maintaining it if necessary in static postures, especially the erect position. These activities are considerably affected by gravity, which aids some movements and opposes others. Hence the muscle groups are unequal in power.

1. The Anterior Femoral Muscles

Included in this group (5.85) are the tensor fasciae latae, sartorius and rectus femoris which can act on the hip and knee joints, and the vasti medialis, lateralis, et intermedius which act only on the latter. The articularis genus, a derivative of the vastus intermedius, a retractor of the synovial capsule of the knee joint, completes the group. The rectus femoris and the vasti extend the knee joint through a common tendon and are hence considered as one muscle, the quadriceps femoris.

The superficial fascia of the thigh, consisting as elsewhere in the limbs of loose areolar tissue containing a variable amount of fat, splits into recognizable layers in places, and between these ramify superficial vessels and nerves. In the inguinal region it is thick and in two layers, between which are the superficial inguinal lymph nodes, great saphenous vein and other smaller vessels. Here, the *superficial* layer is continuous with the abdominal superficial layer; the *deep* layer, a thin fibro-elastic stratum, is best marked medial to the great saphenous vein and inferior to the inguinal ligament, extending between the subcutaneous vessels and nerves and the deep fascia, with which it fuses a little below the ligament. It completes the saphenous opening, blending with its circumference and with the femoral sheath (p. 725). Over the opening it is perforated by the great saphenous vein and other blood and lymph vessels, and is hence termed the sieve-like or *cribriform fascia*. Anterior to the patella the superficial fascia contains a large subcutaneous bursa.

The fascia lata, the **deep fascia** of the region, so termed because of its wide extent, is thicker in the proximal and lateral parts of the thigh where the tensor fasciae latae and an expansion from gluteus maximus are attached to it. It is thin posteriorly, and also in the proximal and medial region over the adductor muscles, but it is thicker around the knee, being strengthened there by expansions from the tendon of biceps femoris laterally, sartorius medially, and quadriceps femoris in front. The fascia lata is attached, above and behind, to the back of the sacrum and coccyx, laterally, to the iliac crest, in front, to the inguinal ligament and to the superior ramus of the pubis, medially, to the inferior ramus of the pubis, the ramus and tuberosity of the ischium, and the lower border of the sacrotuberous ligament. From the iliac crest it descends as a dense layer over gluteus medius to the upper border of gluteus maximus, where it splits into two layers, one passing superficial and the other deep to this muscle; at its lower border the layers reunite. Over the flattened lateral surface of the thigh,

the fascia lata is specially thickened as a strong band, the *iliotibial tract*. At the upper limit of the tract, where it splits into two, the tensor fasciae latae is attached to it and, posteriorly, the greater part of the tendon of gluteus maximus. Of these two layers the superficial ascends lateral to tensor fasciae latae to reach the iliac crest; the deeper layer passes up and medially, deep to the muscle, and blends with the lateral part of the capsule of the hip joint. Distally, the iliotibial tract is attached to the lateral condyle of the tibia, where it is superficial to and blends with an aponeurotic expansion from vastus lateralis. It stands out as a strong, visible ridge anterolateral to the knee, when the leg is extended. (For a clinical, morphological and comparative study of the iliotibial tract, *see* Kaplan 1958.) Below, the fascia lata is attached to all exposed bony points around the knee joint, such as the condyles of the femur and tibia, and the head of the fibula. These remarks apply particularly to the iliotibial tract (Evans 1977). On each side of the patella it is strengthened by transverse fibres from the lower parts of the vasti, which are thereby attached to it; the stronger lateral fibres are continuous with the iliotibial tract. The fascia lata is continuous with two intermuscular septa, which are attached to the whole of the linea aspera and its prolongations above and below: the lateral and stronger septum, extending from the attachment of gluteus maximus to the lateral condyle, is between the vastus lateralis in front and the short head of biceps femoris behind, and provides partial attachment for them; the medial and thinner septum is between the vastus medialis and the adductors and pectineus. Numerous smaller septa pass between the individual muscles ensheathing them and sometimes providing partial attachment for them.

Evans (1977) has drawn attention to the ligamentous arrangement of the iliotibial tract, suggesting that it is more substantial than appears necessary to serve the tensor fasciae latae and gluteus maximus. He regards it as a ligament stretched between the ilium and tibia (*vide infra*).

The saphenous opening (5.86) is an aperture in the deep fascia lateral and a little distal to the medial part of the inguinal ligament, through which pass the great saphenous vein and other smaller vessels. (Its approximate 'centre' is, in the adult, indicated by a point some 3 cm inferior, and 3 cm lateral, to the pubic tubercle.) The cribriform fascia, which is pierced by these structures, fills in the aperture and must be removed to reveal it. The fascia lata in this part of the thigh displays superficial and deep parts (these must not be confused with the superficial and deep layers of the superficial fascia described above).

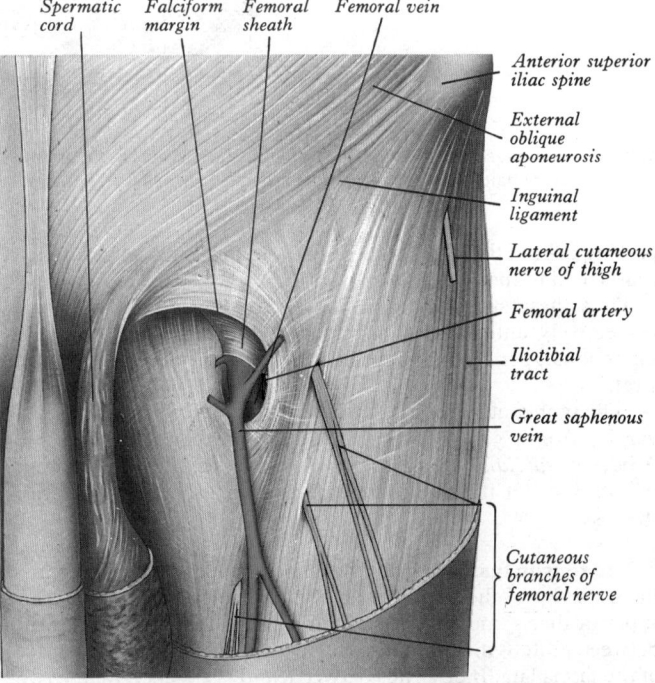

Spermatic cord / Falciform margin / Femoral sheath / Femoral vein

Anterior superior iliac spine

External oblique aponeurosis

Inguinal ligament

Lateral cutaneous nerve of thigh

Femoral artery

Iliotibial tract

Great saphenous vein

Cutaneous branches of femoral nerve

5.86 The left saphenous opening, after the removal of the cribriform fascia.

The *superficial stratum* is lateral to the saphenous opening. It is attached to the crest and anterior superior spine of the ilium, to the whole length of the inguinal ligament, and to the pecten pubis in conjunction with the lacunar ligament (p. 552). From the pubic tubercle it is reflected inferolaterally, as the arched *falciform margin*, which forms the superior, lateral and inferior boundaries of the saphenous opening (**5**.86); this margin is adherent to the anterior layer of the femoral sheath, and the cribriform fascia is attached to it. The falciform margin is considered to have *superior* and *inferior cornua*. The latter is well defined, and is continuous behind the great saphenous vein with the deep portion of the fascia lata.

The *deep stratum* is medial to the saphenous opening, and is continuous with the superficial stratum at its lower margin; traced upwards, it covers pectineus, adductor longus and gracilis, and,

muscle, however, varies in length, and its fibres may even reach the lateral femoral condyle. Its proximal attachment may extend on to the aponeurotic fascia superficial to gluteus medius.

Nerve supply. The superior gluteal nerve, L. 4 and 5.

Actions. The tensor fasciae latae acting through the iliotibial tract, extends the knee with lateral rotation of the leg; it may also assist in abduction and medial rotation of the thigh, but its role as an abductor has been denied (Kaplan 1958). In the erect posture, acting from below, it helps to steady the pelvis on the head of the femur; through the iliotibial tract it steadies the condyles of the femur on the tibia and thus helps to maintain the erect attitude. When the thigh is flexed against gravity and the knee is extended, an angular depression becomes apparent immediately below the anterior superior iliac spine. Its lateral boundary is formed by the tensor fasciae latae, which is active in flexion. The muscle aids

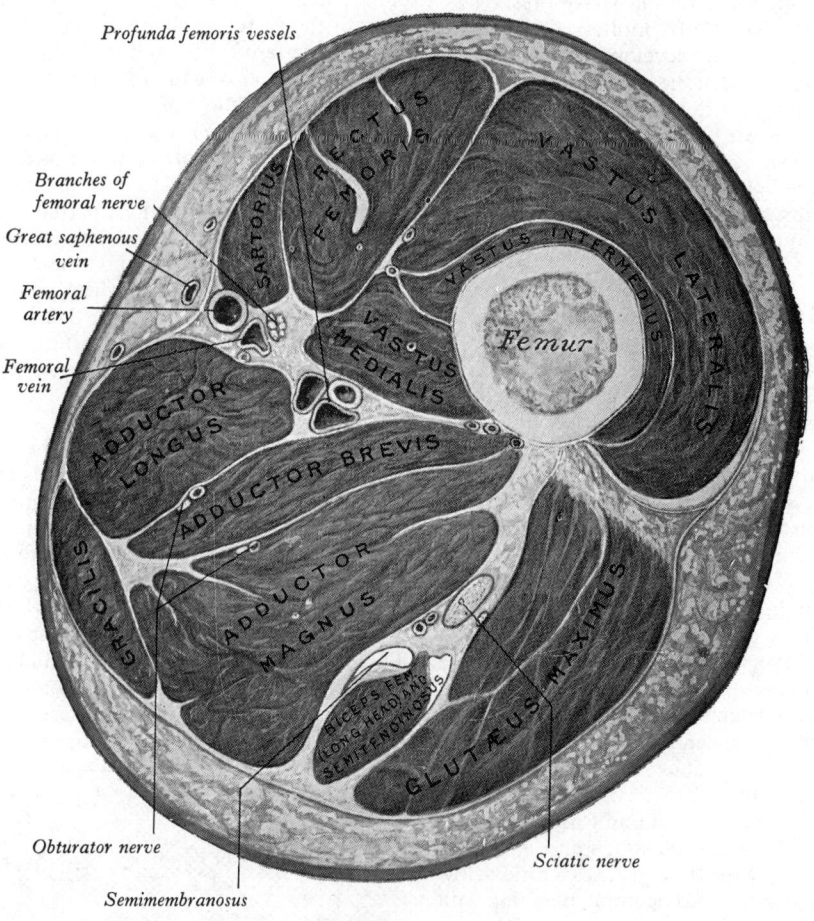

5.87 Transverse section through the right thigh at the level of the apex of the femoral triangle. Proximal (superior) aspect.

passing behind the femoral sheath, to which it is closely united, it is attached to the pecten pubis.

Thus these superficial and deep strata of the fascia lata are respectively anterior and posterior to the femoral sheath, the saphenous opening being formed by the continuity of the two strata.

The saphenous opening varies much in dimensions, its height varying from 1·5 to 8 or 9 cm, its width from about 1 to 3 or 4 cm. Adjacent *subsidiary openings* may exist, to transmit venous tributaries, but these openings are more usually in the floor of the fossa.

The tensor fascia latae (**5**.85) arises from the anterior 5 cm of the outer lip of the iliac crest, the lateral surface of the anterior superior iliac spine, and part of the border of the notch below it, between gluteus medius and sartorius, and from the deep surface of the fascia lata. It descends between, and is attached to, the two layers of the iliotibial tract of the fascia lata and ends about the junction of the middle and the upper third of the thigh. The

gluteus medius in postural abduction at the hip joint (*see* p. 617). Kaplan (1958) and Evans (1977) regard the iliotibial tract as of greater importance in steadying the pelvis than the tensor fasciae latae (*see* p. 595).

The sartorius (**5**.85, 87, 89) is a narrow strap muscle, and incidentally the longest in the body. It arises by tendinous fibres from the anterior superior iliac spine and the upper half of the notch below it. It crosses the thigh obliquely to the medial side, then descends more vertically to the medial side of the knee, where a thin, flattened tendon replaces the muscle fibres and curves obliquely forwards to expand into a broad aponeurosis attached in front of gracilis and semitendinosus to the upper part of the medial surface of the tibia (**5**.98). The upper part of the aponeurosis is curved backwards over the upper edge of the tendon of gracilis to be attached behind it. An offset, from its upper margin, blends with the capsule of the knee joint, and another, from its lower border, with the fascia on the medial side of the leg. This muscle varies little, save in relative bulk. The muscle is, however, occasionally bicipital at its proximal end, the additional slip being attached to the pectineal line or to the femoral sheath; Bhatnagar and Narayan (1959) have described three instances. It is occasionally absent.

Relations. In the upper third of the thigh the muscle's medial border forms the lateral side of the *femoral triangle*, the medial side of which is the medial border of adductor longus, and the base the inguinal ligament; the femoral artery descends through the middle of this triangle from base to apex. In the middle third of the thigh, the femoral artery is contained in the adductor (subsartorial) canal, anterior to which is a strong stratum of deep fascia and sartorius (**5**.89). This fascia bridges the interval between the adductors and quadriceps. It must be incised to expose the vessels.

Nerve supply. The femoral nerve, L. 2 and 3.

Actions. The sartorius assists in flexing the leg on the thigh, and the thigh on the pelvis, and its action is called for particularly when the two movements are carried out simultaneously. It also helps to abduct the thigh and to rotate it laterally. (A combination of these movements, together with inversion of the foot, brings the sole of the foot into direct view.) When acting against gravity, as it usually is, the muscle can be both seen and felt in the living subject.

The quadriceps femoris (**5**.85, 87, 89) is the great extensor muscle of the leg, covering almost all of the front and sides of the femur. It can be divided into four parts, which have distinctive names. One occupies the middle of the thigh, and arises from the ilium; from its straight course it is called the *rectus femoris*. The other three arise from the shaft of the femur, which they surround, apart from the linea aspera, from the trochanters to the condyles; lateral to the femur is the *vastus lateralis*, medial to it the *vastus medialis*, and in front, the *vastus intermedius*.

The rectus femoris (**5**.85, 87, 89) is fusiform, and its superficial fibres are arranged in a bipennate manner, the deep fibres running straight down to an aponeurosis. It arises by two tendons—a straight head from the anterior inferior iliac spine, and a reflected one from a groove above the acetabulum and the fibrous capsule of the hip joint. The two unite at an acute angle, and spread into an aponeurosis which is prolonged downwards on the *anterior* surface of the muscle, and from this the muscular fibres arise. The muscle ends in a broad and thick aponeurosis which occupies the lower two-thirds of its *posterior* surface, and gradually narrows into a flattened tendon, attached to the base of the patella. This is the superficial central part of the quadriceps tendon.

The vastus lateralis (**5**.85, 87, 89) is the largest part of quadriceps femoris. It arises by a broad aponeurosis, attached to the upper part of the intertrochanteric line, the anterior and inferior borders of the greater trochanter, the lateral lip of the gluteal tuberosity, and the upper half of the lateral lip of the linea aspera (**3**.185, **5**.89). This aponeurosis covers the upper three-fourths of the muscle, and from its deep surface many additional fibres arise. A few fibres also arise from the tendon of gluteus maximus and the lateral intermuscular septum between the vastus lateralis and short head of biceps femoris. The muscular mass thus formed is attached to a strong aponeurosis on the deep surface of the lower part of the muscle; this contracts into a flat tendon, attached to the base and lateral border of the patella, and blends with the quadriceps femoris tendon (*vide infra*). It gives to the capsule of the knee joint an expansion which descends to the lateral condyle of the tibia and blends with the iliotibial tract.

The vastus medialis and intermedius appear to be inseparably united, but when rectus femoris is reflected a narrow cleft can be seen extending upwards from the medial border of the patella between the two sometimes as far as the lower part of the intertrochanteric line, where, however, the two muscles are frequently fused.

The vastus medialis (**5**.85, 87, 89) arises from the lower part of the intertrochanteric line, spiral line, medial lip of the linea aspera, upper part of the medial supracondylar line, the tendons

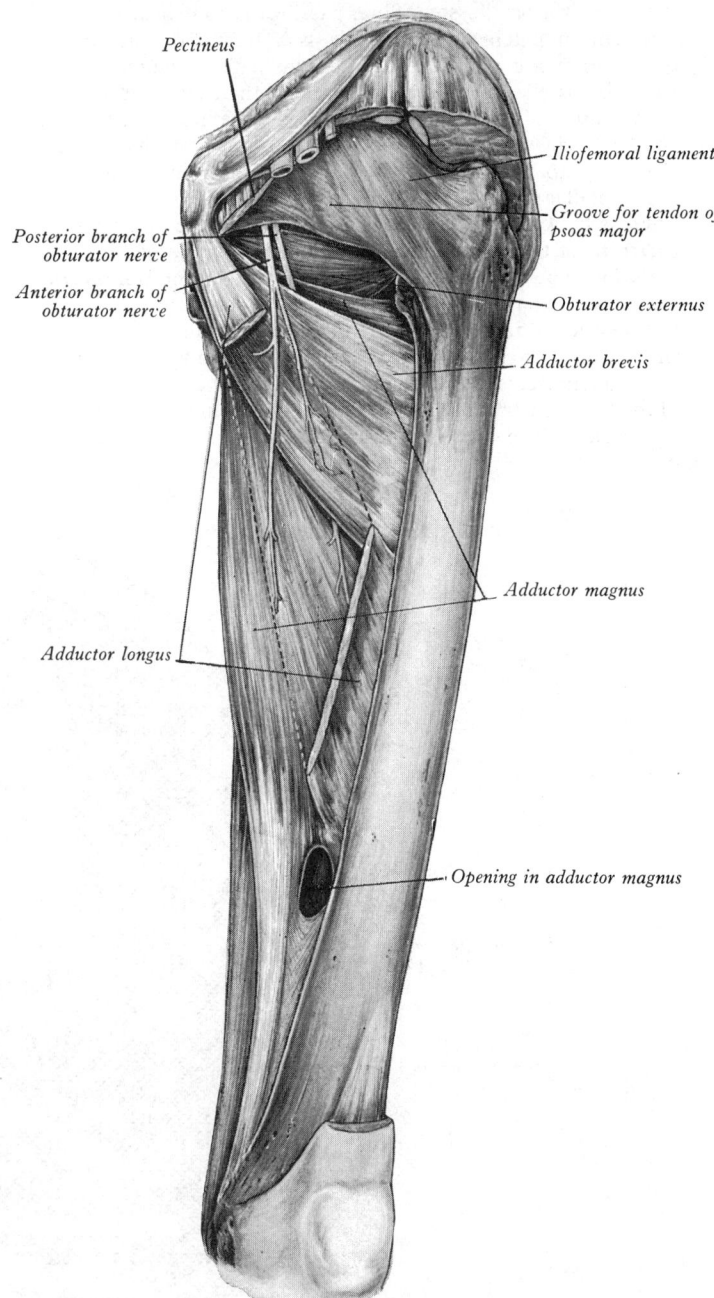

5.88 The adductor muscles of the left thigh. Anterior aspect. Part of the adductor longus has been excised.

Pectineus

Iliofemoral ligament

Groove for tendon of psoas major

Posterior branch of obturator nerve

Anterior branch of obturator nerve

Obturator externus

Adductor brevis

Adductor magnus

Adductor longus

Opening in adductor magnus

of adductor longus and magnus, and the medial intermuscular septum (*see* p. 595). Its fibres pass downwards and forwards, and are chiefly attached to an aponeurosis on the deep surface of the muscle which is attached to the medial border of the patella and the quadriceps femoris tendon (*vide infra*). An expansion from this aponeurosis reinforces the capsule of the knee joint and is attached below to the medial condyle of the tibia. The lowest fibres are almost horizontal and form a bulge in the living subject, medial to the upper half of the patella.

The vastus intermedius (**5**.87, 89) arises from the front and lateral surfaces of the upper two-thirds of the femoral shaft, and from the lower part of the lateral intermuscular septum. Its fibres end in an aponeurosis on the anterior surface of the muscle; this aponeurosis forms the deep part of the quadriceps femoris tendon and, in addition, is attached to the lateral border of the patella and the lateral condyle of the tibia.

The tendons of the four divisions of the quadriceps unite in the lower part of the thigh to form a single strong tendon attached to

597

the base of the patella, some fibres passing over it to blend with the ligamentum patellae. The patella is a sesamoid bone in the quadriceps tendon, and the 'ligamentum' patellae, which extends from the patellar apex to the tubercle of the tibia, is in fact the continuation of the tendon of the muscle, the medial and lateral patellar retinacula (p. 484) being expansions from its borders. The suprapatellar bursa is between the femur and the suprapatellar part of the quadriceps tendon; the deep infrapatellar bursa is between the ligamentum patellae and the upper part of the front of the tibia (4.75). The quadriceps is subject to little variation. The rectus femoris may arise from the anterior *superior* iliac spine, and its reflected head may be absent.

The articularis genus, a small muscle, usually distinct from the vastus intermedius but occasionally blended with it, consists of several muscular bundles which arise from the anterior surface of the lower part of the shaft of the femur and are attached to the upper part of the synovial membrane of the knee joint (DiDio *et al.* 1967).

The articularis genus retracts the synovial membrane of the knee joint proximally during extension of the leg, thereby preventing interposition of redundant synovial folds between patella and femur (c.f. the articularis cubiti or subanconeus at the elbow—p. 573).

2. The Medial Femoral Muscles

This group is derived, as its nerve supply suggests, from both the flexors and extensors between which it lies. All five muscles— gracilis, pectineus, and the adductores longus, brevis et magnus, cross the hip joint, but only gracilis reaches beyond the knee. Collectively they are known as the adductors of the thigh, but their actions are more complex than this.

The gracilis (5.85, 87, 89) is the most superficial of the adductor group. It is thin and flattened, broad above, narrow and

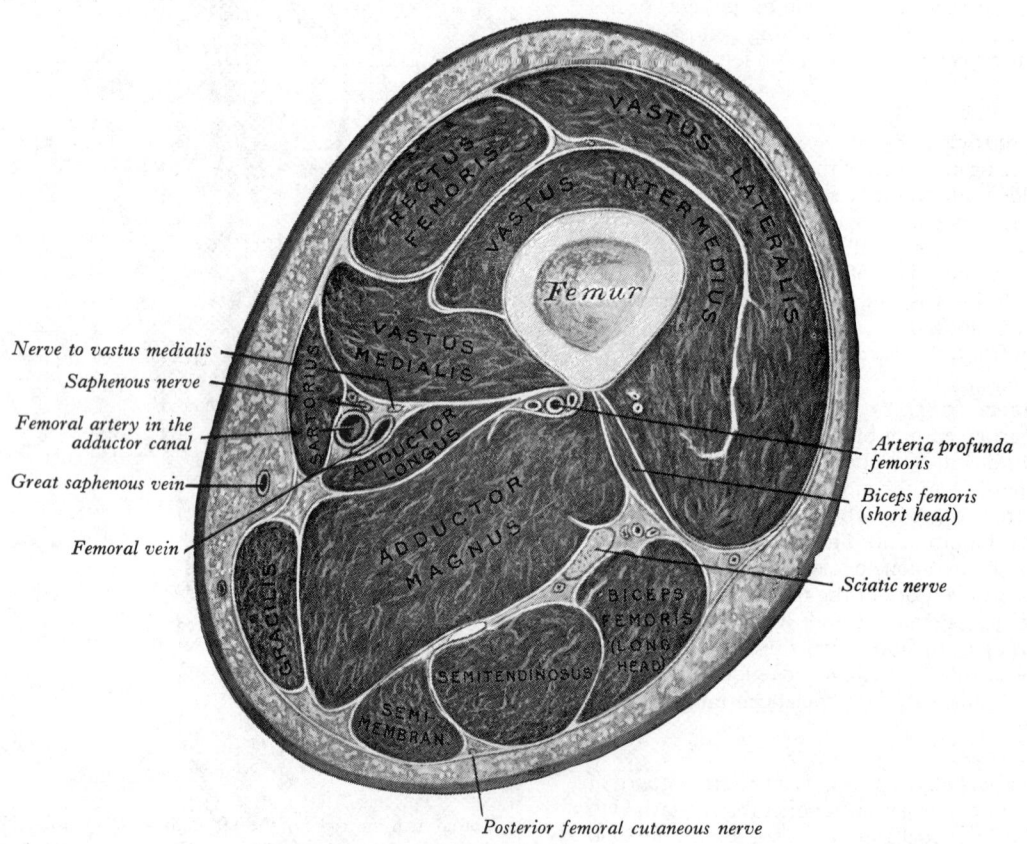

5.89 Transverse section through the middle of the right thigh. Proximal (superior) aspect.

Nerve supply. The quadriceps femoris and the articularis genus are supplied by the femoral nerve, L. 2, **3** and **4**.

Actions. The quadriceps femoris extends the leg upon the thigh. The rectus femoris also assists in flexing the thigh on the pelvis, or, if the thigh be fixed, it helps to flex the pelvis; it is, however, remarkably quiescent in standing (Joseph and Williams 1957). The rectus can, of course, flex the hip and extend the knee simultaneously. The lower fibres of the vastus medialis contract particularly during the terminal phase of extension of the knee joint to retain the patella in its groove on the patellar surface of the femur by counteracting the natural tendency to lateral displacement of the patella, which is attributable to the angulation between the shaft of the femur and the bones of the leg (Lieb and Perry 1968). It has been suggested that the lower fibres of the vasti medialis et lateralis are involved in stabilizing the knee joint. Electromyographic studies indicate that the three vasti are not equally or synchronously active in various phases of extension of the knee or rotation of the thigh, but confirmed details are not yet available.

tapering below. It arises by a thin aponeurosis from the medial margins of the lower half of the body of the pubis, the whole of its inferior ramus and the adjoining part of the ramus of the ischium (3.171). The fibres descend vertically into a rounded tendon which passes across the medial condyle of the femur posterior to the tendon of sartorius. It then curves round the medial condyle of the tibia, where it becomes flattened, and is attached to the upper part of the medial surface of the tibia, below the condyle. A few fibres from the lower part of the tendon are prolonged into the deep fascia of the leg. At its termination the tendon is immediately proximal to that of semitendinosus, and its upper edge is overlapped by the tendon of sartorius, with which it is in part blended. It is separated from the tibial collateral ligament of the knee joint by the tibial intertendinous bursa (p. 487).

Nerve supply. The obturator nerve, L. **2** and 3.

Actions. The gracilis flexes the leg and rotates it medially; it may also act as an adductor of the thigh.

The pectineus (5.85) is a flat, quadrangular muscle, in the femoral triangle. It arises from the pecten pubis, and to a slight extent from the bone in front of it, between the iliopectineal eminence and the pubic tubercle, and from the fascia on its own anterior surface; the fibres pass downwards, backwards and laterally, to be attached along a line leading from the lesser trochanter to the linea aspera. The muscle may be bilaminar, as it is in some other mammals, the two layers receiving separate nerve supplies (*vide infra*). It may be partially or wholly attached to the capsule of the hip joint.

Relations. Its *anterior surface* is related to the fascia lata, which separates it from the femoral vessels and great saphenous vein, its *posterior surface*, to the capsule of the hip joint, the adductor

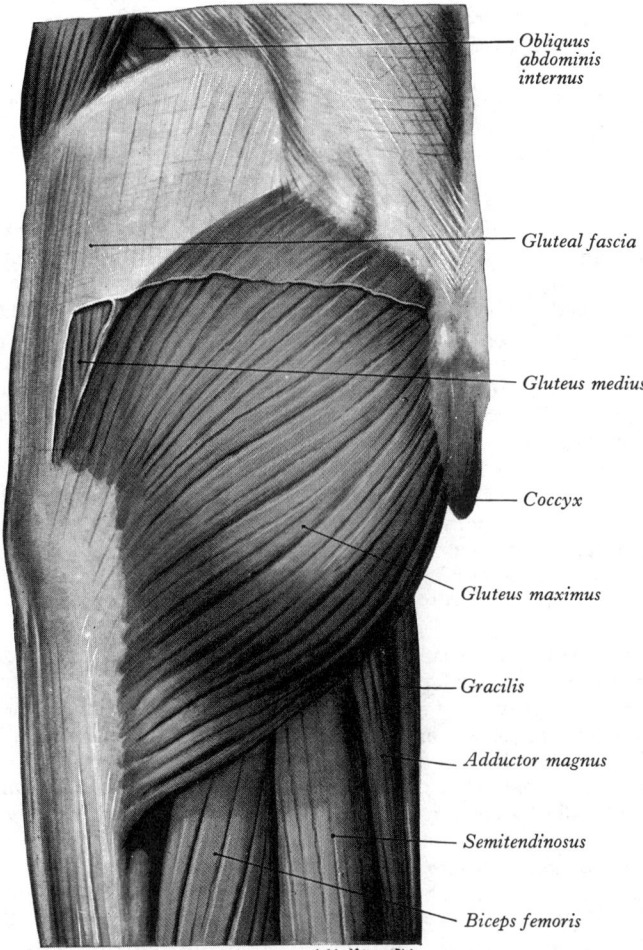

Obliquus
abdominis
internus

Gluteal fascia

Gluteus medius

Coccyx

Gluteus maximus

Gracilis

Adductor magnus

Semitendinosus

Biceps femoris

A·K· MAXWELL

5.90 The left gluteus maximus. Some of the gluteal fascia has been removed to expose part of gluteus medius.

brevis, obturator externus and anterior branch of the obturator nerve, its *lateral border*, to psoas major and the medial circumflex femoral vessels, its *medial border*, to the lateral margin of the adductor longus.

Nerve supply. The pectineus is supplied by the femoral nerve, L. **2** and 3; and by the accessory obturator, L. 3, when present. Occasionally it receives a branch from the obturator nerve. The muscle may be incompletely divided into dorsal and ventral strata, supplied respectively by the obturator and the femoral (or accessory obturator) nerves. In a total of 800 cases investigated, only 69 displayed a partial supply by an accessory obturator nerve (Woodburne 1960).

Actions. The pectineus adducts the thigh and flexes it on the pelvis.

The adductor longus (**5**.88, 89), the most anterior of the three adductors, is triangular and in the same plane as pectineus. It arises by a flat, narrow tendon which is attached to the front of the pubis in the angle between the crest and the symphysis. It soon

expands into a broad fleshy belly which passes downwards, backwards and laterally, and is inserted, by an aponeurosis, on to the linea aspera in the middle third of the femur, between the vastus medialis and the adductores magnus et brevis, with all of which it is usually blended. The adductor longus is occasionally double.

Relations. *Anterior* to the muscle are the spermatic cord, fascia lata, by which it is separated from the great saphenous vein, and, near its attachment, the femoral artery and vein and sartorius. *Posterior* to it are the adductores brevis et magnus, the anterior branch of the obturator nerve, and, near its attachment, the profunda femoris vessels. *Lateral* is pectineus, *medial* is gracilis.

Nerve supply. The anterior division of the obturator nerve, L. 2, **3** and 4.

The adductor brevis (**5**.87, 88) is posterior to pectineus and adductor longus. It is somewhat triangular, and arises by a narrow attachment from the external aspect of the body and inferior ramus of the pubis, between the gracilis and obturator externus. Passing backwards, laterally and downwards, it is attached by an aponeurosis to the femur, along a line leading from the lesser trochanter to the linea aspera, and into the upper part of the latter immediately behind pectineus and the upper part of adductor longus. The adductor brevis is frequently divided into two or three separate parts, or it may be integrated into the adductor magnus.

Relations. Its *anterior surface* is related to pectineus, adductor longus, the profunda femoris artery, and anterior branch of the obturator nerve; *posteriorly*, are adductor magnus and the posterior branch of the obturator nerve; the medial circumflex femoral artery, obturator externus, and conjoined tendon of psoas major and iliacus are related to its *upper border*, and to its *lower border*, gracilis and adductor magnus. Near its attachment the second, or first and second, perforating arteries pierce it.

Nerve supply. The obturator nerve, L. 2, **3** and 4.

The adductor magnus (**5**.87, 88, 89), a massive triangular muscle, arises from a small part of the inferior ramus of the pubis, from the ramus of the ischium, and from the inferolateral aspect of the ischial tuberosity. The fibres from the pubic ramus are short, horizontal, and attached to the medial margin of the gluteal tuberosity of the femur, medial to gluteus maximus; this part of the muscle, anterior in plane to the rest, is sometimes described as *adductor minimus*. The fibres from the ramus of the ischium fan out downwards and laterally, to be attached by a broad aponeurosis, to the linea aspera and the proximal part of the medial supracondylar line. The medial part of the muscle, principally fibres arising from the ischial tuberosity, is a thick mass which descends almost vertically, and ends about the lower third of the thigh in a rounded tendon, palpable proximal to its attachment to the adductor tubercle (p. 394) on the medial condyle of the femur. The tendon is connected by a fibrous expansion to the medial supracondylar line.

The long, linear attachment of the whole muscle is interrupted by a series of osseo-aponeurotic openings, formed by *tendinous arches* attached to the bone. The upper four are small, and give passage to the perforating branches and the termination of the profunda femoris artery. The lowest is large, and the femoral vessels traverse it to the popliteal fossa. The vertical, ischiocondylar part of the muscle varies in its degree of separation from the rest. The upper border of the adductor magnus may be fused with the quadratus femoris.

Relations. *Anterior*, are the pectineus, adductores brevis et longus, the femoral and profunda vessels, and the posterior branch of the obturator nerve; a bursa separates the highest part of the muscle and the lesser trochanter of the femur. *Posterior*, are the sciatic nerve, gluteus maximus, biceps femoris, semitendinosus and semimembranosus. The *superior border* is parallel with the quadratus femoris, the transverse branch of the medial circumflex femoral artery passing between the muscles. The *medial border* is related to the gracilis, sartorius and the fascia lata.

Nerve supply. The adductor magnus is composite and is innervated by the obturator nerve and the tibial division of the sciatic nerve, L. 2, **3** and **4**. The latter supplies the ischiocondylar part. Both nerves are derived from *anterior* divisions in the lumbosacral plexus, indicating a flexor origin for both parts of the muscle.

Actions. Extensive or forcible adduction of the femur is not a common action, and although the adductors can, of course, so act when required, they are essentially synergists in the complex patterns of gait activity, and partly are controllers of posture. For example, they show activity during flexion and extension of the knee (Janda and Stará 1965). The magnus and longus are probably *medial* rotators of the thigh, according to later but as yet unconfirmed data (de Sousa and Vitti 1966). The same authors have pointed out that while the adductors are *inactive* during

adduction of the abducted thigh in the erect posture (where gravity assists), they *are* active in other postures such as the supine position, or during adduction of the flexed thigh when erect. The adductors are also active during flexion (longus) and extension (magnus) of the thigh at the hip joint. In symmetrical easy standing their activity is minimal or absent.

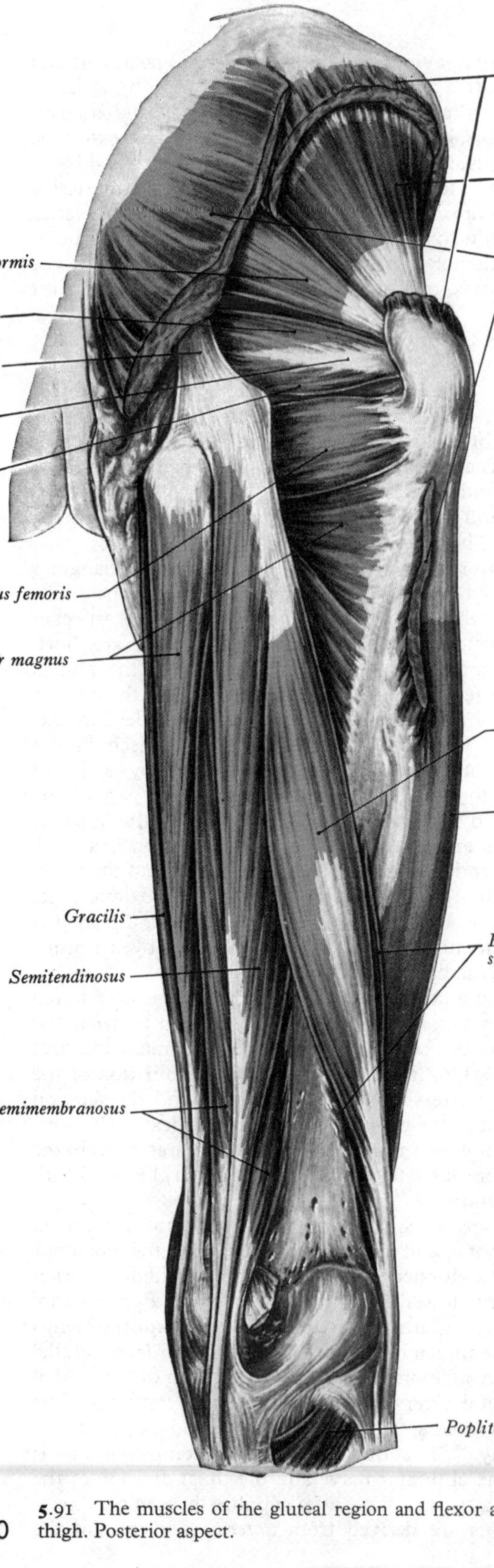

3. The Muscles of the Gluteal Region

The muscles of the region (**5**.90, 91) consist of the glutei, maximus, medius et minimis, which are extensors and abductors at the hip joint, and a more deeply situated group of smaller muscles, lateral rotators at this joint, which includes the piriformis, obturatores internus et externus, gemelli superior et inferior and quadratus femoris.

The gluteus maximus (**5**.90), the largest and most superficial muscle in the region, is a broad and thick quadrilateral mass, forming, together with the superadded characteristic superficial fascial adipose tissue, the prominence of the buttock. Its large size is one of the most characteristic features of the muscular system in man, associated with the frequent need of bringing the trunk upright. The muscle is remarkably coarse in its fascicular architecture, its fibres being collected into large bundles separated by fibrous septa. It arises from the posterior gluteal line of the ilium, and the rough area of bone, including the crest, immediately above and behind it, from the aponeurosis of erector spinae, from the dorsal surface of the lower part of the sacrum and the side of the coccyx, from the sacrotuberous ligament, and from the fascia (gluteal aponeurosis) covering gluteus medius. The fibres descend obliquely and laterally; the upper and larger part of the muscle, together with the superficial fibres of the lower part, end in a thick tendinous lamina which passes lateral to the greater trochanter, and is attached to the iliotibial tract of the fascia lata. The deeper fibres of the lower part of the muscle are attached to the gluteal tuberosity between vastus lateralis and adductor magnus.

Three bursae usually lie deep to this muscle. One, large and generally multilocular, separates it from the greater trochanter (*trochanteric bursa of gluteus maximus*); a second is between the tendon of the muscle and that of the vastus lateralis (*gluteofemoral bursa*); a third, often absent, between the ischial tuberosity and the muscle (*ischial bursa of gluteus maximus*). Additional slips from the lumbar aponeurosis or ischial tuberosity may occur. The muscle may also be bilaminar.

Relations. Its *superficial surface* is related to a thin fascia which separates it from subcutaneous tissue, its *deep surface*, to the ilium, sacrum, coccyx, and sacrotuberous ligament, part of gluteus medius, piriformis, gemelli, obturator internus, quadratus femoris, the tuberosity of the ischium, greater trochanter, the attachments of biceps femoris, semitendinosus, semimembranosus and adductor magnus. The superficial division of the superior gluteal artery reaches the deep surface of the muscle between piriformis and gluteus medius; the inferior gluteal and internal pudendal vessels and the sciatic, pudendal and posterior femoral cutaneous nerves and muscular branches from the sacral plexus issue from the pelvis below piriformis. The first perforating artery and the terminal branches of the medial circumflex femoral artery are also deep to the lower part of the muscle. Its *upper border* is thin, and superficial to gluteus medius. Its *lower border* is free and prominent (directed, as is the upper border, *downwards* and *laterally*) and is *crossed* by the *horizontal gluteal fold*, which is the *posterior flexure line* of the hip joint, and can be taken as marking the upper limit of the back of the thigh on the surface (**5**.39). (*The anterior flexure line* corresponds to the zone of fusion between the membranous layer of superficial fascia and the deep fascia below the medial part of the inguinal ligament.)

Nerve supply. The inferior gluteal nerve, L. 5 and S. **1** and **2**. For details of the intramuscular distribution of the gluteal nerves consult Menning *et al.* (1974).

Actions. When the gluteus maximus acts from the pelvis, it can extend the flexed thigh and bring it into line with the trunk.

Gluteus medius
Gluteus minimus
Gluteus maximus
Piriformis
Gemellus superior
Sacrotuberous ligament
Obturator internus tendon
Gemellus inferior
Quadratus femoris
Adductor magnus
Biceps femoris, long head
Vastus lateralis
Gracilis
Biceps femoris, short head
Semitendinosus
Semimembranosus
Popliteus

5.91 The muscles of the gluteal region and flexor aspect of the right thigh. Posterior aspect.

Taking its fixed point as distal, it may prevent the forward momentum of the trunk from causing flexion at the supporting hip during bipedal gait. In standing, the muscle is inactive and remains so in forward swaying at the ankle joints or during bending forwards at the hip joints to touch the toes. However, in conjunction with the hamstrings, it is active in raising the trunk, after stooping, by rotating the pelvis backwards on the head of the femur (Joseph and Williams 1957). It is active intermittently in the walking cycle and climbing upstairs, and continuously active in strong lateral rotation of the thigh. Its upper fibres are active in powerful abduction of the thigh. It is a tensor of the fascia lata, and through the iliotibial tract it steadies the femur on the tibia, when the knee extensor muscles are relaxed.

The gluteus medius is a broad, thick muscle on the outer surface of the pelvis. Its posterior third is covered by gluteus maximus; its anterior two-thirds is superficial and covered by a strong layer of deep fascia (5.90). It arises from the outer surface of the ilium between the iliac crest and posterior gluteal line above, and the anterior gluteal line below; it also arises from the strong fascia superficial to its upper part. The fibres converge into a flattened tendon, which is attached to the oblique ridge slanting downwards and forwards on the lateral surface of the greater trochanter. A bursa (*trochanteric bursa of gluteus medius*) separates the tendon from the anterosuperior area of the lateral surface of the trochanter, over which it glides. A deep slip of the muscle may be attached to the upper border of the trochanter. The posterior edge of gluteus medius is sometimes blended with piriformis.

The gluteus minimus (5.91), the smallest of the group, is deep to the medius. It is fan-shaped, arising from the outer surface of the ilium between the anterior and inferior gluteal lines and, behind, from the margin of the greater sciatic notch. The fibres converge to the deep surface of an aponeurosis, and this ends in a tendon attached to a ridge laterally situated on the anterior surface of the greater trochanter, and gives an expansion to the capsule of the hip joint. A bursa (*trochanteric bursa of gluteus minimus*) separates the tendon from the medial part of the anterior surface of the greater trochanter. The muscle may be divided into anterior and posterior parts. Separate slips may pass to piriformis, gemellus superior, or vastus lateralis.

Relations. Between the gluteus medius and gluteus minimus are the deep branches of the superior gluteal vessels, and the superior gluteal nerve. The reflected tendon of the rectus femoris and the capsule of the hip joint are deep to the gluteus minimus.

Nerve supply. Both the gluteus medius and minimus are supplied by the superior gluteal nerve, L. 5 and S. 1.

Actions. Both the preceding muscles, acting from the pelvis, abduct the thigh and their anterior fibres rotate it medially. They play an essential part in maintaining the trunk upright when the foot of the opposite side is raised from the ground in walking and running, the body weight tending to make the pelvis sag downwards on the unsupported side. This tendency is counteracted by the gluteus medius and minimus of the supporting side, which, acting from below, exert such powerful traction on the hip bone that the pelvis is actually raised a little on the unsupported side. The muscles are inactive in symmetrical standing. It is, however, necessary to differentiate between different symmetrical standing postures. In standing 'at ease' with the feet somewhat separated, the abductor muscles are electromyographically 'silent' in most individuals. According to Jonsson and Steen (1962), in contrast, the muscles *are* active in the position adopted for Romberg's Test—with the feet placed parallel and close together. Evans (1977) considers that the iliotibial tract is the major factor in preventing descent of the pelvis on the unsupported side.

Applied Anatomy. The supportive effect of the glutei (medius and minimus) on the pelvis, when the foot of the opposite side is raised from the ground, is dependent on the following: (1) the two muscles, and of course their innervation, must be functioning normally; (2) the components of the hip joint, which forms the fulcrum, must be in their usual relation; (3) the neck of the femur

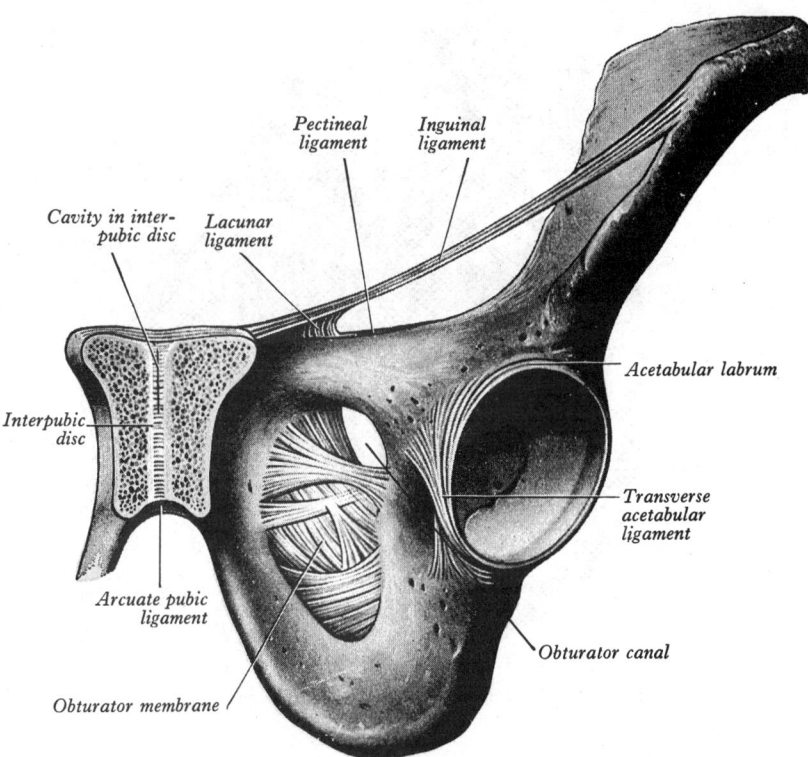

Pectineal ligament
Inguinal ligament
Cavity in interpubic disc
Lacunar ligament
Interpubic disc
Acetabular labrum
Transverse acetabular ligament
Arcuate pubic ligament
Obturator canal
Obturator membrane

5.92 The left obturator membrane. Ventral aspect.

must be intact and must exhibit its usual angulation to the shaft of the bone. When any one of these three conditions is not fulfilled, e.g.—(1) paralysis of the glutei, medius and minimus, (2) congenital dislocation of the hip joint, (3) ununited fracture of the neck of the femur or coxa vara, the supporting mechanism is upset and the pelvis sinks on the unsupported side when the patient stands on the affected limb. This is known clinically as the *Trendelenberg sign.*

Paralysis of the glutei medius et minimus is the most serious muscular disability in the region of the hip, and sufferers have a characteristic lurching gait. On the other hand, when these two muscles remain intact, even though many others acting on the hip joint are paralysed, the patient is able to walk, or even run, with remarkably little disability.

The piriformis (5.91) is almost parallel with the posterior margin of gluteus medius. It is partly within the pelvis, on its posterior wall, and partly posterior to the hip joint. It arises from the front of the sacrum by three digitations, attached to the portions of bone between the pelvic sacral foramina, and to the grooves leading from the foramina (3.43): it also arises from the gluteal surface of the ilium near the posterior inferior iliac spine, from the capsule of the adjacent sacro-iliac joint and sometimes from the upper part of the pelvic surface of the sacrotuberous ligament. The muscle passes out of the pelvis through the greater sciatic foramen, and is attached by a rounded tendon to the upper border of the greater trochanter of the femur, behind and above, but often partly blended with, the common tendon of the obturator internus and gemelli. The muscle may be blended with gluteus medius.

Relations. *Within the pelvis* the *anterior surface* of the piriformis is related to the rectum (especially on the left), the sacral plexus of nerves and branches of the internal iliac vessels, its *posterior surface* to the sacrum. *Outside the pelvis,* its *anterior surface* is in contact with the posterior surface of the ischium and capsule of the hip joint, its *posterior surface,* with gluteus maximus. Its *upper border* is in contact with gluteus medius, and the superior gluteal vessels and nerve, its *lower border,* with the coccygeus and gemellus superior. The inferior gluteal and internal pudendal vessels, the sciatic, posterior femoral cutaneous and pudendal nerves, and muscular branches from the sacral plexus appear in the buttock in the interval between piriformis and gemellus superior. The muscle is frequently pierced by the common peroneal nerve, which may divide it into two parts. The

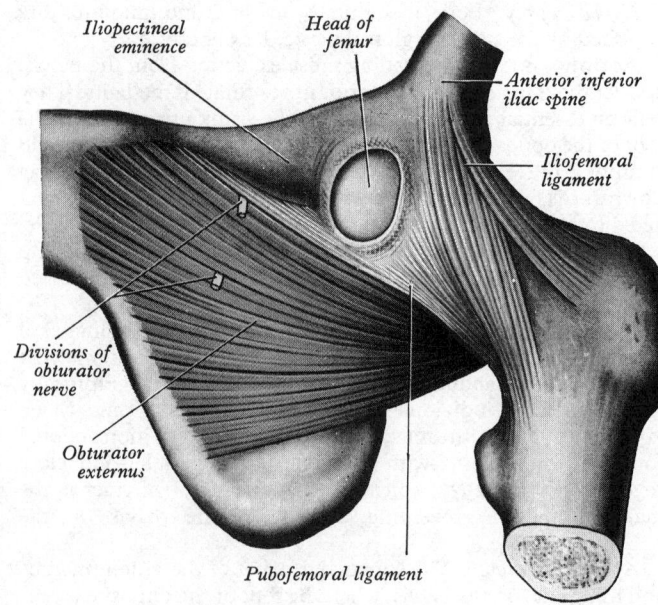

5.93   The left obturator internus. Pelvic aspect.

sciatic nerve in fact varies frequently in its relation to piriformis; in a detailed analysis of 168 dissections Lee and Tsai (1974) observed variation in almost 30 per cent. The undivided nerve emerged *above* the muscle in 2·98 per cent, through the muscle in 1·8 per cent, with its divisions above and below in 4·2 per cent, with one division between the heads of a divided muscle and one division either above (1·2 per cent) or below (19·6 per cent), the last being the commonest variation.

In a larger series (2,250 specimens), reported by Anson (1963), the incidence of variation was only 10·7 per cent (242 specimens); in 10 per cent (222) the common peroneal division separately traversed the muscle, the whole nerve in 0·2 per cent (5), and the two divisions appeared above and below in 0·7 per cent (15); in this series the entire nerve never appeared at the upper border of the muscle.

Nerve supply. Branches from L. 5. S. 1 and 2.

Actions. The piriformis rotates the extended thigh laterally, but abducts the flexed thigh.

The obturator membrane (5.92) is a thin aponeurosis which nearly closes the obturator foramen. Its fibres are arranged in interlacing bundles mainly transverse in direction; the uppermost bundle is attached to the obturator tubercles and completes the obturator canal through which pass the obturator vessels and nerve. The membrane is attached to the sharp margin of the obturator foramen except at its lower lateral angle, where it is fixed to the pelvic surface of the ramus of the ischium, i.e. internal to the margin of the foramen. The obturator muscles are attached on opposite surfaces of this aponeurosis, and some of the fibres of the pubofemoral ligament of the hip joint are attached to its inferior surface. The aponeurosis, of course, *obturates* its foramen.

The obturator internus (5.93) is situated partly within the true pelvis, partly posterior to the hip joint. It arises from the internal surface of the anterolateral wall of the pelvis, where it surrounds the greater part of the obturator foramen, being attached to the inferior ramus of the pubis, the ramus of the ischium and to the pelvic surface of the hip bone below and behind the pelvic brim, reaching from the upper part of the greater sciatic foramen above and behind, to the obturator foramen below and in front (**3**.172). It also arises from the medial part of the pelvic surface of the obturator membrane, from the tendinous arch which completes the obturator canal, and to a slight extent from the obturator fascia, which covers the muscle. The fibres converge rapidly towards the lesser sciatic foramen and

end in four or five tendinous bands in the deep surface of the muscle; these bands make a right-angled bend over the grooved surface of the ischium between its spine and tuberosity. The grooved surface is covered by smooth cartilage, which is separated from the tendon by a bursa, and presents one or more ridges corresponding with the furrows between the tendinous bands. These bands leave the pelvis through the lesser sciatic foramen and unite into a single flattened tendon, which passes horizontally across the capsule of the hip joint, and, after receiving the attachments of the gemelli, is attached to the anterior impression on the medial surface of the greater trochanter anterosuperior to the trochanteric fossa. A narrow and elongate bursa usually exists between the tendon and the capsule of the hip joint; it occasionally communicates with the bursa between the tendon and the ischium.

Relations. *Within the pelvis*, the *anterolateral surface* of the muscle is in contact with the obturator membrane and inner surface of the lateral wall of the pelvis, its *pelvic surface*, with the obturator fascia, and the origin of the levator ani, and with the sheath which surrounds the internal pudendal vessels and pudendal nerve (p. 560). The pelvic surface forms the lateral boundary of the ischiorectal fossa. *Outside the pelvis*, the muscle is covered by gluteus maximus, is crossed by the sciatic nerve, and is posterior to the hip joint. As the tendon of the obturator internus emerges from the lesser sciatic foramen it is overlapped both above and below by the two gemelli, which form a muscular canal for it; near its termination the gemelli pass anterior to the tendon and form a groove in which it lies.

Nerve supply. The nerve to obturator internus, L. 5 and S. 1.

The gemelli (**5**.91) are two small muscular fasciculi, adjacent to, partially enclosing, and finally blending with the tendon of the obturator internus.

The gemellus superior, the smaller of the two, arises from the dorsal surface of the spine of the ischium, blends with the upper part of the tendon of obturator internus, and is attached with it to the medial surface of the greater trochanter. It is sometimes absent.

Nerve supply. The nerve to the obturator internus, L. 5 and S. 1.

The gemellus inferior arises from the upper part of the tuberosity of the ischium, immediately below the groove for the obturator internus tendon. It blends with the lower part of this

Iliopectineal eminence

Head of femur

Anterior inferior iliac spine

Iliofemoral ligament

Divisions of obturator nerve

Obturator externus

Pubofemoral ligament

5.94   The left obturator externus. Antero-inferior aspect. The bursa of psoas major tendon, which in this specimen communicated with the synovial cavity of the hip joint, has been opened to expose the head of the femur.

tendon, and is attached with it into the medial surface of the greater trochanter.

Nerve supply. The nerve to the quadratus femoris, L. 5 and S. 1.

Action. The obturator internus and the gemelli rotate the extended thigh laterally; they abduct the flexed thigh.

The quadratus femoris (**5**.91) is a flat, quadrilateral muscle, between the gemellus inferior and the upper margin of adductor magnus, separated from it by the transverse branch of the medial circumflex femoral artery. It arises from the upper part of the external aspect of the tuberosity of the ischium, and is attached to a small tubercle on the upper part of the trochanteric crest of the femur and for a short distance to the bone below. As it passes to this attachment the muscle is posterior to the hip joint and the neck of the femur, but it is separated from them by the tendon of obturator externus and the ascending branch of the medial circumflex femoral artery. A bursa often exists between the front of this muscle and the lesser trochanter. The muscle may be absent.

Nerve supply. The nerve to quadratus femoris, L. 5 and S. 1. For a detailed analysis of this muscle's arterial supply (chiefly from the medial circumflex and first perforating arteries) consult Leborgne *et al.* (1973).

Action. The quadratus femoris is a lateral rotator of the thigh.

The obturator externus (**5**.94) is flat and triangular, covering the external surface of the anterior pelvic wall. It arises from bone immediately around the medial side of the obturator foramen, viz. from the rami of the pubis, and the ramus of the ischium; it also arises from the medial two-thirds of the outer surface of the obturator membrane, and from the tendinous arch which completes the obturator foramen, viz. from the rami of the pubis for a short distance on to the pelvic surface of the bone, where they obtain a narrow attachment between the margin of the foramen and the obturator membrane (p. 602). The fibres converge and pass backwards, laterally and upwards, and end in a tendon which crosses the back of the neck of the femur and lower part of the capsule of the hip joint to end in the trochanteric fossa of the femur. A bursa, which communicates with the hip joint, may be interposed between this tendon and the hip joint capsule and femoral neck. The obturator vessels lie between the muscle and the obturator membrane; the anterior branch of the obturator

nerve reaches the thigh by passing in front of the muscle, and the posterior branch by piercing it (**5**.94).

Nerve supply. The posterior branch of the obturator nerve, L. 3 and **4**.

Action. The obturator externus is a lateral rotator of the thigh. The short muscles around the hip joint—the pectineus, piriformis, obturatores, gemelli and quadratus femoris—are possibly more important as postural muscles than as prime movers. They may act as adjustable ligaments of the joint in all positions, but they are in any case largely inaccessible to direct observations, and such views are necessarily speculative.

4. The Posterior Femoral Muscles

The posterior femoral muscles (**5**.91), the biceps femoris, semitendinosus, and semimembranosus, often grouped familiarly as the '*hamstrings*', span the hip and knee joints, integrating extension at the former with flexion at the latter.

The biceps femoris (**5**.87, 89, 91) is posterolateral in the thigh and has two proximal attachments—one, the long head, arising from the inferomedial impression on the upper area of the ischial tuberosity (**3**.173) by a tendon common to it and semitendinosus, and from the lower part of the sacrotuberous ligament, the other, the short head, from the lateral lip of the linea aspera, between the adductor magnus and vastus lateralis, extending almost as high as the gluteus maximus, from the lateral supracondylar line to within 5 cm of the lateral condyle, and from the lateral intermuscular septum. The fibres of the long head form a fusiform belly which descends laterally across the sciatic nerve to end in an aponeurosis which covers the posterior surface of the muscle, receives on its deep surface the fibres of the short head, and gradually narrows down into a tendon (the lateral hamstring). The main part of the tendon splits round the fibular collateral ligament and is attached to the head of the fibula. The remainder splits into three laminae, the intermediate of which fuses with the fibular collateral ligament while the others pass superficial and deep to the ligament to be attached to the lateral condyle of the tibia (Sneath 1955). The common peroneal nerve descends along the medial border of the tendon and separates it below from the lateral head of gastrocnemius. The short head may be absent. Additional slips may arise from the ischial tuberosity, linea

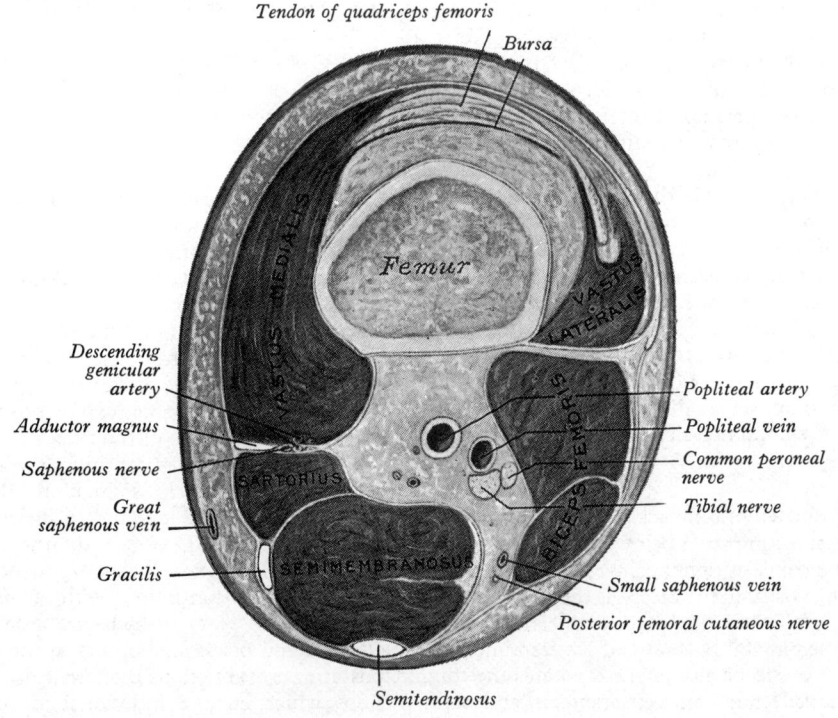

Tendon of quadriceps femoris

Bursa

Femur

VASTUS MEDIALIS

VASTUS LATERALIS

Descending genicular artery

Adductor magnus

Saphenous nerve

Great saphenous vein

Gracilis

SARTORIUS

SEMIMEMBRANOSUS

BICEPS FEMORIS

Popliteal artery

Popliteal vein

Common peroneal nerve

Tibial nerve

Small saphenous vein

Posterior femoral cutaneous nerve

Semitendinosus

5.95   Transverse section through the right thigh, about 4 cm superior to the adductor tubercle of the femur. Proximal (superior) aspect.

aspera, or medial supracondylar line.

Nerve supply. The biceps femoris is supplied by the sciatic nerve; the long head through its tibial part and the short head through the common peroneal part, L. 5, S. 1 and 2, indicating the muscle's composite derivation from flexor and extensor musculature. (See also nerve supply of semimembranosus.)

The semitendinosus (5.89, 91), remarkable for the great length of its tendon, is posteromedial in the thigh. It arises from the inferomedial impression on the upper part of the ischial tuberosity, by a tendon shared with the long head of biceps femoris, and from an aponeurosis connecting the adjacent surfaces of the two muscles to the extent of about 7·5 cm from their origin. The muscle is fusiform and ends a little below the middle of the thigh in a long, rounded tendon, which lies on the surface of semimembranosus; the tendon curves around the medial condyle of the tibia, passes over the tibial collateral ligament of the knee joint, from which it is separated by a bursa, and is attached to the upper part of the medial surface of the tibia behind the attachment of sartorius and below that of gracilis. At its termination it is united with the tendon of gracilis and gives off a prolongation to the deep fascia of the leg. A tendinous interruption is usually present about the mid level of the muscle, which may also receive a muscular slip from the long head of biceps femoris.

Nerve supply. The sciatic nerve through its tibial division, L. 5, S. 1 and 2. (See also nerve supply of semimembranosus.)

The semimembranosus (5.89, 91, 95), so-called because of the flattened membranous form of its upper attachment, is posteromedial in the thigh. It arises by a tendon from the superolateral impression on the ischial tuberosity (3.173). Inferomedially its fibres are interwoven with those of the biceps femoris and semitendinosus and in addition its upper tendon receives, from the ischial tuberosity and ramus, two fibrous expansions which flank the adductor magnus. This upper tendon expands into an aponeurosis which passes downwards deep to the semitendinosus and long head of biceps femoris; from this muscular fibres arise, and converge to a second aponeurosis which is posterior to the lower part of the muscle and narrows into the tendon of lower attachment. The latter divides at the level of the knee into five components, the principal one being attached to the tubercle (sometimes called the *tuberculum tendinis*) on the posterior aspect of the medial tibial condyle (Cave and Porteous 1958). The additional attachments consist of a series of slips to the medial margin of the tibia immediately behind the tibial collateral ligament, a thin fibrous expansion to the fascia over popliteus, a thick cord-like tendon to the inferior lip and adjacent part of the groove on the back of the medial tibial condyle, deep to the tibial collateral ligament, and a variable number of fasciculi, which pass upwards and laterally to the intercondylar line and lateral femoral condyle, forming the oblique popliteal ligament of the knee joint. The muscle overlaps the upper part of the popliteal vessels and is itself partly overlapped by semitendinosus throughout its extent (5.91). (The distal tendons of semitendinosus and semimembranosus are the medial 'hamstring'.)

The semimembranosus varies much in size, and may be absent. It may be double, arising mainly from the sacrotuberous ligament. Slips to the femur or adductor magnus may occur.

Nerve supply. The sciatic nerve from its tibial portion, L. 5, S. 1 and 2. The detailed intramuscular distributions of the nerves supplying the posterior femoral muscles have been described by Himstedt *et al.* (1974).

Actions. The posterior femoral muscles, acting from above, flex the leg on the thigh. Acting from below, as extensors of the hip joint, they draw the trunk upright against gravity when it is raised from the stooping position; the biceps is the most active in this. When the knee is semiflexed, the biceps femoris can act as a lateral rotator, and the semimembranosus and semitendinosus as medial rotators of the leg. Biceps also laterally rotates the thigh when the hip is extended, the semimembranosus and semitendinosus being medial rotators in these circumstances.

As is the case with the quadriceps femoris, the adductors and gluteus maximus, the hamstrings are quiescent in easy symmetrical standing. However, any action which carries the line of body weight in front of the transverse axis of the hip joints, e.g. forward reaching, forward sway at the ankle joints, or forward bending at the hips, is immediately accompanied by strong contraction of the hamstrings. (This is in marked contrast to gluteus maximus which only contracts when powerful extensor activity at the hip joint is required.)

When the knee is flexed against resistance, the tendon of biceps can be felt lateral to the depression over the popliteal fossa. Medial to the fossa, the tendons of gracilis, which is the more medial, and semitendinosus stand out sharply, and in the interval between them the semimembranosus tendon can be felt, though less distinctly owing to its deep location.

There is some evidence that the semimembranosus, semitendinosus and biceps femoris, which pass over both the hip and knee joints, can produce movement at only one of these joints while at the same time offering no resistance to the antagonistic movement at the other joint on which they act (Markee *et al.* 1955). Each of these muscles, however, usually contracts as a whole, and the consequent movement at the hip or knee is determined by other muscles fixing the position of one or other articulation.

Applied Anatomy. In disease of the knee joint, contracture of the flexor tendons is a frequent complication; this causes flexion of the leg, and a partial dislocation of the tibia backwards, with a slight degree of lateral rotation, probably due to biceps femoris.

When the flexor tendons require surgical manipulation the relation of the common peroneal nerve to the medial border of the tendon of biceps femoris must be borne in mind.

The relative length of the flexors, when relaxed, shows considerable variation and in some individuals the muscles are so short that they impose a serious limitation on flexion of the trunk at the hip joints when the knees are kept extended. Such movements as stooping are then largely effected by flexion of the vertebral column or by squatting.

III. The Muscles of the Leg

This group, which consists basically of anterior extensors and posterior flexors, also includes a lateral set, derived from the extensors, the peronei.

1. The Anterior Crural Muscles

The muscles of this group (5.97) consist of a common digital extensor, an extensor of the hallux, and two dorsiflexors of the foot (with other actions), tibialis anterior and peroneus tertius.

The fascia cruris, the deep fascia of the leg, is continuous, of course, with the fascia lata, and is attached around the knee to the patella, the ligamentum patellae, the tubercle and condyles of the tibia and the head of the fibula. Posteriorly, it forms the popliteal fascia, over the popliteal fossa; here it is strengthened by transverse fibres, and perforated by the small saphenous vein. It receives expansions from the tendon of biceps femoris laterally, and from the tendons of sartorius, gracilis, semitendinosus and semimembranosus medially; it blends with the periosteum on the subcutaneous surface of the tibia and the head and malleolus of the fibula; below, it is continuous with the extensor and flexor retinacula (p. 611). It is thick and dense in the proximal and anterior part of the leg, and to its deep surface some fibres of tibialis anterior and extensor digitorum longus are attached; it is thinner behind, where it covers gastrocnemius and soleus. On the lateral side it is continuous with the *anterior* and *posterior crural intermuscular septa*, which are attached respectively to the anterior and posterior borders of the fibula; in the anterior and posterior crural regions the fascia also has several slender extensions which enclose individual muscles. A broad, transverse, intermuscular septum, the *deep transverse fascia of the leg* (5.99), passes between the superficial and deep muscles on the back of the

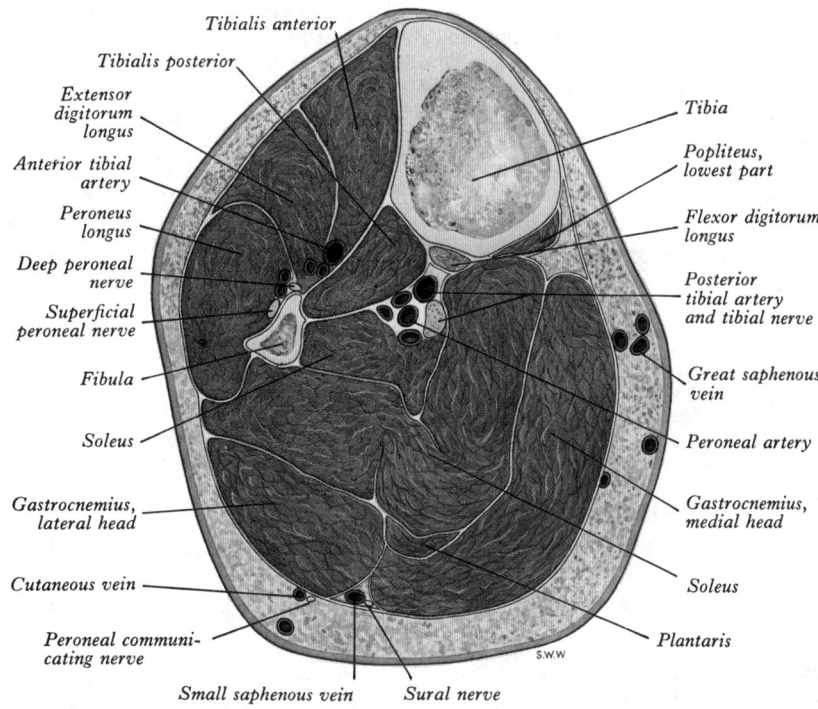

5.96 Transverse section through the right leg, about 10 cm distal to the knee joint. Distal aspect.

leg. These separately named sheets are, of course, all united together, as a kind of auxiliary, connective tissue 'skeleton'.

The tibialis anterior (**5**.96, 97), situated lateral to the tibia, is thick and muscular above, tendinous below. It arises from the lateral condyle and upper half or two-thirds of the lateral surface of the tibial shaft, from the adjoining part of the anterior surface of the interosseous membrane, from the deep surface of the fascia cruris, and from the intermuscular septum between it and extensor digitorum longus. The muscle descends vertically to end in a tendon on its anterior surface in the lower third of the leg; this passes through the medial compartments of the superior and inferior extensor retinacula, inclines medially, and is attached to the medial and inferior surfaces of the medial cuneiform and the adjoining part of the base of the first metatarsal bone. This muscle overlaps the anterior tibial vessels and deep peroneal nerve in the upper part of the leg. Attachments to the talus, first metatarsal head, or the base of the proximal phalanx of the hallux have been recorded.

Actions. The tibialis anterior is a dorsiflexor of the talocrural joint and invertor of the foot. It is most active when both movements are combined, as in walking. It elevates the first metatarsal and medial cuneiform and rotates them laterally. Its tendon can be seen through the skin lateral to the anterior border of the tibia and traced downwards and medially across the front of the ankle to the medial side of the foot.

In standing still the muscle is in most cases quiescent, reflecting the fact that the line of body weight passes anterior to the talocrural joints. It can act from below and assist in maintaining the balance of the body by drawing the leg forwards at the talocrural joint, when there is a tendency to overbalance in a backward direction. It is frequently said to be active during any movement which raises the summit of the longitudinal arch of the foot. Electromyographic experimentation (under somewhat artificial conditions) suggests that activity in this and other 'arch supporting' muscles is minimal in standing, but there appears to be no doubt that tibialis anterior contracts powerfully to increase the arch in the take-off and toeing of walking and running.

The extensor hallucis longus (**5**.97, 99) lies between and partly deep to the tibialis anterior and extensor digitorum longus. It arises from the middle two-fourths of the medial surface of the

fibula, medial to the extensor digitorum longus; it also arises from the anterior surface of the interosseous membrane to a similar extent. The anterior tibial vessels and deep peroneal nerve are between it and tibialis anterior. The fibres pass down into a tendon which occupies the anterior border of the muscle. It passes deep to the superior and through the inferior extensor retinaculum, crosses to the medial side of the anterior tibial vessels near the talocrural joint, and is attached to the dorsal aspect of the base of the distal phalanx of the great toe. Opposite the metatarsophalangeal articulation a thin prolongation from each side of the tendon covers the dorsal surface of the joint. An expansion from the medial side of the tendon to the base of the proximal phalanx is usually present.

The muscle is sometimes united with extensor digitorum longus; it may send a slip into the second toe.

Actions. The extensor hallucis longus extends the phalanges of the great toe and dorsiflexes the foot. When the great toe is actively extended, relatively little external force is required to overcome the extension of the distal phalanx, whereas considerable force must be exerted to overcome the extension of the proximal phalanx. Its tendon can readily be identified on the lateral side of the tendon of tibialis anterior.

The extensor digitorum longus (**5**.96, 97, 99) is a pennate muscle, arising from the lateral condyle of the tibia, the upper three-fourths of the medial surface of the fibula, the upper part of the anterior surface of the interosseous membrane, the deep surface of the fascia cruris, the anterior crural intermuscular septum and the septum between it and tibialis anterior. In the upper part of the leg the anterior tibial vessels and deep peroneal nerve are between the muscle and tibialis anterior and, at a lower level, extensor hallucis longus also intervenes between them. The tendon of extensor digitorum longus passes behind the superior extensor retinaculum and within the loop of the inferior extensor retinaculum (p. 611) with the peroneus tertius (**5**.103). It divides into four slips, which run forward on the dorsum of the foot, and are attached in the same way as the tendons of the extensor digitorum in the hand (p. 579). Opposite the metatarsophalangeal joints the tendons to the second, third, and fourth toes are each joined on the lateral side by a tendon of extensor digitorum brevis. The *dorsal digital expansion* thus formed on the dorsal aspect of the proximal phalanx, like that on the fingers, receives contributions from the lumbrical and interosseous muscles (pp. 615–16). Narrowing, as it approaches a proximal interphalangeal joint, the expansion divides into three slips—an intermediate, attached to the base of the middle phalanx, and two collateral, which reunite on the dorsum of the middle phalanx and are attached to the base of the distal phalanx. The tendons to the second and fifth toes are sometimes doubled. Accessory slips

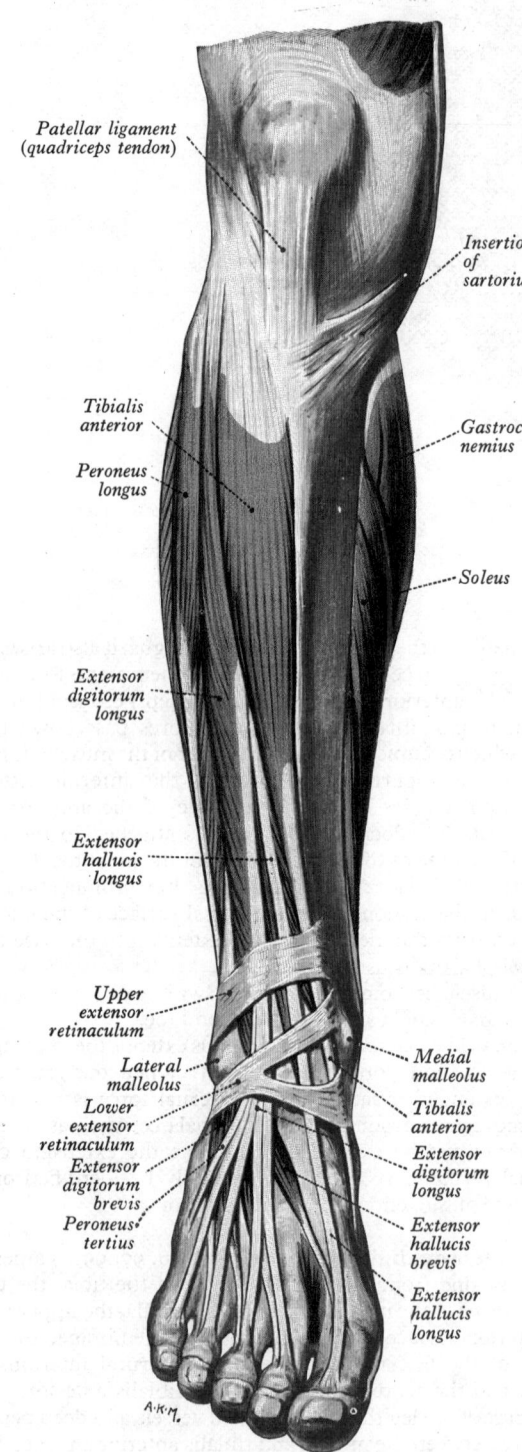

Patellar ligament (quadriceps tendon)

Insertion of sartorius

Tibialis anterior

Gastrocnemius

Peroneus longus

Soleus

Extensor digitorum longus

Extensor hallucis longus

Upper extensor retinaculum

Lateral malleolus

Medial malleolus

Lower extensor retinaculum

Tibialis anterior

Extensor digitorum brevis

Extensor digitorum longus

Peroneus tertius

Extensor hallucis brevis

Extensor hallucis longus

A·K·M.

5.97   Muscles on extensor aspect of right leg. (From Quain's *Anatomy*, 11th edition.)

attached to metatarsals occasionally exist, as may a slip to the hallux.

Actions. The extensor digitorum longus extends the toes, and dorsiflexes the foot in synergy with tibialis anterior and extensor hallucis longus. Acting with the latter it tautens the plantar aponeurosis (pp. 612, 617).

The peroneus tertius (5.97, 104) is a part of the extensor digitorum longus, and might be described as its fifth tendon. The muscular fibres operating on this tendon arise from the lower third or more of the medial surface of the fibula, the adjoining anterior surface of the interosseous membrane and the anterior crural intermuscular septum. The tendon passes behind the superior and within the loop of the inferior extensor retinaculum with extensor digitorum longus (5.103), and is attached to the medial part of the dorsal surface of the base of the fifth metatarsal bone, but a thin expansion usually extends forwards along the medial border of the shaft of the bone. This muscle is sometimes missing.

Actions. The peroneus tertius dorsiflexes the foot, acting as a part of extensor digitorum longus. It may also aid in eversion of the foot.

Nerve supply. All the anterior crural muscles are supplied by the deep peroneal nerve. The tibialis anterior is innervated by L. **4** and 5, but the others receive their supply from L. 5 and S. 1.

2. The Lateral Crural Muscles

The lateral group of leg muscles are the peroneus longus et brevis, evertors of the foot (5.96–101).

The peroneus longus (5.96, 100, 101), the more superficial, arises from the head and upper two-thirds of the lateral surface of the fibula, from the deep surface of the fascia cruris, and from the anterior and posterior crural intermuscular septa and occasionally also by a few fibres from the lateral condyle of the tibia. Between its attachments to the head and body of the fibula, there is a gap through which the common peroneal nerve passes. It ends in a long tendon, which runs behind the lateral malleolus, in a groove shared by the tendon of the peroneus brevis, behind which it lies; the groove is converted into a canal by the superior peroneal retinaculum, and the tendons are contained in a common synovial sheath (5.104). The peroneus longus tendon then runs obliquely forwards across the lateral side of the calcaneus, below the peroneal trochlea and the tendon of peroneus brevis, and beneath the inferior peroneal retinaculum (p. 611); it crosses the lateral side of the cuboid, and then runs under it in a groove converted into a canal by the long plantar ligament (4.90). It crosses the sole of the foot obliquely, and is attached by two slips to the lateral side of the base of the first metatarsal bone and the medial cuneiform; occasionally a third slip is attached to the base of the second metatarsal bone. The tendon changes its direction at two points: (*a*) below the lateral malleolus, (*b*) on the cuboid bone; in both situations it is thickened, and, in the latter a sesamoid fibrocartilage (sometimes a bone) is usually developed in it. A second synovial sheath invests the tendon as it crosses the sole of the foot.

Tendinous slips to the base of the third, fourth or fifth metatarsal bone may occur, or to the adductor hallucis. Fusion of the peroneus longus and brevis is a rare variation.

Nerve supply. The peroneus longus is supplied by the superficial peroneal nerve, L. **5**, S. **1** and 2.

Actions. There is no doubt that the peroneus longus can evert and plantar-flex the foot, and perhaps act on the leg from its distal attachments. The oblique direction of its tendon across the sole also appears likely to be of some value in regard to the longitudinal and transverse arches of the foot. The question is: How do these potentialities contribute to actual movements? With the foot off the ground the eversion is unquestionable, the tendon and muscle being prominent to eye and touch. How far the same action helps to maintain plantigrade contact of the foot in standing is uncertain; however, the electromyographic evidence indicates little or no peroneal activity in standing. But peroneus longus and brevis come strongly into action to maintain the concavity of the foot in take-off and tip-toeing. In deliberate swaying to one side the peronei contract on that side, but how far they enter into postural activity between the foot and leg is not clear.

The peroneus brevis (5.99, 100) arises from the lower two-thirds of the lateral surface of the fibula, anterior to peroneus longus, and from the anterior and posterior crural intermuscular septa. It passes vertically downwards and ends in a tendon which extends behind the lateral malleolus along with, but anterior to, that of peroneus longus, the two tendons being deep to the superior peroneal retinaculum in a common synovial sheath (p. 611). It then runs forwards on the lateral side of the calcaneus above the peroneal trochlea and the tendon of peroneus longus, to

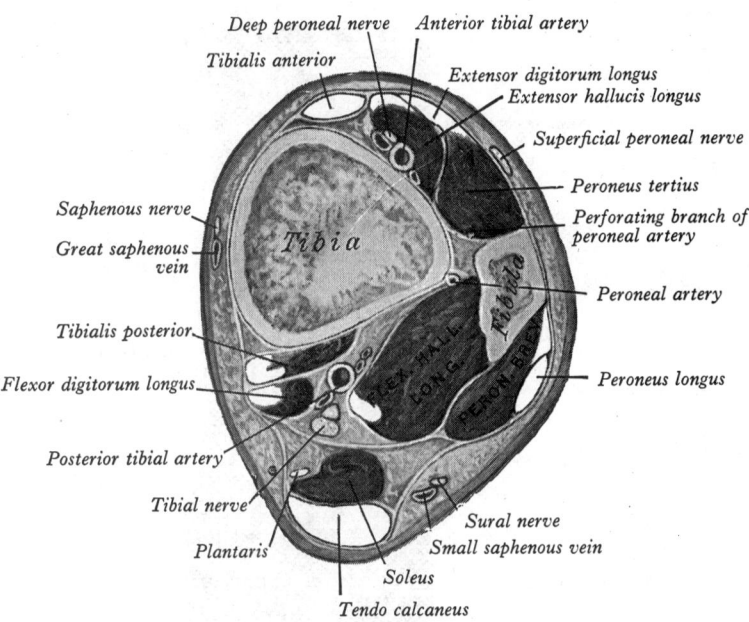

5.99   Transverse section through the right leg, about 6 cm superior to the tip of the medial malleolus. Proximal (superior) aspect.

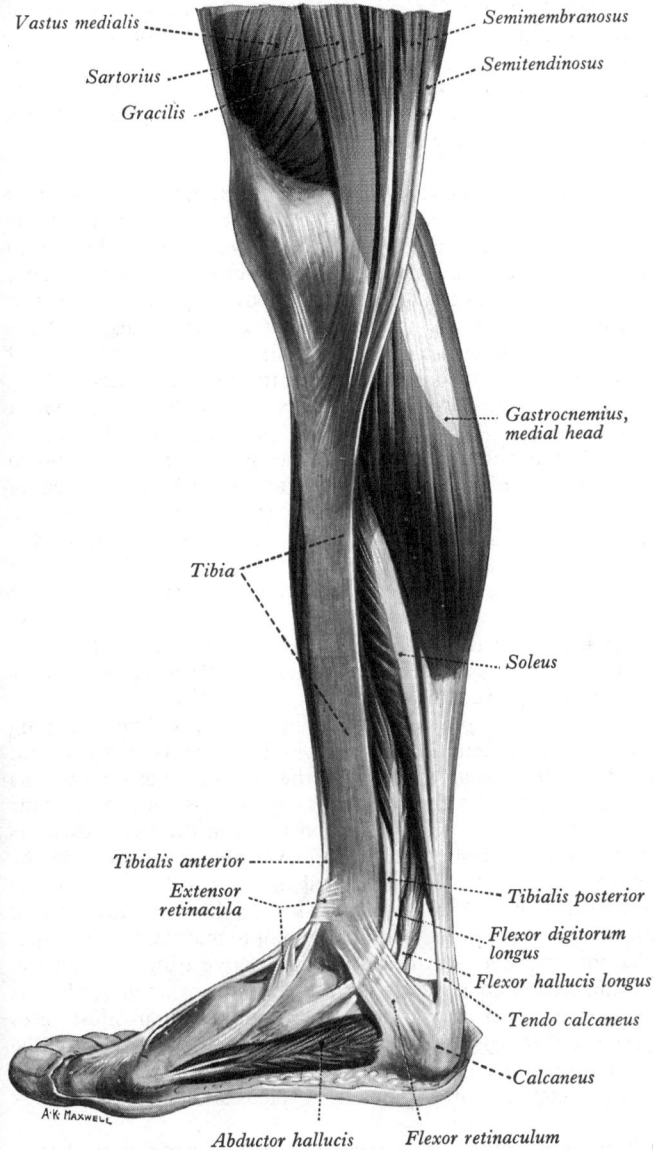

5.98   The muscles of the right leg, medial aspect.

be attached to the tubercle on the base of the fifth metatarsal bone, on its lateral side.

On the lateral surface of the calcaneus the tendons of the peronei longus et brevis occupy separate osseo-aponeurotic canals formed by the calcaneus and the inferior peroneal retinaculum; each tendon is enveloped by a separate forward prolongation of the common synovial sheath (5.104).

Nerve supply. The superficial peroneal nerve, L. **5**, S. **1** and 2.

Action. The peroneus brevis may limit the inversion of the foot and so relieve the ligaments which are tightened by this movement (lateral part of interosseous talocalcanean, lateral talocalcanean and calcaneofibular). It participates in eversion of the foot and may help to steady the leg on the foot (cf. peroneus longus).

3. The Posterior Crural Muscles

The muscles of the back of the leg can be regarded as two groups, superficial and deep, separated by the deep transverse fascia of the leg (pp. 604, 609). Those of the superficial group constitute a powerful muscular mass forming the calf of the leg, and capable of flexing the knee and plantar-flexing the foot. Their large size is one of the most characteristic features of the musculature of man, being directly related to his upright stance and mode of progression.

THE SUPERFICIAL GROUP

In this group (5.100), gastrocnemius and plantaris act on both knee and ankle joints, soleus on the latter alone.

The gastrocnemius (5.96, 98, 100), the most superficial muscle of the group, forms the 'belly' of the calf. It arises by two heads, which are connected to the condyles of the femur by strong, flat tendons. The medial and larger descends from a depression at the upper and posterior part of the medial condyle behind the adductor tubercle, and from a slightly raised area on the popliteal surface of the femur just above the medial condyle. The lateral head arises from a recognizable area on the lateral surface of the lateral condyle and from the lower part of the corresponding supracondylar line. Both also arise from the subjacent part of the capsule of the knee joint. Each head spreads out into a tendinous expansion which covers the posterior surface of its own part of the muscle. From the anterior surfaces of these expansions muscular fibres arise, those of the medial head extending lower than those of the lateral. The two muscular masses remain separate as far as their attachment to a broad aponeurosis on the anterior surface of the muscle. This gradually contracts and unites with the tendon of soleus, to form the *tendo calcaneus* (p. 608). Absence of the lateral head, or the whole muscle, can occur; a more frequent variation is a third head from the popliteal surface of the femur.

Relations. The fascia cruris separates the *superficial surface* of the muscle from the small saphenous vein and the peroneal communicating and sural nerves; the common peroneal nerve crosses the lateral head of the muscle, partly deep to biceps femoris. The *deep surface* is posterior to the oblique popliteal ligament, popliteus, soleus, plantaris, popliteal vessels and tibial nerve. A bursa, which sometimes communicates with the knee joint, is anterior to the tendon of the medial head. The tendon of

607

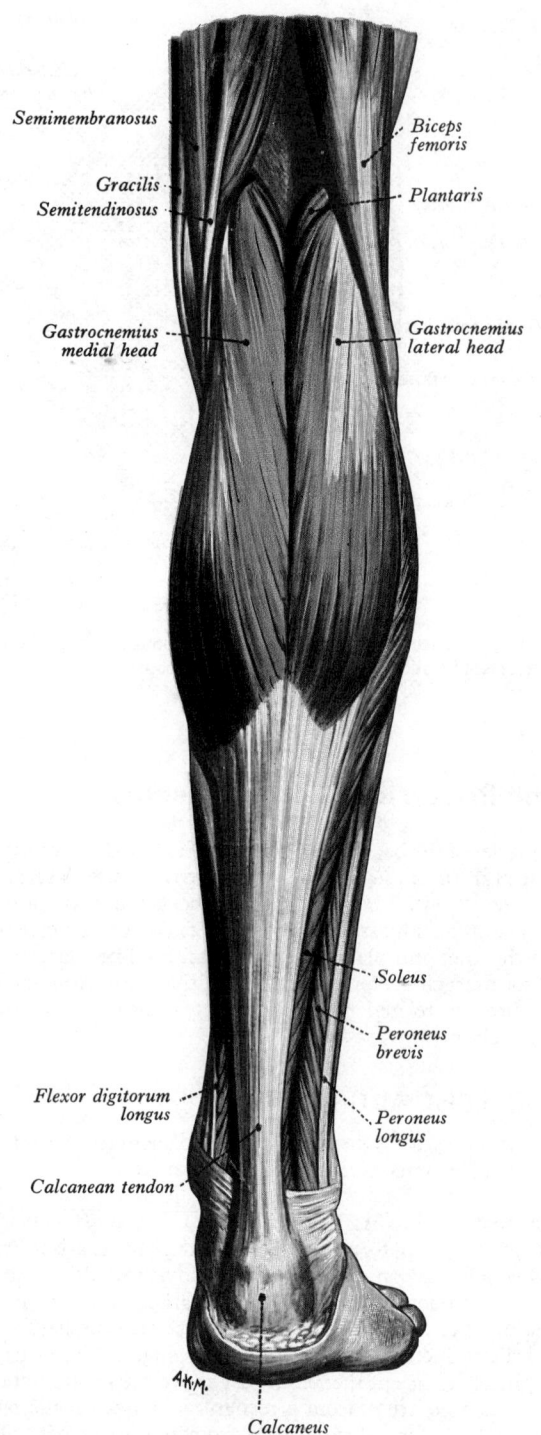

Semimembranosus

Gracilis

Semitendinosus

Biceps femoris

Plantaris

Gastrocnemius medial head

Gastrocnemius lateral head

Soleus

Peroneus brevis

Flexor digitorum longus

Peroneus longus

Calcanean tendon

Calcaneus

5.100 Muscles of the right calf; superficial layer. (From Quain's *Anatomy*, 11th edition.)

the lateral head sometimes contains a sesamoid fibrocartilage or bone where it plays over the lateral condyle; one is also occasionally present in the tendon of the medial head.

Nerve supply. The tibial nerve, S. **1** and 2.

The soleus (5.96, 98) is a broad flat muscle situated immediately deep or anterior to gastrocnemius. It arises from the back of the head and the upper fourth of the posterior surface of the fibula, from the soleal line and the middle third of the medial border of the tibia, and from a fibrous band between the tibia and fibula which arches over the popliteal vessels and tibial nerve. This origin is aponeurotic, most of the muscular fibres arising from its posterior surface and proceeding to the tendon of insertion, which is on the superficial surface. Some muscle fibres also arise from the deep surface of the aponeurosis; they are short,

oblique and bipennate in arrangement, and converge on a narrow, central intramuscular tendon which merges distally with the principal tendon (5.96). The latter, gradually becoming thicker and narrower, joins with the tendon of gastrocnemius to form the *tendo calcaneus*.

Relations. Its *superficial surface* is close to gastrocnemius and plantaris, its *deep surface*, to the flexor digitorum longus, flexor hallucis longus, tibialis posterior, and posterior tibial vessels and tibial nerve, from all of which it is separated by the deep transverse fascia of the leg.

Nerve supply. Two branches from the tibial nerve, S. 1 and **2**. For the intramuscular distribution of the nerve supplies of gastrocnemius, soleus and plantaris consult Schumacher *et al.* (1973).

The gastrocnemius and soleus together form a tripartite muscular mass which shares the tendo calcaneus, and is hence sometimes termed the *triceps surae*.

The tendo calcaneus (5.100) is the thickest and strongest human tendon. It is about 15 cm long, and begins near the middle of the leg, but its anterior surface receives muscle fibres from soleus, almost to its lower end. It gradually becomes more rounded towards a level about 4 cm above the calcaneus; below this it expands and is attached to the posterior surface of the calcaneus at mid level, a bursa separating it from the upper part of this surface. The tendon fibres spiral through 90° in descending, so that, for example, the medial fibres become the most posterior (White 1943). This arrangement permits some degree of elongation and elastic recoil, and hence some storing of energy in the tendon, which can be released at an appropriate phase of locomotion. Alexander and Vernon (1975) have studied such elastic recoil (in kangaroos), and Alexander has supplied a detailed mathematical analysis of the contribution of elasticity in the tendo calcaneus to bipedal gait (Alexander 1976).

Actions. The muscles of the calf are the chief plantar flexors of the foot, and the gastrocnemius is a flexor of the knee; they possess considerable power, and are usually of large size.

Gastrocnemius provides propelling force in walking, running and leaping. Soleus is frequently said to be concerned with steadying the leg on the foot in the standing position and its postural function is emphasized more than its value as a prime mover; but such a rigid distinction between the two muscles is probably an over-simplification. Nevertheless, in standing the centre of gravity lies on a vertical line passing anterior to the talocrural joint (which in this position is loose-packed), and a strong brace is required behind the joint to maintain the posture. Electromyography has demonstrated that during symmetrical standing soleus consistently exhibits continuous activity, whereas the gastrocnemei are active only intermittently in most cases (Joseph *et al.* 1955; Joseph 1960). The phasic activity of the triceps surae in walking has not yet been satisfactorily analysed into the relative contributions from soleus and gastrocnemius.

The plantaris (5.100) arises from the lower part of the lateral supracondylar line and the oblique popliteal ligament. It has a small fusiform belly, from 7 to 10 cm long; this ends in a long slender tendon, which crosses obliquely between gastrocnemius and soleus and runs along the medial border of the tendo calcaneus to be inserted with it. This muscle is sometimes double, and at other times missing. Occasionally, its tendon merges with the flexor retinaculum (p. 611), or the fascia of the leg. Daseler and Anson (1943) found this muscle absent in 160 of 1,545 legs.

Nerve supply. The tibial nerve, S. **1** and 2.

Actions. The plantaris is the rudiment of a large muscle, the tendon of which is inserted into the plantar aponeurosis in some of the lower animals: in man it is accessory to gastrocnemius.

THE DEEP GROUP

The deep crural flexors (5.98, 101), include one, the popliteus, which acts on the knee joint, the rest, flexor hallucis longus, flexor digitorum longus and tibialis posterior, acting on the talocrural joint and other joints of the foot.

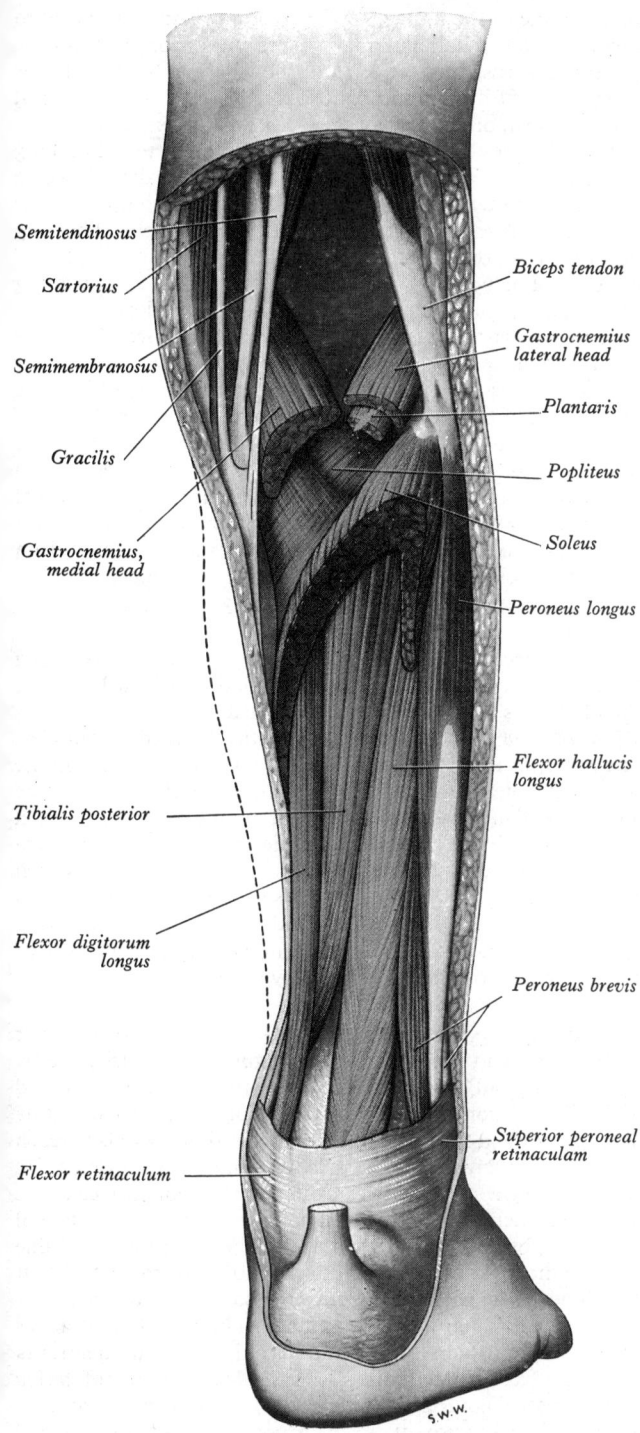

Semitendinosus

Sartorius

Semimembranosus

Gracilis

Gastrocnemius, medial head

Tibialis posterior

Flexor digitorum longus

Flexor retinaculum

Biceps tendon

Gastrocnemius lateral head

Plantaris

Popliteus

Soleus

Peroneus longus

Flexor hallucis longus

Peroneus brevis

Superior peroneal retinaculum

S.W.W.

5.101 The right posterior crural muscles, deep group, in a child, aged 8 years.

The deep transverse fascia of the leg is a fibrous stratum between the superficial and deep posterior crural muscles, extending transversely from the medial margin of the tibia to the posterior border of the fibula. Above, where it is thick and dense, it is attached to the soleal ridge of the tibia and to the fibula, inferomedial to the attachment of soleus. Between these bony attachments it is continuous with the fascia covering popliteus and receives an expansion from the tendon of semimembranosus; it is thin at intermediate levels, but below, where it covers the tendons behind the malleoli, it is thick and continuous with the flexor and superior peroneal retinacula (p. 611).

The popliteus (**5.**101), flat and triangular, forms the deep boundary or floor of the lower part of the popliteal fossa. The lateral, larger part of the muscle arises by a strong tendon, about

2·5 cm long, from a depression at the anterior end of the groove on the lateral condyle of the femur, its medial fibres from the arcuate popliteal ligament (p. 485) where it blends with the fibrous capsule adjacent to the lateral meniscus; fibres also arise from the outer margin of the meniscus itself. The muscle is attached to the medial two-thirds of the triangular area above the soleal line on the posterior surface of the tibia, and into the tendinous expansion which covers its surface. An additional head may arise from the sesamoid in the lateral head of gastrocnemius.

Relations. The popliteal tendon is intracapsular and is overlapped by the fibular collateral ligament of the knee and the tendon of biceps femoris (**4.**76). Invested on its deep surface by synovial membrane, it grooves the posterior border of the lateral meniscus and the adjoining part of the tibia, and emerges inferior to the posterior band of the arcuate ligament (**4.**72). On the floor of the popliteal fossa it is covered by a strong layer of fascia derived mostly from the tendon of semimembranosus.

Nerve supply. The tibial nerve, L. 4 and 5 and S. 1.

Actions. The popliteus rotates the tibia medially on the femur, or, when the tibia is fixed, rotates the femur laterally on the tibia. It is usually regarded as the muscle which 'unlocks' the joint at the beginning of flexion of the fully extended knee (p. 488); electromyographic observations support this view. Its connexion with the arcuate popliteal ligaments, fibrous capsule and lateral meniscus, suggests that it may draw the posterior part of the latter backwards during lateral rotation of the femur and flexion of the knee joint and so protect the meniscus from being crushed between the femur and the tibia during these movements (Last 1950). The muscle shows marked activity in knee flexion during crouching; this may aid the posterior cruciate ligament in preventing forward dislocation of the femur.

The flexor hallucis longus (**5.**99, 101) arises from the inferior two-thirds of the posterior surface of the fibula (save for the lowest 2·5 cm or so), from the lower part of the interosseous membrane, from the posterior crural intermuscular septum, and from the fascia covering tibialis posterior, which it overlaps to a considerable extent. The fibres pass obliquely down to a tendon which occupies nearly the whole length of the posterior aspect of the muscle. This tendon grooves the posterior surface of the lower end of the tibia, the posterior surface of the talus, and the inferior surface of the sustentaculum tali of the calcaneus (**5.**102, 103). The grooves on the talus and calcaneus are converted by fibrous bands into a canal, lined by a synovial sheath. Distal to this it crosses the flexor digitorum longus from lateral to medial side, pursuing an obliquely curved course superior to it; at the crossing point the long digital flexor receives a fibrous slip from the flexor hallucis longus tendon. The latter then crosses the lateral portion of flexor hallucis brevis to reach the interval between the sesamoid bones under the head of the first metatarsal. It continues on the plantar aspect of the great toe running in an osseo-aponeurotic tunnel to be attached to the plantar aspect of the base of the distal phalanx. The tendon is retained in position over the lateral part of flexor hallucis brevis by the diverging stems of the distal band of the medial intermuscular septum. The connecting slip to flexor digitorum longus varies considerably in size; it usually continues into the tendons for the second and third toes, but is sometimes restricted to the second, or occasionally reaches the tendons of the second, third and fourth toes.

Relations. The *superficial surface* of the muscle is related to soleus and the tendo calcaneus, from which it is separated by the deep transverse fascia; its *deep surface* to the fibula, tibialis posterior, the peroneal vessels, the lower part of the interosseous membrane, and the talocrural joint; its *lateral border* to the peronei; its *medial border* to the tibialis posterior, posterior tibial vessels and tibial nerve.

Nerve supply. The tibial nerve, S. **2** and 3.

The flexor digitorum longus (**5.**101), medial to the flexor hallucis longus, is thin and pointed proximally, but gradually widens as it descends. It arises from the posterior surface of the tibia, medial to tibialis posterior, from just below the soleal line to within 7 or 8 cm of the lower extremity of the bone; it also arises from the fascia covering tibialis posterior. The muscle ends in a tendon extended along almost the whole of its posterior surface which gradually crosses tibialis posterior and passes behind the medial malleolus, in a groove shared with tibialis posterior, but separated from the latter by a fibrous septum. Each tendon is in a separate compartment lined by a synovial sheath. It passes obliquely forwards and laterally, in contact with the medial side of the sustentaculum tali (5.102), deep to the flexor retinaculum, and

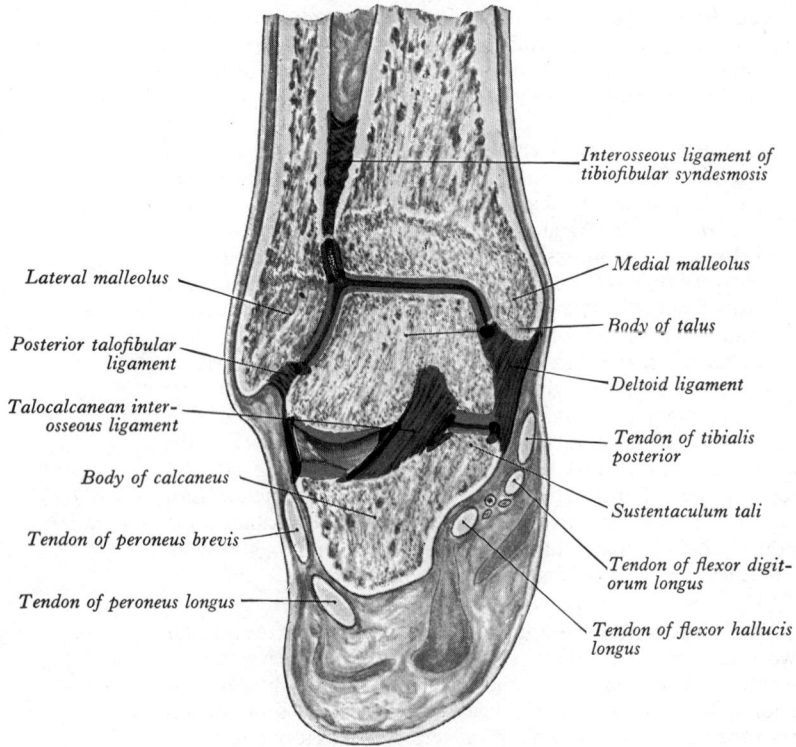

5.102 A coronal section through the left talocrural, talocalcanean and subtalar joints. Ligaments are shown in green, articular cartilage in blue, and synovial membrane in red.

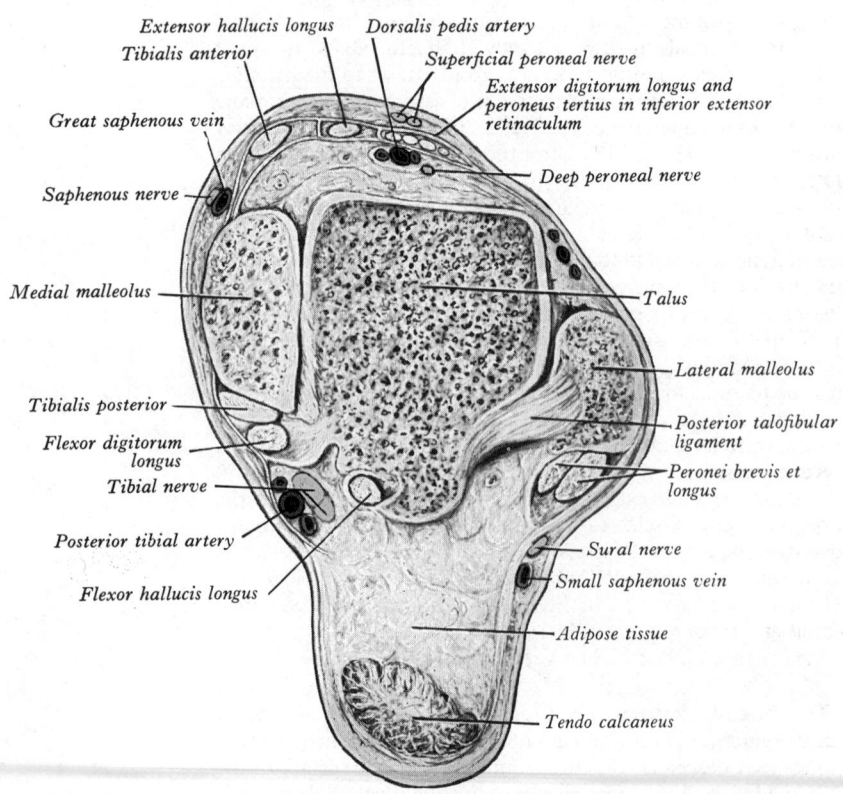

5.103 Transverse section through the inferior part of the talocrural joint.

enters the sole of the foot (5.109), where it crosses superficial to the tendon of flexor hallucis longus, receiving a strong slip from it. It continues across the sole to form the whole of the long flexor tendon of the fifth digit and contributes to those of the second, third and fourth digits. It may also occasionally send a slip to the tendon of the flexor hallucis longus in the great toe. The long flexor tendons of the second, third and fourth digits receive accessions from flexor accessorius and variably from the flexor hallucis through the connecting slip mentioned above. The long flexor tendons of the lateral four digits are attached to the plantar surfaces of the bases of their distal phalanges, each passing through an opening in the corresponding tendon of flexor digitorum brevis at the base of the proximal phalanx. (Compare with the arrangement in the hand, pp. 575, 576.)

The lateral head of flexor accessorius may be inserted into the lateral border of the flexor digitorum longus tendon (p. 615).

Relations. In the leg the muscle's *superficial surface* is in contact with the deep transverse fascia, which separates it from soleus, and, distally, from the posterior tibial vessels and tibial nerve; its *deep surface* is related to the tibia and tibialis posterior. In the foot it is covered by abductor hallucis and flexor digitorum brevis, and crosses superficial to flexor hallucis longus.

Nerve supply. The tibial nerve, S. **2** and 3.

Actions. *When the foot is off the ground*, both the preceding muscles flex the phalanges of the toes, acting primarily on the distal phalanges. They are also plantar flexors.

When the foot is on the ground and under load, these muscles, acting synergically with the small muscles of the foot and—in the case of the flexor digitorum longus—especially with the lumbricals and interossei (p. 617), maintain the pads of the toes in firm contact with the ground, enlarging the weight-bearing area and helping to stabilize the heads of the metatarsal bones, which form the fulcrum on which the body is propelled forwards. Both muscles show little activity in the standing position, contributing little to the longitudinal arch mechanism. During take-off and tip-toe movements, however, their activity is marked.

The tibialis posterior (5.96, 101) arises between flexor hallucis longus and flexor digitorum longus, and is overlapped by both, but especially by the former; it is the most deeply placed muscle of the flexor group. Two pointed processes, separated by an angular interval traversed by the anterior tibial vessels to reach the extensors, characterize the proximal end of the muscle. The medial part of the muscle arises from the posterior surface of the interosseous membrane, excepting its lowest part, from a lateral area on the posterior surface of the tibia between the soleal line above and the junction of the middle and lower thirds of the shaft below. The lateral part arises from a medial strip of the posterior fibular surface in its upper two-thirds. The muscle also arises from the deep transverse fascia, and from the intermuscular septa separating it from adjacent muscles. In the lower quarter of the leg its tendon passes anterior to (i.e. deep to) that of flexor digitorum longus and lies with it in a groove behind the medial malleolus, but enclosed in a separate synovial sheath; it next passes deep to the flexor retinaculum (p. 611) and superficial to the deltoid ligament (5.102) to enter the foot. In the foot it is first inferior to the plantar calcaneonavicular ligament, where it contains a sesamoid fibrocartilage. The tendon then divides into two. The more superficial and larger division, a direct continuation of the tendon, is attached to the tuberosity of the navicular, from which fibres are continued to the inferior surface of the medial cuneiform. A tendinous band also passes laterally and a little proximally to the tip and distal margin of the sustentaculum tali. The deeper, lateral division gives rise to the tendon of origin of the medial limb of flexor hallucis brevis, and then continues between this muscle, and the navicular and medial cuneiform bones, to end on the intermediate cuneiform and the bases of the second, third and fourth metatarsals. The slips to the metatarsals are variable in number, that to the fourth metatarsal being the strongest. Slips to the cuboid and lateral cuneiform have also been described (Lewis 1964; see also **3**.203).

Relations. The *superficial surface* of the muscle is in relation with soleus, from which it is separated by the deep transverse fascia, the flexor digitorum longus, the flexor hallucis longus, the posterior tibial vessels, tibial nerve, and the peroneal vessels; its *deep surface* is in contact with the interosseous membrane, tibia, fibula and talocrural joint.

Nerve supply. The tibial nerve, L. 4 and 5.

Actions. The tibialis posterior is the principal invertor of the foot and may assist powerful plantar flexion. Through its insertions into the cuneiform bones and the bases of the metatarsals it has long been considered to contribute to the elevation of the longitudinal arch of the foot. However, it is quiescent in standing but phasically active in walking, during which, in concert with the intrinsic foot musculature and the lateral calf muscles, it probably affects the degree of pronation of the foot, and the distribution of weight through the metatarsal heads.

It has been suggested that when the body is supported on one leg, the tibialis posterior, acting from below, assists in maintaining balance by resisting any tendency to overbalance to the lateral side. It must be remembered, however, that any act of balancing demands the cooperation of many muscles, particularly groups acting on the hip joint and vertebral column.

The Talocrural Fasciae and Retinacula

As the tendons of the muscles of the leg cross the talocrural joint to the foot, they are bound down by localized thickenings of the deep fascia which constitute retinacular bands, comparable, both in formation and function, with the retinacula of the wrist (pp. 583–584). They comprise superior and inferior extensor, flexor, and superior and inferior peroneal retinacula.

The superior extensor retinaculum (5.97, 104) binds down the tendons of tibialis anterior, extensor hallucis longus, extensor digitorum longus and peroneus tertius, as they descend immediately above the anterior aspect of the talocrural joint; the anterior tibial vessels and the deep peroneal nerve also pass deep to it. It is attached laterally to the lower end of the anterior border of the fibula, and medially to the anterior border of the tibia; above, it is continuous with the deep fascia of the leg and similar dense connective tissue connects its inferior border to the inferior extensor retinaculum. Only the tendon of tibialis anterior has a synovial sheath in this situation (5.104).

The inferior extensor retinaculum (5.97, 104) is a Y-shaped band anterior to the talocrural joint. Its stem is attached to the upper surface of the calcaneus, in front of the sulcus calcanei, and passes medially, forming a strong loop around the tendons of peroneus tertius and the extensor digitorum longus (5.103); from the deep aspect of the loop a band passes laterally behind the interosseous talocalcanean ligament and the cervical ligament to be attached to the sulcus calcanei (Smith 1896; Stamm 1931). From the medial extremity of this loop the two diverging bands pass further medially to complete the Y. The *upper* consists of two layers; the deep one passes deep to the tendons of extensor hallucis longus and tibialis anterior, but superficial to the anterior tibial vessels and deep peroneal nerve, to reach the tibial malleolus. The superficial layer crosses superficial to the tendon of extensor hallucis longus and is then firmly connected to the deep one; it may or may not be continued superficial to the tendon of tibialis anterior to reach the tibia. The *lower band* extends downwards and medially to be attached to the plantar aponeurosis; it is superficial to the tendons of extensor hallucis longus and tibialis anterior, the arteria dorsalis pedis and the terminal branches of the deep peroneal nerve.

The flexor retinaculum (5.98) is attached anteriorly to the tip of the medial malleolus, distal to which it is continuous with the deep fascia on the dorsum of the foot. Its posterior extremity is attached to the medial process of the calcaneus and to the plantar aponeurosis. Its upper border is not clearly demarcated from the deep fascia of the leg, especially the deep transverse layer. Its lower border is continuous with the plantar aponeurosis and many fibres of the abductor hallucis are attached to it. The flexor retinaculum converts the bony grooves in this situation into canals for the tendons and bridges over the posterior tibial vessels and tibial nerve as all these structures enter the sole. From medial

to lateral these structures are: the tendon of tibialis posterior, the tendon of flexor digitorum longus, the posterior tibial vessels, tibial nerve and the tendon of flexor hallucis longus (5.103).

The peroneal retinacula are fibrous bands which retain the tendons of peroneus longus and brevis in position as they curve round the lateral side of the ankle. The *superior retinaculum* (5.101) extends from the back of the lateral malleolus to the deep transverse fascia of the leg and the lateral surface of the calcaneus. The *inferior retinaculum* is continuous in front with the inferior extensor retinaculum; behind it is attached to the lateral surface of the calcaneus; some of its fibres are fused with the periosteum on the peroneal trochlea of the calcaneus, forming a septum between the tendons of peroneus longus and brevis.

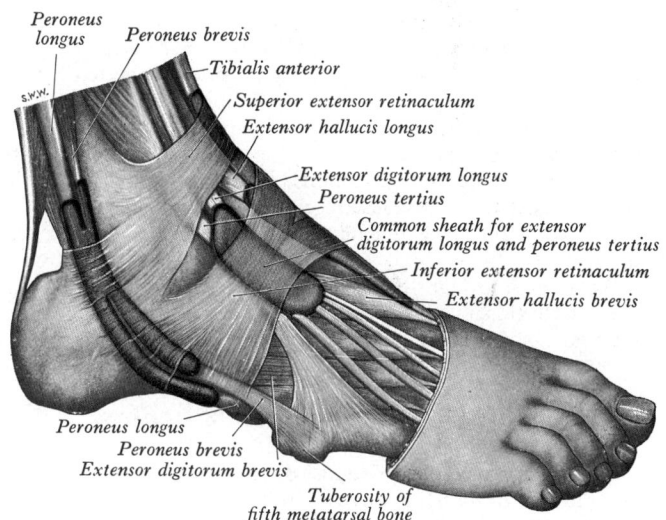

5.104 The synovial sheath of the tendons at the right ankle. Lateral aspect.

The synovial sheaths in the talocrural region. The tendons crossing the talocrural joint are all to some degree deflected from a straight course, and are hence held down by retinacula and enclosed in synovial sheaths.

Anteriorly the sheath for tibialis anterior extends from the proximal margin of the superior extensor retinaculum to the interval between the diverging limbs of the inferior retinaculum (5.105); the sheath for extensor digitorum longus and peroneus tertius and another for extensor hallucis longus start just above the level of the malleoli, the former reaching the higher. The sheath of extensor hallucis longus is prolonged to the base of the first metatarsal bone, while that of extensor digitorum longus reaches only to the level of the base of the fifth metatarsal bone. *Medial* to the ankle (5.105) the sheath for tibialis posterior extends for about 4 cm above the malleolus; below, it ends just proximal to the attachment of the tendon to the tuberosity of the navicular. The sheath for the flexor hallucis longus reaches the level of the malleolus, while that for flexor digitorum longus goes slightly higher; the former is continued to the base of the first metatarsal bone, the latter ends at the navicular. *Lateral* to the ankle (5.104) a sheath, which is proximally single but double below, encloses peroneus longus and brevis. It extends for about 4 cm proximal to the tip of the malleolus, and downwards and forwards for about the same distance.

IV. The Muscles of the Foot

The muscles which are *intrinsic* or confined to the foot follow the primitive limb pattern of dorsal extensors and plantar flexors. Topographically and functionally associated with these are the tendons of the *extrinsic* muscles, already considered. The intrinsic extensor musculature is most limited, but it should be noted that, as in the hand, certain of the intrinsic flexor muscles are also concerned in part with extensor activities.

1. The Dorsal Muscle of the Foot

The deep fascia on the dorsum of the foot (*fascia dorsalis pedis*) is a thin layer, continuous above with the inferior extensor retinaculum; at the sides of the foot it blends with the plantar aponeurosis; anteriorly it ensheathes the dorsal tendons.

The extensor digitorum brevis (5.97, 104) is a thin muscle, arising from the anterior region of the superolateral surface of the calcaneus, in front of the groove for peroneus brevis, from the interosseous talocalcanean ligament, and the stem of the inferior extensor retinaculum. It slants distally and medially across the dorsum of the foot, and ends in four tendons. The medial part of the muscle is usually a more or less distinct slip ending in a tendon which crosses the dorsalis pedis artery superficially to be attached to the dorsal aspect of the base of the proximal phalanx of the great toe; it is sometimes termed the *extensor hallucis brevis*. The other three tendons are attached to the lateral sides of the tendons of extensor digitorum longus of the second, third and fourth toes. This muscle is subject to much variation in the form of accessory slips (from the talus and navicular), extra tendons (to the fifth digit), and suppression of tendons. It may be connected to the adjacent dorsal interosseous muscles.

Nerve supply. The lateral terminal branch of the deep peroneal nerve, S. 1 and 2.

Actions. Through the tendons of extensor digitorum longus the muscle aids in extension of the phalanges of the middle three toes; but, in the hallux, it acts only on the first phalanx.

2. The Plantar Fasciae and Muscles

The plantar fasciae, deep and superficial, show a pattern of arrangement similar to those in the hand (5.106A–D), with some small differences due to absence of opposition and grasping in the foot. They are arranged to limit the mobility of the skin, to hold down muscles and especially tendons in the concavity of the foot and digits, to facilitate the excursions of these tendons, to prevent compression of the plantar and digital vessels and nerves, and perhaps to aid venous return. The plantar skin is repeatedly subjected to forcible shearing and impact stresses in locomotion, particularly in the areas of the posterior calcaneal tubercles, the metatarsal heads, and the pulps of the terminal segments of the toes. As in the hand, pads of adipose tissue occur in these regions, and these are diffusely pervaded by fine but collectively strong connective tissue strands, which tether the skin and limit displacement of the fat, thus augmenting rather than impeding its resilient cushioning effect. The direction of these strands, which extend from the deep fascia through the subcutaneous tissues to the dermis of the skin, is adapted to that of prevailing stresses, as has been clearly demonstrated in the sub-metatarsal region, the '*ball' of the foot*, by Bojsen-Møller and Flagstad (1976) (5.106D).

The parts of the deep fasciae on the inferior aspect of the plantar structures are usually named collectively the *plantar aponeurosis* (*vide infra*); but it is only the central part of this which is extensively aponeurotic, and some anatomists reserve the name for this part alone. The medial and lateral parts are aponeurotic only posteriorly, where they provide muscle attachments (*vide infra*).

The plantar aponeurosis, or fascia (5.106A), is composed of collagen fibres disposed chiefly in a longitudinal manner, but also transversely. It can be conveniently divided for description into medial and lateral parts which overlie the intrinsic muscles of the hallux and minimus, and a central part overlying the long and short digital flexors.

The central part is the strongest and thickest. Posteriorly narrow, where it is attached to the medial process of the calcaneal tuberosity, proximal to the flexor digitorum brevis, it becomes broader and somewhat thinner as it diverges towards the metatarsal heads. Just proximal to these it divides into five bands, corresponding with the toes. These five bands are united by transverse fibres where they begin to diverge (5.106A), below the metatarsal shafts. Proximal to the metatarsal heads the superficial stratum of each band is connected to the dermis by *skin ligaments*

(*retinacula cutis*), which reach the skin of the ball of the foot proximal to the furrow separating toes from sole (5.106D). The deep stratum of each digital band of the aponeurosis divides into *two septa* which flank the digital flexor tendons, separating them from the lumbrical muscles and the digital vessels and nerves. These septa pass deeply to fuse with the interosseous fascia, the deep transverse metatarsal ligaments, the plantar ligaments of the

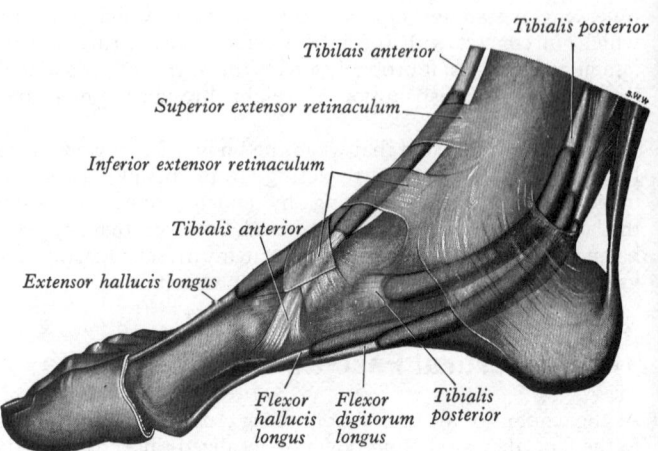

Tibialis posterior
Tibilais anterior
Superior extensor retinaculum
Inferior extensor retinaculum
Tibialis anterior
Extensor hallucis longus
Flexor hallucis longus *Flexor digitorum longus* *Tibialis posterior*

5.105   The synovial sheaths of the tendons at the right ankle. Medial aspect.

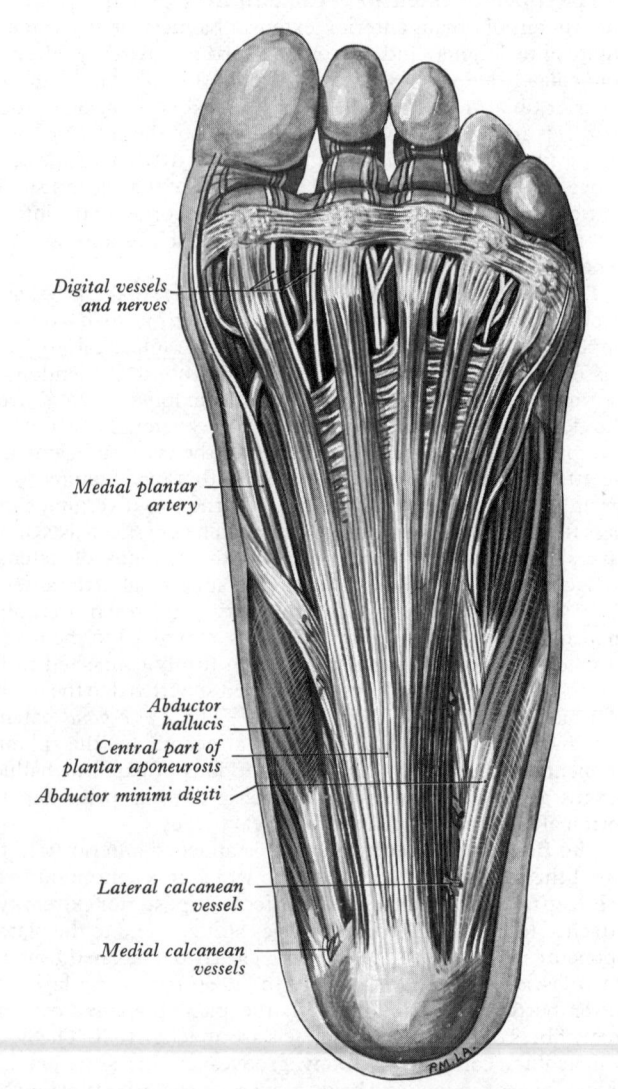

Digital vessels and nerves
Medial plantar artery
Abductor hallucis
Central part of plantar aponeurosis
Abductor minimi digiti
Lateral calcanean vessels
Medial calcanean vessels

5.106A   The plantar aponeurosis of the left foot.

metatarsophalangeal joints, and the periosteum and fibrous flexor sheaths at the base of each proximal phalanx. In the webs between the metatarsal heads and the bases of the proximal phalanges pads of fat are developed, which cushion the digital nerves and vessels from adjoining tendinous structures and extraneous plantar pressures. These four fat pads are stabilized by vertical strands of collagen fibres, which extend from the fibrous flexor sheaths to

the superficial stratum of the plantar aponeurosis and, through the latter, to the skin. Just distal to the metatarsal heads a *plantar interdigital ligament* (officially known as the superficial transverse metatarsal ligament) blends with the *deep* aspect of the superficial stratum of the plantar aponeurosis, where it extends into the toes (**5**.106C, D).

The central part of the plantar aponeurosis thus provides an intermediary between the skin and skeleton of the foot, through numerous *cutaneous retinacula* and *deep septa* extending anteriorly to the metatarsals and phalanges. It is also continuous with its medial and lateral parts, and at these junctions *two vertical intermuscular septa*, medial and lateral, extend deeply between the medial, intermediate and lateral groups of plantar muscles to reach bone (*vide infra*). Thinner *horizontal intermuscular septa*, passing between muscle layers, are derived from the vertical intermuscular septa.

The *lateral part* of the plantar aponeurosis is inferior to abductor digiti minimi; it is thin distally and thick proximally, where it forms a strong band, sometimes containing muscle fibres, between the lateral process of the calcanean tuberosity and the base of the fifth metatarsal bone; it is continuous medially with the central part, and with the fascia on the dorsum of the foot round its lateral border.

The *medial part* of the plantar aponeurosis is thin and inferior to abductor hallucis; it is continuous proximally with the flexor retinaculum, medially with the fascia dorsalis pedis, and laterally with the central part of the plantar aponeurosis.

The *lateral intermuscular septum* is incomplete, especially at its proximal end; distally its deep attachments are to the fibrous sheath of peroneus longus and to the fifth metatarsal bone. The *medial intermuscular septum* is also incomplete and divides into three bands, proximal, intermediate and distal, each of which displays lateral and medial divisions as it approaches its deep

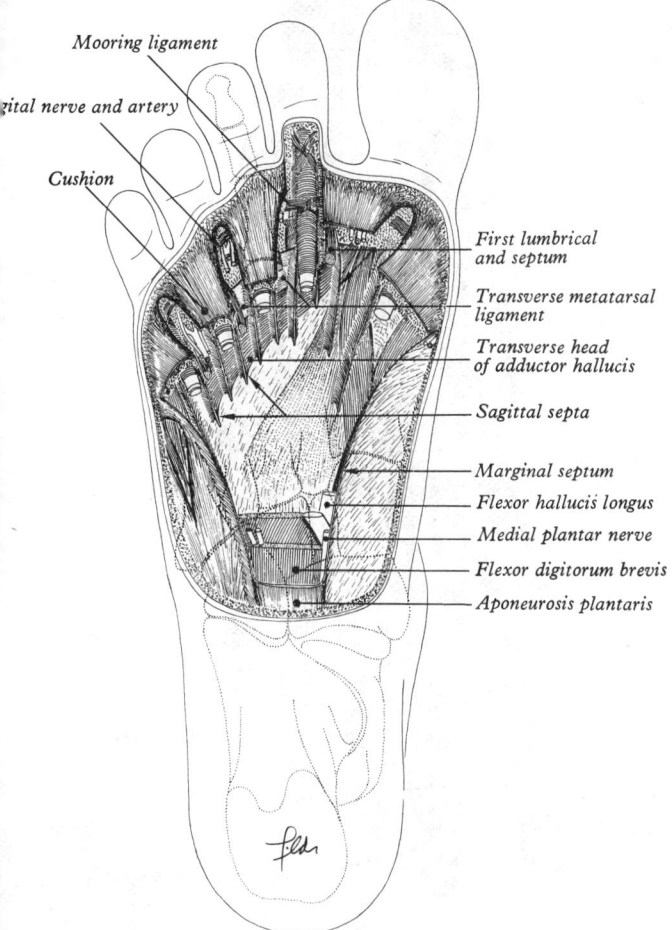

5.106B, C & D Details of the tendinous and fibrous architecture of various regions on the plantar aspect of the foot. (B) Plantar aspect of the central compartment and the structures forming the ball of the foot. Of the plantar interdigital ligament, only the mooring ligament is shown.

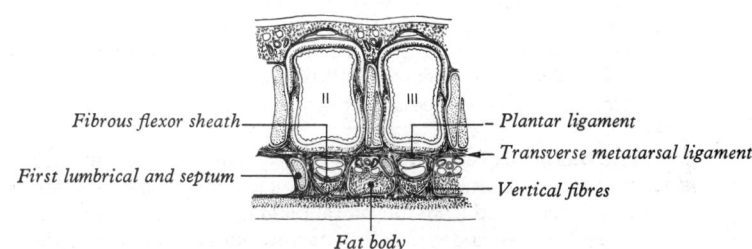

5.106C A transverse section through the heads of the second and third metatarsal bones showing the course of the collagen fibre bundles in the submetatarsal cushions and around the joints. Fat covers the fibrous flexor sheath inside the cushion, and the digital nerves and vessels are lodged between the cushions.

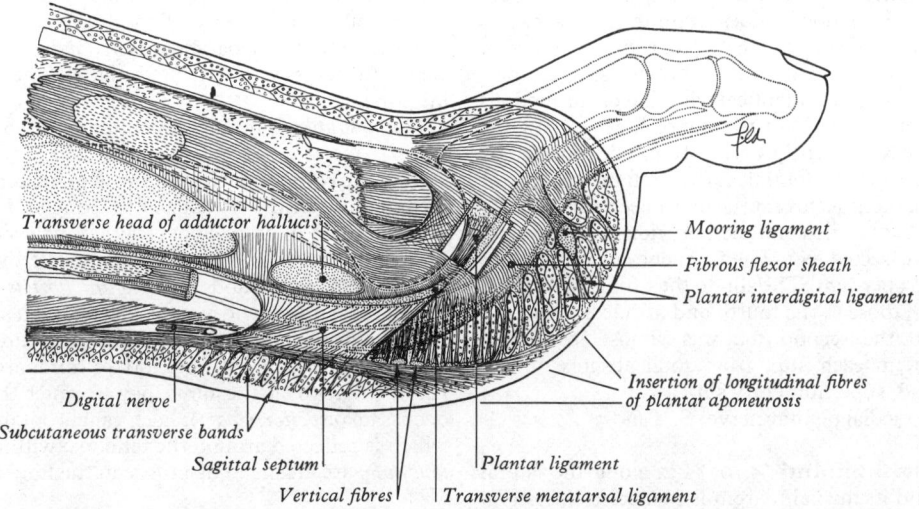

5.106D A sagittal section through the second interosseous cleft showing the internal architecture of the three major areas of the ball of the foot. The sagittal septum is attached to the proximal phalanx through the transverse metatarsal ligament and the plantar ligament of the joint. The vertical

fibres and the lamellae of the plantar interdigital ligament are attached to the proximal phalanx through the fibrous flexor sheath. (From F. Bojsen-Møller and K. E. Flagstad 1976, by courtesy of the authors and the Cambridge University Press.)

attachment. The *proximal band* is attached laterally to the cuboid and blends medially with the tendon of tibialis posterior. The *middle band* is attached laterally to the cuboid and the long plantar ligament and medially to the medial cuneiform bone. The *distal band* divides to enclose the tendon of flexor hallucis longus and is attached to the fascia over flexor hallucis brevis.

The muscles in the plantar region of the foot may be divided into medial, lateral and intermediate groups, as in the hand. The medial and lateral groups comprise the intrinsic muscles of the hallux and fifth digit, while the central or intermediate group includes the lumbricals, interossei and short digital flexors. However, it is customary to group them in four layers, as met with in the course of dissection. These 'layers' can be over-emphasized, and in functional terms the former grouping will often be found more useful. Their actions are discussed together on p. 617.

THE FIRST LAYER

This, the superficial layer (**5**.107), includes the abductores hallucis et digiti minimi and the flexor digitorum brevis. All three extend from the calcanean tuberosity to the toes, and therefore, in this case, comprise a functional group capable of assisting in maintaining the concavity of the foot.

The abductor hallucis (**5**.107) lies along the medial border of the foot and covers the origins of the plantar vessels and nerves. It arises principally from the flexor retinaculum, but also from the medial process of the calcanean tuberosity, the plantar aponeurosis, and the intermuscular septum between it and flexor digitorum brevis. The fibres end in a tendon which is attached, together with the medial tendon of flexor hallucis brevis, to the medial side of the base of the proximal phalanx of the great toe. Some fibres are attached more proximally to the medial sesamoid bone of this toe. A separate tendinous slip may extend to the base of the proximal phalanx of the great toe; the muscle may also derive some fibres from the dermis of the medial border of the foot. It flexes and abducts the hallux.

Nerve supply. The medial plantar nerve, S. 2 and **3**.

The flexor digitorum brevis (**5**.107) is immediately superior or deep to the central part of the plantar aponeurosis. Its deep surface is separated from the lateral plantar vessels and nerves by a thin layer of fascia. It arises by a narrow tendon from the medial process of the calcanean tuberosity, from the central part of the plantar aponeurosis, and from the intermuscular septa between it and adjacent muscles. It divides into four tendons, for the lateral four toes. At the bases of their proximal phalanges, each tendon divides into two slips, around the corresponding tendon of flexor digitorum longus; the two slips then reunite, partially decussate, and form a grooved channel for the tendon of flexor digitorum longus. The short flexor tendon divides again to be attached to both sides of the shaft of the intermediate phalanx. The mode of division of tendons of flexor digitorum brevis, and of their attachments to the phalanges, is identical with that of the tendons of flexor digitorum superficialis in the hand. The part of the muscle for the fifth toe is frequently absent. It may be replaced by a small muscular slip from the long flexor tendon or from flexor accessorius. The variations of this muscle have been surveyed by Nathan and Gloobe (1974) who combined previous findings with observations on 100 dissected feet. The frequency of variation in the latter group was 63 per cent. The slip to the fifth toe was most variable (63 per cent), those to the fourth and fifth less so (10 per cent), while that to the second toe was almost invariable. Variability extended, in each slip, from total absence to the occurrence of a second, supernumerary slip.

Nerve supply. The medial plantar nerve, S. 2 and **3**.

The abductor digiti minimi (**5**.107) lies along the lateral border of the foot, and its medial margin is related to the lateral plantar vessels and nerve. It arises from both processes of the calcanean tuberosity, from the plantar surface of the bone between them, from the plantar aponeurosis, and from the intermuscular septum between it and flexor digitorum brevis. Its

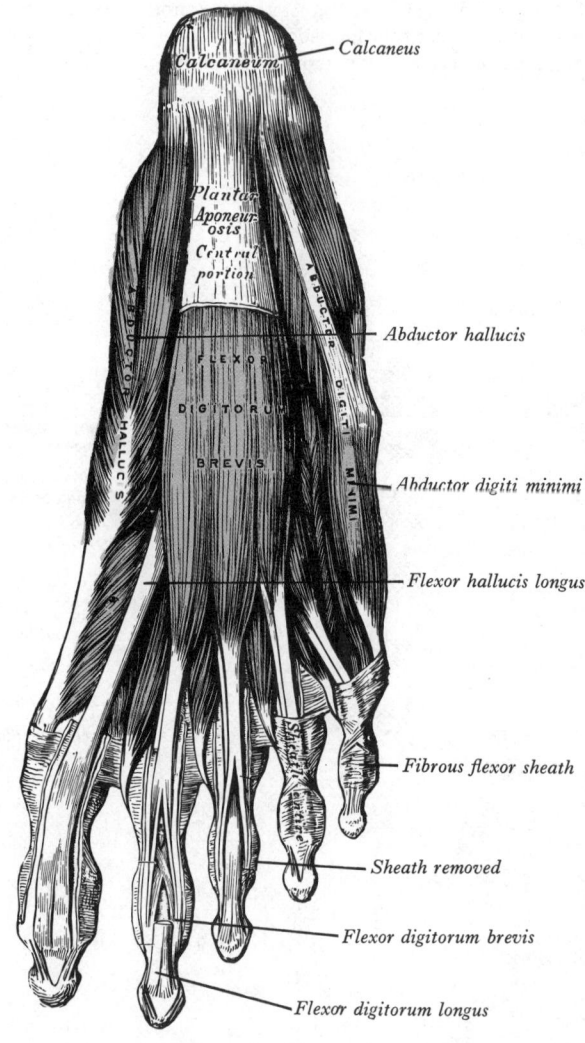

Calcaneus

Abductor hallucis

Abductor digiti minimi

Flexor hallucis longus

Fibrous flexor sheath

Sheath removed

Flexor digitorum brevis

Flexor digitorum longus

5.107　The superficial plantar muscles of the right foot.

tendon glides over a smooth groove on the plantar surface of the base of the fifth metatarsal bone and is attached with flexor digiti minimi brevis, to the lateral side of the base of the proximal phalanx of the fifth toe; hence it is more a *flexor* than an abductor. Some of the fibres arising from the lateral process usually reach the tip of the tuberosity of the fifth metatarsal (**3**.203). These fibres may form a separate muscle, *abductor ossis metatarsi digiti quinti*. An accessory slip from the base of the fifth metatarsal is not infrequent.

Nerve supply. The lateral plantar nerve, S. 2 and **3**.

The fibrous sheaths of the flexor tendons (**5**.107). The terminations of the tendons of the long and short flexor muscles are contained in osseo-aponeurotic canals similar to those in the fingers. These canals are bounded above by the phalanges, and below by fibrous bands, the *digital fibrous sheaths*, which arch across the tendons and are attached on each side to the margins of the phalanges. Along the proximal and intermediate phalanges the fibrous bands are strong, and the fibres are transverse (*annular part*); but opposite the joints they are much thinner, and the fibres decussate (*cruciform part*). Each canal contains a synovial sheath, which is reflected around the tendons; within this sheath *vincula tendinum* are arranged like those in the fingers (p. 577).

THE SECOND LAYER

The muscles of this group are between those of the superficial layer, inferior to them, and those of the 'third layer', which are

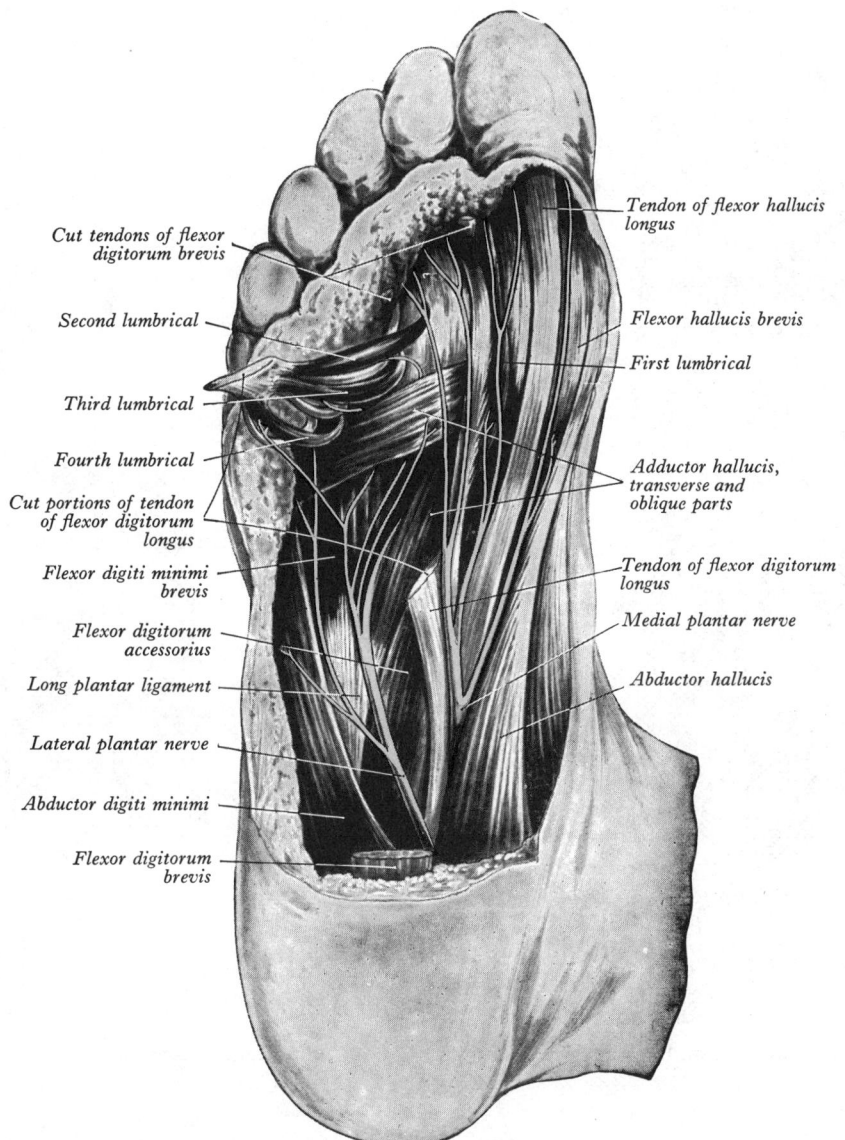

Cut tendons of flexor
digitorum brevis

Second lumbrical

Third lumbrical

Fourth lumbrical

Cut portions of tendon
of flexor digitorum
longus

Flexor digiti minimi
brevis

Flexor digitorum
accessorius

Long plantar ligament

Lateral plantar nerve

Abductor digiti minimi

Flexor digitorum
brevis

Tendon of flexor hallucis
longus

Flexor hallucis brevis

First lumbrical

Adductor hallucis,
transverse and
oblique parts

Tendon of flexor digitorum
longus

Medial plantar nerve

Abductor hallucis

5.108   The plantar muscles and their nerve supply. Most of the flexor digitorum brevis has been removed. The flexor digitorum longus has been divided partially, and its distal end has been turned forwards together with the second, third and fourth lumbricals.

superior. They include the flexor digitorum accessorius and four lumbrical muscles (5.108, 109).

The flexor digitorum accessorius (5.109) arises by two heads separated by the long plantar ligament. The medial, and larger, is muscular, and attached to the medial concave surface of the calcaneus below the groove for the tendon of flexor hallucis longus; the lateral head, flat and tendinous, arises from the calcaneus in front of the lateral process of the tuberosity and from the long plantar ligament. The medial head may end in a fibrous lamina which passes deep to the tendon of flexor digitorum longus dividing into tendinous slips which join the long flexor tendons to the second, third and, sometimes, fourth digits, but about as often it joins the lateral border of the flexor digitorum longus tendon. The lateral head joins this fibrous lamina either on its superficial surface or lateral border, or it may terminate on the lateral border of the flexor digitorum longus tendon (Lewis 1962), thus perhaps 'correcting' the diagonal vector of this muscle. It is sometimes absent. It varies in the number of digits to which its slips proceed. The slip to the fourth is often absent; a slip to the fifth is not infrequent. For details see Novakova and Korbelář (1976).

Nerve supply. The lateral plantar nerve, S. 2 and 3.

The lumbricals (5.109) are four small muscles, accessory to the tendons of flexor digitorum longus, and numbered from the medial side of the foot. They arise from these tendons, as far back as their angles of separation, and, with the exception of the first,

which arises only from the medial border of the first tendon of flexor digitorum longus, each springs from two adjoining tendons. The muscles end in tendons which pass distally on the medial sides of the four lesser toes, to be attached to the dorsal digital expansions on the proximal phalanges.

Nerve supply. The first lumbrical is supplied by the medial plantar nerve and the rest by the deep branch of the lateral plantar nerve, S. 2 and 3.

THE THIRD LAYER

This group comprises the shorter intrinsic muscles of the hallux and the fifth digit—flexor hallucis brevis, adductor hallucis and flexor digiti minimi brevis (5.108, 110). These are the most deeply situated muscles in the sole, except for the interossei, which are superior to them.

The flexor hallucis brevis (5.110) arises by a bifurcate tendon, the lateral limb of which is attached to the medial part of the plantar surface of the cuboid, posterior to the groove for the peroneus longus tendon, and to the adjacent part of the lateral cuneiform; the medial limb has a deep attachment directly continuous with the lateral division of the tendon of tibialis posterior (in the foot), and a more superficial one to the middle band of the medial intermuscular septum (Lewis 1964). The muscle divides into medial and lateral parts whose tendons are attached to the corresponding sides of the base of the proximal

615

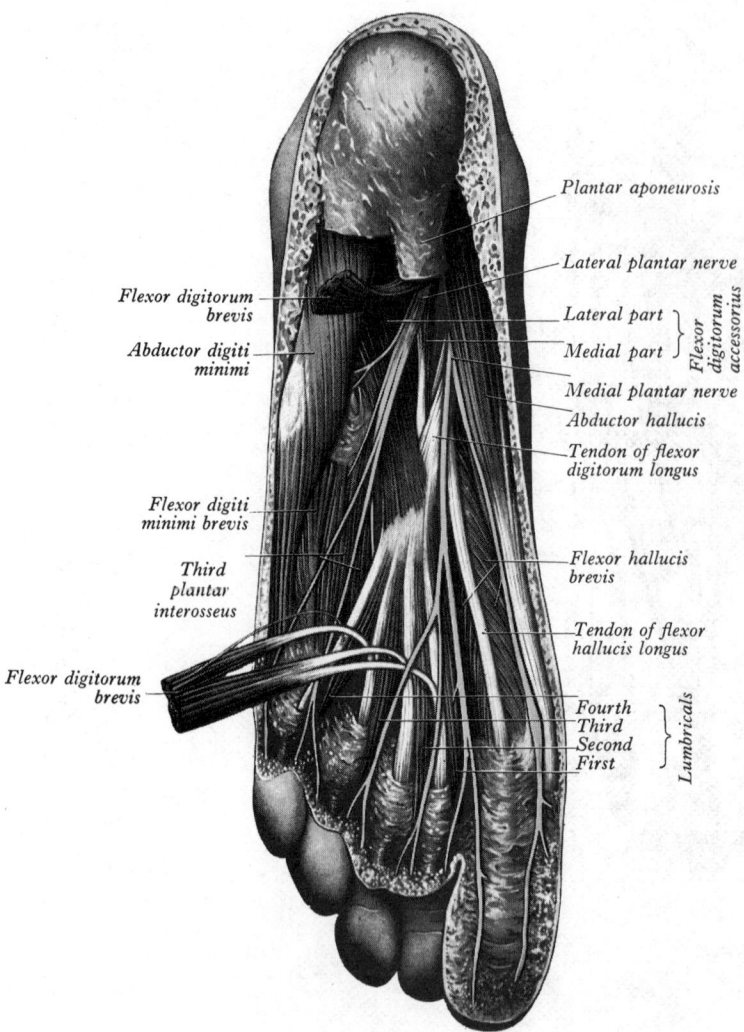

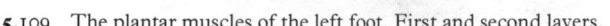

5.109   The plantar muscles of the left foot. First and second layers.

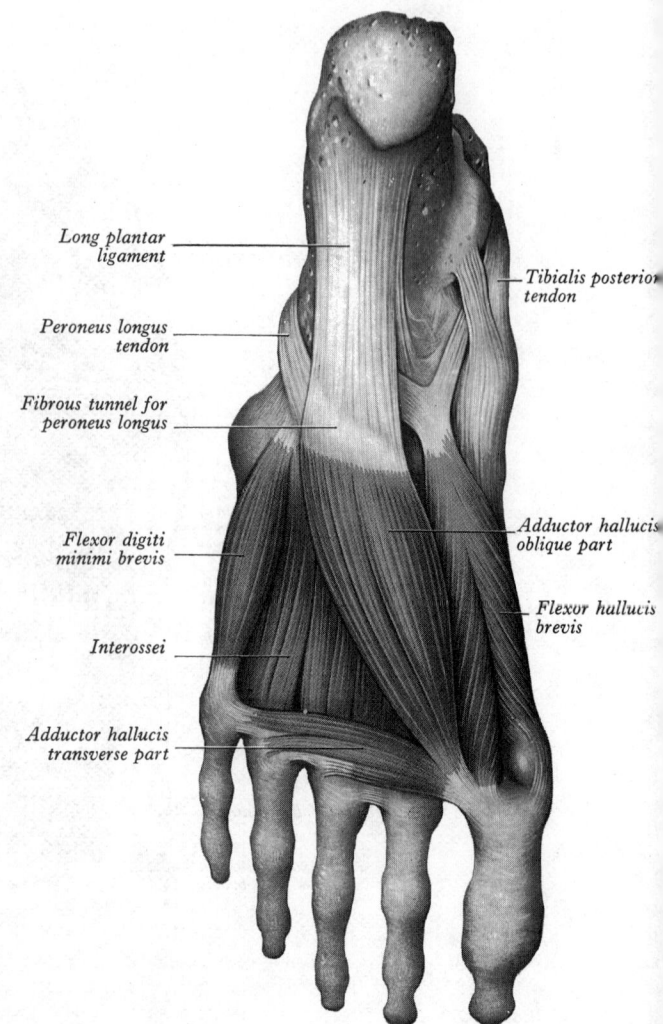

5.110   The plantar muscles of the left foot. Third layer.

phalanx of the hallux, a sesamoid bone usually occurring in each at its attachment. The medial part is blended with abductor hallucis, the lateral with adductor hallucis, as they reach their terminations. Accessory slips from the calcaneus or long plantar ligament may occur, and a tendinous slip may reach the proximal phalanx of the second toe.

Nerve supply. The medial plantar nerve, S. 2 and **3**.

The adductor hallucis (**5**.110) arises by oblique and transverse heads. The *oblique head* springs from the bases of the second, third and fourth metatarsal bones, and from the sheath of the tendon of peroneus longus. The *transverse head*, a narrow, flat fasciculus, arises from the plantar metatarsophalangeal ligaments of the third, fourth and fifth toes (sometimes only from the third and fourth), and from the deep transverse metatarsal ligaments between them. The oblique head has medial and lateral parts; the medial blends with the lateral part of flexor hallucis brevis and is attached to the lateral sesamoid bone of the great toe; the lateral part joins the transverse head and is attached to the lateral sesamoid bone and the base of the first phalanx of the hallux. According to Cralley *et al.* (1976) the transverse part of this muscle is sometimes absent (3 out of 50 feet dissected). These observers failed to identify a phalangeal attachment for the transverse part; those fibres which did not reach the lateral sesamoid bone were attached with the oblique part of the muscle.

Part of the muscle may be attached to the first metatarsal, constituting an *opponens hallucis*. A slip may also pass to the proximal phalanx of the second toe.

Nerve supply. The deep branch of the lateral plantar nerve, S. 2 and **3**.

The flexor digiti minimi brevis (**5**.110) arises from the

medial part of the plantar surface of the base of the fifth metatarsal bone, and from the sheath of peroneus longus; its tendon ends on the lateral side of the base of the proximal phalanx of the fifth toe. Occasionally some of its deeper fibres reach the lateral part of the distal half of the fifth metatarsal bone, sometimes described as a distinct muscle—the *opponens digiti minimi*.

Nerve supply. The superficial branch of the lateral plantar nerve, S. 2 and **3**.

THE FOURTH LAYER

This group comprises the plantar and dorsal **interossei**, which are similar to those in the hand, but they are arranged relative to the *second* and not the *third* digit, as in the hand, the second being the least mobile of the metatarsal bones.

The dorsal interossei (**5**.111A), are situated between the metatarsal bones. They are four bipennate muscles, each arising by two heads from the adjacent sides of two metatarsal bones; their tendons are attached to the bases of the proximal phalanges and to the dorsal digital expansions. The first reaches the medial side of the second toe; the other three pass to the lateral sides of the second, third and fourth toes. In the angular interval between the heads of each of the three lateral muscles one of the perforating arteries passes to the dorsum of the foot; through the space between the heads of the first muscle the terminal part of the dorsalis pedis artery enters the sole.

The plantar interossei (**5**.111B), three in number, are below rather than between the metatarsal bones, and each is connected with only one metatarsal bone. They arise from the bases and medial sides of the third, fourth and fifth metatarsal bones, and are attached to the medial sides of the bases of the proximal phalanges of the same toes, and into their dorsal digital expansions.

Nerve supply. The dorsal and plantar interossei are supplied by the deep branch of the lateral plantar nerve (S. 2 and **3**), except those in the fourth interosseous space, which are supplied by the superficial branch of the same nerve. The first dorsal interosseous frequently receives an extra filament from the medial branch of the deep peroneal nerve on the dorsum of the foot, and the second a twig from the lateral branch of the same nerve.

The Actions of the Intrinsic Muscles of the Foot

As elsewhere in the body the actions of the muscles of the foot, or at least their potentialities, can be deduced from the mere geometry of their attachments, and the names applied to some of them indicate this. Such purely structural deductions must, of course, also take into consideration the modifying effects of contact with the ground, and they are in any case unlikely to provide more than suggestions regarding the actual roles of individual muscles in particular aspects of pedal activity. Direct observations of these muscles, either by eye or touch, is scarcely more informative; even with the aid of electromyography only the crudest of data are available, derived from a highly limited series of experiments. It must therefore be admitted that, despite some attractive speculation (p. 618), much uncertainty persists in these matters, as is attested by the volume of controversial literature on foot function.

There are two major conditions in which the musculature of the foot may be called upon to aid or modify the effects of muscular influences operating upon it from the leg; these are standing still and progression. In standing, with the feet flat on the ground, the feet are used as platforms for the distribution of weight, the centre of gravity of the body being maintained above them by suitable adjustment of tension and length in leg and trunk muscles. It is tempting to suppose that in these conditions the largely longitudinal intrinsic musculature of the sole would be active in sustaining the arched form of the foot against the flattening effect of body weight and this has been, of course, a canon of purely anatomical teaching for many years, despite increasing doubts amongst clinical and experimental workers. All the evidence from such sources, although limited in amount, indicates that the ligamentous ties of the foot, including the plantar aponeurosis, are largely sufficient. It is also often suggested that during standing, the intrinsic muscles, especially the abductors and short flexors, exert a 'tonic' effect in steadying the toe pads on the ground. Electromyographic evidence, admittedly scant, contradicts this and common observation shows that the toes can easily be extended out of contact with the ground with little adverse effect upon the efficiency of the feet in standing. Suzuki (1972), in a detailed electromyographic study of leg and foot muscles of normal and flat-footed individuals, observed that the intrinsic muscles of the foot were relatively inactive in the former individuals when standing, but were active in the flat-footed, as were long arch-supporting muscles such as peroneus longus.

When the heel is lifted from the ground in beginning to take a step, whether in walking or running, the whole of the weight and muscular thrust is transferred to the region of the metatarsal heads and the pads of the toes. This shifts the role of the foot from platform to lever and intensifies the forces acting on the fore part of the foot, especially in running and jumping. So much argument has been extended upon the nature and behaviour of the 'arches' of the foot and of the agents, muscular and ligamentous, which can or do act as 'tie-beams' or trusses across them, that the essential role of the foot as a lever is often overlooked. It is at first sight most ill-suited to act as a lever, being composed of a series of links, though its curvature or arched form has good mechanical precedents. In this situation it is clear that the concavity of the sole is accentuated, and the electromyographic evidence available indicates that the intrinsic muscles become strongly active. This would slacken the plantar aponeurosis, but the dorsiflexion of the toes which must occur tightens it up. The foot is also inverted and supinated, and the position of close-packing of the intertarsal joints is reached as the foot takes the full effects of leverage. In this position the foot loses all its pliancy (Inman 1969). The toes are held extended at the metatarsophalangeal and distal inter-

phalangeal joints and flexed at the proximal interphalangeal including the interphalangeal joint of the hallux. This probably results from contraction of the flexor digitorum brevis and the long flexors of the toes, by drawing the terminal phalanges towards the sole and passively buckling the rest of the toes into the above position. The buckling is perhaps controlled by the action of lumbrical and interosseous muscles by opposing extension at the metatarsophalangeal joints and flexion at the proximal interphalangeal joints. The flexor hallucis brevis could prevent excessive extension at the metatarsophalangeal joint of the hallux. The abductors of the hallux and fifth toe both act as flexors, with probably little abductor effect, though their disposition varies in different feet. It must be emphasized that analyses such as the above are speculative, and in view of the difficulties which impede electromyographic examination of the active foot, the present uncertainties are likely to persist (see also p. 619). Various imbalances in the actions of toe muscles are invoked to explain certain dysfunctions and deformities, such as contracted toes. Actual studies are rare, but an evaluation of the weights of lumbrical muscles has shown no correlation with toe contractures (Challey *et al.* 1975).

Standing, Walking and Running

Although birds are in a sense bipeds, as are some reptiles and eutherian mammals, only man has evolved a completely upright posture as his habitual working position. Some mammals, particularly primates, may adopt such a position for considerable periods, but none displays the degree of specialization associated with human adaptation to continuous upright standing and locomotion. Human erect posture is often held to have been attained at the cost of such a painfully difficult evolution that it is a precarious habit, prone to all manner of drawbacks, breakdowns and pathology. This is probably due to the influence of medically orientated thinking. A completely upright posture was characteristic of most, if not all extinct hominids and has been foreshadowed in other primates for an even greater period of time. The disadvantages of quadrupedal habit do not appear to have attracted similar veterinary attention.

In his *posture* the human primate is singular, if not unique, in having straightened the trunk and lower limbs to such an extent that the centre of gravity of any segment of the body is substantially above the joints upon which its weight impinges; even the head is almost if not wholly balanced upon the cervical spinal column. A vertical from the centre of gravity of the whole

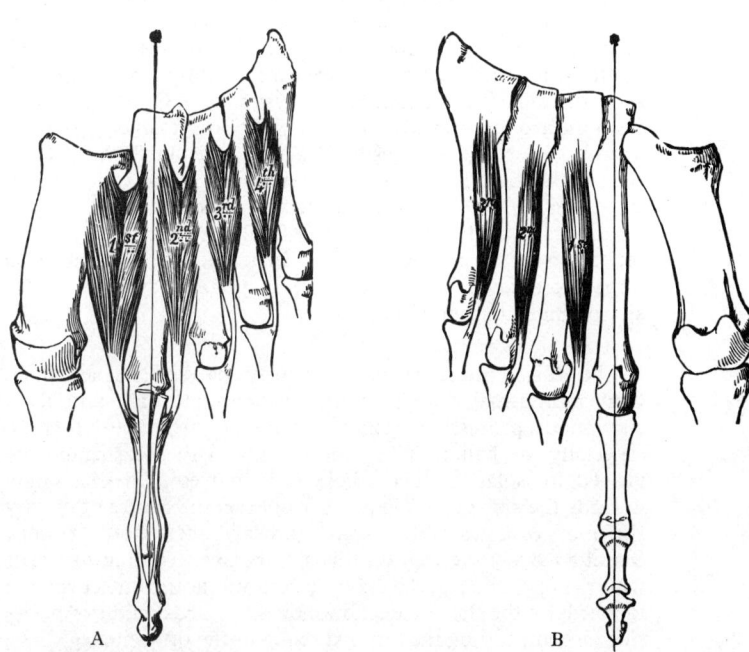

A   B

5.111A and B   The interossei of the left foot. A Dorsal interossei viewed from the dorsal aspect. B Plantar interossei viewed from the plantar aspect. The axis to which the movements of abduction and adduction are referred is as indicated.

body (a point about a centimetre posterior to the sacral promontory) falls anterior to the ankle joint. Strong plantar flexion, particularly due to increased tension in soleus, is hence necessary to prevent forward sway or actual falling. However, in most individuals when standing in an easy, symmetrical manner, the same vertical passes in such a relationship to the hips and knees that gravity tends to increase extension, which is in part limited by the rise of tension in ligamentous structures (associated with the near close-packed position of the joints—*see* p. 438) rather than by muscle activity. Thus, such a posture, often maintained for long periods, requires little muscular activity (Akerblom 1948; Joseph 1960), except for brief bursts in appropriate muscle groups whenever the body sways from this position. It is noticeable that the standing position is often changed, intermittently relieving stress upon ligaments and transferring it to muscles (Smith 1950), the commonest habit being to hyper-extend one knee and to flex the other slightly. Thus one leg after the other is made to take the major stress of body weight, the slightly flexed leg acting for the time being as a kind of outrigger or prop. Evans (1977) has emphasized the role of the iliotibial tract in achieving and sustaining the 'locked' hyper-extended (close-packed) posture of the weight-bearing leg in such asymmetrical standing. The line of gravity is related to the curvatures of the spine in such a way that they are, on the whole, exaggerated by gravity, but can be controlled by small muscular tensions, over the details of which considerable controversy still obtains. Any considerable swaying of the trunk above the pelvis entails not only responsive activity in leg muscles but also in those of the trunk, such as the psoas major when the body sways backwards.

The energy-sparing principle of using gravity to hyperextend (or sometimes flex) joints is applied in some birds, such as the upright penguin, and passerine birds which roost in trees. It is, however, apparently unique in man, among mammals. It should be noted, however, that many quadrupeds also habitually stand for even more prolonged periods of time, apparently without great effort and with much less complaint than the human biped.

In *locomotion*, whether it be walking, running, jumping or climbing, there is an immediate departure from the economical pattern of standing. All muscles of the trunk, arms and legs may be at times involved, some as prime movers, others as fixators or stabilizers. The arms are swung as balancers and also as weighted pendulums to increase the impetus, especially in running and jumping; their use in climbing is obvious.

In studies of locomotion (*vide infra*) most attention has been focused upon analysis of walking movements and the temporal sequence of contraction and relaxation in some muscles involved. As the first step is taken from the standing position, the trunk inclines forwards and the feet are slightly inverted with accentuation of the longitudinal arches (p. 501). One heel is lifted from the ground, transferring the weight progressively along the lateral border of the opposite foot to the metatarsal heads, and then to the toes. This is known as the *stance phase*, which passes into the *swing phase* as soon as the foot leaves the ground, which it meets again in the reverse order—heel first and then again a repetition of the roll forwards on to the toes. From one heel impact to the next is termed a *walking cycle*, the cycle of one leg alternating with that of the other.

The **stance phase** has been variously analysed (Barnett 1956; Gray and Basmajian 1968; Elftmann 1966), but it is essentially a continuous process of weight transference from heel to toes, especially the hallux. These movements of the foot cannot be viewed in isolation. The whole body moves a little laterally towards the side of the leg in stance phase, the centre of gravity shifting to a variable extent (on average about 5 cm). Simultaneously the arm on this side begins to swing forwards with the opposite leg; the pelvis, which was at the outset rotated forwards on the stance side, now moves forwards on the opposite side (accompanying the forward swing of the opposite leg). The foot is, of course, fixed on the ground, and hence, as the knee extends and the ankle dorsiflexes, the tibia rotates medially to a small extent carrying the talus with it, and the femur undergoes an obligatory conjunct medial rotation at the knee joint (p. 488) and

hence at the hip joint. The latter medial rotation of the femur relative to the acetabulum is accentuated by the rotation of the pelvis in the opposite direction mentioned above. This series of congruent medial rotations, combined with dorsiflexion and transient supination of the subtalar skeleton (p. 501) serves to bring all the articulations of the leg and foot near their close-packed positions. The resulting momentary rigidity coincides with the instant of maximum thrust imparted by the powerful plantar flexors of the foot. Thereafter, as the heel is raised from the ground, the foot passes into plantar flexion and pronation and the thrust is now transmitted through a loosely packed pliant lever (under muscular control) to the head of the first metatarsal and the dorsiflexing hallux, which finally leaves the ground to terminate the stance phase. (This brief description should be compared with an analysis of the various movements of the pedal skeleton—pp. 501–503.)

In the **swing phase** the toes leave the ground, as the opposite heel meets it, to transfer support. The leg, relieved of weight bearing, swings forwards with flexion at the hip and knee, dorsiflexion of the foot and lateral rotation of all those elements that had undergone congruent medial rotation during the stance phase. The leg swings forwards with a pendular action which, however, is under the influence of the hip flexors. As the limit of the stride is reached, the momentum of the limb is overcome by the hip extensors to bring the heel to the ground, and to begin another cycle.

These events occur far more rapidly than can be described, but the general analysis is fully supported by cinematographic evidence. The actual role of many of the muscles involved has been investigated at every stage, but the details are beyond the scope of this account and should be sought in original studies.

The speed of walking can, of course, be increased by augmenting the work done by the muscles involved, and by acceleration of their actions. In running, the movements are broadly similar, but the heels do not touch the ground, thrust being taken entirely by the forefoot. Moreover, the body is carried forwards by its momentum entirely unsupported for a brief period between take-off from one foot and descent upon the other. A similar unsupported phase characterizes hopping, whether on one foot or alternate feet. In jumping, take-off may be from one foot or two, either simultaneously or in quick succession; landing occurs in the same manner.

In all forms of progression the forces involved are gravitational, inertial and muscular. The latter may vary from minimal exertion, sustainable over long periods of time, to maximal effort for comparatively brief periods, although reliable estimates of the relative expenditures of energy are few.

Mechanical Analysis of Locomotion

Mathematical analysis of the mechanics and energetics of locomotion, studied in many animals, including invertebrates, has been advancing rapidly in recent years. (For an introduction to the literature in this field consult Alexander and Goldspink, 1977, and Pedley 1977.) Locomotion in many vertebrates has been analysed, and more particularly in mammalian quadrupeds. The basic data are usually cinephotographic and force platform recordings (Alexander *et al.* 1977), but X-ray cinephotography and electromyography (Wentink 1976) have also been used. The mechanics of bipedal locomotion have been studied in amphibia, marsupials, and primates, including mankind. One of the most interesting results of such studies is the recognition of the importance of the storage of recoil energy not only in muscle tissue but also in tendons (Alexander and Bennet-Clark 1977). Cinematic analysis of human walking and running has already contributed to phasic analysis of the movements involved, but the more recent mathematical studies go much further in accounting for the forces concerned. For example, it is possible to predict that there is a limiting optional velocity for walking, beyond which it becomes more economical, in terms of muscle energetics and oxygen consumption, to introduce a 'floating' phase by a change to running. The role of elastic recoil in human tendons and muscles has been studied by combined light and X-ray cinephotography, but reliable data are not yet available. Mathematical interpretation of locomotor mechanics has also

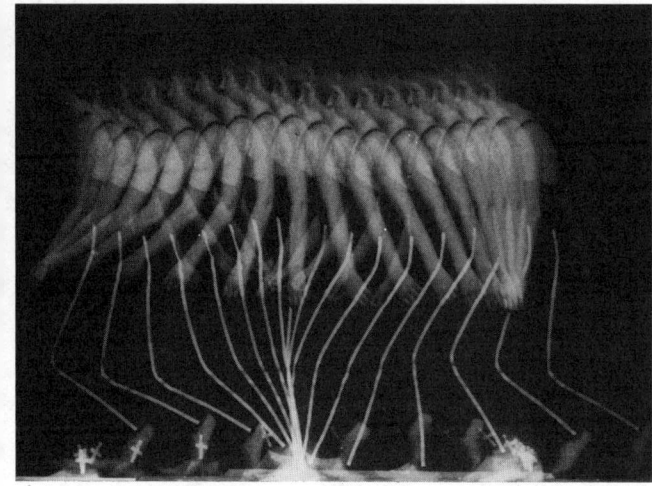 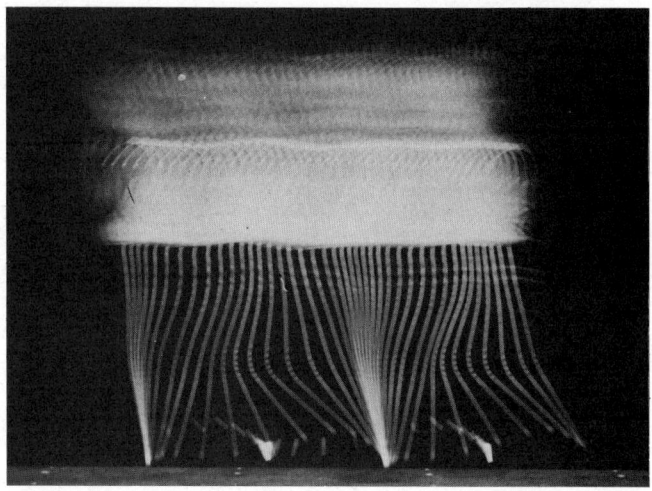

5.112A AND B Chronocyclograms of two individuals walking relatively fast. In A the support phase, with one foot on the ground, occurs at midpoint of the recording, which occupies one full cycle. In B two such support phases are apparent and the recording occupies somewhat less than two cycles. The bright lines are successive images of a stripe attached to the trousered leg proximal to the camera. Note the concentration of lines at the support phase and also the varying angulation at hip, knee, and ankle. (Photographs kindly supplied by Dr. D. W. Grieve, Department of Anatomy, Royal Free Hospital Medical School, London.)

stimulated a renewed interest in bodily proportions, especially in limbs, which provide necessary data for such analyses (McMahon 1977). Not only the dimensions of limb bones, relative to body size and weight, but also the proportions of their osseous segments and those of muscle masses, have become essential data in comparative mechanical studies of gait. The internal arrangements of fibres in muscles have also acquired a renewed significance in quantitative analysis (e.g. Alexander and Vernon 1975). Detailed coordination of all such data into even general concepts of either quadrupedal or bipedal locomotion is inevitably complex; but progress has been rapid in this field and highly interesting and valuable results are to be expected. For a discussion of optimal technique in bipedal walking consult Alexander and Jayes (1978).

The foregoing remarks largely concern analyses of electromyographic and cinematographic recordings in which the data have been considered qualitatively rather than quantitatively. Some workers in this field (e.g. Grieve and Cavanagh 1974) have claimed to be able to quantify electromyograms with considerably improved accuracy by computerized techniques. This has been coupled with an exacting method of measuring cinematic records obtained simultaneously in two or three cardinal planes. Since a temporal parameter is obviously involved, and is of course measurable from the recordings, a very detailed analysis of the angular displacements between the segments of the limbs involved in walking (and the speeds of movement of each segment) can be undertaken. The data obtained can be integrated by computer techniques, and they may also be expressed in graphic forms. Such integrated graphs of the cycle of movements, in space and time, are known as 'chronocyclograms'. The so-called 'heel-strike' and 'toe-off' points of the cycle are commonly chosen as datum points. Such chronocyclography (see 5.112) has already demonstrated changes in pattern with alternation of speed in walking, and comparisons can also be made between the events in 'normal' records with those obtained from patients afflicted with disturbances of locomotion. The latter developments facilitated and refined descriptions of abnormal gaits and thus it was hoped that they would also prove of applied value, but in its original form the technique was somewhat lengthy and tedious. However, the technique has been further improved by Grieve and his coworkers (1979). This depends upon computer evaluation of the cycle of angular changes in the thigh and foreleg segments, the angles being recorded by the impingement of a rotating beam of plane-polarized light upon suitable sensors attached to the limb segments. Since the signals produced by this polarized light goniometer (the Medelec Polygon equipment) are directly passed to a suitably programmed computer, a graphic recording is *immediately* available of the subject's angulation cycle, in stepping or walking. This development now renders the technique suitable for clinical application. (A stroboscopic method of recording has also been used, and the records shown in 5.112A, B illustrate the cycle of angulation in walking.) The researchers quoted, and others, are currently promoting the collection and analysis of data on an international scale.

6

ANGIOLOGY

Introduction

The physico-chemical composition and properties of the fluid micro-environment of the cells comprising the various tissues of the body remain relatively constant, by the operation of an interlocking series of homeostatic control mechanisms, which depend for effectiveness upon adequate circulatory systems. The latter consist of cell-lined arrays of tubes and spaces which permeate or surround the tissues and through which a continuous flow of one of the body fluids occurs. The most extensive circulatory pathways comprise the *blood-vascular system*, consisting of the heart and blood vessels, through which the blood circulates, and the *lymphatic system*, consisting of lymph nodes and vessels which conduct the lymph—although such a restricted definition should perhaps be extended to include the isolated patches of lymphoid tissue, the spleen, thymus and the various pathways additionally involved in the wider concept of 'the *cirulation of the lymphocytes*' (p. 773). Other more restricted circulations are those of the cerebrospinal fluid, the perilymph, the various endocochlear fluids, the aqueous humour of the eye and the synovial fluid. The present section on *Angiology* is restricted to a consideration of the blood-vascular and lymphatic systems.

Repeated contraction of the heart, which consists of a *pair* of hollow, valved, muscular pumps, arranged in *series (vide infra)*, drives the blood through the blood-vascular system to reach all the tissues of the body. Leaving the heart, the blood is distributed through a series of large tubes, the *arteries*, which ramify and continually subdivide until minute vessels, the various grades of *arterioles*, are reached, which again subdivide to form close-meshed networks of microscopic vessels in intimate contact with the tissues. These vary in structure and are called either *capillaries* or *sinusoids* depending upon their particular structural and functional characteristics. From the latter the blood is collected via minute *venules* which repeatedly converge to form increasingly large tubes, the *veins*, which return the blood to the heart.

The terms artery, arteriole, capillary, venule and vein, are essentially *anatomical* names for vessels of particular dimensions, structural characteristics and topographical positions. Another useful *functional* classification groups the blood vessels into *distribution, resistance, exchange* and *capacitance* or *reservoir* vessels. Finally, varieties of *shunt* which provide bypass channels at a pre-capillary or pre-sinusoidal level must be considered.

The *distributing vessels* consist of the low-volume, high-pressure system of large arteries which leave the heart, and the branches of the arterial tree down to arteriolar levels. They show graded structural and functional characteristics with increasing subdivision and distance from the heart. The large vessels leaving the heart and their main branches contain much elastic tissue in their walls (*elastic arteries*), the repeated distension and elastic recoil of which converts the effects of the intermittent contractions of the heart into a more continuous although still pulsatile flow of blood throughout the arterial tree. The smaller arterial branches, while retaining a marked elastic component, contain an ever-increasing amount of non-striated muscle in their walls (*muscular arteries*), the differential contraction of which allows a controlled variation in the overall blood flow to the various large tissue masses, which corresponds to their fluctuating levels of physiological activity. It should also be noted, however, that some authorities prefer to use the term *conducting arteries* for the major outflow vessels, with much elastin in their walls, and *distributing arteries* for muscular arteries lying between these and the resistance vessels provided by the arteriolar channels (Simonescu and Simonescu 1977).

A more intimate control of the blood flow pattern through the microcirculatory units of the various tissues (*vide infra*) is provided by the muscular walls of the arterioles and precapillary sphincters (*resistance* vessels). These provide the principal source of the peripheral resistance to the flow of blood, which, together with the prevailing volume of cardiac output in unit time, determines the blood pressure levels in the arterial tree.

The capillaries, sinusoids and post-capillary venules are collectively termed *exchange vessels* because across their walls occurs the exchange between the blood and the tissue fluid bathing the cells, which constitutes the essential function of the blood circulatory system. This includes the transference of oxygen, carbon dioxide, nutrients, water and inorganic ions, vitamins, hormones, metabolic products, immune substances and finally, immunologically competent and phagocytic cells.

The larger venules and veins comprise an extensive, variable, large-volume, low-pressure system of *capacitance* or *reservoir* *vessels* through which the cardiac return occurs.

The mature heart has four chambers: through widespread usage these are termed the *right atrium, left atrium, right ventricle* and *left ventricle* (some discussion on the somewhat misleading nature of these names will be found on p. 639). Each atrium communicates freely with the corresponding ventricle, but the right and left chambers are separated from one another by partitions, called *septa*. The right and left heart in the adult thus comprise twin pumps *topographically combined* in a single organ but which are interposed *in series* at different points in the blood-vascular system and separate it into a *systemic* and a *pulmonary* circulation (constituting the so-called *double circulation* which characterizes birds and mammals—*vide infra* p. 636). The course of the blood from the left ventricle through the body generally to the right side of the heart constitutes the greater or *systemic circulation*, while its passage from the right ventricle through the lungs to the left side of the heart is termed the lesser or *pulmonary circulation*. The relatively short and restricted pulmonary system offers a much lower peripheral resistance than the systemic circulation and this is reflected in the considerably lower pressures in the pulmonary distribution vessels than in the aorta and in the relatively thin walls of the right ventricle (p. 637). Nevertheless, the average output volume of blood per unit time from the right and left heart must necessarily be the same.

The superior and inferior venae cavae bring to the right atrium blood which has become deoxygenated, taken up carbon dioxide and been otherwise modified during its circulation through tissues of the body. From the right atrium this venous blood passes into the right ventricle, which expels it into the pulmonary trunk to be conveyed to the lungs. As it circulates through the lung capillaries the blood is brought into close relationship with the inspired air and it gives off some of its carbon dioxide and acquires a fresh supply of oxygen. This oxygenated blood is returned by the pulmonary veins to the left atrium and thence passes into the left ventricle. With each beat of the heart the left ventricle pumps its contents into the *aorta* , which distributes the oxygenated blood through its numerous branches to the tissues and organs of the body.

The blood which circulates through the spleen, pancreas, stomach and the intestines is not returned directly from these organs to the heart, but is conveyed by the *portal vein* to the liver. In the liver this vein divides like an artery and ultimately ends in the hepatic blood sinusoids from which the rootlets of the *hepatic veins* arise; the hepatic veins carry the blood into the inferior vena cava, which conveys it to the right atrium. This constitutes the *portal circulation* and it will be understood that the blood supplied to the above-named viscera passes through two sets of minute vessels before reaching the inferior vena cava: (1) the capillaries in the spleen, pancreas, stomach, etc., draining into the portal vein; and (2) the sinusoids in the liver, draining into the hepatic veins. The passage through two sets of capillaries enables the blood to take up products of digestion from the alimentary canal and to convey them to the liver cells. A second venous portal circulation connects the median eminence and infundibulum of the hypothalamus with the pars distalis of the adenohypophysis (p. 1444).

The Structure of the Blood Vessels

Blood is ejected from the left ventricle during systole into the aorta, a single tube of about 2·5 cm diameter in the adult (Wright 1969) and with a wall thickness of about 1·5 mm. After traversing numerous generations of branches from this parent trunk the blood eventually enters the capillary bed in the tissues which, by contrast, consists of many millions of tubes about 8 μm in

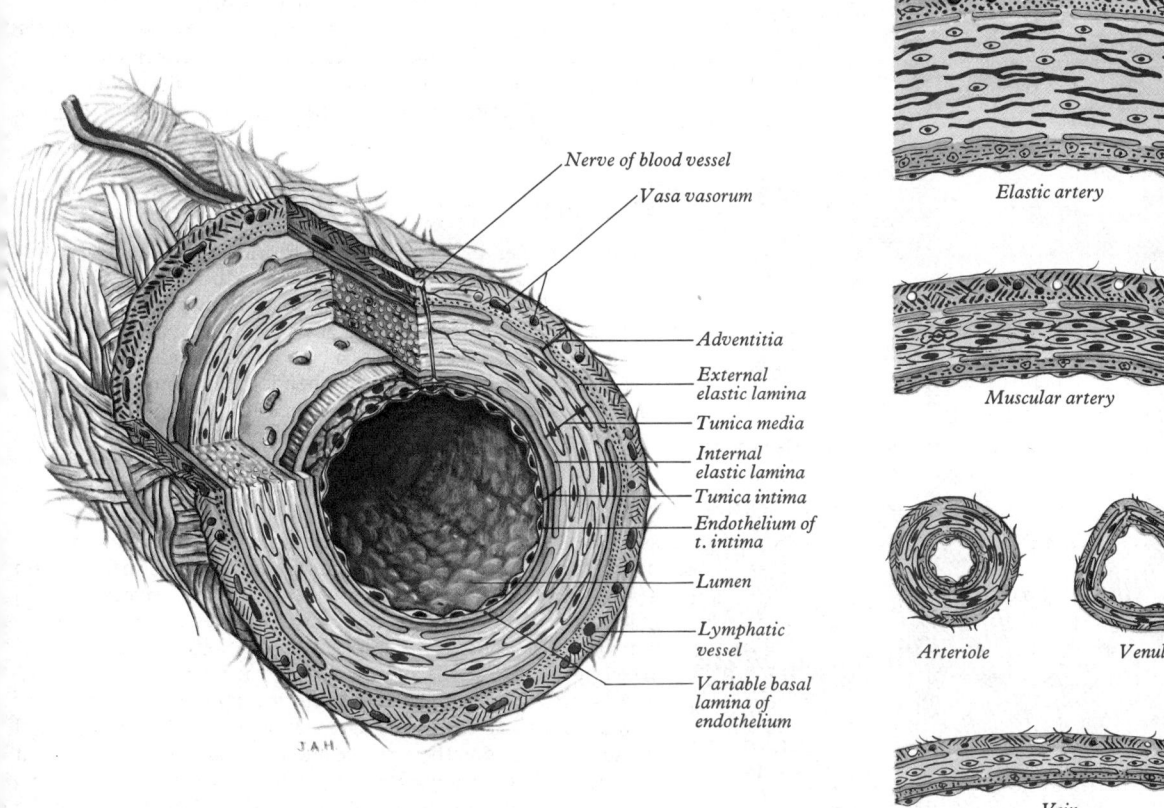

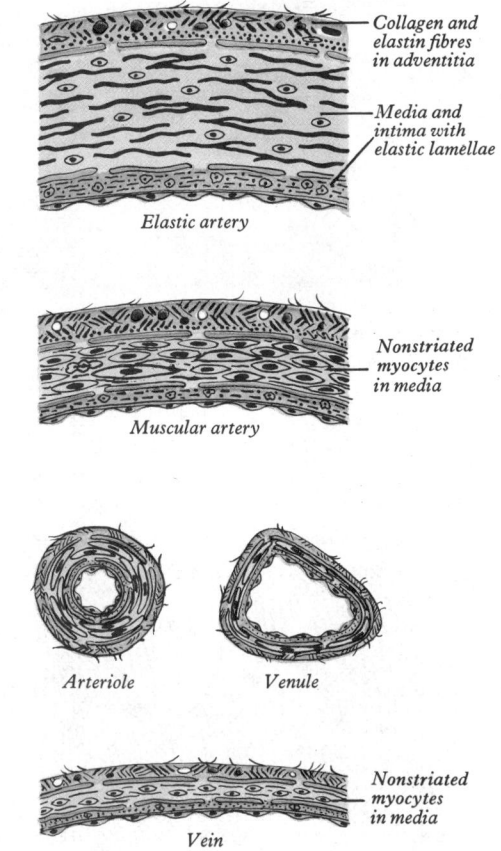

Collagen and
elastin fibres
in adventitia

Media and
intima with
elastic lamellae

Elastic artery

Nonstriated
myocytes
in media

Muscular artery

Arteriole

Venule

Nonstriated
myocytes
in media

Vein

Nerve of blood vessel

Vasa vasorum

Adventitia

External
elastic lamina

Tunica media

Internal
elastic lamina

Tunica intima

Endothelium of
t. intima

Lumen

Lymphatic
vessel

Variable basal
lamina of
endothelium

JAH

6.1   A diagram showing the principal architectural features of the larger blood vessels. On the *left* the major layers and associated structures are depicted. On the *right* the particular features of an elastic artery, muscular artery, arteriole, venule and vein are shown.

diameter with a wall thickness in places of less then an 0·2 μm and a total estimated length in the adult which approaches 60,000 miles (Zweifach 1961). It is often affirmed that the systemic *arterial tree* progressively divides into a larger and larger cross-sectional area. However, the evidence for this in man is by no means clear, particularly in the arterial pathways proximal to the arteriolar level. However, the change in the structural and functional attributes of the arteries is progressive with increasing branching and distance from the heart. There is both a geometric taper of lumen cross-section and an increase in stiffness of the vessel wall. During *arteriolar* and *capillary* branching the total cross-sectional area of the vessels increases greatly and this is accompanied by a progressive reduction in the flow-rate, systolic pressure and pulse pressure of the contained blood. The structural and functional attributes of these vessels change therefore, even more dramatically, with increasing branching and distance from the heart (**6**.1–8). However, the whole blood-vascular system, from the finest vessels up to and including the heart, have one structural component in common, namely, a smooth lining throughout of a single layer of endothelial cells. Even these cells show dimensional and other ultra-structural differences in different parts of the system and particularly in the capillary and sinusoidal networks. Vessels of larger calibre than capillaries or sinusoids possess organized tissue zones which surround the endothelial lining. These zones contain varying amounts of fibrous tissue (reticulin, collagen and elastin) together with fibroblasts and non-striated myocytes. Despite the fact that the radial organization of these components in a vessel wall varies greatly with vessel size and type, closely reflecting the functional roles of the vessel, most classical histologists have attempted to define three analogous zones (coats or tunics) in the walls of all vessels down to arteriolar and venular level (Cliff 1976). These are named from within outwards the *tunica intima, tunica media* and *tunica adventitia*, and because of common usage these terms will be retained in this account. Too strict an adherence to such an arbitrary classifi-

cation should not obscure the overall functional organization of each vessel type. This includes the provision of a low-friction lining, structural elements which balance the longitudinal and circumferential stresses generated by the prevailing pressures of the contained blood, elastic recoil which progressively modifies the pulsatile nature of the pressure and flow variations in the arterial tree, and the ability to control calibre changes, by muscular action, with varying functional demands.

In general, the *tunica intima* consists of a single layer of endothelial cells supported externally by a delicate layer of subendothelial connective tissue with a predominantly longitudinal organization. The fibro-muscular *tunica media* extends from the internal to the external elastic lamina (*elastica interna* to *elastica externa*) and has a circumferential organization. The *tunica adventitia* which surrounds the external elastic lamina is a connective tissue coat with again a largely longitudinal organization and this blends externally with the surrounding fibro-areolar tissue.

Nutrient blood vessels, the *vasa vasorum*, and lymphatics run in the adventitia, and a few in the outer regions of the tunica media; nerve fibres may also supply all three major layers of the vessel wall (*vide infra*). The arteries are enclosed in a thin fibrous connective tissue *sheath*, artery and sheath being connected by delicate areolar tissue. The sheath usually encloses any accompanying veins and perivascular nerves, and sometimes major nerve trunks.

The thickness of vessel walls

The precise form and composition of blood vessel walls varies largely with the blood pressure and, in arteries, the nature of the pressure wave within. Thus, arteries are proportionately much thicker than veins, and both arteries and veins inferior to the heart are in general thicker than their counterparts superior to it, in accordance with the higher hydrostatic pressure of the blood in the lower parts of the body when in the erect posture. Arteries of

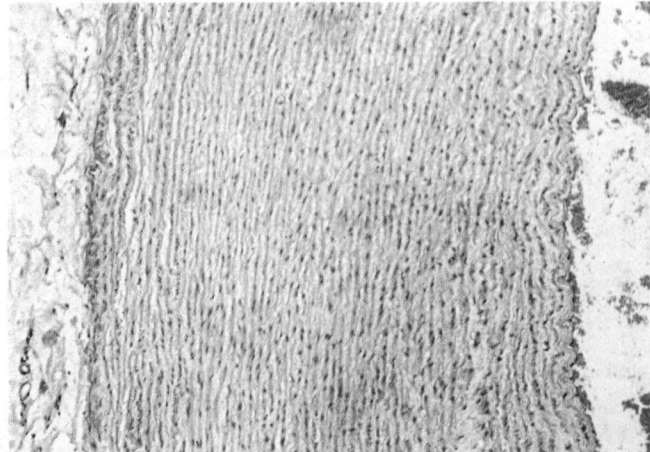

6.2A Part of a transverse section of the aorta of a monkey, stained with haematoxylin and eosin, showing the distribution of cell nuclei.

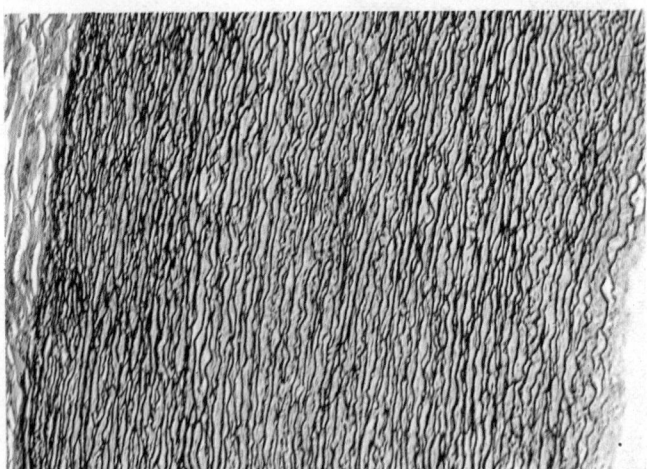

6.2B Part of a transverse section of a young human aorta stained with Verhoeff's stain for elastin. Note the density of the concentric fenestrated elastic laminae. (Kindly supplied by Dr. R. O. Weller, Department of Pathology, Guy's Hospital Medical School.)

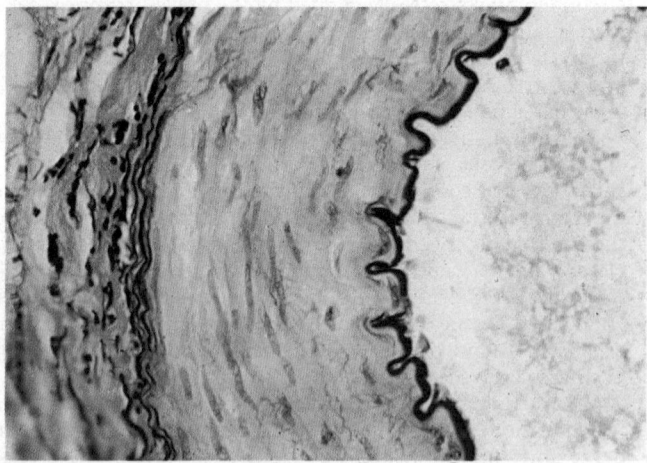

6.2C An oblique section of a medium-sized muscular artery, stained with Verhoeff's stain for elastin (black), and van Gieson's stain for collagen (red). Note the characteristic folded, prominent internal elastic lamina.

the systemic circulation have thicker walls than ones of corresponding diameter in the pulmonary circuit, reflecting their marked differences in pressure. Arteries within the 'closed box' pressure system of the cranium and vertebral canal have very thin walls in relation to their diameter, the external and middle coats being reduced in thickness; similar considerations apply to the vessels within the medullary cavities of the long bones. In contrast, some vessels are exceptionally muscular, particularly where considerable external pressures or other mechanical forces

act upon them, e.g. the coronary and carotid arteries are in possession of especially thick muscular media.

As a general rule, elastic arteries have walls about one tenth the thickness of their interior diameter, muscular arteries about one quarter, arterioles approximately one half, whilst veins, again, have a mural thickness about one tenth that of their luminal diameter, or less.

ARTERIES

Large arteries of the elastic type (6.1, 2, B). These include the aorta and its main branches of distribution, the brachiocephalic, common carotid, subclavian and common iliac arteries, and also the pulmonary arteries and their principal branches.

The *tunica intima* consists of a layer of endothelial cells resting on a basal lamina and a layer of subendothelial connective tissue, together being about 100 μm thick in young adults. The endothelial cells are flattened and polygonal, and are joined by intercellular junctions which include zonulae occludentes and maculae communicantes. The general cytological features of endothelial cells will be considered with capillary structure. The subendothelial connective tissue comprises, first, a layer of interlacing mainly longitudinal collagenous and elastin fibres with a few fibroblasts and occasional nonstriated myocytes, and macrophages. This is succeeded by a fenestrated elastic membrane which corresponds to the internal elastic lamina of the smaller arteries, but is not sharply demarcated from the tunica media, because the latter, in these larger arteries, contains a succession of similar elastic membranes (*see* Dobrin 1978). The intimal thickness varies greatly with age and other factors: in children it may be an almost insignificant layer, but in young adults intimal thickening becomes noticeable due, it is thought, partly to migration and proliferation of nonstriated myocytes from the media, and to a greater or lesser extent to the deposition of lipid material which often forms irregular fatty streaks. In normal young adults the intima is about one sixth of the total wall thickness, but after middle age, lipid deposits may grossly expand this layer.

The *tunica media* consists almost entirely of concentric, circularly disposed, fenestrated elastic lamellae, up to forty layers thick in normal aortae, separated from one another by fibrous tissue containing a basophilic matrix, permeated by bundles of collagen and elastin fibres which surround circularly disposed non-striated myocytes (Somlyo and Somlyo 1968). For the most part these myocytes are unexceptional, but those adjoining the intima often assume highly branched bizarre shapes (Keech 1960), and interconnect adjacent lamellae. There appears to be a good correlation between the internal pressure of elastic arteries and the number of elastic lamellae which, together with their associated nonstriated myocytes, form 'lamellar units' (Wolinsky and Glagov 1967a). The nonstriated myocytes are responsible for the laying down of the lamellae, and also the collagen and glycoprotein components between them (Ross and Klebanoff 1971), an ability also shown to a much smaller degree by non-striated myocytes elsewhere in the body. The numbers of lamellae increase greatly during postnatal development, the aorta of the newborn being almost devoid of them, and therefore, of the muscular type. They also increase in number in hypertension under the influence of the extra mechanical stress (*see* Cliff 1976).

The disposition of the myocytes varies with their overall position in the vascular system. In the elastic arteries nearest the heart, the myocytes interconnect radially adjacent lamellae ('*Spannmuskeln*'), whereas more distally they are orientated in a more circular direction ('*Ringmuskeln*'), being arranged end to end (Benninghof 1927; Wolinsky and Glagov 1964); in both cases they display a slight helical arrangement.

The *tunica adventitia* is composed of fibrous tissue containing a few elastic fibres. It is relatively thin and merges into the surrounding connective tissue.

From middle age onwards, various structural alterations are found in elastic (and other) arteries, in addition to the intimal changes already mentioned; this is particularly true of the aorta, and arteries of the heart and brain. The amount of elastin and collagen increases, muscle decreases, and there is a lower water content leading to an increase in general stiffness. Other associated changes include the desposition of calcium salts and lipid in the media, and the splitting of the internal elastic lamina.

Medium and small arteries of the muscular type (6.1, 2C, 3). The *tunica intima* consists of a single layer of elongate endothelial cells, longitudinally arranged, the outer surfaces of which are covered with a typical basal lamina closely abutting upon the *internal elastic lamina*. The latter is a fenestrated elastic sheet appearing in transverse sections as a bright refractile zone which, because of agonal contraction of the muscular media, is thrown into characteristic wavy folds. As there is less elastic tissue in the media than there is in the larger vessels, the internal elastic lamina stands out in sharp contrast. Here, and in the large arteries, the fenestrations in the elastic layers are considered to act as diffusion paths for metabolites between the various layers of the vessel wall.

In the coronary arteries the internal elastic lamina is bounded by a specialized thick layer of longitudinal muscle and fibrous tissue, which presumably enables the vessels to accommodate to the marked changes in length occurring throughout the cardiac cycle.

The *tunica media* consists mostly of non-striated myocytes with scattered elastic membranes and a few collagen fibres. The myocytes are generally disposed in low-pitched helices except in the smaller vessels where they are circumferential. The myocytes are similar to those elsewhere (p. 518), but are claimed to contain fewer micropinocytotic vesicles and more rough-walled endoplasmic reticulum and ribosomes; the Golgi apparatus is small. Adjacent non-striated myocytes are interconnected with communicating junctions which act to coordinate the activity of the muscle layer, under neural or neuro-humoral control. Longitudinal myocytes exist in both the intima and media of arteries subject to repeated bending, e.g. the carotid, axillary, uterine, cervico-occipital, and palmar arteries. In the umbilical arteries, longitudinal muscle occurs in the inner region, and circular in the outer zone of the tunica media.

The *tunica adventitia* consists of collagenous and elastin fibres which run predominantly in a longitudinal direction. The outer portion is somewhat loosely arranged, merging into the surrounding areolar tissue, and so allows considerable movement between the artery and the neighbouring structures. Immediately adjacent to the smooth muscle of the tunica media the adventitia contains much elastic tissue which here includes the fenestrated *external elastic lamina*.

ARTERIOLES, TERMINAL ARTERIOLES, METARTERIOLES, AND PRECAPILLARY SPHINCTERS (6.4, 5A–E)

The terminal branches of the smallest arteries open into *muscular arterioles* of between 100 and 50 μm in diameter, which in turn branch into *terminal arterioles* of less than 50 μm in diameter and with but one or two layers of non-striated myocytes in their walls (Rhodin 1971). The side branches from the terminal arterioles are 10–15 μm in diameter at their origins, but may decrease over a distance of 50–100 μm to as little as 5 μm at their narrowest point. At their mouth they are surrounded by a prominent layer of circular non-striated myocytes. These vessels are sometimes termed *metarterioles*, and their surrounding muscle rings, *precapillary sphincters*. They open directly into the capillary bed, and act as a final control of nutritive blood flow. In several instances, precapillary sphincters have been seen to undergo periodic opening and closing, each cycle of activity lasting 2–8 seconds.

Both the larger and terminal arterioles are lined by a single layer of flattened endothelial cells which are greatly elongated in the long axis of the vessel. They rest on a basal lamina, separated in places from the lamina of the non-striated myocytes of the media by patches of elastic tissue, homologous with the internal elastic lamina of larger vessels. This elastic tissue becomes scantier and finally disappears in the finest terminal arterioles. The media of the larger arterioles consists of a few layers of non-striated myocytes circularly disposed, but with the outermost layer often

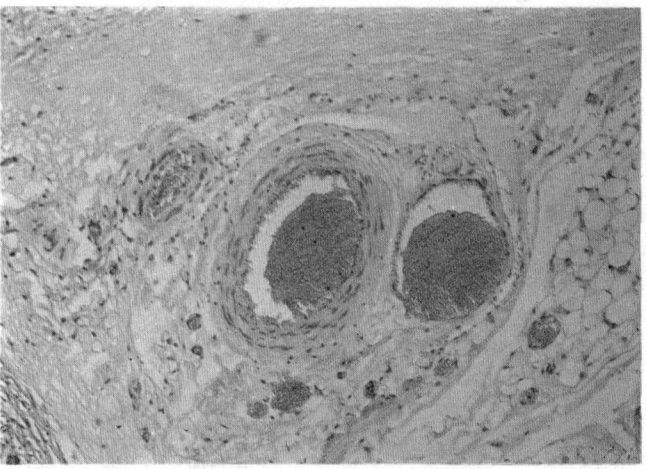

6.3 A transverse section of two small muscular arteries and a small vein, stained with haematoxylin and eosin. Numerous venules and capillaries are included but are indistinct at this magnification. (Kindly supplied by Dr. D. R. Turner of the Department of Pathology, Guy's Hospital Medical School.)

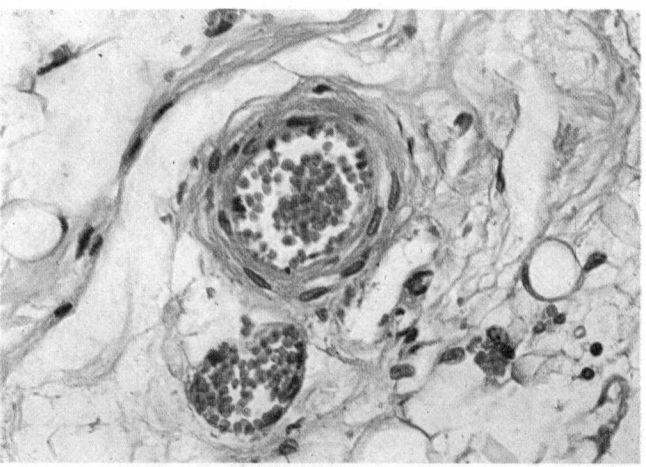

6.4 A transverse section of a large arteriole and venule in loose connective tissue, stained with haematoxylin and eosin. Source as 6·3.

forming a low-pitched helix. The single layer of myocytes in a terminal arteriole has a similar spiralized form and, as in the larger arterioles and metarterioles, they form intercellular (myoendothelial) junctions with the overlying endothelium, at those points where the elastic lamina is lacking.

The adventitial coats of arterioles consist of a fine feltwork of collagen fibres and a few associated fibroblasts.

Arterial Nutrient Supply

The larger arteries are supplied with blood vessels. These nutrient vessels, called the *vasa vasorum*, arise from branches of the artery itself, or of a neighbouring vessel, at some considerable distance from the points at which they are distributed; they ramify in the loose areolar tissue connecting the artery with its sheath, and, after forming a dense capillary network in the adventitia, supply the outer part of the media. The nutritional requirements for the remainder of the vessel wall are obtained by diffusion from the blood in the lumen of the vessel, and it is assumed that the fenestrations in the elastic laminae are important in this regard. Minute veins return the blood from these vessels; they empty themselves into the vein or veins accompanying the artery. Lymph vessels are also present in the outer coat.

Arterial Nerve Supply

Arteries are also supplied with nerves; the majority are non-myelinated, but some are myelinated. The non-myelinated fibres are mostly efferent and constitute the vasomotor nerves to the vessels, along which passes the continuous flow of impulses responsible for the variable 'vasomotor tone' of the vessels. The

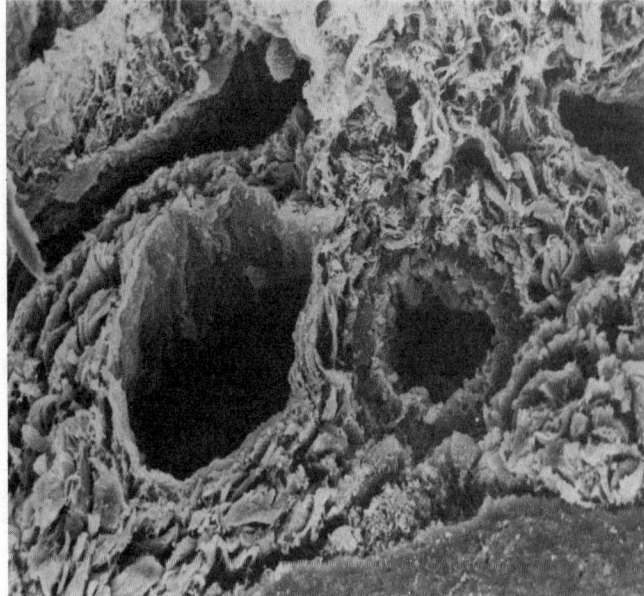

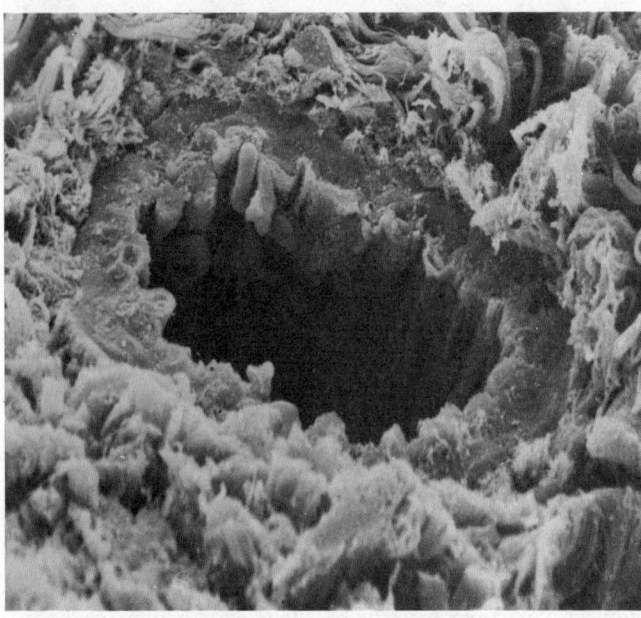

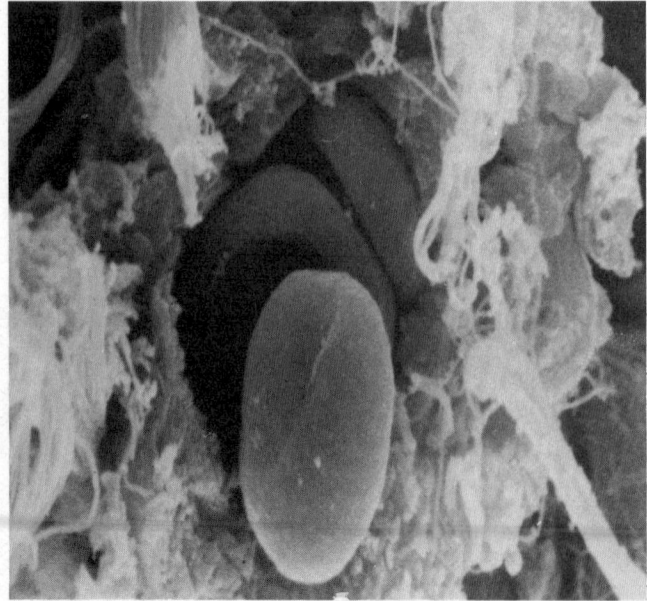

light microscopical evidence of pericellular nerve nets in the media has not been confirmed by electron microscopy, which has emphasized the paucity or absence of nerves within the media. The nerve fibres ramify in the adventitia and then approach the outer layer of medial non-striated muscle cells through the fenestrations in the external elastic lamina. The terminal area of a non-myelinated nerve fibre is extensive; in this area the axon is partly or wholly denuded of Schwann cell processes and contains clusters of mitochondria and numerous vesicles of 30–100 nm in diameter, many of which, in the middle diameter range, are dense-cored and typical of adrenergic synapses. This terminal is separated from the muscle cells by a distance of 60–400 nm, an interval which is considerably larger than that found at the synaptic junctions between non-myelinated fibres and non-striated muscle cells elsewhere, and the still narrower gap at the neuromuscular junctions in striated muscle. One terminal area may be related to several non-striated muscle cells and one smooth muscle cell may be innervated by more than one axon terminal (Appenzeller 1964; Lever 1965, 1968; Burnstock *et al.* 1970). It is assumed that the transmitter substance can diffuse through the basal lamina material which surrounds much of the smooth muscle cell surfaces, and that further spread of the excitatory process into the depths of the media occurs via the low-resistance communicating ('gap') junctions between adjacent muscle cells. The myelinated fibres are believed to be afferent and are distributed to the outer and inner coats where they terminate in expanded and varicose endings.

The significance of afferent nerve fibres in the general systemic arterial tree is uncertain, but some may mediate pain impulses under certain pathological and traumatic states. In some specific sites the afferent nerve fibre terminals are specialized as baroreceptors (pressoreceptors) which monitor the systemic blood pressure levels (e.g. in the carotid sinus and aortic arch), or as chemoreceptors which respond to increases in hydrogen ion and carbon dioxide concentration, or decreases in oxygen tension in the circulating blood (carotid and aortic bodies). Lamellated corpuscles (Pacinian corpuscles), which are mechanoreceptors, are occasionally found in the outer coat of the aorta.

The majority of blood vessels do not receive a vasodilator innervation and, in these, neurogenic vasodilation follows a reduction of sympathetic vasoconstrictor tone. However, nerve-mediated vasodilator action is seen: (1) in skeletal muscle vessels which receive a cholinergic sympathetic innervation (2) in various exocrine glands following secretomotor activity with the secondary release of the vasodilator bradykinin; (3) in the skin following afferent nerve stimulation, the collaterals of which ramify on neighbouring blood vessels and form the structural basis for the so-called 'axon reflex'.

The largest arterial trunks receive branches direct from the sympathetic ganglia, but for the smaller arteries (brachial, femoral, etc.) the supply is carried in the peripheral nerves and diverges in a series of small branches (pp. 1125, 1129). In the splanchnic arteries perivascular plexuses extend along the whole extent of the vessels.

CAPILLARIES AND SINUSOIDS

In many tissues (e.g. muscle, skin, lung, alimentary tract, brain, etc.) the side branches of the terminal arterioles beyond the

6.5A–C Scanning electron micrographs of small blood vessels in the cut surface of the submucosa of the small intestine (mouse).
A Low-power view of a group of vessels showing a small artery (mid-right) with an accompanying vein (mid-left) in transverse section. An obliquely sectioned vein is also visible (extreme right) and a lymphatic vessel is present on the upper left; the extravascular tissue consists mainly of irregular dense connective tissue. Magnification × 320.
B Medium-power view of a transversely sectioned small artery, showing the thick muscular wall and the longitudinally ridged intima with erythrocytes adhering to the endothelial surface. The adventitia blends with the surrounding dense fibrous tissue. Magnification × 1,000.
C High-power micrograph of the cut end of a capillary with three erythrocytes visible. Note the extreme thinness of the endothelial wall (left) and the collagen fibres in the extravascular tissue. Magnification × 6,000 (Preparations by Michael Crowder, Guy's Hospital Medical School.)

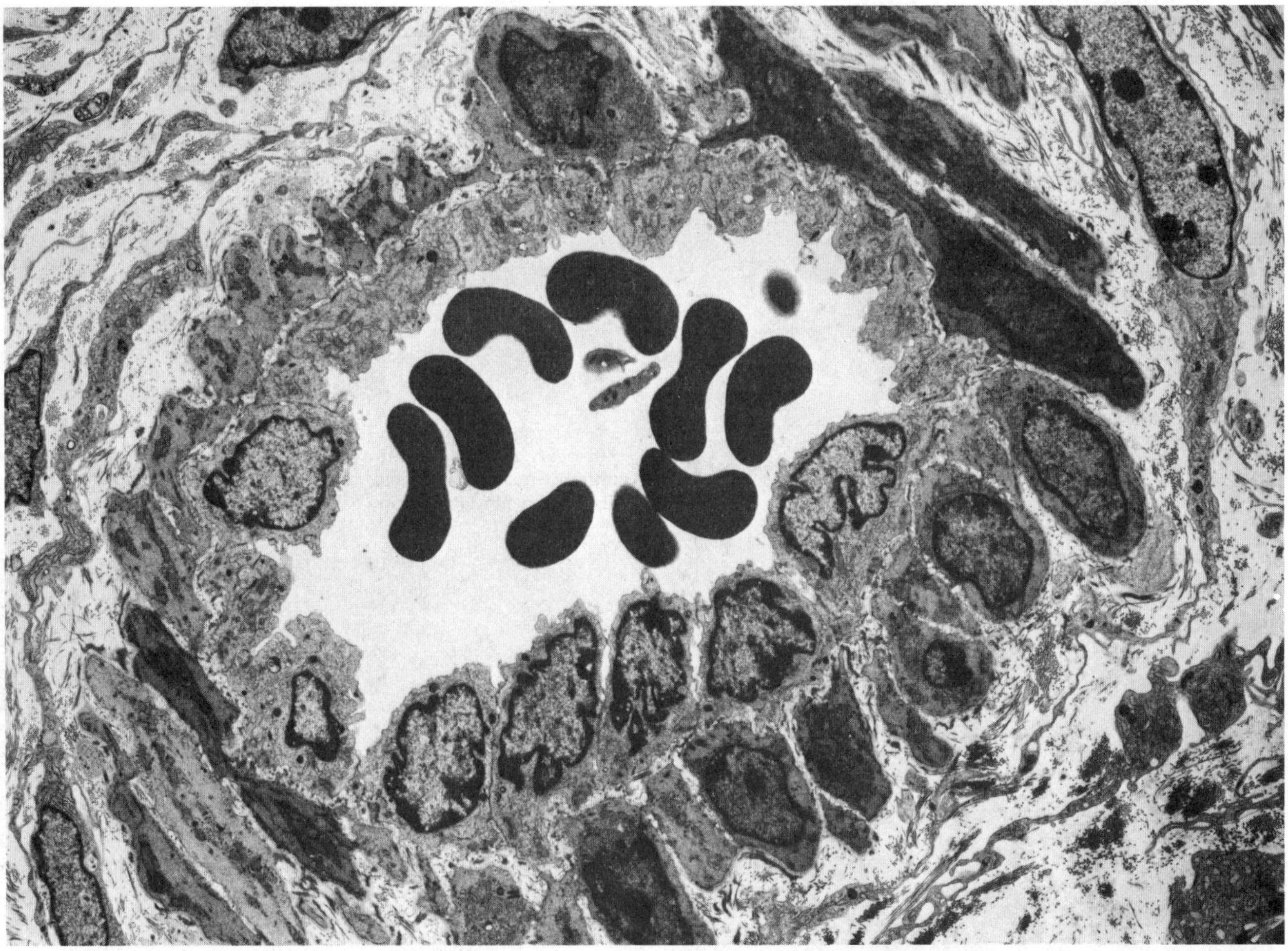

6.5D   Transmission electron micrograph of a partially contracted small arteriole in transverse section, showing an outer zone of non-striated myocytes and an inner lining of endothelial cells. Erythrocytes are visible within the lumen. The specimen is from the uterus of a rat. Magnification × 4,000.

precapillary sphincters, pass through a short transition zone of 50–100 μm where a few isolated, spiralling smooth muscle cells persist external to the endothelium and which then continue as true *capillaries* of fairly uniform diameter. These intercommunicate to form a capillary network, the organization of which varies with the tissue concerned (*vide infra*). In other tissues (e.g. red bone marrow, spleen, liver, suprarenal and parathyroid glands, carotid and coccygeal bodies) the smallest blood vessels differ from true capillaries. They are wider, with an irregular lumen and a very thin wall; little, if any, connective tissue separates the endothelial wall from the neighbouring cells of the organ in which they lie. They are called *sinusoids* and among their lining endothelial cells there may be many actively phagocytic cells of the macrophage type.

Until recent times our knowledge of capillary and sinusoid structure was severely restricted by the relatively low resolution of the light microscope, but electron microscopy has revealed marked variation in the fine structure of these minute vessels in the various tissues (Bennett *et al.* 1957; Rhodin 1962 a & b; Fawcett 1963; Florey 1966; Wisse 1970). In general, capillaries consist of a tube lined by a single layer of polygonal or lanceolate endothelial cells which show variations in their fine structure and intercellular junctions. Surrounding the endothelial tube is usually a typical glycoprotein basal lamina which, at isolated points over the capillary, splits to enclose flattened or branching perivascular cells or *pericytes*. The basal lamina usually merges into an adventitial layer of fine reticular tissue, with occasional fibroblasts and mast cells, which border the extravascular tissue-fluid spaces. Variations in all these features have formed the basis for a variety of classifications, but broadly, a capillary may be considered as either *continuous* or *fenestrated*.

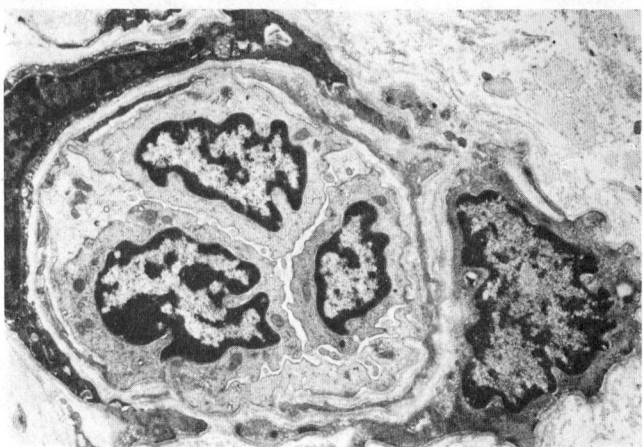

6.5E   Transmission electron micrograph of a metarteriole fully contracted. The endothelial cells almost totally occlude the lumen. The dark cells around the periphery are pericytes. A biopsy specimen of human synovial membrane. Magnification × 6,000.

Continuous capillaries (6.6A, 7, 8) are found in many tissues including the skin, connective tissues, striated and smooth muscle, lung and brain. The cytoplasm of each cell is fairly thick opposite its oval nucleus but becomes attenuated elsewhere and presents a smooth luminal surface with occasional filopodia. The cytoplasm has the usual cell organelles but these are few in number. It contains fine filaments of about 4–7 nm diameter and a small amount of endoplasmic reticulum with attached and free ribosomes; the Golgi apparatus is small. The filaments may represent skeletal or possibly contractile elements. On the luminal and basal surfaces the plasma membrane is the site of numerous flask-shaped invaginations or *caveolae*, and the underlying cytoplasm contains many vesicles of 50–70 nm diameter. These are considered to form on one surface, to detach, cross the

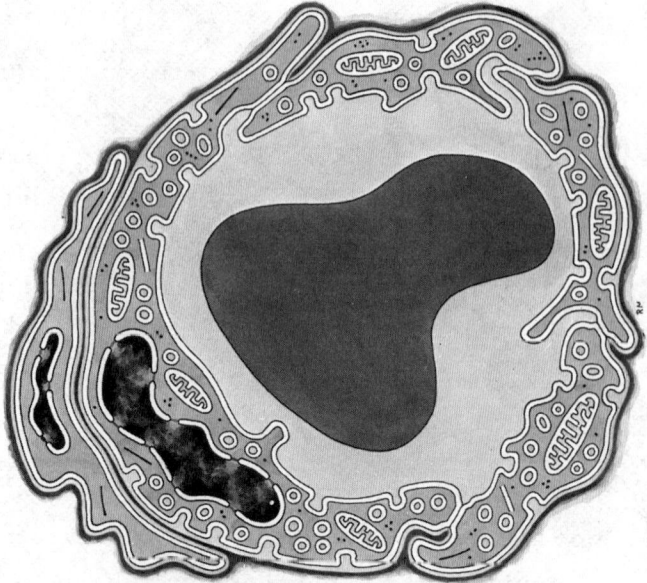

6.6A A 'continuous' capillary in transverse section—a scheme of its ultrastructural appearance.

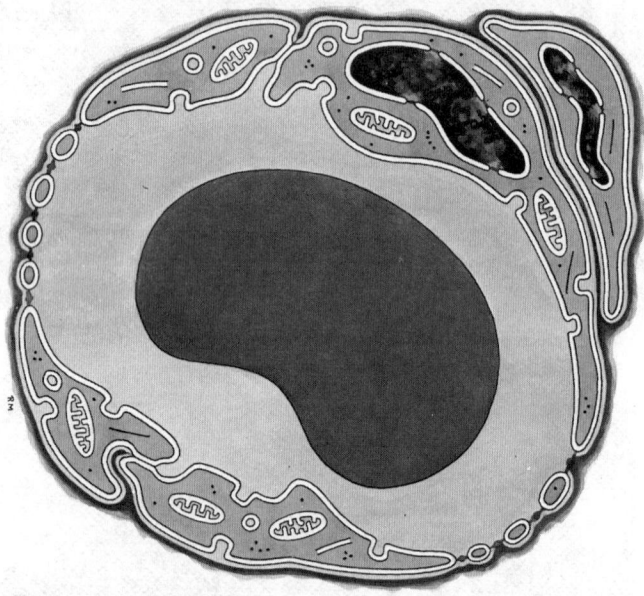

6.6B A 'fenestrated' capillary in transverse section—a scheme of its ultrastructural appearance.

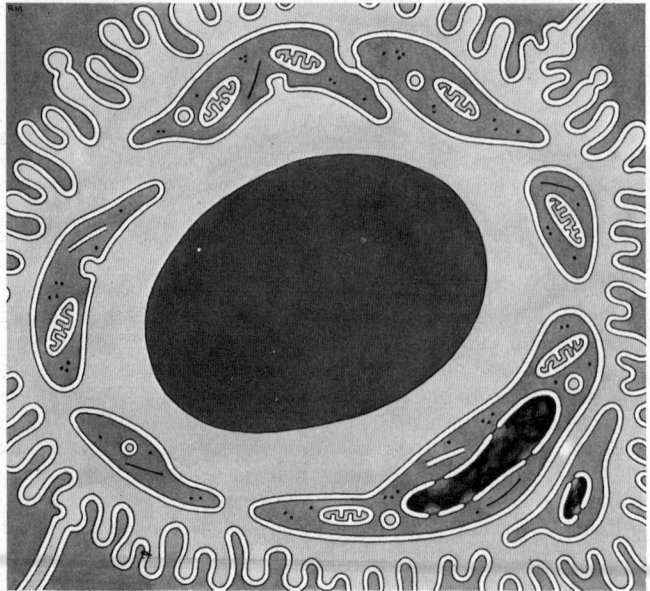

6.6C A 'discontinuous' sinusoid in tranverse section—a scheme of its ultrastructural appearance.

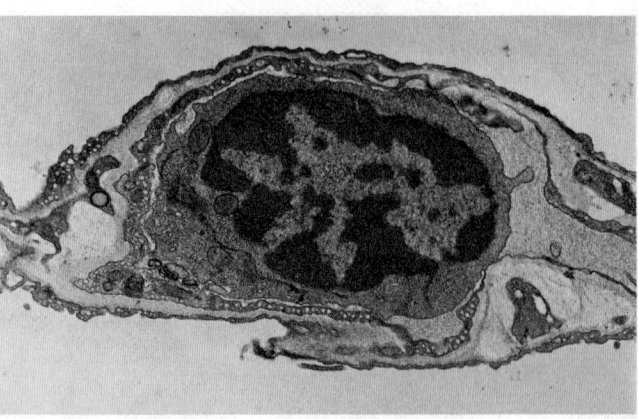

6.7 Transmission electron micrograph of a capillary in the wall of a pulmonary alveolus. Note the presence of numerous pinocytotic vesicles in the attenuated endothelial cell cytoplasm. A small lymphocyte almost fills the lumen. Magnification × 5,500

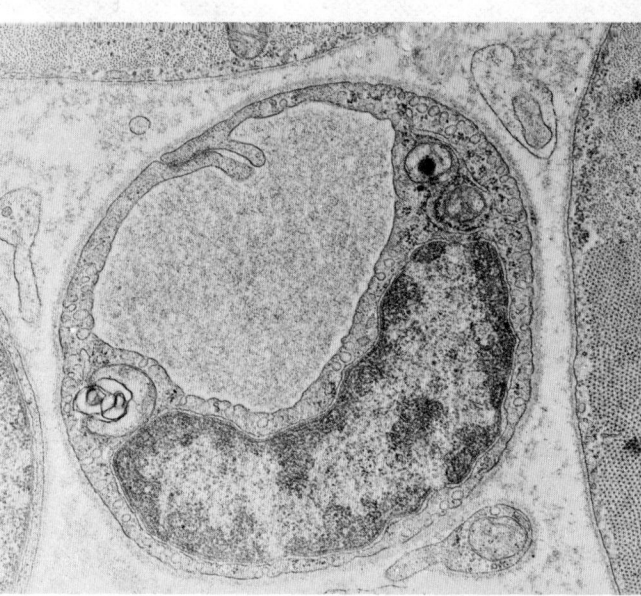

6.8 Transmission electron micrograph of a capillary in transverse section lying in the endomysial space between skeletal muscle fibres. Note the large endothelial cell nucleus, two intercellular tight junctions and conspicuous pinocytotic vesicles. A prominent basal lamina surrounds the capillary. Human biopsy specimen. Magnification × 8,000

cytoplasm and fuse with the opposite wall, thus discharging their contents. The cytoplasm around the invaginations is also rich in phosphatases. At the junctions of the endothelial cells the plasma membrane may be simple, or imbricated and interdigitated; the cells frequently overlap. In most regions a gap of about 20 nm containing electron-dense material separates the plasma membranes, but in other sites, the opposed plasma membranes either approach each other closely, or fuse to form tight junctions (*see* p. 7). True desmosomes are infrequent in man.

Fenestrated capillaries (6.6B) are found in the renal glomeruli, the intestinal villi, endocrine glands and the pancreas. Here, the cytoplasm on each side of the nuclear region is extremely thin, often approaching 40–60 nm and is apparently perforated at intervals by a system of 'pores' which vary greatly in number and size (their diameters may range from 30 to 100 nm). Each pore, however, appears 'closed' by a thin electron-dense diaphragm which often appears to be a condensation of basal lamina material. However, the molecular status of the diaphragm and the permeability of the pores remains uncertain. Whilst the use of electron-dense tracers (e.g. ferritin) and electron microscopy has added to our knowledge, many aspects of the structural basis of variations in capillary permeability are unresolved and under active investigation (*vide infra*). Another unresolved question concerns the mechanism of variations in capillary diameter and blood flow. Many workers believe that flow variations are exclusively determined by the state of contraction of the arterioles and precapillary sphincters, whilst others claim that the endothelial cells themselves, and some types of pericyte, possess poorly developed cytoplasmic contractile systems (Epling 1966).

The sinusoids (6.6c) also show regional variations in structure. Their calibre is larger and more irregular than that of capillaries. They are lined by a variety of cell types and they may be grouped as either *discontinuous, closed* or *fenestrated*. The fenestrated sinusoids are typical of the endocrine glands and possess many of the fine structural features seen in the fenestrated capillaries, including a diaphragm and a delicate, complete basal lamina. Intercellular gaps are absent and the lining cells are not prominently phagocytic. In contrast the discontinuous sinusoids of the liver are lined by flattened, often branched cells, some of them highly phagocytic; large intercellular gaps occur at many points. Further details of these sinusoids and the other specialized types found in the bone marrow and spleen will be considered under the individual organs.

VENULES

With the confluence of a number of capillaries, a *venule*, some 20–30 μm in diameter is formed, consisting essentially of a tube of flattened, oval or polygonal endothelial cells supported by a basal lamina, and a delicate adventitial coat of longitudinal fibres with interspersed fibroblasts (Rhodin 1968). Some authorities (e.g. Simonescu and Simonescu 1977) distinguish between the initial postcapillary venules which are surrounded by pericytes (*pericytic venules*) and the somewhat larger vessels partially enclosed in a muscular tunica media (*muscular venules*). Venules in the immediate postcapillary position are important sites of fluid exchange and leucocyte migration; in the lymphoid tissue of the gut and bronchial walls, lymph nodes and thymus, the endothelial cells of the postcapillary venules are cuboidal in form, and have incomplete intercellular junctions, permitting the ready passage of lymphocytes to and from the blood (*vide supra*, p. 773). In other tissues these vessels are believed to be a major site of transendothelial migration into the extravascular spaces, for neutrophils and other motile cells, as well as a region in which neutrophils can temporarily attach to the endothelium as a *marginated pool* of cells.

The intercellular junctions of venules are particularly sensitive to inflammatory agents which promote the leakiness of these vessels to fluids and defensive cells in the inflammatory response (*see* e.g. Marchesi 1961, 1962).

PERMEABILITY OF CAPILLARIES AND VENULES

The capillary bed, together with the postcapillary venules, forms an enormous area for the *exchange* of nutrients, gases, metabolites and water, between the blood and surrounding tissues, and also constitutes the chief route of leucocyte migration from the vascular system to the extravascular spaces. These capacities are reflected in the capillary bed density, which varies with the general metabolic level of the tissues they supply. Thus, it has been calculated that in the highly demanding myocardium there are about 2,000 capillaries per cubic millimetre of tissue, about 1,000 in skeletal muscle, but only some 50 in the connective tissue component of the skin (*see* Simonescu and Simonescu 1977). The total surface area of such capillaries may be relatively large, up to 2·4 metres^2 per 100 cm^3 in skeletal muscle, and about 4 metres^2 per 100 cm^3 in lung, giving a value of about 60 metres^2 for the capillaries of the systemic circulation and 40 metres^2 for the pulmonary circulation (*see* Krogh 1959; Intaglietta and Zweifach 1971).

The permeability of capillaries and venules has been extensively studied with various methods, ranging from physiological measurements of fluid composition, to structural investigations with tracers, e.g. Evans' blue, horseradish peroxidase, and ferritin, introduced into the bloodstream or extravascular spaces (Karnovsky 1967, 1968, 1970). These techniques have demonstrated that although, in general, the endothelium of blood vessels is a semi-permeable barrier acting as a dialysing membrane, i.e. allowing water, gases, and ions to pass across the wall by diffusion whilst preventing the passage of large molecules, there is also some movement of colloids and even quite large water-soluble molecules both into and out of the bloodstream across the endothelial cells. Physiological measurements of rates of large molecules, e.g. dextrans, have indicated the existence of two distinct categories of diffusion, corresponding to hypothetical '*large pore*' and '*small pore*' channels (Grotte 1956; Landis and Pappenheimer 1963), although the precise structural basis of these diffusion routes is not certain. Tracer studies have shown that large molecules, or molecular aggregates, can be taken across endothelia in the numerous pinocytotic (endocytic) vesicles, and indeed in some cases vesicles from both sides may fuse to form continuous channels (Bruns and Palade 1968). The vesicular transport system probably corresponds to the 'large pore' channels. 'Small pore' diffusion, it has been suggested, might represent leakage through intercellular junctions, and (where they are present) perhaps through fenestrae and their associated septa, as in the renal glomeruli (*see* Maul 1971).

However, the permeability of capillaries and postcapillary venules varies greatly in different tissues, and this can, to some extent, be correlated with the organization of the endothelium in those regions. In tissues where ready exchange of large molecules occurs, e.g. the alimentary tract, endocrine glands, fenestrated or incomplete endothelia, with numerous endocytic vesicles are present, and intercellular junctions are either incomplete or of the 'leaky' variety (Simonescu *et al.* 1972, 1973, 1975). In other areas, where known physiological barriers to large molecule diffusion occur, e.g. the brain, thymic cortex, testis and exocrine pancreas, endothelia are complete and non-fenestrated, with complete and efficient zonulae occludentes between cells, and there are relatively few endocytic vesicles. Most tissues, however, e.g. skeletal muscle, occupy intermediate positions between these two extremes.

VEINS

The walls of the veins, like those of the arteries, are composed of three coats, the internal being endothelial, the middle muscular, and the external of connective tissue (6.1).

The main difference between the veins and the arteries is in the comparative weakness of the middle coat in the former; its much smaller content of muscle and elastic fibres is related to the much lower venous blood pressure.

In the smallest veins the three coats are difficult to distinguish. The endothelium is supported on a membrane separable into two layers, the outer of which is the thicker, and consists of a delicate membrane of collagen fibres and fibroblasts, while the inner is composed of a network of longitudinal elastic fibres. In the veins next above these in size (0·4 mm in diameter) a layer of white fibres with a few non-striated muscular cells circularly disposed forms the middle coat, while the elastic and connective tissue elements of the outer coat are more distinct. In the middle-sized veins, the endothelium is of the same character as in the arteries, but its cells are shorter and broader. It is supported by a connective tissue layer, consisting of a delicate network of branched fibroblasts, and external to this there is a layer of elastic fibres disposed in the form of a network in place of the definite fenestrated membrane seen in arteries. This constitutes the *tunica intima*. The *tunica media* is composed of a thick layer of connective tissue with elastic fibres, intermixed in some veins with non-striated muscular fibres arranged circularly. The white fibres predominate, and the elastic fibres are in much smaller proportion than in the arteries. The *tunica adventitia* consists, as in the arteries, of areolar tissue with longitudinal elastic fibres. In the largest veins it is very much thicker than the tunica media, and contains many longitudinal muscular fibres. These are most distinct in the inferior vena cava, especially at the termination of this vein in the heart, in the trunks of the hepatic veins, in all the large trunks of the portal vein, and in the external iliac, renal and azygos veins. In the inferior vena cava, renal and portal veins they extend through the whole thickness of the outer coat, but in the

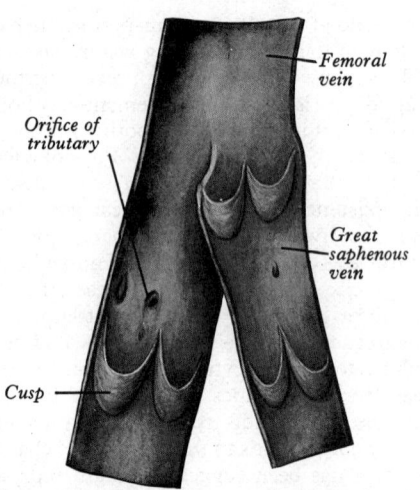

Femoral vein

Orifice of tributary

Great saphenous vein

Cusp

6.9 The upper portions of the femoral and great saphenous veins laid open to show valves. About two-thirds of natural size.

other veins mentioned a layer of connective and elastic tissue is found external to the muscular fibres. The white connective tissue fibres in the tunica adventitia of the inferior vena cava are disposed in bundles which form a network of right- and left-hand spirals, and this arrangement, together with the elastic fibres, facilitates the lengthening and shortening of the vessel which occurs with ascent and descent of the diaphragm (Franklin 1937). The large veins which open into the heart are covered for a short distance with a layer of cardiac muscle continued on to them from the heart and in the case of the coronary sinus the covering is complete (Coakley and King 1959). Muscular tissue is absent: (1) in the veins of the maternal part of the placenta; (2) in the venous sinuses of the dura mater and the veins of the pia mater; (3) in the veins of the retina; (4) in the veins of the spongy substance of bones; (5) in the venous spaces of the corpora cavernosa and corpus spongiosum. These veins consist of an endothelial lining supported by connective tissue.

Most veins are provided with *valves* which serve to prevent the reflux of the blood (**6**.9). Each valve is formed by a reduplication of the intima, strengthened by connective tissue and elastic fibres, and is covered on both surfaces by endothelium, the arrangement of which differs on the two surfaces. On the surface of the valve next the wall of the vein, the cells are arranged transversely; while on the other surface, over which the current of blood flows, the cells are arranged longitudinally in the direction of the current. Most commonly two such valves are found opposite one another, more especially in the smaller veins or in the larger trunks at the point where they are joined by smaller branches; occasionally there are three, and sometimes only one. The valves are semilunar and they are attached by their convex edges to the wall of the vein; the concave margins are free, directed in the course of the venous current, and lie in close apposition with the wall of the vein as long as the current of blood is centripetal; if, however, regurgitation takes place, the valves become distended, their opposed edges are brought into contact, and the backflow is interrupted. The wall of the vein on the cardiac side of the attachment of each valve is expanded into a pouch or *sinus*, which gives to the vessel, when injected or distended with blood, a knotted appearance. The valves are very numerous in the veins of the extremities, especially in the veins of the lower extremities, these vessels having to conduct the blood against the force of gravity in addition to being subjected to intermittent pressure due to muscular contractions. They are absent in the very small veins and very large veins and are absent or few in number in many other tissues. Particular features will be noted with the description of the individual veins.

The return of blood along the veins is under the influence of a number of factors. The smaller veins are filled by the blood overflowing into them from the capillary bed. The deep veins of the limbs are subjected to pressure due to contractions of the surrounding muscles, and the venae comitantes are affected in a similar way by arterial pulsations. This squeezing of the veins would tend to drive the blood in both directions along the vessels,

but the valves prevent flow towards the periphery and so the blood must flow towards the heart so long as the valves are competent. The force of gravity in the veins of the head and neck and the suction due to negative intrathoracic pressure in the veins near the heart are also important factors in venous return.

The rate of flow of the blood in veins being slower than in the arteries, veins are larger and more numerous than the corresponding arteries so that the blood delivered to tissues can be adequately returned to the heart.

The walls of the larger veins, like the arteries, are supplied with nutrient vessels, termed *vasa vasorum*. Unlike those in the arteries these vessels may penetrate as far as the intima probably reflecting the low venous blood pressure and oxygen tension. Nerves (postganglionic sympathetic efferent, and primary afferent) also are distributed to the veins in the same manner as to the arteries, but in much lesser numbers.

ASSOCIATIONS BETWEEN ARTERIES AND VEINS

Arteries and veins usually, though not invariably, run in close relationship to each other within fascial planes, the pedicles of viscera, and neurovascular bundles to muscles and other structures. The vein may be single, but often an artery is accompanied by a pair of veins, one lying on each side, having numerous cross-connexions, and both contained within a connective tissue sheath: these are termed *venae comitantes*.

As noted above, intimate apposition of artery and vein is probably an important determinant aiding the return of venous blood towards the heart, due to the interplay of the arterial pulsations and the valved interior of the veins. In some cases such close associations between vessels are also important in the *countercurrent exchange* of heat or ions from the incoming to the outgoing blood. An excellent example is in the blood supply of the testis where the branches converging to form the testicular vein comprise a complex pampiniform plexus surrounding the testicular artery, which arrangement promotes heat transfer from the arterial to the venous blood, thus permitting the testis to remain at a lower temperature than that of the general abdominal viscera (*see* Evans 1949; Grant and Payling Wright 1971; also p. 1415). An analogous arrangement of arterial and venous sinusoids exists in the vasa recta of the renal medulla; in this situation countercurrent exchange ensures that sodium ions are retained at a high concentration within the medulla (p. 1395) as efferent venous blood gives up its sodium ions to the afferent arterial supply.

Vascular Patterns

The distribution of the systemic arteries is like a highly branched tree, the common trunk of which, formed by the aorta, commences at the left ventricle, while the smallest ramifications extend to the viscera and to the peripheral parts of the body.

There is considerable variation in the mode of division of the arteries: occasionally a short trunk subdivides into several branches at the same point, as in the coeliac and thyrocervical trunks; more usually the vessel gives off several branches in succession, and still continues as the main trunk, as in the arteries of the limbs and those of the head and neck.

A branch of an artery is smaller than the trunk from which it arises; but if an artery divides into two branches, the combined sectional area of the two vessels is, in nearly every instance, somewhat greater than that of the trunk, and the combined sectional area of all the arterial branches greatly exceeds that of the aorta.

ANASTOMOSES

Arteries do not always end in arterioles and capillaries; in many cases they unite with one another, forming what are called *anastomoses*. Anastomosis between trunks of nearly equal size is found in the brain, where the two vertebral arteries unite to form the basilar artery, and the two anterior cerebral arteries are

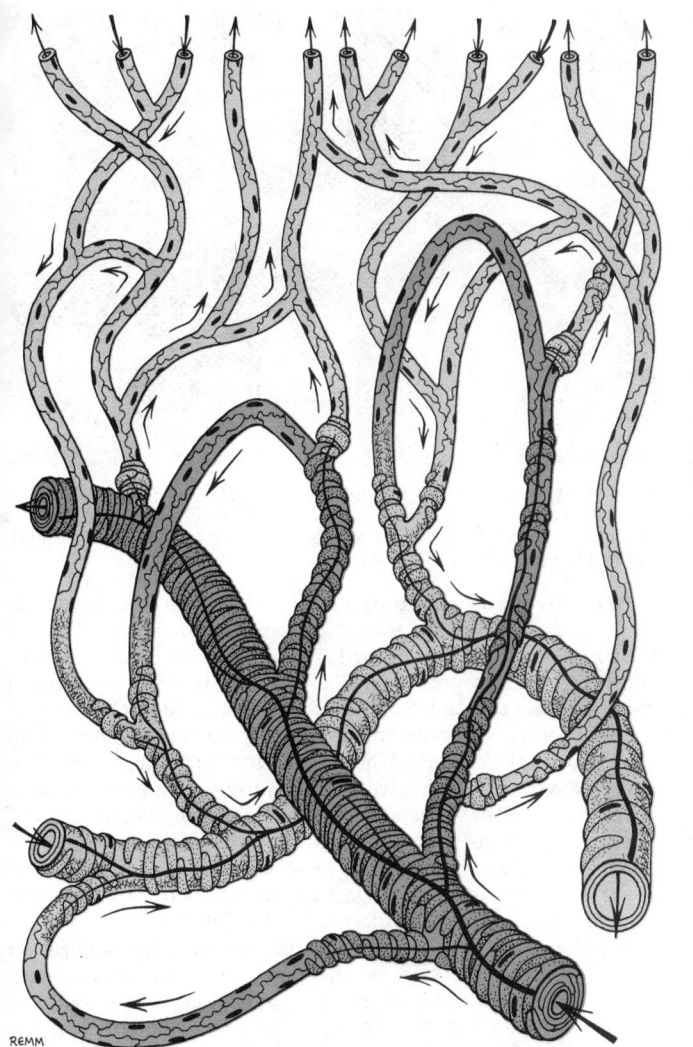

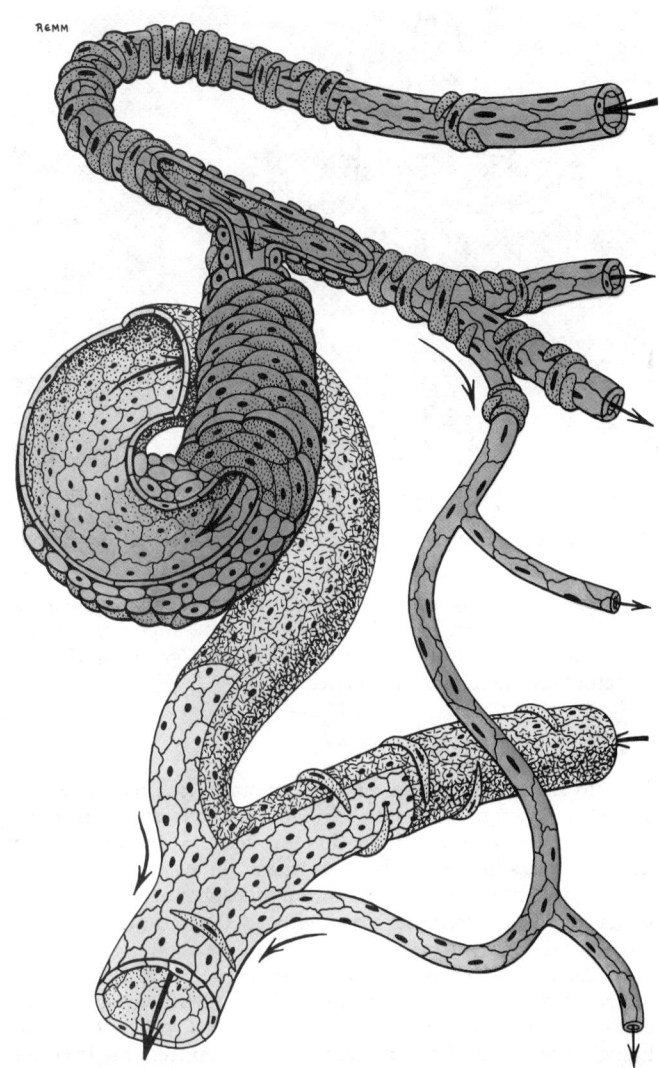

6.10 Diagram of a microcirculatory unit based upon descriptions in Zweifach (1937, 1959, 1961) and Reynolds and Zweifach (1959). Note the terminal arteriole, thoroughfare channels, true capillaries and the collecting venule. The distribution of nonstriated muscle cells and precapillary sphincters is shown. See text for further description.

6.11 Diagram of an arteriovenous anastomosis. Note the thick wall of the anastomotic channel composed of many layers of characteristically modified 'epithelioid' nonstriated muscle cells. See text for further description.

connected by the anterior communicating artery, and in the abdomen, where the intestinal arteries have free anastomoses between their larger branches. In the limbs, the anastomoses are largest and most numerous around the joints; the branches arising from an artery above a joint unite with branches from the vessels below it, and lateral anastomoses complete peri-epiphysial circles (p. 259). Arterial anastomoses increase in frequency with the remoteness of the vessels from the heart and the smaller branches of the arteries anastomose more frequently than the larger; between the smallest twigs these anastomoses may be so numerous that they constitute a close network. Anastomoses allow of an equalization of pressures over the territories which they connect and provide alternative channels of supply to a particular area; hence their frequency around joints where the circulation may be temporarily impeded during movement. If the arteries join end to end (e.g. the palmar and plantar arches, intercostal arteries, arteries of the stomach and intestine), cutting the anastomosis results in bleeding from both ends. An anastomotic channel may be so enlarged as to replace the normal and constitute an aberrant supply to a part (e.g. 721). From the clinical point of view anastomoses provide, by their enlargement, the basis for a *collateral circulation* when a vessel is interrupted by accident, disease or by ligation. The collateral circulation often develops rapidly and if the increased flow through and pressure within the collateral channels are maintained, the vessels enlarge, often become tortuous and structural changes appropriate to the function occur in the walls and new blood vessels may develop.

The resulting collateral circulation is, however, never in excess of the functional needs of the part. Sudden occlusion of a vessel may be followed in some situations by death of the part supplied, whilst gradual occlusion may allow time for the dilatation of the anastomosing channels and adequate nourishment of the tissue. The collateral circulation develops more rapidly in the young. The mechanisms involved in the dilatation of the collateral channels are not fully understood; haemodynamic changes consequent on the fall of the peripheral resistance, anoxia, nervous factors and the effects of accumulating metabolites have all been implicated.

END-ARTERIES

In certain regions of the body there are arteries which have no anastomoses with their neighbours and are therefore called *end-arteries*. If an artery of this type be occluded, serious nutritional disturbances resulting in death (necrosis) will occur in the tissues supplied by the vessel. The central artery of the retina is the best example of an end-artery, and its occlusion is followed by permanent blindness. Whilst the large arteries at the base of the brain form effective anastomoses, the central arteries which enter the brain substance communicate with each other only through the capillary bed, and, from a practical point of view, are end-arteries. The same applies to the arteries of the spleen, kidneys, lungs and the metaphyses of long bones. The branches of the coronary arteries do in fact anastomose to some extent, but all too

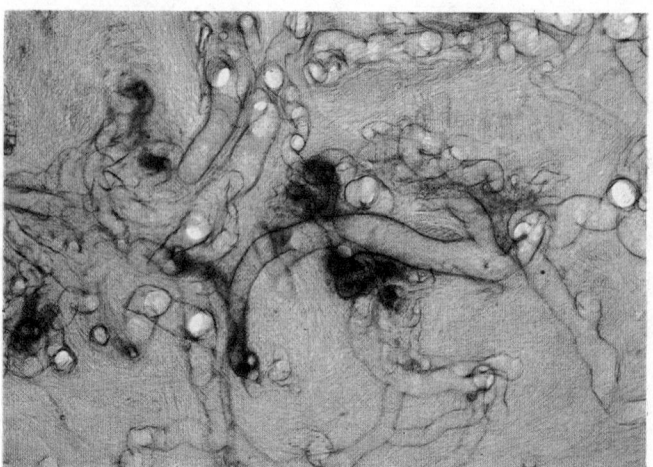 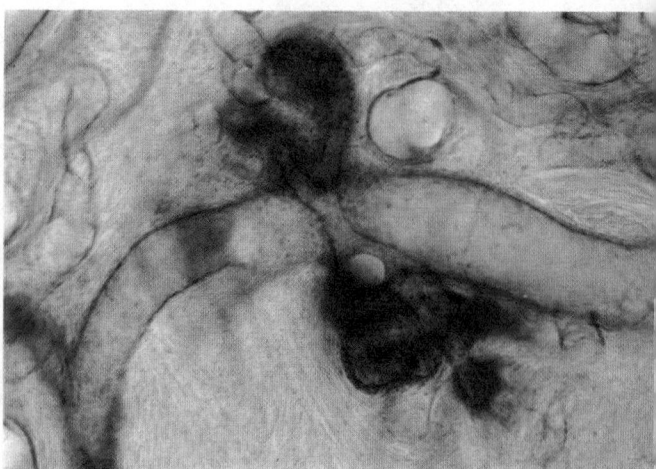

6.12A and **B** Human digital arteriovenous anastomoses prepared by intravascular perfusion of haematoxylin and subsequent clearing, of a full thickness specimen of skin. The heavily stained, thick-walled, tortuous, anastomotic channels contrast with the central arterial stem and the thin-walled venous outflow channels. See text for further details. (The specimens were prepared and kindly provided by Dr. R. T. Grant, Guy's Hospital Medical School.)

frequently this is insufficient to provide an adequate collateral circulation. Occlusive arterial disease in these important organs having an end-arterial blood supply is thus of great importance in pathology and clinical medicine.

VASCULAR SHUNTS

In many tissues there exist vessels of communication between the arterial and venous sides of the circulation which, when open, by-pass the nutritive tissue circulation through the capillaries. They may be classified according to their dimensions, site and complexity as: (1) preferential thoroughfare channels; (2) 'simple' arteriovenous anastomoses; (3) highly specialized arteriovenous anastomoses or *glomera*.

Preferential 'thoroughfare channels'. It has been demonstrated in many tissues that the true capillaries arise not only as direct side branches of terminal arterioles, but also as side branches of a thoroughfare channel which connects the terminal arteriole and the venule (Zweifach 1937, 1959, 1961; Reynolds and Zweifach 1959; Maggio 1965; Grant and Payling Wright 1968, 1970). This *thoroughfare channel* has a larger calibre than the true capillaries, and a fine structure which closely resembles typical continuous capillaries, but with occasional, widely spaced, smooth muscle cells spiralled around the endothelial lining. Each capillary side branch is surrounded at its origin by a precapillary sphincter. It is considered that a thoroughfare channel and its associated true capillaries form a functional *microcirculatory unit* (6.10). In times of low functional demand the blood flow is largely limited to the by-pass channel with the majority of the precapillary sphincters closed. Periodic opening and closing of different sphincters irrigates different parts of the capillary net. With increasing functional demand, the nutritive blood flow to the tissues may increase greatly following the opening of many sphincters. The size of the microcirculatory unit varies greatly with the particular tissue, e.g. in striated muscle each channel may give rise to 20–30 true capillaries, whereas in some glandular tissues only to 1–2. However, detailed investigations of the cremaster muscle and biceps femoris tendon of the rat (Grant and Payling Wright 1968, 1970) have emphasized that in these situations the thoroughfare channels are confined to the perimuscular and peritendinous connective tissues and are absent from striated muscle itself. The form of the capillary net also varies with the tissue, the meshes being either round or elongated. Round or angular meshes are most common and prevail where there is a dense network, as in the lungs, mucous membranes and in the skin. Elongated meshes occur in muscles and nerves, the long axis of the mesh running parallel with that of the muscle or nerve fibres. Sometimes the capillaries have a looped arrangement as in the papillae of the skin and tongue. The number of the capillaries and the size of the meshes determine the degree of vascularity of a part; the smallest meshes are found in the lungs and in the choroid coat of the eye.

Arteriovenous anastomoses. In a number of situations in the body direct connexions exist between the smaller arteries and the corresponding veins termed *arteriovenous anastomoses* (Grant and Bland 1931, Popoff 1934, Clark 1938). The connecting vessel may be straight or coiled and it often possesses a thick muscular coat of peculiar structure and a relatively fine lumen, measuring 10–30 μm on the average. Under the influence of the sympathetic nervous system, which gives a rich supply of non-myelinated fibres to the wall of the vessel, it is capable of complete closure and in that event circulation passes through the capillary bed in the ordinary way. When patent, the vessel carries blood directly from the artery to the vein and so partially or completely excludes the capillary bed from the circulation for the time being.

Arteriovenous anastomoses of relatively simple organization (6.11) occur in the skin of the nose, lips and external ear; the mucous membrane of the nose and alimentary canal; the coccygeal body; the erectile tissue of the sexual organs; the tongue; the thyroid gland and the sympathetic ganglia. They probably occur also in other situations.

In the skin of the hands and feet (especially the digital pads and nail beds), the anastomoses exhibit a special arrangement and form a large number of small units which have been termed 'glomera'. They are situated in the deeper layer of the corium and each 'glomus' has one or more afferent arteries. These vessels arise from branches of the cutaneous arteries which run towards the skin surface (6.12). The afferent arteries come off at right angles from their parent vessels, which thereafter are continued into the papillary layer of the corium, where they end in a capillary plexus. A short distance from its origin the afferent artery of a glomus gives off a number of fine 'periglomeral' branches and at once becomes considerably enlarged. It makes an S-shaped curve and then narrows down to become continuous with a short funnel-shaped vein which opens, at right angles, into a collecting vein. This vein commences on the deep aspect of the glomus, and curves round its outer surface. Having gained the superficial aspect of the glomus the collecting vein retraces its course, receiving venules from the papillary layer of the skin as it does so. Finally, it joins one of the deeper cutaneous veins.

In the child at birth arteriovenous anastomoses are few in number and poorly differentiated; they develop rapidly in the early years of life. In old age they undergo atrophy and sclerosis and are reduced in number.

Structurally the vessels immediately concerned in the digital arteriovenous anastomoses are unusual. At the point where it enlarges into the connecting vessel, the afferent artery presents a number of small endothelio-muscular elevations which project into its lumen, but proximal to this point its structure is typical. The connecting vessel has a lining of endothelium, supported by some fine collagenous fibres, but is devoid of an internal elastic lamina. An inner longitudinal and an outer circular muscle coat are not sharply differentiated from one another. The muscular wall is thick and on section the muscle fibres appear pale and swollen, with centrally situated nuclei, so that they have been described as 'epithelioid'. The emerging vein has a thin wall, the muscular coat being entirely wanting. Numerous elastic fibrils are present and are continued into the tunica adventitia of the collecting vein.

Arteriovenous anastomoses regulate the regional blood flow and their functional state varies. They are capable of contraction, but the mechanism is uncertain. Where circular muscle tissue is present in the walls, the epithelial cells may act as cushions which narrow the lumen. Where such muscle is absent, closure of the lumen may result from swelling of the epithelioid cells. Cutaneous arteriovenous anastomoses play an important part in regulating body temperature both generally and locally. When the local temperature of the rabbit's ear is raised above 40 °C the connecting vessel relaxes and an increased flow of blood at body temperature results, with a consequent cooling effect. When the local temperature is lowered below 15 °C the connecting vessel again relaxes. In this case the increased flow of blood at body temperature will help to raise the local temperature unless the process of artificial cooling is intensified. Again, when the body temperature of a rabbit is raised artificially there is a general opening up of all the subcutaneous arteriovenous anastomoses, a considerable increase in heat radiation and a consequent drop in the body temperature (Grant 1930). The cooling effect of panting in the dog is associated with the opening up of the arteriovenous anastomoses in its tongue. The paucity and underdeveloped character of arteriovenous anastomoses in the newborn, and the marked reduction in number of the subcutaneous arteriovenous anastomoses with advancing age, may be associated with less efficient temperature regulation.

The arteriovenous anatomoses present in the mucous membrane of the alimentary canal fulfil a different function (Spanner 1931). In man the arteriole to a villus has a direct connexion with its corresponding venule and, during periods when absorption is not occurring, the connexion becomes patent and serves to raise the pressure in the portal vein. On the other hand, during absorption the connexion is closed and the circulation passes through the capillary plexus at the apex of the villus.

Other suggested functions of arteriovenous anastomoses include the regulation of the blood pressure, secretion by the epithelioid cells and pressor reception.

THE THORACIC CAVITY AND HEART

The skeletal walls of the thorax are described on pp. 291–292. Its capacity does not correspond with the osseous thorax, because the lower part of the region enclosed by the ribs is encroached upon by the diaphragm and the upper abdominal viscera. The capacity varies with the phase of respiration, which also affects to some extent the positions and relations of the thoracic viscera. Its arbitrary upper limit is usually taken as the plane of the thoracic inlet, but the apices of the lungs extend above this into the neck.

THE THORACIC INLET (6.62, 63, 118)

The bony boundaries of the inlet are described on p. 291, 3. 60. The structures which pass through the upper opening of the thorax fall naturally into two groups, viz.: those in or near the median plane and those placed more laterally on each side in close relationship with the cervical parts of the pleurae and lungs.

In or near the median plane, behind the manubrium, the lowest parts of the sternohyoid muscles just enter the thorax. Immediately behind these are the sternothyroid muscles, the remains of the thymus and the inferior thyroid veins on their way to their termination in the brachiocephalic vein. In the child, in particular, the left brachiocephalic vein itself may be so highly placed as to cross over to the right actually in the thoracic inlet. More posteriorly, the trachea and the oesophagus, together with the recurrent laryngeal nerves, occupy the median region and, behind the left margin of the oesophagus, the thoracic duct enters the neck. Just in front of the vertebral column are the longus colli muscles and the anterior longitudinal ligament.

On each side the upper part of the pleura and the apex of the lung occupy the thoracic inlet. Between the pleura and the neck of the first rib, from medial to laterial side, lie the sympathetic trunk, superior intercostal artery and most of the ventral ramus of the first thoracic nerve passing obliquely upwards and laterally to join the lowest trunk of the brachial plexus. Anteriorly, between the pleura and the first costal cartilage the internal thoracic artery enters the thorax.

On the right side (6.62), the brachiocephalic artery leaves the chest between the trachea and the pleura. The vagus nerve, having passed at a higher level between the subclavian artery and vein, lies at the thoracic inlet between the pleura and the brachiocephalic artery. The right brachiocephalic vein enters the thorax anterolateral to the brachiocephalic artery and the right phrenic nerve crosses the internal thoracic artery and comes to lie lateral to the brachiocephalic vein as these vessels pass behind the first costal cartilage.

On the left side (6.63), the left common carotid and subclavian arteries leave the thorax between the pleura laterally and the trachea medially; the left vagus nerve is running downwards lateral to the interval between the two vessels. Anterolateral to this is the left brachiocephalic vein. The left phrenic nerve crosses the internal thoracic artery at a higher level than it does on the right side so that, at the thoracic inlet, it is running in the interval between the left brachiocephalic vein anterolaterally and the subclavian and common carotid arteries posteromedially,

THE THORACIC OUTLET

This is wider transversely than from before backwards. It slopes obliquely downwards and backwards, so that the thoracic cavity is much longer behind than in front. The diaphragm (p. 549) closes the opening and forms a convex floor for the thorax. The floor is flatter at the centre than at the periphery, and higher on the right side than on the left; in the dead body the right side of the floor reaches the level of the upper border of the fifth costal cartilage, while the left extends only to the corresponding part of the sixth costal cartilage. From the highest point on each side the floor slopes suddenly downwards to the costal and vertebral attachments of the diaphragm; this slope is more marked and longer behind than in front, so that only a narrow space is left between the diaphragm and the posterior wall of the thorax.

DIVISIONS OF THE THORACIC CAVITY

The thoracic cavity is divided by the *mediastinum*, the region between the lungs which extends from the back of the sternum to the vertebral column, and descends from the thoracic inlet above to the diaphragm below. The heart lies in the mediastinum, enclosed within a fibroserous sac, the *pericardium*; the lungs occupy the right and left parts of the thoracic cavity. Each lung is covered with a serous membrane, called the *pleura*, which also lines the walls of the corresponding half of the chest, and forms the lateral boundary of the mediastinum (6.13B).

For the purposes of description the mediastinum is divided into a superior and an inferior part. The *superior part* extends downwards from the thoracic inlet as far as an oblique plane passing through the lower edge of the manubrium sterni in front and the lower border of the fourth thoracic vertebra behind. The *inferior part*, below this plane, is subdivided into three portions, viz. an *anterior* in front of the pericardium, a *posterior* behind the pericardium and diaphragm, and a *middle*, which contains the pericardium, the heart and the large vessels entering or leaving

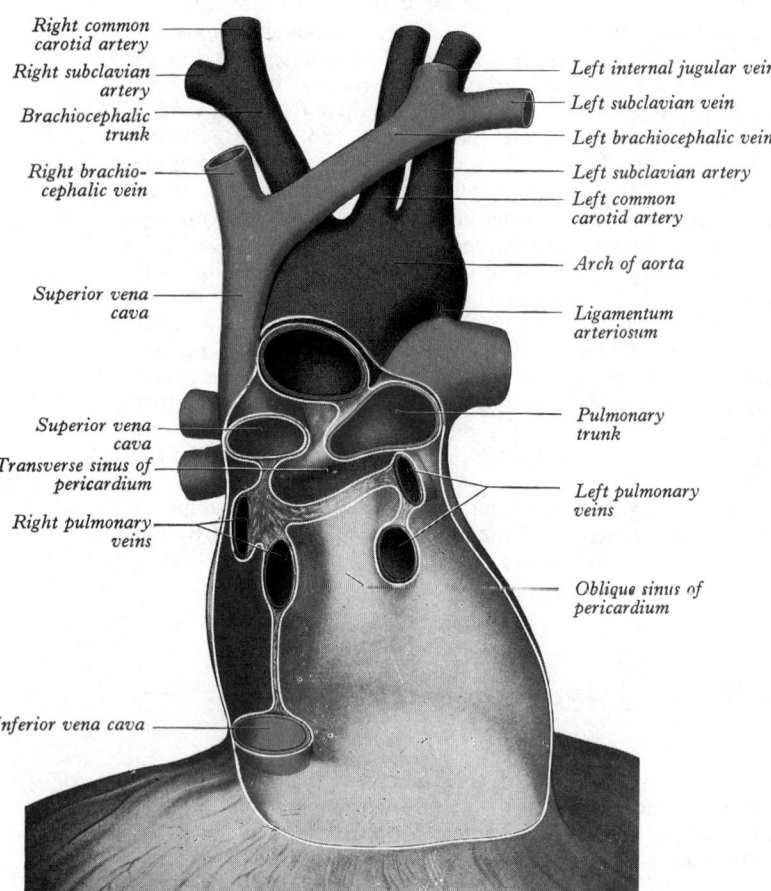

Right common carotid artery
Right subclavian artery
Brachiocephalic trunk
Right brachio-cephalic vein
Superior vena cava
Superior vena cava
Transverse sinus of pericardium
Right pulmonary veins
Inferior vena cava

Left internal jugular vein
Left subclavian vein
Left brachiocephalic vein
Left subclavian artery
Left common carotid artery
Arch of aorta
Ligamentum arteriosum
Pulmonary trunk
Left pulmonary veins
Oblique sinus of pericardium

6.13A The interior of the pericardial sac after removal of the heart, ventral aspect.

the latter (**6.13B**). Details of the different component parts of the mediastinum are given with the description of the respiratory organs.

The Pericardium

The pericardium (**6.13A, B**) contains the heart and the roots of the great vessels. It is placed in the mediastinum behind the body of the sternum and the cartilages of the ribs from the second to the sixth inclusive, and in front of the thoracic vertebrae, from the fifth to the eighth inclusive. It is essentially composed of two opposed layers of serous membrane, separated by a mere film of fluid, which nevertheless provides a complete cleavage between the heart and its surroundings, allowing it freedom of movement within the pericardium.

The pericardium consists of an outer sac, known as the *fibrous pericardium*, consisting of fibrous tissue and an inner, double-layered sac, the *serous pericardium*, a delicate membrane which lines the fibrous sac and covers the heart. The heart invaginates the wall of the serous sac from above and behind, and practically obliterates its cavity, the space being a potential one.

The *fibrous pericardium* is a cone-shaped bag, the apex of which is considered to end where it is continuous with the external coats of the great vessels, while its base is attached to the central tendon and to a small part of the muscular substance of the left half of the diaphragm. In some of the lower mammals the base is either completely separated from the diaphragm or joined to it by some loose fibrous tissue; in man much of its diaphragmatic attachment consists of loose fibro-aerolar tissue which can be readily broken down, but over a small area the central tendon of the diaphragm and the pericardium are fused. Above, the fibrous pericardium not only blends with the external coats of the great vessels, but is continuous with the pretracheal fascia (p. 537). The fibrous pericardium is also attached to the posterior surface of the sternum by *superior* and *inferior sternopericardial ligaments*, the superior passing to the upper end of its body, and the inferior to its lower end. (It should be noted that in the experience of modern thoracic surgeons, the development of these 'ligaments' is extremely variable; in particular the superior one is often indetectible.) By means of all these connexions it is securely anchored within the thoracic cavity and therefore maintains the general position of the heart within the chest. Some have also assumed that it prevents 'over-distension' of the heart, but recognizing the great range of variation that occurs in cardiac dimensions accompanying various pathologies, such a view is entirely speculative.

Anteriorly, the fibrous pericardium is separated from the front wall of the thorax, in the greater part of its extent, by the lungs and pleurae; but a small area, usually corresponding with the left half of the lower part of the body of the sternum and the sternal ends of the cartilages of the fourth and fifth ribs of the left side, is in direct relationship with the chest wall. Until it retrogresses at puberty or adolescence, the lower end of the thymus is in contact with the front of the upper part of the pericardium. *Posteriorly*, the fibrous pericardium rests upon the principal bronchi, the oesophagus, the

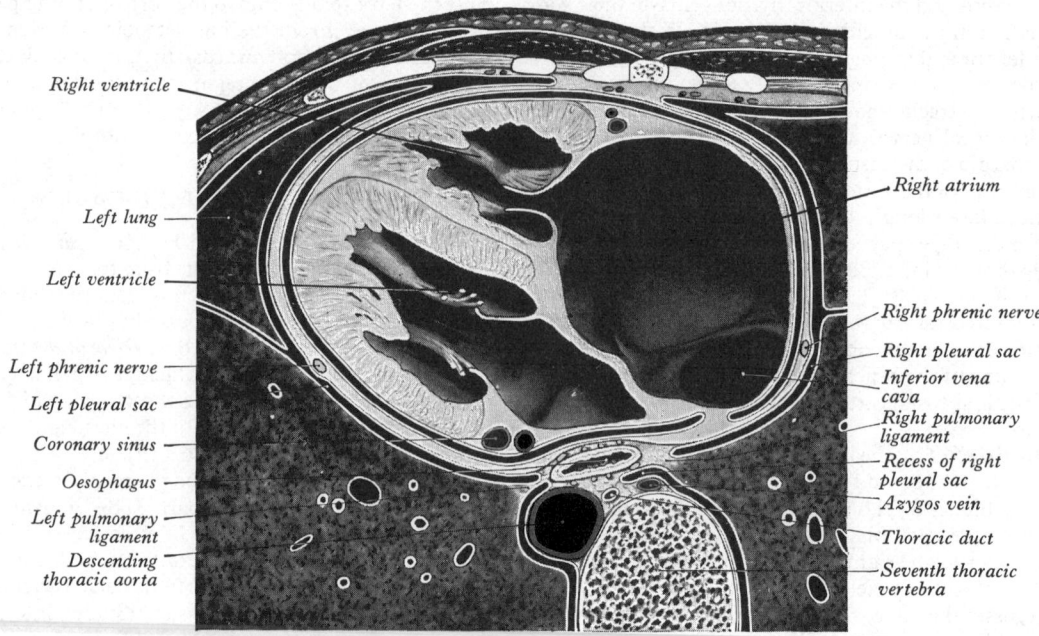

Right ventricle
Left lung
Left ventricle
Left phrenic nerve
Left pleural sac
Coronary sinus
Oesophagus
Left pulmonary ligament
Descending thoracic aorta

Right atrium
Right phrenic nerve
Right pleural sac
Inferior vena cava
Right pulmonary ligament
Recess of right pleural sac
Azygos vein
Thoracic duct
Seventh thoracic vertebra

6.13B A transverse section through the mediastinum at the level of the body of the seventh thoracic vertebra, viewed from above.

oesophageal plexus of nerves, the descending thoracic aorta and the posterior part of the mediastinal surface of each lung. *Laterally*, it is separated by the pleura, from the mediastinal surfaces of the lungs; the phrenic nerve, with its accompanying vessels, descends between the fibrous pericardium and the mediastinal pleura on each side. *Inferiorly*, it is separated from the liver and the fundus of the stomach by the diaphragm.

The vessels receiving prolongations from the fibrous pericardium are: the aorta, the superior vena cava, the right and left pulmonary arteries, and the four pulmonary veins. The inferior vena cava, which enters the pericardium through the central tendon of the diaphragm, receives no covering from the fibrous layer.

The *serous pericardium* is a closed sac which lines the fibrous pericardium and is invaginated by the heart; it therefore consists of a *visceral* and a *parietal* layer. The visceral layer, or *epicardium*, covers the heart and the great vessels, and from the latter it is reflected to form the parietal layer, which lines the fibrous pericardium. The extension which covers the vessels is arranged in the form of two tubes. The aorta and pulmonary trunk are enclosed in one tube; the superior and inferior venae cavae and the four pulmonary veins are enclosed in a second tube, the attachment of which to the parietal layer is Γ-shaped (**6.13**A). The cul-de-sac within the curve of the Γ lies behind the left atrium and is known as the *oblique sinus*, while the passage between the aorta and pulmonary trunk, in front, and the atria, behind, is named he *transverse sinus* (**6.19**B, 37). The sinus lies between the two pericardial tubes described above. The upper boundary of the transverse sinus is shown in **6.13**A.

A small triangular fold of the serous pericardium is reflected from the left pulmonary artery to the subjacent upper left pulmonary vein; it is known as the *fold of the left vena cava*. It contains a fibrous strand, called the *ligament* of the left vena cava, which is a remnant of the obliterated *left common cardinal vein* (or left duct of Cuvier, p. 183) and which extends downwards in front of the root of the left lung from the upper part of the left superior intercostal vein to the back of the left atrium, where it is continuous with the oblique vein of the left atrium (p. 184). The fold often forms the anterior wall of a small blind recess, the mouth of which is directed to the left. In some cases the left,

instead of the right, common cardinal vein persists and forms a left superior vena cava which runs downwards, in front of the root of the left lung and on the back of the left atrium, to enter the right atrium, replacing in the lower part of its course both the oblique vein of the left atrium and the coronary sinus. Sometimes both common cardinal veins persist to form a right and a left superior vena cava, and the transverse anastomosis between the common cardinal veins, which normally forms the left brachiocephalic vein, may be small or absent. When the left common cardinal vein persists it is joined by the left superior intercostal vein.

Vessels and nerves. The *arteries* of the pericardium are derived from the internal thoracic arteries and their musculophrenic branches, and from the descending thoracic aorta; its *veins* are tributaries of the azygos system. Its nerve supply is derived from the vagus and phrenic nerves, and the sympathetic trunks.

Structure. The fibrous pericardium consists of a compacted network of a collagenous fibrous tissue. The serous pericardium consists of a single layer of flattened cells resting on a layer of subserous aerolar tissue which, in the case of the parietal layer, blends with the fibrous pericardium. The areolar tissue under the visceral layer is continuous with the interstitial tissue of the myocardium and contains fat which is greatest in amount along the ventricular border of the coronary sulcus, along the inferior border of the heart and in the interventricular grooves. The main trunks of the coronary vessels and their larger branches are embedded in this fat, the amount of which is related to that of the general body fat and gradually increases as age advances.

Applied Anatomy. Paracentesis of the pericardium (i.e. draining of fluid accumulating in the pericardial cavity due to disease, by means of a hollow needle) may be performed in the fifth or sixth left intercostal space near the sternum, with care to avoid wounding the internal thoracic artery. Alternatively, the exploring needle may be entered at the left costoxiphoid angle, and passed upwards and backwards into the pericardial sac.

It should be remembered, in relation to aortic surgery, that the serous pericardium, which clothes the pulmonary trunk on the wall of the transverse sinus, extends as far as the ligamentum arteriosum (p. 666).

THE HEART

General Introduction

All triploblastic organisms, invertebrate or chordates, possess some form of circulatory system to overcome impractically long diffusion distances by circulating a fluid from regions of high oxygen tension and high concentrations of nutrient substances to the immediate surroundings of mesodermal and other cells remote from the external environment. The constitution of the fluid and the methods by which it is circulated vary much, especially in the many different phyla of the invertebrates. The majority of these are, however, coelomates and hence contain a closed system of blood vessels and some localized means of propelling 'blood' through them in a true circulation. (There are exceptions to this: some very small annelid worms lack a blood-vascular system, and in some acoelomates, e.g. nematode worms, a body cavity (regarded as a *pseudocoel*) does occur, though it has no endothelial lining and merely permits the to and fro movements of the contained fluid under the influence of bodily movements. One nematode, *Ascaris*, possesses a haemoglobin in this fluid.)

In some coelomates, e.g. annelid worms, pulsatile sections of the vascular tubes, usually multiple, occur in many parts of the body, and the larger of such 'hearts' may be valved and may respond to neurogenic 'pacemakers'. In arthropods the heart is usually singular and dorsal in position and the circulation is 'open', i.e. the efferent vessels from the heart open into wide 'haemocoels' which surround most organs. (Some small arthropods, e.g. mites, have no circulatory organs.) Aquatic

arthropods, such as the class *Crustacea*, respire through various forms of gill-like structures, the blood acting as the intermediary; but in the class *Insecta*, most of the species of which are emancipated from water, respiration is served by intercellular networks of tubes, independent of the circulation. If one takes into account other invertebrate phyla, such as the *Mollusca*, *Brachiopoda*, and *Echinodermata*, an even wider variety of circulatory pattern is apparent. But these facts are mentioned briefly as a contrast to the phylum *Chordata*, which displays a much more uniform circulatory pattern. Even in the most 'primitive' of the phylum, the *Hemichordata* (acorn worms) and the *Urochordata* (tunicates) the heart is single, and in the *Cephalochordata* (represented by the familiar *Amphioxus*) there is a ventral tubular 'heart' and a circulatory system basically vertebrate in type. The main, midventral pulsatile tube is assisted by pulsatile vessels in other parts of the circulation.

The *vertebrate* heart is ventral to the gut—to the pharynx in fishes and the embryos of land tetrapods, to the oesophagus in adult forms of the latter. Being developed from a tube, in terms of phylogeny, it is a single pump in earlier forms. However, from the beginning (as evidenced by the simpler extant forms) it shows a specialization into a succession of three or four enlarged segments of the tube—a collecting reservoir for the principal veins, the *sinus venosus*, a pulsatile but relatively thin-walled antechamber, the *atrium*, a thick-walled, highly muscular pumping chamber, the *ventricle*, and a fourth, the *bulbus cordis* which may be muscular or elastic and is absent in some forms. From the conical end of the ventricle (*conus arteriosus*), or from the bulbus cordis, when present, the major arteries commence. Valves, usually consisting of several flaps or leaves (sometimes serially repeated), are interposed between successive cavities. The heart possesses its own *coronary* circulation (except in *Agnatha*), and is contained in a pericardial coelom, which is a separated part of the general 'body cavity'. The pulsatile rhythm of the vertebrate heart is basically *myogenic*, but is coordinated with other systemic demands by a regulating nerve supply. With an increasing degree of specialization from fishes to birds and mammals, particular parts and tracts of the cardiac muscle differentiate to form foci of initiation of muscular contraction and rapidly conducting pathways for dissemination of these cyclic stimuli. In almost all vertebrates the heart tube outgrows the cranio-caudal dimensions of the pericardial sac, developing a sinuous bend. Because of this the 'venous end' (the sinus venosus and atrium) becomes dorsal to the 'arterial end' (ventricle and conus); moreover, the S-shaped heart tends to take up an asymmetric position associated with a transition from symmetrical cardinal venous systems to asymmetric caval veins. The pericardial cavity is semi-rigid in some fish (e.g. elasmobranchs), due to its position dorsal to the pectoral girdle, and the resultant constancy of its volume aids the filling of the atrium by suction as the ventricle empties; in terrestrial and aerial vertebrates this effect is largely lost, because of the caudal 'migration' of the heart.

In discussing the phylogeny of the heart, and especially in deriving the mammalian organ from the cardiac patterns of earlier forms, it is to be emphasized that no direct palaeontological data are available. Comparison of the arrangements at different grades of vertebrate development as they exist in extant creatures provides the only source of information. This comparison can be misleading unless allowance is made for several factors which complicate considerations. Firstly, hearts must function at a very early stage of embryonic life, and they must continue to do so throughout development and into the usually greatly changed environment after birth. Moreover, the environmental demands upon the adult creature vary greatly; and if these demands are less in the case of a more recent (and presumably 'more highly evolved' group), actual retrogression of cardiac structure may occur. For example, in the *Dipnoi*, or lungfish (and also in crossopterygians, if the extant coelacanth fish, *Latimeria*, is representative) the atrium is almost completely divided and the ventricle partly so, whereas in the *Amphibia* the ventricle is uniformly single and the atrium fully divided only in the anuran amphibians. The incomplete division of the atrium in urodele amphibians is considered a retrogression in structural terms, but it must be of physiological advantage to this, the more active of these two sub-classes. Although it is orthodox, and indeed almost inescapable, to derive the mammals from the reptiles, there is not an extant representative of the latter from which the mammalian heart can be directly derived. Although some reptiles (e.g. the order *Crocodilia*) show complete division of the heart, the relation of the major arteries, pulmonary and systemic, to the ventricles precludes such a derivation. The reptiles ancestral to the mammals must have possessed different arrangements. (It is interesting to note, however, that the avian heart and its outflow vessels are much closer in these arrangements to those of extant reptiles.)

It is obvious that a completely satisfactory phylogenetic history is not possible in the case of the vertebrate heart, and for this reason alone an elaborate comparison will not be attempted here. But some particular trends of evolution must be mentioned.

The sinus venosus, typically symmetrical in fishes, already displays asymmetry in the *Cyclostomata*: its right duct alone persists in lampreys, the left in the hagfish. In the *Dipnoi* the sinus opens into the right half of a partly divided atrium. This condition persists in subsequent vertebrate classes; in *Anura*, many *Reptilia* and all *Mammalia* this asymmetry is coupled with some retrogression of the sinus and absorption of its vestiges into the right atrium. With the complete separation of the atria a right-hand systemic venous return is thus established.

The arterial 'end' of the heart, represented by a *bulbus cordis* in vertebrate embryos (and some adult forms) or by a contractile chamber, the *conus arteriosus*, in adult elasmobranch fishes, commonly exhibits valve flaps, often arranged serially in rows.

Probably derived from the latter are the spiral valves of the *Dipnoi*, and these become more elaborate and efficient in the *Amphibia*. This development is closely associated with what must be regarded as the greatest era of transformation in the evolution of the heart—the long series of adaptive changes which accompanied and made possible the spread of vertebrate life from the aquatic to the terrestrial habitat. This transformation is accepted as occurring in the Devonian Period of the Palaeozoic Era, a period of 50 to 60 million years, approximately 300 million years ago. At the beginning of this transformation were dipnoan fish, chiefly dependent upon branchial respiration but also able to use, for aerial breathing, a pharyngeal diverticulum, the so-called 'swim bladder' (p. 1228); at the end of the transformation were amphibians, mainly dependent on pulmonary respiration, with lungs developed from some form of pharyngeal air sac, but also able to respire through a wet skin. These changes of respiratory habit, inseparably associated, of course, with cardiac and circulatory modifications, also demanded locomotor changes, the evolution of limbs, adapted from aquatic fins to terrestrial progression. (It is for such reasons that the *Amphibia* are regarded as a development from crossopterygians related to the extant *Dipnoi*.)

At the time of this great transformation, which may have occupied many millions of years, there already existed an inherent duality in the circulation, both the dipnoan and the amphibian possessing a *systemic* 'portal', arterial circulation, through gills in one case and subcutaneous networks in the other, and in addition a parallel *pulmonary* 'portal' circulation through the walls of either an air sac or true lungs. These separate circuits already return their blood to separate atria in the dipnoan; and although the ventricle is only partly divided, mixing of the two atrial inputs is probably diminished by the highly trabecular internal surface of the ventricle. Of course, the dipnoan ventricular output, though also largely separated into two streams by spiral valves in the large conus arteriosus, is conducted entirely to the gills; but the blood returning to the right atrium is conducted by a separate, dorsal division of the ventral aorta to the more caudal gill clefts and hence largely to the air sacs or 'lungs'.

In the amphibian circulation, with the disappearance of gill structures, the capillary beds of the systemic vessels supplying them disappear, and the main weight of respiratory function is pulmonary. The undivided ventricle (and partial division only of the atrium in some forms) may, however, permit some degree of mixture of the systemic and pulmonary flows through the heart; and presumably this is an advantage when the animal returns to water for breeding, cutaneous respiration then becoming predominant. There is thus no sudden change, in structure or function, during the slow cardiac adaptation from an aquatic to a terrestrial mode of life, but rather a change of emphasis between two, co-existent methods of respiration. The separation of the two flow routes through the heart is carried further in *Reptilia*. Despite the fact that some forms have returned to aquatic existence, respiration is now entirely pulmonary. This is coupled with a complete interatrial septum and at least a partial division of the ventricle, which is actually complete in *Crocodilia*. The latter are unusual in having a triple arterial outflow from the heart—a pulmonary artery from the right ventricle and also right *and* left aortic (systemic) arches, each arising from the contralateral ventricle. Because of this curious arrangement, the aortic arteries cross each other and communicate at this point. In the cayman this allows an almost complete shunt of blood into the left aorta, thus foreshadowing the mammalian condition.

In mammals, including mankind, the division of the heart (into atrium, ventricle, bulbus and conus) is complete and need not be described here. In the embryo, however, a series of stages occurs (p. 181) which closely resemble the final arrangements in earlier classes of vertebrate, and an arrest of development at various stages leads to a congenital defect which may resemble conditions in an earlier vertebrate form. However, the resemblance is modified by the fact that the heart must actually function at all but its earliest stages of development. The foramen ovale is a feature of mammalian pre-natal development, a necessary shunt mechanism rather than an atavistic indication of incomplete atrial septation in earlier forms. Hence, a persistently patent foramen ovale is a manifestation of disturbed *mammalian* rather than *vertebrate* development. Similarly, many other cardiac abnormalities may be tentatively 'explained' as divergencies from ontogenetic development rather than by reference to an earlier vertebrate state.

The fact that every vertebrate heart must function within its particular circulation from a very early stage of development, and long before either has reached its final arrangement, accords with the continuous picture of cardiac phylogeny which appears in retrospect. Although a relatively simple heart of single chambers, serially arranged, has finally led to a fully septated double heart consisting of two muscular pumps arranged 'in parallel', but in fact pumping in series, the transitional stages have been smooth and at no stage dramatic. This has been achieved despite the great change from purely aquatic respiration through gills to a state of exclusively pulmonary respiration. The smooth changeover may be ascribed to the existence of *both* methods of respiration in a long series of dipnoan and amphibian species over a long period of time. The double circulation, systemic through gills and pulmonary through lungs, has gradually evolved from a condition of *parallel* supply by incompletely septated hearts to one of *serial* supply on completion of septation. In this gradual change, between two highly different circulations, evolutionary changes in the vessels served by the heart have, of course, occurred in a harmonious succession of adaptations. For some details of these changes, from the ontogenetic point of view, consult the section on *Embryology* in this volume. The literature of cardiac evolution is large; extensive bibliographies exist in the works of Bolk *et al.* (1931–9); Grassé (1954); Foxon (1955); Robb (1965); Romer (1970); and Foxon and Bannister (1974).

In the following pages it will become evident that there is no entirely satisfactory and logical progression to be adopted in a description of the heart—whatever standpoint is approached first it necessarily assumes some knowledge contained in more detail in later sections. Nevertheless, it is hoped that the sequence presented represents a good compromise. An account of the general organization of the heart precedes considerations of its external features, surface anatomy and radiological appearances; thereafter the chambers are described sequentially in terms of their internal features, flow tracts, valve architecture, with brief functional allusions; finally the fine structure of the myocardium, the 'fibrous skeleton', and both the well-accepted and more controversial aspects of the specialized conducting tissues are combined with an overall illustration of the coordinated cycle of cardiac contraction and relaxation.

General Cardiac Organization

As noted above, the mammalian, including the human, heart consists of a pair of valved muscular pumps topographically combined in a single organ, and whilst their fibromuscular framework and conducting tissues are inextricably interwoven in a single complex functionally interdependent unit, each pump (the so-called 'right' and 'left' hearts) is physiologically *separate*, being interposed *in series* at different points in the double circulation. Thus, despite this *functional* 'in series' disposition (*vide infra*) it is often stated (*v. supra*) that the two pumps are *topographically* 'in parallel': however, as we shall see, even the latter comment is an over-simplification since the two pumps have completely contrasting haemodynamic roles, and their outflow channels exhibit a *mutual spiral* of almost 180° as they discharge towards their primary destinations.

The chambers of the heart are the right and left atria and right and left ventricles, the atria providing the receiving reservoirs for venous blood and a relatively weak contractile system which effects final filling of the ventricles, whilst the ventricles provide the much more powerful expulsive contraction forcing the blood into the main arterial trunks.

The 'right heart' of clinical parlance consists of the *right atrium* which receives the superior and inferior caval and main coronary inflows of venous blood from the systemic circulation: this is passed through the right *atrioventricular orifice*, guarded by

the *tricuspid valve*, into the *inflow tract* of the *right ventricle*. Contraction of the latter causes final closure of the tricuspid valve and, with increasing pressure within the ventricle, a level is reached when ejection of blood through the right ventricular *outflow tract* (the *infundibulum or conus arteriosus* and the *pulmonary valve*) is achieved, ejection obviously necessitating temporary opening of the valve. (The pressure profiles, time relationships, and sequences of valve opening and closure, are treated more fully below.) It should be noted here, however, that the right ventricular outflow tract discharges into the pulmonary trunk, arteries, and thereafter is restricted to perfusion of the *relatively* low-resistance pulmonary arteriolar and capillary vascular bed, and many of the structural features of the 'right heart' including the overall geometry of its chambers, the vectors of its inflow and outflow tracts, its myocardial architecture (in particular the thickness of its ventricular walls), and the construction and relative strengths of the tricuspid and mitral valves, accord with such a role, involving comparatively low and leisurely pressure fluctuations.

The 'left heart', in complete contrast, consists of the *left atrium* which receives the pulmonary (plus a little coronary) venous inflow, and by its contraction completes the filling of the *left ventricle* through the *left atrioventricular orifice*, guarded by the *mitral valve*. The latter provides the portal of entry to the inflow tract of the left ventricle, contraction of which raises its luminal pressure rapidly, causing closure of the mitral valve (*vide infra*) followed, with minimal delay, by opening of the aortic valve and ejection of blood through the left ventricular *outflow tract* via the aortic orifice into the aortic sinuses, ascending aorta, and thence throughout the whole of the non-pulmonary systemic arterial, arteriolar and capillary tree, including the coronary vessels themselves. This vast vascular bed provides a relatively high peripheral resistance, and this, together with the metabolic demands of the tissues, in particular the high and persistent requirements of the cerebral tissues, is the clue to an understanding of the structural organization of the 'left heart' including again, the geometry of its cavities, flow vectors, planes and strengths of the mitral and aortic valves, and the thickness of the left ventricular myocardium. Thus, the ejection phase of the left ventricle is of shorter duration than that of the right ventricle, but its pressure fluctuations are very much greater (**6.**35).

Because of the prolonged and complex phylogenetic history (outlined above), ontogenetic history (p. 181), and contrasting functional demands, the human heart is far from being a simple pair of parallel pumps, topographically combined, despite the obvious fact that (on average) the right and left ventricles deliver the same volume of blood with each heart beat. The heart, in fact, presents a complicated, spiralized, three-dimensional arrangement which is not obviously related directly to the cardinal planes of the body, and hence terms such as 'left' and 'right', 'apex' and 'base', 'surface', 'aspect' and 'border' may prove misleading to those embarking on a study of cardiac anatomy. Another source of confusion is the common practice of studying isolated specimens of whole or dissected hearts with subsequent difficulty in relating the observations to the heart *in situ*. The following few paragraphs are included to highlight some of these difficulties and, it is hoped, circumvent certain misconceptions before proceeding to a more detailed description of individual chambers.

Some of the principal features of cardiac anatomy can, as an introduction, be helped by a consideration of corrosion casts of a normal heart in which the left heart and right heart have been filled with resins of contrasting colours (**6.**14A, B, C), and by a survey of a horizontal section through the mediastinum, including the heart, at the level of the seventh thoracic vertebra (**6.**15).

In illustrations **6.**14A, B, C the left heart (red) and the right heart (blue) are viewed from the *ventral aspect* in their normal mutual relationship (**6.**14A), followed by the right heart (**6.**14B) and left heart (**6.**14C) in isolation (*see* caption for further details). It is seen immediately that the general form of the two is in complete contrast and that the unmodified terms 'right' and 'left' are singularly inappropriate. The right heart, whilst forming the right aspect or border of the heart, then follows a gentle wide-open curve (almost 'stomach-like' in form) and covers most of the

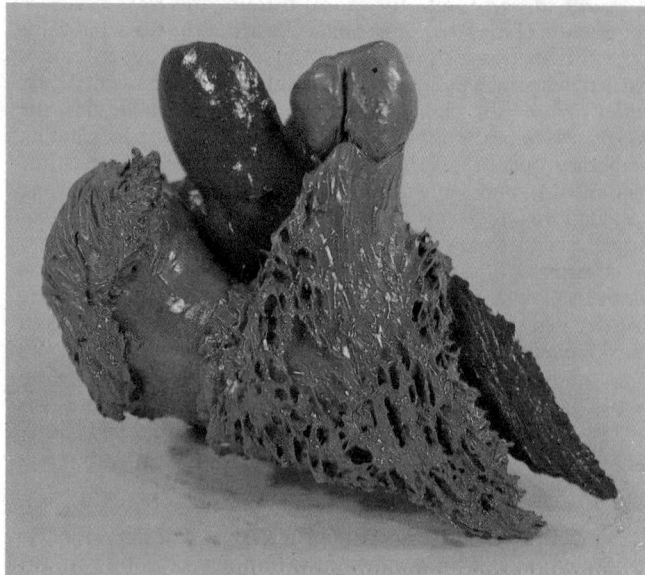

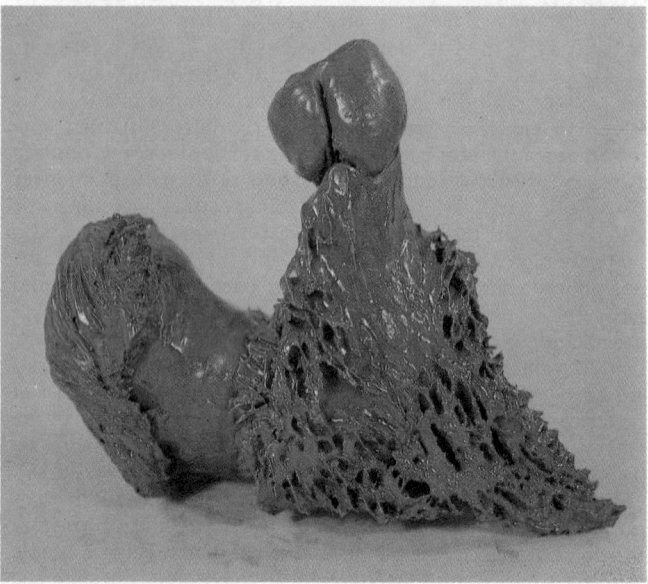

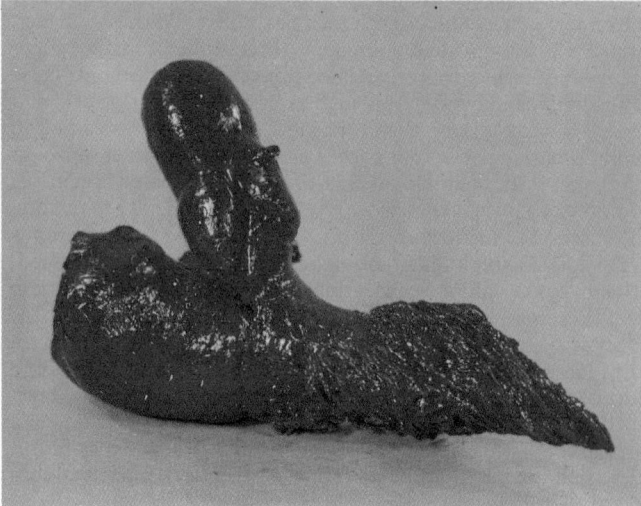

6.14A–C Coloured polyester resin luminal casts of a human heart; right atrium, ventricle, infundibulum, pulmonary root and valvular sinuses—*blue*; left atrium, ventricle, aortic vestibule, root and sinuses—*red*. A Sternocostal surface with right and left hearts *in situ*; B isolated cast of right heart; C isolated cast of left heart. Note the impressions made by the musculi pectinati, trabeculae carneae (spiralized in the left ventricle), the smooth areas related to the atrioventricular valve leaflets, and the dilated sinuses in the walls of the aortic and pulmonary roots. (Casts prepared by Dr. M. C. E. Hutchinson, photography by Mr. Kevin Fitzpatrick, Department of Anatomy, Guy's Hospital Medical School.)

anterior aspect of the left heart (with the exception of a left-sided strip including the apex). Thus, much of the so-called right heart in fact forms a great part of the *ventral surface* of the heart, its outflow tract proceeding upwards and to the left until it even terminates on the *left* side of the outflow tract of the left ventricle. The impressions of the leaflets of the tricuspid valve and those of the sinuses and cusps of the pulmonary valve are *widely separated*, occupying completely different planes, and the flattened cavity of the right ventricle (which is crescentic in section—*vide infra*) is splayed out between them. Conversely, the so-called left heart (except for the narrow left-sided strip and apex mentioned above) is largely *posterior* in position and, as seen in 6.14A, when viewed from the ventral aspect, is virtually obscured by the chambers of the right heart. It should also be noted at this stage that the left ventricular inflow orifice of the mitral valve is *very closely related* to its outflow orifice (i.e. the aortic valve), the two being embraced by the wide divarication of the 'right' inflow and outflow orifices. The planes of the left ventricular orifices, even though somewhat inclined to each other, are much more nearly coplanar than those of the right; the left ventricular cavity also shows marked differences—it is narrow, conical, and the spiral nature of its trabeculated interior is obvious. The tip of the left ventricular cone occupies the cardiac *apex*, whilst most of the *anatomical base* of the heart is provided by the left atrium.

Clearly, the heart as a whole is placed *obliquely* in the thorax (*see further below*) and this is easily confirmed by reference to the horizontal section shown in illustration 6.15. Note particularly that the interatrial septum and interventricular septum are virtually in line but are inclined *forwards and to the left* at an angle of about 45° to the sagittal plane, whilst the planes of the mitral and tricuspid valves, although not precisely coplanar and vertical, are broadly in a plane at right angles to the septal plane. In brief, therefore, the right atrium lies not only to the right but extends *anterior* and *inferior* to the left atrium (and also, to a small extent, anterior to part of the left ventricle, the *atrioventricular septum* intervening). The right ventricle forms most of the *anterior aspect* of the ventricular part of the heart; only its right inferior extremity lies to the *right* of the left ventricle, whilst its upper left extremity (outflow orifice) lies to the *left* and superior to the aortic valve. The left atrium forms most of the *posterior aspect* of the heart, supplemented by a small part of the right atrium, whilst the left ventricle is only prominent *inferiorly*, along the left margin, and at the cardiac apex. The atria are essentially to the right and posterior to their respective ventricles.

Despite some inevitable repetition, it is hoped that the foregoing remarks will assist an elementary understanding of the heart before pursuing a more detailed appraisal of its size, position, shape and parts. Additionally, such general dispositions as these are of the greatest importance when planning or interpreting radiographs, angiocardiograms, and echocardiograms.

Cardiac Size, Shape and External Features

The heart is a compact, hollow, fibromuscular organ of a somewhat conical or pyramidal form possessing, therefore, a base, apex, and a series of surfaces and 'borders' (6.14, 16, 17); it occupies the middle mediastinum between the lungs and pleurae (6.15, 16, 17), and is enclosed in the pericardium (p. 634). It is placed *obliquely* in the chest behind the body of the sternum and adjoining costal cartilages and (left) ribs; approximately one-third lies to the right of the median plane and two-thirds to the left.

An average adult heart measures about 12 cm from base to apex, 8 to 9 cm transversely at its broadest part, and 6 cm anteroposteriorly. Its weight, in the male, varies from 280 to 340 gm (average 300 gm), while in the female, from 230 to 280 gm (average 250 gm). It is held that cardiac weight is about 0·45 per cent of body weight in the male and 0·40 per cent in the female (Hudson 1965); adult weight is achieved between 17 and 20 years.

The oblique disposition of the heart has been mentioned in the previous section and may be emphasized by continuing the analogy of a rather deformed pyramid the base of which faces

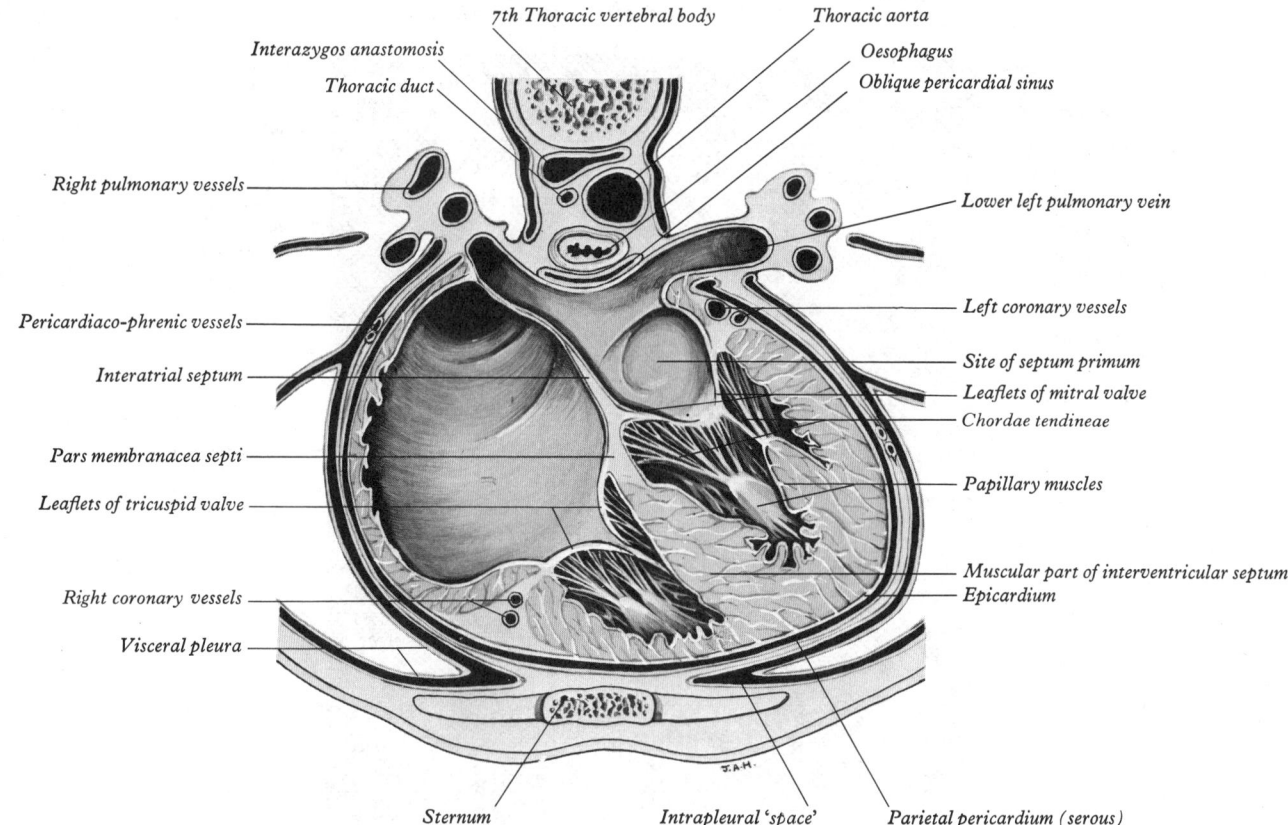

7th Thoracic vertebral body — *Thoracic aorta*
Interazygos anastomosis — *Oesophagus*
Thoracic duct — *Oblique pericardial sinus*

Right pulmonary vessels

Lower left pulmonary vein

Pericardiaco-phrenic vessels

Left coronary vessels

Interatrial septum

Site of septum primum
Leaflets of mitral valve
Chordae tendineae

Pars membranacea septi

Papillary muscles

Leaflets of tricuspid valve

Right coronary vessels

Muscular part of interventricular septum
Epicardium

Visceral pleura

Sternum   *Intrapleural 'space'*   *Parietal pericardium (serous)*

6.15A   Transverse section through the heart, pericardium and related structures at the level of the lower border of the body of the seventh thoracic vertebra to show the principal dispositions of the cardiac chambers and septa, and surrounding structures, viewed from below.

posteriorly and to the right, the apex pointing anteriorly and to the left. Thus a line joining the apex and approximate centre of the base, if projected anterolaterally would pass through the ventral chest wall near the left mid clavicular line; projected postero-laterally it would emerge in the neighbourhood of the right mid-scapular line. Some aspects of the cardiac 'pyramid' are flattened, others presenting a gentle or more pronounced convexity, these various aspects often merging into each other along rather ill-defined boundaries. Clearly, such dispositions render the precise definition of the extent of named surfaces and intervening 'borders' difficult: in the following pages account will be taken not only of the official nomenclature (*Nomina Anatomica* 1977) but in many instances, terms stemming from widespread clinical practice will be given as alternatives. The heart is described as possessing an (anatomical) base and apex; its *surfaces* are designated—sternocostal (anterior), diaphragmatic (inferior), right, and left (pulmonary); its *borders* are termed—upper, inferior (the 'acute' margin or border), and the left (the 'obtuse' margin or border). (It may be noted here that some authorities persist in naming the right surface as a 'border', despite its extensive nature; furthermore, the right surface is unquestionably as 'pulmonary' as the left surface. A further avoidable source of confusion is the use of the term 'posterior' surface for the unambiguous terms diaphragmatic or inferior surface; if posterior is to be used to denote a cardiac surface it should be reserved for its anatomical base. Finally, as will be discussed more fully below, there are a number of different usages of the term 'cardiac base' by different authorities.)

CARDIAC SURFACE GROOVES AND SULCI

The division of the heart into its principal four chambers, i.e. the right and left atria and the right and left ventricles, is indicated externally by a series of grooves or sulci; some of these are deep and obvious, lodging other prominent structures; others are less

distinct and sometimes barely perceptible, whilst some of the grooves or sulci are completely obscured from surface view in part of their extent by major structures which cross them.

The atria are separated from the ventricles by the *coronary (atrioventricular) sulcus* (**6.**16A, B, 17, 27, 28, 39) which contains the principal trunks of the coronary vessels; the sulcus is obliquely set and a study of its position and plane are quite fundamental to any understanding of cardiac anatomy. On the sternocostal surface it proceeds downwards and to the right separating the right atrium (and its auricular appendage) from the oblique right margin of the right ventricle and its infundibulum; the upper left part of the sternocostal course of the coronary sulcus is, however, obscured from the surface where it is crossed by the root of the pulmonary trunk (and deep to this the root of the aorta). Continuing to the left and posteriorly, it curves around the 'obtuse margin' of the heart and proceeds downwards and to the right separating the atrial base of the heart from its diaphragmatic surface. Finally, the posterior and sternocostal aspects of the coronary sulcus curve around the 'acute margin' of the heart at the latter's lower right extremity, where they become confluent. Thus, both the sternocostal and posterior parts of the coronary sulcus pass from high on the left aspect of the heart, downwards and to the right, meeting as indicated, at its lower right extremity: however, the posterior part of the sulcus is a little to the left of its sternocostal element. It will be clear, therefore, that a section through the heart which includes the coronary sulcus will be obliquely set at some 45° to the sagittal plane and at a rather less acute and variable angle to the transverse and coronal planes. Such a section, as will be detailed further below, will pass through the general locations of the lines of attachment of the atrioventricular (mitral and tricuspid) valves, and also (less precisely) those of the aortic and pulmonary valves (**6.**23, 24); further, a line at right angles to the centre of the plane of section will pass obliquely forwards to the left, and a little downwards, towards the cardiac apex.

Common carotid arteries Internal jugular vein

Internal jugular vein

Vagus nerve {

Inferior thyroid vein

Phrenic nerve

Internal thoracic
artery

Superior vena cava

Fibrous pericardium

Serous pericardium

Ascending aorta

Right atrium

Horizontal fissure
of lung

Diaphragm

Jugular lymph
trunk

Subclavian vein

Subclavian artery
Left superior
intercostal vein
Vagus nerve

Phrenic nerve

Pulmonary trunk

Right ventricle

6.16A   A dissection to display the heart, great vessels and lungs *in situ*. The sternum and the sternal ends of the costal cartilages, together with the parietal pleura on each side, have been excised, and the mediastinal pleura and parietal layer of the pericardium over the sternocostal surface of the heart have been removed. The lungs have been displaced to expose the heart, and the epicardium dissected off the heart and roots of the great vessels.

On the right side, the inferior cardiac branch of the vagus nerve descends between the brachiocephalic artery and the right brachiocephalic vein. On the left side, a communication descends from the left superior intercostal vein and crosses the aortic arch and the left pulmonary artery to become continuous with the oblique vein of the left atrium.

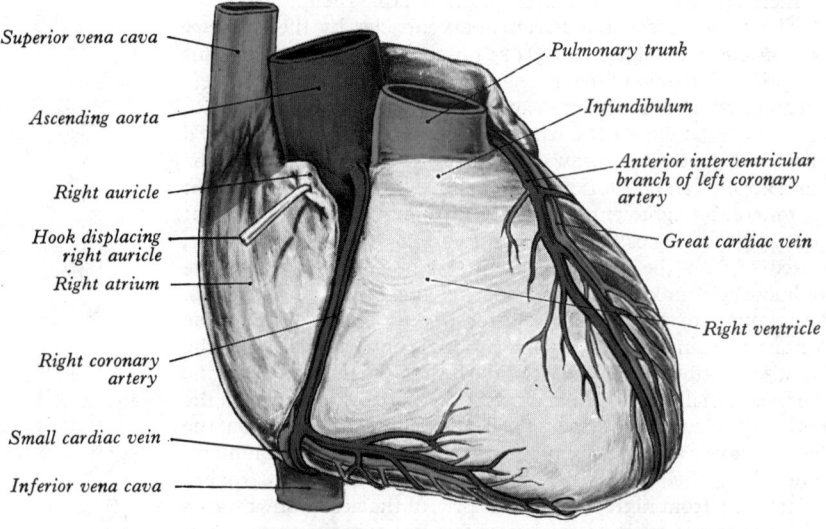

Superior vena cava

Ascending aorta

Right auricle

Hook displacing
right auricle

Right atrium

Right coronary
artery

Small cardiac vein

Inferior vena cava

Pulmonary trunk

Infundibulum

Anterior interventricular
branch of left coronary
artery

Great cardiac vein

Right ventricle

6.16B   The anterior or sternocostal surface of the heart.

The ventricles are separated internally by the interventricular system (p. 649) and its mural margins are indicated superficially by two linear depressions—the *anterior* and *inferior (diaphragmatic) interventricular grooves*. The former is situated on the sternocostal surface of the heart quite near, and almost parallel to, its left ventricular obtuse margin; the latter is placed on the diaphragmatic surface nearer its acute border. The interventricular grooves extend from the coronary sulcus (i.e. the ventricular 'base') to a notch, the *apical incisure*, sited on the acute margin a little to the right of the true cardiac apex.

CARDIAC BASE, APEX, SURFACES, BORDERS

The *posterior aspect* or *base* (**6.17**) is somewhat quadrilateral in form centrally, but with curved lateral extensions beyond the quadrangle; it faces backwards and to the right, and is separated from the thoracic vertebrae (fifth to eighth in the recumbent, sixth to ninth in the erect posture) by the pericardium, right pulmonary veins, oesophagus and aorta. It is formed mainly by the left atrium, and, to a small extent, by the posterior part of the right atrium. Above, it extends as high as the bifurcation of the pulmonary trunk; below it is bounded by the posterior part of the coronary sulcus, containing the coronary sinus and rami of the coronary arteries (p. 669). On the right it is limited by the rounded right surface of the right atrium, and on the left by the rounded left surface of the left atrium. The four pulmonary veins, two on each side, open into the left atrium, whilst the superior vena cava opens into the upper part, and the inferior vena cava into the lower part of the right atrium. The portion of the left atrium between the openings of the right and left pulmonary veins constitutes the anterior wall of the oblique sinus of the pericardium (p. 635). It should be noted that the foregoing description applies to the true *anatomical base* of the whole heart *in situ*. Some confusion arises, however, because of alternative current usages of the term 'base'. Certain authorities apply the term to that aspect of the ventricular heart viewed after section through the coronary sulcus (**6.29A**): the qualified term *base of the ventricles* should be reserved for this aspect. Finally, in clinical practice, it is common to refer to auscultation or other forms of examination in or near the parasternal parts of the second intercostal spaces as applied to the *clinical 'base'* (as opposed to the *clinical 'apex'*). These multiple applications of the term, whilst unfortunate, seem likely to persist.

The cardiac apex, formed by the true apex of the conical left ventricle, is directed downwards, forwards and to the left, and the left lung and pleura overlap it anteriorly. It most commonly lies deep to the fifth left intercostal space, near or a little medial to, the mid-clavicular line.

The anterior or *sternocostal surface (6.16B)* is directed forwards, upwards and to the left. It consists of an atrial portion above and to the right, and a ventricular portion below and to the left of the oblique anterior part of the coronary sulcus. The atrial portion is almost entirely formed by the right atrium; the greater part of the left atrium is hidden by the ascending aorta and pulmonary trunk (6.16B), and only a small part of its auricle projects forwards on the left side of the pulmonary trunk. Of the ventricular portion about one-third is formed by the left, and two-thirds by the right ventricle, the line of separation between the ventricles being marked by the anterior interventricular groove. The sternocostal surface is separated by the pericardium from the body of the sternum, the sternocostalis muscles and the third to sixth costal cartilages; owing to the bulging of the heart towards the left side, the part of the surface which lies behind the left costal cartilages is much larger than the part which lies behind the right. The sternocostal surface is also covered by the pleurae and the thin, anterior parts of the lungs, with the exception of a small, triangular area corresponding with the cardiac incisure in the left lung; the latter and its associated pleura are, however, quite variable in extent.

The inferior or *diaphragmatic surface (6.17)*, largely horizontal, is, however, directed downwards and forwards towards the apex. It is formed by the ventricles (chiefly by the left ventricle), and rests upon the central tendon and a small part of the left muscular portion of the diaphragm. It is separated from the base by the posterior part of the coronary sulcus, and is traversed obliquely by the inferior interventricular groove.

The left surface is directed upwards, backwards and to the left. It is formed almost entirely by the left ventricle, but a small part of the left atrium and its left auricle contribute to its formation superiorly. Convex from before backwards and from above downwards, it is widest above, where it is crossed by the coronary sulcus, and narrows to the apex. It is separated by the serous pericardium from the left phrenic nerve and its accompanying vessels, and by the left pleura, in addition, from the deep hollow on the left lung, below and in front of the hilus.

The right surface is formed by the wall of the right atrium; it is gently convex from above downwards and anteroposteriorly. This surface is separated from the mediastinal aspect of the right lung by the serous and fibrous pericardium and by the mediastinal and visceral pleura. The convexity merges below into the short intrathoracic part of the inferior vena cava, and above, into the superior vena cava. The sulcus terminalis curves approximately along the transitional zone between sternocostal and right surfaces.

The *upper border* of the heart is formed by the atria, principally the left; when the heart is viewed from the front this border is hidden by the ascending aorta and pulmonary trunk. The superior vena cava enters the right atrium at its right extremity.

The so-called *right 'border'*, formed by the convex profile of the right atrium, is rounded and almost vertical.

The *inferior border* or *'acute margin of the heart'* is sharp and thin and is nearly horizontal; it extends from the lower limit of the right border to the apex. It is formed mainly by the right ventricle but by the left ventricle near the apex.

The *left border* or *'obtuse margin of the heart'* separates the sternocostal from the left surface; it is rounded, and is formed mainly by the left ventricle, but to a slight extent, above, by the left auricle. It extends from the left auricle obliquely downwards, with a convexity to the left, to the apex of the heart.

The borders of the heart, as indicated above, apart from the inferior, are rounded, indefinite and difficult to define precisely, they are more in the nature of 'aspects' than borders. The right and left surfaces are together referred to as the pulmonary surfaces but it should be noted that much of the sternocostal surface is also related to the lungs and pleurae with the pericardium intervening. The right and left margins of the cardiac *shadow* as seen in a postero-anterior radiograph of the thorax constitute the right and left borders of the radiologist.

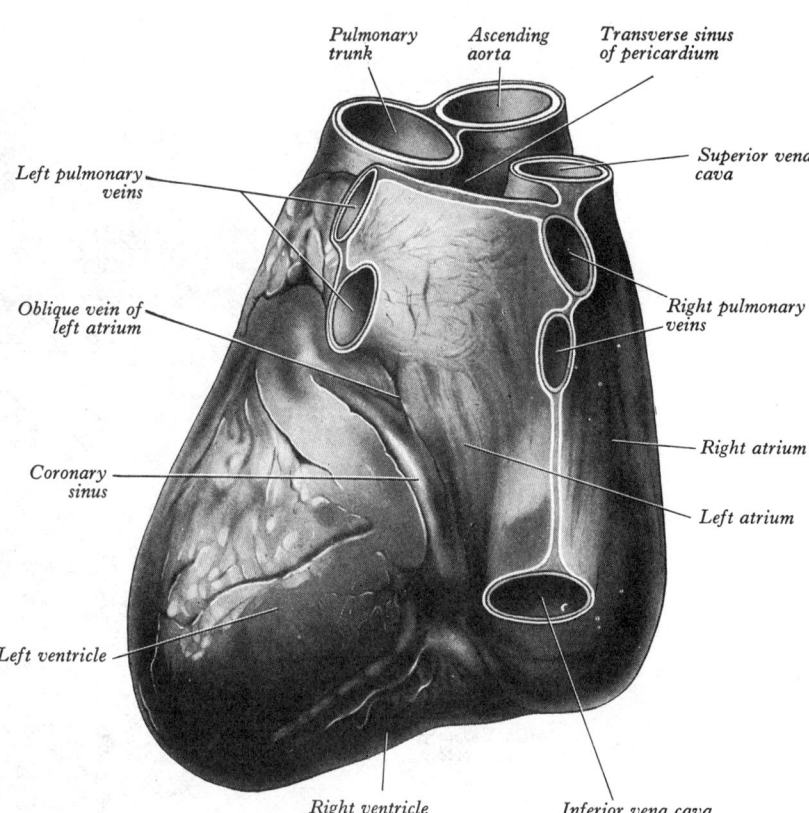

6.17 The base and the diaphragmatic surface of the heart. The serous pericardium is *in situ* and its cut edge is seen around the great vessels.

Cardiac Chambers and Internal Features

In the following sections the chambers of the right heart and the left heart will be described seriatim in terms of their general form, walls and principal internal topographical features. Some aspects of the two hearts, however, have much in common; for example, the structural makeup of the valve cusps or leaflets, chordae tendineae, and papillary muscles of the atrioventricular valves, and cusp architecture of the outflow tract valves, i.e. the pulmonary and aortic valves. Where possible repetition will be avoided.

THE RIGHT ATRIUM (6.16, 17, 18)

As noted above, the *interatrial*, or *atrial septum* is placed obliquely, so that the right atrium, a roughly quadrangular chamber, lies not only to the right of the left atrium, but anterior to it and also extends more inferiorly. Its walls form the right upper part of the sternocostal surface, the convex right (pulmonary) surface, and a small part of the right side of the base of the heart.

The superior vena cava opens into its upper and posterior part and the inferior vena cava into its lower and posterior part. A small, conical, muscular pouch, termed the *auricle*, projects towards the left from its upper and anterior part and overlaps the right side of the root of the ascending aorta. The margins of the auricle are notched, and its interior is encroached on by an irregular muscular reticulum.

In well-fixed hearts the outer surface of the lateral wall of the atrium is marked by a shallow groove, termed the *sulcus terminalis*, which extends between the right sides of the orifices of the superior and inferior venae cavae.

Anteriorly, the right atrium is related to the anterior part of the mediastinal surface of the right lung and is separated from it by the pleura and the pericardium. *Laterally*, it is related to the mediastinal surface of the right lung in front of the hilus but is separated from it by the pleura, the right phrenic nerve and pericardiacophrenic vessels and the pericardium. *Posteriorly* and to the left (6.13B, 15), the right atrium is related to the left atrium

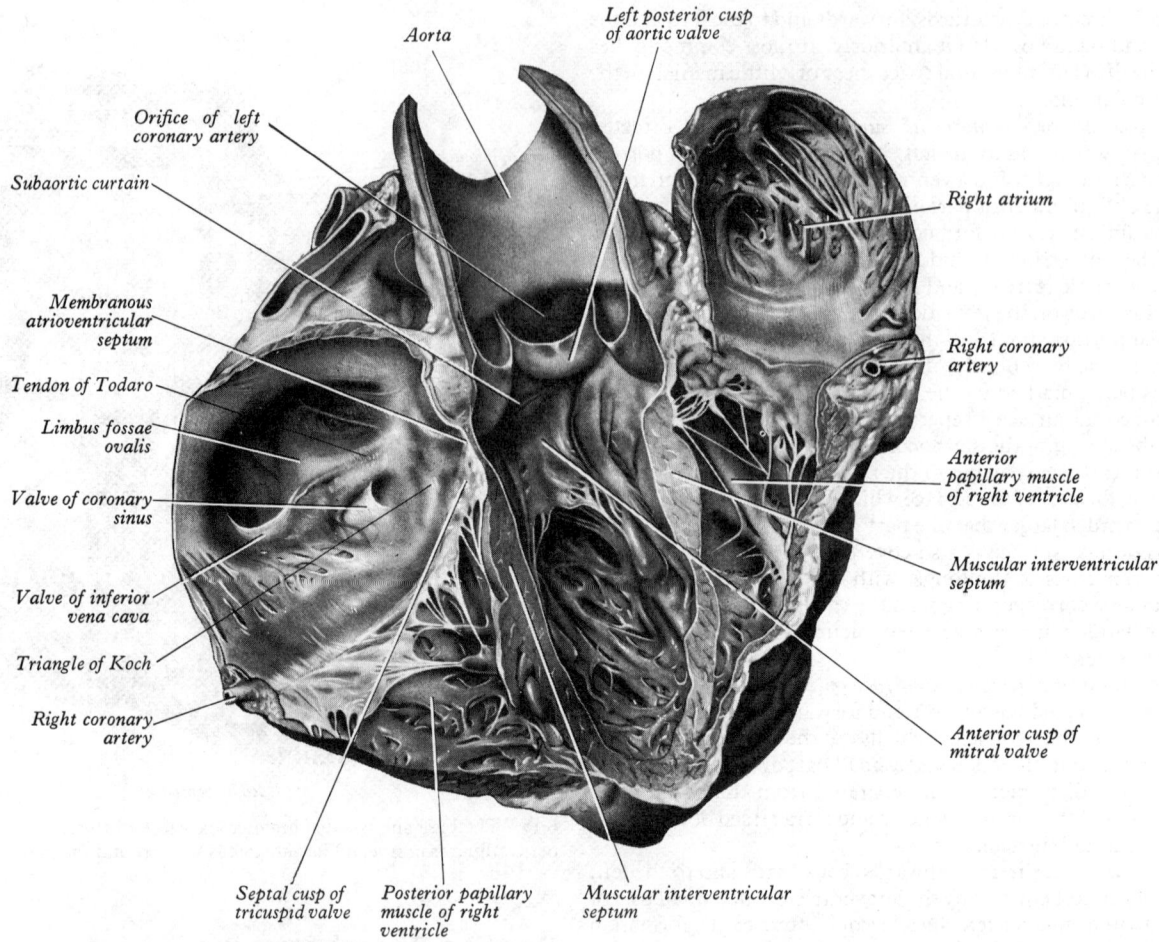

6.18A The interior of the heart revealed by incising it along its right and
lower surfaces and excising the pulmonary trunk and infundibulum. The
rest of the front of the heart has been turned over to the left.

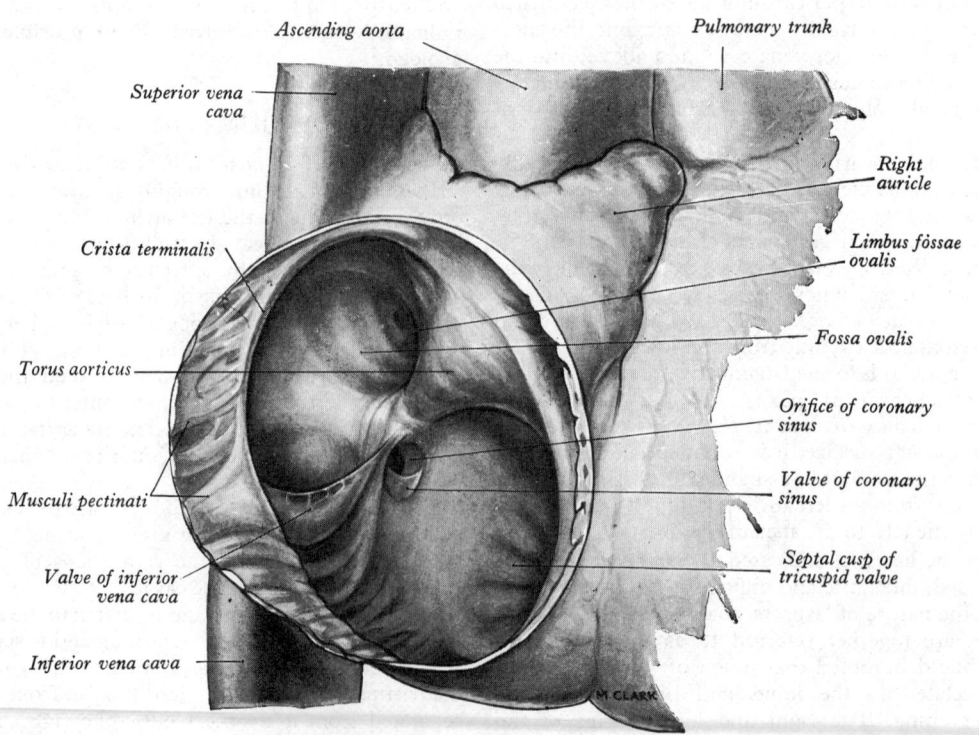

and is separated from it by the interatrial septum; posteriorly and to the right, it is related to the right pulmonary veins. *Medially*, it is related to the commencement of the ascending aorta and, to a lesser extent, to the root of the pulmonary trunk.

The interior of the right atrium (6.18B) presents two main parts for examination: posterior and anterior.

The posterior part, into which the great veins discharge, is derived embryologically from the absorbed right horn of the sinus venosus (p. 183); it has smooth walls and is termed the *sinus venarum*. *Anterior* to this is a portion with ridged walls derived embryologically from the *atrium* proper and which is in continuity anteriorly with the *auricle*. The atrium communicates with the right ventricle through the atrioventricular opening.

The sinus venarum. This region includes the posterior portion and the lateral wall of the cavity as far forwards as the crista terminalis (p. 184). Opening into it are the following vessels:

The *superior vena cava* (6.16B, 18B) returns the blood from the upper half of the body, and opens into the upper and posterior part of the atrium. Its orifice is directed downwards and forwards, and has no valve.

The *inferior vena cava* (6.18B), larger than the superior, returns the blood from the lower half of the body, and opens into the lowest part of the atrium near the interatrial septum. The orifice is skirted anteriorly by a rudimentary semilunar valve, the *valve of the inferior vena cava*. Its convex margin is attached to the anterior margin of the orifice; its concave margin, which is free, ends in two *cornua*, of which the left is continuous with the anterior edge of the limbus fossae ovalis, while the right is lost on the lateral wall of the atrium where it is continuous with the lower end of the crista terminalis. The valve is formed by a duplication of the endocardium of the atrium, enclosing a few muscular fibres. During intrauterine life this valve is of large size, and serves to direct the flow from the *inferior vena cava* into the *left atrium* through an opening, the *foramen ovale*, in the interatrial septum. The valve varies considerably in size and is sometimes cribriform or filamentous; occasionally it is absent.

The *coronary sinus* (6.17) returns the greater part of the blood from the substance of the heart. Its opening is between the orifice of the inferior vena cava and the atrioventricular opening, and is protected by a thin, semicircular *valve of the coronary sinus* (6.18B), which covers the lower part of the orifice. It prevents the regurgitation of blood into the sinus during the contraction of the atrium. This valve may be double or cribriform.

The *foramina venarum minimarum* are the orifices of minute veins (*venae cordis minimae*), which return a small quantity of blood directly from the substance of the heart (p. 737). They are more numerous on the septal wall than elsewhere. Other small orifices opening into the atrium are those of the anterior cardiac veins and sometimes of the right marginal vein (p. 737).

The *intervenous tubercle* is a small variably developed projection on the posterior wall of the atrium, just below the orifice of the superior vena cava. It is distinct in the hearts of quadrupeds, but in man is often scarcely visible. During intrauterine life it is much more prominent and may direct the blood from the superior vena cava towards the right atrioventricular opening.

The atrium proper and the auricle. This region is separated from the sinus venarum by the *crista terminalis*, a smooth, muscular ridge, placed mainly on the lateral wall of the right atrium. It begins on the upper part of the septum and, after passing anterior to the orifice of the superior vena cava, skirts the right margin of that orifice and then extends to the right side of the orifice of the inferior vena cava, where it is connected to the right end of the valve of the latter vessel. It occupies the site of the right venous valve of the embryo (p. 184) and corresponds in position with the sulcus terminalis (p. 641) on the outside of the heart. It indicates the junction between the part of the heart derived from absorption of the right horn of the sinus venosus and the part derived from the original atrium. The upper part of the crista terminalis accommodates the sinuatrial (sinus) node (*see* p. 661), whilst the remainder of the crista is regarded by some to be the route of the posterior internodal tract of specialized conducting tissue (p. 663).

The *musculi pectinati* are nearly parallel muscular ridges which run forwards from the crista terminalis across the lateral and

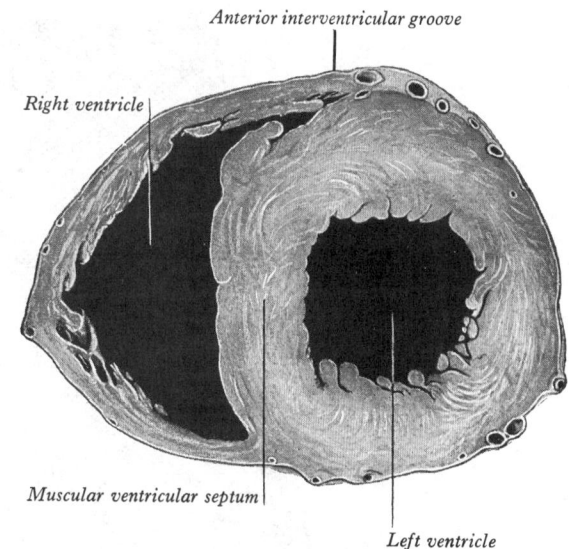

6.19A   A transverse section through the ventricles of the isolated heart, viewed from below. Note that in this illustration the heart is *not* positioned as it would be *in situ*: in the latter position the crescentic 'right' ventricle overlaps most of the anterior surface of the 'left' ventricle.

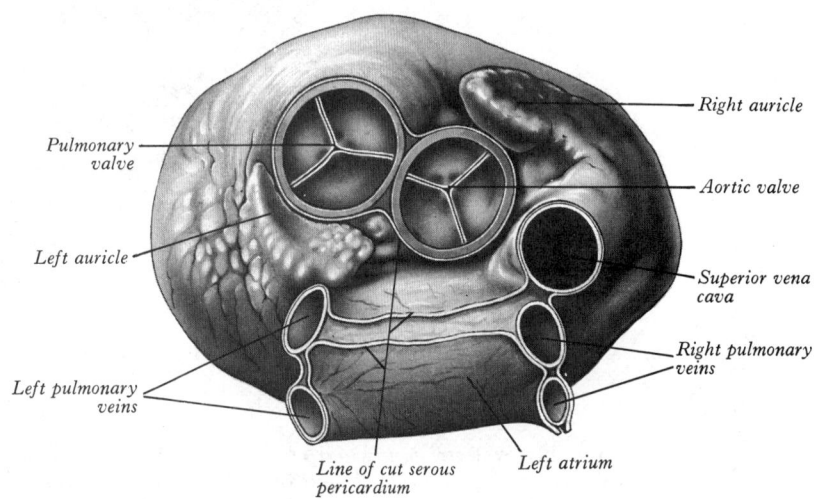

6.19B   The heart, viewed from above. The two continuous white lines which enclose the pulmonary trunk and aorta on the one hand, and the pulmonary veins and the superior vena cava on the other, indicate the continuation of the parietal layer of the serous pericardium with the serous epicardium. The floor of the transverse sinus is seen from above, with the left coronary artery running in it. This diagram, from an earlier edition, has been retained for its pericardial details. However, in some respects it is misleading—the aortic and pulmonary valves are *not* coplanar. The pulmonary valve is distinctly higher than the aortic valve; furthermore the planes of the valves 'face' approximately at right angles to each other.

anterior walls of the right atrium, inclining towards the atrioventricular orifice. In the auricle they are connected to one another so as to form a muscular network.

The septal wall presents the following features:

The *fossa ovalis* (6.18B) is an oval depression on the lower part of the septal wall of the atrium, above and to the left of the orifice of the inferior vena cava. Its floor is formed originally by the septum primum of the fetal heart (p. 184). The *limbus fossae ovalis* is the prominent margin of the fossa ovalis and represents the free edge of the septum secundum (p. 186) of the embryonic heart. It is most distinct above and at the sides of the fossa; below, it is deficient. Its anterior edge is continuous with the left horn of the valve of the inferior vena cava. A small, slit-like valvular opening is occasionally found at the upper margin of the fossa, leading upwards, beneath the limbus, into the left atrium; it is the remains of the foramen ovale between the two atria.

The antero-inferior part of the right atrium is occupied by the

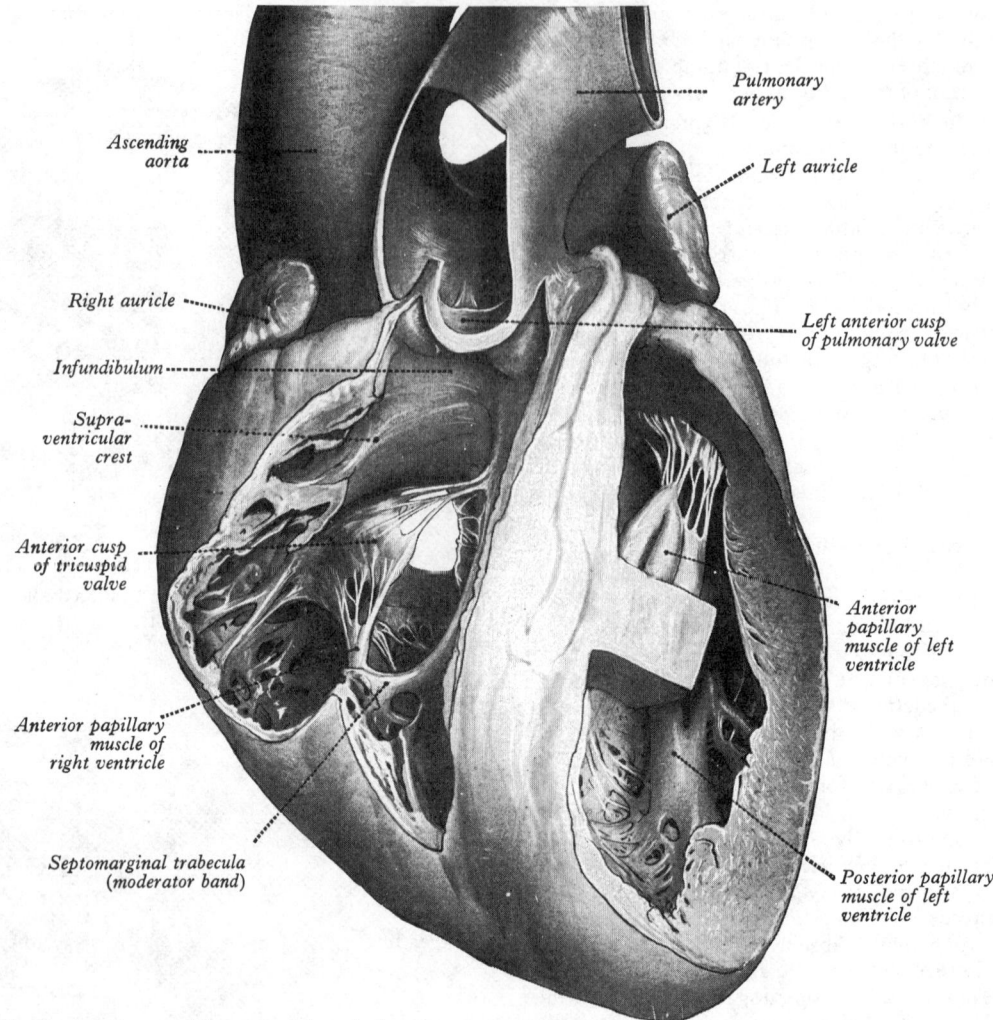

Ascending aorta

Right auricle

Infundibulum

Supra-ventricular crest

Anterior cusp of tricuspid valve

Anterior papillary muscle of right ventricle

Septomarginal trabecula (moderator band)

Pulmonary artery

Left auricle

Left anterior cusp of pulmonary valve

Anterior papillary muscle of left ventricle

Posterior papillary muscle of left ventricle

6.20A   A dissection opening the ventricles, viewed from in front.

large roughly oval *atrioventricular orifice*, the margins of which approximate to the line of attachment of the leaflets of the *tricuspid valve*; both orifice and valve are considered further below.

The roughly triangular zone (*triangle of Koch*—see **6**.18A) bounded by the base of the septal leaflet of the tricuspid valve, the anteromedial margin of the orifice of the coronary sinus, and the rounded collagenous, palpable, subendocardial *tendon of Todaro* which curves from the '*right fibrous trigone*' (*central fibrous body*— see p. 657) towards the medial extremity of the valve of the inferior vena cava, is a landmark of surgical importance, indicating the septal site of the atrioventricular node and its associated juxtanodal bundles of conducting tissue (p. 663). Immediately dorsal to the septal leaflet of the tricuspid valve the septal wall is formed by an extension of the pars membranacea septi which is *atrioventricular* here, since it intervenes between the right atrium and left ventricle.

The anterosuperior part of the septal wall of the right atrium bulges to a variable degree into the atrial cavity as the *torus aorticus*: the torus is caused by the proximity of the right posterior (non-coronary) aortic sinus and cusp (p. 651).

THE RIGHT VENTRICLE

The right ventricle extends from the right atrioventricular (tricuspid) orifice nearly to the cardiac apex and then continues upwards and to the left as a conical pouch, the *conus arteriosus* (*infundibulum*) reach the pulmonary orifice from which the pulmonary trunk arises. On this view, therefore, the right ventricle possesses an inflow orifice and tract, succeeded at an obtuse angle by an outflow tract and orifice; (although some authorities prefer to include an intermediate 'body' of the ventricle, this practice will not be followed here).

In the following paragraphs the external features of the right ventricle will be followed by considerations of the topographical internal features of its inflow and outflow tracts, the tricuspid valve complex, the pulmonary valve, and finally brief functional allusions related to valve operation and flow vectors.

External features: The *anterosuperior surface* of the right ventricle is convex and forms a large part of the sternocostal surface of the heart. In the greater part of its extent it is separated from the chest wall only by the pericardium, but the left pleura and, to a lesser extent, the anterior margin of the left lung are interposed both above and to the left side (**6**.23). Its *inferior surface* is rather flat and is related to the central tendon and the adjoining part of the diaphragm, but is separated from it by the pericardium. Its *left* and *posterior* wall is formed by the ventricular septum, which bulges into the right ventricle, so that when a section is made that is transverse to the midbase-apex axis of the heart, the cavity of the right ventricle presents a crescentic outline (**6**.19A). A collagenous band (**6**.29A, B), the *tendon of the infundibulum* (*conus ligament*) connects the posterior surface of the pulmonary annulus and the muscular infundibulum to the root of the aorta where it blends with the wall and one margin of its right posterior (non-coronary) sinus (p. 651); inferiorly this tendon is continuous with the membranous part of the ventricular septum (p. 649). The wall of the right ventricle varies greatly in thickness (*vide infra*), but is much thinner (3–5 mm on average) than that of the left, the proportion between them being about 1 to 3; it is thickest at its atrial end and gradually becomes thinner towards the apex of the ventricle.

Internal features: The internal surfaces of the inflow and outflow tracts have a completely contrasting topgraphy and the transition between the two occurs superiorly at the supraventricular crest and anteroinferiorly near the apex of the ventricle at the septomarginal trabecula (*vide infra*). The *supraventricular crest* is a massive muscular arch placed between the right atrioventricular and pulmonary orifices (**6.**20A). It is obliquely set, curving forwards and to the right from the interventricular septum (its '*septal limb*') to the right anterolateral ventricular wall (its '*mural or parietal limb*'). The concavity of the arch lies between these limbs in the long axis of the crest; at right angles to this, i.e. between the valve orifices, the crest is highly convex. Posteriorly the crest gives partial attachment to the anterior leaflet of the tricuspid valve. Its septal limb is often continuous with the septal extremity of the septomarginal trabecula (*vide infra*).

The inflow tract has rough walls due to the presence of the *trabeculae carneae* which are round or irregular muscular ridges, columns, bands or protrusions, covered with endocardium, which project into the ventricular cavity: the elevations are, of course, surrounded by or have intervening valleys, grooves or depressions; hence the great variation in thickness of the ventricular myocardium. There are three main types of trabeculae: some, as indicated, are mere ridges, others are fixed to the ventricular wall or septum at their ends but are free in the middle, while a third set, the *papillary muscles* are continuous by their bases with the wall of the ventricle, while their apices project into the cavity and are continuous with the (true) *chordae tendineae*—fine, glistening collagenous cords which, together with the muscles, are part of the tricuspid valve complex (and are further described with it). The *septomarginal trabecula*, a curved muscular band which extends from the lower part of the interventricular septum to the base of the anterior papillary muscle, although variably developed, deserves special mention. Forming the lower limit of the trabeculated inflow tract, it carries the continuation of the right bundle-branch of the specialized conducting tissue (p. 663). Its alternative name of 'moderator band', implied a tenet of the earlier anatomists, namely that it prevented 'over distention' of the ventricle; the latter is of dubious validity.

The outflow tract, conus arteriosus (or *infundibulum*) has smooth walls and leads upwards, to the left, inclining posteriorly to reach the *pulmonary orifice*. In the cadaver it is impossible to recognize the lower limit of the infundibulum on the surface of the heart but it is formed by the free rounded border of the supraventricular crest (**6.**20A) and the upper border of the septomarginal trabecula near the ventricular apex. The infundibulum represents a persistent part of the bulbus cordis which has been incorporated in the right ventricle, and its persistence as the outflow channel of this ventricle is attributable, in part, to the support it provides for the pulmonary valve during ventricular diastole. It has been shown experimentally (in dogs) that, when subjected to increased backward pressure, the pulmonary valve becomes incompetent much more readily than the aortic valve, and it has been inferred that, in life, the muscular walls of the infundibulum retain their contractile tonus throughout ventricular diastole thus providing support for the valve cusps (Brock 1955).

It has been proposed that the trabeculated walls of the inflow tract assist in controlling and slowing the inflow of blood during early diastole, and by their contraction and interdigitation increase the *volumetric efficiency* of ventricular emptying during systole: the smooth walls of the outflow tract are presumed to increase the *velocity* of systolic ejection. (Similar considerations apply to the left ventricle.)

THE TRICUSPID VALVE COMPLEX

It is undesirable that the tricuspid and mitral valves should be described as isolated structural entities, but rather as parts of functionally integrated complexes of structures. The tricuspid valve complex comprises: (*a*) the so-called *tricuspid atrioventricular 'orifice'* and the tissues that make up its associated *tricuspid valve annulus*; (*b*) the *valve leaflets* or *cusps* and their subdivisions; (*c*) the *chordae tendineae* of various types; and (*d*) the *papillary muscles*. Further, it will be clear to the reader that effective action of the foregoing also depends upon harmonious interplay with the activities of the atrial, ventricular and septal myocardia (p. 657), the specialized conducting tissues (p. 659), and the mechanical cohesion provided by the fibro-elastic 'skeleton' of the heart. It will be appreciated that all of these elements undergo substantial

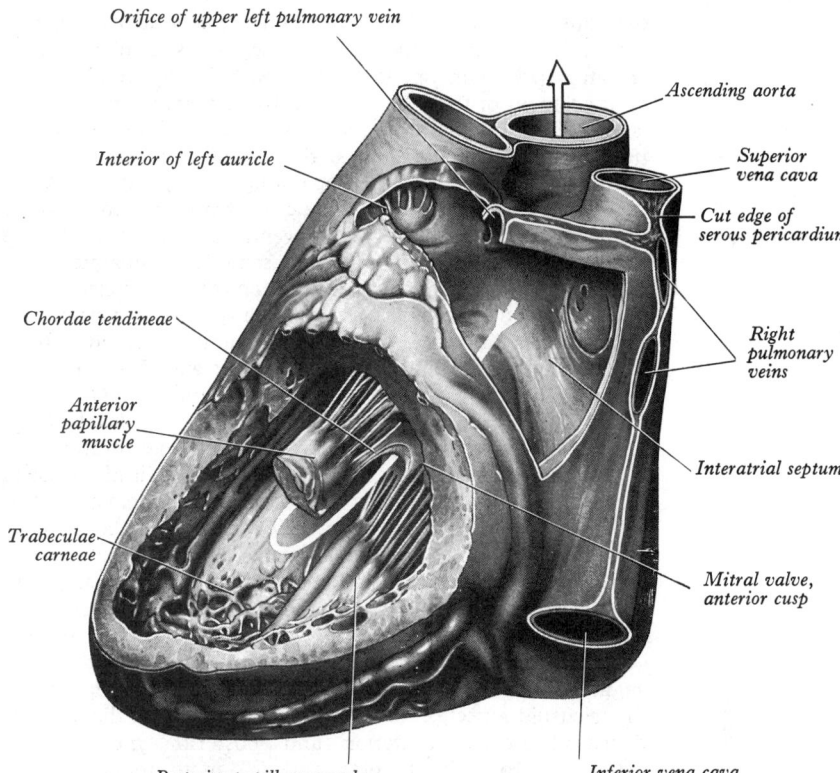

Orifice of upper left pulmonary vein

Ascending aorta

Interior of left auricle

Superior vena cava

Cut edge of serous pericardium

Chordae tendineae

Right pulmonary veins

Anterior papillary muscle

Interatrial septum

Trabeculae carneae

Mitral valve, anterior cusp

Posterior papillary muscle

Inferior vena cava

6.20B A dissection showing the interior of the left side of the heart. The white arrow indicates the course of blood flow from the left atrium through the left ventricle to the aorta.

changes in position, shape, angulation and dimensions throughout the various phases of the cardiac cycle; the *in vivo* condition being the resultant, at any instant, of the state of myocardial contraction or relaxation balanced against the internal pressure profiles and flow patterns of the blood. Thus, either qualitative or quantitative observations made on cadaveric hearts, whether fixed or unfixed, removed or *in situ*, may prove quite misleading; similar strictures apply to many observations made during open heart surgery where the (abnormal) heart has been subjected to severe haemodynamic interference.

The *tricuspid atrioventricular 'orifice'*, the largest of the valve orifices (circumference $11\cdot4 \pm 1\cdot1$ cm in males and $10\cdot8 \pm 1\cdot3$ cm in females according to Silver *et al.* 1971), is best viewed from the atrial aspect (**6.**29A), and is the clearly recognizable line of transition from atrial wall or septum into the bases of the valve leaflets. Thus is it not a true orifice, only becoming so after leaflet removal by dissection; deep to its endocardium lie the various components of the valve annulus. The orifice, as defined above, has been variously described as almost circular, oval, or roughly triangular with rounded angles, doubtless reflecting differences in age, normality, method of preparation and inspection. As detailed below, the margins of the orifice do not lie precisely in a single plane, but for our present purposes, the nearest approximation to such a plane is placed almost vertically at about 45° to the sagittal plane, but also slightly inclined to the vertical so that it 'faces' (on its ventricular aspect) anterolaterally to the left, and also somewhat inferiorly. When roughly triangular in shape, the margins of the orifice are described as presenting anterior, posterior, and septal aspects, corresponding to the bases of the valve leaflets. (It may be noted that the name septal is unambiguous; however, the terms anterior and posterior although having the advantage of simplicity are, because of the near-vertical plane of the orifice, rather misleading—right superolateral and right inferolateral would be more strictly accurate, although cumbersome.)

The *tricuspid valve annulus* is a rather ill-defined term used in different ways by different authorities. In many elementary accounts the view is advanced that all four valve orifices are surrounded by relatively *uniform rings* of collagenous tissue (of different strength and dimensions in the case of each valve), and these rings are interconnected by dense masses of collagen; they are supposed to correspond precisely with the topographical line of attachment of the bases of the valve leaflets, fusing with the lamina fibrosa of the latter, and in the case of the mitral and tricuspid valves, to be situated throughout their extent at the atrioventricular junction (i.e. separating the myocardia of these cavities). However, it has become clear that only some of these propositions are true: with the atrioventricular valves, whilst it is undeniable that connective tissue in some form separates atrial and ventricular myocardium around the complete periphery of the associated 'orifice', it varies greatly in its density and disposition in different regions of the circumference, and with the sex and age of the individual (Walmsley and Watson 1978). Further, the topographical line of 'attachment' of the *free* valve leaflet does not correspond *at all points* to the level of the atrioventricular connective tissue (although it does throughout a considerable part of its extent). Some authorities limit the use of the term valve annulus simply to the line of attachment of the free leaflet: it should be appreciated, however, that the collagenous central stratum of the leaflet may, in some regions, pass basally and then subendocardially for some distance before merging with the atrioventricular connective tissues concerned. The latter will be considered to constitute the valve annulus in the present account, and in the case of the tricuspid valve they include: (*a*) an angulated line crossing the membranous part of the septum, (*b*) the tricuspid aspect of the right fibrous trigone, (*c*) the tapering dense fibro-elastic 'anterior' and 'posterior' *fila coronaria* extending from the trigone, and (*d*) more tenuous sulcal connective tissue completing the annular circumference between the tips of the fila. These elements will be discussed further below and are illustrated in **6.29B**.

The *tricuspid valve leaflets*, customarily regarded as three in number and hence the name of the valve, are termed *septal*, *anterior* and *posterior*, corresponding to the marginal sectors of the atrioventricular orifice similarly named, and each consists of a

reduplication of the endocardium within which is sandwiched a collagenous *lamina fibrosa*. The latter is continuous marginally and on its ventricular aspect, with the diverging fascicles of the chordae tendineae (*vide infra*), whilst basally is confluent with the connective tissue of the valve annulus. Numerous earlier anatomists described the occurrence of small 'accessory leaflets' or 'islands of leaflet tissue' between the major ones, but recent critical quantitative reviews of tricuspid valve architecture have largely rejected the need for such a view (*see* Silver *et al.* 1971). The leaflet tissue of the tricuspid (and mitral—*see* p. 649) valve may be regarded as a *continuous curtain* arising from or near the valve annulus, descending into the ventricular cavity, but presenting three principal grades of indentation in its free border, dividing the tissue into definable subregions. The deepest indentations are bays projecting towards the basal attachment of the principal, smoothly arched, *valve commissures* which separate the main leaflets and are therefore termed anteroseptal, posteroseptal, and anteroposterior. (Thus, a particular region of a leaflet may be designated, for example, 'the anteroseptal half of the septal leaflet'.) The commissures can be identified with precision in most cases (although not as precisely as in the case of the mitral valve) by their characteristic fan-shaped chordae tendineae, and other adjacent anatomical features (Silver *et al.* 1971). Intermediate marginal indentations or '*clefts*', possessing smaller fan-shaped chordae divide the posterior leaflet into three *scallops*, while one small '*notch*' is usually present on the margin of both the septal and the anterior leaflet. Thus viewed from either the atrial or ventricular aspect, the true valve orifice presents a highly irregular profile.

Simple palpation, transillumination, and histology reveal that (except for those parts extending on to the membranous septum) all the tricuspid leaflets possess three *zones*, named from the free margin, rough, clear and basal. As the name implies, the *rough zone* is relatively thick, opaque, uneven on both, but particularly the ventricular aspect; the latter receives the insertion of the majority of the chordae tendineae, whilst the obverse 'atrial' or

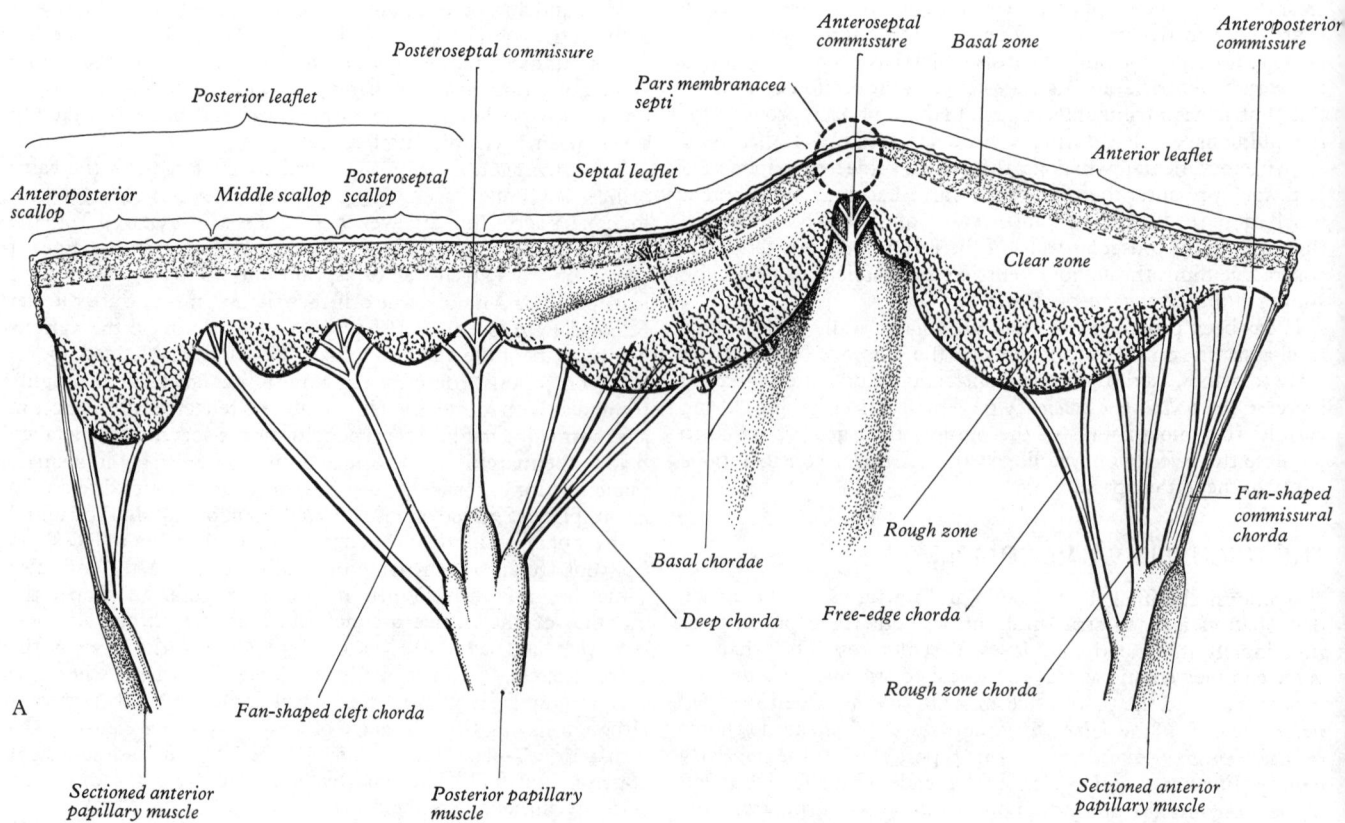

6.21 A, B Diagrams of the tricuspid (A) and mitral (B, facing page) valves, laid flat after division of their fibrous annuli with the atrioventricular axis placed vertically, showing the relative shapes, dimensions and zones of the leaflets and their subdivisions; also the varieties of chordae tendineae. Drawn to scale. A—after Silver *et al.* (1971); B—after Lam *et al.* (1970). Consult text for further details.

'inflowing' aspect of the zone is that area making contact with another leaflet during full valve closure. The *clear zone* is smoother, translucent, receives relatively few chordae, and has a more delicate lamina fibrosa. The *basal zone*, some 2–3 mm in extent from the peripheral attachment, is again thicker, possesses more connective tissue, is vascularized and frequently contains tails of atrial myocardium.

The *anterior leaflet* is the largest and in the main is attached to the atrioventricular junction on the posterolateral aspect of the supraventricular crest, but it extends along its septal limb to reach the membranous septum, ending at the anteroseptal commissure: one or more notches indent its margin. The *septal leaflet* is the smallest, its attachment passing from the posterior ventricular wall across the muscular septum and then angulating over the membranous septum to the anteroseptal commissure. The *posterior leaflet* is wholly mural in its attachment. Diagram **6**.21 A, based on the researches of Silver *et al.* (1971), summarizes some of the main features of the tricuspid valve leaflets, their attachments and chordae: the heart has been incised along its acute margin, opened, and the valve annulus cut through the anteroposterior commissure and laid flat, with the base-apex ventricular axis placed vertically. Note that the posterior leaflet and posteroseptal half of the septal leaflet are horizontal, whereas the anteroseptal half of the septal leaflet and anterior leaflet slope upwards to meet on the surface of the membranous septum at the anteroseptal commissure. The distribution of commissures, clefts, notches, zones, and the pleated form of the septal leaflet are clearly shown, and are drawn to scale.

The *chordae tendineae* are, as noted, glistening white fibrous collagenous cords: as in the left ventricle the majority are *true chordae* associated with the atrioventricular valve complex. The *false chordae* that merely pass between papillary muscles, from papillary muscles to ventricular wall, or between two points on the ventricular wall, are irregular in numbers, dimensions and of no known functional significance – they will not be further discussed here. The *true chordae* usually arise from small projections from the tips or margins of the apical thirds of the papillary muscles, but on occasion from the papillary bases or general ventricular (including septal) walls, and are inserted into some aspect of the valve leaflets, their clefts, commissures or bases. They were early classified by Tandler (1913) into first, second and third order chordae according to their distance of insertion from the free margins of the leaflets: most subsequent authors followed this tripartite classification which, however, possessed little functional or morphological merit. Following an extensive quantitative study of the mitral valve chordae and leaflets, Lam *et al.* (1970) proposed a new and much more generally useful classification, and this was soon applied, with additions, by Silver *et al.* (1971) to the tricuspid valve. This will be followed briefly here: five principal classes of chordae are described in relation to the tricuspid valve complex and have been named *fan-shaped, rough zone, free edge, deep,* and *basal* chordae.

Fan-shaped chordae, as the name implies, possess a short single stem from which a series of branches radiate to attach, in the case of the stouter fans, to the margins (or a short distance on the ventricular aspect) of the main interleaflet commissures (i.e. *commissural chordae*) and also to the extremities of the adjacent leaflets; the more delicate fans have a similar attachment to the inter-scallop clefts of the posterior leaflet (*cleft chordae*). *Rough zone chordae* again arise from a single stem which splits into three soon after its origin, one cord inserting into the free margin of a leaflet; another passes on to the ventricular aspect of the rough zone to insert at its outer limit which corresponds to the line of leaflet contact in full valve closure; the third cord terminating at some intermediate point. *Free edge chordae* are single, threadlike, often long chordae, arising either from the apex or base of a papillary muscle and having a marginal leaflet attachment usually near the mid-point of a major leaflet or one of its scallops. *Deep chordae*, also long, pass deep to the free edge and, sometimes bi- or tri-furcating just before insertion, they reach the more peripheral part of the rough zone, or even the clear zone. *Basal chordae*, the most variable in shape, being either round cords, flattened

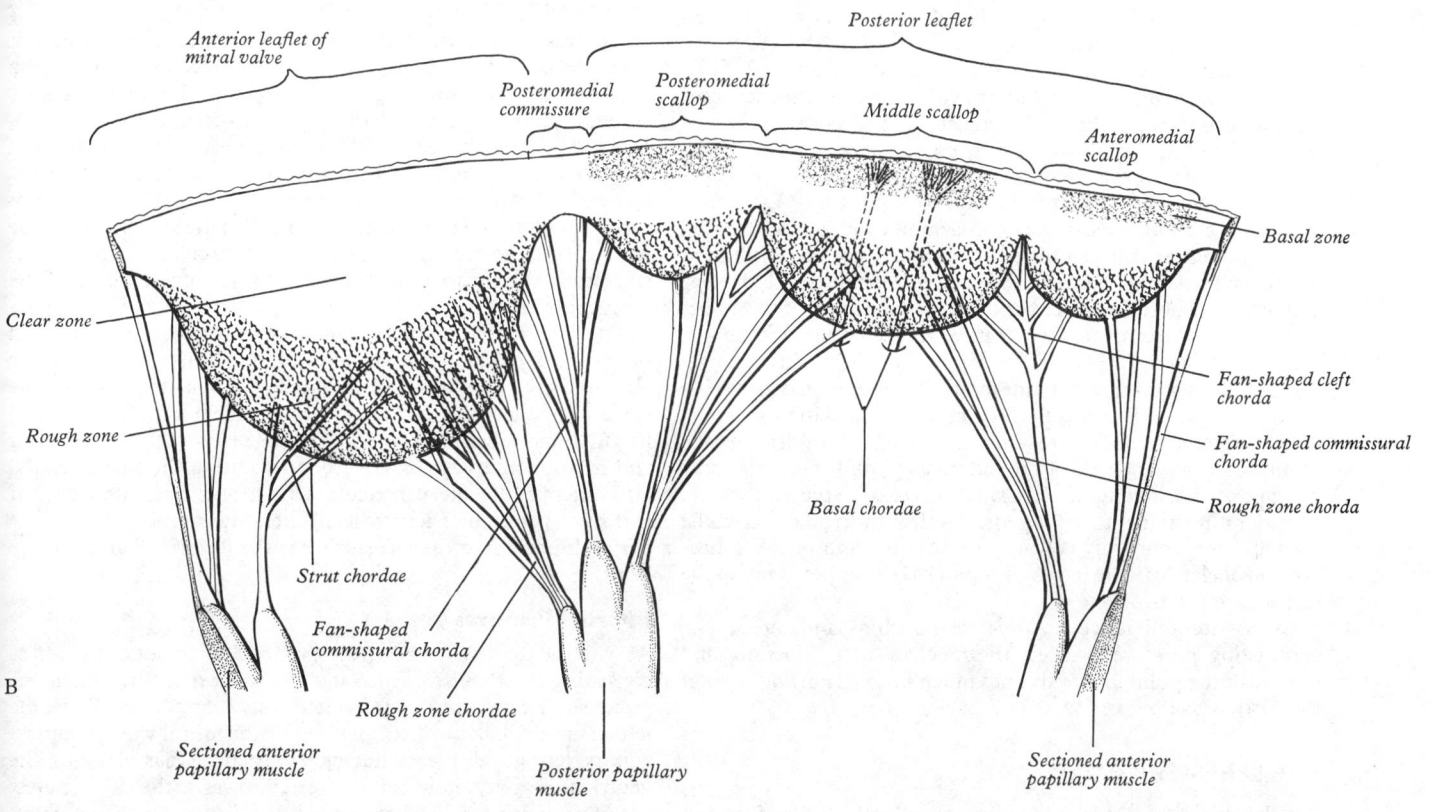

647

ribbons, long and slender, or short and muscular, arise directly from either smooth or trabeculated ventricular wall and insert into the basal 2 mm of the leaflet. (Consult the references quoted above for quantitative details and bibliographies and appertaining to the tricuspid leaflets and chordae tendineae.)

The *papillary muscles* of the right ventricle comprise two principal ones, *anterior* and *posterior*, and a less prominent, variable set of *septal* muscles. The *anterior papillary muscle* is the largest; its base arises from the right anterolateral ventricular wall below the anteroposterior commissure and blends with the right extremity of the septomarginal trabecula (*vide supra*); its apex, usually single but occasionally bifid, gives origin (both apically and subapically) to chordae tendineae which insert into *corresponding zones* of the anterior and posterior tricuspid leaflets. The *posterior papillary muscle* arises from the ventriculo-septal myocardium below the posteroseptal commissure; it is frequently bifid or trifid and its chordae are distributed to the septal and posterior leaflets and their associated commissure and clefts. 'Accessory' *septal papillary muscles* are occasionally absent or merely represented by irregular fibrous cords; more commonly a group of small papillary projections arise from the septal wall of the infundibulum below the level of the crista supraventricularis and are located on the apical aspect of the anteroseptal commissure—their chordae tether the adjoining parts of the anterior and septal leaflets, and the highest muscle of the group is often termed the *papillary muscle of the conus*.

The mechanism of closure of the tricuspid valve has much in common with the mitral valve and since there is firmer evidence with regard to the latter, this will be considered later (p. 650).

THE PULMONARY VALVE

The pulmonary valve, the outflow valve of the right ventricle, surmounts the infundibulum and is somewhat separated from the other three principal cardiac valves, lying anterior and superior to them; its general plane faces superiorly, to the left, and slightly posteriorly. It consists of three semilunar segments or *cusps* (*valvules*), which are attached by their convex margins to the triple-scalloped fibrous thickening in the wall of the pulmonary trunk at its junction with the ventricle and constitutes the valve annulus (*see* p. 657 and **6.**29B); their free borders are directed upwards into the lumen of the vessel. Two of the cusps are situated anteriorly (right and left), the third is posterior. (The cusps of the pulmonary and aortic valves are, in this account, named according to their approximate *in situ* positions in the normal adult heart; this differs from the *Nomina Anatomica* where they are named according to their positions in the fetus before rotation of the heart to its definitive position.) Each cusp consists of a duplication of the endocardium, with an intervening *lamina fibrosa* of collagen fibres variably developed in different locations; the lamina fibrosa is well developed along both the free and attached borders, and at the latter blends with the fibrous 'annulus'. In the centre of the free margin a localized thickening of collagen, the *nodulus* (Arantii) occurs, from which fibres radiate through the lamina fibrosa to the attached margin; on each side of the nodule, however, the collagen is much reduced in the narrow, more delicate, crescentic regions, the *lunules*. Opposite to the semilunar cusps the wall of the pulmonary trunk presents three dilatations, the *sinuses* of Valsalva; the walls of the sinuses are considerably thinner than the general wall of the trunk, and whilst they are predominantly collagenous near the annulus, the proportion of elastic tissue increases rapidly as the upper zones of the sinuses are reached.

Except for differences in time relationships and, of course, contrasting pressure profiles, the mechanism of opening and closure of the pulmonary valve has much in common with that of the aortic valve (*see* p. 652, **6.**33).

THE LEFT ATRIUM

Smaller than the right atrium volumetrically, the walls of the left atrium are, however, thicker (some 3 mm on average); it will be recalled that the cavity and walls of the left atrium are formed to a large extent by the proximal parts of the pulmonary veins which

are incorporated during its development (p. 184). The left atrium is roughly cuboidal and extends to the right *behind* the right atrium from which it is separated by the obliquely set interatrial septum (*see* p. 643 and **6.**15). Thus the right atrium forms its right anterolateral relation; anteriorly, and to the left, it is concealed by the roots of the pulmonary trunk and aorta, part of the transverse sinus of the pericardium intervening between the atrium and the arterial trunks. Antero-inferiorly and to the left, it abuts against the base of the left ventricle where it presents the left atrioventricular orifice (*vide infra*). Its posterior aspect forms most of the anatomical base of the heart; it is approximately quadrangular in shape, receives the terminations of (usually) two pulmonary veins on each side, and provides the anterior wall of the oblique sinus of the pericardium (p. 635).

The *auricle* is somewhat constricted at its junction with the atrium; it is longer, narrower and more curved than that of the right atrium, and its margins are more deeply indented. It is directed forwards on the left side of the pulmonary trunk, and overlaps the commencement of this vessel. (The intimate relation of the auricle to the stem and initial branches of the left coronary artery are mentioned on p. 671).

The interior of the left atrium (**6.**20B) presents several features for examination:

The *pulmonary veins*, four in number, open into the upper part of the posterior surface of the atrium—two on each posterolateral aspect; their orifices are smooth, oval and are not provided with valves. The two left veins frequently end via a common opening.

The *left atrioventricular orifice* is described below.

The *foramina venarum minimarum* are the orifices of minute veins (venae cordis minimae) which return blood from the muscular substance of the heart.

The *musculi pectinati*, fewer and smaller than those in the right atrium, are confined to the inner surface of the auricle.

On the atrial septum a lunate impression may be seen, bounded below by a crescentic ridge, the concavity of which is directed upwards; it corresponds to the site of the ostium secundum of the developing heart (p. 184).

THE LEFT VENTRICLE

General and External Features

In contrast to the right ventricle, the construction of the left ventricle accords with its role as a powerful expulsive pump capable of sustaining a pulsatile flow in the high-pressure systemic arterial system. Variously described as a half-ellipsoid, or roughly cone-shaped, it is longer and narrower than the right ventricle, extending from its ventricular base which lies in the plane of the coronary sulcus, to the cardiac apex. The long axis of the cone points downwards, forwards and to the left. In transverse section its cavity presents an oval or nearly circular outline (**6.**19A) with walls about three times as thick (8–12 mm) as those of the right ventricle.

The left ventricle forms part of the sternocostal, left, and inferior (diaphragmatic) surfaces of the heart. Except where obscured by the aorta and pulmonary trunk, the base of the ventricular cone is superficially separated from the left atrium and its auricular appendage by part of the coronary sulcus; anterior and inferior interventricular grooves indicate the line of mural attachment of the interventricular septum and the limits of the left and right ventricular territories. Where the sternocostal surface curves bluntly into the left surface is the 'obtuse' margin of the heart.

Internal Features (**6.**20A, B)

As with the right ventricle, the left ventricle may be considered as consisting of an *inflow orifice* and *tract* succeeded by an *outflow orifice* and *tract*, with their attendant valve complexes. Thus, the left atrioventricular orifice, guarded by the mitral valve complex, admits left atrial blood during ventricular diastole into the ventricular cavity, flow being directed towards the cardiac apex (i.e. the *inflow vector*). After mitral valve closure, and throughout the ejection phase of ventricular systole (**6.**35) blood is expelled along a path from the apex towards and through the aortic orifice, thus defining the *outflow vector*. It should be noted here, that in complete contrast to the orifices of the right ventricle, the inflow and outflow orifices of the left ventricle are in *close contact*; in part of their extent they are merely separated by the subaortic curtain intervalvar septum—*vide infra*) and by the anterior leaflet of the mitral valve; further, the inflow and outflow vectors are acutely angled with respect to each other.

The anterolateral wall of the left ventricle is formed by the interventricular septum (i.e. the septal wall as opposed to the 'free' ventricular wall) which is thick and muscular throughout most of its extent. The muscular septum is concave towards the left ventricular cavity, its convexity forming the posteromedial profile of the (sectioned) right ventricle: it thus completes the myocardial circular outline of the left ventricle mentioned above (6.19A). However, as the aortic orifice is approached, the interventricular septum becomes thin and collagenous, the *pars membranacea septi*, an oval to round area situated below and partly confluent with the fibrous supports of the anterior (right coronary) and right posterior (non-coronary) cusps of the aortic valve (for further comments *see* pp. 651, 653, 657 and 6.29B).

Between the lower limits of the free margins of the mitral valve leaflets and the ventricular apex, including the apical two-thirds of the interventricular septum, the muscular walls are intensely trabeculated. These *trabeculae carneae* are stouter and more intricate than, but of the same three varieties as, those of the right ventricle (mere ridges, terminally attached but free centrally, or papillary muscles). They form a particularly dense labyrinthine interlacement towards the apex of the ventricle. The endocardial walls of the outflow tract (as in the right ventricular infundibulum) are, however, smooth; additionally, the subendocardial tissue in the walls of the final part of the outflow tract, the *aortic vestibule*, becomes increasingly fibrous with a corresponding reduction in myocardial tissue.

THE MITRAL VALVE COMPLEX

Many of the *general* comments concerning the tricuspid valve apply equally to the mitral valve, and to avoid excessive repetition, remarks will be confined to particular features of the following components of the mitral complex: (*a*) the *mitral atrioventricular 'orifice'* and its associated *valve annulus*; (*b*) *valve leaflets* and their subdivisions; (*c*) varieties of *chordae tendineae*; and (*d*) the *papillary muscles*.

The *mitral atrioventricular orifice*, the well-defined transitional zone between atrial (or aortic) wall into the bases of the leaflets and best viewed from the atrial aspect, is smaller than the tricuspid orifice (mean circumference—9·0 cm in males, 7·2 cm in females according to Ranganathan *et al.* 1970, based on a study of 50 Caucasian hearts from 15 to 85 years: other literature was reviewed and much variation recorded). In diastole the orifice, approximately circular in outline, is placed almost vertically at 45° to the sagittal plane, but with a slight forward inclination: thus, its ventricular aspect 'faces' anterolaterally to the left and also slightly inferiorly, i.e. towards the left ventricular cardiac apex. It is virtually in the same plane as the tricuspid orifice but posterosuperior to it, whereas it lies postero-inferior and a little to the left of the aortic orifice. However, as we shall see, all three orifices are intimately interconnected by the various constituents of the fibrous skeleton of the heart (p. 654).

The *mitral valve annulus* is, again, *not* a simple circumferential fibrous ring, but comprises those elements that vary greatly in consistency, with which the laminae fibrosae of the valve leaflets become continuous. Variations in different parts of the annulus are of the greatest functional importance, allowing major changes in shape and dimensions at different stages of the cardiac cycle; this dynamism ensures optimal efficiency of both the true valvular actions of the mitral complex, and its equally important role in controlling the inflow/outflow patterns of blood passage through the left ventricle. It may also be noted that structural variations are not only related to different locations around the periphery of the valve, but that these are accentuated in some specimens, particularly with advancing years.

The constitution of the mitral annulus is illustrated in 6.19B. It consists of the inner aspects of parts of the left and right fibrous trigones, and extending from the latter the anterior and posterior

fila coronaria—long, tapering, fibrous, whip-like subendocardial tendons which partially encircle the valve orifice at the atrioventricular junction. Between the tips of the fila the atrial and ventricular myocardia are separated by a more tenuous sheet of deformable fibro-elastic sulcal connective tissue. Spanning the zone between the trigones, the lamina fibrosa of the central part of the anterior mitral leaflet is a continuation of the fibrous *subaortic curtain* (*intervalvar septum*) which descends from the adjacent halves of the left posterior (left coronary) and right posterior (non-coronary) cusp regions of the aortic valve annulus; there is no specialized thickening of the curtain where it passes into the base of the free valve leaflet. (It should perhaps be stated that since the line of connective tissue attachment of the valve consists of a malleable fibrous sheet, two dense collagenous masses, two tapering 'tendons', and delicate fibro-areolar tissue at different locations, the official term annulus fibrosus seems singularly inappropriate; however, its use is so deeply engrained in anatomical and cardiological literature that it is retained in this account.)

The *mitral valve leaflets*, since the earliest descriptions, have been widely held as two in number, and hence the name *bicuspid valve* as a more explicit but less picturesque alternative to the familiar term mitral valve. However, considerable confusion, controversy, and difficulties in quantitation have arisen because some authors described occasional or frequent occurrence of small accessory leaflets between the two major ones. These problems have now been resolved by an excellent quantitative study using strict criteria to define the leaflet areas and their subdivisions (for details and a bibliographic review *see* Lam *et al.* 1970; and Ranganathan *et al.* 1970). It is now proposed that the mitral valve "consists of a continuous veil inserted around the entire circumference of the mitral orifice" (Harken *et al.* 1952); the free edge of the veil shows several indentations. Two of the latter are deep and regularly positioned which receive the insertions of unique fan-shaped commissural chordae tendineae (*vide infra*); additionally, the tips of the papillary muscles may be used as guides to identify them (Rusted *et al.* 1952). These deep indentations are the *anterolateral* and *posteromedial commissures* and the valvular tissue may therefore be allocated into two commissural areas, and two leaflet areas termed *anterior* and *posterior*. (It should be noted that naming of the tricuspid commissures was derived from the names of their adjacent cusps, whereas the mitral commissures are named according to their overall position. Further, the official terms anterior and posterior for the valve leaflets, whilst having the advantage of simplicity, are rather misleading because of the oblique positioning of the whole valve; as will be indicated below many authors prefer alternative names.) The general anatomy of the mitral valve, opened and laid flat is diagrammed in 6.21B (after Ranganathan *et al.* 1970).

The *anterior leaflet* (also called the aortic, septal, 'greater', or anteromedial leaflet—Chiechi *et al* 1956) is large, semicircular or triangular, with a free margin which bears few or no indentations. Peripherally, its lamina fibrosa is continuous centrally with the fibrous subaortic curtain, and at the margins of the curtain with the mitral aspect of the right and left fibrous trigones, and beyond these with the substantial trigonal roots of the fila coronaria (6.29B). The leaflet presents a deep crescentic opaque *rough zone* which receives the insertions of various types of chordae tendineae (*vide infra*), and the ridge limiting the outer margin of the rough zone indicates the maximal extent of surface contact with the posterior leaflet in full valve closure during ventricular systole; from the rough zone to the valve annulus is a translucent *clear zone*, devoid of chordal insertions, but its lamina fibrosa carries extensions from the chordae inserting into the rough zone. The anterior leaflet possesses no basal zone. It will be appreciated that the anterior leaflet, hinging on its substantial annular attachment, and continuous centrally with the deformable subaortic curtain, is critically placed between the inflow and outflow tracts of the left ventricle, the whole mechanism having been called the *aortic baffle*. During passive filling of the ventricle and during atrial systole, the smooth atrial surface of the leaflet functions as an important surface directing a relatively turbulence-free flow of blood towards the highly trabeculated ventricular body and apex. After the onset of ventricular systole and mitral valve closure, the ventricular aspect of its clear zone merges into the smooth surface of the subaortic curtain which, together with the remaining fibrous walls of the subvalvular aortic vestibule, form the low-friction boundaries of the terminal part of the outflow tract. The approximate dimensions of the anterior leaflet are—mean vertical height centrally 2·4 cm in males and 2·2 cm in females, mean basal width being 3·6 cm in males and 2·9 cm in females; the mean vertical height of the rough zone is 0·9 cm in males and 0·8 cm in females (these data are derived from Ranganathan et al. 1970, who give many further quantitative details and ranges, and also analyse the data of Rusted et al. 1952, and Cheichi and Lees 1956; the results are too extensive to be quoted here and the interested reader should consult the original papers).

The *posterior leaflet* (also called the ventricular, mural, 'smaller', or posterolateral leaflet) possesses minor indentations or 'clefts', usually two, in its free border and because for many years there was a lack of precise definition of what features distinguished a major interleaflet commissure from an intraleaflet cleft, much disagreement and confusion existed concerning the territorial extent of the posterior leaflet and the existence, or otherwise, of small accessory leaflets. Lam et al. (1969) in a detailed analysis of the chordae tendineae of the mitral valve provided precise criteria for distinguishing commissural chordae (*vide infra*) and showed that the posterior leaflet is best regarded as all the valvular tissue lying posterior to the anterolateral and posteromedial commissures. When defined thus, the posterior leaflet has a wider attachment to the atrioventricular valve annulus than the anterior leaflet (mean values are 5·4 cm in males and 4·3 cm in females). The clefts receive the insertion of characteristic cleft chordae tendineae (*vide infra*) and divide the posterior leaflet into a relatively large *middle scallop*, and smaller *anterolateral commissural* and *posteromedial commissural* scallops. Each scallop bears a crescentic, opaque, rough zone, which receives chordal attachments on its finely serrated margin and ventricular aspect and defines the area of leaflet apposition in full valve closure. From the rough zone to within 2–3 mm of the annular attachment is a translucent, membranous or *clear zone* devoid of chordal insertions, whilst the basal 2–3 mm is thickened, vascularized, and receives the insertions of basal chordae tendineae. The rough zone to clear zone ratio in the anterior leaflet is about 0·6 whilst in the middle scallop of the posterior leaflet it is 1·4; thus a much greater proportion of the posterior leaflet is in apposition with the anterior leaflet during mitral valve closure.

The *chordae tendineae* of the mitral valve have the same general structure and characteristics as those of the right ventricle; false chordae are found irregularly distributed in the left ventricle as in the right, and will not be further considered.

The *true chordae* of the mitral complex may first be divided into *interleaflet* or *commissural chordae* and into varieties of *leaflet chordae*; of the latter those of the anterior leaflet are classified as *rough zone chordae* including specialized *strut chordae*; those of the posterior leaflet include *rough zone chordae*, *cleft chordae*, and *basal chordae*. Most true chordae either divide into a number of branches from a single stem soon after their origin from the tip or apical third of a papillary muscle, or they proceed as single cords only radiating into multiple fine cords immediately before their insertion.

The anterolateral and posteromedial *commissural chordae* arise near the tips of their corresponding papillary muscle as a single stem that branches almost immediately into radiating strands like the struts of a fan and insert into the smooth free margin of the commissure; some pass into the lamina fibrosa of the immediately adjacent leaflet tissue and continue towards the annular attachment. The branches of the posteromedial commissural chorda are longer, thicker and have a wider spread than the anterolateral ones.

Rough zone chordae typically divide into three cords soon after their origin from the papillary muscle; one inserts into the free margin, one beyond the free margin on the ventricular surface of the leaflet at the line of closure, and an intermediate cord inserts between these. Chordae of this type are found in relation to both the posterior and anterior leaflets. Peculiar to the anterior leaflet are, however, two that are much the thickest and strongest ten-

dinous chordae of the mitral valve; they originate from the tips of the anterolateral and posteromedial papillary muscles and insert near the line of valve closure between the '4 & 5 o'clock' positions posteromedially, and between the '7 & 8 o'clock' positions anterolaterally. They are termed *strut chordae* and were present in more than 90 per cent of hearts examined according to Lam et al (1970); in a much earlier publication their zone of attachment was termed the *critical point of tendinous insertion* of the anterior leaflet (Brock 1952).

Cleft chordae are miniature fan-shaped structures, their tiny radial branches attaching to the free margins of the inter-scallop clefts, whilst some deeper branches pass on to the ventricular aspect of the adjacent rough zones.

Basal chordae are confined to the posterior leaflet (most commonly the large middle scallop and one of the two smaller scallops); they arise directly from the mural ventricular myocardium as a single strand which flares into minute branches before insertion into the ventricular aspect of the basal zone of the leaflet. Sometimes the basal chordae are partly or wholly muscular.

The *papillary muscles* of the left ventricle are two in number and vary considerably in their formation from long and finger-like to short and stubby; occasionally their tips are bifid. The *anterior* (anterolateral) *papillary muscle* arises from the sternocostal mural myocardium and the *posterior* (posteromedial) *papillary muscle* has its base attached to the diaphragmatic wall of the ventricle. Chordae tendineae arise most commonly from the tip and margins of the apical third of each muscle, but occasionally they emerge near its base. The chordae from each muscle diverge and are inserted into corresponding points (i.e. points in close apposition in valve closure) on *both* cusps of the mitral valve.

Opening of the mitral valve at the onset of diastole is an entirely passive but rapid process, the leaflets parting and projecting into the ventricular cavity as the pressure of the left atrial venous blood returning from the lungs exceeds the left ventricular diastolic pressure. Passive ventricular filling proceeds as the atrial blood pours towards the ventricular apex, directed by the pendant anterior leaflet of the mitral valve. With increased filling, however, the valve leaflets begin to float passively towards each other, hinging on their annular attachments, and partially occluding the ventricular inflow orifice. Atrial systole now supervenes, directing a powerful jet of blood towards the ventricular apex and causing a transient re-opening of the valve. As the atrial jet-stream approaches maximal ventricular filling, the valve leaflets again float rapidly towards each other; ventricular systole now ensues commencing with papillary muscle and followed, extremely rapidly, by general mural and septal contraction. Coordinated papillary activity causes a progressive increase in tension in the chordae tendineae and coaptation of corresponding points on the opposing valve cusps. With general mural and septal myocardial excitation and contraction left ventricular blood pressure rises with great speed (6.35); the valve leaflets 'balloon' towards the atrial cavity and the atrial aspect of their rough zones come into tight apposition and maximal areal contact. Precisely graded continuing papillary contraction and increasing tension in the various grades of chordae tendineae prevents retropulsion of the leaflets into the atrial cavity, and maintains valvular competence. The essentially *biphasic* nature of mitral valve closure just described is vividly demonstrated in a normal echocardiogram recorded in an appropriate plane (6.28). With the fast drop in ventricular pressure following the end of the ejection phase of systole and the onset of diastole, the valve leaflets part and the next cycle of ventricular filling restarts.

It is assumed that, despite slight differences in time scale, and marked differences in pressure profiles, the general forms of valve opening and closure are similar in the tricuspid valve.

In addition to the foregoing it should be noted that both atrioventricular orifices and their valves undergo large changes in position, form and surface area at different stages of the cardiac cycle. Thus, both valves move anteriorly and to the left during systole and retrace this path during diastole. Also during systole the mitral valve (for which the data are more complete) reduces its orificial (annular) *area* by as much as 40 per cent; furthermore, its shape changes from the circular profile characteristic of diastole to a *crescentic form* at the height of systole. The annular attachment of the anterior leaflet provides the concavity of the crescent, whilst the attachment of the posterior leaflet, although remaining convex, contracts towards the former.

THE AORTIC VALVE (6.21 A, B, 22, 29B)

The smooth-walled outflow tract of the left ventricle, the aortic vestibule, terminates at the aortic orifice and its associated aortic valve, the attached margins of which continue into the root of the aorta. Although considerably stouter in its general construction, the aortic valve has many structural features in common with the pulmonary valve (p. 648), consisting therefore of a fibrous 'annulus' of complex form, three semilunar cusps or valvules attached basally to the annulus, and three dilatations of the aortic wall termed aortic sinuses corresponding to the cusps.

Because of the complex form of the annulus, it is difficult to define a 'plane' in which the aortic valve lies (some authorities choose the apical intercuspid commissures as reference points, whilst others prefer the deepest and most proximal points on the attached margins of the cusps, i.e. the three nadirs of the fibrous annulus). Nevertheless, in general the valve faces superiorly, to the right, and slightly anteriorly; in overall position it lies anterosuperior and slightly to the right of the mitral atrioventricular orifice; in overall shape it approximates to a truncated cone or *frustum*.

The *aortic fibrous annulus* consists of three almost semicircular condensations of collagen forming three 'scallops' which are so arranged that they encircle the vestibulo-aortic junction in the form of a three-pronged coronet; the 'prongs' point distally and apically form the fibrous framework of the commissural regions between adjacent cusps (*vide infra* and 6.29B). The luminal aspect of each scallop fuses with the cuspid lamina fibrosa through the attached border of a semilunar cusp. The proximal aspect of the scallops are thickened at their nadirs (*see* p. 657 for further details) whilst the triangular intervals between the proximal aspects of their ascending limbs are filled, and fuse with, extensions of the largely fibrous walls of the aortic vestibule. As will be noted further below, specialized regions of the latter include the subaortic curtain (intervalvar septum) and part of the *pars membranacea septi*. The distal deeply concave aspect of each scallop forms the lower limit of and continues into the fibro-elastic wall of an aortic sinus. Some observers maintain that although largely fibrous, the walls of the aortic vestibule contain myocardial strands, some of which approach and cross the line of the fibrous annulus, continuing as fine irregular strands into the walls of the sinuses.

The *semilunar cusps* or *valvules* of the aortic valve, three in number (6.22), are again reduplications of the endocardium enclosing a central lamina fibrosa which shows thickened specializations in certain locations. With the valve in the half-open position, each cusp would cover the surface of slightly more than a quarter of a sphere (or ellipsoid), the approximate hemisphere (or hemi-ellipsoid) being completed by the wall of the corresponding aortic sinus. Each cusp possesses a thickened basal attached border which is deeply concave when viewed from the aortic aspect, and an approximately horizontal free border. Throughout most of its extent the latter shows only a slight thickening of its fibrous core, but at the mid-point occurs a substantial aggregation of fibrous tissue forming the valvular *nodule* (Arantii); fine collagenous bundles radiate from the nodule to the attached border where the basal fibrous thickening blends with the corresponding scallop of the fibrous annulus. At the apical commissures, few collagen fibres pass between the adjacent cusps, but strong longitudinal fascicles pass into the ascending limbs of the annular fibrous scallops (*see* p. 657 and 6.29B). On each side of the nodule and its collagenous radiation the lamina fibrosa is tenuous, and in these regions, the *lunules* of the cusp, the valvular tissue is thin, translucent, and occasionally fenestrated. The aortic surface of each cusp is rougher than its ventricular (outflow) surface and it has been shown that the endocardial cells are elongated with respect to the direction of blood flow (Clarke and Fink 1974).

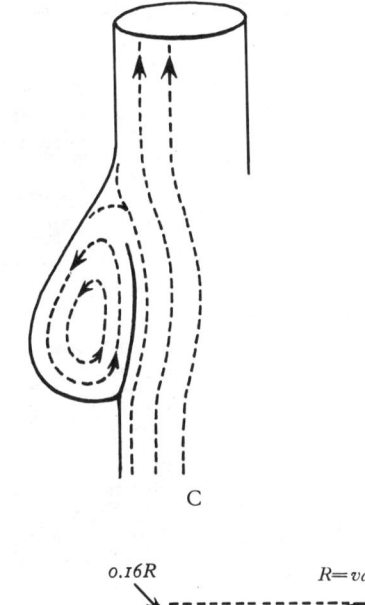

C

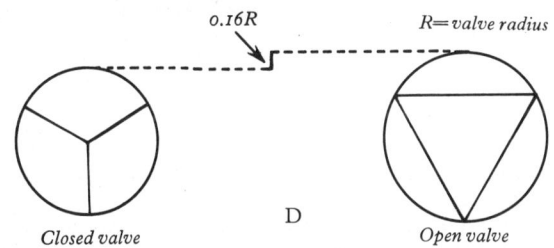

0.16R   *R= valve radius*

D

Closed valve   *Open valve*

6.21 C, D.   C.   A diagram of the vortices formed within an aortic sinus during the ejection phase of left ventricular systole. Note the position of the valve cusp and sinus wall. **D.**   *Left:* the general position of the apposed valve cusps during diastolic closure of the aortic (and pulmonary) valve. *Right:* the open position of the aortic (and pulmonary) valve during the ejection phase of ventricular systole. Note the increase in diameter of the arterial trunk root at the level of the valve commissures and the *triangular* form of the open valve. Consult text for further discussion.

It is unfortunate and the source of considerable confusion, that *three* different systems of naming the aortic cusps are currently in use. The official names advanced by the *Nomina Anatomica* (1977) refer to the embryonic/fetal positions of the cusps before full and definitive cardiac rotation has occurred; hence they are termed posterior, right and left. In this volume it has been preferred to retain the widely used system which refers to the approximate positions of the cusps in the mature heart and in which they are named the *anterior*, *left posterior*, and *right posterior*, corresponding names applying also to the sinuses. However, in clinical literature the practice is widespread of basing the terminology on the relation of cusps and sinuses to the origins of the coronary arteries; thus the anterior is termed *right coronary*, the left posterior the *left coronary*, and the right posterior the *non-coronary*. (In the present volume the mature positional names are often followed by the clinical terms in parenthesis.)

The *aortic sinuses* (of Valsalva) are more prominent than those of the pulmonary trunk, but similarly, are dilatations of the aortic root wall above the attached margin of each cusp. The upper margin of each sinus reaches considerably beyond the level of the free border of the cusp and is limited by a well-defined circumferential *supravalvar ridge* when viewed from the luminal aspect (6.18A). The ostia of the coronary arteries usually open near this ridge from the *upper* part of their sinus of origin, the ostium of the left artery often being a little lower than the right (*see* p. 669). The walls of the sinuses are largely collagenous near the fibrous annulus but the proportion of collagen diminishes and the amount of lamellated elastic tissue increases in the upper part of the sinus; as mentioned above, strands of myocardium may extend into this fibro-elastic wall. At mid-sinus level its wall is about one-half the thickness of that of the general wall of the supravalvar aorta, and less than one-quarter the thickness of the supravalvar ridge; also at this level the mean luminal diameter of

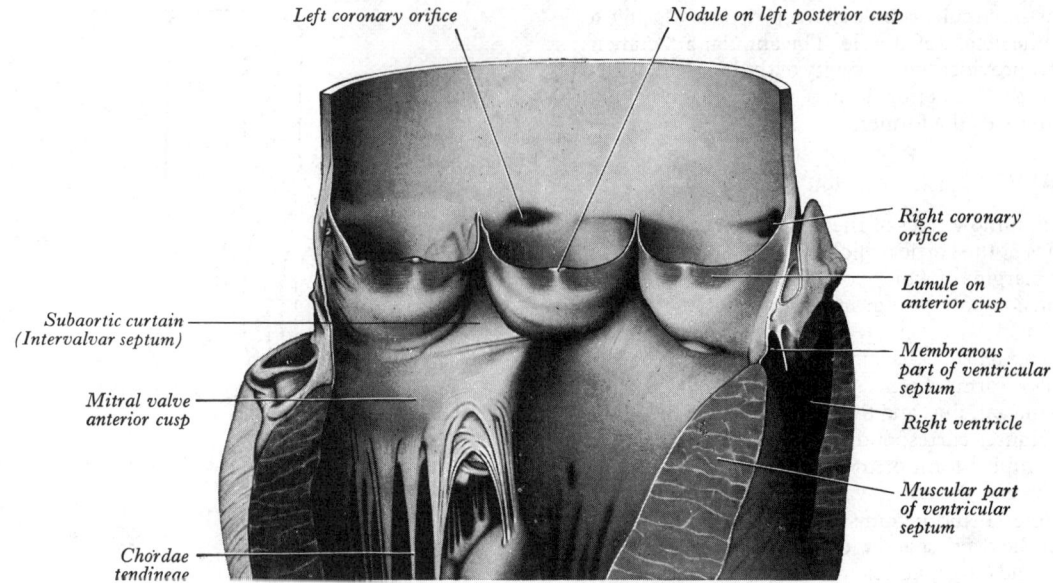

Left coronary orifice

Nodule on left posterior cusp

Right coronary orifice

Lunule on anterior cusp

Membranous part of ventricular septum

Right ventricle

Muscular part of ventricular septum

Subaortic curtain (Intervalvar septum)

Mitral valve anterior cusp

Chordae tendineae

6.22 The aortic orifice opened from the front to show the cusps of the aortic valve.

the aortic root is almost double that of the commencement of the ascending aorta (Reid 1970).

The topographical features, dimensions, and mechanical properties of both cusps and sinuses are widely considered to be of great functional significance in valve opening and effective closure at appropriate points during the cardiac cycle. Throughout diastole the aortic valve is closed, maintaining the aortic column of blood which is at high but slowly diminishing pressure (**6**.33); each cusp and its associated sinus form a roughly hemispherical chamber, the three nodules are in tight apposition centrally, and the free borders and lunular parts of adjacent half-cusps are in tight apposition on their ventricular aspects. Thus, viewed from the aortic aspect the closed valve has a tri-radiate appearance, three pairs of closely compressed lunular segments passing in straight lines out to the peripherally sited commissures. With the onset of ventricular systole and full mitral valve closure, left ventricular pressure rises rapidly and when this exceeds aortic pressure, the aortic valve cusps open passively. The fibrous wall of the sinuses nearest to the ventricular aortic vestibule (i.e. at the nadirs of the annular fibrous scallops) is relatively dense, inextensible, and undergoes little dimensional change. However, in the upper parts of the sinuses, at the level of the valve commissures, the sinus wall is fibro-elastic and deformable; under the influence of full left ventricular ejection pressure, the *radius* of this part increases some 16 per cent in systole compared with its dimension in diastole. As a result, the commissures move apart and the fully open aortic orifice is *triangular*, the free margins of the cusps passing in virtually straight lines between the commissures. It must, however, be emphasized again that the aortic cusps do *not* flatten against the sinus walls, even at the height of ventricular systole, and it is considered that this cuspid position together with the shape of the sinus wall are important factors as a prelude to efficient valve closure. During systolic ventricular ejection most of the blood passes directly into the ascending aorta, but a proportion is thought to enter each sinus centrally forming vortices within the sinus and then leaving through its marginal angles. The vortices help to maintain the triangular 'mid-position' of the cusps throughout most of ventricular systole, and probably initiate the floating together of the cusps at the extreme end of systole; tight approximation and full valve closure ensues with the rapid drop of ventricular pressure accompanying the onset of diastole—the commissures move together, the nodules aggregate centrally, and the valve reassumes its tri-radiate form. It may be noted that experimental results indicate that about 4 per cent of the ejected blood regurgitates through a valve possessing normal aortic sinuses, whereas 23 per cent regurgitation occurs through a valve without

sinuses (Bellhouse 1968). Additionally, it has been suggested that normal aortic sinuses promote a non-turbulent flow of blood into the ostia of the coronary arteries.

It is generally thought that considerations similar to those just outlined apply also to the opening and closure of the pulmonary valve even though the events are more leisurely, the pressure profiles much less extreme (**6**.35), and the structure of the valve complex more delicate in all its features.

CARDIAC SURFACE ANATOMY (**6**.23, 24)

It should be emphasized that the surface projection data given apply to an average adult, and that they are considerably modified

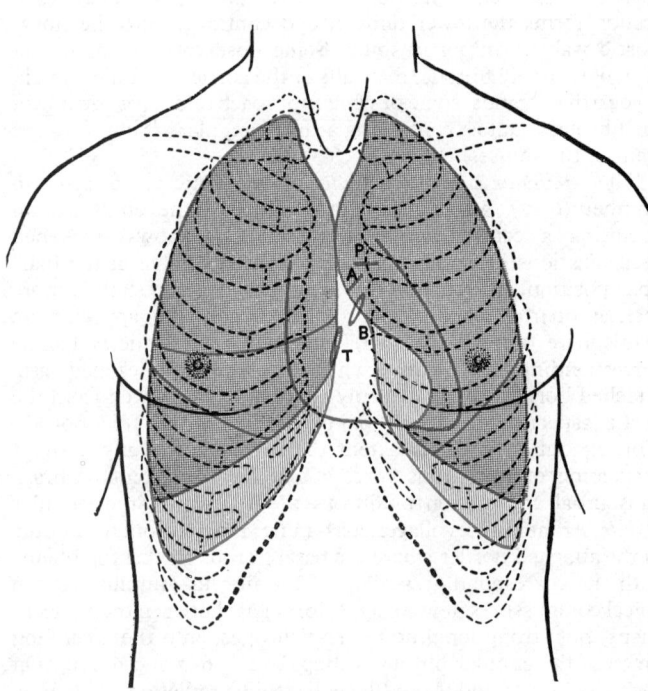

6.23 The front of the thorax, showing the surface relations of the bones, lungs (purple), pleurae (blue), and heart (red outline).
A = Orifice of aorta.
B = Left atrioventricular (mitral) orifice.
P = Orifice of pulmonary trunk.
T = Right atrioventricular (tricuspid) orifice.

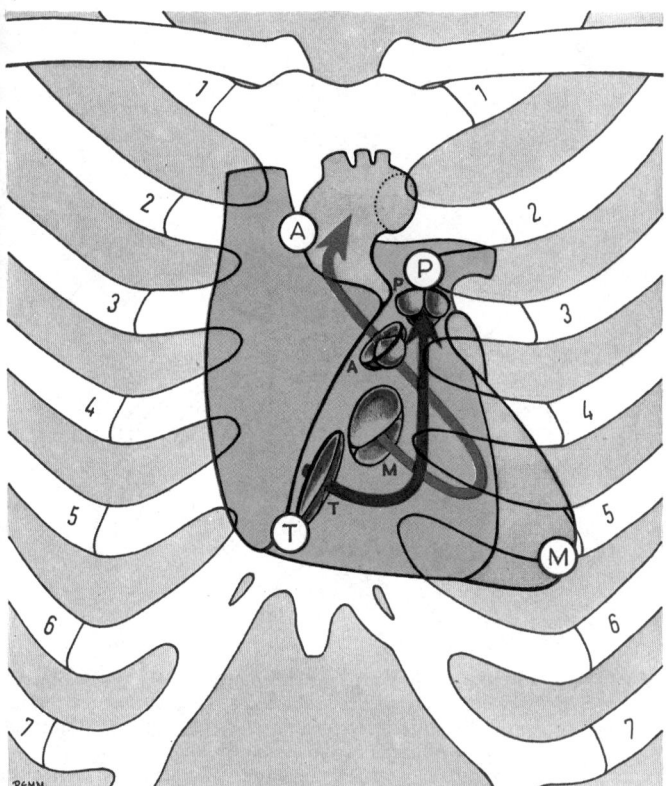

6.24 A diagram illustrating the relation of the sternocostal surface and valves of the heart to the thoracic cage. The right heart is blue, the arrow denoting the inflow and outflow channels of the right ventricle; the left heart is treated similarly in red. The positions, planes and relative sizes of the cardiac valves are shown. The position of the letters A, P, T and M indicated the aortic, pulmonary, tricuspid and mitral *auscultation* areas of clinical practice.

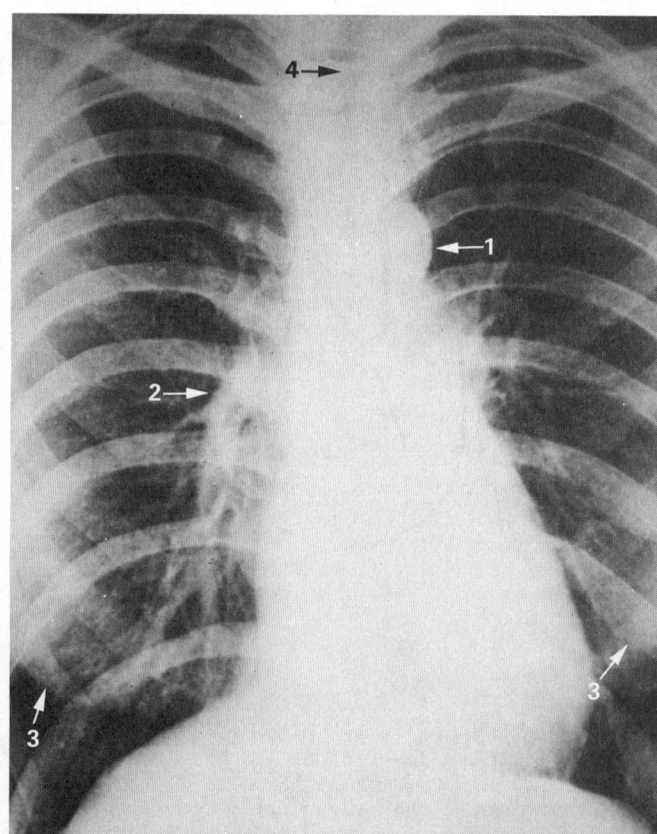

6.25 Radiograph of chest, postero-anterior view. Adult female. Note the difference in level of the right and left halves of the diaphragm. 1. Aortic 'knuckle'. 2. Pulmonary vessels of right side. 3. Edge of shadow caused by breast. 4. Position of trachea.

by age, sex, bodily size and proportions, by respiration and by position. Moreover, the surface projections of the valves do *not correspond* to the best sites for *auscultation* of valvar sounds (**6.24**).

The *apex* of the heart almost corresponds in position to the *apex beat*, which can usually be seen and can always be felt a little below and medial to the left (male) nipple. As a rule it is in the fifth intercostal space, slightly medial to the mid-clavicular line, or about 9 cm from the midline in an average adult male.

The sternocostal surface of the heart can be projected on to the anterior chest wall and forms an irregular, quadrangular area. The *right border* (**6.23, 24**) corresponds to a line drawn from the upper border of the right third costal cartilage, 1·2 cm from the margin of the sternum, downwards to the sixth costal cartilage. This line is gently convex to the right and is at its maximum distance from the median plane—about 3–4 cm—in the fourth intercostal space. It represents the lateral aspect of the right atrium. The continuation of this line in an upward direction, marks the lateral border of the superior vena cava and, in a downward direction the lateral border of the inferior vena cava.

The *lower border* of the cardiac projection can be represented by a line joining the lower end of the right border to the apex beat; it passes through the xiphisternal joint, and corresponds for the most part to the lower margin of the right ventricle. The *left border* is represented by a line drawn from the apex beat upwards and medially to a point on the lower border of the left second costal cartilage, 1·2 cm from the sternal margin. This line is convex upwards and to the left; with the exception of its upper part, which demarcates the left auricle, it corresponds to the left ventricle.

The four-sided area may be completed by a line which joins the upper ends of the right and left borders, and corresponds roughly to the upper limits of the atria. The left and right borders can be mapped out by heavy percussion.

The *pulmonary orifice* lies partly behind the upper border of the left third costal cartilage and partly behind the sternum. It can be represented by a horizontal line, 2·5 cm long. Two parallel lines, drawn from the extremities of this line upwards to reach the left second costal cartilage, map out the *pulmonary trunk*.

The *aortic orifice* lies below and a little to the right of the pulmonary orifice. It corresponds to a line, 2·5 cm long, drawn from the medial end of the left third intercostal space downwards and to the right. Two parallel lines, drawn from the extremities of this line upwards and to the right as far as the right half of the sternal angle, outline the *ascending aorta*.

The *right atrioventricular*, or *tricuspid*, orifice can be represented by a line, 4 cm long, commencing in the median plane opposite the fourth costal cartilage, and passing downwards and slightly to the right. The centre of this line should be opposite the middle of the fourth intercostal space. The *left atrioventricular*, or *mitral*, orifice lies behind the left half of the sternum opposite the fourth costal cartilage and can be represented by a line, 3 cm long, passing downwards and to the right.

The *area of superficial cardiac dullness* is a roughly triangular area, which can be mapped out by light percussion, and corresponds to the portion of the heart which is not covered with lung (p. 1254).

RADIOLOGICAL APPEARANCE OF THE HEART (6.25, 26, 27)

The heart, being full of blood, casts a shadow, occupying the lower part of the mediastinum, which is in sharp contrast with the clearer areas occupied by the air-filled lungs. The pulsatile movements of this shadow are evident on screening. In full inspiration the shadow of the apex is clear of the diaphragm, and presents, in radiographs, a rather blurred outline due to its movement. The shadow of the convex right border of the heart is continuous above and below with those of the superior and inferior venae cavae. Owing to the attachment of the pericardium to the diaphragm, during inspiration the heart becomes longer and narrower and during expiration shorter and broader. The shape of the heart also varies with the bodily proportions and attitude of the individual (*see* p. 1321). In lateral radiographs the

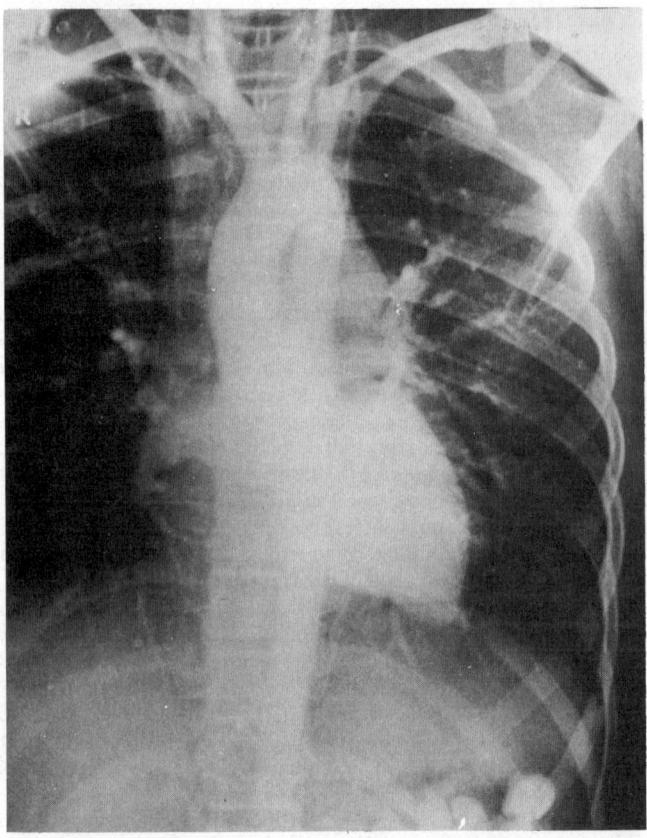

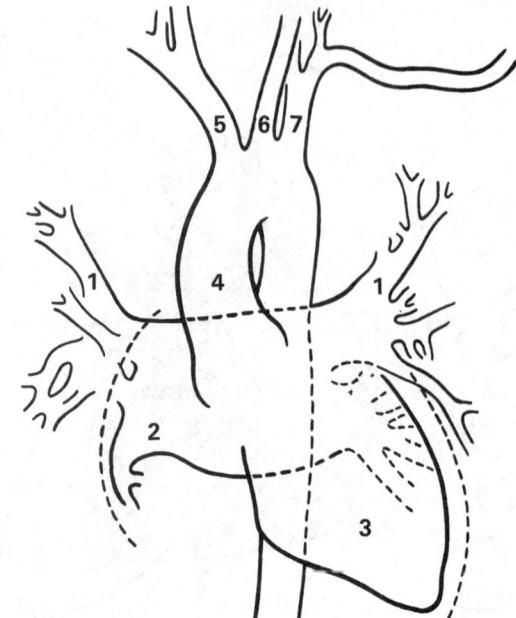

6.26 Angiocardiogram (provided by Dr. Frances Gardner), showing the left side of the heart in a child of 11 years; anteroposterior view. 1. Upper pulmonary vein. 2. Left atrium. (Note that owing to the great obliquity of the atrial septum, the left atrium extends to the right behind the right atrium.) 3. Left ventricle. 4. Ascending aorta. 5. Brachiocephalic trunk. 6. Left common carotid artery. 7. Left subclavian artery. (The arms of the patient are raised above the head and, as a result, the distal end of the artery passes upwards.)

6.27 Angiocardiogram (provided by Dr. Frances Gardner), showing the right side of the heart in a child of 12 years; anteroposterior view. 1. Superior vena cava. 2. Right atrium. 3. Right ventricle. 4. Pulmonary trunk. 5. Right pulmonary artery. 6. Left pulmonary artery.

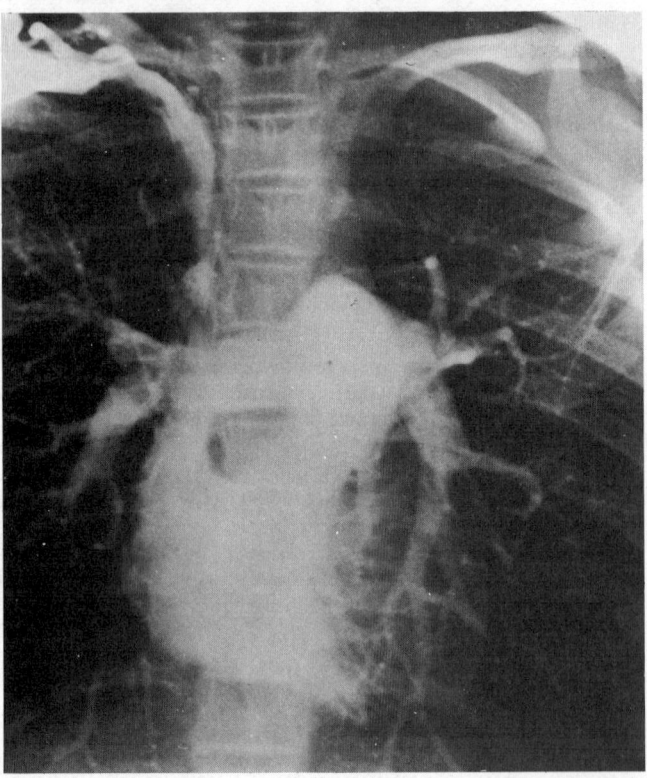

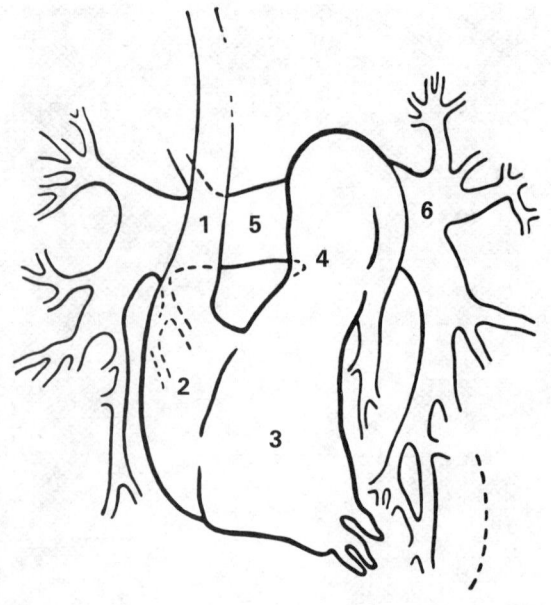

retrocardiac space may be recognized as a translucent area between the heart and the vertebral column. This space is occupied by the descending aorta and the oesophagus (**8**.90). For a detailed study of the cavities of the heart and the large blood vessels the method of angiocardiography can be used (Robb and Steinberg 1939, Gardner 1949). A suitable contrast medium miscible with blood is injected intravenously and the course of the intravascular injection is followed in serial X-ray exposures, usually in anteroposterior or oblique views (**6**.26, 27).

CARDIAC CONNECTIVE TISSUES AND FIBROUS SKELETON (**6**.29A, B)

From epicardium to endocardium and from the orifices of the entrant great veins to the roots of the outflowing great arterial trunks, the intercellular crevices, intra- and inter-tissue planes, between the essential contractile and conducting elements, are everywhere permeated by connective tissue, but this varies greatly in its organization and constitution in different locations.

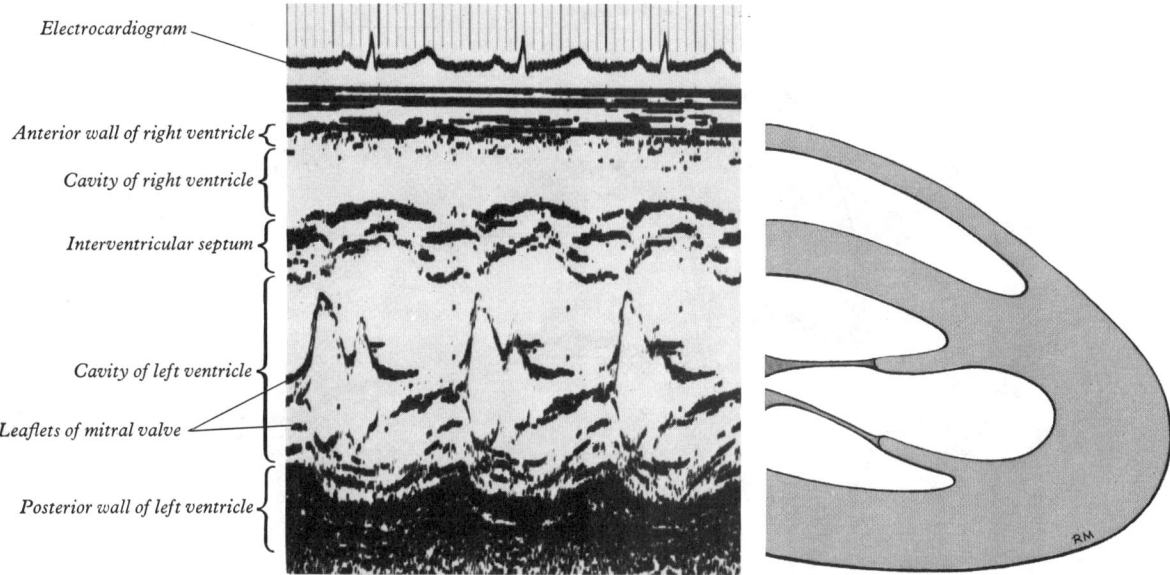

Electrocardiogram

Anterior wall of right ventricle

Cavity of right ventricle

Interventricular septum

Cavity of left ventricle

Leaflets of mitral valve

Posterior wall of left ventricle

6.28 A standard echocardiogram of the heart recorded from the anterior aspect as increasingly used in clinical practice. The outline diagram shows the corresponding reflecting surfaces. Note the biphasic closure of the mitral valve leaflets.

Beneath the squamous cells of the serous visceral epicardium lies, over much of the heart, a fine layer of subserous areolar connective tissue; but this accumulates adipose cells, the *subepicardial fat*, often considerable in quantity and gradually increasing with age, which is concentrated along the acute margin of the heart, in the coronary sulcus, interventricular grooves and their side channels. The principal coronary vessels and their main branches are embedded in this fat. Similarly, the squamous endocardium lies on a fine subendocardial areolar tissue which has a particularly well marked content of elastic fibre. The fibrocellular components of these subepicardial and subendo-cardial layers blend on their mural aspects with the inter-cellular endomysial and perimysial connective tissue of the myocardium. Each cardiac myocyte is invested by a delicate endomysium consisting of fine reticulin, collagen and elastin fibres embedded in ground substance; this investment is, of course, absent at the specialized desmosomal and gap junction contacts that constitute the intercalated discs (p. 517). Similar considerations apply to Purkinje myocytes and their extensive intercellular contacts. It is widely held that cardiac myocytes are structurally and functionally polarized, forming tracts of cells, with interconnexions laterally with adjacent tracts, and that substantial aggregations of these tracts form bundles, strands or sheets of macroscopic proportions that show a high degree of complex geometric patterning (*see* p. 661). Further, it has been claimed that these larger myocardial aggregates are 'surrounded' or 'separated' by, and perhaps attached to, stronger perimysial condensations of connective tissue. However, much detailed research will still be needed before a comprehensive view of the three-dimensional array of tracts and aggregations of cardiac myocytes becomes available, and this applies even more strongly to the disposition and properties of their associated connective tissues.

Lying approximately in the plane of the coronary sulcus, i.e. at the 'ventricular base' of the heart, and intimately related to the atrioventricular inflow orifices and the arterial trunk outflow orifices and their associated valves, is a complex framework of dense collagen with membranous, tendinous, and fibro-areolar extensions, the whole being sufficiently distinct to be termed the *fibrous skeleton of the heart*. (For early literature on this topic consult Wolff 1781, Henle 1876, Mall 1911, and Walmsley 1929, an excellent modern survey is provided by Walmsley and Watson 1978; for a particularly penetrating structural and functional

analysis *see* Zimmerman 1959, Zimmerman and Bailey 1962, and Zimmerman 1966.)

It will have been apparent in the previous pages that the four cardiac valves in fact depart considerably from a single plane, and that the fibrous skeleton is a complicated, deformable, three-dimensional *continuum*. Thus, study (and illustration) of a single section near the ventricular base of the heart (**6**.29A) provides incomplete data, and led some of the earlier anatomists to introduce relatively simple two-dimensional terms such as 'trigones' and 'annuli' which are, in some respects, misleading if taken too literally. The terms, however, are officially recognized and in such widespread use that they will be retained in the present account, but wherever possible, alternatives and additions to the terminology will be provided. Further difficulties are encountered when attempts are made to illustrate such a system, but it is hoped that figure **6**.29B will assist the reader in understanding its main features and interrelationships.

As noted previously, the annuli of the mitral and tricuspid valves are almost coplanar and at 45° to the sagittal plane with

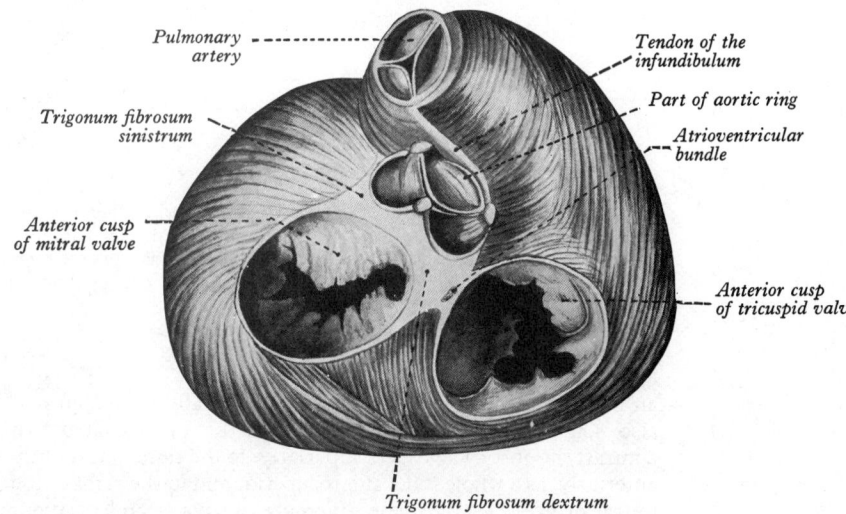

Pulmonary artery

Trigonum fibrosum sinistrum

Anterior cusp of mitral valve

Tendon of the infundibulum

Part of aortic ring

Atrioventricular bundle

Anterior cusp of tricuspid valve

Trigonum fibrosum dextrum

6.29A The base of the ventricles, after removal of the atria and the pericardium. (From Quain's *Elements of Anatomy*, vol. iv, part iii. *The Heart*, by T. Walmsley, 1929.) Contrast with **6**.29B.

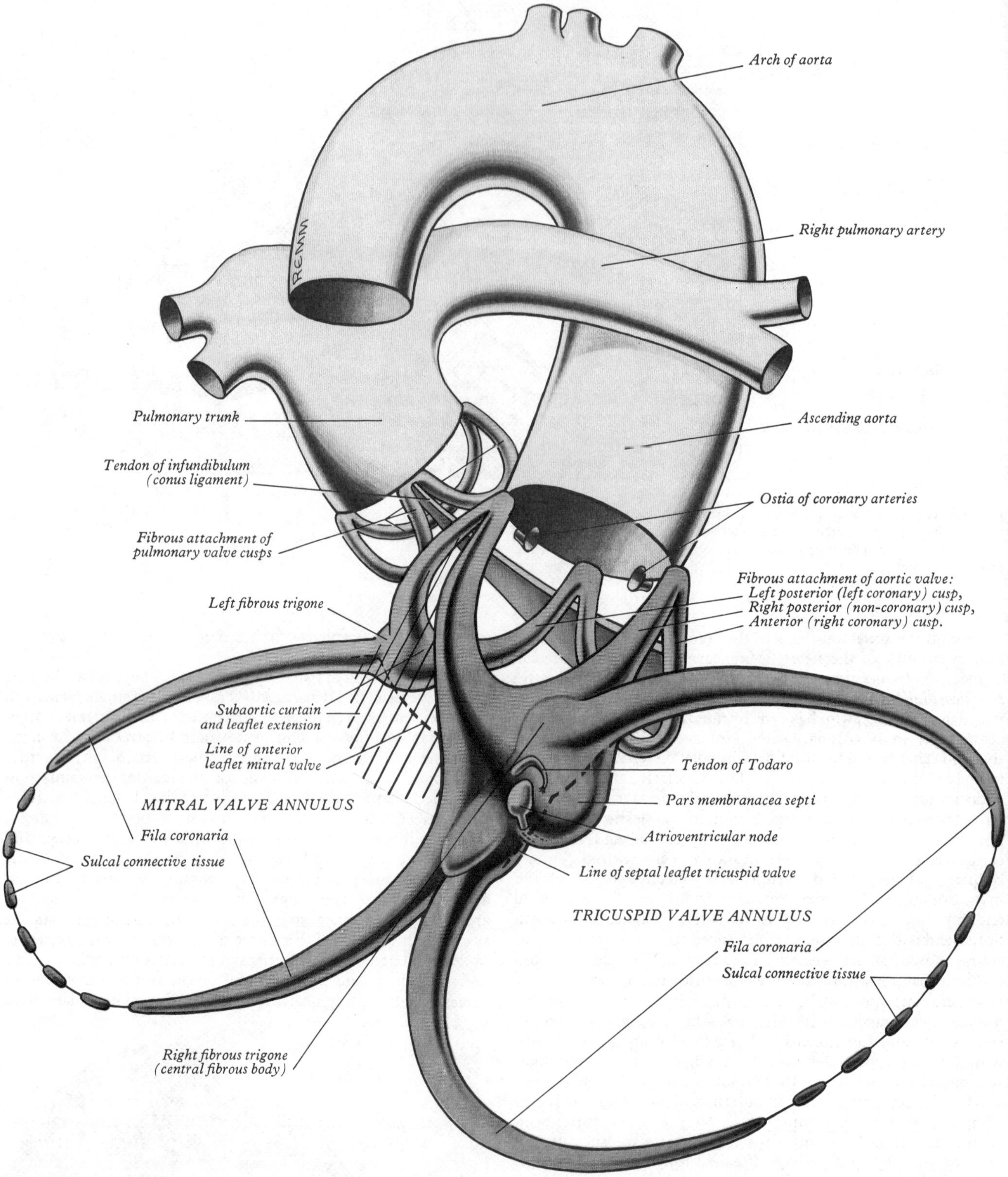

Arch of aorta

Right pulmonary artery

Pulmonary trunk

Ascending aorta

Tendon of infundibulum
(conus ligament)

Ostia of coronary arteries

Fibrous attachment of
pulmonary valve cusps

Fibrous attachment of aortic valve:
Left posterior (left coronary) cusp,
Right posterior (non-coronary) cusp,
Anterior (right coronary) cusp.

Left fibrous trigone

Subaortic curtain
and leaflet extension

Line of anterior
leaflet mitral valve

Tendon of Todaro

MITRAL VALVE ANNULUS

Pars membranacea septi

Fila coronaria

Atrioventricular node

Sulcal connective tissue

Line of septal leaflet tricuspid valve

TRICUSPID VALVE ANNULUS

Fila coronaria

Sulcal connective tissue

Right fibrous trigone
(central fibrous body)

6.29B　The principal elements of the fibrous skeleton of the heart: _red—_ mitral and aortic 'annuli'; _blue_—tricuspid and pulmonary 'annuli'; _green_—tendon of the infundibulum. For clarity the view is from the _right posterosuperior aspect_. Note that due to perspective the pulmonary annulus _appears_ smaller than the aortic annulus, whereas, in fact, the reverse obtains. Consult text for an extended discussion. Based in part on the work of Zimmerman and Bailey (1962) and Zimmerman (1966).

an inclination such that they face anterolaterally to the left and also somewhat inferiorly, i.e. towards the cardiac apex. In contrast the aortic valve faces superiorly, to the right and slightly anteriorly; as a whole it lies anterosuperior and to the right of the mitral orifice. Despite these differences in plane and position, however, these _three_ valvar orifices, mitral, tricuspid and aortic, are _intimately interconnected_ through the medium of their basal collagenous framework. In contrast, the pulmonary valve is considerably removed from the other three, being placed anterior, superior, and virtually at right angles to the aortic valve, its only connexion with the latter being through the relatively lengthy, deformable tendon of the conus arteriosus (infundibulum).

The essential functions of the fibrous skeleton are: (1) to ensure electrophysiological discontinuity between atrial and ventricular myocardium except via the specialized conducting tissues; (2) to provide a strong mechanical attachment for the ventricular myocardium, and a more tenuous one for the atrial myocardium; (3) to maintain the overall position of the heart within the fibrous pericardium; (4) to establish a stable but dynamically deformable base for the attachments of the fibrous cores of the valve leaflets.

It is apposite to start with a consideration of the specializations of the aortic annulus because it is centrally placed, interconnected with all the surrounding fibrous structures, and the aortic root constantly supports a high-pressure column of blood. Furthermore, it is exposed to the incessant bombardment of the full systolic ejection pressure of the left ventricle with each cardiac cycle, which, as Mall pointed out, without such stout fibrous intermediaries would undoubtedly tear the arterial trunk from the ventricular chamber.

The *aortic fibrous annulus* has been briefly described previously (p. 651); it is constituted by the coaptation of three fibrous scallops which together take the form of a simple three-pronged 'coronet'. Thus, reference to illustration **6**.29B will make it clear that each scallop has two lateral 'high points' or zeniths, and a central 'low point', or nadir, each with its particular specialization. The *distal margin* of each scallop merges with the fibro-elastic wall of an aortic sinus; some myocardial strands have been described as intermixed with the fibro-elastic tissue, and near the annulus collagen predominates but is gradually replaced by increasing quantities of lamellated elastic tissue more distally. The *luminal aspect* of each scallop blends with the lamina fibrosa of a valve cusp or valvule. Adjacent *high points* meet and fuse, with some interchange of collagen fasciculi at the three *intercuspid commissures*. The latter do not reach the upper limit of the sinuses which, as indicated, is marked by a circumferential thickening of the aortic wall, the *supravalvar ridge*. Between the commissures the walls of the sinuses are quite elastic, allowing the commissures to retreat from each other, with concomitant changes in the shape and disposition of the cusps in systolic valve opening, and the converse of these movements in diastolic closure. The *nadirs* of the three scallops show collagenous thickenings to very different degrees. The least prominent is that of the anterior (right coronary) cusp, but its margins blend with the aortic end of the tendon of the infundibulum (*vide infra*). The nadir of the left posterior (left coronary) cusp is thickened to an intermediate extent, and because of its appearance in section in the plane of the atrioventricular valves, it is termed the *left fibrous trigone*. It provides an attachment for part of the anterior leaflet of the mitral valve, and posterolaterally to the left it continues as one of the (mitral) whip-like, tapering, collagenous bundles (the *fila coronaria* of Henle) which partially encircle the atrioventricular orifices. The nadir of the right posterior (non-coronary) scallop is the most massively developed; because of its appearance in section the earlier anatomists termed it the *right fibrous trigone*. In the adult heart it is a dense irregularly ellipsoidal mass about 20 × 10 × 5 mm which provides powerful structural and functional links between the aortic, mitral and tricuspid orifices, and in accord with its site and multiple circumvalvar extensions, it is now termed by many authors the *central fibrous body* of the heart. It continues into the second filum coronarium of the mitral orifice, and provides the base for both fila coronaria of the tricuspid orifice. Anteriorly the central body blends with the *pars membranacea septi* (*vide infra*); posterosuperiorly the *tendon of Todaro*, a palpable subendocardial bundle of collagen about 1 mm in diameter extends into the right atrial wall curving across the *torus aorticus* (p. 644) towards the medial extremity of the valve of the inferior vena cava where its identity is lost. The tendon is one of the boundaries of the *triangle of Koch* (p. 644, **6**.18A) and provides an important surgical guide to the position of the atrioventricular node.

The *proximal* (i.e. ventricular) *aspect* of the triple scalloped fibrous condensation which constitutes the aortic annulus, presents three roughly triangular areas, the *subaortic spans* (see Zimmerman 1959; Zimmerman and Bailey 1962). The spans, originally termed *spatia intervalvularia* by Henle (1876), occupy the intervals between adjacent sinuses of Valsalva and their triangular apices correspond to the intercuspid commissures; their walls either consist of collagen, or admixed muscle strands and fibro-elastic tissue, and they encompass the immediately subvalvar recesses of the aortic vestibule. The interval between the right posterior (non-coronary) and left posterior (left coronary) sinuses is filled with the deformable *subaortic curtain* (*intervalvar septum*), which continues into the base of the central part of the anterior leaflet of the mitral valve; down to the leaflet base it is clothed by a thin layer of left atrial myocardium. Its critical functional role as an *aortic baffle* placed between the inflow and outflow tracts of the left ventricle is mentioned on p. 649. The span lying between the right posterior (non-coronary) and anterior (right coronary) sinuses is continuous with the anterior surface of the right fibrous trigone, and is a variably developed collagenous sheet, the *pars membranacea septi*. The latter is an oval area measuring about 1 cm from the aortic annulus to the muscular interventricular septum, about 1.2 cm in the antero-posterior direction, and it is normally some 1 mm in thickness. As far posteriorly as the angulated attachment of the anteroseptal commissure of the tricuspid valve (p. 646) the pars membranacea intervenes between the right and left *ventricles*, whereas posterior to the commissure it separates the left ventricle and right atrium (i.e. here it is an *atrioventricular septum*). After traversing the right fibrous trigone, the atrioventricular bundle of specialized conducting tissue courses along the postero-inferior margins of the membranous septum to reach the muscular interventricular septum where it divides into its initial bundle branches. It will be recalled that this region of the heart has a complex developmental history related to the final stages of septation (p. 185), and it is one of the commonest sites of congenital cardiac anomalies (p. 666). Thus, since it lies at the very centre of the heart interconnecting three valves (aortic, mitral and tricuspid) and three chambers (right and left ventricle and right atrium), and since it is so intimately related to the atrioventricular bundle, and also frequently subject to maldevelopment, the membranous septum is of the greatest interest to the investigative cardiologist and cardiac surgeon.

The third subaortic span, namely that between the anterior (right coronary) and left posterior (left coronary) aortic sinuses, is filled with fibro-elastic tissue which blends along its anterior cusp margin with the *tendon of the infundibulum*; the remainder of the span is clothed with left ventricular myocardium.

The *pulmonary valve fibrous 'annulus'*, situated anterosuperior and almost at right angles to the aortic annulus is, however, of the same general construction as the latter, and although of greater diameter, it is considerably less robust. It is triple scalloped, with commissures at the intercuspid apices, and barely perceptible thickenings at the nadirs of the scallops. The subvalvar spans are filled with fibro-elastic tissue which is well-supported by interlacement with myocardial extensions from the muscular wall of the infundibulum. The *apex* of the subvalvar span nearest to the aorta becomes confluent with the pulmonary end of the tendon of the infundibulum. It has been suggested (Zimmerman and Bailey 1962) that the latter acts as a powerful bond between the two great arterial trunks, and that it "permits a certain degree of torsional movement between them, while preventing them from being torn asunder by the differently directed ejaculatory forces of the ventricles."

The *mitral* and *tricuspid fibrous annuli* have been adequately described on pp. 649, 645: suffice it to re-emphasize here that they are not simple, rigid, collagenous rings, but dynamic, deformable lines of valve leaflet attachment that vary greatly at different points around their periphery, and also change consderably with increasing age.

MYOCARDIAL ARCHITECTURE

It has long been held that the muscular walls of the cardiac chambers consist of cardiac muscle 'fibres' which are transversely and longitudinally striated (p. 515), and present an exceedingly intricate interlacement. They can be conveniently divided into (*a*) the fibres of the atria, (*b*) the fibres of the ventricles and (*c*) the fibres of the conducting system of the heart (p. 659). It is functionally significant that atrial and ventricular muscle fibres are completely separated by the fibrous 'annuli', the only connexion being by elements of the conducting system (p. 659). (It should be noted that in modern terminology such fibres are presumed to consist of polarized rows, bundles, or other

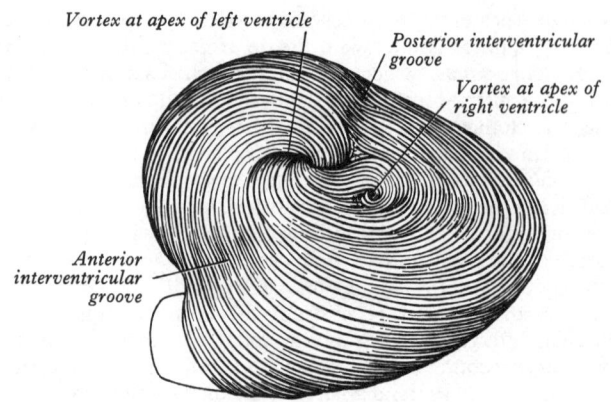

6.30   The two vortices in the myocardium at the apex of the heart. (After Mall.)

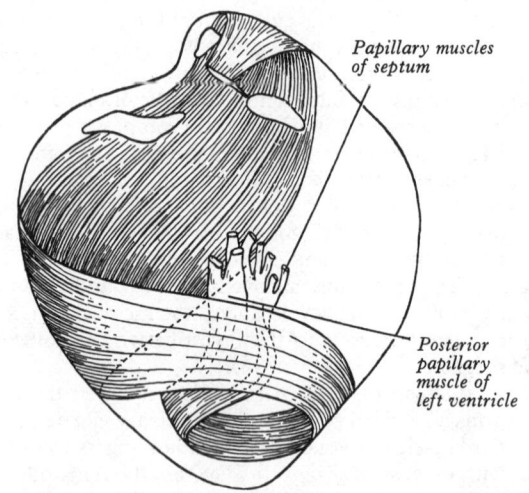

6.31   The superficial muscular fibres of the ventricles of the heart originating in the tendon of the infundibulum. (After MacCallum.)

aggregations of cardiac myocytes with their associated connective tissues—*see* p. 654, and below.)

The *atrial fibres* are in two layers, superficial, common to both atria, and deep, proper to each. *Superficial fibres* are most distinct anterior to the atria, crossing their bases transversely as a thin, incomplete layer; some pass into the atrial septum. *Deep fibres* are looped and annular, the former passing over each atrium to the corresponding atrioventricular ring, in front and behind, while the annular fibres surround the auricles and encircle the vena caval openings and fossa ovalis.

The *ventricular fibres* are complex in arrangement, and few full descriptions except by MacCallum (1897) and Mall (1911, 1912) have been available until recently (*vide infra*). Superficial and deep layers exist, all but two bands of which enter the papillary muscles. In the infundibulum the superficial fibres are transverse to its axis, the deep being longitudinal. Hence, vertical incisions into the infundibulum may be preferable. The *superficial*

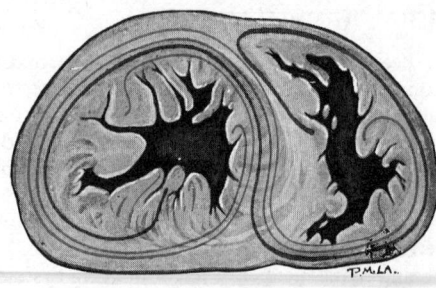

6.32A   The arrangement of the deep layers of the muscular fibres of the ventricles, as seen in cross-section. (After MacCallum.)

ventricular layers include: (1) Fibres which start from the infundibular tendon (**6.**31), curve across the diaphragmatic surface, and sweep to the left across the anterior interventricular groove to form a vortex (**6.**31) round the cardiac apex, finally passing up into the papillary muscles of the left ventricle. Fibres from the anterior half of the tendon pass to the anterior and posterior papillary muscles, those from its posterior half to the anterior alone. (2) Fibres from the *right* atrioventricular annulus which cross the diaphragmatic surface of the right ventricle reach its sternocostal surface. There they pass beneath the fibres just described and cross the *anterior* interventricular groove to wind round the apex, ending in the posterior papillary muscle of the left ventricle. (3) Fibres from the *left* atrioventricular annulus, which cross the *posterior* interventricular groove and pass into the right ventricle, end in its papillary muscles. There are three *deep ventricular* layers, starting in the papillary muscles of one ventricle and curving sinuously into the anterior interventricular groove to end in the papillary muscles of the other (**6.**32A). The most superficial layer in the right ventricle is deepest in the left, and *vice versa*. The first layer almost encircles the right ventricle, crosses in the septum to the left, and unites with the superficial fibres from the right atrioventricular ring to form the posterior papillary muscle. The second layer is less extensive in the right ventricular wall and correspondingly greater in the left, where it joins with superficial fibres from the anterior half of the infundibular tendon to form the septal papillary muscles. The third layer almost encircles the left ventricle to unite with superficial fibres from the posterior half of the tendon to form the anterior papillary muscle. The arrangement of these three layers is held to synchronize ventricular systole with atrioventricular valve closure. Two bands of fibres which do not end in the papillary muscles are also described; one, from the *right* atrioventricular ring, crosses the septum (p. 645), encircles the deep left ventricular layers, to end in the *left* ring. The second band is confined to the left ventricle, passing from the *left* atrioventricular ring to encircle the aortic orifice (Mall 1911, 1912). Fibres of the left ventricular musculature and the lower part of the left atrium reach the base of the aorta, close to its valvular cusps. These fibres are said to rotate the aortic base with the general ventricular twist of systole.

The foregoing account of myocardial architecture agrees, in its broadest features, with the views of a majority of the workers in this field, although MacCallum (1900) and Mall (1911) are the most frequently quoted. However, the concepts of obliquely spiralling laminae of fibres, overlapping at differing angles in the ventricular wall, the existence of vortices at the apex of the ventricles through which superficial fascicles spiral inwards to form internal layers—these and some other features were all described in earlier papers by Lower (1669), Borelli (1681), von Haller (1764), Gerdy (1823), and Pettigrew (1864). All these earlier investigators, like their successors, used blunt dissection of hearts which had been boiled (sometimes even roasted!), usually with the addition of varying concentrations of acetic or nitric acid. This is not the place to discuss the validity of this maceration technique, but it is important to realize that, while it is generally assumed that the loosening of myocardial tissue is considered to reside in changes in the endomysial connective tissue, the branching of the cardiac myocytes inevitably entails that a multitude of minute tears of actual muscle cells must occur in such blunt dissection. Theoretically, at least, it might therefore be as easy to tear across the long axis of muscle cells; but all who have used the technique extensively, from Lower to the most recent exponent, Torrent Guasp (1970), appear to have been satisfied that, with careful dissection, they could trace the predominant orientation of a particular layer or fascicle throughout its supposed extent. MacCallum's *tour de force* was the 'unrolling' of the ventricular muscle of a fetal pig's heart; in his short study, accomplished in three weeks, he dissected no human hearts; this aspect of the work was extended by Mall after the former's sadly premature demise. In his much more extensive investigation Mall (1911) particularly emphasized the helical spreading out of superficial fibres after they had plunged deeply into the ventricular walls. In the pig's heart he also described arrangements of deep fibres, which in his illustrations have a rope-like appearance and which encircle each ventricle. He regarded these bands as capable of imparting a twisting compression on the ventricles, producing a kind of *wringing* action, as had been suggested long before by Borelli (1681). Although they are not describing exactly the same fascicles, other workers have used the analogy of a rope in their descriptions, and particularly in those of Torrent Guasp, whose extensive studies of human, bovine and other vertebrate hearts are somewhat inaccessible, being published privately as monographs (1959, 1970). It is at this level of detail that many accounts become somewhat difficult to comprehend, as noted by Robb (1934), and Thomas (1957). Robb's brief historical survey points to the difficulties of equating the descriptions of many workers, except in their most general statements. Even in major details there exist contradictions. Thomas (1957), who dissected canine and porcine hearts (and contributed one of the few detailed studies of the *atrial* musculature), states that all tracts of myocardial muscle begin and end by actual attachment to the fibrous rings of the atrioventricular openings, except for those which extend into papillary muscles. This is the usual view, but one denied by Pettigrew (1864), who considered the myocardial 'fibres' to be continuous. This latter view, with certain differences, is also inherent in the descriptions by Torrent Guasp (1970) of continuous arrangements of helical bands, which turn on themselves like 'figures of eight' in encircling both ventricles, an analogy used by Lower (1669). Pettigrew (1864) claimed to have revealed a complex seven-layered structure, whereas most observers are content with a thin superficial layer, an irregular internal layer, concerned in the formation of trabeculae and papillary muscles, and a thick intermediate layer, partly interventricular and partly proper to each ventricle. The potential artificiality inherent in the technique of myocardial unravelling was admitted by Mall (1911), who emphasized that 'the fasciculi and sheets are never fully separated from adjacent fasciculi and sheets'. Torrent Guasp (personal communication, 1978) also admits that in his experience the number of strata which can be dissected out is indefinable. This has not prevented him from contributing perhaps the most complicated account of the arrangement of the myocardium in mammals, including mankind. This elaboration, the details of which cannot be included here, is nevertheless completely justifiable if its architectural details can be shown to accord with the mechanics of cardiac contraction and, equally, with the conducting pathways in the myocardium itself. Unfortunately, a comprehensible harmony between the architectural details of the myocardium and the dynamics of cardiac activity has not been achieved, except in the crudest terms. However, it is significant that in the most recent account of myocardial architectonics (Streeter 1979, 1980) the work of Torrent Guasp is in its major details confirmed. Streeter's own approach to the problem is a searching mathematical analysis of the angular orientation between myocardial 'fibres' at different depths through the ventricular wall. He finds his results much in accord with Torrent Guasp's concept of 'nested' spiral laminae

(6.32B), which he believes was also foreshadowed by the work of Krehl in 1891. Streeter's account offers the most up-to-date examination of the literature, which is far more extensive than can be cited here. Although it must be stated that all these efforts have so far failed to impart a clear picture of the *total* architectonic arrangement of the myocardium, there is perhaps already sufficient agreement to satisfy physiological concepts. The accepted existence of 'fibre-pathways' of varying obliquity from almost circular to almost longitudinal provides at least the powers required to decrease the ventricular cavities in all dimensions; the passage of very obliquely helical bands of fibres into the papillary muscles also provides an integrating mechanism between the atrioventricular valves and the cyclical changes in longtidunal dimensions of the ventricles. Some outstanding problems require further investigation, in particular the relation or 'attachment' of the ventricular myocardium to the misnamed atrioventricular 'rings' or annuli of the cardiac 'fibrous skeleton' (*see* further on p. 657).

COORDINATION OF CARDIAC ACTIVITIES
THE CONDUCTING SYSTEM

On average the human heart beats ceaselessly at about 60 to 100 cycles per minute for many decades, maintaining a perfusion of blood through the pulmonary and systemic tissues, the most critical maintenance being through the cerebral tissues, where even transient interruption leads to irreversible neuronal changes. The actual rate of the heart beat and the 'stroke volume' of blood delivered with each cycle, of course, fluctuates considerably from time to time in accord with prevailing physiological demands.

In the previous pages emphasis has been placed on the contrasting roles of the 'right' and 'left' hearts, the topography of the cardiac chambers, the construction and operation of the valves, the main elements of the fibrous skeleton, and a review of the general architecture of the myocardium. The principal events in one cardiac cycle, including the electrocardiogram, mechanical sequences of diastole, atrial systole, the isovolumetric contraction, ejection, and isovolumetric relaxation phases of ventricular systole, the main elements of the phonocardiogram, pressure profiles of the right and left hearts and arterial trunks, and the sequences of valve opening and closure are summarized in figure 6.33. There has now accumulated a vast literature on all aspects of cardiovascular physiology and no attempt will be made to pursue these further in the present volume. However, a brief survey of those structural elements that are involved in the *coordinated sequence* of events in the cardiac cycle, together with some indication of features that remain the subject of much controversy will be attempted. Clearly, the *efficiency* of the pumping action of the heart is heavily dependent upon the precise timing of operation of the interdependent structural complexes. Passive diastolic filling of the atria and ventricles is followed by atrial systole which completes ventricular filling. Excitation and contraction of both atria must therefore be complete before the onset of ventricular contraction and this is effected by the introduction of an *atrioventricular conduction delay*. Further, ventricular contraction is not haphazard but proceeds in a precise pattern; mechanisms exist that ensure the onset of graded papillary muscle contraction and hence control of the atrioventricular valves is followed rapidly by a wave of excitation and contraction that spreads from the ventricular apices, basally towards their outflow tracts and orifices, generating a rapid acceleration of blood during the ejection phase.

It is now generally accepted that the vertebrate heart beat is *myogenic*, originating in the myocytes themselves, neural influences serving to modify the intrinsic rhythm as functional demands vary. *All* varieties of myocyte in the heart (p. 515) are excitable cells and show the properties of autonomous rhythmic depolarization and repolarization of their cell membranes, conduction of the wave of depolarization via nexuses (gap junctions—p. 8) to neighbouring myocytes, and excitation-contraction coupling to their actomyosin complexes (p. 516). However, these properties are developed to very different degrees in different sites and in the various types of myocyte. The *rate* of rhythmic depolarization and repolarization is slowest in

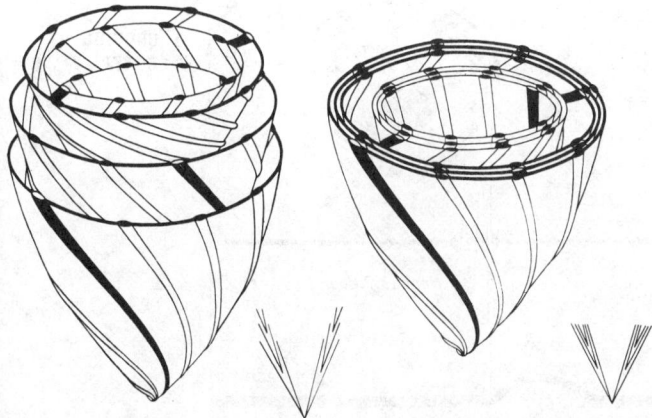

6.32B Diagrammatic representation of one concept of the arrangement of ventricular myocardial 'fibre bundles' from the work of F. Torrent Guasp (1970).

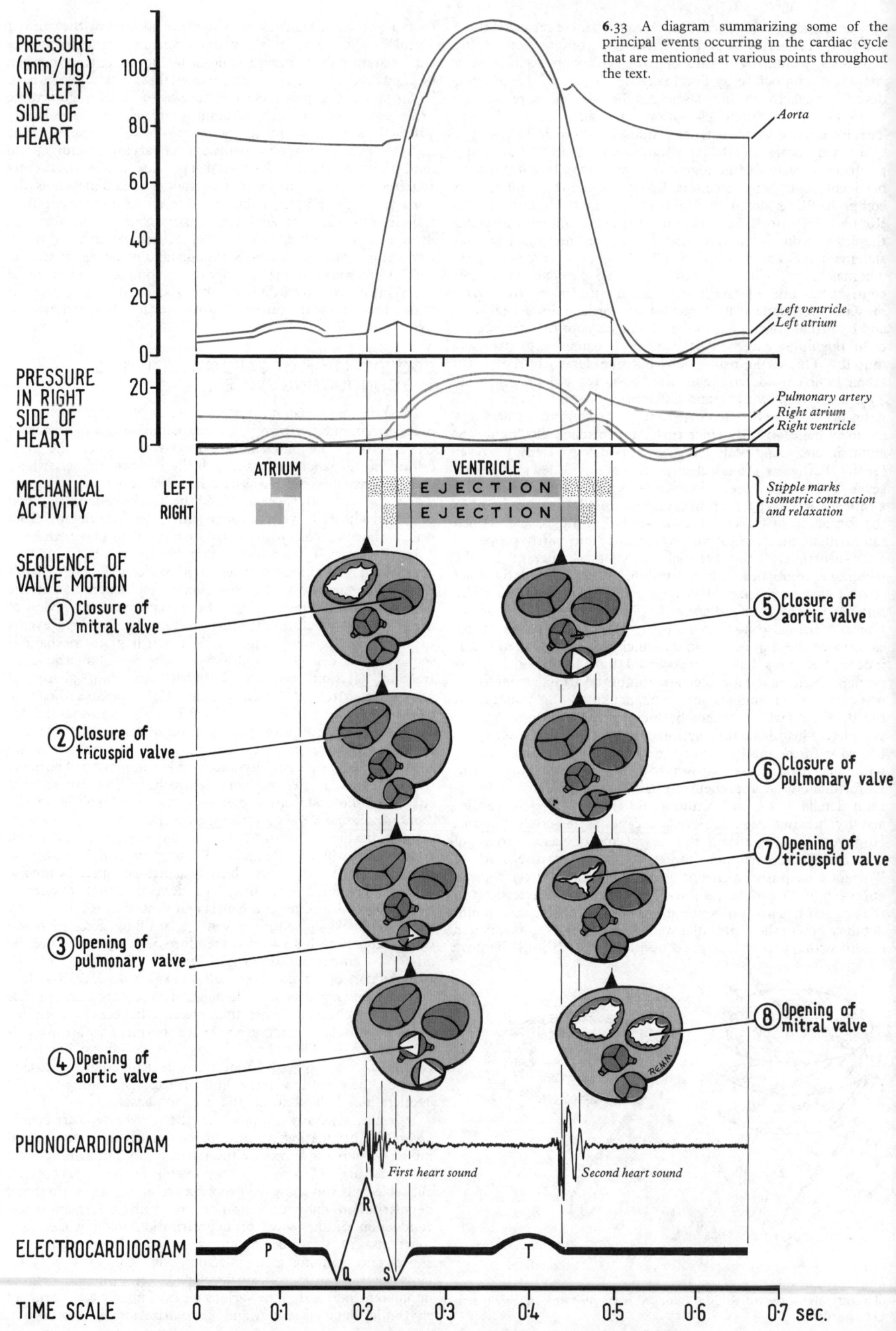

6.33 A diagram summarizing some of the principal events occurring in the cardiac cycle that are mentioned at various points throughout the text.

PRESSURE (mm/Hg) IN LEFT SIDE OF HEART

Aorta

Left ventricle
Left atrium

PRESSURE IN RIGHT SIDE OF HEART

Pulmonary artery
Right atrium
Right ventricle

MECHANICAL ACTIVITY

LEFT
RIGHT

ATRIUM

VENTRICLE

EJECTION
EJECTION

Stipple marks isometric contraction and relaxation

SEQUENCE OF VALVE MOTION

① Closure of mitral valve

② Closure of tricuspid valve

③ Opening of pulmonary valve

④ Opening of aortic valve

⑤ Closure of aortic valve

⑥ Closure of pulmonary valve

⑦ Opening of tricuspid valve

⑧ Opening of mitral valve

PHONOCARDIOGRAM

First heart sound

Second heart sound

ELECTROCARDIOGRAM

P

R

Q S

T

TIME SCALE

0   0·1   0·2   0·3   0·4   0·5   0·6   0·7 sec.

ventricular myocardium, intermediate in atrial myocardium, and fastest in sinuatrial nodal myocytes derived from sinus venosus. The latter, therefore, override the tissues with slower rhythms and, in the normal heart, provide the locus for the repeated initiation of each cardiac cycle. Conversely, conduction *velocity* is slow in nodal and transitional myocytes, intermediate in general 'working' cardiac myocytes and fastest in Purkinje myocytes.

The accumulations, tracts and networks of these various cell types (nodal, transitional and Purkinje myocytes) constitute the *conducting system* or *specialized conducting tissues* of the heart, and while all authorities are in agreement concerning the main constituent parts of the system, their precise cellular makeup and the existence of certain other elements remain debatable.

Generally accepted parts of the conducting system are the sinuatrial (sinus) node, the atrioventricular node, the common atrioventricular bundle, the left and right bundle branches, and the subendocardial network of Purkinje 'fibres'. Strongly advocated by some workers, but rejected by others, are interatrial and internodal tracts of specialized conducting tissue, atrioventricular 'bypass' fibres, and accessory atrioventricular bundles. These various constituents are illustrated in figures **6**.34, 35. For excellent reviews and extensive bibliographies concerning the cardiac cycle the reader should consult Rushmer (1976), and Schlant (1978); similarly, comprehensive (and sometimes controversial) treatments of the cardiac conducting system are to be found in Hudson (1965), Anderson (1974, 1975, 1980), James (1978), James and Scherf (1978).

The sinuatrial node (or sinus node) of Keith and Flack (1907), often called the 'pacemaker' of the heart, because certain of its specialized cells are generally considered to be the site of initiation of excitation in each cardiac cycle, is, as its name suggests, situated at the junctional zone between parts of the mature right atrium derived from the embryonic sinus venosus and primitive atrium. The histological features of the core of the node are quite distinctive, but its periphery is much less obvious and the latter doubtless accounts for some lack of agreement concerning its precise dimensions, shapes and boundaries. Contrary to earlier views the nodal tissue does not extend through the full thickness of the atrial wall from epicardium to endocardium but lies, throughout most of its extent, some 1 mm from the epicardium and rather more from the underlying endocardium. It is often covered by a plaque of subepicardial fat. The node is elongated, extending from the right part of the groove between the root of the right atrial appendage and the anterolateral aspect of the termination of the superior vena cava, then continuing posteroinferiorly to invade the upper part of the crista terminalis. Variously described as a flattened ellipsoid, an 'Indian war club', or possessing a 'head', 'body' and 'tail', in shape, it is generally conceded to be 10–20 mm in length, some 1 mm thick and approximately 3 mm wide at its maximum lateral convexity. (Here, and later, when mentioning the architecture of the conducting tissues, the classification of cardiac cells proposed by James and his co-workers in a series of publications, and summarized in James (1978), will be followed: a brief account of this architecture is given on p. 663. Their classification was based upon exhaustive series of careful dissections, light and electron microscopic studies, and attempts to correlate these with electrophysiological evidence. However, the reader should note that by no means all authorities recognize the full range of cardiac myocyte types proposed by James as being present in all the sites mentioned.)

A most characteristic feature of the sinuatrial node is that it is traversed throughout its length by a centrally placed *artery of the sinuatrial node* which has an unexpectedly large lumen (*see* p. 669 for the variable origin of this vessel). The adventitia of the vessel merges into a dense interlacing collagenous framework, firm to palpation, which permeates the node, and in the interstices of which the myocytes are embedded. Only a few small lateral branches leave the nodal artery as the nutrient supply to the nodal tissue itself; the main vessel continues beyond the node to ramify in the general atrial myocardium. The interesting suggestion has been made that the enlarged adventitia of the nodal artery may subserve a sensory function monitoring aortic pressure and pulsation, and forming part of a feedback loop which stabilizes an appropriate sinus rhythm; pharmacological studies in general support such a view. The myocytes of the node have, on the basis of light microscopy, for long been described as 'primitive' cardiac cells, slender and fusiform or branching, and characteristically pale-staining; the majority of the earlier authors considered that such nodal cells made specialized contacts *directly* with ordinary atrial wall myocytes at the margins of the node. James, however, distinguishes the following cell varieties in the sinuatrial nodal complex and its immediate environment: rounded, polygonal, fusiform or occasionally stellate *nodal myocytes* (pale-staining or P-cells), a heterogeneous and structurally more complex group of short slender *transitional myocytes*, *Purkinje myocytes*, and finally general atrial 'working' *cardiac myocytes*.

The nodal myocytes are confined to the central regions of the node and are circumferentially arranged in the immediate vicinity of the nodal artery, but more irregularly grouped external to this. These cells are now considered to be the origin of pacemaker activities; they make functional contacts with each other and with adjacent transitional cells. The latter are both shorter and of smaller diameter than general myocardial cells and their internal organization varies from the simple almost organelle-free type (as in the P-cell) to one with a well-developed contractile system and complex sarcoplasmic reticulum. They are slow-conducting cells and are interposed between the P-cells and either general myocardium or Purkinje cells. The latter are shorter, and wider than general myocardial cells (their ultrastructure is mentioned on p. 518) and they are thought to have a rapid conduction velocity. It should be noted that in the human heart and the hearts of small mammals in general, the cells in question are not markedly wider than those of the general myocardium (in contrast to the prominent Purkinje cells found in the larger ungulates): accordingly, some authorities do not consider Purkinje cells to be present in the atrial walls and reserve the term for the more distal parts of the ventricular conducting system. James, however, holds the view that Purkinje myocytes are numerous around all the margins of the sinuatrial node (and also extending into the interatrial and internodal tracts—*vide infra*). Thus, he considers that excitation of atrial myocardium follows the sequence: pacemaker activity of nodal myocytes, slow conduction along transitional myocytes, fast conduction along Purkinje myocytes, and finally spreading excitation-contraction between general atrial myocytes.

The atrioventricular node of Tawara (1906) is roughly oval or elliptical in outline, its long (anteroposterior) axis is some 7–8 mm, its maximal vertical dimension being 3 mm, and transversely it is about 1 mm thick. It lies beneath the endocardium of the septal wall of the right atrium in the triangle of Koch (p. 644) immediately dorsal to the basal attachment of the septal leaflet of the tricuspid valve, 1 cm or less above the septal margin of the orifice of the coronary sinus, and encircled by the arching subendocardial collagenous tendon of Todaro. Its right atrial surface is convex whilst its left surface is concave abutting on the base of the mitral valve annulus. Its dorsal extremity projecting into the atrial septum narrows forming the nodal crest whilst its antero-inferior extremity which is closely related to the central fibrous body of the heart continues into the common atrioventricular bundle. The node is everywhere permeated by an irregular collagenous framework enmeshing the myocytes; the framework is, however, much less dense than that of the sinuatrial node. The arterial supply to the node is given on p. 671. *Nodal myocytes* form a relatively small population of cells confined to the deeper parts of the node, particularly near the central fibrous body, and the principal cellular elements are *transitional myocytes*: the latter are irregularly arranged in the dorsal part of the node but become increasingly polarized into longitudinal tracts where they extend into the common atrioventricular bundle. All surfaces of the node are also held to be encrusted with Purkinje myocytes which (1) receive the fine terminals of internodal conduction paths and 'bypass' fibres (*vide infra*) and (2) continue as one of the cellular components of the atrioventricular bundle.

It is probable that the large population of slender, slow-conducting transitional myocytes in the atrioventricular node is responsible for the functionally essential *atrioventricular conduction delay* mentioned above.

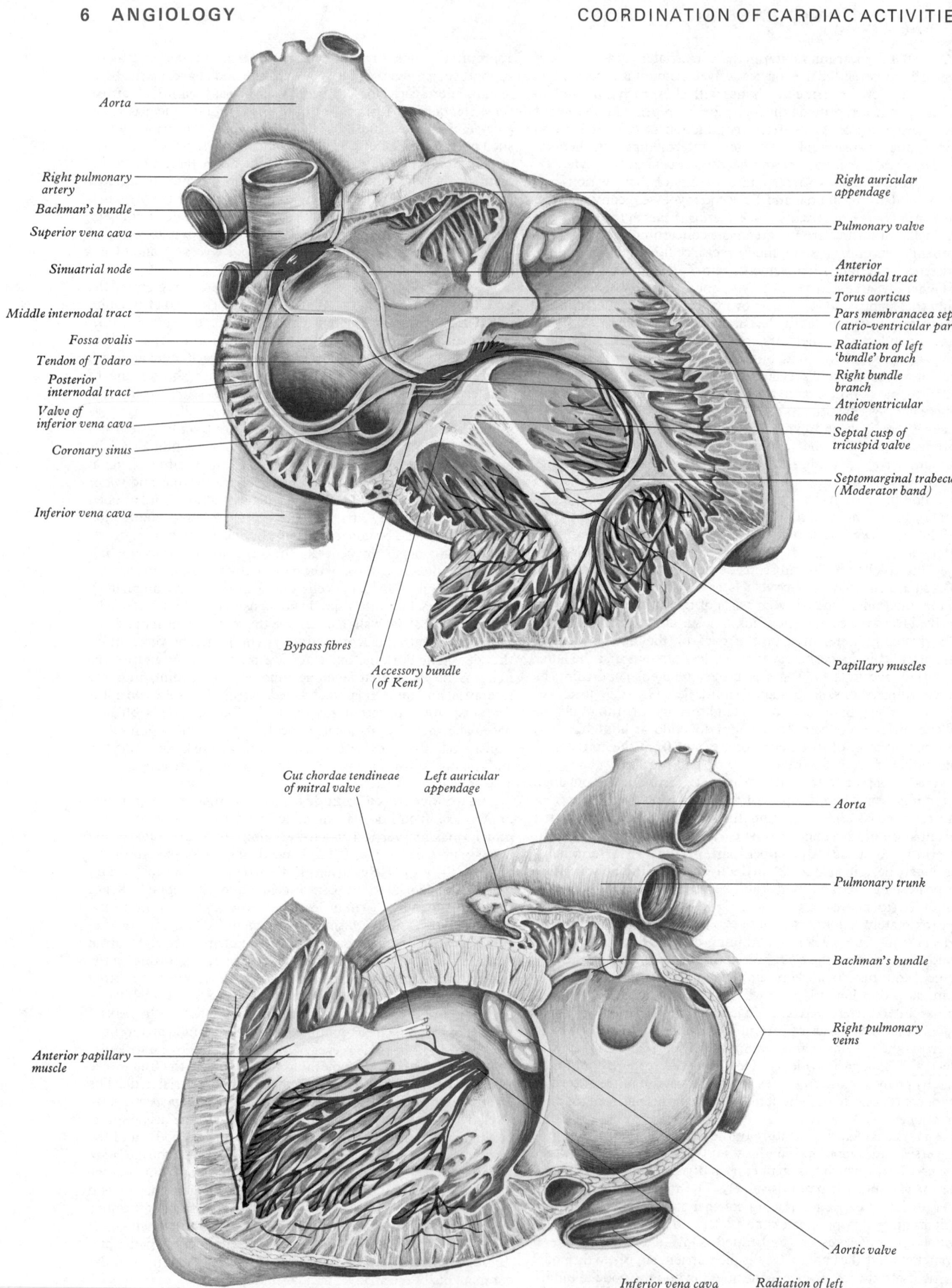

Aorta

Right pulmonary artery

Bachman's bundle

Superior vena cava

Sinuatrial node

Middle internodal tract

Fossa ovalis

Tendon of Todaro

Posterior internodal tract

Valve of inferior vena cava

Coronary sinus

Inferior vena cava

Bypass fibres

Accessory bundle (of Kent)

Right auricular appendage

Pulmonary valve

Anterior internodal tract

Torus aorticus

Pars membranacea septi (atrio-ventricular part)

Radiation of left 'bundle' branch

Right bundle branch

Atrioventricular node

Septal cusp of tricuspid valve

Septomarginal trabecula (Moderator band)

Papillary muscles

Cut chordae tendineae of mitral valve

Left auricular appendage

Aorta

Pulmonary trunk

Bachman's bundle

Right pulmonary veins

Anterior papillary muscle

Aortic valve

Inferior vena cava

Radiation of left 'bundle' branch

6.34, 35   A diagram of the conducting tissues of the heart as seen from the right (**6**.34) and left (**6**.35) aspects. Generally accepted elements are shown in *red*; accessory atrioventricular paths and 'bypass' bundles in *blue*. The internodal and interatrial pathways shown in *yellow* are subject to some controversy. Consult text for further discussion.

The atrioventricular bundle of His (1893) is formed by the convergence of longitudinal tracts of myocytes at the antero-inferior pole of the atrioventricular node. Narrowing rapidly, the bundle, ensheathed in a delicate tunic of vascularized connective tissue, becomes oval, quadrangular, or triangular in transverse section and enters a channel in the central fibrous body (right fibrous trigone—*see* p. 657). Traversing the latter, the undivided (common) bundle reaches the dorsal margin of the pars membranacea septi; thereafter it courses along its postero-inferior margin to reach the crest of the muscular interventricular septum. It is commonly stated that at the crest the common bundle divides into *right* and *left bundle branches* implying a simple bifurcation, but this is misleading in a number of respects. The *right bundle branch* continues as a relatively narrow, discrete rounded aggregation of fascicles which runs at first in a myocardial channel and then subendocardially towards the right ventricular apex; it then enters the septomarginal trabecula which conveys it to the base of the anterior papillary muscle. In its septal course towards the trabecula the bundle gives off relatively few side branches to the ventricular walls, but having reached the papillary muscle it breaks up into a profuse network of fine subendocardial fibres, which first embrace the papillary muscles and then pursue a *recurrent* course towards the ventricular base reaching all parts of the ventricular parieties. The *left bundle branch* is something of a misnomer because numerous fine fascicles, each with its delicate connective tissue sheath, leave the left margin of the common bundle throughout much of its course along the membranous septum: these fascicles form a wide *flattened sheet* which arches over the crest of the muscular septum. The diverging sheet passes apically and subendocardially across the left ventricular aspect of the septum and after 2–3 cm it then separates into anterior and posterior sheets destined for the bases of the papillary muscles bearing the same names. Fine lateral branches leave the margins of the sheaths, but their terminations again form complex subendocardial networks which encompass first the papillary muscles, and then pursue recurrent paths to all parts of the subendocardial ventricular walls.

The common bundle at its outset contains extensions of *transitional myocytes* from the atrioventricular node, but in its distal part, followed by the bundle branches, and terminal subendocardial network, the cells gradually increase in diameter and assume the characteristics of polarized tracts of *Purkinje myocytes*. Between the latter, however, are intermingled tracts of what ultrastructurally appear to be general cardiac myocytes; their conduction characteristics are unknown.

It should be noted that the principal bundle branches are largely insulated from the surrounding myocardium by their sheaths of connective tissue, functional contacts between Purkinje myocytes and the ventricular myocardium first appearing in substantial numbers with the emergence of the profuse subendocardial network. Thus, not only do the papillary muscles contract first followed by a wave of excitation-contraction that proceeds *from* the ventricular *apex* towards the basally situated outflow tract and orifice, but, because of the subendocardial position of the Purkinje network, mural ventricular excitation proceeds from the endocardial to the epicardial surface. However, little is yet known concerning the details of the three-dimensional array of Purkinje myocyte terminals in the ventricular walls; it is even uncertain to what depth they penetrate the walls.

Interatrial and internodal conduction paths. For many years it has been standard teaching in physiological textbooks that sinuatrial pacemaker activity caused *direct* excitation of general atrial cardiac myocytes, and that this spread radially via their intercellular nexuses to encompass both atria, the wave of excitation eventually reaching the atrioventricular node. It was further assumed that the pathways followed by the wave merely reflected the complex topography of the atrial walls and in particular the positions of their orifices. Indeed, a number of modern authorities dismiss the view that any specialized interatrial and internodal conduction pathways exist (*see* for example Anderson *et al.* 1974, 1975; Anderson 1975; Rushmer 1976), or that if they exist they have no demonstrable physiological role in the normal human heart (*see* Marriot and Myerberg 1978).

However, since the earlier years of this century a number of authors have postulated the existence of such specialized routes (e.g. Wenckebach 1908; Thorel 1909, 1910; Bachmann 1916). Further interest was stimulated when electrophysiological studies demonstrated that impulses originating in the sinuatrial node arrived at the atrioventricular node more rapidly than if they had travelled through ordinary myocardium (Hoffman and Cranefield 1960). A strong protagonist of the existence of interatrial and internodal pathways is James (1978) who claims that they may even be demonstrated by careful dissection, and who has examined their histology and ultrastructure. The following pathways have been described, each consisting of an admixture of Purkinje myocytes and ordinary cardiac myocytes.

The *anterior internodal tract* (including the *interatrial bundle of Bachmann*) leaves the anterior extremity of the sinuatrial node, skirts the anterior aspect of the termination of the superior vena cava and passes posteromedially to reach the anterior margin of the interatrial septum. Here, it divides, some fibres penetrating the septum to splay out in the walls of the left atrium (Bachmann's bundle); the remaining fibres descend posteriorly across the right side of the septum behind the torus aorticus to divide into terminal fascicles that approach the atrioventricular node (*vide infra*).

The *middle internodal tract* (of Wenckebach) emerges from the posterosuperior aspect of the sinuatrial node, curving posterior to the orifice of the superior vena cava to reach the interatrial septum, which it crosses, and breaks up in the neighbourhood of the atrioventricular node.

The *posterior internodal tract* (of Thorel) continues from the postero-inferior part of the sinuatrial node, continues through the substance of the crista terminalis and valve of the inferior vena cava, then proceeds to the posterior margin of the atrioventricular node; variable relationships of this tract to the ostium of the coronary sinus have been described.

The fine terminal branches of these internodal pathways are held to enter and make functional contact with transitional myocytes at many levels over the atrioventricular node, including its posterosuperior crest, throughout its convex right atrial surface, and a small number may even reach the root of the common atrioventricular bundle beyond the node. Such radicles thus constitute '*bypass fibres*' circumnavigating much or all of the nodal cells—they provide a possible anatomical basis for a *dual form* of atrioventricular transmission first proposed by Moe *et al.* (1956).

It must be re-emphasized here that the existence, routes, architecture and functional roles of interatrial and internodal conduction pathways remain the subject of vigorous debate; the results of future researches will be awaited with great interest.

Finally, mention must be made of the question of **accessory atrioventricular bundles** the existence of which is often invoked to explain the clinico-pathological background of various cardiac arrhythmias. It is widely held that such bundles may be present near the principal common atrioventricular bundle, or at any point around the circumference of either the mitral or tricuspid fibrous 'annuli'. They are generally thought to be of slender dimensions, of irregular frequency, and of no functional significance in the *normal* cardiac cycle. Whether such bundles consist of specialized conducting tissue or of general myocardium is unresolved; much current research is being directed towards an anatomico-physiological analysis and an assessment of the applied surgical relevance of such connexions.

CARDIAC NERVE SUPPLY

The initiation of each cardiac cycle of activity may be myogenic or neurogenic in invertebrates, but in vertebrates, including mammals, it is myogenic. In the vertebrate heart, with increasing specialization and elaboration of the organ, initiating foci and conducting pathways have become progressively more differentiated to integrate the cycle of successive contraction and relaxation in the cardiac chambers. This 'conducting system' has already been described (p. 659). However, the cardiac cycle is harmonized, in its rate, force, and output, by an extrinsic nerve supply, which operates upon the cardiac nodal tissue and its fascicular prolongations, upon the coronary vessels, and perhaps upon cardiac muscle. This nerve supply is autonomic and comprises both efferent and afferent fibres. The parasympathetic fibres reach the heart through branches of the vagus (p. 1123) and the sympathetic by rami of the sympathetic trunk (p. 1127). The vagal preganglionic fibres proceed from central origins in the brainstem, particularly the medulla, and these include the nucleus ambiguus (p. 1076), reticular nuclei (p. 946), and possibly the dorsal nucleus of the vagus (p. 1076). These pre-ganglionic axons leave the vagus as its cardiac branches and reach the cardiac plexuses without interruption (*see*, however, p. 1078). The sympathetic pre-ganglionic neurons are situated in the lateral grey column of the thoracic spinal cord, in its cranial five or six segments (Kuntz 1953). The pre-ganglionic fibres from these sources end in the cervical and third and fourth thoracic sympathetic ganglia (Mitchell 1953), from all of which bilateral cardiac nerves, consisting of post-ganglionic fibres, proceed to the heart (pp. 1127, 1132).

The central connexions of the cardiac pre-ganglionic neurons, parasympathetic and sympathetic, are described elsewhere (*see* Reticular System, Hypothalamus, and Cerebral Cortex). The existence and behaviour of these integrating influences is better understood in terms of function, the precise locations of connecting pathways in spinal cord, brainstem, and cranial to this, being in general somewhat uncertain, perhaps because they do not form closely aggregated tracts.

As they reach the heart the sympathetic nerves (post-ganglionic fibres) and the parasympathetic cardiac branches of the vagus (pre-ganglionic fibres) together form a mixed *cardiac plexus* (p. 1132), usually considered divisible into a ventral (superficial) part, inferior to the aortic arch and ventral to the pulmonary artery, and a dorsal (deep) part situated between the aortic arch and the tracheal bifurcation. These plexuses contain ganglia, which also occur in the heart itself, along the distribution of the branches of the cardiac plexus, and their neurons are considered to be largely, if not exclusively, post-ganglionic parasympathetic in nature. With the advent of reliable staining techniques for cholinergic and adrenergic nerve cells, axons, and axon terminals, the intrinsic distribution of autonomic elements in the heart has been subjected to detailed examination, although, of course, methylene blue staining had permitted elucidation of many details prior to this. Even a brief historical review of such contributions is impossible here; Kuntz (1953), Mitchell (1956), and Pick (1970) have all examined earlier reports, and the more recent observations cited here are merely representative of an active field of study.

Cholinergic and adrenergic fibres, originating in or passing through the cardiac plexus, are distributed to the nodal tissue (sinuatrial and atrioventricular), to the atrial and ventricular myocardium, and to the coronary arteries and cardiac veins, in rodents and dogs, according to many observers (*see* Glabella 1976 for references). Whole mounts of the right atrial wall, in pigs, demonstrate the rich supply of cholinergic nerve fibres to this chamber, at least (Bojsen-Møller and Tranum-Jensen 1971); and the same technique shows extensions of the cardiac plexus to the nodes and to the atrioventricular bundle. These cholinergic fibres presumably originate in the small ganglia frequently described in the cardiac plexus and its prolongations. In man a single ganglion in the 'superficial' part of the plexus has been described; but in other mammals numerous small ganglia have been identified in the atrial walls, especially in the vicinity of the sinuatrial and atrioventricular nodes, and along the coronary arteries. No adrenergic neurons have yet been identified in such ganglia, but they contain chromaffin cells of the small intensely-fluorescent type (SIF-cells, *see* p. 1454), according to Nielsen and Owman (1968). Adrenergic fibres occur in all chambers of the heart; both cholinergic and adrenergic axons are said to be more abundant in the atria and adjacent to nodal tissue than elsewhere, but in the cat adrenergic fibres appear to be more densely distributed in the ventricles. According to Chiba and Yamauchi (1970) cholinergic fibres are more numerous in the atrial myocardium of the human heart and adrenergic axons in the ventricular wall.

In most reports nerve fibres are merely described as being in close relation to nodal tissue or myocardial cells; but Thaemart (1969) has reported neuromuscular junctions, with a gap of 78–100 mm, in the atria, ventricles and septa of the murine heart. Such ultrastructural studies have been extensively reviewed by Yamauchi (1973). To summarize, adrenergic and cholinergic terminals occur in the nodes, atrial and ventricular myocardium, the atrioventricular bundle, and the coronary vessels of a wide variety of laboratory mammals. The functional significances of such supplies are only in part understood, but are not difficult to speculate upon. Confirmation of such innervations in the human heart is still awaited, but it seems improbable that similar arrangements do not exist.

Afferent cardiac nerves have attracted less interest in anatomical studies, although their existence is physiologically undeniable. Some of the fibres which accompany the coronary arteries have long been considered to be visceral afferents, and afferents from the atria and the terminations of the caval and pulmonary veins are generally accepted. The precise routes of return of such visceral afferents are not fully established (p. 1136), but the cardiac branches of the vagus and of the middle and inferior cervical sympathetic ganglia probably provide the major pathways. Most recent work has concerned the actual 'terminations' of visceral afferents in the heart. Ultrastructural

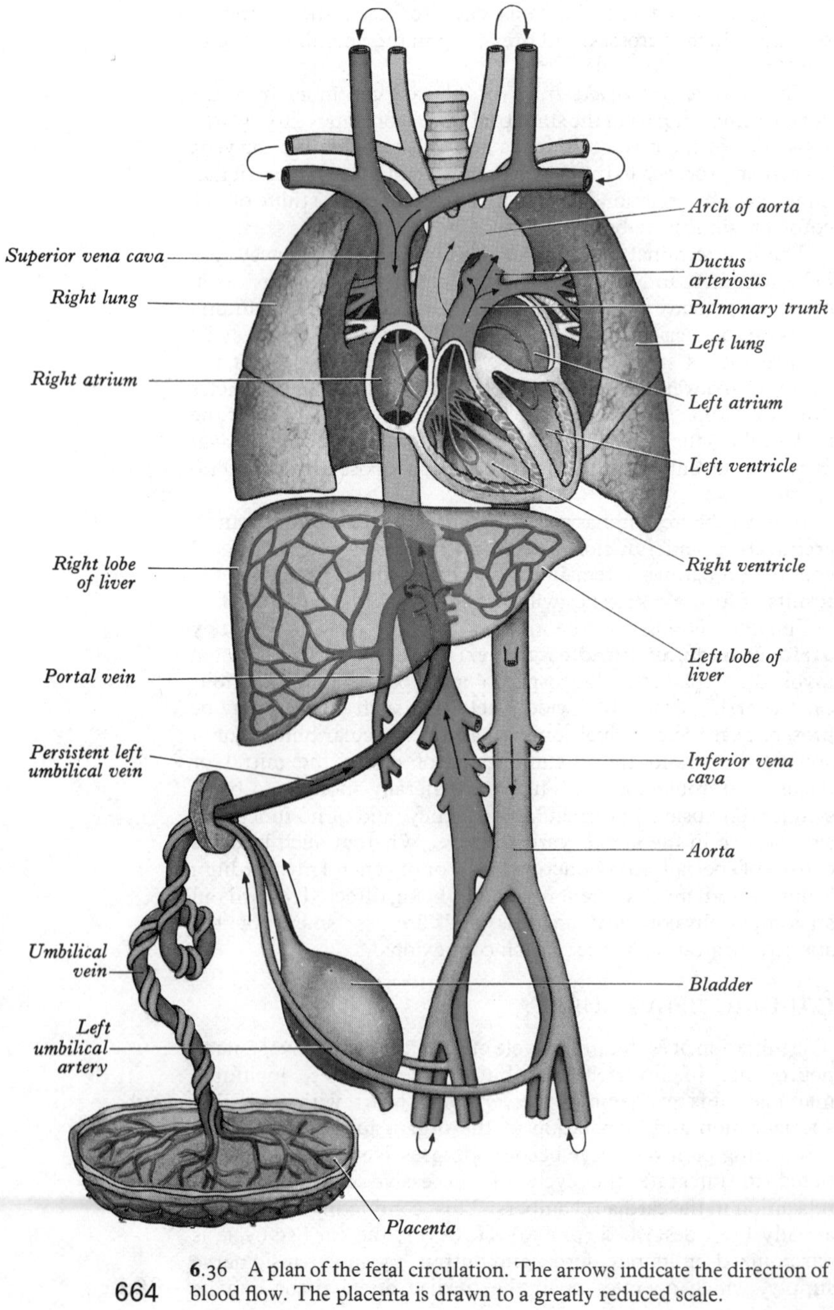
Arch of aorta
Superior vena cava
Ductus arteriosus
Right lung
Pulmonary trunk
Left lung
Right atrium
Left atrium
Left ventricle
Right lobe of liver
Right ventricle
Left lobe of liver
Portal vein
Persistent left umbilical vein
Inferior vena cava
Aorta
Umbilical vein
Bladder
Left umbilical artery
Placenta

6.36 A plan of the fetal circulation. The arrows indicate the direction of blood flow. The placenta is drawn to a greatly reduced scale.

identification of such endings is largely dependent upon mitochondrial accumulations, which are considered to be characteristic. By such criteria Kisch (1958) reported subendo-thelial afferent terminals in the atria and ventricles of mammals, including man; he considered that these mediated the pain phenomena of *angina pectoris*. Similar endings have been demonstrated also in the pulmonary valve (Chiba and Yamauchi 1970). Further developments in this field are awaited with interest.

The Vessels of the Heart

The arteries supplying the heart, the coronary branches of the aorta, are described on pp. 669–673. The cardiac veins and coronary sinus are described on pp. 737–738. For the lymphatic drainage of the heart *see* p. 860.

The Fetal Circulation

The fetal blood is carried to the placenta by the two umbilical arteries, and returned from the placenta to the early fetus by two umbilical veins, though in later fetal life the right umbilical vein disappears (p. 192). The persisting left vein enters the abdomen at the umbilicus and passes in the free margin of the falciform ligament to the visceral surface of the liver, where it gives off two or three branches to the left lobe, including the quadrate lobe. At the porta hepatis it joins the left branch of the portal vein, from which opposite this point, a large vessel arises and ascends on the posterior aspect of the liver to join the left hepatic vein immediately before that vessel opens into the inferior vena cava. This vessel is termed the *ductus venosus*. During fetal life the portal vein is small compared with the umbilical vein, and the parts of its left branch, proximal and distal to where the umbilical vein joins it, function as branches of the latter vessel carrying oxygenated blood to the right and left parts respectively of the liver (**6**.36). It will be seen therefore that the blood conveyed by the left umbilical vein passes to the inferior vena cava in three different ways. Some enters the liver directly and is carried to the inferior vena cava by the hepatic veins; a considerable quantity circulates through the liver with the portal venous blood, before entering the inferior vena cava by the hepatic veins; the remainder passes into the inferior vena cava through the ductus venosus.

In the inferior vena cava the blood carried by the ductus venosus and hepatic veins mixes with that returning from the lower limbs and from the abdominal wall. It enters the right atrium, and, guided by the valve of the inferior vena cava, passes for the most part through the *foramen ovale* into the left atrium, where it mingles with a small quantity of blood returned from the lungs by the pulmonary veins. A small amount of blood returned to the heart by the inferior vena cava, however, instead of passing through the foramen ovale, goes through the right atrioventri-cular orifice with the blood from the superior vena cava. From the left atrium blood passes into the left ventricle, and from that cavity into the aorta, through the branches of which it is probably distributed almost entirely to the heart itself and to the head and the upper limbs, only a small quantity being carried into the descending aorta. The blood from the head and the upper limbs is returned by the superior vena cava to the right atrium and all of it passes through the right atrioventricular orifice, carrying with it a small amount of the blood returned by the inferior vena cava. Having reached the right ventricle this blood passes into the pulmonary trunk. The lungs of the fetus being inactive, only a small quantity of the blood conveyed by the pulmonary trunk is distributed to them by the right and left pulmonary arteries, and returned by the pulmonary veins to the left atrium; the greater part passes through the *ductus arteriosus* into the aorta, where it mixes with the small quantity of blood transmitted by the left ventricle into this part of the aorta. It descends through the aorta and is in part distributed to the lower limbs and to the viscera of the abdomen and pelvis, but most of it is conveyed by the umbilical arteries to the placenta.

This account of the circulation of the blood in the fetus has been confirmed by radiographic observations after the injection of radio-opaque substances, and timing the appearance of tracer material in different parts of the circulation after injection of isotopically labelled serum albumin into the bloodstream in fetal sheep (Barclay *et al.* 1944; Dawes 1961). The following points, which are supported by evidence provided by blood-gas analysis can be inferred: (1) The placenta serves the purposes of nutrition and excretion, receiving deoxygenated blood from the fetus and returning it charged with oxygen and nutritive material. (2) Some of the blood of the left umbilical vein traverses the liver before entering the inferior vena cava; this is correlated with the relatively large size of the liver, especially at an early period of fetal life. (3) Only the pulmonary veins open directly into the left atrium, and the volume of blood which enters it from this source is very small. On the other hand, the volume of the blood entering the right atrium is much greater, and the pressure within that chamber is much higher than the pressure in the left atrium. As a result, the flap-like septum primum (p. 184) is thrust over to the left (**2**.103) and the passage of blood from the right to the left side of the heart is effected easily. The valve of the inferior vena cava is placed so as to enable it to direct nearly all the blood which issues from that vessel to the foramen ovale and so to the left atrium, whereas the blood entering the right atrium from the superior vena cava passes directly into the right ventricle. (4) The refreshed blood carried from the placenta to the fetus, mixed with the blood from the portal vein and inferior vena cava, passes almost directly to the arch of the aorta, and is distributed by the branches of that vessel to the head and the upper limbs. (5) The blood contained in the descending aorta, chiefly derived from that which has already circulated through the head and the upper limbs, together with a small quantity from the left ventricle, is distributed to the abdomen, the lower limbs, and placenta.

The Changes in the Vascular System at Birth

At birth, when pulmonary respiration begins, an increased amount of blood from the pulmonary trunk passes through the pulmonary arteries to the lungs, and a correspondingly increased amount returns by the pulmonary veins to the left atrium. At the same time there is a fall in pressure in the inferior vena cava due to reduction of the venous return by occlusion of the umbilical vein. The pressures within the two atria become equalized, and the foramen ovale, which is valve-like in character, is closed by the apposition, and later by the fusion, of the septum primum to the septum secundum (**2**.103). Possibly, contraction of septum primum musculature, synchronized with that in the wall of the superior vena cava, may play some part in this closure (which occurs some considerable time after *functional* closure of the ductus arteriosus—*vide infra*, and Barclay and Franklin 1938). Not infrequently the fusion is incomplete and a potential communication between the two atria may persist throughout life. Such a communication, unless large, has no functional significance; because of the equality of the intra-atrial pressures and the valve-like arrangement of the opening, no blood can pass from one side to the other.

When the umbilical cord is ligated and the placental circulation is cut off, the umbilical vein becomes thrombosed (*see* p. 764), and is gradually converted into a fibrous vestige, the so-called *ligamentum teres* of the liver. The umbilical vessels have thick muscular walls but are devoid of a nerve supply in their extra-abdominal portions. They constrict in response to a variety of stimuli, such as handling, stretching, cooling and altered oxygen and carbon dioxide tensions in the blood. The ductus venosus also becomes obliterated but the mechanism is not known; there is evidence that it is absent or already closed in about 30 per cent of newborn infants (Rudolph *et al.* 1961). Its fibrous remnant is termed the '*ligamentum*' *venosum* of the liver. Following ligation of the umbilical cord, the umbilical (hypogastric) arteries become thrombosed from the point at which they give off their last branches—the superior vesical arteries—to the umbilicus. They are subsequently converted into fibrous cords which lie in the extra-peritoneal fatty tissue of the abdominal wall and produce the *medial umbilical folds* of peritoneum.

Obliteration of the ductus arteriosus is also essential. This channel contracts rapidly at first. Blood probably flows intermittently through it for a week or two after birth. The direction of blood flow in this vessel is reversed, due to a rise in the systemic vascular resistance resulting from exclusion of the placental circulation and a fall in the pulmonary resistance with expansion of the lungs. When blood flows through the narrowed ductus arteriosus loud murmurs may be heard over the left chest (Burnard 1959). Anatomical closure of the ductus arteriosus occurs by proliferation of the lining endothelium but takes some months to complete. The initial constriction of the ductus arteriosus at birth has been attributed to a direct effect of the raised oxygen tension in the blood on its muscular wall. A nervous factor may also be involved, for the muscular wall is provided with both afferent and efferent nerve endings and responds to adrenalin and noradrenalin (Barcroft 1941; Franklin 1939; Barclay *et al.* 1942).

Before birth the ductus arteriosus is the direct continuation of the pulmonary trunk, and has a similar calibre. Ultimately it becomes an impervious cord which connects the left pulmonary artery (near its origin) with the arch of the aorta, the *ligamentum arteriosum*. For a general review of perinatal vascular changes *see* Dawes (1969).

Abnormalities of the Heart

Malformations of the heart are relatively common and probably represent about one-quarter of all developmental abnormalities. Their incidence is estimated at 6 per 1,000 live births and about 2 per cent of stillbirths. Owing to a heavy early mortality, the incidence in older children is about a half of that in live births. A few cardiac anomalies can be attributed to genetic or environmental factors; the majority are believed to be multifactorial and genetic factors may be effective only in certain environmental conditions (Abbot 1936; Brown 1950; Taussig 1961; Hudson and Wendell-Smith 1966).

Complete absence of the heart, *acardia*, is a rare anomaly. The most recent surveys account for 155 cases (Napolitani and Schreiber 1960; Severn and Holyoke 1973), with an estimated incidence of 1 in 34,600 births (Gillian and Hendricks 1957). Afflicted individuals are almost always monochorionic twins, depending for their development upon the other twin's heart.

1. Abnormalities of Position

(*a*) The position of the heart may be completely reversed so that the apex is directed towards the right instead of the left. This condition is associated with mirror image positioning of the great vessels and of the aortic arch. It may be a part of general transposition of the viscera or *situs inversus*, or the condition may affect only the heart (*dextrocardia*).

(*b*) *Ectopia cordis.* The heart may project on the surface of the thorax through a gap in the lower part of the chest wall. This is associated with a breakdown of the thin body wall and anterior part of the pericardium at a very early stage of development.

2. Abnormalities Due to Failure of Normal Developmental Processes

This group includes the majority of anomalies, of which the commonest is a complete or partial failure of division of the common chambers of the heart into right and left. In the complete form there is a persistence of a common atrium or common ventricle or both (*cor triloculare* and *cor biloculare*). When incomplete, the defects involve the interatrial septum, the interventricular septum or both these septa, or failure of division of the bulbus cordis and truncus arteriosus. The defects in each case tend to occur at specific sites.

(*a*) **Atrial septal defects.**

(i) *Defects of the ostium secundum.* The persistence of a slit-like communication between the right and left atria is common and results from failure of fusion, after birth, of the septum primum with the margin of the septum secundum. It is rarely of functional importance. The communication may be large if the septum primum is insufficient to occlude the foramen ovale.

(ii) *Persistence of an ostium primum.* The normal occlusion of the ostium primum by the union of the free edge of the septum primum with the fused atrioventricular cushions fails to occur. This is often associated with malformation of the septal cusp of the tricuspid valve or of the posterior cusp of the mitral valve, and sometimes with underdevelopment of the septum secundum.

(iii) *Other atrial septal defects.* Rarely defects occur dorsal to the fossa ovalis. These may be associated with drainage of pulmonary veins into the right atrium. They are difficult to explain in terms of normal developmental processes.

(*b*) **Ventricular septal defects.**

(i) *Defects of the membranous part.* The commonest defect is situated in the right wall of the outflow channel of the left ventricle, the aortic vestibule, below the commissure between the right and anterior cusps of the aortic valve. Viewed from the right ventricle the defect lies beneath the septal cusp of the tricuspid valve and below the *supraventricular crest* (p. 645). It results from the failure of development of the membranous part of the interventricular septum. This defect rarely occurs alone and is usually associated with an overriding of the muscular part of the interventricular septum by the aortic orifice, pulmonary stenosis or atresia and hypertrophy of the right ventricle (*Fallot's tetralogy*). Rarely the pulmonary trunk is normal or even dilated (*Eisenmenger's complex*). This defect in the interventricular septum may also be associated with transposition of the great vessels (*vide infra*) in which it is the pulmonary trunk which overrides the muscular part of the interventricular septum (*Tausig-Bing syndrome*). In these interventricular septal defects the atrioventricular bundle and its right and left limbs are in most cases found along the postero-inferior margin of the defect; occasionally they lie on the anterior margin (Campbell 1965).

(ii) *Defects of the proximal bulbar septum.* Less commonly a defect occurs in that part of the interventricular septum which is formed from the bulbar ridges. The defect is found below the commissure between left and posterior cusps of the pulmonary valve and between the anterior and right cusps of the aortic valve. It constitutes a communication between the outflow channels of the two ventricles.

(iii) *Defects of the muscular part.* Rarely, a defect occurs in the muscular part of the interventricular septum (*Maladie de Roget*).

(*c*) **Combined atrial and ventricular septal defects.**

These are usually associated with a persistent common atrioventricular orifice with defects in the adjacent parts of the septa. The defect is bounded below and behind by the free margin of the muscular part of the interventricular septum. The atrioventricular valves are abnormal; there is frequently a superior and inferior cusp common to both sides of the heart and draped across the free border of the muscular interventricular septum.

(*d*) **Persistent truncus arteriosus** (Truex and Bishof 1958).

In this condition a large undivided channel lies above and astride the free margin of the muscular interventricular septum and communicates with both ventricles. The right and left pulmonary arteries arise independently from this channel which continues beyond their origin as the ascending aorta. The persistent truncus arteriosus usually possesses a valve with four semilunar cusps. The defect is due to a failure of development of the bulbar ridges and the aorto-pulmonary septum.

3. Abnormalities Resulting from Defective Progress in some Developmental Process (Truex and Bishof 1958).

(*a*) **Abnormal bulbar and truncal septation.**

(i) *Complete transposition*, in which the aorta arises from the right ventricle and the pulmonary trunk from the left ventricle. This may be associated with an incomplete interventricular septum, or with a defective interventricular septum of the type described above (Tausig-Bing syndrome).

(ii) *Varying degrees of partial transposition of the great arteries*, usually associated with a defective interventricular septum of the type described in (2.*b*. i). In the majority of cases in this category the pulmonary trunk arises from the right ventricle but the position of the aorta varies from one in which it arises wholly from the right ventricle to varying degrees of overriding of the ventricular septum.

(*b*) **Anomalous entry of the great veins.**

(i) *A persistent left superior vena cava* may drain into the coronary sinus or into the superior aspect of the left atrium.

(ii) *The inferior vena cava* may be duplicated.

(iii) *The right pulmonary vein*, or, less frequently, all the pulmonary veins, may drain into the right atrium. Any of the four pulmonary veins may drain into some part of the systemic venous system in the thorax, or less frequently, in the abdomen.

4. Abnormalities of the Visceral Arches

(*a*) Right aortic arch (*see* p. 187).

(*b*) Patent ductus arteriosus (*see* p. 666).

(*c*) Coarctation of the aorta (*see* p. 710).

Congenital malformations of the heart are often multiple, probably occur more frequently in sibs and in children of consanguineous marriages, but there is a rather low concordance between monozygotic twins. About 20 per cent of all cardiac abnormalities are ventricular septal defects; persistent ductus arteriosus, atrial septal defects, coarctation of the aorta, pulmonary stenosis, Fallot's tetralogy and transposition of the great vessels each account for about 10 per cent and aortic stenosis for about 5 per cent. For a discussion of possible causal mechanisms consult Shaner (1962).

THE ARTERIAL SYSTEM

The Pulmonary Trunk

The pulmonary trunk (**6**.37, 38, 41) conveys deoxygenated blood from the right ventricle of the heart to the lungs (*see* Fishman and Hecht 1969). It is about 5 cm in length and 3 cm in diameter, and arises from the base of the right ventricle above and to the left of the supraventricular crest. It runs upwards and backwards, at first in front of the ascending aorta, and then to its left side. In the concavity of the aortic arch it divides, at the level of the fifth thoracic vertebra, into right and left pulmonary arteries, which are of nearly equal size.

Relations. The whole of the pulmonary trunk is contained within the pericardium. Together with the ascending aorta it is enclosed in a common tube of the visceral layer of the serous

of the oesophagus and the right bronchus, to the root of the right lung, where it divides into two branches. The lower and larger of these is distributed to the middle and lower lobes of the lung; the upper and smaller accompanies the upper right lobar bronchus.

The left pulmonary artery, a little shorter and smaller than the right, runs horizontally in front of the descending aorta and left bronchus to the root of the left lung, where it divides into two branches, one for each lobe of the lung. Above, it is connected to the concavity of the aortic arch by the ligamentum arteriosum, on the left of which is the left recurrent laryngeal nerve, and on the right the superficial part of the cardiac plexus. The pericardial fold of the left vena cava (p. 635) passes from its lower border to the upper left pulmonary vein.

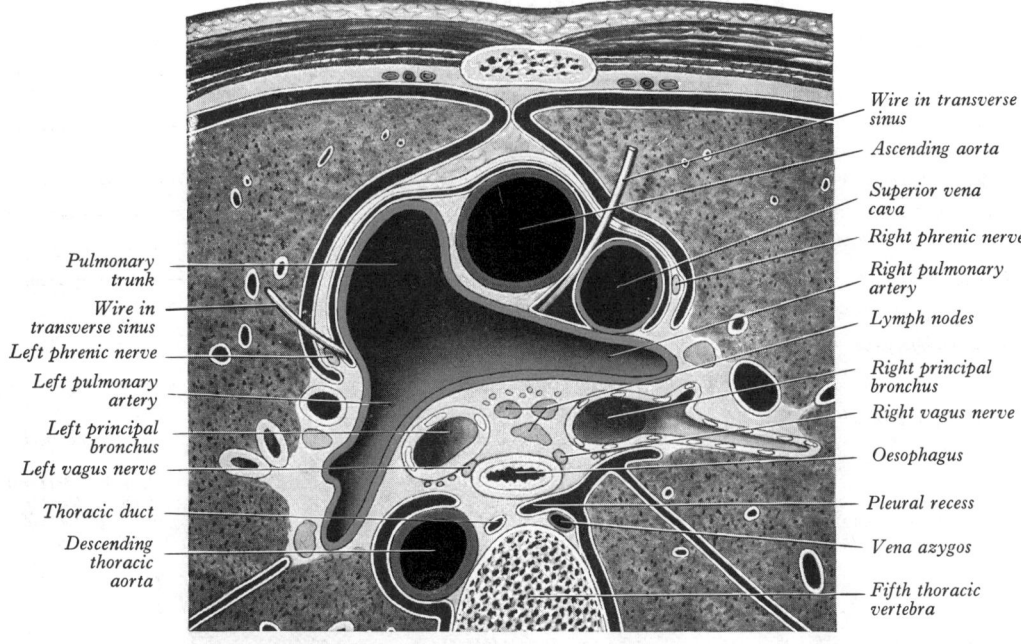

Pulmonary trunk

Wire in transverse sinus

Left phrenic nerve

Left pulmonary artery

Left principal bronchus

Left vagus nerve

Thoracic duct

Descending thoracic aorta

Wire in transverse sinus

Ascending aorta

Superior vena cava

Right phrenic nerve

Right pulmonary artery

Lymph nodes

Right principal bronchus

Right vagus nerve

Oesophagus

Pleural recess

Vena azygos

Fifth thoracic vertebra

6.37 A transverse section through the mediastinum at the level of the upper border of the fifth thoracic vertebra. Superior aspect.

pericardium. The fibrous layer of the pericardium is gradually lost upon the external coats of the two pulmonary arteries. *In front*, the pulmonary trunk is separated from the sternal end of the left second intercostal space by the pleura, the left lung and the pericardium. *Behind*, are at first the ascending aorta and the left coronary artery; at a higher level the pulmonary trunk lies in front of the left atrium, and the ascending aorta is on its right side. The auricle of the corresponding atrium and a coronary artery lie on each side of its origin. The superficial part of the cardiac plexus lies between the division of the pulmonary trunk and the arch of the aorta.

The right pulmonary artery, slightly longer and larger than the left, runs horizontally to the right, behind the ascending aorta, superior vena cava and upper right pulmonary vein, and in front

The pattern of branching of the right and left pulmonary arteries shows little variation (except in cases of dextrocardia, where the lobation of the lungs may also be reversed). Cory and Valentine (1959) found the above pattern constant in all 34 right and all 80 left of 114 pulmonary arteries observed at operation (*see* also p. 1265).

The treatment of congenital stenosis of the pulmonary trunk, both with and without an associated 'right to left intracardiac shunt' has advanced rapidly in recent years from the indirect 'systemic/pulmonary vessel anastomosis' (Blalock 1947), to the direct blind approach to the narrow pulmonary orifice and right ventricular outflow tract (Brock 1949), and finally via the open dry heart techniques employing hypothermia or suitable pump/ oxygenator circulatory replacement machines.

THE AORTA

The aorta is the main trunk of the system of vessels which convey the oxygenated blood to the tissues of the body. It begins at the upper part of the left ventricle, where it is about 3 cm in diameter, and after passing upwards and to the right for about 5 cm, arches backwards and to the left, over the root of the left lung; it then descends within the thorax on the left side of the vertebral column, gradually inclining towards the median plane, and enters the abdominal cavity through the aortic hiatus in the diaphragm. Considerably diminished in size (about 1·75 cm in diameter), it ends a little to the left of the median plane, at the level of the lower border of the fourth lumbar vertebra, by dividing into the right and left common iliac arteries. For convenience it is described in several portions, viz. *ascending*, the *arch*, and *descending*, the last being divided into *thoracic* and *abdominal* parts.

small dilatations called the *aortic sinuses*. At the continuation of the ascending aorta as the aortic arch the calibre of the vessel is slightly increased, owing to a bulging of its right wall. This dilatation is termed the *bulb of the aorta*, and, on transverse section at this level, the vessel presents a somewhat oval outline.

Relations. The ascending aorta is contained within the fibrous pericardium, and is enclosed in a tube of the serous pericardium, common to it and the pulmonary trunk (**6.**13). *Anteriorly*, its lower part is related to the infundibulum of the right ventricle (p. 640), the commencement of the pulmonary trunk and the auricle of the right atrium; higher up, it is separated from the sternum by the pericardium, the right pleura, the anterior margin of the right lung, some loose areolar tissue and the remains of the thymus; *posteriorly*, it is related successively to the left atrium, the

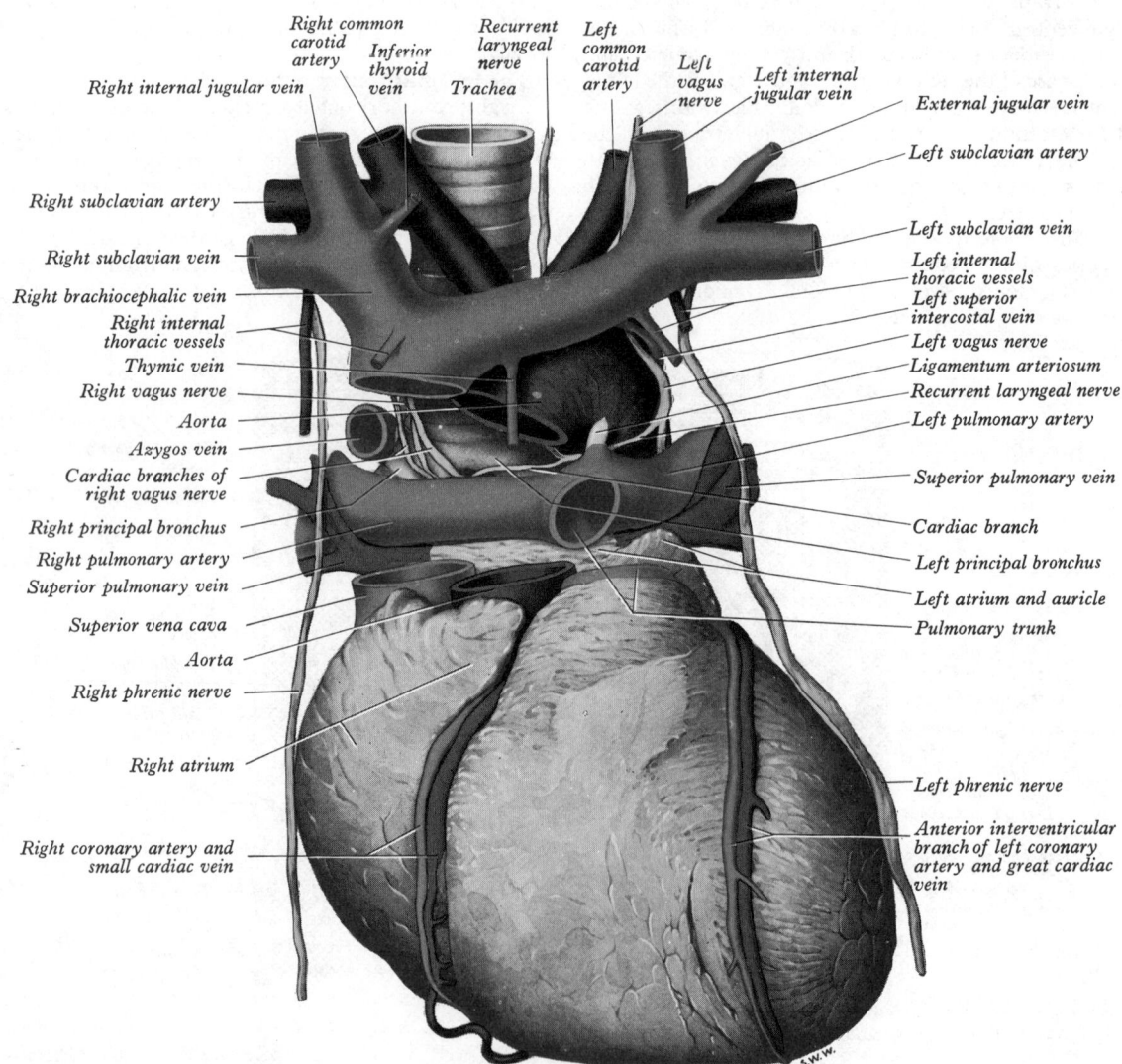

6.38   The relations of the pulmonary arteries and primary bronchi seen from the front. Parts of the ascending aorta, pulmonary trunk and superior vena cava have been removed in the dissection. The right vagal trunk is uncoloured to avoid confusion.

The Ascending Aorta

The ascending aorta (**6.**14, 37, 38, 40) is about 5 cm long. It begins at the base of the left ventricle, on a level with the lower borders of the third costal cartilage, behind the left half of the sternum; it passes obliquely upwards, forwards and to the right, behind the sternum, as high as the upper border of the second right costal cartilage, describing a slight curve in its course. At its origin, immediately distal to the cusps of the aortic valve, there are three

right pulmonary artery and the right principal bronchus; on its *right side*, to the superior vena cava and right atrium, the former lying partly behind it; on its *left side*, first to the left atrium and, at a higher level, to the pulmonary trunk.

At least two small bodies, believed to be chemoreceptors and similar in structure to the carotid body (p. 1462), lie between the ascending aorta and pulmonary trunk. One, sometimes termed the *inferior aorticopulmonary body*, lies close to the heart and anterior to the aorta; the other, sometimes termed the *middle aorticopulmonary body*, lies higher up and towards the right side of the ascending aorta (Boyd 1961).

Branches. The branches of the ascending aorta are the right and left coronary arteries (**6**.39), which supply the heart itself.

THE CORONARY ARTERIES (**6**.39A–F)

The right and left coronary arteries open off the ascending part of the aorta from the anterior and left posterior aortic sinuses. (In clinical terminology these are called the right and left coronary sinuses, respectively, whilst the official right posterior aortic sinus is termed the non-coronary sinus.) Variations in this pattern of origin are rare, but the two vessels may start, separately or by a common opening, from the same aortic sinus. Sometimes three or even four coronary arteries have been observed. The most common variation concerns the ramus of the right coronary artery which suppies the arterial conus (*arteria coni arteriosi* or 'conus artery'), which in the majority of cases (64 per cent) is the first branch of the right coronary; frequently, however, the conus artery arises by a separate opening in the external wall of the anterior aortic sinus (36 per cent), and thus constitutes a third coronary artery. The opening of the left coronary artery may be double, the two orifices leading into the major initial branches of the left coronary artery, usually the circumflex and anterior interventricular rami, one occasionally leading into a stem common to one of these and a diagonal ventricular ramus. The levels of the orifices of the coronary arteries are variable. Thebesius (1708) appears to have initiated a view that the aortic cusps obstruct the openings when the aortic valve is fully patent in systole; but subsequent observers have found the coronary orifices at a higher level, at or above the margins of the cusps in most specimens, but below in about 10 per cent (right coronary) and 15 per cent (left).

The general arrangement of the two coronary arteries is indicated by their name, for they form an obliquely inverted *crown*, consisting of an anastomotic *circle* in the atrioventricular sulcus connected by marginal and interventricular *loops* intersecting at the apex of the heart (**6**.39). This is, of course, only an approximate description, and the degree of anastomosis is most variable and usually functionally insignificant (*vide infra*). These main arteries and their major rami are usually sub-epicardial in position, but especially those which are placed in the atrioventricular and interventricular sulci are often quite deeply situated, being occasionally hidden by overlapping lips of myocardium and even actually embedded in it. Strands of myocardium may also cross atrial or ventricular branches of the coronary system; Polacek (1961) described them as present in the ventricles of more than 80 per cent of hearts, and Bloor and Lowman (1963) have emphasized their importance in the interpretation of coronary arteriograms.

The diameters of the coronary arteries, both of their main stems and larger branches, have been recorded by a number of observers; even ignoring individual and sexual variations in calibre, such figures are of limited value, since the technique involved is not always stated, physiological state is often ignored as a factor, and observers do not always indicate whether they have measured external diameters (including wall thickness) or actual luminal calibres. Calibre is usually the basis, since most measurements have been made on arterial casts or angiograms. The maximum ranges recorded in major studies are 1·5 to 5·5 mm for the left and right coronary arteries at their origins. Baroldi and Scomazzoni (1967) give means of 4·0 and 3·2 mm for the left and right vessels. The left artery exceeds the right in diameter in about 60 per cent of hearts, and the right is the larger in 17 per cent, the vessels being approximately equal in 23 per cent. Vogelberg (1957) considered that the coronary diameters show some increase up to the 30th year.

The Right Coronary Artery

Arising from the anterior ('right coronary') aortic sinus, the right coronary artery passes at first anteriorly and slightly to the right between the right auricular appendage and the pulmonary trunk, between which the sinus usually bulges. Reaching the atrioventricular (coronary) sulcus it descends in this in an almost vertical direction to the right (acute) margin of the heart, curving around this into the posterior part of the sulcus. In this it passes towards the junction of the sulcus with the interatrial and interventricular grooves, a region conveniently named the *crux* of the heart. In about 60 per cent of hearts the artery reaches the crux and terminates a little to the left of this in a variable degree of anastomosis with the circumflex branch of the left coronary. In a minority of instances (*c.* 10 per cent) the right coronary artery ends near the right cardiac margin or between this and the crux (*c.* 10 per cent); more frequently (*c.* 20 per cent) it may even reach the left border of the heart, taking the place of the terminal part of the circumflex artery.

The branches of the right coronary artery supply the right atrium and ventricle and, to a variable extent, parts of the left chambers and the atrioventricular septum. The first branch (which arises separately from the anterior aortic sinus in 36 per cent of cases) is the *conus artery* or 'third coronary artery'; since a similar vessel is derived from the left coronary artery, this branch of the right is more correctly named the *right conus artery*. It ramifies on the anterior aspect of the lowest part of the pulmonary conus and the upper part of the right ventricle. It commonly anastomoses with a similar branch of the left coronary to form the '*annulus of Vieussens*', a somewhat tenuous anastomotic 'circle' around the commencement of the pulmonary trunk. Various observers differ greatly in their descriptions of the conus artery (Baroldi and Scomazzoni 1967), some regarding the right conus artery of particular significance in coronary arterial disease. Some consider it the first ventricular branch of the right coronary, supplying a variable region from the pulmonary conus to the apex.

The anterior atrial and ventricular rami branch from the so-called *first segment of the right coronary*, which extends from its origin to the right margin of the heart. Both sets of vessels diverge from the parent artery at a wide angle, approaching a right angle in the case of the ventricular arteries, in contrast to the more acute origin of the ventricular rami of the left coronary. The **right anterior ventricular rami**, usually 2 or 3 in number, ramify towards the apex, which they rarely reach, unless the **right ('acute') marginal branch** is included in the group, as it is by some observers. The latter is then the largest anterior ventricular ramus of the right coronary artery, being greater in calibre and long enough to reach the apex in most hearts (93 per cent according to Baroldi and Scomazzoni 1967). When the right ('acute') marginal artery is particularly large, the remaining anterior ventricular rami may be reduced to one or even to complete absence. From the *second segment of the right coronary artery* (i.e. between the right border and the crux) 1 to 3 small **right posterior ventricular rami**, commonly two, supply the diaphragmatic aspect of the right ventricle. Their degree of development is reciprocal to that of the right marginal artery, as in the case of the anterior right ventricular supply, the right marginal usually extending on to the diaphragmatic surface. The posterior right ventricular rami may be absent. As the right coronary approaches the crux of the cardiac sulci, it produces 1 to 3 posterior interventricular rami, only one of which occupies the interventricular sulcus. The **posterior interventricular artery** is singular in about 70 per cent of hearts, otherwise being accompanied by parallel branches of the right coronary, which may occur to the right or left of it, or be present on both sides of the sulcus. When these flanking vessels are present, the branches of the posterior (descending) interventricular artery are small and sparse; when it exists alone it gives off a few branches particularly to the right ventricle but also one or two to the left. Its place is taken in about 10 per cent of individuals by a branch of the left coronary artery.

The **atrial rami** of the right coronary artery are sometimes described as forming anterior, lateral (right or marginal), and posterior *groups* of vessels, but in fact they are most frequently single, small vessels of an average diameter of about 1 mm. The right anterior and lateral atrial arteries are occasionally double and, very rarely, triple; they are distributed chiefly to the right atrium. The posterior ramus is usually single, and is distributed to the right and left atria; but in a large minority of hearts (40 per cent or more) there is also a left posterior atrial branch of the right

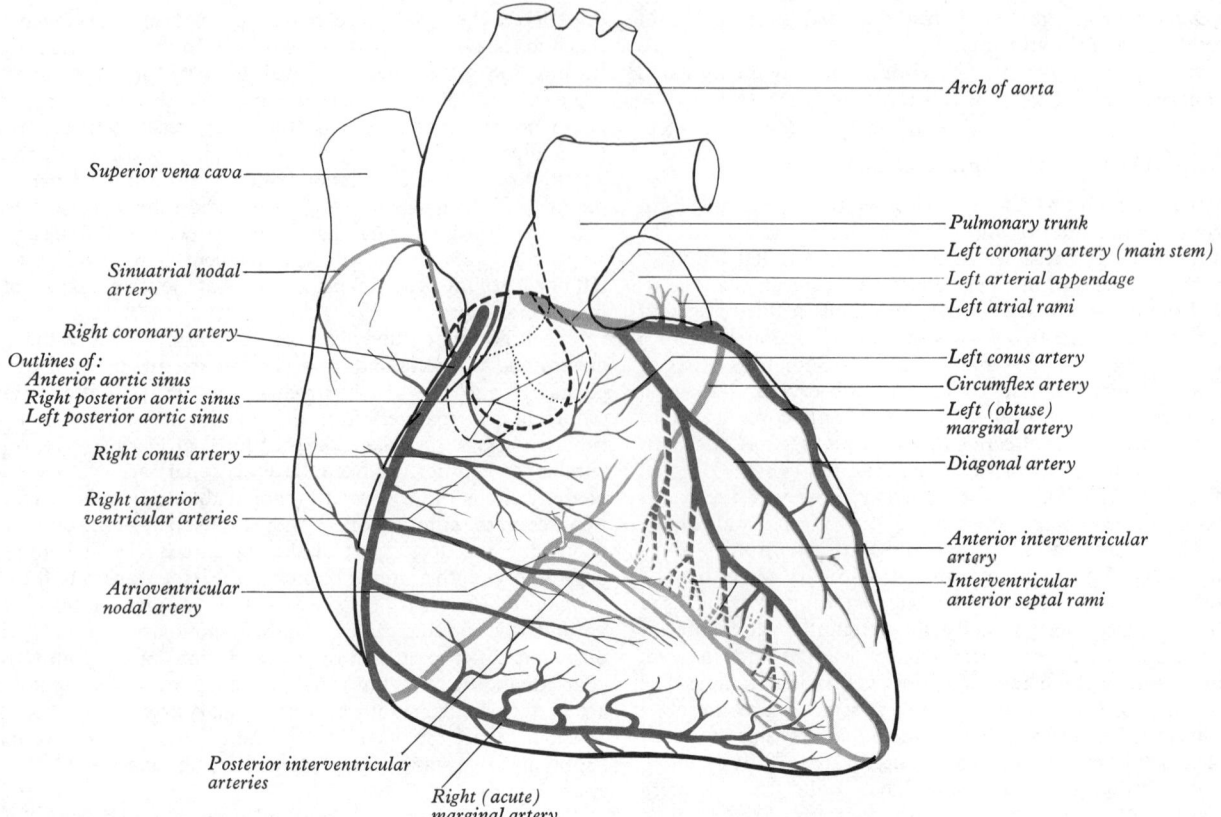

Arch of aorta

Superior vena cava

Pulmonary trunk

Left coronary artery (main stem)

Left arterial appendage

Sinuatrial nodal
artery

Left atrial rami

Right coronary artery

Left conus artery

Outlines of:
Anterior aortic sinus
Right posterior aortic sinus
Left posterior aortic sinus

Circumflex artery

Left (obtuse)
marginal artery

Right conus artery

Diagonal artery

Right anterior
ventricular arteries

Anterior interventricular
artery

Atrioventricular
nodal artery

Interventricular
anterior septal rami

Posterior interventricular
arteries

Right (acute)
marginal artery

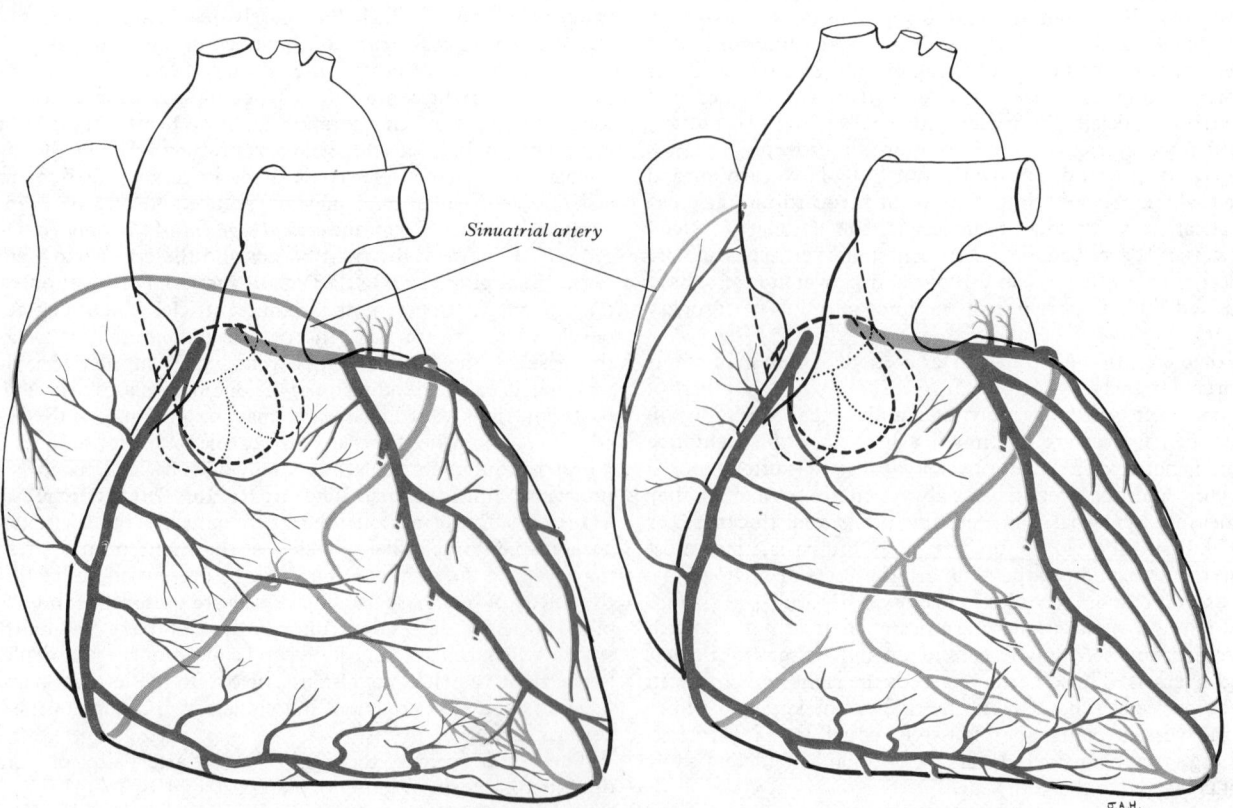

Sinuatrial artery

6.39A–C Anterior views of the coronary arterial system, with the principal variations.

A The commonest arrangement.

B A common variation in the origin of the sinuatrial nodal artery.

C An example of *left* 'dominance' by the left coronary artery, showing also an uncommon origin of the sinuatrial artery.

Note that in 39A to E the right coronary arterial tree is shown in *magenta*, the left in *full red*. In both cases posterior distribution is shown in a paler shade.

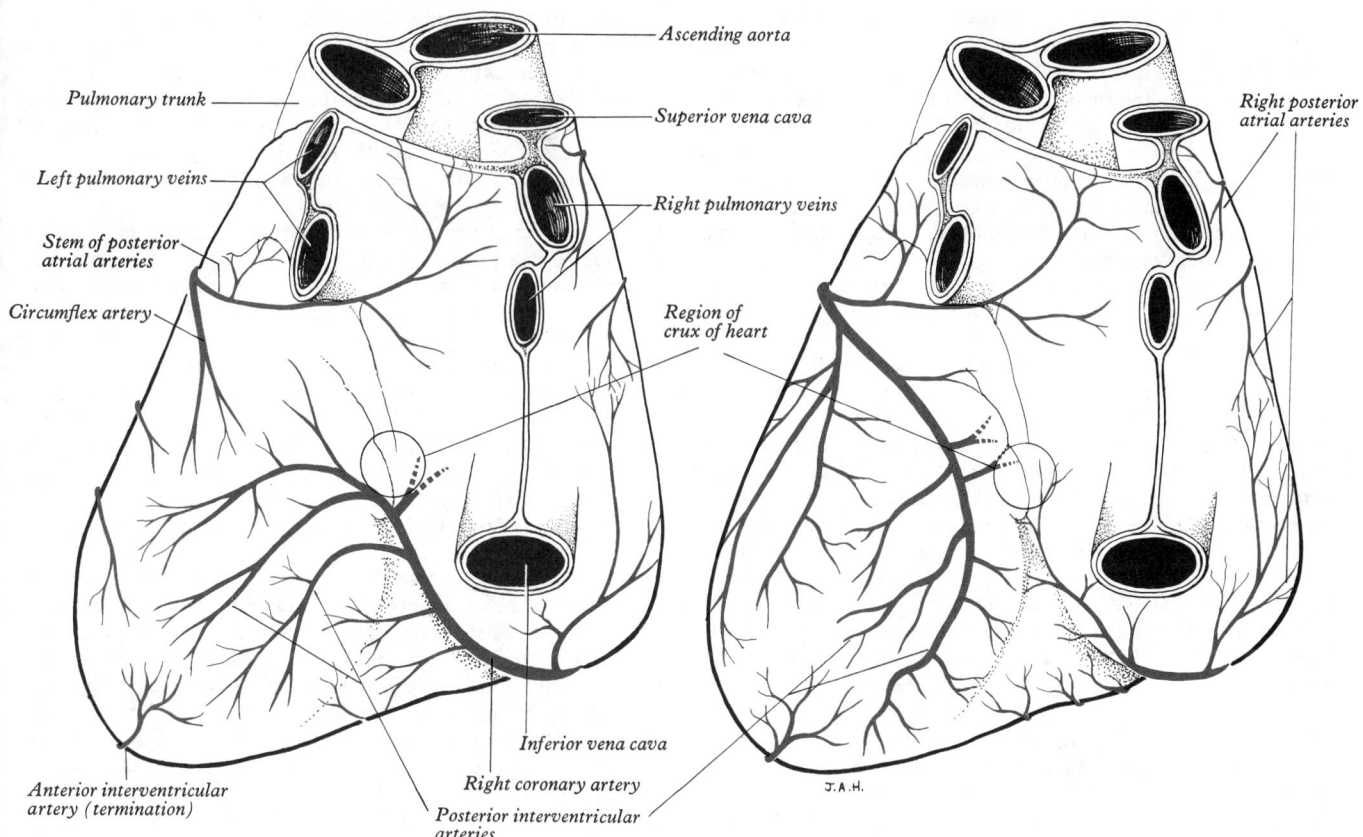

Ascending aorta

Pulmonary trunk

Superior vena cava

Left pulmonary veins

Right pulmonary veins

Stem of posterior atrial arteries

Circumflex artery

Region of crux of heart

Right posterior atrial arteries

Inferior vena cava

Anterior interventricular artery (termination)

Right coronary artery

Posterior interventricular arteries

J.A.H.

6.39D, E   Postero-inferior views of the coronary arterial system.
D   An example of the more normal distribution in *right* 'dominance'.
E   A less common form of *left* 'dominance'.
N.B. In these 'posterior' views the diaphragmatic (inferior) surface of the

ventricular part of the heart has been artificially displaced, and foreshortening ignored, to clarify the details of the so-called posterior (inferior) distribution of the coronary arteries.

---

coronary. The **artery of the sinuatrial node** may be regarded as an atrial artery, since it is distributed largely to the myocardium of both atria, principally the right. The site of origin of this vessel is variable; it is a branch of the *left* coronary artery in about 35 per cent of hearts (Hutchinson 1978), arising from its circumflex branch (*vide infra*). When a branch of the right coronary, it stems from its anterior, initial segment most frequently, less often from the right lateral part of its course, and with least frequency from the posterior atrioventricular part of its course. Most commonly this 'nodal' artery passes backwards from the proximal trunk of the right coronary in the sulcus between the right auricular appendage and the aorta, a disposition of obvious surgical importance. Whatever its origin the artery of the sinuatrial node usually branches around the base of the superior vena cava, commonly forming an *arterial loop* around the vein, from which small rami branch off to supply the right atrium. A large branch of the vessel, the so-called '*ramus cristae terminalis*' (Spalteholz 1924), passes into the sinuatrial node and transverses it (**6**.39); it is this ramus which should, perhaps, be named the artery of the sinuatrial node, since the rest of the vessel supplies the atria. A more appropriate name, the '*main atrial branch*' of the right coronary, has been suggested by Baroldi and Scomazzoni (1967).

The **septal rami** of the right coronary artery are relatively short vessels, which pass from the posterior interventricular ramus into the posterior region of the interventricular septum. They are numerous, but they do not usually reach the apical part of the septum (which is supplied by the septal branches of the termination of the anterior interventricular artery). The first, and usually the largest of these posterior septal arteries, is the first such branch, commonly arising from the *inverted loop* which is said to characterize the right coronary artery at the crux of the heart, where the posterior interventricular artery is derived from it. It is this **large posterior septal artery** which usually

supplies the **atrioventricular node**—in 80 per cent of hearts, according to Hutchinson (1978).

DiDio (1967) has described *atrioventricular rami* of the right coronary artery. These are small recurrent branches from either ventricular arteries, crossing the atrioventricular sulcus to supply the adjoining atrial myocardium, or ventricular twigs from atrial arteries.

The Left Coronary Artery

As stated above, the left coronary artery is distinctly larger in calibre than the right; it supplies a much greater *volume* of myocardium, being distributed to almost the whole of the left ventricle and left atrium, except in those instances of so-called 'right dominance' previously mentioned, where the right coronary supplies a variable part of the posterior region of the left ventricle. Most commonly, also, the left coronary artery supplies most of the interventricular septum.

The initial trunk of the artery, between its opening from the left posterior ('left coronary') aortic sinus and the vessel's first branches, varies from a few millimetres to a few centimetres in length. It lies between the pulmonary trunk and the left auricular appendage, emerging to reach the atrioventricular sulcus, in which it turns to the left. This part of the artery is loosely embedded in subepicardial fat. In general it has no branches, but a small atrial twig sometimes occurs, and very occasionally, according to James (1961), the *artery of the sinuatrial node* may be derived from the left coronary trunk. However, the latter branch, when a ramus of the left coronary, almost always arises from its circumflex branch and will be described with this artery below. As it reaches the atrioventricular or coronary sulcus, the left coronary artery divides into two or three main branches, one of which, the **anterior interventricular (descending) ramus** is commonly described as the continuation of the left coronary artery. It descends obliquely forward and to the left in the interventricular sulcus, sometimes deeply embedded in it, and crossed by occasional bridges of myocardial tissue and by the great cardiac vein and its tributaries. The artery reaches the apex of the heart almost always, terminating there in about one third of specimens, but most commonly turning around the apex into the posterior (inferior) interventricular sulcus. It passes along the sulcus for a few centimetres, usually for a third to a half of its length, to meet the terminal twigs of the corresponding ramus of the right coronary artery.

The **anterior interventricular artery** produces right and left anterior ventricular and anterior septal branches, and, to a variable extent, corresponding posterior rami. The *right anterior ventricular rami* of the left coronary artery are small, and rarely more than one or two in number, the right ventricle being supplied almost exclusively from the right coronary artery.

The **left anterior ventricular arteries** are large vessels, from 2 to 9 in number, which branch at acute angles from the anterior interventricular to course *diagonally* across the anterior aspect of the left ventricle, their larger terminals reaching the rounded left border of the heart. One of these vessels is often especially large and prominent and it may arise separately from the trunk of the left coronary artery (which then ends by trifurcation rather than the more usual bifurcation), and it is then known as the **left diagonal artery**. A left diagonal artery occurs in 33–50 per cent or more of hearts, according to various observers; very occasionally (20 per cent) it is duplicated. A small *left conus artery* frequently branches from the anterior interventricular artery, near its beginning, and anastomoses on the conus with the conus branch of the right coronary, and with the vasa vasorum of the pulmonary artery and aorta. The **anterior septal rami** of the anterior interventricular artery branch off almost perpendicularly, passing backwards and downwards in the substance of the septum, of which they usually supply about the ventral two-thirds. Small *posterior septal rami* from the same source supply the posterior third or so of the septum for a variable distance from the apex.

The **circumflex artery**, usually comparable with the anterior interventricular artery in calibre, curves to the left border of the heart in the atrioventricular sulcus, continuing round into the

posterior part of the sulcus to terminate a little to the left of the crux in the majority of hearts, but sometimes continuing as a posterior interventricular artery. Its proximal part is commonly overlapped by the left auricular appendage. In about 90 per cent of hearts a large ventricular branch, the **left ('obtuse') marginal artery**, arises perpendicularly from the circumflex and ramifies over the rounded left margin of the heart, supplying much of that region of the left ventricle—it usually reaches the apex of the heart. Smaller anterior and posterior ventricular rami of the circumflex artery also supply the left ventricle. The **anterior ventricular branches** (1 to 5 in number, but usually limited to 2 or 3) course over the left ventricle parallel to the diagonal artery, when present, and replacing it when absent. The **posterior ventricular branches** are somewhat smaller and less numerous, since the left ventricle is partly supplied by rami of the posterior interventricular artery. When the latter is small or absent, it is accompanied by, or replaced by, an interventricular continuation of the circumflex artery. Such a **left posterior interventricular artery** is frequently double or triple. **Atrial rami**, both anterior, lateral, and posterior, spring from the circumflex artery to supply the left atrium.

Three inconstant branches of the circumflex artery require description. The **artery to the sinuatrial node** is a branch of the circumflex in about 35 per cent of hearts (Hutchinson 1978), usually from its anterior segment, less often from its circum-marginal part. This artery passes over the left atrium, supplying it and then encircling the termination of the superior vena cava, as in the case of a nodal artery from the right coronary. Similarly, also, it sends a large branch into the sinuatrial node, but is largely an atrial vessel in distribution. The **artery to the atrioventricular node** is a terminal ramus of the circumflex artery in 20 per cent of hearts, arising near the crux, and in such specimens the circumflex usually supplies the posterior interventricular ramus, providing an example of so-called 'left dominance' (*vide infra*). The third branch of the circumflex to be mentioned here is *Kugel's anastomotic artery*, the '*arteria anastomotica auricularis magna*'. Kugel (1927) described this as a constant branch of the circumflex artery, usually from its anterior part, which traverses the interatrial septum (near its ventricular border) to establish some degree of direct or indirect anastomosis with the right coronary artery. The existence of this anastomosis is a matter of controversy, being apparently accepted by James (1974), but denied by Baroldi and Scomazzoni (1967). James regards it as an auxiliary supply to the atrioventricular node.

The foregoing details of distribution of the separate coronary arteries require some integration to provide a total concept of cardiac supply. In the commonest arrangement the right coronary artery supplies the whole of the right ventricle (except for a very small region on its own side of the anterior interventricular groove), a variable part on the diaphragmatic aspect of the left ventricle, the postero-inferior third of the intraventricular septum, the right atrium and part of the left, and the conducting system as far as the proximal parts of the right and left bundle branches. The distribution of the left coronary artery is, of course, reciprocal and includes most of the left ventricle (apart from the diaphragmatic region referred to above), a narrow strip of right ventricle (*vide supra*), the anterior two thirds of the interventricular septum, and most of the left atrium. As already mentioned above, and depicted in illustration **6**.39, variations from the distributions just outlined chiefly affect the diaphragmatic aspect of the ventricles; and these reside in the relative '*dominance*' of supply by the left or right coronary artery. The term is, of course, misleading, since it is clear that the left artery almost always supplies a greater *volume* of tissue. In '*right dominance*' the posterior interventricular is derived from the right coronary and in '*left dominance*' from the left. In the so-called '*balanced*' type of pattern branches of both arteries run in or parallel to the interventricular sulcus. Less is known of the variations in atrial supply, because the small vessels involved are not easily preserved and studied in corrosion casts. From Hutchinson's results (1978) it is apparent that in over 50 per cent of hearts the right atrium is supplied by the right coronary alone, the remainder receiving a dual supply. More than 62 per cent of left atria are supplied by the left coronary and about 27 per cent by

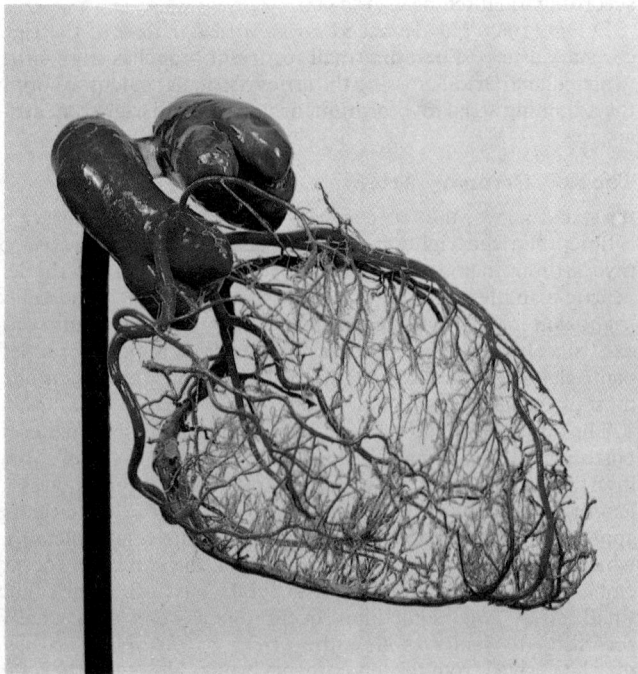

6.39F A composite coloured resin cast; in *red* are the aortic root, aortic sinuses, and coronary arteries; in *blue* are the pulmonary trunk and sinuses, and the coronary sinus and its tributaries. Note reduplication of the right coronary artery; the smaller of these vessels is known as the artery of the conus. Note particularly the septal arterial and venous rami. (See text for detailed account.)

the right, but in each group there is a small accessory supply from the other coronary artery, the remainder being supplied almost equally by both arteries. The sinuatrial and atrioventricular nodes also vary in their sources of supply. According to James (1961) the right and left coronary arteries are the source of vessels to the sinuatrial node in 55 per cent and 45 per cent of hearts, the corresponding values from Baroldi and Scomazzoni's study (1967) being 51 per cent and 41 per cent (8 per cent receiving a bilateral supply), whilst figures from Hutchinson (1978) are 65 per cent and 35 per cent. For the atrioventricular node James's values are 90 per cent (right coronary) and 10 per cent (left coronary); Hutchinson's figures are 80 per cent (right coronary) and 20 per cent (left coronary), while Baroldi and Scomazzoni merely state that a right coronary supply is common and a left supply rare.

Coronary Anastomosis

The existence of anastomoses between the branches of the two coronary arteries, whether subepicardial or myocardial in location, and between these arteries and extracardiac vessels, is obviously a matter of prime medical importance. Because the clinical evidence has largely suggested that such anastomoses are not of much demonstrable functional value, i.e. in providing rapidly available collateral routes to circumvent relatively sudden coronary obstruction, it has become widely customary to regard the coronary circulation as end-arterial in character. Nevertheless, the existence of anastomoses, particularly between finer branches of the subepicardial rami has long been recognized, and according to Gross (1921), such anastomoses may improve through the advancing life of the individual. Moreover, those who have actually investigated the coronary arteries—whether by injection of radio-opaque perfusants (Vastesueger et al. 1957, using post mortem hearts, Laurie and Woods 1958, using in vivo coronary radiography), by perfusion with calibrated spherules (Prinzmetal et al. 1947), or by injection of plastic resins with subsequent production of corrosion casts (Baroldi et al. 1956; James 1961)—have almost all recorded the existence of anastomoses, and in arterial vessels of a calibre as much as 100 to 200 μm. Baroldi and Scomazzoni (1967) have tabulated the results of a large number of investigations carried out since 1880, and it is noticeable in their analysis that no study denying the occurrence of anastomoses has been recorded since 1957. Some workers, however, describe anastomoses as occurring only between the branches of individual coronary arteries; nevertheless, the majority have recorded intercoronary anastomoses. James (1974) considers the evidence for anastomoses at all levels, subepicardial (extramural), myocardial (mural), and subendocardial, to be conclusive. He considered the most frequent sites of extramural anastomoses to be the apex, anterior aspect of the right ventricle, the posterior aspect of the left ventricle, the crux, along the interatrial and interventricular sulci, and between the artery of the sinuatrial node and other atrial vessels. It is generally conceded that the functional value of such anastomoses must be extremely variable, though the evidence suggests that they may become more effective in pathological conditions of comparatively slow onset. The precise structure of coronary anastomoses is not fully clarified; most observations depend upon appearances in corrosion casts. These suggest that the anastomotic vessels are relatively straight in normal hearts, but may be much coiled in hearts previously subjected to episodes of coronary occlusion, indicating, perhaps, that they are exposed to greater pressures in pathological states. Little has been recorded of their microscopic structure, but they appear to be little more than endothelial tubes, unsupported by non-striated muscle or elastic tissue.

Extracardiac anastomoses are connexions between various branches of the coronary arteries and other vessels in the thorax, through the medium of pericardial arteries and the arterial vasa vasorum of veins and arteries linking the heart with the systemic and pulmonary circulations. The classic study of Hudson et al. (1932) demonstrated that coronary injections of India ink could reach as far as the diaphragm through the aortic vasa vasorum, thus connecting with pericardial, mediastinal, and diaphragmatic arteries. Similar connexions along the pulmonary trunk reach mediastinal and bronchial arteries, and connexions also exist along the pulmonary veins and the venae cavae. These results

have been confirmed (Baroldi and Scomazzoni 1967), but in this case also, the effectiveness of these connexions as collateral routes in supplying the heart is unpredictable.

Arteriovenous anastomoses were considered to be present in the coronary circulation by Nussbaum (1912); his evidence was indirect, but Hirsch (1960) described structures like glomera, with a characteristic sphincteric appearance (p. 632), as present in the cardiac sulci. These findings have aroused much controversy and the occurrence of such structures must be regarded as at present uncertain. Various other forms of 'arteriovenous' connexions have been described by many workers. Wearn et al. (1933) recorded the existence of numerous connexions through very thin-walled 'arterial' vessels between the coronary circulation and the heart cavities, naming them 'myocardial sinusoids' and 'arterio-luminal' vessels. Their existence has been frequently confirmed (see, for example, Watanabe 1960), and indirect evidence of them, derived from perfusion experiments, dates back as far as Vieussens (1706). The value of these connexions in coronary disease is uncertain.

The Arch of the Aorta

The arch of the aorta (6.37–42) begins behind the manubrium sterni at the level of the upper border of the second right sternocostal articulation, and runs at first upwards, backwards and to the left in front of the trachea; it is then directed backwards on the left side of the trachea, and finally passes downwards on the left side of the body of the fourth thoracic vertebra, at the lower border of which it is continuous with the descending aorta. Its termination corresponds to the sternal extremity of the second, left costal cartilage (6.26). It thus has two curvatures: one with its convexity upwards, the other with its convexity forwards and to the left. Its upper border is usually about the level of the middle of the manubrium sterni.

Relations. Anteriorly and to the left the vessel is covered with the left mediastinal pleura, under cover of which it is crossed by four nerves; in order from before backwards these are: the left phrenic, the lower of the cervical cardiac branches of the left vagus, the

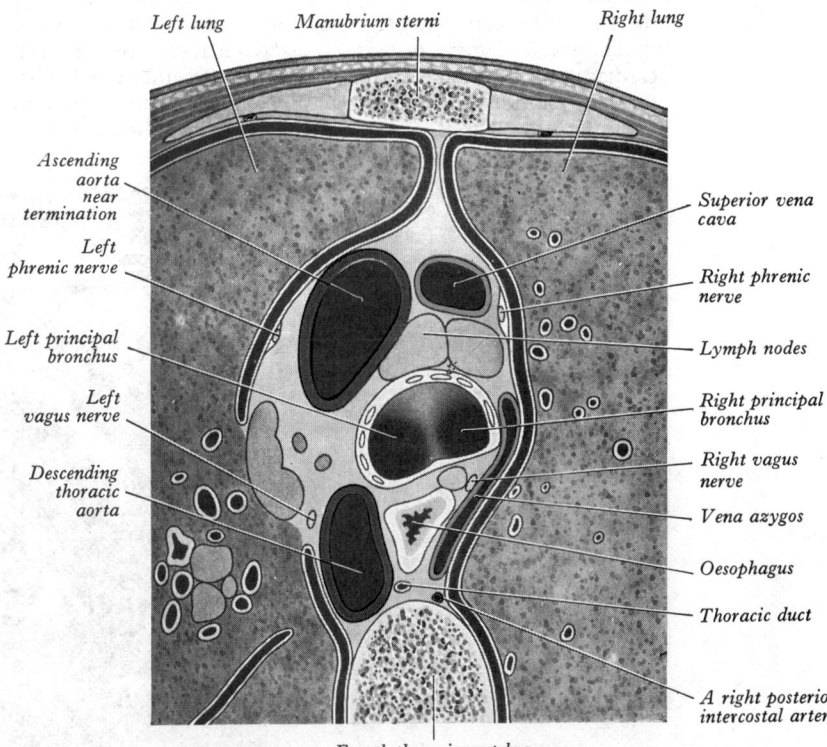

Left lung   Manubrium sterni   Right lung

Ascending aorta near termination

Left phrenic nerve

Left principal bronchus

Left vagus nerve

Descending thoracic aorta

Superior vena cava

Right phrenic nerve

Lymph nodes

Right principal bronchus

Right vagus nerve

Vena azygos

Oesophagus

Thoracic duct

A right posterior intercostal artery

Fourth thoracic vertebra

6.40 A transverse section through the mediastinum at the level of the lower part of the body of the fourth thoracic vertebra, viewed from above.

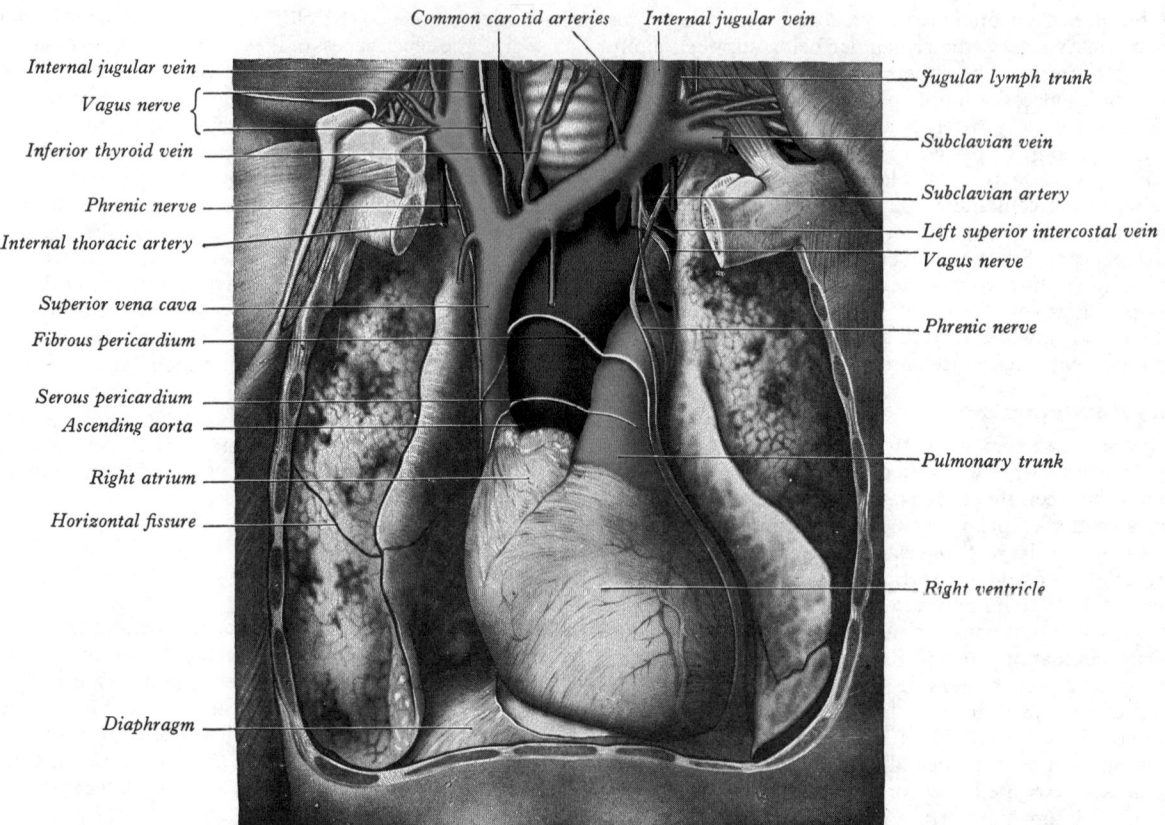

Common carotid arteries   Internal jugular vein

Internal jugular vein —
Vagus nerve {
Inferior thyroid vein —
Phrenic nerve —
Internal thoracic artery —
Superior vena cava —
Fibrous pericardium —
Serous pericardium —
Ascending aorta —
Right atrium —
Horizontal fissure —
Diaphragm —

— Jugular lymph trunk
— Subclavian vein
— Subclavian artery
— Left superior intercostal vein
— Vagus nerve
— Phrenic nerve
— Pulmonary trunk
— Right ventricle

6.41 A dissection to display the heart, great vessels and lungs *in situ*. The sternum and the sternal ends of the costal cartilages, together with the parietal pleura on each side, have been excised, and the mediastinal pleura and parietal layer of the pericardium over the sternocostal surface of the heart have been removed. The lungs have been displaced to expose the heart, and the epicardium dissected off the heart and the great vessels.

superior cervical cardiac branch of the left sympathetic, and the trunk of the left vagus. As the left vagus crosses the arch it gives off its recurrent laryngeal branch, which hooks round below the vessel to the left of the ligamentum arteriosum and then passes upwards on its right side. The left superior intercostal vein runs obliquely upwards and forwards on this surface of the arch, crossing superficial to the vagus but deep to the phrenic nerve (**6.41**). The left lung and pleura separate all these structures from the chest wall. *Posteriorly and to the right*, the trachea and the deep cardiac plexus, the left recurrent laryngeal nerve, the oesophagus, the thoracic duct and the vertebral column are successively in relation to the vessel. *Above*, the brachiocephalic, left common

carotid and left subclavian arteries arise from the convexity of the arch and are crossed in front close to their origins by the left brachiocephalic vein. *Below*, the arch is related to the bifurcation of the pulmonary trunk, the left principal bronchus, the ligamentum arteriosum (p. 666), the superficial part of the cardiac plexus and the left recurrent laryngeal nerve.

In the fetus the lumen of the aorta is considerably narrowed between the origin of the left subclavian artery and the attachment of the ductus arteriosus, forming what is termed the *aortic isthmus*, while immediately beyond the ductus arteriosus the vessel presents a fusiform dilatation (*aortic spindle*)—the point of junction of the two parts being marked in the concavity of the arch

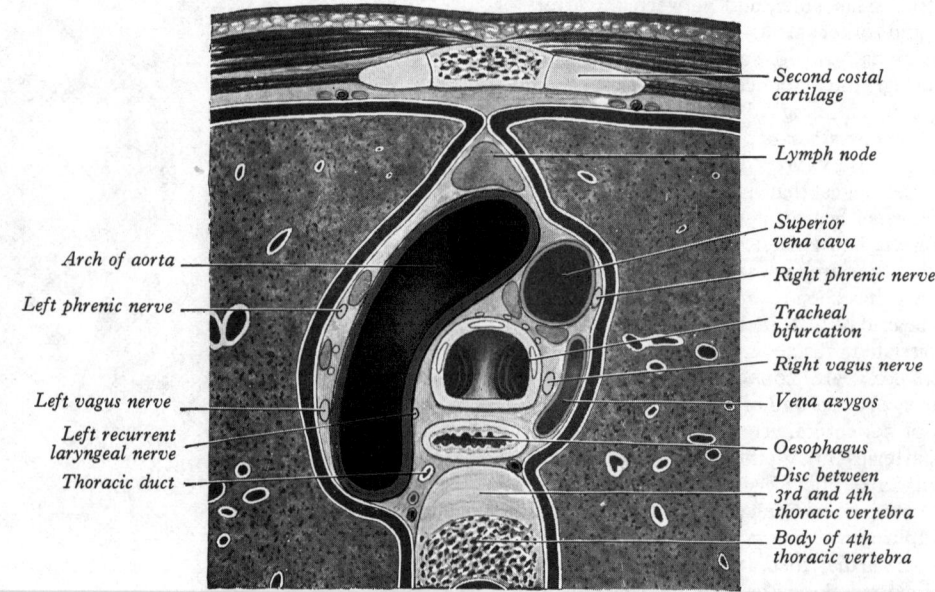

Arch of aorta —
Left phrenic nerve —
Left vagus nerve —
Left recurrent laryngeal nerve —
Thoracic duct —

— Second costal cartilage
— Lymph node
— Superior vena cava
— Right phrenic nerve
— Tracheal bifurcation
— Right vagus nerve
— Vena azygos
— Oesophagus
— Disc between 3rd and 4th thoracic vertebra
— Body of 4th thoracic vertebra

6.42 A transverse section through the mediastinum at the level of the upper part of the body of the fourth thoracic vertebra, viewed from above.

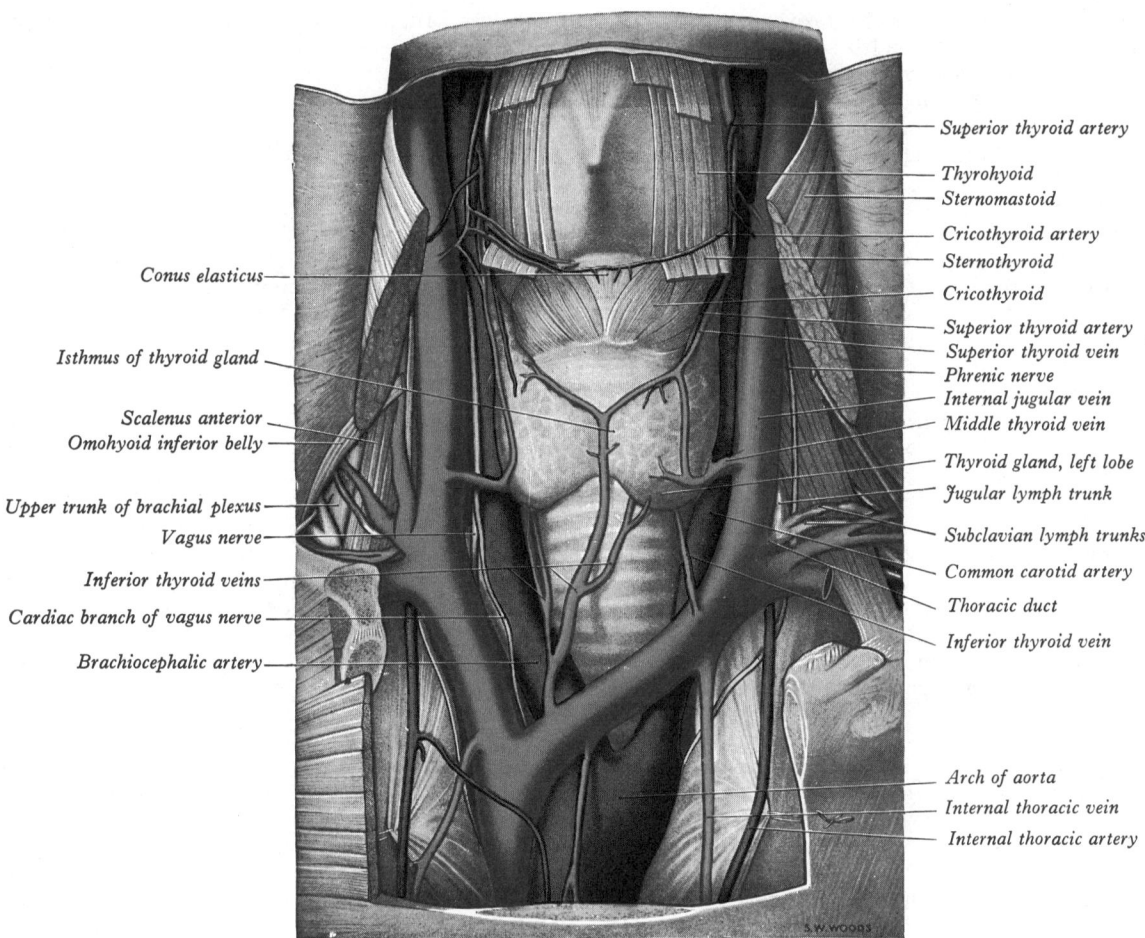

Conus elasticus

Isthmus of thyroid gland

Scalenus anterior
Omohyoid inferior belly

Upper trunk of brachial plexus

Vagus nerve

Inferior thyroid veins

Cardiac branch of vagus nerve

Brachiocephalic artery

Superior thyroid artery

Thyrohyoid
Sternomastoid

Cricothyroid artery
Sternothyroid

Cricothyroid

Superior thyroid artery
Superior thyroid vein
Phrenic nerve
Internal jugular vein
Middle thyroid vein

Thyroid gland, left lobe

Jugular lymph trunk

Subclavian lymph trunks

Common carotid artery

Thoracic duct

Inferior thyroid vein

Arch of aorta
Internal thoracic vein
Internal thoracic artery

6.43 A dissection of the lower part of the front of the neck and of the superior mediastinum. The manubrium sterni and the sternal ends of the clavicles and the first costal cartilages have been removed, and the pleural sac and lung have been retracted on each side. In this specimen each superior thyroid artery arose from the common carotid artery.

by an indentation. These conditions persist, to some extent, in the adult.

Variations. The summit of the arch of the aorta is usually about 2·5 cm below the upper border of the sternum; but it may be considerably higher or lower than this. Sometimes the aorta arches over the root of the right lung (right aortic arch) instead of over that of the left, and passes down on the right side of the vertebral column, a condition which is normal in birds. In such cases there is usually a transposition of the thoracic and of the abdominal viscera. Less frequently the aorta, after arching over the root of the right lung, passes behind the oesophagus to gain its usual position on the left side of the vertebral column; this peculiarity is not accompanied by transposition of the viscera. The aorta occasionally divides, as in some quadrupeds, into an ascending and a descending trunk, the former of which is directed vertically upwards, and subdivides into three branches, to supply the head and upper limbs. Sometimes the aorta subdivides near its origin into two branches, which soon reunite; in these cases the oesophagus and trachea usually pass through the interval between the two branches; this is the normal condition of the vessel in the reptilia and is due to persistence of a part of the right dorsal aorta which usually disappears (p. 187).

Radiological appearances. The shadow cast by the terminal part of the arch in anteroposterior radiographs (**6.27**) and is sometimes called the 'aortic knuckle'. The arch may also be seen in left anterior oblique radiographs enclosing a translucent space, 'the aortic window', in which may be seen the shadows of the pulmonary trunk and its left branch.

Branches (**6.41**, **44**). Three branches are given off from the upper aspect of the arch of the aorta: the brachiocephalic trunk, the left common carotid and the left subclavian.

Variations. The branches may spring from the commencement of the arch or upper part of the ascending aorta; or the distance between them at their origins may be increased or diminished, the most frequent change in this respect being the approximation of the left common carotid artery to the brachiocephalic trunk (Wright 1969).

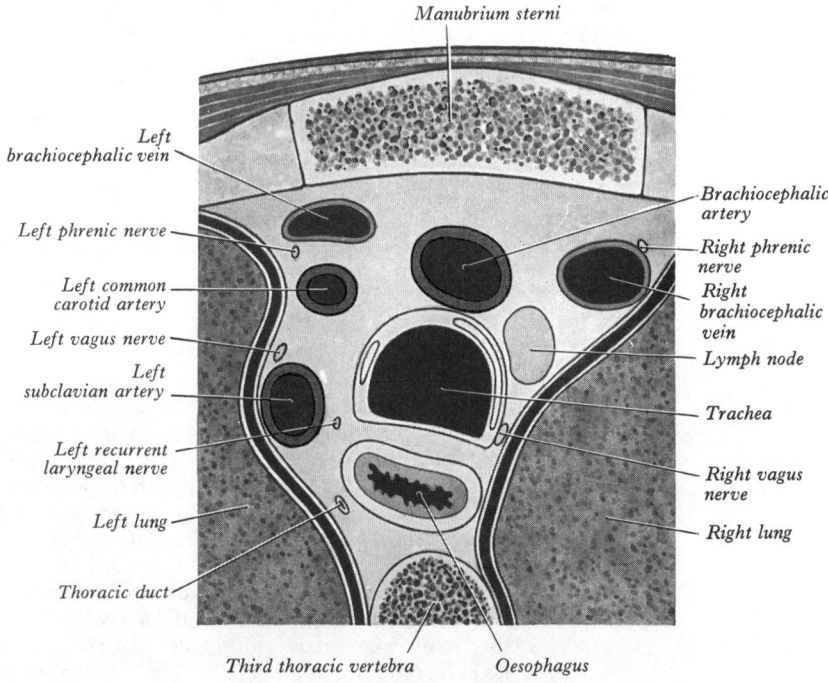

Manubrium sterni

Left
brachiocephalic vein

Left phrenic nerve

Left common
carotid artery

Left vagus nerve

Left
subclavian artery

Left recurrent
laryngeal nerve

Left lung

Thoracic duct

Brachiocephalic
artery

Right phrenic
nerve
Right
brachiocephalic
vein

Lymph node

Trachea

Right vagus
nerve

Right lung

Third thoracic vertebra   Oesophagus

6.44 A transverse section through the superior mediastinum at the level of the body of the third thoracic vertebra, viewed from above.

675

The number of the primary branches may be reduced to one; more commonly there are two, the left common carotid arising from the brachiocephalic trunk (7 per cent), or (more rarely) the common carotid and subclavian arteries of the left side arising from a left brachiocephalic trunk. But the number may be increased to four, through the right common carotid and subclavian arteries arising directly from the aorta; in most of these cases the right subclavian arises from the left end of the arch and passes to the right behind the oesophagus (see also p. 187). Another common variation in which there are four primary branches is that where the left vertebral artery arises from the arch of the aorta between the left common carotid and subclavian arteries. Lastly, the number of trunks may be increased to five or six; very rarely, the external and internal carotid arteries arise separately, the common carotid being absent on one or both sides. In some few cases six branches have been found, and this condition is associated with the origin of both vertebral arteries from the arch.

When the aorta arches over to the right side, the arrangement of the three branches is reversed: there is a left brachiocephalic trunk, and the right common carotid and right subclavian arteries arise separately. In other cases, where the aorta takes its usual course, the two common carotids may be joined in a single trunk, and the subclavians arise separately from the arch, the right subclavian generally arising from the left end of the arch.

Other arteries may spring from the arch of the aorta. Of these the most common are the bronchial (one or both), and the thyroidea ima.

An analysis of variation in the branches of 1,000 aortic arches has been recorded by Anson (1963). In 65 per cent the usual pattern of three major branches was observed; in 27 per cent the left common carotid shared the brachiocephalic trunk (contrast the percentage quoted above). The four large arteries branched separately from the aortic arch in 2·5 per cent. The remaining 5 per cent also showed a great variety of forms of branching, the commonest (1·2 per cent) being the occurrence of symmetrical right and left brachiocephalic trunks.

The Brachiocephalic Artery

The brachiocephalic (innominate) artery is the largest branch of the arch of the aorta, and is from 4 to 5 cm in length (**6.**41, 43, 44, 118). It arises from the convexity of the arch of the aorta, posterior to the centre of the manubrium sterni; it passes obliquely upwards, backwards and to the right, lying at first in front of the trachea and then on its right side. At the level of the upper border of the right sternoclavicular joint it divides into the right common carotid and right subclavian arteries.

Relations. It is separated *anteriorly* from the manubrium sterni by the sternohyoid and sternothyroid, the remains of the thymus, the left brachiocephalic and right inferior thyroid veins, which cross its root, and sometimes the cardiac branches of the right vagus nerve. *Posteriorly* it is related to the trachea below, and the right pleura above, where also the right vagus nerve lies posterolateral to the vessel before leaving it to pass on to the side of the trachea; on the *right side*, are the right brachiocephalic vein, the upper part of the superior vena cava, and the pleura; and on the *left side*, the remains of the thymus, the origin of the left common carotid artery, the inferior thyroid veins, and at a higher level the trachea.

Branches. The brachiocephalic artery is usually devoid of branches other than its terminal ones, but occasionally the *thyroidea ima* arises from it, and sometimes it gives off a *thymic* or a *bronchial branch*.

The thyroidea ima, small and inconstant, ascends in front of the trachea to the isthmus of the thyroid gland, in which it ends. It occasionally arises from the aorta, or from the right common carotid, subclavian or internal thoracic arteries.

THE CAROTID SYSTEM OF ARTERIES

The principal arteries of the head and neck are the two common carotids; they ascend in the neck as far as the level of the upper border of the thyroid cartilage, where each divides into two branches: (1) the external carotid, supplying the exterior of the head, the face and the greater part of the neck; (2) the internal carotid, supplying the parts within the cranial and orbital cavities.

The common and internal carotid arteries, together with the veins and nerves which accompany them, are situated in a cleft on each side of the neck. This cleft may be said to possess three walls: a posterior, formed mainly by the transverse processes of the cervical vertebrae with their attached muscles; a medial, consisting of the trachea, oesophagus, thyroid gland, larynx and the constrictor muscles of the pharynx; and an anterolateral, made up of the sternocleidomastoid with, at different levels, the omohyoid, sternohyoid, sternothyroid, and the digastric and stylohyoid.

The Common Carotid Arteries

The common carotid arteries differ in length and in their mode of origin. The *right* artery begins at the bifurcation of the brachiocephalic trunk behind the right sternoclavicular joint and is confined to the neck. The *left* artery springs from the highest part of the arch of the aorta immediately behind and to the left of the brachiocephalic trunk, and therefore consists of a thoracic and a cervical portion.

The thoracic part of the left common carotid artery (**6.**43, 44) ascends from the arch of the aorta to the level of the left sternoclavicular joint, where it is continuous with the cervical portion. It lies at first in front of the trachea, but later inclines to its left side.

Relations. It is separated *in front* from the manubrium sterni by the sternohyoid and sternothyroid, the anterior portions of the left pleura and lung, the left brachiocephalic vein and the remains of the thymus; *behind*, it is related first to the trachea, then to the left subclavian artery, the left edge of the oesophagus, the left recurrent laryngeal nerve and the thoracic duct. On its *right side* it is related below to the brachiocephalic trunk, and above to the trachea, the inferior thyroid veins and the remains of the thymus; at its *left side* are the left vagus and phrenic nerves, the left pleura and lung.

The cervical parts of the common carotid arteries have similar courses (**6.**43, 44, 46). Each passes obliquely upwards and slightly laterally, from behind the sternoclavicular joint, to the level of the upper border of the thyroid cartilage, where it divides into the external and internal carotid arteries (**6.**47). At its point of division the vessel shows a dilatation, termed the *carotid sinus*, which usually involves, and may be restricted to, the proximal part of the internal carotid artery. In this situation the tunica media is thinner than elsewhere and the tunica adventitia, which is relatively thick, contains a large number of sensory nerve endings, derived from the glossopharyngeal nerve (p. 1075). The nerve fibres and terminals show evidence of degeneration and regeneration, suggesting a continuous turnover which may be more marked in the elderly (Rüsager and Weddell 1962). The structure of the walls of the sinus enables it to react readily to changes in the arterial blood pressure and to bring about appropriate modifications reflexly. Owing to its situation on the main artery of supply to the brain, its function as a baroreceptor mechanism enables it to exercise control over intracranial pressure. The *carotid body*, which lies behind the point of division of the common carotid artery, is a small, reddish-brown structure; it acts as a 'chemoreceptor' (*see* Adams 1958 for an earlier comparative account, and p. 1462 for modern views concerning its fine structure and possible modes of operation).

At the lower part of the neck the two arteries are separated from each other by a narrow interval which contains the trachea; but at the upper part, the thyroid gland, the larynx, and the pharynx project forwards between the two vessels. The common carotid artery is contained in the carotid sheath (p. 537), which is continuous with the deep cervical fascia, and is composed of loose cellular tissue, but the part surrounding the artery is thicker and denser than the rest. This sheath encloses also the internal jugular vein and vagus nerve, the vein lying lateral to the artery, and the nerve between the artery and vein on a plane posterior to both. The superior root of the ansa cervicalis is embedded in its anterior wall.

Relations. The common carotid artery is crossed *antero-laterally*, at the level of the cricoid cartilage, by the superior belly of the omohyoid. Below this muscle the artery is very deeply seated, being covered with the skin, superficial fascia, platysma, deep cervical fascia, sternocleidomastoid, sternohyoid and sternothyroid. Above the omohyoid it is more superficial, being covered merely by the skin, the superficial fascia, platysma, deep cervical fascia and the medial margin of the sternocleidomastoid; this part of the artery is crossed obliquely, from its medial to its lateral side, by the sternocleidomastoid branch of the superior thyroid artery. In front of, or embedded in, its sheath is the superior root of the ansa cervicalis, this branch being joined by the inferior root of the ansa, from the second and third cervical nerves, which crosses the vessel obliquely. The superior thyroid vein usually crosses the artery near its termination, and the middle thyroid vein a little below the level of the cricoid cartilage; each anterior jugular vein crosses the artery just above the clavicle, but is separated from it by the sternohyoid and sternothyroid. *Behind*, the artery is separated from the transverse processes of the fourth, fifth and sixth cervical vertebrae by the longus colli and longus capitis and the origin of the scalenus anterior, the sympathetic trunk and the ascending cervical artery being interposed between the artery and the muscles. Below the level of the sixth cervical vertebra the common carotid artery lies in the angle between the scalenus anterior and the longus colli (**6**.118), anterior to the vertebral vessels, the inferior thyroid and subclavian arteries, the sympathetic trunk, and, on the left side, the thoracic duct. *Medial* to it are the oesophagus, trachea, the inferior thyroid artery and recurrent laryngeal nerve; at a higher level, the larynx and pharnyx are medial to the artery; the lobe of the thyroid gland overlaps it anteromedially. *Lateral* to the artery is the internal jugular vein; *posterolaterally*, in the angle between it and the internal jugular vein is the vagus nerve.

On the right side, at the lower part of the neck, the recurrent laryngeal nerve crosses obliquely behind the artery; the right internal jugular vein diverges from the artery below, but the left vein approaches and often overlaps the lower part of the vessel.

Variations. In about 12 per cent of subjects the *right* common carotid artery arises above the level of the upper border of the sternoclavicular joint. It may arise as a separate branch from the arch of the aorta, or in conjunction with the left common carotid. The *left* common carotid artery varies in its origin more frequently than the right. In the majority of abnormal cases it arises in common with the brachiocephalic artery; if that artery be absent, the two common carotids arise usually by a single trunk. It is rarely joined with the left subclavian, except in cases of transposition of the aortic arch.

Division of the common carotid artery may occur higher than

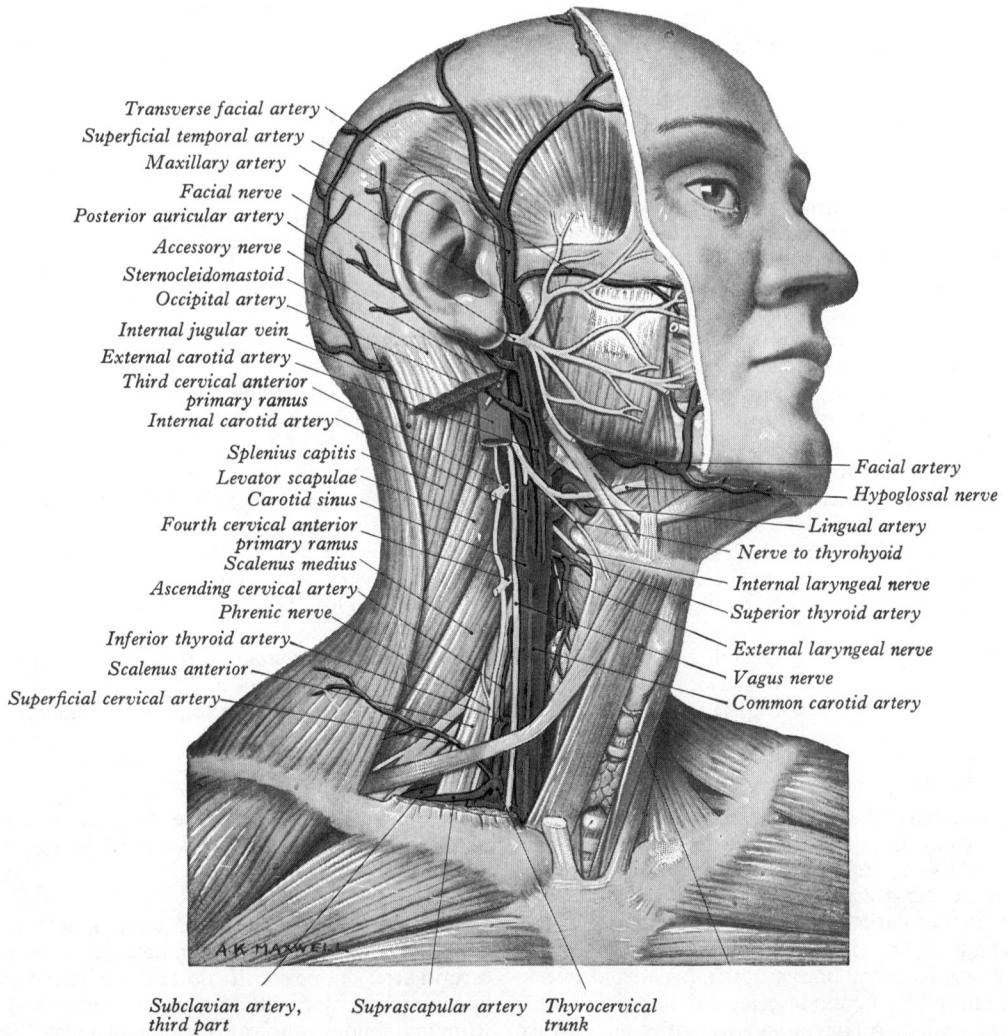

Transverse facial artery
Superficial temporal artery
Maxillary artery
Facial nerve
Posterior auricular artery
Accessory nerve
Sternocleidomastoid
Occipital artery
Internal jugular vein
External carotid artery
Third cervical anterior primary ramus
Internal carotid artery
Splenius capitis
Levator scapulae
Carotid sinus
Fourth cervical anterior primary ramus
Scalenus medius
Ascending cervical artery
Phrenic nerve
Inferior thyroid artery
Scalenus anterior
Superficial cervical artery

Facial artery
Hypoglossal nerve
Lingual artery
Nerve to thyrohyoid
Internal laryngeal nerve
Superior thyroid artery
External laryngeal nerve
Vagus nerve
Common carotid artery

Subclavian artery, third part   Suprascapular artery   Thyrocervical trunk

6.45   A dissection of the right side of the neck, showing the carotid and subclavian arteries and their branches. The parotid and submandibular glands have been removed together with the lower part of the internal jugular vein, most of the sternocleidomastoid, and the upper parts of the stylohyoid and posterior belly of the digastric.

usual, at or about the level of the hyoid bone; more rarely it occurs below the usual level, opposite the middle of the larynx, or the lower border of the cricoid cartilage. Very rarely the artery ascends in the neck without undergoing division, either the external or the internal carotid being absent. In a few instances the artery has been found absent, the external and internal carotid arteries arising directly from the arch of the aorta; this perculiarity exists on both sides in some instances, on one side in others.

The common carotid artery usually has no branches; but it may give origin to the vertebral, the superior thyroid (**6**.43) or its laryngeal branch, the ascending pharyngeal, the inferior thyroid, or the occipital.

The External Carotid Artery

The external carotid artery (**6**.45, 53) begins opposite the upper border of the thyroid cartilage, at the level of the disc between the third and fourth cervical vertebrae, and, taking a slightly curved course, passes upwards and forwards, and then inclines backwards to a point behind the neck of the mandible midway between the tip of the mastoid process and the angle of the jaw, where, in the substance of the parotid gland, it divides into the superficial temporal and maxillary arteries. It diminishes rapidly in size, owing to the number and large size of its branches. In the child, it is a little smaller than the internal carotid artery; but in the adult, the two vessels are of nearly equal size. At its origin, where its pulsations are easily felt, it is contained within the carotid triangle (p. 585), and lies anterior to and nearer the median plane than the internal carotid artery; higher up it is situated lateral to this artery.

Relations. Within the carotid triangle the external carotid artery is covered by the skin, the superficial fascia, the loop between the cervical branch of the facial nerve and the transverse (anterior) cutaneous nerve of the neck, the deep fascia, and the anterior margin of the sternocleidomastoid; it is crossed by the hypoglossal nerve and its vena comitans, by the lingual and (common) facial veins and sometimes by the superior thyroid vein. Leaving the carotid triangle it is crossed by the posterior belly of the digastric and the stylohyoid and then ascends between the latter muscle and the posteromedial surface of the parotid gland. Finally, it enters the gland, where it lies deep to the facial nerve and the junction of the superficial temporal and maxillary veins. Deeply, it is related at first to the wall of the pharynx, the superior laryngeal nerve, and the ascending pharyngeal artery. At a higher level the internal carotid artery is deep to it, but is separated from it by the styloid process, the styloglossus and stylopharyngeus, the glossopharyngeal nerve, the pharyngeal branch of the vagus nerve, and a part of the parotid gland (**6**.46). The relation of the external carotid artery to the parotid gland is, in fact, a matter of some controversy, many clinicians asserting that it is often medial to, rather than within the gland. The most recent study (Guffarth and Graumann 1975), based on a short series of dissections, suggests that both relationships occur and at about equal frequency.

Branches. The branches of the external carotid artery (**6**.45, 49) are:

1. Superior thyroid.
2. Ascending pharyngeal.
3. Lingual.
4. Facial.
5. Occipital.
6. Posterior auricular.
7. Superficial temporal.
8. Maxillary.

1. The Superior Thyroid Artery (**6**.45)

This arises from the front of the external carotid artery just below the level of the greater cornu of the hyoid bone and divides into terminal branches at the apex of the lobe of the gland. It may arise from the common carotid artery (**6**.43).

Relations. From its origin beneath the anterior border of the sternocleidomastoid it runs downwards and forwards in the carotid triangle along the lateral border of the thyrohyoid, and is covered by the skin, platysma and fasciae; it then passes under cover of the omohyoid, sternohyoid and sternothyroid. To its medial side are the constrictor pharyngis inferior and the external branch of the superior laryngeal nerve, but the nerve frequently lies on a more posterior plane.

Branches. It supplies the adjacent muscles and the thyroid gland; it anastomoses with its fellow of the opposite side and the inferior thyroid arteries. The branches to the gland are: the anterior branch, which follows the medial border of the upper pole of the lobe and supplies principally the anterior surface, sending a branch across the upper border of the isthmus to anastomose with the artery of the opposite side; the posterior branch descends on the posterior border of the gland, supplying its medial and lateral surfaces, and anastomoses with the inferior thyroid artery. Sometimes a lateral branch is distributed to the lateral surface of the gland.

Besides arteries to muscles and the thyroid gland, the superior thyroid artery has the following branches:

Infrahyoid.   Superior laryngeal.
Sternocleidomastoid.   Cricothyroid.

The infrahyoid artery is small, runs along the lower border of the hyoid bone deep to the thyrohyoid, and anastomoses with the opposite vessel.

The sternocleidomastoid branch, which frequently arises from the external carotid artery, runs down and laterally across the carotid sheath.

The superior laryngeal artery accompanies the internal laryngeal nerve and passes deep to the thyrohyoid; it pierces the lower part of the thyrohyoid membrane, and supplies the larynx, anastomosing with the opposite artery and with the inferior laryngeal artery, a branch of the inferior thyroid.

The cricothyroid branch is small and runs transversely across the upper part of the cricothyroid ligament communicating with the artery of the opposite side.

2, The Ascending Pharyngeal Artery (**6**.56)

This, the smallest branch of the external carotid, is a long, slender vessel. It arises close to the origin of the external carotid, and ascends vertically between the internal carotid and the side of the pharynx to the base of the skull, being crossed by the styloglossus and the stylopharyngeus, with longus capitis posterior to it; it anastomoses freely with the ascending palatine branch of the facial artery.

Branches. The named branches of the ascending pharyngeal artery are:

Pharyngeal.   Inferior tympanic.   Meningeal.

The pharyngeal branches, three or four in number, supply the constrictors and the stylopharyngeus. A branch of variable size is distributed to the palate, and may take the place of the ascending palatine branch of the facial artery; it runs downwards and forwards between the superior border of the superior constrictor and the levator veli palatini, and accompanies the latter muscle to the soft palate; it gives minute branches to the tonsil, and supplies a twig to the auditory tube.

The inferior tympanic artery is a small branch which traverses the temporal bone, in the canaliculus for the tympanic branch of the glossopharyngeal nerve, to supply the medial wall of the tympanic cavity.

The meningeal branches are several small vessels which supply the dura mater and adjacent bone. They enter the cranium through the foramen lacerum, the jugular foramen and the hypoglossal canal, and supply nerves traversing them.

Numerous small vessels supply the longi capitis et colli, the sympathetic trunk, the hypoglossal, glossopharyngeal and vagus nerves, and cervical lymph nodes; they anastomose with branches of the ascending cervical and vertebral arteries.

3. The Lingual Artery (**6**.45)

This is the principal supply to the tongue and the floor of the mouth. It arises from the anteromedial surface of the external carotid artery opposite the tip of the greater cornu of the hyoid bone, and between the superior thyroid and facial arteries. Running obliquely upwards and medially at first, it then curves downwards and forwards to the greater cornu of the hyoid bone, forming a characteristic loop; it passes medial to the posterior border of the hyoglossus, runs horizontally forwards under cover of that muscle, and finally, ascending almost perpendicularly, courses tortuously forwards on the under surface of the tongue as far as its tip (**6**.48).

Course and relations. The *first part* of the lingual artery is in the carotid triangle; superficial to it are the skin, fascia and platysma; medial to it, the middle constrictor. It runs upwards and medially for a short distance, and then descends to the level of the hyoid bone, forming a loop, which is crossed by the hypoglossal nerve. The *second part* of the artery traverses the upper border of the hyoid bone, deep to the hyoglossus, the tendon of the digastric or its fascial retinaculum, the stylohyoid, the lower part of the submandibular gland and the posterior part of the mylohyoid; the hyoglossus separates the artery from the hypoglossal nerve and its vena comitans; in this part of its course it lies on the middle constrictor and crosses the stylohyoid ligament. It is accompanied by the lingual veins (p. 742). The *third part* of the artery is named the *arteria profunda linguae*. It bends sharply upwards near the anterior border of the hyoglossus, and then runs forwards close to the inferior surface of the tongue near the frenulum, accompanied by the lingual nerve. Medially, it is related to the genioglossus, laterally, to the longitudinalis linguae inferior, below, to the mucous membrane of the tongue. At the tip of the tongue it anastomoses with the lingual artery of the opposite side.

Branches. The branches of the lingual artery are:

The suprahyoid branch, which is very small; it runs along the upper border of the hyoid bone, and anastomoses with the contralateral artery.

The dorsal lingual branches, consisting usually of two or three small vessels, arise medial to the hyoglossus muscle, ascend to the posterior part of the dorsum of the tongue, and supply its mucous membrane, the palatoglossal arch, the tonsil, soft palate and epiglottis; they anastomose with the vessels of the opposite side.

The sublingual artery which arises at the anterior margin of the hyoglossus, travels forward between the genioglossus and mylohyoid to the sublingual gland. It supplies the gland and gives branches to the mylohyoid and neighbouring muscles, and to the mucous membrane of the mouth and gums. One branch runs behind the alveolar part of the mandible in the substance of the gum to anastomose with a similar artery from the other side; another pierces the mylohyoid and anastomoses with the submental branches of the facial artery.

The lingual artery often shares a common trunk with the facial or, less frequently, with the superior thyroid artery. It may be replaced by a branch of the maxillary artery.

4. The Facial Artery (**6**.49, 56)

This arises from the front of the external carotid artery in the carotid triangle above the lingual artery, and immediately above the greater cornu of the hyoid bone. Medial to the ramus of the mandible, it arches upwards and grooves the posterior border of the submandibular gland. It next turns downwards and forwards between the gland and the medial pterygoid, and, reaching the lower border of the mandible, hooks round it at the anterior edge of the masseter, and enters the face. On the face it passes forwards and upwards across the mandible and buccinator almost to the angle of the mouth where its pulsations can be felt if the cheek be grasped lightly between finger and thumb. It then ascends the side of the nose, and ends at the medial palpebral commissure, to supply the lacrimal sac and anastomose with the dorsal nasal branch of the ophthalmic artery. The facial artery is remarkably tortuous—in the neck to accommodate itself perhaps to the movements of the pharynx during deglutition and on the face to the movements of the mandible, lips and cheeks.

Distal to its superior branch (**6**.49) the facial artery is occasionally termed the *angular artery*.

Relations. In the neck, at its origin the artery is superficial, being covered merely by skin, platysma, and fasciae, and is often crossed by the hypoglossal nerve. It runs upwards and forwards, deep to the digastric and stylohyoid and the posterior part of the submandibular gland. At first on the surface of the middle constrictor of the pharynx, it may reach as high as the lateral surface of the styloglossus, and it is then separated from the tonsil only by that muscle and the lingual fibres of origin of the superior

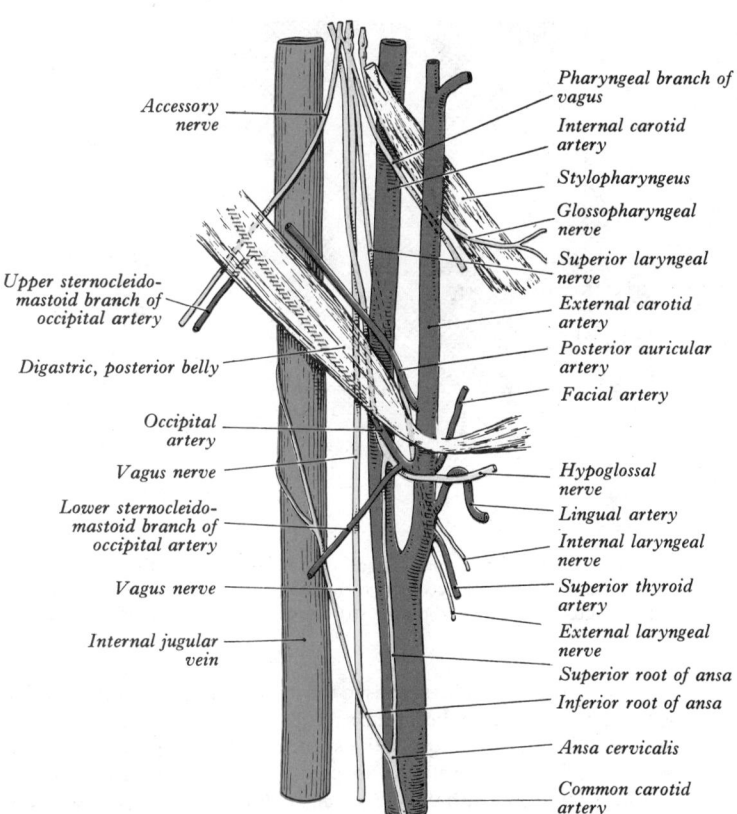

6.46 A diagram showing the structures crossing the internal jugular vein and carotid arteries, and those intervening between the external and internal carotid arteries.

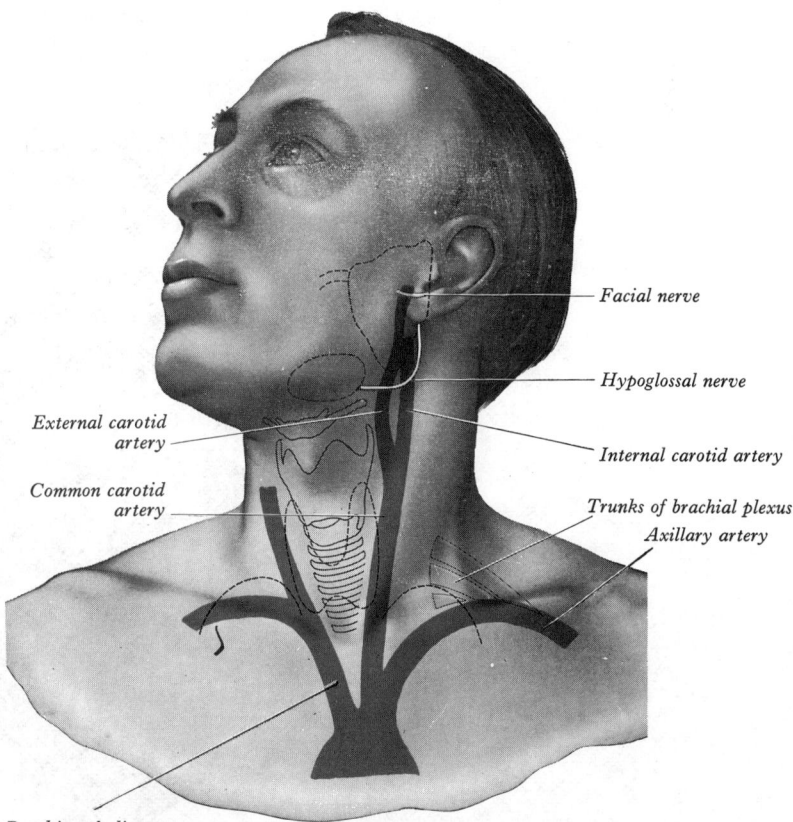

6.47 The surface projection of some of the larger structures in the face and neck. *Note.*—The parotid gland and its duct, the submandibular and thyroid glands, and the apices of the lungs are shown as interrupted outlines. The hyoid bone and the thyroid, cricoid and tracheal cartilages are indicated by continuous outlines.

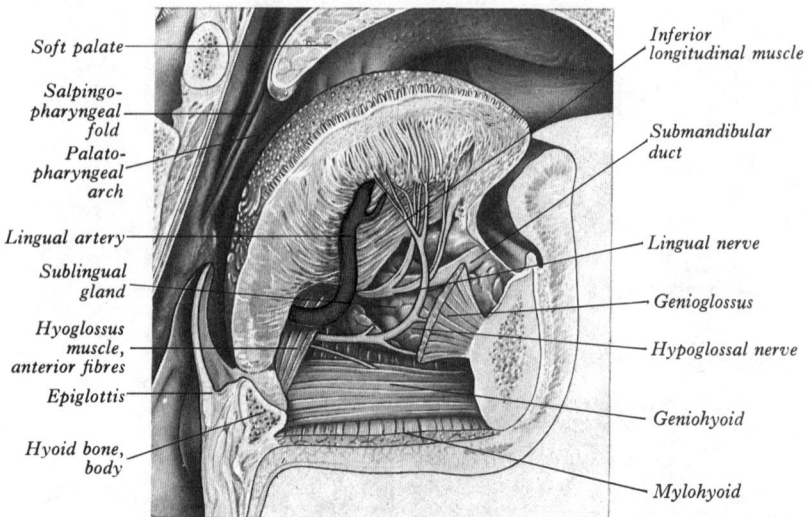

6.48 A dissection of the left half of the tongue from the medial side, exposing the end of the second part and the beginning of the third part of the left lingual artery and adjoining structures, in an edentulous subject.

constrictor. Next, it descends to the lower border of the mandible, lying in a groove on the lateral aspect of the submandibular gland. *In the face*, where its pulsations can be felt as it passes over the base of the mandible, it is comparatively superficial, lying immediately beneath the platysma. In its course over the face, it is under cover of the skin, the fat of the cheek, and, near the angle of the mouth, the platysma, risorius and zygomaticus major. It rests on the buccinator and levator anguli oris, and passes either over or through the levator labii superioris. Its terminal part is embedded in the fibres of the levator labii superioris alaeque nasi. The facial vein lies posterior to the artery, and, taking a more direct course across the face, is at some distance from the artery. At the anterior border of the masseter the two vessels are in contact; in the neck the vein is superficial to the artery. The branches of the facial nerve cross the artery from behind forwards.

The facial artery supplies the muscles and tissues of the face, the submandibular gland, the tonsil and the soft palate. Its branches can be divided into cervical and facial groups.

Cervical Branches

The ascending palatine artery (**6**.56) arises close to the origin of the facial artery and passes up between the styloglossus and stylopharyngeus to the side of the pharynx, along which it ascends between the superior constrictor and the medial pterygoid towards the base of the skull. Near the levator veli palatini it divides into two branches: one follows this muscle, and, winding over the upper border of the superior constrictor, supplies the soft palate, anastomosing with its fellow of the opposite side and with the greater palatine branch of the maxillary artery; the other pierces the superior constrictor and supplies the tonsils and the auditory tube, anastomosing with the tonsillar and ascending pharyngeal arteries.

The tonsillar artery is the principal supply to the tonsil. Sometimes derived from the ascending palatine artery, though usually a separate branch, it ascends between the medial pterygoid and styloglossus, and at the upper border of the latter muscle it perforates the superior constrictor and ramifies in the tonsil and the root of the tongue.

The glandular branches, three or four large vessels, supply the submandibular salivary gland and lymph nodes, the neighbouring muscles and overlying skin.

The submental artery, the largest cervical branch of the facial artery, branches off as it quits the submandibular gland: it

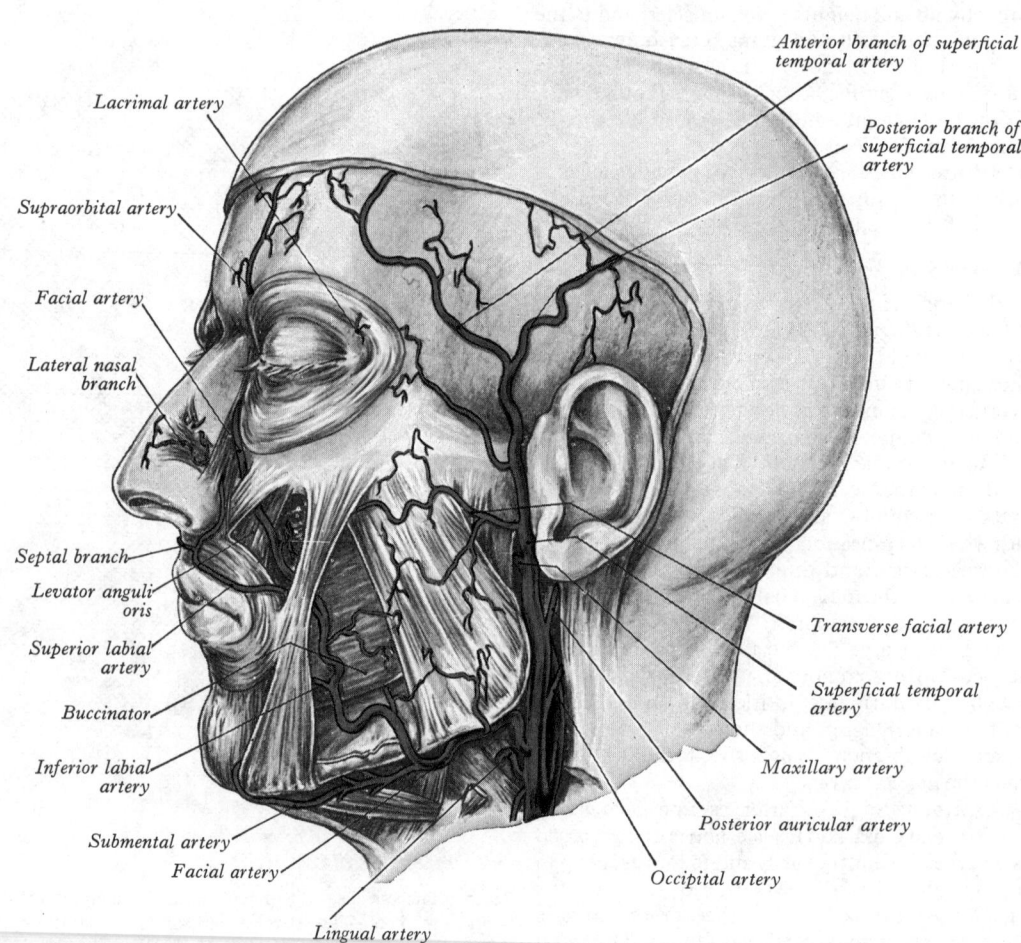

6.49 The arteries of the left side of the face and their main branches. Note the less usual origin of the lateral nasal branch in this specimen.

runs forwards upon the mylohyoid (**6**.49), below the body of the mandible. It supplies the surrounding muscles, and anastomoses with the sublingual branch of the lingual artery and the mylohyoid branch of the inferior alveolar artery; at the chin it turns upwards over the base of the mandible and divides into superficial and deep branches which anastomose with the inferior labial and mental arteries, supplying the chin and lower lip.

Facial Branches

The inferior labial artery (**6**.49) arises near the angle of the mouth; it passes upwards and forwards under cover of the depressor anguli oris and, penetrating the orbicularis oris, runs in a tortuous course near the edge of the lower lip between this muscle and the mucous membrane. It supplies the glands, mucous membrane and muscles of the lower lip and anastomoses with the artery of the opposite side, and with the mental branch of the inferior alveolar artery.

The superior labial artery (**6**.49) is larger and more tortuous than the inferior. It follows a similar course near the edge of the upper lip, lying between the mucous membrane and the orbicularis oris, and anastomoses with the artery of the opposite side. It supplies the upper lip, and gives off a *septal branch*, which ramifies on the lower and front part of the nasal septum, and an *alar branch*, to the ala of the nose.

The lateral nasal branch (**6**.49) is derived from the facial artery as that vessel ascends along the side of the nose. It supplies the ala and dorsum of the nose, anastomosing with its fellow, with the septal and alar branches of the superior labial artery, with the dorsal nasal ramus of the ophthalmic artery, and with the infraorbital branch of the maxillary artery. The lateral nasal artery may be represented by several small rami; it may alternatively arise from the superior labial artery, diverging from its septal branch (as in **6**.49).

The anastomoses of the facial artery are very numerous, not only with the branches of the vessel of the opposite side, but, *in the neck*, with the sublingual branch of the lingual, with the ascending pharyngeal and with the palatine branch of the maxillary; *on the face*, with the mental branch of the inferior alveolar, the transverse facial branch of the superficial temporal, the infraorbital branch of the maxillary and the dorsal nasal branch of the ophthalmic artery. The anastomoses across the midline between the superior and inferior labial arteries are by their main trunks, an important fact in the treatment of split lip.

Variations. The facial artery not infrequently arises in common with the lingual artery, constituting the *linguo-facial trunk*. It varies in size, and in the extent to which it supplies the face: it occasionally ends by forming the submental artery, and not infrequently extends only as high as the angle of the mouth or nose. The deficiency is then compensated for by enlargement of one of the neighbouring arteries. In a series of 110 human fetuses a common linguo-facial trunk occurred in 43 per cent. In 42 per cent it did not reach the medial orbital angle, terminating as the superior (20 per cent) or inferior (22 per cent) labial artery (Kozielec and Joźwa 1977).

5. The Occipital Artery (**6**.50)

This artery arises from the back of the external carotid artery, opposite to the facial artery and, running at first on the medial surface of the posterior belly of the digastric, ends in the posterior part of the scalp.

Course and relations. At its origin, it is crossed superficially by the hypoglossal nerve, which winds round it from behind forwards. The artery passes backwards and upwards deep to the lower border of the posterior belly of the digastric, crossing in its course the internal carotid artery, the internal jugular vein, and the hypoglossal, vagus and accessory nerves (**6**.50). Reaching the interval between the transverse process of the atlas and the mastoid process of the temporal bone, it comes into contact with the lateral border of the rectus capitis lateralis. It then runs in the occipital groove on the temporal bone, and here is medial to the mastoid process and the attachments to it of the sternocleidomastoid, splenius capitis, longissimus capitis and digastric, and lies successively on the rectus capitis lateralis, obliquus superior and semispinalis capitis. Finally, it turns upwards and pierces the

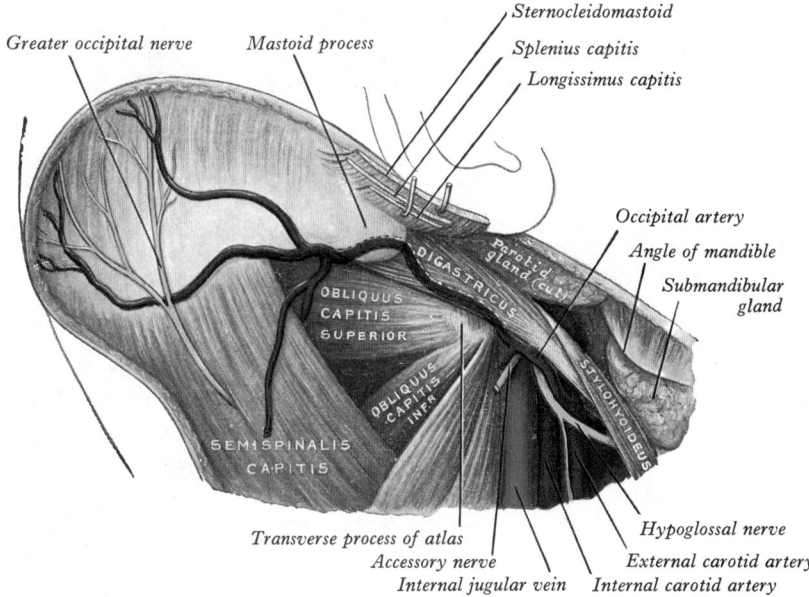

Greater occipital nerve — Mastoid process — Sternocleidomastoid — Splenius capitis — Longissimus capitis — Occipital artery — Angle of mandible — Submandibular gland — DIGASTRICUS — Parotid gland (cut) — OBLIQUUS CAPITIS SUPERIOR — OBLIQUUS CAPITIS INFR. — STYLOHYOIDEUS — SEMISPINALIS CAPITIS — Transverse process of atlas — Accessory nerve — Internal jugular vein — Hypoglossal nerve — External carotid artery — Internal carotid artery

6.50 A dissection to show the course of the occipital artery.

fascia connecting the cranial attachments of the trapezius and the sternocleidomastoid, and ascends in a tortuous course in the superficial fascia of the scalp, where it divides into numerous branches. The terminal portion of the occipital artery is accompanied by the greater occipital nerve.

Branches. The following branches arise from the occipital artery:

The sternocleidomastoid branches are usually two in number. The *lower branch* usually arises from the beginning of the occipital artery, but sometimes comes directly from the external carotid. It passes downwards and backwards over the hypoglossal nerve and the internal jugular vein, and enters the substance of the sternocleidomastoid; it anastomoses with the sternocleidomastoid branch of the superior thyroid artery. The *upper branch* arises from the occipital artery as it crosses the accessory nerve, and runs downwards and backwards superficial to the internal jugular vein. It enters the deep surface of the sternocleidomastoid in company with the accessory nerve.

The mastoid branch, small in size and sometimes absent, enters the cranial cavity through the mastoid foramen; it gives branches to the mastoid air cells and ramifies in the dura mater.

The stylomastoid artery in two thirds of subjects (*see* p. 682).

The auricular branch supplies the back of the auricle and anastomoses with the posterior auricular artery.

Unnamed **muscular branches** supply the digastric, stylohyoid, splenius and longissimus capitis.

The descending branch (**6**.50) arises from the occipital artery as the latter lies on the obliquus superior, and divides into superficial and deep branches. The *superficial branch* passes deep to the splenius, and anastomoses with the superficial branch of the transverse cervical artery; the *deep branch* descends between the semispinales capitis et cervicis, and anastomoses with the vertebral artery, and with the deep cervical artery, a branch of the costocervical trunk (**6**.56).

The meningeal branches enter the skull through the jugular foramen and the condylar canal, to supply the dura mater and bone in the posterior fossa, and the caudal four cranial nerves.

The occipital branches, which are the terminal rami, are distributed to the scalp, and reach as high as the vertex of the skull; they are very tortuous, and lie between the skin and the occipital belly of the occipitofrontalis, anastomosing with the artery of the opposite side and with the posterior auricular and temporal arteries, and supplying the occipital belly of the occipitofrontalis, the skin and the pericranium. One of the terminal branches may give off a meningeal twig, which passes through the parietal foramen.

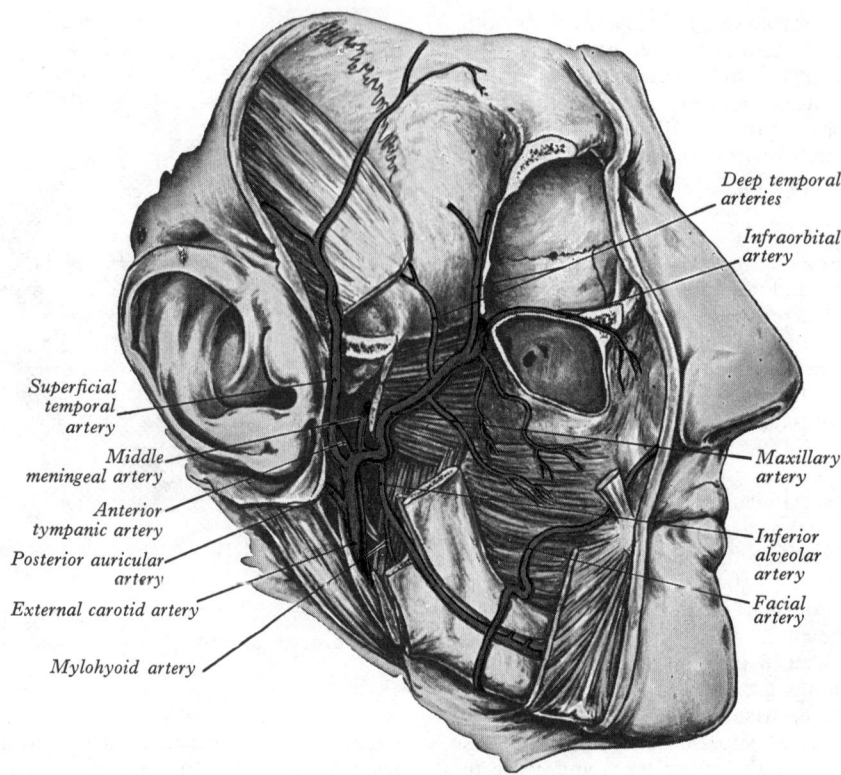

6.51   The right maxillary artery. An extensive dissection has been
carried out, involving the removal of the parotid gland, the zygomatic
arch, part of the ramus of the mandible, the lateral walls of the orbit and
maxillary sinus and the orbital contents.

6. The Posterior Auricular Artery (6.49)

This is small and branches from the posterior aspect of the
external carotid immediately above the digastric and stylohoid. It
ascends between the parotid gland and the styloid process of the
temporal bone to the groove between the cartilage of the auricle
and the mastoid process, where it divides into auricular and
occipital branches.

Branches. As well as supplying the digastric, stylohyoid,
sternocleidomastoid, and parotid gland, the posterior auricular
artery gives off:

The stylomastoid artery, an indirect branch of the posterior
auricular in about a third of subjects (Blunt 1954), enters the
stylomastoid foramen to supply the facial nerve, tympanic cavity,
mastoid antrum and air cells, and semicircular canals. In the
young its posterior tympanic ramus forms an encircling
anastomosis to supply the tympanic membrane with the anterior
tympanic artery (p. 683).

The auricular branch, which ascends under cover of the
auricularis posterior, ramifies on the cranial surface of the auricle;
some of its branches pierce the auricle, and others curve round its
margin to supply its lateral surface.

The occipital branch, which passes laterally across the front
of the mastoid process, then turns backwards over the
sternocleidomastoid to the occipital belly of the occipitofrontalis
and to the scalp above and behind the ear; it anastomoses with the
occipital artery.

7. The Superficial Temporal Artery (6.45)

This, the smaller terminal branch of the external carotid, begins
in the parotid gland, behind the neck of the mandible, and crosses
over the posterior root of the zygomatic process of the temporal
bone; about 5 cm above this process it divides into anterior and
posterior branches.

Relations. As it crosses the zygoma, it is covered by the
auricularis anterior; it is crossed in the substance of the parotid
gland by the temporal and zygomatic branches of the facial nerve,
and is accompanied in the scalp by the corresponding veins and by
the auriculotemporal nerve, which lies immediately behind it.

Branches. The superficial temporal artery supplies the parotid
gland, the temporomandibular joint, the masseter, and certain
named branches:

The transverse facial artery (6.49) arises from the
superficial temporal artery before that vessel emerges from the
parotid gland; it runs forwards through the gland, passes across
the masseter between the parotid duct and the zygomatic arch,
accompanied by one or two branches of the facial nerve. It divides
into numerous branches, which supply the parotid gland and
duct, the masseter, and the skin, and anastomose with the facial,
masseteric, buccal, lacrimal, and infraorbital arteries.

The anterior auricular branches are distributed to the
lobule and anterior portion of the auricle, and to the external
acoustic meatus.

The zygomatico-orbital artery, sometimes a branch of the
middle temporal artery, runs along the upper border of the
zygomatic arch between the two layers of the temporal fascia to
the lateral angle of the orbit. It supplies branches to the
orbicularis oculi, and anastomoses with the lacrimal and palpebral
branches of the ophthalmic artery.

The middle temporal artery arises immediately above the
zygomatic arch, and, perforating the temporal fascia, gives
branches to the temporalis; it anastomoses with the deep temporal
branches of the maxillary artery.

The frontal (anterior) branch runs tortuously upwards and
forwards towards the tuberosity of the frontal bone; it supplies the
muscles, skin and pericranium in this region, and anastomoses
with its fellow of the opposite side, and with the supraorbital and
supratrochlear arteries.

The parietal (posterior) branch, larger than the preceding
vessel, curves upwards and backwards on the side of the head,
lying superficial to the temporal fascia, and anastomosing with the
opposite artery and with the posterior auricular and occipital
arteries.

Variation in the superficial temporal artery is largely limited to
the relative sizes of its frontal, parietal, and transverse facial
branches; either of the first two may be absent, while the
transverse facial may be large enough to replace a shortened facial
artery. Variations in fetal material have been described by
Kozielec and Jóźwa (1976).

Applied Anatomy. As the superficial temporal artery crosses the zygomatic process it lies beneath the skin and fascia, and its pulsations may be readily felt, for example, during an anaesthetic, or in circumstances where the radial pulse is not available; it may be compressed easily against the bone in order to check bleeding from the temporal region of the scalp. It should be noted that this vessel and the other arteries to the scalp enter from below and are well protected by dense tissue. It is rarely found that they are all implicated in scalping injuries and their anastomoses are so free that so long as one is intact the detached scalp may be replaced with reasonable hope of its survival. When a flap is raised from this part of the head, for craniotomy, the incision should be shaped like a horseshoe, with its convexity upwards, so that the flap shall contain the superficial temporal artery, which ensures a sufficient supply of blood.

In angiograms of the external carotid artery the branches of the superficial temporal and middle meningeal arteries are superimposed. They may be distinguished by the straighter courses, lack of anastomoses, and narrower calibre of the meningeal rami (Domnić-Stošić and Jeličić 1974).

8. The Maxillary (Internal Maxillary) Artery (6.51)

This, the larger terminal branch of the external carotid, arises behind the neck of the mandible and is at first embedded in the parotid gland; it passes forwards medial to the neck of the mandible and then runs either superficial or deep to the lower head of the lateral pterygoid to enter the pterygopalatine fossa between the two heads of that muscle. It may be divided into mandibular, pterygoid and pterygopalatine portions associated successively with bone, muscle, and bone, a useful indication of the branches of each part.

The *first* or *mandibular part* passes horizontally forwards, between the neck of the mandible and the sphenomandibular ligament, where it lies parallel with and a little below the auriculotemporal nerve; it crosses the inferior alveolar nerve, and runs along the lower border of the lateral pterygoid.

The *second* or *pterygoid part* runs obliquely forwards and upwards medial to temporalis, and superficial to the lower head of the lateral pterygoid; very frequently it lies deep to the latter, between it and the branches of the mandibular nerve, and in this case, before entering on the third part of its course, it often forms a wide loop which projects laterally between the two heads of the lateral pterygoid.

The *third* or *pterygopalatine part* passes between the upper and lower heads of the lateral pterygoid, and through the pterygomaxillary fissure into the pterygopalatine fossa, where it lies in front of the pterygopalatine ganglion.

Branches. The maxillary artery is distributed to the upper and lower jaws, the teeth, muscles of mastication, palate, nose and cranial dura mater. Its branches may be divided into three groups, corresponding with its three parts.

Branches of the First Part (6.51)

The deep auricular artery, a small branch, often arises with the anterior tympanic. It ascends in the substance of the parotid gland, behind the temporomandibular joint, pierces the cartilaginous or bony wall of the external acoustic meatus, and supplies its cuticular lining and the outer surface of the tympanic membrane; it gives a branch to the temporomandibular joint.

The anterior tympanic artery, a small branch, ascends behind the temporomandibular joint, and enters the tympanic cavity through the petrotympanic fissure; it ramifies upon the medial aspect of the tympanic membrane, and forms a vascular circle around it with the posterior tympanic branch of the stylomastoid artery; it anastomoses with small branches of the artery of the pterygoid canal and with caroticotympanic branches from the internal carotid artery in the tympanic cavity.

The middle meningeal artery is the largest of the meningeal arteries. It arises from the maxillary artery between the sphenomandibular ligament and the lateral pterygoid, and, passing between the two roots of the auriculotemporal nerve, may lie on the lateral surface of the tensor veli palatini just before it enters the cranial cavity through the foramen spinosum of the sphenoid bone; it then runs forwards and laterally for a variable distance in a groove on the anterior part

of the squamous part of the temporal bone, and divides into a frontal and a parietal branch. The *frontal (anterior) branch*, the larger, crosses the greater wing of the sphenoid bone, reaches the groove, or canal, in the sphenoidal angle of the parietal bone, and then divides into branches which spread out between the dura mater and internal surface of the cranium, some passing upwards as far as the vertex, and others backwards to the occipital region. One branch runs upwards, grooving the parietal bone about 1·5 cm behind the coronal suture. It corresponds, in a general way, to the line of the precentral sulcus of the brain. The *parietal (posterior) branch* curves backwards on the squamous part of the temporal bone, and, reaching the lower border of the parietal bone some distance in front of its mastoid angle, divides into branches which supply the posterior part of the dura mater and cranium. The branches of the middle meningeal artery anastomose with the arteries of the opposite side, and with the anterior and posterior meningeal arteries.

The middle meningeal artery gives off the following branches within the cranial cavity: (1) Numerous small *ganglionic branches* supply the trigeminal ganglion and the roots of the trigeminal nerve. (2) A *petrosal branch* enters the hiatus for the greater petrosal nerve, gives twigs to the facial nerve and the tympanic cavity, and anastomoses with the stylomastoid artery (of variable origin—*see* pp. 681, 682). (3) A *superior tympanic artery* runs in the canal for the tensor tympani, and supplies this muscle and the lining membrane of the canal. (4) *Temporal branches* pass through minute foramina in the greater wing of the sphenoid, and anastomose in the temporal fossa with the deep temporal arteries. (5) An *anastomotic branch* with the lacrimal artery (p. 688) runs forwards and enters the orbit through the lateral part of the superior orbital fissure. It anastomoses with a recurrent meningeal branch of the lacrimal artery, and an enlargement of this anastomosis explains the occasional origin of the lacrimal from the middle meningeal artery. Apart from these small branches and a supply to the dura mater, the middle meningeal artery is predominantly a periosteal artery supplying bone and red bone marrow.

Surface Anatomy (6.119). The *middle meningeal artery* enters the skull opposite a point immediately above the middle of the zygoma (6.119) and divides 2 cm above this. From this situation the frontal branch runs first upwards and slightly forwards to the pterion (3.126) and then turns upwards and backwards towards the mid-point between the inion (3.69) and the nasion (3.126). The parietal branch runs upwards and backwards towards the lambda (3.61).

Applied Anatomy. The middle meningeal artery is of considerable surgical importance, for it may be torn in fractures of the temporal region of the skull, or, indeed, by injuries causing separation of the dura mater from the bone without fracture. The injury may be followed by considerable haemorrhage between the bone and dura mater, which produces symptoms of compression of the brain, and requires trephining for its relief. As the compression implicates the motor region of the cortex, paralysis on the opposite side of the body forms the prominent phenomenon of the lesion.

The accessory meningeal branch may arise from the maxillary artery or from the middle meningeal artery. It enters the cranial cavity through the foramen ovale, and supplies branches to the trigeminal ganglion, dura mater and bone. Its main distribution, however, is *extracranial* (Baumel and Beard 1961), passing principally to the medial pterygoid, lateral pterygoid (upper head), tensor veli palatini, parts of the sphenoid bone (greater wing and pterygoid processes), the mandibular nerve and the otic ganglion, and it is sometimes represented by a number of separate small arteries.

The inferior alveolar (dental) artery descends posterior to the inferior alveolar nerve to the mandibular foramen on the medial surface of the ramus of the mandible. Here the artery lies between the bone on the lateral side, and the sphenomandibular ligament on the medial side. Before it enters the mandibular foramen, it gives off a *mylohyoid branch*, which pierces the sphenomandibular ligament, and descends with the mylohyoid nerve in the mylohyoid groove on the ramus of the mandible; it ramifies on the superficial surface of the mylohyoid and anastomoses with the submental branch of the facial artery. The inferior alveolar artery then runs in the mandibular canal, accompanied by the inferior alveolar nerve, and, opposite the first premolar tooth, divides into two branches, incisor and mental. The *incisor* branch is continued forwards below the incisor teeth as far as the median plane, where it anastomoses with the artery of the opposite side. Within the canal the inferior alveolar artery and its incisor branch supply the mandible, tooth sockets, and a series of branches which correspond in number to the roots of the teeth; these enter the minute apertures at the extremities of the roots, and supply the pulp of the teeth. The *mental branch* escapes at the mental foramen, supplies the chin, and anastomoses with the submental and inferior labial arteries. Near its origin the inferior alveolar artery gives off a *lingual* branch, which descends with the lingual nerve and helps to supply the mucous membrane of the mouth.

Branches of the Second Part (6.51)

The deep temporal branches, an anterior and a posterior, ascend between the temporalis and the skull; they supply the muscle, and anastomose with the middle temporal artery; the anterior communicates with the lacrimal artery by means of small branches which perforate the zygomatic bone and greater wing of the sphenoid bone.

The pterygoid branches, irregular in their number and origin, supply the pterygoid muscles.

The masseteric artery is small and passes with the corresponding nerve behind the tendon of the temporalis and through the mandibular notch to the deep surface of the masseter. In the substance of that muscle it anastomoses with the masseteric branches of the facial and with branches of the transverse facial artery.

The buccal artery is small and runs obliquely forwards with the buccal nerve, between the medial pterygoid and the insertion of the temporalis, to the outer surface of the buccinator, to which it is distributed, anastomosing with branches of the facial and infraorbital arteries.

Branches of the Third Part

The posterior superior alveolar (dental) artery is given off from the maxillary artery as that vessel enters the pterygopalatine fossa. Descending upon the infratemporal surface of the maxilla, it divides into branches, some of which enter the alveolar canals, and supply the molar and premolar teeth and the lining of the maxillary sinus, while others are continued forwards on the alveolar process to supply the gums.

The infraorbital artery often arises in conjunction with the posterior superior alveolar artery. It enters the orbital cavity through the posterior part of the inferior orbital fissure, runs along the infraorbital groove and canal with the infraorbital nerve, and emerges with the nerve on the face through the infraorbital foramen, deep to the levator labii superioris. In the canal, it gives off (*a*) *orbital branches*, which assist in supplying the rectus inferior, the obliquus inferior and the lacrimal sac, and (*b*) *anterior superior alveolar (dental) branches*, which descend through the anterior alveolar canals to supply the upper incisor and canine teeth and the mucous membrane of the maxillary sinus. On the face, some branches ascend to the medial angle of the eye and the lacrimal sac, anastomosing with the terminal branches of the facial artery; others run towards the nose, anastomosing with the dorsal nasal branch of the ophthalmic artery; others descend between the levator labii superioris and the levator anguli oris, and anastomose with the facial, transverse facial and buccal arteries.

The remaining branches of the maxillary artery arise in the pterygopalatine fossa.

The greater palatine artery descends through the greater palatine canal with the greater palatine nerve from the pterygopalatine ganglion, and gives off two or three *lesser palatine arteries*, which are transmitted through the lesser palatine canals to supply the soft palate and tonsil, and to anastomose with the ascending palatine artery. The greater palatine artery emerges on the oral surface of the palate through the greater palatine foramen, runs forwards, in a groove near the alveolar border of the hard palate, to the incisive canal; its terminal part passes upwards through this canal, and anastomoses with a branch of the sphenopalatine artery. Branches are distributed to the gums, the palatine glands and the mucous membrane of the roof of the mouth.

The pharyngeal branch is very small; it runs backwards through the pharyngeal (palatovaginal) canal with the pharyngeal branch of the pterygopalatine ganglion, and is distributed to the mucosa of the roof of the nose, pharynx, sphenoidal air sinus and auditory tube.

The artery of the pterygoid canal, frequently a branch of the greater palatine artery, passes backwards along the pterygoid canal with the corresponding nerve supplying its walls and contents. It is also distributed to the mucous membrane of the upper part of the pharynx, the auditory tube and the tympanic cavity.

The pharyngeal branch is medial, and the artery of the pterygoid canal lateral, to the pterygopalatine ganglion, while the trunk of the maxillary artery is anterior to it.

The sphenopalatine artery, the terminal part of the maxillary artery, passes through the sphenopalatine foramen into the cavity of the nose at the posterior part of the superior meatus. Here arise its *posterior lateral nasal branches*, which ramify over the conchae and meatuses, anastomose with the ethmoidal arteries and the nasal branches of the greater palatine artery, and assist in supplying the frontal, maxillary, ethmoidal and sphenoidal sinuses. Crossing the anterior part of the inferior surface of the sphenoid bone, the sphenopalatine artery ends on the nasal septum as the *posterior septal branches*, which anastomose with the ethmoidal arteries; one branch descends in a groove on the vomer to the incisive canal and anastomoses with the terminal ascending branch of the greater palatine artery, and with the septal branch of the superior labial artery.

Collateral Circulation. After ligature of the common carotid, the collateral circulation can be established in many instances by the free communication which exists between the carotid arteries of opposite sides, both outside and inside the cranium, and by enlargement of the branches of the subclavian artery. The chief communications outside the skull take place between the superior and inferior thyroid arteries, and between the deep cervical and descending branch of the occipital; the vertebral takes the place of the internal carotid within the cranium.

Nevertheless, wounds of the common carotid should be treated by suture whenever possible, because, after ligature of the vessel, symptoms of cerebral disturbance supervene in about 25 per cent of cases.

After ligature of the external carotid artery the circulation is re-

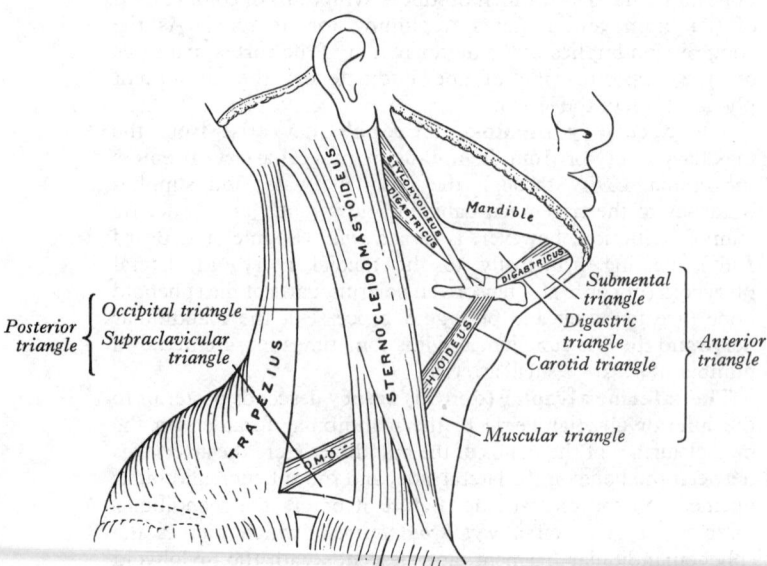

6.52 The triangles of the right side of the neck.

established by the free communication between most of the large branches of the artery (facial, lingual, superior thyroid, occipital) and the corresponding arteries of the opposite side, and by the anastomosis of its branches with those of the internal carotid artery, and of the occipital artery with branches of the subclavian, etc.

The Triangles of the Neck

The side of the neck (**6**.52) presents a somewhat quadrilateral outline, limited above by the base of the mandible, and a line drawn from the angle of the mandible to the mastoid process, below, by the upper border of the clavicle, in front, by the anterior median line of the neck, behind, by the anterior margin of the trapezius. This space is subdivided by the sternocleidomastoid, which passes obliquely across the neck, from the sternum and clavicle below, to the mastoid process and occipital bone above. The area in front of this muscle is called the *anterior triangle* of the neck, that behind it, the *posterior triangle*. While these and their subdivisions are purely arbitrary topographical regions, they have a certain value in description and as surface landmarks.

THE ANTERIOR CERVICAL TRIANGLE

The anterior triangle of the neck is bounded anteriorly by the anterior median line of the neck, posteriorly, by the anterior margin of the sternocleidomastoid; its base, directed upwards, as the inferior border of the mandible, and a line from the angle of the mandible to the mastoid process: its apex is below, at the sternum. This triangle may be subdivided into muscular, carotid, digastric and submental triangles.

The muscular triangle is bounded, in front, by the median line of the neck from the hyoid bone to the sternum, behind and below, by the anterior margin of the sternocleidomastoid, behind and above, but the superior belly of the omohyoid.

The carotid triangle is limited, behind, by the sternocleidomastoid, in front and below, by the superior belly of the omohyoid, above, by the stylohyoid and the posterior belly of the digastric. (It should be noted here that in the living non-obese neck, the carotid triangle is usually an obvious *surface feature*, presenting as a relatively small triangular depression, best seen with the head and cervical vertebral column slightly extended, and the head rotated a little to the contralateral side.)

It is covered by the skin, superficial fascia, platysma and deep fascia, ramifying in which are branches of the facial and the cutaneous cervical nerves. Its floor is formed by parts of the thyrohyoid, hyoglossus, and the inferior and middle constrictor muscles of the pharynx. When this space is dissected it is seen to contain the upper part of the common carotid artery, which divides opposite the superior border of the thyroid cartilage into the external and internal carotid arteries. These vessels are overlapped by the anterior margin of the sternocleidomastoid. The external and internal carotid arteries lie side by side, the external being the more anterior. The following branches of the external carotid artery are also encountered: the superior thyroid, running forwards and downwards, the lingual, forwards with an upward loop, the facial, forwards and upwards, the occipital, upwards and backwards and the ascending pharyngeal, directly upwards on the medial side of the internal carotid. The veins encountered are those corresponding to the above-mentioned branches of the external carotid artery—viz. the superior thyroid, the lingual, facial, ascending pharyngeal, and sometimes the occipital—all of which end in the internal jugular vein. The hypoglossal nerve crosses both the internal and external carotid arteries, curving round the origin of the lower sternocleidomastoid branch of the occipital artery; in this position it gives off the superior root of the ansa cervicalis, and this small nerve runs down in the anterior part of the carotid sheath. On the medial side of the external carotid artery, below the hyoid bone, is the internal laryngeal nerve and, still more inferiorly, the external laryngeal nerve.

It should be noted that many important structures in this

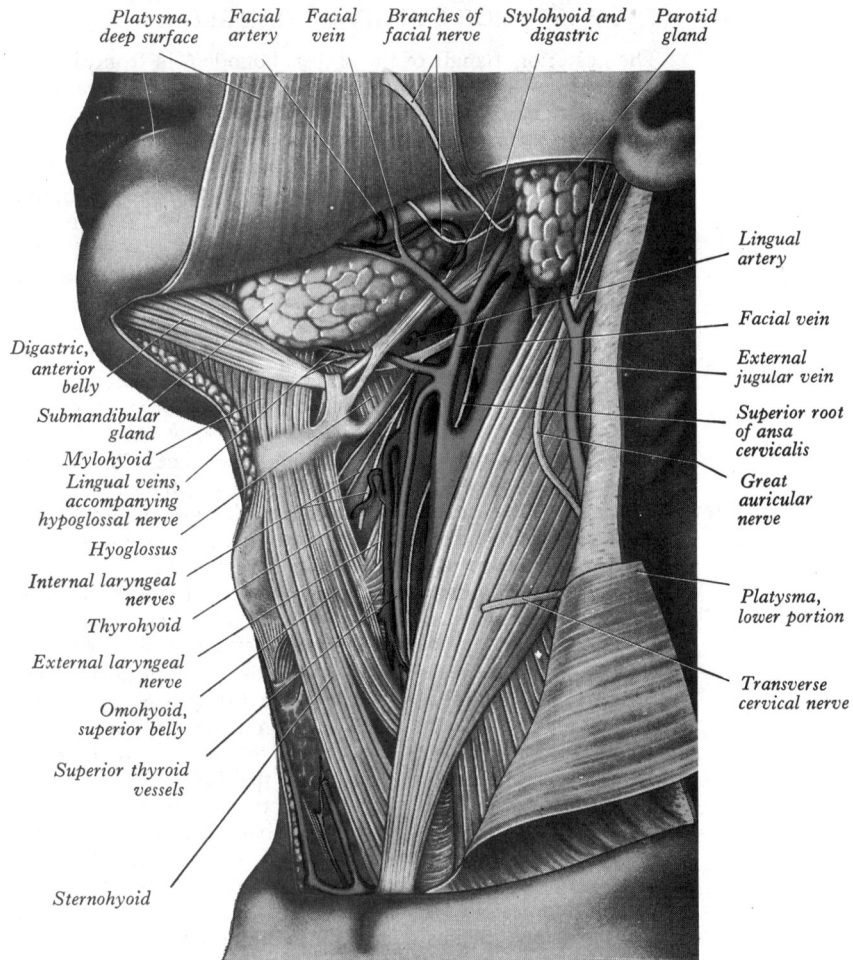

Platysma, deep surface | *Facial artery* | *Facial vein* | *Branches of facial nerve* | *Stylohyoid and digastric* | *Parotid gland*

Digastric, anterior belly
Submandibular gland
Mylohyoid
Lingual veins, accompanying hypoglossal nerve
Hyoglossus
Internal laryngeal nerves
Thyrohyoid
External laryngeal nerve
Omohyoid, superior belly
Superior thyroid vessels
Sternohyoid

Lingual artery
Facial vein
External jugular vein
Superior root of ansa cervicalis
Great auricular nerve
Platysma, lower portion
Transverse cervical nerve

6.53 A dissection of the left anterior triangle. The platysma has been divided transversely; its upper part has been turned upwards on to the face, its lower part turned backwards, exposing the lower part of the sternocleidomastoid.

region, such as the internal jugular vein, the vagus nerve, etc., lie entirely subjacent to the sternocleidomastoid and are therefore excluded from the triangle.

The digastric triangle is bounded, above, by the base of the mandible and a line drawn from its angle to the mastoid process, below and behind, by the posterior belly of the digastric and the stylohyoid, below and in front, by the anterior belly of the digastric.

It is covered by the skin, superficial fascia, platysma and deep fascia, ramifying in which are branches of the facial and transverse cutaneous cervical nerves. Its floor is formed by the mylohyoid and hyoglossus. The anterior part contains the submandibular gland, superficial to which is the facial vein, while deep to it is the facial artery, which crosses the lower border of the mandible at the anterior edge of the masseter; on the surface of the mylohyoid are the submental artery and the mylohyoid artery and nerve. The posterior part of this triangle contains the lower part of the parotid gland; the external carotid artery, having passed deep to the stylohyoid, curves over the upper border of the muscle so as to overlap to some extent its superficial surface where it ascends deep to the parotid gland before entering its substance. In this triangle the external carotid artery lies superficial to the internal carotid and crosses lateral to it; more deeply placed, and separated from the external carotid artery by the styloglossus, the stylopharyngeus and the glossopharyngeal nerve, are the internal carotid artery, the internal jugular vein and the vagus nerve.

The submental triangle is limited on each side by the anterior belly of the digastric; its apex is at the mandible; its base is formed by the body of the hyoid bone, and its floor by the mylohyoid. It contains one or two lymph nodes and some small veins; the latter unite to form the anterior jugular vein.

685

THE POSTERIOR CERVICAL TRIANGLE

The posterior triangle of the neck is bounded, in front, by the sternocleidomastoid, behind, by the anterior margin of the trapezius, its base is formed by the middle one-third of the clavicle; its apex is at the occipital bone between the attachments of the sternocleidomastoid and the trapezius. The 'apex' is often blunted by considerable separation between the occipital attachments of the latter two muscles, in which case the triangle is more of an elongated parallelogram between them. The triangle is crossed, about 2·5 cm above the clavicle, by the inferior belly of the omohyoid, which divides the triangle into an occipital and a supraclavicular triangle.

The occipital triangle, the upper and larger division of the posterior triangle, is bounded in front by the sternocleidomastoid, behind, by the trapezius, below, by the inferior belly of the omohyoid. Its floor is formed from above downwards by the splenius capitis, levator scapulae, and the scaleni medius et posterior; sometimes a small part of the semispinalis capitis is seen at the apex of the triangle. It is covered by the skin, the superficial and deep fasciae, and below by the platysma. The accessory nerve pierces the sternocleidomastoid and courses on the levator scapulae obliquely backwards across the space to reach the deep surface of the trapezius; the cutaneous and muscular branches of the cervical plexus become more superficial at the posterior border of the sternocleidomastoid; below, the supraclavicular nerves, the transverse cervical vessels and the upper part of the brachial plexus cross the space. A chain of lymph nodes is also arranged along the posterior border of the sternocleidomastoid, from the mastoid process to the root of the neck.

The supraclavicular triangle, the lower and smaller division of the posterior triangle, is bounded, above, by the inferior belly of the omohyoid, below, by the clavicle; its base is formed by the lower part of the posterior border of the sternocleidomastoid. Its floor consists of the first rib, the insertion of the scalenus medius and the first digitation of the serratus anterior. The size of this triangle varies with the extent of attachment of the clavicular portions of the sternocleidomastoid and trapezius, and also with the level at which the inferior belly of the omohyoid crosses the neck; this level is lowered when the arm is raised, and raised when the arm is depressed. The triangle is covered by the skin, the superficial and deep fasciae, and the platysma, and crossed by the supraclavicular nerves. Just above the level of the clavicle, the third portion of the subclavian artery curves laterally and downwards from the lateral margin of the scalenus anterior, across the first rib, to the axilla. The subclavian vein lies behind the clavicle, and is not usually seen in this space; but in some cases it rises as high as the artery, and has even been seen to accompany that vessel behind the scalenus anterior. The brachial plexus of nerves lies partly above and partly behind the artery, and in close contact with it. The suprascapular vessels pass transversely behind the clavicle and are not, strictly speaking, in the triangle; running in the same direction, but at a slightly higher level, are the transverse cervical artery and vein. The external jugular vein descends behind the posterior border of the sternocleidomastoid to terminate in the subclavian vein; it receives the transverse cervical and suprascapular veins, which form a plexus in front of the third part of the subclavian artery, and occasionally it is joined by a small vein which crosses the clavicle from the cephalic vein. The small nerve to the subclavius also crosses this triangle about its middle, and some lymph nodes are contained within the space.

The Internal Carotid Artery

The internal carotid artery (6.54, 55, 56) supplies the greater part of the cerebral hemisphere, the eye and its accessory organs, and sends branches to the forehead and nose. It begins at the bifurcation of the common carotid artery, where it usually presents a localized dilatation, termed the carotid sinus (p. 676). It ascends to the base of the skull, and enters the cranial cavity through the carotid canal of the temporal bone. It then runs forward through the cavernous sinus, lying in the carotid groove on the side of the body of the sphenoid bone, and ends below the anterior perforated substance of the brain by dividing into anterior and middle cerebral arteries.

The internal carotid artery may for convenience be divided into four parts: cervical, petrous, cavernous and cerebral.

The Cervical Part

This section of the internal carotid artery begins at the carotid bifurcation, at the upper border of the thyroid cartilage, and ascends in front of the transverse processes of the upper three cervical vertebrae to the lower end of the carotid canal in the petrous portion of the temporal bone. Comparatively superficial at its commencement, it is contained in the carotid triangle, but after passing medial to the posterior belly of the digastric it lies on a much deeper plane. Except at the base of the skull, the internal jugular vein and the vagus nerve lie on its lateral side; the external carotid artery is at first anterior and medial to the internal carotid but on leaving the carotid triangle it curves backwards so as to lie superficial to it.

The cervical part of the internal carotid artery has many additional relations. *Posteriorly*, it adjoins the longus capitis, but the superior cervical sympathetic ganglion intervenes and the superior laryngeal nerve crosses obliquely behind the vessel. *Medially*, the artery is related to the wall of the pharynx, from which it is separated by an interval containing some fat, connective tissue and the pharyngeal veins, the ascending pharyngeal artery and the superior laryngeal nerve. *Antero-laterally*, the internal carotid artery is covered throughout by the sternocleidomastoid. In addition, *below the digastric*, the hypoglossal nerve and the superior root of the ansa cervicalis, the lingual and facial veins are superficial to the artery. *At the level of the digastric* the vessel is crossed by the stylohyoid and by the occipital and posterior auricular arteries. *Above the digastric*, it is separated from the external carotid artery by the styloid process, the styloglossus and stylopharyngeus, the glossopharyngeal nerve, the pharyngeal branch of the vagus and the deeper part of the parotid gland. At the base of the skull the glossopharyngeal, vagus, accessory and hypoglossal nerves are between the internal carotid artery and the internal jugular vein, which here lies posterior to the artery.

The Petrous Part

The internal carotid artery at first ascends in the carotid canal in the petrous temporal bone, and then curves forwards and medially. As it leaves the canal to enter the cranial cavity, it runs upwards and medially above the fibrocartilage which in life fills the foramen lacerum. Finally, it passes between the lingula and petrosal process of the sphenoid bone. The artery is at first in front of the cochlea and tympanic cavity; it is separated from the latter and from the auditory tube by a thin, bony lamella, cribriform in the young subject, often partly absorbed in old age. Further forwards it is separated from the trigeminal ganglion by a thin plate of bone, the floor of the trigeminal impression and the roof of the horizontal portion of the carotid canal; frequently this bony plate is more or less deficient. The artery is surrounded by a plexus of small veins and the carotid plexus of nerves, which is derived from the internal carotid branch of the superior cervical ganglion of the sympathetic trunk.

The Cavernous Part

When the internal carotid artery is in the cavernous sinus, it is covered by the lining endothelium of the sinus. It at first ascends towards the posterior clinoid process, then passes forwards on the side of the body of the sphenoid, and again curves upwards on the medial side of the anterior clinoid process, traversing the dura mater forming the roof of the sinus; occasionally the anterior and middle clinoid processes form a bony ring round the artery. The cavernous portion of the artery is surrounded by a sympathetic plexus, and the oculomotor, trochlear, ophthalmic and abducent nerves (6.127) are lateral to it.

The Cerebral Part

After perforating the dura mater medial to the anterior clinoid process, the internal carotid artery turns backwards below the

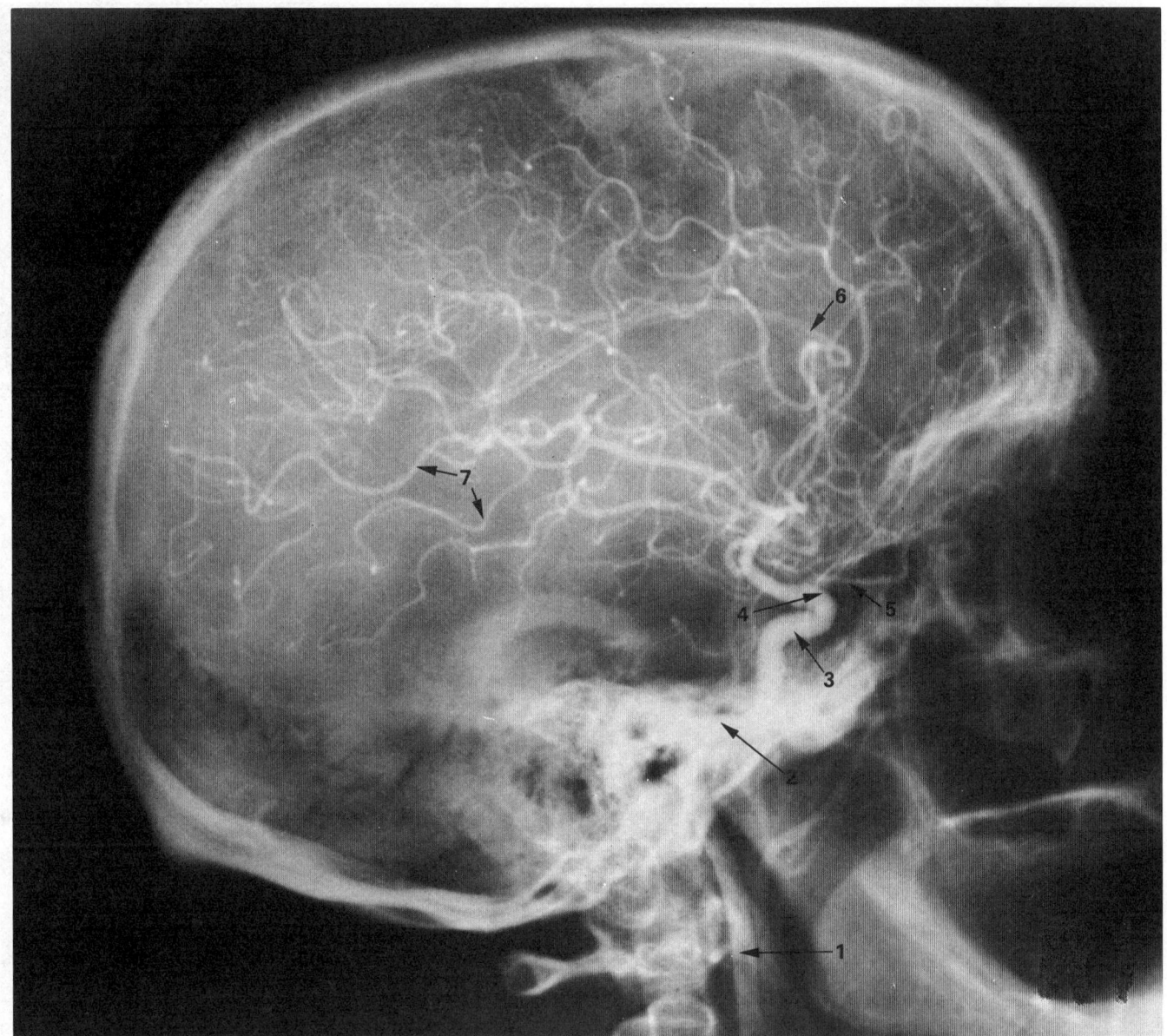

6.54 Internal carotid arteriogram (right). Lateral view. (Adult male of 33 years.) The following can be identified: Parts of internal carotid artery and individual vessels—1. Cervical. 2. Intrapetrous. 3. Cavernous. 4. Terminal. 5. Ophthalmic artery. 6. Anterior cerebral artery. 7. Branches of middle cerebral artery. Note absence of radio-opaque injectant from cerebellar vessels.

optic nerve, and then passes between the optic and oculomotor nerves to the anterior perforated substance at the medial end of the lateral cerebral sulcus, where it divides into the anterior and middle cerebral arteries.

Variations. The length of the internal carotid artery naturally varies with the length of the neck, and with the point of carotid bifurcation. It arises occasionally from the arch of the aorta, and then has been found medial to the external carotid as far as the larynx, where it crosses behind the latter vessel to reach its usual position. The course of the cervical part of the artery, instead of being straight, may be very tortuous. When this occurs the vessel approaches nearer to the pharynx than usual, and may lie very close to the tonsil. Its absence has been occasionally recorded.

Surface Anatomy. The internal carotid corresponds to a broad line drawn upwards from the termination of the common carotid to the posterior border of the neck of the mandible (6.47).

Branches. The cervical portion of the internal carotid has no branches. Those from the other parts are:

From the petrous part	1. Caroticotympanic.
	2. Pterygoid.
From the cavernous part	3. Cavernous.
	4. Hypophysial.
	5. Meningeal.

From the cerebral part	6. Ophthalmic.
	7. Anterior cerebral.
	8. Middle cerebral.
	9. Posterior communicating.
	10. Anterior choroid.

1. The caroticotympanic branch (or branches) is small; it enters the tympanic cavity through a foramen in the wall of the carotid canal, and anastomoses with the anterior tympanic branch of the maxillary artery, and with the stylomastoid artery.

2. The pterygoid branch, small and inconstant, enters the pterygoid canal with the nerve of that name, and anastomoses with a branch of the greater palatine artery.

3. The cavernous branches are numerous small vessels which supply the trigeminal ganglion, the walls of the cavernous and inferior petrosal sinuses and their contained nerves. Some anastomose with middle meningeal rami.

4. The hypophysial branches are a group of small but important vessels; for details *see* p. 1443.

5. The meningeal branch is a minute branch which passes over the lesser wing of the sphenoid to supply the dura mater and bone of the anterior cranial fossa; it anastomoses with the meningeal branch from the posterior ethmoidal artery.

6. The ophthalmic artery (6.57) is a branch from the internal

687

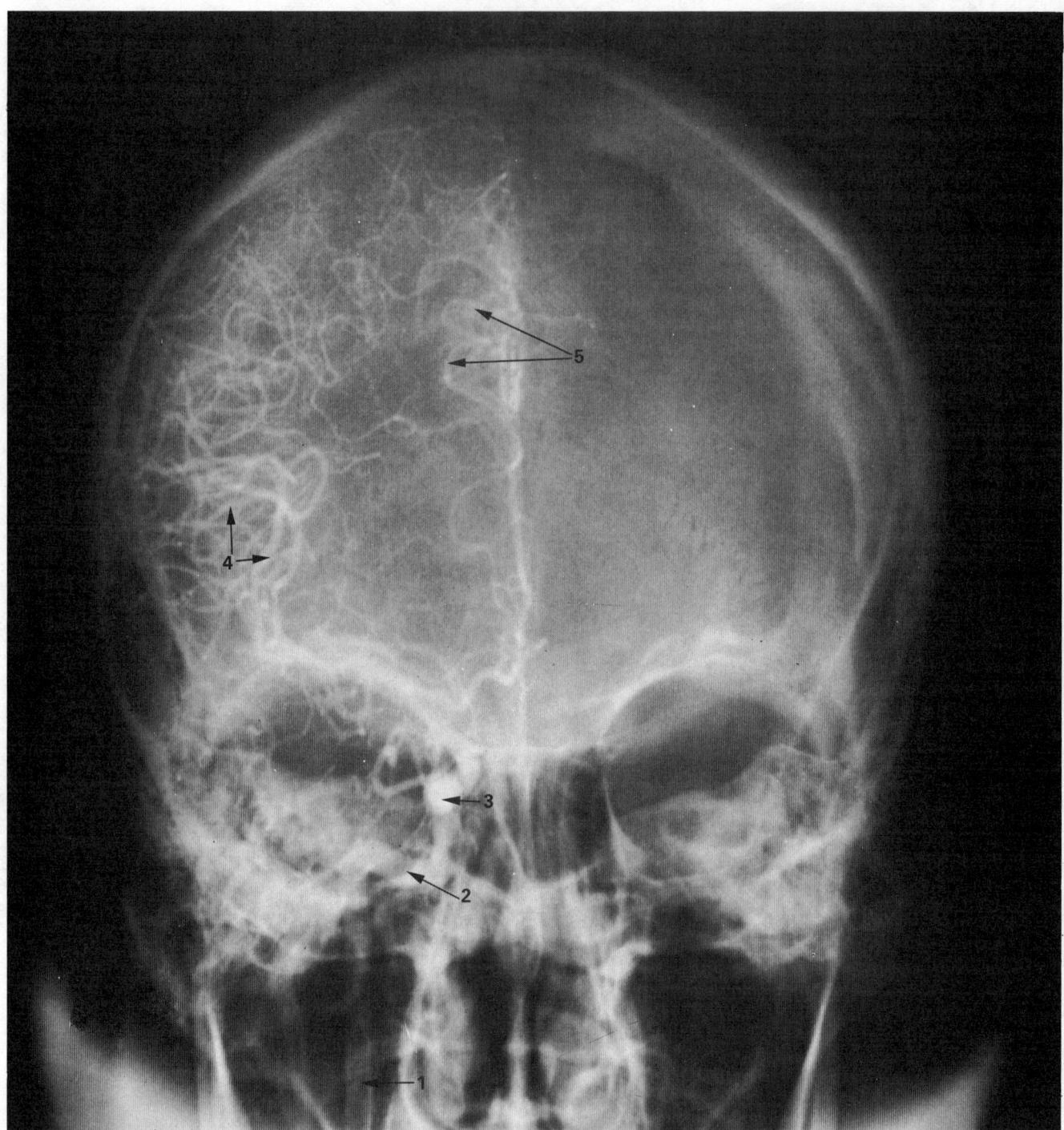

6.55   Internal carotid arteriogram (right). Anteroposterior view. (Same subject as **6**.54.) Parts of internal carotid artery—1. Cervical. 2. Intra-petrous. 3. Cavernous. 4. Branches of middle cerebral artery. 5. Branches of anterior cerebral artery. Note lack of contrast medium on left side.

carotid as it emerges from the cavernous sinus medial to the anterior clinoid process; it enters the orbital cavity through the optic canal, inferolateral to the optic nerve. In the orbital cavity it runs for a short distance lateral to the optic nerve and medial to the oculomotor and abducent nerves, the ciliary ganglion and the rectus lateralis. It next crosses obliquely above the optic nerve and below the rectus superior to reach the medial wall of the orbit. It then runs forwards between the obliquus superior and the rectus medialis, and, at the medial end of the upper eyelid, divides into two branches, named *supratrochlear* and *dorsal nasal*. As the artery crosses the optic nerve it is accompanied by the nasociliary nerve, and is separated from the frontal nerve by the rectus superior and levator palpebrae superioris; the terminal part of the artery is accompanied by the infratrochlear nerve. In about 15 per cent of subjects the ophthalmic artery crosses below the optic nerve. Its branches are as follows.

The *central artery of the retina*, the first and one of the smallest branches of the ophthalmic artery, commences below the optic nerve. It runs for a short distance within the dural sheath of the optic nerve, and about 1·25 cm behind the eyeball it pierces the inferomedial surface of the nerve, and runs forward to the retina along the centre of it. Its distribution is described elsewhere with the eye (p. 1175).

The *lacrimal artery* arises from the ophthalmic artery close to its exit from the optic canal, and is one of its largest branches; sometimes it is given off before the ophthalmic artery enters the orbit; occasionally its place is taken by a branch of the middle meningeal artery (p. 683). It accompanies the lacrimal nerve along the upper border of the rectus lateralis, and supplies the lacrimal gland. Its terminal branches, after traversing the gland, are distributed to the eyelids and conjunctiva: of those supplying the eyelids, two are of considerable size and are named the *lateral palpebral arteries*; they run medially in the upper and lower lids respectively and anastomose with the medial palpebral arteries. The lacrimal artery gives off one or two *zygomatic branches*, one of which passes through the zygomaticotemporal foramen to the temporal fossa, and anastomoses with the deep temporal arteries; another appears on the cheek through the zygomaticofacial foramen and anastomoses with the transverse facial and zygomatico-orbital arteries. A *recurrent meningeal branch* passes backwards through the lateral part of the superior orbital fissure and anastomoses with a branch of the middle meningeal artery. It is enlargement of this anastomosis which may provide an alternative origin of the lacrimal artery.

The *muscular branches* frequently spring from a common trunk. They consist of a superior and an inferior group, and most of them accompany the branches of the oculomotor nerve. The inferior group, more constantly present, gives off most of the anterior ciliary arteries. Additional muscular branches are derived from the lacrimal and supraorbital arteries, or from the trunk of the ophthalmic artery.

The *ciliary arteries* are divisible into three groups, long and short posterior, and anterior.

The *long posterior ciliary arteries*, two in number, pierce the posterior part of the sclera a short distance from the entrance of the optic nerve. Their distribution is described on p. 1162.

The *short posterior ciliary arteries*, about seven in number, pass forwards around the optic nerve to the posterior part of the eyeball, and, after dividing into from fifteen to twenty branches, pierce the sclera around the entrance of the nerve, and supply the choroid coat and the ciliary processes. At the entrance of the optic nerve they anastomose with twigs of the central artery of the retina, and at the ora serrata with branches of the long posterior and anterior ciliary arteries.

The *anterior ciliary arteries* are derived from the muscular branches of the ophthalmic artery; they run to the front of the eyeball on the tendons of the recti, form a circumcorneal vascular zone beneath the conjunctiva, and then pierce the sclera a short distance from the sclerocorneal junction and end in the greater arterial circle of the iris (p. 1162).

The *supraorbital artery* leaves the ophthalmic artery as the vessel crosses the optic nerve. It ascends medial to the rectus superior and levator palpebrae superioris and, meeting the supraorbital nerve, accompanies it between the periosteum and levator palpebrae superioris to the supraorbital foramen or notch; passing through this it divides into superficial and deep branches, which supply the skin, muscles and pericranium of the forehead, anastomosing with the supratrochlear artery, the frontal branch of the superficial temporal artery, and the artery of the opposite side. In the orbit it supplies twigs to the rectus superior and the levator palpebrae, and sends a branch across the trochlea of the obliquus superior, to the parts at the medial angle of the eye. At the supraorbital margin it frequently sends a branch to the diploë of the frontal bone, which may supply the mucoperiosteum of the frontal sinus.

The *posterior ethmoidal artery* runs through the posterior ethmoidal canal, supplies the posterior ethmoidal air sinuses, and, entering the cranium, gives off a meningeal branch to the dura mater and nasal branches which descend into the nasal cavity through the cribriform plate of the ethmoid bone, to anastomose with branches of the sphenopalatine artery supplying bone.

The *anterior ethmoidal artery* accompanies the anterior ethmoidal nerve through the anterior ethmoidal canal, supplies the anterior and middle ethmoidal and frontal air sinuses, and, entering the cranium, gives off a meningeal branch to the dura mater, and nasal branches; the latter descend into the nasal cavity with the anterior ethmoidal nerve and, running along the groove on the inner surface of the nasal bone, supply twigs to the lateral wall and septum of the nose, and a terminal branch which appears on the dorsum of the nose between the nasal bone and the upper nasal cartilage. Angiographic studies (Kuru 1967) have shown that the meningeal branch may extend in to the falx; Müller (1977, 1978) has confirmed this finding in fetal and adult material and has shown that these 'falciate arteries' may also be derived from the recurrent meningeal branch of the lacrimal artery.

The *meningeal branch* is a small artery which passes backwards

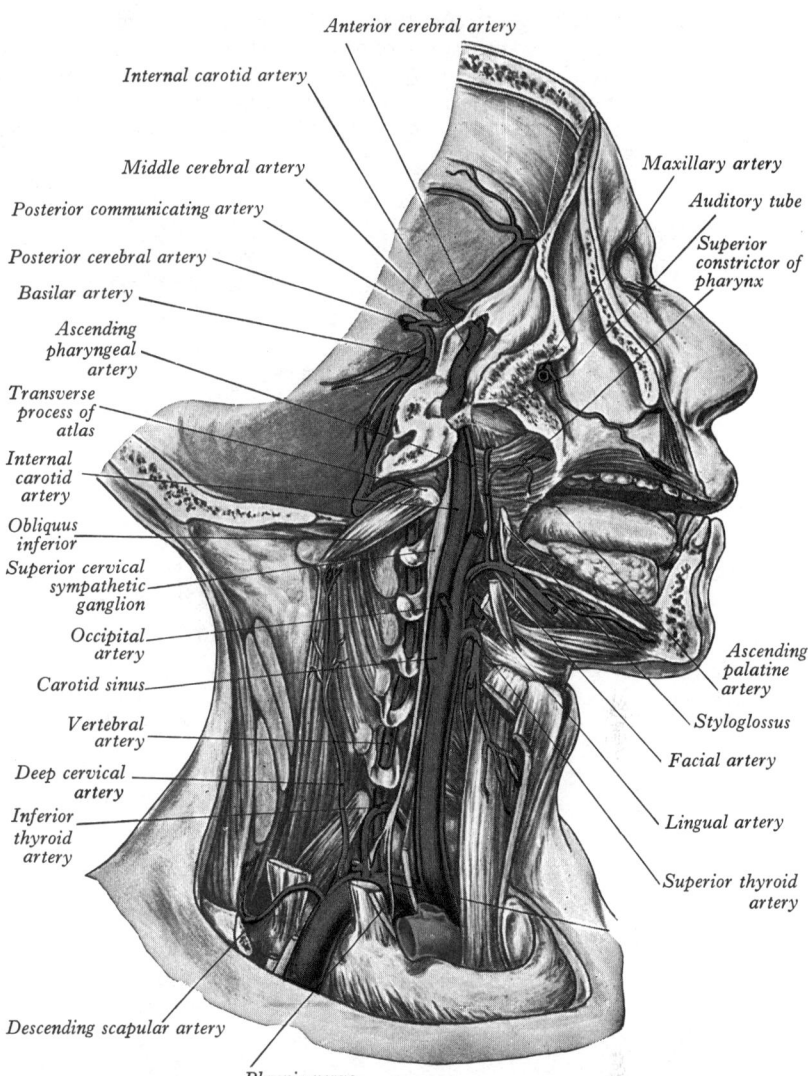

Anterior cerebral artery
Internal carotid artery
Middle cerebral artery
Posterior communicating artery
Posterior cerebral artery
Basilar artery
Ascending pharyngeal artery
Transverse process of atlas
Internal carotid artery
Obliquus inferior
Superior cervical sympathetic ganglion
Occipital artery
Carotid sinus
Vertebral artery
Deep cervical artery
Inferior thyroid artery
Descending scapular artery
Phrenic nerve
Maxillary artery
Auditory tube
Superior constrictor of pharynx
Ascending palatine artery
Styloglossus
Facial artery
Lingual artery
Superior thyroid artery

6.56 A dissection to show the course of the right vertebral and internal carotid arteries and some of their branches.

through the superior orbital fissure to the middle cranial fossa, and anastomoses with the middle and accessory meningeal arteries. It supplies bone.

The *medial palpebral arteries*, two in number, superior and inferior, arise from the ophthalmic artery below the trochlea of the obliquus superior. They descend behind the lacrimal sac, and enter the eyelids, where each divides into two branches which course laterally along the edges of the tarsi, thus forming two arches (a superior and an inferior) in each eyelid. The superior palpebral artery anastomoses with the supraorbital artery, and, at the lateral part of the eyelid, with the zygomatico-orbital branch of the superficial temporal artery, and with the upper of the two lateral palpebral branches of the lacrimal artery. The inferior palpebral artery anastomoses at the lateral part of the eyelid with the lower of the two lateral palpebral branches of the lacrimal artery and with the transverse facial artery and, at the medial part of the eyelid, with a twig from the facial artery; from this last anastomosis a branch passes to supply the nasolacrimal duct, ramifying in its mucous membrane.

The *supratrochlear artery*, one of the terminal branches of the ophthalmic artery, leaves the orbital opening at its superomedial angle, with the supratrochlear nerve, and, ascending on the forehead, supplies the skin, muscles and pericranium, anastomosing with the supraorbital artery and with the artery of the opposite side.

The *dorsal nasal artery*, the other terminal branch of the ophthalmic artery, emerges from the orbit between the trochlea of the obliquus superior and the medial palpebral ligament, and,

689

after giving a twig to the upper part of the lacrimal sac, divides into two branches, one of which anastomoses with the terminal part of the facial artery; the other runs along the dorsum of the nose, supplies its outer surface, and anastomoses with the artery of the opposite side, and with the lateral nasal branch of the facial artery.

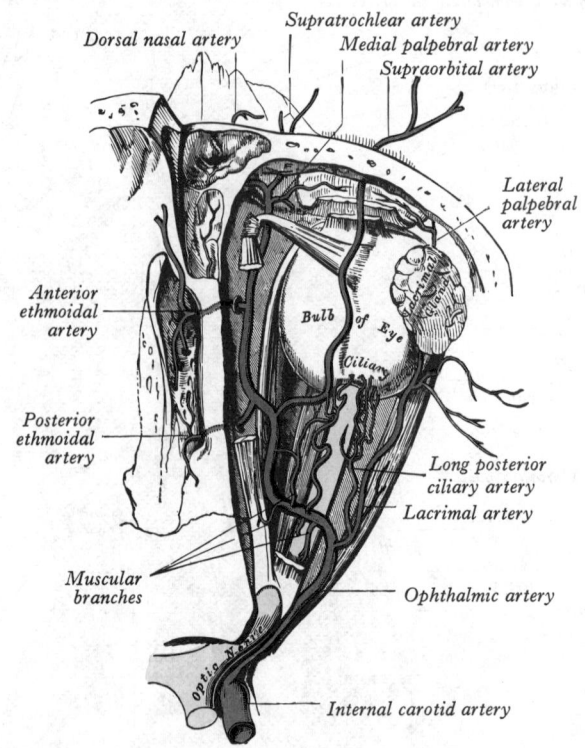

6.57   The ophthalmic artery and its branches in the right orbit, as seen from above.

7. The anterior cerebral artery (**6**.58, 59, 60), the smaller of the two terminal branches of the internal carotid artery, starts at the medial end of the lateral cerebral sulcus. It passes forwards and medially above the optic nerve, to the commencement of the longitudinal fissure. Here it comes into close relationship with the opposite artery and is joined to it by a short transverse trunk (sometimes duplicated) named the anterior communicating artery. From this point, the two anterior cerebral arteries run side by side in the longitudinal cerebral fissure, curving round the genu of the corpus callosum, and running backwards along the upper surface of this structure to its posterior extremity, where they end by anastomosing with the posterior cerebral arteries. Occasionally the two arteries join to form a single vessel.

In its course the anterior cerebral artery gives off central and cortical branches.

The *anterior communicating artery* has an average length of about 4 mm and connects the two anterior cerebral arteries across the commencement of the longitudinal fissure; sometimes it is double. It gives off a few anteromedial central branches (p. 1642). According to Crowell and Morawetz (1977) the branches of the anterior communicating artery are constantly present, varying from 3 to 13 in number, and can be seen to supply the optic chiasma, lamina terminalis, hypothalamus, parolfactory areas of Broca, fornix (anterior columns), and the cingulate gyrus.

The *central branches* are a group of small arteries which arise from the commencement of the anterior cerebral artery; they pierce the anterior perforated substance and lamina terminalis, and supply the rostrum of the corpus callosum, the septum pellucidum, the anterior part of the putamen of the lentiform nucleus and the head of the caudate nucleus. The *cortical branches* are named according to their distribution. The *orbital branches*, two or three in number, are distributed to the orbital surface of the frontal lobe, where they supply the olfactory lobe, gyrus rectus and medial orbital gyrus. The *frontal branches* supply the corpus callosum, the cingulate gyrus, the medial frontal gyrus and the paracentral lobule, and send twigs over the superomedial border of the cerebral hemisphere to the superior and middle frontal gyri and the upper part of the precentral gyrus. It is to be noted that

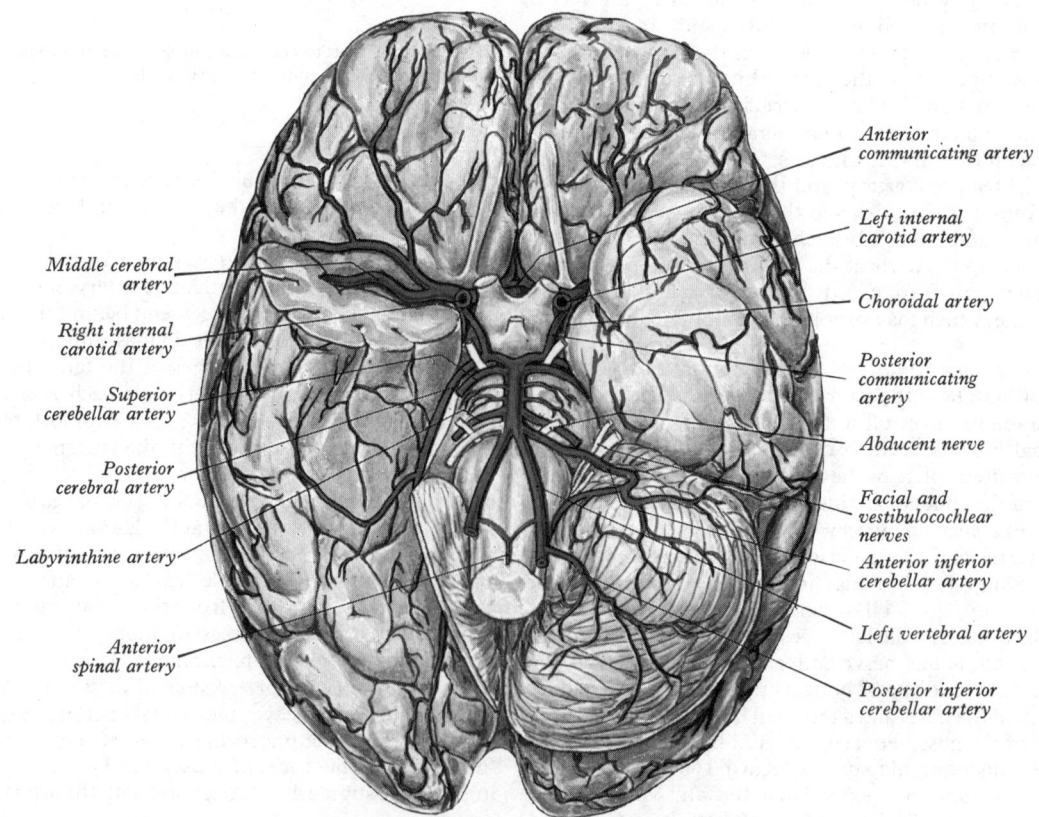

6.58   The arteries at the base of the brain. The right temporal pole and most of the right hemisphere of the cerebellum have been removed. Variations in the pattern of these vessels is common.

these branches supply the 'leg area' of the motor cortex (p. 1013). The *parietal branches* supply the precuneus and adjacent lateral surface of the hemisphere.

8. The middle cerebral artery (**6**.58, 59, 60), the larger terminal branch of the internal carotid artery, runs first laterally in the lateral cerebral sulcus and then backwards and upwards on the surface of the insula, where it divides into branches which are distributed to the insula and to the lateral surface of the cerebral hemisphere (**6**.59).

The *central branches* are small arteries from the commencement of the middle cerebral artery which enter the substance of the brain through the anterior perforated substance. They are arranged in two sets: one, *medial striate arteries*, ascend through the lentiform nucleus, and supply it, the caudate nucleus, and the internal capsule; the other, *lateral striate arteries*, ascend over the lower part of the lateral aspect of the lentiform nucleus (in the external capsule) and then, bending medially, traverse this nucleus and the internal capsule to supply the caudate nucleus. One of this group is usually larger than the rest, and of special importance as the artery in the brain most susceptible to rupture; it was termed by Charcot the '*artery of cerebral haemorrhage*'. The *cortical arteries* supply *orbital branches* to the inferior frontal gyrus and the lateral part of the orbital surface of the frontal lobe. The *frontal branches* supply the precentral and the middle and inferior frontal gyri. The *parietal branches*, two in number, are distributed to the postcentral gyrus, the lower part of the superior parietal lobule, and the whole of the inferior parietal lobule. The *temporal arteries*, two or three in number, are distributed to the lateral surface of the temporal lobe. It is to be noted that the cortical branches of the middle cerebral artery supply the motor area (excluding the 'leg area') of the cerebral cortex (p. 1013), its corresponding somaesthetic area (p. 1015), and the auditory area (p. 1021).

9. The posterior communicating artery (**6**.58, 61) runs backwards from the internal carotid above the oculomotor nerve, and anastomoses with the posterior cerebral, a branch of the basilar artery. It is usually a small vessel, but is occasionally so large that the posterior cerebral may be considered as arising from the internal carotid rather than from the basilar. It is frequently larger on one side than on the other. From its posterior half are given off several small *central branches*, which, with similar vessels from the posterior cerebral artery, pierce the posterior perforated substance and supply the medial surface of the thalamus and the walls of the third ventricle (p. 1043).

10. The anterior choroidal artery, a small but constant ramus, leaves the internal carotid, near the posterior communicating artery (Abbie 1933, 1934). Passing backwards above the medial part of the uncus, it crosses inferior to the optic tract and reaches the crus cerebri, to which it gives several branches. It then turns laterally, recrossing the optic tract, and arrives at the lateral aspect of the lateral geniculate body, to which it supplies a number of branches. Finally it enters the inferior horn of the lateral ventricle through the choroidal fissure and ends in the choroidal plexus. It also supplies branches to the globus pallidus, caudate nucleus, and amygdaloid body. Other branches supply the hypothalamus, tuber cinereum, red nucleus, substantia nigra, posterior limb of the internal capsule, optic radiation, optic tract, hippocampus and fimbria of the fornix.

The Circulus Arteriosus (of Willis)

A considerable part of the brain is supplied by the two vertebral arteries (p. 694) and a remarkable anastomosis, the *circulus arteriosus*, exists between these and the two internal carotid arteries. This circle (which is really more polygonal than circular), is situated in the cisterna interpeduncularis at the base of the brain, and encloses the optic chiasma and the structures in the interpeduncular fossa (p. 966). It is formed as follows: in front, the two anterior cerebral arteries are joined to each other by the anterior communicating artery; behind, the basilar artery (p. 696) divides into the two posterior cerebral arteries, each of which is joined to the internal carotid artery of the same side by the posterior communicating artery (**6**.61).

The arteries forming the arterial 'circle' evince much variation in calibre, being frequently hypoplastic and sometimes absent.

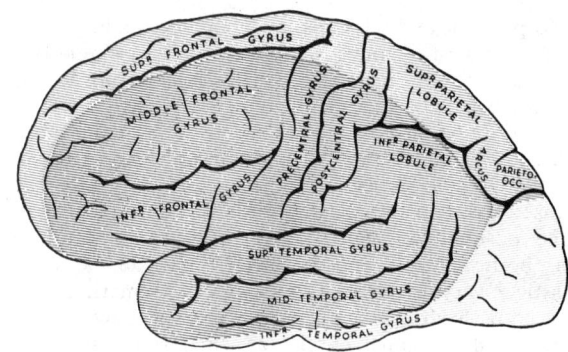

6.59   The lateral surface of the left cerebral hemisphere, showing areas supplied by the cerebral arteries. In this and the next figure the area supplied by the anterior cerebral artery is coloured *blue*, that by the middle cerebral artery, *pink*, and that by the posterior cerebral artery, *yellow*.

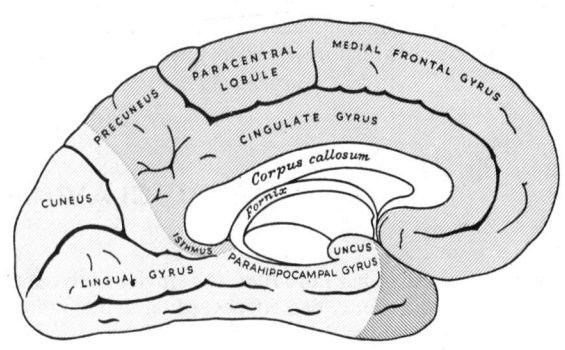

6.60   The medial surface of the left cerebral hemisphere, showing the areas supplied by the cerebral arteries (see description of **6**.59).

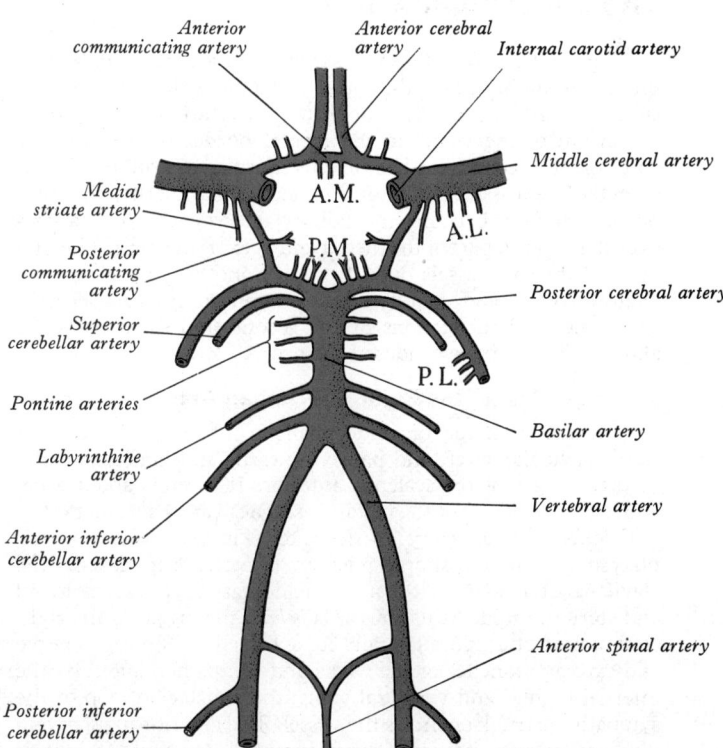

6.61   A diagram of the arteries at the base of the brain, showing the constitution of the arterial circle. A.L.—Anterolateral central branches, A.M.—Anteromedial central branches, P.L.—Posterolateral central branches, P.M.—Posteromedial central branches. The arteries constituting this so-called arterial 'circle' are commonly asymmetric, and sometimes a constituent vessel is missing.

About 60 per cent of circles display 'anomalies', and hence the arrangement described above is applicable to only a minority of cases. These variations have been much studied, from Windle's account in 1888 of 200 specimens, to that of Puchades-Ort *et al.* (1976) based on 62 dissections, the largest series being those of Fawcett and Blachford (1906)—700 dissections, and Riggs and Rupp (1963)—994 dissections. Fields *et al.* (1965) have summarized the anomalies in such studies, and these authorities should be consulted for details. The cerebral and communicating arteries, both anterior and posterior, may all be absent, hypoplastic to varying degrees, double or even sometimes triple. In about 90 per cent or more of circles there is, nevertheless, a complete 'circular' channel, but in a large majority of these one or other of the formative vessels is sufficiently narrowed in calibre to cast doubt on its functional value as a collateral route. The haemodynamic 'balance' of the circle is more often than not disturbed by variation in the calibres of the communicating arteries, and this is often coupled with variation in the proximal segments of the anterior and posterior cerebral arteries, that is, the segments extending from their origins to their junctions with the corresponding communicating artery. This segment is, in the case of the posterior cerebral artery, particularly 'labile', being often much reduced in calibre or even absent. Where this is coupled, as it often is, with an enlarged posterior communicating artery on the side in question, the posterior cerebral artery is derived, in effect, from the internal carotid. As Abbie (1933), Williams (1936), and Kaplan (1956) have pointed out, this anomaly is in harmony with the ontogenetic and phylogenetic association of the posterior cerebral with the internal carotid artery. In the anterior region of the arterial circle agenesis or hypoplasia of the initial segment of the anterior cerebral artery is more frequent than similar anomalies in the anterior communicating, and is hence a commoner cause of defective collateral circulation. Angiographic evidence suggests that defective or absent collateral circulation of this kind may be encountered in about one third of individuals (Sedzmir 1959). It is therefore clear that the existence of an effective arterial circle cannot be assumed in any patient, and that surgical procedures involving its 'feeders' must depend upon adequate angiographic investigation.

For diagnostic purposes, radio-opaque substances may be injected into the internal carotid or vertebral arteries in the neck and the condition of their intracranial branches studied radiologically (**6**.54, 55, 65, 66).

Further details of the distribution of the arteries and veins of the brain appear on pp. 1042–1045. For the intracranial venous sinuses *see* pp. 744–749.

THE SUBCLAVIAN SYSTEM OF ARTERIES

The stem artery of the upper limb runs as a single trunk as far as the elbow; but it is differently named, according to the regions it traverses. From its origin to the outer border of the first rib it is termed *subclavian*; from this point to the lower border of the tendon of the teres major it is named *axillary*; and from here to its division opposite the neck of the radius it is called *brachial*.

The Subclavian Arteries

The right subclavian artery arises from the brachiocephalic trunk, the left, from the arch of the aorta. To facilitate description, each subclavian artery is divided into three parts; the first extends from the origin of the vessel to the medial border of the scalenus anterior, the second lies behind this muscle, and the third runs from the lateral margin of the muscle across the first rib to its outer border, where it becomes the axillary artery; each artery arches over the cervical part of the pleura and apex of the lung. The first parts of the two vessels differ from one another in their origin, length, course and relations, and therefore require separate descriptions. The relations of the second and third parts are almost alike on the two sides of the neck.

The First Part of the Right Subclavian Artery (**6**.45, 62)

This arises from the brachiocephalic trunk, behind the right sternoclavicular joint, and passes upwards and laterally to the medial margin of the scalenus anterior. It ascends about 2 cm above the clavicle, but the height it reaches varies considerably.

Relations. The artery is deep to skin, superficial fascia, platysma, anterior supraclavicular nerves, deep fascia, the clavicular attachment of the sternocleidomastoid, the sternohyoid and sternothyroid. At its start it is *behind* the origin of the right common carotid; more laterally it is crossed by the vagus nerve and cardiac branches of the vagus and sympathetic, and by the internal jugular and vertebral veins; the subclavian loop of the sympathetic trunk encircles the vessel. The anterior jugular vein diverges *laterally* in front of the artery, but is separated from it by the sternohyoid and sternothyroid. *Below* and *behind*, the artery is related to the pleura and the apex of the lung, but it is separated from them by the suprapleural membrane (p. 1249), the ansa subclavia, a small accessory vertebral vein (p. 742), and the right recurrent laryngeal nerve (which winds round the lower and posterior part of the vessel).

The First Part of the Left Subclavian Artery (**6**.41, 44, 63, 118)

This arises from the arch of the aorta, behind the left common carotid, usually at the level of the disc between the third and fourth thoracic vertebrae; it ascends to the root of the neck and then arches laterally as far as the medial border of the left scalenus anterior.

Relations. Within the thorax it is related, *in front*, to the left common carotid artery and the commencement of the left brachiocephalic vein, from which it is separated by the left vagus, cardiac (p. 1128) and phrenic nerves. Superficial to these structures, the anterior margin of the left lung and pleura and the sternothyroid and sternohyoid intervene between the vessel and the upper, left part of the manubrium sterni. *Behind*, it lies, successively, on the left edge of the oesophagus, the thoracic duct and the longus colli, and it is in contact posterolaterally with the left lung and pleura. *Medially* it is related, successively, to the trachea, the left recurrent laryngeal nerve, the oesophagus and the thoracic duct. *Laterally* the artery grooves the mediastinal surface of the left lung and pleura, and these structures, as already indicated, tend to encroach on its anterior and posterior aspects.

In the neck near the medial border of the scalenus anterior the artery is crossed *anteriorly* by the left phrenic nerve and the terminal part of the thoracic duct. Otherwise the relations are the same as those previously described for the first part of the right subclavian artery. *Posteriorly* and *inferiorly*, the relations of the two vessels are identical, but the left recurrent laryngeal nerve, which is related to the left subclavian artery in the thorax, is not directly related to its cervical part.

The Second Part of the Subclavian Artery (**6**.45, 64)

This lies behind the scalenus anterior; it is very short, and forms the highest part of the arch described by the vessel.

Relations. Deep to skin, superficial fascia, platysma, deep cervical fascia, sternocleidomastoid and scalenus anterior. On the right the phrenic nerve is often said to be separated from the second part by scalenus anterior, on the left side crossing the first part of the artery at the muscles' edge. Qvist (1977) holds that *both* nerves lie anterior to the muscles at this level. *Behind* and *below* are the suprapleural membrane, pleura and lung, and the lower trunk of the brachial plexus; *above*, the upper and middle trunks of the brachial plexus. The subclavian vein lies below and in front of the artery, separated by scalenus anterior (**6**.64).

The Third Part of the Subclavian Artery

This runs slightly downwards and laterally from the lateral margin of the scalenus anterior to the lateral border of the first rib, where it becomes the axillary artery. This is the most superficial portion of the vessel, and is contained in the supraclavicular triangle (p. 686).

Relations. It is deep to the skin, the superficial fascia, the platysma, the supraclavicular nerves and the deep cervical fascia. The external jugular vein crosses its medial part and receives the suprascapular, transverse cervical and anterior jugular veins, which frequently form a plexus in front of the artery. The nerve to the subclavius descends behind the veins and in front of the artery. The terminal part of the artery lies behind the clavicle and the subclavius, and is crossed by the suprascapular vessels. The subclavian vein is in front of, and at a slightly lower level than, the artery. The lower trunk of the brachial plexus lies *behind* and *below* the artery and intervenes between it and the scalenus medius. *Above*, and to its *lateral side*, are the upper and middle trunks of the brachial plexus, and the inferior belly of the omohyoid. *Below*, it rests on the superior surface of the first rib.

Surface Anatomy. The subclavian artery can be represented by a broad line, convex upwards, drawn from the sternoclavicular joint to the middle of the lower border of the clavicle (**6**.47). The third part of the vessel can be felt pulsating by deep pressure in the lower and anterior angle of the posterior triangle.

Variations. The subclavian arteries vary in their origin, course, and the height to which they rise in the neck.

The right subclavian may arise from the brachiocephalic above or below the level of the sternoclavicular joint. It may be a separate trunk from the arch of the aorta, and may then be either its first or last branch. When it is the first branch, it occupies the ordinary position of the brachiocephalic trunk; and when the last, it arises from the left extremity of the arch, and ascends obliquely towards the right side behind the trachea, oesophagus, and right common carotid to the inner border of the first rib, whence it follows its ordinary course. In these cases, the proximal part of the artery represents a persistent part of the right dorsal aorta, and the right fourth aortic arch takes no part in its formation (p. 188), and hence the right recurrent laryngeal nerve hooks round the lateral side of the common carotid, which is derived from the artery of the third arch. Sometimes, when it arises as the last branch of the arch of the aorta, the right subclavian artery passes between the trachea and the oesophagus.

Occasionally the subclavian artery perforates the scalenus anterior; very rarely it passes in front of that muscle. Sometimes the subclavian vein passes with the artery behind the scalenus anterior. The artery may ascend as high as 4 cm above the clavicle or may only reach the level of the upper border of the bone.

The left subclavian is occasionally joined at its origin with the left common carotid. It is more posterior in position than the right subclavian in the first part of its course, and, as a rule, does not reach quite as high a level in the neck.

Applied Anatomy. The third part of the subclavian artery is the most accessible. As the posterior border of the sternocleidomastoid corresponds closely to the lateral border of the scalenus anterior, the artery lies immediately lateral to the former and, with firm pressure, can be felt pulsating in the lower, anterior angle of the posterior triangle.

Effective compression of the subclavian artery can be attained only where it passes across the upper surface of the first rib. To compress the vessel here, the shoulder should be depressed and pressure exercised downwards, backwards and medially in the angle formed by the posterior border of the sternocleidomastoid with the upper border of the clavicle.

BRANCHES OF THE SUBCLAVIAN ARTERY

1. Vertebral.	4. Costocervical.
2. Internal thoracic.	5. Dorsal scapular.
3. Thyrocervical.	

On the left side all branches except the dorsal scapular arise close together from the first part of the artery; on the right side the costocervical trunk usually springs from the second part. The

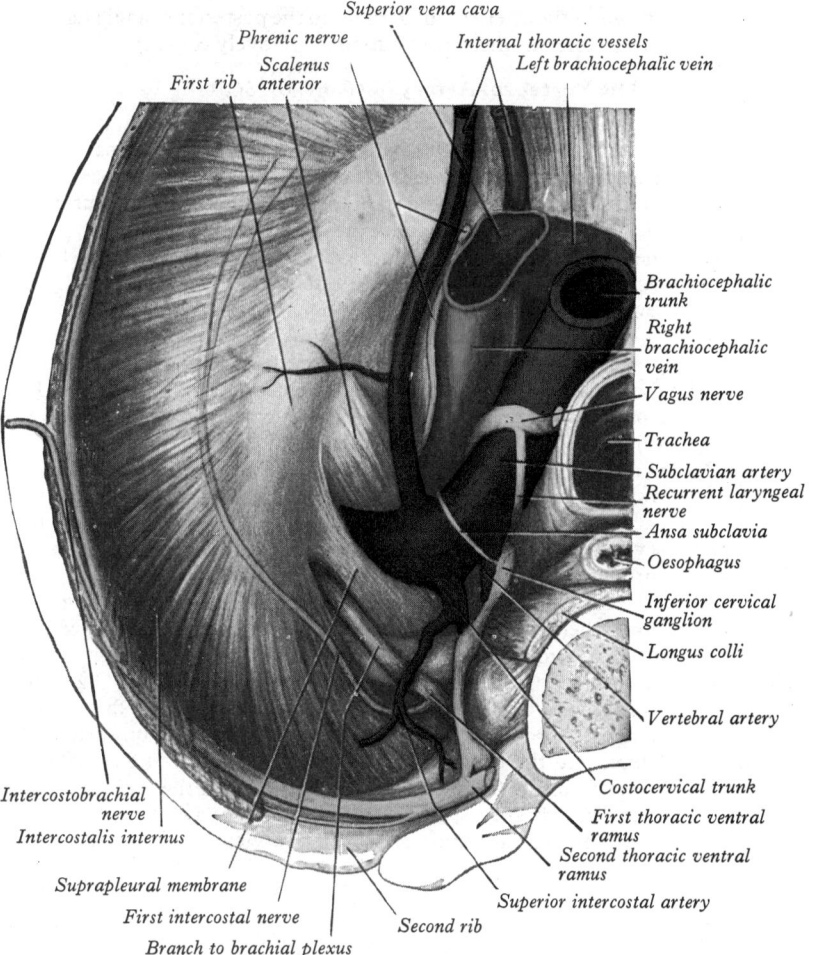

6.62 Structures related to the right cervical pleura, as seen from below.

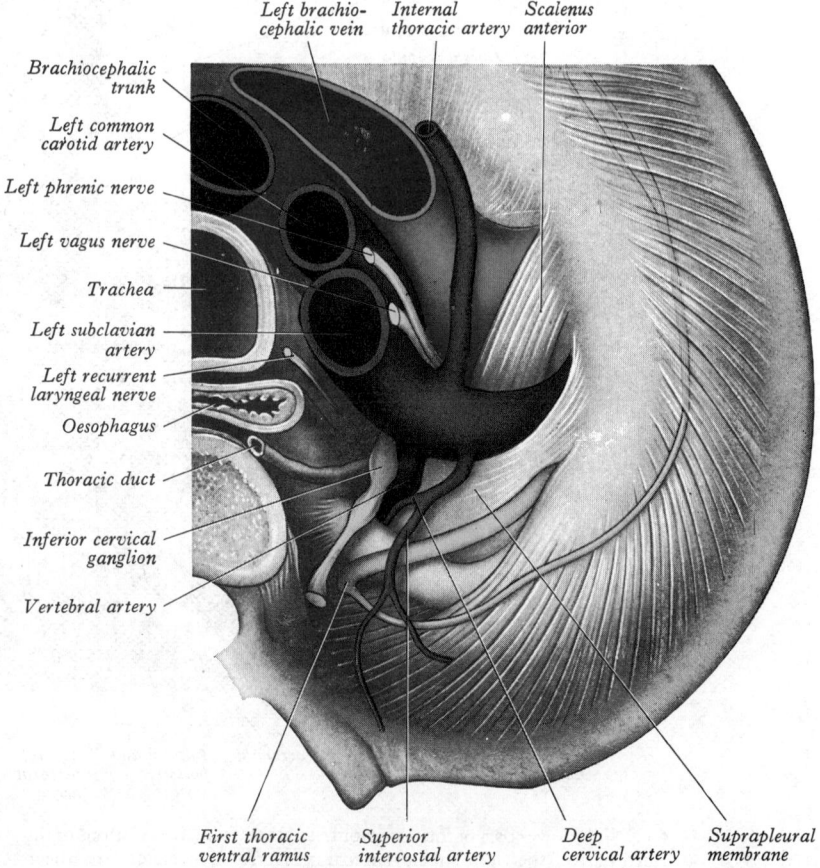

6.63 Structures related to the left cervical pleura, as seen from below. 693

origins of the arteries which run into the posterior triangle tend to be variable but their distribution is relatively constant.

1. The Vertebral Artery (6.56, 61, 65, 66, 118)

This arises from the upper and posterior part of the first part of the subclavian artery. It ascends through the foramina in the transverse processes of all the cervical vertebrae save the seventh, winds behind the lateral mass of the atlas, enters the skull through the foramen magnum, and, at the lower border of the pons, joins the vessel of the opposite side to form the basilar artery. Occasionally the vertebral artery may enter the fifth, fourth, or seventh cervical vertebra. (For other variations *see* p. 675.)

Relations. The vertebral artery consists of four parts. The *first part* runs upwards and backwards between the longus colli and the scalenus anterior and behind the common carotid artery. In front, it is related to the common carotid artery and the vertebral vein, and is crossed by the inferior thyroid artery; on the left side it is crossed also by the thoracic duct. Behind, it is related to the transverse process of the seventh cervical vertebra, the cervicothoracic ganglion (6.62, 63) and the ventral rami of the seventh and eighth cervical nerves. The *second part* ascends through the transverse foramina of the upper six cervical vertebrae, with a large branch derived from the cervicothoracic sympathetic (stellate) ganglion, and by a plexus of veins which unite to form the vertebral vein at the lower part of the neck. It lies in front of the ventral rami of the cervical nerves (C.2–C.6) (6.118), and pursues an almost vertical course as far as the transverse process of the axis, through which it runs upwards and laterally to the transverse foramen of the atlas. The *third part* issues from the foramen on the medial side of the rectus capitis lateralis, and curves backwards behind the lateral mass of the atlas, the ventral ramus of the first cervical nerve being on its

medial side; it then lies in the groove on the upper surface of the posterior arch of the atlas, and enters the vertebral canal by passing below the lower, arched border of the posterior atlanto-occipital membrane. This part of the artery is covered by the semispinalis capitis and is contained in the suboccipital triangle. The dorsal ramus of the first cervical nerve lies between the artery and the posterior arch of the atlas (3.30). The *fourth part* pierces the dura and the arachnoid mater, ascends in front of the roots of the hypoglossal nerve (p. 1083) and inclines medially to the front of the medulla oblongata where, at the lower border of the pons, it unites with the opposite artery to form the basilar artery (6.58).

Winckler (1972) has described variations in the elastic and muscular tissues in different parts of the vertebral artery. In its first and third parts it appears to be adapted, by increased elasticity, to the greater mobility and lack of support in these regions.

The branches of the vertebral artery may be divided into two sets—those given off in the neck, and those within the cranium.

Cervical Branches of the Vertebral Artery

Spinal branches enter the vertebral canal through the intervertebral foramina, and each divides into two branches. Of these, one passes along the roots of the nerves to supply the spinal cord (*see* p. 896, 7.54) and its membranes, anastomosing with the other arteries of the spinal cord; the other divides into an ascending and a descending branch, which unite with similar branches from the arteries above and below, so that two lateral anastomotic chains are formed on the posterior surfaces of the bodies of the vertebrae, near the attachment of the pedicles. From these anastomotic chains branches are supplied to the periosteum and the bodies of the vertebrae, and others communicate with similar branches from the opposite side; from these com-

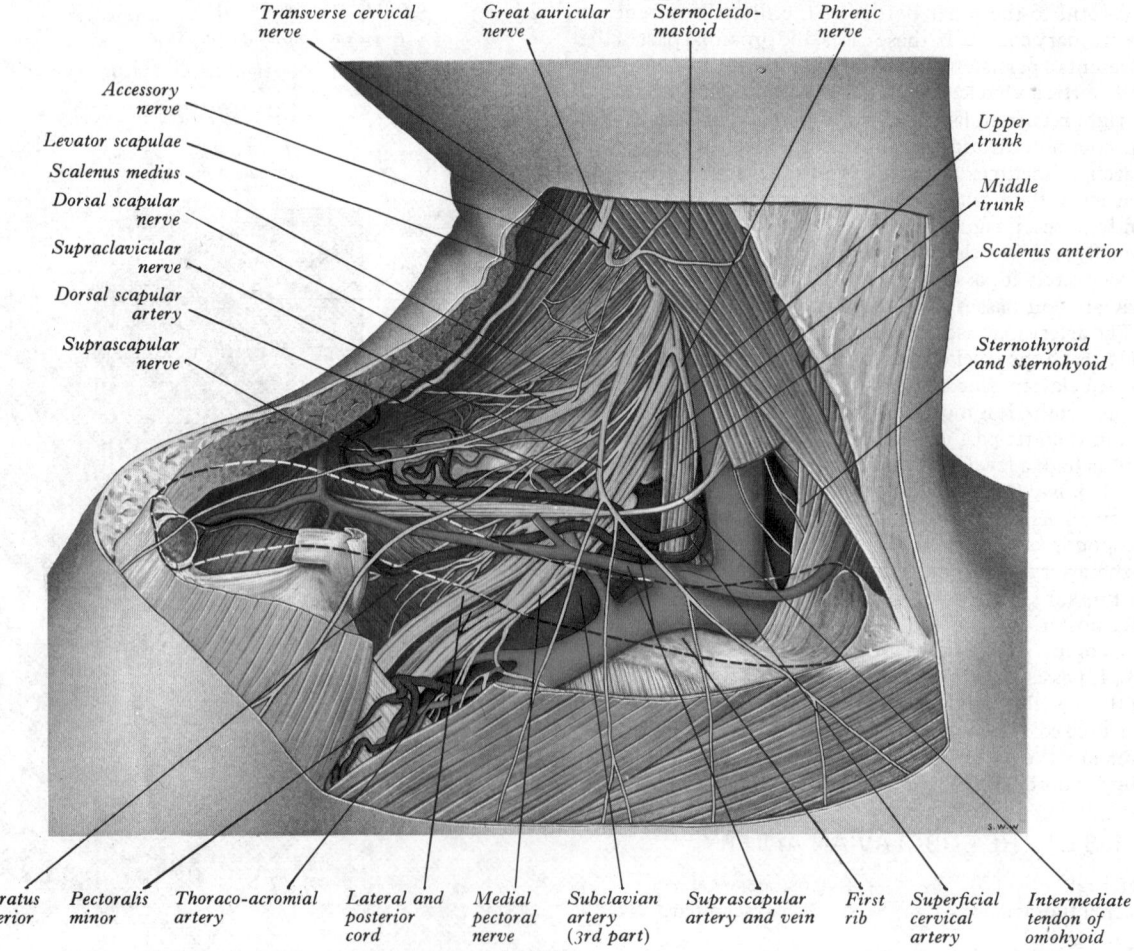

6.64 The lower part of the posterior triangle showing the relations of the third part of the right subclavian artery. *Note.*—The clavicle has been removed but is shown as a dotted outline. The middle trunk of the brachial plexus gives an unusual contribution to the medial cord.

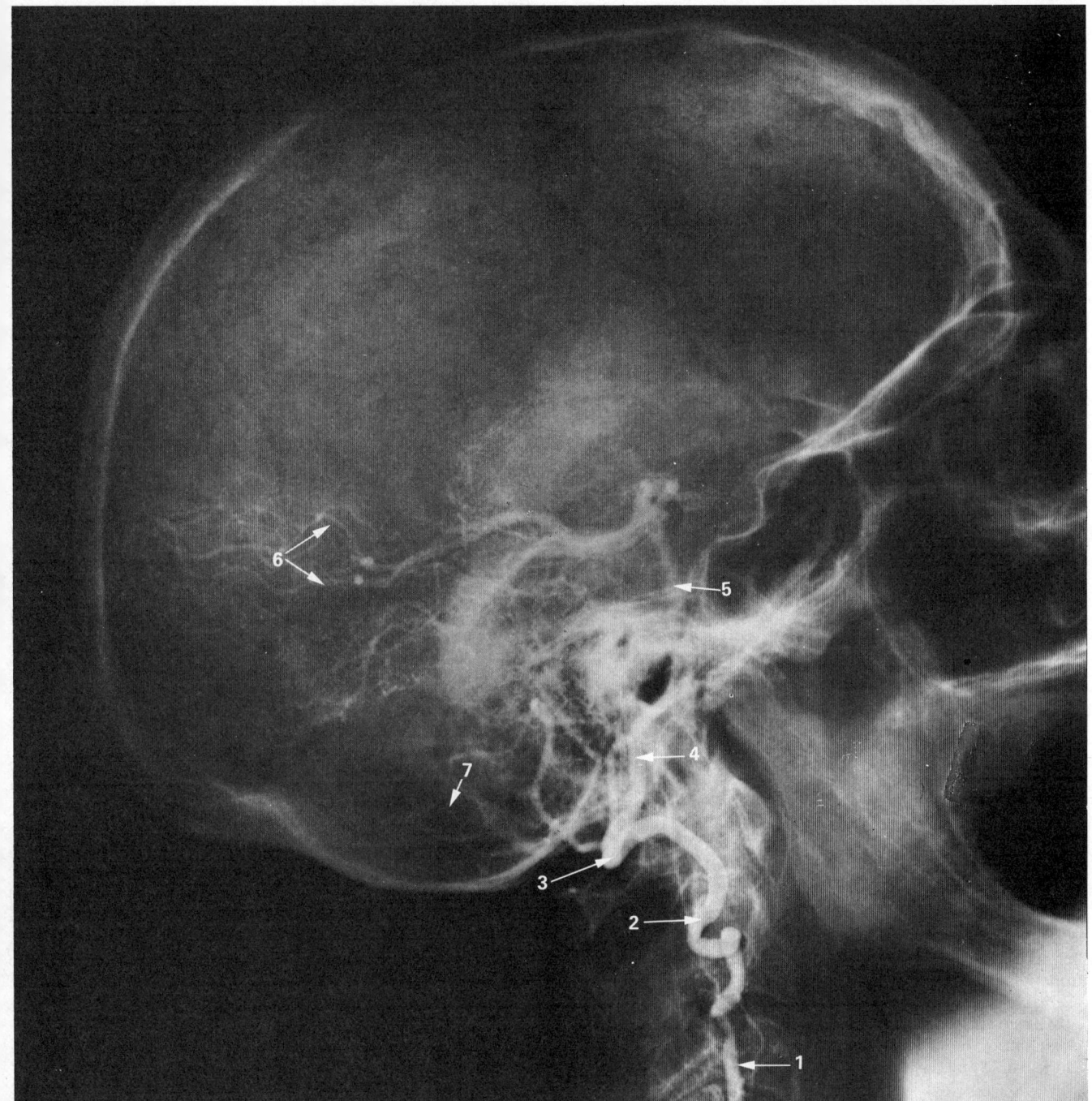

6.65   Vertebral arteriogram (left). Lateral view. 1. Vertebral artery, ascending part. 2. Loop between transverse foramina of axis and atlas. 3. Suboccipital part. 4. Intracranial part. 5. Basilar artery. 6. Posterior cerebral branches. 7. Inferior cerebellar branches.

munications small twigs arise which join similar branches above and below, to form a central anastomotic chain on the posterior surfaces of the bodies of the vertebrae.

Muscular branches arise from the vertebral artery as it curves round the lateral mass of the atlas. They supply the deep muscles of this region and anastomose with the occipital artery, and with the ascending and deep cervical arteries.

Cranial Branches of the Vertebral Artery

One or two **meningeal branches** spring from the vertebral artery opposite the foramen magnum; they ramify between the bone and dura mater in the cerebellar fossa and supply bone and the falx cerebelli.

The posterior spinal artery may arise from the vertebral artery at the side of the medulla oblongata, but is most frequently derived from the posterior inferior cerebellar artery. It passes

backwards, and then descends as two branches, one in front of and the other behind the dorsal roots of the spinal nerves; these are reinforced by a succession of spinal twigs which arise from the vertebral, ascending cervical, posterior intercostal and first lumbar arteries, and enter the vertebral canal through the intervertebral foramina; by means of these branches the posterior spinal arteries are continued to the lower part of the spinal cord. (For details of the blood supply of the spinal cord *see* p. 896 and 7.54.)

The anterior spinal artery is a small branch, which arises near the termination of the vertebral artery; it descends in front of the medulla oblongata and unites with its fellow of the opposite side near the level of the lower end of the olives of the medulla oblongata. The single trunk thus formed descends on the front of the spinal cord, and is reinforced by a succession of small spinal branches which enter the vertebral canal through the inter-

695

vertebral foramina; these branches are derived from the vertebral, the ascending cervical, posterior intercostal, and first lumbar arteries. They unite, by means of ascending and descending branches, to form a single anterior median artery, which extends as far as the lower part of the spinal cord, and is continued as a slender twig on the filum terminale. This vessel is placed in the pia mater along the anterior median fissure; it supplies the substance of the spinal cord, and sends off branches at its lower part to be distributed to the cauda equina. Branches pass from the anterior spinal arteries, and from the beginning of the trunk formed by their union, to the medulla oblongata, where they are distributed to its central portion, being sharply limited dorsally to the region of the trigonum hypoglossi. (*See* p. 1042 and 7.54.)

The posterior inferior cerebellar artery (6.58) is the largest branch of the vertebral artery, but is not infrequently absent. It arises near, and winds backwards round, the lower end of the olive of the medulla oblongata; it then ascends behind the roots of the glossopharyngeal and vagus nerves to the lower border of the pons, where it turns downwards along the inferolateral border of the fourth ventricle. Finally, it runs laterally into the vallecula of the cerebellum, where it divides into a medial and a lateral branch. The *medial branch* runs backwards between the cerebellar hemisphere and the inferior vermis, supplying branches to both; the *lateral branch* supplies the under surface of the hemisphere, as far as its lateral border, and anastomoses with the anterior inferior cerebellar and superior cerebellar branches of the basilar artery. The trunk of the artery supplies branches to the medulla oblongata and to the choroid plexus of the fourth ventricle, and sends a branch upwards lateral to the tonsil of the cerebellum to supply the dentate nucleus of the cerebellum. The area supplied in the medulla oblongata lies dorsal to the olivary nucleus and lateral to the nucleus and emerging fila of the hypoglossal nerve.

Applied Anatomy. It is to be noted that the posterior inferior cerebellar artery supplies the lateral part of the medulla. Thrombosis of this artery will therefore cause a loss of function of the following structures ('*lateral medullary syndrome*'): nucleus ambiguus, nuclear solitarius, vestibular and cochlear nuclei, spinocerebellar tracts, lateral spinothalamic tract, spinal nucleus and tract of the trigeminal nerve. The anterior spinal artery supplies the medial part of the medulla and thrombosis of this vessel causes loss of function of the following structures ('*medial medullary syndrome*'): hypoglossal nucleus and nerve, medial lemniscus, the corticospinal (pyramidal) tract. (For the normal functions of the above structures consult the section on the Central Nervous System.)

The medullary arteries are several minute vessels which spring from the vertebral and its branches, and are distributed to the medulla oblongata.

The Basilar Artery

This is formed by the junction of the two vertebral arteries; it extends from the lower to the upper border of the pons, and is in the cisterna pontis. It lies in a shallow, median groove on the ventral surface of the pons. It is placed between the two abducent nerves at the lower border and the two oculomotor nerves at the upper border of the pons, where it divides into the two posterior cerebral arteries.

Branches:

The pontine branches are numerous small vessels which come off from the front and sides of the basilar artery, and supply the pons and adjacent parts of the brain.

The labyrinthine (internal auditory) artery, a long slender branch, may arise from the lower part of the basilar artery but is more often derived from the anterior inferior cerebellar artery (see also below); it accompanies the facial and the vestibulocochlear nerves into the internal acoustic meatus, and is distributed to the internal ear (p. 1214).

In a radiographic study of the labyrinthine artery Wende *et al.* (1975) were able to identify the origin of 238 examples. In only 38 (16 per cent) was the origin from the basilar artery, while 108 (45·4 per cent) were derived from the anterior inferior cerebellar. The superior cerebellar artery accounted for 58 (24·4 per cent) and the posterior inferior cerebellar for 13 (5·4 per cent). The remaining 21 (8·8 per cent) examples were reduplicated and were branches, in various combinations, from the basilar artery and one or other of the three cerebellar arteries. The same observers found a labyrinthine artery on one side only in 24 of 316 basilar angiographic examinations. Other workers have recorded different incidences of these variations. For example, Cavatori (1908) observed a basilar origin for the labyrinthine artery in about 70 per cent in Italians, whereas Stopford (1916) and Adachi and Hasche (1928) recorded it as most commonly arising from a common trunk with one of the cerebellar arteries. Gillilan (1972) has suggested that racial variations may exist in this regard.

The anterior inferior cerebellar artery (6.58) arises from the lower part of the basilar artery. It runs backwards and laterally, usually ventral to the abducent nerve, the facial and vestibulocochlear nerves, and commonly forms a loop which penetrates for a variable distance into the internal acoustic meatus below the nerves (Sunderland 1945). The labyrinthine artery frequently arises from the summit of the loop. On emerging from the meatus the artery is distributed to the anterolateral parts of the inferior surface of the cerebellum, where it anastomoses with the posterior inferior cerebellar branch of the vertebral artery. A few branches are supplied by the anterior inferior cerebellar artery to the lower and lateral parts of the pons, and sometimes to the upper part of the medulla oblongata.

The superior cerebellar artery (6.58) arises near the termination of the basilar. It passes laterally immediately below the oculomotor nerve, which separates it from the posterior cerebral artery, winds round the cerebral peduncle close to and below the trochlear nerve, and, arriving at the superior surface of the cerebellum, divides into branches which ramify in the pia mater, supplying this aspect of the cerebellum and anastomosing with branches of the inferior cerebellar arteries. In addition, branches are given to the pons, the pineal body, the superior medullary velum and the tela choroidea of the third ventricle.

The posterior cerebral artery (6.58, 59, 60), frequently double, is larger than the superior cerebellar artery, from which it is separated near its origin by the oculomotor nerve, and on the side of the midbrain by the trochlear nerve. Passing laterally, parallel with the superior cerebellar artery, and receiving the posterior communicating branch from the internal carotid artery, it winds round the cerebral peduncle, and reaches the tentorial surface of the cerebrum, where it breaks up into branches for the supply of the temporal and occipital lobes.

The branches of the posterior cerebral artery are divided into two sets, central and cortical.

Central branches. Several small *posteromedial central branches* (6.61) arise at the commencement of the posterior cerebral artery; these, with similar branches from the posterior communicating, pierce the posterior perforated substance, and supply the anterior part of the thalamus, the lateral wall of the third ventricle and the globus pallidus of the lentiform nucleus. The *posterior choroidal rami* vary in number and arrangement (Abbie 1933). One, or more, courses over the lateral geniculate body and helps to supply it, before entering the posterior part of the inferior horn of the lateral ventricle through the lower part of the choroidal fissure. The others curl round the posterior end of the thalamus and pass through the transverse fissure, some to the tela choroidea of the third ventricle, and some to pass through the upper part of the choroidal fissure; they supply the choroid plexuses of the third and lateral ventricles, and give some twigs to the fornix. (For further information on the distribution of the choroidal arteries consult Percheron 1977, and see also p. 1042.) The *posterolateral central branches* are small arteries which arise from the posterior cerebral artery after it has turned round the cerebral peduncle; they supply the cerebral peduncle, the posterior part of the thalamus, the colliculi and the pineal and medial geniculate bodies.

Cortical branches. The *temporal branches*, usually two in number, are distributed to the uncus, the parahippocampal, the medial and lateral occipitotemporal gyri; the *occipital branches* supply the cuneus, lingual gyrus and the posterior part of the lateral surface of the occipital lobe; the *parieto-occipital* supply the cuneus and the precuneus. It is to be noted that the posterior

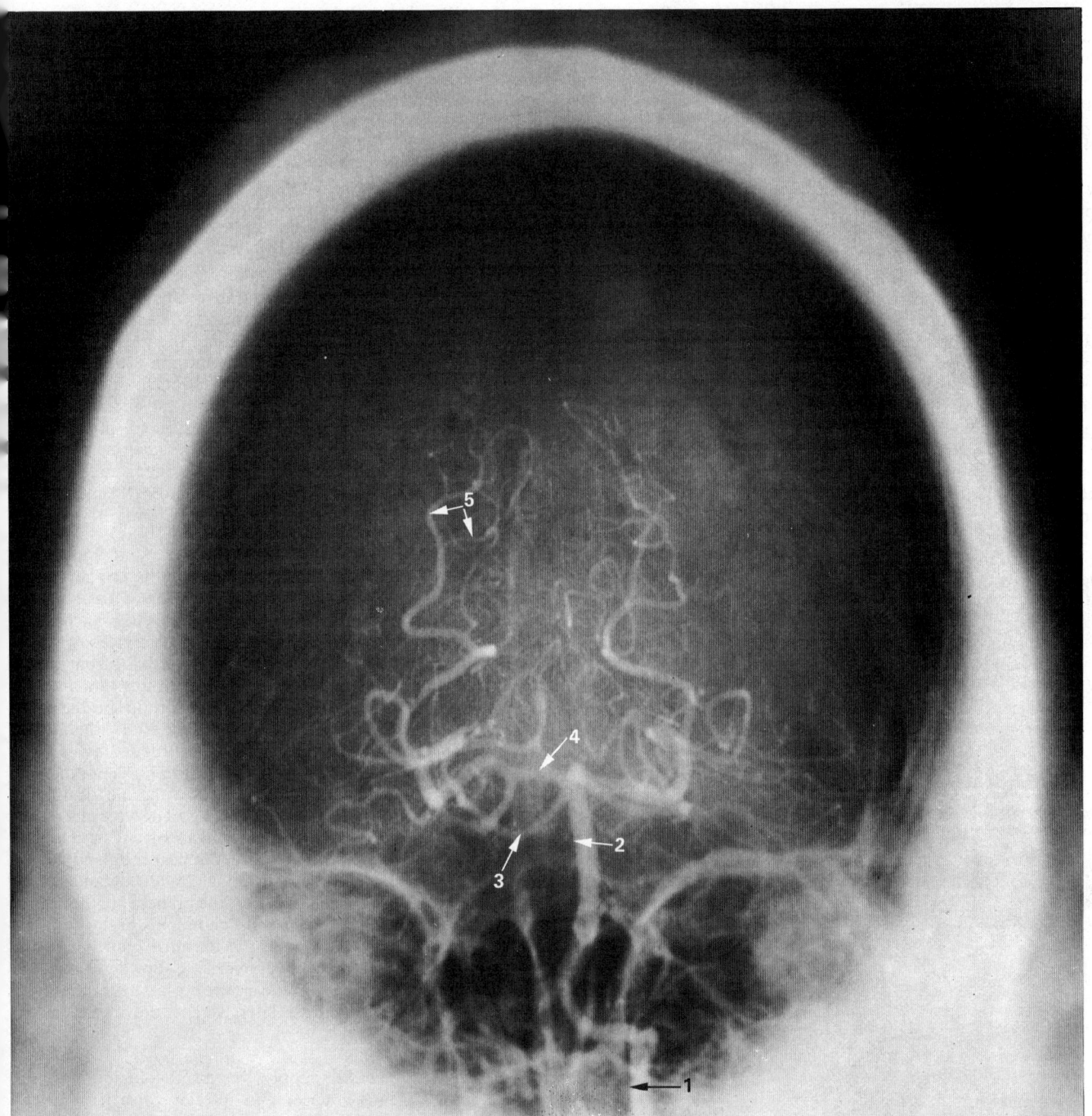

6.66 Vertebral arteriogram (left). Anteroposterior view. Circular shape of field is due to use of a cone to improve resolution. 1. Left vertebral artery. 2. Basilar artery. 3. Right superior cerebellar artery. 4. Right posterior cerebral artery. 5. Branches of right posterior cerebral artery.

cerebral artery supplies the visual area of the cerebral cortex (p. 1042), and many other structures associated with the visual pathway.

2. The Internal Thoracic (Mammary) Artery (6.67)

This arises about 2 cm above the sternal end of the clavicle from the inferior surface of the first part of the subclavian artery, opposite the thyrocervical trunk. It descends behind the cartilages of the upper six ribs about 1·25 cm from the lateral border of the sternum, and at the level of the sixth intercostal space divides into the musculophrenic and superior epigastric arteries.

Relations. It runs at first downwards, forwards, and medially behind the sternal end of the clavicle, the internal jugular and brachiocephalic veins, and the first costal cartilage. As the artery enters the thorax, the phrenic nerve crosses it obliquely from the lateral to the medial side, the nerve usually passing in front of the artery. Below the first costal cartilage it descends almost vertically to its bifurcation. It is covered in front by the pectoralis major, the cartilages of the upper six ribs and the intervening external intercostal membranes and internal intercostals, and is crossed by the terminal parts of the upper six intercostal nerves. It is separated from the pleura, as far as the second or third costal cartilage, by a strong layer of fascia, below this level, by the transversus thoracis. It is accompanied by a chain of lymph nodes and by venae comitantes which unite at the third costal cartilage to form a single vessel, medial to the artery.

Branches:

The pericardiacophrenic artery is a long, slender branch which accompanies the phrenic nerve, between the pleura and pericardium, to the diaphragm, and anastomoses with the musculophrenic and phrenic arteries.

697

The mediastinal arteries are small vessels, distributed to the areolar tissue and lymph nodes in the anterior mediastinum, and to the remains of the thymus.

The pericardial branches supply the upper part of the anterior surface of the pericardium.

The sternal branches are distributed to the transversus thoracis and to the posterior surface of the sternum.

The mediastinal, pericardial and sternal branches, together with some twigs from the pericardiacophrenic, anastomose with branches from the posterior intercostal and bronchial arteries, and form a *subpleural mediastinal plexus*.

The anterior intercostal branches are distributed to the upper six intercostal spaces. Two in each space, they pass laterally, one lying near the lower margin of the upper rib, and the other near the upper margin of the lower rib, and anastomose with the posterior intercostal arteries. They are first situated between the pleura and the internal intercostals, and then between the intercostales intimi and the internal intercostals. They supply the intercostals and send branches through them to the pectoral muscles, the breast, and to skin.

The perforating branches emerge through the upper five or six intercostal spaces, with the anterior cutaneous branches of the corresponding intercostal nerves. They pierce the pectoralis major and, curving laterally, supply that muscle and the skin. In the female, those of the second, third and fourth spaces give branches to the breast, and during lactation are of large size.

The musculophrenic artery passes obliquely downwards

and laterally, behind the seventh, eighth and ninth costal cartilages; it perforates the diaphragm near the ninth costal cartilage, and ends opposite the last intercostal space. It anastomoses with the inferior phrenic artery, the lower two posterior intercostal arteries and the ascending branch of the deep circumflex iliac artery. It gives off two anterior intercostal branches to each of the seventh, eighth and ninth intercostal spaces; these are distributed in a manner similar to the anterior intercostals from the internal thoracic. The musculophrenic also supplies the lower part of the pericardium and the abdominal muscles.

The superior epigastric artery descends through the interval between the costal and xiphoid origins of the diaphragm, anterior to the lower fibres of the transversus thoracis and the upper fibres of the transversus abdominis. It enters the sheath of the rectus abdominis, at first lying behind the muscle, and then perforating and supplying it, and anastomosing with the inferior epigastric artery from the external iliac. Branches perforate the sheath of the rectus, and supply the skin of the abdomen, and a small branch passes in front of the xiphoid process and anastomoses with the opposite artery. The superior epigastric artery gives some twigs to the diaphragm, while from that of the right side small branches extend into the falciform ligament of the liver and anastomose with the hepatic artery.

3. The Thyrocervical Trunk (6.45, 118)

This short, wide vessel arises from the front of the first part of the subclavian artery, close to the medial border of the scalenus anterior, and divides almost immediately into three branches, the inferior thyroid, suprascapular and superficial cervical.

The inferior thyroid artery runs upwards in front of the medial border of the scalenus anterior; it then turns medially in front of the vertebral vessels and behind the carotid sheath and its contents, and usually behind the sympathetic trunk, the middle cervical ganglion of which generally rests upon the vessel; it finally descends on the longus colli to the lower border of the lobe of the thyroid gland. As it approaches the thyroid gland, the relation of the artery to the recurrent laryngeal nerve is of surgical importance (see also p. 1080). At a little distance from the gland the artery usually passes behind the nerve, but near the gland, on the right side the nerve may lie with equal frequency anterior or posterior to, or intermingled amongst, the branches of the artery, while on the left side the nerve is most commonly posterior to the artery. The relationship between the terminal branches of the artery and those of the nerve is very variable (Bowden 1955). On the *left side*, close to its origin, the artery is crossed anteriorly by the thoracic duct as it curves laterally and downwards to its termination.

Branches:

The *muscular branches* supply the infrahyoid muscles, the longus colli, scalenus anterior and inferior constrictor of the pharynx.

The *ascending cervical artery* is a small branch which arises from the inferior thyroid artery as that vessel turns medially behind the carotid sheath; it ascends on the anterior tubercles of the transverse processes of the cervical vertebrae in the interval between the scalenus anterior and longus capitis. It gives twigs to the muscles of the neck, and sends one or two *spinal branches* into the vertebral canal through the intervertebral foramina to be distributed to the spinal cord and its membranes, and to the bodies of the vertebrae, in the same manner as the spinal branches of the vertebral artery. It anastomoses with branches of the vertebral, ascending pharyngeal, occipital and deep cervical arteries.

The *inferior laryngeal artery* ascends upon the trachea in company with the recurrent laryngeal nerve; it enters the larynx deep to the lower border of the inferior constrictor, and supplies its muscles and mucous membrane, anastomosing with the artery from the opposite side, and with the superior laryngeal branch of the superior thyroid artery.

The *pharyngeal branches* supply the lower part of the pharynx.

The *tracheal branches* are distributed to the trachea and anastomose below with the bronchial arteries.

The *oesophageal branches* supply the oesophagus and ana-

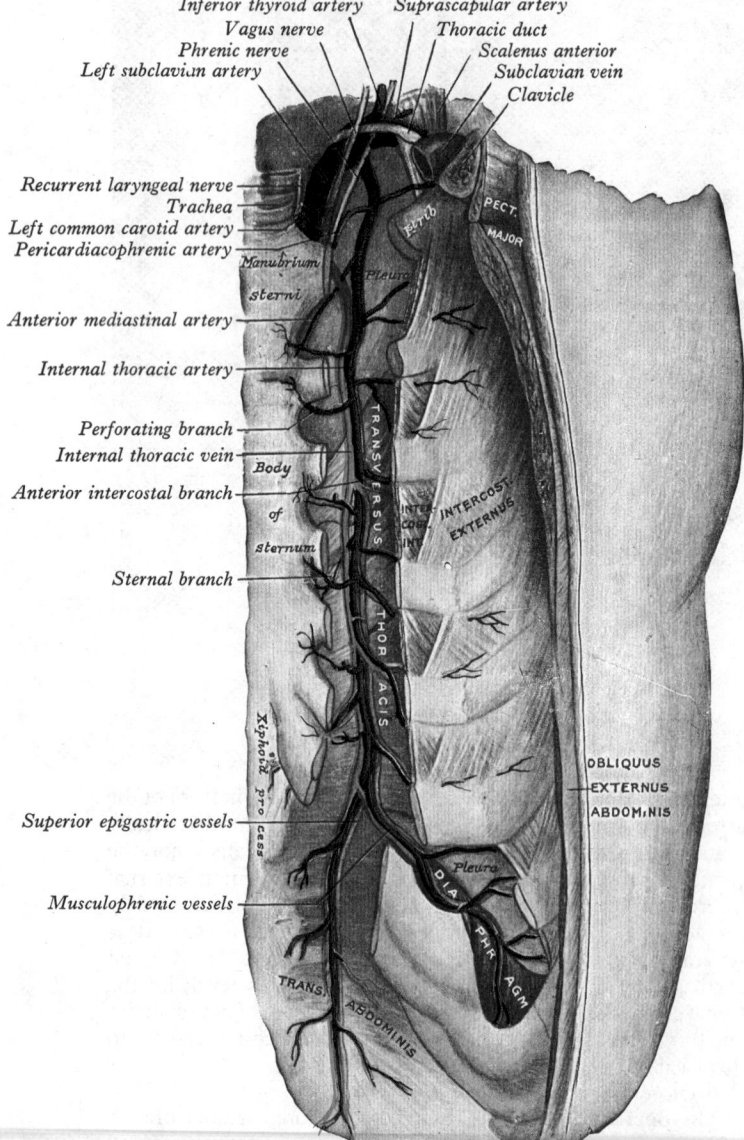

Inferior thyroid artery
Vagus nerve
Phrenic nerve
Left subclavian artery
Suprascapular artery
Thoracic duct
Scalenus anterior
Subclavian vein
Clavicle

Recurrent laryngeal nerve
Trachea
Left common carotid artery
Pericardiacophrenic artery
Anterior mediastinal artery
Internal thoracic artery
Perforating branch
Internal thoracic vein
Anterior intercostal branch
Sternal branch
Superior epigastric vessels
Musculophrenic vessels

6.67   The left internal thoracic artery and vein and their main branches.

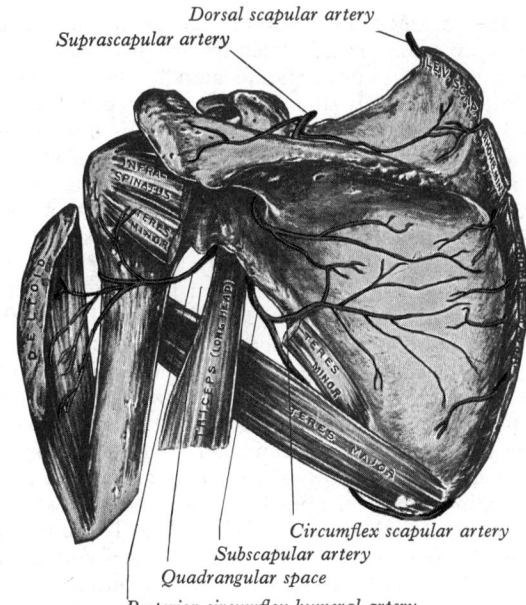

Dorsal scapular artery
Suprascapular artery

Circumflex scapular artery
Subscapular artery
Quadrangular space
Posterior circumflex humeral artery

6.68   The scapular anastomosis of the left side, dorsal aspect.

stomose with the oesophageal branches of the thoracic aorta.

The large *glandular branches* comprise an inferior and an ascending branch; they are distributed to the posterior and inferior parts of the thyroid gland, and anastomose with the superior thyroid artery and with the opposite inferior thyroid artery; the ascending branch supplies the parathyroid glands.

The suprascapular artery (**6**.64, 68) passes at first downwards and laterally across the scalenus anterior and the phrenic nerve, behind the internal jugular vein and the sternocleidomastoid; it then crosses in front of the subclavian artery and the brachial plexus, and runs behind and parallel with the clavicle and subclavius, and deep to the inferior belly of the omohyoid, to the superior border of the scapula; here it passes above (occasionally below) the superior transverse ligament, which separates it from the suprascapular nerve, and enters the supraspinous fossa (**6**.68). In this situation it lies on the bone, and supplies branches to the supraspinatus. It then descends behind the neck of the scapula, through the great scapular notch and deep to the inferior transverse ligament, to reach the deep surface of the infraspinatus, where it anastomoses with the circumflex scapular artery, and the deep branch of the transverse cervical artery. Besides distributing branches to the sternocleidomastoid, subclavius, and neighbouring muscles, it gives off a *suprasternal branch*, which crosses over the sternal end of the clavicle to the skin of the upper part of the chest and, an *acromial branch*, which pierces trapezius and supplies the skin over the acromion, anastomosing with the thoraco-acromial and posterior circumflex humeral arteries. As the suprascapular artery passes above the superior transverse ligament, it sends a branch into the subscapular fossa, where it ramifies beneath the subscapularis, and anastomoses with the subscapular artery and with the deep branch of the transverse cervical artery. The suprascapular artery also sends articular branches to the acromioclavicular and shoulder joints, and nutrient arteries to the clavicle and scapula. Not infrequently the suprascapular artery arises from the third part of the subclavian artery.

The superficial cervical artery (**6**.64) lies at a higher level than the suprascapular artery; it crosses in front of the phrenic nerve and the scalenus anterior, and in front of the brachial plexus, and is covered by the internal jugular vein, sternocleidomastoid and platysma. It crosses the floor of the posterior triangle of the neck, to reach the anterior margin of the levator scapulae. It then ascends deep to the anterior part of the trapezius, distributing branches to it and the neighbouring muscles, and to the lymph nodes in the neck; it anastomoses with the superficial branch of the descending branch of the occipital artery.

4. The Costocervical Trunk (**6**.62)

This arises from the back of the second part of the subclavian artery on the right, but from the first part of the artery on the left. It arches backwards above the cervical pleura to the neck of the first rib, and divides into the superior intercostal and deep cervical arteries.

The superior intercostal artery descends between the pleura and the necks of the first and second ribs, and anastomoses with the third posterior intercostal artery. As it crosses the neck of the first rib it is medial to the ventral ramus of the first thoracic nerve, which it crosses at a lower level (**6**.62), and lateral to the cervicothoracic ganglion of the sympathetic trunk. In the first intercostal space, it gives off the first posterior intercostal artery, which is arranged in a manner similar to the distribution of the lower posterior intercostals. It then descends to become the second posterior intercostal artery and usually joins with a branch from the third. The second posterior intercostal artery is not constant, but is more commonly found on the right side; when absent, its place is supplied by a branch from the aorta.

The deep cervical artery (**6**.56) arises, in most cases, from the costocervical trunk, and is analogous to the posterior branch of a posterior intercostal artery: occasionally it is a separate branch from the subclavian artery. Passing backwards above the eighth cervical nerve and between the transverse process of the seventh cervical vertebra and the neck of the first rib (sometimes between the transverse processes of the sixth and seventh cervical vertebrae) it ascends in the back of the neck, between the semispinales capitis et cervicis, as high as the second cervical vertebra. It supplies the adjacent muscles, and anastomoses with the deep division of the descending branch of the occipital artery (p. 681) and with branches of the vertebral artery. It gives off a spinal twig which enters the vertebral canal through the foramen between the seventh cervical and first thoracic vertebrae.

5. The Dorsal Scapular Artery (**6**.64, 68)

This arises from the third or, less frequently, the second part of the subclavian artery. It passes laterally through the brachial plexus and in front of the scalenus medius. It then goes deep to the levator scapulae to the superior angle of the scapula where it descends in company with the dorsal scapular nerve under cover of the rhomboids along the medial border of the scapula as far as the inferior angle of the bone. It supplies branches to the rhomboids, latissimus dorsi and trapezius, and anastomoses with the suprascapular and subscapular arteries, and with the posterior branches of some of the posterior intercostal arteries.

Variations. In about one-third of cases the superficial cervical and dorsal scapular arteries have a common origin from the thyrocervical trunk termed the *transverse cervical artery*. This gives rise to a *superficial branch* (superficial cervical artery) and a *deep branch* (dorsal scapular artery); the latter passes laterally in front of the brachial plexus before going deep to the levator scapulae.

The Axilla

The axilla is a pyramidal region, situated between the upper parts of the chest wall and of the medial side of the arm.

The blunted *apex* of the axilla is directed upwards into the root of the neck (sometimes termed the *cervico-axillary canal*), and corresponds to the interval between the outer border of the first rib, the superior border of the scapula, the posterior surface of the clavicle, and the medial aspect of the coracoid process; through it the axillary vessels and nerves enter the space from the neck. The *base*, directed downwards, is broad at the chest but narrow and pointed at the arm; it is formed by the skin and a thick layer of fascia, termed the *axillary fascia*, extending between the lower border of the pectoralis major in front, and the lower border of the latissimus dorsi behind. It is of course convex upwards, conforming to the concavity of the armpit. The *anterior wall* is formed by the pectorales major et minor, the former covering the whole of this wall, the latter only its intermediate part. The space between the upper border of the pectoralis minor and the clavicle is occupied by the clavipectoral fascia. The *posterior wall* is formed by the subscapularis above, the teres major and latissimus dorsi below. On the *medial side* are the first four ribs with their corresponding intercostal muscles, and the upper part of the serratus anterior. This 'wall' displays the convexity of the thorax. On the *lateral side*, where the anterior and posterior walls converge, the space is narrow, and is bounded by the intertubercular sulcus of the humerus, the coracobrachialis and the biceps.

The axilla contains the axillary vessels, the infraclavicular part of the brachial plexus with its branches, the lateral branches of some of the intercostal nerves, a large number of lymph nodes, and a quantity of fat and loose areolar tissue. The axillary vessels and the brachial plexus of nerves run from the apex to the base along the lateral wall of the axilla; they are nearer to the anterior than the posterior wall, the axillary vein lying anteromedial to the axillary artery. Owing to the obliquity of the upper ribs, the neurovascular bundle, just after it emerges from behind the clavicle, crosses the first intercostal space. The relations of the bundle are therefore somewhat different in the upper portion of the axilla from those in the lower parts. The thoracic branches of the axillary artery are in contact with the pectoral muscles, and along the lateral margin of the pectoralis minor the lateral thoracic artery passes to the side of the thorax. The subscapular vessels descend on the posterior wall in contact with the lower margin of the subscapularis, and the subscapular nerves and the thoraco-dorsal nerve (to latissimus dorsi) cross the anterior surface of the muscle with different degrees of obliquity; the circumflex scapular vessels wind round the lateral border of the scapula, and the posterior circumflex humeral vessels and the axillary (circumflex) nerve curve backwards close to the surgical neck of the humerus. No vessel of any importance lies on the medial or thoracic side, the upper part of the space being crossed merely by a few small branches form the superior thoracic artery. The long thoracic nerve (to serratus anterior) descends on the surface of the muscle which it supplies; and the intercostobrachial nerve perforates the upper and anterior part of this wall, and passes across the axilla to the medial side of the upper arm.

The position and arrangement of the lymph nodes are described on p. 789.

Applied Anatomy. When suppuration occurs in the axilla, the arrangement of the fascia plays an important part in the direction in which pus may spread. As described on p. 568, the clavipectoral fascia, after filling the space between the clavicle and the medial border of the pectoralis minor, splits to enclose it and blends at its lateral border with the axillary fascia at the anterior fold of the axilla. Suppuration may take place either superficial or deep to this layer of fascia; that is, either between the pectoral muscles or behind the pectoralis minor; in the former case, the abscess would point either at the border of the anterior axillary fold, or in the groove between the deltoid and the pectoralis major; in the latter case, pus would tend to surround the vessels and nerves, and ascend into the neck, that being the direction in which there is least resistance. Instances have been recorded where the pus found its way along the vessels into the arm.

When an axillary abscess is opened, the knife should be entered in the floor of the axilla, midway between the anterior and posterior margins and near the thoracic side of the space so as to avoid the lateral thoracic, subscapular, and axillary vessels which are in contact respectively with the anterior, posterior and lateral walls of the axilla.

The relations of the vessels and nerves in the several parts of the axilla are important, for it is still a frequent procedure to remove the lymph nodes from the axilla in operating for cancer of the breast. When such an operation is performed, it is necessary to proceed with much caution in the direction of the lateral wall and apex of the space, because of the axillary vessels.

The Axillary Artery

The axillary artery (**6**.69), the continuation of the *subclavian*, begins at the outer border of the first rib, and nominally ends at the lower border of the teres major, where the artery becomes the *brachial*. Its direction varies with the position of the limb: thus

the vessel is nearly straight when the arm is at right angles with the trunk, concave upwards when the arm is elevated above the shoulder, and convex upwards and laterally when the arm lies by the side. At first deeply situated, the artery becomes superficial, covered only by skin and fasciae. The pectoralis minor crosses the vessel and divides it into three parts; the first part is proximal, the second posterior, and the third distal, to the muscle.

Relations of the first part. The first part of the axillary artery is covered *in front* by the skin, superficial fascia, platysma, supraclavicular nerves, deep fascia, clavicular fibres of the pectoralis major and the clavipectoral fascia. This part of the artery is crossed anteriorly by the lateral pectoral nerve, the loop of communication between it and the medial pectoral nerve, and by the thoraco-acromial and cephalic veins. *Behind*, the artery is related to the first intercostal space and external intercostal, the first and second digitations of the serratus anterior, the long thoracic nerve and medial pectoral nerve, and the medial cord of the brachial plexus. On the *lateral side* it is related to the lateral and posterior cords of the brachial plexus, on the *medial side*, to the axillary vein, which also overlaps in front. The first part of the artery is enclosed, together with the axillary vein and the brachial plexus, in the fibrous *axillary sheath*, continuous above with the prevertebral layer of the deep cervical fascia.

Relations of the second part. In *front* of the second part are skin, superficial and deep fascia, and pectoralis major and minor, *behind*, the posterior cord of the brachial plexus and some areolar tissue, both of which intervene between it and the subscapularis, on the *medial side*, the axillary vein, separated from it by the medial cord of the brachial plexus and the medial pectoral nerve, on the *lateral side*, the lateral cord of the brachial plexus, which separates it from the coracobrachialis. The cords of the brachial plexus thus surround the second part of the artery on three sides, and separate it from direct contact with the vein and adjacent muscles.

Relations of the third part. The third part of the axillary artery extends from the lateral border of the pectoralis minor to the lower border of the teres major. Its upper part is covered *in front* by th lower part of the pectoralis major; its lower part by the skin and fasciae only. *Behind* it are the lower part of the subscapularis and the tendons of the latissimus dorsi and teres major. On its *lateral side* is the coracobrachialis, and *medially*, the axillary vein. The nerves of the brachial plexus bear the following relations to this part of the artery; *laterally* are the lateral root and the trunk of the median, and, for a short distance, the musculocutaneous; *medially*, the medial cutaneous nerve of the forearm lies between the axillary artery and vein anteriorly, and the ulnar nerve between the artery and vein posteriorly; the medial cutaneous nerve of the upper arm is on the medial side of the vein; *in front* is the medial root of the median nerve, and *behind*, the radial and axillary nerves, the latter only as far as the lower border of the subscapularis.

Branches:

Superior thoracic.	Subscapular.
Thoraco-acromial.	Anterior circumflex humeral.
Lateral thoracic.	Posterior circumflex humeral.

The superior thoracic artery (**6**.69), a small vessel, arises from the first part of the axillary artery near the lower border of the subclavius, but it may branch from the thoraco-acromial artery. Running forwards and medially above the medial border of the pectoralis minor, it passes between it and the pectoralis major to the thoracic wall. It supplies branches to these muscles, and to the thoracic wall, and anastomoses with the internal thoracic upper and intercostal arteries.

The thoraco-acromial (acromiothoracic) artery (**6**.64, 69) is a short branch from the second part of the axillary artery, at first overlapped by the upper edge of the pectoralis minor. Skirting the medial border of this muscle, it pierces the clavipectoral fascia and divides into four branches—pectoral, acromial, clavicular and deltoid.

The *pectoral branch* descends between the two pectoral muscles, and is distributed to them and to the breast, anastomosing with the intercostal branches of the internal thoracic artery and with the lateral thoracic artery. The *acromial*

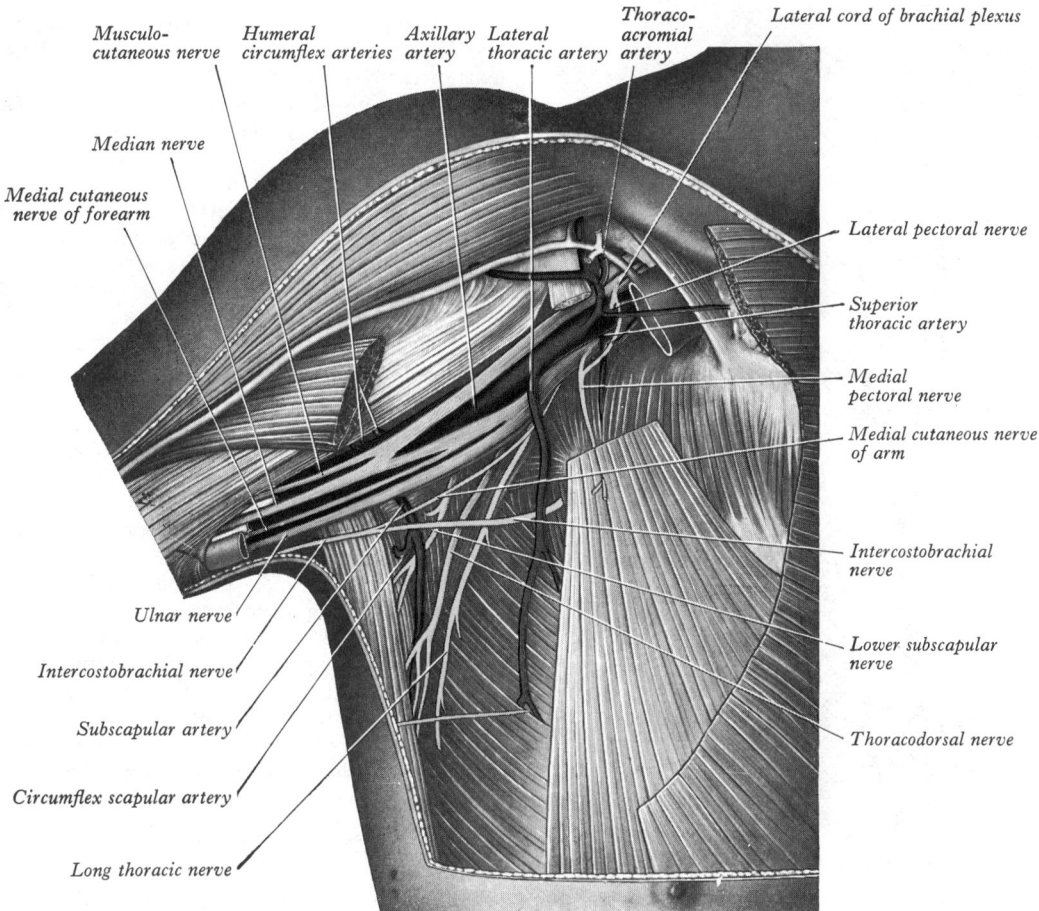

6.69   The right axillary artery and its branches. The pectoralis major and part of the pectoralis minor have been removed.

branch runs laterally over the coracoid process and under the deltoid, to which it gives branches; it then pierces that muscle and ends on the acromion, where it anastomoses with the branches of the suprascapular, the deltoid branch of the thoraco-acromial and the posterior circumflex humeral arteries. The *clavicular branch* runs upwards and medially between the clavicular part of the pectoralis major and the clavipectoral fascia; it gives branches to the sternoclavicular joint, and to the subclavius. The *deltoid branch* often arises with the acromial branch; it crosses over the pectoralis minor and runs with the cephalic vein in the interval between the pectoralis major and deltoid, supplying both muscles.

The lateral thoracic artery (**6**.69) follows the lateral border of the pectoralis minor to the side of the chest, supplies the serratus anterior and the pectoral muscles, the axillary lymph nodes, and subscapularis; it anastomoses with the internal thoracic, subscapular, and intercostal arteries, and the pectoral branch of the thoraco-acromial artery. In the female, the lateral thoracic artery is large, and gives off *lateral mammary branches*, which turn round the lateral border of the pectoralis major to reach the mamma.

The subscapular artery (**6**.69) is the largest branch of the axillary artery; usually it arises at the lower border of the subscapularis, which it follows to the inferior angle of the scapula, where it anastomoses with the lateral thoracic and intercostal arteries and with the deep branch of the transverse cervical artery; finally it ends in the neighbouring muscles and adjacent part of the chest wall. In the lower part of its course it is accompanied by the nerve to the latissimus dorsi; about 4 cm from its origin it gives off the *circumflex scapular artery*. This is generally larger than the continuation of the subscapular. It curves round the lateral border of the scapula, traversing the triangular space between the subscapularis above, the teres major below, and the long head of the triceps laterally (**6**.68); it enters the infraspinous fossa under

cover of the teres minor, and gives off two branches: one (*infrascapular*) enters the subscapular fossa deep to the subscapularis, and anastomoses with the suprascapular artery and the dorsal scapular (or deep branch of the transverse cervical) artery; the other is continued along the lateral border of the scapula between the teres major and the teres minor, and at the dorsal surface of the inferior angle anastomoses with the deep branch of the transverse cervical artery. In addition, small branches are distributed to the posterior part of deltoid and the long head of triceps, and anastomose with an ascending branch of the arteria profunda brachii.

The anterior circumflex humeral artery (**6**.69) is a small branch from the lateral side of the axillary artery at the lower border of the subscapularis. It runs horizontally, behind coracobrachialis and short head of biceps, in front of the surgical neck of the humerus. On reaching the intertubercular sulcus (bicipital groove), it gives off a branch which ascends in it to supply the head of the humerus and the shoulder joint. The artery then continues laterally under the long head of biceps and deltoid, and anastomoses with the posterior circumflex humeral artery.

The posterior circumflex humeral artery (**6**.68) is considerably larger than the anterior. It branches from the third part of the axillary artery at the lower border of subscapularis, and runs backwards with the axillary nerve through the quadrangular space, which is bounded by subscapularis, the capsule of the shoulder joint and teres minor above, teres major below, the long head of triceps medially, and the surgical neck of the humerus laterally. It winds round the surgical neck of the humerus and distributes branches to the shoulder joint, deltoid, teres major and minor, and the long and lateral heads of triceps, and gives off a descending branch which anastomoses with the deltoid branch of the arteria profunda brachii. It also anastomoses with the anterior circumflex humeral, and with the acromial branches of the suprascapular and thoraco-acromial arteries.

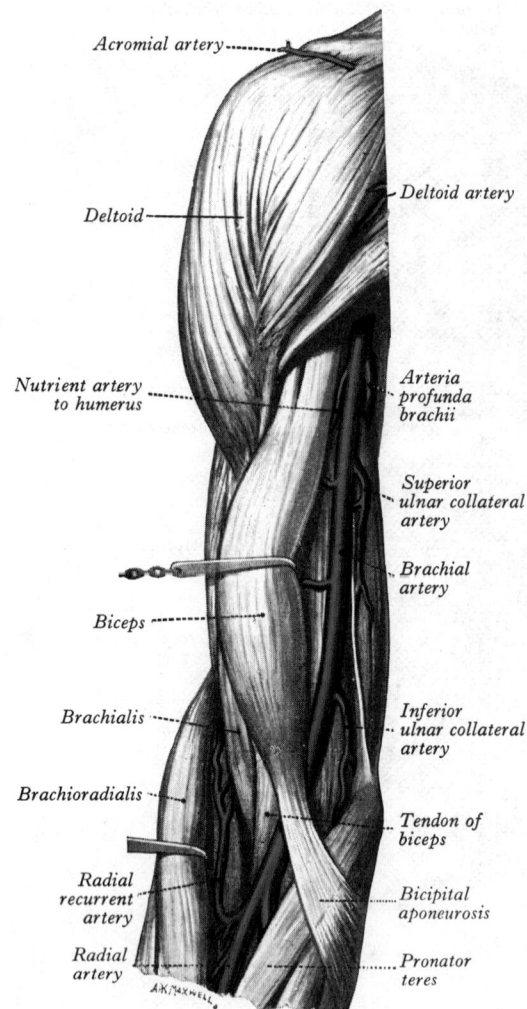

6.70 The right brachial artery and its branches.

Acromial artery

Deltoid

Deltoid artery

Nutrient artery to humerus

Arteria profunda brachii

Superior ulnar collateral artery

Brachial artery

Biceps

Brachialis

Inferior ulnar collateral artery

Brachioradialis

Tendon of biceps

Radial recurrent artery

Bicipital aponeurosis

Radial artery

Pronator teres

Surface Anatomy. The axillary artery can be felt pulsating against the lateral wall of the lower part of the axilla. Its upper portion can be mapped out when the arm is raised to a right angle with the trunk, if the point of pulsation indicating the lower portion of the vessel is joined to the midpoint of the lower surface of the clavicle.

Variations. The branches of the axillary artery vary considerably in different subjects. One, named *alar thoracic*, and frequently derived from the second part of the artery, is distributed to the fat and lymph nodes in the axilla. Occasionally the subscapular, circumflex humeral and profunda arteries arise from a common trunk, and when this occurs the branches of the brachial plexus surround this trunk instead of the main vessel. The posterior circumflex humeral artery may arise from the arteria profunda brachii; it then passes backwards below the teres major, instead of accompanying the axillary nerve through the quadrangular space. Sometimes the axillary artery divides into the radial and ulnar arteries, and occasionally it gives off the anterior interosseous artery of the forearm.

Applied Anatomy. Compression of the axillary artery may be required in injuries, in the removal of tumours, or in amputations. Compression can be made most effectually in the lower part of its course, by pressing the artery against the humerus.

With the exception of the popliteal, the axillary artery is perhaps more frequently lacerated by violent movements than any other artery in the body, being particularly susceptible when diseased. It has occasionally been ruptured in attempts to reduce old dislocations of the shoulder joint, especially when the artery has become fixed to the capsule of the joint.

The Brachial Artery

The brachial artery (6.70, 71, 72, 73) is the continuation of the axillary artery. It begins at the lower border of the tendon of the teres major, runs down the arm, and ends about a centimetre or so below the elbow joint by dividing into radial and ulnar arteries. At first it lies medial to the humerus, but it gradually spirals to the front of the arm and the midpoint between the humeral epicondyles at the elbow. Its pulsations can be felt throughout its length.

Relations. The artery is superficial throughout, covered only by skin and superficial and deep fasciae; the bicipital aponeurosis lies in front of it at the elbow and separates it from the median cubital vein; the median nerve crosses the artery from lateral to medial side near the insertion of coracobrachialis. *Posteriorly*, it lies at first on the long head of the triceps, separated by the radial nerve and the profunda brachii artery. It then lies successively on the medial head of the triceps, the insertion of the coracobrachialis, and brachialis. *Laterally*, it is related above to the median nerve and coracobrachialis, below to biceps, the two muscles overlapping the artery to some extent. *Medially*, its upper half is next to the medial cutaneous nerve of the forearm and ulnar nerve, its lower half to the median nerve. The basilic vein lies medially, but is separated from it in the lower part of the arm by the deep fascia. The artery is closely accompanied by two venae comitantes, connected at intervals by short transverse branches.

At the bend of the elbow the brachial artery sinks deeply into a triangular intermuscular depression which is named the *cubital fossa*. The base of the triangle is represented by a line connecting the two humeral epicondyles; the sides are formed by the medial edge of brachioradialis and the lateral margin of pronator teres; the 'floor' consists of brachialis and supinator. This fossa contains the tendon of biceps, the terminal part of the brachial artery, and its accompanying veins, the commencement of the radial and ulnar arteries, and parts of the median and radial nerves. The brachial artery occupies the middle of the fossa, and divides opposite the neck of the radius into its radial and ulnar branches. In *front* of it are the skin, superficial fascia, and median cubital vein, the last separated from the artery by the bicipital aponeurosis. *Behind*, brachialis separates it from the elbow joint. The median nerve lies close to the *medial* side of the artery above, but is separated from its ulnar branch below by the ulnar head of pronator teres. The tendon of biceps is *lateral* to the artery; the

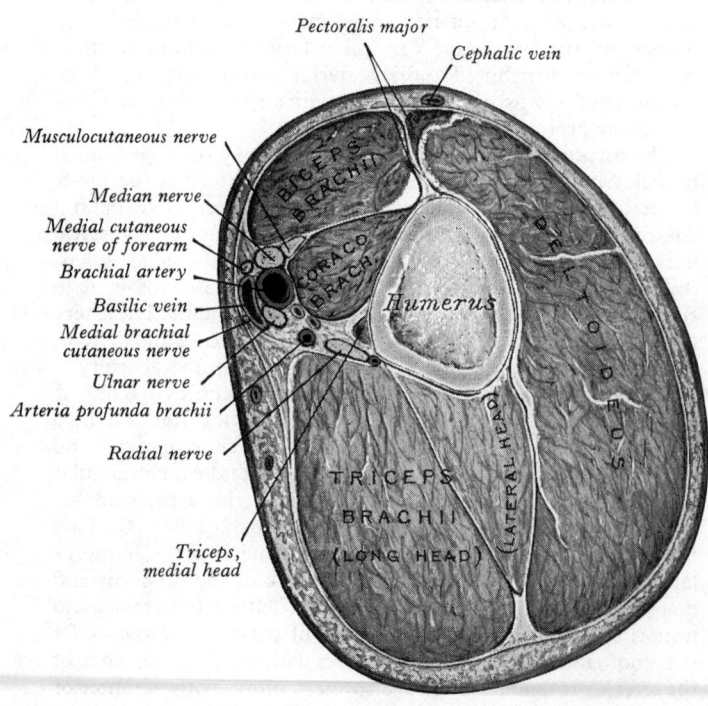

Pectoralis major

Cephalic vein

Musculocutaneous nerve

Median nerve

Medial cutaneous nerve of forearm

Brachial artery

Basilic vein

Medial brachial cutaneous nerve

Ulnar nerve

Arteria profunda brachii

Radial nerve

Triceps, medial head

BICEPS BRACHII

CORACO BRACH.

Humerus

DELTOIDEUS

(LATERAL HEAD)

TRICEPS BRACHII (LONG HEAD)

6.71 A transverse section through the right arm at the junction of the proximal and middle thirds of the humerus. Proximal aspect.

radial nerve is concealed, more laterally, between supinator and brachioradialis.

Variations. The brachial artery, accompanied by the median nerve, may leave the medial border of the biceps, and descend towards the medial epicondyle of the humerus; in such cases it usually passes behind a *supracondylar process* of the humerus, from which a fibrous arch is most often thrown over the artery; it then runs behind or through the substance of the pronator teres, to the bend of the elbow. This variation resembles the normal condition in some carnivores (p. 363). Occasionally, the artery splits at a high level into two trunks which re-unite. Frequently it divides at a higher level than usual, and the vessels resulting from this high division are three: viz. radial, ulnar and common interosseous arteries. Most frequently the radial is given off high up, the other limb of the division consisting of the ulnar and common interosseous; in some instances the ulnar originates above the ordinary level, and the radial and common interosseous form the other limb of the division; occasionally the common interosseous alone arises high up.

Sometimes long, slender vessels, termed *vasa aberrantia*, connect the brachial to the axillary artery or with one or other of the arteries of the forearm. These vessels usually join the radial.

The brachial artery is occasionally crossed in some part of its course by muscular or tendinous slips derived from the coracobrachialis, biceps, brachialis or pronator teres.

Branches:

Profunda brachii.	Superior ulnar collateral.
Nutrient.	Inferior ulnar collateral.
Muscular.	Ulnar and radial.

The arteria profunda brachii (6.70, 71, 74) is a large branch from the medial and posterior aspect of the brachial artery, just below the lower border of teres major. It follows the radial nerve closely, running at first backwards between the long and medial heads of triceps, then along the groove for the radial nerve where it is covered by the lateral head of triceps; here it divides into its terminal branches (6.74). Apart from muscular branches it supplies the following named vessels: nutrient, deltoid, middle collateral and radial collateral arteries. The *nutrient artery* enters the humerus just posterior to the deltoid tuberosity; it may be absent. The *deltoid (ascending) branch* ascends between the lateral and long heads of triceps and anastomoses with the descending branch of the posterior humeral circumflex artery. The *middle collateral (posterior descending) branch*, the larger of the two terminal branches, arises behind the humerus and descends in the substance of the medial head of triceps to the elbow where it anastomoses with the interosseous recurrent artery behind the lateral epicondyle; it often gives a fine vessel which accompanies the nerve to anconeus. The *radial collateral*, the other terminal branch, is the continuation of the profunda brachii. It accompanies the radial nerve through the lateral intermuscular septum and then descends between brachialis and brachioradialis to the front of the lateral epicondyle where it anastomoses with the radial recurrent artery.

The main nutrient artery of the humerus arises usually about the middle of the arm; it enters the nutrient canal near the insertion of coracobrachialis, and is directed downwards.

The superior ulnar collateral artery (6.70, 72, 74), a small vessel, arises a little below the middle of the arm; it frequently springs from the upper part of the arteria profunda brachii. It accompanies the ulnar nerve, pierces the medial intermuscular septum, descends between the medial epicondyle and the olecranon, and ends deep to flexor carpi ulnaris by anastomosing with the posterior ulnar recurrent and inferior collateral arteries. It sometimes sends a branch in front of the medial epicondyle, to anastomose with the anterior ulnar recurrent artery.

The inferior ulnar collateral (supratrochlear) artery (6.70, 74, 75) starts about 5 cm above the elbow. It passes medially between the median nerve and brachialis behind and, piercing the medial intermuscular septum, winds round the back of the humerus between triceps and bone, forming, by its junction with the middle collateral branch of the arteria profunda brachii, an arch above the olecranon fossa. As the vessel lies on brachialis, it

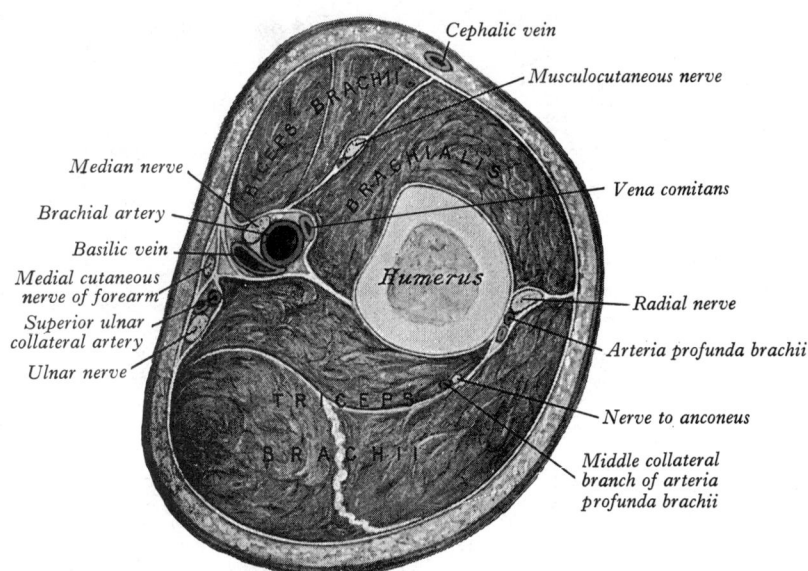

6.72   A transverse section through the right arm, a little below the middle of the shaft of the humerus. Proximal aspect.

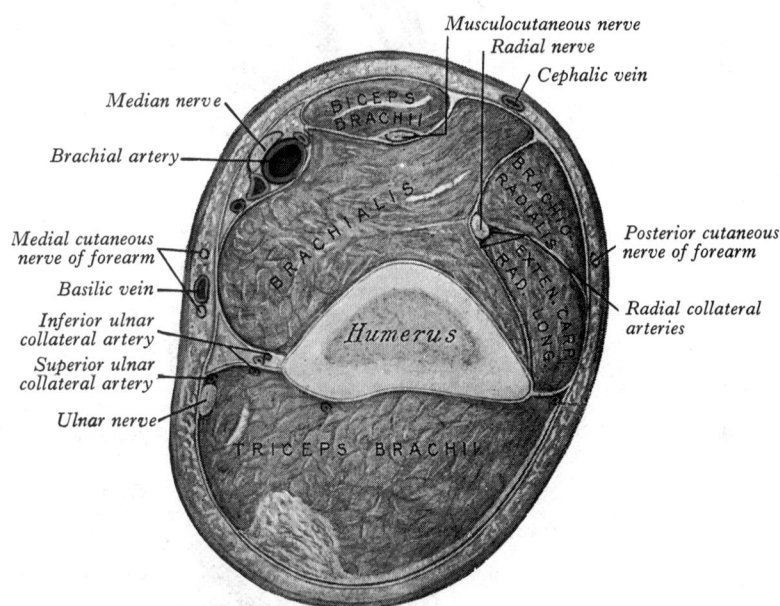

6.73   A transverse section through the right arm, 2 cm above the medial epicondyle of the humerus. Proximal aspect.

gives off branches which descend in front of the medial epicondyle, to anastomose with the anterior ulnar recurrent artery. Behind the medial epicondyle a branch anastomoses with the superior ulnar collateral and posterior ulnar recurrent arteries.

Muscular branches, three or four in number, are distributed to coracobrachialis, biceps and brachialis.

Applied Anatomy. Compression of the brachial artery in case of injury or operation may be effected in almost any part of its course in the arm. If pressure be made in the upper part of the arm, it should be directed laterally, if in the lower part, backwards, because the artery lies medial to the humerus above, and in front below. The most favourable situation is about the middle of the arm, where the artery lies on the tendon of the coracobrachialis medial to the humerus.

The Radial Artery

The radial artery (6.75, 76, 77), though smaller than the ulnar, appears to be the more direct continuation of the brachial trunk. It begins at the division of the brachial, about 1 cm below the bend of

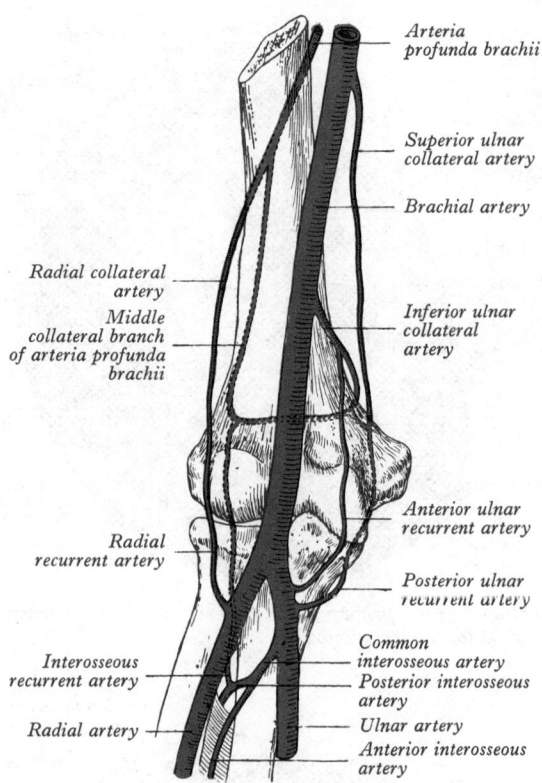

Arteria
profunda brachii

Superior ulnar
collateral artery

Brachial artery

Radial collateral
artery

Middle
collateral branch
of arteria profunda
brachii

Inferior ulnar
collateral
artery

Radial
recurrent artery

Anterior ulnar
recurrent artery

Posterior ulnar
recurrent artery

Interosseous
recurrent artery

Common
interosseous artery

Posterior interosseous
artery

Radial artery

Ulnar artery

Anterior interosseous
artery

6.74 The arterial anastomoses around the elbow joint.

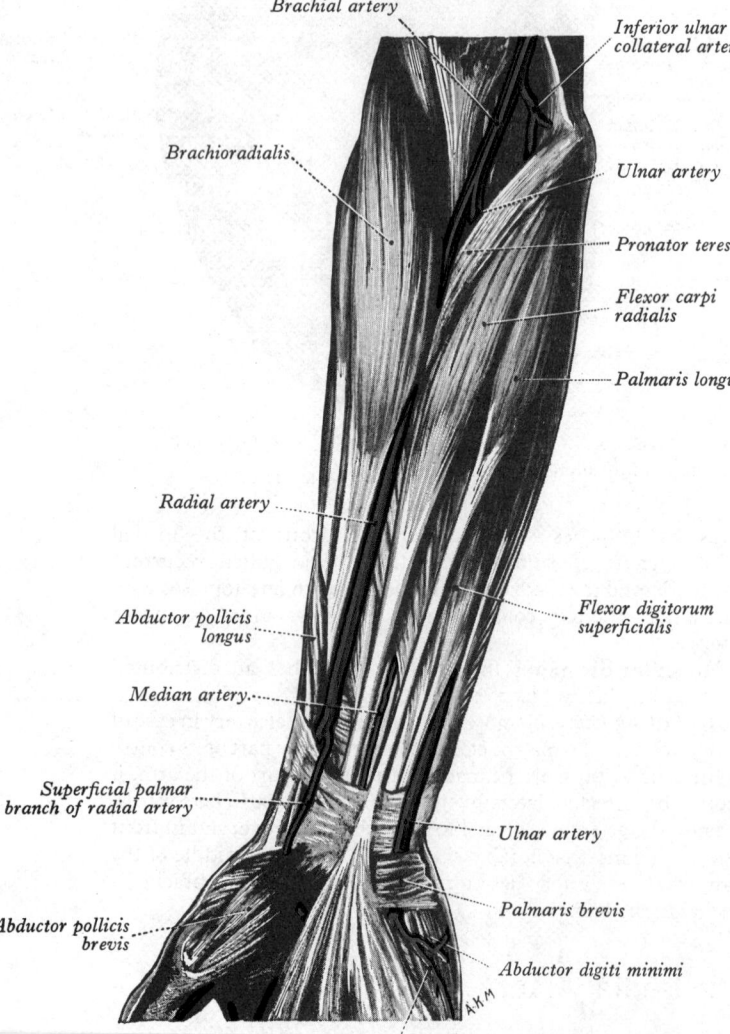

Brachial artery

Inferior ulnar
collateral artery

Brachioradialis

Ulnar artery

Pronator teres

Flexor carpi
radialis

Palmaris longus

Radial artery

Flexor digitorum
superficialis

Abductor pollicis
longus

Median artery

Superficial palmar
branch of radial artery

Ulnar artery

Abductor pollicis
brevis

Palmaris brevis

Abductor digiti minimi

Palmar digital artery

6.75 The right radial and ulnar arteries, superficial dissection.

the elbow, and passes along the radial side of the forearm to the wrist, where its pulsation can readily be felt in the interval between flexor carpi radialis tendon medially and the salient lower part of the anterior border of the radius laterally. It then winds backwards, round the lateral side of the carpus, beneath the tendons of abductor pollicis longus and extensores pollicis brevis et longus, to the proximal end of the space between the first and second metacarpal bones, where it swerves medially between the two heads of the first dorsal interosseous into the palm of the hand; it crosses towards the ulnar side and forms the deep palmar arch by uniting with the deep branch of the ulnar artery. The radial artery is therefore divisible into three parts, one in the forearm, a second at the wrist, and a third in the hand.

Relations:

In the forearm (**6**.75, 76, 77), the radial artery extends from the medial side of the neck of the radius to the front part of its styloid process, being medial to the shaft of the bone above, and in front of it below. Its upper part is overlapped by the fleshy belly of the brachioradialis; the rest of the artery is covered only with the skin, and the superficial and deep fasciae. It lies successively upon the tendon of biceps, supinator, the insertion of pronator teres, the radial origin of flexor digitorum superficialis, flexor pollicis longus, pronator quadratus, and the lower end of the radius. The pronator teres is medial, and brachioradialis lateral, to the upper one-third of the artery; the tendon of flexor carpi radialis is medial, and that of brachioradialis lateral to its lower two-thirds. The superficial branch of the radial nerve is close to the lateral side of the middle third of the vessel; and some filaments of the lateral cutaneous nerve of the forearm run along the lower part of the artery as it winds round the wrist. Throughout its course the vessel is accompanied by a pair of venae comitantes. The part of the radial artery which lies in front of the lower end of the radius and on the lateral side of the tendon of the flexor carpi radialis is used clinically for observations on the pulse.

At the wrist (**6**.78, 80), the radial artery passes on to the dorsal aspect of the carpus between the lateral ligament of the wrist and the tendons of the abductor pollicis longus and extensor pollicis brevis. It then crosses the scaphoid bone and the trapezium, where its pulsation is obvious, and, before disappearing between the heads of the first dorsal interosseous, it is crossed by the tendon of the extensor pollicis longus. In the interval between the two extensores pollicis it is crossed by the origin of the cephalic vein, and by the digital branches of the radial nerve which go to the thumb and index finger.

In the hand (**6**.77), the radial artery, having travelled through the proximal end of the first interosseous space between the heads of the first dorsal interosseous, runs transversely across the palm; it passes at first deep to the oblique head of adductor pollicis and then between its oblique and transverse heads, or through the transverse head. At the base of the fifth metacarpal it anastomoses with the deep branch of the ulnar artery, to complete the *deep palmar arch* (**6**.77).

Variations. Sometimes the origin of the radial artery is higher than usual; it then branches more often from the axillary or upper part of the brachial artery than from the lower part of the latter. In the forearm it is sometimes superficial to the deep fascia instead of beneath it; in turning round the wrist, it is occasionally superficial instead of deep to the extensor tendons of the thumb.

Branches:

The radial recurrent artery (**6**.74, 77) arises immediately below the elbow. It passes between the superficial and deep branches of the radial nerve and ascends behind brachioradialis, in front of supinator and brachialis; it supplies these muscles and the elbow joint, and anastomoses with the radial collateral branch of the arteria profunda brachii.

The muscular branches are distributed to the muscles on the radial side of the forearm.

The palmar carpal branch (**6**.77) is a small vessel which arises near the lower border of pronator quadratus, and, running medially across the palmar surface of the carpus, anastomoses behind the flexor tendons with the palmar carpal branch of the ulnar artery. This anastomosis is joined by a branch from the anterior interosseous artery, and by recurrent branches from the deep palmar arch, thus forming a *palmar carpal arch*, which supplies the articulations of the wrist and carpus and the bones involved.

The superficial palmar branch (6.81) arises from the radial artery where this vessel is about to wind round the lateral side of the wrist. It passes through, occasionally over, the muscles of the thenar eminence, which it supplies, and sometimes anastomoses with the terminal part of the ulnar artery, completing the *superficial palmar arch*.

The dorsal carpal branch (6.80) is small and arises deep to the extensor tendons of the thumb; running medially across the dorsal carpal surface under these tendons, it anastomoses with the dorsal carpal branch of the ulnar artery, and with the anterior and posterior interosseous arteries, to form a *dorsal carpal arch*. The dorsal and palmar carpal arches are both close to bone and supply the lower, epiphysial parts of the radius and ulna. Arising from the dorsal arch, three slender *dorsal metacarpal arteries* descend on the second, third and fourth dorsal interosseous muscles and bifurcate into dorsal digital branches for the adjacent sides of the index, middle, ring and little fingers; they anastomose with the palmar digital branches of the superficial palmar arch; near their origins they anastomose with the deep palmar arch by the *proximal perforating arteries*, and, near their points of bifurcation, with the palmar digital vessels of the superficial palmar arch by the *distal perforating arteries*.

The first dorsal metacarpal artery (6.80) arises just before the radial artery passes between the two heads of the first dorsal interosseous, and divides almost immediately into two branches, which supply the adjacent sides of the thumb and index finger; the radial side of the thumb receives a branch directly from the radial artery (*vide infra*).

The arteria princeps pollicis (6.77) arises from the radial artery, as it turns medially into the palm; it descends on the palmar aspect of the first metacarpal bone under the oblique head of the adductor pollicis and lateral to the first palmar interosseous muscle. At the base of the proximal phalanx, where it lies deep to the tendon of the flexor pollicis longus, it divides into two branches. These make their appearance between the medial and lateral insertions of the oblique head of adductor pollicis, and run along the sides of the thumb, forming, on the palmar surface of the distal phalanx, an arch from which branches are distributed to the skin and subcutaneous tissue of the thumb. The arteria princeps pollicis is the usual source of supply of the nutrient artery to the first metacarpal.

The arteria radialis indicis (6.77, 81), which frequently arises from the proximal part of the arteria princeps pollicis, descends between the first dorsal interosseous and transverse head of the adductor pollicis, running along the lateral side of the index to its extremity; it anastomoses with the digital artery supplying the medial side of the finger. At the distal border of the transverse head of the adductor pollicis this vessel anastomoses with the arteria princeps pollicis, and gives a communicating branch to the superficial palmar arch.

The arteriae princeps pollicis and radialis indicis may start in common as the *first palmar metacarpal artery*.

The Deep Palmar Arch (6.77)

This is formed by anastomosis of the terminal part of the radial artery with a deep palmar branch of the ulnar. It lies on the proximal ends of the metacarpal bones and the interossei, and is covered by the oblique head of adductor pollicis, flexor tendons of the fingers and lumbricals. In its concavity, but running towards the lateral side of the hand, is the deep branch of the ulnar nerve. Some variation occurs in the formation of this anastomotic arch, which is sometimes incomplete—in 6 of 200 arches, according to Coleman and Anson (1961). The variation is chiefly in the size of the contribution from the ulnar artery.

Surface Anatomy. The deep palmar arch can be represented by a horizontal line about 4 cm long drawn from a point just distal to the hook of the hamate bone; it lies about 1 cm proximal to the superficial arch (6.82).

The branches of the deep palmar arch are palmar metacarpal, perforating and recurrent.

The three palmar metacarpal arteries (6.77), arise from the convexity of the deep palmar arch; they run distally upon the

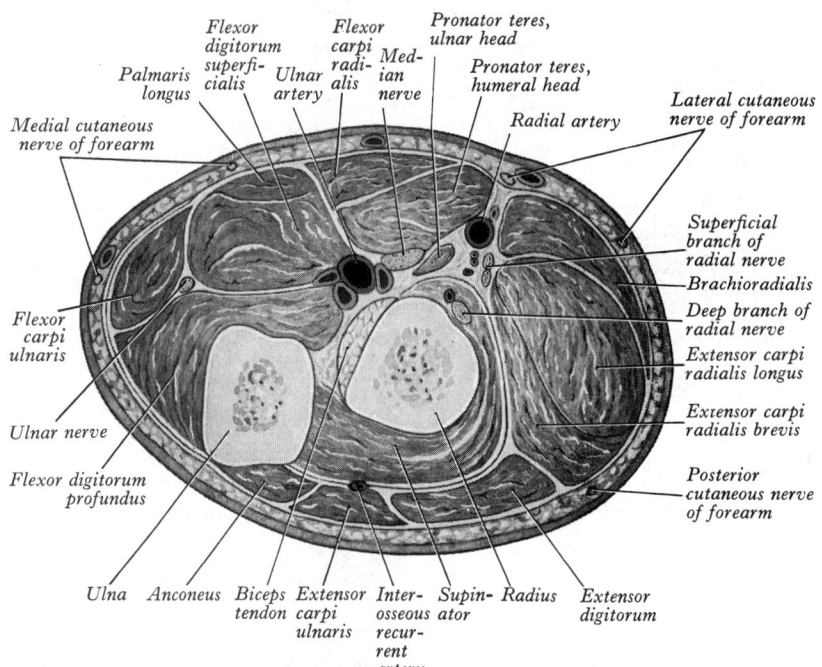

6.76 A transverse section through the forearm at the level of the radial tuberosity.

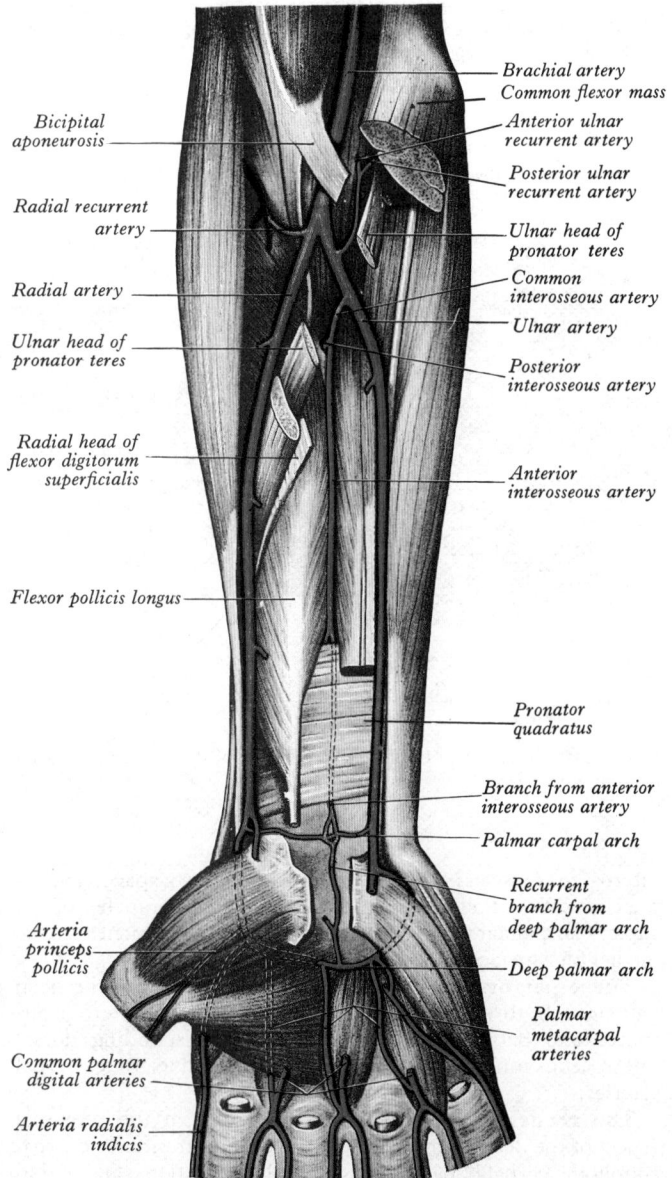

6.77 The arteries of the right forearm and hand, deep dissection.

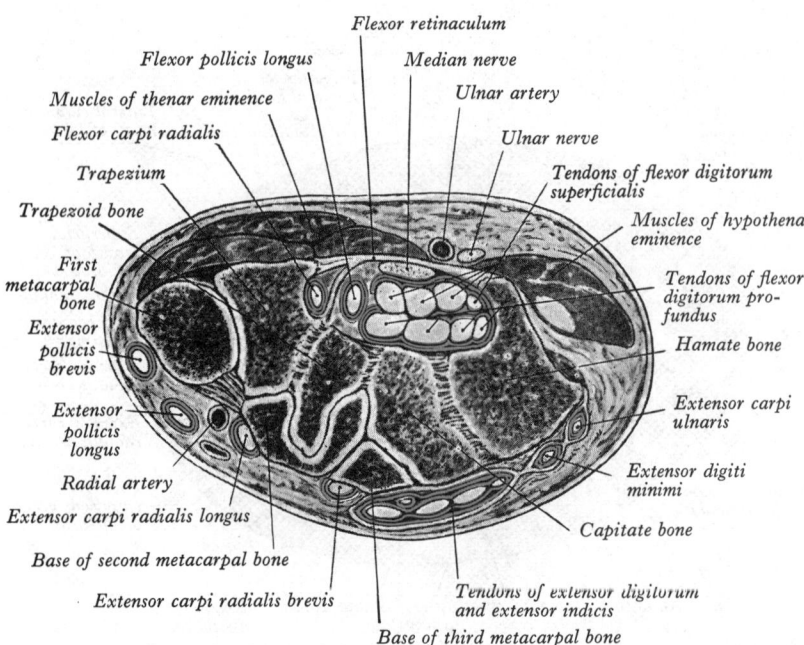

Flexor retinaculum
Flexor pollicis longus
Median nerve
Muscles of thenar eminence
Ulnar artery
Flexor carpi radialis
Ulnar nerve
Trapezium
Tendons of flexor digitorum superficialis
Trapezoid bone
Muscles of hypothenar eminence
First metacarpal bone
Tendons of flexor digitorum profundus
Extensor pollicis brevis
Hamate bone
Extensor pollicis longus
Extensor carpi ulnaris
Radial artery
Extensor digiti minimi
Extensor carpi radialis longus
Capitate bone
Base of second metacarpal bone
Tendons of extensor digitorum and extensor indicis
Extensor carpi radialis brevis
Base of third metacarpal bone

6.78 A transverse section through the left wrist, superior aspect. The section is slightly oblique and divides the distal row of the carpus, and the bases of the first, second and third metacarpal bones. The arrangement of the tendons of the flexors of the fingers shown in the figure represents the actual condition in the specimen. Observe that the carpometacarpal joint of the thumb is separate from the joint between the trapezium and the base of the second metacarpal bone.

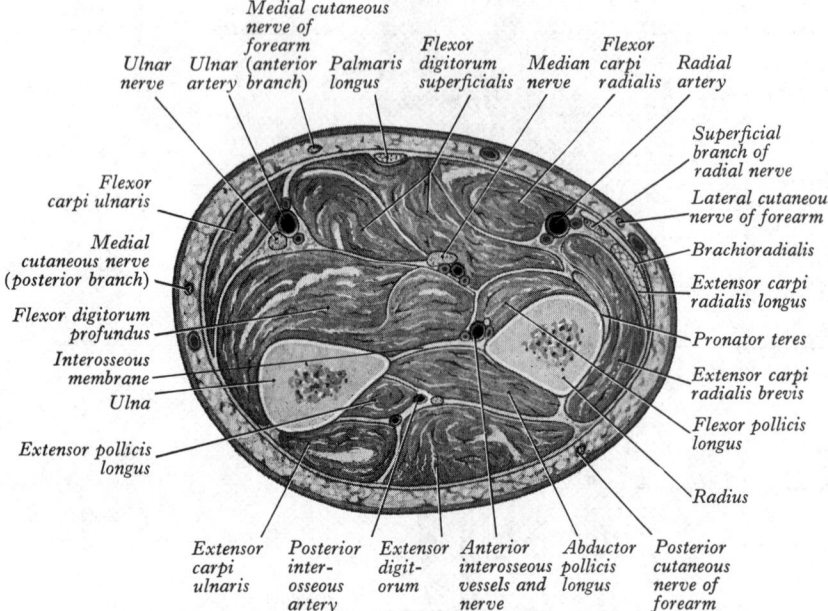

Medial cutaneous nerve of forearm
Ulnar nerve
Ulnar artery (anterior branch)
Palmaris longus
Flexor digitorum superficialis
Median nerve
Flexor carpi radialis
Radial artery
Superficial branch of radial nerve
Flexor carpi ulnaris
Lateral cutaneous nerve of forearm
Brachioradialis
Medial cutaneous nerve (posterior branch)
Extensor carpi radialis longus
Flexor digitorum profundus
Pronator teres
Interosseous membrane
Extensor carpi radialis brevis
Ulna
Flexor pollicis longus
Extensor pollicis longus
Radius
Extensor carpi ulnaris
Posterior interosseous artery
Extensor digitorum
Anterior interosseous vessels and nerve
Abductor pollicis longus
Posterior cutaneous nerve of forearm

6.79 A transverse section through the middle of the left forearm, seen from proximal aspect.

interosseus muscles of the second, third and fourth spaces, and, at the clefts of the fingers, join the common digital branches of the superficial palmar arch. They supply nutrient branches to the medial four metacarpals.

Three perforating branches pass dorsally from the deep palmar arch, through the second, third and fourth interosseous spaces and between the heads of the corresponding dorsal interosseous muscles, to anastomose with the dorsal metacarpal arteries.

The recurrent branches (**6**.77) arise from the proximal aspect of the deep palmar arch; they ascend in front of the wrist, supply the carpal bones and intercarpal articulations, and end in the palmar carpal arch.

The Ulnar Artery

The ulnar artery (**6**.75–81), the larger of the two terminal branches of the brachial artery, begins level with the neck of the radius, about 1 cm below the bend of the elbow, and, passing downwards and medially, reaches the medial side of the forearm at a point about midway between the elbow and the wrist. It then runs straight to the medial side of the wrist, and crosses the flexor retinaculum lateral to the ulnar nerve and pisiform bone. Immediately beyond this bone it gives off a deep branch, and is then continued across the palm as the superficial palmar arch.

Relations.

In the forearm the *upper half* of the vessel (**6**.75, 76, 79) passes obliquely deep to pronator teres, flexor carpi radialis, palmaris longus and flexor digitorum superficialis to the medial side of the forearm, where it is overlapped in its middle third by the flexor carpi ulnaris; it lies in front of brachialis and the flexor digitorum profundus. Below the elbow the median nerve is medial to the artery for about 2·5 cm and then crosses it, but is separated from it by the ulnar head of pronator teres. The *lower half* of the vessel (**6**.75, 81) lies upon flexor digitorum profundus, covered by the skin, superficial and deep fasciae, and is between flexor carpi ulnaris and flexor digitorum superficialis.

The ulnar artery is accompanied by two venae comitantes, the ulnar nerve is adjacent and medial to the lower two-thirds of the artery, and the palmar cutaneous branch of this nerve descends on the lower part of the vessel to the palm of the hand.

At the wrist (**6**.77, 78, 81) the ulnar artery is covered by skin and fasciae and palmaris brevis, and lies between the superficial and the main part of the flexor retinaculum (p. 583). The ulnar nerve and the pisiform bone are on its medial side.

Surface Anatomy. The vessel commences in the midline of the limb opposite the neck of the radius. From this point the upper part passes to a point at the junction of the upper and middle thirds of a line drawn from the base of the medial epicondyle of the humerus to the lateral edge of the pisiform bone; this line represents the lower two-thirds of the ulnar artery.

Variations. The ulnar artery varies in its origin; it sometimes arises above the elbow, the brachial being more often the source of origin than the axillary. When its origin is normal, the course of the vessel is rarely changed. When the artery arises high up, it is usually superficial to the flexor muscles in the forearm, lying commonly beneath the deep fascia, more rarely between it and skin; the brachial artery then supplies the common interosseous artery, and the latter, the anterior and posterior ulnar recurrent arteries. Occasionally it is subcutaneous in the upper part of the forearm, and subfascial in the lower part.

Branches. The ulnar artery supplies the muscles on the medial side of the forearm and hand, the common flexor synovial sheath and the ulnar nerve (Blunt 1959). It supplies the following named branches:

The anterior ulnar recurrent artery (**6**.74, 77), a small branch, arises immediately below the elbow joint, runs upwards between the brachialis and pronator teres, supplies these muscles, and, in front of the medial epicondyle, anastomoses with the inferior ulnar collateral artery.

The posterior ulnar recurrent artery (**6**.74, 77) is larger, and arises lower than the anterior artery. It passes backwards and medially between flexor digitorum profundus and flexor digitorum superficialis, and ascends behind the medial epicondyle of the humerus. In the interval between this process and the olecranon, it lies deep to flexor carpi ulnaris, and ascends between the two heads of this muscle, in contact with the ulnar nerve; it supplies the neighbouring muscles, bone and elbow joint, and anastomoses with the two ulnar collateral and the interosseous recurrent arteries (**6**.74).

The common interosseous artery (**6**.77), about a centimetre long, arises immediately below the tuberosity of the radius, and, passing backwards to the upper border of the interosseous membrane, divides into two branches, the *anterior* and *posterior interosseous arteries.*

The anterior interosseous artery (**6**.77, 80) descends on the anterior surface of the interosseous membrane, accompanied by the anterior interosseous branch of the median nerve, and is

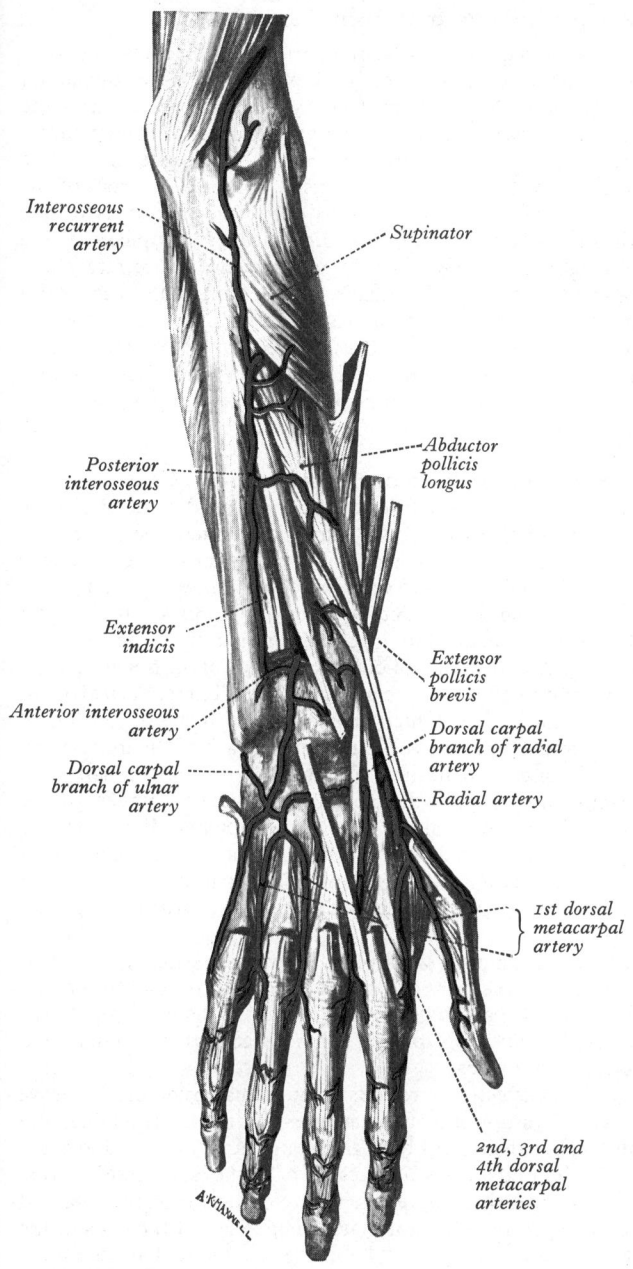

6.80   The arteries of the posterior surface of the right forearm and hand.

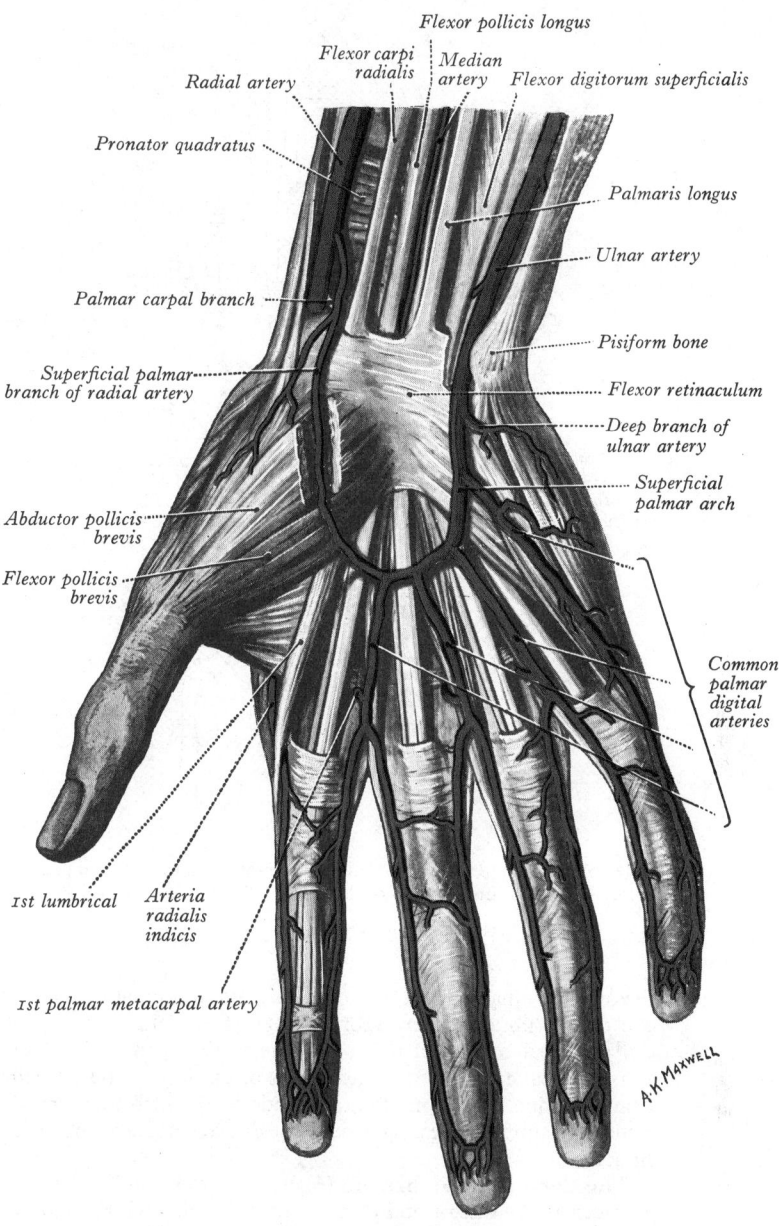

6.81   The superficial palmar arch and its branches. A part of the abductor pollicis brevis has been excised to expose the superficial palmar branch of the radial artery.

overlapped by the contiguous margins of the flexor digitorum profundus and flexor pollicis longus; it gives off *muscular* branches, and the *nutrient* arteries of the radius and ulna. As it lies on the interosseous membrane, the artery gives off branches which pierce the interosseous membrane and pass to the deep extensor muscles in the forearm. At the upper border of the pronator quadratus it pierces the interosseous membrane and reaches the back of the forearm, where it anastomoses with the posterior interosseous artery and descends on the back of the wrist in the compartment of the extensor retinaculum containing the tendons of extensor digitorum and extensor indicis to join the dorsal carpal arch. Before the artery pierces the interosseous membrane, a branch descends behind pronator quadratus to join the anterior carpal arch. The *median artery*, a long, slender branch, starts from the beginning of the anterior interosseous artery; it accompanies and supplies the median nerve. It often arises from the common interosseous artery. Sometimes it is much enlarged and runs with the nerve into the palm of the hand (p. 192), where it may join the superficial palmar arch or end as one or two of the palmar digital arteries.

The posterior interosseous artery (**6**.77, 80), usually much smaller than the anterior, passes backwards between the oblique

cord and the upper border of the interosseous membrane. It appears on the back of the forearm between supinator and abductor pollicis longus, and descends between the superficial and deep layers of muscles, to which it distributes branches. As it lies upon abductor pollicis longus, it is accompanied by the deep branch of the radial nerve. At the lower part of the forearm it becomes very small and anastomoses with the termination of the anterior interosseous artery, and with the dorsal carpal arch. It gives off, near its origin, the *interosseous recurrent artery*, which ascends to the interval between the lateral epicondyle and olecranon, on or through the fibres of the supinator, but deep to anconeus, and anastomoses with the middle collateral branch of the arteria profunda brachii, and with the posterior ulnar recurrent and the ulnar collateral arteries.

The muscular branches of the ulnar artery are distributed to the muscles in the ulnar region of the forearm.

The palmar carpal branch (**6**.77) is a small vessel which crosses the front of the carpus behind the tendons of flexor digitorum profundus; it anastomoses with the palmar carpal branch of the radial artery to form the palmar carpal arch (p. 704).

The dorsal carpal branch (**6**.80) arises immediately above the pisiform bone, and winds backwards deep to the tendon of

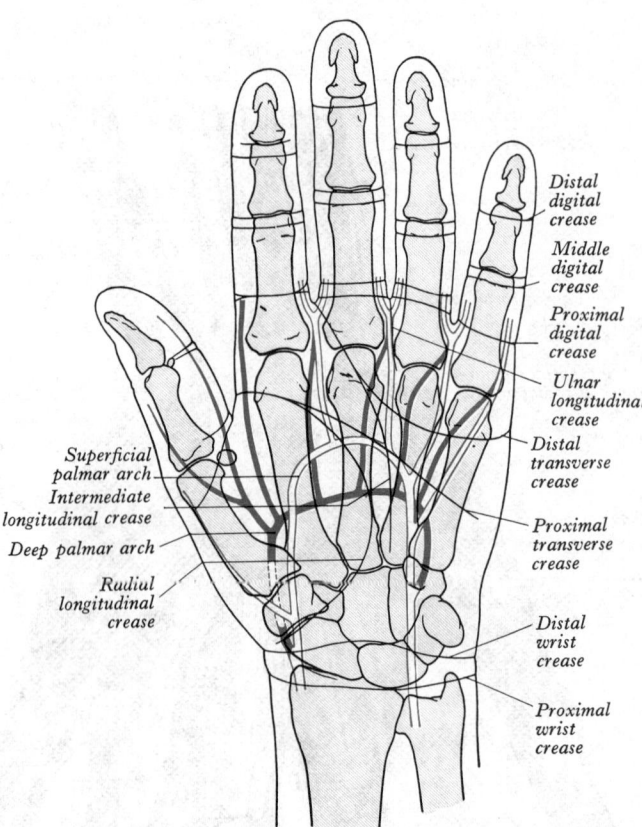

Distal
digital
crease

Middle
digital
crease

Proximal
digital
crease

Ulnar
longitudinal
crease

Distal
transverse
crease

Proximal
transverse
crease

Distal
wrist
crease

Proximal
wrist
crease

Superficial
palmar arch
Intermediate
longitudinal crease

Deep palmar arch

Radial
longitudinal
crease

6.82 A diagram to show the relation of the skin flexure lines and palmar arterial arches to the bones of the left hand.

flexor carpi ulnaris; it passes laterally across the dorsal surface of the carpus under the extensor tendons, anastomoses with the dorsal carpal branch of the radial artery, thus completing the dorsal carpal arch (p. 705). Near to its origin it gives off a small branch which runs along the ulnar side of the fifth metacarpal bone, and supplies the ulnar side of the dorsal surface of the little finger.

The deep palmar branch (**6.**77, 81), often double, passes between the abductor and flexor digiti minimi, and through or deep to the origin of the opponens digiti minimi; it anastomoses with the radial artery, thus completing the deep palmar arch; it is accompanied by the deep branch of the ulnar nerve.

The Superficial Palmar Arch (**6.**81, 82)

This is formed mainly by the ulnar artery, which enters the hand with the ulnar nerve in front of the flexor retinaculum, and lateral to the pisiform bone. It then passes medial to the hook of the hamate and crosses the palm, forming the superficial palmar arch, which is convex towards the fingers, and is at the level of a line drawn across the hand from the distal border of the extended thumb. In about one-third of subjects the arch is formed by the ulnar artery alone; in a further third it is completed by the superficial palmar branch of the radial artery. More rarely it is completed by the arteria radialis indicis, a branch of the arteria princeps pollicis or the median artery. (For variations *see* Coleman and Anson 1961.) It is covered by the palmaris brevis and the palmar aponeurosis, and lies on the flexor digiti minimi, the branches of the median nerve, which it supplies, the flexor tendons and the lumbricals.

Branches:

Three common palmar digital arteries (**6.**81) arise from the convexity of the superficial palmar arch and proceed distally on the second, third and fourth lumbricals. Each is joined by the corresponding palmar metacarpal artery from the deep palmar arch, and then divides into a pair of *proper palmar digital arteries*, which run along the contiguous sides of the index, middle, ring and little fingers, dorsal to the corresponding digital nerves; they anastomose freely in the subcutaneous tissue of the finger tips and by smaller branches near the interphalangeal joints. Each gives off two dorsal branches, which anastomose with the dorsal digital arteries, and supply the soft parts on the back of the middle and distal phalanges, including the matrix of the finger nail. The *palmar digital artery* for the medial side of the little finger springs from the arch under palmaris brevis. The palmar digital arteries supply nutrient branches to the phalanges and to the metacarpophalangeal and interphalangeal joints. The palmar digital arteries are the main supply to the digits, the dorsal digital arteries (p. 705) being very small.

A free anastomosis takes place between the radial and ulnar arteries, (*a*) on the front and back of the wrist through the palmar and dorsal carpal arches, and (*b*) in the hand through the superficial and deep palmar arches, and their digital and metacarpal branches.

Applied Anatomy. In wounds of the palmar arches it may prove useless to ligature one of the arteries of the forearm alone, and simultaneous ligature of both radial and ulnar arteries above the wrist is often unsuccessful, because of of the anastomotic carpal networks. Therefore, upon the failure of local pressure to arrest haemorrhage in the hand it may be expedient, if the ends of the bleeding vessels cannot readily be exposed and tied, to compress, or even ligate, the brachial artery.

THE ARTERIES OF THE TRUNK

The Thoracic Aorta

The thoracic aorta (**6.**83, **8.**88) is that part of the *descending aorta* which is confined to the posterior mediastinum. It begins at the lower border of the fourth thoracic vertebra, where it is continuous with the aortic arch, and ends in front of the lower border of the twelfth thoracic vertebra at the aortic opening in the diaphragm. At its origin it is situated on the left of the vertebral column; as it descends it approaches the median plane, and at its termination lies in front of the column.

Relations. It is in relation, *anteriorly*, from above downwards, with the root of the left lung, the pericardium, separating it from the left atrium, the oesophagus and diaphragm; *posteriorly*, with the vertebral column and the hemiazygos veins; on the *right side*, with the azygos vein and thoracic duct and, in the lower part of its course, with the right pleura and lung; on the *left side*, with the left pleura and lung. The oesophagus, with its accompanying plexus of nerves, lies on the *right side* of the aorta above; but in the lower

part of the thorax it is placed in front of the vessel, and, close to the diaphragm, is placed anteriorly and to its left side.

Surface Anatomy. The vessel can be represented by a band 2·5 cm broad extending from the sternal end of the second left costal cartilage to a point in the median plane about 2 cm above the transpyloric plane (p. 1319).

Branches:

The thoracic aorta gives off *visceral branches* to the pericardium, lungs, bronchi, and oesophagus, and *parietal branches* to the walls of the thoracic cavity.

The pericardial branches consist of a few small vessels which are distributed to the posterior surface of the pericardium.

The bronchial arteries vary in number, size and origin. There is as a rule one *right bronchial artery*, which arises from the third posterior intercostal artery, or from the upper left bronchial artery. It runs on the posterior surface of the right bronchus, dividing and subdividing along the bronchial tubes, supplying them, the areolar tissue of the lung and the bronchopulmonary

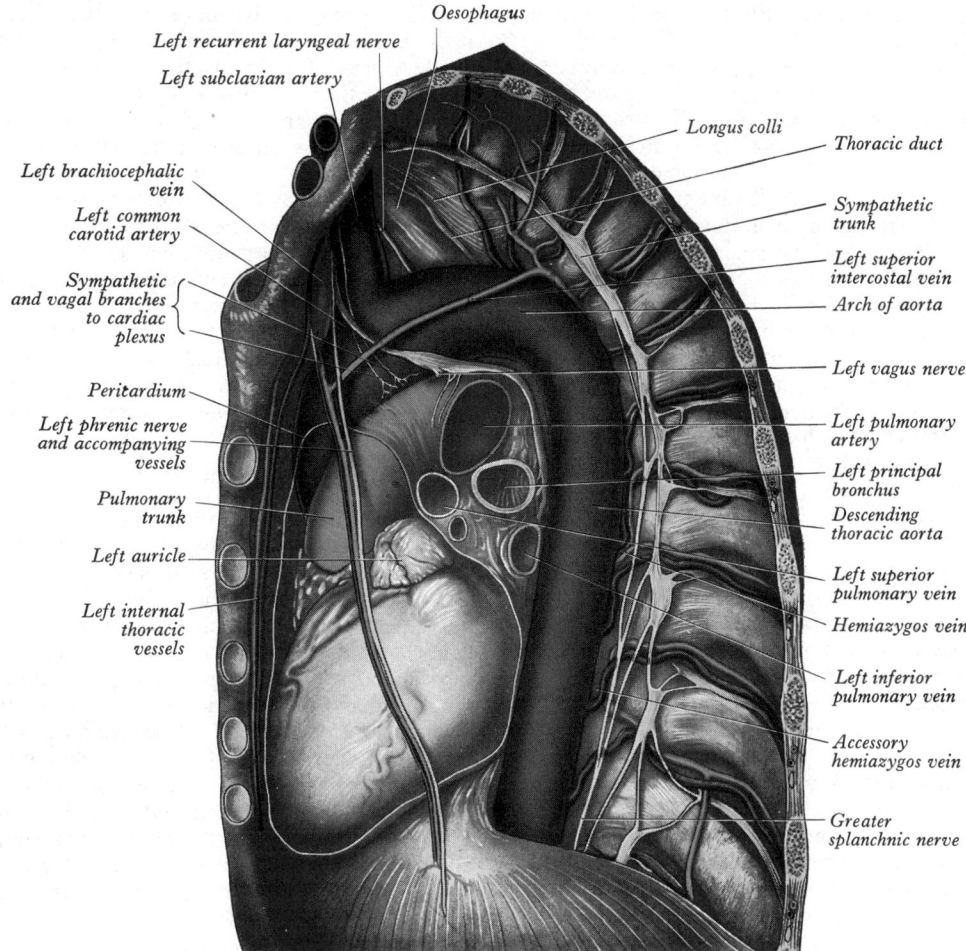

Oesophagus

Left recurrent laryngeal nerve

Left subclavian artery

Left brachiocephalic vein

Left common carotid artery

Sympathetic and vagal branches to cardiac plexus

Pericardium

Left phrenic nerve and accompanying vessels

Pulmonary trunk

Left auricle

Left internal thoracic vessels

Longus colli

Thoracic duct

Sympathetic trunk

Left superior intercostal vein

Arch of aorta

Left vagus nerve

Left pulmonary artery

Left principal bronchus

Descending thoracic aorta

Left superior pulmonary vein

Hemiazygos vein

Left inferior pulmonary vein

Accessory hemiazygos vein

Greater splanchnic nerve

6.83   The left aspect of the mediastinum. The left lung and pleura have been removed, and an extensive opening has been made into the pericardial sac to expose the heart.

lymph nodes; it also sends branches to the pericardium and the oesophagus. The *left bronchial arteries*, usually two in number, arise from the thoracic aorta, the upper opposite the fifth thoracic vertebra and the lower just below the left bronchus. They run on the posterior surface of the left bronchus and have a distribution similar to that of the right bronchial artery. (See also p. 1024.)

Cauldwell *et al.* (1948) found the above arrangement in 40 per cent of 150 cadavers. The next most frequent arrangements (at about 20 per cent each) were two left and two right bronchial arteries, or one on each side, all as direct branches of the descending thoracic aorta arising near the levels of the third and fourth intercostal arteries. In about 10 per cent of instances there were *one* left and *two* right bronchial arteries. Other, more complex variations consisted chiefly of more numerous vessels arising directly from the descending aorta. Very rarely one or other of the bronchial arteries may arise from the aortic arch.

The oesophageal arteries, four or five in number, arise from the front of the aorta, and pass obliquely downwards to the oesophagus; on this they form a *vascular chain*, which anastomoses above with the oesophageal branches of the inferior thyroid arteries, and below with ascending branches from the left phrenic and left gastric arteries. (It is often forgotten that the aortic oesophageal arteries are persisting ventral splanchnic arteries to the foregut, in addition to the subdiaphragmatic coeliac, superior and inferior mesenteric arteries).

The mediastinal branches are numerous small vessels which supply the lymph nodes and the areolar tissue in the posterior mediastinum.

The phrenic branches are small, and arise from the lower part of the thoracic aorta; they are distributed to the posterior part of the upper surface of the diaphragm, and anastomose with the musculophrenic and pericardiacophrenic arteries.

The Posterior Intercostal Arteries

There are usually nine pairs of posterior intercostal arteries derived from the thoracic aorta. They arise from the back of the vessel, and are distributed to the lower nine intercostal spaces, the first and second spaces being supplied by the superior intercostal artery (p. 699). The *right* posterior intercostal arteries are longer than the left, on account of the position of the aorta on the left side of the vertebral column; they cross the bodies of the vertebrae behind the oesophagus, thoracic duct and azygos vein, and are covered by the right lung and pleura. The *left* posterior intercostal arteries run backwards on the sides of the vertebrae and are in contact with the left lung and pleura; the upper two vessels are crossed by the left superior intercostal vein, the lower vessels by the hemiazygos and accessory hemiazygos veins. The further course of the posterior intercostal arteries is practically the same on both sides. Opposite the heads of the ribs the sympathetic trunk passes downwards in front of them, and the splanchnic nerves also descend in front of the lower arteries.

Each artery (**6**.83) crosses its intercostal space obliquely towards the angle of the rib above, and thence is continued forward in the costal groove. It is placed at first between the pleura and the internal (posterior) intercostal membrane, as far as the angle of the rib; from this onward it runs between the internal intercostal and the intercostalis intimus (p. 547), and anastomoses in front with the anterior intercostal branch of the internal thoracic or musculophrenic artery. Each artery is accompanied by a vein and a nerve, the former above and the latter below the artery, except in the upper spaces where the nerve is at first above the artery. The third posterior intercostal artery anastomoses with the superior intercostal artery, and may form the chief supply of the second intercostal space. The lower two posterior intercostal arteries are continued anteriorly from the intercostal spaces into

709

the abdominal wall and anastomose with the subcostal, superior epigastric and lumbar arteries.

Each posterior intercostal artery gives off a number of branches.

The dorsal branch runs backwards through a space which is bounded above and below by the necks of the ribs, medially by the body of a vertebra and laterally by a superior costotransverse ligament. It gives off a *spinal branch* which enters the vertebral canal through the intervertebral foramen, and is distributed to the vertebrae, spinal cord (*see* p. 896 and 7.54) and its membranes, anastomosing with the spinal arteries above and below and with the opposite artery. The dorsal branch then courses over the transverse process with the dorsal ramus of the corresponding thoracic spinal nerve, supplying offshoots to the muscles of the back, and a cutaneous twig which accompanies the cutaneous branch of the dorsal ramus of the nerve.

The collateral intercostal branch leaves the posterior intercostal artery near the angle of the rib, and descends to the upper border of the rib below, along which it courses to anastomose with an anterior intercostal branch of the internal thoracic or musculophrenic artery.

Muscular branches are given to the intercostal and pectoral muscles and to the serratus anterior; they anastomose with the superior and lateral thoracic branches of the axillary artery.

The lateral cutaneous branches accompany the lateral cutaneous branches of the thoracic nerves.

Mammary branches are given off by the vessels in the second, third and fourth spaces; they increase considerably in size during lactation.

The right bronchial artery may arise from the right third posterior intercostal artery (p. 109).

Applied Anatomy. In puncturing the thorax the needle should not be introduced nearer the posterior median line than the angle of the rib, for the artery crosses the space medial to this point. In the lateral portion of the chest, where the puncture is usually made, the main intercostal artery lies at the upper part of the intercostal space, and therefore the puncture should be made just above the rib forming the lower boundary of the space.

The subcostal arteries, the last pair of branches from the thoracic aorta, are in series with the posterior intercostal arteries, but are named *subcostal* because they are situated below the twelfth rib and are, of course, not 'intercostal'. Each artery runs laterally over the body of the twelfth thoracic vertebra, and behind the splanchnic nerves, the ganglionated trunk of the sympathetic, the pleura and diaphragm. The right artery also passes behind the thoracic duct and the azygos vein, and the left behind the accessory hemiazygos vein. Each subcostal artery then enters the abdomen under the lateral arcuate ligament, and courses with the twelfth thoracic nerve along the lower border of the twelfth rib, anterior to quadratus lumborum and posterior to the kidney. The right artery runs behind the ascending, and the left behind the descending, colon. Each then pierces the aponeurosis of origin of the transversus abdominis, and, passing forward between this muscle and the obliquus internus, anastomoses with the superior epigastric, lower posterior intercostal and lumbar arteries. Each subcostal artery gives off a dorsal branch, which is distributed like the dorsal branch of a posterior intercostal artery.

A small *aberrant artery* sometimes branches from the right side of the thoracic aorta near the origin of the right bronchial. It passes upwards and to the right behind the trachea and the oesophagus, and may anastomose with the right superior intercostal artery. It represents the remains of the right dorsal aorta (p. 187), and occasionally is enlarged to form the first part of the right subclavian artery (p. 693).

Variations. The lumen of the aorta is occasionally found to be partly or completely obliterated, either above (preductal or infantile type), or opposite to or just beyond (postductal or adult type), the site of entry of the ductus arteriosus. The condition is congenital and is known as *coarctation* of the aorta. The ductus arteriosus may remain patent after birth, but rarely plays any compensatory role, for in most cases the systemic blood pressure is much higher than that in the pulmonary circulation.

In the preductal type the length of the coarctation is very variable and may involve the origin of the left subclavian and even of the brachiocephalic artery, thereby providing little scope for the development of an effective collateral circulation to regions supplied by the aorta distal to the lesion. Therefore many cases of this type are incompatible with more than a few months of life. Furthermore, the surgical problems posed are considerable. Sometimes, however, the coarctation in the preductual type is restricted to that part of the aorta between the brachiocephalic and left subclavian arteries. In this instance the blood and pulse pressures in the left arm are considerably lower than those in the right, and a collateral circulation may develop through branches of the brachiocephalic artery.

The postductal type of coarctation has been attributed to an abnormal extension of ductal tissue into the aortic wall, giving rise to a simultaneous stenosis of both vessels as it contracts after birth. This form of coarctation is compatible with many years of normal life, and leads to the establishment of an extensive collateral circulation to carry blood to the aorta immediately below the stenosis.

The extreme vascularity of the whole thoracic wall is an important feature, for many arteries which arise indirectly from the aorta, above its obliterated portion, anastomose with vessels connected with the aorta below the obliteration and the connecting channels become greatly enlarged. In the anterior thoracic wall the thoracoacromial, lateral thoracic and the subscapular arteries from the axillary, the suprascapular from the subclavian and the first and second posterior intercostal arteries from the costocervical trunk anastomose with the third, and lower, posterior intercostal arteries, while the internal thoracic artery and its terminal branches anastomose with the lower posterior intercostal arteries and the inferior epigastric arteries. The posterior intercostal arteries are always especially involved. Enlargement of their dorsal branches may eventually cause notching of the inferior margins of the ribs with which they are in contact. The X-ray shadow of the enlarged left subclavian artery is increased. The enlargement of the scapular vessels and anastomoses may lead to pulsation in the interscapular region ('pulsating scapula').

The Abdominal Aorta

The abdominal aorta (**6**.84, 92) begins in the median plane at the aortic hiatus of the diaphragm, in front of the lower border of the last thoracic vertebral body and it descends in front of the vertebral column, ending at the level of the body of the fourth lumbar vertebra, a little to the left of the midline, by dividing into the two common iliac arteries. It diminishes rapidly in size, because such large branches arise from it. Recent measurements of casts of the abdominal aorta in 100 individuals, ranging from 16 to 70 years, showed an increase with age. In males the superior and inferior extremities measured 9·8 to 14·1 mm and 8·1 to 14·6 mm; in females the luminal diameters were 9·7 to 15·7 mm and 9·1 to 14·6 mm (Aleksandrowicz *et al.* 1974). These values contrast with those derived from radiological observations of 61 adults (17–41 years) contributed by Leithner *et al.* (1975), who recorded diameters of 26 mm and 19 mm (averages) for the extremities of the abdominal aorta. These latter workers also attributed a mean value of 37° to the angle of aortic bifurcation. These dimensional observations are of interest in connexion with current attempts to assess a suspected hydrodynamic ('haemodynamic') factor in the genesis of atherosclerosis in the abdominal aorta (*see* Newman *et al.* 1971; Lallemand and Newman 1973). Theoretically it can be shown that the pressure pulse wave in arteries is reflected at any junction at certain values of the combined arterial *luminal* areas of the branch or branches to that of the parent vessel; this relation is the *area ratio* of the junction. At an equal bifurcation, such as the aortic, with an area ratio of 1:15, reflexion of the pressure pulse wave is near to zero, and the vessels are said to be 'matched'. Oscillations and possibly turbulence set up by 'mis-matching' (at other ratios), perhaps also influenced by asymmetry of the bifurcation, may be causal in intimal damage and hence predispose to aortic atheroma. The luminal and other dimensions of the bifurcation may, therefore, assume a special significance, as may the changes in these values which occur during life. Measurement of the aortico-iliac junction in humans, dogs and domestic fowls (free from vascular disease) has shown that the area ratio is usually close to the theoretical value for 'matching', and that the ratio is independent of age in the dog and fowl (Gosling *et al.* 1971). However, in mankind the aortic bifurcation appears to be 'matched' only in early infancy; it is 1·11 ±0·02 at birth, becoming progressively lower with advancing age and reaching a value of about 0·7 in the 5th decade, at which theory predicts a 'mismatch' producing a reflection of the pulse pressure wave at about one-third of its oscillatory amplitude. These studies are continuing, and their implications make the latest report on the geometry of the aortic bifurcation (Shah *et al.* 1978) of particular interest. This embodied the most extensive dimensional data so far recorded, including not only diamters but also angles of deviation, iliac lengths and curvatures, and data for the

proximal parts of its testicular (or ovarian) branches, and is crossed by the horizontal part of the duodenum. In its lowest part the aorta is covered by the posterior parietal peritoneum and is crossed by the parietal attachment of the mesentery and its contents.

Posteriorly the abdominal aorta descends in front of the upper four lumbar vertebrae, the intervening intervertebral discs and corresponding part of the anterior longitudinal ligament. The lumbar arteries, which arise from its dorsal aspect, and the third and fourth (and sometimes the second) left lumbar veins, which cross behind it to reach the inferior vena cava, separate the aorta from the ligament. The vessel may overlap the anterior border of the left psoas major to a slight extent.

On the *right* the aorta is related above to the cisterna chyli and the thoracic duct, the azygos vein, and the right crus of the diaphragm, which overlaps it and separates it from the inferior

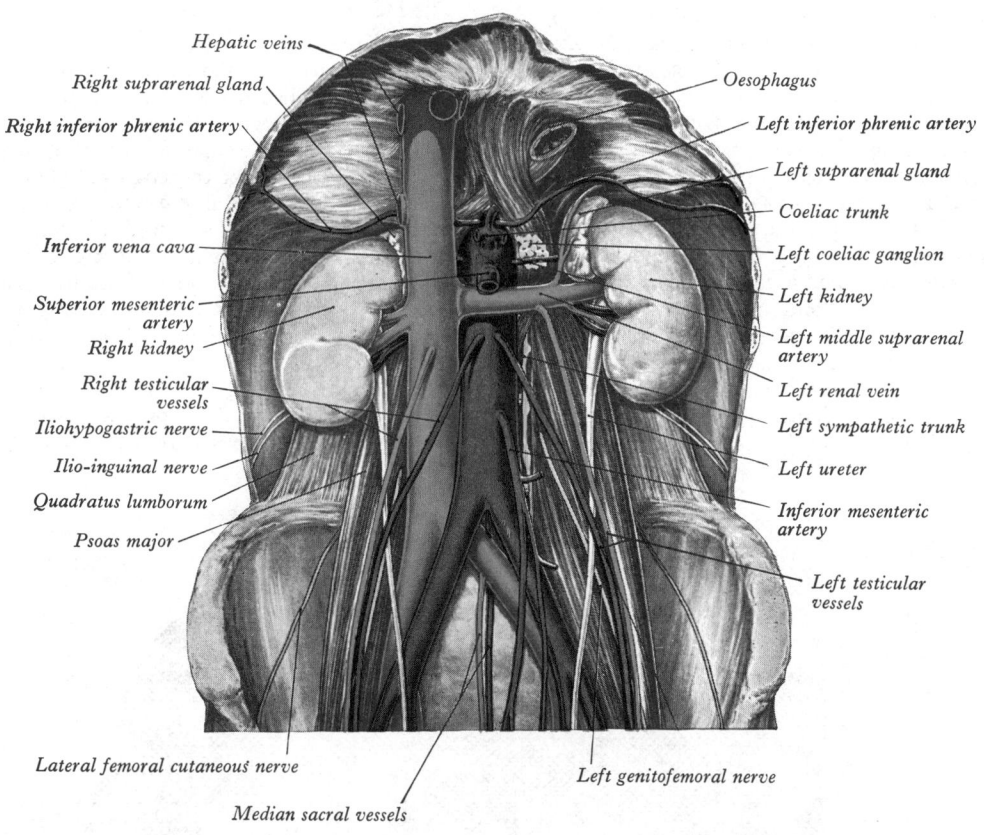

6.84   The abdominal aorta and its branches in the male.

dorsal angulation of these vessels as they enter the pelvic cavity. Unfortunately, the diameters were measured externally by calipers and in a relatively small series of cadavers at autopsy. These data cannot, therefore, be compared with those cited above, of which these recent workers appear to have been unaware. It is to be hoped that these interesting observations will be carried further, with improved techniques and greater cohesion between the different groups involved.

Relations. The abdominal aorta is at first related *anteriorly* to the coeliac trunk and its branches, the coeliac plexus of nerves and the omental bursa, which intervene between it and the papillary process of the liver and the lesser omentum. Immediately below this level the superior mesenteric artery leaves the aorta and crosses anteriorly to the left renal vein. The body of the pancreas, with the splenic vein applied to its posterior aspect, extends obliquely to the left and slightly upwards across the abdominal aorta, but is separated from it by the superior mesenteric artery and left renal vein. Below the pancreas, the aorta is related to the

vena cava and the right coeliac ganglion. Below the level of the second lumbar vertebra it is in contact with the inferior vena cava.

On the *left*, it is related above to the left crus of the diaphragm and the left coeliac ganglion. Opposite the second lumbar vertebra, it is related to the duodenojejunal flexure and the sympathetic trunk, which continues downwards along the left side of the vessel. The ascending part of the duodenum and the inferior mesenteric vessels constitute additional relations.

Surface Anatomy. The vessel can be represented by a band about 2 cm wide extending from a point in the median plane 2·5 cm above the transpyloric plane to the point where it divides about 1 cm below and to the left of the umbilicus. When the abdominal wall is relaxed the lowest portion of the aorta may be felt pulsating just above this point. The pulsations can even be seen in thin people when recumbent.

The branches of the abdominal aorta (**6**.84) may be divided into four sets: ventral, lateral, dorsal and terminal. The ventral and lateral branches are distributed to viscera, while the dorsal branches supply the body wall. The terminal branches supply the pelvis and lower limbs.

Ventral	Lateral
Coeliac.	Inferior phrenic.
Superior mesenteric.	Middle suprarenal.
Inferior mesenteric.	Renal.
	Testicular or Ovarian

Dorsal	Terminal
Lumbar.	Common iliac.
Median sacral.	

(For a detailed account of the upper abdominal vasculature consult Michels 1956.)

The Coeliac Trunk

The coeliac trunk (**6**.85, 86) is a wide branch, about 1·25 cm long, from the front of the aorta, just below the aortic hiatus of the diaphragm; it passes nearly horizontally forwards and slightly to the right above the pancreas and the splenic vein, and divides into three: (1) *left gastric*; (2) *hepatic*, (3) *splenic*. It may give off one or both of the inferior phrenic arteries; the superior mesenteric artery may arise in common with the coeliac trunk from the aorta, or the usual branches of the coeliac trunk may branch independently from the aorta.

Relations. The coeliac trunk is behind the omental bursa (lesser sac of the peritoneum), and is surrounded by the coeliac plexus of nerves, which sends extensions along the three divisions of the artery. On its *right side* it is related to the right coeliac ganglion, the right crus of the diaphragm, and the caudate process of the liver; on its *left* are the left coeliac ganglion, the left crus of the diaphragm, and cardiac end of the stomach. *Below*, it is related to the upper border of the pancreas and the splenic vein. The suspensory muscle of the duodenum (p. 1344) may split to encircle the artery, although it usually lies on its left.

1. The Left Gastric Artery (**6**.85)

This, the smallest branch of the coeliac trunk, passes upwards and to the left, behind the omental bursa, to the cardiac end of the stomach. In its course, the left gastric artery lies close to the left inferior phrenic artery and medial to or in front of the left suprarenal gland. At or near the cardiac end of the stomach it gives off two or three *oesophageal branches*, which ascend through the oesophageal opening of the diaphragm and anastomose with the aortic oesophageal arteries; others supply the cardiac part of the stomach and anastomose with branches of the splenic artery. The artery then turns forwards and downwards in the left gastro-pancreatic fold, and runs (frequently as two branches) along the lesser curvature of the stomach to the pylorus, between the layers of the lesser omentum; it gives branches to both surfaces of the stomach and anastomoses with the right gastric artery. An *accessory left gastric artery* may arise from the left branch of the hepatic artery and traverse the lesser omentum to gain the lesser curvature of the stomach.

2. The Hepatic Artery (**6**.85, 86)

This is intermediate in size between the left gastric and splenic branches, in *later* fetal and early postnatal life it is the largest branch of the coeliac trunk. It is accompanied by the hepatic plexus of nerves and is first directed forwards and to the right, to the upper surface of the superior part of the duodenum, passing below the medial end of the epiploic foramen (**6**.85). It then crosses in front of the portal vein, and ascends, between the layers of the lesser omentum and in front of the epiploic foramen, to the porta hepatis, where it divides into right and left branches to supply the corresponding lobes of the liver, accompanying the ramifications of the portal vein and hepatic ducts. In the lesser omentum the hepatic artery lies in front of the portal vein and on the left of the bile duct, and its right branch crosses behind (occasionally in front of) the common hepatic duct (**6**.87). For purposes of description, the hepatic artery is subdivided into (*a*) the *common hepatic artery*, extending from the coeliac trunk to the

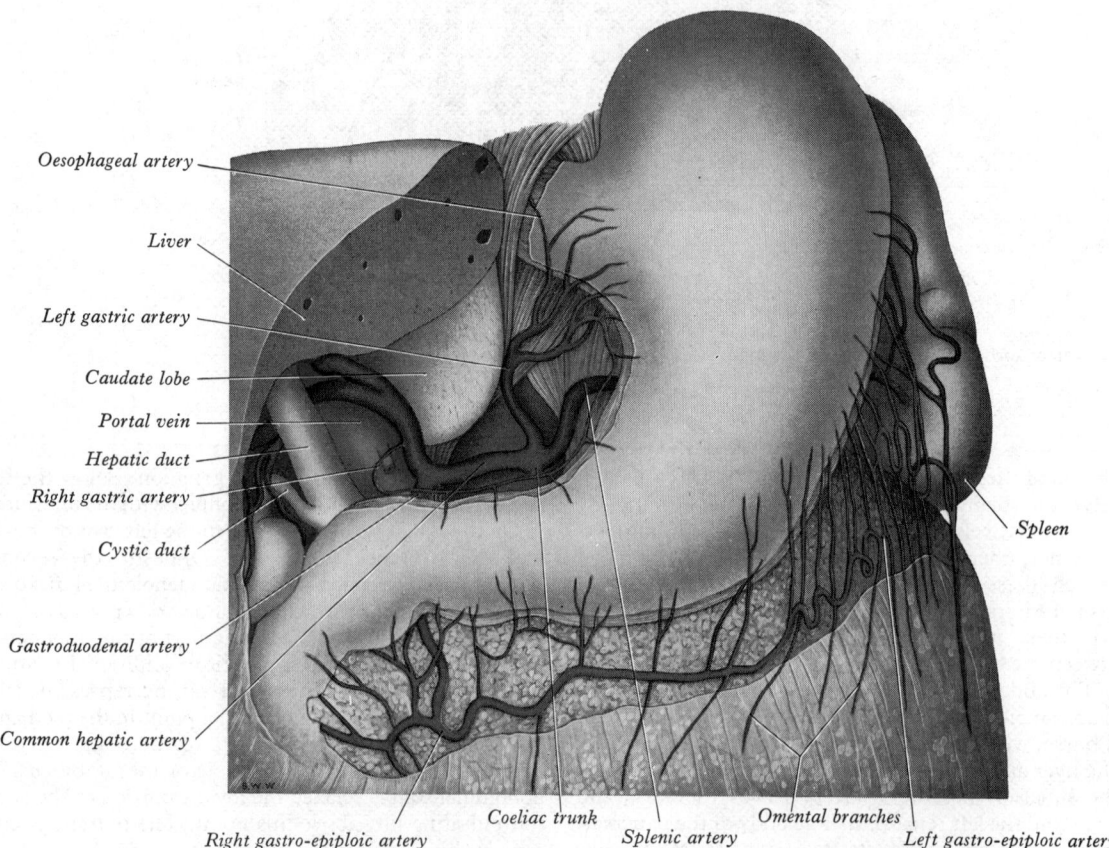

Oesophageal artery

Liver

Left gastric artery

Caudate lobe

Portal vein

Hepatic duct

Right gastric artery

Cystic duct

Gastroduodenal artery

Common hepatic artery

Spleen

Right gastro-epiploic artery

Coeliac trunk

Splenic artery

Omental branches

Left gastro-epiploic artery

6.85 The coeliac trunk and its branches. Part of the liver and all the lesser omentum have been removed, as well as the posterior wall of the omental bursa and part of the anterior layer of the greater omentum.

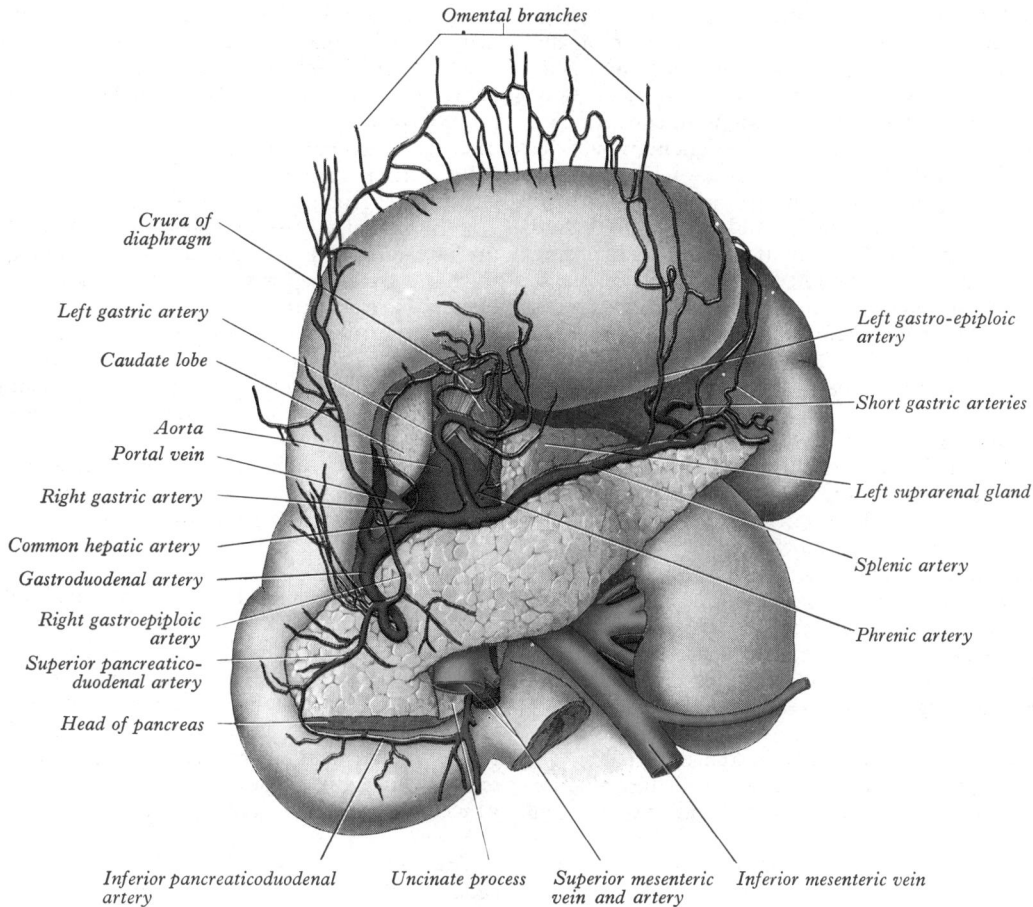

Omental branches

Crura of diaphragm

Left gastric artery

Caudate lobe

Aorta
Portal vein

Right gastric artery

Common hepatic artery

Gastroduodenal artery

Right gastroepiploic artery

Superior pancreatico-duodenal artery

Head of pancreas

Left gastro-epiploic artery

Short gastric arteries

Left suprarenal gland

Splenic artery

Phrenic artery

Inferior pancreaticoduodenal artery   *Uncinate process*   *Superior mesenteric vein and artery*   *Inferior mesenteric vein*

6.86   The coeliac trunk and its branches exposed by turning the stomach upwards and removing the peritoneum on the posterior abdominal wall.

origin of the gastroduodenal artery, and (*b*) the *hepatic artery proper*, extending from that point to the bifurcation of the artery into its right and left branches.

In embryonic and *early* fetal life, the hepatic artery arises from the left gastric (in 67 per cent of 56 individuals, according to Godlewski *et al.* 1975). This arrangement rarely persists, but the hepatic artery may arise from the superior mesenteric; alternatively the right or left hepatic branches may be derived from other vessels, the former from the superior mesenteric, the latter from the left gastric artery. For these and other variations consult Quain (1865, 1899), and Woodburne (1962); see also p. 715 and p. 1383.

The hepatic artery gives off the right gastric, the gastro-duodenal and the cystic arteries, branches to the bile duct from the right hepatic artery, and sometimes the supraduodenal artery (*vide infra*).

The right gastric artery (**6**.85) arises above the superior part of the duodenum, usually before, but sometimes beyond the gastroduodenal artery. It descends in the lesser omentum to the pyloric end of the stomach, and passes from right to left along the lesser curvature, supplying the upper parts of the anterior and posterior surfaces of the stomach, and ends by anastomosing with the left gastric artery. It may give off the supraduodenal artery (*vide infra*).

The gastroduodenal artery (**6**.85, 86), which arises behind, or sometimes above, the superior part of the duodenum, is a short but large branch. It descends between the superior part of the duodenum and the neck of the pancreas, lying immediately to the right of the line along which the peritoneum is reflected from the posterior surface of the first half-inch (1·25 cm) of the duodenum (**6**.85). The artery usually lies to the left of the bile duct, but sometimes is in front of it. At the lower border of the superior part of the duodenum it divides into the *right gastro-epiploic* and *superior pancreaticoduodenal* arteries. Prior to this division it supplies small branches to the pyloric end of the stomach and to

the pancreas, small retroduodenal branches to the back of the superior part of the duodenum, and sometimes the supraduodenal artery (*vide infra*). The first branch of the common hepatic artery is more frequently the gastroduodenal artery than the right gastric artery. The gastroduodenal artery may occasionally come from the superior mesenteric, the coeliac trunk, or an aberrant right hepatic artery (p. 715). The most invariable feature of the gastroduodenal artery is its intermediate position between the pancreas and duodenum, a point of surgical importance in view of the common involvement of this artery in duodenal ulceration (Bradley 1973).

The *supraduodenal artery*, which may be double, is an inconstant vessel. It arises from either the gastroduodenal, the hepatic artery (common hepatic, hepatic artery proper, or the right or left branches of the latter), or the right gastric artery. It supplies the superior aspect and upper parts of the anterior and posterior surfaces of the proximal half or more of the superior part of the duodenum, though in many cases the proximal part of the duodenum is invaded by branches from the right gastric artery (see also p. 1344).

The right gastro-epiploic artery (**6**.85, 86), which is the larger terminal branch of the gastroduodenal, skirts the right margin of the omental bursa and then runs to the left along the greater curvature of the stomach, between the layers of the greater omentum. It ends by anastomosing with the left gastro-epiploic branch of the splenic artery. Except at the pylorus, where it is in contact with the stomach, it lies about a finger's breadth from the greater curvature. This vessel has numerous branches, some of which ascend to supply both surfaces of the stomach, while others descend into the greater omentum (omental or epiploic branches). It also gives off branches to the lower surface of the superior part of the duodenum.

The superior pancreaticoduodenal arteries (**6**.86), are usually two in number. The *anterior* descends on the front of the groove between the duodenum and the head of the pancreas. It

713

supplies both of these organs, and anastomoses with the anterior division of the inferior pancreaticoduodenal branch of the superior mesenteric artery. The *posterior superior pancreaticoduodenal artery*, which generally arises independently from the gastroduodenal at the upper border of the superior part of the duodenum, runs downwards and to the right in front of the portal vein and the bile duct. It then passes downwards on the back of the head of the pancreas, supplying branches to the gland and to the duodenum, and crosses behind the bile duct, just before that structure pierces the duodenal wall. It ends by anastomosing with the posterior division of the inferior pancreaticoduodenal artery. (The pancreaticoduodenal arteries form vascular arcades of very variable disposition before their final branches of distribution—*see* Michels 1962.) The superior artery supplies several branches to the lower part of the common bile duct (*see* p. 1383).

The cystic artery (**6**.87) usually comes from the right branch of the hepatic artery, and passes behind the common hepatic and cystic ducts to gain the upper surface of the neck of the gall bladder, on which it runs downwards and forwards before dividing into *superficial* and *deep* branches. The former ramifies on the lower, and the latter on the upper surface. occasionally the cystic artery arises from the hepatic artery itself (or rarely from the gastroduodenal artery) and crosses in front of, or behind, the bile duct or common hepatic duct to reach the gall bladder. Such variations in the *origin* of the cystic artery are of considerable surgical interest. In a large series of 800 specimens Anson (1963) observed the following incidences: from the right hepatic—63·9 per cent, the hepatic trunk—26·9 per cent, the left hepatic—5·5 per cent, the gastroduodenal—2·6 per cent, the superior pancreaticoduodenal—0·3 per cent, the right gastric—0·1 per cent, the coeliac trunk—0·3 per cent, and the superior mesenteric—0·8 per cent. The direct origins from the hepatic artery varied from the latter's start to its bifurcation into right and left hepatic branches. An *accessory cystic artery* may arise from the common hepatic artery or from one of its branches. The cystic artery supplies branches to the hepatic ducts and to the upper part of the common bile duct (*see* p. 1383). A comparative study of the *distribution* of the cystic artery in a variety of reptilian, avian and mammalian species included 74 injected and cleared

human gall bladders (Gordon 1967). Briefly, the cystic artery of man reached the gall bladder at its neck and in no specimen was it at any point in contact with the cystic duct. Coursing over the neck, one or two pairs of vessels encircled the neck, anastomosing posteriorly. Division into its two principal longitudinal branches occurred at any point along the neck, and each then gave off 4–8 pairs of lateral branches at right angles to the parent vessels, anastomoses completing 4–8 vascular circles. The principal vessels bifurcated and formed a diamond-shaped anastomosis at the junction of body and fundus, the centre of the diamond being traversed by a fine anastomotic network. The overall arterial pattern was classified as *bipinnate*.

The terminal, intrahepatic branches display a pattern of branching which, in its major detail (p. 1376), is relatively constant from individual to individual. This ramification justifies a **segmental** description of the liver, and it is the result, as in other organs, of the mode of growth and branching of an original epithelial blastema, its pattern being accompanied by arterial rami (and innervation). Arterial hepatic segmentation is noted on p. 1379. Consult also Woodburne (1962).

3. The Splenic Artery (**6**.85, 86)

This, the largest branch of the coeliac trunk, is remarkable for its tortuosity. Surrounded by the splenic plexus of nerves, and accompanied by the splenic vein, which lies behind the pancreas, it passes horizontally to the left, behind the stomach and the omental bursa, and along the upper border of the pancreas; it crosses in front of the left suprarenal gland and the upper part of the left kidney, and enters the lienorenal ligament. On arriving near the spleen it divides into five or more *segmental* branches which enter the hilum of the spleen (*see* p. 778).

Branches:

The pancreatic branches (**6**.86) are numerous small vessels supplying the neck, body and tail of the pancreas; they are derived from the splenic artery as it runs along the upper border of the pancreas. A *dorsal branch*, which may arise from the superior mesenteric, middle colic, hepatic or, more rarely, the coeliac artery, descends behind the pancreas and divides into right and left vessels. The former, usually two in number, run between the neck and uncinate process of the gland and form *a prepancreatic arterial arch* with a branch from the anterior superior pancreaticoduodenal: the latter runs to the left along the inferior border of the gland to reach the tail. It anastomoses with the branches (*arteria pancreatica magna* and *arteria caudae pancreatis*) of the splenic artery which supply the left of the body and tail of the pancreas.

Short gastric arteries (**6**.86), five to seven in number, arise from the end of the splenic artery and its terminal divisions or from the left gastro-epiploic artery. They pass between the layers of the gastrosplenic ligament, and are distributed to the fundus of the stomach, anastomosing with branches of the left gastric and left gastro-epiploic arteries.

The posterior gastric artery, derived from any part of the splenic artery, but most commonly from its intermediate section, has been described by many authorities (e.g. Quain 1844), but recent texts have omitted mention of this vessel as pointed out by Susuki *et al.* (1978), who have surveyed reports on this vessel. In their observations 38 (62·3 per cent) of 61 adult cadavers displayed this vessel; incidences from 14 other reports (appearing from 1904 to 1968) varied from 12·7 to 77 per cent, with an average of about 58 per cent in a total of 870 cadavers (three studies cited percentages only). The authors quoted describe the vessel as ascending behind the peritoneum of the omental bursa towards the fundus and reaching the posterior gastric wall through the gastrophrenic fold of peritoneum. In their series it was usually about 2 mm in diameter, and hence of considerable significance in surgical manoeuvres in this region.

The left gastro-epiploic artery (**6**.85, 86), the largest of the branches of the splenic artery, arises near the hilum of the spleen and runs obliquely downwards, forwards and to the right. It sends several branches through the gastrosplenic ligament to be distributed to about the upper third of the greater curvature, and these are appreciably longer than the gastric branches of the right gastro-epiploic artery and may be 8 to 10 cm long. Its terminal

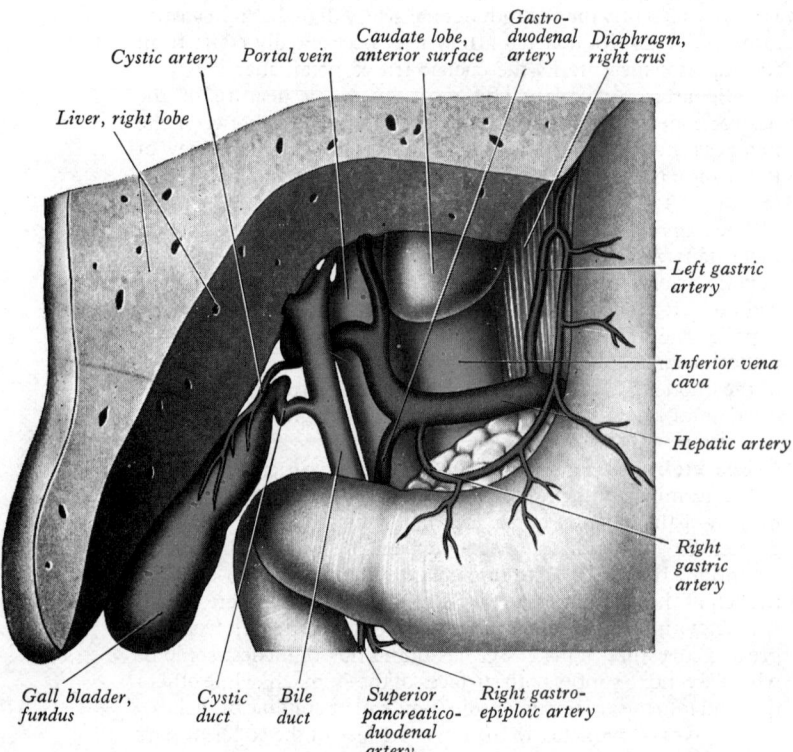

Cystic artery   *Portal vein*   *Caudate lobe, anterior surface*   *Gastroduodenal artery*   *Diaphragm, right crus*

Liver, right lobe

Left gastric artery

Inferior vena cava

Hepatic artery

Right gastric artery

Gall bladder, fundus   *Cystic duct*   *Bile duct*   *Superior pancreaticoduodenal artery*   *Right gastro-epiploic artery*

6.87 The relations of the hepatic artery, bile duct and portal vein exposed by removal of the lesser omentum and the peritoneum on the posterior abdominal wall.

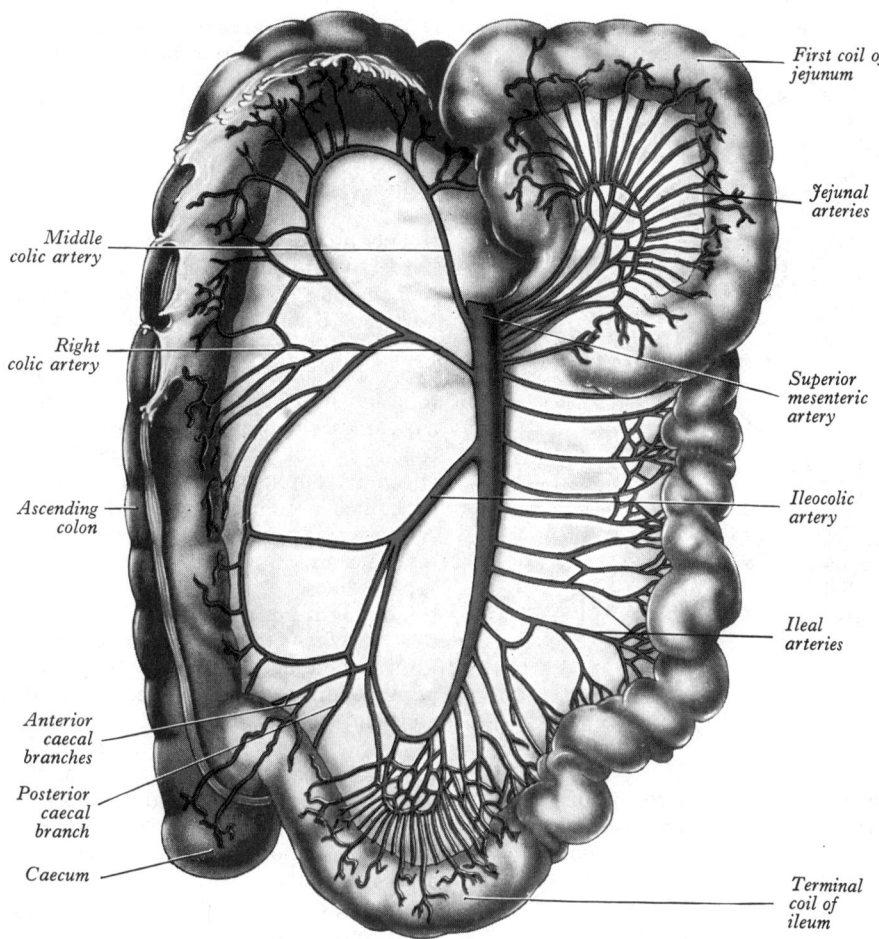

Middle
colic artery

Right
colic artery

Ascending
colon

Anterior
caecal
branches

Posterior
caecal
branch

Caecum

First coil of
jejunum

Jejunal
arteries

Superior
mesenteric
artery

Ileocolic
artery

Ileal
arteries

Terminal
coil of
ileum

6.88 The superior mesenteric artery and its branches. The first coil of
the jejunum and the terminal coil of the ileum have been spread out to
show the arrangement of their arteries.

part gives off a large omental branch which runs downwards and
to the right in the greater omentum, and itself curves forwards at a
higher level to join the right gastro-epiploic. The looped course of
the terminal part of the left gastro-epiploic leaves a portion of the
greater curvature devoid of branches in its upper part. In the
operation of partial gastrectomy the greater omentum is divided
below the right gastro-epiploic artery, cutting all its omental
branches. The greater omentum survives because its supply from
the large omental branch of the left gastro-epiploic artery has not
been damaged (Horton 1952).

The arteries that supply the greater omentum are the epiploic
(omental) branches of the right and left gastro-epiploic arteries.
The right, middle and left colic arteries do not give branches to
the greater omentum, and it is to be noted that the transverse
mesocolon, though adherent to the greater omentum, is separable
from it (*see* p. 1330).

The terminal splenic branches enter the hilum of the spleen
between the two layers of the lienorenal ligament. Their
distribution within the spleen is described with the anatomy of
that organ (p. 778).

Variations of the hepatic artery. Variations in the arrangement
of the hepatic artery and its branches are common and of surgical
importance. They include the following. (1) The common hepatic
artery may arise from the superior mesenteric artery or, less
commonly, directly from the aorta. In these cases it usually passes
to the right behind the portal vein to enter the lesser omentum. (2)
An accessory left hepatic artery arises most frequently from the
left gastric artery and passes to the right in the lesser omentum to
the porta hepatis. (3) An accessory right hepatic artery arises most
commonly from the superior mesenteric artery and generally runs
behind the portal vein and bile duct in the lesser omentum to gain
the porta hepatis.

The accessory left and right hepatic arteries may arise from
vessels other than the superior mesenteric artery, e.g. the
gastroduodenal or the aorta. They may exist in conjunction with
'normal' branches of the hepatic artery and constitute additional
sources of blood supply to the liver; on the other hand they may
replace the normal branches and constitute the sole supply to the
appropriate parts of the liver and they are then called '*aberrant
replacing arteries*'.

Collateral circulation after ligature of the hepatic artery.
Obstruction of the hepatic artery may lead to necrosis of the liver.
The effects on the liver depend on the site of the block. Occlusion
of the common hepatic artery, proximal to the origin of the right
gastric artery, will allow a collateral circulation to the liver
through the left and right gastric, left and right gastro-epiploic,
pancreaticoduodenal and gastroduodenal arteries, and hepatic
necrosis is unlikely to occur. If, however, the obstruction of the
hepatic artery proper is beyond the origin of the gastroduodenal
artery, the only collateral circulation is by means of the small
inferior phrenic arteries (p. 719).

The Superior Mesenteric Artery

The superior mesenteric artery (**6**.88) supplies the whole of the
small intestine except the superior part of the duodenum; it also
supplies the caecum, the ascending colon and most of the
transverse colon. It diverges from the front of the aorta about 1 cm
below the coeliac trunk, and is crossed anteriorly at its origin by
the splenic vein and the body of the pancreas. It is separated from
the front of the aorta by the left renal vein. It next passes
downwards and forwards, anterior to the uncinate process of the
pancreas and the horizontal part of the duodenum, and descends

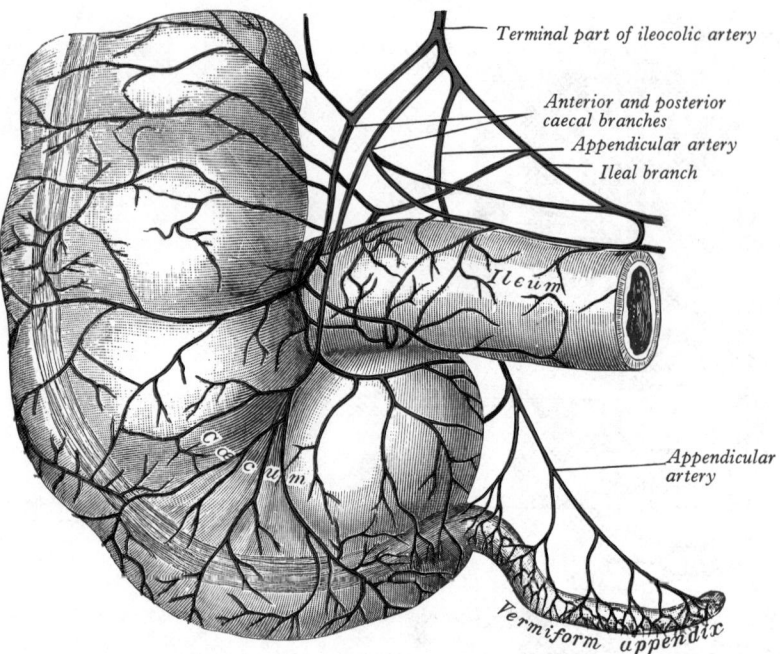

Terminal part of ileocolic artery

Anterior and posterior caecal branches

Appendicular artery

Ileal branch

Appendicular artery

6.89 The arteries of the caecum and vermiform appendix.

between the layers of the mesentery near its root until it reaches the right iliac fossa. Here, considerably diminished in size, it anastomoses with one of its own branches, the ileocolic artery. As it descends it crosses in front of the inferior vena cava, the right ureter and psoas major, and forms an arch, its convexity forwards, downwards and to the left. It is accompanied by the superior mesenteric vein, on its right, and is surrounded by the superior mesenteric plexus of nerves. Occasionally a fibrous strand passes to the umbilicus from the terminal part of the superior mesenteric artery, or from one of its lower ileal branches; it is a remnant of the embryonic artery which runs along the vitelline or yolk duct to the yolk sac.

Branches:

The inferior pancreaticoduodenal artery (**6**.86) leaves the superior mesenteric artery or its first jejunal branch, level with the upper border of the horizontal part of the duodenum. Usually it divides at once into anterior and posterior branches. The *anterior* goes to the right in front of the head of the pancreas and then ascends to anastomose with the anterior superior pancreaticoduodenal artery. The *posterior* passes upwards to the right behind the head of the pancreas, which it sometimes pierces, to anastomose with the posterior superior pancreaticoduodenal artery. Both branches supply the head of the pancreas, including its uncinate process, and adjoining parts of the duodenum.

The jejunal and ileal branches (**6**.88) arise from the left side of the superior mesenteric artery, usually twelve to fifteen in number, and are distributed to the jejunum and ileum, except for the terminal part of the latter, which is supplied by the ileocolic artery. They run nearly parallel with one another between the layers of the mesentery, each vessel dividing into two, which unite with adjacent branches to form a series of arches (**6**.88). The branches from the arches unite to form a second series of arches, and the process may be repeated three or four times. In the short, upper part of the mesentery only one set of arches exists, but as the mesentery increases in depth, second, third, fourth, and even fifth series are present. From the terminal arches numerous small straight vessels supply the intestine. These terminals are distributed, roughly alternately, to opposite surfaces of the gut, and it has been claimed that neighbouring vessels do not anastomose with one another (Cokkinis 1930). As a rule the jejunal arteries are longer and less numerous than the ileal arteries. From both groups small branches are given off to the lymph nodes and other structures between the layers of the mesentery. It should be noted that the changes in mode of distribution from jejunal to ileal levels is gradual. (*See* **8**.120 C, D.)

The ileocolic artery (**6**.88) is the lowest of the branches from the right of the superior mesenteric artery. It passes downwards and to the right behind the peritoneum, towards the right iliac fossa, where it divides into superior and inferior branches; the *superior* branch anastomoses with the right colic artery, the *inferior* with the end of the superior mesenteric artery. In its course, the ileocolic artery crosses in front of the right ureter, testicular (or ovarian) vessels and psoas major.

The inferior branch of the ileocolic runs towards the upper border of the ileocolic junction and supplies the following branches (**6**.89): (a) *ascending* (*colic*) which passes upwards on the ascending colon; (b) *anterior* and *posterior caecal*, distributed to the front and back of the caecum; (c) an *appendicular artery*, which descends behind the termination of the ileum and enters the mesoappendix; after giving off a recurrent branch, which anastomoses with a branch from the posterior caecal artery, it runs at first close to and later in the free margin of the meso-appendix, but its terminal part is in actual contact with the appendicular wall; (d) *ileal*, which runs upwards and to the left on the lower part of the ileum, supplying this part of the small intestine, and ends by anastomosing with the termination of the superior mesenteric artery.

The right colic artery (**6**.88) arises from near the middle of the concavity of the superior mesenteric, or from a stem common to it and the ileocolic artery. It passes to the right behind the peritoneum, and in front of the right testicular (or ovarian) artery and vein, the right ureter, and psoas major, towards the ascending colon. Sometimes the vessel lies at a higher level, and crosses the descending part of the duodenum and the lower end of the right kidney. At the colon it divides into a descending branch, which anastomoses with the ileocolic artery, and an ascending branch, which anastomoses with the middle colic. These branches form arches, from the convexity of which vessels are distributed to the ascending colon, supplying about its upper two-thirds and the right colic flexure.

The middle colic artery (**6**.88) departs from the superior mesenteric artery just below the pancreas and, passing downwards and forwards in the transverse mesocolon, divides into a right and a left branch; the former anastomoses with the right colic artery, the latter with the left colic artery, a branch of the inferior mesenteric artery. The arches thus formed are 3 or 4 cm from the transverse colon, to which they distribute branches.

Occasionally the superior mesenteric artery may give origin to the common hepatic artery, the gastroduodenal artery, or an accessory right hepatic artery. It may also provide accessory pancreatic or splenic arteries. The superior mesenteric artery has been observed to arise from a common coeliaco-mesenteric trunk (Mangoushi 1975).

The Inferior Mesenteric Artery

The inferior mesenteric artery (**6**.90) supplies the left third of the transverse colon, the whole of the descending colon, the sigmoid colon and the rectum. It is smaller than the superior mesenteric artery, and arises 3 or 4 cm above the aorta's division into the common iliac arteries, just above the lower border of the horizontal part of the duodenum. It descends behind the peritoneum, at first in front of the aorta, and then on its left. It crosses the left common iliac artery medial to the left ureter, and is continued in the sigmoid mesocolon into the lesser pelvis as the *superior rectal artery*. In the lower part of its course, the inferior mesenteric vein lies on its lateral side. The inferior mesenteric artery gives off left colic, sigmoid and superior rectal branches.

The left colic artery (**6**.90) runs upwards and to the left behind the peritoneum and in front of the psoas major, and, after a short but variable course, divides into an ascending and a descending branch. The trunk or the branches of the artery cross in front of the left ureter and left testicular (or ovarian) vessels. The ascending branch passes in front of the left kidney into the transverse mesocolon, where it anastomoses with the middle colic artery; the descending branch anastomoses with the highest sigmoid artery. From the arches formed by these anastomoses, branches are distributed to the left half of the transverse colon,

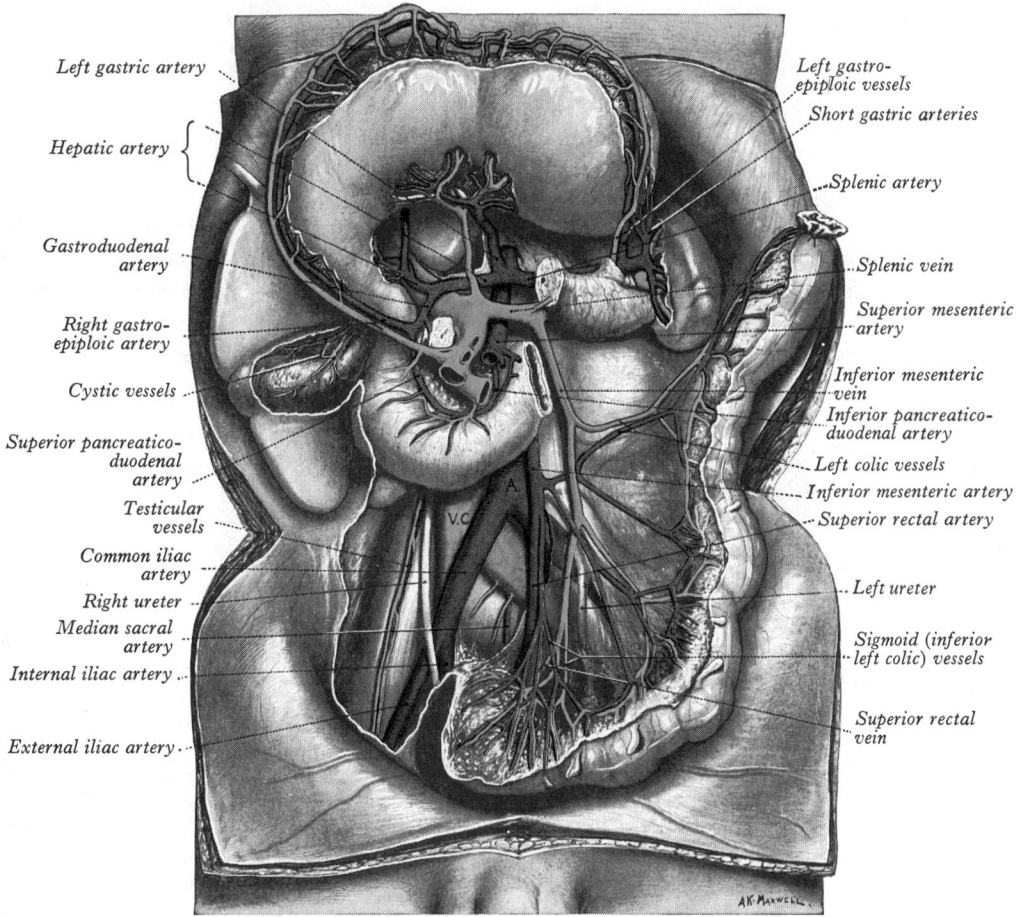

6.90 The inferior mesenteric vessels and their branches (male subject). *Note.*—The stomach has been turned upwards and the whole of the jejunum and ileum, the caecum, ascending colon and transverse colon have been removed, together with part of the pancreas.

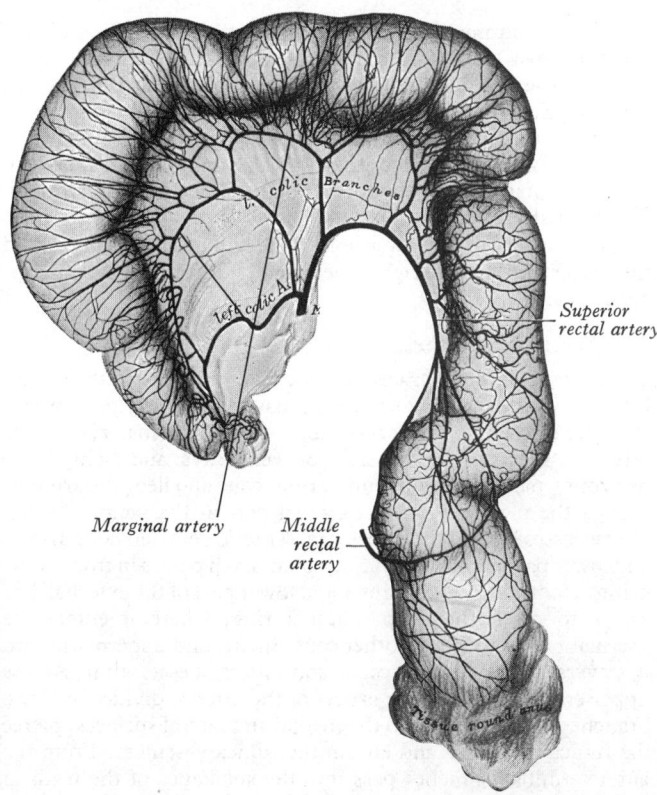

6.91 The sigmoid colon and rectum, showing the distribution of the branches of the inferior mesenteric artery and their anastomoses. (From a preparation by Hamilton Drummond.)

and to the descending colon. The territories of supply of the middle and left colic arteries show some reciprocal variation; the left branch of the middle colic may take over supply of the splenic flexure (in 19 of 100 cadavers according to investigations by Sierociński 1975).

The sigmoid (inferior left colic) arteries (6.90, 91), two or three in number, run obliquely down to the left behind the peritoneum in front of the left psoas major, ureter, and testicular or ovarian vessels. Their branches supply the lower part of the descending and the sigmoid colon, anastomosing above with the left colic artery, below with the superior rectal. The anastomoses between the branches of the left colic and sigmoid arteries form what is sometimes a continuous 'marginal artery' near the colon. The continuity of the marginal artery (of Drummond 1914) is, in fact, rarely interrupted, even in the vicinity of the sigmoido-rectal junction. Anastomosis at this level is sometimes effected by tenuous vessels which may be overlooked in dissection; injection experiments have demonstrated filling of the superior rectal vessels after high aortic injection, and the most recent series of dissections (100 individuals), by Sierociński (1976), indicated a constant anastomosis between the last sigmoid and the superior rectal arteries. Sudeck (1907) introduced the concept of uncertainty in this anastomosis, suggesting it as a '*critical point*' in colico-rectal surgery, and many authorities have repeated this view. (*See* Hollinshead, 1971, for a review of the literature.)

The superior rectal artery (6.90, 91), the continuation of the inferior mesenteric, descends into the pelvis in the sigmoid mesocolon, crossing the left common iliac vessels. It divides, opposite the third sacral vertebra, into two branches; these descend one on each side of the rectum; about halfway down they break up into several small branches. These pierce the muscular coat of the rectum and run straight down between the muscular and mucous coats, to the level of the sphincter ani internus; here,

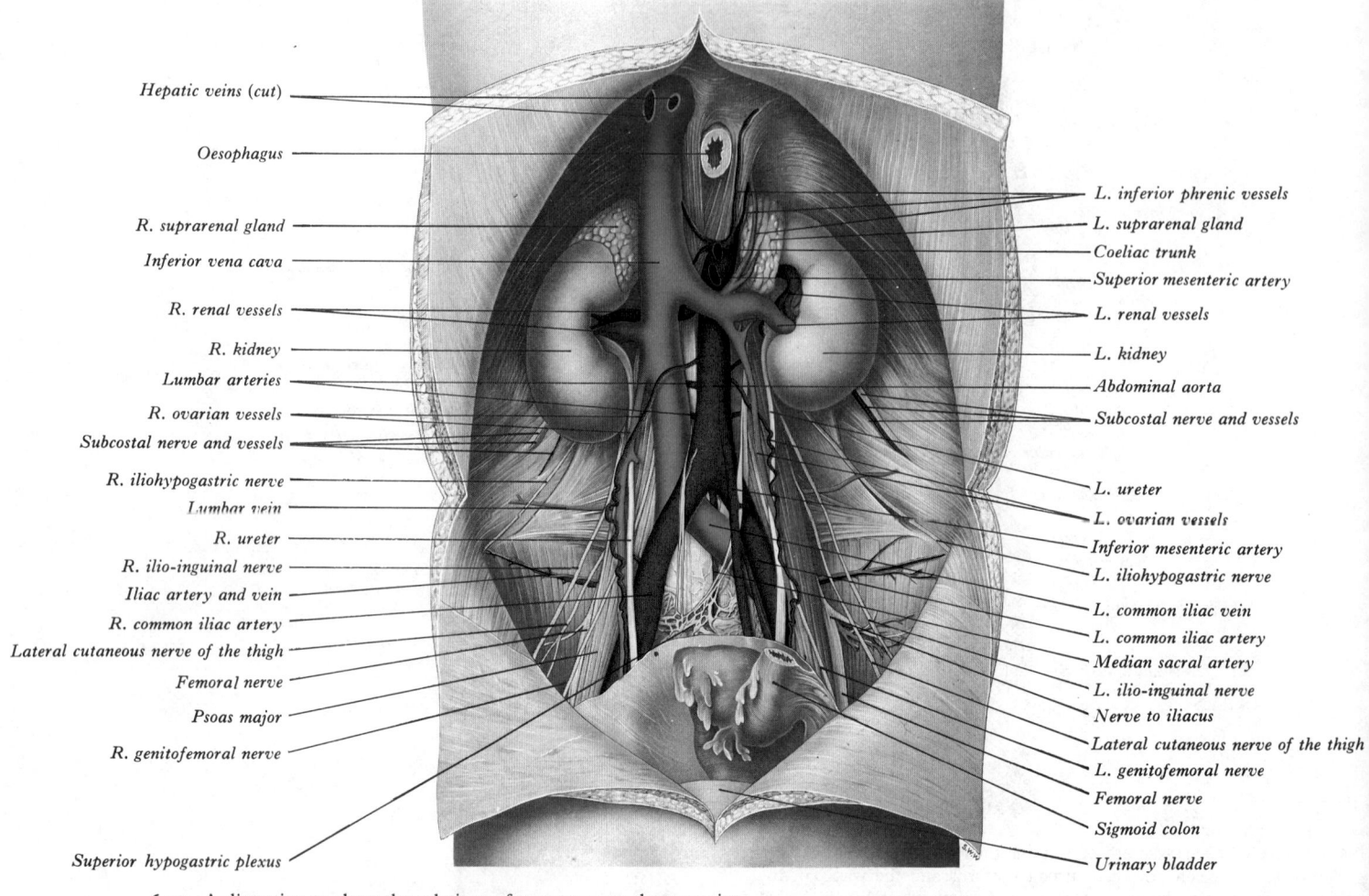

Hepatic veins (cut)

Oesophagus

R. suprarenal gland

Inferior vena cava

R. renal vessels

R. kidney

Lumbar arteries

R. ovarian vessels

Subcostal nerve and vessels

R. iliohypogastric nerve

Lumbar vein

R. ureter

R. ilio-inguinal nerve

Iliac artery and vein

R. common iliac artery

Lateral cutaneous nerve of the thigh

Femoral nerve

Psoas major

R. genitofemoral nerve

Superior hypogastric plexus

L. inferior phrenic vessels

L. suprarenal gland

Coeliac trunk

Superior mesenteric artery

L. renal vessels

L. kidney

Abdominal aorta

Subcostal nerve and vessels

L. ureter

L. ovarian vessels

Inferior mesenteric artery

L. iliohypogastric nerve

L. common iliac vein

L. common iliac artery

Median sacral artery

L. ilio-inguinal nerve

Nerve to iliacus

Lateral cutaneous nerve of the thigh

L. genitofemoral nerve

Femoral nerve

Sigmoid colon

Urinary bladder

6.92   A dissection to show the relations of structures on the posterior abdominal wall (female subject).

by anastomoses with one another, they form loops around the lower end of the rectum, and communicate with the middle rectal artery, a branch of the internal iliac artery, and with the inferior rectal, which arises from the internal pudendal (*see* p. 721).

The Middle Suprarenal Arteries

These two small vessels arise, one from each side of the aorta, level with the superior mesenteric artery. Each passes laterally and slightly upwards over the crus of the diaphragm to the suprarenal gland, where it anastomoses with suprarenal branches of the phrenic and renal arteries. The right artery passes behind the inferior vena cava and is close to the coeliac ganglion. The left is related to the coeliac ganglion, splenic artery and upper border of the pancreas.

The Renal Arteries (**6**.84, 92)

These two large trunks branch from the sides of the aorta immediately below the superior mesenteric, and both cross the corresponding crus of the diaphragm, nearly at right angles to the aorta. The *right* is longer, on account of the position of the aorta; it passes behind the inferior vena cava, right renal vein, head of the pancreas, and descending part of the duodenum. The *left* is a little higher; it lies behind the left renal vein, the body of the pancreas and splenic vein, and may be crossed anteriorly by the inferior mesenteric vein. Just before the hilum of the kidney, each artery divides into four or five branches; most of these lie between the renal vein and the pelvis of the ureter, the vein in front and the pelvis behind; but one or more are usually behind the pelvis. Each renal artery gives off some small *inferior suprarenal branches* (p. 191), and supplies the ureter and surrounding cellular tissue and muscles. The distribution of the renal arteries in the kidneys is described on p. 1397.

Surface Anatomy. The renal arteries can be represented by

broad lines running laterally for 4 cm from the lateral margins of the aorta just below the transpyloric plane. In the case of the artery of the left side the line should incline upwards across the transpyloric plane.

One or two *accessory renal arteries* frequently occur, especially on the left side: usually from the aorta, they may come off above or below the main artery, the former being slightly more frequent. They usually enter the upper or lower part of the kidney; an accessory artery to the lower pole of the kidney crosses in front of the ureter and, on the right side, usually in front of the inferior vena cava.

The Testicular Arteries (**6**.84, 92)

These two long, slender vessels arise from the front of the aorta a little below the renal arteries. Each passes obliquely downwards and laterally behind the peritoneum, on the psoas major; the right artery is in front of the inferior vena cava and behind the horizontal part of the duodenum, right colic and ileocolic arteries, root of the mesentery, and terminal part of the ileum; the left artery passes behind the inferior mesenteric vein, left colic artery, and lower part of the descending colon. Each passes in front of the genitofemoral nerve, the ureter and lower part of the external iliac artery to reach the deep inguinal ring, where it enters the spermatic cord. With the other constituents of the spermatic cord it traverses the inguinal canal and enters the scrotum. At the upper end of the posterior aspect of the testis it divides into two branches which pass on to the medial and lateral surfaces, pierce the tunica albuginea and end in the tunica vasculosa. From the latter terminal branches pass into the substance of the testis at various points over the free surface. Some pass into the mediastinum testis and loop back again before reaching their distribution (Harrison and Barclay 1948). In the abdomen the testicular artery supplies the fat around the kidney, the ureter and iliac lymph nodes; in the inguinal canal it gives one or two twigs to the cremaster muscle.

Sometimes the right testicular artery passes behind the inferior vena cava. Both arteries represent persistent lateral splanchnic branches of the aorta (p. 191), which enter the mesonephros and cross ventral to the supracardinal vein but dorsal to the subcardinal. Normally the lateral splanchnic artery, which persists as the right testicular artery, passes caudal to the particular suprasubcardinal anastomosis which persists in the formation of the inferior vena cava. When it passes cranial to this anastomosis the right testicular artery is behind the inferior vena cava in the adult.

The Ovarian Arteries

In the female these correspond to the testicular arteries but enter the pelvis to supply the ovaries (6.95). The details of the first part of each artery are the same as those of the testicular, but on arriving at the brim of the lesser pelvis the artery crosses the external iliac artery and vein, and enters the pelvic cavity. It then runs medially in the suspensory ligament of the ovary and gains the broad ligament of the uterus, below the uterine tube. At the level of the ovary it passes backwards in the mesovarium and breaks up into branches to the ovary. Small branches are given to the ureter and uterine tube, and one passes to the side of the uterus to unite with the uterine artery. Others run on the round ligament of the uterus, through the inguinal canal, to the skin of the labium majus and the inguinal region.

Early in intrauterine life, when the testes or ovaries flank the vertebral column below the kidneys, the testicular and ovarian arteries are short; but with the descent of the gonads into the pelvis, the arteries gradually lengthen.

The (Inferior) Phrenic Arteries (6.84, 92)

These are two small vessels which supply the diaphragm. They may arise separately from the front of the aorta, immediately above the coeliac trunk, or by a common stem, either from the aorta or the coeliac trunk; sometimes one is derived from the aorta, the other from one of the renal arteries. From its origin the artery runs upwards and laterally in front of the crus of the diaphragm and close to the medial border of the suprarenal gland. The left passes behind the oesophagus, and runs forwards on the left side of the oesophageal opening. The right phrenic artery passes behind the inferior vena cava, along the right side of the opening for this vein. Near the posterior border of the central tendon of the diaphragm each vessel divides into a medial and lateral branch. The medial curves forwards, and anastomoses with its fellow of the opposite side in front of the central tendon, and with the musculophrenic and pericardiacophrenic arteries. The lateral branch passes towards the side of the thorax, and anastomoses with the lower posterior intercostal arteries and with the musculophrenic artery. The lateral branch of the right artery gives off a few twigs to the inferior vena cava; the left artery sends some branches to the oesophagus. Each vessel gives off two or three small *superior suprarenal branches* to the suprarenal gland of its own side. The liver (*see* p. 715), Collateral Circulation) and the spleen also receive a few twigs from the phrenic arteries.

The Lumbar Arteries (6.92)

These are in series with the posterior intercostal arteries. Usually four in number on each side, they arise from the back of the aorta, opposite the bodies of the upper four lumbar vertebrae. A fifth pair, smaller in size, occasionally arise from the median sacral artery; but the lumbar branches of the iliolumbar arteries usually take their place. The lumbar arteries run laterally and backwards on the bodies of the lumbar vertebrae, behind the sympathetic trunks, to the intervals between adjacent transverse processes, and then continue into the abdominal wall. The right arteries pass deep to the inferior vena cava, and the upper two on the right, the first only on the left, run deep to the corresponding crus of the diaphragm. The arteries of both sides pass under the tendinous arches providing attachment for psoas major, and travel on behind this muscle and the lumbar plexus. They then cross quadratus lumborum, the upper three arteries running behind, the last usually in front of that muscle. At the lateral border of

quadratus lumborum they pierce the posterior aponeurosis of transversus abdominis, and pass forwards between this and internal oblique. They anastomose with one another and with the lower posterior intercostal, subcostal, iliolumbar, deep circumflex iliac and inferior epigastric arteries.

Branches. Each lumbar artery gives off a *dorsal ramus* which passes backwards between adjacent transverse processes to the muscle and skin of the back. The posterior ramus also furnishes a *spinal branch*, which enters the vertebral canal to supply its contents and adjacent vertebra, anastomosing with the arteries above and below it, and with the artery of the opposite side. The spinal branch of the first lumbar artery gives off branches to the lower part of the spinal cord. Branches of the lumbar arteries and their dorsal rami also supply the neighbouring muscles.

The Median Sacral Artery (6.84, 92)

This is a small branch from the back of the aorta a little above its bifurcation. It descends in the midline in front of the fourth and fifth lumbar vertebrae, the sacrum and coccyx, and ends in the coccygeal body. At the level of the fifth lumbar vertebra it is crossed by the left common iliac vein, and it frequently gives off on each side a small lumbar artery (*arteria lumbalis ima*). Minute branches pass from it to the posterior surface of the rectum. On the last lumbar vertebra it anastomoses with the lumbar branch of the iliolumbar artery; in front of the sacrum it anastomoses with the lateral sacral arteries, and sends offsets into the pelvic sacral foramina.

The Common Iliac Arteries

The abdominal aorta divides, on the left side of the body of the fourth lumbar vertebra, into the right and left common iliac arteries (6.84, 92). These diverge as they descend to divide opposite the intervertebral disc between the last lumbar vertebra and sacrum into two branches, the *external* and *internal iliac arteries*; the former supplies the greater part of the lower limb, the latter, the viscera and walls of the pelvis, perineum and gluteal region. The bifurcation is anterior to the sacroiliac joint.

The Right Common Iliac Artery (6.84, 92)

This is about 5 cm long and passes obliquely across the body of the last lumbar vertebra. *In front*, it is crossed by sympathetic fibres passing to the superior hypogastric plexus, and, at its point of division, by the ureter; throughout its course it is covered with the parietal peritoneum, by which it is separated from the coils of the small intestine. *Behind*, it is separated from the bodies of the fourth and fifth lumbar vertebrae and the intervening intervertebral disc by the sympathetic trunk, the terminal parts of the two common iliac veins and the commencement of the inferior vena cava. The obturator nerve, the lumbosacral trunk and the iliolumbar artery are situated more deeply and traverse the fatty tissue which occupies the interval between the last lumbar vertebra and the psoas major. *Laterally* it is in relation above with the inferior vena cava and the right common iliac vein, below, with the psoas major. *Medially* its upper part is related to the left common iliac vein.

The Left Common Iliac Artery (6.84, 92)

This is about 4 cm long, and is in relation, *in front*, with the peritoneum, ileum, sympathetic fibres passing to the superior hypogastric plexus, and the superior rectal artery, and is crossed at its bifurcation by the ureter. *Behind*, are the sympathetic trunk and bodies of the fourth and fifth lumbar vertebrae, and intervening intervertebral disc. The obturator nerve, lumbosacral trunk and iliolumbar artery are placed more deeply. The left common iliac vein lies partly *medial* to, and partly *behind*, the artery; *laterally* the artery is close to the psoas major.

Surface Anatomy. The vessel corresponds on each side to the upper third of a broad line from the bifurcation of the aorta (p. 710) to the point midway between the anterior superior iliac spine and the pubic symphysis. The *external iliac artery* corresponds to the lower two-thirds of this line, which should be slightly convex laterally.

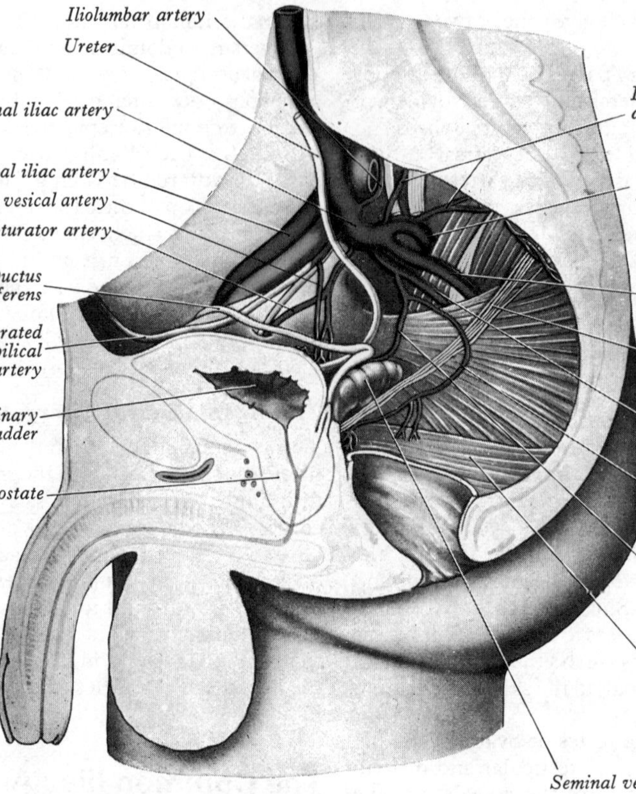

6.93   The arteries of the male pelvis. Right side. The internal iliac vein and its tributaries have been removed; the rectum has been divided just above the anal canal and its upper part has been taken away.

Branches. In addition to the terminal branches each common iliac artery gives small branches to the peritoneum, psoas major, ureter, and surrounding areolar tissue; occasionally the iliolumbar and accessory renal arteries are branches.

The Internal Iliac Arteries

Each internal iliac artery (**6**.93), about 4 cm long, arises at the bifurcation of the common iliac artery, level with the lumbosacral intervertebral disc and in front of the sacroiliac joint; it descends to the upper margin of the greater sciatic foramen, where it divides into an *anterior trunk*, which continues in the line of the parent vessel towards the spine of the ischium, and a *posterior trunk*, which passes backwards towards the foramen. (*See* Braithwaite 1952.)

In front are the ureter and, in the female, the ovary and the fimbriated end of the uterine tube; *behind*, is the internal iliac vein, the lumbosacral nerve trunk and sacroiliac joint; *laterally*, near its origin, is the external iliac vein, between it and psoas major; lower down, is the obturator nerve; and *medially* it is next to the parietal peritoneum, separating it from the terminal part of the ileum on the right and the sigmoid colon on the left, and to some of the tributaries of the internal iliac vein.

In the fetus, the internal iliac artery is twice as large as the external iliac, and is the direct continuation of the common iliac. It ascends on the back of the anterior wall of the abdomen to the umbilicus, converging on its fellow. Having passed through the umbilical opening, the two arteries, now termed *umbilical*, enter the umbilical cord, where they are coiled round the umbilical vein, and ultimately ramify in the placenta.

At birth, when the placental circulation ceases, only the pelvic portion of the artery remains patent, as the internal iliac artery and the first part of the superior vesical artery of the adult; the remainder of the vessel becomes a fibrous cord, termed the *medial umbilical ligament*, raising the peritoneal *medial umbilical fold*, which extends from the pelvis to the umbilicus. In the male, the

patent part of the umbilical artery usually gives off the artery to the ductus deferens.

BRANCHES FROM THE ANTERIOR TRUNK OF THE INTERNAL ILIAC ARTERY

The superior vesical artery (**6**.93) supplies numerous branches to the upper part of the bladder (Braithwaite 1951). From one of these the *artery to the ductus deferens* occasionally starts and accompanies the ductus to the testis, where it anastomoses with the testicular artery. Other branches supply the ureter. The first part of the superior vesical artery is the proximal, patent section of the fetal umbilical artery (*vide supra*).

The inferior vesical artery (**6**.93) frequently arises in common with the middle rectal artery, and is distributed to the fundus of the bladder, prostate, seminal vesicles and lower part of the ureter. The prostatic branches communicate with corresponding vessels of the opposite side. The inferior vesical artery may sometimes provide the artery to the ductus deferens.

The middle rectal artery (**6**.91, 93) usually arises with the preceding vessel. It vascularizes the muscular tissue of the lower rectum, anastomosing with the superior and inferior rectal arteries. It supplies the seminal vesicles and prostate by branches which join those of the inferior vesical artery.

The uterine artery (**6**.95) runs medially on the levator ani towards the cervix uteri; about 2 cm from this it crosses above and in front of the ureter—to which it supplies a small branch—and above the lateral vaginal fornix. Reaching the side of the uterus, it ascends along it tortuously in the broad ligament to the junction of uterine tube and uterus. It then runs laterally towards the hilum of the ovary, and ends by joining with the ovarian artery. It supplies the cervix uteri and branches descend on the vagina; the latter anastomose with branches of the vaginal arteries forming two median longitudinal vessels, the *azygos arteries of the vagina*; one descends in front of, the other behind, the vagina. It supplies numerous branches to the body of the uterus, and from its termination, twigs reach the uterine tube and round ligament of the uterus. The terminal branches in the uterine muscle are exceedingly tortuous (helicine arteries).

The vaginal artery, frequently double or triple, usually corresponds to the inferior vesical in the male; it descends upon the vagina, supplying its mucous membrane, and sends branches to the bulb of the vestibule, the fundus of the bladder and the adjacent part of the rectum. It assists in forming the azygos arteries of the vagina (*vide supra*).

The obturator artery (6.93) passes forwards and downwards on the lateral wall of the pelvis, to the upper part of the obturator foramen, and, leaving the pelvic cavity through the obturator canal, divides into an anterior and a posterior branch. In the pelvic cavity this vessel is related laterally to the obturator fascia, separating it from the obturator internus; medially, it is crossed by the ureter and ductus deferens, which separate it from the parietal peritoneum; in the nullipara the ovary is a medial relation. The obturator nerve is above and the obturator vein below.

Branches. Inside the pelvis, the obturator artery provides (*a*) *iliac branches* to the iliac fossa; these supply the bone and iliacus, and anastomose with the iliolumbar artery; (*b*) a *vesical branch* which runs medially to the bladder, and may replace the inferior vesical branch of the internal iliac; and (*c*) a *pubic branch* just before the obturator leaves the pelvic cavity; this branch ascends the pubis and anastomoses with the opposite vessel, and the pubic branch of the inferior epigastric.

Outside the pelvis, the obturator artery divides, at the upper margin of the obturator foramen, into an anterior and posterior branch, which encircle the foramen between the obturator externus and obturator membrane.

The *anterior branch* runs forwards on the outer surface of the obturator membrane and then curves downwards along the anterior margin of the foramen. It supplies branches to the obturator externus, pectineus, adductors and gracilis, and anastomoses with the posterior branch and with the medial circumflex femoral artery.

The *posterior branch* follows the posterior margin of the foramen and turns forwards on the ramus of the ischium where it anastomoses with the anterior branch. It supplies the muscles attached to the ischial tuberosity and anastomoses with the inferior gluteal artery. It also has an *acetabular branch* which enters the hip joint through the acetabular notch, ramifies in the fat of the acetabular fossa, and sends a twig along the ligament of the head of the femur.

Variations. In 20 to 30 per cent of people, the obturator artery is replaced by an enlarged pubic branch of the inferior epigastric (p. 723); this descends almost vertically to the upper part of the obturator foramen. Such an *abnormal obturator artery* is usually in contact with the external iliac vein, and lateral to the femoral ring; in such cases it would not be endangered in the operation for strangulated femoral hernia. Occasionally, however, it curves along the free margin of the lacunar part of the inguinal ligament, and may partly encircle the neck of a hernial sac; moreover, it may be inadvertently incised at operation. Rarely, the artery passes across the femoral ring and may be pushed either medially or laterally by a femoral hernia. According to Pick *et al.* (1942), who reported on 640 dissections of the obturator artery, the vessel is most commonly a branch of the internal iliac (24 per cent), or from this artery's anterior (21 per cent) or posterior (3 per cent) divisions. They put the incidence of epigastric origin at 27 per cent (*vide supra*); apart from the superior gluteal (11 per cent) and the inferior gluteal artery (9 per cent); rare origins include the external iliac, internal pudendal, and iliolumbar arteries.

The Internal Pudendal Artery (6.93, 94, 96, 97)

This, the smaller of the two terminal branches of the anterior trunk of the internal iliac, supplies the external genitalia. Though its course is the same, it is smaller in the female and its distribution somewhat different.

In the male the artery runs downwards and laterally to the lower part of the rim of the greater sciatic foramen, leaves the pelvic cavity between piriformis and coccygeus, and enters the gluteal region. Here it curves round the back of the ischial spine to enter the perineum through the lesser sciatic foramen, and thence traverses the pudendal canal, in the lateral wall of the ischiorectal

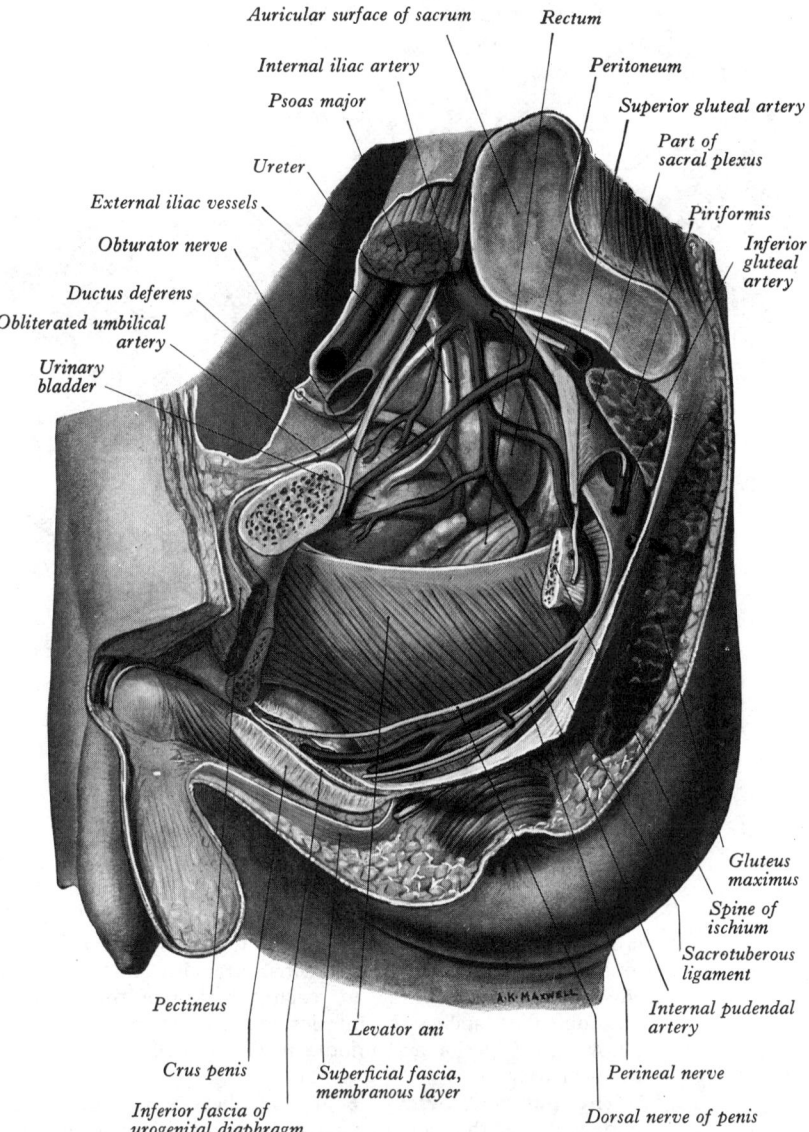

6.94 Structures of the male pelvic contents from the left side. Most of the left innominate bone has been removed together with the obturator internus. The sciatic nerve has been cut away close to its origin from the sacral plexus. All the vessels and nerves exposed are those of the *left* side. Note superior vesical, obturator, inferior vesical and middle rectal arteries which are, for technical reasons, without leader lines or tabbing labels.

fossa, medial to the obturator internus and about 4 cm above the lower limit of the tuberosity. Gradually approaching the margin of the ischial ramus, it proceeds forwards above, or deep to, the inferior fascia of the urogenital diaphragm, skirts the medial margin of the inferior pubic ramus, and ends a little behind the inferior pubic ligament by dividing into the *deep and dorsal arteries of the penis*. It may descend through the inferior fascia of the diaphragm before its bifurcation. (It is perhaps worthy of note that the part of the internal pudendal artery distal to its perineal branch has sometimes been known as the *artery of the penis*, an appropriate description in view of its distribution—*vide infra*.)

Relations. Within the pelvis the internal pudendal artery crosses anterior to piriformis, the sacral plexus, and the inferior gluteal artery. Crossing the ischial spine it is covered by gluteus maximus, lying between the pudendal nerve medially and the nerve to obturator internus lateral to it. In the perineum, as noted, it clings to the lateral wall of the ischiorectal fossa. Here, in the pudendal canal (p. 562), it travels at first with two companion veins and the pudendal nerve; beyond this, the latter's terminal branches flank the artery, the dorsal nerve of the penis above, the perineal nerve below.

Branches (**6**.96, 97):

Muscular branches leave the internal pudendal both in the pelvis and in the gluteal region to supply related muscles.

721

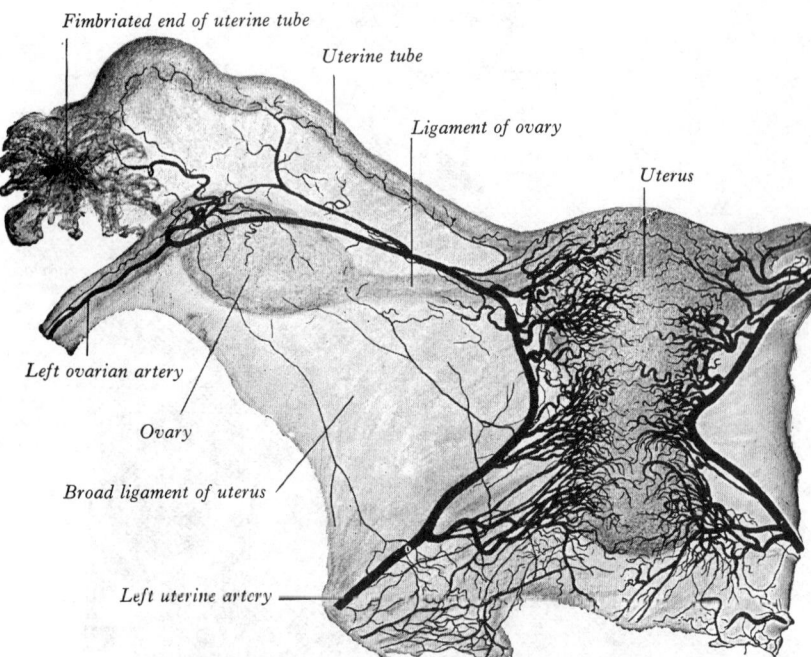

Fimbriated end of uterine tube

Uterine tube

Ligament of ovary

Uterus

Left ovarian artery

Ovary

Broad ligament of uterus

Left uterine artery

6.95   The left uterine and ovarian arteries of a nullipara of 17½ years. Posterior aspect. (From a preparation by Hamilton Drummond.)

The inferior rectal artery branches off above the ischial tuberosity. Escaping through the connective tissue wall of the pudendal canal (p. 562), it divides into two or three stems which cross the ischiorectal fossa medially to supply the skin and musculature of the anal region. Small branches skirt the lower edge of gluteus maximus to reach the skin of the buttock. Through its branches the inferior rectal anastomoses with its fellow, the superior and middle rectal arteries, and with the perineal artery.

The perineal artery (**6**.96) branches from the internal pudendal near the anterior end of the pudendal canal; it turns down through the inferior fascia of the urogenital diaphragm (p. 560) and then runs forwards towards the scrotum in the superficial perineal space, or region, between bulbospongiosus and ischiocavernosus. Just after its exit from the diaphragm, and near the latter's base, it has a small *transverse branch*, which passes medially on the inferior surface of the superficial transverse perineal muscle to anastomose with the opposite vessel and with the posterior scrotal and inferior rectal arteries. This small branch supplies tissues between the anus and penile bulb. The *posterior scrotal arteries*, small vessels distributed to the scrotal skin and dartos muscle, are usually the terminals of the perineal artery, but they may also be derived from its transverse branch. They also supply the musculature of the perineum.

The artery of the bulb of the penis, a short vessel of considerable calibre, runs medially through the deep transverse perineal muscle and the inferior fascia of the urogenital diaphragm to reach the bulb of the penis. Penetrating into this, it supplies the erectile tissue of the bulb and posterior part of the corpus spongiosum. It sends a branchlet to the bulbo-urethral gland.

The urethral artery also pierces the inferior fascia to enter the corpus spongiosum, travelling therein as far as the glans penis. It supplies the urethra and surrounding erectile tissue.

The deep artery of the penis, a terminal branch of the internal pudendal, passes through the inferior fascia of the urogenital diaphragm to enter the crus penis of its own side. It traverses the corpus cavernosum in a central position and supplies its erectile tissue.

The dorsal artery of the penis is the other terminal branch of the internal pudendal. Leaving the inferior aspect of the urogenital diaphragm, it ascends between the crus penis and the

pubic symphysis, passes through the suspensory ligament of the penis, and runs along its dorsum as far as the glans, where it forks into a branch to the glans itself and one to the prepuce. In the penis it is between the dorsal nerve and deep dorsal vein, the latter being most medial of the three structures. The artery supplies penile skin and the fibrous sheath of the corpus cavernosum, sending anastomotic branches through the sheath to the deep penile artery.

In the female the internal pudendal artery is naturally smaller, but its origin, course and branches are similar. *Posterior labial branches* supply the labia; the *artery of the bulb* is distributed to the erectile tissue of the vestibular bulb and vagina; a *deep artery of the clitoris* supplies the corpus cavernosum; and a *dorsal artery* supplies the dorsum, glans and prepuce of the clitoris.

Variations. Occasionally certain of the branches of the internal pudendal artery are derived from an *accessory pudendal*, usually a branch of the former before its exit from the pelvic cavity.

The inferior gluteal artery (**6**.94, 98), the larger terminal branch of the anterior trunk of the internal iliac, chiefly supplies the buttock and back of the thigh. It descends in front of the sacral plexus and piriformis, behind the internal pudendal artery. Passing between the first and second, or second and third sacral anterior rami, and then between piriformis and coccygeus, it traverses the lower part of the greater sciatic foramen to gain the gluteal region. It then descends between the greater trochanter and the ischial tuberosity with the sciatic and posterior femoral cutaneous nerves, covered by the gluteus maximus; it continues down the back of the thigh, supplying skin, and anastomosing with branches of the perforating arteries. Commonly, the inferior gluteal and internal pudendal arteries are a common stem from the internal iliac; the stem sometimes also includes the superior gluteal artery.

Surface Anatomy. The inferior gluteal artery leaves the pelvis at about the midpoint of a line joining the posterior superior iliac spine and the ischial tuberosity.

Branches:

Inside the pelvis: (*a*) to piriformis, coccygeus and levator ani; (*b*) to the fat around the rectum, and occasionally to take the place of the middle rectal artery; (*c*) vesical branches to the fundus of the bladder, seminal vesicles and prostate.

Outside the pelvis. Muscular branches supply gluteus maximus, obturator internus, the gemelli, quadratus femoris and the upper parts of the hamstring muscles; they anastomose with the superior gluteal, internal pudendal, obturator and medial circumflex femoral arteries.

Coccygeal branches run medially, pierce the sacrotuberous ligament, and supply the gluteus maximus and structures attached to the back of the coccyx.

The *artery to the sciatic nerve*, a long slender vessel, runs on the sciatic nerve for a short distance, penetrates it, and descends in its substance to the lower part of the thigh.

An *anastomotic branch* descends obliquely across obturator internus, the gemelli and quadratus femoris, to join the so-called *cruciate anastomosis* (p. 727) linking with the first perforating and the medial and lateral circumflex femoral arteries.

An *articular branch*, generally from the anastomotic, is distributed to the hip joint.

Cutaneous branches are distributed to the buttock and back of the thigh.

BRANCHES FROM THE POSTERIOR TRUNK OF THE INTERNAL ILIAC ARTERY

The iliolumbar artery (**6**.93) ascends laterally in front of the sacroiliac joint and lumbosacral trunk, behind the obturator nerve and external iliac vessels, to the medial border of psoas major, behind which it divides into a lumbar and iliac branch.

The lumbar branch supplies psoas major and quadratus lumborum, anastomoses with the fourth lumbar artery, and sends a small *spinal branch* through the intervertebral foramen between the fifth lumbar vertebra and the base of the sacrum, into the vertebral canal to supply the cauda equina.

The *iliac branch* supplies the iliacus; between the muscle and the bone, it anastomoses with iliac branches of the obturator; one branch enters an oblique canal to supply the ilium, others run along the iliac crest supplying the gluteal and abdominal muscles and anastomosing with the superior gluteal, circumflex iliac and lateral circumflex femoral arteries.

The lateral sacral arteries (**6**.93) arise from the posterior trunk of the internal iliac; there are usually a superior and an inferior. The *superior*, and larger, passes medially, enters the first or second pelvic sacral foramen, supplies the sacral vertebrae and the contents of the sacral canal, and escapes by the corresponding dorsal sacral foramen to reach the skin and muscles on the dorsum of the sacrum, anastomosing with the superior gluteal artery. The *inferior* crosses obliquely in front of the piriformis and the sacral anterior rami, descends lateral to the sympathetic trunk, and anastomoses with the median sacral and opposite lateral sacral artery over the coccyx. Branches from this vessel enter the pelvic sacral foramina to be distributed like those of the superior lateral sacral.

The superior gluteal artery (**6**.93, 98) is the largest branch of the internal iliac artery, and is the short continuation of its posterior trunk. It runs backwards between the lumbosacral trunk and first sacral ramus, or between the first and second rami, passing out of the pelvis through the greater sciatic foramen above the piriformis, and dividing into a *superficial* and *deep branch*. Within the pelvis it supplies piriformis and obturator internus and a nutrient artery to the innominate bone.

The *superficial branch* enters the deep surface of gluteus maximus, and has numerous branches; some supply the muscle and anastomose with the inferior gluteal artery; others perforate the tendinous origin of the muscle, to supply skin posterior to the sacrum, anastomosing with the posterior branches of the lateral sacral arteries.

The *deep branch* lies between gluteus medius and bone, soon forking into a superior and inferior division. The *superior* runs along the upper border of gluteus minimus to the anterior superior iliac spine, anastomosing here with the deep circumflex iliac and ascending branch of the lateral circumflex femoral artery. The *inferior division* crosses gluteus minimus obliquely, supplies it and the gluteus medius, and anastomoses with the lateral circumflex femoral artery; one branch enters the trochanteric fossa to join the inferior gluteal and the ascending branch of the medial circumflex femoral artery; other branches pierce the gluteus minimus and supply the hip joint.

The superior gluteal artery may arise from the internal iliac by a stem common to it and the inferior gluteal, and sometimes the internal pudendal.

Surface Anatomy. The superior gluteal artery emerges from the pelvis at the junction of the upper and middle thirds of a line joining the posterior superior iliac spine to the apex of the greater trochanter of the femur.

The External Iliac Arteries

The external iliac arteries (**6**.93, 99) are larger than the internal. Each runs obliquely down and laterally along the medial border of psoas major from the bifurcation of the common iliac artery to a point midway between the anterior superior iliac spine and symphysis pubis, where it enters the thigh behind the inguinal ligament to become the femoral artery.

In front and *medially* the external iliac artery is related to parietal peritoneum and extraperitoneal tissue, which separate the right artery from the terminal ileum and frequently the vermiform appendix, and the left artery from the sigmoid colon and some coils of small intestine. The beginning of the artery may be crossed by the ureter; in the female it is crossed by the ovarian vessels. The testicular vessels lie for some distance upon it near its termination, and here it is crossed by the genital branch of the genitofemoral nerve, the deep circumflex iliac vein, the ductus deferens or the round ligament of the uterus. *Posteriorly* it is separated from the medial border of psoas major by the iliac fascia. The external iliac vein is partly behind the upper part of the artery, but is medial to its lower part. *Laterally* it is related to psoas major, the iliac fascia being between them. Numerous

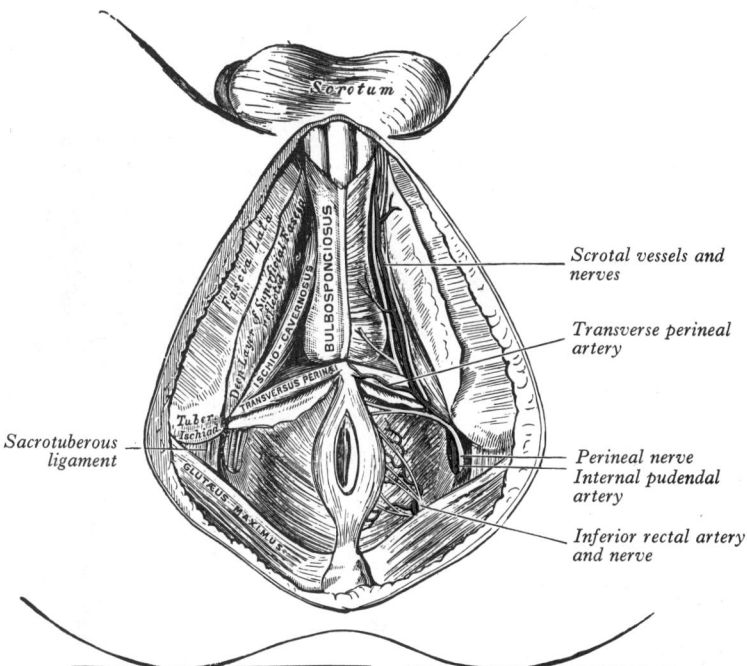

6.96　The superficial branches of the internal pudendal artery, in the male.

lymph vessels and nodes lie on the front and sides of the vessel.

Branches. Besides supplying small branches to psoas major and neighbouring lymph nodes, the external iliac artery gives off inferior epigastric and deep circumflex iliac branches.

The inferior epigastric artery (**5**.52, **6**.99) leaves the external iliac immediately above the inguinal ligament. It curves forwards in the extraperitoneal tissue, and ascends obliquely along the medial margin of the deep inguinal ring; continuing upward, it pierces the transversalis fascia, passes in front of the arcuate line, and ascends between rectus abdominis and the posterior lamina of its sheath. It finally divides into numerous branches, which anastomose, above the umbilicus, with those of the superior epigastric artery and the lower posterior intercostal arteries. The inferior epigastric artery thus skirts the lower and medial margins of the deep inguinal ring, and is behind the

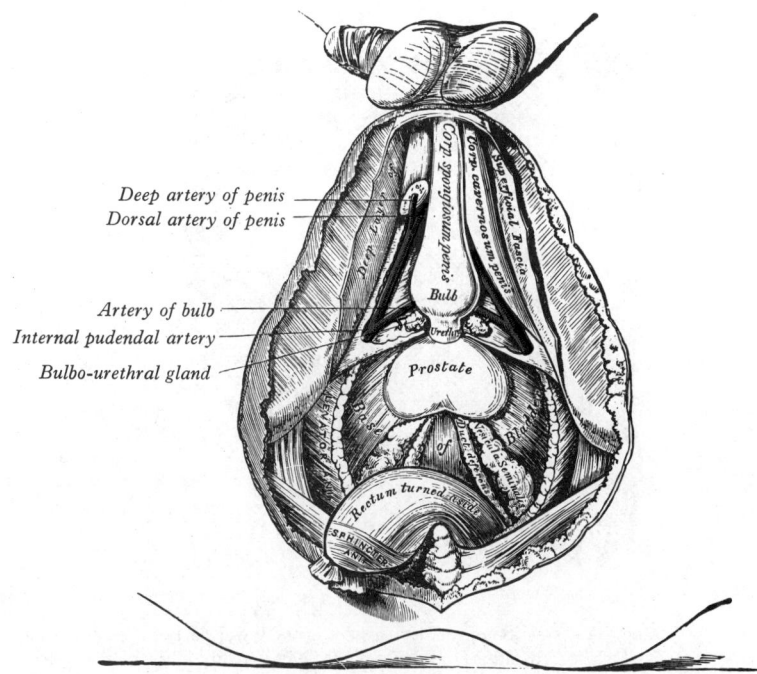

6.97　The deeper branches of the internal pudendal artery, in the male.

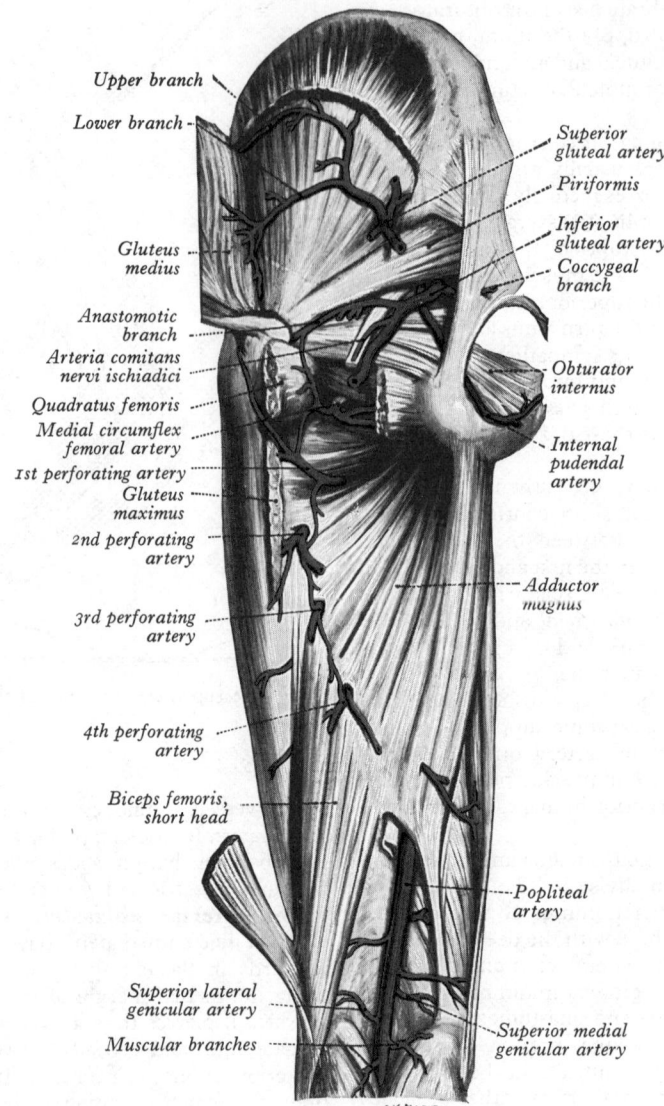

Upper branch

Lower branch

Gluteus medius

Anastomotic branch

Arteria comitans nervi ischiadici

Quadratus femoris

Medial circumflex femoral artery

1st perforating artery

Gluteus maximus

2nd perforating artery

3rd perforating artery

4th perforating artery

Biceps femoris, short head

Superior lateral genicular artery

Muscular branches

Superior gluteal artery

Piriformis

Inferior gluteal artery

Coccygeal branch

Obturator internus

Internal pudendal artery

Adductor magnus

Popliteal artery

Superior medial genicular artery

A·K·MAXWELL.

6.98   The arteries of the left gluteal and posterior femoral regions.

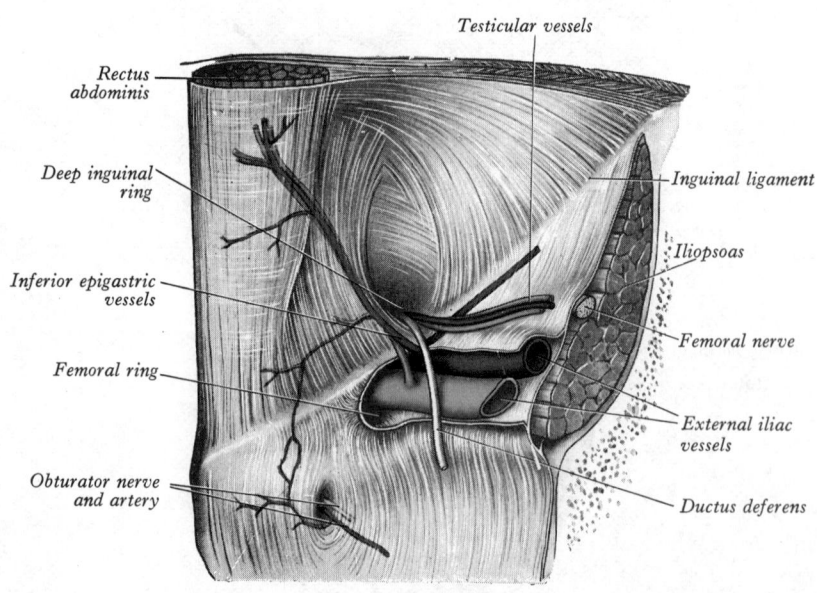

Testicular vessels

Rectus abdominis

Deep inguinal ring

Inferior epigastric vessels

Femoral ring

Obturator nerve and artery

Inguinal ligament

Iliopsoas

Femoral nerve

External iliac vessels

Ductus deferens

6.99   A dissection of the deep aspect of the lower part of the abdominal wall of the right side below the level of the arcuate line of the rectus sheath. The femoral and deep inguinal rings are displayed, together with the vessels and other structures in relation to them.

commencement of the spermatic cord, separated from it by the transversalis fascia. It raises the parietal peritoneum on the posterior surface of the anterior abdominal wall, forming the *lateral umbilical fold* (p. 1324). The ductus deferens, or the round ligament of the uterus, winds round its lateral surface. The inferior epigastric artery supplies the following branches:

The *cremasteric artery* accompanies the spermatic cord, supplies cremaster and other coverings of the cord, and anastomoses with the testicular artery. In the female the artery is very small and accompanies the round ligament of the uterus.

A *pubic branch* descends, close to the femoral ring behind the pubis, and there anastomoses with the pubic branch of the obturator. Between 20 and 30 per cent of pubic branches are large, and take the place of the obturator (p. 721).

Branches are distributed to the abdominal muscles and peritoneum, and anastomose with the circumflex iliac and lumbar arteries.

Cutaneous branches perforate the aponeurosis of the external oblique, supply the skin and anastomose with branches of the superficial epigastric artery.

Variations. The inferior epigastric artery may arise from the femoral artery, and then it ascends in front of the femoral vein to enter the abdomen. It frequently comes from the external iliac artery by a common trunk with an abnormal obturator and, rarely, it arises from the obturator artery.

Applied Anatomy. The inferior epigastric artery is one of the principal means, through its anastomosis with the internal thoracic, of establishing the collateral circulation after ligature of either the common or external iliac arteries. It is *medial* to the neck of the sac of an oblique inguinal hernia, but *lateral* to that of a direct inguinal hernia, as these emerge from the abdomen (p. 1367).

The deep circumflex iliac artery branches from the lateral side of the external iliac nearly opposite the inferior epigastric. It ascends laterally to the anterior superior iliac spine behind the inguinal ligament, in a sheath formed by junction of the transversalis and iliac fasciae. There it anastomoses with the ascending branch of the lateral circumflex femoral artery. It then pierces the transversalis fascia and passes along the inner lip of

the crest of the ilium to about its middle, where it perforates the transversus abdominis and runs back between this muscle and the internal oblique, to anastomose with the iliolumbar and superior gluteal arteries. At the anterior superior iliac spine it gives off a large *ascending branch* (**5**.50), which runs between the internal oblique and transversus, supplying them, and anastomosing with the lumbar and inferior epigastric arteries.

Collateral Circulation. A collateral circulation may be established in young adults after ligature of the common iliac artery; when the arterial walls are degenerated in older subjects it is unlikely to supply the limb efficiently.

THE ARTERIES OF THE LOWER LIMBS

The chief artery of the lower limb is the continuation of the external iliac. It extends from the level of the inguinal ligament to the lower border of popliteus, where it divides into the anterior and posterior tibial arteries. Its proximal part is the femoral artery, its distal, the popliteal.

The Femoral Artery

The femoral artery (**6**.100–104) is the continuation of the external iliac. It begins behind the inguinal ligament, midway between the anterior superior iliac spine and symphysis pubis, and passes down the front and medial side of the thigh. It ends at the junction of middle and inferior thirds of the thigh, where it passes through an opening in adductor magnus to become the popliteal. Above, the femoral artery is in the *femoral triangle*, below in the *adductor (subsartorial) canal*. The first 3 or 4 cm of the vessel are enclosed, with the femoral vein, in the *femoral sheath*.

The Femoral Sheath (**6**.100)

This is a downward prolongation, behind the inguinal ligament, of the transversalis fascia in front of the femoral vessels and the iliac fascia behind them. It has the form of a short funnel, wide end upwards, the lower, narrow end fusing with the fascial investment of the vessels, about 3 or 4 cm below the ligament. In the newborn the sheath is much shorter, becoming elongated after birth when extension of the thigh becomes habitual. The lateral wall of the sheath is vertical and perforated by the femoral branch of the genitofemoral nerve; the medial wall is directed obliquely down and laterally, and is pierced by the great (long) saphenous vein and lymphatic vessels. Like the carotid sheath the femoral sheath consists of a mass of connective tissue in which the femoral vessels are embedded. It is customary to describe three compartments; the lateral contains the femoral artery, the intermediate the femoral vein, while the medial and smallest compartment is named the *femoral canal*, and contains some lymph vessels and a lymph node, embedded in a small amount of areolar tissue; it is said to permit expansion of the vein. The canal is conical and measures 1·25 cm in length; its base, above, is the *femoral ring*, oval in form, its long or transverse diameter measuring 1·25 cm. The femoral ring (**6**.100) is bounded in front by the inguinal ligament, behind by pectineus and its fascia, medially by the crescentic edge of the lacunar ligament, and laterally by the femoral vein (*see*, however, p. 1367). The spermatic cord, or the round ligament of the uterus, lies immediately above the anterior margin of the ring, while the inferior epigastric vessels are close to the junction of its anterior and lateral boundaries. The ring is larger in women than in men due partly to the greater breadth of the female pelvis and partly to the smaller size of the femoral vessels. The femoral ring is filled by a condensed portion of the extraperitoneal tissue, the *femoral septum*, containing a small lymph node and covered by parietal peritoneum (see also p. 1367). The femoral septum is traversed by numerous lymph vessels connecting the deep inguinal to the external iliac lymph nodes.

The femoral triangle (**5**.85) underlies the depression immediately below the fold of the groin. Its apex is below, and its

limits are, laterally, the medial margin of sartorius, medially, the medial margin of adductor longus, above the inguinal ligament. The triangle is gutter-like and is floored laterally by iliacus and psoas major, and medially by pectineus and adductor longus. The femoral vessels, extending from near the middle of its base to its apex, lie in the deepest part of the gutter. Lateral to the femoral artery the femoral nerve divides into branches. The triangle also contains some fat and lymph nodes.

The adductor (subsartorial) canal (**6**.102) is an aponeurotic tunnel in the middle third of the thigh, from the apex of the femoral triangle to the opening in adductor mangus, through which the femoral vessels pass to the popliteal fossa. It is triangular on transverse section, bounded, in front and laterally, by vastus medialis, behind, by adductor longus above and adductor magnus below, anteromedially by a strong aponeurosis extending between these muscles, across the femoral vessels, to the vastus medialis. The sartorius lies anterior to them. The canal contains the femoral artery and vein, the saphenous nerve, and the nerve to the vastus medialis proximally until it enters its muscle.

The relations of the femoral artery. In the femoral triangle (**6**.103) the artery is covered with skin and superficial fascia, the superficial inguinal lymph nodes, the fascia lata and the femoral sheath, and it is crossed by the superficial circumflex iliac vein in the superficial fascia. The femoral branch of the genitofemoral nerve travels a short distance in the lateral part of the femoral sheath, at first lateral to and then in front of the artery. Near the apex of the triangle the medial cutaneous nerve of the thigh crosses the artery from its lateral to its medial side.

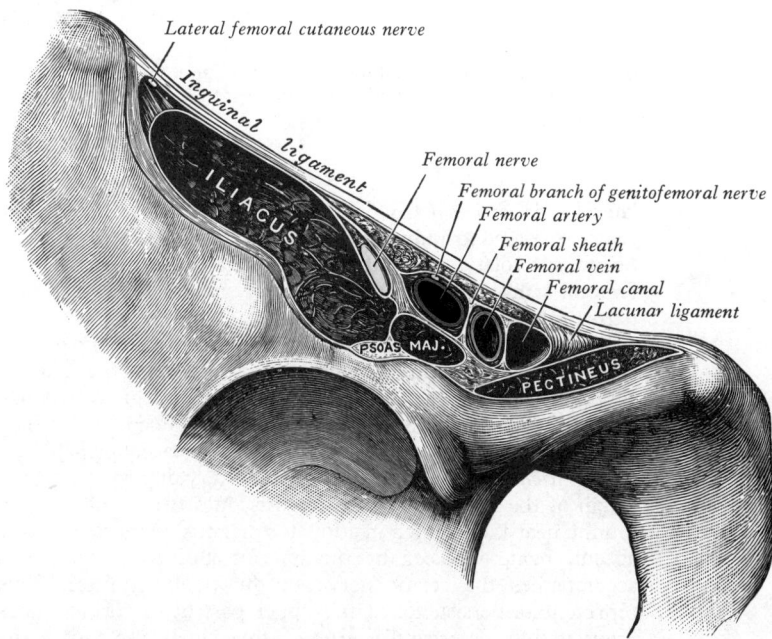

Lateral femoral cutaneous nerve

Inguinal ligament

ILIACUS

Femoral nerve

Femoral branch of genitofemoral nerve

Femoral artery

Femoral sheath

Femoral vein

Femoral canal

Lacunar ligament

PSOAS MAJ.

PECTINEUS

6.100 The structures passing posterior to the right inguinal ligament. Inferior (distal) aspect.

Behind, the artery is separated by the femoral sheath from the tendon of psoas major, pectineus and adductor longus, in this order distally. The artery is separated from the capsule of the hip joint by the tendon of psoas major, from pectineus by the femoral vein and profunda vessels, and from adductor longus by the femoral vein—the vessels having passed posterior to the adductor longus. The nerve to pectineus passes medially behind the upper end of the artery. Lateral to the artery is the femoral nerve. The femoral vein is medial to the artery in the upper part of the femoral triangle, and posterior in the lower part.

In the *adductor canal* (**6**.102, 103, 104) the femoral artery lies deeper, covered with skin, superficial and deep fasciae, sartorius and the fibrous roof of the canal. The saphenous nerve is at first

effectual immediately below the inguinal ligament. Here the artery is superficial and separated from the superior pubic ramus merely by the tendon of psoas major.

Branches:

The superficial epigastric artery (**6**.103) arises from the front of the femoral about 1 cm below the inguinal ligament. Piercing the femoral sheath and cribiform fascia it ascends in front of the ligament, and between the two layers of the superficial fascia of the abdominal wall nearly to the umbilicus. It supplies the superficial inguinal lymph nodes, superficial fascia and skin; it anastomoses with branches of the inferior epigastric and with the opposite vessel.

The superficial circumflex iliac artery (**6**.103), smallest of

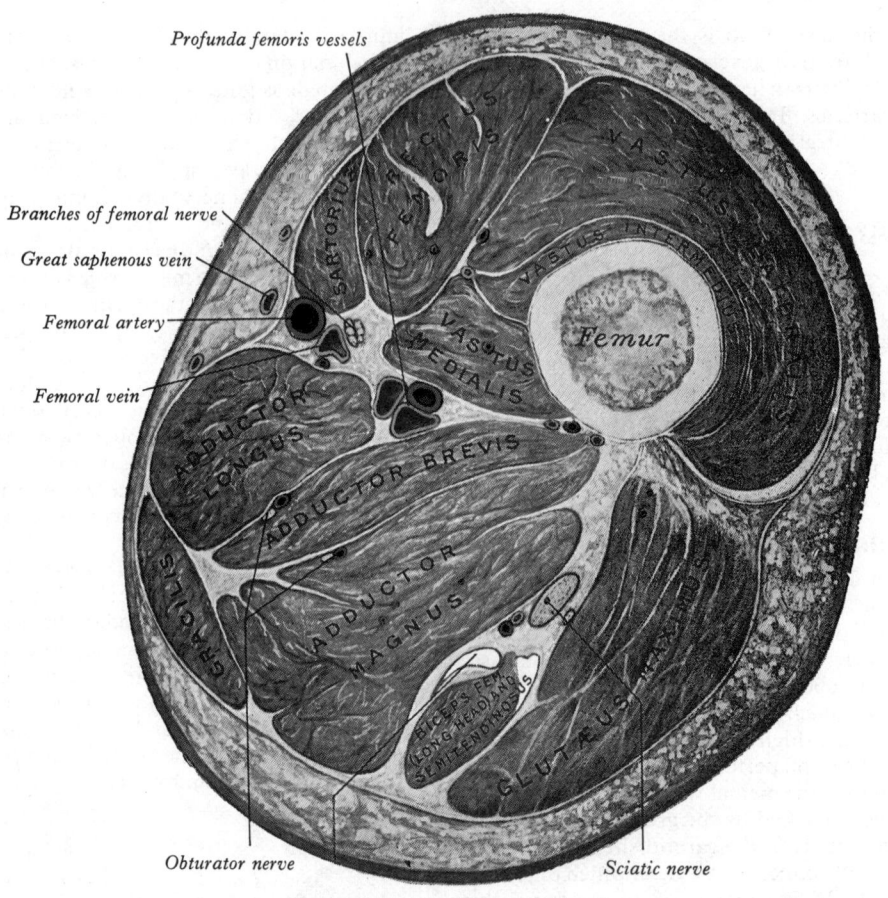

Profunda femoris vessels

Branches of femoral nerve

Great saphenous vein

Femoral artery

Femoral vein

Femur

Obturator nerve

Sciatic nerve

6.101　A transverse section through the right thigh at the level of the apex of the femoral triangle. Superior (proximal) aspect. About four-fifths of the natural size.

lateral and then lies in front of it, and is medial below. Behind, the artery is related to adductor longus above, and adductor magnus below; anterolateral to it are vastus medialis and its nerve. The femoral vein lies posterior to the upper, and lateral to the lower parts of the artery.

Surface Anatomy. The artery corresponds to the upper two-thirds of a line joining a point midway between the anterior superior iliac spine and pubic symphysis to the adductor tubercle when the thigh is semiflexed, abducted and rotated laterally. The pulsations of the upper part of the vessel are readily palpable.

Variations. In rare cases the femoral artery divides, below the origin of the arteria profunda femoris, into two trunks, which reunite near the opening in adductor magnus. Occasionally it is absent, being replaced by the inferior gluteal artery, which accompanies the sciatic nerve to the popliteal fossa. This represents a persistence of the upper part of the original axis artery and the external iliac artery is then small, and ends as the arteria profunda femoris.

Applied Anatomy. Compression of the femoral artery is most

the superficial branches of the femoral, arises close to, or in common with, the superficial epigastric artery. It usually emerges through the fascia lata lateral to the saphenous opening and turns laterally just below the inguinal ligament toward the anterior superior iliac spine; it gives branches to skin, superficial fascia and superficial inguinal lymph nodes, and anastomoses with the deep circumflex iliac, superior gluteal and lateral circumflex femoral arteries.

The superficial external pudendal artery (**6**.103) arises from the medial side of the femoral artery, close to the preceding vessels. Emerging from the femoral sheath and cribriform fascia, it goes medially, usually deep to the great saphenous vein, across the spermatic cord (or round ligament of the uterus), to the skin on the lower abdomen, the penis and scrotum or the labium majus, anastomosing with branches of the internal pudendal artery.

The veins accompanying the superficial epigastric, superficial circumflex iliac and external pudendal arteries join the great saphenous vein before it enters through the saphenous opening.

The **deep external pudendal artery** (6.103) passes medially across pectineus and either in front of or behind adductor longus; it is covered by the fascia lata, which it pierces at the medial side of the thigh, and is distributed to the skin of the scrotum and perineum or the labium majus; its branches anastomose with the posterior scrotal (or labial) branches of the internal pudendal artery.

Muscular branches are supplied to the sartorius, vastus medialis and the adductor muscles.

The Arteria Profunda Femoris (6.102, 103, 104)

This is a large branch from the lateral side of the femoral about 3·5 cm below the inguinal ligament. At first lateral to the femoral, the

The arteria profunda femoris is the principal supply to the adductor, extensor and flexor muscles, and in addition it establishes a number of anastomoses with the internal and external iliac arteries above and the popliteal below.

Branches:

The lateral circumflex femoral artery arises from the lateral side of the profunda, passes laterally between the divisions of the femoral nerve, behind sartorius and rectus femoris, and divides into ascending, transverse and descending branches. Occasionally it arises from the femoral.

The *ascending branch* passes up along the intertrochanteric line, under the tensor fasciae latae, to the lateral side of the hip; it anastomoses with the superior gluteal and deep circumflex iliac

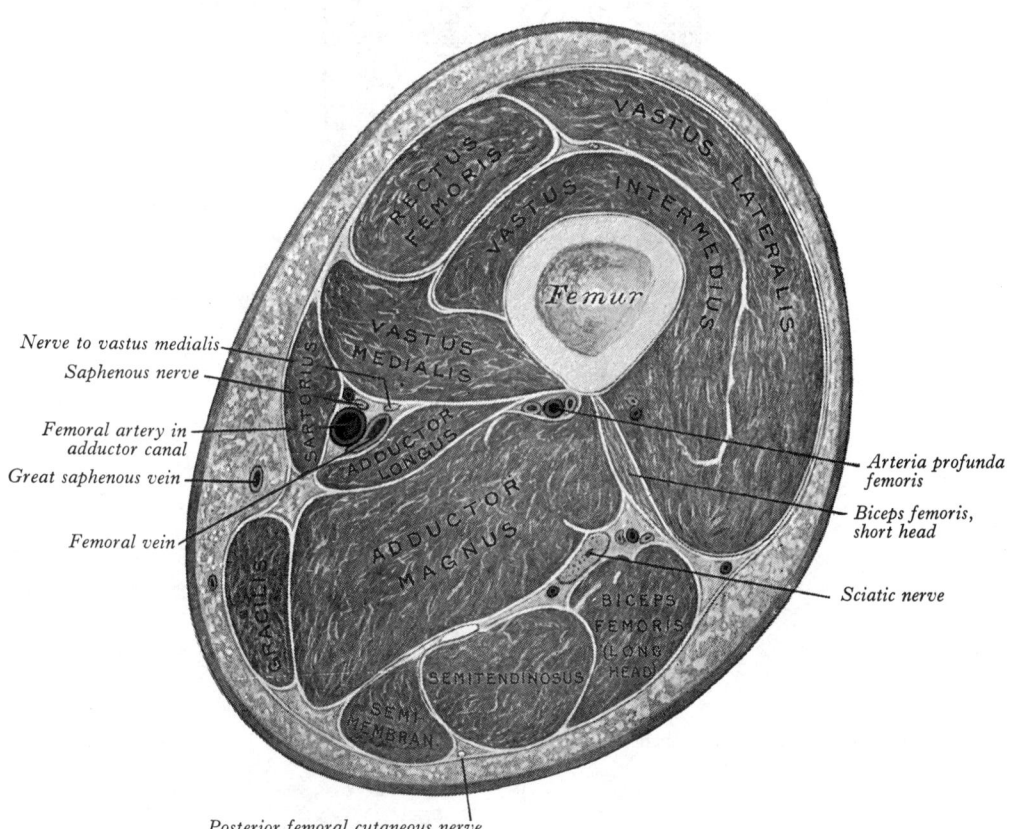

Nerve to vastus medialis

Saphenous nerve

Femoral artery in adductor canal

Great saphenous vein

Femoral vein

Posterior femoral cutaneous nerve

Arteria profunda femoris

Biceps femoris, short head

Sciatic nerve

6.102 A transverse section through the middle of the right thigh. Superior (proximal) aspect. About four-fifths of natural size.

profunda runs behind it and the femoral vein to the medial side of the femur, where it passes between pectineus and adductor longus and then between the latter and adductor brevis. It continues to descend between adductor longus and adductor magnus and ends by piercing the latter to anastomose with the upper muscular branches of the popliteal artery. This terminal part of the profunda is sometimes named the *fourth perforating artery.*

Relations. Behind, from above downwards, it is related to iliacus, pectineus, adductor brevis, and adductor magnus. *In front,* it is separated from the femoral artery by the femoral and profunda veins above, and by adductor longus below. *Laterally,* the origin of vastus medialis intervenes between the proximal part of the artery and the femur.

Variations. This vessel sometimes arises from the medial side, more rarely, from the back of the femoral artery. In the former condition it may cross in front of the femoral vein and then pass backwards round its medial side. The distance from the inguinal ligament of the origin of the vessel is variable. It is usually between 2·5 and 5 cm below it.

arteries. It supplies the greater trochanter, and with branches of the medial circumflex femoral artery forms an anastomotic ring round the base of the neck of the femur, from which the femoral head and neck are supplied.

The *descending branch,* sometimes an independent branch direct from the profunda, or from the femoral, runs down behind the rectus femoris, along the anterior border of the vastus lateralis, which it supplies: one long branch descends in the latter muscle as far as the knee, and anastomoses with the lateral superior genicular branch of the popliteal artery. It is accompanied by the nerve to vastus lateralis.

The *transverse branch,* the smallest, passes laterally in front of vastus intermedius, pierces vastus lateralis, and winds round the femur, just below the greater trochanter, anastomosing on the back of the thigh with the medial circumflex femoral, inferior gluteal, and first perforating arteries (*cruciate anastomosis*).

The medial circumflex femoral artery usually arises from the posteromedial aspect of the profunda, but frequently from the femoral artery. It supplies the adductor muscles and then winds

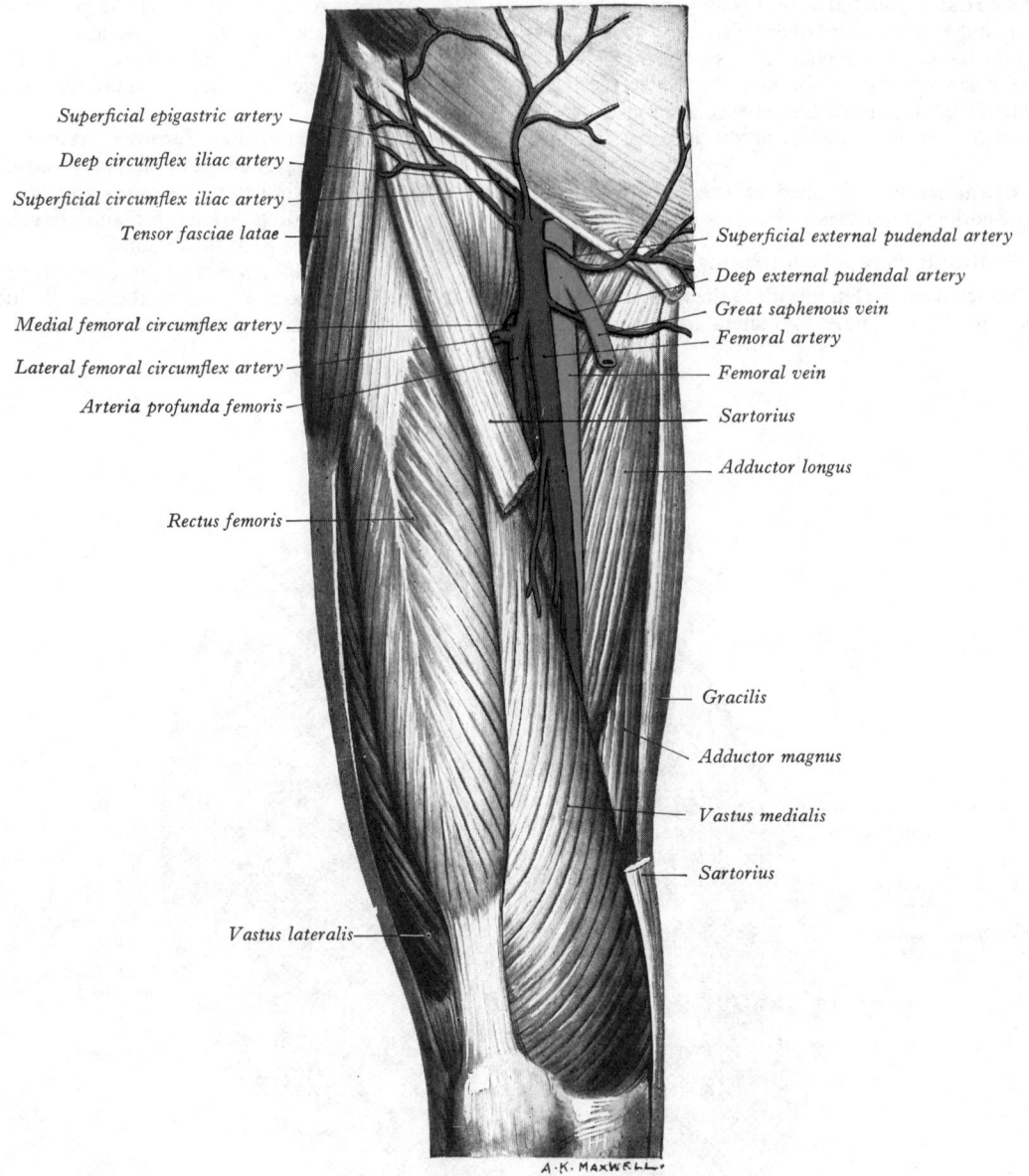

Superficial epigastric artery
Deep circumflex iliac artery
Superficial circumflex iliac artery
Tensor fasciae latae

Medial femoral circumflex artery
Lateral femoral circumflex artery
Arteria profunda femoris

Rectus femoris

Vastus lateralis

Superficial external pudendal artery
Deep external pudendal artery
Great saphenous vein
Femoral artery
Femoral vein
Sartorius

Adductor longus

Gracilis

Adductor magnus

Vastus medialis

Sartorius

A·K·MAXWELL·

6.103   The right femoral vessels and some of their branches.

medially round the femur, first between pectineus and psoas major and then between obturator externus and adductor brevis, and finally appears between quadratus femoris and the upper border of adductor magnus, where it divides into transverse and ascending branches. The *transverse branch* takes part in the cruciate anastomosis. The *ascending branch* runs obliquely upwards upon the tendon of obturator externus and in front of quadratus femoris towards the trochanteric fossa, where it anastomoses with branches of the gluteal arteries and lateral circumflex femoral artery. An *acetabular branch* from the medial circumflex femoral at the upper border of adductor brevis enters the hip joint below the transverse acetabular ligament with the acetabular branch of the obturator; it supplies fat in the acetabular fossa, and reaches the head of the femur along its ligament. For details of blood supply of the proximal extremity of the femur consult Crock (1965).

The perforating arteries (**6**.98), usually three, are so named because they perforate the insertion of the adductor magnus to reach the back of the thigh. They pass back close to the linea aspera of the femur under small tendinous arches in the insertion of the muscle, and give off muscular, cutaneous and anastomosing branches. Reduced in size, they pass deep to the short head of biceps femoris (the first usually pierces the insertion of gluteus maximus), go through the lateral intermuscular septum and enter the vastus lateralis. The first perforating artery is above the

adductor brevis, the second in front of it, and the third immediately below it.

The *first perforating artery* passes back between pectineus and adductor brevis (sometimes it perforates the latter); it then pierces adductor magnus close to the linea aspera. It supplies adductor brevis, adductor magnus, biceps femoris and gluteus maximus, and anastomoses with the inferior gluteal, medial and lateral circumflex femoral, and second perforating arteries.

The *second perforating artery*, larger than the first, frequently arising in common with it, pierces the insertions of adductor brevis and adductor magnus, and divides into ascending and descending branches, which supply the posterior femoral muscles, anastomosing with the first and third perforating arteries. The *nutrient artery* of the femur usually comes from it; when two nutrient arteries exist, they usually come from the first and third perforating vessels.

The *third perforating artery* starts below adductor brevis, pierces the insertion of adductor magnus, and divides into branches to the posterior femoral muscles; it anastomoses above with the higher perforating arteries, and below with the termination of the profunda and the muscular branches of the popliteal. The nutrient artery of the femur may arise from this branch. Offshoots from the principal diaphysial nutrient arteries and other branches of the profunda, also provide subsidiary cortical arteries of the femur (Crock 1967).

The termination of the profunda, as already described, is sometimes called the *fourth perforating artery*.

The perforating arteries form a *double chain of anastomoses:* (*a*) in the adductor muscles and (*b*) close to the linea aspera.

Numerous muscular branches arise from the arteria profunda femoris; some end in the adductors, others pierce adductor magnus, give branches to the hamstrings, and anastomose with the medial circumflex femoral artery and with the superior muscular branches of the popliteal artery. The profunda femoris artery is the *main* supply to the thigh muscles.

The *anastomosis on the back of the thigh*. This important chain of anastomoses stretches from the gluteal region to the popliteal fossa, and is formed from above downwards, as follows: (*a*) the gluteal arteries anastomose with the terminal branches of the medial circumflex femoral artery, (*b*) the circumflex femoral arteries with the first perforating artery, (*c*) the perforating arteries with each other, and (*d*) the fourth perforating artery with the superior muscular branches of the popliteal artery.

The descending genicular artery (**6**.108) arises from the femoral just before it passes through adductor magnus. It immediately gives off a *saphenous branch*, and then descends in the substance of vastus medialis, and in front of the tendon of adductor magnus, to the medial side of the knee, where it anastomoses with the medial superior genicular artery. *Muscular branches* supply vastus medialis and adductor magnus, and give off *articular branches*, which take part in the anastomosis round the knee joint. One of the articular branches crosses above the patellar surface of the femur, forming an arch with the lateral superior genicular artery, and supplying the knee joint.

The *saphenous branch* emerges through the lower part of the roof of the adductor canal and accompanies the saphenous nerve to the medial side of the knee. It passes between sartorius and gracilis, and is distributed to the skin of the upper and medial part of the leg, anastomosing with the medial inferior genicular artery.

Collateral Circulation. After ligation of the femoral artery above the origin of the profunda femoris artery, the main channels for carrying on the circulation are the following anastomoses: (1) the superior and inferior gluteal branches of the internal iliac artery with the medial and lateral circumflex femoral and first perforating branches of the arteria profunda femoris; (2) the obturator branch of the internal iliac artery with the medial circumflex femoral of the arteria profunda femoris; (3) the internal pudendal branch of the internal iliac artery with the superficial and deep external pudendal branches of the femoral artery; (4) the deep circumflex iliac branch of the external iliac artery with the lateral circumflex femoral branch of the arteria profunda femoris and the superficial circumflex iliac branch of the femoral artery; (5) the inferior gluteal branch of the internal iliac artery with the perforating branches of the arteria profunda femoris.

The Popliteal Fossa (**5**.100, **6**.106, **7**.216, **7**.217)

This is a diamond-shaped region at the back of the knee joint only apparent as a depression when disturbed by dissection. *Laterally*, it is bounded by biceps femoris above, the plantaris and lateral head of gastrocnemius below; *medially*, it is limited by semitendinosus and semimembranosus above and by the medial head of gastrocnemius below. *Anteriorly* the region is limited by the popliteal surface of the femur, the oblique popliteal ligament of the knee joint, the back of the upper end of the tibia and the fascia covering the popliteus, which collectively form a so-called floor. The fossa is covered *posteriorly* by the popliteal fascia.

Contents (**6**.106; **7**.216). Until its boundaries are disturbed, the popliteal fossa is only about 2·5 cm wide, and very little can be seen of its contents, especially so in the lower part of the space, where the two heads of gastrocnemius are in contact. When, however, the boundaries are drawn apart the fossa is seen to contain the popliteal vessels, tibial and common peroneal (medial and lateral popliteal) nerves, termination of the small saphenous vein, lower part of the posterior femoral cutaneous nerve, an articular branch from the obturator nerve, a few small lymph nodes and fat. The tibial nerve descends through the middle of the region lying anterior to the popliteal fascia, and crossing the vessels posteriorly from lateral to medial side. The common

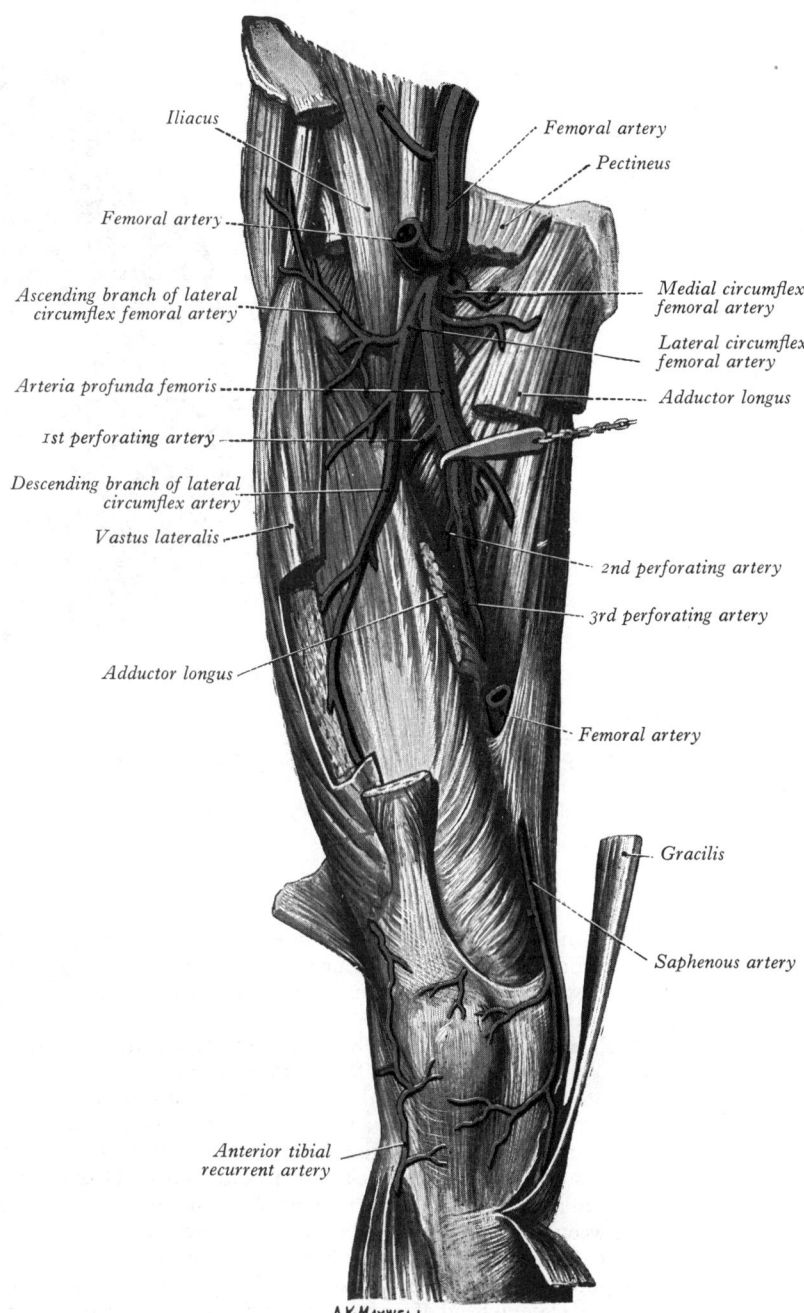

Iliacus

Femoral artery

Pectineus

Femoral artery

Ascending branch of lateral circumflex femoral artery

Medial circumflex femoral artery

Lateral circumflex femoral artery

Arteria profunda femoris

Adductor longus

1st perforating artery

Descending branch of lateral circumflex artery

Vastus lateralis

2nd perforating artery

3rd perforating artery

Adductor longus

Femoral artery

Gracilis

Saphenous artery

Anterior tibial recurrent artery

A.K.MAXWELL.

6.104   The right profunda femoris artery and its branches.

peroneal nerve descends laterally in the fossa, close to the tendon of biceps femoris. The popliteal vessels are on the floor of the fossa, the vein superficial to the artery and united to it by dense areolar tissue. The vein is thick-walled, and lateral to the artery above, and then crosses it posteriorly to its medial side below; sometimes it is double, the artery lying between the two veins, usually connected by short transverse branches. The articular branch from the obturator nerve descends upon the popliteal artery to the knee joint. The popliteal lymph nodes, six or seven in number, are embedded in the fat; one lies beneath the popliteal fascia near the termination of the small saphenous vein, another between the popliteal artery and the back of the knee joint, while the others flank the popliteal vessels.

The Popliteal Artery

The popliteal artery (**6**.105, 106, 107) is the continuation of the femoral, and traverses the popliteal fossa. It commences at the opening in adductor magnus, at the junction of middle and distal

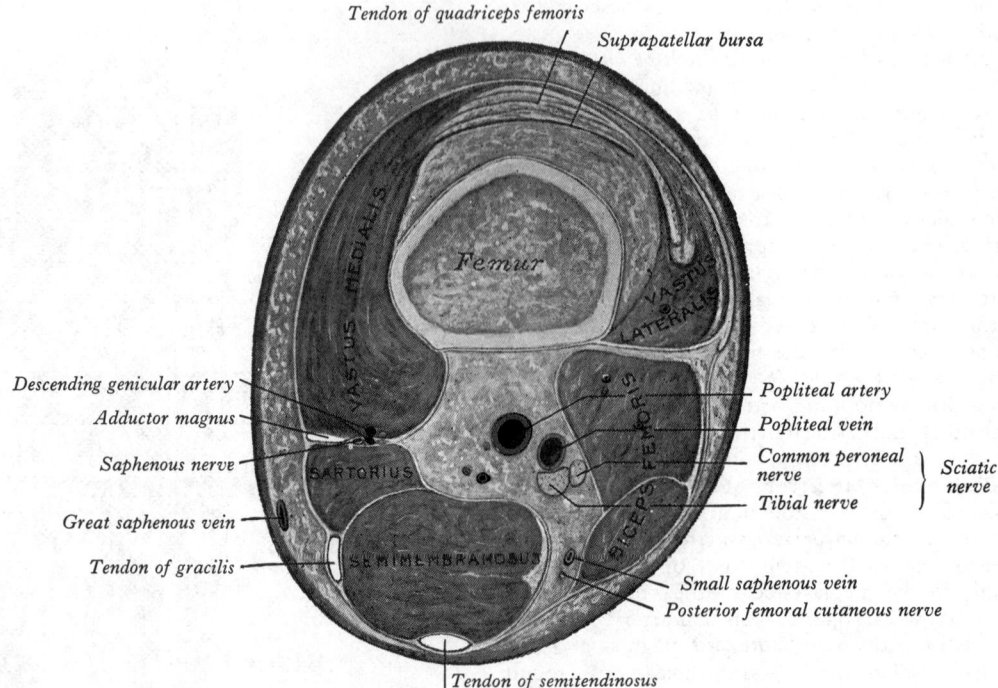

Tendon of quadriceps femoris

Suprapatellar bursa

Femur

VASTUS MEDIALIS

VASTUS LATERALIS

Descending genicular artery

Adductor magnus

Saphenous nerve

SARTORIUS

Great saphenous vein

Tendon of gracilis

SEMIMEMBRANOSUS

BICEPS FEMORIS

Popliteal artery

Popliteal vein

Common peroneal nerve

Tibial nerve

Sciatic nerve

Small saphenous vein

Posterior femoral cutaneous nerve

Tendon of semitendinosus

6.105   A transverse section through the right thigh, 4 cm above the adductor tubercle of the femur. Superior (proximal) aspect. About four-fifths of natural size.

thirds of the thigh, and extends downwards and laterally to the intercondylar fossa of the femur. Thence it continues *obliquely* to the lower border of popliteus, where it divides into *anterior* and *posterior tibial arteries* (*see* **6**.107).

Relations. In *front*, from above downwards, are the fat covering the popliteal surface of the femur, capsule of the knee joint, and the fascia of popliteus. *Behind* (superficially), it is overlapped by semimembranosus above, and is covered by gastrocnemius and plantaris below. In the middle part of its course the artery is separated from skin and fasciae by fat, and is crossed from lateral to medial side by the tibial nerve and popliteal vein, the vein being between the nerve and the artery and closely adherent to the latter. *Laterally*, above, are biceps femoris, the tibial nerve, popliteal vein and lateral condyle of the femur, below, plantaris and the lateral head of gastrocnemius. On its *medial* side, above, are semimembranosus and the medial condyle of the femur, below, the tibial nerve, popliteal vein and medial head of gastrocnemius. The relations of the popliteal lymph nodes to the artery are described on p. 791.

Variations. Occasionally the popliteal artery divides into terminal branches above the level of popliteus, and then the anterior tibial artery descends *anterior* to it. The popliteal artery sometimes divides into the anterior tibial and peroneal arteries, the posterior tibial artery being wanting or rudimentary; occasionally it divides into three branches, the anterior and posterior tibial, and peroneal arteries.

Surface Anatomy. The popliteal artery is represented by a line from the junction of the middle and lower thirds of the thigh 2·5 cm medial to the middle line of the back of the limb, running down and laterally to the midline between the femoral condyles. It continues obliquely inferolaterally until the level of the tibial tuberosity.

Branches. The popliteal artery has cutaneous, muscular and genicular branches, reaching the tibiofibular interosseous gap.

The cutaneous branches leave either the popliteal artery or some of its branches; they descend between the two heads of gastrocnemius, and pass through the deep fascia to the skin of the back of the leg; one usually accompanies the small saphenous vein.

Superior muscular branches (two or three) arise from the

upper part of the artery, and pass to adductor magnus and the hamstring muscles, anastomosing with the terminal part of the arteria profunda femoris.

Sural arteries (two in number and large) arise behind the knee joint to supply gastrocnemius, soleus and plantaris.

The superior genicular arteries (**6**.106, 108) diverge from the popliteal artery, and wind round the femur immediately above both condyles to gain the front of the knee joint. The *medial superior genicular artery* runs under semimembranosus and semitendinosus, above the medial head of gastrocnemius, and passes deep to the tendon of adductor magnus. It divides into a branch which supplies vastus medialis and anastomoses with the descending genicular and medial inferior genicular arteries, and one which ramifies on the surface of the femur and anastomoses with the lateral superior genicular artery. The size of the medial superior genicular artery varies inversely with that of the descending genicular. The *lateral superior genicular artery* passes under the tendon of biceps femoris, and divides into a superficial and a deep branch; the superficial supplies vastus lateralis, and anastomoses with the descending branch of the lateral circumflex femoral and the lateral inferior genicular artery; the deep branch anastomoses with the medial superior genicular artery and forms an arch across the front of the femur with the descending genicular.

The middle genicular artery, a small branch, arises from the popliteal artery opposite the back of the knee joint; it pierces the oblique popliteal ligament, and supplies the cruciate ligaments and the synovial membrane of the knee joint.

The inferior genicular arteries (**6**.106, 108) arise from the popliteal artery under gastrocnemius. The *medial* lies deep to the medial head of gastrocnemius and descends along the upper margin of popliteus, which it supplies; it then passes below the medial condyle of the tibia and under the tibial collateral ligament of the knee; at the anterior border of this ligament it ascends to the front and medial side of the joint, supplies it and the upper end of the tibia, and anastomoses with the lateral inferior and medial superior genicular arteries. It also anastomoses with the anterior tibial recurrent artery and the saphenous branch of the descending genicular artery. The *lateral inferior genicular artery* runs laterally across popliteus, and then forwards above the head

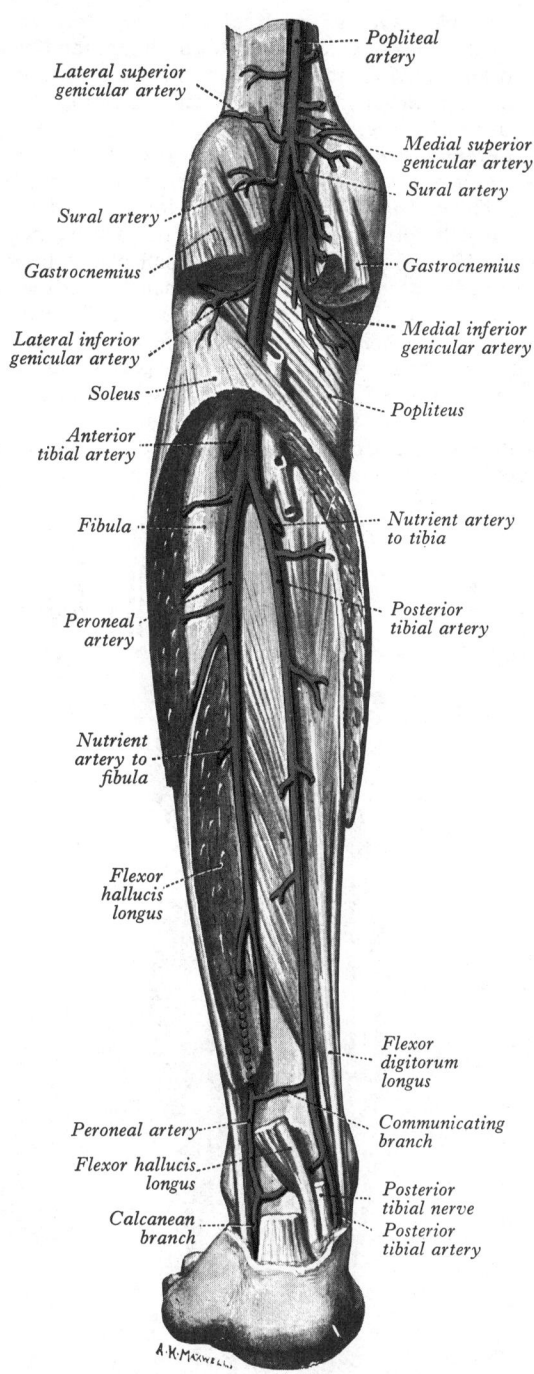

6.106   The left popliteal, posterior tibial and peroneal arteries. Dorsal aspect.

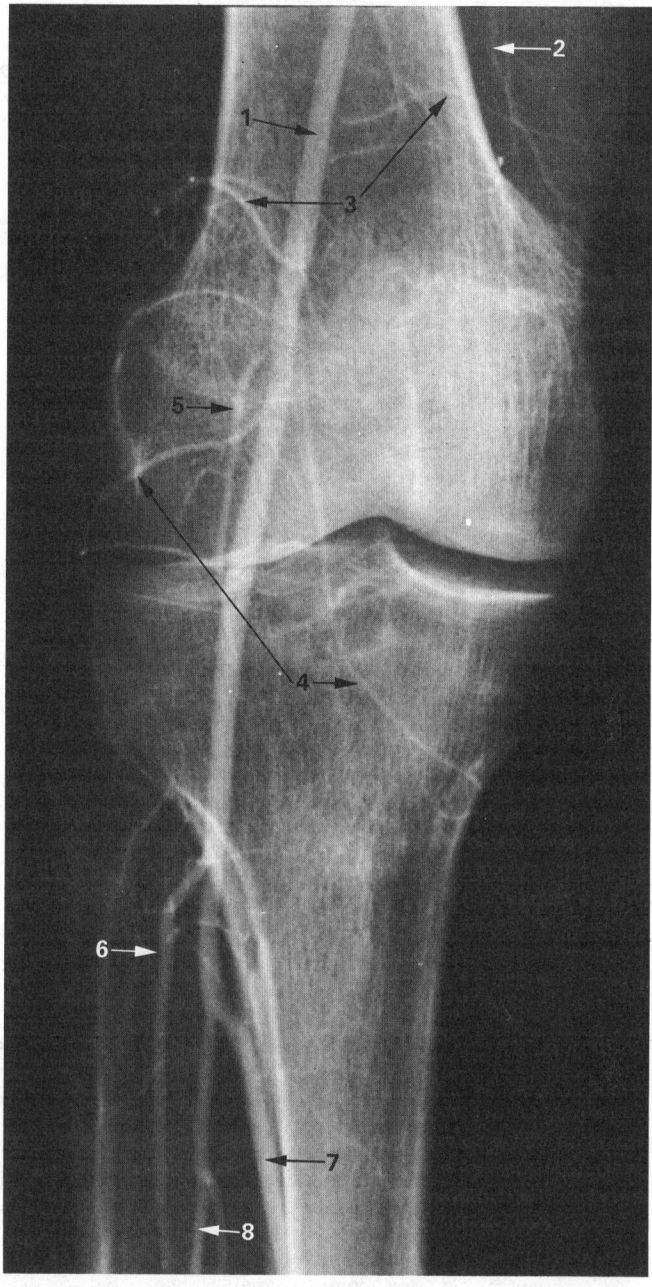

6.107   Popliteal arteriogram. Anteroposterior view. (Adult male of 63 years.) The following arteries can be identified—1. Popliteal. 2. Descending genicular. 3. Superior medial and lateral genicular. 4. Inferior medial and lateral genicular. 5. Middle genicular. 6. Anterior tibial. 7. Posterior tibial. 8. Peroneal. Note the obliquity of the popliteal artery.

of the fibula to the front of the knee joint, passing under the lateral head of gastrocnemius, the fibular collateral ligament of the knee, and the tendon of biceps femoris. It divides into branches which anastomose with the medial inferior genicular, lateral superior genicular and anterior and posterior tibial recurrent, and circumflex fibular arteries.

The Genicular Anastomosis (6.108)

Around the patella, and the contiguous ends of femur and tibia, an intricate arterial anastomosis exists. A *superficial network* is situated between the fascia and skin around the patella and in the fat behind the ligamentum patellae. A *deep network* lies on the femur and tibia round the adjoining articular surfaces, supplying bone and marrow, the capsule and synovial membrane of the joint. The vessels involved are the medial and lateral genicular, the descending genicular, the descending branch of the lateral circumflex femoral, the circumflex fibular, and the anterior and posterior tibial recurrent arteries.

The Anterior Tibial Artery (6.106, 107, 109, 110)

This is a terminal branch of the popliteal, arising at the lower border of popliteus. At first in the back of the leg, it passes forwards between the two heads of tibialis posterior and through the upper part of the interosseous membrane to the front of the leg, medial to the neck of the fibula. Descending on the front of the interosseous membrane, it gradually approaches the tibia, lying on it lower in the leg (6.111). In front of the ankle it is midway between the malleoli, and is continued on the dorsum of the foot as the *dorsalis pedis artery*.

Relations. In its upper two-thirds, the anterior tibial artery descends upon the interosseous membrane, in the lower third, upon the front of the tibia and ankle joint. In its upper third, it is between tibialis anterior and extensor digitorum longus; in the middle third between tibialis anterior and extensor hallucis longus. At the ankle it is crossed superficially from the lateral to the medial side by the tendon of extensor hallucis longus, and is then between this and the first tendon of extensor digitorum

731

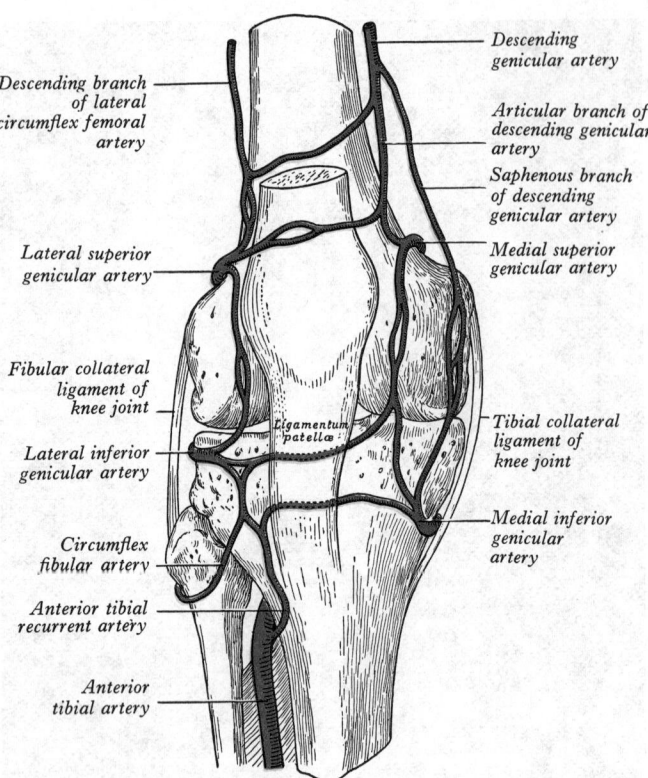

6.108   The arterial anastomosis around the knee joint. (Schematic.)

longus. Its upper two-thirds are covered by the muscles which flank it and by deep fascia, its lower third by the skin and fasciae, and the extensor retinacula.

Venae comitantes lie on each side of the artery. The deep peroneal nerve, coursing round the lateral side of the neck of the fibula, comes into relation with the lateral side of the artery shortly after it reaches the front of the leg; about the middle of the leg the nerve is in front of the artery; lower, it is generally lateral.

Surface Anatomy. In surface projection the anterior tibial artery begins 2·5 cm below the medial side of the head of the fibula and ends at the midpoint between the malleoli. The vessel can be felt pulsating lateral to the tendon of extensor hallucis longus at the ankle.

Variations. This vessel may be smaller than usual, or even absent, being replaced by perforating branches from the posterior tibial, or by the perforating branch of the peroneal artery. The artery occasionally deviates towards the fibular side of the leg, regaining its usual position at the front of the ankle.

Branches:

The posterior tibial recurrent artery—an inconstant branch—is given off from the anterior tibial before it reaches the front of the leg. It ascends in front of popliteus with the nerve to the muscle, anastomoses with the inferior genicular branches of the popliteal, and supplies the superior tibiofibular joint.

The anterior tibial recurrent artery (6.109) arises as soon as the anterior tibial has reached the front; it ascends in tibialis anterior, ramifies on the front and sides of the knee joint, and joins the patellar network, anastomosing with the genicular branches of the popliteal and the circumflex fibular artery.

Numerous muscular branches are distributed to the muscles on each side of the anterior tibial; some pierce deep fascia to supply skin, others pass through the interosseous membrane to anastomose with branches of the posterior tibial and peroneal arteries.

The anterior medial malleolar artery (6.109) starts about 5 cm above the ankle, passing behind the tendons of extensor hallucis longus and tibialis anterior to the medial side of the joint, where it joins branches of the posterior tibial and medial plantar arteries.

The anterior lateral malleolar artery (6.109) passes behind the tendons of extensor digitorum longus and peroneus tertius to supply the lateral side of the ankle, anastomosing with the perforating branch of the peroneal and ascending twigs from the lateral tarsal artery.

The arteries around the ankle joint anastomose freely (6.112) to form networks around and below the corresponding malleoli. The *medial malleolar network* is formed by the anterior medial malleolar branch of the anterior tibial artery, the medial tarsal branches of the dorsalis pedis, the malleolar and calcanean branches of the posterior tibial, and branches from the medial plantar artery. The *lateral malleolar network* is formed by the anterior lateral malleolar branch of the anterior tibial, the lateral tarsal branch of the dorsalis pedis, the perforating and the calcanean branches of the peroneal, and twigs from the lateral plantar artery.

The dorsal artery of the foot (arteria dorsalis pedis) (6.109) is the extension of the anterior tibial distal to the ankle joint. It

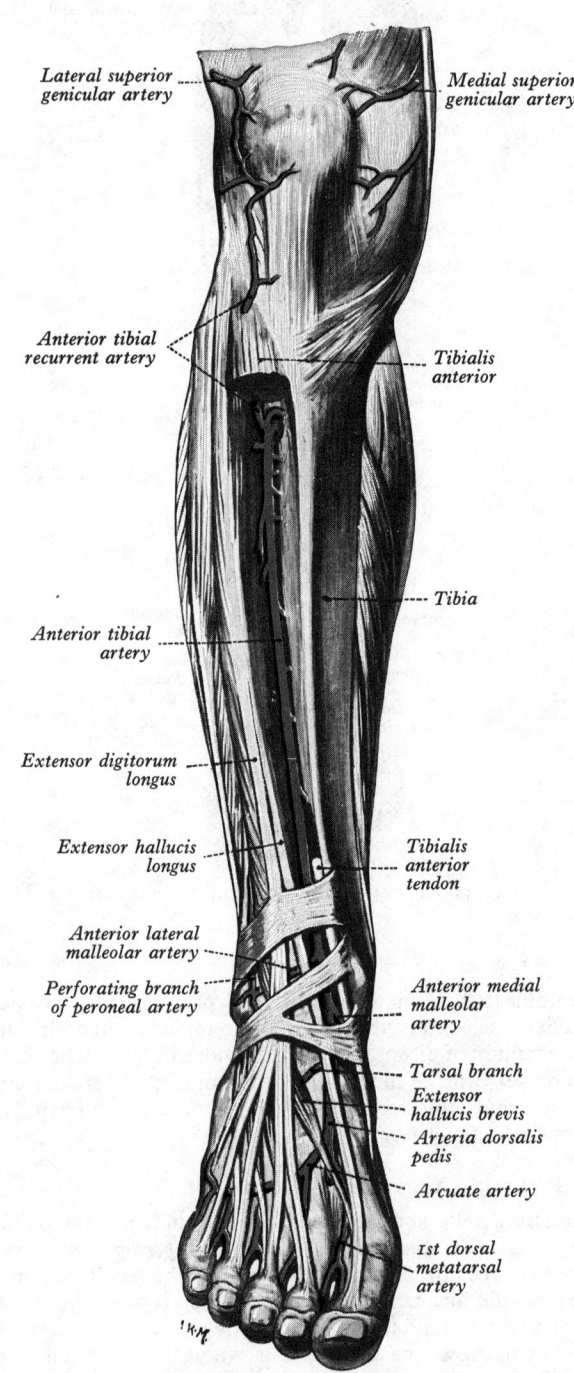

6.109   The right anterior tibial and dorsalis pedis arteries. A large part of the tibialis anterior has been excised and the extensor hallucis longus retracted laterally to expose the anterior tibial artery.

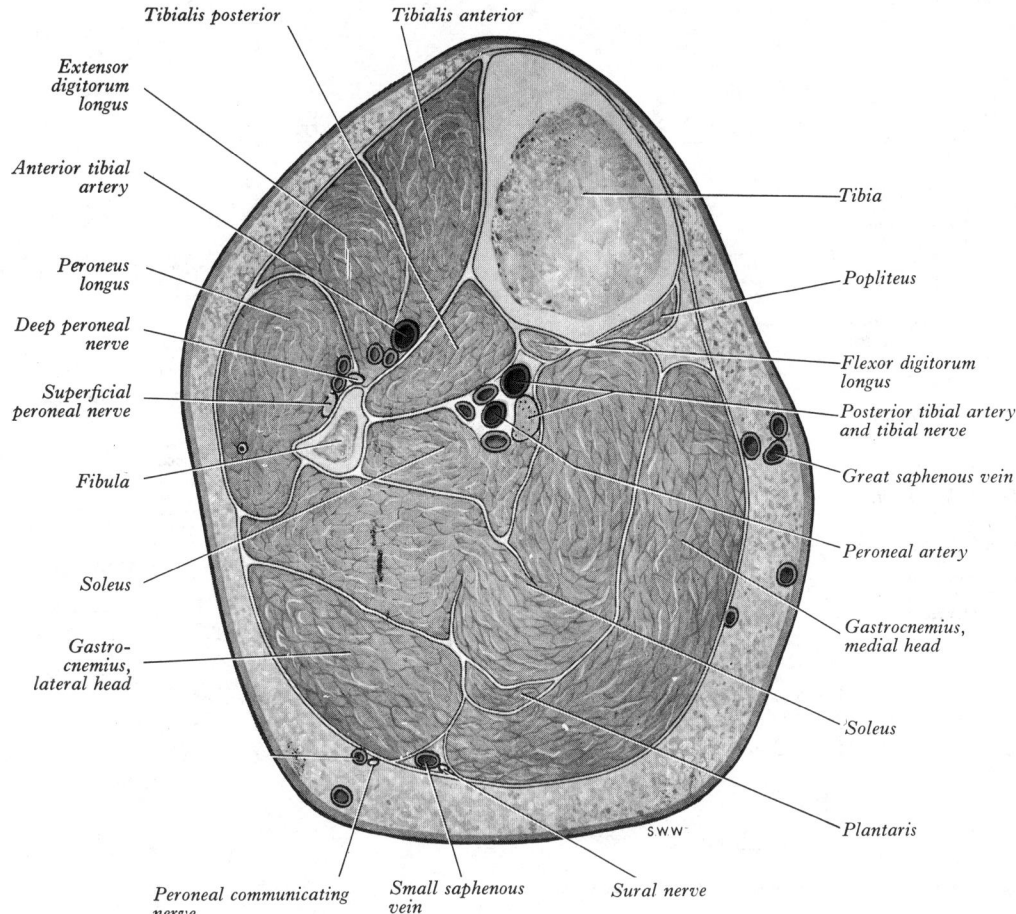

Tibialis posterior　*Tibialis anterior*

Extensor digitorum longus

Anterior tibial artery

Peroneus longus

Deep peroneal nerve

Superficial peroneal nerve

Fibula

Soleus

Gastro-cnemius, lateral head

Tibia

Popliteus

Flexor digitorum longus

Posterior tibial artery and tibial nerve

Great saphenous vein

Peroneal artery

Gastrocnemius, medial head

Soleus

Plantaris

Peroneal communicating nerve　*Small saphenous vein*　*Sural nerve*

SWW

6.110　A transverse section through the right leg, about 10 cm below the knee joint. Inferior (distal) aspect. At a slightly lower level the flexor digitorum longus intervenes between the soleus and the fascia on the posterior surface of the tibialis posterior.

follows the tibial side of the dorsum of the foot to the proximal end of the first intermetatarsal space, where it turns into the sole of the foot between the two heads of the first dorsal interosseous muscle, to complete the plantar arch. At this junction it provides the *first plantar metatarsal artery*.

Relations. The dorsal artery of the foot successively crosses in front of the articular capsule of the ankle, the talus, navicular and intermediate cuneiform, and their ligaments. Superficial to it are the skin, fasciae and inferior extensor retinaculum and, near its termination, the extensor hallucis brevis. *Medially* is the tendon of extensor hallucis longus; *laterally* are the first tendon of extensor digitorum longus, and the medial terminal branch of the deep peroneal nerve.

Surface Anatomy. Being superficial, the vessel can be felt pulsating along a line from the midpoint between the malleoli to the proximal end of the first intermetatarsal space.

Variations. The dorsal artery of the foot may be larger than usual, to compensate for a small lateral plantar artery; or it may be replaced by a large perforating branch of the peroneal artery. It frequently curves laterally, diverging from its usual direct route.

Branches. The dorsal artery of the foot has tarsal, arcuate and first dorsal metatarsal branches.

The tarsal arteries, lateral and medial (**6.**109), come from the arteria dorsalis pedis as it crosses the navicular bone. The lateral branches outwards under extensor digitorum brevis; it supplies this and the articulations of the tarsus, and anastomoses with branches of the arcuate, anterior lateral malleolar, and lateral plantar arteries, and with the perforating branch of the peroneal.

Two or three medial tarsal branches ramify on the medial border of the foot and join the medial malleolar network.

The arcuate artery (**6.**109) arises from the dorsal artery of the foot opposite the medial cuneiform; it passes laterally over the bases of the metatarsal bones deep to the tendons of the extensors

digitorum longus et brevis, and anastomoses with the lateral tarsal and lateral plantar arteries. It gives off the *second, third* and *fourth dorsal metatarsal arteries*, which run distally upon the corresponding dorsal interosseous muscles; in the clefts between the toes each divides into two dorsal digital branches for the sides of adjoining toes. At the proximal parts of the interosseous spaces

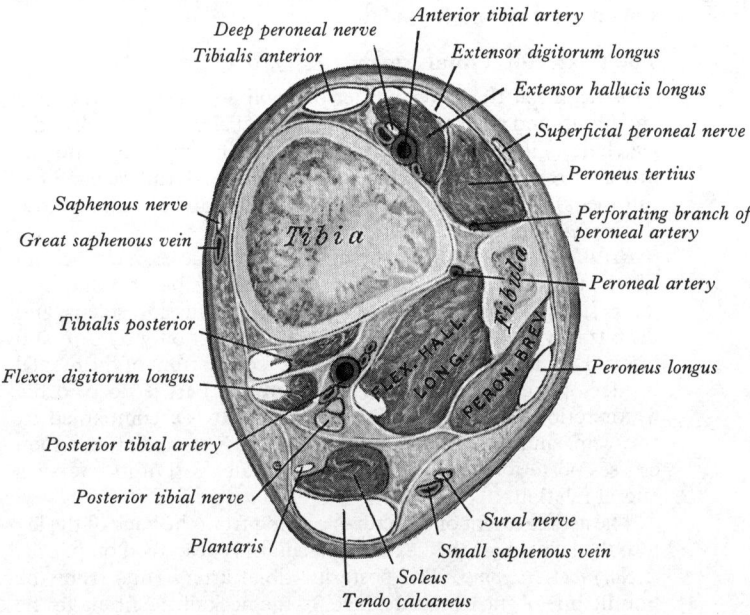

Deep peroneal nerve　*Anterior tibial artery*

Tibialis anterior　*Extensor digitorum longus*

Extensor hallucis longus

Superficial peroneal nerve

Peroneus tertius

Perforating branch of peroneal artery

Peroneal artery

Saphenous nerve

Great saphenous vein

Tibia

Fibula

Peroneus longus

Tibialis posterior

FLEX. HALL. LONG.

PERON. BREV.

Flexor digitorum longus

Posterior tibial artery

Posterior tibial nerve

Plantaris

Sural nerve

Small saphenous vein

Soleus

Tendo calcaneus

6.111　A transverse section through the right leg, about 6 cm above the tip of the medial malleolus. Superior (proximal) aspect.

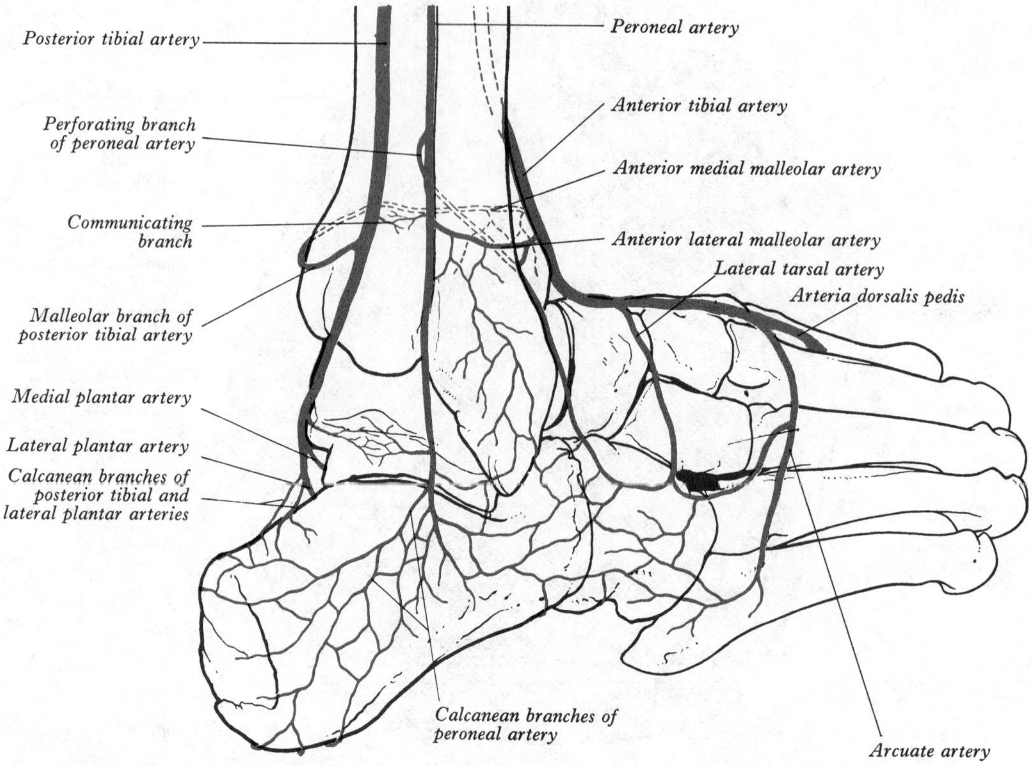

6.112   The arterial anastomoses of the ankle and tarsus and metatarus.

the dorsal metatarsal arteries receive the *proximal perforating branches* from the plantar arch, and at the distal parts of the spaces they are joined by the *distal perforating branches* from the plantar metatarsal arteries. The fourth dorsal metatarsal artery gives off a branch to the lateral side of the fifth toe.

The first dorsal metatarsal artery arises just before the arteria dorsalis pedis passes into the sole; it runs distally on the first dorsal interosseous muscle, and at the cleft between the first and second toes divides into two, a branch passing beneath the tendon of extensor hallucis longus and distributed to the medial border of the great toe, and one which bifurcates to supply the adjoining sides of the first and second toes.

The Posterior Tibial Artery (**6**.106, 107, 110)

This begins at the lower border of popliteus, between the tibia and fibula, and passes downwards and medially on the back of the leg. Below, it is midway between the medial malleolus and the medial process of the tuber calcanei (medial tubercle of the calcaneus). It divides under the origin of abductor hallucis into the *medial* and *lateral plantar arteries*.

Relations. The posterior tibial artery lies successively behind tibialis posterior, flexor digitorum longus, the tibia and the ankle joint. Its upper part is covered by gastrocnemius and soleus, and deep transverse fascia of the leg, its lower part only by skin and fascia. It runs parallel with and about 2·5 cm in front of the medial border of the tendo calcaneus; its terminal part is deep to the flexor retinaculum and abductor hallucis. It is accompanied by two veins and the tibial nerve, which is at first medial, but soon crosses posteriorly, to become, in the greater part of its course, a lateral relation.

The arrangement of structures passing from the back of the leg into the sole under the flexor retinaculum is described on p. 611.

Surface Anatomy. The posterior tibial artery runs from the middle line of the calf at the level of the neck of the fibula to the midpoint between the medial malleolus and the prominence of the heel, where its pulsations can be felt.

Branches:

The circumflex fibular artery, sometimes a branch of the anterior tibial artery, passes laterally round the neck of the fibula, through soleus, to anastomose with the lateral inferior genicular, the medial genicular and anterior tibial recurrent arteries. It supplies bone and articular structures.

The peroneal artery (**6**.106) arising from the posterior tibial about 2·5 cm below the lower border of popliteus passes obliquely towards the fibula, and descends along its medial crest in a fibrous canal between tibialis posterior and flexor hallucis longus, or in the substance of the latter muscle. It is then behind the tibiofibular syndesmosis, and divides into calcanean branches, which ramify on the lateral and posterior surfaces of the calcaneus. Its *upper* part is covered by soleus and the deep transverse fascia, between this and the deep muscles; its *lower* part is overlapped by flexor hallucis longus.

Variations. The peroneal artery may spring from the posterior tibial higher than usual, or may even branch from the popliteal; sometimes it arises 7 or 8 cm below the inferior border of popliteus. It is more frequently increased than diminished in size; it either joins and reinforces the posterior tibial artery, or takes its place when large, in the lower leg and foot.

Branches:

Muscular branches supply soleus, tibialis posterior, flexor hallucis longus and the peronei. A *nutrient artery* is directed downwards into the fibula. A *perforating branch* traverses the interosseous membrane about 5 cm above the lateral malleolus, to reach the front of the leg, where it anastomoses with the anterior lateral malleolar artery; it then descends in front of the tibiofibular syndesmosis, supplies the tarsus, and anastomoses with the lateral tarsal artery. The perforating branch is sometimes enlarged, and may take the place of the dorsalis pedis artery. A *communicating branch* connects the peroneal artery about 5 cm above the lower end of the tibia to the communicating branch of the posterior tibial artery. The *calcanean* or terminal branches of the peroneal artery communicate with the anterior lateral malleolar artery

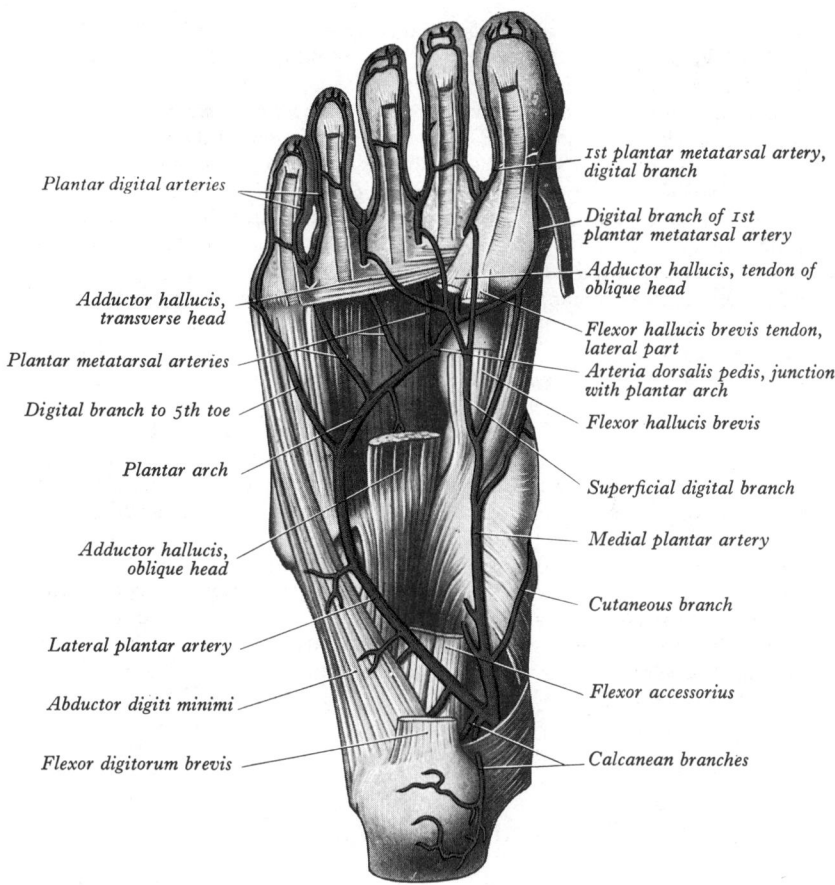

Plantar digital arteries

Adductor hallucis, transverse head

Plantar metatarsal arteries

Digital branch to 5th toe

Plantar arch

Adductor hallucis, oblique head

Lateral plantar artery

Abductor digiti minimi

Flexor digitorum brevis

1st plantar metatarsal artery, digital branch

Digital branch of 1st plantar metatarsal artery

Adductor hallucis, tendon of oblique head

Flexor hallucis brevis tendon, lateral part

Arteria dorsalis pedis, junction with plantar arch

Flexor hallucis brevis

Superficial digital branch

Medial plantar artery

Cutaneous branch

Flexor accessorius

Calcanean branches

6.113   The plantar arteries of the right foot.

lateral to the heel and with the calcanean branches of the posterior tibial artery behind it.

The nutrient artery of the tibia arises from the posterior tibial artery near its origin; after supplying a few minute muscular branches, it descends into the nutrient canal in the bone, immediately below the soleal line. It is one of the largest nutrient arteries in the body.

Muscular branches of the posterior tibial artery are distributed to soleus and the deep muscles on the back of the leg.

The communicating branch runs transversely across the back of the tibia about 5 cm above its lower end, deep to flexor hallucis longus, and joins the communicating branch of the peroneal.

The medial malleolar branches wind round the tibial malleolus to the medial malleolar network.

The calcanean branches arise from the posterior tibial just above its division; they pierce the flexor retinaculum to supply fat and skin behind the tendo calcaneus and about the heel, and the muscles on the tibial side of the sole; they anastomose with the medial malleolar arteries and calcanean branches of the peroneal arteries.

The medial plantar artery (6.113, 114), the smaller terminal branch of the posterior tibial, passes distally along the medial side of the foot with the medial plantar nerve, which is lateral to it. At first deep to abductor hallucis, it runs distally between this and flexor digitorum brevis, supplying both. At the base of the first metatarsal bone, much diminished in size, it passes along the medial border of the first toe and anastomoses there with a branch of the first metatarsal artery. It supplies three small superficial digital branches which accompany the digital branches of the medial plantar nerve and join the first, second and third plantar metatarsal arteries.

Surface Anatomy. The trunk of the medial plantar artery begins midway between the medial malleolus and the prominence of the heel and runs forwards in the direction of the first interdigital cleft as far as the navicular bone.

The lateral plantar artery (6.113), the larger terminal branch of the posterior tibial, passes laterally and distally to the base of the fifth metatarsal bone with the lateral plantar nerve on its medial side. Turning medially, with the deep branch of the nerve, to the interval between the bases of the first and second metatarsal bones, it unites with the dorsalis pedis artery, thus completing the *plantar arch*. As the lateral plantar artery passes

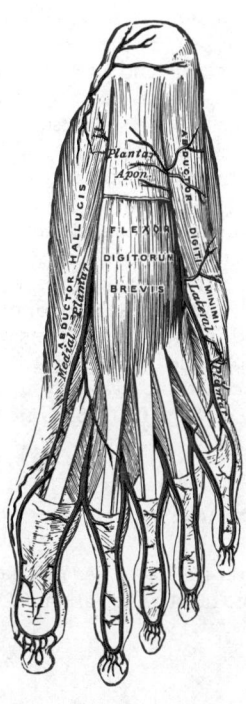

6.114   The plantar arteries. Superficial dissection.

laterally, it is at first between the calcaneus and abductor hallucis, and then between flexor digitorum brevis and flexor accessorius; as it runs distally to the base of the fifth metatarsal it lies between flexor digitorum brevis and abductor digiti minimi, and is covered by the plantar aponeurosis, superficial fascia and skin.

Branches. The lateral plantar artery has muscular, superficial, and anastomotic branches. The *muscular branches* supply adjoining muscles; the *superficial branches* emerge along the line of the lateral intermuscular septum and supply skin and subcutaneous tissue in the lateral part of the sole; the *anastomotic branches* run to the lateral border, where they anastomose with branches of the lateral tarsal and arcuate arteries. In addition, the lateral plantar artery sometimes has a *calcanean* branch, which pierces the origin of abductor hallucis to supply the skin of the heel.

The Plantar Arch

This is deeply situated and extends from the base of the fifth metatarsal bone to the proximal part of the first interosseous space. Convex distally, it is plantar to the bases of the second, third and fourth metatarsal bones and corresponding interossei, and dorsal to the oblique part of adductor hallucis.

Branches. The plantar arch has three perforating and four plantar metatarsal branches, and distributes numerous small vessels to skin, fasciae and muscles in the sole.

The three perforating branches ascend through the proximal ends of the second, third and fourth interosseous spaces, between the heads of the dorsal interosseous muscles, and anastomose with the dorsal metatarsal arteries.

The four plantar metatarsal arteries (**6**.113) extend distally between the metatarsal bones in contact with the interossei. Each divides into two *plantar digital arteries*, which supply the sides of adjacent toes. Near its division each plantar metatarsal artery sends dorsally a *distal perforating branch* to join the corresponding dorsal metatarsal artery. The first plantar metatarsal artery springs from the junction between the lateral plantar and the dorsalis pedis arteries, and sends a digital branch to the medial side of the first toe. The digital branch for the lateral side of the fifth toe arises from the lateral plantar artery near the base of the fifth metatarsal bone.

Surface Anatomy. Beginning at the termination of the posterior tibial, the lateral plantar artery crosses the sole obliquely to a point 2·5 cm medial to the tuberosity of the fifth metatarsal bone. From here, a line drawn with a slight forward convexity to the proximal end of the first intermetatarsal space indicates the course of the plantar arch.

Applied Anatomy. Wounds of the plantar arch are always serious, on account of the depth of the vessel and the important structures which must be interfered with in an attempt to ligature it. They must be treated on similar lines to those of wounds of the palmar arches (p. 708).

THE VENOUS SYSTEM

The veins consist of three sets—pulmonary, systemic, and portal (p. 710).

The pulmonary veins contain oxygenated blood, which they return from the lungs to the heart.

The systemic veins return venous blood to the heart. The *superficial veins* lie in the superficial fascia and are very variable in their disposition. The *deep veins* are usually enclosed in the same connective tissue sheaths with the arteries that they accompany, an arrangement which helps venous return (p. 630). Smaller arteries are generally accompanied by a pair of veins, lying on either side of the artery (*venae comitantes*), while larger arteries have usually only one accompanying vein. Some arteries, however, have no companion veins.

In general the systemic veins are more variable than the corresponding arteries and anastomoses occur more frequently and between larger vessels in the venous system. In many situations, such as the pelvis and in and around the vertebral column, the veins form extensive anastomosing plexuses and are characteristically devoid of valves. These plexuses provide the basis of free anastomosis between the veins of the trunk; they may also act as reservoirs of variable capacity in the vascular system. At a number of points, often at topographic junctional regions, and in the neighbourhood of principal joints, valved *connecting veins* interconnect the superficial and deep systemic veins.

The portal vein receives the radicles which drain venous blood from almost the whole of the subdiaphragmatic intestinal tract and its associated intrinsic and extrinsic glands, and from the spleen: having approached the porta hepatis it divides and, entering the substance of the liver, continually subdivides, ultimately discharging its contained blood into the hepatic venous sinusoids.

The Pulmonary Veins

The pulmonary veins return the oxygenated blood from the lungs to the left atrium of the heart. They are most commonly four in number, two from each lung, and are destitute of valves. They commence in the capillary network in the walls of the alveoli of the lungs, and, joining together, form a single trunk from each lobe, three from the right lung and two from the left. The vein from the middle lobe of the right lung generally unites with that from the upper lobe, so that ultimately two veins, a superior and an inferior, leave each lung; they perforate the fibrous layer of the pericardium and open separately into the upper and posterior part of the left atrium (**6**.16). Occasionally the three veins on the right side remain separate. Sometimes the two left pulmonary veins unite to form a single trunk before entering the heart. Occasionally the two left pulmonary veins, each draining a single lobe, may be augmented by *accessory lobar veins*, one from each lobe, and these may unite to form a *third* left pulmonary vein. (Consult Corry and Valentine 1959 for further details.)

In the root of the lung (pp. 1255, 1266), the superior pulmonary vein lies in front of and a little below the pulmonary artery; the inferior is situated at the lowest part of the hilum of the lung and on a plane posterior to that of the superior vein. The principal bronchus is behind the pulmonary artery. On the right side the superior pulmonary vein passes behind the superior vena cava, and the lower behind the right atrium. On the left side both pulmonary veins pass in front of the descending thoracic aorta. Within the pericardium, their anterior surfaces are invested by the serous layer of this membrane. Between the veins of the right and left side is the oblique sinus of the pericardium.

The Systemic Veins

The systemic veins may be arranged into three groups: (1) the veins which drain and discharge into *the heart*; (2) the veins of the upper limbs, head, neck and thorax, all of which end in *the superior vena cava*; (3) the veins of the lower limbs, abdomen and pelvis, all of which end in *the inferior vena cava*.

THE CARDIAC VEINS

The veins which drain the heart may be systematized in three groups: (*a*) *The coronary sinus* and its tributaries, which return the blood to the right atrium from all the heart (including its septa) except the anterior region of the right ventricle, and variable, but small parts of both atria and the left ventricle. (*b*) *The anterior cardiac veins*, whose tributaries drain the anterior part of the right ventricle, and a region extending round the 'acute' right border of the heart when the right marginal vein joins this group. (*c*) *The venae cordis minimae* (Thebesian veins), which open directly into the right atrium and ventricle and, to a lesser extent, into the left atrium and sometimes the left ventricle.

THE CORONARY SINUS

Most of the cardiac veins drain into the coronary sinus. This wide venous channel, about 2 or 3 cm long, is in the posterior part of the coronary sulcus (atrioventricular groove), between the left atrium and left ventricle (**6.**115). It ends in the right atrium between the opening of the inferior vena cava and the right atrioventricular orifice, its opening displaying a semilunar flap, the *valve of the coronary sinus* (**6.**18).

Its tributaries are the great, small, and middle cardiac veins, posterior vein of the left ventricle and oblique vein of the left atrium, all of which, except the last, have valves at their orifices.

The great cardiac vein (**6.**115) begins at the apex of the heart and ascends in the anterior interventricular sulcus to reach the coronary sulcus. It follows this groove round to the left and to the back of the heart, entering the beginning of the coronary sinus. It receives tributaries from the left atrium and both ventricles, including the *left marginal vein*, which ascends along the left aspect ('obtuse border') of the heart and is of considerable size.

The small cardiac vein (**6.**115) runs in the coronary sulcus between the right atrium and ventricle posteriorly, and opens into the end of the coronary sinus, i.e. near its termination. It receives blood from the back of the right atrium and ventricle; the *right marginal vein* passes to the right along the lower margin ('acute border') of the heart and may join the small cardiac vein in the coronary sulcus, more often it opens directly into the right atrium.

The middle cardiac vein (**6.**115) begins at the apex of the heart, runs backwards in the posterior interventricular groove, and ends in the coronary sinus near its termination.

The posterior vein of the left ventricle (**6.**115) runs on the diaphragmatic surface of the left ventricle a little to the left of the middle cardiac vein; it usually opens into the coronary sinus, but may end in the great cardiac vein.

The oblique vein of the left atrium, a small vessel, descends obliquely on the back of the left atrium to join the coronary sinus near its left extremity; it is continuous above with the *ligament of the left vena cava* (p. 193), and the two structures are remnants of the left common cardinal vein (**2.**119).

THE ANTERIOR CARDIAC VEINS

The *anterior cardiac veins* draw their tributaries from the anterior part of the right ventricle. Usually two or three in number, though as many as five may occur (Baroldi and Scomazzoni 1967), they ascend in the subepicardial stratum to cross the right part of the atrioventricular sulcus, where they pass deep or superficial to the right coronary artery. They terminate in the right atrium, close to the sulcus, either separately or by variable junctions. A subendocardial collecting channel, into which all may open, has been described (James 1961). **The right marginal vein**, which courses along the right 'border' of the heart, draining adjacent parts of the right ventricle, usually opens independently into the right atrium, but it may join the anterior cardiac veins or even, less frequently, the coronary sinus. Because of its usual independence it is sometimes grouped with the venae cordis minimae, but it is considerably larger in calibre, being comparable in size with the anterior cardiac veins, or even larger. It is perhaps better considered as one of the latter, which also sometimes drain with it into the coronary sinus. Mechanik (1934) described all the

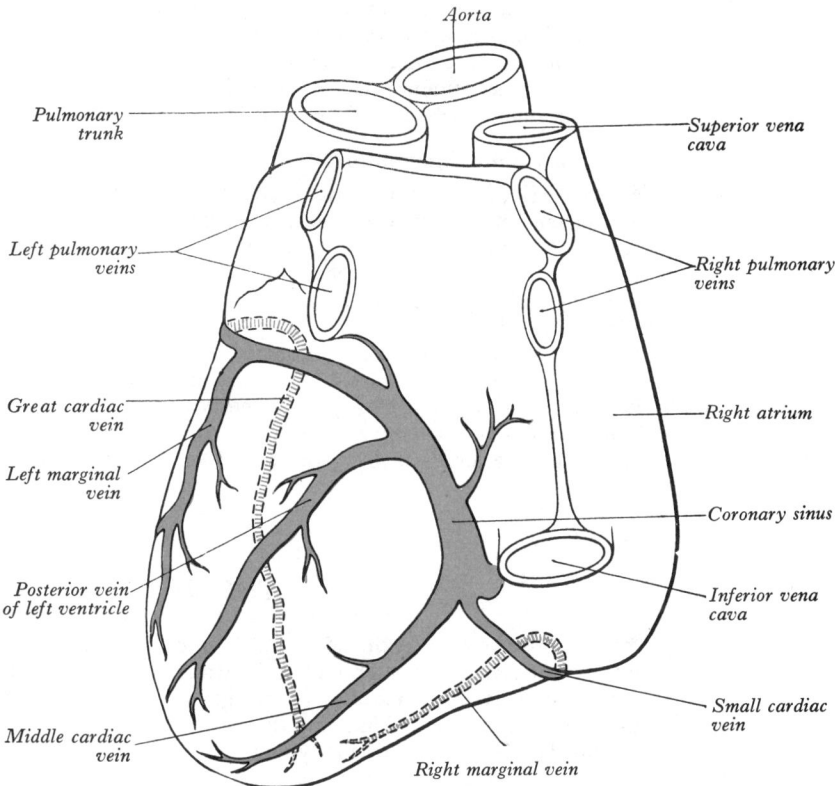

6.115A   The principal veins of the heart.

cardiac veins as draining into the coronary sinus in early fetal stages.

THE VENAE CORDIS MINIMAE

The existence of 'the smallest veins' or *venae cordis minimae*, opening into all the cavities of the heart, has been confirmed by many experimenters subsequent to their first recorded description by Thebesius (1708). They are, however, much more difficult to demonstrate than the larger cardiac vessels. Their numbers are highly variable, as is their size. Aho (1950)

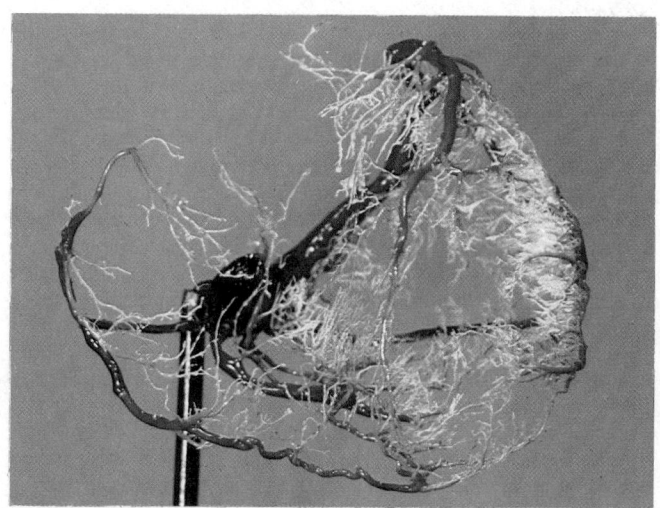

6.115B   A coloured resin cast of the coronary sinus and its tributaries. Of course, the venae cordis minimae and anterior cardiac veins are not filled. Compare with **6.**115A. (Specimen by Dr. D. H. Tompsett, Department of Anatomy, Royal College of Surgeons of England.)

demonstrated 'minimal' veins of up to 2 mm in diameter opening into the right atrium, and of about 0·5 mm into the right ventricle. He considered venae minimae as numerous in the right atrium and ventricle, occasional in and often absent from the left atrium, and rare in the left ventricle. Grant and Regnier (1926) considered that the venae minimae are derived from the pronounced intertrabecular spaces of the developing mammalian heart.

Cardiac Venous Anastomoses

As in the case of the coronary arteries, most investigators agree to the existence of widespread anastomosis at all levels of the cardiac venous circulation, and on a scale exceeding that of the arteries, amounting to a veritable venous network according to some authorities (Baroldi and Scomazzoni 1967). Not only are adjacent veins frequently interconnected, but there are also connexions between the tributaries of the coronary sinus and those of the anterior cardiac veins (see e.g. Mierzwa and Kozielec 1975). Regions of particularly marked anastomosis are the apex of the heart and its anterior and posterior aspects. As also in the case of the coronary arteries (p. 673), the cardiac veins connect with extracardiac vessels, chiefly with the vasa vasorum of the large vessels entering and leaving the heart.

Variation in the Cardiac Veins

Attempts have been made to categorize the individual variations encountered in the cardiac venous circulation (see Aho 1950) into a number of 'types', but these have not produced any agreed pattern. The major variations concern the general directions of drainage. The coronary sinus may receive all the cardiac veins (except the venae minimae), including the anterior cardiac veins (33 per cent). The latter may be reduced, by diversion of some of them into the small cardiac vein and thus to the coronary sinus (28 per cent). The percentages represent the views of Aho (1950), the remainder (39 per cent) representing the 'normal', most usual pattern, as described above. Baroldi and Scomazzoni (1967) distinguished two major variants: a majority (70 per cent), in which the small cardiac vein is independent, small or even absent, and a less frequent pattern (30 per cent) in which the small cardiac vein, though variable in size, connects with both the coronary and anterior cardiac 'systems'.

THE VEINS OF THE HEAD AND NECK

The venous channels of the head and neck may be subdivided into: (1) the veins of the exterior of the head and face; (2) the veins of the neck; (3) the diploic veins, the meningeal veins, the veins of the brain and the venous sinuses of the dura mater.

This classification is particularly significant at the cranial level, where the veins, like the arteries, are arranged as a three-layered system: (a) the vessels of the scalp, (b) the dural vessels, and (c) the cerebral and cerebellar vessels. There are, however, some differences between the arteries and veins, largely in respect of the interconnexions between the three levels. The arteries of the scalp and dura, both derived from the external carotid system, intercommunicate to a limited extent, as do the corresponding veins; the latter although quite variable usually communicate to a more extensive degree (see Emissary Veins, p. 749). The dural or meningeal arteries, on the other hand, are completely independent of the cerebral and cerebellar arteries, derived from the internal carotid, whereas the dural venous sinuses share a common drainage into the internal jugular vein, with the veins of the cerebrum and cerebellum. To this concept it is possible to add a fourth venous tier—the diploic veins; since these, however, drain directly into the dural veins, they are here grouped with the latter, following the schema of Browder and Kaplan (1976). It must be added at once that the intracranial veins communicate at many points with extracranial vessels by emissary and other veins (p. 749).

In usual accounts of the venous sinuses these channels are for the most part described as if they were simple arrangements with a single lumen. In embryological development, of course, these sinuses emerge as venous *plexuses*; and it is clear, from angiographic studies and examination of corrosion casts, that most of the sinuses preserve a *plexiform* arrangement to a variable degree. Browder and Kaplan (1976) have studied human venous sinuses in hundreds of corrosion casts, and they have observed complex vascular plexuses adjoining, in particular, the superior and inferior sagittal and straight sinus, and with a lesser incidence in the transverse sinuses. The details of these arrangements show much individual variation, and departures from 'average' patterns are specially frequent in the earlier years of life. For example, the falx cerebelli may, in infancy, contain large plexiform channels and even venous lacunae, augmenting the occipital sinus. Such individual variations obviously cannot be minutely considered in a general text, and must in any case be established for the individual by angiography when clinical necessity arises. It is important to recognize, however, the wide variations possible in the structure of the venous sinuses within the cranium, and also their frequent plexiform nature and wide connexions with cerebral and cerebellar veins. One other kind of connexion may be mentioned here. Experimental evidence (Rowbotham and Little 1962; Browder and Kaplan 1976) shows that some parts of the sinus system (and even the diploic veins) can be filled by injectants forced into the internal carotid artery, suggesting the existence of arteriovenous shunts. Browder and Kaplan, by injection of the middle meningeal artery, established a connexion between this vessel and the superior sagittal sinus. The precise location of such anastomoses requires further investigation.

External Veins of Head and Face (6.116)

The supratrochlear vein starts on the forehead from a venous network which communicates with the frontal tributaries of the superficial temporal vein. Veins converge from the network into a single trunk, which descends on the forehead near the midline parallel with its fellow. At the root of the nose the two supratrochlear veins are joined transversely by the *nasal arch*, draining small veins from the dorsum of the nose. The supratrochlear veins then diverge, and at the medial angle of the orbit, each joins with a *supraorbital vein* to form the (*anterior*) *facial vein*. Occasionally the supratrochlear veins unite in a single trunk, which divides at the root of the nose into the two anterior facial veins.

The supraorbital vein begins near the zygomatic process of the frontal bone, where it communicates with the superficial and middle temporal veins. It courses medially along the upper margin of the orbital opening under orbicularis oculi, and, near the medial angle of the eye, pierces this muscle to form, with the supratrochlear, the *facial vein*. It sends a branch through the supraorbital notch into the orbital cavity to join the superior ophthalmic vein; as this branch traverses the supraorbital notch it is joined by the frontal diploic vein and by a vein draining the frontal air sinus..

The facial vein, formed by junction of the supratrochlear and supraorbital veins, runs obliquely downwards on the side of the root of the nose to the level of the lower margin of the orbital opening. It then runs downwards and backwards behind the facial artery, but follows a less tortuous course. It passes under zygomaticus major, risorius and platysma, descends along the anterior border of and then on the surface of masseter, crosses the body of the mandible, and runs obliquely backwards deep to platysma and superficial to the submandibular gland, digastric and stylohyoid. A little below and in front of the angle of the mandible it is joined from above and behind by the anterior division of the retromandibular vein, and then, descending across the loop of the lingual artery, hypoglossal nerve and external and internal carotid arteries, it enters the internal jugular vein near the

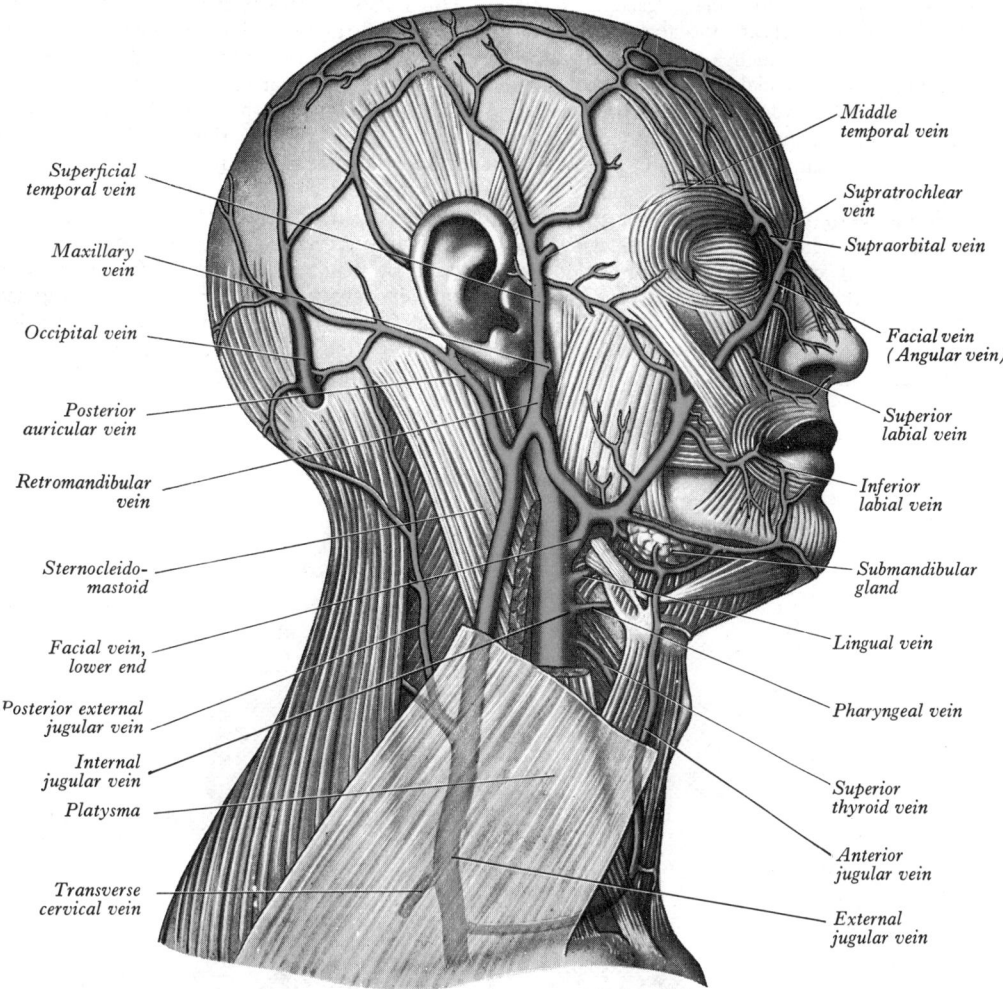

6.116   The veins of the right side of the head and neck. Parts of the right sternocleidomastoid and platysma have been excised to expose the trunk of the internal jugular vein. The external jugular vein is visible through the lower part of the platysma.

greater cornu of the hyoid bone. Near the termination of the facial vein a branch of considerable size often diverges down the anterior border of sternocleidomastoid to the lower part of the anterior jugular vein.

Above its junction with the superior labial vein (*vide infra*) the facial vein is often termed the *angular vein*.

Tributaries. At the commencement the facial vein connects with the superior ophthalmic vein both directly and also through the supraorbital vein (*vide supra*). The facial vein is thus connected with the cavernous sinus through the superior ophthalmic vein. It receives the veins of the ala nasi and, at a lower level, a large branch, the *deep facial vein*, from the pterygoid venous plexus. It is also joined by the inferior palpebral, superior and inferior labial, buccinator, parotid and masseteric veins. Below the mandible it receives the submental, tonsillar, external palatine (paratonsillar) and submandibular veins. The facial vein may be joined by the vena comitans of the hypoglossal nerve, and often also the pharyngeal and superior thyroid veins.

Applied Anatomy. The facial vein has no valves and it communicates with the cavernous sinus by two routes; firstly, through connexions directly with the ophthalmic vein or with its tributary, the supraorbital; secondly, by the deep facial vein, which links it to the pterygoid plexus and hence also to the cavernous sinus. Thus infective thrombosis of the facial vein may extend into the intracranial venous sinuses.

The superficial temporal vein (6.116) begins in a network which extends widely over the scalp. Through this it is joined to the corresponding vein of the opposite side, and with the supratrochlear, supraorbital, posterior auricular and occipital veins, all of which drain the same network. Anterior and posterior tributaries

unite above the zygomatic arch to form the superficial temporal vein, which is joined here by the *middle temporal vein*. It then crosses the posterior root of the zygomatic arch, enters the parotid gland, and unites with the maxillary to form the *retromandibular vein*.

Tributaries. The superficial temporal vein receives some veins from the parotid gland, articular veins from the temporomandibular joint, anterior auricular veins from the external ear, and the *transverse facial* from the face. The middle temporal vein, after receiving the *orbital vein*, formed by lateral palpebral branches, passes back between the layers of the temporal fascia and then becomes superficial to join the superficial temporal vein.

The pterygoid plexus is of considerable size, and is placed partly between the temporalis and lateral pterygoid, and partly between the two pterygoids. The sphenopalatine, deep temporal, pterygoid, masseteric, buccal, dental, greater palatine and the middle meningeal veins, and a branch or branches from the inferior ophthalmic vein are all tributaries. The pterygoid plexus anastomoses with the facial vein through the *deep facial*; it is also connected with the cavernous sinus by veins which pass through the sphenoidal emissary foramen, foramen ovale and foramen lacerum. The deep temporal veins frequently connect with the tributaries of the anterior diploic vein (p. 743) and, through these, with the middle meningeal veins.

The maxillary vein, a short trunk, accompanies the first part of the corresponding artery and is formed by confluence of the veins of the pterygoid plexus. It passes back between the sphenomandibular ligament and the neck of the mandible, and unites with the superficial temporal to form the retromandibular vein.

The retromandibular vein (posterior facial), descends in the parotid gland, superficial to the external carotid artery but deep to the facial nerve. It divides into two branches, an anterior which passes forwards and unites with the facial vein, and a posterior which joins the posterior auricular to form the external jugular vein. Occasionally the retromandibular vein is not connected with the external jugular, and then the latter is small and the anterior jugular vein often very large.

The posterior auricular vein (6.116) begins in the posterior part of the scalp in a network which communicates with the tributaries of the occipital and superficial temporal veins. It descends behind the auricle, and joins the posterior division of the retromandibular vein in or just below the parotid gland to form the external jugular vein. It receives the stylomastoid vein and tributaries from the cranial surface of the auricle.

The external jugular vein (6.116) receives blood mostly from the scalp and face, including its deeper parts. It is formed by union of the posterior division of the retromandibular vein with the posterior auricular vein. It begins level with the mandibular angle just below, or in the parotid gland, and runs down the neck towards the middle of the clavicle. It crosses sternocleidomastoid obliquely, and in the subclavian triangle perforates the deep fascia to end in the subclavian vein, lateral to, or in front of, the scalenus anterior; the wall of the vein is adherent to the circumference of the opening in the deep fascia. It is covered by platysma, superficial fascia and skin, and separated from sternocleidomastoid by the investing layer of the deep cervical fascia; it crosses the transverse cervical nerve, and its upper half runs parallel with the great auricular nerve, which ascends behind it. The vein varies in size, bearing an inverse

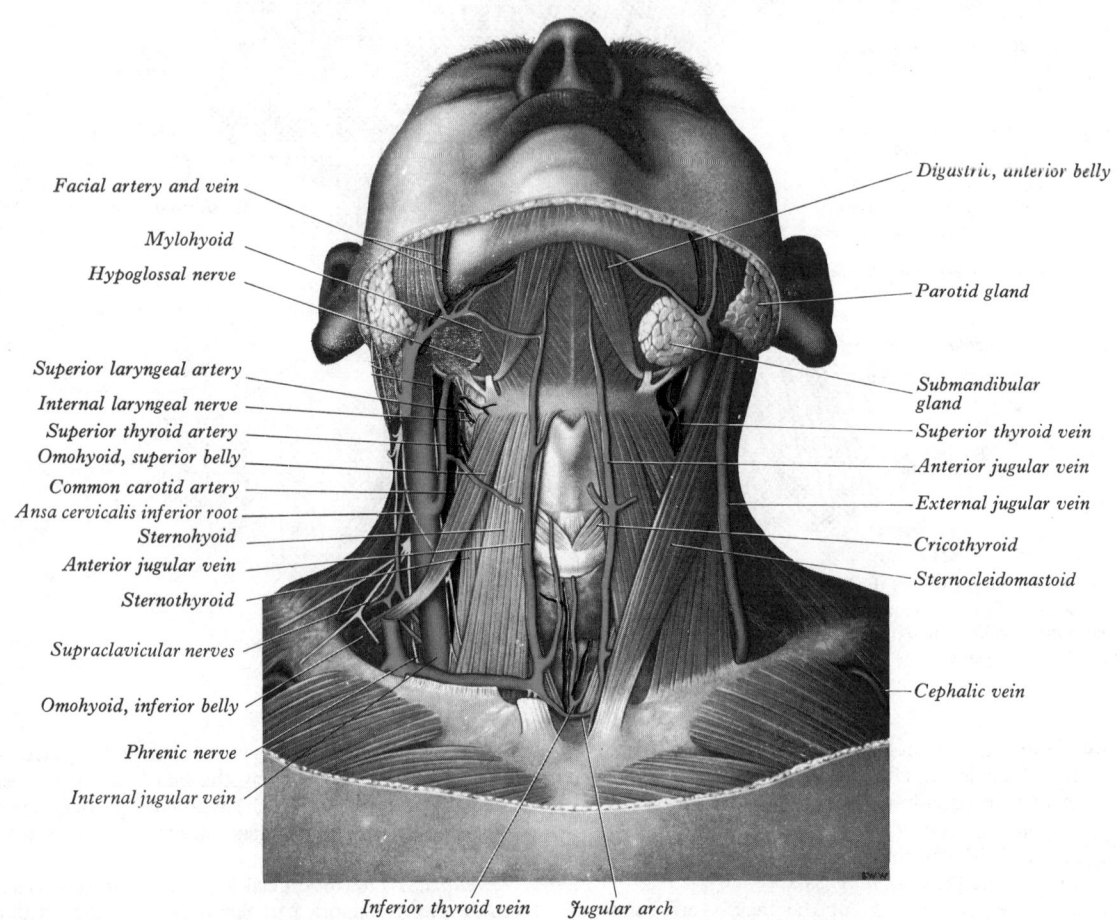

Facial artery and vein

Mylohyoid

Hypoglossal nerve

Superior laryngeal artery

Internal laryngeal nerve

Superior thyroid artery

Omohyoid, superior belly

Common carotid artery

Ansa cervicalis inferior root

Sternohyoid

Anterior jugular vein

Sternothyroid

Supraclavicular nerves

Omohyoid, inferior belly

Phrenic nerve

Internal jugular vein

Digastric, anterior belly

Parotid gland

Submandibular gland

Superior thyroid vein

Anterior jugular vein

External jugular vein

Cricothyroid

Sternocleidomastoid

Cephalic vein

Inferior thyroid vein   Jugular arch

6.117   Anterior view of the veins of the neck.

The occipital vein (6.116) begins in a venous network at the posterior part of the scalp. It pierces the cranial attachment of trapezius, dips into the suboccipital triangle and joins the deep cervical and vertebral veins. Occasionally it follows the occipital artery and ends in the internal jugular vein; sometimes it joins the posterior auricular vein and, through it, opens into the external jugular vein. The parietal emissary vein connects it with the superior sagittal sinus, and the mastoid emissary vein connects it with the transverse sinus. The occipital diploic vein sometimes joins it.

Veins of the Neck (6.116, 117, 131)

The veins of the neck may be divided into those superficial to the deep fascia and those deep to it; but the two groups are not entirely independent. The superficial veins, tributaries of the external jugular, drain a much smaller volume of tissue than the deep veins, which collect blood from all except subcutaneous structures, returning it largely into the internal jugular vein.

proportion to the other veins of the neck; it is occasionally double. It has two pairs of valves, a lower at its entrance into the subclavian vein, an upper about 4 cm above the clavicle. The part of the vein between the two sets of valves is often dilated, and is sometimes termed the *sinus*. These valves do not prevent regurgitation of blood.

Tributaries. In addition to its formative tributaries, the external jugular vein receives the posterior external jugular and, near its end, the transverse cervical, suprascapular and anterior jugular veins; in the parotid gland it is frequently joined by a branch from the internal jugular vein. The occipital vein occasionally drains into it.

The posterior external jugular vein begins in the occipital region and returns the blood from the skin and superficial muscles in the upper and posterior part of the neck. It opens into the middle part of the external jugular vein.

The anterior jugular vein (6.116, 117) starts near the hyoid bone by the confluence of several superficial veins from the submandibular region. It descends between the anterior median

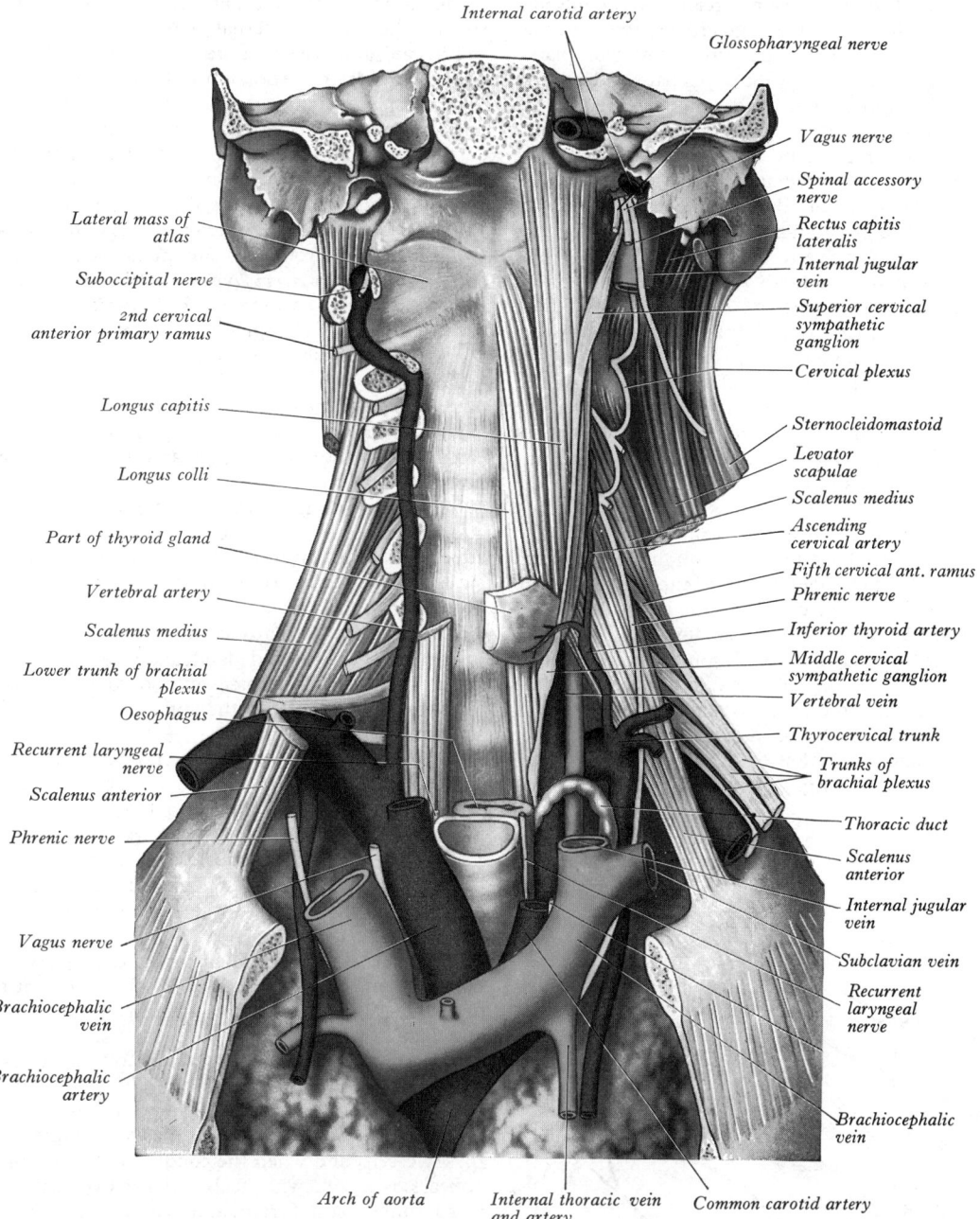

Internal carotid artery

Glossopharyngeal nerve

Vagus nerve

Spinal accessory nerve

Lateral mass of atlas

Rectus capitis lateralis

Suboccipital nerve

Internal jugular vein

2nd cervical anterior primary ramus

Superior cervical sympathetic ganglion

Cervical plexus

Longus capitis

Sternocleidomastoid

Levator scapulae

Longus colli

Scalenus medius

Ascending cervical artery

Part of thyroid gland

Fifth cervical ant. ramus

Vertebral artery

Phrenic nerve

Scalenus medius

Inferior thyroid artery

Middle cervical sympathetic ganglion

Lower trunk of brachial plexus

Vertebral vein

Oesophagus

Thyrocervical trunk

Recurrent laryngeal nerve

Trunks of brachial plexus

Scalenus anterior

Thoracic duct

Phrenic nerve

Scalenus anterior

Internal jugular vein

Vagus nerve

Subclavian vein

Recurrent laryngeal nerve

Brachiocephalic vein

Brachiocephalic artery

Brachiocephalic vein

Arch of aorta

Internal thoracic vein and artery

Common carotid artery

6.118 A dissection to show the prevertebral region and the superior mediastinum. On the right the costal elements of the upper six cervical vertebrae have been removed to expose the cervical part of the vertebral artery. On the left most of the deep relations of the common carotid artery and the internal jugular vein are exposed.

line and anterior border of sternocleidomastoid; low in the neck it turns laterally posterior to this muscle, but superficial to the depressors of the hyoid bone, and opens into the termination of the external jugular veins, or directly into the subclavian vein. Its size varies considerably, and is usually inverse to that of the external jugular vein. It communicates with the internal jugular and receives some laryngeal veins, and occasionally a small thyroid vein. There are usually two anterior jugular veins, a right and a left; just above the sternum they are united by a large transverse trunk, the *jugular arch*, which receives tributaries from the inferior thyroid veins. The anterior jugular veins have no valves. They may be replaced by a single trunk which descends in the anterior midline of the neck.

Surface Anatomy. Usually the *external jugular vein* is easily seen crossing the sternocleidomastoid obliquely. When it is not obvious, it can be brought into view by the effort of blowing with the mouth closed. The *anterior jugular vein* can often be rendered visible in the upper two-thirds of the neck in a similar manner. The terminal part of the facial vein runs from a point on the lower border of the mandible at the anterior border of masseter to join the internal jugular just below the greater cornu of the hyoid bone.

The internal jugular vein (**6.**117) collects blood from the brain, superficial parts of the face, and the neck. It begins at the base of the skull in the posterior compartment of the jugular foramen, as a direct continuation of the sigmoid sinus. At its origin is a dilatation, the *superior bulb*, below the posterior part of the floor of the tympanic cavity. The vein runs downwards through the neck within the carotid sheath (p. 537), and, behind the sternal end of the clavicle, it unites with the subclavian to form the brachiocephalic vein. The internal jugular vein is also dilated near its termination as the *inferior bulb*; directly above this the vein contains a pair of valves. *Posterior* to the internal jugular vein, from above downwards, are the rectus capitis lateralis, the transverse process of the atlas, levator scapulae, scalenus medius and the cervical plexus, then scalenus anterior, the phrenic nerve, the thyrocervical trunk, the vertebral vein and first part of the subclavian artery; on the left side the vein passes in front of the thoracic duct (6.118). *Medially* the vein is related to the internal and common carotid arteries and the vagus nerve, the last lying between the vein and the arteries but on a posterior plane. *Superficially* the vein is overlapped by the upper part, and covered by the lower part, of sternocleidomastoid, and is crossed by the posterior belly of digastric and the superior belly of omohyoid. Above the digastric, the parotid gland and the styloid process are superficial, and the accessory nerve and the posterior auricular and occipital arteries cross the vein. Between the digastric and the omohyoid, the sternocleidomastoid arteries and the inferior root of the ansa cervicalis cross the vein, but the nerve often passes between it and the common carotid artery. Below the omohyoid, it is covered by the infrahyoid muscles, in addition to sternocleidomastoid, and crossed, superficial to the infrahyoid muscles, by the anterior jugular vein. The deep cervical lymph nodes lie along the course of the vein, mainly on its superficial aspect. At the root of the neck the right internal jugular vein is a little distant from the common carotid artery, while the left vein usually overlaps its artery. At the base of the skull the internal carotid artery is in front of the internal jugular vein, and is separated from it by the last four cranial nerves.

Surface Anatomy. The internal jugular vein is represented in surface projection by a broad band drawn from the lobule of the ear to the medial end of the clavicle; its lower bulb lies behind the lesser supraclavicular fossa, projecting into the interval between the sternal and clavicular heads of sternocleidomastoid (6.47).

Tributaries. The internal jugular vein is joined by the inferior petrosal sinus, the facial, lingual, pharyngeal, superior and middle thyroid veins, and sometimes the occipital vein. In the upper neck it may communicate with the external jugular vein. The thoracic duct opens into the angle of union of the left subclavian and internal jugular veins, and the right lymphatic duct at the same site on the right.

The inferior petrosal sinus leaves the skull through the anterior part of the jugular foramen and, crossing either lateral or

medial to the ninth, tenth and eleventh cranial nerves, joins the superior bulb of the internal jugular.

The veins of the tongue follow two routes. (1) The *dorsal lingual veins* drain the dorsum and sides of the tongue and join the *lingual veins*, which accompany the lingual artery, lying in the interval between hyoglossus and genioglossus. Near the greater cornu of the hyoid bone the lingual veins join the internal jugular. (2) The *deep lingual vein* commences near the tip of the tongue and runs backwards close to the mucous membrane on its inferior surface (8.74). Near the anterior border of hyoglossus it joins the *sublingual vein*, from the sublingual salivary gland, to form the *vena comitans nervi hypoglossi* (the vein accompanying the hypoglossal nerve), which runs backwards in the interval between the mylohyoid and hyoglossus with the hypoglossal nerve to end by joining the facial, the internal jugular, or the lingual vein.

The pharyngeal veins begin in the *pharyngeal plexus* on the outer surface of the pharynx, and, after receiving some meningeal veins and the vein corresponding to the artery of the pterygoid canal, end in the internal jugular vein. They occasionally open into the facial, the lingual, or the superior thyroid vein.

The superior thyroid vein (6.116, 117) is formed by deep and superficial tributaries corresponding to the branches of the superior thyroid artery. It accompanies the artery, receives the superior laryngeal and cricothyroid veins, and ends in the internal jugular or facial vein.

The middle thyroid vein (6.117) collects blood from the lower part of the thyroid gland, and receives some veins from the larynx and trachea. It crosses anterior to the common carotid, and joins the lower part of the internal jugular vein behind the superior belly of omohyoid.

The facial and occipital vein have been described (pp. 738, 740), and the inferior thyroid veins on p. 754.

Applied Anatomy. When thrombosis occurs in the superior bulb of the internal jugular vein (as a complication of otitis media), the glossopharyngeal, vagus and accessory nerves may cease to conduct. The hypoglossal nerve is also sometimes affected by extension of the thrombus to the veins of the hypoglossal canal.

The internal jugular vein is surrounded by deep cervical lymph nodes; when these are enlarged in tuberculous or malignant disease, their adherence to the vessel may render removal difficult and dangerous.

Venous pulsation is often observable in the external jugular veins at the root of the neck. There are no valves in the brachiocephalic veins or superior vena cava; consequently, the systole of the right atrium causes a wave of distension to pass up these vessels, and when the conditions are favourable this wave appears as a somewhat feeble flicker over the external jugular veins, quite distinct from, and just preceding, the more forcible impulse transmitted from the underlying common carotid artery and due to ventricular systole. This atrial systolic venous impulse is much increased in conditions in which the right atrium is abnormally distended with blood or is hypertrophied, as is often the case in disease of the mitral valve.

The vertebral vein is formed in the suboccipital triangle from numerous small tributaries from the internal vertebral plexuses which issue from the vertebral canal above the posterior arch of the atlas. They unite with small veins from the deep muscles in the upper part of the back of the neck, and form a vessel which enters the foramen in the transverse process of the atlas, and descends, as a dense plexus around the vertebral artery, through the transverse foramina of successive cervical vertebrae. This plexus ends in the vertebral vein, which emerges from the transverse foramen of the *sixth* cervical vertebra, and runs downwards, at first anterior and then anterolateral to the vertebral artery, to open into the upper and posterior part of the brachiocephalic vein, the opening guarded by a pair of valves. In its course the vertebral vein descends behind the internal jugular, and in front of the first part of the subclavian artery (6.118). A small vein, termed the *accessory vertebral vein*, usually descends from the plexus around the vertebral artery, passes through the transverse foramen of the *seventh* cervical vertebra, and curves forwards between the subclavian artery and the cervical pleura to join the brachiocephalic vein.

6.119 The relations of the brain, the middle meningeal artery and the transverse and sigmoid sinuses to the surface of the skull. 1. Nasion. 2. Inion. 3. Lambda. 4. Lateral cerebral sulcus. 5. Central sulcus. AA.=Frankfurt plane, which traverses the lower margin of the orbital opening and the upper margin of the external acoustic meatus. B. Point for trephining over the frontal branch of the middle meningeal artery. C. Suprameatal triangle. D. Sigmoid sinus. E. Point for trephining over the transverse sinus, exposing dura mater of both cerebrum and cerebellum. The outline of the cerebral hemisphere and its major sulci are indicated in blue; the course of the middle meningeal artery is in red.

Tributaries. The vertebral vein communicates with the sigmoid sinus inside the skull by a vein which passes through the posterior condylar canal, when this canal exists. It also receives branches from the occipital vein, from the prevertebral muscle and from the internal and external vertebral plexuses. It is joined by the anterior vertebral and the deep cervical veins (*vide infra*); close to its termination it is sometimes joined by the first intercostal vein.

The anterior vertebral vein commences in a plexus around the transverse processes of the upper cervical vertebrae, descends in company with the ascending cervical artery between the scalenus anterior and longus capitis, and opens into the terminal part of the vertebral vein.

The deep cervical vein accompanies its artery between the semispinales capitis et cervicis. It begins in the suboccipital region by communicating branches from the occipital vein and small veins from the deep muscles at the back of the neck. It receives tributaries from the plexuses around the spines of the cervical vertebrae, and passes forwards between the transverse process of the seventh cervical vertebra and the neck of the first rib to end in the lower part of the vertebral vein.

Cranial and Intracranial Veins

DIPLOIC VEINS

The diploic veins occupy channels in the diploë of the cranial bones and are devoid of valves. They are large, and exhibit pouch-like dilatations at irregular intervals; their walls are thin, and formed merely of endothelium supported by a layer of elastic tissue. In radiographs of the skull the diploic veins may show as relatively transparent bands some 3 or 4 mm wide. They are absent from the skull of the newly born and begin to develop with the diploë at the age of about two years.

They communicate with the meningeal veins, the sinuses of the dura mater, and the veins of the pericranium. Recognizably regular channels are (1) the *frontal diploic vein*, which emerges from the bone at the supraorbital foramen and opens into the supraorbital vein; (2) the *anterior temporal (parietal) diploic vein*, which is confined chiefly to the frontal bone and pierces the greater wing of the sphenoid bone to end in the sphenoparietal sinus or the anterior deep temporal vein; (3) the *posterior temporal (parietal) diploic vein*, which is situated in the parietal bone; it descends to the mastoid angle of the parietal bone and joins the transverse sinus through an aperture placed at that angle or through the mastoid foramen; and (4) the *occipital diploic vein*, the largest of the four, which is confined to the occipital bone and opens into the occipital vein, or into the transverse sinus near the confluence of the sinuses, or into the occipital emissary vein. Numerous small diploic veins (**7.166**) pierce the inner table close to the margins of the superior sagittal sinus and terminate in the venous lacunae (p. 745).

MENINGEAL VEINS

The meningeal veins begin from plexiform vessels in the dura mater and drain into efferent vessels which lie in the outer layer of the dura. The efferents communicate with the lacunae of the superior sagittal sinus, with other cranial sinuses, including those which accompany the middle meningeal arteries (p. 683), and with the diploic veins.

CEREBRAL AND CEREBELLAR VEINS

The veins of the brain (for details *see* p. 1044) have no valves, and their walls, from which muscular tissue is absent, are extremely thin. They pierce the arachnoid mater and the inner layer of the dura mater, and open into the cranial venous sinuses. They comprise the cerebral and cerebellar veins and the veins of the brain stem.

The cerebral veins (6.121) are divisible into external and internal groups, draining the surfaces or the inner regions of the hemispheres.

The external cerebral veins form *superior*, *middle* and *inferior* groups.

The *superior cerebral veins*, eight to twelve in number on each hemisphere, drain the superolateral and medial surfaces of the hemispheres, and mainly follow the sulci between the gyri, but some run across the gyri. They ascend to the superomedial border of the hemisphere, where they receive small veins from the medial surface of the hemisphere, and open into the superior sagittal sinus; the anterior veins run nearly at right angles to the sinus; the posterior and larger veins are directed obliquely forwards, and thus open into the sinus against the current of the blood within it. This arrangement may prevent the collapse of the thin-walled cerebral veins which might otherwise result from a rise of intracranial pressure, but the causative factor is the backward growth of the cerebral hemispheres and the consequent displacement of the vessels during development.

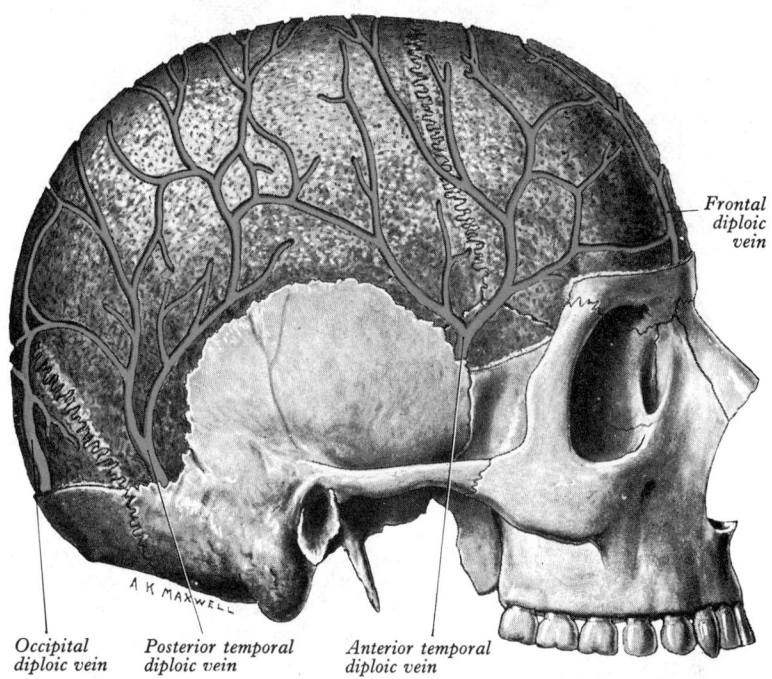

Frontal diploic vein

Occipital diploic vein *Posterior temporal diploic vein* *Anterior temporal diploic vein*

6.120 The veins of the diploë, displayed by the removal of the outer table of the skull.

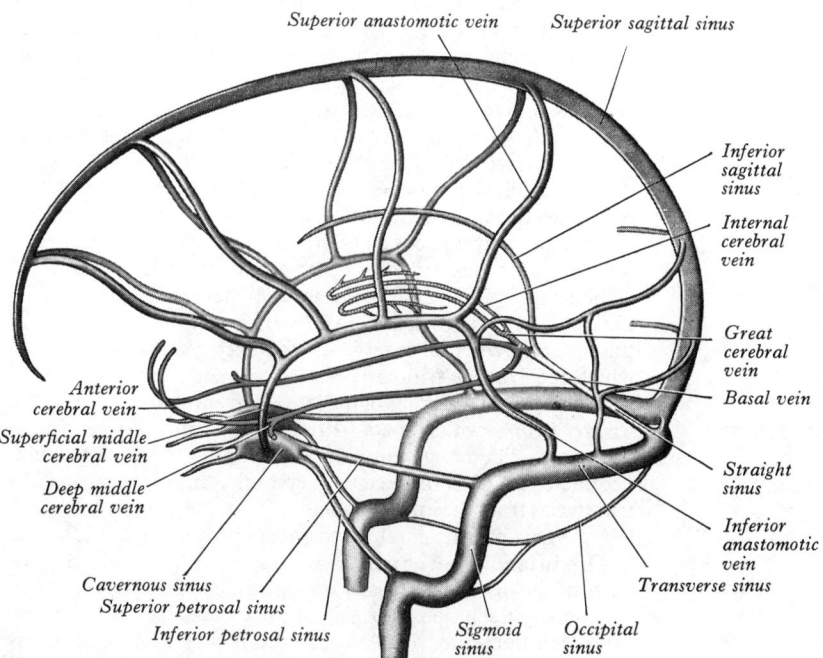

Superior anastomotic vein *Superior sagittal sinus*

Inferior sagittal sinus

Internal cerebral vein

Great cerebral vein

Basal vein

Anterior cerebral vein

Superficial middle cerebral vein

Deep middle cerebral vein

Straight sinus

Inferior anastomotic vein

Cavernous sinus
Superior petrosal sinus

Transverse sinus

Inferior petrosal sinus *Sigmoid sinus* *Occipital sinus*

6.121 A schema of the venous sinuses of the dura mater and their connexions with the cerebral veins. The more deeply placed cerebral veins are shown in *blue*, and those inside the brain are shown in *interrupted blue*.

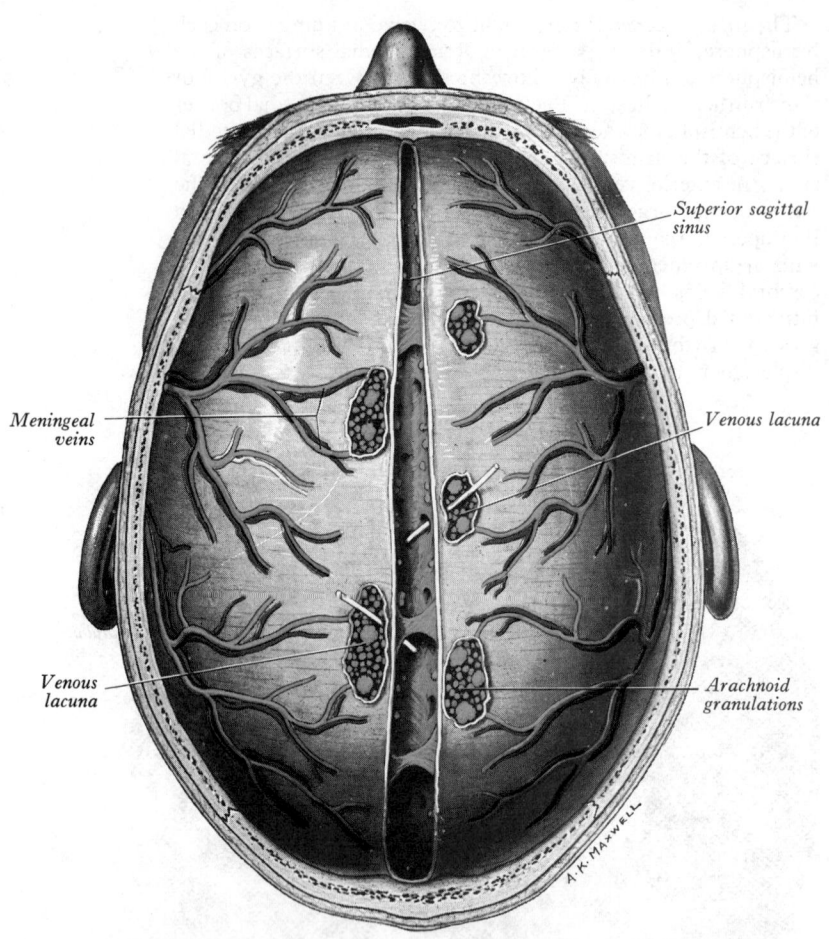

6.122 The superior sagittal sinus laid open after removal of the cranial vault. Some of the fibrous bands which cross the sinus are clearly seen; from two of the venous lacunae, bristles are passed into the sinus.

The *superficial middle cerebral vein* begins on the lateral surface of the hemisphere, and, following the posterior ramus and the stem of the lateral sulcus, ends in the cavernous sinus. The *superior anastomotic vein* runs backwards and upwards between the middle cerebral vein and the superior sagittal sinus, and thus a communication is established between the superior sagittal and cavernous sinuses. A second vein, named the *inferior anastomotic vein*, courses over the temporal lobe, and connects the middle cerebral vein to the transverse sinus.

The *inferior cerebral veins* are of small size, and drain the inferior surface of the hemisphere. Those on the orbital surface of the frontal lobe join the superior cerebral veins, and through these open into the superior sagittal sinus; those of the temporal lobe anastomose with the basal and middle cerebral veins, and empty into the cavernous, superior petrosal and transverse sinuses.

The *basal vein* begins at the anterior perforated substance by the union of (*a*) a small *anterior cerebral vein*, which accompanies the anterior cerebral artery, (*b*) the *deep middle cerebral vein*, which receives the tributaries from the insula and neighbouring gyri, and runs in the floor of the lateral cerebral sulcus, and (*c*) the *striate veins*, which pass through the anterior perforated substance. The basal vein passes backwards round the cerebral peduncle and ends in the great cerebral vein (**6.**121); it receives tributaries from the interpeduncular fossa, the inferior horn of the lateral ventricle, the parahippocampal gyrus and the midbrain.

The internal cerebral veins, right and left, drain the deep parts of the hemisphere; each is formed near the interventricular foramen by the union primarily of the *thalamostriate* and *choroid veins* (but numerous smaller veins from surrounding structures also converge becoming confluent near here). They run backwards parallel with each other, between the layers of the tela choroidea of the third ventricle, and below the splenium of the corpus callosum, where they unite to form the *great cerebral vein* (**6.**121).

The *thalamostriate vein* runs in the groove between the caudate nucleus and the thalamus, receives numerous veins from both of these structures, and unites behind the anterior column of the fornix with the choroid vein, to form the internal cerebral vein. The *choroid vein* runs along the whole length of the choroid plexus, and receives veins from the hippocampus, the fornix, and the corpus callosum and numerous adjacent structures.

The great cerebral vein starts from the union of the two internal cerebral veins, as a short median trunk which curves sharply upwards around the splenium of the corpus callosum and opens into the anterior end of the straight sinus, after receiving the right and left basal veins.

The cerebellar veins are placed on the surface of the cerebellum, and consist of superior and inferior sets. Some of the *superior cerebellar veins* pass forwards and medially, across the superior vermis, to end in the straight sinus or in the great cerebral vein; others run laterally to the transverse and superior petrosal sinuses. The *inferior cerebellar veins* include a small median vessel, which runs backwards on the inferior vermis to enter the straight or one of the sigmoid sinuses, and laterally coursing vessels which join the inferior petrosal and occipital sinuses.

The *veins of the brain stem* form a superficial venous plexus deep to the arteries. The veins of the midbrain may drain upwards into the great cerebral vein or basal vein. Over the pons the veins tend to form a lateral channel on each side which, with the upper medullary veins, may drain into the petrosal sinuses, the transverse sinus, cerebellar veins or the venous plexus of the foramen ovale. Sometimes a median pontine vein drains the pons and may join one of the basal veins above. The veins of the lower part of the medulla oblongata communicate with the veins of the spinal cord and drain into the adjacent venous sinuses, or along very variable radicular veins following the last four cranial nerves to the inferior petrosal or occipital sinuses, or to the upper part of the internal jugular vein. Anterior and posterior median veins may run along the anterior medial fissure or the posterior median sulcus respectively, continuous with the veins in the corresponding positions on the spinal cord (see also pp. 896, 1045).

CRANIAL DURAL VENOUS SINUSES

The sinuses of the dura mater are venous channels which drain the blood from the brain and the bones of the cranium; they are situated between the two layers of the dura mater and are lined by endothelium continuous with that which lines the veins; they contain no valves, and their walls are devoid of muscular tissue. (It should be noted here that although most accounts present the sinuses as, in most cases, simple smooth channels, recent researches have emphasized their complex 'cavernous' or plexiform nature in many sites—*see* Browder and Kaplan 1976; also p. 747.) They may be divided into two groups: (1) a posterosuperior, in the upper and posterior parts of the cranial cavity, and (2) an antero-inferior, on the base of the skull.

1. *The posterosuperior group of venous sinuses*: superior sagittal, inferior sagittal, straight, two transverse, two petrosquamous, two sigmoid, and occipital sinuses.

The superior sagittal sinus (**6.**121, 122, 123) occupies the attached, convex margin of the falx cerebri. It commences in front of the crista galli, where it receives a vein from the nasal cavity on the rare occasions when the foramen caecum is patent. Kaplan *et al.* (1973) found no such venous tributary to the superior sagittal sinus in 201 specimens, in only 9 per cent of which did the sinus begin as far forward as the foramen. In most instances the first tributaries were cortical veins from the adjacent frontal poles of the cerebral hemispheres, the *ascending frontal veins* of Krayenbühl (1967). The sinus usually begins a few millimetres posterior to the foramen caecum. It runs backwards, grooving the inner surface of the frontal bone, the adjacent margins of the two parietal bones, and the squamous part of the occipital bone. Near the internal occipital protuberance it deviates to one or other side (usually the right), and is continued as the corresponding transverse sinus. It is triangular in cross-section, and gradually increases in size as it passes backwards. Its inner surface presents the openings of the superior cerebral veins, projecting arachnoid

Meningeal veins

Venous lacuna

Superior sagittal sinus

Venous lacuna

Arachnoid granulations

granulations, and numerous fibrous bands which cross the inferior angle of the sinus; the sinus also communicates through small openings with irregularly shaped *venous lacunae*, which are situated in the dura mater near the sinus. There are usually three lacunae on each side of the sinus: a small frontal, a large parietal, and an occipital, intermediate in size. In elderly subjects these lacunae tend to become continuous with one another as one elongated lacuna on each side. Many fine fibrous bands cross the lacunae, and numerous arachnoid granulations project into them from below. The superior sagittal sinus receives the superior cerebral veins, and, near the posterior extremity of the sagittal suture, veins from the pericranium which pass through the parietal foramina; the venous lacunae drain the diploic and meningeal veins.

The complexity of the lateral lacunae of the superior sagittal sinus and of the sinus itself have been somewhat obscured by over-simplification in general texts of anatomy; but these complexities have often been emphasized (*see*, for example, LeGros Clark 1920, and Baló 1950), and corrosion cast studies (Browder and Kaplan 1976) and cerebral angiography have more recently revived these earlier views. The lateral lacunae are frequently so complex as to be almost plexiform and are rarely the simple venous spaces usually depicted. All the observers cited above have described plexiform arrangements of small veins in association with the sagittal, transverse, and straight sinuses, and LeGros Clark and Baló regarded these venous masses as cavernous tissue. Masses of this kind commonly adjoin all the sinuses which inter-communicate at the confluence of the sinuses. Ridges of such spongy venous tissue frequently project into the lumen of the superior sagittal and transverse sinuses. The function of these arrangements can only be conjectured (*see* p. 1050). The lumen of the superior sagittal sinus is also invaded, in its intermediate third, by variable bands and projections of its dural walls, which may even extend as horizontal shelves, dividing its lumen into superior and inferior channels for considerable distances. All these variable factors make it impossible to give a simple description of this or other venous sinuses. Their variations have been noted in detail by Browder and Kaplan (1974) in a large series of corrosion casts; even with this information individual variations can only be established by angiography in the living.

The confluence of the sinuses (6.124) is the term applied to the dilated posterior extremity of the superior sagittal sinus. It is one side (generally the right) of the internal occipital protuberance, and from it the transverse sinus of the same side is derived. It receives also the blood from the occipital sinus, and connects with the commencement of the opposite transverse sinus. The size and degree of inter-communication of the venous channels which meet at the confluence are highly variable (consult Browder and Kaplan 1974, for the statistics of 215 specimens). In more than half of specimens examined all the venous channels which converge towards the internal occipital protuberance do in fact interconnect, and this includes the straight and occipital sinuses. In many instances communications are lacking or tenuous. Any of the sinuses involved may be reduplicated, narrowed, or widened as it approaches the confluence. The variety of arrangements is too great to make particularization useful until much larger series than that quoted above become available.

Applied Anatomy. The communications between the superior sagittal sinus and the veins of the nose, scalp, and diploë explain the occasional spread of infective thrombosis from suppurative processes in these parts.

The inferior sagittal sinus (6.123) is in the posterior half or two-thirds of the free margin of the falx cerebri. It increases in size posteriorly, and ends in the straight sinus. It receives several veins from the falx, and occasionally a few from the medial surfaces of the cerebrum.

The straight sinus (6.123, 124) is in the junction of the falx cerebri and tentorium cerebelli. It is triangular in cross-section and displays a few transverse bands. It runs backwards and downwards, continuing the inferior sagittal sinus to the transverse sinus of the side opposite to that into which the superior sagittal sinus is prolonged. Its terminal part communicates with the confluence of the sinuses. Besides the inferior sagittal sinus, it receives some of the superior cerebellar veins, and, at its commencement, the great cerebral vein, the site of the opening of this vein being marked by a dilatation. A small body, resembling an arachnoid granulation, projects into the floor of the sinus at its angle of union with the great cerebral vein. This body, which contains a sinusoidal plexus of blood vessels, is believed to become engorged from time to time and may then act as a ball-

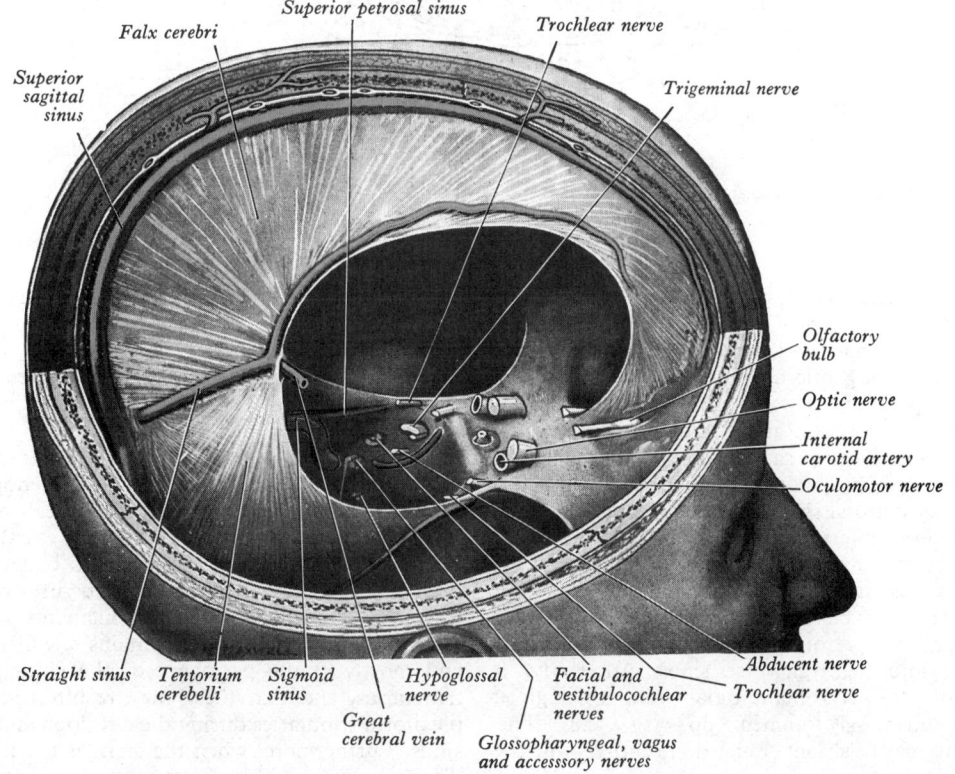

6.123   The dura mater, its processes, and venous sinuses. Right aspect.

valve mechanism controlling the outflow from the great cerebral vein and thus affecting the secretion of cerebrospinal fluid by the choroid plexuses of the lateral ventricles. As noted, other masses of cavernous tissue have been described in relation with many dural sinuses. Engorgement possibly influences the blood flow through them: structural data make this unlikely (*vide supra*).

The transverse sinuses (**6**.124, 126) are of large size and begin at the internal occipital protuberance, one, generally the right, being the direct continuation of the superior sagittal sinus, the other of the straight sinus. Each transverse sinus passes laterally and forwards to the posterolateral part of the petrous part

transverse sinuses, beginning where the latter leave the tentorium cerebelli. Each sigmoid sinus curves downwards and medially in a deep groove on the mastoid part of the temporal bone, crosses the jugular process of the occipital bone, and then turns forwards to become the superior bulb of the internal jugular vein in the posterior part of the jugular foramen. Anteriorly, only a thin plate of bone separates the upper part of the sigmoid sinus from the mastoid antrum and mastoid air cells. Each sinus communicates with the veins of the pericranium by means of the mastoid and condylar emissary veins.

The occipital sinus (**6**.126), the smallest of the cranial

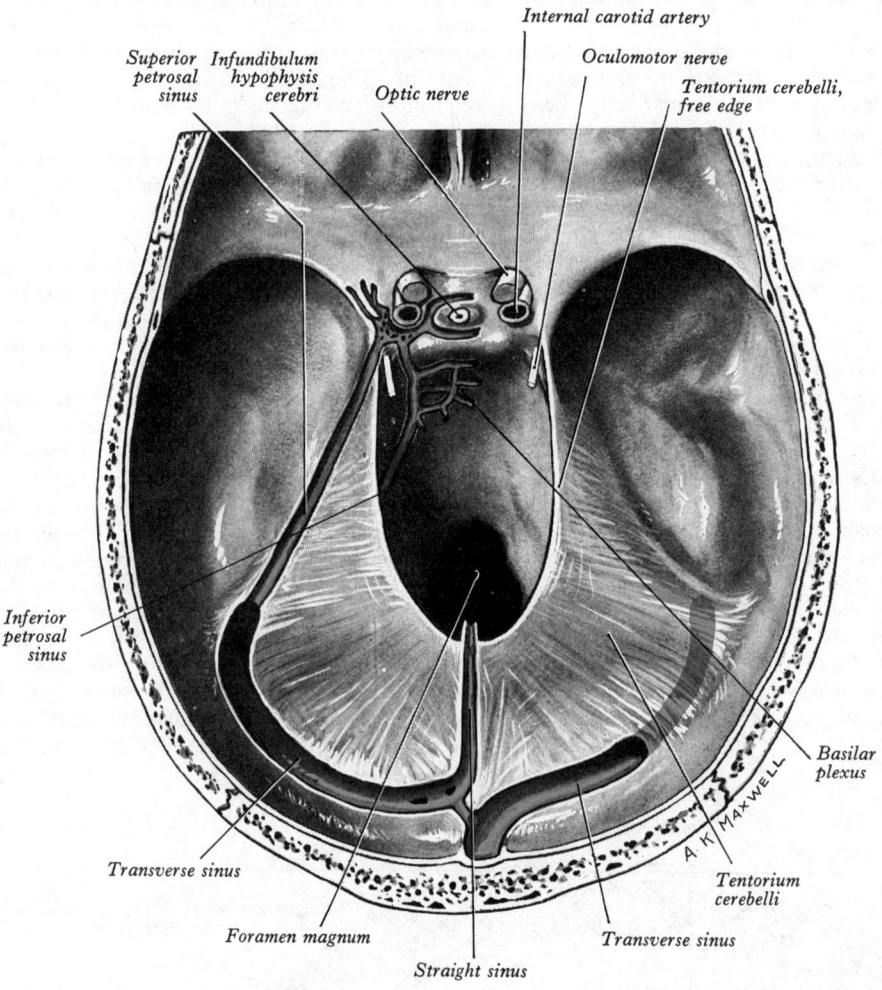

Internal carotid artery

Superior   *Infundibulum*   *Oculomotor nerve*
petrosal   *hypophysis*
sinus   *cerebri*   *Optic nerve*   *Tentorium cerebelli,*
                                     free edge

Inferior
petrosal
sinus

Basilar
plexus

Transverse sinus

Tentorium
cerebelli

Foramen magnum   *Transverse sinus*

Straight sinus

6.124   The tentorium cerebelli and venous sinuses. Superior aspect.

of the temporal bone, where it curves down as the sigmoid sinus. It lies in the attached margin of the tentorium cerebelli, at first on the squama of the occipital bone and then on the mastoid angle of the parietal. It describes a gentle curve, convex upwards, and increases in size as it proceeds forwards. The transverse sinuses are triangular on transverse section, and are frequently of unequal size, the one draining the superior sagittal sinus being the larger. Where they become continuous with the sigmoid sinuses, they are joined by the superior petrosal sinuses; and in their course they receive inferior cerebral, inferior cerebellar, and diploic veins, and the inferior anastomotic vein (p. 744).

The petrosquamous sinus runs backwards in a groove, the posterior part of which may be converted into a canal, along the junction of the squama and petrous portion of the temporal bone, and opens behind into the transverse sinus. Anteriorly, it communicates with the retromandibular vein through a postglenoid or a squamosal foramen (pp. 327, 328). The petrosquamous sinus may be absent or may drain entirely into the retromandibular vein.

The sigmoid sinuses (**6**.126) are direct continuations of the

sinuses, is in the attached margin of the falx cerebelli, and is occasionally paired. It commences near the margin of the foramen magnum in several small venous channels, one of which joins the terminal part of the sigmoid sinus; it communicates with the internal vertebral plexuses, and ends in the confluence of the sinuses.

2. *The antero-inferior group of venous sinuses*: cavernous, intercavernous, inferior petrosal, sphenoparietal, superior petrosal, basilar, and the middle meningeal veins.

The cavernous sinuses (**6**.121, 126, 127) are placed one on each side of the body of the sphenoid bone, and are so named because they present a spongy structure, due to their being traversed by numerous interlacing filaments. It is claimed that the distended sinus in the adult contains few filaments (trabeculae), and mostly at the periphery of the sinus near the entry of tributaries, and that these may result from incorporation of plexiform tributaries during the developmental expansion of the sinus. Furthermore, when the sinus is collapsed, as is usual in dissecting-room cadavers, its cavity is encroached upon by the nerves and arachnoid granulations in its wall which thus give a

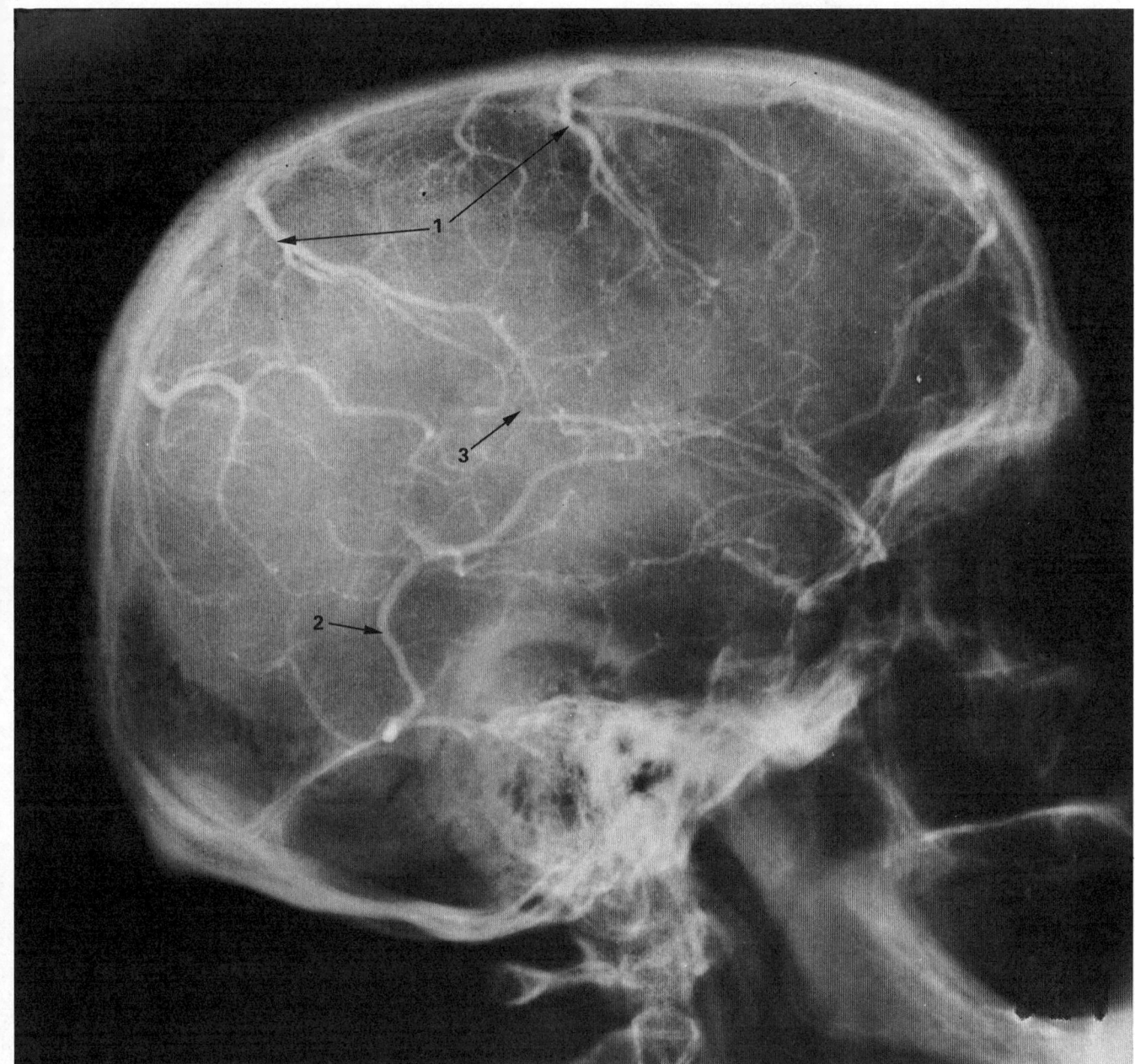

6.125 Internal carotid arteriogram (right), venous phase. Lateral view. (Same subject as **6.**54 and **6.**55, p. 633.) 1. Superior cerebral veins. Note anterior course at entry into superior sagittal sinus. 2. An inferior cerebral vein ending in the straight sinus. 3. Region of venous anastomoses.

spurious resemblance to cavernous tissue (Butler 1957). From a study of corrosion-cast preparations Parkinson (1973) concluded that the sinus is, in fact, usually a plexus (as during its development), a finding which is in accord with some earlier descriptions. Pernkopf (1963), for example, depicted the 'sinus' as a venous plexus (*see* **6.**127B, C). Browder and Kaplan (1976) have also examined a large series of corrosion casts of cavernous sinuses, prepared in human cadavers; they described the sinus as 'reticulated' in external view. It is not clear whether this signified *plexiform* or *cavernous*, but the latter seems more probable. It extends from the superior orbital fissure in front, to the apex of the petrous part of the temporal bone behind, and has an average length of 2 cm and width of 1 cm. The internal carotid artery, surrounded by a plexus of sympathetic nerves, passes forwards through the sinus; the abducent nerve, in this part of its course, lies inferolateral to the artery; the oculomotor and trochlear nerves, and the ophthalmic and maxillary divisions of the trigeminal nerve (**6.**127) are frequently described as being *in the thickness* of the lateral wall of the sinus. In fact, they are of such diameters that they project considerably into the sinus itself

(**6.**127A); and while they may be surrounded by dural connective tissue, in some cases, they are usually covered medially by little more than endothelium. (Consult McGrath (1977), for a survey of this and other details of this region.) The sphenoidal air sinus and the hypophysis cerebri are medial to the cavernous sinus. The trigeminal cave is close to the lower and posterior part of its lateral wall, and extends backwards beyond the sinus, enclosing the trigeminal ganglion. The uncus forms an additional relation of the lateral wall.

The tributaries of the cavernous sinus are the superior ophthalmic vein, a branch from the inferior ophthalmic vein (or the whole vessel), the superficial middle cerebral vein, inferior cerebral veins, and the sphenoparietal sinus; the central vein of the retina and the frontal tributary of the middle meningeal vein sometimes open into it. The cavernous sinus drains into the transverse sinus through the superior petrosal sinus, into the internal jugular vein by way of the inferior petrosal sinus and a plexus of veins on the internal carotid artery, into the pterygoid venous plexus by veins which pass through the emissary sphenoidal foramen, foramen ovale, and foramen lacerum, and

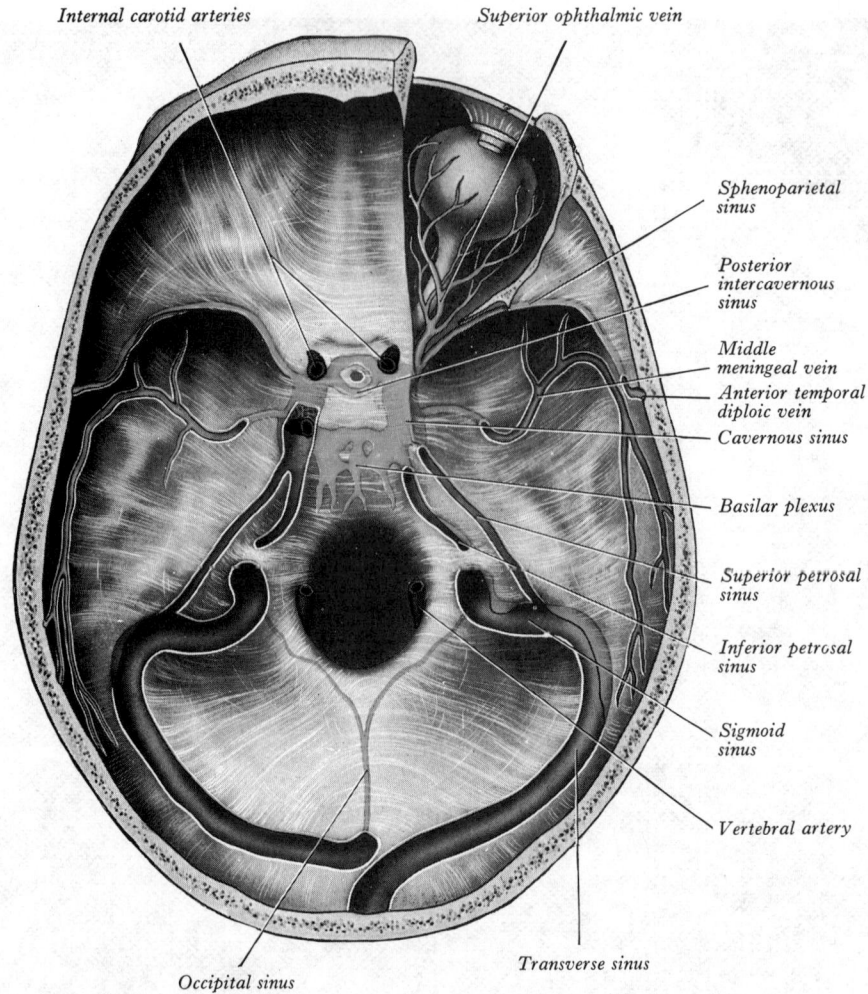

Internal carotid arteries

Superior ophthalmic vein

Sphenoparietal sinus

Posterior intercavernous sinus

Middle meningeal vein

Anterior temporal diploic vein

Cavernous sinus

Basilar plexus

Superior petrosal sinus

Inferior petrosal sinus

Sigmoid sinus

Vertebral artery

Transverse sinus

Occipital sinus

6.126 The sinuses at the base of the skull. The sinuses coloured dark blue have been opened up.

into the facial vein through the superior ophthalmic vein. The two sinuses also communicate with each other by means of the anterior and posterior intercavernous sinuses and the network of the basilar plexus. All these communications are valveless, and therefore the direction of flow in them is reversible.

The expulsion of blood from the cavernous sinus is due partly to the expansile pulsation of the internal carotid artery in its interior. It is also influenced to some extent by gravity and the position of the head.

Applied Anatomy. An arteriovenous communication may occur between the cavernous sinus and the internal carotid artery, giving rise to a pulsating swelling in the orbit. This may result from various injuries, such as a bullet wound, or a fracture of the base of the skull. Ligation of the internal or common carotid artery has been performed in these cases with considerable success.

Suppuration in the upper nasal cavities and certain paranasal sinuses may lead to septic thrombosis of the cavernous sinuses, with subsequent meningitis.

The ophthalmic veins (**6**.126, 128) superior and inferior, are devoid of valves.

The *superior ophthalmic vein* is formed behind the medial part of the upper eyelid by two tributaries which communicate anteriorly with the facial and supraorbital veins (p. 738). It runs with the ophthalmic artery, receives tributaries corresponding to its branches, passes through the superior orbital fissure, and ends in the cavernous sinus.

The *inferior ophthalmic vein* begins in a venous network near the forepart of the floor and medial wall of the orbit; it receives veins from the rectus inferior, obliquus inferior, lacrimal sac and eyelids, and runs backwards above the rectus inferior. It frequently joins the superior ophthalmic vein, but may open

separately into the cavernous sinus. It communicates with the pterygoid venous plexus by small veins passing through the inferior orbital fissure.

The central vein of the retina first traverses the optic nerve behind the lamina cribrosa and then leaves it to pursue a long course in the subarachnoid space before piercing the dura mater to enter either the cavernous sinus or the superior ophthalmic vein. Within the nerve it receives a central vein draining its proximal part.

The sphenoparietal sinuses (**6**.126) are inferior to the lesser wings of the sphenoid bone, next to periosteum and near their posterior edges. Each receives small veins from the adjacent part of the dura mater and sometimes the frontal trunk of the middle meningeal vein; it opens into the anterior part of the cavernous sinus. The sinus frequently receives connecting rami, in the middle part of its course, from the superficial middle cerebral vein, and its tributaries may also include veins from the temporal pole of the cerebrum, and the anterior temporal diploic vein. When these cerebral venous connexions are well developed the sphenoparietal sinus is a particularly large channel.

The intercavernous sinuses, an anterior and a posterior, connect the cavernous sinuses across the median plane, and are in the anterior and posterior attached borders of the diaphragma sellae; with the cavernous sinuses they thus form a venous circle (*circular sinus*, **6**.126). Small, irregular venous sinuses which lie below the hypophysis cerebri drain into the intercavernous sinuses. The *inferior intercavernous sinuses* were studied by casts in 27 specimens by Kaplan *et al.* (1976), who emphasized their size and plexiform arrangement, factors of importance in the trans-nasal surgical approach to the hypophysis cerebri.

The superior petrosal sinuses (**6**.126), small and narrow, drain the cavernous into the transverse sinuses. Leaving the posterosuperior part of the cavernous sinus, each superior petrosal sinus runs backwards and laterally in the attached margin of the tentorium cerebelli. It crosses above the trigeminal nerve and then lies in a groove on the superior border of the petrous part of the temporal bone. Finally, it terminates by joining the transverse sinus where the latter curves downwards to become the sigmoid sinus. It receives some cerebellar and inferior cerebral veins, and veins from the tympanic cavity. It also has connexions with the inferior petrosul sinus and basilar venous plexus.

The inferior petrosal sinuses drain the cavernous sinuses into the internal jugular vein. Each (6.126) begins in the postero-inferior part of the corresponding cavernous sinus, runs backwards in the groove between the petrous part of the temporal and the basilar part of the occipital bone. Passing through the anterior part of the jugular foramen it ends in the superior bulb of the internal jugular. It receives *labyrinthine veins* from the cochlear canaliculus, from the aqueduct of the vestibule, and tributaries from the medulla oblongata, pons, and inferior surface of the cerebellum. According to Browder and Kaplan (1976) the inferior petrosal sinus is more often a venous plexus than a single vessel. They also reported that it sometimes drains by a vein traversing the hypoglossal canal to end in the suboccipital vertebral plexus.

The relations of the structures transmitted through the jugular foramen are as follows: the inferior petrosal sinus lies medially and anteriorly with the meningeal branch of the ascending pharyngeal artery, and is directed obliquely downwards and backwards; the sigmoid sinus is situated at the lateral and posterior part of the foramen with a meningeal branch of the occipital artery; between the two sinuses are the glossopharyngeal, vagus, and accessory nerves (*see* p. 1082).

The basilar venous plexus (6.126) consists of several interconnecting venous channels situated between the layers of the dura mater over the clivus of the skull; it connects the two inferior petrosal sinuses, and communicates with the internal vertebral venous plexus. The plexus usually connects with the cavernous sinuses and the superior petrosal sinuses at its anterior (rostral) extremity. When the marginal sinuses are large they communicate anteriorly with the basilar venous plexus. There

may then be an almost complete venous channel encircling the foramen magnum and forming communications between the basilar plexus and the inferior petrosal, sigmoid, and occipital sinuses at the intracranial level, and variable connexions with the extracranial vertebral plexuses in the suboccipital region.

The middle meningeal veins (6.126) communicate above with the superior sinus through the adjoining venous lacunae, and unite to form two principal trunks, a frontal and a parietal, which accompany the branches of the middle meningeal arteries more or less closely in the grooves on the inner surface of the parietal bone; the veins are closer to bone than the arteries and sometimes they occupy separate grooves. The grooves on the inner surfaces of the parietal bones are in reality impressed not by the arteries but by the middle meningeal veins which are liable to be torn when the bones are fractured (Wood Jones 1911). Their ending is subject to variation. The parietal trunk may pass through the foramen spinosum into the pterygoid plexus. The frontal trunk may reach the pterygoid plexus through the foramen ovale, or it may end in the sphenoparietal or cavernous sinus. Besides their meningeal tributaries they receive some small inferior cerebral veins, and communicate with the diploic veins and the superficial middle cerebral vein. Browder and Kaplan (1976) state that the middle meningeal 'veins' are histologically *sinuses*, which in places almost surround the middle meningeal artery; they also report the frequent occurrence of arachnoid granulations in these channels.

Surface Anatomy of the Venous Sinuses. The *superior sagittal sinus* runs from the glabella (p. 298) to the inion (3.126). Narrow anteriorly, it widens posteriorly until it is about 1 cm wide. The *transverse sinus* begins at the inion and runs laterally, with a slight upward convexity to the base of the mastoid process. From this the *sigmoid sinus* passes downwards just in front of the posterior border of the mastoid process to a point about 1 cm from its tip. These sinuses are usually a little more than 1 cm wide.

The Emissary Veins

The emissary veins pass through apertures in the cranial wall and establish communications between the venous sinuses inside the skull and the veins external to it. Some are constant, but others sometimes absent. (1) A *mastoid emissary vein* runs through the

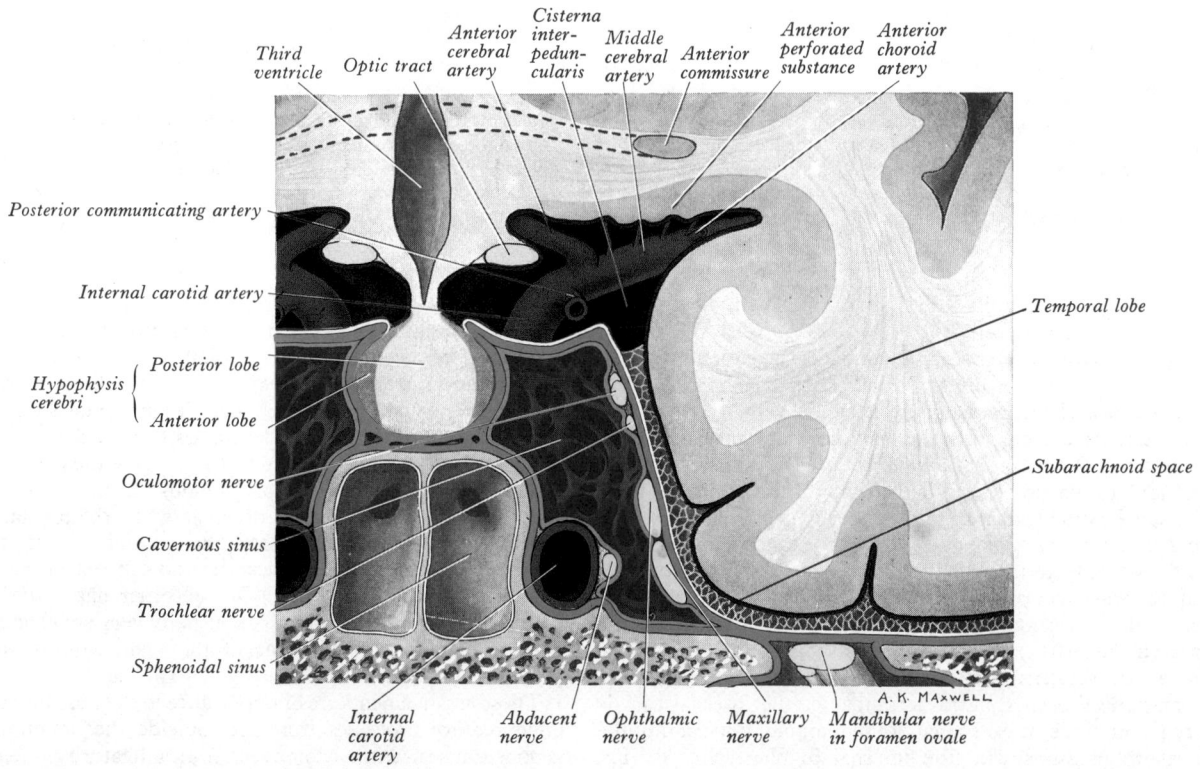

6.127A A coronal, slightly oblique section through the middle cranial fossa, showing the cavernous and cerebral portions of the internal carotid artery and the cavernous sinus. Pia mater—mauve; arachnoid mater—white; layers of dura mater—green; mesothelium of dura mater not indicated, endothelium of cavernous sinus—blue.

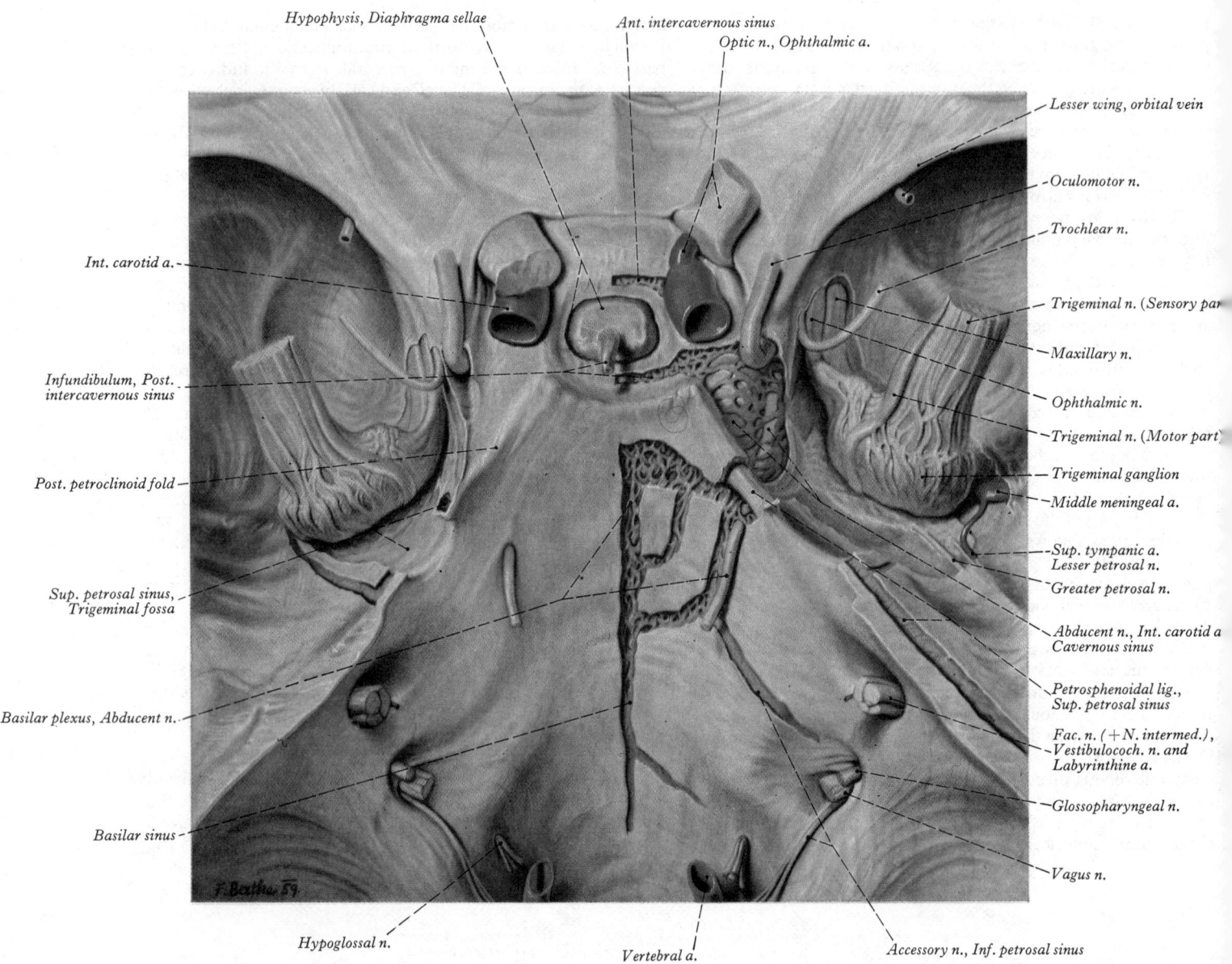

Hypophysis, Diaphragma sellae

Ant. intercavernous sinus

Optic n., Ophthalmic a.

Lesser wing, orbital vein

Oculomotor n.

Trochlear n.

Int. carotid a.

Trigeminal n. (Sensory part)

Maxillary n.

Infundibulum, Post. intercavernous sinus

Ophthalmic n.

Trigeminal n. (Motor part)

Post. petroclinoid fold

Trigeminal ganglion

Middle meningeal a.

Sup. tympanic a.
Lesser petrosal n.

Greater petrosal n.

Sup. petrosal sinus,
Trigeminal fossa

Abducent n., Int. carotid a.
Cavernous sinus

Petrosphenoidal lig.,
Sup. petrosal sinus

Basilar plexus, Abducent n.

Fac. n. (+ N. intermed.),
Vestibulococh. n. and
Labyrinthine a.

Glossopharyngeal n.

Basilar sinus

Vagus n.

F. Batke 59

Hypoglossal n.

Vertebral a.

Accessory n., Inf. petrosal sinus

6.127B The middle cranial fossa, viewed from above to show the termination of the internal carotid artery, its branches and the cavernous sinus. Note the plexiform nature of the 'sinus', which communicates with similar venous plexuses in the hypophysial fossa and over the clivus. These have been exposed by partial removal of the dura mater.
c An oblique vertical section through the cranial base to display in lateral view the right internal carotid artery and the continuity of the venous plexus around the intra-osseous and cavernous parts of the artery. (Both figures from Pernkopf's *Atlas of Topographical and Applied Human Anatomy*, English Edition 1963, by kind permission of W. B. Saunders and Urban and Schwarzenberg.)

mastoid foramen and unites the sigmoid sinus with the posterior auricular or occipital vein. (2) A *parietal emissary vein* passes through the parietal foramen and connects the superior sagittal sinus with the veins of the scalp. (3) The *venous plexus of the hypoglossal canal*, or occasionally a single vein, traverses the hypoglossal canal and joins the sigmoid sinus to the internal jugular vein. (4) A *posterior condylar emissary vein* passes through the condylar canal and connects the sigmoid sinus with the veins in the suboccipital triangle. (5) A network of emissary veins (the *venous plexus of the foramen ovale*) unites the cavernous sinus with the pterygoid plexus through the foramen ovale. (6) Two or three small emissary veins run through the foramen lacerum and connect the cavernous sinus with the pharyngeal veins and pterygoid plexus. (7) A vein traverses the emissary sphenoidal foramen (of Vesalius) and connects the same vessels. (8) The *internal carotid venous plexus* accompanies the internal carotid artery through the carotid canal of the temporal bone and unites the cavernous sinus with the internal jugular vein. (9) The petrosquamous sinus (p. 746) connects the transverse sinus with the external jugular vein. (10) A vein may pass through the foramen caecum, which, however, is patent in only a little over

1 per cent of adult skulls; it connects veins in the nose with the superior sagittal sinus. (11) An *occipital emissary vein* connects the confluence of sinuses with the occipital vein through the occipital protuberance. It also receives the occipital diploic vein. This emissary vein may be absent. (12) The occipital sinus usually communicates with variably developed venous channels around the rim of the foramen magnum (the so-called *marginal sinuses*) and through these with vertebral venous plexuses, providing an alternative escape for venous drainage when the jugular vein is blocked or deliberately tied. (13) The ophthalmic veins must be regarded as potential emissaries; they do connect intracranial to extracranial venous channels. On the other hand, the parietal emissary veins, included here, are usually very small and do not appear to connect with veins of the scalp in corrosion cast preparations.

These arrangements are of importance in the possible spread of inflammatory processes from foci outside the cranium to the venous sinuses within. Moreover, surgical ligature of the internal jugular vein, as a measure to limit oral and pharyngeal malignant disease, is necessarily dependent upon the adequacy of collateral drainage.

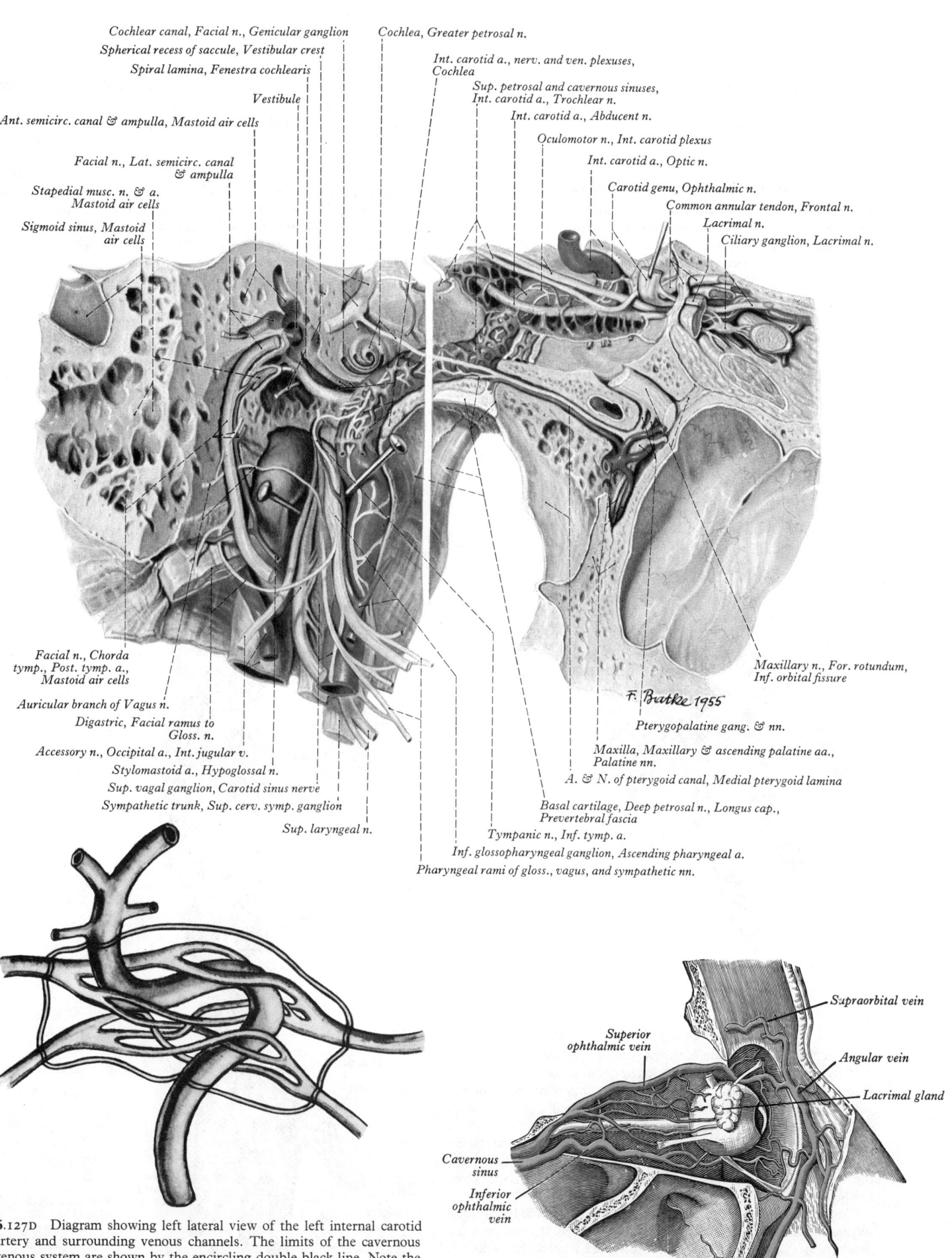

Cochlear canal, Facial n., Genicular ganglion
Spherical recess of saccule, Vestibular crest
Spiral lamina, Fenestra cochlearis
Vestibule
Ant. semicirc. canal & ampulla, Mastoid air cells
Facial n., Lat. semicirc. canal & ampulla
Stapedial musc. n. & a., Mastoid air cells
Sigmoid sinus, Mastoid air cells

Cochlea, Greater petrosal n.
Int. carotid a., nerv. and ven. plexuses, Cochlea
Sup. petrosal and cavernous sinuses, Int. carotid a., Trochlear n.
Int. carotid a., Abducent n.
Oculomotor n., Int. carotid plexus
Int. carotid a., Optic n.
Carotid genu, Ophthalmic n.
Common annular tendon, Frontal n.
Lacrimal n.
Ciliary ganglion, Lacrimal n.

Facial n., Chorda tymp., Post. tymp. a., Mastoid air cells
Auricular branch of Vagus n.
Digastric, Facial ramus to Gloss. n.
Accessory n., Occipital a., Int. jugular v.
Stylomastoid a., Hypoglossal n.
Sup. vagal ganglion, Carotid sinus nerve
Sympathetic trunk, Sup. cerv. symp. ganglion
Sup. laryngeal n.

F. Batke 1955

Maxillary n., For. rotundum, Inf. orbital fissure
Pterygopalatine gang. & nn.
Maxilla, Maxillary & ascending palatine aa., Palatine nn.
A. & N. of pterygoid canal, Medial pterygoid lamina
Basal cartilage, Deep petrosal n., Longus cap., Prevertebral fascia
Tympanic n., Inf. tymp. a.
Inf. glossopharyngeal ganglion, Ascending pharyngeal a.
Pharyngeal rami of gloss., vagus, and sympathetic nn.

Superior ophthalmic vein
Supraorbital vein
Angular vein
Lacrimal gland
Cavernous sinus
Inferior ophthalmic vein
Facial vein

6.127D Diagram showing left lateral view of the left internal carotid artery and surrounding venous channels. The limits of the cavernous venous system are shown by the encircling double black line. Note the *plexiform* nature of this system, as described by some authorities. This is an artist's impression founded upon numerous dissections and corrosion casts by Dr. Dwight Parkinson (University of Manitoba) to whom we are indebted. See text for further details.

6.128 The veins of the right orbit. Lateral aspect.

THE VEINS OF THE UPPER LIMBS

The veins of the upper limb can be divided into two sets, *superficial* and *deep*, which anastomose freely with each other. The superficial veins are immediately under the skin, in the superficial fascia; the deep veins accompany the arteries. Both sets are provided with valves, which are more numerous in the deep than in the superficial veins.

The Superficial Veins of the Upper Limb (6.129, 130)

The superficial veins of the upper limb are the cephalic, basilic, and median cubital and antebrachial veins, and their tributaries.

The *dorsal digital veins* pass along the sides of the fingers, joined by oblique communicating branches. Those from adjacent sides of the fingers unite into three *dorsal metacarpal veins* (6.129),

anterior surface, receiving tributaries from both surfaces. Below the front of the elbow it gives off the *median cubital vein*, which receives a communicating branch from the deep veins of the forearm and passes medially to join the basilic vein. The cephalic vein then ascends subcutaneously in front of the elbow superficial to the groove between the brachioradialis and the biceps. It crosses superficial to the lateral cutaneous nerve of the forearm and runs upwards lateral to biceps. In the upper one-third of the arm it lies in the interval between pectoralis major and deltoid, where it is accompanied by the deltoid branch of the thoraco-acromial artery. Entering the infraclavicular fossa, it passes behind the clavicular head of pectoralis major. It then pierces the clavipectoral fascia, crosses the axillary artery, and ends in the

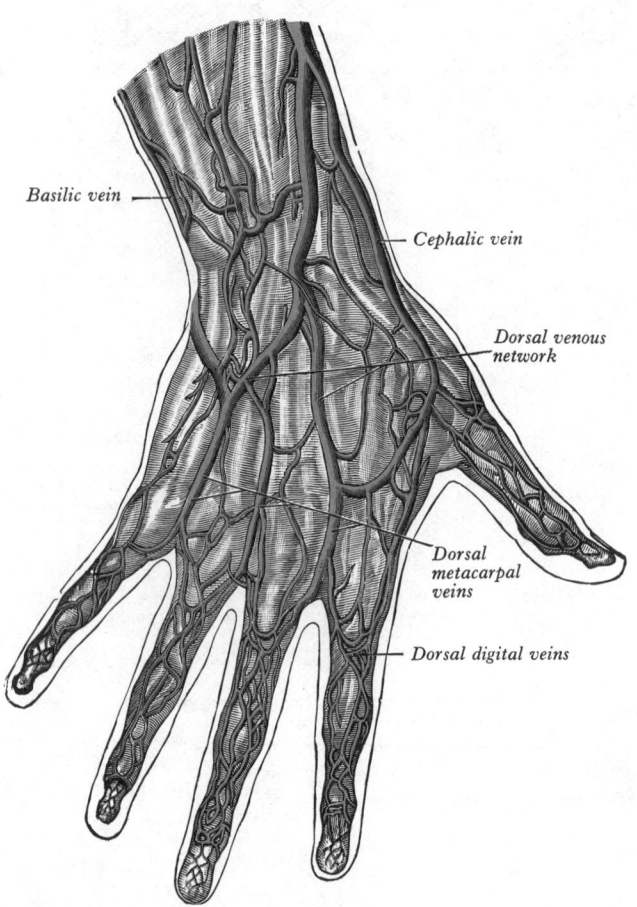

Basilic vein

Cephalic vein

Dorsal venous network

Dorsal metacarpal veins

Dorsal digital veins

6.129   The veins of the dorsum of the hand.

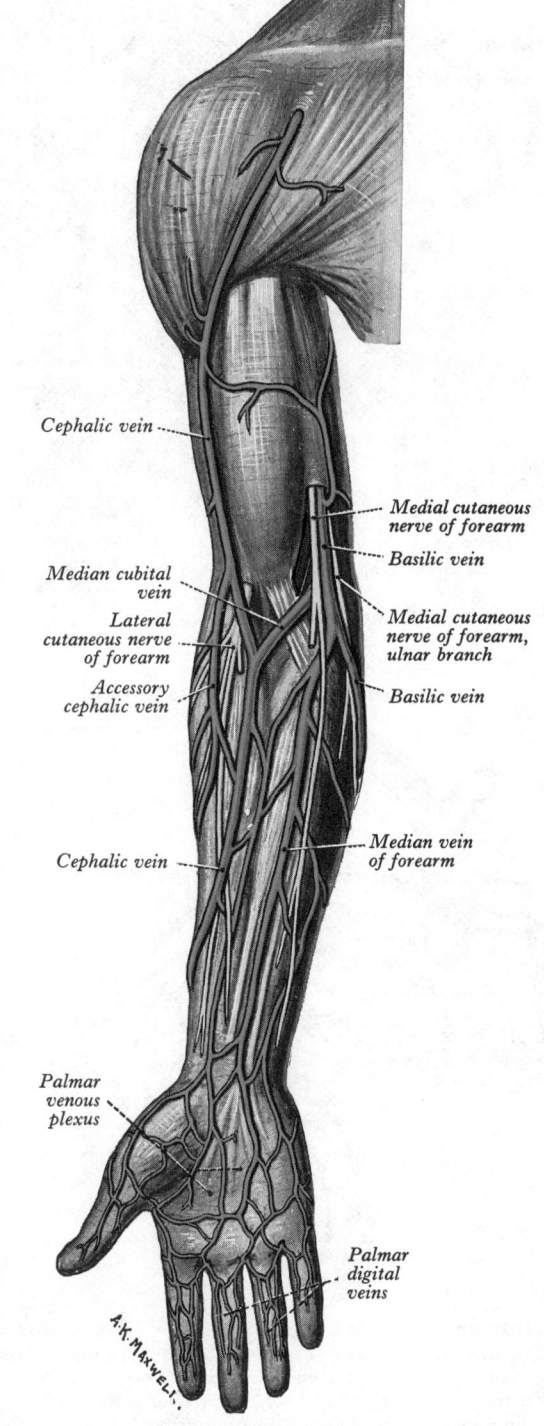

Cephalic vein

Median cubital vein

Lateral cutaneous nerve of forearm

Accessory cephalic vein

Medial cutaneous nerve of forearm

Basilic vein

Medial cutaneous nerve of forearm, ulnar branch

Basilic vein

Cephalic vein

Median vein of forearm

Palmar venous plexus

Palmar digital veins

which end in a *dorsal venous network* opposite the middle of the metacarpus. Laterally the network is joined by the dorsal digital vein from the radial side of the index finger and both dorsal digital veins of the thumb, and is prolonged proximally as the cephalic vein. Medially the network receives the dorsal digital vein of the ulnar side of the little finger, and is continued upwards as the basilic vein. A communicating vein frequently connects the central part of the dorsal venous network with the cephalic about the middle of the forearm (see also accessory cephalic vein, below).

The *palmar* digital veins are connected to the *dorsal* by oblique *intercapitular veins*, which pass back between the metacarpal heads. They also drain into a venous plexus superficial to the palmar aponeurosis and extending over the thenar and hypothenar eminences.

The cephalic vein (6.130) winds upwards from the dorsal venous network round the radial border of the forearm to its

6.130   The superficial veins of the right upper extremity. Anterior aspect.

axillary vein just below the level of the clavicle. Sometimes it communicates with the external jugular vein by a branch ascending anterior to the clavicle.

Sometimes the median cubital vein is large and carries all or most of the blood from the cephalic into the basilic vein, the proximal half of the cephalic vein being either absent or proportionately diminished.

The *accessory cephalic vein* arises from a small tributary plexus on the back of the forearm or from the ulnar side of the dorsal venous network; it joins the cephalic below the elbow. In some cases it springs from the cephalic vein above the wrist and joins it again higher up. A large oblique branch frequently connects the basilic and cephalic veins on the back of the forearm.

The basilic vein (6.130) begins in the ulnar part of the dorsal venous network of the hand. It ascends for some distance on the posterior surface of the ulnar side of the forearm, but inclines forward to the anterior surface below the elbow. It is joined by the median cubital vein and ascends obliquely and superficially, in the groove between the biceps and pronator teres; filaments of the medial cutaneous nerve of the forearm pass in front of and behind this section of the vein. It then runs upwards medial to biceps, perforates the deep fascia a little below the middle of the arm, and ascends medial to the brachial artery to the lower border of teres major, continuing as the axillary vein.

The median vein of the forearm (6.130) drains the superficial palmar venous plexus. It ascends on the front of the forearm and ends in the basilic or median cubital vein; occasionally it divides below the elbow into two branches, one to the basilic, the other to the cephalic vein.

Surface Anatomy. All the superficial veins can be seen, especially in spare individuals, until they pierce the deep fascia. They become more obvious when the limb is dependent and its muscles are contracted to drive blood from deep to superficial veins. This is most effective with a light encircling pressure in the upper arm.

Applied Anatomy. Blood sampling and transfusion, and intravenous injections in general, are often performed at the bend of the elbow, and the largest vein, usually the median cubital, is commonly selected. The cubital veins are also the site for the introduction of cardiac catheters to secure blood samples from the chambers of the heart and great vessels, to assess pressure profiles, and for cardio-angiography. Equally useful for sampling and transfusions, is the cephalic vein overlying the 'anatomical snuffbox'.

The Deep Veins of the Upper Limb

These veins follow the arteries as their companions. They are generally in pairs, flanking the corresponding artery (venae comitantes), and connected at intervals by short transverse branches. Since most of the blood draining the upper limb is *returned* by the the *superficial veins*, the deep veins are relatively small.

The deep veins of the hand. The superficial and deep palmar arterial arches are each accompanied by venae comitantes, respectively the *superficial* and *deep palmar venous arches*, receiving veins corresponding to the branches of the arterial arches. Thus the *common palmar digital veins* open into the superficial, and the *palmar metacarpal veins* into the deep, palmar

venous arches. The deep veins accompanying the dorsal metacarpal arteries receive perforating branches from the palmar metacarpal veins, and end in the radial veins and the dorsal venous network.

The deep veins of the forearm are companion vessels of the radial and ulnar arteries and constitute respectively the upward drainage of the deep and superficial palmar venous arches; they unite in front of the elbow as the brachial veins. The radial veins are smaller than the ulnar, and receive the deep veins of the dorsum of the hand. The ulnar veins receive tributaries from the deep palmar venous arch and communicate with the superficial veins at the wrist; near the elbow they receive the anterior and posterior interosseous companion veins and send a large branch to the median cubital vein.

The brachial veins are on each side of the brachial artery, and receive tributaries corresponding with its branches; near the lower margin of subscapularis, they join the axillary vein; the medial one frequently joins the basilic vein before becoming the axillary.

These deep veins have numerous anastomoses, not only with each other, but also with the superficial veins.

The axillary vein begins, by definition, at the lower border of the teres major, as the continuation of the basilic, increases in size as it ascends, and ends at the outer border of the first rib, where it becomes the subclavian vein. Near the lower border of subscapularis it receives the brachial veins and, close to its termination, the cephalic vein; its other tributaries correspond with the branches of the axillary artery. It lies medial to the axillary artery, which it partly overlaps; between the two vessels are the medial pectoral nerve, the medial cord of the brachial plexus, the ulnar nerve and the medial cutaneous nerve of the forearm. Medially it is accompanied by the medial cutaneous nerve of the arm, and both on its medial and posterior aspects it is intimately related to the lateral group of the axillary lymph nodes. It is provided with a pair of valves opposite the lower border of subscapularis; valves are also found in the ends of the cephalic and subscapular veins.

The subclavian vein (6.64), continuing the axillary, extends from the outer border of the first rib to the medial border of scalenus anterior, where it unites with the internal jugular to form the brachiocephalic vein. It is in relation, in front, with the clavicle and subclavius, behind and above, with the subclavian artery, separated from it by scalenus anterior and the phrenic nerve. Below, it rests in a shallow groove on the first rib and upon the pleura. It usually has a pair of valves, situated about 2 cm from its termination.

Its tributaries are the external jugular, the dorsal scapular and sometimes the anterior jugular veins, and occasionally a small branch, which ascends in front of the clavicle, from the cephalic. At its angle of junction with the internal jugular, the left subclavian vein receives the thoracic duct and the right subclavian the right lymphatic duct.

Surface Anatomy. The subclavian vein can be represented by a broad band convex upwards, from a point just medial to the mid-clavicular point to the medial edge of the clavicular attachment of sternocleidomastoid.

THE VEINS OF THE THORAX

The brachiocephalic (innominate) veins are two large trunks in the root of the neck and the uppermost part of the thorax; each is formed by union of the internal jugular and subclavian veins of its side, and both are devoid of valves.

The right brachiocephalic vein (6.131), about 2·5 cm long, begins behind the sternal end of the right clavicle, and passes almost vertically downwards to join the left brachiocephalic vein and form the superior vena cava, behind the lower border of the first right costal cartilage, close to the right border of the sternum. It lies anterolateral to the brachiocephalic artery and the right vagus nerve. The right pleura, phrenic nerve, and internal thoracic artery are posterior to its upper part, and become lateral below this.

Its tributaries are the right vertebral, internal thoracic and inferior thyroid veins, and sometimes the first right posterior intercostal vein.

The left brachiocephalic vein (6.131), about 6 cm long, begins behind the sternal end of the left clavicle, in front of the cervical pleura. It runs obliquely downwards to the right behind the upper half of the manubrium sterni to the sternal end of the first right costal cartilage, where it unites with the right brachiocephalic vein to form the superior vena cava. It is separated from the left sternoclavicular joint and the manubrium sterni by sternohyoid and sternothyroid, the thymus or its remains, and some areolar tissue; at its termination it is overlapped by the right pleura. In its course it crosses in front of

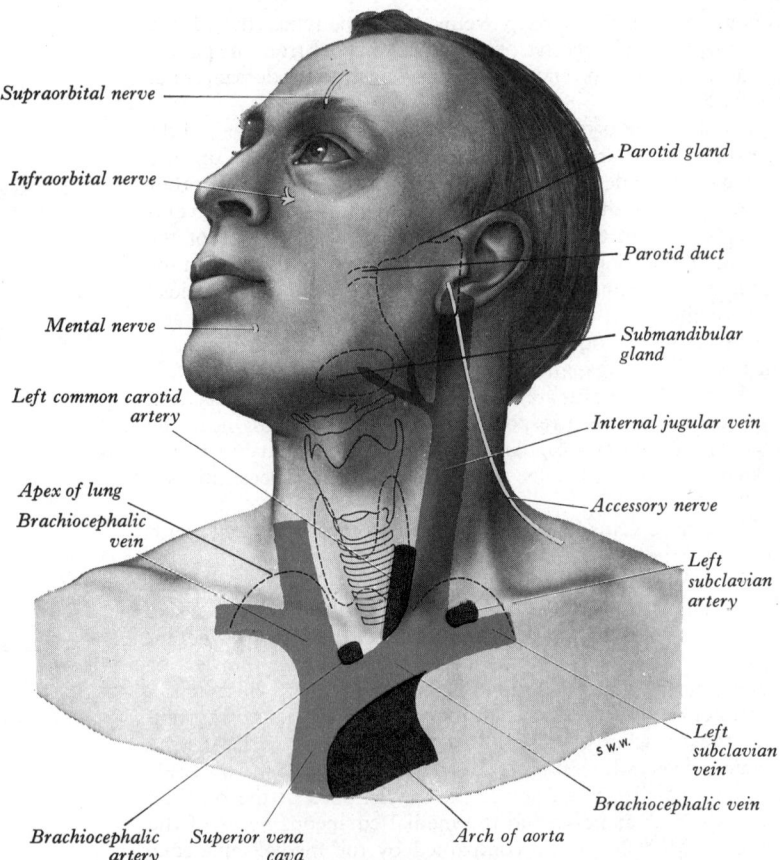

Supraorbital nerve

Infraorbital nerve

Mental nerve

Left common carotid artery

Apex of lung

Brachiocephalic vein

Brachiocephalic artery

Superior vena cava

Parotid gland

Parotid duct

Submandibular gland

Internal jugular vein

Accessory nerve

Left subclavian artery

Left subclavian vein

Brachiocephalic vein

Arch of aorta

S.W.W.

6.131   The surface projections of some of the important structures in the face, neck and upper part of thorax. The apices of the lungs, the thyroid, submandibular and parotid glands, and the parotid duct are indicated in interrupted dotted outline: the hyoid bone, the thyroid and cricoid cartilages and the rings of the trachea are shown in continuous outline.

the left internal thoracic, subclavian and common carotid arteries, the left phrenic and vagus nerves, the trachea and the brachiocephalic artery. The arch of the aorta lies inferior to it.

Its tributaries are the left vertebral, internal thoracic, inferior thyroid, and superior intercostal veins, sometimes the first left posterior intercostal vein and some thymic and pericardial veins.

Surface Anatomy. The brachiocephalic veins can be projected on to the surface as broad bands 1·5 cm wide from the sternal end of the clavicle on each side to the formation of the superior vena cava at the lower border of the first right costal cartilage.

Variations. Sometimes the brachiocephalic veins open separately into the right atrium: when this is so the right vein takes the ordinary course of the superior vena cava; the left vein—*left superior vena cava,* as it is then termed—may communicate by a small branch with the right one, and crosses the left side of the arch of the aorta, passes in front of the left pulmonary hilum, and, turning to the back of the heart, enters the right atrium. It replaces the oblique vein of the left atrium and the coronary sinus and receives all the tributaries of the latter. This abnormality is due to the persistence of the early fetal condition (p. 193), and is normal in birds and some mammals.

The left brachiocephalic vein sometimes projects above the level of the manubrium sterni (more frequently in childhood), crossing the suprasternal fossa, in front of the trachea.

The internal thoracic (mammary) veins (**6.**43, 67) are venae comitantes to the lower half of the internal thoracic artery, and are provided with a number of valves. About the level of the third costal cartilage the companion veins unite into a single trunk, which runs up medial to the artery and ends in the corresponding brachiocephalic vein. Veins accompanying branches of the artery (p. 697), and the pericardiacophrenic vein, are tributaries.

The inferior thyroid veins (**6.**43) arise in the thyroid gland in a venous network which also communicates with the middle and superior thyroid veins. They form a plexus in front of the trachea.

From this plexus the left vein descends and joins the left brachiocephalic vein, and the right passes obliquely downwards to the right across the brachiocephalic artery to open into the right brachiocephalic vein, at its junction with the superior vena cava; frequently the two veins open by a common trunk either in the vena cava or the left brachiocephalic vein. They drain oesophageal, tracheal and inferior laryngeal veins, and have valves at their terminations.

The left superior intercostal vein (**6.**132) receives the second and third (and sometimes the fourth) left posterior intercostal veins; it runs obliquely upwards and forwards on the left side of the aortic arch, passing lateral to the left vagus and medial to the left phrenic nerve, and opens into the left brachiocephalic vein. It usually receives the left bronchial veins, and sometimes the left pericardiacophrenic vein; it communicates below with the accessory hemiazygos vein.

The superior vena cava (**6.**38, 40, 41r, 42, 132) collects blood from the upper half of the body. It measures about 7 cm in length, is formed by the junction of the two brachiocephalic veins, and has no valves. It begins behind the lower border of the first right costal cartilage close to the sternum, descends vertically behind the first and second intercostal spaces, ends in the upper part of the right atrium opposite the third right costal cartilage; the lower half of the vessel is within the fibrous pericardium, which it pierces at the level of the second costal cartilage. It is covered in front and on each side with the serous pericardium. In its course it describes a slight convexity to the right.

Relations. In *front,* the superior vena cava is related to the anterior margins of the right lung and its pleura, with the pericardium intervening below; these structures separate it from the internal thoracic artery and the first and second intercostal spaces, and from the second and third costal cartilages; the trachea and the right vagus nerve are *posteromedial* and the right lung and pleura *posterolateral* to its upper part, while the root of the right lung is a direct posterior relation below. On its *right* it is related to the right phrenic nerve and right pleura; on its *left* to the commencement of the brachiocephalic artery and the ascending aorta, the latter overlapping it.

Surface Anatomy. The superior vena cava is 2 cm wide and partly behind the right margin of the sternum. It extends from the lower border of the first to the lower border of the third right costal cartilage. The shadow of its lateral border is visible in anteroposterior radiographs of the chest.

Tributaries. These include the azygos vein and several small veins from the pericardium and other structures in the mediastinum.

The azygos vein (**6.**132, 133) is inconstant in origin (as is the hemiazygos vein—*see* Gladstone 1929). On developmental grounds it may be expected to arise from the posterior aspect of the inferior vena cava, at or below the level of the renal veins. Such a vein (*lumbar azygos*) is indeed frequently present, and ascends in front of the upper lumbar vertebrae. The azygos vein may pass behind the lateral border of the right crus of the diaphragm, or it may pierce the crus. Occasionally it may pass through the aortic opening of the diaphragm, on the right side of the cisterna chyli. In front of the body of the twelfth thoracic vertebra, it is joined by a large vessel which, formed by the union of the right ascending lumbar with the right subcostal vein, passes forwards on the right side of the body of the twelfth thoracic vertebra behind the right crus of the diaphragm. This common trunk may, in the absence of a lumbar azygos vein, form the azygos itself. Whatever its mode of origin, the azygos vein ascends in the posterior mediastinum to the level of the fourth thoracic vertebra, where it arches forward above the root of the right lung, and ends in the superior vena cava, just before it pierces the pericardium. In its course it lies anterior to the bodies of the lower eight thoracic vertebrae (*see* comments below), the anterior longitudinal ligament, and the right posterior intercostal arteries. On its right are the right greater splanchnic nerve, lung and pleura; on its left, throughout the greater part of its course, are the thoracic duct and aorta, and higher up, where it arches forward above the root of the right lung, the oesophagus, trachea and right vagus. In the lower part of the thorax it is covered anteriorly by a recess of the right pleural sac and by the oesophagus, but it emerges from behind the right edge of the latter and ascends behind the root of the right lung (**6**.133). For a variable extent in the lower thorax, the azygos vein is quite *closely applied* to the right posterolateral aspect of the descending thoracic aorta (Hutchinson, personal communication), and thus often reaches and sometimes crosses to the left of the midline of the vertebral bodies. It seems probable that aortic pulsations will, by intermittent compression, assist venous return in both the azygos and hemiazygos veins (cf. comments on venous return in venae comitantes in general on p. 630).

Tributaries. The azygos vein collects from the posterior intercostal veins of the right side, with the exception of the first; the veins from the second, third and fourth intercostal spaces usually open by a common stem called the right superior intercostal vein. It receives also the hemiazygos and accessory hemiazygos veins, several oesophageal, mediastinal, and pericardial veins, and, near its termination, the right bronchial veins. When it begins as a lumbar azygos vein, the common trunk formed by the union of the right ascending lumbar and subcostal veins is its largest tributary. A few imperfect valves are found in the azygos vein, but its tributaries are provided with complete valves.

The hemiazygos vein starts on the left like the azygos vein on the right side. Ascending on the front of the vertebral column as high as the eighth thoracic vertebra, it passes across the column, behind the aorta, oesophagus, and thoracic duct, to end in the azygos vein. Its tributaries are the lower three posterior intercostal veins and the common trunk formed by the union of the ascending lumbar and the subcostal veins of the left side, and some oesophageal and mediastinal veins. Its lower end often communicates with the left renal vein.

The accessory hemiazygos vein (**6**.132) descends on the left side of the vertebral column. It receives the veins from the fourth (or fifth) to the eighth intercostal spaces inclusive of the left side, and sometimes the left bronchial veins. It crosses the body of the seventh thoracic vertebra and joins the azygos vein. The accessory hemiazygos vein sometimes joins the hemiazygos vein, and the common trunk thus formed opens into the azygos vein.

The azygos veins exhibit much variation, not only in their mode of origin, but also in their course, tributaries, anastomoses and termination. For a recent survey of the literature consult Grzybiak *et al.* (1975). According to this study the accessory hemiazygos is the most variable; it may drain into the left brachiocephalic, directly into the azygos vein, or into the hemiazygos. The arrangement illustrated in **6**.132 represents in its general features the commonest pattern (except for the position and calibre of the azygos vein). In a few instances (about 1 or 2 per cent according to Anson, 1963) there are two completely independent azygos veins (left and right), which is the early embryonic form, and in an equally small group of instances a single azygos vein, without any hemiazygos tributaries, which occupies an approximately midline position. In more than 95 per cent of cases there is a main 'right-sided' azygos vein and at least some representative of the hemiazygos veins. The latter, however, show much variation, one or other being absent or poorly developed. The number of retro-aortic trans-vertebral connexions from the hemiazygos and accessory hemiazygos veins, connecting them to the azygos vein, is also extremely variable; they may number from one to five, or more; when either hemiazygos vein is absent, the intercostal veins which would have drained into it themselves cross their vertebral bodies to end in the azygos vein. These trans-vertebral routes are, however, frequently extremely short, since the azygos vein is, as noted above, more commonly situated anterior to the vertebral column rather than on its right lateral aspect (Anson 1963, and Dr. M. C. E. Hutchinson, personal communication), and quite often passes to the *left* of the midline in a variable part of its vertical extent.

The posterior intercostal veins (**6**.83, 132, 133) run with the posterior intercostal arteries and are eleven in number on each side. As they approach the vertebral column each vein receives a tributary which accompanies the posterior branch of the corresponding artery returning blood from the muscles and skin of the back and from the vertebral venous plexuses.

On both sides of the thorax the first posterior intercostal vein ascends in front of the neck of the first rib. It arches forwards

above the pleura to end in the corresponding brachiocephalic or vertebral vein.

On the right the second, third and, often, the fourth posterior intercostal veins unite to form the *right superior intercostal vein*, which joins the terminal part of the azygos vein. The veins from the intercostal spaces below the fourth open separately into the vena azygos.

On the left the second and third and sometimes the fourth posterior intercostal veins unite to form the left superior intercostal vein (p. 754). The veins from the fourth (or fifth) to the eighth intercostal spaces inclusive end in the accessory hemiazygos vein, and the veins from the lower three spaces in the hemiazygos vein.

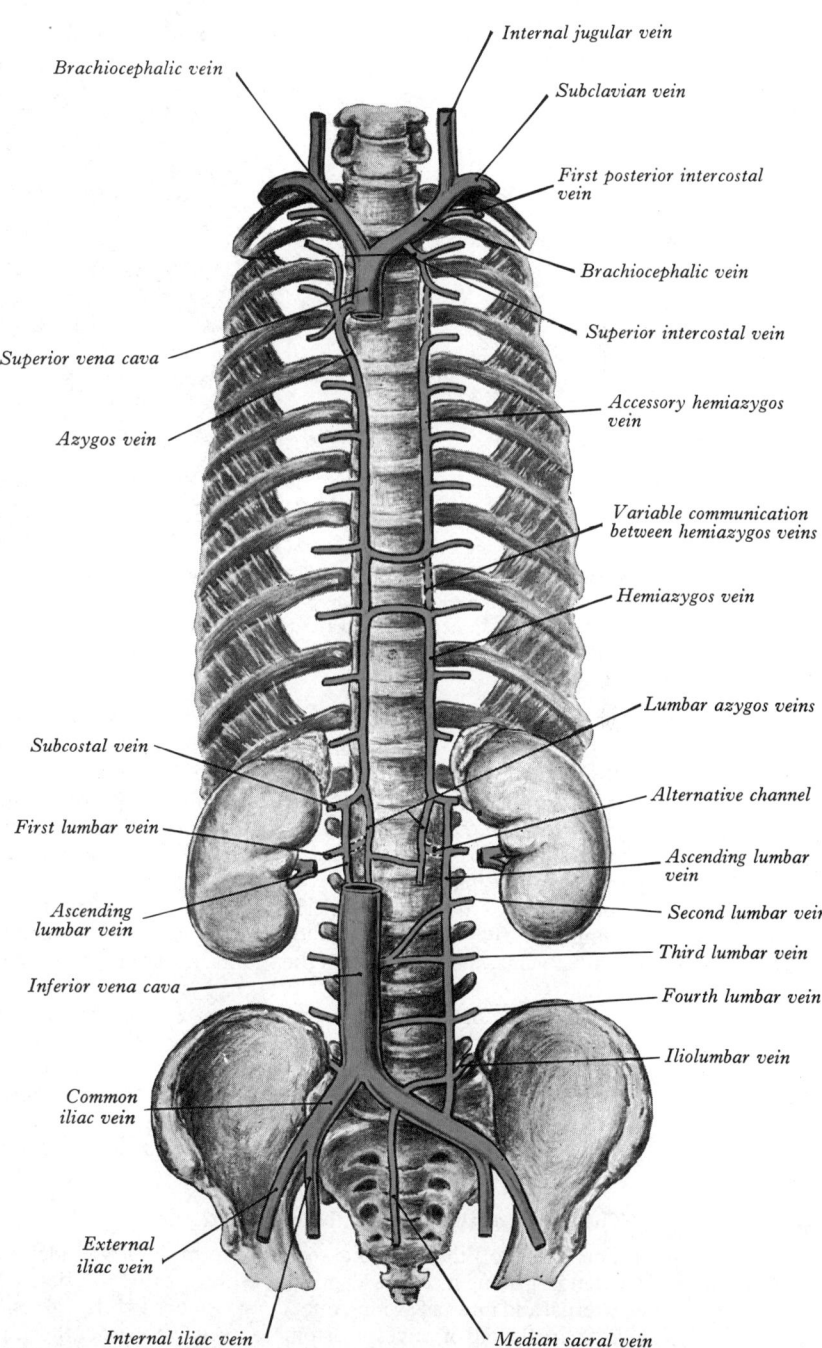

6.132 A highly schematic diagram of the azygos, hemiazygos and ascending lumbar veins, their tributaries and connexions. On both sides the first lumbar vein is terminated in the ascending lumbar, its occasional termination in the lumbar azygos vein is also shown. The hemiazygos and accessory hemiazygos veins both cross the midline one vertebra lower than usual. It should be noted that the azygos vein is usually much greater in calibre than the hemiazygos veins, and seldom follows a vertical course on the right anterolateral aspect of the vertebral bodies as diagrammed here. *See* text for more detailed comments on course and variations.

Oesophagus

Brachiocephalic artery

Trachea

Sympathetic trunk

Superior intercostal vein

Azygos vein

Right superior lobe bronchus

Right pulmonary artery

Right principal bronchus

Right phrenic nerve

Greater splanchnic nerve

Diaphragm

Right vagus nerve

Right brachiocephalic vein

Left brachiocephalic vein

Internal thoracic vessels

Pericardium

Ascending aorta

Superior vena cava

Right pulmonary veins

Right atrium

6.133   The right aspect of the mediastinum. The right lung and most of the right pleura have been removed and a large opening made into the pericardial sac to expose the heart. In this specimen the fourth right posterior intercostal vein did not join the superior intercostal vein.

The posterior intercostal veins are called 'posterior' to distinguish them from the small *anterior intercostal veins*, which are tributaries of the internal thoracic and musculophrenic veins.

Applied Anatomy. In obstruction of the venae cavae, the azygos and hemiazygos veins and the vertebral venous plexuses are important means by which the venous circulation is carried on, connecting, as they do, the superior and inferior venae cavae, and communicating with the common iliac veins by the ascending lumbar veins, and with many of the tributaries of the inferior vena cava.

The bronchial veins, usually two on each side, return blood from the larger bronchi, and from the structures at the roots of the lungs. The bronchial veins of the right side open into the terminal part of the azygos vein, those of the left side, into the left superior intercostal vein or the hemiazygos vein. Some of the blood carried to the lungs through the bronchial arteries returns to the heart through the pulmonary veins (*see* p. 1266).

The Veins of the Vertebral Column (6.134, 135)

The veins of the vertebral column form intricate plexuses extending along its entire length; these plexuses are divisible into external and internal groups, outside or inside the vertebral canal. Both are devoid of valves, anastomose freely with each other, and end in the intervertebral veins. Interconnexions are established on a wide scale between the plexuses and longitudinal veins very early in fetal life (Loginova 1972).

The external vertebral venous plexuses, most developed in the cervical region, consist of anterior and posterior plexuses, which anastomose freely. The *anterior external plexuses* lie in front of the bodies of the vertebrae, communicate with the basivertebral and intervertebral veins, and receive tributaries from the vertebral bodies. The *posterior external plexuses* are on the

posterior surfaces of the laminae and around the spines and the transverse and articular processes. They anastomose with the internal vertebral venous plexuses, and end in the vertebral, posterior intercostal and lumbar veins.

The internal vertebral venous plexuses lie within the vertebral canal between the dura mater and vertebrae, and receive

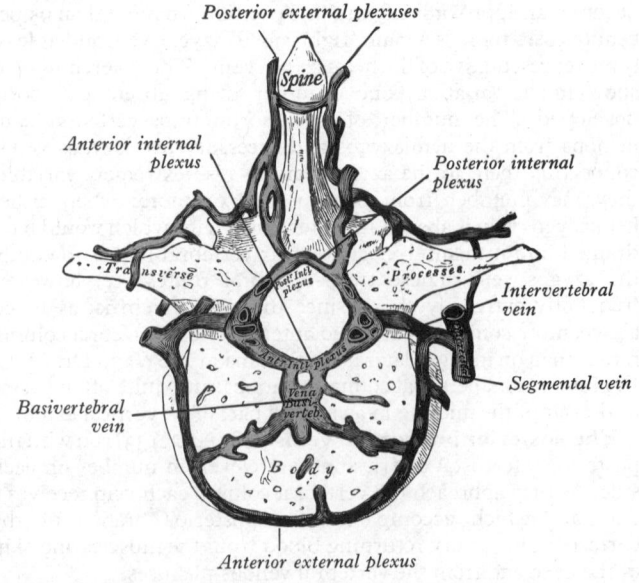

Posterior external plexuses

Spine

Anterior internal plexus

Posterior internal plexus

Transverse

Processes

Intervertebral vein

Basivertebral vein

Segmental vein

Body

Anterior external plexus

6.134   A transverse section through the body of a thoracic vertebra, showing the vertebral venous plexuses and basivertebral veins.

tributaries from the bones and from the spinal cord. They form a closer network than the external plexuses, and are arranged vertically as four longitudinal veins, two in front and two behind; therefore they may be divided into anterior and posterior groups. The *anterior internal plexuses* consist of large veins which lie on the posterior surfaces of the vertebral bodies and intervertebral discs, on each side of the posterior longitudinal ligament; under cover of this ligament they are connected by transverse branches into which the basivertebral veins open. The *posterior internal plexuses* are on each side of the median plane in front of the vertebral arches and ligamenta flava, and anastomose, by veins passing through and between those ligaments, with the posterior external plexuses.

The anterior and posterior internal plexuses communicate freely with one another by a series of venous rings, one opposite each vertebra. Around the foramen magnum they form an intricate network, which connects with the vertebral veins, occipital and sigmoid sinuses, the basilar plexus, the venous plexus of the hypoglossal canal and the condylar emissary veins.

The basivertebral veins emerge from foramina on the posterior surfaces of the vertebral bodies. They are large, tortuous channels in the substance of the bones, similar in every respect to those found in the diploë of the cranial bones. It is to be noted that the cancellous bone in the bodies of the vertebrae contains much red haemopoietic bone marrow. They also drain into the anterior external vertebral plexuses through small openings on the front and sides of the bodies of the vertebrae, and converge behind to form single (sometimes double) veins, which open into the transverse branches uniting the anterior internal vertebral plexuses. The basivertebral veins become enlarged in advanced age.

The intervertebral veins accompany the spinal nerves through the intervertebral foramina; they drain veins from the spinal cord, the internal and external vertebral plexuses and end in the vertebral, posterior intercostal, lumbar and lateral sacral veins. There is some difference of opinion as to whether the basivertebral or intervertebral veins contain effective valves; but

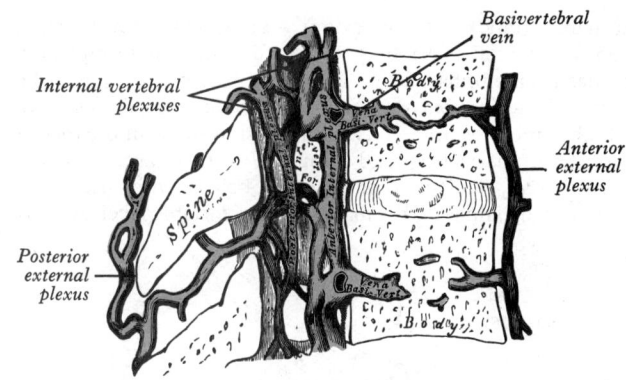

6.135   A median sagittal section through two thoracic vertebrae showing the vertebral venous plexuses and basivertebral veins.

experimental studies strongly suggest that the blood flow in them may be reversed (Batson 1957). This may explain how pelvic neoplasms in particular may lead to metastases in vertebral bodies, tumour cells spreading into the internal vertebral venous plexuses through connexions with the pelvic veins when the direction of blood flow is temporarily reversed by raised intra-abdominal pressure.

The veins of the spinal cord are situated in the pia mater and form a tortuous venous plexus in this membrane. In this plexus there are: (*a*) two *median longitudinal veins*, one in front of the anterior median fissure, and the other behind the posterior median septum of the spinal cord; (*b*) two *anterolateral* and two *posterolateral longitudinal veins*, which run behind the ventral and the dorsal nerve roots respectively. They drain into the internal vertebral venous plexuses, and thence the intervertebral veins. Near the base of the skull they unite to form two or three small trunks which communicate with the vertebral veins, and end in the inferior cerebellar veins, or in the inferior petrosal sinuses.

THE VEINS OF THE LOWER LIMBS

The veins of the lower limbs can be subdivided, like those of the upper limbs, into two sets, *superficial* and *deep*; the superficial veins are immediately under the skin in the superficial fascia; the deep veins accompany the arteries. Both sets are provided with valves, which are more numerous in the deep than in the superficial veins. Valves are also more plentiful in the veins of the lower than in those of the upper limbs.

The Superficial Veins of the Lower Limb (6.136, 137)

The principal named superficial veins are the great and small saphenous veins. Most of their tributaries are unnamed. (For variations in the superficial veins consult Kosinski 1926.)

The *dorsal digital veins* receive, in the clefts between the toes, communications from the plantar digital veins, and then join to form *dorsal metatarsal veins*, which unite across the proximal parts of the metatarsal bones in a *dorsal venous arch*. Proximal to this arch there is an irregular dorsal venous network, which receives tributaries from the deep veins and is continuous with the venous network on the front of the leg. At the sides of the foot this network communicates with a *medial* and a *lateral marginal vein*, both of which are formed mainly by the union of veins from the superficial parts of the sole of the foot.

In the sole of the foot the superficial veins form a *plantar cutaneous arch*, which extends across the roots of the toes and opens at the sides of the foot into the medial and lateral marginal veins. Proximal to this arch there is a *plantar cutaneous venous network*, which is especially dense in the fat beneath the heel; this network communicates with the plantar cutaneous venous arch and with the deep veins, but is chiefly drained into the medial and lateral marginal veins.

The great (long) saphenous vein (**6**.136), the longest in the body, begins in the medial marginal vein of the foot, and ends in the femoral vein about 3 cm below the inguinal ligament. It ascends about 2·5 to 3 cm in front of the tibial malleolus, crosses the lower part of the medial surface of the tibia obliquely to its medial border, about a finger's breadth behind which it ascends to the knee. It runs upwards postero-medial to the medial condyles of the tibia and femur and along the medial side of the thigh and passes through the saphenous opening (p. 595), into the femoral vein. The saphenous opening is about 2·5 to 3·25 cm below and lateral to the pubic tubercle, and the vein may be represented by a line from here to the adductor tubercle of the femur. In the thigh it is accompanied by some branches of the medial femoral cutaneous nerve, at the knee by the saphenous branch of the descending genicular artery, and in the leg and foot by the saphenous nerve, which is in front of the vein. The great saphenous vein is often duplicated, especially below the knee. The valves in it number from ten to twenty and are more numerous in the leg than in the thigh. One valve lies just before the vein pierces the cribriform fascia and another at its junction with the femoral vein. In practically its entire extent the vein is in the superficial fascia, but it has many communications with the deep veins, especially in the leg (*vide infra*).

Tributaries. At the ankle it drains the sole of the foot through the medial marginal veins. In the leg it communicates freely with the small saphenous vein and deep veins. Just below the knee it usually has three large tributaries, one from the front of the leg, one from the region of the tibial malleolus which communicates with the 'perforating' veins and a third from the calf which communicates with the small saphenous vein. The second of these tributaries, ascending on the medial aspect of the calf, was called the '*posterior arch vein*' by Dodd and Cockett (1956), and the clinical importance of its connexions with the venae comitantes of the posterior tibial artery by 3 to 6 perforating (communicating) veins has been emphasized by Platz and Adelmann (1976), who proposed the term '*vena arcuata cruris posterior*' for this vessel. In the thigh it receives numerous tributaries; those from the medial and posterior parts frequently unite to form a large *accessory saphenous vein*, which communicates below with the small saphenous vein and joins the main vein at a variable level. A fairly

wall of the trunk, connecting the superficial epigastric vein, or the femoral vein, with the lateral thoracic veins and establishing a communication between the femoral and axillary veins. This communication is an important connecting channel between the superior and the inferior vena caval fields of drainage. It is held by some to reflect the line of the primitive mammary ('milk') ridge which extends from axilla to pubic region (p.156).

The small (short) saphenous vein (6.137) begins behind the

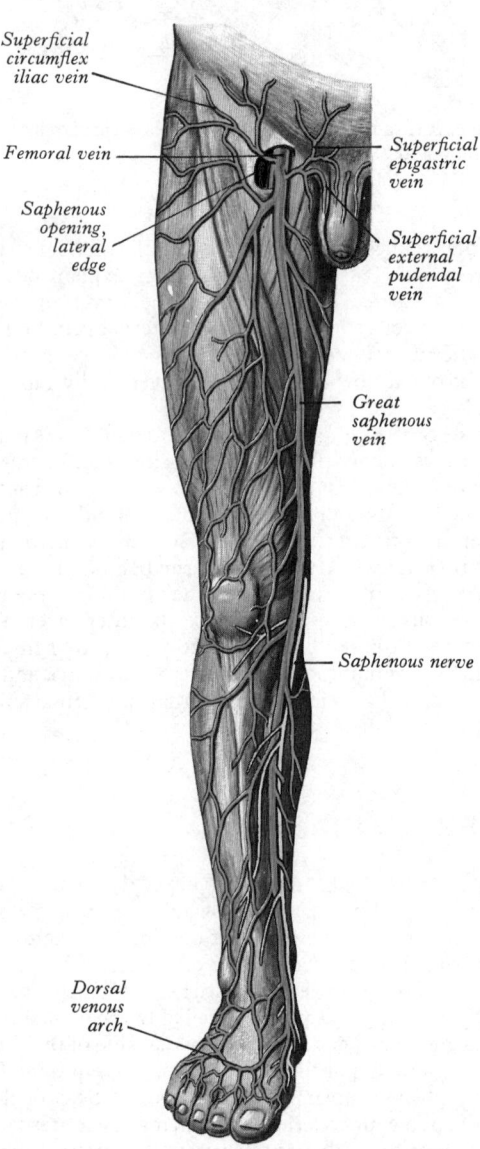

6.136   The great saphenous vein and its tributaries.

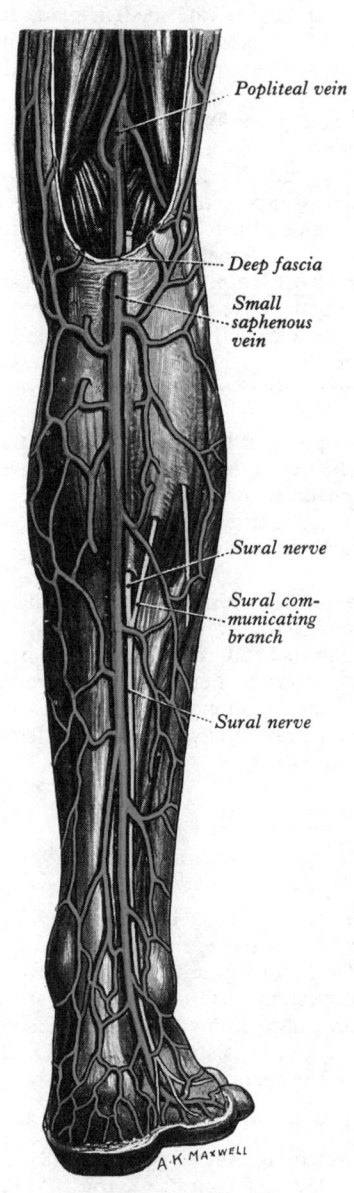

6.137   The small saphenous vein and its tributaries.

constant large vessel, sometimes called the *anterior femoral cutaneous vein*, commences from a network of veins on the lower part of the front of the thigh and crosses the apex of the femoral triangle to enter the great saphenous vein in the upper part of the thigh. Just before it pierces the saphenous opening (6.136) it is joined by three veins, the superficial epigastric, superficial circumflex iliac and superficial external pudendal. Their precise mode of union with the great saphenous varies. The superficial epigastric and superficial circumflex iliac veins drain the lower part of the abdominal wall, the latter veins also receiving tributaries from the upper and lateral parts of the thigh; the superficial external pudendal veins drain part of the scrotum and one is joined by the superficial dorsal vein of the penis. The deep external pudendal vein joins the great saphenous vein at its termination in the saphenous opening.

A vein, named *thoraco-epigastric*, runs along the anterolateral

lateral malleolus as a continuation of the lateral marginal vein of the foot; it first ascends lateral to the tendo calcaneus, and then along the middle of the back of the leg. It perforates the deep fascia and passes between the two heads of gastrocnemius in the lower part of the popliteal fossa, and ends in the popliteal vein from 3 to 7.5 cm above the level of the knee joint. It communicates with the deep veins on the dorsum of the foot, receives numerous cutaneous tributaries from the back of the leg, and sends several branches upwards and medially to join the great saphenous vein. Sometimes a communicating branch from the small saphenous vein just before it pierces the deep fascia passes upwards and medially to join the accessory saphenous vein. This communication may occasionally be the main continuation of the small saphenous. In the leg it is close to the sural nerve. The small saphenous vein possesses from seven to thirteen valves, one near its termination in the popliteal vein.

The mode of ending of the small saphenous vein is variable. It may join the great saphenous vein in the upper third of the thigh, or may bifurcate, one branch joining the great saphenous vein, the other the popliteal or the deep posterior veins of the thigh; occasionally it ends, below the knee joint, in the great saphenous vein or in the deep muscular veins of the calf.

Applied Anatomy. In the upright position, the venous return from the lower limb depends almost entirely on muscular activity (*see* p. 630), especially the contraction of the calf muscles, known as the 'calf pump', whose efficiency is aided by the tight sleeve of deep fascia. A number of '*perforating*' *veins* connect the great saphenous vein with the deep veins, particularly near the ankle and in the lower medial part of the leg. In these perforating veins the valves are arranged so that normally they prevent the flow of blood from the deep to the superficial veins. At rest, the pressure in a superficial vein of the leg is equal to the height of the column of blood extending therefrom to the heart. When the calf muscles contract, the blood is pumped upwards in the deep veins; it is normally prevented from flowing outwards into the superficial veins by the valves in the perforating veins; and during relaxation of the calf muscles blood is actually aspirated from the superficial into the deep veins. If the valves in the perforating veins become incompetent, these veins become 'high pressure leaks' during muscular contraction, and this transmission of the high pressure in the deep veins to the superficial veins results in dilatation and venous stagnation in the superficial veins, producing varicosities, devitalization of the tissues, and in some cases, varicose ulceration. In the operative treatment of severe varicose veins and ulcers it is essential that these perforating veins be ligatured. Similar perforating venous connexions occur on the anterolateral aspect of the leg and varicosities of the superficial veins may occur in this region (Cockett 1956; Green *et al.* 1958). Veins connecting the great saphenous vein to the femoral vein, as the latter lies in the adductor canal, may also become varicose (Dodd and Cockett 1956; Dodd 1959).

The Deep Veins of the Lower Limb

These accompany the arteries and their branches; they possess numerous valves.

The *plantar digital veins* arise from plexuses on the plantar surfaces of the digits. They communicate with the dorsal digital veins and unite to form four *plantar metatarsal veins*; which run back in the metatarsal spaces, communicate, by means of perforating veins, with the veins on the dorsum of the foot, and unite to form the *deep plantar venous arch*, accompanying the plantar arterial arch. From the deep plantar venous arch the *medial* and *lateral plantar veins* run backwards close to the corresponding arteries and, after communicating with the great

and small saphenous veins, unite behind the medial malleolus to form the posterior tibial veins.

The posterior tibial veins accompany the posterior tibial artery. They receive veins from the calf muscles, especially from the venous plexus in soleus, communicating vessels from the superficial veins of the leg and the *peroneal veins*. The latter, which accompany the peroneal artery, also receive veins from soleus and communications from the superficial veins.

The anterior tibial veins are the upward continuations of the venous companions of the dorsal artery of the foot. They leave the front of the leg by passing between the tibia and fibula, through the upper part of the interosseous membrane of the leg, and unite with the posterior tibial veins to form the *popliteal vein* at the lower border of popliteus.

The popliteal vein ascends through the popliteal fossa to the aperture in the adductor magnus, where it becomes the femoral vein. In the lower part of its course it is medial to the popliteal artery; between the heads of the gastrocnemius it is superficial to it; above the knee joint it is posterolateral to it. Its tributaries are the small saphenous vein, the veins corresponding to the branches of the popliteal artery and muscular veins, including a prominent vessel from each head of the gastrocnemius. There are usually four valves in the popliteal vein.

The femoral vein accompanies the femoral artery, beginning at the opening in adductor magnus as the continuation of the popliteal vein, and ending at the level of the inguinal ligament, by becoming the external iliac. In the lower part of the adductor canal it is posterolateral to the femoral artery; in the upper part of the canal, and in the lower part of the femoral triangle, it is behind the artery. At the base of the femoral triangle it is medial to the artery (**6**.100, 103); here it occupies the middle compartment of the femoral sheath, between the femoral artery and the femoral canal, the fatty tissue in the latter allowing considerable expansion of the vein. It receives numerous muscular tributaries, and about 4 to 12 cm below the inguinal ligament is joined posteriorly by the vena profunda femoris, and a little higher by the great saphenous vein, which enters its anterior aspect. In addition, the lateral and medial circumflex femoral veins are usually tributaries. There are usually four or five valves in the femoral vein, the most constant being just below the entry of the profunda femoris vein; there is also commonly a valve below the inguinal ligament.

The *vena profunda femoris* is anterior to the arteria profunda femoris; it receives tributaries corresponding to the muscular and perforating branches of that artery, and through these establishes communications with the popliteal vein below and the inferior gluteal vein above. It sometimes drains the medial and lateral circumflex femoral veins. It has a pair of valves just below its termination.

THE VEINS OF THE ABDOMEN AND PELVIS

The External Iliac Vein

The upward continuation of the femoral vein constitutes the external iliac vein: it begins behind the inguinal ligament, and ascends along the brim of the lesser pelvis, to end in front of the sacro-iliac joint, by joining the internal iliac vein to form the common iliac vein. On the right, it lies at first medial to the external iliac artery; but, as it passes upwards, gradually inclines behind it. On the left, it lies altogether on the medial side of the artery. On its medial aspect, it is crossed by the ureter and the internal iliac artery; elsewhere it is covered with parietal peritoneum. In the male it is crossed by the ductus deferens and in the female by the round ligament of the uterus and the ovarian vessels. Laterally it is related to the psoas major, except where the external iliac artery intervenes. It is usually valveless but may contain one valve.

Tributaries. These are the inferior epigastric, deep circumflex iliac, and pubic veins.

The inferior epigastric vein is formed by union of the venae comitantes of the inferior epigastric artery, which communicate

above with the superior epigastric vein; it joins the external iliac about 1 cm above the inguinal ligament.

The deep circumflex iliac vein is derived from the venae comitantes of the deep circumflex iliac artery, and joins the external iliac vein about 2 cm above the inguinal ligament after crossing in front of the external iliac artery.

The pubic vein, which connects the external iliac with the obturator vein in the obturator foramen, ascends on the pelvic surface of the pubis alongside the pubic branch of the inferior epigastric artery. It is sometimes enlarged and replaces the normal obturator vein.

The Internal Iliac Vein

A confluence of veins occurs near the upper part of the great sciatic foramen and constitutes the internal iliac vein. It ascends behind and slightly medial to the internal iliac artery, and joins with the external iliac vein to form the common iliac vein at the brim of the lesser pelvis, in front of the lower part of the sacro-iliac joint and is covered with parietal peritoneum on its anteromedial aspect.

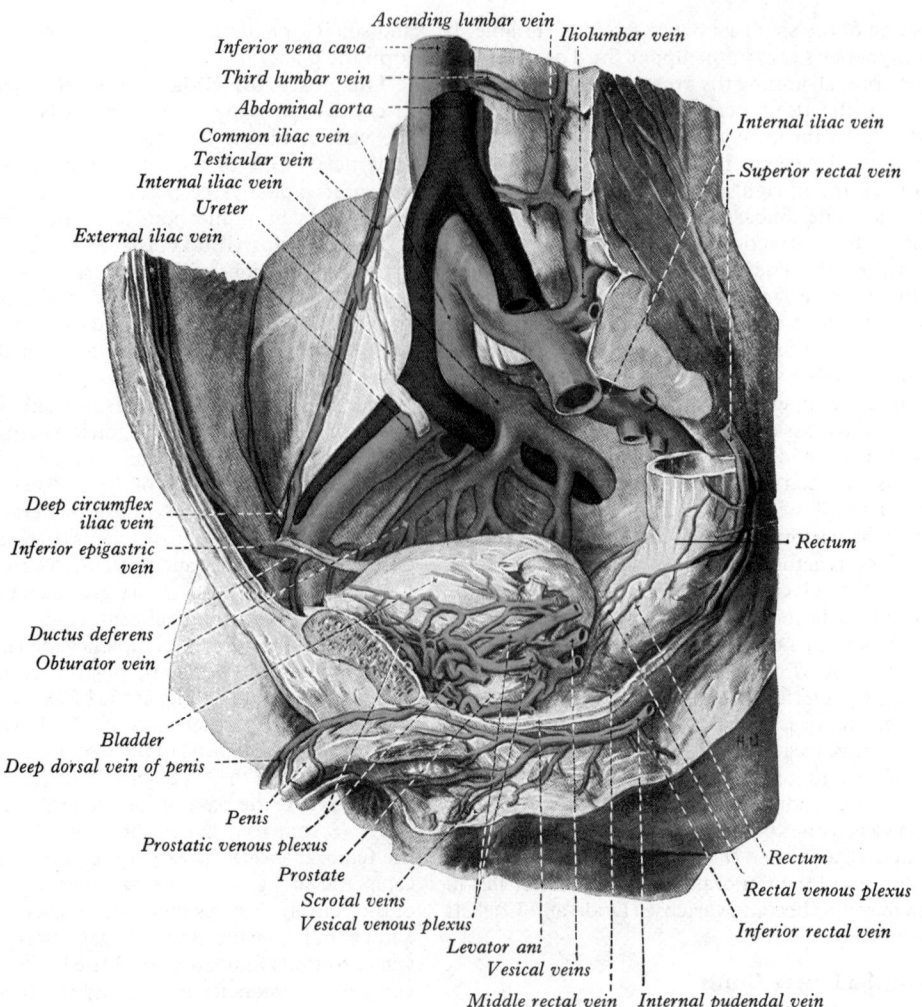

Ascending lumbar vein
Inferior vena cava
Third lumbar vein
Abdominal aorta
Common iliac vein
Testicular vein
Internal iliac vein
Ureter
External iliac vein

Iliolumbar vein
Internal iliac vein
Superior rectal vein

Deep circumflex iliac vein
Inferior epigastric vein

Ductus deferens
Obturator vein

Bladder
Deep dorsal vein of penis

Rectum

Penis
Prostatic venous plexus
Prostate
Scrotal veins
Vesical venous plexus
Levator ani
Vesical veins
Middle rectal vein

Rectum
Rectal venous plexus
Inferior rectal vein
Internal pudendal vein

6.138   The veins of the right half of the male pelvis (after Spalteholz).

Tributaries. With the exception of the iliolumbar vein, which usually joins the common iliac, the tributaries of the internal iliac vein correspond with the branches of the internal iliac artery. These are (*a*) the gluteal, internal pudendal, and obturator veins, which have their origins outside the pelvis; (*b*) the lateral sacral veins, which lie in front of the sacrum; (*c*) the middle rectal, the vesical, uterine, and vaginal veins, which originate in venous plexuses connected with the pelvic viscera.

The superior gluteal veins are the venae comitantes of the superior gluteal artery; their tributaries from the buttock correspond to the branches of the artery; entering the pelvis through the greater sciatic foramen, above piriformis, they end in the internal iliac vein, frequently uniting to form a single trunk before this.

The inferior gluteal veins are the venae comitantes of the inferior gluteal artery; they begin in the upper part of the back of the thigh, where they anastomose with the medial circumflex femoral and first perforating veins; they enter the pelvis through the lower part of the greater sciatic foramen and join to form a stem which opens into the lower part of the internal iliac vein.

The gluteal veins communicate with the superficial veins of the buttock by a considerable number of perforating veins (Doyle 1970), as in the calf (p. 759). These *gluteal perforating veins* are, indeed, more numerous. In addition to their possible venous 'pumping' role, in association with the activities of the gluteal musculature, these venous connexions provide collaterals between the femoral and internal iliac veins.

The internal pudendal veins are the venae comitantes of the internal pudendal artery. They begin in the prostatic venous plexus (p. 761), accompany the internal pudendal artery, and unite to form a single vessel, which ends in the internal iliac vein. They receive the veins from the bulb of the penis, and the scrotal

(or labial) and inferior rectal veins. The deep dorsal vein of the penis communicates with the internal pudendal veins, but ends mainly in the prostatic plexus.

The obturator vein begins high in the adductor region of the thigh, and enters the pelvis through the upper part of the obturator foramen. It runs back and upwards on the lateral wall of the pelvis below the obturator artery covered by peritoneum; it passes between the ureter and the internal iliac artery, and ends in the internal iliac vein. Sometimes it is replaced by an enlarged pubic vein, which joins the external iliac vein (p. 759).

The lateral sacral veins accompany the lateral sacral arteries. The veins are interconnected by the *sacral venous plexus*.

The middle rectal vein varies in size; it begins in the rectal venous plexus, with tributaries from the bladder, prostate and seminal vesicle; it runs laterally on the pelvic surface of levator ani and ends in the internal iliac vein.

The rectal venous plexus surrounds the rectum, and communicates in front with the vesical plexus in the male, and the uterovaginal plexus in the female. It consists of two parts, an *internal* beneath the epithelium of the rectum and anal canal, and an *external* outside their muscular coats. In the anal canal the internal plexus presents a series of longitudinal dilatations, connected by transverse branches, which are arranged in circles around the tube, immediately above the anal valves. These dilatations are more obvious in the left lateral, right anterolateral, and right posterolateral sectors of the tube. The plexus drains mainly into the superior rectal vein but communicates freely with the external plexus. The lower part of the external plexus is drained by the inferior rectal veins into the internal pudendal, middle part by the middle rectal vein into the internal iliac vein, and the upper part by the superior rectal vein, which is the commencement of the inferior mesenteric, a tributary of the portal vein. A free communication between the portal and systemic venous systems is thus established through the rectal plexus.

Applied Anatomy. The veins of the internal rectal plexus are apt to become further dilated and varicose, and form *internal* haemorrhoids (piles), most commonly at one or more of the three sites of concentration mentioned above. This is due to several factors: the vessels are contained in very loose connective tissue, so that they get less support from surrounding structures than most other veins, and are less capable of resisting increased blood pressure; the condition is favoured by the fact that the superior rectal and portal veins have no valves; the veins pass through muscular tissue and are liable to be compressed by its contraction, especially during the act of defaecation; finally they are affected by every form of portal obstruction.

The veins of the subcutaneous part of the external plexus may suffer thrombosis, the basis of one type of *external* haemorrhoid; alternatively, rupture of these vessels leads, on occasion, to an acute perianal haematoma.

The prostatic venous plexus lies behind the arcuate pubic ligament and the lower part of the symphysis pubis and in front of the bladder and prostate (**6.**138). Its chief tributary is the deep dorsal vein of the penis, but it also receives tributaries from the front of the bladder and prostate. It communicates with the vesical plexus and the internal pudendal vein, and drains into the vesical and internal iliac veins. The veins of the plexus are embedded in the lateral part of the fascial sheath of the prostate (*see* p. 1422).

The vesical plexus envelops the lower part of the bladder and, in the male, the base of the prostate. It communicates with the prostatic plexus in the male, and with the vaginal plexus in the female. It is drained by means of several vesical veins which usually form a single trunk before entering the internal iliac vein.

The dorsal veins of the penis are superficial and deep: The *superficial dorsal vein* drains the prepuce and skin of the penis, and, running backwards in the subcutaneous tissue, inclines to the right or left, and opens into the corresponding external pudendal vein, a tributary of the great saphenous. The *deep dorsal vein* lies within the fibrous envelope of the penis; it receives blood from the glans penis and corpora cavernosa penis, and courses backwards in the median plane between the dorsal arteries; near the root of the penis it passes deep to the suspensory ligament and then through an aperture between the arcuate pubic ligament and the anterior margin of the perineal membrane, and divides into two branches (right and left), which enter the prostatic plexus after communicating below the symphysis pubis with the internal pudendal veins. The *dorsal vein of the clitoris*, after a similar course, ends in the vesical plexus.

The uterine plexuses extend along the lateral side of the uterus within the broad ligament, and communicate with the ovarian and vaginal plexuses. They are drained by two uterine veins on each side; these arise from the lower parts of the plexuses, opposite the external opening of the cervix of the uterus, and open into the corresponding internal iliac vein.

The vaginal plexuses flank the vagina; they communicate with the uterine, vesical and rectal plexuses, and are drained by the vaginal veins, one each side, into the internal iliac veins.

The common iliac veins (**6.**138) result from the union of the external and internal iliac veins, in front of the sacro-iliac joint; passing obliquely upwards, they end on the right side of the fifth lumbar vertebra by uniting at an acute angle to form the inferior vena cava. The *right common iliac vein*, shorter than the left, is nearly vertical, and ascends behind, and then lateral to its artery. The right obturator nerve passes behind it obliquely, as it runs downwards and forwards to the obturator foramen. The *left common iliac vein*, longer and more oblique than the right, is at first medial to its artery, and then behind the right common iliac artery. It is crossed anteriorly by the root of the sigmoid mesocolon and the superior rectal vessels. In the rest of its course it is covered only with peritoneum. Each common iliac vein receives the iliolumbar, and sometimes the lateral sacral veins; the left common iliac drains the median sacral vein. There are no valves in these veins.

The median sacral veins accompany the corresponding

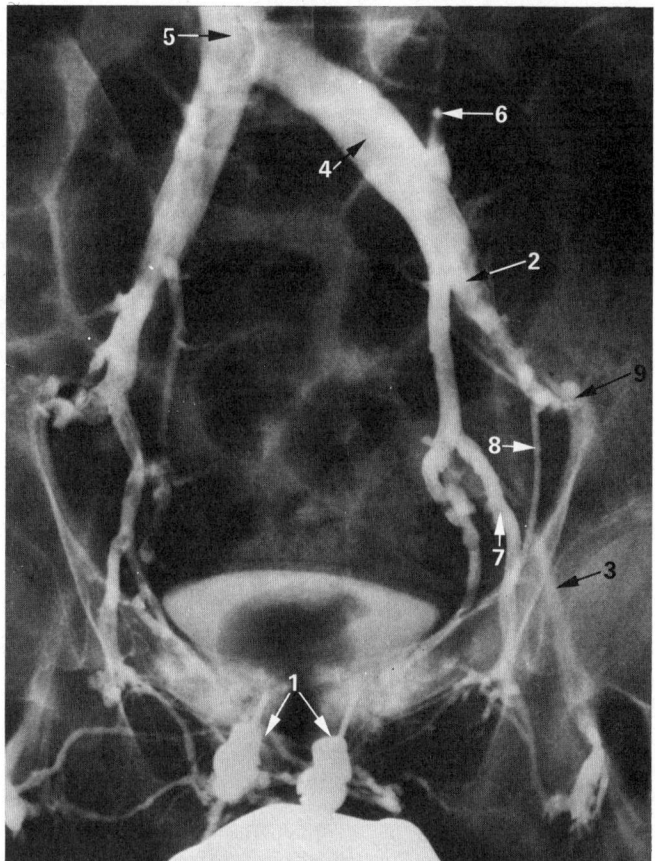

6.139 Venogram showing the veins of the pelvis and groin. The contrast medium has been injected into the bodies of the pubic bones. 1. Injected contrast medium in pubic bones. 2. Internal iliac vein. 3. External iliac vein (faintly outlined). 4. Common iliac vein. 5. Inferior vena cava. 6. Ascending lumbar vein. 7. Obturator vein. 8. Internal pudendal vein. 9. Gluteal vein. (Radiograph kindly supplied by Dr. M. Lea Thomas.)

artery along the front of the sacrum, and join to form a single vein, which usually ends in the left common iliac, but sometimes in the junction of the two common iliac veins.

Variations. The left common iliac vein, instead of joining the right in its usual position, occasionally ascends on the left side of the aorta as high as the kidney, where, after receiving the left renal vein, it crosses in front of the aorta to join the inferior vena cava. This anomalous vessel represents the persistent caudal half of the left postcardinal or supracardinal vein (p. 194).

The Inferior Vena Cava (**6.**84, 183)

This conveys blood to the right atrium of the heart from all the body below the diaphragm. It is formed by junction of the two common iliac veins, anterior to the body of the fifth lumbar vertebra, a little to the right. It ascends in front of the vertebral column, on the right of the aorta. Reaching the liver it is contained in a deep groove on its posterior surface—a groove which is occasionally converted into a tunnel by a band of liver substance. It then perforates the tendinous part of the diaphragm between the median and right portions of its tendinous centre, inclining slightly forwards and medially. After passing through the fibrous pericardium, it passes behind the serous pericardium to open into the lower posterior part of the right atrium. In front and to the left of its atrial orifice there is a semilunar valve, the *valve of the inferior vena cava*; this is rudimentary in the adult, but large and functionally important in the fetus (p. 665). The inferior vena cava is devoid of valves.

Relations of the abdominal part:

Anteriorly the inferior vena cava is overlapped at its commencement by the right common iliac artery and is covered, below the horizontal part of the duodenum, with the posterior parietal peritoneum. It is crossed obliquely by the root of the mesentery and its contained vessels, and by the right testicular (or ovarian) artery. Passing behind the horizontal part of the duodenum, it loses its peritoneal covering and ascends behind the head of the pancreas and then behind the superior part of the duodenum, from which it is separated by the bile duct and the portal vein. Above the duodenum it is again covered for a short distance with peritoneum in the posterior wall of the epiploic foramen (aditus to the lesser sac) (**8**.99, 100), by which it is separated from the right free border of the lesser omentum and its contents. Above this level the liver is anterior to it.

Posterior to the lower part of the inferior vena cava are the bodies of the lower three lumbar vertebrae and the anterior longitudinal ligament, the right psoas major and the right sympathetic trunk, while the third and fourth right lumbar arteries pass behind to its medial border. In its upper part, it is anterior to the right crus of the diaphragm, from which it is partially separated by the medial part of the right suprarenal gland and the right coeliac ganglion; the right renal, suprarenal and inferior phrenic arteries cross behind it.

On its *right side* the inferior vena cava is related to the right ureter, though not in immediate contact, to the descending part of the duodenum, the medial border of the right kidney and the right lobe of the liver.

On its *left side* it is related to the aorta, below, and to the right crus of the diaphragm and the caudate lobe of the liver above.

Relations of the thoracic part:

This part of the inferior vena cava is very short, and is partly inside and partly outside the pericardial sac. The *extrapericardial part* is separated from the right pleura and lung by the right phrenic nerve. The *intrapericardial part* is covered on the front and sides by the serous layer of the pericardium.

Surface Anatomy. The vein begins in, or just below, the transtubercular plane, its centre 2·5 cm from the midline; about 2·5 cm wide, it ends above behind the sternal end of the sixth right costal cartilage. A line, slightly convex laterally, from its lower end to a point 1 cm medial to the mid-inguinal point, indicates the course of the *common* and *external iliac veins* on each side.

Variations. Numerous anomalies of the inferior vena cava have been recorded and are attributable to arrests or errors in the complicated series of developmental changes which result in its formation. The vessel is sometimes represented, below the level of the renal veins, by two more or less symmetrical vessels. This is often associated with failure of the cross anastomosis connecting the two common iliac veins, and is due to persistence on the left of one of the longitudinal channels (usually supra- or subcardinal) which normally disappear in early fetal life (p. 194). In complete transposition of the viscera, the inferior vena cava lies to the left side of the aorta.

Applied Anatomy. Thrombosis of the inferior vena cava leads to oedema of the legs and back, without ascites. An extensive collateral venous circulation is soon established by enlargement of either the superficial or deep veins, or both. In the first case, the epigastric, the circumflex iliac, the lateral thoracic, the thoraco-epigastric (p. 758), the internal thoracic, the posterior inter-costals, the external pudendal and the lumbovertebral anastomotic veins effect the communication with the superior vena cava; in the second, the deep anastomosis is made by the azygos, hemiazygos and lumbar veins. It has also been suggested that the vertebral venous plexuses may provide effective collateral circulation between the venae cavae (Batson 1957).

Tributaries. These include in addition to the two common iliac veins the following: lumbar, right testicular or ovarian, renal, right suprarenal, inferior phrenic, and hepatic veins.

The lumbar veins, four on each side, collect the blood by dorsal tributaries from the muscle and skin of the loins, and by abdominal tributaries from the walls of the abdomen, where they communicate with the epigastric veins. Near the vertebral column they drain veins from the vertebral plexuses, and they are here connected with one another by the *ascending lumbar vein*—a longitudinal vessel in front of the roots of the transverse processes of the lumbar vertebrae. The *third* and *fourth lumbar veins* pass forwards on the sides of the bodies of the corresponding vertebrae and enter the posterior aspect of the inferior vena cava. Those of

the left side pass behind the abdominal aorta and are longer than the right. The *first* and *second lumbar veins* may end in the inferior vena cava, the ascending lumbar, or lumbar azygos veins. Mostly, the first lumbar vein does not pass directly into the inferior vena cava. It may turn downwards to join the second and so open into it indirectly, but more frequently it ends in the ascending lumbar vein or passes forwards over the side of the body of the first lumbar vertebra and terminates in the lumbar azygos vein (p. 754). The second lumbar vein may join the inferior vena cava at or below the level of the renal veins. Sometimes it joins the third lumbar vein or it may terminate in the ascending lumbar vein. The first and second lumbar veins are frequently connnected to each other, to the vessels of the opposite side, and to the right and left lumbar azygos veins by a plexiform network which lies on the bodies of the upper lumbar vertebrae.

The *ascending lumbar vein* connects the common iliac, iliolumbar and lumbar veins. It is behind psoas major and in front of the roots of the transverse processes of the lumbar vertebrae. At its upper end it joins the subcostal vein, and the trunk so formed turns forwards over the side of the body of the twelfth thoracic vertebra and, passing deep to the crus of the diaphragm, ascends in the thorax as the azygos vein on the right, and as the hemiazygos vein on the left. There is an angled bend on the vessel as it turns upwards, and it is usually joined at this point by a small vessel from the back of the inferior vena cava (or from the left renal vein, on the left side). This little vessel represents the azygos line (p. 193), and has already been described as the *lumbar azygos vein* (p. 754). Not infrequently the ascending lumbar vein ends in the first lumbar vein, which then turns forwards round the first lumbar vertebra in company with the first lumbar artery and joins the lumbar azygos vein. In this case the subcostal vein joins the azygos vein on the right and the hemiazygos vein on the left.

The testicular veins (**6**.84) emerge from the back of the testis, and receive tributaries from the epididymis; they unite and form a convoluted mass, the *pampiniform plexus*, the chief constituent of the spermatic cord, which ascends in the cord, in front of the ductus deferens. Below the superficial inguinal ring the veins of the plexus are drained by three or four veins which pass along the inguinal canal, enter the abdomen through the deep inguinal ring and coalesce into two veins, which run upwards in front of psoas major and the ureter, behind the peritoneum, one on each side of the testicular artery. These two veins become a single vessel, which on the right side opens into the inferior vena cava at an acute angle a little below the level of the renal veins; on the left side it opens into the left renal vein at a right angle. The testicular veins contain valves. The left vein passes behind the lower part of the descending colon and the lower margin of the pancreas and is crossed by the left colic vessels; the right passes behind the terminal part of the ileum and the horizontal part of the duodenum and is crossed by the root of the mesentery, the ileocolic and right colic vessels.

Applied Anatomy. The testicular veins are frequently varicose, a condition known as *varicocele*. Varicocele almost invariably occurs on the left, and this has been accounted for by the fact that the left testicular vein joins the left renal at a right angle; also it is overlaid by the lower part of the descending colon, and when this portion of the gut is full of faecal matter, in cases of constipation, its weight impedes the venous return. After the removal of a varicocele the venous return is carried out by the small veins of the ductus deferens, the cremaster, and the scrotal tissues.

The ovarian veins correspond with the testicular veins; each forms a plexus in the broad ligament near the ovary and uterine tube, and communicates with the uterine plexus. Two veins issue from this plexus and ascend across the external iliac artery, on each side of the ovarian artery. Their further course and mode of termination are like those of the testicular veins. Valves are occasionally found in the ovarian veins. Like the uterine veins, they are much enlarged during pregnancy.

The renal veins, of large size, are anterior to the renal arteries, and open into the inferior vena cava almost at right angles. The *left* is thrice the length of the right (7·5 cm to 2·5 cm), and crosses the posterior abdominal wall, behind the splenic vein and the body of the pancreas. Near its termination it passes in front of the aorta, just below the origin of the superior mesenteric artery. The left testicular (or ovarian) vein enters it from below, and the left suprarenal vein, which generally receives one of the left inferior phrenic veins, enters it above a little nearer the median plane. The left renal vein opens into the inferior vena cava at a slightly higher level than the right. The *right renal vein* is behind the descending part of the duodenum and sometimes the lateral part of the head of the pancreas.

Variations. Occasionally the left renal vein may be double, then one vein passes behind and one in front of the aorta to join the inferior vena cava—persistence of the renal collar (p. 194)—or the anterior vessel may be entirely absent. The latter condition represents persistence of the posterior limb of the renal collar combined with absence of the intersubcardinal anastomosis.

The suprarenal veins issue from the hilum of each suprarenal gland. The right is very short and passes directly into the posterior aspect of the inferior vena cava; the left runs downwards and medially, in front of or just lateral to the left coeliac ganglion, and passes behind the body of the pancreas to the left renal vein.

The inferior phrenic veins follow the course of the corresponding arteries on the inferior surface of the diaphragm; the right ends in the inferior vena cava; the left is often represented by two branches, one of which ends in the left renal or suprarenal vein, while the other passes in front of the oesophageal opening in the diaphragm and joins the inferior vena cava.

The hepatic veins drain the liver, and commence in the *intralobular veins*, which collect blood from the sinusoids of the liver lobules (*see* p. 1381). The intralobular veins open into *sublobular veins*, and these in turn unite to form the hepatic veins, which emerge from the posterior surface of the liver and open *immediately* into the inferior vena cava as it lies in the groove on the posterior surface of the liver. The hepatic veins are arranged in two groups, upper and lower. The *upper group* usually consists of three large veins, right, left and middle, the last emerging from the caudate lobe; those of the *lower group* vary in number; they are of small size and come from the right and caudate lobes. The hepatic veins are in direct contact with hepatic tissue and are destitute of valves. The occurence of large 'accessory' hepatic veins belonging to the lower group, and draining a variable volume of the intermediate to inferior part of the right lobe has been studied in 93 adult livers using corrosion cast techniques. Such veins are usually single, occasionally double and have an incidence of 15 per cent (*see* Sledziński and Tyszkiewicz 1975).

The Hepatic Portal System

The portal system (**6.**140, 141) includes all veins collecting blood from the abdominal part of the digestive tube (with the exception of the lower part of the anal canal) and from the spleen, pancreas, and gall bladder. From these viscera blood is conveyed into the liver by the *portal vein*. In the liver this vein ramifies like an artery and ends in capillary-like vessels termed sinusoids, from which blood is conveyed to the inferior vena cava by the hepatic veins. The blood of the portal system therefore passes through two sets of 'exchange' vessels, (*a*) the capillaries of the digestive tube, spleen, pancreas, and gall bladder, (*b*) the sinusoids of the liver. In

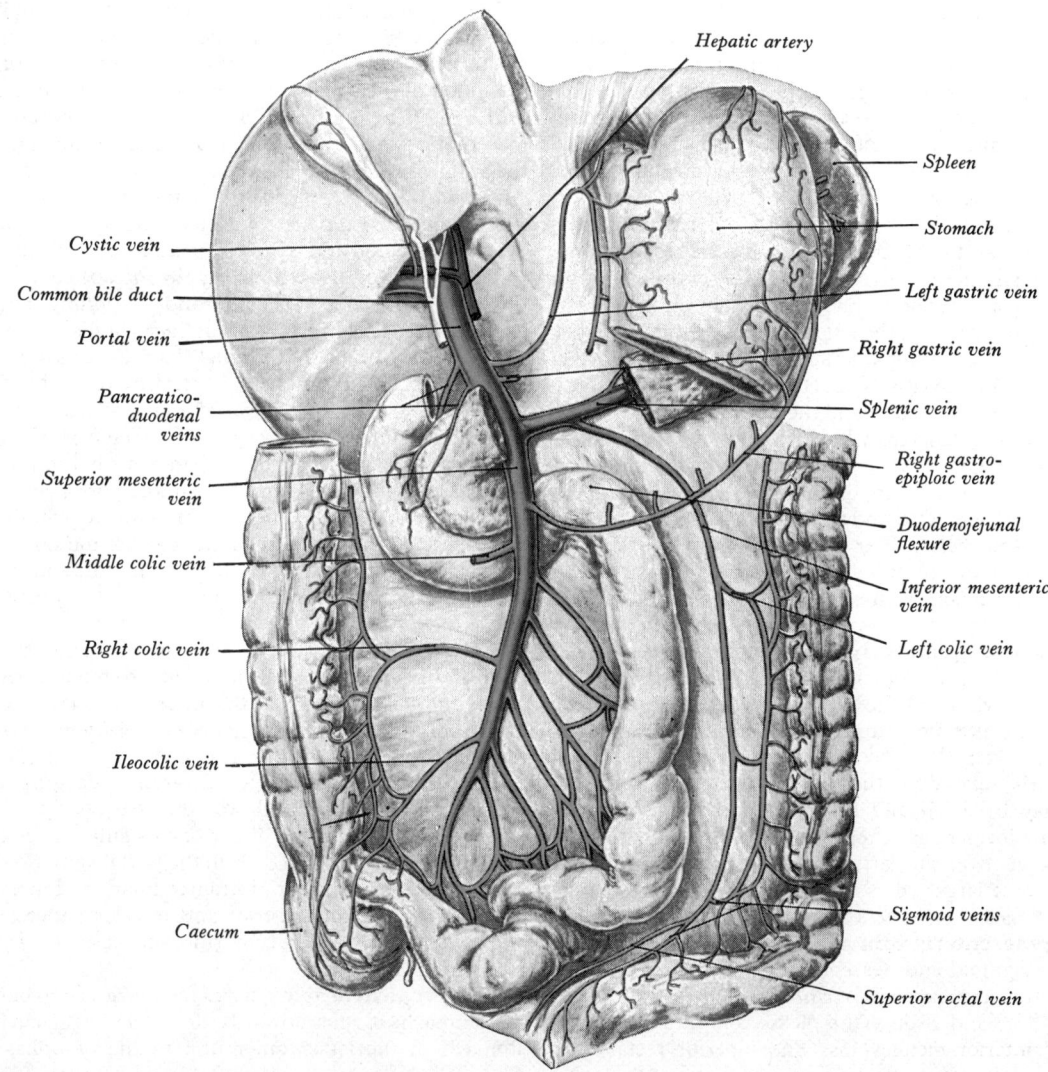

6.140   The portal vein and its tributaries. Semi-diagrammatic. Portions of the stomach, pancreas and left lobe of the liver, and the transverse colon have been removed.

the adult the portal vein and its tributaries have no valves; in the fetus and for a short time after birth valves can be demonstrated in its tributaries; as a rule they atrophy and disappear, but some may persist in a degenerate form.

The **portal vein** (**6**.87, 140) is about 8 cm long, and starts at the level of the second lumbar vertebra from the junction of the superior mesenteric and splenic veins, in front of the inferior vena cava and behind the neck of the pancreas. The vein inclines slightly to the right as it passes upwards behind the superior part of the duodenum, the bile duct and the gastroduodenal artery, and in front of the inferior vena cava; it then ascends in the right border of the lesser omentum in front of the epiploic foramen to reach the right end of the porta hepatis, where it divides into right and left stems, which accompany the corresponding branches of the hepatic artery into the substance of the liver. In the lesser omentum it is behind the bile duct and the hepatic artery, the former to the right of the latter; it is surrounded by the hepatic plexus of nerves, and is accompanied by numerous lymph vessels and some lymph nodes. The *right branch* of the portal vein enters the right lobe of the liver, but before doing so generally receives the cystic vein. The *left branch*, longer but of smaller calibre than the right, gives branches to the caudate and quadrate lobes and then enters the left lobe of the liver. As it does so, it is joined in front by the para-umbilical veins (p. 765) and by a fibrous cord, the *ligamentum teres*, the remains of the obliterated left umbilical vein. It is connected to the inferior vena cava by a second fibrous cord, the *ligamentum venosum*, a vestige of the obliterated ductus venosus, and ascends in a fissure on the posterior aspect of the liver. The small extrahepatic portion of the left branch, from which the vessels to the quadrate and left lobes arise, is a persistent part of the left umbilical vein.

The tributaries of the portal vein are: splenic, superior mesenteric, left gastric, right gastric, para-umbilical, and cystic.

The **splenic vein** (**6**.140) of large size, but not tortuous like the artery, commences from five or six tributaries issuing from the spleen. These unite into a single vessel, which traverses the lienorenal ligament with the splenic artery and the tail of the pancreas. It then descends to the right across the posterior abdominal wall, lying at a lower level than the splenic artery and immediately posterior to the body of the pancreas (which it grooves), receiving numerous short tributaries from the gland. In its course it crosses anterior to the left kidney and its hilar structures (or the lower pole of the left suprarenal gland), and it is separated from the left sympathetic trunk and crus of the diaphragm by the left renal vessels, and from the abdominal aorta by the superior mesenteric artery and the left renal vein. It ends behind the neck of the pancreas where it unites at a right angle with the superior mesenteric vein, to form the portal vein.

Tributaries. The splenic vein drains the short gastric, left gastro-epiploic, the pancreatic, and the inferior mesenteric veins.

The **short gastric veins**, four or five in number, drain the fundus and left part of the greater curvature of the stomach, traversing the gastrosplenic ligament to reach the splenic vein or one of its large tributaries.

The **left gastro-epiploic vein** drains both surfaces of the stomach and adjacent greater omentum; it runs from right to left along the greater curvature of the stomach, between the anterior two layers of the greater omentum, and ends in the commencement of the splenic vein.

The **pancreatic veins** drain the body and tail of the pancreas. These veins may be small and numerous or larger and less numerous. In the former case they empty more or less directly into the splenic vein; in the latter case, superior and inferior arcades receive these larger veins, their ultimate drainage being into the splenic vessel (Sow *et al.* 1976; see also p. 1372).

The **inferior mesenteric vein** (**6**.140) returns blood from the rectum, and the sigmoid and descending parts of the colon. It begins in the rectum as the *superior rectal vein*, which starts in the rectal plexus (p. 760), and through this plexus communicates with the middle and inferior rectal veins. The superior rectal vein leaves the lesser pelvis, crosses the left common iliac vessels on the medial side of the left ureter together with the superior rectal artery, and is continued upwards as the inferior mesenteric vein. This lies to the left of its artery and ascends behind the

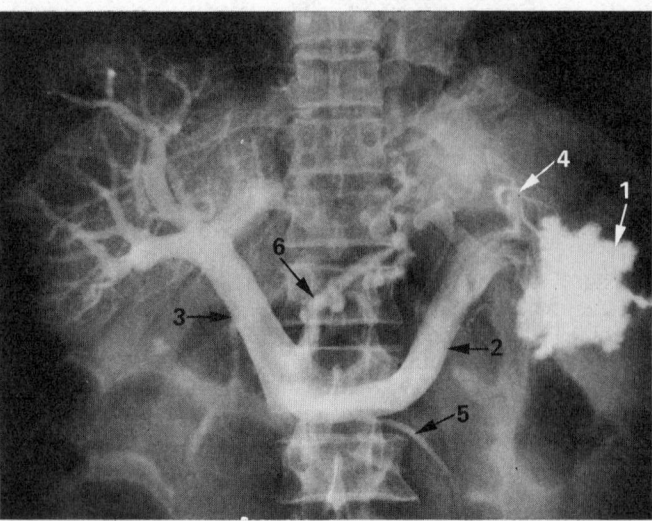

6.141 Portal venogram. A contrast medium has been injected percutaneously into the spleen. There has been some retrograde filling of some of the gastric veins and of the inferior mesenteric vein. The ramifications of the portal vein in the liver are especially prominent in the right lobe. 1. Injection mass in spleen. 2. Splenic vein. 3. Portal vein. 4. Short gastric vein. 5. Inferior mesenteric vein. 6. Left gastric vein. (Radiograph kindly supplied by Dr. M. Lea Thomas.)

peritoneum and in front of the left psoas major; it may cross the testicular (or ovarian) vessels or lie medial to them; it then passes above, or behind, the duodenojejunal flexure and opens into the splenic vein behind the body of the pancreas; sometimes it ends at the union of the splenic and superior mesenteric veins.

If a duodenal or a paraduodenal fossa is present, the inferior mesenteric vein usually lies in the fold of peritoneum which forms the anterior wall of the fossa (*see* p. 1331).

Tributaries. The inferior mesenteric vein receives the *sigmoid veins* from the sigmoid colon, and the *left colic vein* from the descending colon and left colic flexure.

The **superior mesenteric vein** (**6**.140) collects blood from the small intestine, caecum, and ascending and transverse portions of the colon. It begins in the right iliac fossa by the union of tributaries from the terminal ileum, caecum and vermiform appendix. It ascends in the mesentery on the right of the superior mesenteric artery; in its upward course it passes in front of the right ureter, inferior vena cava, horizontal part of the duodenum, and the uncinate process of the pancreas. Behind the neck of the pancreas it joins the splenic to form the portal vein (**6**.141).

Tributaries. The superior mesenteric vein's tributaries correspond to the branches of the superior mesenteric artery, viz. jejunal, ileal, ileocolic, right colic, and middle colic veins; it is also joined by the right gastro-epiploic and the pancreaticoduodenal veins.

The right gastro-epiploic vein drains the greater omentum and the lower part of the stomach; it runs from left to right along the greater curvature, between the anterior two layers of the greater omentum and joins the superior mesenteric vein below the neck of the pancreas.

The **pancreaticoduodenal veins** accompany their corresponding arteries; the lower one frequently joins the right gastro-epiploic vein; the upper one usually passes up to the left behind the bile duct and ends in the portal vein. Sow *et al.* (1975) have described anterior, intraglandular, and posterior *venous arcades* between the superior and inferior pancreaticoduodenal veins in about 70 per cent (in each case) of 157 pancreatic corrosion preparations.

The **left gastric vein** derives tributaries from both surfaces of the stomach; it runs upwards along the lesser curvature of the stomach in the lesser omentum, to the oesophageal opening, where it receives some oesophageal veins. It then turns backwards and passes downwards and to the right behind the omental bursa (lesser sac), and ends in the portal vein at the upper border of the superior part of the duodenum.

The right gastric vein, of small size, runs from left to right along the pyloric portion of the lesser curvature of the stomach in the lesser omentum, and ends in the portal vein. It is joined by the *prepyloric vein* which ascends in front of the pylorus and usually marks the site of the pyloric opening.

The para-umbilical veins, which establish an anastomosis between the veins of the anterior abdominal wall and the portal vein, extend along the ligamentum teres of the liver and the median umbilical ligament (p. 1405). The best marked of these small veins is one which begins at the umbilicus and runs backwards and upwards in, or on the surface of, the ligamentum teres in the falciform ligament, to end in the left branch of the portal vein.

The cystic veins. The veins draining the gall bladder vary considerably. Those from its upper surface lie in the areolar tissue between the gall bladder and liver and usually run directly into the liver through the fossa for the gall bladder to join the hepatic veins. Those from the rest of the gall bladder join to form one or two cystic veins on its neck, and these commonly enter the liver, either directly or after joining with veins draining the hepatic ducts and the upper part of the bile duct. Only rarely does a single or double cystic vein drain directly into the right branch of the portal vein.

Applied Anatomy. Obstruction to the portal vein may produce ascites, whether the site of the obstruction is intra- or extra-hepatic. In cirrhosis of the liver, the radicles of the portal vein are compressed by the contraction of the fibrous tissue in the portal canals. In valvular disease of the heart backward pressure on the

hepatic veins, and so on the whole circulation through the liver, exerts a similar effect. In addition the portal vein may be compressed by tumours of the liver, or by enlarged lymph nodes in the lesser omentum or carcinoma of the head of the pancreas. Thrombosis of the portal vein may accompany a variety of conditions. In cases of obstruction of the portal vein, the *anastomoses between the portal and systemic circulations*, which may collectively offer an effective collateral circulation, are as follows. (1) In the abdominal part of the oesophagus tributaries of the left gastric vein (portal drainage) anastomose with the oesophageal tributaries of the azygos and accessory hemiazygos (systemic) veins. Enlargement of the anastomoses may result in varicosity of these veins, leading even to a fatal haematemesis. (2) In the wall of the anal canal the opening up of communications between the inferior and middle rectal (systemic) and superior rectal (portal) veins may result in varicosity of these venous connexions. (3) At the umbilicus the veins running with the ligamentum teres of the liver to the left branch of the portal vein (p. 764) anastomose with the epigastric veins (systemic); enlargement of these connexions may produce a varicose condition of veins radiating from the umbilicus, known clinically as the *caput Medusae*. (4) Retroperitoneal veins of the abdominal wall communicate with venous radicles of the colon and bare area of the liver. (5) Very rarely the ductus venosus remains patent and connects the left branch of the portal vein directly to the inferior vena cava.

Some success has been achieved in case of portal obstruction by anastomosis of the portal vein to the inferior vena cava, or of the splenic vein to the left renal vein, after removal of the spleen.

MONONUCLEAR PHAGOCYTE AND LYMPHATIC SYSTEMS

Introduction

Scattered widely throughout the body are systems of tissues, or fluids and their cellular contents, which are concerned with a variety of interrelated functions. The latter include the circulation and modification of some of the tissue fluid formed in the blood capillary beds; phagocytosis of foreign matter and cell débris; long-term specific immune responses of the body, both humoral and cellular and, finally, the production and maintenance of the circulating cell populations in the blood, lymph and tissue fluid. Since these activities are interdependent, often having a common cellular basis, precise definition of various so-called 'systems' presents difficulties of terminology and may obscure rather than clarify important concepts. However, certain terms, widely used, will be retained in this account, even though they may reflect historical stages in the scientific analysis of the phenomena, or a preoccupation with a particular functional attribute, rather than any exclusive basis for their separate classification.

The term *reticulo-endothelial system* included all those cells and tissues with well-marked phagocytic properties and whilst the term is adequately descriptive for certain tissues, in others, the cells are neither reticular nor endothelial. Accordingly, this term has been generally replaced by the designation *mononuclear phagocyte* (or *macrophage*) *system* to include all major phagocytic cell types, except haemal polymorphonuclear leucocytes.

Most of the tissue fluid formed at the arterial end of the capillary bed returns to the blood circulation via the venous ends of the capillaries and the post-capillary venules. Some 10–20 per cent, however, is transported through a system of fine lymphatic capillaries, traverses one or more groups of lymph nodes, and finally larger lymph vessels, before returning to the venous system. These channels and nodes together with other masses of lymphoid tissue in the walls of the alimentary tract, spleen, thymus and bone marrow are often grouped as the *lymphatic system*. However, in addition to transporting and modifying lymph, parts of the lymphatic system are essentially involved in phagocytosis, effecting immune responses and contributing to the cell populations of the blood and lymph, and are accordingly often classed as *lympho-reticular organs*.

The term *haemopoietic system* is used when the origins of circulating blood cells are being considered; this topic has been discussed earlier (p. 63).

Finally, the term *immunocyte system* or *complex* is often used when considering the origins, transformations and activities of the cell types directly or indirectly concerned in mounting an immune response (e.g. certain phagocytes, plasma cells, varieties of lymphocyte, their precursors, and transformed progeny).

In this section an account of the mononuclear phagocyte system and an outline of the immune response is followed by general consideration of a lymph node and its vessels, the spleen, alimentary lymphoid tissue, thymus, and finally, a review of the topographical distribution of lymphatic vessels and nodes.

Mononuclear Phagocyte or Macrophage System

The macrophage, mononuclear phagocyte or reticulo-endothelial system (Aschoff 1924) is a collective term for a widespread system of highly phagocytic cells (*see* p. 45). Its cells are numerous and are characteristically present in large numbers in certain situations in the body. In postnatal life they are believed to arise primarily in the bone marrow, where they are formed as monocytes which migrate into connective and peripheral lymphoid tissues, and the liver, where they may form a relatively fixed or 'resting' component. In some cases, e.g. the liver, the cells are frequently highly branched, the ends of the branches coming into close apposition (but not syncytial fusion) to form a three-dimensional meshwork or *reticulum*, often supported by a complementary meshwork of fine reticulin fibres, a product of the cells themselves. Under appropriate stimulation they are capable of detaching themselves from the tissues, which, when they are static, they help to form, and of acquiring mobility which is amoeboid in character. In ordinary stained histological preparations they display few characteristics by which they can be distinguished with certainty. They can be identified by their marked affinity for inert particles and certain colloidal dyes, when these are injected into the living animal. Particulates such as India ink, when injected into the bloodstream of the living animal, are taken up by the intravascular components of the macrophage system as well as by the granulocytes in the blood. Diffusible colloidal dyes, such as trypan blue, when injected into the vascular system, are taken up by both the intravascular and extravascular components of the mononuclear phagocyte system; for electron microscopic identification, cells may be visualized in the same manner by injecting electron-dense tracers such as ferritin or thorium dioxide, or enzymic markers (e.g. horseradish peroxidase) into the circulation. The basis for the inclusion of particular cells in the mononuclear phagocyte system is rather arbitrary and depends upon the readiness of uptake of particulate matter. Thus, the general endothelial cells lining most of the blood-vascular and lymphatic systems, fibroblasts and most leucocytes take up only small amounts of a test injection and are excluded from the system by most workers in the field.

Cells of the mononuclear phagocyte type occur in considerable number and varying morphology in the following situations: (1) Connective tissues, where they have already been described as macrophages. In nervous tissue the microglial cells are considered to belong to this system. In the subserous connective tissue of the pleura and peritoneum the macrophages are frequently aggregated and appear as white streaks, known as *milky spots*, close to the small lymphatic trunks where they adhere; many also wander freely in the fluid of the pleural or peritoneal cavities (*pleural* or *peritoneal macrophages*). (2) Blood, where they are represented by monocytes. There is much evidence that these migrate into the tissues from the bloodstream, to become tissue macrophages, and are distinct from other circulating white blood cells. (3) Lining the blood sinuses of the liver where they are termed Kupffer's cells. (4) The *reticular tissue* of the spleen, lymph nodes, solitary and aggregated lymph follicles, tonsils and bone marrow where they adhere to the fibres or migrate within the cavities which permeate the reticulum. (5) The meninges, where they are known as *meningocytes*. (6) The alveoli of the lung, which are constantly patrolled by *alveolar macrophages* migrating from the interalveolar connective tissue through the intercellular junctions between the epithelial cells. These cells engulf inhaled particles and bacteria, and in the case of congestive heart failure, extravasated red blood corpuscles; they are accordingly sometimes termed *dust cells* or *heart failure cells*, and may be found in the sputum bearing their ingested material. (7) In certain infections of the tissues, where macrophages may fuse together to form multinucleate *foreign body giant cells* around foci of bacterial, viral, or other types of damage. (8) In resorbing bone, where some authorities consider the *osteoclast* to be a cell similar in origin and function to the foreign body giant cell, although others view the osteoclast as a distinctive cell type (p. 261); the same considerations apply to the *chondroclast*.

Because of their amoeboid and phagocytic properties the cells of the macrophage system form one of the most important short-term defences of the body against micro-organisms and under appropriate stimulation, the cells proliferate freely. Repeated injections of particulate matter into an animal can 'block' the normal function of the macrophages, so that subsequent destruction of micro-organisms is hindered. These cells are involved in the initial phagocytosis and destruction of aged red blood cells in the spleen and liver, although it is not certain whether such erythrocytes may be, to some extent, lysed prior to their uptake by macrophages and Kupffer cells. The haemoglobin is split into globin and an iron-porphyrin ring compound, which is opened and the iron removed. The released iron is re-used for haemoglobin synthesis and the remaining group degraded to bilirubin which is excreted by the liver. There is evidence also that the macrophage system is concerned in the metabolism of lipids; it is extensively involved in certain disorders of lipid metabolism.

In addition to providing a first-line defence system against invading micro-organisms, the macrophages present in lymphoid tissue are now considered to be intimately concerned with the establishment of specific immune responses by neighbouring cells (*see* p. 45).

Many of the prominent reticulo-endothelial tissues are also important sites of haemopoiesis, both during mid-fetal life (spleen, liver and some connective tissues), and later fetal and postnatal life (lymphoid tissues and bone marrow).

The Lymphoid System

The other major defensive arm, which operates in close association with the system of phagocytes, is provided by the lymphocytes which are constantly present in the bloodstream, lymph, and other body fluids, as well as in the connective tissues and in their major areas of lymphocyte proliferation, the specialized lymphoid tissues and organs. The lymphocytes constitute several populations of cells which have in common their ability to produce chemicals such as antibodies which can inactivate foreign substances, microbes and neoplastic cells when these arise in the body, or are introduced to the tissues (*see* p. 61).

Lymphocytes undergo their initial differentiation in the **central lymphoid tissues**, i.e. the bone marrow, where B-lymphocytes capable of synthesizing antibodies are initially produced, and the thymus, where the T-lymphocytes able to kill virus-infected cells, fungi and neoplastic cells, etc., are first formed. The progeny from these central tissues then pass in the circulation to the **peripheral lymphoid organs**—the *lymph nodes*, *epithelio-lymphoid tissue* ('lymphoid nodules' in Peyer's patches, vermiform appendix, bronchial tree, etc.) and similar tissues in the bone marrow and spleen; in these peripheral sites lymphocytes may proliferate, mature, and migrate into the surrounding tissues, or they may return to the circulation. This system of tissues is quite labile, and a given component may become extremely active in lymphocyte production and maturation when subjected to effective antigenic stimulation. The lymphocytes produced within them can travel freely in the haemal and lymphatic circulation, populating lymphoid tissue temporarily, and providing a circulating pool which can be called upon during antigenic emergencies (*see* Roitt 1977).

DIFFERENTIATION OF LYMPHOCYTES

As in all cell lineages, the cells of the lymphoid series undergo a succession of stages in their development from 'pluripotent' lymphoid stem cells before they finally emerge as fully competent defensive elements of the immune system. (However, *see* discussion on cell 'potentiality' on pp. 83–85). During early fetal life this process occurs in the liver, but in later fetal and postnatal life the site is transferred to the bone marrow, where the stem cells proliferate giving rise to uncommitted lymphocytes which must then pass through another process of differentiation before they are established as either B- or T-lymphocytes. This step occurs in the thymus in the case of T-lymphocytes, whilst the B-lymphocytes apparently undergo this stage within the bone marrow itself, although in birds the wall of the bursa of Fabricius, a hind-gut diverticulum, is the site of B-cell commitment.

B-lymphocytes are able to produce an enormous range of antibodies, each able to recognize and inactive a narrow range of antigens, and it is generally established that each B-lymphocyte is able to synthesize antibody with only one type of specificity.

According to the selective hypothesis, only those cells capable of reacting by cell division and the synthesis and secretion of an appropriate antibody are activated when a particular antigen is introduced into the body, so there must be a large 'resting' population of cells each with a distinctive capacity to produce antibodies of different types. The manner in which this variability is initially achieved has been the subject of much debate. It is known that the ability to produce antibodies of different kinds is under genetic control, and therefore it seems likely that antibodies are coded for on the chromosomes in the normal manner. However, as there may be many thousands of antibodies formed within an individual, it has been argued that the amount of genetic material devoted to antibody production would be a major proportion of the whole genome, and that instead it may be possible that the genes in the lymphocyte precursors could undergo somatic mutation, thus generating a wide diversity of genes in the lymphoid series.

Recent evidence, although not decisive, favours the more

usual genetic view that mutation does not take place and that the genes of the germ line are all that is necessary for antibody diversity. Further, since antibody molecules are composite structures of two light and two heavy chains, with variable regions in each, it is feasible that genes code for small parts of antibody molecules, which could thereafter be assembled in different ways to create the observed diversity of the final antibody molecules. (For further discussion consult Williamson 1976; Cunningham 1976; Roitt 1977.)

The Lymph Nodes and Vessels

The lymphatic system comprises (a) plexuses of minute vessels (*lymph capillaries*) that commence blindly in the tissue spaces in many tissues of the body and ultimately empty their contents (*lymph*) into certain veins; (b) *lymph nodes*, consisting of small solid masses of lymphoid tissue, into which the lymph vessels at some part of their course pour their lymph, so that generally the lymph from any tissue or organ traverses one or more lymph nodes before it eventually reaches the venous bloodstream; (c) collections of lymphoid tissue situated in the walls of the alimentary canal and respiratory tract (*epitheliolymphoid tissue*), and in the spleen and thymus; (d) the circulating *lymphocytes* (*vide infra*). (For general accounts of the lymphatic system consult Rouviére 1932; Yoffey and Courtice 1956; Rusznyák et al. 1960; Allen 1967.)

LYMPH VESSELS

The lymph capillaries form networks in the tissue spaces, the meshes of which are larger than those of the neighbouring blood capillaries. They often commence with a dilated, bulb-like, blind extremity and their calibre is greater and less regular than that of capillaries. An important feature of the endothelial wall of the lymph capillaries is that it is permeable to substances of much greater molecular size than those which can pass through the endothelial wall of blood capillaries (Allen 1967). The lymph capillaries form the pathway for absorption of colloidal material (and particular matter including cell débris and micro-organisms) from the tissue spaces, whereas the blood capillaries are more concerned with the absorption of soluble crystalloid substances. Thus, if the lymph vessels become obstructed, the tissues drained by them become oedematous and distended with a fluid containing much protein. Experimental evidence suggests that the absorption of macromolecular and particulate substances takes place through intercellular fenestrations between the endothelial cells or by micropinocytosis across the cells. Whereas the lymph draining from most tissues is a clear colourless fluid, that draining from the small intestine is milk-white in appearance, due to the absorption of fat, as chylomicra, from the gut; the fluid is called *chyle* and the lymph vessels are known as *lacteals*. Lymph capillaries are present in many tissues of the body, but are absent from avascular structures (epidermis, hair, nails, cornea, and articular and some other cartilages), the brain and spinal cord, the splenic pulp and bone marrow.

The lymph capillaries join to form larger trunks that pass to neighbouring, or sometimes remote, lymph nodes. The lymph nodes are for the most part arranged in *regional groups*, as described later. Most of these groups are sufficiently regular to receive names. Each group has a characteristic region of drainage, but the immediately local group is often by-passed. The nodes in a particular group are also frequently interconnected by short trunks (*see* Kubik 1974, for literature). In general, most of the lymph appears to traverse a series of nodes before reaching the major collecting ducts. There are, however, exceptions to the general rule that lymph traverses one or more lymph nodes before reaching the bloodstream. These are some of the lymph vessels of the thyroid gland, oesophagus, and the lymph vessels running in the coronary and triangular ligaments of the liver, all of which enter directly into the thoracic duct (Rusznyák et al. 1960). The larger superficial lymph vessels of the skin lie near the deep fascia and tend to accompany the superficial veins, though some run independently; they have very few connexions with the deep lymphatics. The deep lymphatic trunks in most cases closely accompany arteries or veins. Eventually, practically all the lymph from the body is collected into two channels, the *thoracic duct* (p. 784 and the *right lymphatic duct* (p. 785), which pour their lymph into the left and right brachiocephalic veins respectively. Most lymphatic vessels anastomose freely and those of the two sides of the body are in communication *across the midline*. The larger lymph vessels are supplied with their own vasa vasorum and are accompanied by a plexus of fine blood vessels. If the walls of the lymph vessels are acutely infected (lymphangitis) this plexus becomes congested and the paths of superficial lymph vessels are often marked by painful red lines visible through the skin. Nerve networks are described both within and around the walls of the larger lymph vessels.

Lymph vessels have a great capacity for repair and for the formation of new vessels after damage. The new vessels are formed first as solid cellular sprouts produced by the mitotic division of the endothelial cells of the persisting vessels, the sprouts later becoming canalized.

The structure of the lymph vessels. The wall of a lymph capillary consists of a single layer of endothelial cells. It resembles that of a blood capillary, but the basement membrane is often lacking and specialized attachments between endothelial cells are few, though fenestrae between adjacent cells have been demonstrated in subserous lymphatics. The latter are absent in well-fixed subcutaneous lymphatics, but are very evident after trauma to neighbouring tissue. Filopodia are more frequent on their luminal surfaces. In the lacteals similar projections may also occur on the outer surfaces of the cell and pericytes are absent (Fraley and Weiss 1961). Thus, as with blood capillaries, there may exist considerable structural variation between lymph capillaries in different tissues (Allen 1967; Leak and Burke 1968). As the capillaries unite to form larger vessels, a thin connective tissue coat is added outside the endothelium. The larger collecting lymph trunks have three tunics, comparable with those of small veins. The tunica intima consists of endothelial cells and a thin layer of fibrous tissue; the tunica media contains some smooth muscle cells, most disposed circularly, and the tunica adventitia is composed mainly of fibrous tissue with a slight admixture of smooth muscle.

An optical and electron microscopic study of human lymphatic trunks by Boggon and Palfrey (1973) confirmed this trilaminar structure, though these authors did not observe muscle fibres in the adventitia. Elastic fibres were sparsely distributed in the subendothelial connective tissue, but were numerous enough to constitute an external elastic lamina in the tunica adventitia. The lymph channels differ from small veins in that they possess many more valves. These are semilunar and are generally arranged in pairs; they are formed by a reduplication of the endothelium with a delicate fibrous tissue core. Their free edges are directed along the course of the lymph current, and the wall of the lymph vessel immediately proximal to the attached edges of the valves is expanded into a sinus or pouch which gives the vessel, when distended, a knotted or beaded appearance. The valves are of considerable importance in connexion with the mechanism of the flow of lymph. The thoracic duct is similar in structure to a medium-sized vein, though the non-striated muscle in its tunica media is more prominent.

Satiukova and Rassokhina-Volkova (1972) have studied the regeneration of lymph capillaries in dogs after auto-transplantations of hind legs and lungs; they observed an early formation of buds from severed lymph channels in the junctional scar tissue, and they considered that lymph flow was largely restored by such new collateral routes.

Movement of lymph. Several factors are concerned in propelling lymph from the tissue spaces towards lymph nodes and the venous bloodstream. (a) The 'filtration pressure' in the tissue spaces, generated by the filtration of fluid from the blood capillaries. (b) Contraction of surrounding muscles compresses the lymph vessels and the lymph moves in the direction determined by the valves; that this is an important factor is shown by the fact that extremely little lymph flows along the lymphatics of a limb that has been rendered immobile, whereas the flow is considerably increased when the limb is moved (actively or

6.142A Diagram of a lymph node. (Modified from Maximow and Bloom.) In part of the diagram, lymphocytes have been omitted to show the reticulum. This diagram from an earlier source has been retained for historical interest. Greater detail and modern concepts have been displayed in **6.142B**.

passively). Clinically, immobilization of a limb is used to diminish or prevent dissemination of toxic material from infected tissues. Massage or movement of a part in which there is excess of fluid (oedema) in the tissue spaces promotes the flow of lymph from the affected regions. (c) Where a lymph trunk is close to an artery, pulsation of the artery probably compresses the lymphatic vessel and thus assists the lymph flow. (d) Respiratory movements and the negative pressure in the brachiocephalic veins are also factors promoting lymph flow. (e) The smooth muscle in the walls of the lymphatic trunks is most marked just proximal to the valves; stimulation of sympathetic nerves accompanying the trunks results in contraction of the vessels; the intrinsic muscle of the vessels thus probably aids the flow of lymph. What form of muscle contraction occurs naturally has, however, not been determined. The valves are of considerable importance in determining the direction of the flow of lymph. If, however, the lymph vessels become markedly dilated, the valves may become incompetent and a retrograde flow may occur. This may explain some instances of the retrograde spread of malignant tumours.

Methods of study of lymph vessels. Infected material and malignant growths may be spread from an affected site by the lymph vessels. It is therefore of considerable clinical importance to know the precise lymph pathways from the various regions and organs of the body. Ordinary dissection methods are not suitable for investigating these pathways because, apart from the large terminal collecting lymph ducts, the vessels are very slender and difficult to see. Knowledge of these pathways has been obtained by the following methods. (1) *Experimental.* Injection of substances into the organs or tissues of living or dead animals, including man. In successful experiments, the substances pass into the lymph vessels draining the site and render visible the vessels and the lymph nodes into which they drain. The substances most commonly used are suspensions of India ink, Neoprene latex or Prussian blue. The latter was used by Jamieson and Dobson (1907–20), for their extensive studies of the lymph pathways in man. In living animals methylene blue and radio-opaque substances, such as ultrafluid lipiodol, have been injected; with the latter the lymph vessels are visualized by radiography. Lymphangiography in the human subject after cannulation of appropriate peripheral lymphatic channels and the injection of lipiodol, is now adding considerably to our knowledge of lymphatic channels and is being increasingly used as a diagnostic aid (Kinmonth 1964; Kinmonth and Taylor 1964). (2) *Clinical.* By noting the lymph nodes involved by the spread of inflammatory or malignant disease of an organ or tissue, the lymph path from that site is inferred. However, the phenomenon of retrograde spread of tumour cells after blockage of a lymphatic channel limits the usefulness of such observations when attempting to determine *normal* directions of lymph flow.

LYMPH NODES

These are small, oval or reniform bodies (0·1–2·5 cm long), situated in the course of lymph vessels so that the lymph passes through them on its way to the blood. Generally each presents on one side a slight depression, termed the *hilum* through which *blood* vessels enter and leave the node. The *efferent lymph vessel* (usually single) also emerges from the node at this spot, while the *afferent vessels* enter it at different parts of the periphery. A lymph node has a highly cellular *cortex* and a darker *medulla* containing numerous cavities, with an indefinite line of demarcation between the two. The cortex does not form a complete investment but is deficient at the hilum, where the medulla reaches the surface of the node; thus the efferent lymph vessel is derived directly from the medulla, while the afferent vessels empty themselves into the cortex. Lymph nodes are most numerous in the thoracic mediastinum, on the posterior abdominal wall, in the abdominal mesenteries, and in the pelvis, neck and proximal ends of the limbs.

STRUCTURE OF LYMPH NODES

A lymph node consists essentially of a continuous framework which includes a capsule, trabecula and reticular fibres, and the cells entangled in this framework (**6.142–144**).

The capsule and trabeculae. The capsule is composed mainly of collagen fibres, a few fibroblasts and some elastin fibres, especially in its deeper layers. In some animals many smooth

6.142B Schema of the general architecture and cellular organization of the lymph node. Particular reference is made to the differential distribution of lymphatic spaces and cell masses. Coloured arrows indicate the circulatory pathways of T- and B-lymphocytes. For details see text.

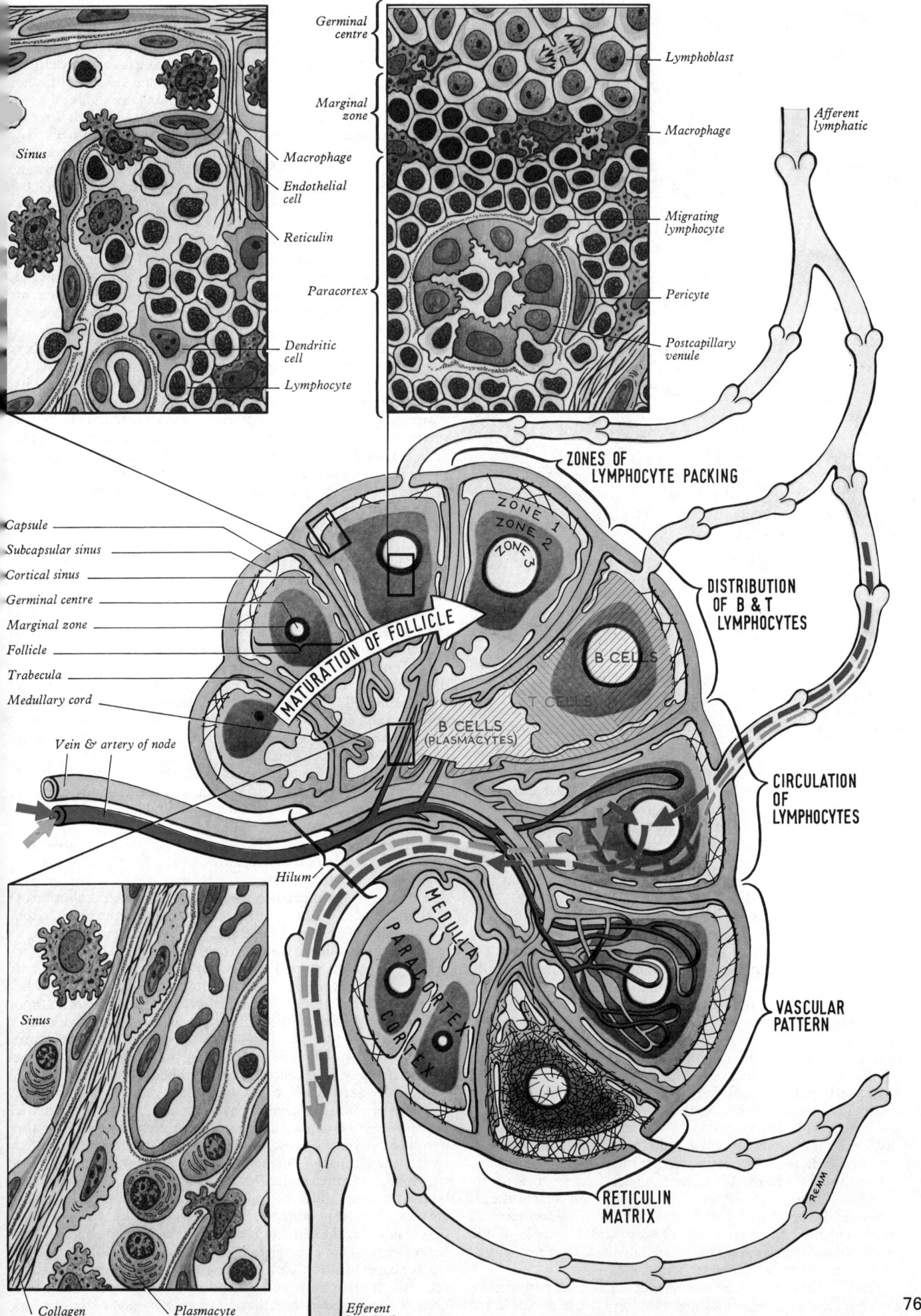

Germinal
centre

Marginal
zone

Sinus

Macrophage

Endothelial
cell

Reticulin

Paracortex

Dendritic
cell

Lymphocyte

Lymphoblast

Macrophage

Migrating
lymphocyte

Pericyte

Postcapillary
venule

Afferent
lymphatic

ZONES OF
LYMPHOCYTE PACKING

ZONE 1
ZONE 2
ZONE 3

Capsule

Subcapsular sinus

Cortical sinus

Germinal centre

Marginal zone

Follicle

Trabecula

Medullary cord

MATURATION OF FOLLICLE

B CELLS

T CELLS

B CELLS
(PLASMACYTES)

DISTRIBUTION
OF B & T
LYMPHOCYTES

CIRCULATION
OF
LYMPHOCYTES

Vein & artery of node

Hilum

MEDULLA

PARACORTEX

CORTEX

VASCULAR
PATTERN

Sinus

RETICULIN
MATRIX

Collagen

Plasmacyte

Efferent
lymphatic

REMM

769

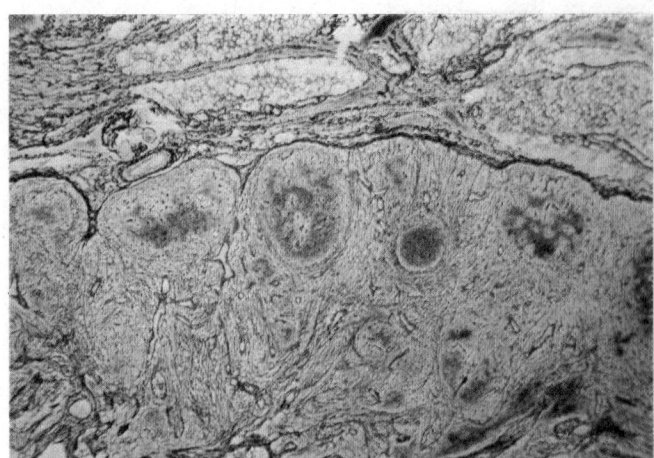

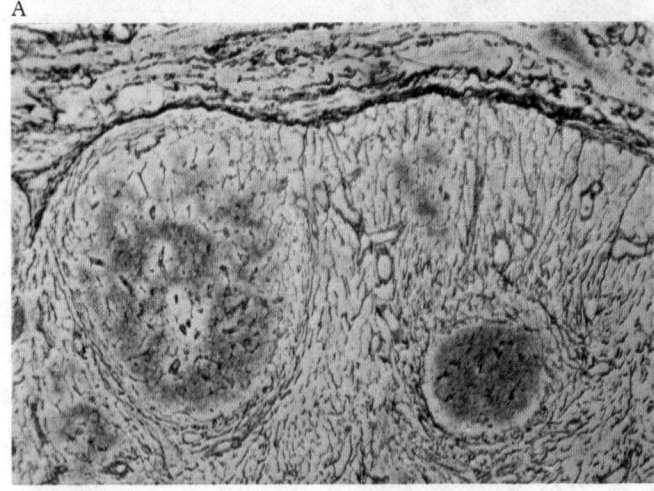

6.143A and B Sections of a lymph node stained by the method of Glees and Marsden for reticulin—note the heavy concentration of fibres in the capsule and trabeculae. A fine network permeates the rest of the node with a concentric accumulation surrounding the cortical lymphatic follicles.

muscle cells are found but these are few in man. The capsule covers the outside of the node and from its deep surface, trabeculae of a similar structure extend radially into the interior of the node, where they are continuous with the fine *reticulum* that forms a supporting framework for the lymphoid tissue. At the hilum, dense fibrous tissue may extend some distance into the medulla and the efferent lymphatic vessel is embedded in this before it leaves the node.

The reticulum. This is a meshwork of fine reticulin fibres and attendant cells which permeates the spaces outlined by the capsule and trabeculae (**6.142**) providing mechanical support for the cell masses lying in their vicinity. The fibres are histologically demonstrable with special reticulin stains (**6.143**) which show that the bundles of fibres branch and interconnect freely, forming a particularly dense network in the cortex, although the germinal centres have fewer fibres than the areas surrounding them. Fibres of this type, clothed with endothelial cells, criss-cross the sinuses of the node, providing attachment sites for various cells, particularly macrophages and lymphocytes, entangled in their mesh. The reticulin fibres together with the glycoprotein matrix associated with them appear to be laid down by cells in-distinguishable from fibroblasts, although various names (e.g. 'reticular cells') have been appended to them in the past.

Lymphatic channels. Lymph nodes are permeated by lymphatic channels through which the lymph percolates after it has entered from the afferent lymphatic vessels, an arrangement which ensures the maximum exposure of the lymph to the macrophages and lymphocytes which line them or are entangled between the fibres which cross them.

The afferent vessels enter at different points on the periphery of the node, and after branching and forming a dense plexus in the

substance of the capsule, open into the *subcapsular sinus*, a cavity running around the periphery of the cortex except in the region of the hilum (**6.142**). From this space numerous radial *cortical sinuses* lead towards the medulla, eventually coalescing to form the larger *medullary sinuses* which are confluent at the hilum with the efferent lymphatic vessel or vessels draining the node. These spaces are everywhere lined by endothelial cells, although there is a constant passage of lymphocytes, macrophages and other cells, through the sinus walls, in both directions. Numerous trabeculae (*vide supra*) cross the sinuses, converting their lumina into an almost labyrinthine system, and providing widespread attachment areas for the various cell types present in the spaces (Nopajaroonsri *et al.* 1971; Luk *et al.* 1973).

Lymphatic blood system. Arteries and veins serving the interior of lymph nodes enter through the hilum and give off straight branches which traverse the medulla in company with the connective tissue trabeculae, giving off a few minor vessels *en route*. On reaching the cortex, the arteries break up into dense arcades of arterioles and capillaries which form numerous anastomosing loops, eventually to pass back into the highly branched venules and veins. The capillaries are particularly profuse around the periphery of the follicles, with fewer vessels in the germinal centres (Herman *et al.* 1973; Blau 1976); the postcapillary venules are abundant in the paracortical zones where they are an important route of lymphocyte migration (*vide infra*). This pattern of vascularization is considerably altered when nodes undergo antigenic stimulation to produce large numbers of lymphocytes, with consequent alterations in general architecture, and the density of the capillary beds greatly increases (Herman *et al.* 1972).

The structure of these various blood vessels is not unusual except in the case of the *postcapillary venules*, which are lined by tall cuboidal endothelial cells, between which colloidal materials pass quite readily into the perivascular space (Mikata and Niki 1970) and which allow extensive movements of lymphocytes from the bloodstream into the paracortex of the lymph node, and probably also in the reverse direction (Marchesi and Gowans 1964; Gowans and Knight 1964). At one time it was thought that these cells actually migrated through the cytoplasm of the endothelial cells, but in view of subsequent data this appears unlikely, and the lymphocytes would seem merely to migrate through the rather easily dislocated endothelial intercellular junctions—in the usual manner of leucocytes (Schoefl 1972). In addition to the vessels of the nodal medulla, some may also leave the lymph node through its principal trabeculae and general capsule, supplying those structures, and the surrounding connective tissue.

The entangled cells. The majority of these are B- and T-lymphocytes, though macrophages which have become freed from the reticulum are also frequent, particularly in the sinuses and around the germinal centres (*vide infra*). The distribution of the lymphocytes is different in the various parts of the node. In the lymph sinuses are found some free cells which have been swept into the lymph as it circulates through the node. In the cortex the cells are very densely packed and may form more or less isolated masses called *lymphatic follicles* (nodules) (**6.142–144**). The number and the degree of isolation of the follicles vary from time to time according to the prevailing level of antigenic stimulation. The central part of each follicle is composed of cells which are larger, less deeply staining and dividing more rapidly, than those at the periphery. The central areas are called *germinal centres*; the cells are mainly *lymphoblasts* which by their prolific mitotic divisions produce small lymphocytes accumulating initially in the *marginal zones* around the germinal centres. These leave the peripheral parts of the follicles to enter the lymph sinuses, which convey them through the medulla to the efferent lymph vessel at the hilum of the node. In the medulla of the node the lymphocytes are much more loosely packed than they are in the cortex. They constitute irregular branching *medullary cords* between which the reticulum of the medullary lymph sinuses is easily seen. The entangled cells include, in addition, some macrophages which are more numerous in the medulla, plasma cells which often surround individual macrophages, and a few granulocytes. Under conditions of intense antigenic stimulation,

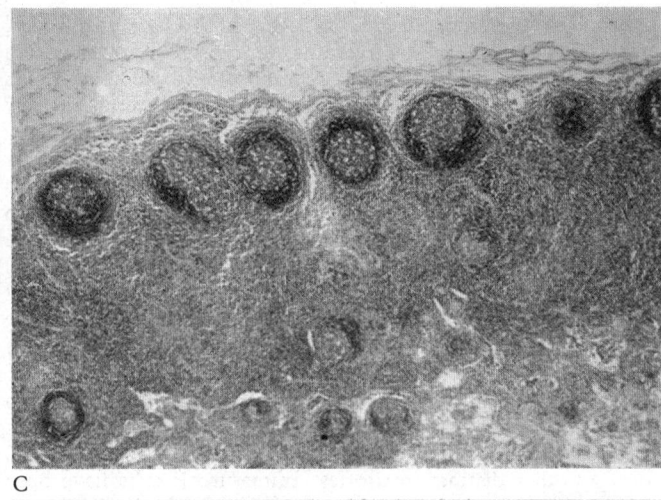

C

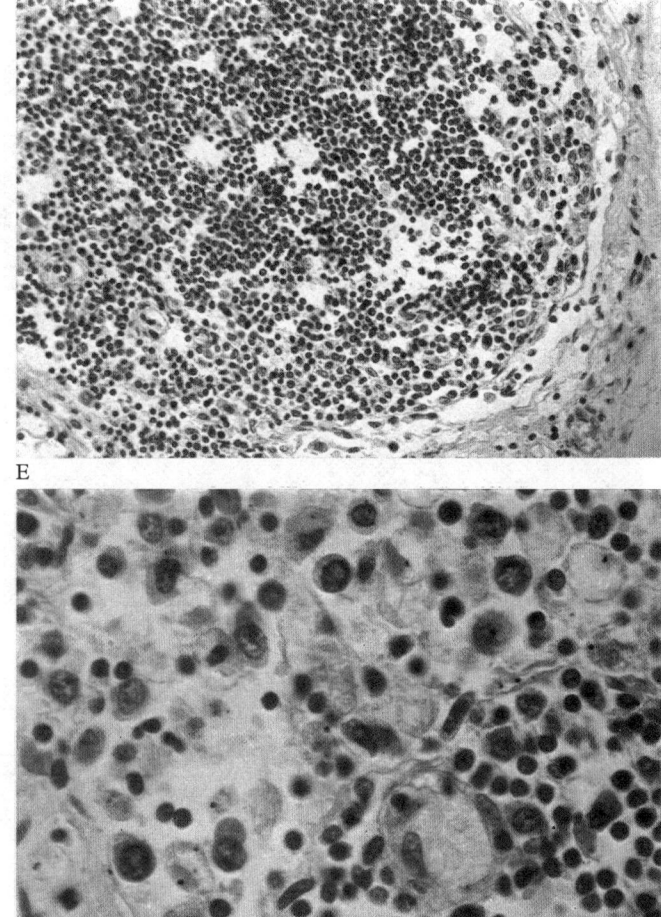

E

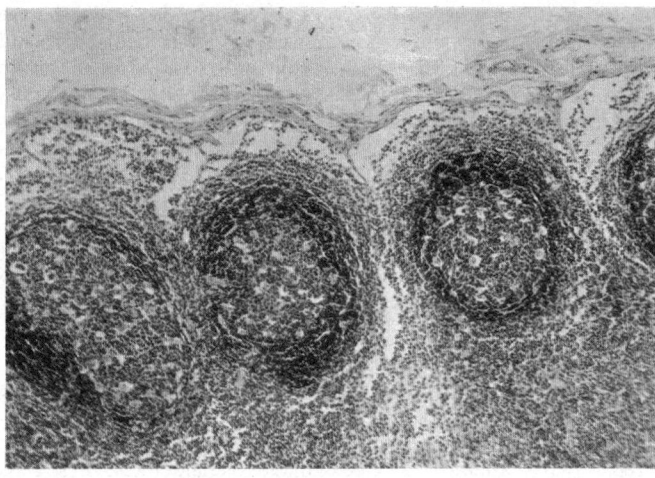

D

F

6.143C–F Sections of lymph nodes stained with haematoxylin and eosin.
C Note the round cortical lymphatic follicles with their dense, dark periphery and pale germinal centres, and the irregular medullary tissue; very low power survey micrograph.
D A low-power view of germinal centre showing the variation in cell density.

E Higher-power view of the peripheral zone of a follicle showing the densely packed small lymphocytes.
F Higher-power view of the medulla showing a variety of cell types including small and large lymphocytes and prominent rounded plasma cells.

e.g. when the node drains a focus of infection or an immunization site, the whole node increases in size and vascularity and becomes a *reactive node*. The number and size of its germinal centres increases, with a raised rate of lymphopoiesis, the proliferation of macrophages, and the differentiation of numerous plasma cells

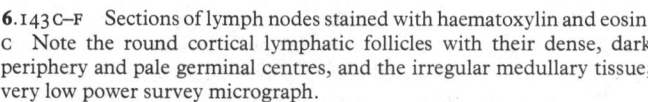

6.143G Lymph node (guinea pig) in section, after blood vessels have been injected with indian ink. Note the large vessels in the medulla, ramifying to form a capillary plexus in the cortex. (Kindly provided by Dr. N. Blau, Guy's Hospital Medical School.)

within the sinuses. (The experimental methods employed, and the theories concerning the cellular origins of lymphocytes, the distinction between B- and T-lymphocytes, and the views concerning their respective roles in the cellular functioning of the lymphocytic system is discussed in some detail on pp. 59–63, in addition to the following pages.)

Cell Zones in Lymph Nodes

Although for many years it was useful to divide the lymph node into medullary and cortical areas, recent studies have indicated the value of further subdivisions. As already mentioned, the cortex is divided into a number of rather indistinct lobules by connective tissue trabeculae passing inwards from the capsule, and within each of these territories lie several *lymphoid follicles* or rounded masses of cells, continuous centrally with the medullary cords. Variations occur in the cell packing and cell types in these divisions (**6.**142B). Nopajaroonsri *et al.* (1971) have suggested that these populations can be divided into three zones. *Zone 1* is a region of loosely packed cells consisting predominantly of small lymphocytes, macrophages and occasional plasma cells, found around the extreme periphery of each follicle and extending into the medullary cords. *Zone 2* is a more densely packed region within zone 1, limited to the cortical and paracortical areas of the node, consisting mainly of small lymphocytes and macrophages. *Zone 3* comprises the germinal centres forming the core of the lymphoid follicles, and they are prominent in antigenically stimulated lymph nodes; cells within these include large lymphoblasts, some undergoing mitosis, together with macrophages. Fibroblasts, reticulin fibres, and other associated cells

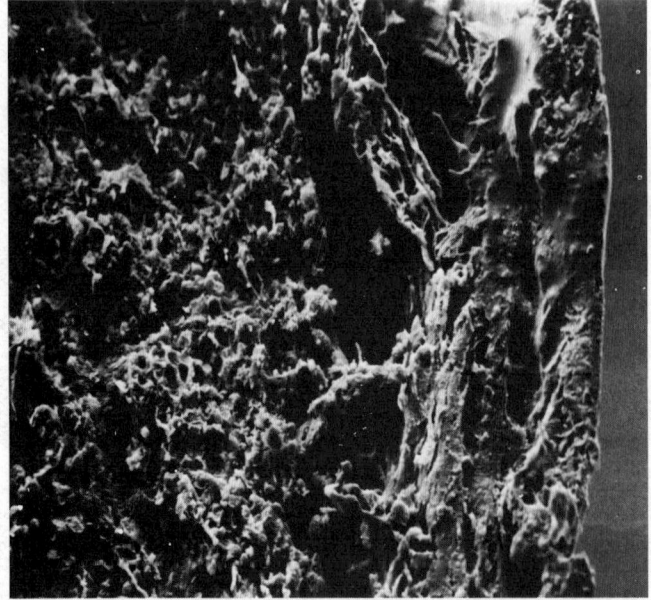

A

B

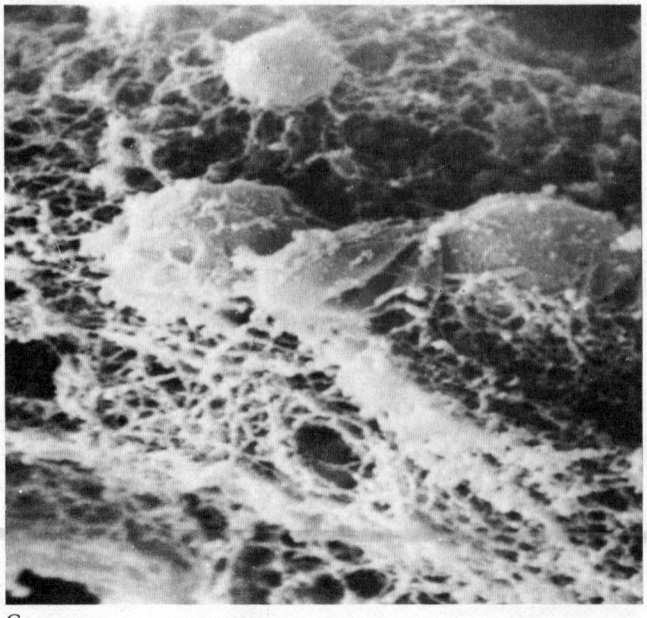

C

of the node, mentioned below, are scattered throughout all zones. It seems likely that these three zones represent, in part at least, a maturational sequence, and that lymphocytes formed by division of cells within the germinal centres (zone 3) pass to the densely packed zone 2, becoming smaller in this process, and finally migrating to the loosely packed zone 1, from which they may pass on through the endothelium into the sinuses. However, this sequence is complicated by the immigration of other cells from the lymphatic and haemal vascular systems, and the precise relation between the cellular architecture and maturational processes are not altogether clear. In any case, although helpful for purposes of description, such purely structural schemes are incomplete without a consideration of the sub-classes of lymphocytes occurring within these zones.

Another type of subdivision is concerned with the distribution of B- and T-lymphocytes in the node, and immunofluorescent staining techniques have demonstrated that these two types of cell occupy quite distinct territories. Immature B-cells have been detected in the outermost parts of the follicles (6.142B) whereas T-cells are sited in the region between the germinal centres and the medulla, i.e. the *paracortex* or *thymus-dependent zone*. Mature B-cells (plasma cells) are present mainly in the medullary cords, some also being found at the edges of the follicular masses. This distribution of T-lymphocytes is demonstrated clearly in animals with congenital absence of a thymus; these fail to develop a paracortex. Whether the germinal centres contain T- or B-cells, or both, is arguable as there is some conflicting evidence, and it is also possible that the cytological markers allowing detection of these cells only become effective after the lymphocytes have left the germinal centres.

Non-lymphocytic Cell Types in Lymph Nodes

The nature of the varieties of non-lymphocytic cells, like the names given to them by the many investigators of nodal histology, is a most confused and confusing subject, and it is only quite recently that attempts have been made to characterize these cells using adequate techniques. Following Steinman *et al.* (1974) we can distinguish the following cell varieties: (1) endothelial cells, (2) fibroblasts, (3) macrophages, (4) perivascular cells including pericytes, and (5) 'dendritic cells'. The endothelial cells of the nodal sinuses appear to be similar both in structure and properties to those elsewhere, and, contrary to the older suggestions, do not show any particular phagocytic ability which had prompted such terms as 'endothelial macrophages', 'reticuloendothelial cells', 'reticular cells', or 'littoral cells' by various authors (*see* Nopajaroonsri *et al.* 1971). Fibroblasts (corresponding to the 'reticular cells' of some authors) lay down reticulin and collagen, forming the framework of the node, including its capsule and trabeculae. Macrophages are identifiable as typical of that genre in both structure and behaviour (*see* p. 45). Perivascular cells (pericytes and smooth muscle cells), as elsewhere, surround the larger blood vessels. Another class of cell has recently been identified as a distinct entity, named the 'dendritic' cell because of its irregular outline (Steinman *et al.* 1974); it possesses few organelles, and may possibly represent a precursor for some other cell type. Whether it is similar to the presumed 'reticulum' cell of the earlier literature is uncertain, although the concept of a multipotent stem cell for the various defensive cell lines present in lymph nodes now seems unlikely, since it appears that the various cells are derived separately from stocks differentiating in other ('central') lymphoid tissues.

6.144A–C Scanning electron micrographs of the cut surface of a lymph node (guinea pig).
A Low-power micrograph of the outer cortex showing the capsule (right) together with the subcapsular sinus traversed by reticular fibres. Part of a germinal centre is visible on the left. Magnification × 400.
B Medium-power micrograph of part of a germinal centre, showing lymphocytes clustered around a capillary. Magnification × 2,000.
C High-power micrograph of the wall of the subcapsular sinus, showing the fine network of reticular fibres with some attached cells. Magnification × 6,000.

Circulation of the Lymphocytes

The lymph node, being a peripheral lymphoid organ, is constantly receiving lymphocytes from the vascular system and from the afferent lymphatic supply, and in turn, continually provides lymphocytes to its efferent lymphatic drainage which eventually leads back to the haemal circulation. In fetal life, and to a decreasing extent through postnatal life, these lymphocytes arise from the central lymphoid organs of the bone marrow and thymus where the initial differentiation of B- and T-lymphocytes takes place respectively. The cells then pass in the bloodstream to the lymph nodes and other peripheral lymphoid organs where they may proliferate when suitably stimulated, providing new lymphocytes which may rejoin the circulatory system and move to other peripheral organs, sites of inflammation, and other tissues.

As already mentioned, the lymphocytes migrate from the bloodstream into the lymph node through the postcapillary venules, and may then move to the germinal centres, in the case of B-lymphocytes, or into other regions of the cortex and medulla. Some cells appear to be quite transient visitors, whereas others may remain for considerable periods. Antigenically stimulated B-lymphocytes transform into *plasma cells*, migrating as they do so to the medullary cords and sinuses where they may remain, secreting antibodies into the lymph as it flows by, although some may also pass into the lymphatic channels, and ultimately the haemal circulation.

FUNCTIONS OF THE LYMPHOID SYSTEM

Lymph capillaries are primarily concerned with absorption from the tissue spaces. The absorbed material is conveyed to lymph nodes and in its passage through these, particulate matter (e.g. dust material inhaled into the lungs) is to a large extent filtered off by the phagocytic activity of cells in the nodes and prevented from entering the bloodstream. The same applies to bacteria and other micro-organisms in lymph draining from an affected area. Lymph nodes thus have a *protective function* in preventing to a considerable extent the entrance of noxious material from tissues into the bloodstream. Some observers, however, have stressed that the lymphatic system, under certain conditions, operates adversely, providing a pathway along which the *spread* of infection or malignant disease from tissues may occur.

The essential significance of the lymph node includes, therefore: (1) The provision of an intricate network of spaces, of large volume and surface area, through which the lymph percolates slowly. (2) The exposure of any contained foreign material to the phagocytic action of macrophages in the sinuses. (3) The trapping of antigen by the phagocytes. (4) The provision of centres for lymphocyte production and a pool of stem cells potentially capable of transforming into antibody-producing B-lymphocytes and mature T-lymphocytes. (5) The interaction between antigen-laden phagocytes and the lymphoid tissues with the mounting of an immune response both cellular and humoral. (6) A portal of entry of blood-borne lymphocytes back into the lymphatic channels. (7) Humoral antibody production.

As indicated above further details concerning the origins, varieties, and functional roles of lymphocytes are found on pp. 59–63.

Connexions of the lymphatic system with the bloodstream. It is generally believed that the main, if not the only, direct connexions between the lymph vessels and the venous bloodstream are by means of the thoracic duct and the right lymphatic duct with the veins at the root of the neck. Various observers have, however, reported additional connexions with the inferior vena cava, the renal, azygos, suprarenal and iliac veins. As the lymph vessels are very closely associated with veins in their development (2.124), such additional connexions would not be surprising, though they may be very variable. In this context, the importance of the post-capillary venules of the lymph nodes should again be stressed (p. 770).

Haemal nodes. In typical lymph nodes, some red blood cells may sometimes be found in the lymph sinuses. In some animals a variable number of small bodies are found, mainly related to the abdominal and thoracic viscera, which are called haemal nodes. In structure they resemble lymph nodes but the sinuses are filled with blood, which gives them a red colour, and they have no afferent or efferent lymph vessels. They appear to be more closely related to the blood-vascular system than to the lymphatic system. Investigations on the haemal nodes of the rat (Turner 1969) show that they possess no afferent, but one large efferent lymphatic vessel. Fast and slow components of the blood circulation are claimed, the former through arterioles, capillaries and venules and the latter through a tortuous sinusoidal system. Specialized post-capillary venules are present as in true lymph nodes. It is suggested that haemal nodes may represent an intermediate stage between a lymph node and the spleen, and phylogenetically, may be a basis from which these have evolved. However, their existence, structure and distribution in man is still uncertain. In some animals, *haemolymph nodes* have been described, with a structure intermediate between lymph nodes and haemal nodes, possessing lymphatic as well as vascular connexions. Some observers maintain that they are stages in the transformation of lymph nodes into haemal nodes, but others are strongly of the opinion that no such intermediary structures exist.

Applied Anatomy. The lymph vessels and lymph nodes draining any infected area of the body are very liable to become inflamed, resulting in acute or chronic lymphangitis and lymphadenitis. Chronic lymphangitis, together with the blocking of numerous lymphatic vessels by the escaped ova of the minute parasitic worm, *Microfilaris nocturna*, is the cause of elephantiasis, a condition common in the tropics and subtropics, and characterized by enormous enlargement and thickening of the skin of some parts of the body, most frequently of the leg and scrotum. Blockage of lymphatic vessels may also occur as a result of extensive spread of malignant disease or widespread removal of lymphatic glands.

The present view is that cancer spreads both by minute emboli and by permeating the lymph vessels as a solid cell growth. Operations for the removal of cancer are therefore planned to take away *in one mass* the cancer, the intervening lymph vessels, and the lymph nodes. In this connexion it is important to note that the lymph vessels from a given region may drain *directly*, not to neighbouring lymph nodes, but to nodes quite far removed, thus rendering operative removal of the primary site, the intervening vessels and the nodes extremely difficult, if not impracticable.

The Spleen

The spleen (6.145–149) is situated for the most part in the left hypochondriac region of the abdomen, but its posterior edge extends into the epigastric region; it lies between the fundus of the

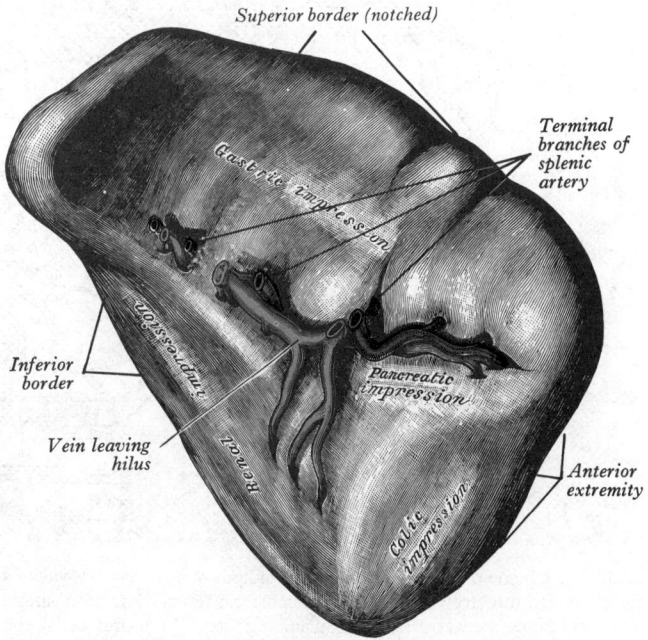

6.145   The visceral surface of the spleen.

stomach and the diaphragm. It varies in shape, according to the size of the colic impression, from that of a slightly curved wedge (if the colic impression is small) to a tetrahedron (if the colic impression is large). Its long axis lies in the line of the tenth rib, its posterior extremity being about 3·5 to 4·0 cm from the mid-dorsal line opposite the spine of the tenth thoracic vertebra, and its anterior extremity reaching as far as the mid-axillary line. It is soft, of very friable consistence, highly vascular, and of a dark purplish colour.

Relations. It presents diaphragmatic and visceral surfaces, upper and lower borders, anterior and posterior extremities. The *diaphragmatic surface* is convex, smooth and faces upwards, backwards and to the left, except at its posterior edge, where it faces slightly medially. It is in relation with the abdominal surface of the diaphragm, which separates it from the lowest parts of the left lung and pleura and the ninth, tenth and eleventh ribs of the left side. The costodiaphragmatic recess of the pleura extends down as far as the inferior border of the spleen.

The *visceral surface* (**6**.145) is directed towards the abdominal cavity, and presents gastric, renal, pancreatic and colic impressions.

The *gastric impression*, directed forwards, upwards and medially, is broad and concave. It is in contact with the posterior

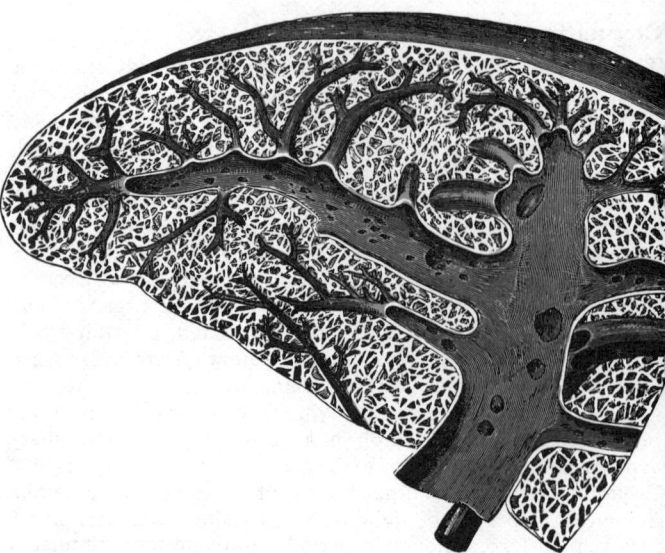

6.146 A transverse section through the spleen, showing the trabecular tissue and the splenic vein and its tributaries. From the first edition of *Gray's Anatomy* (1858).

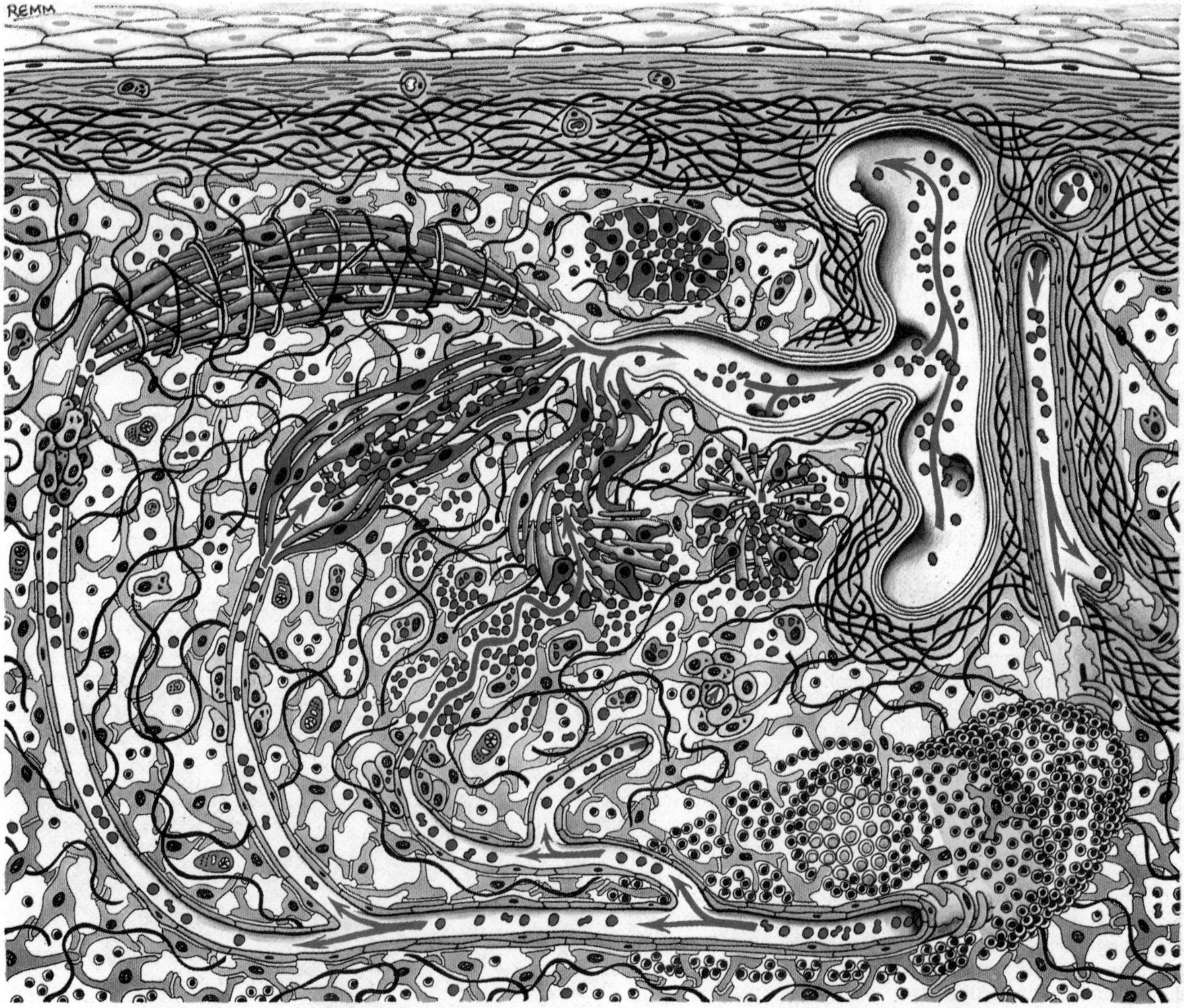

6.147 A scheme of the main features of splenic structure—the various elements are not drawn to scale, to enable representation on a single diagram. Note the capsule, trabeculae, reticular fibres and cells, the perivascular lymphatic aggregations (white pulp), the ellipsoids, and cell cords and venous sinusoids of the red pulp. The 'open' and 'closed' theories of splenic circulation are illustrated. The venous sinusoids are shown in two states (*a*) with their lining of 'stave' cells (bright blue) in close apposition, and (*b*) with intercellular gaps—these have been over-emphasized for clarity. Consult text for further details.

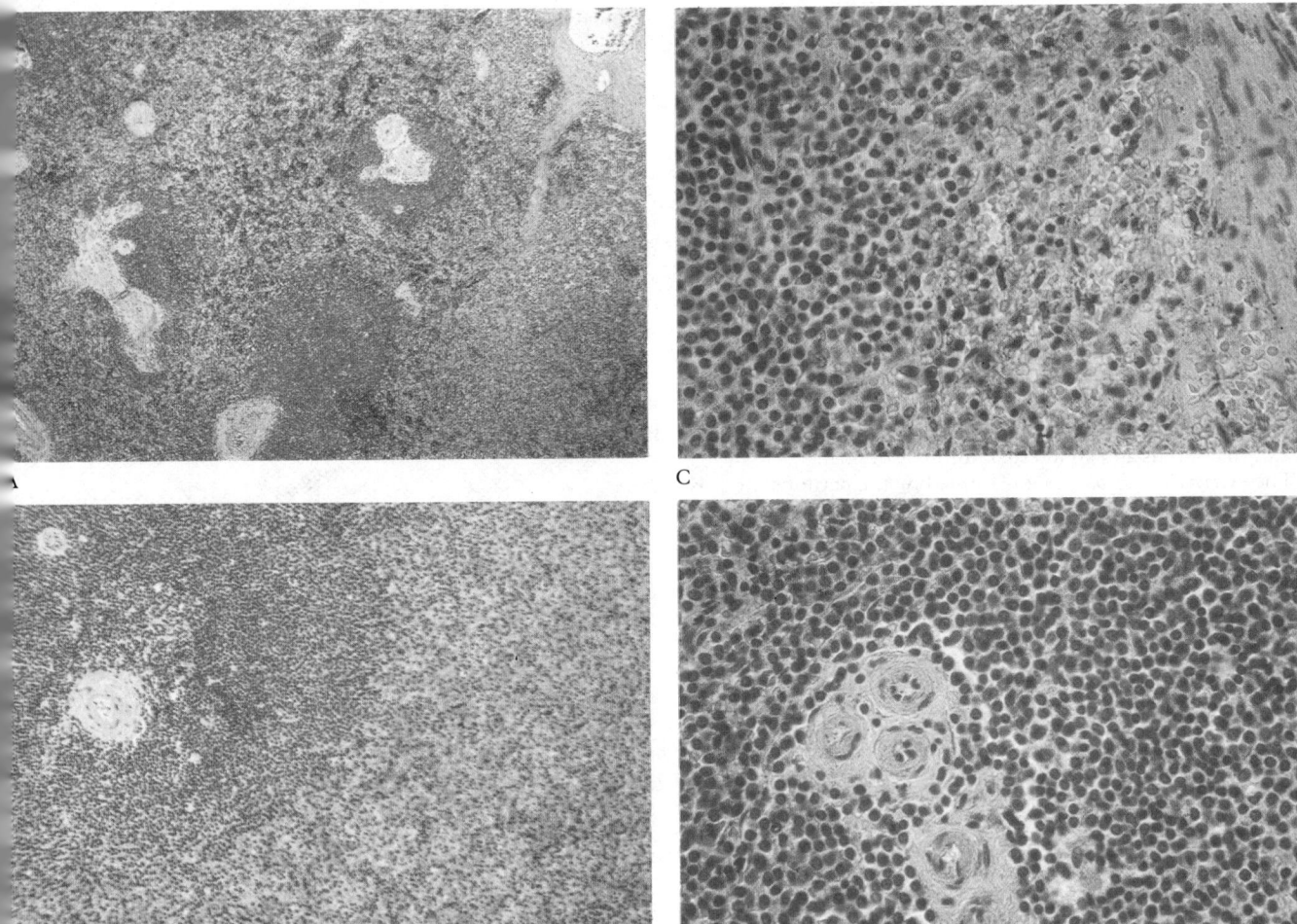

A

B

C

D

6.148 Sections of human spleen stained with haematoxylin and eosin. (Kindly supplied by Dr. D. R. Turner of the Department of Pathology, Guy's Hospital Medical School.)

A and B Survey photographs at low power showing the general contrast between the white pulp (perivascular lymphatic aggregates, stained blue) and the red pulp (venous sinusoids and intervening cellular cords, stained reddish purple).

C High-power view of the junctional zone between the densely packed lymphocytes of the white pulp and the blood-filled sinusoids and cell cords of the red pulp.

D A group of small (penicillar) arteries ensheathed by densely packed small lymphocytes.

wall of the stomach, from which it is separated by a recess of the greater sac of the peritoneum. It presents near its lower limit a long fissure, termed the *hilum*. This is pierced by several irregular apertures for the entrance and exit of vessels and nerves.

The *renal impression*, which is very slightly concave, is placed on the lower part of the visceral surface and is separated from the gastric impression above by a raised margin. It is directed medially, downwards and a little backwards, and is related to the upper and lateral part of the anterior surface of the left kidney and, sometimes, to the upper pole of the left suprarenal gland.

The *colic impression* is placed at the lateral extremity of the spleen and is usually flattened. It is related to the left colic flexure and phrenicocolic ligament (**8.111**).

The *pancreatic impression*, when present, is small and is placed between the colic impression and the lateral part of the hilum. It is directly related to the tail of the pancreas which lies in the lienorenal ligament (**8.111**).

The *superior border* separates the diaphragmatic surface from the gastric impression. It is usually convex upwards and is marked, near its lateral end, by one or two notches of variable depth. The notches indicate the lobulated character of the spleen in early fetal life (p. 207). The *inferior border* separates the renal impression from the diaphragmatic surface and lies between the diaphragm and the upper part of the lateral border of the left kidney. It is blunter and more rounded than the superior border and corresponds to the lower margin of the eleventh rib.

The *posterior extremity* of the spleen is blunt and rounded in most cases. It is directed towards the vertebral column. The *anterior extremity* is more expanded, and commonly takes the form of a margin connecting the lateral ends of the upper and lower borders. It is related to the left colic flexure and to the phrenicocolic ligament.

The spleen is almost entirely surrounded by peritoneum, which is firmly adherent to its capsule. Recesses of the greater sac intervene between the spleen and the stomach, and between the spleen and the left kidney. It develops in the upper part of the dorsal mesogastrium (p. 205) and remains connected with the stomach and the posterior abdominal wall by two folds of peritoneum. One, termed the *lienorenal ligament*, is derived from the peritoneum where the wall of the general peritoneal cavity comes into contact with the omental bursa between the left kidney and spleen; the splenic vessels pass between its two layers (**8.99A**). The other fold, termed the *gastrosplenic ligament*, also consists of two layers, and is formed by the meeting of the walls of the greater sac and omental bursa between the spleen and stomach (**8.99A**); the short gastric and left gastro-epiploic branches of the splenic artery run between its two layers. The lateral part of the lateral end of the spleen is in contact with the phrenicocolic ligament.

The *size and weight* of the spleen vary at different periods of life, in different individuals, and in the same individual under different conditions. *In the adult* it is usually about 12 cm in length, 7 cm in breadth, and 3 or 4 cm in thickness, but it tends to diminish in size and weight with advancing age. Its average weight in the adult is about 150 gm, though it may range between 80 and 300 gm, largely according to the amount of blood it contains.

The size of the spleen slowly increases during digestion, and varies according to the state of nutrition of the body, being large in highly fed, and small in starved animals.

Frequently in the neighbourhood of the spleen, and especially in the gastrosplenic ligament and greater omentum, small encapsulated nodules of splenic tissue may be found, either isolated or connected to the spleen by thin bands of splenic tissue. They are known as *accessory spleens* or *spleniculi*, and they may be very numerous and widely scattered in the abdomen. The spleen may retain its fetal lobulated form, or there may be deep notches on the diaphragmatic surface and inferior border, in addition to those usually present on the superior border.

Surface Anatomy. The position of the spleen in the living can be determined by percussion. The dull area extends over the ninth, tenth and eleventh ribs in vertical extent and should not extend forwards further than the mid-axillary line. The normal spleen is not palpable.

Structure (**6**.146–149). The spleen is invested by two coats: an external serous and an internal fibro-elastic coat.

The *external*, or *serous*, *coat* is formed by the peritoneum; it is thin, smooth, and in the human subject intimately adherent to the fibro-elastic coat. It invests the entire organ, except at the hilum and along the lines of reflection of the lienorenal and gastrosplenic ligaments.

The *fibro-elastic coat*, or *capsule*, invests the organ and from it *trabeculae* pass into the spleen and branch to form a network which constitutes the framework of the spleen (**6**.146). The largest trabeculae pass in from the hilum and ensheath the splenic vessels, which divide into branches that run in the subdivisions of the trabeculae (**6**.147, 148). The capsule and trabeculae consist of collagenous white fibrous tissue and yellow elastic fibres, the latter being more numerous in the trabeculae. In many mammals the capsule and trabeculae contain many non-striated muscle cells and the rhythmical contraction of the spleen is attributed to these. In man, few muscle cells are present in the capsule and trabeculae, and contraction and distension of the spleen are attributed to constriction or relaxation of the blood vessels (**6**.147) with consequent alteration in the amount of blood in the organ. Increase in the blood content distends the spleen and stretches the elastic fibres in the capsule and trabeculae; contraction of the spleen is due to the recoil of these fibres.

The subdivisions of the trabeculae are continuous with a fine reticulum which pervades the remainder of the organ. Electron microscopy reveals that the reticulum consists of dense amorphous matrix, probably mainly proteoglycan in nature, with only occasional fine collagen fibres. Closely associated with the reticulum are reticular cells, or fibroblasts, into which strands of the matrix may be invaginated, and macrophages. The interstices of the reticulum are occupied by the *splenic pulp*, of which there are two kinds, *red pulp* and *white pulp*, which are both related to the blood vessels permeating the spleen. The reticulum is particularly dense around the borders of the white pulp where it constitutes a *marginal zone*.

Circulation of blood inside the spleen. (*See:* Knisely 1936; MacKenzie *et al.* 1941; Snook 1950; Peck and Hoerr 1951; Lewis 1957; Wennberg and Weiss 1969, and Li-Tsun Chen and Weiss 1972, 1973.) The large, tortuous splenic artery, before it reaches the spleen, divides in the lienorenal ligament into five or more branches which enter the hilum of the organ and ramify throughout its substance in the trabeculae. The splenic vein is formed in the lienorenal ligament by the junction of five or more tributaries which emerge from the hilum. Small arteriolar branches of the splenic arteries leave the trabeculae, and their adventitial coat becomes replaced by a *periarteriolar sheath* of lymphatic tissue which accompanies the vessels and their branches almost as far as their division into capillaries (**6**.147, 148). These lymphatic sheaths constitute the *white pulp* of the spleen, and here and there the sheaths are enlarged containing rounded masses of lymphocytes, the *splenic lymphatic follicles* (Malpighian bodies), which vary in diameter from 0·25 mm to 1 mm and are visible to the naked eye on the freshly cut surface of the organ as whitish semi-opaque dots contrasting with the dark red colour of the surrounding tissue, which constitutes the *red pulp* of the spleen. These lymphatic or lymphoid follicles are centres of lymphocyte production and when antigenically

stimulated, show germinal centres similar to those of lymph node follicles. The follicles atrophy with increasing age and may be absent in old age. The arterioles usually occupy an eccentric position in the follicles and give off fine side branches to supply the follicles. Before the arterioles finally lose their sheaths of lymphatic tissue and enter the surrounding red pulp, they divide into a number of straight vessels termed *penicilli*. After running a course of about 0·5 mm, entering or penetrating through the marginal zone around the white pulp, each of these vessels develops a slight thickening of its coat known as an *ellipsoid* or *sheath of Schweigger-Seidel*, which is formed by an aggregation of macrophages and fibroblastic (reticular) cells, and the lumen of the vessel is considerably narrowed. The ellipsoids are well

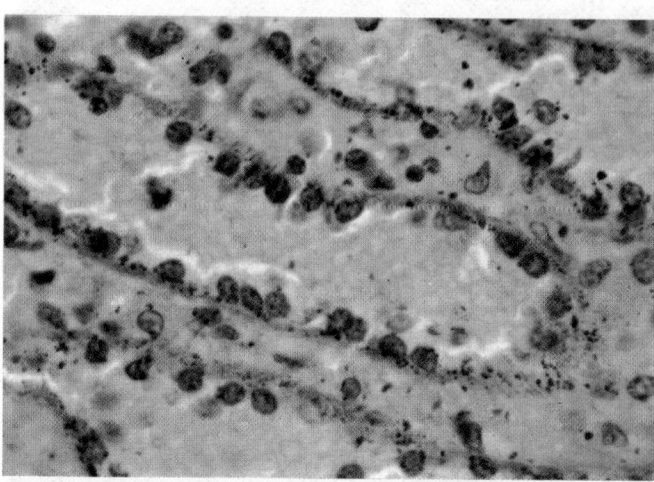

A

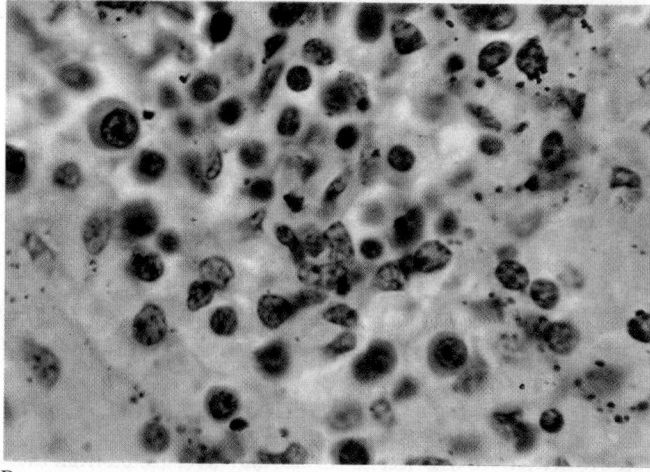

B

6.149   A section of monkey spleen following intravascular perfusion with a suspension of carbon particles followed later by perfusion fixation, stained by Weigert's haematoxylin and Van Gieson's stain.

A   Showing empty, dilated venous sinusoids and intervening cell cords. The 'stave' cells lining the sinusoids are prominent.

B   High-power view of cellular region between venous sinusoids. The cell types seen include reticular macrophages with carbon particles in their cytoplasm, small and large lymphocytes and a number of prominent rounded plasma cells.

developed in some mammals, e.g. pig, cat and dog, but are only feebly developed in man. Beyond the ellipsoids each vessel continues as a capillary-like arteriole, or may divide into two vessels, but thereafter its precise course is somewhat uncertain, although it is known that the blood passes into the red pulp and eventually drains into the veins of the spleen. The red pulp, in man (as in rabbit) consists of numerous *venous sinusoids* which are large and complex cavities containing blood, separated by areas of tissue rich in macrophages attached to the reticulum of the spleen (reticulin fibres and fibroblasts), termed *splenic cords* or *cords of Billroth*.

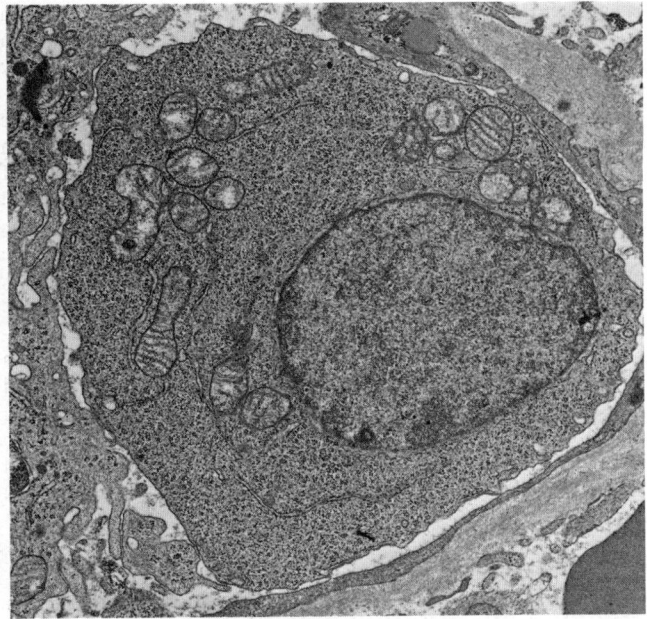

C

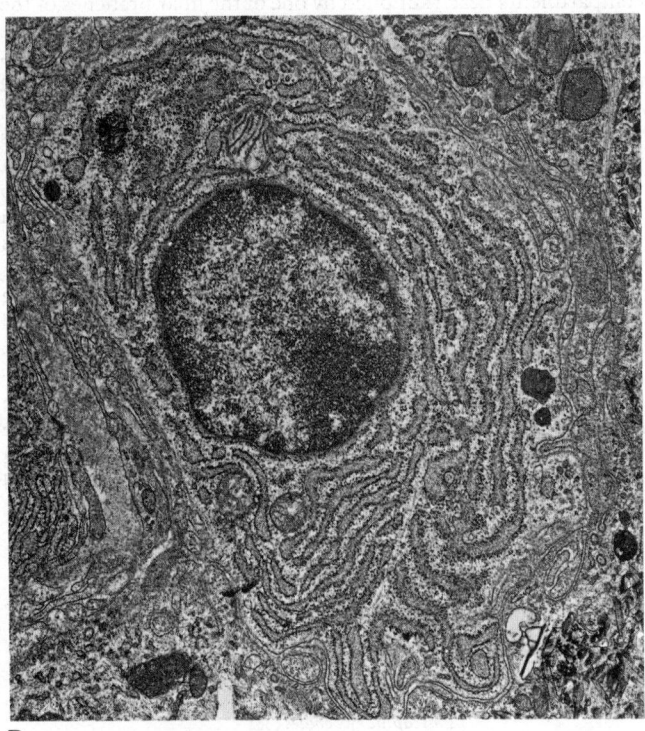

D

6.149C Transmission electron micrograph of a stimulated T-lymphocyte in the spleen (monkey). Note the numerous free ribosomes and relative paucity of granular endoplasmic reticulum. Magnification × 10,000.

6.149D Transmission electron micrograph of a stimulated B-lymphocyte (plasmocyte) in the spleen (monkey). Note the copious granular endoplasmic reticulum containing newly synthesized antibodies, seen here as finely granular material. Magnification × 10,000.

The splenic venous sinusoids are lined by flattened, greatly elongated endothelial cells often called *stave cells* because of their shape, reminiscent of the planks of a barrel; when the sinusoids are inflated by a rise in internal pressure (as, for example, during perfusion in preparation for routine electron microscopy), numerous gaps appear between the stave cells which allow blood to pass in and out of the surrounding splenic cords. It should be noted, however, that the dimensions and dynamic state of such intercellular crevices in the *in vivo* condition remain conjectural. The luminal surfaces of the endothelial cells are uneven and beset with protrusions and the cells are moderately phagocytic; they are

supported externally by strands of, in the main, circularly arranged reticulum. Further ultrastructural details of the red pulp of the human spleen (Chen and Weiss 1972), and experimental results after modification of rats' splenic vasculature with phenylhydrazine (Chen and Weiss 1973) have been added. The fine architecture of the predominantly circumferential ('ring') and occasional longitudinal fibrillary components of the surrounding basal lamina of the reticulum were analysed, together with details of the neighbouring stave cell cytoplasm. The endothelial (stave) cells of the venous sinusoids possessed three distinctive structural features. Numerous pinocytotic vesicles about 0·1 μm in diameter were present beneath the plasma membrane on both luminal and basal aspects of the cell. Loosely organized filaments some 7–8 nm in diameter coursed longitudinally through the cell, whilst bands of finer filaments about 3–5 nm in diameter arched through the basal cytoplasm. The latter filaments appeared to attach to the cell membrane at points adjacent to the fibrillary ring component of the reticular basal lamina, and were possibly in structural continuity with the rings. It was proposed that the filaments dynamically maintained the shape of the stave cells, stabilized the relation of the cells to their basal lamina, and controlled the dimensions of the intercellular crevices—thereby also controlling cellular passage through the crevices.

The splenic cords are areas of intense phagocytic activity because of their numerous macrophages, and also contain innumerable lymphocytes derived from the white pulp and elsewhere. The term 'cord' is perhaps misleading, since these areas of perivascular tissue form a continuous network throughout the spleen, and have numerous cavities between the indigenous cells through which blood can pass. In some animals, including mouse and cat, there do not appear to be any sinusoids, and so the majority of the red pulp is formed by splenic cord tissue.

The passage of blood from the penicillar arterioles to the veins of the trabeculae has been a subject of much controversy, stemming from the variations in techniques of observation and species of animal studied. According to the 'open' theory of splenic circulation, blood passes from the arterioles directly into the splenic cord tissue, eventually collecting in the venous sinusoids to enter the veins. In the 'closed' theory, the blood passes from the arterioles directly into the venous sinusoids and does not at any point enter the extravascular tissue of the cords. Electron microscope studies do not appear yet to have settled this issue unequivocally, but in observations of spleens in living animals, blood appears to flow directly from the arteries into the sinusoids, then to escape through their walls into the surrounding cordal tissue, where it can certainly be demonstrated by electron microscopic and other methods. It appears probable that the flow of blood through these routes depends on the local control of blood flow by the intermittent opening and closing of the blood vessels at either end by perivascular smooth muscle contraction, which allows the tidal flow of blood between sinusoids and the spaces outside them, but this has yet to be demonstrated unambiguously. However, in those animals which apparently lack sinusoids, there must be some other type of circulation, presumably of the 'open' type.

After entering the venous tributaries which pass along the trabeculae towards the hilum, the blood eventually flows into the main splenic vessels.

FUNCTIONS OF THE SPLEEN

These are essentially phagocytosis, cytopoiesis, erythrocyte storage, and the mounting of immune responses.

Phagocytosis. The splenic macrophages form an important part of the mononuclear phagocyte system. They consist of the macrophages of the marginal zones of the white pulp, and those of the ellipsoids and splenic cords. Particulate matter such as carbon is also taken up to some extent by the stave cells lining the venous sinusoids and some authorities have considered these cells to be part of the macrophage system. However, since all endothelial cells have a minor ability to phagocytose particles, the stave cells do not appear to warrant separate classification, and certainly have various features which indicate that they are not true macrophages.

The phagocytes remove cell débris and the products of effete erythrocytes, leucocytes, platelets, and micro-organisms from the circulating blood. It is especially important in the processing of aging or damaged red cells, and all stages of erythrophagocytosis, from disintegrating erythrocytes to granules of haemosiderin, may occur in the cytoplasm of the splenic macrophages. Ultimately, bilirubin is produced, which is conveyed to the liver for excretion, and the iron is largely re-used by the bone marrow in the production of further erythrocytes. As in the other peripheral lymphoid organs, the spleen is important in the phagocytosis of circulating antigens and in the initiation of humoral and cellular immune responses.

Lymphocyte activities. Various studies have shown that T- and B-lymphocytes are sited in different parts of the white pulp, the periarteriolar layers of cells being mainly T-lymphocytes and the follicles, with their germinal centres, being B-lymphocytes. Cells from these regions can migrate into the splenic cord tissue and other areas of the spleen, to perform their various characteristic functions. Some of the B-lymphocytes transform into plasma cells which secrete antibodies into the fluid-filled cavities, when suitably stimulated, and the T-lymphocytes carry out cytotoxic killing of infected circulating cells, or cooperating with the B-cells in complex immune reactions. Lymphocytes are also provided for the general defence of the body, some passing into the haemal circulation via the venous sinusoids, but the great majority apparently moving into the lymphatic drainage to rejoin the blood system indirectly.

Cytopoiesis. In the human fetus, from the fourth month onwards, the spleen is an important *haemopoietic* organ and the red pulp contains groups of myelocytes, erythroblasts and megakaryocytes. In some anaemias and myeloid leukaemia in post-natal life, as yet unidentified cell types in the red pulp may revert to a haemopoietic function.

In the mature spleen, lymphopoiesis in the white pulp contributes to the circulating pool of immunologically competent lymphocytes and it also contributes to the large mononuclear cell population of the blood and connective tissue. (*See* p. 45.)

Immune response. Under conditions of adequate antigenic stimulation there is a proliferation of splenic macrophages, increased lymphopoiesis and an increased growth and differentiation of antibody-producing plasma cells. (See also pp. 59–63,

and above.) In individuals who have suffered many episodes of a blood-borne infection, for example malaria, the splenic tissues may be permanently hypertrophied and the spleen greatly enlarged (*splenomegaly*).

Erythrocyte storage. The storage mechanism involves the separation of erythrocytes from plasma, but the site of separation and storage is uncertain; according to the 'closed' theory of splenic circulation it is in the venous sinusoids, the filling and emptying of which are controlled by sphincters at each end of the sinusoids, whereas according to the 'open' theory it is in the reticular meshes of the red pulp. In states of emergency, especially those associated with anoxia, the erythrocytes are discharged into the circulation, thus increasing the oxygen-carrying capacity of the blood. In most mammals, the discharge is due to contraction of the unstriped muscle in the capsule and trabeculae under the influence of the sympathetic nervous system and adrenalin; in man it is due to recoil of the stretched elastic fibres of the distended spleen, though the blood storage and discharge function of the human spleen is not nearly so marked as in other species.

Splenic Vascular Segmentation

There is evidence (Dreyer and Budtz-Olson 1952) that the spleen in man and other animals consists of separate 'segments' or 'compartments', each supplied by one of the hilar branches of the splenic artery and drained by one of the hilar tributaries of the splenic vein (Braithwaite and Adams 1957). Adjacent compartments, it has been claimed, are connected by an intersegmental vein so that if one compartment becomes congested with blood, excess blood can pass by these channels to adjacent compartments (but *vide infra*). It is suggested that the splenic segments may act in normal circumstances as separate units when the blood flow to them is not excessive. (For a review of the relevant literature on splenic vascular segmentation since its first proposal over a century ago by Kyber 1870, consult Gupta *et al.* 1976.) The latter authors studied corrosion casts of 50 adult human spleens. In 42 specimens (84 per cent), only two segments (superior and inferior) were present; in 8 specimens (16 per cent), three segments (superior, intermediate, and inferior) were demonstrated. There was no apparent anastomosis between adjacent segments. A comparable segmental arrangement of the splenic veins was claimed by Fuld and Irwin (1954). This finding of only two or three segments (arterial or venous) was supported

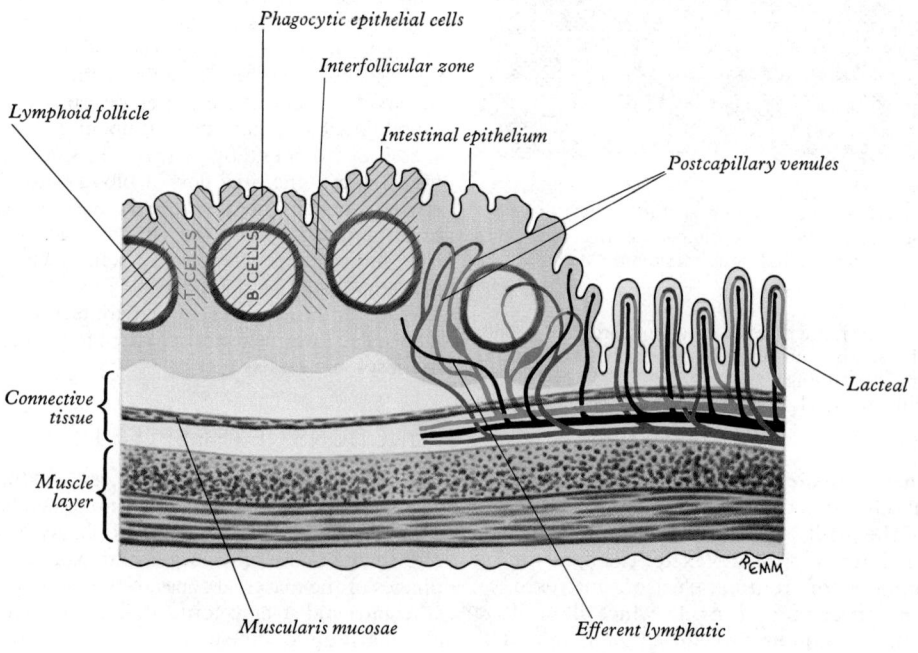

6.150 A diagram of the organization of an epitheliolymphoid complex in the wall of the small intestine (a Peyer's patch), showing the distribution of B-lymphocytes associated with the follicles and T-lymphocytes in the interfollicular zones; the arrangement of the vascular supply of the lymphoid tissue is also shown (right).

by a number of the investigators quoted by Gupta *et al*. However, it should be noted that this does not accord with the usual description of the splenic artery breaking up into five or more major branches in the lienorenal ligament before entering the splenic hilum: no explanation of this discrepancy is advanced by the authors. (Braithwaite and Adams described 5 to 7 vascular splenic segments in the rat.) Clearly, further investigations are necessary.

The effects of splenectomy. Partial removal of the spleen is followed by rapid regeneration of the lost tissue, but even total splenectomy has few serious effects, as its functions can be carried out fairly effectively by the remaining reticulo-endothelial organs. However, splenectomy, particularly in the early years of life, may be accompanied by a general reduction in the rate of mounting an immune response and consequently an increased susceptibility to infection. Splenectomy in later life is soon attended by a leucocytosis with increased lymphocyte, neutrophil, eosinophil and platelet counts in the peripheral blood. This has been interpreted as the result of removal of humoral factors produced by the spleen which oppose the formation and release of cells from the haemopoietic tissues. These effects gradually disappear over the course of a few weeks.

The *lymphatic vessels* of the spleen, once thought to be an insignificant component merely draining the capsule and thickest trabeculae, has recently been shown (in at least some species) to be an important feature of *all* parts of the tissue. Extensive blind-ending (efferent) lymphatic vessels are present in the white pulp, running in tortuous fashion alongside the arterioles and arteries, the flow of lymph being counter to that of the arterial blood. The lymphatic vessels eventually emerge at the hilum. No afferent lymphatic vessels are present. This lymphatic pathway is a drainage area for fluid infiltrating the white pulp, and also provides a major route for lymphocytes to pass from the spleen into the general circulation, in a manner similar to that of lymph nodes (*see* Weiss 1979).

The *nerves* are derived from the coeliac plexus and are chiefly nonmyelinated. In man their distribution is confined mainly to the branches of the splenic arteries; in other mammals they also supply the non-striated muscle in the capsule and trabeculae.

Applied Anatomy. Enlargement of the spleen, or *splenomegaly*, may accompany a variety of conditions, e.g. increased phago-cytosis in certain generalized infections, either bacterial or viral; in conditions of increased erythrocyte breakdown, and in various lipoidoses. Any circumstance leading to a massive immune response may be accompanied by enlargement of the spleen, which is also a characteristic of many of the reticuloses. A full review of the causes of splenomegaly is outside the scope of this volume, but the general features of the enlarging spleen will be considered. It is the anterior extremity, the anterior part of the diaphragmatic surface and the notched superior border that become palpable below the left costal margin; the notches become exaggerated and may be easily palpable. The left part of the transverse colon and the left colic flexure are displaced downwards, so that there is no area of colonic resonance on percussion over the enlarged organ; this is in contrast to a retroperitoneal tumour (e.g. of the kidney) which does not displace the gut and in which, therefore, an area of colonic resonance may sometimes be elicited on percussion over the palpable tumour. There is no anastomosis between the smaller branches of the splenic arteries, so that obstruction leads to infarction of the spleen. In surgical removal of the spleen, damage to the tail of the pancreas must be avoided when ligating the splenic vessels near the hilum.

Gut-Associated Lymphoid Tissue

In addition to the encapsulated peripheral lymphoid organs of the lymph nodes and spleen, there are large amounts of unen-capsulated lymphoid tissue, the *lymphoid nodules* in the walls of the alimentary and respiratory tracts, termed collectively the *epithelio-lymphoid* or *gut-associated lymphoid tissue*. These are located chiefly in the lamina propria just beneath the adjacent epithelium, although when active, they may extend into deeper layers, e.g. the submucosa of the intestine, and their cells may also be disseminated diffusely throughout the neighbouring tissues. Such nodules include the pharyngeal, tubal, palatine, and lingual tonsils, various nodules in the wall of the oesophagus, the large groups in the small intestine (Peyer's patches) and vermiform appendix, smaller aggregates in the colon and rectum, and nodules in the trachea and bronchial tree.

The precise form of the nodules depends on their location, but in all sites it is possible to distinguish numerous rounded follicles similar to those of lymph nodes, which develop noticeable germinal centres when antigenically stimulated (*see*, for example, Anderson 1977). Between the follicles (**6**.150) lie wedge-shaped masses of somewhat less closely packed (parafollicular) lympho-cytes. The whole mass of cells, together with numerous macrophages, is supported by a fine meshwork of reticulin fibres with their associated fibroblasts, and occasional coarser con-nective tissue trabeculae in some of the larger nodules, such as those of the pharyngeal tonsil. The adjacent epithelium of the gut covers the luminal surface of the tissue and in most places gives off glandular or other diverticula which penetrate into the lymphocyte aggregate. In some areas (e.g. the palatine tonsils) such crypts may become loci of infection leading to a general inflammatory response of the whole surrounding area. In many sites, dome-like areas of specialized epithelium cover the lymphoid nodule (*vide infra*). The vascular pathways consist of blood vessels branching from the surrounding connective tissue to supply the follicles with a capillary plexus which drains into post-capillary venules possessing cuboidal endothelial cells similar to those of lymph nodes, and which allow the frequent migration of lymphocytes to and from the bloodstream into the nodules. The lymphatic vessels associated with lymph nodules are exclusively efferent, draining into the general network of lymphatic channels serving the organ in which they are embedded.

Immunofluorescence studies have shown that the B- and T-lymphocytes of lymphoid nodules are segregated, again in a manner resembling the lymph node, the follicles being composed of B-lymphocytes and the parafollicular areas of T-lymphocytes. These cells can migrate either into the lymphatic drainage and so rejoin the hemal circulation, or may migrate out into the adjacent tissues, including the intercellular spaces of the neighbouring epithelia. In the case of non-stratified epithelia, they may eventually pass into the lumen of the alimentary or respiratory tract. In the lamina propria of these tracts migrating B-lymphocytes are often seen to have transformed into plasma cells (**6**.150).

FUNCTIONS OF LYMPHOID NODULES

There has been much speculation about the precise role of these structures in the total lymphoid system of the body. This interest has, in part, stemmed from the condition in birds in which the B-lymphocytes differentiate in lymphoid tissue in the wall of a hind-gut diverticulum, the *bursa of Fabricius*, suggesting that the lymphoid tissues of the gut in mammals may have a similar function. However, various experiments have so far failed to support this idea, and it seems more likely that these regions provide areas in which B- and T-lymphocytes can proliferate and act as reservoirs of defensive cells which can infiltrate the surrounding tissues providing local defences. The B-lymphocytes are particularly important in the synthesis of secretory antibodies of the IgA class which are present in the secretions of the alimentary tract. These antibodies are secreted first by plasma cells in the lamina propria and intercellular spaces of the unicellular epithelia and sub-epithelial glands. The latter are of considerable importance where the general epithelial lining is of stratified squamous type. The antibodies appear to be taken up by certain varieties of glandular cell (although not by goblet cells, at least in the gut) and then modified by the addition of carbohydrate to form the final secretory IgA (sometimes termed sIgA) which is secreted into the lumen of the gut or bronchial tree. These antibodies are of great importance in eliminating pathogenic organisms in the various tracts in which they are found, although other types of antibody (IgM and IgG) secreted by the plasma cells of the lamina propria are also vital to the destruction of pathogens which have already passed through the epithelial surface.

For this system to operate efficiently there must presumably be some way in which lymphocytes can detect antigens present on the luminal side of the epithelium covering them. Recently, specialized phagocytic cells have been demonstrated in the epithelium lying over lymphoid follicles, which are capable of passing particulate material to the lymphoid tissue beneath, thus providing a route for antigens to reach the immune system (Bockman and Cooper 1973; Chamberlain et al. 1973; Owen and Jones 1974).

The Thymus

The thymus (**6**.151A) varies in size with age. Until recent years the significance of the thymus was completely unknown, but evidence from intensive and elegant experimentation has now placed the thymus as a primary *central organ* of the lymphoid system. At birth it commonly weighs between 10 and 15 gm; it continues to grow up to the age of puberty, when its weight ranges between 30 and 40 gm. Thereafter it generally progressively diminishes in size, undergoing gradual atrophy and replacement by fat, so that after mid-adult life it may weigh only about 10 gm, though it may remain large and weigh between 28 and 50 gm (Young and Turnbull 1931; Keynes 1954). Certain disease processes greatly accelerate the involution and this may account for much of this variation.

In early life it is pinkish grey in colour, soft and finely lobulated on its surfaces, and consists of two unequally sized, pyramidal lobes connected together by areolar tissue. Although the thymus is customarily described as a single unpaired organ, each lobe is developed from the third pharyngeal pouch of its own side (p. 198), and strictly there are two separate thymic bodies, right and left. The thymus lies in the anterior and superior mediastina of the thorax, extending as far inferiorly as the fourth costal cartilages, and its upper tapering parts extend into the neck, sometimes as far as the lower part of the thyroid gland, or even higher. Its shape is largely determined by the structures related to it and upon which it is moulded. Anteriorly, it is covered by the sternum and the adjacent parts of the upper four costal cartilages, and the sternohyoids and sternothyroids. Posteriorly, it is moulded on the pericardium, the aortic arch and its branches, the left brachiocephalic vein, and the front and sides of the trachea. After mid-adult life it becomes yellowish in colour due to its gradual replacement by fat. Small accessory nodules of thymic tissue may occur in the neck as detached parts of the thymic diverticula in their developmental descent (p. 198), or the main thymus may be continued upwards as thin strands into the neck along this path, reaching the thyroid cartilage or even slightly higher. Sometimes the cervical part of the thymus is represented by strands of connective tissue which connect the thymus to the inferior parathyroid glands. The nodules may be closely associated with the parathyroids, those related to the superior parathyroids (parathyroids IV) being developed from the fourth pharyngeal pouch, and to the inferior parathyroids from the third.

Structure (**6**.151–154). Each lobe of the thymus is surrounded by a delicate fibrous capsule, from which septa penetrate for a short distance to divide the lobe into irregular lobules, each of 0·5–2·0 mm in diameter, which consist of an outer, darkly staining, densely cellular *cortex* and an inner, lightly staining, less dense *medulla*. The lobules are not completely separated from each other by these septa, and the medullary parts of the lobules are continuous with each other in the central parts of the lobe, running through which is a central parenchymatous cord derived from the embryonic thymic diverticulum. The lobules are connected to this cord through prolongations of their medullary substance, and if the interlobular connective tissue is dissected away (in the young child), the lobules are seen to have an irregular 'necklace' arrangement.

Essentially, throughout both the cortical and medullary parts of the thymus, there are two principal tissue components, which differ quantitatively in the two regions. These are: (1) a framework of irregularly branched epithelial 'reticular' cells, quite distinct from those found in the other lymphoid tissues; and (2) lymphocytic cells, macrophages and occasional other cells in the interstices of the epithelial framework.

The epithelial framework. The cells vary in their size and shape in the different regions of the thymus, but in all regions they retain epithelial characteristics, more immediately obvious in the fetal thymus or after the loss of thymic lymphocytes following X-irradiation or in early stages after thymic grafting. The cells form flattened, incomplete sheets which extend as a continuous layer on the surface of the thymus, beneath the connective tissue capsule (**6**.151B) and trabeculae of the lobules, and ensheathe the branches of the thymic blood vessels. The medullary framework is formed by a system of incomplete, fairly tightly packed, anastomosing sheets of similar, but larger, eosinophilic epithelial cells. The cortex is permeated by a more reticular arrangement of epithelial cells which are more highly branched with larger interstices, but the cells of this reticulum retain intercellular contacts, both with each other, and with the sheets of medullary cells and those lining the capsule and covering the blood vessels.

These epithelial cells are almost certainly derived from the original endodermal thymic diverticulum from the embryonic pharynx. They have oval, pale-staining nuclei and intercellular junctions with many typical desmosomes having well-marked tonofibrillae. Many of the cells show, in addition to typical lysosomes and vacuoles, other prominent smooth, round or oval, electron-dense granules in their cytoplasm, possibly secretory in nature. The epithelial cells of the cortex are quite uniform in their fine structure, whereas those of the medulla take many varied forms, some having very irregular surfaces and others being quite smooth in outline; the cytoplasm of the medullary cells is likewise pleomorphic, suggesting that these cells may have different functions. Where their surfaces abut on the trabecular, capsular or perivascular reticulin fibres, they are separated from them by a prominent basement membrane. Many investigators believed that the outer layers of the epithelial framework formed a partial *haemothymic barrier* which limited access of certain circulating materials, including antigens, into the tissue spaces of the thymus. Paradoxically, the interruptions in the cell layers are sufficient to allow passage of nutrients, metabolites and a small population of stem cells and lymphocytes to and from the thymus. (However, *see* further more recent discussion below concerning the blood–thymus barrier.) The thymic epithelial reticular cells are not members of the reticuloendothelial system in that they do not phagocytose diffusable dyes or particulate matter used in investigation of this system, but they are widely thought to be the origin of the thymic humoral factors important in lymphopoiesis and in the establishment of immunologic competence (*vide infra*). Macrophages are also present throughout the thymus, although

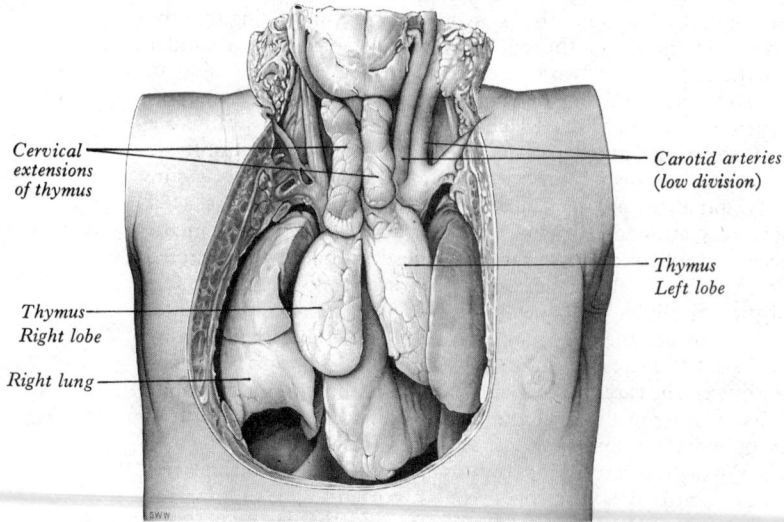

Cervical extensions of thymus

Carotid arteries (low division)

Thymus Left lobe

Thymus Right lobe

Right lung

6.151A A dissection to display the neonatal thymus.

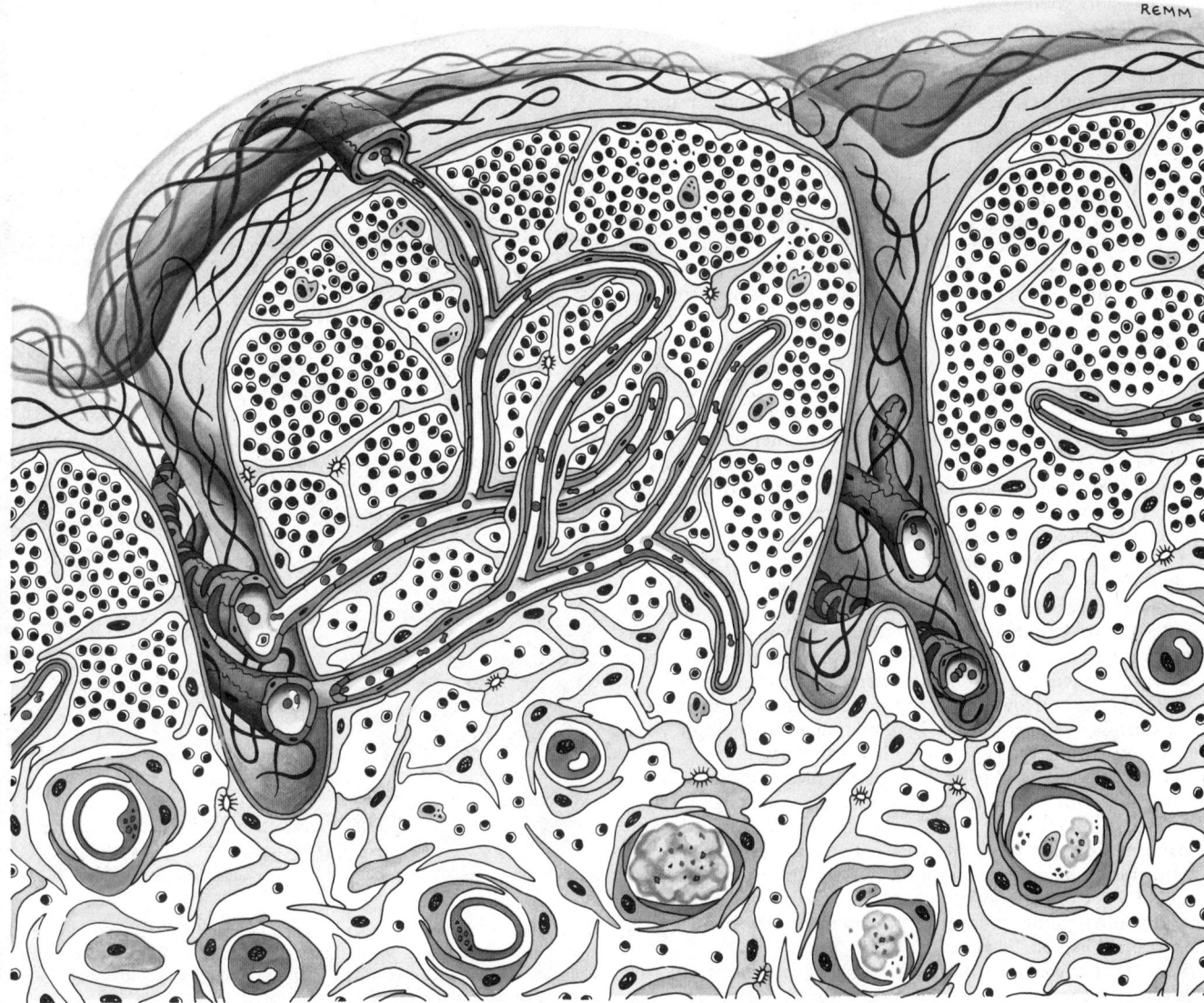

6.151B A scheme of thymic structure—the various elements are not drawn to scale to enable representation in a single diagram. Note the lobular outlines, capsule, delicate interlobular septa, cortical lympho-cytes, the epithelial 'reticular' cells and their junctions, and the medullary corpuscles of Hassall showing a graded series of increasing maturity. Note transcortical circulation. See text for further discussion.

they are much fewer than in many other lymphoid organs (e.g. the spleen).

The cortical cells. Densely packed small lymphocytes pre-dominate in the cortex, occupying the interstices of the epithelial reticulum, which they largely obscure, and are so numerous that they form about 90 per cent by weight, of the total organ. Two *cortical zones* are present, an outer region containing numerous large lymphoblasts, some in stages of mitosis, and an inner cortex occupied mainly by small lymphocytes. Macrophages occur in both zones (Hwang *et al.* 1974). Thymic lymphocytes are quite distinct from those of the peripheral lymphoid organs and circulating pool of lymphocytes (*vide infra*). They arise mainly or entirely in the outer cortex by division of stem cells which have originally migrated to the thymus from the bone marrow. The need for a continual supply of stem cells, however, is quite low despite the lymphopoietic level of the thymic cortex which is the highest of any lymphoid tissue. This is because the stem cells divide asymmetrically, one resultant cell retaining the potential for further divisions, the other, being a potential lymphocyte. The majority of the thymic lymphocytes have a short life span (3–5 days) and undergo degeneration still within the thymus, whilst the remainder leave the thymus to form a part of the circulating pool of lymphocytes which are not at first immunologically mature.

However, in the perivascular regions and near the capsule and trabeculae are found a number of true macrophages which are large, vacuolated and largely concerned with phagocytosis of the products of the massive lympholysis occurring in the cortex. The fate of these phagocytes may also be intimately associated with the natural history of further interesting specializations of the thymic medulla, i.e. the concentric corpuscles. In a similar position, plasma cells are also present, although their numbers are never large. These cells are thought to enter the medulla from the bloodstream through the post-capillary venules, rather than differentiating within the thymus itself, and their functional significance is uncertain, although their presence has intriguing possibilities for interactions between B- and T-lymphocytes in the thymus (Hwang *et al.* 1974).

The concentric corpuscles of Hassall are first formed in fetal life and are then continuously formed throughout the life of the thymus. A medullary epithelial cell enlarges, develops an intense eosinophilia and then follows a train of further degenerative changes. Progressive vacuolation of its cytoplasm and fragmen-tation of its nucleus is followed by engulfment of the nuclear débris by the vacuoles which then become confluent. Further epithelial cells increase in size, approach the degenerating central cell, and become arranged as a series of eosinophilic concentric cellular lamellae around the central hyalinizing mass. As the corpuscle grows, further degenerate epithelial cells are added to the central mass together, it has been claimed, with effete macrophages which carry the products of cortical lympholysis. They pass between the epithelial cell coverings of the corpuscle to

781

reach its centre, where they undergo hyalinization or are made soluble. The corpuscles vary from 30 to 100 μm in diameter and they increase in size and number during periods of intense lympholysis and during thymic involution.

Thymic involution. From the age of 5–6 years until puberty there is a slow, relative, reduction in the number of cortical lymphocytes even though the organ continues to grow. After puberty there is a more massive, progressive reduction in both cortical lymphocytes and epithelial cells, and these are replaced by fibro-adipose tissue, and at the same time the number of Hassall's corpuscles increases. There is an absolute reduction in size and weight so that in the later years of life the gland is identified only with difficulty, and consists of nests of corpuscles surrounded by a thin coat of fibrous and adipose tissue which entraps a few degenerating lymphocytes.

Vessels and nerves. The *arteries* are derived from the internal thoracic and inferior thyroid arteries, and the branches of these main vessels pass into the depths of the interlobular trabeculae where they give off a series of arterioles which enter the substance of the lobule. Here, they course along the junction of cortex and medulla where they remain at least partly ensheathed by epithelial cells, macrophages, pericytes, and other perivascular elements. From the arterioles a series of radial capillaries pass into the cortex and less regular vessels into the medulla. The returning post-capillary venules of the cortico-medullary junction present a thickened endothelium similar to that of the post-capillary venules of the lymph node (p. 770), across which lymphocyte passage is a frequent event. Some venules retrace their arteriolar path and converge to form principal veins again in the interlobular septa. However, a proportion of venous return occurs via a capsular venous plexus. In the latter case, therefore, the circulation from arteriole to radial capillary continues its *transcortical centrifugal* path as radial venules and small veins, finally draining into the capsular veins (**6.151B**). These contrasting microvascular routes may be important in the cell dynamics of the thymus (Blau 1976). The main *veins* end in the left brachiocephalic, internal thoracic and inferior thyroid veins; frequently one or more veins emerge from the medial side of each lobe and join together to form a common trunk which opens into the left brachiocephalic vein.

The blood–thymus barrier is a concept arising from the observation that materials injected intravascularly do not penetrate to the extravascular spaces of the thymic cortex where the lymphocytes are proliferating, suggesting that the walls of thymic blood vessels may act as efficient barriers to the passage of antigens and other materials into the thymic tissue which would thus be an *immunologically sequestered site*, allowing the untroubled differentiation of lymphocytes (*vide infra*). This subject was investigated by Raviola and Karnovsky (1972) who found that in rodents, electron-opaque tracers passed readily through the intercellular junctions of the medullary vascular endothelia, particularly those of the post-capillary venules, and diffused extensively into the medullary tissue. However, the endothelium of the cortical capillaries forms a much better barrier, and the small amount of tracer which passed through by vesicular transport was immediately phagocytosed by the perivascular macrophages. Presumably macrophage action also prevents diffusion of particulate substances from the medulla into the cortex. The lymph vessels are described on p. 800. The *nerves* are small and are derived from the sympathetic (cervicothoracic ganglion or ansa subclavia) and the vagus; branches from the phrenic and descendens cervicalis are distributed mainly to the capsule of the thymus.

FUNCTIONS OF THE THYMUS

Our understanding of thymic function is by no means complete and still under intensive investigation, but in recent years dramatic advances have occurred. The evidences upon which recent views of thymic function are based stem from a variety of experimental approaches. These include studies of the effects of thymectomy or radiation-induced thymic destruction in neonatal and mature animals, and the study of certain mutant mouse

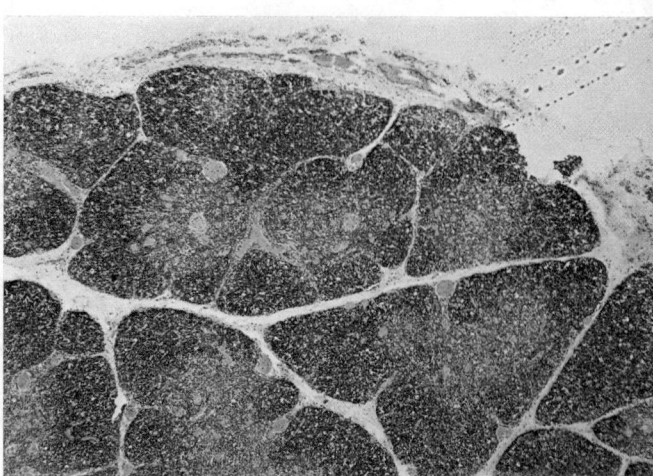

A

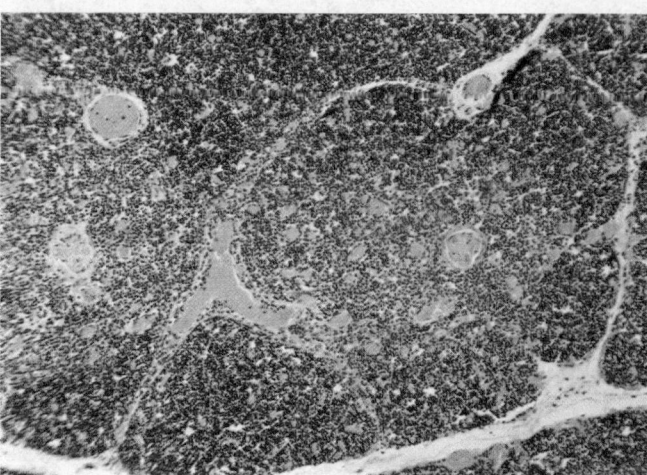

B

6.152A and B Survey photographs of a neonatal human thymus stained with haematoxylin and eosin. The general lobular architecture is seen, each lobule contains a relatively pale medullary core surrounded by a densely cellular, dark, heavily stained cortex. (**6.**152 A and B, 153 and 154 are from a specimen kindly prepared and provided by Dr. R. O. Weller of the Department of Pathology, Guy's Hospital Medical School.)

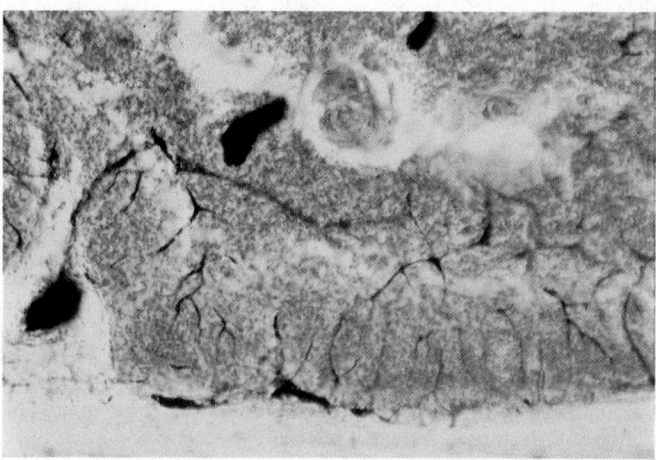

C

6.152C Section of a thymus gland (guinea pig) after injection of the arterial system with indian ink. Note the presence of fine capillaries in the cortex, some of them forming recurrent loops, and others passing through to the capsular drainage. (Kindly provided by Dr. N. Blau, Guy's Hospital Medical School.)

strains, e.g. *nude*, in which the thymus is vestigial or absent in homozygous individuals. Attempts have been made to compensate for the results of thymectomy by administering cell suspensions from a variety of lymphoid tissues or bone marrow; also with normal or irradiated thymic tissue, either as a free graft, or surrounded by a cell-impermeable barrier; finally, with cell-free thymic extracts.

The stem cell sources concerned in the repopulation of cell-depleted lymphoid tissues or thymic grafts, have been followed using chromosome-marker or autoradiographic techniques. Further studies have followed the restoration of immune competence in animals rendered immunologically tolerant to certain antigens, by exposing them to these antigens, in the early neonatal period.

A full review of this complex and highly important field of experimentation lies outside the scope of this volume. What follows is a short summary of some of the principal concepts which are emerging.

The essential functions of the thymus are the differentiation of lymphocytes into immunologically competent T-cells (thymocytes), and the maintenance of an effective pool of these cells in the circulation and peripheral lymphoid organs, able to react to a wide and increasing array of antigenic stimuli, and to modify the activities of B-lymphocytes. Thus, during the neonatal period and in early postnatal life, thymic function is essential for the normal development of the lymphoid tissues. Thymectomy at this time leads to a progressive, eventually fatal disease, with hypoplasia of the peripheral lymphoid organs, wasting, and infective processes which reflect an inability to mount a cellular immune response. By puberty, the main lymphoid tissues are fully developed, and thymectomy in the mature animal is less dramatic, but in time, there follows a reduction in the effectiveness of responses to unfamiliar antigens.

A second major function is the *control of lymphopoiesis* in the *central* and *peripheral lymphoid organs*, including the thymus itself, both with respect to the rate of cell division and the differentiation of lymphocytes into immunologically competent cells. These effects appear to be mediated by humoral factors, e.g. *lymphopoietin* probably secreted by the epithelial cells of the thymus, which stimulates lymphoblast division.

A third possible role is the production of hormones not necessarily directly connected with lymphocyte proliferation (*thymosins*), which have been considered by some authors to be concerned with neuromuscular junction maintenance amongst other activities. However, recent evidence suggests that these hormones may be similar to, or identical with, the factors controlling lymphocyte production, and that neuromuscular disease of the myasthenia gravis type is basically an auto-immune condition (*vide infra*).

Origin and fate of thymic lymphocytes: The cortical lymphocytes of the thymus arise from stem cells of the bone marrow, and in earlier fetal life, the spleen, which migrate in the circulatory system to the thymus, passing into its tissues through the post-capillary venules. Such cells become lymphoblasts in the outer cortex, and proliferate there extensively, but about 90 per cent of the progeny die within the thymus in the course of a few days; the remainder apparently become immunologically competent T-lymphocytes, and can pass back into the circulatory system to join the circulating lymphocyte pool, and populate the thymus-dependent areas of lymph nodes and other peripheral lymphoid organs.

The observed death of lymphocytes in the thymus is an intriguing problem, and its massive scale suggests that it may have considerable biological importance. Some authors have suggested that it may be part of a mechanism enabling lymphocytes to distinguish between *foreign* antigens and chemicals indigenous within the normal body. In this theory (the *clonal hypothesis* of McFarlane Burnett) it has been proposed that within the population of differentiating lymphocytes, individual stem cells develop the ability to detect and respond immunologically to a restricted range of antigens, or even to a single antigen, so that within the body, any antigen being introduced to the tissues, e.g. the cell walls of bacteria or the coats of viruses, will already be matched by a group of lymphocytes (a *clone*) which has descended from an appropriate stem cell, and which is able to inactivate the antigen. On meeting the matching antigen, this clone would itself proliferate and so provide an overwhelming defensive force. In the case of B-lymphocytes this initial differentiation into separate classes of stem cells presumably occurs in the bone marrow, to

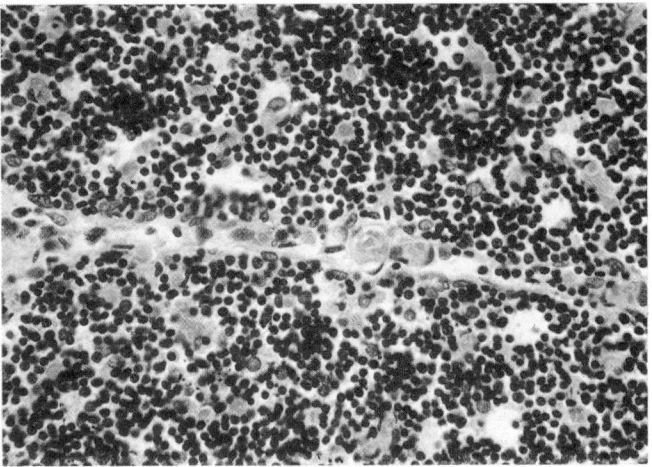

6.153   Neonatal thymus. A cortical field showing a delicate interlobular connective tissue septum, densely packed cortical lymphocytes and scattered profiles of larger, pale, eosinophilic, epithelial, reticular cells.

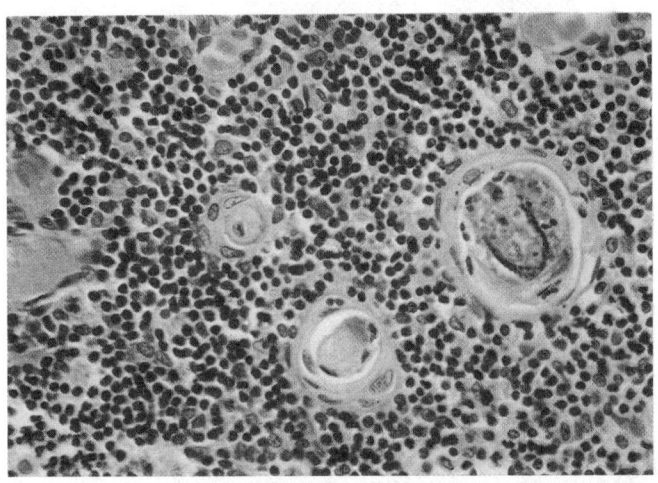

6.154   Neonatal thymus. A medullary field showing three concentric corpuscles of Hassall of varying degrees of maturity, surrounded by many closely packed lymphocytes and a number of reticular cells.

provide an extensive battery of cells capable of producing an enormous range of antibodies. With the T-lymphocytes, differentiation of stem cells committing them to become T-cells takes place in the thymus, and results in cells collectively able to recognize a wide range of antigens, and to respond to them by cytotoxic cell killing, 'arming' of macrophages, triggering of large mononuclear cells ('killer cells'), 'helper' activities with respect to B-lymphocyte reactions, and so forth.

Recently it has been shown that T-lymphocytes require 'priming' by macrophages which have ingested antigenic material, adding significance to the numerous macrophages of the thymus. However, it is also known that T-lymphocytes respond by killing mainly, or perhaps only, the cells of the body bearing normal HLA antigens together with a 'foreign' or altered antigen (e.g. of viral origin), so the process of 'education' in the thymus must presumably include the recognition of 'self' by surviving T-lymphocytes; clearly the picture is more complex than we yet understand. However, this having been said, it is obviously important that lymphocytes can distinguish foreign antigens from the substances natural to the body, and certain disease conditions are known where this distinction breaks down and the lymphocytes attack the normal tissues (*auto-immune diseases*). In the clonal selection theory it is further proposed that the great many types of stem lymphocyte which the body produces include those directed against the body, but that these are eliminated to leave only those directed against external antigens. The massive destruction of lymphocytes observed in the thymus may be part of this 'weeding-out' process. The actual site of destruction is a matter of some debate; certainly the numerous macrophages of the thymus ingest many dead or dying lymphocytes, and it has been suggested that such macrophages subsequently enter the Hassall's corpuscles and degenerate.

The cortical thymic lymphoid tissue of normal mammals does not form germinal centres, but these are formed in a number of types of auto-immune disease, perhaps reflecting a defect in the normal role of the thymus of recognizing and destroying auto-allergic cells.

Applied Anatomy. Castration or adrenalectomy delays involution of the thymus, whereas hypertrophy of the suprarenal cortex (or injection of cortisone) or the injection of androgenic hormones causes atrophy of the thymus.

In young children, a large thymus may, in rare cases, press on the trachea and cause attacks of respiratory stridor, or noisy and difficult breathing. Tumours of the thymus may press on the trachea, the oesophagus, and the large veins at the root of the neck, causing hoarseness; cough, dysphagia and cyanosis.

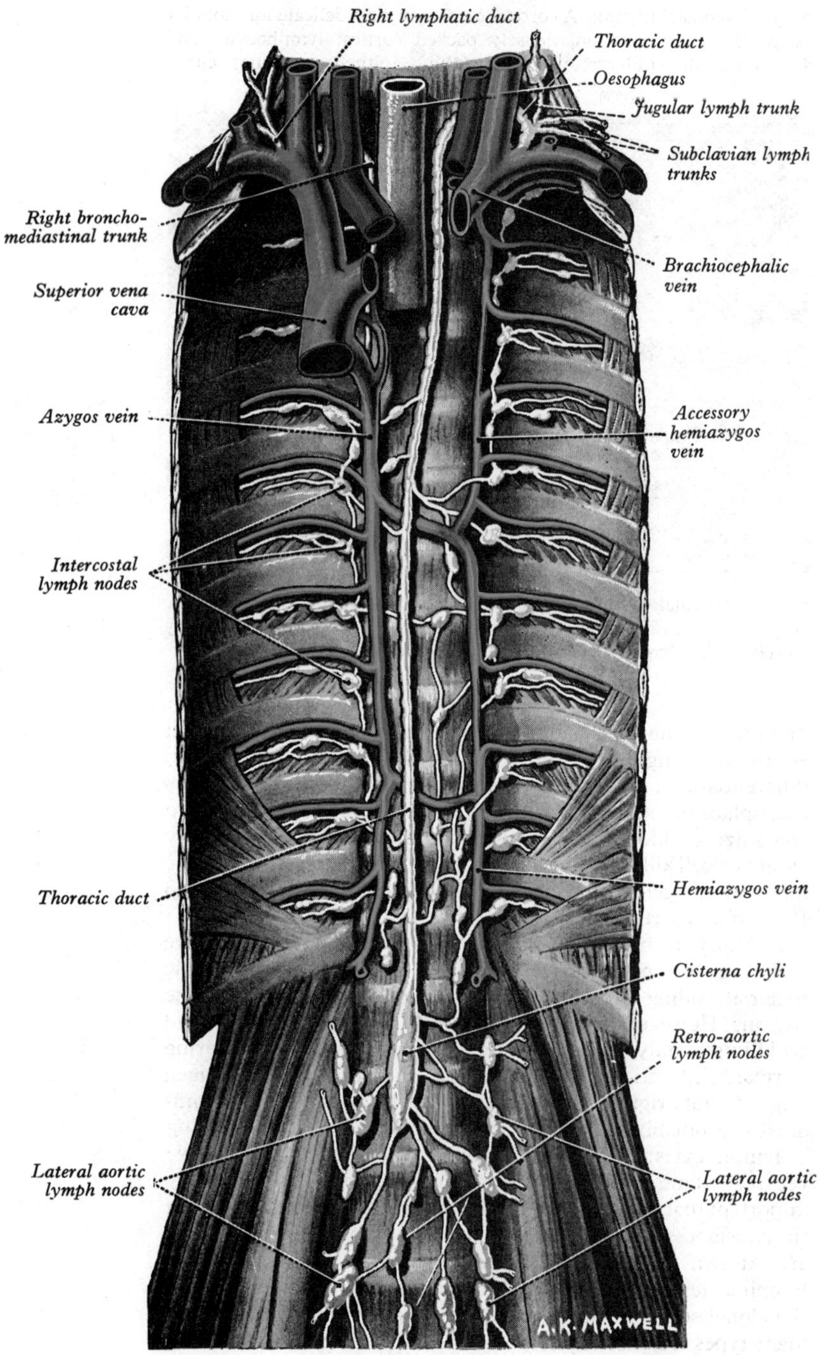

Myasthenia gravis is a chronic auto-immune disease in adults (Castleman 1966) characterized by diminution in the power of contraction of certain voluntary muscles after repetitive contractions have been carried out. The muscles most commonly involved are the levator palpebrae superioris (leading to ptosis or drooping of the upper eyelid) and the orbital muscles (leading to diplopia or double vision). Other muscles (facial, masticatory, neck and limb muscles) may be involved, including in severe cases, the respiratory muscles. In many of these cases some abnormality of the thymus has been observed, such as a general enlargement (hyperplasia) or a tumour (thymoma, composed of epithelial cells derived from the reticular cells or possibly from the lymphocytes), and in a number of these cases transient or more lasting improvement may follow removal of the thymus. Although there may be more than one condition bearing somewhat similar symptoms, it is now thought that myasthenia gravis is an auto-immune disease in which the acetylcholine receptor proteins of the neuromuscular junctions are attacked by the body's defensive systems; the coincidence of this disease with thymic disorders may reflect the part which the latter organ plays in modulating the immune responses of the body as a whole.

Topography of the Lymph Nodes and Vessels

The thoracic duct (**6**.155, 156) conveys the greater part of the lymph back into the circulating blood. It is the common trunk of all the lymph vessels of the body, except those of the right side of the head, neck, and thoracic wall, the right upper limb, right lung, right side of the heart, and part of the convex surface of the liver. In the adult it varies in length from 38 to 45 cm and, including the cisterna chyli, extends from the second lumbar vertebra to the root of the neck. The duct proper begins at the upper end of the cisterna chyli (p. 785) near the lower border of the twelfth thoracic vertebra and enters the thorax through the aortic opening of the diaphragm. It then ascends through the posterior mediastinum with the aorta on its left and the azygos vein on its right. In this part of its course the vertebral column and anterior longitudinal ligament, the right aortic intercostal arteries, and the terminal parts of the hemiazygos and accessory hemiazygos veins are behind it. The diaphragm and the oesophagus are in front of it, but a recess of the right pleural cavity may intervene between the duct and the oesophagus. Opposite the fifth thoracic vertebra the thoracic duct inclines to the left, enters the superior mediastinum, and ascends to the thoracic inlet along the left side of the oesophagus. Crossed anteriorly by the aortic arch, it then lies behind the commencement of the left subclavian artery in close contact with the mediastinal pleura of the left side.

Passing into the neck, it arches laterally at the level of the transverse process of the seventh cervical vertebra, the arch rising about 3 or 4 cm above the clavicle. Here the duct runs anterior to the vertebral artery and vein, the sympathetic trunk and the thyrocervical trunk or its branches. It also passes in front of the phrenic nerve and the medial border of scalenus anterior, but is separated from these by the prevertebral fascia. In this situation it is placed behind the left common carotid artery, vagus nerve, and internal jugular vein. Finally, it descends in front of the first part of the left subclavian artery and ends by opening into the junction of the left subclavian and internal jugular veins (**6**.157). Sometimes the duct joins the adjacent parts of the subclavian or internal jugular veins and sometimes it breaks up into a variable number of smaller vessels before terminating (*see* also below). At its commencement it is about 5 mm in diameter, but it diminishes considerably in calibre in the middle of the thorax, and in about half the cases is again slightly dilated just before its termination. It is slightly sinuous, constricted at intervals and appears varicose. Not infrequently it divides in the middle of its course into two unequal vessels which soon unite, or into several branches which form a plexiform interlacement. It occasionally divides at its upper levels into two branches, right and left, the left ending in the usual manner, while the right opens into the right subclavian vein with the right lymphatic duct. The thoracic duct has several valves, and these correspond to sites of exposure to pressure. At

6.155 The thoracic and right lymphatic ducts. The accessory hemiazygos vein is crossing the median plane lower and the hemiazygos higher than usual. The reader should also note the comments concerning the more common course of the azygos vein made in the caption to illustration **6**.132 and on p. 755.

Labels on figure:
Right lymphatic duct
Thoracic duct
Oesophagus
Jugular lymph trunk
Subclavian lymph trunks
Right broncho-mediastinal trunk
Brachiocephalic vein
Superior vena cava
Azygos vein
Accessory hemiazygos vein
Intercostal lymph nodes
Thoracic duct
Hemiazygos vein
Cisterna chyli
Retro-aortic lymph nodes
Lateral aortic lymph nodes
Lateral aortic lymph nodes
A.K. MAXWELL

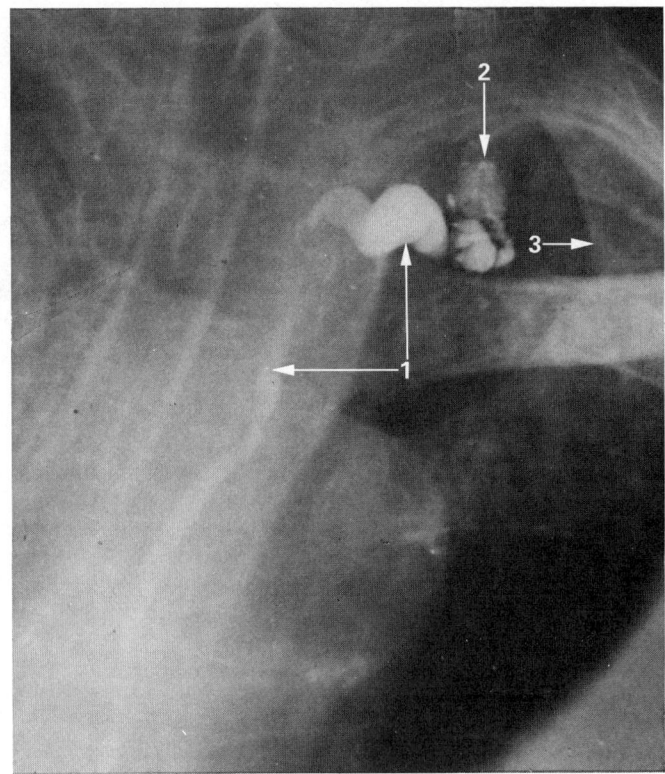

6.156 Lymphangiogram showing the upper part of the thoracic duct. There has been some filling of a cervical lymph node. 1. Thoracic duct. 2. Lower deep cervical lymph node. 3. First rib. (Kindly supplied by Professor J. B. Kinmonth.)

its entry into the venous system a bicuspid valve faces into the vein to prevent reflux of blood into the duct. After death, however, blood regurgitates into the terminal part of the duct, which therefore looks like a vein. Kinnaert (1973) has collected the recorded accounts of 480 dissections of the cervical termination of the thoracic duct adding 49 further cases. He gives full statistics of all previous studies. In 0 to 4·5 per cent, in different studies, no thoracic duct was found on the left. Multiple terminal openings were noted frequently (in 10 to 40 per cent, according to different observers). In Kinnaert's series the duct was multiple near its termination in 66 per cent of dissections, but in only 21 per cent were actual terminal openings multiple. Patterns of termination varied greatly in different studies, but in the two largest series, those of Jdanov (1959) and Kinnaert (1973), the percentages for sites of termination were respectively 48 and 36 per cent into the internal jugular vein, 9 and 17 per cent into the subclavian, 35 and 34 per cent into the jugulo-subclavian junction. Termination in the left innominate vein occurred in 8 per cent of Jdanov's series, but in none of Kinnaert's.

The cisterna chyli (**6**.155) is commonly described as a saccular dilatation in the lymphatic route from the abdomen and lower limbs, extending for 5 to 7 cm in front of the first and second lumbar vertebral bodies, immediately to the right of the abdominal aorta (but see also below). The upper two right lumbar arteries and the right lumbar azygos vein (p. 754), when it is present, are between the cisterna chyli and the vertebral column. Anteriorly the cisterna is covered by the medial edge of the right crus of the diaphragm. It is joined by right and left lumbar and intestinal lymphatic trunks. The *lumbar trunks* are formed from the efferent vessels of the lumbar (lateral aortic) lymph nodes; they receive the lymph from the lower limbs, the walls and viscera of the pelvis, the kidneys and suprarenal glands, the testes or ovaries, and the deep lymph vessels of the greater part of the abdominal wall. The *intestinal trunks* comprise the large lymph vessels which receive the lymph from the stomach, intestine, pancreas and spleen, and from the lower and anterior part of the liver. They join the cisterna chyli, lumbar lymph trunks or thoracic duct.

Although this description of the cisterna as a single, simple receptaculum is repeated by most textbooks, a large statistical study of the formation of the cisterna and thoracic duct in mankind appears to be lacking. Anson (1963) depicted a number of variations, and in many of these no cisterna was shown; when present it appeared to be often multilocular. No statistics of incidence were given. Kubik (personal communication, 1978) has observed a true cisterna in only 14 of 70 dissections. In 7 of these a single cisterna existed, while there were two cisternae in 5 specimens and three in 3 of them. In the remaining 56 dissections no clearly distinguishable cisterna was observed; in about half of these the collecting channels formed a thoracic duct without cisternal formation; in the other half intercalated lymph nodes (as also depicted by Anson 1963) simulated cisternae by their external appearance. If these findings are confirmed, the customary description of the cisterna chyli, and the concept of it as a usual feature of the human lymphatic system, will require modification.

Tributaries. At its start the thoracic duct receives a *descending trunk* from the intercostal lymph nodes of the lower six or seven intercostal spaces, on each side. In the thorax the thoracic duct is joined by two *ascending trunks* draining the upper lumbar lymph nodes which pierces the crura of the diaphragm. It also receives efferents from the posterior mediastinal lymph nodes and from the intercostal lymph nodes of the upper six spaces of the left side. In the neck it is joined usually by the *left jugular trunk* from the left side of the head and neck, and the *left subclavian trunk* from the left upper limb, but these vessels may open independently into the internal jugular or subclavian vein, respectively; sometimes it is joined by the *left bronchomediastinal trunk*, but this usually opens independently into the junction of left subclavian and internal jugular veins. The tributaries of the thoracic duct are usually described as possessing terminal valves which prevent reflux of lymph. Sapin and Boryiak (1974) studied the behaviour of injected masses in the thoracic ducts of 180 cadavers, and they observed that reflux into several groups of mediastinal and paravertebral lymph nodes was a usual occurrence, a fact of great importance in the spread of metastases.

The right lymphatic duct (**6**.155, 157), about 1 cm in length, courses along the medial border of the scalenus anterior at the root of the neck, and ends by opening into the junction of the right subclavian and internal jugular veins. At its orifice two semilunar valves prevent the passage of venous blood into the duct.

Tributaries. The duct receives lymph from the right half of the head and neck collected into the *right jugular trunk*, from the right upper limb through the *right subclavian trunk*; from the right side of the thorax, right lung, right side of the heart, and part of the convex surface of the liver, through the *right bronchomediastinal trunk*. A single right lymph duct occurs in about one-fifth of subjects; more frequently its three tributaries open separately at the union of the two veins (**6**.157).

Applied Anatomy. Blockage of the thoracic duct, usually the result of malignant disease, leads to no signs or symptoms; hence alternative routes for the passage of lymph to the venous circulation must exist. Widespread blockage of multiple pathways

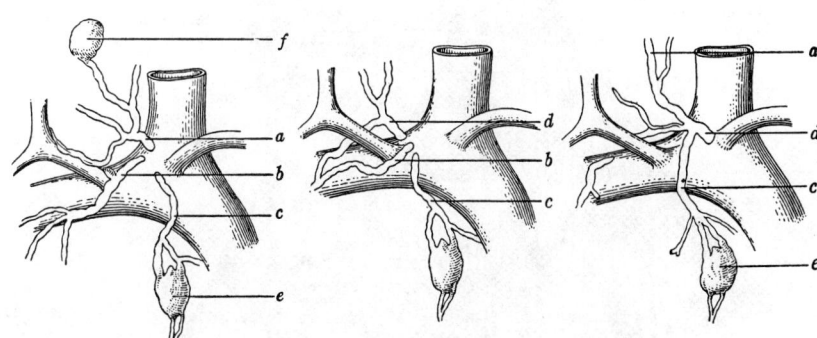

6.157 Variations in the terminal lymph trunks of the right side. (After Poirier and Charpy.) *a*. Jugular trunk. *b*. Subclavian trunk. *c*. Bronchomediastinal trunk. *d*. Right lymphatic duct. *e*. Lymph node of parasternal chain. *f*. Lymph node of deep cervical chain.

may occasionally occur, e.g. in filariasis, and cause effusions into the peritoneal or pleural cavities; nevertheless such effusions are more likely to result from rupture of the duct from trauma, surgical accidents, or a malignant disease.

The Lymphatic Drainage of the Head and Neck

The lymph nodes of the head and neck comprise a terminal group and a number of intermediary, outlying groups. The terminal group is closely associated with the carotid sheath and is named the *deep cervical group*. All the lymph vessels of the head and neck drain into this group, either directly from the tissues themselves, or indirectly after passing through one of the outlying groups. The efferents of the deep cervical lymph nodes form the *jugular trunk*, which, on the right side, may end in the junction of the internal jugular and subclavian veins or may join the right lymphatic duct; on the left side, it usually enters the thoracic duct, although it may join either the internal jugular or the subclavian vein.

The Deep Cervical Lymph Nodes

These nodes lie along the carotid sheath and may be divided into superior and inferior groups.

The superior deep cervical lymph nodes (6.158) adjoin the upper part of the internal jugular. Most of them are deep to sternocleidomastoid, but a few extend beyond it. One group, consisting of one large and several small nodes, is in the triangular region bounded by the posterior belly of digastric, the facial vein, and the internal jugular vein, and is termed the *jugulodigastric group*; it is concerned particularly with lymph drainage of the tongue.

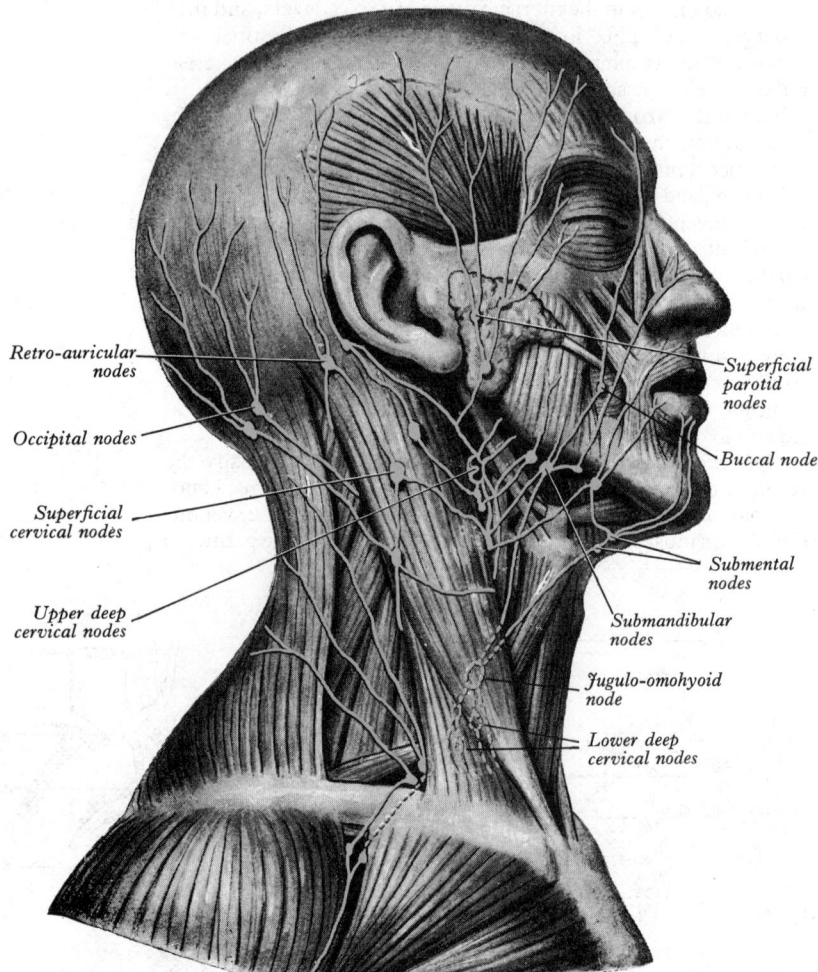

Retro-auricular nodes
Occipital nodes
Superficial cervical nodes
Upper deep cervical nodes
Superficial parotid nodes
Buccal node
Submental nodes
Submandibular nodes
Jugulo-omohyoid node
Lower deep cervical nodes

Efferents from the upper deep cervical lymph nodes pass to the lower deep cervical group and direct to the jugular trunk.

The inferior deep cervical lymph nodes are partly deep to the lower part of sternocleidomastoid, and extend also into the subclavian triangle, and are closely related to the brachial plexus and the subclavian vessels. One node of this group lies on, or just above, the intermediate tendon of the omohyoid; it is called the *jugulo-omohyoid lymph node* and is concerned especially with the drainage of the tongue (p. 788). The efferents from the lower deep cervical lymph nodes join the jugular trunk.

In their lymphatic drainage the tissues of the head and neck, like other regions, can conveniently be considered in two groups: (*a*) the superficial tissues and (*b*) the deeper structures, including the viscera.

Lymphatic Drainage of the Superficial Tissues of the Head and Neck (6.158)

Most of the superficial tissues are drained by vessels which go first to locally sited groups of lymph nodes and the efferents from these drain into the deep cervical lymph nodes. Some, however, may pass directly to the deep cervical nodes.

Regional groups concerned in the drainage of the superficial tissues are:

In the head.	*In the neck.*
Occipital.	Submandibular.
Retro-auricular (mastoid).	Submental.
Parotid.	Anterior cervical.
Buccal (facial).	Superficial cervical.

Lymphatic drainage of the scalp and ear. Lymph vessels from the frontal region just above the root of the nose drain into the submandibular group (6.158) and are considered with the face (*vide infra*).

The vessels from the rest of the forehead, the temporal region, and the upper half of the lateral surface of the auricle and the anterior wall of the external acoustic meatus drain into the *superficial parotid lymph nodes*. These are immediately in front of the tragus, on, or deep to, the fascia of the parotid gland. These nodes also drain the lateral lymph vessels from the eyelids and the skin over the zygomatic bone. Their efferent vessels pass to the upper deep cervical lymph nodes.

A strip of scalp above the auricle, the upper half of the cranial surface and margin of the auricle, and the posterior wall of the external acoustic meatus are drained by vessels which pass, some to the upper deep cervical lymph nodes and others to the retro-auricular group.

The *retro-auricular lymph nodes* (6.158) are superficial to the mastoid attachment of the sternocleidomastoid, and deep to the auricularis posterior. Their efferents pass to the upper deep cervical lymph nodes.

The lobule of the auricle, the floor of the meatus, and the skin over the angle of the jaw and the lower part of the parotid region are drained by vessels which may pass to the superficial cervical lymph nodes or the upper deep cervical group. The *superficial cervical lymph nodes* extend along the external jugular vein superficial to sternocleidomastoid. Some of the efferents from this group pass round the anterior border of sternocleidomastoid to reach the upper deep cervical lymph nodes; others follow the external jugular vein and join the lower deep cervical lymph nodes in the subclavian triangle.

The occipital region of the scalp is drained partly to the occipital group of lymph nodes, and partly by a trunk along the posterior border of sternocleidomastoid ending in the lower deep cervical lymph nodes. The *occipital lymph nodes* are in the upper angle of the posterior triangle and superficial to the attachment of trapezius.

Lymphatic drainage of the face. The lymph vessels draining the eyelids and conjunctiva commence in a superficial plexus beneath the skin and a deep plexus in front of and behind the tarsi; these communicate with one another, and medial and lateral sets of vessels arise from them. The lateral vessels drain the whole thickness of the upper lid with the exception of the skin over its medial part; they also drain the whole thickness of the lateral half of the lower lid and all the ocular conjunctiva. They

6.158   The superficial lymph nodes and lymph vessels of the head and neck.

pass laterally from the lateral commissure to end in the superficial parotid lymph nodes and also in the *deep parotid lymph nodes*, which are embedded in the parotid salivary gland. The deep parotid lymph nodes receive vessels also from the middle ear (p. 787). The lymph vessels of the medial set drain the skin over the medial part of the upper eyelid, the whole thickness of the medial half of the lower lid, and the caruncula lacrimalis. Following the course of the facial vein, they terminate in the submandibular group of lymph nodes.

The *submandibular lymph nodes* (6.158) are internal to deep cervical fascia in the submandibular triangle. There are usually three, one at the anterior end of the submandibular gland, one in front of and another behind the facial artery where it reaches the mandible. Additional members of this group are often embedded in the gland or on its deep surface. The submandibular nodes receive afferents from a wide area, including vessels from the submental, buccal and lingual groups of lymph nodes; their efferents pass to upper and lower deep cervical lymph nodes.

The external nose, cheeks, upper lip, and lateral parts of the lower lip send their lymph to the submandibular nodes. These vessels may have along their course a few *buccal lymph nodes* lying in relation to the facial vein. The mucous membrane covering the oral surfaces of the lips and cheeks is drained by vessels which also end in the submandibular nodes. Lymph from the lateral part of the cheek drains into the parotid nodes, whilst that from the skin over the root of the nose and the central part of the forehead just above this drains partly into the parotid nodes, and partly, along the facial lymphatics, into the submandibular nodes.

The central part of the lower lip, the floor of the mouth and the tip of the tongue drain to the submental group of nodes. The *submental lymph nodes* are placed on the mylohyoid between the anterior bellies of the two digastrics (6.159). They receive afferents from both sides of the median plane, some of the vessels decussating over the symphysis of the mandible; their efferents pass to the submandibular and jugulo-omohyoid nodes.

Lymphatic drainage of the neck. Many of the vessels draining the superficial tissues of the neck pass round the borders of sternocleidomastoid to either the upper or the lower deep cervical lymph nodes. Some, however, pass from the region over the upper part of the sternocleidomastoid and the posterior triangle of the neck to the superficial cervical and occipital lymph nodes. The lymph from the upper part of the anterior triangle of the neck is drained into the submandibular and submental nodes; vessels from the skin of the anterior part of the neck below the hyoid bone pass to the *anterior cervical lymph nodes*, which are associated with the anterior jugular veins. The efferents from this group pass to the deep cervical lymph nodes of both sides of the neck: they pass to infrahyoid, prelaryngeal, and pretracheal nodes (*vide infra*). One of the anterior cervical lymph nodes often occupies the suprasternal space (p. 537).

Lymphatic Drainage of the Deeper Tissues of the Head and Neck

The deeper tissues of the head and neck drain to the deep cervical lymph nodes either directly, or indirectly through one of the outlying groups. In addition to the outlying groups which have been considered already, the following also are concerned with drainage of the deeper tissues, viz.: retropharyngeal, paratracheal, lingual, infrahyoid, prelaryngeal and pretracheal.

The *retropharyngeal lymph nodes* comprise a median and two lateral groups, the former near the midline, the latter in front of the lateral mass of the atlas along the lateral border of the longus capitis. They are all between the fascia covering the pharynx and the prevertebral fascia. They receive afferents from the nasopharynx, the auditory tube and the atlanto-occipital and atlanto-axial joints. Their efferents pass to the upper deep cervical lymph nodes.

The *paratracheal lymph nodes* flank the trachea and oesophagus, along the recurrent laryngeal nerves. Efferents from both groups pass to the corresponding deep cervical nodes.

The *infrahyoid, prelaryngeal and pretracheal lymph nodes* are deep to the investing layer of the deep cervical fascia. Some of their afferents are from the anterior cervical nodes and their efferents join the deep cervical groups. The nodes of the

infrahyoid group are on the front of the thyrohyoid membrane; those of the prelaryngeal group lie on the conus elasticus and its cricovocal membrane; the pretracheal lymph nodes are in front of the trachea in close relation with the inferior thyroid veins.

The *lingual lymph nodes* are small and inconstant. They are situated on the external surface of hyoglossus and between the genioglossi.

Lymphatic drainage of the nasal cavity, nasopharynx and middle ear. The lymphatics of the nasal cavity can be injected from the subarachnoid space, through communications along the course of the olfactory nerves. The lymph vessels from the anterior part of the nasal cavity pass superficially to join those draining the skin covering the external nose and end in the submandibular nodes. The remainder of the nasal cavity, paranasal sinuses, nasopharynx, and pharyngeal end of the auditory tube drain to the upper deep cervical nodes, either directly or through the retropharyngeal lymph nodes. It is probable that the posterior part of the floor of the nasal cavity is drained by vessels which enter the parotid groups of nodes.

The lymph vessels of the mucous lining of the tympanic cavity and mastoid antrum pass to the parotid or upper deep cervical lymph nodes; those from the tympanic end of the auditory tube probably end in the deep cervical nodes. Lymphatic channels have been identified in the submucosa of the tube by injection and electron microscopy (Pulec *et al.* 1975).

Lymphatic drainage of the larynx, trachea, and thyroid gland. The lymph vessels of the larynx comprise upper and lower groups; on the lateral wall the two systems are distinct, the line of division being the vocal fold; the two systems anastomose on the posterior wall. The vessels of the upper set pierce the thyrohyoid membrane, and accompany the superior laryngeal vessels to end in the upper deep cervical nodes. The vessels of the lower set either pass between the cricoid cartilage and the first tracheal ring

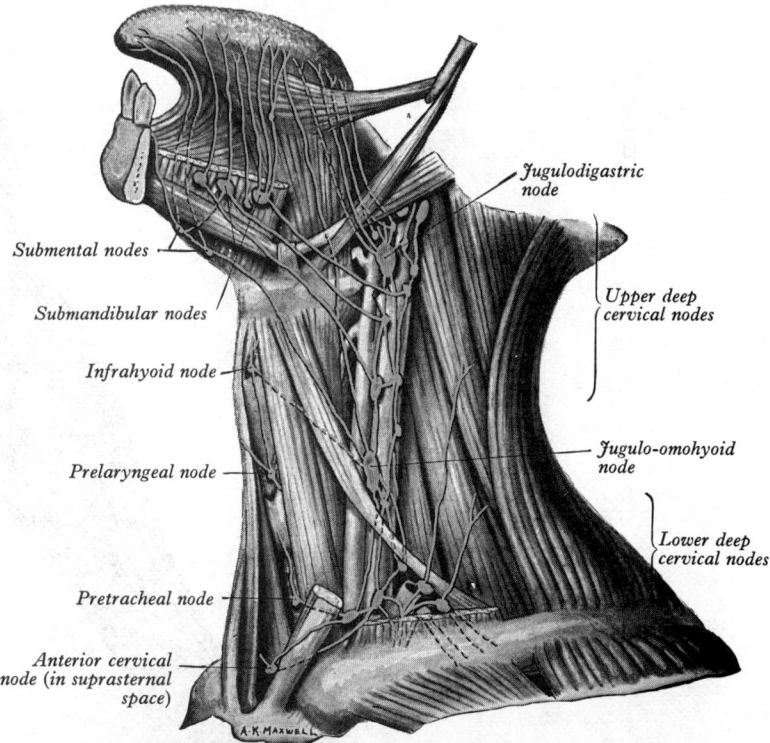

Submental nodes

Submandibular nodes

Infrahyoid node

Prelaryngeal node

Pretracheal node

Anterior cervical node (in suprasternal space)

Jugulodigastric node

Upper deep cervical nodes

Jugulo-omohyoid node

Lower deep cervical nodes

A·K·MAXWELL

6.159 The lymphatic drainage of the tongue. (After Jamieson and Dobson.) Removal of sternocleidomastoid has exposed the whole chain of deep cervical lymph nodes.

to go directly to the lower deep cervical lymph nodes, or, piercing the cricovocal membrane, pass to the pretracheal and pre-laryngeal groups which drain into the deep cervical lymph nodes.

There is a dense network of lymph vessels in the wall of the trachea. The cervical part of the trachea is drained to the pretracheal and paratracheal nodes, or directly to the nodes of the lower deep cervical group.

The lymph vessels of the thyroid gland communicate freely with the tracheal plexus. They pass to the prelaryngeal nodes just above the isthmus and to the pretracheal and paratracheal nodes; some may drain into the brachiocephalic lymph nodes, which are related to the thymus in the superior mediastinum. Laterally, the gland is drained by vessels which accompany the superior thyroid vein to the deep cervical lymph nodes. Some lymph vessels from the thyroid gland may enter directly into the thoracic duct (p. 784).

Lymphatic drainage of the mouth, teeth, tonsil and tongue. The *mouth*. The vessels of the gums end in the submandibular lymph nodes; those of the hard palate are continuous in front with those of the upper gum, but run backwards to pierce the superior constrictor, and end in the upper deep cervical and retropharyngeal lymph nodes; those of the soft palate pass backwards and laterally and end partly in the retropharyngeal, and partly in the upper deep cervical nodes. The vessels of the anterior part of the floor of the mouth go to the inferior nodes of the upper deep cervical group, either directly or indirectly through the submental lymph nodes; the vessels from the rest of the floor of the mouth pass to the submandibular and upper deep cervical nodes.

The *teeth*. Lymph vessels pass to the submandibular and deep cervical lymph nodes.

The *tonsil*. The vessels from the tonsil pierce the buccopharyngeal fascia and the superior constrictor and pass between the stylohyoid and the internal jugular vein to the upper deep cervical lymph nodes. Most of them end in the jugulodigastric nodes; occasionally one or two additional vessels run to small nodes on the lateral side of the internal jugular vein, deep or medial to the sternocleidomastoid.

The *tongue* (**6.**159). The lymphatic plexus in the mucous membrane of the tongue is continuous with the intramuscular plexus. The part of the tongue in front of the vallate papillae is drained into marginal and central lymph vessels. The part of the tongue behind the vallate papillae drains into a set of dorsal lymph vessels.

(1) *Marginal vessels*. The vessels from the tip of the tongue and the region of the frenulum descend under the mucous membrane and end in widely distributed lymph nodes.

(*a*) Vessels pierce the origin of the mylohyoid in contact with the periosteum of the mandible; one or two of these vessels enter the submental lymph nodes, and one descends over the hyoid bone to the jugulo-omohyoid node. It should be noted that vessels arising in the plexus on one side of the tongue may cross under the frenulum and end in the lymph nodes of the opposite side, and

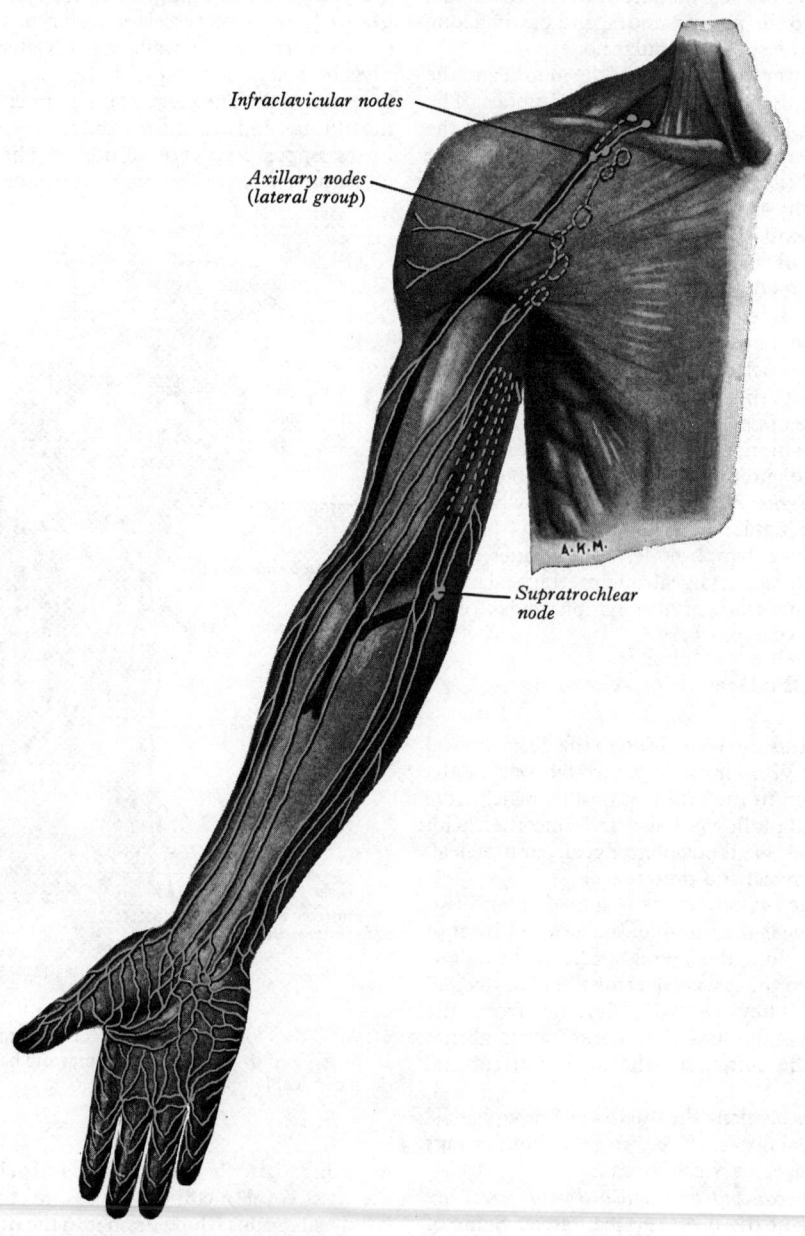

Infraclavicular nodes

Axillary nodes
(lateral group)

A.K.M.

Supratrochlear
node

6.160 The lymphatic drainage of the superficial tissues of the upper limb. Anterior aspect. Semi-diagrammatic.

that the efferent vessels of the submental nodes, which are placed in or near the median plane, pass impartially to either side.

(*b*) Some vessels pierce the origin of the mylohyoid, and enter the anterior or middle submandibular lymph node.

(*c*) Some vessels pass deeply under the sublingual salivary gland and, accompanying the companion vein of the hypoglossal nerve, end in the jugulodigastric nodes. One vessel often descends over or deep to the intermediate tendon of the digastric to reach the jugulo-omohyoid lymph node.

Some of the lymph vessels from the lateral margin of the tongue pass over the sublingual salivary gland, pierce the mylohyoid, and end in the submandibular nodes; others pass under the sublingual salivary gland and end in the jugulodigastric or jugulo-omohyoid nodes.

The vessels from the posterior part of the margin of the tongue make their way through the pharyngeal wall to the jugulodigastric lymph nodes.

(2) *Central vessels.* There is no clear line of demarcation between areas on the surface of the tongue draining into the marginal or central vessels. The central lymph vessels descend in the median plane between the genioglossi. Some turn laterally through the muscles, but the majority appear between their free borders and diverge to the right or left, i.e. the vessels from one side of the tongue may run to the lymph nodes of the opposite side. They follow the lingual blood vessels, and end in the deep cervical lymph nodes, especially in the jugulodigastric and jugulo-omohyoid lymph nodes. Some pierce the mylohyoid and enter the submandibular nodes.

(3) *Dorsal vessels.* The vessels draining the area of the vallate papillae, and the part of the tongue behind these papillae, run downwards and backwards—those near the median plane may divide and run to both sides. They turn laterally to join the marginal vessels, and all pierce the pharyngeal wall, passing in front of or behind the external carotid artery, to reach the jugulodigastric and jugulo-omohyoid lymph nodes, or the nodes between them. One vessel may descend behind the hyoid bone, perforate the thyrohyoid membrane, and end in the jugulo-omohyoid node.

Lymphatic drainage of the pharynx and cervical part of the oesophagus. The collecting vessels from the pharynx and cervical part of the oesophagus pass to the deep cervical nodes either directly, or indirectly through the retropharyngeal or paratracheal nodes. From the region of the epiglottis the lymph vessels run to the infrahyoid nodes.

The Lymphatic Drainage of the Upper Limbs

All the lymph vessels of the upper limb drain into a terminal group of lymph nodes in the axilla, either directly, or after passing through an outlying group of lymph nodes. Vessels internal to the deep fascia follow the principal vascular and neurovascular bundles, while the superficial vessels, except in the hand and on the back of the forearm, converge towards the superficial veins, which they closely accompany.

The axillary lymph nodes (**6.**161, 163), which are the terminal group for the whole of the upper limb, are of large size; they vary from twenty to thirty in number, and may be divided into five groups, which are not, however, sharply demarcated:

(1) A *lateral group* (**6.**160) of from four to six lymph nodes lies medial to, and behind, the axillary vein, the afferents of this group drain the whole limb except for the part whose lymph vessels accompany the cephalic vein. The efferent vesssels pass partly to the central and apical groups of axillary nodes, and partly to the lower deep cervical nodes.

(2) An *anterior* or *pectoral group* of four or five lymph nodes lies along the lower border of the pectoralis minor, in relation to the lateral thoracic vessels. Its afferents drain the skin and muscles of the anterior and lateral walls of the body, above the level of the umbilicus, and the central and lateral parts of the mammary gland (p. 790); its efferents pass partly to the central, and partly to the apical groups of axillary lymph nodes.

(3) A *posterior* or *subscapular group* of six or seven lymph nodes extends along the lower margin of the posterior wall of the axilla in the course of the subscapular vessels. The afferents of this group drain the skin and muscles of the lower part of the back of the neck and of the dorsal aspect of the trunk, as low down as the iliac crest; their efferents pass to the apical and to the central group of axillary lymph nodes.

(4) A *central group* of three of four large lymph nodes is embedded in the fat of the axilla. It has no special area of drainage, but it receives afferents from all the preceding groups of axillary lymph nodes: its efferents pass to the apical group.

(5) An *apical group* of six to twelve lymph nodes is situated partly posterior to the upper portion of the pectoralis minor and partly above its upper border, and extends upwards into the apex of the axilla along the medial side of the axillary vein. The only *direct* territorial afferents of this group are those which accompany the cephalic vein and one or two which drain the upper and peripheral part of the mammary gland, but it receives the efferents of all the other axillary lymph nodes. The efferent vessels of this group unite to form the *subclavian trunk*, which opens either directly into the junction of the internal jugular and subclavian veins or into the jugular lymphatic trunk; on the left side it may end in the thoracic duct. A few efferents from the apical group usually pass to the lower deep cervical lymph nodes.

The outlying lymph nodes in the upper limb are few in number. They comprise (1) the supratrochlear group, (2) the infra-clavicular group (both interposed on the path of the superficial vessels), and (3) a few isolated lymph nodes occasionally present along the course of the principal blood vessels of the arm and forearm.

(1) The *supratrochlear lymph nodes*, merely one or two in number, are superficial to the deep fascia above the medial epicondyle of the humerus medial to the basilic vein. Their efferents accompany the basilic vein and join the deep lymph vessels.

(2) The *infraclavicular lymph nodes*, one or two in number, are found beside the cephalic vein, between pectoralis major and deltoid, immediately below the clavicle. Their efferents pass through the clavipectoral fascia to the apical group of axillary nodes, or, more rarely, they may pass anterior to the clavicle to reach the lower deep cervical (supraclavicular) group.

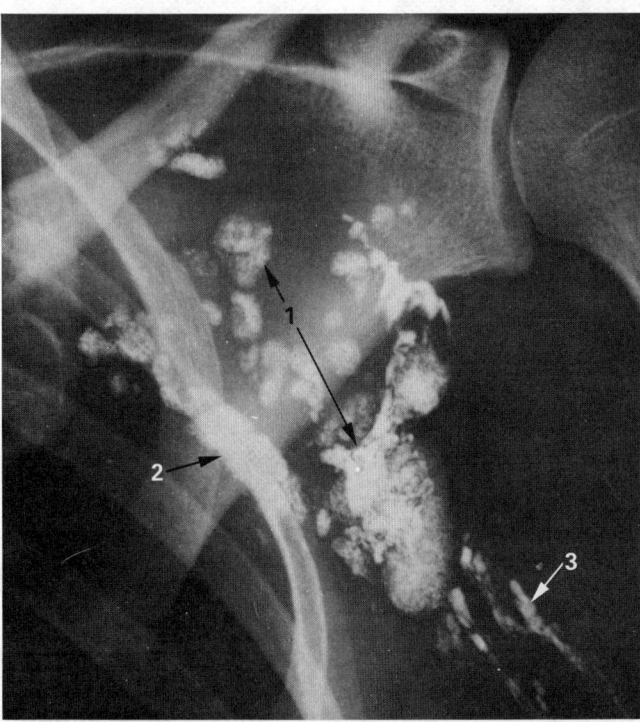

6.161 Normal axillary lymphangiogram, four days after injection of ultrafluid lipiodol into a lymph vessel on the dorsum of the hand. 1. Lateral group of lymph nodes. 2. Pectoral group of lymph nodes. 3. Brachial lymph vessels. (Kindly supplied by Professor J. B. Kinmonth.)

(3) Isolated lymph nodes, small in size, sometimes occur in the forearm along the radial, ulnar and interosseous vessels, in the cubital fossa near the bifurcation of the brachial artery, and in the arm along the medial side of the brachial vessels.

Lymphatic drainage of the superficial tissues of the upper limb. The superficial lymph vessels being in the lymphatic plexuses in the skin.

In the hand, the meshes of the plexus are much finer on the palmar than on the dorsal surface. The digital plexuses are drained by vessels which run along the borders of the digits to the webs, where they join vessels from the distal part of the palm and then pass back to the dorsal aspect of the hand (**6**.160, 162). The rest of the palm is drained by vessels which pass proximally towards the wrist, medially to join vessels along the ulnar border of the hand, and laterally to joint those on the thumb. Several collecting vessels from the central part of the palmar plexus unite to form a trunk which winds round the metacarpal bone of the index finger to join the dorsal vessels from the same finger and from the thumb.

In the forearm and arm, the superficial vessels mostly run with the superficial veins. Collecting vessels from the hand pass into the forearm on all aspects of the wrist.

The vessels on the dorsum, after running upwards parallel with one another, finally pass successively round the borders of the limb to join vessels on the front (**6**.162). The vessels on the front of the wrist pass up the forearm parallel with the median vein of the forearm to the cubital region. Beyond this point they follow the medial border of the biceps and pierce the deep fascia at the anterior

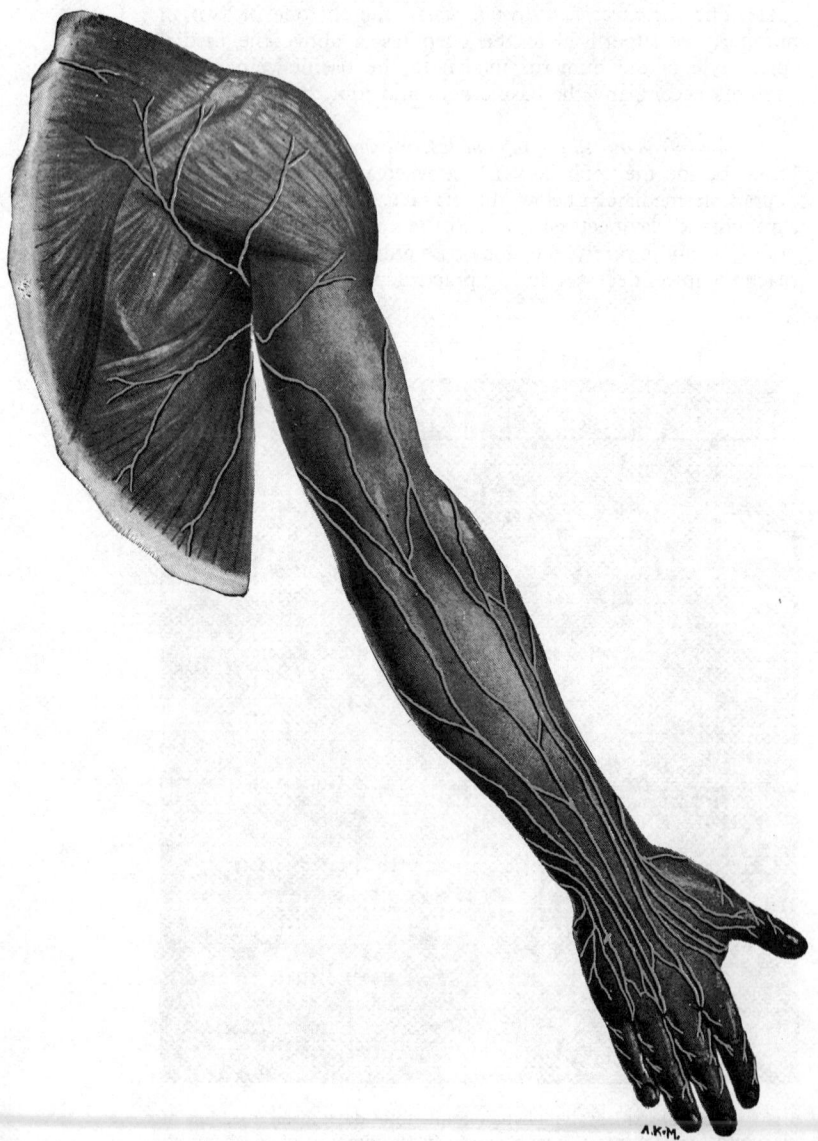

axillary fold, to end in the lateral group of axillary lymph nodes

The lateral vessels become associated in the forearm with the cephalic vein. They follow it to the level of the insertion of deltoid where most incline medially to enter the lateral group of axillar nodes; a few continue with the cephalic vein and end in th infraclavicular nodes. These lateral lymph vessels receive thos which wind round the lateral border of the limb from th posterior surface.

The medial vessels follow the basilic vein in the forearm. Jus above the elbow some of them end in the supratrochlear lymp nodes. The efferents from these, together with the medial vessel which have bypassed the supratrochlear nodes, pierce the dee fascia with the basilic vein, and end in the lateral group of axillar nodes or the deep lymph vessels. The vessels which wind roun the medial border of the limb from the back join those of th medial set.

The collecting vessels from the front and back of the deltoi region pass respectively round the anterior and posterior axillar folds to end in the axillary nodes. The skin over the scapular regio drains either into the subscapular group of axillary nodes, o following the transverse cervical vessels into the lower dee cervical lymph nodes.

Lymphatic drainage of the deep tissues of the uppe limb. The deep lymph vessels follow the main neurovascula bundles (radial, ulnar, interosseous and brachial) and end in th lateral axillary nodes. They are less numerous than the superficia vessels, with which they communicate at intervals. Along thei course a few lymph nodes occur.

The muscles of the scapular region are drained by vessels which pass mainly to the subscapular group of axillary lymph nodes Lymph from the pectoral muscles is drained into the pectoral central and apical groups.

Lymphatic drainage of the breast (**6**.163). The lymph vessels of the mammary gland originate in a plexus in the interlobular connective tissue and in the walls of the lactiferous ducts. This communicates with the overlying cutaneous lymphatic plexus especially around the nipple in the *subareola plexus.* It is also claimed to communicate with the plexus o minute vessels on the deep fascia underlying the breast; this anastomosis may play no part in the normal lymphatic pathway from the breast nor in the early spread of cancer from the organ (Turner-Warwick 1959). It offers an alternative route when the usual pathways are obstructed.

Efferent lymph vessels arising directly from the breast pass around the anterior border of the axilla, pierce the axillary fascia and end in the pectoral lymph nodes. Some lymph may pass directly to the subscapular nodes. From the upper part of the breast a few lymph vessels pass to the apical nodes; this pathway may be interrupted in the infraclavicular nodes or in small and inconstant interpectoral nodes. The axillary lymph nodes commonly receive more than 75 per cent of the lymph from the breast. Most of the remainder enters the parasternal nodes which receive lymph vessels from both the medial and lateral halves of the organ; these vessels accompany the perforating branches of the internal thoracic artery. Lymph vessels from the breast occasionally follow the lateral cutaneous branches of the posterior intercostal arteries to the intercostal lymph nodes.

The drainage of the skin over the breast is dealt with on p. 798.

Applied Anatomy. Enlargement of the axillary lymph nodes is frequent in malignant disease and also in infective processes implicating the upper part of the back and shoulder, the front of the chest and mammary gland, the upper part of the front and side of the abdomen, or the upper limb. In operations for carcinoma of the breast the pectoralis major and the deep fascia overlying it and the surrounding muscles are usually removed on account of the wide ramifications of the lymphatics. The axillary lymph nodes, the sternocostal head of pectoralis major and, frequently, the pectoralis minor are also taken away, in an endeavour to ensure, as far as possible, the complete removal of affected lymph vessels and nodes. (Some surgeons, however, now advocate less radical extirpation.) It should be noted that, in disease, the pathological blockage of some of the normal lymph channels can result in a flow of lymph and consequently the spread of disease along other than the usual pathways.

6.162 The lymphatic drainage of the superficial tissues of the upper limb. Posterior aspect. Semi-diagrammatic.

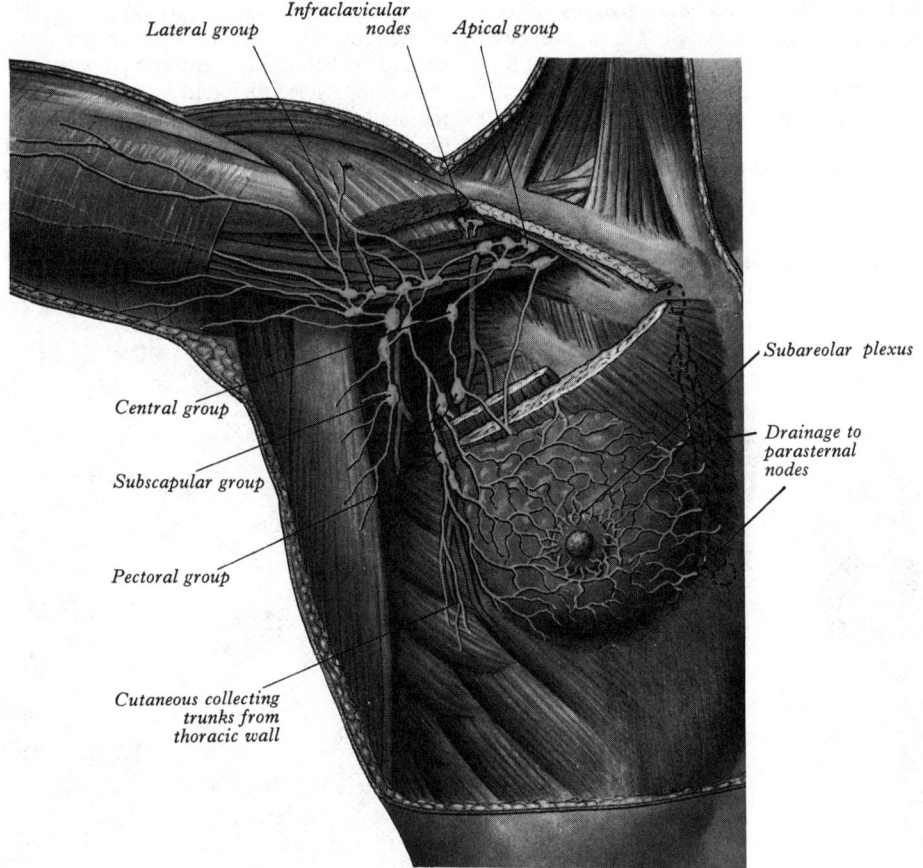

6.163   The lymph vessels of the mammary gland and the axillary lymph nodes.

The Lymphatic Drainage of the Lower Limbs

Most of the lymph from the lower limb traverses a terminal group of lymph nodes in the groin. Before reaching these terminal nodes the lymph may pass through outlying, intermediary nodes, which, however, are less numerous in the lower limb than elsewhere.

The terminal nodes are the inguinal groups, superficial and deep to the deep fascia.

The superficial inguinal lymph nodes (**6.**164, 166, 167) are arranged in upper and lower groups. The *upper group*, usually of five or six nodes, forms a chain immediately below the inguinal ligament. The lateral members of the group receive afferent vessels from the gluteal region and from the adjoining part of the anterior abdominal wall below the umbilicus. The medial members of the group receive the superficial vessels from the external genitalia (including in the female the vagina below the hymen), from the lower part of the anal canal and the perianal region, from the adjoining part of the anterior abdominal wall the umbilicus and uterine lymph vessels which run along the round ligament of the uterus.

The *lower group*, usually four or five in number, is disposed vertically along the terminal part of the great saphenous vein. It receives all the superficial lymph vessels of the lower limb, except those from the back and lateral side of the calf of the leg.

All the superficial inguinal nodes send their efferents to the external iliac lymph nodes, some traversing the femoral canal, others passing in front of, or lateral to the femoral vessels. In addition numerous vessels interconnect the individual nodes.

The deep inguinal lymph nodes vary from one to three, and are on the medial side of the femoral vein. When three are present, the lowest is situated just below the junction of the great saphenous and femoral veins, the middle in the femoral canal, and the highest in the lateral part of the femoral ring. The middle one is the most inconstant, but the highest is also frequently absent.

They receive as afferents the deep lymph vessels which accompany the femoral vessels, the lymph vessels from the glans penis (or glans clitoridis), and a few of the efferents from the superficial inguinal lymph nodes; their efferents pass through the femoral canal to the external iliac lymph nodes.

The outlying, intermediary lymph nodes are few in number and are all deeply placed. Except for a single node, sometimes present on the upper part of the interosseous membrane of the leg in relation with the anterior tibial vessels, they are restricted to a single group found in the popliteal fossa.

The popliteal lymph nodes (**6.**165), small in size and six or seven in number, are embedded in fat in the popliteal fossa. One lies near the termination of the small saphenous vein, and drains the superficial region from which this vein derives its tributaries. Another is between the popliteal artery and the posterior surface of the knee joint; it receives the lymph vessels from the knee joint together with those which accompany the genicular arteries. The remainder lie at the sides of the popliteal vessels, and receive as afferents the trunks which accompany the anterior and posterior tibial blood vessels. The efferents of the popliteal lymph nodes ascend in close relation with the femoral blood vessels to the deep inguinal nodes, but a few may accompany the great saphenous vein and end in the superficial inguinal nodes.

Applied Anatomy. Inflammation and suppuration of the popliteal lymph nodes are most commonly due to a lesion on the lateral side of the heel.

The superficial inguinal lymph nodes frequently become enlarged in diseases implicating the parts from which their lymph vessels originate. Thus in malignant or syphilitic affections of the prepuce and penis, or labia majora, in scrotal carcinoma, in abscess in the perineum, anus and lower part of the vagina, or in similar diseases affecting the skin and superficial structures in those parts, or the infra-umbilical part of the abdominal wall, or the gluteal region, the upper group of lymph nodes is almost invariably enlarged, the lower group being implicated in diseases affecting the lower limb.

Lymphatic drainage of the superficial tissues of the lower limb. The superficial lymph vessels begin in lymphatic plexuses beneath the skin. Collecting vessels leave the foot in a medial set, whose vessels follow the general course of the great saphenous vein, and a lateral set, associated with the small saphenous vein.

The vessels of the *medial group* are larger and more numerous than those of the lateral, and begin on the tibial side of the dorsum of the foot. They ascend, some in front of the medial malleolus and others behind it, and accompany the great saphenous vein to the groin, where they end in the lower group of the superficial inguinal nodes. The vessels of the *lateral group* begin on the fibular side of the dorsum of the foot. Some of them cross the front of the leg to join the vessels of the medial group and so reach the

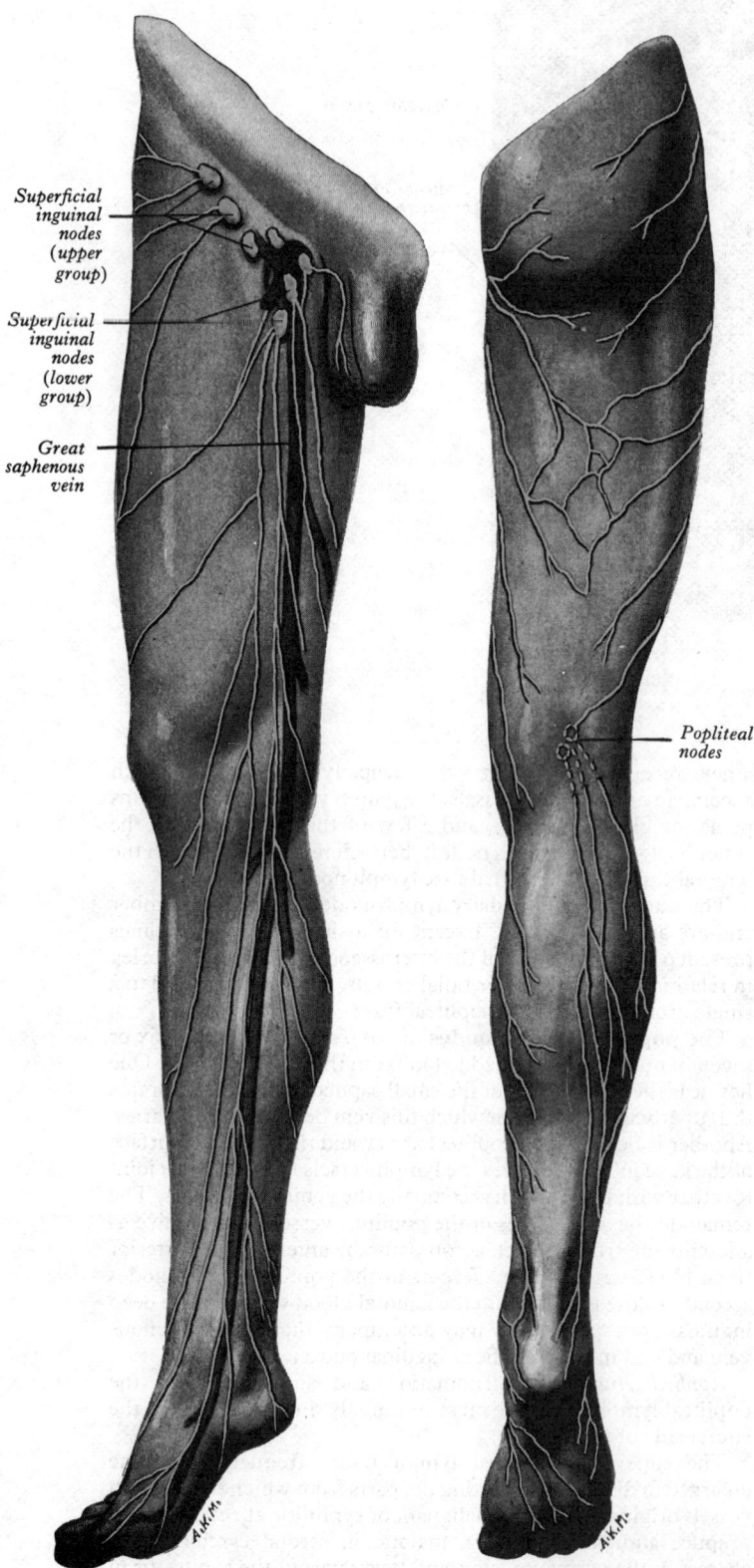

6.164   The lymphatic drainage of the superficial tissues of the lower limb. Anteromedial aspect. Semi-diagrammatic.

6.165   The lymphatic drainage of the superficial tissues of the lower limb. Posterior aspect. Semi-diagrammatic.

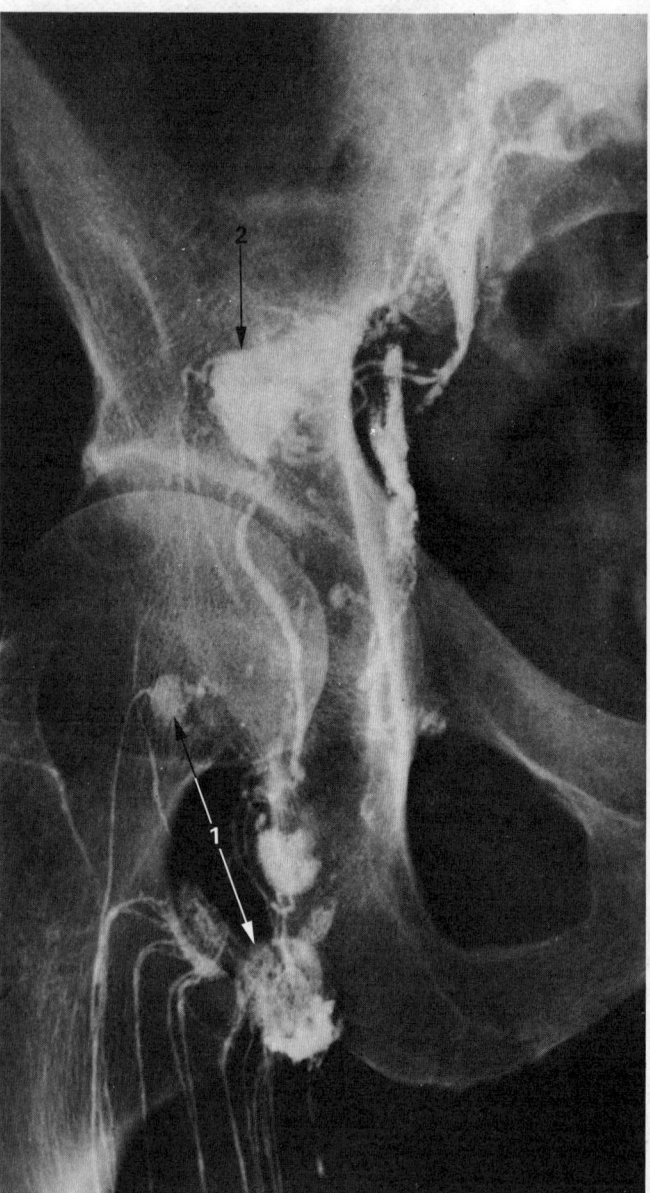

6.166   Lymphangiogram showing the inguinal lymph vessels and nodes taken immediately following injection of ultrafluid lipiodol into a lymph vessel on the dorsum of the foot. 1. Inguinal lymph nodes. 2. External iliac lymph node. (Kindly supplied by Professor J. B. Kinmonth.)

lower group of the superficial inguinal lymph nodes; others accompany the small saphenous vein and end in the popliteal nodes.

The superficial lymph vessels of the buttock pass round to the front of the limb and terminate in the upper group of the superficial inguinal lymph nodes.

Lymphatic drainage of the deeper tissues of the lower limb. The deep lymph vessels accompany the main blood vessels of the limb and so comprise anterior tibial, posterior tibial, peroneal, popliteal and femoral sets. The deep vessels of the foot and leg are interrupted by the popliteal nodes, but those from the thigh pass direct to the deep inguinal nodes.

The deep lymph vessels of the gluteal and ischial regions follow the course of the corresponding blood vessels. Those accompanying the superior gluteal vessels end in a lymph node which lies on the intrapelvic portion of the superior gluteal artery, near the upper border of the greater sciatic foramen. Those following the inferior gluteal vessels traverse one or two small nodes which lie below the piriformis, and end in the internal iliac lymph nodes.

The Lymphatic Drainage of the Abdomen and Pelvis

The lymph from most of the abdominal wall and from all the abdominal viscera (except a small part of the liver) is returned to the bloodstream via the thoracic duct. The lymph vessels run with the corresponding arteries. The lymph nodes comprise a large number of outlying, intermediary groups which are placed along the arteries concerned, and a smaller number of terminal groups which are all in close relation with the abdominal aorta.

The lumbar lymph nodes (6.167) include three principal groups of terminal nodes and one subsidiary group, each of which, though not sharply separated from the others topographically, nevertheless has its own particular area of drainage. These four groups are *pre-aortic*, *lateral aortic* (right and left), and *retro-aortic*. The pre-aortic group drains the viscera supplied by the *ventral branches* of the aorta, i.e. the abdominal part of the alimentary canal and its derivatives. The lateral aortic groups drain the viscera and other structures supplied by the *lateral* and *dorsal branches* of the aorta and receive the efferents from the large, outlying groups associated with the iliac arteries. They therefore constitute the terminal groups for the suprarenal glands, kidneys, ureters, testes, ovaries, pelvic viscera (apart from the gut) and the posterior abdominal wall. The retro-aortic group has no particular area of drainage; although it may have been primarily associated with the drainage of the posterior abdominal wall, it may be regarded as being formed by peripheral members of both lateral aortic groups.

The Pre-aortic Lymph Nodes and their Region of Drainage

The pre-aortic lymph nodes are directly anterior to the abdominal aorta. They receive lymph from the regional, intermediary nodes associated with the subdiaphragmatic part of the alimentary canal, the pancreas, liver and spleen. Their efferents form the *intestinal trunks* which enter the cisterna chyli (p. 785). They are divisible into coeliac, superior mesenteric and inferior mesenteric groups, intimately related to the origins of these arteries.

In the alimentary canal the lymph vessels begin as minute subepithelial radicles, blind at one end but opening at the other into a fine *periglandular plexus*. In the small intestine each villus contains a central lymphatic vessel, known as a *lacteal* from the milky nature of its contents. From the periglandular plexus vessels pierce the muscularis mucosae and join the *submucous plexus*, efferents from which pass through the muscular coat, where they anastomose with or sometimes by-pass vessels draining the muscular coat. The submucous plexus is also joined by vessels from the lymph spaces at the bases of the solitary lymphatic follicles. The lymphatics of the intestinal muscle drain into a plexus of vessels found mainly between the longitudinal and circular coats. The collecting vessels from both sets leave the gut by piercing the muscle and entering larger vessels which follow the arteries in the mesentery.

The collecting vessels from the alimentary canal pass through local groups of lymph nodes before reaching the pre-aortic group.

The coeliac lymph nodes and their areas of drainage. The coeliac lymph nodes lie on the front of the abdominal aorta close to the origin of the coeliac artery. They are the terminal group for the stomach, duodenum, liver, gall bladder, pancreas and spleen, and their afferents are derived from the regional nodes along the branches of the coeliac artery. Of these there are, therefore, three main sets, gastric, hepatic and pancreatico-splenic.

The gastric lymph nodes (6.168, 169) consist of left gastric, right gastro-epiploic and pyloric groups.

The *left gastric lymph nodes* lie along the left gastric artery and are themselves divisible into groups, *upper*, on the stem of the artery, *lower*, associated with the descending branches of the artery along the cardiac half of the lesser curvature of the stomach, in the lesser omentum, and *paracardial*, a chain around the cardiac orifice of the stomach. They receive their afferents from the stomach and the abdominal part of the oesophagus; their efferents pass to the coeliac group of pre-aortic nodes.

The *right gastro-epiploic lymph nodes*, four to seven in number, are in the greater omentum along the pyloric half of the greater curvature in relation with the vessels of the same name. They receive afferents from the stomach; their efferents mostly pass to the pyloric lymph nodes.

Four or five *pyloric lymph nodes* lie close to the bifurcation of the gastroduodenal artery, in the angle between the superior and descending parts of the duodenum; an outlying member of this group is sometimes found above the duodenum on the right gastric artery. They receive afferents from the pyloric part of the stomach, the first part of the duodenum, and the right gastro-epiploic nodes; their efferents end in the coeliac group.

The hepatic lymph nodes (6.168) extend in the lesser omentum along the common hepatic artery and the hepatic artery proper, together with its right and left branches, as well as the bile duct. They vary in number and position, but two are fairly constant: one at the junction of the cystic and common hepatic ducts, known as the *cystic node* (or node of the neck of the gall bladder), the other alongside the upper part of the bile duct is sometimes called the *node of the anterior border of the epiploic foramen*. The hepatic nodes receive afferents from the stomach, duodenum, liver, gall bladder, bile ducts and the pancreas; their efferents pass through the coeliac nodes. Pathologically enlarged hepatic nodes may press on and obstruct the portal vein.

The pancreaticosplenic lymph nodes (6.169) accompany the splenic artery, and are related to the posterior surface and upper border of the pancreas; one or two of this group are in the gastrosplenic ligament. Their afferents are derived from the stomach, spleen, and pancreas; their efferents join the coeliac group.

Lymphatic drainage of the stomach and duodenum. The

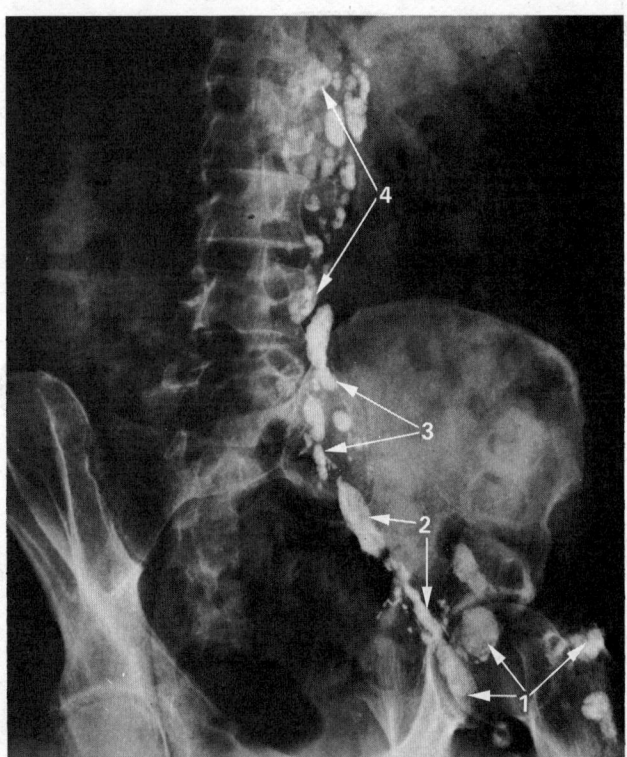

6.167 The lymph nodes in the inguinal region and abdomen seen after the injection of a contrast medium into a lymph vessel on the dorsum of the foot. 1. Inguinal lymph nodes. 2. External iliac lymph nodes. 3. Common iliac lymph nodes. 4. Lateral aortic lymph nodes. (Kindly supplied by Dr. S. J. R. Reynolds.)

lymph vessels of the stomach (**6.**168, 169) are continuous at the cardiac orifice with those of the oesophagus, and at the pylorus with those of the duodenum. They follow the blood vessels for the most part, and may be described in four sets. Vessels of the first set accompany the branches of the left gastric artery, receive tributaries from a large area on both surfaces of the stomach, and terminate in the left gastric lymph nodes. Those of the second set drain the fundus and body of the stomach to the left of a vertical line from the oesophagus; they accompany the short gastric and left gastro-epiploic vessels and end in the pancreaticosplenic

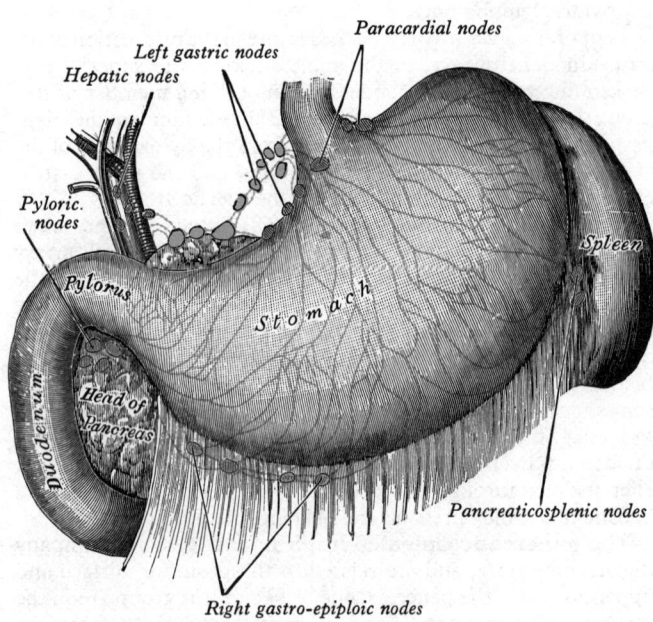

6.168   The lymphatic drainage of the stomach and duodenum. (After Jamieson and Dobson.)

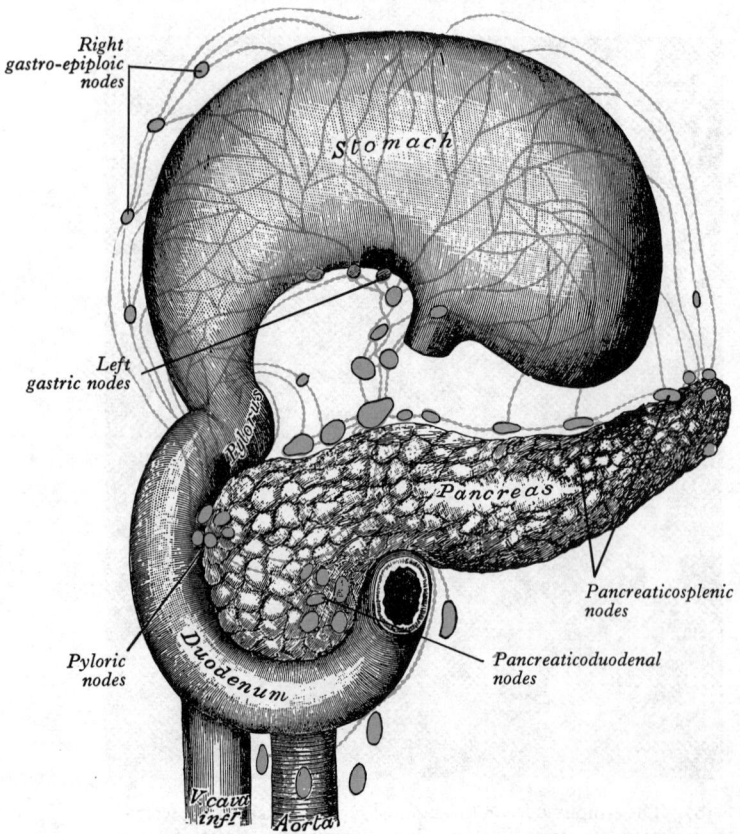

6.169   The lymph vessels and nodes of the stomach, duodenum and pancreas. The stomach has been turned upwards. (After Jamieson and Dobson.)

nodes. Vessels of the third set drain the right part of the greater curvature as far as the pylorus, and end in the right gastro-epiploic lymph nodes, the efferents of which pass to the pyloric group. Those of the fourth set drain the pyloric part of the stomach and pass to the hepatic, pyloric, and left gastric nodes. Although the vessels of these regions communicate, their valves are arranged to direct lymph from the right part of the stomach towards the lesser curvature and from the left towards the greater curvature.

The *lymph vessels of the duodenum* run anteriorly and posteriorly into a series of small *pyloric lymph nodes*, on the anterior and posterior parts of the groove between the head of the pancreas and duodenum. Their efferents run upwards to the hepatic nodes, and downwards to the pre-aortic nodes, around the origin of the superior mesenteric artery.

Lymphatic drainage of the liver. The collecting vessels from the liver are divisible into superficial and deep systems.

The *superficial lymph vessels of the liver* run in the subserous areolar tissue over the whole surface of the organ. They drain in four directions: (1) From the middle part of the posterior surface, from the caudate lobe, from the posterior part of the convex surfaces of both lobes near the attachment of the falciform ligament, and from the posterior part of the inferior surface of the right lobe the vessels accompany the inferior vena cava to nodes round its terminal part. The lymph vessels in the coronary and right triangular ligaments may enter the thoracic duct. (2) The vessels from the remainder of the inferior surface, and from the anterior part of the convex surface of both lobes near the attachment of the falciform ligament, converge on the porta hepatis and end in the hepatic group of lymph nodes. (3) From the posterior part of the left lobe a few vessels pass towards the oesophageal opening in the diaphragm and end in the paracardial lymph nodes. (4) From the rest of the convex surface of the right lobe one or two trunks accompany the inferior phrenic artery across the right crus of the diaphragm to the coeliac nodes.

The *deep lymph vessels* of the liver join to form ascending and descending trunks. The ascending trunks accompany the hepatic veins and pass through the vena caval opening in the diaphragm to end in the nodes round the end of the inferior vena cava. The descending trunks emerge from the porta hepatis and end in the hepatic lymph nodes. (See also p. 1379.)

Lymphatic drainage of the gall bladder and bile ducts. Numerous lymph vessels run from the submucous and subserous plexuses on all surfaces of the gall bladder and cystic duct, those on the upper surface communicating sparsely with vessels in the liver. They pass to the hepatic nodes, especially the cystic node and the node of the anterior border of the epiploic foramen (p. 793). These nodes also collect the lymph vessels of the hepatic ducts and the upper part of the bile duct, while those of the lower part of the bile duct drain into the lower hepatic nodes and the upper pancreaticosplenic nodes.

Lymphatic drainage of the pancreas. The lymph capillaries commence around the acini and the lymph vessels follow the course of the blood vessels; there are no lymph vessels in the interalveolar cell islets. Most of the vessels end in the pancreaticosplenic nodes, but some end in lymph nodes along the pancreaticoduodenal vessels and others in the superior mesenteric group of pre-aortic nodes.

Lymphatic drainage of the spleen. The collecting vessels from the capsule of the spleen end in the pancreaticosplenic lymph nodes.

The superior and inferior mesenteric lymph nodes and their area of drainage. The superior and inferior mesenteric lymph nodes lie on the front of the abdominal aorta close to the points of origin of the arteries so named. They are the terminal groups for the alimentary canal, from the duodenojejunal flexure to the upper part of the anal canal, and collect from the outlying groups, which include the lymph nodes of the mesentery, the ileocolic nodes, nodes of the colon and the pararectal nodes.

The lymph nodes of the mesentery number about one hundred to one hundred and fifty and comprise three series, viz.: one close to the wall of the intestine amongst the terminal twigs of the jejunal and ileal arteries; a second, among the loops and primary branches of the vessels; and a third, along the upper part of the trunk of the superior mesenteric artery. Lymph vessels from the terminal few inches of the ileum follow the ileal branch of the ileocolic artery to the ileocolic nodes.

Applied Anatomy. Enlargement of mesenteric lymph nodes occurs in many diseases of the intestine, and is well marked in typhoid fever, tuberculous ulceration and malignant tumours of the region. The enlarged nodes can often be palpated through the wall of the abdomen.

The ileocolic lymph nodes (6.160, 171) form a *chain* of ten to twenty around the ileocolic artery, but show a tendency to

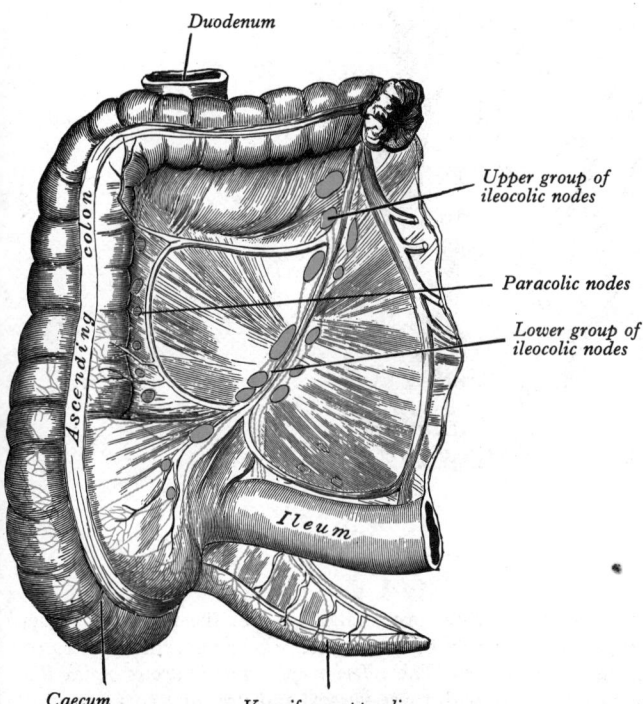

6.170 The lymph vessels and nodes of the caecum and vermiform appendix. Anterior aspect. (After Jamieson and Dobson.)

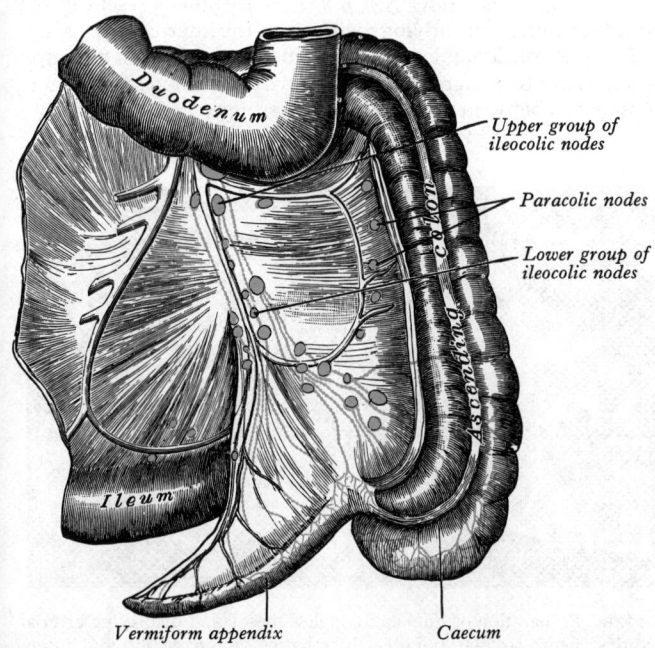

6.171 The lymph vessels and nodes of the caecum and vermiform appendix. Posterior aspect. (After Jamieson and Dobson.)

subdivision into two groups, one near the duodenum and another along the lower part of the artery. Where the vessel divides into its terminal branches, the chain is broken up into several groups, (*a*) *ileal*, in relation to the ileal branch; (*b*) *anterior ileocolic*, usually three in number, in the ileocaecal fold, near the wall of the caecum; (*c*) *posterior ileocolic*, mostly in the angle between the ileum and the colon, but partly behind the caecum at its junction with the ascending colon; (*d*) *appendicular*, a single lymph node in the meso-appendix.

The lymph nodes of the colon are in four groups: (*a*) epicolic, (*b*) paracolic, (*c*) intermediate colic and (*d*) terminal colic.

The *epicolic nodes* are merely minute nodules situated on the wall of the gut, sometimes in the appendices epiploicae. The *paracolic nodes* lie along the medial borders of the ascending and descending colon, and along the mesenteric border of the transverse and sigmoid colon. The *intermediate colic nodes* lie along the right, middle and left colic arteries. The *terminal colic nodes* are near the main trunks of the superior and inferior mesenteric arteries, and are continuous with the corresponding pre-aortic nodes.

The pararectal lymph nodes lie in contact with the muscular coat of the rectum. Their efferents pass to an intermediate group around the superior rectal artery and thence to the nodes at the origin of the inferior mesenteric artery. Others pass to the lymph nodes at the bifurcation of the common iliac artery.

Lymphatic drainage of the jejunum and ileum. The lacteals pass between the layers of the mesentery but, before reaching the superior mesenteric nodes, the lymph passes through the nodes in the mesentery.

Lymphatic drainage of the vermiform appendix and caecum (6.170, 171). The lymph vessels are numerous, since there is a large amount of lymphoid tissue in the walls of these parts of the digestive tube. From the body and tip of the vermiform appendix eight to fifteen vessels ascend in the meso-appendix, one or two being interrupted by nodes which lie in this peritoneal fold. They unite to form three or four vessels, which end in the lower and upper lymph nodes of the ileocolic chain. The vessels from the root of the vermiform appendix and from the caecum comprise anterior and posterior groups. The anterior vessels pass in front of the caecum, to the anterior ileocolic nodes and the upper and lower nodes of the ileocolic chain; the posterior vessels ascend over the back of the caecum to the posterior ileocolic nodes and the lower nodes of the ileocolic chain.

Lymphatic drainage of the colon (6.172). The lymph vessels of the ascending and transverse parts of the colon end in the superior mesenteric lymph nodes, after traversing the groups of nodes along the right and middle colic arteries and their branches. Those of the descending and sigmoid parts of the colon are interrupted by the small nodes on the branches of the left colic arteries, and ultimately end in the pre-aortic lymph nodes around the origin of the inferior mesenteric artery.

Lymphatic drainage of the rectum and anal canal. From the upper half, or more, of the rectum the lymph vessels emerge from its wall and ascend along the superior rectal vessels, through the pararectal lymph nodes to reach the nodes in the lower part of the sigmoid mesocolon and those associated with the inferior mesenteric artery. From the lower half, and from the anal canal *above the mucocutaneous junction* the lymph vessels pass upwards through the wall of the gut and accompany the middle rectal vessels to the internal iliac nodes. Some of these are said to pierce the insertion of levator ani and to gain the ischiorectal fossa where they accompany the inferior rectal and internal pudendal vessels to reach the internal iliac nodes.

The lymph vessels of the anal canal *below the mucocutaneous junction* descend to the anal margin and then pass laterally to reach the most medial of the superficial inguinal lymph nodes.

The Lateral Aortic Lymph Nodes and their Area of Drainage (6.167, 173)

The lateral aortic lymph nodes lie on each side of the abdominal aorta in front of the medial margin of the psoas major, the crus of the diaphragm and the sympathetic trunk. On the right side some

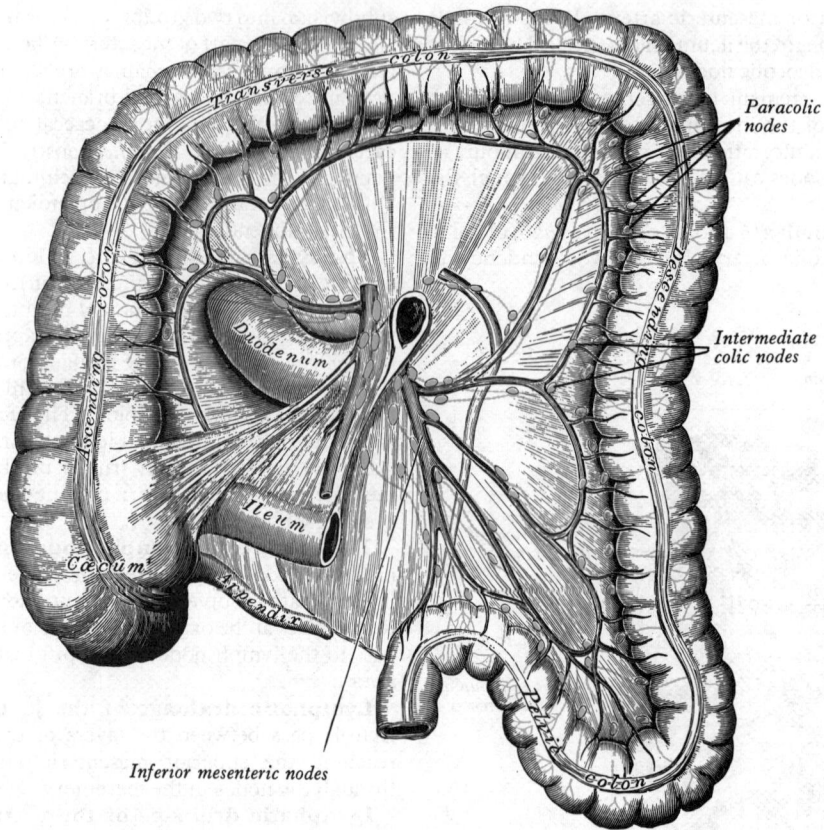

Paracolic nodes

Intermediate colic nodes

Inferior mesenteric nodes

6.172A   The lymph vessels and nodes of the colon. (After Jamieson and Dobson.)

of the group lie lateral to the inferior vena cava and in front of the vessel near the termination of the right renal vein. Afferents reach these nodes from the structures supplied by the lateral and dorsal branches of the aorta and from outlying nodes associated with the iliac arteries and their branches. Efferents from the lateral aortic lymph nodes form a *lumbar trunk* on each side, and both terminate in the cisterna chyli. A few efferents may pass to the pre-aortic and retro-aortic groups. Efferents from the right lumbar lymph trunk and its nodes usually cross over the midline to their left counterparts.

The lymph vessels from the kidney, adrenal gland, abdominal part of the ureter, posterior abdominal wall, testis and ovary, uterine tube and upper part of the uterus, all pass directly to the lateral aortic lymph nodes. Lymph vessels from the pelvis, most of the pelvic viscera, and from the lateral and anterior parts of the abdominal wall, pass first through outlying groups of lymph nodes associated largely with the internal iliac arteries and their numerous branches.

They include the following groups:

Common iliac.	Inferior epigastric.
External iliac.	Circumflex iliac.
Internal iliac.	Sacral.

The common iliac lymph nodes (four to six), are grouped around the common iliac artery, one or two below the bifurcation of the aorta in front of the fifth lumbar vertebra or the sacral promontory. They receive efferents from the external and internal iliac nodes, and send their efferents to the lateral aortic group. They are usually arranged in medial, lateral and intermediate (anterior) chains, the lateral being the main drainage route.

The external iliac lymph nodes (6.167) (eight to ten), lie along the external iliac vessels. They are usually in three groups, one lateral, another medial, and a third anterior to the vessels, but the anterior group is inconstant. The medial chain or group is considered to be the main channel of drainage. They collect from the inguinal lymph nodes (p. 791), from the deeper layers of the infra-umbilical part of the abdominal wall, from the adductor region of the thigh, from the glans penis or clitoridis, the

membranous urethra, prostate, fundus of the urinary bladder, cervix uteri and part of the vagina. Their efferents pass to the common iliac nodes. The *inferior epigastric* and *circumflex iliac nodes*, associated with these vessels and draining corresponding areas, are outlying members of the external iliac group, inconstant in number.

The internal iliac lymph nodes surround the internal iliac vessels. They receive afferents from all the pelvic viscera, from the deeper parts of the perineum, and from the muscles of the buttock and back of the thigh. Their efferents pass to the common iliac nodes. The *sacral lymph nodes*, along the median and lateral sacral vessels, and the *obturator lymph node*, sometimes present in the obturator canal, are outlying members of the internal iliac group.

There is considerable by-passing in the iliac groups of lymph nodes. Lymphangiographic studies have also demonstrated inter-connexions between the right and left groups.

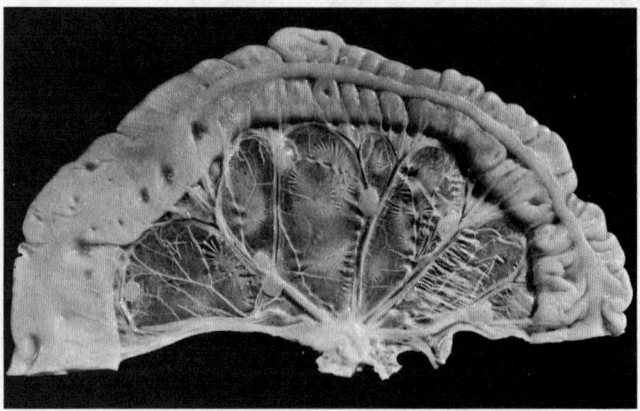

6.172B   Preparation of human colon and mesocolon displaying arterial arcades, neurovascular bundles, lymphatics and paracolic and inter-mediate lymph nodes. (Kindly provided by Prof. S. Kubic, University of Zürich.)

Lymphatic drainage of the urinary tract. (*a*) The *kidney*. The lymph vessels of the kidney begin in three plexuses: one between and around the renal tubules, a second beneath the fibrous capsule, and a third in the perirenal fat which communicates freely with the sub-capular plexus. The collecting vessels from the intrarenal plexus form four or five trunks which follow the renal vein and end in the lateral aortic nodes; as they issue from the hilum they are joined by the collecting vessels from the sub-capsular plexus. The plexus in the renal fat drains directly into the same nodes.

(*b*) The *ureter*. The lymph vessels begin in submucous, intramuscular and adventitial plexuses which communicate with each other. The collecting vessels from the upper part of the ureter may join the renal collecting vessels or may pass directly to the lateral aortic nodes near the origin of the testicular or ovarian artery; those from the succeeding part pass to the common iliac nodes; and those from the pelvic part may end in the common, the external or the internal iliac lymph nodes.

(*c*) The *bladder*. The vesical lymph vessels (**6.**174) take origin in three plexuses—mucous, intramuscular and extramuscular. The collecting vessels, nearly all of which end in the external iliac nodes, are arranged in three sets: the vessels from the region of the trigone emerge on the base of the bladder and run upwards and laterally; those from the superior surface converge on the posterolateral angle and then pass upwards and laterally across the lateral umbilical ligament to reach the external iliac lymph nodes (one of these vessels may go to the internal or common iliac group); those from the inferolateral surface pass upwards to run

with those from the superior surface. Minute nodules of lymphoid tissue may be found along the course of the lymph vessels of the bladder.

(*d*) The *urethra*. (i) The vessels derived from the prostatic and membranous parts in the male, and from the whole urethra in the female, pass mainly to the internal iliac nodes; a few may end in the external iliac lymph nodes. The vessels from the membranous part accompany the internal pudendal artery.

(ii) The vessels of the male spongiose urethra accompany those of the glans penis and end in the deep inguinal nodes. Some may terminate in the superficial nodes, and others may pass along the inguinal canal to reach the external iliac group.

Lymphatic drainage of the male reproductive organs. (*a*) The *testis*. The lymph vessels of the testis commence in two plexuses—superficial, under the tunica vaginalis, and deep, in the substance of the testis and in the epididymis. Four to eight collecting trunks ascend in the spermatic cord and accompany the testicular vessels as they lie on the psoas major; they end in the lateral aortic and pre-aortic lymph nodes.

(*b*) The *ductus deferens, seminal vesicle and prostate gland*. The collecting vessels from the ductus end in the external iliac nodes, those from the seminal vesicle go to both the internal and external iliac groups.

The prostatic vessels terminate chiefly in the internal iliac and sacral nodes; a trunk from the posterior surface passes with the lymph vessels of the bladder to the external iliac nodes, and another from the anterior surface gains the internal iliac group by joining the vessels of the membranous urethra.

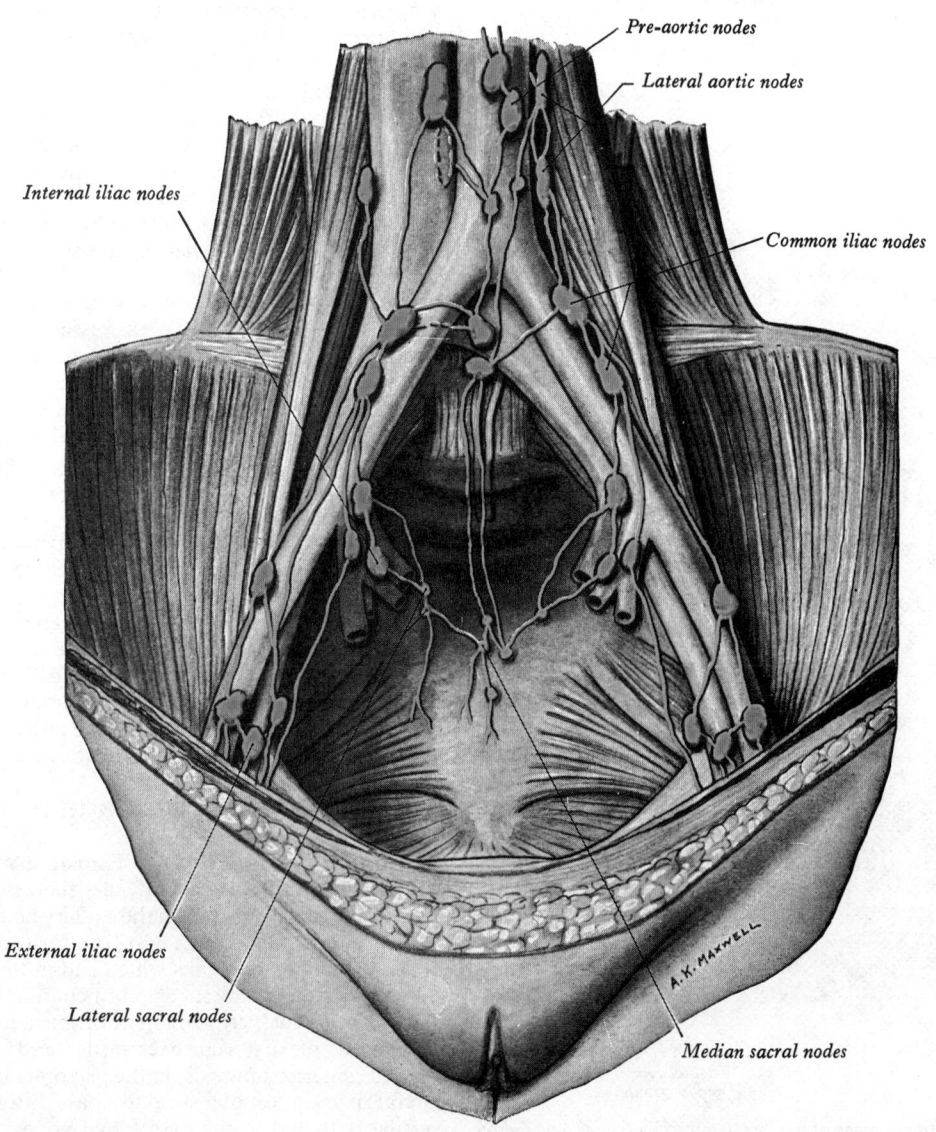

6.173   The lymph vessels and nodes of the pelvis.

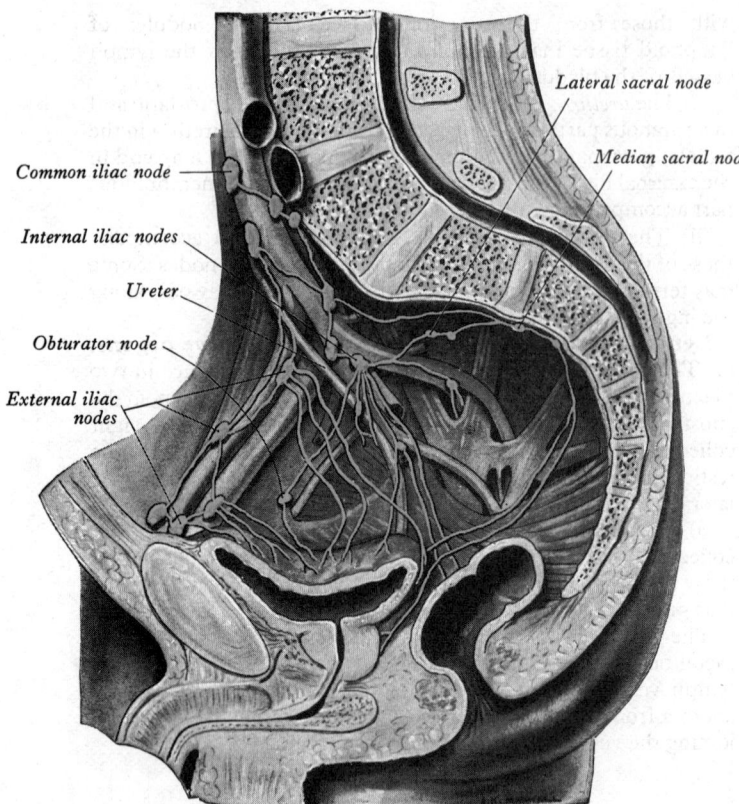

Common iliac node

Internal iliac nodes

Ureter

Obturator node

External iliac nodes

Lateral sacral node

Median sacral node

6.174   The lymphatic drainage of the urinary bladder. Semi-diagrammatic.

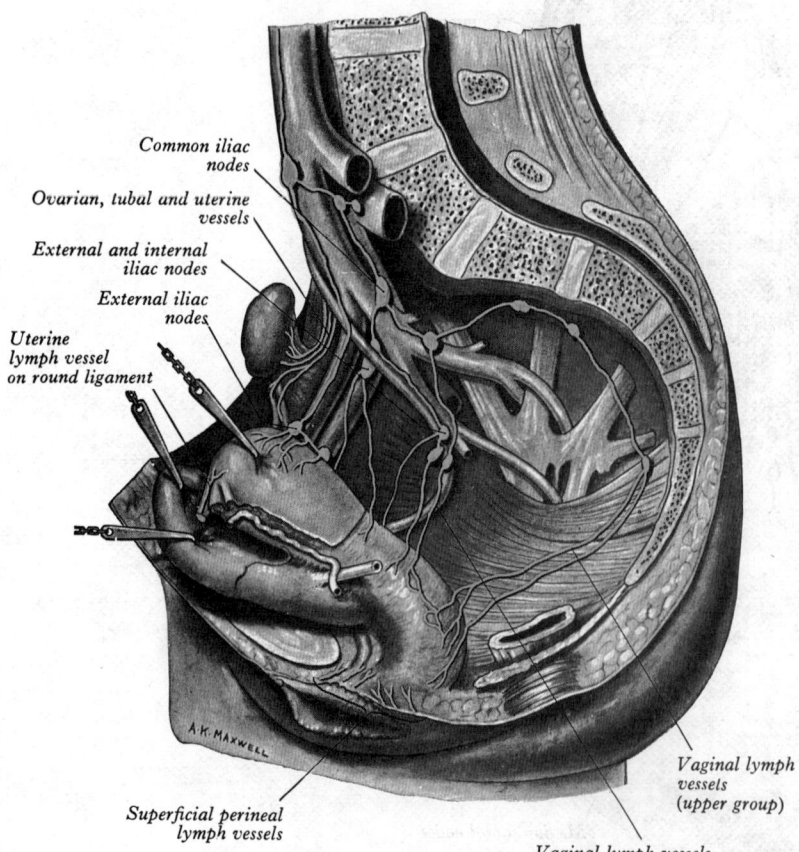

Common iliac nodes

Ovarian, tubal and uterine vessels

External and internal iliac nodes

External iliac nodes

Uterine lymph vessel on round ligament

Superficial perineal lymph vessels

Vaginal lymph vessels (upper group)

Vaginal lymph vessels (middle group)

6.175   The lymphatic drainage of the female reproductive organs. Semi-diagrammatic. (After Cunéo and Marcille.)

(c) The *scrotum and penis.* The skin covering these parts is drained by vessels which, together with those of the whole of the perineal skin, pass along the course of the external pudendal blood vessels to the superficial inguinal nodes. The lymph vessels of the glans penis pass to the deep inguinal and external iliac groups. The lymph drainage of the erectile tissue and the penile urethra is into the internal iliac lymph nodes.

Lymphatic drainage of the female reproductive organs (**6.**175). (*a*) The *ovary.* The vessels, like those of the testis, ascend along the ovarian artery to the lateral aortic and pre-aortic lymph nodes.

(*b*) The *uterus and uterine tube.* The lymph vessels of the uterus comprise two sets: superficial, beneath the peritoneum, and deep, in the substance of the uterine wall. Collecting vessels from the cervix pass in three directions: laterally in the parametrium to the external iliac nodes; posterolaterally to the internal iliac nodes; and backwards in the sacrogenital fold to the rectal and sacral nodes. Some cervical efferents may also reach the obturator or gluteal nodes. The vessels from the lower part of the body pass mostly to the external iliac lymph nodes, accompanying those from the cervix. From the upper part of the body, the fundus and the uterine tube the vessels accompany those of the ovary to the lateral aortic and pre-aortic nodes; a few, however, pass to external iliac nodes. The region near the point of entry of the uterine tube is drained by vessels which accompany the round ligament and so reach the superficial inguinal nodes. The lymph vessels of the uterus enlarge greatly during pregnancy.

(*c*) The *vagina.* The lymph vessels link up with those of the cervix uteri, the rectum and the vulva. They are in three groups, but the areas drained are not sharply demarcated. The upper vessels accompany the uterine artery to the internal and external iliac nodes. The middle part is drained by vessels which accompany the vaginal artery to the internal iliac lymph nodes. The vessels of the vagina below the hymen, those of the vulva and the skin of the perineum as a whole pass to the superficial inguinal nodes, but the clitoris and labia minora drain to the deep inguinal nodes, and direct efferents from the clitoris may pass to the internal iliac lymph nodes (Kubik 1967).

Lymphatic drainage of the abdominal wall. These vessels are in two sets: superficial, in the superficial fascia, and deep, draining with muscles, etc.

The *superficial lymph vessels* accompany the superficial blood vessels. The lumbar and gluteal vessels, run with the superficial circumflex iliac vessels, those from the skin of the anterior wall below the umbilicus with the superficial epigastric vessels. Both sets drain into the superficial inguinal nodes. The region above the umbilicus is drained by vessels which chiefly run obliquely upwards to the pectoral and subscapular groups of the axillary nodes; a few end in the parasternal nodes.

The *deep lymph vessels* accompany the deep arteries. Those from the posterior abdominal wall pass without interruption along the course of the lumbar arteries, to the lateral aortic and retro-aortic lymph nodes; those from the upper part of the anterior abdominal wall run with the superior epigastric vessels to the parasternal nodes, those of the lower part end in the circumflex iliac, inferior epigastric, and external iliac nodes. The vessels of the pelvic wall follow the internal iliac artery and its parietal branches and end in the iliac or lateral aortic nodes.

The Lymphatic Drainage of the Thorax

Lymphatic Drainage of the Thoracic Walls
Superficial lymph vessels of the thoracic wall ramify subcutaneously and converge on the axillary nodes. Those superficial to trapezius and latissimus dorsi run forwards and unite to form about ten or twelve trunks which end in the subscapular group. Those over the pectoral region, including the vessels from the skin covering the peripheral part of the mammary gland and the subareolar plexus, run backwards, and those superficial to serratus anterior upwards, to the pectoral nodes. Others near the lateral margin of the sternum pass inwards between costal cartilages to end in the parasternal nodes, while the vessels of both sides anastomose across the front of the sternum. A few

vessels from the upper part of the pectoral region ascend over the clavicle to the inferior deep cervical lymph nodes.

Lymph vessels from the *deeper* tissues of the thoracic walls drain mainly into three sets of nodes—parasternal, intercostal, and diaphragmatic.

(*a*) **The parasternal (internal thoracic) lymph nodes**, four or five on each side, are at the anterior ends of the intercostal spaces, alongside the internal thoracic artery. They derive afferents from the mammary gland, the deeper structures of the anterior abdominal wall above the umbilicus, the upper surface of the liver through a small group of lymph nodes which lies behind the xiphoid process, and from the deeper parts of the anterior region of the thoracic wall. Their efferents usually unite with those of the tracheobronchial and brachiocephalic nodes to form a single trunk, the *bronchomediastinal trunk*; this may open directly into the junction of the internal jugular and subclavian veins, but on the right side it may join the right subclavian trunk, and on the left, the thoracic duct.

(*b*) **The intercostal lymph nodes** are in the posterior parts of the intercostal spaces in relation to the heads and necks of the ribs. They receive the deep lymph vessels from the posterolateral aspect of the chest and from the mammary gland; some of these vessels are interrupted by small lateral intercostal lymph nodes. The efferents of the nodes in the lower four or five spaces unite to form a trunk which *descends* and opens either into the cisterna chyli or into the commencement of the thoracic duct (p. 784). The efferents of the lymph nodes in the upper spaces of the left side end in the thoracic duct, those of the corresponding right spaces in the right lymphatic duct.

(*c*) **The diaphragmatic lymph nodes** are on the thoracic surface of the diaphragm, and consist of anterior, right and left lateral, and posterior groups.

The *anterior* group consists of two or three small nodes behind the base of the xiphoid process, which receive afferents from the convex surface of the liver, and one or two nodes on each side near the junction of the seventh rib with its cartilage, which receive lymph vessels from the front of the diaphragm. The efferent vessels of the anterior set pass to the parasternal nodes.

The *lateral* groups consist of two or three nodes on each side close to where the phrenic nerves enter the diaphragm. On the right side some of the nodes of this group lie within the fibrous wall of the pericardium on the front of the termination of the inferior vena cava. Their afferents are from the middle part of the diaphragm, those on the right side also receiving afferents from the convex surface of the liver. Their efferents pass to the posterior mediastinal, parasternal and brachiocephalic nodes.

The *posterior* group consists of a few nodes on the back of the crura of the diaphragm, connected on the one hand with the lateral aortic nodes, and on the other with the posterior mediastinal nodes.

Lymphatic drainage of the deeper tissues. The collecting lymph vessels from the deeper tissues include the following:

(*a*) Lymph vessels of the muscles attached to the ribs: most of these end in the axillary lymph nodes, but some from pectoralis major pass to the parasternal nodes. (*b*) Intercostal vessels which drain the intercostal muscles and parietal pleura; those from the anterior half of the thoracic wall and pleura end in the parasternal nodes; those from the posterior half, in the intercostal nodes. (*c*) Vessels of the diaphragm, which form two plexuses, one on its thoracic and another on its abdominal surface; these anastomose freely with each other, and are best marked in areas covered respectively by the pleurae and peritoneum. The *plexus on the thoracic surface* unites with the lymph vessels of the costal and mediastinal parts of the pleura, and its efferents are: *anterior*, passing to the anterior diaphragmatic nodes, which lie near the junction of the seventh rib with its cartilage, *middle*, to the nodes on the oesophagus and around the termination of the inferior vena cava, and *posterior*, to the nodes which surround the aorta at the point where it leaves the thoracic cavity. The *plexus on the abdominal surface* is composed of fine vessels, and anastomoses with the lymph vessels of the liver and, at the periphery of the diaphragm, with those of the subperitoneal tissue. The efferents from the right half of this plexus terminate partly in a group of nodes on the trunk of the corresponding inferior phrenic artery, while others end in the right lateral aortic nodes. Those from the left half of the plexus pass to the pre-aortic and lateral aortic

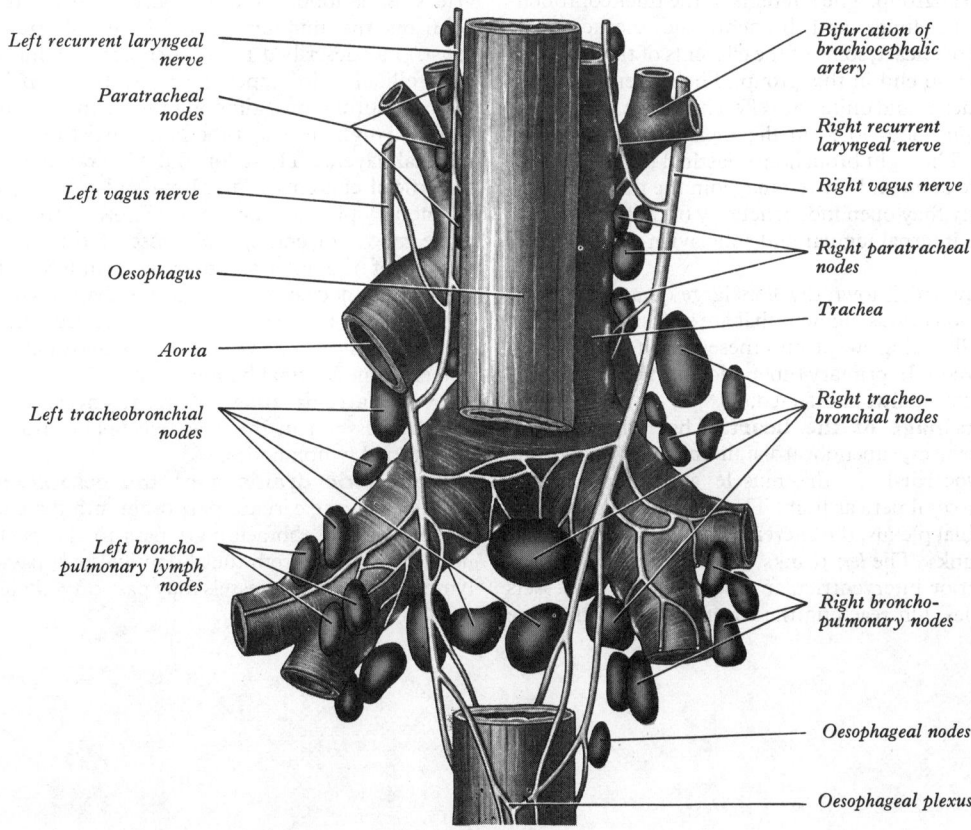

Left recurrent laryngeal nerve
Paratracheal nodes
Left vagus nerve
Oesophagus
Aorta
Left tracheobronchial nodes
Left broncho-pulmonary lymph nodes

Bifurcation of brachiocephalic artery
Right recurrent laryngeal nerve
Right vagus nerve
Right paratracheal nodes
Trachea
Right tracheo-bronchial nodes
Right broncho-pulmonary nodes
Oesophageal nodes
Oesophageal plexus

6.176 The lymph nodes of the trachea, bronchi and lungs. Note the large 'carinate' node lodged between the bifurcation of the principal bronchi.

lymph nodes and to the nodes on the terminal portion of the oesophagus.

Lymphatic Drainage of the Thoracic Contents

Lymph from the thoracic viscera traverses one or other of three sets of lymph nodes—brachiocephalic, posterior mediastinal or tracheobronchial—before entering the thoracic duct, the right lymphatic duct or some other vessel which itself enters one of the great veins at the root of the neck.

The brachiocephalic lymph nodes are placed in the anterior part of the superior mediastinum, in front of the brachiocephalic veins and the large arterial trunks which arise from the aortic arch. They receive afferents from the thymus, thyroid, pericardium, and from the lateral diaphragmatic lymph nodes; their efferents unite with those of the tracheobronchial lymph nodes, to form the right and left bronchomediastinal trunks.

The posterior mediastinal lymph nodes are behind the pericardium, close to the oesophagus and descending thoracic aorta. Their afferents are derived from the oesophagus, the posterior part of the pericardium, diaphragm, lateral and posterior diaphragmatic nodes, and, occasionally, the left lobe of the liver. Their efferents mostly end in the thoracic duct, but some join the tracheobronchial nodes.

The tracheobronchial lymph nodes (6.176) are arranged in five main groups, and include some of the largest nodes in the body: (*a*) *paratracheal*, at the sides of the thoracic part of the trachea; (*b*) *superior tracheobronchial*, in the angles between the lower part of the trachea and bronchi; (*c*) *inferior tracheobronchial* in the angle between the two bronchi (in clinical and pathological practice these are commonly termed *carinate nodes* because of their close relationship to the *carina* or 'keel-shaped' cartilage sited between the two principal bronchi); (*d*) *bronchopulmonary*, in the hilum of each lung (in clinical practice often simply called *hilar lymph nodes*); (*e*) *pulmonary*, in the lung substance, on the larger branches of the principal bronchi. These groups are not sharply demarcated. The pulmonary lymph nodes become continuous at the hilum of the lung with the bronchopulmonary nodes, and they in turn are continuous with the inferior and superior tracheobronchial nodes, while the latter are continuous with the paratracheal group. The afferents of the tracheobronchial nodes drain the lungs and bronchi, the thoracic part of the trachea, and the heart; some of the efferents of the posterior mediastinal nodes also end in this group. Their efferent vessels ascend upon the trachea and unite with efferents of the parasternal and brachiocephalic nodes to form the *right* and *left bronchomediastinal trunks*. The right bronchomediastinal trunk may join the right lymphatic duct, and the left may join the thoracic duct, but more frequently they open independently of these ducts into the junction of the internal jugular and subclavian veins of their own side.

Applied Anatomy. In all town dwellers large quantities of the dust and black carbonaceous pigment that are so freely inhaled in cities are continually being swept into these lymph nodes, from the bronchi and alveoli. In primary tuberculosis of the lungs these lymph nodes are almost always infected.

Lymphatic drainage of the heart. The lymph vessels consist of three plexuses, subendocardial, immediately under the endocardium, myocardial, in the muscle and subepicardial, subjacent to the visceral pericardium. The deeper plexuses open into the subepicardial plexus, the efferents of which form left and right collecting trunks. The *left* trunks, two or three in number, ascend in the anterior interventricular groove, receiving vessels from both ventricles. On reaching the coronary sulcus they are joined by a large trunk from the diaphragmatic surface of the left ventricle which first ascends in the posterior interventricular groove and then turns to the left along the coronary sulcus. The single vessel formed by the union of these trunks ascends between the pulmonary trunk and the left atrium, and ends, usually, in one of the inferior tracheobronchial nodes. The *right* trunk receives its afferents from the right atrium and from the right border and diaphragmatic surface of the right ventricle. It runs upwards in the coronary sulcus, close to the right coronary artery, and then ascends in front of the ascending aorta to terminate in one of the brachiocephalic nodes, usually to the left of the median plane.

Lymphatic drainage of the lungs and pleurae. The lymph vessels of the lungs originate in superficial and deep plexuses. The superficial is beneath the pulmonary pleura, the deep accompanies branches of the pulmonary vessels and the ramifications of the bronchi. In the case of the larger bronchi the deep plexus consists of two networks—submucous, and peribronchial, outside the walls of the bronchi. In the smaller bronchi there is but a single plexus, which extends as far as the bronchioles, but fails to reach the alveoli, in the walls of which there are no traces of lymph vessels. The superficial efferents turn round the borders of the lungs and the margins of their fissures, and converge to end in the bronchopulmonary lymph nodes; the deep efferents are conducted to the hilum along the pulmonary vessels and bronchi, and end, for the most part, in the same nodes. No free anastomosis occurs between the superficial and deep lymph vessels of the lungs, except in the hilar region. In the peripheral parts of the lung small connecting channels exist between the superficial and the deep lymph vessels, capable of becoming dilated so as to direct lymph from the deep to the superficial vessels when the outflow from the deep vessels is obstructed by disease of the lung or pulmonary lymph nodes. At the bottom of the fissures the lymph vessels of the adjoining lobes communicate with one another, so that, although there is a general tendency for the lymph vessels from the upper lobes of the lungs to pass to the superior tracheobronchial nodes and for those from the lower lobes to join the inferior tracheobronchial group, these connexions are not exclusive. At the level of pulmonary lobulation the arrangement of lymphatic vessels is said (Kubik 1970) to follow both the central artery of a lobule and its peripherally situated veins. This confirms the findings of Celtis and Porter (1952). Policard (1950) has described the occurrence of lymphoid aggregations, non-follicular in appearance, in peribronchial sites and in '*placoid*' formations adjoining the pulmonary pleura.

The lymph vessels of the pleura exist in both the visceral and parietal layers. Those of the visceral pleura drain into the superficial efferents of the lung and form a plexus beneath the pulmonary pleura (*vide supra*). Those of the parietal pleura have three modes of ending: (*a*) those of the costal region join the vessels of the internal intercostal muscles and so reach the parasternal nodes; (*b*) those of the diaphragmatic pleura form a plexus on the thoracic surface of the muscle and are dealt with on p. 799, while (*c*) those of the mediastinal pleura end in the posterior mediastinal lymph nodes.

Lymphatic drainage of the thymus. The lymph vessels of the thymus end in the brachiocephalic, tracheobronchial, and parasternal lymph nodes.

Lymphatic drainage of the oesophagus. The efferent vessels from the cervical part drain into the deep cervical nodes, those from the thoracic part pass to the posterior mediastinal nodes, and those from the abdominal part pass to the left gastric lymph nodes. Some vessels may pass directly to the thoracic duct (p. 784).

7

NEUROLOGY

Introduction

The dependence of living organisms upon environmental energy sources and the essentially dynamic nature of their life processes have been emphasized elsewhere in this volume (pp. 2–4). Many aspects of this dynamic state stem from the ability of organisms continually to interact with a fluctuating external environment whilst preserving their own structural integrity. Such effective adaptations to a varying environment can, as we have seen, be considered in relation to either long or brief time scales (p. 72). Thus, over many generations, the forces of natural selection

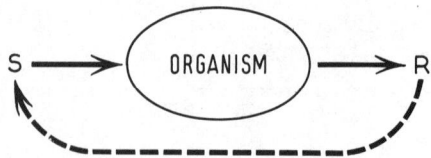

7.1 A diagram of the relationship between a stimulus (S) and a response (R) emphasizing its closed-loop nature.

operate upon the genetic variants introduced by sexual reproduction and occasional mutation, resulting in species with an enhanced ability to adapt to changing or different environments. Alternatively, within the lifetime of a complex organism, such adaptive responses range from complicated behaviour patterns, associated with mating, rearing of young, capturing of prey, avoidance or combat with predators, etc. to the innumerable transient readjustments, e.g. of posture, or the reaction and composition of the body fluids which constantly occur. The foregoing are examples of *homeostatic responses* which are a central feature of the behaviour of living organisms and which may be simply summarized as in illustration **7.1**.

When an environmental change (S or *stimulus*), to which the

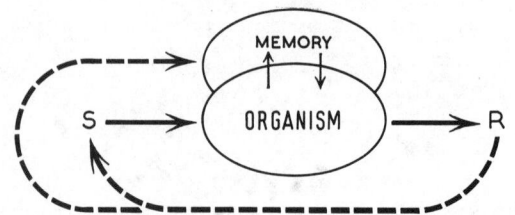

7.2 The stimulus-response loop with the added flexibility of a memory store.

organism is capable of reacting, occurs in its vicinity, an appropriate response (R) follows. In general, the nature of the response is such that it tends to preserve the internal constancy of the organism within the fairly narrow limits necessary for the continued life of its cells and hence ultimately to preserve the whole organism and the species. It should be noted that because of the response, the relationship of the organism to the stimulus is altered (the broken arrow in the diagram), and that in this sense, the whole sequence of events is a *loop*, and not an *arc* with open ends. Further, in the simple case illustrated, the structural organization of the organism (the result of its phylogenetic history) limits the array of stimuli to which it can react, and also restricts the repertoire of its responses, which may be mechanical, chemical, photic or electrical. Accordingly, there are many environmental disturbances to which it is *unable* to respond, or the disturbances may be of such an intensity that any responses are inadequate; the organism 'dies' and gradually undergoes dissolution (i.e. it merges with its environment). In the example quoted the *genetic memory* of the organism, which emerged by natural selection over countless generations, has determined its receptivity of stimuli and its range of responses, and hence its degree of fitness for survival within some particular environment. However, such mechanisms lead to rather stereotyped sets of responses, and in more complex organisms, selective processes

have led to the addition of more flexible *memory stores* which operate throughout the life of the individual (**7.2**). In the latter the general sequence of the stimulus–response loop is similar to the previous example, but in this case the information derived from a particular stimulus pattern is transferred to a memory store, where it is compared with a record of previous experiences. As a result of this comparison, a choice of a particular response is made (out of the many alternatives available), as the one with the highest probability of effectiveness under the circumstances prevailing. It is considered that once having occurred, the effectiveness of the response is somehow assessed, and the results of this assessment are transferred to the memory store, which is modified accordingly; the probability of the same response being chosen under similar conditions in the future being either raised or lowered. In this manner the memory store (or hierarchy of memory stores) is continually modified, readjusted and refined as, with time, further experiences occur, i.e. the organism *learns*.

As implied above, homeostatic responses are a characteristic attribute of *all* living organisms, from unicellular to compound multicellular forms. However, with increasing size and complexity of structure there is a corresponding increase in the range and flexibility of the responses, and this has paralleled the emergence of a *nervous system*—the concern of the present section of this volume.

The human nervous system is, without question, the most complex, widely investigated, and least well understood system known to mankind. Its structure and activities are inseparably interwoven with every aspect of our lives, physical, cultural and intellectual. Accordingly, investigators of many different disciplines, methodologies, motivations and persuasions converge in its study. Similarly, depending upon the context, there are many more or less appropriate ways of embarking upon a study of nervous systems; for example the approach may be developmental, phylogenetic, physicochemical, energetic, structural—gross or cellular, cybernetic, behavioural or ethological.

For our present purposes, the detailed neuroanatomy of the various arbitrary but convenient divisions of the human nervous system is preceded by an account of some relevant experimental methods, the biology of its components in cellular terms, a brief review of some simple nervous systems, and introduced by comparing nervous systems with the information-processing, communication, and control systems of homeostatic machines. (For an extended discussion of the latter approach, the interested reader should consult Young 1964; 1978.)

A SIMPLE HOMEOSTAT

It will, perhaps, prove helpful at this point to draw certain analogies with the main features of a simple man-made homeostatic device such as a thermostat (**7.3**).

In the thermostatic system illustrated, the *energy level* (temperature) of the water bath is continuously *monitored* by an appropriate *sensor* or *receptor* (in this case a thermocouple), and variations in this level are converted into variations in the flow of a *pattern of signals* or *coded information* (in this case current flow) along an *afferent communication channel* to a *controller*. The physical design of the latter is such that it combines or *integrates* the information flow along its afferent channel with that provided by the temperature reference level, and makes a *decision* between alternative patterns of activity (*responses*) of the *effectors* (steam or cold water). If, for example, there is a fall of temperature in the bath (of sufficient magnitude to be detected by the sensor), this change is transformed into signals in the afferent channel which, after comparison and computation in the controlling centre results in an increased flow of steam. It may be noted that it is inherent in the design of such a *control loop system* that there will not be a simple, smooth, immediate return to the reference level of temperature; some overshoot occurs and, with time, there follows a series of oscillations of decreasing amplitude, which gradually converge on the reference level. All the features described in this man-made homeostat have their counterparts in primitive nervous systems. These include the possession of a variety of sensors each capable of monitoring the rate and magnitude of some particular type of environmental change, communication

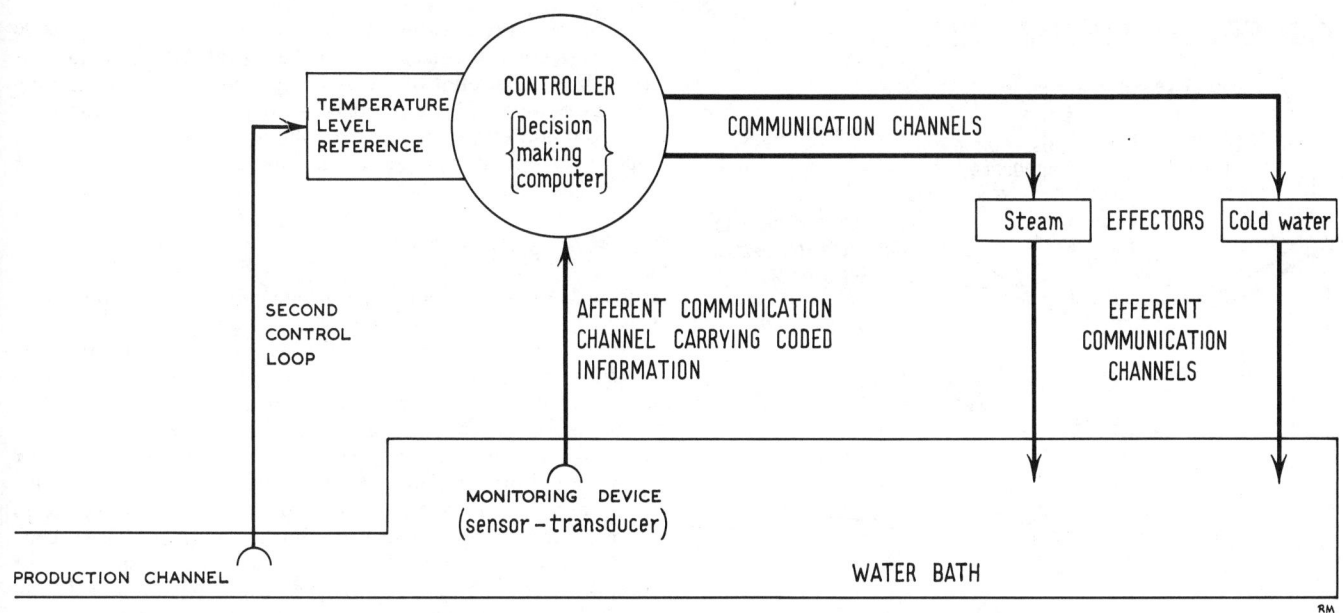

7.3   A diagram of the essential features of a thermostat coupled to an industrial production line. This is an example of a man-made homeostatic device; *see* text for further details and discussion. Modified from an illustration in: *A Model of the Brain*, by J. Z. Young, published by the Clarendon Press, with the permission of the author and the publishers.

channels carrying coded information (nerve fibres bearing temporo-spatial patterns of nerve impulses), integrative computing and decision-making centres (geometric patterns of contact between nerve cells), further communication channels (nerve fibres) and finally effectors (gland or muscle cells) which may be thrown into appropriate patterned responses. All these features will be considered in greater detail in subsequent sections.

As also indicated in illustration **7.3**, simple control loops of the type described, which are a characteristic feature of all organisms (and widely used in engineering design), are frequently *coupled* with *further loops*, thus giving more complex responses. For example, the reference level of such a factor as the bath temperature may be continually readjusted to maximize the output of an industrial production line. As we shall see, many features of more complex nervous systems are conveniently considered in terms of an *integrated hierarchical system of control loops*.

Cell Communication

Many aspects of intracellular and intercellular communication (**7.4**A, B) have been mentioned elsewhere in this volume. Thus, consideration has been given to the selective interaction between cell surfaces, when in close apposition, and the phenomena of contact guidance and contact inhibition (p. 91). The structure and informational role of nucleic acids in cells has been discussed in some detail (p. 21), and brief reviews of current theories concerning how local cytoplasmic factors, or intercellular chemical messengers (embryonic induction factors, hormones, etc.) may effect changes in genotropic control loops, and result in the enhancement or repression of different patterns of gene activity (pp. 24, 25). Where cells are in particularly close apposition (e.g. at maculae communicantes or 'gap junctions') a variety of ions and molecules are able to pass selectively between the cytoplasms of the adjacent cells at such sites (p. 7). (However, much work remains to be carried out concerning the precise nature and variety, mechanisms of transfer, and informational role of the substances involved at these sites, *see* p. 87.) Alternatively, chemical messengers, such as the secretory products of endocrine glandular cells, pass into the blood circulation and then diffuse through the tissue fluid to reach groups of cells which may be quite remote from their site of production. Necessarily, such mechanisms operate upon relatively slow time scales.

EXCITABLE TISSUES

More rapid mechanisms of both intracellular and intercellular communication have evolved in certain specialized *excitable tissues* (including receptor, neural, muscular and certain glandular cells). Some general features of an excitable cell are illustrated in **7.4**B. All cells possess a plasma membrane which separates the surrounding tissue fluid from the various intracellular compartments (p. 5, **1.1**). By virtue of the permeability characteristics of this membrane, and its associated energy-dependent ionic pumping mechanisms, the ionic composition of the intracellular fluid contrasts sharply with that outside the cell, and consequently a large recordable difference in electrical potential exists between the two compartments across the membrane. Upon the receipt of an appropriate form of *stimulus* at some point upon its surface, a transient, reversible wave of change spreads over the surface of the excitable cell from this point. Whilst the details of the change in molecular terms (presumably involving conformational alterations in the membrane substructure) are not yet understood, changes in the permeability (conductance) of the membrane result. Accordingly there is a rapid local redistribution of ions across the membrane, as they flow down their concentration gradients; this soon ceases as the permeability returns to its resting state. These ionic fluxes may be recorded as variations in electrical potential using suitable equipment, and they are also accompanied by characteristic energetic (thermal) exchanges between the cell and its environment. These properties of excitable cells are discussed in greater detail elsewhere, as are the features which lead to *trains* or *volleys* of such *impulses* upon prolonged stimulation, and the relationship of stimulus strength to impulse pattern (number and frequency). The impulse, having spread over the surface of the cell, then causes, by some imperfectly understood process, either the release of a specific secretory product or the contraction of actin-myosin complexes, depending upon the type of cell involved. The different cells which cooperate in the functioning of nervous systems are modified in various ways, and exhibit these properties of excitable tissues to greater or lesser degrees (**7.5**A, B, C). Thus, whilst in unicellular forms, and some primitive multicellular forms, *individual cells* are capable of reacting to stimuli, in the majority of multicellular forms possessing nervous systems, a *division of labour* occurs between the excitable cells which make up the system. Thus, we see the emergence of *receptor* cells specialized particularly for the reception of stimuli, *neurons* for the

GENERAL TISSUE CELLS

Selective interaction
between surfaces
(Reaggregation of dissociated
tissues, contact guidance,
contact inhibition, etc.)

Genotropic Loops
altering patterns
of gene activity

Cytoplasmic
factors

Intercellular 'messengers'
(e.g. embryonic organizers
hormones, chalones,
neural transmitters)

Passage of ions
and other molecules
between cells across
specialised junctions

Homeostatic Loops
involving local
cytoplasmic machinery
e.g. mitochondrial activity

7.4A A diagram illustrating various forms of cellular and intercellular communication systems.

EXCITABLE TISSUE CELL

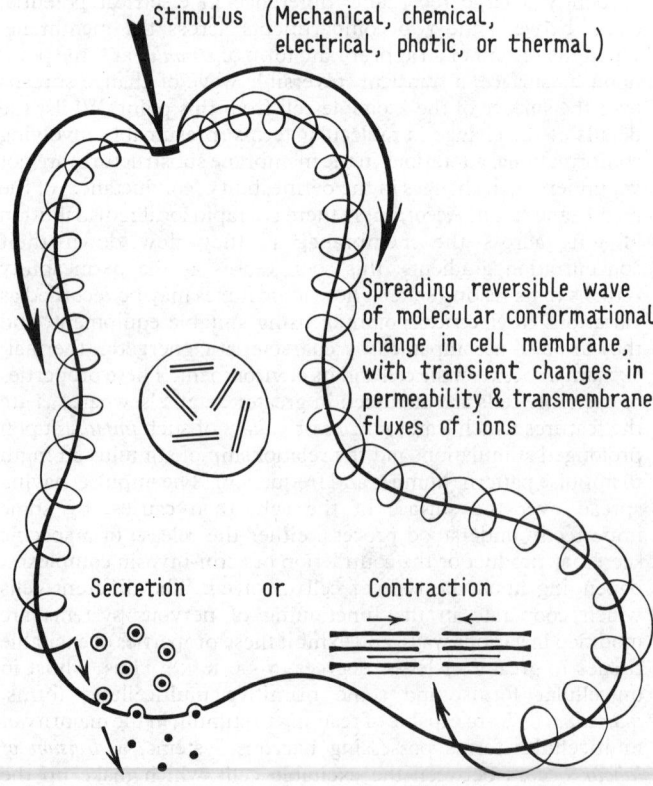

Stimulus (Mechanical, chemical,
electrical, photic, or thermal)

Spreading reversible wave
of molecular conformational
change in cell membrane,
with transient changes in
permeability & transmembrane
fluxes of ions

Secretion   or   Contraction

7.4B A summary of the main events which may follow the application of an effective stimulus to the surface of an excitable tissue cell.

integration and conduction of information, and *effectors* (contractile or glandular cells) for the operation of responses. A more detailed account of the cytology, and theories concerning the functioning and evolutionary history of these cell varieties, will be found elsewhere in this volume (pp. 820 *et seq.*); what follows is a brief introduction to features common to many types of nervous system.

RECEPTOR CELLS

Within the various epithelial layers which clothe the body surfaces, and their associated sub-epithelial connective tissue, within the walls of some blood vessels, incorporated into the structure of many of the solid and hollow viscera and into muscles, tendons, ligaments, etc. are arrays of *receptor cells*. The latter possess specialized regions of their surfaces which enable them to monitor selectively changes in either the internal or external environment. Some receptors are particularly responsive to mechanical deformation whilst others are sensitive to either thermal, chemical, electrical or photic variations. Views concerning the *degree of specificity* of action of individual 'types' of receptor cell have been modified on a number of occasions during the last one hundred and fifty years. Thus, since the first enunciation of a law of *specific nerve energies* by Johannes Müller in 1840, many investigators have held that any particular receptor reacted preferentially to only one type of energetic change, and this constituted its usual physiological role. A receptor was considered to exhibit a *low threshold* with respect to this form of stimulation, i.e. it reacted to small variations in energy level, whereas to all other types of change it exhibited a *high threshold*, only reacting if the stimulus strength was of relatively great, perhaps abnormal, intensity. Following experiments on cutaneous innervation in mammals, however, other investigators reached the opposite conclusion—that the cutaneous receptors were *non-specific* and that sensory perception depended entirely upon the *spatio-temporal patterns* of nerve impulses in a *common set* of neural pathways (Weddell 1941). With more refined methods of recording from individual nerve fibres and nerve cell bodies, however, there has been a partial return to the specificity theory, but with many complexities, subtleties and grades of interaction, not envisaged in the earlier theories (Mountcastle and Powell 1959; Sinclair 1967; Wall 1967). Thus, some nerve fibres have been identified which stem from receptors reacting to more than one class of environmental change (*polymodal receptors*). Further, receptors may be *fast* or *slowly adapting*: the former respond to a brief stimulus with a sharp volley of nerve impulses which ends as the stimulus ceases, but if the stimulus is maintained the frequency of nerve impulse firing rapidly drops to the resting level. Slowly adapting receptors may continue to generate a volley of impulses for prolonged periods if the stimulus strength persists. Yet other varieties of receptor show a resting level of impulse discharge which may be either raised or lowered with different changes in the environment. These variants will be further considered with the anatomy of the individual receptor cells (p. 849).

The intimate molecular mechanisms whereby an environmental change, whether mechanical, chemical or thermal, causes an alteration in the receptor cell are imperfectly understood (although more is known concerning the activation of photosensitive pigments in retinal receptors—p. 1166). Nevertheless, electrophysiological recording from some varieties of receptor cell has demonstrated that, when the cell is exposed to a stimulus of adequate strength, there follows a change of permeability to ions in its specialized receptor surface. Consequently, some of the ions which are differentially concentrated in the intra- and extracellular compartments flow down their concentration gradients across the specialized receptor surface; a microelectrode inserted into the cell records these ionic fluxes as variations in electrical potential—the so-called *receptor potential*. The circuit for this flow of ionic current is completed by a flow of ions in the opposite direction across the neighbouring non-specialized (but excitable) regions of the receptor cell membrane, where the resulting *graded variations* in potential are termed the *generator potential*. The detailed consequences of fluctuations in the level of the generator potential have only been analysed in a few sites, but vary with the type and geometry of different receptors. In some situations, for example, the gustatory cells of the tongue, the generator potential may *directly* affect the level of activity of the points of synaptic contact between the receptor cell and the peripheral process of a neuron (*vide infra*). In other situations (the Pacinian corpuscle is a particularly well investigated example, p. 855), where the specialized receptor surface is the peripheral part of a long cytoplasmic conducting process (nerve fibre) which proceeds to the central nervous system, other events occur. If the variation in the generator potential reaches a critical *threshold* level, it causes a dramatic but rapidly reversible series of changes of ionic permeability in the regions of cell membrane which border the specialized receptor surface. The consequent ionic fluxes are

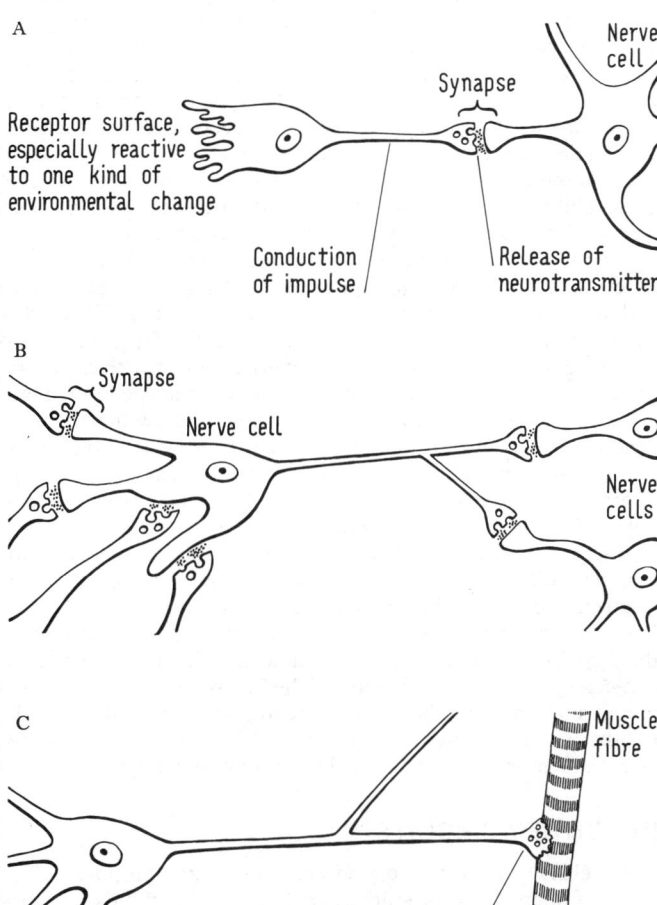

7.5A–C The three main avenues of differentiation which may be followed by a primitive nerve cell: A a receptor neuron; B an interneuron; and C an effector neuron. See text for further discussion.

recordable as an *action potential*, and they set in train a self-generating wave of similar changes which spreads along the conducting process as a *nervous impulse*. Maintenance of the generator potential at an adequate level above threshold results in a *volley of impulses* which passes centrally along the fibre. The *frequency* of the impulses constituting the volley is related (logarithmically) to the energy level, or *intensity* of the stimulus. In this manner, receptors act as *transducers*, whereby fluctuations in various environmental energy levels are transformed into coded volleys of nerve impulses. Much work is currently in progress attempting to analyse the detailed mechanisms and quantitative aspects of their responses.

SYNAPTIC CONTACTS

Excitable cells (receptors, neurons, and effectors) can communicate at specialized points of intercellular contact termed

synapses and these are of two principal kinds; by far the most common are *chemical synapses* where chemicals (neurotransmitters) released by one cell stimulate the plasma membrane of an adjacent cell so that its permeability to certain ions, and hence its electrical state is altered. Such junctions are found throughout the animal kingdom wherever a nervous system exists. A second type is the *electrical synapse*, similar to the *communicating junctions* or *nexuses* between many classes of cell including cardiac myocytes, non-striated myocytes, etc. (*see* pp. 515, 518). At such junctions the adjacent cells establish channels of direct ionic communication between their cytoplasms so that one cell can excite another without the intermediary step of transmitter release.

As noted above at most mammalian synapses (as in the majority of sub-mammalian and invertebrate synapses) chemical transmission occurs. The cell membranes are, in this case, separated by a *synaptic cleft*, and in the simplest synapses transmission occurs in one direction only, i.e. the synapse is *polarized* and is bounded by the membranes of a *presynaptic* and a *postsynaptic* cell. As the nerve impulse which has spread over the surface of the presynaptic cell (either a receptor cell, or a neuron) approaches the presynaptic membrane, it causes the release of a *neurochemical transmitter* which rapidly diffuses across the confines of the synaptic cleft and approaches the postsynaptic membrane, where it causes a change in its ionic permeability. The action of the transmitter is short-lived, as it is removed or broken down almost immediately by the action of a specific enzyme. The character of the permeability change in the postsynaptic membrane, however, varies with the chemical nature of the transmitter substance at a particular synapse, and also with the nature of the membrane. In some cases the permeability change is such that the resulting flow of ions tends to *depolarize* the postsynaptic membrane, i.e. to *reduce* the level of the electric potential difference which reflects the asymmetric distribution of ions across the membrane of the 'resting cell'. Such synapses and their transmitters are said to be *excitatory* with respect to the postsynaptic cell (*vide infra*). In contrast, other varieties of neurotransmitter tend to *hyperpolarize* the postsynaptic membrane and are said to be *inhibitory*. These distinctive effects, as recorded by a microelectrode impaling the postsynaptic cell, are known as *excitatory postsynaptic potentials* (EPSP's) and *inhibitory postsynaptic potentials* (IPSP's) respectively. (Further information concerning the detailed architecture and modes of operation of different varieties of synapse, and groups of synapses, will be found on pp. 826–834.) To understand some features of their interaction, the morphology of a common type of nerve cell must now be outlined.

NEURONS

Neurons or *nerve cells* are essentially excitable cells which are specialized for the *reception, integration, transformation* and *onward transmission* of coded information.

Vertebrate neurons consist of a *cell body* or *soma*—a localized mass of specialized cytoplasm which carries a diploid nucleus and is bounded by an excitable membrane, from the surface of which project one or more *neurites*. The latter are delicate, branching, cytoplasmic processes, each enclosed by an excitable plasma membrane, which extend varying distances from the cell body. Accordingly, neurons may be classified as either *unipolar, bipolar* or *multipolar* (7.6A–C). The neurites in each case may be divided into those which conduct towards or directly influence the cell body, namely one or more *dendrites*, and another one (which may branch more or less profusely), and conducts away from the cell body, the *axon*.

Neurons present a great range of shapes, sizes and inter-relationships with other cells, both in different species and in different regions of the nervous system. Many of these variations will be treated in greater detail in subsequent parts of the present section. For our present introductory purposes it should be noted that the dendrites of a bipolar or unipolar neuron are either in synaptic contact with specialized receptor cells in the tissues of the body, or, by their dendritic terminals, themselves function as specialized peripheral receptor surfaces. In contrast, virtually all the nerve cells of the vertebrate *central* nervous system (brain and spinal cord) are some form of the multipolar variety (7.7).

It is clear that the surfaces of the dendritic tree and of the cell body provide a large area, which receives the synaptic terminals of other neurons. Such terminals may be relatively few in number, but usually they number many thousands, being derived from the axonal branches of some hundreds of neurons which are presynaptic relative to the one on to which they converge. Such presynaptic cells may be widely scattered in different regions of the nervous system; evidently there is a great *convergence* of information paths in such a case.

The synaptic contacts illustrated may thus be classified as *axodendritic*, or *axosomatic*, according to their site on the

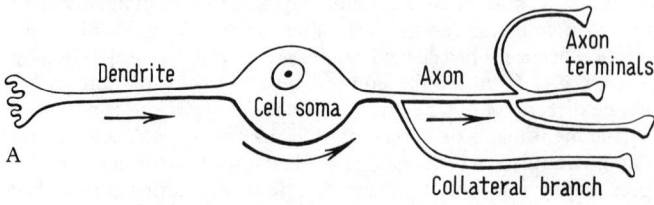

BIPOLAR NEURON

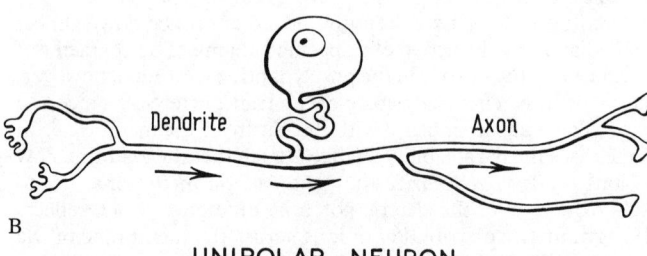

UNIPOLAR NEURON

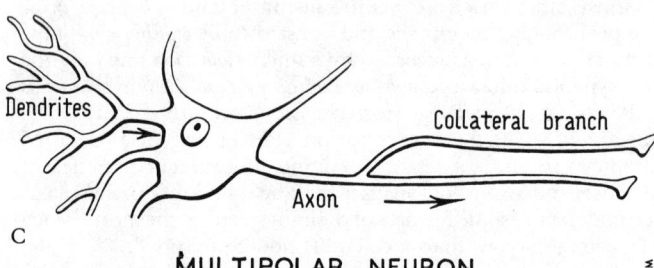

MULTIPOLAR NEURON

7.6A–C Three general morphological groups of neurons classified according to the number of neurites which arise from the surface of the cell soma: A bipolar neuron; B unipolar neuron; and C multipolar neuron. For a more detailed analysis of the branching patterns of neurites, see 7.15.

postsynaptic membrane; other less common varieties include *axoaxonic* synapses involving either the initial segment of the axon or the presynaptic terminals of the cell concerned, *dendrodendritic* synapses, *reciprocal* and *serial* synapses, and finally synaptic *cartridges* and *glomeruli*, in which a number of neurites are interrelated in complex geometrical patterns. The latter provide the *higher integrative units* or *microcircuits* of the functioning nervous system. The cytological details and possible modes of operation of these and other varieties will be described in later pages (pp. 827, 829).

Of the numerous synaptic terminals clustered over the dendrites and cell soma of a multipolar neuron, those from certain sources are *excitatory* whilst the remainder, from other sources, are *inhibitory*. Depending upon the states of relative activity or quiescence of these different sources, the proportion of excitatory and inhibitory synapses which are active also varies continuously with the passage of time. Accordingly, their overall effect is summated (*integrated*) and the *excitatory state* of the cell, i.e. the relative level of depolarization or hyperpolarization of its cell membrane also fluctuates. Only when the depolarization reaches a critical *threshold level* are the local electronic current flows, which encircle the dendritic tree and cell soma, sufficient so to alter the permeability characteristics of the *initial segment* of the axon, that an action potential is generated and spreads down the axon as a *nervous impulse*. If the threshold excitatory state continues to be exceeded, a volley of such impulses is discharged along the axon and its collateral branches. It is to be noted that, since the excitatory level of the neuron *continuously* reflects the innumerable variations in pattern of the information paths which converge upon it, the nerve cell bears some similarity to the operation of an *analogue computing device*. In contrast, the outflow information channel, the axon, exhibits an *all or none* response. Thus, it is either quiescent, or carries one or more discrete, identical, intermittent nerve impulses—only their frequency varies. In this regard the axon is more like a *digital computer*.

The axon terminals of a particular neuron may be few in number and concentrate their effect on one or only a small localized group of postsynaptic cells. In other situations *collateral branches*, which may be numerous, arise from an axon at points throughout its course, and each of these may themselves arborize and *diverge* to many destinations. The 'many to one' *convergence* of information channels with the integration of their effects, and, in other situations, the 'one to many' *divergence* of channels, are essential features of all advanced forms of nervous system.

INHIBITORY CIRCUITS

Some elementary inhibitory circuits, currently proposed as a result of microelectrode studies, are illustrated in 7.8A–C.

In **feed-forward inhibition** (7.8A) an excitatory neuron (X) is functionally linked with two others (Y and Z). However, it is

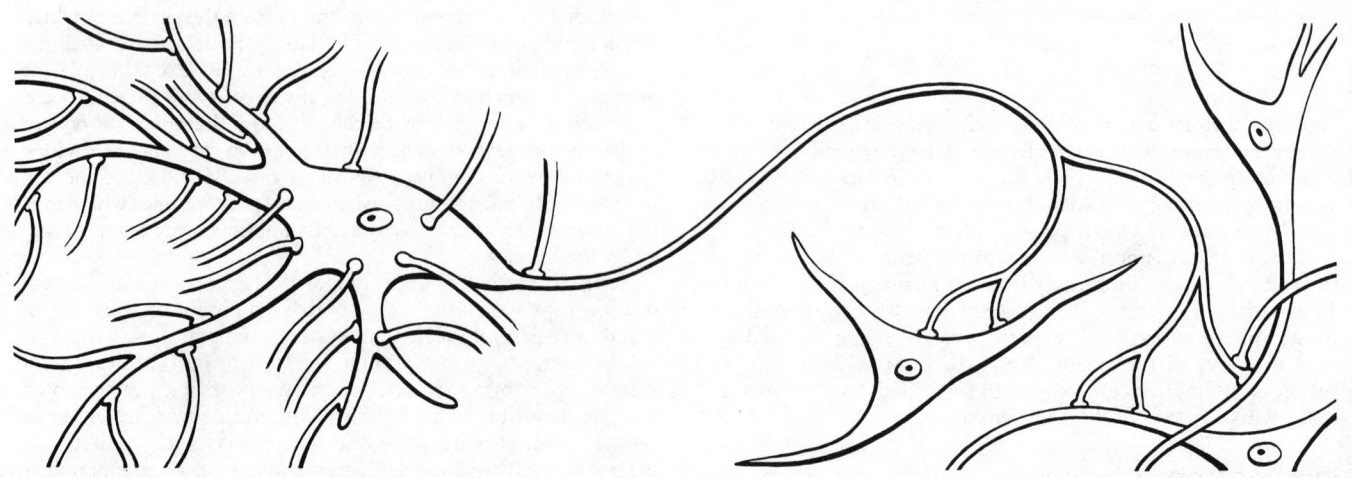

7.7 A diagram to illustrate a generalized multipolar neuron and the parts of its surface upon which synaptic terminals from other neurons may converge. These include the surfaces of the dendrites, the cell soma, the initial segment of the axon, and the proximal surface of its own axonal synaptic terminals. For a more detailed representation of the cytology and surface contacts of a multipolar neuron see 7.17.

FEED-FORWARD INHIBITION

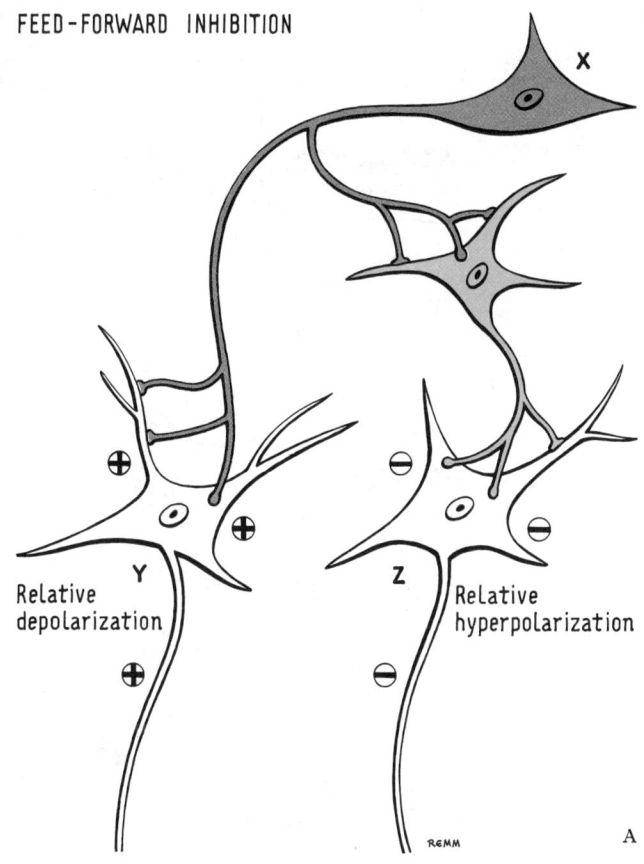

Relative
depolarization

Relative
hyperpolarization

A

clear that the interposition of an inhibitory neuron on the second path causes the excitatory level of Z to fall whilst that of Y rises, during the period that X is active. Of course, in practice, many other local circuits are simultaneously active; nevertheless, the principle illustrated probably accounts in part for such examples as the increase in force of contraction of a muscle group during a particular activity, whilst, concurrently, the contraction level of a cooperative group of muscles is progressively reduced.

In **feed-back inhibition** (7.8B) a volley of impulses initiated by a multipolar neuron (A) proceeds down its axon to its principal group of synaptic terminals (B). However, the persistence of the volley is limited in time by the operation of a feed-back loop. A collateral branch leaves the axon and its excitatory terminals synapse with a single inhibitory neuron (X), the terminals of which pursue a recurrent course and synapse with the dendrites and soma of the original cell (A), where their hyperpolarizing effect terminates the volley. Such feed-back circuits may vary greatly in complexity, and in some situations two inhibitory neurons (Y and Z) may be interposed at some point in a recurrent loop. Evidently, some inhibitory effect on A by neuron Z, perhaps initiated by an alternative source (C), is diminished by the

FEED-BACK
INHIBITION

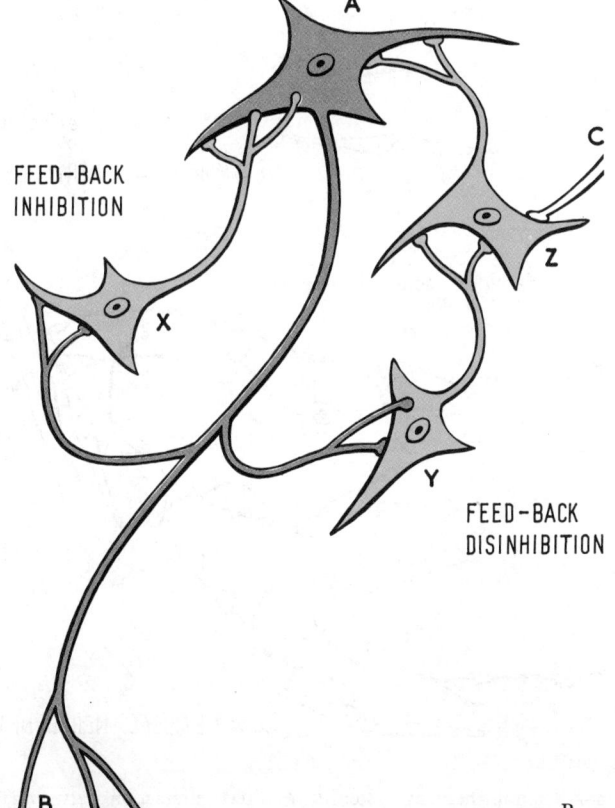

FEED-BACK
DISINHIBITION

B

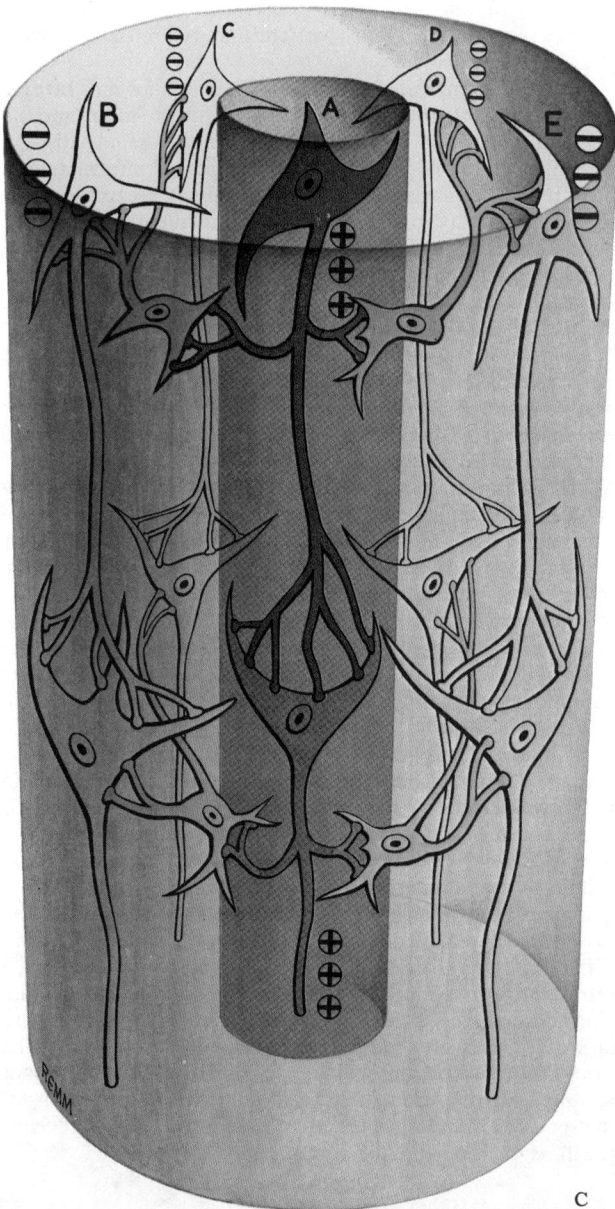

C

7.8A–C Examples of elementary inhibitory circuits; excitatory neurons shown in red, inhibitory neurons in blue: A a direct excitatory circuit (on the left) is compared with a simple feed-forward inhibitory circuit (on the right); B feed-back circuits of two orders of complexity with one (on the left) and two (on the right) inhibitory interneurons interposed on the

recurrent pathway; C lateral inhibition. A central excitatory train of neurons is surrounded by a hollow cylindrical zone of inhibition, mediated by shells of inhibitory interneurons which are activated by the central column. See text for further discussion.

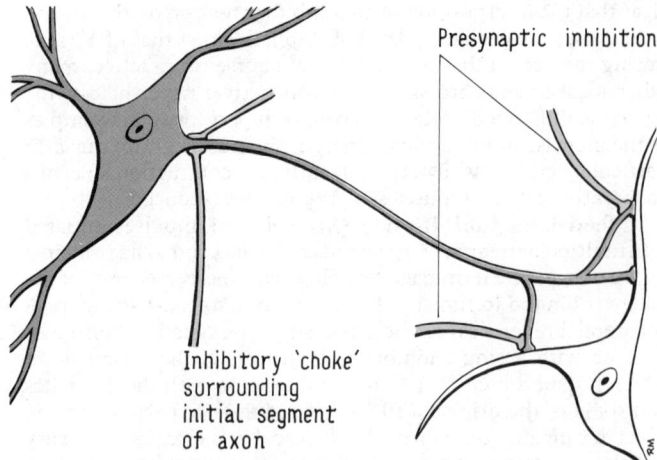

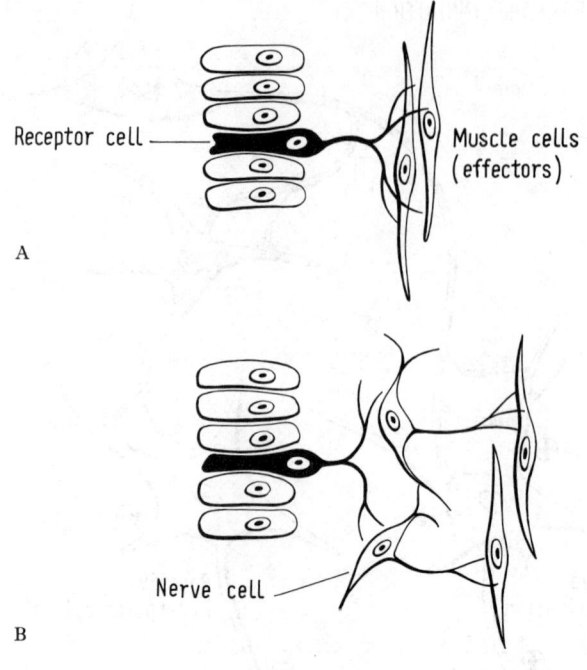

7.9 The multipolar neuron, shown in red, has inhibitory synaptic terminals (blue) applied to the initial segment of its axon, and others to the surfaces of its own axonal synaptic terminals.

operation of inhibitory neuron Y. This release from an inhibitory effect by a second one, in series with the first, is termed, somewhat inelegantly, *disinhibition*.

In **7.8**C one form of feed-forward inhibition known as **lateral inhibition** or **inhibitory surround**, of great importance in many neural pathways, is illustrated. A series of parallel pathways A–E which carry functionally similar types of information is shown. It may be imagined that all these channels are transmitting sporadically with a rather low information content. If, however, the excitatory state of one (e.g. channel A) rises significantly above the others, by the operation of neighbouring shells of inhibitory neurons, the activity in B–E is reduced. Consequently, the centrally activated channel becomes surrounded by a quiescent zone; there is less chance of confusion by random activity in the neighbouring channels, and the discriminative value of the central channel is greatly increased—a phenomenon termed *neural sharpening*. Many examples of this important process will be referred to in the subsequent sections.

Mention should also be made at this point of two further sites of inhibitory interaction between neurons (**7.9**). These forms are of less common occurrence, but they evidently have completely contrasting roles. An inhibitory terminal surrounding the initial segment of the axon is advantageously placed to *prevent* rapidly and completely an outflow from the neuron. On the other hand, *presynaptic inhibitory terminals* may *selectively diminish* the activity or effectiveness of some terminals whilst others continue to act unimpeded—a number of examples will be encountered in later sections.

It should be stated here, however, that our knowledge of the geometry of dendritic trees, cell somata and axonal branching, in quantitative terms, of the three-dimensional interconnexions between neurons, and the spatial distribution of synaptic types, is still in its infancy. So, too, is our knowledge of how individual neurons and cooperative groups of neurons transform patterns of information flow, and how these relate to the operation of the whole nervous system and the overall economy of the organism. For the same reason, much research is currently concentrated upon investigations into the causal mechanisms which operate during neurogenesis, to determine which features of neurons, and neuron assemblies, are specified by the genome, and on what is the nature of the modifications which occur during the establishment of acquired memory traces during the lifetime of the organism. Some specific examples, where illuminating research is proceeding, will be mentioned in the appropriate sections.

Some General Features of Neural Organization

What follows is an attempt to present some aspects of the organization of nervous systems, of varying intricacy, which may

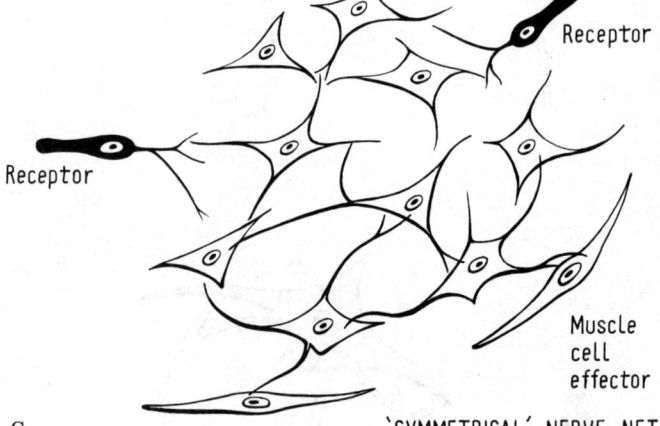

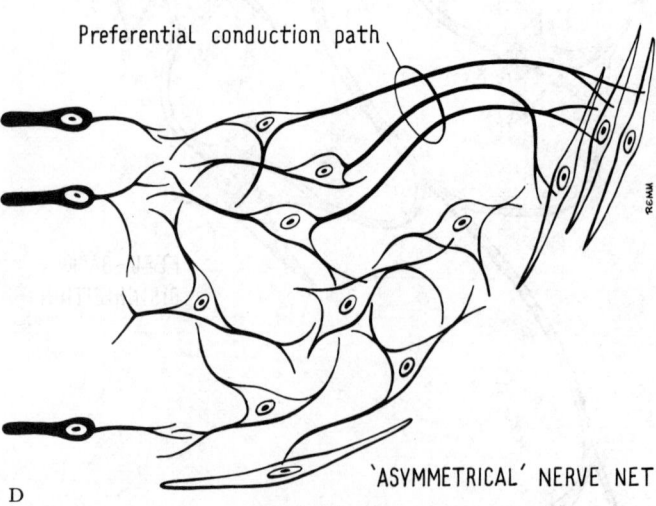

7.10A–D Simple nervous systems: A direct interaction between a receptor cell and the muscle cell effectors. B interneurons which make functional contacts with each other are interposed between the receptor cell terminals and the muscle cell effectors. C receptor cell terminals feed into a fairly symmetrical network of interneurons which in turn makes contact with the effectors. D the interneurons between the receptors and effectors are organized where 'diffuse' multisynaptic conduction pathways predominate, and others where more direct routes are present.

be of assistance before embarking upon a study of the complex nervous arrangements of vertebrates, and in particular mankind. It is in no sense intended to reflect a phylogenetic history, and no consideration will be given here to the distribution of inhibitory and excitatory neurons, to synaptic morphology, or chemical or behavioural studies; comments are limited to the elemental aspects of the arrangement of neurons.

In the unicellular protozoans, of course, the single cell constitutes the whole organism, and its various parts can act as primitive receptor, conductor and effector mechanisms. Some ciliated protozoans, for example, show coordinated beating of their cilia, so that the organism may advance towards or retreat from particular sources of stimulation (p. 18).

Simple multicellular forms possess many *independent effectors*, such as the contractile cells which surround the pore-system of sponges, or the stinging organs of coelenterates, which can respond to direct mechanical or chemical stimulation of their surfaces.

Coelenterates are the simplest multicellular forms which possess a true nervous system: and in various species and bodily regions different orders of complexity can be recognized (7.10). In some sites specialized receptor cells situated between the general epithelial cells of the surfaces of the organism send a conducting process from their bases which is in direct functional contact with underlying effectors (muscle cells). Elsewhere, however, the processes from the receptor cells discharge into a sub-epithelial '*network*' of nerve cells which are functionally interconnected in a tangential plane. From the latter, processes pass to the effectors. The interconnexions or 'synapses' between the cells forming the network are, however, not functionally polarized as they are in more complex systems; a strong stimulus applied to any point on the surface of the organism causes motor effects that spread widely. A superficial examination might indicate that the network of neurons is random (7.10C), but closer study shows that some regions have a greater density of receptor cell inputs, others have closer meshes in the network, and in yet others the conducting fibres are longer and have a distinct orientation. These features presumably subserve the more complex behavioural patterns associated with the capturing of prey, avoidance reactions, and other locomotor activities of these simple organisms. It should also be noted that these structural characteristics foreshadow similar fundamental arrangements which persist in more complex nervous systems, including those of Mammalia.

In simple forms which exhibit *bilateral symmetry* we see the first appearance of a *central nervous system* (7.11). The receptor cells are still largely situated in the surface epithelial layers, but their basal conducting processes are longer, since they traverse the tissue layers to approach the region on each side of the plane of symmetry. Here, many neuronal somata and their associated neurites are concentrated as the central nervous system, and the synaptic terminals of the receptor cells end in relation to some of them. The somata of motor (effector) neurons are sited within this concentration, and their axonal processes leave the central nervous system to reach the effectors. The remaining nerve cells are interposed between some of the receptor cell terminals and the motor neurons. They may possess short or longer axonal processes and are termed *interneurons* (*intercalated* or *internuncial* neurons are less frequently used synonyms). Thus, paths of varying complexity connect the receptor cell input and the motor neuron output; some paths are *polysynaptic*, with a number of neurons successively interposed, whilst in others a direct receptor-effector *monosynaptic* path exists. It is in such diversity of *connectivity*, that the variations in speed and complexity of behavioural responses to different environmental conditions reside.

In the higher invertebrates, which exhibit *metametric segmentation* of the body, many of the same considerations apply. The receptor cells are still largely intra-epithelial in position and their centrally directed nerve fibres collect into bundles, or *sensory nerves*, within each segment, as they approach the central nervous system. The latter consists of masses of interneurons and motor neurons—usually a ventrally situated pair of masses (or 'ganglia') in each segment. Each pair is interconnected by axonal processes which *cross the midline*, and the masses are also interconnected

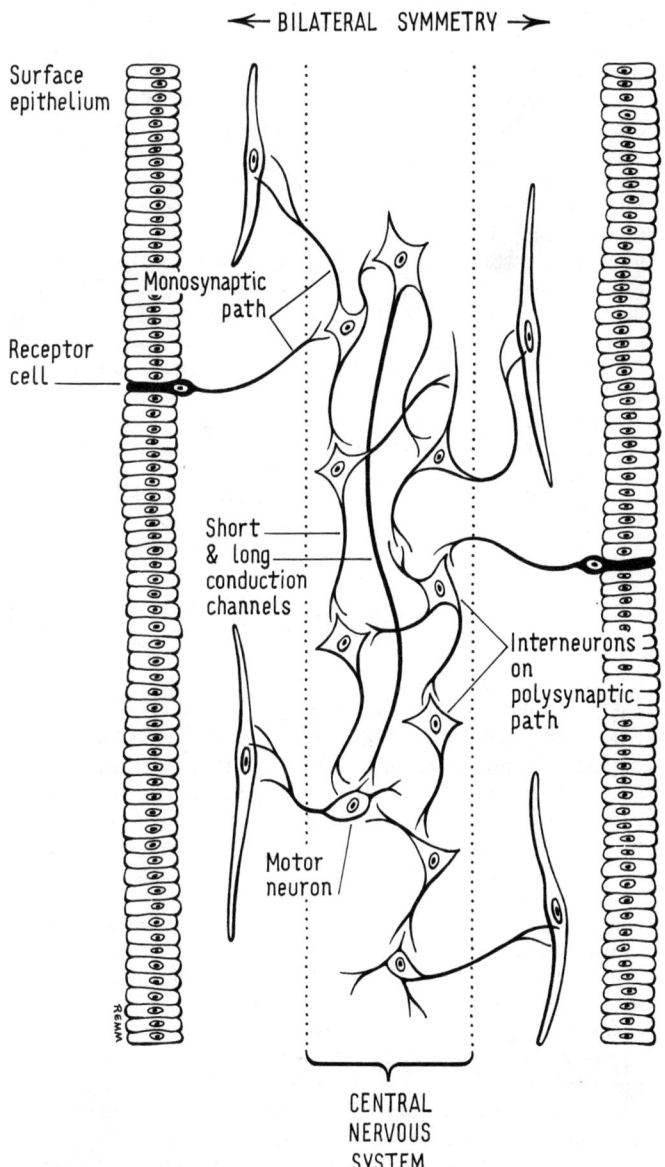

7.11   Some features of the nervous system in a primitive organism with bilateral symmetry. Note the intraepithelial positions of the receptor cells, and the aggregation of interneurons and motor neurons near the plane of symmetry as a central nervous system. Note also the different orders of complexity of the conduction pathways, and the possibility of cooperative actions involving both sides of the body.

longitudinally on both sides of the midline, so that a structure often likened to a rope-ladder is formed. The axons of the motor neurons emerge from the central nervous system as a series of bundles, *motor nerves*, within each body segment. The transverse communications also carry axonal processes of interneurons and are responsible for cooperative responses involving both sides of the body. Similarly, short-axoned neurons interconnect adjacent body segments longitudinally in both directions. Others possessing longer axons interconnect more distant segments, including the increasingly important 'head' end, whose specialized receptors often adopt a dominant role in behaviour.

As we have seen elsewhere (p. 112), during embryogenesis in *Chordata*, including *Mammalia* and mankind, the central nervous system develops in the dorsal midline, as a hollow tube-like structure which contains a *central canal*, and expands cranially to form a *brain* containing extensions of the central canal termed *ventricles*. Throughout the vertebrates, the walls of the brain and spinal cord again consist of interneurons and, less numerous, motor neurons, together with associated non-nervous glial cells and blood vessels; but two quite distinctive tissue varieties can be recognized, namely *grey matter* and *white matter*, when freshly cut nervous tissue is examined.

Grey matter consists of the cell somata, dendritic trees and initial axon segments of the neurons, the terminal segments and synaptic endings of axons which are associated with them, together with varieties of glial cell and blood vessels. Some interneurons have short axons that remain wholly within the grey matter. Many, however, possess longer axons and these leave the grey matter, where they course in functionally associated bundles termed *tracts*. Outside the grey matter each acquires a laminated lipo-protein sheath of *myelin* derived from certain varieties of neighbouring glial cells. The sheaths are essentially concerned with the functional efficiency of the nerve fibres during impulse conduction, in terms of an enhanced conduction velocity for a given diameter, and a reduction in their energy requirements (p. 843). They impart a pinkish-white opalescence to the fresh tissue—hence the term **white matter**.

The vertebrate central canal is largely lined by a ciliated columnar epithelium termed *ependyma*, and during early embryogenesis the grey matter develops in a zone immediately surrounding the ependymal lining, whilst the white matter forms external to this. At first these relationships are present throughout the length of the neural tube, and, in the spinal cord, they persist to maturity. With the development and elaboration of hindbrain, midbrain and forebrain vesicles, however, whilst some masses of grey matter are retained in the primitive para-ependymal position, others migrate outwards to insinuate themselves between the developing tracts of white matter on the ventrolateral aspects of the tube. Most importantly, profuse secondary migrations of developing grey matter occur in the roof of each of the three vesicles, where they become arranged as a subsurface series of fibrocellular laminae, which are of the greatest significance in the functioning of the system.

Before embarking upon a detailed account of human neuroanatomy, methods which have been employed in neuroanatomical research and the cytology of nervous tissue will be reviewed.

TECHNIQUES AND DEVELOPMENT OF NEUROANATOMY

The previous section has been largely a consideration of the analytical philosophy of nervous systems—only possible on the results of a long historical process of accumulation of data, hypotheses, and explanations. These results, due to the efforts of a large succession of observers and experimenters (*see*—Anglo-American Symposium 1958; Brazier 1959; Singer and Underwood 1962; Clarke and O'Malley 1968; Meyer 1971; Bellairs and Gray 1974; Neurosciences Study Programs 1967, 1970, 1974) have been primarily dependent upon perceiving relationships between observed phenomena, especially in the most effective contributions. Even these, however, were conditioned by the technical means available at the time. It hence seems appropriate to examine the *methods* by which major advances leading to current levels of knowledge have been achieved. Inevitably such a survey must be brief, and while historical in approach, it is not intended as an account of neurological discovery, but of the neuroanatomical and other techniques involved. Hence, it is impossible here to cite more than a few examples of the fruits of the various techniques and methods included, nor can treatment of these subjects be exhaustive. In any case, other references to technique, and to the advances made by them, will appear in the general description of the nervous system which follows. Ideally, the names of innovators should be kept before us by habitual citation, but this is scarcely practicable in the textbooks of today. Perhaps the ensuing pages will help to correct this deficiency, though necessarily in a limited and irregular manner.

The development of neuroanatomy, as with any other corpus of knowledge, has been linked at every stage to contemporary means of observation and experiment. Hence, it was initially slow: a period of about two millennia separated the earliest recorded observations from the introduction of even simple magnifying lenses. The great accessions to knowledge during the nineteenth and twentieth centuries have coincided with the acceleration of technical innovation and improvement during this period. Until recent times physiology lagged centuries behind anatomy, a frustration to all who find function the chief interest in structure; but in the same explosion of technical progress functional studies have overtaken the purely structural in many fields, and have indeed carried concepts of function beyond structural observation in some. In neurology this has provided a much needed intellectual stimulus to morphologists.

From the Greeks to the Renaissance

Granted the importance of technique, and hence the extremely simple means available, the paucity of neurological observation, or even speculation, recorded by earlier civilizations, prior to classical Greek times, is surprising. Opportunities presented by injury, sacrifice and mummification are obvious; yet the earliest records are limited to a description of the optic nerves by Alcmaeon (*c.* 500 B.C.) and a few statements in the Hippocratic Collection (*c.* 400 B.C.), amounting to little more than recognition of the cleft between the cerebral hemispheres and a belief that the brain is associated with intelligence. Even Aristotle (384–322 B.C.) confused peripheral nerves with tendons (with lasting effects visible in the term 'aponeurosis'); and he unaccountably set intelligence in the heart, unlike Plato (429–347 B.C.), who sited sensation and thought in the brain. Of course, biological thought was already permeated by the Four Elements of Empedocles (*c.* 493–433 B.C.) and the Four Humours of Polybus (*c.* 390 B.C.). The humoral concept in particular was to dominate and, indeed, obstruct biological thought for almost two thousand years, however improbable it now seems or how extraordinary that simple contradictory observations were ignored.

Human dissection began (as far as records support) with the rise of the Alexandrian School, whose leaders in biology, Herophilus and Erasistratus (*c.* 300–250 B.C.), are often regarded as the 'father figures' of anatomy and physiology. (The possible debt of such dissectors to the embalmers of Egypt should not be overlooked, though how far this age-old source of practical knowledge influenced deliberate dissection cannot be assessed; *see* Cave 1958.) Herophilus distinguished cerebrum and cerebellum, describing the fourth ventricle and even its calamus scriptorius; he also differentiated motor and sensory nerves. Erasistratus detailed the ventricles further, ascribing them functions entangled in humoral concepts, as was his entire physiological teaching, complicated further by the *pneuma*, or vital spirit, an intellectual impediment lingering long after the Renaissance. However, he also equated the more elaborate gyral pattern of the human brain with mankind's pre-eminent intelligence, a remarkably prophetic speculation.

With the Roman annexation of Egypt (50 B.C.) the Alexandrian School decayed; but Galen (A.D. 129–199) left his native Pergamon to study there, his most productive years being spent in Rome. His significance in medicine and biology, as a dissector and experimentalist, calls for no reiteration (Sarton 1954; Singer 1956). He classified the cranial nerves, omitting the olfactory and trochlear and confusing others, though his description was not improved until some 1,500 years later. Much of Galen's voluminous and dogmatic writings was theoretical, but his work on spinal cord function was directly based on experiment, and was not extended until Bell and Magendie added their observations early in the nineteenth century. Section between the first two vertebrae, he concluded, led to immediate death, between the third and fourth to respiratory arrest, and at lower levels to paralysis of bladder, intestines and legs. In contrast, most of Galen's prolific output was couched in humoral and vitalistic ideas which—in view of the veneration with which his teaching was held for many centuries—were a serious brake on progress.

After Galen, biological and medical thinking entered a long suspense. The classical legacy was all but extinguished in Europe during the ensuing 'Dark Ages'. Greek texts survived chiefly in Arabic translations, with little improvement in knowledge; in accord with the general intellectual lethargy, neuroanatomy marked time for a thousand years; indeed, anatomical accuracy degenerated. The rising popularity of astrology and the insistence of early Christian authority upon the unimportance and even degradation of man's physical estate, displaced interest from his own true fabric.

From the eleventh century Greek medical manuscripts were being translated into Latin, at first from Arabic versions. Though this was the start of the rewakening of European science, progress was very slow, the ensuing centuries remaining barren of fresh anatomical or physiological advance. The opinions of Hippocrates, Aristotle, and especially Galen were the approved sources of anatomical knowledge in medieval times, with almost unchallenged authority. Even in the rising universities of the twelfth and thirteenth centuries, observation of natural phenomena was neglected; achievements were literary. Dissection reappeared, but at first as an exercise to confirm established dogma. Even Mondino of Bologna, whose *Anathomia* of 1316 is regarded as the first text in the modern period, ascribed the functions of antique teaching to the brain ventricles, and in sum he added nothing of significance to neuroanatomy. Though dissection continued in Bologna, spreading to other Italian centres and to Montpellier, the fourteenth and fifteenth centuries brought little more than translations from Greek, rather than Arabic texts. The anatomical Humanists of the sixteenth century were preoccupied with the replacement of Arabic anatomical terms by Greek and Latin, but great exceptions from such drab polemics were at work, as we shall now see.

From the Renaissance to the Nineteenth Century

The movement towards naturalism amongst artists of the Renaissance in the late fifteenth and early sixteenth centuries culminated in Leonardo da Vinci (1452–1519)—*see* Belt (1956). Like his great contemporaries, Dürer, Michelangelo and Raphael, he turned to dissection for direct information on human structure. The extent of his contemporary influence is uncertain, but he brilliantly epitomizes the revival of an inquiring spirit in Europe, reacting powerfully against traditionalism and the excesses of scholastic orthodoxy. His contribution to neuroanatomy was limited but impressive—some remarkable figures of the brain, depicting wax casts of the ventricles, a technique invented by him. The movement represented by Leonardo, together with the introduction of improved techniques of engraving, paved the way for Vesalius.

Vesalius (1514–64) has received perhaps excessive medical adulation; his plates, starkly structural when compared with the functional inspiration of Leonardo's, nevertheless introduced an accuracy unrivalled in his time, marking a new era in medicine (*see* O'Malley 1964). Dissatisfied with Galenic inaccuracies, and with the near sanctity accorded them despite the ascertainable facts, and yet unable fully to emancipate himself from Galen's physiology, he revised the whole field of anatomy by his own dissections. In neuroanatomy he struck a new high level in detail and precision, though with lesser success in the peripheral nervous system. He illustrated the brain partly in a sectional manner (a technical innovation), going far beyond existing knowledge. He clearly indicated the basal nuclei, hippocampus, fornix, internal capsule, pulvinar, colliculi, fourth ventricle, and many other details, with almost the accuracy of a modern atlas.

Eustachius (1550–74), a contemporary of Vesalius, has received less than just acclaim; his work, like Leonardo's, only became generally known long after his death. He left a brilliant depiction of the sympathetic nervous system (7.12), and he portrayed the cranial nerves more accurately than Vesalius and the pons (Varolii) than Varolius.

Vesalius' successors at Padua were less eminent, but exerted much influence on visiting scholars. Thus, Coiter of Montpellier (1534–76), a notable early comparative anatomist, studied under Fallopius, and he improved the work of Vesalius and Eustachius on the cranial nerves, describing the two roots of spinal nerves and distinguishing white and grey matter in the spinal cord. Fabricius, also a comparative anatomist, taught Harvey (1578–1657), whose celebrated discovery of the circulation was strangely combined with a veneration for Galen's views in other matters, sometimes almost servile. Nevertheless, Harvey rejected the flow of 'spirits' in hollow nerves, a break with dogma perhaps not unassociated with the influence of his contemporary, Bacon (1561–1626). Bacon's contempt for scholastic orthodoxy, and his inductive approach to experimentation and factual verification of hypothesis, have proved the first clear advance beyond the logic of Aristotle. This new scientific attitude, though a working philosophy rather than a practical technique, has perhaps influenced subsequent scientific endeavour as profoundly as Harvey's experimental genius. Thereafter the tempo of discovery has slowly augmented, with the focus upon humours in nerves or the blood, as avenues of control and communication, gradually transferring to the nervous system.

Until the times of Vesalius and Harvey the only techniques for study of the nervous system were dissection and the simplest experimentation, both observed with the unaided eye. No sudden changes occurred; but during the seventeenth and eighteenth centuries discoveries in parallel fields, especially in electrical phenomena and optics, have proved highly significant to nervous system investigation. The experiments of Gilbert (1540–1603) in electricity and magnetism foreshadowed the development of electrical stimulation of nerves by Galvani (1737–98) and Volta (1745–1827), which heralded the evolution of neurophysiology. Simple microscopes began to appear, enabling van Leeuwenhoek (1632–1723) to dispute the hollowness of nerves; the compound instruments which rapidly followed, though notably exploited by observers such as Hooke (1635–1703), were not applied extensively to neuroanatomy until the early nineteenth century.

Meanwhile, topographical neuroanatomy prospered, and such exponents as Willis (1621–75) and Vieussens (1641–1716) hardened the brain, either by soaking it in wine or by boiling. They were hence able to dissect it, an advance over the slicing methods of their predecessors, enabling them to achieve a new level of excellence. Willis in particular was a pioneer of macroscopic dissection of nerve fibre bundles in the central nervous system, the only technique available for such investigation for nearly two centuries after him. Naturally disposed towards experiment, he was also approaching modern concepts of reflex activity.

During the latter part of the eighteenth century the speculations of Willis, von Haller, Whytt and others regarding involuntary reactions to stimulation were beginning to emerge as realities in the experiments of the 'galvanists'. At the same time the compound microscope was coming into wider use; and during the first half of the nineteenth century the accessory supporting techniques of fixation, microtomy, and staining were also evolving. Neuroanatomists were turning from external features to internal structure, though the former diverged into a wide interest in comparative morphology. Many workers were becoming more specialized, and it is perhaps less confusing to consider further technical progress as a number of arbitrarily separated topics.

Comparative Anatomy and Embryology

Though Erasistratus is sometimes regarded as the first comparative neuroanatomist, the modern movement stems from the observations of Willis (1664); see also Cole (1944), Needham (1959). At first the accent was on functional rather than evolutionary concepts; and in this Willis laid the foundations of nineteenth-century preoccupation with localization of function, especially in the cerebral cortex. Gall (1758–1828) and his pupil, Spurzheim (1776–1832), though achieving a notoriety undeserved by the former through their erroneous theory of phrenology, nevertheless were initially responsible for the

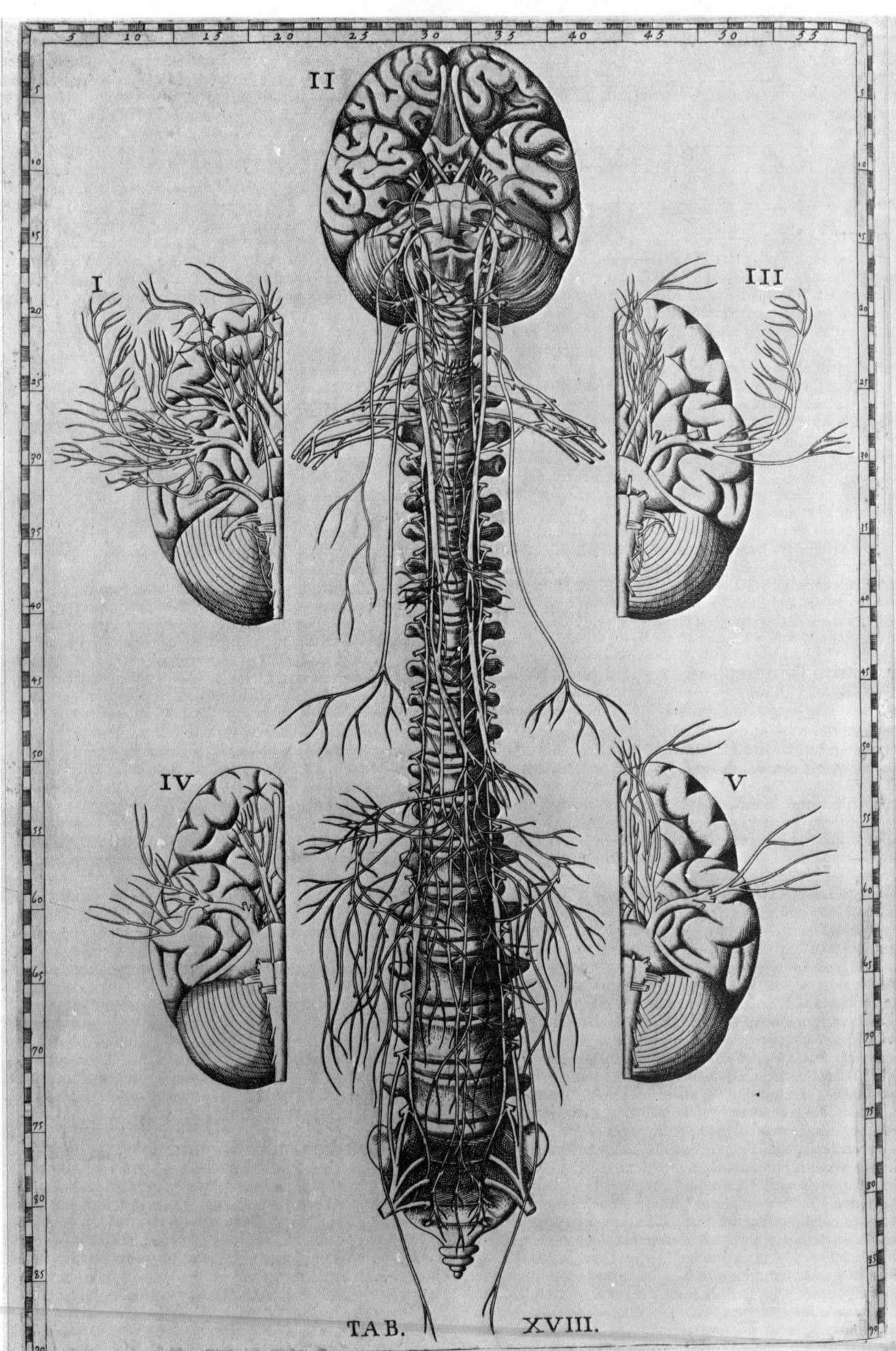

TAB. XVIII.

principle of cortical localization. In their massive publications (Gall and Spurzheim 1810–19) they much improved on Willis's comparative observations on cerebral gyri, and they also carried further his method of brain dissection. Tiedemann (1781–1861) was a pioneer in description of the human fetal brain, and Reil (1759–1813) concentrated on the development and comparative anatomy of the cerebellum and insula, advocating the technical value of combining such studies in functional interpretation (Tiedemann 1860; Reil 1807). Owen (1868) added the concept of brain weight and volume as a comparative technique, and he also linked the early developmental appearance of certain gyri with the regularity of their occurrence. Many others extended the series of animals studied, especially in respect of primates, and in this the advent of Darwinism was an additional spur. Throughout the nineteenth century a wealth of comparative and embryological observations were adduced in evidence of homologies in cortical pattern to support the polemics of the 'localizationists' and their opponents, a now almost forgotten controversy. All these contributions were based on macroscopic technique; but with Campbell, the Vogts and Brodmann (1909), at the end of the century and extending into the twentieth, microscopic investigation of the cortex on a comparative basis became established. The vast cytoarchitectonic investigations, familiar in the 'Brodmann maps' of many current texts, have never been equalled except by Conel's studies of developing cortex (1939–59).

With the great advances in neurohistology and neurophysiology in the late nineteenth and twentieth centuries, interest in topographical comparative neuroanatomy has somewhat waned. Nevertheless, with the new morphological techniques classical work such as that of Herrick (1931), of Larsell (1937) on the cerebellum, and Polyak (1957) on the visual system, continued to appear. The great compendia of comparative neuroanatomy compiled by Ariëns Kappers and his collaborators must also be mentioned (Kappers 1920–1; Kappers et al. 1936). (References to the work of others, including Tilney, Conolly, Bailey and von Bonin can be found in the recent text of Crosby et al. 1962.)

The linkage between embryology and comparative anatomy was early established, as already noted. Arnold (1842), for example, described the early development of the hippocampal gyrus in primates; His (1874) distinguished 'complete' fissures, which appear first, and secondary cortical sulci; and Broca (1878) supported his concept of a speech centre as much by such considerations as by functional data.

More significant, however, have been embryological studies at the microscopic level. His (1887) demonstrated stained preparations showing that axons grow from neuroblasts, thus providing positive evidence for the neuron doctrine, and in a sense foreshadowing the tissue culture studies of Harrison (1907). The work of His was confirmed by Held (1893), who nevertheless concluded that synaptic gaps, as we would call them today, disappeared during maturation of neurons—evidence against the individuality of nerve cells. Flechsig (1876) discovered that myelination proceeds—before and after birth—at different rates in various tracts, thus introducing a *myelogenetic technique* for investigation of fibre connexions.

Neurohistological Technique and Light Microscopy

(*See* Clay and Court 1932; Brachet and Mirsky 1960; Clark 1961.)

Leeuwenhoek (1674) identified nerve fibres with a simple high-powered lens, and with a similar 'microscope' Swammerdam (1675) even dissected out the nervous system of the mayfly larva, a

brilliant anticipation of microdissection. Jansen (1590) had already constructed the first compound microscope, put to practical but not biological use by Galileo. Kepler is also credited with a similar invention, carried into effect by Scheiner (1611). Huygens (*c.* 1684) introduced the compound eyepiece. Malpighi (1666) has been supposed to have been first to visualize nerve cells with a somewhat crude compound microscope, but a reconstruction of his observations with a similar instrument has neatly demolished this claim (Clarke and Bearn 1968). Fontana (1781) described nerve fibres clearly, but failed to distinguish axon from sheath. Further advances were in fact impossible without improvement in microscopy. Lister (1827) and Amici (1827) manufactured *achromatic* objectives; however, Lister's pioneer efforts were sustained by no British successor, the centre of further development passing to Germany, where Abbé, working for Carl Zeiss, was able to construct *apochromatic* objectives in 1886, using the newly discovered Jena glasses. He also improved the condenser and devised compensating eyepieces. Darkfield condensers (Wenham *c.*1853) and immersion objectives (Tolles 1874 and Abbé 1878) further accelerated the refinement of the late nineteenth-century microscope.

The immediate result of even achromatic objectives was a flood of descriptions of animal tissues. Botanical *cells* had been portrayed long before, of course, by Hooke (1665) and so named. Nerve cells, a discovery at one time accorded to Schwann when enunciating the cell theory with Schleiden in 1839, are now considered to have been first satisfactorily recognized by Ehrenberg (1833). However, there seems to be little doubt that Schwann first described the neurolemmal cell. These early achievements were rapidly surpassed with the new optical resources, but this progress rested equally upon development of adequate methods of preparation and staining of animal tissues.

PREPARATION OF MATERIAL FOR MICROSCOPY

Early microscopists used teased or squashed preparations of either fresh or, at most, partially preserved biological material. Since fixation and hardening are essential to adequate sectioning, especially of animal tissues, Reil's introduction of alcohol for this purpose in 1809 was an important advance. Hannover added chromic acid as a fixative in 1840; and Stilling (1842) used both methods, sometimes coupled with freezing, in accomplishing his classical work on the spinal cord (1846), cutting it both transversely and in longitudinal section. He also advocated serial sections, as did Virchow (1846), in describing and naming neuroglia. Formalin fixation was introduced much later by Blum (1893). Stilling and Wallach (1842) devised a simple hand microtome, improved by Welcher (1856); machines of increasing accuracy were invented and improved throughout the second half of the nineteenth century, still linked with names such as Jung, Thoma, Cambridge Rocker, etc. Von Gudden and Catsch (1875) were able to cut whole sections of the cerebral hemisphere, and Rutherford and Cathcart (1873) introduced the first freezing microtome. By the 1880s semi-automatic microtomes, such as the Cambridge Rocker, were capable of producing ribbons of thin serial sections from wax embedded material, paraffin wax having been introduced by Klebs in 1869. Embedding of tissue in collodion (Duval 1879), celloidin (Schiefferdecker 1882) and other materials followed, leading to such modern innovations as water-soluble waxes, freeze-drying, and cryostat techniques.

TISSUE STAINING METHODS IN NEUROANATOMY

It is a source of surprise and admiration to note how much was achieved by the early nineteenth-century microscopists without staining techniques, though it would be misleading to overlook the sporadic use of colouring methods from the beginnings of microscopy (Baker 1960; Gurr 1965; Pearse 1968). Nevertheless, Remak's early description of axons and their sheaths in 1836 was based on unstained embryonic material (Remak 1836), and under similar difficulties Purkinje, a year later, confirmed these observations and identified the cerebellar nerve cells named after him. Incidentally, Purkinje adopted acetic alcohol as a clearing agent and Canada balsam for permanent mounting of sections. It

7.12 The brain and autonomic nervous system depicted by Bartolomeo Eustachio (1550–74), a contemporary and rival of Vesalius (1514–64), from *Tabulae Anatomicae*, published posthumously in 1714 by Lancisi. The copperplates of Eustachio (or Eustachius) are not only anatomically more accurate than the woodcuts of Vesalius, but also technically superior. (By courtesy of the Trustees of the British Museum.)

is nowadays difficult to believe that Waller's study of degeneration in nerve fibres reported in 1850 was conducted on teased nerves, unfixed and unstained; or that Kühne's discovery of motor endings in 1862 followed mere treatment with weak hydrochloric acid to clarify the sarcoplasm.

The first neurohistological stain was probably the carmine and gold method of Gerlach (1858), which enabled him to advance a *reticular* or *net* theory of neuronal continuity. The controversy between the protagonists of this and the rival *neuron theory* reverberated down the rest of the nineteenth into the twentieth century. Though Faraday had discovered benzene in 1825, and Perkins had prepared the first aniline dye, mauveine, in 1856, the industrial development of dyes was quickly taken up in Germany, a by-product being a rapidly increasing range of such dyes available for histological use from 1862 onwards (*see* Conn 1948). Carmine staining was applied by Deiters (1865) in differentiating between dendrites and axons. Ranvier's (1871) classical delineation of nerve fibres and their nodes depended upon a silver impregnation of chromic acid fixed tissue; and Golgi (1878) further elaborated silver staining, thus rendering occasional whole neurons visible, from which preparations he described the

cell types still known today by his name. (While Golgi's technique is usually regarded as applicable only to small volumes of nervous tissue, Golgi himself and some recent workers have stained whole human brains in this manner. *See* Kemali *et al.* 1977.) Golgi emphatically supported the reticular theory, but Ramon y Cajal (1908), who adopted Golgi's metallic impregnation methods and added his own refinements, was thus able to accumulate such an overpowering array of observations on neurons in all parts of the human nervous system, and in many other animals, that the neuron theory has prevailed. The accuracy and volume of this work was the chief factor in advancing the individuality of nerve cells, although the theory is often accorded to Waldeyer. Not until the advent of the electron microscope was it possible to adduce more convincing evidence in favour of Cajal's teaching; however, even as recently as 1933, a year before the appearance of the first electron microscope, he considered it necessary to re-emphasize his views. Cajal also used Nissl's methylene-blue method for staining 'chromatin granules' (*vide infra*). Ehrlich (1886), applying the same dye to living tissues and thus initiating vital staining, was thereby able to define peripheral nerve cell processes to their terminations; being an axonal stain, this technique has

7.13A–C Diagram of the sequelae of interruption of the fibres of central nervous system neurons. As shown, regeneration or atrophy may follow the response to injury, both in the neuron primarily damaged and in those affected transneuronally. Transneuronal effects may occasionally supervene in neurons more remotely associated with the damaged neuron. See text for further details.

proved most valuable, in various modifications, for the demonstration of peripheral non-myelinated autonomic nerve fibres.

Nissl (1892) discovered the *acute response* of a nerve cell body to damage of its axon, characterized by *chromatolysis*, a breakdown of the *Nissl* or *chromatin granules*, as well as other manifestations, such as swelling of the cell and eccentricity of its nucleus. (These, and other features of retrograde changes are described more fully elsewhere—pp. 861–863.) He thus provided further evidence of continuity of nerve cell and axon and also a most valuable means of delineating the arrangement of neurons. Weigert (1882), somewhat earlier, had elaborated a staining technique for *normal* myelin sheaths based on pretreatment with potassium bichromate, followed by acid fuchsin (haematoxylin is now more often used). It is interesting to note that he was already using collodion for embedding material, xylol for clearing, and balsam as a mounting medium. This innovation was quickly followed by Marchi's osmic acid method for staining *degenerating* myelin, for which it is specific and hence an early example of histochemical technique (Marchi and Algeri 1885–6). A few years before Nissl's description of the early response of neurons to damage, Gudden (1870) had studied the later sequelae of these changes, finding that the injured cells commonly atrophied and ultimately disappeared; he also noted that the changes were more rapid and marked in young animals.

Both Gudden and Nissl made considerable personal contributions using their own techniques. The former's original experiments were an investigation of thalamocortical connexions, which his pupil von Monakow (1882) carried further, also discovering the visual function of the lateral geniculate body. It is noteworthy that both problems have been reinvestigated by the same technique half a century later (Le Gros Clark and Penman 1934; Le Gros Clark 1936), an example of the continuing value of such degeneration techniques; these have continued in use throughout the twentieth century, and their application, together with certain other allied methods, is of such general importance in neuroanatomy that they must now be considered in more detail.

Neuron Tracing Techniques

The researches of Waller, Gudden, Golgi, Weigert, Marchi, Nissl and Cajal were all published within the last thirty years of the nineteenth century, except for Waller's basic observations in 1850. His demonstration that damage to nerve fibres induces microscopically identifiable changes proved a most fruitful starting point in this brilliant period and, indeed, in neuro-anatomy and neuropathology in general. His immediate successors established the broad basis of neuronal morphology and the major techniques for an enormous volume of investigation into the organization of the nervous system. Modifications and some additions have followed, as we shall see, but their methodology remains substantially unchanged, persisting up to the present in parallel with subsequently evolved techniques in neurophysiology, histochemistry and electron microscopy (*vide infra*).

The essential features of the use of degeneration techniques in tracing out the arrangements of nerve cells and their neurites are shown in illustrations 7.13, 14.

When the cell body is actually ablated or damaged to such an extent that it dies, all its neurites, including the whole of the axon and its sheath of myelin, degenerate and finally disappear (for further details *see* p. 862), and during this period the changes described by Waller can be distinguished by Marchi staining of the degenerating myelin. The axon's terminal branchings, including their terminal synaptic buttons (*boutons terminaux*), are equally involved in their degenerative process, though this was not at first appreciated. Being devoid of myelin these degenerating terminals did not stain by Marchi's technique, and this often led to errors when interpreting their presumed destination. More recently, however, methods introduced by Glees (1946), and by Nauta and Gygax (1951), with various modifications by other workers, do stain these degenerating preterminal and terminal structures. More recently still, ultrastructural changes in *boutons*

terminaux during degeneration have been utilized for fibre tracing.

However, as we shall see, there are many technical difficulties associated with the placing of small discrete experimental lesions within a mass of nervous tissue. Further, there is no known method of selectively ablating *cell bodies alone* within the central nervous system; in all experimental situations, their dendritic trees and the synaptic terminals from other functionally associated neuron groups, blood vessels, and often other structures, are involved simultaneously. In some respects a clearer experiment results when axonal processes are interrupted at some point along their course, and various terms have been introduced to indicate the different situations in which degenerative changes occur following such focal damage. These are merely summarized here.

Firstly, the effects confined to the neuron that has been damaged—those affecting that part of the axon, and associated structures, which has been detached from the parent cell, i.e. *distal* to the lesion, are termed *anterograde*; those affecting the dendritic tree, parent cell body and that part of the axon still attached to the cell body, i.e. *proximal* to the lesion, are *retrograde*. Secondly, changes may follow in neurons which have suffered no direct damage, but are functionally associated with the damaged cell. These are *transneuronal* (or *transynaptic*) changes, and they may be *primary* or *secondary*, etc. depending upon their degree of separation from the damaged cell, and further they may be *anterograde* or *retrograde*, depending upon their position relative to the damaged cell in a functional train of neurons. These varieties are detailed further in the following pages, and elsewhere in the present section.

Following a focal lesion of the axon, anterograde degeneration (of the Wallerian type involving again both the axon, its terminals, and myelin sheath) occurs distal to the lesion. The changes do not occur simultaneously at all points, but spread as a 'wave-front' distally from the lesion. After suitable time intervals they may be followed experimentally using the methods of Marchi, Glees, Nauta, Fink-Heimer, and histochemical and ultrastructural methods.

The proximal part of the fibre often does not show any changes when stained by Marchi's method, and for long it was believed that no degeneration occurred there although Cajal had described a process of retrograde degeneration of Wallerian type which extended backwards along the fibre, but only as far as the first node of Ranvier proximal to the site of damage. However, it has now been established that more extensive changes sometimes do occur in the fibre proximal to the injury, and that is when there occurs a *retrograde atrophy* of the nerve cell body in question (which is of course accompanied by degeneration of *all* its neurites). Such an atrophy of the cell soma is a later sequel of the *acute retrograde* (chromatolytic) *response* studied by Nissl (but as we shall see the latter is not followed in every case by the more chronic manifestations of atrophy). The degenerative changes in the nerve fibres are thus termed *retrograde* in the sense that they occur proximal to the lesion, but they proceed in a *centrifugal* direction, from the cell soma towards the lesion. Again, at suitable time intervals they may be studied using the methods of Marchi, Glees, Nauta and Fink-Heimer.

These various phenomena develop more rapidly and are more marked in degree in newborn or at least very young animals, as Gudden noted, whereas Nissl apparently did not. Brodal (1940), who has worked extensively with these methods, has introduced a so-called *modified Gudden technique*, which depends upon using very young experimental animals, in which both the acute and chronic types of retrograde change occur much more rapidly, the former being more usually relied upon.

The application of such methods, and the selection of a particular one in a given problem, require careful consideration; and the control of timing between infliction of a lesion and inspection of the consequent degeneration is critical. Moreover, workers in this field have reported much variation in the extent and rate of degenerative sequelae in different kinds of neurons and in the same cells in different species. Critiques of these experimental difficulties should be consulted (e.g. Glees 1961; Brodal 1969), as also the many publications of workers by whom these techniques have been employed.

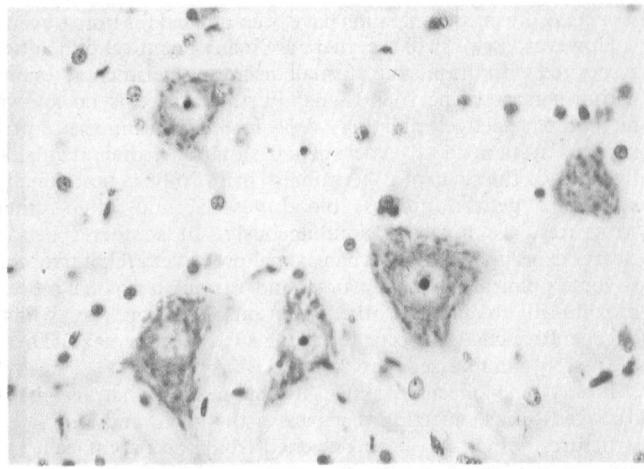

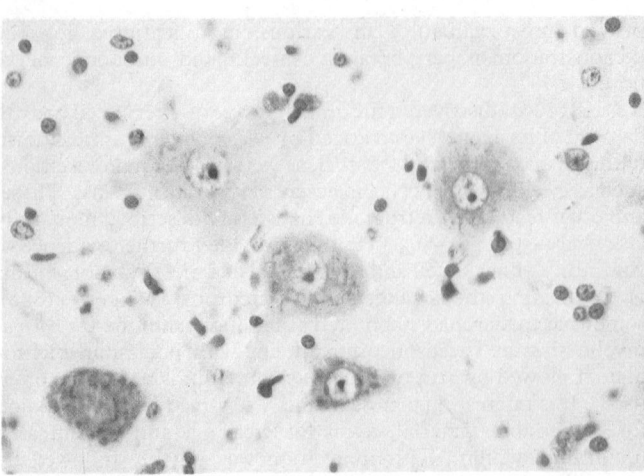

7.14A–D A Large multipolar neuronal perikarya in the magnocellular part of the feline red nucleus, showing prominent Nissl's granules, bases of dendrites and axon hillocks. The nuclei are euchromatic and vesicular, with prominent nucleoli. The small nuclei scattered in the surrounding neuropil are characteristic of the various categories of neuroglial cells.

B A field similar to that depicted in A, but after previous contralateral hemisection of the spinal cord at the level of the fifth cervical segment, thereby severing the rubrospinal tract. The section shows characteristic chromatolytic retrograde changes in the cytoplasm of three large neurons. Two smaller neurons are unaffected.

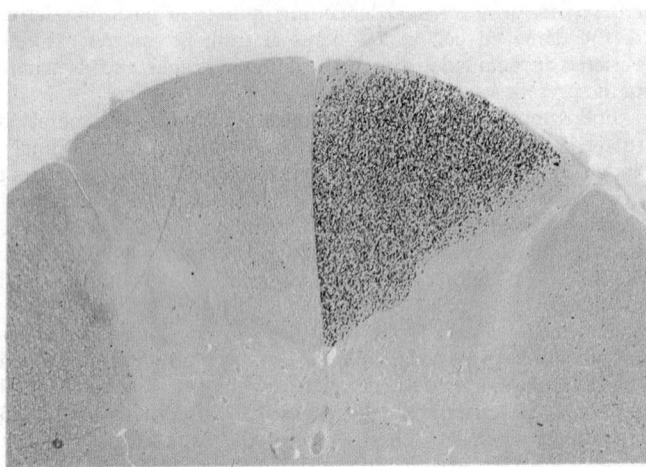

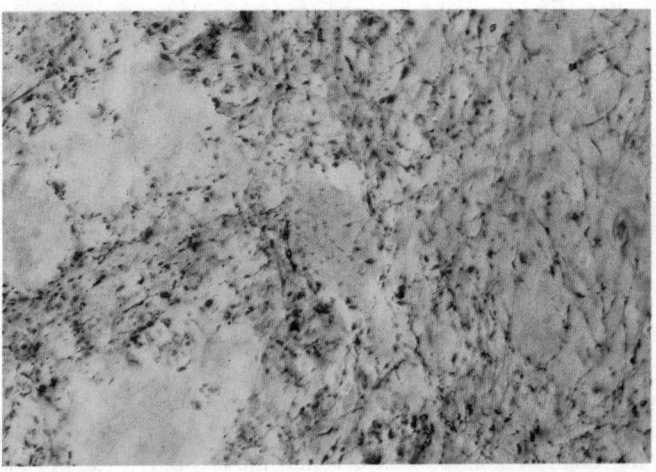

C Transverse section through dorsal funiculi of feline cervical spinal cord, after unilateral dorsal column section at a more caudal level. Note anterograde Wallerian degeneration of ascending nerve fibres: degenerating myelin sheaths are stained black by Marchi technique. (Provided by Dr. E. W. Baxter, Dept. of Biology, Guy's Hospital Medical School.)

D A high-power light micrograph showing preterminal degeneration of afferent axons ending in relation to neurons of the red nucleus of a rat, after the previous placing of a cerebellar lesion. The preparation was stained by the Nauta-Gygax method. (Kindly supplied by Prof. K. E. Webster, King's College, University of London.)

The ultimate fate of neurons damaged deliberately for experimental purposes is not always the same, and the following remarks apply also to the long-term results of naturally occurring injury. As stated above, when the nerve cell is ablated or coagulated, or otherwise severely damaged as in the infliction of focal lesions (*vide infra*), the whole neuron, including all its processes, degenerates and disappears. However, where axonal injury is inflicted, the acute retrograde reaction which supervenes may be followed either by a gradual recovery extending over some weeks, or by chronic retrograde atrophy, a much slower process with ultimate disappearance of the cell and, of course, its processes. Quantitative aspects of these phenomena have been investigated in some respects by Turner (1943) and Bodian and Mellors (1945), amongst others. Where the nerve cell survives, its characteristic pattern of chromatin granules returns and the proximal, attached part of its axon recovers, whereas its severed part is finally absorbed, after passing through the changes of Wallerian degeneration. The time relations of the latter vary in different instances, and they may be very slow. In the spinal cord it has been possible to apply Marchi staining to identify degenerating tracts up to a year or so after injury (Smith 1951).

So far the retrograde and anterograde degenerative responses to injury *within a single neuron* have been considered. Early in the present century, however, it was recognized that experimentally induced degeneration in one set of neurons may be followed by similar changes in others, arranged in the same functional sequence, the phenomenon being termed *transneuronal* or *transsynaptic degeneration*.

The best known example of *anterograde transneuronal degeneration* occurs in the visual pathway: selective lesions of the retina result in degeneration not only of the nerve fibres of the injured ganglion cells which travel thence to the lateral geniculate body, but also of the geniculate neurons with which these fibres form synapses. The process may extend even further, involving a second group of functionally associated neurons in the striate cortex. It is thus possible to speak of *primary* and *secondary* transneuronal degeneration. The potentiality of this finding for elegant experimentation has not been overlooked, but this kind of 'polysynaptic' degenerative response has not been discovered in more than a few instances. There are indications that it may be limited to situations where neurons receive their afferent impulses predominantly, if not exclusively, from a single source. However, ventral grey column cells in the spinal cord receive afferents from multiple sources, and yet it has been claimed that they exhibit transneuronal degeneration when appropriate dorsal spinal roots are cut; these claims have been severely criticized. However that may be, the technique offers obvious possibilities in tracing pathways, and it may be more widely applicable than is so far expected.

Retrograde transneuronal degeneration has received less attention, though its occurrence in Gudden's cortical ablation experiments is now regarded as undoubted. The damage inflicted caused not only degeneration in anterior thalamic neurons, whose axons reach the cortex, but also in other thalamic nuclei now known to be in a primary and secondary functional relation to the anterior thalamus (Cowan 1970). In Gudden's investigations the destructive technique was surgical ablation, a method still used when applicable, and when no more accurate means is required. A number of other methods of making highly selective lesions have been devised.

SELECTIVE LESIONS AND STEREOTAXIS

In using fibre-tracing techniques an appropriately accurate method of killing nerve cells or dividing their fibres is essential. Simple division may suffice in peripheral nerves or in superficial central tracts; crushing or repeated injury has been advocated as more effective in the former. Superficial targets, such as cerebral or cerebellar cortex, can be damaged by ablation or by simple removal of pia mater, which naturally devascularizes a considerable depth of tissue. Most targets, however, are at some depth, and the surgical approach may invalidate some experiments by undesired damage to other regions. This difficulty has prompted alternative techniques, such as the introduction of fine electrodes, which can be used for stimulation as well as electrocoagulation, or of a fine probe capable of producing intense cold at its tip. Implantation of destructive substances, including alcohol, hydrocyanide, carbon dioxide snow, or radioactive yttrium, has also been practised. Irradiation by focused ultrasound (Warwick and Pond 1968) or proton beam (Malis *et al.* 1957) offers the advantage of causing damage within the nervous system without entering it, thus obviating even the usually trivial damage entailed by the passage of an electrode or probe. The latter complication becomes more serious when large or irregular structures, such as nuclei or ganglia, are to be destroyed, because repeated overlapping lesions must be made, requiring repeated insertions. The use of antibody-antigen reactions has been suggested, and it is claimed that the caudate nucleus has been selectively destroyed in this manner.

All the methods commonly employed to make selective lesions of the central nervous system, usually some part of the brain, entail the complication of *stereotaxis*—some means of aligning an electrode, cold probe, transducer, or the like with respect to the target selected (Carpenter and Whittier 1952). The usual method is to fix the head of the experimental animal in a rigid frame in relation to which the electrode, etc. can be adjusted in three planes according to the coordinates of a particular target. These coordinates are established by measurements of the position of the target structure in serial sections of the whole head prepared in three planes, or by the use of an X-ray guidance system, using such features as air-filled ventricles as reference points. Preliminary dissection down to the target for direct estimation of coordinates in preserved material has also been advocated. Since all such coordinate data vary a little from brain to brain, and with the method employed, the degree of accuracy obtainable is limited, but is acceptable for all but very small structures. Horsley and Clarke (1908) constructed the prototype stereotactic apparatus, and a considerable succession of modifications have improved on this, particularly under the stimulus of neurosurgery.

The Rise of Histochemistry, Electron Microscopy, and Neurophysiology

The morphological techniques so far discussed continue to be used and refined, and in parallel with them are electrophysiological methods which must also be mentioned.

With improving instrumentation in the early part of the twentieth century, the study of action potentials in nerve fibres has become increasingly accurate, and with continuous refinement in electrodes, smaller and smaller groups of fibres or nerve cells can be subjected to electrical recording. The technique of *evoked potentials* consists in controlled stimulation at one point with recording of the induced action potentials at another (Chang 1959). The stimulus may be applied at receptor organs and, for example, the recording terminal may be in the cerebral cortex, and the pathway may in fact involve a number of synapses. Another technique—*strychnine neuronography*—is less applied at the present time; strychnine solution applied to small areas of the cortex, for example, causes rapid firing in subjacent neurons, and the recording of distant responses can be correlated with the area stimulated (Dusser de Barenne *et al.* 1942).

These two kinds of approach to fibre tracing are often differentiated as anatomical and physiological, but in fact each inescapably contains an element of the other, and they often provide complementary advantages and an important check on each other. With either, whenever relatively lengthy pathways are under investigation, precise details of the intermediate route between the points of stimulation or destruction and the sites of consequent distant effects may require repeated experiment for full elucidation. The distance between such experimental terminals may be contracted or extended, or intermediate blockage at suspected sites may abolish a response and thus reveal the pathway situation at that level. A useful combined technique consists in the use of the same electrode to produce electrocoagulation in neurons to set in train degeneration along fibres from which electrophysiological activity has just been recorded.

By combination of data from all these various methods a most impressive volume of knowledge of the interconnexions of the parts of the central and peripheral nervous system has now accumulated; and by further application of the methods most likely to give accurate and reliable information, and by combining them where appropriate, this knowledge will doubtless continue to grow. Of late, however, attention has been directed most intensively towards the finer details of interneuronal connexions (e.g. Horridge 1968). By insertion of microelectrodes or micropipettes directly into individual neurons the activity of single cells has been recorded in a wide variety of situations in the central nervous system. By simultaneous recording from numbers of cells in small 'integrative units' or 'microcircuits' of interconnexion, and by combining the results of such experimentation with recent highly detailed studies of the morphology of the neurons involved and their synapses, a new era of analysis of nervous system function is beginning (e.g. Lissák 1967; Shepherd 1974; Rakic 1975; Schmitt *et al.* 1976). Earlier concepts of excitation, inhibition and facilitation are being expanded by disinhibition, disfacilitation, lateral, pre-, and post-synaptic inhibition, and the like, a somewhat inelegant neology, but perhaps necessary to the rapidly growing information on activity in the repetitive interneuronal patterns so characteristic of much of the central nervous system. The role of the synapse in these studies is basic, and in this regard the techniques of histochemistry and electron microscopy have mediated great progress.

PHYSIOLOGICAL TECHNIQUES FOR TRACING NERVE FIBRES

By the end of the nineteenth century microscopes and supporting histological techniques were well advanced. The general appearance of nerve cells and their fibres were established, their diverse types had been described in some detail, and the neuron theory was widely accepted. The gross and microscopic arrangement of the nervous system, central and peripheral, including its autonomic moiety, had been defined in much detail and in sufficiently explanatory functional terms to be highly useful in clinical diagnosis, even though understanding of the basic mechanisms might still be elementary. It is appropriate to note here the considerable contributions to neuroanatomy made by neurologists, both in the nineteenth century, when a combination of clinical and laboratory activities was more usual and perhaps easier, and in the twentieth, despite the effects of specialization and the increasing technical complexity of experimentation. By this stage also, neurophysiological approaches were proving particularly successful in localization of function and reflex activity. The spectacular evolution of electronics had provided a continually increasing array of accurate instrumental techniques for the measurement of nerve fibre activities, the behaviour of synapses, and the electrical phenomena of integrated cerebral activity. The study of synapses has perhaps been even more potently stimulated by the discovery of transmitter substances (by Loewi, Dale, and others), such as acetylcholine and adrenalin, with immense physiological, biochemical and pharmacological consequences. In company with these brilliant advances morphological investigation has continued under the particular impetus of new light microscope techniques, new histochemical developments and the invention of the electron microscope.

DEVELOPMENTS IN LIGHT MICROSCOPY

General development of the light microscope has produced a highly flexible instrument with refined optics, and with such alternatives as dark-field illumination for examination of living cells in tissue culture, polarization accessories allowing the definition of birefringent substances such as myelin, and the addition of a host of auxiliary equipment, including varieties of moving stage, warmed stages for tissue culture, recording cameras for still, ciné and time-lapse photography, devices for micromeasurement including microdensitometry, microdissection and micro-injection.

The introduction of the phase contrast microscope of the Zernicke type in 1935, and of its congener, the interference microscope by Francis Smith in 1947, has ushered in a new era in the examination of living cells. Phase microscopy has become a normal adjunct of tissue culture, which has been extensively applied in the study of nerve cells and neuroglia (Geiger 1963; Murray 1965). Like the phase instrument, the interference microscope can be used to assess refractive index, even in different parts of the same cell, and the latter also makes possible measurements of concentration, water content and dry mass in living cells (Ross 1967).

Recently, incident illumination optics have been introduced as a method for studying myelinated nerve fibre populations *in vivo* by Williams and Hall (1970, 1971); the technique permits examination of the dynamics of the axon and its satellite cells under various experimental conditions such as osmotic stress, severance, varieties of trauma, and after the micro-injection of, for example, electron-dense tracers and substances which cause demyelination.

Ultraviolet microscopy offers two advantages—firstly, about twice the resolving power of the ordinary light microscope (now rendered less important by the advent of the electron microscope, especially since the latter's lowest magnification overlaps the highest in light microscopy) and, secondly, the selective absorption of particular frequencies in the ultraviolet spectrum by various substances and especially nucleoproteins. Caspersson (1936) has evolved a technique of ultraviolet microspectography applicable to single cells, and this has been widely applied in quantitative assessments of DNA and RNA. The amount, distribution and behaviour of these and other substances in various kinds of nerve cells have been studied extensively in normal, experimental and pathological situations (Hydén 1960).

NEUROHISTOCHEMISTRY

Phase, interference, and ultraviolet microscopy may be considered as adjuncts to histochemistry and, more especially, to cytochemistry. Histochemistry stretches back at least a century; Raspail, a botanist, published in this field as long ago as 1830; but like other innovations of the nineteenth century, histochemistry has developed slowly and by an accelerating multiplication of individual identification methods for particular biological compounds, whose distribution in tissues and at the intracellular level has been recorded in innumerable studies, some of outstanding functional significance (*see* Pearse 1968). A wide spectrum of such tests is available for nervous tissues and neurons

both in normal and pathological states (Adams 1965). Of particular interest are the methods elaborated for the identification of transmitter substances and their associated enzymes, in nerve cells, fibres and synapses. These extend from the original chromaffin reaction of Henle (1865) and its modern modifications (Coupland 1965; *see* p. 1454) to the Gomori method (1943) and the more recently devised fluorescence techniques, including immunofluorescence (p. 819). The application of fluorescence techniques in neuroanatomy (Falck and Hillarp 1959; Falck 1962) have been recently surveyed (e.g. Fuxe *et al.* 1970; Fuxe and Jonsson 1974; see also Nieuwenhuys *et al.* 1978 for a convenient bibliography). Such methods have led to a new classification of nerve cells, and a useful, if inelegant, jargon. Thus, such techniques have allowed the mapping of the mammalian brain in terms of the distribution of noradrenergic (NA), dopaminergic (DA), and serotonergic (ST) neurons (see also p. 947).

In the more general field of neurohistology other comparatively recent innovations must be mentioned. New staining methods such as those of Bodian (1936) for nerve fibres, Glees (1946), Nauta (1950), and Fink-Heimer for terminal degeneration in experimentally divided axons, Einarson (1932) and Klüver (1953) for 'chromatin granules' (Klüver's method also stains axons)—all these and others augmented the neurohistologist's repertoire.

Of special interest, in view of the resurgence in the use of Golgi's technique, are its various modifications (e.g. Fox *et al.* 1951), which have been employed widely of late in working out the morphological interrelationships of small groups of neurons, particularly interneurons. The Golgi type of staining has an unexplained peculiarity in that only a fraction of the cells are actually stained, and this makes the unravelling of their intricacies practicable, and often allows the application of quantitative histological methods, including those of microreconstruction. Studies of this kind, carried out in a wide variety of vertebrates and invertebrates, have led to a considerable *rapprochement* between current electrophysiological and anatomical investigation of small-scale neuron networks and individual neurons (e.g. Eccles *et al.* 1967). An equally potent technique in drawing together workers in separated disciplines in neuroanatomical studies has been the advent of electron microscopy.

ELECTRON MICROSCOPY

Although practicable electron microscopes have been in use since the Ruska prototype in 1933, years elapsed before sufficiently thin sections of biological material could be reliably prepared. Pearse and Baker succeeded in doing this in 1948, using an adapted Spencer rotary microtome. Sjöstrand (1967) has described these early struggles and the emergence of the ultramicrotome in its modern varieties. It must also be emphasized that biological electron microscopy would have been equally impracticable without the introduction of glass and diamond knives to cut the hard embedding media required. From the 1950s all the basic requirements were met, and from then onwards biological preparations could be submitted to the increasingly high resolutions and magnifications of continuously improving electron microscopes, the use of which has transformed the morphological approach to cytology. Three decades of application have shown great improvement and innovation in supporting techniques. Heavy metal shadowing, replication methods, the production of homogenate films, heavy metal staining including negative contrast, autoradiography, densitometry, microreconstruction, cell fractionation methods depending upon differential centrifugation (of fundamental importance in neurochemistry), electron diffraction and electron phase contrast optics, freeze-fracture and freeze-etch methods, cytochemical staining, and many other techniques have enriched the basic potentialities of electron microscopy. Perhaps even more necessary has been the rapid improvement in the interpretation of the appearances of the greatly enlarged intracellular detail, and the growth of an experimental attitude and of quantitative methods in the use of electron microscopy; otherwise the singularly productive approximation of cytology, biochemistry and molecular biology, which is fast occurring, would have been improbable. In neuroanatomical research such bonds are developing with much effect. Elucidation of the structure of the myelin sheath and of synapses have been notable achievements. With improvement in microtomes and in the orientation of material, it is becoming possible to counter the two main criticisms of electron microscopy—the small scale of sampling and the difficulty in equating some of the findings of light and electron microscopy. Techniques have thus been developed by which adjoining thick and thin sections of the same block can be prepared, stained by contrasting techniques, and examined using the light and the electron microscopes respectively. Even short runs of serial ultra-thin sections can be produced, with the possibility, for example, of tracing the full complexities of part of a nerve cell process using reconstruction techniques. A number of further developments are being elaborated or projected, and more than the most elementary potentialities of the scanning and high-voltage electron microscopes have yet to be exploited by neuroanatomists.

OTHER MODERN TECHNIQUES

The tracing of detailed neuronal circuitry, both central and peripheral, has received a renewed impetus during the last decade by the introduction of several new techniques. Two of these have proved to be particularly valuable: autoradiographic identification of labelled substances, absorbed by nerve cells and transported along their processes, and the retrograde transport of identifiable 'marker' substances (usually horseradish peroxidase). Both techniques may be regarded as dependent upon physiological activities, in contrast to the older 'degeneration' techniques, which involve the pathological reactions of neurons to deliberate injury. It is claimed that they are also more reliable (see Jones and Hartman 1978), but each has its peculiar advantages and disadvantages. The autoradiographic technique was introduced by Lasek (1968) and Hendrickson (1969) and was popularized by Cowan et al. (1972). It consists of the injection of radio-actively labelled amino acids or sugars around nerve cells, which take up these substances and transport them along their axons (usually combined in various macromolecules), where their presence is revealed by suitable autoradiographic technique and examined by light or electron microscopy (Rogers 1973, Hendrickson 1975). The method is *anterograde*, with respect to the axon, and transport extends to terminals, where the labelled molecules ultimately accumulate. The transport is relatively rapid, but the technique is much lengthened by the necessity of prolonged autoradiographic exposure. It has already been very widely applied; and while most studies using it have merely confirmed previous findings, it has also mediated a number of discoveries, such as the revelation of previously unidentified connexions (e.g. hypothalamo-spinal autonomic and olivo-cerebellar paths). It has been shown to produce transneuronal effects (e.g. in the visual system).

The horseradish peroxidase technique depends upon *retrograde* axonal transport, a phenomenon reported by Matsumoto (1920) and confirmed by Lubinska (1964) and others (p. 826). The usefulness of inducing a traceable material to pass along an axon, from its remote distribution to its cell of origin, is obvious, but it was not until 1971 that Kristensson and Olsson showed that labelled albumin and horseradish peroxidase could be used as markers in this manner. The LaVails (1972) quickly confirmed the use of horseradish peroxidase, and the technique has subsequently been used most extensively. It consists essentially in the injection of solutions of the enzyme into nervous tissues or in the immersion of their cut surfaces (e.g. a severed peripheral nerve) in such solutions. Mechanical application has been supplanted by iontophoretic injection in some cases, to reduce local damage. The material is taken up by axons, transported along them in residues and appears in their neuronal perikarya in identifiable form. The method is applicable to all types of nerve fibre and can be employed at light and electron microscopic levels. A large literature in regard to the technique itself has already accumulated (see reviews by Winer 1977, and Jones and Hartman 1978). It has been utilized in innumerable researches and its introduction has stimulated extensive efforts to find other suitable markers, such as fluorescent dyes, tetanus toxin, labelled cobalt, etc.

Some of the most exciting developments that have occurred in recent years in neuroanatomical research have been associated with the widespread application of standard immunological techniques to neurobiological problems. For example, immuno-histochemical methods have resulted in the localization of neurotransmitters, neurotransmitter-related enzymes, pituitary hormones and their releasing factors and other nervous system-specific proteins with a spatial resolution frequently superior to that obtained with either customary staining or biochemical regional assays (see Jones and Hartman 1978). Basically these techniques involve the production and subsequent visualization of sites of antibody-antigen interaction within tissue sections (the quantity of antigen can in some cases be increased by varying the physiological or pharmacological status of the tissue prior to examination). As might be expected, there are a number of factors which can influence the final distribution of these sites, and therefore complicate interpretation of staining patterns. These may be grouped as follows: (i) the production and purification of specific antibodies (the preparation of pure antigen is difficult when tissue concentrations are low). (ii) Adequate tissue preparation, especially during fixation: most fixation regimes are designed to achieve the optimum compromise between preservation of morphology and antigenicity. The commonest fixative in use for this purpose is formaldehyde; other aldehydes, with their stronger cross-linking abilities have proved less popular, although a recent improvement involving periodate-lysine-paraformaldehyde has been reported (Nakane and Kawaoi 1974). (iii) Selection of an appropriate visual marker, which in neuroanatomical work is commonly either fluorescein iso-thiocyanate (FITC) or horseradish peroxidase (HRP). FITC emits a high yield of green fluorescence (517 nm) when excited with blue light (peak excitation 490–500 nm)—properties which in combination with modern fluorescence microscopes permit the examination of, for example, fine adrenergic fibres (Hartman 1973). The disadvantages are (a) the lack of permanence of the preparation, as a consequence of quenching in mountant or photodecomposition with time; and (b) fluorescent markers cannot be used in electron microscopy. HRP is used as a marker molecule in two completely different ways in the nervous system. As a marker in immunohistochemical studies it can be conjugated directly to antibodies by a number of bifunctional cross-linking reagents, although this results in a partial inactivation of peroxidase activity. A recent development, the peroxidase-antiperoxidase method (PAP) in which antibodies to HRP are bound to peroxidase without inactivating it, has had an enormous impact on immunohistochemistry, especially at the ultrastructural level (Sternberger 1974). The relatively large size of the PAP complex (c. 420,000 daltons) can, however, produce problems of penetration in certain situations (e.g., fine nerve terminals). A recent extension of the general method outlined above has involved the use of suitably tagged antigenic markers in the identification of different neurons and satellite cells, either in suspension after isolation from whole tissue or growing and differentiating in culture prior to fixation. Thus, *in vitro*, oligodendrocytes can be 'marked' with anti-galactocerebroside; astrocytes with anti-glial fibrillary acidic protein (GFAP); Schwann cells with anti-RAN I and certain neurons with anti-tetanus toxin (Raff et al. 1978). Moreover, antisera raised against certain components of the myelin sheath and against nerve growth factor are being used in studies of the factors controlling myelinogenesis and other ontogenetic and pathological mechanisms in the nervous system.

A number of other relevant techniques can be barely mentioned—autoradiography (Rogers 1967, 1973), ciné and time-lapse photography (Rose 1963), microreconstruction (Gaunt and Gaunt 1978), microphotometry, micromanipulation (Kopac 1960), neuron counting techniques (Konigsmark 1970; Corsellis et al. 1975), morphometric analysis (Weibel and Elias 1969), and the elaboration of models to provide functional analogues (Weiner and Schadé 1963). Enough has been said, however, to illustrate the ever-growing dependence of neuroanatomical investigations upon technical development. Despite widespread misapprehension, neuroanatomy—and indeed anatomy in general—has at no stage been an entirely morphological, merely descriptive discipline, though it can descend, and frequently has so fallen, to this somewhat sterile level. However, there have always been those, such as Willis, Harvey, or Sherrington, who have disregarded the artificial separation of nervous structure and function. Unfortunately, levels of knowledge of one or the other have often been imbalanced. After a considerable period, early this century, of separation of disciplines concerned with morphology and function (perhaps chiefly due to the exigencies of teaching organization in medical schools), the consequent isolationism now shows signs of dissolution. This is particularly true of studies of the nervous system. Preoccupation with the minutiae of shape and of the nervous impulse have borne their

separate fruit, and we can now pass on to more hopeful prospects. As morphologists become more experimental and the functionalists look more attentively at the arena of their experiments, instrumentation and outlooks are evincing an increasing similarity, and purposes a new cohesion. Doubtless we are still far from concepts of integrated brain activity; but, at simpler divisions or units of the whole, electrophysiological and anatomical inquiries into the behaviour of the individual neuron and synapse, and their associations in small functional networks, seem to afford a prospect of joint success. We may at least be much nearer to comprehending the 'miniaturized' basic units in central nervous mechanisms. It is to the individual elements at this level of organization that we must now turn our attention.

CYTOLOGY OF THE NERVOUS SYSTEM

The foregoing sections have been concerned with the organization and investigation of the nervous system. We shall now consider the cellular elements of the nervous tissue in more detail, since it is impossible to create a model of brain function without an appreciation of how the individual units work (Schmitt *et al.* 1967, 1970, 1977).

As already stated, nervous tissue is composed of two distinct sets of cells, the excitable cells or *neurons*, and the non-excitable cells which constitute the *neuroglia* and *ependyma* in the central nervous system, and the *Schwann cells* and associated elements in the periphery. Since our interest in terms of neural behaviour centres on the neurons, more space will be given to outlining their structure and behaviour.

The Neuron

Neurons, although diverse in appearance and size, have many characteristics in common. All possess a large surface area compared with most other cell types, and this surface is specialized for the reception, conduction and transmission of neural information by virtue of its molecular composition, and by the topographic specialization of the neuron. The branch-like extensions of its periphery are termed *neurites*, and these may be divided into receiving processes or *dendrites*, and efferent processes termed *axons*. Usually there are many dendrites but only one axon extending from each cell. The protoplasm massed around and including the nucleus of the cell is termed the *soma*, *perikaryon*, or *cell body*; the cytoplasm of this structure is characterized by the presence of numerous strongly basophilic inclusions, the *Nissl bodies* or *granules* (7.14). The soma is often angular in shape (7.15), the projection which leads to the axon being the *axon hillock*, and its bare continuation the *initial segment* of the axon. Sometimes the axon emerges from the base of one of the main dendrites rather than directly from the soma. From the other angles of the soma spring a number of main or *primary dendrites* which branch repeatedly to form a *dendritic tree*. Each axon may also possess side branches or *collaterals*, usually much finer than the main axonal process. Axons at their termination usually break up into several fine branches, the *axon terminals* or *telodendria*, which end in apposition, variously, with other neurons to form zones of interneuronal transmission or *synapses*, with effectors such as muscles (*neuromuscular junctions*) or with glands; adipose tissue also receives efferent terminals (*see* p. 851). The plasma membrane of the axon is known as the *axolemma* and its internal cytoplasm the *axoplasm*.

Neurons vary considerably in size, from microneurons with cell bodies 7 μm or less in diameter, to some of the largest cells in the body, e.g. spinal ganglion cell bodies 120 μm across. In some of the larger cells (e.g. Purkinje cells of the cerebellum, Betz cells of the cerebral cortex) the cell nuclei, which are proportionately large, are tetraploid in their chromosomal content (Herman and Lapham 1969). The size of the cell body reflects the need to sustain the large volume of cytoplasm present in dendrites and axon; in a large human motor neuron, the axon, which can be a

metre long and possesses numerous branches, may be more than one hundred times the cell body in volume.

HISTOLOGICAL DEMONSTRATION OF NEURONS

As stated previously (p. 813), what we know of neuronal structure is dependent upon the special methods which have been elaborated to show various aspects of neuronal cytology, and it is useful to consider these methods further here. Although the standard methods of light microscope histology are valuable in showing their cell bodies, the neurites of neurons do not stain sufficiently well to allow them to be traced with any degree of confidence. Various types of impregnation of *blocks* of tissue with metallic salts, particularly those of silver, do, however, stain some or all parts of the cytoplasm. One of the most useful class of techniques, invented by Golgi (*see* p. 813), results in the precipitation of silver chromate or other metal salts within individual neurons, which can therefore be demonstrated in their entirety in thick sections; and, since only a small percentage of the total cell population is affected, they are visible as black spidery profiles against a clear background (7.15B). Although the fidelity of the method has in the past been questioned, electron microscopy supports the view that the results obtained with the Golgi techniques are an accurate representation in even the finest of light microscopic details, so that now there is a resurgence of interest in this means of investigation. However, there has, so far, been no critical evaluation of the distortion introduced by either of these preparative techniques. Recently, some other methods for demonstrating entire neuronal structure have been introduced, including horseradish peroxidase, cobalt nitrate, and various micro-injection methods (*see* p. 819).

Next, there are the various techniques of staining *sections* with silver salts, some of which are specific for the neurofibrils within neurons (Gray and Guillery 1966), and others which seem to stain the entire cytoplasm (7.15E). In this category are the ammoniacal silver nitrate methods of Bielschowsky and their subsequent modifications, the silver proteinate methods of Bodian, and the various silver nitrate techniques such as that of Holmes. The block staining methods of Cajal also give good results by a similar process. Various modifications of the silver reactions have also been used to demonstrate anterograde degeneration of axons (the Nauta methods) and of their terminals (such as the techniques of Glees) in sections, by chemically suppressing the staining of normal as opposed to degenerating axons (7.14, see also p. 817). More recently, important modifications of these degeneration techniques for use with the electron microscope (e.g. the Fink-Heimer method) have been introduced, but these are being superseded by the horseradish peroxidase methods for anterograde analysis.

Numerous methods have been developed to demonstrate the sheaths of myelinated fibres (*vide infra*, p. 844), some based on osmium staining, and some on haematoxylin and allied stains such as Luxol fast blue (7.15). A modification employing chromium salts has also been developed to demonstrate *degenerating* myelin sheaths (the method of Marchi) although this is now less widely used because of various difficulties of interpretation (p. 838). Added to these, there are modifications of normal cytological stains which demonstrate various organelles within the cytoplasm. The most useful of these, the Nissl stains, are basic substances which bind strongly to the nucleic acids which are abundant in neuronal nuclei and cytoplasm. Such stains as cresyl fast violet (7.14) and toluidine blue are widely employed by neuroanatomists for this reason. A number of stains also demonstrate mitochondria, particularly in axon terminals, and modifications show the *degeneration of axon terminals* after experimental trauma.

Electron microscopy has in many cases revolutionized our concepts of neuronal structure or, in some instances, has determined which of several alternative models is the correct one. However, since this discipline is restricted to the examination of ultra-thin sections, three-dimensional reconstruction of nervous tissue is laborious and limited in scope. It is therefore of great importance to correlate light and electron microscopy to obtain an understanding of neuronal organization.

All of the above methods involve the examination of dead material, but much may also be learnt from studies of neurons stained by supra-vital methods. Methylene blue was found by early neuroanatomists to stain the entire cytoplasm of neurons, giving a picture of individual cells as striking as those obtained with the Golgi methods. Because methylene blue is a vital stain, however, its use is restricted largely to thin or small whole tissue situations such as the peripheral and embryonic nervous systems, where penetration and staining are rapid. Many other techniques are also available for specific purposes.

GENERAL STRUCTURE OF NEURONS

Neurons can be described in several ways according to the patterns of their neurites, their physiological action, the transmitter which they release and so on (Hydén 1967; Bourne 1968). Of these various descriptions, only the first is generally possible, since our understanding of the physiological and biochemical properties of neurons is limited to relatively few instances, whereas the morphology of many neuron populations is well established. Even within the field of general structure, several types of classification are possible, mostly based on the patterns of ramification of the neurites. As mentioned previously (p. 805) a commonly used scheme distinguishes: (1) *unipolar neurons*, in which the cell body has a single extension giving rise to both dendritic and axonal branches, examples being dorsal root ganglion cells (sometimes termed *pseudo-unipolar* because this form is arrived at secondarily in development), and granule cells of the olfactory bulb; (2) *bipolar neurons*, with an extension at each end of the cell body, for example retinal bipolar cells, and cells of the cochlear and vestibular ganglia; (3) *multipolar neurons*, with several extensions of the cell body; most cells of the central nervous system are of this last type. Although this scheme of classification (7.15) is useful for descriptive purposes, it appears to have limited physiological significance. Of more importance is the classification of central neurons into relatively large, Golgi type I neurons which have long axons connecting different parts of the nervous system (Cajal 1911), and small Golgi type II neurons (microneurons) in which the axon is short and terminates in the neighbourhood of the cell body, or else is entirely absent (7.15). The Golgi type II neuron is often present in inhibitory situations, an example being the periglomerular cells of the olfactory bulb (p. 992). In a special category of microneurons, lacking an obvious axon, nervous conduction is apparently possible in either direction along their dendrite-like processes. Such neurons have long been known in the retina (*amacrine cells*), where they lie in synaptic contact with ganglion and other cells (p. 1171), but their presence is also indicated in other parts of the central nervous system, including the olfactory bulb (granule cells) and, possibly, the lateral geniculate body. It is probable that they constitute a characteristic element of all the main sensory pathways. In some situations, at least, microneuron cell processes make *reciprocal synapses* with adjacent dendrites, so that *dendrodendritic transmission* occurs (*see* p. 806).

The small inhibitory neurons appear responsible for an important type of interaction between neurons, such as those of the sensory projection pathways, in which maximally excited cells can inhibit the activity of less excited, adjacent neurons, by means of inhibitory interneurons, thus 'sharpening' the sensory patterns at various levels in the pathway (7.8 and p. 808), a process termed *lateral inhibition* (*see* Horridge 1968, and pp. 808, 889).

Central neurons can also be classified according to the branching patterns and shape of their dendritic fields (7.15). Those with dendrites extending equally in all directions away from the cell body, to fill a roughly spherical volume with their smaller branches, are termed *stellate cells* (7.15), a type common in the cerebral cortex, and in certain nuclei of the brainstem and spinal cord, amongst other locations. In *pyramidal cells*, the cell body is pyramidal or conical in shape, and *basal dendrites* emerge from the angles made by the base and the walls of the pyramid to fill a roughly hemispherical volume (7.15); an *apical dendrite* emerging from the apex of the cell may also be present to give a second dendritic field at some distance from the cell body, as in the pyramidal cells of the cerebral cortex (7.15). Spindle-shaped cell bodies with dendrites emerging at one or both ends are termed *fusiform*.

Many variations of these basic patterns exist, of course, throughout the nervous system, depending upon the pattern of afferent fibres, their synaptic sites, and the mechanical constraints imposed by other cells and so on. One of the most remarkable dendritic fields is that of the cerebellar Purkinje cell (p. 919), in which a primary dendrite emerges apically and branches to form a complex two-dimensional fan-like array (7.15, 79, 80). *Glomerular neurons* are also known, in which there may be relatively few dendrites with highly convoluted branches at their tips (7.15, 79, 81) where most synaptic contact occurs, as in the mitral and tufted cells of the olfactory bulb (p. 992), in the lateral geniculate nucleus, in certain 'relay' nuclei of the thalamus (p. 957), and in the cerebellum (p. 924).

Several attempts have been made to classify dendritic patterns by their geometry, but most of these schemes, although useful for descriptive purposes, seem to have limited functional significance. More important seems to be the mathematical analysis of the branching patterns of dendrites in terms of the frequency, angles and dimensions of the branches, the spatial distribution of cell surface and volume, and the frequency, type, and spatial location of synaptic endings, since all of these parameters are known to be significant in the electrical activity of neurons. Although such studies are still in their infancy, a number of important contributions to the field have already been made. The use of computers to collate such information holds out considerable promise for the future. Other biometric parameters, such as the ratio between nuclear and cytoplasmic volume have also been examined.

STRUCTURAL PARAMETERS OF DENDRITES

When dendritic fields are examined, as for instance with the Golgi methods, the dendritic branching is seen to be largely dichotomous (Sholl 1956; Aitken and Bridger 1961; Gelfan *et al.* 1970). Where dendrites spread and branch symmetrically around the cell body, the *total number* of dendritic branches is, of course, dependent upon the number of primary dendrites and the number of sub-orders of branches. The *surface area* of cell membrane associated with the *dendrites* can be calculated from simple, although laborious, measurements of the diameters and lengths of the branches, and this can be compared with the *surface area* of the *cell body*. In spinal cord motor neurons and interneurons, up to 80 per cent of the total neuronal surface area (excepting the axon) is associated with the dendrites, emphasizing their important role as a receptive area for afferent stimuli (see also p. 883). Various methods of describing the dendritic tree have been used; the frequency of branching can be expressed by imagining a series of equally-spaced concentric spheres centred on the cell body; as the dendrites emerge and branch, the number of dendritic segments passing through the surface of each imaginary sphere can be counted and expressed as a function of distance from the cell body. In neurons of the cerebral cortex and spinal cord, the number of branches rises quickly to a maximum, and then tails off asymptotically or logarithmically, with increasing distance from the base of the primary dendrite (Sholl 1956). Counts of the thorn-like extensions of the dendrite surfaces, the *dendritic spines*—an important index of numbers of incoming synapses (p. 826)—show that in some types of neuron there are often many thousands equally spaced over much of the dendritic tree (Gelfan and Rapisarda 1964; Gelfan *et al.* 1970). Thus it follows that the number of synapses probably first increases, and then decreases asymptotically with distance from the cell body in the same manner as the number of dendritic branches. Careful measurements indicate that different populations of neurons, although showing the same general mode, have their own peculiar relationships between cell body and dendritic surface areas, volumes, spine counts, and so on (Lindsay and Scheibel 1974).

The importance of these parameters, of course, lies in the numbers of afferent synapses made with the cell, their position, and the spreading patterns of electrical disturbance at the cell membrane which follow their activation (Rall 1977). A vital feature is the relative distribution of excitatory and inhibitory synapses, but, as yet, this can only be estimated from electron micrographic studies, which are often statistically invalid (*see*, however, p. 922). Microelectrode studies are a possible means of analysis of dendritic action, but in most cases the dendrites are too small to be monitored individually, and elaborate mathematical techniques are necessary to extract any meaning from the records. There are, however, some functional observations indicating the general significance of dendritic branching patterns; for example when neonatal mice are made thyroid hormone-deficient to simulate the condition of cretinism in man, they show behavioural retardation, and the dendritic trees of their cerebral cortical neurons are also smaller, with fewer branches, than those in control animals (Eayrs 1955).

PHYSIOLOGICAL PROPERTIES OF NEURONS

Neurons, as we have seen (p. 805), are characterized by their ability to receive, conduct and transmit *information*, which is

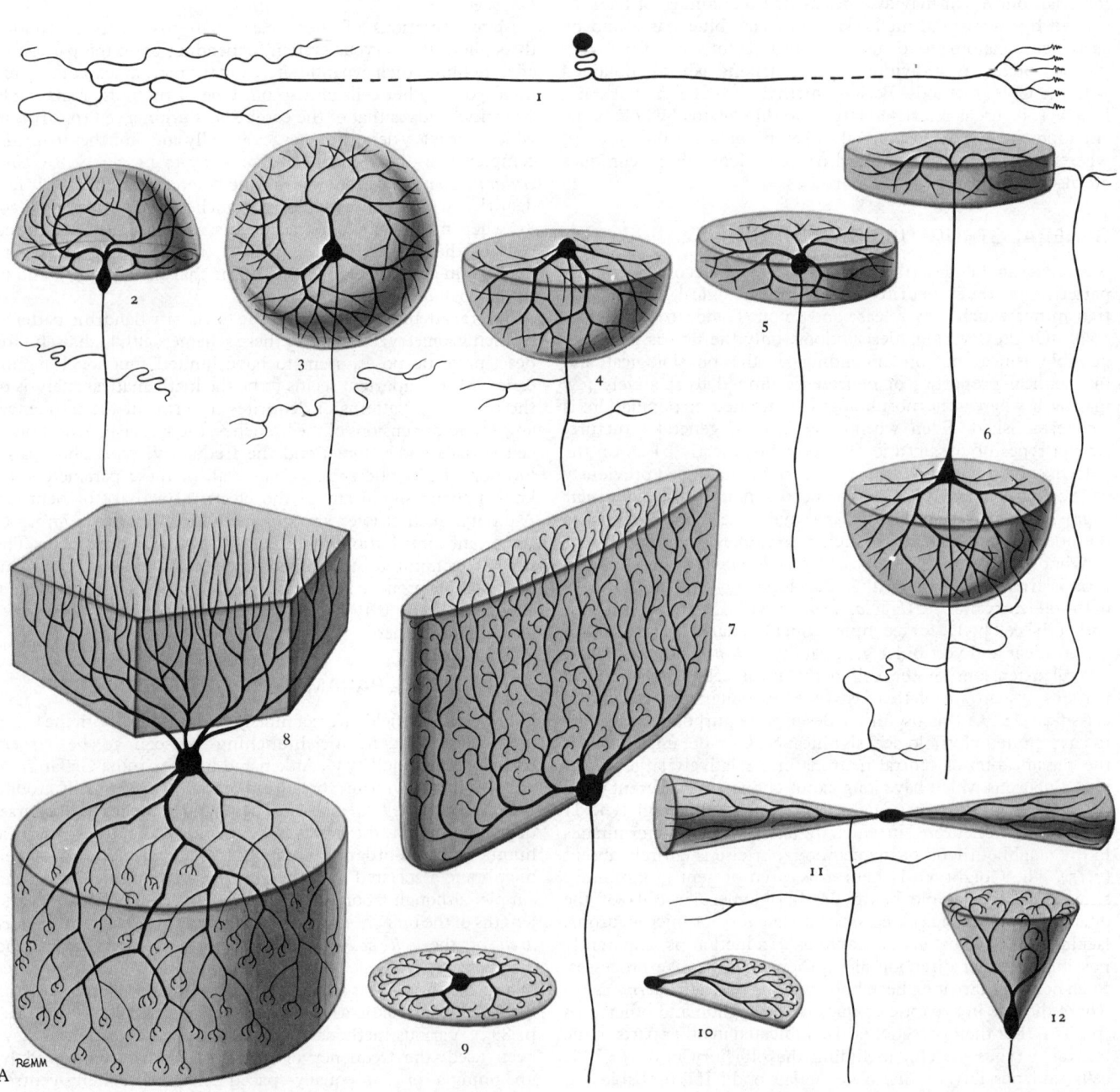

7.15A–E A Schema showing pattern variations of neuronal geometry: (1) unipolar, sensory ganglionic neuron; (2) bipolar neuron; (3) stellate (isodendritic) neuron, with (4), (5), and (11) which are modifications of this pattern; (6) pyramidal neuron with an apical and a series of basal dendrites, and recurrent axon collaterals, from the cerebral cortex; (7) Purkinje neuron from the cerebellar cortex; (8) Golgi neuron from the cerebellar cortex; (9) and (10) amacrine cells lacking axons; (12) glomerular neuron (mitral cell) from the olfactory bulb, showing recurved dendritic tips.

coded in terms of transient electro-chemical changes in their plasma membranes (Katz 1966; Kuffler and Nicholls 1977). These changes are associated with rapid fluxes of ions in and out of the cytoplasm, which occur against the background of a steady electrical potential difference maintained across this surface. The measurement of these changes by means of various electro-physiological instruments, and their analysis, provide a means of understanding the structure of the nervous system in functional terms, and progress in these two fields is coming more and more to depend on a close working relationship between them. A detailed survey of electrophysiology is beyond the scope of the present account and only a brief summary will be attempted.

The resting potential (7.16) of a neuron is in many respects similar to the membrane potential of non-excitable cells, consisting of a steady potential difference of about 80 mV which can be measured across the plasma membrane, the inside negative with respect to the outside. This bioelectrical potential is the result of differences in concentration of various anions and cations inside and outside the cell, which is further dependent upon the permeability properties of the external membrane, and the active transport of ions, particularly sodium and potassium, by the membrane. These various factors result in a high concentration of potassium, and of negatively charged, non-diffusible ions (proteins, etc), and a low concentration of sodium and chloride ions, *inside* the cell, relative to their concentrations in the tissue fluid which bathes the cell externally. Calcium is also present

within the cell at much lower concentrations than in the extracellular fluids. The resting potential can change in two distinct ways, giving rise to *graded potentials* or *action potentials* respectively. Graded potentials involve the membrane of *dendrites* and neuron *cell bodies*, and consist of fluctuating increases or decreases in the resting potential, i.e. their membranes are either relatively *hyperpolarized* or *depolarized*. In contrast, the action potential involves a transient complete reversal of polarity across the membrane of *axons* (and also of some large dendrites which resemble axons in their structure).

Graded potential variations may be either *excitatory* or *inhibitory*. When excitatory, there occurs an increase in permeability of the membrane to sodium ions, which therefore flow down their concentration gradient to enter the cell, and thereby progressively *depolarize* the membrane. This sudden influx results from the transient opening of special ionic channels present in certain membrane proteins (*sodium* and *potassium gates*). The size and time course of the depolarization at a particular membrane locus, to which an effective stimulus has been applied, is related (logarithmically) to the stimulus strength. Furthermore, surrounding zones of membrane also suffer a depolarization, but with decreasing amplitude, at increasing distances from the initiating locus. Thus, the larger the initial depolarization, the greater the spread of effect to surrounding areas of membrane. Such a mechanism allows the *summation* of effect of multiple excitatory stimuli applied at numerous points

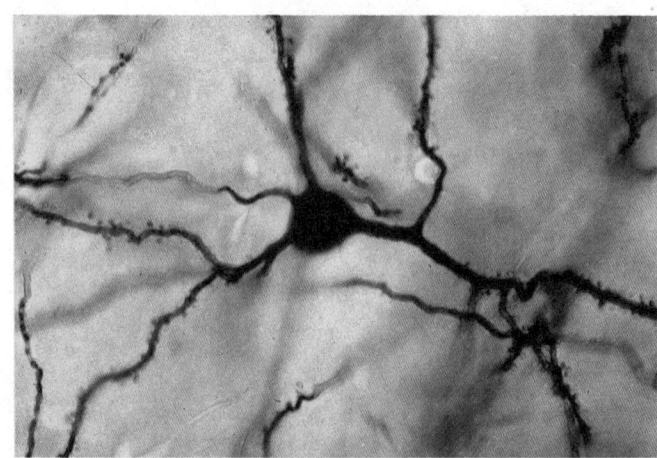

B A high-power light micrograph showing part of a pyramidal neuron from the cerebral cortex of a rat, prepared by the rapid method of Golgi. The bases of several dendrites covered with dendritic spines, and a thin axon (left) are visible. (Preparation kindly supplied by Dr. A. R. Lieberman of University College, London.)

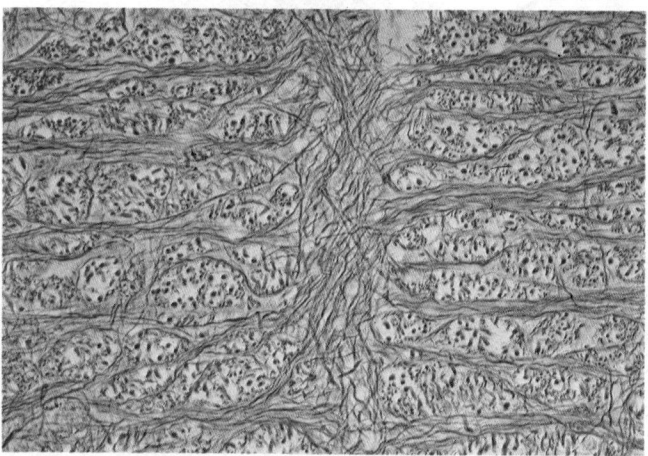

C A low-power light micrograph of groups of interweaving axons from the medulla oblongata of a cat, stained by the Holmes' silver method. (Preparations C–E kindly supplied by Dr. E. L. Rees of the Anatomy Dept., Guy's Hospital Medical School, London.)

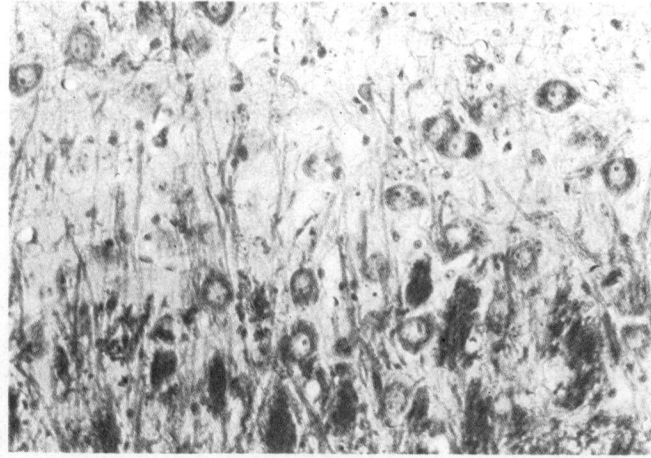

D A group of neuronal perikarya (purple) amongst bundles of myelinated fibres (blue), from the brainstem of a rat. Stained by the cresyl fast violet-Luxol fast blue method.

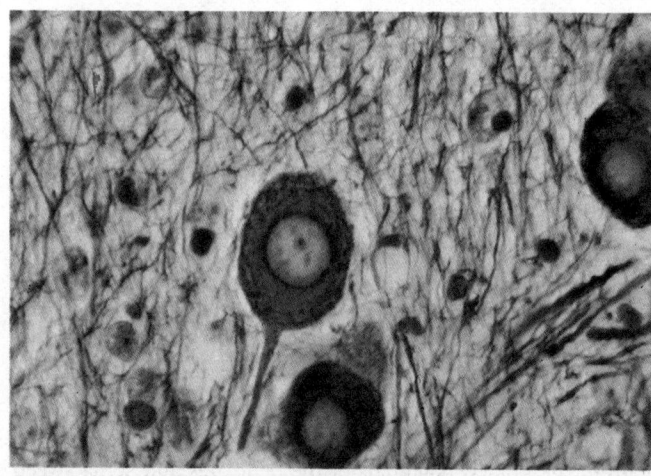

E A section through the mesencephalic nucleus of the trigeminal nerve (rat), stained with the Holmes' silver nitrate method to show large unipolar neurons from one of which a dendro-axonal process is emerging.

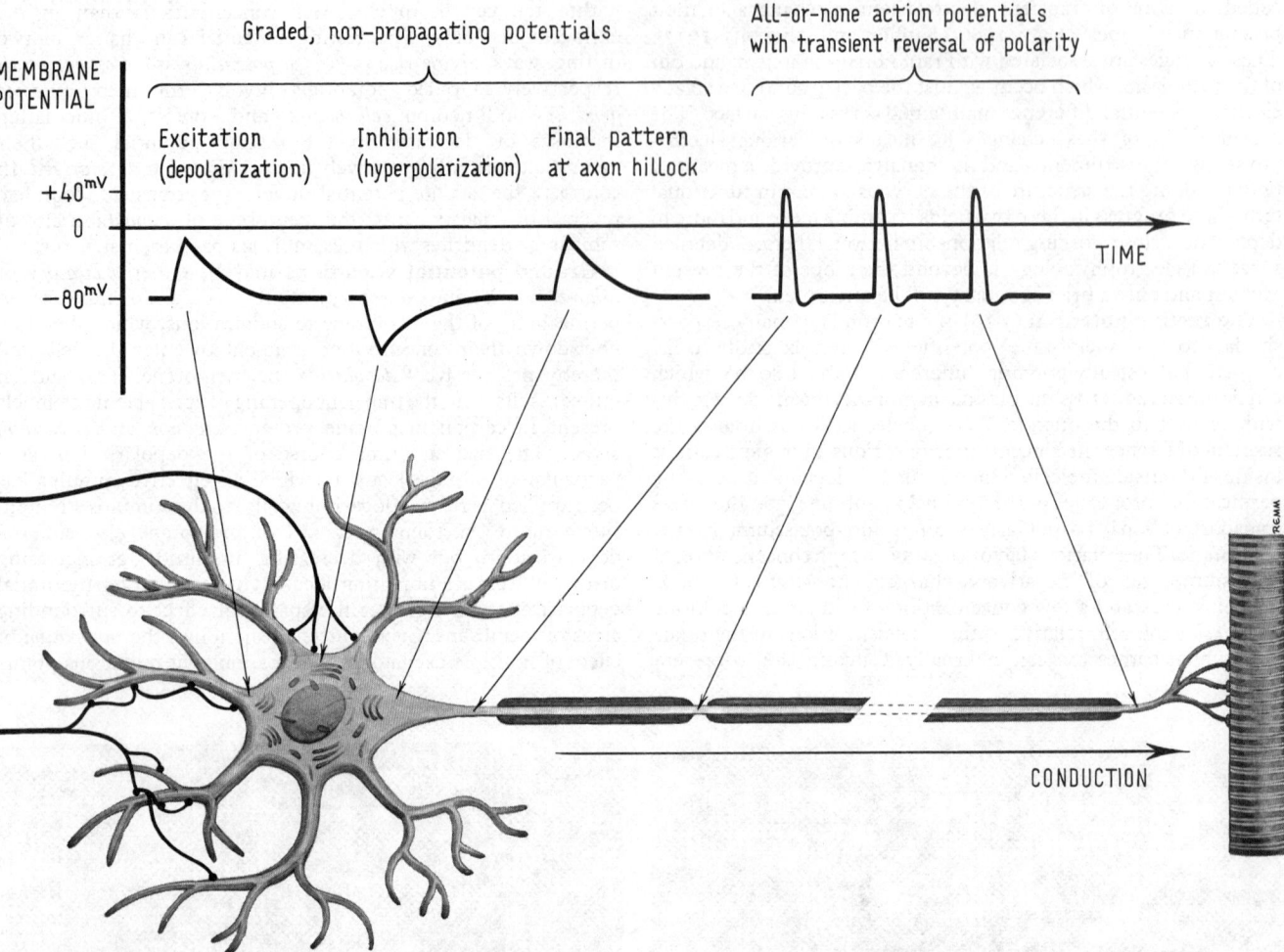

7.16   A diagram showing the types of change in electrical potential which can be recorded across the cell membrane of a motor neuron at the points indicated by the arrows. Excitatory and inhibitory synapses on the surfaces of the dendrites and soma cause local graded changes of potential which summate at the axon hillock, and may initiate a series of all-or-none action potentials, which in their turn are conducted along the axon to the effector terminals.

over the receiving surface of the neuron. Inhibitory stimuli, thought to act by increasing the inflow of negatively charged chloride ions, therefore tend to increase the membrane potential (*hyperpolarization*), and thus they oppose or subtract from, the *total excitatory state* of the neuron.

The action potential, nerve impulse or **spike potential** in contrast, is characterized by a brief *reversal* of polarity caused by the sudden influx of sodium ions, followed by a return to the resting potential as potassium ions flow out; these processes take about 5 msecs in all (Hodgkin and Huxley 1952). For any system, the *amplitude* and *duration* of each successive action potential is identical, no matter how large the stimulus is above a certain *threshold value*, so the response is an *all-or-none* event. The wave of excitation associated with the action potential spreads away from the point of initiation, but, unlike the graded potential, its size and time course do not alter, so the wave sweeps rapidly over the membrane at a constant rate. The action potential is therefore said to be *propagated* and *non-decremental*. After a region of membrane has been stimulated, however, there is an irreducible *refractory period* during which another action potential cannot be elicited, thus fixing a maximum frequency of their formation. When a series of action potentials is generated, they are rapidly conducted along the neurite, maintaining their original frequency pattern precisely. This phenomenon is the basis of nervous communication, which is therefore said to be *frequency-coded*.

The graded and action potentials are related to one another at specific points on the neuronal surface. For example, in a motor neuron, the graded potentials of the dendrites and soma sweep over the axon hillock to initiate action potentials at the *initial segment* of the axon. The greater the graded depolarization of the soma, the more frequent the action potentials, so that the *state of*

excitation of the cell is reflected by the *frequency pattern* of the *volleys of action potentials* which flow along its axon.

When an action potential reaches the axon terminals, it causes the release of a discrete amount of *neurotransmitter substance* which further changes the excitability of the receiving cell, either another neuron, a muscle cell, or some other element. A closely grouped series of action potentials (*spike trains* or *volleys*) naturally causes a greater effect than single ones, since extracellular enzyme action, or other processes, usually limits the time during which the neurotransmitter can act.

Although the most common cell region for the change from graded to action potential to occur is the initial segment of the axon, in many peripheral receptors with an axon-like dendrite (for example cutaneous sensory neurons), it occurs in a region close to the sensory ending, often the first node of Ranvier, in myelinated fibres.

In amacrine cells (p. 1171) and some other very small neurons lacking true axons *all* electrical activities may be of the graded type, and such cells are able to spread these changes in any direction depending on the point of stimulation. In cells with an axon, however, conduction occurs under physiological conditions in one direction only, that is, from the dendrites and cell body to the axon (*orthodromic conduction*); although when artificially stimulated at the periphery, axons can conduct in the opposite direction (*antidromic conduction*), and electrical changes then invade the cell body.

SIGNIFICANCE OF NEURONAL MORPHOLOGY

Because of the mode of synaptic excitation and decremental conduction of the dendrites and perikarya, the shape of these structures and the distribution of excitatory and inhibitory synapses over their surfaces, undoubtedly play a vital part in determining the relation between the input and output of a neuron. This provides a cogent reason for analysing the geometry of neurons in as precise a manner as possible, in order to understand their role in different physiological processes.

ULTRASTRUCTURE OF NEURONS

Since the pioneering studies of Palay and Palade (1955), and of de Robertis and Bennett (1955), much information on neuronal ultrastructure has been amassed (see, for example, Peters *et al.* 1976; Palay and Chan-Palay 1977). Following the terminology of light microscopy, it is useful to distinguish between those regions of the nervous system consisting mainly of the cell bodies of neurons, or of regular arrays of their axons which constitute relatively discrete *nuclear cell masses* and *fibre tracts* respectively, and those parts made up of complex meshworks of axon terminals, dendrites and neuroglial processes which form the *neuropil* of the nervous system (e.g. Peters *et al.* 1970). The different components of the neurons will now be considered in detail.

The soma consists of cytoplasm rich in granular and agranular endoplasmic reticulum (**7.17, 18**) with attached and free polyribosome clusters, often forming large aggregations; these are visible with the light microscope, by virtue of their constituent RNA, as basophilic *Nissl* or *chromatin bodies* or *granules*. These are more obvious in highly active cells, such as the motor neurons of the anterior grey column of the spinal cord, where *stacks* of granular endoplasmic reticulum occur (**7.18**).

The *nucleus* is usually large, rounded and quite pale, that is, euchromatic, with one or more prominent *nucleoli*. All of these features are typical of cells actively engaged in protein synthesis, as too is the presence of numerous mitochondria as a source of energy, available also for other cellular activities. *Lysosomes* are another prominent cell component, as are also stacks of agranular cytomembranes comprising the *Golgi complex* which often form several distinct groups in different parts of the cell body. A Golgi complex-associated endoplasmic reticulum in which lysosomal enzymes are collected is also a prominent feature of many neurons. *Neurofilaments* and *microtubules* are present in abundance, bundles of the former corresponding to the 'neurofibrils' of light microscopists. Microtubules and neurofilaments extend into and throughout the lengths of **dendrites** and **axons**, the relative proportion of these two components varying with the type of neuron and the type of cell process involved. In general, dendrites are richer in microtubules than are axons, which may be almost completely filled with neurofilaments; neither microtubules nor neurofilaments appear to be continuous from the cell body to the tips of its neurites, and although bundles of them diverge at points of neurite branching, individual tubules and filaments do not branch. Actin microfilaments have also been described in neurons, as indeed in most cells (see, for example, Lasek and Hoffman 1976).

Centrioles, formerly believed to be absent from mature neurons, have been described in every neuronal type so far investigated, and may be associated with the generation or maintenance of the microtubular apparatus. In some neurons, centrioles and ciliary projections from the cell surface have been reported; although cilia appear to be common on the surface of neuroblasts, they have also been found in mature cells; their significance, except at some special sensory terminals, is obscure.

Various other inclusions are also commonly present in the cytoplasm of neurons. *Pigment granules* of various types characterize certain parts of the brain (Barden 1969); the substantia nigra (p. 936), for example, has a *neuromelanin* whose presence is probably related to the catecholamine-synthesizing ability of these neurons, these neurotransmitters being closely related chemically to the pigment. In the locus ceruleus, likewise, a similar pigment, rich in copper, gives a bluish colour to the neurons. Other neurons are unusually rich in certain metals, zinc for example in the hippocampus, and iron in the oculomotor nucleus. These metals may be part of the prosthetic groups of various special enzymes. As neurons age, all of them tend to accumulate granules of a yellowish pigment termed *lipofuscin*

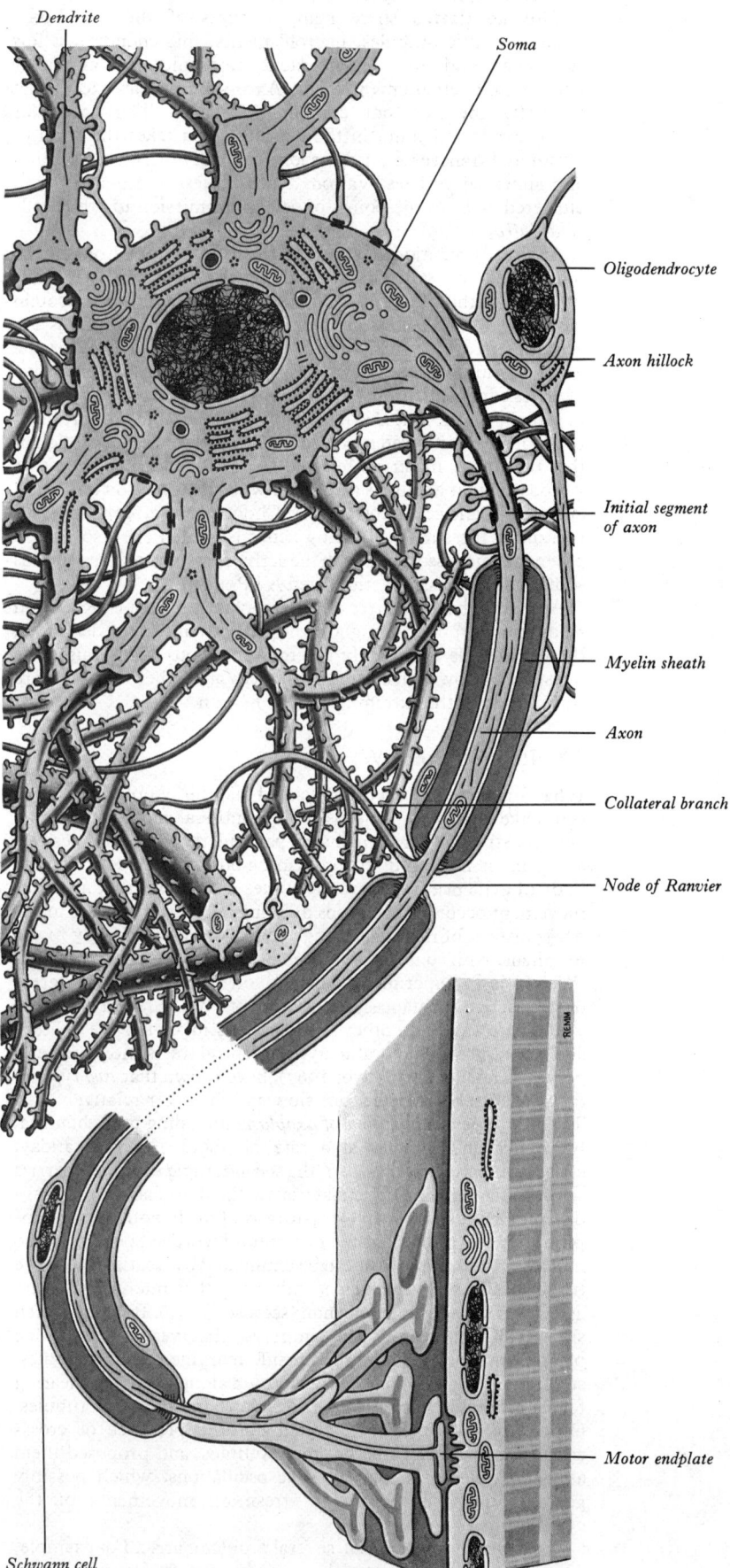

Dendrite

Soma

Oligodendrocyte

Axon hillock

Initial segment of axon

Myelin sheath

Axon

Collateral branch

Node of Ranvier

Motor endplate

Schwann cell

7.17 A schematic drawing of the ultrastructure of a motor neuron, showing part of its dendritic field (above left); the dendrites are studded with spines which are contacted by different types of synaptic terminal. The cytoplasm of the neuronal soma contains stacks of rough endoplasmic reticulum, and other organelles. See text for a detailed description.

(senility pigment), which becomes particularly prominent in the cells of the spinal ganglia.

The dendrites share many features of the cell body, containing microtubules, neurofilaments, mitochondria, a few lysosomes, and, to a lesser extent, ribosomes and agranular endoplasmic reticulum (7.18, 20). **Axons** are similar, except that normally they do not contain ribosomes. The agranular endoplasmic reticulum in the neurites often takes the form of perforated transverse septa, or longitudinally orientated tubules and spherical vesicles. Various types of vesicle are also found clustered close to the points of neurotransmission to other cells (*vide infra*, p. 829).

As already mentioned, several of the features of neurons can be interpreted as indicating a high level of protein synthesis, prompting the question, For what activity is this protein destined? Maintenance and repair of cytoplasmic proteins are activities in which all living cells engage, and in view of the relatively huge volume of cytoplasm in the cell body and extensive neurites of many neurons, it is perhaps not surprising that the rate of protein synthesis is also high. However, a sizable proportion of the protein synthesized may also be involved in the elaboration of the transmitter materials released at nerve endings, and also in various cellular activities associated with the reception of stimuli at dendritic and cell body surfaces. Cholinesterase, for example, is present at the surface of many neurons, which are also rich in other enzymes associated with the active transport of ions, such as sodium and potassium stimulated ATP-ase.

Although the machinery of protein synthesis, that is, RNA and ribosomes, is found throughout the cell body and in the dendrites, these materials are usually absent from axons. This raises the question of how transmitter substances and other materials are transported to the extremities of the neuron.

AXOPLASMIC FLOW

It has for some time been appreciated that the cytoplasm of nerve cells, like that of other cells, is in continual motion; in tissue culture, streaming movements of particulate matter and vesicles along axons in both directions can be observed. A considerable body of evidence from several sources now shows that a similar movement occurs in the axons of living neurons, resulting in the net transport of materials along axons from the cell body to the terminals, with some lesser movement also in the other direction (Weiss 1970). Experiments which involve following the progress of radioactive substances (for example, from the retina along the visual pathway), and other investigations involving the ligature and subsequent structural analysis of central tracts and peripheral nerves (Kapeller and Mayor 1967), have shown that *two types of transport* can be detected, one slow and the other relatively fast. The first type is a *bulk flow of axoplasm* including mitochondria, lysosomes and vesicles at a rate of about 1–3 mm a day, constituting the *slow transport*; the second, *rapid transport*, carries selected proteins and other materials at the rate of about 100 mm a day, and in the neurosecretory pathway of the hypothalamohypophyseal tract (p. 969), at the maximum recorded speed of 2,800 mm a day. This rapid flow can be eliminated by treating the nerve trunk locally with colchicine, indicating that microtubules are associated with the phenomenon (see also p. 970). It has also been shown ultrastructurally, in lampreys, that vesicles with side projections line up along the outside margins of microtubules, suggesting that they may be transported along axons by shearing forces generated between the projections and the microtubules. Other investigators have pointed out the presence of cross-connexions between adjacent microtubules, and proposed them as a tentative basis for wave-like oscillations, which possibly generate a motile force for streaming movements of the cytoplasm.

Whatever the mechanism, several problems arise. For example, what happens to the materials transported to the periphery—do they pass out of the neuron, and if so, into what tissue compartment? Ligature experiments show that there is a build-up of vesicles on *both* sides of the ligature, so that any mechanism must account for *bi-directional movement* of cytoplasmic organelles. Many unsolved questions remain to be answered in this field.

Morphology of Synapses

The concept that neural pathways are interrupted at specific junctional points, was derived first from the physiological observation that there is an irreducible delay period inherent in reflex responses to sensory stimuli, and that neural conduction occurs only in certain directions (Sherrington 1947; Eccles 1964). When it was established histologically that the central nervous system consisted of large numbers of individual neurons, rather than a continuous syncytial network, it became possible to attempt the localization of the junctional zones or synapses morphologically. Studies with the silver impregnation methods of Golgi (*vide supra*, p. 820) showed various structural specializations at axon terminals in the central nervous system, and other methods demonstrated rather similar peripheral endings in muscles (7.19A, B, 20, 26). Later studies with the electron microscope, correlated with a considerable body of physiological and pharmacological knowledge, have helped to establish the morphological basis of synapses (see 7.19A–C, 20, 21), and their molecular organization. Since individual chemical synapses conduct only in one direction, some type of asymmetry might be expected in their structure; and, being regions of apposition between adjacent neurons, they may involve the junction of almost any parts of the two neuronal surfaces. The most common type of synaptic junction (7. 19A, B, 20) is that between an axon and a dendrite or a soma, the afferent fibre being expanded to form a small bulb or *bouton*; this may either be the terminal expansion of an axonal branch (*bouton terminal*), or it may be one of a series of expansions which form a row of bead-like endings, each of which makes synaptic contact (*bouton de passage*). Boutons may synapse with: (1) *dendritic spines* or with the *flat surface* of a *dendrite*; (2) the *spines* or *flat surface* of the *soma*; (3) the *initial segment* of an *axon*; (4) the *boutons* of other *axons*. The patterns of axon termination in different populations of neurons vary considerably; a single axon may synapse with only one neuron, for example the climbing fibres ending on cerebellar Purkinje cells (p. 921), or more commonly, with a number of cells, the parallel fibres of the cerebellum being an extreme instance of this type (p. 922). An axon may synapse primarily with the dendritic tree or it may enwrap and arborize around the soma. Afferent axons of different origins may synapse only with certain parts of the neuron: for example, in the pyramidal cells of the visual cortex, optic radiation afferents synapse chiefly with the basal segments of the apical dendrite. (See also the account of the differential distribution of afferent terminals on the pyramidal cells of the cornu ammonis on p. 1000.)

In *synaptic glomeruli* groups of axons make contact with the dendrites of one or more neurons in localized regions encapsulated by neuroglial cells (7.21); often complex interactions take place which will be further described below (p. 829).

SYNAPTIC SPINES

Thorn-like extensions of dendrites and other parts of the neuron surface (7.15B, 17, 19A, B, C) are termed *spines* or *gemmules*; they form the receptive points of contact with many of the incoming boutons, and may take several forms. Most frequently they are slender extensions (7.15B, 17, 19A, B) not more than about 2 μm long, with one or more expansions at their free ends, but they can also be short and stubby, branched, and bulbous. Their fine structure and distribution will be described below.

ULTRASTRUCTURE OF SYNAPSES

Ultrastructurally, chemical synapses can be defined as regions of structural specialization between two or more neurons, which possess some type of asymmetrical organization (7.19A, B, C, 20). They can be classified in various ways, for example by the type of neuronal processes involved and the direction of transmission. Thus synapses may be *axodendritic* (most common), *axosomatic* (also quite common), and less commonly, *axoaxonic*, *dendroaxonic*, *dendrodendritic*, *somatodendritic*, or *somatosomatic* (see Shepherd 1974; Gray 1974).

Axodendritic and axosomatic synapses are found in all regions

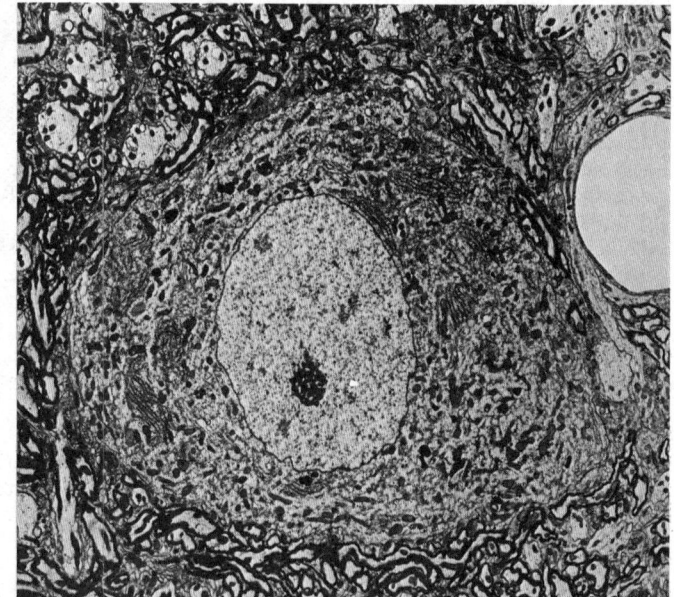

A

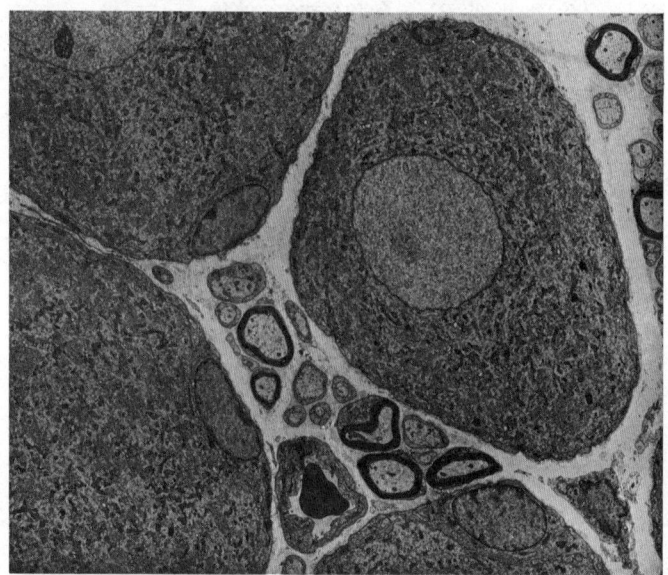

B

or attaching to the presynaptic dendrites (Gray 1975). Each microtubule is surrounded by a cluster of synaptic vesicles connected to it by thin filaments; although it is not understood why these details are revealed by the rather unusual method of preparation, it is feasible that the complex of structures represents a means of conveying transmitters close to the presynaptic membrane, to which they may attach and fuse, releasing their contents at specific sites ('*synaptopores*') into the synaptic cleft when the bouton is depolarized. Membrane added by this fusion of vesicles with the presynaptic surface is taken back into the bouton for recycling in the form of endocytic vesicles at the lateral margins of this structure (Heuser and Reese 1975).

Considerable cytochemical investigation has also revealed interesting details of presynaptic organization (*see*, for example, the review by Pfenninger and Rees 1976), mainly of the 'type I' synapse (*vide infra*). Staining with ethanolic phosphotungstic acid has shown a hexagonal grid pattern to occur on the cytoplasmic side of the presynaptic membrane, probably consisting of proteinaceous pillars guiding the vesicles to attachment points on the membrane (7.19A). Various other methods, such as the bismuth iodide staining technique coupled with enzyme digestion, have also demonstrated glycoproteins of various types in the synaptic cleft and postsynaptic proteins within the neighbouring dendrite. Freeze-fracture techniques have further shown the presence of rings of membrane particles within the presynaptic membrane, corresponding to vesicle attachment sites, and also of numerous particles in the postsynaptic membrane, believed to be related to receptors for neurotransmitter substances which are responsible for activating the postsynaptic structures electrically (Akert *et al.* 1972).

In some sites, the neurofilaments of the afferent axon enter the bouton and form a loop which takes a variety of forms visible under the light microscope with silver stains. On the postsynaptic side there are frequently membranous structures within the cytoplasm, such as the parallel cisternae which constitute the *spine apparatus* of some mammalian dendritic spines. Glial cell processes commonly enwrap these various structures, but they do not extend into the synaptic cleft. The synaptic surface of the bouton is flat where it ends on the smooth surface of a dendrite or soma, but where postsynaptic spines are involved it may be highly curved to enwrap the spine or several spines belonging to adjacent neurons (7.19A, B, 20). In some situations the dendrite surface

of the central nervous system and in autonomic ganglia. Axoaxonic synapses occur between the boutons of two axon terminals (p. 806) and also between axon terminals and the initial segments of other axons (7.19B, C, 20). The other types of synapse appear to be restricted to regions of complex interaction between the larger neurons of sensory pathways and microneurons.

For clarity, the **axodendritic arrangement** will first be described in some detail (7.19, 20), since considerable effort has been made by various investigators to understand its structure, e.g., Gray (1959, 1961, 1969, 1974); Uchizono (1965); Pappas and Purpura (1972); Shepherd (1974); Rakic (1975); Jones (1978). Each synapse involves the apposition of a *presynaptic bouton* or *synaptic bag* with a postsynaptic process from which it is separated by a *synaptic cleft*. On both sides of this cleft there are zones of dense cytoplasm, usually broken on the presynaptic side into several groups, and on the postsynaptic side often extended into a filamentous meshwork, the *subsynaptic web*. The cleft shows indications of fine transverse filaments. The presynaptic expansion is typified by numerous small membranous *synaptic vesicles* clustered in groups against the edge of the synaptic cleft and often, in a particular sectional profile, filling the whole bouton; mitochondria, membranous sacs and occasional lysosomes are also present. Recently it has been shown that after pretreatment with albumen, fixed boutons may demonstrate numerous microtubules stemming from the attendant axon and extending in a regular manner close to the presynaptic membrane

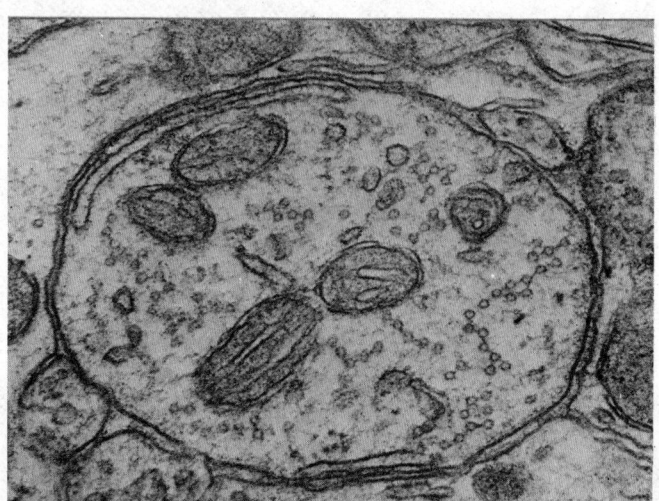

C

7.18A–C Electron micrographs of neuronal perikarya of the rat. A Typical ventral grey column multipolar neuron placed amongst numerous profiles of dendrites and axons, including myelinated nerve fibres. Note the prominent nucleolus, and in the cytoplasm the clusters of rough endoplasmic reticulum. B A section through the somata or a number of unipolar spinal ganglionic neurons, associated with which are small flattened capsular (satellite) cells. Myelinated and non-myelinated nerve fibres and a capillary are also present. C An electron micrograph of a transverse section through the initial segment of an axon, showing the rows of linked microtubules which characterize this region. (Specimens 7.18A–C kindly provided by Dr. A. R. Lieberman of University College, London.)

itself forms a large spike, which interdigitates with the surface of the bouton. Conversely, dendritic expansions may be invaginated by one or more axonal terminals.

FUNCTIONAL CLASSIFICATION OF SYNAPSES

Since we know that synapses may be either inhibitory or excitatory and that different transmitters are released in their locality, it is important to look for corresponding structural differences between synapses. These can be classed morphologically on the basis of the shape of their synaptic vesicles and the arrangement of their pre- and post-synaptic cytoplasmic densities, into a number of distinct types. Thus, synapses which are rich in *catecholamines* (noradrenalin, adrenalin, dopamine), contain large (40–60 nm) *dense-cored vesicles* (7.20), similar to those of chromaffin cells in the adrenal medulla, and their action in some sites is inhibitory, whilst in others it is excitatory. *Neurosecretory endings* such as those of the posterior pituitary contain relatively huge (50–200 nm) dense-cored vesicles of an irregular shape (7.20). Of the other types of central synapse, it was first recognized by Gray (1961) in osmium-fixed material that two

classes could be distinguished—a *type I* in which the zone of subsynaptic dense cytoplasm is much thicker than that on the presynaptic side, and a *type II* in which the two zones are more symmetrical but thinner (7.19). The widths of the synaptic clefts are also different, being about 30 nm in type I synapses against 20 nm in type II synapses. With the advent of aldehyde perfusion as a method of fixation, these two types of synapse were found to be associated with two distinct classes of synaptic vesicles, type I endings containing small spherical vesicles about 50 nm in diameter, and type II boutons showing a variety of flattened forms (7.20). Where the electrophysiology was known, as in the cerebellum, *type I synapses* could be correlated with *excitation* and *type II endings* with *inhibition* (Uchizono 1965). Other reports, however, indicate that the situation is more complicated than this simple classification implies, since it is possible to distinguish between large (25 × 60 nm) and small (15 × 40 nm) flattened vesicles (Pinching and Powell 1971), and also between discoidal and cigar-shaped forms (Dennison 1971). Irregular (*pleomorphic*) shapes are also reported. Other studies have shown that in fresh specimens all vesicles may be spherical, but that some flatten on exposure to the buffers used in the preparative procedures

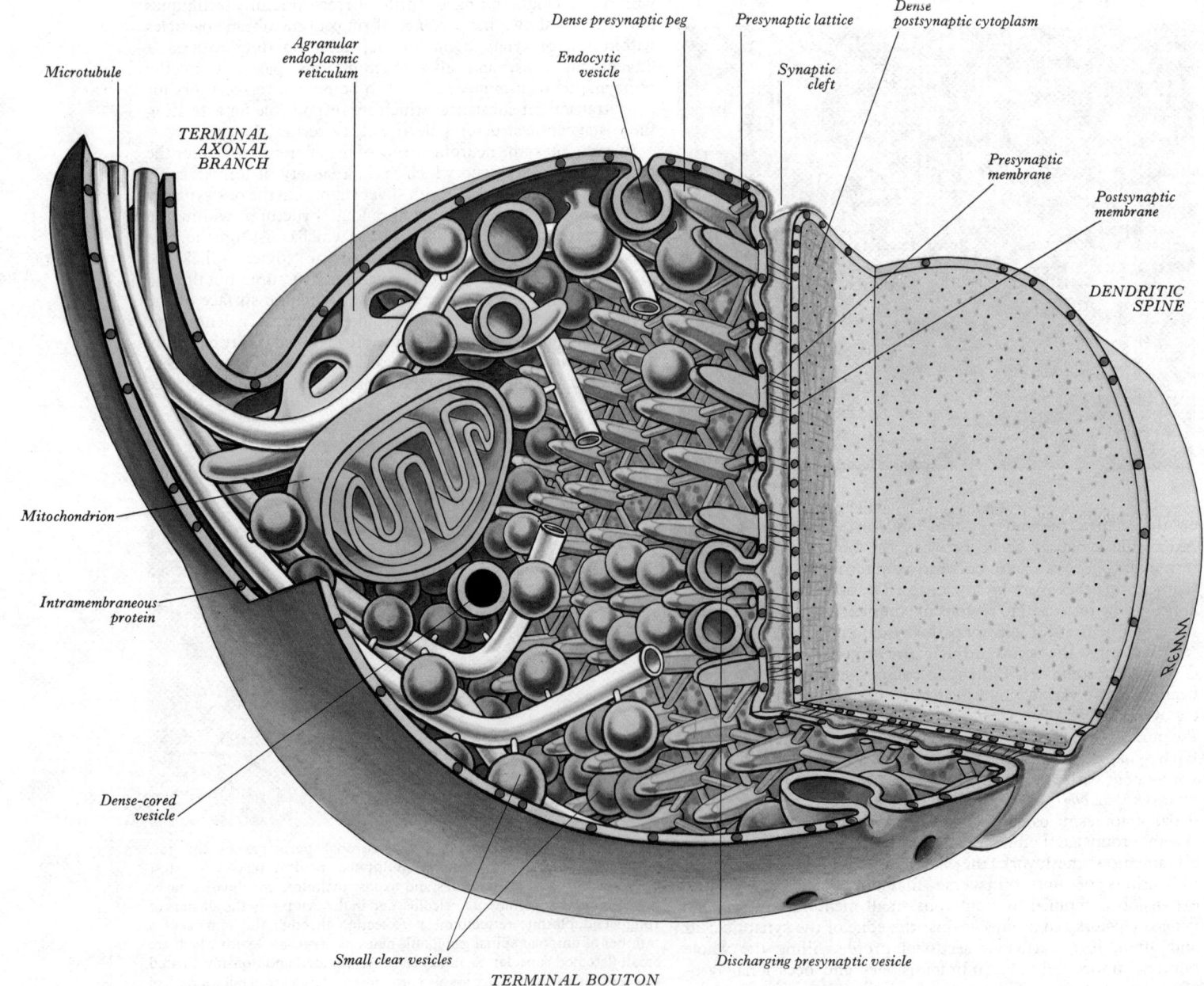

7.19A The general organization of an excitatory synapse (asymmetrical, Gray type I) showing a synaptic bouton (left) cut away to show presynaptic structures, and an associated dendritic spine (right).

(Bodian 1970). Different synaptic endings also show *mixed populations* of flat and spherical vesicles of varying proportions, depending upon the osmotic strength of the fixative. Although these various appearances are undoubtedly artefactual, they nevertheless point to real differences in the chemistry of the synapses. There have been several attempts at classifying different forms of synaptic boutons in different regions of the central nervous system, according to such criteria as vesicle shape and size, distribution of pre- and post-synaptic cytoplasmic thickenings, the overall size of the bouton, presence of mitochondria, and so forth. One commonly used notation (Guillery 1969; Partlow *et al.* 1977) designates the type I synapse of Gray as a round vesicle-asymmetrical thickening (R.A.) class, and Gray's type II as flat vesicle-symmetrical thickening (F.S.) class; various subdivisions of these categories have also been suggested (*see*, for example, Robson and Hall 1977). Whilst these schemes have some descriptive value, our knowledge of the transmitters present at the various types of synapse, and of their functional status, is so slender that any serious effort beyond a crude classification would appear to be premature. Mixed populations of small, clear vesicles and large, dense-cored ones are also commonly found; in some cases such synapses have been correlated with the presence of the peptide transmitter substance 'P', of serotonin, or of the enkephalins (Beaudet and Descarries 1979; Pickel *et al.* 1979); (*see* also below).

Some central synapses apparently lack the specialized contact zones mentioned above, and seem to be formed by areas of release of transmitter in varicosities of nonmyelinated axons, so that synaptic action may in these cases be somewhat diffuse and could affect several neighbouring structures, as, for example, described in the aminergic pathways of the basal nuclei (*see* also p. 948).

It is also interesting that freeze fracture studies of type II synapses have failed to demonstrate the numerous pre- and post-synaptic intramembrane particles typical of the type I variety, suggesting that there may be fundamental differences in the method of release and of action of the transmitter on the postsynaptic membrane (Landis *et al.* 1974).

The junction between the retinal receptors and bipolar neurons, and between cochlear hair cells or vestibular sensory cells and their afferent endings is also marked by a synapse-like structure; this differs from the usual pattern in that the vesicles are clustered around an internal cytoplasmic rodlet, rather than being situated at the margin of the synaptic cleft (Boycott and Kolb 1973; Flock 1971).

NEUROCHEMICAL TRANSMISSION

Transmission at all of the synapses described above is *neurochemical*, involving the release of *transmitters* associated with, or contained within, the synaptic vesicles, into the synaptic cleft. Here, the transmitter causes changes in potential at the postsynaptic surface, tending to depolarize it in the case of excitatory synapses, or to hyperpolarize the membrane in inhibitory ones. The cue for transmitter release is the depolarization of the presynaptic membrane by the arrival of a nerve impulse, and the consequent release of calcium ions in the cytoplasm of the terminal.

Synaptic bags ('*synaptosomes*') can be separated by cell fractionation techniques from brain homogenates, and further subfractionated into various components. Where acetylcholine is the transmitter, enzymes capable of synthesizing this substance from its precursor, choline, are present in synaptosomes, and the transmitter is stored in association with the synaptic vesicles. Catecholamines, and other possible transmitters, have been demonstrated in the same way in synaptosomes (*vide infra*).

ELECTRICAL SYNAPSES

Although in mammals chemical transmission is usual, in lower vertebrates, and in several groups of invertebrates, electrically acting synapses are abundant in certain pathways where speed or synchrony of action is important. Examples are the 'spoon' endings in the chick ciliary ganglion, 'club' endings on the giant Mauthner cells of the medulla in bony fishes, electromotor

synapses in electrogenic fishes, and the giant fibre 'escape' systems of crayfish and earthworms. In structure and physiology, these synapses are essentially the same as the electrical junctions in cardiac muscle (p. 517), and non-striated muscle (p. 520), and are much more rapid than chemical synapses. Some of the invertebrate types can operate equally well in either direction, though usually the transmission is undirectional.

More recently, similar junctions have been shown in mammals at various sites, including the superior olivary and vestibular nuclei, cerebellar and cerebral cortices, although electrophysiological corroboration of their transmission properties are available in only a few cases (*see*, for example, Gwyn *et al.* 1977).

ORGANIZATION OF SYNAPTIC GROUPS

The way in which synaptic endings affect other neurons is partly dependent on their detailed arrangement on the cell surface, and their interrelations with other synapses. At *serial synapses* (7.19B, C, 20E), boutons end on other boutons, to affect their ability to respond to an afferent volley of nerve impulses; this could be the basis of *presynaptic inhibition* observed physiologically in the spinal cord, although of course it could also be responsible for *presynaptic facilitation*, depending upon the types of synapse involved. In *reciprocal synapses* (7.19C, 20F) found initially in the olfactory bulb and in the lateral geniculate body, and subsequently in many other sites, transmission between two processes can occur in either direction by way of staggered synaptic zones placed on each side of the synaptic cleft. Commonly the zones appear to be excitatory in one direction and inhibitory in the other; they are believed to be the basis of lateral inhibition, at least at the first of these sites (Rall *et al.* 1966; Shepherd 1974, 1978). Other arrangements involving serial synapses at axonal or dendritic junctions have been described; some of these are depicted in illustration 7.19C (*see* also Gray 1974; Colonnier 1974; Shepherd 1974, 1978).

In **synaptic glomeruli** (7.20, 21) several boutons may synapse with dendrites, and with each other, in localized regions of the neuropil which are usually encapsulated by layers of glial cells (Szentágothai 1970). Sometimes, where microneurons are involved, the synaptic relationships are quite complex, involving both excitatory and inhibitory interactions between many cell processes; these may occupy extensive zones or **synaptic clusters**. In **synaptic** or **neuropil cartridges**, a local region of a dendrite is enclosed partially by a glial sheath to isolate a cylindrical zone of synaptic endings both on spines and the smooth dendritic surface between spines (7.21B). Glial cells may also be so situated as to isolate or allow juxtaposition between various groups of interacting cells in various ways which may be of some physiological importance. Although much is known about the arrangement of synapses in limited areas of the nervous system, much remains to be learnt about the disposition of different types of synapse on the surfaces of neurons. It has been suggested, for example, that inhibitory synapses might be particularly effective in blocking excitation if situated on the flat surfaces of the stems of dendrites, and on the soma and initial segment of the axon, where they could block excitation caused by other types of endings occurring on the spines of dendrites or other more peripheral positions, an example being the terminals of the basket cell axon which synapse with the 'pre-axon' of the Purkinje cell (p. 921).

DEVELOPMENT AND PLASTICITY OF SYNAPSES

When first formed embryonically, synapses are recognizable by inconspicuous zones of density on each side of the synaptic cleft, and they only gradually mature into their fully differentiated structure (Bodian 1970; Pfenninger and Rees 1976). Similar profiles are often seen in postnatal nervous systems, prompting the suggestion that synapses may be labile structures capable of being recruited for transmission, and perhaps dispersing when redundant. A change such as this is implicit in some theories of memory, which require synapses to be subject to modification according to the frequency of their use, thus establishing

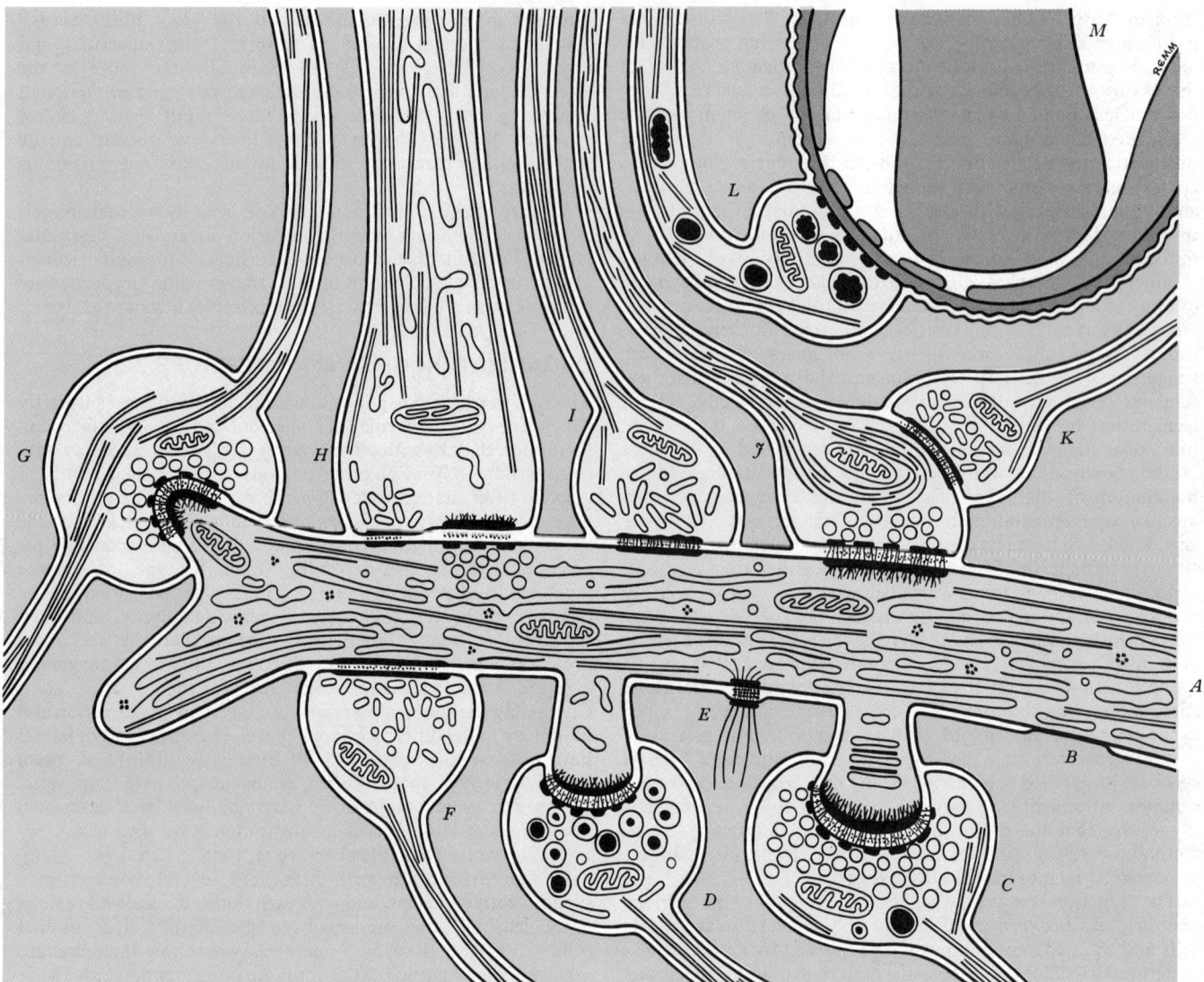

7.19B A scheme of the ultrastructural morphology of synapses, showing various junctional structures, grouped around a dendrite (A). The tight junction (B) and the desmosome (E) are without synaptic significance. Excitatory synaptic boutons are shown (C, G) containing small spherical translucent vesicles. D: a bouton with dense-cored, catecholamine-containing vesicles; F: an inhibitory synapse containing small flattened vesicles; H: a reciprocal synaptic structure between two dendritic profiles, inhibitory towards dendrite A and excitatory in the opposite direction; I: an inhibitory synapse containing large flattened vesicles. J and K: two serial synapses; J is excitatory to the dendrite; K is inhibitory to J. L: a neurosecretory ending adjacent to a vascular channel (M), surrounded by a fenestrated endothelium. All the boutons in this diagram are of the terminal type, except G which is a *bouton de passage*.

preferential conduction pathways in the brain. Unfortunately, what little evidence is available from training experiments, and from artificial stimulation, either does not support this hypothesis, or is equivocal. Neurophysiological recording of cortical neurons, however, suggests that synapses may become less easy to excite with repeated use, and they may change size. However, perhaps it is not necessary to postulate structural changes on this scale, since an increase of molecular receptor sites on the postsynaptic membrane, for example, would have the same effect as a change in the size of the presynaptic *bouton*. It is interesting that a somewhat controversial body of evidence has accumulated, suggesting that long-term memory formation is associated with increased protein synthesis; any change in the proteins associated with the synapse, even of receptor proteins, might require such an increase (*vide infra*).

There is, however, much evidence that during early postnatal development there is a considerable increase in the number of synapses, correlated with a parallel increase in dendritic spines, and that this phenomenon is dependent upon the degree of neural activity. This is seen particularly well in the visual cortex in young animals made temporarily blind in one eye, when the dendritic spines of related cortical neurons show greatly impaired development (Rothblat and Schwarz 1979).

NEUROTRANSMITTERS

Up to the present time the major classes of transmitter substance, which have been authenticated as such and extensively examined, are *acetylcholine*, various *monoamines*, the catecholamines *noradrenalin* and *dopamine*, the indoleamine *5-hydroxytryptamine* (serotonin), and the amino acid *gamma aminobutyric acid*, or GABA (*see* the reviews by Krnjevic 1974; Snyder and Bennett 1976; Burgen and Mitchell 1978). The physiological effects of these vary considerably from one site to another and are beyond the scope of the present account. However, their common characteristics are that they are released from nerve endings when a nerve impulse arrives, and that they affect the resting potential of the cells which they influence, either to raise or lower it, for only a limited period of time. With acetylcholine, this temporal restriction is achieved by enzymic destruction of transmitter in the synaptic cleft, and the enzyme *acetylcholinesterase* can be demonstrated by various means, including cytochemical localization, in that region. With the catecholamines, as with other possible transmitters (*vide infra*), the transmitter is taken back into the nerve ending, to be stored for subsequent use, so terminating its extracellular action.

The mechanism of action of neurotransmitters is not yet fully

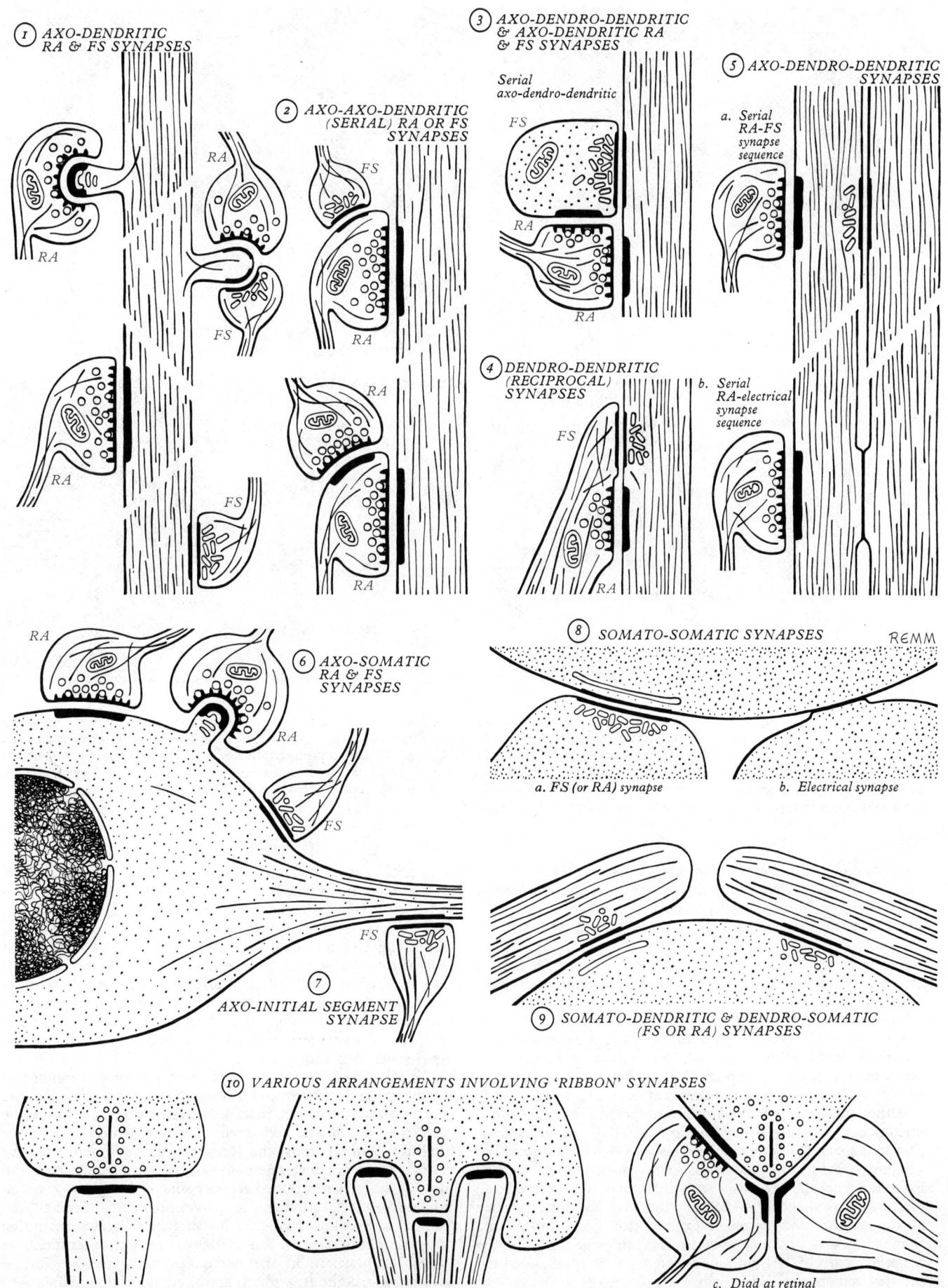

① *AXO-DENDRITIC RA & FS SYNAPSES*

② *AXO-AXO-DENDRITIC (SERIAL) RA OR FS SYNAPSES*

③ *AXO-DENDRO-DENDRITIC & AXO-DENDRITIC RA & FS SYNAPSES*

Serial axo-dendro-dendritic

④ *DENDRO-DENDRITIC (RECIPROCAL) SYNAPSES*

⑤ *AXO-DENDRO-DENDRITIC SYNAPSES*

a. Serial RA-FS synapse sequence

b. Serial RA-electrical synapse sequence

⑥ *AXO-SOMATIC RA & FS SYNAPSES*

⑦ *AXO-INITIAL SEGMENT SYNAPSE*

⑧ *SOMATO-SOMATIC SYNAPSES*

REMM

a. FS (or RA) synapse

b. Electrical synapse

⑨ *SOMATO-DENDRITIC & DENDRO-SOMATIC (FS OR RA) SYNAPSES*

⑩ *VARIOUS ARRANGEMENTS INVOLVING 'RIBBON' SYNAPSES*

a. Cochlear hair cell

b. Triad at base of retinal rod

c. Diad at retinal bipolar cell terminal

7.19C A diagram illustrating various types of simple and multiple arrangements of synapse involving 'asymmetrical' synapses with rounded vesicles (R.A.), 'symmetrical' synapses with flattened vesicles (F.S.), and electrical synapses. For further details, see text.

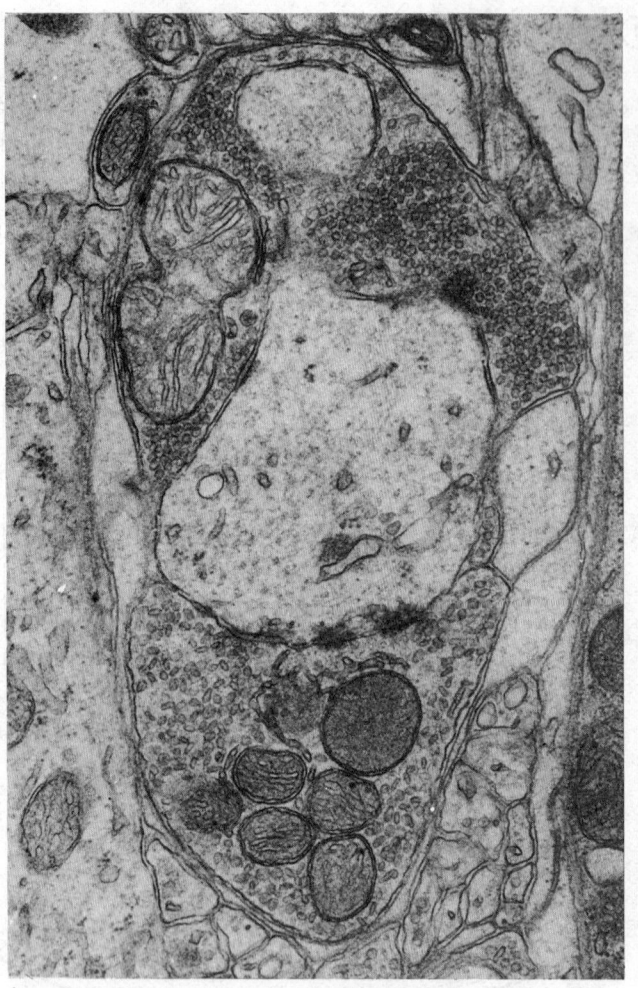

A

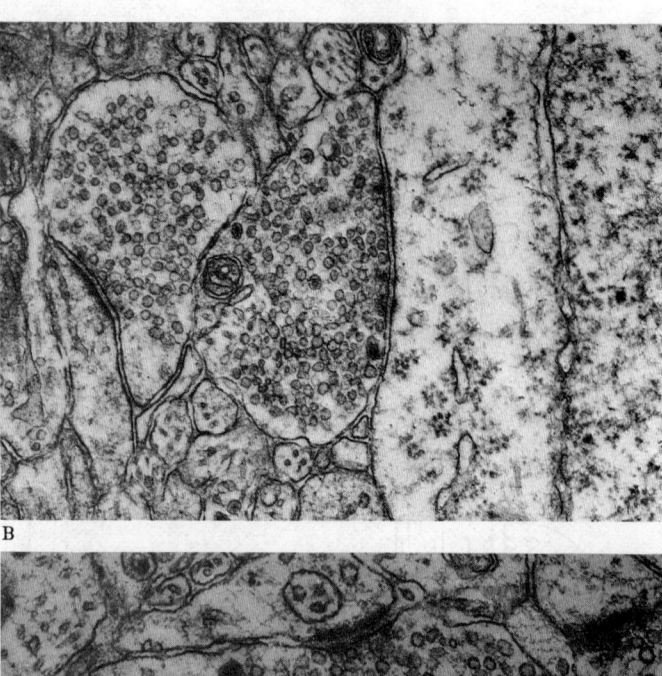

B

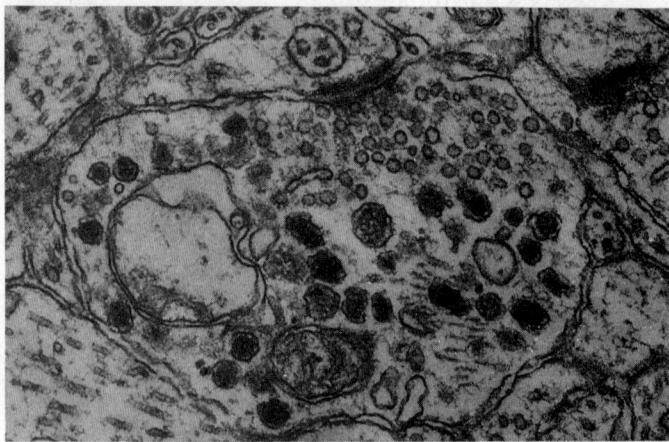

C

7.20A–C Electron micrographs demonstrating various types of synapse. A This shows a pale cross-section of a dendrite upon which end two synaptic boutons. One of them (above) contains round vesicles, and the other (below) contains flattened vesicles of the small type. A number of pre- and post-synaptic thickenings mark the specialized zones of contact. B Two types of synaptic structures are shown; one of them (left), a type I synapse between an axon terminal containing round vesicles, and a dendritic spine, the other (right), a type II synapse between an axon terminal containing pleomorphic vesicles, and the surface of a neuronal soma. C A type I synapse containing both small, round, clear vesicles, and also large dense-cored vesicles of the neurosecretory type.

understood, and most of our knowledge comes from investigations of peripheral motor endings. In the cholinergic endings in muscle, the subsynaptic sarcolemma of the muscle cells bears special *receptor molecules*, proteins, which bind the transmitter strongly, and presumably initiate the changes in the membrane which alter its permeability to ions. A similar mechanism appears to exist with other transmitters, although details may vary.

The distribution of particular neurotransmitters within the brain has been studied with a variety of methods, including pharmacological techniques, radioactive tracer methods, fluorescence microscopy with respect to the monoamines and recently choline acetyl transferase, cytochemical localization and so on.

Although relatively few neurotransmitters have been identified unequivocally, many of the synapses of the central nervous system appear to operate by means of other transmitters of unknown chemistry. Numerous substances which can be extracted from the nervous system are known to have profound pharmacological effects at synapses, but have not so far satisfied all the classical pharmacological criteria for neurotransmitters. Some of these agents are present in relatively large concentration in specific parts of the brain, and are known to be synthesized and metabolized locally. They may be precursors or breakdown products of transmitters, rather than actual transmitters themselves; however, some of them have been localized in synaptic endings, and are taken up from extracellular media in the same manner as catecholamines.

Morphological and pharmacological evidence points to a multiplicity of transmitters within the central nervous system; why so many are needed is not clear, but obviously much remains to be learnt about the physiology of central synapses which will have a direct bearing upon concepts of nervous organization. One simplifying concept, first formulated by Dale *et al*. (1936), is that each neuron synthesizes only one transmitter substance, which is then released at all of its axon terminals including those of its collaterals. This concept has been upheld in many peripheral and central locations, although in the latter it is perhaps at variance with the occasional observation of mixed populations of synaptic vesicles within boutons.

Acetylcholine is present at the terminals of somatic motor and parasympathetic axons, at synapses in autonomic ganglia, and centrally has been located in the corpus striatum, some ascending pathways, and in the collaterals of motor neuron endings on inhibitory interneurons, the Renshaw cells. (For an extensive review of the 'cholinergic nervous system' *see* De Feudis 1974.) In most of the sites mentioned acetylcholine is thought to exert an excitatory effect, although at parasympathetic terminals inhibitory actions may, in some instances, also occur depending upon the type of receptor site acted upon. In effector terminals, as probably elsewhere in the nervous system, acetylcholine is associated structurally with synaptic vesicles of the small (40–50 nm), round, clear type, and with the presence of acetylcholinesterase in the neighbouring synaptic gap.

Gamma aminobutyric acid (GABA) is an important transmitter which has a postsynaptic inhibitory (hyperpolarizing) action on many central neurons, although in the spinal cord it also may

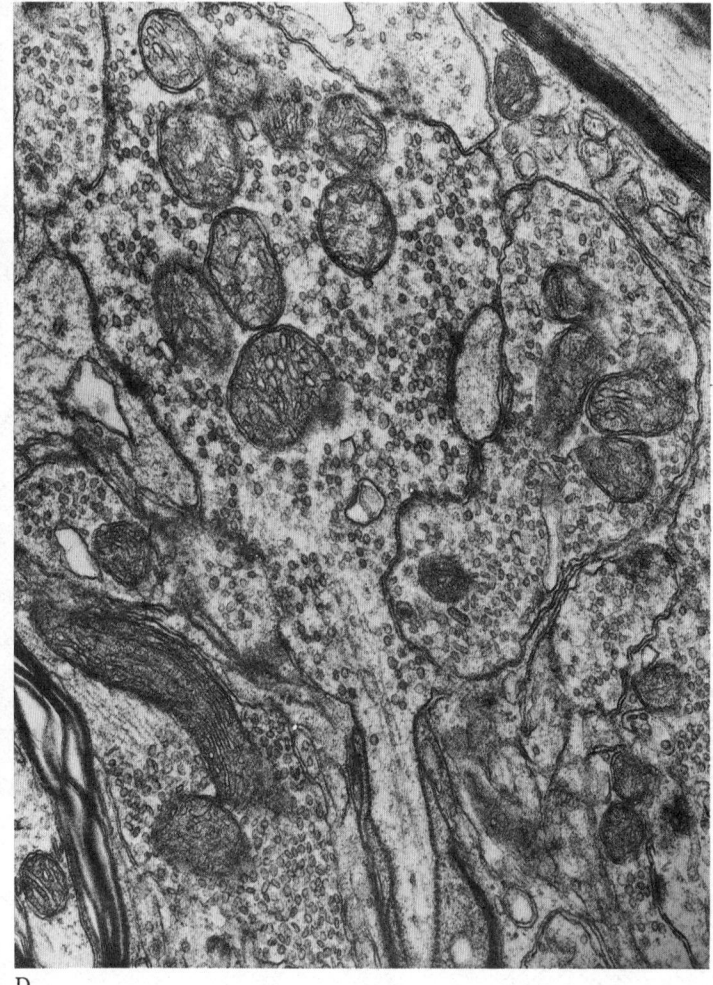

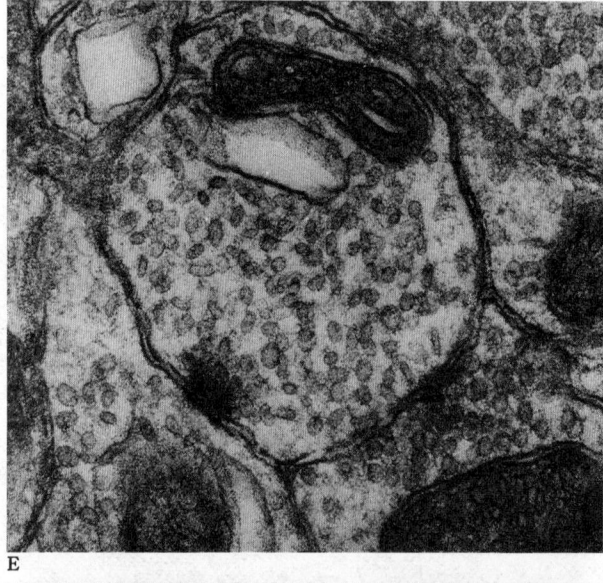

E

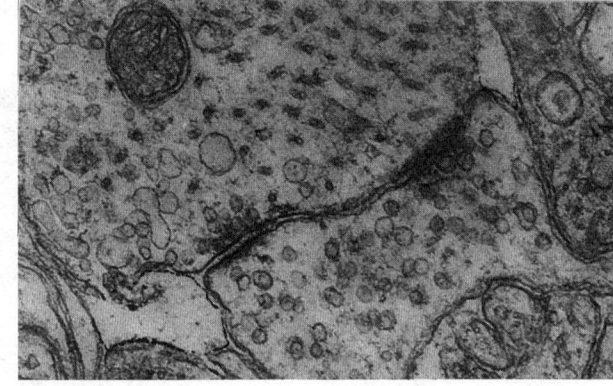

D

F

7.20D–F Electron micrographs of complex arrangements of synapses. D This shows a large terminal bouton of an optic nerve afferent fibre, which is making contact with a number of postsynaptic processes, in the dorsal lateral geniculate nucleus of the rat. One of the postsynaptic processes (right), also receives a synaptic contact from a bouton containing flattened vesicles. E Three neuronal processes in serial contact. On the lower right, a process, containing round vesicles, synapses with a second process (centre) containing flattened vesicles; in turn the latter makes contact with a third process (lower left): specimen from the dorsal lateral geniculate nucleus of the rat. F This demonstrates reciprocal synapses between two neuronal processes in the olfactory bulb of the rat. (7.20A, D and E provided by Dr. A. R. Lieberman of University College, London.)

cause presynaptic inhibition by depolarization of the axon terminals of excitatory fibres. Its action has been particularly well established at the terminals of the cerebellar Purkinje neurons synapsing in the lateral vestibular nucleus.

Glycine is another probable inhibitory transmitter, and is found abundantly in the spinal cord and brainstem where it appears to act by increasing chloride conductance, thus hyperpolarizing the postsynaptic membrane. Various studies have suggested that it is contained in the flattened vesicles of the type II synapse (*vide supra*). Other amino acids which have been proposed as central neurotransmitters because of their presence and known pharmacological actions are *glutamic* and *aspartic acids*, both excitatory in the spinal cord, and *taurine*, which has an inhibitory effect.

The *monoamines* form a class of substances widely distributed in the nervous system. These include chemicals having excitatory effects in the central nervous system, namely, *noradrenalin* (a catecholamine), *5-hydroxytryptamine* or *serotonin* (an indoleamine), and *histamine*; *dopamine* (also a catecholamine) is an inhibitory substance. Of these putative transmitters there is considerable evidence that noradrenalin, serotonin, and dopamine are major central nervous transmitters; noradrenalin is, of course, also the classical transmitter at peripheral sympathetic nerve endings, where its precise action depends upon the type of receptor site present in the innervated cell surface.

The precise structure of synapses associated with the monoamines is not yet fully understood, although noradrenalin and dopamine are known to be present where moderate-sized or large dense-cored vesicles abound. Serotonin has been correlated with clear rounded vesicle populations, often mixed with dense-cored vesicles (Beaudet and Descarries 1979).

Recently, another class of possible transmitters has been discovered in the brain, all of them having polypeptide chains of 5 or more amino acid residues. These include *substance P*, an undecapeptide, which occurs extensively in the axon terminals of primary sensory neurons, amongst other sites, where excitatory synapses are known to contain mixed dense-cored and small clear vesicles at 'asymmetrical' junctions (Hökfelt 1977; Pickel *et al.* 1979).

The *enkephalins* are pentapeptides which have various inhibitory actions resembling those of morphine when injected into the brain and have been shown to bind to 'opiate receptor sites' widely distributed in the spinal cord and brain, particularly throughout many of the subdivisions of the so-called limbic system (*see* p. 990). Synapses rich in the enkephalins apparently resemble those containing substance P (Pickel *et al.* 1979).

The *endorphins* have somewhat longer molecules, derived from the pituitary peptide *β-lipotropin*, but have certain actions similar to those of the enkephalins.

Finally, the *purine* derivative *adenosine triphosphate* has been suggested as an important transmitter in a group of peripheral ('*purinergic*') postganglionic autonomic neurons supplying gut musculature and various other tissues (Burnstock 1975), the terminals of such cells containing large dense-cored vesicles (for further comments *see* p. 520).

It should be borne in mind that not all of these substances may be actual transmitters themselves, but may instead be metabolites

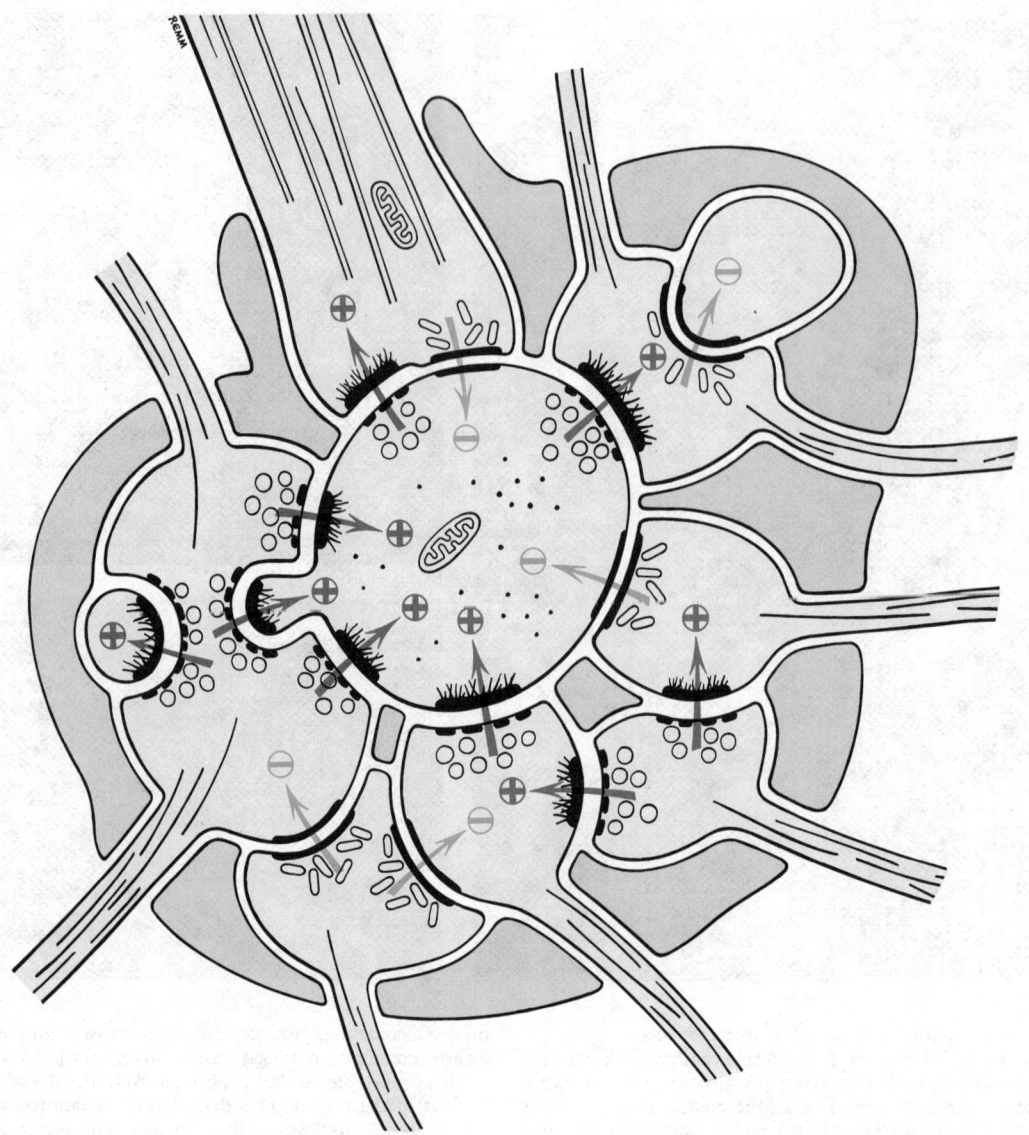

7.21A and B Specialized multiple groups of synaptic contacts. A A synaptic glomerulus, showing various arrangements of synapses grouped around a centrally placed terminal dendritic expansion, seen in cross-section. Both excitatory (+) and inhibitory (−) synapses are shown; the direction of transmission is indicated by arrows (red for excitation, blue for inhibition). A glial capsule surrounds the whole complex.

of transmitters; alternatively, they may exert their effects in a general, diffuse manner, thus modulating the firing patterns of large numbers of neurons.

(For information concerning the distribution of a number of these substances in the mammalian brain and a general discussion consult the section on 'The Reticular Formation', pp. 943–953.)

Neuroglia

In addition to neurons, several varieties of non-excitable cells are present in the nervous system (de Robertis and Carrera 1965), forming a major component in its total composition (7.22, 23). In the central nervous system these are the *glial cell types* in the parenchyma of the brain and spinal cord, and the *ependymal cells* lining their internal cavities. In the ganglia of the spinal nerve roots and autonomic nervous system, are the *capsular cells* which surround the cell bodies in those masses and peripherally the *Schwann cells* ensheathing the axons (*see* p. 843), terminal *lemmal cells* surrounding sensory capsules, *teloglial cells* ensheathing motor terminals, and the *supporting cells* of the sensory epithelia. Although the precise roles of the various cell types is not yet clear, their importance in the activities of the nervous system is becoming more and more obvious. One barrier to study has been their small size and variable appearance as seen with the special

staining methods necessary to demonstrate them with light microscopy. Combined electron and light microscopic studies have helped to clarify some of these details, and a picture is gradually emerging, at least of their appearances.

NEUROGLIAL CELL TYPES

In the central nervous system the chief non-nervous cells are the glial cell types (7.22, 23). These vary in numbers and type from one part of the nervous system to another, but two basic classes have long been distinguished (del Rio Hortega 1924) by their size and embryonic origin, namely: *macroglia*, relatively large cells derived from the neural plate, and the smaller *microglia*, stemming from the mesodermal tissues surrounding the nervous system, which they enter at a relatively late stage of development (*see* pp. 157, 861).

The *macroglia* (7.22, 23) comprises three cell types, the *astrocytes* (*astroglial cells*), the *oligodendrocytes* (*oligodendroglial cells*) and the *glioblasts*.

Special methods have to be used to demonstrate different neuroglial cells, since only their nuclei are made visible by the commonly used histological stains, and these lack obvious distinguishing features. Apart from various metallic stains, which are capricious in their results, the recent use of $0.5–2.0$ μm epoxyresin sections stained for light microscopy, and correlated with electron microscope studies, have enabled clear distinctions

individual astrocytes, and sometimes between astrocytes and neurons. Cells with these features may take a number of different forms, and it appears likely, on the basis of isotopic labelling studies, that astrocytes, which are able to divide in mature animals, pass after mitosis through a series of structural transformations before finally disintegrating. In areas of brain injury they proliferate (gliosis), and also act as phagocytes to clear cellular débris.

Oligodendrocytes (7.22, 23), as their name implies, have fewer cell processes, and occur in two distinct areas, namely as *intrafascicular* cells in myelinated tracts, and as *perineuronal* oligodendrocytes where their processes come into apposition with the surfaces of the somata of neurons.

Since the advent of electron microscopy, the earlier suggestion that these cells may be connected with myelin sheaths has been amply confirmed (*see* Bunge 1968); indeed it is evident that they are responsible for laying down myelin and that these cells are the central nervous counterpart of the peripheral myelinating Schwann cell. There are some interesting differences between central and peripheral myelin, however, including a distinctive chemical composition (*see* Gregson 1975), and a different relationship between the axon and myelinating cell: the oligodendrocyte may enclose several neighbouring axons in separate myelin sheaths, whereas Schwann cells of myelinated peripheral nerve fibres are each associated with only one axon. Ultrastructurally, these cells are characterized by a rounded nucleus and a cytoplasm rich in mitochondria, microtubules and glycogen. As with astrocytes, there is a spectrum of appearances, ranging from cells with relatively large euchromatic nuclei and a pale cytoplasm to those with heterochromatic nuclei and a dense cytoplasm. Tritium labelling, again, shows that oligodendrocytes can proliferate in normal *mature* animals, and that they pass through stages of maturation and degeneration.

Glioblasts are embryonic or postnatal stem cells capable of differentiating into macroglial cells. They have structural features typical of such cellular elements, namely a pale nucleus and cytoplasm, free ribosomes, and in general a rather rounded shape. They are particularly numerous, even in adults, in a diffuse layer beneath the ependyma (the sub-ependymal zone). It is not known whether, in adults, such glioblasts are multipotent, i.e. able to form either astrocytes or oligodendrocytes, or if individual glioblasts for each cell line exist. However, they appear to be an important source of macroglial cells formed by mitotic division throughout postnatal life as well as in the fetus (Mori and Leblond 1970; Ling *et al.* 1973).

Pituicytes of the neurohypophysis are similar to astrocytes in some respects, except that their processes end mostly on the endothelial cells of vascular sinuses.

Müller cells of the retina (p. 1171) also have many features in common with astrocytes, and are of a similar origin.

Bergmann glial cells of the cerebellum (p. 921) have a special structure; their cell bodies and nuclei are situated in a row some distance below the pial surface. A single apical process passes superficially and then branches into terminal pial expansions. In many respects they have the appearance of a rather primitive glial cell type.

Ependymal cells are arranged as an epithelial layer one cell thick which lines the ventricles of the brain and the central canal of the spinal cord. The cells vary from squamous to columnar according to their locality. At the surface they are in contact with each other by gap junctions and occasional desmosomes, and their free faces bear numerous microvilli and cilia which are often motile, their movements contributing to the flow of cerebrospinal fluid. Ultrastructurally (Brightman and Palay 1963; Bleier 1977) the nucleus is rather heterochromatic, with an indented outline; the cytoplasm is rich in mitochondria, lysosomes, microtubules and microfilaments (7.22, 23). Secretion bodies are also sometimes present. Great regional variations in surface specializations occur in the ventricular linings of the brain, some areas being heavily ciliated, others being lined by microvilli or by secretory structures (Scott *et al.* 1974). In embryonic nervous systems, and in lower vertebrates, ependymal cells possess one or more long, radially orientated processes which may branch and give off short side extensions, and are termed *tanycytes, ependymal*

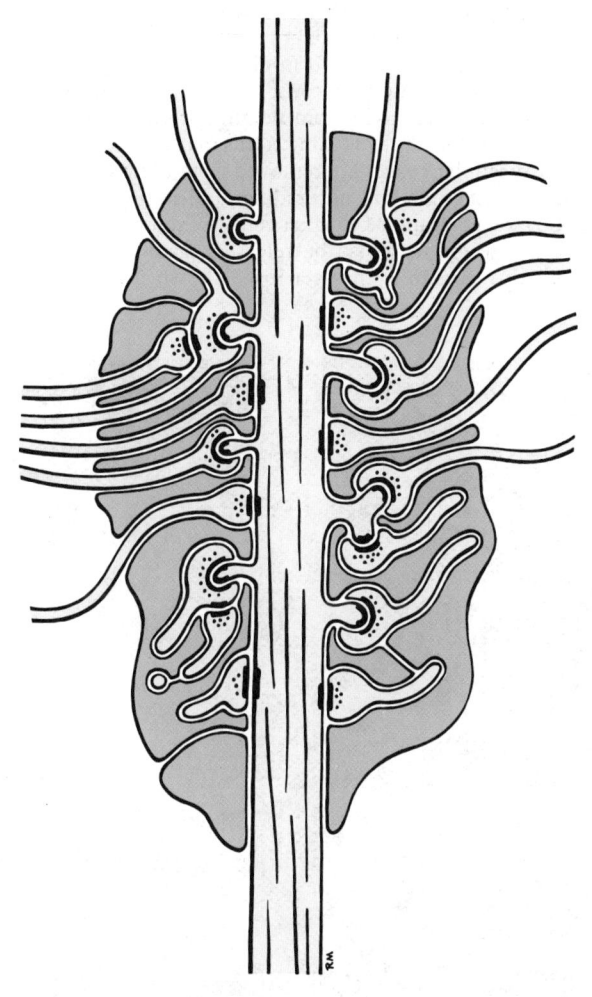

B A synaptic cartridge, with synapses grouped around a segment of a dendrite, and enclosed within a glial capsule (green).

to be made routinely, at least in some mammalian species (Ling *et al.* 1973).

Astrocytes possess small cell bodies (the nucleus is about 8 μm in diameter in man) with ramifying dendrite-like extensions. These cells can be further divided into star-like *protoplasmic astrocytes* with broad symmetrically spreading processes, confined to the grey matter, and *fibrous astrocytes* situated chiefly in the white matter, with asymmetrically spreading processes ramifying amongst the nerve fibres. The cytoplasmic processes of astrocytes carry fine, foliate extensions which partly engulf and separate neurons and their neurites, and often end in plate-like expansions on blood vessels, ependyma (*vide infra*), and on the pial surface of the central nervous system. Ultrastructurally, the cells are typified by a somewhat pale nucleus lined by a narrow rim of heterochromatin (7.23), a pale cytoplasm which is rich in glycogen, lysosomes (corresponding to the *granules* or *gliosomes* of earlier light microscopy), Golgi complexes, and in fibrous astrocytes, bundles of filaments extending throughout the cell processes (Mori and Leblond 1970). Glial filaments have received considerable attention recently because of their notable contribution to the total protein content of the brain. The 'glial fibrillary acidic protein' composing much of each filament, appears to be similar to that of tonofibrils, intermediate filaments, and similar structures elsewhere; it is likely that the astrocytic filaments are skeletal in function (*see* Schacher *et al.* 1977). Desmosomes and gap junctions (*see* pp. 5–6) form special contact zones between *astrocytes,* or *ependymoglial cells.* Later, in most, but not all regions of mammalian brains, the basal process is absorbed and a rapidly increasing number of functional roles are being ascribed to them (*see,* for example, pp. 1442, 1443; **8**.197). Some authorities consider these cells as important in providing 'guides' for early neuroblast migration (*vide infra*).

The functions of glial and ependymal cells in general appear to be numerous although not fully explored (*see* Kuffler and Nicholls 1976): (1) Undoubtedly they act mechanically as a supporting component of the nervous system, and their various filaments, microtubules, and surface contact zones fit them for this task. (2) They act as insulators, separating neurons and their processes from each other, or grouping interacting regions, and limiting the spread of ions from neurons during electrical activity. Electrical

studies of glial cells in leeches, where they are particularly large, show that they can act in this way, and further that they lack the ability to conduct surface depolarizations for any distance (Nicholls and Kuffler 1964). (3) They can act defensively by phagocytosis of foreign material or cell débris, and can provide a means of limited repair to form glial scar tissue, or to fill the gaps left by degenerated neurons. (4) They take up and store neurotransmitters released from neighbouring synapses and in some cases may metabolize them; in this manner the transmitters in the extracellular spaces are removed from their sites of action, as seen for example, by autoradiographic studies of gamma aminobutyric acid uptake by glial cells (Schon and Kelly 1974). Such transmitters may also be released again from glial cells, although this possibility has yet to be shown in normal neural

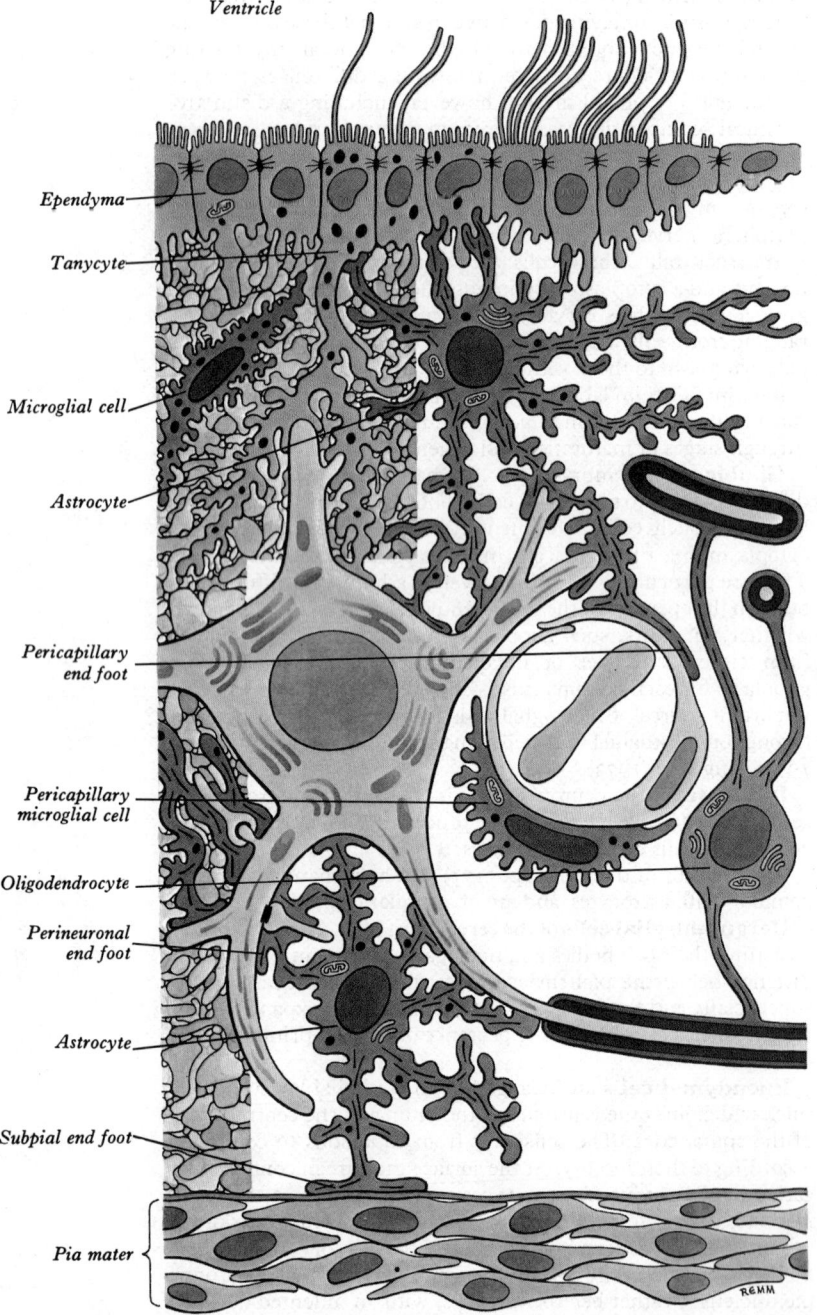

Ventricle

Ependyma

Tanycyte

Microglial cell

Astrocyte

Pericapillary end foot

Pericapillary microglial cell

Oligodendrocyte

Perineuronal end foot

Astrocyte

Subpial end foot

Pia mater

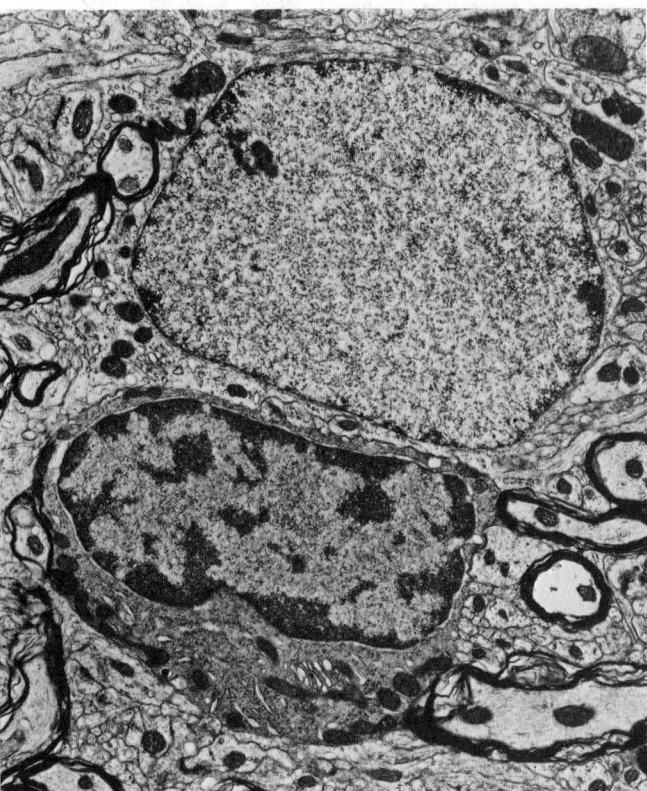

7.23 An electron micrograph of two neuroglial cells situated amongst myelinated and non-myelinated nerve fibres, in the rat thalamus. An oligodendrocyte containing numerous mitochondria, a well-developed endoplasmic reticulum, and an indented nucleus is shown below; a larger astrocyte with a vesicular nucleus and scanty cytoplasm, is demonstrated above. (Kindly provided by Dr. A. R. Lieberman of University College, London.)

7.22 A schema showing the types of non-neuronal cells in the central nervous system. The ependymal and glial cells are shown in green. The ependyma includes examples of ciliated and non-ciliated cells, and one tanycyte, with a centrally directed basal process. Two astrocytes are shown apposed to a neuronal soma and dendrites; one (above) also contacts a capillary, the other (below) expands on the pial surface. An oligodendrocyte (middle right) provides myelin sheaths for two axons. Two flattened microglial cells, one adjacent to a capillary (middle right), and the other within the neuropil at the top left, are also illustrated.

activity. Certainly changes in external potassium, such as may occur when neighbouring axons are active, cause GABA release (Iversen and Kelly 1975) and this may have some kind of modulating action on local synaptic processes (*see* discussion by Kuffler and Nicholls 1976). (5) Oligodendrocytes form and maintain the myelin sheaths of the larger neuronal processes of the central nervous system, in a manner similar to that of the Schwann cell in the periphery. (6) Ependymal and related cells are associated with secretion into, and uptake and transport from, the cerebrospinal fluid; these functions have been explored in regions associated with the hypophysis cerebri where they may be involved in transport of hormone-controlling factors in the median eminence and hypophysial stalk (*see* Bleier 1977). (7) We can speculate that macroglial cells may play an important part in determining levels of physiological activity in the neuronal populations which they infiltrate by regulating the metabolic and ionic environment of the neuronal population. Although there is no evidence that glial cells supply ions to neurons during their activity, alterations in the metabolism of glial cells may have profound effects on the ionic environment of neurons and therefore on their patterns of excitability (*see* Kuffler and Nicholls 1976). Such alterations may be involved experimentally where anions such as chloride, applied topically to the brain surface, cause spreading electrical changes, the *spreading depression of Leao* (p. 1023); it is conceivable that various metabolic changes within the brain may effect changes of electrical behaviour in its various parts by this means, either as a normal biological mechanism, or as a pathological one. Pathological disturbances of glial function may equally result in profound disturbances of neuronal function. (8) It was first suggested in the last century by Golgi that macroglial cells because of their contacts with blood vessels and ventricular surfaces as well as with neurons, could provide an intracellular route for nutrient diffusion from the vascular system and cerebrospinal fluid to the neuronal population. Whilst such an idea has its attractions, particularly in the case of neuronal processes remote from their cell bodies, there is little firm evidence in favour of this view. Against this theory we may adduce the lack of any consistent or frequent neuroglial connexion between nutrient supply and neuronal surface, the availability of reasonably rapid intercellular diffusion pathways from the ventricles and central canal to neurons, and the ability of material to pass from cell bodies to axon terminals by axonal flow mechanisms. It would appear, therefore, that whilst glial nutrient pathways cannot be excluded, they do not need to be invoked to explain neuronal nutrition.

Microglia

The microglial cells (7.22) are the smallest of the glial elements, and have flattened outlines, with fine, rather short, dendritic processes inserted between neurons, or applied to the external surfaces of capillaries. Ultrastructurally, the cells contain rather heterochromatic, flattened or indented nuclei and, beside the usual organelles, large numbers of lysosomes. These features are identical with those of connective tissue macrophages, which indeed they are thought to be. Recent experimental studies with labelled bone marrow or circulating monocytes have shown that these blood cells migrate as macrophages into damaged or degenerating areas of the central nervous system to phagocytose and eliminate cellular débris, and are then indistinguishable from indigenous microglial cells which are activated to carry out a similar role. Indeed, macrophages may remain afterwards as sedentary microglial cells, although the majority of this type of cell apparently migrate into the brain and spinal cord during late fetal and early postnatal life (Imamoto and Leblond 1977, 1978; Ling 1978). However, it should be mentioned that some authors still regard the microglial cell as being derived from neural plate tissue in the same way as other neuroglial cells.

ORIGINS OF NEUROGLIA AND EPENDYMA

Early schemes of cellular differentiation in the neural tube such as that of His envisaged germinal cells giving rise, *simultaneously*, to *neuroblasts* and to glial stem cells or *spongioblasts*, which differentiate further into glioblasts and ependymoglial cells. Glioblasts subsequently differentiate into astrocytes and oligo-dendrocytes and the ependymoglial cells into mature ependymal elements (p. 1037). With the coming of autoradiographic methods of analysis, this scheme was modified in the sequence of events (*vide infra*), the germinal (matrix) cells *first* giving rise to neuroblasts, and *later* to the other cell elements (p. 860). As the matrix cell of later stages can only form non-nervous elements, it corresponds in principle to the spongioblast. However, it has been established that astrocytes and oligodendrocytes are also able to multiply once they have reached their position in the parenchyma of the nervous system (p. 860), even in mature animals. Ultrastructurally, the maturation of the glial and ependymal cells is marked by a gradual increase of intracellular membranes, to which ribosomes become attached, an increasing density of the nucleus and eventually the laying down of characteristic organelles such as filaments and microtubules (Mori and Leblond 1970; Vaughan and Peters 1974; Sturrock 1974, 1975, 1976). These events accompany the cell migration and the contacting of

the neuronal and vascular structures with which they establish their particular functional relationships. Microglia migrate in from the meninges, along the blood vessels, during embryonic development; they also appear able to divide in postnatal life.

THE BLOOD-BRAIN BARRIER

In early investigations of the brain it was noticed that certain dyes, when injected into the circulatory system, failed to stain the parenchyma of the brain and spinal cord, although they passed easily into non-nervous tissues. Many subsequent studies confirmed that access from the circulation to the extracellular space of the central nervous system is severely limited for many substances, as though a physiological barrier existed at this junction. The structural nature of this '*blood-brain barrier*' has been the subject of much debate in view of its importance in the therapy of nervous disorders, anaesthesia, and so on (Davson 1970). However, many materials when injected into the *ventricles* can enter the brain with ease, indicating that the barrier is probably present at the level of the capillary. Experiments with peroxidase as an ultrastructural colloidal tracer (Brightman and Reese 1969; Brightman *et al.* 1970) show that diffusion can occur quite freely in the interstices between the cells of the central nervous system. The tracer passes from the ventricles, through the intercellular junctions of the ependymal layer, and across the nervous tissues to reach the pial surface. Diffusion is, however, stopped at the endothelial lining of capillaries, and it appears that central nervous capillaries are unusually impermeable to *large molecules* by virtue of their well-formed tight junctions and lack of fenestrations. Likewise, diffusion of substances through the choroid plexus from the vascular system is restricted by the impermeability of both vascular endothelial and choroid epithelial junctions, so that the cerebrospinal fluid contains only those materials secreted selectively by the choroidal epithelium (Cserr 1971).

EXTRACELLULAR SPACE IN THE BRAIN

When considering the movements of water and solutes in any tissue it is convenient to think of the system in terms of a number of functional barriers which create a series of physiological compartments, namely the capillary bed, the lymphatic drainage, the intracellular space and the extracellular space (Davson 1970). The central nervous system lacks an intrinsic lymphatic drainage and the position is therefore simplified. At one time it was considered that large extracellular spaces were present between the cell processes, on the basis of light microscopy; but fine structural studies soon showed that the central nervous parenchyma consists of a complex three-dimensional meshwork of neurons and their processes, glial cells and capillaries, all apparently tightly packed with little more than a 20 nm gap between their surfaces. However, tracer studies show that in spite of the narrow spaces between cells, there is relatively rapid diffusion of substances from the ventricles and central canal through the nervous tissue, and a similar rapid diffusion of *small* molecules from intrinsic blood vessels (*vide supra*).

Peripheral Nerve Fibres

As in the central nervous tracts, two types of nerve fibre exist in the peripheral nervous system, namely myelinated and non-myelinated fibres. Peripherally, however, they are ensheathed by the *cells of Schwann* instead of the central oligodendrocytes. Both central and peripheral nerve fibres present a number of difficulties to the neurohistologist, either because of their small size in the case of non-myelinated fibres, or because the sheaths of myelinated fibres are severely disrupted by the lipid solvents employed for histological preparation. A number of special methods have been devised to overcome these problems; with myelin sheaths these are based chiefly on means of fixing and staining the myelin either in teased material which demonstrates individual fibres, or in blocks of tissue which can be later processed and sectioned in the usual way. The osmium-based methods are particularly valuable in this respect, either used as osmium tetroxide in aqueous solution, or with chromate and haematoxylin solutions, both methods fixing and staining the sheaths simultaneously. Other methods include the use of frozen sections stained with lipid-soluble dyes, examination of material prepared by any of these methods by polarization, phase or interference microscopy, and observation of freshly teased or living nerves *in situ* by various means including incident light techniques. The extension of time-lapse cinemicrography, and videotape recording techniques in conjunction with closed-circuit television, to this field of investigation have aided the analysis of living nerve fibres (Williams and Hall 1970, 1971 *a* and *b*). The various means of demonstrating central neurons already described (p. 820), have also been valuable in demonstrating peripheral nerve fibres including non-myelinated ones; as with other parts of the nervous system, electron microscopy has contributed greatly to our understanding of peripheral nerve structure.

Structure of the Peripheral Nervous System

The peripheral nervous system comprises the cerebrospinal and autonomic systems of nerves and their associated ganglia, containing nerve cell somata, together with the cellular and connective tissue elements which ensheathe them. All these structures, of course, lie peripheral to the pial envelope which covers the central nervous system, through which the central and peripheral nerve fibres are continuous at many points; but the histology of the two contrasts in a number of ways, and that of the peripheral structures will now be considered (7.24, 25, 26, 27, 28).

The sensory ganglia of dorsal roots of spinal nerves (7.25), and corresponding ganglia on the trunks of the trigeminal, facial, glossopharyngeal and vagal cranial nerves, are ensheathed by periganglionic connective tissue which is similar to perineurium in composition (p. 1052). The neurons are *unipolar*, possessing spherical or ovoid somata of varying size, and these are aggregated in groups, interspersed with fasciculi of myelinated and non-myelinated nerve fibres. The fine structure of neuronal cytoplasm is described elsewhere (p. 825) and will not be pursued further here, but it should be noted that a single non-myelinated neurite (sometimes termed a 'dendro-axonal' process) leaves each cell soma. It often follows a highly convoluted course near the parent soma before bifurcating at a T-junction to become continuous with the central and peripheral processes of the sensory nerve fibre. In the case of myelinated fibres, such a junction occurs at a node of Ranvier (p. 845). The peripheral process terminates in a sensory ending and since it conducts towards the cell body it is functionally an *elongated dendrite*, but it possesses all the structural and physiological characteristics of a *peripheral axon* and, following common usage, will be so termed in this account. Each unipolar cell body is closely enveloped by a *nucleated capsule* consisting of flattened, epithelioid *capsular cells*, also variously termed *amphicytes* or *satellite cells*. (It should be noted that the name satellite cell is used in a number of different ways. Some authors use the term for small rounded extracapsular ganglionic cells, others include ganglionic capsular cells and Schwann cells, and yet others include all non-neuronal cells both central and peripheral which are perineuronal in position. Further, the name is also applied to cells associated with the surfaces of striated muscle fibres—p. 513). The cytoplasm of the capsular cells in many ways resembles that of Schwann cells (*vide infra*), and their deep surfaces are irregular, interdigitating with reciprocal irregularities on the surface of the subjacent nerve cell body. The layer of capsular cells is continuous with a layer of similar cells which encloses the convoluted part (*initial glomerulus*) of the dendro-axonal process, and this, in turn, is continuous with the Schwann cell layer of the peripheral and central processes of the nerve fibre. Outside these cell layers lies a delicate vascular connective tissue which is continuous with the endoneurium of the peripheral nerve and nerve root.

Sensory ganglionic neurons are not confined to the discrete ganglia of the cerebrospinal nerves, but either singly or in small

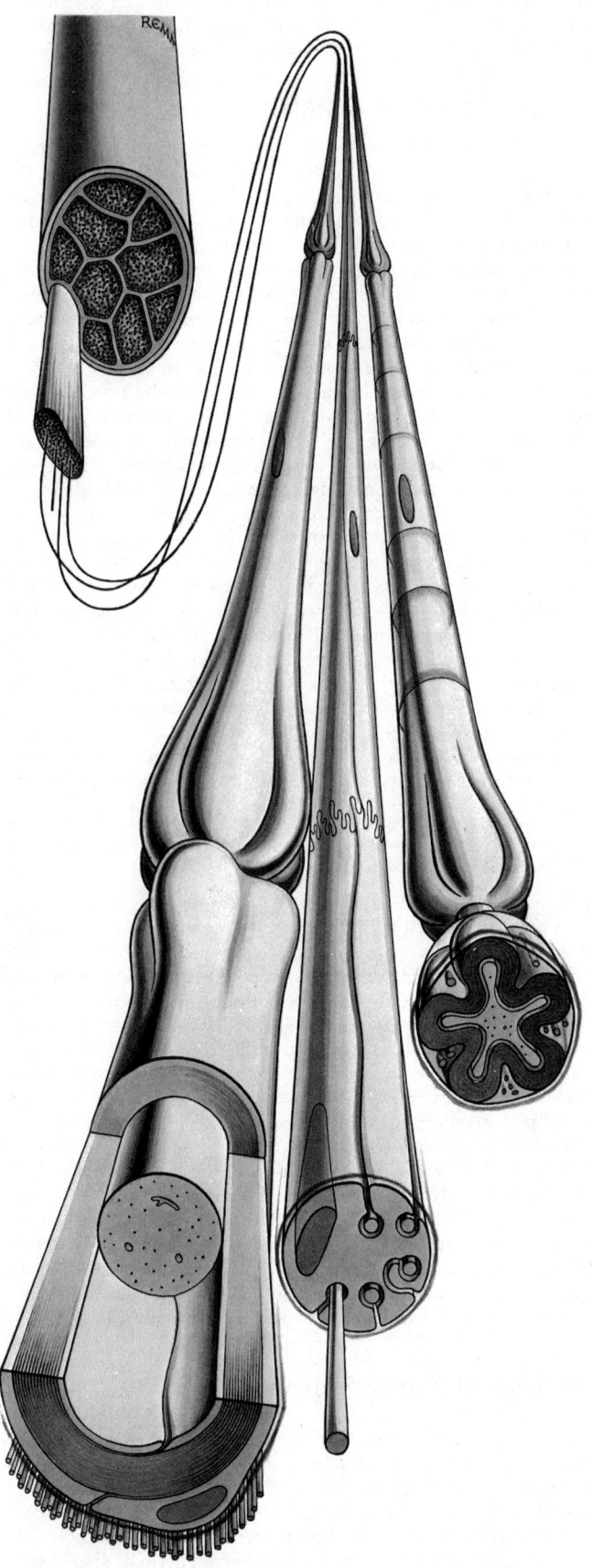

7.24A This diagram shows some structural features of peripheral nerve fibres. A nerve trunk (top left) is cut away to expose a single fasciculus, from which three fibres are indicated in detail. These include two myelinated axons, one on each side of a group of non-myelinated axons enclosed within a Schwann cell sheath. The myelinated fibre on the left has been cut away at various points to demonstrate the relationship between the axon, the Schwann cell, and its sheath of myelin.

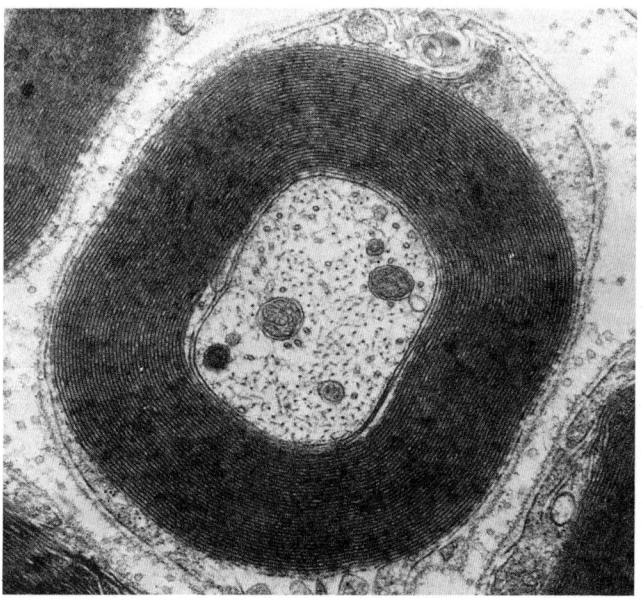

7.24C An electron micrograph of a myelinated peripheral nerve fibre, showing an axon containing neurofilaments, microtubules and mitochondria, surrounded by a myelin sheath, enclosed in turn by Schwann cell cytoplasm and endoneurial spaces. Magnification × 25,000.

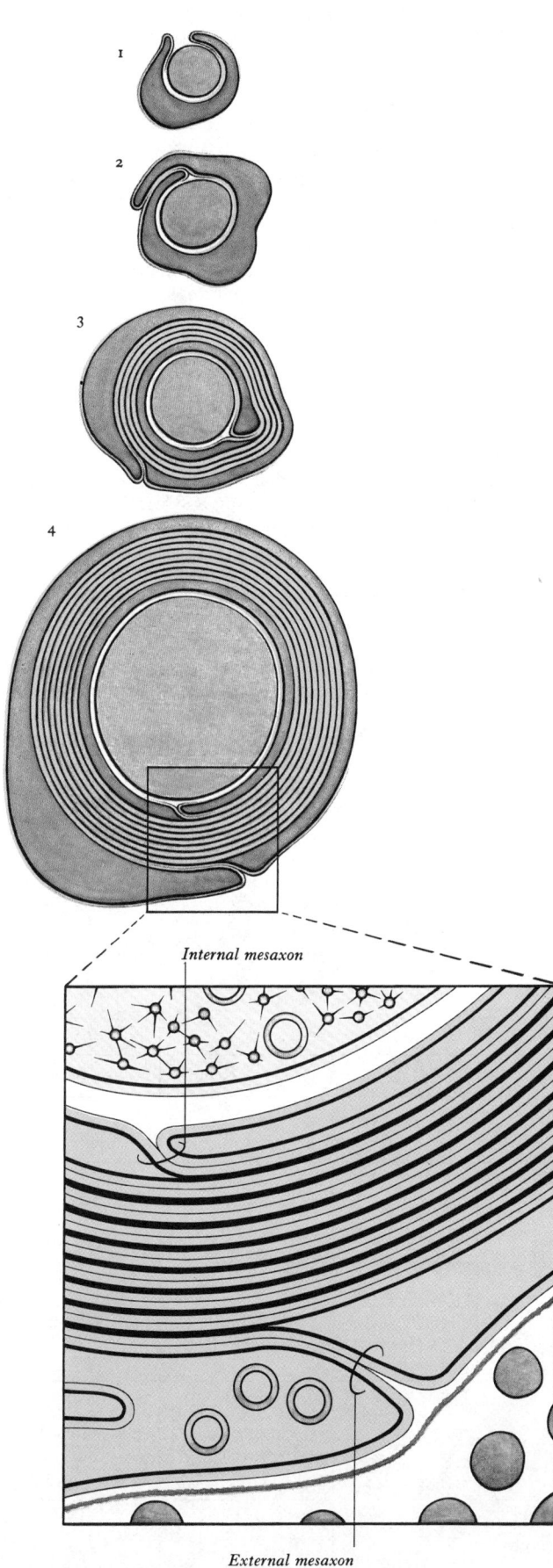

Internal mesaxon

External mesaxon

7.24B A diagram showing the development and organization of the myelin sheath of a peripheral nerve fibre. In stages 1–4 myelin formed by a Schwann cell (blue) progressively envelops the growing axon (yellow), to form the final pattern of spirally disposed myelin lamellae: see enlarged detail at the base of the diagram.

groups, they often occupy 'heterotopic' positions either peripheral to the ganglia.

The autonomic ganglia have a different structure, since their cell bodies are multipolar, with dendritic trees which receive synapses from incoming preganglionic visceral motor fibres, and sections show a mixed neuropil of afferent and efferent fibres, dendrites, synapses and cell bodies (see p. 1124).

The nerve trunks and their principal branches are composed of roughly parallel bundles of nerve fibres including efferent and afferent axons, their ensheathing Schwann cells, which in some cases elaborate *myelin sheaths*, surrounded by connective tissue sheaths at different levels of organization. The fibres are grouped together within the trunk in a number of *fasciculi* each of which may contain from relatively few to many hundreds of nerve fibres. The size, number and pattern of the fasciculi vary greatly in different nerves and at different points along the length of a nerve. The number of fasciculi increases and their size decreases some distance before there is macroscopic evidence of *branching* of a nerve trunk. Similarly, where the nerve is subjected to considerable increases of pressure, as when passing deep to a fibrous retinaculum, the fasciculi are again increased in number and reduced in size; further, their volume of associated connective tissue and their vascularity also increase. Consequently, the nerve shows a pink, fusiform dilatation of contour at this point, which is sometimes termed a *pseudoganglion* or *gangliform enlargement*.

A dense irregular connective tissue sheath, the *epineurium*, surrounds the whole trunk, and a similar but less fibrous *perineurium* encloses each fasciculus, within which the spaces between nerve fibres are penetrated by a loose delicate connective tissue network, the *endoneurium*. These connective tissue sheaths serve as convenient planes of access for the vasculature of peripheral nerves, the capillary bed and associated lymphatics running in the main parallel to the nerve fibres in the endoneurial spaces, but short oblique cross-connexions pass between them.

The epineurium is a collagenous adventitial coat with little regular organization (Thomas 1963); the perineurium in contrast has a regular structure of highly flattened laminae of fibroblasts alternating with fine collagenous sheets running in various directions. Macrophages are also present. The fibroblasts, surrounded by dense intercellular material, form junctional complexes and it appears from tracer experiments that these limit the diffusion of large molecules across the sheath (Waggener and

Beggs 1967). The fibroblasts also show numerous pinocytotic vesicles, an indication of active transport mechanisms.

THE CLASSIFICATION OF PERIPHERAL NERVE FIBRES

Several schemes of classification have been proposed for peripheral nerve fibres, based on relative conduction velocities, functional nature of fibres, total fibre diameter and other considerations. Of these, two are in common use (*see* accompanying table); the first scheme, proposed initially by Erlanger and Gasser (1937) in relation to the different conduction speeds in frog nerves, divides fibres into three major categories, designated A, B and C, corresponding to three peaks in the distribution of velocities. The group A fibres could be further divided into three subgroups termed a, β, and γ; B group fibres are preganglionic autonomic efferents, and C fibres are non-myelinated. Since total diameter and conduction velocities are in most fibres roughly proportional to each other (in mylinated fibres conduction, expressed in metres per second is approximately six times the fibre diameter in micrometers), the group Aa fibres are both the widest and the most rapidly conducting, and the C fibres the narrowest and slowest. In mammals it was found that Aβ fibres were negligible, but also that a subclass of non-autonomic fibres similar in conduction speeds to B fibres exists; these are termed Aδ fibres.

In the Erlanger and Gasser scheme, the *afferent (sensory) fibres* are confined to the Aa, Aδ, and C groups. The largest of these (the Ad fibres) include the axons of the encapsulated cutaneous, joint and muscle receptors and some large alimentary or related interoceptors. The Aδ fibres belong to various nociceptors, including those of the dental pulp, of skin, and of various other connective tissue sites, and the C fibres have thermoreceptive, nociceptive and interoceptive functions.

The somatic efferent (motor) fibres include Aa, Aβ, and Aγ fibres. The Aa fibres innervate extrafusal muscle only, being up to 22 μm in diameter and conducting at a maximum of 120 m/sec. The fibres to the 'fast' twitch muscle are slightly larger than those to

'slow' muscle. The Aβ fibres are restricted to collaterals of Aa fibres, forming plate endings on intrafusal muscle fibres, and the Aγ fibres are exclusively fusimotor nerves to plate and trail endings on intrafusal muscle fibres (*vide infra*).

Autonomic efferents comprise the preganglionic B fibres and postganglionic sympathetic and parasympathetic axons of the C (non-myelinated) group.

As can be seen, this scheme can be applied to all fibres of spinal nerve trunks and all cranial ones except perhaps olfactory nerve fibres which form a uniquely small diameter and slow conducting fibre population.

Another type of classification, proposed for use in designating the afferent fibres of somatic muscles, was instituted by Lloyd in 1943. In this scheme, the myelinated fibres are divided into three groups, I, II, and III, the non-myelinated fibres forming group IV.

Group I fibres are large in diameter (12–22 μm), and include the primary sensory fibres of muscle spindles (Group Ia) and the slightly smaller fibres of the Golgi tendon organs (Ib).

Group II comprises the fibres of the secondary sensory terminals of muscle spindles, with diameters of about 6–12 μm.

Group III fibres, 1 6 μm in diameter, have free endings in the connective tissue sheaths surrounding and within muscles, and appear to be nociceptive, related to the experience of 'pressure-pain' in externally stimulated muscles. The paciniform endings of muscle sheaths may also contribute fibres to this class.

Group IV includes non-myelinated fibres below 1·5 μm again with 'free' endings in muscle sheaths and interiors, chiefly with nociceptive functions.

For a full discussion of the sensory nerves of muscle, the reader is referred to the excellent reviews by Matthews (1973), Barker (1974), and Hunt (1974).

Whilst such general forms of classification have proved of great assistance in neurophysiology, it must be emphasized that caution should be exercised before assuming that a single structural parameter, such as total fibre diameter, measured at one point along a nerve trunk, adequately categorizes the nerve fibre population along the length of the nerve. Likewise, many of the

CLASSIFICATION OF PERIPHERAL NERVE FIBRES

CHARACTERISTICS OF FIBRE						
Type of sheath	Myelinated					Non-myelinated
Fibre diameter	22 μm				1·5 μm	2·0–0·1 μm
Conduction speed (Metres/second)	120	60	50	30	4	0·5
CLASSIFICATION 1. Erlanger and Gasser *All fibres*			A		B	C
Subclasses: Efferent	Aa Skeletomotor	Aβ Fusimotor collaterals of A fibres	Aγ Fusimotor		B Preganglionic autonomic	C Postganglionic autonomic
Afferent	Aa and smaller. Muscle and tendon; cutaneous				Aδ Cutaneous, muscle visceral, etc.	C Cutaneous, muscle visceral, etc.
2. Lloyd Afferent— skeletal muscle and articular	I *a* primary spindle ending *b* tendon ending		II Secondary spindle ending		III Free ending (nociceptor, etc.) paciniform ending?	IV Free ending (nociceptor, etc.)

It should be noted that the scale for conduction velocities is not arithmetic.

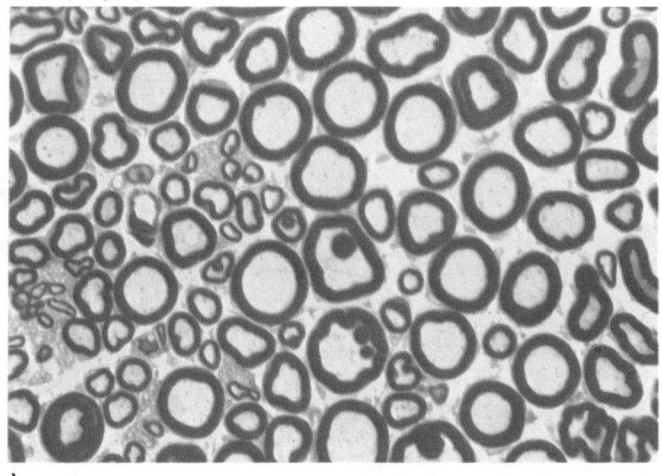

A

7.25A Transverse section of part of a mixed peripheral nerve from a mouse, showing a wide range of diameters of myelinated nerve fibres. Myelin: dark blue; axoplasm: pale blue, embedded in which are slightly darker blue dots, the axonal mitochondria. Note particularly the wide range of *axon* diameters; the smaller axons possess thinner myelin sheaths. The fibre with a particularly thick, double-contoured myelin sheath (bottom left) is sectioned through an incisure of Schmidt-Lanterman. The large fibres with more complex profiles (centre and bottom) are sectioned near the commencement of paranodal bulbs. On the left, between the myelinated fibres, groups of non-myelinated axons, enclosed by Schwann cells (medium blue) are just visible. A 1 μm epoxy resin section, stained with osmium tetroxide solution, and toluidine blue, photographed using oil immersion optics.

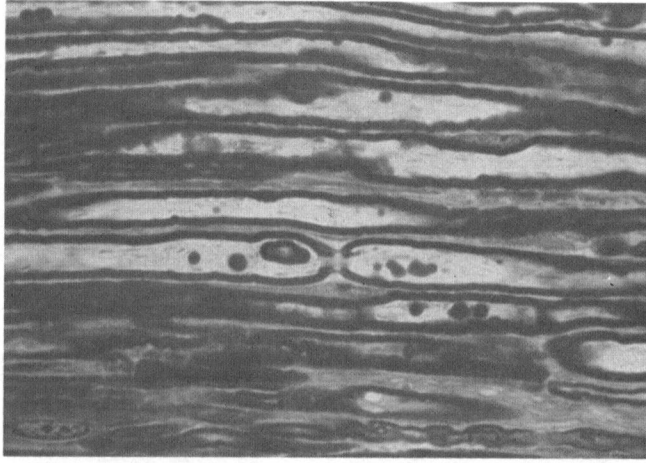

B

B A longitudinal section of material prepared in a similar manner to that in 7.25A. Note particularly the fibre just below the centre of the field which is sectioned through a node of Ranvier. As the internodal myelin sheath approaches the node, its diameter increases to a maximum (the paranodal bulb) and the myelin sheath then curves sharply inwards, to terminate at the limits of the 'nodal gap'. The myelin profiles, apparently lying free within the paranodal bulbs, are demonstrated to be paranodal shelves of myelin, when serial sections are examined (compare with 7.25D2). Note also the constriction of the nodal axon; long, narrow axonal mitochondria are also visible.

7.25C A transverse section of an immature peripheral nerve of a rat, taken one week after birth, showing the profiles of numerous axons. Some of the latter are non-myelinated and, either singly or in groups, are invaginated into the surfaces of cytoplasmic processes of adjacent Schwann cells. Other axons are in the process of myelination. Note the Schwann cell nucleus (right) to one side of a myelin sheath, and the ultrastructure of the Schwann cell cytoplasm at this stage of development.

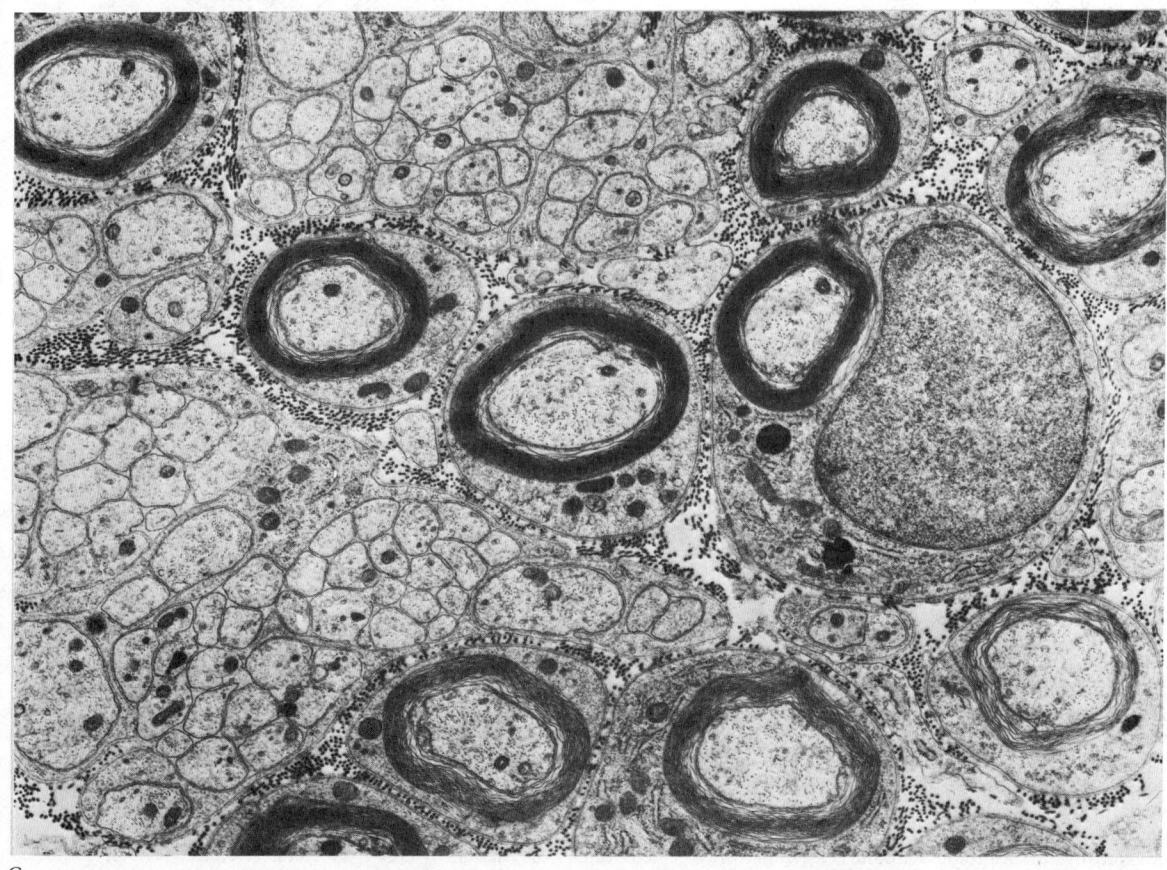

C

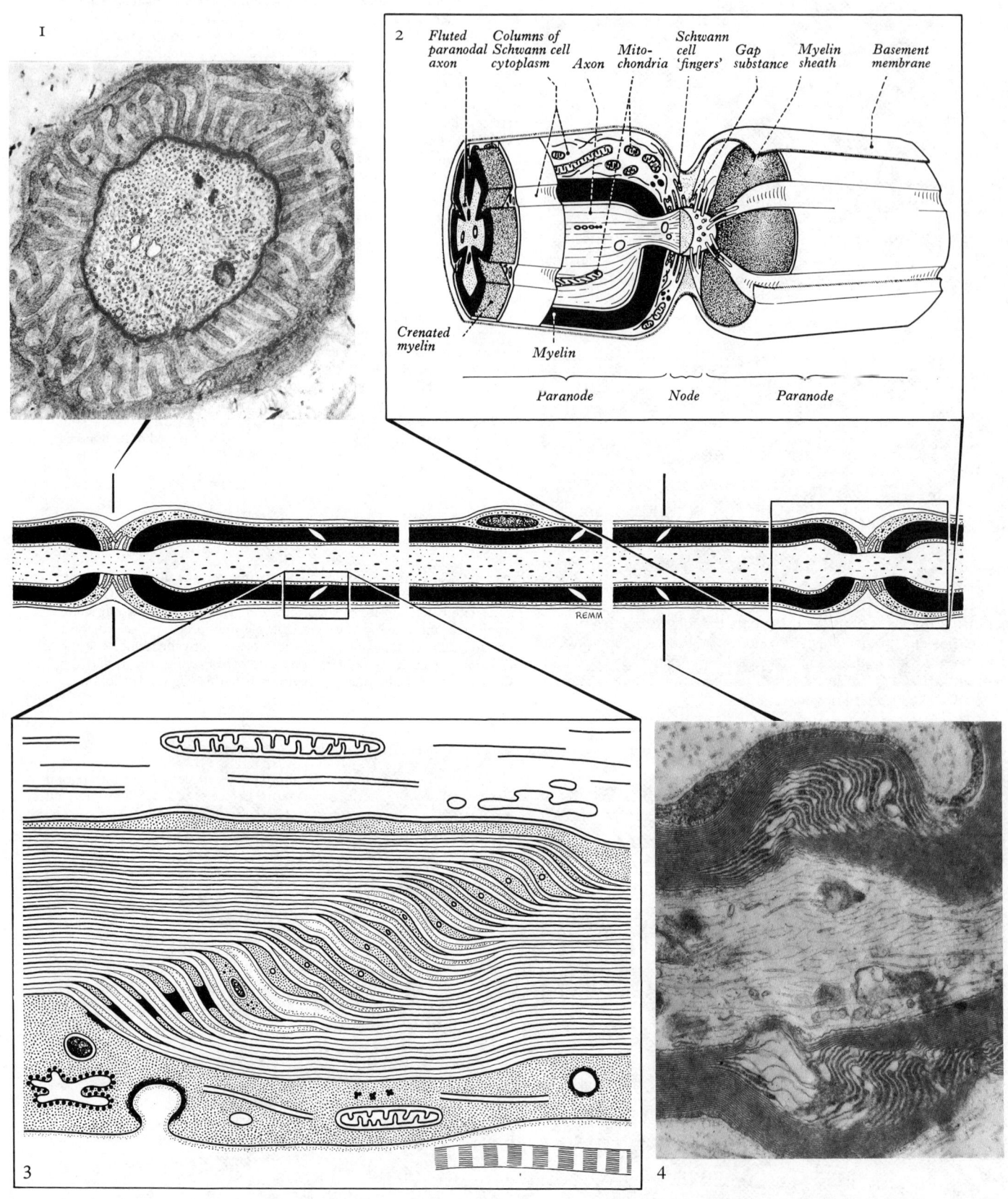

7.25D A general plan of a myelinated nerve fibre in longitudinal section including one complete internodal segment and two adjacent paranodal bulbs, used as a key for the more detailed microarchitecture of specific subregions. 1: a transverse section through the centre of a node of Ranvier, with numerous finger-like processes of adjacent Schwann cells converging towards the nodal axolemma. Many microtubules and microfilaments are visible within the axoplasm. 2: a diagram showing the arrangement of the axon, myelin sheath and Schwann cell cytoplasm at the node of Ranvier in the paranodal bulbs. (Supplied by Prof. P. L. Williams and Dr. D. N. Landon.) 3: detailed substructure of one-half of an incisure of Schmidt-Lanterman; consult text for further information. (Supplied by Dr. Susan M. Hall and Prof. P. L. Williams.) 4: longitudinal section of part of a myelinated nerve fibre (mouse), including an incisure of Schmidt-Lanterman; this appears as oblique zones in the myelin sheath on both sides of the fibre. Consult text for structural details. (Nos. 1 and 4 kindly supplied by Dr. Susan M. Standring (née Hall) of Department of Anatomy, Guy's Hospital Medical School, London University.)

values quoted are only related to peaks in histograms of conduction velocity or fibre diameter frequency measured only in a few sites in certain nerves of a few species of mammal. However, if it is borne in mind that the figures quoted are relative, and that considerable overlap occurs between the functional classes of fibres, these schemes enable us to detect a general pattern of organization in the peripheral nerves of mammals including man, which is of considerable interest.

CONDUCTION IN PERIPHERAL NERVES

When a nerve trunk is stimulated, the *sum* of the action potentials in its many different nerve fibres, the *compound action potential*, can be recorded with extracellular electrodes. Since these potential variations travel at different speeds along different fibres, and since large fibres create larger electrochemical fluxes than small ones, a compound action potential recorded at a distance from the point of stimulation has a complex form, with at least four successive waves of different amplitude and velocity. These are known as the *a*, *β*, *γ* and *δ* waves. Just as each nerve trunk has its own characteristic spectrum of nerve fibres types, so the precise wave pattern is also distinctive. In the past, the fibre types have been classified by their conduction velocities, but it is now recognized that fibres of different physiological roles may have similar conduction velocities so that a more complex terminology taking into account size, conduction velocity and functional connexions has emerged (*vide supra*).

It is becoming clear that there is a series of factors which govern the conduction velocity of nerve fibres. In non-myelinated fibres, the action potential sweeps continuously over the axolemma as ionic eddy currents (electrotonus) which progressively excite neighbouring areas of membrane. The rate of spread is proportional to the cross-sectional area of the axon (Hodgkin 1964). In myelinated fibres, however, the presence of an insulating layer of myelin prevents the membrane from being excited at any point except the nodes of Ranvier, but once an action potential is initiated at one node, the resultant ionic eddy currents spread to excite the next node and others in sequence. Since the longitudinal flow of ions in tissue fluid and axoplasm is much more rapid than a continuous electrotonic spread along the axonal membrane, much higher conduction velocities are possible by this mechanism, which is known as *saltatory conduction* or *saltation* (Stämpfli 1954).

During excitation at a node of Ranvier, there is initially an increased permeability of the short, exposed, segment of nodal axolemma to sodium ions, which therefore flow down their concentration gradients to enter the nodal axon. Consequently, longitudinal ionic currents are generated which flow along the *internodal axon* to leave at the next node, the circuit being completed by longitudinal ionic currents in the reverse direction in the endoneurial tissue fluid. When the outflowing current density at the *second node* rises to a critical *threshold level*, its depolarizing effect is sufficient to initiate similar permeability changes in the nodal axolemma, and the whole process is repeated, with excitation of the third and subsequent nodes. However, myelin is an imperfect insulator and, as a result, the longitudinal currents are not completely confined to the axoplasm and external tissue fluid. Some radial leak of current occurs between the axoplasm and tissue fluid, with a resulting loss of efficiency. The details of these relationships will not be pursued here, but it may be noted that for a given total fibre diameter, maximum efficiency is achieved when the diameter of the axon and the thickness of its myelin sheath are such that the *ratio* of axon diameter to total fibre diameter is 0·6. Only the largest mammalian myelinated nerve fibres reach this ratio during growth.

Despite the general usefulness of the classification of nerve fibres described above, based in part upon total nerve fibre diameter variations, it should be emphasized that other dimensions are equally essential in determining the conduction characteristics of a myelinated nerve fibre. As just mentioned, these include the *diameter* of the axon *and the relative thickness* of its *myelin sheath*; in addition, the *internodal distance* between successive nodes of Ranvier, and the *area, characteristics* and *micro-environment* of the *nodal axolemma* are important.

These structural parameters vary between species, between different functional categories of fibre, and at different points along the length of individual fibres and their branches (Williams and Wendell-Smith 1971).

SCHWANN CELLS

Schwann cells are the satellite cells of the peripheral nervous system. All peripheral axons lie for most of their course in invaginations of Schwann cell plasma membrane, separated from the endoneurial compartment by a Schwann cell basal lamina. Schwann cells are also sometimes known as neurolemmocytes, a not altogether fortunate term.)

In view of the large cytoplasmic volume and surface area of many neurons (which possess but a single diploid nucleus) it has long been inferred that this structural association is indicative of a

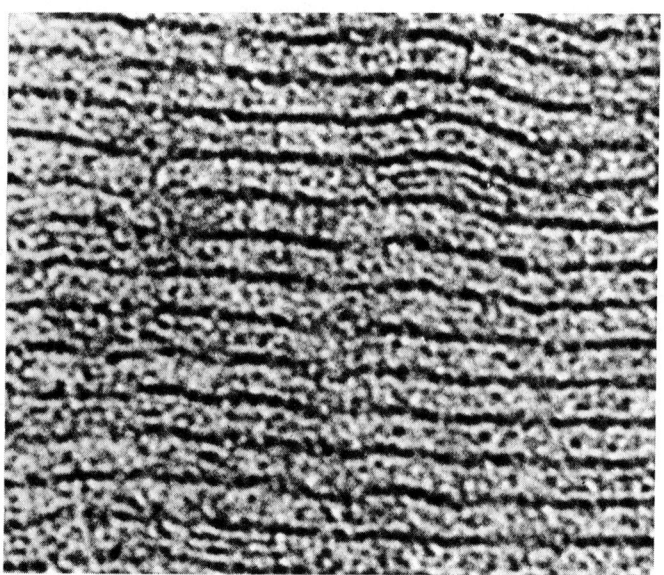

7.25E A high-power electron micrograph of part of a myelin sheath in section, showing dense period lines (the apposition of the cytoplasmic aspects of the cell membranes), alternating with less dense intraperiod lines (the apposition of the external aspects of the Schwann cell membranes). The preparative techniques used have caused the intraperiod line to split into two distinct lines in some places.

functional interdependence. In recent years considerable experimental evidence has accrued to support this idea. Thus, Schwann cells have been implicated in the local support of the axon, supplying metabolites and trophic factors (Varon & Bunge 1978); influencing the ionic milieu of the periaxonal space and possibly the distribution of neuro-transmitters (Villegas 1975), and even in determining the siting of sodium channels along the axolemma (Ritchie & Rogart 1977). It is also clear that Schwann cells are influenced by the neurons they surround. The physiologically important parameters of myelin thickness and internodal length (i.e., expressions of Schwann cells' territories) are related to axonal size, possibly reflecting the spacing of antigenic determinants on or in the axolemma. Experiments using tissue culture of homogenous populations of neurites and Schwann cells, and allo- and xeno-grafting of nerve fascicles have demonstrated (*a*) that axons are mitogenic for Schwann cells, and (*b*) that Schwann cell differentiation (into myelinated or non-myelinated states) is at least in part dependent upon neuronally derived signals (Wood & Bunge 1975; Aguayo, Perkins, Duncan & Bray 1978). Schwann cells can even form myelin around central axons under certain experimental conditions.

The life history of the Schwann cell is divisible into two distinct phases: a perinatal *migratory* and *proliferative* stage, which is succeeded by the definitive *axon-associated* state when, if appropriately stimulated, myelin formation and maintenance may occur.

The immature Schwann cell is large and rounded with an oval nucleus and dense cytoplasm. As migration from the neural crest proceeds, the cells are at first fusiform and subsequently become irregular in outline, putting out processes which insinuate between bundles of outgrowing neurites, effecting their subdivision. In the mature myelinated fibre, Schwann cell cytoplasm is evident as thin, often apparently discontinuous, strips lying external to the myelin sheath and occasionally between the inner aspect of the sheath and the periaxonal space. Only at the paranodes, in the nodal gap fingers, throughout the spiral of the Schmidt-Lanterman incisure and in the perinuclear region, does any substantial volume of Schwann cell cytoplasm persist.

The relationship between axon and Schwann cell is emphasized in pathological material. It is well known that any disturbance of either axon or Schwann cell results in breakdown of the normal morphology of the nerve. Crushing or cutting a peripheral nerve produces Wallerian degeneration. All affected axons distal to the injury degenerate and there is a period of intense Schwann cell division during which myelin and axonal débris is phagocytosed by the Schwann cell population. Successful axonal regrowth into the persistent basal laminal tubes of the remaining Schwann cells in the distal stump is accompanied by a re-establishment of Schwann cell/axon associations and remyelination where appropriate. An insult (metabolic or chemical) which is primarily directed against the Schwann cell results in either demyelination or in hypomyelination but not necessarily in axonal degeneration.

Recent experimental trends have been directed towards elucidating the mitogenic and myelinogenic signals, preparing antigenically homogenous populations of Schwann cells for immunological studies, and examining abnormalities of Schwann cell behaviour in various human neuropathies by grafting abnormal nerves into normal (non-human) recipient nerves.

All of the larger axons of the nervous system of mammals are enwrapped in *myelin sheaths* and are termed *myelinated* or *medullated* fibres. The fatty composition of the myelin is responsible for the glistening whiteness of the peripheral nerve trunks and of the white matter centrally. Axons less than about 0·5–1·0 μm in diameter generally lack these sheaths and are therefore termed *non-myelinated*, *non-medullated* or *grey* fibres. These two types of fibre have great physiological significance and will be described in some detail.

NON-MYELINATED FIBRES

In mammals, approximately 75 per cent of axons are non-myelinated in cutaneous nerves and dorsal spinal nerve roots, 50 per cent in muscle nerves and 30 per cent in ventral spinal nerve roots. In the autonomic nervous system, post-ganglionic axons are almost exclusively non-myelinated, and pre-ganglionic fibres contain significant numbers of non-myelinated axons. A non-myelinated 'fibre' consists of a group of small axons (0·15–2 μm in diameter) associated with a longitudinal chain of Schwann cells. In mature nerve, the way in which the Schwann cells enclose the axons within each group shows inter-species variation, and even varies in different nerves in the same species. In general, individual axons are segregated from their neighbours by tongues of Schwann cytoplasm, but are sometimes further isolated by separate processes of cytoplasm that only come together in the perinuclear region of the Schwann cell (Gamble and Eames 1964). A more primitive organization, resembling that of fetal nerves, persists in some adult nerve, notably the olfactory nerve, where bundles of many small axons are engulfed within common invaginations of Schwann cell cytoplasm.

Three-dimensional reconstruction based on analyses of semi-serial ultra-thin sections of somatic and autonomic nerves has revealed that the spatial relationships between axons and satellite cells alters continuously and in a complex fashion within each cell, and along the chain (Aguayo *et al.* 1973). Transfer of axons between Schwann cells usually occurs at the extremities of the engulfing Schwann cells where their elongated cytoplasmic processes interdigitate (Gamble *et al.* 1978).

Thus, non-myelinated fibres comprise the smaller axons of the central nervous system, in addition to peripheral post-ganglionic axons of the autonomic nervous system, and the several types of fine sensory fibres such as some of those responsible for signalling pain (i.e. from nociceptors, but see also below and pp. 852, 856), the olfactory nerve axons, and others. As indicated above, these axons are engulfed either singly or in small bundles into longitudinal invaginations of a series of Schwann cells which therefore separate them from the endoneurial space (7.25c). The original line of invagination during development of non-myelinated axons which are deeply embedded in Schwann cells is marked by a *mesaxon* consisting of a double layer of the Schwann cell plasma membrane, the external surfaces of which are parallel but separated by a gap of 15–20 nm. Deeply, the two layers separate to enclose the axon from which they are again separated by a *periaxonal space* of similar dimensions; superficially the mesaxonal layers separate to become continuous with the general outer plasma membrane of the Schwann cell. The endoneurial tissue fluid has access to that in the periaxonal space by way of the gap between the two layers of the mesaxon. Although narrow, these intercellular spaces are of sufficient dimensions to allow the mobility of ions necessary for the *continuous* electronic (ionic) current fluxes which characterize non-myelinated fibre conduction. Thus, in the absence of a myelin sheath and nodes of Ranvier, saltatory conduction does not occur, and the progressive uninterrupted passage of impulses in non-myelinated fibres is a relatively slow process, conduction velocities being in the range of about 4·0–0·5 m/s.

On the classification of Erlanger and Gasser, non-myelinated fibres are classified as C fibres; on that proposed by Lloyd non-myelinated afferents are designated Group IV (*see* p. 840).

Where synaptic or motor transmission occurs, the Schwann cell is reflected back from the axonal surface to expose the active regions. The terminal branches of myelinated fibres where they taper, are also provided with Schwann cell sheaths of a similar nature.

MYELINATED FIBRES

Myelinated fibres (7.24, 25) form the bulk of the elements of the somatic nervous system. To understand the structure of the myelin sheath it is convenient to describe its developmental sequences. The formation of myelin begins some time before birth in many tracts and nerve trunks although it is not completed until a considerable and variable time postnatally. The pattern of myelination varies in its timing in different mammalian groups and in different parts of the same nervous system.

In the peripheral nervous system axons grow out from the central and ganglionic cell bodies along with Schwann cells which initially multiply as the nerve grows in length, but later they cease dividing and increase only in size. At first, all axons are non-myelinated, but later a flap-like extension of the Schwann cell cytoplasm begins to spiral around the neurite, with the bulk of the cytoplasm and nucleus remaining outside the spirals (7.24B, 25c). As development proceeds, more and more turns are made, and the spirals become transformed into compacted layers of cell membrane (Robertson 1955). In electron micrographs, the myelin sheath, when mature, is seen as a laminated structure with regular dense *period lines* (representing the plane of apposition between the internal surfaces of the plasma membrane) alternating with less conspicuous *intraperiod lines*, representing appositions between the external surfaces of the plasma membrane spiral (7.24B). Suitable osmotic treatment which results in swelling of the sheath causes the intraperiod line to split, showing that the external surfaces are not fused to form a specialized junctional zone, but are apparently able to move on one another (Napolitano and Scallen 1969). On the inside and outside of the myelin sheath is a delicate zone of Schwann cell cytoplasm which, in many parts of the internode in mature fibres, is only about 20–30 nm thick. The internal layer is covered deeply by the inner plasma membrane of the Schwann cell which adjoins the 15–20 nm wide *periaxonal space*, and which is continuous through an *internal mesaxon* with the deepest lamella of myelin. Similarly an *external mesaxon* connects the outermost myelin lamella with the superficial part of the plasma membrane of the Schwann cell. Larger accumulations of Schwann cell cytoplasm are found in the external layer related to the grooves in the paranodal myelin sheath, in the perinuclear region, where the nucleus dimples the external aspect of the myelin sheath near the mid-point of the cell, and also on the internal and external aspects of the Schmidt-Lanterman incisures (*vide infra*, p. 846). These cytoplasmic regions often contain mitochondria, lysosomes and a variety of vesicles and granules. The inner and outer zones of cytoplasm are also connected with each other by helical channels containing granular cytoplasm and microtubules at the nodes of Ranvier and Schmidt-Lanterman incisures (*vide infra*). The outer plasma membrane of the Schwann cell and its adjacent basement membrane correspond to the neurolemma of light microscopists.

When the lipids are removed from the myelin sheath chemically, as in the processing of histological material for wax embedding, without prior fixation and stabilization of the lipid components, a network of proteinaceous filaments, termed *neurokeratin*, is seen by light microscopy, but this has not yet been confirmed by other means.

CHEMICAL COMPOSITION OF MYELIN

Myelin is formed by the compaction of Schwann or glial cell plasma membranes. It is therefore not surprising to find that myelin resembles the plasma membranes of any cell in that it contains lipids, proteins and water (the bound water in myelin accounts for at least 20 per cent of the wet weight). However, the relative proportions of the various components and their disposition within the sheath are unique to myelin.

Myelin can be isolated from central or peripheral nervous systems by techniques involving the centrifugation of tissue suspensions against sucrose solutions, followed by osmotic shock of the bulk material and subsequent sedimentation or isolation on a sucrose gradient. Recently, immunological and autoradiographical methods have been used to 'dissect' further the constituents of the sheath. The proportions of cholesterol, total phospholipid and glycolipid in myelin are different from those found in other membranes, and differ between central and peripheral myelin. On a molar basis, cholesterol represents the major lipid of the myelin sheath (approximately 40 moles per cent of the total lipid). Although it is not synthesized within the sheath, it has been demonstrated that interference with cholesterol synthesis affects myelin synthesis as a whole. Of the glycolipids, the most important are galactocerebroside, sulphatide and ganglioside. Peripheral myelin has a lower concentration of galactocerebroside and a higher concentration of choline glycerophosphatide and sphingomyelin than central myelin. The choline glycerophosphatides have a higher proportion of unsaturated fatty acids than their central counterparts, although long chain fatty acids (C_{25} and C_{26}) constitute less than 2 per cent of the total, whereas in the central nervous system they account for up to 20 per cent of the total fatty acids. It is interesting that most of the organ-specific complement-fixing or precipitating activity of certain demyelinating antisera appears to be anti-galactocerebroside. Galactocerebroside has been found to be a cell-surface marker for cultured rat oligodendroglia, the satellite cells responsible for myelinogenesis in the central nervous system.

About 30 per cent by weight of myelin in the adult human central nervous system is protein. The myelin proteins have been examined by means of polyacrylamide gel electrophoresis of isolated myelin which has been made soluble by either phenol or sodium dodecylsulphate. Whatever solvent system is used, at least three major proteins have been described in central and peripheral myelin; as with the lipid moiety, there are considerable differences between myelin proteins from these sites. Peripheral myelin contains a low molecular weight glycoprotein P_0, and two basic proteins P_1 (which constitutes *c.* 5 per cent of the total protein) and P_2 (which constitutes *c.* 20 per cent of the total protein). P_1 is considered to be analogous to the encephalitogenic basic protein of central myelin. Analysis of the immune responses mounted against various regions of the basic proteins has provided information about the position and relationships of these proteins within the myelin sheath in various species.

It seems likely that very few enzymes are associated with myelin. Only one, 2:3′ cyclic AMP 3′-phosphohydrolase, has been examined in any detail, and is used experimentally as a myelin marker. The significance of this enzyme remains obscure.

FORMATION AND GROWTH OF MYELIN

Although the geometry of myelin is now understood in some detail, exactly how it is laid down during myelination is still debatable. Early models, supported by tissue culture observations, involved the formation of a loop of Schwann cell cytoplasm around the enclosed axon, and the subsequent migration of the perinuclear region of the cell around it in a spiral manner (Geren 1954). In the central nervous system at least, such a mechanism appears to be ruled out since a single oligodendrocyte can simultaneously form myelin sheaths around several separate axons which would be impossible by the above method of growth (Bunge 1968). Other possibilities are the interstitial deposition of myelin within the lamellae of the sheath, or ingrowth of the spiral by slipping of the turns on one another, new myelin being formed externally (Webster 1971). Alternatively, it was suggested that the enclosed axon may rotate to wind myelin upon itself; recent reports that myelin in different internodes of the same fibre has different directions of spiralization would appear to rule this possibility out, however. Whilst perhaps excessive ingenuity has been expended in devising models which may explain 'spiralization' of the myelin sheath, much less attention has been paid to the other dimensional changes which are occurring. Thus, whilst the myelin sheath increases in *thickness* from the initial few lamellae to one containing perhaps 300, in a large mature peripheral fibre, the *axon* also increases in *diameter* from about 1 to 15 μm, and each internodal segment increases in length from about 150 μm to perhaps 1,500 μm (Williams and Wendell-Smith 1971). After the onset of myelination, no further multiplication of Schwann cells occurs and each elongates in proportion to the overall growth in length of the part of the body in which the nerve lies (Vizoso and Young 1948; Williams and Kashef 1966). Clearly, much remains to be learnt about the initiation, mechanisms, and cessation of myelination and the continued maintenance of the mature sheath.

In late fetal and early postnatal development, myelination does not occur simultaneously in all parts of the nervous system, different nerves and tracts having their own particular myelination characteristics which can be related to the functional maturity of various functional pathways (Ochoa and Mair 1969 *a* and *b*). Amongst other factors which are largely unknown, the onset of myelination is associated with the growth in diameter of the non-myelinated axons, and in those animals which have been investigated, the critical point appears to be reached when the axon is about 1.0 μm in diameter, although some larger non-myelinated fibres are sometimes found (Matthews 1968) and, with the advent of electron microscopy, many fibres <1 μm in diameter have been found to possess myelin sheaths. Once begun, myelination continues at a steady rate with the growth of the axon until the latter stops, so that there is a predictable mathematical relationship between axon diameter and sheath thickness. The numerical value of this relationship, however, varies with the type of fibre (i.e. in terms of the tissue it innervates), and its distance from the cell body of the parent neuron (Williams and Wendell-Smith 1971).

THE NODE OF RANVIER

At the junction between adjacent Schwann cells or oligodendrocytes of myelinated nerve fibres is a gap where the axolemma is exposed, termed the node of Ranvier (*see* Hess and Young 1952, for a review of the earlier literature, and Landon 1976 for a comprehensive recent survey). These gaps are placed at regular intervals, the myelinated segments between being known as *internodes*. The distance between nodes varies directly with the diameter of the fibre from about 150 to 1,500 μm (Kashef 1966). When fibres branch, they do so at nodes.

In peripheral nerves, the myelin sheaths on both sides of each node are usually expanded into *paranodal bulbs* (7.25B, D) which   845 often show an asymmetry related to the growth patterns of the surrounding topographical region. The surfaces of the bulbs and of the underlying axolemma are fluted as they approach the node, and numerous Schwann cell mitochondria are present in the external cytoplasm-filled grooves so formed (Landon and Williams 1963; Berthold 1968; Landon 1976). The axon itself also narrows considerably at the node. Where each myelin lamella ends it makes contact with the paranodal axolemma as an expanded loop in which are situated spirally orientated microtubules and dense cytoplasm. The external paranodal cytoplasm of the Schwann cell sends a number of finger-like processes which curve towards, and their tips abut on, the otherwise naked axolemma of the node (7.25D). The fingers are numerous and form a regular hexagonal array in large fibres, but are few and irregular in small ones. The depression formed at the node by the narrowing of the axon at this point is filled between the nodal fingers with an acidic mucosubstance (*gap substance*) which may form a reservoir of ions, or a selective barrier to their flow in the conduction of nerve impulses (*vide supra*, p. 843, and Landon and Langley 1971). The narrowing of the nodal axon probably increases the transmembrane current density and hence the efficiency of nodal excitation.

Central nodes have been less extensively investigated, but appear to be of a more simple construction, with few nodal fingers, and small or absent paranodal bulbs.

INCISURES OF SCHMIDT-LANTERMAN

Originally described as cytological features, the clefts or incisures of Schmidt-Lanterman were subsequently long regarded as artefacts of preparation. Ultrastructural and *in vivo* observations have restored them as characteristic features of central and peripheral myelinated fibres (Hall and Williams 1970; Landon 1976). The incisures are fairly regular oblique discontinuities in the close packing of the myelin lamellae (7.25D), forming funnel-shaped zones of cytoplasm spiralling between the internal and external layers of Schwann cell cytoplasm. Thus, at the incisure, the major dense line separates to enclose a helical band of granular Schwann cell cytoplasm, bearing one or more microtubules which follow this spiral pathway, microfilaments, and occasionally other organelles. The dense cytoplasm is modified in places, where it is similar to that found in the zonula adherens of epithelial and other cells, forming an oblique row or *stack* of 'desmosomoid' attachments at one or more points near the external surface of the incisure. The intraperiod line also splits at the incisure to create an extracellular space interconnecting the periaxonal and endoneurial spaces.

The incisures possibly provide conduction channels for metabolites into the depths of the myelin sheath and to the subjacent axon; they have been claimed to be importantly involved in the initial stages of Wallerian degeneration (Williams and Hall 1971) and focal demyelination.

The *classical view* of Wallerian degeneration was that the first week after crush or section of a peripheral nerve was largely a period of *morphological disruption* of the part of the nerve distal to the injury, but it was also a period of *biochemical stability*. The morphological disturbances included the formation of myelin *ovoids* containing 'digestion chambers' with disruption of the axon. Only in the second week was there considered to be any biochemical change, and this consisted principally of the degradation of phospholipid in the sheath with the liberation of cholesterol esters.

More recently, *in vivo* examination of nerves undergoing Wallerian degeneration (Williams and Hall 1971), has shown that the time scale of events is considerably faster. The initial morphological changes include retraction of the paranodal myelin sheath, wide dilatation of the incisures of Schmidt-Lanterman, with the subsequent collapse and rounding off of the myelin to form ovoids at these points. These changes start within minutes in zones near the lesion, and occur progressively over the next 36–48 hours in more distal parts of the nerve. Further, it has been shown that there is an increase in the concentration of hydrolytic enzymes within 12 hours of injury; this is associated with a loss of trypsin-digestible *basic protein* from the sheath which can be demonstrated as a loss of *trypanophilia* in histological sections.

Electron microscopically the changes associated with Wallerian degeneration have been described as an accumulation of membrane-bound bodies in the axoplasm proximal and distal to the site of injury. Distally, there follows a degradation of the axoplasmic organelles, and at the previous sites of dilated incisures, there is a collapse of the previously compact myelin lamellae, with alterations in the repeat distances in their radial structure. Within 4 days, lipid droplets accumulate around the degrading myelin within the Schwann cell cytoplasm, and these are then extruded into the endoneurial space, where they are subsequently phagocytosed by invading haematogenous macrophages. The early increase in acid phosphatase activity has been correlated with the appearance of numerous lysosomes within the Schwann cell cytoplasm at this time.

As degradation of the myelin proceeds, the Schwann cells begin to proliferate rapidly; in some nerves which contain many large diameter myelinated fibres, the total number of nuclei per transverse section of the nerve rises to about 16 times that of a normal nerve, by the end of the third week of degeneration. These proliferating Schwann cells become aligned in parallel longitudinal chains within the persistent basement membrane, where they are of great importance in any subsequent regenerative phenomena which may ensue (p. 861).

CENTRAL-PERIPHERAL NERVOUS SYSTEM TRANSITIONAL ZONE

Crossing the transitional zone between the central (CNS) and peripheral (PNS) nervous system the sections of axons that comprise a nerve root are enclosed within a short glial segment lying close to the surface of the spinal cord or brainstem. In man the zone lies more peripherally in sensory nerves than in motor nerves: in both cases, the apex of the transitional region has been described as the '*glial dome*', with its convexity directed toward the periphery. Electron microscopy has shown that the centre of the dome consists of fibres showing typical central organization, surrounded by an outer mantle of astrocytes (corresponding to the external glial limiting membrane). From this mantle numerous processes, the '*glial fringe*', project into the endoneurial compartment of the peripheral nerve and interdigitate with the Schwann cells. The astrocytes form a loose meshwork through which the axons pass. There is currently some debate as to whether the basement membrane which surrounds the astrocytes is capable of preventing central Schwann cell migration. In general, peripheral myelinated fibres cross the transitional zone at a node of Ranvier, termed a *PNS-CNS compound node* by Carlstedt (1977). At such a node, on the peripheral side, the axon has a corona of Schwann cell microvilli and mitochondria-laden paranodal Schwann cell cytoplasm, while on the central side it is related to a few astrocyte processes, which typically make specialized contacts with the axolemma (Berthold 1978). In the transitional region of cat first sacral spinal nerve roots, an average of 4–5 nodes were associated with the processes of a single astrocyte, and it has been suggested that these node-astrocyte relationships are specific to the borderline nodes. In some myelinated fibres, however, the central–peripheral transition occurs along an internode, the central myelin being telescoped within an external peripheral myelin sheath.

Considerable rearrangement of axons occurs in the rootlets; moreover, many of the largest non-myelinated peripheral axons become invested with a thin myelin sheath as they pass through the transitional region.

Peripheral Endings of Effector Neurons

Of the various types of effector endings, the most intensively studied are those which innervate muscles, and particularly those of skeletal muscle, which are quite distinct from the efferent nerve endings in non-striated muscle. All such endings, termed *neuromuscular* or *myoneural junctions*, are, however, similar in being specialized regions of the neuronal cytoplasm from which neurotransmitters are released on to the surface of an adjacent muscle unit, causing a change in its electrical state. Because of the similarity between these junctions and synapses, and the relative ease with which neurophysiological studies can be carried out at the former, much general knowledge of neurotransmission stems from ultrastructural, physiological, biochemical and pharmacological analysis of neuromuscular junctions rather than synapses.

NEUROMUSCULAR JUNCTIONS IN SKELETAL MUSCLE

The general structure of these has been outlined elsewhere (p. 512) and will be described only briefly here. Essentially, the terminals consist of end branches of somatic motor fibres, each of which innervates from a few to many hundreds of muscle fibres, depending upon the precision of muscle control. The structural specialization of the motor terminal varies with the type of muscle innervated. Two major types of ending have been recognized, the 'en plaque' terminals typical of extrafusal muscle fibres together with the plate endings of intrafusal fibres; the second type includes the 'en grappe' terminals found on certain extraocular muscle fibres and also present in the 'trail' endings on intrafusal fibres (Harker 1972; Barker 1974). In the former type each terminal branch of the axon ends midway along a muscle fibre in a discoidal expansion, the motor end-plate (7.26A–C). In the latter, the axon gives numerous subsidiary branches which form a cluster of small expansions which may extend for some distance along the muscle fibre (7.26A). The en plaque endings usually initiate action potentials in the muscle fibre which are rapidly conducted to all parts of the fibre, whereas in the en grappe ending, in the absence of propagated muscle excitation, the muscle fibre is excited at several points by the branched axon, ensuring activation of the entire fibre. Both types of ending are characterized by a specialized region of the muscle fibre surface, the sole-plate, in which a number of muscle fibre nuclei are grouped in a mass of granular sarcoplasm, and ultrastructurally, by the presence in the muscle of numerous mitochondria, endoplasmic reticulum, and by the folding of the sarcolemma into a series of parallel grooves or gutters (7.26B, C). The terminal expansion of the efferent fibre is separated by a gap of about 30–50 nm from this complex subneural apparatus, and contains mitochondria and large numbers of clear spherical vesicles similar to those of presynaptic boutons, which are clustered against the membrane throughout the zone of apposition. Ensheathing the motor terminal are Schwann cells sometimes termed teloglia, with fingers sometimes projecting into the synaptic cleft. It is interesting that en plaque endings of 'fast' and 'slow' twitch muscle fibres differ in their detailed structure, the sarcolemmal gutters being deeper and the presynaptic vesicles more numerous in the former (Padykula and Gauthier 1970). Recent freeze-fracture studies have shown the presence of numerous intramembranous particles corresponding to acetylcholine receptors in the crests of the sarcolemmal folds where they lie nearest to the presynaptic membrane (Heuser et al. 1974). In the motor end-plate opposite these crests the membrane is slightly invaginated, and vesicles are clustered along the borders of the groove so formed, which is considered to be the main site of their release (i.e. the 'active zone' of the synapse). Microtubule arrays bearing synaptic vesicles, similar to those in central synapses, have also been described (Gray 1978).

Physiological and pharmacological studies (Katz and Miledi 1965) have shown that the neuromuscular junctions with skeletal muscle fibres are cholinergic, that is, they act by the release of acetylcholine from the nerve terminal on to the specialized surface of the underlying sarcolemma, and this changes its ionic permeability. The resulting depolarization of the sarcolemma, if sufficiently intense, initiates an action potential which is propagated over the surface of the muscle fibre and results in its contraction (p. 511). As with synapses, the quantity of neurotransmitter released depends upon the number of nerve impulses arriving at the motor terminal, and since the transmitter is inactivated at a constant rate by the enzyme acetylcholinesterase, the amount present at a particular time is dependent upon its rate of release, which in turn reflects the frequency of arrival of action potentials along the nerve fibre. In this way the contraction of a muscle is controlled by the frequency of firing of its efferent

nerves. Even in quiescent muscle fibres, small sporadic depolarizations have been detected (miniature end-plate potentials), which are too small to cause contraction. When recorded electrophysiologically, their amplitudes cluster at a series of preferential levels, indicating that multiples of small quanta of transmitter are occasionally released in the resting stage; calculations suggest that the amount of transmitter necessary to cause these unitary changes probably occupies the volume of a single synaptic vesicle (Katz and Miledi 1965).

Acetylcholinesterase can be demonstrated cytochemically in the myoneural cleft, and experiments with the specialized motor endings in the electric organ of the rayfish Torpedo show the enzyme to be located in the sarcolemma close to its site of action. Activity of the neurotransmitter can be blocked by various substances including curare and its derivatives, a sea-snake venom substance a-bungarotoxin and black widow spider venom, all of which bind irreversibly with receptor molecules so causing paralysis at the neuromuscular junction (Miledi et al. 1971). If the activity of the enzyme acetylcholinesterase is inhibited with eserine (physostigmine) or other poisons, transmitter action is cumulative and the muscle responds with a tetanic spasm. The neuromuscular junction is also partially blocked by high concentrations of lactic acid, and this is responsible for some types of muscle fatigue after prolonged exercise during which lactic acid accumulates (Miledi et al. 1971).

AUTONOMIC MOTOR ENDINGS

Unlike those in skeletal muscle, autonomic nerve-endings are not closely applied to the non-striated myocytes but end at variable distances from their surfaces (Richardson 1962). Non-myelinated autonomic axons branch to give many tapering, varicose collaterals (7.27A). At zones where transmission occurs, clusters of ultramicroscopic vesicles are present within an expanded region of axoplasm, and the Schwann cell, which closely enwraps the axon elsewhere, is retracted at this point so that the axon lies in a shallow groove, thus forming a free diffusion path between the axon and the muscle cells. In sympathetic nerves the vesicles are usually of the dense-cored type (p. 828), and these can be correlated with the characteristic fluorescence of catecholamines using light microscopy, when preparations which have been fixed in formalin vapour are viewed with ultraviolet illumination (Falck and Owman 1965). Besides noradrenalin, these vesicles also contain certain enzymes involved in their synthesis. Larger vesicles, 100 μm across are often also present.

Generally, noradrenalin is the transmitter at sympathetic postganglionic endings, whilst some adrenalin is also present chiefly in the chromaffin cells of the adrenal medulla. It is possible to distinguish ultramicroscopically between the synaptic vesicles which contain the two types of catecholamine by using appropriate fixatives. Catecholamines can be released experimentally with various drugs, such as reserpine, and this constitutes a valuable research tool, in addition to its clinical usefulness. After release a proportion of the transmitter is taken back into the ending by endocytosis in vesicles. Sympathetic endings are also rich in monoamine oxidases involved in degrading catecholamines, and the control of these enzymes is important in clinical regulation of sympathetic function.

At parasympathetic endings, clear spherical vesicles, similar to those in the motor end plates of skeletal muscle, are present. Several lines of evidence indicate that these are cholinergic in nature; similar terminals are also found in the neighbourhood of sympathetic endings, and it has been suggested that in addition to their action on non-nervous tissues, some cholinergic endings may cause the release of catecholamines from sympathetic endings (Burn 1968).

Recently a third category of autonomic neurons with properties quite distinct from those of noradrenergic or cholinergic cells has been described by Burnstock and his colleagues (see the review by Burnstock 1975). Since there is evidence that the conjugated purine adenosine triphosphate or ATP (a nucleoside), or a related compound, is the neurotransmitter at their axon terminals, these cells have been classed as purinergic in their properties. Typically, their axons contain large, dense vesicles some 80–200 nm across

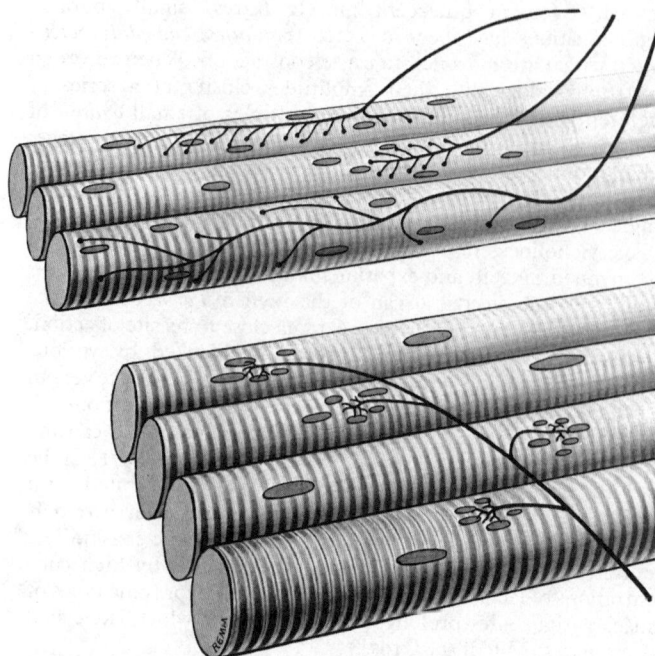

7.26A Diagram showing some types of innervation of striated muscle, including the 'en plaque' terminals of α efferents (below), and the more widely spread 'trail' and 'en grappe' endings of γ efferents (above).

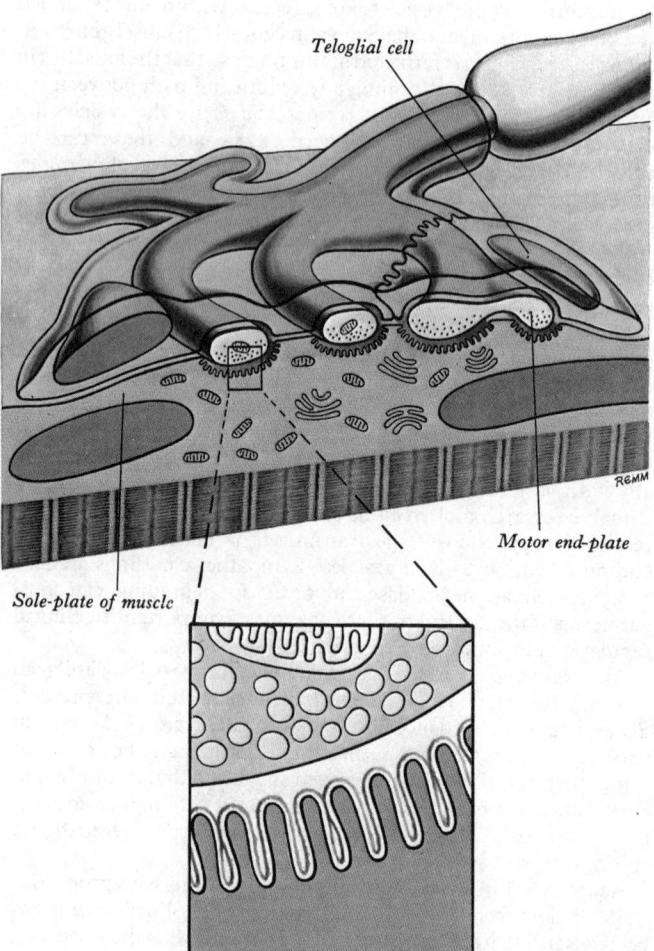

Teloglial cell

Motor end-plate

Sole-plate of muscle

7.26B A diagram of the detailed structure of an 'en plaque' neuromuscular junction. An enlarged portion of the terminal is shown below; note the folding of the sarcolemma to form subsynaptic gutters, the disposition of the basal lamina, and the synaptic vesicles within the axon terminal.

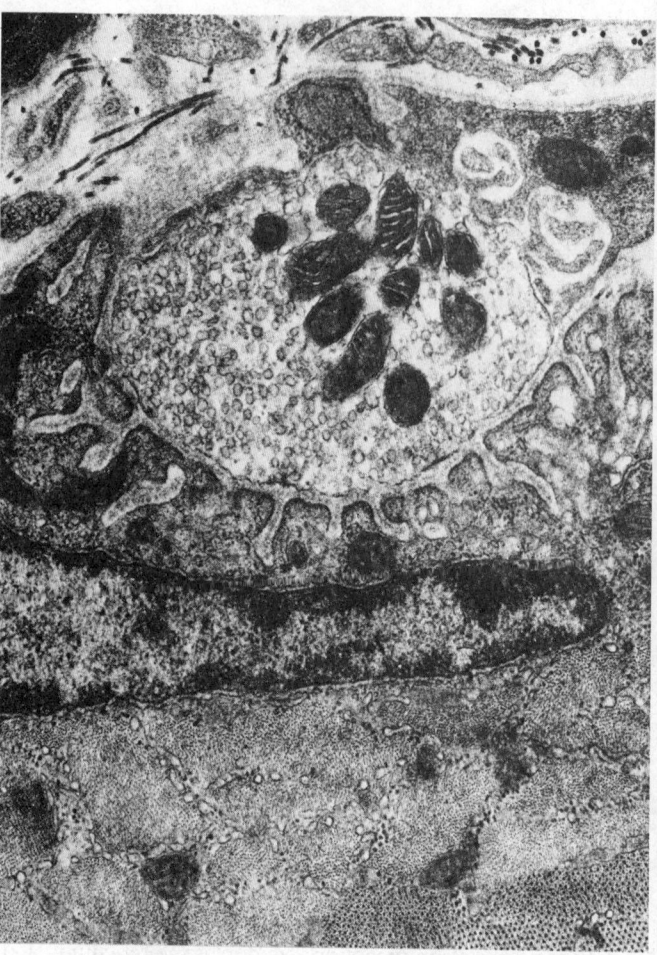

7.26C An electron micrograph of a neuromuscular junction in striated muscle, showing part of a motor end-plate containing synaptic vesicles (above). The latter is situated in a groove in the sarcolemma; this is further convoluted to form the subsynaptic gutters of the sole plate (below). (Kindly provided by Dr. D. N. Landon, Neurobiological Unit, National Hospital for Nervous Diseases, London.)

('*large opaque vesicles*') when seen in electron microscopic sections, congregated particularly in varicosities which occur at intervals along the nerve fibres. Such nerves have been demonstrated in many sites in the mammalian body, including the external muscle layers and sphincters of the alimentary tract, lungs, blood vessel walls, in the urogenital tract and central nervous system. In the wall of the gut, the cell bodies are in the myenteric plexus, and their axons branch and spread caudally for a few millimetres, chiefly to innervate the circular non-striated muscle (7.27B). The purinergic cells are themselves under cholinergic control from preganglionic sympathetic efferent nerve fibres.

The main action of purinergic endings is to hyperpolarize the muscle cells they innervate, causing their relaxation, for example, in front of peristaltic waves in the alimentary tract, opening sphincters and, in the case of the stomach, probably causing reflex distension on the receipt of food. Release of transmitter appears to be rather similar to that in other autonomic endings, that is, from axon varicosities at up to 100 nm away from the target cell, although details are not yet available.

The precise pattern of axonal branching and termination of autonomic efferents is closely related to the nature of the controls which they exert (Burnstock 1970); in non-striated muscle, with a slow widespread action, the branching may be extensive, a single neuron innervating a large number of muscle cells, and the distance between synaptic varicosity and muscle fibre surface being relatively large (50 nm or more). In quick-acting muscle which has a greater precision of control, such as that in the iris or ductus deferens the innervation is more localized, with much less branching of the axons, and the apposition is usually close (15–20

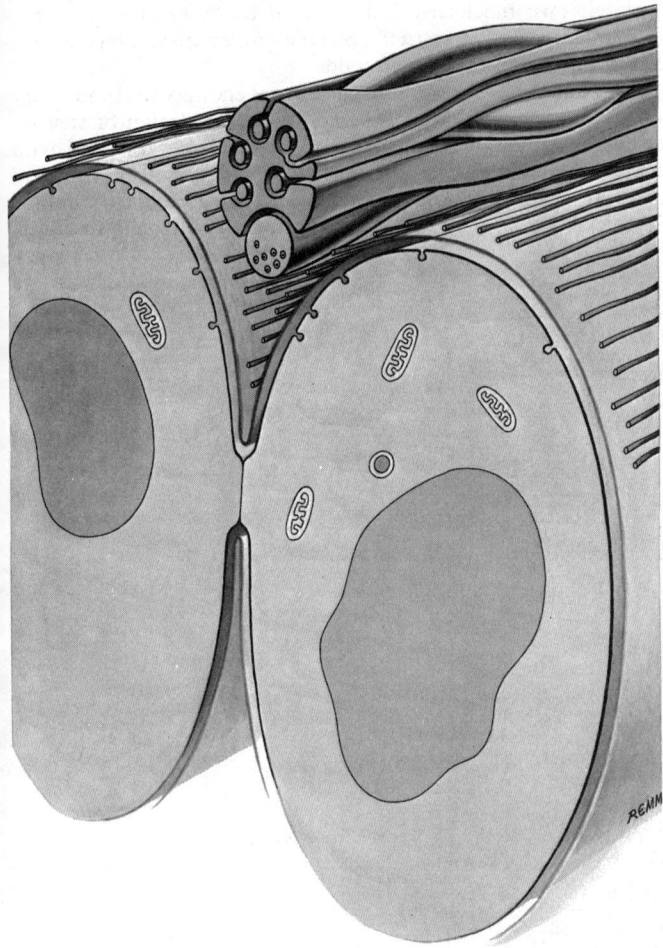

known example of this type of receptor in mammals is the sensory cell of the olfactory epithelium, although in many invertebrate groups it is the chief type of sensory receptor and must be regarded as phylogenetically primitive. A second arrangement is for the sensory cell to be a modified epithelial cell, an **epithelial receptor**, derived from the non-nervous tissue at the sensory surface and innervated by the peripheral process of a primary sensory neuron, the cell body of which is situated close to the central nervous system. Activity in this type of receptor involves the passage of excitation from the sensory cell across a synaptic gap; examples are gustatory and cochlear auditory receptors. It is known that in gustatory sensory receptors there is a high turnover of individual cells which are constantly being renewed from the pool of surrounding epithelial cells, a phenomenon which may

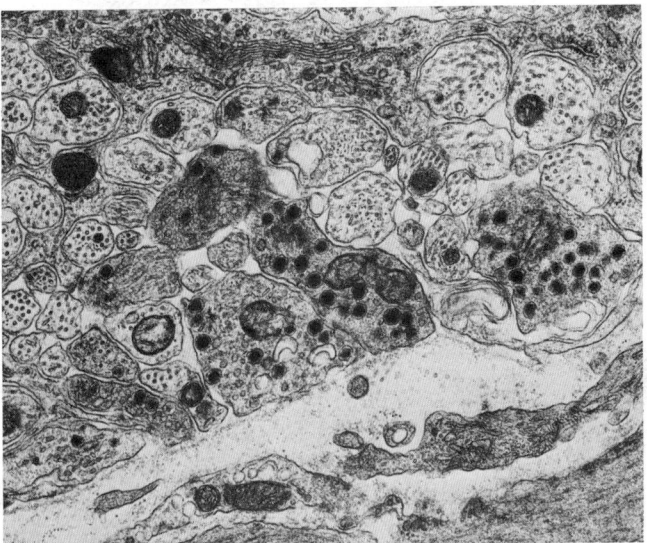

7.27A A diagram of an autonomic neuromuscular junction between a group of non-myelinated axons (above) and smooth muscle cells (below). The Schwann cell (blue) is reflected at intervals to expose enlargements of the axons (yellow) which contain synaptic vesicles.

7.27B Electron micrograph of a group of autonomic axons in the myenteric plexus, showing profiles with large dense-cored vesicles typical of the putative purinergic system (see text); smaller, catecholamine-type vesicles are also visible (lower left). Magnification × 40,000.

nm), the ending sometimes being invaginated in the sarcolemma (Burnstock 1975). In addition to non-striated muscle, many other tissues are innervated by autonomic efferents, including glands, myoepithelial cells, and adipose tissue. Many physiological actions depend upon autonomic control of the cardiovascular system whereby the activities of the tissues are in part regulated by varying the perfusion of blood through them.

Sensory Receptors

As has already been stated, the manner in which an organism reacts to changes in its environment depends upon the presence of suitable sensory receptors which can monitor various parameters important to its survival. Although in unicellular organisms the whole cell may be able to detect mechanical deformation, changes of temperature, osmotic pressure, concentrations of various chemicals, light, and so on, in more elaborate animals individual cells have become differentiated to sense such changes with great acuity, different cells being responsive to their own particular range of physico-chemical variables. Such cells are termed sensory receptors; they take various forms in the animal kingdom, but in mammals there is, with minor exceptions, a common pattern which will be outlined below.

Structurally, sense organs take one of three basic forms, depending upon the relationship of the nervous system to the sensory surface. In **neuroepithelial receptors** the sensory cell is itself a neuron of which the perikaryon is peripherally situated near the sensory surface and the axon extends back into the central nervous system to connect with second order neurons. The only

also occur in other sensory systems of this general type. Visual receptors are in many ways similar in their organization, since they are formed from the lining of the ventricle of the fetal brain; but no cell replacement occurs. Thirdly, a **neuronal receptor** is itself a primary sensory neuron with a perikaryon situated in a craniospinal ganglion and a long peripheral process, the ending of which constitutes the actual sensory terminal. All cutaneous sensors and many proprioceptors are thought to be of this type, although the sensory terminals may in some cases be encapsulated or lie in association with specialized mesodermal or ectodermal elements which form an integral part of the sensory apparatus; however, it appears that these non-neuronal elements are not themselves excitable but rather create the right environment for the excitation of the neuronal dendrite.

RECEPTOR RESPONSES

The various events occurring between the presentation of a stimulus and the signalling of this to the central nervous system have been extensively studied (e.g. Granit 1962), but are not yet entirely understood. It is convenient to divide these processes into a number of stages: initially there is the presentation of an effective stimulus, then its transduction at the receptor surface into a graded change of electrical potential (*generator potential*), and the initiation of an all-or-none action potential which passes to the central nervous system. All of these processes may occur within the confines of the receptor where this is a neuron, or some of them may take place partly in the receptor and partly in the dendrite of the neuron innervating it in the case of epithelial receptors.

The nature of the transduction process varies, of course, with the modality of the stimulus, and involves a change in the permeability to certain ions in the receptor membrane, usually to cause a depolarization (or in retinal cells, a hyperpolarization).

How a stimulus causes such a change is not yet known in any sensory cell. In mechanoreceptors it may involve a physical deformation in membrane structure; in chemoreceptors the action may correspond to that postulated for the action of acetylcholine at the neuromuscular junction (p. 514), involving

by their own bioelectric processes, or by those of other fishes (Lissmann and Machin 1958). So far a similar sensory mechanism has not been reported in tetrapods.

The quantitative responses of sensory endings to stimuli vary greatly and give additional flexibility to the design of sensory systems. Although an increase in the level of excitation with increase of stimulus strength is a common pattern (the 'on' response), some receptors respond instead to a decrease in stimulus strength (the 'off' response). Even when not stimulated, receptors

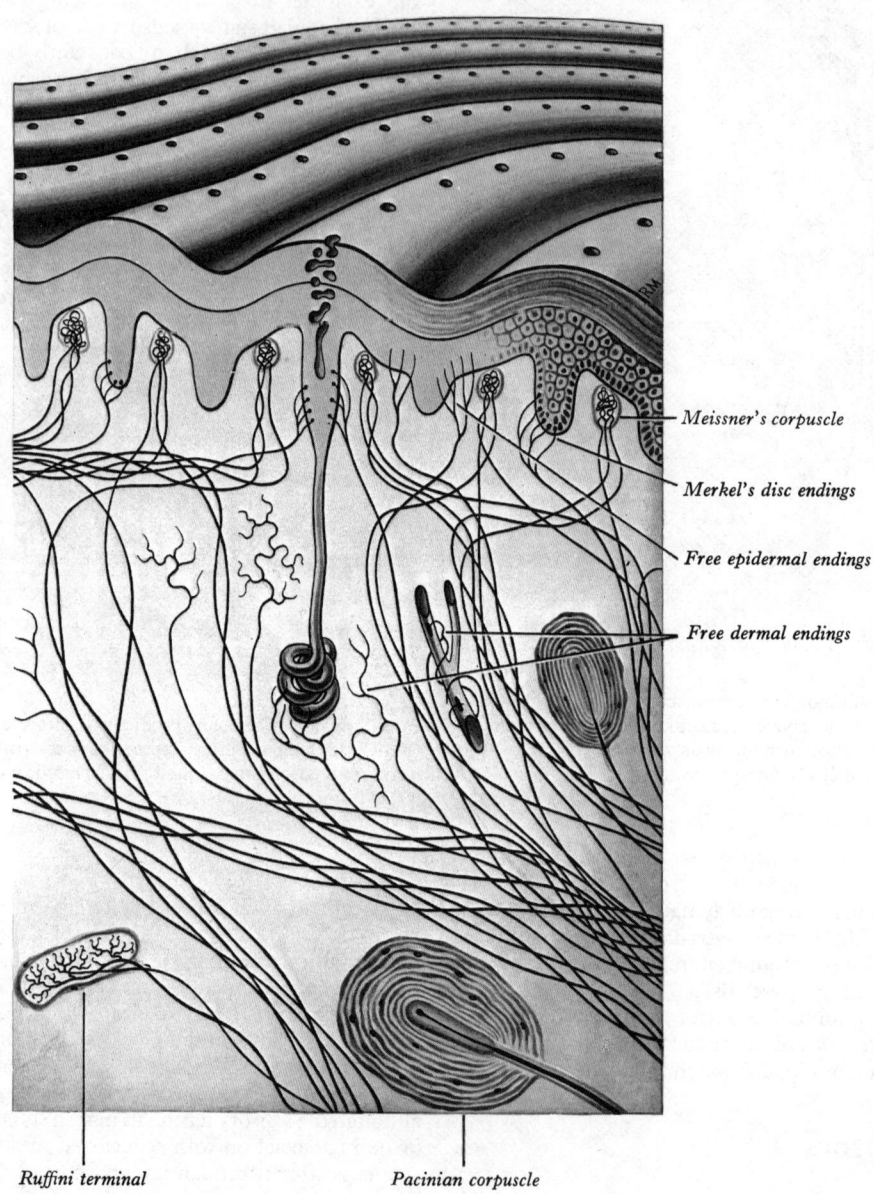

Meissner's corpuscle

Merkel's disc endings

Free epidermal endings

Free dermal endings

Ruffini terminal   Pacinian corpuscle

7.28A Schematic illustration of some of the principal sensory endings of thick (hairless) skin and associated structures, including various types of encapsulated and 'free' endings, of the epidermis, dermis and subcutaneous connective tissue.

specific receptor proteins which undergo alterations when stimulated, leading to the opening of ion channels with consequent depolarization of the sensory ending. Visual receptors appear to be rather similar to chemoreceptors; chemical changes occur within the cell as a result of illumination and so presumably act on the *internal* aspect of its membrane. Osmoreceptors may be similar to mechanoreceptors except that they react to mechanical deformation of their cell surface resulting from osmotic inflow or outflow of water with respect to the cytoplasm.

It is interesting that in certain fishes a quite distinct type of transduction occurs in electroreceptors which are capable of sensing changes in the electric fields set up in the external medium

show varying degrees of spontaneous activity, sometimes of a high level, against the background of which an increase or decrease in activity occurs with changing levels of stimulus strength.

When a steady stimulus is presented, there is, in all receptors yet studied, an initial burst of activity (the *dynamic phase*) followed by a gradual slowing or *adaptation* to the steady level of the stimulus (the *static phase*). Although all receptors appear to show these two phases, either one or the other may predominate, so that it is possible to distinguish between *rapidly-adapting* endings giving accurate estimates of changes in stimulus strength, and *slowly-adapting* endings which signal relatively constant stimuli such as those associated with position sense. Dynamic and static phases occur in both the amplitude of the generator potential and the frequency of action potentials set up in the sensory fibre. The stimulus strength necessary to elicit a response in a receptor (i.e. its *threshold level*) varies considerably. Such high and low threshold responses gives added information on stimulus strength.

GENERAL CLASSIFICATION OF RECEPTORS

Receptors can be classified in various ways, for example, by the particular energy forms or '*modalities*' to which they are especially sensitive. We can therefore group them into *mechanoreceptors* which are particularly responsive to mechanical disturbances (touch, pressure, sound waves, etc.), *chemoreceptors* sensitive to chemical changes, *photoreceptors* responsive to electromagnetic waves in the visual frequency range, *thermoreceptors* which sense changes in temperature, *osmoreceptors* which react to changes in osmotic pressures (in contrast to chemoreceptors which are activated by specific chemical groups in their environment). Some receptors are selectively sensitive to more than one mode of stimulation (*polymodal receptors*), these usually being endings with high thresholds, responding to noxious or damaging stimuli, and associated with irritation or pain (*i.e.* they are *nociceptors*).

Another widely used (but rather arbitrary) classification divides receptors on the basis of their distribution and role in sensory activities of the body, into three main groups, namely *exteroceptors*, *proprioceptors* and *interoceptors*. Exteroceptors and proprioceptors are the receptor end-organs of the somatic afferent components of the nervous system and the interoceptors constitute the receptor end-organs of the visceral afferent components.

Exteroceptors respond to stimuli from the external environment and are placed at or close to the surface of the body (Sinclair 1967); they can be divided into the *general* or *cutaneous sense organs* and the *special sensory organs*. The general sensory organs include the non-encapsulated and encapsulated terminals in the skin and around hairs; the special sensory organs are olfactory, visual, acoustic, and taste receptors, which will be described elsewhere (p. 1138).

Proprioceptors respond to stimuli arising in the deeper tissues particularly in the locomotor system. They are concerned with movement, position and pressure, and include the neurotendinous organs of Golgi, the neuromuscular spindles, deeply placed Pacinian corpuscles and perhaps other endings such as those of joints, and the vestibular receptors in the membranous labyrinth. The proprioceptors are stimulated by the activity of the muscles, movements of joints and changes in the position of the body as a whole or its various parts, and are essential to the coordination of muscles, the grading of muscular contraction and the maintenance of its equilibrium.

Interoceptors include the receptor end-organs in the walls of the viscera, glands, and blood vessels, (*see* e.g. p. 1136), where a variety of fibre terminations and end-organs including naked nerve endings, loops and encapsulated terminals, has been described. Nerve terminals are found in all layers of the visceral walls including the lining epithelium, and they are numerous in the adventitia of blood vessels, although the nature of many of these endings is open to doubt. Lamellated corpuscles have been described in the heart, adventitia of blood vessels, pancreas and mesenteries; free terminal arborizations are also present in many situations including the endocardium, loose connective tissue, the endomysium of all types of muscle, and connective tissue generally.

The nerve terminals in the viscera are not, in general, responsive to the same stimuli which act on the exteroceptors placed at the surface of the body and with certain exceptions do not respond to localized mechanical and thermal stimuli. However, tension produced by overstretching or excessive muscular contraction, often gives rise to visceral pain, particularly in pathological conditions, which is frequently poorly localized and of the deep-seated variety.

Amongst the interoceptors are included various vascular chemoreceptors and baroceptors (pressure receptors) the activity of which is important in the regulation of blood flow and pressure

in the whole cardiovascular system, the control of respiration and other important aspects of homeostasis.

Irritant receptors responding polymodally to various noxious chemical or mechanical stimuli are also widely distributed in the epithelial linings of the alimentary and respiratory tracts where they may initiate protective reflexes.

It should be pointed out that this scheme of classification is in many cases arbitrary, since many types of end-organ are present in all three classes, and also their activities may be closely linked

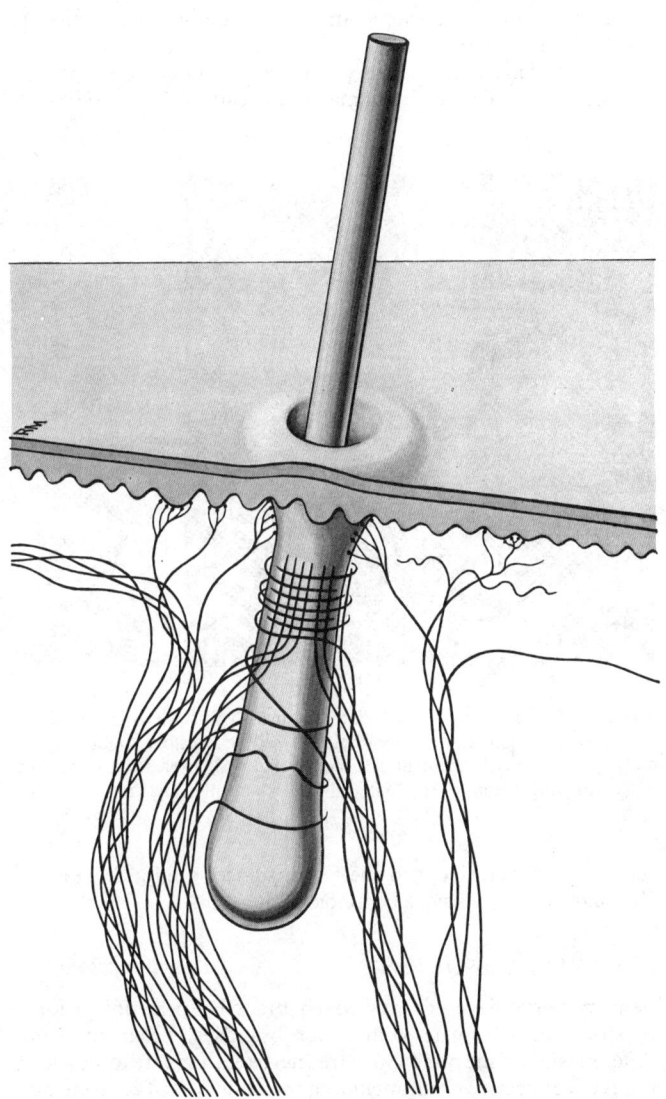

7.28B The innervation of a hair follicle; fine and coarse axons terminate around the intermediate and superficial regions, some branching in a circular direction, and others pursuing a longitudinal course. Nerve terminals associated with Merkel's cells are also shown surrounding the neck of the follicle.

in the central nervous system. For convenience, therefore, the sensory terminals will be classified in the following description on the basis of their structural characteristics.

STRUCTURAL CLASSIFICATION OF SENSORY ENDINGS

In addition to the foregoing division of receptors into broad functional categories, the general receptors have been grouped on morphological grounds (7.28 to 7.32). How much functional significance can be attached to some differences in detail in such a structural classification is debatable. Complicating factors include changes in receptor ending morphology associated with ageing, regeneration, local topography and also species differences. The terminology is further confused, as there is no general agreement on the structural definition of many types of sensory ending. Accordingly, the scheme which will be outlined below is not exhaustive. Some receptor endings are limited to a particular tissue or combination of tissues; others are present in many situations in the body. Generally, however, we may distinguish between *free nerve endings* which form plexuses or are otherwise spread freely without any particular association with other cell types, *encapsulated endings* where specialized non-nervous cells completely invest the neural process with several or many layers, and *epidermis-associated endings*, where the sensory fibre is attached to, but not enclosed in, specific non-nervous cells or tissues (Bannister 1976).

Such specialized non-sensory cells are, of course, additional to the Schwann cells which normally ensheath all three types of

(*unimodal nociceptors*) or to damaging stimuli of all kinds (*polymodal nociceptors*). Similar fibres in deeper tissues may also signal extreme conditions, experienced, as with all nociceptors, as pain. The free endings in some regions, e.g. those of the cornea, dentine and periosteum may be exclusively of the nociceptor type.

ENDINGS ASSOCIATED WITH EPIDERMAL STRUCTURES

Terminals on hairs arise from myelinated fibres derived from the deep dermal cutaneous plexus of nerves; their number, size and form are related to the size and type of hair follicle innervated (Cauna 1966). In *palisade endings* the fibres approach the hair follicle from different directions to reach it just below the duct of the sebaceous gland where they divide into branches which run parallel to the hair in the outer coat of the follicle (7.28B). Some of these fibres pass into the middle of the outer coat where they give

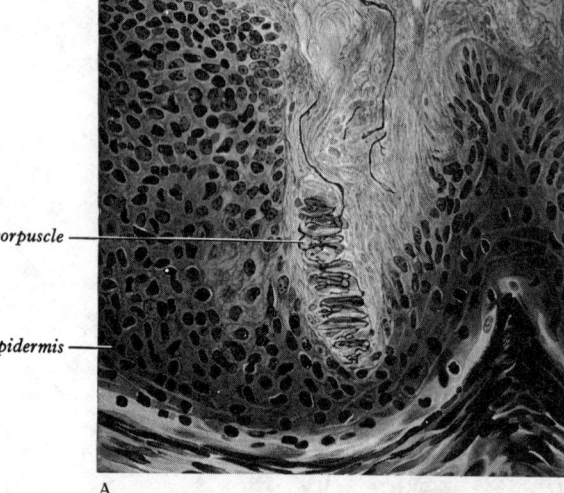
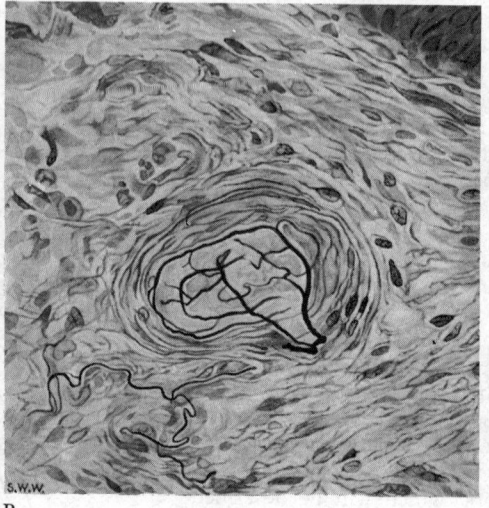
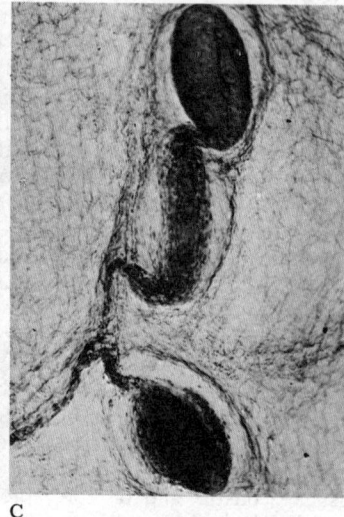

Tactile corpuscle

Epidermis

A B C

7.29A–C Specialized sensory end organs. A Tactile corpuscle of Meissner in human skin. (Gros-Bielschowsky technique, about ×250.) B Bulbous corpuscle from human anal canal. (Gros-Bielschowsky and haematoxylin, about ×480. Material kindly supplied by Prof. M. J. T.

FitzGerald, University College, Galway.) C Whole mount of developing Pacinian corpuscles in feline mesentery (Gros-Bielschowsky technique, about ×120. Specimens A and C kindly provided by Prof. N. Cauna, University of Pittsburgh.)

sensory terminal except in the few instances mentioned below, although they may have a common origin.

FREE NERVE ENDINGS

Sensory nerve fibre endings which branch repeatedly to form plexuses, or terminate with fewer branches, occur in many different sites of the body. Such free nerve endings are found in all types of connective tissue including the dermis, fascia, ligaments, tendons, sheaths of blood vessels, meninges, joint capsules, periosteum, perichondrium, Haversian systems of bone, and the endomysial spaces of all types of muscle. They also innervate the epithelium of the integument, cornea, buccal cavity, and the alimentary tract with its associated glands, although in the latter examples they are devoid of Schwann cell sheaths; they are enveloped instead by the cells of the epithelium and are not truly 'free' endings (Cauna 1966).

The afferent fibres from free terminals are both myelinated and non-myelinated, but are always of small diameter and low conduction speeds, being of the group III (Aδ) or IV (C) afferent types. Where the fibre is myelinated, terminal arborizations are non-myelinated and it is possible that they are devoid of Schwann cell sheaths altogether at their tips.

Electrical recordings from their afferent fibres show that such fine terminals subserve a number of different modalities. In the dermis, some of these narrow terminals belong to fibres responsive to moderate cold or heat (*thermoreceptors*), to light mechanical touch (low threshold, rapidly adapting *mechanoreceptors*), to damaging temperature changes or mechanical trauma

rise to axon branches encircling the hair and terminating as free nerve endings amongst the collagen bundles. Others pass into the hyaline layer between the outer and inner coats and, after losing their myelin sheaths, break up into fine filaments which course in a series of parallel branches along the axis of the follicle, each fibre being ensheathed by two Schwann cell cuffs. The precise pattern of innervation depends upon the type of hair; in many mammals three forms of hair are found, the fine *down hairs*, coarser *guard hairs*, and *whiskers* or *vibrissae* which have erectile vascular tissue around their follicles and which are the most complex, with at least three types of nerve ending (see 7.28B). Human hair is probably of the first two of these types. The palisade hair endings respond mainly to rapid movements of the hair shaft when the hair is deformed mechanically, and belong to the rapidly-adapting mechanoreceptor group. (See also 7.29H.)

Merkel cell endings or *tactile menisci* are found immediately below the epidermis or around the apical ends of certain hair follicles (7.29F). The nerve fibre of these structures expands into a disc or flattened sole applied closely to the base of a specialized non-nervous cell (the *Merkel cell*) which is inserted into the base of the epithelium, and bears numerous protrusions in the form of spikes which interdigitate with surrounding cells. Such cells contain numerous large (50–100 nm) dense-cored vesicles which are congregated towards the junction with the nerve fibre. Often many such nerve-Merkel cell units are assembled at the base of a dome-like epidermal disc (the *touch dome*) supplied by a single branched nerve fibre, and are also frequently associated with specialized hairs (*tylotrichs*). Electrophysiology shows these endings to be slowly-adapting (Type I) mechanoreceptors

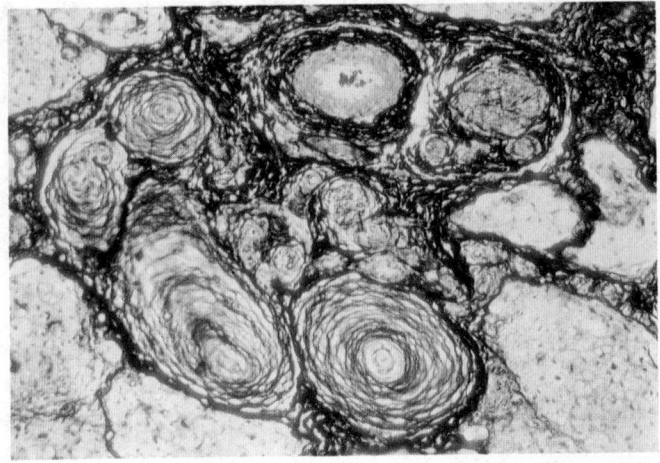

7.29D A cluster of Pacinian corpuscles stained with the Glees and Marsland silver technique to show the capsules and the central axons, sectioned in various planes. Rhesus monkey finger; magnification × 150. Kindly provided by Dr. R. Bilous, Guy's Hospital Medical School, University of London.

7.29E Electron micrograph of a Pacinian corpuscle in transverse section, showing the central core region with lamellar cells surrounding the axon. Note the presence of large intercellular spaces between the lamellar cells and the numerous mitochondria in the axon. Rhesus monkey finger. Magnification × 5,000.

7.29F Electron micrograph showing a Merkel cell (M) associated with a type I slowly-adapting cutaneous mechanoreceptor nerve terminal (N) from a rabbit touch-dome. Note the numerous dense vesicles clustered near the sensory nerve ending (arrowheads) and epidermal keratinocytes (K). Material kindly provided by Dr. W. Hamann, Dept. of Physiology, Guy's Hospital Medical School, University of London. × 5,000.

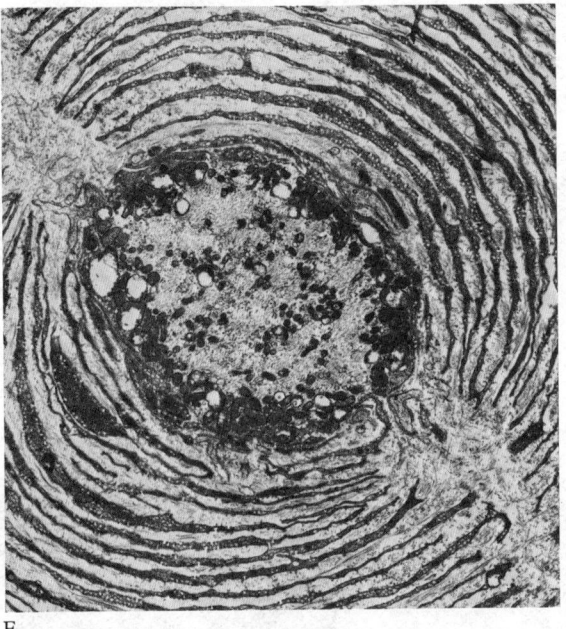

E

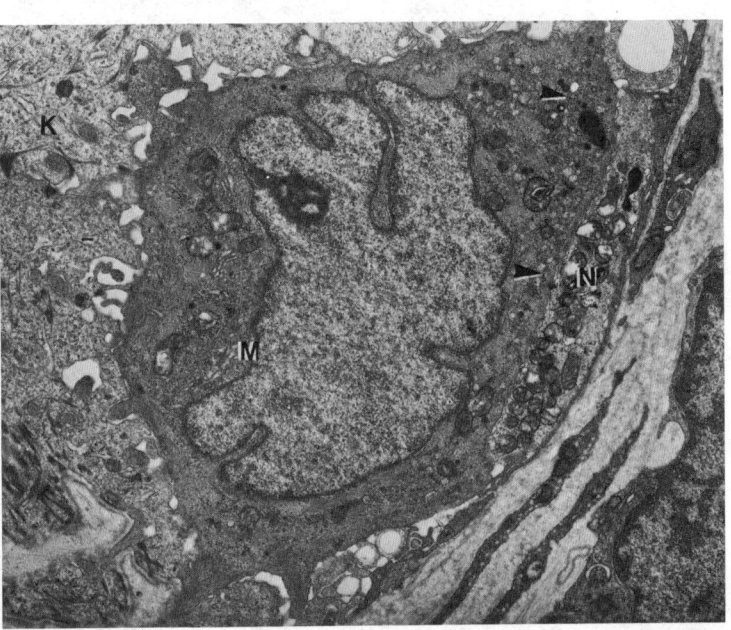

F

responsive to vertical pressure and served by large myelinated (Aa) afferents. The structure of the ending suggests some form of synaptic transmission, but all attempts to demonstrate this have so far been unsuccessful (see Iggo and Muir 1969; English 1978).

ENCAPSULATED NERVE ENDINGS

These are special end organs (or *corpuscular endings*) which exhibit great variety in size and shape, but have one feature in common, that is, the termination of the nerve is enveloped by a capsule. Included in this group are the lamellated corpuscles together with the neurotendinous endings, neuromuscular spindles, and Ruffini endings, all of which are well-defined morphological entities. Many other types of nerve endings have been described such as bulbous and hederiform corpuscles, articular corpuscles, genital corpuscles, and Golgi-Mazzoni endings, but there is considerable confusion as to their nature and function, contributed to by several factors, for example, technical difficulties, species and regional variations and age changes. (*See* 7.28, 29.*)

The tactile corpuscles (of Meissner) are sited in the papillae of the skin of all parts of the hand and foot, in the front of the forearm and lips, palpebral conjunctiva and the mucous membrane of the tip of the tongue. They occur mainly in glabrous skin. Mature corpuscles are cylindrical in shape, orientated with their long axes perpendicular to the deep surface of the epidermis and about 80 μm long and 30 μm broad. The corpuscle consists of a capsule and a central core (7.28A, 29A). The capsule is only loosely attached to the core of the corpuscle and is absent at the

extremities. Light microscopists have maintained that the capsule consists of fine elastic fibres orientated along the long axis of the corpuscle and interspersed with fibrocytes and possibly other cell types. It has been claimed that the elastic fibres anchor the corpuscle to the epidermis (Quilliam 1966). Electron microscopy shows the capsule to be continuous with the perineurium of the nerves supplying the corpuscle and to consist of a variable number of lamellae of greatly flattened cells with their associated basement membranes. Between successive lamellae there is a substantial amount of collagen but no elastic fibres have been identified. Extensions of the capsule, often only one lamella in thickness, may form complete or incomplete septa dividing the corpuscle into lobules, particularly at the epidermal extremity. The core of the corpuscle consists of cells and nerve fibres. At the epidermal end the cells are described as discoid, lightly staining, transversely disposed and about 2–4 μm thick and 30–40 μm in diameter; these cells are claimed to be epidermal in origin. At the opposite end of the core the cells are irregular in shape with small oval or round nuclei devoid of nucleoli and with deeply staining chromatin. These are believed to be Schwann cells. In addition to Schwann cells, electron microscopy reveals occasional fibroblasts and collagen fibres, disposed in bundles or singly and mainly orientated parallel to the nerve fibres; the flattened apical cells have been studied in electron micrographs (7.29G). Each tactile corpuscle is supplied by a few heavily myelinated nerve fibres derived from the deep corial plexus and possibly also by additional non-myelinated fibres which are branches of myelinated fibres in the deep corial plexus. Within the corpuscle the nerve fibres ramify profusely and decrease in size; large numbers

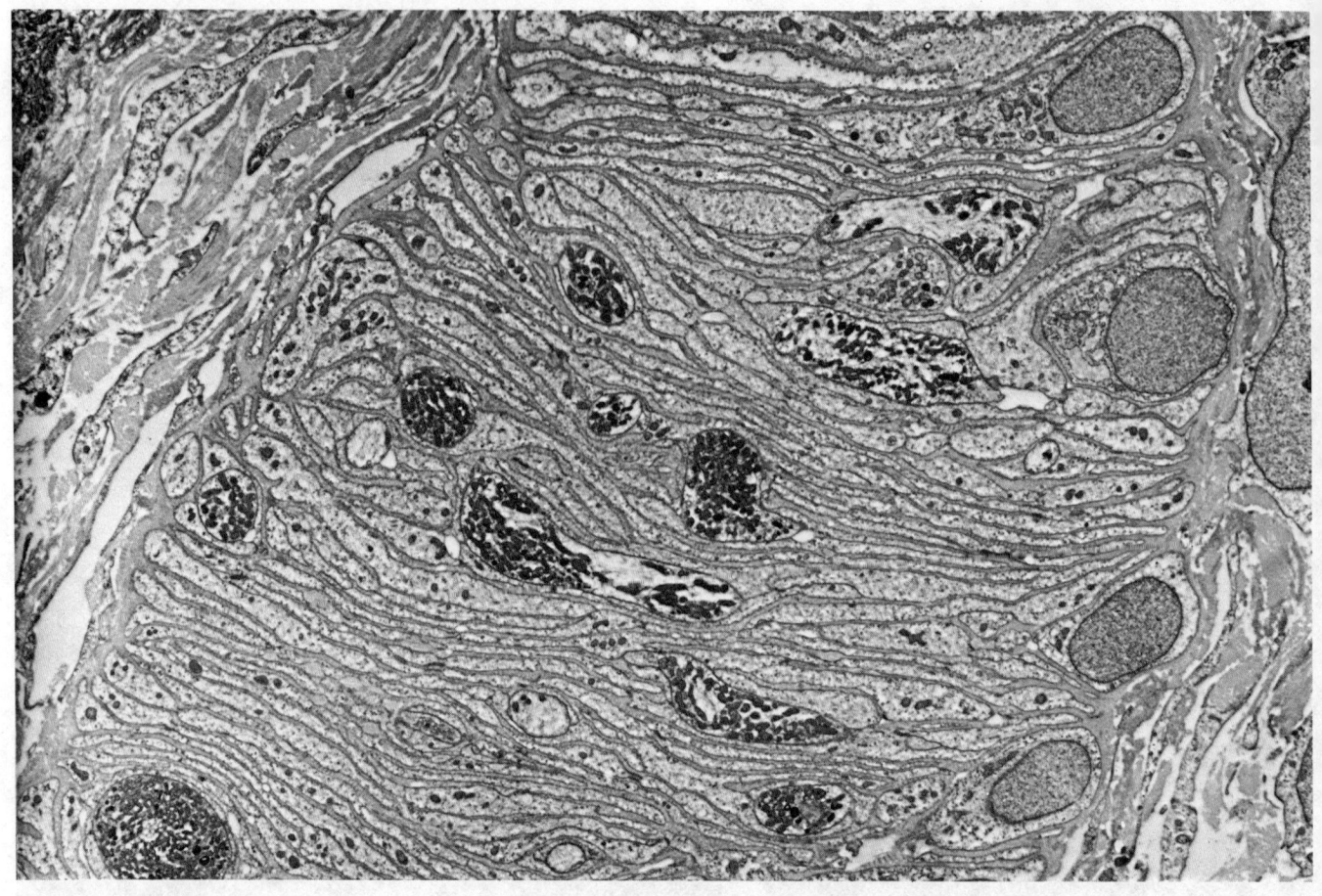

G

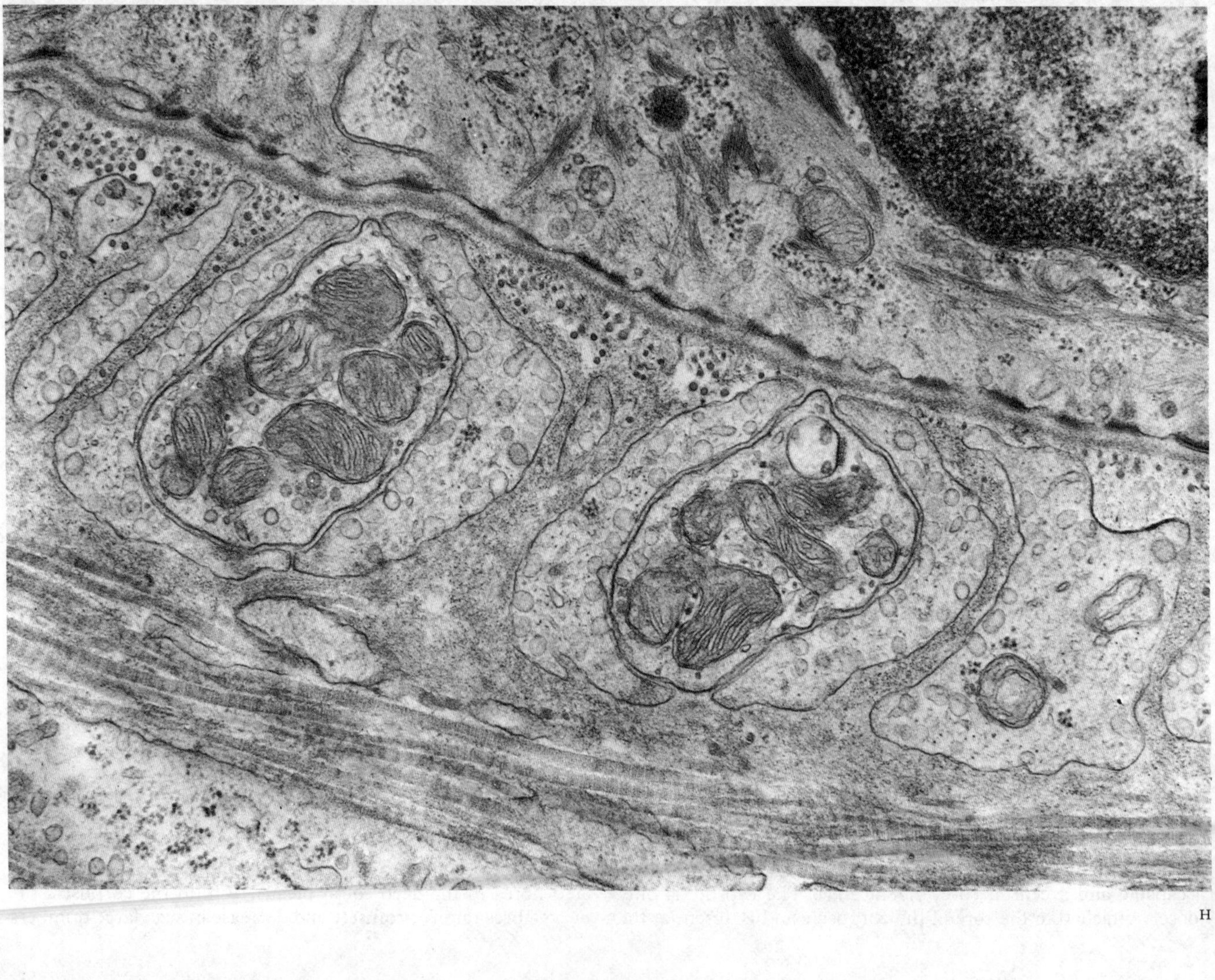

H

of both myelinated and non-myelinated branches are visible in ultra-thin sections, the former being most numerous at the deep extremity of the corpuscle and the latter constituting the majority at the epidermal extremity (7.29G) and in the smaller lobules. In all cases the nerve fibres are associated with Schwann cells or their processes and no naked axons are present. Some of the non-myelinated fibres in the core of the corpuscle are less than 0·1 μm in diameter. The cytoplasm of the non-myelinated fibres may contain small vesicles but no special significance can be attached to them.

Tactile corpuscles develop just before or just after birth. There is a reduction of about 80 per cent in their numbers between birth and old age when the number of nerves supplying each corpuscle is also reduced and the nerve endings are confined to the deep end of the core. These endings function as low threshold, rapidly adapting mechanoreceptors, providing information about fluxes in mechanical forces acting on the volar surfaces of fingers, toes and other areas of glabrous skin (Valbo *et al.* 1979).

The lamellated (Pacinian) corpuscles of Vater-Pacini (7.28A, 7.29C, D, E, 7.30) are found subcutaneously in the palmar aspect of the hand and plantar aspect of the foot and their digits, in the genital organs of both sexes, the arm, neck, nipple, periostea, interosseous membranes of the forearm and leg, near joints and in the mesentery and pancreas of the cat. The corpuscles are oval in the fetus and oval, spherical or even irregularly coiled in the adult. They are relatively huge, being up to 2 mm in length and 100–500 μm across, the larger ones being visible to the naked eye. Each corpuscle consists of a capsule, an intermediate growth zone and a central core containing the nerve terminal. The capsule comprises about 30 concentrically arranged lamellae of flattened cells each about 0·2 μm thick. The neighbouring cells overlap at their edges and successive lamellae are separated by intervals containing some amorphous material and collagen fibres. The latter tend to be disposed circularly and to be closely applied to the surfaces of the lamellae, in particular to their outer surfaces. The amount of collagen increases with age. The intermediate zone is a cellular layer between the capsule and core. Occasional mitoses are seen in this zone and with growth of the corpuscle the cells are incorporated either into the capsule or the core, and the intermediate zone is not a conspicuous feature of the mature corpuscle. The core of the corpuscle consists of about 60 bilaterally arranged, closely packed lamellae, placed on both sides of the central nerve terminal and separated by two longitudinally running clefts. Nucleated cell bodies are situated in the outermost part of the core, at its junction with the intermediate zone. From these cells there arise cylindrical cytoplasmic arms which fold into the longitudinal clefts. Here they give rise to flattened sheet-like processes which pass to one or both sides to form the core lamellae and interdigitate with processes from other cytoplasmic arms. Adjacent lamellae do not arise from the same arm.

Each corpuscle is supplied by one or, rarely, two thick myelinated nerve fibres (Aα) derived directly from peripheral nerves without the intervention of the corial plexus. The fibre first loses its myelin sheath and at its junction with the core the Schwann cell sheath terminates. The naked axon traverses the

7.29G Ultrastructure, near the apex, of a tactile corpuscle of Meissner in vertical section, showing flattened lemmal cells arranged horizontally, with their nuclei to the right of the field. The sectional profiles of the terminal nerve fibres, appearing between the lemmal cells, contain numerous dense mitochondria. (Magnification about × 5000.)

7.29H Ultrastructure of terminal nerve fibres in close association with a hair follicle in the auricle of a rat. Note, above, part of the nucleus and cytoplasm of a keratinocyte; the plasma membrane adjoins a well-defined basal lamina, and presents a series of hemidesmosomes. The two nerve fibres contain numerous mitochondria and are enveloped by Schwann cell processes. Surrounding the latter are basal laminae, reticulin fibres (above) and collagen fibres (below). Magnification × 35,000. (G and H respectively from Ciba Foundation Symposium on *Touch, Heat and Pain*, 1966, and *J. comp. Neurol.*, **136**, 1969, by kind permission of Prof. N. Cauna and the publishers, J. & A. Churchill and the Wistar Press.)

length of the central axis core, usually without division or branching and terminates in an expanded end bulb. It is in contact with the innermost core lamellae and is oval in transverse section, with the longer axis in the plane of the longitudinal clefts between the core lamellae. It contains numerous large mitochondria, the more superficial of which are usually arranged radially in palisade fashion beneath the axolemma. Minute vesicles of about 5 nm diameter are also present; these tend to be aggregated opposite the longitudinal clefts. The cells comprising both the capsule and core lamellae are believed to be modified fibroblasts and appear to be distinct from Schwann cells (Pearse and Quilliam 1957). Butyryl cholinesterase has been demonstrated in the core of the lamellated corpuscle. Lamellated corpuscles are said to be closely associated with glomeral arteriovenous anastomoses and to derive their blood supply from the capillaries accompanying the nerve fibre to the capsule at the site of entry. A condensation of the surrounding fibrous tissue forms an external capsule to the whole corpuscle.

Lamellated corpuscles commence to develop in the third month of fetal life; the nerve terminal becomes surrounded by capsular lamellae which continue to be laid down into adult life, thus increasing the size of the corpuscle (Cauna and Mannan 1959).

Lamellated corpuscles are believed to possess considerable turgidity due to fluid pressure between the capsular lamellae. Extensive electrophysiological studies have shown this type of ending to be a very rapidly adapting mechanoreceptor, responding only to sudden mechanical disturbances and particularly sensitive to vibration (Loewenstein 1971; Gray and Sato 1973). It is thought that the extreme rapidity with which the ending adapts to stimuli is partly due to the lamellated capsule, which acts like a frequency 'filter', damping the effects of slow distortions by virtue of the fluid movements between the lamellar cells which only allow rapid movements to be transmitted to the central axon. If the capsule is removed, the selectivity of the response, although still quite rapidly adapting, is much less extreme.

Other lamellated endings include a wide variety of encapsulated receptor terminals, described in various species and different sites in the body. Since there are very few electrophysiological studies of such endings, it is difficult to establish whether they represent many distinctive categories or varieties of a few. However, the structures of some of them have been described in detail, and have been shown to possess a circumferential, concentric, lamellar pattern similar to that in the Pacinian endings, though perhaps less regular, and certainly much smaller. Amongst such terminals are the paciniform endings present in the connective tissue sheaths of vibrissae in cats and some other mammals (Gottschaldt *et al.* 1973), in joint capsules, and other connective tissue sites (*see* Poláček and Halata 1970); other endings with a similar organization are the *end bulbs of Krause, bulbous corpuscles, genital corpuscles*, and probably the '*innominate corpuscles*' and *Golgi-Mazzoni endings* (*see* Bannister 1976).

Ruffini terminals (Type II slowly-adapting cutaneous mechano-receptors) are found in the dermis of hairy skin, and each consists of the highly branched, unmyelinated endings of an Aα (or II) myelinated afferent, enclosing and invading a bundle of collagen fibres and enclosed in a cellular capsule (Chambers *et al.* 1972). Although less tightly organized, the structure has many characteristics in common with the neurotendinous ending of Golgi (*vide infra*), and appears to be somewhat similar electrophysiologically, being responsive to stresses set up in the dermal collagen. Similar structures appear to be present in joint capsules.

Neurotendinous endings of Golgi (Golgi tendon organs) are chiefly sited near the junctions of tendons and muscles (7.31). Each is about 500 μm long and 100 μm in diameter and consists of small bundles of tendon fibres enclosed in a delicate capsule. The tendon fibres (*intrafusal fasciculi*) are less compact than those elsewhere in the tendon, the collagen fibres are smaller and the associated tendon cells are larger and more numerous. The sheath or capsule consists of concentric sheets of cytoplasm, each about 100–300 nm thick, belonging to capsular cells. Within each layer the capsular cells are closely opposed and successive layers are

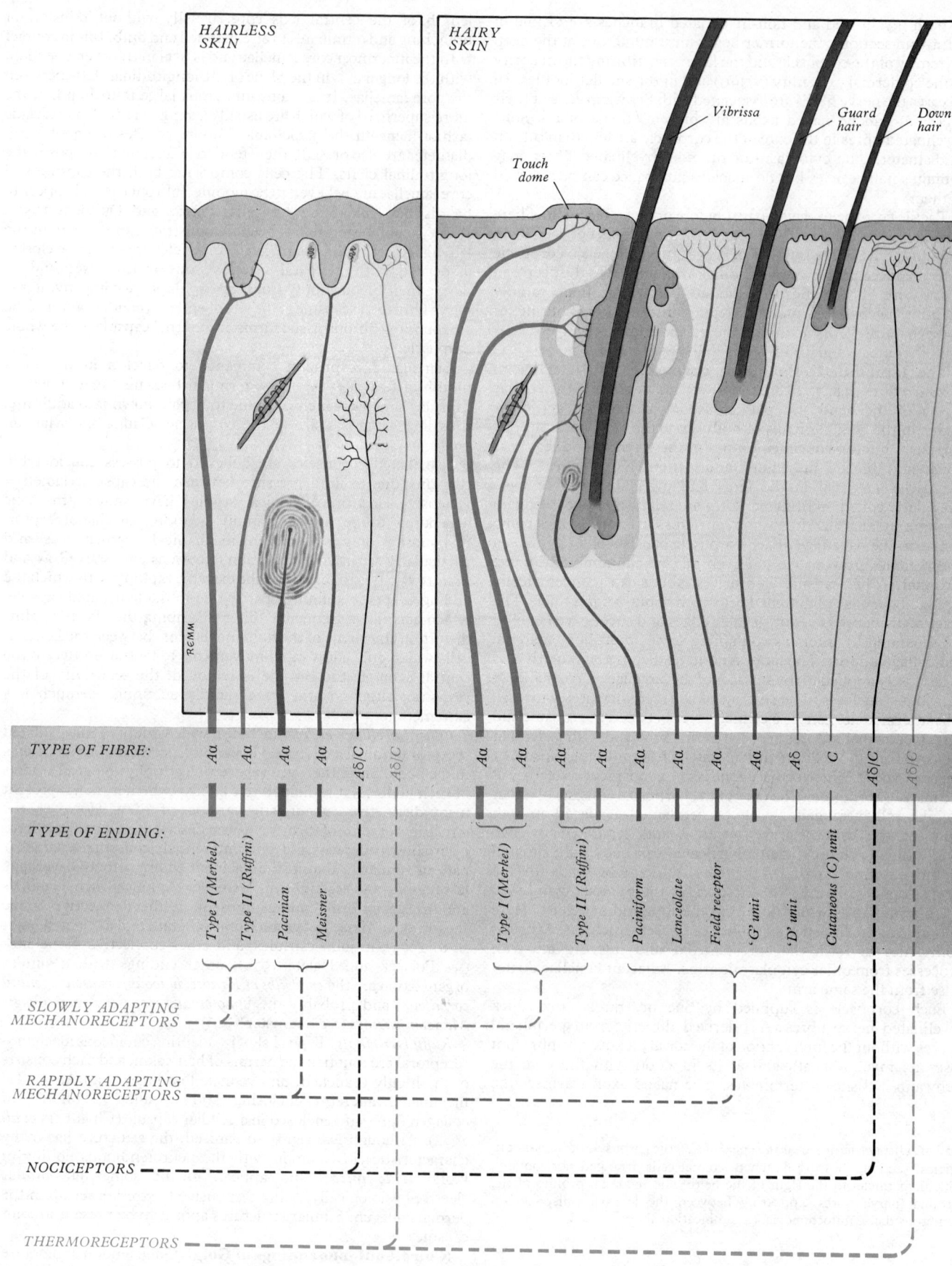

7.30 A schematic survey of the major types of mammalian cutaneous sensory endings and their afferent fibres. The endings are depicted symbolically according to their structural features, and their mode of response to specific types of stimulus are indicated by colour coding, i.e. slowly-adapting mechanoreceptors, green; rapidly-adapting mechano-receptors, red; nociceptors, black; thermoreceptors, blue. The class of afferent fibre is labelled according to the classification of Erlanger and Gasser referring to relative conduction velocities and fibre diameter (see text and accompanying table). The endings typical of hairless skin are shown on the left, and those of hairy skin on the right.

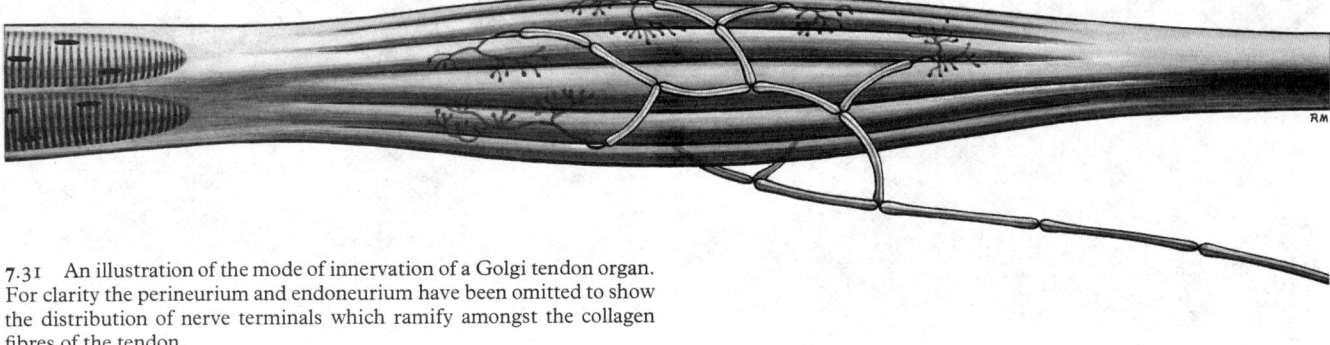

7.31   An illustration of the mode of innervation of a Golgi tendon organ. For clarity the perineurium and endoneurium have been omitted to show the distribution of nerve terminals which ramify amongst the collagen fibres of the tendon.

separated by intervals of varying width containing basal laminae. Numerous minute endocytic vesicles are present, suggesting that the capsular cells may be concerned with the maintenance of the internal environment of the ending. Outside the capsule is a thin layer of collagen fibres. One or more heavily myelinated group I*b* nerve fibres pierce the capsule and divide in a spray-like manner. The ramifications, which may lose their Schwann cell sheaths, terminate in leaf- or clasp-like enlargements, rich in small vesicles and mitochondria, which are wrapped around the tendon fasciculi with the basal lamina or Schwann cell cytoplasm intervening (Schoultze and Swett 1972, 1974).

The endings are highly active when tendons are stretched either actively or passively, and initiate myotactic reflexes which inhibit the development of excessive tensions during muscular contraction. Their responses are of the slowly adapting type, signalling maintained tension particularly well (*see* Matthews 1973). Structural and physiological evidence indicates that these sensory endings may be deformed by the surrounding collagen fibres which tend to lie more parallel and closer together as tension develops in the tendon.

Neuromuscular spindles are, somewhat arbitrarily, treated below under: Special configurations of sensory endings—p.858.

Special Configurations of Sensory Endings

The different varieties of sensory endings have now been discussed individually. Physiologically, however, it is necessary to consider *arrays* of several types of ending, which act in concert, to provide information about the forces and influences acting at a particular locality, to be analysed together in the central nervous system. Although our knowledge of such multiple patterns is as yet only partial, at least three situations have been explored in some detail. These are cutaneous and joint receptors, and neuromuscular spindles.

CUTANEOUS ARRAYS

The skin is a major sensory organ, containing the receptor endings of multitudes of sensory nerve fibres which report on the mechanical and thermal state of the body surface, including the presence of noxious or damaging stimuli. The way in which this array of sensory endings is able to discriminate between different types of stimulus has been a matter of constant debate; in the second half of the last century and the first half of the present one it was widely supposed that the different types of end-organ made visible by the current histological methods corresponded to specific sensory modalities, e.g. touch, pressure, and cold, although conclusive physiological evidence was lacking. However, in the 1950s it was pointed out that various areas of the body which are sensitive to different stimuli lacked complex end-organs (e.g. Weddell, Palmer and Pallie 1955). It was then suggested that each sensory ending might be responsive to a range of stimuli, and that small differences in the optimum sensitivities of a large number of fibres would be detectable as a pattern of activity by the

central nervous system, each pattern being characteristic for each stimulus category (*see* Sinclair 1967).

Although this idea has considerable attractions, subsequent work with improved electrophysiological and electron microscopic techniques has reinstated the concept of a wide range of specific nerve endings each highly sensitive to a particular type of stimulus (*see* e.g. the reviews by Brown 1973; Burgess and Perl 1973; Iggo 1978), although these are not always separable by their structural features. Although the types of ending fall into three major categories, that is mechanoreceptors, thermoreceptors and nociceptors (responding to potentially or actually damaging stimuli), several subdivisions of these have also been discovered (*see* 7.30). Thus within the mechanoreceptors are rapidly adapting endings which respond only, or chiefly, to *changes* in mechanical stimulation, and slowly adapting endings which report on steady deformations of the skin. Examples of the first class are the palisade endings close to the follicles of hairs, Pacinian endings in the subcutaneous tissues, and Meissner or 'field' endings adjacent to the epithelium; the second class of receptors includes 'Type I' slowly adapting cutaneous mechanoreceptors, or Merkel endings which are generally associated with thickened patches of epidermis ('touch domes') giving information concerning steady pressures operating perpendicularly to the skin surface, and the 'Type II' slowly adapting cutaneous mechanoreceptors sensitive to steady stresses in the dermis. Another rather different type of mechanoreceptor is the ending of a class of fine non-myelinated fibres which have a low threshold but require prolonged stimulation.

Thermoreceptor fibres with branched 'free' endings can be separated into those which increase their firing frequency on warming and those which are more active on cooling ('warm' and 'cold' fibres, respectively, supplying restricted areas of the skin corresponding to the 'warm' and 'cold' spots first described by Blix in 1882).

Nociceptor fibres have rather similar free endings, responding to various types of damaging stimulus signalled centrally as pain, discomfort or irritation of the skin. Some fibres only respond to extreme cold, others to extreme heat or gross mechanical insult; yet others are stimulated by combinations of noxious agents. Often such fibres may also be activated by the release of inflammatory substances in the tissues through which they pass, giving the experience of prolonged pain or itching (Iggo 1974).

The precise form of the sensory endings varies depending upon their location at the body surface, and indeed with the species of mammal studied, so that in man and other primates the special skin of the hairless areas of the volar and plantar surfaces of the hands and feet has a different pattern of endings from the rest of the body. Likewise the density of endings and therefore of the fineness of discrimination varies. In general, however, we can distinguish similar arrays of sensory ending in each, ranging from the complex encapsulated or tissue-associated endings of large diameter myelinated nerve fibres capable of carrying rapid information about mechanical stimuli of various kinds, to the free endings of fine myelinated or non-myelinated fibres conveying, more slowly, signals of thermal or damaging effects or of light mechanical stimulation. The entire complex clearly provides a

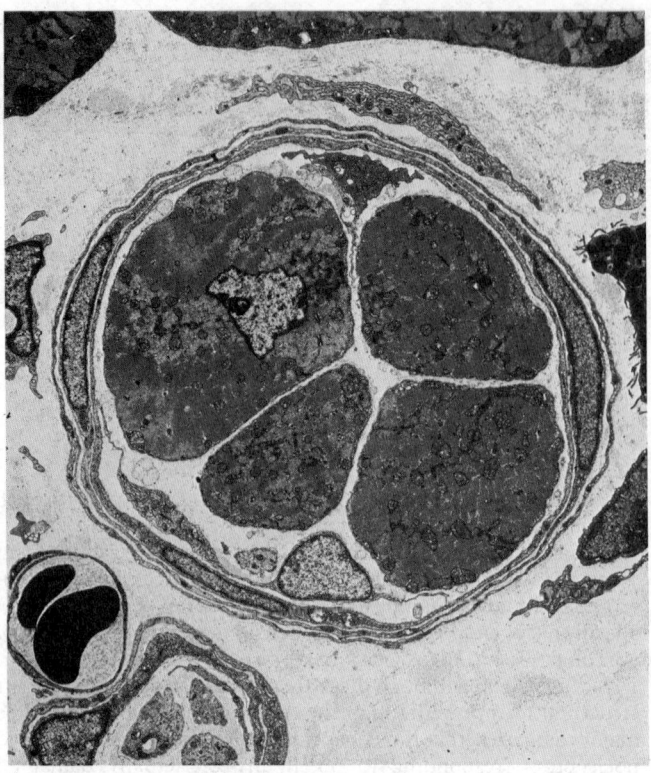

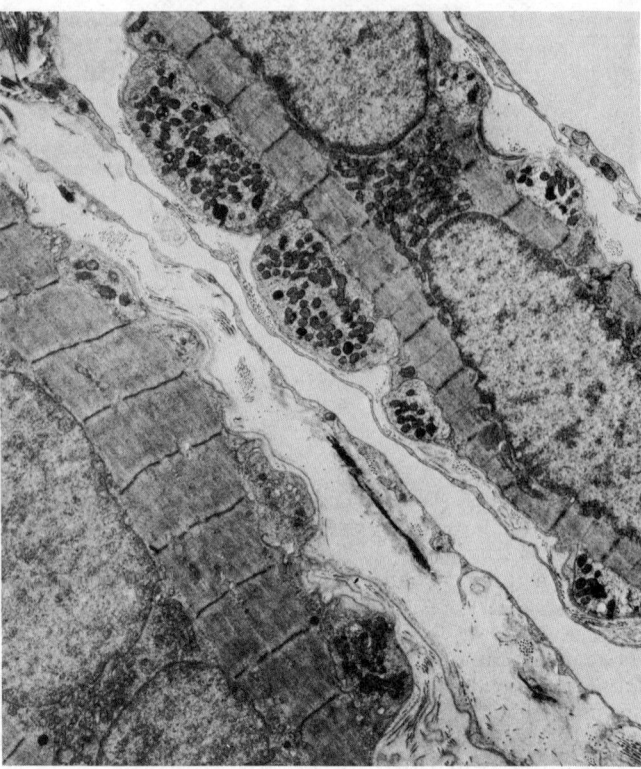

A

B

7.32A and B A An electron micrograph of a neuromuscular spindle of a rat, in transverse section, showing the capsule, capsular space and four intrafusal muscle fibres, one with a centrally positioned nucleus. B An electron micrograph of a longitudinal section through two intrafusal muscle fibres from a neuromuscular spindle of a rat. Note the primary (annulospiral) afferent nerve fibre endings cut in cross-section as they spiral around the equatorial region of a nuclear chain fibre (top right) and of a nuclear bag fibre (lower left). Note also the large numbers of mitochondria present in the sensory fibres. See text for further description. (A and B were kindly provided by Dr. D. N. Landon, National Hospital for Nervous Diseases, London.)

wealth of data about the state of the body surface, to be further analysed by the central nervous system.

JOINT RECEPTORS

The arrays of sensory endings placed within and around the articular capsules of synovial joints are of great importance in providing information about the position, movements and stresses acting on these structures (Wyke 1967). Both structural and physiological studies have shown that there are at least four classes of receptor terminals present, the proportion of each varying with the topographical position of the joint. Three of these classes comprise encapsulated endings and the fourth free terminal arborizations. Wyke has named these terminals types I to IV, and this classification will be followed here.

Type I endings are encapsulated spray terminals of the Ruffini type (*vide supra*) present in the superficial layers of the fibrous capsule in small clusters, supplied by myelinated afferent fibres of the group II category. Physiologically they are slow-adapting receptors, giving conscious awareness of joint position and joint movement, responding, it is thought, to patterns of stress in the articular capsule (Skoglund 1973). This type of ending is particularly common in articulations such as the hip joint, where static positional sense is of importance in the control of posture.

Type II endings are lamellated (Paciniform) receptors, similar to, but smaller than, Pacinian receptors found elsewhere in connective tissue. They are present in small groups throughout the joint capsule but particularly in the deeper layers and also, in the temperomandibular joint, in the posterior articular fat pads. They are rapidly adapting, low-threshold receptor endings, highly sensitive to movement and to pressure change, responding to joint movement and transient stresses in the capsule. They are supplied by group II or III myelinated afferent fibres; but they are not considered to be associated with conscious awareness of joint sensation.

Type III endings are identical to Golgi tendon organs in structure and physiology (p. 855), being present in the specialized ligaments of joints, though not in their capsules. They are high-threshold, slowly adapting receptors which apparently serve to prevent excessive stresses being developed at joints by the reflex inhibition of the adjacent muscular activity. Their innervation is from the large myelinated group Ib afferent nerve fibres.

Type IV endings are free, unencapsulated terminals of group III myelinated fibres and type IV non-myelinated fibres which ramify within the capsule, the adjacent fat pads, and around blood vessels particularly of the synovial layer. These are also high-threshold, slowly adapting endings which are thought to sense excessive joint movements and also provide a basis for the signalling of joint pain.

NEUROMUSCULAR SPINDLES

The *muscle* or *neuromuscular spindles* form an important sensory element in the control of muscle contraction, and as might be expected from the complicated nature of this control, the detailed structure and behaviour of spindles are also complex and not entirely understood. The precise anatomy of spindles varies greatly in submammalian species, and hence the present account applies only to mammals (Bowden 1966; Landon 1966; Matthews 1971; Barker 1974).

Basically, each spindle consists of a few, small, specialized *intrafusal muscle fibres*, which are innervated by both sensory and motor nerve endings (7.32A–C). This complex is surrounded in the equatorial region of the intrafusal fibres by an expanded *spindle capsule* formed of connective tissue, divisible into an *outer sheath* of flattened fibroblasts and collagen similar to those of the perineurium (p. 839), and also elements of an *inner axial sheath* (of Sherrington) which form delicate tubes around the individual intrafusal muscle fibres. The space between internal and external sheaths is occupied by fluid which is rich in hyaluronic acid.

The intrafusal fibres number from six to fourteen, varying from one muscle to another, and also with species. Two distinct types of fibre are present in most mammalian spindles, the *nuclear bag* and *nuclear chain fibres*, distinguishable by the arrangement of nuclei within the sarcoplasm in the equatorial region of the spindle. In nuclear bag fibres the equatorial nuclei are gathered together in a cluster to give a slight expansion of the fibre profile, whereas in the nuclear chain type the nuclei form a single longitudinal row in the centre of the fibre. The nuclear bag fibres are greater in diameter and much longer, extending beyond the capsule to attach to the endomysium of the surrounding extrafusal muscle fibres; the nuclear chain fibres are attached at their poles to the capsule or to the sheaths of the nuclear bag fibres.

The contractile apparatus of the intrafusal fibres is similar to that of extrafusal ones except that the zone of myofibrils is rather thin at the equator where they ensheath the nuclei. Ultrastructurally, the nuclear bag fibres are similar to 'slow' extrafusal fibres of amphibians, lacking the M lines, possessing little sarcoplasmic reticulum, but containing abundant mitochondria and oxidative enzymes: nuclear chain fibres are more like immature 'fast' twitch muscles of mammals, containing M lines, well-formed sarcoplasmic reticulum and T-tube elements, fewer mitochondria and lower oxidative enzyme levels. There is also evidence, of a histochemical and ultrastructural nature, of subtypes of muscle fibre in at least some species (Banks *et al.* 1975), termed 'second bag' or 'intermediate' intrafusal muscle fibres; this indicates that the organization of spindle fibres is even more complex than originally thought, and may provide even more subtle information concerning muscle activity. Nuclear bag fibres, when stimulated, contract more slowly than the nuclear chain type, supporting the ultrastructural findings.

Sensory innervation of muscle spindles is of two types, both of which are the non-myelinated terminal branches of two distinct classes of myelinated somatic sensory nerve fibres. The *primary* or *annulospiral* endings are placed centrally at the equator, and form spiral and sometimes annular terminals enwrapping the nuclear bag or chain portions of the intrafusal muscle fibre. A single large myelinated sensory fibre (group Ia), branches to provide the innervations of a number of intrafusal fibres in this way. Ultrastructurally, each terminal runs in a deep groove in the sarcolemma beneath the basement membrane, and it contains numerous mitochondria, vesicles, microfilaments, microtubules, and a pervading flocculent material. The *secondary* or *flower-spray* endings, which are largely confined to nuclear chain fibres, are branched terminals of slightly narrower myelinated group II afferents; these endings are beaded in appearance and spread in a narrow band on both sides of the primary endings near the equator. These beads, ultrastructurally, are expansions similar in content to the primary endings, and they lie in close apposition to the sarcolemma, although not in grooves. Electrophysiological studies have shown the primary endings to be relatively rapidly adapting whereas the secondary endings have a strongly regular, slowly adapting response to static stretch (*see* Matthews 1972; Hunt 1974).

Motor Endings in Muscle Spindles

Three types of motor terminals have been distinguished, two being the endings of fine fusimotor (γ-) efferents and one of a β-efferent nerve fibre. The first two comprise the *trail ending* nearest to the equatorial region, ramifying to form unspecialized cholinergic motor terminals with no obvious end plate or sole, and further pole-wards the P_2 *ending*, with a typical 'en plaque' end plate and sole. At the extreme ends of nuclear bag fibres is the P_1 ending, likewise forming a typical 'en grappe' end plate (p. 514), whose efferent nerve fibre arises as a collateral branch of a myelinated fibre supplying the extrafusal slow twitch muscle in the neighbourhood of the spindle. Stimulation of the efferent

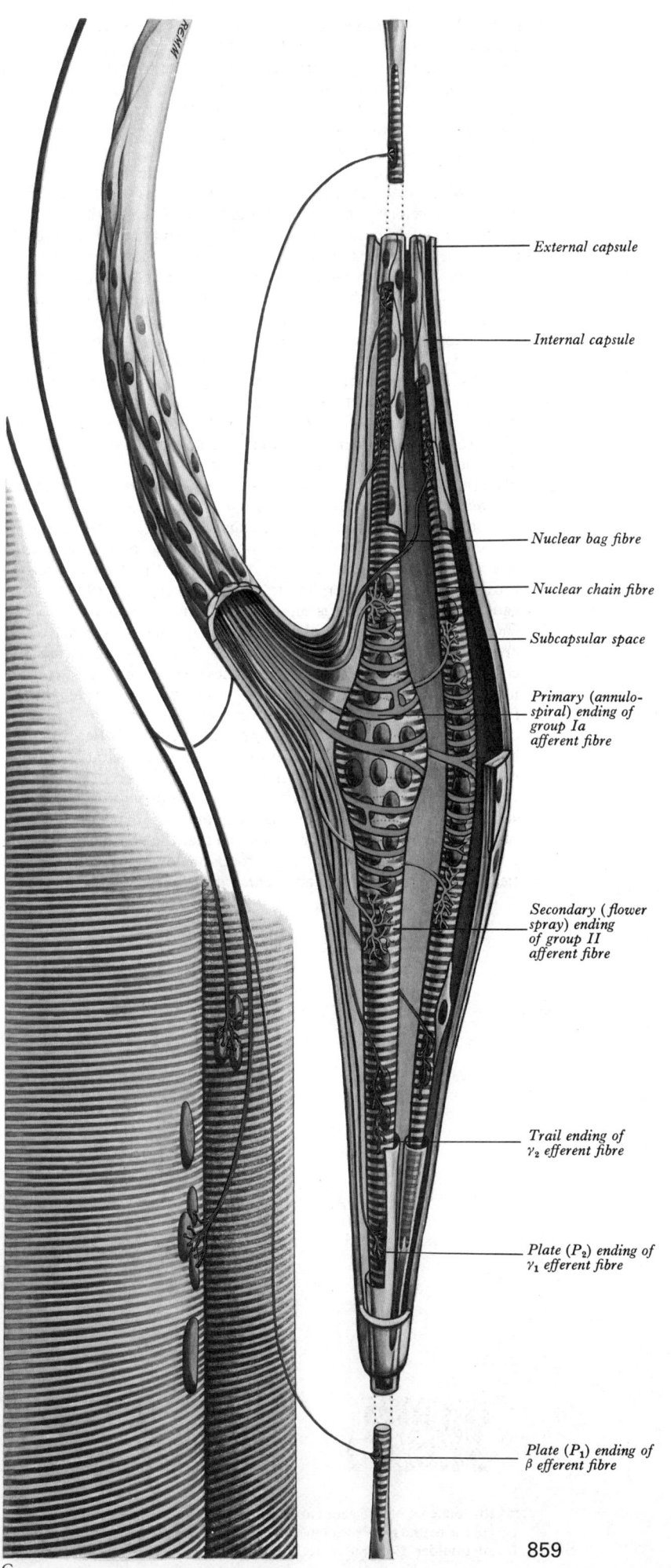

External capsule

Internal capsule

Nuclear bag fibre

Nuclear chain fibre

Subcapsular space

Primary (annulo-spiral) ending of group Ia afferent fibre

Secondary (flower spray) ending of group II afferent fibre

Trail ending of γ_2 efferent fibre

Plate (P_2) ending of γ_1 efferent fibre

Plate (P_1) ending of β efferent fibre

7.32C Schematic three-dimensional reconstruction of a mammalian neuromuscular spindle, showing nuclear bag, and nuclear chain fibres; these are innervated by the sensory annulo-spiral and 'flower-spray' terminals (blue) and by the γ and β fusimotor terminals (red). For further details consult text.

C

endings causes contraction of intrafusal fibres and corresponding activation of their sensory endings. There is some evidence that bag and chain fibres are stimulated to contract in rather different ways, and their fusimotor endings reflect this difference in the characteristic excitation of the two types of fibre. However, this duality is not entirely distinct, and further evidence of the functions of the bag and chain fibre is awaited.

The activities of the spindle appear to provide information on the length of the extrafusal muscle, its velocity of contraction, and changes in velocity, and it is thought that these modalities are related to the different behaviours of the two types of intrafusal fibres when stretched, either actively by their motor endings, or passively by stretching of the extrafusal muscle surrounding them. It is probable that because of the composition of sarcoplasm in the equatorial nuclear zone of the intrafusal fibres a passive stretch causes greater elongation in that region than in the fibres on each side where the myofibrils are more numerous, causing the response to deformation in the primary sensory endings to be different from that of the secondary endings which are placed further pole-wards. The equatorial region of nuclear bag fibres will also stretch more than that of the nuclear chain types because of the lower viscosity of the former, giving differential responses from the two. From physiological studies it is thought that *nuclear bag fibres* are concerned with *position*, *velocity*, and *acceleration* responses of a rapidly adapting, *dynamic* (phasic) nature, and the *nuclear chain fibres* with static, slow adapting ones (the *static response*). The fusimotor fibre can adjust the length of the intrafusal fibre, and therefore the activity of its sensory fibres by causing the polar regions to contract. This can compensate for shortening of the extrafusal fibres during normal muscle activity, or, as was formerly suggested, could be the normal means of causing extrafusal fibres to contract, these being stimulated by the intrafusal muscle. It was at one time envisaged that the γ fusimotor fibres, by causing contraction in the intrafusal fibres, could stimulate the sensory endings, which would then initiate contraction in the extrafusal muscle by reflex excitation of α-motor neurons (Granit 1970; Boyd 1962). More recent evidence, however, indicates that activity in the α-motor neurons may, in some cases of voluntary movement, precede that in γ-motor neurons during muscle contraction, so that the spindle probably monitors the *extent* of muscle contraction allowing *comparisons* to be made between *intended* and *actual movements* (Matthews 1971;

(Valbo 1974). It should be noted that although spindles have generally been considered to play no role in the conscious appreciation of limb position, this role being subserved by tendon, ligament and articular nerve endings, there is mounting evidence that spindles do indeed play a part in our awareness of muscle and limb movements and that spindle receptor activity is represented in the sensory cortex (Matthews 1977).

Development of neuromuscular spindles has been studied in detail by various investigators (*see* Zelená 1957; Landon 1972; Milburn 1973). The intrafusal fibres form by fusion of myoblasts in the usual manner, and the primary sensory endings are well established (in rodents) before birth, followed by the formation of fusimotor terminals, secondary sensory fibres, and the development of the capsular space. The early development of the spindle is absolutely dependent on the formation of the primary sensory ending (Zelená 1957), but later, spindles may survive denervation, although they may undergo marked changes even if crushed nerves are allowed to re-innervate them (Schiaffino and Pierobon Bormioli 1976).

The Natural History of Neurons

The structure of mature neurons considered in the previous paragraphs covers only a limited aspect of the total biology of these cells. Their mature structure must be considered in the context of the development, maturation, reactions to morphogenetic and pathological influences, senescence, and final death of these cells.

ORIGIN AND DEVELOPMENT OF NEURONS

Neurons originate from two main embryonic regions. Central nervous neurons stem from the neural plate and tube, whilst those of the cerebrospinal and autonomic ganglia arise from the neural crest (p. 161). The neural plate also gives rise to the ependymal and macroglial cells, whilst the neural crest produces the Schwann cells of the peripheral nervous system, and to the chromaffin cells of the suprarenal medulla.

The development of the cells of the central nervous system has attracted much attention since the first detailed descriptions by His (p. 157), and many conceptual schemes have been advanced to

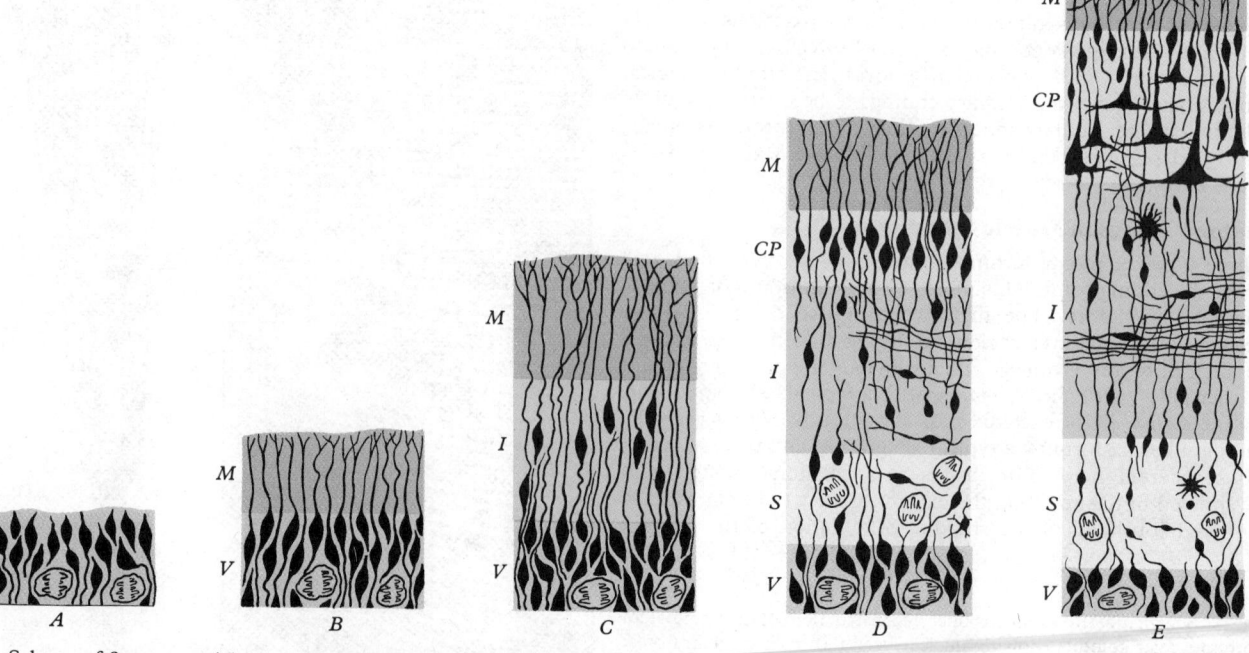

7.32D Schema of five sequential stages (*A to E*) in the development of part of the neural tube's wall in section, to form cerebral cortex. (Modified from: Boulder Committee recommendations, *Anat. Rec.*, **166**, 1970.)

Abbreviations: *V*—ventricular zone; *M*—marginal zone; *I*—intermediate zone; *S*—subventricular zone; *CP*—cortical plate. See accompanying text and also pp. 157 *et seq.*, for details.

explain the complex series of changes involved. Recently many of the difficulties of interpretation have been resolved by the use of autoradiographic methods, which demonstrate the origin and subsequent fate of the various cell types involved in the development of the nervous system.

In the central nervous system, the earliest observers considered the wall of the neural tube to be divided into ependymal, mantle and marginal zones (p. 157). This was later modified to include the concept of a matrix cell layer (p. 157). More recently (Sidman 1970), however, the latter has been further diversified into a fourfold division of the wall of the early neural tube, into *ventricular*, *subventricular*, *intermediate* and *marginal* zones (7.32D).

As described previously (pp. 157 *et seq.*), the neural plate and early neural tube consists of a single layer of pluripotent epithelial stem cells which give rise to all the cell types of the central nervous system except the microglia. The nuclei of these stem cells are situated near the ventricular surface in the *ventricular zone* of the tube, with an anuclear cytoplasmic marginal zone consisting of the laterally extending 'tails' of the cells. As these cells undergo mitotic division they change shape, contracting during the act of mitosis so that the nucleus moves first towards, and then subsequently away from the ventricular surface within the ventricular zone. Since the mitoses do not occur synchronously the nuclei appear at varying levels on histological section. Later, some of the progeny of these cell divisions cease mitosis permanently, and move deep to the ventricular zone to form an *intermediate* (mantle) *zone* where they differentiate into neuroblasts. Others, however, form a *sub-ventricular zone* between the ventricular and intermediate zones, where they continue to multiply, giving rise to further generations of neuroblasts and, later, glioblasts. Both of these cell types subsequently migrate into the intermediate and marginal zones. However, in particular regions (e.g. the cerebellar cortex, *see* p. 919), some of these mitotic subventricular stem cells migrate across the wall of the neural tube to assume a sub-pial position, where they establish a new zone of cell division and differentiation. Many of the cells' formed remain in this sub-surface position, but others migrate, yet again, through the developing nervous tissue to their various deeply-placed definitive positions where final differentiation into neurons and macroglia occurs.

During the genesis of these various cell types, neuroblasts are the first to be formed, followed later by glioblasts. The timing of these events differs in the various parts of the central nervous system and also varies between species. The majority of neuroblasts are formed before birth in mammals, although a number of examples of postnatal neurogenesis are also known, namely, the granule cells of the cerebellar cortex, olfactory bulb, and hippocampus. In contrast gliogenesis continues after birth both near the ventricular surface and in other more deeply placed regions.

Autoradiographic studies have shown that particular classes of neurons develop at specific times. Usually large neurons differentiate before small ones; however, the subsequent pattern of migration appears to be independent of the time of initial formation. For example, the cerebellar Purkinje cells form before the main body of granule cells, which arise in the sub-pial zone and later migrate towards the ventricle via the zone of maturing larger neurons. Other studies have shown that neurons can migrate for considerable distances through populations of maturing, relatively static cells, before arriving at their final destination. It is considered that the early migration and differentiation of neurons is determined by intrinsic factors, or factors in their immediate vicinity, since at this stage they have no afferent connexions which might determine their paths of migration. Later, however, the final form of their neurites, cell volume, and indeed the continuance of the cell itself, depends upon the establishment of the correct patterns of connexion with other neurons. (*See* also pp. 802–816.)

STRUCTURE OF DEVELOPING NEURONS

Initially the neuroblast is a rotund or fusiform cell, the cytoplasm of which bears a prominent Golgi apparatus, many lysosomes, glycogen deposits and numerous unattached ribosomes (Tennyson 1969). As maturation proceeds the cell sends fine cytoplasmic processes (neurites) into the surrounding regions. The latter are rich in microfilaments and microtubules, and centrioles frequently occur at the bases of the neurites where microtubules are being formed (p. 16). Internally, further maturation is marked by the development of the membranes of the endoplasmic reticulum, and the proliferation of attached ribosomes and mitochondria; glycogen is progressively reduced in maturing cells. With development one of the neurites becomes differentiated as the axon, whilst the other cell processes extend to establish the early dendritic tree. The growth of axons has been extensively studied, particularly in tissue culture; the rate of longitudinal growth is initially quite rapid—up to 1 mm per day. The tip of growing axons and dendrites is marked by a bulbous enlargement, the *growth cone*, containing many microfilaments (p. 162) and membranous vesicles, with small surface filopodia ('microspikes') growing out in an 'exploratory' manner into the surrounding intercellular spaces (Tennyson 1970; Bunge 1976; Pfenninger and Rees 1976). Thus, like many other morphogenetic processes, axonal growth appears to depend upon the presence of microfilaments (*see* p. 16).

The direction taken by a growing axon appears to be governed in part by mechanical factors such as the micellar architecture of neighbouring intercellular matrix, or the presence of other axons and Schwann cells along which further generations of axons grow to form bundles, and hence central nerve fibre tracts, or peripheral nerve trunks (*see* p. 161). The final direction and termination of axons, however, appear to be related to the overall fundamental plan of the nervous system, and much evidence is available to indicate that their growth patterns are determined by embryonic chemical concentration gradients, 'field effects', the chemical recognition of the surfaces of other cells, and so forth (*vide infra*). Experimental work, particularly on lower vertebrates, in which it is possible to graft or manipulate parts of the embryonic nervous system such as the visual pathway from the retina to the optic tectum, indicates that axons are able to establish precise relationships with various target regions of the brain in spite of mechanical interference, as though they are guided both by their relationship to surrounding axons, and also by a chemical recognition in their areas of termination. Similar considerations apply to the motor innervation of skeletal muscle. The mechanisms whereby such specific connexions are established are as yet unknown in detail, but may involve the sensing of metabolites such as 'nerve growth factor' produced by the target tissues, and the recognition of particular cells by the composition of their surfaces (Sidman 1974; Moscona 1976), impulse firing characteristics, and so on (but see also pp. 161, 162). There is also evidence (Hughes 1968) that if, during embryonic development, the axon fails to establish the correct contacts, its parent perikaryon atrophies and finally disintegrates; such mechanisms may explain the close numerical relationship between the number of neurons in a given motor pool, and the number of muscle cells which it innervates (Tennyson 1969).

The final growth of the dendritic tree is also influenced by its pattern of afferent connexions and their activity, and, if deprived of these experimentally, the dendrites fail to develop fully, and may be permanently affected, e.g. in the visual system of young animals lacking a normal type of visual environment (Blakemore 1974). Various metabolic factors are also known to affect the final branching pattern of dendrites; for example, thyroid deficiency in perinatal rats results in cerebral cortical neurons which are smaller and have less profuse branching of their dendrites than those of control animals. This is probably analogous to the mental retardation of cretinism (Eayrs 1955).

Once established, however, the dendritic tree in broad outline appears to be remarkably stable, and partial deafferentation affects only the detailed organization of dendritic spines or similar surface projections. If, however, cells lose their afferent connexions completely or are deprived of the normal sensory input (*see* Guillery 1974), atrophy and degeneration of much of the dendritic tree, and even of the whole neuron may ensue, although different regions appear to vary in the quantitative nature of such deprivative responses (*anterograde transneuronal degeneration—see* p. 815). Similar effects are also often seen in retrograde transneuronal degeneration. Thus, as development proceeds there is a gradual loss of plasticity so that soon after birth the neuron is a stable structure with a reduced rate of growth. After this two main types of alteration occur. The first follows physical or chemical trauma to any part of the neuron, the second involves the ill-understood changes which accompany the establishment of temporary or permanent memory traces.

Reaction of neurons to physical trauma has been studied most extensively in motor neurons with peripheral axons, which are convenient of access, but also centrally where their axons form well-defined tracts, and incidentally the latter provides the basis for the various degenerative techniques for exploring neuronal connectivity (*see* p. 815). When an axon is crushed or severed, changes occur on both sides of the lesion (Nauta and Ebesson 1970). Distally, the axon initially swells, and subsequently breaks up into a series of membrane-bound spheres; the process begins near the point of damage and progresses distally. These *anterograde* changes, which also involve the axon terminals (*terminal degeneration*), continue to total degeneration and removal of the cytoplasmic débris (see also p. 820). Proximally, a similar series of changes may occur close to the point of injury, followed by a number of sequential (*retrograde*) changes in the cell body (Cragg 1970). The latter are firstly directed towards the removal of much of the original protein-synthesizing apparatus by autophagic lysosomal action. There is first a rise in cytoplasmic RNA which is associated in part with the increased synthesis of lysosomal acid phosphatases; this is followed by a dispersion of the large Nissl granules with a loss of affinity for stains in the cytoplasm (*chromatolysis*), reflecting a reduction of cytoplasmic RNA (*see* the review by Lieberman 1974). Subsequently, this is followed by the formation of new protein-synthesizing organelles which produce distinctive proteins, many of which are destined for the regrowth of the axon. This is further indicated by the movement of the nucleoli to the periphery of the nucleus, and the restoration of polysomal clusters in the cytoplasm, and the consequent return of staining affinity in the cytoplasm when prepared with Nissl's technique for light microscopy. During this process the synapses on the cell body may be temporarily retracted away from the neuronal surface membrane and glial processes invade the gap so formed (Matthews and Nelson 1975).

Where regrowth of the axon is possible, as in the peripheral nervous system, the presence of an intact endoneurial sheath near to and beyond the region of injury, is important if the axon is to re-establish satisfactory contact with its previous end-organ, or a closely adjacent one. The degeneration of the myelin sheath which occurs distal to the point of injury has already been mentioned (p. 815), and this is accompanied by mitotic proliferation of the Schwann cells, their progeny filling the space inside the basal lamina of the old endoneurial tube (Webster 1964). Further, where a gap is present between the severed ends of the nerve, proliferating Schwann cells emerge from the stumps (mainly the distal stump) and form a series of nucleated cellular cords (the *bands of Bungner*) which bridge the interval. These may persist for a long time, even in the absence of satisfactory nerve regeneration. The proximal part of the axon develops a terminal swelling from the surface of which many small axonal sprouts develop which grow into the surrounding tissues. The majority of these are ultimately abortive, but the successful one enters the proximal end of the endoneurial tube and grows distally in close contact with the surfaces of its contained Schwann cells. A process of *contact guidance* (p. 91) is involved between the growing tip of the axon and the Schwann cell surfaces in the endoneurial tube and, when present, those which form Bungner's bands. When the axon tip has reached and successfully reinnervated an end-organ, the surrounding Schwann cells commence to synthesize myelin sheaths, including typical junctional nodes of Ranvier and internodal Schmidt-Lanterman incisures. However, if regeneration is occurring in a mature animal in which general growth of the body and limbs has virtually ceased, there is no increase in the lengths of the internodal segments of myelin, which remain uniformly short (often about 200 μm). Before full *functional regeneration* can occur, however, a considerable period of growth of both axonal

diameter and myelin sheath thickness is necessary. When a high proportion of effective peripheral connexions has been established, the nerve fibre population eventually recovers a virtually normal nerve fibre diameter spectrum, but this does not occur with a lack of appropriate peripheral connexions or if the onset of regeneration is long delayed. Regeneration of central axons does not normally occur in mammals, perhaps because of the absence of definite endoneurial tubes, and the disorganized proliferation of macroglial cells which follows central nervous injury.

A second series of changes occurs in nerve cells throughout the life of the individual. In addition to the normal turnover and replacement of the cytoplasm and surface of the neuron which occurs continuously, there are the constantly changing patterns of afferent nerve fibre activity, circulating hormone levels and other factors which are important to changes in neuronal behaviour in different parts of the brain. In the normal individual, cyclic or intermittent processes such as those involved in feeding, reproduction, aggression and so on, are obviously connected with such alterations of activity. It may be envisaged that changes in the chemical micro-environment of the cell can alter its electrical firing pattern in a probabilistic manner and thus change the characteristic responses of whole cell populations. More subtle changes are probably involved at the biochemical level so that the neuron modulates its responses (and structure) with reference to its previous history. This is a probable basis of certain aspects of memory formation which have received much attention in recent years. Changes in RNA content and composition have been reported to accompany the learning of conditioned responses, and claims have been made that experimental inhibition of protein synthesis also specifically inhibits certain types of memory formation (Barondes 1969). Despite the interest of such findings, it is by no means clear what relevance they have to learning processes in the intact nervous system.

Finally, more drastic changes may occur; it is known that during fetal development many neurons degenerate and die. This process continues into post-natal life so that in senescence it has been estimated that up to 20 per cent of the original population of neurons is lost. In the embryo the loss of neurons can be seen as a process of 'thinning out' those which fail to establish proper functional connexions (Prestige 1970), or lose the competition with other neurons to innervate structures such as muscles (Hamburger 1975). In post-natal life the loss of neurons is not so easily understood; it has been suggested that it may be a similar process, perhaps related to the stabilization of behaviour patterns formed during the early years. Certainly, with increasing age, the logical efficiency of the nervous system and its reaction speeds deteriorate, and it seems probable that this is related in some way to the loss of neurons, although whether as a direct result is unknown.

TROPHIC INFLUENCE OF NEURONS

In addition to the functions of neurons in conduction and transmission, they have other important functions in the integration of the body during development, and in the continued post-natal maintenance of the other tissues. Of course, nervous tissue is not unique in this respect since most tissues influence the metabolism of surrounding cell masses in some fashion, but the effects of nervous tissue are perhaps more dramatic and far reaching than that of many other cellular components.

The most obvious relationship of this kind is the mutual dependence of motor neurons and the muscles which they innervate. If, during the course of development, a nerve fails to connect with a muscle, the nerve degenerates and likewise, if a muscle is denervated the muscle degenerates. If, however, the innervation of a slow (red) and a fast (white) skeletal muscle, each of which has distinct metabolic and physiological properties, is exchanged, the muscles change their structure and properties in accordance with the innervation, showing that the nerve determines the type of muscle and not vice versa (Buller 1970). However, in this case, the type of muscle is apparently determined chiefly by the firing pattern in its efferent nerve fibre, rather than by the release of a trophic chemical (Lomo and Westgaard 1974).

Trophic influences are best known in lower vertebrates, however, which are capable of regenerating whole limb structures after amputation, but only if the nerves to the limb are not destroyed (Hamburger 1968). In higher vertebrates, also, the trophic influences of axonal growth on dendritic trees is well known (p. 861), and it seems likely that many phenomena of sensory cell development, such as the specificity of gustatory sensory cells in the tongue, are under direct trophic influence from the afferent nerves (Rosenthal 1977).

It is interesting that in many invertebrates, also, regeneration of metameric segments is under direct neurosecretory control, and it has been suggested that the nervous system may have originated in the ancestral metazoans chiefly as a system for coordinating the regeneration and development of the body as a whole. Seen in this light, the neurosecretory activity of the hypothalamohypophyseal neurons may be a primitive survival from the distant past, other neurons being a subsequent specialization for more rapid types of communication.

MAJOR DIVISIONS OF THE NERVOUS SYSTEM

Although in essence a continuum, the nervous system may be divided for convenience of study into a number of parts, regions, or sub-systems. The *encephalon* or brain (7.33) and the *medulla spinalis* or spinal cord together form the *central nervous system*. Extended from this in pairs are twelve cranial and thirty-one spinal nerves, constituting a *peripheral nervous system*. This itself includes not only all the ramifications of these nerves, which mediate *somatic sensory* and *motor* functions, but also the entire complex of *visceral* or *splanchnic* nerves, connected to the central nervous system through the somatic channels, thus forming a *peripheral autonomic nervous system*. This division into central and peripheral parts of the nervous system is basically justifiable, for the latter consists of relatively simple conductors connecting the peripherally situated receptor and effector organs to each other through the complex intermediation of the brain and spinal cord. Though the latter also contains certain further extensions of the long afferent and efferent pathways deployed through the peripheral nerves, the particular significance of the whole central nervous system resides in extremely complex networks of interconnected neurons in which arise the appropriate patterns of response to the stimuli of both the external and the internal environment. This same highly intricate area of intermediation, between incoming patterns of afferent information and the emergence of suitable arrays of 'commands' to the effectors, is also the domain in which learning, memory and consciousness are intrinsic, each to a degree dependent upon the level of development of the central apparatus. Indeed, the elaboration of these activities, which clearly has increased along the lines of evolution, and especially along that leading ultimately to the human animal, is equally clearly related to a vastly increased population of central interneurons, rather than to changes in the peripheral afferent and efferent conductors. Nevertheless, the essential continuity and interdependence of all parts of the nervous system should never be overlooked.

The elements of which the nervous system is composed have already been described (p. 820). Whereas the peripheral nerves largely consist, except in their somatic and autonomic ganglia, of parallel and unconnected fibres, the central nervous system contains not only the terminations or beginnings of these peripheral axons but also very large numbers of nerve cells, together with their dendritic processes, the ramifications and synapses of which make up the greater part of the volume of the brain and spinal cord. Both parts of the nervous system also contain, of course, amounts of either connective tissue and Schwann cells or neuroglia, and numerous blood vessels. The central nerve fibres and cells are distributed in an organized manner, and one or the other frequently predominates in particular regions. The somata of neurons are usually (but not always) gathered together into masses, which are called *nuclei* and sometimes *ganglia*; since they are devoid of any covering of myelin they appear darker to the unaided eye than do collections of nerve fibres, except where the latter are non-myelinated. This difference is accentuated by fixation, and with alcohol treatment nerve cells aggregations appear somewhat grey, and myelinated nerve fibres almost white, giving rise to the crude but useful terms 'grey' and 'white matter' or substance, although their colours are more nearly buff and cream in material fixed by the more commonplace formalin technique. It is not to be supposed that 'grey matter' contains no myelinated fibres; a large proportion of central axons are myelinated, and since every group of nerve cells will contain at least their own fibres, the grey or buff colour is merely due to the relatively low proportion of myelin. Similarly, tracts of myelinated nerve fibres, identified by the unaided eye as 'white matter', may nevertheless contain small numbers of nerve cells, and where the fibres are largely or entirely non-myelinated the colour will be 'grey'. With these provisos the two terms are usefully employed in the grosser descriptive topography of the central nervous system.

The spinal cord occupies approximately the cranial two-thirds of the vertebral canal, and it is arbitrarily considered to be continuous with the lowest part of the brain, the medulla oblongata, just below the level of the foramen magnum. The first pair of spinal nerves emerge from the spinal cord immediately caudal to this. The walls of the cord are thick, the cavity of the

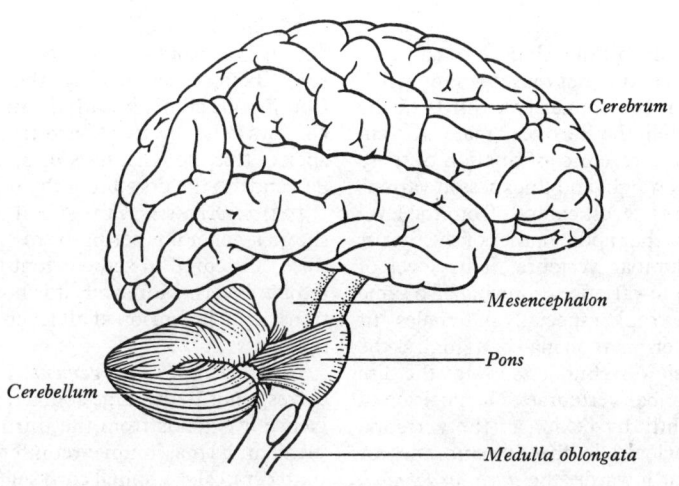

7.33   Semi-diagrammatic scheme of the main divisions of the brain.

central nervous system being here reduced to an almost microscopic *central canal*, extending nearly the full length of the cord.

The encephalon or brain (7.33), which is wholly within the cranial cavity, is itself described as consisting of a number of regions, which are of considerable morphological and functional significance. These are—in ascending order from the spinal cord—the **rhombencephalon** or hindbrain, the **mesencephalon** or midbrain, and the **prosencephalon** or forebrain. The rhombencephalon includes the **myelencephalon** or medulla oblongata, the **metencephalon** or pons, and the **cerebellum**. The prosencephalon is also subdivided into the **diencephalon** ('between brain'), which is the central connecting part of the forebrain, corresponding approximately to the thalamus and hypothalamus, and the **telencephalon**, which comprises the two so-called cerebral 'hemispheres' or cerebrum. The midbrain, pons, and medulla oblongata are collectively termed the *brainstem*, connecting the forebrain and spinal cord. The relation of these divisions of the brain to each other will be appreciated more clearly if their development is considered (p. 170). The pattern of fore-, mid-, and hind-brain is not only an expression of ontogenetic growth in the individual, it is also phylogenetic, inasmuch as it represents the basic central nervous hierarchy of vertebrates. These three successive levels are sometimes regarded as the 'segments' of the brain, and in extension of this idea the telencephalic part of the forebrain, the cerebrum, and particularly the cerebral cortex and its connexions, is described as 'suprasegmental', a term also applied to the cerebellum. Both are phylogenetically later outgrowths upon the basically elongated form of the primitive vertebrate brain.

The *medulla oblongata* is the most caudal part of the brainstem. It is immediately above the basilar region of the occipital bone, and is continuous with the pons above and the spinal cord below. The *pons* is also related to the basi-occiput and to the dorsum sellae of the sphenoid bone. It is markedly greater in transverse and anteroposterior dimensions than the medulla, from which it is easily distinguished by the large lamina of transverse nerve fibres protruding from its ventral apsect. The *cerebellum*, consisting of paired lateral parts, or *hemispheres*, united by a median *vermis*, is dorsal to the pons and medulla, occupying most of the posterior cranial fossa. Ventrally it is continuous with the midbrain, pons, and medulla. The cavity of the rhombencephalon or hindbrain is expanded as the *fourth ventricle*, which occupies the dorsal parts of both the pons and the cranial half of the medulla, being continuous with a fine canal in the caudal part of the medulla, and through this with the central canal of the spinal cord. The fourth ventricle contracts at its cranial end into a narrow channel in the mesencephalon, somewhat inappropriately termed the *cerebral aqueduct*. The *midbrain* is a relatively short part of the brainstem and somewhat constricted in comparison with the pons. The *diencephalon*, which is almost completely embedded in the cerebrum and thus largely hidden from surface view, contains a narrow, vertical *third ventricle*, which communicates caudally with the aqueduct.

The *cerebrum* is the most cranial or rostral part of the brain, and it accounts in mankind for much the major fraction of its volume. It occupies the anterior and middle cranial fossae and is directly related to almost the whole concavity of the vault of the skull. It is described as consisting of two large convoluted *cerebral hemispheres*. In fact, the halves of the cerebrum are *not* hemispherical; *together* they do form roughly a hemisphere, and perhaps the term was originally used in the singular, but the plural usage, however unsuitable as regards shape, is customary. Each hemisphere is roughly equal to a quarter of a sphere in shape and contains a large, crescentic *lateral ventricle*, continuous medially with the third ventricle in the diencephalon. Each hemisphere has an external layer of grey matter, the *cerebral cortex*, and a central core of white matter, or *medullary substance*, in which are several large masses of grey matter, the *basal nuclei* or *'ganglia'*.

As already mentioned, the *peripheral nervous system* contains a somatic component, the *cerebrospinal nervous system*, and a visceral component, the *autonomic nervous system*. In the former the efferent nerve fibres pass directly from their cells of origin in the central nervous system to the effector organs, muscles. The efferent autonomic fibres, however, do not do so, but terminate in ganglia outside the central nervous system, where they form synapses with neurons whose axons pass onwards to innervate non-striated muscle and glandular tissue. Autonomic efferent pathways thus consist of two orders of neurons termed, for obvious reasons, *preganglionic* and *postganglionic* (although the variety of neuron types in automatic ganglia is now known to be much greater than this simple dualistic terminology implies, *see* p. 1124). The arrangement of afferent fibres is, on the other hand, similar in both cerebrospinal and autonomic systems.

The autonomic nervous system consists of *sympathetic* and *parasympathetic* divisions. Preganglionic sympathetic efferent fibres issue from a region of the spinal cord limited to the segments between the first thoracic and second or sometimes third lumbar levels. Preganglionic parasympathetic efferent fibres emerge from the central nervous system only in certain cranial nerves (oculomotor, facial, glossopharyngeal, vagus and accessory) and in the second, third and fourth sacral spinal nerves. These two groups of autonomic efferents are therefore usually designated the *thoracolumbar* (sympathetic) and *craniosacral* (parasympathetic) *outflows*. A detailed description of the autonomic system appears on p. 1121.

THE CENTRAL NERVOUS SYSTEM

THE SPINAL MEDULLA OR CORD

The spinal cord (medulla spinalis) is the elongated, approximately cylindrical part of the central nervous system which occupies most of the diameter of the cranial two-thirds of the vertebral canal. Its average length in the European male is 45 cm and its weight about 30 gm. (For a recent confirmation of these values, and comparison of length, weight and thickness at various levels, consult Barson and Sands 1977.) It extends from the level of the cranial border of the atlas to the caudal border of the first or cranial border of the second lumbar vertebra. This level of termination is, of course, subject to variation, showing also some correlation with the length of the trunk, especially in females (Jit and Charnalia 1959). The cord's termination may be as high as the lower third of the twelfth thoracic vertebra or as low as the disc between the second and third lumbar vertebrae. The position of the termination is elevated slightly by flexion of the vertebral column. The spinal cord is enclosed in three membranes or *meninges*; these are, from without inwards, the *dura, arachnoid*, and *pia maters*, which are separated from each other by *subdural* and *subarachnoid spaces*, the former being merely potential, the latter being occupied by the cerebrospinal fluid (p. 1050). Continuous cranially with the medulla oblongata, the spinal cord narrows caudally to a sharp tip, the *conus medullaris*. From the apex of this the *filum terminale*, a fine connective tissue filament, descends to the dorsum of the first coccygeal segment (7.34).

In *transverse width* the spinal cord varies from level to level. It shows a general tapering from cranial to caudal extremities, but this is obscured to some extent by enlargements at cervical and lumbar levels. Moreover, it is not cylindrical, being greater in its transverse dimension at all levels, and especially so in its cervical segments.

The *cervical enlargement* is the more pronounced and corresponds to the large spinal nerves supplying the upper limbs. Hence it extends from the third cervical to the second thoracic segment, its maximum circumference (about 38 mm) being in the sixth cervical. (A spinal cord segment is the region of attachment of one pair of spinal nerves.)

The *lumbar enlargement* similarly corresponds in level to the segmental innervation of the lower limbs, beginning at the first lumbar segment and extending to the third sacral, the equivalent *vertebral* levels being ninth to twelfth thoracic. Its greatest circumference (about 35 mm) is level with the lower part of the twelfth thoracic vertebra, beyond which it rapidly dwindles into the conus medullaris.

Fissures and *sulci* mark the external surface of the spinal cord through most of its length. An anterior median fissure and a posterior median sulcus and septum almost completely divide the cord into symmetrical right and left halves, joined across the midline by a commissural band of nervous tissue, in which is the central canal (7.35, 36).

The *anterior median fissure*, traversing the whole length of the ventral surface of the spinal cord, has an average depth of 3 mm, being deeper than this at more caudal levels. It contains a reticulum of pia mater, and immediately dorsal to it is a lamina of nerve fibres, the *anterior white commissure*. Perforating branches of the spinal vessels pass from the fissure into the commissure to supply the central region of the spinal cord.

The *posterior median sulcus* is much shallower, and from it a *posterior median septum* of neuroglia penetrates somewhat more than half way into the substance of the cord, reaching almost to the central canal. The septum varies in its anteroposterior extent from 4 to 6 mm, diminishing caudally as the canal becomes more dorsal in position and the cord itself contracts.

A *posterolateral sulcus* exists on each side of and a short distance from the posterior median sulcus, and along it the dorsal spinal nerve roots enter the cord. The white substance of the cord between the posterior median and posterolateral sulci on each side is the *posterior funiculus*. Through the cervical and upper thoracic segments the surface of this funiculus presents a further longitudinal furrow, the *postero-intermediate sulcus*, which marks the position of a septum extending into the posterior funiculus

and dividing it into two large tracts of fibres, the *fasciculus gracilis*, which is medial, and the *fasciculus cuneatus* which is lateral.

The region of the spinal cord between the posterolateral sulcus and anterior median fissure is the *anterolateral funiculus*, and this is further subdivided into *anterior* and *lateral funiculi* by the issuing anterior roots of the spinal nerves. The anterior funiculus lies medial to (and includes) the zone of emergence of the ventral roots, whilst the lateral funiculus lies between the roots and the posterolateral sulcus (7.36, 38). In the upper cervical segments a series of nerve roots emerges through the lateral funiculus on each side to form by their union the spinal part of the accessory nerve, which ascends in the vertebral canal lateral to the spinal cord to enter the posterior cranial fossa through the foramen magnum (7.59). It is composed of fibres which supply the sternocleidomastoid and trapezius muscles.

The *filum terminale* (7.34, 37), a fine filament of connective tissue about 20 cm long, descends from the apex of the conus medullaris. Its cranial 15 cm, the *filum terminale internum*, is surrounded by tubular extensions of the dural and arachnoid meninges and reaches as far as the lower border of the second sacral vertebra. Beyond this its final 5 cm, the *filum terminale externum*, is closely united with the investing sheath of dura mater, descending to an attachment to the dorsum of the first coccygeal vertebral segment. The filum, consisting mainly of fibrous tissue, is continuous at its cranial end with the pia mater of the spinal cord; adherent to the upper part of its surface are a few strands of nerve fibres which probably represent the roots of rudimentary second and third coccygeal spinal nerves. The central canal of the cord is also continued into the filum terminale for 5 or 6 mm. A particularly roomy part of the subarachnoid space surrounds the internal part of the filum; it is the site of election for spinal (lumbar) puncture.

Continuous with the cord at intervals along it are the paired dorsal and ventral roots of the spinal nerves (7.38). These cross

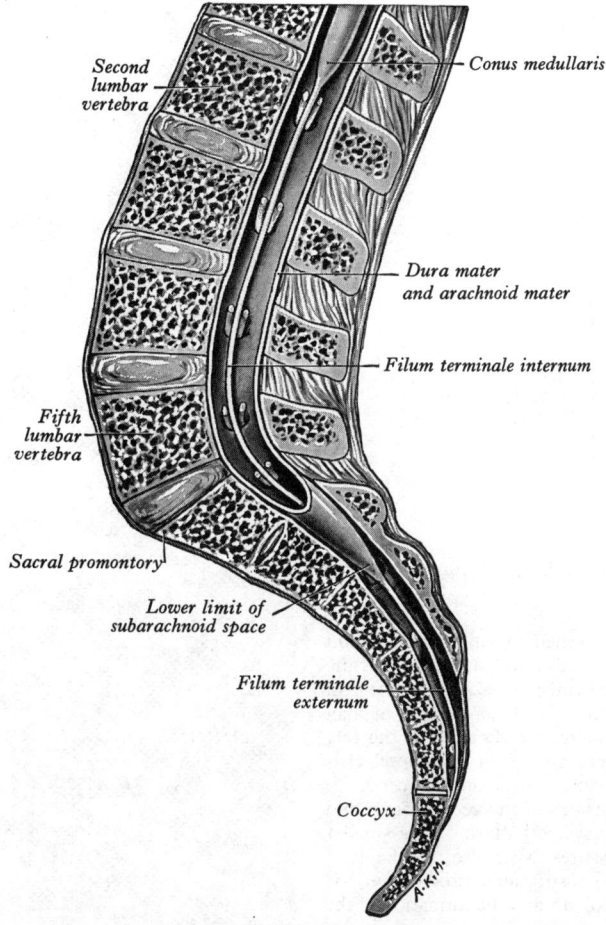

Second lumbar vertebra

Conus medullaris

Dura mater and arachnoid mater

Filum terminale internum

Fifth lumbar vertebra

Sacral promontory

Lower limit of subarachnoid space

Filum terminale externum

Coccyx

7.34 Median sagittal section of the lumbosacral part of the vertebral column to show the conus medullaris and filum terminale. The section has opened up the subarachnoid space as far as the first sacral vertebra. Note

the difference in levels between the inferior limits of the spinal cord and its meninges.

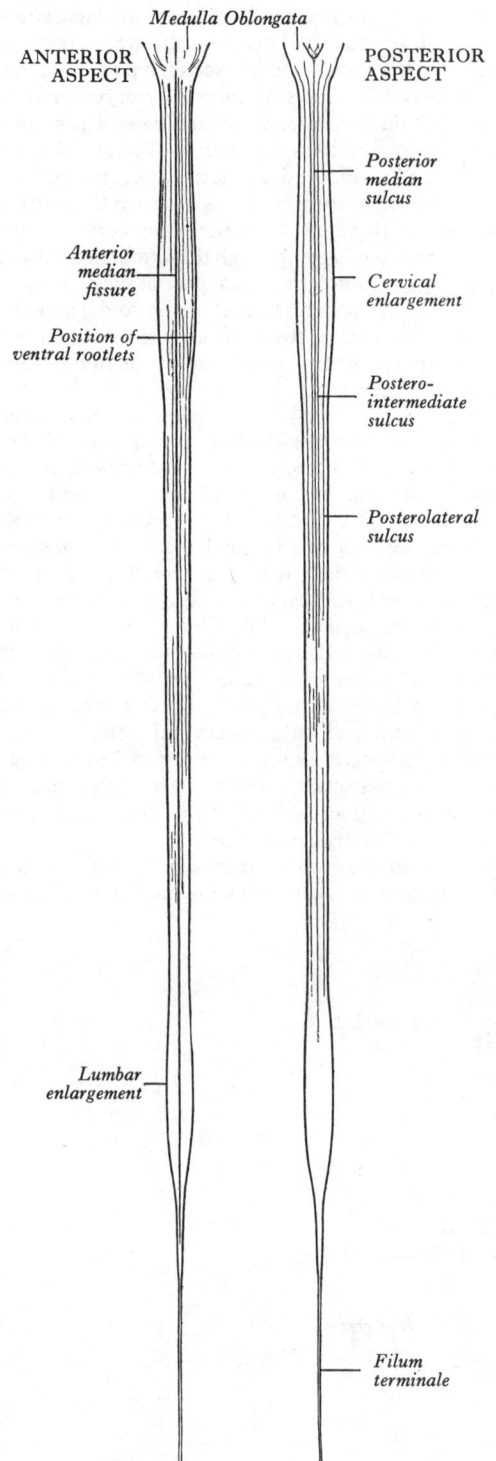

Medulla Oblongata

ANTERIOR
ASPECT

POSTERIOR
ASPECT

Posterior median sulcus

Anterior median fissure

Cervical enlargement

Position of ventral rootlets

Postero-intermediate sulcus

Posterolateral sulcus

Lumbar enlargement

Filum terminale

7.35A Diagram to show the main features of the spinal cord.

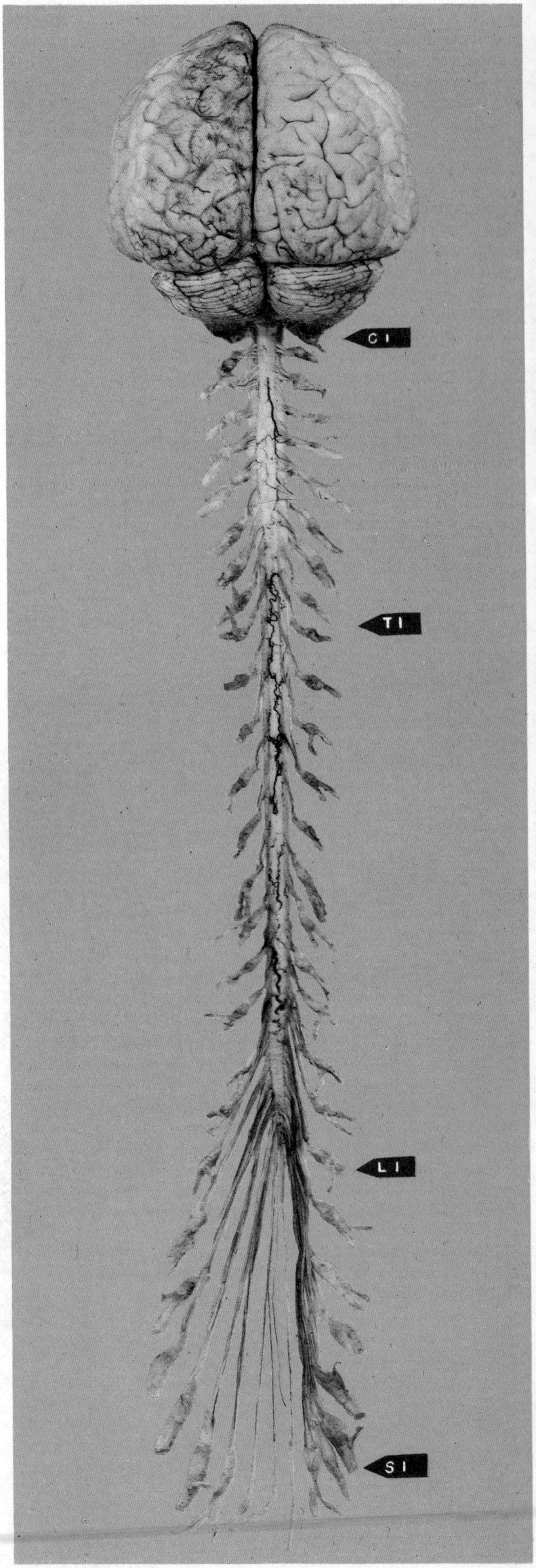

7.35B The brain and spinal cord with attached spinal nerve roots and dorsal root ganglia, photographed from the dorsal aspect. Note the relative sizes of the cerebral and cerebellar hemispheres, and the fusiform cervical and lumbar enlargements of the spinal cord. The median longitudinal fissure between the hemispheres which receives the falx cerebri and falx cerebelli is visible, together with the horizontal cleft between cerebrum and cerebellum which receives the tentorium cerebelli. Contrast the irregular pattern and dimensions of the cerebral gyri and sulci with the more regular, largely transverse pattern of the smaller cerebellar folia, and their intervening fissures. Note also the changing obliquity of the spinal nerve roots in their rostrocaudal progression, the stouter roots attached to the limb enlargements and the formation of the cauda equina and filum terminale. The cauda is undisturbed on the right, and has been fanned out on the left to facilitate identification of its individual components. (Dissection by Dr. M. C. E. Hutchinson, Department of Anatomy, Guy's Hospital Medical School.)

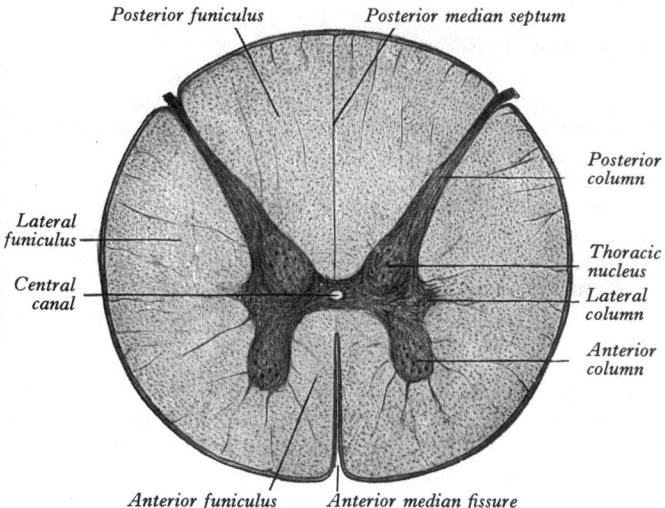

7.36 Typical transverse section of the spinal cord at a mid-thoracic level. Magnification about × 8.

the subarachnoid space, traverse the dura mater separately, and then unite in or close to their intervertebral foramina to form the spinal nerves. Since the spinal cord is markedly shorter than the vertebral column, the more caudal spinal roots descend for varying distances around and beyond the cord to reach their corresponding foramina; and in so doing they form, largely caudal to the apex of the cord, a divergent sheaf of spinal roots, the *cauda equina*, gathered around the filum terminale in the spinal theca. (For the developmental changes which occur in the human

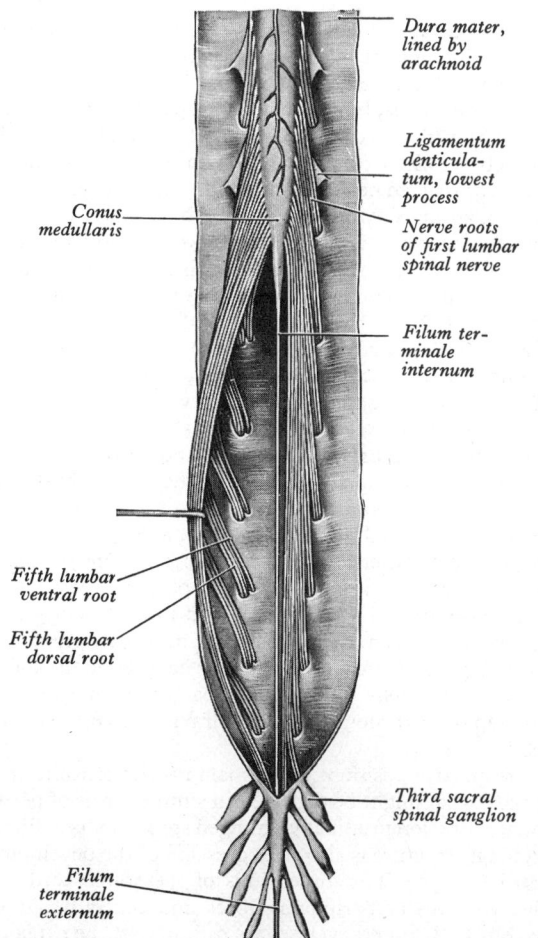

7.37 The lower end of the spinal cord, the filum terminale and the cauda equina exposed from behind. The dura mater and the arachnoid have been opened and spread out.

embryo's spinal cord consult p. 157 and Pearson and Sauter 1971.)

The *ventral spinal roots* consist of efferent somatic and, at certain levels, visceral (sympathetic) nerve fibres, the axons of which are emerging from their sources in the spinal cord. The *dorsal spinal roots* are characterized by ovoid swellings, the *spinal ganglia*, one on each root just proximal to its junction with a corresponding ventral root in its intervertebral foramen. Each dorsal root fans out into six to eight rootlets which enter the cord in a vertical series along the posterolateral sulcus. The dorsal roots are generally regarded as consisting entirely of afferent nerve fibres derived from unipolar nerve cells in the spinal root ganglia, but it has been suggested that a small number of them (3 per cent) are efferent (Young and Zuckermann 1936), and that autonomic vasodilator fibres may issue in dorsal roots (p. 1121). These views have not received wide support. Each ganglionic

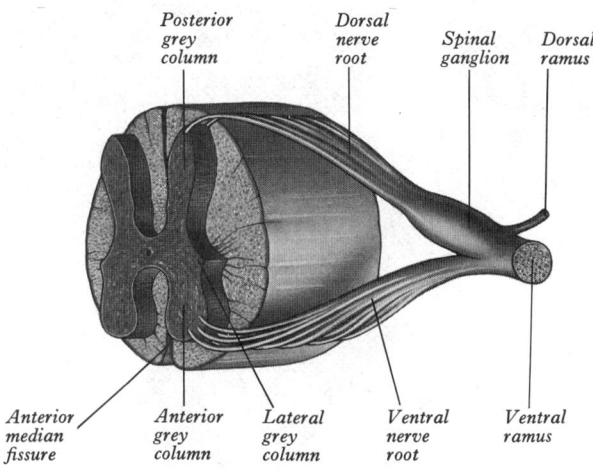

7.38 Diagram of a spinal cord segment showing mode of formation of a typical spinal nerve, and the gross relationships of the grey and white matter.

neuron has a single very short process which almost immediately divides into a medial process entering the spinal cord through the dorsal root and a lateral one which passes peripherally to some form of sensory end-organ. The central process is morphologically an axon, whereas the lateral or peripheral extension is derived from a dendrite. The region of the spinal cord associated with a given pair of spinal nerves is described as a *spinal segment*, but there is no clear indication of segmentation apart from this.

Internal Structure of the Spinal Cord

The internal structure of the spinal cord may be considered from several different but complementary points of view. Ignoring for the present the all-pervading supply and support tissues, blood vessels and neuroglia, the arrangement of the nerve cells and their processes can be studied by a variety of means, including dissection and unaided observation of sections, by light and electron microscopy, and by experimental techniques, and, of course by combinations of these. The first method reveals little more than the general layout of nerve fibres and cells; microscopy provides much more detailed information of cell types and fibre calibres and of their intimate disposition in the substance of the cord. While both light and electron microscopy have contributed greatly to knowledge of the interconnexions between neuronal elements, the potential of these techniques can only be fully exploited in conjunction with experimental manœuvres, such as degeneration, microelectrode and many other methods. We shall therefore consider spinal organization in a series of steps: (*a*) the large-scale arrangement of the grey and white matter, (*b*) the distribution of nerve cells in the grey matter, (*c*) the white substance and the deployment of tracts of fibres in it, and (*d*) the more detailed organization of the spinal neurons.

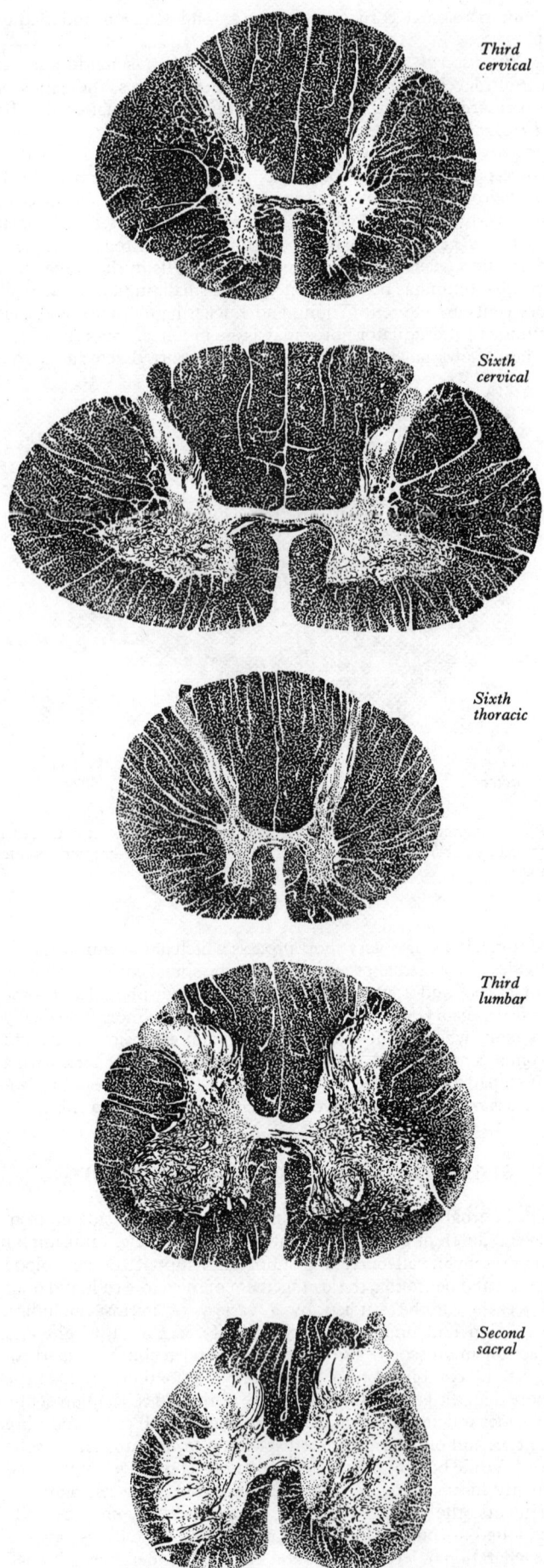

Third cervical

Sixth cervical

Sixth thoracic

Third lumbar

Second sacral

7.39 Transverse sections through the spinal cord at representative levels. Magnification about ×5. Note changes in overall profile, and the relative changes in grey and white regions.

GENERAL ARRANGEMENT OF GREY AND WHITE MATTER

The grey matter of the spinal cord is central in situation and has the form of a fluted column throughout, except where this pattern is modified at its continuation into the medulla oblongata and in its tapered caudal extremity, the conus medullaris. In transverse sections (7.39) this column consists of symmetrical right and left comma-shaped masses connected by a transverse *grey commissure*, the whole somewhat resembling a letter H. The commissure is traversed by the central canal, which may be just visible by unaided vision. Each of the lateral crescentic masses has a concavity directed laterally and can be regarded as having *anterior* and *posterior* parts or *columns* relative to the grey commissure. At some levels a small *lateral column* projects from the intermediate region of the concavity. In transverse sections of the spinal cord these various columns appear as more or less pointed projections and are hence commonly called 'horns'. This picturesque nomenclature is misleading and unnecessary, and the elongated divisions of the grey matter will be referred to as *columns* in this account. The central grey matter is surrounded by the white matter of the cord, the latter consisting largely of nerve fibres, many of which are longitudinal in direction and are therefore grouped appropriately into *funiculi* (little cords) or *white columns*. The general arrangement of these as dorsal, lateral and ventral funiculi has already been described above, and further details of the types and arrangements of nerve fibres in these white columns will be the concern of a subsequent section (pp. 874 *et seq.*).

The anterior or ventral grey column projects ventrally and somewhat laterally with respect to the grey commissure. It is comparatively short and broad and does not reach the surface of the spinal cord, being separated from this by the lateral part of the anterior funiculus (7.39). Its anterior and posterior limits are sometimes named its head and base—arbitrary terms of little value.

The posterior or dorsal grey column projects dorsolaterally, and in contrast to the anterior column it is transversely narrow and also extends almost to the surface near to the posterolateral sulcus, from which is it separated by a thin lamina of nerve fibres, the *dorsolateral tract* (p. 881). It also is considered to have a base, where it is continuous with the intermediate grey region, a constricted neck which expands into an oval or fusiform head, and an apex, which is capped by a mass of somewhat translucent nervous tissue, the *substantia gelatinosa (Rolandi)*. This crescentic mass of small nerve cells and fibres is intimately ·concerned with the connexions of incoming afferent nerve fibres, further details of which appear later (p. 887).

The lateral grey column is a small, angular projection extending from the second thoracic to the first lumbar segment of the cord and does not appear at other levels.

The boundary between white and grey matter is in most places definite, but in the cervical region of the cord strands of grey matter invade the lateral funiculus from the base of the dorsal grey column; these strands are separated by interlacing groups of nerve fibres, and this gives the arrangement the appearance of a loosely interwoven net, whence its name, the *reticular formation*, or formatio reticularis. Similar regions appear, in a less developed form, at more caudal spinal levels; and more recently reticular formations have been identified in the brainstem, in connexion with which physiological investigations have led to the concept of an extensive *reticular 'system'*, of great functional importance, widely deployed throughout the neuraxis. See pp. 943–953 for details.

The respective positions of the main masses of white and grey matter are what might be expected in simple terms of peripheral conductors and longitudinally arranged spinal tracts of fibres, but this general structure is also an expression of the development of the cord (p. 157). The dimensions of the spinal cord and the relative volumes of peripheral fibres and centrally aggregated nerve cells at different levels (7.39) can, in part, be explained by the amounts of muscle, skin and other tissues innervated by different segments, and by consideration of the fact that the fibres in the longitudinal tracts are on the whole inevitably more numerous as the cord is traced in a cranial direction. In the thoracic region the grey matter is absolutely and relatively small in volume, while the white substance shows a progressive increase in the ascending direction. In the cervical and lumbar enlargements the amount of grey matter, especially in the ventral columns, is much increased by the presence of large accumulations of nerve cells concerned in the innervation of the limbs; but, while the amount of nerve fibres at cervical levels is marked, in the lumbar enlargement and particularly the conus medullaris the white funiculi contain very many less fibres passing through these segments; 7.39 shows these details in several representative segments of the spinal cord, and it is obvious that various levels can easily be distinguished from each other. It is, however, more important to recognize the explanation of these differences in terms of the criteria noted above.

The central canal exists throughout the spinal cord and also extends into the caudal half of the medulla oblongata, opening out above this into the fourth ventricle (p. 932). In the caudal part of the conus medullaris it expands into a fusiform *terminal ventricle*, triangular in section with its base ventrally directed. It is about 8 to 10 mm in length, but tends towards obliteration above the age of forty years. At cervical and thoracic levels the canal is slightly ventral to the midpoint of the cord, central in the lumbar enlargement, and more dorsal in position in the conus medullaris. It extends for 5 or 6 mm into the filum terminale. During life it contains cerebrospinal fluid, and it is lined by a columnar, ciliated epithelium, the *ependyma*, which is encircled by a zone consisting principally of neuroglia but also containing a few nerve cells and fibres, the *substantia gelatinosa centralis*. This is traversed by processes spreading centrifugally from the basal aspects of the ependymal cells. The grey matter surrounding the central canal external to the gelatinous substance is the *grey commissure*. Ventral to the canal the commissure is thin, and further ventral still is a slender lamina of decussating nerve fibres, the *ventral white commissure* (p. 875). The grey commissure is traversed by two longitudinal veins (p. 896). The part of it dorsal to the canal is contiguous with the ventral edge of the posterior median septum; it is thinnest in the thoracic region of the cord and thickest in the conus medullaris; it is permeated by a variable number of transverse white myelinated nerve fibres which, collectively, are sometimes termed the *dorsal white commissure*.

Internal Structure of the Grey Matter

Like other parts of the central nervous system the spinal grey matter (7.40) consists of a complex intermingling of nerve cells and their processes with neuroglia and blood vessels. The predominance of the somata of neurons in relation to their myelinated processes, which are also present, is responsible for the so-called grey appearance, as we have seen above. The neuroglia (p. 834) forms a most intricate lattice among the nerve cells and their neurites, being particularly condensed in the central gelatinous substance around the central canal. The processes of the nerve cells will be described in detail later in this section in connexion with the tracts of the spinal cord (p. 875) and with the organization of its interneuronal networks (p. 882); they include axons arriving from or departing to the fibre tracts of the white funiculi, the commencement of efferent and the termination of afferent peripheral nerve fibres, together with collaterals from these sources and a most complex neuropil, the latter being composed of the innumerable neurites of nerve cells which are largely confined to the grey matter or at least to the spinal cord itself. Many cell processes cross the midline in the commissures, and the right and left halves of the cord, including its grey matter, are a functional continuum. The nerve cells in the grey substance are multipolar, varying much in size and other characteristics, and particularly in the length and morphology of their axons and dendrites. Many are Golgi types I and II nerve cells (p. 821), the axons of the former being long and passing out of the grey matter into the ventral spinal roots or the fibre tracts of the white matter. The axons and dendrites of the Golgi type II cells are largely confined to the neighbouring grey matter. Some neurons are *intrasegmental*, being deployed within the limits of a single segment; others spread through several segments, being

intersegmental in distribution. Details of this kind will, however, be described later (p. 895). We shall here be concerned only with the mode of arrangement of the actual nerve cell bodies or somata in the different regions of the grey matter.

In many parts of the central nervous system nerve cells are gathered together in groups, often in large numbers; and this usually indicates involvement in some particular common function. In some instances a large group exhibits division into smaller subgroups sufficiently constant in occurrence in different individuals to justify description and the application of specific names. The existence of such constant patterns of nerve cell distribution inevitably suggests functional implications, though the influence of developmental or growth processes may be of even greater signifance in many places. The nerve cells of the spinal grey matter are not distributed evenly or at random; they also occur in major and minor aggregations, some of which have obvious broad functional significance, while the meaning of others remains obscure or a subject of controversy. The following details of the cytoarchitecture of the spinal grey substance are described in the first place as purely topographical arrangements, and in doing this it is convenient to divide this account into sections dealing with the ventral and dorsal columns and with the intermediate region between them, including with the latter the lateral grey column. After this the functional interpretations of some of the subgroups in these columns will be considered in the light of the available evidence.

NERVE CELL GROUPS OF THE ANTERIOR GREY COLUMNS

The nerve cells of the ventral columns vary greatly in size; most prominent are large multipolar elements exceeding 25 μm in average dimensions of their somata, and with axons emerging as the fibres of ventral roots which innervate striated skeletal muscles as a-efferents. Large numbers of smaller nerve cells, of the order of 15 to 25 μm, also occur and the axons of some of these are the γ-efferents which innervate the intrafusal muscle fibres of neuromuscular spindles. On the evidence of retrograde degeneration experiments and other considerations, not all the smaller nerve cells in this column produce such efferents, and most in fact are interneurons (see also p. 883).

As in other parts of the spinal grey matter, the ventral column cells are basically arranged in elongate groups, a number of separate longitudinal columns extending through characteristic series of segments in the cord. This arrangement is perhaps seen most easily in transverse sections (7.40A) and is usually so studied; longitudinal sections are rarely described or depicted. In the latter, however, it is clear that these columns of nerve cells are themselves not uniformly continuous through the cord (7.40B), but are clustered in small aggregations, which are too diminutive to bear any segmental significance (Laruelle and Reumont 1933). The basic division of the ventral grey region is into three columnar groups—*medial*, *central*, and *lateral*—but all of these exhibit further subdivision at certain spinal levels, usually into dorsal and ventral parts. Thus a considerable nomenclature has accumulated, and while there is a satisfactory measure of agreement regarding the organization of the ventral column into cell groups, some confusion persists in the naming of these. The terms employed here are derived from an authority widely consulted (Crosby *et al.* 1962) but only recently (1977) have these terms been officially established in *Nomina Anatomica*. As will be more easily appreciated from illustration 7.41, the **medial group** extends through most segments of the cord, being perhaps absent from the fifth lumbar and first sacral. Through all the thoracic and the upper three or four lumbar segments it is divided into subsidiary **ventromedial** and **dorsomedial groups**, and in segments cranial and caudal to this the medial group is represented only by its ventromedial moiety, except in the first cervical segment, where only the dorsomedial group is considered to exist.

The **central group** of the ventral grey column is the least extensive and is identifiable only in some cervical and lumbosacral segments. In the cervical cord, through the third to seventh segments, is a centrally situated columnar group termed the phrenic nucleus; there is abundant experimental and clinical evidence that these nerve cells innervate the diaphragm, probably constituting the least controversial individual motor pool in the entire spinal cord. For experimental evidence in various mammals, including primates, *see* Kohnstamm (1898); Sharrard (1955); Keswani and Hollinshead (1956); Warwick and Mitchell (1956); Ullah (1978). The **lumbosacral nucleus**, traversing the second lumbar to first sacral segments, is also central in the ventral column, the distribution of its axons being unknown. Nerve cells whose axons are considered to enter the spinal part of the accessory nerve form an irregularly shaped **accessory group**, in the upper five or six cervical segments, occupying the ventral border of the anterior grey column in an intermediate or central position. They are, however, lateral to the dorsomedial group in the first cervical segment; the ventral siting of this nucleus may be coupled with the absence of the lateral groups from the first three cervical segments (7.41).

The **lateral group** of the ventral column can be further divided into **ventral, dorsal, and retrodorsal groups**, all of which are largely confined to spinal segments innervating the limbs. Their individual extents are indicated in illustration 7.41, and the significance of their arrangements will be discussed below (p. 872).

Onufrowicz (1899) described a ventrolateral group of motor neurons, in the first and second sacral segments, which he considered to innervate the perineal striated muscles. This so-called *nucleus of Onufrowicz* has been confirmed in mankind by Mannen *et al.* (1977) and by Konishi *et al.* (1978) in the cat.

NERVE CELL GROUPS OF THE POSTERIOR GREY COLUMNS

The cell groups of the dorsal region of the spinal grey matter comprise two which extend through the whole length of the cord and two limited to the thoracic and upper lumbar segments.

The **substantia gelatinosa** (of Rolando), present at all levels, consists chiefly of small Golgi type II neurons with some larger nerve cells. The connexions of these with incoming fibres from dorsal roots and with spinothalamic tract fibres have long been accepted, but somewhat drastic review of this teaching in its details is occasioned by recent experimental work (pp. 888, 890). Also extending throughout the cord is a column of rather large cells, ventral to the gelatinous substance, the **dorsal funicular group** or *nucleus proprius*. As in the case of the anterior column, a thin lamina of nerve cells, distinguishable from those of the substantia gelatinosa by their larger size, is described by some authorities as a *marginal zone*, dorsal to the substantia (7.40, 41). (For further structural details and earlier postulated roles of the substantia gelatinosa see pp. 887, 890.)

The **nucleus dorsalis** or *thoracicus* (of Clarke) occupies the basal region of the posterior grey column, immediately dorsal to the intermediate zone in laminae VI–VII (but *see* also p. 883). At most levels it is close to the dorsal white funiculus and may project slightly into it. There are variable accounts of its extent, but in the human spinal cord it can usually be identified from the eighth cervical caudally to the third or fourth lumbar. Similarly situated groups of nerve cells have been described as occurring at cervical

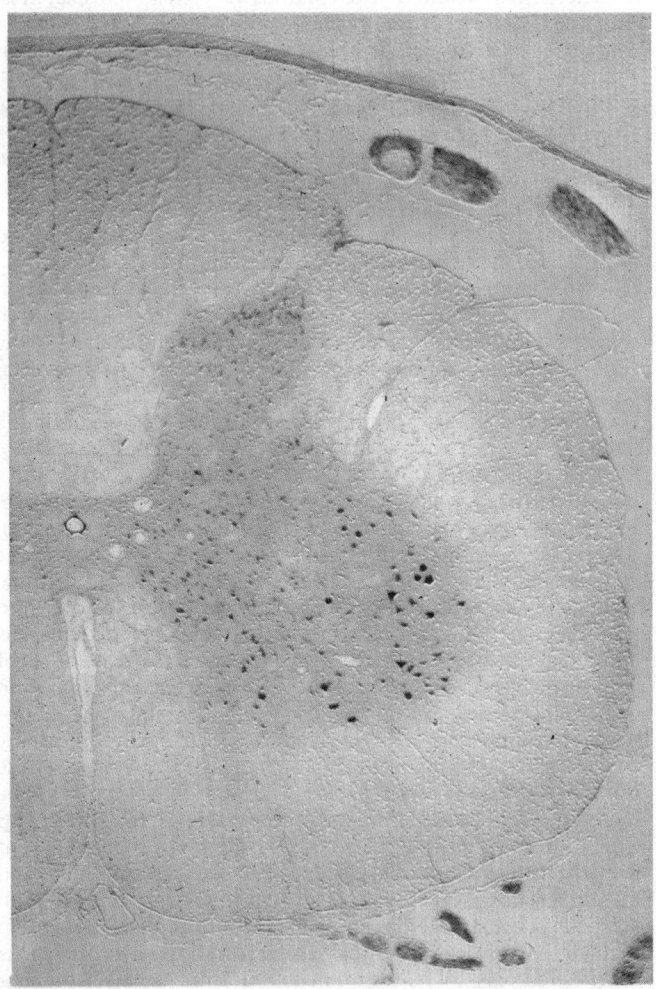

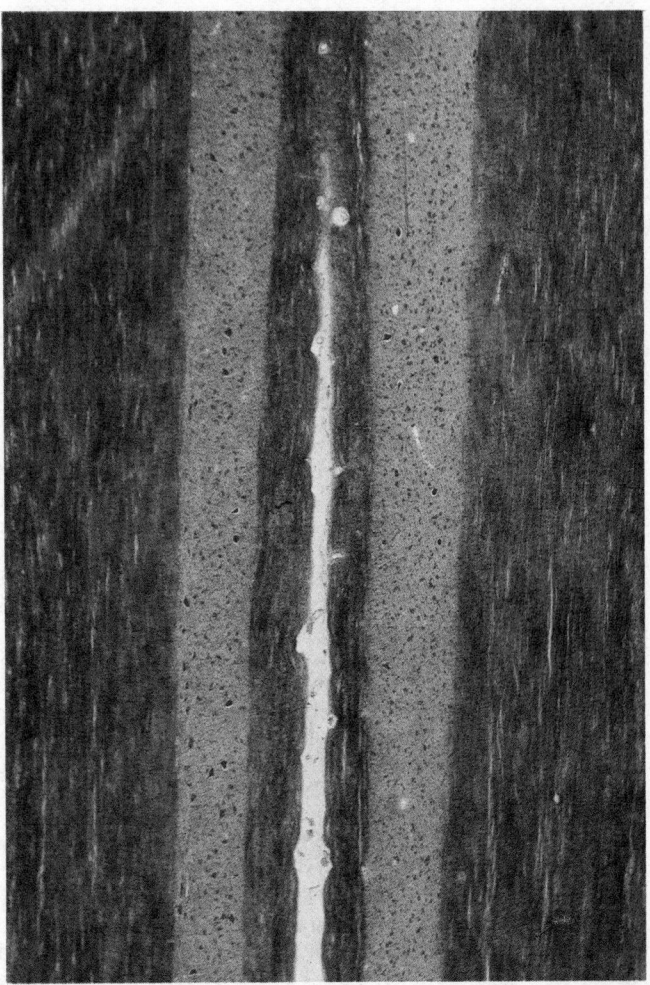

A   B

7.40A and B Transverse and longitudinal sections of spinal cord. A Transverse section of left half of human spinal cord at a mid-lumbar level. Note dorsal and ventral grey columns and commissural grey mass. The larger motor neurons in the ventral grey column are visibly grouped. For details see text. (Stained with cresyl fast violet.) B Longitudinal section of feline spinal cord showing the anterior median fissure, and anterior white columns, and lateral to these the ventral grey columns, in which motor neurons show some degree of grouping into longitudinal columns. (Material prepared by the late Professor L. Laruelle and kindly supplied by Professor J. André-Balisaux, Institut Neurologique Belge, Brussels.)

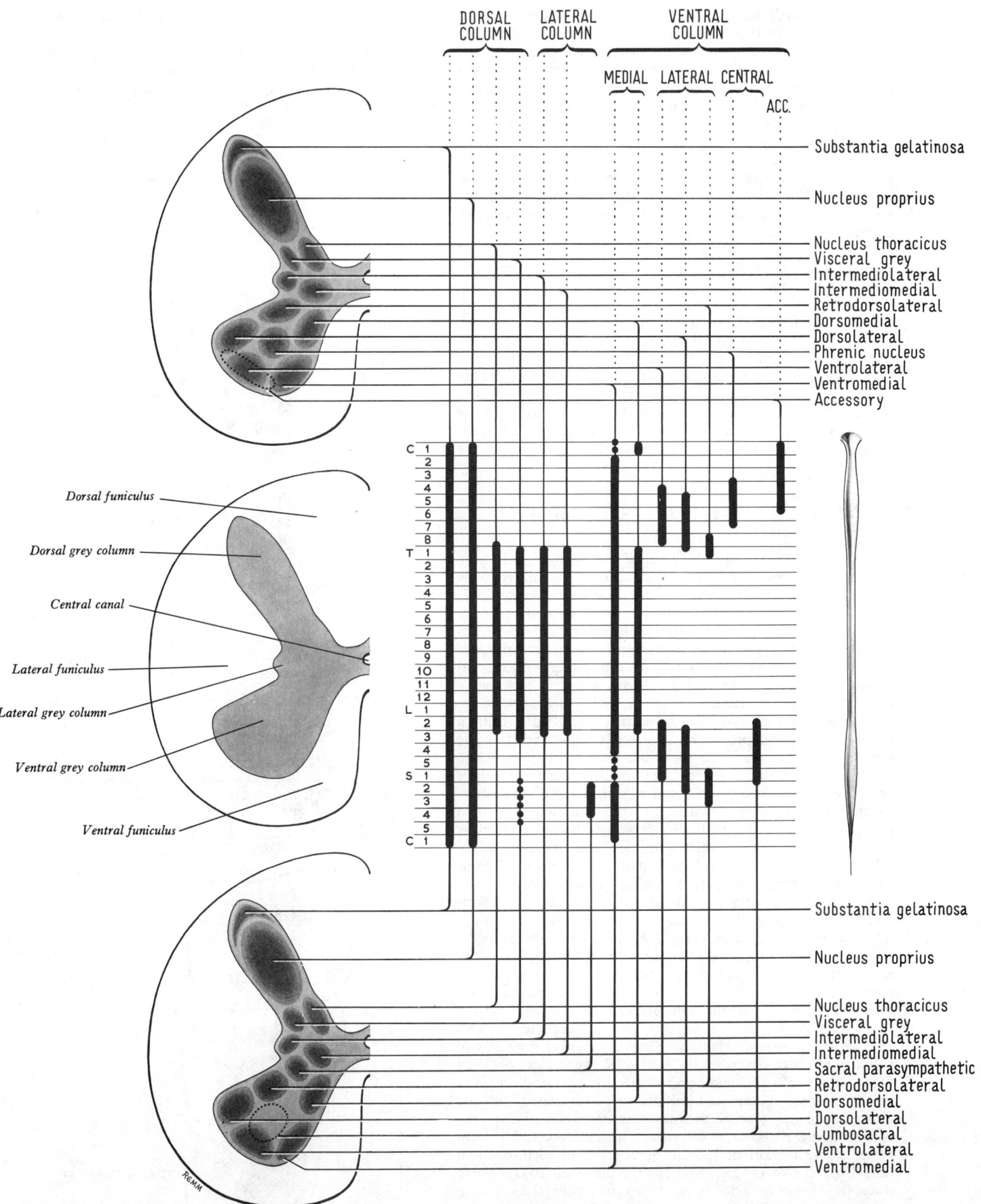

DORSAL COLUMN — **LATERAL COLUMN** — **VENTRAL COLUMN** (MEDIAL · LATERAL · CENTRAL · ACC.)

Substantia gelatinosa

Nucleus proprius

Nucleus thoracicus
Visceral grey
Intermediolateral
Intermediomedial
Retrodorsolateral
Dorsomedial
Dorsolateral
Phrenic nucleus
Ventrolateral
Ventromedial
Accessory

Dorsal funiculus

Dorsal grey column

Central canal

Lateral funiculus

Lateral grey column

Ventral grey column

Ventral funiculus

Substantia gelatinosa

Nucleus proprius

Nucleus thoracicus
Visceral grey
Intermediolateral
Intermediomedial
Sacral parasympathetic
Retrodorsolateral
Dorsomedial
Dorsolateral
Lumbosacral
Ventrolateral
Ventromedial

7.41 The groups or nuclei of nerve cells in the grey columns of the human spinal cord as generally accepted. Relative positions of these columnar groups, as well as their extension through varying series of spinal segments, are as indicated. (Adapted from data in *Correlative Anatomy of the Nervous System*, by E. C. Crosby, T. Humphrey and E. W. Lauer, Macmillan, 1962, with the permission of the authors and publishers.)

levels cranial to the nucleus dorsalis, and extensive prolongations in caudal levels of the cord appear to exist in some long-tailed monkeys (Chang 1951). But since these 'cervical' and 'sacral' nuclei consist of cells of considerably different characteristics, and have only been described in other mammals, it is premature to extrapolate such observations to the human spinal cord (*see*, however, p. 891). The cells of the dorsal nucleus itself vary in size, most being comparatively large, especially in the lower thoracic and lumbar segments. Some of these cells send axons into the dorsal spinocerebellar tract (p. 879), some are interneurons. Petras and Cummings (1977) have described the cells of this nucleus and their connexions in some detail in neonatal dogs, also confirming its role as a 'relay' group on the dorsal spinocerebellar pathway. The same workers demonstrated a similar role for the **nucleus centrobasalis**, which occupies a similar position in the dorsal grey column to the nucleus thoracicus but appears in the lower cervical and lumbosacral segments of the spinal cord. These groups have not been clearly established in the human cord.

Lateral to the nucleus dorsalis, and dorsal to the intermediolateral column (7.41), is a small region of nerve cells of medium size, extending throughout the same segments (approximately first thoracic to third lumbar) as the intermediate columns (Takahashi 1913). This columnar group is identifiable in the human cord, but its functional status is uncertain, though it has naturally been associated with the neighbouring autonomic nerve cells.

NERVE CELL GROUPS OF THE INTERMEDIATE GREY MATTER

The intermediate region of the spinal grey matter (7.40, 41), which includes the lateral grey column, is composed of relatively small nerve cells, many with the features of autonomic preganglionic cells, which develop from elements in the embryonic cord at first dorsolateral to the central canal. Many of these migrate to a position lateral to it and at some little distance from it; these nerve cells constitute the **intermediolateral group**. An **intermediomedial column** is formed from nerve cells which remain nearer to the central canal. The intermediolateral group forms the projecting lateral grey column proper, and a large proportion of its cells send axons into the ventral spinal roots and via the white rami communicantes to reach the sympathetic trunk (p. 1123); preganglionic nerve fibres are similarly derived from some of the cells of the intermediomedial group (the remainder being interneurons). Both groups extend from the eighth cervical or first thoracic segment as far as the second or third lumbar, thus corresponding approximately to the thoracolumbar outflow. In the second, third and fourth sacral segments a similar group of nerve cells, intermediate in position, is the source of the pelvic or sacral outflow of parasympathetic preganglionic nerve fibres (p. 1123). This **sacral parasympathetic grey column** is lateral to the central canal and substantia gelatinosa centralis, in the junctional zone between the bases of the anterior and posterior grey columns. It shows no division into medial and lateral parts, nor does it project from the intermediate grey zone like the thoracolumbar lateral grey column. The emergence of parasympathetic preganglionic nerve fibres from other segments of the cord has been described by Kuré *et al.* (1930), and Sheehan (1933), their cells of origin being ascribed to the basal region of the dorsal grey column and perhaps to be associated with the intermediate grey zone. Such fibres were stated to issue from the cord in *dorsal* roots, to be vasodilator in function, and to form synapses with small multipolar nerve cells in corresponding dorsal spinal root ganglia (Kiss 1932; Kuré *et al.* 1934). These interesting views have not, however, received general acceptance nor any substantial confirmation.

The foregoing description of the arrangement of nerve cells in the spinal cord, which is largely dependent upon the study of material specifically stained to show the *somata* of neurons rather than their processes, has of late been considerably amplified by a *laminar concept* of spinal grey matter organization, an account of which follows below (p. 882). This concept is more widely based upon the interconnexions of nerve cells, and its structural data

have been correlated with the results of degeneration experiments to a much greater degree than in the case of the older mode of description outlined above, combining also more aptly with the observations of microelectrode studies. The laminar pattern of organization has thus helped to establish a more precise definition of spinal cord activities, but the two modes of description are not exclusive and, as will become apparent, the older scheme of

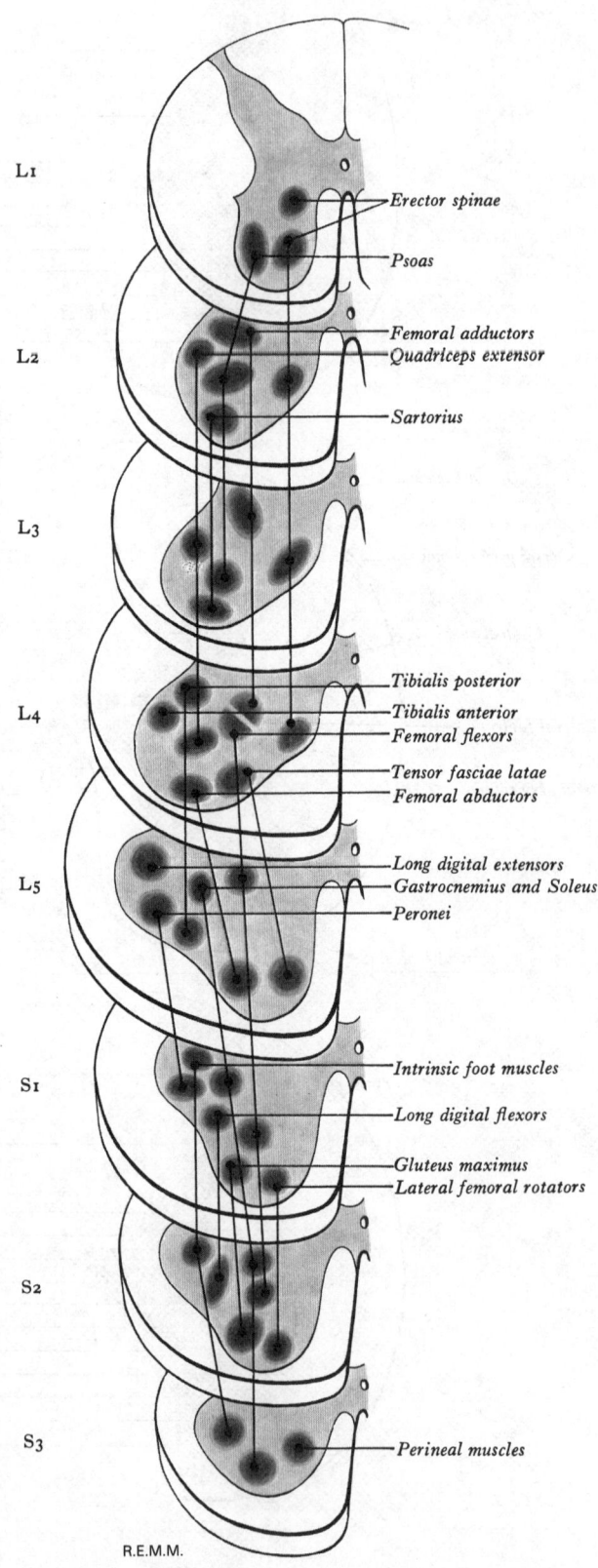

7.42A Diagram of the approximate location in the transverse plane, and in longitudinal extent, of the nerve cell groups innervating muscles, chiefly in the leg, in the lumbosacral segments of the human spinal cord. (Based on clinicopathological studies of poliomyelitis, by W. J. W. Sharrard, *J. Bone Jt Surg.*, 1955.)

columnar grouping of nerve cells is in most of its features adaptable into the newer laminar pattern. The latter does, nevertheless, involve some important modifications in regard to the structural relationships between the nerve cells of the dorsal grey column and the fibres of the dorsal roots and spinal tracts, and also involves more precise concepts of functional implications.

There is one aspect of the spatial relationships of nerve cells in the spinal grey matter which so far has scarcely been mentioned; this concerns the significance of cell grouping in the ventral grey columns in relation to individual muscles and movement, and to this attention must now be directed.

THE FUNCTIONAL IMPLICATIONS OF ANTERIOR GREY COLUMN CELL GROUPS

Even in the earliest accounts of the columnar arrangement of nerve cells in the spinal cord—most of which were based on Nissl stained material studied only in transverse sections, and derived from a truly extraordinary miscellany of animals (including tapoles, an ostrich, a gorilla, and occasionally man!)—a somatotopic interpretation of cell grouping in the ventral column was advanced with some confidence (Elliot 1942). Thus, it was an early tenet of such speculations that the medial groups innervate axial musculature (supplied by posterior primary rami), the limbs being innervated from the lateral groups. This attractive hypothesis, originally based entirely on structural data, has in fact been confirmed to some degree by subsequent experimental work, as will appear below. The criticism has been made that few of the investigators in this field have observed the grouping of ventral grey column cells in longitudinal sections, and that the errors which may arise in tracing elongated aggregations of cells through transverse series of sections probably account for at least some of the disagreements and variations between the topographical results of many earlier workers. Obviously an agreed pattern of cell groups, free at least from major discrepancies, is a necessary prelude to any attempts to assess how far such grouping might represent, for example, the neurons innervating an individual muscle. Considerable agreement in this regard is apparent in the more recent work on this problem, particularly among observers who have examined the distribution of ventral grey column cell groups in fetal material, where it is agreed that the groups are more discrete and hence more easily identified (Romanes 1941; Elliot 1943). It is also of importance to note that results of this kind were markedly similar in cat and human material, though only one human fetus was examined. On the other hand, the degree of motor cell grouping in the ventral column is much less developed in amphibians and reptiles, or even mammals such as the bat and mole, all of which possess a fairly complex forelimb musculature, than in the whale, in which the same muscle system is much simplified (Romanes 1953). It must be added that these comparative observations are of limited value, having been carried out on a somewhat heterogeneous collection of vertebrates, usually in extremely small series and often single animals, with little reference to taxonomic relationships. Moreover, little new information appears to have been recorded during the last few decades and certainly no major series of observations.

These difficulties must be mentioned in some detail to introduce some element of caution into further consideration of the functional interpretations of the motor neuron groups which undoubtedly occur in the ventral grey column, and not to deny the existence of such patterns. Pattern is always significant, and a topographical pattern certainly exists in the ventral column. The question in the present context is—What is the physiological meaning of cell grouping in this region? That it concerns the innervation of skeletal muscles seems inescapable, and it would appear a relatively simple, if tedious, undertaking to cut the nerves supplying individual muscles and to observe the distribution of the resultant retrograde degeneration in the motor neurons of the ventral columns. Few experimental studies of this kind have in fact been carried out—at least, on an extensive scale—and since they have involved the use of different species of animal and sometimes different levels of the spinal cord, they provide only limited checks upon each other. In an investigation

of the lumbosacral region of the cord in cats, in which various peripheral nerves were divided in the hind limb (Romanes 1951), the affected groups of nerve cells were all in the lateral part of the anterior column, and the cells innervating the more distal muscles in the limb were dorsal to the 'motor columns' for proximal muscles. In general terms these findings corroborated the speculations of earlier topographical observers, and limited

THE INNERVATION OF THE LOWER LIMB MUSCLES

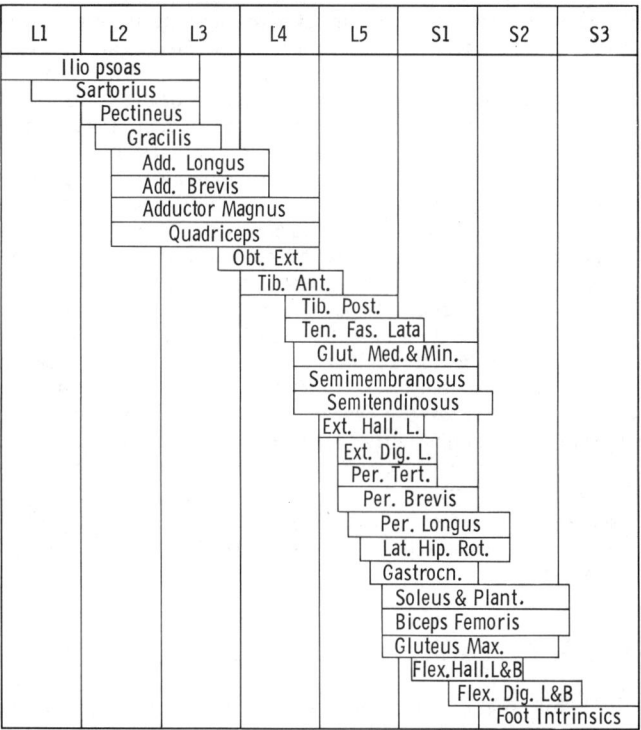

L1	L2	L3	L4	L5	S1	S2	S3
Ilio psoas							
Sartorius							
Pectineus							
Gracilis							
Add. Longus							
Add. Brevis							
Adductor Magnus							
Quadriceps							
Obt. Ext.							
Tib. Ant.							
Tib. Post.							
Ten. Fas. Lata							
Glut. Med. & Min.							
Semimembranosus							
Semitendinosus							
Ext. Hall. L.							
Ext. Dig. L.							
Per. Tert.							
Per. Brevis							
Per. Longus							
Lat. Hip. Rot.							
Gastrocn.							
Soleus & Plant.							
Biceps Femoris							
Gluteus Max.							
Flex. Hall. L&B							
Flex. Dig. L&B							
Foot Intrinsics							

7.42B   The segmental arrangement of the innervation of the lower limb muscles. (According to W. J. W. Sharrard, *Ann. R. Coll. Surg.*, **35**, 1964.)

experimental confirmation, also in the cat, has been recorded with regard to the anterolateral group of leg muscles (Balthasar 1952). In the former investigation it was concluded that the cell groups identifiable topographically were not usually individual *motor pools*, but more often (but not always) represented a somatotopic grouping of the motor nerve cells innervating muscles involved together in some *common effect on a joint*.

Another investigator has interpreted his experimental results in monkeys as indicating that the ventral column cell groups can be accounted for not only on the above basis, but also and equally well in terms of peripheral nerves, limb segments, and muscles grouped on a morphological or developmental basis (Sprague 1948). In this investigation the effects of division of dorsal and ventral primary rami of spinal nerves were studied both in the segments innervating limbs and at thoracic levels; and this led to a denial of the differentiation of the ventral grey column into medial and lateral groups in thoracic segments. Even in the limb enlargements of the cord, where both groups were admitted to be present, both appeared to be sources of fibres entering dorsal and ventral rami, a direct confutation of the classical view assigning the innervation of axial and limb musculature respectively to the medial and lateral groups (Kaiser 1891). The same worker, in a later communication, re-emphasized his view that the concept of columnar organization of the ventral grey region had become so ingrained as to be a serious source of error in functional interpretation of this part of the spinal cord. While this is perhaps an extreme view, it does at least underline the unsatisfactory state of knowledge in these matters, which require a wider re-investigation in primate animals, a formidable undertaking. More precise and reliable description of the motor neuron groups in the ventral grey column might, of course, prove to be of rather limited interest in functional interpretation, though it would clearly be of some value when applied in clinical situations involving spinal lesions in mankind. In this connexion it is interesting to note that the effects of poliomyelitis on motor neurons in the ventral column have been correlated by several clinical observers with the distribution of the resultant paralysis. The most detailed report of this kind (Sharrard 1955, 1956) is of special interest, being concerned with human conditions and also a confirmation to a considerable extent of some of the experimental findings in cats referred to above (7.42). It is much to be hoped that further data will be forthcoming from such clinicopathological correlation.

To summarize, it must be admitted that, despite a copious literature reporting a large number of topographical studies and a much smaller succession of experimental investigations, very considerable uncertainties remain to be clarified. That the topographical arrangement of the nerve cells of the ventral grey columns is basically columnar is well established, but many points of imprecision require further study; and perhaps the columnar mode of organization—to some extent inevitable in an elongated structure such as the spinal cord—has been somewhat over-emphasized. Somatotopic organization with respect to muscles, either in individual representation or in functional groupings, appears to be confirmed in general terms; but the evidence is against the occurrence of discrete motor pools for individual muscles, of which few have been satisfactorily demonstrated. Overlapping of the nerve cell groups of associated muscles seems to be a commoner pattern of functional organization. It may be added that most of the workers in this field of study and experiment have apparently allowed their concentration upon the *somata* of neurons to divert consideration too far from the *dendritic regions* between them, which in simple ratio occupy a much greater volume in the spinal grey matter. It is to this aspect of spinal organization that we shall return after consideration of the 'white matter' in the next section.

Nerve Fibre Tracts of the Spinal Cord

The 'white matter' of the spinal cord consists of nerve fibres, neuroglia and blood vessels. It surrounds the fluted column of grey matter and its whiteness is, of course, due to the large proportion of myelinated nerve fibres. Its arrangement into anterior, lateral and posterior funiculi has already been described (7.36). Its constituent fibres vary much in calibre, large numbers being small and non-myelinated. Some tracts are characterized by fibres of a small diameter, for example the dorsolateral tract, the fasciculus gracilis and the central part of the lateral funiculus. The nerve fibres in the fasciculus cuneatus, anterior funiculus, and the peripheral zone of the lateral funiculus all contain many large

diameter fibres. Most regions of the white substance of the spinal cord contain a considerable *spectrum of fibre diameters*, extending from 1 μm or less up to about 10 μm. Fibres with a diameter of 3 μm or less predominate, and those at the upper end of the spectrum (a few of which may exceed 10 μm) form only a small fraction of the total. Detailed studies of the distribution of fibres of differing diameter in the human spinal cord have been few (Häggqvist 1936; Giok 1956), but it has been claimed that many tracts can be identified on this basis alone. The proportion of fibres of particular diameters has been estimated in a few instances (Szentágothai–Schimert 1941), but in general precise data of this kind are lacking. In any case, delineation of tracts by such observations in transverse sections of normal material can only be regarded as valid when confirmed by the results of experimentation. Most of the available information regarding the tracts of the spinal cord is derived from the results of controlled selective damage of fibres in the cord itself, in experimental animals, but the effects of damage to dorsal nerve roots and of lesions placed in the brainstem, cerebellum and cerebrum have also provided much information. In all such experiments retrograde degeneration indicates the nerve cells from which particular fibres proceed, while anterograde terminal degeneration provides the evidence for their sites of termination. Suitable staining to reveal Wallerian degeneration of the fibres at intermediate levels demonstrates the position of a tract and the degree to which its fibres are compacted together, dispersed, or overlapped with others. It must be emphasized at once that, while certain tracts are relatively discrete and the central regions of most are located regularly in definable parts of the funiculi, reciprocal overlapping at least of their fringes (and often much more) is usual (*vide infra*). This accounts in part for the variation in their extent, as seen in diagrams of transverse sections of the cord according to different authorities; in all such diagrams arbitrary boundaries must necessarily be set to delineate the supposed limits of tracts. This is especially true in regard to the human spinal cord; deliberate experimentation is here impossible, and the only data available are the results of disease and injury, neither of which produce the selective and clear-cut kind of lesions possible in animal experimentation. The results of such investigations, particularly when derived from experiments on other primates, can be regarded as likely to be closely similar to the human arrangements; but it would obviously be unwise to accept them as being identical. The evidence collected from examination of the human spinal cord for degenerating nerve fibres resulting from disease or injury does in general support the presumption that the layout of tracts in the human primate is very much the same as in others. The amount of information accumulated from this source is surprisingly large, though usually less precise; it was in fact the

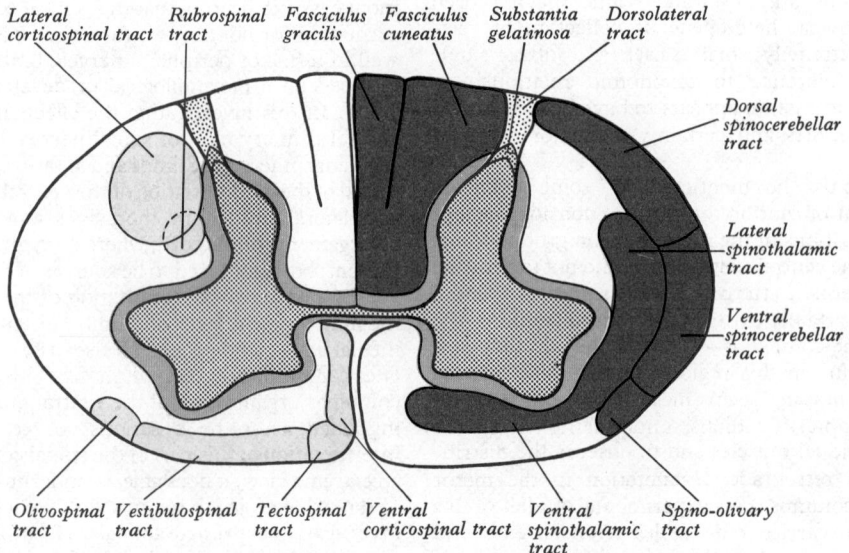

Lateral corticospinal tract   Rubrospinal tract   Fasciculus gracilis   Fasciculus cuneatus   Substantia gelatinosa   Dorsolateral tract

Dorsal spinocerebellar tract

Lateral spinothalamic tract

Ventral spinocerebellar tract

Olivospinal tract   Vestibulospinal tract   Tectospinal tract   Ventral corticospinal tract   Ventral spinothalamic tract   Spino-olivary tract

7.43 Simplified diagram of the main tracts of the spinal cord. The ascending tracts are shown in red on the right side of the figure; the descending tracts are shown in yellow on the left side; the 'intersegmental' tracts are in orange on both sides. Many tracts are omitted (see 7.44).

principal source of information at first (p. 817), and a very large number of reports of this kind are scattered through neurological literature (*see* Nathan and Smith 1955, 1959).

The very predominance of data derived from clinical sources has produced a curious vicious circle in associating an observed grouping or *syndrome* of sensory and motor disturbances with injury to or disease in particular regions of the cord—the exact extent of the latter being not invariably assessed with precision. Because some tracts are undeniably concentrated in certain parts of the white funiculi, their delineation has become artificially crystallized. The fringe overlap mentioned above may be quite extensive; some tracts are in fact almost completely mingled, and others are merely a concept, rather than a circumscribed reality, their fibres being scattered far and wide. Experimental observations, pursued in parallel with clinical deductions, have now far outstripped the latter in their recognition of the intermingling of spinal tracts, whose supposed discrete nature is still widely accepted in clinical practice, but this can become an impediment rather than an aid to accurate diagnosis.

The account of spinal tracts which follows is primarily concerned with their arrangement in the human spinal cord, but inevitably reference to findings in animal experimentation must be made at many points where adequate clinicopathological or other data based on human material are not available. In the succeeding section (p. 882), dealing with the finer details of neuronal organization and activity in the spinal cord, evidence derived from animals other than man predominates, but it is improbable that the deductions from this are not largely applicable to the human spinal cord.

As a convenient simplification, the nerve fibres in the white matter of the spinal cord may be assigned to five groups: (*a*) afferent fibres from the cells in the dorsal root ganglia which have entered by the dorsal roots and extend for longer or shorter distances in the cord; (*b*) long ascending fibres, derived from nerve cells in the cord and conducting afferent impulses to supraspinal levels; (*c*) long descending fibres from supraspinal sources which synapse with cord cells; (*d*) fibres effecting intrasegmental and intersegmental connexions; and (*e*) fibres from the motor neurons in the ventral and lateral grey columns which issue in the ventral spinal nerve roots. Fibres in all these categories except the last form the *longitudinal tracts* of the spinal cord. Arrangements are not, however, as simple as this. Examination of the cord in different planes reveals that many fibres proceed for considerable distances in oblique and even horizontal directions, particularly when crossing the midline, as many do in the grey and white commissures. Many of these are *decussating* fibres, that is to say they are crossing to the opposite side of the cord to continue to some more distant point. Many others are *commissural*, being intrasegmental connexions which link nerve cells in the grey columns on one side of the cord to cells on the other. Moreover, some of the longitudinal tracts are polysynaptic, that is, composed of a train of neurons. Most, if not all, of the fibres entering by the dorsal root divide into ascending and descending branches which both give off a series of collateral branches extending into the grey matter. In the latter the number of nerve cells which either send axons into the ventral roots or into the tracts in the white funiculi is a mere fraction of the total, the majority being local interneurons (p. 885).

In the following account of the spinal tracts an arbitrary order has been adopted, those in the anterior, lateral and posterior funiculi being described in that order; in each region the descending, ascending and intersegmental tracts are successively considered. This is done primarily to present a relatively orderly topographical picture of the arrangement of tracts which is useful in the context of medical diagnosis, and in phylogenetic studies. However, tracts vary somewhat in their relative positions at different levels in the cord, and in different species. For a detailed analysis of tracts in the cat and monkey, and a critical review of some of the difficulties in their determination; *see* Verhaart (1953, 1954); van Beusekom (1955); Verhaart and van Beusekom (1958). The general positioning of the major tracts is illustrated in a simplified form in **7.43** and in greater detail at two spinal levels in **7.44**A and B. Some features are further summarized in **7.47**A, B and C.

TRACTS IN THE ANTERIOR FUNICULUS

1. Descending Tracts

(*a*) **The anterior corticospinal tract** is usually small, but varies in size inversely with the lateral corticospinal tract. It lies alongside the anterior median fissure, from which it is separated by the sulcomarginal fasciculus (*vide infra*). It is present in the upper part of the spinal cord, gradually diminishing in size as it descends; it cannot be traced below the middle of the thoracic region. The origin and termination of its constituent fibres are considered with those of the lateral corticospinal tract (p. 877). This tract is found only in primates; its precise significance is unknown; it is subject to much variation (Nyberg-Hansen and Rinvik 1963); it may be absent or it may, very rarely, contain all the corticospinal fibres. Usually it is composed of about 10–30 per cent of them. Its variations have been said to accord with its late phylogenetic and ontogenetic development (Humphrey 1960).

(*b*) **The vestibulospinal tract**, which is principally derived from both the large and small cells of the lateral vestibular nucleus (p. 908), descends along the periphery of the anterior funiculus; its fibres end around cells in the anterior grey column. This tract is uncrossed and brings the anterior column cells under control of the vestibular nuclei of the same side, serving as an efferent pathway for equilibratory control. The most medially situated fibres of the vestibulospinal tract have, however, been described as starting from cells in the medial and perhaps also the lateral and inferior vestibular nuclei (Rasmussen 1932), though more recent work suggests that only the medial nucleus is a source (Pompeiano and Brodal 1957). These fibres, which pursue a different course in the medulla oblongata, descending in the medial longitudinal fasciculus (p. 939), constitute the *medial vestibulospinal tract*. This tract is partly crossed and probably does not reach lumbar segments of the cord, and may therefore be concerned with the upper extremity and neck alone. In the white funiculus it is dorsal to the tectospinal tract and immediately adjacent to the anterior median fissure (**7.43, 44**) in the position of the so-called *sulcomarginal fasciculus*. The remainder of the vestibulospinal fibres (i.e. the majority, which are uncrossed and descend from the lateral nucleus) may hence be regarded as a *lateral vestibulospinal tract*, which extends to all levels in the cord. The fibres of both tracts terminate in the medial part of the anterior grey column (laminae VII and VIII, *see* p. 883), and they influence both *a* and *γ* motor neurons through interneurons (Gernandt 1959), but there is physiological evidence that some fibres end monosynaptically on motor neurons. Both tracts have been widely regarded as uncrossed, but, as we have seen, decussating fibres have been described in the medial tract. There is some degree of somatotopic representation in the lateral vestibular nucleus.

(*c*) **The tectospinal tract** is variably disposed lateral to the anterior median fissure. Its fibres arise in the superior colliculus of the opposite side (p. 941) and end by forming synapses with cells in the anterior grey column, especially in the cervical segments of the cord. These fibres influence motor neurons through interneurons (Szentágothai 1948), their terminations being probably in laminae VI–VIII (p. 883). The contralateral origin of this tract now seems well established; suggestions of a bilateral origin from the superior colliculi may have arisen through confusion with the *lateral tectotegmentospinal tract* of some authors. When the latter is recognized, the tectospinal fibres which descend in the anterior funiculus are often termed the *medial tectospinal tract*. Autoradiographic fibre-tracing experiments in primates (Frankfurter *et al.* 1976) suggest that only a minority of the fibres in this descending tectal pathway actually reach the spinal cord, the majority terminating in relation to various brainstem nuclei. For further discussion of the tectospinal systems *see* pp. 879 and 941.

(*d*) **Reticulospinal fibres** are also widely scattered through the anterior funiculus, chiefly in its medial part. They arise from the nerve cells of the ipsilateral pontine reticular nuclei (p. 945), but some may decussate shortly before their spinal terminations (Torvik and Brodal 1957). They descend throughout the cord in the anterior funiculus and probably end as synapses with interneurons in the medial part of the ventral grey column (laminae VII and VIII; *see* Nyberg-Hansen 1965, and p. 888), through which they possibly exert a facilitatory effect on motor neurons. This tract is commonly known as the *medial* (pontine) *reticulospinal tract*, its origin in the pons corresponding in level to the 'facilitatory area' of many physiologists. (Contrast the lateral reticulospinal tract—p. 879.) These details are entirely derived from observations in non-human tissues; for a discussion of the reticulospinal tracts in man *see* Nathan and Smith (1955), Brodal (1957), Webster (1978); further details are also furnished on pp. 946, 947.

(*e*) **The interstitiospinal tract** has its origin in the interstitial nucleus (p. 939) and its fibres descend without crossing into the medial longitudinal fasciculus (p. 939), and thence into its spinal continuation, the ipsilateral fasciculus proprius (anterior intersegmental tract). In the cat the tract is claimed to be traceable

to sacral levels (Nyberg-Hansen 1966), but no such details are available for the human spinal cord, although the tract as an entity has long been recognized (Muskens 1914).

(*f*) **The solitariospinal tract** (of Cajal 1909) is a small group of descending fibres which may originate in the neurons of the caudal part of the solitary nucleus (p. 903); it has been best demonstrated in the cat (Torvik 1957), and has been equated with part of the medial reticulospinal tract by another authority (Crosby *et al.* 1962). It was named by Cajal and may be involved in visceral reflexes involving the oesophagus and stomach.

2. Ascending Tracts

The anterior spinothalamic tract is in fact continuous with the lateral tract of the same name, but may be considered to be the part of this complex in the anterior funiculus, medial to the fibres

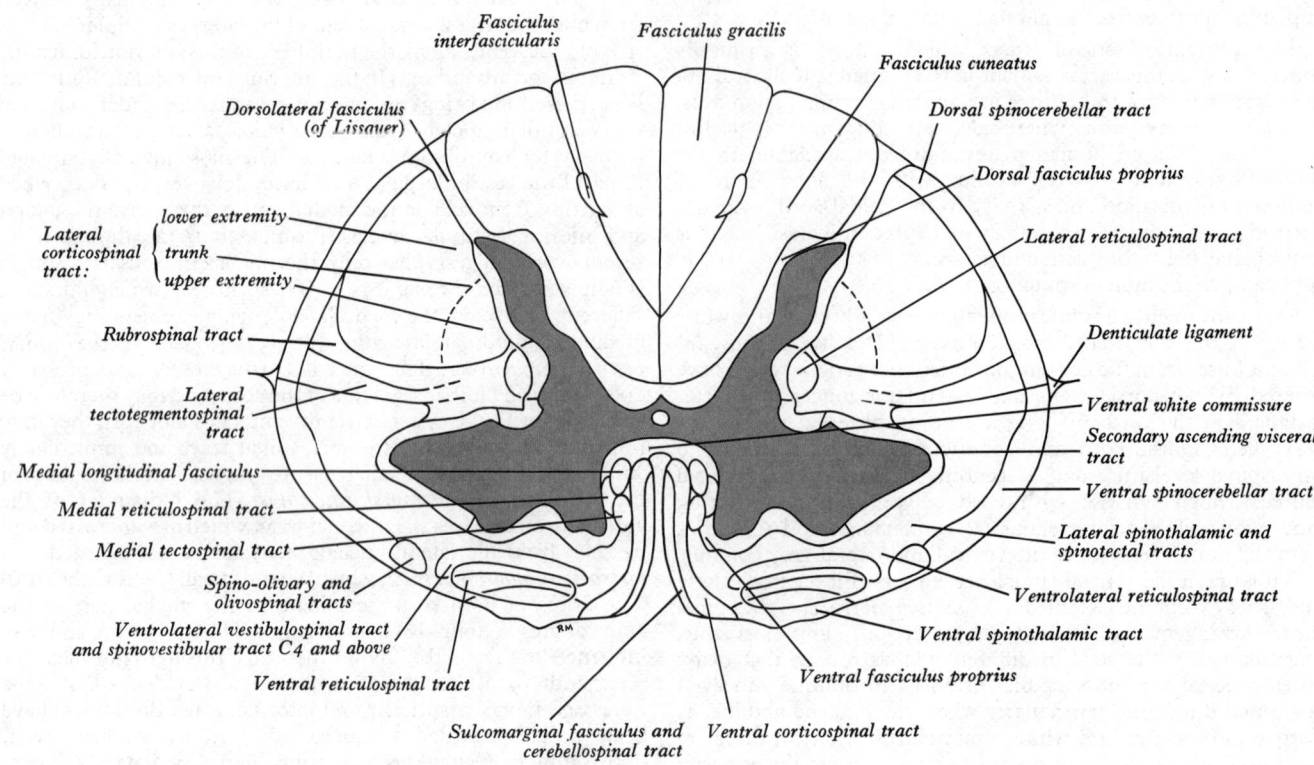

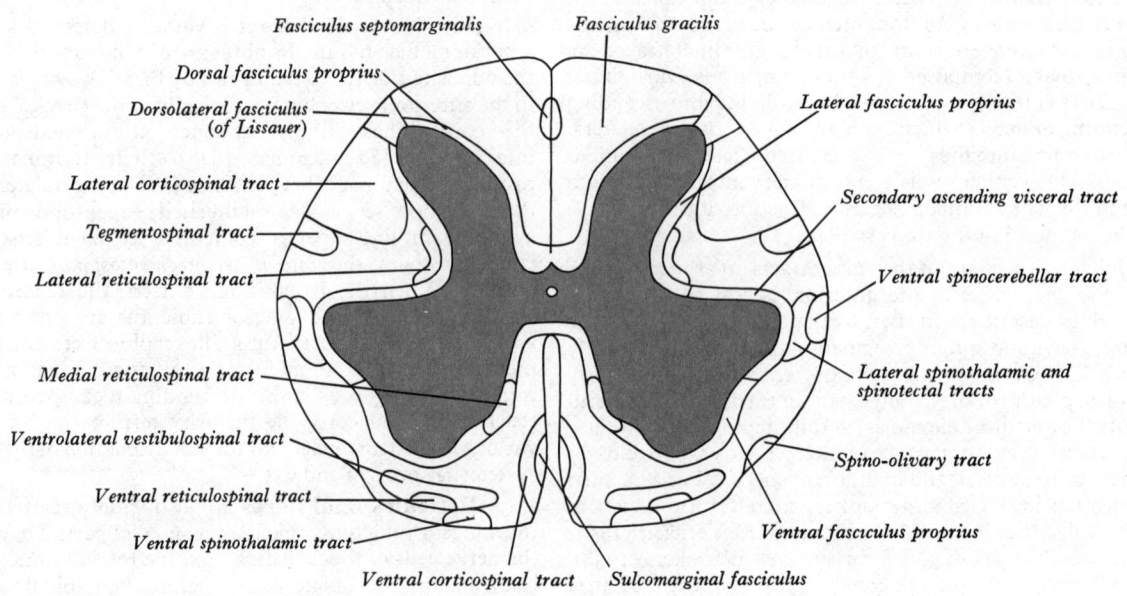

7.44A and B Diagrams of the approximate relative positions of nerve fibre tracts of the human spinal cord at midcervical (A) and lumbar (B) levels. (Adapted from *Correlative Anatomy of the Nervous System*, by E. C. Crosby, T. Humphrey and E. W. Lauer, Macmillan, 1962.)

of the ventral nerve roots and dorsal to the vestibulospinal tract (7.43), which it overlaps, as it does all its neighbours. It has been claimed, largely on clinical evidence, that it is chiefly concerned with crude tactile and with pressure sensibility (but this does not wholly accord with recent physiological investigations—pp. 890, 958). According to Applebaum *et al.* (1975) segregation upon a modality basis, within the anterior/lateral spinothalamic complex is not supported by experimental data, and the two tracts are to be considered as a continuum, both structurally (as stated above) and also in the functional sense. However, as noted below, the neurons from which spinothalamic axons are derived are diverse in distribution and in the modalities which they subserve. It is therefore no longer possible to consider simple segregations of sensory mediation either in the spinothalamic system or other ascending tracts. Nevertheless, the physiological characteristics associated with the neurons in the laminae of origin of spinothalamic projections provide some indications. Thus, cells in laminae I and V respond to 'noxious' mechanical and thermal stimuli, applied cutaneously, and some lamina V cells are associated with non-noxious cutaneous stimuli. Laminae VII and VIII are concerned with muscular and articular receptors, and some show responses on cutaneous stimulation. The precise locations of the nerve cells of origin of this tract are subject to much controversy; they have usually been described as 'secondary neurons' situated in the dorsal grey column which receive direct synapses from the axons which enter via the dorsal roots from the primary sensory neurons in the dorsal root ganglia. Since the investigations of Cajal it has been known that some neurons in all regions of the grey matter (dorsal and ventral columns, and the intermediate zone) have axons which cross the midline and ascend in the anterolateral white funiculus. More recently, investigators using Golgi techniques have claimed that these are particularly concentrated in spinal cord laminae IV–VII and especially lamina V; but there are also contributions from the pericornual cells superficial to the substantia gelatinosa (lamina I) and also from lamina VIII (*see* Webster 1977, for literature). Physiological experiments have corroborated some of these details in the cat and the rat (Dilly *et al.* 1968). Electrical stimulation of the region where the medial lemniscus and spinothalamic systems enter the thalamus was accompanied by intracellular recording of antidromic responses from the various grey matter laminae in the cord. Responsive cell somata were identified in laminae V and VI, and this provides some confirmation of the origin of the spinothalamic tracts; but it should be noted that with the limited sampling inherent in such a technique, responding cells in other laminae may well have been missed. Further, it is by no means certain that all spinothalamic fibres are the axons of 'secondary neurons'; in some cases one or more interneurons may be interposed between the primary dorsal root afferents concerned and the spinothalamic tract neuron.

The evidence from cordotomies in man and from other sources suggests that most of the fibres cross the cord, probably within a single segment above their origin, the decussation being in the anterior white commissure. Although physiological studies (p. 890) suggest some spatial separation of neuronal elements concerned with different combinations of the sensory modalities associated with the spinothalamic tracts in the dorsal grey column, the arrangement of the ascending fibres in these tracts has long been regarded as somatotopic (Nathan 1963; Morin *et al.* 1951). Fibres crossing at any particular level join the medial aspect of those already ascending on the contralateral side of the cord, so that both tracts are segmentally laminated (7.45), fibres originating in lower segments, and therefore of greater total length, being most superficial. In addition a slight spiral twist is said to occur, the more superficial fibres being progressively more dorsal in position as the tracts ascend the cord. This arrangement is continued through the medulla and pons to the nucleus ventralis posterior lateralis of the thalamus (p. 958). However, the precise mode of arrangement of these fibres, whether somatotopic, dermatomic, or otherwise is by no means certain. Intermingling with the anterior spinothalamic fibres are not only descending *reticulospinal fibres*, already mentioned, but also numbers of ascending *spinoreticular fibres*. It is possible that the spinothalamic fibres give off collaterals to the brainstem reticular

nuclei, but it is now considered that the majority of their afferents form separate spinoreticular pathways, some of which are probably polysynaptic in nature. As in the case of other ascending tracts, there is definite physiological evidence that the spinal cells of origin of the spinothalamic tracts, or their associated interneurons, may be selectively inhibited by fibres descending from nerve cells in the sensory and motor parts of the cerebral cortex, in the anterior cerebellar lobe, and certain regions of the brainstem reticular formation; but the spinal distribution and sites of termination of these inhibitory fibres have not yet been clarified. Consult Carpenter *et al.* (1963) and Webster (1977) for details. Finally, mention should be made of a small *spinovestibular tract* which is intermingled with the vestibulospinal system, and which conveys both exteroceptive and cutaneous information to the vestibular nuclei. This tract may be confined to cervical levels of the cord; the degree of crossing of its fibres is uncertain, but they are thought to terminate in restricted regions of the lateral and descending vestibular nuclei. The cells of origin of the tract probably correspond to those of the spinocerebellar system.

3. Intersegmental Tracts

The remaining fibres of the anterior funiculus constitute the **anterior intersegmental tract**. It consists of intersegmental fibres, some of which have crossed from the opposite side. Intermingled with them are believed to be reticulospinal and descending autonomic fibres. The tract is continuous above with the medial longitudinal fasciculus. Although it is perhaps preferable to combine together all the intersegmental or propriospinal fibres in the anterior funiculus under one term, a number of separate entities have been described in this region by the classical neuroanatomists of the past, as in the case of the propriospinal tracts elsewhere in the cord (Nathan and Smith 1959), and the finer details of distribution of such fibres will be briefly considered with the dorsolateral tract (p. 881).

TRACTS IN THE LATERAL FUNICULUS

1. Descending Tracts

(*a*) **The lateral corticospinal tract** extends throughout nearly the whole length of the spinal cord. It gradually diminishes in size as it descends, and it ends about the level of the third or fourth sacral segment. On transverse section it occupies an oval area anterolateral to the posterior grey column and medial to the posterior spinocerebellar tract (7.44); in the lumbar and sacral regions, where the posterior spinocerebellar tract is absent, the lateral corticospinal tract reaches the surface of the spinal cord. The anterior and lateral corticospinal tracts are widely accepted as important motor pathways in the spinal cord, but in recent years physiological experiments have focused attention on other important functional associations of these pathways (pp. 890, 891). They contain about a million fibres of varying diameter, 70 per cent of which are myelinated. About 90 per cent of the fibres have a diameter of 1–4 μm, about 9 per cent are from 5 to 10 μm and less than 2 per cent range from 11 to 22 μm. However, it should be noted that here, as at many other points in the nervous system, these figures, and particularly those concerning the small-diameter fibres, may well need substantial revision in the future, when systematic quantitative studies using the electron microscope have been carried out. Many of the fibres arise from cells in the motor area (area 4 of Brodmann) in the frontal lobe; the thicker fibres (11–22 μm in diameter) are believed, from cell and fibre counts, to be the axons of the giant pyramidal cells of Betz (p. 1007). Some of the fibres come from other layers in area 4, but experimental evidence shows that at least two-thirds of the corticospinal fibres arise elsewhere. A number start from other cortical areas in the frontal and parietal lobes, especially areas 3, 1, 2 and 6 of Brodmann. However, the origin of about 50 per cent only of the total number of fibres can be thus explained (Lassek 1954). All the fibres descend through the internal capsule of the cerebrum, traverse the cerebral peduncle and pons and enter the pyramids of the medulla oblongata. In the lower part of the medulla oblongata a variable number, usually about two-thirds, cross the median plane and turn caudally in the lateral funiculus of the spinal cord as the *crossed lateral cortiospinal tract*, while the

7.45 Diagram to illustrate the general plan of segmental organization of the fibres in the posterior funiculus, the lateral corticospinal tract and the lateral and anterior spinothalamic tracts. The probable cross-sectional areas of these tracts are enlarged to provide adequate space. (After Dr. O. Foerster, by kind permission of Springer Verlag, Berlin.)

remainder do not cross but are continued into the same side of the spinal cord, where they form the *anterior corticospinal tract*. The lateral corticospinal tract also contains some fibres which are derived from the cerebral hemisphere of the same side (*uncrossed lateral corticospinal fibres*)–see Fulton and Sheehan (1935).

The corticospinal tracts, anterior and lateral, thus defined, have commonly been described as the *pyramidal* pathway, a usage which is likely to persist. It is true that they consist of the fibres which pass through the pyramids of the medulla oblongata (p. 898), but the same expression is sometimes also made to include *corticobulbar* fibres which diverge cranial to this level to terminate in association with the motor nuclei of the cranial nerves (p. 996). But the chief objection to the continued use of the term 'pyramidal' lies elsewhere. The corticospinal tracts have a single distinction, in that they descend without interruption from cerebral cortex to a termination in the spinal grey matter; and in this they have become contrasted with a group of other descending tracts, such as the rubrospinal and vestibulospinal and others, the fibres of which form intermediate synapses with nerve cells in the cerebrum and brainstem, and which have been collectively labelled as 'extrapyramidal'. The functional distinction between these so-called 'systems' is rapidly becoming blurred, and clinical concepts based upon them are in process of gradual dissolution. The more recent revelation of the reticular system (p. 943) has added further confusion, because in both structural and functional details it overlaps much of the 'extrapyramidal' system. The latter concept is probably to be regarded as obsolescent, and for this reason the term 'pyramidal' is perhaps also better avoided.

Throughout its course in the cerebrum and brainstem, with the possible exception of the pons, somatotopic arrangement of its fibres has been established, chiefly as a result of clinical observations in man (7.45). The details of this arrangement will be described in the appropriate sections below. Whether the same kind of distribution obtains in the medullary pyramids and spinal cord is a matter of controversy (Foerster 1936; Barnard and

Woolsey 1956), but in the cord the longer fibres are said to be most superficial, the shorter ones being internal to them. The majority of the corticospinal fibres, in both ventral and lateral tracts, probably terminate in synapses with interneurons situated in the basal part of the dorsal grey column and spreading ventrally from this through the intermediate zone into the ventral grey column, laminae IV–VII (*see* p. 883; and Hoff and Hoff 1934; Chambers and Liu 1957; Nyberg-Hansen 1969). In these experimental investigations, mostly in cats, no corticospinal fibres were traced as far as motor neurons; but monosynaptic connexions of this kind have been described in monkeys (Liu and Chambers 1964), and it is presumed that such findings indicate similar arrangements in the human spinal cord, perhaps in greater numbers. Nevertheless, it appears certain that the majority of corticospinal fibres, which influence motor neurons innervating skeletal muscle, do so through polysynaptic chains of interneurons. Physiological evidence suggests that this influence is facilitatory with regard to flexors and inhibitory in the case of extensors—the reverse of the effects mediated by the lateral vestibulospinal tract (p. 875). Both a and γ motor neurons are influenced by corticospinal fibres. There is some evidence that corticospinal fibres from the precentral or motor cortex terminate in a more ventral position than those from the postcentral or sensory area (Scheibel and Scheibel 1966), and a considerable number of them (perhaps up to one quarter in some cases) reach their termination without crossing the midline. The influence of corticospinal fibre activity on the transmission of sensory information has been mentioned elsewhere (pp. 890, 892). The nature of the modifications of sensory input thus effected have been discussed *in extenso* by Schmidt (1973), Gordon (1973), Tone (1973), and McCoskey and Torda (1975).

For the most detailed review of the evolutionary status, neuroanatomy, electrophysiology and possible functional roles of corticospinal neurons in general consult Phillips and Porter (1977).

Ascending spinopontine and *spinocortical fibres* have been described in both the anterior and the lateral corticospinal tracts in the cat (Walberg and Brodal 1953) and in man (Nathan and Smith 1955), derived from all levels of the spinal cord, but especially from cervical segments. They may be concerned in the transmission of cutaneous afferent information (Brodal and Walberg 1952). These circumstances add further confusion to the traditional concepts of 'pyramidal' and 'extrapyramidal' systems.

(b) **The rubrospinal tract** is composed of nerve fibres of variable diameter which descend from large and small cells in the red nucleus; immediately caudal to this they decussate, and in the lateral funiculus they form a relatively compact band ventral to the lateral corticospinal tract (7.43, 44). Similar fibres have been said to be derived from other nerve cells in the tegmentum of the midbrain caudal to the red nucleus, constituting a *tegmentospinal tract* associated with the ventral border of the rubrospinal tract (Woodburne *et al.* 1946). The latter was first established in the cat, in which animal it has been most extensively studied; but it is also well authenticated in the macaque monkey, and in both these species it descends through the whole cord to lumbosacral levels. It is important to note that fibres from the smaller nerve cells of the red nucleus reach the most caudal segments of the cord. Because an erroneous view was at one time current that only the largest cells in the nucleus (p. 938) were the source of such long fibres, and since there are few large cells in the human red nucleus, a persuasion has gradually developed that the rubrospinal tract is negligible in man, an opinion doubtless aided by the dearth of information from clinical observation (Nathan and Smith 1955). Since the tract is well attested in the monkey and chimpanzee, it may exist in mankind; and although the details stated here apply to other mammals, it is wiser to maintain an open verdict as regards the human condition.

In their origins from the red nucleus the rubrospinal fibres show, in the cat (Pompeiano and Brodal 1957), a somatotopic arrangement on a dorsoventral plan, the shortest fibres arising in the most dorsal cells and proceeding to cervical levels in the cord to influence, presumably, the muscles of the neck and upper limb—probably through intermediary neurons. The axons of the most ventral rubral nerve cells are considered to be concerned with the movements of the lower limbs. Some of the fibres leave the tract in the brainstem to influence the motor neurons of cranial nerves, possibly through interneurons in the reticular formation, and some of these *rubrobulbar* axons may reach the ipsilateral inferior olivary nucleus (Walberg *et al.* 1958).

The distribution of the terminals of the rubrospinal fibres is similar to that of the corticospinal tract, largely to laminae V–VII (**7.47**B), except that the rubrospinal terminals do not extend quite so far ventrally, none reaching the motor neurons. This spatial overlap of the terminals of the two tracts suggests a close interaction of their effects on the motor neurons through a common interneuron pool. It is worthy of note here (see also p. 938) that the red nucleus receives afferent fibres not only from the cerebellum but also from the cerebral cortex, including the 'motor' cortex (area 4) and other areas which are concerned in the origins of the corticospinal pathway. This corticorubral projection displays a somatotopic arrangement in the cat (Mabuchi and Kusama 1966) and in primates (Kupyers and Lawrence 1967) similar in its details to the corticospinal pathway. Both tracts appear to exercise somewhat similar effects on the spinal motor neurons, such as facilitation of those concerned with flexor musculature; and it is therefore probably misleading to separate them, either in physiological or clinical thinking, as 'pyramidal' and 'extrapyramidal'. However, it would be premature to formulate a revised outlook upon these tracts and their funicitoning *in man*, despite the highly suggestive data available from investigations in other primates.

It is appropriate to mention here that the cells of origin of the *tegmentospinal fibres* mentioned above are thought to receive the synaptic terminals of *tectotegmental fibres* from the superior colliculi of both sides. Further, some direct tectospinal fibres from both colliculi may accompany the tegmentospinal fibres. This complex of descending fibres (which are closely associated topographically with the rubrospinal system) are accordingly often grouped as the *lateral tectotegmentospinal tract*.

(*c*) **The lateral reticulospinal tract** is usually described as medial to the rubro- and corticospinal fibres, but its constituent fibres are to some degree dispersed among those of the neighbouring tracts, and in some experimental mammals, reticulospinal fibres have been demonstrated descending lateral to the rubro- and corticospinal fibres (*see* **7.47**C). This description applies to experimental animals, for although there is little doubt that reticulospinal pathways exist in the human spinal cord, there is as yet little direct evidence of this. The lateral reticulospinal tract differs in several characteristics, anatomical and physiological, from the medial tract (p. 875). Its axons are largely derived from large nerve cells in the medullary zone of the brainstem reticular formation (*nucleus reticularis gigantocellularis*), but also from smaller cells, this origin according with the varying diameters of the reticulospinal fibres. In contrast with the pontine reticulospinal fibres in the anterior funiculus, the lateral tract is largely crossed, but some fibres also remain on their side of origin. Each half of the medullary reticular formation hence exerts a distinctly bilateral effect on the spinal cord. The terminals of the lateral, like the medial reticulospinal tract, form synapses with interneurons in the intermediate grey region and the medial zone of the anterior grey column, but they are more dorsal in position (chiefly lamina VII), though some may reach motor neurons (lamina IX). These details apply chiefly to the cat (Nyberg-Hansen 1965), and in the same study it was established that reticulospinal fibres descend to all levels of the cord, contrary to earlier views (Torvik and Brodal 1957). The medullary origin of the lateral reticulospinal tract approximates in extent to the brainstem 'inhibitory area', and this again contrasts with the facilitatory effect of the medial tract. In addition to these motor influences, which extend to both α and γ neurons in the activities of the medial and lateral tracts, the reticulospinal axons may also modify the transmission of afferent impulses. (See also **7.47**C and pp. 943–953 for details and bibliography concerning the pontobulbar origins of reticulospinal tracts, their aminergic/non-aminergic status, and further comments concerning possible functional roles.)

(*d*) **Descending autonomic fibres**, mediating both direct and crossed connexions between brainstem visceral 'centres' and the preganglionic nerve cells of the spinal cord, must obviously exist, as is attested by an abundance of physiological evidence. Nevertheless, accurate knowledge of their positions and terminations in the spinal cord is exiguous and conflicting. That such fibres are predominantly in the lateral funiculus is generally agreed, but it is likely that a smaller number also descend in the anterior funiculus. The failure of experimental efforts to define any compact groups of degenerating nerve fibres in the spinal cord as a result of hypothalmic or brainstem lesions has naturally led to the concept that descending autonomic connexions are diffusely dispersed through the cord. But individual investigations have associated them with the lateral corticospinal, reticulospinal and intersegmental tracts, and have also assigned some of them to a superficial position in the lateral funiculus (Enoch and Kerr 1967). The descending pathways are considered to consist for the most part of small-diameter fibres, probably arranged in polysynaptic chains. Since lesions which were supposed to interrupt reticulospinal fibres have been associated with disturbances of visceral function, such as vasomotor and sudomotor control (Johnson *et al.* 1952), it is probable that some of these fibres are to be regarded as autonomic. Vasomotor changes can be elicited by stimulation of the precentral area of the cerebral cortex, and there is thus a possibility that some corticospinal fibres are autonomic in function. In sum the evidence, much of it negative in character, suggests that the descending autonomic fibres are dispersed through much of the anterolateral region of the cord; hence lesions in any part of this are likely to be associated with disturbance of visceral control.

(*e*) **The olivospinal tract** (of Helweg) is described in considerable detail in most textbooks, and these descriptions are markedly uniform in assigning it to a small triangular area, as seen in transverse spinal sections, immediately lateral to the most lateral of the issuing ventral root fibres. It is said to be confined to upper cervical segments, and to contain fibres which end in the ventral grey column, and also fibres which form an ascending *spino-olivary tract* (*see* p. 881, and Jansen and Brodal 1954). The olivospinal tract was identified in the human spinal cord, but there appears to be no experimental evidence that its axons are in fact derived from the nerve cells of the inferior olivary nucleus (Brodal 1969). Because of this uncertainty it has been suggested that the tract be named *bulbospinal*, but this merely displaces the uncertainty elsewhere. These conflicting views serve to emphasize the need of further evidence in this matter.

2. Ascending Tracts

(*a*) **The posterior spinocerebellar tract** is a flattened band situated at the periphery of the posterior region of the lateral funiculus; medially it is in contact with the lateral corticospinal tract, dorsally with the dorsolateral tract. It begins about the level of the second or third lumbar nerve, increases in size as it ascends, and finally passes into the cerebellum through the inferior cerebellar peduncle. Its fibres are the axons of the cells of the ipsilateral thoracic nucleus. This nucleus receives *afferent* impulses from two sources. (1) Many of the long ascending fibres of the posterior funiculus give off collaterals which form synaptic connexions with the cells of the thoracic nucleus. (2) Terminals of the intermediate ascending fibres of the posterior column may behave in the same way, especially in the thoracic segments of the spinal cord. It is uncertain which of these two sources preponderates in the formation of the posterior spinocerebellar tract.

(*b*) **The anterior spinocerebellar tract**, as identified in transverse section, is a crescentic, flattened tract which occupies the periphery of the lateral funiculus, in front of the area occupied by the posterior spinocerebellar tract. The precise origin of its constituent fibres cannot be regarded as settled, but they are usually said to be derived from the large cells of the posterior grey column. They are therefore secondary neurons on the spinocerebellar pathway. The primary neurons concerned are probably similar to those described above in connexion with the posterior spinocerebellar tract, but the ascending axons of the secondary neurons are mostly derived from the opposite side of the spinal cord, only a small proportion ascending on the ipsilateral side (Jansen and Brodal 1954). Experimental evidence indicates that the 'secondary neurons' are in fact sited dorsolaterally in the *ventral* grey column (Ha and Liu 1968). The tract commences in the upper lumbar region and extends upwards to cranial pontine levels, where it turns to descend dorsally in the superior cerebellar peduncle to reach the cerebellum.

The fibres in the spinocerebellar tracts have a laminated arrangement, with those carrying impulses from the lower limb being placed most superficially (Yoss 1952/3; Smith 1957). They convey to the cerebellum both exteroceptive and proprioceptive impulses arising in receptors of the skin and locomotor apparatus which are essential for the adjustments of muscle and for synergic control during the performance of voluntary movements. As already stated (p. 870), the thoracic nucleus diminishes in size as it is traced upwards, and does not extend cranial to the first thoracic or last cervical segments. It would appear, therefore, that the posterior spinocerebellar tracts are concerned chiefly with the trunk and lower limbs, and evidence has been adduced to show that the corresponding proprioceptive impulses from the upper limbs travel by the posterior external arcuate fibres which originate in the accessory cuneate nucleus in the medulla oblongata (p. 899), forming a *cuneocerebellar tract*. In addition to the classical spinocerebellar pathways there is considerable evidence for at least two indirect projections, interrupted by synapses in the inferior olivary and lateral reticular nuclei of the medulla, both of which have been shown to project to the cerebellum, receiving their afferents respectively from the dorsal funiculi and the spinothalamic tracts. Consult Brodal (1969) for further details and discussion.

It has long been established that the spinocerebellar tracts convey proprioceptive information; more recently physiological evidence has demonstrated that the posterior, or dorsal, tract also mediates impulses from touch and pressure endings. Both tracts display a marked somatotopic termination in the cerebellum in experimental animals, including monkeys, and in man an arrangement which will be considered in greater detail later (p. 928). The posterior spinocerebellar tract ends in an area of cerebellar cortex concerned with the hind limb, while its associated cuneocerebellar tract reaches a fore-limb area. Ventral spinocerebellar fibres end only in the hind-limb area; but physiological evidence (Oscarsson and Uddenberg 1964; Oscarsson 1965) indicates the existence of ascending fibres functionally associated with the anterior spinocerebellar tract and serving tendon stretch receptors in fore-limb muscles; this has been tentatively named the *rostral spinocerebellar tract*. A spinocerebellar projection concerned with proprioceptive information from neck muscles has been described as originating in the *nucleus cervicalis*. This entity, chiefly identifiable in young animals and not satisfactorily described in the human spinal cord, exists at high cervical levels only. According to Cummings and Petras (1977) cerebellar ablations in newborn dogs are followed by a retrograde response in cells of this nucleus, and in particular in the *nucleus cervicalis centralis*, and not in the *nucleus cervicalis lateralis*, as was claimed by Rexed and Brodal (1951). Petras (1977) has investigated the spinal origins of spinocerebellar tracts in monkeys, by parallel observations of retrograde degeneration and retrograde horseradish peroxidase transport; these studies yielded similar results. (See also below.)

Some of the neuronal somata which contribute to the spinocerebellar tracts have been studied using intracellular recording techniques. Some receive the termination of only one type of exteroceptive or proprioceptive afferent, whilst others show a convergence and interaction of different types of afferent upon single cells. Their impulse transmission characteristics are modified by activity in tracts which descend from the brainstem.

(*c*) **The lateral spinothalamic tract**, which lies medial to the anterior spinocerebellar tract in the lateral funiculus of the spinal cord, as noted above (p. 879), is continuous with the anterior spinothalamic tract. As in the case of the ventral part of the tract, in the anterior white column (p. 876), it should be noted that the simple zonal, 'modality-specific' pattern of fibres, still widely ascribed to be spinothalamic tracts, contrasts sharply with the great variation in physiological responses of the cell types investigated in the grey matter laminae IV–VII, some of which probably give rise to spinothalamic fibres. As we shall see (pp.

890, 891), these variations include marked differences in the size, characteristics, specificity and somatotopic mapping of the receptive fields of the different cells, and in the degree of convergence of different functional types of primary afferent fibre on to single interneurons. In view of this, it seems likely that some, at least, of the spinothalamic fibres transmit complex patterns of information, and the assumption that they are all modality-specific is certainly an oversimplification. (Reference is made to the possible role of the substantia gelatinosa in the pain pathway on p. 890.)

The somatotopic lamination of the anterior spinothalamic tract is also evident in the lateral. This arrangement is of considerable practical importance to the neurosurgeon, and it is maintained throughout the passage of the lateral tract through the medulla oblongata and the pons. In the midbrain, however, the fibres from the lower limbs extend dorsally, and in this part of their course it is possible for the surgeon to divide the pain and temperature fibres from the upper limb and trunk without injury to the corresponding fibres of the lower limb (Earl Walker 1942).

Although it is generally accepted that the lateral spinothalamic tract is the predominant pathway for somatic pain and thermal sensibilities, it has not infrequently been suggested that an alternative pathway may exist and is provided by a series of intersegmental fibres with their neuronal bodies situated in the grey matter of the spinal cord (p. 946). In addition, the spinotectal tract has been regarded by others (e.g. Earl Walker 1943) as an alternative pathway for painful and thermal sensibility.

It is convenient to add at this juncture a brief reference to the *spinocervical tract* and an associated *spinocervico-thalamic system* of connexions. The *lateral cervical nucleus*, though small in man, is an accepted entity and is situated in the lateral funiculus lateral to the dorsal grey column in the upper two cervical segments of the cord (*see* Truex *et al.* 1970). The tract is derived from nerve cells in the dorsal grey column, particularly from laminae IV to VI, among those giving rise to axons of the spinocerebellar tracts. It extends upwards from lumbosacral levels. The further projection from the lateral cervical nucleus was at first considered to enter the cerebellum, but the most recent anatomical and physiological studies indicate that the axons from the cells in this nucleus ascend with the medial lemniscus to the nucleus ventralis posterior lateralis of the contralateral thalamus, that they mediate light tactile and pressure modalities of sensation (and perhaps also joint sensibility), and are hence for the present to be associated with the spinothalamic system (Morin and Catalano 1955; Ha and Liu 1966). Some of these neurons may also respond to noxious stimuli (Brown and Gordon 1977). Interestingly, a column of nerve cells, considered to be identical with the lateral cervical nucleus, has been described in the rat as extending throughout the spinal cord (Gwyn and Waldron 1968).

Little definite information is available concerning the pathway followed by impulses arising in connexion with painful pathological conditions of viscera. It has been clearly shown that the first neuron fibres travel in the splanchnic nerves, and it seems certain that they enter the spinal cord via white rami communicantes and dorsal spinal roots. Whatever pathway they follow in the spinal cord they are all interrupted by the operation of bilateral cordotomy of the lateral funiculi carried out at the level of the first thoracic segment (Hyndman and Wolkin 1943) and on this account it has been suggested that they travel in the lateral spinothalamic tract. The reader should note that, although hitherto defined as being formed by the union of the lateral and anterior spinothalamic tracts, the spinal lemniscus may now be defined more accurately as the brainstem continuation of the lateral spinothalamic tract alone.

Owing to the fact that the fibres which form the lateral spinothalamic tract cross at once, decussating with the corresponding fibres of the opposite side, lesions affecting the commissural area, such as occur in the disease, *syringomyelia*, produce a bilateral loss of pain and thermal sensibilities for the areas represented in the particular segments involved.

(*d*) **The spinotectal tract** is medial to the anterior spinocerebellar tract and anterior to the lateral spinothalamic tract, the three tracts being intimately related throughout their ascending course in the spinal cord and brainstem. Its constituent fibres start in laminae of the deeper contralateral grey matter and soon cross the median plane to reach the lateral funiculus. The tract is most easily identified at cervical levels. Its fibres ascend to the midbrain where they terminate in the superior colliculus of the tectum. They provide an ascending pathway for spinovisual reflexes. In this connexion it is to be remembered that the superior colliculi constitute a reflex centre in the visual path and are not concerned with the transmission of visual impulses to the cerebral cortex. Afferent impulses passing up the spinotectal tract result in movements of the head and eyes towards the source of stimulation. However, some fibres topographically associated with the tract pass onwards to the ventral thalamus, and have been considered to mediate cutaneous pain and perhaps other modalities.

(*e*) **The dorsolateral tract** (of Lissauer) is a small strand of fine myelinated and non-myelinated fibres situated between the tip of the posterior grey column and the surface of the spinal cord, close to the dorsal roots. It is formed in part by fibres of the lateral bundle of the dorsal roots which bifurcate into ascending and descending branches. The ascending branches travel one or two segments in the tract; they give collaterals to and terminate around cells in the posterior grey column (pp. 887, 890). Although the dorsolateral tract is present throughout the cord, no fibres travel more than a few segments in it, and it is hence more appropriate to regard it as a *fasciculus*. This is all the more desirable in view of the experimental confirmation (Poirier and Bertrand 1955; Earle 1952; Szentágothai 1964) of an old suggestion (Flechsig 1876) that the tract contains many propriospinal fibres, the contribution from entering dorsal root fibres amounting to no more than 25 per cent of the total at any particular level (Earle 1952). (It is interesting to note that Flechsig's suggestion considerably predated Lissauer's descriptions in 1886.) Many of the propriospinal fibres are axons of the small neurons of the substantia gelatinosa (laminae II and III, *see* p. 883).

(*f*) **Spinoreticular fibres** have been described in the lateral funiculus as a result of experiments in the cat (Brodal 1949), these are intermingled with spinothalamic fibres and hence do not form a discrete identifiable tract. The majority of these ascending fibres do not cross, but there appears to be a generalized and bilateral projection to nuclei at all levels of the brainstem reticular formation, though particularly to those at pontine and medullary levels. The existence of such spinoreticular connexions has also been demonstrated in the human brainstem (Mehler 1962).

(*g*) **The spino-olivary tract** is described as arising from neurons in the deeper laminae of the spinal grey matter, the axons of which cross the midline and then ascend near the surface of the cord at the junction of the anterior and lateral white funiculi, to terminate in specific 'spinal' regions of the dorsal and medial accessory olivary nuclei (Mizuno 1966). The tract conveys information derived from cutaneous as well as proprioceptive receptors—both muscle and tendon organs (Grant and Oscarsson 1966). A functionally similar pathway which constitutes a *dorsal spino-olivary system* has been suggested on physiological grounds as ascending with the dorsal white funiculi and then relaying in the dorsal column nuclei before projecting on to the accessory olivary nuclei (Oscarsson 1967). Interest in these pathways has heightened in company with the increased awareness of the importance of the olivocerebellar system of climbing fibres with their specific localized excitatory effects on the cerebellar Purkinje cells (pp. 916, 919).

3. Intersegmental Tracts

The lateral intersegmental tract constitutes the remainder of the lateral funiculus, separated ventrally from the anterior intersegmental tract by the emerging fibres of the ventral nerve roots. It consists of intersegmental fibres, some of which have passed from the opposite side of the spinal cord and, probably, reticulospinal and descending autonomic fibres.

TRACTS IN THE POSTERIOR FUNICULUS

1. Ascending Tracts

This funiculus comprises two large ascending tracts, the fasciculi gracilis and cuneatus, which are separated from each other by the postero-intermediate septum. Essentially the dorsal funiculi mediate projections from the dorsal root ganglia *and* the dorsal grey columns upon the gracile and cuneate nuclei; but this simple description must be modified, because many of the fibres do not reach the nuclei, and many are *propriospinal*. Moreover, some ascending fibres reach the nuclei by other routes.

The fasciculus gracilis commences at the caudal limit of the spinal cord and is composed mainly of the long ascending branches of the medial bundle of fibres of the dorsal nerve roots; but a considerable fraction (15 per cent according to Rustioni and Dekker 1974) are axons of *secondary* neurons, whose perikarya are in the ipsilateral dorsal grey column (laminae III and IV). These remarks also apply to the cuneate fasciculus. They run upwards in the posterior funiculus, and as the tract ascends it receives accessions from each dorsal root. The fibres which enter in the coccygeal and lower sacral regions are thrust medially by the fibres which enter at higher levels. The fasciculus gracilis, which contains fibres derived from the lower thoracic, lumbar, sacral and coccygeal segments, occupies the medial part of the posterior funiculus in the upper part of the spinal cord (7.45).

The fasciculus cuneatus commences in the mid-thoracic region and derives its fibres from the dorsal roots of the upper thoracic and cervical nerves and, in consequence, is situated lateral to the fasciculus gracilis (7.45, 46).

Both fasciculi are myelinated, and the fibres are larger in the fasciculus cuneatus than they are in the fasciculus gracilis, probably due to the finding that proprioceptive fibres (which are of greater diameter) remain in the cuneate fasciculus, but leave the gracile fasciculus at lumbar levels (*vide infra*). Both fasciculi contain the central processes from cells in the spinal ganglia, i.e. receptor or *primary afferent* neurons, and some of these pass without interruption or decussation to the medulla oblongata, where they end in the *gracile* and *cuneate nuclei*, in which the second neurons of this pathway begin. The majority of the fibres of the second neurons sweep ventrally round the central grey matter (7.61) as the *internal arcuate fibres*, and take part in the *decussation of the lemnisci*. Thereafter, as the *medial lemnisci*, they ascend on each side to the ventral nucleus of the thalamus (p. 958) and are there relayed to the cortex of the post-central gyrus (areas 3, 1 and 2). Some of the second neurons form *posterior external arcuate fibres* (p. 902) which pass to the cerebellum. Presynaptic dendrites, which effect dendro-dendritic connexions, have been described in the cuneate nucleus of the macaque (Wen *et al.* 1977), but the origin of the fibres from which they branch remains obscure.

These two tracts, which occupy nearly the whole of the posterior funiculus, convey proprioceptive sensibility, including vibration and pressure and some elements of exteroceptive tactile sensibility. The fibres concerned all pass up to the medulla oblongata in the ipsilateral posterior funiculi, together with the fibres which convey sensations of posture and of movements, both active and passive. It should also be noted, however, that during their passage in the cord a number of the long ascending dorsal column fibres give rise to a series of collateral branches which enter the dorsal column of spinal grey matter.

The laminar somatotopic pattern of the gracile and cuneate tracts has been shown to be more intricate than the segmental arrangement described above; there is also a segregation of fibres on the basis of modality, those conducting impulses from hair receptors being most superficial, followed by fibres mediating tactile and vibratory sensibility in successively deeper layers (Uddenberg 1968). Physiological experimentation has also indicated that the somatotopic arrangement is carried through to the nuclei gracilis and cuneatus, and that these contain numerous interneurons; unit recording in the nuclei shows a high specificity in fibres arriving along the two tracts. Stimuli immediately outside the receptive field from which a single unit can be activated may show an inhibitory effect. There is evidence of a projection of descending fibres from the pre- and post-central cerebral cortex which travel in the corticospinal projection and exert facilitatory and inhibitory influences on the interneurons of the gracile and cuneate nuclei. Similar projections also originate in the brainstem reticular formation (pp. 943–953). It is to such modulating effects that the highly discriminative nature of activity in the dorsal funicular pathways is probably due, and these tracts, like others, can no longer be regarded as a simple through route for sensory information. (For further discussion see p. 891).

2. Descending Tracts

A somewhat confusing number of small tracts of descending fibres have been described in the dorsal funiculi by a variety of observers, who worked chiefly in the late nineteenth century and based their accounts on pathological appearances in human spinal cords. Thus, extending through cervical and upper thoracic levels, in the medial part of the cuneate tract, is the *comma tract* (of Schultze), also known as the *semilunar tract* or, more recently, the *interfascicular fasciculus*. In lower thoracic segments a thin superficial strand of fibres has been described, but this is almost

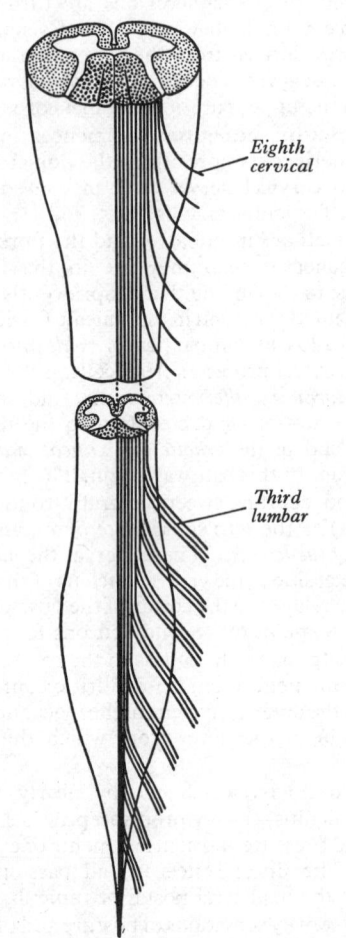

Eighth cervical

Third lumbar

7.46 Diagram showing the lamination of the fibres in the posterior funiculus. The spinal cord is viewed from the dorsal aspect. The drawing shows that the posterior funiculus is formed (in part) by the long ascending fibres of the dorsal roots and that the sacral fibres adjoin the median plane, the lumbar to their lateral side, the thoracic more laterally, and the cervical most lateral of all. For dorsal fibres arising elsewhere *see* p. 881.

certainly the cranial end of the *septomarginal tract*, which is ascribed to a deeper situation bordering the posterior median septum. The so-called '*oval field of Flechsig*' and the '*triangular field of Gombault and Philippe*' are also merely different levels of the septo-marginal tract, which appears to alter its shape and position remarkably as it descends in the cord from lower thoracic segments. It is tempting to assume that the interfascicular fasciculus and septomarginal tract are also continuous, but this remains a matter of doubt. They are reputed to differ in one respect; the former is said to consist only of the descending branches of entering dorsal root fibres, whereas the septomarginal tract is considered to contain also large numbers of intersegmental connexions. It must be emphasized that almost all the literature concerning these descending dorsal column fibres is based on

clinical observation, experimental evidence being scant. For an exhaustive review consult Nathan and Smith (1959).

3. Intersegmental Tracts

Occupying the anterior or deepest part of the posterior funiculus is a small strand of fibres named the **posterior intersegmental tract**. It is somewhat crescentic on transverse section, and is just posterior to the grey commissure (7.43, 44); it is best marked in the lumbar region, but can be traced into the thoracic and cervical regions. Its fibres, which are intersegmental, are derived from the cells of the posterior grey column; they divide into ascending and descending branches which re-enter and ramify in the grey matter.

Further Aspects of Spinal Cord Organization

The more traditional and somewhat simplified account of spinal cord organization given in the preceding pages has been considerably expanded, complemented, and in a number of respects modified, in recent years following the intensive application by many investigators of both the classical and newer methods of neuroanatomy and neurophysiology (Eccles and Schadé 1964a and b; Lissák 1967; Ralston 1974). Their researches include more detailed comparative studies in a wide array of vertebrates, alternative schemes for the classification of the spinal cord grey matter based upon cytoarchitectonics, and more precise analyses of the dendritic patterns and axonal arborizations using modified Golgi techniques. The more recent degeneration techniques have added considerably to our knowledge of the sites of termination of dorsal spinal nerve root afferent fibres and their collaterals, the siting of terminals of fibre systems which descend from supraspinal levels, and some approach has been made to unravel the exceedingly complex intrinsic organization of the cord by placing minute focal lesions within the grey matter and following subsequent degenerative events in neighbouring regions. Many of these neuro-anatomical investigations, based upon light microscopical techniques, have been improved progressively by the application of quantitative histological methods, whilst detailed ultrastructural studies of the synaptic arrangements have provided new data; putative transmitter distribution studies progress rapidly.

This intensified interest in spinal cord structure has greatly assisted, and equally been assisted by, parallel studies in neurophysiology. In particular, the electrophysiological analyses of the dorsal and ventral spinal nerve root fibres under different conditions, recordings from different points within the grey matter, and unit recordings made with microelectrodes which have impaled single neuron somata, can with some success be correlated with the anatomical findings. As a result, whilst some of the classical views have been substantiated, others have needed a drastic revision and new concepts have been proposed.

LAMINAR ARCHITECTURE

For many years the *general outline* of the grey matter of the spinal cord as seen in transverse section provided an arbitrary terminological basis used by morphologists and experimentalists alike. In this manner, as we have seen elsewhere (p. 869), dorsal, lateral and ventral columns were recognized. The ventral column was considered to consist of a 'base' and a ventrolateral 'head', whilst the dorsal column was dignified with an 'apex', 'head', 'neck' and 'base' proceeding from dorsal to ventral. Between these, a rather imprecisely defined intermediate zone was recognized, whilst more easily defined subdivisions included the substantia gelatinosa, the thoracic nucleus (Clarke's column) and the various subgroupings of large motor neurons which, in considerable variety, characterize a part of the ventral column (p. 870). Whilst much valuable pioneering work was carried out using this scheme, the increasing volume of experimental analysis, both structural and functional, has revealed a relative lack of precision, proving a hindrance and prompting further attempts to classify cord structure (Rexed 1952, 1954, 1964). The most widely adopted scheme stemmed initially from extensive studies using thick as well as the more usual thin sections, stained by the method of Nissl for cell somata, of new-born, young and adult specimens of the feline spinal cord. Based upon observations on the size, shape, packing density, and cytological features of the neurons in different regions of the grey matter, nine *cell layers* or *laminae* have been distinguished which are roughly parallel with the dorsal and ventral surfaces of the grey matter, and which extend throughout the length of the cord, together with a region surrounding the central canal. As an example, the disposition of the laminae as seen in a transverse section through the fifth lumbar segment of the cord is shown in illustration 7.48A. General confirmation of the laminar pattern in *human* material has subsequently been provided by Schoenen (1973).

Briefly the constitution of the cell layers is as follows:

Lamina I is an extremely thin layer with an ill-defined boundary adjoining the white matter (within which outlying cell groups occur). It presents a reticular appearance because of the presence of many coarse and fine nerve fibre bundles intermingling in a variety of directions. It contains small, intermediate and fairly large neuron somata, many being spindle-shaped. Alternative names proposed are *lamina marginalis* or the *layer of Waldeyer* (who recognized a similar zone in 1888).

Lamina II consists of tightly-packed small cells crossed, especially medially, by numerous fibre bundles which enter from the dorsal funiculus. An outer (dorsal) zone, which appears darkly stained because of its high density of the smallest cells, stands in contrast to an inner (ventral) pale zone.

Lamina III consists of neuron somata which are in general larger, more variable in shape and less closely packed than those in lamina II.

Lamina IV, thicker than the preceding layers, is a loosely packed, heterogeneous zone permeated by many fibres. The cell somata vary greatly in size and shape, from small and round, through intermediate and triangular, to very large star-shaped profiles.

Laminae I–IV correspond to the general term *head* of the dorsal column of previous workers. Lamina II (and some workers consider in addition part or all of lamina III) corresponds to the *substantia gelatinosa* of earlier accounts, whilst the imprecisely defined *nucleus proprius* of the dorsal column roughly corresponds to some of the cell constituents of laminae III and IV.

Lamina V, a thick layer which includes the *neck* of the dorsal column, is divisible into a lateral one-third and a medial two-thirds. Both have a mixed population of cell somata, but the former contains a large number of prominent well-stained cells interlaced by numerous bundles of nerve fibres running transversely, dorsoventrally and longitudinally—hence the restricted use of the term 'formatio reticularis' for this region, particularly well seen in the cervical region, and recognized for over a century (Deiters 1865). (It should be noted, however, that in modern neurology the term *reticular formation* is used much more widely—pp. 943–953.)

Lamina VI, most easily recognized in the limb enlargements, particularly of young animals, consists of a medial one-third of small tightly-packed cells and a lateral two-thirds which possesses larger triangular or star-shaped cells, more loosely packed. Accordingly, in lamina VI the medial zone stains more heavily than the lateral zone, in contrast to lamina V where the converse applies. Lamina VI corresponds roughly to the topographical *base* of the dorsal column.

Laminae VII–IX present a variety of complex forms in the limb enlargements (7.48A), and to assist understanding the simpler arrangement found at thoracic levels is included for comparison (7.48B).

Lamina VII—this corresponds to much of the intermediate grey column or *zone* of previous authors. Within its confines it includes the prominent cells of the *thoracic nucleus* (Clarke's column) and the *intermediomedial* and *intermediolateral cell columns* at appropriate levels in the spinal cord—these have been detailed elsewhere (p. 872). The remaining large areas of lamina VII (i.e. between these cell columns and, in the limb enlargements, extending ventrally between lamina VIII and the constituent groups of IX) consist of a rather homogeneous population of medium-sized triangular or star-shaped cell somata.

Lamina VIII spans the base of the thoracic ventral column, but in the limb enlargements it is restricted to its medial aspect. It consists of a heterogeneous mixture of cell sizes and shapes from small to moderately large.

Lamina IX comprises the complex array of cell columns (p. 870) which include the very large somata of the *a* motor neurons, and numerous smaller cells.

The smaller cells include motor neurons which give rise to the small-diameter efferent fibres (*γ* efferents) to the muscle spindles, and numerous interneurons some of which may be the inhibitory Renshaw cells (*vide infra*). The location of the *γ* motor neurons was long in doubt, but studies of the retrograde changes in the cell somata following section of peripheral nerves (Nyberg-Hansen 1965) and intracellular recording with microelectrodes (Eccles *et al.* 1960) have demonstrated that these cells are dispersed between the *a* motor neurons within the motor cell columns. The precise location and morphology of the Renshaw cells, however, is still admittedly uncertain. Intracellular microelectrode recordings have indicated the presence of inhibitory interneurons in the ventral extension of cell lamina VII, where it is insinuated between laminae VIII and IX (Willis and Willis 1966). Studies of this region with the Golgi technique (Scheibel and Scheibel 1966) has failed to demonstrate typical Golgi type II neurons (with profuse, short, local arborizations of their axons)—the cell type long assumed to be the basis of Renshaw loop inhibition. Nevertheless, another view has been expressed (Szentágothai 1967), namely that neurons with longer axons are not incompatible with such an inhibitory function, and further, that the ventral extension of lamina VII receives the greatest density of initial collateral branches from the axons of the *a* motor neurons (and which are assumed to synapse with the Renshaw cells).

The remaining area of grey matter (**lamina X**) surrounds the central canal, and consists of the *dorsal* and *ventral grey commissures* and the *substantia gelatinosa centralis*.

FURTHER ASPECTS OF SPINAL LAMINAR ARCHITECTURE

It must be emphasized that the preceding description is the barest outline of the proposed scheme of classification and for further details the extensive original papers quoted should be consulted.

It should also be noted that the original scheme only applied in full to the spinal cord of the *cat* and no comparably detailed study was for long carried out on the spinal cord of man or any other primate (*see*, however, the more recent analysis by Schoenen 1973). Nevertheless, the same general principles of laminar organization (with doubtless appreciable variation in detail) are *considered* to apply to the spinal cords of *all* higher mammals (Rexed 1964). *The originator* of the scheme *proposed* the following tentative functional analysis of the laminar pattern (although a number of his conclusions have been *revised* by subsequent workers— *vide infra*).

Laminae I–IV were considered the main receiving areas for the cutaneous exteroceptive primary afferent fibre terminals and collateral branches. (For further details of this sensory input, *vide infra*.) From this region are initiated many complex polysynaptic reflex paths, both ipsilateral and contralateral, and both intrasegmental and intersegmental. It was also considered that from this region arise many of the long ascending tracts to higher centres (but *vide infra* for further discussion and alternative points of view).

Laminae V and VI were thought to receive most of the terminals from proprioceptive primary afferents and also profuse connexions from corticospinal fibres descending from the motor and sensory regions of the cerebral cortex, and descending systems from other sub-cortical centres. Accordingly, these laminae were regarded as being of great importance in the detailed regulation of movement patterns.

Lamina VII in its lateral part has extensive ascending and descending connexions with midbrain centres and the cerebellum (e.g. via spinocerebellar, spinotectal, spinoreticular, tectospinal, reticulospinal and rubrospinal tracts), and is hence of importance in the regulation of posture and movement. The medial part of lamina VII has a wealth of propriospinal reflex connexions with neighbouring regions of grey matter and adjacent segments

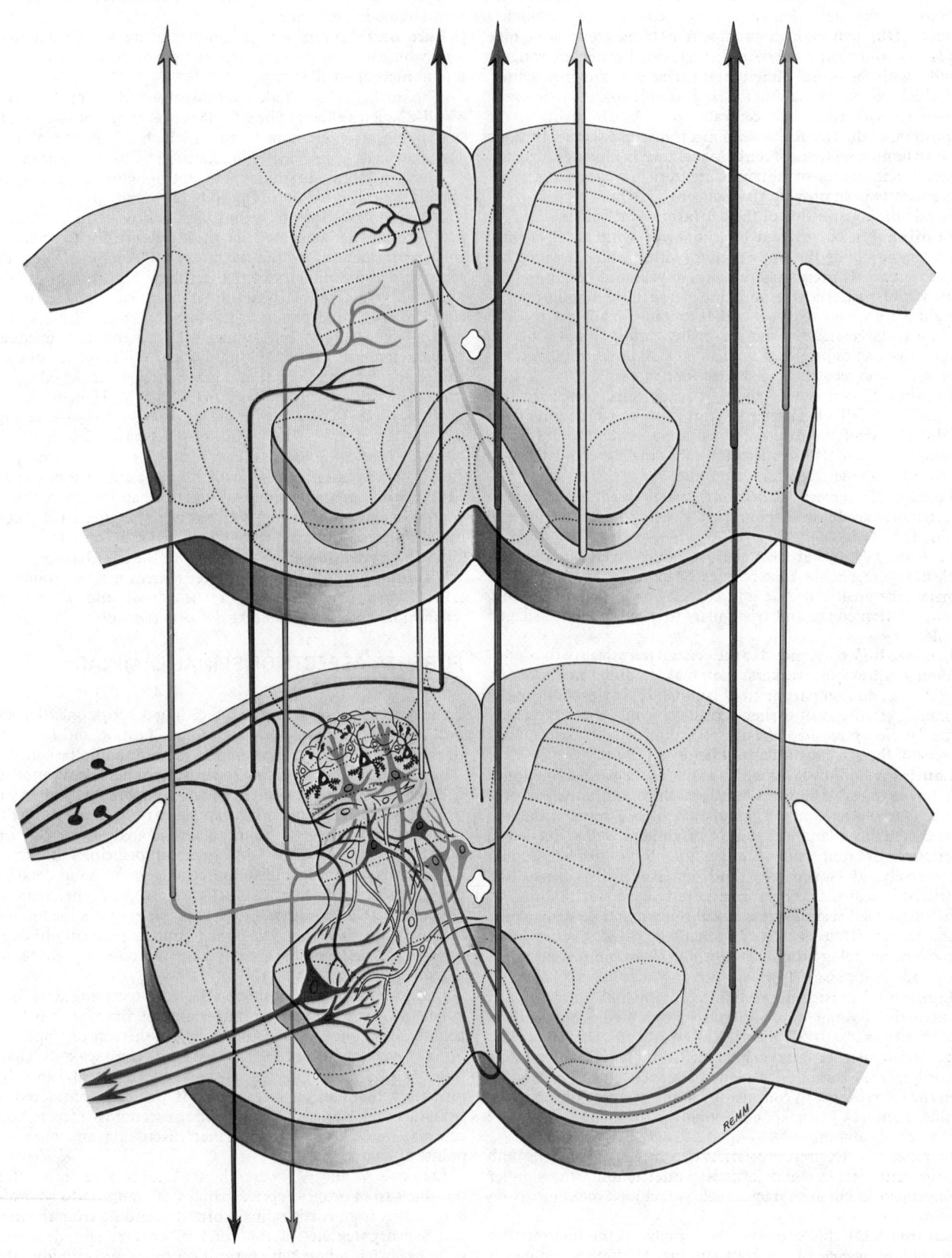

7.47A Simplified scheme of some of the major ascending tract systems of the spinal cord and some features of grey matter organization. Within the grey matter the dotted lines show the laminar pattern, within the white matter they are an approximate guide to the topography of the tracts. Attempts have been made to indicate in a simplified manner the overlapping of dendritic fields described in the text. An alpha and a gamma motor neuron (grey) are included, together with some of the structural features of the substantia gelatinosa which are described and illustrated more fully in 7.49. Some of the small substantia gelatinosa neurons are uncoloured, as are some interneurons in the deeper laminae. The larger substantia gelatinosa neurons are solid black. Large lamina IV cells, and associated ascending and descending intersegmental fibres are green. Primary sensory afferent fibres, including a fibre in the fasciculus gracilis, are purple. Spinotectal fibre—orange; anterior spinothalamic fibre—yellow; lateral spinothalamic fibre—magenta; and dorsal and ventral spinocerebellar fibres—blue. Compare with 7.47C.

concerned both with movement patterns and autonomic functions (p. 970). As we have seen, the ventral extension of lamina VII may be in part occupied by inhibitory interneurons (Renshaw cells—pp. 807, also *see* Renshaw 1941, 1946).

Lamina VIII again consists of a mass of propriospinal interneurons receiving terminals from adjacent laminae, profuse commissural terminals from the contralateral lamina VIII, and descending pathway terminals from interstitiospinal, reticulospinal, and vestibulospinal tracts, and the medial longitudinal fasciculus. Their axons influence both contralateral and ipsilateral motor neuron pools, perhaps directly, but more probably by an excitatory action on the small motor neurons which supply γ-efferent fibres to the muscle spindles.

Lamina IX consists of an admixture of a and γ motor neurons and also many interneurons.

The large a motor neurons supply the motor end plates of the extrafusal muscle fibres in the *motor units* of striated muscle (p. 514). They vary in size and physiological recording techniques have demonstrated two varieties, *tonic* and *phasic a* neurons (Granit *et al.* 1956). The former have a lower rate of impulse firing, a lower conduction velocity in their axons, and are assumed to be dimensionally smaller. Attempts have also been made to correlate these varieties with different structural and functional types of striated muscle fibre, e.g. *slow* and *fast* muscle—*see* p. 512. However, as yet there is no histological means of differentiating these types of a motor neuron. Additionally some of the large motor neurons (sometimes termed β *motor neurons*) have been considered to supply both extrafusal and intrafusal muscle fibres.

Similarly, there are at least two physiologically distinct types of γ *motor neuron*, the axons of which (fusimotor fibres) innervate the intrafusal fibres of the muscle spindles. It has been shown that the 'static' and 'dynamic' responses of muscle spindles (p. 860) are subject to separate control mechanisms mediated by *static* and *dynamic fusimotor fibres* which are distributed to *nuclear chain* and *nuclear bag* intrafusal fibres respectively. However, again it has proved impossible to differentiate the types of γ motor neuron somata histologically, although two types of neuromuscular junction—*plate endings* and *trail endings*—have been recognized, as have two varieties of γ-efferent myelinated nerve fibre (γ *1* and γ *2*) in peripheral nerves to muscle (Boyd 1962). How closely these varieties can be correlated has been disputed. For a detailed review of the motor innervation of muscle spindles and an extensive bibliography consult Barker (1974).

THE GEOMETRY OF SPINAL NEURONS

Since the unrivalled pioneering investigations of Cajal (1908), which foreshadowed so much of what was to follow and, indeed, remain today a primary source of information, there has gradually accumulated a volume of increasingly precise data on the form of spinal neurons. As we have seen, much has been learned from a study of the sizes, shapes, distribution, packing density and cytological characteristics of *cell somata* following staining by the Nissl technique or one of its modifications. However, as has been emphasized, such techniques provide only limited information and they must be supplemented by alternative techniques which allow analyses, preferably in quantitative terms, of the patterns of dendritic ramifications, and of the courses, arborizations and sites of termination of axons and their collateral branches. Furthermore, degenerative and other techniques (p. 815) are necessary for a precise determination of their connexions with other cells.

Excellent examples of such quantitative approaches may be found in (Aitken and Bridger 1961; Schadé 1964), in which the lumbosacral regions of the spinal cord of the cat were studied. Whilst the results cannot be examined in detail here, they contained the following kind of information. In the ventral zone of the ventral column of one complete spinal segment (lumbar 6) there was a total neuron population of about 7,000; of these 700 were considered to be the small cells from which the γ-efferent fibres to the muscle spindles originated; of the larger cells 3,276 were propriospinal (interneurons), 126 were classified as 'spinal border' cells, whilst 2,898 were a motor neurons. Thus, in this region about half the cells were interneurons and half motor

neurons; if the whole ventral column is assessed, the ratio is about 7 interneurons to 1 motor neuron, and if the intermediate zone and base of the dorsal column are included also, the ratio rises to about 13 to 1. Neuron packing density was estimated and was highest in the intermediate zone of grey matter (7 cells per 100 μm cube of tissue) and lowest in the ventral column (1–2 cells per 100 μm cube). The *total* surface area of individual cell somata and associated dendrites ranged from 11,000 to 97,000 μm², whereas the surface area of the *cell somata* alone ranged up to 25,000 μm²; the number of stem dendrites varied up to a maximum of 13,

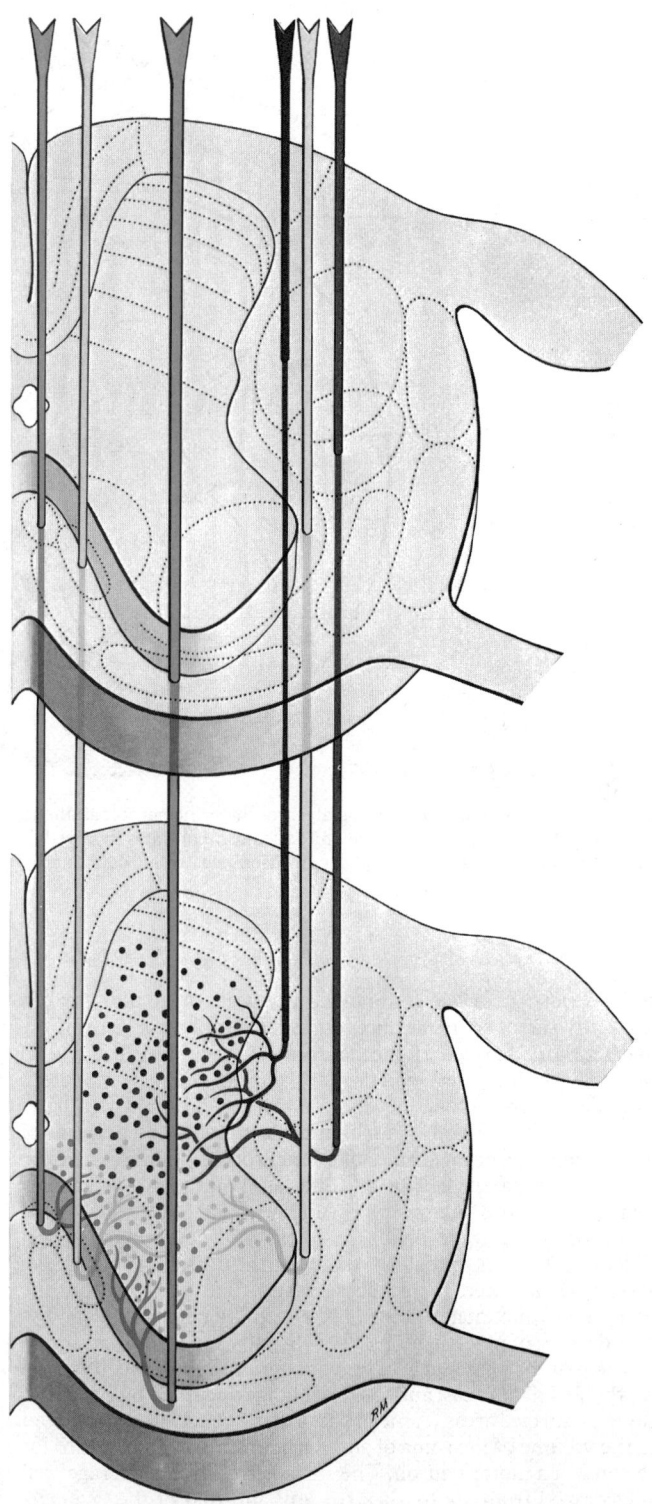

7.47B A simplified scheme of some of the major descending tract systems of the spinal cord including their overlapping zones of termination in the grey matter. The significance of the dotted lines is the same as for 7.47A. Corticospinal tract—mauve; rubrospinal tract—magenta; reticulospinal tracts—yellow; vestibulospinal tracts—blue.

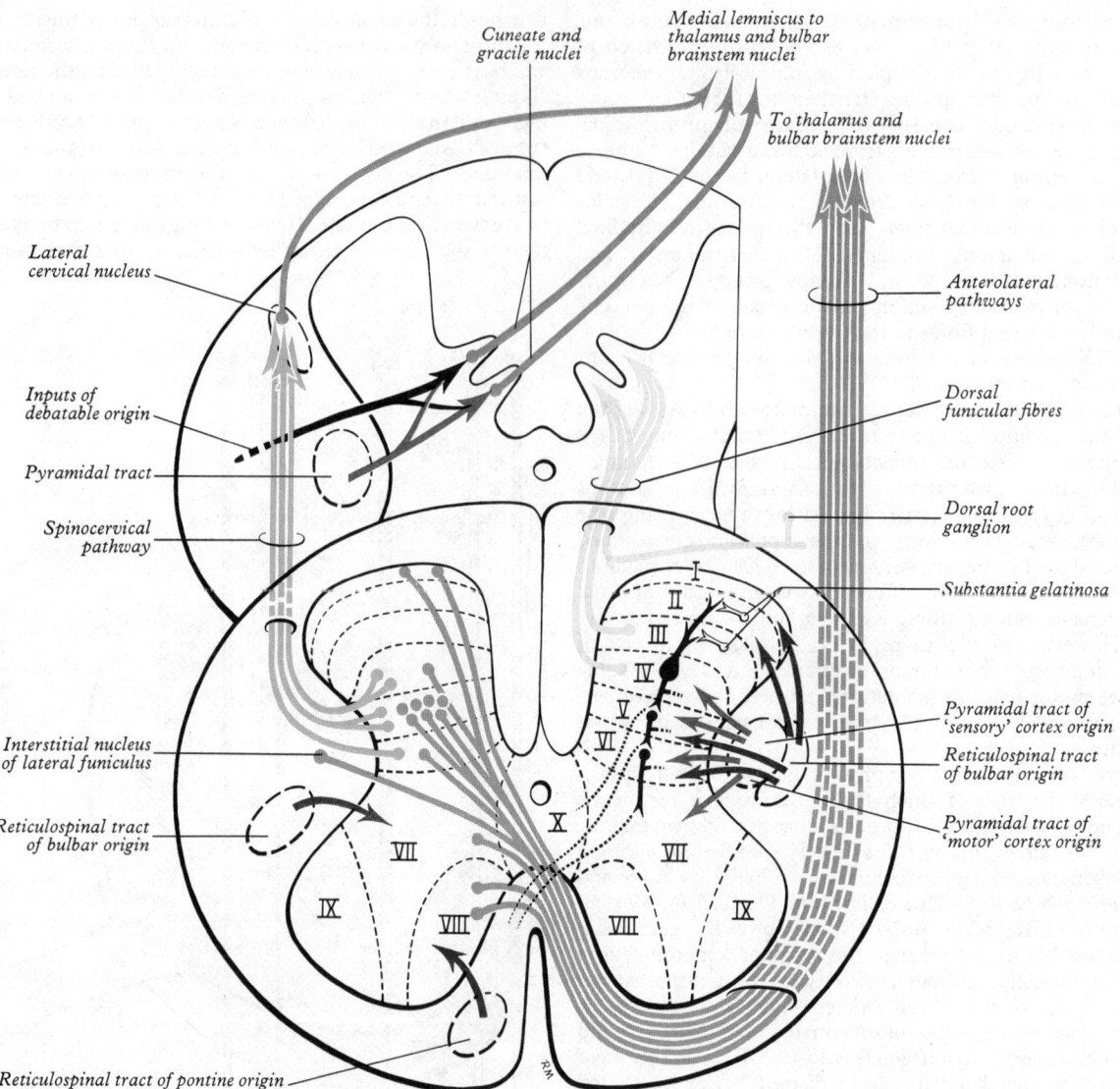

7.47C A recent and more detailed analysis of the principal somaesthetic pathways. Descending corticospinal and reticulospinal tracts involved in sensory modulation are also indicated. Modified from data kindly provided by Professor K. E. Webster, King's College, University of London.

whilst *dendritic* surface area ranged up to about 76,000 μm^2. In this study the dendrite surfaces formed up to 80 per cent of the total *receptive* surface of the neuron (i.e. excepting the axonal surface), the dendrites sometimes extending as far as 1,000 μm from the parent soma, in which cases the terminal dendrites passed into the adjacent white matter, as much as 50 per cent of the estimated receptor surface of the neuron being more than 300 μm from the parent cell body. The latter figure is particularly interesting, since many types of neuron have very wide arborizations of their dendritic trees; but neurophysiological evidence (Eccles 1957) has cast doubt on the possible effectiveness of synapses situated at distances greater than 300 μm from the soma; their functional role still remains to be determined. In the second series of investigations cited (Schadé 1964) estimates were provided of the percentage of grey matter occupied by cell somata and their surface areas, and the percentage occupied by dendrites and *their* surface areas. A mean value of 7,500 μm^3 was estimated for the volume of the soma of an interneuron, and 29,000 μm^3 for the soma of a motor neuron. The surface area of the 'average' cell body ranged from 4·4 to 5·9 × 10^3 μm^2 and that of the 'average' dendritic tree from 59 to 73 × 10^3 μm^2 in different specimens.

Other investigations (Romanes 1964; Sprague and Ha 1964) have also been concerned with the complex dendritic patterns of ventral column motor neurons and of interneurons located in adjacent cell laminae. Emphasis is placed upon the wide arborization of the dendrites of the motor neurons, seen to best advantage when both transverse and particularly longitudinal sections are examined. Not only do the dendrites of adjacent *cells* interlace horizontally in complicated patterns, but they also spread even more widely, overlapping the territories of other motor neuron cell *columns*; some penetrate adjacent cell laminae (VII and VIII), interweaving with their dendrites and cell somata, whilst still others pass quite deeply into the superjacent white matter, weaving between the longitudinal myelinated nerve fibres and terminating at varying depths (sometimes one-third to one-half the thickness of the white matter is traversed). Even more prominent is the longitudinal spread of dendritic branches from *a* motor neurons which occurs within their cell column of origin, and which may overlap the territories of many hundreds of neighbouring motor neurons. Again, these observations raise the question of the functional role of axodendritic synapses far removed from the cell soma. Clearly, the causal mechanisms responsible for the relatively circumscribed grouping of motor neuron *somata* into cell columns contrasts sharply with those related to the development of interlacing *dendritic fields*. The significance of the latter is by no means clear but attempts have been made to interpret the patterns of overlap in terms of the geometry of the terminals of primary afferent fibres from muscle nerves, and those of the interneurons in the adjacent cell laminae, both of which spread longitudinally for distances of three or four spinal cord segments (for details consult Sprague and Ha 1964; Sterling and Kuypers 1967).

A further detailed study of the dendritic fields in spinal cord grey matter has been published (Scheibel and Scheibel 1968), whilst particular regions which have been the subject of intensive study include the *thoracic nucleus* or Clarke's column (Szentá-gothai and Albert 1955; Böhme 1962) and the *substantia gelatinosa* (Szentágothai 1964; Ralston 1965). For a more recent general survey of the neuronal organization of the spinal cord consult Ralston (1974).

The general arrangement of neurons in the **substantia gelatinosa** is as follows. The neurons of the thin lamina I usually possess large dendritic networks which spread tangentially across the dorsal aspect of the dorsal grey column, where they interdigitate with the deepest fibres of the dorsolateral (Lissauer's) tract and some of the dorsal nerve root afferents; in the main their axons ramify in the subjacent laminae, but some are more lengthy, decussating and thereafter ascending to brainstem and diencephalic levels in the ventrolateral white columns of the contralateral side (p. 880). The small neurons of lamina II and the somewhat larger ones of lamina III have predominantly radially disposed dendrites (i.e. they occur in longitudinal 'sheets' which are perpendicular to the surface of the dorsal column); possibly a few of their axons may pass ventrally to the deeper laminae but the vast majority remain within the substantia gelatinosa (*vide infra*). These fine axons ramify immediately, either after ascending or descending a short distance within lamina II or after passing briefly into the dorsolateral tract and then returning to either lamina II or III. Between the radial dendrites of the small cells are two other principal components also radially disposed—the terminal parts of the long dendrites of the larger, more deeply situated cells of lamina IV and the terminal branches of many primary afferent cutaneous dorsal nerve root fibres. The small-diameter non-myelinated afferents approach the substantia gelatinosa from its dorsal aspect and their terminal branches point ventrally, whilst the larger diameter afferents curve around the substantia and approach it from its ventral aspect, their terminals point dorsally.

Another detailed study of the substantia gelatinosa was carried out, based upon the Golgi technique and ultrastructural methods, combined with either dorsal root transection or chronic surgical isolation of the dorsal grey column. The morphology of the neurons in laminae II and III were studied, and particular reference was made to a larger pyramidal type of neuron sited at the junctional zone between laminae III and IV. The latter possess recurrent axons which pass into lamina II, where they expand to form the core of large *glomerular synaptic complexes*. These glomeruli contain axodendritic synapses between the

pyramidal cell axons and the dendrites of the small gelatinosa cells, and axoaxonic contacts between the pyramidal cell axons and the terminals of primary dorsal root afferent fibres. The principal structural features of these synapses, together with some synaptic details of the input from primary dorsal root afferents, the axonal ramifications and terminals of the small gelatinosa neurons, the main output channel via lamina IV neurons, and terminals of fibres descending from supraspinal sources, are summarized in illustration **7.49** which is taken from (Réthelyi and Szentágothai 1969); this should be consulted for further details. Consideration of *earlier* interesting proposals concerning the possible functional significance of this arrangement is given in subsequent paragraphs. However, it must be emphasized that in the light of further researches in the intervening years, it is evident that at present the detailed role of the substantia gelatinosa remains quite obscure (*see* comprehensive reviews by Nathan 1976, and by Wall 1978).

THE SITES OF TERMINATION OF DORSAL ROOT AFFERENT FIBRES

The mode of formation, general topography, and division into a lateral fine-fibred and a medial coarse-fibred bundle of a dorsal spinal nerve root have been described elsewhere (p. 887). This distribution has been confirmed in man by Sindou *et al.* 1974. Consideration has also been given to the manner in which these fibres divide into ascending and descending branches, each of which gives rise to a series of collateral branches which approach the grey matter of the dorsal column at various levels over a substantial length of the cord, the longest ascending branches of the medial division proceeding as far as the gracile and cuneate nuclei in the medulla. In the present section, the zones of termination of these dorsal root fibres in the various regions of the spinal grey matter are discussed.

There have been two principal approaches to this problem: anatomical and neurophysiological. In the former, the fields of termination of fibres at different levels have been followed by application of such techniques as those of Glees, Nauta, and Fink-Heimer after severance of individual dorsal roots and the elapse of appropriate time intervals to allow terminal degeneration to occur. In the latter, the focal electrical potentials generated in the spinal grey matter have been explored by using microelectrodes to record the results of stimulation of muscle and cutaneous nerves and dorsal nerve roots.

In relation to earlier degeneration studies a number of points should be noted. Extensive and informative studies of dorsal root terminals (Sprague 1958; Sprague and Ha 1964) were carried out on the spinal cord of the *cat* (although similar general principles

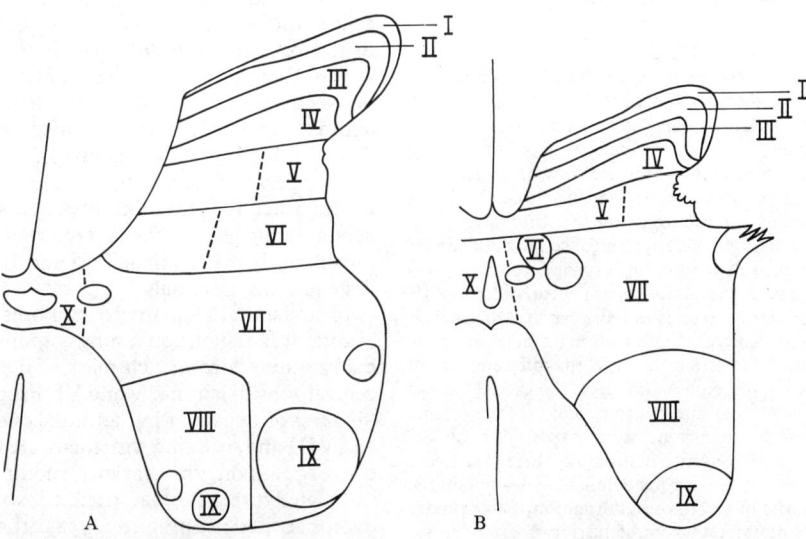

7.48A and B   The pattern of lamination proposed by B. Rexed (*Progress in Brain Research*, Vol. II) for the spinal cord grey matter of the cat, viewed in transverse section. A the fifth lumbar segment. B the third thoracic

segment. (With the permission of Prof. Rexed and the Elsevier Publishing Company.)

probably apply to all higher mammals, *details* cannot be applied without reserve to the spinal cord of man). The authors related their findings to the laminar architecture of the grey matter described previously (p. 882), and included a critical review of the limitations and advantages of the Nauta technique. It is pointed out that the degenerating terminals seen are only those in association with cell somata and dendrite trunks, whereas the branches of the dendrites of many neurons radiate widely from the cell (p. 887). Since the dendrites are encrusted with synaptic contacts throughout their length (Armstrong *et al.* 1956; Wyckoff and Young 1956; Rasmussen 1957; Young 1958; Illis 1964; Gelfan and Rapisarda 1964), numbering some thousands in large neurons, and since these are not revealed by the degeneration techniques mentioned, it is essential to correlate such studies with others carried out using the Golgi techniques.

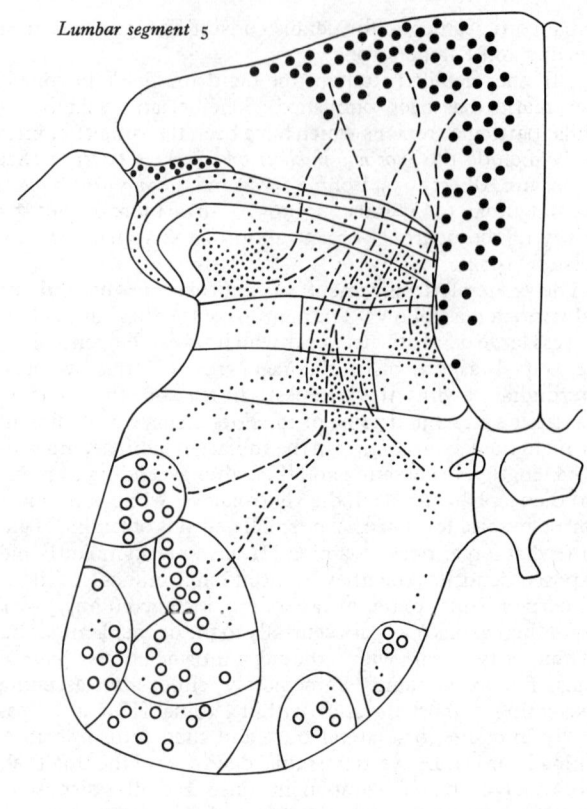

A

7.50A–C The pattern of degeneration of nerve fibres and their terminals, demonstrated with the Nauta-Laidlaw technique in the ipsilateral half of the spinal cord at various segmental levels, five days after surgical division of the *sixth* lumbar dorsal spinal nerve root of the cat. A the fifth lumbar

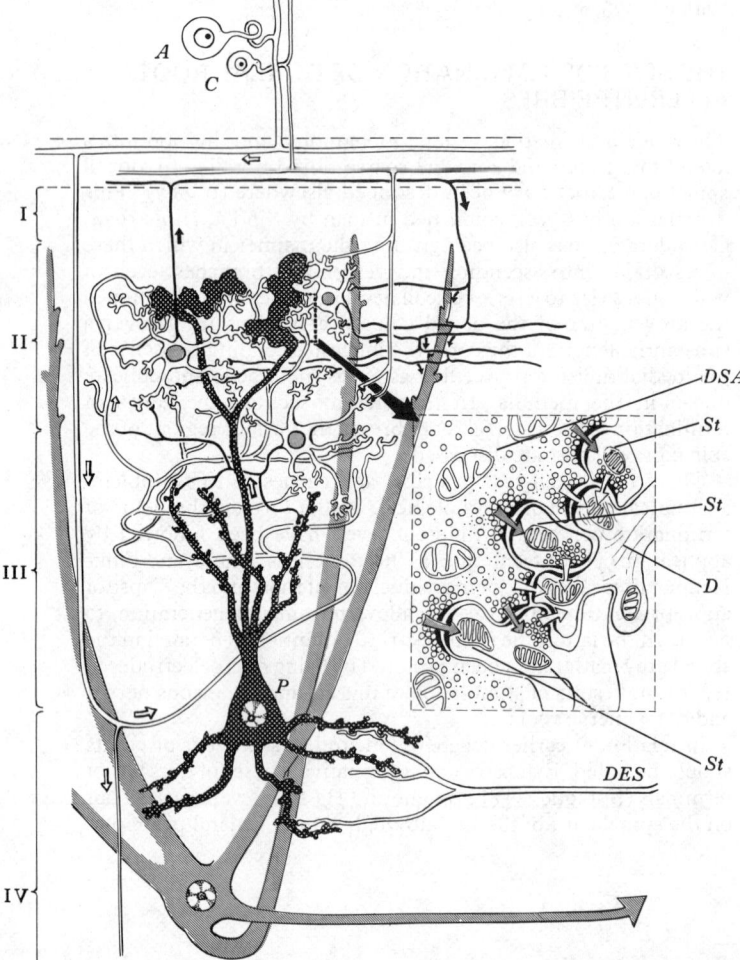

7.49 Diagram of the arrangement of neurons and their interconnexions in a longitudinal section through the dorsal grey column of the spinal cord, which includes the substantia gelatinosa and adjacent neuronal laminae (Roman numerals I–IV). Inset (right) shows synaptic detail of the area indicated. Two primary sensory afferent fibres are shown: a cutaneous afferent of large calibre (A), and a small calibre non-myelinated afferent (C). Small substantia gelatinosa interneurons are in white with black outline; their axons are single black lines. A pyramidal cell with dendrites and spines, and a recurrent axon expanding into synaptic complexes is shown in black with white dots. A large multipolar neuron of lamina IV with long radial dendrites and initial axon is in cross-hatch. DES—axon descending from a supraspinal source. Arrows on main diagram show presumed direction of impulse conduction. Note (1) the different sites of synaptic termination of primary afferents A and C, (2) the axonal pattern and terminals of substantia gelatinosa interneurons, and (3) the synaptic complexes formed between the recurrent axon expansions of the pyramidal cell, the small gelatinosa interneuron dendrites, and the primary sensory axon terminals. See inset for details: DSA—pyramidal cell axon terminal; D—dendrite of gelatinosa interneuron; St—primary afferent fibre axon terminal; white arrows—axo-dendritic synapses; cross-hatched arrows—axoaxonic synapses. Consult text for a discussion of the 'gate' theory and structural details. (By courtesy of Professor J. Szentágothai and *Experimental Brain Research*.)

The fields of termination of fibres examined after section of the sixth right lumbar dorsal spinal nerve root in the cat is shown at three levels (spinal cord segments, lumbar 5, 6 and 7) in illustration **7.50A, B, C**. These findings are briefly summarized as follows.

All the large calibred fibres of the dorsal funiculus (excepting some of the medially placed ones) have collaterals which pass through the medial two-thirds of laminae I, II and III, many curving around the medial aspect of these laminae and these form a dense plexus of degenerating fibres of passage (in transit to other regions) and degenerating terminals around the majority of the cells in the broad lamina IV. Many of the fibres of passage recurve to approach the substantia gelatinosa from its ventral aspect, into which they pass, forming numerous degenerating terminals between the radially disposed dendrites of the small cells of laminae II and III and the terminal segments of the long dorsally directed dendrites from lamina IV.

Degeneration occurs in some of the fine fibres of the dorso-lateral tract for distances of three segments both cranial and caudal to the level of the severed root, and from these, collaterals pass directly into laminae I, II and III, in all of which they form degenerating terminals.

From lamina IV many larger fibres pass to the medial zones of V and VI (a region containing commissural interneurons), whilst many others form a rich mass of degenerating terminals in the central zone of laminae V and VI. From this central concentration 'fingers' of degenerating terminals radiate through laminae VII and VIII and, running with them, are fibres of passage which then converge upon the various motor neuron pools and their associated interneurons, which constitute lamina IX. It has been demonstrated (Sprague 1958) that degenerating synaptic terminals from dorsal root afferents terminate both on the cell somata and the dendritic surfaces of the large multipolar motor neurons where they are interspersed between the numerous

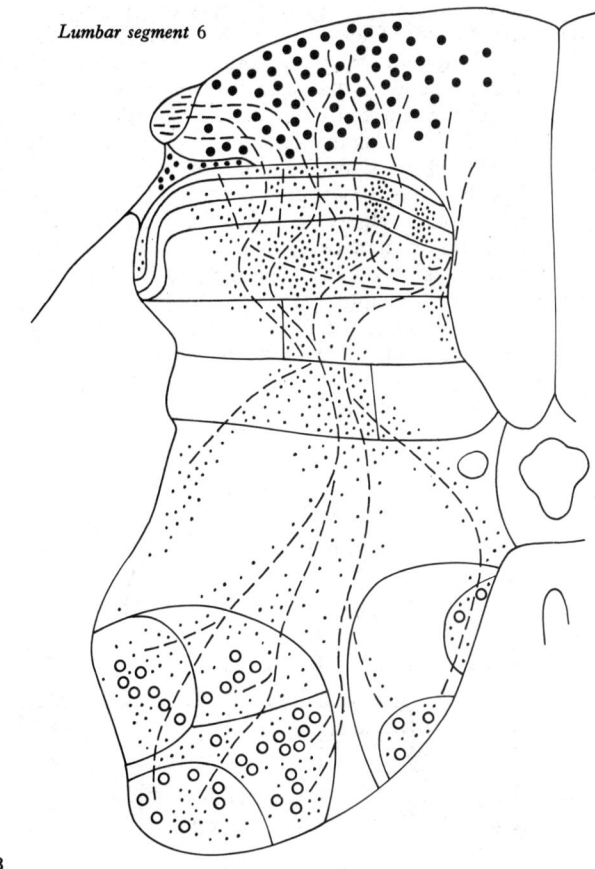

Lumbar segment 6

B

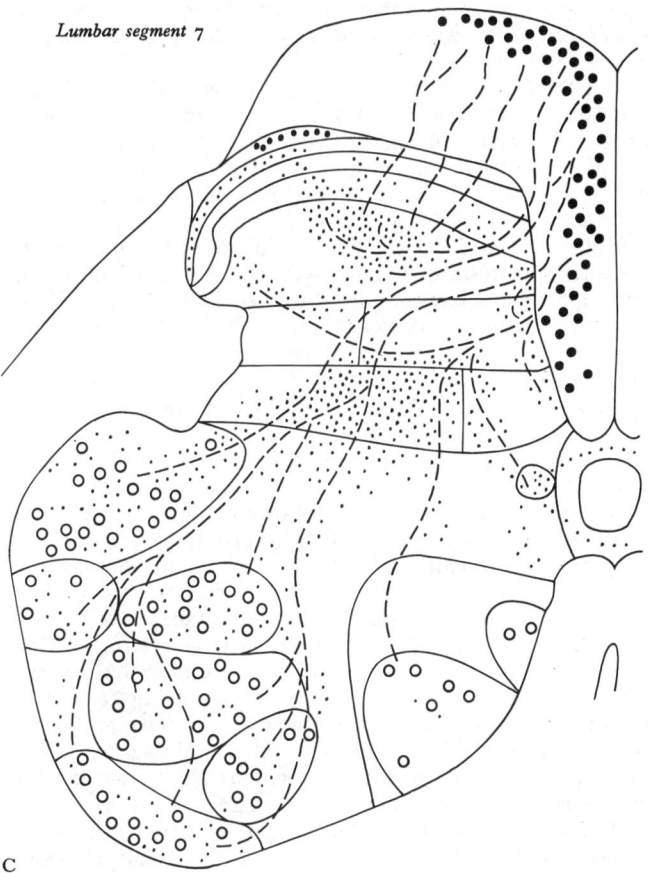

Lumbar segment 7

C

segment; B the sixth lumbar segment; and C the seventh lumbar segment. The large dots indicate degeneration of fibres in the dorsal funiculus; smaller dots—degenerating fibres in the dorsolateral tract of Lissauer; smallest dots—degenerating nerve terminals; dashed lines—

degenerating fibres of passage; circles—neuronal somata of motor neurons. (From J. M. Sprague and H. Ha, *Progress in Brain Research*, Vol. II, with the permission of the authors and the Elsevier Publishing Company.)

terminals derived from interneurons in other laminae and at other levels in the cord. In the cat such monosynaptic terminals of dorsal root afferents on motor neurons were found to extend for two segments cranial and caudal to the severed root, whereas terminals on interneurons in other laminae were also identified one or two segments even more cranial or caudal to this. Similar extensive studies have also been carried out on the spinal cord of the rhesus monkey (Shriver *et al.* 1968; Carpenter *et al.* 1968). In a number of respects the findings correspond with those in the cat, but a number of differences in detail have also emerged, but will not be discussed here; the interested reader should consult the references quoted.

In recent years there have been numerous attempts to correlate the architectural features of the spinal cord grey matter derived from Nissl, Golgi, and degeneration techniques, supplemented by the newer powerful methods employing tracers, of the kind detailed in the preceding pages (p. 819), with more critical analyses of the sites of origin and termination of the tract systems in the white matter, with the exceedingly complex array of propriospinal interneurons, and with electrophysiological studies of their functional activities. In relation to the latter a voluminous literature has accumulated which lies outside the scope of the present volume. Only a few points can be mentioned briefly and superficially here, and further details may be sought in the readily available review articles and original publications in this field, for example: Eccles (1957); Eccles and Schadé (1964 *a* and *b*); Ralston (1974).

SOME NEUROPHYSIOLOGICAL CORRELATIONS

In addition to providing a most convenient site in which to study the sequence of cytological changes which follow the severance of their axons (p. 862), the large multipolar neurons of the motor columns in lamina IX of the spinal cord formed the basis for

Sherrington's classical study of the characteristics of reflex responses (Sherrington 1906). During this emerged the first clear evidence of *integrative* activities occurring at the contact points between neurons, and the foundations for the concepts of opposing *excitatory* and *inhibitory* action were laid. Subsequently the large size and position of the somata of these cells enabled investigators to impale them with microelectrodes. The additional fact that they can be caused to fire volleys of nerve impulses (or to be inhibited from so doing) *orthodromically* by the stimulation of appropriate peripheral nerves or dorsal nerve roots, and that they can be invaded *antidromically* by stimulation of ventral nerve roots was of great assistance to experimentalists. These means have made possible detailed analyses of the electrical and ionic events which occur at synapses during the generation of *excitatory* and *inhibitory postsynaptic potentials* (EPSPs and IPSPs). Similarly, the early recognition of inhibitory inter-neurons (Renshaw 1941, 1946), in the neighbourhood of the large motor neurons, has led to an increasing volume of enquiry into *lateral* and *feedback* inhibitory phenomena, not only in the spinal cord but at many other cell stations throughout the nervous system.

Since 1940 much evidence has accumulated concerning another form of inhibitory action, namely *presynaptic inhibition*, in which spinal neurons have, again, provided a major experimental source (for reviews *see* Eccles 1964). In presynaptic inhibition the synaptic terminals (A) which impinge upon a neuron (B) are themselves subjected to the action of synaptic terminals (C) from a neighbouring interneuron, i.e. there exist axoaxonic synapses. When the inhibitory terminals (C) are active they are held to cause a relative depolarization of terminals (A), thus reducing the effectiveness of the latter in causing a postsynaptic change in neuron (B). Usually two or more interneurons are interposed on a presynaptic inhibitory pathway. Further, in common with postsynaptic excitatory and inhibitory phenomena, presynaptic inhibition may be mediated by fibre system terminals which have descended from supraspinal sources.

Once thought to be uncommon, *presynaptic inhibitory effects* have now been demonstrated in relation to many, possibly all varieties of primary afferent fibre terminals. In this manner the flow of sensory information into the nervous system does not simply and directly reflect environmental changes; it is continually readjusted and modified at the first synapse depending upon local conditions in the grey matter. A particularly interesting and intensively investigated site at which presynaptic effects have been held to occur is in laminae II and III—the substantia gelatinosa (Wall 1964; Mendell and Wall 1964; Melzack and Wall 1965; Mendell 1966; Heimer and Wall 1968; Ralston 1974; Nathan 1976; Wall 1978). *One view* of the anatomy of this region has been described elsewhere (pp. 870, 887) and it *was* proposed as a mechanism whereby inflowing impulses from cutaneous (and other) afferents are subjected to a *tonic control mechanism* involving relative levels of depolarization and hyperpolarization of the primary afferent fibre terminals. The effect was thought to be mediated by the small and pyramidal cells of the substantia gelatinosa, but the precise synaptic mechanisms that exist are still the subject of much research (for earlier references consult Réthelyi and Szentágothai; but see also below). A most interesting and provocative *gate control theory* first proposed by Melzack and Wall (1965) concerning the possible mode of operation of the gelatinosal system in providing a 'gate' in relation to the inflow of impulses along nociceptive and other afferent pathways has promoted intense discussion. The *mechanisms initially* envisaged by the authors (as opposed to other fundamental elements of the theory) are summarized in 7.51.

Briefly, it *was* proposed that large-diameter afferents (e.g. from hairs and touch corpuscles) are excitatory to both small gelatinosa interneurons (SG) and to the larger neurons (T-cells) of the deeper lamina IV (from which, directly or indirectly, spinothalamic fibres arise). In contrast, small-diameter and non-myelinated afferents are excitatory to the T-cells but inhibitory to the SG cells, whilst the axons of the latter were presumed to exert presynaptic inhibition upon the terminals of all afferents which synapse with the T-cells. Within such a system, a low level of activity in the small-diameter afferents inhibits the SG cells, which are therefore prevented from inhibiting the T-cells, and thus the 'gate' to the lamina IV T-cells is open and will transmit intermittent small volleys of impulses from the large fibres. A prolonged high-frequency volley of impulses in the large-diameter afferents, however, will be transmitted to the lamina IV T-cells initially, but this soon ceases as activity in the SG cells closes the gate. Conversely, a persistent high level of activity in the small-calibre afferents will open the gate wide resulting in a massive bombardment of the lamina IV cells. The latter include some high threshold cells which are only activated by such an intense bombardment, the presumption being that onward transmission from such cells over the lateral spinothalamic tract will be appreciated as pain in supraspinal centres. In this concept, therefore, pain would result from the *imbalance* between varieties of inflowing impulses which occurs when there is a disproportionately large traffic along the small-diameter afferents. The latter, perhaps, should thus not be regarded as '*specific pain afferents*' but as a fibre system which continually monitors the *state of the tissues* innervated. Thus, in addition to *interaction* and *comparison* between the information flowing along different fibre types, other essential ingredients of the gate control theory (as initially stated) involve *presynaptic inhibition* of primary afferent terminals; furthermore, the overall 'sensitivity' of the gate may be varied by *descending control systems* from supraspinal sources.

Since the original publication of the theory many additional data have accumulated, and it has been progressively modified, modernized, and refined (Melzack 1973; Wall 1973, 1974, 1976). Nevertheless, Nathan (1976) was prompted to write a lengthy critical review of the theory, which the interested reader should undoubtedly consult. In his most recent publication at the time of writing, however, Wall (1978) has presented a *re-examination* and *re-statement* of the gate control theory of pain mechanisms. In summary he states: 'In 1965, we proposed that the transmission of

information about injury from the periphery to the first central cells was under control. The setting of this control or gate was influenced by peripheral afferents, other than those which signalled injury. The gate was also influenced by impulses descending from the brain. Subsequent work has fully supported and enlarged this view. All the cells so far discovered which transmit information from nociceptors are inhibited by low threshold afferents and by descending controls. The mechanism by which this control is achieved remains completely unknown. Pre-synaptic inhibition as a phenomenon isolated from post-synaptic inhibition is in doubt. Whether the inhibitions and facilitations are pre-synaptic or post-synaptic or both is unknown. The role of the substantia gelatinosa in any function is unknown. That a gate control exists is no longer open to doubt but its functional role and its detailed mechanism remain open for speculation and for experiment.'

Other varieties of electrophysiological investigation have

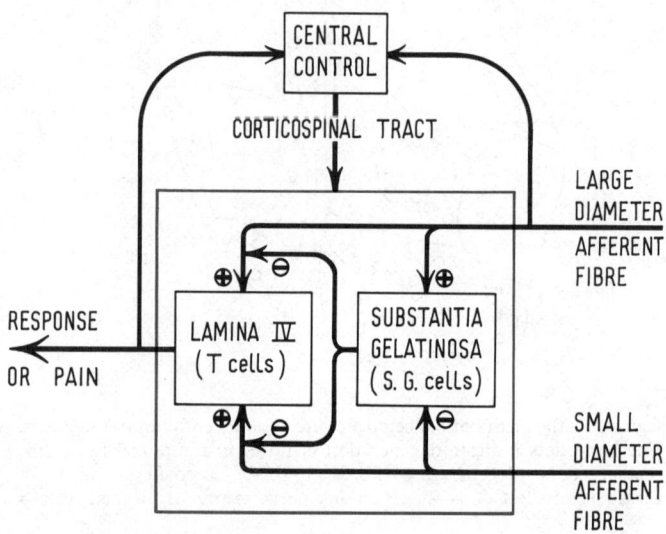

7.51 The sensory 'gate' mechanism as *originally* proposed by Dr. R. Melzack and Prof. P. D. Wall for the *modus operandi* of the dorsal laminae of grey matter of the spinal cord. See text for a discussion of the effects of an imbalance in the sensory inflow along the large and small diameter afferent fibres. (From an illustration in *Science*, Vol. 150, with the permission of the authors and publishers. Copyright, 1965, American Association for the Advancement of Science.) Consult references for a critical review of the gate theory by Nathan (1976), and a comprehensive reappraisal of the evidence and modern re-statement of the theory by Wall (1978).

involved the exploration of the spinal grey matter with microelectrodes during the stimulation of muscle and cutaneous nerves (Eccles *et al.* 1954; Coombs *et al.* 1956) and during the 'natural' stimulation of peripheral receptors (Kolmodin 1957; Kolmodin and Skoglund 1960). The focal extracellular potentials evoked by such means correspond well with certain specific regions of the grey matter laminae and with the areas of termination of dorsal root afferent fibres determined by degeneration techniques. For further details the references quoted should be consulted.

UNIT RECORDING WITH INTRACELLULAR MICROELECTRODES

Physiological recordings made with microelectrodes inserted into individual cell somata deep within the spinal cord grey matter have, in recent years, provided much information, often of an unexpected nature, which has necessitated a radical reappraisal of many widely held views. Excellent examples of this experimental approach are a series of investigations of the activities of cells in spinal laminae IV–VI during natural forms of skin stimulation, passive limb movements, stimulation of cutaneous, muscle and visceral nerves in various combinations, and on occasion, simultaneous stimulation of various fibre tracts which descend from supraspinal sources (or after the removal of such influences).

Consult Wall (1967); Pomeranz et al. (1968); Wall (1970; 1978).

In the earlier experiments it was confirmed that a *functional lamination* exists which corresponds in large measure to that proposed on cytoarchitectonic grounds (p. 883). It was demonstrated that cells in all three laminae would respond to cutaneous stimulation, whereas movement only elicited a response in lamina VI. Cells in lamina IV possessed small receptive fields and responded as if many different types of specific cutaneous afferents *converged* upon an *individual cell* (e.g. it responded to hair movement, touch, skin cooling, etc.). Lamina V cells responded as if many cells of lamina IV converged upon each of them; they possessed much larger and more diffuse receptive fields. Response to movement was limited to cells in lamina VI, but many of the latter responded to both passive movement and cutaneous stimulation. Simultaneous stimulation of corticospinal fibres affected cells in all three laminae, sometimes facilitating and sometimes inhibiting their responses. It was also demonstrated that activity of certain brainstem centres inhibited cutaneous stimulation responses but enhanced responses to movement and could even *switch the 'modality'* of lamina VI cells from cutaneous to proprioceptive. In later investigations in which cutaneous, muscle, and visceral nerves were stimulated, the findings were expanded and refined. Particular emphasis was placed upon lamina V in which *fine myelinated afferents* from the *three* sources converged on *individual cells* and interacted in various ways. They possessed very large, non-discriminative receptive fields and it was proposed that in some way they signalled information concerning the general 'state' of the various tissues. These results will not be examined in detail here, but they are well summarized in a subsequent publication (Wall 1970) in which the characteristics of the spinal cord cells in laminae IV–VI are compared with those of the dorsal column nuclei (nucleus gracilis and nucleus cuneatus). The former are now considered to provide an important source of the crossed spinothalamic tract (Dilly et al. 1968) which is, as we have seen (p. 876), one of the principal ascending sensory pathways, whilst the latter receive some terminals of the fasciculi gracilis et cuneatus, the second major ascending sensory pathway. Their respective roles have long been a subject of controversy and a comparison emphasizes some important features of sensory physiology.

COMPARISON OF THE DORSAL COLUMN—MEDIAL LEMNISCUS AND THE SPINOTHALAMIC TRACT PATHWAYS

It is instructive to compare the size of the receptive fields, the degree of anatomical localization (somatotopical mapping), the maintenance of specificity of channels or the convergence and interaction among them, and the forms of control of transmission of impulses in the two systems.

The cells of the dorsal column nuclei (Mountcastle 1968; Norton 1968) receive the synaptic terminals of the long, ascending, uncrossed, primary afferent fibres of the fasciculus gracilis and cuneatus (p. 881) which carry information concerning deformation of the skin, movement of hairs, joint movement, and rhythmical vibration of the tissues. In the cat, the fibres from hair receptors are superficial, whilst touch and vibration receptors are more deeply placed. The somatotopical localization of fibres in the dorsal white columns has already been described (p. 881) and further consideration of the structure and connexions of the dorsal column nuclei will be considered with the medulla oblongata (p. 881).

Unit recording with microelectrodes implanted in the cells of the dorsal column nuclei has shown that their *receptive fields* for touch (that is the skin area from which a response can be elicited) vary in size in different regions, but are generally small (and smallest in the digits). Some receptive fields have an *excitatory centre* and an *inhibitory surround*; thus a stimulus applied just outside the excitatory receptive area will inhibit the cell in question. The cells of the nuclei are spatially organized into an accurate somatotopic map of the periphery (in accord with the similar localization in the dorsal columns). *Specificity* in the cells is high, each responds to *only one type* of afferent fibre stimulation;

thus a cell may respond to hair movement alone, or to joint movement, or to an applied sinusoidal vibration, but never to two of these. Indeed, a number of fibres converge upon a single cell but they are always of the same functional type.

The transmission of impulses from the dorsal columns to the medial lemniscus are subjected to a variety of control mechanisms (Jabbur and Towe 1961; Andersen et al. 1964). Concomitant activity in neighbouring dorsal column fibres may result in presynaptic inhibition by depolarization of the presynaptic terminals of one of them. Stimulation of the sensorimotor cortex also modulates the transmission of impulses by both pre- and post-synaptic inhibitory mechanisms, and sometimes by facilitation. These descending influences are mediated by the corticospinal tract. Modulation of transmission also follows reticular formation stimulation.

Clearly, the dorsal column nuclei are not simple 'relay nuclei' as was long supposed. They have been described as having the characteristics of 'a private and highly reliable telephone system in which afferent information is separated in channels which are discrete both for spatial origin and stimulus specificity of afferent fibres'. These features contrast strongly with those of the cells of origin of the spinothalamic tracts (see also p. 887).

The cells in laminae IV–VI of the spinal cord have very different receptive fields. Those of lamina IV have small fields and there is a high degree of somatotopic localization of the cells, whereas those of lamina V have extremely large diffuse fields, and respond to a wide range of stimulus strengths. *Specificity* of separate channels as it exists in the dorsal column nuclei is absent in the laminae of the cord. *Convergence of different functional types* of afferent fibre on to the *same cell* is the rule in the cord, the convergence varying with the lamina. Lamina IV cells receive large cutaneous afferents, lamina V cells receive fine afferents from skin, muscle, and splanchnic nerves in various combinations, whilst lamina VI cells receive large muscle and cutaneous afferents. The converging synaptic terminals from different sources result in an *interaction* between excitatory and inhibitory states, which summate to determine the output from the cell. The control of impulse transmission is modulated in a variety of ways. Firstly, the cutaneous afferents are possibly influenced by a tonic regulating mechanism in the substantia gelatinosa described previously. In addition, impulse transmission is greatly influenced by a variety of descending tracts from the sensorimotor cortex and brainstem centres (*vide supra*). The degree of control varies from simple changes in excitability of the cells in lamina IV to a complete reversal of modality in lamina VI.

The roles of the two major sensory pathways—dorsal columns and spinothalamic tract—have aroused considerable controversy. The classical view (Mountcastle 1968) holds that the dorsal column is the *essential discriminatory pathway* and that in its absence a mechanical stimulus is recognized as having occurred, but that it is impossible to specify it exactly as to location, intensity and shape. This view has been strongly challenged (Wall 1970), and it is pointed out that experimental animals with complete section of a dorsal column can in fact discriminate for weight, texture, two-point stimulation, vibration and position. It is proposed that environmental stimuli may be classified into two types: those which are 'passively impressed on an animal' and those that 'must be actively explored by motor movement or sequential analysis before they can be successfully discriminated'. Reception of the former is considered the role of the spinothalamic system, whilst that of the dorsal column system is to *initiate and programme an exploration* of a source of stimulation, and to carry the resultant flow of information. Further evidences in this field, perhaps derived from intracellular microelectrodes carried by freely moving animals—a technique currently being developed—will be awaited with great interest.

THE SPINAL TERMINATIONS OF DESCENDING TRACTS FROM SUPRASPINAL SOURCES

In recent years degeneration technique such as that of Nauta have added considerably to our knowledge of the sites of termination within the spinal cord grey matter of many of the principal

7.52A–C The spinal terminations of various descending tracts of the spinal cord determined experimentally in the cat, and referred to the laminar pattern of the grey matter which is described elsewhere in the present section. (Redrawn from *Neurological Anatomy* by A. Brodal, Oxford University Press, 1969, by courtesy of the author and publishers.)

The original papers and illustrations stemmed from the numerous publications by R. Nyberg-Hansen and his collaborators to whom we are grateful—consult bibliography under 'Spinal Terminations of Descending Tracts from Supraspinal Sources'.

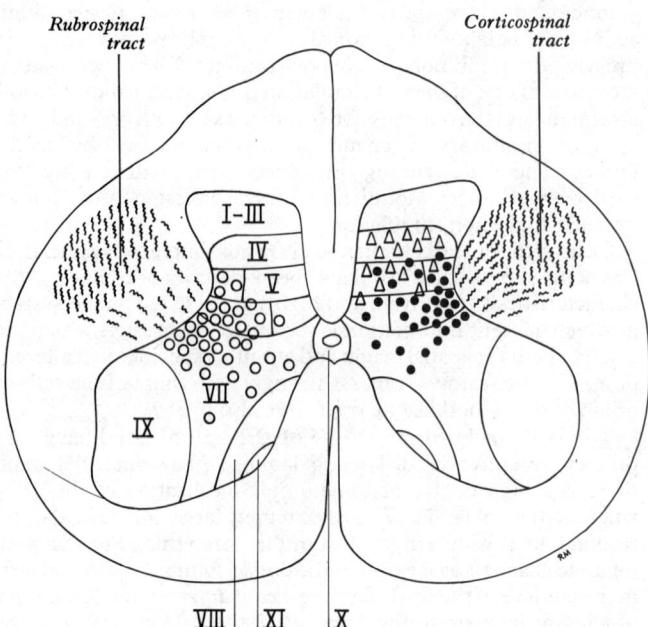

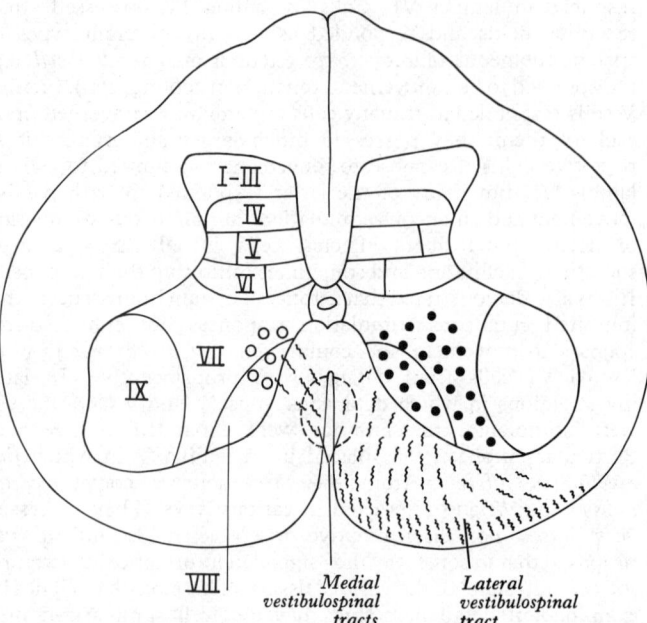

A   Terminations of corticospinal fibres from 'motor' areas of the cerebral cortex—black dots; corticospinal fibres from 'sensory' areas of cerebral cortex—white triangles; rubrospinal fibres—white dots.

B   Terminations of vestibulospinal fibres; those from the lateral vestibulospinal tract—black dots; those from the medial vestibulospinal tract are shown on the opposite side of the cord as white dots.

descending fibre tract systems (Nyberg-Hansen 1964a and b, 1965, 1966 a and b, 1969; Nyberg-Hansen and Brodal 1963, 1964; Nyberg-Hansen and Mascitti 1964; Brodal 1969). The majority of these investigations, again, involve the spinal cord of the cat and occasionally primates other than man. The results, which may well embody general principles applying to many higher mammals, must not, however, be extrapolated in detail directly to the situation in the human cord. Some of the principal results are summarized in illustration **7.52A**, B and C based upon the above references, in which the distributions of the terminals are related to the laminar architecture of the grey matter. The following salient points, some of which have already received brief mention with the human tracts, may be noted.

The corticospinal tract fibres in the cat terminate almost exclusively on interneurons in laminae IV–VII. Perhaps all such fibres terminate in this manner, but because of the very extensive ramifications of the dendrites of the multipolar motor neurons of lamina IX, some of which penetrate lamina VII, the presence of a few axodendritic contacts with motor neurons cannot be entirely excluded. Further experiments demonstrated that the corticospinal fibres which arise from 'sensory' cortical regions terminate chiefly in laminae IV and V, whereas those from 'motor' regions end in V–VII with the greatest concentration in the lateral part of lamina VI. Thus, although some overlap occurs, the two types of corticospinal fibre (from different cortical regions) are fairly distinct in their areas of termination. These findings are of particular interest in relation to the increasing volume of evidence concerning the corticospinal (pyramidal) tract as one of the principal supraspinal systems that *modulate the inflow of sensory information*. As we have seen, the latter may be mediated either by presynaptic inhibition of the terminals of the primary afferent fibres, or by postsynaptic inhibition or facilitation of the second (or subsequent) neurons in the afferent pathway. It should also be pointed out, however, that in contrast to the cat there is both anatomical (Hoff and Hoff 1934; Kupyers 1960; Liu and Chambers 1964), and physiological (Bernhard *et al.* 1953; Preston and Whitlock 1961; Landgren *et al.* 1962) evidence that in monkeys a small proportion of corticospinal fibres end monosyn-

aptically on large α motor neurons. Much less is known with quantitative precision about the synaptic terminals in man; the majority undoubtedly end upon interneurons, but some evidence (Kupyers 1958) again indicates a small population of fibres ending directly on motor neurons.

Briefly, it may again be mentioned that physiological studies (Corazza *et al.* 1963) in the cat indicate that the corticospinal tract influences both α and γ motor neurons, and that in both cases the effect is mediated via interneurons. In general an *increased* flux of impulses along corticospinal axons has an excitatory effect on motor neurons which supply flexor musculature and an inhibitory one to those supplying extensor musculature. It must be stressed here and elsewhere, however, that the *converse* effects accompanying a *decreased* flux of impulses are equally important in functional terms.

The rubrospinal tract fibres in the cat arise from both the large and small celled parts of the contralateral red nucleus, which shows evidences of some degree of somatotopic organization. Degeneration studies after stereotactic lesions of the nucleus indicate that the tract descends throughout the cord to lumbosacral levels and that its terminals end upon interneurons in spinal cord laminae V–VII. Its zones of termination correspond rather closely with those of the corticospinal fibres which arise in 'motor' regions of the cerebral cortex. Again there are physiological evidences that increases in activity of the rubrospinal projection influences both α and γ motor neurons, via interneurons, facilitating flexor groups and inhibiting extensor ones. Since in some experimental animals a somatotopically organized *corticorubral projection* of fibres has been demonstrated from the sensorimotor cortex, some authors have stressed the presence of dual pathways from the cortex to the cord, one a *direct corticospinal projection*, the other an *indirect corticorubrospinal projection*. Both have similar terminations in the cord and many physiological attributes in common. Little precise knowledge is available, as we have seen, concerning the origin, localization, termination and functional significance of the rubrospinal system in man. Although often stated to be rudimentary and of little significance, clear evidence for such a view is lacking.

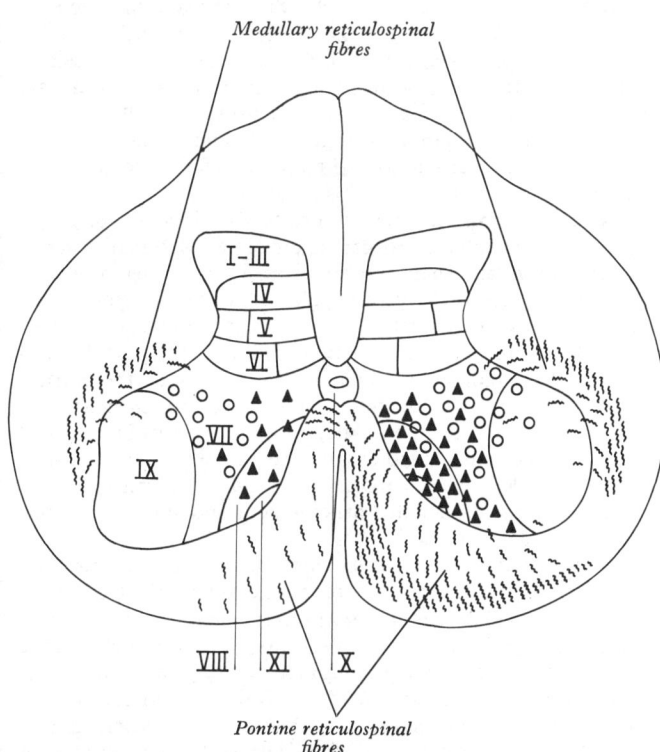

Medullary reticulospinal
fibres

Pontine reticulospinal
fibres

C   Terminations of reticulospinal fibres; those originating in the medulla oblongata—white dots—are in general more dorsally placed than those originating in the pons—black triangles.

The vestibulospinal tracts, medial and lateral, have also been quite intensively investigated in experimental animals, but are much less well understood in man. As stated previously, the lateral vestibulospinal tract originates from both large and small cells of the *lateral vestibular nucleus* of the same side, descends in the anterolateral white column throughout the cord, its fibres terminating at successive levels largely in laminae VII and VIII, only a trivial number of terminals entering lamina IX. The medial vestibulospinal tract arises mainly from the medial vestibular nuclei of both sides. The fibres run in company at brainstem levels with other components of the medial longitudinal fasciculi and then descend, perhaps to mid-thoracic levels only, on both sides of the anterior median fissure of the cord. They, too, terminate in a more restricted zone which includes parts of laminae VII and VIII. Activation of the main lateral vestibulospinal tract results in excitation of extensor motor neurons and inhibition of flexor motor neurons. The excitation is a direct monosynaptic effect and it must be concluded that the terminals of the vestibulospinal fibres make synaptic contact with the extensive dendrites of some of the motor neurons that penetrate laminae VII and VIII. Gamma motor neurons are also considered to be facilitated whilst the inhibitory effect to the flexors is presumably mediated by inhibitory interneurons of the laminae in which the fibres terminate.

The reticulospinal tracts have been notoriously difficult to evaluate in the spinal cord of man and, again, much more precise information is available in the cat, from studies using the Nissl method for studying the acute retrograde responses in cell somata after spinal cord section, and also the Nauta technique for anterograde terminal degeneration after the placing of stereotactic lesions in the brainstem. Both pontine and medullary reticulospinal fibres from one side of the brainstem apparently pass to *both sides* of the cord, but the pontine fibres are much more densely concentrated on the ipsilateral side, whereas the medullary fibres also send a substantial population to the contralateral side. The spinal cord course of the two differs, the medullary fibres are claimed to pass in the lateral funiculi whilst the pontine fibres are concentrated ventromedially. Their zones

of termination are summarized in 7.52C. (For further information concerning the pontine and medullary origins and the possible non-aminergic, noradrenergic, serotonergic, and enkephalin-containing nature of these reticulospinal fibres, consult pp. 847–853.) Briefly it may be mentioned that similar experiments indicate that the **tectospinal tract** from the contralateral superior colliculus and the **interstitiospinal tract**, which arises mainly in the ipsilateral interstitial nucleus of Cajal in the cranial midbrain, both terminate in spinal cord laminae VI–VIII. The latter and the reticular formation will be further discussed in subsequent sections.

SUMMARY OF SPINAL CORD ORGANIZATION

It has been proposed by Szentágothai (1967) that in broad outline it is convenient to consider the grey matter of the spinal cord as composed of a *central core* with paired *dorsal* and *ventral appendages*, each with some distinctive characteristics (7.53). (The unfamiliar terms core and appendage are retained in this summary since they do not correspond precisely with the dorsal and ventral grey columns of topographical neuroanatomy.)

The core, with our present methods of investigation, appears to present a diffuse, non-discriminative, reticular type of organization, within which there is both a great divergence and also convergence of information paths; the majority of its interneurons interconnect with many hundreds, perhaps thousands of other core interneurons distributed over a substantial length of the cord. In contrast the dorsal and ventral appendages have a greater degree of precise discriminative organization in terms of somatotopic mapping and various orders of functional localization. To give precise topographical limits to core and appendages is probably an intellectual abstraction and is certainly impossible at the present rudimentary state of our knowledge; indeed, the concept itself will undoubtedly need much revision (and perhaps lose much of its relevance) as knowledge and understanding increase. Roughly the central core is considered to encompass the interneurons of laminae VII and VIII (i.e. the intermediate zones of previous workers and the areas between the motor neuron columns). However, it may well be that many of the interneurons of the deeper laminae of the dorsal column and those within the motor cell columns should also be included under the concept of a reticular core. The ventral appendage corresponds to the cell columns of lamina IX whilst the dorsal appendage includes laminae I–VI.

The dorsal appendage is, as we have seen (pp. 870, 887), the principal receiving zone of the various exteroceptive, proprioceptive and interoceptive dorsal nerve root afferent fibres. The dorsally situated laminae (I–IV) are the main cutaneous afferent receiving areas, lamina V receives small-diameter afferents from skin, muscle and viscera, whilst lamina VI receives proprioceptive and some cutaneous afferents. It should be emphasized, however, that relatively few investigations have as yet been carried out; the functional boundaries are certainly not as clear-cut as the above description suggests, and undoubtedly a great variety of further functional 'types' of interneuron with many other grades and subtleties of interaction between them remain to be discovered. Further, it is inherent in the technique of intracellular microelectrode recording that because of their situation, size and shape, some cells can be investigated with relative ease, others with difficulty, and a large (and indeterminate) number are beyond our current technical reach; for these reasons a bias in our view of cord dynamics is inevitable.

Nevertheless, it is becoming clear from the existing recordings that the dorsal appendage cells abstract information about the internal and external environment in many different forms. Cells and cell groups in different regions show great variety in their degree of somatotopic mapping, the size and response characteristics of their receptive fields, in their specificity or in the degree of convergence and interaction of different types of afferent channel upon a single cell. A brief comparison between the behaviour of the cells in the dorsal appendage of the cord with those in the dorsal column nuclei of the medulla oblongata has been given elsewhere (p. 891). Clearly, the simple view of spinal cord cells as providing merely relays in a series of invariant, discrete, 'unimodality' channels which transmit an elementary punctate view of the environment, either ventrally to motor neurons or cranially to brain centres, as they are often presented in neurological textbooks, must now be firmly rejected. A series of complex transformations of the patterns of inflowing information have already occurred within the dorsal appendage. The outflow along the axons of the dorsal appendage neurons passes to many destinations, either directly to motor neuron columns in the ventral appendage, indirectly via more ventrally situated laminae including the exceedingly complex arrangements of interneurons in the reticular core, on both sides of the cord and cranial and caudal to the level of the input in question. Cells with long axons in both the dorsal appendage and the reticular core contribute to the long ascending tracts, which reach a multiplicity of centres in the brainstem.

Further, transmission of impulses to and beyond the dorsal appendage cells may be modified in various ways by mutual interaction and the operation of controlling mechanisms. Mention has already been made (p. 890) of the facilitatory and inhibitory effects which may follow simultaneous activity in different categories of afferent fibre which converge on the same cell, and of the postulated tonic modulating system of the substantia gelatinosa, which affects the inflow of impulses from cutaneous afferents. In addition, much evidence is now accumulating which demonstrates that impulse transmission in all the laminae of the dorsal appendage (as in the remainder of the spinal cord grey matter) are strongly influenced by fibre systems which descend from supraspinal sources such as the sensorimotor cortex and the reticular formation of the brainstem; such effects may be mediated by either pre- or post-synaptic contacts and be either facilitatory or inhibitory depending upon the site and the circumstances. The biological significance of these mechanisms which control sensory input channels are still far from clear; they may be concerned with eliminating redundancy, reducing confusion by excessive bombardment of central networks, or be linked to 'states of readiness' of central mechanisms, or to the temporary 'preoccupation' of such centres with more immediately significant transformations. The subject is under intense and continuing analysis. It is increasingly clear, however, that the majority of fibre tract systems that descend from the brain and influence motor patterns of behaviour, do so by modifying the state and impulse transmission characteristics of interneurons in the deeply placed laminae of grey matter and much less frequently by influencing the motor neurons directly.

The reticular core interneurons form an exceedingly complex network in which each cell receives inputs from and transmits to large numbers of others (c.f. 'The Reticular Formation'—pp. 943–953). The core receives an input from the axonal terminals of some cells of the dorsal appendage, from proprioceptive dorsal root afferent fibres, and from some of the long descending tract systems. Because of the wealth of interneuronal connexions in this region, analysis has proved of the greatest difficulty, but the earlier view that it almost approaches a *random nerve network* in form is receding with the extensive application of the newer methods of analysis. The latter include the placing of minute stereotactic lesions within the core and observing subsequent degenerative phenomena, intracellular recording, elegant researches on thousands of specimens using the Golgi technique (for an extensive review of this field consult the collection of essays in Brazier 1969), and the newer techniques of mapping of somata or axonal terminals according to their content of putative neurotransmitters (*see* p. 950). One type of organization that has emerged concerns the quantitative analysis of connectivity or 'transmitting power' of different types of interneuron and some of the descending tract systems. Details cannot be given here, but briefly some axonal terminals pursue an extremely long course through the grey matter giving off perhaps only two or three synaptic end bulbs to each of the many hundreds of interneurons encountered in transit. Others concentrate large numbers of terminals on one or a small group of neurons. Thus, diffuse or '*non-discriminative*' connexions have been contrasted with '*discriminative*' ones. Further, it has been shown that contrary to earlier opinion many proprioceptive afferents have terminals in the core that are strictly 'segmentally' localized to

narrow *transverse* 'sheets' of grey matter (in contrast to the dorsal appendage cutaneous afferents which terminate in extensive narrow *longitudinal* sheets). It is largely through such complex interneuronal aggregates that inflowing sensory information interacts with that descending from higher centres to set in train the endless variety of locomotor responses; and although our understanding of them is still in its infancy, the continued application of modern powerful methods offers a considerable prospect of further success in this region.

The ventral appendage as described elsewhere (pp. 869, 873, 883) consists of a columnar organization of the cell somata of a and γ motor neurons and again neighbouring interneurons. Attention was also directed to the physiological evidence for 'tonic' and 'phasic' types of a cell related to different varieties of striated muscle fibre, and to the existence of different types of γ cell whereby the 'static' and 'dynamic' responses of the muscle spindles are under independent central control mechanisms. Unfortunately, the detailed synaptology of these different cell types is uncertain. Further, it was shown that whilst a considerable degree of somatotopic localization exists in the ventral columns, its detailed arrangement and significance also remain to be determined.

The principal synaptic connexions of the *motor neurons* are derived from: (1) direct monosynaptic terminals from proprioceptive dorsal root afferents within the same or neighbouring spinal cord segments; (2) terminals from collateral branches of the axons of dorsal appendage interneurons; (3) terminals from interneurons of the reticular core which are of high density and 'discriminative' when derived from the same spinal segment, and diffuse or 'non-discriminative' when derived from neighbouring segments; (4) a few direct monosynaptic terminals from the vestibulospinal and (in various primates including man) the corticospinal tracts, although in the main these tracts end on interneurons.

How these various channels which converge on the motor neurons interact to produce integrated motor behaviour is still far from clear. However, a few generalizations can be made. The principal descending pathways can be grouped: those wherein impulse volleys are excitatory to flexor musculature and inhibitory to extensors (the corticospinal, rubrospinal and medullary reticulospinal tracts) and those which have the converse effect (the vestibulospinal and pontine reticulospinal pathways). However, this simple view of the broad dualistic action of certain descending tracts does little justice to the large volume of detailed investigation that has been reported on the complex modifications of reflex activities by descending systems; but, as yet, it has proved difficult to fit these into an overall behavioural picture.

Secondly, there are two distinct pathways by which a muscle may be thrown into a state of contraction (or relaxation); the first or a *pathway* involves an immediate and direct change in the excitatory level of the a motor neurons innervating the motor units. It is considered that such a mechanism operates only infrequently, and that is when a sudden forceful response is appropriate. In the majority of instances a γ *pathway* is operative in which the sequence appears to be—activity in local interneurons, followed by activity (or inhibition) of the γ efferents to the muscle spindles, which in turn via the local muscle servo-loop mechanism (p. 859) causes an appropriate change in the tonic and phasic a motor neurons. The detailed mechanism whereby this so-called a-γ *linkage* is maintained or broken under different conditions is still under investigation (Granit 1970). It should also be noted, however, that some recent researches have indicated that during voluntary actions, initiating activity in the a system may be more frequent than was previously recognized (*see* p. 860).

Applied Anatomy. In injury to the spinal cord the segmental level of the interference may be determined with some precision from clinical data and the application of accurate anatomical knowledge, where available.

Complete division of the spinal cord above the fourth cervical segment causes death by respiratory failure resulting from paralysis of the phrenic and intercostal nerves. Lesions between C_5 and T_1 produce paralysis in all four limbs (quadriplegia); the degree of paralysis in the upper limb varies with the site of the

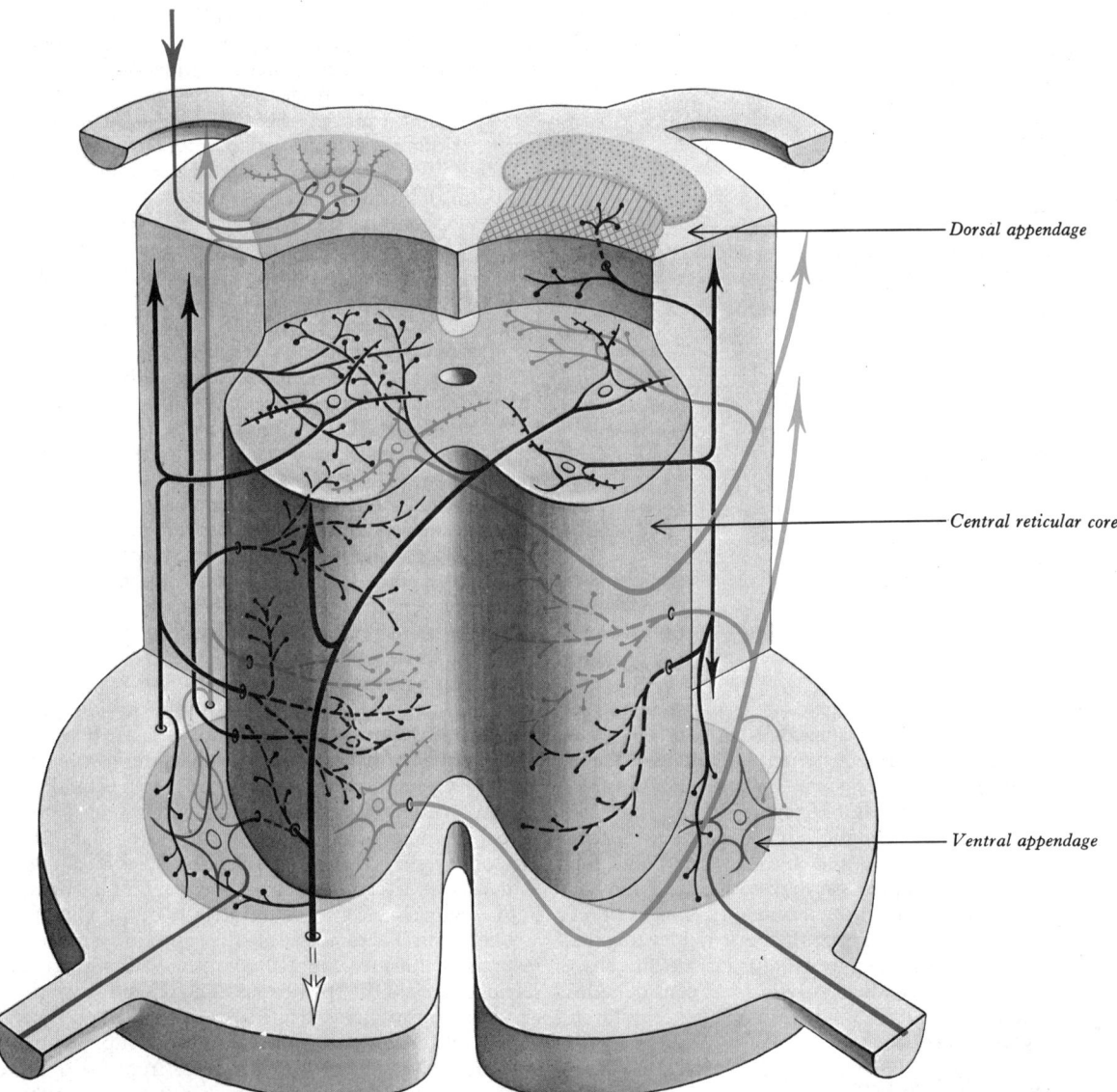

Dorsal appendage

Central reticular core

Ventral appendage

7.53 A highly simplified stereodiagram illustrating the concept of the spinal cord as consisting of a central 'reticular core' of grey matter, with related dorsal and ventral 'appendages' of grey matter. Many structural features are omitted, and only a few examples, relevant to the concept, are included. A dorsal column neuron, and others, more ventrally placed, which give rise to descending, long ascending, and local collateral branches, are shown in blue. Varieties of interneuron are in black. Two motor neurons are shown in red; also in red is a single example of a fibre descending from a supraspinal source. See text for a more detailed description; see also 7.47A and B for the origin and termination of some tracts, and 7.49 for one view of the fine structure of the substantia gelatinosa. (Redrawn from J. Szentágothai in: *Recent Development of Neurobiology in Hungary*, I. L. Lissák (ed.). By courtesy of the author and publishers, Akadémiai Kiadó, Budapest, 1967.)

lesions. At the fifth segment the upper limb paralysis will be complete; at the sixth the arms adopt a position of abduction and lateral rotation, with the elbows flexed and the forearms supinated due to the unopposed activity of the deltoids, spinati, rhomboids, bicipites and brachiales, which are supplied by the fifth cervical nerve. In cervical lesions at progressively lower levels the innervation of correspondingly more muscles of the upper limb is retained. Lesions of the first thoracic segment result in paralysis of the small muscles of the hand together with interference with sympathetic outflow, resulting in contraction of the pupil, recession of the eyeball, narrowing of the palpebral fissure and absence of sweating on the face and neck (Horner's syndrome). Sensation will be retained in areas deriving their innervation from segments above the site of the lesions. In particular, cutaneous sensation will be retained in the neck and chest down to the second intercostal space, because this area is innervated by the supraclavicular nerves (C_3 and C_4).

In the thoracic region, division of the spinal cord results in paralysis of the trunk below the segmental level of the lesion and of both lower limbs (paraplegia).

The first sacral neural segment is opposite the junction of the thoracic and lumbar vertebrae and injury, which commonly occurs here, results in paralysis of the urinary bladder and rectum and of the muscles supplied by the sacral segments. Cutaneous sensibility is lost in the 'saddle' area in the perineum and buttocks, back of the thigh, leg and sole of the foot. The lumbar nerve roots which pass distally to join the cauda equina at this level may be divided and then the result is complete paralysis of both lower limbs. Lesions below the first lumbar vertebra may divide or damage the nerves of the cauda equina, but severe nerve damage is uncommon and is usually confined to the nerve root at the level of the bony injury.

Neurological symptoms may also arise from interference with the blood supply of the spinal cord, particularly in the lower thoracic and upper lumbar regions.

The Vertebral Levels of Spinal Cord Segments

Of clinical importance is the position of the spinal segments relative to the vertebrae. A useful approximation is: in the cervical region the tip of the spine of a particular vertebra corresponds to the level of the succeeding cord segment (i.e., the sixth cervical spine is opposite the seventh cord segment); in the upper thoracic

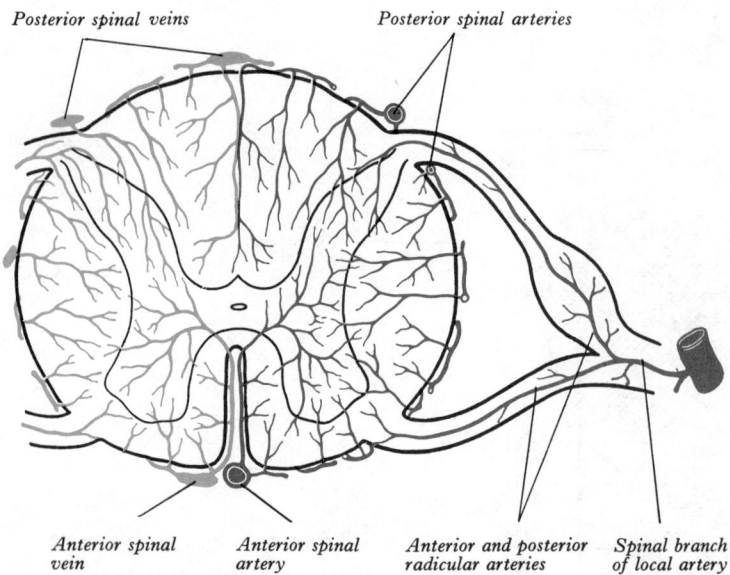

Posterior spinal veins   *Posterior spinal arteries*

Anterior spinal vein   *Anterior spinal artery*   *Anterior and posterior radicular arteries*   *Spinal branch of local artery*

7.54 Diagram of the intrinsic blood vessels of the spinal cord. The position of the veins is quite variable.

region the apex of a vertebral spine corresponds to two segments lower in number (i.e., the fourth spine is level with the sixth segment); in the lower thoracic region there is a difference of three segments (i.e., the tenth thoracic spine is level with the first lumbar segment). The eleventh thoracic spine overlies the third lumbar segment and the twelfth thoracic spine is opposite the first sacral segment. In the newborn child, on the other hand, the spinal cord extends to the upper border of the third lumbar vertebra. Barson (1970) examined the terminal level in 258 pre- and postnatal subjects, and stated that the perinatal level is the third lumbar vertebra, rising somewhat during the first two months of postnatal life. Individual variation was considerable: first to fourth lumbar at birth, first to third in a number of children (3 months to 15 years of age).

Blood vessels of the Spinal Cord

Blood reaches the spinal cord along the spinal branches of the vertebral, deep cervical, intercostal and lumbar arteries which, with the anterior and posterior spinal arteries, contribute to the formation of longitudinal anastomotic channels along the cord (*see* pp. 695, 710, and Gillilan 1972). These spinal branches give rise to anterior and posterior radicular arteries which approach the spinal cord along the ventral and dorsal nerve roots. Most of the anterior radicular arteries are small and terminate within the ventral nerve roots or in the plexus in the pia around the cord. The posterior radicular arteries supply the dorsal root ganglia; according to Somogyi *et al.* (1973) these ganglionic ramules enter at both poles of each ganglion, forming distinctive patterns of distribution around ganglion cells and amongst nerve fibres. The same authors have described the arterial supply of spinal roots (Undi *et al.* 1973). A small but variable number of radicular arteries (usually four to nine), mainly situated in the lower cervical, lower thoracic and upper lumbar regions, are larger than the remainder and in addition reach the anterior median sulcus of the spinal cord, where they divide into slender ascending and large descending branches. These branches anastomose with one

another and with the anterior spinal arteries above to form a single, or in places paired, longitudinal vessel of uneven calibre along the anterior median sulcus. Frequently one of these anterior radicular arteries is considerably larger than the remainder and is often termed the *arteria radicularis magna* (of Adamkiewicz). Its exact position varies, but it arises from one of the intersegmental branches of the descending aorta in the lower thoracic or upper lumbar vertebral levels. In two-thirds of cases it arises on the left-hand side. Reaching the spinal cord, this vessel sends one branch to join the anterior spinal artery below and another to anastomose with the division of the posterior spinal artery lying in front of the dorsal roots. The arteria radicularis magna may sometimes be responsible for most of the blood supply of as much as the lower two-thirds of the spinal cord.

From the anterior spinal artery central branches pass into the anterior median fissure. Here each one passes either to the right or left to supply the anterior grey column, base of the posterior grey column, including the dorsal nucleus, and the adjacent white matter (7.54).

Each posterior spinal artery gives rise to a pair of longitudinally running channels which anastomose frequently, one lying in front of, the other behind the attachment of the dorsal roots. These longitudinal channels are reinforced at intervals by posterior radicular arteries, which are variable both in number and size but are more numerous than the anterior radicular arteries. In addition the vessel in front of the dorsal nerve roots receives an anastomotic ramus from the descending branch of the arteria radicularis magna. Lazorthes *et al.* (1971) have largely confirmed the above description, but they emphasize the uneven calibre and interruptions in the longitudinal spinal arteries which do not form completely continuous channels. At the level of the conus they communicate by anastomotic loops. They also point out the potential importance of anastomoses other than those between pial, or peripheral spinal arterial branches, indicating a posterior spinal series of anastomoses between rami of the dorsal divisions of the segmental arteries in the vicinity of the spinous processes.

The central branches of the anterior spinal artery are responsible for the supply of as much as two-thirds of the cross-sectional area of the spinal cord. The remainder of the posterior grey columns, posterior white columns and the peripheral parts of the lateral and anterior white columns are supplied by numerous small radially directed vessels derived from the posterior spinal arteries and from the vessels forming the plexus in the pia mater (Torr 1957; Gillilan 1958). In a detailed microangiographic investigation of the human spinal cord at cervical levels (Turnbull *et al.* 1966) there were from 1 to 6 anterior, and 0 to 8 posterior radicular spinal arteries. Each centimetre of the anterior spinal artery gave rise to 5 to 8 central branches. No anastomoses were observed within the cord itself. Overlapping of the territories of the central spinal arteries was confirmed, and similar overlapping in the longitudinal direction was emphasized.

The veins of the spinal cord drain into six tortuous, often plexiform and longitudinal channels, one along the anterior median fissure and a second along the posterior median sulcus and the others, often incomplete, situated on each side, one pair just behind and the other in front of the line of attachment of the ventral and dorsal nerve roots. These six vessels communicate freely with one another and above with the veins of the cerebellum and the cranial venous sinuses. The veins accompanying the anterior spinal artery receive sizable venules from the central grey matter, while much of the blood draining the periphery of the spinal cord enters the veins of the pia mater. Further details of these veins are found on p. 756.

THE RHOMBENCEPHALON OR HINDBRAIN

The rhombencephalon comprises the medulla oblongata, pons and cerebellum; its cavity is the fourth ventricle. The medulla oblongata and pons are traversed by fibre tracts which interconnect these and other parts of the central nervous system and contain, amongst others, collections of nerve cells which constitute the nuclei of several of the cranial nerves. The following cranial nerves have their superficial origins from the pons and medulla oblongata: trigeminal, abducent, facial, vestibulocochlear, glossopharyngeal, vagus, cranial roots of the accessory, and hypoglossal. Scattered among the nuclei and tracts is the *reticular formation* (p. 943) which consists of intermingled grey and white matter and also includes nuclei of nerve cells concerned with the control of the heart, the respiratory apparatus and the alimentary tract.

THE MEDULLA OBLONGATA

The medulla oblongata extends from the lower margin of the pons to a transverse plane passing above the first pair of cervical nerves; this plane corresponds with the upper border of the atlas behind, and the middle of the dens of the axis in front; at this level the medulla oblongata is continuous with the spinal cord. However, it must be emphasized that the plane of junction is arbitrary; the internal structure of the spinal cord changes *gradually* to that of the medulla oblongata. The anterior surface of the medulla oblongata is separated from the basilar part of the occipital bone and the upper part of the dens by the membranes of the brain and the occipito-axial ligaments. Posteriorly it is received into the notch between the hemispheres of the cerebellum, and the upper portion of this surface forms the lower part of the floor of the fourth ventricle.

The medulla oblongata is somewhat piriform in shape (7.55, 56), its broad extremity being directed upwards to merge with the pons, while its narrow lower end is continuous with the spinal cord. It measures about 3 cm longitudinally, 2 cm transversely at its widest part, and 1·25 cm anteroposteriorly. The central canal of the spinal cord is prolonged into its lower half, and then expands as the cavity of the fourth ventricle; the medulla oblongata may therefore be divided into a lower, *closed part* containing the central canal, and an upper, *open part* corresponding with the lower half of the fourth ventricle. Its anterior and posterior surfaces are marked by median fissures.

The anterior median fissure contains a shallow fold of pia mater, and extends along the entire length of the medulla oblongata; below, it is continuous with the anterior median fissure of the spinal cord; above, it ends at the lower border of the pons in a small triangular expansion termed the *foramen caecum*. Its lower part is interrupted by bundles of fibres which cross obliquely from one side to the other, the *decussation of the pyramids*. Some fibres, the *anterior external arcuate fibres*, emerge from the fissure above this decussation and curve laterally over the surface of the medulla oblongata.

The posterior median sulcus is a narrow groove which exists only in the closed part of the medulla oblongata; it is continuous below with the posterior median sulcus of the spinal cord, but becomes rapidly shallower at cranial levels, and ends about the middle of the medulla oblongata, where the central canal expands into the cavity of the fourth ventricle.

Many of the cranial nerves emerge from or enter the substance of the medulla oblongata, and they appear at the surface in line with the roots of spinal nerves. The fibres of the hypoglossal nerve correspond in position with ventral spinal roots and emerge in linear series from a furrow termed the *anterolateral sulcus*.

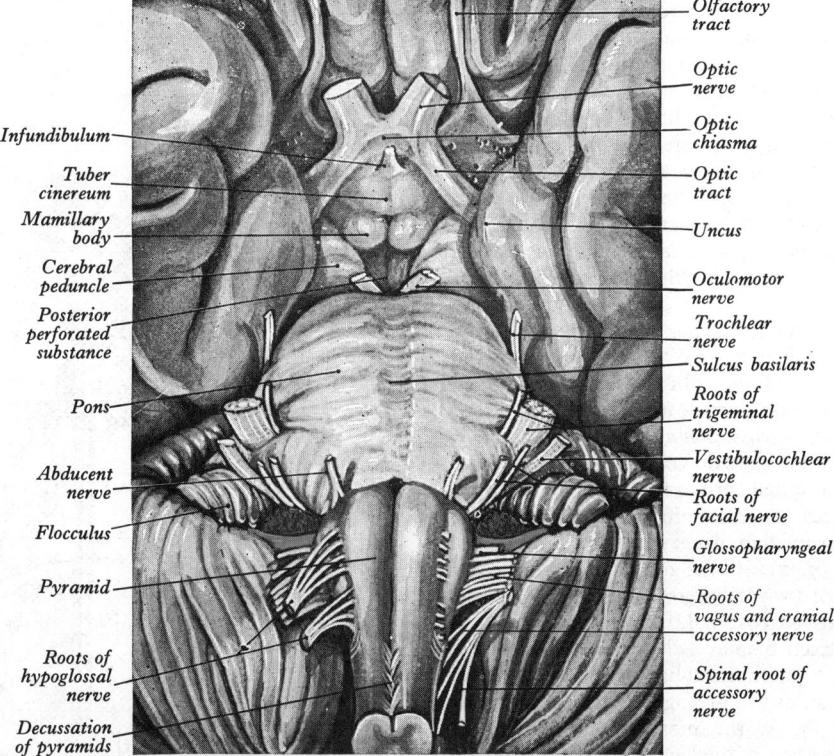

Infundibulum
Tuber cinereum
Mamillary body
Cerebral peduncle
Posterior perforated substance
Pons
Abducent nerve
Flocculus
Pyramid
Roots of hypoglossal nerve
Decussation of pyramids

Olfactory tract
Optic nerve
Optic chiasma
Optic tract
Uncus
Oculomotor nerve
Trochlear nerve
Sulcus basilaris
Roots of trigeminal nerve
Vestibulocochlear nerve
Roots of facial nerve
Glossopharyngeal nerve
Roots of vagus and cranial accessory nerve
Spinal root of accessory nerve

7.55 The ventral aspect of the brainstem and the interpeduncular fossa. The wall of the lateral recess of the fourth ventricle is shown in blue, and the choroid plexus, which protrudes through the foramen of the lateral recess into the subarachnoid space, is coloured crimson. Note that the lateral recess covers the medial part of the flocculus and is itself partially obscured by a rootlet of the glossopharyngeal nerve.

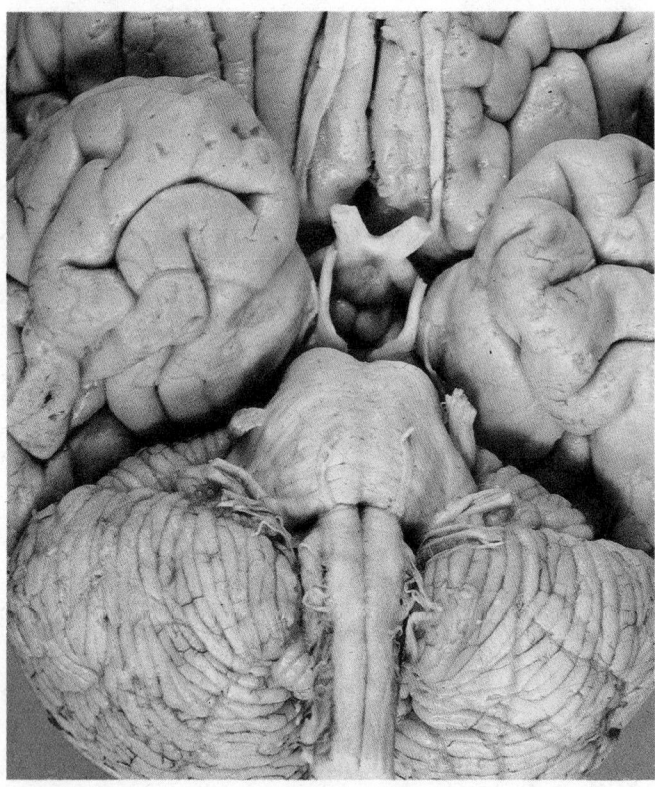

7.56 The ventral aspect of the brainstem, interpeduncular fossa and adjacent parts of the cerebellar and cerebral hemispheres. For identification of the various structures, compare with 7.55. (Dissection by Dr. E. L. Rees, Department of Anatomy, Guy's Hospital Medical School.)

Similarly, the accessory, vagus, and glossopharyngeal nerves are in line with dorsal spinal roots (p. 867) and enter or leave through the bottom of a sulcus named the *posterolateral sulcus*. Advantage is taken of this arrangement to subdivide each half of the medulla oblongata into anterior, middle and posterior regions. Although these three regions *appear* to be directly continuous with the corresponding funiculi of the spinal cord, they do not contain precisely the same nerve fibres, since some of the fasciculi of the spinal cord end or begin in the medulla oblongata, while others alter their course in passing through it.

The anterior region of the medulla oblongata (7.55, 56) lies between the anterior median fissure and the anterolateral sulcus, forming an elongated surface elevation which is named, somewhat inappropriately, the *pyramid*. Its upper end is slightly constricted and between it and the pons the fibres of the abducent nerve emerge; below, it tapers into the anterior funiculus of the spinal cord, with which it is superficially continuous.

The two pyramids contain descending fibres which pass from the cerebral cortex to the spinal cord. When traced downwards, approximately 70–90 per cent of these fibres leave the pyramids in successive bundles, and decussate in the anterior median fissure, forming what is termed the *decussation of the pyramids*. Having crossed the median plane, they pass down in the posterior part of the lateral funiculus of the spinal cord as the lateral corticospinal tract. The remaining fibres—i.e. those in the lateral part of the pyramid—do not cross the median plane; some descend as the anterior corticospinal tract (7.57) into the anterior funiculus of the same side of the spinal cord while others incline backwards and laterally to join the lateral corticospinal tract of the same side (p. 877). The corticospinal tracts display a clear segregation of their descending fibres upon a topographical basis at almost all levels; and in the medullary pyramids this arrangement is similar to that which exists at cranial levels, the most lateral fibres being concerned with the innervation of the legs, the most medial with the arms and neck. How far the same pattern is carried on into the ventral corticospinal tracts or through the decussating cortico-spinal fibres is not fully clarified, but a similar somatotopic sorting

of fibres is usually described in the lateral corticospinal tracts as they extend into the spinal cord.

The lateral region of the medulla oblongata (7.58) is limited in front by the anterolateral sulcus and the roots of the hypoglossal nerve, and behind by the posterolateral sulcus and the roots of the accessory, vagus and glossopharyngeal nerves. Its upper part consists of a prominent oval mass which is named the *olive*, while its lower part is of the same width as the lateral funiculus of the spinal cord, and appears on the surface to be a direct continuation of it. Only a portion of the lateral funiculus of the spinal cord is continued upwards into this region, because the lateral corticospinal tract is derived mainly from the contralateral pyramid, and most of the fibres of the posterior spinocerebellar tract leave the funiculus to enter the inferior cerebellar peduncle in the posterior region of the medulla. The lateral intersegmental tract and the anterior spinocerebellar tract are continued upwards in the lateral region of the medulla oblongata.

The olive is a smooth, oval elevation between the anterolateral and posterolateral sulci and lateral to the pyramid. It is caused by underlying groups of nerve cells forming the *inferior olivary nuclei* (p. 904). It is lateral to the pyramid, separated by the anterolateral sulcus and the fibres of the hypoglossal nerve. It is about 1·25 cm long, and dorsolateral to its cranial end there is a slight depression at the lower border of the pons in which the roots of the facial nerve appear. The anterior external arcuate fibres emerge from the anterior median fissure, and wind backwards across the pyramid and the olive to enter the inferior cerebellar peduncle (7.67).

The posterior region of the medulla oblongata (7.58A, B; 59) lies behind the posterolateral sulcus and the roots of the accessory, vagus and glossopharyngeal nerves, and, like the lateral region, is divisible into caudal and cranial levels.

The *caudal part*, limited behind by the posterior median sulcus, consists of the upward continuation of the *fasciculus gracilis* and the *fasciculus cuneatus* of the spinal cord. The fasciculus gracilis flanks the posterior median sulcus, and is separated from the fasciculus cuneatus by the cranial continuation of the postero-intermediate sulcus and septum of the cervical spinal cord (p.

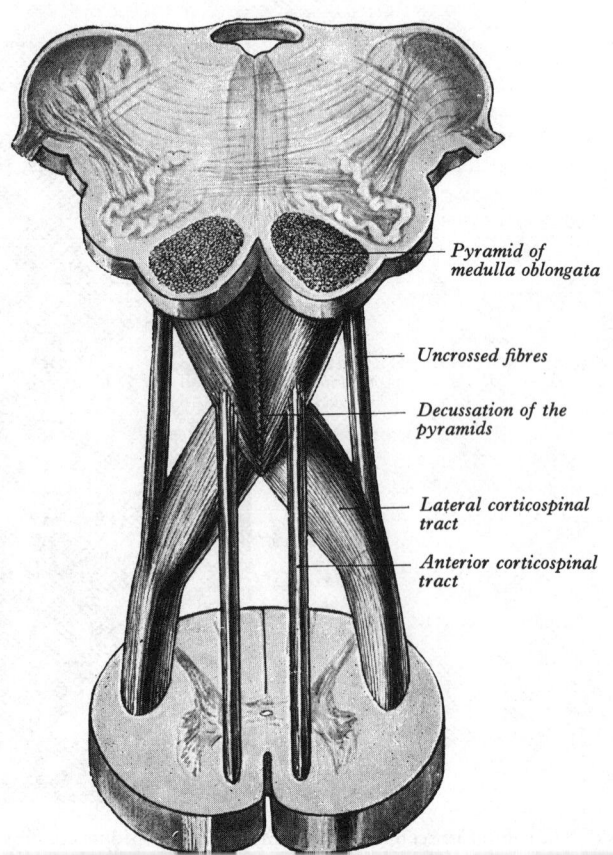

Pyramid of medulla oblongata

Uncrossed fibres

Decussation of the pyramids

Lateral corticospinal tract

Anterior corticospinal tract

7.57 Schematic dissection to show the decussation of the pyramids.

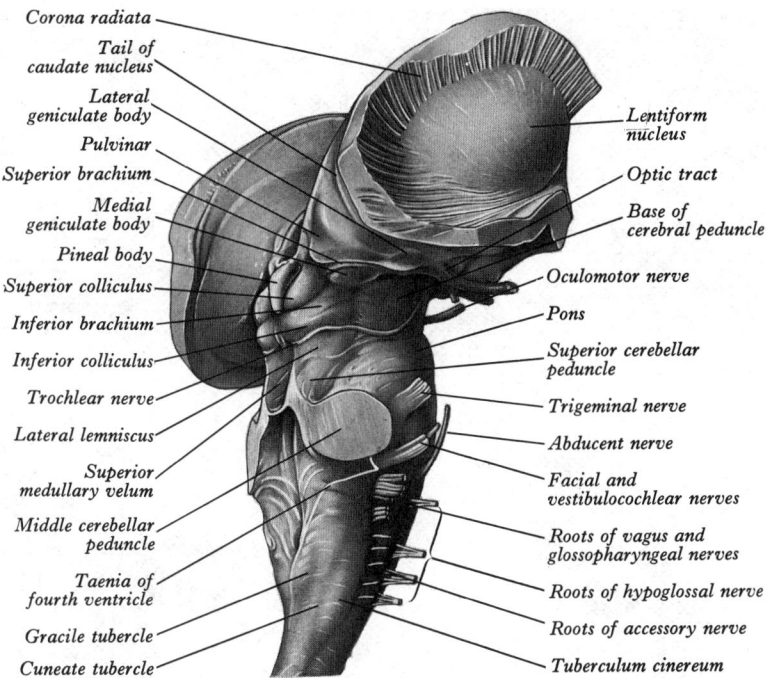

Corona radiata
Tail of caudate nucleus
Lateral geniculate body
Pulvinar
Superior brachium
Medial geniculate body
Pineal body
Superior colliculus
Inferior brachium
Inferior colliculus
Trochlear nerve
Lateral lemniscus
Superior medullary velum
Middle cerebellar peduncle
Taenia of fourth ventricle
Gracile tubercle
Cuneate tubercle

Lentiform nucleus
Optic tract
Base of cerebral peduncle
Oculomotor nerve
Pons
Superior cerebellar peduncle
Trigeminal nerve
Abducent nerve
Facial and vestibulocochlear nerves
Roots of vagus and glossopharyngeal nerves
Roots of hypoglossal nerve
Roots of accessory nerve
Tuberculum cinereum

7.58A   The brainstem, posterolateral aspect.

865). These two fasciculi are at first vertical; but at the caudal end of the fourth ventricle they diverge from the median plane, and each presents an elongated swelling. The swelling on the fasciculus gracilis is the *gracile tubercle*, and is produced by the upper end of a subjacent nucleus of grey matter termed the **nucleus gracilis**; that on the fasciculus cuneatus is termed the *cuneate tubercle*, and is caused similarly by the **nucleus cuneatus**. Most of the fibres of these two fasciculi end by forming synapses with the cells in their respective nuclei. A third elevation, the *tuberculum cinereum*, can sometimes be recognized in the caudal part of the posterior region of the medulla (**7.58**). It is located between the fasciculus cuneatus and the roots of the accessory nerve, and is narrow below but wider above. It is produced by a nucleus which is continuous below with the substantia gelatinosa, and in which the fibres of the spinal tract of the trigeminal nerve end; these fibres separate the nucleus from the surface of the medulla oblongata (p. 911). For further details of this region, including the obex and the taeniae of the fourth ventricle, *see* p. 932.

The *cranial part* of the posterior region of the medulla oblongata is occupied by the *inferior cerebellar peduncle*, a thick rounded ridge situated between the lower part of the fourth ventricle and the roots of the glossopharyngeal and vagus nerves. The two inferior cerebellar peduncles incline away from the dorsolateral aspect of the medulla oblongata towards the cerebellum. As they ascend, they diverge from each other, and form the lower parts of the lateral boundaries of the fourth ventricle; higher up, they are directed backwards, each passing into the corresponding cerebellar hemisphere. Near their entrance into the cerebellum they are crossed by several strands of fibres, the *striae medullares*, which run to the median sulcus of the floor, or anterior wall, of the fourth ventricle (**7.88, 90**). The inferior cerebellar peduncle is not the upward continuation of the fasciculus gracilis and fasciculus cuneatus, although it appears to be so, for the fibres of these fasciculi end in the gracile, cuneate and accessory cuneate nuclei. The composition of the inferior cerebellar peduncle is described on p. 916.

INTERNAL STRUCTURE OF THE MEDULLA OBLONGATA

The internal structure of the medulla oblongata, like that of other parts of the brainstem, has been studied chiefly in transverse sections by a combination of histological and experimental methods, three-dimensional reconstructions also being largely

based upon serial transverse sections. It is hence customary and convenient to reconstruct such regions in description by reference to appearances in sample transverse sections at a variety of levels. In the account which follows, both of the medulla and other brainstem levels, this method will be employed, the medulla being considered at four successive levels, the first being the most caudal. The disposition of the principal brainstem nuclei is shown in profile in illustrations **7.62** and **7.64**, which should be consulted frequently when reading the subsequent sections.

(1) **A transverse section through the caudal end of the medulla oblongata** shows details very similar to those at the cranial end of the spinal cord, as would be expected (**7.60**). The posterior, lateral and anterior funiculi can be identified, and they contain the same nerve tracts. The grey matter shows two very

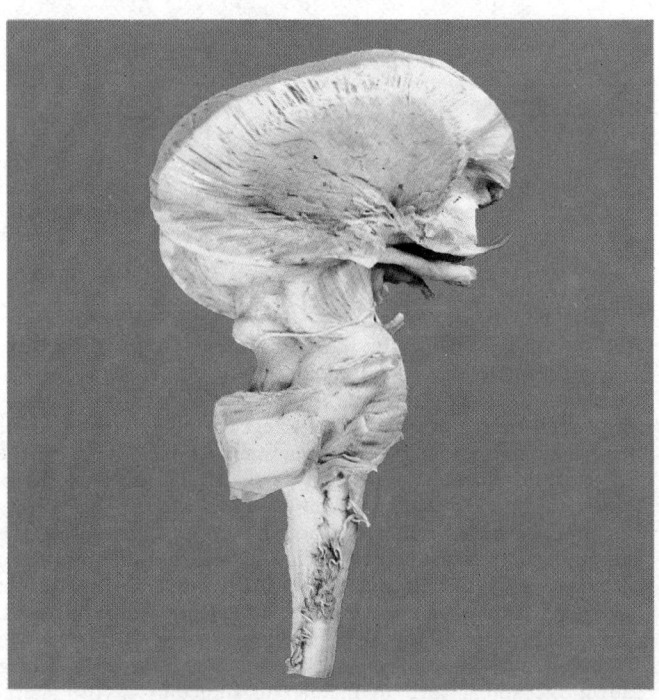

7.58B   The right lateral aspect of the brainstem, lentiform nucleus and corona radiata. For identifications compare with **7.58A**. (Dissection by Dr. E. L. Rees, Dept. of Anatomy, Guy's Hospital Medical School.)

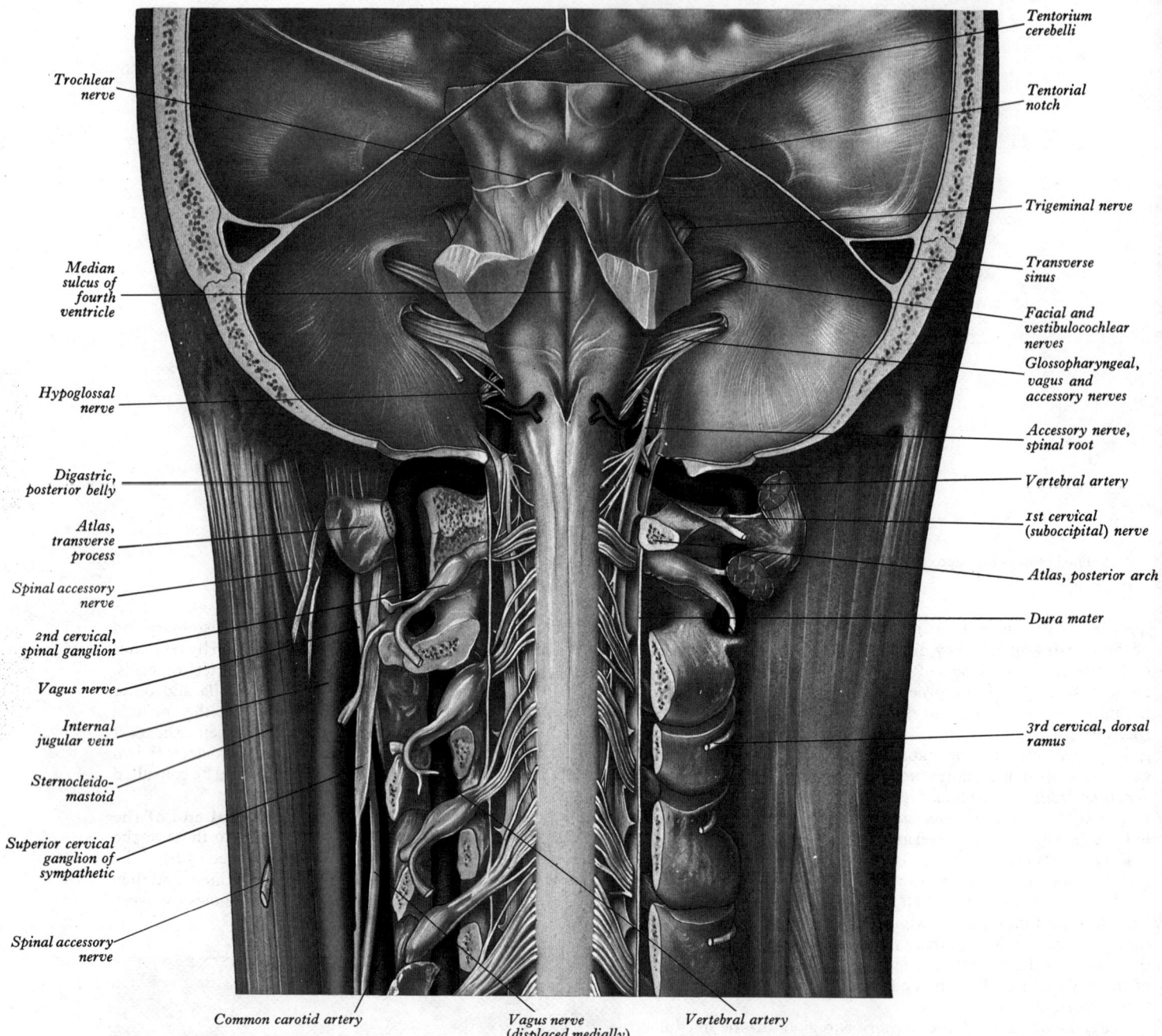

Trochlear nerve

Median sulcus of fourth ventricle

Hypoglossal nerve

Digastric, posterior belly

Atlas, transverse process

Spinal accessory nerve

2nd cervical, spinal ganglion

Vagus nerve

Internal jugular vein

Sternocleido-mastoid

Superior cervical ganglion of sympathetic

Spinal accessory nerve

Tentorium cerebelli

Tentorial notch

Trigeminal nerve

Transverse sinus

Facial and vestibulocochlear nerves

Glossopharyngeal, vagus and accessory nerves

Accessory nerve, spinal root

Vertebral artery

1st cervical (suboccipital) nerve

Atlas, posterior arch

Dura mater

3rd cervical, dorsal ramus

Common carotid artery   *Vagus nerve (displaced medially)*   *Vertebral artery*

7.59 Dissection exposing the brainstem and upper five cervical spinal segments after removal of large portions of the occipital and parietal bones and the cerebellum together with the roof of the fourth ventricle. On the left the foramina transversaria of the atlas and the third, fourth and fifth cervical vertebrae have been opened to expose the vertebral artery. On the right the posterior arch of the atlas and the laminae of the succeeding cervical vertebrae have been removed.

striking alterations. The anterior grey column is separated from the central grey matter by the decussating corticospinal fibres, which are coursing backwards and laterally to reach the lateral funiculus of the opposite side. In the upper part of the medulla oblongata the corticospinal fibres occupy the pyramid in its ventromedial portion, but in the lower part the majority of the corticospinal fibres cross the median plane, inclining backwards, as they do so (**7.57**), and decussating anterior to the central grey matter. The decussation takes place in an orderly manner with the fibres which terminate in the cervical segments of the spinal cord decussating first. This *decussation of the pyramids* is the most striking feature in sections of the medulla oblongata at this level. The actual proportion of the fibres which take part is subject to variation, but, as a rule, at least three-quarters do so and continue down the spinal cord in the lateral funiculus as the lateral corticospinal tract. Of the remaining fibres some retain their ventromedial position and descend in the anterior funiculus of the spinal cord as the anterior corticospinal tract; others descend with

the crossed fibres in the lateral funiculus of the same side (**7.57**). As a result of this decussation the anterior intersegmental tract of the spinal cord is displaced nearer to the central grey matter, which also takes up a more dorsal position so that the central canal inclines backwards as it ascends. The continuity between the anterior grey column and the central grey matter, maintained throughout the whole length of the spinal cord, is severed. The detached anterior grey column rapidly diminishes in size as it ascends; it is subdivided into the *supraspinal nucleus*, the source of efferent fibres of the first cervical nerve, and the *spinal nucleus of the accessory nerve*, which lies dorsolaterally and gives origin to the upper fibres of the spinal part of that nerve. The supraspinal nucleus is continuous above with that of the hypoglossal nerve. The nucleus of the spinal part of the accessory nerve is continued into the upper five segments of the spinal cord where it lies in the dorsolateral part of the anterior grey column. Above it merges with the nucleus ambiguus (p. 903).

The outline of the posterior column of grey matter can still be made out, but it, too, has undergone some modification. A narrow, strip-like portion of grey matter appears in the middle of the fasciculis gracilis, continuous ventrally with the base of the posterior horn. This is the inferior end of the *nucleus gracilis*, which extends upwards as far as the caudal limit of the fourth ventricle and forms an elevation on the posterior surface of the medulla oblongata, already described as the gracile tubercle (p. 899). A second cuneiform projection from the base of the posterior horn, beginning at a slightly higher level, invades the ventral part of the fasciculus cuneatus, and constitutes the *nucleus cuneatus*.

The *substantia gelatinosa* is a prominent feature in the direct upward continuation of the apex of the posterior grey column of the spinal cord. Here this apex is continuous with the lower end of the *nucleus of the spinal tract of the trigeminal nerve*, and the fibres of the tract itself are interposed between the nucleus and the surface of the medulla oblongata (7.60). It is considered in detail on pp. 911, 1086.

(2) **A transverse section just rostral to the decussation of the pyramids** shows an accentuation of the differences already noted and the appearance of certain new elements (7.61).

The *nucleus gracilis* has increased in breadth and the fibres of its corresponding fasciculus are grouped together on its dorsal, medial and lateral surfaces; the *nucleus cuneatus* has undergone a similar change. At first both retain their continuity with the central grey matter, but this is lost at higher levels. The fibres of the fasciculus gracilis and cuneatus have ascended uncrossed through the spinal cord, and the majority terminate in their respective nuclei at different levels by forming synapses with their contained nerve cells. New fibres arise in the nuclei and constitute further neurons on the pathway of tactile and proprioceptive sensibilities. These *internal arcuate fibres* emerge from the ventral aspects of the nuclei and, curving forwards and laterally at first round the central grey matter, they bend medially to reach the median plane, where they decussate with the corresponding fibres of the opposite side (7.61). Thereafter, they turn upwards and ascend on the opposite side close to the median raphe, constituting the *medial lemniscus*. The *decussation of the lemnisci* occurs in the area dorsal to the pyramids and in front of the central grey matter, which is in this way displaced still more dorsally towards the dorsal surface of the medulla oblongata. As the internal arcuate fibres sweep forwards they intervene between the spinal tract of the trigeminal nerve and the central grey matter. From the foregoing remarks it might appear that the gracile and cuneate nuclei are simply relay stations on a main sensory pathway widely regarded as the major route for impulses concerned with the more discriminative aspects of tactile and locomotor sensibility. During the last decade abundant anatomical and physiological evidence has led to a reappraisal of these views. Anatomically, the neuronal population of these nuclei has long been considered somewhat uniform, despite the early definition of many cell types within them by Cajal (1900). Although many of the neurons are multipolar, they vary much in size and are by no means uniformly distributed. Taber (1961) and Kuypers and Tuerk (1964) have described a variety of neurons in the nuclei and Biedenbach (1972) has defined two zones, of contrasting cytoarchitecture. In all species examined, including monkeys, the rostral region of each nucleus is *reticular* in nature, consisting of small and large multipolar neurons with long dendrites, and large but rounded neurons with short and profusely branching dendrites; this third type of neuron predominates in the caudal part of each nucleus, in which the neurons are *clustered* in various formations. These two zones differ in their connexions. Both receive dorsal root fibres, which terminate at all levels of the gracile and cuneate nuclei; but the caudal zone receives a denser input than the rostral (reticular) zone. Dorsal funicular fibres which arise from neurons in the spinal grey substance terminate only in the reticular zone. In all these connexions there is variable somatotopical distribution of terminals on the basis of spinal roots, which follows that in the dorsal funiculi; but considerable and variable degrees of overlap have been observed (*see* for example, Millar and Basbaum 1975). In monkeys, the hind leg and tail are represented medially in the nuclei, the trunk ventrally, and the digits dorsally. There is also

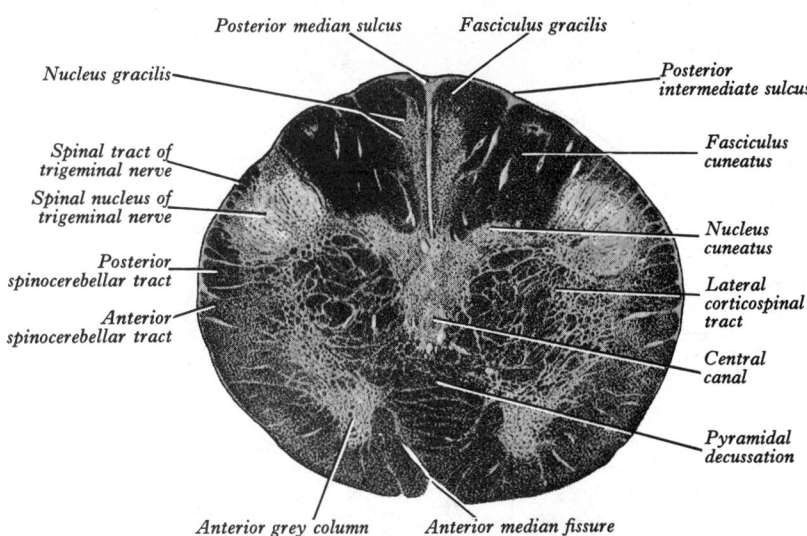

7.60 Transverse section through the medulla oblongata at the level of the pyramidal decussation. Weigert Pal preparation. Magnification × 7.

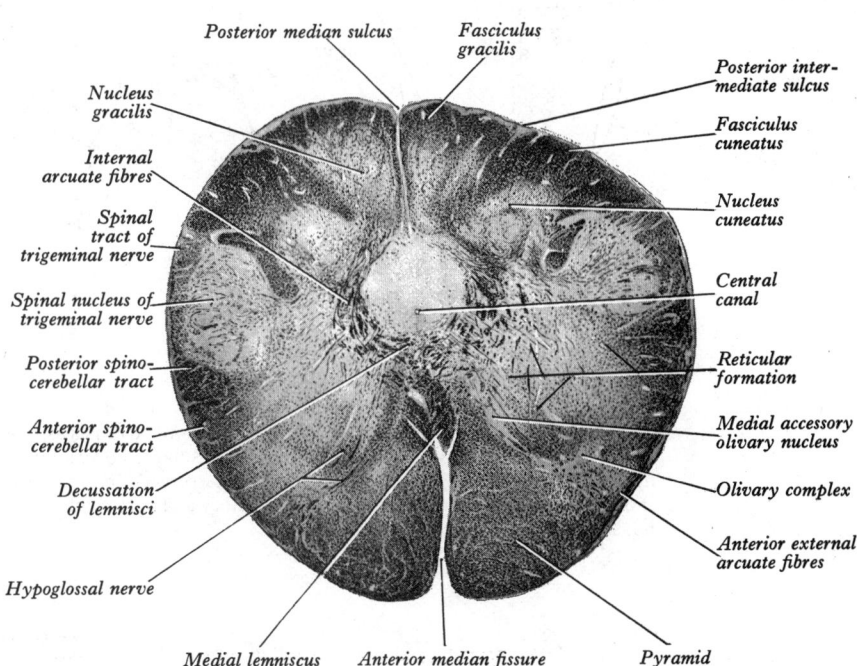

7.61 Transverse section through the medulla oblongata at the level of the decussation of the lemnisci. Weigert Pal preparation. Magnification *c.* × 6.

modality specificity; the caudal zone responds to low threshold cutaneous stimuli, while the neurons of the reticular (rostral) region respond to inputs from fibres subserving cutaneous and proprioceptive stimuli from joints and muscles. Both nuclei contain intrinsic cells or interneurons (Andersen *et al.* 1964; Rustioni and Sotelo 1974), and many of these are inhibitory in their effects. However, primary spinal afferents form synapses with the large rounded multipolar neurons which constitute the major source of efferent projection from the nuclei, thus forming a simple relay between spinal cord and higher levels. The descending afferents from the somatosensory cortex (p. 958), which reach the nuclei through the corticospinal tracts appear to be restricted in termination to the reticular zones; they may inhibit or enhance activity at the nuclear level, which is clearly a region of complex modulation of incoming sensory information. The same reticular zones also receive a projection from the reticular formation proper (Špaček and Lieberman 1974). It appears probable, for example, that a 'feed-back' mechanism from the gracile and cuneate nuclei to the spinal cord also exists. By antidromic stimulation (Dart 1971), retrograde horseradish peroxidase studies (Kuypers and Maisky 1975), and anterograde autoradiographic transport techniques (Burton and Loewy 1977), it has been shown in cats and monkeys that cells in these nuclei project to various regions of the ipsilateral dorsal appendage of the spinal grey matter. These descending fibres are presumably those involved in the depression of dorsal column neuronal activity which follows stimulation of the column itself (Hillman and Wall 1969).

The *accessory cuneate nucleus* lies dorsolateral to the cuneate nucleus. It contains large cells similar to those of the thoracic nucleus of the spinal cord and gives origin to the *posterior external arcuate fibres* (p. 916); these enter the cerebellum through the ipsilateral inferior cerebellar peduncle. The accessory cuneate nucleus receives its afferents from the lateral fibres in the fasciculus cuneatus which are derived from the cervical segments. It provides a pathway, the *cuneocerebellar tract*, for proprioceptive impulses from the upper limb, destined for the cerebellum, which enter the spinal cord above the upper limit of the thoracic nucleus. A group of neurons, currently known as '*nucleus Z*', was first identified in the cat by Brodal and Pompeiano (1957) between the rostral pole of the nucleus gracilis and the termination of the inferior vestibular nucleus and it is said to be present in the human medulla (Webster 1977). Its principal input appears to be from the dorsal spinocerebellar tract (Rustioni 1973), and it may constitute a separated part of the reticular zone of the gracile nucleus, which it resembles in cytoarchitecture. It is considered to be concerned in proprioception from the ipsilateral hind limb (Landgren and Silfvenius 1971).

The *nucleus of the spinal tract of the trigeminal nerve* (7.62) is at this level separated from the central grey matter by the internal arcuate fibres. It is separated from the lateral surface of the medulla oblongata only by the spinal tract of the trigeminal nerve, the fibres of which terminate in the nucleus, and by some of the fibres of the posterior spinocerebellar tract, which is beginning to incline dorsally to enter the inferior cerebellar peduncle (p. 916).

Two additional collections of grey matter occur at this level. One is dorsal to the lateral part of the pyramid, while the other is placed to its medial side and not far from the median plane. These are parts of the *medial accessory olivary nucleus* and will be considered together with the inferior olivary nucleus (p. 904).

The central grey matter, now occupying a position near the dorsal surface of the medulla oblongata, contains three important nuclei. A prominent group of large motor nerve cells, interspersed

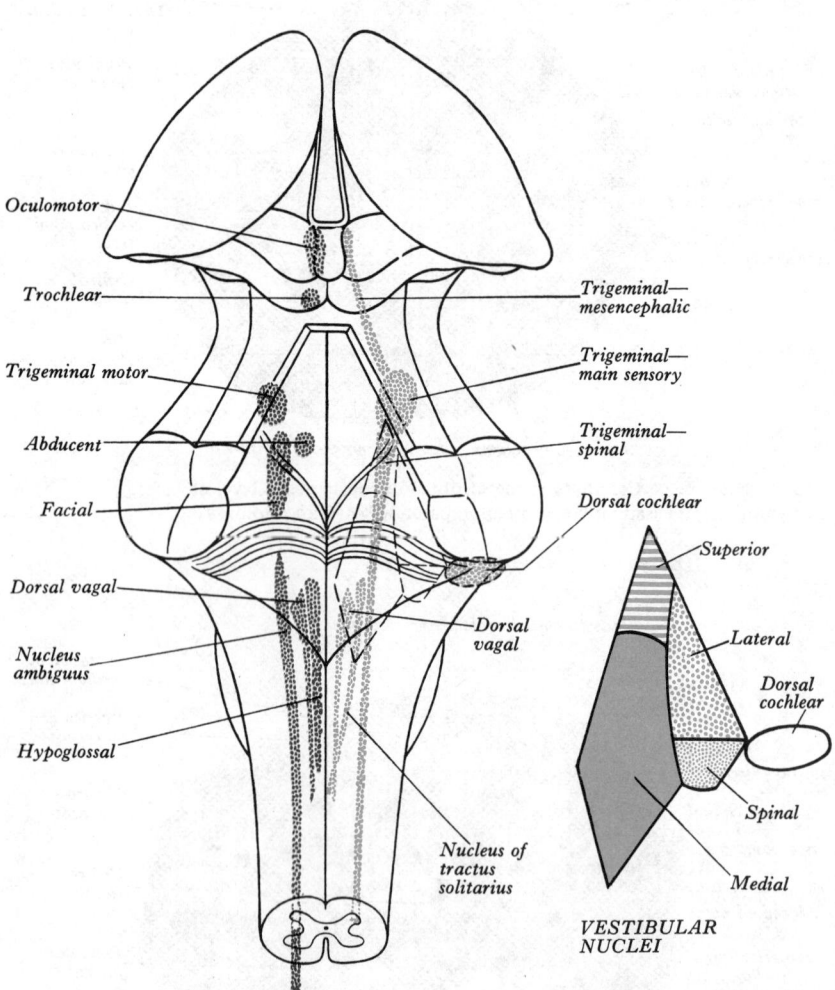

7.62 *Surface projection* of some cranial nerve nuclei on the dorsal aspect of the brainstem. Motor nuclei in red; sensory in blue. The vestibular nuclei are indicated in the main diagram by interrupted lines and are shown in detail in the small diagram. The olfactory and optic centres are not shown.

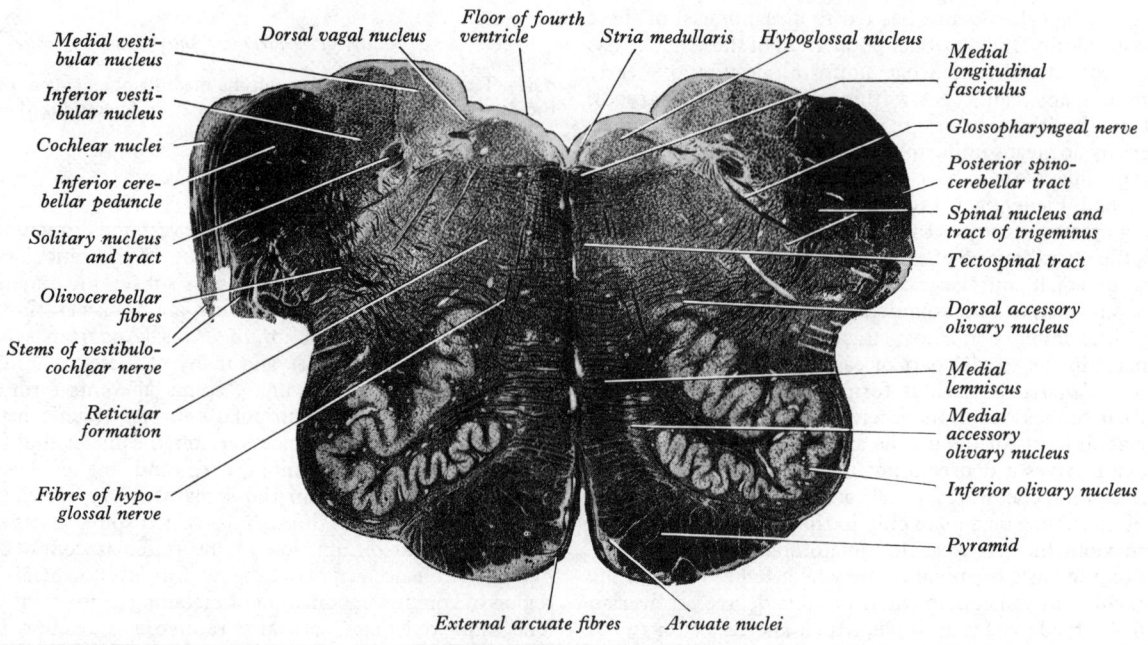

7.63 Transverse section through the medulla oblongata at mid-olivary level. Weigert Pal preparation. Magnification *c.* ×4·5.

with myelinated fibres, is situated in the ventromedial part of the central grey matter. This is the *nucleus of the hypoglossal nerve*. It extends upwards into the open part of the medulla oblongata, where it lies under the medial part of the trigonum hypoglossi in the floor of the fourth ventricle. Immediately adjacent to the hypoglossal nucleus are several other smaller groups of nerve cells collectively spoken of as the '*perihypoglossal complex*' or '*perihypoglossal grey*'. Neither term has more than topographical significance, for none of them is known to be in any way connected with the hypoglossal nerve or its nucleus. Included in the complex

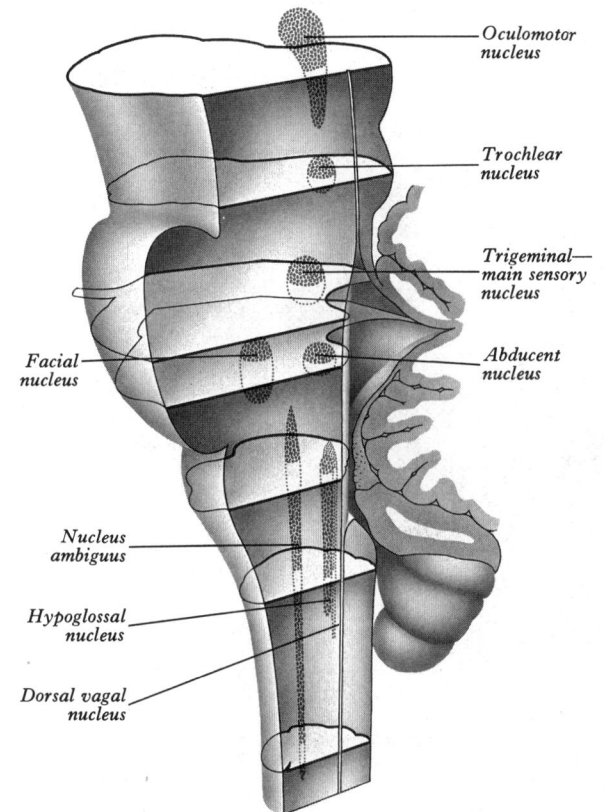

Oculomotor nucleus

Trochlear nucleus

Trigeminal— main sensory nucleus

Abducent nucleus

Facial nucleus

Nucleus ambiguus

Hypoglossal nucleus

Dorsal vagal nucleus

7.64   Diagram of motor nuclei of the cranial nerves. The sectional planes shown correspond to those depicted elsewhere in the text.

are the nucleus intercalatus (of Staderini), the 'sublingual' nucleus (of Roller), the nucleus prepositus hypoglossi, and the nucleus paramedianus dorsalis (reticularis), all of which contain cells suggestive in their characteristics of reticular connexions, which have been definitely ascribed to the paramedian nucleus (Brodal 1957). Gustatory and visceral afferent connexions have been attributed to the intercalated nucleus, but there is more convincing evidence that the perihypoglossal nuclei in general project to the cerebellum, at least in the cat (Torvik and Brodal 1954) and the monkey (Mehler *et al.* 1960). A topographical representation of lingual musculature has been described in the hypoglossal nucleus (p. 1083).

Dorsolateral to the hypoglossal nucleus, there is a second group of cells, the *dorsal nucleus of the vagus*. It is a mixed nucleus, containing cells of at least two types. The larger cells give rise to the fine fibres which innervate non-striated muscle; the smaller spindle-shaped cells may possibly be concerned with visceral afferent impulses. On the other hand many authorities believe that all the vagal visceral afferent fibres terminate in the nucleus of the tractus solitarius (*vide infra*). At a higher level the dorsal vagal nucleus is situated lateral to the hypoglossal nucleus in the floor of the fourth ventricle and corresponds in position to the trigonum vagi.

A third group of cells lies dorsolateral to the dorsal nucleus of the vagus at this level. It is the *nucleus of the tractus solitarius* (7.65), and it is intimately related to a group of descending fibres which constitute the *tractus solitarius* itself. At the caudal end of

the medulla oblongata these two nuclei fuse dorsal to the central canal. As the nucleus of the tractus solitarius is traced upwards it comes to lie more deeply in the medulla oblongata, on the ventrolateral aspect of the dorsal nucleus of the vagus, with which it is practically coextensive. The tractus solitarius receives afferent fibres from the facial, glossopharyngeal and vagus nerves, and they enter the nucleus in that order from above downwards, conveying to it gustatory sensibility from the mucous membrane of the tongue and palate (facial, glossopharyngeal and vagus), and, according to many authorities, visceral sensibility from the pharynx (glossopharyngeal and vagus) and from the oesophagus and the abdominal part of the alimentary canal (vagus). In this rostrocaudal representation there is some degree of overlap (Schwartz *et al.* 1951; Kerr 1962). The nerve cells of the solitary nucleus are smaller than those of the dorsal vagal nucleus. It is presumed that their axons project to the thalamus and perhaps thence to the cerebral cortex, though experimental attempts to establish this in the cat have failed (Morest 1967). The same study demonstrated connexions with the dorsal vagal nucleus and nucleus ambiguus. There is evidence that the solitary nucleus projects to the upper levels of the spinal cord through a solitariospinal tract (p. 876) both in man (Collier and Buzzard 1903), and the cat (Torvik 1957). The nucleus is considered to receive fibres from the spinal cord, cerebral cortex and cerebellum (Angaut and Brodal 1967). Nerve cells aggregated in a position ventrolateral to the nucleus of the solitary tract have sometimes been termed the *nucleus parasolitarius* (Crosby *et al.* 1962).

Numerous scattered islets of grey matter occur in the centre of the ventrolateral portion of the medulla oblongata. They occupy an area which is freely intersected by nerve fibres running in all directions and which is therefore termed the *reticular formation*. It is present at all levels of the medulla oblongata and extends upwards into the tegmentum of the pons and midbrain. There is no clear demarcation between the reticular formations in these various regions and they are considered collectively on p. 943.

The white matter has undergone an important rearrangement above the corticospinal decussation. The *pyramids* contain corticospinal and corticonuclear fibres, the latter being distributed to the many nuclei of the cranial nerves. They form two large bundles in the ventral part of the section, on each side of the anterior median fissure. Dorsally they are related to the accessory olivary nuclei and the decussation of the lemnisci.

The fibres of the *medial lemniscus* (p. 901), after emerging from the lemniscal decussation, turn upwards on each side in the form of a flattened tract, closely applied to the median raphe. In this

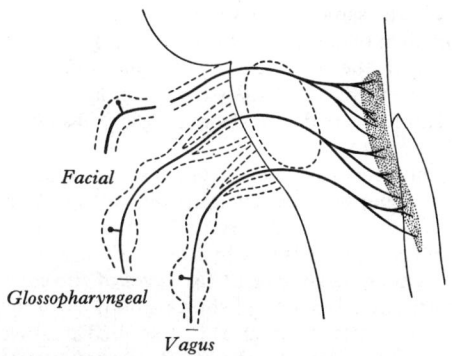

Facial

Glossopharyngeal

Vagus

7.65   Diagram of afferent fibres of the facial, glossopharyngeal and vagus nerves, conveying gustatory impulses to the nucleus of the tractus solitarius.

position they ascend to the pons, increasing in number as additional fibres join them from the upper levels of the decussation. Ventrally they are related to the pyramidal fibres, and dorsally to the medial longitudinal fasciculus and tectospinal tract. In the decussation the fibres undergo a rearrangement whereby those derived from the gracile nucleus come to lie ventral to those derived from the cuneate nucleus; at a higher level where the disposition of the medial lemniscus in the brainstem becomes altered (p. 938), the gracile fibres are lateral and the cuneate fibres medial. The fibres of the medial lemnisci at this level have been shown to display in the monkey (Ferraro and Barrera 1936) and chimpanzee (Walker 1937) a laminar somatotopic arrangement on a segmental basis.

The *medial longitudinal fasciculus* forms a small compact tract of nerve fibres, situated close to the median plane and ventral to the hypoglossal nucleus. Below, it is continuous with the anterior intersegmental tract of the spinal cord, but at this medullary level it is displaced dorsally by the decussations of the pyramids and

medially, intersecting the fibres of the medial lemniscus (7.63, 66). They cross the median plane and sweep dorsal to or traverse the olivary nucleus of the opposite side, intersecting the lateral spinothalamic tract, rubrospinal tract and nucleus of the spinal tract of the trigeminal nerve to enter the inferior cerebellar peduncle, by which they are conveyed to the cerebellum. Afferent connexions to the nucleus can be divided into ascending and descending fibres. The ascending fibres, mainly crossed, reach the nucleus from all levels of the spinal cord travelling in one or, possibly, two *spino-olivary tracts*. Some ascending connexions also reach the inferior olivary complex via the dorsal white

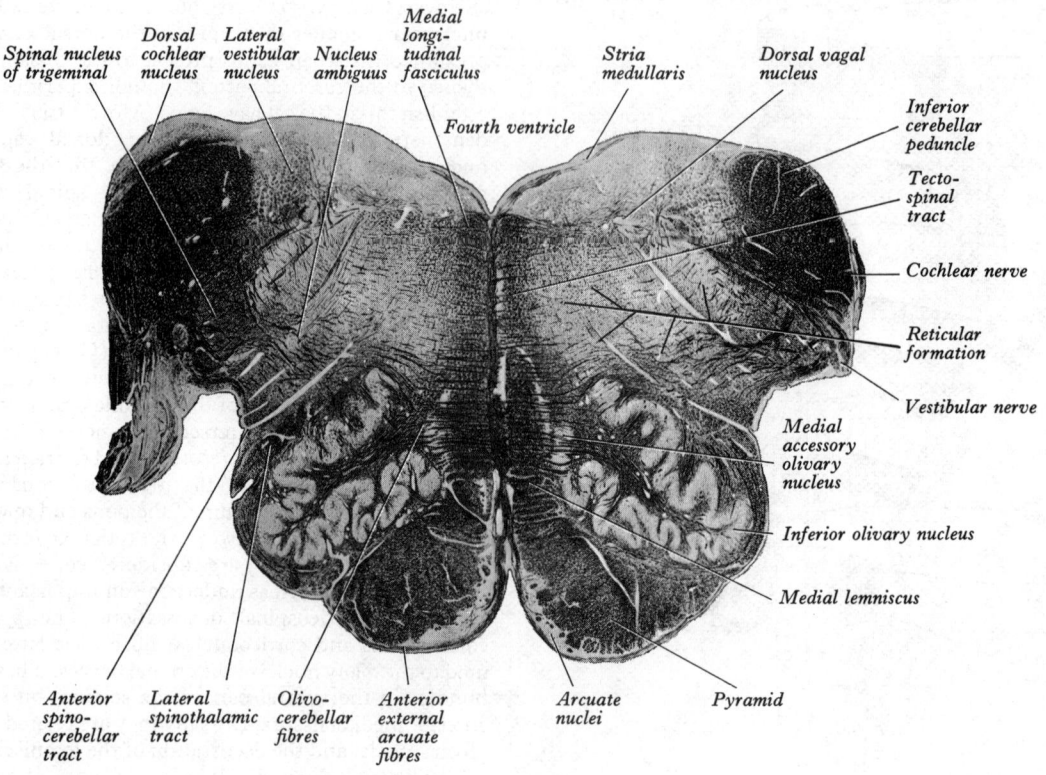

7.66 Transverse section through the superior half of the medulla oblongata. Weigert Pal preparation. Magnification *c*. ×4·5.

lemnisci. It is continued upwards through the pons and the midbrain in the same position relative to the central grey matter and the median plane, and therefore is closely related throughout its course with the somatic efferent column of the grey matter. The constituent fibres of the tract run relatively short courses within it, for they are derived from a variety of sources, which are detailed on p. 939.

The spinocerebellar, spinotectal, vestibulospinal, rubrospinal and lateral spinothalamic (spinal lemniscal) tracts are all in the anterolateral area, limited dorsally by the nucleus of the spinal trigeminal tract and ventrally by the pyramid.

(3) **A transverse section at the level of the caudal limit of the fourth ventricle** (7.63), shows a number of new elements, together with most of those already described at a lower level. The total amount of grey matter shows a distinct increase owing to the presence of the large olivary nucleus, the arcuate nucleus and nuclei associated with two divisions of the vestibulocochlear, the glossopharyngeal, vagus and accessory nerves.

The *inferior olivary nucleus* is a large hollow mass of grey matter, with irregularly crenated walls and a longitudinal hilum placed on its medial side. It is surrounded by a capsule of myelinated fibres forming the *amiculum of the olive*. Situated dorsolateral to the pyramid, the nucleus corresponds to the surface elevation of the olive, but extends upwards almost to the pons. The olivary nucleus consists of small cells, a large number of their axons forming the *olivocerebellar tract*. These axons emerge from the hilum or through the adjacent wall and run

columns (Hand and Liu 1966). The descending fibres arise from the cortex of the cerebrum, thalamus, basal nuclei, red nucleus and central grey matter of the midbrain (Walberg 1960). Some of these are said to travel in a bundle called the *central tegmental fasciculus* (Jansen and Brodal 1950).

The *medial accessory olivary nucleus* is a curved lamina of grey matter which is found at this level. The concavity of the curve is directed laterally and the nucleus is interposed between the medial lemniscus and the pyramid, on the one hand, and the medial and ventral aspects of the inferior olivary nucleus on the other.

The *dorsal accessory olivary nucleus* is a second lamina of grey matter, placed dorsal to the medial part of the inferior olivary nucleus.

Both the inferior and the accessory olivary nuclei are intimately associated with the cerebellum. Phylogenetically, the accessory olivary nuclei are older than the inferior nucleus, and they send their fibres to the paleocerebellum (p. 916). The inferior olivary nucleus occurs only in mammals and, in the course of evolution, it has enlarged in a caudal direction. In all their connexions, whether with higher levels in the cerebral hemispheres, the spinal cord, or the cerebellum, the olivary nuclei show a marked and often highly specific organization upon a somatotopic basis. This is particularly so in the case of cerebellar connexions, which are considered in detail later (pp. 916, 918, 924).

The *arcuate nuclei* form curved, interrupted bands of grey matter which are closely applied to the anterior and medial aspects of the pyramids; they appear to be caudally displaced nuclei pontis (Rasmussen and Peyton 1946). They give origin to the anterior external arcuate fibres.

The central grey matter, which, at this level, is spread out over the floor of the ventricle, contains the *hypoglossal nucleus* and the *dorsal nucleus of the vagus*, the *nucleus of the tractus solitarius* lying ventrolateral to the last-named; lateral to these, and on the medial side of the inferior cerebellar peduncle, the lower part of the *inferior and medial nuclei of the vestibular nerve* may be recognized (p. 908). Between the hypoglossal nucleus and the dorsal nucleus of the vagus is the *nucleus intercalatus* (p. 903).

A small isolated group of large motor nerve cells, termed the *nucleus ambiguus*, is placed deeply in the reticular formation. It extends upwards as far as the upper limit of the dorsal nucleus of the vagus. The fibres which emerge from its upper end join the glossopharyngeal nerve, and those which emerge at a lower level join the fila of the vagus and cranial part of the accessory nerves. Inferiorly it is continuous with the spinal nucleus of the accessory nerve (p. 900). It consists of large motor neurons, the fibres of which are distributed to striated muscle of branchial origin (p. 1074). These first pass dorsally and medially for a short distance and then curve laterally to join the emerging fila of the cranial portion of the accessory, the vagus and glossopharyngeal nerves. Histologically, the nucleus ambiguus can be divided into several groups of cells in man, as in other mammals; experimentally some degree of representation of the muscles innervated has been established (Lawn 1966, and p. 1076). At the rostral extremity of the nucleus ambiguus, between this and the facial nucleus, is a small group of cells named the *retrofacial nucleus*. Though thus apparently in line with special visceral efferent nuclei, it is reputed to be a source of general visceral efferent fibres of the vagus nerve (*see* p. 1076).

The nucleus gracilis and the nucleus cuneatus, now diminishing in size and irregular in outline, occupy the dorsolateral portion of the section, and ventral to them the *nucleus of the spinal tract of the trigeminal nerve* can be recognized without difficulty.

The *cochlear nuclei* may be observed on the surface of the inferior cerebellar peduncle (pp. 909, 910).

The white matter of the medulla oblongata shows little change at this level apart from the development of the inferior cerebellar peduncle on the lateral side of the fourth ventricle. The pyramid, the medial lemniscus, the tectospinal tract and the medial longitudinal bundle occupy the same relative positions as they did at a lower level. The fibres of the olivocerebellar tract, sweeping across the median plane and turning dorsally to join the inferior cerebellar peduncle, have already been described in connexion with the olivary nucleus (p. 902). The *anterior external arcuate fibres* have their cells of origin in the arcuate nuclei of both sides and, emerging from the anterior median fissure, they run laterally, backwards and upwards over the surface of the pyramid, the olive and the spinal tract of the trigeminal nerve. Reaching the posterior spinocerebellar tract, they ascend with it to enter the inferior cerebellar peduncle (**7.67**).

The emerging fila of the hypoglossal nerve leave the ventral aspect of its nucleus and run forwards through the reticular formation. Passing lateral to the medial lemniscus and medial to, or sometimes through the wall of, the olivary nucleus, they curve laterally to emerge from the anterolateral sulcus. A relatively small lesion in the ventral part of the medulla oblongata at this level may therefore involve both the corticospinal tract and the hypoglossal nerve, causing a characteristic crossed paralysis. The muscles of the tongue are paralysed on the same side as the lesion, but it is the limbs of the opposite side of the body that are affected, for the lesion is situated above the level of the pyramidal decussation.

More dorsally, the reticular formation is traversed by the fibres of the vagus, travelling from their origin in the dorsal nucleus, the nucleus ambiguus and the nucleus of the tractus solitarius to the posterolateral sulcus, where they emerge.

The *spinal lemniscus* (p. 880), lies dorsal to the olivary nucleus and separated from the surface of the medulla oblongata by the anterior spinocerebellar and the spinotectal tracts. There is evidence from surgical procedures in man, and experiments in other animals, that the fibres are arranged somatotopically; those

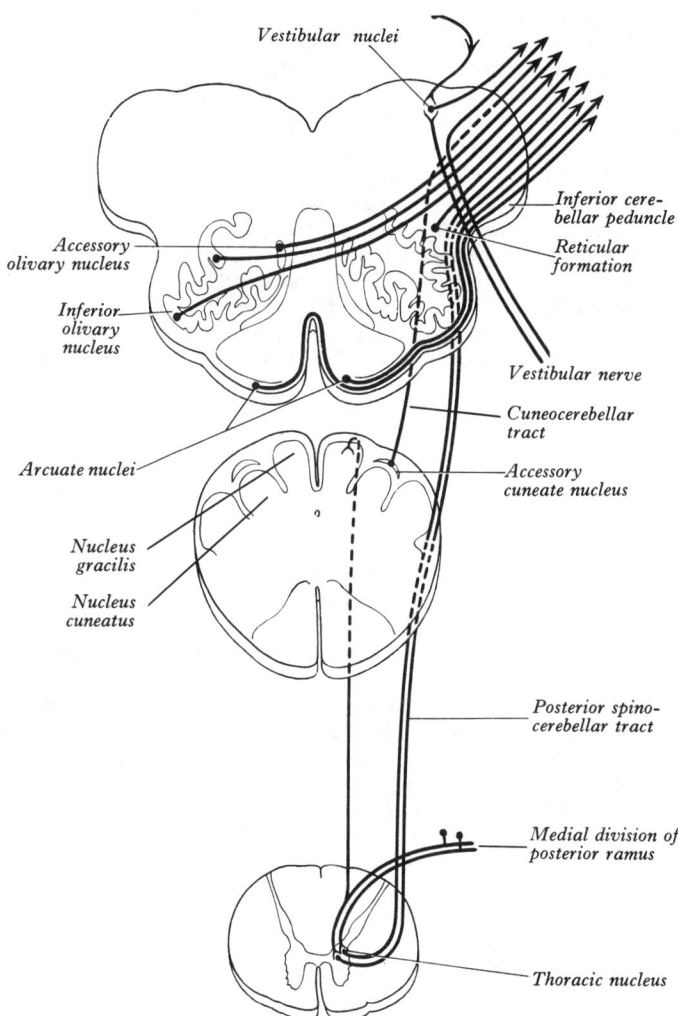

7.67 Diagram to show some of the afferent components of the inferior cerebellar peduncle; the efferent components have been omitted. See text for further details.

conveying impulses from the lower limb are superficial, those from the upper limb deep and those from the trunk intermediate. As it ascends through the upper part of the medulla oblongata, the spinal lemniscus is closely related to the nucleus ambiguus, and a small lesion in the ventral part of the reticular formation may cause paralysis of the vocal fold and of the soft palate of the same side, but a loss of sensibility to pain and temperature on the opposite side of the body.

(4) **A transverse section of the most rostral part of the medulla oblongata** shows little change. The dorsal surface of the medulla oblongata here is relatively flat as compared with the preceding level and may show a few fibres of the stria medullaris just under this surface (p. 910). The inferior olivary nucleus occupies the same relative position, but the accessory olivary nuclei are broken up and diminishing (**7.66**).

The medial nucleus of the vestibular nerve has widened and lies on the lateral and dorsal sides of the dorsal vagal nucleus, which is now depressed below the floor of the rhomboid fossa. The inferior nucleus of the vestibular nerve intervenes between the medial nucleus and the inferior cerebellar peduncle. At the pontomedullary junction the *lateral vestibular nucleus* replaces the inferior nucleus and the lower parts of the cochlear nuclei are usually seen. (*See* p. 908 for details of the vestibular nuclei.)

The nucleus of the tractus solitarius, the nucleus of the spinal tract of the trigeminal nerve and the nucleus ambiguus show little alteration in position.

The arrangement of the white substance at this level shows no conspicuous alteration. The lateral spinothalamic tract, or spinal lemniscus, ascends dorsal to the olivary nucleus, its fibres retaining the somatotopical arrangement already described (p. 880). The *inferior cerebellar peduncle* has increased in size and forms a well-marked elevation on the dorsolateral aspect of the medulla oblongata. The extensive array of fibres, from many sources, which compose the peduncle are described on p. 916.

The disposition of the *medial lemniscus* becomes altered in the upper part of the medulla oblongata. Its ventral region widens and becomes insinuated between the dorsal aspect of the pyramid and the narrowing cranial end of the olivary nucleus (7.71). At the same time its dorsal part recedes from the tectospinal tract and the medial longitudinal bundle. This alteration is continued so that, as it enters the pons, the medial lemniscus is extended in a coronal

plane (7.70) in the ventral part of the tegmentum (p. 911). The medial lemniscus comprises the projection fibres on the pathway for proprioceptive and tactile sensibility. It is believed that, in its course through the medulla oblongata, it is joined by the fibres of the anterior spinothalamic tract (p. 876). On entry into the pons, therefore, the medial lemniscus contains many of the proprioceptive, tactile and pressure fibres from the lower limb, the trunk and the upper limb of the *opposite* side and there are good grounds for believing that the lower limb fibres are placed most laterally and adjoin those from the upper limb, while those from the neck lie most medially.

The medullary reticular formation: in the preceding pages no attempt has been made to describe the reticular 'columns', 'nuclei' and their subdivisions, classification and connectivity. These matters are reviewed in some detail on pp. 943–953.

THE PONS

The pons is ventral to the cerebellum. Rostral to it is the midbrain. Inferiorly the pons is continuous with the medulla oblongata, but is demarcated from it in front and on each side by a transverse furrow in which the abducent, facial and vestibulo-cochlear nerves appear.

The *ventral* or *anterior surface* of the pons (7.55, 56) is prominent, being markedly convex from side to side, less so from above downwards. It consists of transverse fibres arched like a bridge across the median plane, and converging on each side into a compact mass which forms the middle cerebellar peduncle. It adjoins the dorsum sellae of the sphenoid bone and the adjacent basilar part of the occipital bone, and is limited above and below by well-defined borders. The anterior surface of the pons is marked by the shallow median *sulcus basilaris*, which usually lodges the basilar artery; this sulcus is bounded on each side by an eminence caused by the descent of the corticospinal fibres through the substance of the pons. Lateral to these eminences, a little above the mid level of the pons, the trigeminal nerves make their exit, each consisting of a smaller, superomedial, motor root, and a larger, inferolateral, sensory root; lines immediately lateral

to the superficial origins of the trigeminal nerves, may be taken as arbitrary boundaries between the ventral surface of the pons and the middle cerebellar peduncles.

The *dorsal* or *posterior surface* of the pons is hidden by the cerebellum. It contributes to the upper half of the rhomboid fossa, with which it is described (p. 933).

Transverse sections through the pons show that it is divided into a dorsal region or *tegmentum* and a *ventral (basilar) part*. The tegmentum is the direct upward continuation of the medulla oblongata, excluding the pyramids. The ventral part contains bundles of longitudinal fibres, some of which are continued into the pyramids of the medulla oblongata, numbers of transverse fibres and scattered collections of grey matter which constitute the *nuclei pontis*. (It should be noted that the term 'pons' is commonly used in two senses: firstly, to denote the externally visible protuberance of the ventral part of the region and, secondly, to designate both this and the tegmental part, i.e. the whole of the brainstem between medulla oblongata and mesencephalon.)

In mammals a correlation exists between the degree of

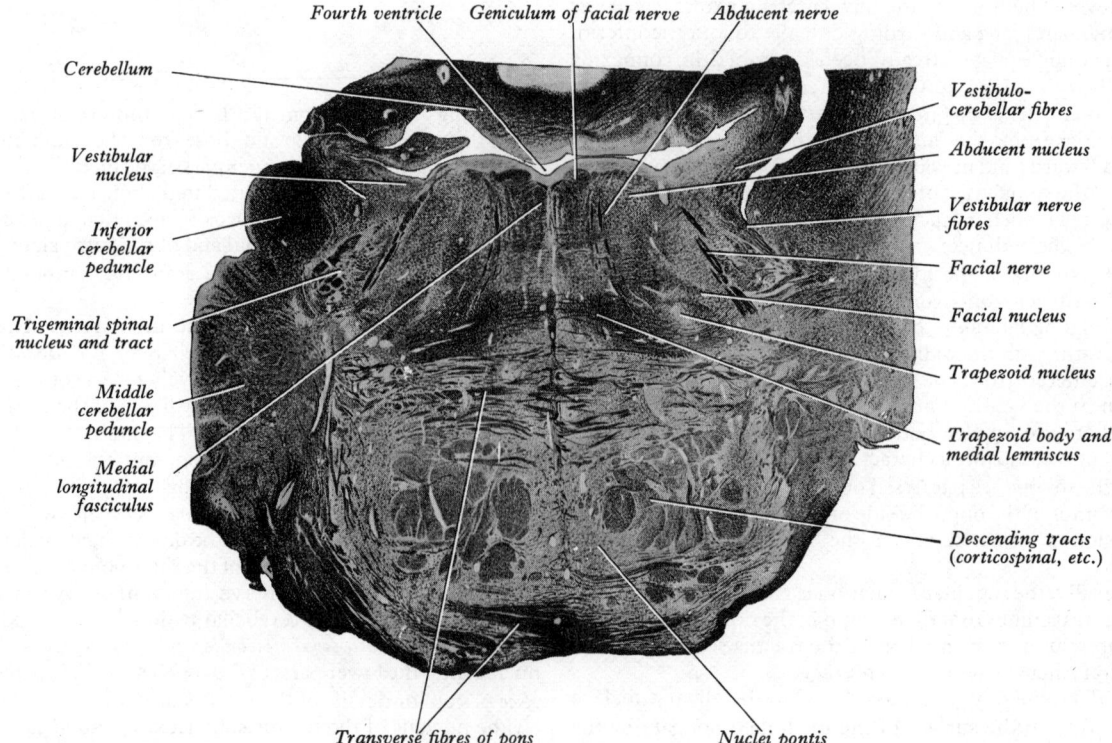

Fourth ventricle   *Geniculum of facial nerve*   *Abducent nerve*

Cerebellum

Vestibular nucleus

Inferior cerebellar peduncle

Trigeminal spinal nucleus and tract

Middle cerebellar peduncle

Medial longitudinal fasciculus

Vestibulo-cerebellar fibres

Abducent nucleus

Vestibular nerve fibres

Facial nerve

Facial nucleus

Trapezoid nucleus

Trapezoid body and medial lemniscus

Descending tracts (corticospinal, etc.)

Transverse fibres of pons   *Nuclei pontis*

7.68 Transverse section through the pons at the level of the facial colliculus. Weigert Pal preparation. Magnification c. ×3·5.

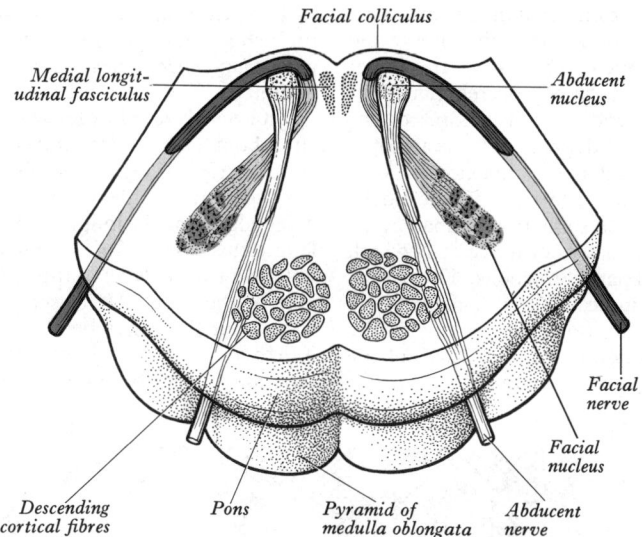

7.69 Diagram of central course of the fibres of the facial nerve, superior aspect, in a transverse section of the pons.

development of the cerebral hemispheres, ventral part of the pons and cerebellum (neocerebellum, p. 914). The ventral part of the pons is not present in submammalian forms; it is present in marsupials and higher mammals and possibly in the monotremes. As the mammalian scale is ascended, the ventral part increases in size *pari passu* with cerebrum and cerebellum.

INTERNAL STRUCTURE OF THE PONS

The ventral part of the pons presents a similar arrangement of its grey and white matter at all levels. The longitudinal bundles (**7.71**) comprise corticopontine, corticonuclear and corticospinal fibres, which are continued downwards from the crus cerebri (p. 935). As they enter the upper limit of the ventral part of the pons, they form a compact collection of fibres, but they rapidly become dispersed into numerous smaller bundles, separated from one another by the *nuclei pontis* and the transverse fibres of the pons. The *corticospinal fibres* descend through the whole length of the pons and enter the pyramids of the medulla oblongata, where they converge into compact tracts (p. 898). They are accompanied by *corticonuclear fibres*, some of which pass to the various contralateral nuclei of the cranial nerves in the pons; the remainder continue into the medulla oblongata to end in a similar manner. Clinically there is evidence that the facial nucleus and certain other nuclei (p. 910) also receive ipsilateral corticonuclear fibres. The *corticopontine fibres*, which are derived from the cerebral cortex of the frontal, temporal, parietal and occipital lobes, terminate at different levels in the nuclei pontis (**7.68**). The axons of the cells of the nuclei pontis form the *transverse fibres of the pons* (pontocerebellar fibres), and constitute the middle cerebellar

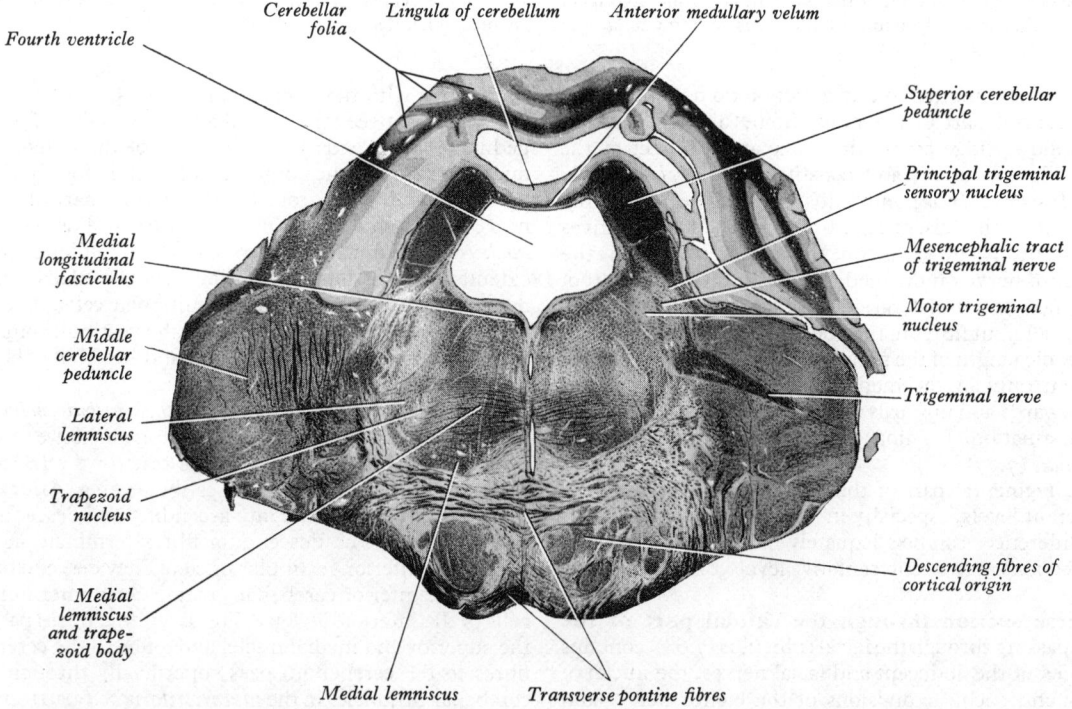

7.70 Transverse section of the pons at the level of the trigeminal nerve. Weigert Pal preparation. Magnification *c*. ×2·5.

peduncle. The frontopontine fibres terminate in the nuclei pontis above the level of the emerging roots of the trigeminal nerve and are relayed to the opposite half of the cerebellum as the upper transverse fibres of the pons. A few of the pontocerebellar fibres do not cross the midline; they all end as mossy fibres of the cerebellar cortex (p. 924). Some degree of somatotopic organization is carried through this system of connexions. Axons from the tectum of the midbrain may also relay in the nuclei pontis (Pearce 1960), as may in addition some from spinal levels.

The *nuclei pontis* comprise all the masses of nerve cells which are scattered throughout the ventral part of the pons. They are of various sizes and shapes. As already indicated, they constitute cell stations on the pathway from the cerebral cortex to the cerebellum. The cells of the nuclei pontis are derivatives of the rhombic lip which migrate ventrally and cranially.

The *medial nucleus* of the vestibular nerve is continued upwards for a short distance into the tegmentum of the pons. The *lateral vestibular nucleus* lies between it and the inferior cerebellar peduncle.

The vestibular nuclei lie subjacent to the vestibular area in the rhomboid fossa (p. 933) and comprise the medial, lateral, inferior and superior. They receive fibres from the vestibular division of the vestibulocochlear nerve and send their axons to the cerebellum, medial longitudinal fasciculus, spinal cord and lateral lemniscus. The *medial vestibular nucleus* extends from the medulla oblongata at the level of the upper end of the olive into the lower part of the pons. As it ascends it broadens, so that the dorsal nucleus of the vagus becomes depressed below the floor of the fourth ventricle. It is crossed by the striae medullares, which separate it from the floor of the fourth ventricle. Caudally it is

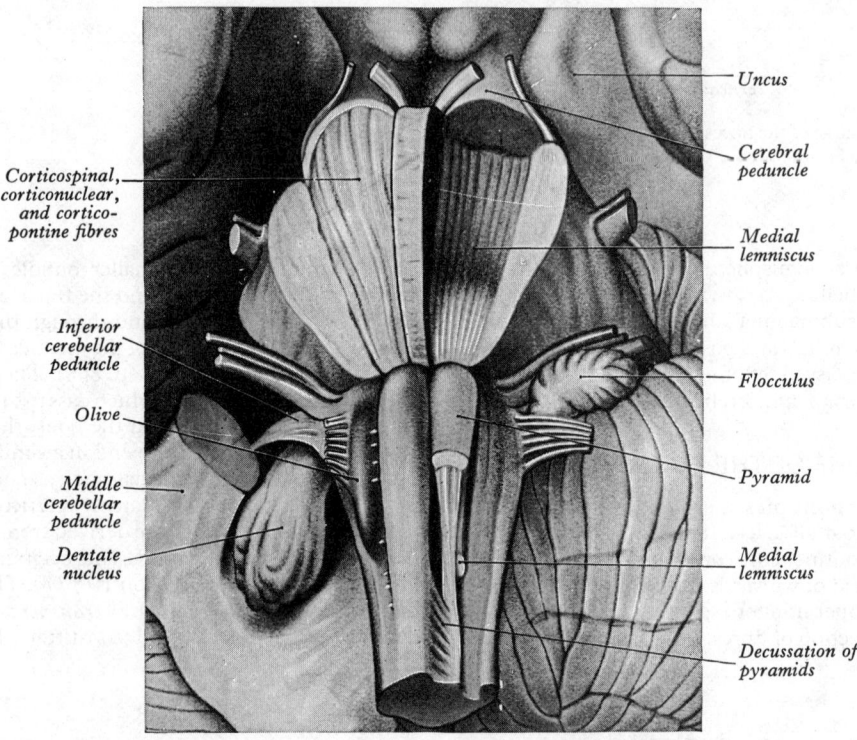

Corticospinal, corticonuclear, and cortico-pontine fibres

Inferior cerebellar peduncle

Olive

Middle cerebellar peduncle

Dentate nucleus

Uncus

Cerebral peduncle

Medial lemniscus

Flocculus

Pyramid

Medial lemniscus

Decussation of pyramids

7.71 Ventral aspect of a dissection of the pons, the medulla oblongata and the right cerebellar hemisphere. In the pons and medulla, the dissection is deeper on the left (right side of figure). Note the spiralling of the medial lemniscus and compare with 7.103.

All the cells which migrate in this direction do not succeed in reaching the ventral part of the pons. Some of them remain, forming an oblique ridge across the dorsolateral aspect of the inferior cerebellar peduncle, and constitute the *nucleus of the circumolivary bundle* (corpus pontobulbare). It has been held that the fibres, to which this discrete part of the nuclei pontis gives origin, run vertically upwards on the surface between the emerging seventh nerve on the medial side, and its sensory root and the eighth nerve on the lateral side (but for an alternative view *see* p. 910). The afferent fibres to the nucleus have been claimed to traverse the whole length of the pons with the corticospinal fibres and leave them only in the medulla oblongata. They course obliquely backwards and upwards over the surface of the olive, to reach their destination, forming part of the *fasciculus circum-olivaris pyramidis* (**7.77**).

The dorsal, tegmental part of the pons varies in its internal details at different levels, especially in regard to its cytoarchitecture. These differences can be adequately illustrated for general purposes by sections at two representative levels, one cranial, one caudal.

A transverse section through the caudal part of the tegmentum passing through the facial colliculus (**7.68**), contains the motor nuclei of the abducent and facial nerves, the nuclei of the vestibular and cochlear divisions of the eighth nerve, and certain isolated collections of grey matter which will be described below.

continuous with the nucleus intercalatus. The *inferior vestibular nucleus* lies between the medial nucleus and the inferior cerebellar peduncle. It extends from the level of the cranial limit of the nucleus gracilis to the pontomedullary junction. It is interspersed by bundles of fibres from the descending part of the vestibular nerve and from the vestibulospinal tract. The *lateral vestibular nucleus* lies immediately cranial to the inferior nucleus and extends upwards almost to the level of the nucleus of the abducent nerve. It is composed of large multipolar cells. The cells of this nucleus give origin to the fibres of the vestibulospinal tract. The *superior vestibular nucleus* is small and lies above the medial and lateral nuclei.

The *fibres of the vestibular part of the vestibulocochlear nerve* can be seen entering the medulla oblongata between the inferior cerebellar peduncle and the spinal tract of the trigeminal nerve and are directed towards the vestibular area. After entering the brainstem, they separate into ascending and descending bundles and branches. The descending fibres terminate in the medial, lateral and inferior vestibular nuclei. They descend on the medial side of the inferior cerebellar peduncle and intermingle with the cells of the inferior nucleus. The ascending fibres pass either into the superior and medial nuclei and a few to the cerebellum. The fibres to the cerebellum pass superficially through the inferior cerebellar peduncle (in the *juxtarestiform body*) to terminate in the nucleus fastigii, the flocculonodular lobe and the uvula (p. 913). The vestibular nuclei not only project extensively to the cerebellum (p. 916), but also receive axons from parts of the cerebellar cortex and the fastigial nuclei (p. 917). Their projections to the spinal cord, through the vestibulospinal tract, have already been described (p. 875). Vestibular connexions also reach the cord through the medial longitudinal fasciculus (p. 885). Efferent axons are also projected to higher levels in the cerebrum, probably mediating a bilateral cortical representation, but evidence for this is scanty. The vestibular complex sends fibres in addition to the pontine reticular nuclei (p. 946) and via the medial longitudinal fasciculus to the motor nuclei of the ocular muscles.

A number of minor groups of neurons in the vicinity of the named vestibular nuclei have been identified in experimental animals; only one of these, the *interstitial nucleus* of the vestibular nerve, is known to receive axons from it. The *nucleus parasolitarius* (p. 903) may be associated with the vestibular complex on the basis of afferent connexions from the fastigial nucleus of the cerebellum (p. 926). The total number of nerve cells in the vestibular nuclei much exceeds the complement of afferent fibres in the vestibular part of the eighth cranial nerve, and this is linked with the observation that vestibular afferents reach only limited regions of the vestibular nuclei, many of whose cells may be interneurons, many also being the sources of the varied projections mentioned above. There is considerable evidence of spatial representation of the vestibular apparatus in the nuclei (p. 1073). For an appraisal of the large volume of experimental research in connection with the vestibular pathway *see* Rasmussen and Windle (1960); Brodal *et al.* (1962).

The *fibres of the cochlear division of the vestibulocochlear nerve* partially encircle the lateral surface of the inferior cerebellar peduncle to terminate in the *dorsal* and *ventral cochlear nuclei*, which project slightly from the surface of the peduncle. The *dorsal cochlear nucleus*, which forms the auditory tubercle (p. 934), is on the posterior aspect of the peduncle and is continuous medially with the vestibular area in the rhomboid fossa. The *ventral cochlear nucleus* is on the anterolateral side of the peduncle in the interval between the cochlear and vestibular fibres of the vestibulocochlear nerve. The two parts of the vestibulocochlear nerve and the cochlear nuclei are usually seen in sections at the pontomedullary junction.

The ventral cochlear nucleus is highly complex in its cytoarchitecture, consisting of many cell types, variously described as giant, large and small spherical, multipolar, granular, and so on, the variants being on the whole grouped in separate regions of the nucleus (Cajal 1909; Lorente de Nó 1933). (*See* also p. 1074.) Degeneration experiments demonstrate a marked degree of topographical order in the termination of cochlear nerve fibres in the nucleus, in that different parts of the spiral organ of the cochlea and differing frequencies of stimulation appear to be related to specific cells serially arrayed in the ventrocaudal part of the ventral nucleus (Schuknecht 1960; Whitfield 1960). All the cochlear nerve fibres enter the nucleus, bifurcating as they do so into ascending branches, which terminate in the ventral nucleus, and descending branches traversing it to reach the dorsal cochlear nucleus (*vide infra*). The total number of cochlear nerve fibres has been estimated at about 25,000 in man, and since the ventral nucleus alone contains at least three times as many nerve cells as this in the cat, whose cochlear fibres are about half the human complement, it is clear that the cells in the cochlear nuclei greatly outnumber the fibres in the nerve. (Ferraro and Minckler 1977 have assessed the fibre count of the lateral lemniscus at a much higher value. *See* p. 1074.) Thus, a mere fraction of the cochlear nerve cells receive terminals from the nerve, though it is said that each fibre forms connexions with several cells (Lewy and Kobrak 1936). These terminals are limited to the ventral and caudal region of the ventral cochlear nucleus, some of whose neurons are probably local interneurons, the remainder being the sources of axons leaving the nucleus, as described below.

The dorsal cochlear nucleus is topographically almost continuous with the ventral nucleus, the two being separated by a thin stratum of nerve fibres. The dorsal nucleus also shows a complex organization, based upon a laminar pattern and containing a considerable spectrum of nerve cell types, including giant cells and other types like those in the ventral nucleus and also a fusiform or pyramidal type peculiar to it. Again, the terminals in the dorsal nucleus appear to be limited in their distribution, being derived, of course, from the descending branches of the cochlear nerve mentioned above (Osen 1969).

While details of the types of cells in the cochlear nuclei which project from them are not yet available, such axons are probably derived from several types, since a large proportion of the total number of cells have efferent axons, a few only being confined to the nuclei. These efferents all end at pontine levels in the superior olivary, trapezoid, and lateral lemniscal nuclei (*vide infra*), and they leave the cochlear nuclei by at least three routes. (1) The most ventral contingent of efferent fibres are also the most numerous, forming by their decussation the *trapezoid body* at the level of the pontomedullary junction (7.70). These axons ascend slightly as they approach the midline to decussate, but a minor number do not cross; they form connexions in the ipsilateral superior olivary nuclei. Those which decussate relay in the contralateral nuclei, and in both cases the next order of axons ascend in the corresponding *lateral lemniscus*. A few of the decussating fibres pass through the contralateral superior olive into the lateral lemniscus, where they relay in the lemniscal nuclei. (2) Some of the axons derived from ventral cochlear nucleus cells pass dorsally superficial to the descending spinal fibres of the trigeminal nerve and to the cerebellar fibres in the inferior peduncle, together with axons of the dorsal cochlear nucleus, which form the third group. These ventral cochlear fibres, which are smaller in calibre than those forming the trapezoid decussation, swerve ventromedially to cross the midline ventral to the medial longitudinal fasciculus as the *intermediate acoustic striae* (of Held 1893); their further progress is uncertain, but they probably pass rostrally within the opposite lateral lemniscus. (3) The most dorsal group of cochlear projection fibres, derived from the dorsal cochlear nucleus, turn medially around the dorsal aspect of the inferior cerebellar peduncle and continue towards the midline as the *dorsal acoustic striae* (of Monakow 1905). They are not to be confused with the striae medullares of the fourth ventricle (p. 910), to which they are ventral and thus more deeply situated. As the dorsal acoustic striae incline ventromedially they approach and cross the midline to reach and ascend in the opposite lateral lemniscus, in the nuclei of which they probably relay.

The *superior olivary complex of nuclei* is located laterally in the reticular formation at the level of the pontomedullary junction. Medial to the complex are the trapezoid nucleus and body, inferior to it the much larger inferior olivary group of nuclei (p. 904). The superior olive consists of a number of named nuclei and other small groups of nerve cells (Irving and Harrison 1967). The *lateral superior olivary nucleus* (the S-shaped segment of the feline olivary complex) is relatively small in primates, including man. Medial to it is the *medial (accessory) superior olivary nucleus* (the para-olivary nucleus of Minckler), which is large in man. Medial again to this is the nucleus of the trapezoid body (*vide infra*). Dorsally situated in the complex is a *retro-olivary group* of nerve cells, reputed to be the origin of the efferent cochlear fibres described below. Interconnexions between the individual nuclei of the superior olivary complex have been described. Unit recording from the lateral nucleus in the cat (Warr 1966) has established the existence of a tonotopical mode of organization, adjoining groups of nerve cells being related to ipsilateral cochlear fibres concerned with different but related acoustic frequencies. The medial superior olivary nucleus receives impulses from both cochlear spiral organs, and physiological evidence suggests that it forms part of a pathway concerned in auditory localization. Together with the trapezoid nuclei, the superior olivary groups are the main relay stations of the ventral and largest projection of the cochlear nuclei. Their complex connexions, as demonstrated in experimental animals, are not yet well authenticated in the human brainstem. Neuronal counts of certain of these nuclei have been recorded (for man) by Ferraro and Minckler (1977).

The *trapezoid nucleus* is sometimes described as consisting of two parts—a ventral component of cells scattered among the fascicles of the trapezoid body and a more compact dorsal nucleus located medial to the superior olivary complex. Although usually regarded as relay stations on the auditory pathway, the trapezoid nuclei in man remain in this regard an uncertain issue. Some of the axons of the trapezoid nuclei may enter the medial longitudinal fasciculus, ascending therein to terminate in the trigeminal and facial nuclei and also those of the motor nerves of the extraocular muscles. Such connexions could mediate reflexes involving the muscles in the middle ear (stapedius and tensor tympani) and the eye muscles.

The *nucleus of the lateral lemniscus* consists of small aggregations of nerve cells dispersed among the fibres of this fasciculus. Lateral and medial groups have been described. They receive bilateral afferent axons from both cochlear nuclei, and their efferents pass cranially into the midbrain through the lateral lemniscus; their further connexions will be considered later (pp. 941, 976). Total neuronal counts for these nuclei were of the order of 18,000–24,000 in human brainstems (Ferraro and Minckler 1977).

Efferent cochlear nerve fibres travel centrifugally in the cochlear nerves to innervate the spiral organ (Held 1893; Rasmussen 1967). Though comparatively few in number (about 500 in the cat), experiment suggests that they play an important part in hearing, perhaps being involved in both inhibitory and excitatory reflexes through the cochlear nuclei (Allanson and Whitfield 1955). Lateral inhibition has also been demonstrated by the unit recording technique applied at the trapezoid level of the auditory pathway. The efferent cochlear fibres appear to be derived from retro-olivary cells in the superior olivary complex, fibres from each side proceeding to both cochleae—hence the term 'olivocochlear' fibres. Borg (1973) has confirmed a projection to the cochlea from neurons not only dorsal but also ventral to the superior olivary nucleus, in the rabbit; he has also described a descending connexion from the inferior colliculus to the olivary complex (confirming a much older finding by Münzer and Wiener 1902). Since there is some evidence of a descending projection from the medial geniculate body to the inferior colliculus, a complete efferent corticocochlear pathway appears to be established (p. 976).

The *striae medullares of the fourth ventricle*, as already stated above, are *not* formed of nerve fibres involved in the auditory pathway. There is considerable evidence on the contrary that they are part of an *aberrant cerebropontocerebellar* connexion, in which the arcuate nuclei (p. 905) pontobulbar body (p. 908), and the external arcuate fibres (p. 905) are all concerned. Some confusion exists among the accounts of various authorities, especially regarding the role of the pontobulbar body. The following description merely presents the consensus of these views, some of which appear to be more reliably established than others. Embryological evidence (p. 164) suggests that some of the nerve cells which migrate ventrally towards the pons from the rhombic lip to form the nuclei pontis largely fail to do so and remain near the fourth ventricle as the pontobulbar body or nucleus. Others migrate further and are scattered superficially over the ventral aspect of much of the extent of the medullary pyramids as the arcuate nuclei, a sizable congregation of them occurring immediately caudal to the pons. Both groups of cells are considered to receive corticobulbar projections, those to the arcuate nuclei descending in the pyramids with the corticospinal fibres. Axons from the arcuate nuclei of both sides pass dorsally round the side of the medulla, cranial, superficial and caudal to the inferior olive as a thin stratum, fascicles of which are usually visible on the surface. All these fibres pass into the inferior cerebellar peduncle, being known collectively as *external arcuate fibres*. Some external arcuate fibres (so-called) follow an *internal* course passing dorsally from the arcuate nuclei through the substance of the medulla close to the median raphe. Reaching the floor of the fourth ventricle, these fibres decussate and turn laterally subjacent to the ependymal lining of the ventricle, entering the cerebellum by the inferior peduncle. In this latter part of their course these fibres form the striae medullares; from their connexions they are also known as the *arcuatocerebellar tract*. There is some evidence that this ends in the flocculus (Szentágothai 1955). Some of these fibres are said to end in the pontobulbar nucleus, but this may be a confusion with projection fibres of the nucleus. Some of the latter may travel ventrally forming part of a compact bundle, the *circumolivary fasciculus,* which skirts the caudal pole of the inferior olive, usually appearing as a surface feature of this part of the medulla. The

fasciculus and the pontobulbar nucleus have been found absent in aplasia of the pons, an interesting confirmation of their affiliation with the pontocerebellar projection (Baumgarten 1959). The efferent circumolivary fibres which pass ventrally join the arcuato-cerebellar tract, with which they reach the striae medullares and enter the contralateral inferior cerebellar peduncle. However, as pointed out previously (p. 908), afferent fibres to the pontobulbar nucleus have also been considered to run in the circumolivary fasciculus, and the precise contribution of efferent and afferent fibres to the fasciculus cannot yet be regarded as settled.

The nucleus of the abducent nerve is in the central grey matter a short distance from the median plane, and in line with the nuclei of the third and fourth cranial nerves, above, and the hypoglossal nerve, below, thus forming a somatic motor column (p. 164). It is close to the medial longitudinal fasciculus, which is placed to its ventromedial side. In this way fibres from the vestibular and cochlear nuclei and the nuclei of other cranial nerves, especially the oculomotor, communicate with the abducent, which also bears an intimate relation to the emerging fibres of the facial nerve (*vide infra*). The outgoing fibres of the sixth nerve pass ventrally downwards through the reticular formation, intersecting the trapezoid body and the medial lemniscus and traversing the basilar part of the pons to emerge at its lower border. (See also p. 1069.)

The facial nucleus lies in the ventrolateral part of the reticular formation of the pons, immediately behind the dorsal nucleus of the trapezoid body. Dorsal to it, and somewhat to its lateral side, is the spinal tract of the trigeminal nerve and its associated nucleus. The facial nucleus receives fibres from the corticonuclear tract of the opposite side, a smaller number from that of the same side (p. 1070), and also fibres from the ipsilateral rubroreticular tract. Its large motor cells give origin to the fibres of the facial nerve. These fibres do not pass directly from their origin to the surface of the brainstem, but pursue a very remarkable course. At first they incline dorsally and medially towards the rhomboid fossa, passing below the abducent nucleus (7.69). They then course upwards on the medial side of this nucleus, coming into close relationship with the medial longitudinal fasciculus, through which the facial nerve may be brought into communication with the other cranial nerves. Finally, the fibres of the facial curve forwards and laterally around the upper pole of the abducent nucleus forming the *genu of the facial nerve* and pass forwards, laterally and downwards through the reticular formation. In the last part of their course to the surface they pass between their own nucleus medially and the nucleus of the spinal tract of the trigeminal nerve lateral to them.

The unusual behaviour of the emerging fibres of the facial nerve provided apparent evidence in favour of the theory of neurobiotaxis (p. 164). In the 10 mm human embryo the facial nucleus lies in the floor of the fourth ventricle, occupying the position of the branchial (special visceral) efferent column, and at this stage it is placed at a level rostral to the abducent nucleus. As growth proceeds, the facial nucleus migrates relative to surrounding structures at first caudally, dorsal to the sixth nucleus, and then ventrally to reach its adult position. As it migrates the axons to which its cells give rise elongate, and their subsequent course maps out the pathway along which the facial nucleus has travelled.

It must be remembered that the facial nucleus not only receives fibres from the corticonuclear tracts for volitional control, but (in addition to many other sources) it also receives afferents from its own sensory root (through the nucleus of the tractus solitarius) and from the nucleus of the spinal tract of the trigeminal nerve. These latter sources of stimulation complete local reflex loops, in every way similar to the segmental reflex loops in the spinal cord. It was considered that to retain its proximity to the nucleus of the tractus solitarius and to the nucleus of the spinal tract of the trigeminal nerve the facial nucleus migrated 'neurobiotactically' from its original position in the basal lamina.

The nucleus of the facial nerve is divided into several parts. The cells which give rise to the axons innervating the muscles in the scalp and upper part of the face are placed towards the dorsal part of the nucleus and are believed to receive corticonuclear fibres from both sides (Papez 1927; Buskirk 1945). The subgroups of neurons which form the nucleus may represent discrete motor pools (p. 1070).

The salivatory nucleus is near to the rostral end of the dorsal nucleus of the vagus, just above the junction of the medulla oblongata and pons (Lewis and Shute 1959). It is in close relation to the caudal end of the nucleus of the facial nerve. It is customary to divide it into *superior and inferior salivatory nuclei* which send their secretomotor fibres to the salivary and, perhaps, the lacrimal glands through the facial and glossopharyngeal nerves respectively (pp. 1070, 1074).

The nucleus of the spinal tract of the trigeminal nerve is continued up through the lower part of the pons, the fibres of the tract still being closely applied to the lateral aspect of the nucleus. It is placed ventral to the lateral vestibular nucleus and is intersected by the fibres of the vestibular nerve destined for that nucleus. The inferior cerebellar peduncle lies to its lateral side below, but inclines dorsally as it ascends to the cerebellum, and the spinal tract of the trigeminal nerve and its nucleus are subsequently related laterally to the middle cerebellar peduncle. Above, the nucleus becomes continuous with the superior sensory nucleus of the trigeminal. The spinal 'tract' is unfortunately named, for it does not consist of fibres originating in the central nervous system but of the descending axons of the trigeminal nerve derived from nerve cells in the trigeminal ganglion. From these fibres collaterals and terminals enter the spinal nucleus, which is continuous at its caudal end with the substantia gelatinosa of the dorsal grey column in the spinal cord (p. 1060). Rostrally it merges with the main sensory nucleus of the nerve. It is mainly concerned with the mediation of pain and thermal sensibilities in the trigeminal area. There is a well-established topographic organization within the nucleus. (For further details *see* p. 1060.)

In addition to tracts already studied at a lower level the white matter of the lower part of the tegmental region of the pons contains the trapezoid body, the lateral lemniscus and the emerging fibres of the sixth and seventh cranial nerves, which are new elements not present in the upper part of the medulla oblongata.

The *medial lemniscus* occupies the ventral part of the tegmentum. Its outline, on transverse section, is a flattened oval, extending laterally from the median and paramedian raphe (7.68). The vertically running fibres of the medial lemniscus are intersected by the horizontal fibres of the trapezoid body. Laterally they are related to the *lateral spinothalamic tract* and to the *trigeminal lemniscus*. The fibres of the latter are derived from the cells of the nucleus of the spinal tract of the trigeminal nerve of the opposite side, and convey painful and thermal impressions from the skin of the face, the mucous membranes of the conjunctiva, tongue, mouth, nose, etc. The lemnisci are now arranged as a transverse band composed, from the medial to the lateral side, of the medial and trigeminal lemnisci, lateral spinothalamic tract and lateral lemniscus.

The *trapezoid body* is formed by fibres derived from the cochlear nuclei (mainly the ventral) and from the nuclei of the corpus trapezoideum. They run transversely and rostrally in the ventral part of the tegmentum, and, having intersected or passed ventral to the vertical fibres of the medial lemniscus, they cross the median raphe, decussating with the corresponding fibres of the opposite side. Before they reach the emerging fibres of the facial nerve the fibres of the trapezoid body turn upwards to contribute to the *lateral lemniscus*, as the *ascending* part of the auditory pathway.

The course of the outgoing fibres from the nuclei of the abducent and facial cranial nerves has already been examined.

The *medial longitudinal fasciculus* is sited close to the midline, immediately ventral to the floor of the fourth ventricle. It is closely related to the nucleus of the abducent nerve and to the emerging fibres of the facial nerve, as they ascend on the medial side of that nucleus. The proximity of the fasciculus suggests that it may receive fibres from and transmit fibres to both structures (p. 939). As it lies in the lower part of the pons, the medial longitudinal fasciculus receives fibres from the vestibular nuclei and possibly from the dorsal nucleus of the trapezoid body (p. 909), through the peduncle of that nucleus. These contributions from the eighth nerve form the greater part of the fasciculus (p. 939) which is the main 'intersegmental' tract in the brainstem, particularly concerned with interactions between the nuclei innervating the ocular muscles, and between these and the vestibular system.

A transverse section at a rostral level in the tegmentum of the pons contains new elements in connexion with the trigeminal nerve (7.70), but otherwise shows no very notable alteration.

The motor nucleus of the trigeminal nerve lies in the reticular formation of the pons, deep to the lateral part of the floor of the fourth ventricle in line with the fibres of the trigeminal nerve traversing the ventral part of the pons (7.70).

The principal (superior) sensory nucleus of the trigeminal nerve lies on the lateral side of the motor nucleus, intervening between it and the middle cerebellar peduncle, and is continuous below with the nucleus of the spinal tract. The second neuron fibres from the principal sensory nucleus cross the median plane and ascend with the medial lemniscus to the thalamus (p. 1060).

The *nucleus of the lateral lemniscus* is a small collection of cells placed on the medial aspect of the tract in the cranial part of the pons. It receives synaptic terminals from some fibres of the lateral lemniscus; some of its efferent fibres enter the medial longitudinal fasciculus, whilst others return to the lemniscus. It is to be associated, as a relay station, with the nucleus of the trapezoid body.

The white matter of the tegmentum at this level is marked by the absence of the trapezoid body, which is now replaced by the lateral lemnisci, and the invasion of its dorsolateral part by the superior cerebellar peduncles.

The *medial lemniscus* (7.71) occupies a position in the ventral part of the tegmentum, but it has moved laterally a short distance from the median raphe. Here it is joined medially by the projection fibres from the principal sensory nucleus of the trigeminal nerve, which convey proprioceptive, tactile and pressure impulses from the receptive area covered by it. More laterally it is related dorsally to the trigeminal lemniscus and lateral spinothalamic tract, conveying pain and thermal impulses, and to the lateral lemniscus and its nucleus. As the lateral lemniscus ascends, it passes dorsally and lies close to the surface. It will be seen subsequently to send its fibres into the inferior colliculus and the medial geniculate body. The *medial longitudinal fasciculus* retains its paramedian position.

The *superior cerebellar peduncle* is a large collection of fibres many of which take origin in the dentate nucleus of the cerebellum (p. 926), and pass upwards and forwards to enter the lateral part of the roof of the fourth ventricle. As it ascends in this position it inclines forwards and medially and enters the dorsolateral part of the tegmentum. The *anterior spinocerebellar tract* is intimately associated with the foregoing. It has already been traced up through the medulla oblongata, where it lies dorsal to the olivary nucleus and separated from the surface only by the anterior external arcuate fibres. In the lower part of the pons it inclines dorsally between the sensory nucleus of the trigeminal nerve and the middle cerebellar peduncle until it reaches the lateral aspect of the superior peduncle. Its fibres then descend dorsally in a curve to enter the cerebellum.

The *reticular formation* is continued through all levels of the pons and its named nuclei, columns, and aggregates of aminergic neurons, and what is known of their connectivity is summarized on pp. 946–947.

THE CEREBELLUM

The cerebellum, the largest part of the hindbrain, lies behind the pons and medulla oblongata, and its median portion is separated from these structures by the cavity of the fourth ventricle. It lies in the posterior cranial fossa and is covered by the tentorium cerebelli (p. 986). It is somewhat ovoid in form, but constricted in its median part, and flattened from above downwards, its greatest diameter being from side to side. Its average weight in the male is about 150 grams. In the adult the proportion between the cerebellum and cerebrum is about 1 to 8, in the infant about 1 to 20.

GENERAL FORM OF THE CEREBELLUM

The cerebellum consists of two *cerebellar hemispheres* joined by a narrow median strip, the *vermis*. On the *superior surface*, however, there is no deep grooving in the paramedian planes, so that the surface of the superior vermis, which is raised into a slight median ridge, is directly continuous with the hemisphere on each side (7.72). Anteriorly the *superior vermis* projects beyond the free margin of the tentorium cerebelli, and from there it slopes downwards and backwards, related above to the straight sinus. The upper surface of each hemisphere is in contact with the tentorium cerebelli, and slopes downwards and laterally from the superior surface of the vermis. It is bounded, in front, by an anterolateral margin, which corresponds to the attachment of the tentorium cerebelli to the superior border of the petrous part of the temporal bone, and behind, by a curved posterior margin, which abuts against the transverse sinus as it lies in the attached margin of the tentorium cerebelli.

On the *inferior surface* the cerebellar hemispheres are separated from each other by a deep hollow, which is termed the *vallecula* (7.75). The inferior surface of the hemisphere is irregularly convex and lies in contact with the posterior surface of the petrous part of the temporal bone, the sigmoid sinus, the mastoid part of the temporal bone and the lower part of the squamous portion of the occipital bone. The *inferior vermis* projects into the floor of the vallecula and is limited on each side by the *sulcus valleculae*.

Anteriorly the cerebellum presents a wide, shallow *anterior cerebellar notch*, adapted to the pons and the upper part of the medulla oblongata, but these portions of the brainstem are separated from it by the fourth ventricle. In the floor of the anterior cerebellar notch the penduncles pass into the white centre of the cerebellum.

Posteriorly the hemispheres are separated from each other by the posterior cerebellar notch, which is a deep and narrow interval containing the falx cerebelli of the dura mater.

SURFACE TOPOGRAPHY OF THE CEREBELLUM

The surface of the cerebellum is everywhere marked by closely set transverse but somewhat curved fissures which give it a laminated appearance and separate its constituent *folia*. Some of the fissures are deeper than others and divide the organ into several *lobules*. The most conspicuous of these fissures is the *horizontal fissure*. This extends around the lateral and posterior borders of each cerebellar hemisphere from the middle cerebellar peduncle in front to the posterior cerebellar notch behind; it marks the junction of the superior and inferior surfaces of the cerebellum.

Superior surface (7.72, 85). The most conspicuous fissure of the superior surface is the *fissura prima*. This is somewhat V-shaped with its apex directed dorsally, and cutting into the superior surface of the vermis at the junction of its anterior two-thirds with the posterior third. The lines of the fissure are directed anterolaterally around the superior surfaces of the cerebellar hemispheres to meet the horizontal fissures close to their anterior ends.

The superior surface of the vermis is divided by short, deep fissures into the *lingula, central lobule, culmen, declive* and *folium vermis* in that anteroposterior order. Each of these divisions, excepting the lingula, is continuous laterally with the adjoining lobule of the cerebellar hemispheres (7.73, 85). The fissura prima indents the superior surface of the vermis between the culmen and declive.

The lingula consists of a single lamella which presents four or five poorly marked folia on its surface; its white matter is directly continuous with that of the superior medullary velum. The lingula is separated from the central lobule by the *postlingual fissure*. The central lobule is continuous laterally with the *alae of the central lobule*. These are limited behind by the *postcentral fissure*. Between this fissure and the fissura prima lies the culmen medially and laterally the *quadrangular lobule*.

The superior surface of the cerebellar hemispheres and vermis behind the fissura prima is divided by the curved *postlunate fissure* into an anterior portion which consists of the declive with its lateral extensions, the *lobuli simplices* and a posterior portion, the *folium vermis* with the adjoining parts of the cerebellar hemisphere termed the *superior semilunar lobules* which are limited behind by the horizontal fissures.

Inferior surface (7.75, 85). This includes the inferior surface of the vermis and the inferior aspect of each cerebellar hemisphere. The inferior vermis is divided into four smaller portions named from behind forwards the *tuber vermis, pyramid, uvula* and *nodule*. The tuber vermis is continuous laterally with

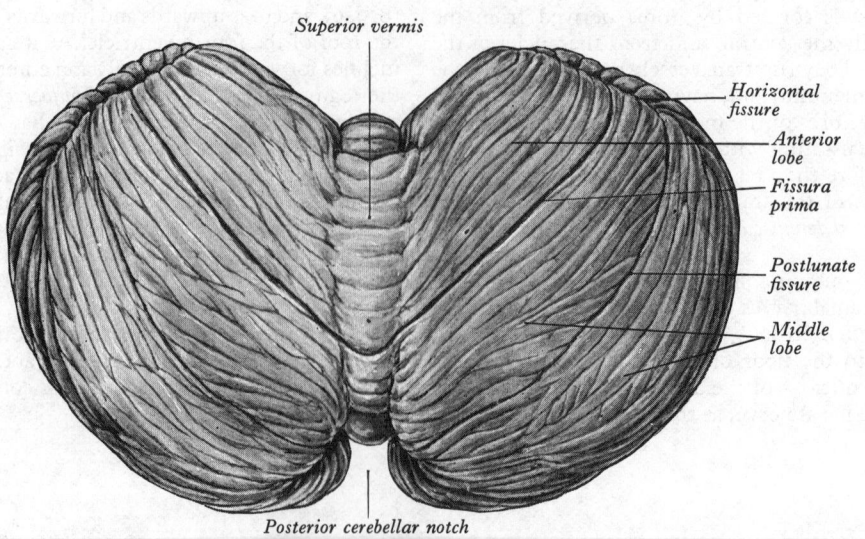

Superior vermis

Horizontal fissure

Anterior lobe

Fissura prima

Postlunate fissure

Middle lobe

Posterior cerebellar notch

7.72 Superior aspect of the cerebellum to show major fissures and lobes. Compare with 7.74.

the *inferior semilunar lobules*. These parts are bounded, behind, by the horizontal fissure and, in front, by the *prepyramidal fissure*. The *pyramid* is separated from the *uvula* by the *postpyramidal fissure* or *fissura secunda* and is continuous laterally with the *biventral lobule* on the inferior surface of each hemisphere. In front of the uvula and separated from it by the median portion of the *posterolateral sulcus* is the nodule (**7**.73, 75).

On the inferior aspect of the cerebellar hemisphere, anterior to the biventral lobule, is a deep fissure, the *retrotonsillar*, which passes laterally from the sulcus valleculae opposite the fissura secunda and then curves forwards to gain the anterior part of the inferior surface of the hemisphere. Together with the anterior part of the sulcus valleculae it bounds a circumscribed portion of the cerebellum, termed the *tonsil*, which is connected to the uvula across the floor of the sulcus valleculae by a strip of cortex, termed the *furrowed band* (**7**.75). Superiorly the tonsil lies in intimate relation with the inferior surface of the inferior medullary velum.

The *nodule* is the most anterior part of the inferior surface of the vermis. Behind, it is separated from the uvula by the posterolateral fissure, and on each side it is connected to the flocculus and the white core of the hemisphere by the inferior medullary velum. Its anterosuperior aspect is directed towards the fourth ventricle. Anteriorly it is covered with grey matter and crossed by two or three shallow fissures. In this situation it is separated from the ventricular cavity by a double layer of pia mater and its associated choroid plexus, and the ventricular ependyma (**7**.89). Posteriorly the grey matter is deficient, and the white matter is on the surface covered only with a layer of neuroglia and the ependyma (**7**.89). The lateral aspect of the nodule is free anteriorly and is covered with grey matter; posteriorly, it presents a narrow strip, where its white core would be exposed, were it not directly continuous with the nervous layer of the inferior medullary velum.

The *flocculus* is a small, partially detached portion of the cerebellum which lies immediately below the vestibulocochlear nerve as it enters the brainstem, and is crossed anteriorly by the fila of the glossopharyngeal and vagus nerves as they pass laterally to reach the jugular foramen. It is somewhat oval in outline, with a crenated margin, and from its medial end a narrow band of white fibres emerges, which constitutes the *peduncle* of the flocculus; it is covered anteriorly by the lateral recess of the fourth ventricle and the part of the choroid plexus which projects from the aperture of the recess (**7**.55). The peduncle contains both afferent and efferent fibres. At the lateral angle of the floor of the fourth ventricle it divides into dorsal and ventral parts. Through the dorsal part the flocculus establishes connexions with the nodule and the uvula. The ventral part passes medially and turns upwards close to the lateral border of the pontine part of the floor of the ventricle. Many of these fibres are afferent and are derived from the

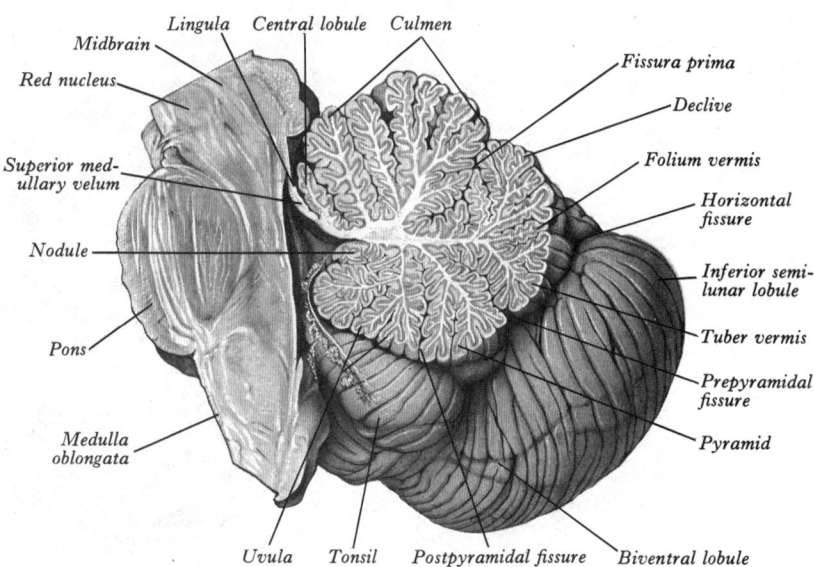

7.73 Median sagittal section of the cerebellum and brainstem.

vestibular nuclei and also, according to some authorities, from the medial accessory olivary nucleus, but others are efferent to the vestibular nuclei and some appear to ascend to a higher level.

THE LOBES OF THE CEREBELLUM

The cerebellum can be divided into two fundamental parts termed the flocculonodular lobe and the corpus cerebelli, the latter comprising an anterior and middle lobe (**7**.74, 85A). These subdivisions possess functional as well as morphological and embryological significance. The *flocculonodular lobe* consists of both flocculi, their peduncles and the nodule. The *corpus cerebelli* comprises the remainder of the cerebellum and is separated from the flocculonodular lobe by the *posterolateral fissure*, which is the first to appear on the cerebellum both in phylogeny and ontogeny. The corpus cerebelli is subdivided by the *fissura prima* into anterior and middle lobes. The *anterior lobe* lies in front of the fissure and comprises the lingula, central lobule, culmen, alae of the central lobules and quadrangular lobules. The remainder of the corpus cerebelli is termed the *middle lobe* and comprises the declive, folium vermis, tuber vermis, pyramid, uvula, lobuli simplices, biventral lobules, semilunar lobules and tonsils.

Certain sectors of the cerebellum are phylogenetically older than the rest. The flocculonodular lobe, which is predominantly

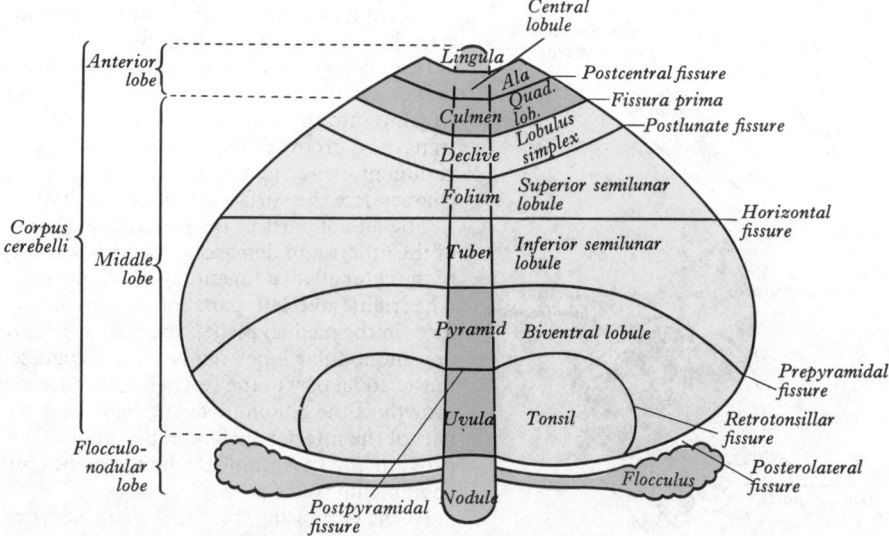

7.74 Diagram to show the morphological and functional subdivisions of the cerebellum. *Blue*: archicerebellum. *Green*: paleocerebellum. *Yellow*: neocerebellum. See also illustration **7**.85 for further details.

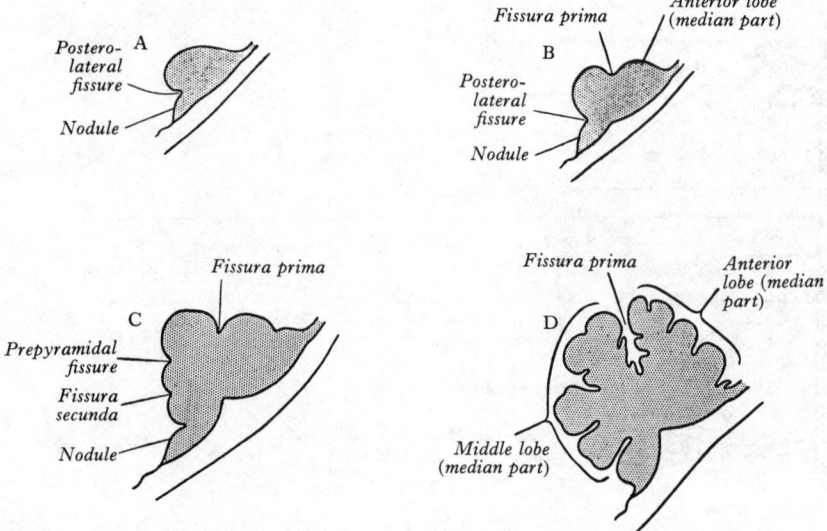

7.75   Inferior aspect of the cerebellum. The tonsil and the adjoining part of the biventral lobule of the right side have been removed.

vestibular in its connexions, together with the lingula, which receives spinocerebellar in addition to vestibular connexions, constitute the oldest part of the cerebellum, the *archicerebellum*. The anterior lobe, excluding the lingula, but together with the pyramid and uvula of the middle lobe, is phylogenetically the next part to appear and is predominantly *spinocerebellar* in its afferent connexions, constituting the *paleocerebellum*. At this stage in phylogeny, this newly acquired lobe separates the archicerebellum into two parts, the lingula in front and the flocculonodular lobe behind. With the evolution of the neopallium in the mammal, there is a further expansion of the cerebellum with the addition of the middle lobe, excepting the pyramid and uvula. This addition constitutes the *neocerebellum* and is predominantly *corticoponto-cerebellar* in its afferent connexions. Like the paleocerebellum, the neocerebellum intervenes between the anterior and flocculo-nodular lobes (**7.74**, 85A).

The superior medullary velum is a thin lamina of white substance, which stretches between the superior cerebellar peduncles (brachia conjunctiva), and with them forms the roof of the cranial part of the fourth ventricle; its deep surface is covered with the ventricular ependyma. The velum is narrow antero-superiorly, where it extends into the interval between the inferior colliculi, and broader postero-inferiorly, where it is continuous with the white substance of the superior part of the vermis. The folia of the lingula are prolonged on to the dorsal surface of its lower half, and a median ridge, termed the *frenulum veli*, descends upon its superior part from between the inferior colliculi. The trochlear nerves emerge at the sides of the frenulum (**7.87**).

The inferior medullary vela are two thin, somewhat crescentic, sheets placed one on each side of the nodule. Each consists of a thin layer of white matter and neuroglia, surfaced over internally by the ventricular ependyma, and externally by the pia mater. Its internal surface forms the lower wall of the lateral dorsal recess of the fourth ventricle (p. 932); its external surface is related to the superior aspect of the tonsil. Its convex peripheral margin is continuous with the white core of the cerebellum and with the sides of the pyramid, uvula and nodule; its anterior (sometimes inferior) border is free (**7.87**) and from it the ventricular ependyma is prolonged downwards in close apposition with the pia mater to form the thin part of the roof of the ventricle and to reach the taeniae. At its anterolateral corner the velum is continuous with the dorsal part of the peduncle of the flocculus, from which most, if not all, of its nerve fibres are derived (**7.75**).

GROSS DEVELOPMENT OF THE CEREBELLUM

For the initiation of cerebellar development *see* p. 165. Early in the third month the cerebellum is represented by a mass which stretches across the roof of the cranial part of the hindbrain vesicle, becoming bilobar like an hour-glass. Its narrow median part is destined to form the vermis, and its enlarged extremities develop into the hemispheres. As growth proceeds a number of transverse grooves appear on the dorsal aspect of the cerebellar rudiment, and give rise to the numerous fissures which characterize the surface of the cerebellum (**7.75**, 76).

The lateral parts of the *posterolateral fissure* appear before any of the others and demarcate the most caudal region from the rest of the cerebellar rudiment, by which the flocculi can be identified. The right and left parts of this fissure extend medially and meet in the median plane, where they demarcate the nodule. The flocculonodular lobes can now be recognized and constitute the most caudal part of the cerebellum at this stage; but, owing to the growth of the adjoining areas, they come to occupy the anterior part of the inferior surface in the adult. They are formed in close proximity to the line of attachment of the epithelial roof, i.e. to the rhombic lip (p. 164).

At the end of the third month a transverse furrow appears on the cranial slope of the cerebellar rudiment, and deepens to form the *fissura prima*, which cuts into the vermis and both hemispheres, separating off the most cranial region of the rudiment to form the anterior lobe.

7.76   Median sagittal sections through the developing cerebellum at four successive stages.

About the same time two short transverse grooves appear on the inferior surface of the vermis behind the postnodular fissure. The first of these is the *fissura secunda*, which demarcates the uvula, and the second is the *prepyramidal fissure*, which demarcates the pyramid (**7.76**). The whole cerebellum grows in a dorsal direction, and the caudal, or inferior, aspects of the hemispheres undergo much greater enlargement than the inferior surface of the vermis, which therefore becomes buried at the bottom of a deep hollow—the *vallecula*. While these changes are taking place numerous additional fissures develop, but they have little morphological significance. The most extensive of them forms the *horizontal fissure*.

In many mammals a portion of the hemisphere immediately cranial to the flocculus becomes demarcated, and in some it forms a very prominent part of the cerebellum. Owing to its relation to the flocculus, it is termed the *paraflocculus*, but the relationship is purely topographical, and, in contradistinction to the flocculus, the paraflocculus derives its afferent connexions mainly, if not entirely, from the cerebral cortex. It is uncertain whether any homologue of the paraflocculus exists in the human cerebellum, or whether it is represented by some small patches of grey matter which are found not infrequently on the inferior surface of the middle cerebellar peduncle. (For a résumé of *cellular neurogenesis* of the cerebellum consult pp. 166, 167.)

INTERNAL STRUCTURE OF THE CEREBELLUM

The cerebellum exhibits a profound difference in structure from the spinal cord, the medulla oblongata and the pons, for the grey and white matter of which it is comprised are arranged in the opposite manner. The grey matter or cortex covers the whole surface of the cerebellum, folding into all the various fissures which cross its surface. Certain aggregations of grey matter are found in its interior, but that does not in any way alter the prominence of the peripheral distribution of the grey matter and the central arrangement of the white matter. In this way the cerebellum resembles the cerebrum, and it is this modification of the disposition of the grey matter which has rendered possible the enormous degree of expansion which these two parts of the nervous system have undergone during the process of evolution.

The White Matter of the Cerebellum

The white matter forms a central core, which is much thicker in the lateral parts than it is in the median area, where it forms a flattened strip connecting the enlarged lateral portions with each other. From its surfaces a series of nearly parallel plates or laminae project towards the surface, and these give off secondary laminae, usually more or less at right angles to the primary laminae. In turn the secondary laminae may give off still shorter laminae, all of which are covered with grey matter. When a section is made through the cerebellum parallel with the median plane it divides the primary laminae at right angles, and the cut surface presents a characteristic branched appearance which is termed the *arbor vitae* (**7.73**).

The white matter consists of (1) fibrae propriae, (2) projection fibres, and (3) the myelinated axons of the Purkinje cells (*vide*

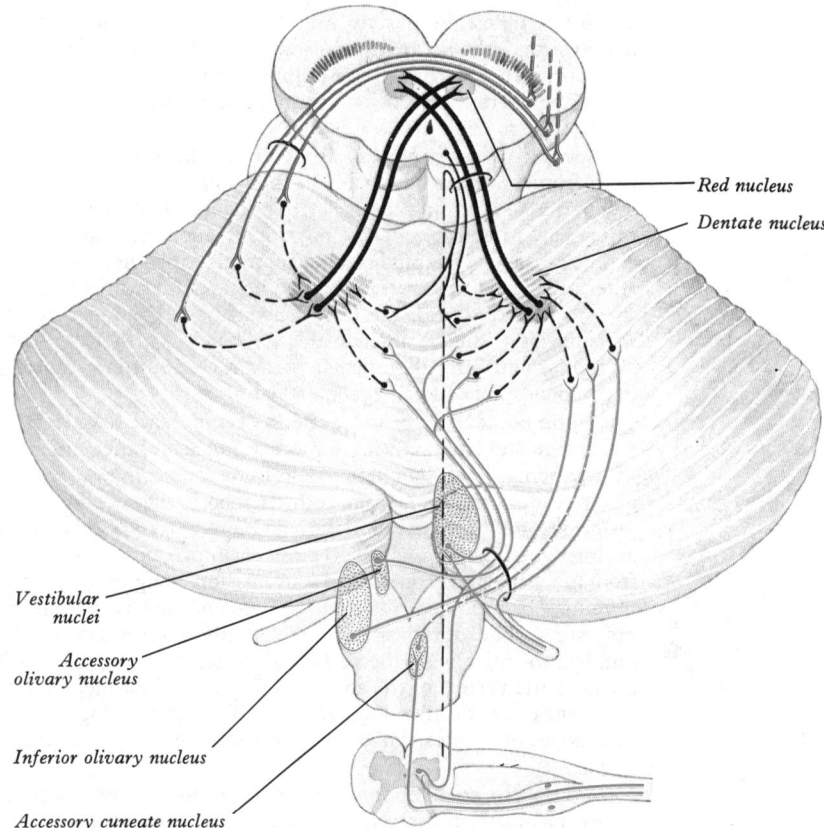

7.78 A simplified diagram of the connexions of the cerebellum showing some components of the peduncles—inferior (blue), middle (magenta) and superior (black). The important cerebellothalamic fibres and other efferent systems, have been omitted.

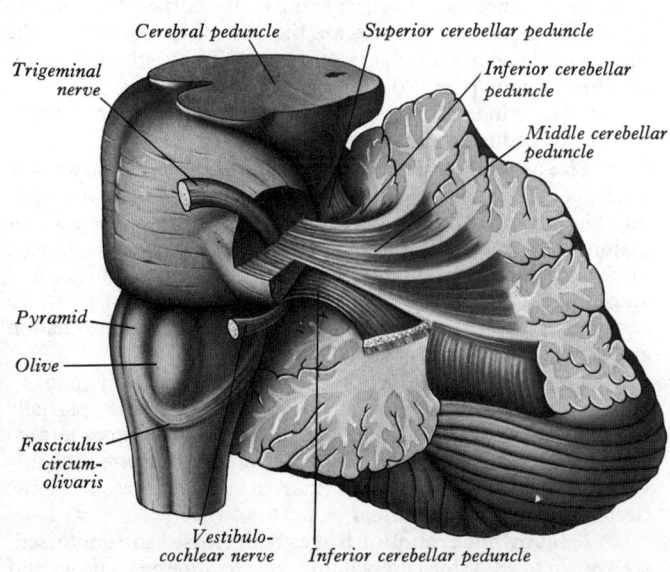

7.77 Dissection of the left cerebellar hemisphere and its peduncles (by the late Dr. E. B. Jamieson, University of Edinburgh).

infra), (4) 'climbing' afferent fibres (*vide infra*), and (5) 'mossy' afferent fibres (*vide infra*).

The fibrae propriae or intrinsic fibres, as their name suggests, do not leave the cerebellum, but interconnect different regions of the organ. The *association fibres* connect adjacent or more distant folia of the cerebellar cortex, including those of the vermis. They do not cross the midline and the majority are relatively short bundles. Their detailed anatomy has not been extensively investigated (but see also p. 922). Numerous *commissural fibres* interconnect the two hemispheres; many are grouped into an *anterosuperior commissure*. A *postero-inferior commissure* also exists and crosses the midline near the fastigial nuclei (*vide infra*). In addition to interhemispheric cortical connexions, and running with them, are *decussating* spinocerebellar and cerebellovestibular fibres.

The projection fibres connect the cerebellum with other parts of the brain and spinal cord. They are grouped together into three large bundles or *peduncles* (**7.77**) on each side and these issue from the anterior cerebellar notch. The superior peduncles connect the cerebellum to the midbrain, the middle peduncles to the pons, and the inferior peduncles to the medulla oblongata. Some of these connexions are shown in a highly simplified manner in 7.78.

THE INFERIOR CEREBELLAR PEDUNCLE

The inferior cerebellar peduncle is a thick bundle of white fibres, both entering and leaving the cerebellum from many sources and to many destinations. As described elsewhere (p. 899) it is formed on the dorsolateral aspect of the superior half of the medulla oblongata. The two peduncles diverge as they continue to ascend, but upon reaching the anterior cerebellar notch, each bends sharply posteromedially to enter into its respective hemisphere. As it bends and curves backwards it is insinuated between the middle peduncle on its lateral aspect and the superior peduncle medially. Some authorities consider it convenient to divide the inferior peduncle, for descriptive purposes, into a small medially placed strand, the *juxtarestiform body*, and a large lateral bundle, the *restiform body*. The relation of the oblique *pontobulbar body* to the dorsolateral aspect of the peduncle has already been mentioned (p. 908).

The tracts entering the cerebellum through the inferior peduncle stem from a variety of sources in the spinal cord and medulla. They include the olivocerebellar, parolivocerebellar, vestibulo-cerebellar, reticulocerebellar, posterior spinocerebellar, cuneo-cerebellar and trigeminocerebellar tracts, together with the anterior external arcuate fibres and other arcuatocerebellar fibres (the striae medullares). The tracts *leaving* the cerebellum through the inferior peduncle include the cerebello-olivary, cerebello-vestibular and cerebelloreticular pathways, with minor bundles of cerebellospinal and cerebellonuclear fibres.

(1) **The posterior spinocerebellar tract**, as we have seen (p. 879), originates from cells in the thoracic nucleus of the same side, shows a somatotopic lamination of its fibres, and conveys a variety of types of information from both cutaneous exteroceptors and proprioceptors from the hind limb and lower trunk. Some fibres are 'modality-specific' whilst others carry information converging from more than one source; the transmission along both types of fibre may be modified by activity in descending tracts from the brainstem. The dorsal spinocerebellar fibres terminate in the 'hind-limb' regions of the cerebellar cortex (*vide infra*) which includes the vermis of the anterior lobe (central lobule, culmen and some of the declive) together with the neighbouring parts of their adjacent lobules, whilst other fibres end in the pyramid and uvula.

(2) **The cuneocerebellar tract** (posterior external arcuate fibres) passes from the cells of the external (accessory) cuneate nucleus to gain the anterior and posterior regions of the vermis of the same side. This tract and its nucleus of origin are somatotopically organized, and are functionally similar to the posterior spinocerebellar tract, but in relation to the fore-limb and upper trunk. The anterior and posterior areas of the cerebellar cortex which receive spinal afferents, and the localization of afferent terminals within these areas, are illustrated in 7.85A, B.

(3) **The olivocerebellar tract** is composed of the axons of neurons in the inferior olivary nucleus, whilst (4) **the parolivocerebellar tracts** originate from the dorsal and medial accessory olivary nuclei. It had long been suspected that a close correspondence existed between points on the surface of the inferior olive and ones on the cortex of the contralateral cerebellar hemisphere. Subsequently, studies of human cases with de-generative disease, and both physiological and anatomical experimentation in animals, have confirmed this. Further, it has been shown that a remarkably precise point-to-point inter-connexion exists between all points of the olivary complex of nuclei and the opposite hemisphere. Broadly, a dorsoventral and a mediolateral correspondence exists with the inferior olive projecting to the larger and more lateral parts of the hemisphere, whilst the accessory olives project to the vermian and para-vermian regions. Illustration 7.85E is a simplified summary of these findings based largely on clinical cases (for further details *see*

Jansen and Brodal 1954; Brodal 1969). Whilst the olivocerebellar fibres largely cross the midline, some authorities have described a few fascicles which terminate ipsilaterally. The olivocerebellar fibres end in the cortex as *climbing fibres* which have, in recent years, been recognized as of great functional significance—they are discussed further below. In association with this it should be emphasized that afferents converge upon the olivary nuclei both from the spinal cord, the cerebral cortex and other subcortical centres. The spino-olivary tracts already described (p. 881) convey cutaneous and proprioceptive afferents to the accessory olivary nuclei, which in turn project largely on to the 'spinal regions' of the contralateral cerebellar hemisphere completing a *spino-olivocerebellar* system. Cortico-olivary fibres descend largely from motor regions of the cerebral cortex and terminate in localized areas of the inferior and accessory olivary nuclei (Sousa-Pinto and Brodal 1969) completing a *cortico-olivocerebellar pathway*. Additionally, other subcortical nuclei, including the corpus striatum, red nucleus and brainstem reticular formation, all project on to restricted regions of the olivary complex and may thus differentially affect cerebellar cortical activity.

Not all olivocerebellar fibres, however, terminate as climbing fibres in the cortex—some end in the contralateral fastigial, emboliform, globose and dentate intracerebellar nuclei, and the projections are to some degree somatotopically organized. These various nuclei will be further discussed below.

(5) **The reticulocerebellar tract** arises mainly from both large and small cells of the *lateral reticular nucleus* of the medulla (*nucleus of the lateral funiculus—see* p. 946), but other *peri-hypoglossal* and *paramedian* reticular nuclei also contribute—p. 903. Parts of the latter, in particular neuron group A2, the *nucleus commissuralis* part of the perihypoglossal group are noradrenergic (p. 949) and almost certainly contribute such fibres via the inferior peduncle to both the intracerebellar nuclei and cortex. The reticulocerebellar fibres end in the 'spinal' regions of the cerebellar vermis mentioned above.

It should be noted at this point that various indirect channels to the cerebellum thus exist through the intermediary of the medullary reticular nuclei, because of the convergence on them of spinoreticular, vestibuloreticular and corticoreticular pathways which are described elsewhere in this section.

(6) **The vestibulocerebellar tract** is in the main composed of primary afferent (root) fibres of the vestibular part of the vestibulocochlear nerve, but some secondary fibres stem from cells of the medial and inferior vestibular nuclei. The fibres terminate ipsilaterally in the cortex of the archicerebellum, principally in the flocculus and nodule, but a small number are considered to end also in the uvula and lingula. Other vestibulocerebellar fibres are distributed to the fastigial nuclei of both sides. It is held (Stein and Carpenter 1967) that the primary vestibulocerebellar fibres are mainly those with sensory endings in the ampullae of the semicircular ducts. In contrast, the nuclei of origin of the secondary fibres are the sites of termination of the rather small spinovestibular tract (p. 877). The vestibulocerebellar fibres occupy part of the juxtarestiform body (*vide supra*).

The remaining pathways which enter the cerebellum through its inferior peduncle are less well understood.

(7) **The anterior external arcuate fibres** arise from arcuate nuclei of both sides and, together with some superficial reticulocerebellar fibres, course over the surface of the inferior peduncle to enter the cerebellum. As stated above, the arcuate nuclei are possibly homologous with the pontine nuclei, and accordingly, the anterior external arcuate fibres form part of a *cortico-arcuatocerebellar* pathway, but the cerebellar termination and significance of the latter is unclear.

(8) **The striae medullares**, encountered previously (p. 910), are also thought to originate in the arcuate nuclei; they partially decussate and pass to the flocculus. They are sometimes termed the *arcuatofloccular tract* (fibres of Piccolomini) and are probably to be grouped with the anterior external arcuate fibres, and those derived from the pontobulbar body (p. 908).

(9) **Trigeminocerebellar fibres** both crossed and uncrossed, are known to enter the cerebellum from the superior sensory and descending (spinal) nuclei of the trigeminal nerve. Their distribution, and the status of other nucleocerebellar tracts which

have been described are uncertain in the human cerebellum.

The tracts leaving the cerebellum in the inferior peduncle include:

(10) **Cerebello-olivary fibres**, as yet of uncertain origin, which leave in the inferior peduncle to terminate on cells of the inferior olivary nucleus.

(11) **Cerebellovestibular fibres**, which issue from the ipsilateral flocculus, nodule, and fastigial nucleus (and to some extent the opposite fastigial nucleus). They run with other constituents of the juxtarestiform body to reach the four vestibular nuclei, in which they terminate. Other cerebellovestibular fibres are the axons of Purkinje cells in the anterior and posterior regions of the vermis which do not synapse in the intracerebellar nuclei, but pass directly to the lateral vestibular nucleus. The anterior vermian projection shows a somatotopic localization which corresponds with that of the origin of the lateral vestibulospinal tract (p. 875). Further vermian projections pass to the fastigial nucleus, and thence to the lateral vestibular nucleus in the fastigiovestibular tract—again these pathways are somatotopically organized (Brodal *et al.* 1962).

(12) **Cerebelloreticular fibres**, which start in both, but mainly the contralateral, fastigial nuclei. They pursue an indirect course, partially decussating and then proceeding as the *hook bundle of Russell*, which curves around the cerebellar end of the superior peduncle before joining the inferior peduncle. In the main these fascicles are distributed to the pontine and medullary reticular formation. Accordingly, different functional zones of the nucleus fastigii (*vide infra*) and *fastigiovestibular* and *fastigiobulbar* tracts have been recognized. However, the two tracts are partially admixed in both the direct juxtarestiform body pathway and in the less direct hook bundle described above. Some observers believe that a proportion of the hook bundle fibres pass through the medulla oblongata to enter the anterior funiculus of the spinal cord as a *cerebellospinal tract*. The evidences with regard to the latter, and those concerning direct *cerebellonuclear tracts* to brainstem motor nuclei remain meagre and unconvincing.

THE MIDDLE CEREBELLAR PEDUNCLE

The middle cerebellar peduncle (brachium pontis) is more massive than the superior and inferior peduncles, and passes lateral to them as it continues from the dorsolateral region of the pons, curving dorsally and then radiating to become continuous with the laminae of white matter within the cerebellar hemisphere (7.77). It is composed almost exclusively of the axons from the second neurons on the extensive *corticopontocerebellar pathway*. These neurons constitute the nuclei pontis which are scattered throughout the substance of the ventral (basilar) part of the pons. The fibres arising from the nuclei in one half of the pons almost all cross the midline, traversing the opposite middle peduncle to reach the cortex of the contralateral hemisphere. However, much of the vermis probably receives pontocerebellar fibres from both sides of the pons, whilst the lingula, some small regions of the anterior lobe and the flocculonodular lobe receive no pontocerebellar projection.

The fibres of each middle peduncle are arranged in three fasciculi, superior, inferior and deep. The *superior fasciculus* is derived from the upper transverse fibres of the pons; it is directed backwards and laterally superficial to the other two fasciculi, and is distributed mainly to the lobules on the inferior surface of the cerebellar hemisphere, and to the parts of the superior surface adjoining the posterior and lateral margins. The *inferior fasciculus* is formed by the lowest transverse fibres of the pons; it passes inferomedial to the superior fasciculus and is continued downwards and backwards more or less parallel with it, to be distributed to the folia on the under surface close to the vermis. The *deep fasciculus* comprises most of the deep transverse fibres of the pons. It is at first covered by the superior and inferior fasciculi, but crossing obliquely it appears on the medial side of the superior, from which it receives a bundle; its fibres spread out and pass to the upper anterior cerebellar folia. The fibres of this fasciculus cover those of the inferior peduncle.

Despite a number of investigations the precise distribution of pontocerebellar fibres from different regions of the pons in terms of the localization of their cerebellar terminals has not yet been satisfactorily explored. In general, the most lateral and also the most medial pontine nuclei project to the vermis, whilst the intermediate ones project to the contralateral hemisphere.

The pontine nuclei receive the terminals of corticopontine fibres which arise from many, perhaps all, parts of the cortex of the ipsilateral cerebral hemisphere, and they also receive collateral branches from some of the descending corticospinal fibres. While the origin of corticopontine fibres is undoubtedly extensive, the quantitative contribution from different cortical regions, and any precise somatotopic pattern of the projection from the different cerebral lobes on to specific groups of pontine nuclei, cannot yet be regarded as established in primates, including man. However, experimental investigations demonstrate that a fairly precise somatotopic pattern of corticopontine fibres extends from the sensorimotor cortical regions in the cat (Brodal 1968 *a* and *b*). Similarly, reports of *spinopontine* and *tectopontine* fibre systems in the cat have, as yet, received no definite confirmation in primate brains, although they are probably present. Recently another class of afferent fibres to the cerebellum that traverses the middle peduncle to reach both the intracerebellar nuclei and cortex has been demonstrated. These are aminergic and constitute the *cerebellar serotonergic pathway* derived from cell groups B5 (part of the nucleus raphe pontis) and B6 (part of the central pontine reticular grey matter)—*see* p. 950. Their precise terminations and functional roles are, as yet, undetermined.

THE SUPERIOR CEREBELLAR PEDUNCLE

The superior cerebellar peduncles (7.77, 90) emerge from the cranial part of the anterior cerebellar notch and are hidden from view by the anterior lobe of the cerebellum. When that structure is pulled aside they can be seen connected with one another by the superior medullary velum, and ascending in the lateral part of the roof of the fourth ventrical to disappear just caudal to the inferior colliculus.

A number of fibres *enter* the cerebellum through the superior peduncle (*vide infra*), but the great majority of the fibres which constitute the strand are *leaving* the cerebellum, and take origin for the most part in the cells of the nucleus dentatus. **The efferent fibres** emerge from the hilum of this nucleus and, having been joined by efferent fibres from the emboliform, globose and fastigial nuclei, they pass upwards, forwards and medially, covered over at first by the medial fibres of the inferior and the deep fibres of the middle peduncle. As they ascend in the roof of the fourth ventricle the fibres gradually incline anteriorly and sink into the tegmental region of the midbrain medial to the lateral lemniscus. They then sweep medially and the majority of the fibres decussate with the corresponding contralateral fibres. However, a proportion do not decussate and accordingly, considering the contribution from both peduncles, there is now, on both sides of the midline, a small uncrossed component and a much larger crossed component of the superior cerebellar peduncles. Both the crossed and uncrossed components now separate into ascending and descending bundles. Each of these will be considered briefly, but it must be emphasized that the *crossed ascending bundle* is by far the most prominent, and is often the only one referred to in elementary accounts.

The uncrossed component of the superior cerebellar peduncle which arises in part from the fastigial nucleus, is mainly distributed to the brainstem reticular formation. *Ascending fibres* terminate in the reticular nuclei of the midbrain tegmentum and the grey matter surrounding the aqueduct of the midbrain, whilst the *descending fibres* reach reticular nuclei of the pons and medulla.

The crossed descending fibres of the superior cerebellar peduncle are distributed mainly to the inferior and accessory olivary nuclei and neighbouring reticular formation. Some authorities describe additional fascicles which pass into the spinal cord and others which run in company with the medial longitudinal fasciculus to the motor nuclei of the nerves supplying the extrinsic eye muscles.

The crossed ascending fibres of the superior cerebellar peduncle are quantitatively the major outflow from the cerebellum. They arise from the neurons of the dentate, globose and emboliform nuclei (p. 926) and form the *cerebellorubral* and *dentatothalamic tracts*. **The cerebellorubral fibres** are principally axons from the globose and emboliform nuclei (of the opposite side) and they terminate on the cells of the red nucleus which lies in the midbrain tegmentum (p. 938). On the basis of animal experimentation, a somatotopic localization of the cells in the globose and emboliform nuclei and in the issuing cerebello-rubral tract, which corresponds with that in the red nucleus and rubrospinal tract, has been described.

The dentatothalamic fibres, having issued from the cells of the opposite dentate nucleus and crossed the midline, now by-pass the red nucleus and terminate in synaptic relationship with the neurons of the nucleus ventralis intermedius (lateralis) and, to a lesser degree, the nucleus ventralis anterior of the thalamus (*see* p. 958). From these thalamic centres fibres radiate to end in the 'motor' regions of the cerebral cortex. In this manner large areas of the cortex of the cerebellar hemispheres, through their connexions with the dentate nucleus, and then via the dentothalamic tract and the thalamocortical radiation, are able to influence the activities of the motor areas of the cerebral cortex.

In addition, some authors have described fibres in the crossed ascending component of the superior peduncle which terminate in other thalamic nuclei, including the nucleus centromedianus, the reticular nucleus, the nuclei ventralis posterior and lateralis posterior, and also in a variety of nuclei in the subthalamus and hypothalamus (Hassler 1950; Cohen 1958; Nimi *et al.* 1962; Snider 1967).

The afferent fibre systems which enter the cerebellum through the superior cerebellar peduncle include the anterior spinocerebellar, tectocerebellar, ceruleocerebellar tracts, and descending hypothalamocerebellar tracts.

The origin and constitution of the **anterior spinocerebellar tract** was considered previously (p. 879). As this superficial laminated tract ascends the cord it consists of some fibres which originated on the same side, but a majority which crossed from neurons on the opposite side of the cord. It continues to ascend superficially in the brainstem and then curves posteriorly along the lateral aspect of the superior peduncle to enter the cerebellum, within which many of the fibres which crossed the midline in the cord recross in the inferior cerebellar commissure to regain their side of origin. The fibres terminate in the 'hind-limb' receiving areas of the anterior and posterior vermian and paravermian regions of the cerebellar cortex (7.85A, B). As mentioned previously, the existence of a **rostral spinocerebellar tract** which is the fore-limb counterpart of the anterior spinocerebellar tract, has been recognized by physiological recording techniques in experimental animals. Its anatomical verification, and the demonstration of its existence in the primate brain, including that of man, is still awaited.

The tectocerebellar tracts descend from the tectum of the midbrain to the cerebellum through the neural substance of the superior medullary velum close to the midline. They probably arise from the inferior and superior colliculi of both sides, and their fibres terminate in the intermediate vermian and para-vermian regions of the cerebellar cortex, including the posterior part of the anterior lobe, the declive and lobulus simplex, the folium, tuber and pyramid (7.85D). The precise origin and constitution of these tracts await further analysis. While it is widely assumed that they convey both auditory and visual information directly to the cerebellum, alternative routes have been proposed; these include *tectopontocerebellar* and *occipitopon-tocerebellar* pathways.

The ceruleocerebellar tract consists of noradrenergic fibres stemming from some of the aminergic neurons constituting brainstem reticular group A6 (the central core of the nucleus ceruleus—*see* pp. 933, 949). Having traversed the superior cerebellar peduncle they are distributed to both the intra-cerebellar nuclei and the cerebellar cortex. Their precise modes of termination and functional roles still await analysis.

Hypothalamocerebellar tracts composed, at least in part, of cholinergic fibres, have been described by Shute and Lewis

(1965) as descending from the caudal hypothalamic nuclei, in company with the dorsal longitudinal fasciculus (of Schütz) to enter the cerebellum through the superior peduncle; their destination is uncertain. (For a review of possible cholinergic mechanisms in the cerebellum, and elsewhere, *see* De Feudis 1974.)

A Summary of Cerebellar Connexions

The foregoing account of the cerebellar peduncles and their contained tracts is rendered particularly lengthy, complex, and the overall pattern somewhat obscured, by the multiple nature of the topographic pathways involved and the plethora of tract names. In the following paragraphs some details of the topography of the peduncles and tracts will be ignored, and certain groupings of connexions emphasized, which would otherwise perhaps be less apparent.

THE CEREBELLAR INPUT

The input to the cerebellum may be *direct* or *indirect*. *Fibres passing into the cerebellum directly* without intervening synapses, to terminate predominantly in the cerebellar cortex, stem from neuron groups in the spinal cord, medulla oblongata, pons and midbrain. All other tracts which influence cerebellar activity do so *indirectly* by converging on one or more of the direct sources just mentioned.

The main direct cerebellar inputs are: (1) from the spinal cord via the various spinocerebellar pathways (including functionally the cuneocerebellar tract); (2) from the olivary, reticular, vestibular and arcuate nuclear complexes of the medulla oblongata; (3) from the pontine nuclei in the ventral part of the pons, and from the raphe nuclei and other reticular nuclei of the pontine tegmentum; (4) from the core of the nucleus ceruleus; and (5) from the tectum of the midbrain via the tectocerebellar tract. These are detailed in previous pages.

Reference has already been made to a useful approximate subdivision of the cerebellar cortex on phylogenetic and ontogenetic grounds into archicerebellum, paleocerebellum and neocerebellum (*see* 7.85A). Ignoring for our present purposes minor inconsistencies and some degree of overlap, the *archicere-bellum* or 'vestibulocerebellum' receives principally vestibular connexions; the *paleocerebellum* or 'spinocerebellum' receives direct terminals from the spinal cord, the medullary reticular formation and the accessory olivary nuclei; the *neocerebellum* receives, laterally in the large hemispheres of the middle lobe, the massive *pontocerebellar* connexions and those from the inferior olive, whilst nearer the median plane, in its vermian and paravermian regions, it receives *tectocerebellar* connexions. (Perhaps to correspond to the rather imprecise but useful terms 'vestibulocerebellum' and 'spinocerebellum', the addition of 'pontocerebellum' and 'tectocerebellum' would seem appropriate.) This crude parcellation of the cerebellar cortex into the receiving areas for the main groups of *direct* afferent inputs, despite some overlap, can aid further discussion of the complexities added when indirect pathways are included, and it has also proved useful in clinicopathological description (*vide infra*).

The indirect routes by which cerebellar cortical activity can be modified must, as pointed out above, necessarily involve the convergence of pathways on to the direct pathways just summarized. In particular, other routes from the spinal cord, various subcortical nuclei and the extensive projections from the cerebral cortex must be considered.

In addition to the array of direct spinocerebellar pathways previously described, it will be recalled that both the accessory olivary nuclei and the medullary reticular formation also project to the spinocerebellum, and it is interesting that both the latter receive spinal cord projections through the spino-olivary and spinoreticular tracts. Further, circumscribed regions of the vestibular nuclei and the tectum both receive spinal connexions through the spinovestibular and spinotectal tracts, although the degree to which information flowing into the cerebellum along the vestibulocerebellar and tectocerebellar tracts is modified by influences from the spinal cord remains uncertain. Nevertheless, it is clear that information from the cord may reach the cerebellum in many different orders of complexity. The direct spinocerebellar pathways convey relatively elementary forms of information; single fibres may carry impulse patterns which follow stimulation of a single type of cutaneous or proprioceptive peripheral receptor, whilst others may carry patterns derived from the convergence and interaction of fibres from two or more types of peripheral receptor—even here, however, impulse transmission at the first synapse in the pathway may be modulated by activity in descending pathways such as the corticospinal. In the indirect spinal pathways there are presumably mechanisms for the further integration of impulses from the cord with those flowing along the other pathways which converge on the medullary reticular formation, accessory olivary nuclei, etc., before transmission into the cerebellum (i.e. these are sites of 'cerebellar preintegration').

Similar considerations apply to the numerous routes by which the cerebral neocortex may modify cerebellar cortical activity. As we have seen, the most massive is the corticopontocerebellar pathway with which many observers would link a cortico-arcuatocerebellar pathway. In addition it must be remembered that other cell stations from which axons reach the cerebellar cortex are also in receipt of connexions from the cerebral cortex. Thus, there are corticotectal, corticoreticular, and cortico-olivary tracts; and as mentioned above, even the cells of origin of the spinocerebellar tracts are influenced by descending corticospinal fibres. Accordingly, it will be appreciated that cerebral cortical activity may well influence the information flow into the cerebellum along most of the pathways that enter the organ.

However, the various brainstem centres which project to the cerebellum are not homogeneous systems with respect to their connexions; on the contrary, there is much evidence that, for example, the olivary complex is not only somatotopically organized in relation to its projections to the cerebellar cortex, but varies much in its different parts in terms of their afferent connexions. Thus, some parts may be largely concerned with particular through routes, e.g. the spino-olivocerebellar pathway; other regions may provide sites for the pre-integration of the information flow along various afferent channels which converge on the olivary nuclei. The latter include some of the afferents from the spinal cord and cerebral cortex just mentioned, and others from the corpus striatum, the red nucleus and the brainstem reticular formation.

Without elaborating further here, it is clear that for a fuller understanding of cerebellar mechanisms, much more needs to be known about all the forms in which information is presented to the cerebellar cortex. The relatively simple direct spinocerebellar and vestibulocerebellar pathways have already been analysed quite extensively, but the significance of the cerebral cortical control systems, and the degree to which complex transformations occur by pre-integration in the various brainstem centres remain obscure, but are the subject of much current research.

THE CEREBELLAR OUTPUT

The output from the cerebellum commences with the axons of the Purkinje cells of the cerebellar cortex and involves the four intracerebellar nuclei (fastigial, globose, emboliform and dentate), on both sides of the midline. Both the Purkinje cells and the nuclei will be further detailed in subsequent sections, but for our present purposes it must be stated that the vast majority of the Purkinje cell axons converge upon and synapse with the neurons of the nuclei just mentioned; the axons of the latter form the main outflow tracts from the cerebellum. However, some of the Purkinje cells situated in the cortex of the flocculonodular lobe and others in the anterior and posterior parts of the vermis send axons which by-pass the intracerebellar nuclei, to extend directly to their destinations outside the cerebellum.

The most obvious effects of cerebellar dysfunction are disturbances in patterns of muscle contraction, and it is instructive to consider the cerebellar outflow in relation to the regions known to be intimately concerned, at least in part, with locomotor control. Whilst a few cerebellospinal and cerebellonuc-

lear fascicles have occasionally been mentioned in the research literature, it seems that direct cerebellar connexions to motor neuron pools in either the spinal cord or brainstem are either non-existent, or so sparse as to be of little functional significance. In contrast, all the major motor control regions of the brainstem and cerebrum receive, directly or indirectly, profuse connexions from the cerebellum. These include the vestibular nuclear complex, the brainstem reticular formation, the red nucleus, the tectum, and, via the thalamus, the corpus striatum and motor regions of the cerebral cortex. From these nuclei arise all the main descending pathways, corticospinal, tectospinal, vestibulospinal, rubrospinal, reticulospinal, etc. by which—usually through the intermediary of local interneurons—the patterns of activity in the a and γ motor neurons are varied. These pathways will not be further detailed here, but it is evident that for a comprehensive view of cerebellar control to emerge, it will be necessary to consider the control which is exerted at all these levels simultaneously. Further, despite the prominence of locomotor sequelae in states of deranged cerebellar function, there may well be many other activities in which the cerebral and cerebellar cortices cooperate and which are not revealed by current methods of investigation. Certainly a number of aspects of autonomic regulatory functions may be modified in cases of cerebellar disease or after experimental stimulation or ablation. There is also an increasing weight of evidence that the cerebellum may play a role in modifying the electrical activity of the sensory areas of the cerebral cortex; and through its connexions with the reticular formation and thalamus it may modulate transmission in ascending sensory pathways (Snider 1967).

The Grey Matter of the Cerebellum

The cerebellar grey matter is distributed in two locations—as an extensive surface coat, the cerebellar cortex, and as independent masses deep within the substance of the organ, the intracerebellar nuclei.

THE CEREBELLAR CORTEX

The surface grey matter of the cerebellar cortex is, as we have seen (p. 912), folded by a large number of predominantly transverse but somewhat curved fissures. The latter are closely arranged, approximately parallel, and vary considerably in depth, dividing the organ into a series of lobes, lobules and folia. These subdivisions correspond to the primary, secondary, tertiary and sometimes quaternary laminae of the central cerebellar white matter. The smallest terminal laminae and their curved caps of grey matter constitute individual cerebellar folia (7.73, 84). So deeply grooved is the surface of the cerebellum, that if 'unfolded' the anteroposterior length of the cortex would be more than one metre, while its transverse dimension is only one-seventh of this at its maximum. Throughout its whole extent the cerebellar cortex shows a virtually uniform microscopic structure. Local differences which are so characteristic of the cerebral cortex do not occur in the cerebellum, so that it is impossible to distinguish between sections taken from different areas. Not only is the cerebellar cortex of man homogeneous structurally in all regions, but also a similar structure obtains throughout the mammalia and with only minor differences throughout the vertebrates.

The cerebellar cortex shows a high degree of geometric order, many of its elements being precisely arranged with respect to the surface and to the longitudinal and transverse axes of individual folia (7.79). It consists of: (a) the terminations of fibres entering the cortex ('climbing' and 'mossy' fibres); (b) five varieties of neuron—granule cells, outer stellate cells, basket cells, inner stellate (Golgi) cells, and finally, the Purkinje cells, whose axons leave the cerebellar cortex; (c) specialized neuroglial cells and blood vessels.

The cortex of the cerebellum consists of two main strata, external molecular and internal granular layers. Some authorities consider that a third Purkinje cell layer should be distinguished which is intermediate in position, but in this account it is treated as the deepest part of the molecular layer. In the following paragraphs the constituents of each layer are listed and briefly summarized, and subsequently individual features are treated in greater detail. Nevertheless, the detailed researches on the anatomy and physiology of the cerebellar cortex are now so numerous and contain such a wealth of data, that all cannot be treated in depth here. For further details and extensive bibliographies the interested reader should consult Dow (1942); Jansen and Brodal (1954); Dow and Moruzzi (1958); Eccles *et al.* (1967); Fox and Snider (1967).

In brief, the exceedingly numerous granule cells of the inner layer are one of the two principal *input* stations of the cerebellar cortex, since they receive the synaptic terminals of the mossy afferents (i.e. *all* the fibre systems converging on the cerebellar cortex other than the olivocerebellar fibres). The large Purkinje cells form the *output* channel from the cerebellar cortex, most of

their axons converging upon the deeply placed intracerebellar nuclei, from which the various outflow tracts leave the cerebellum for the numerous destinations described in the previous pages.

The dendritic trees of the Purkinje cells receive synaptic terminals from the profuse array of *parallel fibres* derived from the granule cell axons, and from the *climbing fibre* input to the cortex derived from the olivocerebellar tract. The remaining neurons, i.e. outer stellate cells, basket cells and Golgi cells, are essentially *inhibitory interneurons*, which interconnect the various elements in the cerebellar cortex in complex geometrical patterns (*vide infra*).

In broad outline, therefore, the cerebellar cortex receives *two lines of input*, the climbing olivocerebellar fibres which synapse directly with the Purkinje cells, and the mossy fibres which do so through the intermediary of numerous granule cells and their

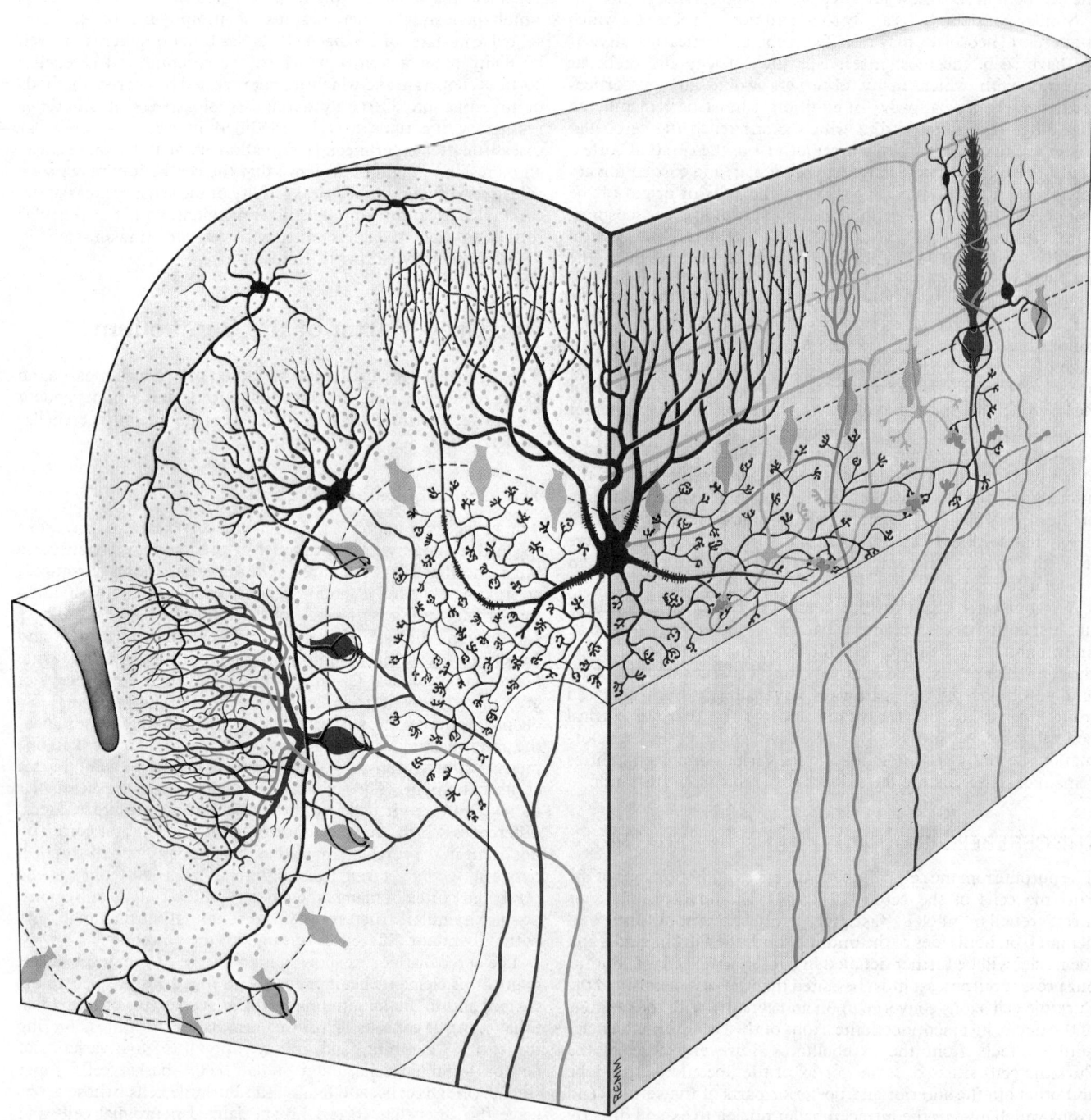

7.79 The general organization of the cerebellar cortex: a single cerebellar folium has been sectioned vertically, both in its longitudinal axis (right part of the diagram) and transversely (on the left). Note: (1) Purkinje cells (red); (2) inhibitory interneurons (black) including outer stellate, basket and Golgi cells; (3) granule cells and their ascending axons which bifurcate into longitudinally disposed horizontal fibres (yellow); (4)

climbing fibres and mossy afferents (blue). Note also the synaptic glomeruli formed between the terminals of the mossy afferent fibres, the complex dendrite tips of the granule cells and the ramifications of the Golgi cell axon. (Redrawn from: *The Cerebellum as a Neuronal Machine* by J. C. Eccles, M. Ito and J. Szentágothai. With the permission of the authors, and the publishers, Springer, 1967.)

parallel-fibred axon terminals. Both are excitatory to the Purkinje cells but are completely contrasting, as we shall see, in their structural arrangement and presumed modes of operation. The *single line of output* consists of the axons of the Purkinje cells; the latter are inhibitory to the cells of the intracerebral nuclei, which in turn modify the activities of all the major motor control centres of the brainstem and cerebrum. Purkinje cell activity is further complicated by the operation of complex surrounding 'shells' of inhibitory interneurons.

THE LAYERS OF THE CEREBELLAR CORTEX

The molecular layer is some 300–400 μm in thickness and consists of a rather sparse population of different types of neuron, numerous dendritic arborizations, non-myelinated axons and neuroglial cell processes. In conventional histological preparations it appears rather featureless, hence its name, in contrast to the immense tightly-packed nuclear population of the granular layer. The molecular layer contains (7.79):

(1) In its deepest part a single layer of the flask-shaped cell somata of Purkinje cells which are flattened in the transverse axis of the folium, and appear narrow when the folium is sectioned longitudinally.

(2) The rich dendritic trees of the Purkinje cells which reach towards the surface of the cortex and are also flattened in the transverse plane of the folium.

(3) Recurrent collaterals from Purkinje cell axons.

(4) The dendritic trees of the Golgi cells whose somata occupy the outer granular zone; these dendritic trees also reach towards the surface of the cortex, but are not flattened, and they span the territories of a number of Purkinje cells both transversely and longitudinally.

(5) The cell somata, dendrites and axonal branches of the outer stellate cells which are superficially placed.

(6) The cell somata, dendrites and axonal branches of the more deeply placed basket cells.

(7) The axonal terminals of the more deeply placed granule cells which pass radially into the molecular layer, where each divides into two branches which proceed in opposite directions in the longitudinal axis of the folium, giving, with others, the parallel fibre bundles which intersect with the various dendritic trees just mentioned.

(8) The terminals of the olivocerebellar climbing fibres, which ascend from the deeper parts of the hemisphere through the granular layer to become closely applied to the surfaces of the Purkinje cell dendrites in the molecular layer.

(9) The radially disposed branches of large neuroglial cells which are sited in the granular layer; these branches surround or intervene between the foregoing neuronal elements, except at points of synaptic contact, and at the surface of the cortex form conical expansions which meet neighbouring ones to make an external limiting membrane for the cerebellar cortex.

The general disposition of these different elements is summarized in illustration 7.79. **The granular layer** of the cerebellar cortex (7.79) varies in thickness, being about 100 μm deep in the furrows and 400–500 μm at the apex of the folia. It contains an almost unbelievably large population of 'microneurons', the *granule cells*, which are about $2 \cdot 4 \times 10^6$ per cubic millimetre for the cerebellar cortex of the monkey and $3–7 \times 10^6$ per cubic millimetre in the human cortex (Fox and Barnard 1957; Braitenberg and Atwood 1958). In summary the granular layer consists of:

(1) The cell somata of the granule cells just mentioned and the initial parts of their axons which ascend to the molecular layer to form the parallel fibres.

(2) The dendritic extensions of the granule cells, with their claw-like terminal expansions.

(3) The branching terminals of the mossy afferent fibres to the cerebellar cortex.

(4) The climbing fibres passing through to the molecular layer.

(5) The cell somata, basal dendrites and the profuse and complex axonal arborization of the Golgi cells.

(6) The cerebellar *glomeruli* which are synaptic complexes involving four types of neurite; the mossy fibre terminal establishes synaptic contacts with both granule cell and Golgi cell dendrites, whilst the granule cell dendrites also receive synaptic contacts from Golgi cell axon terminals (p. 922).

(7) The cell bodies of the large neuroglial cells mentioned above.

The cells and synaptic contacts of the cerebellar cortex will now be described in greater detail.

THE PURKINJE CELL

As mentioned above the Purkinje neurons are a highly differentiated type of cell, with a specific geometry, and throughout the vertebrates are characteristically found in the cerebellar cortex and in no other part of the nervous system.

Their flattened cell bodies are flask-shaped when viewed in a transverse section across a folium and appear as a vertical narrow strip in longitudinal section. They are arranged in a single stratum in the deepest part of the molecular layer of the cortex (7.79), immediately adjacent to the granular layer; individual Purkinje cells are separated by about 50 μm transversely and 50–100 μm longitudinally. Their cell bodies have a vertical diameter of 50–70 μm and a transverse diameter of 30–35 μm.

One, sometimes two large, smooth primary dendrites arise from the 'neck' of the flask, that is, the superficial pole of the cell, but its precise pattern of branching varies with the position of the cell. In the depths of the furrows between folia the primary dendrite branches almost immediately into two large smooth stems which recede from each other at nearly 180° in the transverse plane of the folium, and from these a remarkably rich arborization of second, third and subsequent order sub-branches arise and pass towards the surface of the cortex. Near the apex of a folium the primary dendrite passes superficially for some distance before branching at a more acute angle. In both cases, however, the branches of the dendritic tree are confined to a narrow 'sheet' of tissue which lies precisely in the transverse plane of the folium. The Purkinje cells are rather more tightly packed together at the apices of the folia than in the depths of the fissures, and this, together with the variations in dendrite branching, have been interpreted as Cartesian transformations, which correspond to the convexities and concavities of the surface, consequent upon the 'folding' of the tissue planes which occurs during development.

The primary dendrite and its first and second order branches have smooth surfaces, as noted above, but the third order branches and subsequent ramuli are densely covered over much of their surfaces by short, thick *dendritic spines* which form synaptic contacts with the parallel fibres derived from the granule cell axons (*vide infra*). The terminal *spiny branchlets* of the dendritic tree carry about 45 spines on each 10 μm of their length, and it has been computed (Fox and Barnard 1957) that on average each Purkinje cell carries about 180,000 spines on the whole of its dendritic tree—this is further discussed below.

The base of the cell body of a Purkinje cell, which lies near the granular layer, gives origin to the axon, which passes through the granular layer into the subjacent white matter. The initial 30 μm or so of the process, however, is narrow, non-myelinated, has ultrastructural features which correspond with those of the soma rather than the true axon, and its surfaces make specialized synaptic contacts with basket cell axon terminals. For these reasons some authorities prefer to call this initial region the *preaxon*; beyond this point it suddenly increases in diameter, acquires a myelin sheath and gives rise to a number of collateral branches. The principal part of the axon continues through the white matter and eventually breaks up into a basket-work of terminals which form synapses with cells of one of the intracerebral nuclei, but a small proportion of the axons by-pass these nuclei and pass directly to the vestibular nuclei of the brainstem (p. 917). The collateral branches of the Purkinje axon interweave to form supra- and infra-ganglionic plexuses internal and external to the level of the Purkinje cells. Their termination is still rather an open question, but any significant number of these recurrent collaterals ending upon neighbouring Purkinje cells seems doubtful. The majority probably pass to make synaptic contacts with the dendrites of basket and Golgi cells in the same, neighbouring and even quite distant folia, and may well form the so-called cortical association fibres, the significance of which has long remained obscure.

Synaptic contacts are made with Purkinje cells (7.80) by: (1) the parallel fibre bundles of the granule cell axons, which synapse with the dendritic spines of the Purkinje cell; (2) the climbing fibres from the olivocerebellar system; one climbing fibre is largely restricted to a single Purkinje cell and its numerous branches follow the subdivisions of the Purkinje dendrites and establish hundreds, perhaps thousands of synaptic contacts with the non-spiny (i.e. smooth interspine) areas of the spiny branchlets; (3) synaptic terminals from the outer stellate cells on the smooth surfaces of the large primary and secondary dendrites; (4) synaptic contacts from collaterals of basket cells, which end interspersed between those of the stellate cells; (5) complex 'beard-like' terminals of the basket cell axons which surround the Purkinje cell preaxon described above. These various contacts are summarized in 7.80.

The ultrastructure of the Purkinje cell has been well studied (Herndon 1963; Hámori and Szentágothai 1965, 1966 *a* and *b*; Léránth and Hámori 1970); it contains all the elements common to neurons (p. 825) and particular emphasis has been placed upon distinctive arrays of compressed lamellar systems of membranes which assist microscopic identification. They will not be further detailed here. Other than at points of synaptic contact with other neuronal processes, all parts of the cell surface are clothed with the processes of Bergmann glial cells.

THE GRANULE CELL

The varying thickness of the granular layer and its immense population of granule cells (up to 7,000,000 per cubic millimetre in man) has been mentioned above. Each granule cell has a spherical nucleus of only 5–8 μm diameter with a mere veil of cytoplasm 0·5–1·0 μm thick containing a few small mitochondria, ribosomes and a diminutive Golgi complex. From the deep aspect of the granule cell pass usually 3–5 dendrites (occasionally 1–7), and each dendrite is some 10–30 μm in length and often remains single, but sometimes branches before breaking up into claw-like terminal expansions. The latter receive the synaptic terminals of the mossy fibre input and the other components adjoining them to form the *cerebellar glomeruli*, which occupy the small clear crevices between the granule cells, seen when they are examined with conventional light microscope techniques.

The small-diameter axon of each granule cell passes from its superficial aspect into the molecular layer and then dichotomizes at a T-junction, the branches diverging parallel to the long axis of the cerebellar folium. The total length of the two branches is 2–3 mm, and together with those of other granule cells, they form the system of *parallel fibres* (7.79, 80). Near the point of bifurcation the parallel fibres are smooth, but throughout much of their course they develop intermittent fusiform dilatations of outline, and then near their termination a series of short hook-like or club-like surface projections. The dilatations and projections are sites of synaptic contact between the parallel fibre and the numerous types of dendrite in the molecular layer. Some quantitative aspects of these connexions will be mentioned below, but by far the most numerous are those with the dendritic spines of the Purkinje cells, less frequent are those with the spines of Golgi cell dendrites, and with the dendrites of the outer stellate and basket neurons. It may be noted that quantitative histological studies suggest that 150,000–300,000 parallel fibres cross, and probably synapse with, the dendrites of a *single* Purkinje cell.

THE BASKET NEURONS

These neurons are of stellate form and occupy the deeper half of the molecular layer of the cerebellar cortex. Their dendritic trees have no special geometry in their details of branching, but again, like the Purkinje cells, their dendrites are confined to thin 'sheets' of tissue which lie strictly in the transverse plane of the cerebellar folia. The dendritic surfaces carry rather sparse, irregular, long, thin dendritic spines which make synaptic contact with the parallel fibres that intersect the dendritic tree at right angles. The

cell body is also densely beset with axosomatic synapses, derived it is thought from climbing fibre collaterals and from recurrent collaterals of the Purkinje cell axons.

The basket cell axons course in the deeper part of the molecular layer just superficial to the layer of Purkinje cells, and again, the principal axon passes in the transverse plane of the folium. It continues for a distance of about 1 mm, covering the territories of about 10–12 Purkinje cells in its course. From the second Purkinje cell onwards, descending collateral branches pass towards the cell somata of the Purkinje cells, where they interweave with collaterals from other basket cells, forming the pericellular networks of fibres from which the cells derive their name. Side branches pass from each descending collateral and from the main axon, weaving longitudinally in the folium to reach an additional 3–6 rows of Purkinje cells on both sides of the main axon. Accordingly, 100–200 Purkinje cells may be reached by the synaptic terminals of a single basket cell. The general cell body surfaces of Purkinje neurons are devoid of synaptic contacts and those derived from the descending collaterals of the basket cell axons are clustered in a complex fashion around the point of origin and throughout the length of the preaxon mentioned above (7.80). The basket cell axons also give rise to a much less profuse system of ascending collateral branches which probably synapse with the smooth bases of the Purkinje cell dendrites.

Thus, the basket cells provide a unique synaptic system. Activity in a longitudinal bundle of parallel fibres will activate a longitudinal row of basket cells whose dendrites are intersected by the bundle. Because the basket cell axons pass transversely in the folium, they will therefore exert a powerful synaptic action on the origins of axons from Purkinje cells which lie in longitudinal strips on both sides of the parallel fibre bundle (*see* also below).

THE OUTER STELLATE NEURONS

Smaller than the basket cells, the outer stellate neurons are scattered through the superficial half of the molecular layer of the cerebellar cortex. Their dendritic trees, although smaller than those of basket cells, are similar in their general form, possession of spines and their transversal disposition in the folium. Again, they are intersected at right angles by bundles of parallel fibres, with which their spines and interspinous smooth areas establish synaptic contacts. The origin of the sparse axosomatic synapses which have been seen on these cells is uncertain.

The smallest stellate neurons have profuse local axonal arborizations, which terminate on Purkinje cell dendritic spines, whilst the longer axons of the larger stellate cells again pass transversely in the folium to similar destinations on more distant Purkinje cells. The deepest and largest stellate neurons are similar but possess a number of descending collaterals which accompany those of the typical basket cells.

THE GOLGI NEURONS

These are the largest neurons of a stellate form in the cerebellar cortex and they occupy the superficial zone of the granular layer, immediately below the Purkinje cell bodies. A series of large dendrites radiate from every aspect of the cell, but many eventually curve to enter the molecular layer, within which they arborize, their course within the latter being predominantly at right angles to the surface of the cortex. Their dendritic trees are not compressed into the transverse plane of the folium, and they appear much the same in both transverse and longitudinal folial section. In both these planes they overlap the territories of three Purkinje cells. Thus, the whole dendritic tree of a Golgi cell approaches, but does not overlap appreciably, those of adjacent Golgi cells, and each is intermingled, within the molecular layer, with the dendritic fields of about ten Purkinje cells. Some of the Golgi dendrites, however, do not enter the molecular layer, but divide within the granular layer and make a contribution to the cerebellar synaptic glomeruli (*vide infra*).

The Golgi cell axon arises from the deep aspects of the cell soma or from the base of one of its deeper dendrites, and immediately breaks up into a profuse arborization which permeates the full thickness of the granular layer, occupying a volume of tissue

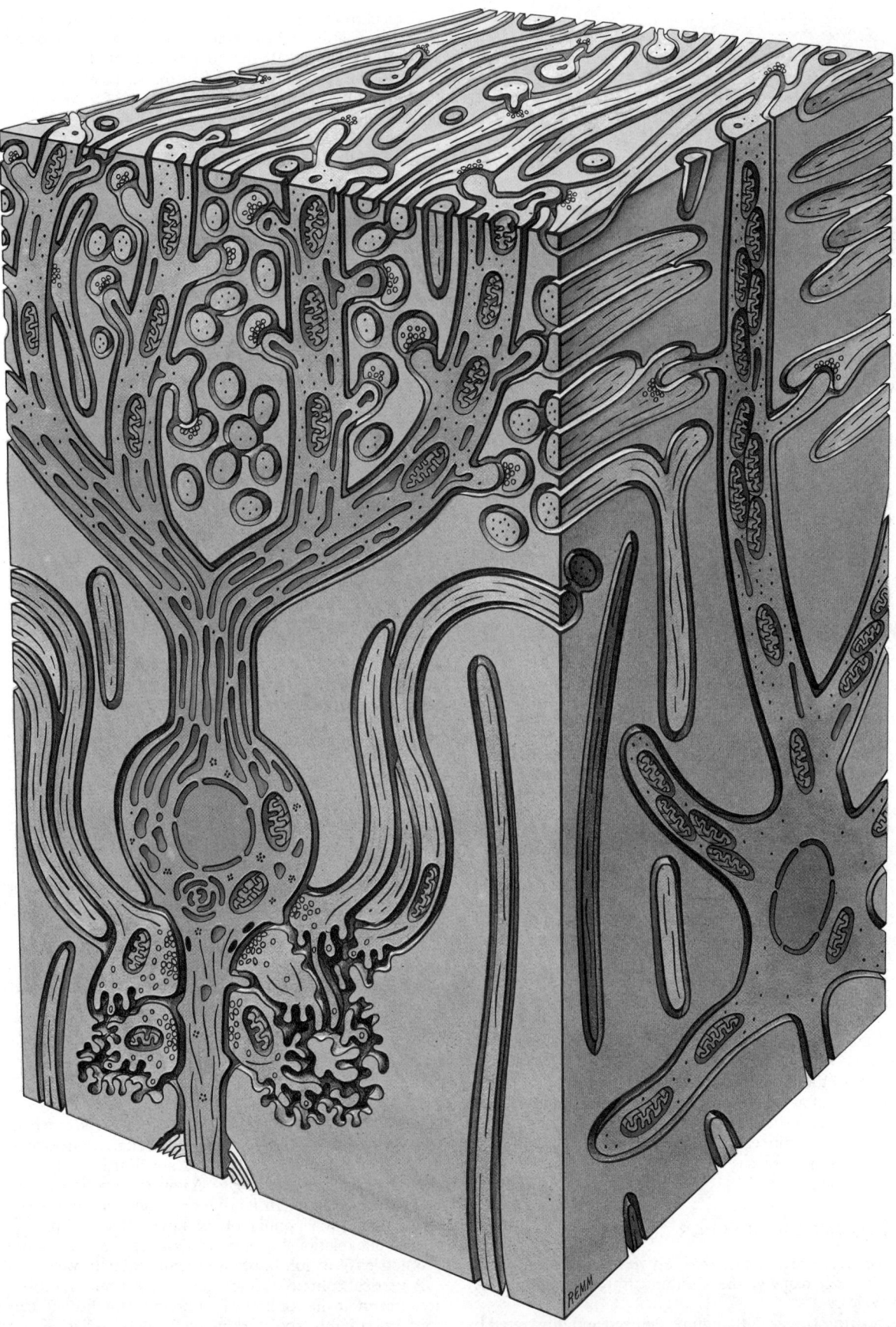

7.80 Detailed cytoarchitecture of part of the cerebellar cortex which includes the layer of Purkinje cell somata and the zones immediately superficial and deep to this. The vertical block face to the right is in the longitudinal axis of the cerebellar folium; the vertical face to the left is in the transverse plane of the folium; the upper face is tangential with respect to the convex summit of the folium. The cell details are shown in the following colours: red—the soma of the Purkinje cell, its 'preaxon' passing from the lower pole, and its apical stem dendrite with its first-order branches bearing dendritic spines. Blue—climbing fibres. Orange/yellow—horizontal fibres derived from the bifurcation of the ascending axons of granule cells. Mauve—Golgi cell soma, dendrites with spines and initial segment of its axon. Pale brown—the descending basket cell axons forming complex synapses on the preaxon of the Purkinje cell. Green—areas occupied by glial cell processes. Note the synapses between the horizontal axons and the dendritic spines of the Purkinje cell and Golgi cell. (Compounded and redrawn from information and illustrations in: *The Cerebellum as a Neuronal Machine* by J. C. Eccles, M. Ito and J. Szentágothai.)

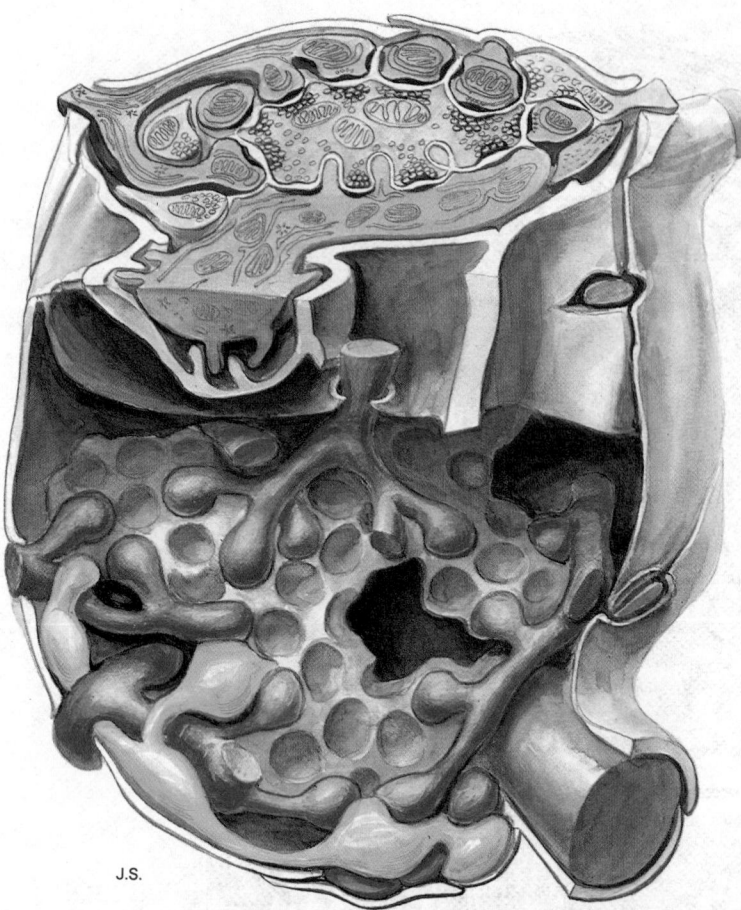

J.S.

7.81 A stereodiagram illustrating the structure of a cerebellar synaptic glomerulus. *Blue:* mossy afferent fibre rosette. *Red:* granule cell dendrites. *Yellow:* terminals of Golgi cell axon. *Green:* Golgi cell dendrite. *Grey:* neuroglial capsule. Note that the essential synaptic contacts are axodendritic between: mossy afferent fibres and granule cell dendrites; mossy afferents and Golgi cell dendrites; Golgi cell axons and granule cell dendrites. (From: *The Cerebellum as a Neuronal Machine*, by J. C. Eccles, H. Ito and J. Szentágothai, by courtesy of the authors, and the publishers, Springer, 1967.)

which corresponds on a deeper plane, to that of its dendritic tree in the molecular layer, i.e. does not overlap that of adjacent Golgi cells.

The main synaptic input to the Golgi neuron is from the parallel fibres of the molecular layer which synapse with the rather sparse spines on its dendrites. The axonal terminals of the Golgi cell take part in the formation of the cerebellar synaptic glomeruli.

In relation to the Golgi cells, therefore, the cerebellar cortex may be thought of as consisting of a series of tissue units, each roughly hexagonal in surface view, which abut on each other but do not overlap. Within each unit are about 10 Purkinje cell territories and deep to these an enormous population of granule cells and synaptic glomeruli.

INPUTS TO THE CEREBELLAR CORTEX

As we have seen elsewhere, there are two lines of input to the cerebellar cortex, namely, the climbing fibres and the mossy afferents (7.79).

The climbing fibres, although recognized and illustrated by Cajal, were for long in doubt, as to origin, and only recently has it been established that many are olivocerebellar fibres. The precise point-for-point correlation between the olivary complex of nuclei and the cerebellar cortex has already been mentioned (p. 916). It may be noted that comparatively recently evidence has accumulated demonstrating electrotonic coupling (through nexuses or gap-junctions) between the inferior olivary neuronal origins of the climbing fibres (see Llinas *et al.* 1974, Sotelo *et al.* 1974).

Each climbing fibre passes straight without branching through the white matter and the granular layer to become attached to a single Purkinje cell. It then divides repeatedly and its terminal arborization is closely applied to the surfaces of the branching Purkinje cell dendrites throughout much of their length, making large numbers of synaptic contacts with the smooth areas of dendrites between the dendritic spines.

A few collateral branches leave the climbing fibre and some of these descend to synapse with the cell somata and dendrites of Golgi neurons, whilst others make synaptic contacts with neighbouring basket or outer stellate cells. Physiological studies have now established that both mossy and climbing fibres are excitatory in their effects. Thus, climbing fibres are able to exert a powerful, localized, excitatory effect upon single Purkinje cells and also a much weaker effect upon the different varieties of interneuron in its neighbourhood.

The mossy afferent fibres are now widely considered to include all the afferent systems other than those of the olivocerebellar tract described above. They also are excitatory in their effects, but in all other respects they contrast markedly with the sharply localized climbing fibre input.

As each mossy fibre traverses the white matter of the cerebellum it gives off a series of collateral branches which diverge to a number of adjacent folia, and within each folium the branch runs through the central lamina of white matter, again giving off numerous sub-branches to the granular layer on both aspects of the folium. Each sub-branch then enters the granular layer and divides yet again into two or three terminal branchlets which expand into a group of grape-like synaptic terminals, or mossy fibre *rosettes*, each of which occupies the centre of a cerebellar *glomerulus* or *islet*.

The cerebellar glomeruli are complex synaptic arrangements involving a central mossy fibre rosette, the dendrites of a number of granule cells, the synaptic terminals of Golgi cell axons; in addition, certain glomeruli are also invaded by a Golgi cell dendrite. Each glomerulus is roughly spherical or ovoid, about 20 μm in its greatest dimension, and the ratio of glomeruli to granule cells is about 1 to 5.

All synaptic contacts in the glomerulus are of the axodendritic type. The central rosette establishes contact with the internal surfaces of a surrounding leash of up to 20 granule cell dendrites and, when present, with the spine-studded surface of a Golgi cell dendrite. The terminals of the Golgi cell axons establish synaptic contact with the lateral or external surfaces of a number of the granule cell dendrites. These arrangements are most easily appreciated by reference to a diagram (7.81).

The mossy fibre terminals are excitatory and can thus excite the granule cell dendrites and, if included, a Golgi cell dendrite. Conversely, the Golgi cell, which is an inhibitory interneuron, can inhibit the granule cell through its axodendritic contacts.

SOME QUANTITATIVE ASPECTS OF CEREBELLAR CORTICAL STRUCTURE

Because of its regularly repetitive geometry, the cerebellar cortex has been much studied using quantitative neurohistological methods (Eccles *et al.* 1967; Fox and Barnard 1957; Fox *et al.* 1967; Braitenberg 1967, 1977). A few prominent features only can be mentioned here. In its different zones, it shows extremes in terms of neuronal population, in the convergence, divergence and one-to-one relationship of transmission paths, and in the complex geometries of its inhibitory interneuronal pathways.

A vertical column of cerebellar cortex 1 mm in cross-sectional area, taken at the summit of a folium in the human cerebellum, contains roughly 500 Purkinje cells, 600 basket cells, 50 Golgi cells and perhaps 3,000,000 granule cells, with some 600,000 synaptic glomeruli.

On the input side each climbing fibre synapses with *only one* Purkinje cell (but through its collaterals with an undetermined number of neighbouring interneurons). In contrast a single mossy afferent fibre shows an enormous divergence—it may make synaptic contact with 400 or more granule cells within a single folium, and if its branches to neighbouring folia are included, the number probably rises to several thousand. Conversely, each granule cell probably receives synaptic contacts from 4 to 5 different mossy fibre terminals.

The ascending axons of the granule cells, as we have seen, bifurcate to form parallel fibres which run longitudinally in the molecular layer and enter into synaptic contact with the various dendritic trees in their paths, i.e. those of Purkinje cells, Golgi cells, basket cells and outer stellate cells. The total length of the parallel fibre from a single granule cell axon is about 2–3 mm and in its path it makes a synaptic contact with 300–450 Purkinje cells. Thus the divergence from a single mossy afferent fibre through the intermediary of granule cells is to perhaps hundreds of thousands of Purkinje cells. The uncertainty in the latter calculation stems from our lack of knowledge concerning the degree of overlap in parallel fibre territories. Finally, there is an enormous convergence of paths on to individual Purkinje cells, the dendritic tree of each cell receiving some 200,000 synaptic contacts from different parallel fibres.

Since the foregoing was written, when discussing the 'module concept' of *cerebral* cortical architecture Szentágothai (1975) has emphasized the *evolution* of quantitative knowledge of the cerebellar cortex. Its first recognition as a three-dimensional rectangular lattice (Braitenberg and Atwood 1958) was followed by the intellectual subdivision of the network into assumed functional spaces (Szentágothai 1963, 1965); later to be defined physiologically (Eccles *et al.* 1967) and structurally corroborated (Uchizono 1967) in terms of excitatory and inhibitory zones of interaction. Finally extensive and detailed analyses on the cerebellum of the cat (Palkovits *et al.* 1971 *a*, *b*, *c*, 1972) made it possible 'to describe the cerebellar cortical network statistically in numerical and geometric terms, giving considerable refinement to the earlier rather crude qualitative guesses on the spatial and numerical distribution both of excitatory and inhibitory interactions'.

MECHANISMS OF THE CEREBELLAR CORTEX

The large volume of structural research on the cerebellum ranging from comparative morphology, developmental analysis, and gross connectivity studies, to the quantitative cytology, cell geometry and synaptic relationships pursued by ultrastructural techniques, has been paralleled by an equally impressive volume of neurophysiological research (for example: Dow 1942; Dow and Moruzzi 1958; Eccles *et al.* 1967; Ito *et al.* 1964; Ito and Yoshida 1966; Snider 1967; Szentágothai 1967; Eccles 1970; *see* also references on p. 920). The increasing interdependence of these approaches has led to the postulation of a number of elementary mechanisms which probably interlock during the normal functioning of the cerebellar cortex; only some brief comments indicating these kinds of approach will be included here.

It is clear, as repeated in previous sections, that the cerebellar cortex has two highly distinctive lines of input—the climbing and the mossy fibres—and only one line of output, the axons of the Purkinje cells (7.82). Both the climbing fibres and the mossy fibres ultimately convey information to the cerebellar cortex derived, at least in part, from similar sources, namely the exteroceptors and proprioceptors, the brainstem reticular formation and the cerebral cortex (*see* previous paragraphs). Further, both input channels are excitatory in their effects; but the climbing fibre exerts a one-to-one, powerful, all-or-none excitation on individual Purkinje cells, whilst each mossy afferent diffuses its excitatory effect through hundreds of excitatory granule neurons to thousands of Purkinje cells. The Purkinje cells are exclusively inhibitory in their action and their outflowing axons exert varying inhibitory patterns upon the intracerebellar and vestibular nuclei.

The remaining cells of the cerebellar cortex—basket, outer stellate and Golgi cells—each with its distinctive cell geometry, input and output, all function as inhibitory interneurons, the particular shape, site and kind of inhibitory effect varying with the cell type in question.

Whatever means of neurophysiological recording is employed, and whatever type of tissue preparation used, from the cerebellar cortex of the unanaesthetized animal to the chronically isolated slab of cortex, most of the cellular elements show a low level of 'background' impulse discharge even under conditions of minimal sensory input. Some researchers have used the term 'spontaneous' for this type of activity, but whether it represents an intrinsic characteristic of the individual neurons and their assemblies remains unclear; nevertheless, it is upon such a background that the more dramatic and orderly responses to specific patterns of input occur.

It has been suggested (Szentágothai 1967) that the climbing fibres and their terminal synapses, which are largely concentrated on individual Purkinje cells, do not provide an integrative mechanism, and it is presumed that for the climbing fibres their information flow has been *pre-integrated* from the various channels which converge and synapse with the cells of the olivary nuclear complexes. Further, it is suggested that the important *integrative units* of the cerebellar cortex involve the mossy afferent input, granule cells, parallel fibres, Purkinje cells and the varieties of interneuron. Some of their possible elementary interactions are summarized in 7.82, 83.

It may be imagined that activity in some of the mossy afferents has excited a small locus of granule cells with resultant activity in a narrow bundle of parallel fibres in the molecular layer of the cortex. The bundle of active parallel fibres, about 3 mm in length, courses longitudinally within the folium and exerts an excitatory effect upon the dendritic fields in its path, i.e. those of Purkinje, Golgi, basket and outer stellate cells. If the bundle is very narrow, or its activity of a low order, none of these cells will respond.

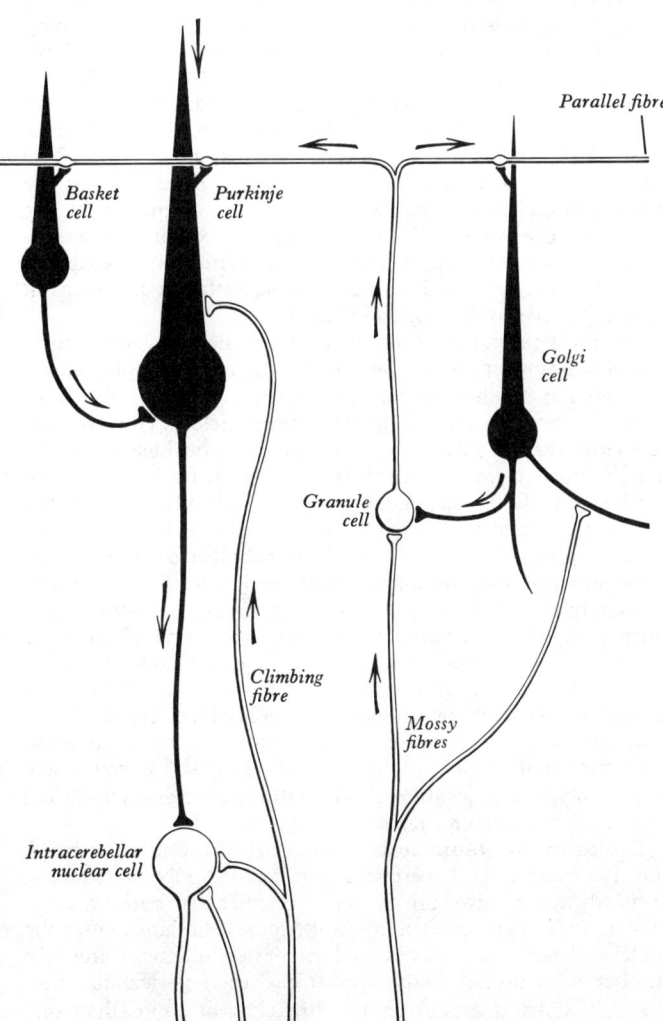

Parallel fibre

Basket cell

Purkinje cell

Golgi cell

Granule cell

Climbing fibre

Mossy fibres

Intracerebellar nuclear cell

7.82 An analysis of the essential circuitry and synaptic contacts between the climbing and mossy afferent fibres, the main neuronal elements of the cerebellar cortex, and the neurons of the intracerebellar nuclei, based upon cytological and microelectrode studies. Excitatory cells, neurites and terminals are white surrounded by a black line; inhibitory elements are solid black. (By courtesy of Professor J. C. Eccles.)

However, as the size of the bundle increases until it roughly corresponds to the width of the Purkinje cell dendritic fields, excitatory responses follow in a row of Purkinje cells and also in the related basket and outer stellate cells. Under these conditions the Golgi neurons do not respond, because only a small proportion of their much larger dendritic fields are receiving an input from the parallel fibre bundle. The net result, therefore, is that a single row of Purkinje cells about 3 mm long and parallel to the long axis of the folium respond with volleys of impulses along their axons. A crucial point to note is that because of the relatively slow conduction velocity of an impulse along the small diameter parallel fibres (a few decimetres per second) and the regular spacing of the Purkinje cells forming a row, the latter are excited *sequentially*. The difference in the time of arrival of a parallel fibre impulse between *successive* Purkinje cell dendritic trees being of the order of a tenth of a millisecond; this forms the basis for the concept of the cerebellar cortex as a *timing device* mentioned below. Because of the transverse disposition of the basket and outer stellate cell axons and their terminals, simultaneous activation of these inhibitory interneurons by the bundle results in longitudinal strips of cortex containing inhibited Purkinje cells which flank the active row on both sides. It is considered that with the continual fluctuations of activity in the mossy fibre input, the essential response of the cerebellar cortex is one of a constantly changing pattern of innumerable 'unit rows' of excited Purkinje cells flanked by inhibitory zones. The latter provide a process of neural sharpening by which the more active bundles of parallel fibres are selected from the general background of cortical activity. Should the active bundle of parallel fibres become too wide (i.e. much greater than the width of a single Purkinje cell row), another mechanism comes into play mediated by the Golgi cells. If sufficiently wide, the parallel fibres excite the Golgi cells via their dendrites and these, in turn, cause a reduction in the mossy fibre input via the granule cells by means of their inhibitory synaptic terminals on the granule cell dendrites in the synaptic glomeruli (p. 922). Hence, the principal function of the Golgi neurons appears to be in providing a 'choke' whereby excessive broadening of the active bundle of parallel fibres is prevented.

It is upon this incessantly changing pattern of *sequentially* excited, inhibited or relatively quiescent rows of Purkinje cells, that the sharply localized, powerful, excitatory effect of the climbing fibres is exerted, the resultant effect upon any individual Purkinje cell depending upon the strength of the climbing fibre stimulus and the state of the Purkinje cell upon its receipt. By mechanisms such as these, fluctuating inhibitory patterns are transmitted by the Purkinje cells to the intracerebellar nuclei, and thence to the motor control centres in the cerebrum and brainstem. It must be emphasized, however, that the manner in which cooperative interaction between localized climbing fibre activity and the time-base provided by active parallel fibres acts as a coordinator of striated muscle dynamics remains an enigma.

There are, of course, many other orders of complexity in the intracortical circuits of the cerebellum. These include the added complication of the effects of Purkinje cell and climbing fibre collateral branches on the various inhibitory interneurons, the problems posed by the presence of Golgi cell dendrites in the synaptic glomeruli, and differential 'on' and 'off' responses in the different categories of interneuron. Further interesting mathematical analyses of the structural arrangements in the cerebellum have, as intimated above, led to proposals concerning its possible mode of operation as an accurate timing device or *biological clock*, which incorporates critical *delay paths* in its design; these may be important in controlling the correct temporal sequence of events in, for example, rapid 'voluntary' movements (Braitenberg 1967). In the following decade these ideas and approaches were expanded and refined, the most recent exposition of the concept of the cerebellar cortex as a timing device being advanced by Braitenberg (1977) in his elegant monograph *On the Texture of Brains* which the interested reader should consult. Further details will not be pursued here, but perhaps enough has been said to indicate the general lines along which our knowledge of the *higher integrative units* of the cerebellar cortex is increasing. Despite such increases in understanding, they still fall far short of a comprehensive view of how the cerebellum as a whole integrates its total complex inflow of information, and utilizes it to regulate the overall locomotor patterns of the complete organism.

THE INTRACEREBELLAR NUCLEI

Four independent accumulations of grey matter are embedded in the white matter of the cerebellum of man on each side of the midline. These are the *intracerebellar nuclei*, sometimes also rather imprecisely termed the 'roof' nuclei because of the close relationship of *some* of them to the roof of the fourth ventricle.

Most laterally placed and the largest is the nucleus dentatus, on the medial aspect of which are the smaller nucleus emboliformis and nucleus globosus, whilst nearest the midline lies the nucleus fastigii. In most mammals, however, only three nuclei are recognized—a nucleus lateralis corresponding in part to the nucleus dentatus, a nucleus interpositus which corresponds largely to the globose and emboliform nuclei of man, and finally a nucleus medialis which corresponds to the nucleus fastigii. However, the precise homologies of the nucleus interpositus have been the subject of some debate (Jansen and Brodal 1954).

The nucleus fastigii is phylogenetically the oldest and is largely associated with the archicerebellum (vestibulocerebellum); the globose and emboliform nuclei are more recent and associated with the paleocerebellum (spinocerebellum), whilst the nucleus dentatus is the most recent and is associated with the neocerebellum (pontocerebellum). These associations are, however, approximate only, and are further discussed below.

The nucleus dentatus (7.84) sited near the white centre of the hemisphere, consists of an irregularly folded grey lamina containing white matter centrally, the fibres of which are derived in large measure from the axonal processes of the cell somata in the grey lamina. The grey lamina is deficient anteromedially and through this so-called 'hilum' of the nucleus, the white fibres stream to form a large part of the superior cerebellar peduncle (p. 917). **The nucleus emboliformis** lies close to the medial side of the nucleus dentatus and partially covers its hilum, whilst the **nucleus globosus** is still more medially placed and despite its name is elongated anteroposteriorly. **The nucleus fastigii** is larger than the latter and is situated close to the median plane in the anterior part of the superior vermis.

The most prominent cells of the cerebellar nuclei are fairly typical large multipolar neurons possessing rather simple, stellate but irregular dendritic arborizations, the numbers of dendrites increasing from the fastigial to the dentate nucleus. However, the dendritic trees of adjacent neurons overlap considerably in the fastigial nucleus and are much more distinct, having their own territories, in the dentate nucleus. These large multipolar neurons give rise to axons which make one or two characteristic loops before leaving the nuclei to form the **cerebellar outflow tracts** in the superior and inferior cerebellar peduncles for the various destinations in the brainstem detailed in the previous pages. During the looped part of their course the axons give off a number of recurrent collateral branches which terminate locally in the nuclei of origin. The intracerebellar nuclei also contain numerous small neurons, some of which are probably local Golgi type II interneurons, whilst others contribute to the peduncles. (Experimentally some of the small cells show retrograde chromatolytic changes after section of the peduncles—Jansen and Jansen 1955; Flood and Jansen 1966.)

The afferent connexions to the cells of the intracerebellar nuclei are from both cerebellar and extracerebellar sources. Primarily they receive both axodendritic and axosomatic synapses from the terminals of Purkinje cell axons. The latter enter the nuclei and each axon branches to form pericellular nests around a number of neurons, whilst conversely, each pericellular nest receives contributions from the branches of more than one Purkinje cell axon, i.e. there is both an element of divergence and of convergence in the construction of the nucleus.

It is now widely held that the Purkinje cells are exclusively inhibitory in their action and this fact has focused attention on other afferents, many of which are presumably excitatory to the intracerebellar nuclei. Fibres have been described as reaching these nuclei from the red nucleus, i.e. rubrocerebellar fibres

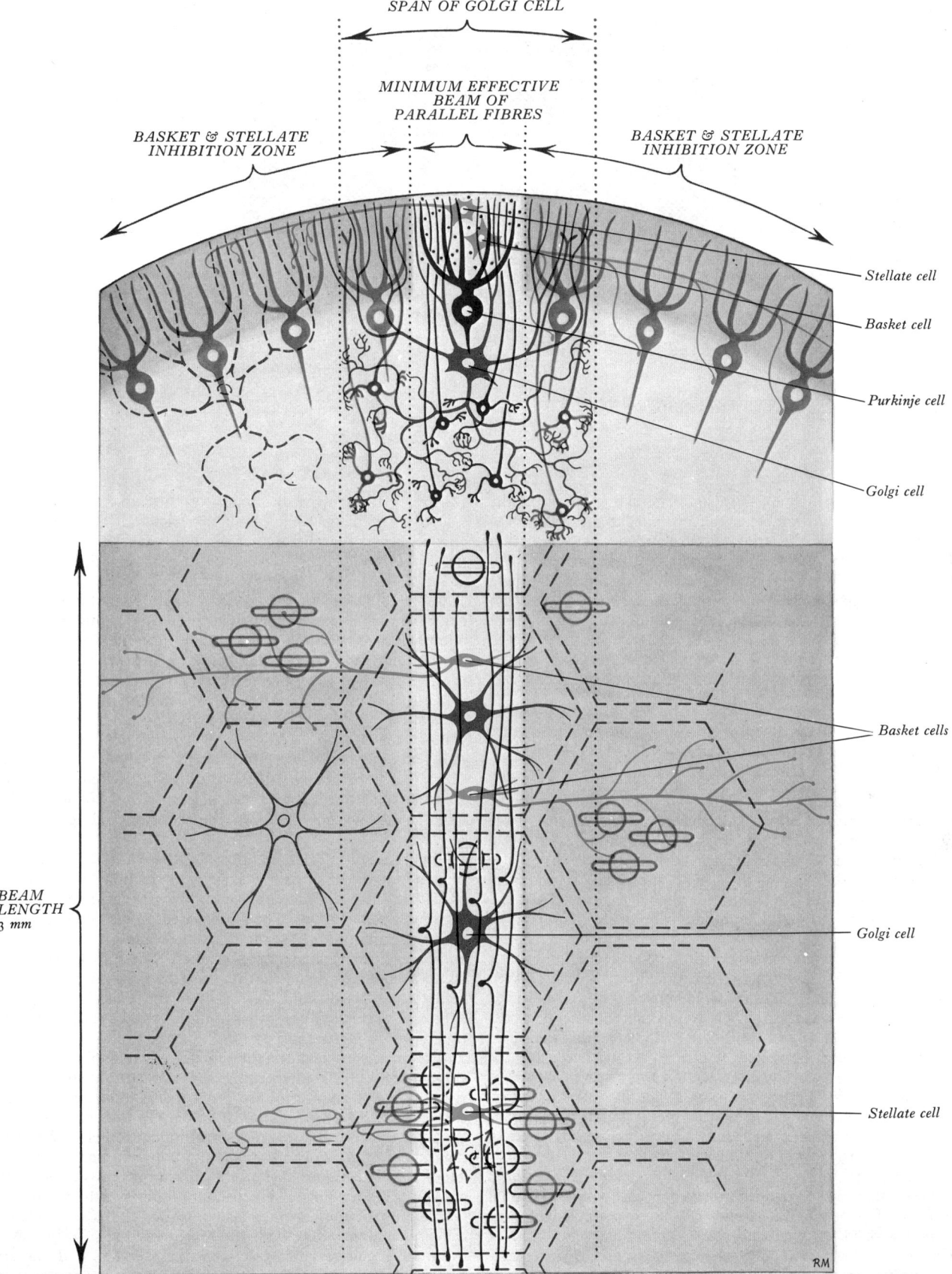

SPAN OF GOLGI CELL

MINIMUM EFFECTIVE
BEAM OF
PARALLEL FIBRES

BASKET & STELLATE
INHIBITION ZONE

BASKET & STELLATE
INHIBITION ZONE

Stellate cell

Basket cell

Purkinje cell

Golgi cell

Basket cells

Golgi cell

Stellate cell

BEAM
LENGTH
3 *mm*

RM

7.83 A diagram illustrating the concept of the complex neuronal integrative units of the cerebellar cortex. The neurons and their processes are shown in a transverse section of a cerebellar folium (above) and in surface view (below). Neurons in full colour (black, red or blue) or in full coloured outline, are in a state of excitation. Neurons in grey or interrupted outlines are inactive. *Full black or grey, black or grey outlines:* Purkinje cells, granule cells and their parallel fibre bundles; *blue:* basket and stellate cells; *red:* Golgi cells. The white longitudinal strip indicates activation of a locus of granule cells, their bundle of parallel fibres, and the associated longitudinal row of Purkinje cells. Simultaneous activation of the basket and stellate inhibitory interneurons by the parallel fibre bundle causes zones of lateral inhibition (grey shading) to flank the activated strip. These zones of inhibited Purkinje cells are in reality about ten rows wide—only four rows are illustrated here (grey profiles). If the active bundle of parallel fibres reaches a critical width, inhibitory Golgi interneurons become active, and reduce or stop mossy fibre input in the cerebellar glomeruli. The inhibitory territories of the Golgi cells are indicated by the array of red hexagons. See text for further description. (Redrawn from J. Szentágothai in: *Recent Development of Neurobiology in Hungary,* I. K. Lissák (ed.). By courtesy of the author and publishers, Akadémiai Kiadó, Budapest, 1967.)

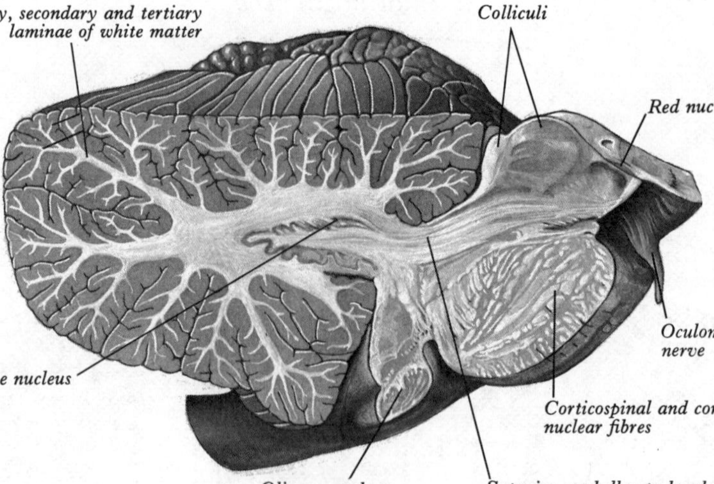

7.84 Oblique vertical section through the right cerebellar hemisphere and right half of the brainstem.

(Courville and Brodal 1966), and also collateral branches from the spinocerebellar, pontocerebellar, olivocerebellar and reticulocerebellar tracts (Eccles *et al.* 1967). Recent evidence has shown that both noradrenergic and serotonergic fibres from particular cell groups in the brainstem reticular formation reach and innervate the intracerebellar nuclei. The outflow of information through the cerebellar peduncles, therefore, is thought to stem from an integration within the intracerebellar nuclei of the patterns of excitation delivered from these various extracerebellar sources with the inhibitory patterns delivered by Purkinje cell axons, and modified by the action of local interneurons.

STRUCTURAL AND FUNCTIONAL LOCALIZATION IN THE CEREBELLUM

Since the cerebellar cortex shows an identical microscopic structure and microcircuitry in all parts of the cerebellum, it may be assumed—and physiological investigations support this—that its operational characteristics in terms of this local circuitry is also identical in all regions. In this sense, therefore, a functional localization does not exist within the cerebellar cortex *itself* as it does, for example, in the different laminae of grey matter in the spinal cord.

Structural and functional localization in the cerebellum is thus a reflexion of the manner in which functionally dissimilar afferent tracts end preferentially in different regions of this homogeneous cortex. Equally, it is a reflexion of how the different areas of cortex project to particular intracerebellar nuclei, and therefore influence different brainstem centres. A vast number of investigations have been directed towards the problem of cerebellar localization and the foundations have been laid by comparative morphological, developmental and gross connectivity studies (Jansen and Brodal 1954; Smith 1903; Riley 1930; Larsell 1937, 1953; Nieuwenhuys 1967). These were supplemented by more detailed analysis of connectivity patterns using degeneration techniques in experimental animals (Brodal 1969). Neurophysiological studies of the changes in electrical potential evoked by the 'natural' or 'artificial' stimulation of peripheral receptors including the special sense organs, peripheral nerves, and different regions of the cerebrum, brainstem and spinal cord, have been carried out (Snider 1936, 1940, 1945; Snider and Eldred 1951, 1952; Dow and Moruzzi 1958). The type of recordings have varied from rather crude monitoring of the surface of the exposed cerebellum, through recording of focal potentials by electrodes inserted into the cerebellar tissues, to unit recording with intracellular microelectrodes. Stimulation of the exposed cerebellar cortex in animals has been used to determine whether discrete movements are elicited, and which cerebellar regions influence movements that have been initiated reflexly or by stimulation of the cerebral cortex. Another extensive field of

research has involved the study of the behavioural effects of selective cerebellar ablation in experimental animals and the effects of cerebellar disease in man. For details of these important fields the interested reader should consult the articles quoted, some of which are devoted exclusively to these topics. A few examples only can be mentioned and illustrated here; further, since the various experiments have been carried out on a range of mammals, and because their considerable variations in cerebellar morphology cannot be treated here, the illustrations are referred to simplified plans of the cerebellum, and the areas indicated may well vary considerably between species and in mankind.

Reference has already been made to the primary division of the cerebellum into *archicerebellum*, *paleocerebellum* and *neocerebellum*, and how these correspond approximately, with some degree of overlap, to the major patterns of afferent cerebellar connexions summarized in the terms 'vestibulocerebellum', 'spinocerebellum' and 'pontocerebellum' (7.85A), the latter including the 'tectocerebellum'.

Experimental determinations, in the cat, of the areas of termination of the primary vestibulocerebellar fibres, and of the spinocerebellar, cuneocerebellar and spino-olivocerebellar tracts are illustrated in 7.85B. The double areas of distribution of these various spinal tracts to separate anterior and posterior, vermian and paravermian zones should be noted. These areas make up the spinocerebellum noted above, and within them a general somatotopic separation of hind-limb and fore-limb regions is shown. The primary vestibular fibres are shown ending mainly in the flocculus, nodule and part of the uvula.

Illustration 7.85C shows the somatotopic map of the cerebellar cortex produced by recording the evoked potentials during stimulation of cutaneous receptors distributed over the body surface, based on the findings of Snider (1952), and Hampson *et al.* (1952). There is a marked correspondence between this and diagram 7.85B, which was based upon neuroanatomical degeneration techniques. Again, the anterior and posterior parts of the spinocerebellum are evident, but in this case the responses to stimulation of the face are also included. It is seen that the different regions of the body are represented in reverse order in the anterior and posterior receiving areas; further, in the anterior area the representation is ipsilateral, whereas in the posterior area it is bilateral. As would be expected, a rather similar map is produced on stimulation of proprioceptive endings, and therefore need not be reproduced here.

In illustration 7.85D is shown the area from which responses are recorded during visual and auditory stimulation, and also during stimulation of the occipital areas of the cerebral cortex. The visual and auditory areas are virtually coincident and include the central regions of the vermis and their adjacent paravermian zones, overlapping the face area of the anterior spinocerebellum determined by tactile stimulation. How far these visual and auditory responses are due to direct tectocerebellar pathways, or to the involvement of other less direct pathways is uncertain (p. 918).

Repeated reference has been made in the previous pages to the strict point-to-point correlation that exists between the nuclei of the olivary complex and the opposite cerebellar cortex—the general correspondence of zones is illustrated in a simplified form in 7.85E.

The foregoing examples of cerebellar localization are based upon the differential distribution of afferent tracts to the cerebellar cortex, and broadly, the main receiving areas are arranged as a sequence of *transverse* strips which proceed from the (morphologically) anterior to posterior ends of the cerebellum. Equally clear evidences exist for a localization based upon the different destinations of Purkinje cell axons which leave the cerebellar cortex, but this is a *longitudinal* form of organization (Jansen and Brodal 1954; Eager 1966; Goodman *et al.* 1963; Voogd 1964; Korneliussen 1968).

It will be recalled that the majority of the Purkinje cell axons terminate in the intracerebellar nuclei, but some by-pass the latter and pass directly to the lateral vestibular nucleus. The main features of the corticonuclear projections in the cerebellum are summarized in illustration 7.85F. It is seen that direct cerebellovestibular fibres pass to the vestibular complex from the flocculus, nodule and much of the remainder of the vermis, but particularly from its anterior and posterior regions. The projections to the intracerebellar nuclei show an orderly longitudinal arrangement, and on both sides of the midline the cerebellar cortex is considered to be divided into *medial*, *intermediate*, and *lateral zones*. The fibres which leave the medial zone (vermis) converge upon the nucleus fastigii and their rostro-caudal sequence is preserved during this convergence and in their termination in the nucleus. The intermediate zone (a longitudinal paravermian strip) projects to the nucleus globosus and nucleus emboliformis, whilst the lateral zone (the cerebellar hemisphere) projects to the nucleus dentatus. In each case the orderly rostro-caudal sequence of the cortical origin of the axons is preserved through to their terminations in the nuclei. The vast majority of these corticonuclear projections are ipsilateral, but the fastigial nuclei receive some axons from the vermian cortex on both sides of the midline.

Recalling the main cerebellar outflow tracts and their nuclei of origin (pp. 919, 926), it will thus be appreciated that the vermis principally exerts a control over the vestibular nuclei and the brainstem reticular formation. The intermediate paravermian cortex controls the red nucleus and midbrain tegmentum, whilst the cerebellar hemisphere, through the nucleus dentatus and thalamus, mainly influences activity in the corpus striatum and cerebral cortex.

Interestingly, some parts of these outflow pathways have also been shown to be somatotopically organized. These include the direct and indirect vermian projections to the lateral vestibular nucleus (7.85G), and the intermediate projections to the red nucleus; this correlates well with the organization of the lateral vestibular and red nuclei themselves, and also that of the lateral vestibulospinal and rubrospinal tracts (Brodal *et al.* 1962; Courville 1966). Further, a similar localization has been shown in the nucleus ventralis lateralis of the thalamus which receives the dentatothalamic projections, and itself projects to the motor regions of the cerebral cortex (Walker 1934).

It should be emphasized at this point that the foregoing descriptions are based upon a range of experimental animals, but it is most likely that the general principles will apply to the cerebellum of man. Similarly, in experimental animals, interesting physiological parallels have been demonstrated. Cerebellar stimulation at different points often modifies movements which are being generated by simultaneous reflex or cerebral cortical stimulation, whilst cerebellar stimulation in the decerebrate animal elicits discrete movements (7.85H). Again, anterior and posterior areas which correspond to the spinocerebellum have been shown to display a somatotopic order in respect to such movement patterns (Hampson *et al.* 1952). The anterior area when stimulated gives ipsilateral responses whereas the posterior area gives bilateral responses.

Such investigations, combined with the effects of ablation (Chambers and Sprague 1955 *a* and *b*), suggest that the vermis controls the posture, tone, locomotion and equilibrium of the *whole body*, whilst the intermediate zone controls the posture and discrete movements of the *ipsilateral limbs*. The lateral zone is much less well understood.

CEREBELLAR DYSFUNCTION

There have been innumerable reports on the behavioural effects of cerebellar disease in man and on those of selective or total cerebellar ablations in experimental animals. Only the briefest mention can be made here and the interested reader should consult the excellent reviews which are readily available (Holmes 1939; Wyke 1947; Brown 1949; Dow and Moruzzi 1958; Dow 1969).

In essence, the cerebellum receives an informational input from cutaneous receptors, proprioceptors, the eyes, ears, brainstem reticular formation and cerebral cortex, which is integrated and then discharged to the motor controlling centres of the cerebrum and brainstem. Its normal functioning is necessary for smooth, coordinated, effective locomotor responses to occur.

Without reference to particular cerebellar regions the more obvious effects of cerebellar dysfunction may include: (*a*) disturbances in equilibrium of the whole body; (*b*) disturbances in muscle 'tone' or their resistance to stretch, tendon reflexes and ability to stabilize joint positions; (*c*) incoordination of movements (*ataxia*) due to irregularities in the timing of onset, rate and force of contraction of synergistic muscle groups.

Disequilibrium in man is manifest by a tendency to fall forwards, backwards or laterally when standing, and by an unsteady staggering gait; it may be accompanied by sensations of spinning, nausea, etc.

There is usually a 'softness' in the affected muscle bellies on palpation, diminished tendon reflexes and the muscles tire easily (*asthenia*). The lowering of joint control may progress to pendular swinging of a dependent limb segment after displacement, or to the condition of 'flail joints'.

Muscular incoordination is the essential feature of most varieties of cerebellar dysfunction. It may affect different regions of the body and to varying degrees; for this reason a variety of symptoms may occur, and a wide array of clinical tests have been devised to demonstrate its presence. Briefly, the term *asynergia* is often used to denote the diminished capacity for smooth, cooperative, sequential action between a series of muscle groups. A complex movement may be carried out as a sequence of irregular disjointed episodes—*decomposition of movements*. There may develop an inability to carry out rapid movements which alternate in direction, e.g. supination and pronation of the forearm and hand, *dysdiadochokinesis*; control of the range of movement may be lost with either 'undershoot' or 'overshoot'—*dysmetria*. Disturbances of locomotion with a *staggering gait* and a tendency to fall, and— particularly with closed eyes—deviations from the intended direction of progression, are common. Similarly, the inability to point with the finger in a particular direction—*past-pointing*—or to trace a specified course with either finger or heel may be present. *Tremor* is usually absent when the arm and hand are at rest, but a coarse transverse *intention tremor* may appear, which intensifies as the movement nears completion; tremor may also affect movements of the head and trunk. Muscular incoordination may lead to characteristic *defects of speech*, with a slowness of onset, slurring, jerky intermittent sound production which sometimes is intensified and has an explosive nature—so-called *scanning speech*. A *cerebellar nystagmus* may be present; there is an inability to fixate an object with the eyes, resulting in a repetitive conjugate drift of the visual axes away from the object, followed by a rapid return. The nystagmus is sometimes *positional*, i.e. more pronounced when the body adopts particular postures, or it may be *directional*, i.e. increasing when the subject attempts to gaze in a particular direction.

Attempts to correlate particular groups of clinical signs and symptoms of cerebellar disease in man with the different regions of the organ involved, or with the results of animal experimentation, have met with only partial success.

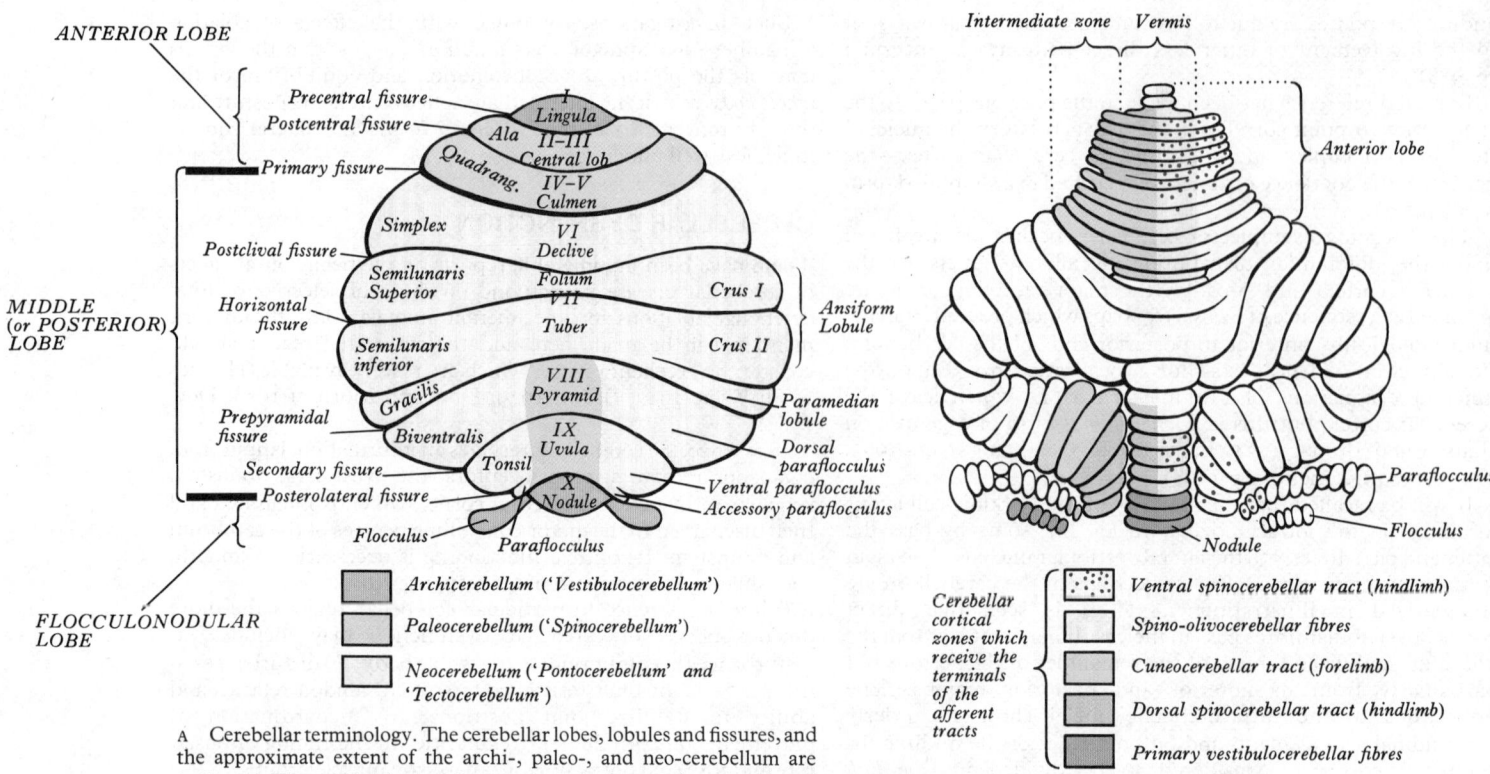

A   Cerebellar terminology. The cerebellar lobes, lobules and fissures, and the approximate extent of the archi-, paleo-, and neo-cerebellum are indicated. The subdivisions of the vermis are both named and numbered (after Larsell). On the left are terms widely used in human neuroanatomy; on the right are additional terms often used in general mammalian neuroanatomy.

B   The cerebellar cortical areas of termination of the afferent tracts indicated, derived from experimental studies.

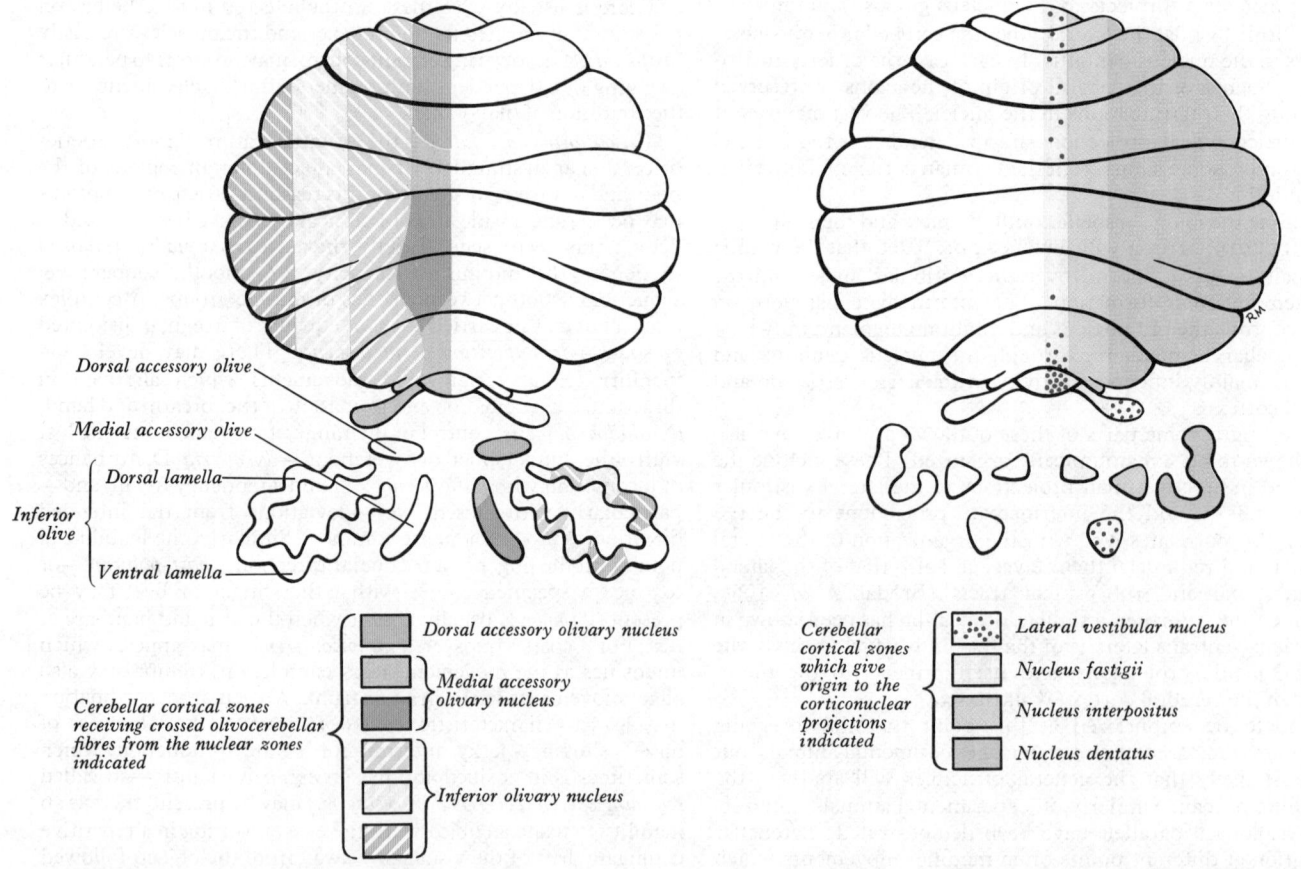

E   An analysis of the topical organization of the cortical zones of the cerebellum which receive crossed olivocerebellar (climbing) fibres from the different parts of the inferior olivary complex of nuclei.

F   An analysis of the topically organized projections of cerebellar cortical Purkinje cell axons on to the lateral vestibular nucleus, and the intracerebellar nuclei.

7.85A–H   This series of diagrams illustrates certain features of the localization of structure and function in the cerebellum. G is a median sagittal section of the feline cerebellum; in all the remaining diagrams it is assumed that the cerebellum has been flattened and viewed from the dorsal aspect, so that its whole rostro-caudal extent can be seen. A, E and F are approximate outlines of the human cerebellum; B, C, D and H are of the feline cerebellum. Detailed descriptions of the information in these diagrams are in the text, and in the references quoted.

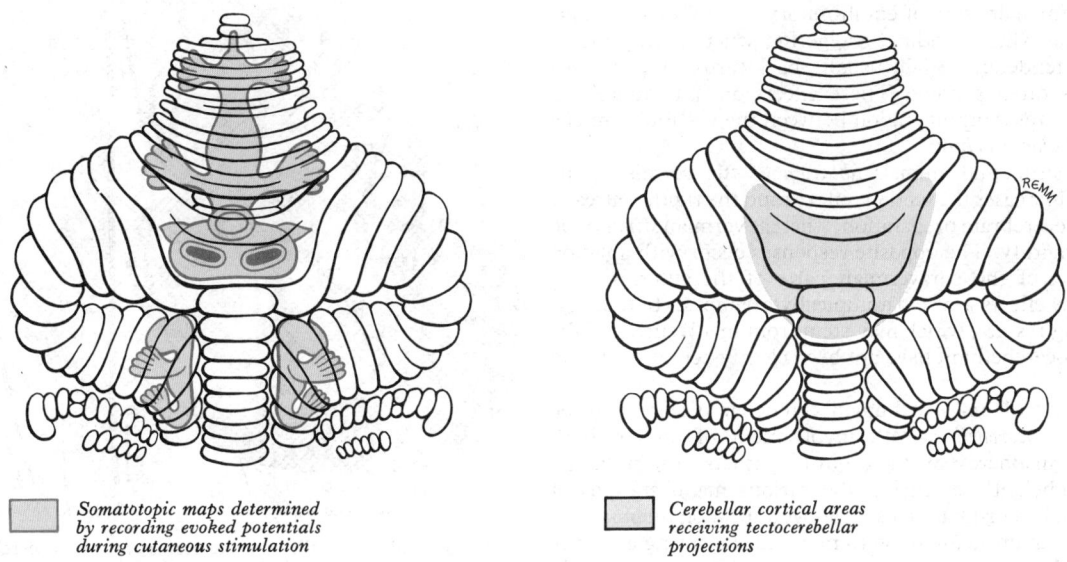

Somatotopic maps determined
by recording evoked potentials
during cutaneous stimulation

Cerebellar cortical areas
receiving tectocerebellar
projections

C   The somatotopic arrangement of the evoked potentials recorded from the cerebellar cortex during cutaneous stimulation.

D   The cerebellar cortical areas receiving tectocerebellar projections.

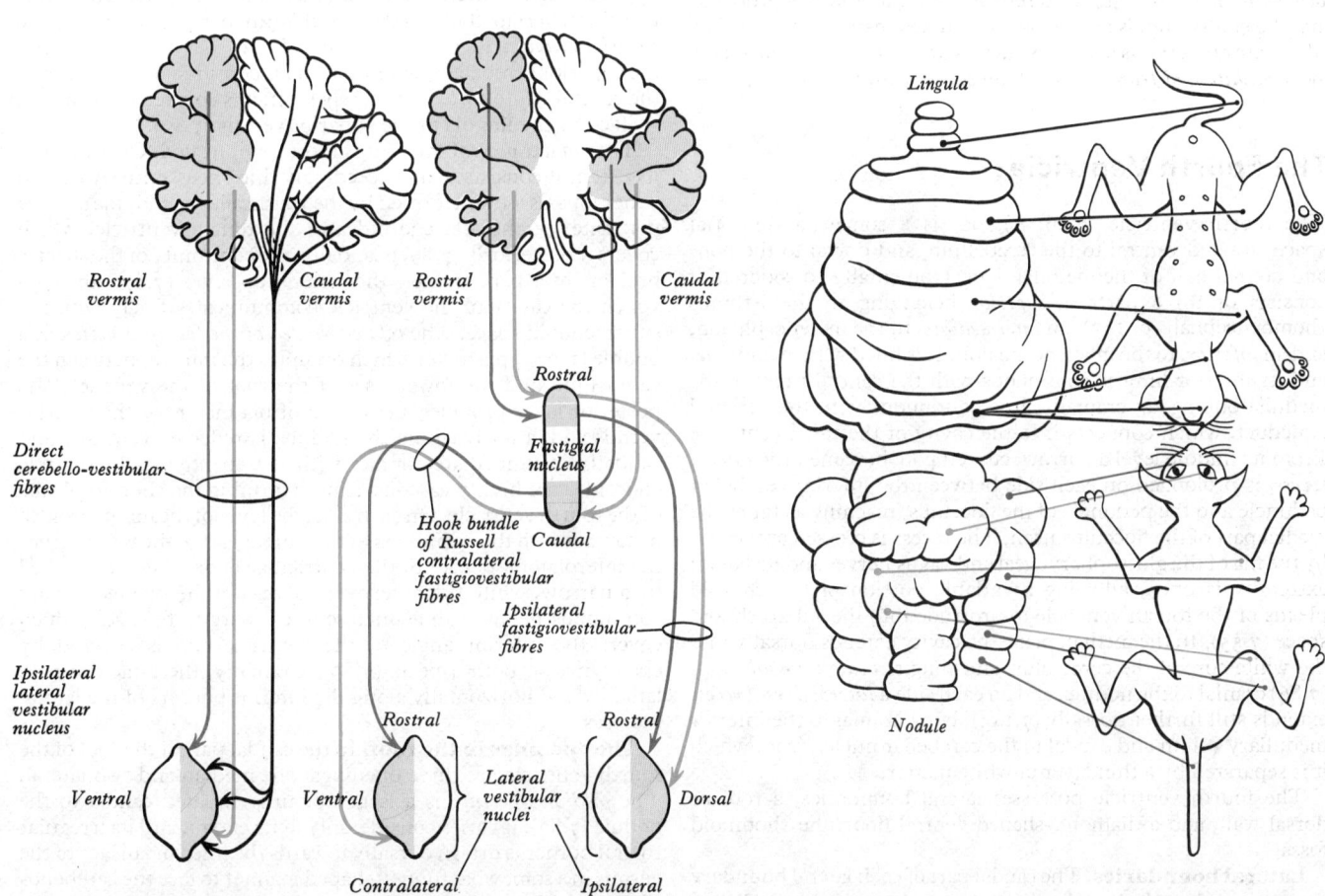

G   The arrangement of direct cerebello-vestibular and indirect cerebello-fastigio-vestibular projections from the cerebellar vermis of the cat.

H   The somatotopic pattern of movements elicited in different parts of the body during stimulation of the cerebellar cortex in the decerebrate cat.

(Illustrations B, F and G were redrawn and modified from *Neurological Anatomy*, 1969, by courtesy of the author Professor A. Brodal, and the publishers, Oxford University Press; C, D and H were redrawn and modified from *Publications of the Association for Research in Nervous and* *Mental Disease*, **30**, 1952, by courtesy of the authors Professor R. Snider (C and D) and of Professors J. L. Hampson, C. R. Harrison and C. N. Woolsey (H) and the publishers, The Williams and Wilkins Company)

In the *flocculonodular syndrome*, with damage to the nodule, uvula and flocculus, both in man and experimental animals, the principal features are loss of equilibratory control of the whole body, swaying when standing, staggering when attempting to walk, and a tendency to fall, usually backwards. A positional nystagmus is often present. These effects are attributable in general to the upset of integration between the vestibular nuclei and the *vestibulocerebellum*.

In the experimental animals *ablation* of the vermis of the anterior lobe increases the tendon reflexes and the rigidity already present in a decerebrate preparation, whereas vermian *stimulation* reduces the rigidity. The opposite responses occur with ablation or stimulation of the paravermian parts of the anterior lobe. Further, these effects are somatotopically organized; but, so far, corresponding results, which may stem from involvement of the anterior spinocerebellum, have not been clearly demonstrated in man.

The majority of cases of human cerebellar disease may be classified as *neocerebellar*, with involvement of one or both cerebellar hemispheres or their outflow tracts. In unilateral disease, if sufficiently extensive, the various manifestations of hypotonia and incoordination listed above appear on the *same side* as the lesion, but gross intention tremor and staggering gait only supervene if the dentate nucleus or superior cerebellar peduncle are involved. It should be emphasized that relatively small cortical cerebellar lesions cause little obvious effect, and even quite extensive disease, although causing a transient derangement of function, is followed by rapid and marked improvement in locomotor control. The mechanism of this *cerebellar compensation* remains obscure.

Finally, as mentioned above, despite the prominence of *locomotor* effects in states of cerebellar dysfunction, it is probable that the cerebellum is less obviously, but extensively, involved in other central nervous activities such as *autonomic homeostasis* and the *modulation of transmission* along sensory input channels.

The Fourth Ventricle

The fourth ventricle (7.86, 89, 90) is a somewhat tentorial space situated ventral to the cerebellum, and dorsal to the pons and cranial half of the medulla. Developmentally considered, it consists of three parts: a *superior* belonging to the isthmus rhombencephali (p. 163), an *intermediate*, to the metencephalon, and an *inferior*, to the myelencephalon. It is lined with ependyma, and its inferior limit is continuous with the central canal of the medulla oblongata; cranially, it is continuous with the cerebral aqueduct, which connects it to the cavity of the third ventricle. From its middle level a narrow, curved pouch, named the *lateral recess*, is prolonged on each side between the inferior cerebellar peduncle and the peduncle of the flocculus, reaching as far as the medial part of the flocculus itself. The recess is crossed anteriorly by the fila of the glossopharyngeal and vagus nerves and its lateral extremity is open, allowing a variable portion of the choroid plexus of the fourth ventricle to protrude into the subarachnoid space (7.55). In the median plane the cavity extends dorsally into the white core of the cerebellum, forming a *median dorsal recess* (7.86) cranial to the nodule; and on each side a *lateral dorsal recess* extends still further dorsally (7.143), lying cranial to the inferior medullary velum and caudal to the cerebellar nuclei, from which it is separated by a thin layer of white matter.

The fourth ventricle possesses lateral boundaries, a roof or dorsal wall, and a diamond-shaped ventral floor, the rhomboid fossa.

Lateral boundaries. The caudal part of each lateral boundary is constituted by the gracile and cuneate tubercles, the fasciculus cuneatus, and the inferior cerebellar peduncle, the cranial part, by the superior cerebellar peduncle.

The roof or dorsal wall (7.87). This extends backwards into the median and lateral dorsal recesses. The cranial portion of the roof is simple, and is formed by the superior cerebellar peduncles and the superior medullary velum. The *superior cerebellar peduncles* (p. 917), on emerging from the central white matter of the cerebellum, pass cranially and ventrally, forming at first the

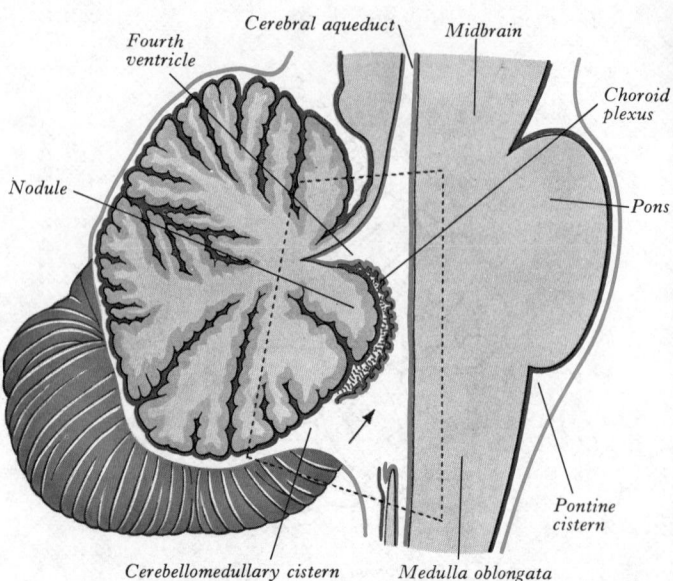

7.86 Sagittal section through the brainstem and the cerebellum close to the median plane. The black arrow is placed in the median aperture of the fourth ventricle. The area enclosed by the interrupted lines is shown enlarged in 7.88. *Blue:* arachnoid mater. *Red:* pia mater. *Green:* ependyma.

lateral boundaries of the upper part of the ventricle; on approaching the inferior colliculi, they converge, and their medial margins overlap the ventricle and form part of its roof. The *superior medullary velum* (p. 914) fills the angular interval between the superior cerebellar peduncles, and is continuous dorsally with the central white core of the cerebellum; it is covered on its dorsal surface by the lingula of the superior vermis (7.89).

The *caudal part* of the roof is more complicated. Over most of its extent it consists of an exceedingly thin sheet, entirely devoid of nervous tissue and formed by the ventricular ependyma and the pia mater of the tela choroidea of the fourth ventricle, which covers it posteriorly (7.89). Caudally, the continuity of the sheet is broken by a gap, termed the *median aperture* (7.87), through which the cavity of the ventricle communicates freely with the subarachnoid space. The *tela choroidea of the fourth ventricle* is a double layer of pia mater which occupies the interval between the cerebellum and the lower part of the roof of the ventricle. Its posterior layer provides a covering of pia mater for the inferior vermis and, after reaching the nodule, is reflected ventrally and caudally in immediate contact with the ependyma. In the tela choroidea are highly vascular fringes forming the choroid plexus of the fourth ventricle. On each side, the layer of the tela choroidea in contact with the ependyma of the caudal part of the roof reaches the inferolateral border of the ventricular floor, which is marked by a narrow, white ridge, termed the *taenia*; the two taeniae are continuous below with a small, curved margin, the *obex*, which covers the inferior angle of the ventricle and is covered by ependyma on both aspects (7.89). Cranially, these taeniae pass laterally and horizontally along the inferior borders of the lateral recesses.

The openings in the roof. In the caudal part of the roof of the fourth ventricle are three openings, one median and two lateral. The *median aperture* is a large opening, situated caudal to the nodule (7.87); it varies considerably in its extent, and its irregular rostral border is drawn dorsally towards the inferior surface of the vermis in a somewhat funnel-shaped manner to face the cerebellomedullary cistern (7.89). The *lateral apertures* are situated at the ends of the lateral recesses and are partly occupied by parts of the choroid plexus, which protrude into the subarachnoid space (7.55). The ependyma and pia mater are continuous at the margins of these openings. Through these openings alone the ventricular cavity communicates with the subarachnoid space. Occasionally one of the lateral recesses may fail to open into the subarachnoid space, but the median aperture is constantly present.

The choroid plexuses. Two highly vascular fringe-like processes of the tela choroidea contain the choroid plexuses of the fourth ventricle; they invaginate the caudal part of the roof of the ventricle and are everywhere covered by ependyma, which is modified to form a true secretory epithelium. Each consists of a vertical and a horizontal region. The former consists of two longitudinal fringes adjacent to the median plane, fusing at the cranial margin of the median aperture and frequently prolonged beyond this margin on to the ventral aspect of the vermis to which they adhere (Hewitt 1960). The horizontal portion projects into the fourth ventricle, passes into the lateral recess and emerges through the lateral aperture still covered by ependyma. The entire structure presents the form of the letter T, the vertical limb of which, however, is double. Wide variation in this form has been recorded by Lang and Schäfer (1977). The arterial supply of these plexuses is from the inferior cerebellar arteries (p. 696). A detailed study of these arterial rami and the venous drainage has been contributed by Maillot *et al.* (1976).

The rhomboid fossa (7.90). The floor of the fourth ventricle is rhomboidal in shape; it is formed by the posterior surface of both the pons and the cranial, open part of the medulla oblongata. It is covered by a layer of grey matter continuous with that surrounding the central canal of the medulla oblongata and spinal cord; superficial to this there is a thin lamina of neuroglia covered with ependyma. The floor consists of three parts, superior, intermediate and inferior. The *superior* part is triangular in shape and limited laterally by the superior cerebellar peduncles; its cranial apex is directly continuous with the wall of the cerebral aqueduct; its base is represented by an imaginary line at the level of the cranial ends of two small depressions, named the *superior foveae*. The *intermediate* part extends from this level to that of the horizontal sections of the taeniae of the ventricle and is prolonged into the lateral recesses. The *inferior* part is triangular, and its caudally directed apex is continuous with the wall of the central canal of the closed part of the medulla oblongata.

The rhomboid fossa is divided into symmetrical halves by a *median sulcus*, which reaches from its upper to its lower apex and is deeper below than above. On each side of this sulcus there is an elevation, the *medial eminence*, bounded laterally by a sulcus, the *sulcus limitans*. In the superior part of the floor the medial eminence has a width equal to that of the corresponding half of the floor, but opposite the superior fovea it forms an elongated swelling, the *facial colliculus*, which overlies the nucleus of the abducent nerve, and is in part produced by the ascending section of the root of the facial nerve. In the inferior part of the floor the medial eminence forms a triangular area, termed the *hypoglossal triangle* or *trigonum hypoglossi*. When examined closely with a lens the hypoglossal triangle is seen to consist of medial and lateral areas separated by a series of oblique furrows; the medial area corresponds with the cranial part of the nucleus of the hypoglossal nerve, the lateral with the *nucleus intercalatus* (see p. 903).

The *sulcus limitans* forms the lateral boundary of the medial eminence. Its superior part corresponds with the lateral limit of the floor and presents a bluish-grey area, named the *locus ceruleus*, which owes its colour to a group of pigmented nerve cells as seen through overlying tissue. The *nucleus ceruleus* corresponds to some extent with the locus of the same name in the floor of the fourth ventricle, but it ascends cranially beyond this as far as the caudal end of the mesencephalic trigeminal nucleus; it may even overlap the latter, appearing in the same sections in the most caudal level of the midbrain. Its cells are of medium size and some contain neuromelanin, especially so in the human midbrain, and it is this which is responsible for the colour of the locus ceruleus (Foley and Baxter 1958). At pontine levels the nucleus spreads into the neighbouring reticular formation and indeed has for long been, and is currently, regarded as a reticular element by many observers (see, for example, Russell 1955; Webster 1978). It has been demonstrated (Dahlström and Fuxe 1964; Fuxe *et al.* 1970) that all the centrally placed neurons of the nucleus are noradrenergic; these neurons are designated A6 (using the classification of the authors—p. 949) and they constitute by far the largest aggregation of noradrenergic cells within the mammalian brains so far examined. Stereotactic lesions of both nuclei coerulei (in rats) demonstrated that they are the main and

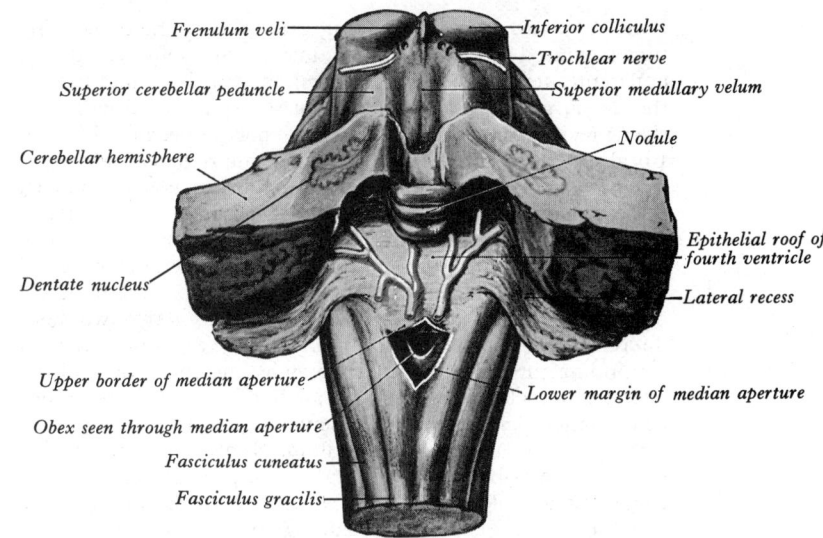

7.87 Dorsal aspect of the roof and the lateral recesses of the fourth ventricle, exposed by removal of parts of the cerebellum.

probably sole source of fibres constituting the *dorsal noradrenergic bundle*, part of the central tegmental fasciculus through which they approach some of their many destinations (Ungerstedt 1971). Ramón-Moliner and Dansereau (1974), however, have demonstrated in the feline nucleus ceruleus high concentrations of acetylcholinesterase, a finding not necessarily inconsistent with the above view, as they note in discussion. The afferent connexions of the nucleus are not particularly well authenticated but probably include projections from ascending spinal fibres, from neighbouring brainstem reticular nuclei, from trigeminal nuclei, and possibly descending systems from forebrain centres. The distribution of efferent projections from the nucleus ceruleus have, in contrast, been subjected to intensive investigation (see, for example, Maeda *et al.* 1973; Lindvall and Björkland 1974; Freedman *et al.* 1975; Kievit and Kuypers 1975; Gatter and Powell 1977; Pickel *et al.* 1977; Bowden *et al.* 1978); these included both histofluorescence and autoradiographic studies. Briefly, the nucleus ceruleus presented descending, local

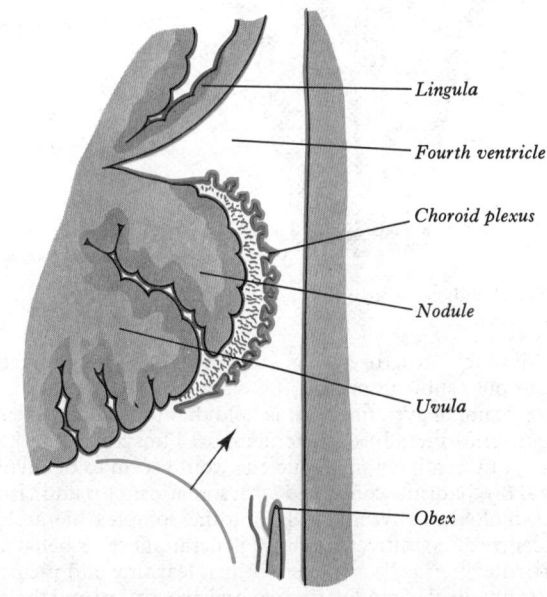

7.88 An enlargement of the part of 7.86 which is enclosed by an interrupted line. *Blue:* arachnoid mater. *Red:* pia mater. *Green:* ependyma. The black arrow traverses the median aperture of the roof of the fourth ventricle.

brainstem, and profuse ascending projections. The descending fibres entered the ventral funiculus of the spinal cord and apparently terminated in the ventral, intermediate, and base of the dorsal grey columns throughout the spinal cord, although their precise laminar destinations have not yet been established. Local fibres projected to many 'specific' and reticular bulbopontine nuclei; others pass through the superior cerebellar peduncle, some ending in the intracerebellar nuclei whilst the majority diverge in the cerebellar cortex either arborizing around Purkinje cell somata or continuing into the molecular layer. As indicated above the ascending fibres constitute the dorsal noradrenergic bundle; in the upper midbrain the bundles of the two sides interchange a proportion of their fibres, the bundles then continuing through the diencephalon as one constituent of the lateral hypothalamic medial forebrain bundle (p. 968), its further continuation being through the septal region into the supracallosal longitudinal stria and cingulum. Numerous complex side branches, described elsewhere, leave these principal channels for a truly remarkable diversity of destinations which can merely be listed here (for details and bibliography consult references quoted). Midbrain destinations include the periaqueductal grey matter, other reticular nuclei, and the colliculi; diencephalic arborizations occur in all the principal groups of thalamic nuclei, the geniculate bodies and the habenular nuclei (a direct supply to hypothalamic nuclei, although claimed on occasion, is less certain). Telencephalic destinations, widely accepted, are the

inferior fovea. Lateral to the foveae there is a rounded elevation named the *vestibular area,* which extends into the lateral recess, where it forms the *auditory tubercle* produced by the underlying dorsal cochlear nucleus and cochlear part of the vestibulocochlear nerve (p. 909). Winding round the inferior cerebellar peduncle, and crossing the vestibular area and the medial eminence to pass deeply at the median sulcus, are a number of white strands termed the *striae medullares* (*see* p. 910). Caudal to the inferior fovea, and between the hypoglossal triangle and lower part of the vestibular area, a triangular dark field, the *vagal triangle or trigonum vagi,* overlies the dorsal nucleus of the vagus nerve (p. 1076). The lower part of the vagal triangle is crossed by a narrow translucent ridge, named the *funiculus separans,* and between this funiculus and the gracile tubercle is a small tongue-shaped area, termed the *area postrema.* On section it is seen that the funiculus separans is covered by a strip of thickened specialized ependyma including tanycytes, and the area postrema has a similar lining over a loose, highly vascular, neuroglial tissue containing nerve cells of moderate size. In these regions the usual blood-brain barrier may be modified. The specialised ependyma is analogous to similar distinctive areas in the third ventricle and cerebral aqueduct (pp. 935, 981). The columnar ependymal cells and tanycytes may be involved in (1) secretory activities into the cerebrospinal fluid; (2) transport of neurochemical substances from underlying neurons, glia, or vasculature to the cerebrospinal fluid; (3) transport of neurochemical substances from the cerebrospinal fluid to

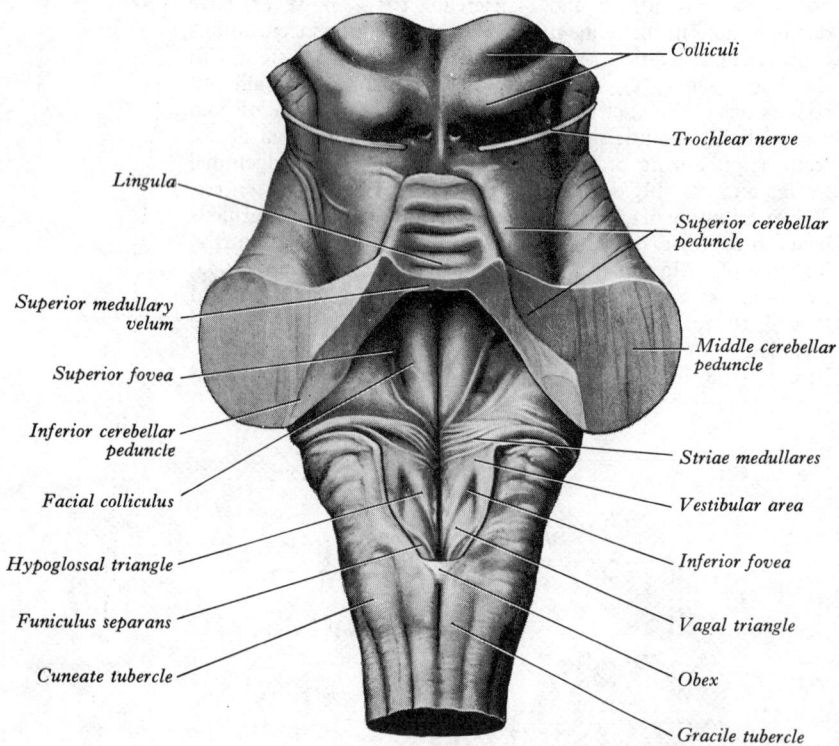

7.89   The rhomboid fossa, or 'floor' of the fourth ventricle.

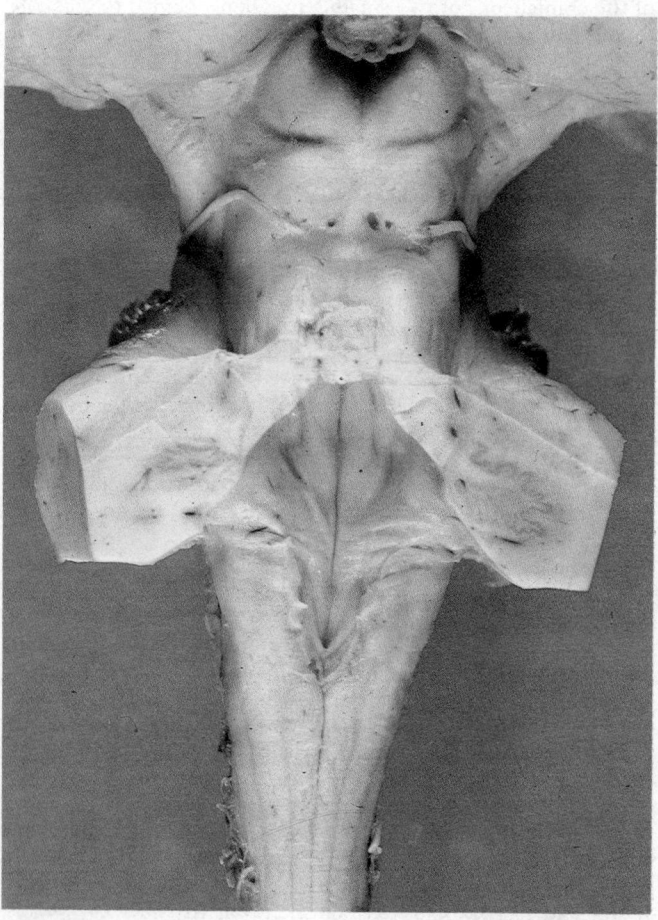

7.90   A dorsal view of the brainstem including the floor of the rhomboid fossa—for identification of structures compare with 7.89. In addition to the structures shown in the latter, note: (1) the crenated outlines of the right and left dentate nuclei in the sectioned surface of the cerebellar white matter opposite the widest part of the rhomboid fossa; (2) the midline pineal gland cranial to the superior colliculi; (3) the rounded pulvinar of the dorsal thalami which encroach on the uppermost part of the photograph; (4) the right and left habenular trigones immediately lateral to the base of the pineal; and (5) the medial geniculate bodies, lateral to the superior colliculi. (Dissection by Dr. E. L. Rees, Department of Anatomy, Guy's Hospital Medical School.)

major 'limbic' structures, e.g. the septal nuclei, amygdala, hippocampus, subiculum, and the cingulate, retrosplenial, and parahippocampal gyri; finally, it is held that the whole neocortex receives a uniform, diffuse, bilateral input. Thus, this remarkable, compact, but relatively small nucleus would seem to be involved in neural mechanisms concerned with somatic motor and visceral control, biological rhythms including the complex hierarchy of sleep 'centres', cognitive mapping, general affective behaviours (in particular 'reward' or 'pleasure' loci, learning and memory), and to be involved in modulating general neocortical and thalamic activity.

At the level of the facial colliculus the sulcus limitans widens into an angular depression, termed the *superior fovea,* and in the inferior part of the floor presents a distinct dimple termed the

underlying neurons, glia, or vasculature; and (4) it has been proposed that some cells of the area postrema may function as chemoreceptors (Brizzee and Neal 1954; Borison and Borison 1973).

The inferior part of the floor of the fourth ventricle towards its apex, presents the appearance of a pen nib, and is hence called the *calamus scriptorius*. (For the phylogenetic and ontogenetic development of the fourth ventricle consult Kier 1977.)

THE MESENCEPHALON OR MIDBRAIN

The mesencephalon or midbrain is derived from the intermediate of the three primary cerebral vesicles. In the course of its development in man and in its phylogenetic history it retains, from some standpoints, a rather simpler form than either the forebrain or the hindbrain. In lower vertebrates the leading feature of the midbrain is the development in its roof plate of higher visual (p. 941), and, later, higher auditory centres. In the mammals these become reflex centres and their original function as 'higher' or sensory centres becomes transferred to the cerebral cortex. As this change, *telencephalization*, increases, the midbrain is traversed by an increasing number of axons subserving cortico-spinal and spino-cortical pathways. Midbrain mechanisms also contribute extensively to the reticular system (pp. 943–953).

It should be noted that the foregoing statements are given merely as an aid to those embarking on an elementary course of neurology. Whilst some midbrain activities may justifiably be termed reflex, for example, the 'light', 'accommodation', 'visuospinal', and 'auditory' reflexes amongst many others, it would be an *over-simplification* to regard as reflex, the complex activities of such subdivisions as the colliculi, substantia nigra and the numerous entities constituting the midbrain reticular formation. In such regions all grades of convergent/divergent interaction, and subtleties of information processing occur, that affect many other regions of the nervous system. Thus the midbrain acts *in concert with*, and not merely as reflex side-channels of, the forebrain and hindbrain. *See*, in particular, p. 951.

EXTERNAL FEATURES

The midbrain lies athwart the hiatus in the tentorium cerebelli and connects pons and cerebellum with the forebrain. It is the shortest segment of the brainstem, being not more than 2 cm in length. On each side it is related to the parahippocampal gyrus, which hides its lateral aspect from view when the inferior surface of the brain is examined. Its long axis inclines ventrally as it ascends.

The midbrain can for description be divided into right and left halves, the *cerebral peduncles*, each of which is further subdivided into a ventral part, the *crus cerebri*, and a dorsal *tegmental part*, by a lamina of pigmented grey matter, the *substantia nigra*. The two crura are separate, whereas the tegmental parts are united. The tegmentum is traversed by the *cerebral aqueduct*, which connects the third and fourth ventricles. The region of the tegmentum dorsal to the cerebral aqueduct is called the *tectum* and comprises the *colliculi* which consist of four rounded elevations, symmetrically arranged in superior and inferior pairs; these include visual and auditory reflex centres respectively (but *vide supra*).

The *crura cerebri* are two white, superficially corrugated structures which emerge from the cerebral hemispheres, one on each side of the median plane. They converge as they descend and meet where they enter the pons; here they form the posterior boundaries of the *interpeduncular fossa* (p. 966). The surface of the posterior part of the interpeduncular fossa is formed by a greyish area, the *posterior perforated substance* (p. 966), through which pass the central branches of the posterior cerebral artery.

The ventral surface of each crus is crossed close to the pons, from medial to lateral, by the superior cerebellar and posterior cerebral arteries; near to the point of entry of the crus into the cerebral hemisphere, the optic tract winds backwards around it. Over the surface of the crus, also close to the pons, a thin white band, the *taenia pontis*, is frequently seen as it enters the cerebellum between the middle and superior peduncles.

The medial surface of each crus bears a longitudinal groove, the *medial sulcus*, from which the roots of the oculomotor nerve emerge (7.92). The lateral surface of each peduncle is in relation with the parahippocampal gyrus of the cerebral hemisphere and is crossed in a ventral direction by the trochlear nerve (7.97). This surface is marked by a longitudinal groove, the *lateral sulcus*; the fibres of the lateral lemniscus come to the surface in this sulcus, and then turn dorsally, some to enter the inferior colliculus, the rest passing into the brachium of the inferior colliculus.

The *colliculi (corpora quadrigemina)* (7.97) are four rounded eminences, situated cranial to the superior medullary velum, and caudal to the pineal gland and posterior commissure, the whole region inclining ventrally as it ascends. They are inferior to the splenium of the corpus callosum, and are partly overlapped on each side by the pulvinar of the thalamus. The colliculi are arranged in pairs (superior and inferior), and are separated from one another by a cruciform sulcus. The vertical part of this sulcus expands superiorly to form a slight depression in which the *pineal gland* (7.97) lies. From the inferior end of the vertical sulcus a white ridge, the *frenulum veli*, is prolonged caudally to the superior medullary velum; at the sides of this ridge the trochlear nerves emerge, pass ventrally on the lateral aspects of the cerebral peduncles and traverse the interpeduncular cistern to reach the posterior end of the cavernous sinus. The *superior colliculi* are larger and darker in colour than the inferior, and constitute centres for visual responses (p. 941). The *inferior colliculi*, though smaller, are somewhat more prominent than the superior and are associated with the auditory pathway (p. 941). The difference in colour is due to the greater accumulation of nerve cells near the surface of the superior colliculus (p. 941).

From the lateral aspect of each colliculus a ridge, termed the *brachium*, ascends in a ventrolateral direction. The *brachium of the superior colliculus* passes inferior to the pulvinar. It partly overlaps the medial geniculate body and is partly continued into the lateral geniculate body (p. 977), and partly into the optic tract. It conducts fibres from the retina and from the optic radiation to the superior colliculus. The *brachium of the inferior colliculus* ascends ventrally from the inferior colliculus; it conveys fibres from the lateral lemniscus and the inferior colliculus to the medial geniculate body.

INTERNAL STRUCTURE OF THE MIDBRAIN

On transverse section, each cerebral peduncle is seen to consist of a dorsal and a ventral part, separated by a deeply pigmented lamina of grey matter, termed the *substantia nigra* (7.91, 92). The dorsal part is named the *tegmentum*; the ventral, the *crus cerebri*. The crura are separated from each other, but the tegmental parts are continuous with one another across the midline, and the term *tegmentum* usually implies the bilateral mass between both nigral masses and the *tectum*. The latter is the part dorsal to the cerebral aqueduct and contains the colliculi.

The crus cerebri is semilunar on transverse section, and consists of corticospinal, corticonuclear, and corticopontine fibres (7.91). The *corticospinal* and *corticonuclear* fibres occupy the middle two-thirds of the crus; they descend through the pons and medulla oblongata, where the corticonuclear fibres end in the nuclei of the various cranial nerves, mainly of the opposite side; corticospinal fibres are continued into the pyramid of the medulla oblongata. The corticopontine fibres arise in the cerebral cortex and terminate in the nuclei pontis, where they are relayed mainly to the opposite cerebellar hemisphere. They are subdivided into two main groups: (a) *frontopontine* and (b) *temporopontine*. (a) The frontopontine fibres arise in the frontal lobe, principally areas 6 and 4 of Brodmann, traverse the anterior limb of the internal

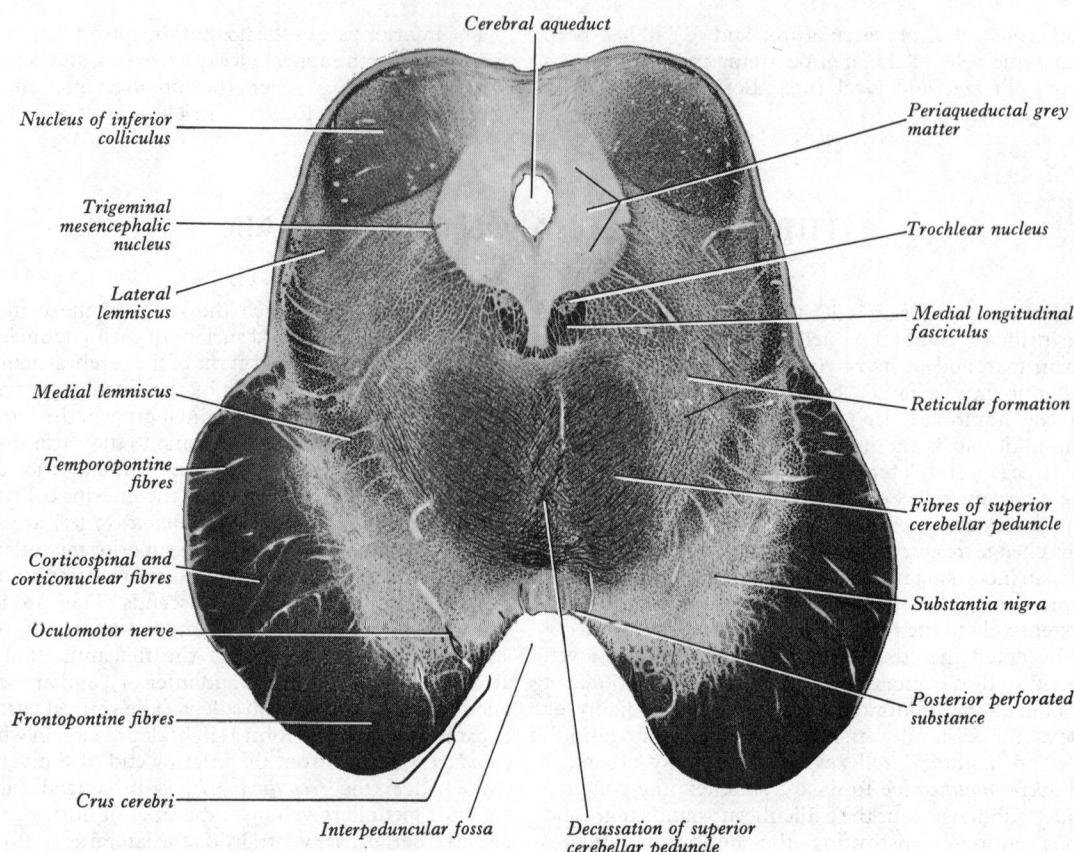

Cerebral aqueduct

Nucleus of inferior colliculus

Trigeminal mesencephalic nucleus

Lateral lemniscus

Medial lemniscus

Temporopontine fibres

Corticospinal and corticonuclear fibres

Oculomotor nerve

Frontopontine fibres

Crus cerebri

Interpeduncular fossa

Decussation of superior cerebellar peduncle

Periaqueductal grey matter

Trochlear nucleus

Medial longitudinal fasciculus

Reticular formation

Fibres of superior cerebellar peduncle

Substantia nigra

Posterior perforated substance

7.91 Transverse section of midbrain through the inferior colliculi. Weigert Pal preparation. Magnification c. × 4·1.

capsule, and occupy the medial sixth of the crus cerebri. (b) The temporopontine fibres arise in the temporal lobe, traverse the posterior limb of the internal capsule and occupy the lateral sixth of the crus cerebri; they end in the nuclei of the pons. *Parieto-* and *occipito-pontine* fibres are also described in the crus cerebri, medial to the temporopontine fibres. The temporopontine projection is largely derived from the posterior (caudal) region of the temporal lobe, and the occipitopontine fibres do not include a projection from the striate area (Verhaart and Mechelse 1954).

A band of fibres, named the *tractus peduncularis transversus*, is sometimes seen emerging from the optic tract on the lateral aspect of the cerebral peduncle; it passes round the ventral surface of the peduncle about midway between the pons and the optic tract, and disappears by entering the interpeduncular fossa behind and lateral to the corpus mamillare, where it terminates in a small *nucleus of the transverse peduncular tract* which lies medial to the substantia nigra. The tract is a constant structure in many lower mammals, but is only identifiable in 30 per cent of human brains. Since it undergoes atrophy after enucleation of the eyeballs, it may be considered as being associated with the visual pathway; the nucleus projects to the oculomotor nuclei in some mammals (Gillilan 1941; but *see* also p. 975).

The substantia nigra is a lamina of grey matter containing numerous, deeply pigmented (neuromelanin-containing), multipolar nerve cells and extending throughout the whole length of the midbrain. Owing to its pigmentation it can readily be recognized with the naked eye in transverse or coronal sections (7.98) through the midbrain. Its pigment is related to the aminergic status of many of the mesencephalic reticular neurons; it increases with age, is greater in amount in primates, maximal in man (Marsden 1961), and is even present in albinos.

The substantia nigra is semilunar on transverse section, its concavity being directed towards the tegmentum; from its convex surface, processes extend between the fibres of the crus cerebri. Thicker medially than laterally, it reaches from the medial to the lateral sulcus, and extends from the upper surface of the pons to the subthalamic region; its medial part is traversed by the fibres of

the oculomotor nerve as they stream forwards to reach the oculomotor sulcus. It is divided into a dorsal *pars compacta* containing medium-sized cells, many of which are pigmented, and a ventral *pars reticularis* intermingled with fibres of the crus cerebri and containing fewer cells, only some of which contain a small amount of pigment. The reticular part extends into the subthalamic region, where it is considered to be continuous with the globus pallidus, which it resembles structurally; both nuclei contain iron in unusual amounts. In vertebrates below mammals a well-developed *pars lateralis* is present; this is recognizable but has been claimed to be rather insignificant in man (Huber and Crosby 1933). It has nevertheless been shown to project to the tectum in primates (Woodburne *et al* 1946); see also below and p. 941.

The substantia nigra is connected with the cerebral cortex, spinal cord, hypothalamus and basal nuclei. Corticonigral fibres arise from the precentral and, probably, the postcentral gyri and terminate on cells in the reticular region, but they are few in number, many being fibres of passage to the red nucleus and reticular formation. Collaterals from fibres in the sensory tracts ascending from the spinal cord are also said to terminate in the substantia. Fibres from the mamillary peduncle and subthalamic nucleus have also been traced on to its cells. The connexions with the basal nuclei are efferent nigrostriatal fibres which pass to the caudate nucleus, putamen and possibly also the globus pallidus (although the latter are strongly debated by many authors); afferent strionigral fibres also exist. Efferent fibres also pass into the tegmentum of the midbrain, probably to terminate on cells of the reticular formation, whence impulses arising in the substantia nigra are relayed to anterior column cells of the spinal cord. The nigrotegmental projection was observed to be a major efferent connexion in cats (Afifi and Kaelber 1965). Lesions in the medial part of the substantia, also in cats (Marcós 1969), suggest that some fibres pass to the ventral thalamus; nigrocortical connexions have been described in the same experimental animal (Negro 1969). A nigrothalamic projection has been confirmed in monkeys by autoradiographic tracing techniques (Carpenter *et al.* 1976);

erminals were identified in the medial part of the nucleus ventralis intermedius, magnocellular part of the nucleus ventralis anterior, and paralaminar part of the nucleus medialis dorsalis (*see* p. 959). Jayaraman *et al.* (1977), using the same techniques in monkeys, have described a projection to the tectum; the fibres are derived largely from the cranial and lateral region of the substantia nigra, and they terminate in intermediate strata of the caudal levels of the ipsilateral inferior colliculus.

It may be noted here that the pars lateralis of the substantia nigra includes the dopaminergic cell group A9, while the pars compacta constitutes group A10 (for further details *see* p. 949).

The tegmentum of the midbrain presents appearances which differ according to the level of the section examined. Caudally, it is directly continuous with the tegmentum of the pons and contains the same fibre tracts.

At the level of the inferior colliculi, the grey matter is restricted to the immediate environs of the cerebral aqueduct, to the scattered collections in the reticular formation (7.91) and to the tectum, which is described separately (p. 940).

The *nucleus of the trochlear (fourth cranial) nerve* lies in the ventral region of the central grey matter, close to the midline. It occupies a position homologous with the abducent and hypoglossal nuclei at lower brainstem levels. Closely related throughout to the medial longitudinal fasciculus, which is on its ventral aspect, the nucleus extends through the caudal half of the midbrain, and it is just caudal to the oculomotor nucleus. In some primates the nuclei are merged and only distinguishable by the arrangement of their cells, which are also smaller in the trochlear nucleus. Its *outgoing fibres* pass laterally and dorsally round the central grey matter. Descending dorsally on the medial side of the mesencephalic nucleus of the trigeminal nerve, they reach the cranial end of the superior medullary velum, wherein they decussate to emerge at the lateral side of the frenulum veli (7.90). A few fibres probably do not decussate. The trochlear nerve fibres emerge dorsally and largely decussate in all vertebrates. It has been suggested that the nerve originally supplied the muscles of the pineal eye, which would account for its dorsal course. Embryological evidence (Pearson 1943) and phylogenetic data support a more dorsal origin of this nucleus.

The *mesencephalic nucleus of the trigeminal nerve* is in the lateral part of the central grey matter. It extends from the cranial end of the main sensory nucleus of the trigeminal nerve in the pons to the level of the superior colliculus in the midbrain. It is accompanied by a tract composed of the peripheral and central branches of the axons of its cells. The large ovoid nerve cells of this nucleus are unipolar and resemble those of sensory ganglia. They occur in many small groups each containing only a few cells; these groups extend as curved laminae on each side, in the lateral margins of the periaqueductal grey matter. The greatest number of cells is apparent at the caudal end of the mesencephalic nucleus. Attempts to categorize the cells of the nucleus have been largely unsuccessful. These cells have very few dendrites and are usually so closely clustered together that somato-somatic contacts have been suggested. (For a discussion of nerve cells of the mammalian nucleus consult Hinrichsen and Larramendi 1969.) For connexions of this nucleus *see* pp. 1060, 1085.

Apart from these nuclei the tegmentum at this level contains large numbers of scattered nerve cells, most of which are included in the reticular formation (p. 945).

The *white matter* at this level of the midbrain contains all the tracts which have already been mentioned in the tegmentum of the pons, and it is characterized by the great decussation of the fibres of the superior cerebellar peduncles.

The *superior cerebellar peduncle*, which has already been described (p. 917), enters the dorsolateral part of the tegmentum and passes ventromedially round the central grey matter to reach the median raphe, where the majority of its fibres form with their fellows the *decussation of the superior cerebellar peduncles*. Having crossed the median plane, the fibres separate into ascending and descending bundles. Some of the *ascending* fibres and branches terminate in or give collaterals to the red nucleus, which they encapsulate and penetrate. Numerous others are continued cranially to end in the nucleus ventralis lateralis of the thalamus. Some of the fibres, however, are uncrossed, and branches are believed to end in the periaqueductal grey matter and reticular formation of the midbrain, in the interstitial nucleus (of Cajal) and the nucleus of the posterior commissure (sometimes, perhaps mistakenly—Ingram and Ranson 1953—called the nucleus of Darkschewitsch). The latter is considered to send efferent fibres to the medial longitudinal fasciculus and posterior commissure, the significance of which is uncertain. The *descending* branches end in the reticular formation of the pons and medulla oblongata, the olivary complex, and possibly the motor nuclei of certain cranial nerves (see also pp. 917, 919).

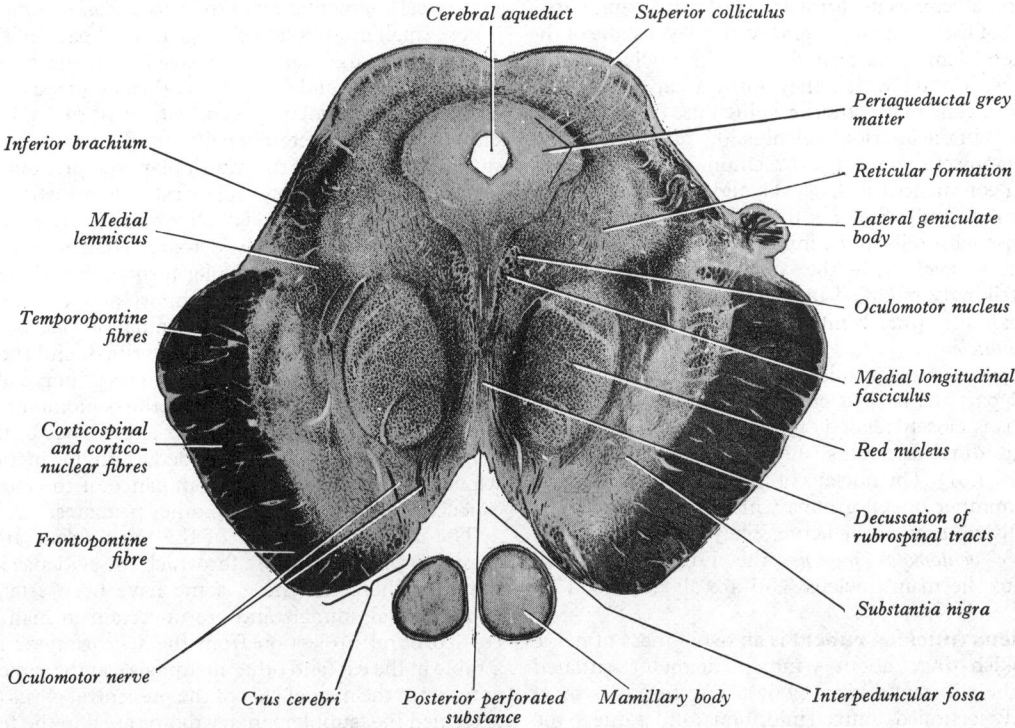

Cerebral aqueduct   Superior colliculus

Inferior brachium

Medial lemniscus

Temporopontine fibres

Corticospinal and cortico-nuclear fibres

Frontopontine fibre

Oculomotor nerve   Crus cerebri   Posterior perforated substance   Mamillary body   Interpeduncular fossa

Periaqueductal grey matter

Reticular formation

Lateral geniculate body

Oculomotor nucleus

Medial longitudinal fasciculus

Red nucleus

Decussation of rubrospinal tracts

Substantia nigra

7.92   Transverse section of the midbrain through the superior colliculi. Weigert Pal preparation. Magnification *c.* × 3.

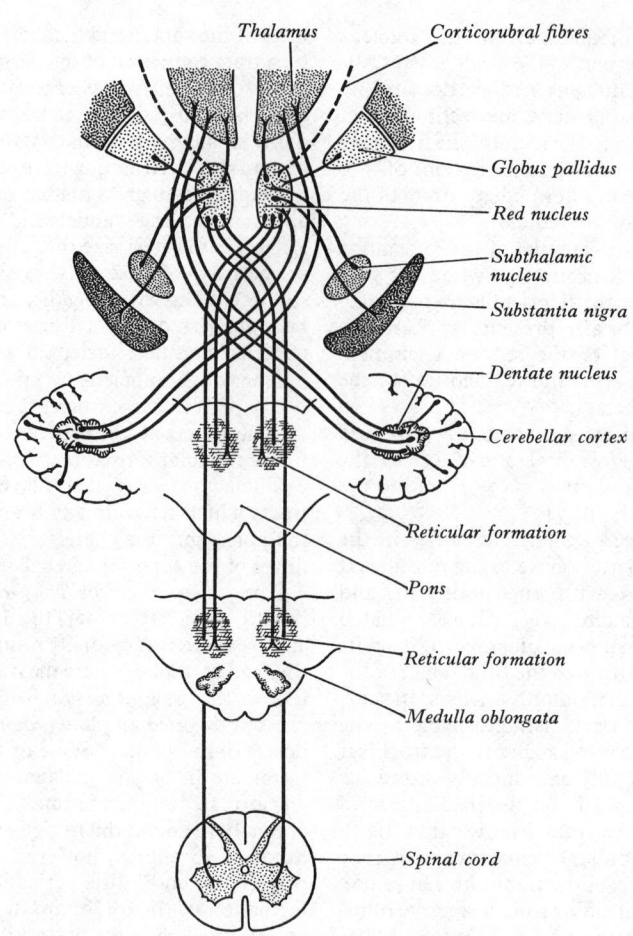

7.93　Simplified diagram of the principal connexions of the red nucleus.

The *medial longitudinal fasciculus* retains its intimate relationship to the somatic efferent column, and therefore lies dorsal to the decussating fibres of the superior cerebellar peduncles.

The *medial*, *trigeminal* and *lateral lemnisci* and *lateral spinothalamic tract* (spinal lemniscus) form a curved band lying dorsal to the lateral part of the substantia nigra. At this level some of the fibres of the lateral lemniscus terminate in the nucleus of the inferior colliculus, around which they form a capsule. These fibres synapse with cells in the inferior colliculus. Accompanied by fibres from cells in the inferior colliculus, the remaining fibres of the lateral lemniscus enter the brachium of the inferior colliculus, which commences at this level, and conducts them to the medial geniculate body. Some of these fibres give collaterals which also synapse with cells in the inferior colliculus.

Rostrally, on a level with the superior colliculus, the tegmentum is strikingly modified by the appearance of a large nucleus, which extends upwards into the subthalamic region and is termed the *red nucleus*.

The *central grey matter* surrounds the aqueduct and contains in its ventromedial part the *nucleus of the oculomotor nerve*. This elongated nucleus is closely related on its ventrolateral aspect to the medial longitudinal fasciculus; inferiorly it extends to the trochlear nucleus (7.91). The nucleus of the oculomotor nerve is divisible into a number of cell groups which can be correlated with the motor distribution of the nerve. They include a visceral efferent *accessory oculomotor nucleus*, the Edinger-Westphal nucleus, dorsal to the main nucleus (and are all described on p. 999).

The red nucleus (nucleus ruber) is an ovoid mass of nerve cells with a pinkish tinge, about 5 mm in diameter, situated dorsomedial to the substantia nigra (7.92). The colour is most apparent in freshly sectioned, unfixed midbrains, and is due to the presence of an iron-containing pigment which occurs in many, but not all, of the cells in the nucleus. These are mostly multipolar

and of varying sizes, rather large and small categories being predominant. The proportions and distribution of these types differ in various mammalian groups; in primates, including man, the *magnocellular* element is relatively decreased, with a reciprocal augmentation of the *parvocellular* complement of cells. These small multipolar cells occur in all parts of the nucleus, but in man the large cells are restricted to the caudal part of the nucleus. Condé and Condé (1973) categorized the rubral cells of the cat as multipolar, pyramidal, fusiform and spherical; they emphasized the heterogeneity of the cytoarchitecture of the nucleus, finding that the multipolar type predominated at caudal levels, the pyramidal and spherical cells at rostral levels. All types were, however, described at all levels. Rostrally, the boundary of the red nucleus is poorly demarcated from other cell groups, where it fades into the reticular formation and the caudal pole of the interstitial nucleus (Davenport and Ranson 1930). Some division of the red nucleus into caudal *compact* and rostral *diffuse* architectonic regions has been described, and the whole mass is traversed and encapsulated by fascicles of nerve fibres, including many of the fibres emerging from the oculomotor nucleus (7.92). These bundles impart a reticular appearance to the nucleus. Its magnocellular element is regarded as phylogenetically the older, which accords with the predominance of the parvocellular part especially in man, but also in other primates.

The *afferent connexions* of the red nucleus are complex and numerous, and only those for which the evidence is strong will be described here. Of these, some have been established only in experimental animals and are uncertain in man. An uncrossed corticorubral projection from the sensorimotor area is demonstrable in the cat (and other mammals); in the same animal an area anterior to the medial part of the precentral gyrus (p. 957), which is termed the 'supplementary motor area' in the feline cerebrum, projects bilaterally upon the red nuclei. A somatotopic arrangement of these corticorubral fibres and their terminations in the nucleus has been demonstrated; this conforms with the localization of the cells of origin of the rubrospinal tract (*vide infra*). This projection is derived from neurons in the primary somatomotor and somatosensory areas—the sensorimotor cortex. Brown (1974) has described the fibres as terminating largely in the periphery of the red nucleus and as being axodendritic at their terminals. There are also bilateral connexions, probably in both directions, with the superior colliculi. The red nucleus also receives nerve fibres from the contralateral nucleus interpositus in the cat; this corresponds to the nuclei globosus and emboliformis of the human cerebellum. Afferents are also derived from the contralateral dentate nucleus. All these cerebellar connexions, and perhaps others (p. 917), pass through the superior cerebellar peduncle. They are said to display some degree of somatotopic organization. Connexions from other sources have been reported—from the globus pallidus, subthalamic and hypothalamic nuclei, substantia nigra, and spinal cord—but these are only putative in regard to the human red nucleus. (Some of the connexions of the red nucleus are outlined in **7.93**.)

The main *efferent outflow* of axons is into the **rubrospinal tract** (p. 878), which is derived from both the large and small cells in the nucleus in other mammals, and probably also in mankind. As mentioned elsewhere (p. 878), this tract is currently dismissed as unimportant in man by many authorities, clinical and academic, who state that it is derived only from the magnocellular neurons in the nucleus, and that even their axons do not travel far down the spinal cord. The evidence for this negative view is exiguous; all the studies carried out on the tract in other mammals, chiefly the cat and monkey, indicate that it reaches lumbosacral levels. The tract and its nuclear origins are somatotopically localized, the axons ending at cervical levels being from cells segregated in the dorsomedial region of the nucleus, those for lumbosacral segments in its ventrolateral part, and those with a destination in the thoracic cord are in an intermediate position. This mode of organization has been confirmed by recording antidromic impulses arriving in the nucleus as a result of stimulation of different levels of the tract. Some of the efferent axons terminate in the brainstem, as a *rubrobulbar tract*, which projects upon the motor nuclei of the trigeminal and facial nerves. Fibres are also claimed to pass to the inferior olive, and the nuclei of the oculomotor, trochlear and abducent nerves; many of the 'rubrobulbar' fibres mentioned above are thought to terminate at many points throughout the brainstem reticular formation, thus constituting a *rubroreticular system*; in the cat *rubrocerebellar* connexions to the dentate nucleus have been described (Courville and Brodal 1966). A projection to the ventrolateral thalamic nuclei from the caudal third of the red nucleus has been described.

As it descends, the rubrospinal tract decussates at once in the ventral decussation of the tegmentum, ventral to the decussation of the tectospinal tracts, and then passes caudally, ventral to the decussation of the superior cerebellar peduncles, and continues through the reticular formation of the pons and medulla to reach the spinal cord. Since its spinal terminals are distributed to almost the same laminae of grey matter in the cord as those of the corticospinal tract, the *cortico-rubro-spinal* projection can be regarded as a kind of indirect corticospinal pathway (p. 878).

Closely associated with the rubrospinal fibres are *tegmento-spinal* fibres which arise in the tegmental reticular formation lateral and caudal to the red nucleus. These are probably to be grouped with the other (medullary and pontine) reticulospinal tracts (p. 879).

The tectospinal and the **tectobulbar tracts** also take origin at this level. Their fibres arise in the grey matter of the superior colliculi and sweep ventrally round the central grey matter to decussate with one another in the median raphe ventral to the oculomotor nucleus and medial longitudinal fasciculus, forming the *dorsal part of the tegmental decussations*. Emerging from this the tectospinal tract descends on the ventral aspect of the medial longitudinal fasciculus until the decussation of the medial lemniscus in the medulla oblongata. Thereafter it diverges ventrolaterally and in the spinal cord it lies in the anterior column adjacent to the ventral end of the anterior median fissure. The tectobulbar tract is mainly composed of crossed fibres and

descends close to the tectospinal tract. It ends in the pontine nuclei and motor nuclei of the cranial nerves, particularly those concerned with the innervation of the orbital muscles. It serves as a pathway for reflex movements of the eyes in response to visual stimuli. (See also references to the lateral tectotegmentospinal tract, p. 875).

The medial longitudinal fasciculus (7.94) is a heavily myelinated composite tract lying on the ventrolateral aspect of the oculomotor nucleus. At this level its fibres are more spread out than they are at lower levels in the brainstem, but the intimate relationship to the efferent nuclei is retained. The fasciculus extends cranially to the *interstitial nucleus* (of Cajal)—a collection of cells situated in the lateral wall of the third ventricle immediately above the cranial end of the cerebral aqueduct—which contributes fibres to it. (Closely related to the cranial ends and decussating fibres of the fasciculi are the *interstitial* and *dorsal* groups of *nuclei of the posterior commissure* which probably both receive fibres from, and contribute fibres to these complex bundles. Also closely related are the *nuclei of Darkschewitsch*, but their status in relation to the fasciculi remains uncertain.) As has already been seen, the medial longitudinal fasciculus retains its position relative to the central grey matter throughout the midbrain, pons and cranial part of the medulla oblongata. It is displaced ventrally by the successive decussations of the medial

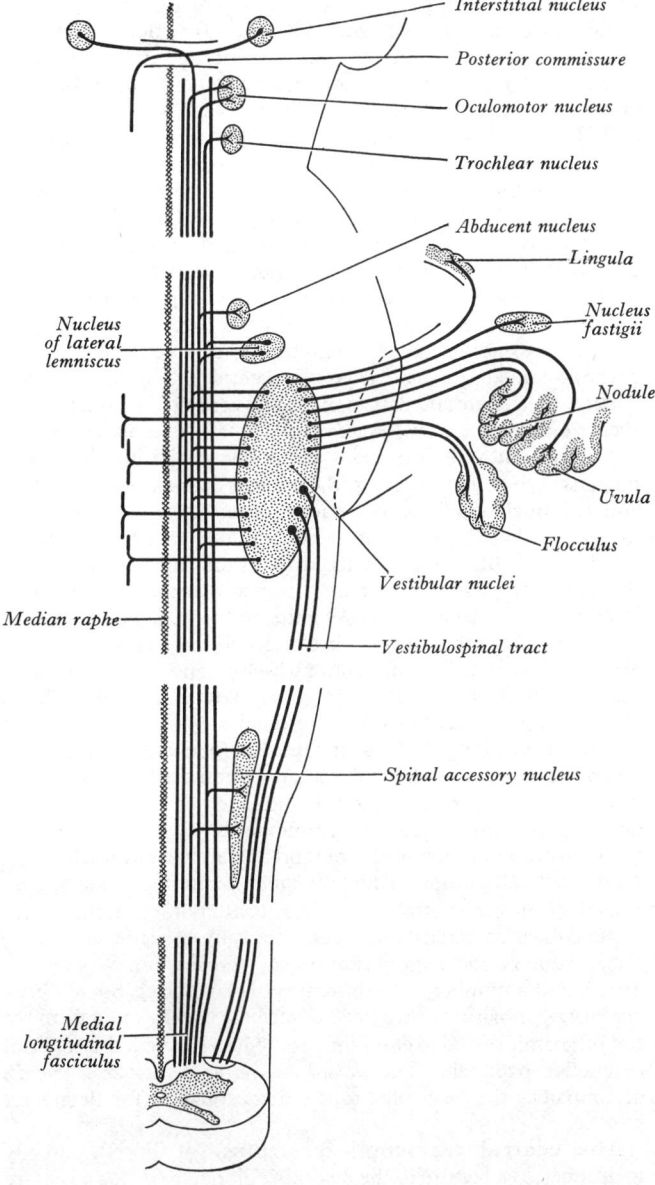

Interstitial nucleus

Posterior commissure

Oculomotor nucleus

Trochlear nucleus

Abducent nucleus

Lingula

Nucleus fastigii

Nodule

Uvula

Flocculus

Vestibular nuclei

Nucleus of lateral lemniscus

Median raphe

Vestibulospinal tract

Spinal accessory nucleus

Medial longitudinal fasciculus

7.94 Simplified diagram of *some* of the components of the *medial longitudinal fasciculus* and of the distribution of its fibres to cranial nerve nuclei.

lemnisci and the lateral corticospinal tracts, and becomes continuous with the anterior intersegmental fasciculus of the spinal cord.

The intimate relationship which the fasciculus bears successively to the nuclei of the oculomotor, trochlear and abducent nerves, to the emerging fibres of the facial, the fibres from the dorsal cochlear nucleus in the floor of the fourth ventricle and the nucleus of the hypoglossal nerve, renders it a very convenient pathway for the passage of fibres from one nucleus to another in the brainstem. Its continuity with the anterior intersegmental tract of the spinal cord provides a route for connexions between these nuclei and the cervical anterior grey column cells, principally those innervating neck musculature. The harmonious cooperation obviously existing between the facial and hypoglossal nerves in movements of the lips and tongue in speech is frequently attributed to connexions between their nuclei conveyed by the medial longitudinal fasciculus. It is, however, doubtful whether it is the medial longitudinal fasciculus which provides the pathway for these connexions. It has long been established that the most substantial contributions to the fasciculus are made by the *vestibular nuclei*, and to a much smaller extent by fibres derived from the nucleus of the lateral lemniscus, and that its chief function is to ensure the coordinate movements of the eyes and head in response to stimulation of the vestibulocochlear nerve. (*See* Maciewicz *et al.* 1977, and Yamamoto *et al.* 1978; also p. 1073.) Lesions of the fasciculus are associated with partial ophthalmoplegia. Fibres from the vestibular nuclei, both of the same and of the opposite side, join the fasciculus, where they ascend, descend or divide into ascending and descending branches. These vestibular fibres send collaterals to, or they may end in, the nuclei of the third, fourth, fifth and sixth cranial nerves and the spinal nucleus of the eleventh nerve (7.94). All four vestibular nuclei contribute ascending fibres to the fasciculus, those from the superior nucleus remaining uncrossed (McMasters *et al.* 1966), the others crossing in part. Some of these fibres, as stated, pass to the nuclei of the nerves of the eye muscles, some pass further to reach the interstitial nucleus and the nuclei of the posterior commissure, through which a proportion decussate before ending in these two nuclei (Matano 1970; see also pp. 963, 1073). Descending axons, from the medial vestibular nuclei, perhaps joined by some from the lateral and inferior nuclei, partially decussate and pass to the spinal cord, in the fasciculus, as the *medial vestibulospinal tract* (Brodal *et al.* 1962, and *see* pp. 875, 893). In addition, fibres join the fasciculus from the dorsal nucleus of the trapezoid body, the nucleus of the lateral lemniscus and the nucleus of the posterior commissure. It is probable, therefore, that the cochlear as well as the vestibular nerve is capable of influencing the movements of the eyes and head through the fasciculus. Some fibres conveying vestibular impulses may ascend in the fasciculus to the thalamus.

A number of other longitudinal nerve fibre systems have been described in the midbrain, some of which extend through much of the entire brainstem. It is not possible to detail all of these here, but two in particular must be mentioned.

The dorsal longitudinal fasciculus (of Schütz) runs in the central grey matter of the midbrain, pons and part of the medulla, in a position ventrolateral to the aqueduct. It is a pathway for descending and ascending connexions, largely uncrossed, between numerous hypothalamic and dorsal thalamic nuclei and a number of cell groups in the brainstem, including the accessory nuclei (Edinger-Westphal) of the oculomotor complex, the super colliculus, nucleus ambiguus, the salivatory nuclei, and the facial, solitary and hypoglossal nuclei (Crosby and Woodburne 1951), and a number of brainstem reticular nuclei. Some fibres, *cholinergic* in nature, have been described as *descending* from the hypothalamic nuclei to enter the cerebellum through the superior cerebellar peduncle. The *dorsal ascending serotonergic bundle* accompanies the fasciculus to the diencephalon (for details *see* p. 950).

The central tegmental fasciculus (of Forel), already mentioned as a feature of the medulla oblongata (p. 904), is more ventral in the midbrain, being at first lateral to the medial longitudinal fasciculus and dorsolateral to the red nucleus and the decussation of the superior cerebellar peduncles. As it descends

through the pons and into the medulla it swerves further laterally to end in the inferior olivary complex of the same side (Bebin 1956). Although few of its fibres are in fact derived from the thalamus, it was sometimes called the thalamo-olivary tract. Most of its descending fibres were said to proceed from nerve cells in the central grey zone surrounding the cerebral aqueduct at the level of the superior colliculus, but fibres from the basal ganglia and red nucleus also travel in it. Through the red nucleus, ansa lenticularis (p. 965) and the lenticular nucleus, the central tegmental fasciculus mediates a pathway from the motor cortex via these structures to brainstem reticular nuclei and the inferior olivary nuclear complex.

While these earlier described constituents of the bundle are still, in the main, accepted, much additional information has accumulated in recent years. Many of its fibres have proved to be the ascending, descending (or both) longitudinal axons of neurons intrinsic to the brainstem reticular formation, their collaterals and final terminals innervating other 'reticular' or neighbouring 'specific' nuclei (p. 943). Additionally, the newer staining techniques have demonstrated that within the confines of the central tegmental fasciculus are the *dorsal* and *ventral ascending noradrenergic bundles*, the *ventral ascending serotonergic bundle*, and some, at least, of the *dorsal* and *ventral ascending cholinergic bundles* (for further details *see* pp. 948, 949, 950).

The brachium of the inferior colliculus forms a rounded strand on the lateral aspect of the upper part of the midbrain (7.107). Its fibres are derived from the inferior colliculus and from the lateral lemniscus and ascend to reach the medial geniculate body. In their course they separate the dorsolateral fibres of the medial lemniscus from the surface. Quantitative examination of the brachium by Ferraro and Minckler (1977) in 28 autopsy specimens (from individuals aged from birth to 97 years) revealed a nerve fibre density of about 68,000/mm², with little variation between left and right sides and no significant diminution until the ninth decade. The total number of fibres in an average brachium was estimated as 354,314 (right) and 346,810 (left) near the inferior colliculus, rising to 557,122 (right) and 561,359 (left) near the medial geniculate body. These total counts must be treated with some reserve, since the brachium was dissected away from the brainstem. However, the actual counting was done with great care. Ramon-Moliner and Dansereau (1974) have examined the 'peribrachial' region of the pontomesencephalic junction in detail, and have proposed an arbitrary topological division of this part of the tegmental brainstem.

THE TECTUM OF THE MIDBRAIN

Although the tectal part of the midbrain exhibits dorsal swellings from the earliest vertebrate stages, its full differentiation into superior and inferior pairs of colliculi (corpora quadrigemina) appears only in the mammals (7.90). Even in the first mammals, the prototherian monotremes, only a single pair of elevations, the *optic lobes* (corpora bigemina), can be distinguished. The optic lobes are highly developed in fishes, being actually larger than the olfactory lobes in Osteichthyes. It is interesting to note that at this and later levels of vertebrate evolution the olfactory lobes are in volume the major development of the forebrain. With the emergence of the cerebral hemisphere, i.e. a forebrain development with more than olfactory connexions, the optic lobes become relatively smaller, but remain substantial structures in amphibians, reptiles and birds. Their connexions and functions show a progressive cranial shift from reptiles through to mammals, a process called *telencephalization*. In this process the dominant position of the optic lobes, which in earlier vertebrates are probably concerned with *all* sensory modalities other than olfactory, shows a gradual diminution. This may be coupled with a decrease in the complexity of cytoarchitectonic organization in the optic lobes which is considered to reach its acme in primates, whose highly developed cerebrum has taken over much of the activity of the primitive tectum. The differentiation of the optic lobes into superior and inferior colliculi appears in the eutherian mammals, the inferior being perhaps a derivative of the optic lobes. Though it has been usual to link the superior colliculi with visual and the inferior pair with auditory behaviour, to accept this as an *exclusive* interpretation is likely to mislead; the superior colliculi in particular appear to be concerned in other ascending afferent pathways from cutaneous receptors and perhaps in auditory connexions. The inferior colliculi are certainly interconnected with the superior; it will nevertheless be more convenient to consider them separately.

Each inferior colliculus consists internally of a central main nucleus, ovoid in form, derived embryologically from the periaqueductal grey matter, with which it preserves some degree of continuity. This nucleus is surrounded by a laminar zone of nerve fibres, many of which are from the lateral lemniscus (p. 909) and terminate in the nucleus. Small and medium-sized nerve cells, many being stellate or multipolar, predominate in the nucleus, and some of these are scattered in the nerve fibres surrounding it, particularly on its lateral and medial aspects. These are sometimes regarded as constituting a separate nuclear entity, and they almost certainly have a different origin, being continuous through scattered cells with the superficial lamina of the superior colliculus. The nuclear and neuronal organization of the inferior colliculus have been described in considerable detail in the cat (Morest 1966; Geniec and Morest 1971; Rockell and Jones 1973; Meininger and Baudrimont 1977) and to a lesser extent in man. With minor differences, there is a general agreement that the main or *central nucleus* (described above) is divisible into *dorsomedial* and *ventrolateral* parts, chiefly on the basis of neuronal characteristics. The 'laminar zone', also described above, forms a lateral peripheral feature in the cat. Disagreements persist in regard to the distribution of different sizes of neurons; Meininger and Baudrimont (1977) in the most recent analysis, carried out using careful stereological techniques but on a very small series of cats, described the dorsomedial nucleus as consisting of small neurons, those of the ventrolateral nucleus being larger. The two inferior colliculi are connected by a substantial *commissure*, which contains not only axons derived from their own cells but also lateral lemniscal fibres passing across the midline to end in the contralateral colliculus. The chief afferent pathway to the colliculi is the lateral lemniscus, whose origins have been described elsewhere (p. 909); similarly, its major contingent of efferent fibres passes through the inferior brachium to the medial geniculate body, a largely ipsilateral projection, though a crossed connexion, through the intercollicular commissure, may exist (Ades 1944). The lemniscal fibres relay only in the main collicular nucleus (Goldberg and Moore 1967). Some of them pass through without relay to reach the medial geniculate body, and similarly some of the colliculogeniculate fibres do not actually relay in the body, but pass on with those which do into the auditory radiation (p. 1074) to reach the auditory cortex (area 42). As in other sensory pathways there is a *descending* projection from the same cortical area to the inferior colliculus via the medial geniculate body, through which some may pass without relay. This descending pathway may produce effects at various levels, from the medial geniculate body onwards; it is likely that it links with efferent cochlear fibres, through the superior olivary and cochlear nuclei, as previously noted (p. 910). A tonotopical projection of the lateral lemnisci on the inferior colliculi has been described in cats; it is likely that some similar orderly arrangement extends throughout these auditory connexions, ascending and descending (*see* Auditory Cortex, p. 963 and Moore and Goldberg 1966).

Projection pathways from the inferior colliculi to the brainstem and spinal cord appear to pass first to the superior colliculi, through which connexions with the tectospinal and tectotegmental tracts are effected. These collicular projections are relatively small, and it is possible that the customary view of the inferior colliculi as reflex centres for auditory responses may require amplification. There is, however, considerable evidence that they are concerned in the ability to localize the source of sounds (Masterton *et al.* 1967).

The superior colliculus, unlike the inferior, which reaches full development only in the mammals, is generally regarded as much simplified in the higher vertebrates, and particularly so in the primates. In man, however, it still exhibits much of the complex laminar organization of earlier forms. At least six laminae have been described in the human colliculus and

considerably more intricate arrangements in some other mammals (Cajal 1909–11; Angaut and Repérant 1976). In this pattern of organization the superior colliculus closely resembles the cerebellar and cerebral cortex which, indeed, it surpasses in complexity in lower vertebrates. This resemblance has prompted the view that the optic lobes, from which the superior colliculi are derived, were the first suprasegmental development of the vertebrate brain, which mediates as a summating or integrating mechanism on a much wider basis than the term optic lobe indicates. Such considerations are worthy of attention, if only as a corrective to the widely current teaching that the superior colliculus is exclusively concerned with visual activities.

The laminae of the human colliculus have been variously named, and official terms for them are not yet established. From the exterior inwards strata zonale, cinereum, opticum and lemnisci are recognized, the stratum lemnisci itself being divisible into strata griseum medium, album medium, griseum profundum and album profundum. These seven layers have also been termed zonal, superficial grey, optic, intermediate grey, deep grey, deep white and periventricular strata (Crosby *et al.* 1962). The two systems of description do not completely accord, but as a generalization the layers may be considered to be alternately composed of nerve cells or their fibres and dendrites, though some admixture occurs. For example, the most external, the *zonal layer*, consists chiefly of myelinated and non-myelinated nerve fibres derived from the occipital cortex (areas 17, 18 and 19—*see* p. 960), and arriving as the *external corticotectal tract*; but among these fibres are a few small cells, typically horizontal in arrangement. The next layer, the *superficial grey layer* (stratum cinereum) consists of numerous small multipolar interneurons, with which the cortical fibres in part form synapses, the whole mass forming a cap-shaped lamina, thicker at its centre, over the deeper layers. The *optic layer* consists partly of the fibres of the optic tract which reach the colliculus, a much diminished afferent element in higher primates; as they terminate, these fibres invade neighbouring layers with numerous collateral branches. The optic layer also contains nerve cells, some being large multipolar in type; efferent fibres to the retina are said to start in this layer in man (Wolter and Liss 1956) and other mammals. The remaining four layers are sometimes considered together as the stratum lemnisci, but will be briefly outlined here as separate strata. The next two, the *intermediate grey* and *white layers*, are collectively the main reception zone of the colliculus. As the names indicate these layers consist of nerve cells of various sizes intermixed with numerous axons and dendrites. Its main afferent source is the medial corticotectal pathway from the occipital cortex (area 18) and possibly preoccipital cortex (area 7); these areas are concerned with following movements of the eyes (p. 960). These layers also receive afferents from the spinal cord through spinotectal and spinothalamic routes, and probably from the inferior colliculus. Finally, the *deep grey* and *deep white* layers appear most deeply situated, immediately adjacent to the periaqueductal grey substance, again containing a mixture of cell types, with dendrites extending peripherally as far as the optic layer and axons forming many of the efferent connexions of the colliculus. (For recent evaluation of the literature and details of neurons of the neonatal superior colliculus consult Labrida and Laemle 1977.)

The superior colliculus thus receives *afferents* from a wide area (Wolter and Liss 1956; Meikle and Sprague 1964), including fibres from the retina, spinal cord, inferior colliculus and occipital and temporal cortex, the first three conveying visual, tactile, and probably thermal, pain and auditory impulses, the cortical projection acting as a 'command' and possibly modulating pathway. Its *efferents* pass out to the retina and to a wide array of brainstem and spinal nerve cell groups. Thus, *tecto-oculomotor* fibres project to nuclei of the ocular muscles. In the cat (Edwards and Henkel 1978) neurons in the rostral third of the superior colliculus project to the periaqueductal grey substance dorsal to the rostral third of the oculomotor complex, some also making contacts with dendrites of oculomotor neurons. Neurons of this central grey region also project to both abducent nuclei. *Tectospinal* fibres (p. 875) travel caudally to the cervical segments of the spinal cord, *tectotegmental* projections reach various reticular masses in the tegmentum, passing also to the substantia nigra, red nucleus and probably as far as the spinal cord. *Tectopontine* fibres, which probably descend in the tectospinal tract, terminate in the dorsolateral part of the pontine nuclei, whence a relay carries this path into the cerebellum (p. 918). All these descending fibres decussate, mostly in the *dorsal tegmental decussation*, but also to a lesser extent through a commissure which unites the two superior colliculi. For a review of these connexions *see* Altman and Carpenter (1961). The tectoreticular projection has been studied in cats by Kawamura *et al.* (1974). Stereotactic lesions in the superior colliculus produced terminal degeneration in ipsilateral mesencephalic and contralateral pontomedullary reticular nuclei, namely the nucleus reticularis gigantocellularis, the nucleus reticularis pontis caudalis, the nucleus reticularis pontis oralis, and other reticular groups. The distribution of these tectoreticular projections correspond closely to those from cortical, fastigial and vestibular neurons, and thus to the main sources of reticulospinal fibres. A tecto-olivary projection, from the deeper laminae of the superior colliculus to the rostral third of the medial accessory olivary nucleus has been identified in primates (Frankfurter *et al.* 1976). The connexion is crossed and provides a link with the posterior vermis in the execution of eye movements.

The relative paucity of the retinal projection to the superior colliculus in primates throws doubt upon the validity of the commonly accepted view of it as primarily a centre for visual reflexes, though it certainly has this function in many other mammals. But while the projection of the retinal quadrants on the superior colliculi in cats and rodents has been established with a satisfactory measure of agreement, both by degenerative and stimulation techniques, attempts to extend such experiments to monkeys have been much less successful, probably because of the great reduction in the retinocollicular projection in primates. Clinical evidence merely suggests that a similar pattern of organization may exist in man, this pattern being an association of the craniomedial half of each colliculus with the inferior retinal quadrants and the caudolateral with the superior quadrants. Even the connexions to the eye muscle motor nuclei, which are held to accord with the collicular somatotopic pattern, have been put in doubt by experiments on cats, though it is suggested that these may be merely interrupted by a relay in the pretectal area. For the moment these matters must remain *sub judice*.

Much attention has been focused upon behavioural observations in experimental animals subjected to various types of damage to subcortical parts of the visual pathway in recent times (Ingle and Schneider 1969). In addition the visual projection on the optic tectum has been mapped and compared by electrophysiological techniques in a wide series of lower vertebrates from fish to mammals. In general a similar pattern of retinal representation exists in the widely different groups examined, the chief difference in mammals being the degree of ipsilateral projection. This accords with the findings with respect to the movements elicited by stimulation of different parts of the tectum. For example, central collicular stimulation produces a contralateral movement of the head, natural in character, in the cat (Hess *et al.* 1946). A considerable array of similar responses, identical with those occurring in the intact animal, have been observed in a number of experimental species, involving eyes, ears, head, trunk and limbs, suggesting that the superior colliculus is concerned in complex integrations between vision and widespread bodily activity (Schaefer and Schneider 1968). In view of the bilateral representation of the retinae in the colliculi the results of various brain splitting experiments are of special interest. Division of the tegmentum, for example, leads in cats to profound changes in visual behaviour, not so much in reflex responses as in the ability of the animal to interpret its visual environment and to adapt to it (Voneida 1965). Responses to threatening stimuli are permanently abolished, and the ability to locate edges and follow them is also lost. Similar results have been observed in the monkey. These results have something in common with the split brain syndrome in human patients (p. 967).

A descending projection from the *auditory cortex* has been described by several experimentors in various mammals,

including the cat (Diamond *et al.* 1969) and monkey (Whitlock and Nauta 1956). Paula-Barbosa and Sousa-Pinto (1973) have confirmed this finding and explored the projection in greater detail in the cat. In this animal a bilateral projection from each secondary auditory area (AII) reaches the superficial laminae of each superior colliculus. No such projection was found to occur from the primary auditory area (AI). Although the projection is derived from parts of the auditory cortex which do not have, as far as is known, any visual connexions, it is suggested that this corticocollicular pathway may be involved, together with known connexions between the superior and inferior colliculi, in complex integrations between visual and auditory behaviour.

The pretectal nucleus is a somewhat indistinctly demarcated mass of nerve cells in the pretectal area, at the junction of the mesencephalon and diencephalon. It extends from a position dorsolateral to the posterior commissure towards the superior colliculus, with which it is in part continuous. It receives fibres through the superior brachium from the occipital and preoccipital cortex and from the lateral root of the optic tract (*see* Kuhlenbeck and Miller 1949; and p. 975). Its efferent fibres pass to both accessory oculomotor nuclei (of Edinger-Westphal)—*see* Ranson and Magoun (1933). Those which decussate pass ventral to the aqueduct or through the posterior commissure. By the autonomic outflow from these nuclei, with a relay in the ciliary ganglia (p. 999), the sphincter muscles of the iris in *both* eyes are made to contract in response to impulses from either (7.108). This light reflex appears to be the only activity mediated by the pretectal nucleus, but some of its efferent axons are said to enter the tectobulbar and tectospinal tracts. The significance of the descending cortical projection is not yet established.

EXTRAGENICULATE VISUAL PATHWAYS

As a corrective and extension to the above account it is apposite to refer briefly to the volume of literature which has accumulated in regard to the visual pathways in mammals (e.g. Jones 1974) and particularly birds (reviewed *in extenso* by Webster 1974). It is almost a dictum of comparative neuroanatomy that the vertebrate visual system was primarily mesencephalic in its connexions. The concept of 'telencephalization' has also long been established as a part of a greater process of 'encephalization', the evolutionary forward or cranial migration of 'centres' intermediating sensorimotor activities from spinal to *encephalic* levels. The intrusion of visual projections into the *telencephalon* via the thalamus in mammals with a volumetric reduction in *mesencephalic* visual arrangements, has been and still is frequently cited as a major example of *telencephalization*. Moreover, the change is often regarded as in some respects more advanced in mammals, especially primates, than in birds. Compared with the mammalian condition the avian colliculi, or optic lobes, are indeed relatively larger and more complex in their laminar pattern, and, furthermore, the avian corpus striatum is more highly developed than its superincumbent cortex, proportions which appear to be reversed in the mammalian cerebral hemisphere. The generalizations thus engendered have tended towards assigning the avian and mammalian visual systems to opposite balances in a supposed *duality of pathways*—a mesencephalic level vaguely regarded as 'reflex' and a telencephalic level loosely labelled as 'perceptual'. This is oversimplification. Much relatively recent work on avian visual neuroanatomy and physiology has shown that both visual pathways, whether passing through the midbrain tectum or direct to the dorsolateral thalamus, involve thalamic and cortical projection, as well as striatal connexions. In mammals both these and other pathways exist, in particular a pretectal route (in addition to the tectal superior collicular path) via the pulvinar of the thalamus to the occipital cortex. Physiological and behavioural studies have also demonstrated that it is misleading to consider the two, 'major' visual pathways, either in birds or mammals, as being in one case perceptual and the other reflexive. In both vertebrate classes there is evidence that the mesencephalic level is involved in processes of perception, and is not merely at a level of intrinsic and extrinsic ocular musculature reflexes. For example, in the avian brain neurons of the tecto-striate projection exhibit responses not only to movement but also to colour, though not to shape or orientation. In contrast, the more direct route through the thalamus to the cortical region known as the *'wulst'* in birds (the equivalent of the mammalian geniculostriate pathway) has so far not been shown, by unit recording, to evince any special sensitivities, except perhaps to movement. In mammals, including mankind, perhaps even more than in birds, the complexity of the inflow paths of visual information, and the multiplicity of their lateral and descending interconnexions, provide a 'system' of intricate projection and 'feed-back' which at present defies analysis. Despite a great accumulation of data regarding, in particular, the response peculiarities of individual neurons at every level, it is becoming more difficult, rather than easier, to ascribe different aspects of visual information to identifiable neuronal aggregations, and perhaps it is inherently misleading to seek to do so. Even in man, clinical evidence in cases of injury to the visual cortex suggests that the customary ascription of 'conscious' vision to cortical levels may be an obfuscation of the intimate interaction between various parts of the occipital cortex with the tectum and pretectum of the midbrain. Although it is not possible at present to effect a functional synthesis from the mass of structural and physiological findings available, it seems clear that visual function involves very widespread networks of neurons at all levels of the brain.

The cerebral aqueduct is a narrow canal, about 15 mm long, connecting the third with the fourth ventricle. (Since it is in the midbrain, and certainly does not conduct fluid *to* the cerebrum, it is misnamed.) Its form, as seen in transverse sections, varies at different levels, being T-shaped below, triangular above, and oval in the middle, at which level it is slightly dilated (Flyger and Hjelmquist 1957). It is lined with ependyma which is surrounded by a layer of grey matter named the *central grey*; the latter is continuous below with the grey matter in the floor of the fourth ventricle, and above with that of the third ventricle. Remains of the sulcus limitans, apparent on each lateral wall in embryonic development, may persist. Dorsally the aqueduct is partly separated from the grey matter of the colliculi by the fibres of the stratum lemnisci; ventral to it are the medial longitudinal fasciculi, and the reticular formation of the midbrain. Scattered throughout the central grey matter are numerous nerve cells of various sizes, interlaced by a network of fine fibres. Besides these scattered cells it contains three groups, which constitute the nucleus of the mesencephalic tract of the trigeminal nerve and the nuclei of the oculomotor and trochlear nerves.

THE RETICULAR FORMATION

It has long been recognized that scattered amongst the more conspicuous fibre bundles and nuclei of the brainstem are extensive fields of intermingled grey and white matter collectively termed the **reticular formation**. Throughout the present section of this volume there are numerous references, some brief and some in greater detail, to the supposed distribution, connexions and possible functions of various parts of this 'system'. When appropriate, reference should be made to these accounts, and to the detailed reviews which have appeared. *See*, for example Moruzzi and Magoun (1949); Meessen and Olszewski (1949); Olszewski (1954); Olszewski and Baxter (1954); Brodal (1957); Jasper *et al.* (1958); Taber (1961); Valverde (1961 *a* and *b*, 1962); Ramón-Moliner and Nauta (1966); Horridge (1968); Brazier (1969); Scheibel and Scheibel (1970); Webster (1978).

GENERAL CONSIDERATIONS

A strict scientific definition of the criteria to be used when designating a particular region of the nervous system as 'reticular' or not has proved impossible to achieve. Simple inspection of sections of the vertebrate nervous system prepared by the usual neuroanatomical methods, however, allows some measure of distinction. Many areas consist of either predominantly grey or white matter (p. 810) and their contained neurons or fibre bundles have definite and recognizable quantitative cytological variations which are polarized with respect to the three principal planes of the body. Thus, many regions of white matter contain fibre bundles with a characteristic fibre diameter spectrum, and a definable direction, either longitudinal, transverse, oblique or curved, which carries them between a series of more or less well-defined origins and destinations. Similarly, many areas of grey matter contain populations of neurons with their somata, axon collaterals, dendritic trees and synapses organized into fairly precise three-dimensional arrays; further, depending upon the criteria used, they appear to possess fairly distinct 'boundaries'. These constitute the main nuclei, cortical formations and fibre tracts of the nervous system. Obvious examples are the well-defined cell columns or laminae in the dorsal and ventral appendages of the spinal cord (p. 869) and their homologues in the brainstem, many other brainstem, diencephalic and telencephalic nuclei, the cerebral and cerebellar cortices, the tectum, and the named tracts of white matter throughout the nervous system, the majority of which are described individually in the present section. Between these geometrically organized *non-reticular*

nuclei and tracts are found areas where the grey and white matter is admixed, the fibre bundles interlace in many directions, and the scattered neurons have diffuse, ill-defined patterns of connectivity (*vide infra*); these constitute the *reticular regions* of the nervous system. However, although many authorities agree concerning the geometric organization and non-reticular nature of the majority of the main nuclei and tracts, the status of a number of regions remains unclarified, and the following points should be noted.

The structural extremes of a mathematically precise, repeating geometric organization on the one hand, and a truly random nerve network on the other, are abstractions which do not exist in nervous systems. Even the highly regular cerebellar cortex shows considerable quantitative variation in different regions when examined critically (p. 925), and the most 'reticular' parts of the nervous system show some regionally distinctive structural features. Thus, a graded range of levels of structural organization exists and, inevitably, those with a more diffuse structure prove the more difficult to investigate. For this reason, coupled with the fact that detailed knowledge of the quantitative aspects of neuronal morphology and connexions is still in its infancy, many authorities have expressed divergent views about which regions of the nervous system should be embraced by the term *the reticular formation*. Most are agreed about the inclusion of certain deeply placed parts of the medulla, pons and midbrain (*vide infra*), and some would restrict the use of the term to these regions. The status of other cell groups in these parts of the nervous system such as the olivary complex, the parvocellular part of the red nucleus, the pars reticularis of the substantia nigra, the nucleus of the posterior commissure, the interstitial nucleus, the intracerebellar nuclei and others, is uncertain, being included by some and excluded by others. Further disagreement reigns concerning the inclusion of the central regions of grey matter of the spinal cord which have been termed a 'reticular core' by some investigators (p. 894), and whether the non-specific nuclei of the thalamus (p. 960) and certain hypothalamic nuclei should be included. These terminological debates merely serve to emphasize the undesirable nature of rigidly categorizing some centres as reticular or non-reticular, and the need for further researches, in the light of which the necessity for such crude parcellations of the nervous system will gradually recede.

No attempt will be made in the present account to draw precise boundaries between such areas, but it should be appreciated that *all* levels of the neuraxis, including the spinal cord, brainstem and diencephalon, have regions with a relatively high level of ordered structure, which are *interlocked* both structurally and functionally with neighbouring reticular regions where the organization is much more diffuse.

The more obviously reticular regions of the nervous system are often regarded as phylogenetically ancient, representing a supposed 'random nerve network' of early ancestors, upon which background, during subsequent evolution, the more circumscribed, highly organized, discriminative parts of the nervous system have appeared. But, as pointed out elsewhere (p. 808), even the most primitive known nervous systems possess both diffuse and more highly organized regions, which cooperate in their responses to different environmental stimuli. It is undoubtedly preferable to regard *both* these elements as evolving together, each providing *indispensable* and *interdependent* contributions to the total response patterns of the organism.

THE RETICULAR PARTS OF THE NERVOUS SYSTEM

In general terms, the reticular parts of the nervous system consist of: (1) deeply placed, rather poorly defined groups of neurons and fibres with diffuse patterns of connexions; (2) the conduction paths through these regions are difficult and often impossible to define anatomically, but physiological evidence indicates that they are complex and often *polysynaptic*; (3) both *ascending* and *descending* components can be recognized; (4) stimulation of a locus on one side often elicits responses both on the ipsilateral and the contralateral side, i.e. both the ascending and descending systems contain *crossed* and *uncrossed* elements; (5) they mediate both *somatic* and *visceral* functions.

While the earlier investigators discerned little structural variation in different regions, and simply dubbed all regions, other than the long-recognized 'specific' named nuclei and tracts as reticular, more recently, a series of critical studies have emphasized that zonal variations in structure (and presumably 'functional' associations) are present in these reticular areas. These include differences in cell size, number, packing density, geometry of neurites, cytochemistry including putative neurotransmitter production, and in some measure, differences in their connexions. On such bases (and therefore largely dependent upon the technique employed) a series of *reticular nuclei* have been named by some authors in a variety of animals. However, the 'borders' of such nuclei are often defined with difficulty, their dendritic fields interweave over wide areas, and interspecific variation exists, as does the map yielded by different techniques and the employment of different criteria. Thus, the results of electrophysiological or pharmacological exploration, and focal stimulation or ablation of different regions must be subject to the most severe critical interpretive scrutiny; even greater caution is necessary before ascribing functional roles to particular reticular zones or nuclei. The complex geometry of reticular neurons and their neurites has, in the main, been pursued by means of Golgi metallic impregnation techniques, and the latter, in combination with cytoarchitectonic examination of the form and distribution of cell somata, have led to the basic terminologies for reticular zones and nuclei proposed by different groups of workers. The smaller subdivisions, details, and variations cannot be pursued here and for these the references quoted should be consulted; however, some general features, a simplified but useful plan, and brief allusions to overall connectivity and histochemical analyses are given below.

Studies with the Golgi technique (*see* especially Ramón-Moliner and Nauta 1966; Horridge 1968; Brazier 1969; Scheibel and Scheibel 1967, 1970) have shown that in the brainstem few of the reticular neurons are of the classical Golgi type II with short axons which branch locally (*vide infra*). Their dendritic trees are in general spread at right angles to the long axis of the brainstem forming transversely orientated flattened sheets. In the majority of instances the dendrites are long, with a relatively simple branching pattern that has been likened to 'the spokes of a rimless wheel' and classified as the *isodendritic configuration* by Ramón-Moliner and Nauta (1966): they were described as forming an *isodendritic core* for the brainstem, the territory of which encompasses many of the less well-defined, 'diffusely' connected reticular nuclei. Their radiating dendrites may cover up to 50 per

cent of the transverse sectional profile of that half of the brainstem in which the neuron lies, and the dendritic trees are intersected sequentially by, and synapse with, the complex aggregation of ascending and descending fibre bundles that constitute the *central tegmental fasciculus of Forel*. The latter includes components from many sources, one prominent one being the axons of large numbers of reticular neurons themselves: thus, the main stem of a reticular axon usually pursues a longitudinal course, either ascending, descending, or bifurcating to pass in both directions, and in so doing traversing considerable lengths, perhaps the whole, of the length of the brainstem and even beyond these confines. As an extreme example, a reticular neuron of the magnocellular medullary nucleus (*vide infra*) may generate an axon which bifurcates, its ascending component passing through the upper medulla, pons, midbrain tegmentum, thence proceeding to subthalamic, hypothalamic or dorsal thalamic centres in the diencephalon; its descending component traverses the reticular core of the lower medulla and may then continue into the intermediate grisea (laminae V and VI) of the cervical spinal cord. Many reticular neurons may, of course, have unidirectional and considerably shorter axons than the extreme example just described. During their longitudinal course such axons synapse with the radiating dendrites of innumerable other reticular neurons *en route* and also give off a series of collaterals that approach, enter and synapse with cells in neighbouring 'specific' brainstem nuclei or cortical formations such as that of the cerebellum. It is clear, therefore, that the multiple sources of afferent fibres that converge on individual reticular neurons and the myriads of synapses and profusion of destinations of their outflowing axons provide the structural basis for the polymodal responses frequently elicited during physiological investigation, and also the reason for the generation of such terms as 'diffuse' and 'non-specific'. Equally, such arrangements pose perplexing questions for the investigative neurologist and currently it is certainly unwise, and intellectually misleading, to attempt to attach unitary or even complex functional/behavioural labels to individual neurons or small aggregations of neurons; obviously there are myriad opportunities for convergence and divergence of information paths in such a system as the reticular core.

A contrasting type of dendritic pattern is termed the *idiodendritic configuration* in which the profusely branched dendrites are short, highly sinuous or curved, and so arranged that they pursue a re-entrant course at the periphery of a nuclear group, providing a rather precise boundary between the nucleus and surrounding neural tissue. Such neurons characterize many cranial nerve nuclei, pontine nuclei (in the basilar pons), the olivary complexes, nuclei of the 'specific' sensory pathways, the nucleus funiculi lateralis (the so-called lateral 'reticular' nucleus) of the medulla, and a complex of midbrain tegmental nuclei (of Gudden). Neurons with an intermediate degree of complexity of dendritic pattern are found within and around such nuclei, and with varying densities throughout much of the remainder of the reticular formation—these have been termed *allodendritic*: further, in different zones the proportion of neurons with somata of different sizes varies greatly, some regions consisting exclusively of small to intermediate sized multipolar cells (the *parvocellular* regions), while in certain restricted localities these are intermixed with extremely large multipolar neurons—the '*gigantocellular*' or '*magnocellular*' nuclei.

In *general terms*, therefore, the reticular formation has been described as consisting of a continuous isodendritic core which traverses the whole of the medullary, pontine and midbrain brainstem, is continuous caudally with the reticular intermediate grey laminae of the spinal cord, and cranially merges into the subthalamus and parts of the hypothalamus and dorsal thalamus. Either bordering, partially encompassed by, or completely embedded within the core are numerous more easily defined, circumscribed nuclei, only a small proportion of which are considered to merit the name reticular (for discussion and details *see* Leontovich and Zhukova 1963; Ramón-Moliner and Nauta 1966). On such a view the reticular formation may be regarded as consisting of a 'diffuse network' roughly divisible into *three longitudinal columns*, the first consisting of neurons occupying the median plane and a narrow immediately adjacent paramedian

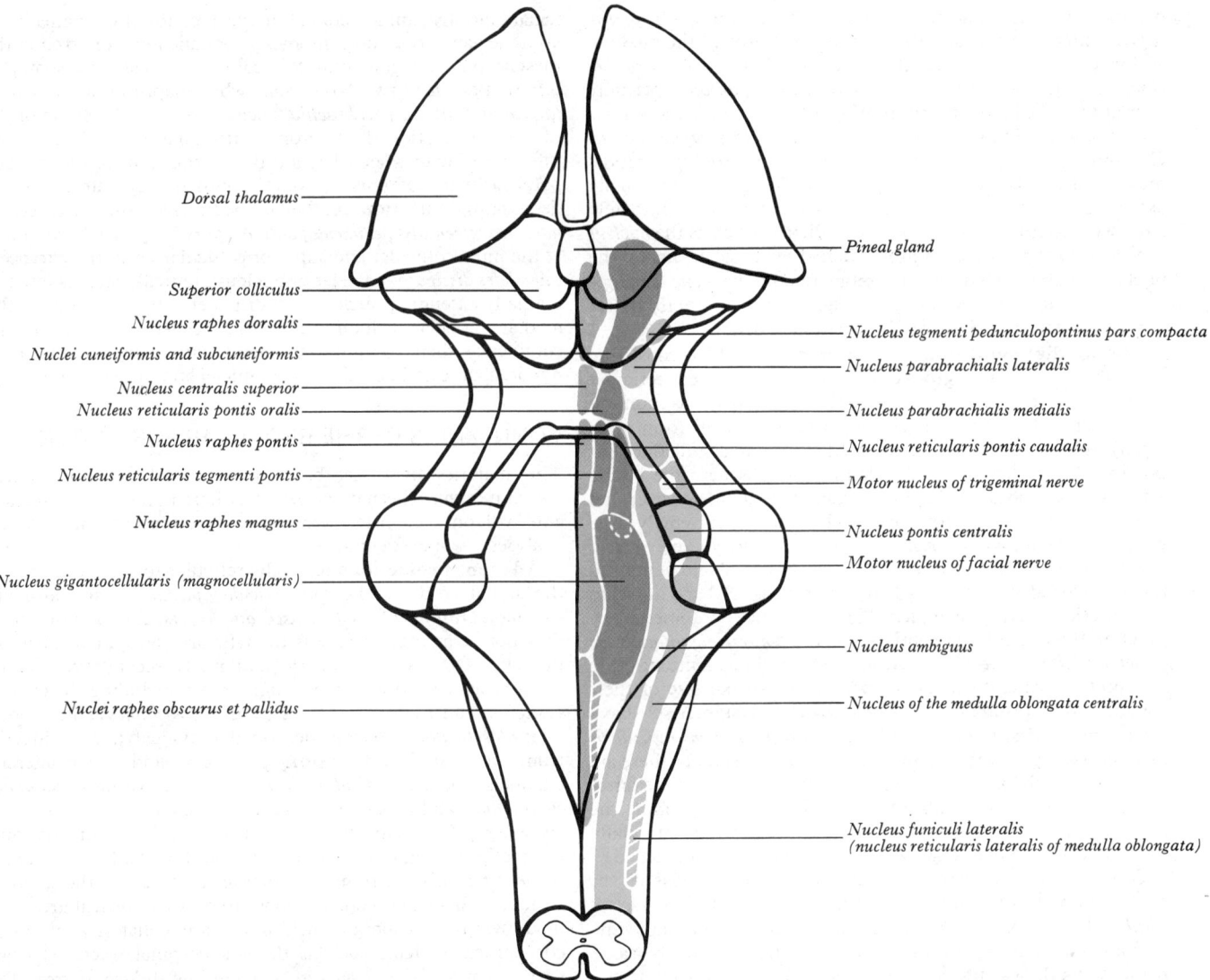

Dorsal thalamus

Superior colliculus

Nucleus raphes dorsalis

Nuclei cuneiformis and subcuneiformis

Nucleus centralis superior

Nucleus reticularis pontis oralis

Nucleus raphes pontis

Nucleus reticularis tegmenti pontis

Nucleus raphes magnus

Nucleus gigantocellularis (magnocellularis)

Nuclei raphes obscurus et pallidus

Pineal gland

Nucleus tegmenti pedunculopontinus pars compacta

Nucleus parabrachialis lateralis

Nucleus parabrachialis medialis

Nucleus reticularis pontis caudalis

Motor nucleus of trigeminal nerve

Nucleus pontis centralis

Motor nucleus of facial nerve

Nucleus ambiguus

Nucleus of the medulla oblongata centralis

Nucleus funiculi lateralis
(nucleus reticularis lateralis of medulla oblongata)

7.95 An outline of the human brainstem (*black*) extending from the caudal end of the medulla to the dorsal thalami: note the margins of the rhomboid fossa, the lateral angles of which indicate the pontomedullary junction; note also the profiles of the transected surfaces of the cerebellar peduncles, the colliculi and pineal gland. The principal nuclear derivatives of the brainstem reticular formation are indicated in approximate outline, those from the median and paramedian nuclear column are in *magenta*; medial column derivatives in *purple*; lateral column derivative in *blue*. In reality, of course, considerable overlap of the nuclear profiles would be present when the third dimension is considered. A number of 'non-reticular' nuclei are also included.

zone, the second or medial column containing the majority of the large reticular neurons, and the third or lateral column containing predominantly small to intermediate neurons. Regional localization is conferred on the system solely by the (overlapping) zones of termination of afferents (*vide infra*) and sources of efferents for their multiple destinations. A further corollary of this view is that unitary functional localization of individual neurons, small neuronal groups, named reticular nuclei, or groups of nuclei is absent, the 'system' largely operating as a whole, enhancing or depressing the 'general background activity' of all the major elements of the central nervous system. (For a stimulating discussion of these and alternative views of the reticular elements of the brain consult Webster 1978.)

In contrast to the foregoing, the parcellation of the reticular formation into named nuclei on cytoarchitectonic grounds was pursued vigorously by a number of observers; perhaps the most devoted exponent of this approach being Olszewski who described over 40 such nuclei in the mammalian brainstem (*see* Olszewski and Baxter 1954; also consult Brodal 1957). Whilst such subdivisions are undoubtedly useful for the experimental neurologist, only a limited number have specific and well-authenticated patterns of connectivity and for the reasons implied above their individual functional significance is obscure.

It would be quite inappropriate to attempt any detailed listing and description, be it concerned with cytoarchitectonics and neurite geometry, overall three-dimensional topography, or what little is known of individual connectivity, of this plethora of named nuclei, in a volume such as this. What follows is the briefest summary of the names of the principal columns and nuclear groups (based on traditional neuroanatomical techniques) accompanied by a diagram of their approximate territories (7.95) as viewed from the dorsal aspect of the brainstem. The reader will also appreciate that their positional relationships can only be studied in a closely spaced, and therefore numerically large, specialized, series of transverse sections: they have therefore not been included *in extenso* in the relatively elementary account of brainstem anatomy attempted in the preceding pages—for such details consult, for example, Olszewski and Baxter (1954); Brodal (1957); Jasper *et al.* (1978); Ungerstedt (1971); Nieuwenhuys *et al.* (1978).

The median column of reticular nuclei extends throughout the medulla, pons and midbrain, and their cell somata are, in the main, aggregated in bilateral, narrow, vertical sheets that blend in the median plane and occupy the immediately adjacent paramedian zone: collectively they are called the *nuclei of the raphe*. Extending through the cranial two-thirds of the medulla and crossing the pontomedullary junction is the *nucleus raphes obscurus* (and closely associated *nucleus raphes pallidus*): partially overlapping the latter and continuing into the pons is the *nucleus raphes magnus*, followed cranially by the *nucleus raphes pontis*. Continuing from upper pontine levels towards the caudal mesencephalon is the *nucleus raphes centralis superior*, and finally expanding at the level of the inferior colliculus and ultimately narrowing to extinction at superior collicular levels is the *nucleus raphes dorsalis (rostralis)*. A proportion of neurons making up the nuclei of the raphe have been demonstrated to be *monoaminergic* (more specifically *serotonergic*) and some of their better authenticated connexions will be mentioned below.

The medial column of reticular nuclei whilst consisting throughout of a background of intermediate-sized reticular neurons, in some regions possesses the largest neurons found in such nuclei; the great majority are isodendritic in form (*vide supra*). In the caudal medulla the medial column is indistinct but possibly represented by a thin lamina of cells near the lateral aspect of the rapheal nuclei; in the cranial medulla, however, the column expands into a prominent nuclear mass of extremely large neurons, the *medullary gigantocellular (magnocellular) nucleus* situated lateral to the rapheal nuclei, ventrolateral and ventral to the hypoglossal and dorsal vagal nuclei respectively, and dorsal to the inferior olivary complex. Crossing the pontomedullary junction the cell column continues as the *pontine gigantocellular (magnocellular) nucleus* situated in the medial tegmentum; quite abruptly the constituent neurons diminish in the size of their somata and, progressing cranially, the systematists have distinguished first the practically co-extensive *nucleus reticularis tegmenti pontis* and the *nucleus reticularis pontis caudalis*; these are succeeded by the *nucleus reticularis pontis oralis*. From the latter the medial column expands into the midbrain tegmentum as the *nucleus cuneiformis* and *nucleus subcuneiformis* which gradually become less distinct as cranial midbrain levels are approached.

Other nuclear masses which some authorities classify as associated with the foregoing column derivatives are the *nucleus ceruleus* (p. 933) and *subceruleus* which cross the pontomesencephalic junction beyond the confines of the locus of the same name (p. 933); near the same junction and closely related to the mesencephalic dorsal and medial longitudinal fasciculi are the *dorsal* and *deep tegmental nuclei* (of Gudden) together with numerous aggregates of neurons in the *mesencephalic periaqueductal grey matter*.

The lateral column of reticular nuclei is almost exclusively formed of well-dispersed *small* neurons and hence many authorities simply term it the *parvocellular reticular column* or *zone*, and its derivatives the *parvocellular 'nuclei'*; in the main, its neurons are of the allodendritic or idiodendritic form (*vide supra*). The column extends throughout the length of the brainstem and forms a rather poorly differentiated 'background' with few systematically named, discernible 'reticular' nuclei; partially embedded or totally enclosed within this background, however, are a series of well-defined 'specific' nuclei which many authorities prefer to exclude from the reticular formation. (In some cases, however, the distinction, as intimated above, appears to be based on quite arbitrary criteria.)

Extending through the caudal two thirds of the lateral part of the pontine tegmentum and upper medulla, the lateral column skirts the lateral aspect of the gigantocellular nucleus and lies medial to the sensory trigeminal nuclei; continuing caudally the column expands to form most of the reticular formation lateral to the rapheal nuclei, and then continues, to blend with the intermediate grey laminae of the cervical spinal cord. In the medullary part it is sometimes considered to constitute a *central medullary reticular nucleus* and a *ventral medullary reticular nucleus*: these are confluent and enclose the distinctive nucleus ambiguus (pp. 905, 1074), whilst bordering them laterally is the *nucleus of the lateral funiculus*—often, despite its cytoarchitecture the latter is named the *lateral medullary reticular nucleus*. Similarly, the *nucleus of the anterior funiculus* is also often called the *medullary nucleus reticularis paramedianus*. The *pontine parvocellular column* or *nucleus* embraces the motor nuclei of the facial and trigeminal cranial nerves; continuing cranially to approach and cross the pontomesencephalic junction to enter the mesencephalic tegmentum the column is held, by some, to differentiate into two large, somewhat inappropriately named *lateral* and *medial parabrachial nuclei*. The lateral column of the reticular formation is not prominent cranial to the level of the inferior colliculi where, beyond the rostral limit of the nucleus parabrachialis lateralis is found, immediately lateral to the decussating superior cerebellar peduncles, the large-celled *nucleus tegmentalis pedunculopontinus, pars compacta*; beyond this at the inter-collicular and superior collicular level, the parvocellular *pars dissipata* of the latter nucleus gradually merges into the cranially extending nuclei cuneiformis et subcuneiformis of the medial reticular column. The small-celled product of this confluence then continues to blend with many subthalamic, hypothalamic and dorsal thalamic nuclei and fibre systems.

CONNEXIONS OF THE RETICULAR FORMATION

The majority of these, based on classical neuroanatomical techniques, have been described elsewhere in the present section and will only be listed here. (For more recent and detailed analyses, *see* subsequent pages.)

Afferent projections reach the reticular formation from: (1) the spinal cord via the *spinoreticular pathways*, and also via *collateral branches* of the *long ascending tracts*, although the latter may not be as profuse as was thought previously; (2) collateral branches from some primary, and many secondary, afferent neurons associated with the *cranial nerves*, including the central *vestibular* and *acoustic* pathways; (3) the *cerebellum* via the various *cerebelloreticular pathways* described previously; (4) indirectly from the *visual* and *acoustic* pathways, and other afferent channels, via *tectoreticular fibres*; (5) from various *thalamic, subthalamic* and *hypothalamic* nuclei, which are further detailed elsewhere; (6) from the *corpus striatum*, both directly and indirectly through its complex of outflow paths; (7) direct *corticoreticular fibres* from the *sensorimotor regions* of the cerebral cortex, with smaller contributions from other cortical areas; an unknown proportion of such fibres are collaterals of other corticofugal systems such as the corticospinal tract; (8) from various parts of the *limbic system*, including the *septal areas*, the *amygdaloid nuclei* and the *hippocampus* through a variety of descending pathways.

Thus, many different afferent pathways converge upon the reticular formation, and although there is some evidence for a degree of zonal localization in the levels of termination of diverse afferents, there is also abundant evidence for much overlap and convergence of different channels. Unit recording with intracellular microelectrodes has provided much information that supports this view. Single units have been studied in experimental animals during multimodal forms of stimulation of cutaneous receptors and proprioceptors in many parts of the body, and during stimulation of a wide variety of peripheral nerves, the special sense organs, and diverse parts of the central nervous system including the spinal cord, the cerebral and cerebellar cortices, the hippocampus, corpus striatum, etc. For example, units are found which respond solely to nociceptive stimuli applied to different body regions; others respond to a wide range of cutaneous stimuli, again from extensive bilateral receptive fields; yet others respond to stimulation of many different regions of the cerebral cortex, and so forth. Further, individual units can be identified which respond to the stimulation of a number of different types of afferent channel, e.g. visual, acoustic, cortical, cerebellar, etc. These different afferent channels also interact in various ways, some with mutual enhancement, others with mutual inhibition. Although the different levels of the reticular formation are not all equivalent in terms of their input and output, there is still a high degree of convergence of afferent channels, and a profusion of interconnexion between its neurons. The question of how such a bewilderingly complex and fluctuating array of information channels is used to direct cooperative responses, remains one of the central and most intractable problems in neurology (Klimer *et al.* 1968; Webster 1978).

The efferent connexions of the reticular formation are: (1) to the autonomic and locomotor control centres, and interneuronal pools of the *spinal cord* via the *reticulospinal tracts*; (2) by short descending pathways to similar centres in the *brainstem*; (3) to the *cerebellum*; (4) to the *red nucleus*, *substantia nigra* and *tectum* of the midbrain; (5) to numerous nuclei in the *subthalamus*, *thalamus* and *hypothalamus*; (6) indirectly, through radiations of the latter diencephalic nuclei, to the corpus striatum and to the cerebral cortex, including most regions of the neocortex and many areas of the limbic system.

Further comments on reticular connectivity. It has already been noted that the majority of neurons forming the *medial reticular column* nuclei together with many *rapheal nuclear neurons* are of isodendritic form (*vide supra*) with radiating, spoke-like, flattened, extensive dendritic trees which are intersected at right angles (i.e. longitudinally) by innumerable ascending and descending fibres which synapse with successive neurons. Their dendrites also receive inputs from ascending spinoreticular fibres (*vide infra*), from cranial nerve sensory nuclei, cerebellar (particularly fastigial) efferents, descending fibres from the limbic forebrain complex and hypothalamus, the neocortex, tectal fibres, and medially directed axons from a proportion of the lateral column parvocellular nuclei. The ascending, descending, or bifurcating axons and their collaterals, of these isodendritic neurons may remain intrinsic to the reticular system completing local reflex loops or acting as intermediaries in longitudinal polysynaptic chains, whilst others project to spinal centres, neighbouring brainstem centres, or ascend to the forebrain, that is, they are the source of the majority of the outflow channels to the more distant destinations. In contrast the allodendritic and idiodendritic neurons of the parvocellular lateral zone nuclei appear to be mainly in receipt of inputs from sensory cranial nerve nuclei and collaterals from corticoreticular fibres of neocortical origin. Some of their axons pass medially presumably modulating medial column nuclear activity, but the majority surround, interlace and synapse with, brainstem cranial nerve motor nuclei thereby completing reflex loops or mediating corticonuclear control.

As examples of zonal or nuclear variations of activity in the reticular formation we may summarize the components assumed to form part of the so-called ascending activating system and also some reticular nuclei with well-defined connexions. The *ascending activating system* (further described below) is widely held to be the structural substrate for the *arousal reaction* and *desynchronization* of the electroencephalogram, prerequisites for wakening, attention, and 'specific' cortical activities (p. 952). It comprises (*a*) spinoreticular inputs, (*b*) collaterals from other long ascending pathways, (*c*) cranial nerve sensory nuclear inputs, (*d*) reticulo-diencephalic projections, and (*e*) hypothalamo-limbic and diffuse 'non-specific' thalamo-cortical radiations; probably the thalamo-striate and pallido-thalamic loop should also be included. The majority of these connexions are considered elsewhere in this section and only a few points will be emphasized here. The spinoreticular projections reach most parts of the medial reticular column nuclei, but at many levels their terminals are sparse, rich populations of terminals being restricted to the medullary gigantocellular (magnocellular) nucleus, the oral part of the central medullary reticular nucleus and the nucleus reticularis pontis caudalis. Similar nuclear zones receive the majority of the terminals from the pontomedullary trigeminal sensory nuclei and the nucleus of the tractus solitarius. In contrast the vestibulocochlear and retinotectal inputs are concentrated on the mesencephalic reticular nuclei. The origins of the reticulospinal tracts and some of the constituent bundles that effect reticulo-diencephalic connexions are considered further on pp. 875, 879, 893, but in general they broadly correspond to the inputs just described.

Certain reticular nuclei have relatively precise boundaries and (doubtless amongst many others) well defined afferent or efferent projections. Thus, for example, the 'lateral reticular nucleus', magnocellular nucleus, and nucleus reticularis paramedianus, all medullary aggregates, and the nucleus reticularis tegmenti pontis all have profuse and precise cerebellar connexions. The nucleus ceruleus is described on p. 933; the nucleus tegmenti pedunculopontinus, pars compacta receives a discrete pallidofugal input.

COMPARTMENTALIZATION OF THE RETICULAR FORMATION

In the preceding pages it will have become apparent that different investigators using widely varying techniques as these become available, have found it appropriate to compartmentalize the reticular formation in a number of contrasting ways, each having its own limited usefulness. These have ranged from:

(1) Considering the formation as a diffuse, 'non-specific', virtually *random nerve network*, with multiple inputs and outputs, polymodal interaction between neurons, and the view that the 'system' should be treated as a whole, with perhaps the application of 'network theory' combined with computer analysis (Scheibel and Scheibel 1967).

(2) The concept of the formation as consisting of a central isodendritic core extending from spinal cord to diencephalon and surrounded by, or surrounding, 'islands' of allodendritic and idiodendritic neurons and their 'specific' nuclei.

(3) The view that the formation, essentially bilateral but with numerous cross-connexions, consists on each side of three longitudinal columns, 'median', 'medial', and 'lateral', each with its characteristic varieties of neuron.

(4) The extreme parcellation of the foregoing columns into named nuclei using traditional cytoarchitectonic techniques.

(5) Attempts to locate columnar 'zones', individual nuclei or parts thereof, as the sites of termination of definable bundles of afferent fibres, or the sources of recognizable efferent bundles with distinct destinations: these have been pursued with the older degeneration techniques (p. 815) and the more recent fibre tracing methods (p. 819).

(6) Efforts to analyse the results of micro-electrophysiological recording, and the structural and behavioural effects of micro-stimulation or ablation of these zones or nuclei. Such manoeuvres often result in dramatic focal effects, the nodal point often being termed a 'centre' by neurophysiologists. However, much coincidence or overlap is often present and the relation to named nuclei has not been established.

(7) More recently it has become fashionable to consider the formation 'to be a collection of more or less specific neuron populations, but populations that are to a large extent mixed together rather than separated into spatially isolated groups' (*see* review by Webster 1978): the criteria employed entails the use of techniques that allow the mapping of the distribution of groups of neurons that produce a common neurotransmitter substance (*see* p. 832). Thus, terms such as '*the cholinergic nervous system*', and '*the aminergic nervous system*', and its variations, are being employed with ever-increasing frequency.

With regard to the latter approach it is, however, essential to appreciate certain general points. Quite a large, and increasing number of substances are being proposed as putative neurotransmitters (*vide infra* and p. 831), but as yet only a limited number have met all the rigid criteria that allow them to be firmly classed as such—whilst some may be true transmitters, others may be precursors or metabolic products of transmitters. When a reasonably complete list of undoubted transmitters becomes available, together with sensitive histochemical or other means for the demonstration of their spatial and quantitative distribution, it is obvious that each will yield its own distinctive 'map'. Doubtless much overlap and interlacement, particularly of neurites, will be present. Furthermore, in any complex neural aggregate, individual neurons will in most instances receive synaptic terminals from several varieties of neuron each producing its characteristic transmitter, the responses of the neurons' receptor sites also being neuron specific. Thus, the intellectual problem arises as to what extent it is valid to group all the neuron aggregates that possess a common neurotransmitter as a 'system'. Certainly cooperative behavioural responses in the intact normal organism result from complex, balanced interactions between many such systems. Equally, however, *disordered* or *modified* behaviour often results from the pathological, experimental, or pharmacologically induced over- or under-activity of a particular transmitter. The most precise localizing histochemical techniques yet available relate to the biogenic amines (*vide infra*) and they have therefore become favourite objects of study; the distribution of cholinergic neurons, and those containing other putative transmitters is less certain.

THE CHOLINERGIC RETICULAR SYSTEM

Despite the elegance of the early researches and innumerable subsequent investigations that firmly established acetylcholine as a neurotransmitter, the spatial distribution of cholinergic neurons throughout the nervous system has proved a much less tractable problem (see for example, the excellent reviews by Shute and Lewis 1967; and De Feudis 1974). Nevertheless, the latter author felt that sufficient biochemical, pharmacological, structural and behavioural evidence had accumulated to write a monograph entitled *Central Cholinergic Systems and Behaviour*. Standard techniques for the localization of acetylcholinesterase have long been employed (Koelle and Friedenwald 1949) and this substance is widespread in the nervous system; however, its presence is not necessarily an indication of the presence of cholinergic neurons since it is often abundant in tissues devoid of neural elements or, for example, in neuron aggregates that are strongly mono-aminergic. However, these earlier techniques have now been supplemented by traditional histochemical techniques for the localization of a synthetic enzyme choline acetyltransferase (Burt 1969, 1970; Kasa et al. 1970 a and b) and more recently immunohistochemical methods for the same substance (Chao and Wolfgram 1973; Eng et al. 1974; McGreer et al. 1975) and these are generally regarded as more critical, particularly when *combined* with biochemical assays.

On these bases it is accepted that acetylcholine acts as a transmitter in virtually all parts of the central nervous system but varies greatly in concentration in different zones. Bearing in mind the difficulties in localization, it appears that cholinergic neurons are found scattered in the majority of the brainstem reticular nuclei. In the medulla and pons they are related to those areas adjoining both sensory and motor cranial nerve nuclei and particularly in those reticular nuclei, noted above, as constituting part of the 'ascending activating system'. In the midbrain and pontomedullary junction all the principal reticular nuclei described above have a cholinergic component (this includes the nuclei raphes dorsalis, cuneiformis, subcuneiformis, deep and dorsal tegmentalis, and tegmentalis pedunculopontinus; also the non-dopaminergic part of the substantia nigra pars compacta). The dorsally sited nuclei have axons forming a loose series of bundles, the *dorsal tegmental cholinergic pathway*, from which they radiate to terminate in the inferior and superior colliculi, the pretectal nuclei, the nuclei of the geniculate bodies and the more dorsally and anteriorly placed nuclei of the dorsal thalamus, including a number of the midline and intralaminar nuclei (particularly the nucleus centromedianus). The reticular nuclei sited in the ventral mesencephalic tegmentum have issuing axons that associate with those emerging from the compact part of the substantia nigra to form the *ventral tegmental cholinergic pathway*. The fibre bundles of the latter then diverge to end in numerous nuclei in the subthalamus, ventral and anterior nuclear groups in the dorsal thalamus, and most of the nuclear complexes of the hypothalamus, both lateral and medial, including the mamillary, supra-optic and pre-optic groups. For details of these pathways, a critical appraisal of methodology, bibliography, and further cholinergic projections from these nuclei to telencephalic structures the reader should consult Shute and Lewis (1967), and for an extensive review of the cholinergic innervation of limbic structures see Kuhar (1975), and Moore (1975).

THE MONOAMINERGIC RETICULAR FORMATION

With the advent of the Falck-Hillarp technique (Falck and Hillarp 1959, Falck 1962; Falck et al. 1962) in which it was demonstrated that following pre-treatment with formaldehyde, neurons and their neurites containing certain biogenic mono-amines exhibited a characteristic fluorescence when examined using ultraviolet light. Those containing the catecholamines dopamine or noradrenalin showed a greenish fluorescence whilst those containing the indoleamine, serotonin, (5-hydroxytryptamine) were a distinct yellow. For a period it was difficult to distinguish between the two catecholamines

(except for slight morphological differences—*vide infra*), but subsequently this became practicable by the application of quantitative photofluorometry combined with pharmacological manipulation. These reliable and reproducible techniques have, in recent years, led to the accumulation of a wealth of data concerning the differential distribution of aggregations of dopaminergic, noradrenergic, and serotonergic neurons in the brainstem and diencephalon of usually rats, cats and monkeys (see, for example, Dahlström and Fuxe 1964, 1965; Pin et al. 1968; Ungerstedt 1971; Björklund and Nobin 1973; Felton et al. 1974; Jacobowitz and Palkovits 1974; Palkovits and Jacobowitz 1974; Lindvall et al. 1974; Webster 1978), but general confirmation has also been obtained on adult and fetal human brains (Nobin and Björklund 1973; Olson et al. 1973 a and b). The reader should note the general points concerning interpretation—the axons of these cells are characteristically extremely fine, often branch profusely, and present minute varicosities along their course, from each of which it is assumed that transmitter is released. Thus, they may form innumerable synapses *en passage*, or perhaps produce miniature 'lakes' of transmitter influencing numerous neurons in their neighbourhood—a 'paracrine' effect. Furthermore, some investigators have emphasized that the varicosities on noradren-ergic fibres are almost spherical, closely and regularly spaced, exhibiting an intense greenish fluorescence, whereas those of dopaminergic fibres are usually fusiform, somewhat smaller, more widely and irregularly spaced, and their fluorescence is less prominent. It should also be noted that the *dendrites* of these monoaminergic neurons are also fluorescent; such dendrites may invade the territories of neighbouring nuclei; and whilst they *may* be involved in forming, say, dendrodendritic synapses, they may also give the spurious appearance of an aminergic innervation to a group that is, in fact, devoid of one. Intensive studies continue, and new data appear almost daily.

The investigators who have studied the rat's brain in detail have recognized 15 catecholaminergic groups (classified as groups A1–15 of which A1–7 are noradrenergic and A8–15 dopaminer-gic), and 9 serotonergic groups (classified as groups B1–9). There has been considerable correspondence between the different groups of animals studied, but it must be realized that the aminergic groups thus classified only rarely correspond spatially to the con-fines of one of the reticular nuclei defined and named on classical cytoarchitectonic grounds: usually they form part of, lie adjacent to, or across the boundaries between, one or more of these nuclei.

THE NORADRENERGIC RETICULAR FORMATION

A brief account of the general location of the main noradrenergic neuron groups (A1–7) will be followed by a summary of their principal projections: the groups, initially identified in the rat, have mostly but not exclusively been identified in primates, are confined to the medullary and pontine parts of the brainstem.

Group A1, in the caudal medulla, lies ventral to the lateral reticular nucleus but has extensions which partly surround the latter; A2 is part of the *nucleus commissuralis*, one of the *para-hypoglossal nuclei*, which encloses the dorsolateral aspect of the hypoglossal nucleus and approaches the ependymal floor of the fourth ventricle. Group A3 is part of, and extends dorsal to, the dorsal accessory olivary nucleus (a subdivision of the inferior olivary nuclear complex); its noradrenergic status in primates is still under investigation. Group A4, the *nucleus pigmentosus tegmentocerebellaris*, lies subependymally embracing the ventric-ular aspect of the superior cerebellar peduncle. Group A5 is situated ventral or deep to the subcerulean nucleus partially invading the territory of the superior olivary complex and closely related to the motor nucleus of the facial nerve. Group A6 is the compact aggregation of neurons that form the *core* of the *nucleus ceruleus* previously described (p. 933). Group A7 constitutes the *nucleus subceruleus*, sited in the parvocellular lateral reticular column of the oral part of the pons and partly confluent with groups 4 and 6, particularly in some primates (Hubbard and Di Carlo 1973).

Concerning the patterns of connectivity of these groups, the majority of the noradrenergic (and some of the serotonergic groups) exhibit both complex *ascending* and *descending* projections in addition to *local* destinations in the brainstem. Because of their contrasting ascending brainstem pathways it is convenient to consider group A6 first, and then the remaining groups separately, although the separation into two may well not be as distinct as was first envisaged.

Group A6—the central noradrenergic cells of the nucleus ceruleus—has been described in some detail previously (p. 946) but for completeness and convenience will again be summarized here. The *descending fibres*, joined by some from A7 neurons provide collaterals to a number of bulbar nuclei including vagal and inferior olivary complex nuclei, then traverse the ventrolateral white column of the spinal cord providing noradrenergic inputs to the ventral, intermediate and base of the dorsal grey columns of the cord; these bundles constitute the *ventrolateral ceruleospinal tract*. The local brainstem terminals are largely collaterals of the principal ascending pathway, but a substantial bundle joins the superior cerebellar peduncle, the *ceruleocerebellar tract* to distribute both to the cerebellar roof nuclei and cortex (p. 918). The ascending fibres form the *dorsal tegmental noradrenergic bundle* (part of the central tegmental fasciculus) which traverses the midbrain ventrolateral to the periaqueductal grey matter, continues through the lateral hypothalamic part of the diencephalon largely as part of the medial forebrain bundle, thence through the septal region to join the cingulum and supracallosal longitudinal stria. Side branches have terminals in the mesencephalic periaqueductal grey matter, reticular nuclei and colliculi; diencephalic terminals include the dorsal thalamic nuclei, geniculate and habenular nuclei; telencephalic terminals include all major limbic structures—the septal and amygdaloid nuclei, hippocampus, subiculum, and the cingulate, retrosplenial and parahippocampal gyri; the neocortex receives a uniform diffuse bilateral input, fibres of the two dorsal bundles having interchanged some fibres across the midline in the upper brainstem. It has even been claimed that the profusely branching axon of a *single* cerulean neuron may diverge so widely as to provide terminals in the spinal cord, brainstem, and forebrain! (For further comments *see* p. 933.)

Groups A1, 2, 5 and 7 (with possible accessions from A3 and 4, but the status of these is less certain in primates) provide ascending branches that converge to form the *ventral tegmental noradrenergic bundle* (part of the central tegmental fasciculus). Proceeding through the ventral reticular regions of the midbrain it approaches the dorsal bundle at upper midbrain levels; both bundles exchange fibres across the midline, the ventral bundle then continuing mainly as a constituent of the medial forebrain bundle through the lateral hypothalamus. *En route* the ventral bundle provides collaterals that terminate at mesencephalic levels in the neighbouring reticular nuclei (p. 945) and periaqueductal grey matter; in the diencephalon it diverges to nuclei throughout the *hypothalamus*, particularly profuse terminals being found in the dorsomedial, periventricular, tuberal, paraventricular, and supra-optic nuclei; it finally appears to terminate as sprays of endings in the bed nucleus of the stria terminalis and the pre-optic area. Thus, the dorsal and ventral tegmental noradrenergic bundles are mainly to be contrasted in their diencephalic zones of termination and the extremely wide continuation of the dorsal bundle into limbic and neocortical areas.

Groups A1 and 2 provide *local* brainstem inputs, particularly to the dorsal vagal nucleus and the nucleus of the solitary tract.

Finally, Groups A1, 2, 5 and 7 provide *descending* fibres to the spinal cord; these have been described (Dahlström and Fuxe 1965) as constituting an *anterior noradrenergic bulbospinal tract* with terminals in the anterior grey column of the spinal cord, and a *dorsolateral noradrenergic bulbospinal tract* with terminals in the intermediolateral grey column and also in the substantia gelatinosa.

THE DOPAMINERGIC NEURONAL GROUPS

No dopaminergic cell aggregations have been identified in the hindbrain; of the eight groups recognized so far (classified as groups A8–15) three (A8, 9 and 10) are mesencephalic, four (A11, 12, 13 and 14) are diencephalic, and a single group (A15) has been found in the telencephalon.

The three mesencephalic groups are interconnected plates or strands of dopamine-positive cells and their limits are therefore difficult to define. Group A8, generally accepted as part of the mesencephalic reticular formation is also known as the *substantia nigra pars lateralis* (but *see* p. 936), whilst A9 constitutes the whole of the *substantia nigra pars compacta*. Group A10 is a median group or complex that abuts ventrally on the nucleus interpeduncularis (the group receives the following terminologies— *nucleus linearis, nucleus parabrachialis pigmentosus, interstitial nucleus of the ventral tegmental decussation*: how far these are synonymous in primates is obscure).

Of the diencephalic aggregations, A11 is in the subependymal grey matter in the caudal part of the third ventricular wall, partially encompassing the fasciculus retroflexus. Group A12 includes the *infundibular (arcuate) nucleus*, and rostral extensions of the latter into the anterior periventricular grey matter are often classified as group A14. Some neurons of the zona incerta (p. 965) make up group A13.

The solitary telencephalic group (A15) is provided by the periglomerular cells of the olfactory bulb (p. 992).

Summarizing briefly the efferent projections of the dopaminergic neurons—those derived from groups A8 and 9 (subdivisions of the substantia nigra) ascend through the caudomedial subthalamus and lateral hypothalamus, the fibre bundles then diverging and, intersecting the internal capsular fibres (as part of the *comb* bundle—*see* p. 978), they join the ansal system and continue to radiate, finally terminating as a profuse input to the whole of the caudate nucleus and putamen of the corpus striatum. Collectively these fibre bundles form the *nigrostriatal dopaminergic system*. In contrast, the fibres derived from the midline third mesencephalic group A10 ascend more medially and traverse the medial forebrain bundle to terminate in a variety of further telencephalic structures. These include the nucleus accumbens, the bed nucleus of the stria terminalis, the olfactory tubercle, nuclei of the septal area, granular areas of frontal neocortex, anterior cingulate cortex, and also the entorhinal cortex. These bundles have been grouped as the *mesolimbic dopaminergic system*. The remaining dopaminergic neuron groups do not belong to the 'reticular formation' as defined by the classical neuroanatomists, but such terminological irrelevances shall not deter us from summarizing them here.

Groups A11, 13 and 14 are considered together since their axons are relatively short and remain within the confines of the diencephalon. Group A11 projects to medial hypothalamic nuclei, the nuclei of the midline of the dorsal thalamus, the habenular nuclei, and the pretectal area. Groups A13 and 14 have axons that ramify to terminate in nuclei of the caudal part of the dorsal thalamus, other parts of the zona incerta, and nuclei of the dorsal and oral parts of the hypothalamus. Collectively the latter fibre bundles have been termed the *incerto-hypothalamic dopaminergic system*.

Neurons forming group A12 with their somata in the infundibular nucleus have axons which pass to the rostroventral part of the median eminence and upper infundibulum; here, they may terminate directly in synaptic relation with, or merely in the neighbourhood of, the axons or terminals of parvocellular releasing-factor neurons, or in relation to specialized elongate ependymal cells (tanycytes) of the infundibular recess of the third ventricle (for further comments and illustration *see* pp. 981, 1447 and **8.**197 A, B). Thus this *tubero-infundibular dopaminergic system* may modulate the transport and/or exocytosis of releasing factors and their uptake by venous portal system capillaries, also secretory activity or transport to and from the third ventricular cerebrospinal fluid by the tanycytes.

Finally, dopaminergic cells of group A15 are the periglomerular cells of the olfactory bulb; some details of their cytology and connectivity in this exceedingly complex part of the brain are given on p. 992.

THE SEROTONERGIC RETICULAR FORMATION

The reticular neurons synthesizing and containing the putative transmitter serotonin (5-hydroxytryptamine) have been identified at medullary, pontine and midbrain levels; they all lie in or near the median plane (thus, they are mainly parts of the rapheal nuclei—*see* above); nine relatively distinct groups have been located and classified as B1–9. Group B1 is formed by some neurons of the nucleus raphes pallidus and B2 by the nucleus raphes obscurus; both groups have extensions ventrolaterally to include neurons of the nucleus paramedianus. Extending cranially from B1, further neurons of the nucleus raphes pallidus, and those of the nucleus raphes magnus approach and reach the pontomedullary junction together constituting group B3. Thus, groups B1–3 are serotonergic groups, median and paramedian in position and largely confined to the medulla. Group B4 is located in the pontine periventricular grey matter of the cranial part of the fourth ventricle; B5 is a circumscribed group in the nucleus raphes pontis near the motor nucleus of the trigeminal nerve, whilst group B6 is found in the cranial part of the central pontine tegmental reticular grey matter and extending into the superior central nucleus of the pons. Aggregate B7, the largest of the serotonergic cell groups, occupies much of the mesencephalic nucleus raphes dorsalis with extensions into the periaqueductal grey matter; B8 is a cranial extension of B6 continuing within the confines of the nucleus centralis superior (of Bechterew); finally group B9 is a lamina of serotonergic neurons embracing the lateral aspect of the nucleus interpeduncularis.

The routes and fields of termination of projection fibres from the serotonergic cell groups have much in common with those of the noradrenergic groups, but of course, with some exceptions. Thus, they may be considered as (*a*) descending, (*b*) local brainstem, (*c*) cerebellar, and (*d*) profuse and widespread dorsal and ventral ascending pathways. Moreover, a single serotonergic reticular neuron may have an axon that diverges to many of these destinations, whilst others appear to have a strictly localized zone of termination.

The *descending bulbospinal serotonergic projections* are contributed by medullary groups B1–3, particularly those located in the nucleus raphes magnus; they form two bulbospinal tracts—one traverses the anterior white funiculus giving off re-entrant terminals to the neurons of the ventral grey columns (laminae VIII and IX), the other descends through the dorsal part of the lateral white funiculus largely lateral to the corticospinal projection, its terminals innervating the preganglionic autonomic neuron-containing intermediolateral cell column, and the remaining laminae of the dorsal horn including the substantia gelatinosa. It has been held that the latter are particularly concerned in modulating nociceptive inputs to the spinal cord grey matter.

The *intrinsic brainstem serotonergic projections* involve groups B3–8; those from B5–8 project to the nucleus ceruleus, tegmental nuclei, and descending fibres to other (probably both aminergic and non-aminergic) zones of the pontine and medullary reticular formation. Groups B3–5 also project to neighbouring reticular nuclei and questionably to the inferior olivary complex. It is noteworthy that the largest aggregate of serotonergic neurons (group B7) sends the most massive bundle of axons yet identified to terminate in the largest aggregation of noradrenergic neurons which form the core of the nucleus ceruleus (group A6).

The *cerebellar serotonergic pathway* consists of axons that emerge from groups B5 and 6, then having entered the cerebellum through the homolateral middle cerebellar peduncle, branch to have terminal fields in the intracerebellar nuclei and then diverge to wide areas of the cerebellar cortex (cf. the noradrenergic cerebellar pathway through the inferior cerebellar peduncle).

The *dorsal ascending serotonergic bundle*, much smaller and with more restricted destinations than the ventral, arises from neurons with their somata in cell groups B3–6. These axons converge dorsocranially and become constituents of the dorsal longitudinal fasciculus (of Schütz) in which they ascend giving off side branches to the mesencephalic reticular nuclei, periaqueductal grey matter, and, entering the diencephalon, diverge to their fields of termination in the caudal zone nuclei of the hypothalamus.

The *ventral ascending serotonergic bundle* is relatively massive and stems from neuron groups B6–8 (rapheal nuclei of the upper pons and midbrain); curving ventrorostrally it traverses the ventral midbrain tegmentum as a constituent of the central

tegmental fasciculus, then entering the diencephalon, courses through the hypothalamus in its lateral part as yet another constituent of the complex of fascicles forming the medial forebrain bundle; numerous side branches and a wide terminal radiation reach a great diversity of destinations. Mesencephalic bundles pass to many parts of the substantia nigra; others enter the interpeduncular nucleus—some terminate in this nucleus and the remainder of this side branch continues in the habenulo-interpeduncular tract to reach the epithalamus and dorsal thalamus, where they terminate in the habenular nuclei, the medial and midline nuclear groups and parafascicular nucleus of the thalamus.

Coursing through the diencephalon as part of the medial forebrain bundle, side branches are given to the mamillary and lateral hypothalamic nuclei; other fibres penetrate the hypothalamic region obliquely and reach the caudate nucleus and putamen of the corpus striatum, whilst some continue to reach parts of the lateral neocortex; a small bundle also enters the postcommissural column of the fornix and then recurves through its body, crus, and fimbria to distribute terminals to the dentate gyrus and cornu ammonis. At the rostral end of the hypothalamus the ventral serotonergic bundle diverges into numerous radicles. Some pass ventrolaterally as constituents of the ansa peduncularis and reach the nucleus of the diagonal band of Broca, the prepiriform and entorhinal cortex, and parts of the amygdaloid complex. Further bundles splay out to reach the nuclei of the preoptic and septal areas, the olfactory tubercle and adjacent grey matter, and some continue rostrally to reach the medial frontal neocortex and along the olfactory tract to end in the olfactory bulb. A third contingent of fibres radiate to the accumbens nucleus and then furnish a second (rostral) input to the anterior zones of the caudate nucleus and putamen of the corpus striatum. Ultimately, what remains of the ventral ascending bundle is still, however, a substantial fascicle, and this passes first rostrally, then dorsally to join the cingulum within which it arches over the corpus callosum and then curves downwards and forwards to run the length of the temporal lobe. Throughout its course it provides side branches and terminals to the cingulate, retrosplenial and entorhinal (parahippocampal) cortices, and from their ventrolateral aspects, sequential side branches to the cornu ammonis and subiculum, thus completing a *dual input* to the hippocampal formation (*vide supra*).

Thus, *in summary*, the serotonergic 'system' of neuron groups in the reticular formation of the brainstem possess axonal bundles that project to (*a*) other brainstem reticular nuclei, (*b*) many hypothalamic nuclei, (*c*) all the major limbic structures, (*d*) some 'limbic' and 'non-specific' nuclei of the dorsal thalamus, and the epithalamic nuclei, (*e*) much of the spinal cord grey matter.

THE MAPPING OF OTHER PUTATIVE NEUROTRANSMITTERS

There is now strong evidence that gamma-amino-butyric acid (GABA) and glycine act as inhibitory central neurotransmitters; data are also being accumulated concerning the transmitter status and possible excitatory roles of the amino acid taurine, and also glutamic and aspartic acids. However, their spatial location is limited to high resolution autoradiography after administration of radio-isotope labelled derivatives, with its restriction to relatively minute pieces of tissue. A further obstruction to progress is that these substances (other than GABA) form part of the metabolic pool of virtually all cells, neural or not, and they are therefore quite ubiquitous in their distribution, whatever their specifically neural functional roles, if any, may turn out to be. Thus, apart from gross biochemical assays, the effects of agonists and antagonists, and severely limited ultrastructural studies, there are no specific histochemical methods yet available for their identification. Accordingly, there has been virtually no progress in the 'mapping' of their sites of production, transport, release and postsynaptic action; whether such maps will become feasible, and if produced, how far they will assist an understanding of the operation of the whole, and parts of, the central nervous system remains quite speculative and must await future researches.

However, the mapping of the distribution of certain other

possible neurotransmitter substances which have recently come into prominence has made, and doubtless will continue to make, rapid and substantial progress (*see*, for example, Hughes and Kosterlitz 1977). These include the peptides substance P, the enkephalins, endorphins, and also the spatial distribution of opiate *receptor* sites are under intensive investigation (for detailed reviews *see* Guillemin 1978; and particularly Chan-Pham 1978).

As pointed out by Webster (1978) 'the distribution of enkephalin-containing fibres and terminals in the forebrain shows remarkable similarity to that of the monoamines'—(the basal nuclei, septal area, hypothalamus, amygdala, and areas of the cerebral neocortex all contain enkephalins—*see* Simantov *et al.* 1977); at the time of writing the siting of their parent cell somata was, in most cases, not established. Nevertheless, Hökfelt *et al.* (1977) held that the somata of met-enkephalin-containing cells occur in zones of the mesencephalic periaqueductal grey matter, in the nuclei raphes magnus et pallidus of the medulla (cf. serotonergic groups B1–3), in the nucleus caudalis segment of the spinal nucleus of the trigeminal nerve, and throughout the spinal cord dorsal grey column laminae I, II, and V. There is additionally a close similarity in the distribution of substance P; it has been proposed, therefore, that these met-enkephalin-, endorphin-, substance P-, and serotonin-containing cells and fibres may form an integrative complex that modulates the mechanisms underlying pain perception (Basbaum *et al.* 1976; Nathan 1977; *see* also review and bibliography by Snyder, in: Reichlin *et al.* 1978).

FUNCTIONAL ASSOCIATIONS OF THE RETICULAR FORMATION

It will be clear from the preceding paragraphs that the reticular formation receives convergent information channels from *all* the principal parts of the nervous system and, in turn, it projects directly or indirectly back to all these regions, modulating their activities through complex and ill understood mechanisms that involve synaptic neurotransmission, and local neuroparacrine as well as neuroendocrine effects. Thus, the comments of some earlier neuro-anatomists and physiologists, and some current elementary texts, that the formation can be dismissed as merely that phylogenetically ancient neural network that remains 'filling in' those crevices that persist when all the well-defined 'specific' nuclei and tracts have been described, should be ignored. It is equally unsatisfactory, and a reflexion of our lack of comprehension of intimate mechanisms, that two completely contrasting and extreme views have been advanced. On the one hand some observers have found it sufficient to feed into the formation a channel from another neural complex, for example, the corpus striatum, and describe effects elsewhere as *mediated* by the formation without any attempt to analyse structural organization within the formation: this treats the latter as a 'black box'; analogous approaches are those that invoke the 'activity of the reticular formation' as sufficient explanation for almost *any* behavioural phenomenon which, as yet, cannot be explained by reference to what is known of the spatially more ordered 'specific systems'. On the other hand, with our present knowledge it is similarly facile to attempt to attach a *simple unitary functional label* to individual neurons, named nuclei, or small neuronal groups (whatever criteria are currently employed for their classification). The methods available for structural/functional evaluation have been outlined elsewhere (pp. 815, 817–820): the various cytoarchitectonic, impregnation, histochemical, histofluorescence, immunofluorescence, degeneration, autoradiographic and other modern tracer techniques having been used to form a picture of structural organization and connectivity patterns, albeit incomplete; this provides (as with other neural complexes) the reference framework for functional probing. The latter entails microelectrode recording under experimental conditions, analysing the behavioural changes or deficits that accompany or follow focal or zonal microstimulation or ablation, pharmacological enhancement or depression of neurotransmitter action, a wide range of neuropathological states, genetic/developmental deviations and studies of psychoses. All such methods have severe limitations, particularly when applied to neural complexes such as

the reticular formation, hypothalamus, dorsal thalamus, limbic 'system' and cerebral neocortex. In the present context any approach towards a satisfactory understanding of the reticular formation would necessitate a knowledge of the form and significance of the flow of information along the numerous channels that converge upon it, including any pre-processing and convergent/divergent interaction that has already occurred; the transformations occurring within the formation itself, and the manner in which the outflow channels modulate the activities of their multiple destinations, and through the latter modify simple to complex behavioural patterns. We are far from such an understanding, and currently it seems appropriate to assume that if one or more of the foregoing techniques (although often highly artificial) produces a clear cut set of behavioural changes, the locus, zone, cell aggregate, or whole 'system' is in some unspecified way, *functionally associated* with the behaviour under study. Some of the more prominent fields currently engaging functional neurobiologists are noted below: they may all, of course, as has been a persistent theme throughout this volume, be allocated more or less obviously to some aspect of long, intermediate, or short term homeostasis.

(*a*) **Somatomotor control:** The reticular formation forms part of the complex of subregions of the nervous system exerting controlling influences on activity patterns of skeletal (striated) muscle ranging from relatively simple reflex loops, coarse to fine postural re-adjustments, positioning and scanning of distance receptors, manipulative skills, locomotion, to complicated patterns associated with intergroup and interpersonal communication and emotional expression involving the complexities of speech, gestures and fluctuations of facial expression. The degree and temporal sequence of involvement of the multiplicity of parts of the central nervous system that may be considered as fairly intimately concerned with influencing the activity levels of alpha and gamma motor neuron pools and their attendant neighbouring interneurons, varies greatly with the speed, force, finesse and complexity of the movement pattern in question. Reticular formation involvement may be direct or indirect through other major neural complexes. Thus, *direct* spinal influences are mediated through the reticulospinal tracts, stemming from the medial two thirds of the pontomedullary reticular formation; that from the pons descends in the ventral white funiculus and is largely ipsilateral; that from the medulla is bilateral and occupies a wide area of the lateral white funiculus. Stimulation and ablation experiments led to the concept of an *inhibitory motor 'centre'* or *'zone'* located in the medulla, and a *pontine facilitatory 'centre'*. It should be appreciated that such names are somewhat misleading—thus, in the case of medullary reticulospinal fibres inhibition results from an *increased* flux of impulses and therefore an increased release of quanta of transmitter from their terminals influencing the complex array of interneurons, alpha and gamma efferents in the spinal cord grey matter with varieties of receptor sites, and involving pre- and postsynaptic mechanisms. However, it is equally clear that, *on balance*, a *reduced* flux of impulses from the medullary origin would have the converse, i.e. a facilitatory effect, and that during normal activities there are *continual fluctuations* of activity making the terminological choice quite arbitrary. The descending fibres involved stem from some multipolar neurons of the medullary and pontine gigantocellular nuclei, the central medullary reticular nucleus and the nucleus reticularis pontis caudalis, and include cell aggregations A6 and 7 and the aggregations B1–2 in the nuclei of the raphe. Thus there are non-aminergic fibres intermixed with inhibitory noradrenergic and serotonergic fibres. It is noteworthy, also, that the nucleus-cerulean origin of bulbospinal fibres receives a serotonergic input from the caudal nuclei of the raphe, and the serotonergic bulbospinal fibres from the latter have been held as mediating the general relaxation of body musculature that characterizes rapid eye-movement (REM) sleep. Similar direct influences are directed to the motor neuron pools of the cranial nerves; also groups of the foregoing reticular neurons controlling the motor neuron pools innervating the diaphragm and intercostal musculature have been said to constitute the hierarchy of respiratory 'centres' described by neurophysiologists.

Indirect influences of the reticular formation on somatomotor control are too numerous to be pursued *in extenso* here, and the relevant sections of this volume should be consulted. Listing them briefly, they include, in virtually all instances, non-aminergic (presumably many of which are cholinergic) and aminergic fibres (*see* individual sections devoted to the noradrenergic, serotonergic and dopaminergic 'systems'), but their relative numbers, precise sites of termination and roles in functional integration remain to be discovered. Thus, the *cerebellum* receives reticular inputs through both the inferior and middle cerebellar peduncles that distribute to the intracerebellar nuclei and the cerebellar cortex; other brainstem complexes including the olivary nuclei, colliculi, red nucleus and substantia nigra; the corpus striatum—in particular the caudate nucleus and putamen through the comb bundle and central tegmental fasciculus; the nuclei ventralis anterior and lateralis of the dorsal thalamus, the zona incerta and other subthalamic nuclei, and finally the somatomotor cortex, either directly from reticular nuclei or indirectly from many of the regions just mentioned.

(*b*) **Somatosensory control:** All the subregions of the nervous system involved in the processing of somatosensory information are subject to influences from the reticular formation. Thus, the spinal and brainstem terminals of primary afferent fibres, their nuclear destinations and attendant interneurons, and the dendrites and somata of neurons projecting long axons to supraspinal centres are all subject to reticular control. (These include the dorsal spinal grey column, dorsal column nuclei gracilis et cuneatus, trigeminal sensory nuclei, nucleus of the solitary tract, amongst others.) In different situations the effects may be pre- or postsynaptic, and facilitatory or inhibitory. Currently, much interest persists in relation to the possible relevance of such controls to theories of sensory 'gating' mechanisms (p. 890); as noted previously, it has been proposed that descending bulbospinal and local bulbar reticular fibres that are serotonergic may cooperate with others that are noradrenergic, and yet others that contain met-enkephalins or substance P in an integrated mechanism that modulates pain perception (p. 962). (In this regard, compare also the diencephalic and limbic distribution of monoaminergic and enkephalin-containing reticular fibre terminals with the distribution of opiate receptors—*see* Guillemin 1978; Chan-Pham 1978.) More cranially sited centres receiving and processing somatosensory information, and also in receipt of a reticular input are the sensory relay nuclei of the dorsal thalamus and the somatosensory cortices.

Similar considerations apply to the special senses, reticular inputs being received by the cochlear and vestibular nuclei, tectal and pretectal laminae or nuclei, the nuclei of the geniculate bodies, pulvinar and other dorsal thalamic nuclei, and finally the visual, auditory and olfactory cortices (*see* appropriate sections).

In addition to influences controlling the activities of the more spatially ordered 'specific' sensory systems, spinoreticular and bulboreticular connexions are, in themselves, held by many to provide bilateral, polysynaptic, ascending, 'slow pain' pathways that discharge into multiple diencephalic nuclei; these convey both somatosensory and viscerosensory information.

(*c*) **Visceromotor control:** Cardiovascular readjustments and the activity of non-striated muscle and many glandular cells in the thoracico-abdominal viscera are, of course, in most cases under controlling influences of postganglionic autonomic neurons. Their preganglionic neuronal somata and dendrites are either directly, or through the intermediary of local interneurons, partially controlled by reticulobulbar and descending reticulospinal fibres. Thus, mainly on the basis of focal stimulation and ablation, or brainstem transection experiments, and also pharmacological manipulation, the concept of, for example, cardiovascular controlling 'centres' has gained popularity in general physiological texts. However (as in the case of the respiratory 'centres'), while the sites of the sympathetic and parasympathetic preganglionic somata are now fairly well established, it has not yet proved possible to locate with precision the functionally related groups of reticular neurons.

Forebrain control of autonomic activities involves areas of the orbitofrontal, cingulate and entorhinal cortices, the major limbic structures (*see* pp. 968, 990), the medial and anterior groups of dorsal thalamic nuclei, the hypothalamic nuclei, and descending fascicles of the various tegmental bundles with destinations in the reticular formation, and thence via the reticulobulbar and reticulospinal tracts mentioned above and elsewhere in this section.

(*d*) **Neuroendocrine transduction:** Through the complex of ascending bundles accompanying the central and dorsal tegmental fasciculi, cholinergic and all varieties of monoaminergic fibres pass in side channels to reach and terminate in most, if not all, hypothalamic nuclei including the infundibular nuclei of the median eminence, and the habenular nuclei. Thus, either directly, through the intermediary of other hypothalamic nuclei, or via the dopaminergic infundibular cells (*vide supra*), the reticular neurons are one source of modification of the synthesis and probably transport and exocytosis of 'releasing' or 'release-inhibiting' factors by the parvocellular neurons of the mediobasal hypothalamus (p. 966), thereby controlling adenohypophysial activity; similar mechanisms probably operate with respect to the neurohypophysis. The neural input to the pineal gland involves the sympathetic nervous system via the upper thoracic spinal cord outflow of preganglionic fibres, the superior cervical ganglion and nervus conarii; there is in addition a strong neural connexion via the habenulopineal tracts. The role of the suprachiasmatic nucleus has been mentioned elsewhere (*see* pp. 966, 1446 for further details). Thus, both main neuroendocrine transduction centres are, in part, under reticular control.

(*e*) **Biological rhythms:** The notion of the existence of '*biological clocks*' has, of recent years, permeated all branches of biological science and only a few of the more obvious ones can be listed here. A time-base is imposed on all living systems by the intrinsic turnover rates of biological macromolecules, including the genetic apparatus of the cell (pp. 23, 27), the annual seasonal variations, and daily environmental fluctuations. Thus biological rhythms attend, for example, the reproductive cycles, developmental horizons, cell division, death and replacement, and so forth, the list being endless. As stated elsewhere (p. 971) many rhythms are dependent upon an intact hypothalamus with its multiple input and output channels including those involving the reticular formation. As mentioned above the latter is also involved in the neural control of the pineal gland and much interest is currently centred upon the neuroendocrine role of the latter in relation to circadian rhythms (pp. 972, 1445).

(*f*) **Sleep, arousal, states of consciousness, perception:** Much philosophical, psychological, neurophysiological, neuropharmacological and neuroanatomical thought, effort and research has been directed, first to attain satisfactory definitions of the varying states of consciousness and their neural substrates, ranging from stuporose unresponsive conditions, through varieties of sleep, arousal and awakening, attention, perception, and the preparation for, and enactment of, active responses. Vigorous researches are being pursued on all these fronts and a vast literature has accumulated; this is beyond the scope of this or any volume (*see*, for example, Jouvet 1969, 1972: and for a concise summary and an extensive bibliography, Ingram 1976). Extremely briefly, it is important to distinguish between different *behavioural* states of unconsciousness, sleep and awakening, and the *electroencephalographic* records that usually, but not invariably, accompany these states. The terminology associated with the latter is widely employed by investigators in this field. Thus, there are two principal forms of sleep: *deep* or *slow wave* (*recuperative*) *sleep* (SWS) in which the person is deeply but reversibly unconscious, random limb and postural changes occur sporadically, there is an elevation of sensory thresholds, and the electroencephalogram records *synchronized*, high-voltage, slow waves; interspersed are periods of *paradoxical* (PS) or *rapid eye movement* (REM) sleep. In the latter dreaming normally occurs, there is relaxation of the general body musculature, but as the name implies, rapid oscillatory movements of the eyeballs. The latter are accompanied by corresponding bursts of electrical activity, similar waveforms being recordable from the pontine reticular formation, the dorsal lateral geniculate nucleus and the occipital visual cortex (PGO activity). Behavioural arousal and awakening is accompanied in the intact animal by a rapid, low amplitude, *desynchronized* waveform. Mention has already been made of the early proposal of an '*ascending activating reticular system*', polysynaptic in nature, with multiple inputs from the principal sensory pathways, and following diencephalic synaptic complexes, bilateral widespread radiations, pass to most regions of the neocortex and limbic structures. Recent investigations indicate that both cholinergic and monoaminergic reticular groups in the upper pons and midbrain are, when stimulated, the most powerful instigators of cortical desynchronization. It is widely assumed that the latter state is a necessary prerequisite for maintaining wakefulness, for selective attention to particular features of the environment, for effective perceptual and cognitive utilization of inflowing information, and preparation for comparator and cooperative responses. The varieties of sleep are currently thought to be the result of activity of a hierarchy of hypnogenic neuron aggregates (balanced against the activating 'system' just mentioned), the majority being monoaminergic in nature. Slow wave sleep is dependent upon serotonin production by the rapheal nuclei, whereas paradoxical sleep after priming by serotonin follows intermittent activity on the part of both cholinergic and noradrenergic cells in the cranial part of the nucleus ceruleus, whilst its medial cells appear to function as a '*pontine pacemaker*', which together with the former cooperate in generating PGO responses. The descending serotonergic bulbospinal fibres from the rapheal nuclei and noradrenergic fibres from the caudal part of the nucleus ceruleus are held responsible for the relaxation of limb and trunk musculature during paradoxical sleep.

(*g*) **Spatio-temporal discrimination, cognitive mapping and exploration, reward, learning and memory, emotional content, long term homeostasis:** All the foregoing complex neural activities are intimately dependent upon an intact hypothalamo-limbic 'system' and will be discussed briefly in the appropriate parts of the present section. However, as indicated in previous paragraphs, in addition to numerous other inputs, all the hypothalamic nuclei and limbic structures receive widespread inputs from both aminergic and non-aminergic elements of the reticular formation, and surgical or pharmacological manipulation of these brainstem reticular elements cause profound alterations in these diverse aspects of complex behaviour.

As stated at the outset, therefore, the reticular core of the brainstem and its attendant cell aggregations, have reciprocal connexions with *all* the major parts of the central nervous system, and in accord with this, influences behavioural activities ranging from the most elementary to the most complex.

THE PROSENCEPHALON OR FOREBRAIN

The prosencephalon or forebrain develops from the most cranial of the three midline primary cerebral vesicles (p. 163). This *forebrain vesicle* and its contained cavity, the future *third ventricle*, is soon divisible into a caudal *diencephalon* and a cranial *telencephalon*. From the side walls of the latter, right and left diverticula form and develop into the cerebral hemispheres. Each hemisphere contains a lateral ventricle, and the sites of evagination are marked by the interventricular foramina, through which the lateral ventricles communicate with each other and with the cavity of the third ventricle. (Survey photographs of the cerebrum in sagittal and coronal section are shown in **7**.95, 96, 110.)

Thus, the *diencephalon* corresponds in large measure to the third ventricle and the structures which bound it, whilst the telencephalon consists of a midline *telencephalon medium* or *impar* containing the cranial extension of the third ventricle, and the massive *bilateral telencephalic cerebral hemispheres* each containing a lateral ventricle. (For the development of these regions *see* p. 169.)

THE DIENCEPHALON OR 'INTERBRAIN'

As stated, the cavity of the diencephalon forms the greater part of the median slit-like third ventricle, which narrows caudally to continue into the aqueduct of the midbrain, while cranially it extends into the telencephalon medium (*vide supra* and p. 966).

More precise topographic limits for the diencephalon are: caudally—a plane which includes the posterior commissure and the caudal margins of the mamillary bodies; cranially—a plane which passes from the interventricular foramina through the posterosuperior border of the midline part of the optic chiasma. Cranial to the latter plane are the structures comprising the telencephalon medium. It should be emphasized, however, that while the boundaries just described have some measure of developmental and phylogenetic significance, in terms of functional anatomy of the brain they are merely arbitrary descriptive aids, since many systems cross the boundaries into adjacent areas.

Thus, the diencephalon is a midline structure with symmetrical right and left halves. Traversing each lateral wall of the third ventricle is a *hypothalamic sulcus* of variable prominence, which extends from the cerebral aqueduct to the interventricular foramen (p. 981). The line of the sulcus is used to divide each lateral half of the diencephalon into a *pars dorsalis* and a *pars ventralis diencephali*.

The pars dorsalis diencephali consists on each side of: (1) the dorsal thalamus, (2) the metathalamus, and (3) the epithalamus; whilst the pars ventralis diencephali includes (1) the hypothalamus, and (2) the ventral thalamus.

Briefly, above the hypothalamic sulcus most of each side wall of the third ventricle is formed by a large egg-shaped mass of grey matter and associated laminae of white matter which together constitute the *dorsal thalamus*. The unqualified term *thalamus* is used by the majority of neuroanatomists as synonymous with dorsal thalamus and will frequently be used in this manner in the following account. As described in greater detail below, the thalamus can be divided into a series of major *parts*, and each of these can be subdivided into several *nuclei*. Inferior to the caudal end of the thalamus, and partly continuous with it, are two further swellings—the *medial* and *lateral geniculate bodies*—which overlie nuclei of the same names, and which together constitute the *metathalamus*. The roof of the diencephalon is largely formed by a layer of ependyma which is continuous with that lining the remainder of the third ventricle. Throughout most of its extent this ependyma is in close apposition with the overlying vascular pia mater, with no nervous tissue intervening. However, in the caudal part of the roof and in the adjoining lateral walls of the diencephalon, are the habenular nuclei and their commissure, the epiphysis cerebri or pineal gland, and the posterior commissure. Together these constitute the *epithalamus*.

Some neuroanatomists include the whole of the pars ventralis diencephali in an 'anatomical' definition of the hypothalamus, but this includes regions which are quite dissimilar functionally, and, accordingly, a more restricted definition is preferred in this account. The *hypothalamus* extends from the lamina terminalis to a vertical plane immediately caudal to the mamillary bodies, and dorsoventrally from the hypothalamic sulcus to include the structures in the side wall and floor of the third ventricle. The latter include the mamillary bodies themselves, the tuber cinereum and infundibulum, and nervous tissue adjacent to the

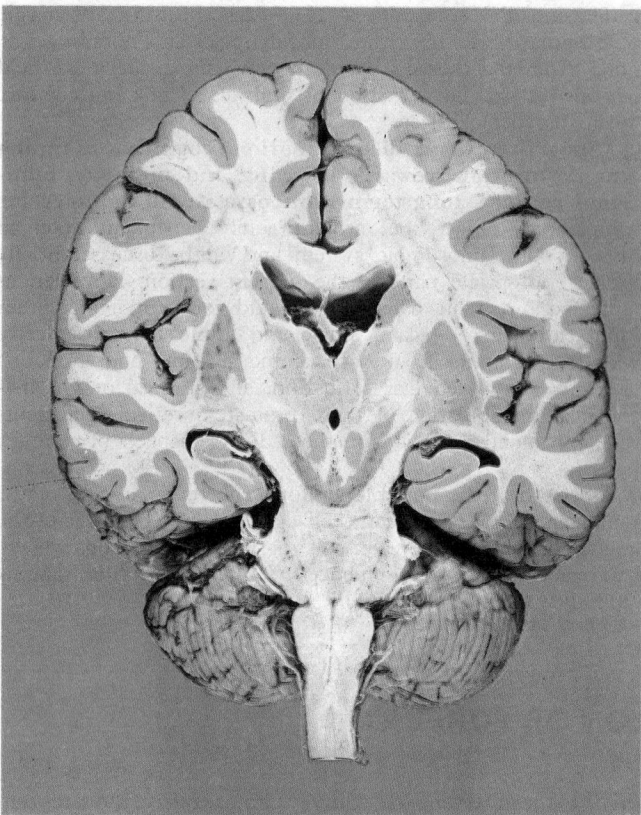

7.96A The dorsal half of a brain sectioned in an oblique coronal plane which passes through the cerebral hemispheres, diencephalon, midbrain, pons and medulla oblongata, to show the general disposition of main structures, many of which are labelled on 7.98. Note: (1) the complex folding of the cerebral cortical gyri and sulci of the frontoparietal, insular and temporal regions; (2) the sectioned surfaces of the corpus callosum, septum pellucidum, body of the fornix, the corona radiata, internal capsule, ventral pons and medulla oblongata; in the latter, part of the decussation of the corticospinal tracts is visible. Note also: (3) the body and inferior horn of the lateral ventricle; (4) the lentiform and caudate nuclei, and the dorsal thalami which are fused across the midline. This illustration includes features referred to at many points in the text which are too numerous to include in a caption; these should be studied as appropriate. (Dissection by Dr. E. L. Rees, Dept. of Anatomy, Guy's Hospital Medical School.)

optic chiasma (7.104). Rostrally, the preoptic region, which lies adjacent to the lamina terminalis and is therefore strictly a telencephalic structure, is for functional reasons usually included in descriptions of the hypothalamus.

The remaining zones of the pars ventralis diencephali are situated in part lateral to the hypothalamus, as a thin sheet immediately ventral to the dorsal thalamus, and as a thick zone which merges caudally with the tegmentum of the midbrain. Collectively these zones constitute the *ventral thalamus* or *subthalamus*. The region includes the rostral extensions of the red nucleus and substantia nigra, the prominent subthalamic nucleus, the prerubral field, the zona incerta, and their associated complexes of nuclei and fibre tracts.

The Thalamus

The thalami (7.96A, B, 97, 98) are two large ovoid masses of grey matter, situated one on each side of the third ventricle and reaching for some distance caudal to that cavity. Each thalamus is about 4 cm long, and has two ends and four surfaces.

The *anterior end* is narrow; it lies near the median plane forming the posterior boundary of the interventricular foramen.

The expanded *posterior end*, the *pulvinar*, is directed dorsally and laterally and overhangs the superior colliculus and its brachium. On the inferior aspect of its lateral part the lateral geniculate body (p. 977) forms a small, oval elevation. Inferiorly,

the pulvinar is separated from the medial geniculate body (p. 977) by the brachium of the superior colliculus.

The *superior surface* (7.97) is free, slightly convex, and covered by a layer of white matter, termed the *stratum zonale*. It is separated laterally from the ventricular surface of the caudate nucleus by a white band, termed the *stria terminalis*, and by the thalamostriate vein (p. 744). It is separated medially from the medial surface by the reflexion of the ependyma of the third ventricle to form its roof. This reflexion is termed the *taenia thalami*. The superior surface is divided into a lateral and a medial part. The lateral part forms a portion of the floor of the lateral ventricle. It is covered with the epithelium of that cavity, and is partly hidden by the vascular fringe of the choroid plexus (7.98). The medial part of this surface is covered with the tela choroidea of the third ventricle, by which it is separated from the body of the fornix which grooves the surface. Between the lateral edge of the fornix and the upper surface of the thalamus the lateral margin of the tela choroidea with its contained plexus is invaginated into the ventricle through the choroidal fissue (7.157). In front, the

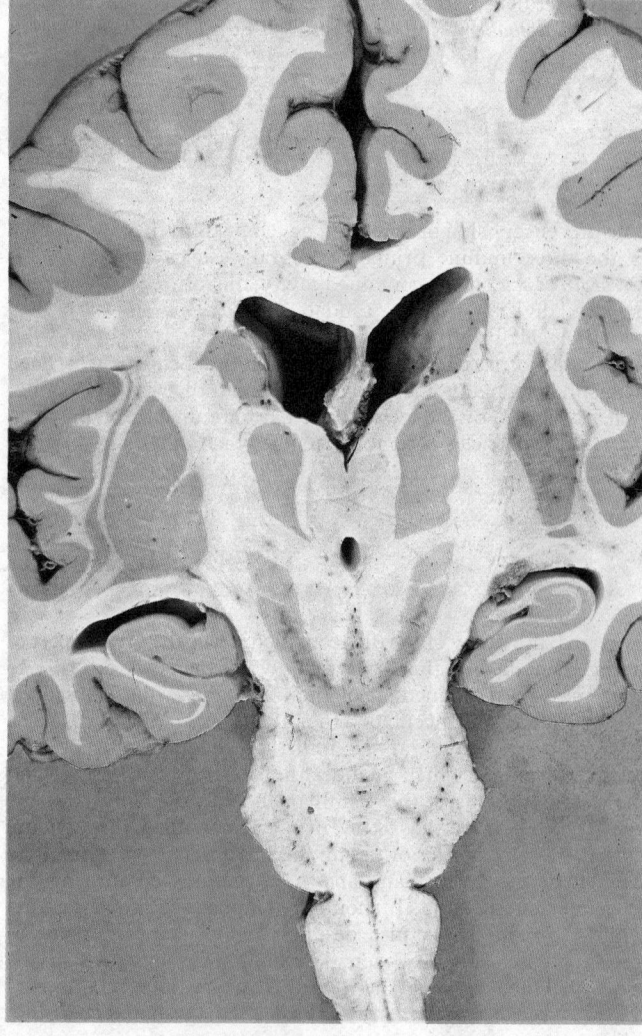

7.96B The central area of the ventral part of the oblique coronal section of the brain shown in 7.95, photographed at higher magnification to show some structural features in greater detail; compare with 7.98 for appropriate labelling. Note in particular: (1) the anterior, medial and lateral parts of the dorsal thalamus; separated by the internal medullary laminae; (2) the relation of the caudate nucleus to the anterior and inferior horns of the lateral ventricle; (3) the lentiform nucleus, divided into an external putamen and an internal globus pallidus, the latter again divided into internal and external parts; (4) the internal capsule, external capsule, claustrum, extreme capsule and insular cortex; (5) the profiles of the sectioned subthalamic and red nuclei and substantia nigra; (6) the hippocampus projecting into the floor of the inferior horn of the lateral ventricle. Other structural features on this section are referred to at many points throughout the text. Compare also with 7.152. (Dissection by Dr. E. L. Rees, Dept. of Anatomy, Guy's Hospital Medical School.)

superior surface is separated from the medial surface by a narrow, raised ridge from which the epithelial lining of the third ventricle is reflected to the under surface of the tela choroidea. This ridge covers a small bundle of white fibres, named the *stria medullaris thalami* (p. 968). Posteriorly it turns medially to form the anterior boundary of the *trigonum habenulae* (7.97), from which the superior surface of the thalamus is separated by the *sulcus habenulae*.

The *inferior surface* rests upon and is continuous with the cranial prolongation of the tegmentum (*subthalamus*). In front of this it rests on that part of the hypothalamus lying in the lateral wall of the third ventricle.

The *medial surface* (8.195) is the superior part of the lateral wall of the third ventricle; it is usually connected to the corresponding surface of the opposite thalamus by a flattened grey band, named the *interthalamic adhesion* (*connexus interthalamicus*). This band is posterior to the interventricular foramen, and averages about 1 cm in its anteroposterior diameter; it sometimes consists of two or even three parts, and occasionally is absent. It contains nerve cells and nerve fibres; some of the latter cross the median plane, but many of them pass towards the median plane and then curve laterally on the same side. This surface of the thalamus is limited below by a curved groove, often ill-defined, termed the *hypothalamic sulcus*. It curves anterosuperiorly from the rostral end of the cerebral aqueduct to the floor of the interventricular foramen and is usually regarded as the cranial continuation of the sulcus limitans of the spinal cord and brainstem, but this view has been strongly challenged (Christ 1969).

On the *lateral surface* there is a thick band of white substance consisting of projection fibres, which form the posterior limb of the internal capsule and separate the thalamus from the lentiform nucleus of the corpus striatum (7.98).

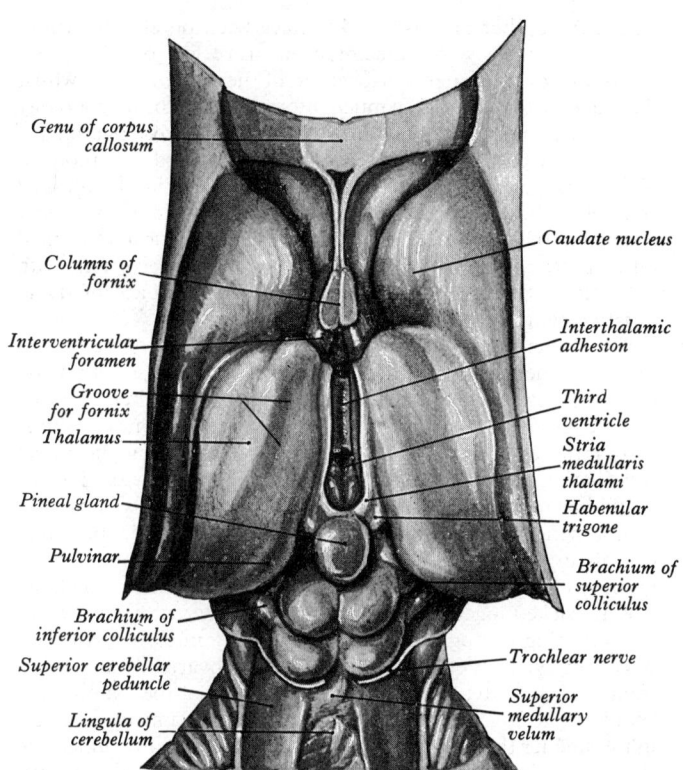

7.97 Dorsal aspect of the caudate nuclei, thalami, pineal gland and tectum, revealed by removal of most of the corpus callosum, the body of the fornix, and of the tela choroidea.

The Gross Structure and Parts of the Thalamus

The thalamus is chiefly grey matter but its superior surface is covered by a layer of white matter, the *stratum zonale* and its lateral surface by a similar layer, the *external medullary lamina*. Internally, the thalamic grey matter is incompletely divided by a vertical sheet of white matter, the *internal medullary lamina* which splits anterosuperiorly in a Y-shaped manner. Accordingly, the main mass of the thalamus is divided into three major *parts*—anterior, medial and lateral; the lateral part is further divided into dorsolateral and ventromedial *parts*. Each of these major parts of the thalamus contains a *group of thalamic nuclei* and, further, these large groups are separated by more restricted groups of nuclei which occur within the internal medullary lamina, and also flank the medial and lateral surfaces of the thalamus.

The main nuclear groups of the thalamus are therefore:

(1) The *anterior group*, which constitutes the anterior part of the thalamus.

(2) The *medial group*, which lies between the internal medullary lamina laterally, and approaches but does not reach, the ependymal lining of the third ventricle medially.

(3) The *lateral group*, which constitutes the dorsolateral moiety of the lateral part of the thalamus, and which expands posteriorly to form the pulvinar.

(4) The *ventral group*, which occupies the ventromedial moiety of the lateral part of the thalamus.

(5) The *intralaminar group*, embedded in the substance of the internal medullary lamina.

(6) The *nuclei of the midline*, which are small, interlaced with a periventricular system of fine nerve fibres and which constitute part of the periventricular grey matter separating the medial part of the thalamus from the ependyma of the third ventricle and in part forming the variable interthalamic adhesion (p. 981).

(7) The *reticular nucleus of the thalamus*, which is a long, thin, curved lamina of cells separating the external medullary lamina from the white matter of the posterior limb of the internal capsule.

The anterior and medial parts of the thalamus, together with some of the smaller groups of nuclei, are often regarded as phylogenetically the older regions, and designated *paleothalamus*

in contrast with the lateral part or *neothalamus*, which reaches its greatest development in anthropoid apes and man. However, such modifications have occurred in the 'older' parts of the thalamus during the course of evolution, and the structural and functional relationships of the various regions of the primate thalamus are so intermingled, that some authorities regard such a division as little more than a convenient aid to elementary teaching.

As we shall detail further below, many parts of the thalamus are connected via relatively well-defined fibre bundles with all the main subcortical parts of the nervous system—the brainstem and spinal cord, cerebellum, hypothalamus and corpus striatum. Further, many parts of the thalamus establish reciprocal connexions with many parts of the cerebral cortex, and classically these two-way radiations of fibres between thalamus and cortex are grouped into four **peduncles** or **stalks of the thalamus**. Taken together, however, these peduncles form an almost continuous fan-like radiation from the ventral, superior, posterior and inferolateral margins of the thalamus to most regions of the cortex, and they form a substantial part of the internal capsule and corona radiata (7.99, 150B, and p. 1030). The *anterior* or *frontal thalamic peduncle* interconnects the anterior and medial parts of the thalamus with much of the cortex of the frontal lobe of the cerebrum. The *superior peduncle* interconnects the ventral and lateral thalamic regions with the pre- and postcentral gyri and the neighbouring parts of the frontal and parietal lobes. The *posterior peduncle* joins the occipital and posterior parietal cortex with the posterior parts of the lateral thalamus including the pulvinar and lateral geniculate body. Finally, the *inferior thalamic peduncle*, smaller than the others, interconnects the posterior thalamus and medial geniculate body with restricted regions of the temporal lobe cortex.

The Thalamic Nuclei and their Connexions

The nuclear groups of the thalamus outlined above have been intensively studied since the close of the nineteenth century in man and a number of other mammalian and submammalian forms. A number of these studies have been purely structural, concerning the cytoarchitectonics as revealed by the Nissl technique, or the general disposition of the white matter, whilst the Golgi method has given much information on the morphology of individual neurons and their processes. Many investigations of connectivity patterns have been carried out with the methods of Nauta and Marchi; the effects of selective stimulation and ablation have been recorded; more recently, in as yet only restricted regions of the thalamus, there have been detailed ultrastructural studies of its synaptology, unit recording with microelectrodes, and analysis of the distribution of accepted and putative neurotransmitters.

What emerges from these studies, which are far too numerous to review here, is that the thalamus is a region of immense structural complexity, composed of a large number of often interdependent, but functionally distinctive zones. Whilst substantial agreement has been reached concerning the broad parcellation of the thalamus into nuclear groups, and often individual nuclei, it is still premature to assume that we possess a comprehensive picture of the structural and functional design of the organ. Difficulties of investigation result from the large number of nuclear groups, some of them very small, which are closely packed together, their afferent and efferent fibre systems often passing through or near neighbouring nuclear territories. Accordingly, stereotactic lesions almost invariably affect unwanted regions in addition to the target area of the experiment. Similarly, retrograde degenerative changes are difficult to assess in the smaller thalamic neurons, and these are equally difficult to study with microelectrode techniques. Nevertheless, despite these problems of technique, a large volume of data has accumulated and the main nuclei of the thalamus and their best substantiated connexions will be summarized here (7.100, 101). For details and an introduction to extensive bibliographies the reader should consult Le Gros Clark (1932, 1933 a and b, 1936, 1937, 1949); Walker (1938); Toncray and Kreig (1946); Sheps (1945); Dekaban (1953); Kuhlenbeck (1954); Hassler (1959); Schaltenbrand and Bailey (1959); Ajmone-Marsan (1965); Purpura and Yahr (1966); Scheibel and Scheibel (1966, 1967); Van Buren and Borke (1972).

In general terms a thalamic nucleus may establish connexions with: (1) other thalamic nuclei; (2) neighbouring subcortical masses of grey matter; (3) long ascending pathways from the brainstem and spinal cord; and (4) the cerebral cortex. The intrinsic connexions with other thalamic nuclei are the most difficult to analyse and thus are the least well understood both structurally and functionally. The general anatomy and thalamic termination of the more prominent long ascending pathways and those from neighbouring centres have been well established for a considerable period, but the finer details of their sites and modes of termination, and those of less substantial pathways are often uncertain and the source of much disagreement. Many thalamic nuclei show marked degenerative changes after experimental resection of regions of the cerebral cortex—hence the term *cortically dependent thalamic nuclei* for such centres. Again, some nuclei, whilst having local and intrinsic connexions, are predominantly in receipt of a discrete, often somatotopically organized input tract, and give rise to an equally well-organized projection to the cortex. These are often termed *relay nuclei* in contrast to *association nuclei* in which multiple subcortical connexions predominate. However, the former are not to be regarded as 'simple' relays since complex transformations of the flow of information are the rule in such centres. Further, based upon focal stimulation of the thalamus, together with recording from the cerebral cortex, thalamic nuclei have often been termed either *specific* or *non-specific*. In the former, a rapid, sharply localized, ipsilateral response is recorded, whilst with the latter the response is widespread, bilateral and occurs after a greater latency. With increasingly refined methods of investigation, however, many nuclei previously thought to have no connexion with the cerebral cortex are now shown to be 'cortically dependent', and the distinction between specific and non-specific nuclei is becoming less clear cut as a considerable measure of interaction between the two is becoming apparent.

SYNAPTIC ORGANIZATION OF THALAMIC RELAY NUCLEI

Studies with the electron microscope have shown that the ultrastructural organization of the principal sensory 'relay' nuclei is extremely complex, and attempts to analyse synaptic

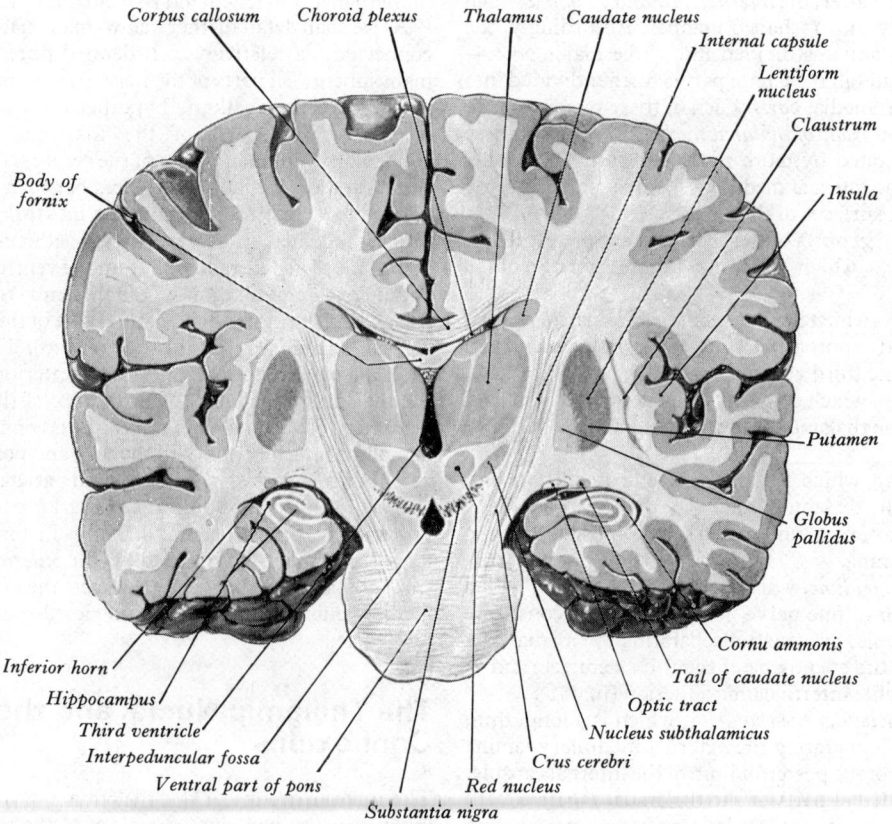

Corpus callosum   Choroid plexus   Thalamus   Caudate nucleus

Internal capsule

Lentiform nucleus

Claustrum

Insula

Body of fornix

Putamen

Globus pallidus

Cornu ammonis

Tail of caudate nucleus

Optic tract

Nucleus subthalamicus

Inferior horn

Hippocampus

Third ventricle

Interpeduncular fossa

Ventral part of pons

Crus cerebri

Red nucleus

Substantia nigra

7.98 Coronal section of the brain through the ventral part of the pons.

relationships are in their infancy. A striking feature of all such nuclei so far studied is the *synaptic glomerulus*, a spherical, ovoid or more irregularly shaped tissue unit from 2 or 3 μm to 20 μm or more along one axis, consisting of closely packed axons and dendrites of different types, establishing with each other a large number and usually a considerable variety of synaptic relationships (Szentágothai 1970; *see* also p. 829 and 7.21). These neural components, which include dendrites or dendritic appendages of the thalamic neurons together with axon terminals of different types, are encapsulated, and separated from the surrounding and generally more simply organized neuropil, by thin, sheet-like astrocyte processes. It is principally within synaptic glomeruli that the specific afferents terminate (for example, the axons of the optic nerve and tract in the lateral geniculate nucleus, and those of the medial lemniscus in the nucleus ventralis posterior lateralis). Because of the synaptic interactions occurring within the environment of the glomeruli, each probably functions to some extent as an independent integrative unit whose output (to the cortically projecting thalamic neuron), following the arrival of an impulse in the specific afferent, is modified according to other variable factors, such as activity in the cortico-thalamic and other afferent systems and the extent of activity in neighbouring specific afferents and synaptic glomeruli.

One of the most interesting recent findings has been the discovery (Morest 1971) in thalamic nuclei of *presynaptic dendrites* morphologically similar to conventional dendrites but containing, in addition to the usual organelles, clusters of synaptic vesicles and establishing synaptic contacts with other dendrites. Some cells with presynaptic dendrites may also have conventional axons; others may resemble the amacrine cells of the retina and olfactory bulb and bear only one type of 'dual-purpose' neurite. More recent ultrastructural analyses (Lieberman and Webster 1972) have shown that the presynaptic dendrites eventually give rise to boutons, packed with distinctive disc-like synaptic vesicles. These dendritic boutons are very similar to axonal boutons, and constitute one of the principal classes of synaptic endings in the thalamic glomeruli. The significance of presynaptic dendrites in the information processing activities of thalamic nuclei, and elsewhere in the nervous system are discussed by Shepherd (1974, 1978); Schmitt *et al.* (1976); Rakic (1975).

THE ANTERIOR GROUP OF NUCLEI

This group is divisible into three nuclei: *anterior dorsalis* (AD), *anterior medialis* (AM) and *anterior ventralis* (AV), the first being the least prominent. The anterior group of nuclei is interconnected with both the medial and lateral groups of thalamic nuclei, and possibly with the contralateral anterior group. It receives some fibres directly from the postcommissural fornix, and the substantial mamillothalamic tract from the mamillary nuclei of the same side, a few fascicles stemming from the opposite side, and it also sends a small contingent of thalamomamillary fibres back to these nuclei. The main radiation from the anterior group are thalamocortical fibres to many parts of the gyrus cinguli (cortical areas 23, 24 and 32), and the anterior group also receives corticothalamic fibres from the same areas. Some authors contend that the anterior nuclei also have reciprocal connexions with the retrosplenial gyrus (areas 29 and 30).

The anterior thalamic nuclei are thus a principal centre linking the activities of the hippocampus and hypothalamus with other thalamic nuclei and with extensive cortical areas which belong to the *limbic system* (p. 990). Stimulation or ablation of the mamillothalamic tract causes alterations in autonomic control and is presumably involved in complex homeostatic cycles involving visceral responses. A normally functioning hippocampus-fornix-mamillary-body/thalamus/limbic-cortex system is now widely considered necessary for the establishment of a *recent memory trace*. In Korsakow's syndrome, characterized by a memory loss for recent events, lesions in the mamillothalamic system are common (Talland 1965; Barbizet 1963). However, experiments in rats (Krieckhaus 1967; Krieckhaus and Randall 1968) failed to

show disturbances of recent memory following such lesions, and the interesting suggestion was made that the system is important in determining the balance between repetitive stereotyped behavioural responses and the initiation of 'novel' exploratory responses under different conditions of environmental stimulation.

THE MEDIAL GROUP OF NUCLEI

This consists of the large *nucleus medialis dorsalis* (MD) and a series of smaller nuclei (*parafascicularis, submedius, paracentralis* and *centralis lateralis*)—the connexions and significance of the latter are uncertain in man and will not be pursued further here.

The *nucleus medialis dorsalis* presents a rostral *pars magnocellularis* and a caudolateral *pars parvocellularis* and is considered to make intrinsic intrathalamic connexions with virtually all the other groups of thalamic nuclei. It receives, in certain regions, fibre connexions from the amygdaloid complex of nuclei (p. 996) and the piriform cortex, and establishes two-way connexions, thalamostriate and striatothalamic, with parts of the corpus striatum (p. 1035), but the precise regions of the latter involved are still rather uncertain.

The most widely described connexions of the nucleus medialis dorsalis of the thalamus are with the hypothalamic nuclei and with the cerebral cortex of the frontal lobe (*vide infra*). The two-way connexions with the hypothalamic nuclei are thought to run in the *periventricular system* of fibres, which are fine myelinated or nonmyelinated fibres and which pursue vertical or oblique courses immediately deep to the ependymal side wall of the third ventricle, where they are intermingled with patches of periventricular grey matter. Some of the periventricular pathways may be polysynaptic, and the system is often described as consisting of a rostrocaudal series of fascicles which connect different parts of the medial thalamus with the preoptic, tuberoinfundibular and mamillary parts of the hypothalamus respectively (p. 966). The most caudal fascicles continue as part of the *dorsal longitudinal fasciculus* (of Schütz) in the periaqueductal grey matter of the midbrain through which they reach numerous brainstem reticular nuclei. It should be noted, however, that not all investigators regard the existence of thalamohypothalamic

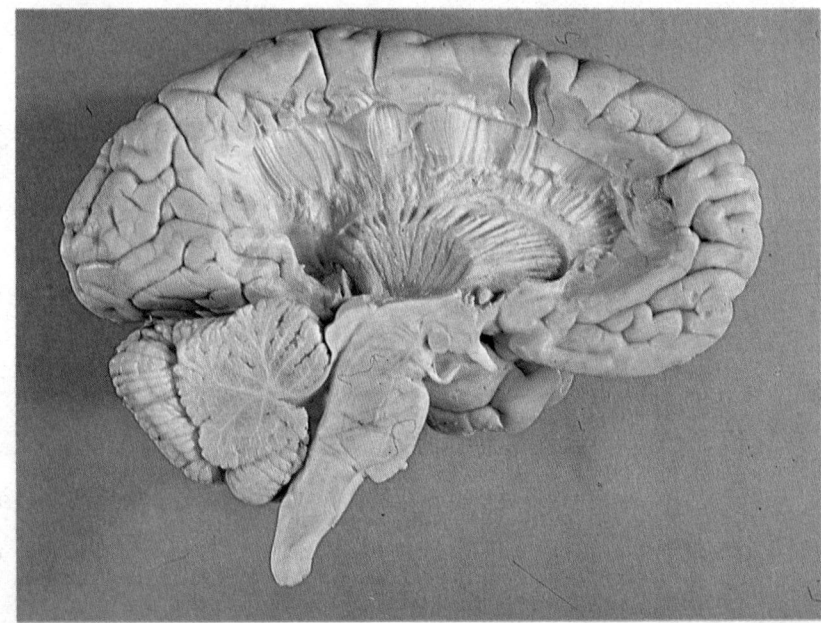

7.99 After median hemisection of the cerebrum, the left cerebral hemisphere has been dissected from its *medial* aspect to display the fibre bundles of the corona radiata and internal capsule. This entailed the removal of the cingulate gyrus and subjacent white matter, much of the paramedian corpus callosum and fornix, the dorsal thalamus, and the head and body of the caudate nucleus. The oval depression previously occupied by the dorsal thalamus can clearly be seen within the curved depression left after removal of the caudate nucleus. (Dissection by Dr. Andrew Seal, Dept. of Anatomy, Guy's Hospital Medical School.)

connexions in the periventricular system as proven (Raisman 1966; Szentágothai *et al.* 1968).

Profuse reciprocal thalamocortical and corticothalamic connexions exist between the nucleus dorsalis medialis and almost the whole of the prefrontal cortex of the frontal lobe of the cerebral hemisphere which lies cranial to cortical areas 6 and 32 (p. 1003). These radiations show an orderly topographic arrangement, specific regions of the nucleus connecting with specific cortical areas.

Clearly the medial part of the thalamus is a region of great functional significance, as evidenced by the wealth and complexity of its interconnexions; but equally it is impossible to attach simple functional labels to such a region of the nervous system. It probably provides mechanisms for the integration of a great variety of information channels, olfactory, visceral and somatic, by the convergence of multiple paths via the hypothalamus, amygdaloid complex, piriform cortex, corpus striatum and prefrontal cortex, and in turn has an output which can affect these and many other regions of the nervous system.

Stimulation of the dorsomedial thalamic nucleus in monkeys elicits movement patterns even after ablation of cortical motor areas, but the patterns disappear on ablation of the corpus striatum. Stimulation, disease or surgical ablation of the nucleus in man results in complex changes in motivational drive, the ability to solve problems and in the consciousness level, general 'personality' and subjective feeling states or 'affective tone' of the individual. The effects of ablation parallel in many ways the results of prefrontal lobotomy

THE VENTRAL GROUP OF NUCLEI

The ventral thalamic group form a rostrocaudal sequence of three main nuclei: *ventralis anterior* (VA), *ventralis intermedius* (VI) and *ventralis posterior* (VP). (Many workers use the term ventralis lateralis (VL) rather than the official term ventralis intermedius.) The *nucleus ventralis posterior* can be further subdivided into the important nuclei *ventralis posterior lateralis* (VPL) and *ventralis posterior medialis* (VPM), together with smaller and less well-understood nuclei which are inferior and oral in position.

The connexions of the main nuclei in this group include some intrinsic ones with adjacent thalamic nuclei such as the centrum medianum and pulvinar (*vide infra*), and also with neighbouring centres in the subthalamus and corpus striatum. However, by far the most prominent connexions of these nuclei are with the large ascending sensory systems, the medial, spinal and trigeminal lemnisci, and gustatory pathways, and also two-way links with the cerebral cortex. The origin, course and constitution of these ascending systems has already been described. The VPL receives the terminals of the medial and spinal lemnisci, whilst the VPM receives the trigeminal and gustatory pathways. (The cells receiving trigeminal terminals are sometimes called the *arcuate* or *semilunar nucleus*, whilst the gustatory fibres are sometimes held to end in an *accessory arcuate nucleus*.) Much research has been directed towards an understanding of the structural and functional localization and the types of interaction which occur during impulse transmission through these thalamic nuclei.

Briefly, there is an overall somatotopic pattern in the ascending pathways, in the nuclei themselves and in their connexions with the cerebral cortex. Both medial lemniscal and spinothalamic fibres from caudal body segments (i.e. from the nucleus gracilis in the case of medial lemniscal fibres) end in the lateral zone of the VPL, while fibres from successively more cranial body segments end in more medially placed zones in the nucleus. Still more medially placed is the VPM, which receives the trigemino-thalamic fibres conducting information from the face and head, whilst most medial are the solitariothalamic gustatory fibres from the nucleus of the solitary tract. Other authors have emphasized that whilst such a lateromedial localization undoubtedly exists, the various tracts are not simply superimposed but end in different rostrocaudal regions. The lateral spinothalamic tract has been described as ending in the caudal nuclear regions followed by the medial lemniscal and anterior spinothalamic fibres, while the trigeminothalamic fibres end in a craniomedial position. The terminations of medial lemniscal and spinothalamic tract

fibres show further contrasting features. The lemniscal fibres are wholly crossed, originating exclusively in the gracile and cuneate nuclei of the opposite side and their terminals are confined to the VPL. Whilst the majority of the spinothalamic fibres are also crossed, an appreciable number ascend on the same side and terminate in the ipsilateral thalamus. Further, although many spinothalamic terminals occur in the VPL, others proceed to a number of alternative destinations, including restricted parts of the medial geniculate nucleus, a small suprageniculate nucleus and various 'non-specific' thalamic nuclei—these are sometimes called the *posterior* (PO) *group* of terminals (Poggio and Mountcastle 1960, 1963). The latter may receive the terminals of high threshold spinal cord neurons which transmit nociceptive information (p. 890). The PO group of nuclei may have reciprocal connexions with the secondary sensory area SmII (p. 1015). Studies of the fine morphology of ascending tract terminals in the VPL have been carried out with the Golgi technique Scheibel and Scheibel (1966). Spinothalamic and reticulothalamic terminals form diffuse interweaving networks, medial lemniscal fibres end as small cone-shaped bushy terminals arranged in regular laminae, whilst corticothalamic fibres have terminal arborizations confined to narrow disc-shaped 'sheets' of tissue which encompass the territories of many lemniscal fibres. Ultrastructural studies have demonstrated the presence of small local interneurons within the VPL (Tombōl 1967), whilst others have started to explore the detailed synaptology of the nucleus (Andersen *et al.* 1964 *a* and *b*; Lieberman and Webster 1972; Shepherd 1974; Rakic 1975). Unit recordings have been taken with microelectrodes inserted into cells of the VPL during medial lemniscal and spinothalamic tract activity in the monkey, by a number of investigators, and also in man, before the placing of stereotactic lesions in the thalamus for various types of locomotor disease. In the monkey, units responding to lemniscal activity retained many of the features described previously for the cells of the gracile and cuneate nuclei. The VPL cells were in this case highly specific for both the type of stimulus and the bodily site of origin. Units responded to contralateral stimulation which consisted of either mechanical distortion of the skin or hairs, or joint movement, or static joint position, or sinusoidal tissue vibration, but never to two of these varieties. Their receptive fields were small and sharply localized, the smallest being recorded by stimulation of the terminal segments of the limbs. Units responding to spinothalamic tract activity in general showed rather larger receptive fields, and whilst some were 'modality specific' others showed interaction due to convergence of impulses from different types of stimulus pattern (p. 891). Impulse transmission in the VPL is now shown to be a most complicated series of events. Many of the units show the phenomenon of lateral inhibition of 'inhibitory surround' discussed previously (p. 808) which greatly enhances their discriminative value. Furthermore, transmission in the nucleus may be modulated by activity in descending corticothalamic fibres and in those which converge on the nucleus from other thalamic and extrathalamic sources. In this regard physiological studies have demonstrated that both presynaptic and postsynaptic inhibitory and facilitatory effects occur in the nucleus.

The recordings made in the nucleus ventralis posterior lateralis of the *human* thalamus, although necessarily more restricted in scope, substantially support the findings concerning somatotopic localization and modal specificity of units demonstrated in the monkey. Furthermore, localized stimulation of the nucleus in conscious subjects evoked sharply circumscribed sensations described as 'tingling', 'numbness', etc. on the opposite side of the body.

The main thalamocortical radiations from the VPL and VPM proceed through the posterior limb of the internal capsule to the primary somatic sensory areas of the cerebral cortex—the postcentral gyrus which includes cortical areas 1, 2 and 3 (p. 1003). Throughout this radiation the precise somatotopic organization of the nuclei of origin is preserved. The same cortical areas project reciprocal corticothalamic fibres back to the nuclei. Unfortunately, the thalamic projections to the secondary somatic sensory areas cannot yet be regarded as settled, although possibly some stem from the PO group (p. 962).

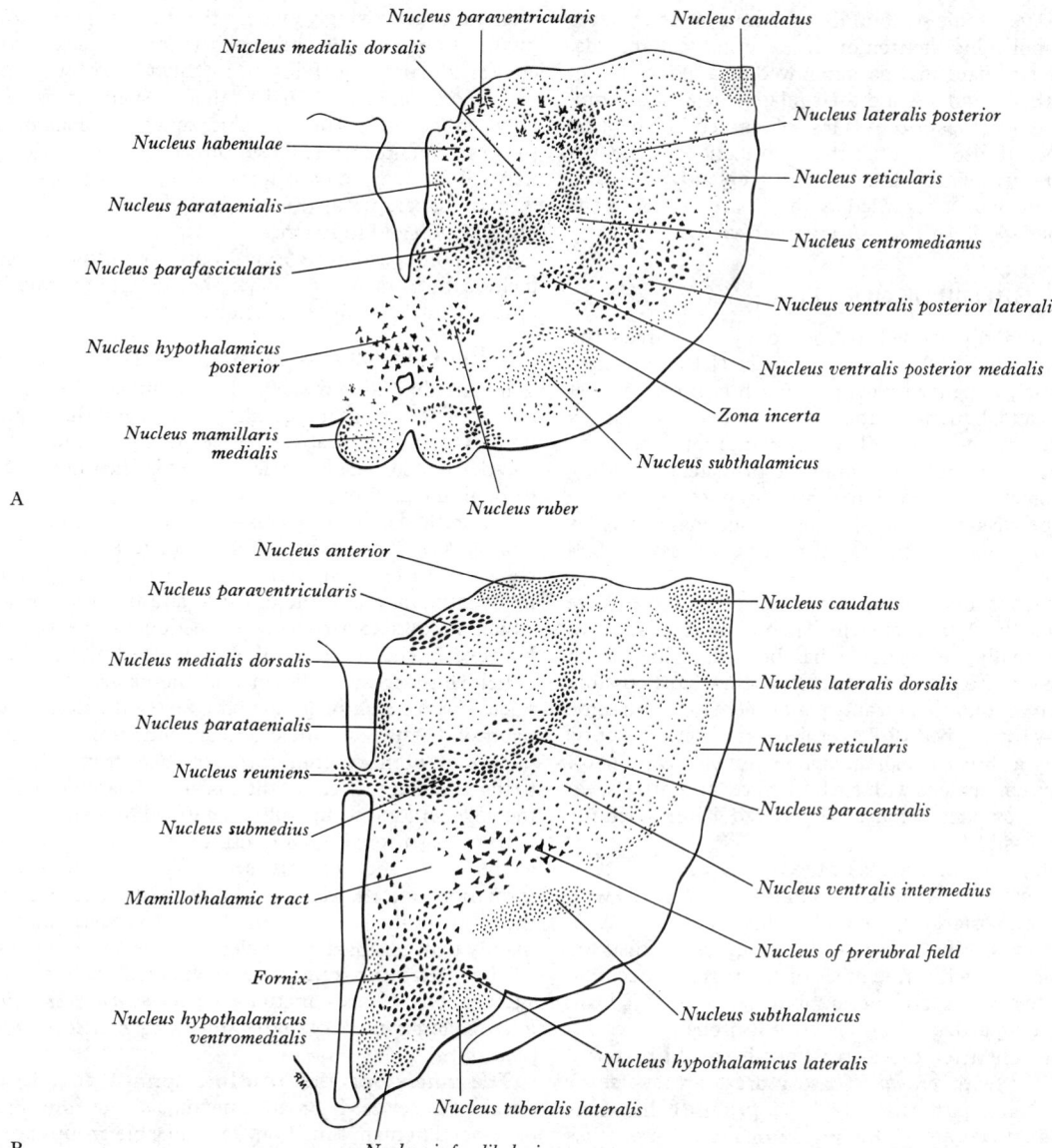

7.100A and B Drawings of coronal sections through the diencephalon, stained with the method of Nissl to show the main nuclear aggregations of nerve cell somata. A at the level of the mamillary bodies; B at the level of the tuber cinereum. Note the variations in cell size, shape and packing density, which characterize the nuclear masses of the dorsal thalamus, subthalamus and hypothalamus at these levels.

The *nucleus ventralis intermedius* (VI, which includes the nucleus ventralis lateralis, VL, of many authors) has been less extensively investigated. It establishes connexions with adjacent thalamic nuclei but its main sources of input are from the contralateral nucleus dentatus of the cerebellum via the dentatothalamic tract and rubrothalamic fibres of the ipsilateral red nucleus (pp. 918, 939). It also receives some fascicles from the globose and emboliform nuclei of the cerebellum and from the globus pallidus. A nigrothalamic projection has been described in monkeys (Carpenter *et al.* 1976). These various afferents are constituents of the *thalamic fasciculus* (*see* p. 965).

The main projection from the VI is somatotopically arranged and passes through the internal capsule to reach the motor and premotor regions of the cerebral cortex (areas 4 and 6). The caudal part of the nucleus has often been selected for neurosurgical ablation in cases of severe Parkinsonism.

The *nucleus ventralis anterior* (VA) has a complex organization and has come into greater prominence in recent research, together with the nucleus ventralis intermedius, because of the therapeutic placing of destructive lesions in or near these nuclei in various types of locomotor disorder.

The nucleus ventralis anterior has numerous interconnexions with other thalamic nuclei, but in particular with the *nucleus centromedianus* (*vide infra*), the other *intralaminar nuclei*, the nuclei of the midline, and the *reticular thalamic nucleus*, i.e. all the so-called 'non-specific' thalamic nuclei. In addition the VA receives an input from the brainstem reticular formation and profuse connexions from the globus pallidus through the *thalamic fasciculus* (*see* p. 965), together with the most rostral of the dentatothalamic fibres of the superior cerebellar peduncle. Carpenter *et al.* (1976) have adduced corroborative evidence for a nigrothalamic projection to VA in monkeys. Thalamocortical fibres radiate from the VA to the premotor cortex (area 6), to a lesser extent to the motor cortex (area 4), and fibres to restricted regions of the insular cortex have also been described. In turn corticothalamic fibres leave areas 6 and 4 and converge upon the VA. (For a detailed Golgi study of the nucleus see Scheibel and Scheibel 1966.)

Thus, the VA is an important focal centre through which the corpus striatum, the ascending reticular formation and non-specific thalamic nuclei, and to a lesser degree the cerebellum, are able to exert a powerful effect on the activities of the motor and premotor cortices and probably on many other regions of the cortex as well. To what degree the information flowing in the various channels which converge on the VA is integrated, before onward transmission to the cortex, is not clear.

Views concerning the possible functional significance of the VA are rapidly changing. Some relationship to the motor control systems is evident—stimulation of the nucleus increases Parkinsonian rigidity and tremor, whilst ablation of the nucleus (often with, however, some surrounding tissue) may be effective in reducing or abolishing the tremor. In experimental animals, stimulation of the caudate nucleus is followed by a reduced level of activity in both VL and VA and their related areas of cerebral cortex, and this is accompanied by a loss of learned behavioural responses. It should also be noted that stimulation of the VA causes *desynchronization* of the electroencephalogram, and increasingly the nucleus is regarded as an important link in the final stages of the '*ascending activating system*' (see also p. 953).

THE LATERAL GROUP OF NUCLEI

This group is commonly divided rostrocaudally into three—the *nuclei lateralis dorsalis* (LD), *lateralis posterior* (LP), and most caudally the greatly expanded *pulvinar*, which occupies almost the whole of the caudal quarter of the thalamus. The pulvinar is a late phylogenetic development, only becoming prominent in the higher mammals and increasingly so in the primates, including mankind. It is sometimes divided into three or more subregions. The separation between the subregions of the pulvinar and LP is imprecise and some authors prefer to group them as the *LP-pulvinar complex*.

The three nuclei of the lateral group are all presumed to have interconnexions with other thalamic nuclei, but the details are not yet clear. Additionally, the *pulvinar* has been described as in receipt of connexions from the lateral geniculate nucleus, possibly the medial geniculate nucleus and amygdaloid complex, and some investigators have described direct projections from the optic tract constituting a direct *retinothalamic projection* (*see* p. 976). Whilst numerous connexions with such subcortical centres seem probable, further evidences must be awaited before definite conclusions are possible.

The best established connexions of the lateral group are two-way links with specific regions of the cerebral cortex, the LD with inferior parietal and posterior cingulate regions, LP with much of the parietal cortex behind the postcentral gyrus, whilst the pulvinar interconnects with wide areas of the parietal, occipital and temporal lobe cortices, each of its subnuclei projecting in an orderly fashion to a particular cortical area; it probably receives reciprocal connexions from these areas, but it should be noted that its connexions with *primary* sensory areas are sparse or absent. The pulvinar in particular establishes profuse connexions with Wernicke's speech area. This region of the thalamus has been implicated by different authors in such diverse activities as perception of chronic pain, oculomotor control, speech pattern control, and as a *polysensory data processor* involved in general '*gnosis*' (Ingram 1976).

THE 'NON-SPECIFIC' GROUPS OF THALAMIC NUCLEI

Since the recognition over a quarter of a century ago (Moruzzi and Magoun 1949) that stimulation or ablation of different regions of the brainstem reticular formation was followed by widespread changes in the state of synchronization or de-synchronization of electroencephalograms, attention has been focused upon possible cranially directed pathways from the reticular formation to the cerebral cortex. The suggestion soon followed that the brainstem reticular centres established profuse connexions with a series of thalamic nuclei from which proceeded a widespread *diffuse thalamocortical radiation* to virtually all parts of the cerebral cortex. Further, since stimulation of a number of these nuclei caused extensive, *bilateral* changes in cortical electrical activity, which were characterized by a slow onset and then a progressive build-up to a maximum, followed by fluctuations in level (recruitment)—the responses, and the nuclei themselves were called *non-specific*. It was widely assumed that the latter effects provided a background level of preparedness of the cortex which is necessary for the effective reception of the rapid, sharply localized ipsilateral, impulse patterns stemming from the larger *specific relay nuclei*, which have precise somatotopically organized thalamocortical and corticothalamic connexions. A large research literature has accumulated around this subject and, whilst many of the general concepts remain valid today, it is increasingly evident that the *reticulocortical pathways* involved are *multiple* and much more complex than originally envisaged, some involving extrathalamic pathways through the ventral diencephalon. Further, there is such extensive interaction between specific and non-specific parts of the thalamus that whilst the terms retain a general usefulness, and are so widely used that they will be retained here, it must not be assumed that they imply a sharp, separable distinction between two aspects of thalamic organization and function.

The non-specific nuclear groups are usually considered to include the intralaminar nuclei, the nuclei of the midline, and the reticular nucleus of the thalamus.

The Reticular Nucleus of the Thalamus

This is a thin curved sheet of cells situated between the white matter of the external medullary lamina and that of the posterior limb of the internal capsule. Accordingly, all the corticothalamic and thalamocortical fibres which run in the internal capsule pass through the territory of the nucleus.

The reticular nucleus was for long regarded as an important final link in the diffuse thalamocortical radiation mentioned above, because of widespread effects on cortical activity which follow stimulation of the nucleus, and the nuclear degeneration which follows experimental resection of the cerebral cortex. However, studies using the Golgi technique (Scheibel and Scheibel 1966) have shown that the axons of the cells in this nucleus run *caudally*, giving off a series of collateral branches to many of the thalamic nuclei and the midbrain reticular formation. The whole of the cerebral cortex projects in an orderly manner on to the reticular nucleus of the thalamus (Carman *et al.* 1964), and the degeneration which follows cortical resection is now thought to be transneuronal and not retrograde in type (Rose 1952). Other afferents to the nucleus are from the brainstem reticular formation, and the globus pallidus. Thus, afferents converge on the reticular nucleus from these various sources and its output is mainly to other thalamic nuclei both specific and non-specific.

The intralaminar nuclei include a number of small aggregations of grey matter (the *nuclei paracentralis, centralis lateralis* and *limitans*) and the much larger *nucleus centromedianus* (*vide infra*).

The nuclei of the midline form a complex and well-developed series in many mammals but are often poorly developed in man. Small but recognizable groups are sometimes to be found near the taenia thalami, the interthalamic adhesion when present, and others scattered in the periventricular wall. They will not be named and described individually here.

Because of their topography the non-specific nuclei are extremely difficult to investigate both anatomically and physiologically. In general, they are thought to receive the terminals of ascending fibres from the brainstem reticular formation, while some establish connexions with the corpus striatum, cerebellum, spinothalamic tracts, the hypothalamus, and many intrathalamic connexions with specific and non-specific nuclei on both sides of the midline. The most cranial of the non-specific nuclei project to the cerebral cortex including the phylogenetically older prepiriform and entorhinal cortices, together with the parieto-occipital and frontal regions of the neocortex. Clearly, much remains to be learned about the arrangement, connexions and significance of these nuclei.

The nucleus centromedianus, which is embedded in the white matter of the internal medullary lamina, only becomes a prominent structure in primates, and it is an easily recognized landmark in the human thalamus. Despite intensive investigation there has been much disagreement concerning its status and connexions.

However, the majority of investigators are now agreed that it is not connected with the cerebral cortex. A variety of ascending pathways have been described as ending partly in this nucleus; these include collaterals or terminals from a small proportion of the fibres of the spinal, medial and trigeminal lemnisci, ascending reticulothalamic fibres, and a contribution from the superior cerebellar peduncle. However, its main connexions are with parts of the corpus striatum, some of which are topically organized, and other profuse interconnexions are with the remaining non-

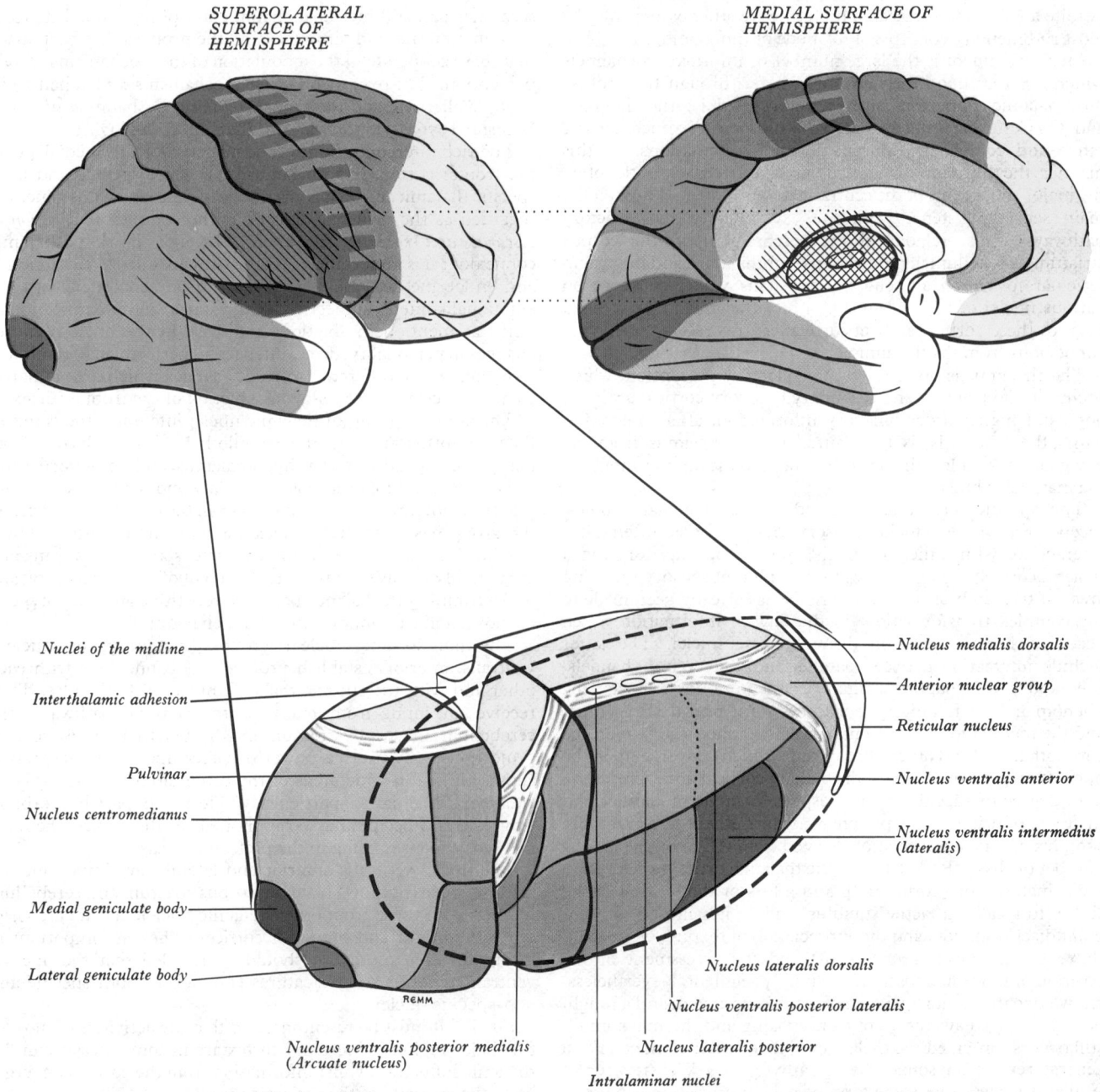

*SUPEROLATERAL
SURFACE OF
HEMISPHERE*

*MEDIAL SURFACE OF
HEMISPHERE*

Nuclei of the midline

Interthalamic adhesion

Pulvinar

Nucleus centromedianus

Medial geniculate body

Lateral geniculate body

Nucleus medialis dorsalis

Anterior nuclear group

Reticular nucleus

Nucleus ventralis anterior

Nucleus ventralis intermedius
(lateralis)

Nucleus lateralis dorsalis

Nucleus ventralis posterior lateralis

Nucleus lateralis posterior

Nucleus ventralis posterior medialis
(Arcuate nucleus)

Intralaminar nuclei

7.101 The main nuclear masses of the dorsal thalamus (below) have been labelled and colour coded, and the same colours have been used to indicate the areas of cerebral neocortex interconnected with these nuclei. The lack of colour in the centromedian, intralaminar and reticular nuclei, and in restricted areas of the frontal and temporal lobes are *not* related to the colour code. The boundaries of the coloured cortical zones may well need revision in the future as experimental and pathological data accumulate.

specific nuclei on both sides of the midline, and also with some 'specific' nuclei, particularly the nucleus ventralis anterior.

Further connexions of the dorsal thalamus. All the connexions of the dorsal thalamus mentioned in the previous pages were established by the older neuroanatomical and neurophysiological techniques. However, data are now accumulating rapidly in a variety of mammals with the advent of the newer tracing techniques (p. 819) and in particular, the methods for locating the somata and neurites of cholinergic and monoaminergic neurons. These were treated in some detail in the section devoted to 'The Reticular Formation' (pp. 943–953) and will merely be mentioned here. The *dorsal tegmental ascending cholinergic bundle* provides terminals for the nuclei of the geniculate bodies, the dorsally and anteriorly situated nuclear groups of the dorsal thalamus, and additionally, terminals are found in the midline and intralaminar nuclei, especially the nucleus centromedianus. The *ventral tegmental ascending cholinergic bundle* also provides terminals for the ventrally and anteriorly situated nuclear groups. The *dorsal ascending*

noradrenergic bundle provides an afferent innervation to many, perhaps all, the nuclei of the dorsal thalamus. *Dopaminergic neuron cell aggregate* A11 sends fine axons that weave through, and terminate in, the nuclei of the midline, whilst aggregates A13 and 14 similarly innervate the caudal nuclei of the dorsal thalamus. The *ventral ascending serotonergic bundle*, in addition to many other destinations, provides axons that innervate the medial nuclear group, the nuclei of the midline, and the parafascicular nucleus. For brief comments on fibres containing met-enkephalins, endorphins, substance P, and the distribution of opiate receptor sites, *see* pp. 833, 951, 962. The functional roles of these multiple varieties of fibre converging on the dorsal thalamus are not clear at present, and progress in this field will be awaited with great interest.

SUMMARY

Clearly, the dorsal thalamus is a region of immense structural complexity and of outstanding functional significance, but despite a voluminous research literature we still possess only the most rudimentary concepts of its mode of functioning.

Most prominent is the large number of information channels which *converge* upon the thalamus where, through the profuse intrathalamic connexions, many are *integrated*, i.e. they interact, and the resulting information patterns, now of much greater range and complexity, *diverge* to many destinations. In this manner, the thalamus is essentially involved in the activities of *all* the major subregions of the central nervous system. Thus, all the main sensory systems (with the exception of the olfactory pathways), most regions of the cerebral cortex, the corpus striatum, cerebellum, hypothalamus, subthalamus and brainstem reticular formation all send fibre systems which converge on various nuclei of the thalamus and metathalamus, and in turn, most of these cortical and subcortical centres receive reciprocal connexions from the thalamus.

The thalamus is not essential for olfactory perception, which occurs in the primary and secondary olfactory cortical areas (p. 995), but higher order olfactory information, after integration with other channels, is transmitted to the thalamus from the amygdaloid complex, the piriform lobe, and the hippocampus via the mamillary body.

The specific relay nuclei, including the geniculate bodies, receive the various major sensory tracts where, often after interaction with other channels, somatotopically organized thalamocortical radiations pass to the different sensory receiving areas of the cerebral cortex. Reference has already been made to the complex transformations in the flow of information which occur during transmission through these nuclei. The latter include interaction between parallel and converging channels, pre- and post-synaptic facilitatory and inhibitory effects, the phenomenon of inhibitory 'surround' with neural sharpening, and the modulation of transmission which follows activity in the corticothalamic and other fibre systems which converge upon the nucleus in question. Some of the thalamocortical fibres preserve a high degree of specificity with respect to single modalities and their site of origin in the periphery, whilst others transmit more complex orders of information derived from the convergence of a number of channels. Ablation of the thalamocortical radiations to the somatosensory cortex results in a loss or diminution of the ability to localize a tactile stimulus, and an impairment of two-point discrimination, and the appreciation of texture, weight and shape of an object (*astereognosis*) and the assessment of the position or movement pattern of a bodily segment. Nevertheless, knowledge that contact with an object has occurred, and a rough, poorly localized awareness of pain-evoking and thermal stimuli, still occurs, provided the thalamic mechanisms are intact. (For a general review on somaesthetic pathways *see* Webster 1976.) Whilst it is clear that some form of pain appreciation can occur after loss of the cerebral cortex, the degree to which the cortex is involved in an intact nervous system has still not been solved (Albe-Fessard and Delacour 1968). Disease in the lateral or central thalamus is sometimes followed by sudden apparently 'spontaneous' attacks of an unexplained type of *thalamic pain*. Paradoxically, in addition to spinal, medullary, or mesencephalic tractotomy as surgical measures for the relief of pain, selective thalamotomy with ablation of either the nucleus ventralis posterior and surrounding regions, or of the intralaminar nuclei including the nucleus centromedianus, has sometimes been effective in pain relief. In summary, therefore, it appears that the VPL and VPM, the so-called PO group (including areas of the medial geniculate, suprageniculate and intralaminar nuclei), the nucleus centromedianus, probably the nucleus dorsalis medialis and pulvinar, together with many hypothalamic nuclei, the globus pallidus and caudate nucleus of the corpus striatum, and limbic structures such as certain amygdaloid nuclei and the bed nucleus of the stria terminalis, are all involved in pain appreciation, but how they cooperate in this activity remains obscure. It may be recalled that many of these regions (including laminae I, II, and V of the spinal cord grey matter, the lateral reticular nucleus, the nucleus ambiguus, and nucleus of the solitary tract in the medulla, areas of the mesencephalic periaqueductal grey and of the pars compacta of the substantia nigra) are all in receipt of monoaminergic (particularly

serotonergic), and met-enkephalin, endorphin, and substance P containing fibres and terminals, and the proposal has been made that these cooperate in the modulation of the mechanisms of pain perception. (For reviews on spinal mechanisms *see* Nathan 1977, 1978; Wall 1978; and for general reviews, Basbaum *et al.* 1976; Webster 1976, 1978; Snyder, in: Reichlin *et al.* 1978.)

The rich interconnexions of the nucleus dorsalis medialis with the frontal cortex, hypothalamus, and other specific and non-specific thalamic nuclei has already been described. It is generally regarded as the principal centre for the complex *integration of visceral and somatic functions*. Through its hypothalamic connexions it is importantly involved in a wide array of autonomic and endocrine activities. It appears to be concerned with the emotional content, subjective feeling states, and identification of 'self'; as mentioned, ablation or disease in the nucleus causes changes in personality, drive, intellectual performance, emotional level etc., and indifference to pain, changes which are similar in some respects to those following ablation of the frontal cortex.

The anterior group of thalamic nuclei integrates the complex flow of information along the mamillothalamic tract derived from many visceral pathways, the hypothalamus, and limbic structures (including the hippocampus, amygdala and septal areas), with that from other thalamic nuclei and from the cingulate cortex, and the group has reciprocal connexions with these centres. These circuits are important in complex homeostatic mechanisms and are possibly involved in the establishment of recent memory, and in determining the balance between 'repetitive and stereotyped', or 'novel and exploratory' forms of behaviour.

The non-specific nuclear groups (including the nucleus ventralis anterior) establish profuse interconnexions with each other and with the specific thalamic nuclei of both sides. They receive an input from many of the sensory pathways, the cerebellum and corpus striatum, and through their diffuse cortical projections they exert a powerful effect upon the background activity levels in wide areas of the cerebral cortex—the *arousal reactions*. The latter particularly affect the parietal, orbital, cingulate and occipital association areas of the cortex, the most marked responses being in the prefrontal areas.

The nuclei ventralis anterior and lateralis are important cell stations where outflows from the corpus striatum and cerebellum interact with those from other thalamic nuclei, and then project on to the motor and premotor cortices. They are important in locomotor control, and it should be recalled that the nucleus ventralis anterior shows features common to both specific and non-specific nuclei.

Finally, it must be re-emphasized that the activities of most of the major regions of cerebral cortex are in some measure under thalamic influence, whilst information from the cortex converges on to the majority of the thalamic nuclei.

The Epithalamus—Pineal Gland and Habenula

The epithalamic structures, as stated previously, occupy the caudal roof of the diencephalon together with adjacent areas on the side walls of the third ventricle (7.95). They include the right and left *habenular nuclei*, each situated deep to the floor of a *habenular trigone*, and each receives the termination of a complex fibre bundle, the *stria medullaris thalami*. The epithalamus also includes the midline *pineal gland* or epiphysis cerebri, and the *habenular* and *posterior commissures*, which cross the midline in the cranial and caudal laminae of the pineal stalk. The development of the pineal gland is briefly mentioned on p. 169, and its structure and possible neuroendocrine roles on p. 1445, together with some proposals concerning its nerve supply.

The *trigonum habenulae* is, as its name suggests, a small triangular surface depression on each side, situated cranial to the superior colliculus and medial to the pulvinar of the thalamus from which it is separated by the *sulcus habenulae*. The trigone is bounded rostromedially by the caudal end of the ridge occupied by the stria medullaris thalami and by the pineal stalk.

The habenular nucleus, sometimes regarded as consisting of medial and lateral parts, is a station on some of the olfactory reflex pathways, and is probably closely concerned with one of the sources of innervation of the pineal gland. A number of connexions have been described but their complete pattern is still uncertain. Many *afferents to the nucleus* run in the *stria medullaris thalami* which is formed near the anterior pole of the thalamus. These include connexions from the amygdaloid complex of nuclei via the stria terminalis (p. 996), and from the hippocampal formation via the fornix (p. 1001). Other components of the stria are from the olfactory tubercle, anterior perforated substance, the preoptic and septal areas and various hypothalamic nuclei. Direct tectohabenular fibres from the superior colliculi have been described.

The stria medullaris courses across the superior part of the medial surface of the thalamus, skirts the medial margin of the trigonum habenulae, and many of the fibres end in the ipsilateral nucleus habenulae. Others, however, cross the midline in the cranial leaflet of the pineal stalk, interlacing with their fellows to form the *habenular commissure*, so reaching the contralateral habenular nucleus. Some of the fibres which also follow this route are truly commissural and interconnect the amygdaloid complexes and the hippocampal cortices of the two sides; crossed tectohabenular fibres accompany them in the commissure. Additional afferent axons that provide terminals for the habenular nuclei are 5-hydroxytryptamine-containing fibres derived from the *ventral ascending tegmental serotonergic bundle* that join the habenulopeduncular tract (*vide infra*) and pass, in the opposite direction to its main contingent of fibres, to reach the nuclei. Thus, these fibres may be involved in exerting some control over the habenular neurons that give origin to the *habenulopineal tract*, thereby influencing the activity of the innervation of the pinealocytes (*see* p. 1445). Similar remarks apply to habenular nuclear innervation by fine fascicles from the *dorsal ascending tegmental noradrenergic bundle* (pp. 940, 949).

Thus, although the habenulae are relatively small in man, they provide a nodal point for the integration of a considerable variety of olfactory, visceral and somatic afferent pathways.

The main outflow paths from the habenular nuclei pass to the interpeduncular nucleus, the nucleus medialis dorsalis of the thalamus, the tectum, and to the reticular formation of the midbrain tegmentum. The largest of these paths constitutes the *habenulopeduncular tract*—often known as the *fasciculus retroflexus* (of Meynert) (**7.**103). The latter courses rostroventrally, skirts the caudal zone of the nucleus medialis dorsalis of the thalamus, and then passes through the rostromedial part of the red nucleus to reach the interpeduncular nucleus from which nerve fibres relay to the midbrain reticular formation. From these centres, descending fibres such as the tectotegmentospinal tracts and the dorsal longitudinal fasciculi ultimately connect with the autonomic preganglionic centres through which control of salivation, gastric and intestinal secretory activity and motility are effected, and others pass to the motor centres for the muscles concerned in mastication and swallowing. It should also be noted that ablation of the habenular complexes in experimental animals leads to extensive changes in metabolism, endocrine regulation and thermo-regulation (Szentágothai *et al.* 1962).

The posterior commissure is a complex fibre bundle which crosses the midline in the caudal lamina of the pineal stalk. Its size is relatively reduced in primates and its fibre constitution is imperfectly known in man. It acquires its myelin sheaths early, and some estimates of the number of fibres in different specimens of the commissure have been made (Tomasch and Malpass 1958). Various nuclei are associated with the commissure—small groups scattered along its length and sometimes called the *interstitial nuclei of the posterior commissure*, accumulations in the periventricular grey matter which constitute *dorsal nuclei of the posterior commissure*, the *nucleus of Darkschewitsch* in the cranial part of the periaqueductal grey matter, and the *interstitial nucleus* (of Cajal) situated near the rostral end of the oculomotor nucleus and closely linked with the medial longitudinal fasciculus (p. 939). Fibres from all these nuclei, and continuations from the medial longitudinal fasciculus all cross the midline in the posterior commissure. Other centres which contribute fibres to the commissure include the posterior thalamic nuclei, the pretectal nuclei, the superior colliculi, and connexions between the tectum

and habenular nuclei. The precise destinations and functional significances of many of these bundles in the human brain are incompletely understood and will not be discussed further here.

Caudal and ventral to the commissure (i.e. in close relation to the caudal wall of the pineal recess), the ependymal cells lining the dorsal aspect of the cerebral aqueduct are specialized, being tall, columnar and ciliated, with a granular basophilic cytoplasm and characteristic histochemical reactions. This patch of cells, which are possibly secretory, with their products passing into the cerebrospinal fluid, is sometimes termed the *subcommissural organ* (Keene and Hewer 1935; Wislocki and Leduc 1953; Møllgård *et al.* 1973). They may also be involved in the transport of materials, gained from neighbouring nerve terminals or blood capillaries, to the cerebrospinal fluid. Alternatively, substances taken up from the cerebrospinal fluid may be transported to more deeply placed nerve cell somata, blood vessels or pinealocytes (*see* p. 969, and cf. the infundibular recess of the third ventricle—pp. 966, 1445). For a review of the possible neuroendocrine role of specialized ependymal cells (tanycytes) *see* Knowles (1974); a number of references throughout Reichlin *et al.* (1978) and especially, Joseph and Knigge (1978). It may also be noted here that in addition to the specialized ependyma of the median eminence and infundibular recess, and that of the subcommissural organ, another patch of similarly specialized ependyma projects towards the cavity of the third ventricle from its anterior wall between the diverging columns of the fornix; it is called the *subfornical organ*, or the *intercolumnar tubercle*. These (and in some vertebrates, other) various specialized regions of the lining of the third ventricle are sometimes collectively termed the **circumventricular organs**.

The Ventral Thalamus or Subthalamus

As mentioned previously, the pars ventralis diencephali, which lies caudoventral to the hypothalamic sulcus, may be divided on phylogenetic, ontogenetic and functional grounds into a *hypothalamus* and a *ventral thalamus* or *subthalamus*. The hypothalamus, which includes the structures forming the floor of the third ventricle as far caudally and including the mamillary bodies, together with the structures embedded in the rostroventral side wall of the ventricle, will be described in a later section. The remainder of the pars ventralis diencephali constitutes the subthalamus. It merges caudally with the tegmentum of the midbrain, and the rostral extensions of the substantia nigra and red nucleus project into the caudal subthalamus, whilst important fibre tracts ascend and descend between the two regions (**7.**103). Dorsally, the subthalamus adjoins the ventral nuclei of the dorsal thalamus, whilst rostromedially it is bounded by the various subdivisions of the hypothalamus (**7.**100). Ventrolaterally and directly laterally, the subthalamus is in contact with the expanding and rotating junctional zone where the cerebral peduncle merges into the internal capsule; and the latter separates the subthalamus from the medial aspect of the globus pallidus of the lentiform nucleus.

The main aggregations of nerve cells in the subthalamus are: (1) the rostral end of the red nucleus; (2) the rostral end of the substantia nigra; (3) the nucleus subthalamicus; (4) the zona incerta; (5) the nucleus of the prerubral or tegmental field; and (6) the entopeduncular nucleus or nucleus of the ansa lenticularis.

The main tracts of nerve fibres in the subthalamus are: (1) the rostral ends of the medial, spinal and trigeminal lemnisci and the solitariothalamic tract, as they approach their terminations in the thalamic nuclei; (2) the dentatothalamic tract from the opposite superior cerebellar peduncle accompanied by ipsilateral rubrothalamic fibres; (3) the fasciculus retroflexus; (4) the fasciculus lenticularis; (5) the fasciculus subthalamicus; (6) the ansa lenticularis; (7) the fibre aggregates of the prerubral field (the H field of Forel); (8) the continuation of the fasciculus lenticularis (in the H_2 field of Forel); and (9) the fasciculus thalamicus (the H_1 field of Forel).

The topographical neuroanatomy of the subthalamus is rather complex and is best understood by reference to *three-dimensional* models and closely graded series of coronal and sagittal sections in

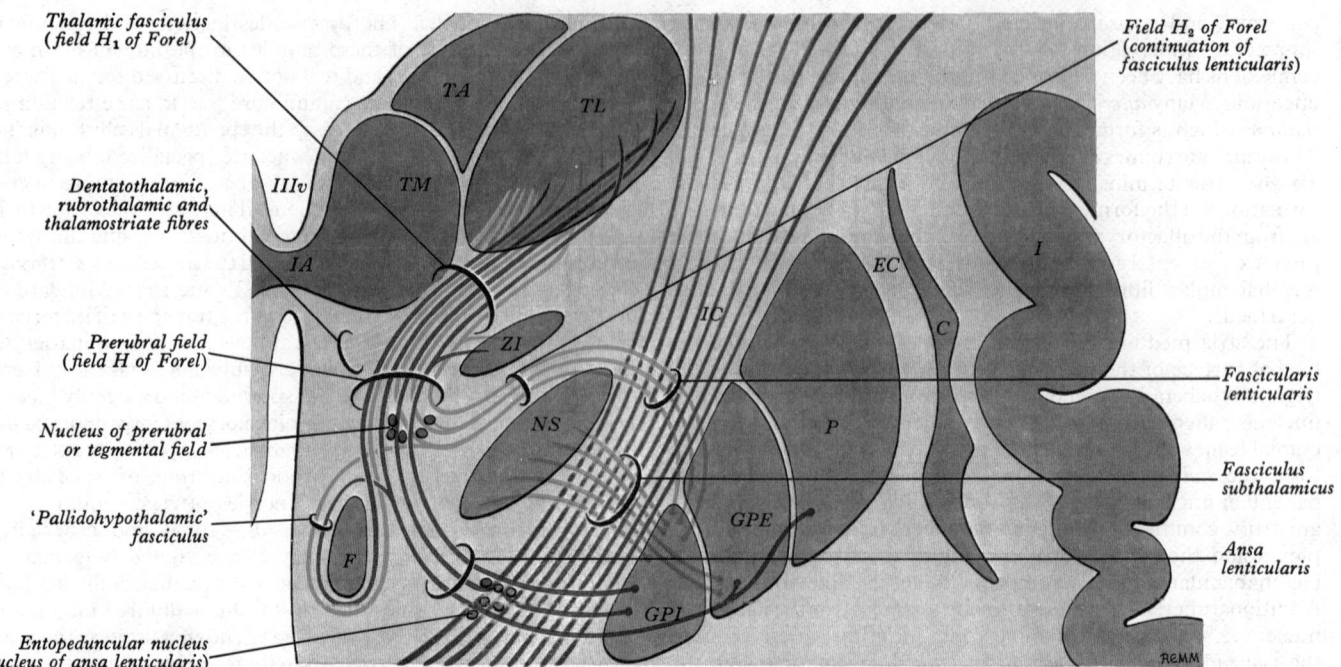

7.102 A diagram of the nuclear masses of grey matter and fibre tract systems associated with, or closely related topographically to, parts of the dorsal thalamus, subthalamus and globus pallidus. The information presented is compounded from a series of closely spaced coronal sections through this region, attempts being made to include what are essentially three-dimensional structures in a two-dimensional diagram; it must be emphasized that all the structures shown would not appear on the same single coronal section. The significance of the letters is as follows: IIIv—third ventricle; M—medial nuclear group of thalamus; TA—anterior nuclear group of thalamus; TL—lateral nuclear group of thalamus; IA—interthalamic adhesion; ZI—zona incerta; F—column of fornix; NS—nucleus subthalamicus; IC—internal capsule; GPI and GPE—internal and external parts of globus pallidus; P—putamen; EC—external capsule; C—claustrum; I—cortex of insula.

the practical laboratory; a parallel study of the corpus striatum (p. 1032) from which a number of the prominent subthalamic tracts are derived is also helpful. Illustration 7.102 may assist with the complicated terminology of the region, but the topography can only be appreciated in part from such a *two-dimensional* diagram.

As the rostral ends of the red nucleus and the substantia nigra pass into the caudal part of the subthalamus they gradually diminish in cross-sectional area to terminate a little caudal to the mamillary bodies. The changing relationship of the lemnisci to the red nucleus and substantia nigra as they ascend through the midbrain has already been described. The lemniscal fibres enter the subthalamus largely lateral to the red nucleus, and as they continue to ascend they pass on to the dorsal aspect of the nucleus to reach the inferior surface of the nucleus ventralis posterior of the thalamus, in which the majority of their fibres terminate. Dentatothalamic and rubrothalamic fibres run in company with pallidothalamic fibres and form part of the thalamic fasciculus, which passes cranial to the termination of the lemnisci noted above to distribute largely to the nuclei ventralis intermedius and anterior of the thalamus (*see* pp. 959, 962, and below for further description).

The nucleus subthalamicus does not occur in sub-mammalian forms, is small in most mammalian groups, and is prominent only in primates. In man the nucleus is an aggregation of medium to large multipolar neurons and classically described as of the general shape of a biconvex lens when seen in coronal sections. It lies in the caudal subthalamus, but extends into the junctional zone of subthalamus and midbrain tegmentum, where it lies dorsolateral to the cranial end of the substantia nigra and lateral to the cranial margin of the red nucleus. The subthalamic nucleus is closely related to the medial aspect of the internal capsule, which separates it from the globus pallidus of the corpus striatum. Medially, the subthalamic nucleus abuts upon the hypothalamic region, whilst dorsally it is separated from the ventral nuclei of the thalamus by a thin strip of grey matter, the zona incerta, which is interposed between complex fibre bundles, the continuation of the fasciculus lenticularis, and the thalamic fasciculus (7.102, and *vide infra*).

The connexions of the nucleus subthalamicus are numerous, some being well-established, whilst considerable uncertainty attaches to the remainder. The principal connexions are with the corpus striatum; two-way fibre systems pass by a number of routes between the globus pallidus and the subthalamic nucleus. Interconnexions with the putamen and caudate nucleus have also been described but are less well documented. The other connexions reported include ones with the opposite subthalamic nucleus and globus pallidus, and with the ipsilateral red nucleus, substantia nigra, the reticular formation of the midbrain tegmentum, the zona incerta and other small nuclei in the subthalamus, various hypothalamic and thalamic nuclei, and possibly with regions of the cerebral cortex. Recently, it has been demonstrated that some of the afferent fibres to the nucleus subthalamicus are derived from the *ventral ascending tegmental cholinergic pathway,* and probably from the *incerto-hypothalamic dopaminergic system*; their specific functional roles remain enigmatic.

The subthalamic nucleus is clearly an important site for the integration of a number of motor control centres, but particularly through its connexions with the corpus striatum and midbrain tegmentum. Relatively discrete lesions of one subthalamic nucleus in man result in the condition of *hemiballismus*, with uncontrollable, violent, torsional movements, choreiform in type, affecting the contralateral side of the body. The movements, which often continue for long periods, usually affect the proximal musculature of one or both limbs and may mimic throwing or kicking; but facial and trunk muscles may also be involved, though with a much lower incidence. Interestingly, unlike most of the other *dyskinesias*, hemiballismus can be reproduced in the experimental animal; in the monkey the condition follows surgical ablation of at least one-quarter of the subthalamic nucleus, provided the surrounding fibre systems associated with the globus pallidus are preserved intact. The hemiballismus so generated is unaffected by ablation of the rubrospinal, vestibulo-spinal and reticulospinal tracts, or area 6 of the cerebral cortex, but the condition *is* abolished by ablation of the globus pallidus or its outflow tracts, the nucleus ventralis anterior of the thalamus, Area 4 of the cerebral cortex or the corticospinal tract. It is assumed, therefore, that the subthalamic nucleus exerts an inhibitory form of control on the globus pallidus, and therefore on its main outflow pathway via the thalamus to the motor cortex, the principal origin of the crossed corticospinal tract (Whittier and Mettler 1949; Carpenter 1950; Carpenter *et al.* 1950, 1951, 1958, 1960).

The zona incerta is a thin lamina of grey matter interspersed with fine fibre bundles which extends throughout most of the diencephalon a little ventral to the thalamus but separated from it by the fibres of the thalamic fasciculus (*vide infra*). Laterally the zone is continuous with the reticular nucleus of the thalamus, whilst ventrally it is in apposition with fibres of the fasciculus lenticularis (*vide infra*). Functionally associated with the zona incerta are scattered groups of neurons along its caudomedial border which constitute the *nucleus of the prerubral* or *tegmental field*, and other groups interspersed between the fibre bundles of the ansa lenticularis (*vide infra*), which some authorities regard as 'detached' parts of the globus pallidus and which constitute the *entopeduncular nucleus*. These various cell groups are to be regarded in the main as stations on discharge pathways from the globus pallidus to the reticular formation in the tegmentum of the midbrain, some of the descending fibres running with the central tegmental fasciculus as far as the inferior olivary complex of nuclei (p. 904); numerous side-fascicles leave these descending bundles to innervate many of the 'reticular' and 'non-reticular' brainstem nuclei. Additionally, fibres descending from the cerebral cortex to terminate in the zona incerta have been described, and these various subthalamic nuclear masses probably also interconnect with the main subthalamic nucleus, the intralaminar and ventral nuclei of the thalamus, and with the red nucleus. The zona incerta has recently been shown to receive terminals from the *ventral tegmental ascending cholinergic pathway*, and to be involved in the *incerto-hypothalamic dopaminergic system* (pp. 948, 949). Much remains to be learned concerning their functional significance.

In addition to the termination of the lemniscal systems and that of the dentatothalamic and rubrothalamic tracts (p. 959), the subthalamus is characterized by a series of bundles of white fibres, often interspersed with small groups of neurons. The fibre constitution of these bundles and their topography is complex, and a number of rather confusing and sometimes conflicting terminologies have been proposed. One approach, adopted here, is to consider the various bundles in relation to the main fibre systems which flow out of the corpus striatum (their relationships to that organ are further discussed on p. 978). The various fibre bundles derived in part from the putamen, but predominantly from the globus pallidus, appear at the surfaces of the latter, from which they diverge medially in a fan-shaped radiation. The dorsal and intermediate fibres of the fan intersect with the fibres of the internal capsule, whilst the ventral fibres curve around the caudoventral border of the capsule. Earlier investigators (von Monakow 1882) termed the whole radiation the ansa lenticularis and considered it to consist of dorsal, intermediate and ventral divisions. The term ansa lenticularis is, however, now restricted to the ventral division, the intermediate division is termed the fasciculus subthalamicus, whilst the dorsal division is termed the fasciculus lenticularis (*see* **7**.102 for an outline of these relationships).

The fasciculus lenticularis consists of the dorsal bundle of pallidofugal fibres which pass through the internal capsule, intersecting with its fibres, from lateral to medial side. They then course medially, closely related to the medial aspect of the internal capsule, and partly intermingled with the dorsal aspect of the subthalamic nucleus and the ventral aspect of the zona incerta; in this part of its course the fasciculus traverses what is sometimes called the *H₂ field of Forel*. Having reached the medial border of the zona incerta, the fibres of the fasciculus lenticularis meet and intermingle with the fibres of the ansa lenticularis, with the scattered cell groups of the nucleus of the prerubal field, and also with bundles of dentatothalamic and rubrothalamic fibres. This zone of merging of diverse fibre pathways and associated cell groups, is variously called the *prerubral, tegmental,* or *H field of Forel*.

The ansa lenticularis has a complex origin from both parts of the globus pallidus, the putamen and possibly other neighbouring centres; its fibres partly relay in scattered cell groups along its course—the *entopeduncular nucleus*. The fibres of the ansa curve

medially around the ventral border of the internal capsule and then continue to curve dorsomedially until they meet and become admixed with the other fibre systems mentioned above in the prerubral field. A number of the fibres in both the fasciculus and ansa lenticularis synapse with cells in the nucleus subthalamicus, the nucleus of the prerubral field and the zona incerta, whilst the remainder continue in company with the other fibre bundles, and then loop laterally and cranially to reach a number of thalamic nuclei, particularly nuclei ventralis anterior, intermedius and centromedianus.

The thalamic fasciculus is the complex bundle of fibres which extends from the prerubral field, passing dorsal to, but also partly traversing, the zona incerta, and related dorsally to the ventral nuclear groups of the thalamus. It contains the continuation of the fasciculus and ansa lenticularis just described, dentatothalamic and rubrothalamic fibres, and, running in the opposite direction, bundles of thalamostriate fibres. Another name often used for the territory occupied by the thalamic fasciculus is the *H₁ field of Forel*.

The 'pallidohypothalamic' fasciculus is the name commonly given to a bundle of fibres which leaves the main pallidofugal system in the prerubral field and then pursues a curious course, curving ventromedially around the column of the fornix to approach the hypothalamus (Bard and Rioch 1937; Vidal 1940; Ingram 1940), where it was for long widely assumed to terminate in its dorsomedial nucleus. However, there is no conclusive evidence for a hypothalamic termination, and more recent experiments in the monkey (Nauta and Mehler 1966) indicate that the bundle recurves yet again, passing laterally below the column of the fornix, and then dorsally to rejoin the main system of fibres in the H₁ field of Forel.

The fasciculus subthalamicus comprises the profuse two-way array of fibres which passes through the internal capsule, interdigitating at right angles with its main fibre systems, to interconnect the nucleus subthalamicus with the globus pallidus, and to a lesser extent with the putamen.

Some of these connexions will be mentioned again in the description of the basal ganglia (pp. 1033 *et seq.*).

The Hypothalamus

The general position and extent of the hypothalamus has been considered previously (p. 953). In the present account it will be assumed to extend from the lamina terminalis to a vertical plane caudal to the mamillary bodies, and from the hypothalamic sulcus, to include the structures in the ventral side-wall and floor of the third ventricle. Strictly, the most cranial part of this region,

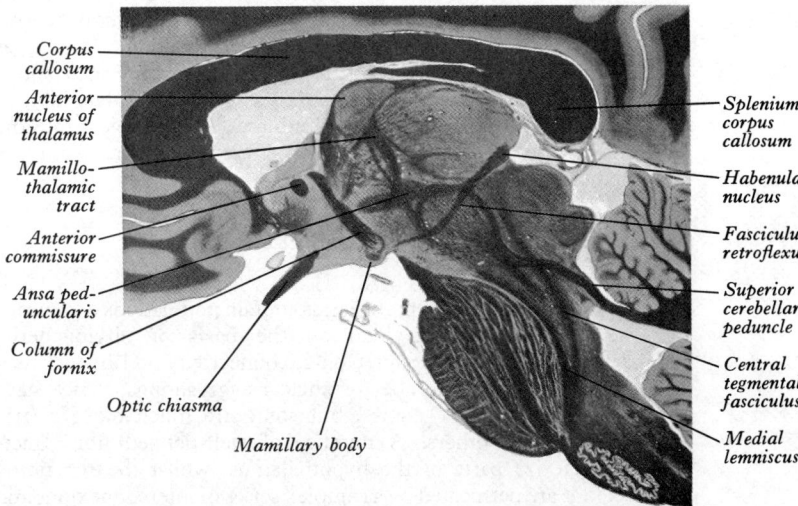

Corpus callosum
Anterior nucleus of thalamus
Mamillo- thalamic tract
Anterior commissure
Ansa ped- uncularis
Column of fornix
Optic chiasma
Mamillary body
Splenium of corpus callosum
Habenular nucleus
Fasciculus retroflexus
Superior cerebellar peduncle
Central tegmental fasciculus
Medial lemniscus

7.103 Sagittal section of the right cerebral hemisphere through the mamillary body, viewed from the left. Myelin stain. The fasciculus retroflexus is seen crossing the medial side of the red nucleus, which is surrounded by a capsule of white fibres derived chiefly from the superior peduncle. (After C. Foix and J. Nicolesco, *Anatomie cérébrale*, Paris, 1925.)

the preoptic area, belongs to the telencephalon impar, but is included for functional reasons. *Lateral* to the hypothalamus lies the rostral part of the subthalamus, the internal capsule and the optic tract; *caudally* the hypothalamus abuts on the tegmental part of the subthalamus which is continuous caudally with the tegmentum of the midbrain; *dorsal* to the hypothalamus lie the various nuclei of the dorsal thalamus; *rostrally* the optic chiasma, lamina terminalis and the anterior commissure separate the preoptic area from the precommissural septum, i.e. the continuation of the diagonal band of Broca into the paraterminal gyrus (p. 995), anterior to which lies the parolfactory gyrus. As mentioned previously, however, these topographical 'boundaries' are arbitrary, and functionally continuous systems cross many of them.

The structures which form the floor of the third ventricle extend to the free surface of the brain in the **interpeduncular fossa** (7.104) and consist, rostrocaudally, of: (1) the optic chiasma; (2) the tuber cinereum with its eminences, and the stalk of the infundibulum; (3) the mamillary bodies; and (4) the posterior perforated substance which is not usually included in the hypothalamus, but is included here for convenience.

The posterior perforated substance is a small depressed area of grey matter which lies in the caudal part of the interval between the diverging crura cerebri. It is pierced by a number of small apertures which transmit the central branches of the posterior arteries. Deep to its floor lies the *interpeduncular nucleus*, a relatively small structure in man which is homologous with a much more extensive nuclear complex in submammalian forms. It receives the curiously looped terminals of the fibres of the fasciculus retroflexus (p. 963) of both sides, and establishes further connexions with the reticular formation of the midbrain tegmentum and with the mamillary bodies.

The mamillary bodies are a pair of smooth, hemispherical masses, each about the size of a small pea, placed side by side in the floor of the interpeduncular fossa just rostral to the posterior perforated substance. Each is enclosed by bundles of white fibres derived in large part from the column of the fornix; internally are a number of aggregations of grey matter; these and their connexions are described further below.

The tuber cinereum lies rostral to the mamillary bodies and caudal to the optic chiasma. It is, as its name suggests, a convex mass of grey matter when viewed from its inferior aspect, but it is not a completely smooth convexity, its surface presenting a series of eminences with intervening grooves which vary in their prominence. In the midline caudal to the optic chiasma, the hollow, conical **infundibulum** proceeds ventro-inferiorly to become continuous with the solid expanded posterior lobe of the hypophysis cerebri (p. 1438). The tuber cinereum around the base of the infundibulum is raised to form a **median eminence** which, however, is superficially marked by a shallow *tubero-infundibular sulcus* along their lines of junction. The tuber cinereum, also presents a pair of **lateral eminences** and a midline **postinfundibular eminence**, whilst caudolaterally, on each side, a row of two or three minute *hillocks* project into the shallow groove between the tuber cinereum and the mamillary bodies.

THE DIVISIONS AND NUCLEI OF THE HYPOTHALAMUS

Most regions of the hypothalamus contain populations of neurons which have been divided on the basis of phylogenetic, developmental, cytoarchitectonic, connectivity and histochemical studies into a number of nuclear aggregations, which have been given specific names, but some are much more clearly defined than others. A number of well-defined fibre tracts characterize parts of the hypothalamus, whilst the remaining regions are permeated by a complex series of interconnexions and more diffuse fibre arrays entering, leaving, passing through, or intrinsic to the region. Although the connexions of some of these fibres are fairly well-established, others are known only in outline. Further complications arise in the study of this small but immensely complex region of the brain; different authorities have, on occasion, used different criteria for its classification and their

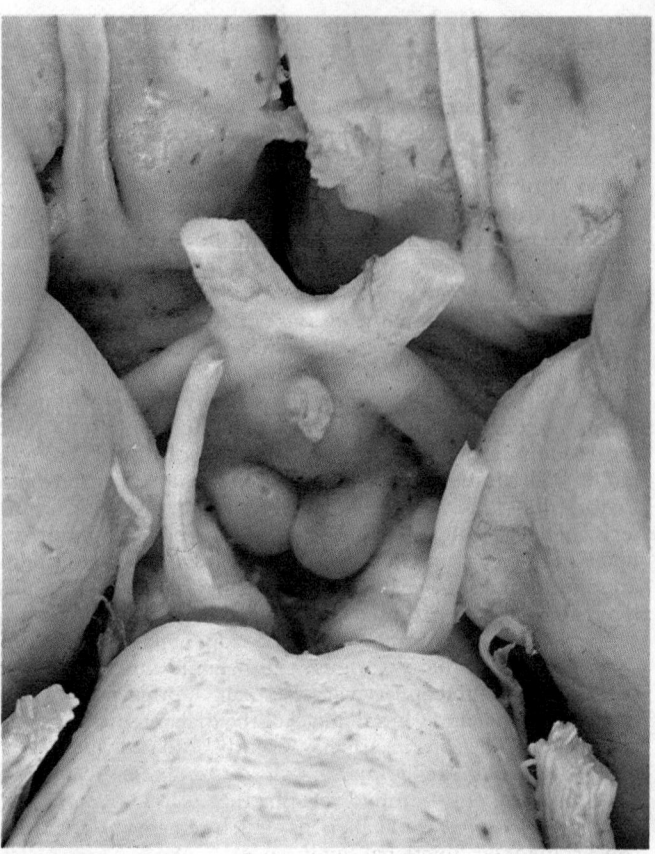

7.104 The interpeduncular fossa and surrounding structures. Note from above downwards: (1) the gyri recti of the frontal lobes, and the olfactory tracts, trigones and anterior perforated substances; (2) the optic nerves, chiasma and diverging optic tracts which disappear beneath the medial borders of the unci hippocampi; (3) the tuber cinereum with the attached superior part of the infundibular stem; (4) the mamillary bodies; (5) the deep recess between the mamillary bodies, the diverging crura cerebri and the cranial border of the pons, the floor of which is the posterior perforated substance; (6) on each side the prominent oculomotor nerves, more laterally the slender trochlear nerves, and bordering the pons, the thick trigeminal nerves are visible. (Dissection by Dr. E. L. Rees, Dept. of Anatomy, Guy's Hospital Medical School.)

terminology does not always correspond. Only the major nuclear groups and principal connexions of the hypothalamus, followed by a brief outline of its main functional associations can be given here. For details, extended discussion and comprehensive bibliographies consult Le Gros Clark *et al.* (1938); Crosby and Woodburne (1940); Harris (1955); Haymaker *et al.* (1959); Szentágothai *et al.* (1962). For an excellent modern review *see* Reichlin, Baldessarini and Martin—editors (1978).

In general, the hypothalamus may be divided into a rostrocaudal sequence of three zones—*supraoptic*, *infundibulotuberal* and *mamillary*. Similarly, a longitudinal division has been proposed into *lateral* and *medial zones* on each side. These are separated by a paramedian plane, which includes on each side the prominent fibre bundles of the column of the fornix, the mamillothalamic tract, and the fasciculus retroflexus. Some authorities prefer to subdivide the medial zone into a thin subependymal *periventricular zone*, and a thicker *intermediate* (or *medial*) *zone*.

Some of the main nuclear groups of the hypothalamus are illustrated in 7.105 and a more complete list is given below for reference purposes. Many of the latter are small and poorly localized, but despite this, some authorities even subdivide these into still smaller aggregations. Each zone may be considered as a rough, rostrocaudal sequence of nuclei, but some overlap occurs dorsoventrally, and the preoptic grey matter is common to all three zones.

The periventricular zone consists of: (1) part of the *preoptic* nucleus; (2) a small *suprachiasmatic* nucleus—but *vide infra*; (3) the large *paraventricular* nucleus; (4) the *infundibular* nucleus;

and (5) the *posterior* nucleus of the hypothalamus. In general, these masses of grey matter are interspersed between the periventricular region of fibres (pp. 957, 968) with which some of them are functionally associated. The most prominent nucleus of this zone, and the best understood in functional terms, is the paraventricular nucleus, which is further discussed below.

The intermediate zone (the medial zone of many authors) consists of: (1) part of the *preoptic* nucleus; (2) the *anterior* nucleus; (3) the *dorsomedial* nucleus; (4) the *ventromedial* nucleus; and (5) small *premamillary* nuclei.

The lateral zone consists of: (1) part of the *preoptic* nucleus;

(2) the well-known *supraoptic* nucleus; (3) the extensive *lateral* nucleus; (4) the *tuberomamillary* nucleus; and (5) the *lateral tuberal* nuclei.

Finally, the mamillary body, with its main *medial* and *lateral mamillary* nuclei and a number of smaller associated aggregations of grey matter, may be considered as lying between the intermediate and lateral zones of the hypothalamus, although in fact they overlap both these zones.

A detailed description of the shapes, sizes, architectonics, and what is known of the connexions of these individual nuclei cannot be given in this volume, and for this, the references quoted should

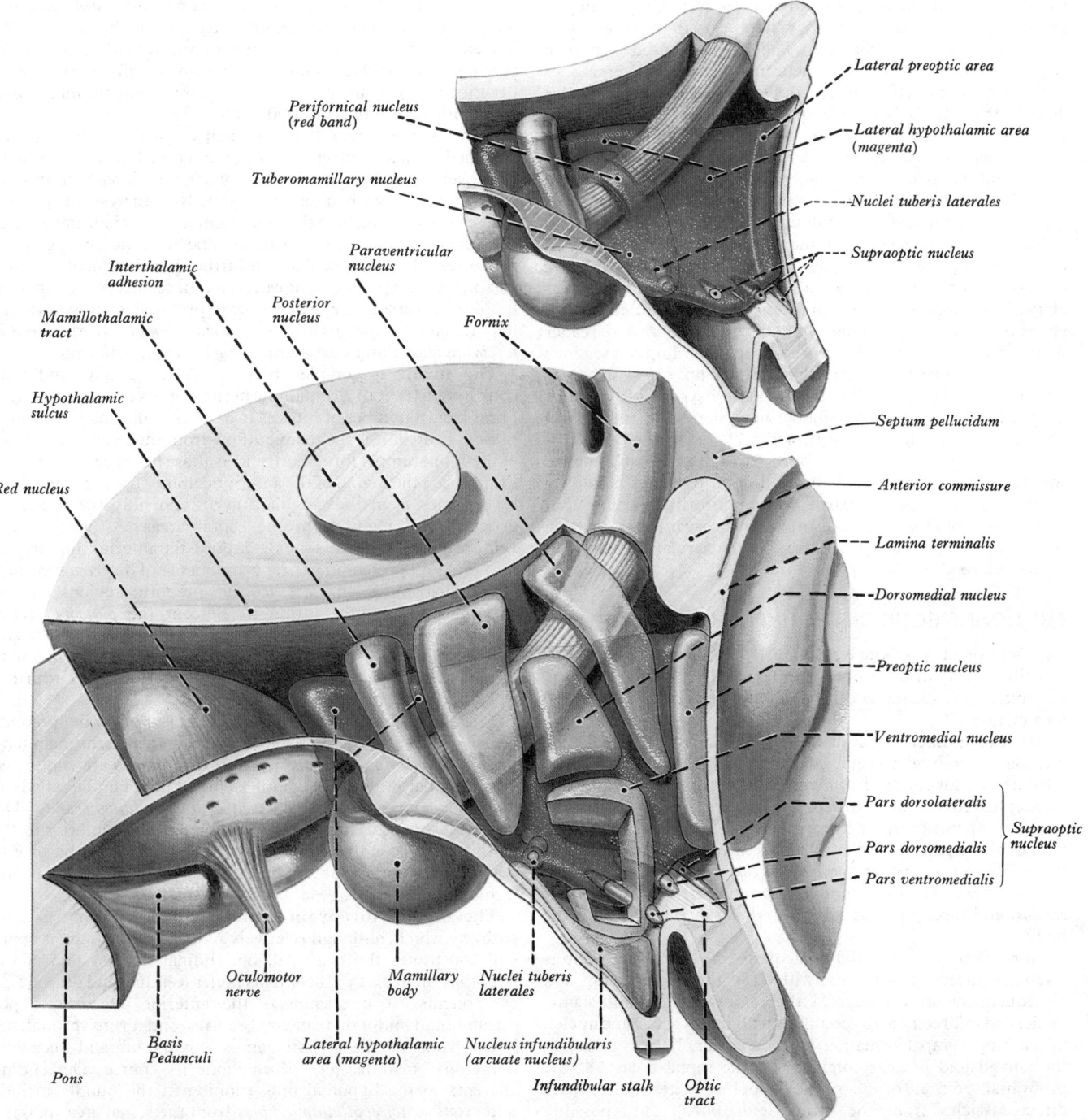

7.105 Schemata of the hypothalamic region of the left cerebral hemisphere from the medial aspect to display the major hypothalamic nuclei. In the upper diagram the medially placed nuclear groups have been removed, whilst in the lower diagram both lateral and medial groups are included. Lateral to the fornix and the mamillothalamic tract is the lateral hypothalamic region (magenta), in which the tuberomamillary nucleus is situated. Situated rostrally in this area is the lateral preoptic nucleus. Surrounding the fornix is the perifornical nucleus (red band), which joins the lateral hypothalamic area with the posterior hypothalamic

nucleus. The medially situated nuclei (yellow) fill in much of the region between the mamillothalamic tract and the lamina terminalis, but also project caudal to the tract. The nuclei tuberis laterales (blue) are situated ventrally, largely in the lateral hypothalamic area. The supraoptic nucleus (green) consists of three parts. See text for further description. (From *The Hypothalamus*, by courtesy of the authors, W. J. H. Nauta and W. Haymaker, and the publishers, Charles C. Thomas, Springfield, Illinois, 1969.)

be consulted; but the following outstanding points may be noted. The preoptic, anterior, dorsomedial, and ventromedial nuclei and part of the posterior nucleus are mainly composed of small or medium-sized neurons and, in man, are rather poorly differentiated from their surroundings. Relatively large scattered neurons surrounded by small ones characterize the rest of the posterior nucleus and much of the lateral nucleus. The lateral tuberal nuclei, although small, are well defined, each being encapsulated by a dense meshwork of fine fibres, and each is located within one of the caudolateral surface hillocks between the tuber cinereum and mamillary body described above. The large size of the medial mamillary nucleus is characteristic of the human brain. The supraoptic nucleus, which covers the lateral part of the optic chiasma, and the large subependymal paraventricular nucleus are richly vascularized and they have certain distinguishing cytological features in common. They are composed in part of large, bipolar or multipolar neurons, which are sometimes multinucleate and which, using Nissl's method, stain much more densely than the cells of surrounding nuclei. Their cytoplasm contains characteristic granules or droplets of neurosecretory material (pp. 969, 1441, and **8**.197).

In addition to the foregoing, however, it should be noted here that the advent of the newer neurite tracing techniques (p. 819), the formaldehyde histofluorescence method of Falck and Hillarp (pp. 819, 948), and the refinement of the powerful immunofluorescence techniques (p. 819) has caused an explosive increase in the detailed research data related to neuroendocrinology. In the following paragraphs only the briefest mention can be made to this voluminous literature; the interested reader should consult the cross-references given, including the sections devoted to the neuro- and adenohypophysis (pp. 1438–1443), the recent proposals concerning the circulatory system of the latter (p. 1443), the 'diffuse neuroendocrine system' and the APUD concept (p. 1454), and the habenulopineal system as a 'model' of neuroendocrine transduction (pp. 963, 1445). Most apposite among recent publications is the collection of essays edited by Reichlin *et al.* (1978), particularly informative being that contributed by Hökfelt and his numerous co-workers (1978) on 'aminergic and peptidergic pathways in the nervous system with special reference to the hypothalamus'.

THE CONNEXIONS OF THE HYPOTHALAMUS

As mentioned previously, some hypothalamic connexions constitute fairly discrete fibre bundles, whilst others are diffuse, difficult to investigate, and hence often of uncertain origin and termination.

In broad outline, fibre systems *converging* on the hypothalamus include ascending visceral and somatic sensory pathways, olfactory pathways, and numerous tracts from the midbrain, diencephalon, 'limbic' structures, and from the neocortex. *Outflowing* paths from the hypothalamus return to many of these regions but, in particular, control paths reach lower loci from which the *peripheral autonomic nervous system* originates; and it also controls the secretory cycles of the *hypophysis cerebri* and *epiphysis cerebri*, and through these activities, the *endocrine system* of the body.

More precisely, the different parts of the hypothalamus establish direct connexions with: (1) the tegmentum and periaqueductal grey matter of the midbrain; (2) subthalamic nuclei and indirectly with the globus pallidus; (3) thalamic nuclei; (4) the hippocampal formation; (5) the anterior olfactory areas; (6) the amygdaloid nuclear complex; (7) the septal areas; (8) the prefrontal cerebral cortex; (9) the hypophysis cerebri; (10) some direct fascicles from the retina (*vide infra*); (11) possibly accessions from the superior cerebellar peduncle and the globus pallidus; and (12) numerous inputs from centres throughout the pontine and medullary reticular formation (*see* pp. 943–953). Some of the more important pathways will now be briefly detailed.

Afferent Connexions

The ascending sensory pathways to the hypothalamus include collateral branches of the lemniscal somatic afferent fibres, but in the main they are considered to be polysynaptic routes conveying visceral and gustatory information from the spinal cord and brainstem. They include the **mamillary peduncle** which is formed by the convergence of a number of fibre systems in the midbrain tegmentum to give a discrete bundle which ends in the mamillary body; ascending fibres run with the other components of the **dorsal longitudinal fasciculus** (*see* p. 840 and below) to reach the hypothalamus, and a number of **ascending pathways** from the **brainstem reticular formation** have also been described (for further details of the latter *see* pp. 943–953, and also below).

Direct projections reach the hypothalamus from the subthalamic nucleus and zona incerta of the same and the opposite side, the latter crossing in the supraoptic decussations. Thalamohypothalamic connexions pass from the dorsomedial nucleus of the thalamus to many of the hypothalamic nuclei through the **periventricular system** of fibres previously described (p. 967) and probably also by traversing the medial part of the **inferior thalamic peduncle**.

A group of forebrain structures, many of which were originally regarded as mainly olfactory in their functional associations, have now been recognized as of much wider functional significance, and are often described together as the **limbic system** (p. 990). These structures include the hippocampal formation, amygdaloid and septal complexes, the piriform lobe and adjacent regions of neocortex which will be detailed further in subsequent sections. For our present purposes, however, it is necessary to state that the limbic structures give rise to a series of prominent pathways to the hypothalamus—the *fornix*, the *stria terminalis*, the *medial forebrain bundle* and the *ventral amygdalofugal pathways*.

The fornix is complex both in its topography and fibre constitution (p. 1001), containing both commissural and projection fibres from a number of sources to a series of destinations. In the present context it is the main outflow from the neuronal laminae of the hippocampal formation, and as these fibres curve ventrally to approach the region of the anterior commissure they are joined by fascicles from the cingulate gyrus, possibly the indusium griseum, and many from the septal areas. As it continues ventrocaudally the fornix divides around the anterior commissure into pre- and postcommissural components. The *precommissural fornix* distributes in part to the preoptic regions of the hypothalamus, whilst during its descent the *postcommissural (column) of the fornix* gives offshoots to the dorsal, lateral and periventricular regions of the hypothalamus before continuing to its main termination in the mamillary nuclei. The fornix is further described on p. 1001.

The amygdaloid complex of nuclei project on to the preoptic area and most of the hypothalamic nuclei cranial to the mamillary bodies. These **amygdalohypothalamic fibres** are conducted along two pathways—the complex curved fibre bundle called the *stria terminalis*, and also by a *ventral amygdalofugal route*. The latter consists of a diffuse array of fibres passing inferior to the lentiform nucleus through the anterior perforated substance to reach the hypothalamus. The amygdaloid complex and stria terminalis are further described below (p. 996).

The medial forebrain bundle is another complex fibre pathway which, although relatively reduced in the human brain, still constitutes the principal longitudinal fibre system of the hypothalamus (**7**.123). It contains both ascending and descending components. It interconnects the anterior olfactory, hypothalamic and midbrain tegmental centres, and it runs through the lateral zone of the hypothalamus, giving off and receiving numerous small fascicles throughout its course. Descending afferents to the hypothalamus running in the bundle include numerous *septohypothalamic fibres* from the septal area (p. 997), *olfactohypothalamic fibres* from parts of the piriform cortex, and possibly *corticohypothalamic fibres* from the orbitofrontal cortex. Ascending visceral afferents reach the hypothalamus through the caudal part of the medial forebrain bundle. It should also be noted that descending fibres from the anterior terminations of the fornix and stria terminalis accompany the medial forebrain bundle and its branches. The details of some of these fibre systems have only been established in subprimate brains, but it is widely assumed that they also exist in the human brain.

The constituents and destinations of certain fascicles in the medial forebrain bundle have been somewhat clarified by the application of the newer fibre-tracing and fluorescence techniques (pp. 819, 820, 825). Some of the fascicles derived from the *ventral tegmental ascending cholinergic pathway* accompany or lie near the bundle and most, if not all, hypothalamic nuclei receive a cholinergic input from their side branches. The *ventral tegmental noradrenergic bundle* runs with the medial forebrain bundle, its hypothalamic distribution being to the dorsomedial, periventricular, paraventricular, supraoptic and lateral nuclei. The *mesolimbic dopaminergic fibre system* (p. 949) passes through the hypothalamus in the medial forebrain bundle. Dopaminergic cell aggregate A12, in the infundibular (arcuate) nucleus, provides terminals in the median eminence and infundibulum; group A11 innervates the medially placed hypothalamic nuclei; groups A13 and 14 distribute axons to the dorsal and rostral hypothalamic nuclei. (*See* p. 949 and **8.**197A.)

Corticohypothalamic fibres project to the hypothalamus from wide areas of the prefrontal cortex by direct and indirect routes, but there has been much debate concerning the relative importance of the two pathways. The indirect fibres converge on the nucleus medialis dorsalis of the thalamus where they relay, and are then continued to the hypothalamus in the periventricular fibre system. In different experimental animals direct corticohypothalamic fibres have been variously described as ending in the lateral, dorsomedial, mamillary and posterior nuclei of the hypothalamus, but these have been denied by some authors. For a detailed review of these and other hypothalamic connexions consult Nauta and Haymaker (1969); Reichlin *et al.* (1978). Finally, as mentioned previously, some hypothalamic nuclei may receive an input from the superior cerebellar peduncle.

Efferent Connexions

In broad outline, the connecting pathways from the hypothalamus include: (1) reciprocal paths to parts of the limbic system; (2) a series of descending polysynaptic paths to lower autonomic and motor centres; and (3) nervous and vascular links with the hypophysis cerebri.

The septal areas and the amygdaloid complex receive hypothalamic fibres which return along the pathways described above.

The prominent medial nucleus of the mamillary body gives rise to a large fibre bundle which ascends and, partly by branching of its individual fibres, diverges into two substantial tracts. The *mamillothalamic tract* (**7.**117) continues to ascend and terminates in all parts of the anterior nuclear group of the thalamus, where, after relay, massive projections radiate to the cingulate gyrus (pp. 960, 1023). The *mamillotegmental tract* curves caudally to enter the midbrain ventral to the medial longitudinal fasciculus and is distributed to tegmental nuclei of the reticular formation. Small bundles of *mamillosubthalamic fibres* pass to the prerubal field of the subthalamus, but their destination is uncertain. In addition to the mamillotegmental tract, descending fibres from the hypothalamus also reach the midbrain tegmentum in the *dorsal longitudinal fasciculus* and in the caudal extension of the *medial forebrain bundle*.

As mentioned above, the **connexions of the hypothalamus with the hypophysis cerebri** are by means of: (1) *nerve fibres* derived from various groups of hypothalamic neurons, which terminate at different levels throughout the median eminence, infundibulum, and *posterior lobe* of the hypophysis; and (2) a system of long and short *portal blood vessels* which interconnect plexuses of sinusoids in the median eminence and infundibulum with other plexuses in the *anterior lobe* of the hypophysis. The anatomy of these nervous and vascular channels is further considered with that of the hypophysis (p. 1438; **8.**197, 198), but certain points must be emphasized here. The main glandular mass of the anterior lobe of the hypophysis receives no nerve supply; and virtually the whole of its blood supply is from the hypothalamohypophysial portal vessels. Direct observation in the living experimental animal has established that the predominant direction of the blood flow in these vessels is from a series of complex sinusoidal tufts in the median eminence and infundibulum, through the portal vessels, to discharge into the blood sinusoids which run between the cords of glandular cells in the anterior lobe. However, as indicated on p. 1443, *flow reversal* may occur in certain parts of this vasculature, with consequently widespread neuroendocrinological implications.

It is now widely accepted that a process of *neurosecretion* is involved both in the production and release of the posterior lobe hormones *oxytocin* and *vasopressin*, their related transport proteins the *neurophysins*, and also in the neurovascular mechanism whereby secretory activity in the various types of glandular cell in the anterior lobe is controlled. The hypothalamic centres containing neurons whose axons terminate in the median eminence or upper infundibulum, or pass through the latter to terminate in the posterior lobe of the hypophysis, include the supraoptic, paraventricular, and infundibular nuclei, and part of the ventromedial nucleus.

The supraoptic nucleus, as mentioned above, is draped over the lateral part of the optic chiasma, and some authorities divide it into three or more subsidiary nuclei which are named according to their relationship to the chiasma. It is composed of a uniform population of large cells. **The paraventricular nucleus**, which causes a visible bulging of the ependyma into the third ventricle, extends from the hypothalamic sulcus and crosses the medial aspect of the column of the fornix, whilst its ventrolateral angle points towards the supraoptic nucleus. A number of its cells are large like those of the latter, but in addition there are many intermediate and small cells of varying shapes. **The infundibular nucleus** (often called **the arcuate nucleus**) occupies the median zone of the postinfundibular part of the tuber cinereum, and it extends into the median eminence and largely encircles the base of the infundibulum. However, it is deficient cranially where the infundibulum adjoins the midline part of the optic chiasma. No glial layer intervenes between the nucleus and the specialized ependymal (tanycyte) lining of the *infundibular recess* of the third ventricle. (For the possible neuroendocrine role of these tanycytes consult Knowles 1974; for a recent review and extensive bibliography *see* Joseph and Knigge 1978; also pp. 836, 1441.) In contrast to the other nuclei just described, it is composed exclusively of numerous small cells, which appear round in coronal section and oval or fusiform in sagittal section.

Thus, the neurons which project into or through the infundibulum may be considered in two groups: **magnocellular**, including the large and intermediate cells of the paraventricular nucleus, the large cells of the supraoptic nucleus and small groups or rows of internuclear cells between them, and **parvocellular**, which includes the small cells of the infundibular nucleus and some in neighbouring nuclei. Whilst the magnocellular group is characterized by the presence of neurosecretory material (*vide infra*), it must be emphasized that both groups consist of true neurons. Each cell possesses a soma, dendrites and an axon with collateral branches, and these contain the usual neuronal organelles (p. 825). Furthermore, they receive axodendritic and axosomatic synapses from unknown sources, and their axons conduct volleys of nerve impulses. Their axonal terminals expand into end-bulbs which contain mitochondria and clear synaptic vesicles about 40–60 nm in diameter, and many of them are closely applied to the basement membranes of capillaries with no intervening neuroglia, in either the infundibulum or the posterior lobe. Whilst it is clear that important neural connexions exist which may influence the activities of these various nuclei—probably stemming from other hypothalamic nuclei and also from a number of the channels which converge upon the hypothalamus from other centres—their precise connexions have not yet been established. Equally, the significance of volleys of nerve impulses in these neurons, and the functional meaning of the presynaptic vesicles is by no means clear: no direct relationship to the process of neurosecretion has been demonstrated. (For further details *see* p. 1441.)

The terminology used in connexion with the neural pathways from the hypothalamus to the hypophysis has undergone several changes with increasing knowledge. This has resulted in an array of terms, some of which lack precision, and often different meanings attach to the same term when used by different authorities.

The term *supraopticohypophysial tract* was often used to denote *all* fibres which entered the infundibulum, whatever their origin or destination, but its use is now often restricted to those originating in the supraoptic and paraventricular nuclei. Since the nuclei are by no means equivalent in either their cytology or functional associations, some workers in the field prefer the more specific terms *supraopticohypophysial* and *paraventriculohypophysial tracts*. However, it must be realized that many of their fibres end in the infundibulum and do not reach the posterior lobe. The term *tuberohypophysial* or more precisely **tubero-infundibular tract** is used to denote the fibres of the small neurons of the infundibular and related nuclei of the mediobasal hypothalamus (*vide infra*) which end in the infundibulum.

The cells and axons of the **magnocellular neurosecretory** pathways from the supraoptic and paraventricular nuclei, scattered internuclear neurons, and some contend occasional neuron groups of the suprachiasmatic nucleus, contain what were originally termed colloid droplets. These vary in size, are sometimes found between the cells and fibres, and they stain selectively with the chrome-alum haematoxylin stain of Gomori, when the fine axons and their branches are seen to contain a row of stained masses, each of which distends the axon, the whole appearance resembling a 'string of beads'. Irregularly distended nerve terminals containing large masses of neurosecretory material are termed *Herring bodies*. Ultrastructurally, the material consists of membrane-bound aggregations of dense granules, the whole mass varying from 200 to 300 nm in diameter. Much intensive research has been directed towards an understanding of neurosecretion, and the methods used include: studies on the effects of transection of the infundibulum; transplantation of the hypophysis; culture of neurohypophysial explants; autoradiography following the administration of labelled cysteine; subcellular fractionation and extraction followed by biological assay or chemical analysis; studies on the effects of dehydration, hydration, various drugs and stressful situations; chemical or electrical stimulation and ablation of focal points in the hypothalamus.

Briefly, it is now accepted that neurosecretory material is synthesized in the large cell somata of the supraoptic, paraventricular, and neighbouring neurons in relation to the endoplasmic reticulum, further elaborated in the Golgi complex, and then transported along the axons and their branches, to be released at the axon terminals, and then to be taken up by the blood which is percolating through adjacent capillaries. Further, there is some evidence that, whilst *both* the supraoptic and paraventricular nuclei are associated with the production of both hormones vasopressin and oxytocin, vasopressin is *predominantly* a product of the supraoptic nucleus and oxytocin of the paraventricular nucleus. Suitable immunofluorescence techniques have allowed a precise three-dimensional analysis of the distribution of oxytocin and vasopressin cells in a number of mammals; marked regional and species differences exist. However, many questions remain unanswered (Donovan 1970; Knowles 1974). For example, the precise chemical constitution of the neurosecretory material remains to be determined, and it appears that the chemically related hormones oxytocin and vasopressin (both are nonapeptides) are transported in a relatively non-active form, each probably conjugated with a specific low molecular weight protein carrier which is termed a *neurophysin*. Secondly, the rate and mechanism of transport, presumably involving some form of fast axoplasmic flow, remains to be defined accurately. Finally, the detailed manner in which the hormones are released exocytotically at the axonal terminals is not understood, although suggestions have been made that both calcium ions and the local release of acetycholine by the synaptic terminals may be involved. The special vascular relationships of the supraoptic and paraventricular nuclei are discussed in some further detail below. (For a detailed review of the magnocellular neurosecretory system consult Defendini and Zimmerman 1978.)

In addition to the production of posterior lobe hormones oxytocin and vasopressin, and their respective neurophysins, neurosecretory pathways are also considered to play an important role in the control of secretory activity by the anterior lobe of the hypophysis. Much evidence supports the view that groups of hypothalamic cells produce a variety of *peptidergic releasing*

hormones which are transported along their axons and then discharged into the upper radicles of the portal system of blood vessels, by which means they are carried to the vascular bed which permeates the anterior lobe, where they reach and influence the appropriate glandular cells. The term *releasing hormone* is not, however, entirely appropriate; some inhibit the release of hormone by anterior lobe cells, whilst others stimulate synthesis of the hormone, as well as promoting its release. They include *releasing hormones* for corticotropin, luteinizing hormone, follicle stimulating hormone; thyrotropin and somatotropin, and *release inhibiting hormones* for prolactin, melanocyte stimulating hormone, and somatostatin. The **parvocellular neurosecretory pathway** provided by the '*tubero-infundibular tract*' has been proposed as the main route for the synthesis and transmission of these releasing hormones to the portal blood flow. Immunofluorescence techniques have now allowed the localization of some of these peptides (*see* p. 1441, table on p. 1455, and Renaud 1978). For example, corticotropin releasing factor has been localized in the small cells of the paraventricular, anterior and posterior hypothalamic nuclei, thyrotropin releasing factor in the nuclei dorsomedialis and ventromedialis, luteinizing hormone releasing factor in some neurons of the nucleus infundibularis (arcuatus), somatotropin releasing factor in the anterior and posterior hypothalamic nuclei, and somatostatin in some of the nuclei periventriculares. In addition to releasing factors some fibres in the tract also have particularly high concentrations of acetylcholine, whilst others are rich in catecholamines, and this is paralleled by the ultrastructural appearances of their end-bulbs (p. 1441). Some contain small clear synaptic vesicles, whilst others contain larger dense-cored vesicles, and yet others contain a mixed population. There are no structural evidences of the typical large, electron-dense, membrane-bound granular masses which characterize the magnocellular pathway. Accordingly, it has been proposed that the fibres of the tubero-infundibular tract carry releasing hormones in a dispersed, active, form, and may also carry catecholamines or acetylcholine. It will have become clear to the reader that with such relatively widespread origins, the name tubero-infundibular may well be too restrictive. Mention has already been made (p. 969) of the possible role of the tubero-infundibular dopaminergic (and other monoaminergic) pathways in modulating the synthesis, transport and exocytosis of releasing or release-inhibiting factors, and also the activities of tancytes (see also p. 1441 and **8**.197A). According to Renaud (1978), it should also be noted that the tubero-infundibular neurosecretory parvocellular neurons are in receipt of neural inputs from the amygdala, preoptic nuclei and hippocampus: in turn their axons provide collaterals back to these regions, to the medial nuclei of the dorsal thalamus, and to other hypothalamic nuclei, in addition to their *infundibular* terminals.

Finally, it must be emphasized that many of the axonal terminals of the magnocellular pathway also end in the upper infundibulum and may also be involved in control of the anterior lobe of the hypophysis. For these reasons, the magnocellular and parvocellular pathways may not be as functionally independent as has been envisaged by some investigators.

FUNCTIONS OF THE HYPOTHALAMUS

Clinically it had long been recognized that lesions in the hypothalamus often lead to widespread and bizarre combinations of symptoms and signs, which stem from endocrine, metabolic, visceral and behavioural disturbances. However, it has been widely assumed that such combinations result from the interruption of control pathways for these various 'functions' which, although largely independent, are closely related topographically either in or near the hypothalamus. In particular, although the general significance of the hypophysis cerebri as an endocrine gland was gradually emerging in the closing decades of the last century and the opening ones of the twentieth (Cushing 1912), it is only in the last forty years that the interrelation of the hypophysis and nervous system has been firmly established (Harris 1955). Thus, the science of *neuroendocrinology* has emerged (Scharrer and Scharrer 1963; Heller and Clark 1962; Gabe 1966; Donovan 1970; Knowles 1974; Reichlin *et al.* 1978) and has made increasingly rapid progress, complementing the widely accepted role of the hypothalamus as the 'head ganglion' of the autonomic nervous system (Sherrington 1947). More recently, intensive investigation has not only confirmed in greater detail the controls exerted by the hypothalamus on the *endocrine system* and the *lower autonomic centres*, but has also emphasized that hypothalamic action depends upon afferent information channels, both nervous and vascular, and that it is interlocked, both structurally and functionally, with higher regions of the nervous system. The latter include the complex of structures which many authors group as the *limbic lobe* or system (p. 990), and the pre-frontal regions of the cerebral cortex.

Reference has been made elsewhere (p. 802) to one possible approach to the study of nervous systems in general, namely, that they operate in more complex animals as controllers of the *homeostatic cycles* which tend to preserve the individual and the species within a fluctuating environment. In mammalia, the frontal cortex, limbic system, hypothalamus, and lower regions of the brainstem and spinal cord are conveniently regarded as forming a hierarchy of controls particularly directed towards those homeostatic cycles, which are mediated by the autonomic nervous system, the endocrine system, and the locomotor patterns associated with them. The concept of such a hierarchy is important because of the widespread use of the terms lower, intermediate and higher 'centres' for a variety of visceral (and locomotor) controls. The term centre is largely a product of the available methods of analysing central nervous activities—usually by the focal stimulation or destruction of more or less discrete patches of nervous tissue, usually by electrical or chemical means. Such experimental manœuvres are then combined with other forms of observation—biochemical or cytological analysis, changes in general metabolic status, growth or behaviour patterns, electrophysiological recording and so forth. Whilst it is undoubtedly true that focal stimulation or obliteration may cause dramatic changes in a particular group of responses, and that some regions exhibit apparently spontaneous and sometimes rhythmic electrical changes when isolated, great caution must be used before regarding such regions as independent autonomous, controlling centres of particular activities. The bulk of the available evidence indicates that such 'centres' are best regarded as *nodal regions* of nervous tissue which, by virtue of their chemical constitution, intrinsic geometry and connexions, and extrinsic connexions, nervous or vascular, are essential for the normal range of certain responses which follow particular patterns of inflowing information. It must be appreciated that the various other regions of the nervous system from which afferent information channels converge upon the 'centre', whilst often less dramatic in their effect upon the responses in question when individually modified, are nevertheless collectively as important as the so-called centre itself. Further, particularly within a confined region such as the hypothalamus, to and from which numerous pathways converge and diverge, and which is composed of many rather ill-defined neuron groups, some of which have diffuse patterns of connexions, it is found that many of the functional 'centres' of different investigators overlap or virtually coincide. For these reasons, many investigators prefer to discuss the effects of stimulation or ablation of general zones of the hypothalamus, e.g. lateral, posterior, preoptic, mamillary, etc., rather than attempting to ascribe particular functional roles to individual hypothalamic nuclei. Partial exceptions to this are the well-recognized neurosecretory nuclei with specific cell products which pass into the bloodstream in the infundibulum or posterior love of the hypophysis, but even in these cases their afferent neural and vascular connexions are incompletely understood.

In general terms, experiments in animals involving transection of the nervous system at various levels demonstrate that as progressively higher levels are included below the plane of section, more complex and effective homeostatic readjustments can be carried out (Anand 1970). Decerebrate preparations with transection at upper midbrain levels can perform individual minor reflex readjustments of the cardiovascular, respiratory and alimentary systems, but these are not integrated and normal body temperature is not maintained. When transection is performed cranial to the hypothalamus, separating it from the limbic system, but retaining the connexions between the hypothalamus, hypophysis, brainstem and spinal cord, a completely different picture emerges. Quite effective homeostasis is maintained over a moderate range of environmental conditions. The various visceral and endocrine control systems are integrated into more ordered patterns of responses; furthermore, 'innate drives' and 'motivated behaviour patterns' make their appearance, including forms of feeding, drinking, apparent satiation, copulatory responses, and so forth. However, the homeostatic mechanisms break down if the environmental stresses involved exceed a certain range, e.g. persistent fairly high or low environmental temperatures. In addition, the motivated behaviour patterns become abnormal under stress, the animal may attack, attempt to eat, drink, or copulate with a bizarre range of objects. If, however, the connexions between the limbic system (p. 990) and hypothalamus are also preserved, and only neocortical regions are ablated, highly effective homeostatic responses occur under a wide range of adverse environmental conditions.

An enormous research literature has accumulated concerning the functional roles of the hypothalamus and the subject of neuroendocrinology, which cannot be reviewed in this volume; and the references quoted should be consulted for details and an introduction to extensive bibliographies. In summary, therefore, animal experimentation, usually involving focal stimulation or ablation of the hypothalamus, has emphasized its intimate relationship to the following activities.

(1) **Endocrine control** by the formation of releasing or release-inhibiting factors related to the production of thyrotropin, corticotropin, somatotropin, somatostatin, prolactin, luteinizing hormone, follicle stimulating hormone, and melanocyte stimulating hormone, by the cells of the *anterior pituitary*. Some of these affect target cells in the general body tissues directly, whilst others do so through the intermediary of a second endocrine organ, e.g. the thyroid gland, adrenal cortex, or the gonads. In the latter instances the sequence of control mechanisms is illustrated diagrammatically in **7**.106. Neural pathways converge upon, and are integrated within, the hypothalamus, and modify its production of releasing or inhibiting factors (p. 970). In turn these modify the activities of endocrine glands A and B, and ultimately the target tissues. Each of the intermediate levels in the chain are also modified by the operation of negative or positive feed-back loops which depend upon the blood concentration of the secretory products of subsequent stages in the chain. (For further details consult Haymaker *et al.* 1969; Donovan 1970; Bargmann and Scharrer 1970; Knowles 1974; Reichlin *et al.* 1978.)

(2) **Neurosecretion** of oxytocin and vasopressin (antidiuretic hormone, ADH) directly into the bloodstream of the infundibulum and posterior lobe of the hypophysis. Despite its name, whether vasopressin exerts any significant effect on vascular non-striated muscle under physiological conditions remains uncertain. However, it has a powerful effect on the reabsorption of water by the kidney, whilst oxytocin causes contraction of uterine and mammary smooth muscle which may be of importance during suckling, coitus and parturition. Stimulation of neural pathways from the nipples and genitalia causes an increase in the rate of oxytocin release. The rate of release of vasopressin is determined, in part, by the tonicity and osmotic pressure of the blood perfusing *osmoreceptor sites* both within the hypothalamus (Verney 1947) and in the peripheral circulation—probably in the hepatic portal vein (Haberich 1968). From the latter, lengthy nervous pathways are involved before the impulse pattern generated reaches the hypothalamus. Similarly, a reduced blood volume is considered to stimulate *volume receptors* in the great veins and the walls of the atria of the heart, thus causing an increased release of vasopressin (Gauer 1968). Finally, pain, and a variety of emotional states and conditions of stress, greatly modify the rate of release of the hormone.

(3) **General autonomic effects.** It has long been held that rostral zones of the hypothalamus mediate general para-sympathetic activity, whilst caudal zones mediate sympathetic activity. Whilst there is undoubtedly a preponderance of one or other effect in these zones, there is much interaction and overlap between the effects in different regions, and a rigid distinction between parasympathetic and sympathetic 'centres' cannot be

maintained. However, focal stimulation or ablation of the hypothalamus shows that it is intimately concerned with *cardiovascular*, *respiratory* and *alimentary* control. A detailed analysis cannot be given here, but such experiments are often accompanied by profound changes in heart rate, cardiac output, vasomotor tone and peripheral resistance, blood pressure, differential blood flow through different organs, and also in the frequency and depth of ventilatory excursions of the thorax, and in the motility and secretory activity of the stomach and intestines.

(4) **Temperature regulation.** In all warm blooded animals a critical balance is achieved between the overall heat production and loss of the body, and the hypothalamus provides a central regulating mechanism for this. A raised body temperature is depressed by vasodilatation, with an increased cutaneous blood flow, sweating, panting, and reduced heat production, whilst with a lowered body temperature the converse conditions apply, with shivering and, if prolonged, an increase in the activity of the thyroid gland. Information concerning body temperature reaches the hypothalamus via neural pathways from peripheral *thermo-detectors* for cold and heat, and from thermodetector neurons within the hypothalamus itself. The latter monitor the prevailing temperature of the blood perfusing the hypothalamus, whilst other hypothalamic neurons are chemoreceptors which react specifically to the presence of blood-borne viruses, toxins, pyrogenic drugs, etc. These converging flows of information are passed to regions of the hypothalamus which then set in train the appropriate autonomic, endocrine and muscular responses. It has been proposed that an *'antirise'* region exists in the cranial hypothalamus which controls vasodilatation, sweating, panting and possibly reduction of heat production, whilst the caudal hypothalamus houses an *'antidrop'* region controlling vasoconstriction, shivering and other means of raising heat production. The topographical limits and degree of separation of this *dual hypothalamic thermostat* cannot yet be regarded as finally settled. Some investigators (Myers 1969) consider that a *dual chemical thermostat* is interposed between the inflow of information and the 'antirise' and 'antidrop' regions of the hypothalamus, consisting of groups of neurons containing 5-hydroxytryptamine, or noradrenalin (and possibly dopamine). Warmer blood, impulses from heat receptors, antipyretic substances or anaesthetics, are thought to cause the release of noradrenalin (or dopamine) which in turn stimulates the 'antirise' mechanisms. Cooler blood, impulses from cold receptors, or pyrogenic substances, stimulate a release of 5-hydroxytryptamine which causes activation of the 'antidrop' mechanisms.

(5) **Regulation of food and water intake.** Many investigations (*see* Anand and Brobeck 1951; Anand 1961) have demonstrated that there exist regions with opposing actions in the medial and lateral zones of the hypothalamus. Ablation of the medial zone leads to over-eating or *hyperphagia*, eventually leading to gross obesity, whilst ablation of the lateral zone leads to *hypophagia* or even complete *aphagia* with death from starvation. Conversely, stimulation of the medial zone causes the animal to eat less, whilst lateral stimulation initiates eating or increases and prolongs the food intake. Thus, a laterally placed *hunger* or *feeding* 'centre' balanced against a medially placed *satiety* 'centre' have been proposed. Neurons which are specifically sensitive to the concentration of glucose in the circulating blood have been demonstrated in the medial zone of the hypothalamus and are probably important in the operation of the satiety centre.

Water balance and the osmolality of the blood are in part controlled by the *osmoreceptor/volume receptor—vasopressin—renal system* described previously, which determines the rate of water loss in the urine. Coupled with this is a *thirst* or *drinking centre* in the lateral zone of the hypothalamus which regulates water intake. Experimental stimulation of this region causes immediate and copious drinking, and, if the stimulation persists, gross hyper-hydration results.

(6) **Sexual behaviour and reproduction.** Through its control of gonadotrophin production by the anterior pituitary, the hypothalamus controls many aspects of reproductive physiology, including gametogenesis, cyclic variations in the reproductive tracts, and the maturation and maintenance of secondary sexual

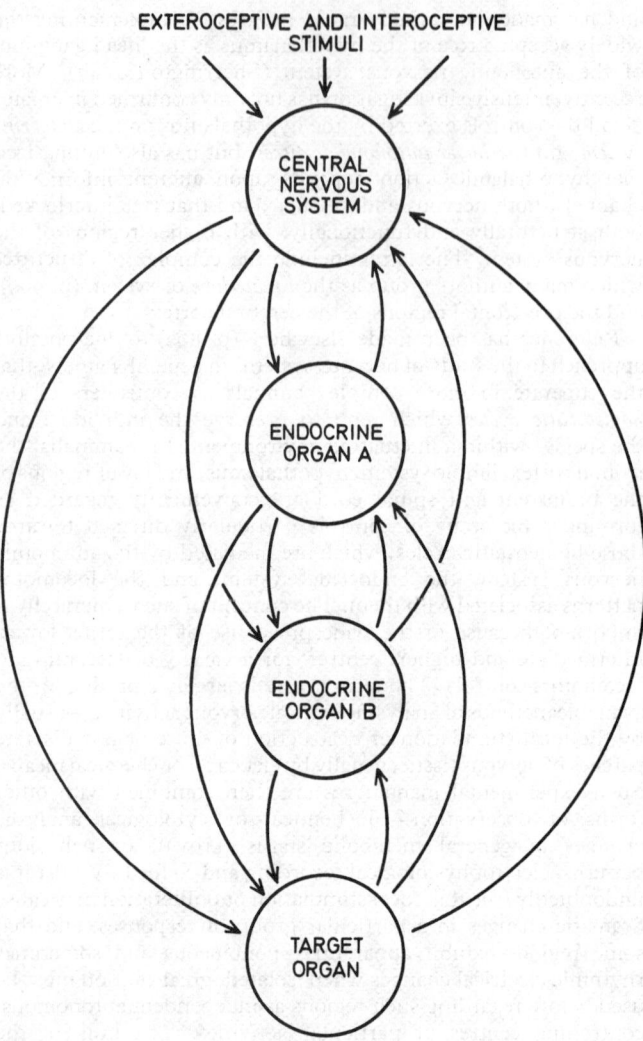

7.106 A flow diagram showing a stimulus pattern impinging on the central nervous system, and the possible interactions of the latter with endocrine organs A and B and a target tissue. Note the different orders of feed-forward and feed-back pathways envisaged. (From *Neuroendocrinology* by E. Scharrer and B. Scharrer, Columbia University Press, 1963. By courtesy of the authors and publishers.)

characteristics. Furthermore, hypothalamic stimulation may induce receptivity of the male in the female, and simple copulatory movements in the male, and it has also been shown that some hypothalamic neurons are sensitive to the circulating level of oestrogen or testosterone.

It should be noted at this point that whilst the elementary *drives* associated with hunger, thirst and sex may be considered to stem in part from an intact hypothalamus, for their full integration into complex behaviour patterns, a two-way commerce between the hypothalamus and limbic system is necessary (p. 997). Such patterns include searching for and procuring a mate, food and drink, home-building, rearing of young, etc.

(7) **Biological clocks.** Many tissues, organs and systems in the body show a cyclic variation in some of their functional activities, the cycle having a periodicity of approximately twenty-four hours. Well-known examples of such *circadian rhythms* include regular fluctuations in the body temperature, the concentration levels of a number of plasma constituents, the eosinophil count, adrenocortical secretory activity, and renal secretory mechanisms. In some cases the rhythm is partly an intrinsic property of the organ or tissue itself, but in many cases an overall control is exerted by an intact hypothalamus. Widespread lesions in other parts of the nervous system leave the biological timing mechanism unaffected, but hypothalamic lesions alone cause serious disturbances of rhythm. The suprachiasmatic nucleus appears in many animals to be an *endogenous neural pacemaker*, a 'biological clock' whose rhythmic activity can respond to regulating cues

provided by changes in environmental lighting (p. 1448). Although of widespread occurrence throughout the vertebrates (Joseph and Knigge 1978) occasional doubts have been expressed concerning the prominence, or even the presence, of the suprachiasmatic nucleus in mankind (Defendini and Zimmerman 1978). Bilateral experimental destruction of this nucleus permanently interferes with the circadian rhythms of, for example, adrenocorticosterone secretion (Moore and Eichler 1972), drinking and locomotion (Stephan and Zucker 1972), sleeping and wakefulness (Ibuka and Kawamura 1975) and pineal N-acetyltransferase activity (Moore and Klein 1974). As far as pineal activity is concerned, it has been demonstrated by Nishino *et al.* (1976) that the neuronal activity of the suprachiasmatic nucleus is increased by visual stimulation, and that this increased activity inhibits the sympathetic neurons which supply the pinealocytes, an action which results in a decrease in their secretory activity (p. 1448). Although still a somewhat controversial proposal, there is now considerable experimental evidence supporting the existence of direct retino-suprachiasmatic connexions which may be involved in the photic regulation of circadian rhythms. Terminations of retinal fibres in the suprachiasmatic nucleus have been demonstrated by a variety of techniques, including autoradiography after intra-ocular injection of tritiated amino acids, in a range of mammals from the tree-shrew to the macaque (Moore 1973)—findings which support the nerve degeneration studies of other workers (Moore and Len 1972; Printz and Hall 1974). Further supporting evidence for the existence of *retino-suprachiasmatic* connexions has been provided by the cobalt precipitation techniques of Mason and Lincoln (1976), by electron microscopy (Wenisch 1976) and by electrophysiological investigation (Sawaki 1977). *Retinal projections* to the rostral part of the infundibular (arcuate) nucleus have also been demonstrated (Mason and Lincoln 1976). Efferent fibres have now been traced from the suprachiasmatic nucleus to the ventromedial, dorsomedial and arcuate nuclei of the hypothalamus, suggesting that the fine control of the endogenous circadian regulator is a mechanism of considerable delicacy and sensitivity.

The periodic conditions of sleeping and wakefulness are an outstanding example of a circadian rhythm. In recent years a large volume of investigation has been directed towards an understanding of the mechanisms of sleep (*see* Jouvet 1962, 1964, 1965; Rossi 1964; Koella 1969 for an introduction to the literature). Briefly, sleep is no longer regarded as a passive 'switching off' process whereby there is simply a reduction in the activity of a cerebral activating system. Two principal states of sleep are now recognized. During the waking state, the electroencephalogram is characterized by a low-amplitude wave pattern with a frequency of about 10 cycles per second, upon which background there are bursts of irregular 'random' activity. With the onset of sleep, the waveform changes to one which is *synchronized*, of a large amplitude and a slow frequency. This state of *deep* or *slow wave sleep* (SWS) is one from which the subject can however be aroused relatively easily; it is a dreamless condition. After some time the condition of slow sleep is interspersed with bouts of *paradoxical sleep* (PS). The electroencephalogram shows fast, low-amplitude, bursts of waves. Muscle tone is in general reduced, particularly in the neck musculature, but there occur characteristic rapid eye movements (so-called REM sleep), during which the subject is difficult to rouse, and experiences dreams. The electroencephalographic waveform corresponds to the pattern of ocular oscillation. Three or four sessions of REM sleep occur each full night. It is generally thought that the level and type of sleep depends upon the *balance* between two opposing systems, which are disposed throughout much of the brainstem and include parts of the thalamus, hypothalamus and limbic system. On the one hand there is the *reticular activating* or *arousal system,* with its cranial extensions to the cerebral cortex, stimulation of which leads to wakefulness and desynchronization of cerebral activity. On the other hand, there has been demonstrated a series of *hypnogenic zones* in the medulla, pons, midbrain, thalamus, hypothalamus and basal limbic structures, the stimulation of which leads to synchronization of cerebral cortical activity, and a *variety* of induced sleep-like states, which differ in their detailed

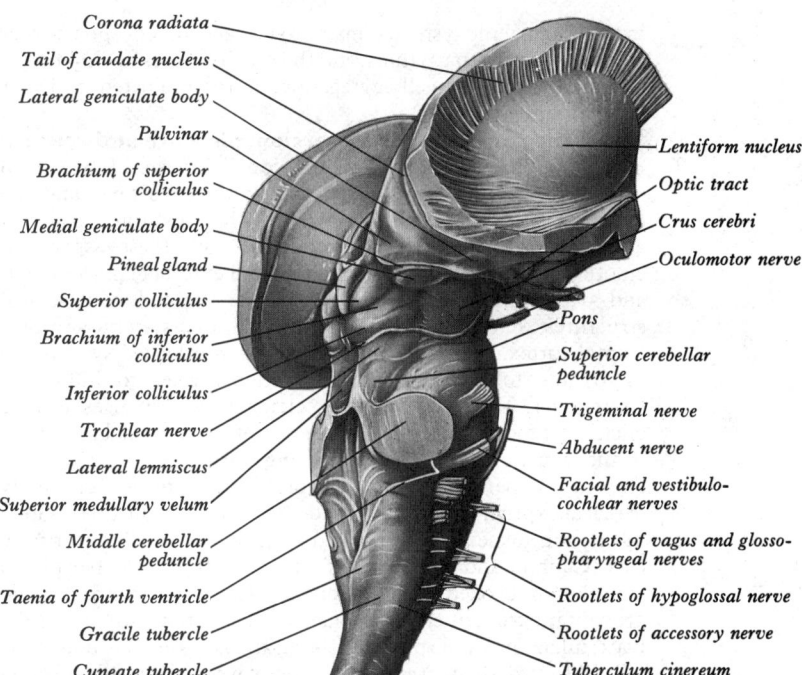

7.107A Posterolateral aspect of the hindbrain and midbrain, exposed by removal of the cerebellum and most of the cerebrum.

Corona radiata
Tail of caudate nucleus
Lateral geniculate body
Pulvinar
Brachium of superior colliculus
Medial geniculate body
Pineal gland
Superior colliculus
Brachium of inferior colliculus
Inferior colliculus
Trochlear nerve
Lateral lemniscus
Superior medullary velum
Middle cerebellar peduncle
Taenia of fourth ventricle
Gracile tubercle
Cuneate tubercle

Lentiform nucleus
Optic tract
Crus cerebri
Oculomotor nerve
Pons
Superior cerebellar peduncle
Trigeminal nerve
Abducent nerve
Facial and vestibulo-cochlear nerves
Rootlets of vagus and glosso-pharyngeal nerves
Rootlets of hypoglossal nerve
Rootlets of accessory nerve
Tuberculum cinereum

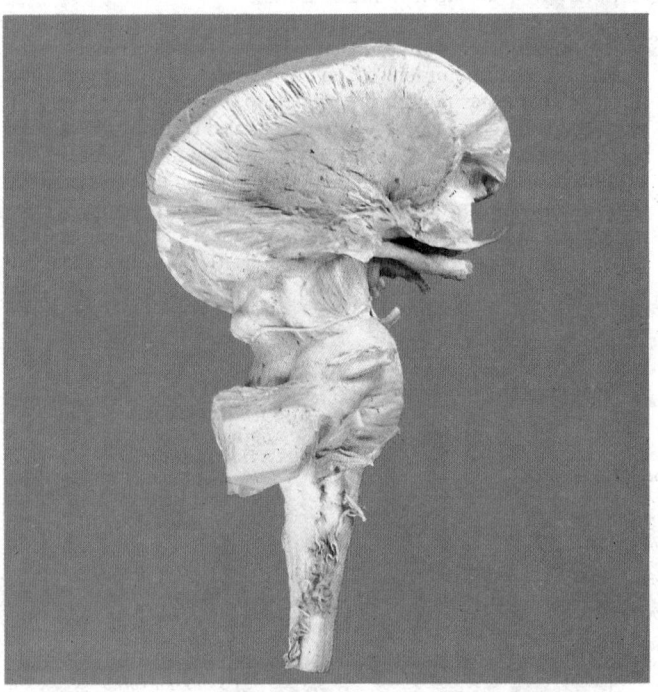

7.107B Right lateral view of the brainstem; compare with 7.107A. (Dissection by Dr. E. L. Rees, Dept. of Anatomy, Guy's Hospital Medical School.)

characteristics. The precise manner in which these opposing, hierarchical systems of sleep-controlling regions interact awaits further investigation, but in general the thalamohypothalamic centres seem dominant whilst those in the medulla, pons and midbrain are subordinate. It is of interest that intact centres in the nucleus ceruleus, rich in both noradrenergic and cholinergic neurons, seem to be important in the generation of paradoxical sleep. Furthermore, the overall circadian rhythm of sleeping and waking is seriously disturbed with lesions in the cranial hypothalamus, and the disturbances being paralleled by corresponding fluctuations of serotonin (5-hydroxytryptamine) concentration in the hypnogenic zones of the brainstem and diencephalon. Whether the serotonin functions as a humoral sleep-inducing and maintaining factor, or as a neural transmitter in the hypnogenic system, remains to be determined. (For further comments on sleep rhythms, and the possible roles of cholinergic and monoaminergic cell aggregations in the brainstem reticular formation, *see* p. 951.)

(8) **Emotion, fear, rage, aversion, pleasure and reward.** The emotional content of an individual consists of two main elements: the *subjective feeling state* or *affective tone*, and the *objective physical accompaniments* which constitute *emotional expression*. For the full integration of both these aspects of emotion, with changes in the internal and external environment, and with other cerebral activities, the essential neurological structures are an intact hypothalamus, limbic system, and pre-frontal cortex. With such a complex and interlocked system it is impossible to ascribe specific functional roles to individual subregions. Nevertheless, some elementary responses can be elicited by focal stimulation or ablation of the hypothalamus both in the intact animal (or human being during neurosurgery), or after the experimental removal of higher centres. The early classical experiments were carried out on decorticate dogs (Goltz 1892), and subsequently analysed in greater detail on decorticate cats (Cannon and Britton 1925). In the latter, mild peripheral stimulation evoked a condition of *sham rage*, with hissing, growling, baring of claws and fangs, piloerection, arching of the back, dilatation of the pupils, striking at objects and lashing of the tail. The term 'sham' was used because it was assumed that feeling states and directed behaviour could not occur in the decorticate animal. Later it was shown that the sham rage was abolished by obliteration of the caudal hypothalamus, but could be elicited by stimulation of the latter in intact animals.

Many subsequent studies have demonstrated that stimulation or ablation of different hypothalamic regions cause changes in the emotional accompaniment of the basic drives of sex, hunger and thirst, whilst other regions are related to the behavioural and (in man) the subjective aspects of rage, fear and pleasure. Particular emphasis has recently been placed upon the existence of centres which are broadly opposed in their actions—*positive* and *negative reward centres*. Stimulation of the former ('*pleasure centres*') leads to 'pleasurable sensations' or the gratification of an intense drive—with an appropriate experimental situation, an animal will

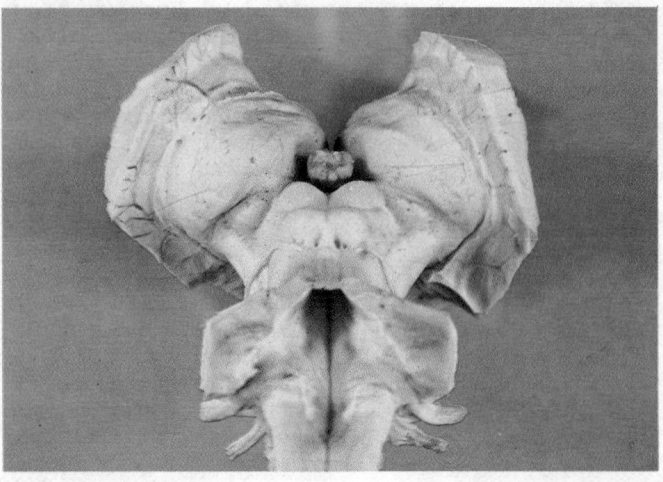

7.108B An oblique view of the dorsal aspect of the brainstem looking cranially. In the foreground is the floor of the rhomboid fossa, bounded laterally by the sectioned white matter of the cerebellum containing the dentate nuclei; the cranial recess of the fourth ventricle passes inferior to the superior medullary velum to continue into the aqueduct of the midbrain. More cranially, the emerging trochlear nerves, the colliculi, superior and inferior brachia, the medial and lateral geniculate bodies, and the pineal gland may be identified. Lateral to the pineal gland on each side is the rounded pulvinar of the dorsal thalamus, skirted laterally by the curving body and tail of the caudate nucleus, whilst most laterally is the cut surface of the corona radiata. (Dissection by Dr. E. L. Rees, Dept. of Anatomy, Guy's Hospital Medical School.)

repeatedly *self-stimulate* its hypothalamus until exhausted, even ignoring food and drink after prolonged periods of deprivation. The principal brain areas from which positive reward (experimental self-stimulation) can be elicited include parts of the catecholaminergic complex of pathways in the brainstem, the medial forebrain bundle of the hypothalamus, the septal areas, the orbitofrontal and parahippocampal (entorhinal) areas of the cerebral neocortex. The most effective region, in the earlier studies, proved to be the medial forebrain bundle (p. 968), in which both ascending and descending fascicles with multiple sources and destinations are intermixed. Furthermore, noradrenergic, dopaminergic, serotonergic and non-aminergic pathways are in close apposition. Readers interested in the history and methods employed in attempts to analyse the functional neuroanatomy of brain stimulation reward should consult Wanquier and Rolls (1976); Olds (1977); Routhenberg and Santos-Anderson (1977). In brief, the latter authors, after highly selective focal stimulation, ablation, and pharmacological manipulation came to the conclusion that intact dopaminergic pathways ascending from the pars compacta of the substantia nigra were essential for 'normal' patterns of self-stimulation. Stimulation in man often causes a general sensation of well-being, and on some occasions it has a strong erotic content. Such *positive reinforcement* may be used to expedite the responses of experimental animals in situations which demand *learning*. In this regard, the forward projection of the *mesolimbic dopaminergic pathways* to their multiple destinations in the limbic system (p. 992), with their possible relevance to the establishment of memory patterns, is perhaps noteworthy. Conversely, stimulation of the negative reward centres presumably causes pain or 'displeasure' and the experimental animal will go to considerable effort and complex patterns of behaviour to avoid repetitive stimulation.

The Metathalamus and Visual Pathway

The optic chiasma (7.108A) is a flattened, somewhat quadrilateral bundle of nerve fibres situated at the junction of the anterior wall of the third ventricle with its floor. Its anterolateral angles are continuous with the optic nerves, the posterolateral angles with the optic tracts. The lamina terminalis (p. 981) is continuous with its upper surface and is crossed, just above the chiasma, by the

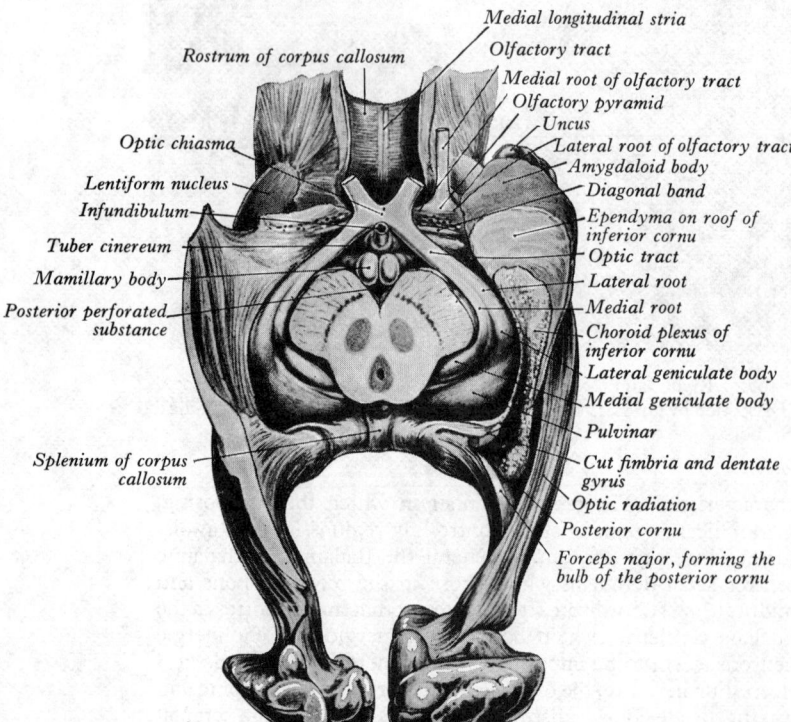

Rostrum of corpus callosum

Medial longitudinal stria
Olfactory tract
Medial root of olfactory tract
Olfactory pyramid
Uncus
Lateral root of olfactory tract
Amygdaloid body
Diagonal band
Ependyma on roof of inferior cornu
Optic tract
Lateral root
Medial root
Choroid plexus of inferior cornu
Lateral geniculate body
Medial geniculate body
Pulvinar
Cut fimbria and dentate gyrus
Optic radiation
Posterior cornu
Forceps major, forming the bulb of the posterior cornu

Optic chiasma
Lentiform nucleus
Infundibulum
Tuber cinereum
Mamillary body
Posterior perforated substance

Splenium of corpus callosum

7.108A A ventral dissection of the brain showing the metathalamus and the optic tracts. On the right side of the figure the inferior horn of the ventricle is exposed. The floor has been removed but the choroid plexus is *in situ* and obscures most of the roof.

anterior communicating artery. Inferiorly the chiasma usually rests on the diaphragma sellae a short distance posterosuperior to the 'optic' groove of the sphenoid bone. In about 10 per cent of subjects it is either more anterior and in the groove, or altogether more dorsal in position. It is always in close relation to the hypophysis (**8**.196). Posteriorly it is related to the tuber cinereum and the infundibulum below, and to the third ventricle above. Laterally it is related to the termination of the internal carotid artery and the anterior perforated substance. A small recess of the third ventricle, the *optic recess*, passes ventrocaudally over its superior surface as far as the lamina terminalis (**8**.196).

Most of the fibres of the optic chiasma start in the retina and reach the chiasma through the optic nerves. In the chiasma the fibres from the nasal half of each retina, including the nasal half of the macula, cross the median plane and enter the optic tract of the opposite side, while the fibres from the temporal half pass backwards in the optic tract of the same side. The crossed fibres loop for a short distance into the ipsilateral optic tract before crossing in the chiasma or into the contralateral optic nerve after crossing. The macular fibres, and those from the central area in the immediate vicinity of the macula, form a flattened band which occupies almost two-thirds of the central region of the chiasma. This bundle is superior to all peripheral decussating fibres. Inferior to this are the fibres from the extramacular parts of the nasal half of the retinae. Most inferior are the nasal fibres concerned with the monocular fringes of the binocular visual field. Dorsal to and within the optic chiasma are bundles of fibres which are not derived from the optic nerve and form no part of the visual pathway. Although these are termed, collectively, the *supraoptic commissures*, they are not really commissures but decussations. One of these is the *supraoptic commissure* (of Gudden), which was formerly, but incorrectly, believed to connect the medial geniculate bodies of the two sides. Its connexions are not fully known, but evidence, from work on mammals, suggests that its fibres arise in the brainstem and spinal cord; its existence, in man, is denied by some authorities. The so-called supraoptic commissure is only one of at least three dorsal connecting pathways described in the optic chiasma. They are reputed to terminate in various hypothalamic and subthalamic

nuclei in some animals, such as the cat, but it appears improbable that they are in any way associated with vision. There is, however, convincing evidence of a retinal projection to the suprachiasmatic, the ventral part of the infundibular (arcuate), and perhaps the supraoptic nucleus in the rat; and whilst these connexions are not concerned with perceptual aspects of vision, they do, through projections to the pineal gland (p. 1448), mediate the influence of light and dark on hypothalamic, pineal and autonomic activity in circadian rhythms. Similarly, detached bands of nerve fibres have been noted as leaving the optic tract to pass dorsally across the cerebral peduncle inferior to the main tract. Such bundles as the *transverse peduncular tract* and various other *accessory optic tracts*, though present in some lower mammals and infra-mammalian forms, have been claimed to be absent in primates by some authorities (but *see* p. 936). According to this view all the myelinated fibres of the optic tracts in man pass either to the lateral geniculate body or to the superior colliculus and pretectal nucleus (Polyak 1957). However, as detailed above and elsewhere (p. 969) there is an impressive body of evidence that the suprachiasmatic and ventral part of the infundibular (arcuate) nuclei, and as indicated below, a thin lamina in the pulvinar, receive optic tract fibres, i.e. forming *retinohypothalamic* and *retinothalamic projections*.

The optic chiasma is supplied from the network of arteries in the investing pia mater. This network receives branches from the superior hypophysial, internal carotid, posterior communicating, anterior cerebral and anterior communicating arteries. The veins drain into the basal veins and the anterior cerebral vein above.

The optic tracts (7.108, 109) are continued dorsolaterally from the posterolateral angles of the chiasma. Each passes between the anterior perforated substance and the tuber cinereum, as the anterolateral boundary of the interpeduncular fossa. The tract becomes flattened and winds round the upper part of the cerebral peduncle, to which it adheres. In this part of its course it is hidden from view on the basal surface of the cerebrum by the uncus and parahippocampal gyrus. Reaching the lateral geniculate body the optic tract divides into medial and lateral 'roots'. The medial of these is believed to contain supraoptic commissural fibres; the lateral root or ramus consists

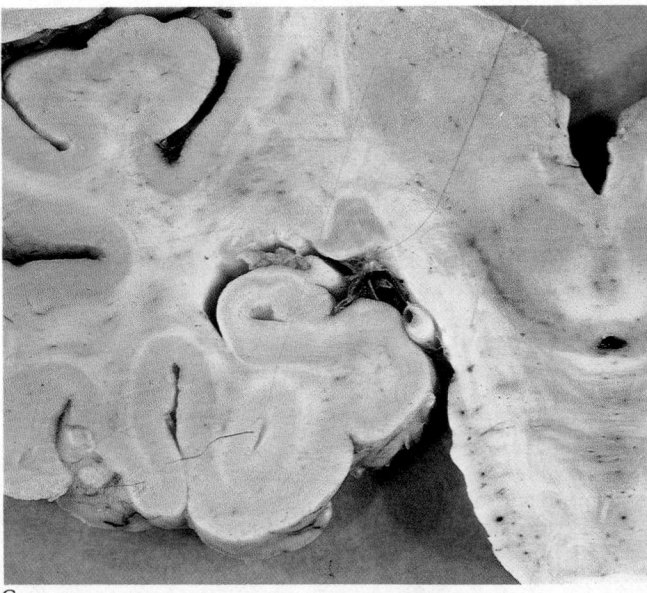

C

7.108c–G (c) The lateral geniculate body, shown in coronal section in the human brain to display its general position and orientation—it is the cap-shaped mass of grey and white matter near the centre of the field. Note its relationship to the inferior horn of the lateral ventricle, the structures visible on the sectioned surface of the midbrain tegmentum, and the dorsal thalamus. Even at this low magnification the lamination of the lateral geniculate body is visible. The four diagrams show the right lateral geniculate nucleus in section: (D) a mature human nucleus in coronal section near its central region and (E) near its posterior pole. Note the reduction in the number of discernible laminae in the latter. The lamination is also visible (F) in an approximately horizontal section at

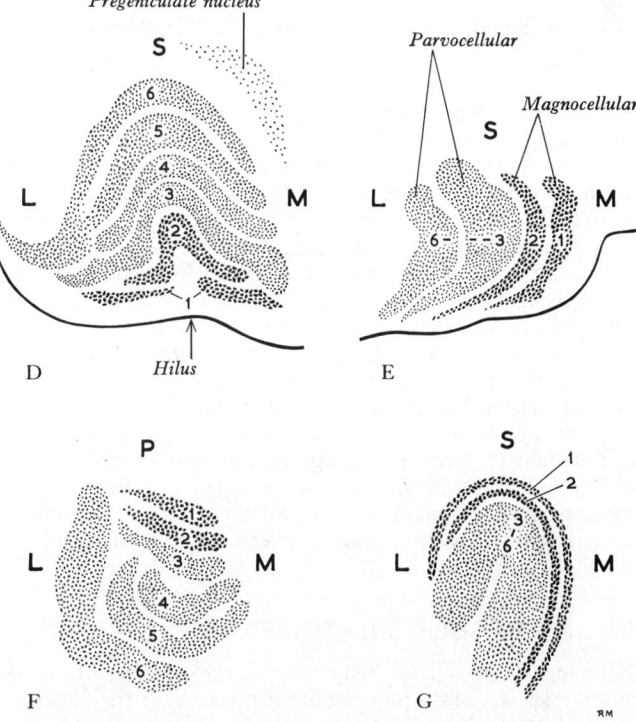

the time of birth. In some primates, for example *Tarsius* (G), the magnocellular layers (1 and 2) are curved externally around the parvocellular layers (3–6); the higher primate arrangement is almost the reverse of this when examined in the coronal plane. See text for further details, and consult Le Gros Clark (1932), Cooper (1945); Chacko (1955).

of afferent fibres which arise in the retina and undergo partial decussation in the optic chiasma, as already described, but it also contains a few fine efferent fibres which are passing forwards to terminate in the retina. Most of the fibres of the lateral ramus are found to end in the lateral geniculate body (p. 977); some sweep medially below the pulvinar and gain the superior colliculus and the pretectal nucleus (pp. 941, 942). The termination of some of these fibres in the pulvinar itself has suggested in various mammals, including primates (Campos-Ortega *et al.* 1972) and cats (Berman and Jones 1977) that they may provide part of a more direct route to cortical areas outside the striate cortex. The latter authors describe the pulvinar projection as ending in a thin zone which borders the lateral aspect of the lateral geniculate nucleus. The detailed arrangement of retinal fibres in the optic chiasma and tract has been worked out in the monkey (Polyak 1957) and clinical observations had already suggested a similar form of organization in human structures (Traquair 1948). Illustration 7.109 shows these details. A notable feature is that nerve fibres derived from the macula, which occupy a central position in the optic nerve and chiasma, assume an eccentric and dorsolateral location in the optic tract, where fibres from both retinae mingle. As will be detailed later, this rearrangement is a prelude to the mode of spatial representation in the lateral geniculate body (p. 941), the superior colliculus and visual cortex (p. 1018).

Further fibres arise from nerve cells in the lateral geniculate body, and pass through the posterior limb of the internal capsule. Emerging from the capsule as a broad bundle termed the *optic radiation*, the fibres of the second visual neurons curve backwards and medially to reach the cortex of the occipital lobe of the cerebrum, where the cortical visual apparatus is situated (p. 1018). On their way they are separated from the posterior horn of the lateral ventricle only by the tapetum of the corpus callosum.

Some of the fibres in the optic radiation take an opposite course, arising from the cells of the occipital cortex and passing to the superior colliculus, which therefore receives cortical in addition to retinal fibres. From the superior colliculus further fibres travel by the tectobulbar tracts to reach the nuclei of the third, fourth, sixth and eleventh cranial nerves, and the anterior grey column of the spinal cord. Recent studies suggest that such 'oculomotor' connexions are doubtful in primates, including man (p. 941).

The Metathalamus—the Geniculate Bodies

The metathalamus (7.108, 109) consists of the *medial* and *lateral geniculate bodies*, bilateral eminences on the inferior aspect of the thalamus lateral to each side of the midbrain. As this collective term suggests, they are relatively late specializations of the vertebrate thalamus in both phylogenetic and ontogenetic senses. Their basic roles as relay nuclei on the acoustic (medial geniculate) and visual (lateral geniculate) pathways have long been recognized; but more recent evidence shows that their connexions, and hence their functions, are much more complex than this. In particular, there exists in both pathways—as in most, if not all sensory projections—a *corticofugal* or efferent component, modulating the patterns of *centripetal* inputs, which can hence no longer be regarded as arrays of simple responses to stimulation of peripheral receptors, proceeding independently along completely separated parallel channels. For example, neuronal mechanisms exist, as elsewhere, for collateral inhibition and probably for interaction between different modalities of sensibility.

THE MEDIAL GENICULATE BODY

The medial geniculate body (7.108, 120) projects from the inferior surface of the pulvinar of the thalamus (7.108), lateral to the superior colliculus. Its contour is ovoid, with the long axis directed anterolaterally. The inferior brachium (p. 935) ascends with a ventrolateral inclination to reach the medial geniculate body, passing between it and the pulvinar; it contains nerve fibres from the ipsilateral lateral lemniscus (derived in part from

the opposite superior olive) and from both inferior colliculi.

In its **internal structure** the medial geniculate nucleus exhibits no clear-cut division into sub-nuclei; the entire mass presents a knee-shaped profile in section and hence its name. On the basis of size its cells are largely segregated into a *dorsal, pars parvocellularis*, or *principal division*, and a smaller *ventromedial division*, the *pars magnocellularis*, which in fact contains an admixture of the smaller cell type. Wedged between the pulvinar and the pretectal region, dorsomedial to the medial geniculate body, is a small mass of nerve cells, the *supragangliculate nucleus* of some forms; its connexions are uncertain, but it probably receives fibres from reticular elements in the lateral tegmental region of the midbrain and pons concerned in other modalities of sensation, including pain (p. 943). Closer study of the constituent nerve cells of the medial geniculate nucleus, particularly of their dendritic organization and their connexions in the cat, has prompted a more complex partition into regions and component nuclei (Morest 1964, 1965). In this animal the medial geniculate nucleus has been allotted *dorsal, ventral* and *medial divisions*, with further subdivisions, one of which—the '*ventral nucleus*' of the ventral division—has a laminar structure. This is a detail of special interest, because this part of the geniculate body receives most of the inferior collicular projection in the cat (Moore and Goldberg 1963); and, as will be described below, the lateral geniculate body is also laminated, this pattern being clearly associated with a spatial organization of sensory terminals. In the same feline schema, the medial division is regarded as the terminus of direct lateral lemniscal fibres, and evidence is growing that these are, in the cat and monkey, spinal afferents concerned in modalities of sensibility other than acoustic, ending in the magnocellular (ventromedial or medial) part of the medial geniculate body. It has hence been suggested that this, and perhaps the supragangliculate nucleus, may be involved in integrations between visual, auditory, general somatic and even visceral inputs.

The role of the medial geniculate body in the acoustic pathway is still in many details obscure, apart from its function as a relay nucleus. This is in large part due to its relative inaccessibility to the recording electrode or destructive probe, neither of which can be accurately placed without marked difficulties of approach. By analogy the laminar appearance of the ventral division, and the existence of a tonotopical organization at cortical level, both suggest similar arrangements in the medial geniculate body; and evidence both in favour of this and against it has been adduced in experimental animals. It is clearly not yet justifiable to suggest tonotopical representation in man. The efferent auditory projection is considered to be derived directly from the small-celled regions of the medial geniculate nucleus, terminating in the temporal lobe, in an area in monkeys corresponding to area 41 in man (p. 1021)—part of the superior temporal gyrus. Other acoustic areas have been delineated by animal experimentation, and they may also exist in man (p. 1021). A primary acoustic area (designated AI in the cat) is directly connected to the medial geniculate body; secondary acoustic area (AII and others) may also receive projections from the same parvocellular parts of the geniculate complex, or from its magnocellular part (Niimi and Naito 1974). All these areas may be concerned in various aspects of acoustic function (p. 1021), and there is abundant evidence of tonotopical representation of the spiral organ of the cochlea in area AI. The role of the medial geniculate body, however, in determination of frequencies of sounds remains *sub judice*, the evidence being largely negative (Whitfield 1967). Lesions of the medial geniculate body do not lead to pronounced deafness unless they are bilateral, since both cochlear organs are represented in each lateral lemniscus (p. 909). As in the case of the first somatosensory (p. 1016) and other receptive areas, the auditory cortex projects to lower centres. Cajal first suggested a cortico-geniculate projection, which was demonstrated in monkeys by Whitlock and Nauta (1956). Pontes *et al.* (1975) have described in detail projections from AI and adjoining areas upon the medial geniculate nuclei (dorsal and magnocellular) in the cat and have also summarized previous work in this field. Their results confirm the findings of Walther and Rasmussen (1960) and hence corroborate the concept of a *descending auditory pathway* (see Rasmussen 1967, and pp. 964, 1074).

THE LATERAL GENICULATE BODY

The lateral geniculate body is a small nuclear complex distinguishable on the surface as a small ovoid projection from the inferior aspect of the posterior region of the thalamus (7.108, 120). Its long axis is approximately sagittal, and its anterior pole blends with the larger, lateral moiety of the optic tract, the medial 'root' or ramus of which separates the geniculate elevation from the lateral aspect of the crus cerebri. The medial geniculate body is, of course, medial to it, but in a more dorsal position. The inferior or superficial aspects of both geniculate bodies are obscured from view by the medial region of the temporal lobe, the para-hippocampal gyrus being the immediate relation (p. 990). Proceeding dorsally from the posteromedial region of the lateral geniculate body, to reach the pretectal area and lateral side of the superior colliculus, is a narrow linear elevation, the *superior brachium* (7.107). This contains uninterrupted retinal nerve fibres which mediate optic reflexes through the pretectal and thence via the oculomotor nuclei and tectospinal tracts (p. 941). The superior brachium is between the medial geniculate body and the pulvinar in much of its extent. From comparison of the lateral ramus of the optic tract and superior brachium, whose relative sizes represent the balance between the forebrain and midbrain destinations, it is obvious that the phylogenetically older, mesencephalic termination is greatly reduced in man, as in other primates, but to a lesser degree in mammals in general. The emergence of the lateral geniculate body was during sub-mammalian evolution, at a stage where the optic tract contains only contralateral nerve fibres from the eye. This decussation is basic in vertebrates, and while it is possible to formulate various explanatory hypotheses to account for it—like other central nervous decussations—these must remain speculative (Polyak 1957; Duke-Elder 1932). The advent of an element of uncrossed retinal connexions in mammals—especially developed in the primates and others with forward looking eyes and overlap of visual fields—can be associated with an increasing specialization of the *dorsal part* of the *primitive lateral geniculate body*, which forms by volume the main part of the structure in higher mammals, including man. It is in this region of the whole nuclear mass that a *laminar organization* appears, perhaps in direct association with the terminals of ipsilateral and contralateral retinal projections (*vide infra*). The *ventral part* of the mammalian lateral geniculate body is also known in many forms as the *pregeniculate nucleus*, which is in man represented by a small but not exiguous group of nerve cells (Le Gros Clark 1932) medial to the main laminar nucleus in its ventral third or so. Medially, the cells of the pregeniculate nucleus extend in a scattered array towards the subthalamic region, marking the evolutionary and embryonic origin of this part of the lateral geniculate complex. It receives a bilateral retinal projection (possibly as collaterals from fibres passing to the tectum and to the main nucleus); this projection is said to be limited to the central region of the retina and to form part of the light reflex pathways through a *pregeniculo-mesencephalic connexion* (Polyak 1957). Such an arrangement has a particular clinical interest to explanations of the Argyll-Robertson phenomenon (p. 1184). However, Pierson and Carpenter (1974) could find no evidence of involvement of this nucleus in the light reflex in monkeys subjected to various lesions of the eye, geniculate and pretectal regions. A cortical connexion with the pregeniculate nucleus has not been established, and it is to be emphasized that some workers deny that optic nerve fibres terminate in this nucleus in primates, in which it is nevertheless a definable entity, though in this order it is not so well developed as in some other mammalian groups, such as ungulates and carnivores (Niimi *et al.* 1963).

The dorsal (or main) lateral geniculate nucleus exhibits a well-known laminar structure, there being six (and perhaps seven) layers of nerve cells discernible in some part of the nucleus in higher primates, such as the rhesus macaque (the subject of much experimental study) and mankind. It should be noted at once, however—particularly in respect of proposals to equate this pattern with functional ideas, such as trichromatic theories of colour vision (Le Gros Clark 1940, 1949)—that the number and arrangement of laminae varies in different mammals, being three

in the cat (another much used experimental animal). The laminar pattern also displays regional degrees of elaboration in the geniculate nucleus in most species, including man. Phylogenetically, an original division into two laminae has been suggested, concerned with separate terminations of ipsi- and contra-lateral retinal fibres (Polyak 1957). Nevertheless, all six layers appear together in human embryonic development (Cooper 1945). With the development of two major classes of neurons, large and small, a splitting of these primary layers may have occurred. In this connexion it is pertinent to note that interlaminar strata of larger cells occur between the three recognized laminae in the feline lateral geniculate nucleus (Thuma 1928).

In describing the human lateral geniculate body, particularly as regards its intrinsic and external interconnexions and their functional interpretation, it is necessary to remember that such details are almost entirely extrapolated from experimental results observed chiefly in cats and monkeys, together with a considerable volume of comparative structural study in mammals, including many primates. Nevertheless, the commonly used experimental animals, especially monkeys, display such a similarity of structural organization in the lateral geniculate body that it is unlikely that data established thus cannot be extended, in most instances, to mankind.

The appearance and orientation of the laminae of the human lateral body are shown in illustration 7.108C–G. This orientation is often confusingly presented in textbooks and other accounts (consult Cooper 1945; Chacko 1955), its relation to other structures being obscured by omission of reference data. Often depicted in coronal sections (7.108C), it is also sometimes shown in sagittal section, and in both views the lamination is apparent, since the laminae are somewhat like a series of caps set one upon another. The lamination is best displayed in coronal sections from the central region of the nuclear mass and in a direction towards its posterior (caudal) pole. Anteriorly or cranially, i.e. near the entry of the optic tract, the layers are blended together into a common mass, and at the posterior extremity the number is reduced to four—in a region of the nucleus which is considered to be monopolized by terminating macular fibres. As illustration 7.108C shows, the laminae are curved, with a ventral concavity forming a kind of 'hilum', at which some of the retinal efferents enter from the adjacent optic tract. This hilum is much more prominent in the monkey than in man. Dorsally, the nucleus is dome-shaped and projects towards the thalamus; from this aspect the majority of the fibres of the optic radiation, or *geniculostriate projection*, emerge from the nucleus. Both medially and laterally the laminae blend, as if at the edge of the cap-shaped mass; and the lateral border is sharp; the blending layers reverse their curvature here, becoming ventrally convex, or upturned like the peak of a cap. It is also noteworthy that concave contours of the laminae are also reversed in the posterior (macular) part of the geniculate nucleus. As a result, the profiles of the laminae, in the coronal plane, are slightly sinuous on the lateral side. (Some of the confusion concerning the lamination of the lateral geniculate body is due to the fact that the order of the layers is reversed in some lower primates, the large celled laminae 1 and 2 being *external* and *dorsal* to the others—on the outer aspect of the curvature of the entire laminar structure. In *Tarsius* the situation is extreme, the layers being an inverted replica of the human situation—or vice versa. In the lemuroids transitional states occur. Much controversy has been entailed by the use of the terms inversion and eversion, which need not detain us here.)

The most careful estimates (Kupfer *et al.* 1967) suggest that there are about 1 million nerve cells in the human lateral geniculate body, corresponding closely to the count of fibres in the optic nerve and tract. (Estimates in the macaque monkey are as much as 1·8 million, Le Gros Clark 1941 and 1·7 million, Rakic 1977, but only 1·2 million in the optic nerve, Bruesch and Arey 1942. *See*, however, Chow *et al.* 1950.) This 1:1 ratio in man does not, however, imply that each optic nerve fibre connects with only one geniculate cell. This relationship may be the case in respect of the macula, but such details are still uncertain. It has long been established (Minkowski 1913; Le Gros Clark 1941; Glees 1941) that each fibre, derived from the retinal ganglion cells, ends by dividing into no more than five or six terminals connecting with a corresponding number of geniculate nerve cells in *one* lamina. (The geniculate cells may connect with terminals from more than one retinal fibre.) The laminae are usually numbered 1 to 6, starting from the concave, ventral, or hilar aspect of the nucleus. Fibres from the *contralateral* optic nerve end in layers 1, 4 and 6, *ipsilateral* fibres in layers 2, 3 and 5. Early recognition that transneuronal degeneration (p. 816) occurs in the lateral geniculate body (Minkowski 1913), coupled with an adequate terminal degeneration technique (Glees 1946), has led to the detailed elucidation of this bilateral representation of the retinae in each lateral geniculate body. Even with the Marchi degeneration technique alone (Brouwer and Zeeman 1926) it had been possible to show in monkeys—by studying the results of small retinal lesions—that fibres from different quadrants of the retinae are located correspondingly in the optic nerve (p. 1055 and 7.109), macular fibres occupying a central position. Similarly, an orderly but somewhat different arrangement was demonstrated in the optic chiasma and tract (7.109), with the difference that in the tract, fibres from corresponding quadrants in both retinae occupy the same location. This pattern is carried into the lateral geniculate body, where the macular fibres terminate in the central and posterior part of the nucleus, the peripheral fibres in its anterior part, the upper retinal quadrants ending laterally, the lower ones medially (7.109). Transneuronal techniques enabled much more precise plotting of the retinogeniculate connexions (Le Gros Clark and Penman 1934), and since the cortical projection fibres of the affected geniculate nerve cells also degenerate, it was demonstrable that the same kind of a precise point-to-point connexion continues through the optic radiation to the striate cortex (p. 1018). The precision of this topical organization in the visual pathway has been repeatedly confirmed in cats and monkeys (Polyak 1957; Meikle and Sprague 1964; Saavedra *et al.* 1969). (Some doubts as to the factors causing transneuronal degeneration in the lateral geniculate nucleus have been expressed. There is some evidence that the effect is not due solely to deafferentation by, for example, optic tract interruption. Consult Garey *et al.* (1976) for discussion.) As will be detailed elsewhere (p. 962), ablation of small parts of the striate cortex in the occipital lobe leads to highly localized patches of degenerating geniculate cells in *six layers* on the same side. These effects appear in a *columnar representation* of corresponding points of the two retinae, each in *three* of the six geniculate laminae. The results of transneuronal degeneration in the lateral geniculate nucleus, consequent upon destruction of the retinal ganglion cells or interruption of their axons, have been confirmed by the substantially equivalent but less dramatic effects of prolonged suturing of eyelids in kittens (Wiesel and Hubel 1963; and Guillery 1973) and infant monkeys (Von Noorden 1973; and Headon and Powell 1973). These experiments have been performed unilaterally and sometimes bilaterally. In the latter case (Headon and Powell 1978) one unexplained finding was that the shrinkage of geniculate neurons observed in such experiments was greater in those parts of the nucleus receiving fibres from the monocular fields of the retinae when *both* eyes had been subjected to prolonged closure. More recently the application of the technique of autoradiographic tracing of labelled amino-acids (which are subject to orthograde axoplasmic flow and even transneuronal transport) to the visual pathway has confirmed these earlier results (*see* Graybiel 1975, Hubel *et al.* 1976, and Rakic 1976), both as regards the laminar terminations of retinal ganglionic neurons in the lateral geniculate nucleus and anatomical orientation of neurons in this nucleus and the visual cortex.

So far this account has considered the visual pathway as a series of parallel conducting pathways, with no mention of possible interaction between them—whether from the same retina or from both. In fact, it has for long been assumed that, apart from inevitable convergence of activities in the far more numerous rods and cones of the retina upon the more limited population of optic nerve neurons (retinal ganglion cells—p. 1170), the remainder of the visual pathway consists of functionally isolated conductors, with a simple relay in the lateral geniculate body. This simple view has been customary textbook teaching, with the modification that the bringing together of fibres from both retinae in the geniculate body constitutes subcortical 'fusion'. The latter would be a naïve assumption in the absence of data beyond those recounted above. The neuronal organization of the retina itself is, of course, complex; there is clear evidence, both anatomical and physiological (p. 1166), of interaction between its neurons. Hence the pattern of impulses in the optic nerve is not the simple result of a mosaic of receptor responses, like the signals in a cable between a television camera and its monitor—even if the effects of corticofugal connexions are ignored (p. 1174). Integrative activities in the cat's lateral geniculate body have also been demonstrated (Hubel and Wiesel 1961) between corticofugal fibres and the incoming sensory projection, as in other sensory pathways. While unit recording has also shown the existence of lateral inhibition and of units responsive to specific modes of stimulation, there is apparently no clear evidence to indicate that individual geniculate cells receive impulses from both retinae (Glees 1961).

An increasing elaboration of cytological detail, particularly in respect to dendritic fields and synaptic characteristics, has been recorded in the cat and monkey. It is not possible even to survey this voluminous literature here, but reference to the reports of Peters and Palay (1966), Campos-Ortega *et al.* (1968), Lieberman (1973, 1974), and Lieberman and Webster (1974) will provide an introduction to the detailed synaptic organization of the dorsal lateral geniculate nucleus. Lieberman in particular has identified a wide variety of synapses, pre- and post-synaptic, inhibitory and excitatory, dendrodendritic, and the triplet synapses (triads) of the glomeruli (*vide infra*). These and other researches have been carried out in rodents, cats, and monkeys, and whilst species differences are apparent, the basic resemblances are predominant. While it is not yet possible to correlate these findings into a fully connected model of geniculate activity, it is obvious that the view of this as a comparatively simple relay mechanism is completely demolished. Three major elements are involved: (*a*) *afferents* from the retinae—to which may be added corticofugal fibres and perhaps fibres from the ubiquitous reticular system; (*b*) the *efferent geniculostriate projection*; and (*c*) the *intrinsic neuronal population*, consisting largely of Golgi type II nerve cells, not all the axons of which are confined to a single lamina, and hence might mediate connexions between bilateral retinal afferents. Most accounts say little of the precise arrangement of the cells (Golgi type I) whose axons stream dorsally through intervening laminae to form the geniculostriate pathway; attention has been predominantly concentrated—both in light and electron microscope studies—upon dendritic arrangements and the distribution and functional attributes of synapses. The profusion of the resultant neuropil is in itself suggestive evidence of highly complex interneuronal mediation between the afferents and efferents of the lateral geniculate body (Szentágothai 1963; Lieberman 1974). One particularly interesting feature is the occurrence of *glomeruli*—conglomerations of synapses between dendritic and axonal terminals, the latter in part derived from optic nerve fibres (Peters and Palay 1966; Campos-Ortega *et al.* 1968; Lieberman 1974). These are similar in some respects to the synaptic glomeruli described in the cerebellum (p. 924), and in other thalamic 'relay' nuclei (p. 956), and subserve the same kind of integrative function. The glomeruli are highly complex structures consisting of a main and peripheral axons, a main and peripheral dendrites, and various other neuronal and glial elements (see also p. 829). Some axons may be involved in the formation of several glomeruli; others, morphologically distinguishable, occur in the periphery of the glomerulus, making

7.109 A diagram of the visual pathway to show the spatial arrangement of nerve cells and their fibres in relation to the quadrants of the retinae and visual fields. The proportions at various levels are not exactly to scale, and in particular the macula is exaggerated in size in the visual fields and retinae. In each quadrant of the visual field, and the parts of the visual pathway subserving it, two shades of the respective colour are used—the paler for the peripheral fields and a darker shade for the macular part of the quadrant. From the optic tract onwards these two shades are both made more saturated to denote intermixture of neurons from both retinae, the palest shade being reserved for parts of the visual pathway concerned with uniocular vision. The path of the light reflex has also been indicated.

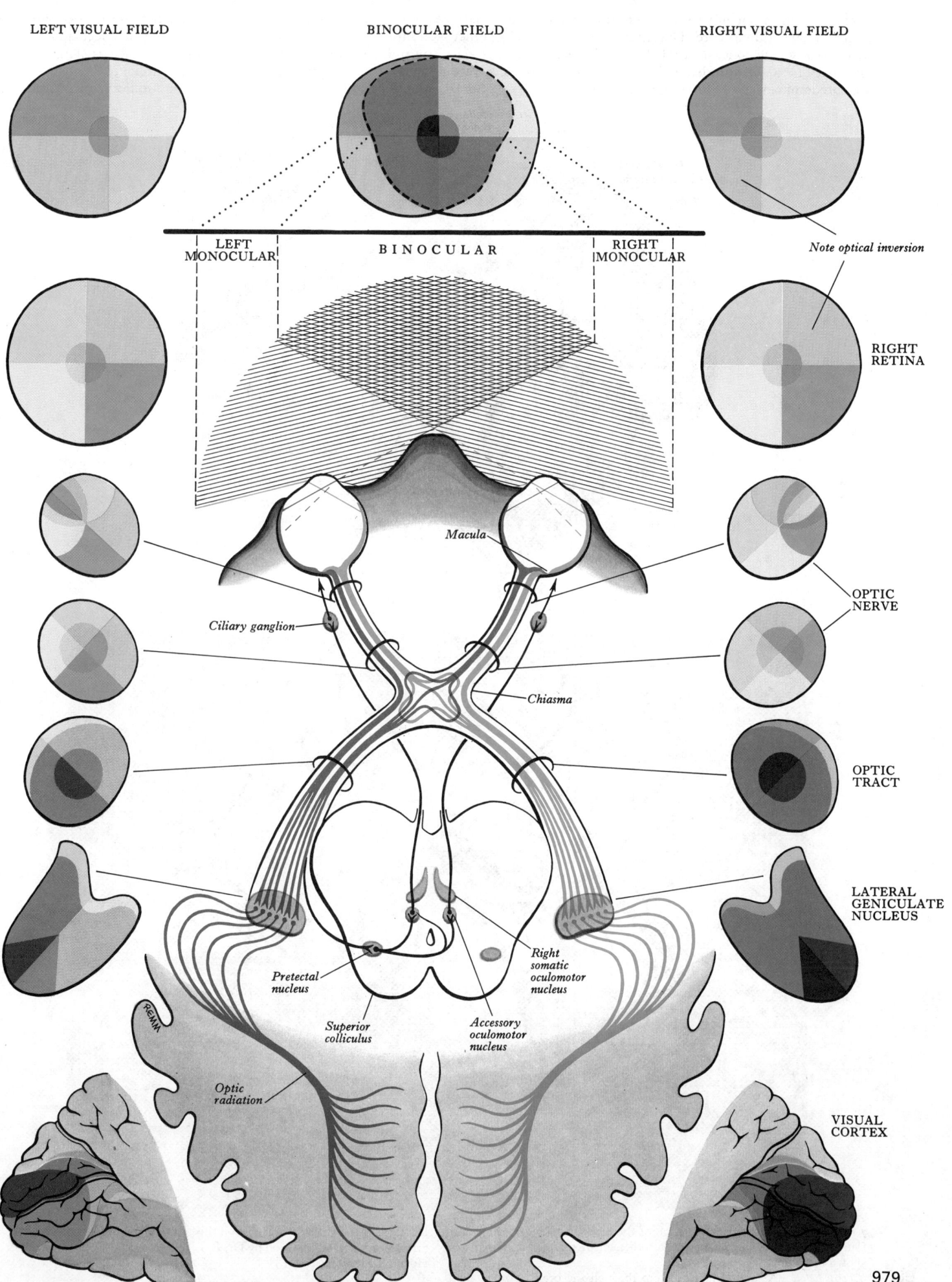

LEFT VISUAL FIELD

BINOCULAR FIELD

RIGHT VISUAL FIELD

LEFT MONOCULAR

BINOCULAR

RIGHT MONOCULAR

Note optical inversion

RIGHT RETINA

Macula

Ciliary ganglion

Chiasma

OPTIC NERVE

OPTIC TRACT

LATERAL GENICULATE NUCLEUS

Right somatic oculomotor nucleus

Pretectal nucleus

Superior colliculus

Accessory oculomotor nucleus

Optic radiation

VISUAL CORTEX

REMM

979

synaptic contacts not only with the dendrites in it but also with the central axon within its core. The latter type of synapse, like other *axoaxonal* connexions, is probably inhibitory (Eccles 1964; Shepherd 1974). Dendrodendritic and axodendritic, however, are the predominant types of synapses (Lieberman 1974). Such observations provide the anatomical substrate for the occurrence of inhibitory as well as excitatory receptive field effects identified by unit recording in the lateral geniculate nucleus (Hubel and Wiesel 1961). Glomerular structures of a similar nature have been identified in the monkey (Colonnier and Guillery 1964). Brauer

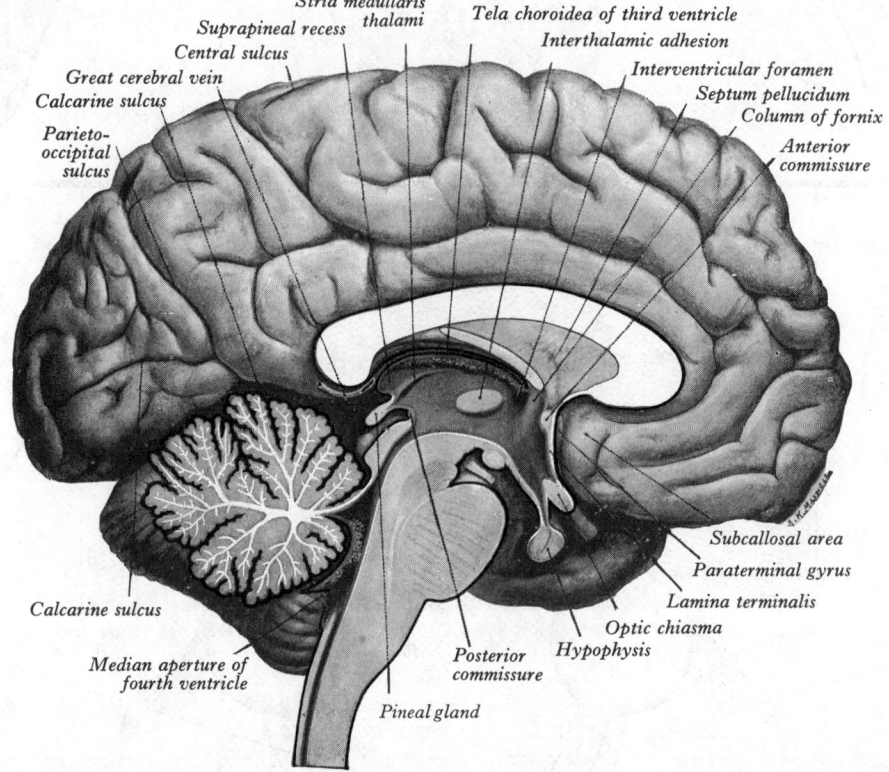

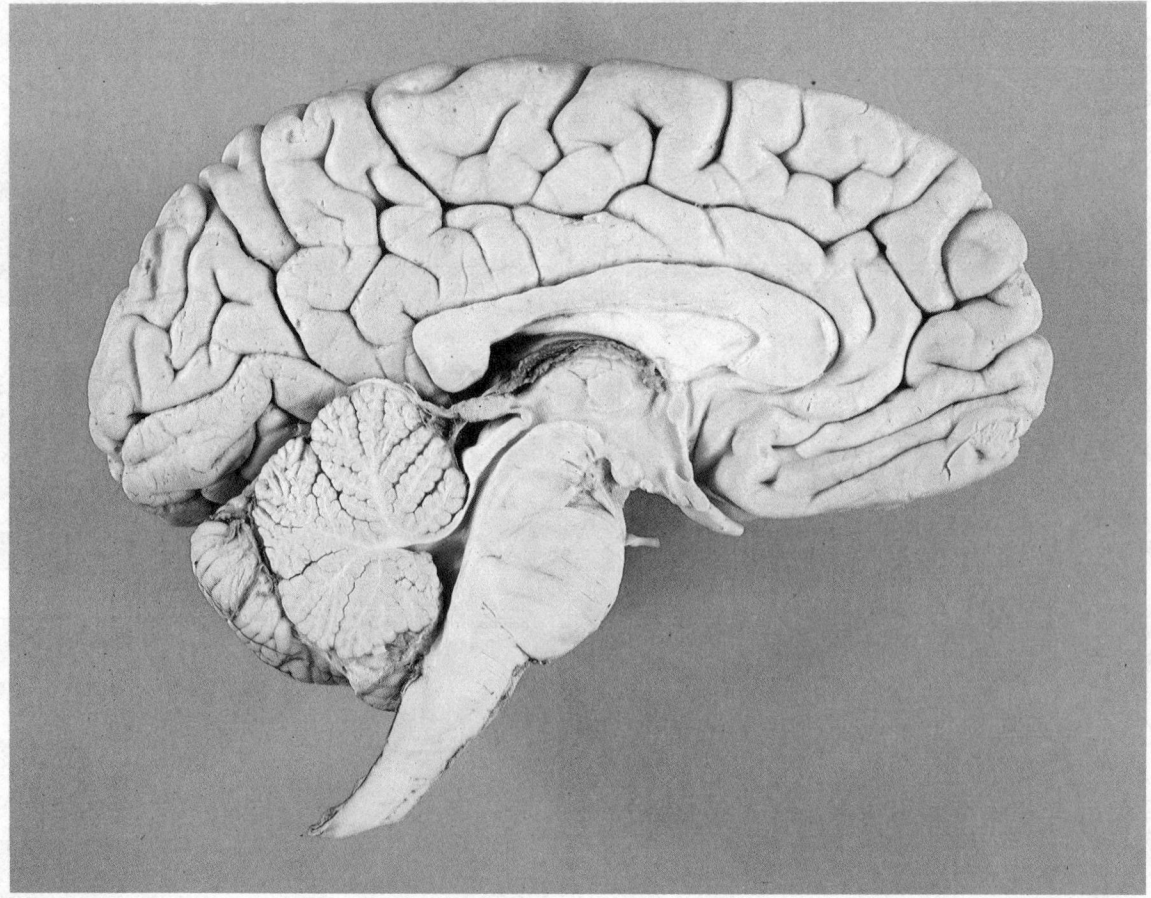

7.110A *above* Median hemi-section of the brain to show the third and fourth ventricles. The pia mater is indicated in red; the ependyma is shown in blue.

7.110B *below* A sagittal hemi-section through the brain. For detailed labelling of the structures visible on this specimen compare with 7.110A, 7.114 and 8.195. (Dissection by Dr. E. L. Rees, Dept. of Anatomy, Guy's Hospital Medical School.)

and Winkelmann (1976) have surveyed views of the functions of interneural participation in mammalian glomeruli. Such *synaptic* glomeruli are not to be confused with clusters of nerve cells, also called glomeruli, described at a much earlier date (Taboada 1927); these exist in all the laminae of the monkey's geniculate nucleus, the number in the clusters increasing from 12–15 in layer 1 to 60–100 in layer 6. It has been suggested (Peters and Palay 1966) that the somata of the neurons whose neurites are involved in a *synaptic glomerulus* may be clustered together as a *cellular glomerulus*, a view which tentatively equates the latter with the nerve cell groups described above. Electron microscope evidence suggests that the synaptic glomeruli may become more complex in the outer layers of the lateral geniculate nucleus. However this may be, the intricate synaptic relationships thus established provide an illuminating prelude to full structural delineation of the integrative mechanisms in the lateral geniculate nucleus.

The findings so far noted on the whole indicate only *intra*laminar integration; the question of *inter*laminar interaction is of special interest with regard to integration of information from both eyes. Since each retinal ganglion cell fibre terminates, as far as is known, merely in one lamina, any interaction between fibres derived, as is presumed, from the same locus in one retina but ending separately in three geniculate laminae, would require some form of interlaminar connexion. Even were such connexions unsupported by anatomical evidence, the results of unit recording at the geniculate level could indicate their existence. The same reasoning appertains to interaction between impulses derived from corresponding loci in *both* retinae. Again, as far as is known, each lamina projects separately to the striate cortex—and not, as some diagrams loosely depict, by a fictitious form of neuron with a soma in each lamina and a single conjoined axon projecting to the cortex! Physiological evidence has tended to corroborate the view that impulses from the two retinae are integrated only at the level of the striate cortex (Hubel and Wiesel 1969). However, it *was* claimed fairly recently that unit recording demonstrated binocular interaction at the geniculate level; and electron microscope observations suggest that in the fibre zones between laminae are synapses between neurons intrinsic to the lateral geniculate nucleus (Guillery and Colonnier 1970), however, *vide infra*. Neurons in the laminae—probably second order interneurons—have also been claimed to mediate interlaminar connexions (Guillery 1966, 1971); these *could* be concerned in channelling effects from both retinae into single geniculostriate fibres. It is precisely upon these problems that most of the intense synaptological research on the geniculate neurons is currently focused. The peculiarities of the intrinsic neurons (interneurons) have attracted special attention (consult Lieberman and Webster 1974). Many, perhaps almost all of these are amacrine (anaxonal) in rodents and some, at least, are without axons in monkeys (LeVay 1971). They are, therefore, integrative in function, as is obvious from their complex pre-synaptic contacts with retinal and cortical connexions in the glomeruli of the lateral geniculate. In the most recent summary of activity in the visual pathways (Hubel and Wiesel 1977) it appears that the weight of evidence is *strongly against* the existence of any neurons in the lateral geniculate nucleus which respond to impulses from both retinae. These two workers, pre-eminent in this field, firmly reject any 'fusion' of information from both retinae at the geniculate level.

The Third Ventricle

The third ventricle (7.97, 110), which is the derivative of the primitive forebrain vesicle, is a median cleft between the two thalami and hypothalami. It communicates posteriorly with the fourth ventricle through the cerebral aqueduct, and anteriorly with the lateral ventricles through the interventricular foramina. It has a roof, a floor, anterior and posterior boundaries, and two lateral walls.

The *roof* is formed by a layer of ependyma which stretches between the upper edges of the lateral walls of the cavity and is continuous with the rest of the ependymal lining of the ventricle. It is covered by, and adherent to, a fold of pia mater, named the *tela choroidea* of the third ventricle, from the inferior surface of which a pair of vascular fringed processes, the *choroid plexuses of the third ventricle*, project downwards, one on each side of the median plane, and invaginate the ependymal roof into the ventricular cavity (7.157).

The *floor* (7.110) descends ventrally and is formed mainly by structures which belong to the hypothalamus; ventrodorsally these are: the optic chiasma, the infundibulum and tuber cinereum, and the mamillary bodies. Posterior to the last-named, the floor is formed by the posterior perforated substance and by the tegmentum of the cerebral peduncles. The ventricle is prolonged downwards into the infundibulum as a funnel-shaped *infundibular recess*. The neurohypophysis is continuous with the apex of the infundibulum.

The *anterior boundary* (7.110) is inferiorly the *lamina terminalis*, which represents the cranial terminal of the primitive neural tube. It forms a thin layer of grey matter stretching from the superior surface of the optic chiasma to the rostrum of the corpus callosum. In its superior part the anterior boundary is formed by the columns of the fornix, which diverge as they descend into the lateral walls of the ventricle, and by the anterior commissure (p. 997), which crosses the median plane anterior to them. At the junction of the floor and anterior wall, immediately above the optic chiasma, the ventricle presents a small angular diverticulum, the *optic recess*. At the junction of the roof with the anterior and lateral limits of the ventricle is the *interventricular foramen*, through which the third and the lateral ventricles communicate with one another. It represents the site of the original diverticular outgrowth from the telencephalon which forms the cerebral hemisphere, and is *relatively* large and circular in a 10 mm human embryo. In the adult, however, it is altered to a somewhat crescentic slit, bounded anteriorly by the curving column of the fornix and posteriorly by the convex anterior tubercle of the thalamus.

The *posterior boundary* (7.110) consists of the pineal gland, the posterior commissure and the cerebral aqueduct. A *pineal recess* projects into the stalk of the pineal gland, while anterosuperior to the latter is a second, *suprapineal recess*, consisting of a diverticulum of the epithelial ventricular roof.

Each *lateral wall* consists of an upper part formed by the medial surface of the anterior two-thirds of the thalamus, and a lower formed by the hypothalamus and continuous with the grey matter of the ventricular floor. These two regions of the lateral wall are separated by the *hypothalamic sulcus*, which extends from the interventricular foramen to the cerebral aqueduct, but is not always an obvious feature. As noted previously (p. 953), the hypothalamic sulcus can be regarded as dividing the diencephalon into two. Above (dorsal to) the sulcus is the *pars dorsalis diencephali* consisting of the dorsal thalamus and the epithalamus, below (ventral) is the *pars ventralis diencephali* which constitutes the hypothalamus and subthalamus. Each lateral wall of the third ventricle is limited superiorly by the ridge covering the stria medullaris thalami (p. 954). The columns of the fornix curve ventrally rostral to the interventricular foramina, and then run in the lateral walls of the ventricle, where, at first, they form distinct prominences, but subsequently sink into them. The lateral walls are joined to each other across the cavity of the ventricle by a band of grey matter, the *interthalamic adhesion* (p. 955). An extensive description of the phylogeny and development of the third ventricle has been contributed by Kier (1977). The zones of specialized ependyma collectively forming the *circumventricular organs* are described on pp. 197, 836, 1442.

The interpeduncular fossa (7.55, 104, 115) is a trapezoid area of the cerebral base, limited anteriorly by the optic chiasma, posteriorly by the anterosuperior surface of the pons, anterolaterally by the converging optic tracts, and posterolaterally by the diverging cerebral peduncles. The structures contained in it have been described elsewhere; caudorostrally they are the posterior perforated substance, mamillary bodies, tuber cinereum pp. 966–973, infundibulum and hypophysis cerebri pp. 969, 1438.

THE TELENCEPHALON OR 'ENDBRAIN'

The expansion of the telencephalon and the development of the two cerebral hemispheres have already been described (pp. 170–174). In primitive vertebrates each cerebral hemisphere is predominantly concerned with olfactory impulses, which enter it rostrally at the *olfactory lobe*. This lobe may be drawn away to form an *olfactory bulb* which remains connected to the rest of the hemisphere by a stalk, the *olfactory tract*. In the basal parts of each hemisphere masses of grey matter, the *basal nuclei*, are present and form an early motor centre. The wall of the hemisphere constitutes the *pallium*, wherein olfactory and other flows of information are presumed to be integrated. In the course of evolution visual, auditory and other conduction paths have been extended, through the thalamus, to the pallium of the cerebral hemispheres, an instance of encephalization. Consequently, each cerebral hemisphere enlarges as a result of the formation of an additional region, the *neopallium*, the predominantly olfactory pallium being confined to a *piriform lobe* inferolateral in site. The medial wall of the hemisphere becomes specialized to form the *hippocampal formation*. This was for long also regarded as primarily part of the olfactory mechanism, but in recent years this view has become untenable (*see* p. 990). In higher mammals the neopallium is greatly enlarged; the piriform lobe, in contrast, is relatively reduced. With the growth of the neopallium motor pathways are developed from this, but the basal nuclei remain as essential parts of the motor control apparatus. The greatly increased growth of the neopallium in mammals is due to the formation of *association areas* concerned with the interaction of its multiplicity of afferent and efferent connexions. (It should be noted that the hippocampal formation is often termed the *archipallium*, or *primal* cortex, and the piriform lobe the *paleopallium*, or *ancient* cortex. Some authorities, however, group both these regions under the term *archipallium*, and also prefer to spell the term archeopallium.)

The telencephalon includes: (1) the cerebral hemispheres, the commissures which connect them, and the cavities which they contain; and (2) the anterior parts of the third ventricle, including the preoptic regions in the telencephalon impar, already described (p. 965). Each cerebral hemisphere consists of an outer layer of neurons, the so-called grey matter, termed the *cortex*, an inner mass of neuronal processes, the white matter (*centrum semiovale*), the deeply situated *basal nuclei* and a cavity, the *lateral ventricle*.

THE CEREBRAL HEMISPHERES

The cerebral hemispheres form the largest part of the brain, and, when viewed together from above, assume the outline of an ovoid mass broader behind than in front, the greatest transverse diameter corresponding with a line connecting the two parietal tuberosities. The hemispheres are incompletely separated by a deep median cleft, named the *longitudinal cerebral fissure*, and each possesses a central cavity, the *lateral ventricle*.

The longitudinal fissure of the cerebrum contains a sickle-shaped process of dura mater, the *falx cerebri*, and the anterior cerebral vessels. Anteriorly and posteriorly, the fissure completely separates the cerebral hemispheres from each other; centrally, however, it only extends down to a great central white commissure, named the *corpus callosum*, which connects the hemispheres across the median plane.

The Surfaces of the Cerebrum

Each cerebral hemisphere presents three surfaces: superolateral, medial and inferior.

The *superolateral surface* is convex in adaptation to the concavity of the corresponding half of the vault of the cranium. The *medial surface* is flat and vertical and is separated from that of the opposite hemisphere by the longitudinal fissure and the falx cerebri. The *inferior surface* is of an irregular form, and may be divided into two parts: orbital and tentorial. The orbital part, being the orbital surface of the frontal lobes, is concave, and is above the orbital and nasal roofs; the tentorial part is concavoconvex, and is the inferior surface of the temporal and occipital lobes; anteriorly it is adapted to the corresponding half of the middle cranial fossa; posteriorly it rests upon the tentorium cerebelli, which intervenes between it and the superior surface of the cerebellum.

The three surfaces are separated by the following borders: (*a*) *superomedial*, between the superolateral and medial surfaces; (*b*) *inferolateral*, between the superolateral and basal surfaces; the anterior part of this border separates the superolateral from the orbital surface of the frontal lobe, and is known as the *superciliary* border; (*c*) *medial occipital*, between the tentorial part of the inferior and the medial surface; and (*d*) *medial orbital*, separating the orbital part of the inferior from the medial surface. The anterior end of the hemisphere is named the *frontal pole*, the posterior, the *occipital pole*; and the anterior end of the temporal lobe, the *temporal pole*. About 5 cm anterior to the occipital pole on the inferolateral border there is an indentation, the *pre-occipital incisure* or notch.

A paramedian line drawn from a point a little superolateral to the inion, forwards to a point just superolateral to the nasion, corresponds to the superomedial margin. The superciliary border follows the curve of the eyebrows at a slightly higher level as far as the zygomatic process of the frontal bone and then ascends to the pterion. The temporal pole can be indicated on the surface of the head by a line drawn, with a forward convexity, from the pterion to the middle of the upper border of the zygomatic arch, and this line, continued backwards just above the zygomatic arch and crossing the auricle a little above the external acoustic meatus, corresponds to the inferolateral margin of the hemisphere, which then curves downwards to reach the posterior end of the superomedial border (6.119).

The surfaces of the hemispheres are moulded into a number of irregular eminences, named *gyri* or *convolutions*, and separated by furrows termed *sulci* or *fissures*.

The irregular character of the surfaces of the cerebral hemispheres is a very prominent feature, but up to the end of the third fetal month these surfaces are smooth and unbroken, like the surfaces of the brains of reptiles and birds. Thereafter localized depressions become apparent, and they deepen and extend over the surfaces to form the sulci (*see* p. 170 and 2.89A–G). Each sulcus corresponds to an infolding of the cortex; thereby the total amount of grey matter is about three times as much as might be inferred from the surface area of the hemisphere. In certain situations the sulci develop along lines separating areas which differ from one another in the details of their microscopic structure and in the functions which they predominantly subserve (Le Gros Clark 1945). Such sulci may therefore be termed *limiting sulci*, since they establish the limits of certain functional areas. The central sulcus is an example of a limiting sulcus, for it is set between two areas of cortex which differ in thickness so notably that this can be appreciated with the naked eye (7.115A). In other situations sulci develop in the long axis of a rapidly growing homogeneous area and are termed *axial sulci*. The posterior part of the calcarine sulcus is in the centre of the striate area and is merely a fold in the visual cortex. In other situations, again, a sulcus may be situated between two surface areas of cortex which are structurally different, but its lip and not its floor may form the dividing line between the two areas. In these cases a third area may be present in the walls of the sulcus without appearing on the surface at all. Such a sulcus is termed an

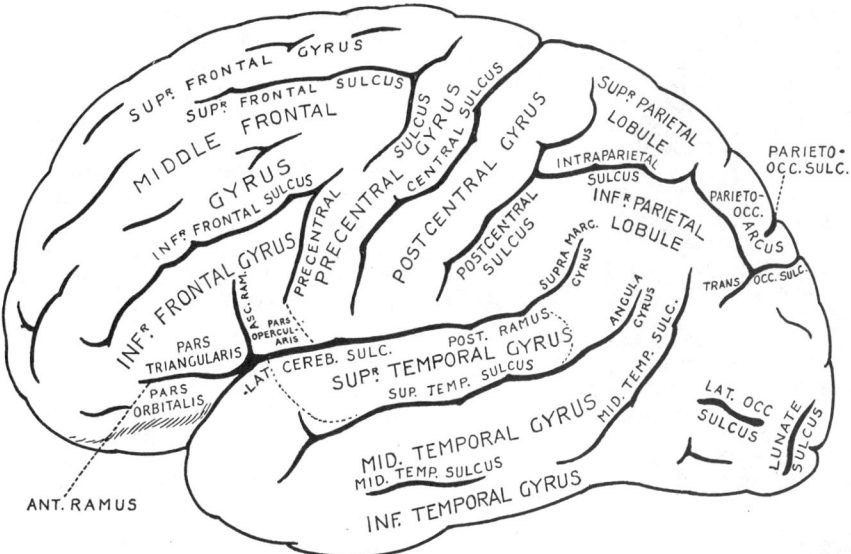

7.111A Lateral aspect of the superolateral surface of the left cerebral hemisphere.

operculated sulcus, and this type is represented in the human brain by the lunate sulcus, which separates the striate from the peristriate areas on the surface, and contains in its wall the submerged parastriate area, which really intervenes between them. These three varieties include all the sulci which develop on the surface of the brain, with the exception of the lateral sulcus and the parieto-occipital sulcus. The former is the result of the slower expansion of the cortex of the insula and its consequent submersion by the adjoining cortical areas, which eventually come into contact with one another so as to delimit the lateral sulcus. The latter is brought about subsequent to the development of the corpus callosum. The posterior end of this great commissure has to convey not only the fibres from the occipital lobes of the brain but also a large number of fibres from its

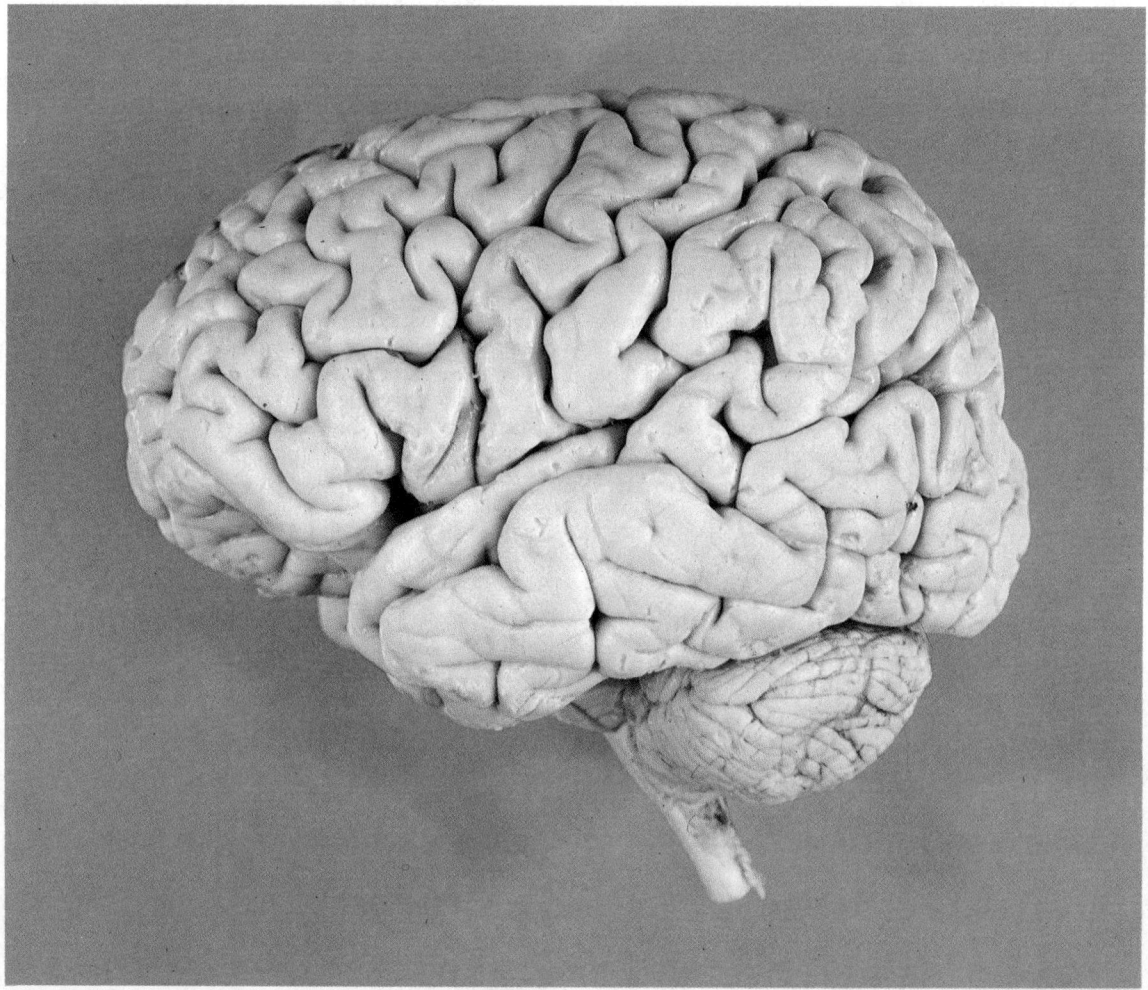

7.111B A left lateral view of the brain to show the pattern of gyri and sulci on the superolateral aspect of the cerebral hemisphere. Compare with **7.111A**, which was drawn from a different specimen, for labelling of the many structures visible. Note also the contrasting cortical patterns of the cerebrum and cerebellum. (Preparation by Dr. E. L. Rees, Dept. of Anatomy, Guy's Hospital Medical School.)

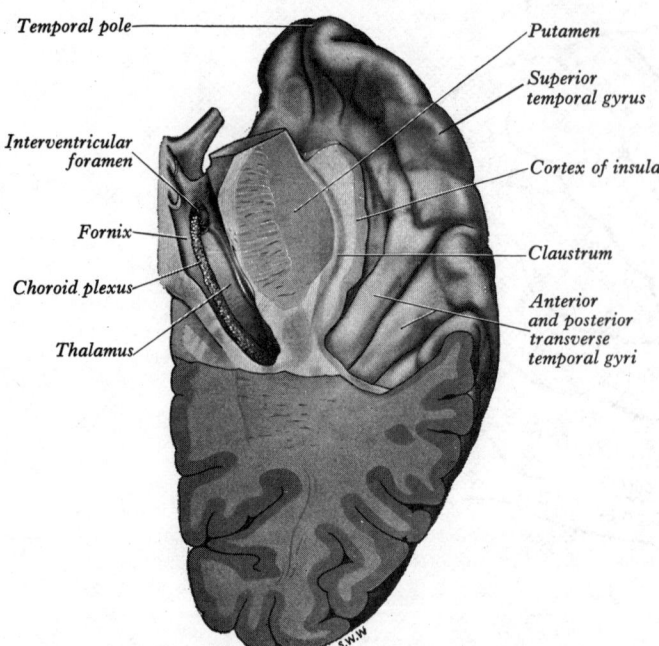

7.112 Horizontal section showing the superior surface of the right temporal lobe.

of the human cortex is about 2,200 cm², and only a third of this is visible on the surface, the rest being obscured from view in the sulci and fissures. By this form of evolution a large increase in cortical area is possible without great change in cranial capacity, but it would be misleading to explain the arrangement in this teleological manner, as is sometimes done. Similarly, any presumed association of high intelligence with great complexity of convolutional pattern is also fallacious. The most intricate arrays of sulci and gyri occur in the cerebra of the elephant and the whale (Hammelbo 1972), both of which have in addition much larger brains than man, though not so in relation to their total size. No close relationship between convolutional complexity and brain size with cerebral abilities has been established in man himself; abundant examples of highly able individuals with relatively small brains and the reverse of this have been attested. Attempts to draw deductions from the endocranial casts of fossil forms of man—as an indication of the development of certain gyri and hence abilities (such as the capacity for speech)—have likewise proved misleading and have been largely abandoned except by anthropologists.

It is convenient for description and reference to separate the surfaces of the cerebral hemisphere into a number of lobes, but it must be remembered that this division is purely one of convenience; moreover the lobes do not precisely correspond in surface extent to the cranial bones from which their names are derived.

THE SUPEROLATERAL SURFACE OF THE CEREBRAL HEMISPHERE

Two sulci, viz. the *lateral sulcus* and the *central sulcus*, take a large part in forming the boundaries of the lobes into which this surface is divided (7.111A, B).

The lateral sulcus (7.111, 115) is a deep cleft situated on the inferior and lateral surfaces of the cerebral hemisphere. It consists of a short stem which ends by dividing into three rami. The *stem* commences on the inferior surface at the anterior perforated substance and extends laterally between the orbital surface of the frontal lobe and the anterior part of the temporal lobe. It is occupied by the posterior border of the lesser wing of the sphenoid bone and the sphenoparietal venous sinus. On reaching the lateral surface it divides into anterior horizontal, anterior ascending and posterior rami. The *anterior ramus* runs forward for 2·5 cm or less into the inferior frontal gyrus, while the *ascending ramus* runs upwards for about an equal distance into the same gyrus. The *posterior ramus* is the longest division. It courses posteriorly and slightly upwards across the lateral surface for about 7 cm before turning upwards to end in the parietal lobe. The floor of this sulcus is formed by the limen insulae and the insula, and it conducts the middle cerebral vessels from the inferior to the superolateral aspect of the hemisphere. It can be

temporal lobes. As a result, a number of smaller axial and limiting sulci become crowded together and some of them become buried within the walls of the parieto-occipital sulcus. These two are really *secondary sulci*, since their occurrence depends on factors other than exuberant growth in closely adjoining areas.

Some of the sulci which indent the cerebral surface are deep enough to produce corresponding elevations in the walls of the lateral ventricles. The anterior part of the calcarine sulcus, which produces the calcar avis of the posterior horn, and the collateral, which produces the collateral eminence in the inferior horn, are therefore termed *complete fissures* or *sulci*. There is, however, no special morphological or functional significance to be attached to the fact that while some sulci are complete others are incomplete. (All the features mentioned in the above account can easily be identified in the illustrations in this section.)

The gyri and their intervening sulci are fairly constant in arrangement; at the same time they vary within certain limits, not only in different individuals, but in the two hemispheres of the same brain.

The convolutional pattern of the cerebral cortex is an inevitable concomitant of the much greater increase in volume of the pallium or mantle of nerve cells in the cortex as compared with the lesser increase in volume of the subjacent white matter. The actual area

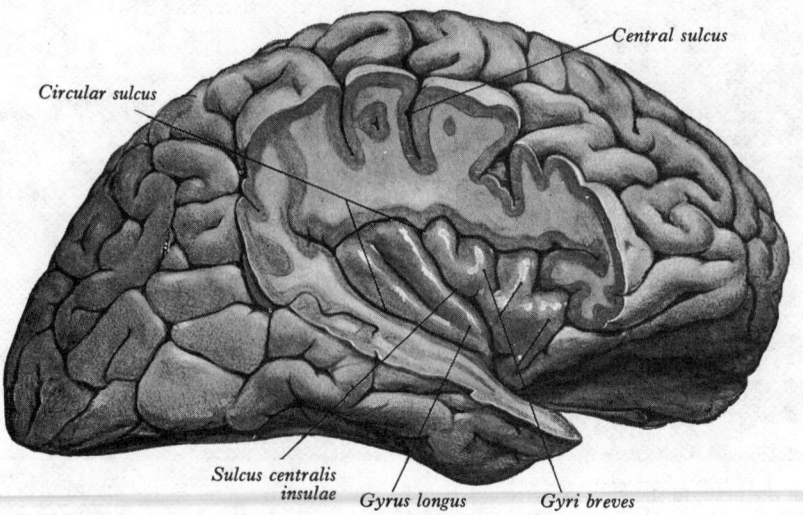

7.113A The right insula, exposed by the removal of its opercula.

represented on the side of the head by a line drawn backwards and slightly upwards for 7 cm from the pterion and then curving upwards to end under the parietal eminence.

The central sulcus (7.111) commences in or near the superomedial border of the hemisphere a little posterior to the mid-point between the frontal and occipital poles (i.e. midway between nasion and inion). It runs sinuously downwards and forwards for about 8 to 10 cm and ends a little superior to the posterior ramus of the lateral sulcus, from which it is always

separated by an arched gyrus. The general direction of the sulcus makes an angle of rather less than 70° with the median plane (7.111). It lies at the junction of the primary motor and somatosensory areas of the cortex (pp. 954 and 958).

When the central sulcus is opened up, the opposed walls are found to be marked by a number of small gyri which interlock with one another after the manner of gears in mesh, and are therefore termed *interlocking gyri*. This arrangement provides additional cortex without any corresponding increase in the

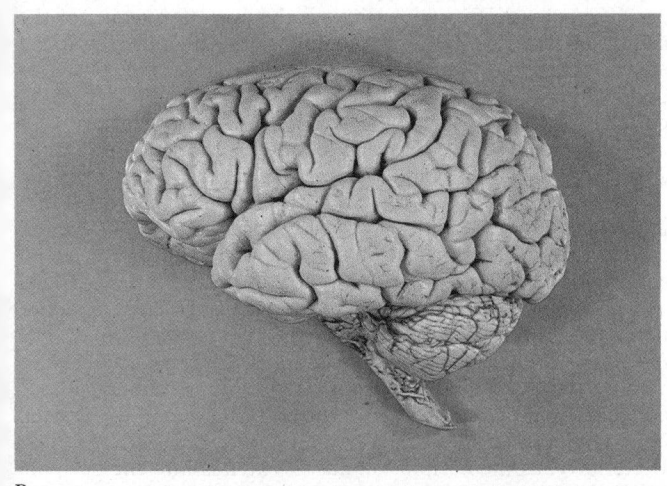

B

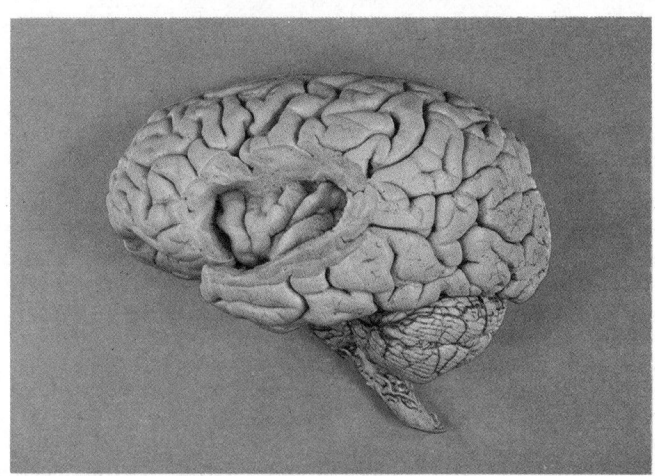

C

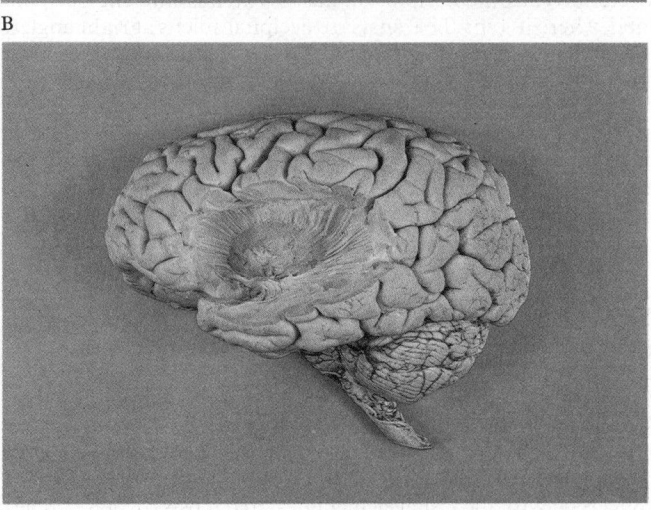

D

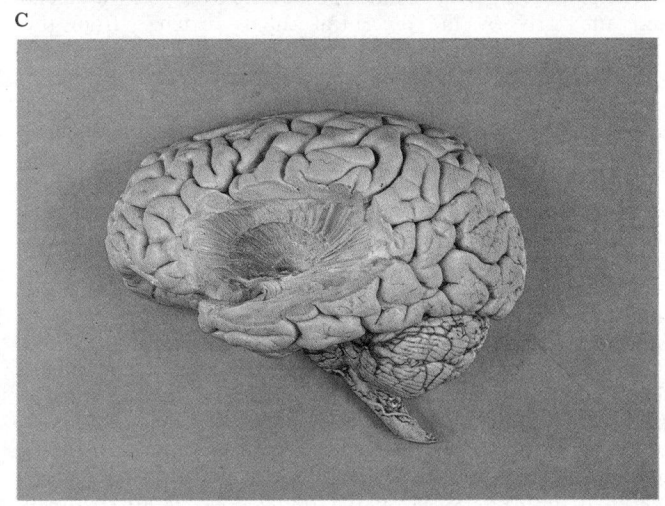

E

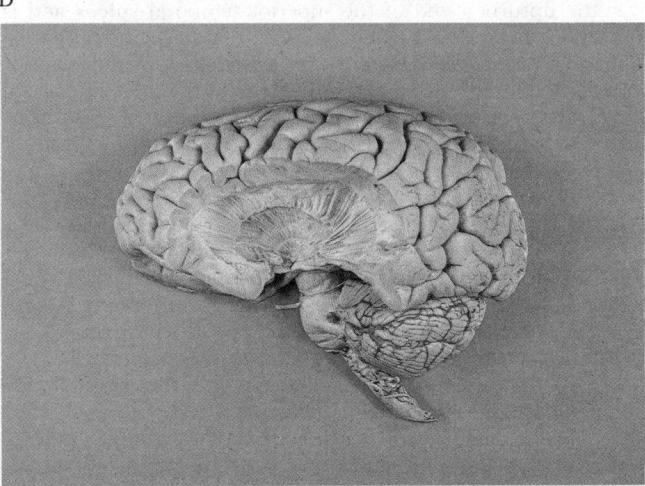

F

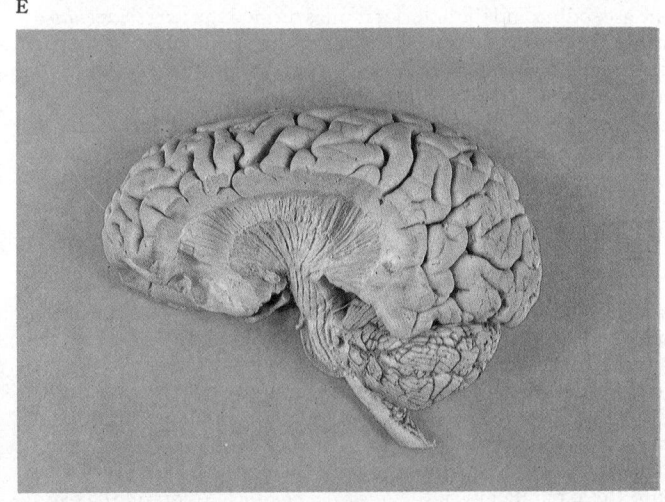

G

7.113B–G A series of dissections of the left cerebral hemisphere at progressively deeper levels to demonstrate the insula and subjacent structures. B the intact brain; not the position of the posterior ramus of the lateral cerebral sulcus on which the dissections are centred; C the cortical gyri of the insula exposed by removal of the frontal, temporal and parietal opercula; D the removal of the insular cortex, extreme capsule, claustrum and external capsule has exposed the lateral aspect of the lentiform nucleus (the putamen); E removal of the lentiform nucleus

displays fibres of the internal capsule coursing across its medial aspect; F removal of part of the temporal lobe shows the internal capsular fibres converging on the crus cerebri of the midbrain; G removal of the optic tract, and superficial dissection of the pons and upper medulla, emphasizing the continuity of the corona radiata, internal capsule, crus cerebri, longitudinal pontine fibres, and the medullary pyramid. (Dissection by Dr. E. L. Rees, Dept. of Anatomy, Guy's Hospital Medical School.)

surface area of the hemisphere. Another feature is brought to light by opening up the sulcus. The floor is not the same depth throughout, for a little inferior to the middle of the sulcus its walls are usually connected to each other by a buried, transverse gyrus. The explanation of this condition is found in the mode of development of the central sulcus. When it makes its appearance in the sixth month, it does so in two distinct parts, superior and inferior, which are at first separated by a transverse gyrus connecting the precentral convolution to the postcentral. The two parts occasionally remain separate, but as a rule they run into each other, and the transverse gyrus becomes buried as a *deep transitional gyrus*.

The frontal lobe is the anterior part of the hemisphere. On the superolateral surface it is bounded behind by the central sulcus, above by the superomedial border and below by the superciliary border and the stem of the lateral sulcus.

The superolateral surface of the frontal lobe is traversed by three sulci which divide it into four gyri. The *precentral sulcus* runs parallel to the central sulcus, and is separated from it by the precentral gyrus. It is usually divided into upper and lower parts, but the two may be confluent. The *superior frontal sulcus* curves forwards from about the middle of the upper part of the precentral sulcus, while the *inferior frontal sulcus* is parallel to it at a lower level. The area of the frontal lobe which lies anterior to the precentral sulcus is thus divided into the superior, middle and inferior frontal gyri. An incomplete sulcus often divides the middle frontal gyrus (7.111).

The *precentral gyrus*, bounded posteriorly by the central sulcus and anteriorly by the precentral sulcus, extends from the superomedial border of the hemisphere, where it is continuous with the paracentral lobule on the medial surface, to the posterior ramus of the lateral sulcus. The cortex of this gyrus is the origin of many of the fibres of large corticonuclear and corticospinal (pyramidal) tracts (p. 877).

The *superior frontal gyrus* is above the superior frontal sulcus, and is continuous over the superomedial margin of the hemisphere with the medial frontal gyrus on the medial surface. It may be divided by an incomplete sulcus (7.111).

The *middle frontal gyrus* lies between the superior and the inferior frontal sulci.

The *inferior frontal gyrus* is below the inferior frontal sulcus and is invaded by the anterior and ascending rami of the lateral sulcus. The areas around these two rami on the left form the *speech area of Broca* (areas 44 and 45) and are associated with the motor element of speech (p. 1014). The region lying below the anterior ramus is termed the *pars orbitalis*, and it curves round the superciliary margin to the orbital surface of the frontal lobe. The area between the ascending and the anterior rami is termed the *pars triangularis*, while the area posterior to the ascending ramus forms the *pars opercularis* (*posterior*) and is continuous posteriorly with the inferior limit of the precentral gyrus.

The temporal lobe is inferior to the lateral sulcus. Posteriorly, it is limited by an arbitrary line from the pre-occipital notch (p. 982) to the parieto-occipital sulcus, where it cuts the superomedial margin about 5 cm anterior to the occipital pole. The lateral surface of the temporal lobe is divided into three parallel gyri by two sulci.

The *superior temporal sulcus* begins near to the temporal pole and runs posteriorly and slightly upwards parallel to the posterior ramus of the lateral sulcus. Its posterior end curves up into the parietal lobe. The *inferior temporal sulcus* is subjacent and parallel to the superior sulcus. It is broken up into two or three short sulci, but its posterior end ascends into the parietal lobe, posterior and parallel to the upturned end of the superior sulcus.

In this way the lateral surface of the temporal lobe is divided into three parallel gyri, the *superior*, *middle* and *inferior temporal gyri*. Along its superior margin the superior temporal gyrus is continuous with the gyri in the floor of the posterior ramus of the lateral sulcus. These vary in number and extend obliquely forwards and laterally from the *circular sulcus* which surrounds the insula. They are termed the *transverse temporal gyri* (7.112). Most commonly there are two transverse gyri, which are then named *anterior* and *posterior*, but on one or other side, occasionally both, the gyrus is singular. (*See* Campain and

Minckler 1976, for the incidence of variation.) The anterior transverse temporal gyrus and the part of the superior temporal gyrus with which it is in continuity are auditory in function (p. 963). The anterior gyrus is approximately equivalent to Brodmann's area 41; the posterior transverse temporal gyrus, and the adjoining superior temporal gyrus (in part), are held to correspond to area 42.

The parietal lobe is bounded anteriorly by the central sulcus and posteriorly by the line joining the pre-occipital incisure to the superomedial margin at the point where it is cut by the parieto-occipital sulcus. Its inferior limit is the posterior ramus of the lateral sulcus and a line drawn to the posterior boundary from the point where the ramus ascends. Thus, both the posterior boundary and the posterior part of the inferior boundary of the parietal lobe on this surface of the hemisphere are arbitrary.

The lateral aspect of the parietal lobe is subdivided into three areas by the *postcentral* and *intraparietal sulci*.

The *postcentral sulcus* (7.111A, B), which may be divided into upper and lower parts, is posterior and parallel to the central sulcus. Inferiorly it ends above the posterior ramus of the lateral sulcus and in front of its upturned end. It divides the parietal lobe into an anterior part, the postcentral gyrus, and a large posterior part which is further subdivided by the intraparietal sulcus. The *intraparietal sulcus* usually commences in the postcentral sulcus about its middle or at the superior end of its lower subdivision. It extends postero-inferiorly across the posterior part of the parietal lobe, dividing it into superior and inferior parietal lobules. Posteriorly, as the occipital ramus, it extends into the occipital lobe, where it joins the transverse occipital sulcus at right angles.

The *postcentral gyrus* lies between the central and postcentral sulci. Its cortex receives somatic sensory impulses (p. 958).

The *superior parietal lobule* is between the superomedial margin of the hemisphere and the intraparietal sulcus. Anteriorly, it is continuous with the postcentral gyrus round the upper end of the postcentral sulcus, while posteriorly it frequently runs into the *arcus parieto-occipitalis*, which surrounds the lateral part of the parieto-occipital sulcus (7.111A, B).

The *inferior parietal lobule* is inferior to the intraparietal sulcus and posterior to the lower part of the postcentral sulcus. It is divided into three parts. The *anterior part* is the *supramarginal gyrus* and arches over the upturned end of the lateral sulcus; it is continuous anteriorly with the lower part of the postcentral gyrus and below and behind with the superior temporal gyrus. Occasionally it is limited posteriorly by a small *sulcus intermedius primus*, which descends from the intraparietal sulcus. The *middle part* or *angular gyrus*, the cortex of which is believed to be concerned with the visual element in stereognosis (p. 966), arches over the upturned end of the superior temporal sulcus and is continuous behind and below with the middle temporal gyrus; sometimes a small *sulcus intermedius secundus* forms its posterior boundary. The anterior and middle parts of the inferior parietal lobule are subjacent to the parietal tuberosity (p. 331). The *posterior part* arches over the upturned end of the inferior temporal sulcus and extends on to the occipital lobe.

The occipital lobe lies behind the line joining the pre-occipital incisure to the parieto-occipital sulcus. The *transverse occipital sulcus* descends from the superomedial margin posterior to the parieto-occipital sulcus and is joined about its middle by the intraparietal sulcus. Its superior part is the posterior boundary of the *arcus parieto-occipitalis*, an arched gyrus which surrounds the end of the parieto-occipital sulcus. The *lateral occipital sulcus* is a short horizontal sulcus on the lateral aspect of the occipital lobe dividing it into *superior* and *inferior occipital gyri* (7.111A, B). The *lunate sulcus*, when present, is just in front of the occipital pole. It is placed vertically and sometimes joins the calcarine sulcus, although the two are more often separated from each other. The lips of the lunate sulcus, which is operculated in type, separate the striate from the peristriate area of the cortex, but the parastriate area is buried within the walls of the sulcus and intervenes between them. The lunate sulcus is the posterior boundary of the *gyrus descendens*, which lies behind the superior and inferior occipital gyri. Two curved sulci, named the superior and inferior polar sulci, are often present near the extremities of the lunate sulcus. The *superior polar sulcus* arches upwards on to the medial

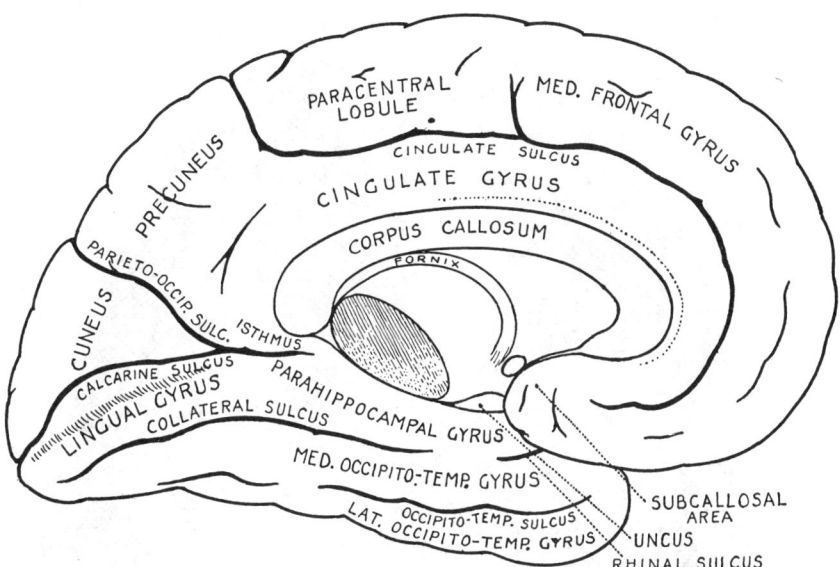

7.114A Diagram of the medial surface of the left cerebral hemisphere, after sagittal section of the brain, and removal of the brainstem.

aspect of the occipital lobe from the neighbourhood of the upper limit of the lunate sulcus; the *inferior polar sulcus* arches downwards and forwards on to the inferior aspect from the lower limit of the same sulcus. These two polar sulci enclose semilunar extensions of the striate area (p. 1018) and indicate the expansion of the visual cortex associated with the formation of its large macular area (Smith 1930).

The insula (7.113A–D) is deep in the floor of the lateral sulcus and is almost surrounded by a *circular sulcus*. It has been overlapped by the overgrowth of the cortical areas which adjoin it, and can only be seen when the lips of the lateral sulcus are widely

separated. These areas of the cortex are therefore termed the *opercula of the insula*, and they are separated from each other by the ascending and posterior rami of the lateral sulcus. The *frontal operculum* or lid lies between the anterior and ascending rami, and is formed by the pars triangularis of the inferior frontal gyrus. It may be of small size in cases where the two rami between which it lies arise by a common stem. The *frontoparietal operculum* is between the ascending and the upturned end of the posterior ramus of the lateral sulcus. It is formed by the pars posterior of the inferior frontal gyrus, by the lower ends of the precentral and postcentral gyri, and by the lower end of the anterior part of the

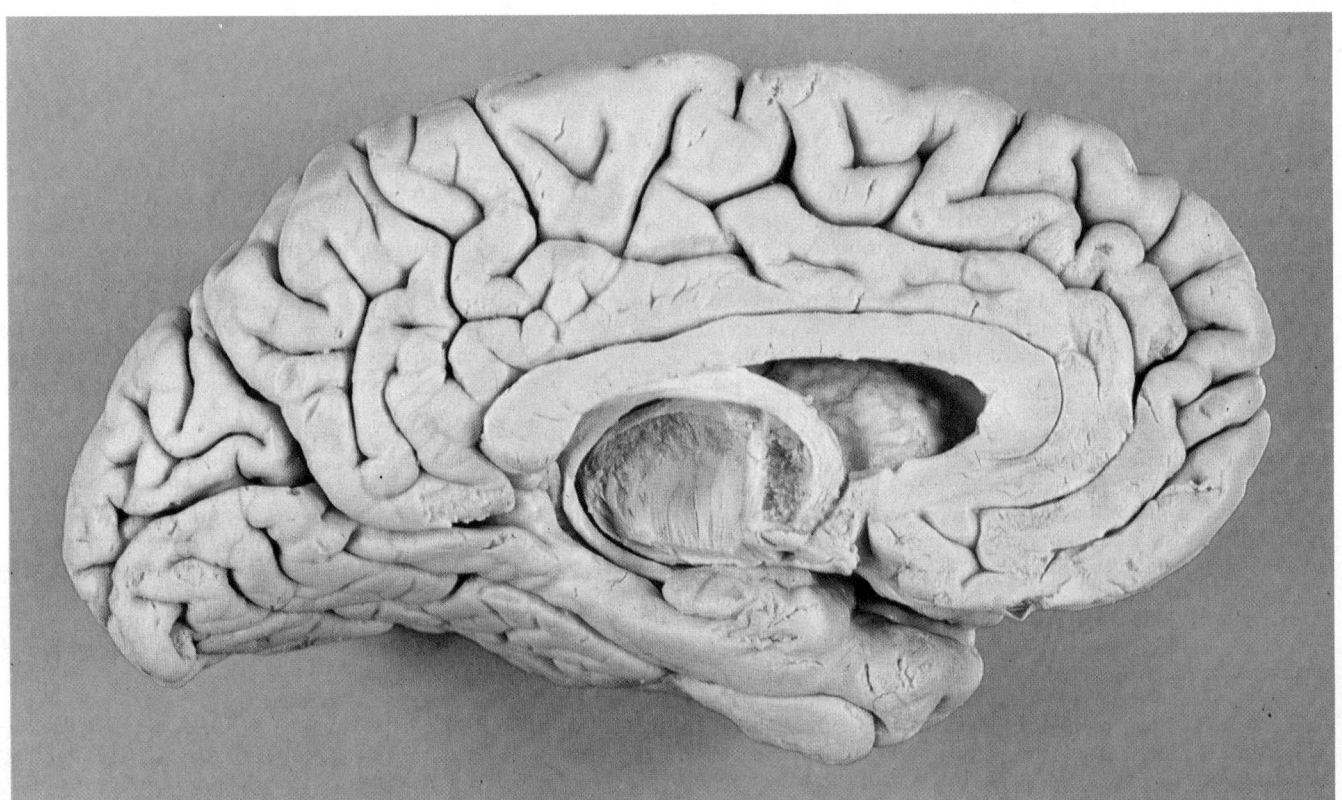

7.114B The medial surface of the left cerebral hemisphere after sagittal section of the brain, followed by removal of the brainstem and septum pellucidum. (For identification of the principal gyri and sulci of the cerebral cortex, compare with 7.114A.) The dissection has been deepened in the region of the dorsal thalamus and hypothalamus to demonstrate the

column, body, crus and fimbria of the fornix, and the mamillothalamic fasciculus (compare with 7.117A and B). The head of the caudate nucleus is visible bulging into the floor of the anterior horn of the lateral ventricle. (Dissection by Dr. E. L. Rees, Dept. of Anatomy, Guy's Hospital Medical School.)

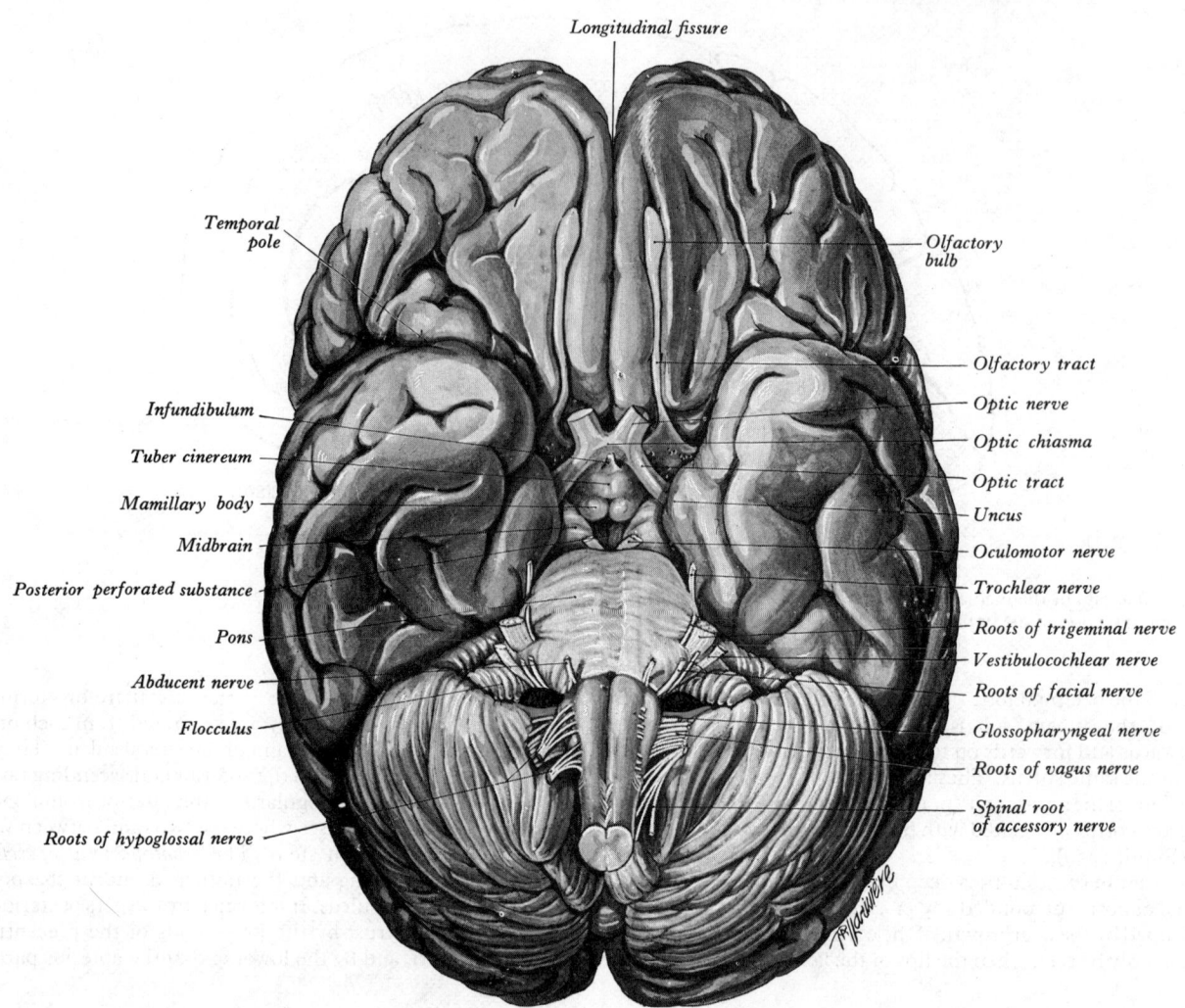

Longitudinal fissure

Temporal pole

Olfactory bulb

Infundibulum

Tuber cinereum

Mamillary body

Midbrain

Posterior perforated substance

Pons

Abducent nerve

Flocculus

Roots of hypoglossal nerve

Olfactory tract

Optic nerve

Optic chiasma

Optic tract

Uncus

Oculomotor nerve

Trochlear nerve

Roots of trigeminal nerve

Vestibulocochlear nerve

Roots of facial nerve

Glossopharyngeal nerve

Roots of vagus nerve

Spinal root of accessory nerve

7.115A   Basal aspect of the brain. The anterior perforated substance (unlabelled) is between the diverging lateral and medial roots of the olfactory tract and anterolateral to the optic tract.

inferior parietal lobule. The *temporal operculum* is below the posterior ramus and is formed by the superior temporal gyrus and the transverse temporal gyri. Anteriorly the inferior part of the insula adjoins the pars orbitalis of the inferior frontal gyrus.

When the opercula are removed, the insula is seen as a pyramidal eminence, the apex of which is directed inferiorly towards the anterior perforated substance (7.113). Here the circular sulcus is deficient and the medial part of the apex is termed the *limen insulae* (*gyrus ambiens*). The surface of the insula is divided into a larger anterior and a smaller posterior part by the *sulcus centralis insulae*, which runs up and back from the apex of the insula. The anterior part is divided by shallow sulci into three or four *short gyri*, while the posterior part is formed by one *long gyrus*, which is often divided at its upper end. The cortical grey matter of the insula is continuous with that of the various opercula in the depths of the circular sulcus. The insula overlies, and is more or less co-extensive with the claustrum and the putamen of the lentiform nucleus (7.113B, C, D).

The angio-architecture of the primate insula (including that of mankind) has been the subject of particular study by Vlahovitch *et al.* (1973). Despite the development of the opercula in man, in contrast to their exiguous size in, for example, the baboon or chimpanzee, these observations indicated a high degree of similarity of vascular pattern in all the primates examined.

THE MEDIAL SURFACE OF THE CEREBRAL HEMISPHERE

The medial surface (7.114A, B) cannot be examined until the two cerebral hemispheres have been separated from each other by the

division of (1) the commissures which connect them and (2) the roof, floor, anterior and posterior walls of the third ventricle (7.114A, B). The most conspicuous feature on this surface is the great commissure which is termed the *corpus callosum*. It forms a broad arched band which lies in the floor of the central part of the longitudinal fissure (7.110). The recurved, anterior end of the corpus callosum is termed the *genu*. Below, it is continuous with the *rostrum*, which narrows rapidly as it passes backwards to become connected to the upper end of the lamina terminalis; above, it is continuous with the *trunk* or main 'body' of the commissure, which arches upwards and backwards to end in a thick, rounded posterior extremity, the *splenium*. To the concave surfaces of the trunk, genu and rostrum are attached the laminae of the septum pellucidum, which occupy the interval between them and the fornix—a curved, flattened band of white fibres inferior to it. Immediately in front of the lamina terminalis and almost co-extensive with it, there is a narrow, triangular field of grey matter, the *paraterminal gyrus* (p. 995). Anteriorly it is separated from the rest of the cortex by a shallow *posterior parolfactory sulcus*. A little anterior to this a second, short, vertical sulcus may be present and is termed the *anterior parolfactory sulcus*. The portion of cortex which lies between these two sulci is the *subcallosal area* (*parolfactory gyrus*) (7.110, 119). The anterior sloping edge of the paraterminal gyrus is sometimes called the *prehippocampal rudiment* (p. 997).

The anterior part of the medial surface of the hemisphere is divided into an outer and an inner zone by the curved *cingulate sulcus*. This commences below the rostrum of the corpus callosum and passes first forwards, then upwards and finally backwards, conforming with the curvature of the corpus callosum. Its posterior end turns up to reach the superomedial margin of the hemisphere, about 4 cm behind the mid-point between the frontal and occipital poles, and is posterior to the upper extremity of the central sulcus (7.110, 114). The outer zone demarcated by the cingulate sulcus forms, with the exception of its extreme posterior end, a part of the frontal lobe. It is subdivided into larger anterior and smaller posterior parts by a short sulcus ascending from the cingulate sulcus above the middle of the trunk of the corpus callosum. The larger anterior region is the *medial frontal gyrus*, while the smaller posterior area is termed the *paracentral lobule*. The superior end of the central sulcus usually cuts into the posterior part of the paracentral lobule and the precentral gyrus is directly continuous with the cortex of the lobule. This area mediates control of movements of the lower limb and perineal region of the opposite side, and clinical evidence suggests that it exercises voluntary control over the defaecation and micturition reflexes (pp. 1357, 1409).

The zone within the curve of the cingulate sulcus is the *cingulate gyrus*. Commencing below the rostrum this gyrus follows the curve of the corpus callosum, from which it is separated by the *callosal sulcus*, and it continues round the splenium on to the inferior surface of the hemisphere to become continuous with the parahippocampal gyrus through the narrow *isthmus* (7.117A).

The line of the cingulate sulcus is interrupted posterior to the paracentral lobule, but is partially continued by a short variable *subparietal (suprasplenial) sulcus*.

The posterior part of the medial surface of the hemisphere is traversed by two deep sulci which converge anteriorly and meet a short distance posterior to the splenium of the corpus callosum. These are the parieto-occipital and the calcarine sulci. The *parieto-occipital sulcus* commences on the superomedial margin of the hemisphere about 5 cm anterior to the occipital pole and is directed downwards and slightly forwards to meet the calcarine sulcus. When the lips of the sulcus are separated it will be found that, although on the surface of the hemisphere the parieto-occipital and the calcarine sulci are apparently continuous, they are in reality separated from each other by the deeply situated *cuneate gyrus*. In addition, the walls of the sulcus show the presence of two or more vertically disposed sulci. These were originally exposed on the medial surface of the hemisphere, but they were included in the parieto-occipital sulcus owing to the growth of the splenium (p. 171). The walls of the parieto-occipital sulcus thus resemble those of the lateral sulcus, although the contained sulci and gyri are fewer in number and smaller in extent.

The *calcarine sulcus* commences near the occipital pole. Although it is usually restricted to the medial surface of the hemisphere its posterior end may extend on to the lateral surface. It runs anteriorly a little above the inferomedial margin of the hemisphere, taking a slightly curved course with an upward convexity, and joins the parieto-occipital sulcus at an acute angle a little posterior to the splenium of the corpus callosum. Continuing forwards the calcarine sulcus crosses the inferomedial margin and gains the inferior aspect of the hemisphere, where it forms the inferolateral boundary of the *isthmus*, which connects the cingulate with the parahippocampal gyrus. At its junction with the parieto-occipital sulcus the floor of the calcarine sulcus is crossed by the buried anterior cuneolingual gyrus. The posterior part of the calcarine sulcus, behind its junction with the parieto-occipital, is an axial sulcus, set in the long axis of the visual cortex (p. 966), but the anterior part is a limiting sulcus and separates the striate, visual cortex from that of the isthmus. The anterior part conforms to the definition of a complete sulcus, since it produces an elevation in the medial wall of the posterior cornu or horn of the lateral ventricle (the *calcar avis*).

The quadrilateral area, bounded anteriorly by the upturned end of the sulcus cinguli, posteriorly by the parieto-occipital sulcus, superiorly by the superomedial margin and inferiorly by the suprasplenial sulcus, is termed the *precuneus* and, together with the part of the paracentral lobule posterior to the central sulcus, constitutes the medial surface of the parietal lobe.

The wedge-shaped area bounded in front by the parieto-occipital sulcus, below by the calcarine sulcus and above by the

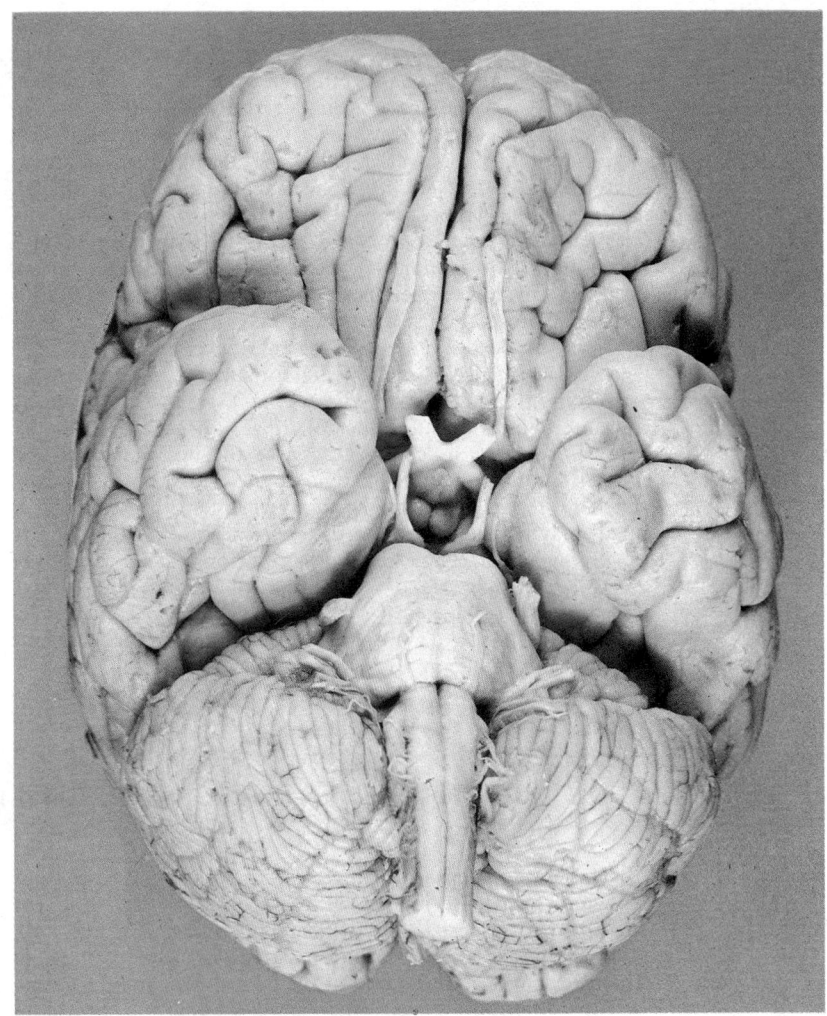

7.115B The base of the brain. For labelling compare with 7.56, 115A. (Dissection by Dr. E. L. Rees, Dept. of Anatomy, Guy's Hospital Medical School.)

superomedial margin, is termed the *cuneus*. Its surface is usually indented by one or two small irregular sulci, and forms the medial surface of the occipital lobe.

THE INFERIOR SURFACE OF THE CEREBRAL HEMISPHERE

The surface is divided into a smaller part, anterior to the stem of the lateral fissure, and a larger part posterior to it (7.115A, B; 116).

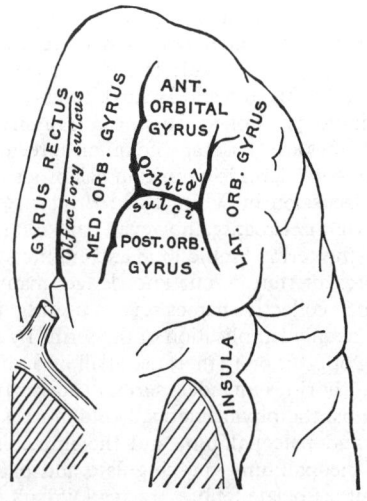

7.116 The orbital surface of the left frontal lobe.

The anterior region forms the orbital part of the inferior surface of the cerebral hemisphere. It is concave from side to side and rests on the cribriform plate of the ethmoid bone, the orbital plate of the frontal and the lesser wing of the sphenoid bone. An anteroposterior sulcus traverses this surface near its medial margin and, since it is overlapped by the olfactory bulb and tract, it is termed the *olfactory sulcus*. The medial strip of cortex which it marks off is the *gyrus rectus*. The rest of this surface is marked by irregular sulci, the *orbital sulci*, generally H-shaped, dividing it into a number of *orbital gyri*. Four can usually be recognized, named, according to their position, anterior, medial, posterior and lateral orbital gyri (7.116).

The larger, posterior region of the inferior cerebral surface is its tentorial part since it is immediately superior in part to the tentorium but also to the middle cranial fossa. It is traversed by two anteroposterior sulci, the collateral and occipitotemporal. The *collateral sulcus* commences near the occipital pole and extends anteriorly, roughly parallel to the calcarine sulcus, from which it is separated by the *lingual gyrus*. Anteriorly the collateral sulcus may be continued into the *rhinal sulcus*, but the two are usually separated. The *rhinal sulcus* runs forwards in the line of the collateral sulcus and separates the temporal pole from a somewhat hook-shaped projection posteromedial to it and termed the *uncus*. This fissure marks the lateral limit of the piriform lobe of the cortex (7.119).

The *occipitotemporal sulcus* is parallel to the collateral sulcus

and lateral to it. As a rule it does not extend as far as the occipital pole, and it is frequently divided into two or more parts.

The *lingual gyrus* lies between the calcarine and collateral sulci. Anteriorly it passes without interruption into the *parahippocampal gyrus*, which commences at the *isthmus*, where it is directly continuous with the cingulate gyrus, and then passes forwards medial to the collateral and rhinal sulci. Anteriorly the parahippocampal gyrus becomes continuous with the uncus, its medial edge being lateral to the midbrain. The uncus is the hook-like, anterior end of the parahippocampal gyrus and forms the posterolateral boundary of the anterior perforated substance. The medial part of the uncus extends laterally above its lateral part and will be described later (7.119); its inferior surface cannot be exposed completely until its lateral and more superficial part has been removed (7.120). The uncus forms part of the *piriform lobe*, which is part of the olfactory system (*vide infra*) and is phylogenetically one of the oldest parts of the pallium. (For further details of the uncal region, and the complex terminology applied to this and surrounding areas, consult illustrations 7.117A and 7.119.)

The *medial occipitotemporal gyrus* extends from near the occipital to the temporal pole. It is limited medially by the collateral and rhinal sulci and laterally by the occipitotemporal sulcus. The lateral part of this area forms the *lateral occipitotemporal gyrus*, which is continuous round the inferolateral margin of the hemisphere with the inferior temporal gyrus.

THE LIMBIC LOBE AND OLFACTORY PATHWAYS

During development the superolateral surfaces of the diencephalon gradually merge with the central areas of the inferomedial surfaces of the two cerebral hemispheres (p. 170). Completely *bordering* the area of fusion on each side, a series of structures develop in the wall of the hemisphere and constitute the **limbic lobe**—a term introduced in 1878 by Broca on comparative anatomical grounds. Many of the structures constituting the limbic lobe (literally the *bordering* lobe) are phylogenetically old, and have a highly arched form; topographically they are interposed between the diencephalon and the massive neopallial areas of the cerebral hemisphere (7.117). Interest in the limbic lobe was heightened when Papez (1937) suggested its role in emotional behaviour. Since that time an ever-increasing volume of research has emphasized that the lobe has profuse interconnexions with the olfactory system, the hypothalamus, thalamus, epithalamus and, to a lesser extent, areas of the neocortex. It is intimately associated with the higher integration of visceral, olfactory and somatic information, and the patterning of complex long and short term homeostatic responses. The latter include seeking and capturing prey, courtship, mating, rearing of young, the subjective and expressive elements in emotional responses, and the balance between aggressive and social behaviour in community living (Maclean 1958, 1969; Livingston 1970). For modern comprehensive reviews including extensive bibliographies consult Isaacson (1974), Isaacson and Pribram (1975), Livingston and Hornykiewicz (1978).

The terminology applied to these regions is complex; alternative schemes are in use, and common agreement in terms of a rational basis for scientific definition has proved impossible to achieve (*see* discussion in White 1965; Brodal 1969; Bargmann and Schadé 1963). Broadly, however, the increasingly widespread use of the terms limbic lobe and limbic system indicate that, at the present state of our knowledge, many investigators believe that such collective names serve a useful purpose, despite differences in detailed application of the terms by some workers, and the strong opposition to their use at all by a minority (Brodal 1969). Most authorities include a series of subcortical nuclei and their connexions, the phylogenetically older areas of cortex, the archipallium and paleopallium, and the immediately adjacent border of the neopallium—the cingulate and parahippocampal gyri, and their associated fibre tracts. Difficulties arise when attempts are made to place precise boundaries between these

regions and adjacent ones, with which some of them are functionally interconnected. For example, opinions differ concerning what areas of frontal cortex should be included, and whether or not the intimately related hypothalamus should be embraced by the term. Of greater practical importance is the need for an agreed international nomenclature for as many orders of subdivision of the brain as possible, based upon the most widely acceptable criteria currently available, with a continuous process of revision as further knowledge accumulates. In the present account the terms limbic lobe and system are included, and the structures described under these headings will be listed below.

Even greater terminologic difficulty attaches to the widely used name **rhinencephalon** (Bargmann and Schadé 1963). Traditionally, the name was used to indicate all those brain structures associated with olfaction, and for many years it was assumed that most of the structures of the limbic lobe were primarily olfactory pathways and centres of integration. However, with advancing researches in comparative and developmental neurology, olfactory physiology, more detailed neuroanatomical studies, together with behavioural and other functional studies on the limbic system and hypothalamus, there has been a considerable change of emphasis. Since the term rhinencephalon has many shades of meaning in the hands of different workers using different approaches, and when investigating different species, it will not be included in this account.

The structures mentioned or described here or elsewhere in this volume, which are included in **the limbic system** are:

(1) The olfactory nerves, bulb and tract (together with the nervus terminalis, and the transient accessory olfactory bulb and vomeronasal nerve).

(2) The anterior olfactory nucleus.

(3) The medial, intermediate and lateral olfactory striae and the medial and lateral olfactory gyri.

(4) The olfactory trigone, the anterior perforated substance with the olfactory tubercle, and the diagonal band of Broca.

(5) The piriform lobe which includes: (*a*) the lateral olfactory gyrus continuing into the gyrus ambiens—together forming the prepiriform cortex; (*b*) the lateral olfactory stria continuing into the gyrus semilunaris (periamygdaloid area); (*c*) the uncus hippocampi including the uncinate gyrus, the tail of the dentate gyrus (band of Giacomini), and the intralimbic gyrus; (*d*) the

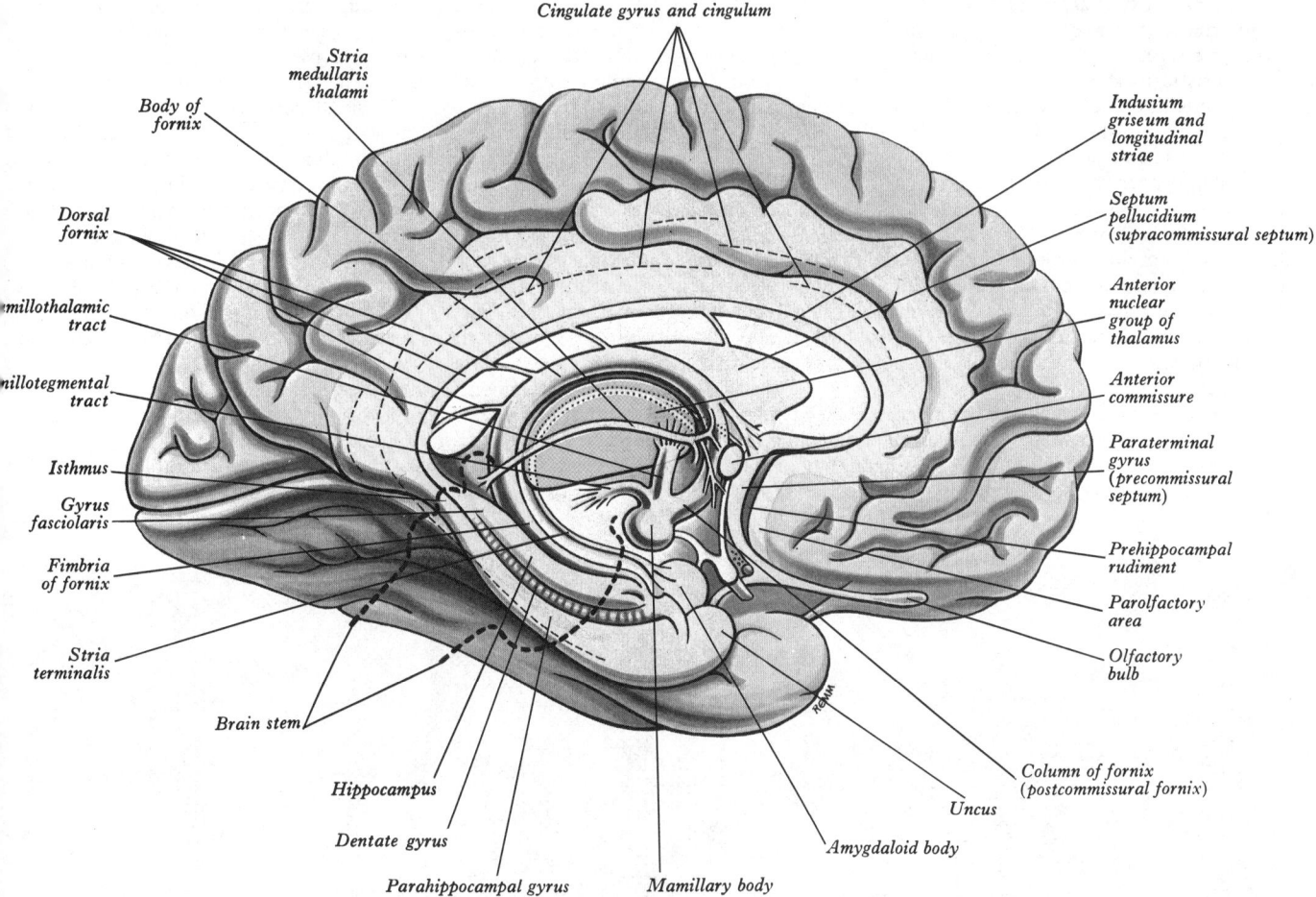

7.117A   A diagram of a dissection of the medial aspect of a cerebral hemisphere to demonstrate the majority of the structures included under the term limbic system in the present account; these are coloured yellow. The anterior nuclear group of the dorsal thalamus is coloured orange and included with the limbic system; the remainder of the dorsal thalamus is magenta. The approximate position of the brainstem which was removed in the course of dissection is outlined in a heavy interrupted line. This diagram should be compared with the colour photograph of a dissection prepared from the *lateral* aspect of a hemisphere (7.117B).

entorhinal area (area 28)—the cranial part of the parahippocampal gyrus.

(6) The amygdaloid complex of nuclei.

(7) The septal areas, including the septum pellucidum, and the septum verum—a nuclear complex much of which is deeply placed but in part corresponds to the paraterminal gyrus.

(8) The hippocampal formation which includes: (*a*) the prehippocampal rudiment, the indusium griseum and longitudinal striae; (*b*) the gyrus fasciolaris, Ammon's horn, the dentate gyrus, the subiculum, and related regions.

(9) The fornix and its various ramifications and divisions.

(10) The stria terminalis.

(11) The stria habenularis.

(12) The cingulate and parahippocampal gyri.

The hypothalamus, medial part of the thalamus, pre-frontal cortex, and the anterior commissure, all of which are intimately related to the foregoing structures, are described elsewhere in the present section. In particular, the hypothalamus is regarded by some authors as the *essential centre* of the limbic system (*see* Isaacson 1974).

THE OLFACTORY BULB

The olfactory bulb (7.115A) is a flattened, oval, reddish-grey mass which lies superior to the medial edge of the orbital plate of the frontal bone near the lateral margin of the cribriform plate of the ethmoid. It is inferior to the anterior end of the olfactory sulcus, on the orbital surface of the frontal lobe of the brain; the gyrus rectus is directly superior to the cribriform plate.

The olfactory nerve fibres which are the central processes of the olfactory cells of the nasal mucosa (p. 1054) converge upon the inferior surface of the cribriform plate and collect into about twenty bundles which pass through the foramina in the plate, and then continue to enter the inferior surface of the bulb.

As described previously (p. 170) the olfactory bulb and tract develop as a hollow diverticulum from the anteromedial part of the floor of the primitive cerebral hemisphere. With growth, the basal part of the diverticulum elongates to form the olfactory tract and, in the human embryo, the cavity of the terminal olfactory bulb (the 'olfactory ventricle') and that of the elongating tract are gradually obliterated by approximation and subsequent fusion of their walls. However, the site of the original cavity is sometimes marked by deeply-placed vestigial aggregations of modified ependymal cells. Thus, the olfactory bulb has a *radial organization* consisting of a number of superimposed layers with increasing depth from the surface. This laminar pattern is particularly well defined in many mammals other than man, and also in the human fetus, but often becomes less distinct as the brain matures. The detailed histology of the olfactory bulb, as revealed by the light microscope after staining with the Nissl and Golgi methods, has been known since the turn of the century (Cajal 1890, 1911, 1955; Blanes 1898; Le Gros Clark 1951, 1957; Valverde 1965), and this has been supplemented by the experimental studies using degeneration techniques, e.g. (Powell and Cowan 1963; Price 1968; Heimer 1968), physiological studies, e.g. (Sem-Jacobsen *et al.* 1956; Døving 1967), more recently, by detailed ultrastructural analyses (Pinching and Powell 1971; Shepherd 1974, 1977), and currently by electron microscopic autoradiography, and immunofluorescence (p. 819).

From the surface, and progressing towards the central core, the olfactory bulb consists of: (1) the olfactory nerve fibre layer; (2) the layer of synaptic glomeruli and interglomerular spaces; (3) the molecular and external granular layer; (4) the mitral cell layer; (5) the internal granular layer; and (6) the fibres of the olfactory tract.

The neurons of the olfactory bulb comprise: (1) the large characteristically shaped *mitral cells* which form a layer only one or two cells thick; (2) internal, middle and external *tufted cells*, analogous to, but smaller than, the mitral cells, and progressively reducing in size throughout the thickness of the molecular layer from the superficial aspect of the mitral layer to the deep aspect of the glomeruli; (3) numerous small round or stellate *internal granule cells*; (4) various types of small *periglomerular cell* distributed around, within, and deep to the synaptic glomeruli. The periglomerular cells and the small external tufted cells constitute the *external granular layer* of the earlier histologists. Modern histofluorescence techniques (p. 819) have demonstrated that the periglomerular cells are *dopaminergic*; they are classified as cell aggregate A15 (*see* p. 949).

It is now clear that previous accounts of the mechanism of the olfactory bulb as a simple relay between incoming olfactory nerve fibres and the dendrites of mitral cells, the axons of which proceed into the olfactory tract, were a gross over-simplification. The bulb is a region of great structural complexity to which only brief reference can be made here (*see* the references quoted for details). In particular the following features are to be noted: (1) there is a great *convergence* of olfactory nerve fibres on to the dendrites of the mitral and tufted cells which occupy much of the synaptic glomeruli (*vide infra*); (2) a number of cell types, mitral, tufted and others, contribute axons to the olfactory tract which proceed to other brain centres; (3) a number of different types of interneuron provide pathways for complex patterns of interaction between the main conduction channels, and also complicated feedback paths; and (4) transmission in the bulb is modulated by centrifugal axons which converge on it from the contralateral olfactory bulb, anterior olfactory nucleus and other centres (*vide infra*). Some of these features are summarized in illustration 7.118.

It is convenient to consider first the main conduction path through the bulb, followed by the complications introduced by local interneurons and centrifugal pathways.

In addition to *basal* dendrites which establish synaptic connexions with granule cell processes, both the large mitral cells

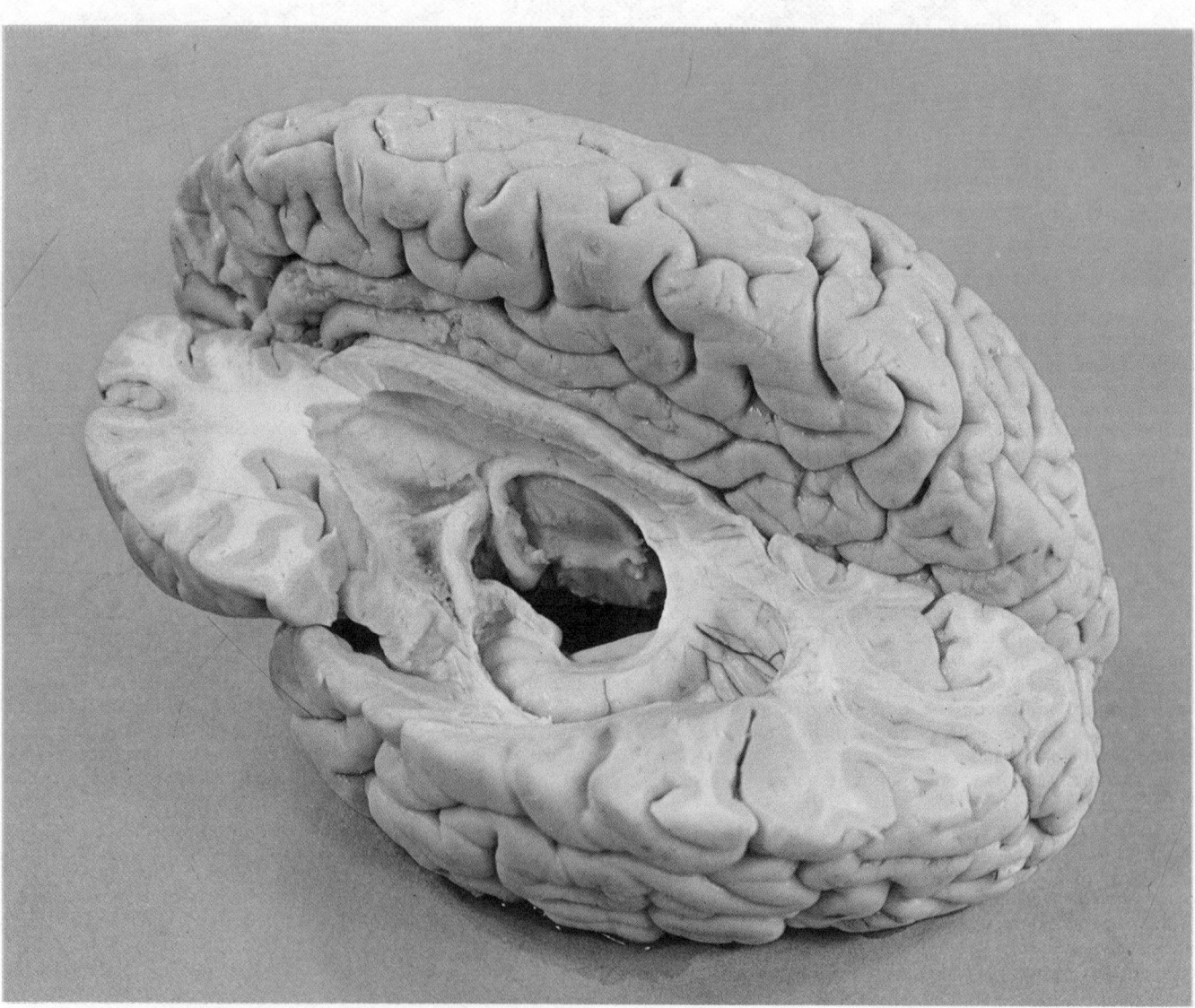

7.117B A dissection of the left cerebral hemisphere from the supero-lateral aspect to demonstrate various structural features of the limbic system. The corpus callosum is divided sagittally in the region of its body only; the frontal, temporal and occipital lobes have been sectioned horizontally and their superior parts removed. The left lentiform nucleus, much of the caudate nucleus and dorsal thalamus have been removed, and the floor of the inferior horn of the lateral ventricle laid open. Note: (1) the horizontally sectioned head of the caudate nucleus; (2) the spiral disposition of the fornix as it curves from the mamillary body through its left column, body, crus and fimbria; (3) the curved elevation of the hippocampus projecting into the floor of the inferior horn of the ventricle, and ending anteriorly as the grooved pes hippocampi; (4) the anterior commissure entering the left hemisphere immediately anterior to the column of the fornix, and passing laterally, to diverge into small anterior and large posterior components; between the latter the deep aspect of the anterior perforated substance is visible; (5) within the curve of the fornix the medial aspect of the *right* thalamus crossed superiorly by the stria medullaris thalami; (6) coursing above the corpus callosum a longitudinal white stria is visible, and above this arches the right gyrus cinguli. Compare with 7.117A. (Dissection by Dr. A. M. Seal, Department of Anatomy, Guy's Hospital Medical School.)

(the shape of their somata resembling a bishop's mitre) and the more prominent external tufted cells possess large *apical* dendrites which pass superficially towards the glomerular layer where they branch and form rich arborizations in a number of synaptic glomeruli. The incoming olfactory nerve fibres converge upon the glomeruli, and form numerous axodendritic synapses with the apical dendrites of the mitral and tufted cells, whilst the axons of the latter pass through the inner granular layer, giving off a number of collateral branches, before continuing into the deep zone of nerve fibres to form the main constituents of the olfactory tract.

The synaptic glomeruli of the olfactory bulb are large, roughly spherical territories, often exceeding 100 μm in diameter, containing many thousands of nerve cell processes from various sources, with a wide variety of types of synaptic interconnexion between them. In addition to the large dendrites of the mitral and internal tufted cells just described, the glomerulus also receives dendrites from various types of *periglomerular cell* (including the external granule cells, the external tufted cells, and the superficial short-axon cells of different authors, *see* Pinching and Powell

dendrites mentioned above, and also with the dendrites of periglomerular cells. The periglomerular cell dendrites also form complex reciprocal dendrodendritic contacts with mitral and tufted cell dendrites; often one or both of the elements in such a double dendrodentric contact also receive an axodendritic synapse from an olfactory nerve axon terminal. Finally, the dendrites of the periglomerular cells receive a number of axodendritic synapses from the axons of other periglomerular cells.

The interglomerular spaces are occupied, in part, by the cell somata of the different categories of periglomerular cell and their processes. Some of these cells have dendrites which ramify both in the spaces and also in the glomeruli as described above. Other interneurons which have been designated short-axon cells have dendritic trees which are wholly extraglomerular. The interglomerular spaces are also penetrated by mitral cell dendrites, many of the axonal ramifications and terminals of the periglomerular cells themselves, and the termination of centrifugal axons from the olfactory tract. Again, reciprocal dendrodendritic contacts between periglomerular cell dendrites and

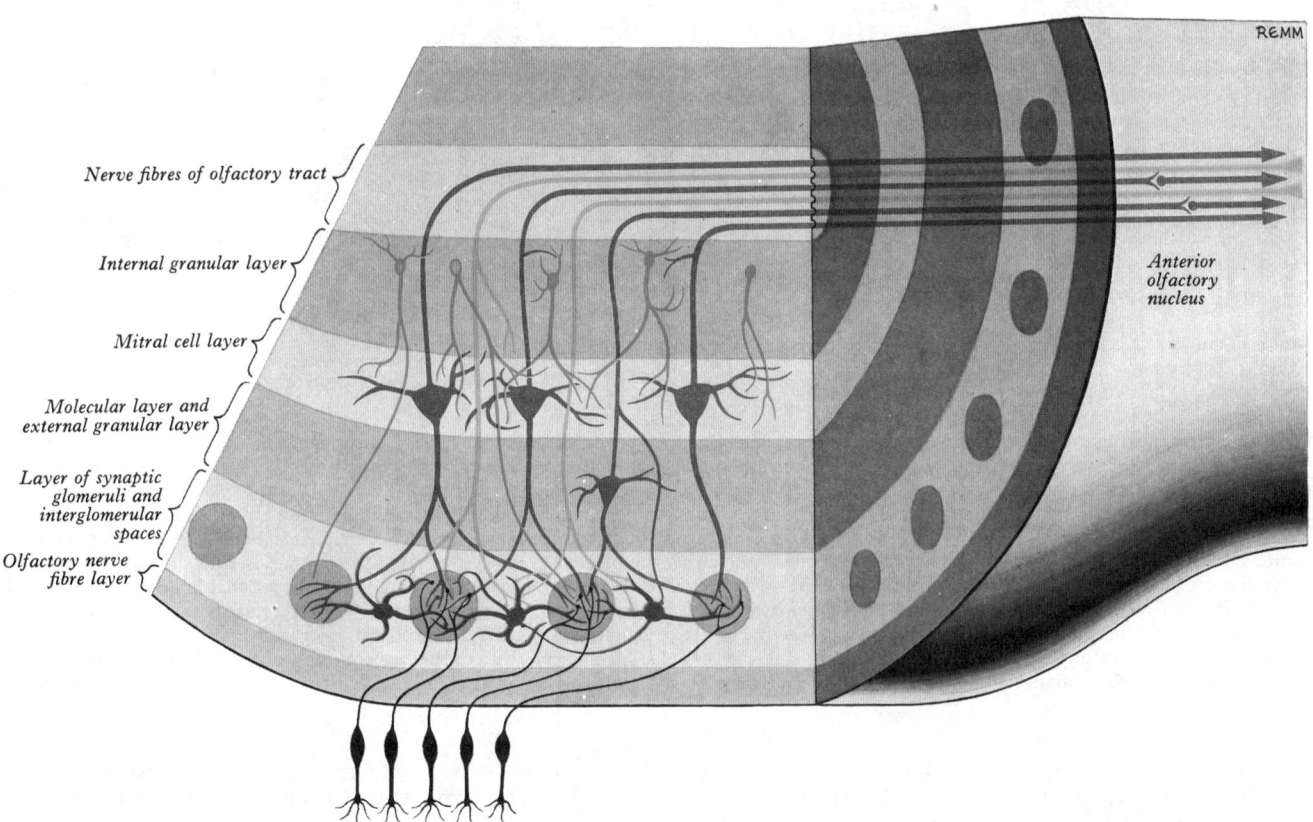

7.118 A scheme of the olfactory bulb based upon neurocytological and experimental studies in a number of mammalian types. The radial organization of the bulb into 'layers', with their principal neuron types, and an approximate indication of their main connectivity patterns, is shown. The 'layers' of the bulb which receive the names indicated are, of course, merely a convenient descriptive frame of reference; they should not obscure the functionally essential interconnexions which cross the boundaries between these zones. The colours used indicate—*red:* mitral and tufted neurons and their processes; *blue:* internal granule neurons; *purple:* periglomerular neurons; *black:* olfactory receptor neurons and their processes. Note that the olfactory tract consists of (1) centripetal axons of mitral and tufted cells, some of which synapse with neurons in the anterior olfactory nucleus, and (2) centrifugal axons (yellow) which terminate in the different zones indicated. Refer to text for a more detailed description of both the organization of the bulb, and the destinations of olfactory tract fibres.

1971). In all, about 100 individual neurons contribute dendrites to each glomerulus, and (in the rabbit) each receives the terminals of some 25,000 olfactory nerve axons. Further, the axonal terminals of a number of periglomerular cells also penetrate the glomerulus. Detailed analyses of the synaptic contacts between these different elements, and their possible modes of operation, are still in progress (Price and Powell 1970; Pinching and Powell 1971; Shepherd 1974, 1977, 1978); in outline, the following synaptic contacts have been described.

The terminal parts of the olfactory nerve axons do not receive axoaxonic (presynaptic) contacts from other sources, but they establish the axodendritic contacts with mitral and tufted cell

those of mitral and tufted cells are found, together with axodendritic contacts between either extrinsic or intrinsic axons and the periglomerular cells. The details of these contacts and the other varieties described will not be pursued further here (for these *see* Pinching and Powell 1971; Shepherd 1974, 1978).

The internal granular layer which lies deep to the mitral cell layer consists of large numbers of microneurons or *amacrine granule cells* (Rall *et al.* 1966; Price and Powell 1970; Shepherd 1974, 1977, 1978) which possess no true axon. Their dendrites branch repeatedly, some spreading locally, whilst others pass deeply into the molecular layer where, once again, reciprocal dendrodendritic synapses are established with the dendrites of

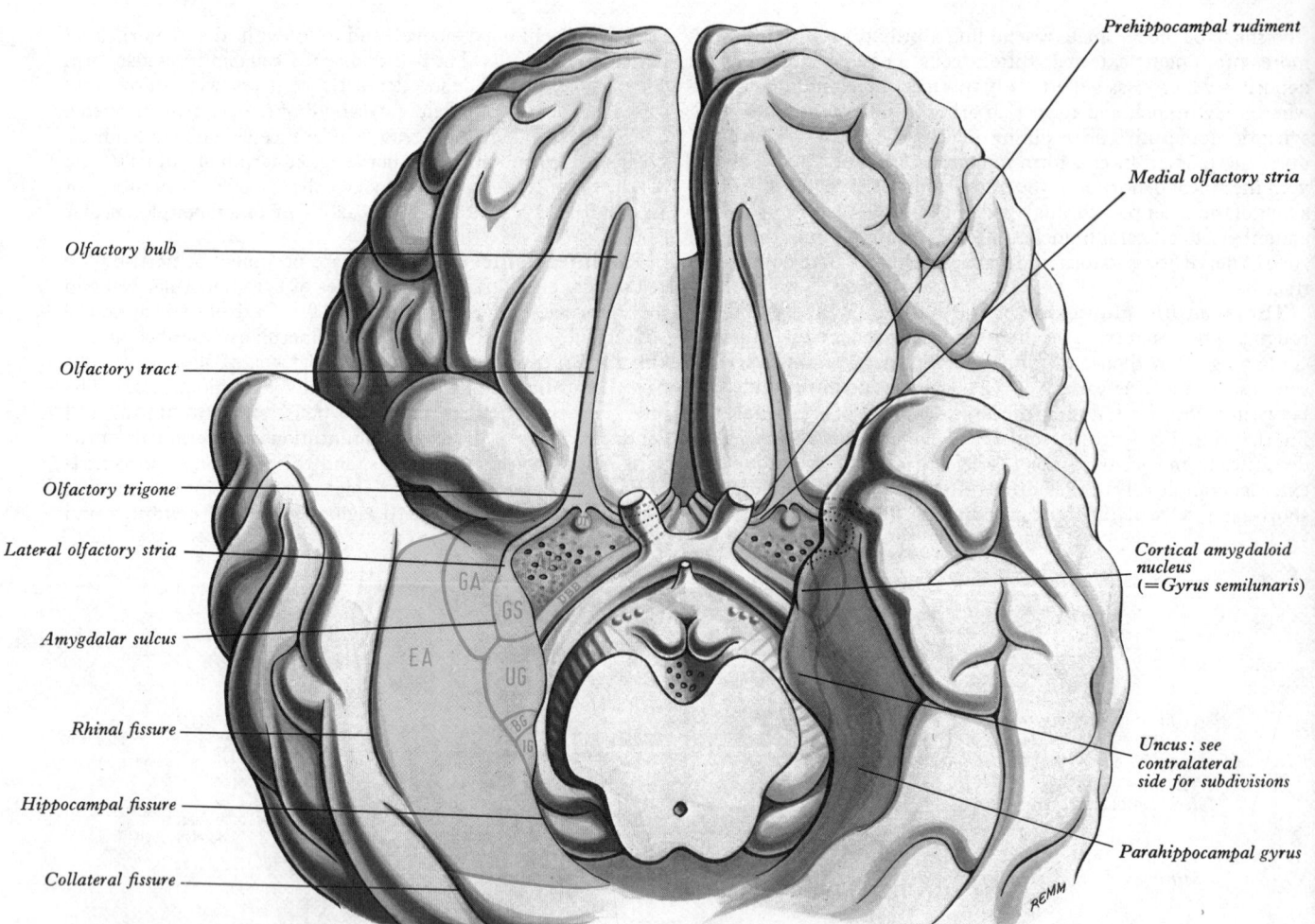

Prehippocampal rudiment

Medial olfactory stria

Olfactory bulb

Olfactory tract

Olfactory trigone

Lateral olfactory stria

Amygdalar sulcus

Rhinal fissure

Hippocampal fissure

Collateral fissure

Cortical amygdaloid nucleus (=Gyrus semilunaris)

Uncus: see contralateral side for subdivisions

Parahippocampal gyrus

7.119 A diagram of structures on the inferior aspect of the human brain in the area immediately surrounding the optic nerves, chiasma, optic tracts and interpeduncular fossa. Many of these structures are intimately related to the olfactory and limbic systems; they are coloured blue. The right temporal pole has been displaced laterally to expose underlying structures. In addition to the features which have been labelled fully, the abbreviations used have the following significance: **OT**—olfactory tubercle; **APS**—anterior perforated substance; **DBB**—diagonal band of Broca. The uncus hippocampi is divided into three areas; **IG**—the intra-limbic gyrus; **BG**—the band of Giacomini; and **UG**—the uncinate gyrus. The lateral olfactory stria continues into the gyrus semilunaris (**GS**); this is bordered laterally by the gyrus ambiens (**GA**); whilst further laterally is the entorhinal area (**EA**) which is the rostral extension of the parahippocampal gyrus. Note the curved extensions of the pre-hippocampal rudiments, medial olfactory striae, and diagonal bands of Broca on to the medial aspect of the hemisphere. The triangular midline zone between the converging diagonal bands, and superior to the optic chiasma, is the lamina terminalis. The occasional intermediate olfactory stria which merges with the olfactory tubercle, is illustrated but unlabelled. (After Kuhlenbeck: redrawn and modified from: *The Hypothalamus*, with the permission of the authors W. J. H. Nauta and W. Haymaker, and the publishers Charles C. Thomas, Springfield, Illinois.)

mitral and tufted cells. The granule cell dendrites also receive axodendritic synapses from recurrent collateral branches of mitral and tufted cell axons, and also from many of the terminals of centrifugal axons in the olfactory tract.

In short, the neurons of the olfactory bulb and their processes present some of the most complex sets of synaptic patterns yet described in the nervous system. The granule and periglomerular cells are considered to be inhibitory interneurons, and by their geometric organization and array of contacts, they provide the substrate for interaction between the principal conduction channels through the bulb. Thus, they mediate feed-forward, lateral and feed-back inhibitory processes, and also inhibitory effects by the centrifugal axons which pass into the bulb from the opposite olfactory tract. Much remains to be determined concerning the detailed operation of these various elements.

In many non-primate mammals, and submammalian forms, a diverticulum lined by specialized epithelium called the *vomero-nasal organ* (of Jacobson) lies on each side of the nasal septum. The nerve bundles from this epithelium run with the ventral groups of olfactory nerve fibres and then converge to form the *vomeronasal nerve*, which ends in an *accessory olfactory bulb*, situated on the dorsomedial aspect of the main bulb. Although an accessory bulb and vomeronasal nerve are present on one or both sides in a proportion of human embryos, both structures degenerate with increasing maturity of the nervous system (McCotter 1915; Hume 1940; Crosby and Humphrey 1941).

THE OLFACTORY TRACT

The olfactory tract (**7.115, 119**) is a narrow white band which issues from the posterior aspect of the olfactory bulb and continues posteriorly on the inferior (orbital) surface of the frontal lobe, where it covers the line of the olfactory sulcus. The tract is triangular in outline in transverse section, its narrow apex being recessed into the olfactory sulcus. It consists of the centrally directed axons of the mitral and tufted cells of the bulb, and also some centrifugal axons which have crossed from the opposite bulb and anterior olfactory nucleus in the anterior commissure, and other centrifugal axons arising from neurons in or near the anterior perforated substance (*vide infra*). The precise contribution of these and other possible sources to the centrifugal outflow is still under active investigation (Cragg 1962; Powell and Cowan 1963; Powell *et al.* 1965; Price 1968; Heimer 1968).

The Anterior Olfactory Nucleus

At the posterior end of the olfactory bulb its various characteristic cell layers disappear, but the deeply placed granule cell layer is replaced throughout the length of the olfactory tract by scattered groups of medium-sized multipolar neurons which constitute the *anterior olfactory nucleus*. Posteriorly, these groups of neurons continue into the olfactory striae and trigone (*vide infra*) to abut on or become continuous with the grey matter of the prepiriform cortex, the anterior perforated substance and the precommissural septal areas. Many of the centripetal axons from mitral and tufted cells which constitute the olfactory tract relay in, or give collaterals to, the anterior olfactory nucleus, and then the axons of the latter continue with the direct fibres from the bulb to form the olfactory striae.

THE OLFACTORY STRIAE AND TRIGONE

Posteriorly, as the olfactory tract approaches the anterior perforated substance, it flattens and becomes splayed out into a smooth *olfactory trigone* (olfactory pyramid) from the caudal angles of which the fibres of the tract continue as separate diverging bundles, the *medial* and *lateral olfactory striae* which border the anterior perforated substance (7.119). In some brains a small *intermediate stria* can be distinguished passing from the centre of the trigone to sink into the anterior perforated substance.

The lateral olfactory stria courses along the anterolateral margin of the anterior perforated substance as a whitish bundle distinguishable by the naked eye. It continues into the limen insulae (p. 988), at which point it makes an abrupt bend posteromedially to merge with an elevated region of grey matter called the gyrus semilunaris which bounds the cranial margin of the uncus hippocampi (7.119). The gyrus semilunaris incorporates part of the *corticomedial* subdivision of the *amygdaloid complex* of nuclei (*vide infra*). A tenuous layer of grey matter which clothes the lateral olfactory stria is called the *lateral olfactory gyrus* and this merges laterally with grey matter of the *gyrus ambiens* which in part forms the limen insulae. The lateral olfactory gyrus and gyrus ambiens, together, constitute the *prepiriform region* of the cortex which passes caudally into the *entorhinal area* of the parahippocampal gyrus.

The prepiriform region, the periamygdaloid region and the entorhinal area (area 28) together comprise **the piriform lobe** of the cerebrum; it is bounded laterally by the *rhinal sulcus* and is relatively more prominent in macrosmatic mammals, and during the early fetal stages of human brain development. The relative positions of these structures and their surroundings are more easily appreciated by reference to illustration 7.119.

The medial olfactory stria, covered by a thin veil of grey matter—the *medial olfactory gyrus*—passes medially along the cranial boundary of the anterior perforated substance and converges towards the medial continuation of the diagonal band of Broca (*vide infra*). Together they curve superiorly on the medial aspect of the hemisphere, anterior to the line of the lamina terminalis (7.119). The diagonal band continues into the paraterminal gyrus, whilst the medial stria becomes indistinct from surface view as it approaches the boundary zone which includes the paraterminal gyrus, parolfactory gyrus, and between them, the narrow prehippocampal rudiment. (*See* illustration 7.117A.)

THE ANTERIOR PERFORATED SUBSTANCE

This is an important topographic landmark on the base of the brain. It lies caudal to the olfactory trigone and the diverging medial and lateral olfactory stria, in the angle between the optic chiasma and tract medially, and the uncus caudally (7.119). Medially, it is continuous above the optic tract with the grey matter of the tuber cinereum, and more anteriorly, with the paraterminal gyrus. Laterally, it reaches the limen insulae, where it is continuous with the prepiriform cortex, and more caudally it merges with the periamygdaloid area (gyrus semilunaris). Superiorly, it is continuous with the grey matter of the corpus striatum and claustrum through the aggregations of grey and white matter which form the *substantia innominata*. Part of the latter, together with fascicles of the ansa lenticularis and anterior commissure separate the anterior perforated substance from the globus pallidus.

The inferior aspect of the anterior perforated substance is related to the termination of the internal carotid artery and the origins of the anterior and middle cerebral arteries; from the latter a number of central arteries pierce the surface of the brain and supply deeper structures (p. 984), and when removed during dissection their paths form the perforations from which the region derives its name.

Immediately caudal to the olfactory trigone the anterior perforated substance presents a small *olfactory tubercle* which varies in its prominence, and into the base of which the occasional intermediate olfactory stria sinks. The tubercle is a large structure in macrosmatic animals but is greatly reduced and sometimes indistinguishable in man. Similarly, the intermediate stria, although sometimes absent, is occasionally represented by two or three fine intermediate striae which radiate into the perforated substance. The caudal zone of the anterior perforated substance where it adjoins the optic tract, is formed by the smooth surface of the *diagonal band of Broca*. Caudolaterally, the band is continuous with the periamygdaloid area, whilst rostromedially it continues above the optic chiasma to merge with the paraterminal gyrus (precommissural septum).

GENERAL ORGANIZATION OF LIMBIC GREY MATTER

The variations in cytoarchitectonic patterns, and the numerous and sometimes divergent views concerning the connexions of these various regions cannot be reviewed in detail here; for this the reader should consult original papers and large works devoted exclusively to neuroanatomy. In summary, the following points may be noted.

The six laminae, which, although varying in their prominence in different regions, are widely held to characterize the neocortex (p. 1006), are not found in the same form in the archipallial and paleopallial regions of the limbic system. In some areas of the latter a laminar pattern is either absent or scarcely distinguishable, whilst in other areas an obvious but distinctive lamination exists, varying from a primitive three-layered to a transitional six-layered variety, where limbic cortex adjoins and merges with neopallial regions. Thus, the medial and lateral olfactory gyri, prepiriform cortex and indusium griseum consist of isolated patches or a thin veil of grey matter and associated fibre bundles, with no laminar pattern. Similarly, the rostral and caudal zones of the anterior perforated substance, the diagonal band of Broca, paraterminal gyrus and the periamgydaloid cortex are poorly differentiated without an obvious laminar pattern, but specific cell aggregations occur in some of them. These include the *nucleus of the lateral olfactory tract* in the periamgydaloid region, and the *nucleus of the diagonal band of Broca*. The central zone of the anterior perforated substance is better differentiated and presents three layers—an outer plexiform, an intermediate pyramidal and a deep polymorphic layer. This region also has a number of characteristic cup-shaped aggregations of granule cells in the pyramidal layer (and these sometimes extend into the plexiform layer). These are the *islands of Calleja*, which vary considerably in size, the medial being the largest—it is closely related to another prominent group of larger pleomorphic cells, the *nucleus accumbens septi*. The structure of the hippocampal formation is further discussed below, but it may be noted at this point that the dentate gyrus and Ammon's horn both consist of a primitive trilaminar type of cortex which shows a gradual transition through the subicular region until a modified six-layered variety is found in the entorhinal area of the parahippocampal gyrus.

THE TERMINATION OF THE OLFACTORY TRACT

The areas of termination of olfactory tract fibres have been widely studied experimentally in mammals, including monkeys, and despite minor differences in some accounts, substantial agreement has been reached. What obtains in the human brain is, of course, less certain.

As we have seen, the olfactory tract fibres are the axons of mitral and tufted cells in the olfactory bulb, some of which synapse with cells in the anterior olfactory nucleus, the axons of the latter then continuing with the direct fibres of the tract. Through the medium of the *lateral olfactory stria*, tract fibres reach and synapse with neurons in the lateral part of the anterior perforated substance, the lateral olfactory gyrus, the prepiriform cortex and in the corticomedial group of amygdaloid nuclei (*vide infra*). These regions are often grouped as the **primary olfactory cortex**, and it should be noted that in contrast to all other sensory pathways, fibres reach these cortical areas directly without a synapse in one of the thalamic nuclei. The entorhinal area (area 28) of the parahippocampal gyrus, which occupies the caudal part of the piriform lobe, receives few or no tract fibres directly, but receives profuse connexions from the primary cortex; accordingly, it is sometimes called the **secondary olfactory cortex**. It is thought that the primary and secondary olfactory cortices are the principal areas responsible for the subjective appreciation of olfactory stimuli.

The variable *intermediate olfactory stria* ends in the anterior perforated substance, but the destination of fibres in the *medial stria* is much less certain. Some of the latter also end in the anterior perforated substance, whilst others have been described as reaching the paraterminal gyrus and adjacent regions, but such claims cannot be regarded as established in the human brain. A number of fibres in the medial olfactory stria cross the midline in the anterior commissure for distribution to the contralateral anterior olfactory nucleus and olfactory bulb.

In addition to the entorhinal area, the primary olfactory cortex also projects to the basolateral part of the amygdaloid complex, the septal areas, the nucleus medialis dorsalis of the thalamus, and to many of the hypothalamic nuclei.

THE AMYGDALA

The amygdala (amygdaloid body, amygdaloid nuclear complex), so named because its general shape resembles that of an almond, consists of a series of neuronal masses and associated nerve fibres in the dorsomedial part of the temporal pole of the cerebrum. It forms the ventral, superior and medial walls of the ventral tip of the inferior horn of the lateral ventricle. Its topographical relationships are complicated. Superiorly, it is partly continuous with the inferomedial margin of the claustrum, and fibres of the external capsule and substriatal grey matter incompletely separate it from the putamen and globus pallidus; it is closely applied to the optic tract. The amygdala is partly deep to the gyrus semilunaris, the gyrus ambiens and the uncinate gyrus (7.119); transitional zones connect it with the anterior perforated substance, prepiriform cortex and parahippocampal gyrus. Caudally, it is closely related to the ventral part of the hippocampus; it fuses with the tip of the tail of the caudate nucleus, which has coursed ventrally in the roof of the inferior horn of the lateral ventricle; the stria terminalis issues from its caudal aspect.

(For a comprehensive review of the neurobiology of the amygdala consult Eleftheriou 1972.)

Divisions of the Amygdala

The amygdaloid complex is divided into two main groups of nuclei, *corticomedial* and *basilateral*, together with junctional zones where these adjoin or partly fuse with adjacent areas. A detailed account of the subdivisions, connexions and homologies of these groups will not be attempted here (*see* Crosby and Humphrey 1941; Crosby *et al.* 1962; Eleftheriou 1972; Isaacson 1974) and only some of the main points will be outlined.

The corticomedial amygdaloid complex consists of the *central, medial*, and *cortical amygdaloid nuclei*, the *nucleus of the lateral olfactory stria*, and a transitional poorly differentiated *anterior amygdaloid area*. The cortical nucleus, as its name suggests, may be regarded as a rudimentary cortex, with irregular groups of pyramidal and granule cells; it occupies the surface elevation of the gyrus semilunaris.

The corticomedial complex is continuous through transitional zones with the anterior perforated substance, the diagonal band of Broca, the substantia innominata, and with the putamen, caudate nucleus, and surrounding cortical areas of the uncus and parahippocampal gyrus. It is relatively small in the human brain.

The basilateral amygdaloid complex, large and well differentiated in the human brain, consists of *lateral, basal*, and *accessory basal amygdaloid nuclei*. It is partly continuous with the claustrum, and, through a corticoamygdaloid transitional zone, with the cortex of the parahippocampal gyrus.

The Connexions of the Amygdaloid Complex

These are incompletely established for the human brain; most of the available information is derived from degeneration and electrophysiological studies in a variety of mammals, including monkeys (*see*, for example Le Gros Clark and Meyer 1947; Allison 1954; Powell *et al.* 1965; Gloor 1960; Wendt and Albe-Fessard 1962; Eleftheriou 1972).

Briefly, **afferent connexions** reach the corticomedial complex via the lateral olfactory stria from the olfactory bulb and anterior olfactory nucleus, whilst the basilateral complex receives many afferents from the cortex of the piriform lobe (p. 990). Other afferents converge on the amygdala from the hypothalamic nuclei, both specific and non-specific thalamic nuclei, the brainstem reticular formation, and probably a number of areas of neocortex (*vide infra*).

The best understood *outflow* from the amygdaloid nuclei occurs via the stria terminalis to the septal areas, the preoptic and adjacent regions of the hypothalamus, whilst some continue into the stria medullaris to reach the habenular nucleus. Other *amygdalofugal fibres* do not run with the stria terminalis, but follow more ventrally placed direct routes (p. 968) to reach many of the hypothalamic nuclei, the cortex of the piriform lobe, the nucleus medialis dorsalis of the thalamus, and the reticular formation of the midbrain tegmentum. Reciprocal connexions probably exist between the amygdaloid complex and the oribitofrontal, cingulate and temporal regions of the neocortex, whilst *interamygdaloid fibres* run with the other components of the anterior commissure (*vide infra*). For a brief review of the more recent views concerning the monoaminergic connexions of the amygdaloid complexes *see* pp. 943–948.

THE STRIA TERMINALIS

This is a small, discrete bundle of fine myelinated nerve fibres, which is visible to the unaided eye throughout much of its course. Fibres pass in both directions within the bundle to a number of different destinations. Topographically, the stria issues from the posterior aspect of the amygdaloid complex and runs caudally in the roof of the inferior horn of the lateral ventricle, on the medial side of the tail of the caudate nucleus. It follows the curve of the nucleus and then passes ventrally in the floor of the body of the ventricle, occupying the groove which separates the caudate nucleus from the thalamus, where it is closely related to the thalamostriate vein. It passes inferior to the interventricular foramen to approach the region of the anterior commissure, where it diverges into a series of components which are *supracommissural, commissural*, and *subcommissural* in position. As the stria nears the anterior pole of the thalamus, within its various subdivisions scattered groups of small neurons are found; these constitute the *bed nucleus of the stria terminalis*, within which a number of its fibres relay. Many of the fibres in the supra- and sub-commissural components are amygdalofugal, and are passing to the septal areas (*vide infra*), the preoptic and anterior hypothalamic nuclei, whilst others descend to the anterior perforated substance and adjacent regions of the piriform lobe. Some of the subcommissural fibres recurve to join the column of the fornix, and yet others pass caudally with the stria medullaris to the habenular nucleus. It is probable that reciprocal connexions are established between some of these regions and the amygdaloid nuclei, by fibres which pass in the reverse direction through the stria. It will be appreciated that interamygdaloid fibres pursue a most intricate topographical course. Leaving one amygdaloid complex they pass through almost the whole length of the highly curved stria terminalis to reach the anterior commissure, where they cross the midline and then curve in the reverse direction through the contralateral stria to reach the opposite amygdaloid complex.

THE ANTERIOR COMMISSURE

The anterior commissure is a compact bundle of myelinated nerve fibres which crosses the midline anterior to the columns of the fornix, embedded in the lamina terminalis, where it forms part of the anterior wall of the third ventricle some 1·5–2·0 cm superior to the optic chiasma (7.110, 114, 141). When seen in sagittal section it is oval in shape with its long diameter (about 2·5 mm) placed vertically. Traced laterally, its fibres are twisted and entwined like the strands of a rope; further laterally, it separates into two principal bundles. The smaller *anterior bundle* curves forwards on each side towards the anterior perforated substance and olfactory tract. The large *posterior bundle* curves backward and laterally on each side and for some distance is lodged in a deep groove on the antero-inferior aspect of the lentiform nucleus. Beyond the latter, the posterior bundle forms a fan-shaped radiation into the anterior part of the temporal lobe including the parahippocampal gyrus. Commissural fibres have been described in various mammals, including primates, as interconnecting the following structures with their fellows—(1) the olfactory bulb and anterior olfactory nucleus; (2) the anterior perforated substance, olfactory tubercle and diagonal band of Broca; (3) the prepiriform cortex; (4) the entorhinal area and adjacent parts of the parahippocampal gyrus; (5) part of the amygdaloid complex—in particular, the nucleus of the lateral olfactory stria; (6) the bed nucleus of the stria terminalis, and the nucleus accumbens septi; (7) the middle and inferior gyri of the temporal lobe in their anterior regions; and (8) possibly other neocortical areas, including small regions of the frontal lobe.

The intertemporal neocortical connexions form the largest component of the anterior commissure in primates, including the human brain. However, in the latter many of the detailed connexions of the fibres in the commissure remain to be elucidated. Further, a proportion of its fibres may not be truly commissural, but decussating pathways between dissimilar centres on the two sides.

THE SEPTAL AREAS

In subprimate mammals, the *septal areas* consist of the thick medial walls of the cerebral hemispheres situated immediately anterior and superior to the lamina terminalis and anterior commissure. They consist of nuclear masses of grey matter together with relatively coarse bundles of white fibres, and according to their relationship to the anterior commissure may be divided into *pre-* and *supra-commissural* parts.

In higher primates and particularly in the human brain, the septal areas are considerably modified in association with the great expansion of the neocortex and corpus callosum in these forms. The supracommissural septum now corresponds in large measure to the bilateral thin laminae of white fibres, sparse grey matter, and neuroglia, which form the right and left halves of the septum pellucidum. The precommissural septum (*septum verum* of some authors) in contrast, consists of relatively well-defined dorsal, ventral, medial and caudal *groups of nuclei*—each of these may be subdivided into a series of individual nuclei, but these will not be detailed here. However, the precise topographical limits of the human septum verum, and the terminology applied to the surface topography and nuclear groups of this region, have been the source of some confusion and disagreement. (For detailed reviews consult Stephan and Andy 1962; Andy and Stephan 1968; Nauta and Haymaker 1969.)

Most authorities agree that the precommissural septum corresponds in part to the paraterminal gyrus. The latter has been mentioned elsewhere (p. 988) as a narrow vertical strip which lies between the anterior surface of the lamina terminalis and the posterior parolfactory sulcus. The anterior slope of the gyrus which passes into the sulcus is sometimes called the *pre-hippocampal rudiment*. Inferiorly, the gyrus and rudiment are continuous with the diagonal band of Broca and with the medial olfactory stria (7.119). Superiorly, they narrow, and spread around the rostrum and genu of the corpus callosum to become continuous with the indusium griseum (p. 998). However, some investigators hold that only the prehippocampal rudiment is

continuous with the indusium griseum, whilst the detailed connexions of the medial olfactory stria are uncertain in the human brain.

Whilst some of the septal nuclei are located within the paraterminal gyrus, others are more deeply placed and are interspersed with fibres of the precommissural fornix (p. 1001). Scattered cell groups interconnect the precommissural septum with the septum pellucidum, the substantia innominata and the anterior perforated substance.

The main afferent connexions to the septal nuclei are from: (1) the amygdaloid complex via the diagonal band and stria terminalis; (2) the anterior perforated substance by fibres which probably run with the medial olfactory stria; (3) the hippocampus via the fornix; (4) the midbrain reticular formation and hypothalamic nuclei through ascending fibres in the medial forebrain bundle (the monoaminergic status of some of the latter afferents are mentioned on pp. 943–948).

The main outflows from the septal nuclei are—(1) fibres which return to the hippocampal formation in the fornix; (2) descending fibres in the medial forebrain bundle which are distributed to many of the hypothalamic nuclei and to the midbrain reticular formation; (3) fibres which pass in the stria medullaris thalami to the habenular nuclei.

Finally, interconnexions are probably established between the septal areas and the vestigial indusium griseum and also with the cingulate gyrus, but details of these remain uncertain.

Thus, the septal areas are important focal zones through which many of the principal limbic and hypothalamic structures are interconnected. For long thought to be reduced, atrophic, and perhaps functionless in primates, recent studies have demonstrated that in fact the septal nuclei have increased in prominence throughout the primates and reach their highest degree of primate development in the human brain (Andy and Stephan 1968).

The Hippocampal Formation

The hippocampal formation develops in the medial pallial fringe of the cerebral hemisphere, immediately adjacent to the outer convex border of the choroidal fissure (p. 980), where, with progressive development, it assumes a highly arched form extending from the interventricular foramen to the ventral extremity of the inferior horn of the lateral ventricle. The modifications in the superior part of the formation, which accompany the expansion of the neopallium and the corpus callosum, have also been discussed previously (p. 171). Essentially, the hippocampal formation consists of a curved band of phylogenetically ancient cortex (the *archipallium*), limited on its concave aspect by the choroidal fissure, and merging on its convex aspect with surrounding areas of *neopallium*. Traced in a radial direction from the choroidal fissure towards the neopallium, three main zones may be distinguished within the archipallium, namely, the *dentate gyrus*, the *cornu ammonis* and the *subiculum*. The dentate gyrus and cornu ammonis show some contrasting structural features, but are generally regarded as the most primitive *trilaminar* types of cortex, whilst the subiculum shows a graded variation in structure from a four, through a five, to a modified six-layered type of cortex where it merges with the surrounding full six-layered neocortex. With further development profound growth changes cause an infolding of these archipallial zones towards the neighbouring cavity of the lateral ventricle. To assist understanding, these zones and their changes in relative position, are shown in a highly simplified manner in illustration 7.121. The general form of infolding is similar throughout Mammalia. However, its *degree*, and the *areas* of hemispheric wall in which a well-differentiated hippocampal formation persists, varies greatly in different groups. These variations will not be detailed here, but they reflect the differences in relative expansion of the neopallium, corpus callosum and temporal lobe in the different groups. In the human brain, that part of the hippocampal arch related to the medial wall and roof of the body of the lateral ventricle becomes greatly reduced in association with the relatively enormous size of the corpus callosum, a well-developed hippocampal formation being confined to the floor and medial wall of the inferior horn of the ventricle.

Antero-inferiorly, the extremities of the archipallial arch are continuous with the septal area including the paraterminal gyrus, the anterior perforated substance and parts of the piriform lobe (p. 990). Further, the topographical limits of the structures to be included in the *hippocampal formation* vary somewhat in the accounts of different authorities. Here, it is considered to include—(1) the indusium griseum, and the longitudinal striae and their extensions; (2) the gyrus fasciolaris; (3) the dentate gyrus, cornu ammonis and subiculum; (4) parts of the uncus. The term *hippocampus* is often used to denote the macroscopic swelling in the floor of the inferior horn of the lateral ventricle consisting of the interlocked dentate gyrus, cornu ammonis and related structures.

The detailed connexions of the striae are uncertain, but in the supracallosal part of their course they probably contribute fibres which pierce the corpus callosum and form part of the *dorsal fornix* (p. 1002).

THE HIPPOCAMPUS

As noted above, the hippocampus consists of the complex interfolded layers of the dentate gyrus and cornu ammonis, the latter being continuous through the subicular region with the cortex of the parahippocampal gyrus. The name hippocampus stems from the supposed resemblance of these cell laminae, when viewed in coronal section, to the outline of a sea-horse. Grossly, the hippocampus is superior to the subiculum and medial part of the parahippocampal gyrus, where it forms a curved elevation about 5 cm long extending throughout the entire length of the

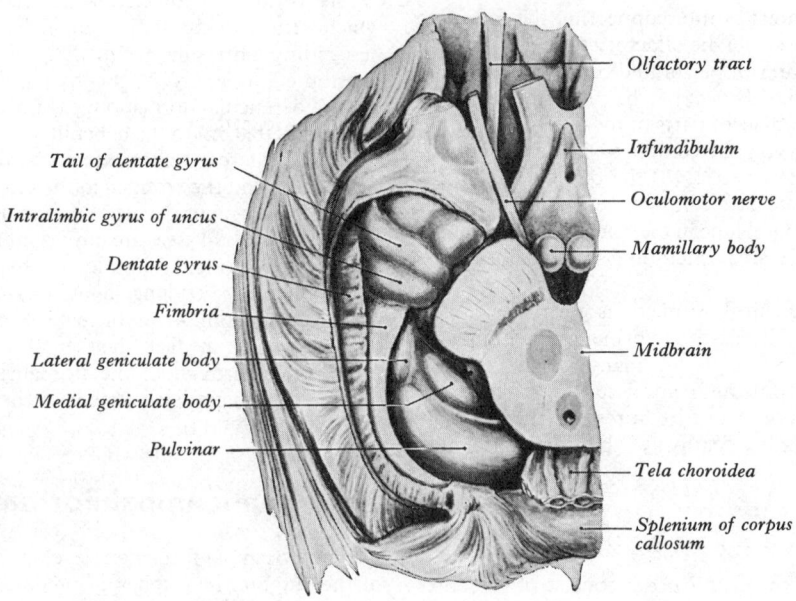

Olfactory tract

Infundibulum

Oculomotor nerve

Mamillary body

Midbrain

Tela choroidea

Splenium of corpus callosum

Tail of dentate gyrus

Intralimbic gyrus of uncus

Dentate gyrus

Fimbria

Lateral geniculate body

Medial geniculate body

Pulvinar

7.120 Basal aspect of part of the brain dissected to display the uncus, dentate gyrus, fimbria, etc.

What follows is a brief topographical account of these regions, some further structural details of the hippocampus, and a description of the functionally related fornix.

THE INDUSIUM GRISEUM

The indusium griseum or *supracallosal gyrus*, is a thin, poorly differentiated veil of grey matter which covers the superior surface of the corpus callosum. Laterally, on each side, it passes into the callosal sulcus to become continuous with the cortex of the cingulate gyrus. Anteriorly, it sweeps around the genu and rostrum of the corpus callosum to merge on each side, with the superior end of the paraterminal gyrus. It will be recalled that the latter is continuously inferiorly with the diagonal band of Broca and, through this, with the anterior perforated substance and periamygdaloid area. Posteriorly, the indusium griseum passes on to the splenium of the corpus callosum where it diverges to become continuous with the right and left *gyrus fasciolaris* (*splenial gyrus*). The latter is a delicate strip of grey matter which curves downwards, forwards and laterally, to blend with the posterior extremity of the *dentate gyrus*. Embedded in the indusium, and causing a ridging of its free surface, are two narrow bundles of white fibres on each side, the *medial* and *lateral longitudinal striae* (of Lancisi) (7.139). The medial striae course near the midline whilst the lateral striae are in the depths of the callosal sulci. The striae are regarded as the reduced white matter of the vestigial indusium. Anteriorly, they pass towards the paraterminal gyri, whilst posteriorly they continue through the gyrus fasciolaris to reach the fimbriae of the fornix (p. 1001).

floor of the inferior horn of the lateral ventricle. Its anterior extremity is expanded, and at this point its margin sometimes presents two or three shallow grooves with intervening elevations giving a paw-like appearance—the so-called *pes hippocampi*. The ventricular surface is convex in coronal section and, of course, covered by ependyma, beneath which pass tangential white fibres of the *alveus* converging medially upon a longitudinal projecting bundle of white fibres, the *fimbria of the fornix*. The general relationships of these various cell layers and fibre bundles as seen in coronal section are best appreciated by preliminary references to a simplified diagram (7.121). It will be noted that passing *medially* from the line of the collateral sulcus, the neocortex of the *parahippocampal gyrus* merges with the transitional cortex of the *subiculum*. The latter curves superomedially to reach the inferior surface of the *dentate gyrus*, and then continuing to curve laterally becomes continuous with the cell laminae of the *cornu ammonis*. The latter continues to curve, first superiorly and then laterally above the dentate gyrus, and finally ends pointing towards the centre of the superior surface of the dentate gyrus. However, it should be appreciated that the degree of curvature of these regions varies somewhat along the length of the hippocampus and also in different specimens.

Topographically, the *dentate gyrus* (7.120) is a crenated strip of cortex which is related inferiorly to the subiculum, laterally to the cornu ammonis, and superiorly to the recurved part of the cornu ammonis, the alveus and, more medially, to the fimbria of the fornix (7.121). However, the distribution and form of the fimbria is quite variable (*vide infra*), but medially it is separated from the notched medial margin of the dentate gyrus by the *fimbriodentate*

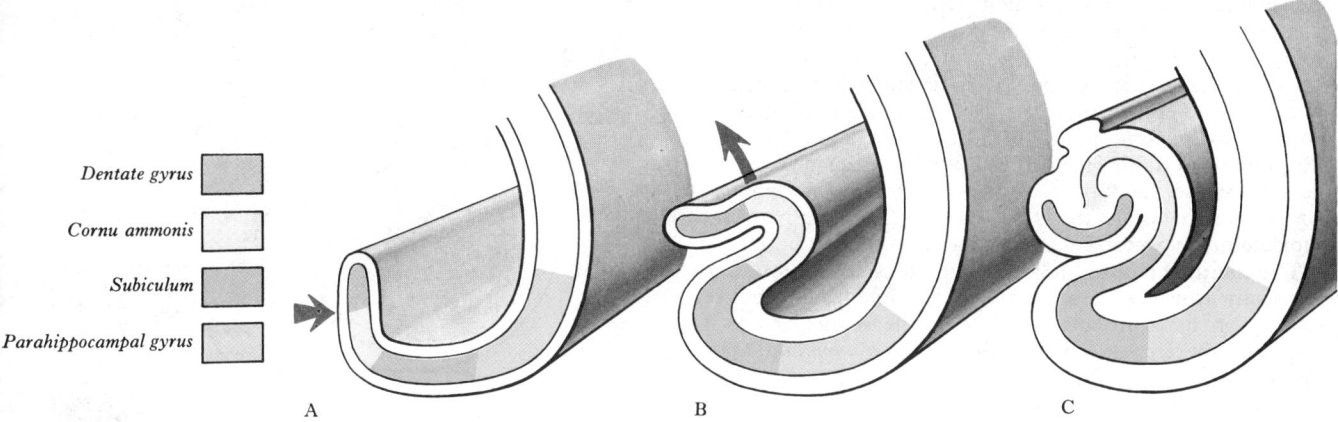

Dentate gyrus

Cornu ammonis

Subiculum

Parahippocampal gyrus

A   B   C

7.121A–C The hippocampus and related structures seen in coronal section A, B and C are a series of diagrams to assist understanding of the assumption of the definitive positions of the dentate gyrus, cornu ammonis, subiculum and parahippocampal gyrus, in the floor of the inferior horn of the lateral ventricle in the human brain. Note that these are *not* tracings from a series of embryonic sections, and that they have been somewhat simplified in the interests of clarity. Note also that the

amount of curvature and infolding which occurs varies along the length of the hippocampus, and in different specimens. It is important to appreciate that following folding, the original *external* surfaces of the dentate gyrus and part of the subiculum are in contact, and that the degree of tissue fusion which occurs along the line of the hippocampal sulcus is quite variable.

sulcus. The *hippocampal sulcus*, of variable depth, intervenes between the dentate gyrus and the subicular extension of the parahippocampal gyrus. Posteriorly, as noted above, the dentate gyrus is continuous with the gyrus fasciolaris, and through this, with the indusium griseum. Anteriorly, the dentate gyrus is continued into the notch of the uncus, where it makes a sharp bend medially across the central part of its inferior surface. This transverse part is smooth and featureless, being termed the *tail of the dentate gyrus* (band of Giacomini), and it becomes indistinguishable on the medial aspect of the uncus. The tail separates the rest of the inferior surface of the uncus into an anterior *uncinate gyrus* and a posterior *intralimbic gyrus* (**7.**119, 120).

THE STRUCTURE OF THE HIPPOCAMPUS

Since the extensive pioneering researches of Cajal (1890, 1911) and Lorente de Nó (1934) a vast literature has accumulated concerning the structure and connexions of the hippocampus. Only a few salient points can be mentioned here, and for details texts devoted wholly to neuroanatomy (e.g. Kappers *et al.* 1936; Crosby *et al.* 1962; Brodal 1969) and original papers should be consulted. Detailed earlier reviews of the complex terminologies proposed for the various laminae and their subdivisions are to be found in the writings of Lorente de Nó (1934), Rose (1927), and Gastaut and Lammers (1960). Examples of the many investigations into the connectivity patterns of the hippocampus using degeneration techniques may be found in (Blackstad 1956, 1958; Raisman *et al.* 1965, 1966; White 1959, 1965), using ultrastructural techniques (Blackstad 1967; Blackstad and Kjaerheim 1967; Blackstad and Flood 1963; Hamlyn 1962), and neurophysiological recording methods (Andersen *et al.* 1964 *a* and *b*; Andersen *et al.* 1966). More recent comprehensive collections of essays devoted to many aspects of the structure, development, neurophysiology, and behavioural aspects, the product of many leading authorities and containing extensive bibliographies are brought together in two volumes entitled *The Hippocampus* and edited by Isaacson and Pribram (1975).

Some aspects of the terminology used and some of the main structural features are shown in outline in illustration **7.**122.

It should be noted that because of the form of cortical folding which occurs during development, the original *external* surfaces of the dentate gyrus and subiculum (the *stratum moleculare*) are closely applied to each other in the depths of the sulcus hippocampi (or along the obliterated line of the sulcus). Throughout much of its extent, the subicular region merges laterally with the modified six-layered cortex of the *entorhinal area* of the parahippocampal gyrus. (Some authorities prefer to include the subiculum with the parahippocampal gyrus, rather

than with the hippocampal formation.) As noted above, during its curved course from the entorhinal area to the beginning of Ammon's horn, the subiculum exhibits a gradual change in its structure from a modified six-layered to a four-layered type of cortex, with accompanying differences in its patterns of connectivity. Accordingly, the subiculum is often divided into four zones, the *parasubiculum*, *presubiculum*, *subiculum* and the *prosubiculum*. Their structural differences will not be detailed here, but it should be appreciated that the prosubiculum merges into the cornu ammonis which, using the same criteria, may be regarded as a primitive trilaminar cortex possessing *molecular*, *pyramidal* and *polymorphic* layers from its original external surface towards its ventricular surface. However, using other methods and criteria, many more sub-layers are often described in Ammon's horn (*vide infra*). Further, although the general

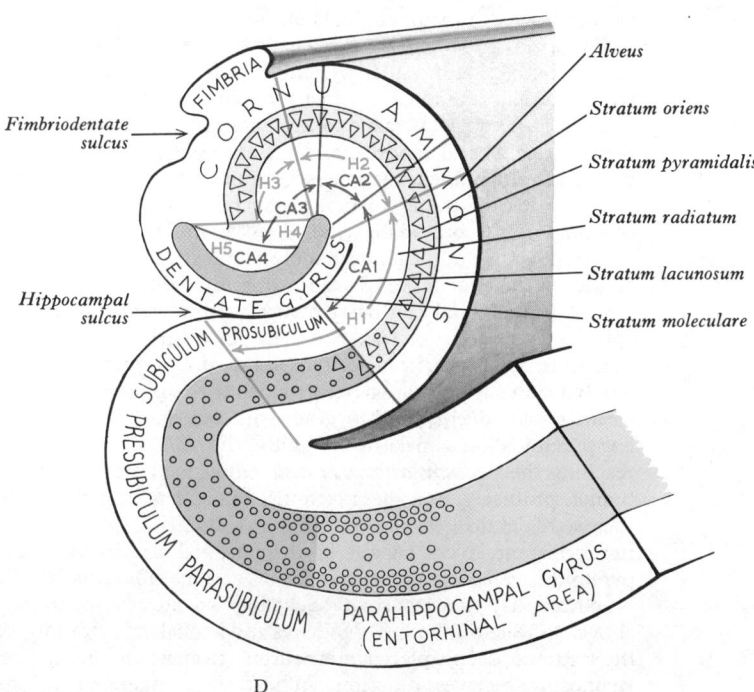

Alveus

Stratum oriens

Stratum pyramidalis

Stratum radiatum

Stratum lacunosum

Stratum moleculare

D

7.121D An analysis of the major topographical zones and the complex terminology applied to the hippocampus and related structures seen in a coronal section of the floor of the inferior horn of the lateral ventricle in a mature human brain. Colour code for tissue zones as in A–C. The approximate limits of the various subdivisions of the subiculum are labelled. CA 1–4 are the cornu ammonis fields of Lorente de Nó (1934), and H 1–5 are the hippocampal fields of Rose (1927).

cytoarchitectonic pattern is roughly similar throughout Ammon's horn, there are regional differences in its detailed structure and connexions. For these reasons, different investigators have proposed various methods and criteria for subdividing this region into a series of radially disposed fields (7.122). These include the **cornu ammonis fields** CA1 to CA4 of Lorente de Nó, and the **hippocampal fields** H1 to H5 of Rose. Unfortunately, these do not correspond; the former, as the name implies, are limited to Ammon's horn, whilst the H1 field of Rose includes part of the subiculum, the prosubiculum, and part of Ammon's horn.

The cornu ammonis, following the intensive studies of Cajal, is usually considered to consist of the following sub-layers starting on its ventricular aspect—(1) the *ependyma*; (2) the *alveus*; (3) the *stratum oriens*; (4) the *stratum pyramidalis*; (5) the *stratum radiatum*; (6) the *stratum lacunosum*; and (7) the *stratum moleculare*. Many authors group the last two sub-layers as a single *stratum lacunosum-moleculare*.

The *alveus* is a subependymal layer of white fibres both entering and leaving the hippocampus. The efferent fibres are predominantly the axons of the large neurons of the stratum pyramidalis, together with a smaller number from some cells of the stratum oriens and dentate gyrus. These axons converge to form the large efferent component of the *fimbria of the fornix (vide infra)* but before joining it they give rise to fine collateral branches which re-enter the hippocampus. Afferent fibres from other regions of the nervous system, including commissural fibres from the opposite hippocampal formation, also approach the cell layers of the hippocampus through the alveus.

The *stratum oriens* is interlaced with the axons and collateral branches of fibres entering and leaving the hippocampus; it is penetrated by some of the basal dendrites of the large pyramidal cells of the adjacent layer, and it contains the cell somata and dendrites of relatively small, irregularly shaped neurons, some of which are termed *basket cells*. These small cells have been shown to be inhibitory interneurons. They receive axosomatic and axodendritic synapses from some of the afferent fibre collaterals to the hippocampus, and also collaterals from efferent fibres. Some of the axons of these interneurons penetrate the radiate and molecular layers establishing axodendritic contacts with pyramidal cells, but the most distinctive terminals are those of the basket cells, which form numerous, crowded, axosomatic synapses on the cell bodies of the pyramidal neurons.

The *stratum pyramidalis* is a particularly well-defined double layer of both large and small pyramidal cells. Their bases face the stratum oriens and alveus, whilst their apices point towards the stratum radiatum. The axon of a pyramidal cell issues from its base or from a basal dendrite, and passes into the alveus, where it gives collateral branches and then continues into the fimbria of the fornix. Some of these collateral branches end on cells in the stratum oriens, but many (the *Schaffer collaterals*) pursue a recurrent course to the stratum moleculare where they terminate on the apical dendrites of adjacent pyramidal cells. The dendritic tree of a pyramidal cell has two main components, basal and apical. The *basal dendrites* radiate into the neighbouring pyramidal layer, but the majority pass into the stratum oriens and overlying alveus; they are beset with dendritic spines. The *apical dendrites* pass deeply, and together with associated axons and a few pyramidal cells, make up the bulk of the *stratum radiatum*. On reaching the *stratum lacunosum-moleculare*, the apical dendrites branch profusely, and these terminal branchlets, together with their stem dendrite, are also covered with dendritic spines. Thus, these deepest layers consist of the terminal dendrites of the pyramidal cells, axonal terminals of various afferents to the hippocampus, the recurrent (Schaffer) collaterals mentioned above, and the cell bodies, dendrites and axonal arborizations of the scattered, deeply placed interneurons, from which the stratum lacunosum receives its name. Whilst the connexions of the hippocampus are treated in a subsequent paragraph, it may be noted here that the various zones of the dendritic tree of the pyramidal cell receive distinctive types of axonal terminal. Thus, commissural fibres from corresponding regions of the opposite hippocampus end upon basal dendrites, whilst those from non-corresponding regions end on the apical dendrites in the strata lacunosum and moleculare. Afferents from the entorhinal cortex

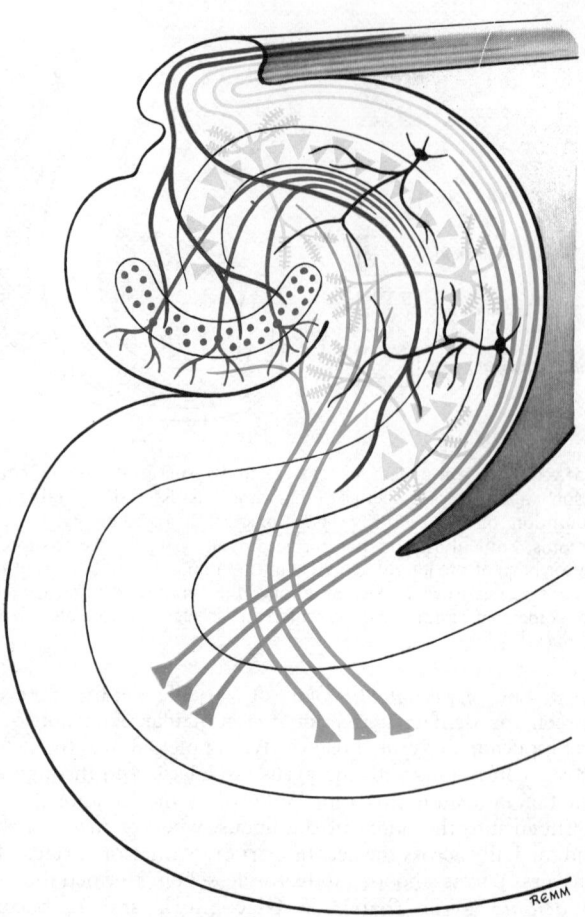

7.122 A diagram of some of the main features of the neuronal organization and connectivity patterns of the dentate gyrus, cornu ammonis, subiculum and parahippocampal gyrus. The cell somata, dendrites and axons of the pyramidal neurons of the cornu ammonis are yellow; their axons form the efferent hippocampal fibres of the alveus and fimbria. Afferent fibres to the cornu ammonis from the fimbria are purple, those following the alvear path are green, whilst those following the perforant path are blue. Basket neurons are in black. The neurons of the dentate gyrus, and their axons which form the mossy fibres of the hippocampus, are in magenta. See text for further details.

form synapses with the most terminal branches of the apical dendrites in the molecular layer, whilst the synaptic endings of the Schaffer collaterals are found in the stratum lacunosum. The mossy fibres derived from dentate gyrus cells form large synaptic terminals which enclose the prominent dendritic spines of the apical dendrites in the stratum radiatum, but as noted above, basket cell terminals form a dense population of axo-somatic synapses. Attempts have been made to correlate these interesting zonal variations in axonal terminals with differences in the character of the electrical recordings made with microelectrodes inserted to varying depths in Ammon's horn (Andersen *et al.* 1964 *a* and *b*; Andersen *et al.* 1966). Thus, many of the afferent systems, and the pyramidal cells themselves, are now known to be excitatory, and some of the interneurons, particularly the basket cells, are known to be inhibitory in their action.

The dentate gyrus is less well understood, but is considered to be a trilaminar cortical structure. Extending deeply from the line of the hippocampal sulcus it consists of: (1) a superficial *molecular layer*; (2) an intermediate *granular layer*; and (3) a deep *polymorphic cell layer*. These will not be considered in detail here. Briefly, the various cell types in the three laminae have spine-studded dendritic trees which either radiate locally or pass in the superficial layer. They receive synaptic terminals from some of the afferent fibres to the hippocampus from extrinsic sources, and also from the axons of neighbouring neurons. Some of the cells are Golgi type II neurons with axons which terminate locally; others have much longer axons which, after collateral branching, run to join the efferent fibres in the fimbria of the fornix. The most distinctive axons, however, are called *mossy fibres*. They arise from many of the cells in the granular layer and then pass through the polymorphic layer giving collateral branches to its neurons. The mossy fibres then proceed along a curved course in the superficial part of the stratum radiatum of the cornu ammonis, making a series of very large ('giant') synaptic contacts with the spines on the initial segments of the apical dendrites of the pyramidal cells.

THE CONNEXIONS OF THE HIPPOCAMPUS

Afferent pathways to the hippocampus arise in: (1) parts of the cingulate gyrus; (2) the septal nuclei; (3) the entorhinal cortex; (4) the indusium griseum; (5) commissural fibres from the opposite hippocampal formation; (6) possibly some fibres from the prepiriform cortex, and (7) a dual monoaminergic fibre input derived from brainstem reticular cell aggregates—*see* pp. 943–953.

The fibres from the cingulate gyrus reach the hippocampus via the cingulum (p. 971) and are distributed both directly to the cornu ammonis, and indirectly after relays in the subicular region. The afferents from the septal nuclei retrace the curved path of the fornix back to the hippocampus. The commissural fibres from one hippocampal formation pass through the fimbria and crus of the fornix on the same side to reach, and cross the midline in the commissure of the fornix (*vide infra*). Thereafter, they recurve back through the opposite crus and fimbria to reach the contralateral hippocampus. Some fibres derived from cells in the indusium griseum run posteriorly in the longitudinal striae, with which they pass through the gyrus fasciolaris to reach the fimbria, which conducts them to the hippocampus.

The most profuse afferent connexions of the dentate gyrus and the cornu ammonis, however, are derived from the entorhinal cortex and subicular regions, and for long were held to pass by two distinct routes. Fibres from the medial part of the entorhinal area and parasubiculum follow an *alvear path* through the pro-subiculum to reach the alveus and stratum oriens of the cornu ammonis. In contrast, the afferents derived from the lateral part of the entorhinal area follow a *perforant path* through the subiculum, which crosses the alvear path, and continues into the stratum lacunosum-moleculare of the cornu ammonis and adjacent parts of the dentate gyrus. These routes, and those entering from the fimbria are shown in illustration **7.122**. It should be noted that the perforant path predominates and is the best attested; some authors doubt the significance or even the existence of an alvear path, at least in some mammals.

THE FORNIX

In addition to the afferent hippocampal fibres and commissural fibres described above, the fornix constitutes the sole **efferent system** from the hippocampus. These efferent fibres are mainly the continuations of the axons of the pyramidal cells of Ammon's horn, but a small proportion are derived from neurons in the stratum oriens and the dentate gyrus. These various axons pass through the alveus and converge upon the medial border of the ventricular surface of the hippocampus to form the *fimbria*. The latter is a flattened band of white fibres which lies superior to the dentate gyrus and forms the inferior boundary of the choroidal fissure. The disposition of the fimbria is variable. It may project above the dentate gyrus with a free medial edge, the *taenia fornicis*, and a lateral border which merges into the alveus; alternatively, its free border may be twisted over towards the lateral side, uncovering the dentate gyrus (**7.140**). Anteriorly, the fimbria continues into the hook of the uncus (**7.120**). Traced posteriorly on the floor of the inferior horn of the ventricle, it ascends towards the splenium and the majority of its fibres now pass anteriorly below the splenium, and then curve forward above the thalamus, forming the *crus of the fornix*. The two crura are closely applied to the inferior surface of the corpus callosum and are connected to each other by transverse fibres which pass between the hippocampal formations of the two sides and form the *commissure of the fornix* (hippocampal commissure). The

commissure is a thin triangular sheet which, together with the converging crura, is sometimes termed the *lyra* or *psalterium* (an instrument resembling a harp). Between the commissure and the corpus callosum, a horizontal cleft (the so-called *ventricle of the fornix*) is sometimes found. Anteriorly, the two crura come together in the median plane to form the *body of the fornix* which is really a symmetrically disposed bilateral structure. The body of the fornix lies above the tela choroidea and the ependymal roof of

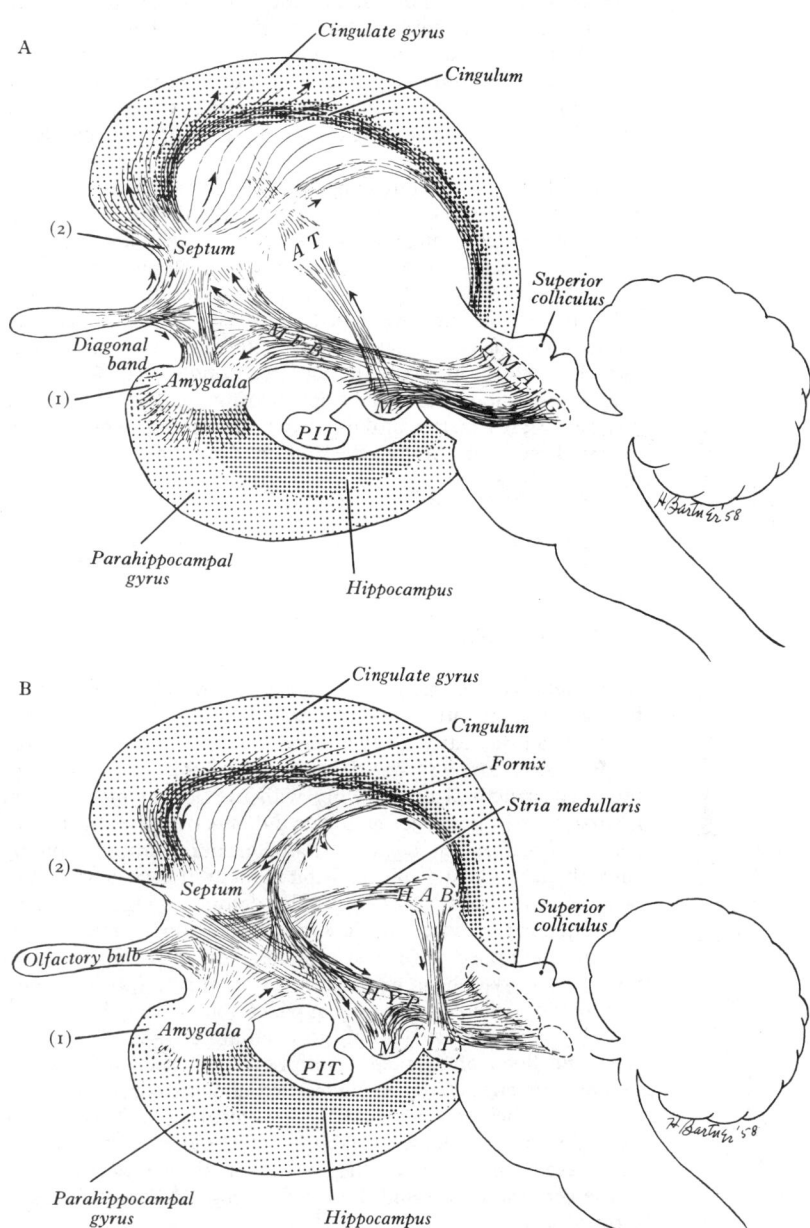

7.123A and B Schematic drawings of one concept of the 'limbic lobe' (after MacLean) in which emphasis is placed upon the medial forebrain bundle (MFB) as a major line of communication between the limbic cortex, the hypothalamus and the midbrain. Note the relationship between the fornix and the cingulum. The limbic cortex, which is considered as a hierarchical system of concentric strips, is indicated by heavy and light stipple. The neocortex is not included. A The ascending pathways to the limbic structures, with emphasis on the divergence of fibres from the medial forebrain bundle to the amygdala (1) and to the septal area (2). Note also the input from the olfactory bulb and tract. B Descending pathways from the limbic system. AT: anterior group of thalamic nuclei; G: tegmental reticular nuclei; HAB: habenular nucleus; HYP: hypothalamus; IP: interpeduncular nucleus; LMA: limbic midbrain area (of Nauta); M: mamillary body; PIT: hypophysis cerebri; (From P. D. MacLean in: *The American Journal of Medicine*, **25**, 1958; by courtesy of the author and publishers.)

the third ventricle (7.157) and is attached to the inferior surface of the corpus callosum and, more anteriorly, to the inferior borders of the laminae of the septum pellucidum. Laterally, the body of the fornix overlies the medial part of the upper surface of the thalamus, and the choroidal fissure is placed below its free lateral edge. Through this fissure the choroid plexus in the lateral margin of the tela choroidea passes into the body of the lateral ventricle (7.157). Anteriorly, above the interventricular foramina, the body of the fornix diverges into right and left bundles which curve inferiorly towards the anterior commissure forming the anterior boundary of the foramina. As each bundle approaches the commissure it divides around it into two components—a *precommissural fornix*, and a *postcommissural fornix* (column of the fornix). Each column continues to curve inferoposteriorly and gradually sinks into the corresponding part of the lateral wall of the third ventricle to reach the superior aspect of the mamillary body.

In addition to the foregoing bundles, some fibres leave the fimbria near the splenium to run anteriorly with the supracallosal medial and lateral longitudinal striae, where they are joined by others arising from the indusium griseum. Some fibres of the striae then pass inferiorly between the fibre bundles of the corpus callosum, where they are joined by other fascicles which pass straight into the corpus callosum from the fimbria. These various fascicles, which intersect the callosal bundles, constitute the *dorsal fornix*. Some fibres of the dorsal fornix relay in the scattered patches of grey matter in the septum pellucidum, whilst others course directly on through the septum; both direct and indirect fibres largely rejoin the main body of the fornix.

Thus, following these different bundles of the fornix, *hippocampal efferents* pass to: (1) the gyrus fasciolaris, indusium griseum, cingulate gyrus, and the septum pellucidum via the *dorsal fornix*; (2) to the precommissural septum, the preoptic and anterior hypothalamic nuclei via the *precommissural fornix*; (3) during its descent the *postcommissural column of the fornix* gives direct fibres to the anterior nuclei of the thalamus and to many of the hypothalamic nuclei (p. 968), before proceeding to its main termination in the medial mamillary nucleus; (4) other fibres curve caudally from the column of the fornix to join the stria medullaris thalami to reach the habenular nuclei, whilst yet others continue into the reticular formation of the midbrain tegmentum.

In summary, illustrations 7.123A and B show the main inflow and outflow pathways associated with the limbic system (based upon MacLean 1958, 1969); but these will obviously require repeated revision in the future.

An immense volume of research has been directed towards a better understanding of the biological roles of the various complex and interconnected structures which are often grouped together under the term limbic system, described in the previous pages. Despite this their significances can, at present, only be described in the most general terms. Even a brief review of the experimental results is beyond the scope of the present volume and the interested reader is directed towards the wealth of information and pertinent bibliographies to be found in Eleftheriou (1972); Isaacson (1974); Isaacson and Pribram (1975); Livingston and Hornykiewicz (1978).

THE CEREBRAL CORTEX

Introduction

The cerebral cortex has been studied over several centuries, with increasing momentum. It was examined by the first microscopists, and as early as 1776 the stria in the occipital cortex named after him was noted by Gennari, this being the first recorded structural detail. Only with the advent of improved microscopes in the 1830s and later did effective investigation of cortical organization begin; it has continued uninterruptedly since then, producing an increasing number of papers which has become an almost unmanageable deluge in this century. Consequently, no more than the most outstanding contributions can be mentioned here, and emphasis must be chiefly upon reports of work done in recent decades. However, it is important to note the basic discoveries from which the intricate knowledge of today has expanded, if only to maintain a wider perspective amid the flood of detailed description and speculation which renders current activity so interesting and yet confusing in its constant sallies and revisions. The accent of endeavour has naturally varied with the availability of techniques; but the major aims have been and remain the elucidation of the *modus operandi* in the cerebral cortex and the localization of different forms of activity in it.

Earlier investigations, inevitably microscopic, were stimulated by the techniques of Nissl, Golgi, Cajal, and Weigert (7.124), amongst others. Even prior to this, with less satisfactory methods, Baillarger (1840) and Meynert (1867–8), had ascribed a *laminar pattern* to the cortex, with particular emphasis upon fibre structure—*myeloarchitecture*. Subsequent workers using newer techniques (Lewis 1878; Campbell 1905; Cajal 1909–11; Brodmann 1909; Vogt and Vogt 1926; von Economo and Koskinas 1929; Lorente de Nó 1949) confirmed this, but with notable variations and disagreements, now largely overlooked. A six-layered schema prevailed (Brodmann), although this was based upon Nissl-stained material alone, in which all the highly significant details of dendrites studied by other workers were invisible; moreover, the data were accumulated from a comparative series which, although including simian brains, did not include the human cortex. These studies, in which a six-layered pattern became almost a dogma, led to the recognition of great variations in this theme in different cortical regions or 'areas', no less than fifty-two being distinguished. These results, embodied in the familiar 'Brodmann maps'—still widely current—were transferred somewhat arbitrarily to man by others, and they gained much credence amongst physiologists, perhaps because they provided a useful numerical reference guide to the cortex (7.125A, B) as they still do today.

Techniques such as Nissl's merely reveal the morphological variations and distribution of nerve cell bodies, and this *cytoarchitectural* mode of study requires *myeloarchitectural* methods to display myelinated nerve cell processes. Only with both methods of examination, coupled with somatic and neurite staining by such techniques as the Golgi method—to which must be added the more recent availability of electron microscopy and refined experimental degeneration techniques—could the finer details of cortical organization be discerned. Unfortunately, the Nissl-based cytoarchitectonic mapping of the cerebral cortex initially attracted more attention than other available techniques, and its continued use, together with myelin staining, has led to an almost excessive parcellation of the cortex into different areas, few of which can be accorded clear physiological significance. One useful outcome of these sometimes rather unfruitful studies was the diversion of interest among younger workers to physiological studies. The rapid advance in this field of cortical investigation gradually lead to much discordance between the results of stimulation studies and architectonic mapping (except in broad details), and this prompted several notable critical reviews (Lashley and Clark 1946; Scholl 1956). These have not entirely deterred the 'architectonicians', and it must be stated that in the *broader* aspects of their schemata there is considerable accord between their structurally established areas and the results of physiological experimentation.

The definition of five or six major types of cortical organization (cf. Campbell 1905, and von Economo and Koskinas 1925, 7.126A, B) has proved valid in functional terms, but the excesses of architectonics are reminiscent, except in their admittedly serious structural basis, of the multi-faculty maps of phrenology. The division of the cortex into such a multiplicity of 'organs' seems basically improbable. As the critics of such concepts have stressed, variations in a single area in different individuals of the

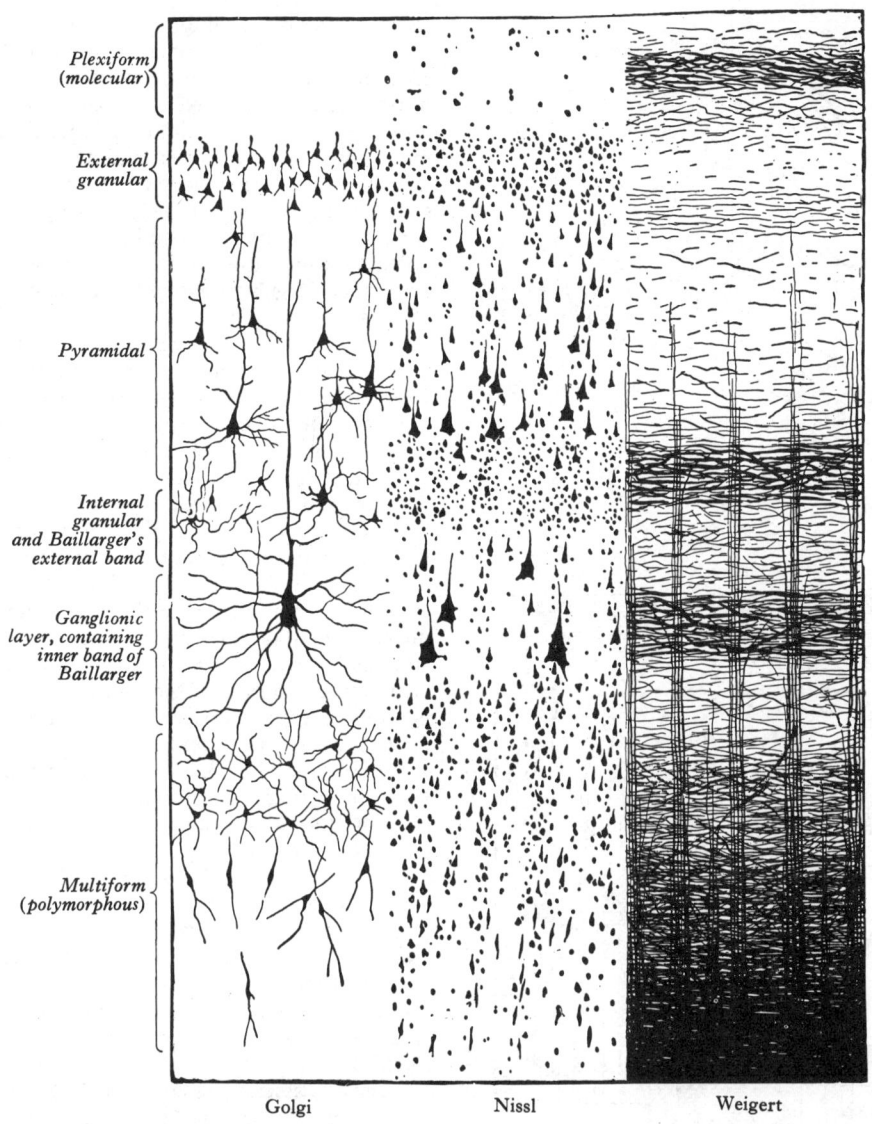

Plexiform
(molecular)

External
granular

Pyramidal

Internal
granular
and Baillarger's
external band

Ganglionic
layer, containing
inner band of
Baillarger

Multiform
(polymorphous)

Golgi   Nissl   Weigert

7.124 Representations of the layers of the human cerebral cortex, as stained by the techniques of Golgi, Nissl and Weigert.

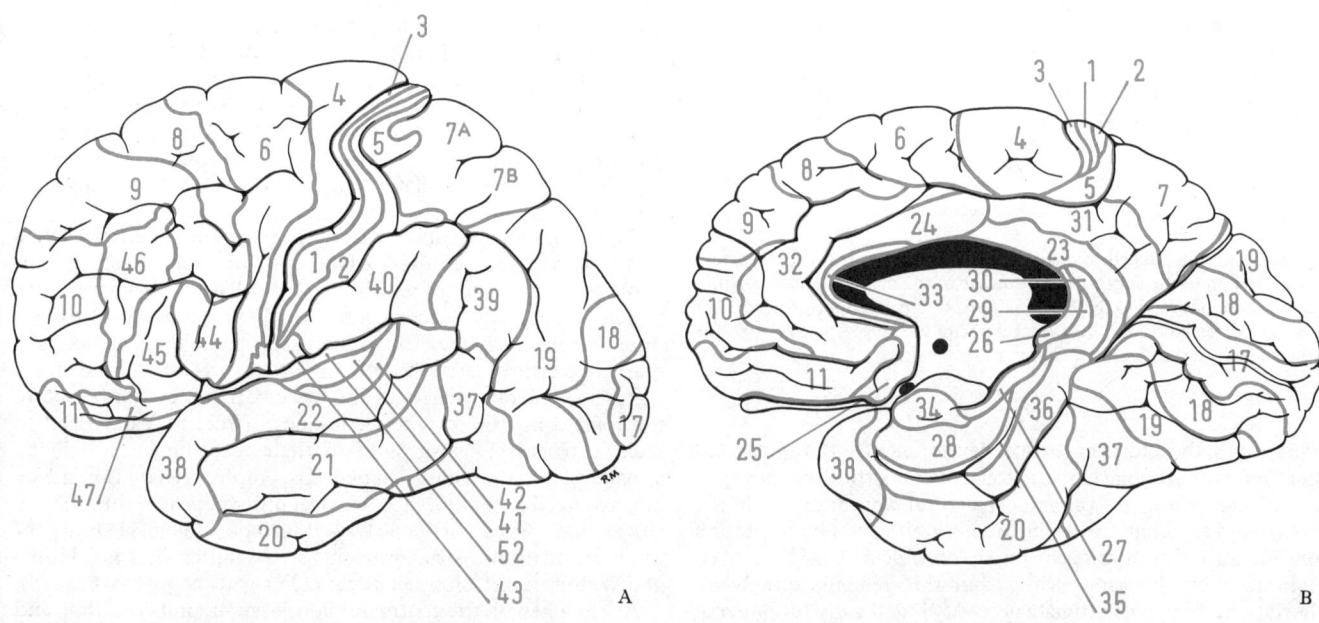

7.125A and B The superolateral (A) and medial (B) surfaces of the human cerebral hemisphere demonstrating the cytoarchitectonic areas identified and designated numerically by K. Brodmann (1909). See text for further details and references. Compare with 7.126.

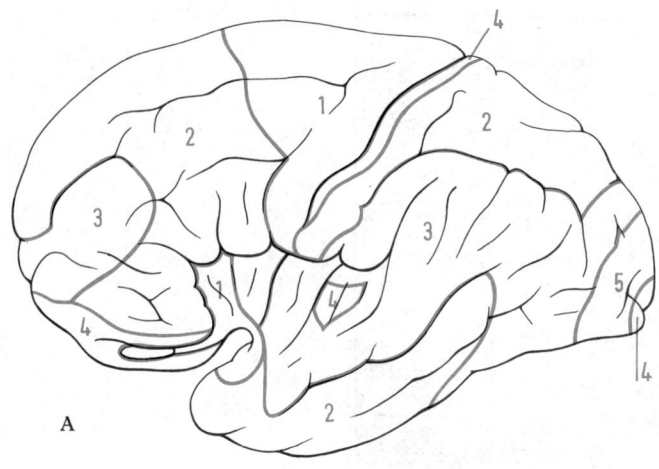

A

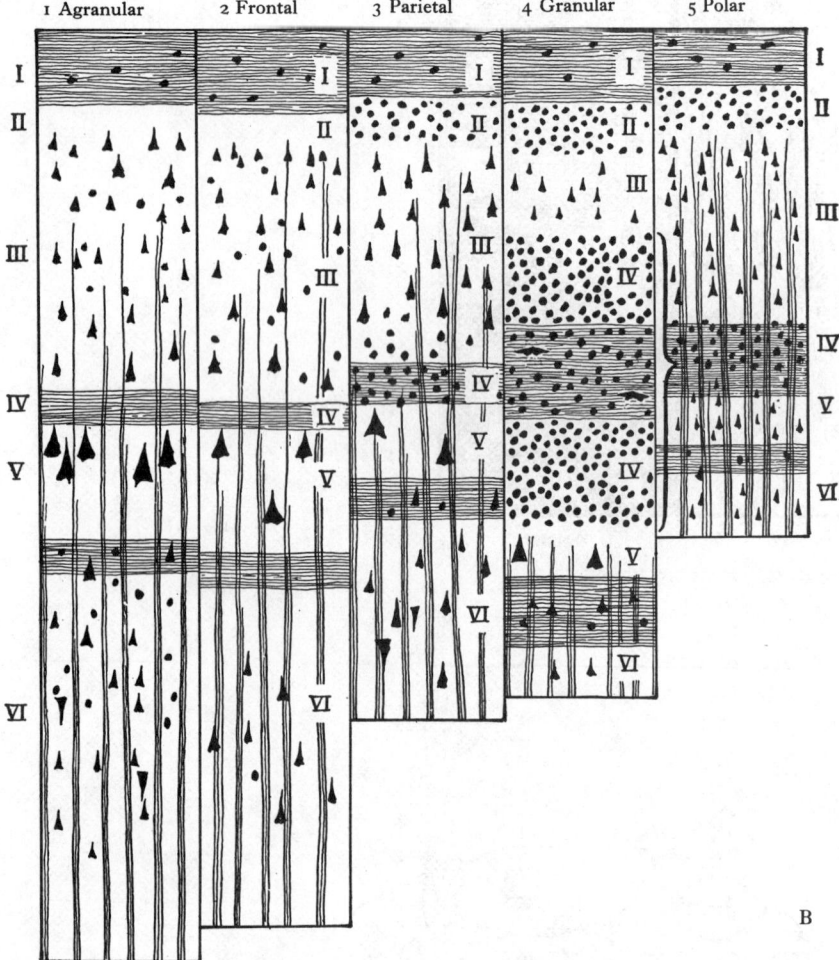

1 Agranular   2 Frontal   3 Parietal   4 Granular   5 Polar

B

7.126A and B. The distribution of the five major types of cerebral cortex, as projected on to the superolateral surface of the hemisphere, according to C. von Economo and G. N. Koskinas (1925). The numbering of cortical areas (A) corresponds to the cytoarchitectonic types (B). See text for further details and references. Compare with **7.125.**

same species, the effects of cortical development and folding, and other factors, particularly the pattern of subcortical connexions, may all contribute to variable structural appearances in the cortex—and to variations which are not uniform. The need for a more quantitative approach to the problem has also been emphasized by the same critics, but this remains largely an exhortation, though its ultimate necessity is generally recognized.

That a laminar appearance characterizes the cerebral cortex cannot be denied, but dogmatic views on the number of layers in

this neuronal continuum, or upon the variations in, and equivalence of, the customarily numbered six layers in different areas, have been shown to be of little or no functional significance, if pushed to excess. Nevertheless, the Brodmann schema of numbered layers is still almost universally employed, perhaps as a mere convenience, and it will perforce be described below. Localization of function to a considerable degree—inasmuch as some areas receive or project large and easily identifiable subcortical connexions—also remains undeniable, and the polemics which this concept occasioned in the late nineteenth century are largely forgotten (Clarke and O'Malley 1968). But the original extents of such 'motor' and 'sensory' areas, as defined (and perhaps artificially confined) by architectonic studies, have been markedly modified and augmented by physiological investigations, and also by the results of degeneration experiments.

Studies of the cortex by Golgi and silver-staining methods, though originally somewhat obscured by work based on Nissl technique, have received a renewed impetus with the recent development of micro-recording from the larger individual nerve cells of the cortex. Prior to this the elaborate structural studies of the neurites of cortical neurons, which have been available for half a century, could not be equated with the results of physiological experimentation; but, as has been mentioned elsewhere (p. 817), during the last decade or so the intimate arrangement of neurons and their individual activities have shown a growing correlation in elaborating concepts of the mode of action of small volumes of nervous tissue, in a variety of sites in the central nervous system, including cerebral cortex.

While classical cyto*architectonic* teaching has become perhaps little more than a useful mode of geographical definition of cortical areas (and a source of some confusion in the 'translation' of areas from species to species of experimental animals), cyto*architectural* details, as revealed by the methods of Golgi and Cajal, are beginning to fit the observed behaviour of individual 'units' in the living cortex. Classical neurite studies of the cerebral cortex—chiefly in lower mammals, such as rodents (*see,* for example Lorente de Nó 1934)—have led to recognition of a large number of nerve cell types. The familiar pyramidal and stellate (granule) cells of Nissl-stained material were chiefly differentiated into a few unsatisfactory categories, subjectively assessed as large, medium, or small, and so on, with few attempts at measurement. The comparatively unilluminating results of this kind of study are well represented in **7.126B**. This should be compared with **7.127B** which illustrates the much more useful information afforded by techniques capable of revealing the interconnexions between cortical neurons. It is upon such criteria that a much larger range of neuronal differentiation can be based; and since it is ultimately interneuronal connexions which must be correlated with their activities in life, this approach to cortical structure is inevitably more appropriate than cytoarchitectonic study of Nissl-stained material. Broadly the results of the two methods do not conflict, but one is overwhelmingly more informative. Even though light microscopy proved unequal to the task of full visualization of synapses in the cerebral cortex, the patterns of interconnexion of cortical neurons were in major details well clarified many years before electron microscopy became available.

The termination of afferent nerve fibres in the superficial layers of the cortex, nearer the surface than the somata of efferent neurons the existence of recurrent axon collaterals from the latter which turn back superficially to form contacts through interneurons with their own dendrites, and the existence of a marked 'vertical' organization in the cortex—these and other features were all available as the basic cortical 'circuitry' before the refined microelectrode techniques of recent years began to seek a structural background to their recordings. As will be apparent later, the 'chains' of neurons, emphasized by Lorente de Nó, arranged in innumerable cortical units, repeated through the cortex, have fitted particularly well into physiological studies of the postcentral and striate regions (*see* Mountcastle 1957; Hubel and Weisel 1962; Colonnier 1974; Zeki 1974; and pp. 1016, 1018).

As elsewhere in the central nervous system, unit recording and more precise physiological study of the transmission processes of excitatory and inhibitory synapses have generated a renewed demand for the finest details of connexions between neurons. Various forms of synapse and their ultrastructural features are given full description elsewhere (p. 826); though identification of these structures in the cerebral cortex was achieved comparatively late (Gray 1959), a large literature concerning cortical synaptology has now accumulated. Axodendritic and axosomatic synapses predominate in the cerebral cortex, and for a period were thought to be the only types present, but dendrodendritic synapses have now been identified in some locations. The profusion of synapses, especially in relation to the dendritic spines of pyramidal cells, obviously provides a structural *mélange* for most complex interactions, however far we may yet be from anything more than the simplest interpretations of the full significance of such arrangements. Nevertheless, structure and function are rapidly progressing together in cortical study; and even if the joint 'models' of cortical activity which are currently evolving evince a somewhat hectic tendency to change, almost from year to year, this is surely a healthier state than the stagnating dogma which has sometimes becalmed neurological research.

In the relatively simple account of the cerebral cortex and its

QUANTITATIVE ASPECTS OF CORTICAL STRUCTURE

Quantitative studies of the cerebral cortex have been comparatively few; the first serious attempt to ascribe numerical values to its features was made by von Economo and Koskinas (1925), who recorded data on variation in cortical depth which remain the most detailed for man. They also computed the total surface areas as 220,000 mm^2, a more recent figure being 285,000 mm^2 (Scholl 1956), with a volume of 300 cm^3. Naturally the total number of cortical nerve cells has attracted much interest and computation, and figures of 14,000 million (von Economo and Koskinas 1925), 6,900 million (Shariff 1953), 5,000 million (Scholl 1956), and 2,600 million (Pakkenberg 1966) are representative, and illustrate a downward trend which may be due to improving technique. Stereological techniques have been applied to the cortex; for example, Foh *et al.* (1973) have estimated the total lengths of neuronal and glial processes in unit volumes of visual cortex, in cats, arriving at the figure of 5000 m/mm^3 for neurites. Such numbers are beyond real comprehension, and it is easier to grasp the magnitude of such a cell population by its proportions in a

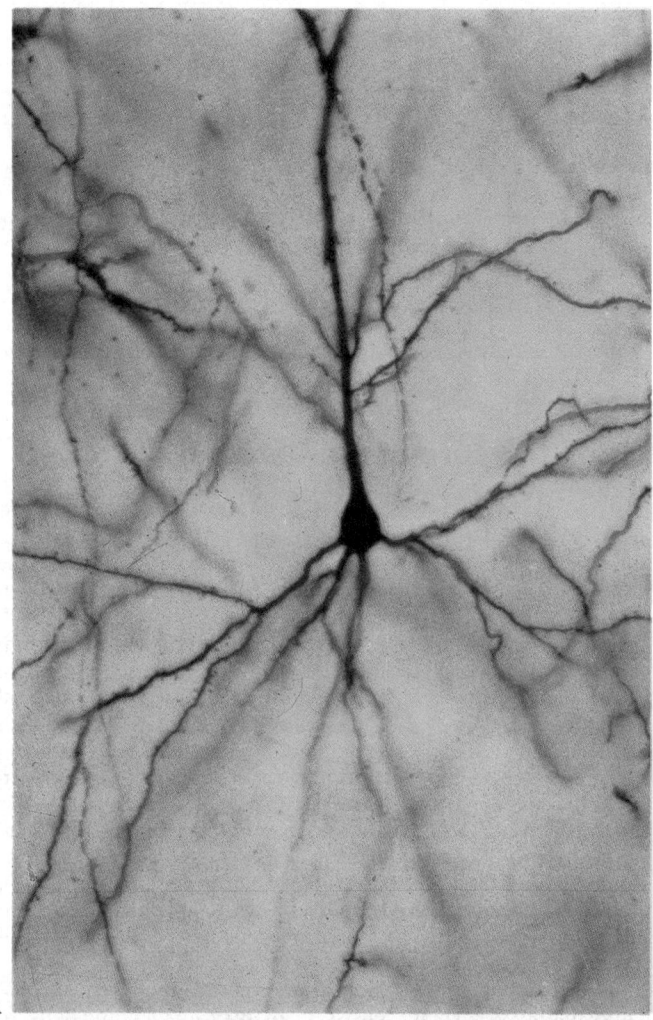

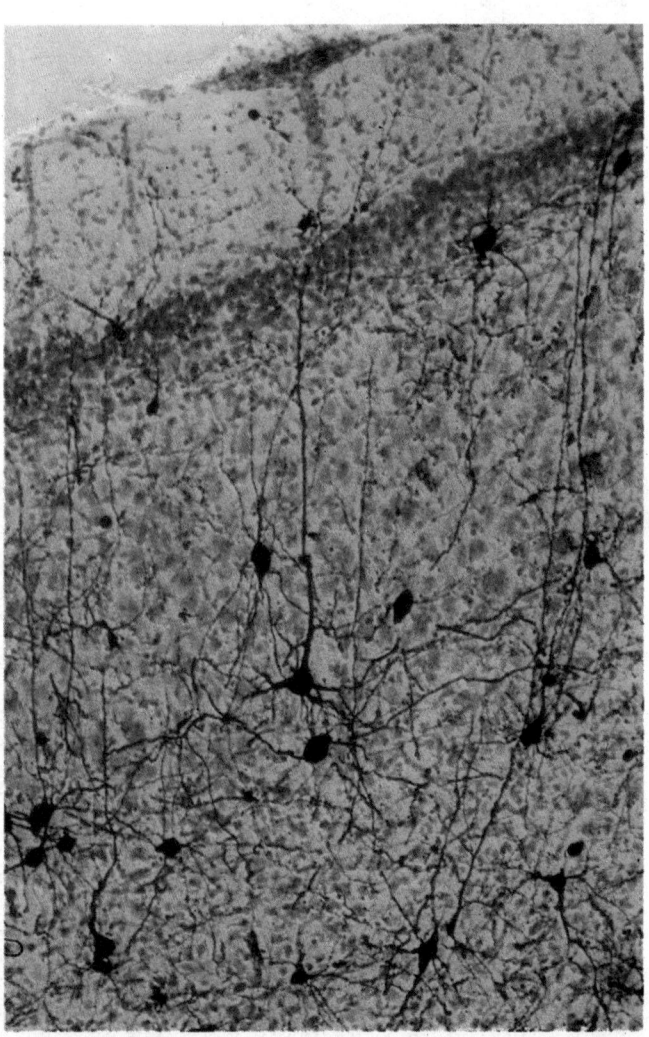

7.127A and B Preparations contrasting the Golgi and Nissl methods of staining nerve cells in the cerebral cortex. In A a single pyramidal cell stands out amongst many unstained elements. In B isolated Golgi-stained neurons are prominent amongst the remaining Nissl-stained cortical elements. (Preparations provided by Dr. A. R. Lieberman, University College, London.)

specific areas which follows, space will not permit an adequate reflexion of the intense activity in current research; for this the reader must consult recent original papers and surveys, e.g. Colonnier (1966, 1968, 1974); Shepherd (1974); Rakic (1975); Szentágothai (1975); Schmitt *et al.* (1976). It may be noted here that vascular patterns (Pfeifer 1940), and distribution of enzymes (e.g. Pope 1967) have also been used as criteria of differentiation of 'areas' in the cerebral cortex.

smaller sample. For example, a column of cells 1 mm square and 2·5 mm deep may contain as many as 60,000 neurons; and in one study of the motor cortex (precentral gyrus) each neuron was considered to take part in about as many synapses and to connect with some 600 other nerve cells (Cragg 1967). Such a small volume of cortex, in itself containing perhaps 3 × 10^9 synapses, might be multiplied a quarter of a million times to represent the whole cortex. The wealth of interconnexions of such huge

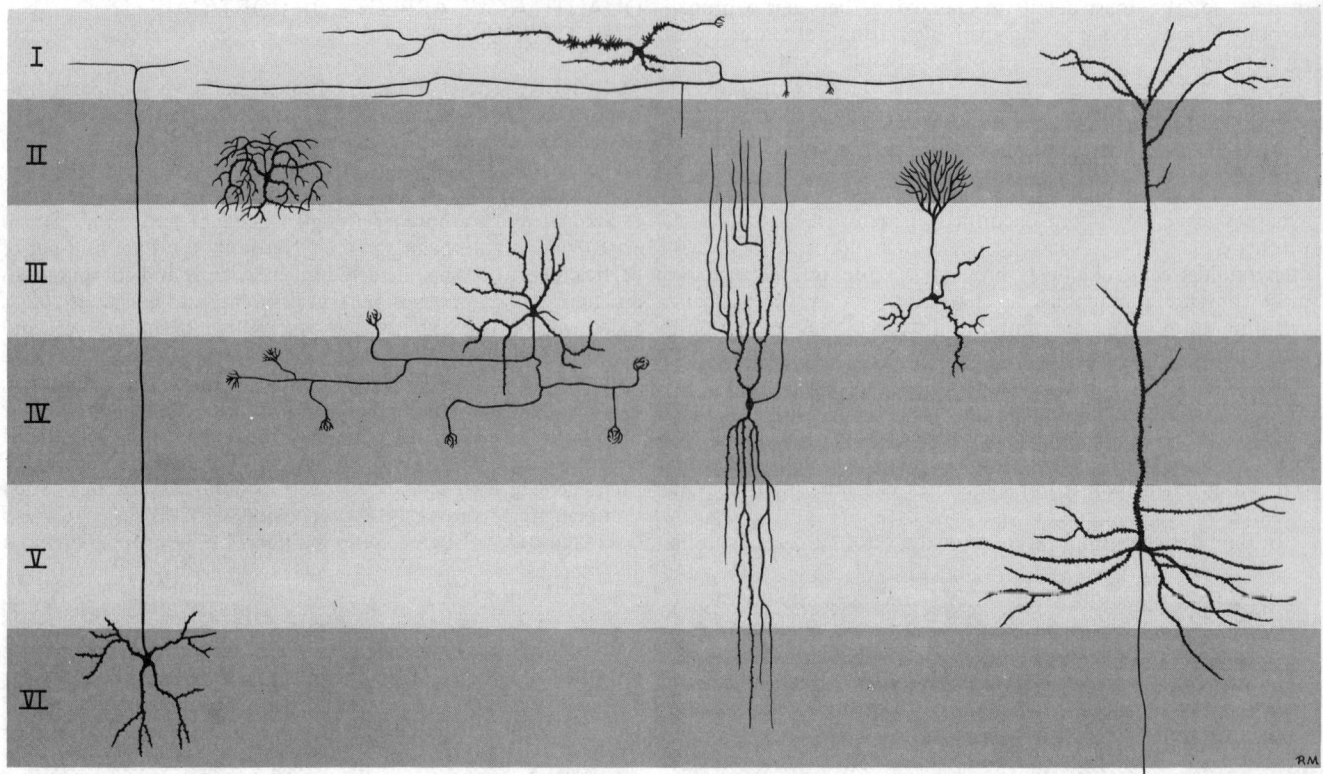

7.128A   Typical outlines of characteristic neocortical neurons as seen in sections prepared by the metallic impregnation techniques introduced by Golgi and Cajal. From left to right are shown Martinotti, neurogliaform, basket, horizontal, fusiform, stellate, and pyramidal types of neuron. Many other forms and variants have been described. See text for literature.

numbers of cells is clearly very great. In the striate area (Scholl 1955) where about one-tenth of the cortical neurons are said to be concentrated, the dendrites of a single neuron may connect with 2,000–4,000 other cells, and an incoming afferent projection fibre may ramify through a volume of cortex containing 5,000 cells. Even the figures given above are, however, dwarfed by the cerebellar neuronal population—*see* p. 924.

The density of packing of nerve cells in different areas and their laminae shows much variation, being most dense in the striate area and perhaps least so in the precentral gyrus. The ratio by volume of the somata of neurons to all other constituents in the cortex (the grey/cell coefficient) has been estimated, and average ratios of 27:1 (von Economo and Koskinas 1925) and 70:1 (Haug 1956) for man have been cited. It is claimed that the ratio shows a phylogenetic increase up to man, due to an increase in the amount of neuroglia.

Various other quantified data regarding the mammalian cortex are scattered through the literature, which should be consulted for further details. In general—apart from an impression of the enormous potentialities for interconnexion and interaction—such figures have no immediate usefulness. The knowledge that each afferent or efferent projection fibre may have upwards of 1,000 neurons associated with it, more or less remotely, merely emphasizes the general truism of an extremely intricate field of interaction between them. The time may be far off when happenings in such large numbers of neurons can be defined in precise spatial and temporal terms, if indeed this can ever be expected in other than generalized statistical approximations. For the present, the limited events in minuscule volumes of cortex offer a more promising and primarily essential field of investigation. Examples of such studies will be mentioned in connexion with the particular cortical areas in which they have been most successful. Arising from such research, a profusion of concepts of neuronal 'circuitry', diagrams to illustrate these, and even models employing artificial 'neurons' have appeared in recent literature, e.g. Walter (1953); Ashby (1960); Weiner and Schadé (1963); Taylor (1964); Young (1964); Arbib (1964); Minsky (1965); Shepherd (1974, 1978); Colonnier (1974); Rakic (1975); Szentágothai (1975). Naturally, these speculations undergo rapid change and modification as knowledge in the very active fields of synaptology and unit recording advances; and hence, despite their great interest, it is as yet premature to include such details here.

The Structure of the Cerebral Cortex

To the unaided eye the cerebral cortex forms a complete mantle or *pallium* covering the hemisphere and obviously variable in thickness (1·5 to 4·5 mm) when seen in section. It is thicker on the exposed convexities of gyri than in the depths of sulci, in which the larger part of the cortex is hidden from surface view. Such variations in thickness might well correspond to structural variations in the pallium; and it has in fact been suggested that the positioning of gyri and sulci is conditioned by such structural differences (Le Gros Clark 1945), but this cannot be claimed with respect to the *functionally* differentiated areas, which in many instances depart in their outlines from the sulcal pattern. In freshly cut cerebral cortex laminar details can often be appreciated even without a simple magnifying lens (e.g. the visual stria of Gennari); and by such means horizontally disposed layers of nerve fibres, the inner and outer bands of Baillarger (p. 1008), can usually be discerned. It has even been claimed that using such simple methods more than a score of structurally distinct areas of cortex can be identified (Smith 1907).

In its **microscopic structure** the cortex of the cerebrum, like 'grey matter' elsewhere, consists of an intricate blending of nerve cells and fibres, neuroglia and blood vessels. Neuroglial cells and the vascular arrangements have been dealt with in other sections (pp. 834 and 837). The features of the neurons and their interconnexions and distributions must now be considered. It should, however, be noted here that variation in the distribution of the blood vessels has also been utilized as a criterion in differentiating cortical areas (Pfeifer 1940).

The neurons of the pallium have been described and categorized in great detail, but they can be assigned to a relatively small number of classes, the great majority in fact falling into two such groups, the *pyramidal* cells and the *stellate* (granule) cells. Both types may be assorted into a variable number of subdivisions on the basis of size and the appearances of their neurites (7.128).

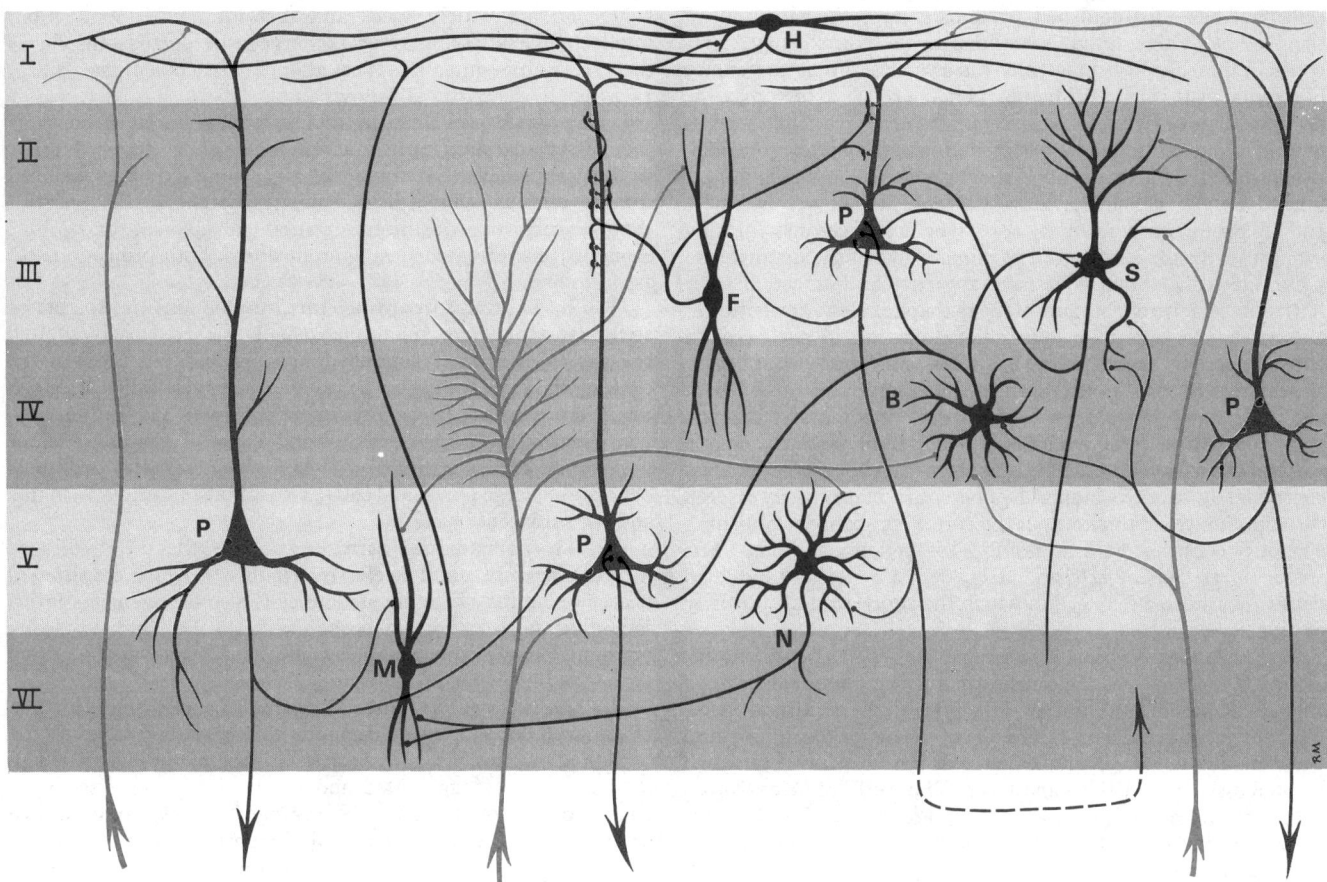

7.128B A diagrammatic representation of the most frequent types of neocortical neuron, showing typical connexions with each other and with afferent fibres (*blue*). Neurons limited to the cortex in their distribution are indicated in *black*. Efferent neurons are in *magenta*. The right and left afferent fibres are association or cortico-cortical connexions, the central afferent is a specific sensory fibre. Neurons are shown in their characteristic lamina, but many types have somata in more than one layer. They are indicated thus: P=pyramidal, M=Martinotti, F=fusiform, H=horizontal, N=neurogliaform, B=basket, S=stellate. See text for details and compare with the stereodiagram in 7.132.

Both types are found at most levels in the cortex and in almost all areas, though their numbers and distribution vary greatly from place to place. (These variations, together with alterations in size, provide the criteria for distinguishing different cortical areas in Nissl-stained material, of course.) Other types of cells commonly distinguished are *fusiform, horizontal,* and *neurogliaform cells,* and the *cells of Martinotti.*

Pyramidal nerve cells, so named from the shape of their somata (**7.15**), vary from small elements measuring about 10 μm across to the giant pyramidal cells (of Betz) which reach 70 μm and more. Their apices are usually orientated towards the surface of the cortex, and·from this region of the cell a thick *apical dendrite* ascends a variable distance, giving off collateral branches and often ending in a complex spray of terminal dendritic twigs. From the other, basal angles of the cell body *basal dendrites* sprout, spreading laterally into the surrounding neuropil to form dendritic fields of varying shapes and extents (Colonnier 1964). A statistical analysis of the branching of such dendrites and those of the stellate cells in the visual cortex has been attempted (Scholl 1953), but this quantitative kind of dendrological study is still a neglected and indeed difficult field. On the whole, the vertical extent of the dendritic extensions of pyramidal cells is markedly greater than it is in the horizontal or tangential direction. The dendrites are studded, especially in their smaller branches, with very large numbers of *dentritic spines* (p. 822), which are now known to be the sites of axodendritic synapses (Gray 1959). Since the number of these synapses, even by the crudest estimates, is very large, it is at once evident that there is a physical basis for the most elaborate interneuronal reactions in the cortex. The axons of pyramidal cells, which are invariably much smaller in calibre than the trunks of their main dendrites, behave in a variety of manners. From larger elements, chiefly in lamina V (*vide infra*), *projection axons* extend centripetally out of the cortex to reach more or less distant subcortical structures, such as the basal nuclei, brainstem nuclei, and the grey matter of the spinal cord. Some may pass back into more or less distant parts of the cortex as *long* or *short association fibres*. The axons of smaller pyramidal cells usually ramify entirely within the cortex, and even those which leave it commonly divide by giving off a small number of collaterals which remain as intrinsic or *intracortical axons*. These latter may extend horizontally, but they more often pursue an obliquely recurrent course towards the more superficial laminae of the cortex. Even from this brief description it is clear that the larger pyramidal cells, whose somata appear in Nissl-stained sections to be sited in a single lamina, in fact extend through most and often all levels of the cortex, since their apical dendrites frequently reach the superficial, *plexiform lamina* (*vide infra*). Thus, each pyramidal neuron forms a species of *columnar unit* extending through the cortex together with its numerous connexions, including other pyramidal elements, many forms of interneuron, and afferent projection fibres. Lateral interconnexions are mediated by various forms of intracortical horizontal neurons and by association neurons.

Stellate nerve cells, often called *granule cells* because of their small size and appearance in Nissl-stained material, appear in variable density of distribution in all the cortical laminae except the most superficial (lamina I); but they are usually concentrated in greater abundance in laminae II and IV (*vide infra*). Like pyramidal cells, stellate neurons would probably have been designated as multipolar if they had not first been described in Nissl-stained sections. They are small, of the order of 6 to 10 μm in diameter, with a rounded soma drawn out at numerous angles by their richly branching dendrites and a single, relatively short axon. They are hence members of the Golgi type II series of neurons. Their dendrites carry numbers of spines, indicating abundant synapses with other neurons. A wide variety of stellate cells have been distinguished, principally upon the behaviour of their axons. These, though confined to the cortex, may travel considerable distances in it, chiefly in a vertical direction but also horizontally in some instances. The vertical axons may be centripetal or centrifugal in direction, the latter reaching as far as lamina I. One particular type of stellate cell, the *basket cell*, which is horizontally extended, has a short vertical axon which almost immediately divides into a horizontal family of collaterals. These end in pronounced terminal sprays or arborizations, forming synapses with the somata and proximal parts of the dendrites of pyramidal cells. Such cells have a particular interest in their horizontal extension, because of the perhaps excessive attention at present concentrated on vertical organization in the cerebral cortex (p. 1018). Another type of stellate cell is in fact *fusiform* in appearance, due to the emergence of two large dendrites which sprout from opposite poles of the soma, dividing at once into elaborate bouquets of branches which extend vertically in the cortex. The axon, sometimes a branch of a stem dendrite, extends centrifugally towards lamina I. The entire neuron may stretch through the whole thickness of the cortex, probably establishing synaptic contacts with a number of pyramidal cells. The *neurogliaform stellate cells* are small, with a dense and localized dendritic arborization, within which the short axon also usually ramifies.

The horizontal cells (of Cajal) are confined to the plexiform lamina I; they are small and fusiform and their dendrites spread short distances in two opposite directions in the plexiform layer. Their axons, often derived from one of the dendrites, divide into two branches which depart from each other to travel to much greater distances in the same layer. **The cells of Martinotti** occur at most levels in the cortex. They are small and multipolar, with a localized dendritic field and a long axon which runs centrifugally to the plexiform lamina, producing a few short horizontal collaterals *en route*. The so-called **pleomorphic cells** are considered to be modified pyramidal cells with axons entering the white matter. Their somata are variously shaped, perhaps in accord with differences in their dendritic sproutings; the dendrites spread widely into the cortex.

Many other forms of nerve cell have been detailed in the cerebral cortex, but since even the array of types mentioned above cannot yet be linked into a fully coordinated scheme of interaction, it is merely confusing to multiply the details. It should never be overlooked that the most intensive studies of cortical neurons have been carried out on subprimate mammals; and although appearances in the primate cortex, as chiefly pursued in material from macaque monkeys, are similar, investigations of even 'cold' structural arrangements in the human cortex have been remarkably few (Economo and Koskinas 1929; Bailey and von Bonin 1951; Conel 1939–1959).

For an account of ultrastructural details of human cerebral cortex consult Braak (1975), Cragg (1976), and Rees (1976).

Laminar Pattern in the Cerebral Cortex

The cortex or pallium of the cerebrum may be conveniently divided into an older and original part, the *allocortex*, consisting of the archicortex and paleocortex (also known as archipallium and paleopallium), which are considered elsewhere (p. 997), and a newer development, the *neocortex* (isocortex or neopallium). The latter may be equated with those systems of sensory and motor activity which originally had little or no connexions with the forebrain, but have acquired these over evolutionary time by the process of prosencephalization. The remarks which follow apply only to the neocortex.

As already stated in the Introduction to this section (p. 1002), the customary description of a six-layered cortical structure successfully promulgated by Brodmann and his followers has, by its general acceptance, obscured earlier disagreements and widespread dissatisfaction with its arbitrary nature. The same strictures must be applied to excessively detailed architectonic mapping of the cortex which has stemmed from the same somewhat dogmatic views. However, Brodmann's numbered layers and areas provide a reference grid of considerable practical

value and are widely used; and therefore, until some more intellectually satisfying system emerges, the details—insofar as they are useful—must be repeated here. The six laminae (7.124, 126) may be described as follows:

I. **The plexiform lamina** (molecular or zonal layer) contains the sparsely scattered horizontal cells (of Cajal), and consists apart from this of a dense net of tangentially orientated fibres, derived from pyramidal cells (apical dendrites), stellate cells (vertical axons), cells of Martinotti (centrifugal axons), and other elements, including cortical afferent fibres, both projection and associational.

II. **The external granular lamina** contains the somata of stellate and small pyramidal cells, usually packed densely, though this varies, like other laminar details, in different areas of the cortex. Passing through the lamina are vertically arranged dendrites and axons from subjacent layers, intermingling with the dense neuropil of local dendrites and axons. Ascending afferent fibres make extensive multiple synaptic contacts with the apical dendrites of large pyramidal cells (with somata in lamina V) in this and the subjacent layer.

III. **The pyramidal lamina** contains the cell bodies of medium-sized pyramidal cells, the smaller of which are situated nearer to lamina II. Some stellate cells also occur in this layer, including horizontally disposed basket cells, and vertically orientated fusiform cells, their dendrites and axons extending far beyond the layer itself.

IV. **The internal granular lamina** is usually narrower than other layers, except lamina I, and is chiefly characterized by the somata of stellate cells, with occasional small pyramidal cells. The cells are densely aggregated and the lamina is traversed by a concentration of horizontally arranged fibres, long known collectively as the *external band of Baillarger*. As in the case of other laminae, the layer also contains large numbers of vertically orientated neurites derived from nerve cells in other parts of the cortex, in subcortical regions, and in adjoining layers.

V. **The ganglionic lamina** contains the largest pyramidal cell somata, but smaller elements of the same type also occur, the actual dimensions of these cells varying in different cortical areas. In any particular area, however, the largest pyramidal cells are in lamina V. Small numbers of stellate cells may also occur. The layer is, of course, permeated by a dense neuropil of dendrites and axons derived both from its intrinsic elements and cells in other laminae. It is also traversed by ascending and descending projection fibres and by association fibres. A considerable complement of horizontally deployed fibres is apparent in lamina V, corresponding, in sections stained to display myeloarchitectural details, to the *internal band of Baillarger*.

VI. **The multiform lamina** contains a considerable range of cell types, as judged by their somata and processes, the variable shape of the former reflecting to some extent the variations in their dendritic arrays. Most of the cells are small and are considered to be modified pyramidal elements, despite the fusiform, triangular, ovoid and other profiles of their somata. The small, multipolar Martinotti cells are often prominent in this lamina. Lamina VI is not always well demarcated from the subjacent cortical zone of fibres approaching or departing from the cortex itself.

The numbering and nomenclature of cortical laminae set out briefly above is in the style of Brodmann. Many synonyms for their names are in circulation, derived from the work of such investigators as Campbell and Cajal; the Vogts also introduced another, somewhat more awkward nomenclature, based on myeloarchitectonic studies, in which the fibre structure is emphasized. Views as to the number of layers have varied widely, and much subdivision has been suggested. For example, lamina VI has been divided into VIa and VIb on the distribution of triangular and fusiform cells, and into no less than four sublaminae (VIa^1, VIa^2, VIb^1 and VIb^2) in material stained to show fibre structure. All layers except lamina II have been further analysed in this manner, as many as sixteen laminae in total being recognized by the Vogts. The usefulness of such minute dissection has been severely criticized, and these details are only mentioned here as examples. Similar remarks are applicable to much of the work in the field of cortical architectonics. Here again the Brodmann maps have achieved most attention. Their early transference (in a form elaborated further by the Vogts but still not worked out in man) to the human cortex by Foerster has overshadowed the contribution of von Economo and Koskinas. In one respect this is fortunate, because the latter chose to designate areas by letters, as has Conel (1939), and these, and other nomenclatures, would have proved much less manageable than Brodmann's simple numbers. However, in recent years, new views on the extent of the major sensory and motor areas of the cortex have generated the need for a more appropriate terminology than Brodmann's (see Woolsey 1964). The new series of symbols suggested will be referred to in connexion with cortical areas involved; many of them are already in wide use in experimental studies, but since a lack of uniformity persists in this, and since such terms are only beginning to penetrate into clinical neurology, the customary numerical designation of cortical areas must be retained for the present.

REPRESENTATIVE VARIANTS OF CORTICAL STRUCTURE

While it is obviously impossible, and probably unprofitable, to detail and discusss here the almost endless nuances of cortical structure in full-blown architectonics, it is necessary to allude to a smaller number of basic variant types of cortex. These were recognized in the classical pioneer studies of Campbell (1905) and consolidated by Economo and Koskinas (1929). *Five* fundamental types are described (7.126) in the neocortex, and while all are considered to develop from the same six-layered or *sesquilaminar pattern*, two of them are regarded as lacking certain laminae when fully differentiated, and are hence described as *heterotypical*; these are the *granular* and *agranular* types. The *homotypical* variants, in which all six laminae are discernible, are called *frontal* (premotor), *parietal* (postcentral), and *polar* (visuopsychic)—names which link them with specific regions in a somewhat misleading manner, as illustration 7.126 shows. For example, the frontal type occurs in the parietal and temporal lobes.

The agranular type of cortex is considered to be lacking in the granular laminae (II and IV), but it does usually display scattered stellate nerve cells. The predominant neuronal type is, however, pyramidal, and it is in this form of cortex that the greatest densities and largest sizes of pyramidal cells occur. Originally identified in the precentral gyrus (area 4), it also occupies areas 6, 8 and 44 (7.125) and occurs in other regions, including parts of the limbic system (p. 990). It is characterized by the projection of large concentrations of efferent fibres from the pyramidal cells, and agranular cortex can thus be equated with the 'motor' areas—with the proviso that such areas are now known to receive afferent projections in addition, as will be detailed in connection with the individual areas.

The granular type of cortex (koniocortex) may be regarded as at the opposite extreme of the main categories of cortical structure. The granular layers are maximally developed in such areas and contain densely packed stellate cells, amongst which are nevertheless a variable but small number of pyramidal cells. Laminae III and IV are poorly developed or unidentifiable. This type of cortex is associated with afferent projections but, here again, there is evidence of a lesser number of efferent fibres,

derived from the few pyramidal cells usually to be found in this otherwise 'granular' cortex. Despite the relative lack of different laminae, the granular and agranular types of cortex exhibit little qualitative distinction, being rather at the opposite extremes of a gradation, in which the pyramidal and stellate series of cell types are reciprocally developed. Typical granular cortex is formed in the postcentral gyrus, striate area, and in the superior temporal gyrus (acoustic area); it also occurs in small parts of the hippocampal gyrus (p. 992). Despite the very large number of stellate cells packed into this form of cortex, especially in the striate area, it is almost the thinnest of the five main types. In the striate cortex the external band of Baillarger (lamina IV) is particularly well defined, as the *stria of Gennari* (or Vicq d'Azyr).

The remaining three types of cortex may be regarded as intermediate forms. In the **frontal type** large numbers of small and medium-sized pyramidal cells occur in laminae III and V, the granular layers (II and IV) being less prominent. The relative numbers of these two major forms of nerve cell vary reciprocally in the different areas in which this form of cortex exists. It is not confined to the frontal region of the cerebrum (7.126). The **parietal type** of cortex contains less pyramidal cells, which are mostly smaller in size than in the frontal type; the granular laminae are, on the contrary, wider and contain more abundant stellate cells. This kind of cortex occupies large areas in the parietal and temporal lobes (7.126). The **polar type** is classically identified with small areas near the frontal and occipital poles of the hemisphere, and hence its name. It is the thinnest form of the five types (7.126). All six laminae are represented in it, but the pyramidal layer (III) is reduced in width, and is not so extensively invaded by stellate cells as in the granular type of cortex. As in the latter, the multiform layer (VI) is more highly organized than in other forms of cortical structure.

While further subdivision of the above five basic types of cortical 'organization' may be useful for specific experimental purposes, it must be emphasized again that in the microscopic sections in which they are customarily distinguished—whether stained to show cell somata or their processes—the finer and more significant details of true organization are not apparent in studies of the whole thickness of the cortex. The functional organization is naturally linked to the spatial distribution of the cells, but it is in their actual patterns of connexion that any real enlightenment as to cortical mechanisms must be sought. Golgi preparations in particular have yielded an immense amount of information indicating the probable designs of neuronal interaction. Functional hypotheses deduced from such merely structural data, however intricate, require confirmation in terms of the precise nature of synapses both in their distribution and mode of action. Details of this kind depend upon electron microscopy and unit recording. Unfortunately, such techniques deal only with much smaller volumes of cortical tissue, but most exciting results are apparent in the research in this field in the last two decades or so, as will be seen in various sections of this account of the central nervous system. The cytoarchitectural approach to the problems of cortical activity was a necessary prelude, and the resultant definition of various forms of organization in structural terms remains as a necessary scheme of orientation to which the finer ultrastructural details must be constantly referred.

THE MAIN CORTICAL AREAS

Before describing the major areas customarily distinguished, on functional and structural data, in the human cerebral cortex— such as the somatomotor, somatosensory, visual and auditory areas—some preliminary general remarks are necessary. As has already been emphasized, the somewhat extreme parcellation of the cortex deriving from the studies of Brodmann has a limited usefulness and is frankly misleading if regarded as more than a reference grid in defining parts of the cortex. Even the simpler differentiation of the cortex into *sensory* areas receiving afferent projection fibres and *motor* areas projecting efferents—the remainder being regarded as 'silent' or *associational*—can no

longer be considered appropriate, being itself an inaccurate over-simplification. Evidence has accumulated during the last three decades to show that the areas receiving or originating projection fibres are much more extensive than the initial classical studies indicated. Furthermore, the division into 'receiving' and 'originating' projection areas is by no means so distinct as at first appeared. Thus, the postcentral gyrus is not the only area to which a somatosensory thalamic projection is directed; at least two other areas of cortex are similarly involved, as will be detailed later in this section. In the same way, the precentral gyrus is supplemented by a second 'motor' area. It is necessary to qualify

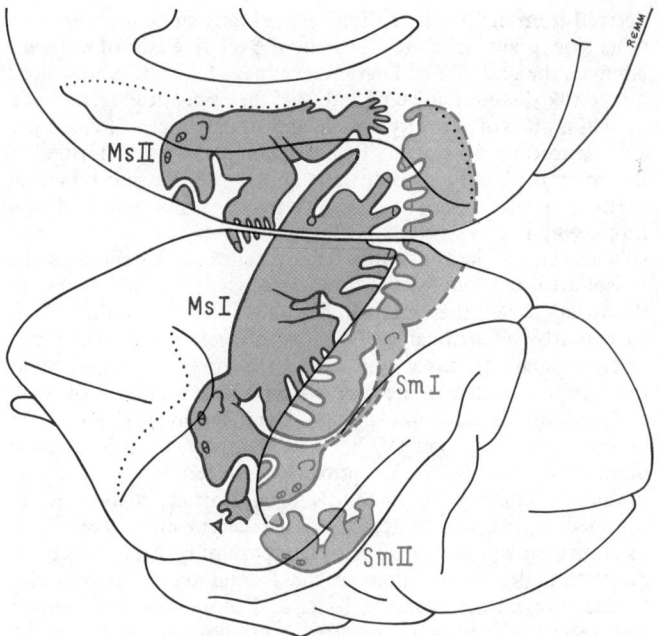

7.129 The main sensorimotor areas projected diagrammatically upon the superolateral surface of the simian cerebral hemisphere. Note the somatotopic arrangement in all four areas. (Adapted from C. N. Woolsey, 1964—see text for details and references.)

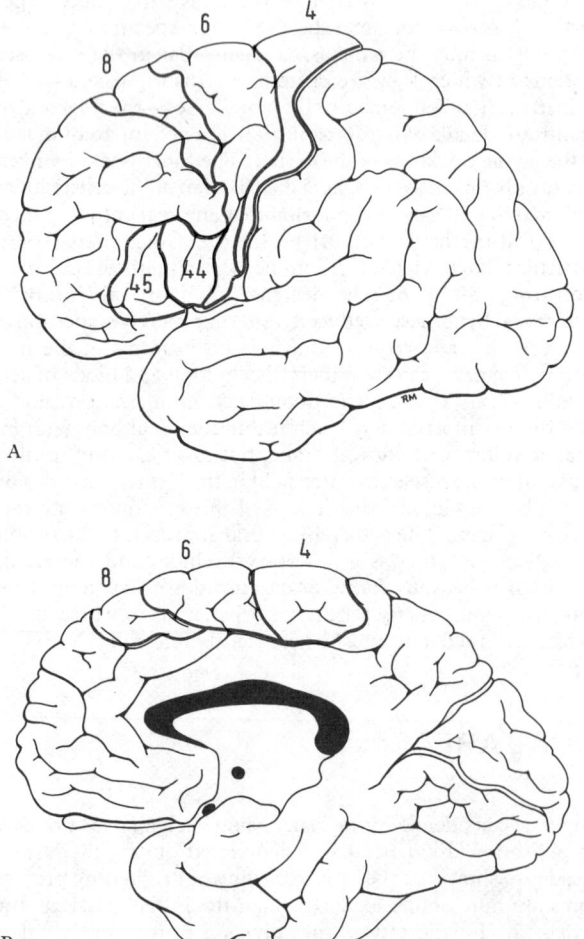

7.130A and B Superolateral (A) and medial (B) surfaces of the cerebral hemispheres, showing approximate correspondence of the Brodmann areas to the main motor area (4) or MsI, the premotor area (6, 8) and motor speech area of Broca (44, 45). See text for details and compare with 7.125, 134, 135A and B.

the term 'motor' because the distinction of motor and sensory areas still customary in simpler accounts of the cerebral cortex is erroneous, and has indeed been known to be so for many years. As long ago as 1933 motor responses to stimulation of the 'sensory' areas were demonstrated (Dusser de Barenne 1933), and projection of efferent ('pyramidal') fibres from the same postcentral area was established shortly afterwards (Levin and Bradford 1938). Since these pioneer studies a mass of confirmatory evidence in various experimental animals has extended these findings to other areas. Moreover, clinical and experimental observations in mankind suggest that similar arrangements obtain. It is hence more appropriate to speak of the pre- and post-central areas as being *sensorimotor*; and since a mixture of afferent and efferent connexions has been shown to exist also in respect to the projection fibres of the acoustic and visual 'sensory' areas, they also are more accurately described as sensorimotor in character.

The recognition that the corticospinal or 'pyramidal' pathway is derived from nerve cells in a much larger area than the precentral gyrus is paralleled by similar findings in regard to the various thalamic and geniculate projections upon the cortex; these terminate in considerably wider regions of the cortex than originally described. For example, the classical somatosensory area in the postcentral gyrus is supplemented by a second area, inferior to it, and by a third, on the medial aspect of the hemisphere, which is also a motor area (*vide infra*). Similarly, the lateral and medial geniculate bodies are now known to project to other regions of the cortex beyond the visual and auditory areas of conventional description. Not only the striate cortex (area 17, visuosensory area), but also the para- and peri-striate areas around it (areas 18 and 19—the 'visuopsychic' cortex) receive projection fibres; and in cats the acoustic radiation has been shown to terminate not only in the *first acoustic area* (41), but also in several other regions in the temporal cortex.

These modifications of the originally simple motor and sensory areas require a revision of terminology. One commonly used (Woolsey 1964), although not entirely satisfactory or as yet accurately adapted to the human cortex, divides the main sensorimotor area into a part in the precentral gyrus termed MsI, because it is the main or *first* predominantly *M*otor area but to a lesser extent also sensory (7.129). Conversely, in the postcentral gyrus, SmI is the primary *S*ensory area, though also partly *m*otor. On the medial surface of the cerebrum is a further sensorimotor area, which, being largely motor, is called the *supplementary motor area*, MsII. Unfortunately, MsII is sometimes known as the *third* somatomotor area, because there is a second sensorimotor area, SmII (7.129), inferior to SmI. It would, of course, have been easier to regard MsII as the second somatomotor area and SmII the third, as some authorities do, but this merely shifts the numerical confusion. Some authors surmount the difficulty by disregarding the useful abbreviations and speaking of first, second and third motor or sensory areas. This terminology is admittedly a little perplexing, but its shortcomings are merely incidental to the important fact that the frontoparietal sensorimotor region is a complex of areas, each with its own degree of somatotopic organization. It should, however, be emphasized that none of these 'areas' is exclusively motor or sensory, though one or the other quality always predominates. In the text which follows ambiguity has been avoided by the use of synonyms wherever necessary.

In the occipital lobe the striate, parastriate and peristriate regions of the cortex (areas 17, 18 and 19) are likewise termed the *first*, *second* and *third visual areas*, or visual areas I, II and III.

A more complex array of terms applied in particular to the feline cerebrum (Rose and Woolsey 1958) is current to describe the *four* temporal areas considered to receive auditory projection fibres in cats. These are the *first acoustic area* (AI)—the auditory area of usual description—the *second acoustic area* (AII), and the *anterior* and *posterior ectosylvian areas* (Ea and Ep). The accessory acoustic areas can only be indicated in a tentative manner in the human cortex (7.138), partly because the acoustic parts of the temporal lobe are entirely superficial in the cat, whereas they are partly folded into the inferior lip of the lateral sulcus in mankind. All the sensory areas named above not only receive projection fibres from particular nuclei of the thalamic complex (though this is not completely established with respect to the ectosylvian acoustic areas), they also project back to the same regions, and perhaps in some instances to other nuclei at lower levels in the central nervous system—connexions which in each modality are probably concerned with inhibitory and sometimes facilitatory effects in their own sensory pathway. It is this afferent-efferent character of most, and probably all, the sensorimotor areas which makes the concept of distinguishable motor and sensory parts of the cortex anatomically invalid and functionally misleading.

It is clear from the above remarks that much less of the cerebral cortex remains to be dubbed as 'associational', in the vague but well-established meaning of the term. Nevertheless, large regions in all the lobes of the cerebrum are not accounted for by the sensorimotor areas. In regard to the human brain, much clinical evidence strongly suggests that such so-called 'silent' areas are indeed concerned in the further elaboration and interpretation of sensory information, and in the combination of its different modalities. Much less is known of the connexions of these parts of the cortex, whether by association or commissural fibres. For these reasons, less can be said of these areas than the sensorimotor domains of the cortex. Nevertheless, ablation experiments coupled with tracing of degenerating axons and with observation of disturbed behaviour, and similar deductions in human patients affected by damage, disease or surgical intervention, are beginning to provide more coherent concepts of the activities of the association areas, as will become apparent below.

Though it might be regarded as functionally appropriate to consider first the predominantly sensory areas of the cortex and then those which are mainly motor, the widespread involvement of many parts of the cortex in even the simplest of activities and the sensorimotor nature of the areas concerned render any such considerations of priority almost meaningless. The main areas will therefore be described on the basis of the lobes in which they are located, and in the arbitrary order—frontal, parietal, occipital and temporal.

THE FRONTAL LOBE

The cortex of the frontal lobe may be divided for convenience into two main regions, *precentral* and the unfortunately named *prefrontal*. The former is largely sensorimotor, the latter 'association' cortex.

The precentral area includes the whole of the precentral gyrus and the posterior (caudal) parts of the superior, middle and inferior frontal gyri (Brodmann areas 4 and 6). The whole of this region (**7.130**) is characterized by the almost complete absence of the granular layers. Intracortical fibres are very numerous and the plexiform layer is particularly densely packed. The precentral area has been divided into posterior and anterior parts, *motor* (area 4) and *premotor* (area 6), but the distinction between these areas differs, according to whether cytoarchitectural or physiological data are applied. As comparison of illustrations **7.125** and **7.130A** shows, the boundary between areas 4 and 6 descends through the precentral gyrus, whereas the entire gyrus is regarded as the *motor area*. Hence it is less confusing to disregard the Brodmann numeration of areas, or to use it as a merely approximate indication of the situation of areas. Thus, the premotor area occupies parts of the three frontal gyri, corresponding largely to area 6, but including parts of areas 8, 44 and 45 (**7.130A**). For the present, it is simpler and more direct to designate the whole precentral area as the *first* or leading **somatomotor area** (MsI), remembering that functional variations occur within it and that these may to some extent coincide with cytoarchitectonic differences.

A feature of the whole precentral area (MsI) is the prominence of pyramidal nerve cells of all sizes. The largest of these, the *giant pyramidal cells of Betz*, vary in height from 30 to 120 μm and in breadth from 15 to 70 μm, being most numerous in the medial part of the first somatomotor area, where it extends over the superomedial border of the hemisphere into the paracentral lobule (**7.130B**). They become progressively less frequent as the precentral gyrus is followed downwards towards the lateral fissure. They also extend less and less forwards in the gyrus in the

same direction, following the boundary of area 4, to which they are largely limited, being absent from the premotor or anterior part of the first somatomotor area. In view of the mode of somatotopical arrangement now to be described in this region, it is tempting to suppose that the size of the larger pyramidal cells is associated with the length of their axons. The number of giant pyramidal cells in the human motor cortex has been estimated as between 25,000 and 30,000, (Campbell 1905; Lassek 1940), a figure which conflicts with the million or so fibres in the medullary pyramid (Lassek and Rasmussen 1939; Lassek 1942). It is clear that the huge majority of corticospinal and probably

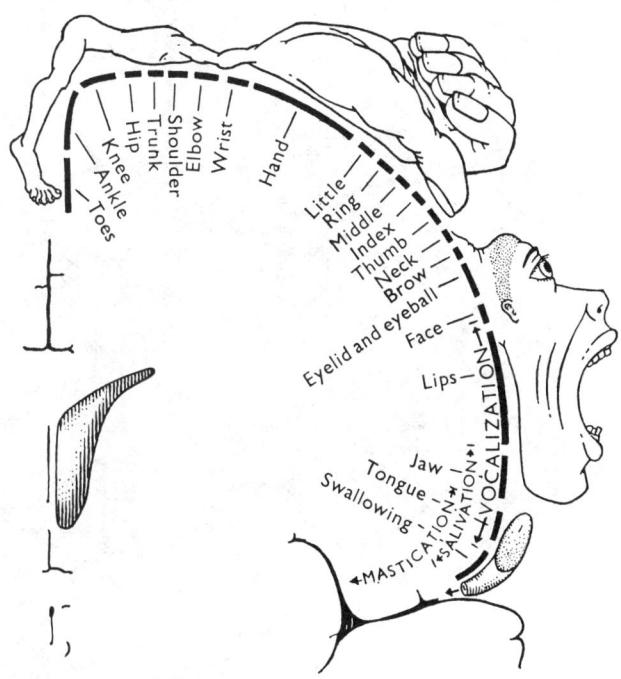

7.131A The *motor homunculus* showing proportional somatotopical representation in the main motor area. (After W. Penfield and T. Rasmussen, *The Cerebral Cortex of Man*, Macmillan, 1950.)

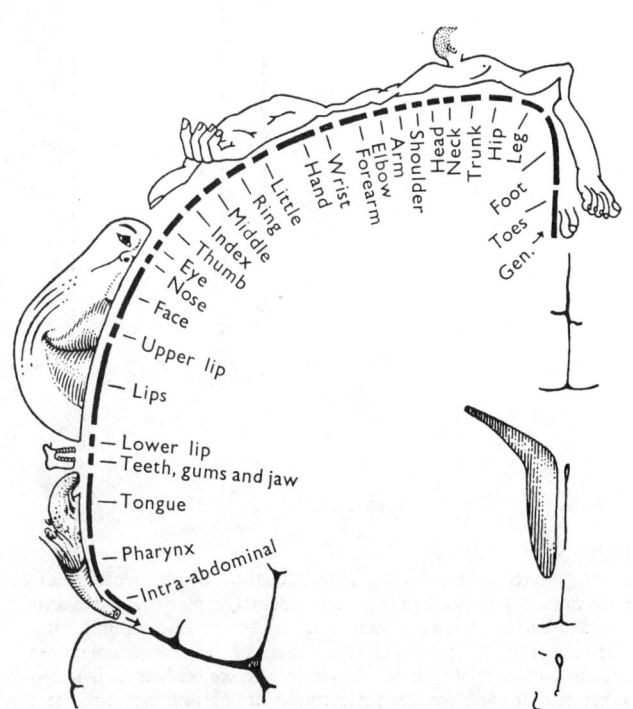

7.131B The *sensory homunculus* showing proportional somatotopical representation in the somesthetic cortex. (After W. Penfield and T. Rasmussen, *The Cerebral Cortex of Man*, Macmillan, 1950.)

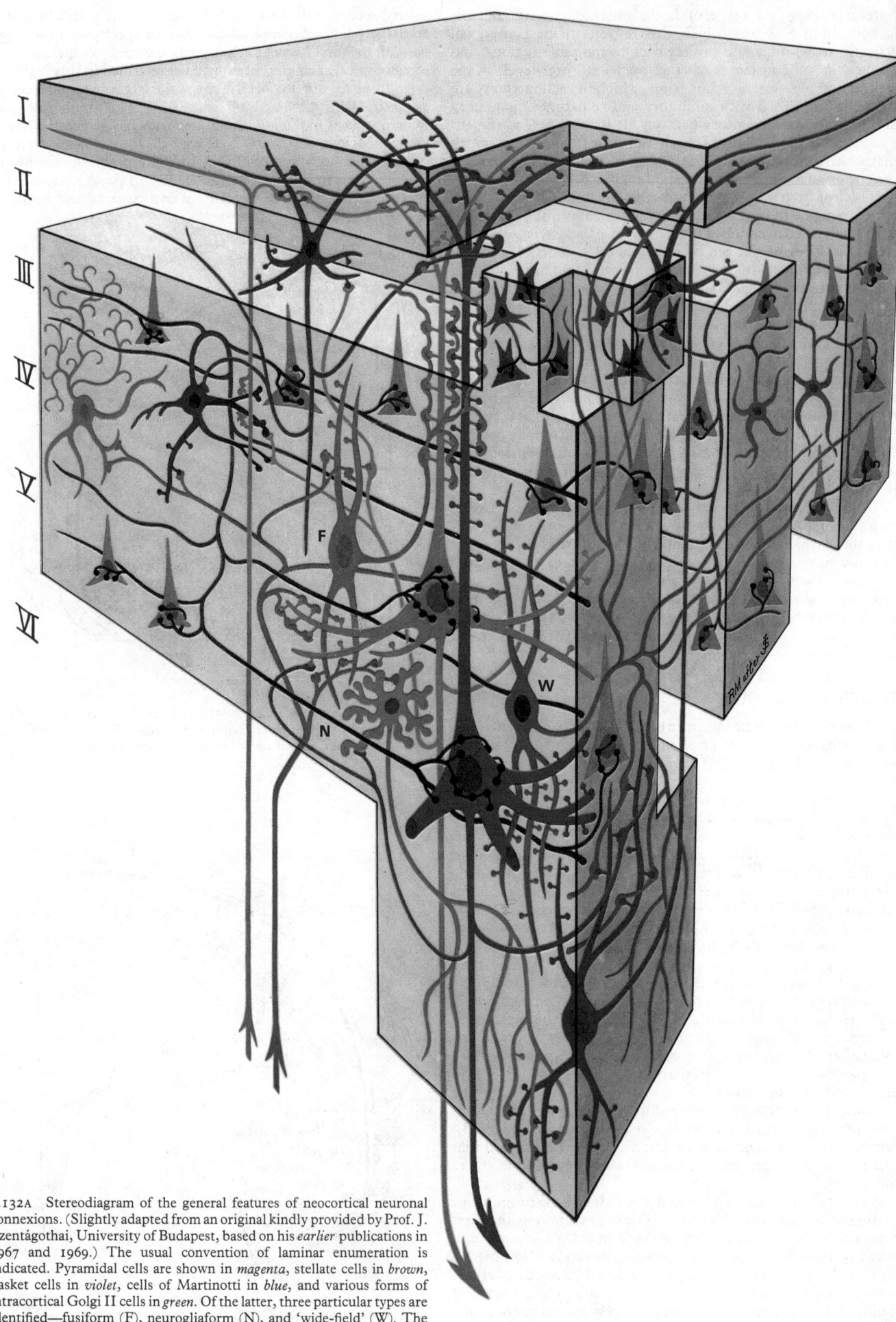

7.132A Stereodiagram of the general features of neocortical neuronal connexions. (Slightly adapted from an original kindly provided by Prof. J. Szentágothai, University of Budapest, based on his *earlier* publications in 1967 and 1969.) The usual convention of laminar enumeration is indicated. Pyramidal cells are shown in *magenta*, stellate cells in *brown*, basket cells in *violet*, cells of Martinotti in *blue*, and various forms of intracortical Golgi II cells in *green*. Of the latter, three particular types are identified—fusiform (F), neurogliaform (N), and 'wide-field' (W). The connexions indicated were in general considered fairly well established, but they were, of course, subject to revision and extension, *see* 7.132B.

corticobulbar fibres are derived from smaller pyramidal cells. As is now known, many such fibres do not originate in area 4 at all, and it is interesting to note here that ablation of this area in monkeys reduces the fibres in the pyramid by a mere 25 per cent (Lassek 1942). Obviously the majority of 'pyramidal' fibres have other origins, as will be detailed below.

The recognition that *contralateral movements*, simple in type, can be elicited in different parts of the body by electrical stimulation of separate loci in the region of the central sulcus dates from the pioneer study of Fritsch and Hitzig (1870) in dogs. This discovery has been confirmed in innumerable subsequent studies in primates (e.g. Leyton and Sherrington 1917) including man (Ferrier 1874; Penfield and Rasmussen 1950), and it has become a tenet of neurology that the area involved is intimately concerned in the mediation of voluntary movements. The order of loci, starting from the paracentral lobule, is associated with the lower limb, the trunk (in the upper part of the precentral gyrus), the upper limb, neck, and head (7.131A). Much greater detail of this bodily 'representation' has accumulated, of course, and it has been established that the amount of cortex mediating movement in any particular region of the body is proportional *not* to the bulk of muscle involved but to the skill with which it is customarily used. All these details can be better appreciated by consulting 7.131A. It is perhaps more important to emphasize that even in the earliest studies of the excitability of the 'motor cortex', it was recognized that movements could be elicited from the postcentral gyrus, as well as from the precentral, and that the same movement could be excited in the same experimental animal's brain from different loci at different times. More careful control of the parameters of stimulation in later investigations has confirmed this apparent 'plasticity' (Liddell and Phillips 1950) which is, of course, difficult to represent in textbook illustrations. This and other factors have obscured to some extent these early findings, and the polemics associated with the great volume of reports on cerebral motor function, especially regarding discrepancies between clinical observations and experimental findings, were only useful insofar as they refocused attention on facts rather than theories (Bucy 1944).

As a result of much painstaking work it is well established that while the whole first somatomotor cortex (MsI—areas 4 *and* 6) is a source of corticospinal fibres, it is not the only one. Both in monkeys (Levin and Bradford 1938) and man (Kuypers 1958) the first somatosensory area (SmI—areas 3, 1 and 2) in the parietal lobe is also a source of corticospinal fibres, and there is some evidence that the second somatosensory area (SmII, approximately equal to areas 40 and 43—*see* p. 1015), below the first area and just above the lateral fissure (posterior limb), also contributes 'pyramidal' fibres. Smaller contributions from the occipital and temporal lobes have also been described in the cat (Walberg and Brodal 1953). Physiological evidence is largely in accord with these anatomical findings. The fibres descending from this extensive origin vary much in calibre. In the human pyramid 90 per cent were estimated to be 1–4 μm in diameter, while only about 2 per cent were 11–22 μm, a fraction corresponding closely with the number of giant pyramidal cells (Lassek 1942). Up to 94 per cent of the fibres are myelinated in man (De Meyer 1959).

The primary, first somatomotor area or MsI receives fibres from the cerebellum, which relay in the nucleus ventralis lateralis of the thalamus (p. 958), and these are distributed particularly to its anterior region (area 6), and to the prefrontal area (area 8). It also receives afferents concerned in other *sensory* modalities, probably via the thalamus also, but in addition from the hypothalamus and from other parts of the cortex. Physiological studies suggest that the corticospinal neurons are influenced not only by somatosensory impulses but also by visual, acoustic and 'nonspecific' thalamic afferents (p. 960). The interactions of all these influences on the pyramidal neurons inevitably demands a most intricate deployment of axons, dendrites and their synapses in the cortex. Illustration 7.132A was an earlier tentative epitomization of the kind of structure involved, provided by Professor J. Szentágothai. It is basically *columnar* in arrangement as in other parts of the cerebral cortex, the pyramidal neurons being extended vertically through the cortex and making abundant vertical associations with afferents and also horizontal

connexions with intracortical interneurons (see also p. 1019). This concept tallies to some extent with recent studies of individual motor neurons by microelectrode techniques in the monkey (Landgren *et al.* 1962) and cat (Asanuma and Sakata 1967). These indicate that some motor neurons at least, though excitable from a considerable area, can be most easily stimulated within a cylindrical section of cortex about 1 mm in diameter. Illustration 7.132B is a more recent and refined summary of the '*module-concept*' in cerebral cortex architecture (Szentágothai 1975). In it: 'An attempt is made to bring earlier circuit models of *primary sensory* cortical areas into better line with recent observations on (1) the distribution of excitatory feedback connexions in cortical tissue volume, (2) putative inhibitory

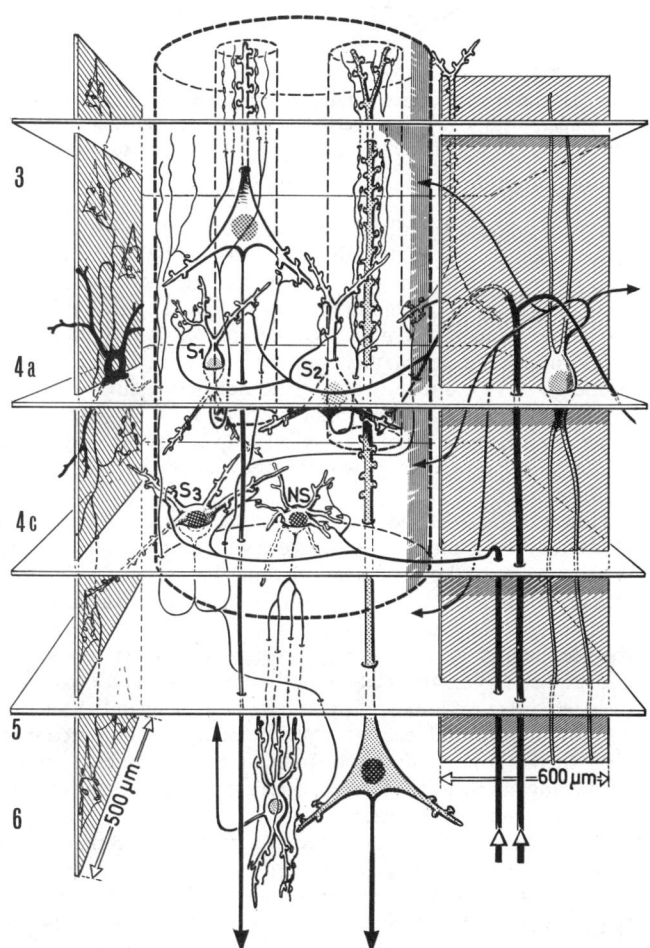

7.132B   A stereoscopic view of the elementary neuron circuit in sensory cortical areas. The horizontal planes are entirely arbitrary and are included to aid stereoscopic visualization; they are unrelated to the cortical lamination (arabic numerals on the left, indicating the 'non-absolute' character of the lamination in such a diagram). Specific sensory afferents (heavy black vertical lines on the right) terminate (separately) on the spiny stellate neurons (S1) and the so-called star pyramids (S2) of upper lamina IV, and in lower lamina IV on another type of spiny stellate neurons (S3) as well as on neurogliaform non-spiny stellate neurons (NS). Ascending relay of sensory afferent impulses is twofold: (1) within *narrow* cylindrical spaces (indicated with dashed outlines) by horsetail-shaped axon arborizations of S1 and S2 type neurons and primarily to selected individual (or small groups of) pyramidal cells, and (2) within *wider* cylindrical spaces (heavier dashed outline) from ascending S3 type neurons, probably mainly to basal dendrites of a much larger group of pyramidal cells. Descending relay is more widely and more indiscriminately distributed by descending branches of the spiny stellate neurons and star pyramid neurons, and occurs in narrow columns by the vertically orientated axonal lacework of neurogliaform non-spiny stellate neurons (NS). The vertical plane on the left shows the strictly orientated axonal arborization of a large basket cell; the vertical plane on the right, orientated at right angles to the basket cell plane, shows a large stellate neuron with vertically orientated dendrites and part of its extended axonal arborization which is strictly confined to this plane. (From Professor Janos Szentágothai and *Brain Research* (1975), by kind permission.)

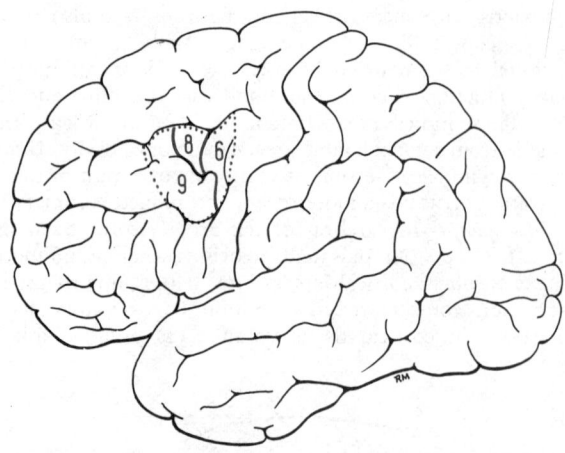

7.133 Superolateral surface of left cerebral hemisphere showing the frontal motor eye field, corresponding to parts of Brodmann areas 6, 8 and 9 (compare with **7.125**). The perimeter of this area is delineated by an interrupted line to indicate uncertainty as to its precise extent.

interneurons and the distribution of inhibition in well defined space modules, and (3) the direct (monosynaptic) cortical target cells of the specific sensory afferents, and the modes of relay to secondary neurons. Even though the concept of cortical circuitry on larger "integrative units" containing smaller modules (or fields) of specific (excitatory and inhibitory) neuronal actions, proposed in 1967 and 1969, had gross deficiencies in the light of newly emerging data, the basic idea of how to look at the functional organization of the cortical neuron network may still be useful as a conceptual framework for the functional interpretation of structural data.'

However the so-called *plasticitity* of the 'motor' cortex may be explained, the anatomical evidence favours a fixed and specific relation between the pyramidal cells and the effectors which they influence. At all levels of the corticospinal pathway a highly specific somatotopical arrangement has been demonstrated. In this connexion it should be emphasized that stimulation of cortical areas is an artificial method, which does not distinguish between direct excitation of pyramidal cells, or even their axons, and of the innumerable intracortical elements which interconnect with them. It is all the more interesting to note that studies employing microelectrode techniques are nevertheless beginning to point in the same direction.

Before dealing with certain specific regions of the somatomotor cortex, it is necessary to point out that stimulation of the cortex in experiment never produces elaborate coordinated movements, but only simple combined activities between synergists and antagonists. The movements elicited bear no relation to the particular skills or motor experience of the individual. This is to be expected, because such stimulation leaves out of account not only other efferent pathways but also the wide range of afferents from the cerebellum, corpus striatum, thalamus, hyothalamus, and other cortical areas mentioned above.

Extending forwards into the middle frontal gyrus from the part of the precentral gyrus concerned with facial movements is the **frontal eye field**; it occupies a considerable part of the Brodmann area 8, invading area 6 behind and probably area 9 in front (**7.133**). Stimulation, which is most effective in area 8, elicits conjugate or binocular movements of the eyes, especially towards the contralateral side, but perhaps also in other directions. Movements of the head and pupillary dilation may also be elicited from the frontal eye field. For a review of work on these responses *see* Crosby (1953). Little is known with certainty concerning the efferent projection from this area; degenerating fibres have been traced from it in monkeys as far as the midbrain, but their ultimate destination is not established. None have been traced with certainty to the nuclei of the motor nerves of the extrinsic ocular muscles, and it has been hypothecated that they influence these through intermediary cell masses or 'centres' in the brainstem. There is, however, no doubt that voluntary and reflex eye movements are mediated by the frontal eye field and its

projection. Like the rest of the first somatomotor area, this region of it also receives association fibres, probably from the occipital cortex, and in addition thalamic afferents from the juxtalaminar part of the dorsomedial nucleus (Scollo-Lavizzari and Akert 1963) to which the central zone of the frontal eye field (area 8) reciprocally projects (Rinvik 1968). A **second frontal eye field**, roughly anterior to the above has been demonstrated in monkeys, and hypothecated in man (Crosby *et al.* 1952).

The motor speech area of Broca is an extension of the primary somatomotor cortex into the inferior frontal gyrus, coinciding approximately with area 44 and part of 45 (**7.134**). Little reliable knowledge is available concerning its anatomical connexions, although subjacent to it are large concentrations of association fibres effecting interconnexions between many other parts of the cortex. Most information as to its functional significance is derived from the results of ablation or stimulation in experimental animals and man (Penfield and Roberts 1959). Ablation may abolish vocalization and usually produces a motor or expressive *aphasia*, or paralysis *of speech* in man, if carried out in the left or 'dominant' hemisphere. In some individuals the leading speech area is, however, on the right, and it is considered that in some there is no such dominance. (See also p. 1024.)

It is to be emphasized that damage to the motor speech area does not entail *paralysis* of the *musculature* involved, all of which is in any case also amenable to stimulation of appropriate parts of the primary motor area in the precentral gyrus. Stimulation of the area may lead to a variety of effects, according to the parameters employed; speech in conscious patients may be actually inhibited, or simple acts of vocalization, such as the utterance of a vowel sound, may ensue. The response in experimental animals consists of simple movements of the face, lips and larynx and perhaps vocalization. Stimulation of the area in macaques has been shown to produce contractions in individual laryngeal muscles (cricothyroid and thyroarytenoid) and in combinations of them (Hast *et al.* 1974). In man also it is to be emphasized that the responses are, as in the case of the somatomotor cortex, always simple, In stimulation studies of the human cortex very similar effects can be elicited from two other areas, one in the frontal lobe, the supplementary motor area (MsII), and another curving round the posterior extremity of the lateral sulcus and occupying a large region in the parietal and, more extensively, temporal lobes. This latter area is sometimes known as the **second motor speech area of Wernicke.** All three regions, which may be named the *anterior, posterior* and *superior motor speech areas*, develop in the dominant hemisphere in each individual; and while the corresponding parts of the cortex in the opposite hemisphere have in some human patients apparently taken over the functions of a damaged area in the dominant hemisphere, they are perhaps in ordinary circumstances uncommitted in the activities of speech.

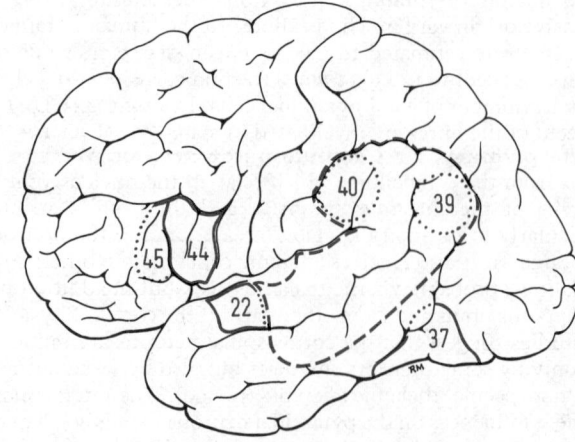

7.134 The superolateral surface of the left cerebral hemisphere showing the motor speech areas of Broca (44, 45) and Wernicke. The latter is variously depicted by different authorities, and is tentatively indicated by the large parieto-temporal area enclosed in an interrupted outline, which itself includes areas 39 and 40. Areas 22 and 37 are considered by some to be respectively auditory and visuo-auditory areas associated with speech and language.

Illustration 7.134 shows the positions of these areas, but it must be said at once that other and often far more complex 'maps' of the cortical areas reputed to be concerned in speech have been propounded. Language, in all its permutations of speaking, listening, reading and writing, is obviously a most intricate activity, and it would be surprising if this, the major form of human communication and perhaps the main distinction of mankind, did not in fact involve widespread and voluminous parts of the brain and central nervous system. In view of the lack of information concerning the connexions of the various cortical areas believed to mediate these activities, it is hardly to be expected that other, subcortical mechanisms will have yet been implicated, despite the intensely active speculation prompted by study of the various forms of aphasia; but the pulvinar of the thalamus has fairly recently been associated with the speech 'centres' (Ojemann *et al.* 1968). However, the inconclusive nature of the evidence in this highly interesting field must limit further pursuit of the subject here, and the reader should consult the excellent reviews available (Brain 1961; Penfield 1966; Millikan and Darley 1967). (See also p. 1024.)

The supplementary motor area (MsII) must now be briefly considered, before passing on to the remainder, the 'prefrontal' region of the frontal lobe. This is, like other projection areas, sensorimotor in nature, but being predominantly motor and in the frontal lobe it is most conveniently described here; its sensory activities, about which less is known, will be merely alluded to in connection with the somatosensory cortex. The situation of this supplementary motor area, on the medial surface of the hemisphere has already been stated (p. 988); it is anterior to, and probably confluent with, the medial extension of the first somatomotor area into the paracentral lobule which mediates leg and perineal movement. It is thus in the medial frontal gyrus in man (area 6 and probably in part area 8). Movements of the contralateral limbs can be elicited from the supplementary motor area in monkeys and in man (Penfield and Welch 1951), but higher thresholds obtain than in the primary somatomotor cortex, as in the case of other accessory areas, and the movements are sometimes ipsilateral or bilateral. A somatotopic pattern (7.129) has been established in monkeys (Woolsey *et al.* 1952), but has not as yet been demonstrated in man. Efferent fibres from this area have been traced into the spinal cord in cats (Nyberg-Hansen 1969), most of these ending contralaterally. There is also a *bilateral* projection to the thalamus, and to the gracile, cuneate and pontine nuclei, contrasting with the similar but unilateral projections from the primary somatomotor cortex. In accord with these variations the contributions of these two areas to the integration of movement are probably different, but the details are not yet sufficiently well defined to warrant discussion here (Travis 1955).

The second somatosensory area (SmII), being preponderantly sensory, will be considered in more detail with the primary sensory cortex in the parietal lobe. Since, however, it is also a *somatomotor area*, its less well-established motor characteristics are included here to complete the picture of the 'motor cortex'. Movements of most parts of the body can be elicited from it and the loci of stimulation show a somatotopic organization both in subhuman primates and in man (Adrian 1941). Its projection fibres have been followed as far as cervical spinal segments in monkeys (Brodal and Angaut 1967), and to the thalamus and dorsal column and pontine nuclei in cats, in which features it resembles the other somatomotor areas. Little can yet be added with regard to its significance in motor function.

The prefrontal area corresponds approximately to all parts of the frontal lobe which have not so far been particularized. It hence includes much of the three frontal gyri, the orbital gyri (orbitofrontal area), most of the medial frontal gyrus, and anterior half, approximately, of the cingulate gyrus, etc. It has been said that the great development of the frontal lobe in particular distinguishes the human from other brains. The so-called *silent areas* have long been regarded as a specially notable feature of the human cortex—at least, in their degree of expansion; and the prefrontal area is usually assigned to this category. If, however, silent or association areas are to be regarded as those parts of the cortex whose connexions are predominantly or exclusively

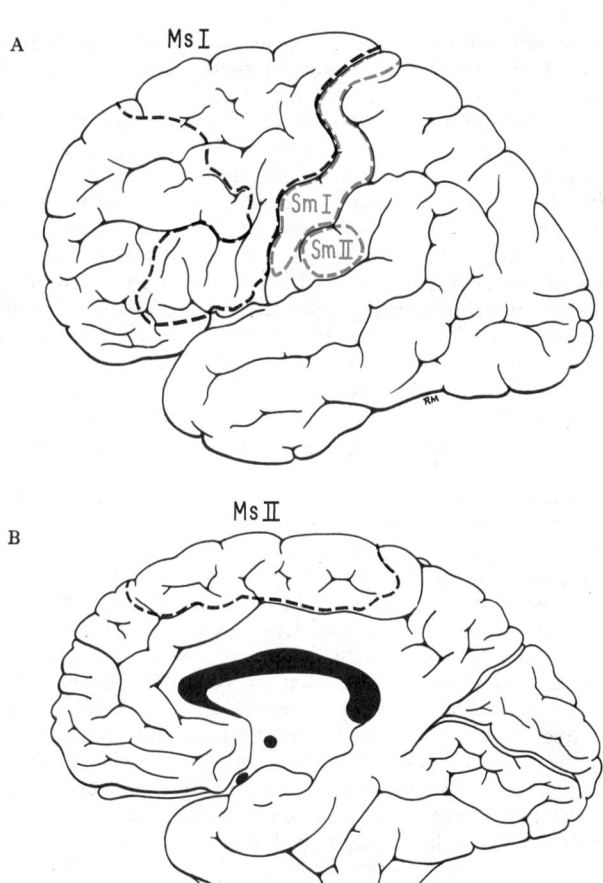

7.135A and B   Superolateral (A) and medial (B) surfaces of the cerebral hemispheres showing the sensorimotor areas. See text for details and compare with 7.125, 130. These areas can only be applied tentatively to the human cortex, and they are hence delineated with interrupted outlines.

mediated by association fibres, the prefrontal region would not qualify. In addition to elaborate afferent and efferent interconnexions with areas of cortex in all lobes of the cerebrum, it also has abundant links with the *thalamus, corpus striatum* and *hypothalamus*, and it projects in addition to the *cerebellum* via the nuclei pontis and perhaps to certain *cranial motor nuclei*.

Through the *superior* and *inferior fronto-occipital fasciculi* the frontal lobe appears to be connected not only with the occipital but also the parietal and temporal lobes. *Commissural* connexions also link corresponding parts of the frontal lobes through the corpus callosum, and there is some evidence that frontal commissural fibres may pass to other lobes. These connexions are said to be largely afferent, but in fact little is known of them in reliable detail, though it is suggested that efferent commissural contralateral fibres may also link the frontal lobe to many other parts of the cortex.

Massive *anterior* and *frontal thalamic radiations*, principally derived from the nucleus medialis dorsalis (MD) of the medial nuclear group of the thalamus (p. 962) pass to the lateral (areas 9 and 10) and orbital regions of the frontal lobe, both of which in turn project back to the same parts of the thalamus. The anterior thalamic nuclei also contribute to these radiations establishing reciprocal connexions with the medial frontal and anterior cingulate cortex. These thalamic nuclei receive many afferents from the mamillary body and thus from the hippocampus (p. 957), which also connects directly with the anterior thalamus. Connexions of the anterior and medial thalamus with the hypothalamus (p. 962) therefore bring this, in addition, into afferent relation with the prefrontal cortex, from which corticohypothalamic projections have also been described. The *prefrontal corticopontine projection*, though less massive than that from the motor areas of the frontal lobe (4 and 6), is equally well established; both relay with considerable specificity in the paramedian components of the nuclei pontis (Nyby and Jansen 1951). The orbitofrontal region in particular projects to the paramedian nucleus of the hypothalamus.

Even this brief survey of the connexions of the prefrontal region suggests highly complex interactions in both *somatic* and *visceral* *activities*. The orbital surface, or orbitofrontal area, has attracted particular attention among experimenters. Stimulation in monkeys (Kaada 1960) depresses the respiratory rate, blood pressure and gastric motility; inhibition of induced cortical and reflex movements and emotional reactions has also been noted. Ablation of orbitofrontal cortex in primates leads to motor hyperactivity, restlessness, loss of attentiveness; and somewhat similar results follow ablation of the anterior cingulate region. The posterior orbitofrontal cortex and anterior cingulate gyrus are often regarded as parts of the limbic system (p. 990). Behavioural studies of monkeys subjected to *anterior* frontal ablations have adduced similar but often conflicting findings (Warren and Akert 1964; Sanides 1964), but the sum total of firmly established fact remains slender.

Some light has been shed on the functional significance of the prefrontal area by the results of lobotomy and leucotomy, performed until recently on human patients in efforts to ameliorate certain mental disorders now treated pharmacologically. The effects of these operations, which were usually bilateral, resembled those caused by extensive disease of the frontal lobes. Abolition of the distressing aspect of illusions and of severe somatic pain were prominent effects, a result which accords with the concept that this part of the cortex is concerned with the 'affective tone' of sensations rather than with any discriminative or localizing aspects of them. Some observers have reported a diminution in intellectual capacity, but limitation of the operation to the orbitofrontal area or the cingulate gyrus is claimed to affect only the emotional balance of personality, suggesting that the superolateral part of the prefrontal area is more concerned with 'intellectual' processes (Lewin 1961). Patients with extensive frontal lobe damage, from whatever cause, almost always exhibit permanent and on the whole undesirable changes. Though perhaps more tractable and docile, they also show less initiative, untidiness, disregard for others and for general tenets of behaviour, and a marked lack of concentration, all of which can for the moment be summed up as a retrogression from the human condition—a somewhat vague statement but excusable and perhaps appropriate in our present dearth of precise information.

THE PARIETAL LOBE

Immediately posterior to the central sulcus, occupying most of the postcentral gyrus and also extending over on to the medial surface of the parietal cortex in the paracentral lobule, is the primary or *first somatosensory area* (SmI, areas 3, 1 and 2). Posterior to it is the large 'silent' area of the parietal lobe, taken up in part below by the second speech area (*vide infra*). Inferiorly, in the lowest part of the postcentral gyrus (and extending forwards a little into the precentral gyrus) is the *second somatosensory area* (SmII).

Anterior to its medial extension is the *third somatosensory area*, the supplementary 'motor' area (MsII). It is appropriate here to note that the region of cortex, centrally situated in the cerebral hemisphere, which forms the great arrival station for somatic afferents and similarly for the departure of somatic efferents, consists of *four separate*, though *juxtaposed sensorimotor areas*. Of these, two, the primary and supplementary motor areas are chiefly efferent projection cortex, MsI and II, while two are preponderantly recipients of afferent projection fibres—the primary and secondary somatosensory areas, SmI and II. As 7.129, 135A, B show, all these four areas evince a somatotopic pattern in cats and monkeys, and evidence that this is largely so in man has already been quoted (p. 1013). The motor aspects of all four areas have been dealt with above, and it remains to consider their involvement in sensory functions.

The first somatosensory area shows interesting cytoarchitectonic variations. Its anterior part (area 3), bordering the central sulcus and extending into its depths to meet the agranular cortex of area 4, is of the granular type but also contains numbers of scattered medium and small pyramidal cells; it is in many respects similar to the striate cortex, the outer band of Baillarger, for example, being broad but not so prominent as the visual stria. The posterior part of the postcentral gyrus (areas 1 and 2) differs particularly in its smaller content of less densely packed granular or stellate cells (for further details see p. 1002). The precise boundary, if such exists, between the pre- and post-central areas in the central sulcus has been a matter of dispute, and may be of some importance in the interpretation of experimental results (Rinvik 1968).

The projection from the *nuclei ventralis posterior lateralis* and *medialis of the thalamus* (VPL and VPM), to the primary somatosensory area is well attested, and it is through this radiation that a wide range of exteroceptive and proprioceptive modalities of sensation are mediated. Unit recording techniques and retrograde degeneration experiments have established that there is a highly specific pattern of localization between the cortex and the thalamus; and this is associated with a somatotopic form of organization in the postcentral gyrus like that in the adjoining first somatomotor area, including an apportioning of cortex in proportion to skill rather than size in the representation of bodily segments. This pattern has been studied in the human cortex by noting the type of sensation and the region to which it is referred when individual loci are stimulated in the exposed cortex at operations (Penfield and Jasper 1954). With carefully moderated stimulation very sharp localization can be established. By such means the localization of vesical, rectal and genital sensations to the lowest part of the medial region of the postcentral area has been established in man. The sensations aroused are for the majority of the body *contralateral*, but may sometimes be solely *ipsilateral* (oral region), or they may be *bilateral* (larynx, pharynx and perineum).

·Experimental studies in animals have largely confirmed the localization demonstrated in the human somatosensory cortex. Responses to natural or artificial stimulation of peripheral receptors have also been shown to evoke potentials in the precentral area (Albe-Fessard and Liebeskind 1966) and by ablation of the somatosensory cortex it is established that there is a *direct* thalamic radiation to the 'motor' area (p. 1013), an elegant demonstration of the latter's sensorimotor nature. Degeneration studies (see, for example, Le Gros Clark and Powell 1953) have provided more refined data as to the specificity of the thalamocortical projection; the nucleus ventralis posterior lateralis exhibits a fairly discrete localizational pattern of connexion with areas 3, 1 and 2, suggesting a functional significance in the structural differences between them, as mentioned earlier in this section (p. 1015). Unit recording observations have confirmed this, revealing that area 3 is activated only by cutaneous stimuli, whereas area 2 is concerned with impulses from proprioceptors (Albe-Fessard and Liebeskind 1966). This implies a transverse deployment of modalities across the long axis of the precentral gyrus (Mountcastle and Powell 1959). To this is to be added a segmental or dermatomic projection to such transverse strips of cortex, as a further refinement of the somatotopic localization in this area (Powell and Mountcastle 1959). Morever, each modality-specific locus appears to be associated with a column of cells vertically arranged through the thickness of the cortex, like the columns described in the visual area (p. 1019). Evidence from anatomical studies in monkeys has demonstrated a cortico-cortical connexion from the first somatosensory area (Brodmann 3, 1, 2) to area 5 in the parietal lobe (Pandya and Kuypers 1969). This connexion has been analysed further (Pearson and Powell 1978); the connexion is isotopological, but reversed in the anteroposterior axis, so that area 3 projects to cells posterior to those receiving fibres from areas 1 and 2. There is some anatomical evidence of modality convergence in area 5, which accords with previous physiological investigations (Mountcastle *et al.* 1975), and provides a basis for the concept of 'association'.

The specificity of the somatosensory neurons concerned with joint receptors appears to be particularly marked, as in the thalamus. Some cortical units respond only to displacement in one direction, and these may continue to discharge even during a statically maintained position of the joint. Such arrangements, if applicable to man, would provide the neuronal basis for the 'joint sense' hypothecated in much earlier studies (Stopford 1922). Fast adapting units activated by the bending of hairs and others adapting slowly to cutaneous deformation have also been identified. Similar studies suggest that afferents from receptors in striated muscle terminate in units sited in the somato*motor* cortex, and that impulses may reach them by pathways independent of the cerebellum, constituting the mediating basis, in part, for consciousness of stretch in muscles. These experimental findings imply a considerable difference in the sensory properties of the pre- and post-central parts of the main sensorimotor area. The two gyri are, however, linked by numerous short association fibres, conducting in both directions (Pandya and Kupyers 1969). As is so in most, if not all, the sensorimotor areas, primary or accessory, the first somatosensory area has abundant reciprocal connexions with the thalamus (p. 962), the corticothalamic projection comparing in its specificity of point-to-point linkage with the thalamocortical radiation (Rinvik 1968). In connexion with this specific and apparently unvarying nature, in the spatial sense, of the cerebral level of some modalities in their pathways, a recent study of the cortex in mice is particularly interesting (Woolsey and van der Loos 1970). Barrel-shaped architectonic groupings of nerve cells were detected in the somatosensory area (SmI) and in the part of this related to the head and muzzle. There is some evidence that these 'barrels' may be units equated with the individual vibrissae of the animal's muzzle (see also Waite 1977). Both this and a study of the spatial distribution of synapses in the same cortical area in neonatal dogs (Molliver and van der Loos 1970), though of course not immediately applicable to the human cortex, illustrate the intricate and quantitative kind of analysis of the cortex being made in current investigations.

The commissural connexions of the primary somatosensory areas of the two sides are restricted to representation of the face, trunk, and proximal parts of the limbs (Pandya and Vignolo 1969; Jones and Powell 1969). These commissural projections, in the monkey, terminate in bands (Jones *et al.* 1975; Shanks *et al.* 1975), as in the visual and motor cortical areas. Shanks *et al.* (1978) have shown that within these bands short intrinsic connexions propagate repeated representations of bodily regions, comparable to the retinotopic repetitions in the visual cortex (p. 1018).

Much less is known of the **second** and **third somatosensory areas** (SmII and MsII). The former is in the superior lip of the posterior limb of the lateral fissure, adjoining the insula in monkeys (7.129), and it occupies a similar position in man. Evoked potentials indicate a somatotopic organization in SmII, with the face area most anterior and the leg at the posterior or caudal end of the area. Single units associated with tactile and vibration senses have been identified in the area, and stimulation of Pacinian corpuscles evokes higher potentials than in the primary somatosensory area (McIntyre *et al.* 1967). The second somatosensory area projects to the thalamus, but its connexions and their reciprocal nature have not yet been studied in detail. It also projects to the dorsal column nuclei, like the other somatosensory areas. The details of such descending projections in monkeys have been studied by many workers (*see* Weisberg and Rustioni 1977), and are of particular interest as an inhibitory modulating influence upon the dorsal column nuclei.

The third somatosensory area, which is the equivalent of the supplementary motor area, is in its sensory functions as yet little understood. It is known to receive thalamocortical fibres and projection fibres leave it, some of which are not motor, but details of these connexions are yet to be worked out. Stimulation in conscious human patients is said to evoke generalized sensations referred to the head and abdominal region, but no exact somatotopic pattern can be ascribed to it, as regards its sensory activities.

The second speech area of Wernicke (7.134), because it is partly in the parietal lobe, must be mentioned here. It occupies a small parietal area, extending more extensively into the temporal lobe (p. 1020). The rest of the parietal lobe, between the main somatosensory and visual areas, is 'silent' cortex. This region corresponds more or less to the inferior parietal gyrus, 'areas' 39 and 40 (Critchley 1953). The connexions of this region, as far as they are known, are indeed in part association and commissural,

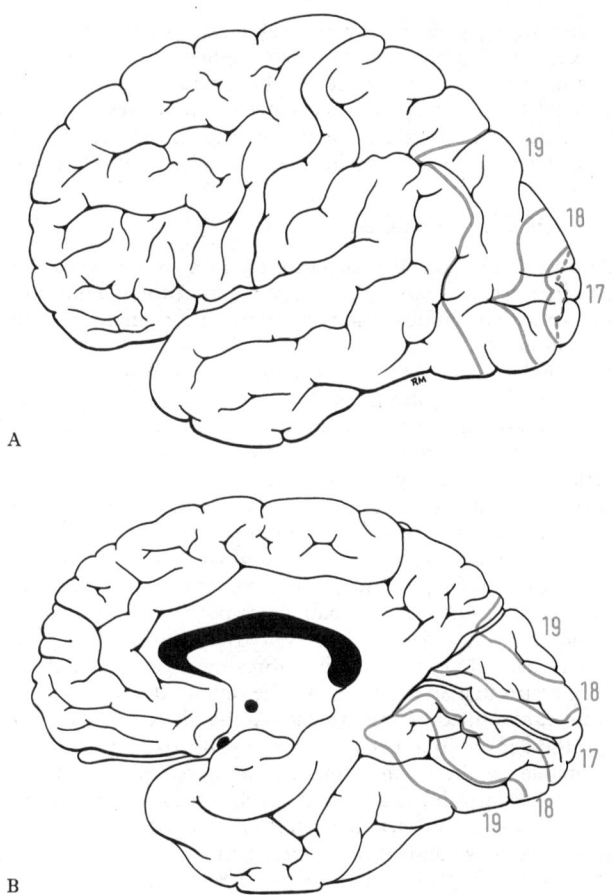

A

B

7.136A and B   Superolateral (A) and medial (B) surfaces of the cerebral hemispheres showing the visual areas in the occipital lobe. The striate (17), parastriate (18) and peristriate (19) areas correspond approximately to the Brodmann areas as indicated, and also to visual areas I, II and III. See text for details.

but little clear detail of them has been established; but there is also a reciprocal interconnexion with all the parts of the lateral nuclear group of the thalamus (p. 891). Corticospinal and corticotegmental projections have also been suggested, the latter influencing motor nuclei in the brainstem, and more particularly those of the nerves to the eye muscles, either directly or through intermediary nerve cells. These latter connexions are specially interesting in conjunction with description of a supplementary motor area in areas 5 and 7 in the macaque monkey (Crosby *et al.* 1959). The strategic position of the parietal association area close to all the cortical regions concerned with general and special modalities of sensation, coupled with its multifarious connexions, is in itself highly suggestive of complex and integrated activities. Most of the evidence in this regard is derived in man from clinical observations of behavioural defects and distortions consequent upon pathological lesions or surgical ablations, and upon extrapolation of similar results in primates submitted to deliberate experimental damage. The reports available and their results are confusingly numerous and varied. Muscular debility, slowing of reflex activity, disorders of speech (aphasias), loss of awareness of bodily parts and of spatial relation to the environment, and other forms of *agnosia* (a state of absent or defective recognition of the significance of objects and situations), have all been observed. A particular form of agnosia, which has long been associated with parietal lobe damage, is *astereognosis*, a failure to interpret the three-dimensional nature of objects when examined by the hand without the help of vision. *Interpretation* has, indeed, been strongly advocated as the prime parietal activity in man (Penfield 1966); in connexion particularly with language on the dominant (left) side of the cerebrum and in a more generalized sense on the other side. The parietal lobe is thus perhaps to be regarded as a region of high activity in learning processes, and of uniquely high development in this respect in mankind. In view of the general nature of the evolutionary process it is perhaps unwise and unnecessary to attribute a qualitatively unique nature to the operations of the human parietal area, but its great importance in our behaviour, especially in discrimination and interpretation, is amply attested by clinical evidence and animal experimentation. (See also p. 1024.)

THE OCCIPITAL LOBE

Almost the whole of the occipital lobe is occupied by the Brodmann areas 17, 18 and 19 (7.136A, B). Although only area 17, the **striate** or **visuosensory cortex**, was originally regarded as containing the terminations of the optic radiation, there is now firm evidence that the lateral geniculate body (p. 977) radiates also to areas 18 and 19 in the cat (Glickstein *et al.* 1967) and monkey (Wilson and Cragg 1967), and that in these, as in area 17, the spatial arrangement of optic radiation terminals is related to the retinae in an orderly manner (Garey and Powell 1967; 1968). Moreover, all three areas project to the thalamus (lateral geniculate body or pulvinar) or brainstem motor nuclei, or both. Short and long association and commissural fibres link the *three visual areas* together, and to their contralateral equivalents and other parts of the cortex in both hemispheres. (For an exhaustive critique of this and other aspects of visual neuroanatomy consult Polyak 1957.) Occipitopontine fibres have been described, but their status in man is uncertain. The actual extent of area 17 is, of course, identified by the presence of the visual stria of Gennari, but the demarcation between areas 18 and 19 is much less easily definable by the usual techniques of cytoarchitectonics. Braak (1977) has introduced a method which depends upon the staining of lipofuscin granules, and this shows a sharp distinction between these two areas and, furthermore, permits the identification of structural subdivisions within them. It is, however, not clear how far such subdivision is functionally significant.

The striate cortex (area 17), or **first visual area**, occupies the upper and lower lips and the depths of the posterior part of the calcarine sulcus (7.136), extending into much of the cuneus and lingual gyrus. As a cytoarchitectonic area it can easily be defined, both macro- and micro-scopically, by the visual stria and by the thinness of the cortex. Posteriorly the area is limited by the lunate sulcus (and by the polar sulci above and below this), and it thus does not extend beyond the occipital pole in man, though it reaches the lateral aspect of the lobe in other primates. Anteriorly the striate area is bounded by the medial parieto-occipital sulcus, but its lower part, below the calcarine sulcus, may extend somewhat further. The stria becomes less obvious as it is followed posteriorly towards the occipital pole, and this change can be correlated with the retinotopical organization of the area (*vide infra*), its prominence being inversely related to proximity to the central retinal area.

Histologically the area is of the *granular* type, or *koniocortex* (p. 1009), in which densely packed stellate cells greatly outnumber the few pyramidal cells present. For full details the reader should consult Polyak (1957); Crosby (1960); Colonnier (1964, 1967, 1968, 1974); only salient features will be mentioned here. (*a*) The deeper stratum of the pyramidal layer (III) contains *large stellate cells*, which almost replace the large pyramidal cells present here in other types of cortex. (*b*) The outer band of Baillarger in lamina IV is accentuated as the *visual stria*, consisting of large numbers of terminations of optic radiation and probably association fibres. (*c*) The ganglionic layer (V), which commonly consists largely of pyramidal cells, including the largest, here contains a few large, scattered *solitary cells* (*of Meynert*); these are of a modified pyramidal shape, about 30 μm in diameter, and distributed in a single row. Their dendrites extend widely through the cortex and their axons form the projection element of the visual cortex, passing through the optic radiation to reach the superior colliculus and possibly the motor nuclei of the extraocular muscles. A study of Meynert's cells in the striate cortex (area 17) of the rhesus monkey (Chan-Palay *et al.* 1974) showed the potential of these neurons for widespread and intricate connexions. It was estimated that each cell's dendrites may display 36,000 spines, and that the apical dendritic field in lamina I is about 400 μm for each cell, being 800 μm for the basal

dendrites in lamina V. The density of distribution of Meynert cells varies in different parts of the striate cortex, reaching its greatest (8000/cm²) in the macular region. (*d*) The external and internal granular layers (II and IV)—especially the latter— contain larger numbers and more densely packed small stellate cells than any other part of the cerebral cortex.

The laminar pattern described above is based upon Brodmann's; other patterns, particularly that of Cajal, have considerable vogue, but a recent review by Billings-Gagliardi *et al.* (1974) advocates a return to the Brodmann scheme.

Although the striate cortex contains only about 3 per cent of the cerebral surface area, approximately one-tenth of all the cortical neurons are said to be concentrated in it. In their cytoarchitectural details, areas 18 and 19 approximate respectively to the polar and parietal types of cortex (p. 1009), in which stellate cells are less prominent and pyramidal cells more so. Myelin and axonal stains have been used to investigate the fibre structure of the visual area, especially the striate cortex (Sholl 1955; Colonnier 1964). In general this resembles arrangements in other cortical areas (p. 1009), but dendritic fields in their quantitative aspects have been studied more intensively in this part of the cortex than perhaps any other excepting the postcentral gyrus. The stimulus to this has been in part the physiological demonstration of a species of columnar organization in the somatosensory area and thereafter in the striate cortex (*vide infra*). So far the physiological techniques in use are providing more illuminating evidence.

The status of area 17 as the primary visual cortical area has long been established, both by electrical stimulation in man and by its connexions with the lateral geniculate body, and through this with the retinae (p. 977). The *retinotopical organization* which exists in the lower parts of the visual pathway has also been shown to be carried through to the cortical level. Each striate area receives impulses from the two ipsilateral half retinae, representing the *contralateral* half of the binocular visual field. As has already been described (p. 977), the patterns of impulses from the retinae do not undergo a simple relay in the geniculate body, where some degree of processing has been shown to occur, excluding any interaction of a biretinal nature (for a more detailed discussion *see* p. 981). These activities have been studied even more intensively in the striate cortex, especially by single unit recording technique. The *geniculate radiation* spreads out as it swerves through the white substance of the occipital lobe, its fibres terminating in a strict point-to-point deployment in the striate area, such that the peripheral parts of the retinae activate the most anterior parts of the area, the macular regions activating a relatively large part of the striate cortex adjoining its posterior extremity. Moreover, the superior and inferior retinal quadrants are thus connected with corresponding divisions of the striate area. In the classical studies of the effects of injuries of the occipital lobe in warfare, similar retinotopic results were obtained. (Holmes and Lister 1916; Teuber *et al.* 1960).

These experimental findings, established in a wide range of mammals and especially primates (*see* p. 977 for references), have been corroborated by stimulation of the human cortex (Penfield and Jasper 1954). The visual impressions thus elicited are simple, such as flashes of light, but they are referred to a specific part of the visual field according to the location of the cortical stimulus. Eye movements are also produced by such stimulation. When similar stimuli are applied to areas 18 and 19 more complex images are reported by the patient, indicating that these regions are concerned in further elaboration of the visual information reaching area 17.

In recent years most interesting results have been obtained by recording the response of single nerve cells to various forms of retinal stimulus; and while the earlier studies have been chiefly pursued in cats (Hubel and Wiesel 1962, 1963, 1965, 1971; Wiesel and Hubel 1963a and b), there was little reason to doubt that in principle they were applicable to the human striate cortex, especially since the cytoarchitectonic areas involved in the cat could be equated with those in primates (Otsuka and Hassler 1962), in which unit cortical recording has also been studied more recently (Hubel and Wiesel 1977). Briefly, each unit from which recordings are made corresponds to a definite receptive field in the retina, and can presumably be excited only from that area. Units responding to a range of modes of stimulation have been identified; some react to white light, especially in the form of stripes or edges between light and dark. This contrasts with the more or less circular receptive fields of excitation demonstrated in the lateral geniculate body. Excitatory units surrounded by inhibitory zones have also been described, but these also evince a strip-like character as well as the circular receptive areas recorded in the geniculate body. Many units are sensitive to the orientation of the stimulus, responding only to vertical, horizontal or intermediate forms of linear stimulation of the retina. More complex units occur, for example, some responding to the particular direction of a moving stimulus. It is possible that these represent the integration of information from the simpler units mentioned above. Further studies by Hubel and Wiesel (1968) have clearly demonstrated the existence of striate cortical neurons, in cats and especially monkeys, which respond preferentially to stimuli from the right or left eye, a phenomenon somewhat misleadingly termed *ocular dominance*. Vertical exploration of the cortex by unit-recording techniques has also shown that these neurons are deployed in vertical columns, extending through all layers of striate cortex, as described below. Certain of these columns may be binocular in response in cats, but in monkeys it is doubtful whether such columns occur, all being right or left orientated. A further evidence of perhaps a greater degree of development of such 'dominance' in the simian cortex resides in the greater facility with which they can be discovered and explored. These workers consider that the columns are of greater diameter in monkeys. It has been possible to refer the majority of the neurons involved to particular laminae of the striate cortex and to equate this distribution with that of termination of the lateral geniculate projection. The full significance of these findings is not yet entirely clear, except that they confirm the segregation of 'information' from each retina to be maintained at least as far as the striate cortex. There is, however, no doubt of the validity of these observations, which have been confirmed by other techniques, such as autoradiographic identification of labelled substances (for example 3H-proline) injected directly into the feline eyeball (LeVay *et al.* 1978). By transneuronal transport in the lateral geniculate body such substances reach the striate cortex, and particularly lamina IV, where they correspond in distribution most clearly with the columnar arrangement identified by other methods. Studies in both cats (LeVay *et al.* 1978) and monkeys (Hubel *et al.* 1977) suggest that these neuronal arrays are subject to some degree of plasticity, since they can be altered by monocular deprivation during early postnatal life.

Further work in this field has led to the identification of columnar unitary loci in the striate cortex which respond to stimulation from one retina, although, as indicated above, some appear to be excited simultaneously from both; a columnar organization of cells coded according to colour responses has also been advanced. Some of these later results have been obtained in monkeys, with an increased likelihood that they correspond to human visual activities.

These studies, carried out with meticulous reference to the laminar pattern of the striate cortex, suggest a *columnar* form of anatomical arrangement corresponding to the vertical physiological organization in 'units' for which there is such undoubted evidence. The concept of 'chains' of neurons linked by synapses through the vertical dimensions of the cortex is comparatively old (p. 1004), and much more recent work, with the advantages of improved staining techniques and electron microscopy, has confirmed and elaborated this view, not only with regard to the striate cortex but also in the case of other areas. Synaptological studies and the results of degeneration experiments indicate clearly that the afferents of such an area as the striate cortex feed their impulses into a limited columnar group of neurons of internuncial type, the resultant of their interactions, excitatory and inhibitory, reaching a pyramidal cell and being thus translated to subcortical neurons or to some other part of the cortex. Such *columnar units* are not, of course, completely independent, being linked to others by recurrent axons, by the tangential fibres in the superficial lamina of the cortex (I), and by actual spatial overlap. (For a survey of the development of concepts concerning the columnar architecture of neocortex compare the publications of Szentágothai in 1967, 1969 and 7.132A with his later, more refined 'module concept' published in 1975 and summarized in 7.132B. The latter relates in particular to the visual cortex. See also below.) It is perhaps appropriate to mention here that within each cortical unit, cells in laminae II and III project to other cortical areas (such as 18 and 19), while lamina V projects to the superior colliculus and lamina VI to the lateral geniculate nucleus.

The effects of lesions of the lateral geniculate body in cats provide particularly detailed patterns of the termination of geniculocortical fibres in all three visual areas (Rossignol and Colonnier 1971). In area 17 degenerating terminals are mainly in lamina IV, but a few fine vertical axons reach lamina I to spread tangentially in it. In area 18 the main termination is in lamina IV, while in area 19 the afferent terminations are much less numerous but are spread through layers IV, V and VI. These results not only confirm earlier findings on the wide region of afferent projection in the occipital lobe, but also accord with the divergence of activity in the three parts of the visual cortex. The above experiments entail the infliction of very large, almost total lesions of the lateral geniculate body. Restricted lesions are naturally required to demonstrate the specific point-to-point relationship between the geniculate projection fibres and the visual areas. The specificity can also be shown by making cortical lesions and observing the resultant retrograde degeneration in the geniculate body, where a columnar series of groups of cells in all six layers are affected (p. 977). It is of special interest here that this effect is only achieved when the cortical lesions involve the whole thickness of the cortex (Rose and Malis 1965). By a similar combined 'attack' the functional details of visual *areas* II and III are being gradually clarified. In both areas evidence of columnar units is accruing, with the difference that the units respond to more complex stimulation and are almost all bi-retinal in respect to their receptive fields.

Hubel and Wiesel (1978), working on the macaque monkey, have been able to carry the above concepts further and to provide at least an initial integration in our understanding of the *modus operandi* of the striate cortex. Briefly, they have been able to show (by transport of radioactively labelled amino-acids from a single eye via the lateral geniculate bodies to the striate cortices) the existence not only of neuronal columns correlated with ocular dominance but also a particular pattern of arrangement of these columns throughout the striate cortex, area 17 (see also LeVay *et al.* 1978). This pattern is based in lamina IV of the cortex, which receives the major part of the geniculate projection and which contains so-called 'simple' neurons responsive to circularly symmetrical retinal stimuli in a strictly uni-ocular manner. The pattern consists of curving and branching stripes, about 0.5 mm in width in the macaque. The stripes are separated by unreactive intervals of similar width and, naturally, identical pattern. The reactive (radioactively labelled) stripes correspond to the cortical 'columns' associated with the injected eye, the intervening, nonreactive, stripes to the unaffected eye. It can be concluded, therefore, that interaction between signals from corresponding areas of both retinae must occur elsewhere in the striate cortex, perhaps in association with the 'complex' cells (in laminae II, III, V or VI), some of which respond to binocular stimuli. A second type of columnar pattern has also been demonstrated in the striate cortex by a combination of unit recording and a radioactive technique revealing sites of high utilization of glucose, the stimulus in both situations being the orientation of linear objects in the visual field. The columns associated with varying angles of object-orientation display a pattern of arrangement of whorled or 'swirling' 0.5 mm wide stripes similar to those concerned in uniocular reception. However, the 'stripes' of the two patterns do not correspond but intersect at frequent points. On the basis of these findings Hubel and Wiesel have hypothecated a cortical arrangement of repeated unitary columns, through the whole thickness of the striate cortex, each unit being able to respond to a stimulus in terms of position and orientation in the visual field, each columnar group of neurons being, however, concerned purely with a uni-ocular input. The theory has not yet been extended to attempt to explain cortical mechanisms concerned in the colour, kinetic, or spatial (stereoscopic) aspects of stimuli. Nevertheless, these studies have provided the most detailed ideas of the manipulation of signals received by any part of the cerebral cortex, though they do not, as the workers cited admit, approach the integratory process which we call perception.

Parts, at least, of the occipital visual area must be considered as sensorimotor, inasmuch as eye movements can be elicited from them. In monkeys this (third) *occipital eye field* (7.136) is confined to visual areas I and II (17 and 18), stimulation of which produces conjugate deviations in various directions, especially to the opposite side. Head turning, facial movements, and sometimes responses in the upper limbs may occur (Rieck 1959; Bender 1964). This occipital motor eye field (III) is connected reciprocally with the *frontal eye fields* I and II. Such connexions are considered to be involved in the mediation of vision in the voluntary ocular movements which are held to characterize the frontal eye fields. The motor responses from area 18, which include accommodation, are on the contrary thought to be, in natural activities, reflex in nature and involved in following and fixation movements. This area may project to the pulvinar, in which connexion a controversial linkage between the lateral geniculate body and the same part of the thalamus should be noted; projection from the pulvinar to areas 18 and 19 is established in principle, but there is disagreement over details

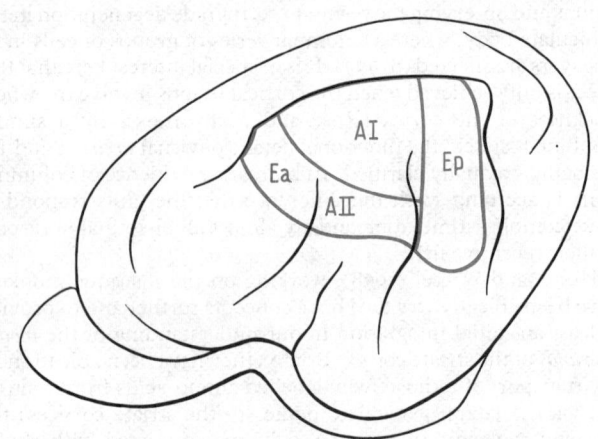

7.137 Lateral aspect of the feline cerebral hemisphere showing the main auditory areas, AI and AII, and accessory areas Ea and Ep. These areas cannot yet be extrapolated with confidence to the human cortex. See text for details. (Adapted from *The Auditory Pathway*, by I. C. Whitfield—see Bibliography.)

(Locke 1961). Efferent projection fibres descend from the occipital motor eye field, particularly from area 19, to the superior colliculus and to the nuclei of the motor nerves of the ocular muscles; but the latter connexion is probably through intermediary nerve cells, such as the para-abducens nucleus (p. 1069), interstitial nucleus of Cajal, nucleus of Darkschewitsch, and the posterior commissural nuclear complex (Carpenter and Peter 1971). These descending fibres are to be distinguished from the corticogeniculate projection (pp. 978 and 1174), the evidence for which is chiefly physiological (Widén and Marsan 1961). A number of workers have now shown in some detail corticofugal connexions from areas 17, 18 and 19 projecting upon the thalamus, tectum, and pretectal region particularly in the cat. In this animal a degree of retinal representation has been demonstrated (Updyke 1977) in the projection from area 19 to the pulvinar.

The three visual areas are intimately linked by short association pathways, which are probably especially specific in the spatial sense between areas 17 and 18. The topographical relation preserved in the connexions between all three visual areas has been demonstrated in monkeys by Cragg (1969) and Zeki (1969, 1974); according to Zeki (1977) the spatial representation displays a 'mirror-image' reversal between adjoining visual areas, as apparently occurs in connexions between other adjacent 'sensory' areas (*see* Somatosensory Area, p. 1016).

In his most recent reviews of the organization of the primate visual cortex, Zeki emphasizes the propagation of '*mosaics*' of retinal representation from the striate to the 'pre-striate' cortex—the latter term including not only peri- and para-striate areas (Brodmann 18 and 19) but perhaps other more remote cortical regions. The latter include in particular the so-called fourth visual area (V4 and V4a), which extends between area 19 and the occipital end of the superior temporal gyrus (7.136A). Zeki's mosaic concept of retinal spatial representation involves the projection of each unitary retinal area at repeated points in all the visual areas, the responsiveness of the cortical loci showing variation in modality, different loci responding to orientation, contour, depth, different colours, and movement. In this concept there appears a possibility of equating such responsive loci with the columnar form of architectonics, already well established, with the results of unit recording. There is considerable evidence that the retinally-linked loci in the major visual areas subserve different modalities of the visual function; for example, area 17 appears to emphasize orientation, whereas columnar loci in area 18 are more concerned with stereoscopic depth. Beyond this, in areas 19 and V4, retinal representation is extended into the modalities of colour and movement. Picturesquely, one may imagine the total pattern of excitation from the retinae as propagated from area to area, the cortical units of each adding its particular modality to the total sensory pattern. Association and commissural fibres connect all visual areas, but more particularly 18 and 19, to many parts of the ipsi- and contra-lateral hemispheres. Vision and ocular movements may therefore be affected by lesions remote from the visual cortex, or from the recognized motor fields. For example, the course of the optic radiation as it swerves through the temporal lobe (Meyer's 'loop') entails that temporal lesions, such as tumours, may lead to homonymous field defects (Falconer and Wilson 1958).

THE TEMPORAL LOBE

The temporal lobe has lateral and inferomedial surfaces, the former presenting superior, middle and inferior temporal gyri, the latter including the medial and lateral occipitotemporal gyri (sometimes regarded collectively as the fusiform gyrus), which are separated from the parahippocampal gyrus by the collateral fissure (7.114 and 119). The temporal lobe is regarded as highly evolved in man and of relatively recent phylogenetic development. In a general way it may be regarded as particularly involved in hearing, language and perception.

Part of the superior temporal gyrus is 'rolled' into the posterior limb of the lateral fissue during human fetal development, and

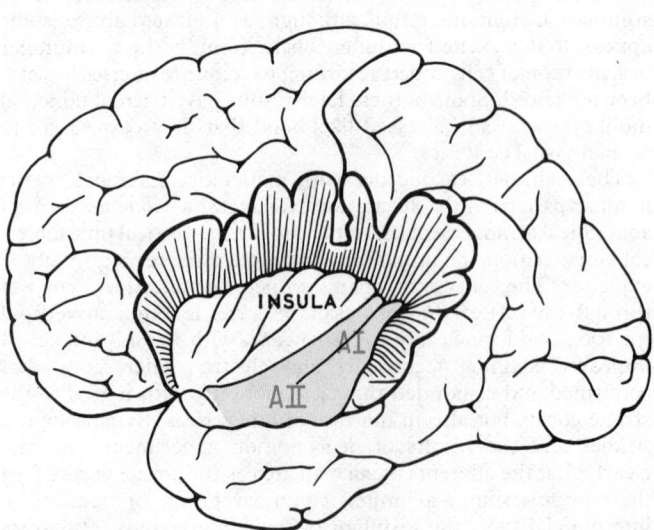

7.138 Superolateral surface of left cerebral hemisphere from which the opercula have been cut away to expose the insula and the adjoining anterior and posterior transverse temporal gyri and their continuity with the superior temporal gyrus. The area shown in blue contains the classical acoustic area, AI (equal to Brodmann area 41 and parts of areas 42 and 52), and the secondary acoustic area, AII (extending into area 22).

this buries two short gyri which ascend obliquely backwards posterior to the limiting sulcus of the insula. These are the anterior and posterior *transverse temporal gyri* (of Heschl), which correspond to areas 52, 41 and 42 approximately (7.112). A large number of other cytoarchitectonic localities have been described in the temporal lobe; but only those areas just mentioned, together with 22, which adjoins them on the lateral aspect of the superior temporal gyrus, can be reliably equated in man. In the original description of Campbell in 1905 the *acoustic area* of the cortex covered rather more than area 41, to which subsequent observers narrowed it down, thus defining what might be called the classical acoustic area, a leading feature of the temporal gyrus (7.126A). As will become apparent, Campbell's view has been revived by more recent experimental work.

Area 41 is histologically a variant of the granular type of cortex, but it is thicker than the cortex in the visual and somatosensory areas (Bailey and von Bonin 1951). Its immediate surroundings, such as area 22 (7.125), are of the parietal type, and beyond this much of the temporal lobe is frontal in type, like the main areas of the parietal and frontal lobes. Both the latter types of cortex show a considerable admixture of pyramidal cells and a diminution in granular elements (stellate cells). For further details consult the authorities referred to on p. 1002, and consult Braak (1978) for an extensive study of temporal areas in the human brain containing large pyramidal cells.

The acoustic area, as classically defined, is approximately equivalent to area 41; for reasons shortly to be related it is now more properly called the **first acoustic area** (AI). It extends a little into the superior temporal gyrus, but is largely obscured in the lateral fissure, forming the anterior transverse temporal gyrus. The earlier evidence on which the acoustic nature of this area was based may be sought in suitable reviews (*see* Whitfield 1967). The cytoarchitecture of the area has been described in detail by Sousa-Pinto (1973), who claims that the area AI thus defined corresponds exactly to the projection of the lateral part of the ventral nucleus of the medial geniculate body (in the cat). Attempts to define more accurately the area in which projection fibres from the medial geniculate body terminate have extended the area (Rose and Woolsey 1949), and the results of cortical stimulation (evoking head and ear movements) and cortical recording of potentials excited by localized stimulation of the cochlea and the cochlear nerve (Woolsey and Walzl 1942) have provided similar evidence. Much of this experimental work has been carried out on cats, and less often on monkeys (consult Sousa-Pinto 1973). Consequently the details of these accessory areas apply primarily to the feline temporal lobe and cannot be confidently translated in their entirety to man (7.138). The **second acoustic area** is often abbreviated to AII and lies inferior to AI in the equivalent of the superior temporal gyrus, in the cat the ectosylvian gyrus. Flanking AII are the *anterior* and *posterior ectosylvian areas*, Ea and Ep (7.137), and even further areas have been advanced. There is some evidence that all the areas named, except Ea, receive projection fibres from the medial geniculate body. Ablation of the first acoustic area, AI, leads to degeneration of cells only in the anterior part of the medial geniculate body, which receives fibres from the inferior colliculus (p. 941); but other parts of the complex of nuclei which make up the body project to all the other areas, except perhaps Ea, but including also part of the insula (Morest 1964). It has been suggested that some of the fibres in the acoustic radiation divide and terminate in more than one area, and that this may be in part the cause of the equivocal results of some experimenters. Short association fibres have been shown to link AI to AII and Ep, but not Ea. (For detailed development *see* Krmpotić-Nemanić *et al.* 1979.)

The geniculocortical projection has been 'mapped' in the first acoustic area in the chimpanzee (Walker and Fulton 1938), and a topical representation in this (Woolsey and Walzl 1942) was considered at an early date to be *tonotopical*, that is, to indicate a localization pattern on the basis of frequency of sound. This has been strongly denied as a feature of the human first acoustic area (Penfield and Jasper 1954), and remains a major controversy (Evans 1968). Stimulation in the conscious patient merely produced poorly localized patterns of sound or noise of simple nature, such as buzzing, humming, or ringing, but in the

neighbouring areas (which may correspond to those identified in the cat), more complex acoustic phenomena were elicited. The topological relationship between the acoustic areas and the various divisions of the medial geniculate nucleus have been investigated in detail by retrograde transport of horseradish peroxidase from small cortical lesions (Winer *et al.* 1977). These observations suggest that only the laminated part of the ventral medial geniculate nucleus projects to a single acoustic area (AI), all other parts of the medial geniculate complex projecting to several of the cortical acoustic areas. It was also noted that the pulvinar projects to acoustic areas. Despite numerous studies of unit recording from the primary acoustic area, this problem has not been settled, but it seems highly likely that there are in fact units which respond to specific frequencies, even if they are not arranged in any particular pattern. Other units which can be excited by *variation* of frequency with time or direction have been described (Whitfield and Evans 1965), and in these specific arrays of units, the acoustic area resembles the visual cortex.

The existence of *efferent fibres* descending from the acoustic areas to the brainstem has been alluded to elsewhere (p. 941). These may not only modulate activities in the medial geniculate body but may also mediate various reflexes by forming connexions with motor nuclei of cranial nerves. Some of the fibres probably influence contraction of the stapedius and tensor tympani in the middle ear cavity; experiments suggest that this is increased in association with high-frequency sounds (Carmel and Starr 1963). *Association fibres*, in addition to the short ones alluded to above, link the acoustic areas to the other lobes in their own hemisphere, and perhaps particularly to the frontal pole and the cortex deep in the superior temporal sulcus, which recent work suggests may be special loci of convergence of association fibres from all the sensory areas. Little is known of commissural fibres, but it should be remembered that the lateral lemniscus is a bilateral tract. As a result of this, lesions in man of the medial geniculate body or the acoustic cortex do not lead to noticeable deafness unless they are bilateral—an unusual occurrence. Quite extensive temporal tumours are associated with only minimal effects on hearing, though they may interfere with the acoustic aspects of interpretation of language, especially if in the 'dominant' hemisphere. In connexion with this it is apposite to note that the posterior motor speech area mentioned in the section on the parietal lobe (p. 1017) extends markedly into the temporal lobe posterior to the acoustic region, practically surrounding the latter according to some authorities (7.134). This proximity doubtless has its implications in the complex cortical organization of speech, hearing and language, but further comment would be at the present juncture mere speculation.

The vestibular area is a matter of uncertainty and disagreement as regards its site, but in the monkey it is said to be located close to the part of the somatosensory area (SmI) concerned with the face in the postcentral gyrus (area 2)—*see* Jones and Powell (1970). Area 2 is a cortical region particularly involved in sensations from deep tissues and joints, and the spatial proximity of a vestibular area to this appears interesting. The area is believed to receive a projection from the vestibular nuclei, but this rests more upon the results of natural or electrical stimulation of the vestibular nerve or its receptors than upon any anatomical demonstration. Unlike the acoustic pathway, the vestibular projection appears to be almost entirely crossed. Indications of some degree of somatotopical specificity in the connexions of the vestibular cortex with parts of the vestibular nuclear complex have been reported (Massopust and Daigle 1960), but in view of the lack of anatomical evidence and the conflicting nature of the physiological data, little more can be said until better evidence is available. Another view of the position of the vestibular area must be mentioned; this depends on the recording of equilibratory responses in man, and implicates part of the superior temporal gyrus, anterior to the acoustic areas (Penfield 1957).

Apart from the acoustic areas, a possible vestibular area, and the extension into it of the posterior motor speech area (of Wernicke), the temporal lobe contains no other functionally designated neocortical areas, but it is not therefore to be assumed that all the remainder can be regarded as association cortex. The existence of large numbers of association fibres connecting parts

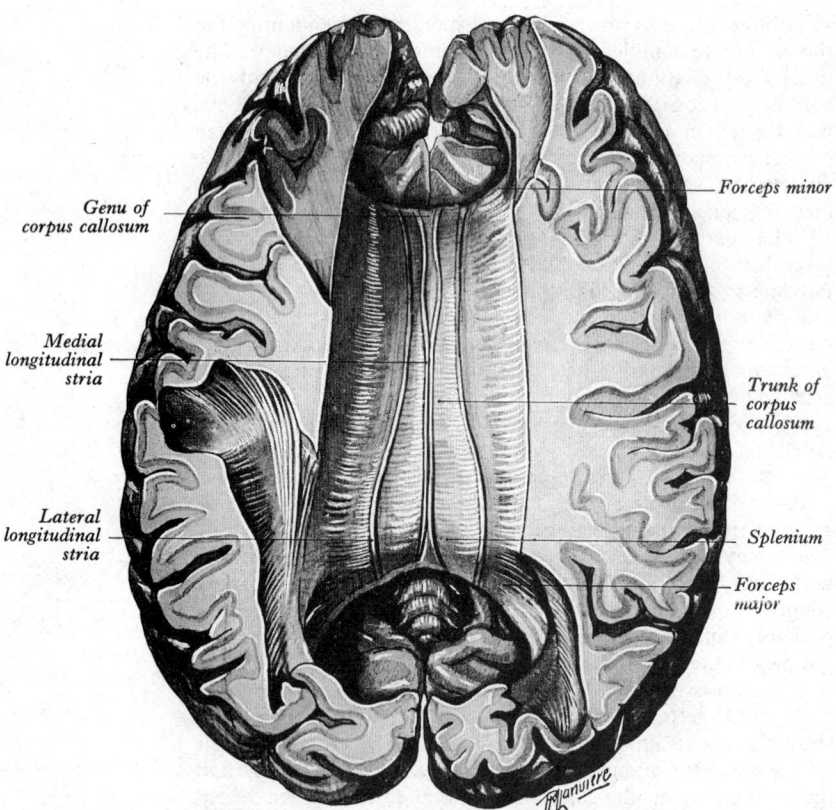

7.139 The corpus callosum, superior aspect, revealed by partial removal and dissection of the cerebral hemispheres.

of the temporal lobe to each other and to areas in other lobes is well established, although precise definitions are often lacking. Such connexions with the visual areas—between areas 18 and 19 and 37—and with the parietal lobe—between areas 7 and 40 and 22—are likely to be involved in complex integration between the sensorimotor areas in the temporal, occipital and parietal lobes. Thus, the more posterior region of the temporal lobe is perhaps more concerned with somatic activities, and especially with linguistic communication in its auditory and visual forms. The results of stimulation studies in man naturally provide the chief basis for such views (Blum et al. 1950). The anterior region of the temporal lobe appears to be more multiform still in its association with somatic and visceral activities. Apart from intrinsic

association connexions, it is linked to the limbic system (p. 990) through the connexions of the parahippocampal gyrus with the hippocampus, through connexions with the piriform lobe (p. 995) and with the amygdala (p. 996), and the temporal pole, area 38, is also involved in these interrelationships (p. 1001). Effects on blood pressure, respiration, and gastric motility—usually depression—have been elicited by stimulation of this part of the anterior temporal lobe; but the major source of evidence in mankind depends on the defects following limited temporal lobe ablations to relieve epileptiform attacks invading this region.

Stimulation prior to temporal lobectomy, which usually does not involve removal of either acoustic or motor speech cortex, may elicit complex memories of auditory, visual or combined content. After actual lobectomy a somewhat confusing plethora of manifestations have been described in the extensive clinical literature on this topic, for which the appropriate textbooks should be consulted. Effects on speech, auditory memory, and the interpretation of both auditory and visual phenomena have been noted (Milner 1967; Falconer 1967). A form of visual agnosia, sometimes called 'psychic blindness', in which patients or experimental animals are unable to recognize the visual significance of objects, coupled with a change in dietary habits, a tendency to examine objects with the oral region, hypersexual behaviour, and loss of emotional responses, together constitute a syndrome named after Klüver and Bucy (1937)—see also Terzian and Ore (1955). But in these instances the lobectomy involved the amygdaloid body, hippocampus, and part of the parahippocampal gyrus, and subsequent experiments have shown that these phenomena are only in part due to temporal lobectomy involving neocortex alone.

In recent years the temporal lobe has been loosely associated with the activities of memory; but except for the possible participation of the parahippocampal gyrus in what has been ascribed to the hippocampus and other associated structures as 'short term' memory, there is little firm evidence of any anatomical entities in the temporal lobe or elsewhere which can be regarded as sites of 'long term' memory.

OTHER CORTICAL AREAS

The insula (7.113A–D) has already been described (p. 987). It is a matter of choice to which lobe it is assigned, since it adjoins and is overlapped by all lobes save the occipital. Histologically it exhibits in its posterior part a type of cortex resembling the parietal (p. 1009), whereas its anterior region is variously described as agranular or like the piriform cortex (Bailey and von Bonin 1951). The fibre connexions of the insula are still largely

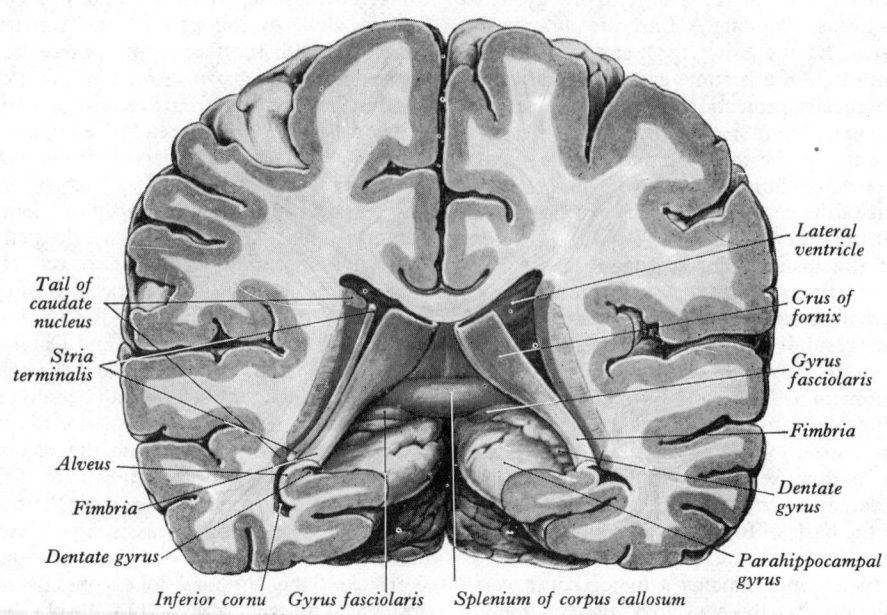

7.140A Anterior aspect of a coronal section of the cerebrum from which the posterior parts of the thalami have been removed to reveal the splenium and parts of the limbic system. (Note that the dentate gyrus is not equally exposed on the two sides.)

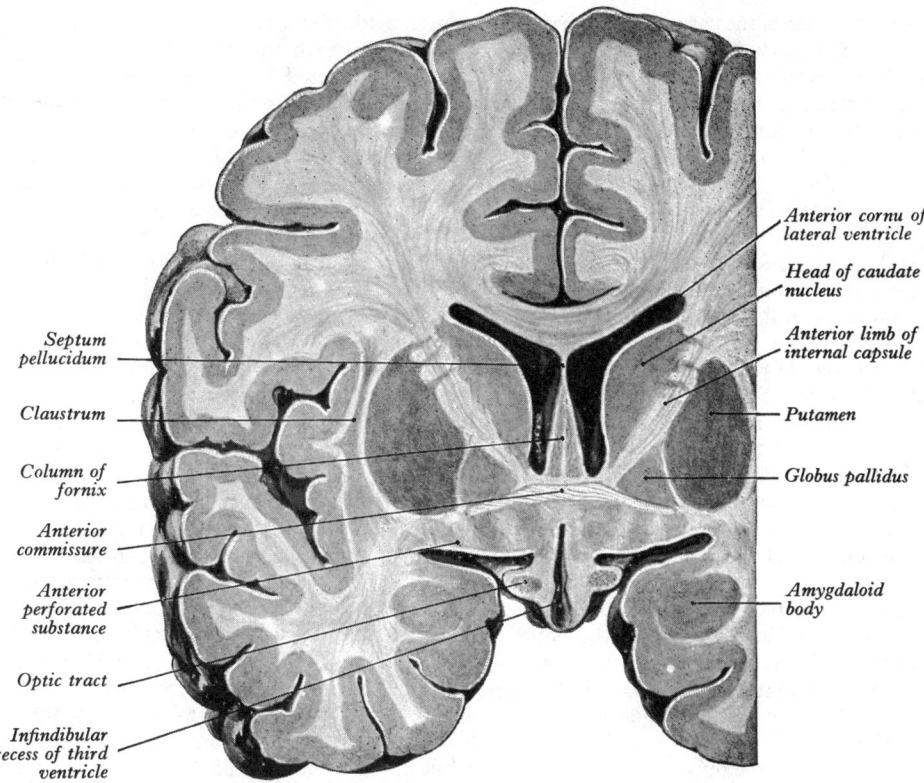

Labels on figure:
- Anterior cornu of lateral ventricle
- Head of caudate nucleus
- Anterior limb of internal capsule
- Putamen
- Globus pallidus
- Amygdaloid body
- Septum pellucidum
- Claustrum
- Column of fornix
- Anterior commissure
- Anterior perforated substance
- Optic tract
- Infindibular recess of third ventricle

7.140B A coronal section of the cerebrum immediately caudal to the optic chiasma and passing through the anterior commissure.

obscure; short association axons between it and all its opercula have been described, and it is said to be connected to the lateral olfactory gyrus and the piriform lobe (p. 995). Thalamic connexions (to the centromedian nucleus) have also been suggested (Le Gros Clark and Russell 1939; Locke 1967). Contralateral motor responses in the face and limbs have been elicited by insular stimulation in macaques and gibbons, and a somatotopic pattern may exist in this possible supplementary field (Showers and Lauer 1961). Stimulation in man excites visceral motor and sensory effects, such as belching, gastric movements, nausea and abdominal 'sensations' (Penfield and Rasmussen 1950). Increased salivation in man and experimental animals has also been elicited, and vasomotor effects in the latter. The insula has also been regarded as a gustatory area by some experimentalists. (For a review see Kaada 1960.)

The 'suppressor areas' reported by earlier experimenters in various parts of the cortex, particularly a vertical strip anterior to area 4, have been the subject of much controversy and adverse criticism (Druckman 1952). It is now considered that the suppression of motor activities from such areas is due to a generalized phenomenon of 'spreading depression' which is not associated with any particular areas of cortex. This phenomenon, which has not yet been completely explained, is characterized by a relatively long depression of neuronal activity, which may last several minutes, during which all responses may be suspended or distorted. It has no special anatomical correlates (Leao 1944).

The cingulate gyrus is discussed elsewhere in connexion with the limbic system (p. 990), but certain aspects of its anterior part, corresponding roughly to area 24, must be mentioned here. In its microscopic structure this area has features in common with the pre- and post-central areas. It receives a projection from the anterior nuclear complex of the thalamus which is reciprocated, and it may have connexions with the corpus striatum, hypothalamus and midbrain tegmentum. Commissural fibres through the corpus callosum have been described, and association fibres appear to link it with the frontal and parietal cortex anterior and posterior to the main sensorimotor region (Showers 1959).

Stimulation in macaques has been held to evoke bodily movements, both contralateral and bilateral, with some form of somatotopic pattern (Showers and Crosby 1958). This supplementary motor area extends into the posterior part of the cingulate gyrus, but the finding requires confirmation. In human patients stimulation of the anterior cingulate area may elicit changes in pulse, respiration and blood pressure, but the most reliable information has come from the results of surgical division or removal of the anterior cingulate gyrus, which is often involved in frontal leucotomy, performed to alleviate certain mental abnormalities. Cingulectomy alone appears to relieve abnormal aggressiveness and obsessional states, resulting in a milder and more placid personality (Tow and Whitty 1953). Similar effects have been recorded in monkeys subjected to ablations of the anterior cingulate gyrus (Glees et al. 1950). More recently, cingulate gyrus ablation has been claimed as a useful manœuvre in relieving intractable pain (Folty and White 1962). This mixture of somatic motor and sensory, emotional and visceral activities is at least in accord with the intermediate position of the anterior cingulate gyrus between the limbic system and the neocortex proper.

The Corpus Callosum

The corpus callosum is the great transverse commissure which connects the cerebral hemispheres and incidentally roofs in the lateral ventricles. Its development in mammals is proportional to the relative volume of the neocortex and is maximal in man (Rakic and Yakovlev 1968). A good conception of its position and size is obtained by examining it in median sections of the brain (7.110, 114). It forms an arched structure about 10 cm in length, its anterior end being about 4 cm from the frontal poles and its posterior end about 6 cm from the occipital poles of the hemispheres.

The *genu*, which forms the anterior end, inclines postero-inferiorly in front of the septum pellucidum and, diminishing rapidly in thickness, is prolonged posteriorly to the upper end of the lamina terminalis as the *rostrum*. The trunk arches back, convex above, to terminate posteriorly in the *splenium*, which is the thickest part of the corpus callosum.

The superior surface of the *trunk* of the corpus callosum (7.139) is covered by a thin grey layer, the *indusium griseum* which extends round the genu to the inferior surface of the rostrum to become continuous with the paraterminal gyrus, and in it are embedded two fine longitudinal bundles of fibres on each side, which are termed the medial and lateral longitudinal striae (p. 998); posteriorly the indusium griseum is continuous with the dentate gyrus and the hippocampus through the gyrus fasciolaris (7.140).

In the median plane the trunk of the corpus callosum forms the floor of the longitudinal fissure, and is related to the anterior cerebral vessels and to the lower border of the falx cerebri, which may come into actual contact with it posteriorly (8.31). On each side of the median plane the trunk is overlapped by the gyrus cinguli, from which it is separated by the slit-like callosal sulcus.

The inferior surface of the trunk is concave in its long axis and convex from side to side. In the median plane, the septum pellucidum is attached to it anteriorly, to an extent which depends on the length of the septum (7.110). Posteriorly it is fused with the body of the fornix and the commissure of the fornix. On each side of the median plane, the inferior surface of the trunk forms the roof of the lateral ventricle (7.141) and is covered with the ventricular ependyma.

The *genu* is continuous above with the trunk and below with the rostrum. Its anterior surface, which is related to the anterior cerebral vessels, is covered with the indusium griseum and the longitudinal striae. To its posterior surface is attached the septum pellucidum in the median plane, and on each side it forms the anterior wall of the anterior horn of the lateral ventricle.

The *rostrum* connects the genu to the upper end of the lamina terminalis. In the median plane its superior surface is attached to the septum pellucidum and, on each side, forms the narrow floor of the anterior horn of the lateral ventricle (7.155). On the inferior surface of the rostrum the indusium griseum and the longitudinal striae are carried backwards to the upper end of the paraterminal gyrus.

The *splenium* overhangs the posterior ends of the thalami, the pineal gland and the tectum of the midbrain. It is, however, separated from them by a number of structures. On each side of the median plane the crus of the fornix and the gyrus fasciolaris (7.140) curve upwards to reach the splenium. The crus of the fornix continues forwards on the inferior surface of the trunk, but the gyrus fasciolaris passes round the splenium, rapidly tapering off and fading away into the indusium griseum. In the median plane, the tela choroidea of the third ventricle passes forwards below the splenium through the transverse fissure, and the internal cerebral veins emerge from between its two layers and unite to form the great cerebral vein. Above, the splenium is covered with the indusium griseum and is related to the falx cerebri and the inferior sagittal sinus in the median plane, and to the cingulate gyrus on each side. Posteriorly the splenium is related to the free margin of the tentorium cerebelli, the great cerebral vein and the beginning of the straight sinus.

The nerve fibres of the corpus callosum radiate into the white core of the hemisphere on each side and disperse to various parts of the cerebral cortex. The rostral fibres extend laterally below the anterior cornua of the lateral ventricles, and connect the orbital surfaces of the two frontal lobes. The fibres of the genu curve forwards and connect the lateral and medial surfaces of the two frontal lobes, as the *forceps minor*. The fibres of the trunk pass laterally and intersect the projection fibres of the corona radiata (7.98). They connect wide cortical areas of the two hemispheres to one another. Those fibres of the trunk and of the splenium which together form the roof and lateral wall of the posterior horn and the lateral wall of the inferior horn of the ventricle constitute the *tapetum* (p. 1029). The remaining fibres of the splenium curve backwards and medially into the occipital lobes as the *forceps major*. This large bundle of fibres bulges into the superior part of the medial wall of the posterior horn of the ventricle and forms a curved elevation which is termed the *bulb of the posterior horn*.

Despite the great size of the corpus callosum and the enormous number of commissural fibres that it contains, limited information is available concerning its functional significance, apart from the obvious inference that it links the two hemispheres together and appears to ensure that they act as a single entity. The number of fibres is unknown in man, but in the cat there are 700,000 in each square millimetre (Myers 1959). The precise linkages effected by these commissural fibres have not been defined, except in the visual area (Zeki 1970, 1974) and the postcentral gyrus (Jones and Powell 1969). In the somatic sensory areas (SmI and SmII) the connexions are limited to the same area on the opposite side, but the right and left loci concerned with the hand and foot appear to lack such commissural connexion. Where commissural linkages occur they are considered to be highly specific, associating 'corresponding' columnar cortical units in bilateral functions.

Cases of complete congenital absence of the corpus callosum are recorded from time to time—the condition is a rare one—but the defect is usually found at autopsy, and the clinical history does not show any characteristic feature which can lead to certain diagnosis during life (Unterharnscheidt *et al.* 1968). In recent decades large portions, and in some cases the whole, of the corpus callosum have been divided as a surgical approach with apparently surprisingly little disturbance of function (Akelaitis 1942). Nevertheless, the accumulated experience of the results of brain damage (*see*, for example, Milner 1974), especially to the temporal lobe and corpus callosum, of experimental studies of commissurotomized cats (Butler 1966) and chimpanzees (Myers and Henson 1960), and particularly from the studies of Sperry (*vide infra* for references) on the effects of division of the human corpus callosum—all this evidence has not only illumined the function of this great commissure in the transfer of information, including memorized data, across the midline of the cerebrum, but has also confirmed a long-suspected asymmetry of function, which has led to a concept of 'dominance', usually by the left hemisphere. This has become a major arena of cerebral research which requires a short section to itself. Though it is impossible in this brief account even to survey more than superficially the profuse, and even at times prolix, literature recorded by clinicians, experimenters (anatomical and physiological), psychologists, behaviourists, and others, it is hoped that the references to major contributions will enable the interested reader to enlarge his knowledge elsewhere.

CEREBRAL ASYMMETRY OR HEMISPHERIC DOMINANCE

The concept of asymmetry of function, and structure, in the cerebrum is by no means modern, although it has attracted intensive study only in recent decades. It arose during the nineteenth century in company with developing views on functional localization in the cerebral cortex. Early protagonists of such asymmetry were Bouillard (1825) and Dax (1834), both concerned with speech, and Weber (1834), who believed he could detect asymmetry in various sensory functions. But it was the observations of Broca (1861), upon a single case of motor aphasia, which not only formed perhaps the first clear indication of cortical localization but also provided the first convincing evidence of *unilateral* representation of function in the cortex. Correlation between pathological changes in the region now known classically as Broca's area (p. 1014), and the observed almost complete defect in speech, clearly pointed to a close association between the initiation of speech and, in this case, the left cerebral hemisphere. Further patients, and similar examples of clinico-pathological deduction from the patients of other clinical observers, confirmed this finding. Moreover, the view was further elaborated into a concept of the left hemisphere as *dominant* and the right as in some, rather vague manner a *minor* hemisphere. This was perhaps partly prompted by an exaggerated identification of language with cognitive processes in general; perhaps also the known high incidence of right-handedness was also a corroborating factor, the more skilful hand being under the command of the apparently more intelligent hemisphere. Others were less dogmatic; Hughlings Jackson (1868), while confirming Broca's clinical findings, was wary of extending the undoubted presence of a unilateral motor speech 'centre' in the left hemisphere (in patients afflicted with vascular lesions of Broca's area) into the concept of left cerebral dominance or an exclusively 'verbal' hemisphere. He considered that the somewhat neglected right or 'minor'

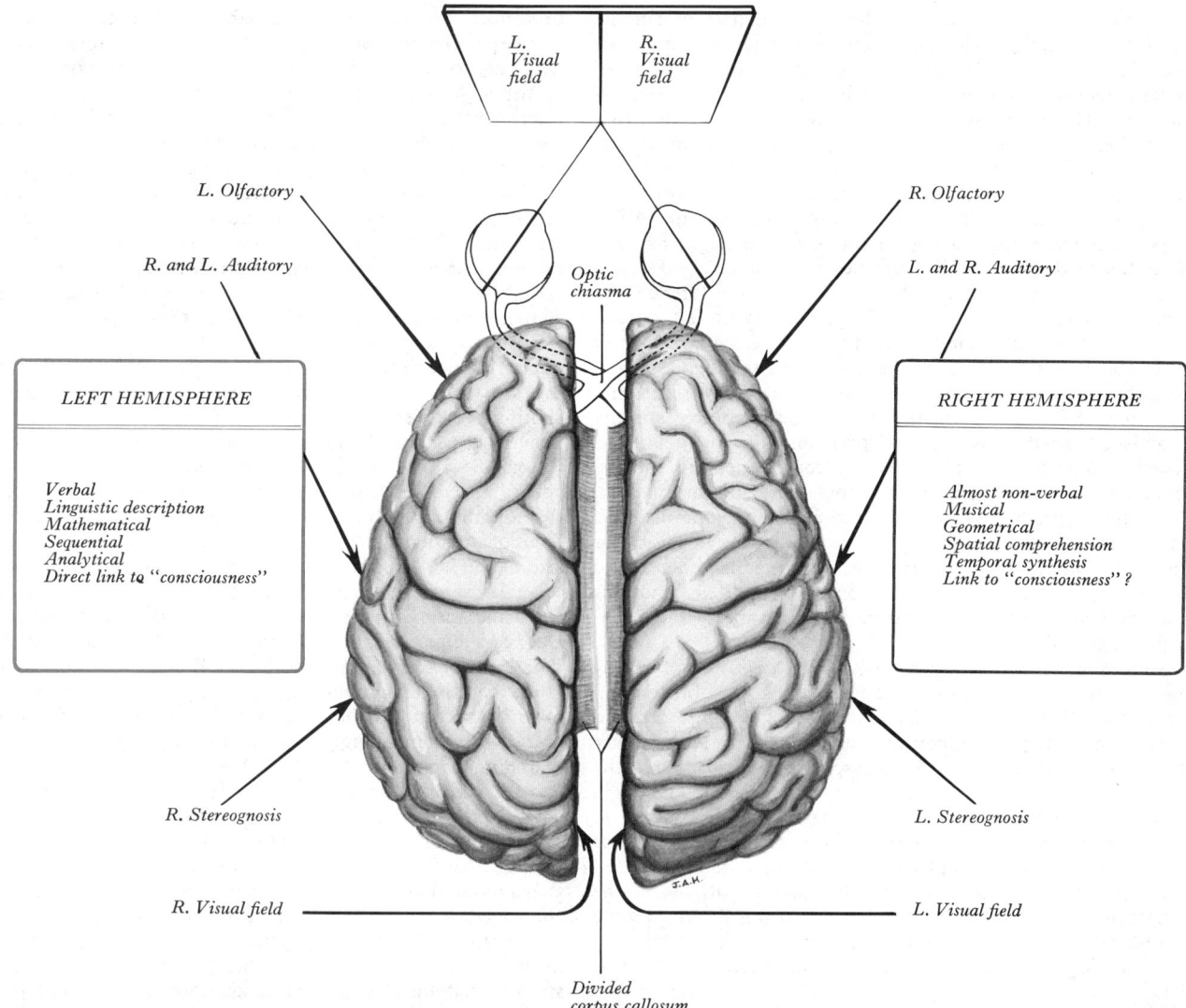

L.
Visual
field

R.
Visual
field

L. Olfactory

R. Olfactory

R. and L. Auditory

L. and R. Auditory

Optic
chiasma

LEFT HEMISPHERE

Verbal
Linguistic description
Mathematical
Sequential
Analytical
Direct link to "consciousness"

RIGHT HEMISPHERE

Almost non-verbal
Musical
Geometrical
Spatial comprehension
Temporal synthesis
Link to "consciousness" ?

R. Stereognosis

L. Stereognosis

R. Visual field

L. Visual field

Divided
corpus callosum

7.141 This diagram is an adaptation (with the author's permission) of the original 'split-brain' schema published by Professor R. W. Sperry in 1964, of hemispheric attributes as revealed by complete division of the corpus callosum. The right and left halves of the total visual field are projected into the contralateral hemispheres and are completely isolated, the right hand being 'controlled' from the left hemisphere. The olfactory and auditory inputs also project to separated hemispheres, but in the former case ipsilaterally and in the latter bilaterally with a contralateral bias. On each side of the diagram the abilities listed refer to the adjoining hemisphere; they are to be taken as extremely generalized attributes, which are likely to be modified as research in this field progresses. The diagram represents, of course, arrangements in an individual with left cerebral dominance or preference.

hemisphere might be of equal functional importance in somewhat vaguely expressed terms, such as an involvement in ideational, non-verbal cognitive activities. As will become apparent, he was closer to modern concepts than his contemporaries, in this as in other matters. However, the idea of dominance persisted with little change into the twentieth century, although it did not become a part of general neurological teaching until much more recent times. Perhaps this stagnation was due to the comparative dearth of further data, especially of an experimental type. Many further studies of cases of motor aphasia were, of course, added in confirmation of Broca's speech 'centre', and the defects consequent upon damage to other cerebral regions in some instances suggested, somewhat vaguely, asymmetry of cortical function and sometimes the reverse. Inevitably the role of the corpus callosum demanded attention—a commissure of such size, containing an estimated 200 million nerve fibres in mankind, cannot be ignored. But, as stated earlier, the observed effects of agenesis or occasional extensive injury proved difficult to assess, and it was easy for many observers to accept that even severe damage to the corpus callosum produced little change in the patient's cerebral performance. Retrospectively it is now easy to appreciate that this was the consequence of the relatively crude and inappropriate tests available. As will be seen, the introduction of adequate testing transformed the scene; but before passing to

this it is appropriate to turn briefly to the subject of '*handedness*', the preference for use of the right or left hand characteristic of mankind.

The close association of cortical areas concerned with the motor 'control' of speech and the right hand in the same, left hemisphere has inevitably evoked much statistical examination of the incidence of right and left hand-preference, a most fascinating problem in itself. A large literature of recorded data, and even more speculation, surrounds this subject (*see* Hicks and Kinsbourne 1978, for a recent brief review). Forelimb preference in mammals other than man is not easy to assess, and while such preferences are considered to exist, they are never so clear and apparently universal as they are in human beings, nor is there the same high frequency of a 'dominant' right hand. This is 'preferred' in about 90 per cent in all races and cultures, however diverse (Chamberlain 1928; Komai and Fukoka 1934; Hécaen and Ajuriaguerra 1964; Annett 1967; Hicks and Kinsbourne 1978). A genetic basis for hand-preference has been confidently postulated for mankind alone (Nagylaki and Levy 1973), but this is still an issue of some but diminishing disagreement. Such a genetic lateralization factor has naturally been extended to the development of the individual's dominant (verbal) cerebral hemisphere, and this influence is widely accepted. Single gene (Annett 1964) and double gene (Levy and Nagylaki 1972) models have been advanced to explain the observed data of familial incidence of hemispheric dominance (often coupled with those regarding hand-preference); but neither of these proposed genetic models accounts completely for the distribution of variants. However, such studies have demonstrated that hemispheric asymmetry is far from simple in its lateral incidence. If one individual is encountered who does not conform to the large majority of left-sided 'verbal' and right-handed people, it does not follow that he is automatically the reverse of this. A few (2 per cent according to Zangwill 1967), who are right-handed, nevertheless display clinical evidence of right cerebral dominance, and similar discordance between hand-preference and cerebral dominance occurs more frequently among left-handed subjects. Occasionally, lateralization of function may be difficult to establish, especially in respect of hand-preference. True ambidexterity may occur, and a limited degree of bilateral hemispheric function is more often suggested but less easily established. In the case of hand-preference in particular, the genetic influence may, of course, be complicated by cultural imitation of a largely right-handed population. How far this may, in childhood, produce secondary effects on cerebral dominance is uncertain; but it is certainly unwise to adopt a dogmatic or inflexible attitude of mind in regard to manual or cerebral asymmetry; and the apparent anomalies in their distributions renders theorization upon the genesis of asymmetry at present highly speculative.

Studies of cerebral dominance have been greatly amplified by the introduction of sodium amytal injections into the cerebral circulation (Wada 1949). Unilateral injection, despite cerebral arterial anastomoses, temporarily suspends or, at least greatly diminishes, function in the corresponding hemisphere; the 'speech' or dominant hemisphere is thus easily deduced in patients or normal volunteers, who may then be observed in much greater detail by appropriate testing techniques. The use of this method has complemented and sometimes extended the opportunities already afforded by injured patients in the investigation of the peculiarities of the two hemispheres; and the recorded data have become almost confusingly profuse, especially with the widespread interest of experimental psychologists in this field.

The contribution, to the study of cerebral asymmetry, of the assessment of functional defects following cerebral lesions, by injury or deliberate surgery, has been collectively large, though scattered through a large number of necessarily limited observations. Historically, of course, this is the oldest mode of study (cf. Broca) and also the newest (cf. Sperry, below), as has been pointed out by Milner (1974), who has herself made a large contribution in this field, particularly upon patients submitted to frontal lobectomies and patients with temporal lobe epilepsy, for which unilateral or bilateral ablations were carried out. The association of the temporal lobes, and more deeply situated hippocampi, with the function of memory has been established for many years; and a particular interest in the effects of unilateral temporal lobectomies resides in the revelation of asymmetrical memory defects (Milner 1958; Milner et al. 1964; Blakemore and Falconer 1967; Milner and Teuber 1968). Removal of the left temporal lobe impairs learning and memorization of verbal material, whether visual or auditory, whereas the right temporal lobe appears to be concerned with the acquisition and retention of visual and auditory experience of a non-verbal nature. Right lobectomy damages the ability to follow mazes, both in terms of vision and proprioception. Some of these lobectomies involve more than cortex, and they may include hippocampal and amygdaloid tissue. The localization of function is therefore sometimes crude and uncertain, but there is no doubt of the asymmetry which is observed.

Although all the avenues of enquiry mentioned above have contributed substantially to unmasking asymmetry of cerebral function, by temporary or permanent partial blockage of function in one hemisphere, the most fruitful approach to the problem has undoubtedly been the isolation of otherwise intact hemispheres from each other, though not from the external environment, by division of the corpus callosum for therapeutic purposes in human patients, and to a lesser extent by performing the same

procedure for experiment in laboratory animals, including monkeys. Before considering this evidence, however, it is necessary to turn briefly to the strictly anatomical contribution.

Although anatomical observation inevitably enters into the interpretation of the effects of disease, injuries, and experimental lesions, it might also be supposed that the marked functional asymmetries, gradually exposed during a century and more of observation, would be associated with clearly ascertainable differences of a structural type, quantitative and perhaps even qualitative. The evidence of such anatomical asymmetry is, however, small and equivocal. Stimulated by the views of Dax (1834) and the discoveries of Broca (1861), some of their contemporaries, and also Broca himself (1875), considered the left hemisphere to be the heavier, but others recorded the reverse opinion. None of the long series of those who have investigated the structure of the cerebral cortex appears to have noted significant differences between right and left sides, and Braitenberg (1977) has re-emphasized this by his own observations. Because of its proximity to Broca's area (on the left side), and its association with hearing, the superior temporal gyrus (and the adjoining speech area of Wernicke (see p. 1014)) has been a focus of attention in the search for asymmetry. A predominance in size of the left superior temporal gyrus has been described by several observers (such as Geschwind and Levitsky 1968; Le May and Culebras 1972; Hyde et al. 1973). The first of those cited claimed a left areal superiority in 80 per cent of human brains. A decade earlier two authorities, Von Bonin (1962) and Bodian (1962) had already pronounced opposite views on the significance of such structural asymmetries in respect of cerebral dominance, even when convincingly demonstrated, Bodian opining that searching for such anatomical differences would be sterile. The general lack of observations in this field, and their uncertainties, corroborates this view so far; but it may be that the structural asymmetries are of a more subtle nature. Di Chiro (1962) noted reciprocal differences in calibre between the superior and inferior anastomotic veins of the two hemispheres, which might denote, if confirmed, some divergence of vascular requirement; but this comparatively crude angiographic study has not been repeated. It must be concluded, therefore, that for the present, at least, a strictly anatomical approach to the problem has not proved useful.

Although the nineteenth-century interest in asymmetric control of speech continued, if somewhat sporadically, into the twentieth (see, for example, the work of Penfield and Roberts 1959), it is not unjust to suggest that the work of Sperry and his collaborators, in the sixties and beyond, has proved the outstanding contribution and stimulus in this field (Sperry et al. 1969; Gazzaniga 1970; Sperry 1970, 1974, 1977). The fact that he was already investigating 'split-brain' preparations in experimental animals and could collaborate with the pioneers (Bogen and Gazzaniga 1965) in complete division of the corpus callosum to alleviate severe, bilateral epilepsy, proved most productive. Sperry (1961) had elaborated, during his work on monkeys, an array of ingenious testing techniques, which could be applied, extended and further refined in observation upon these human patients, with the great advantage of verbal communication, allowing access to subjective phenomena which could merely be hypothecated in monkeys from their motor behaviour. The essence of such techniques is the isolated presentation of a wide variety of sensory cues, visual, auditory, or tactile, to the right or left hemisphere, and also the isolation from each other of such stimuli or cues presented simultaneously or sequentially in these different modes. The most commonly employed stimuli are visual cues, projected with great care in either the right or left half of the total visual field of the two eyes; but auditory and even olfactory cues have been extensively used, and the hands are frequently included, though in the performance of motor tasks, in which tactile and proprioceptive stimuli are significant.

Although corpus callosal division, or commissurotomy, is not a frequent operation, many patients have now been studied postoperatively, and a large array of experimental results has been recorded in such patients, and also in considerable numbers of patients and normal volunteers subjected to amytal injection, both by Sperry and his team and by other workers. Sperry's original findings have been largely corroborated, with some further extensions and modifications, often of a semantic nature (*see* Dimond and Beaumont 1973). They are shown in 7.141. There has, in fact, been so much corroboration of the functional asymmetry of the cerebrum and such detailed analysis of the peculiarities of the two hemispheres, that there is a distinct danger of over-statement. As in any fruitful field of discovery the original impetus is necessarily towards *analysis*, to establish differences and details. As will be emphasized further, below, it is necessary, as a corrective, to preserve also an impulse towards *synthesis*. The two hemispheres are, despite their differences, *complementary*, and they are in normal circumstances undeniably in continuity. The phenomena revealed by their artificial separation have now been studied by a host of workers in much detail, and only examples of those asymmetric qualities which are best established can be summarily recounted here.

In the surgically 'split' brain a suitably brief signal presented in the left half of the binocular visual field is received only by the right hemisphere, with no possibility of transfer of information to the left. (The stimulus must be brief enough, usually about 0·1 second in duration, to counteract momentary ocular movement from midline fixation.) If the right is the 'minor' hemisphere as it usually is, no verbal response can be obtained, and the patient is unaware (so he says) of the occurrence of the signal when he is questioned. (He may, in fact, volunteer statements, but the content of these is unrelated to the experimental situation.) Nevertheless, the patient (or experimentee if sodium amytal is used) must in some manner, apparently outside the consciousness associated with his left hemisphere, recognize signals, especially actual objects, with his right hemisphere. This is shown, for example, by his ability to *remember* an object displayed to him momentarily, and to pick it out subsequently with his *left* hand, unaided by vision, from a small array of dissimilar things presented to his searching hand. Whatever comments his isolated left hemisphere likes to make on his performance, of which he has no experience in this verbalizing, talkative hemisphere, these phenomena demonstrate that the right hemisphere is able to carry out much that is characteristic of a conscious brain: he 'sees', recognizes, employs memory, and directs successfully the corresponding hand. A large series of such experiments, employing a wide variety of visual, auditory, and tactile cues, and recorded by many observers, demonstrates beyond doubt that each hemisphere can, when separated, act to a marked extent as an independent unit, with its own ability for perception, learning, memorization, and motor control; hence, of course, the expression 'split brain' and the assumption that two separate consciousnesses can co-exist in the single individual. However, it is only the verbally skilful hemisphere, usually left, sometimes right, which can communicate directly, explicitly, and intelligently with other consciousnesses by the medium of language, thus uniting its own conscious world with the exterior. The other, the so-called 'minor' hemisphere, is severed from such communication and, moreover, from communication with the 'dominant' hemisphere. Nevertheless, this isolated right hemisphere is even able to recognize the meaning of the names of objects, presented either visually or aurally, and to correlate this recognition with memories of similar experiences and to carry out appropriate motor behaviour, though this half-brain, apparently conscious, is out of touch with any observer, who has no access to its activities except through its motor responses (excluding those of speech). These astonishing findings have persuaded some interpreters to an even firmer belief in the dominance of the verbal hemisphere; and to assign to it a major, if not exclusive role in a supposed brain-consciousness linkage. Some workers have even been led to question the need for two such 'brains' (e.g. Teuber 1974). The enthusiasm for analysis has undoubtedly led to an over-emphasis of the peculiarities of each hemisphere, tending to obscure the normal unity of the two halves of the cerebrum. Whatever may be hypothecated with respect to consciousness, it is clear that in some of the attributes of consciousness, such as spatial conception, geometrical perception, and perhaps 'musicality', the minor hemisphere is the more able partner and must participate in the conscious activities of the united hemispheres. It is easy, of course, to overlook the fact that in the diencephalic

region and caudal to this, the unity persists even after commissurotomy. As Young (1978) has picturesquely stated, the split personality of the commissurotomized patient, dual though it may seem, falls asleep as one, though his two eyes appear to take different views of his environment, when he is questioned on awakening. It is not surprising, perhaps, that such patients exhibit so little deficiency or duality in the ordinary activities of life, in which circumstances they are also able to explore the visual field freely with both eyes, to use both hands, and to rotate the head, thus duplicating in each hemisphere a representative array of signals from the entire sensory field, and not merely from one side or the other. It is only by the careful techniques initially evolved by Sperry and his co-workers that the defects are clearly unmasked. The specialization of the hemispheres in an asymmetrical, unilateral manner, so that in the total cerebrum certain abilities are sufficiently correlated with one hemisphere as to be assignable to it, is at present beyond explanation, despite some interesting theorization. Such arrangements are not, however, wholly unexpected. The existence of two central mechanisms, bilaterally related, for any generalized function is not easy to understand or even accept, without evidence of some linking or coordinating locus. The operation of two such central mechanisms, each dealing with half of what is continuously appreciated as an unbroken visual field has always posed such a problem, and appears to evoke some unitary integrating activity, although Sherrington (1906) was prepared to regard the simultaneous occurrence of the dual sets of neuronal events as a sufficient explanation. But the evidence for the transference of visual and other information across the midline of the cerebrum is impressive, and the memory stores of each hemisphere appear to be accessible to the other and to be probably bilateral to a considerable extent. All such phenomena would in hypothesis demand a large connexion between the two hemispheres, and this of course exists and must provide the anatomical matrix for extensive communication and presumably integration between right and left activities. Although so much of the research in this field has been focused upon functional differentiation of the hemispheres, with a special emphasis at first upon a unilateral mediation of speech, thus leading to the concept of dominance, it is becoming clearer in later investigations that the laterality of many functions is merely greater on one side, and not exclusive to one or other hemisphere. Even in the case of verbalization, some ability has long been assigned to the 'minor' hemisphere (consult Albert and Obler 1978). Moreover, our concepts of what constitutes a discrete 'function' may be in themselves misleading. Though I may *describe* a complex three-dimensional structure to another individual through a symbolic series of sounds called language, which are chiefly organized into an intelligible sequence by left hemispheric activity, it seems equally probable that I am *conceiving* an image or idea of the object I wish to describe in my right hemisphere, and that in the absence of linguistic description the concept may remain largely limited to the 'minor' brain. The too facile ascription of labelled functions or abilities to one or other side has been criticized by Milner (1974), who has redirected attention to the 'parallelism' of right and left activities. Broadbent (1974) has also emphasized the integrative action of the two hemispheres (an interesting reminder of the title of Sherrington's classic text!), regarding speech, for example, as a combined production of both. Such considerations favour the growing tendency to abandon the terms 'dominant' and 'minor'; and it is clearly preferable to think rather of a complex but complementary asymmetry of distribution of ability within the whole cerebrum. Inevitably this will entail a degree of right and left predominance, though perhaps only in so far as we impress our own semantics upon the complete organization. Eccles (1973, and in Popper and Eccles 1977) has professed a preference for the term 'dominance', on the basis of his own interpretation of the dominant, verbalizing hemisphere as the mediator between the external universe and the internal consciousness of the individual. He has resuscitated an old postulate due to Sherrington, whose student he was, that some part of the cerebrum would prove to be a site of what Eccles calls a *liaison* between brain and consciousness, but the latter goes further by designating the *dominant* hemisphere as a particularly *self-conscious* cerebral region. Although we are here being returned to the nineteenth-century association of speech and cognitive processes in a common locus, the evidence derived from Sperry's patients and other sources does also support such a view. But the apparent inaccessibility to verbal communication of the 'minor' hemisphere, and its apparent oblivion in regard to its own activities, are both revealed by the artificial separation. However useful this isolation may be in analysis of cerebral abilities, it may also mislead; for it must be followed by a re-synthesis in accounting for the operations of a normally united and singular brain.

These fascinating studies are currently providing the most fertile common ground between neuroanatomists, neurophysiologists, experimental psychologists, and even philosophers, in a general endeavour towards an understanding of the human brain and hence of what we call ourselves. It is an exciting prospect, particularly because it centres around the great problems of communication between the private world of the individual and his external environment of countless similar individuals, each locked in a private prison except insofar as communication succeeds. Perhaps, as understanding increases, we shall dissolve our differences and disperse our doubts, and then perhaps, in the happily optimistic words of Young (1978) 'a new form of laughter will echo through the system'. However that may be, it is pertinent to conclude with a statement by Braitenberg (1973), another distinguished student of man's brain: 'The *neuroanatomy* of neuronal networks will be for quite some time, and perhaps forever, the context that relates the *physiology* of single neurons and synapses to the macroscopic events of *psychology*.' (Editors' italics.) After analysis must come synthesis.

THE SEPTUM PELLUCIDUM

The septum pellucidum is a thin vertical partition (7.110, 114), consisting of two laminae, separated throughout a greater or lesser part of their extent by a narrow interval, termed the *cavity of the septum pellucidum*, which does not communicate with the ventricles of the brain. The septum is triangular in form, with its base in front and its apex behind. It is attached above to the inferior surface of the trunk of the corpus callosum; below and behind, to the anterior part of the fornix; below and in front to the upper surface of the rostrum of the corpus callosum; anteriorly to the posterior aspect of the callosal genu. The lateral surface of each lamina takes part in the formation of the medial wall of the anterior horn and central part of the lateral ventricle (7.141), and is therefore covered with ependyma.

The laminae contain both grey and white matter, and for what is known of their connexions see the account of the *septal areas* (p. 997) and the *dorsal fornix* (p. 1002). The development of the septum pellucidum is referred to on p. 171.

The anterior commissure is described on p. 997, and the *commissure of the fornix* (hippocampal commissure) on p. 1001.

INTERNAL STRUCTURE OF THE CEREBRAL HEMISPHERES

In the interior of the hemispheres are the lateral ventricles, the basal nuclei and many fibre tracts, both projection and intrinsic.

The Lateral Ventricles

The two lateral ventricles (7.142, 143, 144) are irregular cavities situated in the lower and medial parts of the cerebral hemispheres, one on each side of the median plane. They are almost completely separated from each other by the *septum pellucidum*, but each communicates with the third ventricle and indirectly with the other through the *interventricular foramen* (p. 981). They are lined with ependyma and contain cerebrospinal fluid, which, even in health, is secreted in considerable amounts. Each lateral ventricle consists of a *central part* and three *cornua* or *horns*, anterior, posterior and inferior (7.143, 144, 145).

The central part (7.144) of the lateral ventricle extends from the interventricular foramen to the splenium of the corpus callosum. It is a curved cavity with a roof, floor and medial wall; on transverse section it is triangular anteriorly and rectangular

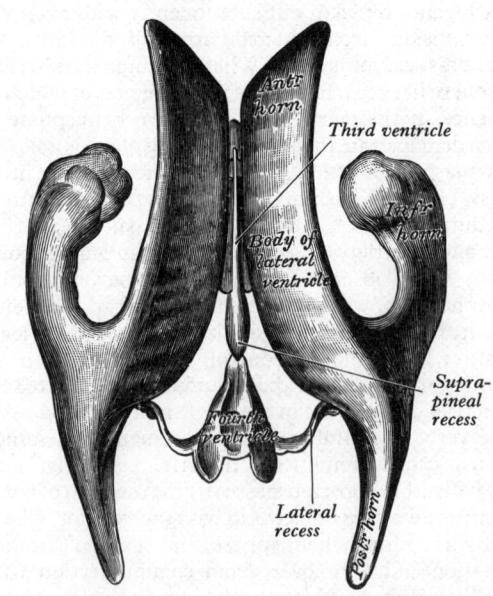

7.143A A drawing of a cast of the ventricular cavities. Superior aspect. (Retzius.) Note that where the lateral recess joins the fourth ventricle, the lateral dorsal recesses of the roof of the ventricle project dorsally on each side beyond the posterior margin of the median dorsal recess. The superior angle of the fourth ventricle and the aqueduct of the midbrain are hidden by the suprapineal recess.

posteriorly. The *roof* is the inferior surface of the corpus callosum; the *floor*, which is concave superomedially, is formed by the following in lateromedial order: the caudate nucleus, the stria terminalis, the thalamostriate vein, and the lateral portion of the upper surface of the thalamus. The caudate nucleus becomes rapidly narrower as it is traced backwards in the floor, and its long axis is directed laterally as well as posteriorly. The stria terminalis, a small bundle of white fibres (p. 996), and the thalamostriate vein occupy a narrow groove which follows the medial border of the caudate nucleus and separates it from the lateral margin of the superior surface of the thalamus. The latter may be almost entirely hidden by the choroid plexus, which invaginates the ependyma into the cavity through the slit-like

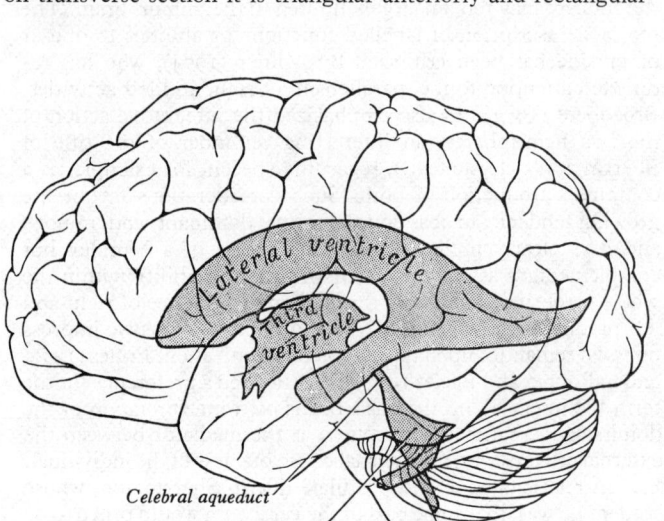

Celebral aqueduct

Fourth ventricle

7.142   Projection of the ventricles on to the left surface of the brain.

interval between the edge of the fornix and the upper surface of the thalamus, the ependymal invagination occupying this *choroid fissure*. The body of the fornix becomes wider as it is traced backwards, and its thin, lateral margin lies parallel with the groove for the stria terminalis.

The *medial wall* is formed by the posterior part of the septum pellucidum with the body of the fornix in its lower margin. Posteriorly, where the septum pellucidum ends, the roof and the floor meet one another on the medial wall.

The anterior cornu (7.141, 142) passes forwards, laterally and slightly downwards, into the frontal lobe. It is continuous behind with the central part at the interventricular foramen. In a coronal section it appears as a triangular slit below the anterior part of the corpus callosum, and it is bounded anteriorly by the posterior surface of the genu and rostrum of the corpus callosum. The *roof* is formed by the anterior part of the trunk of the corpus callosum. The greater part of the *floor* is formed by the rounded head of the caudate nucleus, but, in its medial portion, a small part is formed by the upper surface of the rostrum of the corpus callosum (7.155). The *medial boundary* is formed by the septum pellucidum containing the column of the fornix in its posterior edge.

The posterior cornu curves backwards and medially into the occipital lobe. Its development is very variable and frequently asymmetrical; it may be absent. Its *roof* and *lateral wall* are formed by fibres of the tapetum of the corpus callosum, which separate them from the optic radiation (p. 1031). The splenial fibres which constitute the forceps major pass medial to the posterior horn as they sweep backwards into the occipital lobe. In this part of their course they produce a rounded ridge, the *bulb* of the posterior cornu, in the upper part of the *medial wall*. Below the bulb, a second elevation may be identified on the medial wall. It is the *calcar avis*, and it corresponds to the infolded cortex of the anterior part of the calcarine sulcus (7.146). Posteriorly the lateral and medial walls meet each other. Kier (1977) has shown that the posterior cornu is phylogenetically the most recent extension of the lateral ventricle, being present only in the Anthropoidea. It is symmetrical in the human fetus, but in adults the two cornua are usually unequal in size, a peri-natal change which Kier ascribes to unequal folding of the calcarine fissure.

The inferior cornu (7.144, 147), the largest of the three, traverses the temporal lobe, forming in its course a curve round the posterior end of the thalamus. At first it descends posterolaterally and then curves anteriorly to within 2·5 cm of the temporal pole close to the uncus, its position being usually indicated on the surface of the brain by the superior temporal sulcus. A needle introduced at a trephine hole, the centre of which is placed 3 cm behind and 3 cm above the centre of the external acoustic meatus, and passed in the direction of the tip of the opposite auricle, enters the inferior cornu at a depth of some 5 cm from the surface.

The *roof* of the inferior cornu is formed chiefly by the inferior surface of the tapetum of the corpus callosum, but the tail of the caudate nucleus and the stria terminalis also extend forwards in the roof, at the extremity of which they are continuous with the amygdaloid body. Its *floor* consists of the collateral eminence laterally and the hippocampus medially; the surface of the latter is covered by the alveus, continuing as the fimbria of the hippocampus, which becomes continuous posteriorly with the crus of the fornix (p. 1001). Between the stria terminalis and the fimbria is the inferior, or temporal, part of the choroid fissure, through which the lower part of the choroid plexus of the lateral ventricle invaginates the ependyma closing the fissure. The choroid plexus covers the upper surface of the hippocampus.

The fimbria and the hippocampus have already been considered (pp. 1001 and 997), and a full description of the choroid plexus will be found on p. 1037.

The *collateral eminence* (7.147) is an elongated swelling lying lateral to and parallel with the hippocampus. It corresponds with the middle part of the collateral sulcus, and its size reflects the depth and direction of this sulcus. It is continuous behind with a flattened triangular area, named the *collateral trigone*, which forms the floor of the ventricle between the posterior and inferior cornua. The capacity of the lateral ventricles (and the other

chambers of the system) can be estimated by anatomical and radiological techniques (Weintraub 1953). Levinger and Kedern (1974) have devised a technique for estimation of the surface area of cerebral ventricles from casts of them, but have applied this so far only to cats.

The Nerve Fascicles of the Cerebrum

If the hemisphere is sliced horizontally about 1·25 cm above the corpus callosum, the central white substance of the hemisphere is seen as an oval area surrounded by a narrow convoluted margin of grey matter, and studded with red dots (puncta vasculosa) produced by the escape of blood from divided blood vessels. If the hemispheres be sliced off at the level of the corpus callosum, its nerve fibres will be seen in continuity with the white matter of the hemisphere on each side. The white matter contains many myelinated fibres, of varying size, supported by neuroglia. These fibres may be described according to their course and connexions, as belonging to one of three vast fibre systems: (1) The *commissural fibres* connect the two hemispheres to each other, linking corresponding or *homotopic* loci, and also *heterotopic* loci. (2) The *arcuate (association) fibres* connect different cortical areas of the same hemisphere to one another; some of them are collaterals of the projection and commissural fibres, but the majority are independent axons. (3) The *projection fibres* connect the cerebral cortex with the grey matter of the brainstem and the spinal cord in both directions.

(1) **The commissural fibres** have already been considered (pp. 963, 997, 1001, 1023).

(2) **The arcuate (association) fibres** (7.148, 149), which are all ipsilateral, are of two kinds: short arcuate fibres, connecting adjacent gyri to one another; long arcuate fibres, connecting more widely separated gyri to one another. Details of many individual association pathways have already been given in the section on cortical areas.

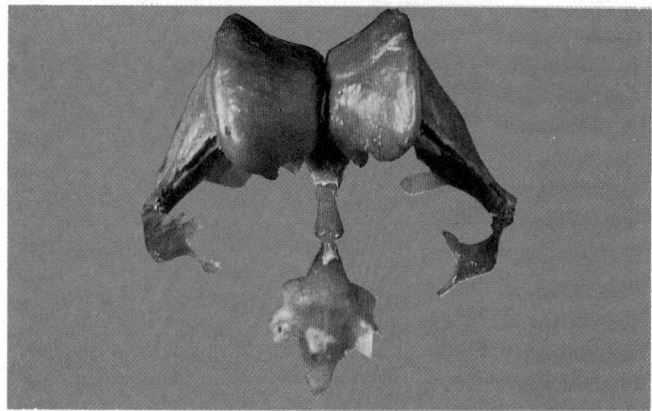

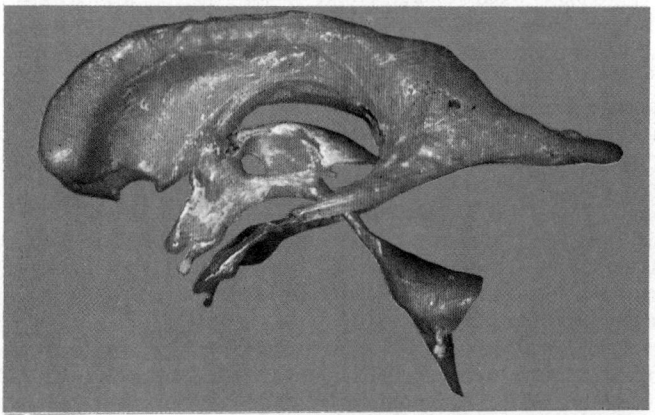

7.143B and C Resin casts of the ventricular system of the human brain prepared by Dr. D. H. Tompsett of the Royal College of Surgeons of England. B Ventral view; C Left lateral view. Compare with the superior aspect shown in 7.143A.

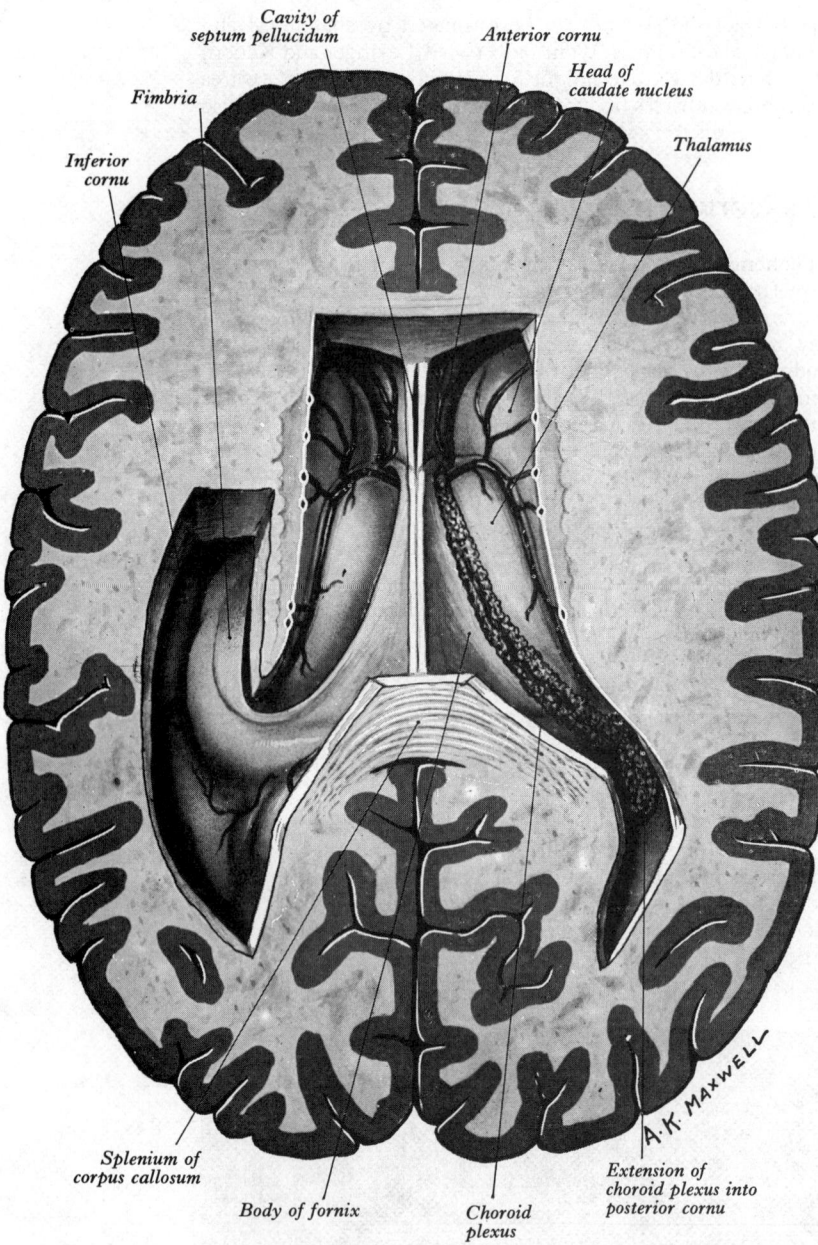

Cavity of septum pellucidum

Anterior cornu

Head of caudate nucleus

Fimbria

Thalamus

Inferior cornu

Splenium of corpus callosum

Body of fornix

Choroid plexus

Extension of choroid plexus into posterior cornu

A.K. MAXWELL

7.144 Horizontal section of the cerebrum dissected to remove the roofs of the lateral ventricles.

The *short arcuate fibres* may be entirely intracortical or they may pass immediately below the cortex to connect adjacent gyri, some merely passing from one wall of a sulcus to the other.

The *long arcuate fibres* group themselves, somewhat indistinctly, into bundles, which can be dissected in the formalin-hardened brain after the cortex and the subjacent short arcuate fibres have been removed. The fibres in each fasciculus show considerable variation in length, and the longest are always situated in the deepest part of the bundle. Concerning the precise connexions of these fibre bundles very little accurate information is at present available, for histological methods are unable to demonstrate them throughout the whole of their length. The following fasciculi can be distinguished: (*a*) the uncinate fasciculus; (*b*) the cingulum; (*c*) the superior longitudinal fasciculus; (*d*) the inferior longitudinal fasciculus; (*e*) the fronto-occipital fasciculus.

(*a*) The *uncinate fasciculus* connects the first motor speech area (p. 1014) and the gyri on the orbital surface of the frontal lobe with the cortex of the temporal pole and the area immediately adjoining. The fibres follow a sharply curved course and cross the floor of the stem of the lateral sulcus. They are related to the antero-inferior part of the insular area (7.148, 149).

(*b*) The *cingulum* is a long, curved bundle which commences on the medial surface of the hemisphere below the rostrum of the corpus callosum. It lies within the gyrus cinguli and so follows the curve of that gyrus. Inferiorly it enters the parahippocampal gyrus and spreads out so as to reach the adjoining parts of the temporal lobe. Along its convexity fibres enter and leave it in groups, giving it a spiked irregular appearance when dissected.

(*c*) The *superior longitudinal fasciculus* is the largest of all the arcuate fibre bundles. It commences in the anterior part of the frontal region and arches backwards above the insular area and lateral to the lower part of the corona radiata (*v. infra*). After giving off a number of fibres to the occipital cortex (areas 18 and 19), it curves downwards and forwards behind the insular area and spreads out into the temporal lobe. Like the other long arcuate fasciculi, it constantly receives new fibres throughout its whole extent and gives off fibres to the adjoining cortex. Its constituent fibres are so intermingled that it is quite impossible to determine their precise connexions by gross methods (7.149), nor is the dissecting microscope of any real help for this purpose.

(*d*) The *inferior longitudinal fasciculus* commences near the occipital pole and its fibres are derived chiefly from areas 18 and 19. They sweep forwards, separated from the posterior horn of the lateral ventricle by the fibres of the optic radiation and the commissural fibres of the tapetum, and after being crossed by the superior longitudinal fasciculus, they are distributed throughout the temporal lobe.

(*e*) The *fronto-occipital fasciculus* commences at the frontal pole and passes backwards on a deeper plane than the superior longitudinal fasciculus and separated from it by the lower part of the corona radiata (*vide infra*). It associates itself with the lateral border of the caudate nucleus, and is therefore closely related to the central part of the lateral ventricle. Posteriorly its fibres radiate into the occipital and temporal lobes in a fan-shaped manner, passing lateral to the posterior and inferior horns, and intersecting and mingling with the fibres of the tapetum of the corpus callosum.

The above details are almost entirely based upon the appearances of blunt dissection of the white substance of the cerebrum. Accurate knowledge of the origins and terminations of association fibres can only be established by experimental studies, which have not been carried out for most regions of the cortex. Studies of the visual areas (Le Gros Clark 1941; Zeki 1970, 1974) and the main somatic sensory areas (SmI)—*see* Jones and Powell (1968)—suggest a high degree of specificity in such connexions.

(3) **The projection fibres** connect the cerebral cortex with the lower parts of the brain (including the diencephalon) and the spinal cord, and include both *corticofugal* and *corticopetal* fibres.

The projection fibres converge from all directions on the corpus striatum (7.150). For the most part they are internal to arcuate fibres, and they intersect the commissural fibres of the corpus callosum and the anterior commissure. At the periphery of the corpus striatum, they form the *corona radiata*. The medial aspect of the corona radiata is separated from the lateral ventricle by the fronto-occipital fasciculus, and its lateral aspect is covered by the superior longitudinal fasciculus. Below, the corona radiata is directly continuous with the internal capsule, a thick, curved zone of white matter which includes all the projection fibres, and which cuts into the corpus striatum, dividing it almost completely into two parts, the lentiform and the caudate nuclei.

The Internal Capsule

In horizontal section through the cerebral hemisphere the internal capsule is a broad band of white fibres, bent with a lateral concavity, which accommodates itself to the convex medial surface of the lentiform nucleus (7.151, 152, 153, 154). It can be divided into an *anterior limb*, a *genu*, a *posterior limb*, a *retrolentiform part* and a *sublentiform part*. The anterior limb is interposed between the lentiform nucleus on the lateral side and the head of the caudate nucleus on the medial side. The posterior limb has the thalamus on its medial side and the lentiform nucleus on its lateral side. The fibres of the internal capsule continue to converge as they pass downwards, and at the same time the frontal fibres tend to pass backwards and medially, while the temporal and occipital fibres pass forwards and laterally. At the lower limit of the lentiform nucleus, they are crossed by the optic tract and enter the midbrain. The corticofugal fibres enter the crus cerebri, where the frontal fibres are placed to the medial side and the temporal, parietal and occipital fibres to the lateral side. What follows is a simplified account of the fibre constitution of the internal capsule. A number of other fibre 'systems' are described at points throughout the section on the telencephalon.

The *anterior limb* of the internal capsule contains *frontopontine fibres*, which arise in the cortex of the frontal lobe and synapse about the cells of the nuclei pontis, the axons of which pass to the cerebellar hemisphere of the opposite side. In addition there are the fibres of the anterior thalamic radiation which interconnect the medial and anterior nuclei of the thalamus and the cortex of the frontal lobe.

The *genu* is usually regarded as containing the *corticonuclear* fibres which arise mainly from area 4 of the cerebral cortex and terminate in the motor nuclei of the cranial nerves to the head, mostly of the opposite side. The most anterior fibres of the superior thalamic radiation interconnecting the thalamus and cerebral cortex also extend into the genu.

The *posterior limb* includes the *corticospinal tract* disposed in scattered bundles, with the fibres concerned with the innervation of the upper limb anteriorly followed by those to the trunk and lower limbs. The location of corticospinal fibres in the anterior part of the posterior limb has been widely accepted since the observations of Charcot (1883) and Déjerine (1901), but more recently evidence derived chiefly from stereotactic lesions in man has accumulated to suggest that these fibres are in fact situated in

the *posterior* part of the posterior limb. This evidence has been reviewed by Smith (1967) and Hanaway and Young (1977). Other descending fibres here include frontopontine fibres from the frontal lobe, in particular areas 4 and 6, *corticorubral* fibres from the frontal lobe to the red nucleus and fibres from the globus pallidus contained in the subthalamic fasciculus. The majority of this portion of the internal capsule contains fibres of the superior thalamic radiation carrying general sensory impulses from the ventral thalamic nuclei to the postcentral gyrus. (For a survey of the posterior limb in the human brain consult Smith 1967.)

In the *retrolentiform part* are *parietopontine* and *occipitopontine* fibres and fibres from the occipital cortex to the superior colliculus and pretectal region. In addition there is the posterior thalamic radiation which includes the optic radiation, and interconnexions between the cortices of the occipital and parietal lobes and the caudal portions of the thalamus, especially the pulvinar.

The fibres of the *optic radiation* arise in the lateral geniculate body and sweep backwards in the angle between the central part and inferior horn of the lateral ventricle. In their course they are intimately related to the superior and lateral surfaces of the inferior horn and lateral surface of the posterior horn of the lateral ventricle, being separated from the latter by the tapetum.

The *sublentiform part* contains the *temporopontine* and some parietopontine fibres, and the acoustic radiation running from the medial geniculate body to the superior temporal and transverse temporal gyri (areas 41 and 42). There are also a few fibres interconnecting the thalamus with the cortex of the temporal lobe and insula.

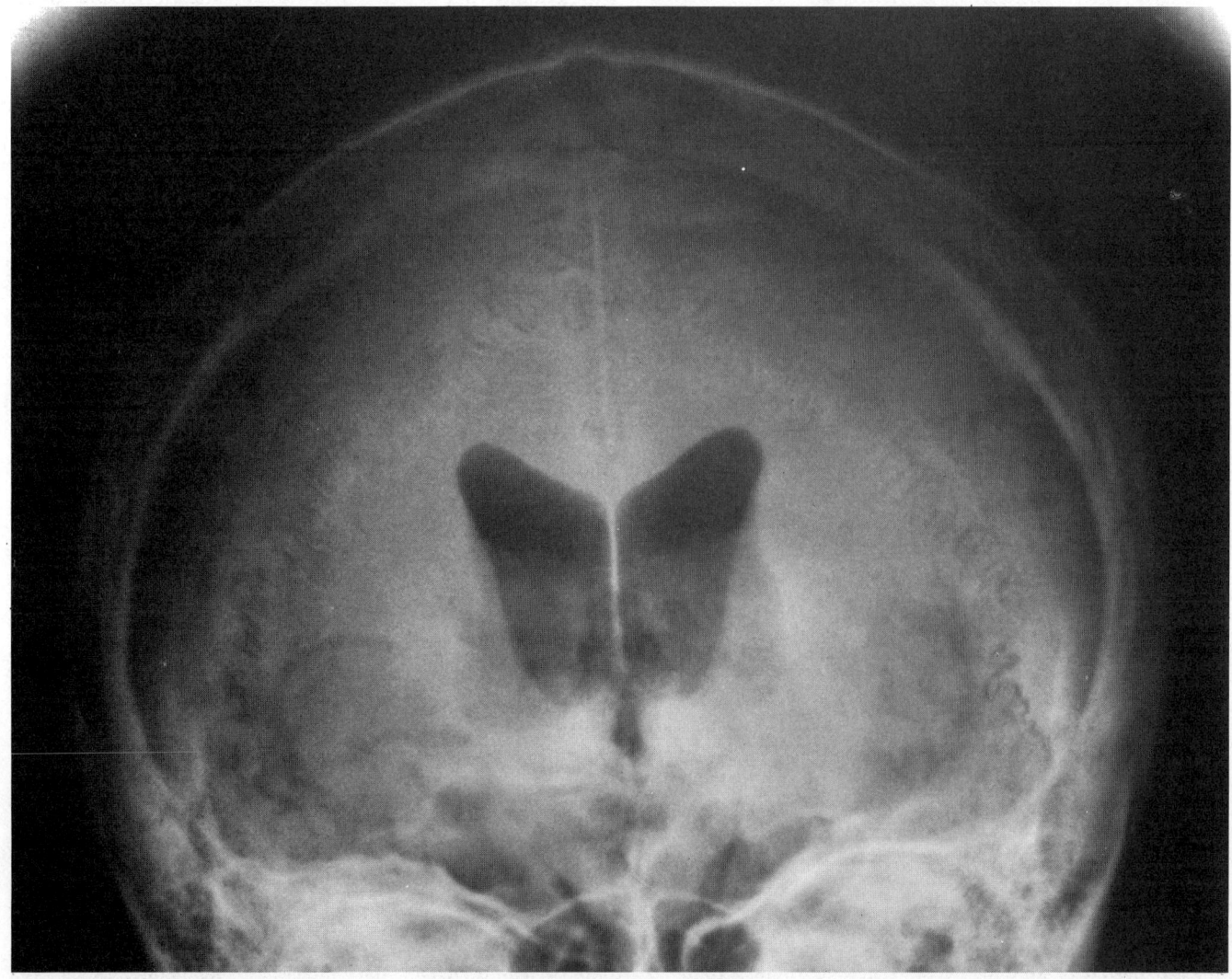

7.145 Anteroposterior radiograph of the head after the introduction of air into the ventricular system of the brain. The outlines of the bodies and anterior horns of both ventricles are separated in the midline by the shadow of the septum pellucidum; directly inferior to the latter, the outline of the third ventricle can be seen. (Kindly provided by Dr. R. D. Hoare, Dept. of Diagnostic Radiology, Guy's Hospital.)

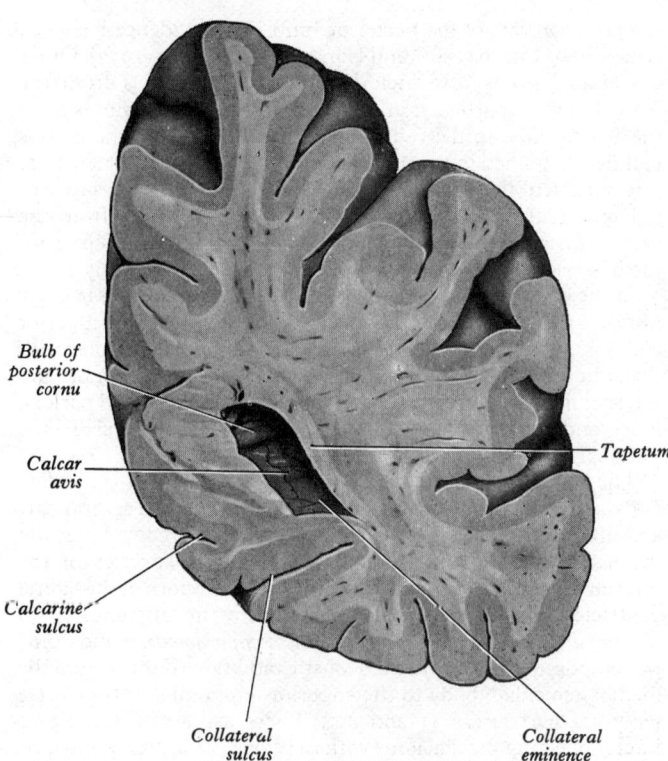

Bulb of posterior cornu

Calcar avis

Calcarine sulcus

Tapetum

Collateral sulcus

Collateral eminence

7.146 Anterior aspect of a coronal section through the posterior cornu of the left lateral ventricle.

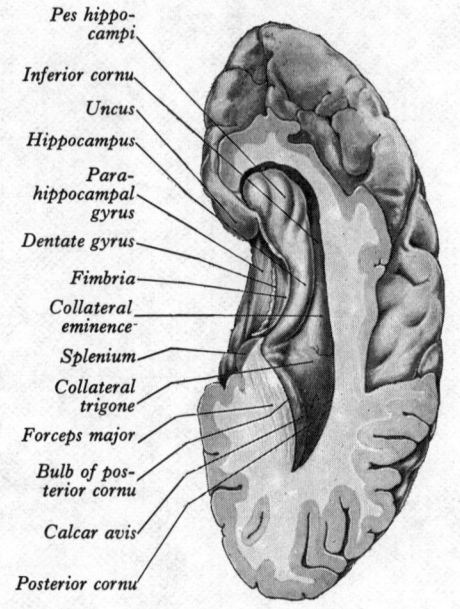

Pes hippocampi

Inferior cornu

Uncus

Hippocampus

Parahippocampal gyrus

Dentate gyrus

Fimbria

Collateral eminence

Splenium

Collateral trigone

Forceps major

Bulb of posterior cornu

Calcar avis

Posterior cornu

7.147 The posterior and inferior cornua of the right lateral ventricle, exposed from above.

The fibres of the *acoustic radiation* sweep forwards and laterally below and behind the lentiform nucleus to reach the cortex.

The connexions between the cortex and the thalamus are discussed on p. 956; corticohypothalamic connexions on p. 957; corticostriate connexions on p. 1034.

The Basal Nuclei

Situated within each cerebral hemisphere is a series of subcortical nuclear masses of grey matter often loosely grouped under the general term *basal nuclei*. Unfortunately, the structures included vary between workers; but most commonly the term includes the amygdaloid complex, the claustrum, the caudate nucleus and the lentiform nucleus (7.141, 151, 152, 153). These topographical structures are a heterogeneous group with respect to their structural and functional associations and phylogenetic history. Accordingly, a large and often confusing terminology has been generated by different investigators, the various structures being subdivided and regrouped in many different ways. The array of terminologies will not be discussed at length, but some of the more commonly used alternatives are mentioned briefly here. (For critical reviews of the basal nuclei *see* Mettler 1968; Webster 1978.)

The *amygdaloid body* (amygdaloid nuclear complex, amygdala, or *archistriatum*) is discussed elsewhere in relation to the limbic system (p. 990).

The *claustrum*, a tenuous sheet of grey matter, surrounded by laminae of white matter which separate it from the deep surface of the insula and the external surface of the lentiform nucleus, is further mentioned on p. 1036.

The *caudate nucleus* and *lentiform nucleus* are commonly grouped together as the *corpus striatum*. The lentiform nucleus is divided into an internal *globus pallidus* and an external *putamen*, which are structurally distinct. The putamen resembles the caudate nucleus in its structure, and together they constitute the *neostriatum*, frequently termed simply the *striatum*. When used in this manner, therefore, striatofugal fibres are those which leave the caudate nucleus or putamen for other destinations, whilst afferents, denoted with the suffix -striate (e.g. thalamostriate), are those passing to the putamen or caudate nucleus, and not to the *corpus* striatum as a whole. (The terms will be used in the manner just described in this account, but it must be realized that some neuroanatomists refer to the whole corpus striatum when they use these terms, and in other cases, the intended meaning is by no means clear.)

The *globus pallidus* or *paleostriatum* is often referred to simply as the *pallidum*, and hence such terms as pallidal afferents or pallidofugal fibres.

TOPOGRAPHY OF THE CORPUS STRIATUM

The caudate nucleus is an arcuate mass of grey matter, which has already been noted in the floor of the anterior cornu and central part of the lateral ventricle and the roof of its inferior cornu (7.141, 144, 151, 152, 153, 156). Its anterosuperior end is massive and termed the *head*. At the interventricular foramen this narrows into the *body* of the nucleus which tapers imperceptibly into the *tail*. As noted the nucleus projects into the *floor* of the anterior cornu and central part of the lateral ventricle and continues round into the roof of the inferior cornu, its tail merging at its anterior and lower extremity (tip) with the amygdaloid body. In all parts of the ventricle it is covered by ependyma. In the central part and inferior cornu of the lateral ventricle the stria terminalis lies along the medial border of the caudate nucleus. In the central part of the ventricle the stria terminalis lies with the thalamostriate vein in the groove between the caudate nucleus and the thalamus. In the inferior cornu it forms the upper border of the lower part of the choroid fissure.

In the anterior cornu and central part of the lateral ventricle the lateral margin of the caudate nucleus is related to the corpus callosum, being separated from it by the fronto-occipital arcuate bundle (p. 1030) and the subcallosal fasciculus (p. 1034). The lateral surface of the nucleus is flat and in contact with the internal capsule. At its anterior end the head of the caudate nucleus is fused below with the putamen of the lentiform nucleus just above the anterior perforated substance. Above this junction strands of grey matter traverse the anterior limb of the internal capsule and connect the putamen of the lentiform nucleus with the head of the caudate nucleus. The striped appearance which this region presents prompted the term corpus striatum.

In the inferior cornu of the lateral ventricle the tail curves forwards behind, below and then lateral to the fibres of the internal capsule entering the crus cerebri; these separate it from the thalamus which lies above and medially. Above, the sublentiform part of the internal capsule and fibres from the external capsule separate the tail from the globus pallidus.

The lentiform nucleus is shaped like a biconvex lens, but the curvature of its medial surface is sharper than the curvature of its lateral surface (7.102, 141, 151, 152, 153, 156). Cut in section, it is seen to consist of two portions, which differ from each other in their colour. The larger, lateral part which is darker in colour, is termed the *putamen*; the smaller, medial portion is of a lighter tint, and is termed the *globus pallidus*.

The lentiform nucleus is completely buried in the substance of the hemisphere. Laterally it is covered by a thin layer of white matter which constitutes the *external capsule*. This sheet is covered on its lateral side by the *claustrum*, which intervenes between it and the subcortical white matter of the insula. Medially the lentiform nucleus is in relation to the internal capsule, which separates it from the thalamus behind, and from the head of the caudate nucleus in front. Round its anterior, superior and posterior margins the nucleus is related to the corona radiata. The inferior part of the lentiform nucleus is deeply grooved by the anterior commissure as it passes backwards and laterally into the temporal lobe (7.141), and anteriorly it is continuous with the head of the caudate nucleus. A little in front of the groove, the grey matter of the corpus striatum is continuous with that of the anterior perforated substance, and the lateral striate arteries, which enter the brain at this site, run laterally and then turn upwards in close contact with the lateral surface of the lentiform nucleus, before they pierce its substance. The lentiform nucleus lies above the inferior cornu of the lateral ventricle and is separated from it by the fibres of the external capsule as they pass medially towards the subthalamic region, the sublentiform part of the internal capsule (p. 1031), the tail of the caudate nucleus and the stria terminalis. More anteriorly, it is separated from the amygdaloid body by the ansa peduncularis.

STRUCTURE OF THE CORPUS STRIATUM

The caudate nucleus and putamen are similar in structure, being highly cellular, well-vascularized zones, permeated by delicate bundles of either finely myelinated or non-myelinated, small-diameter fibres. Hence the pinkish-grey colour of these regions when freshly sectioned, in contrast to the pale colour of the globus pallidus which is encapsulated and traversed by numerous, coarse, heavily myelinated fibres.

The neurons of the caudate nucleus and putamen are mainly small multipolar cells, with round, triangular, or spindle-shaped somata, admixed with a small proportion of large multipolar cells, the ratio of the two being roughly 20:1. The small neurons are considered to be receptive and associative interneurons, which receive the synaptic terminals of many of the striatal afferents. The large neurons possess large and roughly spherical or ovoid dendritic trees, and their principal axons are one source of striatal efferents (*vide infra*), but some of the smaller neurons probably also contribute to the striatal outflow. Although the general pattern of connexions of the striatum with extrinsic regions of the nervous region is becoming clearer (*vide infra*) we have, as yet, little knowledge of the detailed morphology of its neurons, intrinsic connexions, and synaptology. Some progress has, however, been made with both the Golgi technique and electron microscopy in various experimental animals (Fox *et al.* 1966; Kemp 1968*a* and *b*, 1970) but only some general features have emerged. The dendrites of striatal cells bear dendritic spines, and axodendritic synapses both related to spines and smooth interspine dendritic surfaces, and axosomatic synapses have also been identified. Further, both asymmetrical (Type I) and symmetrical (Type II) synapses are present. Lesions placed in the cerebral cortex, thalamus, or midbrain, result in degeneration of asymmetrical synapses, which are presumably excitatory, in relation to both dendrites and (after cortical lesions) to the cell somata as well. In contrast, lesions within the caudate nucleus itself cause degeneration in both type I and type II synapses. The latter, which are possibly inhibitory, are distributed on dendrites, somata and the initial segments of axons. In addition, it was shown that the axons of all striatal neurons studied possessed collateral branches and it was thought that the type II synaptic terminals might be derived not only from intrinsic interneurons, but also from the collaterals of neurons projecting to other regions. It is of considerable interest that histopharmacological

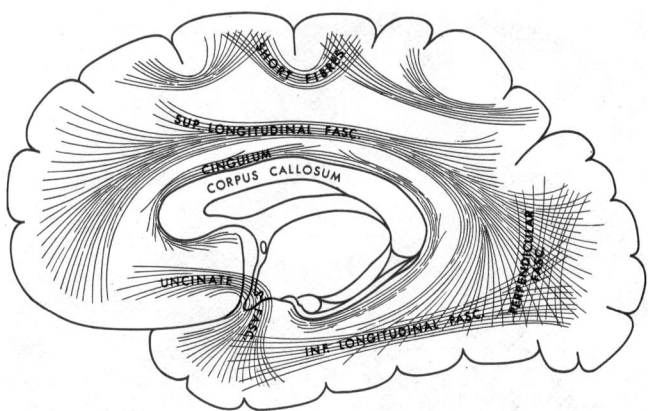

7.148 The principal arcuate (association) fibres in the cerebrum.

studies (Hökfelt 1968) have demonstrated that in the caudate nucleus of the rat many of the synaptic terminals contain dense-cored or 'granulated' vesicles which characterize aminergic neurons (7.19), and are thought, in this site, to represent stores of *dopamine*. The remaining synapses contain the clear spherical small presynaptic vesicles found in cholinergic terminals.

The globus pallidus contains a rather scattered population of large multipolar neurons which in their general cytology closely resemble the lower motor neurons in other situations and also the large cells of the substantia nigra. Their primary dendrites branch infrequently and possess only occasional small dendritic spines. Their synapses are mainly in relation to these dendrites and are of type II, but scattered between these are a small number of type I. The relatively sparse axosomatic synapses are also of both types. After experimental lesions in the caudate nucleus, numerous degenerating symmetrical (type II) terminals were found in relation to both dendrites and somata of neurons in the globus pallidus (and also in the substantia nigra). The axons of the large pallidal neurons give origin to the profuse, well-myelinated, pallidofugal system of fibres (*vide infra*). Topographically, the globus pallidus is separated by a layer of white matter, the *external medullary lamina* or *stria* from the medial aspect of the putamen, whilst an *internal medullary lamina* or *stria* divides it into a smaller, medial, and a larger, lateral part (*pallidum I and II* respectively).

CONNEXIONS OF THE CORPUS STRIATUM

In outline, the neostriatum (putamen and caudate nucleus) constitutes the main receiving station; this projects to the globus

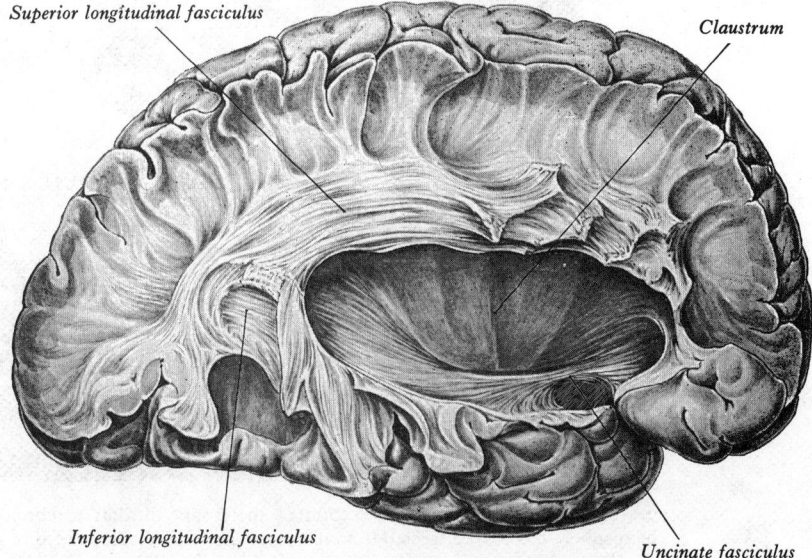

Superior longitudinal fasciculus

Claustrum

Inferior longitudinal fasciculus

Uncinate fasciculus

7.149 A dissection showing some of the long arcuate fasciculi of the right cerebral hemisphere.

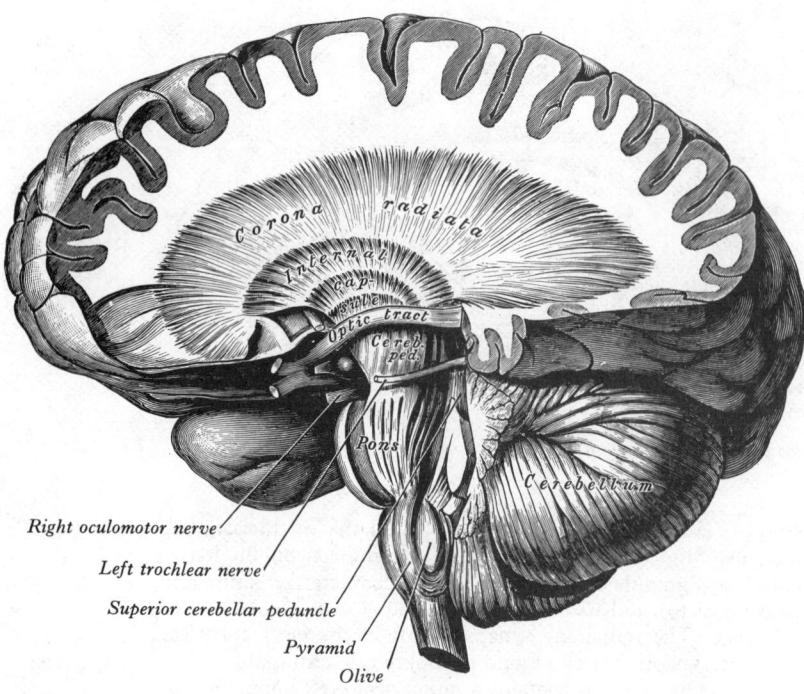

Corona radiata
Internal cap.
Optic tract
Cereb. ped.
Pons
Cerebellum

Right oculomotor nerve

Left trochlear nerve

Superior cerebellar peduncle

Pyramid

Olive

7.150A A dissection showing the convergence of cortical projection fibres through the corona radiata into the cerebral peduncle and pons.

pallidus, which in turn, gives rise to the main outflow pathways. However, in addition some efferent paths leave the striatum directly, whilst the pallidum also receives other afferent connexions.

Afferent connexions to the striatum are derived mainly from the cerebral cortex, thalamus and substantia nigra.

Although their existence was for long in doubt, *corticostriate*

fibres have now been shown to constitute a widespread system which is *topically organized* and converges from almost *all parts of the cerebral cortex* on to the caudate nucleus and putamen. The experimental animals used in these studies included rats, rabbits, cats and monkeys to which the degeneration techniques of Glees, Nauta and more recently, ultrastructural methods, have been applied (Glees 1944; Whitlock and Nauta 1956; Webster 1965; Carman *et al.* 1963; Carman *et al.* 1965; Kemp and Powell 1970). Jones *et al.* (1977), using the retrograde horseradish peroxidase technique, have demonstrated that the corticostriate projection in monkeys is derived from the smaller pyramidal cells of lamina V, distributed sparsely, but widely, in the sensorimotor areas and elsewhere in the frontal and parietal regions in the case of ipsilateral fibres, but only in areas 4, 6 and 8 in the case of contralateral connexions. Yeterian and Hoesen (1978), however, in contrast, used autoradiographic techniques in rhesus monkeys and obtained somewhat differing results. Following the intracortical injection of tritiated leucine or proline they demonstrated the interesting phenomenon that areas having strong reciprocal corticocortical connexions, both projected to common areas of the caudate nucleus. Thus, each part of the cerebral cortex projects to a specific part of the caudate-putamen complex, although some overlap occurs. Most of the projections are derived from the ipsilateral cortex only, but some regions receive bilateral projections from restricted regions of the sensorimotor cortices of both hemispheres. Since the details of corticostriate projection in these different mammals vary somewhat they will not be pursued further here; obviously they should not be extrapolated directly to the human brain, but it is most probable that broadly similar connexions exist. The most profuse projections are from the sensorimotor cortex, whilst the least prominent are from the visual cortex. Corticofugal fibres reach the striatum through both the internal and external capsules, and from the temporal lobe via sublenticular routes, whilst some of the direct and crossed corticocaudate fibres run with the *subcallosal fasciculus* in association with the fronto-occipital arcuate bundle of fibres. It is at present not clear what proportion of corticostriate fibres have

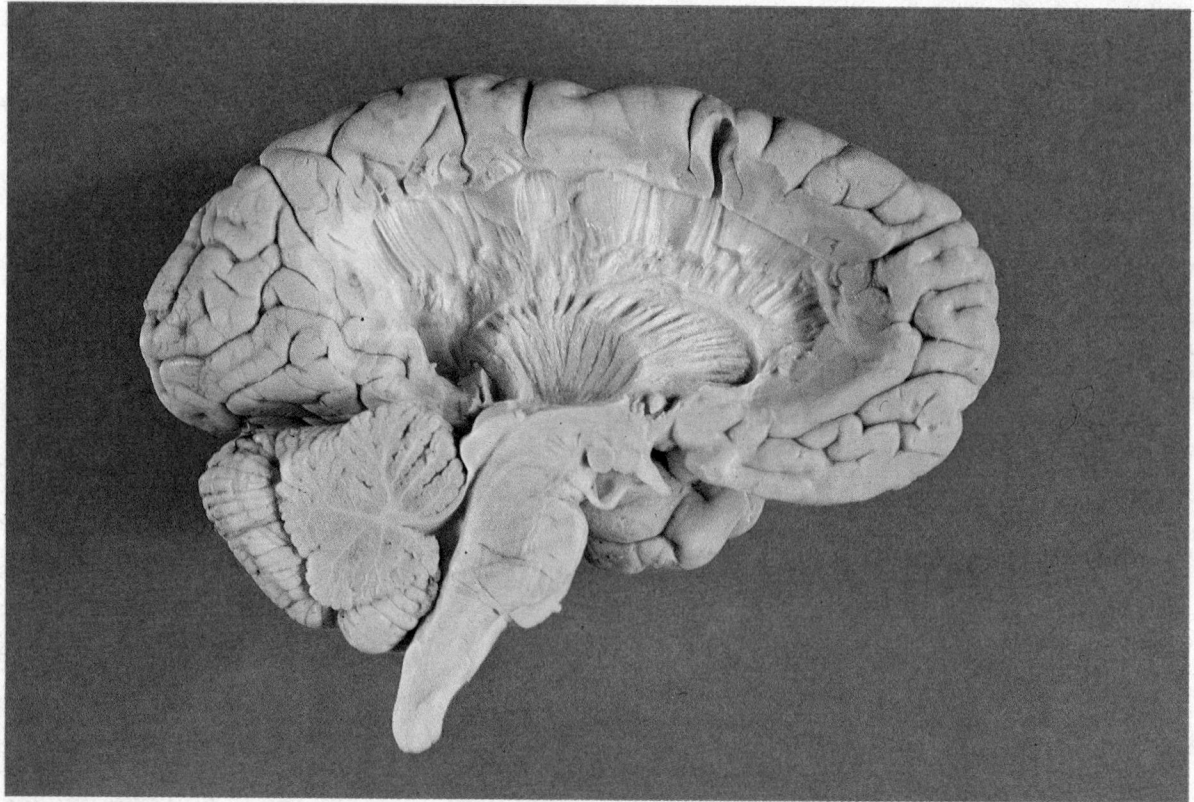

7.150B After median sagittal section of the brain, the left cerebral hemisphere has been dissected from its *medial* aspect to display the fibre bundles of the corona radiata and internal capsule. This entailed the removal of the cingulate gyrus and subjacent white matter, much of the paramedian corpus callosum and fornix, the dorsal thalamus, and the head

and body of the caudate nucleus. The oval depression previously occupied by the dorsal thalamus can clearly be seen within the curved depression left after removal of the caudate nucleus. (Dissection by Dr. Andrew Seal, Dept. of Anatomy, Guy's Hospital Medical School, London.)

their principal terminations in these nuclei, or whether substantial numbers of them are collateral branches of other corticofugal systems, for example, the corticospinal tract. As mentioned above, the corticostriate fibres form type I synapses upon the dendrites and somata of the striatal neurons and thus, on morphological grounds, are probably excitatory.

Thalamostriate fibres form another profuse afferent system to the striatum and are derived from the nucleus centromedianus, various other intralaminar and midline nuclei, and also from the nucleus medialis dorsalis. Some of these fibres pass directly to the caudate nucleus to terminate there; others traverse the caudate, or skirt it, to reach the internal capsule, where they pass between its fibres to reach the putamen. The details of many of these connexions have not been satisfactorily resolved. The nucleus centromedianus has been claimed to project to many parts of the corpus striatum. Studies in the monkey (Powell and Cowan 1956), however, suggested that its projection was topically organized and confined to the putamen, whilst other investigators (Mehler 1966) claimed that its projection was to a band which included the whole mediolateral extent of both putamen and caudate nucleus. The intralaminar and dorsomedial thalamic nuclei are thought to project largely to the caudate nucleus.

Nigrostriate fibres, for long considered by some authorities to provide an important afferent system to the striatum and globus pallidus, although doubted by others, have now gained considerable prominence in relation to theories of the genesis of Parkinsonian tremor (Calne 1970). An increasing body of evidence, derived from electrophysiology (Purpura *et al.* 1967; Feltz and Mackenzie 1969; Connor 1968), neuroanatomy (Adinolfi and Pappas 1968; Nauta and Mehler 1969), and neuropharmacology (Calne 1970), not only supports the existence of such a pathway, but strongly suggests that its neurons, at least in part, utilize *dopamine* as their neurochemical transmitter (*vide infra*).

The *nigrostriate fibres* take origin from cells in both the pars compacta and pars reticularis of the substantia nigra. They ascend through the caudal subthalamus and reach the internal capsule, where separate bundles of nigrostriate fibres interdigitate with the capsular fibres, giving in histological section an appearance like a hair comb, hence the term *comb bundle* for this tract. Some fibres pass to the caudate nucleus, whilst the remainder continue on to reach the putamen and globus pallidus. Carpenter and Peter (1972) have identified degenerating terminals in parts of the caudate nucleus and the putamen following lesions of the pars compacta of the substantia nigra in rhesus monkeys; they have described a topological nigro-caudate relationship.

For further details of the brainstem origins and routes of the monoaminergic pathways to the striatum consult pp. 943–953.

Further afferents may reach the striatum from the subthalamic nucleus and other neighbouring cell groups, but these cannot be regarded as established.

Striatofugal connexions are mainly to the *globus pallidus*, but *striatonigral* and *striatothalamic* fibres, which retrace the pathways just recounted to reach the substantia nigra and thalamus, have also been described. Both the caudate nucleus and the putamen project in a topically organized manner on to the cells of the globus pallidus. The lateral part of the putamen projects only to the external segment of the globus pallidus, whilst the more medial parts of the putamen and the caudate nucleus project to either both segments of the globus pallidus, or to the internal segment alone.

Other striatofugal connexions have been described, but their status is less certain; these include projections to the subthalamic nucleus and to restricted parts of the inferior olivary nucleus.

Afferent connexions to the globus pallidus are principally the topically organized *striatopallidal fibres* from both the putamen and caudate nucleus described above.

Other pallidal afferents include fibres from the subthalamic nucleus, substantia nigra, thalamus and cerebral cortex. The subthalamic fibres reach the globus pallidus (mainly pallidum I) as part of the *subthalamic fasciculus* (p. 965). *Nigropallidal fibres* have been described by a number of authors as part of the comb bundle mentioned above. The neurons forming this tract are dopaminergic, but it is uncertain whether nigrostriate or nigropallidal fibres predominate in the *human* brain. In addition to the extensive, well-organized corticostriate projection mentioned above, some investigators have described fibres which accompany these from many parts of the cerebral cortex, but terminate directly in the globus pallidus. *Thalamopallidal fibres* may also reach the pallidum from the intralaminar, centromedian and dorsomedial nuclei of the thalamus. It must be emphasized, however, that these somewhat variable results have been obtained with different methods in a number of experimental animals, and the existence of some pathways, and the destination to specific parts of the pallidum of others, have not yet been established in the human brain.

The pallidofugal system constitutes by far the quantitatively most important outflow from the corpus striatum. The fibres are thick and well myelinated, and they form a topographically complex series of pathways which diverge to a number of destinations. The main pathways involved include: (1) the *ansa lenticularis*; (2) the *fasciculus lenticularis*; (3) the *fasciculus thalamicus*; (4) the *fasciculus subthalamicus*; and (5) *descending fibres*. The topography of these tracts was described and illustrated (7.102) with the subthalamus (p. 963) and will not be repeated here.

The main destinations of the pallidofugal fibres are to: (1) the *thalamus*, mainly the *nucleus ventralis anterior*, with smaller contributions to the nuclei ventralis intermedius and centro-medianus; (2) the *subthalamic nucleus* and other, smaller, subthalamic centres which include the zona incerta, ento-peduncular nucleus and the nucleus of the prerubral field; (3) the *substantia nigra*; (4) the *red nucleus*; (5) the *midbrain reticular formation*; and (6) the *inferior olivary nucleus*.

Illustration 7.156 summarizes some of the main connexions of the basal nuclei based on an analysis by Webster (1975) which the interested reader should undoubtedly consult.

The following points should also be noted. The external segment of the pallidum projects on to the internal segment, from which most of the pallidofugal fibres originate. The fibres to the subthalamic nucleus, however, arise mainly from the external segment. For long, a *pallidohypothalamic tract* appeared in most accounts of the corpus striatum, but more recent studies using the newer degeneration techniques offered no conclusive evidence that fibres project directly from the globus pallidus to hypothalamic nuclei. The tract described pursues an aberrant course which curves around the column of the fornix; thereafter the majority of its fibres rejoin the main pallidal outflow.

SUMMARY AND FUNCTIONAL STUDIES

Clearly, the corpus striatum and its associated regions of the nervous system constitute an exceedingly complex set of interconnexions, but their details and our understanding of their function role is still elementary.

Broadly, information *converges* on the corpus striatum from most of the cerebral cortex, the thalamus, subthalamus and a number of brainstem centres. After integration within the corpus striatum, information *diverges* again to the thalamus, subthalamus and the various brainstem centres listed above.

For long it was held that the main outflow from the corpus striatum, as part of the classical 'extrapyramidal motor system', *descended*, through various indirect polysynaptic pathways, to the lower motor centres. Whilst such descending pathways as the reticulospinal and rubrospinal tracts undoubtedly carry information derived in part from the corpus striatum, it is increasingly clear that the most prominent outflow from the corpus striatum is to the *nucleus ventralis anterior* of the thalamus (p. 958), where, after integration with other channels, *ascending* pathways radiate to the *motor and premotor areas* (areas 4 and 6). Equally important reciprocal interconnexions are established with the *subthalamic nucleus* and *substantia nigra*.

An impressive list of organic nervous diseases has long been known to affect, in varying measure, the different parts of the corpus striatum and their associated nuclei and fibre pathways. These cannot be reviewed here, and the reader should consult texts devoted to clinical neurology and neuropathology. Briefly,

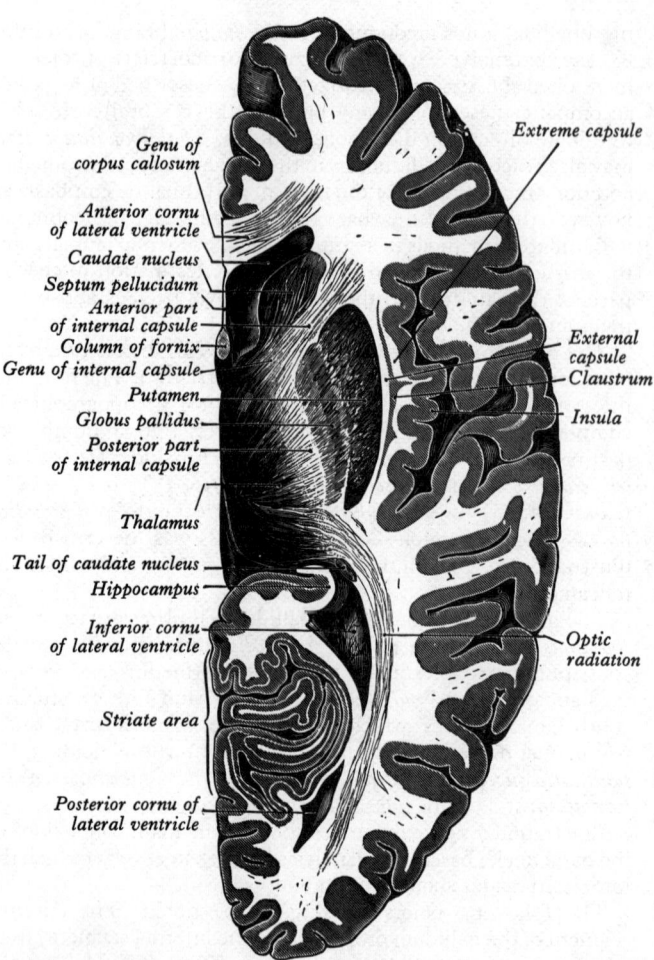

Genu of
corpus callosum

Anterior cornu
of lateral ventricle

Caudate nucleus

Septum pellucidum

Anterior part
of internal capsule

Column of fornix

Genu of internal capsule

Putamen

Globus pallidus

Posterior part
of internal capsule

Thalamus

Tail of caudate nucleus

Hippocampus

Inferior cornu
of lateral ventricle

Striate area

Posterior cornu of
lateral ventricle

Extreme capsule

External
capsule

Claustrum

Insula

Optic
radiation

7.151   Superior aspect of a horizontal section through the right cerebral hemisphere.

in different forms and combinations, these various disease states usually present: (1) *disturbances of muscle 'tone'* or resistance to stretch; sometimes this is reduced, but more commonly some form of *rigidity* is present; (2) diminution or *loss of automatic associated movements* such as arm-swinging during walking, facial expression, etc.; (3) the presence of *unwanted movements* which are uncontrollable and purposeless; these may be *choreiform*, *athetoid*, or *ballistic* (p. 964), or they may take the form of a *tremor* resulting from the fairly rapid alternating contraction of opposing muscle groups. The tremor is usually 'static', that is, present when the limb is otherwise at rest, but occasionally an *intention tremor* is present as in some forms of cerebellar dysfunction (p. 929). Attempts to link the various combinations of symptoms and signs with disease in specific locations have met with limited success. Similarly, experiments with animals have often been uninformative.

Ablation of the putamen, the caudate nucleus, or the globus pallidus, often causes little obvious change in motor behaviour, provided the lesions do not affect surrounding structures. Complete ablation of the globus pallidus on both sides in the monkey, however, results in a general poverty of movement and a reduction in manipulative skills.

Rapid stimulation of the caudate nucleus in unanaesthetized animals sometimes elicits movements of the head and limbs, and some evidences of a somatotopic pattern in the nucleus have been adduced. Low-frequency stimulation, or chemical stimulation, at different points in the corpus striatum, however, usually results in an inhibition of motor responses. It may induce long periods of immobility, an inhibition of cortically induced movements, or 'arrest reactions' in the unanaesthetized animal. Confirmation of this has been obtained by electrophysiological 'unit' recording from the motor cortex and the ventrolateral thalamus—inhibition of unit activity followed stimulation of the caudate nucleus. Other experiments have demonstrated the inhibition of responses which

had been established by learning processes, following caudate nucleus stimulation.

Neurosurgical ablation of the globus pallidus or ventrolateral thalamus has been used, often with considerable success, to diminish contralateral rigidity and tremor in human disease. Thalamic ablations seem more effective in the relief of tremor (including cerebellar tremor) whilst pallidal lesions have a greater effect in reducing rigidity. In addition to their therapeutic value, such manœuvres provide a source of material for histopathological and neuropharmacological study and also allow electrophysiological recording to be carried out. Most interesting in this regard has been the identification of neurons in the ventrolateral thalamus in cases of Parkinsonian tremor, which show a repetitive, rhythmical discharge of impulses, the frequency of which corresponds to that of the tremor (Guiot *et al.* 1962; Guiot *et al.* 1964; Jasper 1966; Bates 1969). Similar findings have been induced experimentally by placing brainstem lesions in the monkey. Intensive experimental analysis involving anatomical, physiological, behavioural and pharmacological methods is now in progress attempting to define the causal mechanism of the rhythmic neuronal discharges. Much interest has been aroused by the demonstration of an *inhibitory dopaminergic nigrostriate pathway* which is probably involved in a proportion of cases of Parkinsonian tremor (Calne 1970). Of the various possible mechanisms proposed, it is currently considered most likely that normal striatal function depends upon a balance between the inhibitory afferent pathway just mentioned, and excitatory afferent pathways which are cholinergic. Reduced activity in the inhibitory pathway leads to an excessive excitatory output from the pallidum, which in turn leads to oscillating bursts of activity in ventrolateral thalamic neurons. Finally, the latter through their cortical radiations are thought to generate rhythmical activity in the various corticofugal descending pathways to the lower motor centres. On this view, dopamine is a normal neurotransmitter in the striatum; there is a considerable body of evidence to show that dopamine levels in the striatum are reduced in many cases of Parkinsonism, and this is paralleled by the encouraging clinical responses to replacement therapy in a substantial proportion of patients who have been given L-dopa, a precursor of dopamine.

Despite this dramatic progress in the pharmacological and clinical sphere, however, beyond the long-held view that the corpus striatum is intimately involved in motor control, and the displacement of emphasis from descending pathways to a control of the cerebral cortex by this region of the brain, its precise role in normal behaviour remains conjectural. (For an excellent 'non-clinical view of the structure and function of the basal nuclei' and an extensive bibliography consult the stimulating paper by Webster 1975.)

The claustrum is a thin sheet of grey matter coextensive with the insula and the putamen of the lentiform nucleus, from which it is separated by the fibres of the external capsule. It is thickest below and in front, where it becomes continuous with the anterior perforated substance, the amygdala, and prepiriform cortex. It has been regarded by some authorities as belonging to the corpus striatum and by others as a detached portion of the insular cortex. However, more detailed studies suggest that it probably consists of at least two structurally and functionally distinct zones. These have been termed the '*insular*' claustrum and the '*temporal*' or '*prepiriform*' claustrum respectively. In experimental animals the insular part of the claustrum has been shown to possess reciprocal, topically organized *corticoclaustral* and *claustrocortical* connexions with many regions of the neocortex. Its detailed connexions and functional significance are unknown in the human brain.

The external capsule is a thin layer of white matter which is interposed between the lateral aspect of the lentiform nucleus and the claustrum. The fibres of the external capsule are derived from the frontoparietal operculum of the insula and, after passing across the lateral surface of the lentiform nucleus, they turn medially below the nucleus and the ansa lenticularis. Their subthalamic connexions are uncertain. Some of the fibres of the anterior commissure are believed to traverse the external capsule.

The Choroid Plexus of the Lateral Ventricle

Projecting into the lateral ventricle on its medial aspect is a highly vascularized fringe composed of pia mater and of the ependymal lining of the cavity (**7.**157, 158). This is the choroid plexus of the lateral ventricle, which is itself only part of a larger structure, the *tela choroidea*, to be described below. The pial basis of the choroid plexus is invaginated during development (p. 169), over a linear region of the medial wall of the hemisphere where no nervous tissue develops. The pia therefore comes into direct contact with the epithelial lining of the ventricle, the *ependyma*, and the two tissues are fused in the structure of the choroid plexus, which otherwise consists chiefly of small blood vessels, capillaries and nerve fibres. The plexus extends anteriorly to the interventricular foramen, where it is continuous across the third ventricle with the choroid plexus of the opposite lateral ventricle. From this point it passes posteriorly above and in contact with the thalamus (**7.**144) to curve round its posterior end into the inferior cornu of the ventricle, reaching as far as the pes hippocampi (**7.**147). When the choroid plexus is torn away from the hemisphere, the line of its

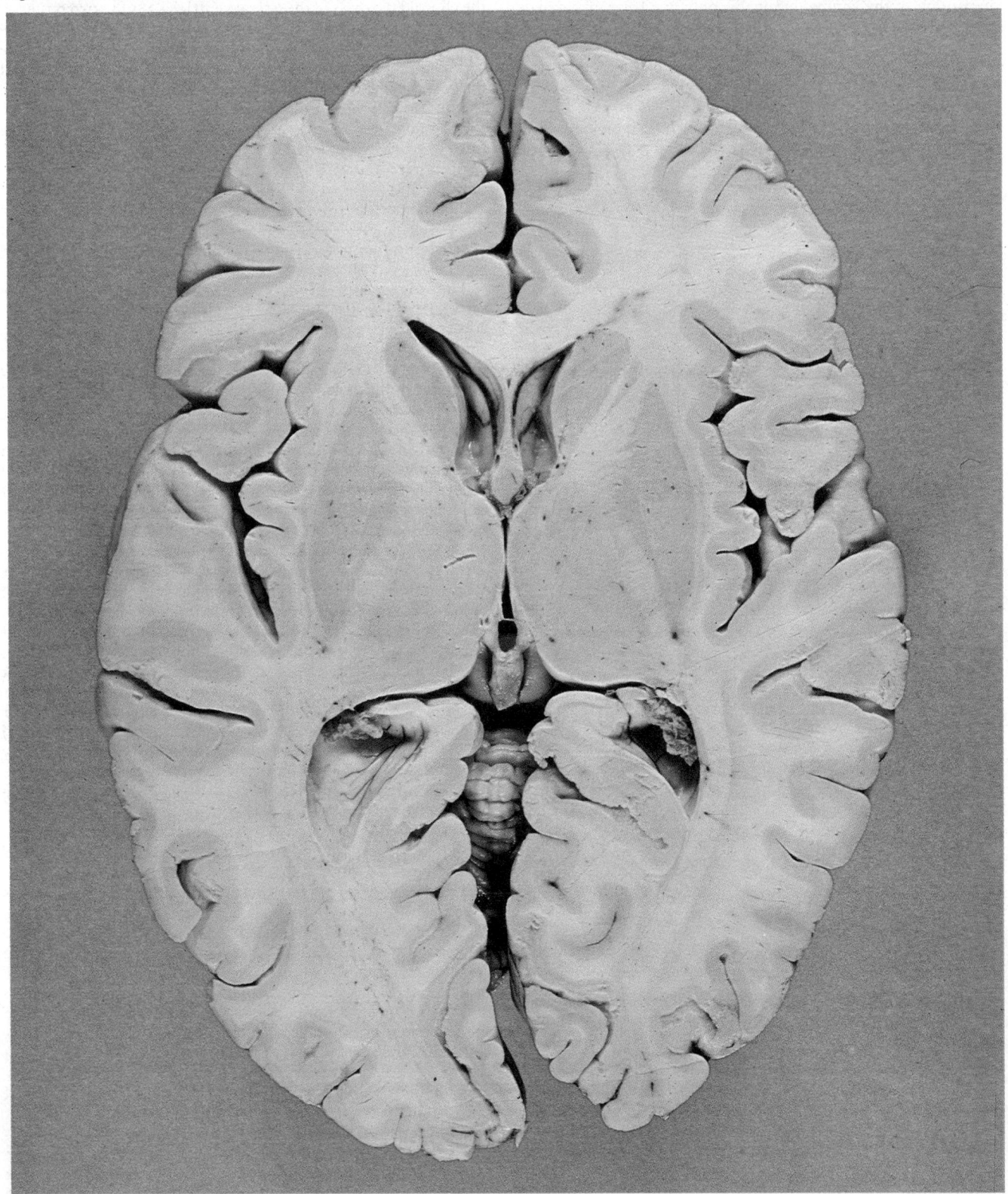

7.152 A horizontal section through the brain including the frontal and occipital poles of the cerebral hemispheres. Features appearing in this section are discussed at many points throughout the text. For appropriate labelling compare with **7.**151. (Dissection by Dr. E. L. Rees, Dept. of Anatomy, Guy's Hospital Medical School.)

invagination appears as a narrow cleft, the *choroid fissure*. Through the main part or body of the ventricle the fissure is limited superiorly by the fornix and inferiorly by the thalamus (**7.**157); in the inferior cornu it is between the stria terminalis above and the fimbria (**7.**158). The choroid fissure is the first groove to appear on the surface of the cerebrum (p. 170). In coronal sections of the brain at eight weeks of embryonic development it can already be seen that the choroid fissure is in direct contact with the ependymal roof of the body of the lateral ventricle, and that vascular pia mater becomes folded into the fissure. At this stage, before the development of commissures and expansion of the lamina terminalis, only one layer of pia mater extends over the roof of the third ventricle (**2.**86A). When the corpus callosum expands posteriorly it does so above the level of the choroid fissure, carrying with it a layer of pia on its inferior surface. This overlies the original single pial layer, fusing with it to form the main part of the tela choroidea, lateral extensions of which, also double layered, form the two choroid plexuses in the lateral ventricles. Posteriorly the two layers separate, the inferior

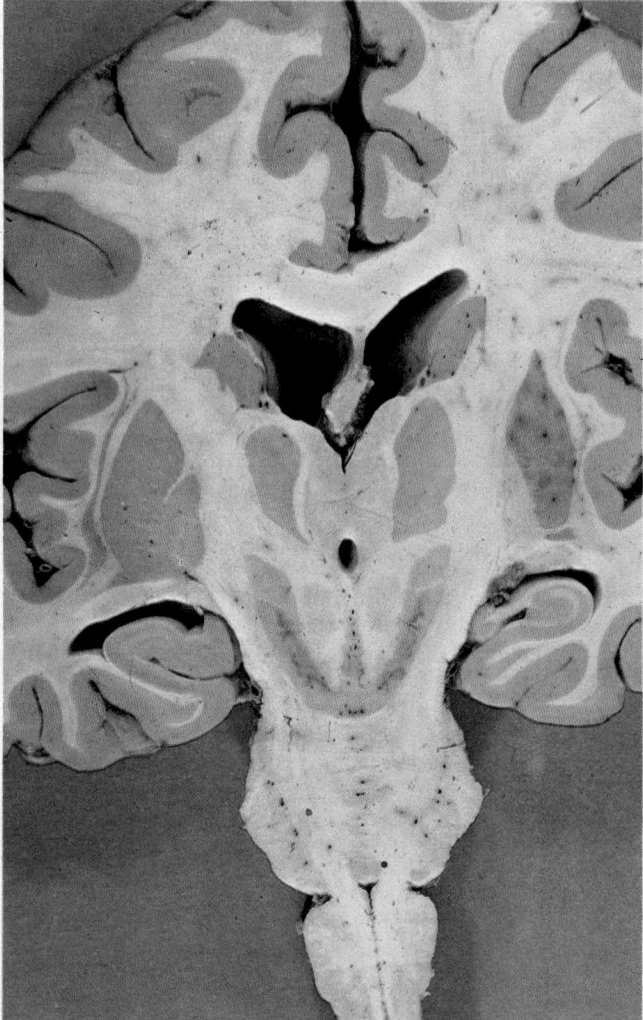

7.153 The central area of the ventral part of the oblique coronal section of the brain shown in **7.**95, photographed at higher magnification to show some structural features in greater detail; compare with **7.**98, 156 for appropriate labelling. Note in particular: (1) the anterior, medial and lateral parts of the dorsal thalamus, separated by the internal medullary laminae; (2) the relation of the caudate nucleus to the anterior and inferior cornua of the lateral ventricle; (3) the lentiform nucleus, divided into an external putamen and an internal globus pallidus; the latter again divided into internal and external parts; (4) the internal capsule, external capsule, claustrum, extreme capsule and insular cortex; (5) the sectioned profiles of the subthalamic and red nuclei, and the substantia nigra; (6) the hippocampus projecting into the floor of the inferior cornu of the lateral ventricle. Other structural features on this section are discussed at many points throughout the text. Compare also with **7.**152. (Dissection by Dr. E. L. Rees, Dept. of Anatomy, Guy's Hospital Medical School.)

(original) layer following the roof of the third ventricle to reach the pineal gland and the tectum, the superior cleaving to the corpus callosum and passing round the splenium to the superior surface of the former (**8.**195).

Where the two layers of pia mater described above are fused they constitute the *tela choroidea of the third ventricle*, as usually defined, although the choroid plexuses of the lateral ventricles are in fact extensions of it. Viewed from above it is a triangular fold, with a rounded apex at the level of the interventricular foramina, often indented by the anterior column of the fornix (**7.**159). Its lateral edges are irregular, due to the contained vascular fringes of the choroid plexuses of the lateral ventricles. At the two posterior or basal angles of the tela choroidea these vascular fringes curve onwards into the inferior cornua of the lateral ventricles, while the wide central part of the base of the tela marks the limit of fusion of the pial layers, which depart from each other here as already detailed. When the tela choroidea has been removed a transverse slit is left between the splenium and the junction of the roof of the third ventricle with the tectum: this is the *transverse fissure*, but it is, of course, not a cerebral fissure in the true sense or usage of the term, for it is not due to a folding of the cerebral cortex. It merely marks the posterior limit of the *extra-cerebral space* enclosed by the posterior growth of the corpus callosum above the third ventricle. In the space, enclosed between two layers of pia mater are the choroid plexuses of the third ventricle (p. 981).

The microscopic structure of a choroid plexus is essentially as follows. The irregular fringes visible to the naked eye are covered with large numbers of microscopic *villous processes*, each containing afferent and efferent vessels, an intervening capillary plexus, a small amount of supporting connective tissue, and nerve fibres. The villi show varying degrees of complexity (**7.**160), and sections demonstrate the large surface area created in this way. This has been estimated to be at least 200 cm² (Voetmann 1949), and this does not allow for the *microvilli* which electron microscopy shows on the ventricular aspect of the ependymal cells (Maxwell and Pease 1956). The ependymal microvilli are not as regular in size and distribution as in some other sites, such as the small intestine (p. 1347), being more akin to those seen in the epithelium of the proximal part of the renal tubule. As in the latter, the basal aspects of the ependymal cells present, to a lesser extent, a series of invaginations (Pappas and Tennyson 1962). (Scott *et al.* 1974 have reported a preliminary scanning electron microscopic study of the surface features of human fetal and infant choroid plexus; their observations largely confirm the above remarks.) These appearances, coupled with an absence of secretory granules, an unspecialized Golgi apparatus, and other features, suggest a role concerned with water transport rather than secretion. However, physiological studies (Davson 1970) provide considerable evidence that the cerebrospinal fluid, largely produced by the choroid plexuses, cannot be regarded as a mere filtrate. Ultrastructural studies have also shown that some ependymal cells possess *cilia*, that tight junctions exist between them, and that a distinct *basement membrane* separates them from the adjacent capillaries (Wislocki and Ladman 1958). The latter are sometimes of the fenestrated type, and experiments (Brightman and Reese 1969) in which the protein peroxidase was used as a tracer, showed that, while the capillary endothelium was readily permeable to the tracer, its passage was blocked by the tight junctions between the choroidal ependymal cells. These findings are probably of importance in relation to the concept of a *blood-brain barrier* which is further discussed on pp. 1052, 837.

The *stroma* of the choroidal villi is derived from the pia mater of the tela choroidea. It consists of pial fibroblasts and few, if any, collagen fibres. The cells are much flattened and do not form complete sheets between the ependyma and the capillary endothelium. Nerve fibres appear to be absent from the actual villi, but in the larger stems from which these branch the amount of connective tissue is greater and myelinated and non-myelinated nerve fibres and small nerve bundles have been identified (Clark 1934; Cooper 1958). The functional significance of these fibres is still uncertain, but the non-myelinated fibres are considered to be vasomotor postganglionic sympathetic elements; others are thought to be derived from the vagus and glosso-pharyngeal nerves. Comparatively few studies of the nerves of the

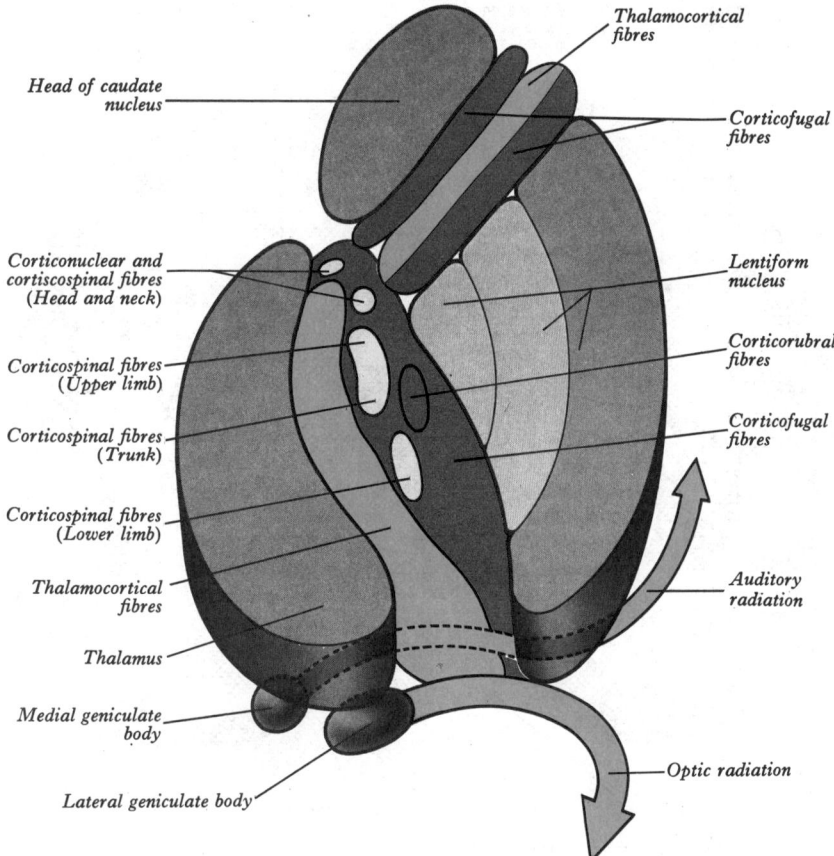

7.154 Diagram of the main components of the internal capsule. Descending motor fibres are shown in yellow, corticofugal fibres to the thalamus and pons, etc. in red, and ascending fibres in blue. (From Strong and Elwyn's *Human Neuroanatomy*—see Bibliography for details). However, see also in text references concerning alternative views about the position of the corticospinal fibres.

choroid plexuses have been reported and much uncertainty of their distribution and role persists. However, histochemical studies (Lindvall *et al.* 1977) have revealed the occurrence of cholinergic nerve fibres in association with the vessels and epithelium of all the choroid plexuses in a variety of mammals, including a primate (baboon).

The *blood supply* of the choroid plexuses in the tela choroidea is derived from the anterior choroidal branch of the internal carotid, and the choroidal branches of the posterior cerebral artery (pp. 691, 698), the former being usually a single vessel, the latter three to five in number (Millen and Woollam 1953). The two sets of arteries anastomose to some extent. The capillaries drain into a rich venous plexus, which is served by a single choroidal vein leaving the tela choroidea. This commences in the vicinity of the basal angle of the tela, at the junction of the parts of the choroid plexus in the body and the inferior cornu of the lateral ventricle. This region also corresponds to the frontier between the territories of the anterior and posterior choroidal arteries, and at this point also a solitary so-called 'glomus' (p. 632) is situated. This structure is near the free edge of the choroid plexus, and many of the tributaries of the choroidal vein converge towards it. Apart from the occurrence of this body nothing appears to have been established regarding its significance in the plexus.

Cerebral Dimensions

The human brain has been weighed and measured by many observers; and it has been compared in its volume, weight, relationship to total body weight, and in the proportionate size of its main divisions with the brains of many other vertebrates. Even without actual quantification, such comparisons permit some interesting generalizations regarding the relation between development of brain size and various abilities in mankind and other animals. But even metrical data, where they are available in reliable quantity, are merely estimates of the bulk of nervous tissue, which largely disregard degrees of organization or even the proportions of particular neuronal elements, such as nerve cells, nerve fibres, neuroglia and vascular elements; and such considerations must obviously be taken into account if valid comparisons are to be made. Clearly, therefore, no more than crude deductions are to be expected from data of this kind. Furthermore, available measurements are usually dependent upon small series of observations in any particular species, even sometimes upon a single example. Hence little allowance for individual variation can have been made, whether due to sex, age, or nutritional state, or indeed any other factors which may influence brain size or its relation to body size and weight; and it is particularly this ratio which is the focus of most attention. Even in respect to the human, for whom much more adequate series of observations have been recorded, the effects of such factors are often overlooked.

An 'average' human male adult's brain is said to weigh about 1,450 gm. and that of his female counterpart about 100 gm. less, but such figures are not useful without a corresponding range, which is of the order of 1,240 to 1,680 in males and 1,130 to 1,510 in females. Proportionately males and females differ rather less in brain weight than total body weight. Even the ranges quoted would exclude numbers of famous people whose brain weights lay well outside such extremes, and in both directions. (Consult Blinkov and Glezer 1968, for examples and much other data.) Different ranges for various races have been stated, but the differences between their means are not significant in general, when the much greater variation in body weight is taken into consideration. Chrzanowska *et al.* (1973, 1975), in a study of 1,670 human brains (896 male and 774 female, ranging from 20 to 89 years in age), concluded that the average difference between the sexes is 130–150 gm. They also stated that brain weight is positively correlated with height, and they recorded progressive diminution in brain weight, in both sexes, from 30 to 40 years onwards. They have computed a loss in weight of one-ninth in the brains of both sexes by the 90th year.

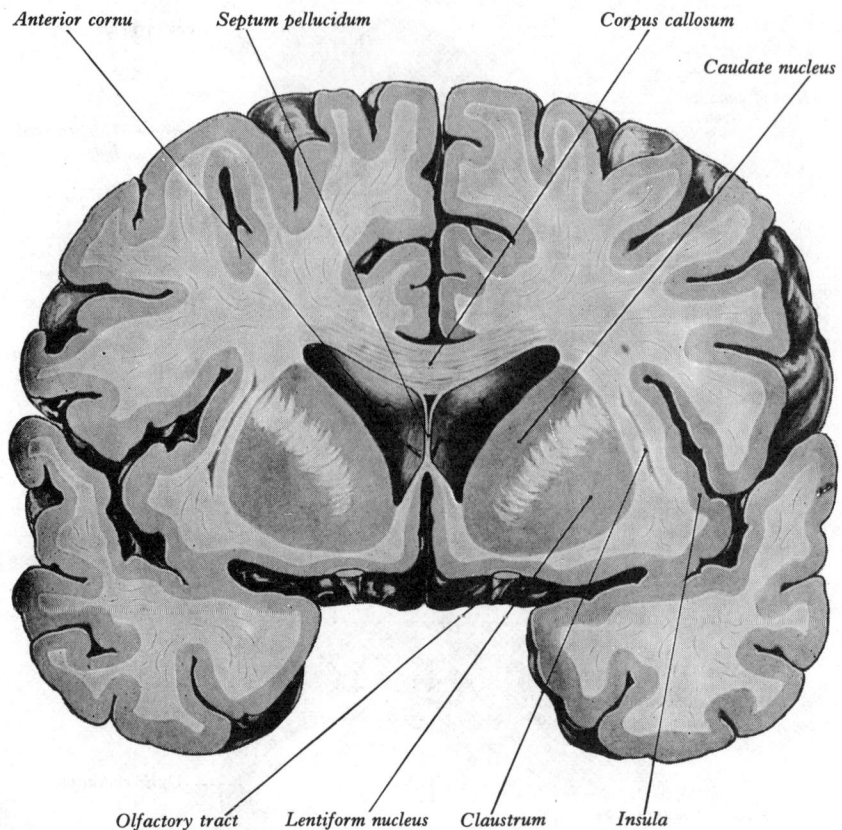

Anterior cornu *Septum pellucidum* *Corpus callosum*

Caudate nucleus

Olfactory tract *Lentiform nucleus* *Claustrum* *Insula*

7.155A Posterior aspect of a coronal section through the anterior cornua of the lateral ventricles.

Caudate nucleus

Thalamostriate vein

Fornix

Thalamus, anterior part

Thalamus, lateral part

Thalamus, medial part

Internal medullary lamina

Nucleus subthalamicus

Substantia nigra

Basis pedunculi

Putamen of lentiform nucleus

Internal capsule

Globus pallidus of lentiform nucleus

Optic tract

Pes hippocampi

Collateral sulcus

7.155B Anterior aspect of a coronal section through the right cerebral hemisphere.

Prior to maturity, of course, the human brain varies markedly in weight and in ratio to total body weight (Purpura and Schadé 1964), showing throughout its growth, however, the same relatively outstanding development which characterizes the adult. Unlike some other mammals, such as rodents, in which the maximum growth rate may be pre- or post-natal, the cerebral growth spurt in mankind and primates in general is peri-natal—in late fetal development and the first year of extrauterine life. During this first year the brain at least doubles in weight (Dogson 1962), and it reaches about 90 per cent of its final weight by the sixth year. Most of this increase is due to factors such as myelination rather than any augmentation in the number of nerve cells, and for this reason alone any comparison of the growing human brain, on the basis of weight, with the brains of adults or of other species, is pointless where correlation with ability is the criterion.

The human brain is obviously large, both absolutely and relatively; but it is surpassed in both respects by those of certain other mammals. Dolphins, elephants (4,000–5,000 gm.) and whales (6,800 gm. has been recorded in the blue whale, *Balaenoptera musculus*) have heavier brains than man, although this is offset by brain:body weight ratios of about 1:600 in elephants and of about 1:850 in large whales. However, in dolphins the ratio is approximately 1:40, which is somewhat 'better' than the human average of about 1:50. In some small mammals, such as mice, the ratio may be as low as 1:35, while in smaller primates ratios as low as 1:12 have been estimated (squirrel monkey). Amongst the primates man occupies an almost average position in brain: body weight ratio, the larger apes and monkeys falling much below him in this regard. In absolute size of brain, of course, he much surpasses any other primate; a male gorilla (Schultze 1969) may have a cranial capacity of 412–752 cm³, and while this is only a rough indication of actual brain weight, it is well below the human range (say, 1,200–1,500 cm³). The largest gorilloid brain recorded weighed 750 gm. (Holloway 1968), which is distinctly less than the usual human range. Nevertheless, extremely small brain weights have been recorded in microcephalic idiots; but what is more significant is that brain weights in dwarfs which were *below* the average upper limit for gorillas have nevertheless been recorded in humans possessing at least elementary speech; and the symbolic abstractions of language are widely regarded as correlated with the large absolute size of the human cerebrum. This view has, however, been questioned (Lenneberg 1964); and while a certain, as yet indefinable minimal size of brain must be associated with the extraordinary potentialities of mankind in cerebration, it is clear that the kind of data so far mentioned cast little illumination on this problem.

As has already been said above, measurements of brain size and brain-to-body weight ratio afford no indication of organization within the brain; nor do they take into account the density of packing or total numbers of nerve cells in any particular brain. While it has been said that this density does not vary greatly in mammalian brains when calculated for the whole organ, estimates of this kind are few and subject to much error. Figures for samples taken from the cerebral cortex are perhaps open to the same strictures, but it is of interest to note here that a cubic millimetre of cortex has been estimated to contain as many as 142,000 cells in the mouse, 21,500 in the macaque, and merely 10,500 in man. However, the area of the human cerebral cortex (p. 1005) is at least twelve times greater than that of the macaque, which may nevertheless have a brain-to-body weight ratio similar to man's.

Another mode of approach to the problems of comparison of mammalian and especially primate brains (Tilney and Riley 1928) has been to estimate the weights of the main divisions of the brain, such as cerebrum, cerebellum, midbrain, medulla oblongata, olfactory bulb, neocortex, paleocortex, archicortex, corpus striatum, and other features, and to express some of these

oblongatae is approximately 9·6:7, despite the fact that the gorilla weighs at least twice as much as the man. The ratios for the mesencephalon, cerebellum and diencephalon are roughly 2:1, but the human telencephalon is three times as heavy as the gorilloid. The olfactory bulb of the gorilla is almost thrice the size of the human, while the latter's hippocampus is nearly two and a half times as heavy as the gorilla's. Incidentally the ratio of olfactory bulb to hippocampus in the gorilla is hence about 1:15 and in man 1:90, an interesting commentary on the status of the hippocampus as part of the 'rhinencephalon' (see p. 990). Legait et al. (1973) have made a volumetric comparison of the hypothalamus and hypophysis relative to total brain volume in a large series of mammals (principally rodents). They observed a constant brain/hypothalamic ratio, but some variation between the comparative volumes of hypothalamus and hypophysis, particularly for the anterior lobe of the latter (adenohypophysis).

Space does not permit further exploration of such data, and it must be said that though considerations of this kind may have marked significance in evolutionary comparisons between man and other mammals, particularly sub-human primates, they do not provide any clear definition in morphological terms of what it is that is responsible in man's brain for the enormous complexity of his behaviour. Perhaps it is rather in this, and his consequently

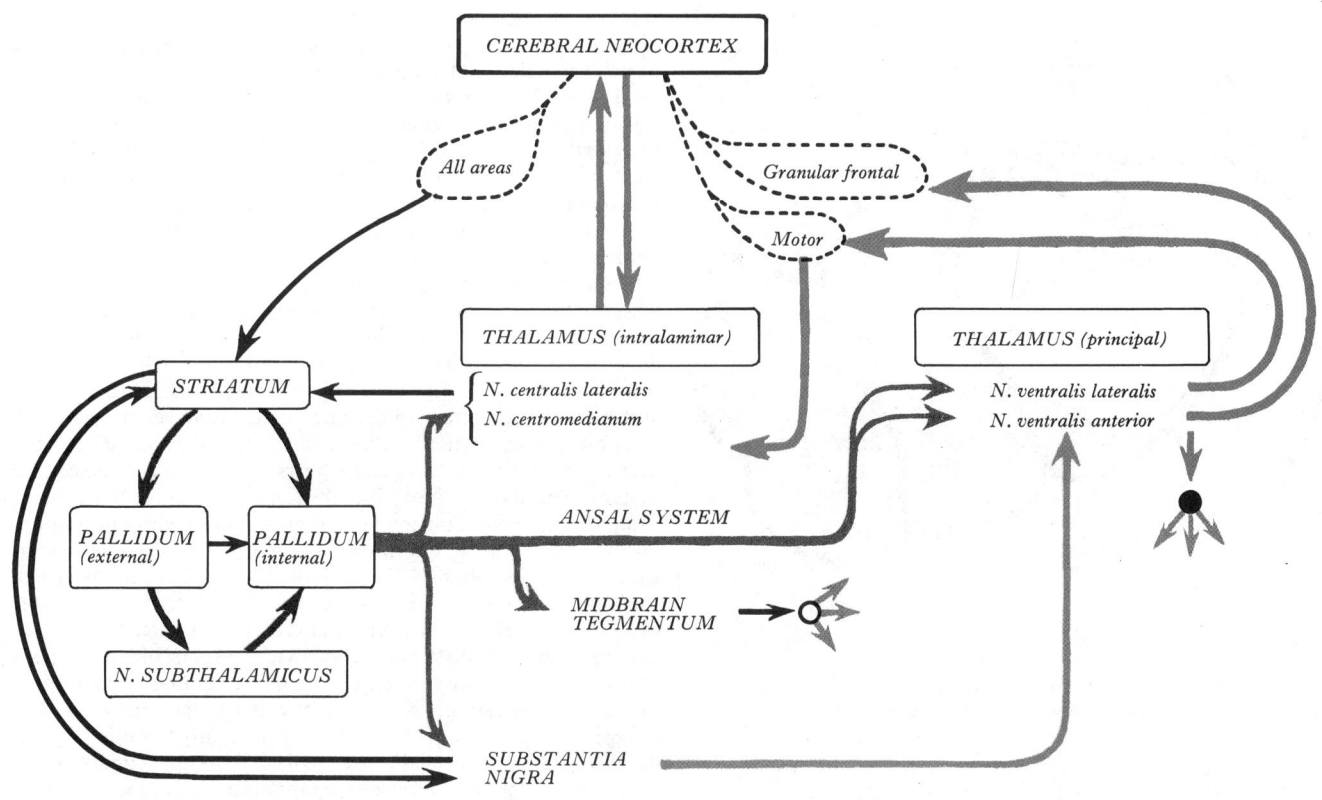

7.156 A scheme of the principal connexions of the basal nuclei. At the points indicated the dorsal thalamic nucleus ventralis anterior projects on to several subcortical regions including the thalamic nucleus ventralis lateralis. The midbrain tegmentum probably establishes polysynaptic relationships with the spinal cord grey matter and the thalamic intralaminar nuclei. Colour has been used to assist visualization of the individual pathways. See text for further comment. (Based on data summarized by Professor K. E. Webster (1975) of the Department of Anatomy, King's College, University of London; by kind permission of the author.)

quantities as indices. (As an example of a more recent such study consult Stephan et al. 1970.) Large numbers of species of insectivores and primates have been compared in this manner, and although in many instances only one sample from each species has been measured, highly interesting comparisons are possible. While it is, of course, easy to appreciate, without any metrical data, that in the human brain the olfactory structures are small, absolutely and relatively, that the cerebrum-pons-cerebellum complex is markedly developed, or that the cerebral cortex is very extensive in the human brain, quantitative assessment of such statements is much more valuable. In comparing man and the gorilla, for example, the ratio by weight of their medullae

ever more complex culture, that the essence of humanness can best be defined. It is noteworthy that in assessing the human paleontological record the accent has gradually passed from comparison of cranial capacities to that of culture. The Australopithecinae, with brains little if at all larger than those of gorillas, may leave us in doubt; but in Homo habilis, with—it is true—a considerably larger but still very modest cranial capacity, doubts evaporate as to his humanity; and perhaps most of all because of his ability, as revealed in surviving artefacts. In more recent ancestors also, and again not merely because of increasing hominoid brain size (in which Homo neanderthalensis even surpassed us, yet is now extinct), but even more because of the

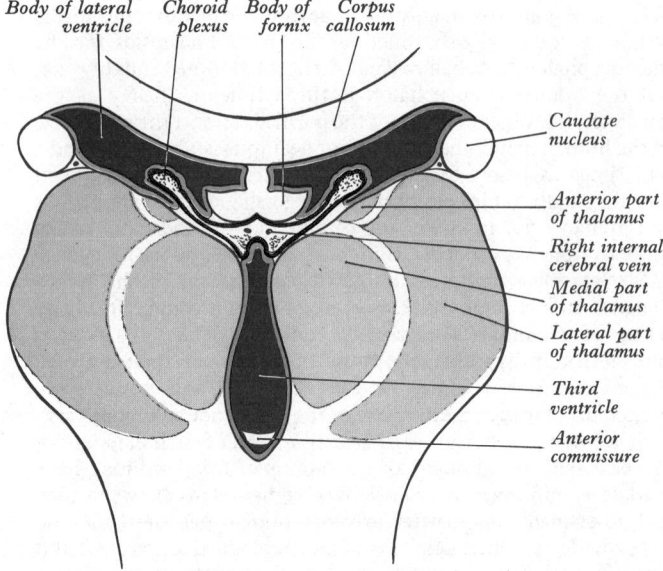

7.157 Diagram of a coronal section through the lateral and third ventricles. The pia mater of the tela choroidea is shown in red and the ependyma in blue.

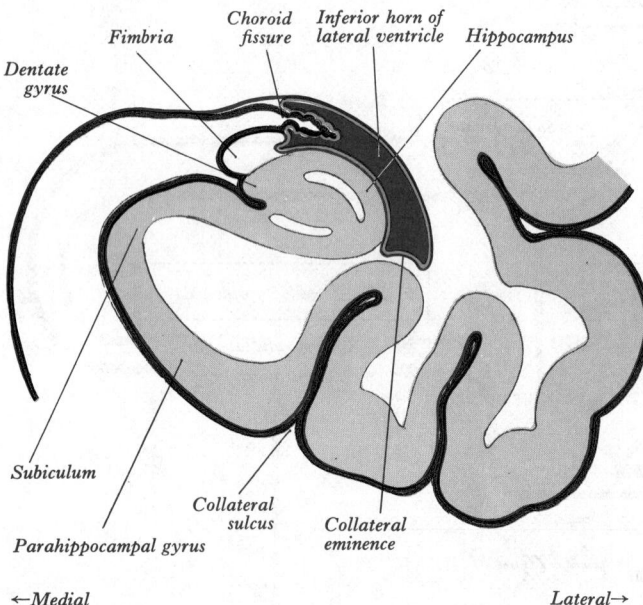

←Medial

Lateral→

7.158 Diagram of a coronal section through the inferior cornu of the lateral ventricle. The pia mater is shown in red and the ependyma in blue.

increasing evidence of their *human* behaviour are we convinced of their close relationship to ourselves. Their abilities, which from such simple beginnings have led to the great achievement and communication of practical and abstract creation in our own era, are no more likely than our own to be explicable in terms of gross cerebral mensuration. (For important discussions of the significance of brain size consult Tobias 1970, 1971; Jerison 1970; Van Valen 1974.)

Blood Vessels of the Brain

The arterial supply of the brain (pp. 686, 695) is derived from the **internal carotid** and **vertebral arteries** which lie in the subarachnoid space (p. 1049). The vertebral and basilar arteries give branches to the spinal cord, brainstem and cerebellum, the basilar artery terminating at the upper border of the pons by dividing into two posterior cerebral arteries. The internal carotid artery terminates by dividing into **anterior** and **middle**

cerebral arteries. The anterior cerebral arteries are interconnected by the **anterior communicating artery**. Just before terminating, the internal carotid artery communicates through the **posterior communicating artery** with the **posterior cerebral artery**, thus completing a vascular circle, the **circulus arteriosus**, around the interpeduncular fossa (**6**.61). For details of collateral circulation following blockage of the main feeders of the circle, *see* Fields *et al.* (1965), and Gillilan (1974). From the circulus arteriosus, or vessels close to it, **central branches** are given off to supply the interior of the cerebral hemisphere and the thalamus. These vessels form six principal groups: (1) an *anteromedial group*, derived from the anterior cerebral and anterior communicating arteries (*see* p. 690); (2) a *posteromedial group*, from the posterior cerebral and posterior communicating arteries; (3 and 4) right and left *anterolateral groups*, from the middle cerebral arteries; (5 and 6) right and left *posterolateral groups*, from the posterior cerebral arteries after they have wound round the cerebral peduncles. (For details consult Kaplan and Ford 1966.) Lang and Brunher (1978) have described a recurrent central ramus arising from the anterior cerebral artery beyond the origin of its anteromedial group of central rami and running back to join these. It was often observed to be double. They name this vessel *arteria recurrens anterior*.

The entire blood supply of the **cerebral cortex** is derived from the **cortical branches** of the anterior, middle and posterior cerebral arteries. They reach the cortex in the pia mater. They divide in its substance, give off branches which penetrate the brain cortex perpendicularly and are divisible into long and short rami (Von Bonin 1950). The **long** or **medullary arteries** pass through the grey matter and penetrate the subjacent white matter to a depth of 3 or 4 cm (Lewis 1957) without intercommunicating and thus constitute many independent small systems. **Deep medullary vessels** extending from the central branches towards the cortex have been described, but these are identified as recurrent branches of the long or medullary vessels (Lewis 1957). The **short arteries** are confined to the cortex, where they form with the long vessels a compact network in the middle zone of the grey matter, the outer and inner zones being sparingly supplied with blood. Lazorthes *et al.* (1968) and de Reuck (1972) have drawn attention to differences in the angio-architecture in iso- and allo-cortical areas, the former being more elaborate, with arterioles terminating in different strata. The vessels of the cortical system are not so strictly 'terminal' as those in the central system (Sunderland 1938) but they approach this type closely, for, although neighbouring vessels anastomose with one another on the surface of the brain, they become end arteries as soon as they pierce its substance. Even the anastomoses on the surface of the brain, however, are in general only between microscopic branches of the cerebral arteries, and there is little clear evidence that they can provide the means of a vicarious circulation in cases of occlusion of larger vessels. Owing to the cellularity of the grey matter, its blood supply is much richer than that of the white. The *lateral surface* of the hemisphere is principally supplied by the *middle cerebral artery*; the area adjacent to the superomedial border as far back as the parieto-occipital sulcus is supplied by the *anterior cerebral artery*, and the occipital lobe and most of the inferior temporal gyrus (excluding the temporal pole) is supplied by the *posterior cerebral artery* (**6**.59, 60). The *medial* and *inferior surfaces* of the cerebral hemisphere are supplied by the *anterior* and *posterior* cerebral arteries. The area supplied by the anterior cerebral artery extends almost to the parieto-occipital sulcus and includes the medial part of the orbital surface. (For a highly detailed study of the distribution of the anterior cerebral artery in 53 hemispheres and 300 angiograms consult Farnarier *et al.* 1977.) The remainder of this surface and the temporal pole are supplied by the middle cerebral artery. The rest of the medial and inferior surface is supplied by the posterior cerebral artery (**6**.59, 60). The junctional zone near the occipital pole between the territories of the middle and posterior cerebral arteries corresponds to the part of the striate cortex concerned with the macula. The phenomenon known clinically as '*sparing of the macula*' is considered to be due to the collateral circulation of blood from branches of the middle cerebral artery into those of the posterior, when the latter vessel is blocked. The middle cerebral artery may in fact itself supply the macular area (Smith and Richardson 1966).

Most of the *corpus striatum* and *internal capsule* is supplied by the medial and lateral striate rami of the central branches of the middle cerebral artery, the remainder being supplied by the central branches of the anterior cerebral artery. One ramus of the middle cerebral is Charcot's 'artery of cerebral haemorrhage' (p. 691).

The *choroid plexuses* of the *third* and *lateral ventricles* are supplied by branches of the internal carotid and posterior cerebral arteries.

The finer details of the vessels of some parts of the diencephalic region have been well explored, particularly in relation to the hypophysis cerebri and its related hypothalamic nuclei (*see* p. 966 and Haymaker *et al.* 1969). Detailed studies of the vascularization of the lamina terminalis (Duvernoy *et al.* 1969) and of the posterior wall of the third ventricle (Plets 1969) have been reported. The arterial supply to the lamina is derived from the anterior cerebral arteries and their communicating vessel, and these are described as forming a superficial capillary plexus in the pia mater which drains into a second, more deeply situated plexus of sinusoidal capillaries, characterized by loops or vortices, these draining in their turn into the veins of the hypothalamus. The physiological significance of these arrangements is unknown. The main artery to the posterior parts of the third ventricle is the medial branch of the posterior choroidal artery, and this supplies the posterior commissure, habenular region, the pineal gland, and medial parts of the thalamus, including the pulvinar. The *thalamus* is supplied chiefly by branches of the posterior communicating, posterior cerebral and basilar arteries. The pattern of branches from these main feeders, and the varying details of the angioarchitecture in the different nuclei of the thalamus, have been described *in extenso* in the human brain, together with a full critique of the literature by Plets *et al.* (1970) and by Percheron (1977). The latter has denied the often described thalamic supply from the anterior choroidal artery; he derives almost the entire thalamic supply from branches of the posterior cerebral and basilar arteries (p. 696).

The *midbrain* is supplied by the posterior cerebral, superior cerebellar and basilar arteries. The crura cerebri are supplied by vessels entering on the medial and lateral sides. The medial vessels pass to the inner side of the crus and also supply the upper and inner part of the tegmentum, including the nucleus of the oculomotor nerve. The lateral vessels supply the lateral part of the crus and the tegmentum. The colliculi are supplied by three vessels on each side from the posterior cerebral and superior cerebellar arteries. There is an additional supply to the crura cerebri and the colliculi and their peduncles from the postero-lateral group of central arteries from the posterior cerebral artery.

The *pons* is supplied by the basilar and the anterior inferior and superior cerebellar arteries. Direct branches from the basilar artery enter along the basilar sulcus. Branches also enter along the trigeminal, abducent, facial and vestibulocochlear nerves and nervus intermedius. There is also a supply from the pial plexus.

The *medulla oblongata* is supplied by the vertebral, anterior and posterior spinal, posterior inferior cerebellar and basilar arteries. Some arteries enter along the anterior median fissure and the posterior median sulcus. Other vessels enter along the rootlets of the last four cranial nerves and intermediately to supply the central substance. In addition there is a supply from the same sources through a pial plexus.

The *cerebellum* is supplied from the three pairs of cerebellar arteries, and the latter, like the cerebral arteries, form superficial anastomoses. Their internal distribution has not been extensively studied, but the possibility of anastomoses of their deeper, medullary branches, as distinct from cortical branches, has been reported (Gomes 1969). The anatomy and development of the cerebellar arteries have been reviewed by Gillilan (1974). According to Kielbasinski (1976) the vermis is supplied by one or both of the inferior cerebellar arteries.

The *choroid plexus* of the *fourth ventricle* is supplied by the posterior inferior cerebellar arteries.

The blood supplies of the optic chiasma, tract and radiation are of marked clinical interest. The chiasma is supplied in part by the anterior cerebral arteries, but its median zone depends upon rami from the internal carotids which reach it by way of the stalk of the

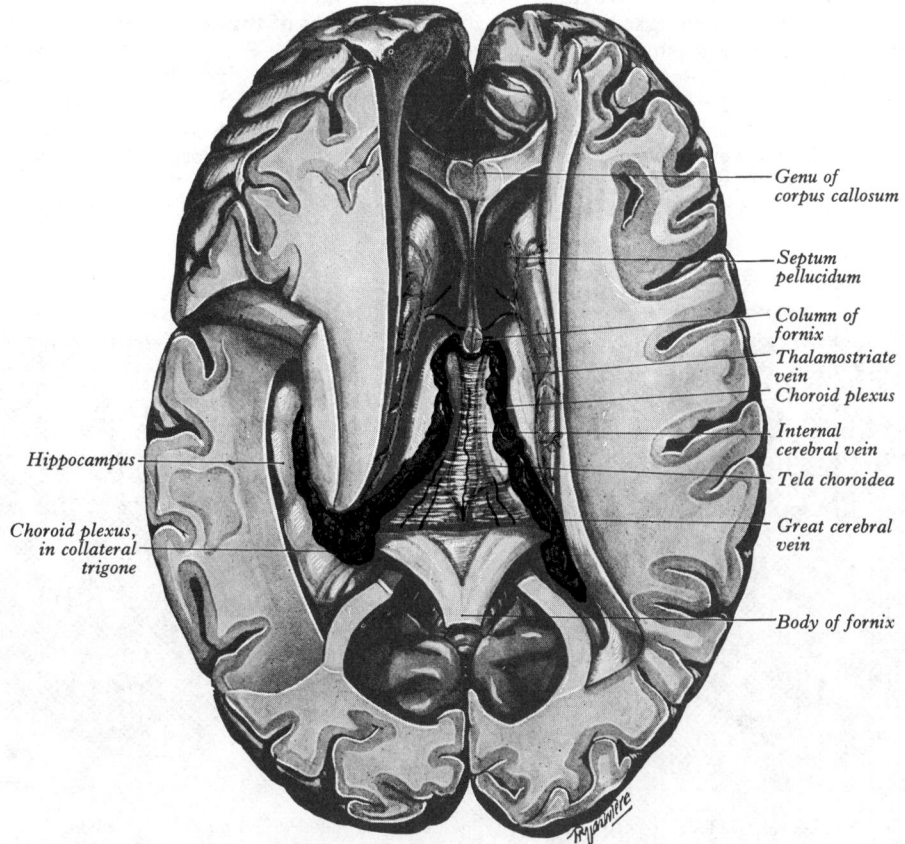

Genu of corpus callosum

Septum pellucidum

Column of fornix

Thalamostriate vein

Choroid plexus

Internal cerebral vein

Tela choroidea

Great cerebral vein

Body of fornix

Hippocampus

Choroid plexus, in collateral trigone

7.159   The tela choroidea of the third ventricle and the choroid plexus of the lateral ventricle.

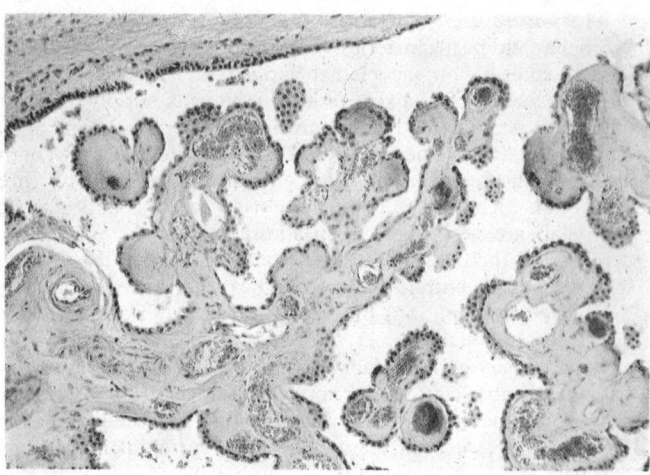

7.160 A section of part of the choroid plexus of the lateral ventricle, stained with haematoxylin and eosin. Note the ependyma lining the ventricular wall above, and covering the loose connective tissue cores of the processes of the plexus, which contain numerous small blood vessels, including many capillaries. A calcareous deposit (dark blue) is present in one process. Owing to the complexity of the ramification of the processes, several appear as disconnected islands of tissue.

hypophysis cerebri. The anterior choroidal and posterior communicating arteries supply the optic tract, and the radiation receives blood through deep branches of the middle and posterior cerebral vessels. For further details consult Abbie (1938); Bergland and Ray (1969).

Cerebral Blood Flow and Functional Localization

That the level of functional activity, metabolism, and volume flow of blood in unit time through most tissues fluctuate in close correspondence has long been known in general physiology. In relation to nervous tissues also, although the mechanisms of selective control of blood flow through particular regions was unknown, as early as 1937, Schmidt and Hendrix using thermocouples demonstrated a marked increase in flow through the feline visual cortex, compared with surrounding regions, during localized retinal illumination. Subsequent years provided many confirmatory experiments in the brains of experimental animals using similar methods, but not only confined to the visual pathway. In recent years, with the advent of a wide and increasing range of appropriate radio-isotopes and scanning cameras there has been a dramatic upsurge of interest in cerebral blood-flow

studies. Much of the initial stimulus stemmed from a new technique, non-invasive in character, potentially able to localize with considerable accuracy intracranial pathological states that modified the blood-flow pattern significantly, for example, neoplasms, post-traumatic loci, intracranial abscess, cerebral infarction and so forth (see Maisey 1978). In large measure, these objectives determined the particular radionuclide and methodology chosen. Some groups, however, soon realized the possible power of the technique in monitoring fluctuating levels of blood flow (and therefore of functional activity) of different regions of, for example, the normal cerebral cortex during a wide range of activities (see Lassen et al. 1978). Briefly, an isotope of the inert gas xenon (xenon 133) dissolved in sterile saline is injected into a carotid artery and the subject's head is subsequently scanned during the arrival and 'washout' of the isotope by means of a gamma-ray camera; the latter carries 254 scintillation detectors each collimated to scan about 1 cm² of the surface of the cerebral cortex. The results from this grid of detectors is processed in a digital computer and then displayed on a colour television screen. The colour of each *pixel*, or picture, denotes a particular range of blood flow through the subjacent cortex (the mean flow rate is green), down to 20 per cent below mean are shades of blue, up to 20 per cent above mean are through orange to shades of red. Illustration 7.161 shows two examples of the technique in use. Already some hundreds of apparently normal cerebral hemispheres have been scanned during a wide range of activities from the 'resting' state, through uncomplicated voluntary movements, sensory stimulation of various modalities, to complex activities such as reading aloud and problem-solving. Broadly, the 'functional areas' mapped by traditional means have been confirmed. Clearly, however, in the forthcoming years there will occur increases in resolution of the technique, both spatial and temporal, and this offers a most exciting prospect for future researches into the localization of cerebral functions.

The venous drainage of the brain (pp. 743–749) can be divided into that serving the cerebral hemispheres and that serving the brainstem and cerebellum. The veins are thin-walled, devoid of valves and the majority cross the subarachnoid space to join the dural venous sinuses.

The veins of the cerebrum can be divided into external and internal groups. The external cerebral veins are grouped in three sets, the superior which drain forwards into the superior sagittal sinus, the inferior which drain principally into the transverse and cavernous sinuses and the middle which are further subdivided into superficial and deep. The *superficial middle cerebral vein* drains the majority of the lateral surface of the hemisphere, follows the lateral sulcus and terminates in the cavernous sinus.

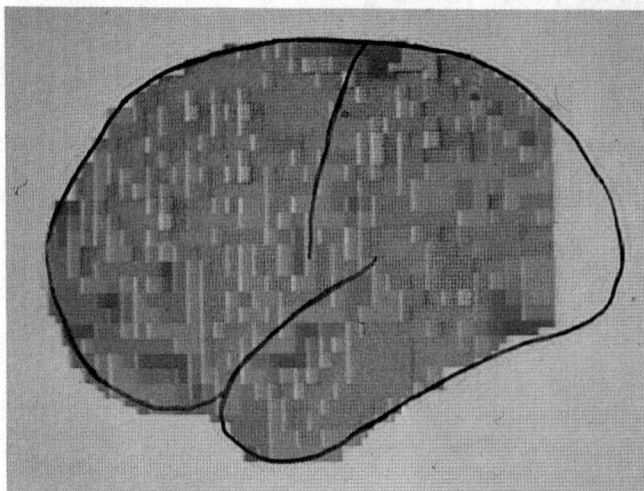

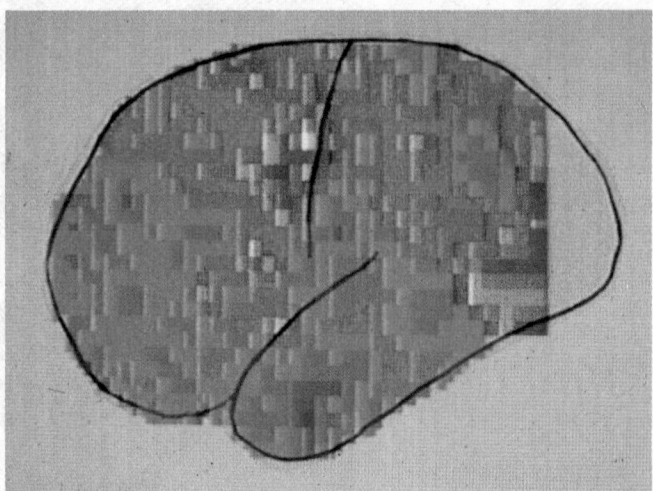

7.161A & B Patterns of presumed relative levels of neuronal activity in different regions of the cerebral cortex as revealed by measurement of regional blood-flow. A shows the pattern which accompanies movement of the contralateral foot, and B shows a pattern which accompanies activity of the eyes engaged in a pursuit movement. The colour scales indicate the percentage flow rate above or below the mean (dark green). (Radioactive xenon scan pictures kindly supplied by Doctors N. A. Lassen, D. H. Ingvar, and E. Skinhoj, Bispebjerg Hospital, Copenhagen. See text.)

The *deep middle cerebral vein* drains the region of the insula and unites with the *anterior cerebral* and *striate veins* to form a *basal vein*. The areas drained by the anterior cerebral and striate veins correspond approximately with those supplied by the anterior cerebral artery and the central branches entering the anterior perforated substance. The striate veins have been described in detail by Rosa and Borzone (1973). The basal veins pass backwards alongside the interpeduncular fossa and midbrain, receive tributaries from this vicinity and join the *great cerebral vein*.

The two **internal cerebral veins** are formed near the interventricular foramen by the union of the *thalamostriate* and *choroidal veins* draining the choroid plexuses of the third and lateral ventricles. They travel backwards parallel to one another between the layers of the tela choroidea of the third ventricle and unite to form the great cerebral vein, which enters the straight sinus.

The veins of the *midbrain* join the basal or great cerebral veins.

The veins of the *pons* drain either into the basal vein, cerebellar veins, the petrosal sinuses, transverse sinus or the venous plexus of the foramen ovale.

The veins of the *medulla oblongata* drain into the veins of the spinal cord, the adjacent dural venous sinuses or along the last four cranial nerves by radicular veins to the inferior petrosal sinus or superior bulb of the jugular vein. For a detailed systematic account of the superficial veins of the brainstem consult Tournade *et al.* (1972) and Duvernoy (1975).

The **veins of the cerebellum** drain mainly into the adjacent sinuses or, from the superior surface, to the great cerebral vein.

Although the *innervation* of the intercranial arteries, including those supplying the brain, remains something of a physiological mystery, there is no doubt that a considerable supply of postganglionic sympathetic fibres is distributed along the arterial trees of the internal carotid and vertebral arteries, and that myelinated fibres accompany these in lesser numbers. A parasympathetic supply is supported by more dubious evidence. For a review of the literature of this topic, consult Nelson and Rennels (1970), and Purves (1972).

There are no lymphatics in the central nervous system. The subarachnoid space is prolonged along the olfactory nerves, thus providing a route between this space and the tissue spaces in the mucoperiosteum of the nasal cavities. The so-called perivascular spaces around brain vessels have been a subject of much controversy (p. 1052). Electron microscope appearances confirm the extension of pial elements into the brain around vessels in various mammals such as the rat (Samarasinghe 1965), and cat (Jones 1970). In the latter study the arachnoid 'space' around the small arterioles entering the cortex of the cerebrum was seen to contain electron-dense material, consisting of collagen bundles, pial cells, and cells resembling macrophages, possibly derived from the leptomeninges. These elements do not, apparently, form a complete sheath to the vessels, large areas being seen in which the basement membranes of the brain and vessel are opposed. The 'space' does not extend around capillaries, where the two basement membranes fuse. These observations also entail that the perivascular spaces are not continuous with whatever space exists around neurons.

The Meninges

The brain and the spinal cord are enveloped by three membranes (meninges), named from without inwards: the dura mater, the arachnoid and the pia mater.

THE DURA MATER

The dura mater is a thick and dense inelastic membrane. The portion of it which encloses the brain (*cerebral dura mater*) differs in several particulars from that which surrounds the spinal cord (*spinal dura mater*), and therefore it is necessary to describe them separately; the two parts, however, form one complete membrane, and are continuous with each other at the foramen magnum.

The cerebral dura mater lines the interior of the skull, and serves the twofold purpose of an internal periosteum to the bones, and a supportive membrane for the brain. It is said to be composed of two layers, an *inner* or *meningeal* and an *outer* or *endosteal*; but these are closely united, except along certain lines where they separate to enclose the venous sinuses which drain the blood from the brain (p. 743). The dura mater adheres to the inner surfaces of the cranial bones, and sends blood vessels and fibrous processes into them, the adhesion being most marked at the sutures, at the base of the skull, and around the foramen magnum. The blood vessels and fibrous processes are torn across when the dura mater is detached from the bones, and consequently the outer surface of the membrane presents a rough and fibrillated appearance; the inner surface is smooth. The endosteal layer of the dura mater is continuous through the sutures and the foramina of the skull with the pericranium, and through the superior orbital fissure with the periosteal lining of the orbital cavity. The meningeal layer provides tubular sheaths for the cranial nerves as the latter pass through the foramina at the base of the skull. Outside the skull these sheaths fuse with the epineurium of the nerves, and the sheath of the optic nerve is continuous with the sclera of the eyeball.

The meningeal layer of the cerebral dura mater is reduplicated inwards as four processes or septa which partially divide the cranial cavity into a series of freely communicating spaces for the lodgement of the subdivisions of the brain.

(1) *The falx cerebri* (7.161), so named from its sickle-like form, is a strong, arched process of dura mater which descends vertically in the longitudinal fissure between the cerebral hemispheres. It is narrow in front, where it is fixed to the crista galli of the ethmoid bone, and broad behind, where it blends in the median plane with the upper surface of the tentorium cerebelli; the narrow, anterior part is thin, and is frequently perforated by numerous apertures. The upper margin of the falx cerebri is convex, and attached to the inner surface of the skull on each side of the median plane, as far back as the internal occipital protuberance; the superior sagittal sinus (p. 744) runs along this margin. Its lower margin is free and concave, and contains the inferior sagittal sinus. The straight sinus runs along its attachment to the tentorium cerebelli.

(2) *The tentorium cerebelli* (7.161, 162) is a crescentic, arched lamina of dura mater which covers the cerebellum, and supports the occipital lobes of the cerebrum. Its concave, anterior border is free, and between it and the dorsum sellae of the sphenoid bone there is a large oval opening, the *tentorial incisure*, which is occupied by the midbrain and the anterior part of the superior surface of the vermis of the cerebellum. Its convex outer margin is attached (*a*) posteriorly, to the lips of the transverse sulci of the occipital bone and the postero-inferior angles of the parietal bones, where it contains the transverse sinuses; (*b*) laterally, to the superior borders of the petrous parts of the temporal bones, where it encloses the superior petrosal sinuses. Near the apex of the petrous part of the temporal bone, the lower layer of the tentorium is pouched forwards and laterally, beneath the superior petrosal sinus, to form a recess between the endosteal and meningeal layers of dura mater of the middle cranial fossa. This recess is called the *trigeminal cave* and envelops the roots of the trigeminal nerve and the posterior part of its sensory ganglion; it is a little more extensive below than above the ganglion. The evaginated meningeal layer terminates by fusing with the anterior part of the ganglion. At the apex of the petrous part of the temporal bone the free border and attached periphery of the tentorium cross each other (7.162); the anterior limits of the free border are fixed to the anterior clinoid processes of the sphenoid bone, whilst those of the attached periphery end on the posterior clinoid processes of

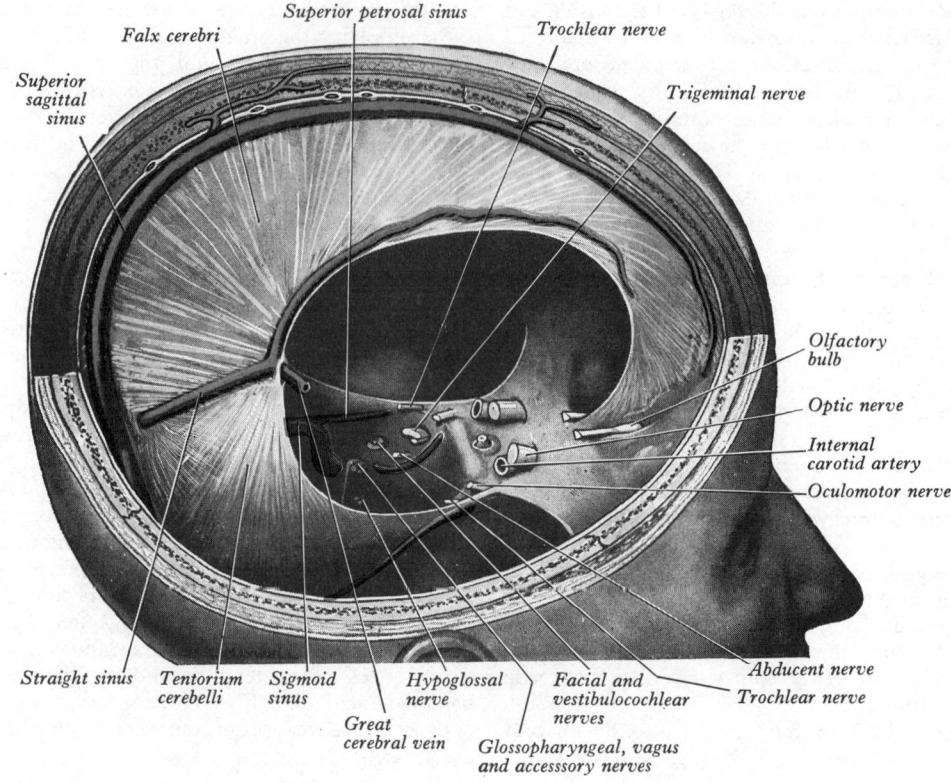

7.162A The cerebral dura mater and its reflexions, exposed by the removal of a part of the right half of the skull and brain.

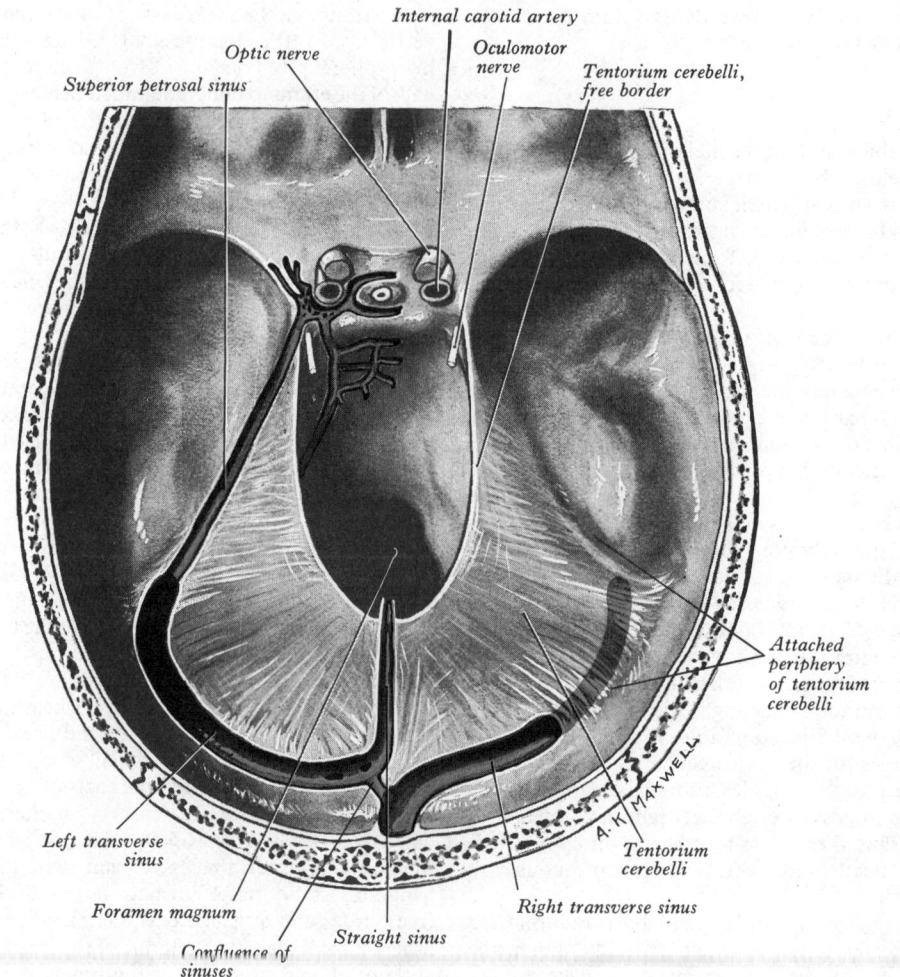

7.162B The superior aspect of the tentorium cerebelli.

that bone. As already described the straight sinus runs in the line of attachment of the posterior part of the inferior border of the falx cerebri to the tentorium cerebelli.

For details of the comparative anatomy and phylogeny of the tentorium cerebelli *see* Klintworth (1968).

(3) *The falx cerebelli* is a small, sickle-shaped process of dura mater which is situated below the tentorium cerebelli, and projects forwards into the posterior cerebellar notch. Its base, directed upwards, is attached to the posterior part of the inferior surface of the tentorium cerebelli, in the median plane; its posterior margin contains the occipital sinus, and is fixed to the internal occipital crest; its apex frequently divides into two small folds, which are lost on the sides of the foramen magnum. Hasan and Das found the falx cerebelli double in 76 of 100 cadavers (*see* p. 746).

(4) *The diaphragma sellae* (7.162) is a small, circular, horizontal fold of dura mater, which forms a roof for the sella turcica and almost completely covers the hypophysis; a small opening in its centre transmits the infundibulum.

The arrangement of the dura mater in the central part of the middle cranial fossa requires further description. As the free rim of the tentorium cerebelli is traced forwards, it converges on the attached periphery, and crosses it near the apex of the petrous part of the temporal bone. It is then continued forwards as a clearly visible ridge on the dura mater as far as the anterior clinoid process, to which it is attached. This ridge marks the junction of the roof and lateral wall of the cavernous sinus (7.161).

The attached periphery of the tentorium cerebelli follows the superior border of the petrous part of the temporal bone and, after being crossed by the free border, continues forwards to the posterior clinoid process as a somewhat rounded and indefinite ridge on the dura mater.

An angular interval exists between the anterior parts of the attached periphery and free border (7.162), and in this interval the dura mater forms part of the roof of the cavernous sinus. Here it is pierced in front by the oculomotor and behind by the trochlear nerves. These two nerves remain in close contact with the dura mater after piercing it and are carried forwards and downwards into the lateral wall of the cavernous sinus (6.127).

From the anteromedial portion of the lateral part of the middle cranial fossa the dura mater ascends, forming the lateral wall of the cavernous sinus. When it reaches the ridge produced by the forward continuation of the free border of the tentorium cerebelli, it is carried medially, forming the roof of the cavernous sinus, and is here pierced by the internal carotid artery (7.162).

Medially, the roof of the cavernous sinus is continuous with the upper surface of the diaphragma sellae. At, or just below, the opening in the diaphragma for the infundibulum of the hypophysis, the dura mater, arachnoid and pia mater blend with one another and with the capsule of the hypophysis (p. 1438), so that within the sella turcica it is impossible to differentiate the individual membranes or to recognize the subdural and subarachnoid spaces (6.127).

Apart from its function as the periosteum of the internal surfaces of the cranial bones, the dura mater may act as a steadying influence, through the agency of the tentorium and falx, upon the movement of the brain within the cranial cavity, but this is merely a speculation. A number of venous sinuses are enclosed within its thickness and certain of these are in actual edges of the dural partitions. The junction of the great cerebral vein with the straight sinus in the tentorium cerebelli is a particularly critical point in the venous drainage of the brain. If the relation between these two venous channels is much altered for anything more than short periods of time, as may occur when 'space-occupying' lesions above the tentorium cause a descent of the brainstem relative to the tentorium, obstruction of the great cerebral vein may ensue, with the sequelae of back-pressure, oedema of the choroid plexuses and over-production of cerebrospinal fluid.

The structure of the dura mater is basically fibrous, white collagen fibres predominating with an admixture of elastic fibres. The collagen fibres are densely arranged in laminae, in which the fibres are often arranged in a parallel manner, with wide angles between these groupings in adjacent laminae, producing a latticed appearance particularly easy to see in the tentorium cerebelli. The cerebral dura mater is often described as consisting of two layers, an *endosteal* layer acting as the periosteum of the cranial bones to which it is attached, and a *meningeal* or cerebral layer internal to this. This description owes more, perhaps, to the separation where venous sinuses occur and to the splitting of the cerebral dura mater at the foramen magnum and optic canals than to any marked histological differences. The smaller branches of the meningeal vessels are, of course, largely in the endosteal region, since they are, despite their name, primarily periosteal in distribution. Fibroblasts occur throughout the dura mater, but osteoblasts are naturally confined to the endosteal level. The elastic fibres separate the laminae of collagen fibres, which are also and more extensively separated by lacunar spaces considered by some to be continuous with the subdural space (*vide infra*). These spaces are mainly confined to the inner part of the dura mater. In sum, therefore, there are a number of features which distinguish the external from the internal levels of the dura, but there is no discontinuity or any other kind of boundary upon which a clear distinction of a bilaminar nature could be based. At all foramina in the cranium the endosteal element is continuous through them with the external periosteum. At the sutures, before their fusion, the endosteal element is continuous with the sutural membrane, and the dura mater is more strongly attached at these locations. Elsewhere it is more easily detached from the cranial bones. The meningeal element is continuous through the appropriate foramina with the dural sheaths of the spinal cord and the optic nerves. At other foramina it is said to be 'pierced' by the nerve or vessel passing through them, but it is perhaps more accurately described as becoming continuous with perineurium or adventitial sheaths. Though in close apposition to the arachnoid mater internal to it, the dura mater is very easily separated from the former, as exemplified by the occurrence of subdural haemorrhages between them. Although contrary views have been recorded, there is little doubt that a layer of flattened 'mesothelial' cells is a constant feature of the internal aspect of the dura mater and that small amounts of fluid, though as a mere capillary film,

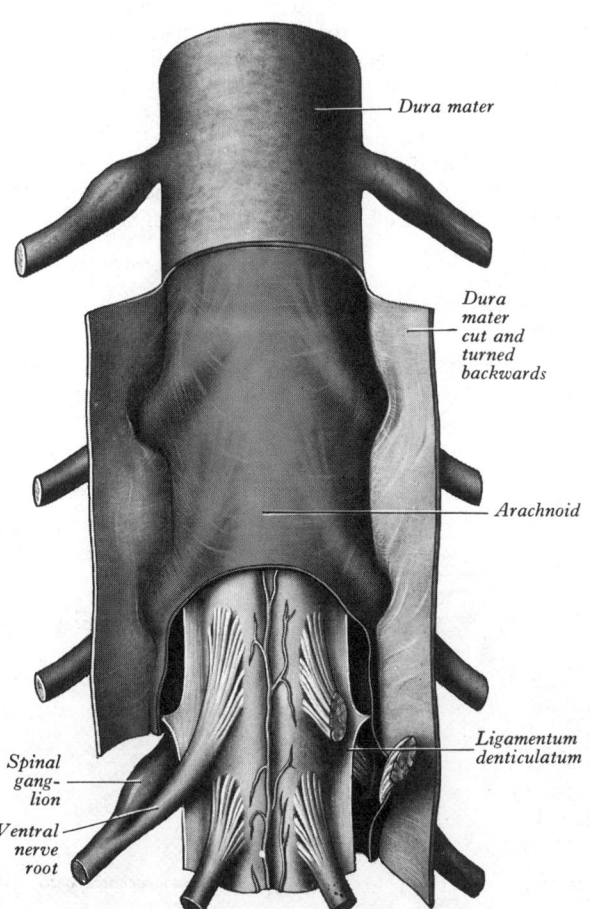

Dura mater

Dura mater cut and turned backwards

Arachnoid

Ligamentum denticulatum

Spinal ganglion

Ventral nerve root

7.163 A part of the spinal cord exposed from the ventral aspect, showing its meningeal coverings.

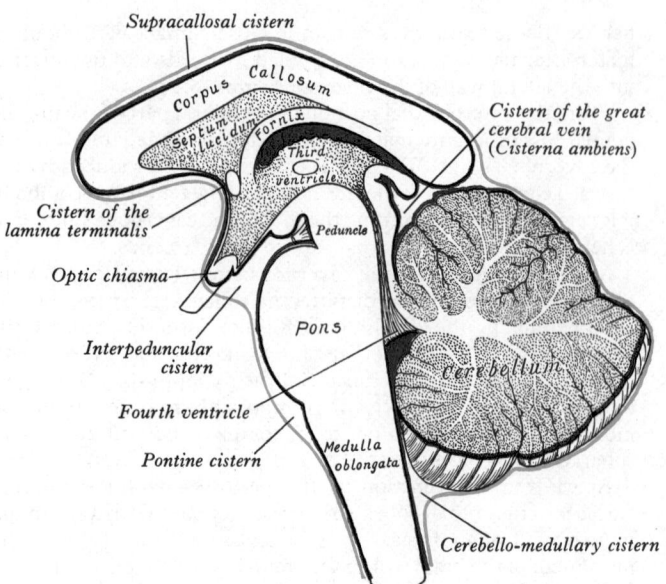

7.164 A diagram showing the positions of the principal subarachnoid cisterns. *Red:* pia mater. *Blue:* arachnoid mater.

usually exist as a zone of potential separation between arachnoid and dura. Rascol and Izard (1974) have confirmed by electron microscopy the existence of epithelial cells in this region, regarding them as 'neurothelium'. They describe them as arranged in several irregular layers, with abundant intercellular spaces, devoid of collagen fibres. The cells display tonofilaments and desmosomes, but these observers consider that the epithelium is fragile and easily torn apart to produce a 'subdural space'. (The structure of the subjacent arachnoid is described on p. 1051.)

The arteries of the cerebral dura mater are very numerous. Those in the anterior fossa of the skull are the anterior meningeal branches of the anterior and posterior ethmoidal and internal carotid arteries, and a branch from the middle meningeal artery. Those in the middle fossa are the middle and accessory meningeal branches of the maxillary artery; a branch from the ascending pharyngeal artery, which enters the skull through the foramen lacerum; branches from the internal carotid artery, and a recurrent branch from the lacrimal artery. Those in the posterior fossa are meningeal branches from the occipital artery, one entering the skull through the jugular foramen, and another through the mastoid foramen; the posterior meningeal branches of the vertebral artery; occasional meningeal branches from the ascending pharyngeal artery, entering the skull through the jugular foramen and hypoglossal canal. The meningeal arteries are chiefly distributed to bone, in contrast to those of the spinal dura mater, and are therefore inappropriately named. Only very fine branches are distributed to the dura itself within the cranium.

The *veins* returning the blood from the cranial dura mater are described on pp. 745–749.

The nerves of the cerebral dura mater have been the subject of a number of investigations over a long period (von Luschka 1850, 1860; Arnold 1851; Hovelacque 1927; Siwe 1931; Penfield and McNaughton 1940; Kimmel 1961*a* and *b*). The best recognized dural innervation arises from the *trigeminal nerve*, including its ganglion and three principal divisions or their branches, from the upper three *cervical nerves*, and from the *cervical sympathetic* trunk. Other less well-established meningeal branches have been described as arising from the vagus and hypoglossal nerves, and also possibly from the facial and glossopharyngeal nerves (but *vide infra*).

In the *anterior cranial fossa* the meningeal nerves are twigs from the anterior and posterior ethmoidal nerves, and anterior filaments from the meningeal branches of the maxillary (*nervus meningeus medius*) and mandibular (*nervus spinosus*) divisions of the trigeminal. The principal area of distribution of the latter two nerves is, however, to the dura of the *middle cranial fossa*, which also receives filaments directly from the trigeminal ganglion. The

tentorium cerebelli receives on each side the recurrent *tentorial nerve*, a branch of the ophthalmic division of the trigeminal. The dura mater of the *posterior cranial fossa* is innervated by ascending meningeal branches from the upper cervical nerves which enter the cranium through the anterior part of the foramen magnum (second and third cervical), and through the hypoglossal canal and jugular foramen (first and second cervical nerves). A number of separate filaments enter by each of these routes. Whilst *direct* meningeal branches from the vagus and hypoglossal, and possibly other cranial nerves, cannot be excluded, strong evidence has been accumulated (Kimmel 1961*a* and *b*) showing that nerves so described are in most cases recurrent filaments from the cervical nerves, which, for a short distance, run within the connective tissue sheath of the cranial nerve concerned.

All the foregoing meningeal nerves contain a postganglionic sympathetic component, either derived directly from the superior cervical sympathetic ganglion, or by receiving a communication from one of its perivascular sympathetic nerve extensions into the cranium. These are probably vasomotor fibres, and experiments have been quoted to show that they exert a vasoconstrictor action on pial vessels. Current physiological teaching, however, accords little importance to autonomic control of intracranial arteries in general (but *see* cerebral blood flow—p. 1044).

Various types of sensory nerve ending, including simple end-bulbs, and Meissner's and Pacinian corpuscles, have been described in the dura mater in various mammals, but little recent information is available concerning man. Thus, the functional role of the sensory and autonomic supply to the dura mater remains uncertain.

The spinal dura mater (7.163, 167) forms a loose sheath around the spinal cord, and represents only the inner, or meningeal, layer of the cerebral dura mater; the outer, or endosteal, layer being represented by the periosteum lining the vertebral canal, which is separated from the spinal dura mater by an interval, termed the *extradural space*. The spinal dura mater is attached to the circumference of the foramen magnum, and to the posterior surfaces of the bodies of the second and third cervical vertebrae; it is also connected by fibrous slips to the posterior longitudinal ligament of the vertebrae, especially near the lower end of the vertebral canal. The subdural cavity ends at the lower border of the second sacral vertebra; below this level the dura mater closely invests the filum terminale of the spinal cord and descends to the back of the coccyx, where it blends with the periosteum. The dura mater has tubular prolongations along the roots of the spinal nerves and the spinal nerves themselves as they pass through the intervertebral foramina (7.167). These

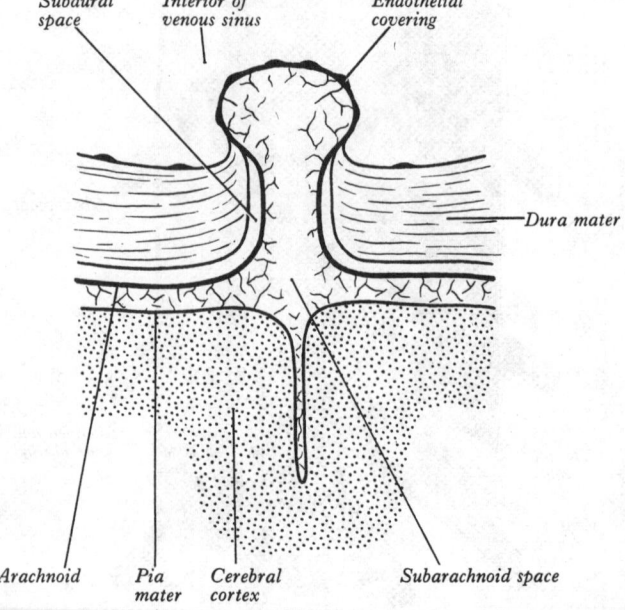

7.165 A diagram to show the structure of a small arachnoid granulation. (After W. E. Le G. Clark.)

prolongations are short in the upper part of the vertebral column, but gradually become longer below, owing to the increasing obliquity of the nerve roots (7.37). The dural sheaths of the spinal nerves fuse with their connective tissue sheaths in or slightly beyond the intervertebral foramina. This continuity may act as a slight check on movements of the nerves when tensed by bodily movement. In the cervical region, where the nerves are short and vertebral movement greatest in range, the nerve sheaths are bound strongly to the periosteum covering the adjacent transverse processes (Sunderland 1974).

The extradural space lies between the spinal dura mater and the periosteum and ligaments within the vertebral canal; it contains a quantity of loose fat and areolar tissue and a plexus of veins. The loose fat and areolar tissue of the space, which is known to clinicians as the *epidural space*, extends laterally for a short distance through the intervertebral foramina along the spinal nerves. Dyes or other fluids injected into the sacral hiatus under pressure can spread upwards to the base of the skull in the extradural space, and local anaesthetics injected in the neighbourhood of one spinal nerve immediately outside the intervertebral foramen may spread either upwards or downwards to affect the nerves of adjoining segments, or may spread to the opposite side. In each instance the spread occurs through the extradural space. (For the nerve supply of the spinal dura mater, *see* p. 1089.)

The subdural space is a potential space between the dura mater and the arachnoid mater. It contains a film of serous fluid which moistens the surfaces of the opposed membranes. It does not appear to communicate with the subarachnoid space, but is continued for a short distance on the cranial and spinal nerves, and is in free communication with the lymph spaces of the nerves. Around the optic nerve it is continued as far as the back of the eyeball. The significance of the subdural space in terms of function remains a matter of argument rather than demonstration. Possible connexions with venous channels on the one hand and with hypothetical lymph spaces in the substance of the dura mater on the other have been claimed and disclaimed. (*See* Hoffmann and Thiel 1956, for a discussion of these views.) The evidence of electron microscope observations is against the occurrence of any specialized cells of epithelial type in the dura mater, apart from its arachnoid surface, all the dural fibroblasts being of similar appearance. The dural lacunae may in fact be artefacts; certainly the evidence available refutes the passage of significant amounts of a lymphatic fluid from the dura into the subdural or subarachnoid spaces (Kaplan and Ford 1966). In the case of the spinal dura mater, however, there is undoubtedly a lymphatic drainage in regard to the extradural adipose tissue, and this may also include the dura itself (Millen and Woollam 1962).

THE ARACHNOID MATER

The arachnoid is a delicate membrane enveloping the brain and spinal cord and lying between the pia mater internally and the dura mater externally. It is separated from the dura mater by the *subdural space*, but here and there this space is traversed by isolated connective tissue trabeculae which are most numerous on the posterior surface of the spinal cord. It is separated from the pia mater by the *subarachnoid space*, which is filled with cerebrospinal fluid.

The arachnoid surrounds the cranial and spinal nerves, and encloses them in loose sheaths as far as their points of exit from the skull and vertebral canal.

The *cerebral part of the arachnoid* invests the brain loosely, and does not dip into the sulci between the gyri, nor into the fissures, with the exception of the longitudinal fissure. On the upper surface of the brain it is thin and transparent; at the base it is thicker, and slightly opaque towards the central part, where it extends between the two temporal lobes in front of the pons, so as to leave a considerable interval between it and the pia mater. It cannot be identified in the hypophysial fossa.

The *spinal part of the arachnoid* (7.163, 167) is a thin, delicate tubular membrane loosely investing the spinal cord. Above, it is continuous with the cerebral arachnoid; below, it widens out, invests the cauda equina, and ends at the level of the lower border

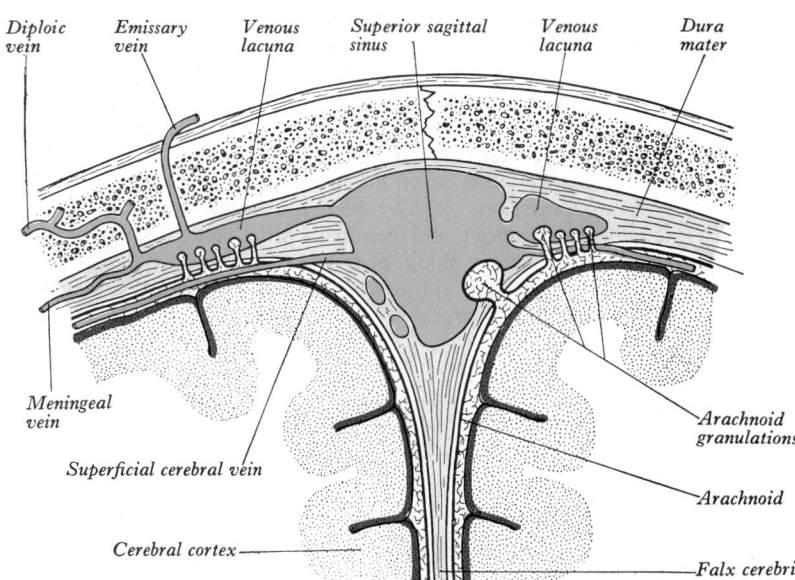

7.166 A coronal section through the vertex of the skull to show the arrangement of the veins and the meninges of the brain and arachnoid granulations.

of the second sacral vertebra. (For the structure of the pia-arachnoid *see* p. 1051.)

The subarachnoid space is the interval between the arachnoid and pia mater. It contains the cerebrospinal fluid and the larger blood vessels of the brain, and is traversed by a network of delicate connective tissue trabeculae, which connect the arachnoid to the pia mater. The pia mater and the arachnoid are in close contact on the summits of the cerebral gyri; but where the arachnoid bridges the sulci, angular spaces are left, in which the subarachnoid trabecular tissue is found. At certain parts of the base of the brain, the arachnoid is separated from the pia mater by wide intervals, which communicate freely with each other and are named *subarachnoid cisterns*; in these the subarachnoid tissue is scanty and may be absent.

The Subarachnoid Cisterns (7.164).

The *cerebello-medullary cistern (cisterna magna)* (7.164) is formed by the arachnoid bridging the interval between the medulla oblongata and the inferior surface of the cerebellum and is triangular on sagittal section; it is continuous below with the subarachnoid space of the spinal cord. The *pontine cistern* (7.164) is an extensive space on the ventral surface of the pons. It contains the basilar artery, and is continuous below with the subarachnoid space of the spinal cord, behind with the cerebello-medullary cistern, and rostral to the pons with the interpeduncular cistern. As the arachnoid extends across between the two temporal lobes, it is separated from the cerebral peduncles and the structures in the interpeduncular fossa by the *interpeduncular cistern*, which contains the circulus arteriosus. Anteriorly the interpeduncular cistern is continued rostral to the optic chiasma and is prolonged over the surface of the corpus callosum; here the arachnoid stretches between the cerebral hemispheres immediately below the free border of the falx cerebri, and this leaves a space in which the anterior cerebral arteries are contained. The *cistern of the lateral fossa* contains the middle cerebral artery, and is formed in front of each temporal lobe by the arachnoid bridging the lateral sulcus. The *cistern of the great cerebral vein (cisterna ambiens, or superior cistern)* occupies the interval between the splenium of the corpus callosum and the superior surface of the cerebellum; it contains the great cerebral vein and the pineal gland. It is a widely used neurosurgical landmark.

Other less prominent cisternae have been described; these include the *prechiasmatic* and *postchiasmatic cisterns* related to the optic chiasma, the *cistern of the lamina terminalis*, and the *supracallosal cistern*, all of which, in the above account, were included as extensions of the interpeduncular cistern, and which contain the anterior cerebral arteries.

The subarachnoid space communicates with the general ventricular cavity of the brain by three openings: the *median aperture* (p. 932) is in the median plane in the inferior part of the roof of the fourth ventricle; the two *lateral apertures* are at the extremities of the lateral recesses of that ventricle (p. 932), behind the upper roots of the glossopharyngeal nerves. There is no direct communication between the subdural and subarachnoid spaces. Communications exist between the tissue spaces in the nasal mucous membrane and the subarachnoid space through channels which are present along the course of the olfactory nerves.

The spinal part of the subarachnoid space is a relatively wide interval, and is largest at the lower part of the vertebral canal,

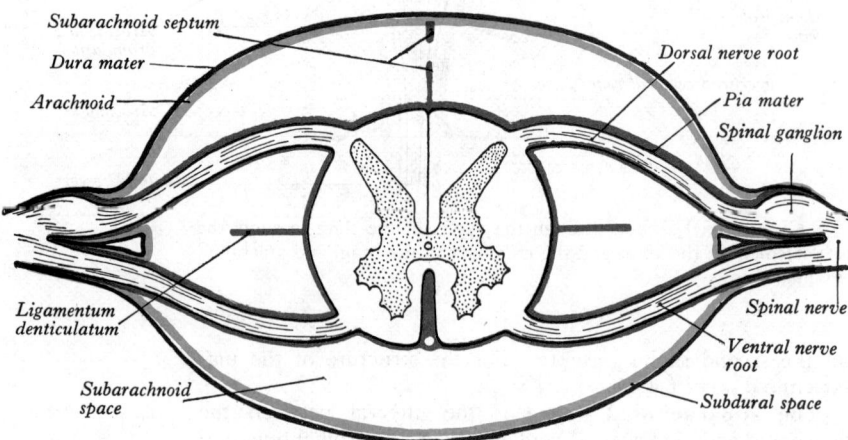

Subarachnoid septum
Dura mater
Arachnoid
Dorsal nerve root
Pia mater
Spinal ganglion
Ligamentum denticulatum
Spinal nerve
Ventral nerve root
Subarachnoid space
Subdural space

7.167 A transverse section through the spinal cord and its membranes. *Black:* dura mater. *Blue:* arachnoid mater. *Red:* pia mater.

where the arachnoid encloses the roots which form the cauda equina. Above, it is continuous with the cranial subarachnoid space; below, it ends at the level of the lower border of the second sacral vertebra. It is partially divided by two septa, termed, respectively, the subarachnoid septum and the ligamentum denticulatum. Both are described later (p. 1051).

The arachnoid granulations (7.165, 166) are small fleshy-looking elevations, usually collected in clusters, which are present in the vicinity of the superior sagittal, transverse, and some other sinuses. When the sagittal sinus and the venous lacunae on each side of it are opened, granulations will be found protruding into their interior (7.166). On close inspection they may be seen at the age of eighteen months, and at the age of three they are disseminated over a considerable area; they increase in number and size as age advances. They cause absorption of the bone, and so produce the pits or depressions on the inner aspect of the skull cap. Arachnoid granulations are macroscopic enlargements, or distensions, of minute projections of the arachnoid mater, termed *arachnoid villi*, which are normally present in great numbers in young subjects.

The growth and structure of the arachnoid villi and granulations have been studied in much detail (Le Gros Clark 1920). Histologically each villus appears as a diverticulum of the subarachnoid space, penetrating into the interstices of the dura mater, and covered by a layer of flattened cells containing large oval nuclei and lightly staining protoplasm. In the subarachnoid space there is a reticulum of fine fibrous tissue, the density of which is as a rule greater at the periphery than at the centre of the granulation; in advanced age it frequently contains calcareous nodules.

At the summit of the villus the mesothelial cells proliferate and form a cap which penetrates the surrounding dura mater, and fuses with the endothelial lining of one of the intradural venous sinuses (7.165); in doing so it pulls out a little stalk of arachnoid containing a diverticulum of the subarachnoid space. Except at the point of fusion with the endothelial lining of the sinus, the villus is surrounded by the subdural space and the dura mater; the latter, covered on its cerebral surface by a layer of mesothelium, is invaginated into the venous sinus by the protrusion of the granulation.

Fluid injected into the subarachnoid space passes into these granulations and villi, and it has been found experimentally that fluid passes by osmosis from the arachnoid villi into the venous sinuses of the dura mater.

The *cerebrospinal fluid* (Davson 1970) is a clear, slightly alkaline fluid, with a specific gravity of about 1007. It contains in solution inorganic salts similar to those in the blood plasma, and also traces of protein and glucose. The cerebrospinal fluid is secreted into the ventricles of the brain by the choroid plexuses and into the subarachnoid space by the plexuses sited in the lateral recesses of the fourth ventricle (p. 932). From the ventricles it passes through the median aperture and the foramina of the lateral recesses of the fourth ventricle and so gains the subarachnoid space in the cerebello-medullary cistern and the pontine cistern. Within the cranium the cerebrospinal fluid flows upwards through the gap in the tentorium cerebelli and then forwards and laterally over the inferior surface of the cerebrum. Finally it ascends the superolateral aspect of each hemisphere to reach the arachnoid villi associated with the superior sagittal sinus, and so is able to pass back again into the bloodstream. It is generally held that within the vertebral canal there is no active flow, but that the process of diffusion and alterations of posture serve to maintain the character of the fluid constant throughout the whole extent of the subarachnoid space. Experimental work (Howarth and Cooper 1949) suggests that the spinal cerebrospinal fluid may drain back locally into the venous system, through the vertebral venous plexuses, the intervertebral veins and the posterior intercostal and upper lumbar veins into the azygos and hemiazygos veins. The cerebrospinal fluid supports the brain and spinal cord, and it maintains a uniform pressure upon them. It has been stated that a brain weighing 1,500 gm. in air, weighs no more than 50 gm. in cerebrospinal fluid, and through the latter, the total weight of the system is evenly distributed to the meningeal parieties, and their mechanical supports. Our knowledge of the circulation of the cerebrospinal fluid and of the arachnoid villi was, to a large extent, built up on the work of Weed (1920, 1938). Electron microscope studies suggest that there is an open communication between the subarachnoid space and the lumen of the superior sagittal sinus by means of fine tubules, lined by an endothelium, traversing the core of arachnoid granulations in the sheep. A valvular action is hypothecated for these tubules, but this form of drainage, if confirmed, does not exclude filtration (Jayatilaka 1965).

Applied Anatomy. Diseases of the central nervous system and its membranes are often reflected in alterations of the cells which are normally found in the cerebrospinal fluid or in alterations in the concentration of its chemical constituents. Interference with the circulation of the fluid is indicated by variations in the pressure within the meninges. The determination of these alterations and variations is often of service in diagnosis.

Specimens of the cerebrospinal fluid may be obtained by the operation of *lumbar puncture*, which is performed through the interval between the laminae or spines of the third and fourth (or fourth and fifth) lumbar vertebrae. A fine trocar and cannula is inserted at the point of intersection of the intertubercular plane with the posterior median line and is passed obliquely upwards and forwards above the upper border of the spine of the fourth lumbar vertebra. It is carried through, or parallel to, the supraspinous and interspinous ligaments into the vertebral canal. The dura mater and the arachnoid are punctured and the instrument is introduced into the subarachnoid space below the lower end of the spinal cord (7.34). When the trocar is withdrawn, the cerebrospinal fluid escapes through the cannula at the rate of one drop per second, under normal conditions, but when the fluid is under increased pressure it escapes in an almost continuous stream. It should also be noted that the introduction of a large needle, such as a trocar, between the vertebral laminae is rather difficult except at lumbar levels. Moreover, the introduction of its point, accurately enough to withdraw fluid, into the narrow subarachnoid space surrounding the spinal cord would be extremely difficult, apart from the danger to the cord itself.

THE PIA MATER

The pia mater closely invests the brain and spinal cord; it is a vascular membrane, consisting of a plexus of minute blood vessels held together by an extremely fine areolar tissue. The *cerebral pia mater* invests the entire surface of the brain, dips between the cerebral gyri and between the cerebellar laminae, and is invaginated to form the tela choroidea of the third ventricle, and the choroid plexuses of the lateral and third ventricles (pp. 1028 and 981); as it passes over the roof of the fourth ventricle, it forms the tela choroidea and the choroid plexuses of this ventricle (p. 932). Upon the surfaces of the hemispheres it gives off from its deep surface a multitude of sheaths around the minute vessels that run perpendicularly for some distance into the cerebral substance. On the cerebellum the membrane is more delicate; the vessels from its deep surface are shorter, and its relations to the cortex are not so intimate. Like the arachnoid, the pia mater cannot be identified in the hypophysial fossa.

The *spinal pia mater* (7.163, 167) is thicker, firmer, and less vascular than the cerebral pia mater, but like the latter it consists of an outer '*epi-pia*' containing the larger vessels, and internal to this a so-called '*pia-glia*' or '*pia-intima*' in direct contact with the nervous tissue (*vide infra*). Between the layers are cleft-like spaces which communicate with the subarachnoid space, and a number of blood vessels. The spinal pia mater covers the spinal cord, and is intimately adherent to it; in front it dips into the anterior median fissure and lines its walls, the two layers involved being connected by a loose open meshwork of fine fibrous strands (7.167). A longitudinal fibrous band, called the *linea splendens*, extends along the median plane anteriorly. The *ligamentum denticulatum* is situated on each side and the *subarachnoid septum* is present posteriorly (7.167). Below the conus medullaris the pia mater is continued as a longer slender filament, named the *filum terminale* (p. 865).

The pia mater forms sheaths for the cranial and spinal nerves; these sheaths are closely connected with the nerves, and blend with their common membranous investments.

The *ligamentum denticulum* (7.163, 167) is a narrow, fibrous sheet situated on each side of the spinal cord, between the ventral and the dorsal nerve roots. Its medial border is continuous with the pia mater at the side of the spinal cord. Its lateral border presents a series of triangular tooth-like processes, the points of which are fixed at intervals to the dura mater. These processes are twenty-one in number, on each side. The first process crosses behind the vertebral artery at the point where that vessel pierces the dura mater, and is separated by the artery from the ventral root of the first cervical nerve; it is attached to the dura mater immediately above the margin of the foramen magnum, a short distance behind the hypoglossal nerve, and the spinal part of the accessory nerve ascends on its posterior aspect (7.59). The last process is between the exits of the twelfth thoracic and first lumbar nerves, and consists of a narrow oblique band running downwards and laterally from the conus medullaris (7.163). Changes in the form and position of the denticulate ligament during spinal movements have been demonstrated by cineradiographic techniques (Epstein 1966).

The *subarachnoid septum* is an interrupted sheet of fibrous tissue, situated in the median plane. It connects the arachnoid to the pia mater opposite the posterior median sulcus (7.167). Incomplete and cribriform in the cervical region, it forms a more complete partition at thoracic levels.

The Microscopic Structure of the Leptomeninges

The finer structure of the arachnoid and pial meninges is more appropriately described in conjunction (Winckler 1960). They have a common phylogenetic history, develop embryologically in continuity, and preserve this intimate relationship in their final differentiation to such an extent that they are frequently regarded as a *pia-arachnoid* rather than two separable entities. They will, however, be described here successively, largely because it is still widely customary to employ separate names for them.

The structure of the arachnoid mater, like that of the pia, is essentially that of loose, irregular connective tissue. Both the leptomeninges, like the dura mater, or pachymeninx, are regarded

as being of mesodermal origin, though there is some doubt with respect to the innermost region of the pia mater, which may be of ectodermal derivation (Millen and Woollam 1961). In the most primitive vertebrates all three meninges are represented by an undifferentiated *meninx primitiva*; but in tetrapods a distinction into the thick external dura mater and a more delicate internal leptomeninx is established, and in mammals and birds this reaches its most differentiated form. All three meninges develop in the mesenchyme surrounding the central nervous system, with the possible proviso mentioned above.

The *arachnoid mater* consists of collagen, elastin, and reticulin fibres, together with flattened cells usually regarded as mesothelial in type, although there is some disagreement as to this latter point. Both on its dural aspect and on its internal surface, where it forms the external boundary of the subarachnoid space, cells with long cytoplasmic processes and a paler cytoplasm than the fibrocytes of the dura mater are arranged in several layers, with an underlying basement membrane. Tight junctions between some of these cells are demonstrable by electron microscopy (Nelson *et al.* 1961), but collagen fibres are interspersed among them and elastic fibres may also appear between the layers of cells. The fibre elements of the arachnoid are of more gracile proportions than those of the dura mater. Fine strands or *trabeculae* of arachnoid tissue extend inwards across the *subarachnoid space*, and these also are usually considered to show a covering of 'mesothelial' cells, within which are collagen fibres and then reticulin fibres surrounding the numerous small vessels which pass through the trabeculae to reach the pia mater and central nervous system. The trabeculae vary much in size and distribution; most are fine strands and they are most numerous in the intracranial part of the subarachnoid space, but are largely absent where this is dilated to form cisternae (p. 1049) and are little evident in the spinal region of the space in man. Descriptions of the trabeculae are frequently illustrated from other mammals and this may convey a false or exaggerated impression of their development in mankind. The spaces between the trabeculae are usually extensive and are described as running together to form a single intercommunicating subarachnoid space; but in the human arrangement large stretches of the space are completely free from trabeculae.

Crossing the subarachnoid space the arachnoid trabeculae fuse with the pia mater, which is itself sometimes regarded as divisible into two layers (*vide infra*), the external being a vascularized connective tissue stratum, covered by a mesothelium bounding the subarachnoid space on its internal or cerebral aspect. Some observers regard this stratum of the pia mater as being arachnoid tissue, considering only the avascular layer of collagen and other fibres and their accompanying cells which are immediately adjacent to nervous tissue to represent the true pia. In this case the subarachnoid space is described as being *in* the arachnoid mater. Others prefer to think of the two leptomeninges as a conjoined pia-arachnoid membrane, as stated earlier.

The arachnoid mater is not itself much vascularized, and is often described as avascular; but it does act as a connective tissue support for large numbers of small vessels and their accompanying nerve supplies, traversing the trabeculae to reach the pia mater. Its mesothelial elements and basement membrane, and particularly the tight junctions between the cells, are a barrier to free diffusion of colloidal substances between the subarachnoid space, subdural space, the central nervous system and its non-capillary vessels in both their extra- and intra-cerebral sites. The return of the cerebrospinal fluid is thus largely, if not entirely, limited to the arachnoid villi (p. 1048), where this permeability barrier is modified and also circumvented by a system of minute valvular canals. The barrier is also absent where the craniospinal nerves pass through fused sleeves of pia and arachnoid as they blend with the perineurium. At these locations cerebrospinal fluid can diffuse through the arachnoid tissue into neighbouring lymphatics.

The structure of the pia mater is also that of a loose connective tissue, containing collagen, elastin and reticulin fibres with flattened 'mesothelial' cells like those of the arachnoid. These cells are considered to be fibrocytic elements by some authorities, especially where they form several layers, associated with reticulin fibres, next to the central nervous system. Here the most internal cells are directly apposed to a convoluted basement membrane, intervening between them and the subjacent end-feet of astrocytes. This relationship is so intimate that the most internal stratum of the pia, together with the basement membrane and the glial processes, is sometimes described as the 'pia-glia'. It is also sometimes distinguished as the 'pia-intima', and thus differentiated from the rest of the membrane, the 'epi-pia', which is the pial element regarded as arachnoid by some, as mentioned above. Sleeves of both pia and arachnoid, with an intervening extension of the subarachnoid space, are carried into the central nervous system around the blood vessels entering or leaving it, forming the perivascular cuffs which have been the subject of so much past disagreement.

The pia mater contains large numbers of blood vessels and their accompanying nerve supplies. On the surface of the spinal cord the more external connective tissue layer of the pia, the epi-pia, is more developed than it is on the cerebral surface, where it is not easily identified. In this stratum the vessels are suspended by trabeculae of pial tissue which meet and blend with arachnoid trabeculae where these are present. Thus many pial vessels project from the surface as they ramify in their pia-arachnoid suspension, and are at least partially surrounded by extensions of the subarachnoid space. Where they turn into the central nervous system prolongations of the space and the two leptomeninges accompany each vessel as far as the commencement of its capillaries. Hence the latter are not separated from nervous tissue by the connective tissue or mesothelial elements which accompany the precapillary vessels; the 'pia-glial' barrier is absent. Current physiological views reject the concept of any significant interchange of fluid between the perivascular extensions of the subarachnoid space and nervous tissue or blood, nor is drainage circulation through them considered likely. It is suggested that these minute fluid-filled spaces merely act as a displaceable medium to accommodate the pulsatile variation in the calibre of the vessels which they surround.

The coverings of the central nervous capillaries are like those in the choroid plexuses (p. 1050) in the complete absence of ordinary connective elements from their surroundings. The neuronal level is separated from the blood only by the vascular endothelium, the end-feet of astrocytes and the basement membrane which they share. While the individual contribution of these structures to the human *blood-brain barrier* cannot yet be stated with certainty, abundant experimental evidence indicates that this is the major and probably the exclusive site of entry of fluid and solutes into the central nervous tissue, the arachnoid villi providing the main exit for cerebrospinal fluid, and the drainage of fluid from the nervous tissue being partly back into the prevenous ends of the capillaries and perhaps through ependyma into the ventricular system. Experiments in the mouse, using horseradish peroxidase as an electron-dense tracer have, however, shed some light on the problem of the blood-brain barrier (Brightman *et al.* 1970). In most of the central nervous situations examined, complete circumferential tight junctions, between adjacent capillary endothelial cells, totally prevent the passage of the tracer. Similar junctions exist between the modified ependymal cells which clothe the choroid plexus. To what extent such findings apply to other sites, animals and substances, remains speculative.

Apart from their functions in the production, circulation and reabsorption of the cerebrospinal fluid, the meninges also serve as a support to the central nervous system, especially in association with the buoyant effect of the fluid. Like connective tissues elsewhere they also provide routes of access and support for the vessels of the central nervous system, and they constitute at least a mechanical barrier against infection.

THE PERIPHERAL NERVOUS SYSTEM

The peripheral nervous system comprises the afferent, or centripetal, fibres which connect the sensory end organs to the central nervous system, and the efferent, or centrifugal, fibres which connect the central nervous system to the effector apparatus. It includes the twelve pairs of cranial nerves which arise from the brain, and the thirty-one pairs of spinal nerves which arise from the spinal cord. The sympathetic trunks with their various ganglia and branches belong to this system, but they will be dealt with in a separate section (pp. 1121–1138).

In the most primitive vertebrates the spinal cord gives rise to a series of ventral nerve roots, arising from the anterior grey column and motor in function, and a series of dorsal nerve roots, connected to the posterior grey column and sensory in function. The ventral and dorsal nerve roots do not unite, and they do not correspond in position. The ventral nerve root is segmental and is distributed to the myotome which corresponds to the neuromere from which it arises. The dorsal nerve root is intersegmental in position and runs in the intersegmental connective tissue to reach its cutaneous distribution. In the majority of fishes and in all higher forms the corresponding ventral and dorsal nerve roots which emerge from the spinal cord unite with one another to constitute the individual spinal nerves. The arrangement of the spinal nerves, therefore, follows a very primitive pattern and has not undergone much modification in the process of evolution.

The arrangement of the cranial nerves, on the other hand, has been very profoundly modified. The development and modification of the branchial system and the suppression of segments owing to the elaborate changes which occur in the region of the head have been largely responsible for this modification. In the brain, corresponding ventral and dorsal nerves *never* fuse, although adjoining ventral or dorsal nerves may and actually do unite. Owing to the complete disappearance of certain myotomes the corresponding ventral nerves become completely suppressed. Further, the dorsal nerves, originally sensory nerves supplying chiefly the skin of the head and the mucous membrane of the mouth and pharynx acquire motor fibres which they distribute to the musculature arising in the branchial region (p. 151). With the growth and modification of the brain and the consequent elaboration of the head region, the cutaneous areas of the head are transferred from one nerve to its neighbour, so that the functions of the individual dorsal nerves become altered.

The incorporation of some of the precervical segments in the head leads to the fusion of the corresponding ventral nerves, and the hypoglossal nerve so formed becomes added to the cranial series.

The cerebrospinal nerves, as we have seen in greater detail elsewhere (p. 838), consist of numerous nerve fibres collected into bundles, which are enclosed in connective tissue sheaths (7.168): a small bundle of fibres is called a *fasciculus*. Each fasciculus is surrounded by a connective tissue sheath, named the *perineurium*. Individual nerve fibres are ensheathed, held together and supported within the fasciculus by delicate connective tissue called the *endoneurium*; it is continuous with septa which pass inwards from the perineurium. The collagen fibres of the endoneurium tend to be longitudinally orientated and bundles are sometimes invaginated into the Schwann cells of non-myelinated nerve fibres (Gamble and Eames 1964). If small, the nerve may consist of only a single fasciculus; but if large, it consists of several fasciculi held together and invested by connective tissue; this investment is known as the *epineurium*. The cerebrospinal nerves consist both of myelinated and non-myelinated nerve fibres, the proportion of each varying with the functional roles of the nerve concerned.

The blood vessels, supplying a nerve, end in a minute plexus of capillaries which pierce the perineurium, and run, for the most part, parallel with the fibres; they are connected together by short transverse vessels, forming narrow, oblong meshes, similar to the capillary system of muscle. Fine, non-myelinated, vasomotor nerve fibres accompany these vessels, and break up into fine fibrils which form a network around them. Myelinated fibres, termed

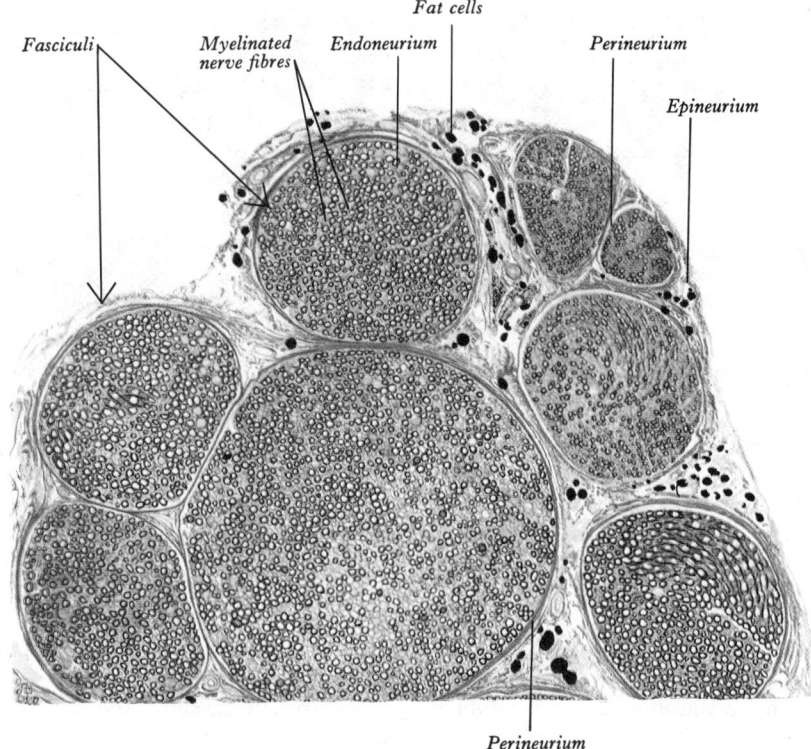

7.168 Transverse section of a peripheral nerve (cat.) Stained with osmic acid and van Gieson's technique. Note the nerve fasciculi, the epi-, peri- and endo-neurial connective tissue sheaths, and the regional variations in the calibre of the myelinated nerve fibres. Magnification about × 55.

nervi nervorum, run in the epineurium and terminate in oval or bulbous corpuscles (p. 855).

The cerebrospinal nerve fibres pursue an uninterrupted course from the centre to the periphery, but in separating a nerve into its component fasciculi, it is found that bundles of fibres from one fasciculus may join, at a very acute angle, another fasciculus proceeding in the same direction.

In their courses, nerves divide into branches, and these frequently communicate with branches of neighbouring nerves; such communications form what is called a *nerve plexus*. Such a plexus is formed by the ventral rami of the trunks of the nerves— as, for example, the cervical, brachial, lumbar and sacral plexuses—or by the terminal fasciculi, as in the plexuses formed at the periphery of the body. In the formation of a plexus, the component nerves divide, then join, and again subdivide in such a complex manner that the individual fasciculi become intricately interlaced. Hence, each branch leaving a plexus may contain filaments from more than one, and even all of the primary nerves entering the plexus. In the formation also of smaller plexuses at the periphery of the body there is a free interchange of fibres. In each case, however, the individual fibres remain separate and distinct.

Through this interchange of fibres, every nerve leaving a plexus gains a more extensive connexion with the spinal cord than if it had proceeded direct to its distribution without joining other nerves.

The origin of a nerve is a phrase usually implying the locus of its emergence from or entry into the central nervous system. This site is sometimes called the *superficial origin*, the *deep origin* being the central group or groups of nerve cells from which the peripheral axons are derived, or to which they are proceeding. The superficial origin is in some cases single—that is to say, the whole nerve emerges from the central nervous system by a single root; in other instances the nerve arises by two or more roots. The *efferent nerve fibres* are the axons of nerve cells situated in the grey matter of the central nervous system. The *afferent nerve fibres* spring from nerve cells in the organs of special sense (e.g. the retina) or from nerve cells in the ganglia on the cerebrospinal nerves. Having entered the central nervous system they branch and send their ultimate twigs to terminate in synaptic association with nerve cells there.

The peripheral terminations of the sensory nerves are dealt with on pp. 849–857, and those of motor nerves on pp. 846–849.

Ganglia are aggregations of nerve cells found on some peripheral nerves. They are present in the dorsal roots of the spinal nerves and the sensory roots of the trigeminal, facial, glossopharyngeal, vagal and vestibulocochlear cranial nerves. They also occur in association with autonomic nerves (p. 1124). They vary considerably in form and size. Each ganglion is invested by a smooth, firm covering of fibrous connective tissue and associated flattened, fibrocyte-like cells, which is continuous with the perineurium of the nerves, and sends numerous processes into the interior of the ganglion.

Ganglia contain nerve cells and their nerve fibres, and fibres derived from cells elsewhere, which pass right through or

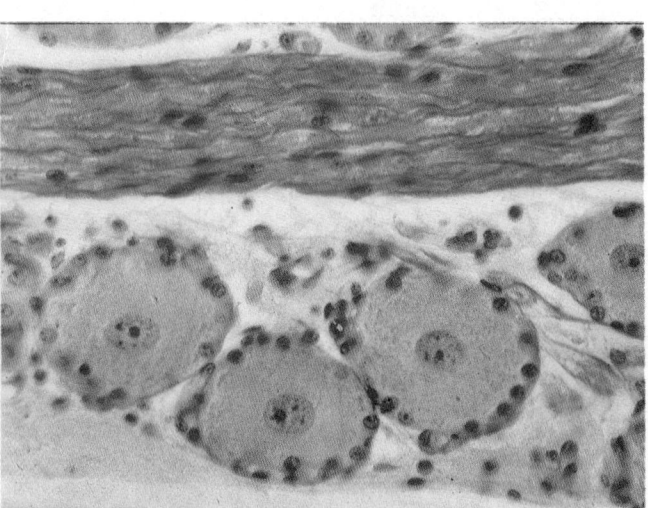

7.169 A typical field in a dorsal spinal nerve root ganglion. Note the characteristic juxtaposition of large ovoid nerve cell somata, and the fascicles of myelinated and non-myelinated nerve fibres. Note also the nuclei of the capsular (satellite) cells which surround each nerve cell. (Grübler's stain—material kindly provided by Dr. Lyn Gregson, Dept. of Anatomy, Guy's Hospital Medical School.)

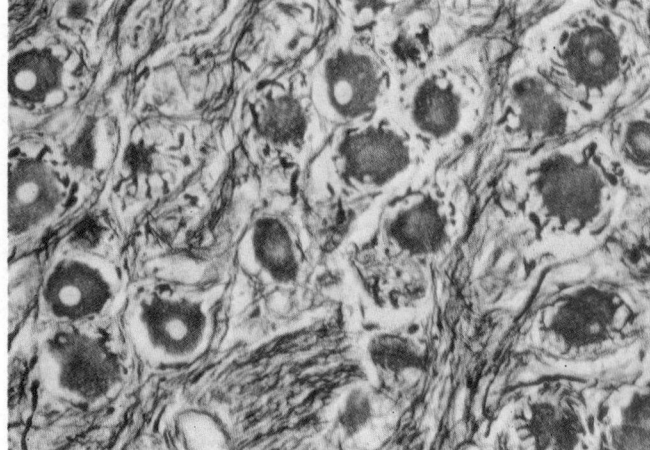

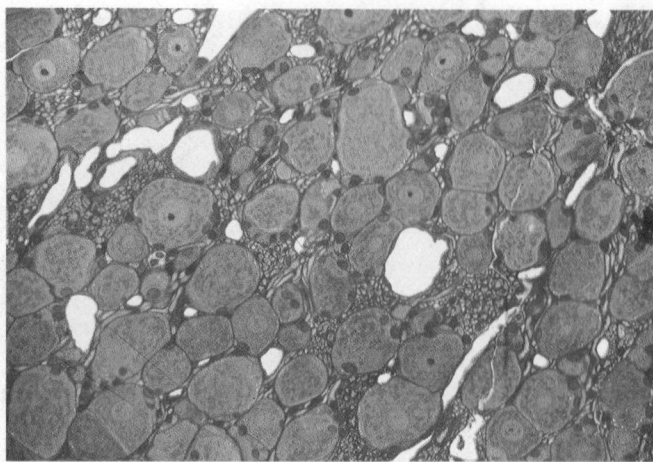

7.170A   Typical field in an autonomic ganglion (human ciliary ganglion), showing nerve cells evenly scattered among fascicles of nerve fibres. (Stained by Bielschowsky's silver and erythrosin technique in material kindly supplied by Dr. N. A. Locket, Institute of Ophthalmology, London.)

7.170B   Cells in superior cervical sympathetic ganglion of rabbit. Thin section (c. 2μm) of araldite-embedded material stained by toluidine blue. (Material kindly supplied by Dr. J. S. Dixon, Department of Anatomy, University of Manchester.)

terminate within the ganglion. Their structure, which is treated briefly here, is considered in greater detail on p. 838. In the *spinal ganglia* the cells are large, unipolar, and occur in groups round the periphery; in the autonomic ganglia they are multipolar and are scattered more or less uniformly (7.169, 170). Each spinal ganglionic nerve cell has a nucleated capsule (p. 838), which is continuous with the Schwann cells of the nerve fibres connected with the cell. The larger nerve cells in the ganglia of the spinal nerves (7.169) are irregularly spherical in shape, and each gives off a single neurite ('dendro-axonal process') which runs towards the centre of the ganglion, and divides in a T-shaped manner; one limb (axon) of the crossbar enters the spinal cord, the other (dendrite) passes outwards to the periphery (but *vide infra*). Near its origin the dendro-axonal process is coiled forming a *glomerulus* (p. 829). The smaller cells, which are more numerous, and give rise to fine non-myelinated fibres in peripheral nerves and their roots, possess delicate initial dendro-axonal processes which do not form complicated glomeruli, and pass straight to their T-bifurcation.

Structurally the peripheral division of the process of a unipolar ganglion cell resembles an axon in every respect, but it functions as a greatly elongated dendrite. (It should be noted that the majority of neuroanatomists term all the elongated neurites in a peripheral nerve 'axons' whether they are afferent or efferent.) The somata of afferent neurons in primitive invertebrates, such as *Hydra*, are situated in the ectoderm, with superficial hair-like dendritic processes projecting from them. Such somata are located much deeper in all vertebrates, being close to the central nervous system. Their peripheral processes are presumably the much elongated homologue of the short 'dendrites' of the primitive nerve cells. The primary olfactory neurons retain this primitive position (p. 1143). In the sensory ganglia of the cranial nerves, the nerve cells are also unipolar, though in the ganglia of the vestibulocochlear nerve they remain bipolar in type.

THE CRANIAL NERVES

The craniocaudal sequence of cranial nerves is as follows:

1 Olfactory	5 Trigeminal	9 Glossopharyngeal
2 Optic	6 Abducent	10 Vagal
3 Oculomotor	7 Facial	11 Accessory
4 Trochlear	8 Vestibulocochlear	12 Hypoglossal

These nerves are continuous with the brain and traverse foramina in the base of the cranium. The **motor**, or **efferent**, parts of the cranial nerves arise within the brain from groups of nerve cells which constitute their *nuclei of origin*. They are connected with the cerebral cortex by the corticonuclear fibres; these arise from the cells of the motor areas of the cortex, descend chiefly in the genicular part of the internal capsule to the brainstem, where many, but not all cross the median plane and end by arborizing round the cells of the nuclei of origin of the motor cranial nerves. The **sensory**, or **afferent**, cranial nerves arise from nerve cells outside the brain; these nerve cells may be grouped to form ganglia on the trunks of the nerves, or may be situated in peripheral sensory organs such as the nose, eye and ear. The centrally directed processes of the cells run into the brain and there end by arborizing around nerve cells which are grouped to form their *nuclei of termination*. Fibres arise from the cells of these nuclei and usually after crossing to the opposite side, run up to connect the nuclei indirectly with the cerebral cortex. (Readers of the relevant parts of the 'Central Nervous System' will be aware that the foregoing remarks are a gross over-simplification.)

The fibres of most of the cranial nerves begin to acquire their myelin sheaths about the fourteenth week of intra-uterine life. The process is delayed until the twenty-second week in the cases of the sensory part of the trigeminal nerve and the cochlear division of the vestibulocochlear. In the case of the optic nerve myelination does not commence until the later stages of gestation.

The Olfactory Nerves

The olfactory nerves, serving the sense of smell (7.171), commence in the mucous membrane of the olfactory region of the nasal cavity; this region comprises the superior nasal concha and the opposed part of the nasal septum. The nerve fibres originate as the central, or deep, processes of the olfactory cells (7.118) of the nasal mucous membrane, and are collected into bundles which cross one another in various directions, thus giving the appearance of a plexiform network in the mucous membrane. They are then collected into about twenty branches, which traverse the cribriform plate of the ethmoid bone in lateral and medial groups, and end in the glomeruli of the olfactory bulb (7.118). Each branch has a tubular sheath of dura mater and pia-arachnoid, the former being continued into the periosteum of the nose, the latter into the delicate trabecular connective tissue perineurium (perineural sheaths) of the nerve bundles. The tissue spaces in these sheaths are continuous with those of the nasal mucous membrane and with the subarachnoid space above.

The olfactory nerves are non-myelinated and consist of bundles of fine axons enfolded within Schwann cells.

The olfactory nerves are unique in that their cells of origin develop in the ectoderm and retain this position throughout life in all vertebrates.

Applied Anatomy. In severe head injuries involving the anterior cranial fossa the olfactory bulb may become separated from the olfactory nerves, or the nerves may be torn, thus producing *anosmia*—loss of smell sensibility. Fractures in this situation may also involve the meninges and cerebrospinal fluid may escape into the nose. Such injuries also open up avenues for infection of the meninges from the nasal cavity. The extensions of the subarachnoid space around the bundles of olfactory nerve fibres have been regarded as a potential lymphatic drainage, and on this basis also certain meningeal infections are considered to spread along these nerves into the cranial cavity. The evidence for this is still equivocal.

Closely associated with the olfactory nerves is a pair of small nerves named the **nervi terminales** (Pearson 1941). These nerves were first seen in lower vertebrates, but their presence has been demonstrated in the human embryo and adult. They consist mainly of non-myelinated nerve fibres, and associated with them are small groups of bipolar and multipolar nerve cells. Each nerve runs along the medial side of the corresponding olfactory tract, and its branches traverse the cribriform plate of the ethmoid bone, and are distributed to the nasal mucous membrane. Centrally the nerve is connected to the brain close to the anterior perforated substance and septal areas; in some animals its fibres have been traced to the lamina terminalis, in others to the hypothalamic region. Its function is unknown; some are inclined to view it as a forward extension of the cephalic part of the sympathetic nervous system, which is distributed to the blood vessels and glands of the nasal cavity. In relation to its possible autonomic nature it is of interest that ganglion cells have been associated with the nerve by Pearson (1941) and others. The presence of such ganglia has been observed in the mouse by Grüneberg (1973), who does not, however, equate this with the ganglion cells scattered along the human terminal nerve. The connexions of the '*ganglion terminale*' are at present unknown. Bojsen-Møller (1975) considered the nerve to be entirely sensory. The **vomeronasal nerve**, with which the nervus terminalis has been frequently confused, is probably absent in adult man. (See also 994, 1114.)

The detailed architecture and central connexions of the olfactory bulb are described on pp. 991–995, and for the detailed structure of the olfactory mucosa *see* p. 1143.

The Optic Nerve

The optic nerve, mediating vision, is distributed to the eyeball. Most of its fibres are afferent and originate in the nerve cells of the ganglionic layer of the retina (p. 1172), but some are efferent, their source of origin being uncertain. Developmentally, the optic nerves and the retinae are parts of the brain (p. 176), and their fibres are provided with glial and not Schwann cell sheaths.

The fibres of the optic nerve form the innermost layer (*stratum opticum*) of the retina and are the axons of the cells in its ganglionic layer; they converge on the optic disc, and there pierce the outer layers of the retina, the choroid coat, and the lamina cribrosa of the sclera at the posterior part of the eyeball, about 3 or 4 mm to the nasal side of its centre. As the fibres traverse the lamina cribrosa they receive their myelin sheaths, and run in bundles which are collected to form the optic nerve.

The optic nerve, about 4 cm long, is directed backwards and medially through the posterior part of the orbital cavity. It then runs through the optic canal into the cranial cavity and joins the optic chiasma.

The intraorbital part of the nerve is about 25 mm long and has a slightly sinuous course, the length of the nerve being about 6 mm more than the distance between the optic canal and the eyeball. Posteriorly it is closely surrounded by the four recti, but anteriorly is separated from them by a quantity of fat, in which run the ciliary vessels and nerves. The ciliary ganglion lies between the nerve and the lateral rectus. The inferomedial surface of the nerve is pierced, at a distance of about 12 mm behind the eyeball, by the central artery and vein of the retina, which are then directed forwards in the centre of the nerve to the optic disc. In the optic canal, which is about 5 mm long, the nerve lies above and medial to the ophthalmic artery, and is separated medially from the sphenoidal and posterior ethmoidal sinuses by a thin lamina of bone; in front of the canal the nasociliary nerve and the ophthalmic artery run forwards and medially, crossing above the optic nerve, whilst the branch to the medial rectus from the inferior division of the oculomotor nerve passes below it (7.183).

The intracranial part of the optic nerve, about 10 mm long, runs backwards and medially from the optic canal to the optic chiasma. The posterior parts of the olfactory tract and gyrus rectus, and, near the chiasma, the anterior cerebral artery lie above it. The internal carotid artery is on its lateral side.

The optic nerve is enclosed in three sheaths, which are continuous with the membranes of the brain (7.258), and are prolonged as far as the back of the eyeball. The *outer sheath*,

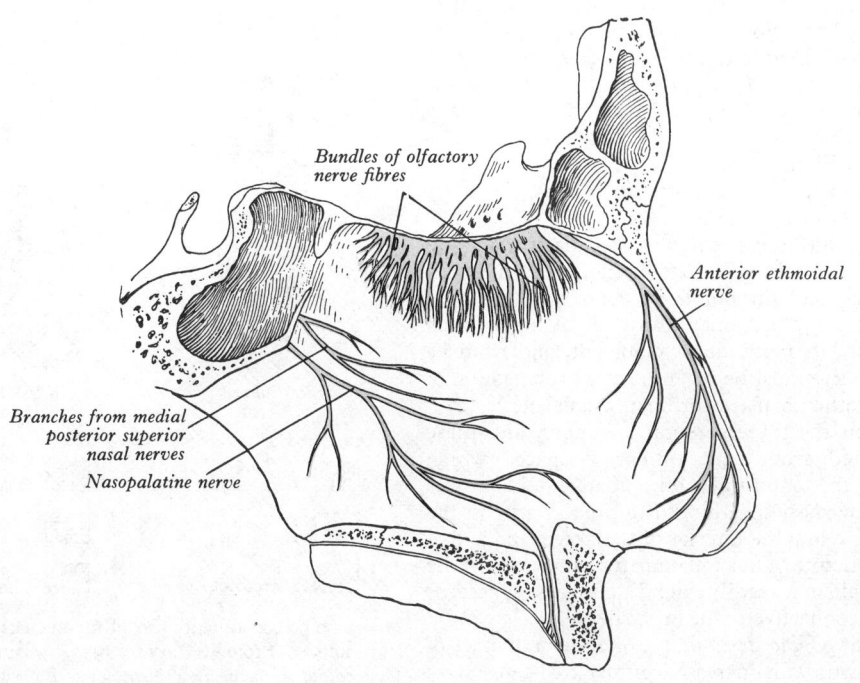

Bundles of olfactory nerve fibres

Anterior ethmoidal nerve

Branches from medial posterior superior nasal nerves

Nasopalatine nerve

7.171   The nerves of the septum of the nose. Right side.

derived from the dura mater, is thick and fibrous, and blends anteriorly with the sclera. The *intermediate sheath*, derived from the arachnoid mater, is thin and delicate. It is separated from the outer sheath by the subdural space, and from the inner sheath by the subarachnoid space. The *inner sheath*, derived from the pia mater, is vascular and closely invests the nerve. From its deep surface septa pass into the nerve and subdivide and reunite to enclose what appear, in transverse sections of the nerve, as polygonal areas, which are occupied by the bundles of nerve fibres. There are about 1,000 such fascicles. From the inner sheath also, an investment is carried on the central vessels of the retina as far as the optic disc.

The ultrastructure of the meninges of the optic nerve resembles that of the meninges elsewhere (Anderson and Hoyt 1969), but the amount of collagen fibres in the pia and arachnoid is greater than in the general intracranial leptomeninges. Here, as elsewhere, the subarachnoid space is lined completely by epithelial cells of the pia-arachnoid which resemble flattened fibroblasts, forming multilaminar surface membranes of 'meso-thelium' or 'meningothelium'.

Close to the eyeball the macular fibres (*papillo-macular bundle*) occupy the lateral part of the nerve, but, as they are traced backwards, they gradually come to lie medially; in front of the chiasma they are close to the medial margin. The fibres from the upper and lower portions of the retina lie above and below respectively; the fibres from the temporal quadrants lie laterally and those from the nasal quadrants medially.

The most recent counts of optic nerve fibres in man (Oppel 1963; Kupfer *et al.* 1967) give a figure of about 1,200,000 myelinated axons, about 92 per cent of which are small (about 1 μm in diameter), the rest varying from 2 to 10 μm. About 53 per cent cross in the chiasma. Most terminate in the lateral geniculate body, but fractions of uncertain size pass to the pretectal nucleus, superior colliculus, and certain hypothalamic nuclei (p. 973). A small proportion of the optic nerve fibres are efferent but of uncertain origin. Centrifugal nerve fibres passing to the retina have attracted much interest in birds, in which their occurrence is well established, being derived from the *isthmo-optic nucleus*, a group of nerve cells situated dorsolaterally at the junction of the mid- and hind-brains. (Consult Cowan and Clarke 1976 for details.)

The optic nerve is supplied by vessels from the plexus in the investing pia and by direct intraneural branches. The pial plexus receives branches from a superior hypophysial artery and the ophthalmic artery intracranially, from recurrent branches of the ophthalmic artery in the optic canal and from the posterior ciliary arteries and the extraneural part of the central retinal artery in the orbit. The intraneural branches to the nerve arise from the central artery, but their actual contribution to the blood supply of the nerve is probably small (Belmonte 1968; François and Neetens 1969). The rich blood supply of the optic papilla and the lamina cribrosa has recently been emphasized (Henkind and Levitsky 1969). The venous drainage is by the central vein of the retina (*see* p. 748, and Steele and Blunt 1956).

The optic chiasma (p. 974), and the **optic tract** (p. 975), have already been described.

Applied Anatomy. The optic nerve is peculiarly liable to neuritis or atrophy in certain affections of the central nervous system, and there are certain points in connexion with the anatomy of this nerve which may throw light upon such associations. (1) From its mode of development, and from its structure, the optic nerve must be regarded as a prolongation of the brain substance, rather than as an ordinary cranial nerve. (2) It receives sheaths from the three cerebral meninges, and these sheaths are separated from each other by spaces which communicate with the subdural and subarachnoid spaces respectively. The innermost sheath sends a process around the arteria centralis retinae into the interior of the nerve, and enters intimately into its structure. Thus inflammatory affections of the meninges or of the brain may readily extend along these spaces, or along the interstitial connective tissue in the nerve.

The *optic neuritis* or *papilloedema* that is often seen in cases of intracranial new growth with increased intracranial tension, is probably caused by increased pressure due to excess of fluid in the general subarachnoid space, since this is in direct communication with a prolongation of the space around the optic nerve as far as the lamina cribrosa (p. 1152).

The Oculomotor Nerve

The oculomotor nerve (**6**.123, 124; **7**.109, 120, 174, 184) supplies all the extraocular muscles, except the obliquus superior and rectus lateralis; it also supplies, through its connexion with the ciliary ganglion, the sphincter pupillae and the ciliaris, which are, of course, intraocular. It contains about 24,000 fibres.

The fibres of the oculomotor nerve arise from a *complex of nuclei* in the grey matter, ventral to the cranial part of the cerebral aqueduct, which extends cranially for a short distance into the floor of the third ventricle. From this the fibres pass forwards through the tegmentum, the red nucleus and the medial part of the substantia nigra, forming a series of curves with a lateral convexity, and emerge from the sulcus on the medial side of the cerebral peduncle (7.62, 64, 92). In its peripheral course the oculomotor nerve contains afferent nerve fibres whose cell bodies are in the trigeminal ganglion. (*See* Bortolami *et al.* 1977, and Manni *et al.* 1978, for recent observations and literature.) These afferents are considered to be proprioceptive. Small neurons, in variable numbers in different species, have been observed in the oculomotor nerve, and these are usually regarded as ectopic

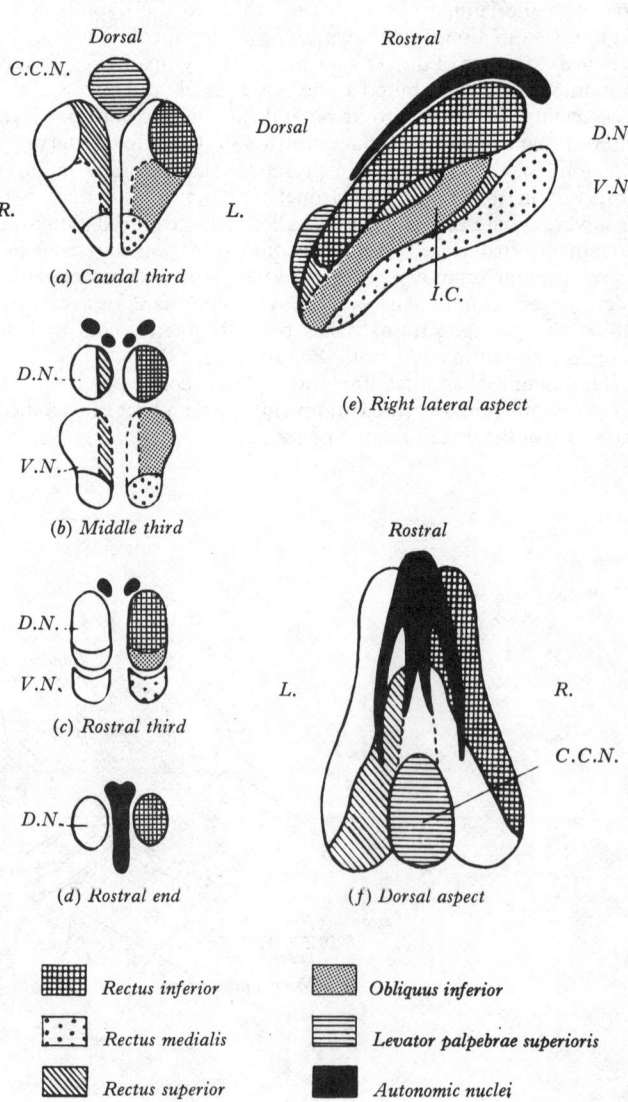

7.172 The constituent parts of the nucleus of the oculomotor nerve in the monkey. (From R. Warwick 1953, by courtesy of the publishers of the *Journal of Comparative Neurology*: for significance of abbreviations consult Warwick 1950 and see text.)

proprioceptive neurons, perhaps displaced from the trigeminal ganglion (Tozer 1912; Bortolami *et al.* 1977).

The nuclear complex from which the fibres of the oculomotor nerve are derived consists of a number of paired groups of large multipolar nerve cells, which can be identified in most mammals, together with paired masses of smaller multipolar nerve cells, which are not so easily identified, though well developed in primates and many other mammals, and which are the source of the parasympathetic outflow in the oculomotor nerves. On each side of the midline the large-celled oculomotor mass is topographically divided into **dorsolateral, intermediate** and **ventromedial nuclei** (7.172), which are extended in a columnar form approximately in the long axis of the midbrain (Crosby and Henderson 1948; Warwick 1950). These sub-nuclei become clearly identifiable in early fetal life in man (Pearson 1944). According to experimental evidence in monkeys they are the motor pools of the *rectus inferior, obliquus inferior* and *rectus medialis*, in that dorsoventral order (7.172). Dorsal to the right and left large-celled masses, and at a level corresponding to their caudal extremities, is a median nucleus, composed of similar large multipolar nerve cells, the **caudal central nucleus**, which in the same experimental study was observed to be the conjoined motor pool of the two *levators* of the upper eyelids (Warwick 1950, 1953). Dorsal to the right and left main oculomotor nuclei are the two **accessory** or **autonomic nuclei** (of Edinger and Westphal), composed of somewhat smaller multipolar nerve cells, whose axons travel out in the oculomotor to relay in the ciliary ganglion (Warwick 1954). These accessory nuclei are best developed at cranial oculomotor levels, where they fuse together and arch ventrally over the cranial end of the main oculomotor nuclei (7.172); in their more caudal, paired parts these autonomic columns frequently show a tendency to further splitting in man and other primates. Recent studies by Sugimoto *et al.* (1978), using the retrograde horseradish peroxidase technique suggest that many of the neurons of the oculomotor parasympathetic outflow may be sited near to but outside these autonomic nuclei.

Although a median group of larger motor nerve cells has been a standard feature of oculomotor topography for many decades, this 'central nucleus of Perlia' is a most variable entity in mammals, and indeed in the same species. The function of convergence was loosely ascribed to it at an early date on inadequate and fallacious evidence. Its degree of development in primates is not commensurate with its supposed function, nor is it possible to equate its size with binocular vision (Le Gros Clark 1926). Topographic observers have described it as frequently absent from the human midbrain (Tsuchida 1906; Crosby and Woodburne 1943). There are always a few scattered large motor cells in the midline raphe between the right and left oculomotor masses, but these never constitute a clear nuclear mass, like the caudal central nucleus; in any case, such nerve cells appear to innervate the superior rectus, and not the medial rectus (Warwick 1955).

Other views of the topographical arrangement of the motor pools of the extraocular muscles have been advanced (Brouwer 1918; Szentágothai 1942; Bender and Weinstein 1943), all of which suggest a craniocaudal pattern rather than the dorsoventral scheme described here; but in both forms of organization, which are very different in other ways, there is general agreement that the issuing oculomotor fibres, somatic and autonomic, are almost completely ipsilateral in their midbrain course. At the most some axons from the median raphe and the caudal central nucleus may cross into the opposite oculomotor nerve to innervate the rectus superior and the levator palpebrae superioris; but the extensive crossing of oculomotor axons which is a usual tenet of textbook accounts is not supported by any of the experimental studies quoted here, almost all of which were carried out in primates. The most recent study of oculomotor organization, carried out in the cat (Tarlov 1972, 1975), describes an array of longitudinal motor pools for individual muscles, but their arrangement differs from that observed in the monkey (Warwick 1955). Tarlov has also defined a projection (*see* p. 1074) to the oculomotor nucleus from the medial and superior vestibular nuclei, the terminals from both of which were found to provide a partially bilateral and overlapping distribution in certain oculomotor muscle 'pools'.

This suggests excitatory and inhibitory roles, and certain physiological data tend to support this concept (High *et al.* 1971).

For a discussion of the claims of other nuclei in the vicinity of the oculomotor complex to be included with it, consult Warwick (1953).

Connexions of the oculomotor nucleus include fibres from: (1) the corticonuclear tracts of *both* sides; some fibres leave the tracts at the level of the oculomotor nucleus but some (aberrant pyramidal fibres) leave at a higher level and thereafter descend in the medial lemniscus; they end either directly on the cells of the nucleus or are linked to them via interneurons; (2) the medial longitudinal fasciculus by which it is connected to the nuclei of fourth, sixth and eighth cranial nerves (p. 939); (3) the tectobulbar tract, by which it is connected to the visual cortex through the medium of the superior colliculus; and (4) the pretectal nucleus of both sides for the light reflex. Horseradish peroxidase studies have indicated a projection from the oculomotor nucleus to the abducent (Graybiel and Hartwig 1974) and in the reverse direction (Maciewicz *et al.* 1975). Such internuclear reciprocal connexions are to be expected from the results of experimental stimulation of or damage to the medial longitudinal fasciculus, and from the clinicopathological data of internuclear ophthalmoplegia. These connexions are part of the 'mechanism' for the control of horizontal eye movements. (For a discussion consult Highstein 1977.) For much information on earlier work in this field consult Bender (1964), and for further comments *see* p. 1085.

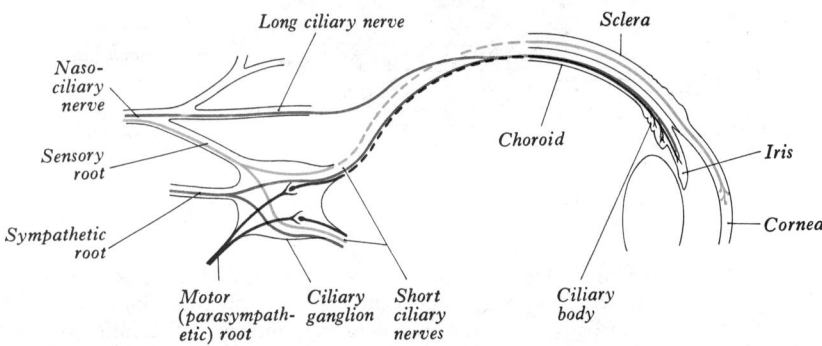

7.173 A diagram showing the ciliary ganglion, with its roots and branches of distribution, *Magenta:* sympathetic fibres. *Heavy black:* parasympathetic fibres. *Blue:* sensory (cerebrospinal) fibres. Alternative pathways are given for the sympathetic fibres to the dilatator pupillae. A schematic sagittal section is shown of the upper lateral quadrant of the eyeball, but the retina has not been included.

Similar experiments (using horseradish peroxidase technique) suggest that many of the neurons in the autonomic nuclei of the oculomotor nerve (*vide supra*) project to the cerebellum and spinal cord (Sugimoto *et al.* 1978).

Course. On emerging from the brain, the nerve, invested by pia mater, lies in the subarachnoid space. It passes between the superior cerebellar and posterior cerebral arteries (**8.196**), and runs forward in the interpeduncular cistern on the lateral side of the posterior communicating artery. It then perforates the arachnoid and lies in the triangular interval between the free and attached borders of the tentorium cerebelli. Piercing the inner layer of the dura mater on the lateral side of the posterior clinoid process the nerve traverses the roof, and further forwards descends into the lateral wall of the cavernous sinus, where it lies above the trochlear nerve (**6.127**). In this situation it receives one or two filaments from the internal carotid plexus of the sympathetic, and communicates with the ophthalmic division of the trigeminal. It then divides into a superior and an inferior ramus, which enter the orbit through the superior orbital fissure, within the anulus tendineus communis to which are attached the four recti; here the nasociliary nerve is placed between the two rami (**7.176, 269**).

The *superior ramus*, the smaller, ascends on the lateral side of the optic nerve, and supplies the rectus superior and levator palpebrae superioris. The *inferior ramus* divides into three

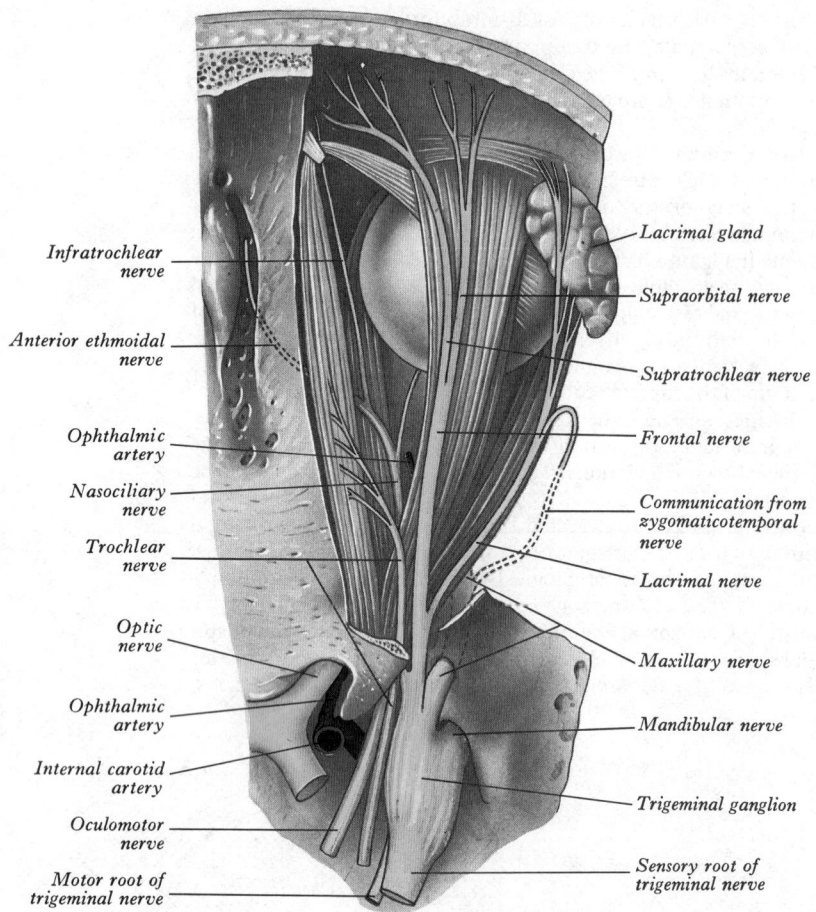

Infratrochlear nerve

Anterior ethmoidal nerve

Ophthalmic artery

Nasociliary nerve

Trochlear nerve

Optic nerve

Ophthalmic artery

Internal carotid artery

Oculomotor nerve

Motor root of trigeminal nerve

Lacrimal gland

Supraorbital nerve

Supratrochlear nerve

Frontal nerve

Communication from zygomaticotemporal nerve

Lacrimal nerve

Maxillary nerve

Mandibular nerve

Trigeminal ganglion

Sensory root of trigeminal nerve

7.174   The nerves of the right orbit, superior aspect.

branches. One passes below the optic nerve to the rectus medialis; another goes to the rectus inferior; the third and longest runs forwards between the rectus inferior and rectus lateralis, to the obliquus inferior. The branches enter the muscles on their ocular surfaces, with the exception of that to the obliquus inferior, which enters the posterior border of the muscle. From the nerve to the inferior oblique a short thick branch (sometimes represented by two or three separate branches) passes to the lower part of the ciliary ganglion and forms its *motor* or *parasympathetic root*. It consists of finely myelinated fibres, derived from the accessory oculomotor nucleus, which synapse about cells in the ganglion, whence postganglionic fibres pass in the short ciliary nerves to supply the sphincter pupillae and ciliaris (pp. 1159, 1161).

The ciliary ganglion (7.173, 176) is a small, flattened ganglion of a reddish-grey colour and the size of a large pin's head; it is situated near the apex of the orbit in some loose fat between the optic nerve and the rectus lateralis, lying usually on the lateral side of the ophthalmic artery. It is a peripheral ganglion of the parasympathetic system and its constituent cells are multipolar. These cells are not typical of autonomic ganglia in general, being much larger in dimensions; a very small number of nerve cells of typical autonomic type are also present (Warwick 1954).

Its *connexions* or *roots* (7.173) enter or leave it posteriorly, but only its parasympathetic function is basic; the sympathetic and sensory fibres merely pass through it, and are absent in some mammals. The *motor* or *parasympathetic root* is derived from the nerve to the inferior oblique and consists of preganglionic fibres which arise from the cells of the accessory (Edinger-Westphal) nucleus (7.109). These fibres are relayed in the ganglion and the postganglionic fibres travel in the short ciliary nerves to supply the sphincter pupillae and ciliaris. More than 95 per cent of these fibres supply the ciliaris, which is much the larger muscle in volume (Warwick 1954); hence this motor pathway is much more concerned with focusing of the eye than with the light reflex. The *sympathetic root* is a branch from the internal carotid plexus. It may pass direct to the ganglion or it may join the sensory root and

reach the ganglion indirectly. It consists of postganglionic fibres from the superior cervical ganglion which traverse the ciliary ganglion without being interrupted, to emerge in the short ciliary nerves. They are distributed to the blood vessels of the eyeball and they may include the fibres which supply the dilatator pupillae when these fibres do not follow their usual course in the ophthalmic, nasociliary and long ciliary nerves. The *sensory root* is formed by a *ramus communicans* to the nasociliary nerve. It contains sensory fibres from the eyeball, which reach the ganglion in the short ciliary nerves and pass through it without being interrupted. It leaves the ganglion posteriorly and runs backwards to join the nasociliary nerve near the point where that nerve enters the orbit.

The *branches* of the ganglion are delicate filaments, eight to ten in number, which emerge from the front of the ganglion in two bundles, of which the lower is the larger. They are termed the *short ciliary nerves*. The postganglionic parasympathetic fibres in them are myelinated. In company with the ciliary arteries they run sinuously forwards, one set above the optic nerve, the other below. They subdivide into about fifteen to twenty branches, which pierce the sclera around the entrance of the optic nerve and pass forwards in delicate grooves on the inner surface of the sclera. They contain both motor and sensory fibres, the former are distributed to the sphincter pupillae and ciliaris, and to the choroidal and iridial blood vessels of the eyeball which provide a nutritive supply to many ocular tissues. The existence of proprioceptive fibres in the oculomotor nerve can no longer be doubted, in view of the occurrence of stretch endings in the extraocular muscles (p. 1085). How far such fibres travel in the oculomotor nerve, and where they terminate centrally are still unsettled problems (but *see* p. 1060). The blood supply of the ciliary ganglion has been investigated in man by Eliscova (1973); it is derived from small rami of the muscular, posterior ciliary and central retinal branches of the ophthalmic artery.

Applied Anatomy. Division of the oculomotor nerve leads, when complete, to (1) ptosis, or drooping of the upper eyelid, consequent on paralysis of the levator palpebrae superioris; (2) lateral strabismus (or squint), stemming from the unopposed action of the rectus lateralis and obliquus superior, which are not supplied by the oculomotor nerve and are therefore not paralysed; (3) dilatation of the pupil, because the sphincter pupillae is paralysed; (4) loss of accommodation, and of pupillary constriction on exposure to light, as the sphincter pupillae and the ciliaris are paralysed; (5) slight prominence of the eyeball, owing to most of its muscles being relaxed; and (6) diplopia, or double vision, the false image being higher than the true. Occasionally paralysis may affect only a part of the nerve—for example, there may be a dilated and fixed pupil, with ptosis, but no other signs. Irritation of the nerve causes spasm of one or other of the muscles supplied by it; thus, there may be medial strabismus from spasm of the medial rectus, accommodation for near objects only, from spasm of the ciliaris, or a contracted pupil owing to irritation of the sphincter pupillae.

The Trochlear Nerve

The trochlear nerve (7.174, 176), the most slender of the cranial nerves, supplies the superior oblique muscle of the eyeball. It contains about 3,400 fibres (Björkman and Wohlfart 1936), but in fetal life the number of fibres is greater, as in some other cranial nerves. According to Mustafa and Gamble (1979), however, the adult human nerve contains approximately 2,400 fibres, whereas fetal counts were 4,000 (at C.R. length 9·2 cm), 6,000 (at C.R. length 10 cm), and 3,200 (at C.R. length 24 cm). Many fibres therefore degenerate during fetal development, after an initial increase in numbers. As mentioned below, the orbital part of the trochlear nerve contains more fibres than its stem. This disparity in numbers is also observed in fetal trochlear nerves.

The trochlear nucleus is situated in the floor of the cerebral aqueduct, opposite the upper part of the inferior colliculus (7.62, 64, 91). This nucleus lies in line with the ventromedial part of the oculomotor nucleus, and occupies the position of the somatic efferent column. The medial longitudinal fasciculus is ventral to the trochlear nucleus. The oculomotor and trochlear nuclei often overlap a little and can only be distinguished by the slightly smaller size of the trochlear nerve cells.

Connexions. The nucleus receives fibres from: (1) the corticonuclear tracts of *both* sides, probably in a manner similar to those of the oculomotor nerve nucleus (p. 1057); (2) the medial longitudinal fasciculus, by which it is connected with the nuclei of the third, sixth and eighth cranial nerves (p. 940); (3) from the tectobulbar tract, through which it receives impulses from the visual cortex through the medium of the superior colliculus (p. 941). (See also p. 1085.) Tarlov and Tarlov (1975) have studied the transmitter substances of the vestibulotrochlear projection in the cat, in which the tract is derived chiefly from the superior vestibular nucleus; they conclude that the transmitter is GABA (gamma-amino butyric acid) at the trochlear terminals of the projection, and they regard them as inhibitory.

Course. After leaving the nucleus the fibres of the trochlear nerve pursue a very unusual course (p. 937). They first run downwards and laterally through the tegmentum and then turn backwards round the central grey matter into the upper part of the anterior medullary velum. Here they decussate with the contralateral fibres, and, having crossed the median plane, emerge from the surface of the velum at the side of the frenulum veli, immediately below the inferior colliculus (7.90) It is the only cranial nerve that emerges from the brainstem on its dorsal aspect.

The nerve is directed laterally across the superior cerebellar peduncle, and then winds forwards round the cerebral peduncle immediately above the pons, and between the posterior cerebral and superior cerebellar arteries. It appears between the upper border of the pons and the temporal lobe, and pierces the inner stratum of the dura mater immediately below the free border of the tentorium cerebelli, a little behind the posterior clinoid process. It then passes forwards in the lateral wall of the cavernous sinus, below the oculomotor nerve and above the ophthalmic division of the trigeminal nerve (6.127). In this part of its course it is closely adherent to the tentorial branch of the ophthalmic nerve, which lies below it. Near the anterior end of the sinus it crosses the oculomotor nerve, and enters the orbit by the superior orbital fissure, above the orbital muscles, and medial to the frontal nerve (7.174). In the orbit it passes medially, above the origin of the levator palpebrae superioris, and finally enters the oribital surface of the obliquus superior (7.176).

In the lateral wall of the cavernous sinus the trochlear nerve communicates with the ophthalmic division of the trigeminal nerve, and with the internal carotid plexus of the sympathetic. In the superior orbital fissure it occasionally gives off a branch to the lacrimal nerve. Although an exchange of fibres through communications in the cavernous sinus has been denied (Sunderland and Hughes 1946), subsequent evidence based on nerve fibre analysis in human material (Zaki 1960) suggests that a substantial component of fairly large nerve fibres, possibly proprioceptive in nature, are present in the trochlear nerve distal to the cavernous sinus. This part of the nerve contained 3,500 fibres, whereas proximal to the sinus the count was only about 2,400. Although it is sometimes assumed, as in the case of the oculomotor nerve, that proprioceptive fibres from the superior oblique muscle enter the brainstem via the trochlear nerve, the most recent study (Manni *et al.* 1970) suggests that these fibres leave the trochlear nerve peripherally and join the ophthalmic division of the trigeminal nerve.

Applied Anatomy. When the trochlear nerve is interrupted the superior oblique is paralysed so that the patient is unable to turn his eye downwards and laterally. Should the patient attempt to do this, the eye of the affected side is rotated medially, producing *diplopia* or double vision. Single vision exists in the whole of the field so long as the eyes look above the horizontal plane; diplopia occurs on looking downwards. To counteract this the patient holds his head forwards, and also inclines it to the sound side.

The Trigeminal Nerve

The trigeminal is the largest cranial nerve. It is the sensory nerve of the face, the greater part of the scalp, the teeth, the oral and the nasal cavities, and the motor nerve of the muscles of mastication and some others. It also contains proprioceptive nerve fibres from the masticatory and probably from the extraocular muscles. It divides into three branches, ophthalmic, maxillary and mandibular. Baumel (1974) has reviewed the functional significance of the widespread peripheral communications between the trigeminal and facial nerves. Despite the voluminous literature on the subject, uncertainties exist; but the preponderant view is that the trigeminal fibres which join from the facial are sensory and largely derived from the facial musculature.

The trigeminal nerve is continuous with the ventral surface of the pons, near its upper border, by a large sensory and a small motor root—the latter being placed medial and anterior to the former.

The fibres of the *sensory root* arise from the cells of the **trigeminal** (*semilunar*) **ganglion**. This ganglion (7.174, 177) occupies a recess (*trigeminal cave*) in the dura mater covering the trigeminal impression near the apex of the petrous part of the temporal bone (p. 311). It lies at a depth of 4·5–5 cm from the lateral aspect of the head, at the posterior extremity of the zygomatic arch. The ganglion is crescentic in shape, with its convexity directed anterolaterally; on its surface are visible a number of interlacing nerve fascicles. Medially it is in relation with the internal carotid artery and the posterior part of the cavernous sinus, inferiorly, with the motor root of the nerve, the greater (superficial) petrosal nerve, the apex of the petrous part of the temporal bone, and the foramen lacerum. It receives filaments from the internal carotid plexus of the sympathetic, and gives twigs to the tentorium cerebelli.

The branches of the unipolar cells of the trigeminal ganglion divide into peripheral and central branches. The former are grouped to form the *ophthalmic* and *maxillary* nerves, and the sensory part of the *mandibular* nerve. The central branches constitute the fibres of the sensory root of the nerve, which leaves the concave margin of the ganglion, runs backwards and medially below the superior petrosal sinus and the tentorium cerebelli, and enters the pons. Some fibres from proprioceptor endings in the masticatory and extraocular muscles may pass through the semilunar ganglion without connexion with nerve cell somata, but electrophysiological observations (Manni *et al.* 1970) strongly support the view that most, if not all, of the neurons innervating proprioceptors in the extraocular muscles have somata situated in the trigeminal ganglion (see also p. 1060).

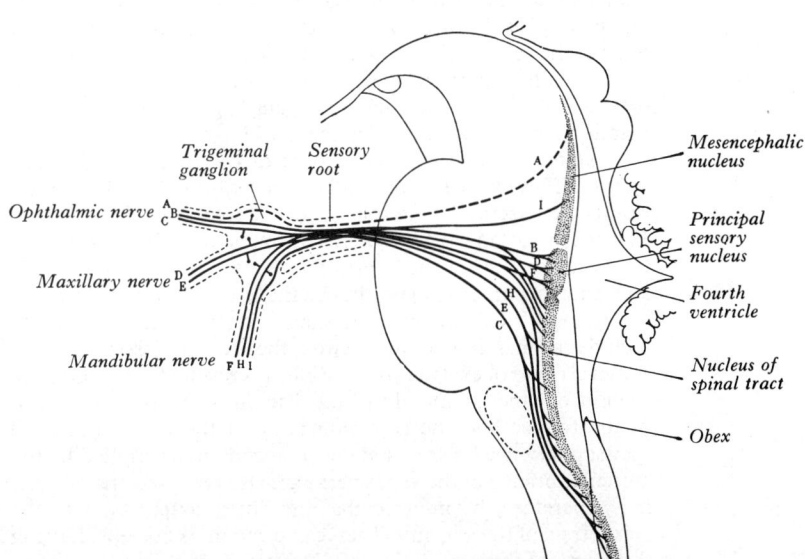

7.175 The nuclei receiving the primary afferent fibres of the trigeminal nerve. A: proprioceptive fibres from ocular muscles. B: tactile and pressure fibres from ophthalmic areas. C: pain and temperature fibres from ophthalmic area. D: tactile and pressure fibres from maxillary area. E: pain and temperature fibres from maxillary area. F: tactile and pressure fibres from mandibular area. H: pain and temperature fibres from mandibular area. I: proprioceptive fibres from muscles of mastication. Proprioceptive fibres are believed to occur in all three divisions of the trigeminal nerve.

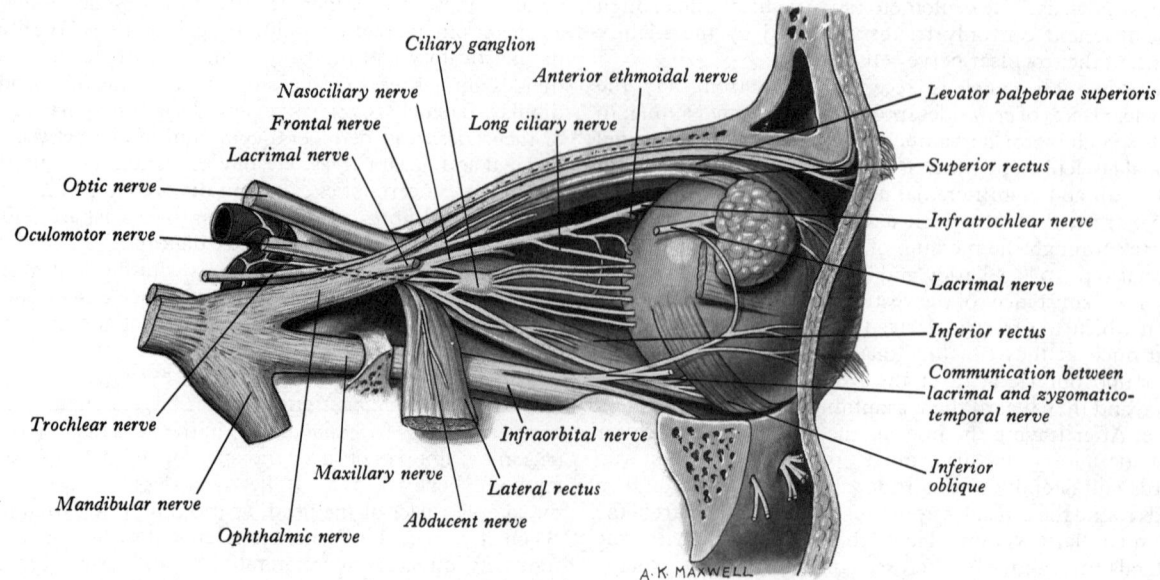

7.176 The nerves of the right orbit, and the ciliary ganglion. Lateral aspect.

On entering the pons the fibres of the sensory root course dorsomedially towards the **principal sensory nucleus** situated at that level (7.70). On approaching the nucleus about 50 per cent of the fibres divide into ascending and descending branches; the others ascend or descend without division. The descending fibres, which are predominantly finely myelinated or non-myelinated, form the **spinal tract of the trigeminal nerve** which descends into the upper cervical part of the spinal cord. As the tract descends terminals and collaterals are given off to synapse with the cells of the **spinal nucleus of the trigeminal nerve** (7.60–63, 66, 68, 175). The nucleus consists of small- and medium-sized cells and is continuous below with the substantia gelatinosa. The tract embraces the nucleus on its dorsolateral aspect. In the lower part of the medulla oblongata it is superficial and lies under the tuberculum cinereum. The fibres which synapse in the nucleus are concerned predominantly (but not exclusively) with painful and thermal sensibility. It has been demonstrated experimentally that (1) the fibres of the ophthalmic nerve are placed ventrally in the tract and descend to the lower limit of the first cervical segment of the spinal cord; (2) the fibres of the maxillary nerve lie in the central part of the tract and do not extend below the medulla oblongata; (3) the fibres of the mandibular nerve are placed in the dorsal part of the tract and do not extend much below the middle of the medulla oblongata (7.175). This experimental evidence is confirmed by the clinical results after section of the spinal tract (Smyth 1939; Brodal 1947; Falconer 1949) in cases of severe trigeminal neuralgia. Section of the tract 4 mm below the level of the obex renders the ophthalmic and maxillary areas analgesic but tactile sensibility, apart from the abolition of 'tickle', is much less affected. When it is desired to include the mandibular area as well, the section must be made at the level of the obex itself. In addition, as a result of this operation, the mucous membrane of the tonsillar sinus, the posterior third of the tongue and the adjoining parts of the pharyngeal wall (glossopharyngeal nerve) and the cutaneous area supplied by the auricular branch of the vagus nerve are also rendered analgesic; it may, therefore, be inferred that the fibres concerned join the spinal tract of the trigeminal nerve and end in its nucleus. During the course of these operations the laminated character of the tract has also been confirmed. A small group of sensory cells, associated with the glossopharyngeal nerve, but related to the spinal trigeminal tract, has been described (Bossy 1968). As well as the trigeminal, vagal and glossopharyngeal afferents to the spinal nucleus, others reach it from the sensory root of the facial nerve, the dorsal roots of the upper cervical nerves, and descending fibres from the sensorimotor regions of the cerebral cortex (see also p. 1086).

It is appropriate to divide the spinal trigeminal nucleus into three levels—*nucleus oralis* (the most rostral part, adjoining the principal sensory nucleus), *nucleus interpolaris* and *nucleus caudalis* (the most caudal part, continuous with the dorsal grey column)—because they differ in structure and to some extent in connexions. The nucleus caudalis closely resembles the neuronal pattern of the dorsal grey column, while the nuclei interpolaris and oralis are devoid of neurons 'equivalent' to the substantia gelatinosa. All levels receive afferents from the whole of the trigeminal sensory field; but all C fibres are believed to end in the caudal nucleus, and its neurons alone respond to noxious stimuli; moreover, its projection is to the contralateral ventroposterior nucleus of the thalamus, as is that of the nucleus oralis. On the other hand, the nucleus interpolaris projects to the anterior lobe of the cerebellum.

Some of the ascending trigeminal fibres, many of which are heavily myelinated, synapse around the small cells of the principal sensory nucleus which lies lateral to the motor nucleus and intervenes between the latter and the middle cerebellar peduncle: it is continuous below with the nucleus of the spinal tract (7.62, 175). It is considered to be concerned mainly with tactile sensibility.

Other ascending fibres enter the **mesencephalic nucleus**. This nucleus is composed of a column of *unipolar* cells. It is believed that their peripheral branches convey proprioceptive impulses from the muscles of mastication and it is also alleged (Corbin and Harrison 1940; Pearson 1949) that proprioceptive impulses travel to the mesencephalic nucleus from the teeth and from the facial and ocular muscles (7.62, 91, 155). The neurons of the mesencephalic nucleus are unique in being the only primary sensory neurons whose cell bodies are within the central nervous system (Johnston 1909). If, however, the primary proprioceptive neurons of the extraocular muscles are in fact situated in their motor nerves or the trigeminal ganglion (*see* pp. 1058, 1085), some at least of the mesencephalic trigeminal neurons may be 'secondary' in status. Small multipolar cells, possibly inter-neurons, occur near the unipolar neurons.

Connexions. The majority of the fibres which arise in the sensory nuclei of the trigeminal nerve cross the median plane and ascend in the trigeminal lemniscus (p. 938) to the nucleus ventralis posterior medialis of the thalamus (p. 958), from which fibres are relayed to the cortex of the postcentral gyrus (areas 3, 1 and 2, p. 1016). Some, however, ascend to the thalamus on the ipsilateral side. It is probable that collateral branches of both primary and secondary afferent trigeminal neurons reach many other central regions such as the cranial nerve nuclei, the reticular formation, cerebellum, tectum, subthalamus, hypothalamus, etc., but such details have not been established in the human brain. (Consult Humphrey 1969; Webster 1977; see also p. 1085.)

The nerve fibres which ascend to the nerve cells of the mesencephalic nucleus may afford collaterals to the motor trigeminal nucleus and the cerebellum. They are, of course, morphologically dendrites; the true axons of the mesencephalic neurons possibly descend in part to the principal trigeminal sensory nucleus, but little is known of them. Physiological evidence suggests that the mesencephalic nucleus is modulated during masticatory reflex activity by connexions with the vagus nerve, but anatomical confirmation is lacking (Manni *et al.* 1977).

The motor nucleus of the trigeminal nerve gives rise to the fibres of the motor root. It is ovoid in shape with typical large multipolar cells interspersed with smaller multipolar cells. It lies in the upper part of the pons on the medial side of the principal sensory nucleus, which is separated from it by the fibres of the trigeminal nerve. It occupies the position of the branchial (special visceral) efferent column (7.62, 64, 70). Detailed analysis has shown that the motor nucleus of the trigeminal consists of a number of relative discrete sub-nuclei, the axons of which innervate individual muscles (Szentágothai 1949).

Connexions. The motor nucleus receives fibres from the corticonuclear tracts of *both* sides; these fibres leave the tracts at the level of the nucleus or at a higher level in the pons (aberrant pyramidal fibres) and descend in the medial lemniscus. They may end on the cells of the nucleus or are linked to them by interneurons. The motor nucleus receives fibres from the sensory nuclei of the trigeminal nerve, and, as stated above, possibly some from the mesencephalic nucleus, forming monosynaptic reflex arcs for the proprioceptive control of the muscles of mastication. It also receives fibres from the reticular formation, red nucleus and the tectum, from the medial longitudinal fasciculus and possibly fibres from the locus ceruleus, by which salivary secretion and mastication are correlated.

THE OPHTHALMIC NERVE

The ophthalmic nerve (7.174, 179, 182), the superior division of the trigeminal nerve, is wholly sensory. It supplies branches to the eyeball, the lacrimal gland and the conjunctiva, to a part of the mucous membrane of the nasal cavity, and to the skin of the nose, eyelids, forehead and scalp. It is the smallest division of the trigeminal nerve, and arises from the anteromedial part of the trigeminal ganglion as a flattened band, about 2·5 cm long, which passes forwards in the cavernous sinus close to its lateral wall and below the oculomotor and trochlear nerves (6.127); just before entering the orbit through the superior orbital fissure, it divides into three branches, viz. *lacrimal, frontal* and *nasociliary*.

The ophthalmic nerve is joined by filaments from the internal carotid plexus of the sympathetic, and communicates with the oculomotor, trochlear and abducent nerves, a route by which the proprioceptive fibres of these nerves may pass into the trigeminal. It supplies a recurrent meningeal branch (*tentorial nerve*), which crosses below and adheres to the trochlear nerve, and is distributed to the tentorium cerebelli (p. 1045).

The lacrimal nerve (7.174) is the smallest of the main branches of the ophthalmic nerve. It sometimes receives a filament from the trochlear nerve, but possibly this filament consists of fibres which have previously passed from the ophthalmic to the trochlear nerve. The lacrimal nerve enters the orbit through the lateral part of the superior orbital fissure (7.177), runs along the upper border of the rectus lateralis with the lacrimal artery, and receives a twig from the zygomatico-temporal branch of the maxillary nerve, often said to contain secretomotor fibres for the lacrimal gland (*see*, however, p. 1123). It enters the lacrimal gland and gives off several filaments to the gland and the conjunctiva. Finally it pierces the orbital septum, and ends in the skin of the upper eyelid, joining with filaments of the facial nerve.

The lacrimal nerve is occasionally absent, and its place is then taken by the zygomaticotemporal branch of the maxillary nerve. Sometimes the latter branch is absent and is replaced by a branch of the lacrimal nerve.

The frontal nerve (7.174, 176) is the largest branch of the ophthalmic division. It enters the orbit through the superior orbital fissure (7.174) above the muscles, and runs forwards between the levator palpebrae superioris and the periosteum. About midway between the apex and base of the orbit it divides into a small supratrochlear and a large supraorbital branch.

The *supratrochlear nerve* runs medially and forwards, passes above the trochlea of the obliquus superior, and gives off a descending filament to join the infratrochlear branch of the nasociliary nerve. The nerve then emerges from the orbit between the trochlea and the supraorbital foramen, curves upwards on the forehead close to the bone in company with the supratrochlear branch of the ophthalmic artery, and sends filaments to the conjunctiva and skin of the upper eyelid; it then ascends under cover of the corrugator and the frontal belly of the occipito-frontalis and divides into branches which pierce these muscles and supply the skin of the lower part of the forehead close to the median plane.

The *supraorbital nerve* runs forwards between the levator palpebrae superioris and the roof of the orbit, passes through the supraorbital notch or foramen, and gives off palpebral filaments to the upper eyelid and conjunctiva. It then ascends upon the forehead with the supraorbital artery, and divides into a smaller medial and a larger lateral branch, which supply the skin of the scalp, reaching nearly as far back as the lambdoid suture. These two branches are at first situated deep to the frontal belly of the occipitofrontalis; the medial branch perforates this muscle, the lateral branch pierces the epicranial aponeurosis. The undivided nerve and both branches supply small twigs to the mucous membrane of the frontal sinus and to the pericranium; some reach the sinus through foramina in the floor of the supraorbital notch.

The nasociliary nerve (7.174, 176) is intermediate in size between the frontal and lacrimal nerves, and is more deeply placed. It enters the orbit through the medial part of the superior orbital fissure within the common tendinous ring, which gives origin to the recti muscles of the eyeball, and here it is situated between the two rami of the oculomotor nerve (7.176). It crosses the optic nerve with the ophthalmic artery, and runs obliquely below the rectus superior and obliquus superior, to the medial wall of the orbital cavity. Here, under the name of the *anterior ethmoidal nerve*, it passes through the anterior ethmoidal foramen and canal and, entering the cavity of the cranium, runs forwards in a shallow groove on the upper surface of the cribriform plate of the ethmoid bone, beneath the dura mater; it then descends through a slit at the side of the crista galli into the nasal cavity, and lies in a groove on the inner surface of the nasal bone. It supplies two *internal nasal branches*—a medial to the mucous membrane of the front part of the nasal septum, and a lateral to the anterior part of the lateral wall of the nasal cavity. Finally it emerges, as the *external nasal branch*, at the lower border of the nasal bone, and, passing down under cover of the transverse part of the nasalis, supplies the skin of the ala, apex and vestibule of the nose.

The nasociliary nerve communicates with the ciliary ganglion, and gives off the long ciliary, the infratrochlear and the posterior ethmoidal nerves.

The *ramus communicans* to the ciliary ganglion (p. 1058, 7.173) usually joins the nasociliary nerve as the latter enters the orbital cavity. It lies on the lateral side of the optic nerve, and emerges from the posterosuperior angle of the ciliary ganglion (7.173); it is sometimes joined by a filament from the internal carotid plexus of the sympathetic, or from the superior ramus of the oculomotor nerve.

The *long ciliary nerves*, two or three in number, are given off from the nasociliary nerve, as it crosses the optic nerve. They accompany the short ciliary nerves from the ciliary ganglion, pierce the sclera near the attachment of the optic nerve, and, running forwards between the sclera and the choroid, are distributed to the ciliary body, iris and cornea. They usually contain the sympathetic fibres for the dilatator pupillae (p. 1161); these are postganglionic fibres and have their cells of origin in the superior cervical ganglion. In view of the susceptibility of the exposed epithelium of the cornea to damage, the corneal distribution of the long ciliary nerves is of pre-eminent importance.

The *infratrochlear nerve* is given off from the nasociliary nerve near the anterior ethmoidal foramen. It runs forwards along the medial wall of the orbit above the upper border of the rectus medialis, and is joined, near the trochlea of the obliquus superior, by a filament from the supratrochlear nerve. It then escapes from the orbit below the trochlea; it supplies branches to the skin of the eyelids and side of the nose above the medial angle of the eye, the conjunctiva, lacrimal sac and lacrimal caruncle.

The *posterior ethmoidal nerve* leaves the orbital cavity through the posterior ethmoidal foramen and supplies the ethmoidal and sphenoidal sinuses. It is frequently inconspicuous.

THE MAXILLARY NERVE

The maxillary nerve (**7**.176, 177), the intermediate division of the trigeminal nerve, is wholly sensory. It is intermediate in position and size between the ophthalmic and mandibular nerves. It begins at the middle of the trigeminal ganglion as a flattened plexiform band, and, passing horizontally forwards along the lower part of the lateral wall of the cavernous sinus (**6**.127), leaves the skull through the foramen rotundum, where it becomes more cylindrical in form and firmer in texture. It then crosses the upper part of the pterygopalatine fossa, inclines laterally on the posterior surface of the orbital process of the palatine bone and on the upper part of the posterior surface of the maxilla, and enters the orbit through the inferior orbital fissure. It is now named the **infraorbital nerve** and, having traversed the infraorbital groove and canal in the floor of the orbit, it appears on the face through the infraorbital foramen. At its termination the nerve lies under

cover of the levator labii superioris, and divides into branches which are distributed to the ala of the nose, the lower eyelid, the skin and mucous membrane of the cheek and upper lip, and which communicate with filaments of the facial nerve.

In view of the fact that the mouth is generally regarded as representing a pair of fused visceral clefts, the maxillary nerve can be described as the pretrematic and the mandibular nerve as the post-trematic branch of the trigeminal nerve. In early fetal life the maxillary nerve primarily supplies the constituent structures of the maxillary process, but later it extends its territory of innervation into the adjoining frontonasal process (*see* p. 149 and 7.182).

The *branches* of the maxillary nerve may be divided into four groups, according to their origin in the cranium, in the pterygopalatine fossa, in the infraorbital canal, or on the face.

In the cranium:	Meningeal
In the pterygopalatine fossa:	Ganglionic, Zygomatic, Posterior superior alveolar
In the infraorbital canal:	Middle superior alveolar, Anterior superior alveolar
On the face:	Palpebral, Nasal, Superior labial

The meningeal nerve (Kimmel 1961 *a* and *b*), sometimes termed the *nervus meningeus medius*, is given off from the maxillary near the foramen rotundum; it receives a communication from the internal carotid sympathetic plexus and then accompanies the

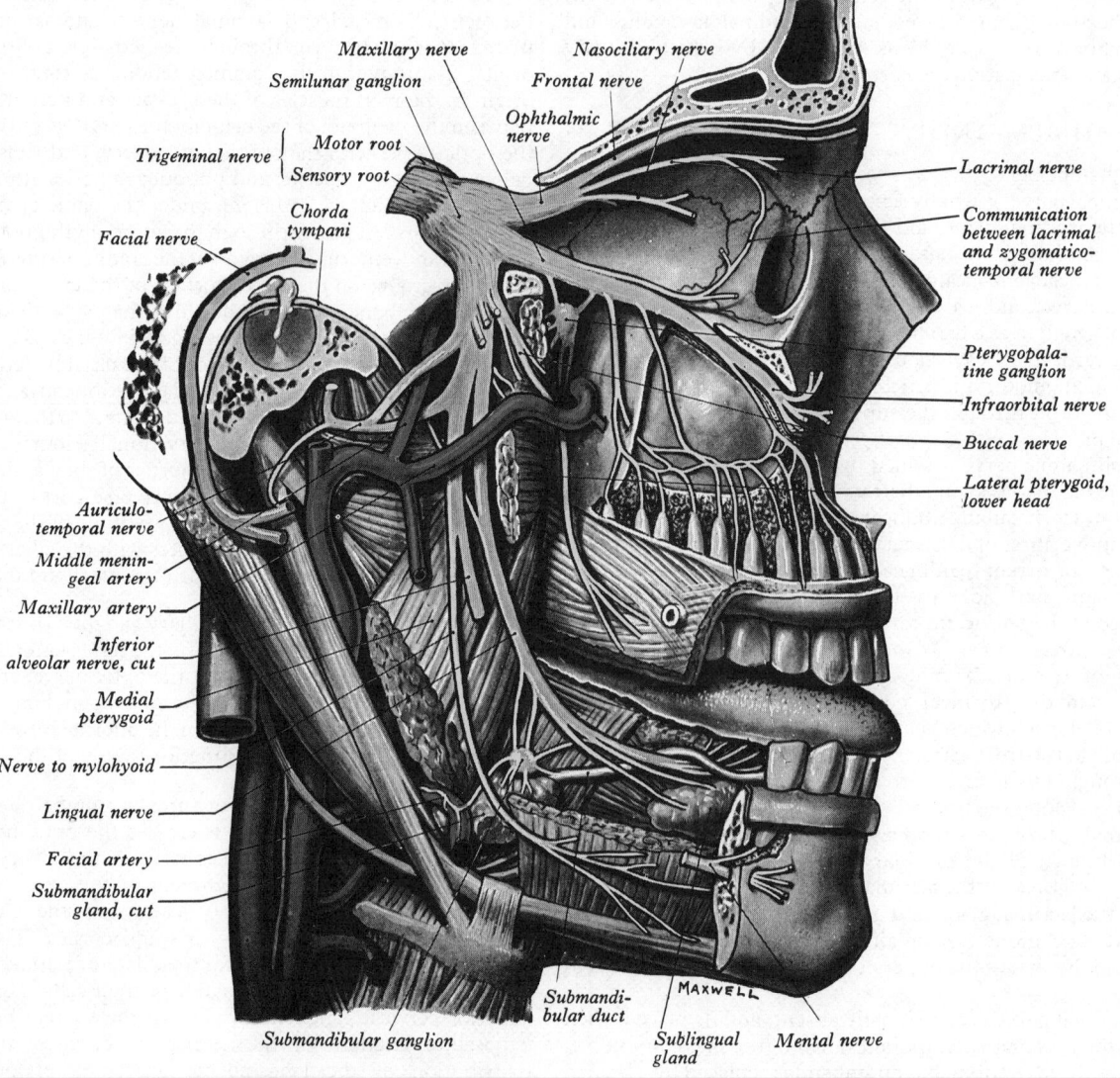

Maxillary nerve

Semilunar ganglion

Nasociliary nerve

Frontal nerve

Ophthalmic nerve

Trigeminal nerve { Motor root / Sensory root

Facial nerve

Chorda tympani

Lacrimal nerve

Communication between lacrimal and zygomatico-temporal nerve

Pterygopala-tine ganglion

Infraorbital nerve

Buccal nerve

Lateral pterygoid, lower head

Auriculo-temporal nerve

Middle menin-geal artery

Maxillary artery

Inferior alveolar nerve, cut

Medial pterygoid

Nerve to mylohyoid

Lingual nerve

Facial artery

Submandibular gland, cut

Submandibular ganglion

Submandi-bular duct

Sublingual gland

Mental nerve

MAXWELL

7.177 The right maxillary and mandibular nerves, and the submandibular ganglion. Note that the zygomatic and superior alveolar nerves are not labelled in this diagram.

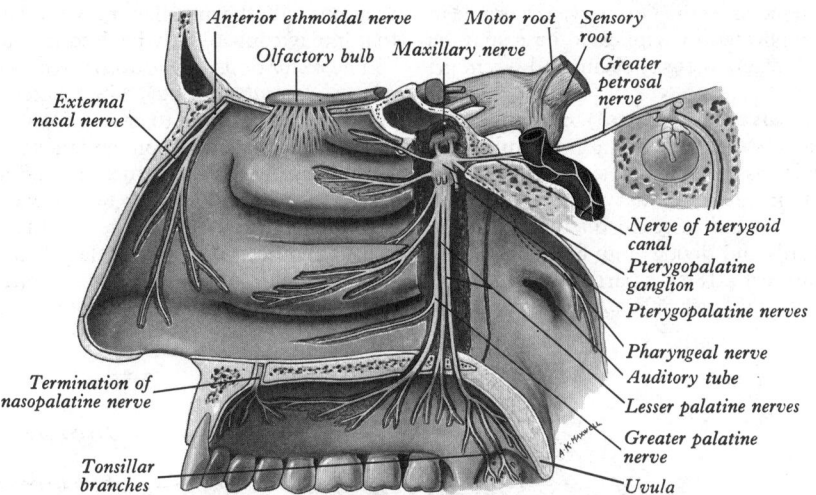

7.178 The right pterygopalatine ganglion and its branches.

frontal branch of the middle meningeal artery and supplies the dura mater of the middle cranial fossa. Its most anterior twigs, however, reach the anterior cranial fossa.

The ganglionic branches, two in number, connect the maxillary nerve to the pterygopalatine (sphenopalatine) ganglion, which lies immediately below it in the pterygopalatine fossa (7.177). They contain the secretomotor fibres for the lacrimal gland (*vide infra*), and sensory fibres from the orbital periosteum and the mucous membranes of the nose, palate and pharynx (p. 1313).

The zygomatic nerve (7.177) arises in the pterygopalatine fossa, enters the orbit by the inferior orbital fissure, courses along the lateral wall of the orbit, and divides into two branches, zygomaticotemporal and zygomaticofacial.

The *zygomaticotemporal branch* passes along the inferolateral angle of the orbital cavity, sends a branch to the lacrimal nerve (p. 1061), and, passing through a canal in the zygomatic bone, enters the temporal fossa. It ascends between the bone and the temporalis, pierces the temporal fascia about 2 cm above the zygomatic arch, and is distributed to the skin of the temple. It communicates with the facial nerve and with the auriculotemporal branch of the mandibular nerve. As it pierces the temporal fascia, it sends a slender twig between the two layers of the fascia towards the lateral angle of the eye. The communication with the lacrimal nerve conveys parasympathetic postganglionic fibres from the pterygopalatine ganglion to the lacrimal gland.

The *zygomaticofacial branch* passes along the inferolateral border of the orbit, emerges upon the face through a foramen in the zygomatic bone, and, perforating the orbicularis oculi, supplies the skin on the prominence of the cheek. It forms a fine plexus with the zygomatic branches of the facial nerve and the palpebral branches of the maxillary nerve. Occasionally this nerve is absent.

Ruskell (1974) has described an *orbito-ciliary* branch, which leaves the maxillary nerve in the pterygopalatine fossa, traverses the inferior orbital fissure, and reaches the eyeball either through the ciliary ganglion or the retro-orbital plexus (p. 1065); dissections in 25 monkeys, together with degeneration experiments, established this pathway and suggested a predominantly sensory role for its nerve fibres.

The superior alveolar (dental) nerves (7.177) arise from the maxillary nerve before it leaves the pterygopalatine fossa, or as it lies in the infraorbital groove or canal. They are termed the posterior, the middle and the anterior superior alveolar (dental) nerves.

The posterior superior alveolar (dental) nerve arises from the maxillary in the pterygopalatine fossa and runs downwards and forwards to pierce the infratemporal surface of the maxilla (p. 1289) and descend under the mucous lining of the maxillary sinus. After supplying the sinus the nerve divides into small branches which link up to constitute the molar part of the *superior dental plexus* supplying twigs to the molar teeth. In addition, it supplies a branch to the upper gum and the adjoining part of the cheek.

The middle superior alveolar (dental) nerve arises from the infraorbital as it passes along the infraorbital groove, and runs downwards and forwards in the lateral wall of the maxillary sinus. Like the posterior, it terminates in a number of small branches which link up with the superior dental plexus, and these give off

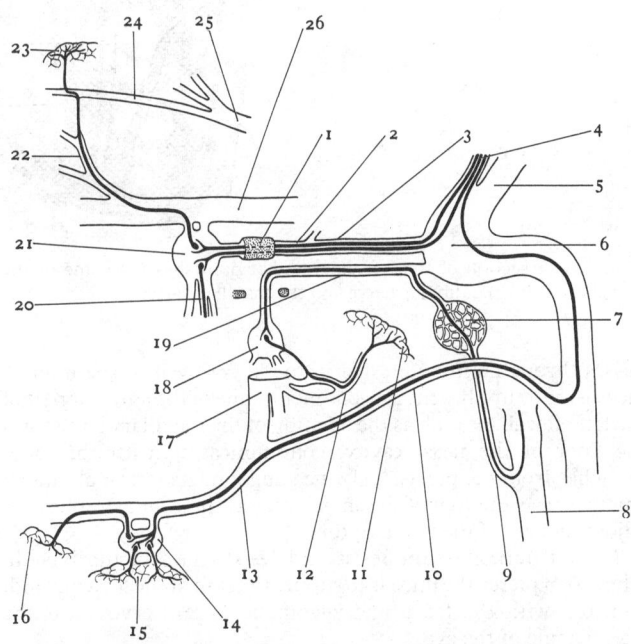

7.179 A diagram to show the parasympathetic connexions of the pterygopalatine, otic and submandibular ganglia. The parasympathetic fibres, both pre- and post-ganglionic, are shown as heavy black lines. The parasympathetic fibres in the palatine nerves (20) are secretomotor to the nasal, palatine pharyngeal glands. Consult text for recent views on the supply to the lacrimal gland.

1. Pterygoid canal.	14. Submandibular ganglion.
2. Nerve of pterygoid canal.	15. Submandibular salivary gland.
3. Greater petrosal nerve.	16. Sublingual salivary gland.
4. Sensory root of facial nerve.	17. Mandibular nerve.
5. Motor root of facial nerve.	18. Otic ganglion.
6. Ganglion of facial nerve.	19. Lesser petrosal nerve.
7. Tympanic plexus.	20. Palatine nerves.
8. Glossopharyngeal nerve.	21. Pterygopalatine ganglion.
9. Tympanic nerve.	22. Zygomaticotemporal nerve.
10. Chorda tympani nerve.	23. Lacrimal gland.
11. Parotid gland.	24. Lacrimal nerve.
12. Auriculotemporal nerve.	25. Ophthalmic nerve.
13. Lingual nerve.	26. Maxillary nerve.

twigs to supply the upper premolar teeth. This nerve is variable in its behaviour. It may be duplicated or triplicated or it may be absent (Wood Jones 1939; Fitzgerald 1956; Pacini and Gremigri 1975).

The anterior superior alveolar (dental) nerve (7.177) leaves the lateral side of the infraorbital nerve near the midpoint of the infraorbital canal, and runs in the canalis sinuosus (p. 1287) in the anterior wall of the maxillary sinus. At first it curves beneath the infraorbital foramen and passes medially towards the nose; it then turns downwards and divides into branches which supply the incisor and canine teeth. It takes part in the formation of the superior dental plexus, and gives off a *nasal branch*, which

relations with the maxillary nerve and its branches are so intimate that it may conveniently be described at this stage.

The *motor* or *parasympathetic root* is formed by the *nerve of the pterygoid canal* (p. 1072), which enters the ganglion posteriorly. Its fibres are believed to arise from a special lacrimatory nucleus in the lower part of the pons and they run in the sensory root of the facial nerve (nervus intermedius) and its greater petrosal branch (p. 1071) before the latter unites with the deep petrosal nerve (7.179) to form the nerve of the pterygoid canal. These preganglionic fibres are relayed in the ganglion and the postganglionic fibres follow a complicated course to gain their destination. Leaving the ganglion in one of the ganglionic

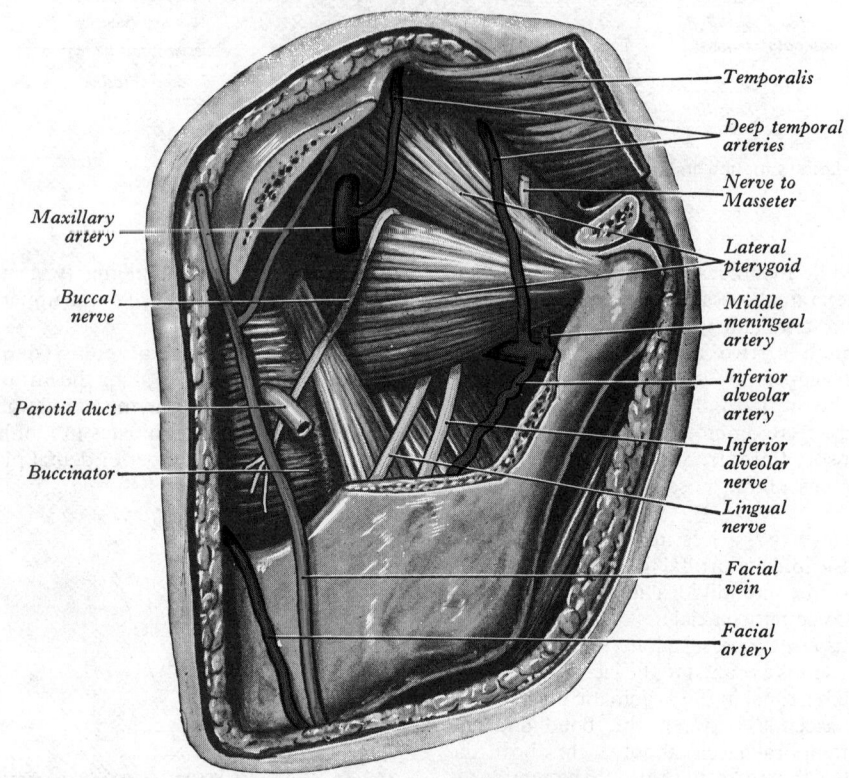

7.180 A dissection of the left pterygoid region, showing some of the branches of the mandibular nerve and the maxillary artery.

passes through a minute canal in the lateral wall of the inferior meatus, and supplies the mucous membrane of the anterior part of the lateral wall (as high as the opening of the maxillary sinus) and the floor of the nasal cavity, communicating with the nasal branches from the pterygopalatine ganglion. Its terminal branch emerges near the root of the anterior nasal spine and supplies the adjoining part of the nasal septum.

The palpebral branches ascend deep to the orbicularis oculi. They soon pierce the muscle to supply the skin of the lower eyelid, and join with the facial and zygomaticofacial nerves near the lateral angle of the eye.

The nasal branches supply the skin of the side of the nose and of the movable part of the nasal septum, and join with the external nasal branch of the anterior ethmoidal nerve.

The superior labial branches are large and numerous; they descend behind the levator labii superioris, and supply the skin of the anterior part of the cheek, the skin of the upper lip, the mucous membrane of the mouth, and the labial glands. They are joined by branches from the facial nerve, and form with them the *infraorbital plexus*.

The pterygopalatine (sphenopalatine) ganglion (7.177, 179) is the largest of the peripheral ganglia of the parasympathetic system. It is deeply placed in the pterygopalatine fossa, close to the sphenopalatine foramen and in front of the pterygoid canal. It is somewhat flattened, of a reddish-grey colour, and is situated just below the maxillary nerve as it crosses the fossa. Although it is *connected functionally with the facial nerve*, its topographical

branches, they join the maxillary nerve and pass into its zygomatic branch. Thence they are usually considered to run in the zygomaticotemporal nerve and later leave it in the communicating branch by which it is connected to the lacrimal nerve (p. 1123). In this way they may reach the lacrimal gland, supplying it with secretomotor fibres (*see*, however, below). In addition, secretomotor fibres—of uncertain origin—for the palatine, pharyngeal and nasal glands are believed to follow a similar route to the ganglion, where they are relayed. Their postganglionic fibres run in the palatine and nasal branches of the ganglion (7.179).

The *sympathetic root* is also incorporated in the nerve of the pterygoid canal. Its fibres, which are postganglionic, arise in the superior cervical ganglion and travel in the internal carotid plexus and the deep petrosal nerve.

The *branches* which appear to arise from the pterygopalatine ganglion (7.178) are, for the most part, derived from the maxillary nerve through its ganglionic branches, and, though intimately related to the ganglion, do *not* establish any synaptic connexions with its cells. They include orbital, palatine, nasal and pharyngeal branches.

The *orbital branches* are two or three delicate filaments which enter the orbit by the inferior orbital fissure, and are distributed to the periosteum and the orbitalis muscle; some twigs pass through the posterior ethmoidal foramen to the sphenoidal and ethmoidal sinuses. The fibres which supply the orbitalis are directly continuous with the fibres of the sympathetic root of the ganglion. Experiments in monkeys and dissections of human material suggest that the orbital rami of the pterygopalatine ganglion form a plexus with branches of the internal carotid sympathetic nerve. This 'retro-orbital' plexus (Ruskell 1970, 1971) is said to supply parasympathetic and sympathetic branches to various orbital structures, including the lacrimal gland (p. 1123).

The *palatine nerves* (7.178) are distributed to the roof of the mouth, the soft palate, the tonsil and the lining membrane of the nasal cavity.

The *greater (anterior) palatine nerve* descends through the greater palatine canal, emerges upon the hard palate through the greater palatine foramen, and runs forwards in a groove on the inferior surface of the bony palate, nearly as far as the incisor

soft palate. The fibres conveying taste impulses from the palate probably pass via the palatine nerves to the pterygopalatine ganglion and thence, without interruption, via the nerve of the pterygoid canal and the greater petrosal nerve to the facial ganglion, where their cells of origin are situated. The central processes of these cells pass through the sensory root of the facial nerve (nervus intermedius) to reach the nucleus of the tractus solitarius (p. 903).

The *nasal branches* enter the nasal cavity through the sphenopalatine foramen. They comprise two sets of nerves. (*a*) About six *lateral posterior superior nasal nerves* innervate the mucous membrane covering the posterior parts of the superior

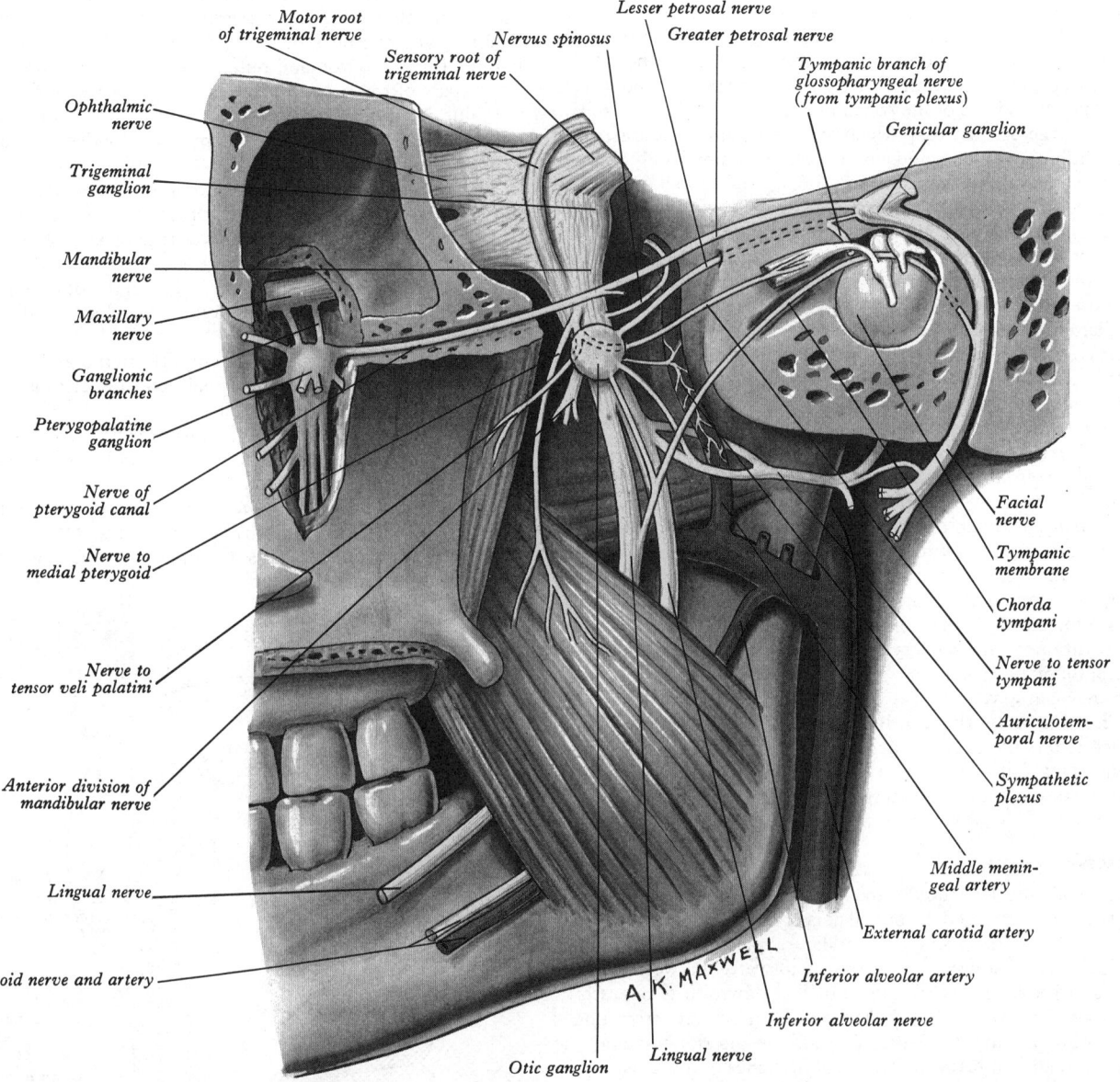

7.181 The right otic ganglion and its branches displayed from the medial side.

teeth. It supplies the gums, and the mucous membrane and glands of the hard palate, and communicates in front with terminal filaments of the nasopalatine nerve. While in the greater palatine canal, it gives off *posterior inferior nasal branches*, which emerge through openings in the perpendicular plate of the palatine bone, and ramify over the inferior nasal concha and the walls of the middle and inferior meatuses; at its exit from the canal, palatine branches are distributed to both surfaces of the soft palate.

The *lesser (middle and posterior) palatine nerves* descend through the greater palatine canal, emerge through the lesser palatine foramina and supply branches to the uvula, tonsil and

and middle nasal conchae and lining the posterior ethmoidal sinuses. (*b*) The *medial posterior superior nasal nerves*, two or three in number, cross the roof of the nasal cavity below the opening of the sphenoidal sinus to supply the mucous membrane of the posterior part of the roof of the cavity and of the nasal septum. The largest of these nerves is the *nasopalatine (long spheno-palatine) nerve*, which runs downwards and forwards on the posterior part of the nasal septum, lying in a groove on the vomer. It descends to the roof of the mouth through the incisive fossa in the anterior part of the hard palate. When an anterior and a posterior incisive foramen (p. 304) are present in this fossa, the left nasopalatine nerve passes through the anterior and the right nerve through the posterior foramen. The nasopalatine nerves furnish a few filaments to the nasal septum and end by supplying the mucous membrane of the anterior part of the hard palate, where they communicate with the anterior palatine nerves.

The *pharyngeal nerve*, a small branch, arises from the posterior part of the ganglion, passes through the palatinovaginal canal with the pharyngeal branch of the maxillary artery, and is distributed to the mucous membrane of the nasal part of the pharynx, behind the auditory tube.

THE MANDIBULAR NERVE

The mandibular nerve (7.177, 181, 182) supplies the teeth and gums of the mandible, the skin of the temporal region, part of the auricle, the lower lip, the lower part of the face, and the muscles of mastication; it also supplies the mucous membrane of the anterior, pre-sulcal part of the tongue and the floor of the mouth. The largest division of the trigeminal nerve, it is made up of two roots: a large, sensory root, which proceeds from the lateral part of the semilunar ganglion and emerges almost immediately through the foramen ovale of the sphenoid bone, and a small motor root (the motor part of the trigeminal) which passes below the ganglion, and unites with the sensory root, just outside the foramen ovale, where the nerve lies between the tensor veli palatini medially and the lateral pterygoid laterally. Immediately beyond the junction of the two roots the nerve sends off from its medial side its meningeal branch and the nerve to the medial pterygoid, and then divides into a small anterior and a large posterior trunk. As it descends from the foramen ovale, the mandibular nerve lies at a depth of 4 cm from the surface and a little in front of the neck of the mandible.

The meningeal branch (*nervus spinosus*) enters the skull through the foramen spinosum with the middle meningeal artery. It divides into two branches, anterior and posterior, which accompany the main divisions of the artery and supply the dura mater of the middle cranial fossa, and to a lesser extent that of the anterior fossa and calvarium; the posterior branch also supplies a twig to the mucous lining of the mastoid air cells; the anterior communicates with the meningeal branch of the maxillary nerve. In addition to its sensory fibres the nervus spinosus carries sympathetic postganglionic fibres from the plexus on the middle meningeal artery.

The nerve to the medial pterygoid is a slender branch which enters the deep surface of the muscle; it gives one or two filaments which pass through the otic ganglion (p. 1076) without being interrupted and emerge from it to supply the tensor tympani and tensor veli palatini (7.181).

Anterior Trunk

The small anterior trunk of the mandibular nerve gives off (*a*) a sensory branch named the buccal nerve, and (*b*) motor branches— the masseteric, deep temporal and lateral pterygoid nerves.

The buccal nerve (7.180) passes forwards between the two heads of the lateral pterygoid, and then downwards beneath or through the lower part of the temporalis; it emerges from under cover of the ramus of the mandible and the anterior border of the masseter, and unites with the buccal branches of the facial nerve. It furnishes a branch to the lateral pterygoid during its passage through that muscle, and may give off the anterior deep temporal nerve. The buccal nerve supplies the skin over the anterior part of the buccinator, and the mucous membrane lining its inner surface and the posterior part of the buccal surface of the gum.

The masseteric nerve (7.180) passes laterally, above the lateral pterygoid, in front of the temporomandibular joint, and behind the tendon of the temporalis; it crosses the posterior part of the mandibular incisure with the masseteric artery, ramifies in the deep surface of the masseter, and gives a filament to the joint.

The deep temporal nerves are usually two in number, anterior and posterior. They pass above the upper border of the lateral pterygoid and enter the deep surface of the temporalis. The *posterior branch*, of small size, is placed at the posterior part of the temporal fossa, and sometimes arises in common with the masseteric nerve. The *anterior branch* is frequently given off from

the buccal nerve, and then ascends over the upper head of the lateral pterygoid. A third, or middle, branch is often present.

The nerve to the lateral pterygoid enters the deep surface of the muscle. It may arise separately from the anterior division of the mandibular nerve, or in conjunction with the buccal nerve.

Posterior Trunk

The posterior trunk of the mandibular nerve is for the most part sensory, but receives a few filaments from the motor root. It divides into auriculotemporal, lingual and inferior alveolar (dental) nerves.

The auriculotemporal nerve generally arises by two roots, which encircle the middle meningeal artery (7.177). It runs backwards under cover of the lateral pterygoid on the surface of the tensor veli palatini and passes between the sphenomandibular ligament and the neck of the mandible. It then passes laterally behind the temporomandibular joint in relationship with the upper part of the parotid gland. Finally, emerging from behind the joint, it ascends, posterior to the superficial temporal vessels, over the posterior root of the zygoma, and divides into superficial temporal branches.

The auriculotemporal nerve communicates with the facial nerve and the otic ganglion. The branches to the facial nerve, usually two in number, pass forwards and laterally behind the neck of the mandible and join the facial nerve at the posterior border of the masseter. The filaments from the otic ganglion join the roots of the auriculotemporal nerve close to their origin (7.184).

The *branches* of the auriculotemporal nerve are the anterior auricular, branches to the external acoustic meatus, articular, parotid, and superficial temporal.

There are usually two *anterior auricular branches*: they supply the skin of the tragus (7.182) and, sometimes, a small part of the adjoining portion of the helix.

The *branches to the external acoustic meatus*, two in number, pass between the bony and cartilaginous parts of the meatus, and supply the skin of the meatus; the upper one sends a twig to the tympanic membrane.

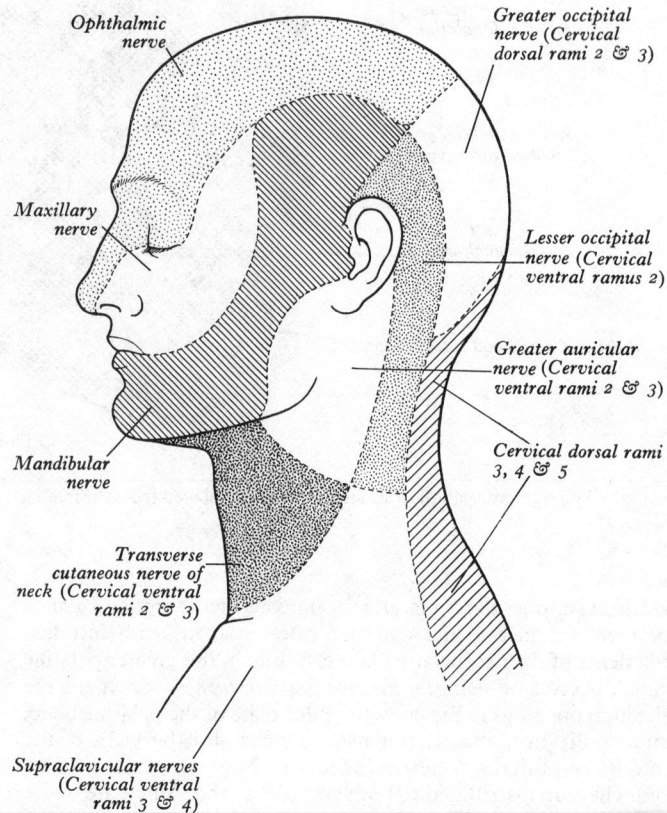

7.182 A diagram showing the cutaneous nerve supply of the face, scalp and neck. Compare with 7.186.

The *articular branches* consist of one or two twigs which enter the posterior part of the temporomandibular joint.

The *parotid branches* convey secretomotor fibres to the parotid gland. The preganglionic fibres are originally derived from the glossopharyngeal nerve through its tympanic branch and travel by the lesser petrosal nerve to the otic ganglion, whence the postganglionic fibres pass to the auriculotemporal nerve and so reach the gland (7.179). They also convey vasomotor fibres to the blood vessels of the parotid gland. These fibres are directly continuous with the fibres of the sympathetic root of the otic ganglion (p. 1076).

The *superficial temporal branches* accompany the superficial temporal artery and its terminal branches; they supply the skin of the temporal region and communicate with the facial and zygomaticotemporal nerves.

The lingual nerve (7.177) is sensory to the mucous membrane of the pre-sulcal part of the tongue, and to the floor of the mouth and the mandibular gums.

It arises from the posterior trunk of the mandibular nerve, and lies at first between the tensor veli palatini and the lateral pterygoid, where it is joined by the chorda tympani branch of the facial nerve, and frequently by a branch of the inferior alveolar nerve. Emerging from under cover of the lateral pterygoid the lingual nerve proceeds downwards and forwards between the ramus of the mandible and the medial pterygoid, lying anterior to, and slightly deeper than the inferior alveolar nerve. It then passes below the mandibular origin of the superior constrictor of the pharynx, and lies against the deep surface of the mandible on the medial side of the roots of the third molar tooth, where it is covered only by the mucous membrane of the gum and can be pressed against the bone by a finger placed inside the mouth. It then leaves the gum and passes on to the side of the tongue, where it crosses the styloglossus, and runs on the lateral surface of the hypoglossus and deep to the mylohyoid; here it is placed above the deep part of the submandibular gland and its duct. It then proceeds forwards on the side of the tongue, lying lateral to the hyoglossus and genioglossus, and divides into its terminal branches, which lie directly under cover of the mucous membrane of the tongue. In the latter part of its course the nerve is in close relation with the submandibular duct; it passes from above downwards and forwards on the lateral side of the duct, and then, winding below it, runs upwards and forwards on its medial side (7.177).

In addition to receiving the chorda tympani and the branch from the inferior alveolar nerve, already referred to, the lingual nerve is connected to the submandibular ganglion (p. 1072) by two or three branches (7.177), and, at the anterior margin of the hyoglossus, forms loops of communication with twigs of the hypoglossal nerve.

The *branches* of the lingual nerve supply the mucous membrane of the floor of the mouth, the lingual surface of the gums, and the mucous membrane of the pre-sulcal part of the tongue, being overlapped to a slight extent by the lingual fibres of the glossopharyngeal nerve (p. 1076); the terminal filaments join, at the tip of the tongue, with those of the hypoglossal nerve. In addition it carries postganglionic fibres from the submandibular ganglion (p. 1072) to the sublingual and anterior lingual glands.

The inferior alveolar (dental) nerve descends deep to the lateral pterygoid, and then, at the lower border of the muscle, it passes between the sphenomandibular ligament and the ramus of the mandible to the mandibular foramen. Here it enters the mandibular canal, and runs below the teeth as far as the mental foramen, where it divides into an incisive and a mental branch. Below the lateral pterygoid the nerve is accompanied by the inferior alveolar artery. Dissection and radiographic studies show that in a majority of mandibles the inferior dental nerve does not occupy a single canal, but is *plexiform* in arrangement. It is also joined, directly, or through its plexiform branches, by rami entering the bone as parts of neurovascular bundles, derived from attached muscles such as masseter. Such 'accessory' dental nerves ramify particularly in a plane lateral to the molar teeth, and their common occurrence accounts for the incomplete abolition of pain by inferior dental nerve block (Carter and Keen 1971).

The inferior alveolar nerve gives off the mylohyoid nerve,

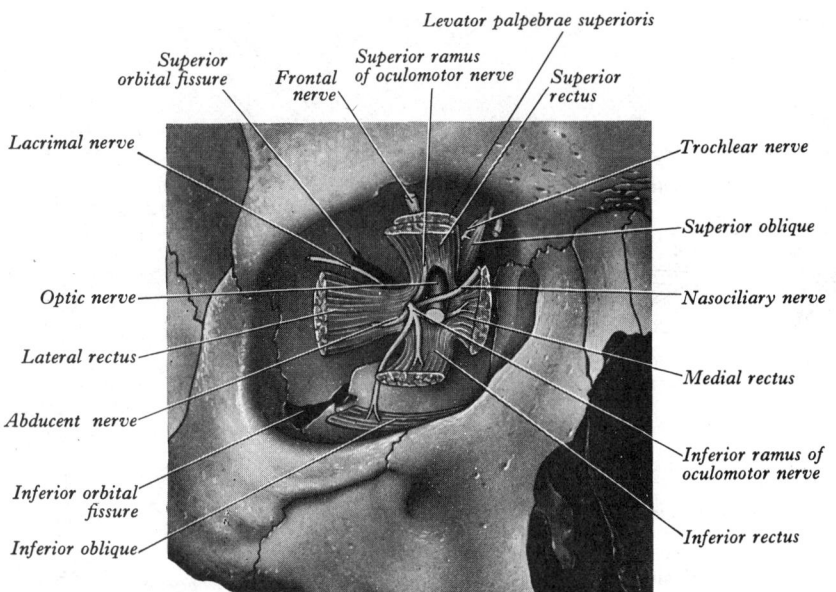

7.183 A dissection of the right orbit viewed from in front, to show the origins of the orbital muscles and the relative positions of the nerves of the orbit.

branches to the molar and premolar teeth of the mandible and the incisive and the mental nerves.

The *mylohyoid nerve* is derived from the inferior alveolar nerve just before the latter enters the mandibular foramen. It pierces the sphenomandibular ligament, descends in a groove on the medial surface of the ramus of the mandible and, passing below the mylohyoid line, it reaches the inferior surface of the mylohyoid, which it supplies together with the anterior belly of the digastric.

The branches to the molar and premolar teeth supply the adjoining gum also. Before they enter the roots of the teeth they communicate with one another and form an inferior dental plexus.

The *incisive nerve* is often described as continuing within the bone to supply the canine and incisor teeth. Alternative views have, however, been expressed: the nerves which supply the incisor teeth form an elaborate plexus on the external aspect of the mandible after emerging from the mental foramen and before they re-enter the bone (Starkie and Stewart 1931). The canine tooth may be supplied either from this incisor plexus or the plexus which innervates the premolars.

The *mental nerve* emerges at the mental foramen, and divides beneath the depressor anguli oris into three branches; one descends to the skin of the chin, and two ascend to the skin and mucous membrane of the lower lip; these branches communicate freely with the facial nerve (mandibular branch).

Applied Anatomy. A lesion of the whole trigeminal nerve causes anaesthesia of the corresponding anterior half of the scalp, of the face (excepting a small area near the angle of the mandible supplied by the great auricular nerve), of the cornea and conjunctiva, and of the mucous membranes of the nose, mouth and pre-sulcal part of the tongue. Paralysis and atrophy occur in the muscles supplied by the nerve, and when the mouth is opened the mandible is thrust over to the paralysed side. Lesions of the divisions of the nerve give a more limited sensory loss and, if affecting the lingual nerve below the point at which it is joined by the chorda tympani, will be accompanied by loss of taste in the corresponding half of the anterior part of the tongue.

Pains referred to various branches of the trigeminal nerve are of very frequent occurrence. As a general rule the diffusion of pain over the various branches of the nerve is at first confined to one only of the main divisions, although in severe cases pain may radiate over the branches of the other main divisions. The commonest example of this condition is the neuralgia which is so often associated with dental caries. Here, although the tooth itself may not appear to be painful, the most distressing referred pains may be experienced, and these are at once relieved by treatment directed to the affected tooth.

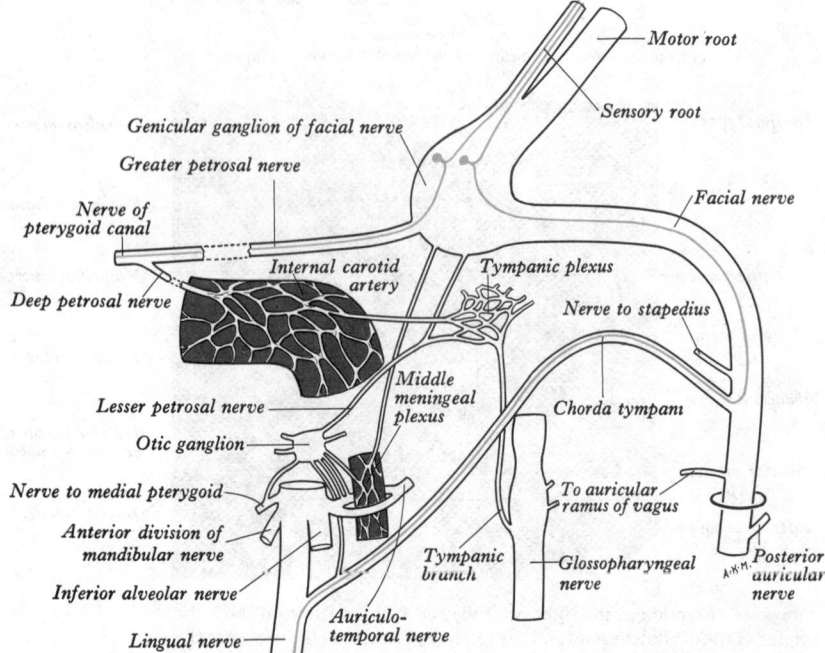

7.184 A plan of the intrapetrous section of the facial nerve, its branches and communications. The course of the taste fibres from the mucous membrane of the palate and from the anterior, oral, or presulcal part of the tongue is represented by the blue lines.

In the area of the ophthalmic nerve, severe supraorbital pain is commonly associated with acute glaucoma or with frontal or ethmoidal sinusitis. Malignant growths or empyema of the maxillary sinus, or unhealthy conditions about the inferior conchae or the septum of the nose, are often found giving rise to 'second division' (maxillary) neuralgia, and should be always looked for in the absence of dental disease in the maxilla. It is in the mandibular nerve, however, that some of the most striking examples are seen. It is quite common to meet with patients who complain of pain in the ear, in whom there is no sign of aural disease, and the cause is usually to be found in a carious tooth in the mandible. Moreover, with an ulcer or cancer of the tongue, often the first pain to be experienced is one which radiates to the ear and temporal fossa, over the distribution of the auriculo-temporal nerve.

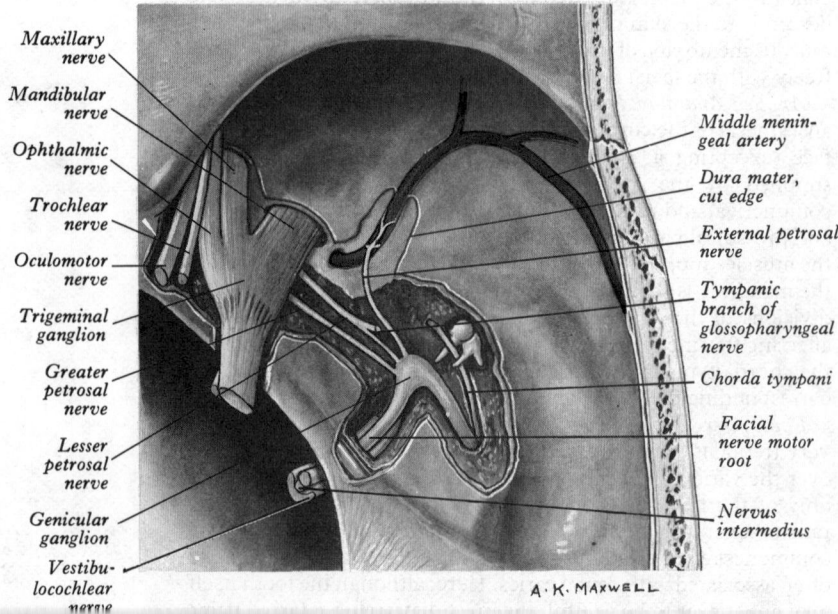

7.185 A dissection of the right middle cranial fossa, showing the course and some of the connexions of the facial nerve within the temporal bone.

The lingual nerve is occasionally divided with a view to relieving the pain in cancer of the tongue. This may be carried out where the nerve lies in direct contact with the mandible below and behind the last molar tooth, covered only by the mucous membrane (p. 317).

In cases of intractable neuralgia of the trigeminal nerve various operative procedures have been introduced from time to time. The trunks of the maxillary and mandibular nerves and the trigeminal ganglion itself have been injected with alcohol with varying degrees of success, and excision of the ganglion, in whole or in part, has frequently been performed successfully, but the last-named operation involves serious risks (laceration of the cavernous sinus, etc.) and is now rarely undertaken. The sensory root of the nerve may be divided behind the ganglion, and this is now the operation of election when the pain is confined to the maxillary and mandibular nerve areas (7.182). Complete division of the sensory root necessarily denervates the cornea completely, and the resulting loss of the corneal reflex leads to neuropathic keratitis. Consequently, in these cases an endeavour is made to preserve the ophthalmic fibres, which lie in the upper and medial part of the root and can be spared if the incision is restricted to the lower and lateral fibres. The motor root of the nerve is left intact.

When the pain is limited to the ophthalmic area or to the ophthalmic and maxillary areas, the operation of election (Falconer 1949) consists in the division of the fibres of the spinal tract of the nerve, where it is most superficial (p. 901) and sometimes forms a recognizable elevation (p. 902) between the lateral margin of the fasciculus cuneatus and the posterior border of the lower part of the olive. Section of the tract 4–5 mm caudal to the obex preserves most, if not all, of the mandibular fibres, because they have entered the cranial part of the nucleus of the tract and so escape injury. Following the operation painful and thermal sensibility are lost over the ophthalmic and maxillary areas, but tactile sensibility is retained and the corneal reflex is not abolished.

The Abducent Nerve

The abducent nerve (7.176) supplies the lateral rectus muscle of the eyeball. Its fibres arise from a small nucleus situated in the cranial part of the floor of the fourth ventricle, close to the median plane and beneath the colliculus facialis (7.62, 64, 68, 69; p. 933). They descend ventrally through the pons, and emerge in the sulcus between the caudal border of the pons and the cranial end of the pyramid of the medulla oblongata (7.55).

The *abducens nucleus* consists of large and small multipolar neurons, the former being sources of the abducent nerve, the latter being known collectively as the *nucleus para-abducens*, which is considered to project to the oculomotor nucleus via the medial longitudinal fasciculus. The total number of cells in the nucleus has been stated to be about 22,000, and only a minority of these can be radicular, since the nerve contains about 6,600 fibres (Konigsmark *et al.* 1969). Claims that all abducent neurons react to division of the nerve (Warwick 1964 and others) are clearly based on inadequate concepts of what cells should be included as 'abducent'. A study in the kitten using the retrograde horseradish peroxidase technique (Gacek 1974) indicated that 50–90 per cent of the cells were radicular. Other studies have clearly demonstrated that some, at least, of the abducent neurons project to the oculomotor and perhaps the trochlear nuclei. (*See* Steiger and Büttner-Ennever 1978, for literature.) Büttner-Ennever (1977) has identified neurons in the pontine reticular formation, just rostral to the abducent nucleus, which project to regions of the oculomotor nucleus reputedly concerned with innervation of the lateral and inferior recti. He has discussed the considerable evidence for the view that these neurons mediate horizontal and perhaps also vertical integration of binocular movement.

Connexions. The nucleus of the abducent nerve receives prominent afferent connexions from: (1) the corticonuclear tract principally, but not wholly, of the contralateral side; some of these fibres are aberrant pyramidal fibres which descend from the midbrain to this level in the medial lemniscus (p. 901), and interneurons may link them to the nucleus; (2) the medial

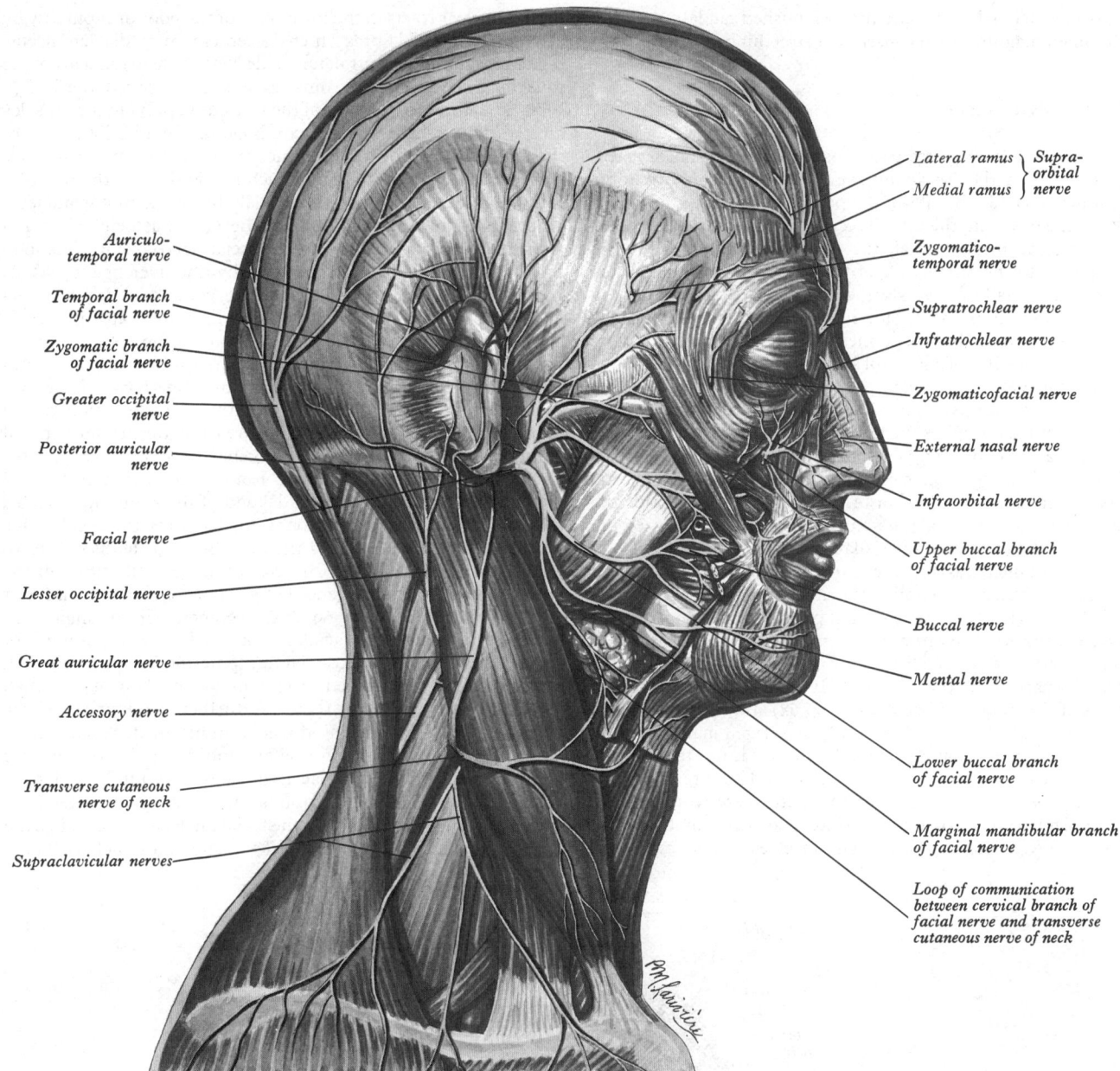

Lateral ramus ⎫ Supra-
              ⎬ orbital
Medial ramus ⎭ nerve

Zygomatico-
temporal nerve

Supratrochlear nerve

Infratrochlear nerve

Zygomaticofacial nerve

External nasal nerve

Infraorbital nerve

Upper buccal branch
of facial nerve

Buccal nerve

Mental nerve

Lower buccal branch
of facial nerve

Marginal mandibular branch
of facial nerve

Loop of communication
between cervical branch of
facial nerve and transverse
cutaneous nerve of neck

Auriculo-
temporal nerve

Temporal branch
of facial nerve

Zygomatic branch
of facial nerve

Greater occipital
nerve

Posterior auricular
nerve

Facial nerve

Lesser occipital nerve

Great auricular nerve

Accessory nerve

Transverse cutaneous
nerve of neck

Supraclavicular nerves

7.186   The nerves of the right side of the scalp, face and neck. Compare with **7.182**.

longitudinal fasciculus, by which it is connected with the nuclei of the third, fourth and eighth cranial nerves; (3) the tectobulbar tract, by which it is connected with the visual cortex and other centres through the medium of the superior colliculus. (See also p. 941).

Course. After leaving the surface of the brainstem, the abducent nerve runs upwards, forwards and laterally through the cisterna pontis, and usually dorsal to the anterior inferior cerebellar artery. It pierces the dura mater lateral to the dorsum sellae of the sphenoid bone and then bends sharply forwards as it crosses the superior border of the petrous part of the temporal bone close to its apex. In this situation it is inferior to the petrosphenoidal ligament—a fibrous band which connects the lateral margin of the dorsum sellae to the upper border of the petrous part of the temporal bone near its medial end. It next traverses the cavernous sinus, lying at first lateral and then inferolateral to the internal carotid artery (**6.**127) and enters the orbital cavity through the medial end of the superior orbital fissure. It passes within the common tendinous ring from which the recti of the eyeball arise, lying inferolateral to the oculomotor and nasociliary nerves, and finally sinks into the ocular surface of the lateral rectus (**7.**183, 269).

In the cavernous sinus the abducent nerve is joined by a considerable branch from the termination of the internal carotid plexus. This communication was originally described by Monro (1746), confirmed by Meckel (1832), but has subsequently been frequently omitted from textbooks and even denied (Sunderland and Hughes 1946). The existence of this connexion has been re-emphasized by Johnston and Parkinson (1974), and Parkinson *et al.* (1978). These workers have also described a branch of similar calibre which leaves the abducent nerve a few millimetres distal to its reception of the sympathetic communication, to join the ophthalmic nerve; they believe that the sympathetic postgang-lionic fibres which thus join the abducent nerve are thereby diverted into the ophthalmic division of the trigeminal. If confirmed, this observation would accord with the passage of sympathetic nerve fibres to the eyeball through the long ciliary branches of the nasociliary nerve, as suggested by some authorities.

Applied Anatomy. The abducent nerve is occasionally involved in fractures of the base of the skull. The result of paralysis of this nerve is medial or convergent squint. Diplopia hence follows. The long course forwards of the sixth cranial nerve through the cisterna pontis and its sharp bend over the superior border of the petrous part of the temporal bone make the nerve particularly liable to damage in conditions producing raised intracranial pressure, as a result of which the brainstem is pushed caudally towards the foramen magnum with consequent stretching of the nerve.

The Facial Nerve

The facial nerve (7.184, 187) possesses a motor and a sensory root, the latter called the *nervus intermedius* (7.55). The two roots appear at the caudal border of the pons just lateral to the recess between the olive and the inferior cerebellar peduncle, the motor part being the more medial; the vestibulocochlear nerve lies immediately to the lateral side of the sensory root. The nervus intermedius usually cleaves to the latter nerve rather than the facial, passing from the eighth to the seventh nerve as they approach the internal acoustic meatus, often as more than one filament. In one-fifth of a series of seventy-three dissections it was not a separate nerve until the meatus was reached, a point of some surgical importance (Rhoton *et al.* 1968).

The *motor root* supplies the muscles of the face, scalp, auricle, the buccinator, platysma, stapedius, stylohyoid, and posterior belly of the digastric. The *sensory root* conveys from the chorda tympani nerve the fibres of taste for the presulcal area of the tongue, and from the palatine and greater petrosal nerves the fibres of taste from the soft palate; in addition, it transmits the preganglionic parasympathetic (secretomotor) innervation of the submandibular and sublingual salivary glands, the lacrimal gland and the glands of the nasal and palatine mucosae.

The motor nucleus from which most of the motor fibres of the facial nerve are derived lies deeply in the reticular formation of the caudal part of the pons (p. 910). It is posterior to the dorsal nucleus of the trapezoid body (7.62, 64, 68) and ventromedial to the nucleus of the spinal tract of the trigeminal nerve. It represents the branchial efferent column, but it lies much more deeply in the pons than might be expected, and its outgoing fibres pursue a most unusual course (7.69). Both these features have been explained by invoking the principle of neurobiotaxis (but *see*, however, p. 164). The nucleus receives fibres from both

corticonuclear tracts in the lower part of the pons or reputedly by aberrant pyramidal fibres which descend in the medial lemniscus. The fibres from the contralateral side contribute to that part of the nucleus which supplies the muscles of the lower part of the face (p. 910). The fibres to that part of the nucleus supplying the muscles around the eyes and forehead are bilateral. In addition, some of the efferent fibres of the facial nerve proceed from the *superior salivatory nucleus* (*see* p. 911), which is said to be in the reticular formation, dorsolateral to the caudal end of the motor nucleus. The cells of the nucleus have been described as being clustered along the intrapontine part of the facial nerve distal to its loop around the abducens nucleus (Crosby and Dejonge 1963). It represents the general visceral efferent column, and it sends its fibres to join the sensory root, by which they are ultimately distributed through the chorda tympani to the submandibular and sublingual salivary glands, and perhaps also to the parotid gland (*see* p. 1123). Further preganglionic fibres having issued via the sensory root are destined for ultimate distribution to the pterygopalatine ganglion which they reach in the greater petrosal nerve and nerve of the pterygoid canal (*vide infra*). From this double origin the fibres of the *motor root* pass dorsally and medially, and, reaching the caudal end of the abducent nucleus, they run rostrally superficial to this nucleus deep to the colliculus facialis. At the cranial end of the nucleus of the abducent nerve they make a second bend, and run caudoventrally through the pons to the point of emergence between the olive and the inferior cerebellar peduncle (7.55, 69) at the cerebellopontine angle.

Topographically the facial motor nucleus is a complex of smaller nerve cell groups—lateral, intermediate and medial, which have been identified in various mammals including man (Marinesco 1899; Papez 1927). A further subdivision of the medial nucleus into ventral, dorsal and intermediate sub-nuclei has also been described. (For details and discussion of the large literature concerning the facial nucleus, consult Vraa-Jensen 1942.) These subsidiary groups of nerve cells extend as craniocaudal columns through the facial nuclear complex, like the columnar groups of the spinal cord or oculomotor nucleus. There

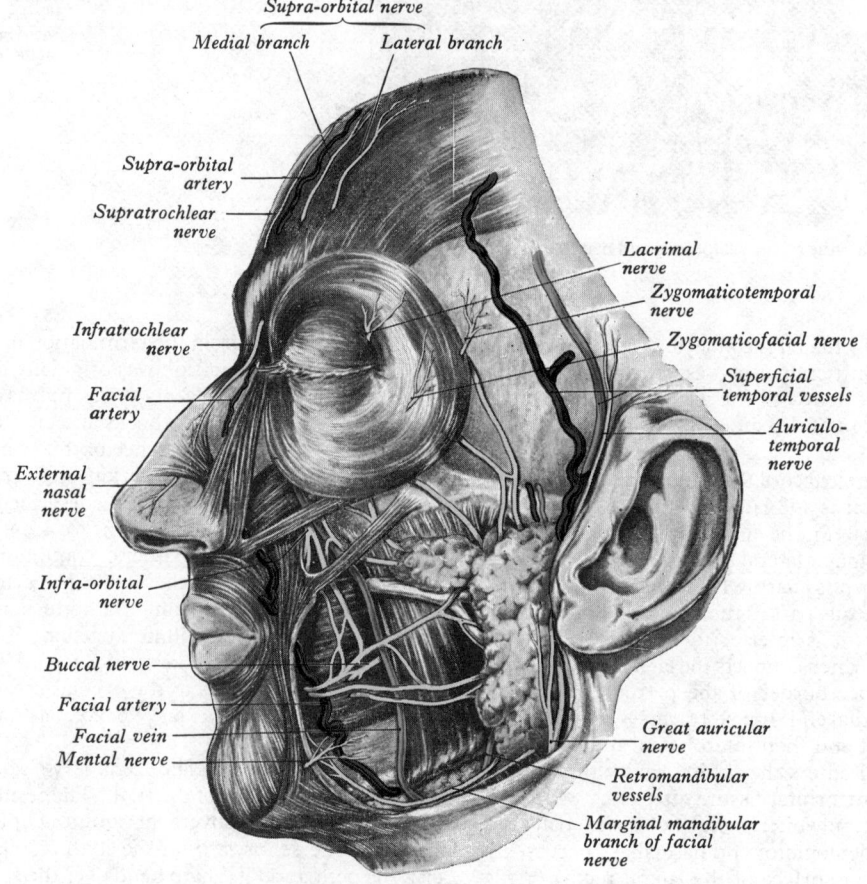

7.187 The terminal branches of the left trigeminal and facial nerves in the face.

is general agreement that the facial subnuclei innervate muscles supplied by individual branches of the facial nerve, or even single muscles, but different observers disagree in details. Retrograde changes due to division of these branches in dogs (Vraa-Jensen 1942; Yagita 1910) or cats (Papez 1927; Courville 1966) have provided most of the evidence, but the effects on motor terminals due to selective lesions of the nucleus have also been studied in cats. General results were that the lateral sub-nucleus innervates the buccal musculature, the intermediate sends axons into the temporal, orbital and zygomatic facial branches, and the medial into the posterior auricular and cervical rami and probably also into the stapedial nerve. Nuclear lesions produced a roughly similar but more detailed schema (Szentágothai 1948).

The sensory nucleus of the facial nerve is the upper part of the *nucleus of the tractus solitarius* of the medulla oblongata (p. 905). It receives afferent fibres from the sensory root and sends efferent fibres to the ventral group of nuclei of the lateral part of the thalamus of the opposite side. As they ascend through the midbrain and subthalamic regions, these fibres are closely related to the median plane (Harris 1952). From the thalamus they are relayed to the inferior part of the postcentral gyrus. (For further details of projections from the nucleus of the solitary tract *see* p. 1085.)

The sensory root (nervus intermedius) consists of the central processes of the unipolar cells of the genicular ganglion, which leave the trunk of the facial nerve in the internal acoustic meatus and pass centrally as one or more slender bundles between the motor root and the vestibulocochlear nerve or adhering to the latter, to enter the brainstem at the lower border of the pons. The peripheral branches from the processes of the ganglion cells are the taste fibres contained in the chorda tympani and the greater petrosal nerves, and also a few somatic afferent fibres from the concha of the auricle (p. 1192). As already stated, the sensory root also contains the efferent preganglionic parasympathetic fibres for the submandibular and sublingual salivary glands, the lacrimal gland, and pharyngeal, nasal and palatine glands.

From their attachments to the brain, the two roots of the facial nerve pass laterally and forwards with the vestibulocochlear nerve to the opening of the internal acoustic meatus. In the meatus the motor root lies in a groove on the upper and anterior surface of the vestibulocochlear nerve, the sensory root being placed between them.

At the bottom of the meatus, the facial nerve enters the facial canal. In this canal the nerve runs at first laterally above the vestibule and, reaching the medial wall of the epitympanic recess, bends sharply backwards above the promontory, and arches downwards in the medial wall of the aditus to the mastoid antrum. Finally it descends to reach the stylomastoid foramen. The point where it bends sharply backwards is named the *geniculum*; it presents a reddish asymmetric swelling named the *genicular ganglion* (**7**.184). On emerging from the stylomastoid foramen, the facial nerve runs forwards in the substance of the parotid gland (p. 1272), crosses the styloid process, the retromandibular vein and the external carotid artery, and divides behind the neck of the mandible into branches which pierce the anteromedial surface of the parotid gland and diverge from one another under cover of it. They form a network (*parotid plexus*) and are distributed to the musculature of the face. As it emerges from the stylomastoid foramen, the facial nerve lies about 2 cm deep to the middle of the anterior border of the mastoid process. Its course through the parotid gland can be represented by a short horizontal line drawn across the upper part of the lobule of the auricle (**6**.47).

The angular measurements of the frequent changes in direction of the facial nerve in the intra-osseous part of its course are of considerable surgical importance. Kudo and Nori (1974) and Guerrier (1975) have recorded highly detailed metrical studies of this part of the nerve.

The facial nerve is supplied by the anterior inferior cerebellar artery in the intracranial part of its course, by the superficial petrosal branch of the middle meningeal artery and the stylomastoid branch of the posterior auricular or occipital arteries in the intrapetrous part, by branches from the stylomastoid, posterior auricular, occipital, superficial temporal and transverse facial arteries extracranially. Its venous drainage is into the venae comitantes of the superficial petrosal and stylomastoid arteries (Blunt 1954).

The branches of communication of the facial nerve may be arranged as follows:

Intracranial:	Vestibulocochlear
Genicular ganglion:	Pterygopalatine ganglion via the greater petrosal nerve
	Otic ganglion via the lesser petrosal nerve
	Middle meningeal sympathetic plexus
Facial canal:	Auricular branch of vagus
At exit from stylo-mastoid foramen:	To glossopharyngeal, vagus, great auricular and auriculotemporal nerves
Post-auricular:	To lesser occipital nerve
Facial:	With trigeminal nerve
Cervical:	With transverse cervical nerve

In the internal acoustic meatus some minute filaments connect the facial nerve with the vestibulocochlear nerve.

The greater petrosal nerve arises from the genicular ganglion, and consists chiefly of taste fibres which are distributed to the mucous membrane of the palate; but it also contains preganglionic parasympathetic fibres which are travelling to the pterygopalatine ganglion; they are said to be relayed through the zygomatic and lacrimal nerves to the lacrimal gland (but *see* p. 1065), and through the nasal and palatine nerves to the nasal and palatine mucosal glands (**7**.179). It receives a twig from the tympanic plexus, passes forwards through the hiatus on the anterior surface of the petrous portion of the temporal bone and runs in a groove on the bone. It passes beneath the trigeminal ganglion and reaches the foramen lacerum. In this foramen it is joined by the **deep petrosal nerve** (**7**.184) from the sympathetic plexus on the internal carotid artery, and forms the **nerve of the pterygoid canal**, which passes forwards through the pterygoid canal and ends in the pterygopalatine ganglion. The taste fibres pass without interruption through or over the surface of the ganglion into the palatine branches which spring from it.

From the trunk of the facial nerve near the genicular ganglion, from the ganglion itself, or from the root of the greater petrosal nerve (Vidić and Young 1967) a branch runs to join the lesser petrosal nerve (**7**.184), and is conveyed through this nerve to the otic ganglion. However, the fibres constituting this small communicating nerve do not stem from the facial nerve itself, but from the auricular branch of the vagus. In twenty-four out of twenty-five human dissections the fibres approached the facial nerve through the stapedius muscle, travelling in the muscle's fascial sheath almost as far as the genicular ganglion, before departing to join the lesser superficial petrosal nerve (Vidić 1968).

The sympathetic plexus on the middle meningeal artery is joined to the genicular ganglion by an inconstant branch, sometimes named the *external petrosal nerve*.

Before the facial nerve emerges from the stylomastoid foramen, it receives another twig from the auricular branch of the vagus.

After its exit from the stylomastoid foramen, the facial nerve receives a twig from the glossopharyngeal nerve, and communicates with the great auricular and auriculotemporal nerves in the parotid gland, with the lesser occipital nerve behind the ear, with the terminal branches of the trigeminal nerve on the face, and with the transverse cutaneous cervical nerve in the neck.

The branches of distribution (7.184, 186) of the facial nerve may be grouped as follows:

In the facial canal:	Nerve to the stapedius
	Chorda tympani
At exit of the stylo-mastoid foramen:	Posterior auricular
	Digastric, posterior belly
	Stylohyoid
On the face:	Temporal
	Zygomatic
	Buccal
	Marginal mandibular
	Cervical

The nerve to the stapedius arises from the facial nerve opposite the pyramidal eminence on the posterior wall of the tympanic cavity; it passes forwards through a small canal to reach the muscle (*vide supra*, and also p. 1199).

The chorda tympani nerve (7.181) arises from the facial nerve about 6 mm above the stylomastoid foramen. It runs upwards and forwards in a canal, and perforates the posterior bony wall of the tympanic cavity through the posterior canaliculus for the chorda tympani nerve, which is situated close to the posterior border of the medial surface of the tympanic membrane and on a level with the upper end of the handle of the malleus. Passing forwards between the fibrous and mucous layers of the membrane, it crosses the medial aspect of the handle of the malleus, and re-enters the bone through the anterior canaliculus for the chorda tympani nerve, which is placed at the medial end of the petrotympanic fissure. The nerve now descends ventrally on the medial surface of the spine of the sphenoid bone (which it sometimes grooves) and passes deep to the lateral pterygoid. In this part of its course the nerve lies lateral to the tensor veli palatini and is crossed by the middle meningeal artery, the roots of the auriculotemporal nerve and the inferior alveolar nerve. Finally it joins the posterior border of the lingual nerve at an acute angle. It contains efferent preganglionic parasympathetic (secretomotor) fibres which enter the submandibular ganglion, and are there relayed as postganglionic fibres to the submandibular and sublingual glands; the majority of its fibres are afferent from the mucous membrane covering the anterior, pre-sulcal part of the tongue, save the vallate papillae; they constitute the nerve of taste for this region of the tongue. Before uniting with the lingual nerve the chorda tympani is joined by a small branch from the otic ganglion.

The posterior auricular nerve arises close to the stylomastoid foramen and runs upwards in front of the mastoid process; here it is joined by a filament from the auricular branch of the vagus nerve, and communicates with the posterior branch of the great auricular nerve, and with the lesser occipital nerve. As it ascends between the external acoustic meatus and the mastoid process it divides into an auricular and an occipital branch. The *auricular branch* supplies the auricularis posterior and the intrinsic muscles on the cranial surface of the auricle. The *occipital branch*, the larger, passes backwards along the superior nuchal line of the occipital bone, and supplies the occipital belly of the occipitofrontalis.

The digastric branch arises close to the stylomastoid foramen, and divides into several filaments which supply the posterior belly of the digastric; one of these filaments joins the glossopharyngeal nerve.

The stylohyoid branch, long and slender, frequently arises in conjunction with the digastric branch; it enters the middle part of the stylohyoid.

The temporal branches cross the zygomatic arch to the temporal region. They supply the intrinsic muscles on the lateral surface of the auricle, the anterior and superior auricular muscles, and join with the zygomaticotemporal branch of the maxillary nerve, and with the auriculotemporal branch of the mandibular nerve. The more anterior branches supply the frontal belly of the occipitofrontalis, the orbicularis oculi and the corrugator, and join the supraorbital and lacrimal branches of the ophthalmic nerve.

The zygomatic branches run across the zygomatic bone to the lateral angle of the eye; they supply the orbicularis oculi, and join with the filaments of the lacrimal nerve and the zygomatico-facial branch of the maxillary nerve.

The buccal branches pass horizontally forwards to be distributed below the orbit and around the mouth. The *superficial branches* run between the skin of the face and the superficial muscles, and supply the latter; some are distributed to the procerus, joining with the infratrochlear and external nasal nerves. The *upper deep branches* pass under cover of the zygomaticus major and the levator labii superioris, supplying them and forming an *infraorbital plexus* with the superior labial branches of the infraorbital nerve; they also supply the levator anguli oris, the zygomaticus minor, the levator labii superioris alaeque nasi and the small muscles of the nose. These branches are sometimes described as lower zygomatic branches. The *lower deep branches* supply the buccinator and orbicularis oris, and join with filaments of the buccal branch of the mandibular nerve.

The marginal mandibular branch runs forwards below the angle of the mandible under cover of the platysma. It lies at first superficial to the upper part of the digastric triangle and then turns upwards and forwards across the body of the mandible to lie under cover of the depressor anguli oris (7.187). It supplies the risorius and the muscles of the lower lip and chin, and joins the mental nerve (p. 1067).

The cervical branch issues from the lower part of the parotid gland, runs forwards and downwards under cover of the platysma to the front of the neck. It supplies the platysma and communicates with the transverse cutaneous cervical nerve.

The submandibular ganglion is a small, somewhat fusiform ganglion which lies on the upper part of the hyoglossus. There are further ganglion cells in the hilum of the submandibular gland. Like the ciliary, pterygopalatine and otic ganglia, these are peripheral ganglia of the parasympathetic system. The submandibular ganglion is superior to the deep part of the submandibular gland and below the lingual nerve, from which it is suspended by an anterior and a posterior filament (7.177). Although so intimately related to the lingual nerve, the ganglion is connected functionally with the facial nerve and its chorda tympani branch.

The *motor* or *parasympathetic root* is formed by the posterior filament connecting the ganglion to the lingual nerve. It conveys preganglionic fibres which leave the superior salivatory nucleus and run in the facial, chorda tympani and lingual nerves to reach the ganglion. There the fibres establish synaptic relations with the cells of the ganglion, and the postganglionic fibres are secretomotor to the submandibular and the sublingual salivary glands. (Some of these fibres may reach the parotid gland. *See* p. 1274). The *sympathetic root* is derived from the plexus on the facial artery. It consists of postganglionic fibres which commence in the superior cervical ganglion and pass through the submandibular ganglion without being interrupted. They are vasomotor to the blood vessels of the submandibular and sublingual glands.

Five or six branches arise from the ganglion and supply the submandibular gland and its duct. Other fibres from the ganglion pass through the anterior filament which connects it to the lingual nerve and are carried to the sublingual and the anterior lingual glands.

The cutaneous fibres of distribution of the facial nerve accompany those of the auricular branch of the vagus and probably innervate skin on both aspects of the auricle, in the conchal depression and over its eminence. Details of this innervation are, however, uncertain, as is the question of whether facial fibres reach the external acoustic meatus and tympanic membrane.

Applied Anatomy. Facial paralysis is commonly unilateral and may be due to: (1) *supranuclear lesions* of the corticonuclear fibres from the frontal lobe, and variably combined with numerous other descending fibres that converge on the facial nucleus; (2) *nuclear* or *infranuclear lesions* involving the lower motor neurons.

Facial paralysis due to a *supranuclear lesion* involving 'upper motor neuron' pathways is usually part of a hemiplegia. The movements of the lower part of the face are usually more severely affected than those of the upper part and voluntary movements are weak or absent whilst emotional and associated movements are little affected. The electrical reactions of the muscles on the affected side are not altered. Occasionally supranuclear lesions result in abolition or weakness of emotional movements with retention of voluntary movements. The dissociation in these forms of paralysis suggests that the supranuclear pathway concerned with emotional movements is distinct from that of the corticonuclear fibres which are concerned with voluntary movements.

The effects of *nuclear* or *infranuclear lesions* vary according to the point on its course at which the facial nerve is injured. If the facial nucleus or facial nerve fibres in the pons are involved, neighbouring structures are inevitably involved as well. The facial muscles are represented in cell groups in the nucleus and their degree of involvement will govern the extent of the paralysis which will be ipsilateral: otherwise the symptoms are identical with those seen in more peripheral lesions of the nerve. The associated lesions may thus include paralysis of the lateral rectus muscle of the eyeball due to involvement of the abducent nerve nucleus, around which the facial nerve fibres loop, paralysis of the muscles of mastication due to involvement of the motor nucleus of the trigeminal nerve, sensory loss on the face from implication of the principal sensory nucleus or of the nucleus of the spinal tract of the trigeminal nerve or spinothalamic tract and paralysis of the upper or lower limbs due to lesions of the corticospinal tracts. Due to the proximity of the sensory root of the facial nerve and of the vestibulocochlear nerve, lesions in the posterior cranial fossa or in the internal acoustic meatus may be accompanied by loss of taste in the anterior two-thirds of the tongue and deafness on the same side as the facial paralysis. When the facial nerve is damaged within the temporal bone the chorda tympani nerve is usually involved and in fractures of the petrous temporal bone the vestibulocochlear nerve is also usually implicated. The most common cause of facial palsy is inflammation of the nerve close to the stylomastoid foramen (Bell's paralysis). The cause of this is uncertain, but it results in oedema of the nerve and compression of its fibres in the facial canal or at the stylomastoid foramen. If the lesion is complete the facial muscles are all equally affected, and voluntary, emotional and associated movements of the face suffer equally. There is asymmetry of the face and the affected side is immobile. The eyebrow droops, the lines on the forehead and nasolabial fold are smoothed out and the palpebral fissure is wider than on the normal side due to the unopposed action of the levator palpebrae. Tears fail to enter the lacrimal puncta because they are no longer in contact with the conjunctiva, the conjunctival reflex is absent and efforts to close the eye merely cause the eyeball to roll upwards until the cornea lies under the upper lid. The ala nasi does not move properly on respiration. The lips remain in contact on the paralysed side, but cannot be pursed for whistling; when a smile is attempted the angle of the mouth is drawn up on the unaffected side but on the affected side the lips remain nearly closed, and the mouth assumes a characteristic triangular form. During mastication food accumulates in the cheek, from paralysis of the buccinator, and dribbles or is pushed out from between the paralysed lips. On protrusion the tongue seems to be thrust over towards the paralysed side, but verification of its position by reference to the incisor teeth will show that this is not really so. The platysma and the muscles of the auricle are paralysed; in severe cases the articulation of labials is impaired. The electrical reactions of the affected muscles are altered (reaction of degeneration), and the degree to which this alteration has taken place after a week or ten days gives a valuable guide to the prognosis. Most cases of Bell's palsy recover completely. For details of variations in the course of the facial nerve in the vicinity of the middle ear, and their surgical importance, consult Durcan et al. (1967). Degeneration studies (Naoris 1968) have shown that the facial nerve contains no fascicular or other pattern corresponding to its peripheral branches, a finding of some significance in suturing the nerve after injury.

The Vestibulocochlear Nerve

The vestibulocochlear nerve (**7.**62, 63, 66, 68) appears in the groove between the pons and medulla oblongata, behind the facial nerve and in front of the inferior cerebellar peduncle (**7.**55). It consists of two sets of fibres, which although differing in their principal central connexions, are both concerned in the transmission of afferent impulses from the internal ear to the brain. One set of fibres forms the vestibular nerve, or nerve of equilibration, and arises from the cells of the vestibular ganglion situated in the outer part of the internal acoustic meatus; the other set constitutes the cochlear nerve, or nerve of hearing, and takes origin from the cells of the spiral ganglion of the cochlea. Both ganglia consist of bipolar nerve cells; from each cell a central fibre passes to the brain, and a peripheral fibre to the internal ear.

The peripheral arrangement of the vestibulocochlear nerve is described on pp. 1202, 1213–1216.

THE VESTIBULAR NERVE

The fibres of the vestibular nerve enter the brain superomedial to those of the cochlear nerve. They pass backwards through the pons between the inferior cerebellar peduncle and the spinal tract of the trigeminal nerve and divide into ascending and descending branches which mostly end in the vestibular nuclei, although some proceed direct to the cerebellum along the inferior cerebellar peduncle (p. 916).

The vestibular nuclear complex comprises the following: (1) The *medial vestibular nucleus* (p. 933), which lies in the vestibular area of the floor of the fourth ventricle, crossed dorsally by the striae medullares. It is the largest subdivision and extends upwards from the medulla oblongata into the pons. (2) The *inferior vestibular nucleus* (p. 905) lies lateral to the medial nucleus and reaches to a lower level in the medulla oblongata. It is placed between the medial nucleus and the inferior cerebellar peduncle, and the descending branches of the incoming vestibular fibres are interspersed among its cells. (3) The *lateral nucleus* (p. 908) lies ventrolateral to the upper part of the medial nucleus, and it is characterized by the large size of its constituent cells. Its upper end becomes continuous with the lower end of (4) the *superior nucleus*, which extends higher into the pons than the other subdivisions and occupies the upper part of the vestibular area. A number of minor subdivisions have been described in the main vestibular nuclei in the cat (Brodal and Pompeiano 1957), and a somatotopic pattern has been tentatively suggested in the lateral vestibular nucleus (Løken and Brodal 1970).

Connexions. All the vestibular nuclei receive *radicular fibres* from the vestibular nerve and it is believed that they receive *afferent cerebellovestibular fibres* through the inferior cerebellar peduncle. These fibres are derived, for the most part, from the flocculus and nodule (posterior lobe), but others have been ascribed to the uvula, the lingula and the fastigial nucleus.

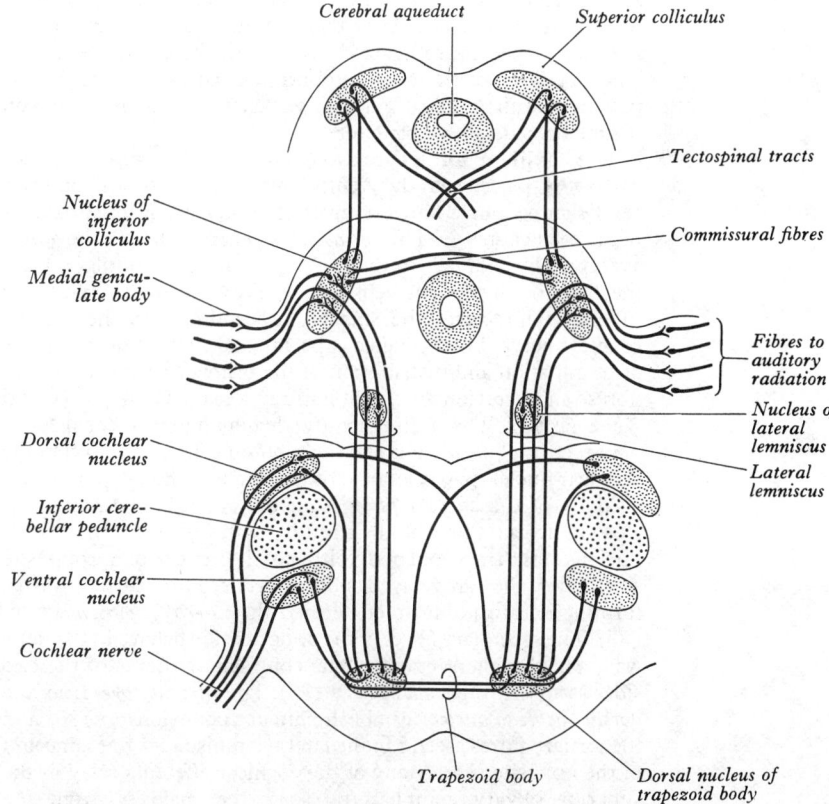

7.188 A simplified diagram to show some of the central connexions of the cochlear nerve and the auditory pathway through the brainstem. Although they are shown in the figure as being of comparable size, the trapezoid body constitutes a much more important and larger commissural bundle than the fibres connecting the nuclei of the two inferior colliculi. *Efferent* fibres in the cochlear nerve, and *descending* fibres in the auditory pathway have been omitted.

From the nuclei *efferent fibres* enter the inferior cerebellar peduncle, most of them being destined for the flocculus and the nodule, though some may pass to the uvula and the lingula, and some have been described as terminating in the fastigial nucleus. As already stated, numbers of fibres of the vestibular nerve 'bypass' the nuclei and traverse the inferior cerebellar peduncle to reach the flocculus and the nodule directly.

As a whole, the vestibular nuclear complex acts as a relay station on an afferent cerebellar pathway, and is in turn a distributing station for efferent cerebellar fibres.

In addition, fibres from the vestibular nuclei enter the medial longitudinal fasciculus (7.94), in which they ascend, or descend, to reach the motor nuclei of the eye muscles and muscles of the neck. This connexion, the *'octavo-oculomotor system'*, has attracted much attention (*see* pp. 909, 946), but much dissension exists between accounts regarding the vestibular nuclei involved, the route of their projections, the degree of decussation, and their particular destinations. Tarlov (1972, 1975) has described a detailed projection from the medial and superior vestibular nuclei to specific motor pools in the oculomotor nucleus of the cat (see also p. 1057). He has also reviewed the literature on this highly interesting topic. It is suggested that an excitatory and an inhibitory projection exists, capable of mediating the complex and subtle integration between vestibular signals and eye movements. From the large cells of the lateral vestibular nucleus, efferent fibres descend to form the vestibulospinal tract (p. 875) and fibres from the other nuclei are believed to join the lateral lemniscus and so may reach the inferior colliculus and the medial geniculate body and, probably, the cerebral cortex.

Speaking broadly, through its connexions the vestibular system is able to influence the movements of the eyes and head and the muscles of the trunk and limbs, so as to maintain equilibrium when loss of balance is threatened.

THE COCHLEAR NERVE

As it reaches the brainstem the cochlear nerve (7.62, 63, 66) is placed on the lateral side of the vestibular nerve, but the two nerves soon become separated by the inferior cerebellar peduncle. The cochlear nerve passes round the lateral aspect of the peduncle, while the vestibular nerve penetrates the brainstem on the medial side of that structure.

The cochlear nuclei are two in number. The *ventral cochlear nucleus* is placed on the ventrolateral aspect of the inferior cerebellar peduncle, and it receives the larger, ascending branches of the cochlear nerve. The *dorsal cochlear nucleus* lies on the dorsal aspect of the peduncle in the lateral part of the vestibular area of the floor of the fourth ventricle, where it forms the auditory tubercle. It receives the smaller, descending branches of the cochlear nerve. For a detailed experimental analysis of the modes of termination and distribution of the primary cochlear afferent fibres, and a review of the literature, consult Osen (1970) and Kane (1974). The former study demonstrated a consistency between the above details and the tonotopical organization in the cochlear nuclei (p. 909); both studies concern the cat.

Many of the *efferent fibres* from the *ventral cochlear nucleus* (second neuron fibres on the auditory pathway) end in the dorsal nucleus of the trapezoid body, either of the same or of the opposite side. There they are relayed and the tertiary fibres turn upwards, forming an ascending tract, termed the *lateral lemniscus* (7.70, 188). The secondary fibres of the opposite side behave in the same way, and the intersections of the contralateral fibres of the two sides form the *trapezoid body* (7.188). The *efferent fibres* from the *dorsal cochlear nucleus* establish similar connexions (7.188) and the tertiary fibres ascend in the lateral lemniscus of the same and of the opposite side. Many of the cochlear efferents relay in the superior olivary complex (p. 904), the majority from the ipsilateral cochlear nuclei. After relay these pathways join both lateral lemnisci.

Each lateral lemniscus, therefore, consists of tertiary neurons from both sides, and on its upward course to the midbrain some of these have a cell station in a small group of nerve cells, intimately related to the tract and termed the *nucleus of the lateral lemniscus*. On reaching the midbrain, some of the fibres end in the nucleus of the inferior colliculus, but others 'bypass' the nucleus and run in the inferior brachium to reach the medial geniculate body, where they are relayed to the auditory cortex (areas 41, 42 and 22). Ferraro and Minckler (1977) have estimated the fibre count of the human lateral lemniscus at about 203,000 (right side) and 185,000 (left side), these being averages from 15 brains (newborn to over 90 years).

Commissural fibres link the two auditory pathways at the level of the inferior colliculi, and auditory reflexes are mediated via the colliculi, tectospinal and tectobulbar tracts, and the medial longitudinal fasciculi (7.94, 188). An *efferent* component in the acoustic pathway has been noted already (p. 910). Its peripheral fibres are derived from nerve cells in or near the superior olivary complex (Rasmussen 1942, 1960), forming in their central course the *olivocochlear fasciculus*. These fibres terminate in relation to the hair cells of the spiral organ of the cochlea. Functionally, they are probably inhibitory in their actions (Whitfield 1967). The existence of sympathetic and possibly parasympathetic fibres in the vestibulocochlear nerve has received some support (Ross 1969).

The vestibulocochlear nerve is soft in texture; the axons are ensheathed in *glial* cells in its proximal part. After leaving the medulla oblongata it passes forwards across the posterior border of the middle cerebellar peduncle, in company with the facial nerve, from which it is partially separated by the labyrinthine artery. It then enters the internal acoustic meatus with the facial nerve. At the outer end of the meatus it receives one or two filaments from the facial nerve, and splits into its *cochlear* and *vestibular* parts, the distribution of which will be described with the anatomy of the internal ear (pp. 1202, 1213–1216).

The Glossopharyngeal Nerve

The glossopharyngeal nerve (7.55, 62, 66, 190, 191) contains motor and sensory fibres. It supplies motor fibres to the stylopharyngeus, secretomotor fibres to the parotid gland, and sensory fibres to the pharynx, the tonsil and the posterior (postsulcal) part of the tongue; it is also the nerve of taste for this part of the tongue. It emerges as three or four rootlets from the cranial part of the medulla, in the groove between the olive and inferior cerebellar peduncle above the rootlets of the vagus nerve.

The sensory nuclei receive the central processes of the unipolar nerve cells in the superior and inferior ganglia of the nerve; the fibres concerned with taste end in the *nucleus of the tractus solitarius* (p. 903) and those concerned with common sensation probably in the *nucleus of the spinal tract of the trigeminal nerve* (Brodal 1947, and *see* pp. 902, 1060).

The motor nucleus is formed by the rostral part of the *nucleus ambiguus* (pp. 905, 1077), which lies deeply in the reticular formation of the medulla oblongata. It is connected with the corticonuclear tracts of *both* sides directly and through interneurons. The corticonuclear fibres leave the tract at the level of the nucleus ambiguus or in the pons whence they descend in the medial lemniscus (aberrant pyramidal fibres, p. 905). The glossopharyngeal part of the nucleus ambiguus sends its efferent fibres to the stylopharyngeus.

In addition, parasympathetic fibres join the motor part of the glossopharyngeal nerve from a representative of the general visceral efferent column which is termed the *inferior salivatory nucleus* (*see* p. 911). This nucleus lies in the reticular formation below the superior salivatory nucleus, and sends its fibres via the tympanic branch of the glossopharyngeal nerve and the tympanic plexus (p. 1075) to the lesser petrosal nerve and the otic ganglion, where they are relayed. The postganglionic fibres pass to the auriculotemporal nerve and so reach the parotid gland (7.179) (but see also pp. 1073, 1274).

From the medulla oblongata the glossopharyngeal nerve passes forwards and laterally towards the triangular depression into which the aqueductus cochleae opens, on the inferior surface of the petrous portion of the temporal bone. It lies at first under cover of the flocculus, and rests on the jugular tubercle of the occipital bone, which is sometimes grooved by it. It leaves the skull by bending sharply downwards through the central part of

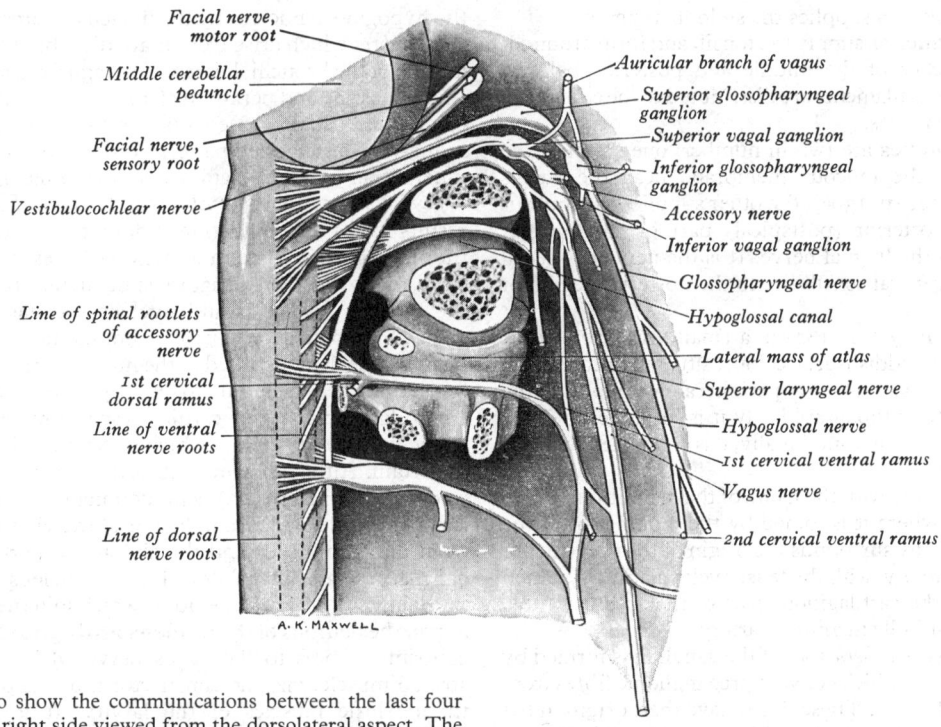

Facial nerve, motor root
Middle cerebellar peduncle
Facial nerve, sensory root
Vestibulocochlear nerve
Line of spinal rootlets of accessory nerve
1st cervical dorsal ramus
Line of ventral nerve roots
Line of dorsal nerve roots

Auricular branch of vagus
Superior glossopharyngeal ganglion
Superior vagal ganglion
Inferior glossopharyngeal ganglion
Accessory nerve
Inferior vagal ganglion
Glossopharyngeal nerve
Hypoglossal canal
Lateral mass of atlas
Superior laryngeal nerve
Hypoglossal nerve
1st cervical ventral ramus
Vagus nerve
2nd cervical ventral ramus

A. K. MAXWELL

7.189   A diagram to show the communications between the last four cranial nerves of the right side viewed from the dorsolateral aspect. The hypoglossal canal has been split in its long axis, and the transverse process of the atlas has been divided close to the lateral mass. The descending branch of the hypoglossal nerve is not shown.

the jugular foramen, anterior to the vagus and accessory nerves, and in a separate sheath of dura mater (7.189). In its transit through the jugular foramen it is lodged in a deep groove leading from the triangular depression for the cochlear aqueduct, and here it is separated by the inferior petrosal sinus from the vagus and accessory nerves. The deep groove is converted into a canal by a bridge which is usually composed of fibrous tissue, but consists of bone in about 25 per cent of skulls. After its exit from the skull it passes forwards between the internal jugular vein and internal carotid artery; it descends in front of the latter vessel, deep to the styloid process and the muscles connected with it, to reach the posterior border of the stylopharyngeus. It then curves forwards, lying upon the stylopharyngeus and either pierces the lower fibres of the superior constrictor of the pharynx or passes between the adjoining borders of the superior and middle constrictors (5.25) to be distributed to the tonsil, the mucous membrane of the pharynx and posterior part of the tongue, and the mucous glands of the mouth.

Two ganglia, named the superior and the inferior, are situated on that part of the nerve which traverses the jugular foramen (7.189).

The superior ganglion is situated in the upper part of the groove in which the nerve is lodged during its passage through the jugular foramen. It is very small, gives off no branches, and is usually regarded as a detached portion of the inferior ganglion.

The inferior ganglion is larger than the superior ganglion and is situated in a notch in the lower border of the petrous portion of the temporal bone (p. 329). Its cells are typical unipolar cells. Their peripheral branches convey taste and general sensibility from the mucous membrane of the posterior third of the tongue, including the sulcus terminalis and the vallate papillae, and general sensibility from the mucous membrane of the oropharynx, soft palate and the fauces.

The glossopharyngeal nerve communicates with the sympathetic trunk, and with the vagus and facial nerves.

The inferior ganglion is connected by a filament with the superior cervical ganglion of the sympathetic. The branches to the vagus consist of two filaments which arise from the inferior ganglion; one joins the auricular branch, and the other the superior ganglion, of the vagus. The branch to the facial arises from the trunk of the glossopharyngeal nerve below the inferior

ganglion; it perforates the posterior belly of the digastric and joins the facial nerve near the stylomastoid foramen.

The *branches of distribution* of the glossopharyngeal nerve are: tympanic, carotid, pharyngeal, muscular, tonsillar and lingual.

The tympanic nerve arises from the inferior ganglion of the glossopharyngeal nerve, and ascends to the tympanic cavity through the inferior tympanic canaliculus (p. 329). In the tympanic cavity it divides into branches which form the **tympanic plexus** and are contained in grooves upon the surface of the promontory. This plexus gives off: (1) a branch to join the greater petrosal nerve (p. 1071); (2) branches to supply the mucous membrane lining the tympanic cavity, the auditory tube and the mastoid air cells; and (3) the lesser petrosal nerve. For a critique of the final distribution and connexions of this nerve, see Winckler and Cochet (1967).

The lesser petrosal nerve contains the secretomotor fibres for the parotid gland (*vide infra*). It enters a small canal inferior to that for the tensor tympani, receives a connecting branch from the ganglion of the facial nerve and reaches the anterior surface of the temporal bone through a small opening on the lateral side of the hiatus for the greater petrosal nerve. It then passes through the foramen ovale or the canaliculus innominatus (p. 323) and joins the otic ganglion.

The carotid branch, often double, arises just below the skull, and descends on the internal carotid artery to be distributed to the wall of the carotid sinus and to the carotid body. It may communicate with the vagus (inferior ganglion or one of its branches) and with a branch from the sympathetic (superior cervical ganglion). Another branch, either from the preceding or from the main trunk, joins a fine plexus which also supplies the carotid body. The other branches to this plexus spring from the sympathetic (superior cervical ganglion) and the vagus (p. 1079). For details of the terminal distribution of the carotid nerve consult Willis and Tange (1959), and also p. 1462.

The pharyngeal branches are three or four filaments which unite, opposite the middle constrictor muscle of the pharynx, with the pharyngeal branch of the vagus nerve and the laryngopharyngeal branches of the sympathetic trunk to form the *pharyngeal plexus*; through this plexus the glossopharyngeal nerve supplies the mucous membrane of the pharynx with sensory branches.

The muscular branch supplies the stylopharyngeus.

The tonsillar branches supply the tonsil, and form around it a plexus with branches of the middle and posterior palatine nerves; from this plexus filaments are distributed to the soft palate and the region of the fauces.

The lingual branches are two in number: one supplies the vallate papillae and the mucous membrane near the sulcus terminalis of the tongue (p. 1306); the other supplies the mucous membrane of the posterior (postsulcal) part of the tongue, communicating with the lingual nerve. It is the nerve of special sense (taste) and of general sensibility to the posterior region of the tongue.

The otic ganglion (7.181, 184) is a small, oval, somewhat flattened ganglion of a reddish-grey colour, situated immediately below the foramen ovale. It is a peripheral ganglion of the parasympathetic system; topographically it is intimately related to the mandibular nerve but, functionally, it is connected with the glossopharyngeal nerve.

It is in relation *laterally* with the trunk of the mandibular nerve at or near the point where it is joined by the motor root of the trigeminal, and it usually surrounds the origin of the nerve to the medial pterygoid; *medially*, with the tensor veli palatini, by which it is separated from the cartilaginous part of the auditory tube; *posteriorly*, with the middle meningeal artery.

The *motor* or *parasympathetic root* of the ganglion is formed by the lesser petrosal nerve, which conveys preganglionic fibres from the glossopharyngeal nerve. These fibres have their origin in the cells of the inferior salivatory nucleus. They are relayed in the otic ganglion and the postganglionic fibres pass by a *communicating branch* to the auriculotemporal nerve. By it they are conveyed to the parotid gland (7.179), to which they supply secretomotor fibres. The *sympathetic root* is derived from the plexus on the middle meningeal artery. It contains postganglionic fibres which arise in the superior cervical ganglion and pass through the otic ganglion without being interrupted. Emerging with the parasympathetic fibres in the communicating branch to the auriculotemporal nerve, they are destined for the supply of the blood vessels of the parotid gland.

Branches. A twig connects the ganglion with the chorda tympani nerve and another ascends from it to join the nerve of the pterygoid canal. According to some neurologists these form an additional pathway by which taste fibres from the anterior, pre-sulcal area of the tongue may reach the facial ganglion without passing through the middle ear (p. 1140). The fibres concerned pass through the otic ganglion without being interrupted. Motor branches are supplied to the tensor veli palatini and the tensor tympani, they are derived from the nerve to the medial pterygoid (p. 1076) and have no synaptic relations with the cells of the ganglion.

The glossopharyngeal nerve is the nerve of the third branchial arch, or it would be more nearly correct to describe it as the post-trematic branch of that arch. The pretrematic branch of the second (hyoid) arch is probably the tympanic branch of the glossopharyngeal nerve, but that is uncertain. Like the trigeminal and the facial nerves, the glossopharyngeal corresponds to a dorsal nerve which has acquired special visceral efferent fibres.

The Vagus Nerve

The vagus nerve (7.55, 62–66, 189–191) is composed of motor and sensory fibres, and has a more extensive course and distribution than any of the other cranial nerves, since it passes through the neck and thorax to the abdomen. It is attached by eight or ten rootlets to the medulla oblongata, below the glossopharyngeal nerve, in the groove between the olive and the inferior cerebellar peduncle.

The fibres of the vagus nerve are connected to four nuclei in the medulla oblongata. (1) **The dorsal nucleus** of the vagus is usually described as a mixed nucleus representing the fused general visceral efferent and general visceral afferent columns. It lies in the central grey matter of the lower, closed part of the medulla oblongata, and extends upwards into the upper, open part, where it is placed under the vagal triangle, separated from

the hypoglossal nucleus by the nucleus intercalatus (p. 903). The *motor fibres* which arise from it are distributed to the nonstriated muscle of the bronchi, heart, oesophagus and stomach, and to the small intestine and part of the large intestine (p. 1123). A number of studies, usually based upon retrograde degeneration techniques, have suggested patterns of localization of visceral motor innervation (*see*, for example, Szentágothai 1952; Mitchell and Warwick 1955; Lewis *et al.* 1970; Rao and Sahu 1974); but the results have been conflicting and, moreover, there is doubt as to the origin of vagal visceromotor nerve fibres. Some observers (e.g. Kerr 1969) considered that most of such fibres are derived from the nucleus ambiguus and nucleus retrofacialis (Szentágothai 1952). Sugimoto *et al.* (1979) claim that neurons innervating the heart can be demonstrated in the nucleus ambiguus (and reticular formation ventrolateral to it) by injection of horseradish peroxidase into right cardiac vagal branches in cats. The particular *sensory fibres* which terminate in the nucleus are uncertain. Although some authorities regard the nucleus of the tractus solitarius (p. 903) as predominantly a vagal nucleus, there is considerable evidence in favour of the view that afferent fibres from the oesophagus and the abdominal part of the alimentary canal terminate in the dorsal vagal nucleus (see also p. 905). (2) Below the origin of the fibres which join the glossopharyngeal nerve, the neurons of the **nucleus ambiguus** (pp. 905 and 1074) contribute fibres to the vagus nerve which are distributed to striped muscle, viz. the constrictor muscles of the pharynx and the intrinsic muscles of the larynx. It is connected to the corticonuclear tracts of *both* sides and to many other brainstem centres (p. 1085). There have been a number of detailed studies on the architecture and regional localization which occurs in the nucleus ambiguus of man and following experimental analysis in a number of mammals (Szentágothai 1943; Getz and Sirnes 1949; Szabo and Dussardier 1964; Lawn 1966). As in the case of many of the other cranial nerve motor nuclei, the nucleus ambiguus can be divided into a number of sub-nuclei. The glossopharyngeal fibres arise from a cranial group of cells, whilst the individual laryngeal muscles are innervated by relatively discrete sub-nuclei in more caudal zones; most caudally the cells of the nucleus ambiguus send axons into the cranial part of the accessory nerve (p. 1082). (3) The lower part of the **nucleus of the tractus solitarius** (pp. 903 and 1071) receives those fibres of the vagus which are distributed through the internal laryngeal nerve to the taste buds of the epiglottis and the vallecula. The middle part of the nucleus receives the visceral afferent fibres from the tongue, tonsil, palate and pharynx (glossopharyngeal nerve). The upper part of the nucleus receives the taste fibres from the anterior two-thirds of the tongue and from the soft palate (facial nerve). (4) The vagus contains somatic afferent nerve fibres, but it is believed that when they enter the medulla oblongata they terminate in the **spinal nucleus of the trigeminal nerve**.

The rootlets of the nerve unite, and form a flat cord which passes below the flocculus of the cerebellum to the jugular foramen, through which it leaves the cranium. In emerging through this opening, the vagus nerve is accompanied by and contained in the same sheath of dura and arachnoid mater as the accessory nerve, a fibrous septum separating them from the glossopharyngeal nerve, which lies in front (7.189). In this situation the vagus nerve presents a well-marked enlargement, named the *superior ganglion*. After its exit from the jugular foramen the vagus nerve enlarges into a second swelling, named the *inferior ganglion*.

The superior ganglion is of a greyish colour, spherical in form, about 4 mm in diameter. It is joined by one or two delicate filaments with the cranial root of the accessory nerve; it is connected by a twig with the inferior ganglion of the glossopharyngeal nerve, and with the sympathetic trunk by a filament from the superior cervical ganglion; the auricular branch of the ganglion gives off an ascending twig which joins the facial nerve (p. 1071).

The inferior ganglion is cylindrical in form, of a reddish colour, and 2·5 cm long. It is connected with the hypoglossal nerve, the superior cervical ganglion of the sympathetic trunk, and the loop between the first and second cervical nerves. The cranial root of the accessory nerve passes over the ganglion, but is

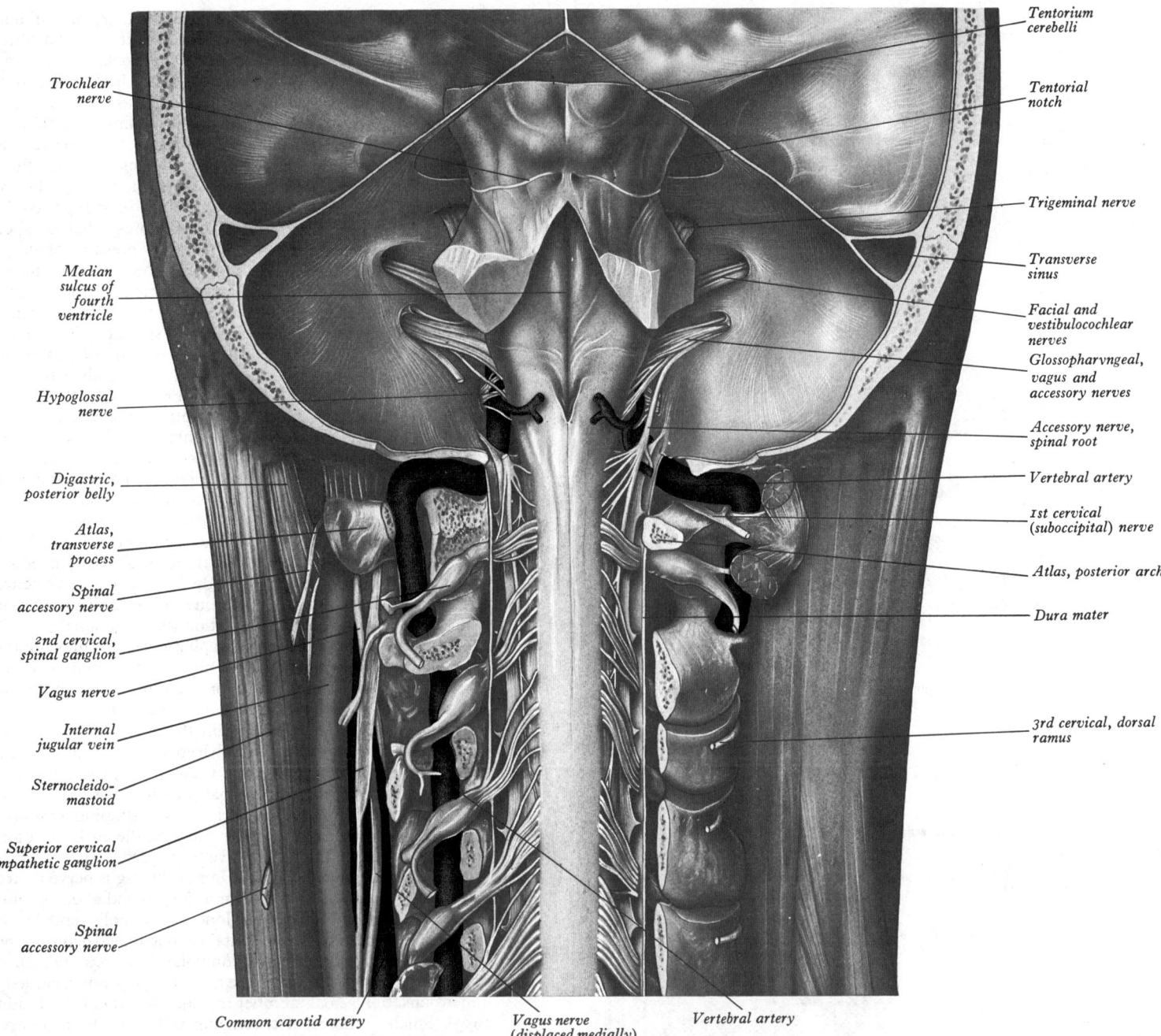

Trochlear nerve

Median sulcus of fourth ventricle

Hypoglossal nerve

Digastric, posterior belly

Atlas, transverse process

Spinal accessory nerve

2nd cervical, spinal ganglion

Vagus nerve

Internal jugular vein

Sternocleido-mastoid

Superior cervical sympathetic ganglion

Spinal accessory nerve

Tentorium cerebelli

Tentorial notch

Trigeminal nerve

Transverse sinus

Facial and vestibulocochlear nerves

Glossopharyngeal, vagus and accessory nerves

Accessory nerve, spinal root

Vertebral artery

1st cervical (suboccipital) nerve

Atlas, posterior arch

Dura mater

3rd cervical, dorsal ramus

Common carotid artery　　　*Vagus nerve (displaced medially)*　　　*Vertebral artery*

7.190　A dissection exposing the brainstem and the upper part of the spinal cord after removal of large portions of the occipital and parietal bones and the cerebellum, together with the roof of the fourth ventricle. On the left side the foramina transversaria of the atlas and the third, fourth and fifth cervical vertebrae have been opened to expose the vertebral artery. On the right side the posterior arch of the atlas and the laminae of the succeeding cervical vertebrae have been divided and have been removed together with the vertebral spines and the laminae of the opposite side. The tentorium cerebelli and the transverse sinuses have been divided and their posterior portions removed.

attached to it by fibrous tissue only.

Beyond the inferior ganglion the cranial root of the accessory nerve blends with the vagus nerve; its fibres are distributed principally to the pharyngeal and recurrent laryngeal branches of the vagus nerve.

The cells of both ganglia are unipolar sensory neurons (Richardson and Hinsey 1933; Evans and Murray 1954). The only evidence to the contrary are some old and unconfirmed electrophysiological experiments and technically unsatisfactory histological observations, both of which have been subsequently refuted. (For details and discussion *see* Mitchell 1956; Lieberman 1968, 1969, 1974.) There is no reliable evidence that any significant number of the nerve cells in the ganglion are motor, nor have synapses between pre- and post-ganglionic neurons been demonstrated in it. Indeed, the preganglionic motor fibres, derived from the dorsal nucleus of the vagus, and the special visceral efferents derived from the nucleus ambiguus, which descend to the inferior vagal ganglion, commonly form a band visible to the unaided eye which skirts the ganglion in some mammals (Hoffman and Kuntz 1957; Mei and Dussardier 1966) without passing through it. The disposition of the sensory or motor fibres which do traverse the ganglion is like that of a dorsal root ganglion. It is necessary to emphasize these facts, because there is current in some textbooks an unsupported view that the parasympathetic component of the vagus relays in its inferior ganglion, based with questionable logic upon the presence of large numbers of thinly myelinated and supposedly non-myelinated fibres in the distal parts of the nerve. Even if it were a safe presumption to regard autonomic fibres as postganglionic purely on the basis of their degree of myelination, to do so in this instance would be merely to ignore another tenet of peripheral autonomic morphology, for such a view produces two relays—one in the

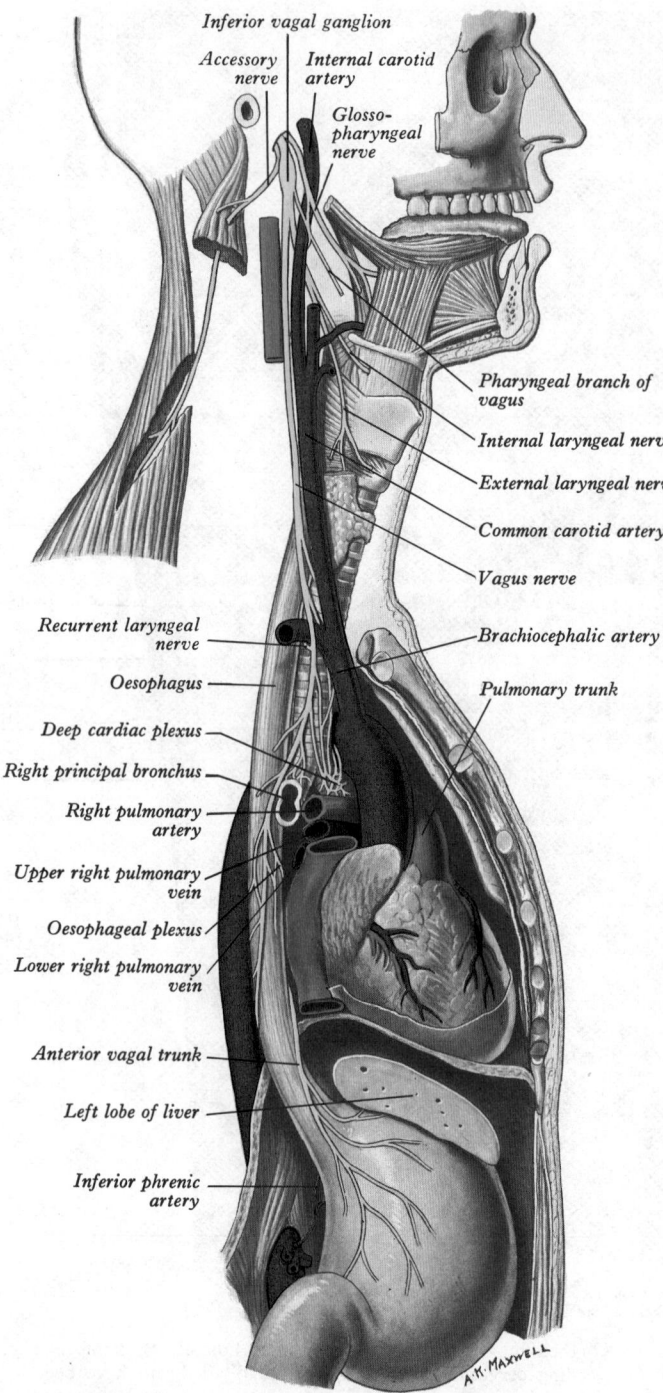

Inferior vagal ganglion

Accessory nerve

Internal carotid artery

Glosso-pharyngeal nerve

Pharyngeal branch of vagus

Internal laryngeal nerve

External laryngeal nerve

Common carotid artery

Vagus nerve

Recurrent laryngeal nerve

Oesophagus

Deep cardiac plexus

Right principal bronchus

Right pulmonary artery

Upper right pulmonary vein

Oesophageal plexus

Lower right pulmonary vein

Anterior vagal trunk

Left lobe of liver

Inferior phrenic artery

Brachiocephalic artery

Pulmonary trunk

A.K. MAXWELL

7.191 The course and distribution of the glossopharyngeal, vagus and accessory nerves.

perhaps some sympathetic nerve fibres. Their cellular populations are unipolar and bipolar (Cajal 1911; Evans and Murray 1954; Gabella 1976). The occurrence of multipolar neurons in the inferior (nodose) ganglion has been suggested in the past, but this has been strongly denied by Mitchell (1956) and Lieberman (1968). Since such motor cells would almost certainly be parasympathetic postganglionic elements, their presence would imply double peripheral relay, which seems unlikely; further investigation is clearly required on this point. Synapses between pre- and post-ganglionic neurons have not been reported. The general consensus is that all efferent fibres whether special visceral (branchial) or parasympathetic preganglionic, pass uninterruptedly through both ganglia. (However, see comments on chromaffin cells below.) The arrangement of the somata of neurons and of nerve fibres resembles that of other sensory ganglia, the fibres being centrally disposed. The superior ganglion has been described as grey, the inferior as pink, a difference which may reflect variation in the distribution of myelinated and non-myelinated fibres or of vessels.

The superior (jugular) ganglion of the vagus nerve is rounded and about 4 mm across in mankind. In the cat (Foley and DuBois 1937) it contains about 8,700 unipolar neurons, 73 per cent of which appear to form the auricular nerve, a branch of the ganglion (p. 1079), while 15 per cent contribute axons to the vagus nerve itself. Similar counts are apparently not available for the human ganglion; but since the human vagus nerve contains, at the mid-cervical level, about 16,500 (Rt.) and 20,000 (Lt.) myelinated fibres, counted in 17 paired nerves, the human ganglionic neurons must be more numerous (Schmitzlein et al. 1958). The ganglion is connected by nerves of communication with the cranial root of the accessory nerve, the inferior glossopharyngeal ganglion, and the superior cervical sympathetic ganglion. The significance of all these connexions is not yet clearly established, but the first probably contains motor fibres from the nucleus ambiguus which take an aberrant course, issuing in the accessory nerve, to be finally distributed to the palatal, pharyngeal, laryngeal and upper oesophageal musculature via the vagus nerve; the sympathetic connexion may be similar to that which exists between this sympathetic ganglion and the *inferior* vagal ganglion (*vide infra*). Shortly after leaving the superior ganglion the auricular branch communicates with the facial nerve (p. 1072).

The inferior (nodose) ganglion of the vagus nerve is larger and elongate, being about 25 mm in length and about 5 mm in maximum breadth. In the cat (Jones 1937; Foley and DuBois 1937) it has been estimated to contain about 30,000 neurons, most of which are unipolar or pseudo-unipolar, but a few are fusiform or bipolar. The majority are in the range of 35–40 μm in perikaryal dimensions, but a small number measure 20–30 μm (Mohiuddin 1953). Small numbers of chromaffin cells have been observed (White 1935; Jacobs and Comroe 1951; Grillo et al. 1974), and these receive agranular synapses but display no efferent synapses. The ganglion is connected with the hypoglossal nerve (p. 1083), the loop between the first and second cervical anterior primary rami (p. 1091), and with the superior cervical sympathetic ganglion (p. 1027). The latter communication may be formed by the peripheral axons of a small number of sensory neurons, resembling nodose ganglion cells, which have been described in this sympathetic ganglion by Terni (1922), as a group of displaced nodose ganglion nerve cells.

While the cells of the vagal ganglia have received much less attention than those of the superior cervical sympathetic ganglion, an extensive series of observations has been contributed in regard to nerve fibre counts of the vagus nerve and its branches in the vicinity of its ganglia (see Gabella 1976).

While overall counts by different workers show general agreement, there are wide discrepancies between estimates of ratios between myelinated and non-myelinated fibres and between fibres of different diameters. There are a large number of thinly myelinated or non-myelinated fibres; and while most of these are pre-ganglionic parasympathetic, a few may be postganglionic sympathetic fibres. Although the cat's superior vagal ganglion contains, according to DuBois and Foley (1937), about 8,700 nerve cells, about 8,800 fibres (chiefly small and non-myelinated) pass into the auricular nerve and 1,700 to 3,500

inferior ganglion, and one in the wall of the target organ. (However, see p. 1124 for alternative modern views in relation to *sympathetic* ganglia.) Retrograde degeneration has been demonstrated in the dorsal nucleus of the vagus by division of vagal branches as remote from their origins as the gastric nerves (Mitchell and Warwick 1955). Reliable counts and analyses of the nerve fibres in the human vagus are not available, but counts in other mammals do not support the view of the inferior ganglion as a relay station, which seems to be based almost entirely upon a misguided adherence to supposed axioms of autonomic morphology. The extensive factual information quoted above with regard to the structure of the inferior vagal ganglion renders such a view untenable, and it is difficult to explain its persistence.

Both vagal ganglia are, as noted, exclusively sensory, containing somatic, special visceral, and general visceral afferent neurons. They are, however, traversed by parasympathetic and

continue into the vagus nerve caudal to the superior ganglion. Many discrepancies and uncertainties of this kind still await clarification, and it would be unprofitable at present to quote further quantitative data.

There is wide agreement that the superior ganglion is chiefly somatic in status, most of its neurons contributing to the auricular nerve. Correspondingly, the neurons in the inferior ganglion are concerned with visceral sensation in connexion with the heart, larynx and lungs, and the alimentary tract from the pharynx to the transverse colon; but some of the neurons transmit impulses from taste endings in the vallecula and epiglottis.

The vagus nerve passes vertically down the neck within the carotid sheath, lying between the internal jugular vein and internal carotid artery as far as the upper border of the thyroid cartilage, and then between the same vein and the common carotid artery until it reaches the root of the neck. The further course of the nerve differs on the two sides of the body.

On the *right side* the vagus nerve continues descending posterior to the internal jugular vein and crosses the first part of the subclavian artery. It enters the thorax and descends through the superior mediastinum, lying at first behind the right brachiocephalic vein, and then to the right of the trachea and posteromedial to the right brachiocephalic vein and the superior vena cava. The right pleura and lung are lateral to the nerve above, but are separated from it below by the azygos vein, which arches forward above the root of the right lung (**6**.133).

The nerve next passes behind the right principal bronchus to reach the posterior aspect of the root of the right lung, and there breaks up into posterior pulmonary (or bronchial) branches, which unite with filaments from the second to fifth or sixth thoracic sympathetic ganglia to form the **right posterior pulmonary plexus**. From the caudal part of this plexus two or three branches descend on the dorsal aspect of the oesophagus, where, with a branch from the left vagus, they form the **posterior oesophageal plexus**; from this a trunk is re-formed which is continued posterior to the oesophagus to enter the abdomen through the oesophageal opening in the diaphragm. This posterior vagal trunk contains fibres from both vagus nerves (*see* p. 1081).

(Note that the origin, disposition and distribution of the **cardiac plexuses** are described on pp. 1123, 1127, 1133.)

In the abdomen the **posterior vagal trunk** divides into a small gastric and a large coeliac branch. The gastric branch supplies the postero-inferior surface of the stomach with the exception of the pyloric canal. The coeliac branch ends chiefly in the coeliac plexus, but sends twigs to the splenic, hepatic, renal, suprarenal and superior mesenteric plexuses. (See also p. 1133.)

On the *left side* the vagus enters the thorax between the left common carotid and left subclavian arteries, and behind the left brachiocephalic vein. It descends through the superior mediastinum, crosses the left side of the aortic arch to pass behind the root of the left lung. Just above the aortic arch the nerve is crossed anterolaterally by the left phrenic nerve, and on the arch by the left superior intercostal vein (**7**.228).

Behind the root of the left lung it divides into posterior pulmonary (or bronchial) branches, which unite with filaments of the second, third and fourth thoracic sympathetic ganglia and form the **left posterior pulmonary plexus**. From this plexus two branches descend on the front of the oesophagus where, with a twig from the right posterior pulmonary plexus, they form the **anterior oesophageal plexus**; from this plexus a trunk, containing fibres from both vagus nerves, is continued in front of the oesophagus, and enters the abdomen through the oesophageal opening of the diaphragm (see p. 549).

In the abdomen the **anterior vagal trunk** supplies twigs to the cardiac antrum, and then divides into right and left groups of branches. The fibres of the left group follow the lesser curvature of the stomach and supply the anterosuperior surface of this viscus. The right group consists of three main branches. The first, which may be duplicated, proceeds between the layers of the lesser omentum towards the porta hepatis, and divides into (*a*) upper branches which enter the porta hepatis, and (*b*) lower rami which supply chiefly the pyloric canal, the pylorus, the superior and the descending parts of the duodenum, and the head of the

pancreas. The second branch is distributed to the anterosuperior surface of the body of the stomach; the third branch follows the lesser curvature of the stomach as far as the angular notch. (See also p. 1340.)

The Branches of the Vagus Nerve

In the jugular fossa	Meningeal
	Auricular
In the neck	Pharyngeal
	Branches to carotid body
	Superior laryngeal
	Recurrent laryngeal (right)
	Cardiac
In the thorax	Cardiac
	Recurrent laryngeal (left)
	Pulmonary
	Oesophageal
In the abdomen	Gastric
	Coeliac
	Hepatic

The meningeal branch or branches appear to spring from the superior ganglion of the vagus nerve and are distributed to the dura mater in the posterior fossa of the skull. However, evidence has been presented (Kimmel 1961 *a* and *b*) which suggests that such meningeal branches are in fact recurrent sensory and sympathetic nerves derived from the upper cervical spinal nerves and the superior cervical sympathetic ganglion, which for a short distance run within the sheath of the upper part of the vagus nerve (pp. 1048, 1089).

The auricular branch arises from the superior ganglion of the vagus nerve, and is joined soon after its origin by a filament from the inferior ganglion of the glossopharyngeal; it passes behind the internal jugular vein, and enters the mastoid canaliculus on the lateral wall of the jugular fossa. Traversing the substance of the temporal bone, it crosses the canal for the facial nerve about 4 mm above the stylomastoid foramen, and here it gives off an ascending branch which joins the facial nerve. (At this point fibres of the nervus intermedius may pass to the auricular branch of the vagus, providing a possible explanation of the cutaneous vesiculation which sometimes accompanies geniculate herpes.) The auricular branch then passes through the tympanomastoid fissure, and divides into two rami; one joins the posterior auricular nerve, the other is distributed to the skin of part of the cranial surface of the auricle and to the posterior wall and floor of the external acoustic meatus and to the adjoining part of the outer surface of the tympanic membrane. The auricular branch of the vagus thus contains *somatic afferent* nerve fibres, but it is believed that when they enter the medulla oblongata they terminate in the spinal nucleus of the trigeminal nerve.

The pharyngeal branch, which is the principal motor nerve of the pharynx, emerges superficially from the upper part of the inferior ganglion of the vagus nerve, and consists principally of filaments from the cranial root of the accessory nerve. It passes between the external and internal carotid arteries to the upper border of the middle constrictor muscle of the pharynx, where it divides into numerous filaments which join with branches from the sympathetic trunk, the glossopharyngeal and external laryngeal nerves, to form the *pharyngeal plexus*. Through this plexus vagal fibres are distributed to the muscles of the pharynx, and the muscles of the soft palate, except the tensor veli palatini. A minute filament joins the hypoglossal nerve as the latter winds round the occipital artery, and is often termed the *ramus lingualis vagi*.

The branches to the carotid body (p. 1462) are variable in number. They may spring from the inferior ganglion or they may travel either in the pharyngeal branch or the superior laryngeal nerve, the latter being most unusual. They form a plexus with rami of the glossopharyngeal nerve and of the cervical part of the sympathetic trunk (*see* Sheehan *et al.* 1941; also pp. 1075, 1128).

The superior laryngeal nerve, which is larger than the pharyngeal branch, issues from the middle of the inferior ganglion of the vagus nerve, and in its course receives a branch from the superior cervical ganglion of the sympathetic trunk. It descends, by the side of the pharynx, first posterior, then medial to the internal carotid artery, and divides into the internal and external laryngeal nerves. The human nerve contains about 15,000 fibres (Ogura and Bello 1952).

The *internal laryngeal nerve* is sensory to the mucous membrane of the larynx as far down as the level of the vocal folds. It also carries afferent fibres from neuromuscular spindles and other stretch receptors in the larynx (Keene 1961; Schever 1964). It descends to the thyrohyoid membrane, pierces this membrane at a higher level than the superior laryngeal artery, and divides into an upper and a lower branch. The upper branch is directed horizontally, and supplies twigs to the mucous membrane of the pharynx, the epiglottis, the vallecula and the vestibule of the larynx. The lower branch descends in the medial wall of the piriform recess, and gives branches to the aryepiglottic fold, and to the mucous membrane on the back of the arytenoid cartilage. It also supplies one or two branches to the arytenoideus, and these branches unite with twigs from the recurrent laryngeal nerve to the same muscle (*see* p. 1239). The internal laryngeal nerve ends by piercing the inferior constrictor muscle of the pharynx, and unites with an ascending branch from the recurrent laryngeal nerve. A number of observers (*see* Latarjet 1947, for citations) have, in the past, described small ganglia on the courses of the superior and internal laryngeal nerves. More recently Ramaswamy (1974) has revived this description in regard to the latter nerve, claiming to have observed an internal laryngeal nerve ganglion in more than 100 human larynges. Cells in the ganglia were described, but without photographic confirmation. Further observations are to be awaited with interest.

The *external laryngeal nerve*, which is the smaller of the two,

descends posterior to the sternothyroid in company with the superior thyroid artery but on a deeper plane; it lies at first on the inferior constrictor of the pharynx, and then, piercing that muscle, winds closely round the inferior thyroid tubercle and reaches the cricothyroid, which it supplies. It gives branches also to the pharyngeal plexus and to the inferior constrictor; behind the common carotid artery it communicates with the superior cardiac nerve, and with the superior cervical sympathetic ganglion (Skórnicki *et al.* 1968).

The recurrent laryngeal nerve differs, as to its origin and course, on the two sides of the body. On the *right* side it arises from the vagus nerve in front of the first part of the subclavian artery; it winds from before backwards, first below and then behind that vessel, and ascends obliquely to the side of the trachea behind the common carotid artery. Near the lower pole of the lobe of the thyroid gland the nerve is always intimately related to the inferior thyroid artery; it may cross either in front of or behind the vessel, or may pass between its branches. On the *left* side, it arises from the vagus nerve on the left of the arch of the aorta, and winds below the arch immediately behind the attachment of the ligamentum arteriosum to the concavity of the arch, and then ascends to the side of the trachea. The nerve on each side ascends in or near the groove between the trachea and oesophagus, and is intimately related to the medial surface of the thyroid gland before it passes under the lower border of the inferior constrictor and enters the larynx behind the articulation of the inferior cornu of the thyroid with the cricoid cartilage. It gives branches to all the muscles of the larynx, excepting the cricothyroid; it communicates with the internal laryngeal nerve, and supplies sensory filaments to the mucous membrane of the larynx below the level of the vocal

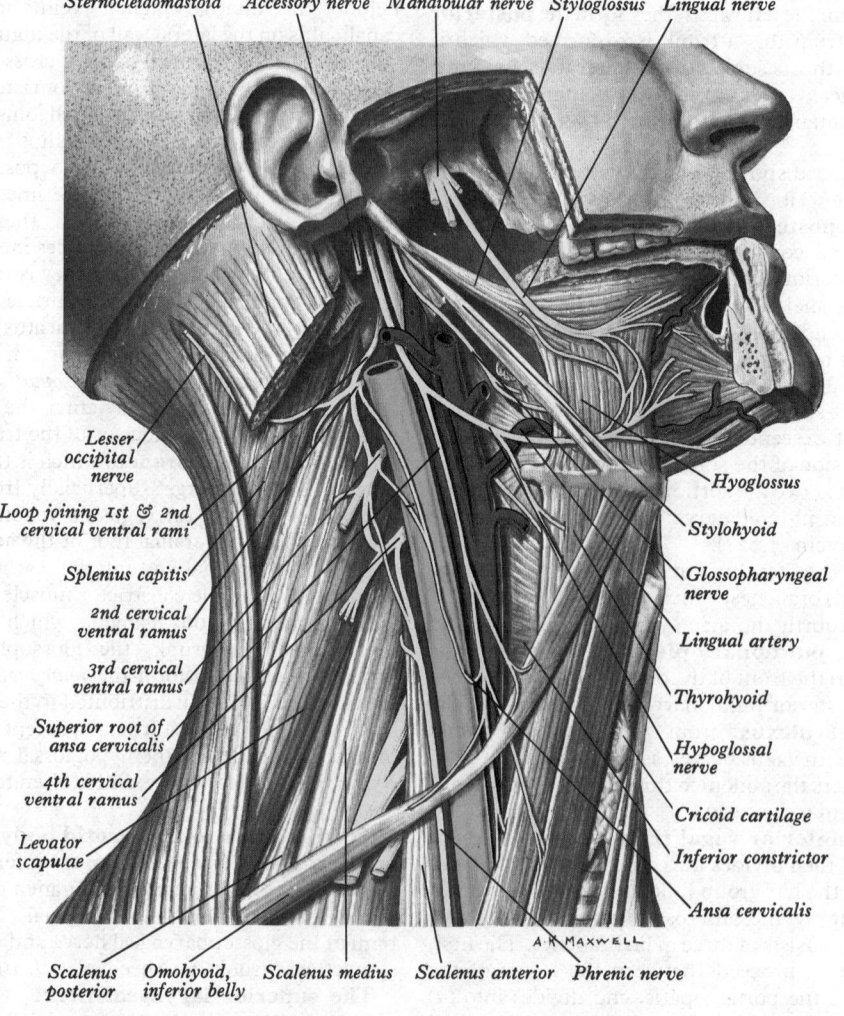

Sternocleidomastoid   Accessory nerve   Mandibular nerve   Styloglossus   Lingual nerve

Lesser occipital nerve

Loop joining 1st & 2nd cervical ventral rami

Splenius capitis

2nd cervical ventral ramus

3rd cervical ventral ramus

Superior root of ansa cervicalis

4th cervical ventral ramus

Levator scapulae

Hyoglossus

Stylohyoid

Glossopharyngeal nerve

Lingual artery

Thyrohyoid

Hypoglossal nerve

Cricoid cartilage

Inferior constrictor

Ansa cervicalis

Scalenus posterior   Omohyoid, inferior belly   Scalenus medius   Scalenus anterior   Phrenic nerve

A·K·MAXWELL

7.192 A dissection to show the general distribution of the right hypoglossal and lingual nerves, and the position and constitution of some parts of the cervical plexus of the right side.

folds. It also carries afferent fibres from stretch receptors in the larynx.

As the recurrent laryngeal nerve curves round the subclavian artery, or the arch of the aorta, it gives several cardiac filaments to the deep part of the cardiac plexus. As it ascends in the neck it gives branches, more numerous on the left than on the right side, to the mucous membrane and muscular coat of the oesophagus; branches to the mucous membrane and muscular fibres of the trachea and some filaments to the inferior constrictor.

The variations in the relations of the recurrent laryngeal nerves as they approach the larynx are especially important in the surgery of the thyroid gland (Bowden 1955; Doyle et al. 1967). The nerve does not always lie in a protected position in the tracheo-oesophageal groove, but may lie a little in front of it (slightly more frequently on the right side of the neck) and it may occasionally be some distance lateral to the trachea at the level of the lower part of the lobe of the thyroid gland. On the right side there are almost equal chances of finding the nerve anterior to, or posterior to, or intermingled with the terminal branches of the inferior thyroid artery, while on the left side the nerve is most likely to be posterior to the artery and least likely to be anterior. The nerve may give off extralaryngeal branches which are distributed to the larynx, arising from the nerve before it passes behind the inferior cornu of the thyroid cartilage. In addition to its true capsule, the thyroid gland is invested with a distinct outer covering formed by the pretracheal fascia (p. 537), which splits into two layers at the posterior border of the lobe of the thyroid gland. One layer clothes the entire medial surface of the lobe and, at and just above the level of the isthmus of the gland, it has a conspicuous thickening, called the *lateral ligament* of the *thyroid gland* (p. 1449), which attaches the gland to the trachea and the lower part of the cricoid cartilage. The other layer is more posterior; it passes behind the oesophagus and pharynx and is attached to the prevertebral fascia. By this splitting of the fascia, a fascial compartment is formed on each side of the neck, just lateral to the trachea and oesophagus, and it is in the fat that occupies this space that the recurrent laryngeal nerve and the terminal parts of the inferior thyroid artery lie. The nerve may lie lateral or medial to the lateral ligament of the thyroid gland; sometimes it may be embedded in the ligament.

The cardiac branches, two or three in number, arise from the vagus nerve at superior and inferior cervical levels. The small, *superior branches* join with the cardiac branches of the sympathetic trunk. They can be traced to the deep part of the cardiac plexus.

The *inferior branches* arise at the root of the neck. That from the right vagus passes in front or by the side of the brachiocephalic artery, and proceeds to the deep part of the cardiac plexus; that from the left runs down across the left side of the arch of the aorta, and joins the superficial part of the cardiac plexus.

Additional cardiac branches arise from the trunk of the right vagus nerve as it lies by the side of the trachea, and from both recurrent laryngeal nerves. They end in the deep part of the cardiac plexus. The cardiac plexus is described on p. 1132.

The anterior pulmonary branches, two or three in number and of small size, are distributed on the anterior surface of the root of the lung. They join with filaments from the sympathetic, and form the *anterior pulmonary plexus*.

The posterior pulmonary branches, more numerous and larger than the anterior, are distributed on the posterior surface of the root of the lung; they are joined by filaments from the second to fifth or sixth thoracic ganglia of the sympathetic trunk, and form the *posterior pulmonary plexus*. Branches from this plexus accompany the ramifications of the bronchi and supply their constrictor muscles and other pulmonary tissues. (See also pp. 1133, 1251, 1267.)

The oesophageal branches are given off both above and below the pulmonary branches; the lower are more numerous and larger than the upper. They form, as already described (p. 1079), the *oesophageal plexus*. From this plexus filaments are distributed to the oesophagus and to the back of the pericardium.

The gastric branches are distributed to the stomach, the anterosuperior surface of which is mainly supplied by the left vagus, and the postero-inferior surface mainly by the right. (*See* pp. 1134, 1340.) Brizzi et al. (1973) investigated the detailed distribution of the gastric nerves, and of the branches of the autonomic nervous system to the stomach. They described separate 'cardiac' branches of the vagus supplying the cardia; but while both gastric branches did supply the fundus and body, the pylorus received a complex innervation mainly from the anterior gastric nerve and the hepatic branches of the vagus, and less frequently from the posterior gastric nerve.

The coeliac branches are derived from the posterior vagal trunks: they join the coeliac plexus.

The hepatic branches arise from both vagus nerves (p. 1134): they join the hepatic plexus and through it are conveyed to the liver.

The renal branches arise from both vagus nerves and join the renal plexus (p. 1134).

For details of the ultimate distribution of the branches of the vagus nerves in the abdomen, see the section on the Autonomic Nervous System, pp. 1123, 1133, 1134.

Applied Anatomy. The trunk of the vagus is not commonly injured, but the functions of the nerve may also be interfered with by damage to its radicular nuclei in the medulla, or during its intracranial course. The symptoms produced by non-functioning of the nerve are palpitation, with increased frequency of the pulse, constant vomiting, slowing of the respiration, and a sensation of suffocation.

'Reflexes' in connexion with the branches of the vagus are not infrequent. The 'ear cough' is perhaps one of the commonest, where a plug of wax in the external acoustic meatus may, by irritating the filaments of the auricular nerve, be responsible for a persistent cough. Syringing the meatus frequently produces cough, and, in children, vomiting is not uncommon. Moreover, syringing of the ear has occasionally been responsible for a sudden fatal reflex cardiac inhibition. Another common example is the persistent cough due to enlarged bronchial lymph nodes in children, which may irritate the recurrent laryngeal nerve.

The anatomy of the laryngeal nerves may also be correlated with some of the morbid conditions of the larynx. When the peripheral terminations of the superior laryngeal nerve are irritated by a foreign body passing over them reflex spasm of the glottis is the result. When the nerve is not functioning there is anaesthesia of the mucous membrane of the upper part of the larynx, so that foreign bodies can readily enter the cavity; since the nerve supplies the cricothyroid, the vocal folds cannot be made tense, and the voice is deep and hoarse. Irritation of the recurrent laryngeal nerves produces spasm of the muscles of the larynx. When both recurrent laryngeal nerves are interrupted, the vocal folds are motionless, in the so-called 'cadaveric position'— in the position in which they also are found in ordinary tranquil respiration—neither closed as in phonation, nor widely open as in deep inspiratory efforts. When one recurrent laryngeal nerve is affected, the vocal fold of the same side is motionless, while the opposite one crosses the median plane to accommodate itself to the affected one; hence phonation is possible, but the voice is altered and weak in timbre. It is generally maintained that in progressive lesions of the recurrent laryngeal nerve the movements of abduction of the vocal cord are abolished before the movements of adduction; conversely, during recovery the movements of adduction are regained before those of abduction (*Semon's law*). There is no direct evidence, however, that the nerve fibres to the abductor muscles have any special grouping in the trunk of the recurrent laryngeal nerve.

The Accessory Nerve

The accessory nerve (7.62, 64, 189–191) is formed by the union of cranial and spinal roots, but these constituent parts are associated with each other only for a very short part of their course before the cranial part joins the vagus to be distributed through its branches. The cranial moiety should be considered as a part of the vagus nerve; it is a branchial or special visceral efferent nerve in constitution. The spinal 'root' may be either somatic, special visceral efferent, or mixed, depending upon the view taken of the evolutionary origin of the sternocleidomastoid and trapezius, which it supplies. This controversy is not likely to be solved, and

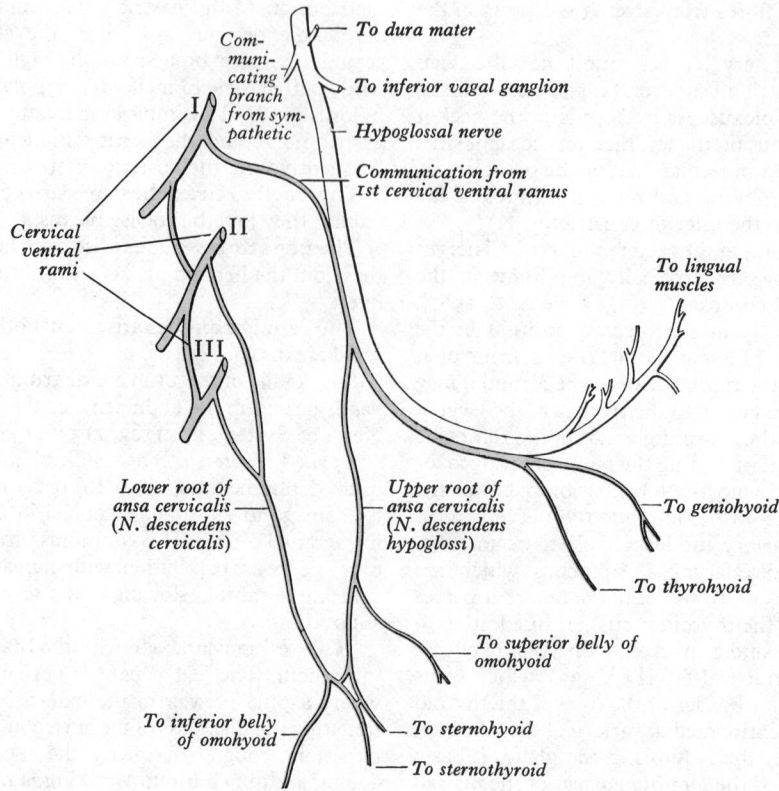

To dura mater

To inferior vagal ganglion

Hypoglossal nerve

Communication from 1st cervical ventral ramus

Communicating branch from sympathetic

I

Cervical ventral rami

II

III

To lingual muscles

Lower root of ansa cervicalis (N. descendens cervicalis)

Upper root of ansa cervicalis (N. descendens hypoglossi)

To geniohyoid

To thyrohyoid

To superior belly of omohyoid

To inferior belly of omohyoid

To sternohyoid

To sternothyroid

7.193 A plan of the right hypoglossal nerve and ansa cervicalis.

hence the custom of describing the two parts as a single cranial nerve has been followed here. It is usually assumed that the spinal part of the accessory nerve is purely motor, but evidence for the presence of afferent fibres is provided by the occurrence of ganglia upon the spinal accessory in prenatal and early postnatal human material (Pearson *et al.* 1964). The nerve may also communicate with the dorsal roots of the upper cervical spinal nerves. Such observations have not been confirmed in adult material.

The cranial root is the smaller; its fibres arise from the cells of the lower end of the **nucleus ambiguus** (p. 1076) and possibly from the **dorsal vagal nucleus**. The former is connected with the corticonuclear tracts of *both* sides. Some of the fibres from this source descend from the midbrain in the medial lemniscus (aberrant pyramidal fibres, p. 905). The fibres of the cranial root emerge as four or five delicate rootlets from the side of the medulla oblongata, below the roots of the vagus. The nerve runs laterally to the jugular foramen, where it is said to interchange fibres with the spinal root, with which it becomes united for a short distance; here it is also connected by one or two filaments with the superior ganglion of the vagus. It passes through the jugular foramen, separates from the spinal portion, and is continued over the inferior ganglion of the vagus, to the surface of which. it is adherent. It is distributed principally in the pharyngeal and recurrent laryngeal branches of the vagus. It is probably the source of the motor fibres which run in the former to supply the muscles of the soft palate, with the exception of the tensor veli palatini. Some filaments from it are continued into the trunk of the vagus below the ganglion, to be distributed with the recurrent laryngeal nerve and possibly also with the cardiac nerves.

The spinal root is firm in texture, and its fibres arise from an elongated column of motor neurons, *the spinal nucleus*, which is situated in the lateral part of the anterior grey column of the spinal cord, and extends downwards as low as the level of the fifth cervical segment (p. 870). Passing through the lateral white column of the spinal cord, they emerge on its surface midway between the ventral and dorsal nerve roots of the upper cervical nerves (7.189), and unite to form a trunk, which ascends between the ligamentum denticulatum and the dorsal roots of the spinal nerves, and enters the skull through the foramen magnum, behind the vertebral artery (7.190). It is then directed upwards and laterally to the jugular foramen, through which it passes in the same sheath of dura mater as the vagus nerve, but separated from

that nerve by a fold of the arachnoid mater. In the jugular foramen, it may receive one or two filaments from the cranial root, or else join it for a short distance and then parts from it again. At its exit from the jugular foramen, it runs laterally and backwards posterior to the internal jugular vein in about two-thirds of subjects, and anterior to it in about one-third; in rare cases it may pass through the vein. In this situation the accessory nerve crosses the transverse process of the atlas and is itself crossed by the occipital artery. The nerve then descends obliquely, passing medial to the styloid process, the stylohyoid and the posterior belly of the digastric. Together with the superior sternocleido-mastoid branch of the occipital artery, it reaches the upper part of the sternocleidomastoid and pierces its deep surface, supplying it and joining with branches from the second cervical nerve. Emerging a little above the middle of the posterior border of the sternocleidomastoid, the nerve crosses the posterior triangle of the neck lying on the levator scapulae (7.186), from which it is separated by the prevertebral layer of the deep cervical fascia and the adipose tissue occupying the triangle. Here it is comparatively superficial, being related to the superficial cervical lymph nodes and receiving communications from the second and third cervical nerves (Raveau 1968). Finally, about 5 cm above the clavicle, the accessory nerve disappears under the anterior border of the trapezius and, together with branches from the third and fourth cervical nerves, forms a plexus on the deep surface of the muscle. From this plexus the trapezius receives its innervation. The course of the accessory nerve in the neck can be represented by a line drawn downwards from the lower and anterior part of the tragus to the tip of the transverse process of the atlas, and then downwards and backwards, across the elevation produced by the sternocleidomastoid and the depression corresponding to the posterior triangle of the neck, to a point on the anterior border of the trapezius 5 cm above the clavicle. For a discussion of the possible significance of the double nerve supply to trapezius and sternocleidomastoid *see* McKenzie (1955), and p. 565.

The cranial and spinal roots, after separating from each other, are also known respectively as the *internal* and *external rami* of the accessory nerves. There is general agreement that the spinal root is the sole motor supply to the sternocleidomastoid, the second and third cervical nerves conveying proprioceptive fibres from the muscle. Whether the spinal root is the sole motor supply to the trapezius is uncertain, some maintaining that the third and fourth cervical nerves are purely proprioceptive, while others believe that they supply motor fibres to the lower part of the muscle.

Applied Anatomy. The functions of the accessory nerve may be interfered with by central changes, or at its exit from the skull, by fractures running across the jugular foramen, or in the neck, by inflamed lymph nodes, etc. Acute torticollis in children is most commonly due to inflamed lymph nodes. Central irritation causes clonic spasm of the sternocleidomastoid and trapezius, or, as it is termed, spasmodic torticollis. In cases of this affection in which all previous palliative treatment has failed, division or excision of a section of the accessory nerve has been resorted to.

In cases where extensive dissections are undertaken in the posterior triangle of the neck for the excision of pathological nodes, it is essential that this nerve should be sought at the outset and isolated from the mass of diseased nodes so as to preserve its continuity.

The Hypoglossal Nerve

The hypoglossal nerve (7.62, 64, 192, 193) is the motor nerve of the tongue. It is in series with the oculomotor, trochlear and abducent nerves and the ventral nerve roots of the spinal nerves, and represents the fused ventral roots of, probably, four precervical or spino-occipital nerves, the dorsal roots of which have disappeared entirely.

The hypoglossal nucleus from which its fibres arise is in line with the modified anterior grey column of the spinal cord. This nucleus is about 2 cm long, and its rostral part corresponds with the hypoglossal triangle of the floor of the fourth ventricle (p. 933). The lower part of the nucleus extends downwards into the closed part of the medulla oblongata, and there lies in the ventral part of the central grey matter, close to the median plane (7.61). The fibres from its cells pass ventrally through the medulla oblongata, and emerge as a linear series of 10–15 rootlets through the anterolateral sulcus between the pyramid and the olive (7.55).

The hypoglossal nucleus displays a longitudinal division into dorsal and ventral laminae, each of which may be further divided into a mediolateral sequence of relatively discrete sub-nuclei. The latter are considered to correspond to the individual muscles innervated (Kosaka and Yagita 1903; Sturman 1916; Barnard 1940). More recently, Jansen and Korneliussen (1977) have described a particularly clear division into four longitudinal columns in Cetacea, and they suggest that these correspond to four occipito-cervical somites and hence, perhaps, to individual lingual muscles.

Connexions. The nucleus of the hypoglossal nerve receives fibres from the precentral gyrus and adjacent areas of mainly the *opposite* cerebral hemisphere through the corticonuclear tract; some of the latter leave the tract in the pons and travel in the medial lemniscus. They are connected to the nucleus directly or through internuncial neurons. Some evidence has been presented, however, that the most medially situated sub-nuclei receive projection fibres from *both* cerebral hemispheres. The hypoglossal nucleus may connect with the cerebellum via neighbouring perihypoglossal nuclei (*see* Torvik and Brodal 1954; also p. 903), and may also be connected with the medullary reticular formation, the sensory nuclei of the trigeminal, and the nucleus solitarius.

The rootlets of the hypoglossal nerve run laterally behind the vertebral artery, and are collected into two bundles, which perforate the dura mater separately opposite the hypoglossal (anterior condylar) canal in the occipital bone, and unite together after their passage through it; in some cases the canal is divided into two by a small bony spicule. The fact that each fascicle acquires a separate sheath from the dura mater is confirmatory evidence of the composite character of the nerve. On emerging from its canal the nerve lies on a deeper plane than the internal jugular vein, the internal carotid artery, the ninth, tenth and eleventh cranial nerves. It passes laterally, with a downward inclination, behind the internal carotid artery and the glossopharyngeal and vagus nerves to gain the interval between the artery and the internal jugular vein. In this part of its course it makes a half-spiral turn round the inferior ganglion of the vagus,

to which it is united by connective tissue. It then descends almost vertically, lying between the vessels and in front of the vagus nerve, to a point corresponding with the angle of the mandible, and becomes superficial below the posterior belly of the digastric, emerging from between the internal jugular vein and the internal carotid artery. The nerve loops round the inferior sternocleido-mastoid branch of the occipital artery (p. 681) and, having crossed the internal and external carotid arteries, it crosses the loop of the lingual artery a little above the tip of the greater cornu of the hyoid bone (7.192), being itself crossed by the facial vein. It inclines upwards as it runs forwards on the hyoglossus, passing deep to the tendon of the digastric, the stylohyoid and the posterior border of the mylohyoid. In the interval between the hyoglossus and mylohyoid the nerve is related above to the deep part of the submandibular gland, the submandibular duct and the lingual nerve. It passes next on to the lateral aspect of the genioglossus and is continued forwards in its substance as far as the tip of the tongue, distributing branches to the muscle.

The hypoglossal nerve communicates with the sympathetic trunk, and with the vagus, first and second cervical, and lingual nerves.

Opposite the atlas the nerve receives branches from the superior cervical ganglion of the sympathetic trunk, and at the same level is joined by a filament from the loop connecting the first and second cervical nerves. This filament soon leaves the hypo-glossal and descends as the *upper root* of the *ansa cervicalis* (7.193).

The communications with the vagus nerve take place close to the skull, numerous filaments passing between the hypoglossal nerve and the inferior ganglion of the vagus nerve through the mass of connective tissue which unites them. As the nerve winds round the occipital artery it receives a filament from the pharyngeal plexus, which is termed the *ramus lingualis vagi* (p. 1079).

Near the anterior border of the hyoglossus it is connected with the lingual nerve by numerous filaments which ascend upon the muscle (p. 1067).

The *branches of distribution* of the hypoglossal nerve are, meningeal, thyrohyoid, descending, muscular.

The meningeal branch or branches leave the hypoglossal nerve as it passes through the hypoglossal canal; they pursue a recurrent course and then ramify to be distributed to the diploë of the occipital bone, to the dural walls of the occipital sinus, and neighbouring structures including the dura of the inferior petrosal sinus and much of the floor and anterior wall of the posterior cranial fossa. As discussed elsewhere (pp. 1048, 1127) these meningeal rami may not be branches of the hypoglossal nerve itself, but ascending, mixed sensory and sympathetic nerves derived from the upper cervical nerves and superior cervical sympathetic ganglion (Kimmel 1961 *a* and *b*).

The descending branch leaves the hypoglossal nerve where the latter turns round the occipital artery, and descends anterior to or in the sheath of the internal and common carotid arteries. It contains no fibres from the hypoglossal nucleus, but only fibres from C1, which constitute the upper root of the ansa cervicalis. After giving a branch to the superior belly of the omohyoid, this nerve is joined by the lower root of the ansa cervicalis from the second and third cervical nerves. The union of the two forms a loop, which is termed the *ansa cervicalis (ansa hypoglossi)*. From the convexity of this loop branches pass to supply the sternohyoid, the sternothyroid and the inferior belly of the omohyoid. Another filament has been described which descends in front of the vessels into the thorax, and joins the cardiac and phrenic nerves.

The nerve to the thyrohyoid arises from the hypoglossal nerve near the posterior border of the hypoglossus; it runs obliquely across the greater cornu of the hyoid bones, and supplies the thyrohyoid. It is derived from the communication with the first cervical spinal nerve.

The muscular branches are distributed to the styloglossus, hyoglossus, geniohyoid and genioglossus. Numerous slender branches pass upwards into the substance of the tongue to supply its intrinsic muscles. Most of these muscular branches are true hypoglossal fibres, but those to geniohyoid stem from the first cervical nerve.

Applied Anatomy. When the hypoglossal nerve is injured or diseased, unilateral lingual paralysis follows, together with hemiatrophy of the tongue; the tongue, when protruded, is directed to the paralysed side owing to the unopposed action of the opposite muscles. On retraction, the wasted and paralysed side of the tongue rises up higher than the other. The larynx may deviate towards the sound side on swallowing, due to unilateral paralysis of the depressors of the hyoid bone. If the paralysis is bilateral, the tongue lies motionless in the mouth; taste and tactile sensibility of the organ are perfect, articulation is slow; swallowing is very difficult.

MORPHOLOGY OF THE CRANIAL NERVES

It is now possible to group the cranial nerves in a manner which conforms with their phylogenetic history, their individual components and their functions.

Group I includes the oculomotor, trochlear, abducent and hypoglossal nerves. These all arise from the cells of the somatic efferent column, and they are distributed to the musculature derived from the *cranial myotomes*. They correspond, therefore, to the *ventral* nerve roots of the spinal nerves, and with the exception of the trochlear, they emerge from the brainstem in line with them. The identification of the individual segments with which each nerve is associated is a matter of considerable difficulty and is not susceptible of proof in the present state of our knowledge, for the precise number of segments represented by the head is still uncertain (p. 117).

Group II includes the trigeminal, facial, glossopharyngeal, vagus and accessory nerves. These nerves are concerned with the innervation of the derivatives of the *branchial arches*. They differ from the spinal nerves in possessing motor roots which are distributed to the musculature derived from the *neural crest* (p. 150) and the *lateral plate mesoderm* of the branchial region. Some of these cranial nerves (the trigeminal, vagus and accessory), are compound nerves and have been formed by the fusion of two or more dorsal nerves (p. 1052). In the process cutaneous branches, originally connected with the facial, glossopharyngeal and vagus nerves, have been taken over by the trigeminal, so that these nerves in man bear but little resemblance to their homologues in the lower forms of vertebrates and still less to the dorsal nerve roots of the spinal nerves.

On account of the complexity of their components, each nerve may possess more than one nucleus of origin and more than one nucleus of termination. It is noteworthy that neurons of the ganglion of the facial, the inferior ganglion of the glosso-pharyngeal and the inferior ganglion of the vagus nerve, though derived to a large extent from the neural crest, owe their origin in part to *ectodermal epibranchial placodes* which develop at the dorsal ends of the first three branchial clefts in close relation to the ganglia (p. 175).

Although there are certain difficulties in the way, the homologies of the nerves in Groups I and II are generally accepted, but the allocation of the three remaining cranial nerves is entirely uncertain. On account of its mode of development, the optic nerve is usually regarded as having nothing in common with any of the other cranial nerves except its function as a *special somatic afferent*. The retinal cells, from which its fibres are derived, really constitute an outlying part of the brain, although it may be urged that they are derivatives of the forerunners of the neural crest cells.

The olfactory and the vestibulocochlear nerves may be grouped together or separately, or the vestibulocochlear nerve may be regarded as being homologous with a dorsal nerve. Both nerves arise, in part at least, from ectodermal cells outside the area of the neural tube and crest, but whereas the olfactory cells remain intercalated amongst the epithelial cells of the nasal mucous membrane, the cochlear cells migrate a short distance away from the otic vesicle. It must be explained, however, that many authorities believe that the contribution made by the neural crest is responsible for the formation of the whole of the vestibulocochlear nerve ganglion, and on this account they prefer to regard the vestibulocochlear as a modified dorsal nerve. In comparing the olfactory and vestibulocochlear nerves it must be remembered that the olfactory nerves are restricted in all forms to the region of the head, whereas the vestibulocochlear nerve in man is the sole survivor of a whole series of nerves of the organs of the lateral line, which in lower forms are distributed not only to the head but also to the whole length of the trunk. There is, therefore, considerable justification for the allocation of the olfactory and vestibulocochlear nerves to separate groups.

For a discussion of the phylogeny of the vertebrate cranial nerves consult Black (1917); Herrick (1922); Bolk *et al.* (1934); Kappers *et al.* (1936); Kappers (1947); Young (1950); Grassé (1954); Goodrich (1958).

General Considerations of the Cranial Nerves

Because of the extremely complex three-dimensional organization of the brainstem, which changes continuously with the level concerned, and the fact that innumerable cell groups and fibre systems exist in relatively small volumes of nervous tissue, the difficulties of investigation are multiplied, and less detail is known of the intimate arrangement and connexions of many cranial nerve nuclei than is the case with the spinal nerves. Accordingly, the account given of the regional localization and connectivity patterns of many of these cranial nuclei, in general neuroanatomical texts, often conveys a simplicity which is undoubtedly far from reality. It may prove helpful, therefore, at this point to summarize briefly some of the general organizational features which are either emerging, or are suspected by analogy with spinal cord organization, but some of which have not yet been established with certainty in the primate brain. The olfactory, optic and vestibulocochlear nerves, however, are considered in some detail elsewhere in the present volume and will not be pursued further here.

The motor nuclei of the cranial nerves, as indicated above, include somatic efferent, special visceral efferent (branchiomotor) and general visceral efferent (autonomic) groups. However, the somatic and special visceral efferent nuclei of the oculomotor, trochlear, trigeminal, abducent, facial glossopharyngeal, vagus, accessory and hypoglossal nerves may, for present purposes, be grouped, since all innervate *striated muscle*. It will be recalled that the ventral grey matter of the spinal cord consists of a series of longitudinal columns of neurons, as judged by the position of their cell somata; and although their significance is not completely understood (p. 873), they exhibit some degree of somatotopic organization. The cells of the columns include three varieties of motor neuron α, β and γ—p. 885) related to the innervation of extrafusal and intrafusal muscle fibres, and further, the α motor neurons to extrafusal muscle may be divided into 'tonic' and 'phasic' types, whilst 'static' and 'dynamic' types of fusimotor γ neurons are also recognized. Between and around these motor neurons are numerous small interneurons, some excitatory, others inhibitory, the latter including the well-known Renshaw cells. Converging on these cell varieties are numerous pathways, completing a wide array of local control loops, both contralateral and ipsilateral, and others descending from supraspinal sources such as the vestibular nuclei, brainstem reticular formation, red nucleus, tectum and tegmentum of the midbrain, and the cerebral cortex. In turn, these parts of the nervous system, which project directly to the lower motor centres, are themselves in receipt of essential control systems from the cerebellum, the corpus striatum, the thalamus and hypothalamus, and many parts of the cerebral neocortex and the limbic system. These have all been considered in detail elsewhere in the present section and no attempt will be made to describe them here. Despite our lack of detailed knowledge of cranial nerve connexions, it seems most probable that comparable control systems operate, and in view of the precision of integrated muscle action involved in phonation, facial expression and ocular movements, the control systems may well be of a *higher* order of complexity than those obtaining at spinal levels.

The trochlear and abducent nerves each innervate a single small muscle, and reflecting this, their motor nuclei are relatively small single groups of neurons. In contrast the oculomotor, trigeminal, facial, glossopharyngeal-vagus-cranial accessory group, and the hypoglossal all innervate complex musculature capable of a highly refined and precise three-dimensional control of their movement patterns. In each case their nuclei can be divided into a series of sub-nuclei which may be related to the major branches of the nerve, or to the innervation of single muscles within the group. In some cases the sub-nuclei show a longitudinal columnar arrangement, the columns varying in their mediolateral and dorsoventral positioning and in their rostro-caudal extent. This somatotopic localization, in terms of the position of cell *somata*, is reminiscent of that existing in the ventral grey column of the spinal cord, and whilst it seems plausible that the delicacy of control may be in some way increased by such grouping, as discussed more fully elsewhere (p. 873), the precise ontogenetic, phylogenetic and functional significance of these groups must await the results of future researches. Also, much less is known in cranial nerve motor nuclei, than in the spinal cord, concerning the quantitative degree of longitudinal and transverse overlap and interlocking of the receptive *dendritic trees* of motor neurons in adjacent nuclei and sub-nuclei. Further, although muscle spindles were for long considered to be absent, or extremely sparse, in facial, masticatory, lingual, laryngeal and extrinsic ocular muscles, their presence has now been amply confirmed in most of these sites (the facial muscles apparently having the least dense population). Nevertheless, the anatomical siting and physiological characteristics of their associated γ efferent neurons has not yet been clarified. Similar uncertainty exists concerning the smaller multipolar neurons often found within or in the neighbourhood of cranial nerve motor nuclei, many of which are presumably excitatory or inhibitory interneurons, including Renshaw cells.

The afferent connexions of the motor nuclei of the cranial nerves include components which correspond to all those which converge upon the spinal grey matter, but some are recognized only in outline. The best recognized tract systems associated with these nuclei are: (1) the *medial longitudinal fasciculus*, interlinking the vestibular nuclear complex, a longitudinal series of cranial nerve nuclei and the cervical spinal cord (*see* p. 939); (2) *tectotegmental, tectopontine* and *tectobulbar* projections from the superior and inferior colliculi (p. 941); (3) projections from the *red nucleus* (p. 938); (4) interconnexions with the *brainstem reticular formation* (p. 946); (5) *corticonuclear* projections from the various sensorimotor areas of the cerebral cortex (p. 1011); (6) projections from the *sensory nuclei* of other cranial nerves.

It must be emphasized, however, that little is known concerning the detailed mode of termination of these afferent systems, for example, whether they terminate directly on α and/or γ motor neurons, or through the intermediary of interneurons, and to what extent they mediate pre- or post-synaptic facilitatory or inhibitory effects. The corticospinal tracts, it will be recalled (p. 877), contain fibres predominantly derived from the contralateral cerebral hemisphere, but they also carry a variable proportion of ipsilateral fibres. Similar considerations apply in varying degree to the cranial nerve motor nuclei. Thus, it is widely held that although the corticonuclear fibres to most parts of the hypoglossal nucleus, and that part of the facial nucleus innervating the lower face, are mainly contralateral in origin, whilst those to the trochlear nucleus are predominantly ipsilateral, the remainder, including the most medial parts of the hypoglossal nuclei, receive a *bilateral* corticonuclear projection. The cerebellum, diencephalic centres, and corpus striatum are considered to influence the cranial nerve nuclei through indirect pathways similar to those proposed for the control of spinal motor centres (see the sections devoted to these regions).

The general visceral efferent nuclei of the oculomotor, facial, glossopharyngeal and vagus nerves are incompletely understood in terms of their central connexions. (The specific connexions of the accessory oculomotor nucleus related to visual reflexes are discussed elsewhere.) It is presumed that the dorsal nucleus of the vagus and salivatory nuclei receive the terminals of ascending tracts, and those from other cranial nerve nuclei (particularly the nucleus of the solitary tract), conveying both somatic and visceral information. It is also considered that they establish interconnexions with the brainstem reticular formation, and receive descending pathways, probably polysynaptic, from the hypothalamus, through which, indirectly, the frontal neocortex, the limbic structures, and the thalamus exert their effects. These descending pathways have not been satisfactorily analysed with the neuroanatomical methods currently available. The same uncertainty attaches to the central connexions of the **general visceral afferent pathways** (p. 1136), which physiological studies have shown to influence many brainstem centres, including particularly the brainstem reticular formation, the hypothalamus, the limbic lobe, and the prefrontal neocortex (consult the sections devoted to the connexions of these regions). The proportion of primary general visceral afferent fibres which terminate in either the dorsal nucleus of the vagus, or in the nucleus of the solitary tract, is undetermined, but the latter are accompanied by the **special visceral afferent** (gustatory) fibres of the facial, glossopharyngeal and vagus nerves. These end in a rostrocaudal sequence of overlapping zones in the solitary nucleus, where, after synaptic relays, secondary gustatory fibres cross the midline and ascend to terminate in the nucleus ventralis posterior medialis of the thalamus, and also in a number of hypothalamic nuclei. The detailed routes taken by these *solitariothalamic* and *solitariohypothalamic* fibres have not been established in the human brain, but it is thought likely that the former accompany the medial fibres of the *medial lemniscus*, whilst the latter may join the *dorsal longitudinal fasciculus* (p. 940), and the *mamillary peduncle* (p. 968). Collateral branches probably leave these ascending fibres to end in association with the brainstem reticular formation and the nuclei of other cranial nerves. From the caudal regions of the nucleus solitarius fibres descend to the spinal cord as a *solitariospinal tract*. The nucleus solitarius, however, is not to be regarded as a simple relay on the visceral afferent pathways. Other systems converge on the nucleus including ascending fibres from the spinal cord, fibres from the vestibulocerebellum, and descending corticonuclear fibres. Accordingly, interaction between these various information channels probably occurs in the nucleus, and its transmission characteristics are likely to be modulated by activity in the descending fibres from the cerebral cortex.

The somatic afferent nuclei of the brainstem constitute the spinal, principal sensory, and mesencephalic nuclei of the trigeminal nerve.

The unique character of the *mesencephalic nucleus* of the trigeminal, which consists of unipolar *primary* sensory neurons, and its probable role as a proprioceptive nucleus have already been described (p. 1060). Electrophysiological investigations support the latter view since rapid responses have been recorded from the nucleus following stretching of the masticatory muscles, during passive jaw movement, and following application of pressure to the teeth. Thus, the peripheral processes of the mesencephalic unipolar cells are considered to innervate the muscle spindles of the masticatory muscles, the articular tissues of the temporomandibular joint and the periodontal tissues. Possibly they are also the source of the fibres which innervate the muscle spindles of the facial, lingual and laryngeal muscles, but this remains uncertain, as does the siting of the neuron somata of the proprioceptive fibres derived from extraocular muscles (*see* p. 1058). Since the cell somata of these primary afferent neurons are deeply embedded in the midbrain, and are arranged as a long, thin, curved lamina of cells on each side, they present particular difficulties to the neuroanatomical investigator, because even the smallest experimental lesions inevitably involve surrounding tissues. Hence, relatively little is known of the central projections of these cells. It has been claimed, however, that some of their central processes descend towards the principal sensory nucleus.

Others are thought to project to the motor nucleus of the trigeminal, completing a masticatory reflex loop, whilst collateral branches have been described as entering the cerebellum in a trigeminocerebellar tract (p. 916). Their other connexions remain conjectural.

The manner in which the primary sensory fibres of the three divisions of the trigeminal nerve, which have their cell somata in the trigeminal ganglion, terminate in the *principal sensory* and *spinal nuclei* of the trigeminal nerve have already been described, and only a few further points need be mentioned here. The view, stemming largely from disease of the brainstem or from neurosurgical manœuvres, that the principal sensory nucleus is a simple relay station for tactile information, and the spinal nucleus for thermal and nociceptive information, is certainly an oversimplification. The observation that a large proportion of the sensory radicular fibres *bifurcate* on entering the pons, one branch entering the principal nucleus, and the other the spinal nucleus, has previously been referred to. The *spinal nucleus*, which is continuous caudally with the substantia gelatinosa of the cervical spinal cord, shows a regional variation in its cytoarchitecture (and is sometimes divided for convenience into a rostrocaudal sequence of three subnuclei termed *oralis, interpolaris* and *caudalis*). Further, the spinal and principal nuclei not only receive the terminals of somatic afferent *trigeminal* fibres, but those of the *facial, glossopharyngeal* and *vagus* nerves, in addition to others which ascend from the cervical spinal cord. Both nuclei establish interconnexions with the brainstem reticular formation, and both receive numerous descending corticonuclear fibres from the sensorimotor cortex.

Thus, many information paths converge on both these nuclei, and unit recording with intracellular microelectrodes has confirmed the *wide range* of different types of cell response in the two nuclei. These vary from rapidly-adapting cells with modality-specific, small receptive fields, to others with multi-modal responses and larger receptive fields. Cells responding to light tactile stimuli, and others to nociceptive stimuli were found in *both* nuclei. Some of the units show the phenomenon of inhibitory 'surround' (p. 808), and detailed analyses have shown that both pre- and post-synaptic inhibition and facilitation occur in the nuclei. Their impulse transmission characteristics are modified by stimulation or ablation of the descending cortico-nuclear tracts, and areas of the brainstem reticular formation. Evidently, the clinical impression of the two nuclei as relatively simple, functionally segregated relay stations, is not supported by experimental evidence, which suggests that they are complex, cooperative, integration centres. In this regard, an interesting analogy has been drawn between the principal sensory nucleus of the trigeminal and the nuclei gracilis and cuneatus, and between the spinal nucleus of the trigeminal and the substantia gelatinosa of the spinal cord, in terms of their structural and functional organization. (*See* pp. 870, 887, 889–890 for further details of these regions, and a proposed theory of action of the substantia gelatinosa as possibly modulating sensory input.) Future researches will determine how far such a comparison is justified.

Finally, the statement on p. 1060 that the ascending efferent connexions from the principal and spinal nuclei of the trigeminal consist of crossed fibres which pass to the thalamus as a 'trigeminal lemniscus' may be amplified. Both animal experimentation and observations on human material have shown that multiple trigeminothalamic pathways exist. From the principal sensory nucleus a substantial bundle of fibres cross the midline, whilst a smaller component remains on the ipsilateral side, and they ascend through the upper pons, midbrain and subthalamus, as the *dorsal trigeminothalamic tract* (dorsal division of the trigeminal lemniscus). In the midbrain this tract lies dorsomedial to the red nucleus and medial lemniscus, and it terminates in the nucleus ventralis posterior medialis (VPM) of the thalamus (p. 958). Other fibres from the principal sensory nucleus and from the length of the spinal nucleus mainly cross the midline and ascend in close company with the fibres of the medial lemniscus; these constitute the *ventral trigeminothalamic tract* (ventral division of the trigeminal lemniscus). Many of these ventral ascending fibres also end in the VPM, but others, mainly from the caudal parts of the spinal nucleus, are described as ending in the medial geniculate body and in the intralaminar nuclei. Collateral branches are considered to leave these ascending pathways to terminate in the brainstem reticular formation, on other cranial nerve nuclei, and some collaterals probably enter into the cerebellum (p. 916).

THE SPINAL NERVES

The spinal nerves are formed by the union of ventral and dorsal spinal nerve roots which are attached in series to the sides of the spinal cord. There are 31 pairs of these nerves grouped as follows: cervical, 8; thoracic, 12; lumbar, 5; sacral, 5; coccygeal, 1. The abbreviations C, T, L, S and Co., followed by the appropriate numeral are commonly used to identify the individual nerves. They emerge through the intervertebral foramina. The first cervical nerve escapes from the vertebral canal between the occipital bone and the atlas, and is therefore called the *suboccipital nerve*; the eighth issues between the seventh cervical and first thoracic vertebrae.

Each nerve is connected with the spinal cord by ventral and dorsal roots, the latter being characterized by the presence of a *spinal ganglion*.

The ventral (anterior) roots contain the axons of cells in the anterior and lateral grey columns of the spinal cord. Each root emerges as a series of rootlets arranged in two or three irregular rows over a distance of about 3 mm across the anterolateral aspect of the spinal cord.

The dorsal (posterior) roots contain the processes of cells in the spinal ganglia which are swellings on the roots. Each *root* consists of two fascicles, medial and lateral, each of which diverges into *rootlets* entering along the posterolateral sulcus. Dissections in a number of different mammals, including man, have shown that the dorsal nerve rootlets of adjacent segments are often interconnected by fine, oblique bundles of nerve fibres, particularly in the lower cervical and lumbosacral regions of the cord (Pallie and Manuel 1968).

The spinal ganglia are collections of nerve cells on the dorsal roots of the spinal nerves. Each ganglion is oval, reddish, and in its size related to that of the nerve root on which it is situated; it is bifid medially where the two bundles of the dorsal nerve root emerge from it to approach and enter the cord. The ganglia are usually in the intervertebral foramina, immediately lateral to the sites where the nerve roots perforate the dura mater (7.37); the ganglia of the first and second cervical nerves, however, lie on the vertebral arches of the atlas and axis, and those of the sacral nerves are inside the vertebral canal, while that of the coccygeal nerve is usually within the dura mater.

The ganglia of the first pair of cervical nerves may be absent, while small *aberrant ganglia* consisting of groups of nerve cells are sometimes found on the dorsal roots of the upper cervical nerves between the spinal ganglia and the spinal cord. (Heterotopic ganglionic neurons are also found in other sites—*see* p. 838.)

Each nerve root receives a covering from the pia mater, and is loosely invested by the arachnoid mater, the latter being prolonged as far as the points where the roots pierce the dura mater. The two roots pierce the dura mater separately, each receiving a sheath from this membrane (7.163, 167); where the roots join to form the spinal nerve this sheath is continuous with the epineurium of the nerve.

The internal structure of sensory ganglia, cranial and spinal, shows a general similarity of pattern (*see* p. 838). (For pertinent literature consult Van Gehuchten 1892; Marinesco 1909; Cajal 1911; De Castro 1932; Scharf 1958; and Lieberman 1976.) Each ganglion has a laminated connective tissue capsule which is continuous with the epineurium of the associated spinal root. An endoneurial stroma permeates the ganglion, surrounding and supporting its neuronal and axonal population, and trabeculae of perineurium extend between groups of these elements from the capsule. This stroma is intimately related to the satellite cells (amphicytes); it contains scattered mast cells and a dense vascular network, which is denser in the vicinity of ganglion cells, into which capillary loops may be actually invaginated (Scharf 1958). These capillaries are non-fenestrated (Lieberman 1968) in rodents, but have been described as commonly fenestrated in primates (Olsson 1971). The vascular permeability in sensory ganglia shows much species variation, amounting to a 'blood-nerve' barrier in some, largely dependent upon the junctional complexes between the endothelial cells, but details are not available for mankind. Ganglionic neurons exhibit a great variation in size (15–110 µm in man—Ohta et al. 1974); most are spheroidal, but smaller cells appear ellipsoidal or display angular profiles in section, the different sizes being randomly distributed in space. In general cytological features ganglion cells resemble other neurons, but they exhibit very marked variation in the distribution of cytoplasmic chromatin from fine dispersal to concentration in the large masses known as Nissl granules. These variations have been used as the chief basis of many classifications, which remain conflicting. Melanin and lipofuchsin pigments occur in some cells, but not as frequently as in some sensory ganglia of cranial nerves. Dense-cored vesicles are commonly observed (Lieberman 1968), despite the apparent absence of catecholamines from such cells. A common feature of agreement in attempts at classification is the occurrence of two extreme types: large 'light' cells (A cells) and small 'dark' cells (B cells). In the light cells the granular endoplasmic reticulum is more dispersed, but highly concentrated in the dark cells. The latter type are sometimes mimicked by artefact, perhaps due to poor fixation. Both types have been identified in prenatal material; and although the functional differences between the two types of nerve cell remain unclarified, it is claimed that the axons of the larger type are myelinated, those of the smaller cells non-myelinated. It has been suggested that the latter are visceral afferent neurons, but this has not been confirmed. So-called 'atypical' cells of various kinds have been described in dorsal root ganglia (and other sensory ganglia) from time to time, the most interesting of which were reputed multipolar neurons; but since synapses have never been satisfactorily identified in dorsal root ganglia, this view is improbable and has not been further confirmed. All, or almost all the ganglion cells are unipolar (see p. 821), many exhibiting marked coiling of the 'stem' process, close to the parent soma—the so-called axonal glomerulus, before branching into peripheral and central parts of the axon (p. 838). Recently, renewed interest in typing of spinal ganglion cells has been engendered by ultrastructural studies (Duce and Keen 1977) and observations of specific intracellular substances (Hökfelt et al. 1976), preferential glutamine absorption (Duce and Keen 1978), and by specific reactions to neurotoxins (Jancso et al. 1977). Collectively, these observations, while confirming the occurrence of 'light' and 'dark' cells, suggest further subdivision into different types, particularly as regards the latter. Small dark cells may occur in at least two populations, differentiated by producing one or other of two substances (substance P and somatostatin). Similarly, different cells respond to different neurotoxins, or differ in their glutamine metabolism.

Counts of ganglion cells in dorsal root ganglia show great specific variation, as would be expected, but few estimates are recorded for human ganglia. There is some evidence that the maximal number is not reached until about three years after birth, but a supposed loss of cells in subsequent life has not been corroborated. The distribution of perikarya and fibres is variable, the former often being concentrated in the periphery of a ganglion; but grouping of the cells between large fascicles of nerve fibres is described as a more usual pattern in human ganglia. This grouping has some interest, for somatotopic organization has been described in dorsal root ganglia, in terms of both neuronal perikarya and processes. Burton and McFarlane (1973), by micro-injection of labelled amino acids into ganglia, were able to associate local groups of ganglion cells, in cats lumbar ganglia,

with particular branches of distribution of the associated spinal nerves and also with particular spinal radicles. Horseradish peroxidase injections near peripheral receptor endings have produced more exact results in certain sensory ganglia of rodents, but no results are yet available for primates.

Size and direction of the spinal nerve roots. The roots of the upper four *cervical* nerves are small; those of the lower four are large. The dorsal roots of the cervical nerves bear a proportion to the ventral of three to one, which is greater than in the other regions. The dorsal root of the first cervical is an exception to this, being smaller than the ventral root; in about 8 per cent of cases it is absent. The roots of the first and second cervical nerves are short, and run nearly horizontally to their points of exit from the vertebral canal. From the third to the eighth cervical they are directed obliquely downwards, the obliquity and length of the roots successively increasing; the distance, however, between the level of attachment of any of these roots to the spinal cord and the points of exit of the corresponding nerves never exceeds the height of one vertebra.

The roots of the *thoracic* nerves, with the exception of the first, are of small size, and the dorsal roots only slightly exceed the ventral in thickness. They increase successively in length, from above downwards, and, in the lower part of the thoracic region,

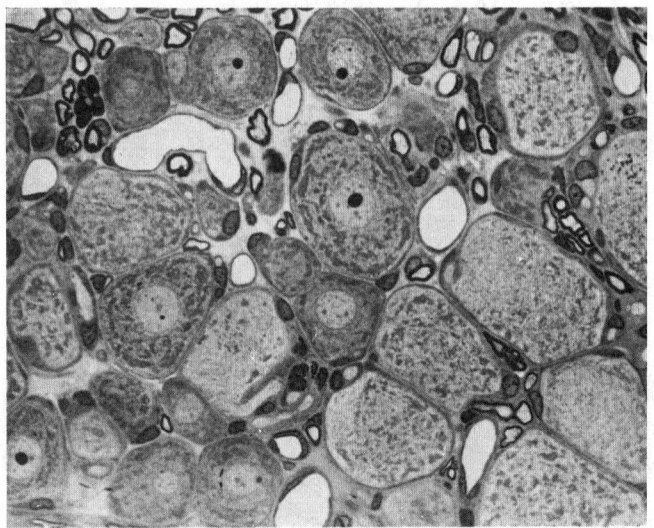

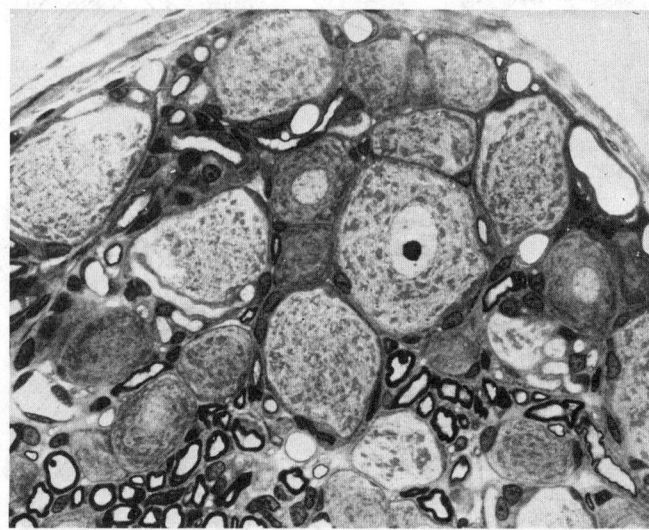

7.194 Two photographs of fields in normal rat cervical dorsal root ganglion to show contrasting features of light and dark neuronal somata. Note capsules of satellite cells; the darkly stained multiple profiles between many of the nerve cells represent glomeruli in repeated transverse section. Interneuronal capillary profiles are also visible. Cresyl fast violet staining of semi-thin sections of material embedded in araldite. (Photographs kindly supplied by Doctor J. M. Jacobs, National Hospital for Nervous Diseases, Queen's Square, London.)

descend in contact with the spinal cord for a distance equal to the height of at least two vertebrae before they emerge from the vertebral canal (but *vide infra*).

Kubik and Müntener (1969) consider that the cervico-thoracic part of the spinal cord grows more in length than other parts, and they thus explain their observations, which differ from the above

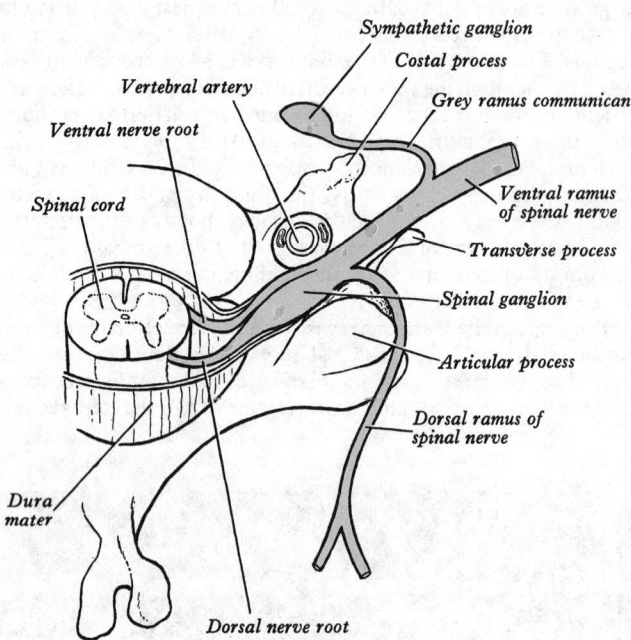

7.195A Scheme showing the relations of a cervical nerve and its ganglion to a cervical vertebra.

Labels on figure 7.195A:
- Sympathetic ganglion
- Costal process
- Grey ramus communicans
- Vertebral artery
- Ventral nerve root
- Spinal cord
- Ventral ramus of spinal nerve
- Transverse process
- Spinal ganglion
- Articular process
- Dorsal ramus of spinal nerve
- Dura mater
- Dorsal nerve root

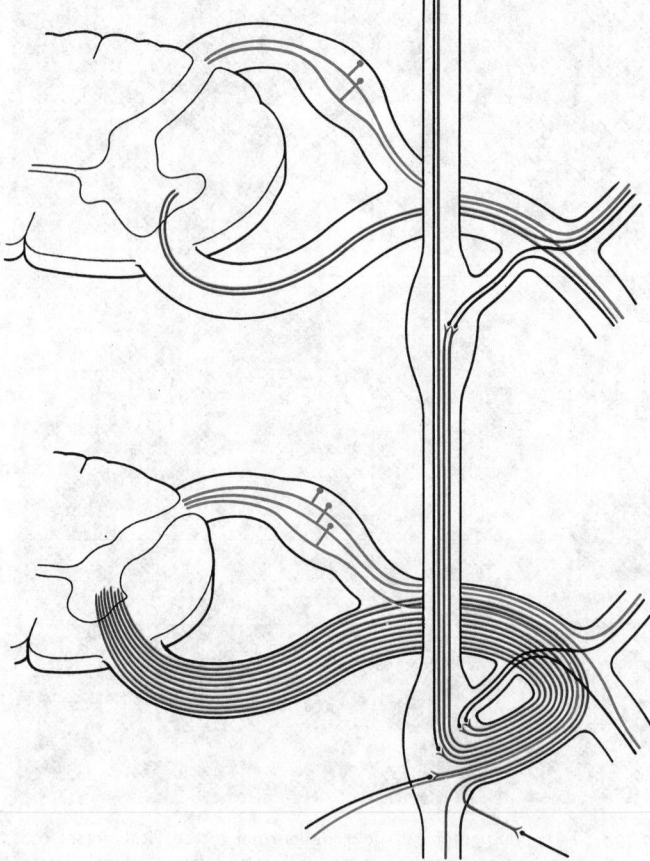

7.195B A scheme showing the constitution of a typical spinal nerve. In the upper part of the diagram the spinal nerve roots show the somatic components; in the lower part of the diagram the spinal roots show the visceral components. *Red:* somatic efferent and preganglionic sympathetic fibres. *Blue:* somatic afferent and visceral afferent fibres. *Black:* post-ganglionic sympathetic fibres.

remarks but confirm the following description of the more caudal spinal nerve roots. They state that the upper cervical roots *descend*, the fifth being horizontal, and that the sixth to eighth actually *ascend*. They describe the first two thoracic roots as horizontal, the next three as ascending, the sixth as horizontal, and the rest as descending.

The roots of the lower *lumbar* and upper *sacral* nerves are the largest, and their individual filaments the most numerous of all the spinal nerves, while the roots of the *coccygeal* nerve are the smallest.

The roots of the lumbar, sacral and coccygeal nerves descend with an increasing degree of obliquity to their respective exits, and since the spinal cord ends near the level of the lower border of the first lumbar vertebra, the lengths of successive roots rapidly increases. As already mentioned (7.35B), the term *cauda equina* is applied to this collection of *nerve roots*.

From the description given it will be seen that the largest nerve roots, and consequently the largest spinal nerves, are attached to the cervical and lumbar swellings of the spinal cord; these nerves are distributed to the upper and lower limbs.

Immediately beyond the spinal ganglia, the ventral and dorsal nerve roots unite to form the *spinal nerve*, which emerges through the intervertebral foramen, gives off recurrent meningeal branches (*vide infra*), and then divides immediately into a *dorsal* and a *ventral ramus* (7.207). (On the data derived from a limited series of dissections to clarify the precise innervation of intertransverse and costal levator muscles Sato (1974) has described a *trifurcation* of the spinal nerves at some cervical and thoracic levels, the third branch being a *ramus intermedius*.) At or immediately distal to its origin the ventral ramus of each spinal nerve is joined by a *grey ramus communicans* from the corresponding ganglion of the sympathetic trunk, while the ventral rami of the thoracic and the first and second lumbar nerves each contribute a *white ramus communicans* joining the corresponding sympathetic ganglion (7.194, 195). The second, third and fourth sacral nerves also give off visceral branches; these, however, are not connected with the ganglia of the sympathetic trunk, but belong to the parasympathetic part of the autonomic system and run directly into the pelvic plexuses (pp. 1123, 1135).

The cervical spinal nerves increase in size from the first to the sixth. The seventh and eighth cervical and first thoracic nerves are similar in size to that of the sixth cervical. The remaining thoracic nerves are relatively small. The lumbar nerves are large and increase in size from the first to the fifth. The first sacral nerve is the largest of all the spinal nerves, which thereafter decrease in size to the coccygeal, which is the smallest of all the spinal nerves.

In the intervertebral foramen the spinal nerves have important relations. *Anteriorly* are the intervertebral discs and adjacent regions of the bodies of the vertebrae. *Posteriorly* are the synovial zygapophysial joints. *Superiorly* and *inferiorly* are the vertebral notches of the pedicles of the adjoining vertebrae. Each nerve accompanied by a spinal artery, a plexus of small veins and its own meningeal branch or branches.

Applied Anatomy. The nerve roots may be compressed or otherwise irritated in their course from the spinal cord to their exit through the intervertebral foramina. In the cervical region, disease of a vertebral body, degeneration of an intervertebral disc or osteoarthrosis of the intervertebral joints may affect nerve roots as they traverse the intervertebral foramen, causing pain, diminished cutaneous sensibility and some muscular weakness in the field of supply. In the lumbar region, posterior protrusion of an intervertebral disc or rupture of its annular fibres with herniation of the nucleus pulposus is very common, affecting particularly the disc between the fifth lumbar and the first sacral vertebrae, or that between the fourth and fifth lumbar vertebrae. The posterior disc lesions may compress one or more nerve roots as they pass towards their intervertebral foramina, resulting in low back pain ('lumbago'), with or without radiation of the pain to one or both lower limbs ('sciatica'), diminished cutaneous sensibility in the area of supply and weakness of the muscles innervated (e.g. tibialis anterior L. 4; extensor hallucis longus L. 5; flexor hallucis longus S. 1). Less commonly, tumours in the vertebral canal, or spina bifida with meningomyelocoele, may cause lesions of one or more nerve roots of the cauda equina.

**Functional Components and Branches of
Spinal Nerves**

Each typical spinal nerve contains both somatic and visceral fibres.

The somatic components consist of efferent and afferent fibres. The *somatic efferent* fibres for the innervation of skeletal muscles are the axons of α, β and γ neurons in the anterior grey column of the spinal cord. The *somatic afferent* fibres convey impulses towards the central nervous system from a variety of receptors in the skin, subcutaneous tissue, muscles, fasciae, joints, etc. (*see* pp. 849, 860), and are the peripheral processes of the unipolar cells in the spinal ganglia.

The visceral components are also afferent and efferent and belong to the autonomic nervous system (p. 1121). They include sympathetic or parasympathetic fibres at different spinal levels. The preganglionic *visceral efferent* fibres of the sympathetic component are the axons of cells in the lateral grey column of the thoracic and upper two or three lumbar segments of the spinal cord; they join the sympathetic trunk through the corresponding white ramus communicans and form synapses with postganglionic neurons that are distributed to non-striated muscle or glands. The preganglionic visceral efferent fibres of the parasympathetic component are the axons of cells in the lateral grey column of the second, third and fourth sacral segments of the spinal cord. They leave the ventral rami of the corresponding sacral nerves to pass to ganglion cells in the pelvis, where they synapse. The postganglionic axons of the latter are distributed principally to muscle or glands in the walls of the pelvic visceral organs. The *visceral afferent* fibres are also derived from the cells of the spinal ganglia. Their peripheral processes are carried through the white rami communicantes, and after passing without synaptic interruption through one or more sympathetic ganglia end in the tissues of the viscera. Some visceral afferent fibres may enter the spinal cord by the *ventral* roots; Coggeshall *et al.* (1973) have claimed that almost 30 per cent of the fibres in the seventh lumbar and first sacral ventral roots in cats are non-myelinated afferents, which possibly project *into* the cord from nerve cells in corresponding dorsal root ganglia. These interesting observations require confirmation.

The central processes of the various ganglionic unipolar cells enter the spinal cord through the posterior nerve roots and form synapses around either somatic or sympathetic efferent neurons, usually through interneurons, thus completing reflex loops, or they synapse with other neurons in the grey matter of the spinal cord or brainstem which give origin to a variety of ascending pathways.

Surprisingly little detail has been recorded in regard to the dendrites of spinal autonomic neurons. Schramm *et al.* (1976), have shown that sympathetic efferent neurons display predominantly horizontal dendritic arrays in newborn rats, which become re-orientated into longitudinal arrangement in the next five or six postnatal weeks. No such data appear to be available for the human spinal cord.

After emerging from the intervertebral foramen, each spinal nerve supplies small *meningeal* branches and then splits almost immediately into a *dorsal* and a *ventral ramus*, each receiving fibres from both spinal nerve roots.

The meningeal branches of the spinal nerves (also known as the *recurrent meningeal nerves* or *sinu-vertebral nerves*) number two to four filaments on each side and are present at all vertebral levels (Kimmel 1961 *a* and *b*). Each receives one or more communications from a neighbouring grey ramus communicans or from a thoracic sympathetic ganglion directly, and the majority then pursue a recurrent (often perivascular) course to re-enter the spinal canal through the intervertebral foramen, passing ventral to the dorsal root ganglion. Here, these mixed sensory and sympathetic nerves divide into transverse, ascending and descending branches which are distributed to the dura mater, the walls of blood vessels, and to the periosteum, ligaments and intervertebral discs in the ventrolateral region of the spinal canal. Occasionally, fine meningeal branches pass dorsal to the spinal ganglion to be distributed to the dorsally situated dura, periosteum and ligaments, whilst others pass ventrally to reach the posterior longitudinal ligament.

The ascending branches of the upper three cervical meningeal nerves are relatively large and they are distributed to the cerebral dura mater of the posterior cranial fossa (p. 1048).

These various meningeal nerves are of importance in relation to the referred pain which characterizes many spinal disorders and also in occipital headache.

DORSAL RAMI OF THE SPINAL NERVES

The dorsal (posterior primary) rami of the spinal nerves are as a rule smaller than the *ventral*. They are directed posteriorly, and, with the exception of those of the first cervical, the fourth and fifth sacral, and the coccygeal, divide into medial and lateral branches for the supply of muscles and skin (**7.**196) of the posterior regions of the neck and trunk.

CERVICAL DORSAL RAMI

The dorsal ramus of each cervical spinal nerve, with the exception of the first, divides into a medial and a lateral branch. All these branches innervate muscles but, in general, only the medial branches of the second, third, fourth and, usually, the fifth, supply cutaneous areas. Except for the first and second, each dorsal ramus passes backwards medial to the posterior intertransverse muscle and winds round the articular process into the interval between the semispinalis capitis and the semispinalis cervicis.

The first cervical dorsal ramus (sometimes termed the **suboccipital nerve**) (**5.**40) is larger than the ventral ramus, and emerges superior to the posterior arch of the atlas and inferior to the vertebral artery. It enters the suboccipital triangle and supplies the muscles which bound this region—the rectus capitis posterior major, and the superior and inferior oblique; it gives branches also to the rectus capitis posterior minor and the semispinalis capitis. A filament from the branch to the inferior oblique joins the dorsal ramus of the second cervical nerve (**5.**40). The nerve occasionally gives off a cutaneous branch which accompanies the occipital artery to the scalp, and communicates with the greater and lesser occipital nerves.

The second cervical dorsal ramus is slightly larger than the ventral and all the other cervical dorsal rami. It emerges between the posterior arch of the atlas and the lamina of the axis, below the inferior oblique. It supplies a twig to this muscle, receives a communicating filament from the dorsal ramus of the first cervical, and then divides into a large medial and a small lateral branch. Its ganglion is said to be extradural in position.

The *medial* branch, called from its size and distribution the *greater occipital nerve* (**7.**196, 198), ascends obliquely between the inferior oblique and the semispinalis capitis, and pierces the latter muscle and the trapezius near their attachments to the occipital bone. It is then joined by a filament from the medial branch of the dorsal ramus of the third cervical, and, ascending in the occipital area with the occipital artery, divides into branches which communicate with the lesser occipital nerve and supply the skin of the scalp as far forward as the vertex of the skull. It gives muscular branches to the semispinalis capitis, and occasionally a twig to the back of the auricle. The *lateral* branch supplies filaments to the splenius, longissimus capitis and semispinalis capitis, and is often joined by the corresponding branch of the third cervical.

The third cervical dorsal ramus is intermediate in size between those of the second and fourth. It courses backwards round the articular pillar of the third cervical vertebra, passing medial to the posterior intertransverse muscle, and divides into medial and lateral branches. Its *medial* branch runs between the spinalis capitis and semispinalis cervicis, and, piercing the

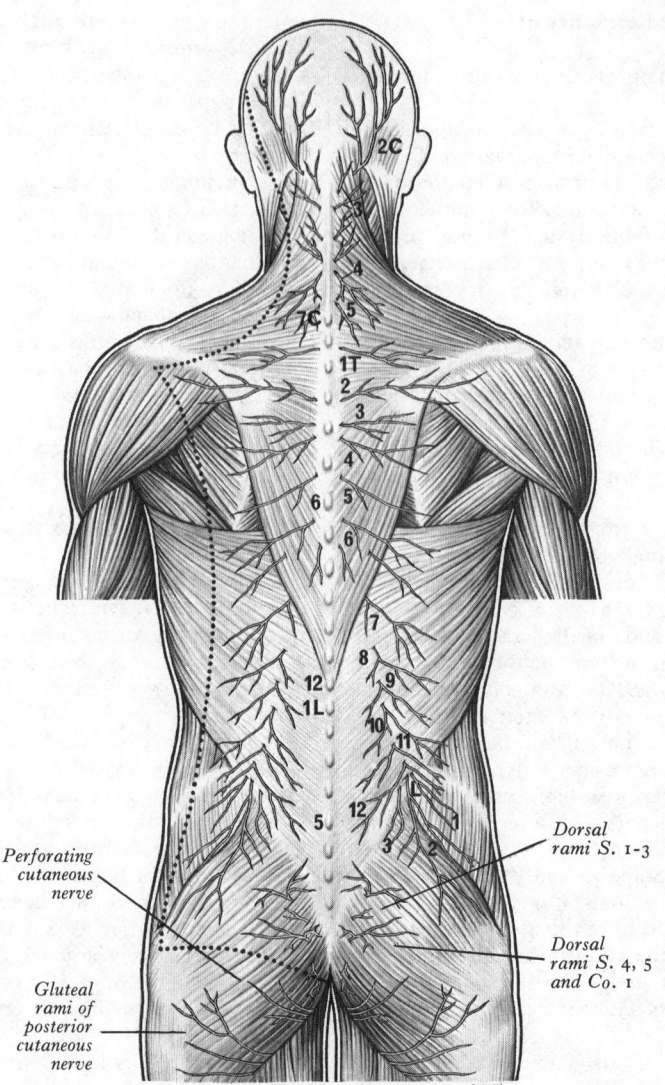

7.196 The cutaneous distribution of the dorsal rami of the spinal nerves. The nerves are shown lying on the superficial muscles; on the left side the limit of the skin area supplied by these nerves is indicated by the dotted line. The nerves are numbered on the right side and the spines of the seventh cervical, sixth and twelfth thoracic and first and fifth lumbar vertebrae are labelled on the left side.

splenius and trapezius, ends in the skin. While deep to the trapezius it gives a branch, called the *third occipital nerve*, which pierces the trapezius and ends in the skin of the lower part of the occipital region (7.196). It is medial to the greater occipital nerve, and communicates with it. The *lateral* branch often joins that of the second cervical dorsal ramus.

The dorsal ramus of the suboccipital, and the medial branches of the dorsal rami of the second and third cervical nerves are sometimes joined by communicating loops to form the *posterior cervical plexus*.

The dorsal rami of the lower five cervical nerves curve backwards round the vertebral articular pillars and divide into medial and lateral branches. The *medial* branches of the fourth and fifth run between the semispinalis cervicis and semispinalis capitis, and, having reached the spines of the vertebrae, pierce the splenius and trapezius to end in the skin (7.196). Sometimes the medial branch of the fifth fails to reach the skin. The medial branches of the lowest three nerves are small, and end in the semispinalis cervicis, semispinalis capitis, multifidus and interspinales. The *lateral* branches of the lower five nerves supply the iliocostalis cervicis, longissimus cervicis and longissimus capitis.

THORACIC DORSAL RAMI

The dorsal rami of the thoracic spinal nerves pass backwards close to the joints between the articular processes of the vertebrae and divide into medial and lateral branches. The medial branch

emerges between the joint and the medial edge of the superior costotransverse ligament and the intertransverse muscle, but the lateral branch runs laterally in the interval between the ligament and the muscle before inclining posteriorly on the medial side of the levator costae.

The *medial* branches of the *upper six* thoracic dorsal rami run between semispinalis thoracis and multifidus, which they supply; they then pierce the rhomboids and trapezius, and reach the skin by the sides of the vertebral spines (7.196). The medial branches of the *lower six* thoracic dorsal rami are distributed chiefly to the multifidus and longissimus thoracis; occasionally they give filaments to the skin near the median plane.

The *lateral* branches increase in size from above downwards. They run through or deep to the longissimus thoracis to the interval between it and the iliocostalis cervicis, and supply these muscles and the levatores costarum; the lower five or six also give off cutaneous branches, which pierce the serratus posterior inferior and latissimus dorsi in a line with the angles of the ribs (7.196). The lateral branches of a variable number of the upper thoracic rami also give filaments to the skin. The lateral branch of the twelfth thoracic, after sending a filament medially along the iliac crest, passes downwards to the skin of the anterior part of the gluteal region.

The medial cutaneous branches of the dorsal rami of the thoracic spinal nerves descend for some distance close to the vertebral spines before reaching the skin, while the lateral branches travel downwards for a considerable distance—it may be as much as the breadth of four ribs—before they become superficial; the branch from the twelfth thoracic, for instance, reaches the skin only a little way above the iliac crest.

LUMBAR DORSAL RAMI

The dorsal rami of the lumbar spinal nerves pass back medial to the medial intertransverse muscles and at once divide into medial and lateral branches.

The *medial* branches run close to the articular processes of the vertebrae and end in the multifidus. (These nerves are related to bone between the accessory and mamillary processes. They may groove the bone, traverse a distinct notch or even a foramen.)

The *lateral* branches supply the erector spinae (sacrospinalis). In addition the upper three give off cutaneous nerves which pierce the aponeurosis of the latissimus dorsi at the lateral border of the erector spinae and cross the posterior part of the iliac crest to reach the skin of the gluteal region (7.196), some reaching as far as the level of the greater trochanter.

SACRAL DORSAL RAMI

The dorsal rami of the sacral spinal nerves are small, and diminish in size from above downwards; with the exception of the fifth, they emerge through the dorsal sacral foramina. The *upper three* are covered at their points of exit by the multifidus, and divide into medial and lateral branches.

The *medial* branches are small, and end in the multifidus.

The *lateral* branches join with one another and with the lateral branches of the dorsal rami of the last lumbar and fourth sacral to form loops on the dorsal surface of the sacrum. From these loops branches run to the dorsal surface of the sacrotuberous ligament and form a second series of loops under the gluteus maximus. From this second series of loops the *gluteal branches*, two or three in number, arise and at once pierce the gluteus maximus along a line drawn from the posterior superior iliac spine to the apex of the coccyx; they supply the skin over the posterior gluteal area (7.196).

The dorsal rami of the *lower two* sacral nerves are small and lie below the multifidus. They do not divide into medial and lateral branches, but unite with each other and with the dorsal ramus of the coccygeal nerve to form loops on the back of the sacrum; filaments from these loops supply the skin over the coccyx. Berthold and Carlstedt (1977) have recorded a highly detailed study of the first sacral nerve of the cat at its junction with the spinal cord.

COCCYGEAL DORSAL RAMUS

The dorsal ramus of the coccygeal spinal nerve does not divide into a medial and a lateral branch, but receives, as already stated, a communicating branch from the last sacral; it is distributed to the skin over the back of the coccyx.

VENTRAL RAMI OF THE SPINAL NERVES

The ventral rami of the spinal nerves supply the limbs and the anterolateral aspects of the trunk; they are for the most part larger than the dorsal rami. In the thoracic region they run independently of one another, retaining, like all the dorsal rami, a more or less segmental distribution. In the cervical, lumbar and sacral regions, however, they unite near their origins to form plexuses. It is to be noted that the dorsal rami of the spinal nerves do not enter into the formation of these plexuses; their distribution has already been described (*vide supra*).

CERVICAL VENTRAL RAMI

The ventral rami of the cervical nerves, with the exception of the first, appear between the corresponding anterior and posterior intertransverse muscles. The ventral rami of the *upper four* nerves unite to form the *cervical plexus*; those of the *lower four*, together with the greater part of the ventral ramus of the first thoracic nerve, join to form the *brachial plexus*.

Each nerve receives at least one grey ramus communicans, the upper four from the superior cervical ganglion, the fifth and sixth from the middle cervical ganglion, and the seventh and eighth from the cervicothoracic ganglion of the sympathetic trunk (*see* p. 1127).

The ventral ramus of the *first cervical nerve* (sometimes termed the *suboccipital nerve*) appears above the posterior arch of the atlas vertebra, and passes forwards lateral to its lateral mass, and medial to the vertebral artery. It supplies a branch to the rectus lateralis, and, emerging on the medial side of that muscle, descends in front of the transverse process of the atlas and behind the internal jugular vein, and joins with the ascending branch of the second nerve.

The ventral ramus of the *second cervical nerve* issues between the vertebral arches of the atlas and axis and runs forwards between the transverse processes of these two vertebrae; passing in front of the first posterior intertransverse muscle and on the lateral side of the vertebral artery it emerges between the longus capitis and levator scapulae, but when the scalenus medius takes origin from the transverse process of the atlas, it intervenes between the nerve and the levator scapulae. It divides into an ascending branch which joins with the first cervical nerve, and a descending branch which unites with the ascending branch of the third cervical nerve.

The ventral ramus of the *third cervical nerve* appears between the longus capitis and scalenus medius. The ventral rami of the remaining cervical nerves emerge between the scalenus anterior and scalenus medius.

The Cervical Plexus

The cervical plexus (7.197) is formed by the ventral rami of the upper four cervical nerves; it distributes branches to some of the muscles of the neck, the diaphragm, and to parts of the integument of the head, neck and chest (7.182). It is at the level of the first four vertebrae, deep to the internal jugular vein and the sternocleidomastoid, in front of the scalenus medius and levator scapulae. The disposition of the nerves in the plexus is as follows: each nerve, except the first, divides into an ascending and a descending part; these are united in communicating loops with the contiguous nerves. From the union of the second and third nerves superficial branches are supplied to the head and neck, and from the junction of the third with the fourth arise some of the cutaneous nerves of the shoulder and chest. Muscular and communicating branches spring from the same nerves.

The branches of the plexus may be divided into two sets—a superficial and deep, the superficial consisting of those which perforate the cervical fascia and supply the integument, the deep comprising branches which are distributed for the most part to the muscles. The superficial nerves may be subdivided into ascending and descending, the deep nerves into a medial and a lateral series.

SUPERFICIAL ASCENDING BRANCHES

The superficial ascending branches of the cervical plexus (7.197–199) include:

Lesser occipital	2 C.
Greater auricular	2, 3 C.
Transverse (anterior) cutaneous	2, 3 C.

The lesser occipital nerve (7.197, 198) arises from the second cervical nerve, sometimes also from the third; it curves around the accessory nerve and ascends along the posterior border of the sternocleidomastoid. Near the cranium it perforates the deep fascia, and is continued upwards on the side of the head behind the auricle, supplying the skin and communicating with the great auricular and greater occipital nerves, and with the posterior auricular branch of the facial nerve. The lesser occipital nerve varies in size, and is sometimes duplicated.

It sends off an *auricular branch* which supplies the skin of the upper third of the cranial surface of the auricle, and communicates with the posterior branch of the great auricular nerve. The auricular branch is occasionally derived from the greater occipital nerve.

The great auricular nerve (7.197, 198) is the largest of the ascending branches. It arises from the second and third cervical nerves, encircles the posterior border of the sternocleidomastoid, and, after perforating the deep fascia, ascends upon that muscle beneath the platysma in company with the external jugular vein. It passes on to the parotid gland, where it divides into an anterior and a posterior branch.

The *anterior branch* is distributed to the skin of the face over the parotid gland, and communicates in the substance of the gland with the facial nerve.

The *posterior branch* supplies the skin over the mastoid process and on the back of the auricle, except at its upper part; a filament pierces the auricle to reach its lateral surface, where it is distributed to the lobule and the concha. The posterior branch communicates with the lesser occipital nerve, the auricular branch of the vagus nerve, and the posterior auricular branch of the facial nerve.

The transverse (anterior) cutaneous nerve of the neck (7.197, 198) arises from the second and third cervical nerves, turns round the posterior border of the sternocleidomastoid about its middle, and runs obliquely forwards, deep to the external jugular vein, to the anterior border of the muscle. It perforates the deep cervical fascia, and divides beneath the platysma into ascending and descending branches, which are distributed to the anterolateral parts of the neck.

The *ascending branches* pass upwards to the submandibular region, and form a plexus with the cervical branch of the facial nerve, beneath the platysma; others pierce that muscle, and are distributed to the skin of the upper and front parts of the neck.

The *descending branches* pierce the platysma, and are distributed to the skin of the side and front of the neck, as low as the sternum.

SUPERFICIAL DESCENDING BRANCHES

These are: Supraclavicular (3, 4 C.)—medial, intermediate and lateral.

The supraclavicular nerves (7.197, 198) arise by a common trunk derived from the third and fourth cervical nerves. This trunk emerges from beneath the posterior border of the sternocleidomastoid, descends under cover of the platysma and deep cervical fascia, and divides into medial, intermediate and lateral (posterior) branches, which diverge from one another and pierce the deep fascia a little above the level of the clavicle.

The *medial supraclavicular nerves* run obliquely downwards and medially, crossing the external jugular vein and the clavicular and sternal heads of the sternocleidomastoid, to supply the skin as far as the median plane and as low down as the second rib. They furnish one or two filaments to the sternoclavicular joint.

The *intermediate supraclavicular nerves* cross the clavicle, and supply the skin over the pectoralis major and deltoid as low down as the level of the second rib, immediately adjoining the area supplied by the second thoracic nerve (7.221). The amount of overlapping in this situation is minimal.

The *lateral (posterior) supraclavicular nerves* descend obliquely across the superficial surface of the trapezius and the acromion, and supply the skin of the upper and posterior parts of the shoulder.

DEEP BRANCHES—MEDIAL SERIES

These include the following communicating and muscular branches:

Communicating branches with	Hypoglossal	1, 2 C.
	Vagus	1, 2 C.
	Sympathetic	1, 2, 3, 4 C.
Muscular branches to	Rectus capitis lateralis	1 C.
	Rectus capitis anterior	1, 2 C.
	Longus capitis	1, 2, 3 C.
	Longus colli	2, 3, 4 C.
	Inferior root of ansa cervicalis	2, 3 C.
	Phrenic	3, 4, 5 C.

The communicating branches consist of several filaments which pass from the loop between the first and second cervical nerves to the vagus, hypoglossal and sympathetic. The branch to the hypoglossal ultimately leaves that nerve as a series of branches, viz. the meningeal, the *superior root of the ansa cervicalis*, the nerve to the thyrohyoid and, probably, the nerve to the geniohyoid (p. 1182). A communicating branch also passes from the fourth to the fifth cervical nerve, while each of the first

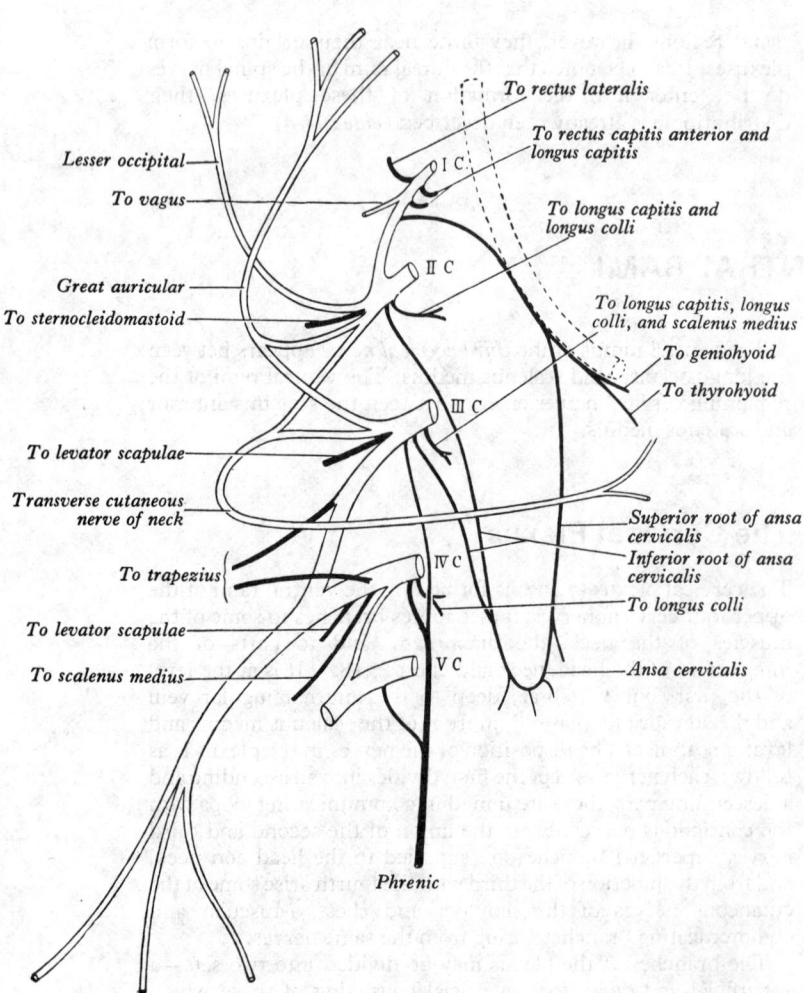

Lesser occipital

To vagus

Great auricular

To sternocleidomastoid

To levator scapulae

Transverse cutaneous nerve of neck

To trapezius

To levator scapulae

To scalenus medius

I C

II C

III C

IV C

V C

To rectus lateralis

To rectus capitis anterior and longus capitis

To longus capitis and longus colli

To longus capitis, longus colli, and scalenus medius

To geniohyoid

To thyrohyoid

Superior root of ansa cervicalis

Inferior root of ansa cervicalis

To longus colli

Ansa cervicalis

Phrenic

Supraclavicular

7.197 A plan of the cervical plexus. The hypoglossal nerve is shown by interrupted lines and the muscular branches by solid black lines. The roman numerals and letters I C to V C indicate the *ventral rami* of these cervical spinal nerves.

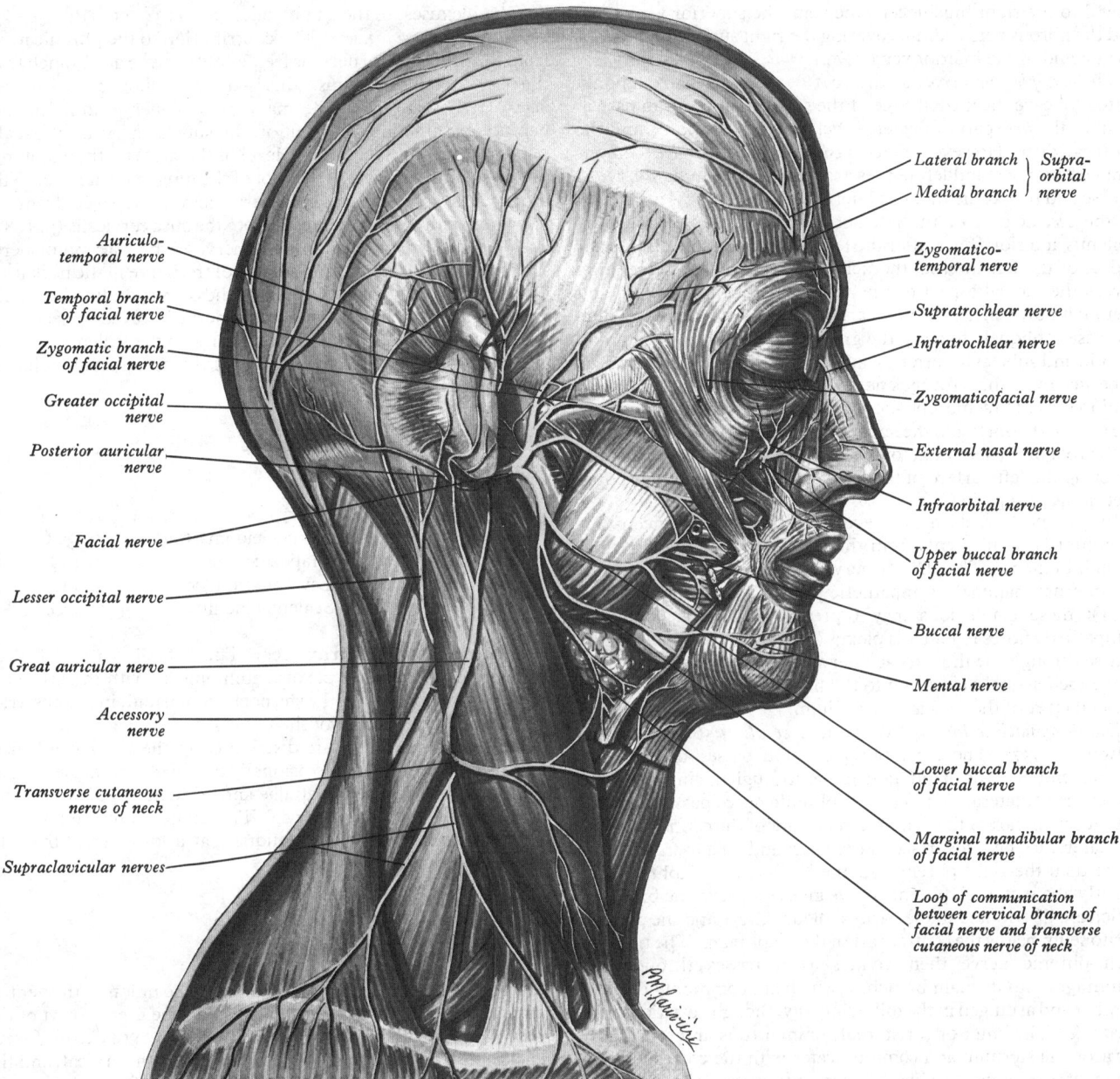

Auriculo-temporal nerve

Temporal branch of facial nerve

Zygomatic branch of facial nerve

Greater occipital nerve

Posterior auricular nerve

Facial nerve

Lesser occipital nerve

Great auricular nerve

Accessory nerve

Transverse cutaneous nerve of neck

Supraclavicular nerves

Lateral branch ⎫ Supra-orbital
Medial branch ⎭ nerve

Zygomatico-temporal nerve

Supratrochlear nerve

Infratrochlear nerve

Zygomaticofacial nerve

External nasal nerve

Infraorbital nerve

Upper buccal branch of facial nerve

Buccal nerve

Mental nerve

Lower buccal branch of facial nerve

Marginal mandibular branch of facial nerve

Loop of communication between cervical branch of facial nerve and transverse cutaneous nerve of neck

7.198   The nerves of the right side of the scalp, face and side of neck.

four cervical nerves receives a grey ramus communicans from the superior cervical ganglion of the sympathetic trunk.

Muscular branches supply the rectus capitis lateralis, rectus capitis anterior, longus capitis and longus colli.

The inferior root of the ansa cervicalis (nervus descendens cervicalis) (**7**.197) is formed usually by the union of two branches, one derived from the second cervical nerve and the other from the third. It passes downwards on the lateral side of the internal jugular vein, crosses in front of this vein a little below the middle of the neck, and continues forwards to join the superior root in front of the common carotid artery, so forming the *ansa cervicalis* (*ansa hypoglossi*). Not infrequently it passes forwards between the internal jugular vein and the common carotid artery to reach the ansa cervicalis (**6**.46), from which all the infrahyoid muscles, with the exception of the thyrohyoid, are supplied. In a series of 160 dissections of the ansa plexus, the inferior root was from the second and third cervical anterior primary rami in 74 per cent, from the second, third and fourth in 14 per cent, from the third alone in 5 per cent, from the second alone in 4 per cent and from the first, second and third in 2 per cent (Poriraer and Chernikov 1965).

The phrenic nerve is the sole motor nerve supply to the diaphragm, but it also contains sensory fibres which have a widespread distribution. It arises chiefly from the fourth cervical nerve but also receives contributions from the third and fifth cervical nerves (**7**.197). Formed at the upper part of the lateral border of the scalenus anterior, the nerve passes downwards almost vertically across the front of that muscle, *behind the prevertebral fascia* covering the anterior surface of the muscle. The phrenic nerve descends through the neck deep to sterno-cleidomastoid, the inferior belly of the omohyoid (near the inter-mediate tendon of that muscle), the internal jugular vein, the transverse cervical and surprascapular arteries (**6**.45) and, on the left side, the thoracic duct. It then runs in front of the subclavian artery and behind the subclavian vein, to enter the thorax by crossing from the lateral to the medial side, and in front of the internal thoracic artery (**6**.67). In the thorax, it descends in front of the root of the lung, between the fibrous pericardium and the mediastinal pleura, to the diaphragm, being accompanied by the pericardiacophrenic vessels. The right and left phrenic nerves differ in their intrathoracic relations.

The *right phrenic nerve*, slightly shorter and more vertical than the left, is separated at the root of the neck from the second part of the right subclavian artery by the scalenus anterior. It then lies lateral to the right brachiocephalic vein, the superior vena cava, and the fibrous pericardium covering the right surface of the right atrium and of the inferior vena cava.

The *left phrenic nerve*, at the root of the neck, is commonly stated to leave the medial edge of the scalenus anterior to pass in front of the *first* part of the left subclavian artery and behind the thoracic duct. However, Quist (1977) claims to have demonstrated that right and left nerves are *symmetrical* in their cervical course and that at the thoracic inlet the left nerve crosses anterior to the *second* part of the left subclavian artery, separated by scalenus anterior. Thereafter the left phrenic crosses the anterior aspect of the left internal thoracic artery obliquely, descending across the medial aspect of the apical part of the left lung and pleura; it now reaches the first part of the subclavian artery which it crosses obliquely to reach the groove between the left common carotid and subclavian arteries, and passes medially and forwards superficially to the left vagus nerve just above the aortic arch and behind the left brachiocephalic vein. It then passes superficially to the arch of the aorta and the left superior intercostal vein, anterior to the root of the left lung, to lie between the fibrous pericardium covering the left surface of the left ventricle and the mediastinal pleura (*see* also p. 692).

In the neck each nerve receives variable and inconstant communicating filaments from the cervical sympathetic ganglia or their branches. The phrenic nerve may also communicate with the internal mammary sympathetic plexus (Pearson and Sauter 1971); these connexions may represent a devious course of sympathetic fibres to join this plexus (see also p. 1129). During its course through the thorax, each nerve supplies sensory branches to the mediastinal pleura and to the fibrous pericardium and the parietal layer of the serous pericardium.

Diaphragmatic relations (Merendino *et al.* 1956; Perera and Edwards 1957). The right phrenic nerve passes through the central tendon of the diaphragm, either through the inferior caval orifice or just lateral to it. The left phrenic nerve passes through the muscular part of the diaphragm in front of the central tendon, just lateral to the left surface of the heart and on a more anterior plane than the right phrenic. At the level of the diaphragm, or slightly above it, each phrenic nerve gives off a few fine branches which are distributed to the parietal pleura above and the parietal peritoneum below the central part of the diaphragm. The trunk of each phrenic nerve then divides as it passes through the diaphragm into its main branches, which are commonly three in number and arranged in the following way, though variations may occur. (*a*) The anterior (or sternal) branch runs anteromedially towards the sternum and communicates with the corresponding nerve of the opposite side; (*b*) the anterolateral branch runs laterally just in front of the lateral leaflet of the central tendon; (*c*) the posterior branch is short and divides into a posterolateral branch which courses just behind the lateral leaflet of the central tendon, and a posterior (crural) branch that passes to the crural part of the diaphragm. The posterolateral and crural branches may arise separately from the phrenic nerve. These branches are often submerged in the muscular substance of the diaphragm, but may lie to some extent below it, and in addition to supplying motor fibres to the diaphragm they give off sensory fibres to the peritoneum and pleura related to the central part of the diaphragm. The branches also contain proprioceptive sensory fibres from the musculature of the diaphragm. The position of these main branches is of surgical importance in planning incisions through the diaphragm without damage to large branches of the phrenic nerve. The right crus of the diaphragm splits to enclose the oesophagus (p. 549). The right phrenic nerve supplies the part of the right crus that lies to the right of the oesophagus, while the left phrenic nerve supplies the left crus and the part of the right crus that lies to the left of the oesophagus (*see* Collis *et al.* 1954; Thornton and Scheweisthal 1969).

On the inferior surface of the diaphragm, rami of the phrenic nerves communicate with phrenic branches of the coeliac plexus (p. 1134); on the right side, at the junction of the plexuses there is a small *phrenic ganglion*. From these plexuses branches are distributed to the suprarenal glands and, on the right side, to the falciform and coronary ligaments of the liver and the inferior vena cava, and, possibly, through communications with the coeliac and hepatic plexuses, to the gall bladder (pp. 1134, 1137).

Accessory phrenic nerve. The contribution to the phrenic nerve from the fifth cervical nerve is frequently derived as a branch from the nerve to the subclavius. This is known as the *accessory phrenic nerve*. It lies lateral to the main phrenic nerve and descends behind, or sometimes in front of, the subclavian vein; it usually joins the main nerve about the level of the first rib, though it may not do so until the level of the root of the lung or even lower in the thorax. An accessory phrenic nerve may be derived from the fourth or sixth cervical nerve, or from the ansa cervicalis (p. 1083).

Applied Anatomy. The phrenic nerve is the sole motor nerve supply to the diaphragm and section of the nerve in the neck leads to complete paralysis and atrophy of the corresponding half of the diaphragm. If an accessory phrenic nerve is present, section or crushing of the main nerve alone as it lies on the scalenus anterior will not produce complete paralysis of the corresponding half of the diaphragm.

DEEP BRANCHES—LATERAL SERIES

These include:

Communicating—Accessory		2, 3, 4 C.
Muscular branches	Sternocleidomastoid	2, (3) C.
	Trapezius	3, 4 C.
	Levator scapulae	3, 4 C.
	Scalenus medius	3, 4 C.

Communicating branches. The lateral series of deep branches of the cervical plexus communicate with the accessory nerve in the substance of the sternocleidomastoid, in the posterior triangle and under cover of the trapezius.

Muscular branches are distributed to the sternocleidomastoid from the second and occasionally the third cervical nerve, and to the trapezius, levator scapulae and scalenus medius from the third and fourth cervical nerves. The branches to the trapezius cross the posterior triangle obliquely at a lower level than the accessory nerve.

The Brachial Plexus

The brachial plexus (7.200) is formed by the union of the ventral rami of the lower four cervical nerves and the greater part of the ventral ramus of the first thoracic nerve (p. 1103); the fourth cervical nerve usually gives a branch to the fifth cervical, and the first thoracic nerve frequently receives one from the second thoracic. The contributions made to the plexus by C. 4 and T. 2 are subject to frequent variation. When the branch from C. 4 is large, the branch from T. 2 is frequently absent and the branch from T. 1 is reduced in size. This constitutes the *prefixed type* of plexus. On the other hand the branch from C. 4 may be very small or entirely absent. In that event the contribution of C. 5 is reduced in size but that of T. 1 is larger and the branch from T. 2 is always present. This arrangement constitutes the *postfixed type* of plexus. These nerves constitute the *roots* of the plexus. The roots are nearly equal in size, but the way in which they form the plexus is subject to some variation. The following is, however, the most constant arrangement. The fifth and sixth cervical nerves unite at the lateral border of the scalenus medius to form the *upper trunk* of the plexus. The eighth cervical and first thoracic nerves unite behind the scalenus anterior to form the *lower trunk* of the plexus, while the seventh cervical nerve itself constitutes the *middle trunk*. These three trunks run downwards and laterally and just above or behind the clavicle, each splits into an *anterior* and a *posterior division*. The anterior divisions of the upper and middle trunks unite to form a cord, which is situated on the lateral side of the axillary artery, and is called the *lateral cord* of the plexus. The anterior division of the lower trunk passes down at first behind and then on the medial side of the axillary artery, and forms the *medial cord* of the brachial plexus; this cord frequently receives fibres from the seventh cervical nerve. The posterior divisions of all three trunks unite to form the *posterior cord* of the plexus, which is situated at first above and then behind the axillary artery.

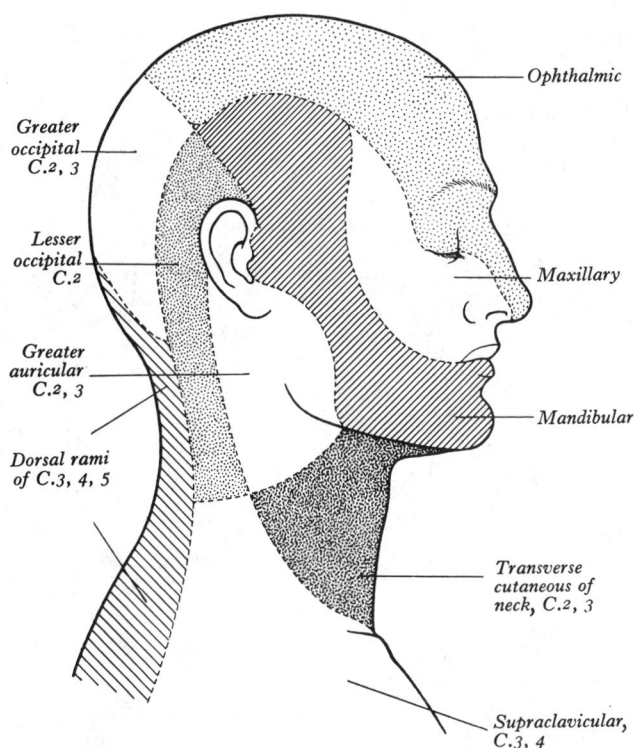

7.199 A diagram showing the cutaneous nerve supply of the face, scalp and neck.

The posterior division of the lower trunk is very much smaller than the others, and contains few fibres, if any, from the first thoracic nerve. It is frequently derived from the eighth cervical nerve before the trunk is formed.

Morphologically the brachial plexus still shows, despite much adaptation to the evolutionary changes in the upper-limb musculature, a clear reflexion of the original flexor-extensor organization of a primitive fin. The posterior cord represents the extensor nerve supply, the medial and lateral cords, the flexor supply. The migration of muscle masses has in some instances modified this basic pattern; for example, the brachialis and the anterior part of deltoid are both supplied (the former only in part) from 'extensor' nerves. For details of the comparative morphology of the plexus consult Harris (1939).

Relations of the Brachial Plexus

In the neck, the brachial plexus lies in the posterior triangle in the angle between the clavicle and the lower part of the posterior border of the sternocleidomastoid, being covered by the skin, platysma and deep fascia. When the arm is by the side, it can be felt in this situation as a bunch of tense cords. The plexus is crossed by the supraclavicular nerves, the nerve to the subclavius, the inferior belly of the omohyoid, the external jugular vein and the superficial ramus of the transverse cervical artery (**6**.64). It emerges between the scalenus anterior and scalenus medius; its proximal part is above the third part of the subclavian artery, while the lower trunk is posterior to the artery; the plexus next passes behind the anterior convexity of the medial two-thirds of the clavicle, the subclavius and the suprascapular vessels, and lies upon the first digitation of the serratus anterior and the subscapularis. *In the axilla* the lateral and posterior cords of the plexus are on the lateral side of the first part of the axillary artery, and the medial cord behind it. The cords surround the second part of the axillary artery on three sides, the medial cord lying on the medial side, the posterior cord behind, and the lateral cord on the lateral side of the artery. In the lower part of the axilla the cords split into the nerves for the upper limb. With the exception of the medial root of the median nerve, the branches of the three cords bear the same relationships to the third part of the axillary artery as the cords from which they spring bear to the second part, i.e. branches from the lateral cord are lateral, branches of the

posterior cord are behind, and branches of the medial cord are medial to the artery.

Close to their exit from the intervertebral foramina the fifth and sixth cervical nerves receive grey rami communicantes from the middle cervical ganglion, and the seventh and eighth cervical similar rami from the cervicothoracic ganglion of the sympathetic trunk (p. 1129). The first thoracic nerve also receives a grey ramus from, and contributes a white ramus to, the cervicothoracic ganglion.

The branches of the brachial plexus are usually considered, for convenience, as two groups—those arising above the clavicle (*supraclavicular*) and those below it (*infraclavicular*).

SUPRACLAVICULAR BRANCHES

The supraclavicular branches may be grouped as follows: (*a*) those arising from the roots, and (*b*) those arising from the trunks of the plexus.

From the roots of the plexus	1. To scaleni and longus colli	5, 6, 7, 8 C.
	2. To join phrenic nerve	5 C.
	3. Dorsal scapular nerve	5 C.
	4. Long thoracic nerve	5, 6, 7 C.
From the trunks of the plexus	1. Nerve to subclavius	5, 6 C.
	2. Suprascapular nerve	5, 6 C.

The branches for the scaleni and longus colli muscles arise from the lower cervical nerves close to their points of exit from the intervertebral foramina.

On the scalenus anterior the phrenic nerve is joined by a branch from the fifth cervical nerve.

The dorsal scapular nerve arises from the fifth cervical nerve, pierces the scalenus medius, passes on to the deep surface of the levator scapulae, to which it occasionally gives a twig, and runs in company with the deep branch of the dorsal scapular

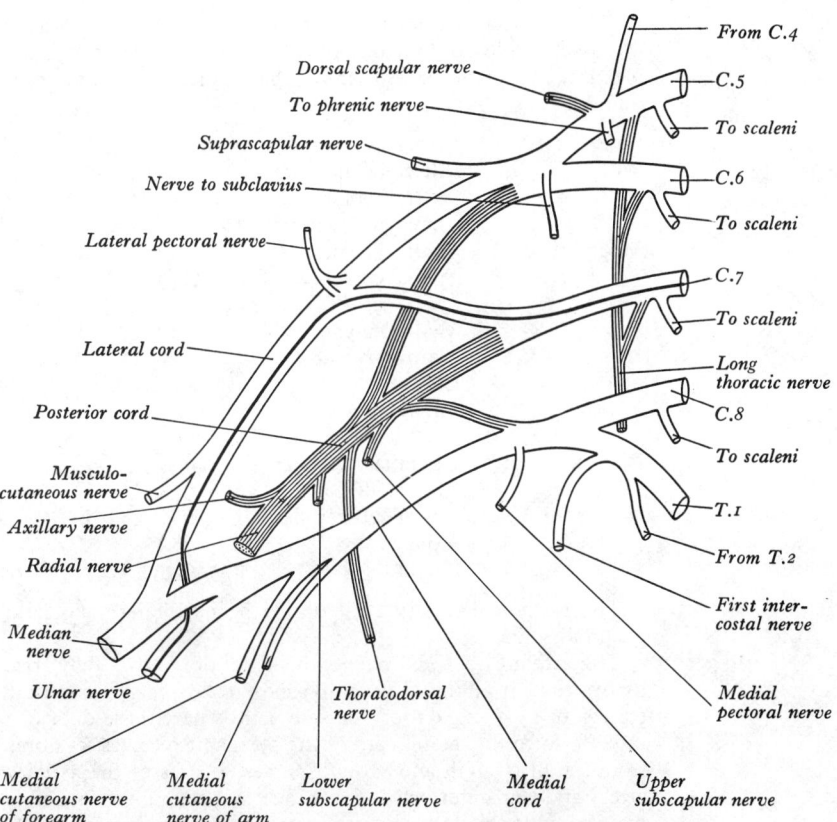

7.200 A plan of the brachial plexus. The posterior divisions of the trunks and their derivatives are shaded and the fibres from C. 7 which enter the ulnar nerve are shown as a heavy black line. Letters and numbers C. 4–C. 8 and T. 1–T. 2 indicate the *ventral rami* of these cervical and thoracic spinal nerves.

artery on the anterior surfaces of the rhomboids; it ends by supplying these muscles.

The long thoracic nerve (7.204) is usually formed by three roots from the fifth, sixth and seventh cervical nerves, but the root from the seventh nerve may be absent. (A study of seventy dissections of the nerve demonstrated all three roots in only 42 per cent of cases; *see* Alexandre *et al.* 1968.) The upper two roots pierce the scalenus medius obliquely, uniting either in the substance of the muscle or on its lateral surface, and the nerve so formed descends dorsal to the brachial plexus and the first part of the axillary artery. Having crossed the upper border of the serratus anterior to gain its outer surface, it is soon joined by the root from C. 7, which emerges from the interval between the scalenus anterior and the scalenus medius at a lower level and descends on the lateral surface of the latter muscle. The nerve is continued downwards to the lower border of the serratus anterior, supplying, in its course, filaments to each of its digitations.

The nerve to the subclavius is small and is derived near the point of junction of the fifth and sixth cervical nerves; it descends in front of the plexus and the third part of the subclavian artery, and is usually connected by a filament with the phrenic nerve. It then passes above the subclavian vein and reaches the subclavius, which it supplies.

The suprascapular nerve (6.64, 7.205) is a large branch from the superior trunk of the brachial plexus. It runs laterally deep to the trapezius and the omohyoid, and enters the supraspinous fossa through the suprascapular notch, inferior to the superior transverse scapular ligament; it then runs deep to the supraspinatus, and curves round the lateral border of the spine of the scapula in company with the suprascapular artery to gain the infraspinous fossa. In the supraspinous fossa it gives two branches to the supraspinatus and articular filaments to the shoulder joint and acromioclavicular joint; and in the infraspinous fossa it gives two branches to the infraspinatus, besides some filaments to the shoulder joint and scapula.

INFRACLAVICULAR BRANCHES

The infraclavicular branches are derived from the three cords of the brachial plexus, but their fibres may be traced through the plexus to the spinal nerves from which they originate. They are as follows:

Lateral cord	Lateral pectoral	5, 6, 7 C.
	Musculocutaneous	5, 6, 7 C.
	Lateral root of median	(5), 6, 7 C.
Medial cord	Medial pectoral	8 C., 1 T.
	Medial cutaneous of forearm	8 C., 1 T.
	Medial cutaneous of arm	8 C., 1 T.
	Ulnar	(7), 8 C., 1 T.
	Medial root of median	8 C., 1 T.
Posterior cord	Upper subscapular	5, 6 C.
	Thoracodorsal	6, 7, 8 C.
	Lower subscapular	5, 6 C.
	Axillary	5, 6 C.
	Radial	5, 6, 7, 8 C., (1 T).

The *pectoral nerves* (7.204) supply the pectoralis major and pectoralis minor.

The lateral pectoral nerve, the larger of the two, may arise by two roots from the anterior divisions of the upper and middle trunks, or by a single root from the point where these divisions unite to form the lateral cord of the plexus; it receives its fibres from the fifth, sixth and seventh nerves. It crosses the axillary artery and vein anteriorly, pierces the clavipectoral fascia, and is distributed to the deep surface of the pectoralis major. It sends a filament to join the medial pectoral nerve and forms with it a loop in front of the first part of the axillary artery (7.204); through this loop the lateral pectoral nerve distributes some fibres to the pectoralis minor.

The medial pectoral nerve receives its fibres from the eighth

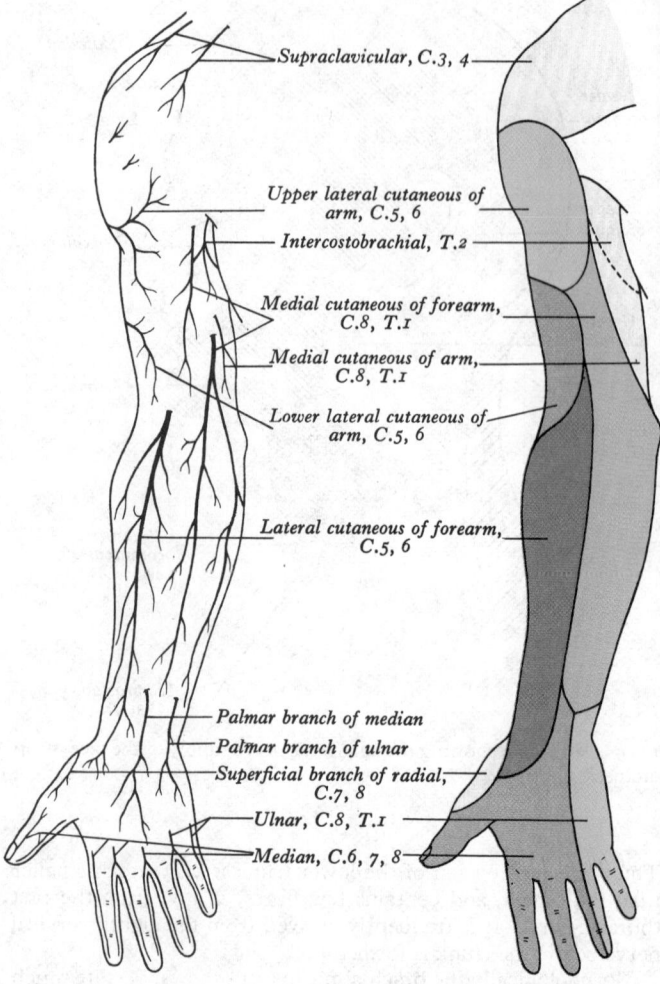

7.201 The cutaneous nerves of the right upper limb, their areas of distribution and segmental origins, viewed from the anterior aspect.

cervical and first thoracic nerves, and arises from the medial cord of the plexus while that cord is still posterior to the axillary artery. It curves forwards between the axillary artery and vein, and unites in front of the artery with a filament from the lateral pectoral nerve. It then enters the deep surface of the pectoralis minor and supplies that muscle. Two or three branches pierce the pectoralis minor, and others may pass round its inferior border, to end in the pectoralis major.

The subscapular nerves, two in number, spring from the posterior cord of the plexus, and through it from the fifth and sixth cervical nerves.

The superior subscapular nerve, the smaller, enters the subscapularis at a cranial level, and is frequently represented by two branches.

The inferior subscapular nerve supplies the caudal part of subscapularis, and ends in teres major; the latter muscle is sometimes supplied by a separate branch.

The thoracodorsal nerve, a branch of the posterior cord of the plexus, derives its fibres from the sixth, seventh and eighth cervical nerves; it arises between the upper and lower subscapular nerves and then accompanies the subscapular artery along the posterior wall of the axilla and supplies the latissimus dorsi, in which it may be traced as far as the lower border of the muscle.

The axillary (circumflex humeral) nerve (7.205) arises from the posterior cord of the brachial plexus, its fibres being derived from the fifth and sixth cervical nerves. It lies at first on the lateral side of the radial nerve and is placed behind the axillary artery, and in front of the subscapularis. At the lower border of that muscle it winds backwards in close relation to the lowest part of the articular capsule of the shoulder joint, and, in company with the posterior circumflex humeral vessels, passes through a quadrangular space bounded *above* by the subscapularis, in front,

Image label annotations: Supraclavicular, C.3, 4 / Upper lateral cutaneous of arm, C.5, 6 / Intercostobrachial, T.2 / Medial cutaneous of forearm, C.8, T.1 / Medial cutaneous of arm, C.8, T.1 / Lower lateral cutaneous of arm, C.5, 6 / Lateral cutaneous of forearm, C.5, 6 / Palmar branch of median / Palmar branch of ulnar / Superficial branch of radial, C.7, 8 / Ulnar, C.8, T.1 / Median, C.6, 7, 8

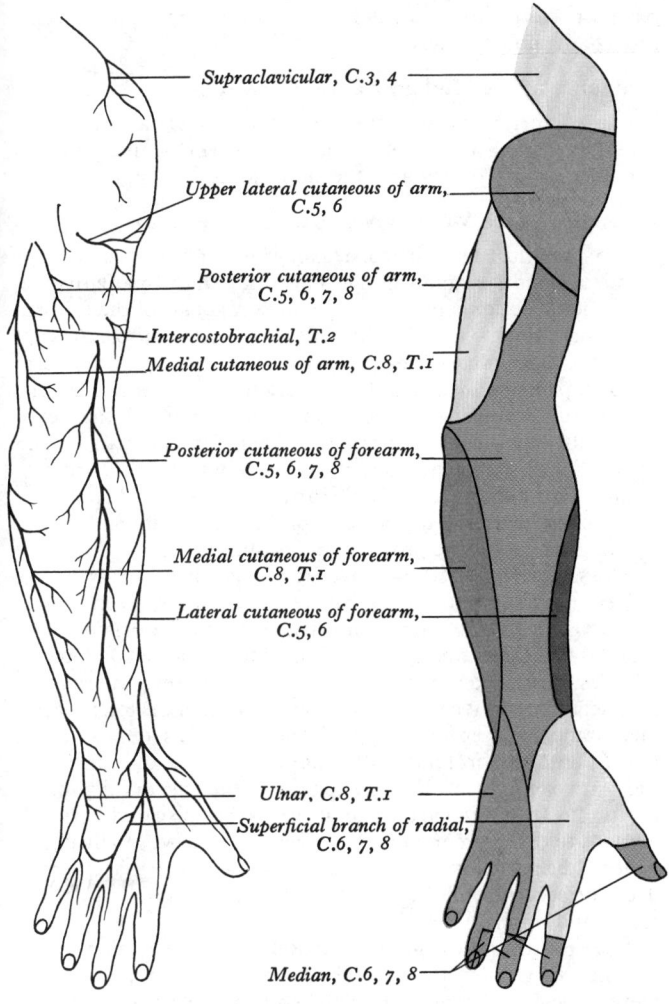

7.202   The cutaneous nerves of the right upper limb, their areas of distribution and segmental origins, viewed from the posterior aspect.

Labels on figure:

Supraclavicular, C.3, 4

Upper lateral cutaneous of arm, C.5, 6

Posterior cutaneous of arm, C.5, 6, 7, 8

Intercostobrachial, T.2

Medial cutaneous of arm, C.8, T.1

Posterior cutaneous of forearm, C.5, 6, 7, 8

Medial cutaneous of forearm, C.8, T.1

Lateral cutaneous of forearm, C.5, 6

Ulnar, C.8, T.1

Superficial branch of radial, C.6, 7, 8

Median, C.6, 7, 8

and the teres minor, behind, *below* by the teres major, *medially* by the long head of the triceps, and *laterally* by the surgical neck of the humerus. The nerve ends by dividing into an anterior and a posterior branch.

The *anterior branch*, accompanied by the posterior circumflex humeral vessels, winds round the surgical neck of the humerus, deep to the deltoid, as far as the anterior border of the muscle, supplying it, and giving a few small cutaneous branches which pierce the muscle and ramify in the skin covering its lower part.

The *posterior branch* supplies the teres minor and the posterior part of the deltoid; upon the branch to the teres minor an oval enlargement (pseudoganglion) usually exists. The posterior branch pierces the deep fascia at the lower part of the posterior border of the deltoid and is continued as the *upper lateral cutaneous nerve of the arm*, which supplies the skin over the lower part of the deltoid and the skin covering the upper part of the long head of the triceps (7.201).

The trunk of the axillary nerve gives an articular filament which enters the shoulder joint below the subscapularis.

The musculocutaneous nerve (7.204) arises from the lateral cord of the brachial plexus, opposite the lower border of the pectoralis minor, its fibres being derived from the fifth, sixth and seventh cervical nerves. It pierces the coracobrachialis and runs downwards and laterally between the biceps and the brachialis to reach the lateral side of the arm; a little beyond the elbow it pierces the deep fascia on the lateral side of the tendon of the biceps and is continued into the forearm as the *lateral cutaneous nerve of the forearm*. A line drawn distally from the lateral side of the third part of the axillary artery, and laterally across the elevations produced by the coracobrachialis and the biceps, to the lateral side of the biceps tendon of insertion, indicates the position of the musculocutaneous nerve relative to the surface; but variation in

the point of entry of the nerve into the coracobrachialis may modify considerably its surface marking (Latarjet *et al.* 1967). In its course through the arm it supplies the coracobrachialis, both heads of the biceps and the greater part of the brachialis. The branch to the coracobrachialis leaves the musculocutaneous before that nerve enters the muscle; it receives its fibres from the seventh cervical nerve, and in some instances arises directly from the lateral cord of the brachial plexus. The branches to the biceps and brachialis leave the musculocutaneous nerve after it has pierced the coracobrachialis; that supplying the brachialis gives a filament to the elbow joint. The nerve also sends a small branch to the humerus; this branch enters the bone with the nutrient artery.

The lateral cutaneous nerve of the forearm (7.201) passes deep to the cephalic vein, and descends along the radial border of the forearm to the wrist. It supplies the skin over the lateral half of the anterior surface of the forearm and distributes branches which turn round the radial border of the forearm to communicate with the posterior cutaneous nerve of the forearm and the terminal branch of the radial nerve. At the wrist joint it is placed in front of the radial artery, and some filaments, piercing the deep fascia, accompany that vessel to the dorsal surface of the carpus. The nerve then passes downwards to the base of the thenar eminence, where it ends in cutaneous filaments. It communicates with the terminal branch of the radial nerve, and with the palmar cutaneous branch of the median nerve.

The musculocutaneous nerve presents frequent variations. It may run behind the coracobrachialis or it may adhere for some distance to the median nerve and then pass behind the biceps instead of through the coracobrachialis. Some of the fibres of the median nerve may run for some distance in the musculocutaneous nerve and then leave it to join their proper trunk; less frequently the reverse is the case, and the median nerve sends a branch to join the musculocutaneous nerve. Occasionally it gives a filament to the pronator teres and, sometimes, it may replace the branches of the radial nerve to the dorsal surface of the thumb.

The medial cutaneous nerve of the forearm (7.204) arises from the medial cord of the brachial plexus. It is derived from the eighth cervical and first thoracic nerves, and at its commencement is placed between the axillary artery and vein. Near the axilla it supplies a filament which pierces the fascia and is distributed to the skin covering the biceps, almost as far as the elbow. The nerve then runs down the arm on the medial side of the brachial artery, pierces the deep fascia with the basilic vein about the middle of the arm, and divides into an anterior and a posterior branch.

The *anterior branch*, the larger, usually passes in front of, but occasionally behind, the median cubital vein. It then descends on the front of the medial side of the forearm, distributing filaments to the skin as far as the wrist, and communicating with the palmar cutaneous branch of the ulnar nerve (7.201).

The *posterior branch* passes obliquely downwards on the medial side of the basilic vein, in front of the medial epicondyle of the humerus, winds round to the back of the forearm, and descends on its medial side as far as the wrist, distributing filaments to the skin. It communicates with the medial cutaneous nerve of the arm, the posterior cutaneous nerve of the forearm, and the dorsal branch of the ulnar (7.202).

The medial cutaneous nerve of the arm is distributed to the skin on the medial side of the arm (7.201). It is the smallest branch of the brachial plexus, and, arising from the medial cord, receives its fibres from the eighth cervical and first thoracic nerves. It passes through the axilla and crosses in front of, or behind, the axillary vein. It then runs on the medial side of this vein, and communicates with the intercostobrachial nerve. It descends along the medial side of the brachial artery and basilic vein (7.204) to the middle of the upper arm, where it pierces the deep fascia, and is distributed to the skin of the medial side of the distal third of the arm, extending on to its anterior and posterior aspects; some filaments reach the skin in front of the medial epicondyle and others over the olecranon. It communicates with the posterior branch of the medial cutaneous nerve of the forearm.

In some subjects the medial cutaneous nerve of the arm and the intercostobrachial nerve are connected by two or three filaments, which form a plexus in the axilla. In others the intercostobrachial nerve is large and may be reinforced by a part of the lateral cutaneous branch of the third intercostal nerve; it then takes the place of the medial brachial cutaneous nerve, receiving from the brachial plexus a communicating filament which represents the latter nerve; occasionally this filament is absent.

THE MEDIAN NERVE

The median nerve (7.204) arises by two roots, one from the lateral cord, C. (5), 6, 7, and the other from the medial cord, C. 8, T. 1, of the brachial plexus; the roots embrace the third part of the axillary artery, uniting either in front, or on the lateral side, of the artery. Often some of the fibres derived from C. 7 leave the lateral root in

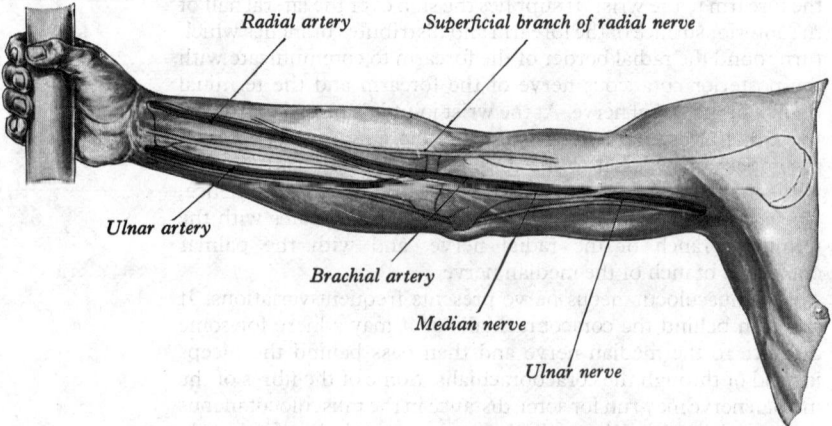

7.203 The anterior aspect of the right upper limb, showing the position of the principal nerves and vessels projected on to the surface.

the lower part of the axilla and pass distally and medially behind the medial root and usually in front of the axillary artery to join the ulnar nerve (7.220). These fibres from C. 7 may arise from the lateral cord or even directly from the seventh cervical nerve root of the brachial plexus. On clinical grounds they are believed to be mainly motor fibres to the flexor carpi ulnaris. If the lateral root is small, the musculocutaneous nerve (C. 5, 6, 7) sends a communicating branch to the median nerve in the arm. The median nerve descends into the arm, lying at first lateral to the brachial artery; about the level of insertion of the coracobrachialis it crosses in front of (rarely behind) the artery and descends on its medial side to the cubital fossa, where it lies behind the bicipital aponeurosis and in front of the brachialis, which separates it from the elbow joint. The nerve usually enters the forearm between the two heads of the pronator teres, crossing from the medial to the lateral side of the ulnar artery, but separated from it by the deep head of the muscle. It then passes deep to the tendinous bridge that connects the humero-ulnar to the radial head of the flexor digitorum superficialis and descends through the forearm deep to, and adherent to, the flexor digitorum superficialis, lying on the flexor digitorum profundus. About 5 cm above the flexor retinaculum it emerges from behind the lateral edge of the flexor digitorum superficialis and, becoming more superficial just above the wrist, it lies between the tendons of the flexor digitorum superficialis and the flexor carpi radialis, projecting laterally from under cover of the tendon of the palmaris longus (if present). The nerve then passes deep to the flexor retinaculum to gain the palm of the hand. In the forearm it is accompanied by the median branch of the anterior interosseous artery and its course can be represented on the surface by a line running approximately along the midline of the front of the forearm from the medial side of the end of the brachial artery in the cubital fossa (7.203).

The relation of the median nerve when entering the forearm is variable. According to Anson (1963) the route described above occurred in 82·8 per cent of 1,000 dissections; 10·8 per cent showed the nerve posterior to the humeral part of the muscle, the ulnar slip being absent. Less frequently (4·6 per cent) the nerve

passed posterior to both heads, and more rarely (1·8 per cent) *through* the humeral head.

Branches of the Median Nerve in the Arm

These comprise vascular branches to the brachial artery and usually a branch to the pronator teres which is given off at a variable distance proximal to the elbow joint.

Branches of the Median Nerve in the Forearm

These are muscular, articular, anterior interosseous, palmar cutaneous and communicating. The **muscular branches** are, with one exception, given off in the proximal part of the forearm (near the elbow) to the superficial flexor forearm muscles (except the flexor carpi ulnaris), namely, the pronator teres, flexor carpi radialis, palmaris longus and flexor digitorum superficialis. The branch of the median nerve supplying the part of the flexor digitorum superficialis that passes to the index finger arises near the middle of the forearm and it may be a branch of the anterior interosseous nerve. **Articular branches** arise from the median nerve at, or just distal to, the level of the elbow joint and supply this joint as well as the proximal radio-ulnar joint.

The anterior interosseous nerve is given off from the posterior surface of the median nerve as it passes between the two heads of the pronator teres and just distal to the origin of the branches of the median nerve to the superficial flexor forearm muscles described above. Accompanied by the anterior interosseous artery, it passes distally on the front of the interosseous membrane, between and deep to the flexor pollicis longus and the flexor digitorum profundus. It supplies a number of branches to these two muscles; those supplying the flexor digitorum profundus are limited to the lateral part of the muscle which sends tendons to the index and middle fingers. The nerve finally passes beneath the pronator quadratus, supplying a branch that enters its deep surface, and ends by supplying the distal radio-ulnar, radio-carpal and carpal joints.

The palmar cutaneous branch of the median nerve commences a short distance above the flexor retinaculum; it pierces the deep fascia or proximal edge of the retinaculum and divides into lateral branches which supply the skin over the thenar eminence and communicate with the lateral cutaneous nerve of the forearm, and medial branches which supply the skin of the central part of the palm and communicate with the palmar cutaneous branch of the ulnar nerve.

The communicating branch of the median nerve, which may take the form of a number of slender nerves, is frequently present. It arises from the median nerve (or sometimes its anterior interosseous branch) in the upper part of the forearm and passes distally and medially between the flexor digitorum superficialis and flexor digitorum profundus and behind the ulnar artery to join the ulnar nerve. This communication is of importance in relation to the anomalous nerve supply of certain hand muscles (p. 1101).

The Median Nerve in the Hand

Just proximal to the flexor retinaculum, the nerve is lateral to the tendons of the flexor digitorum superficialis, but dorsal to the retinaculum it lies immediately deep to the latter and on the anterior aspect of the tendons, in the limited space of the 'carpal tunnel' (*see* p. 583), i.e. the space bounded in front by the retinaculum and behind by the anterior surfaces of the carpal bones; in certain circumstances the nerve may be compressed in this situation. (*See* carpal tunnel syndrome, p. 1102.) Immediately distal to the flexor retinaculum the nerve becomes enlarged and flattened and usually divides into five or six branches, though the exact mode and level of division of the nerve are variable.

The muscular branch is a short stout nerve that arises from the lateral side of the median nerve; it may be the first branch of the median nerve in the palm or it may arise as a terminal branch at the same level as the digital branches. It runs laterally, just distal to the flexor retinaculum, with a slight recurrent curve proximally, lying beneath the lateral portion of the palmar aponeurosis covering the thenar muscles. It winds round the distal border of the flexor retinaculum and comes to lie on the superficial surface of the flexor pollicis brevis, usually giving a branch to this surface of the muscle. It then continues on the superficial surface of the muscle, or it may pass through the muscle itself. It gives a branch to the abductor pollicis brevis, which enters the muscle at its medial edge, then passes deep to this muscle to supply the opponens pollicis, entering the muscle through its medial edge. Rarely, the terminal part of the nerve gives a branch to the first dorsal interosseous, which may be the sole or part nerve supply of this muscle.

The muscular branch may arise in the carpal tunnel and pierce the flexor retinaculum, a point of surgical importance (Papathanassion 1968).

The palmar digital branches (Throughout the following account of the cutaneous innervation of the digits of the hand, note should be taken of the distinction between the proximal, penultimate, as yet undivided rami termed *common palmar digital nerves* and their ultimate branches to individual digits, the *proper palmar digital nerves*. Corresponding terms for the foot are *common plantar digital nerves* and *proper plantar digital nerves*.) The median nerve divides into four or five digital branches, though it often at first divides into two divisions, a lateral which provides the digital branches to the thumb and radial side of the index finger, and a medial, which supplies the digital branches to the adjacent sides of the index, middle and ring fingers. Other variations in the mode of terminal branching of the nerve occur, as well as in the level in the hand at which the branching takes place (Poisel 1974). The arrangement of the digital nerves is commonly as follows. They pass distally, deep to the superficial palmar arch and its digital branches, lying at first in front of the long flexor tendons. Two proper palmar digital nerves, which may arise by a common stem, pass to the sides of the thumb, the one supplying its lateral side crossing in front of the tendon of the flexor pollicis longus. The proper palmar digital nerve to the lateral side of the index finger supplies a branch to the first lumbrical. Two common palmar digital nerves pass distally between the long flexor tendons, the lateral one dividing in the distal palm into two proper palmar digital nerves traversing the adjacent sides of the index and middle fingers, and the medial one dividing into two proper palmar digital nerves which supply the adjacent sides of the middle and ring fingers. The lateral common digital nerve gives a branch to the second lumbrical, and the medial one receives a communicating twig from the common palmar digital branch of the ulnar nerve and may supply the third lumbrical muscle. In the distal part of the palm, the digital arteries pass deeply between the divisions of the digital nerves, so that on the sides of the digits the nerves lie immediately in front of the arteries. In most cases the median nerve supplies palmar cutaneous digital branches to the lateral three and one-half digits (thumb, index, middle and lateral side of the ring finger); in some cases the lateral side of the ring finger is supplied by the ulnar nerve. The proper palmar digital nerves that run along the medial side of the index finger, both sides of the middle finger and the lateral side of the ring finger, enter the bases of these digits in the fat between the slips into which the central portion of the palmar aponeurosis divides (p. 585). They pass, with the lumbricals and palmar digital arteries, dorsal to the superficial transverse metacarpal ligament (p. 586) and ventral to the deep transverse metacarpal ligament (p. 472). In the digits, the nerves run distally on the sides of the long flexor tendons, outside their fibrous sheaths, on the plane of the anterior surfaces of the phalanges, and immediately in front of the accompanying digital arteries. Each nerve gives off several branches to the skin on the front and sides of the digit, many of which end in lamellated corpuscles (p. 855), and branches to the metacarpophalangeal and interphalangeal joints. Branches are also supplied to the fibrous sheaths of the long flexor tendons, to the digital arteries (vasomotor), and to the sweat glands (secretomotor). A little beyond the base of the distal phalanx, the digital nerve gives off a branch passing dorsally to supply the nail bed, while the main nerve divides into branches supplying the skin of the terminal part of the digit and the pulp. Just beyond the base of the proximal phalanx, each *palmar* digital nerve gives off a dorsal branch that runs obliquely, distally and dorsally, to supply branches to the skin over the back of the middle and distal phalanges (7.202). The proper palmar digital nerves to the thumb and lateral side of the index finger run from beneath the lateral

edge of the central part of the palmar aponeurosis, in company with the long flexor tendons to these digits, and in the digits themselves have the same arrangements as described above, but in the case of the thumb small branches from the distal part of the palmar digital nerves are given off to supply the skin on the back of the distal phalanx only.

In addition to the branches of the median nerve described above, variable vasomotor branches pass to supply the radial and ulnar arteries and their branches. Some of the intercarpal, carpometacarpal and intermetacarpal joints are said to receive branches from the median nerve or its anterior interosseous branch, though the precise details of the origin and distribution of these branches are uncertain.

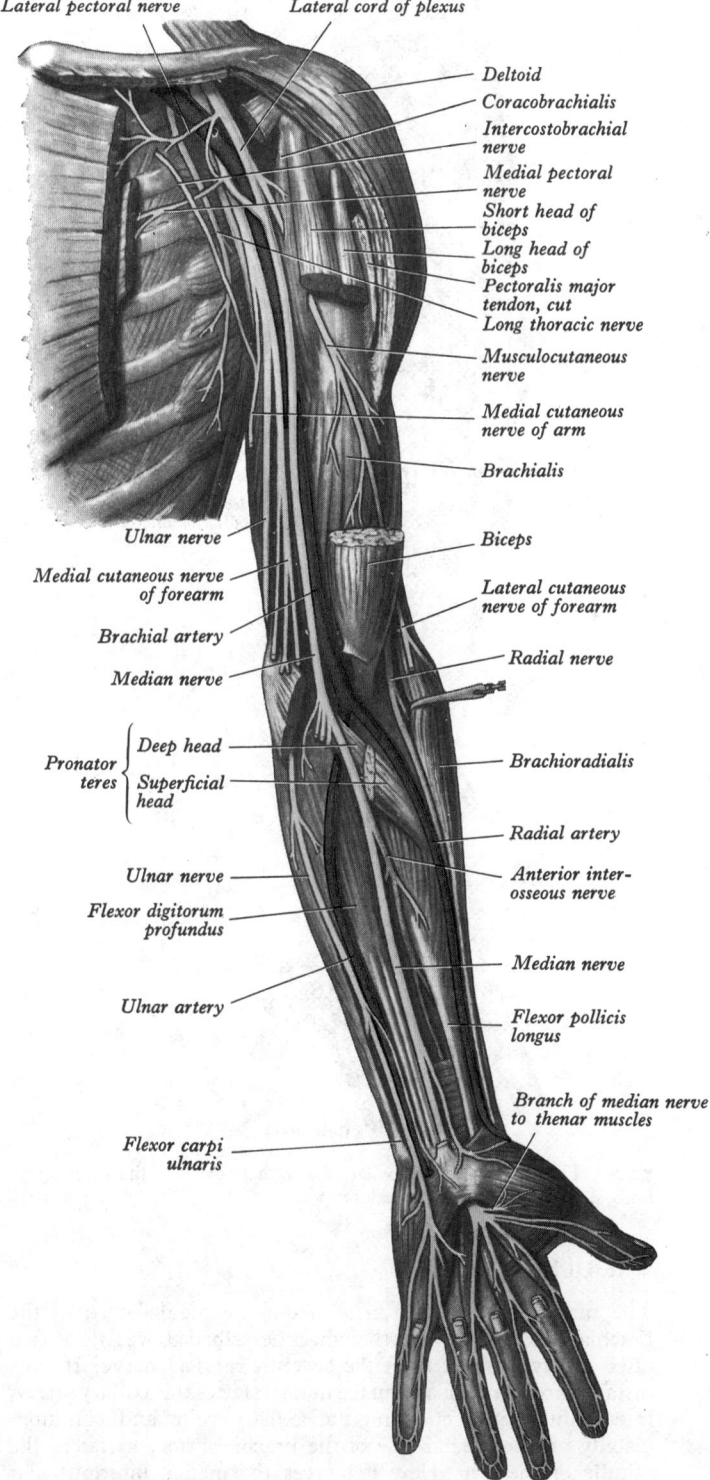

7.204 The nerves of the left upper limb dissected from the anterior aspect.

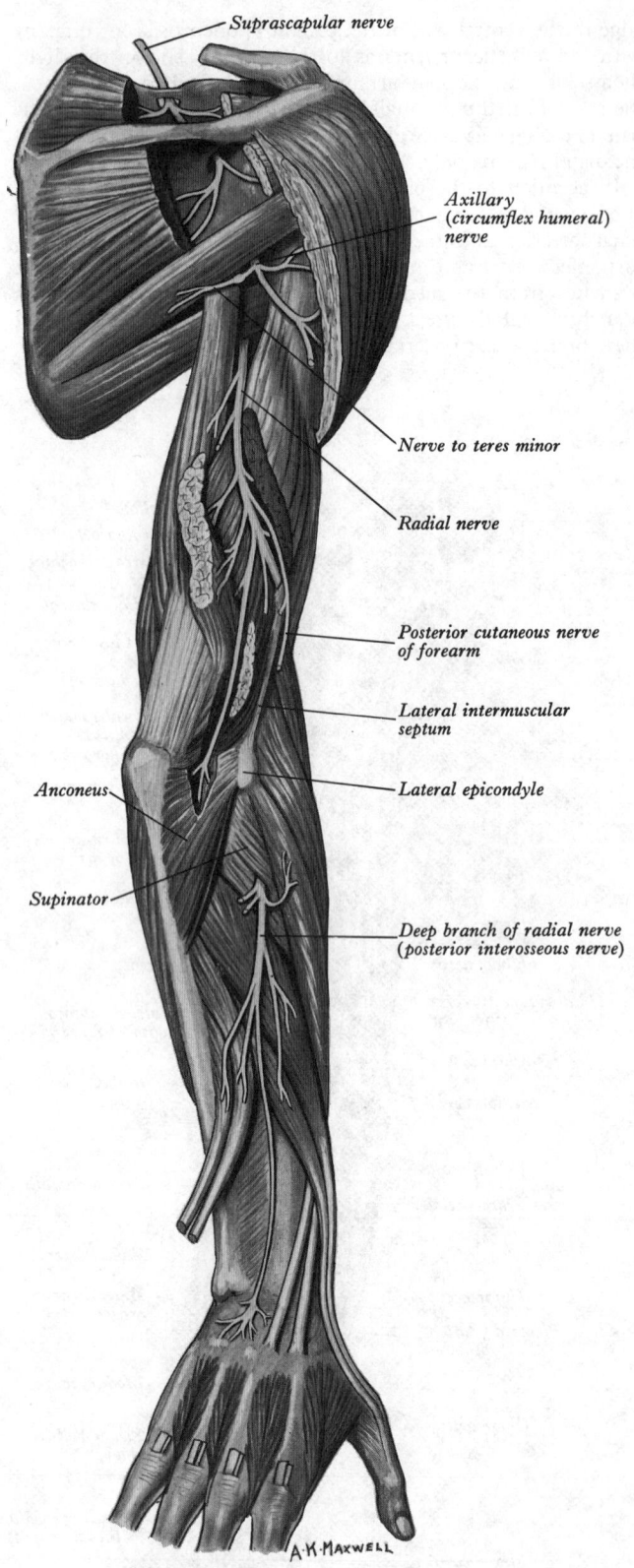

Suprascapular nerve

Axillary
(circumflex humeral)
nerve

Nerve to teres minor

Radial nerve

Posterior cutaneous nerve
of forearm

Lateral intermuscular
septum

Anconeus

Lateral epicondyle

Supinator

Deep branch of radial nerve
(posterior interosseous nerve)

A·K·MAXWELL

7.205   The suprascapular, axillary and radial nerves of the right upper limb, dissected from the posterior aspect.

THE ULNAR NERVE

The ulnar nerve (7.204) arises from the medial cord of the brachial plexus, C. 8, T. 1, though, as described above (p. 1094), it often receives fibres from the seventh cervical nerve. It runs distally through the axilla on the medial side of the axillary artery, intervening between it and the axillary vein, and continues distally on the medial side of the brachial artery as far as the middle of the arm. Here it pierces the medial intermuscular septum, and inclines medially, as it descends in front of the medial head of the triceps to the interval between the medial epicondyle

and the olecranon, accompanied by the superior ulnar collateral artery. At the elbow it lies in a groove on the dorsum of the medial epicondyle, and as it enters the forearm between the two heads of the flexor carpi ulnaris, it lies on the posterior and oblique parts of the ulnar collateral ligament of the elbow joint. It descends along the medial side of the forearm, lying upon the flexor digitorum profundus; its upper half is covered by the flexor carpi ulnaris; its lower half lies on the lateral side of this muscle, and is covered by the skin and fasciae. In the upper one-third of the forearm, the ulnar nerve is separated from the ulnar artery by a considerable interval, but in the rest of its extent it lies close to the medial side of the vessel (7.204). About 5 cm above the wrist it gives off a dorsal branch, and it is then continued downwards into the hand, passing in front of the flexor retinaculum on the lateral side of the pisiform bone and lying medial to and somewhat behind the ulnar artery. In company with the artery the nerve passes behind the superficial part of the retinaculum and ends by dividing into a superficial and a deep terminal branch. Its relationship to the brachial artery in the arm and to the medial epicondyle at the elbow renders the nerve easy to map out in the upper part of its course; a line drawn from the medial epicondyle to the lateral edge of the pisiform bone represents its course through the forearm (7.203).

The branches of the ulnar nerve are: articular, muscular, palmar cutaneous, dorsal, superficial terminal and deep terminal.

The articular branches to the elbow joint are several small filaments which issue from the nerve as it lies between the medial epicondyle and olecranon. Other articular branches are described below.

The muscular branches, two in number, begin near the elbow; one supplies the flexor carpi ulnaris (*see* p. 1098), the other, the medial half of the flexor digitorum profundus.

The palmar cutaneous branch arises about the middle of the forearm, and descends on the ulnar artery (7.204), giving some filaments to the vessel. It perforates the deep fascia and ends in the skin of the palm, after communicating with the palmar branch of the median nerve. It sometimes supplies the palmaris brevis.

The dorsal branch arises about 5 cm above the wrist; it passes distally and backwards deep to the flexor carpi ulnaris, perforates the deep fascia, and, running along the medial side of the back of the wrist and hand, divides into two, frequently three, dorsal digital nerves: one supplies the medial side of the little finger; the second the adjacent sides of the little and ring fingers. The third, when present, supplies the adjoining sides of the ring and middle fingers, but it may be replaced, wholly or partially, by a branch of the radial nerve, with which it always communicates on the dorsum of the hand (7.202). In the little finger the dorsal digital nerves extend only as far as the base of the distal phalanx, and in the ring finger as far as the base of the middle phalanx; the more distal parts of these digits are supplied by dorsal branches derived from the proper digital branches of the ulnar and—on the lateral side of the ring finger—median nerves.

The superficial terminal branch supplies the palmaris brevis and the skin on the medial side of the hand, and divides into two palmar digital nerves, which can be compressed against the hook of the hamate bone (p. 371). One of these palmar digital nerves supplies the medial side of the little finger; the other (a common palmar digital nerve) sends a twig to join the median nerve and then divides into two proper digital nerves for the adjoining sides of the little and ring fingers (7.204). The proper digital branches are distributed to the fingers in the same manner as those of the median nerve. Murakami (1969) has described articular branchlets from the superficial terminal branch of the ulnar nerve.

The deep terminal branch, accompanied by the deep branch of the ulnar artery, passes between the abductor digiti minimi and flexor digiti minimi; it then perforates the opponens digiti minimi and follows the course of the deep palmar arch behind the flexor tendons. At its origin it supplies the three short muscles of the little finger. As it crosses the hand, it gives branches to the interossei and to the third and fourth lumbricals; it ends by supplying the adductor pollicis, the first palmar interosseous and, in most cases (p. 587), the flexor pollicis brevis. It also sends articular filaments to the wrist joint.

It has been pointed out that the medial part of the flexor digitorum profundus is supplied by the ulnar nerve; the third and fourth lumbricals which are connected with the tendons of this part of the muscle, are supplied by the same nerve. In like manner the lateral part of the flexor digitorum profundus and the first and second lumbricals are supplied by the median nerve. The third lumbrical frequently receives an additional twig from the median nerve.

The deep terminal branch of the ulnar nerve is said to give branches to some of the intercarpal, carpometacarpal and intermetacarpal joints, though, as in the case of the median nerve, the precise details of the origin and distribution of these branches are uncertain. Vascular (vasomotor) branches are given off from the ulnar nerve in the forearm and hand to supply the ulnar artery and the palmar arteries.

Anomalous Nerve Supply of Hand Muscles

The nerve supply to the short thenar muscles (flexor pollicis brevis, abductor pollicis brevis and opponens pollicis) is subject to considerable variation. From a study of the results of lesions of the median and ulnar nerves in the forearm (Rowntree 1949; Day and Napier 1961) the following variations have been deduced, expressed here in percentages of 226 hands examined, relating to the innervation of these thenar muscles by the median nerve, the ulnar nerve or both nerves: flexor pollicis brevis—median 36, ulnar 48, both nerves 17; abductor pollicis brevis—median 95, ulnar 2·5, both nerves 2; opponens pollicis—median 83, ulnar 9, both nerves 7·5. The usual description of the nerve supply to these muscles is that they are all supplied by the median nerve; the results of the investigations cited are possibly to be explained, in many cases, by the variable connexions between the median and ulnar nerves in the axilla, arm or forearm, whereby truly median nerve fibres are aberrantly conveyed distally in the ulnar nerve to these muscles.

An arcuate connecting loop has long been known to occur between the median and ulnar nerves in the substance of the flexor pollicis brevis (Cannieu 1886), either part of which, superficial or deep, may be supplied by fibres from both nerves. In a more recent survey (Harness and Sekeles 1971) of the considerable literature on this topic, results of dissections of 35 hands showed a loop in 77 per cent of individuals, indicating that it should be a feature of 'normal' description.

Clinically these variations are important in that even with a complete lesion of the median nerve some of these muscles may not be paralysed, and this may lead to the erroneous conclusion that the median nerve has not suffered a complete lesion. Clinical evidence reveals that the short muscles of the thumb, whether apparently supplied by the median or ulnar nerves, receive their segmental supply from the eighth cervical and first thoracic segments of the spinal cord.

THE RADIAL NERVE

The radial nerve (7.205) arises from the posterior cord of the brachial plexus, C. 5, 6, 7, 8; (T. 1). It is the largest branch of the brachial plexus, and it descends behind the third part of the axillary artery and the upper part of the brachial artery, and in front of the subscapularis and the tendons of the latissimus dorsi and teres major. Accompanied by the arteria profunda brachii and, later, its radial collateral branch, it inclines dorsally between the long and medial heads of the triceps. From here it passes obliquely across the back of the humerus, first between the lateral and medial heads of the triceps and then in a shallow groove deep to the lateral head of the triceps muscle. On reaching the lateral side of the humerus it pierces the lateral intermuscular septum and enters the anterior compartment of the arm. It then descends, lying deeply in the intermuscular furrow, which is bounded on the medial side by the brachialis and on the lateral side by the brachioradialis, above, and the extensor carpi radialis longus, below. On reaching the front of the lateral epicondyle, it divides into terminal rami, *superficial* and *deep*.

In the arm the radial nerve is indicated by a line drawn from the commencement of the brachial artery and carried distally and laterally across the elevations produced by the long and the lateral heads of the triceps to the junction of the upper and middle thirds of a line joining the lateral epicondyle to the deltoid tuberosity. The line of the nerve is then continued on the anterior aspect of the arm to the level of the lateral epicondyle, where it lies 1 cm or less to the lateral side of the biceps tendon.

The branches of the radial nerve are: muscular, cutaneous, articular and superficial and deep terminal branches.

The muscular branches of the radial nerve supply the triceps, anconeus, brachioradialis, extensor carpi radialis longus and brachialis, and are grouped as *medial, posterior* and *lateral*.

The *medial* muscular branches arise from the radial nerve on the medial side of the arm and supply the medial and long heads of the triceps; the branch to the medial head is a long, slender filament, which lies close to the ulnar nerve as far as the distal third of the arm, and is therefore frequently named the *ulnar collateral nerve*.

The *posterior* muscular branch, of large size, arises from the nerve as it lies in the groove. It divides into filaments which supply the medial and lateral heads of the triceps and the anconeus. The branch for the latter muscle is a long nerve which descends in the substance of the medial head of the triceps, and gives numerous branches to it. It is accompanied by the middle collateral branch of the arteria profunda brachii, and passes behind the elbow joint to end in the anconeus.

The *lateral* muscular branches arise from the nerve as it lies in front of the lateral intermuscular septum; they supply the lateral part of the brachialis, the brachioradialis, and the extensor carpi radialis longus.

The cutaneous branches of the radial nerve are the posterior cutaneous and the lower lateral cutaneous nerves of the arm and the posterior cutaneous nerve of the forearm.

The posterior cutaneous nerve of the arm, of small size, arises in the axilla and passes to the medial side of the arm to supply the skin on its dorsal surface nearly as far as the olecranon. It crosses posterior to, and communicates with, the intercostobrachial nerve.

The lower lateral cutaneous nerve of the arm perforates the lateral head of the triceps just below the insertion of the deltoid. It then passes to the front of the elbow, lying close to the cephalic vein, and supplies the skin of the lateral part of the lower half of the arm (7.201).

The posterior cutaneous nerve of the forearm arises in common with the preceding branch. Perforating the lateral head of the triceps, it descends along the lateral side of the arm, and then along the dorsum of the forearm to the wrist, supplying the skin in its course, and joining, near its termination, with dorsal branches of the lateral cutaneous nerve of the forearm (7.201).

The articular branches of the radial nerve are distributed to the elbow joint.

The superficial terminal branch descends from the front of the lateral epicondyle along the front of the lateral side of the upper two-thirds of the forearm, lying at first upon the supinator, lateral to the radial artery and behind the brachioradialis. In the middle third of the forearm it lies behind the brachioradialis, but is now close to the lateral side of the artery. Here it lies first on the pronator teres, next on the radial head of the flexor digitorum superficialis and then on the flexor pollicis longus. It quits the artery about 7 cm above the wrist, passes deep to the tendon of the brachioradialis and, winding round the lateral side of the radius as it descends, pierces the deep fascia and divides into five, sometimes four, dorsal digital nerves.

The dorsal digital nerves are small and four or five in number. The first supplies the skin of the radial side of the thumb and the adjoining part of the thenar eminence, communicating with branches of the lateral cutaneous nerve of the forearm; the second supplies the medial side of the thumb; the third, the lateral side of the index finger; the fourth, the adjoining sides of the index and middle fingers; the fifth communicates with a filament from the dorsal branch of the ulnar nerve and supplies the adjoining sides of the middle and ring fingers, but it is frequently replaced by the dorsal branch of the ulnar nerve. On the dorsum of the hand the superficial branch of the radial nerve usually communicates with the posterior and lateral cutaneous nerves of the forearm. The digital nerves to the thumb reach only as far as the root of the nail; those to the index finger as far as the middle of the middle phalanx and those to the middle and ring fingers not farther than the proximal interphalangeal joints. The remaining distal areas of skin on the dorsal surface of these digits are supplied by the palmar digital branches of the median and ulnar nerves (7.202). The superficial radial nerve may supply the whole of the dorsum of the hand; for this and other variations consult Sayfi (1967).

The deep terminal branch (posterior interosseous nerve) (7.205) winds to the back of the forearm round the lateral side of the radius between the two planes of fibres of the supinator. It gives a branch to the extensor carpi radialis brevis, and another to the supinator before it enters the latter muscle, and as it traverses its substance it supplies additional branches to it. The branch to the extensor carpi radialis brevis may spring from the commencement of the superficial branch of the radial nerve. As soon as it emerges from the supinator on the back of the forearm the deep branch of the radial nerve gives off three short branches—to the extensor digitorum, extensor digiti minimi and extensor carpi ulnaris—and two long branches—a *medial* to the extensor pollicis longus and the extensor indicis, and a *lateral*, which supplies the abductor pollicis longus and ends in the extensor pollicis brevis. The nerve lies at first between the superficial and the deep muscles of the back of the forearm, but, at the distal border of the extensor pollicis brevis, it passes deep to the extensor pollicis longus and, diminished to a fine thread, runs down on the dorsal aspect of the interosseous membrane of the forearm. Finally it reaches the dorsum of the carpus, where it presents a flattened and somewhat expanded termination ('pseudoganglion') from which filaments are distributed to the ligaments and articulations of the carpus (7.205).

Articular branches from the deep branch of the radial nerve are distributed to the carpal and distal radio-ulnar joints and to some of the intercarpal and intermetacarpal joints, while the digital branches of the radial nerve supply branches to the metacarpo-phalangeal and proximal interphalangeal joints of the appropriate digits.

Applied Anatomy. The brachial plexus may be injured by falls from a height on to the side of the head and shoulder, the nerves of the plexus being violently stretched; the upper trunk of the plexus sustains the greatest injury, and the subsequent paralysis may be confined to the muscles supplied by the fifth nerve—the deltoid, biceps, brachialis and brachioradialis, with sometimes the supraspinatus, infraspinatus and supinator. The position of the limb, under such conditions, is characteristic: the arm hangs by the side and is rotated medially; the forearm is extended and pronated. The arm cannot be raised from the side; all power of flexion of the elbow is lost, as is also supination of the forearm. This is known as *Erb's paralysis*, and a very similar condition is occasionally met with in newborn children, either from injury to the upper trunk from the pressure of the forceps used in effecting delivery, or from traction of the head in breech presentations. A second variety of partial palsy of the brachial plexus is known as *Klumpke's paralysis*. In this it is the eighth cervical and first thoracic nerves that are injured, either before or after they have joined to form the lower trunk. The subsequent paralysis affects, principally, the intrinsic muscles of the hand and the flexors of the wrist and fingers.

The brachial plexus may also be injured by direct violence or gunshot wounds, by violent traction on the arm, or by efforts at reducing a dislocation of the shoulder joint; the amount of paralysis will depend upon the amount of injury to the constituent nerves. When the entire plexus is involved, the whole of the upper extremity will be paralysed and anaesthetic. In some cases the injury appears to be rather a tearing away of the roots of the nerves from the spinal cord than a rupture of the nerves themselves, and where this involves the first thoracic nerve the pupil on the same side may be constricted, on account of damage to the preganglionic fibres emerging in it to supply the dilatator pupillae. The brachial plexus in the axilla is often damaged from the pressure of a crutch, producing the condition known as '*crutch paralysis*'. In these cases the radial is the nerve most frequently implicated; the ulnar nerve suffers next in frequency. The median

and radial nerves often suffer from '*sleep palsies*', paralysis from pressure coming on while the patient is profoundly asleep under the influence of alcohol or some narcotic.

Paralysis of the long thoracic nerve throws the serratus anterior out of action, and may occur in porters who have to carry heavy weights on the shoulder, for the nerve is exposed to injury as it lies in the posterior triangle of the neck. The inferior angle of the scapula is drawn towards the median plane, by the unopposed action of the rhomboids and levator scapulae, and tends to project backwards ('winging' of the scapula) when the arm is held horizontally forwards or when forward pushing movements are attempted against resistance. The arm cannot be raised above the horizontal unless the inferior angle of the scapula is pushed anterolaterally for the patient.

The *axillary (circumflex humeral) nerve*, on account of its course round the surgical neck of the humerus, is liable to be injured in fractures of this part of the bone, and in dislocations of the shoulder joint; paralysis of the deltoid, and anaesthesia of the skin over the lower part of that muscle, result. Paralysis of the deltoid renders effective abduction of the arm impossible. The associated paralysis of the teres minor is not easily demonstrated.

The *median nerve* is liable to injury in wounds of the forearm. When it is completely divided proximal to its muscular and anterior interosseous branches, there is loss of flexion of the second phalanges of all the fingers, and of the terminal phalanges of the index and middle fingers. Flexion of the terminal phalanges of the ring and little fingers is effected by the section of the flexor digitorum profundus which is supplied by the ulnar nerve. There is power to flex the proximal phalanges through the interossei. The thumb cannot be opposed or abducted, nor can it be flexed at the interphalangeal joint (*see* p. 1099), and it is maintained in a position of extension and adduction. There is loss in the power of pronating the forearm; the brachioradialis has the power of bringing the forearm into a position of mid-pronation, but beyond this no further pronation can be effected. The wrist can be flexed by the flexor carpi ulnaris, but flexion is combined with adduction of the hand. There is loss or impairment of sensation on the palmar surfaces of the thumb, index, middle, and radial half of the ring fingers, and on the dorsal surfaces of the same fingers over the last two phalanges, except in the thumb, where the loss of sensation is limited to the back of the distal phalanx. Owing to the paralysis of the short muscles of the thumb and the unopposed action of the extensor pollicis longus, an '*ape-like*' hand is produced. Injury of the nerve just above the middle of the forearm may result only in weakness of flexion of the index finger ('*pointing*' index finger), since the branch to the part of the flexor digitorum superficialis distributed to that finger arises at about the middle of the forearm (p. 1098). More commonly, however, the nerve is injured proximal to the flexor retinaculum, when the power of flexion of the fingers and pronation of the forearm remains intact, unless the flexor tendons are also divided. The chief effect of lesions here is usually inability to oppose the thumb, though the intact abductor pollicis longus and adductor pollicis may combine to imitate this action. As it lies deep to the flexor retinaculum, the median nerve is situated in the restricted space (carpal tunnel or carpal canal) between the retinaculum and the carpal bones. Any pathological condition which diminishes the size of this tunnel (e.g. carpal dislocation or arthritis, tenosynovitis of the long flexor tendons, etc.) will cause pressure on the nerve with resultant pain and slight sensory impairment in the digits supplied by the nerve and sometimes slight wasting of the thenar muscles, these clinical features constituting the '*carpal tunnel syndrome*'. Most commonly there is no apparent cause for this syndrome and, in any case, complete division of the retinaculum is curative.

The *ulnar nerve* is also liable to be injured in wounds of the forearm. The commonest cause of complete or partial ulnar nerve lesion is injury to the nerve as it lies behind the medial humeral epicondyle. Such injury leads to impaired power of adduction, and, upon an attempt being made to flex the wrist, the hand is drawn to the radial side by the flexor carpi radialis; there is inability to spread out the fingers owing to paralysis of the dorsal interossei, and for the same reason the fingers, especially the ring and little fingers, cannot be flexed at the metacarpophalangeal joints or extended at the interphalangeal joints, and the hand assumes a claw shape from the action of the opposing muscles; there is loss in the power of flexion in the fourth and fifth digits and inability to adduct the thumb. The muscles of the hypothenar eminence become wasted. Sensation is lost, or impaired, in the skin supplied by the nerve.

The *radial nerve* also is frequently injured. In consequence of its close relationship to the humerus, it may be injured, but is seldom torn, in fractures of this bone. Callus formation in repair of the fracture seldom interferes with the function of the nerve. In fractures of the middle part of the humerus, the triceps is not paralysed, since its supplying nerves arise from the radial nerve more proximally. It is also liable to be contused against the bone by kicks or blows, or to be divided in wounds of the arm. When paralysed, the hand is flexed at the wrist and lies flaccid. This is known as *wrist drop*. The fingers are also flexed, and when an attempt is made to extend them, the last two phalanges only will be extended, through the action of the lumbrical and interosseous muscles; the first phalanges remain flexed. Extension of the wrist is impossible. Supination is completely lost when the forearm is extended on the arm, but is possible to a certain extent if the forearm be flexed to allow effective action of the biceps. The power of extension of the forearm is lost on account of paralysis of the triceps, if the injury to the nerve has taken place near its origin. As the radial nerve has only a very small area of exclusive supply, the extent of the anaesthesia associated with even severe injuries to it is surprisingly small and is confined to a limited region on the lateral part of the dorsum of the hand.

Dislocations and epiphyseal separations of the elbow and supracondylar fractures of the humerus in children frequently lead to ulnar, median or posterior interosseous injury.

THORACIC VENTRAL RAMI

The ventral rami of the thoracic nerves (7.206, 207) are twelve in number on each side. The upper eleven lie between the ribs (*intercostal nerves*), while the twelfth lies below the last rib (*subcostal nerve*). Each nerve is connected with the adjoining ganglion of the sympathetic trunk by grey and white rami communicantes: the grey ramus joins the nerve proximal to the point at which the white ramus leaves it. The intercostal nerves are distributed chiefly to the thoracic and abdominal walls. The first two nerves supply fibres to the upper limb in addition to their thoracic branches; the next four are limited in their distribution to the thoracic wall; the lower five supply the thoracic and abdominal walls (p. 1104); the subcostal nerve is distributed to the abdominal wall and the gluteal skin. Communicating branches link the intercostal nerves in the posterior parts of the intercostal spaces and, in addition, the lower five communicate freely as they traverse the abdominal wall (Davies *et al.* 1932).

The 1st to 6th Thoracic Nerves

The ventral ramus of the first thoracic nerve divides into a large and a small branch. The large branch ascends in front of the neck of the first rib on the lateral side of the superior intercostal artery, and enters the brachial plexus (p. 1094). The small branch is the *first intercostal nerve*; it runs along the first intercostal space, and ends on the front of the chest as the first anterior cutaneous nerve of the thorax. It furnishes a lateral cutaneous branch which pierces the chest wall in front of the serratus anterior and supplies the skin of the axilla; it may communicate with the inter-costobrachial nerve and sometimes joins the medial cutaneous nerve of the arm (Cave 1929). The first thoracic nerve frequently receives a connecting ramus from the second; this twig ascends in front of the neck of the second rib.

The ventral rami of the second, third, fourth, fifth and sixth thoracic spinal nerves pass forwards (7.201) in the intercostal spaces below the intercostal vessels. At the back of the chest they lie between the pleura and the posterior intercostal membranes, but in most of their course they run between the internal intercostals and the subcostales and intercostales intimi (7.207). Near the sternum, they cross in front of the internal thoracic artery and transversus thoracis, pierce the internal intercostals, the external intercostal membranes, and the pectoralis major, and their terminal branches form the **anterior cutaneous nerves of the thorax**; they supply the skin of the front of the thorax; the anterior cutaneous branch of the second nerve may be connected to the medial supraclavicular nerves of the cervical plexus. Twigs from the anterior cutaneous branch of the sixth intercostal nerve supply the abdominal skin in the upper part of the infrasternal angle.

Branches. Numerous slender muscular filaments supply the intercostals, the serratus posterior superior, and the transversus thoracis. At the front of the thorax some of these branches cross the costal cartilages from one intercostal space to another.

Each intercostal nerve gives off a collateral and a lateral cutaneous branch before it reaches the angle of adjoining ribs. The *collateral branch* follows the caudal border of the space in the same intermuscular interval as the main trunk, which it may rejoin before it is distributed as an additional anterior cutaneous nerve. The *lateral cutaneous branch* accompanies the main trunk for a time and then pierces the intercostal muscles obliquely. With the exception of the lateral cutaneous branches of the first and second intercostal nerves, each divides into anterior and posterior branches, which subsequently pierce the serratus anterior. The *anterior branches* run forwards over the border of the pectoralis major and supply twigs to the overlying skin; those of the fifth and sixth nerves supply twigs to a variable number of the upper digitations of the obliquus abdominis externus. The *posterior branches* run backwards, and supply the skin over the scapula and latissimus dorsi.

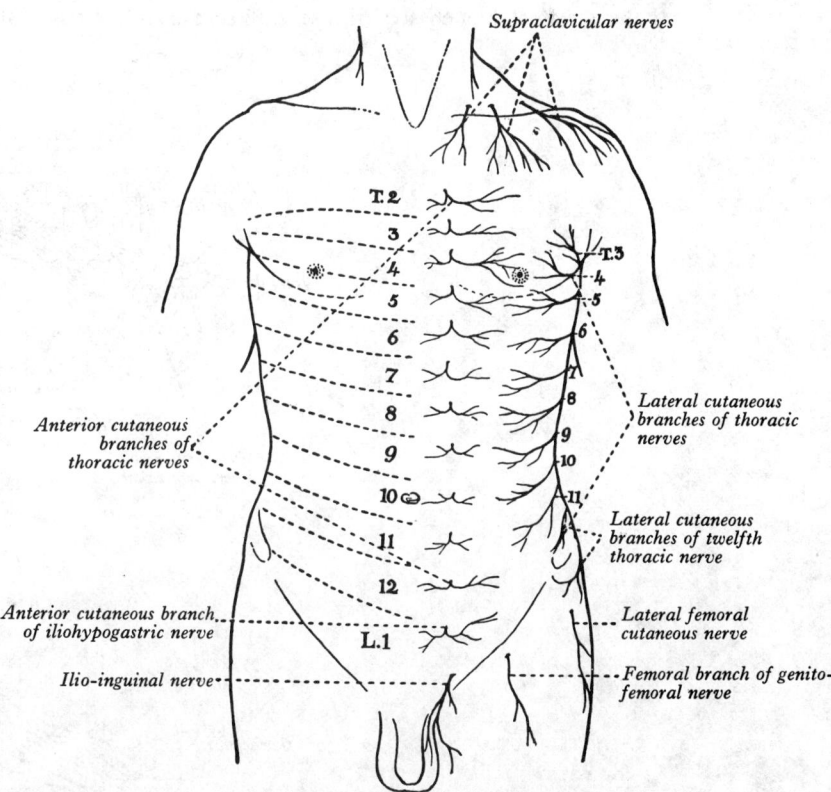

7.206 A diagram showing the approximate segmental distribution of the cutaneous nerves on the front of the trunk. The contribution from the first thoracic spinal nerve is not shown, and the considerable overlap which occurs between adjacent segments is not indicated. For the latter see 7.221.

The lateral cutaneous branch of the second intercostal nerve is named the *intercostobrachial nerve* (7.204). It crosses the axilla to gain the medial side of the arm, and joins with a filament from the medial cutaneous nerve of the arm. It then pierces the deep fascia of the arm, and supplies the skin of the upper half of the medial and posterior parts of the arm, communicating with the posterior brachial cutaneous branch of the radial nerve. The size of the intercostobrachial nerve is in inverse proportion to that of the medial brachial cutaneous nerve. A second intercostobrachial nerve is frequently given off from the anterior part of the lateral cutaneous branch of the third intercostal nerve; it supplies filaments to the axilla and to the medial side of the arm.

The 7th to 12th Thoracic Nerves

The ventral rami of the seventh, eighth, ninth, tenth and eleventh thoracic nerves are continued anteriorly from the intercostal spaces into the abdominal wall.

As they approach the anterior ends of the spaces in which they lie, the seventh and eighth nerves curve *upwards* and medially across the deep surface of the costal margin, insinuating themselves between the digitations of the transversus abdominis to gain the deep aspect of the posterior lamella of the aponeurosis of the internal oblique. Having pierced this layer, they lie behind the rectus abdominis and continue upwards and medially (7.208) for a short distance parallel with the costal margin. Both supply the rectus abdominis and, having passed through the muscle near its lateral edge, pierce the anterior wall of its sheath, to reach and supply the skin. It will be observed that both the seventh and the eighth intercostal nerves cross the costal margin medial to the lateral border of the rectus abdominis and therefore enter its sheath by piercing its posterior wall.

The ninth, tenth and eleventh intercostal nerves pass between the digitations of the diaphragm and transversus abdominis to gain the interval between the latter muscle and the internal oblique. In this intermuscular interval the ninth nerve runs almost *horizontally*, but the tenth and eleventh nerves are inclined caudally and medially. When they reach the lateral edge of the rectus abdominis, they pierce the posterior lamella of the internal oblique aponeurosis and pass behind the muscle. They end like the terminal branches of the seventh and eighth intercostal

nerves. The tenth nerve supplies the band of skin which includes the umbilicus (7.206, 221).

The lower intercostal nerves supply the intercostal, the subcostal and the abdominal muscles, and the last three send branches to the serratus posterior inferior. They also supply sensory fibres to the costal part of the diaphragm and the related parietal pleura and peritoneum. Like the upper intercostal nerves the lower intercostal nerves give off *collateral* and *lateral cutaneous branches* before they reach the angles of the ribs. The collateral branch may rejoin the main trunk, but, if it does so, it leaves it again near the lateral border of the rectus abdominis and runs forwards below it (7.208). It pierces the muscle and the anterior wall of its sheath near the linea alba and supplies the skin. The lateral cutaneous branches pierce the intercostals and the external oblique, in the same line as the lateral cutaneous branches of the upper thoracic nerves, and divide into anterior and posterior branches, which are distributed to the skin of the abdomen and back respectively; the anterior branches also supply twigs to the digitations of the external oblique, and extend downwards and forwards nearly as far as the margin of the rectus abdominis; the posterior branches pass backwards to supply the skin over the latissimus dorsi. Each lateral cutaneous branch descends as it pierces the external oblique and the superficial fascia so that it reaches the skin on a level with the corresponding anterior cutaneous branch and the cutaneous branch of the corresponding dorsal ramus (p. 1090 and 7.196).

The ventral ramus of the **twelfth thoracic nerve** (subcostal nerve) is larger than the others, and gives a communicating branch to the first lumbar nerve (sometimes termed the *dorsolumbar nerve*). Like the intercostal nerves it soon gives off a collateral branch. It accompanies the subcostal artery along the lower border of the twelfth rib and passes behind the lateral arcuate ligament. It then runs behind the kidney (8.148), and in front of the upper part of the quadratus lumborum, perforates the aponeurosis of origin of the transversus and passes forwards between that muscle and the obliquus internus, to be distributed in the same manner as the lower intercostal nerves. It communicates with the iliohypogastric nerve of the lumbar plexus, and gives a branch to the pyramidalis. The *lateral cutaneous branch* of the twelfth thoracic nerve pierces the internal

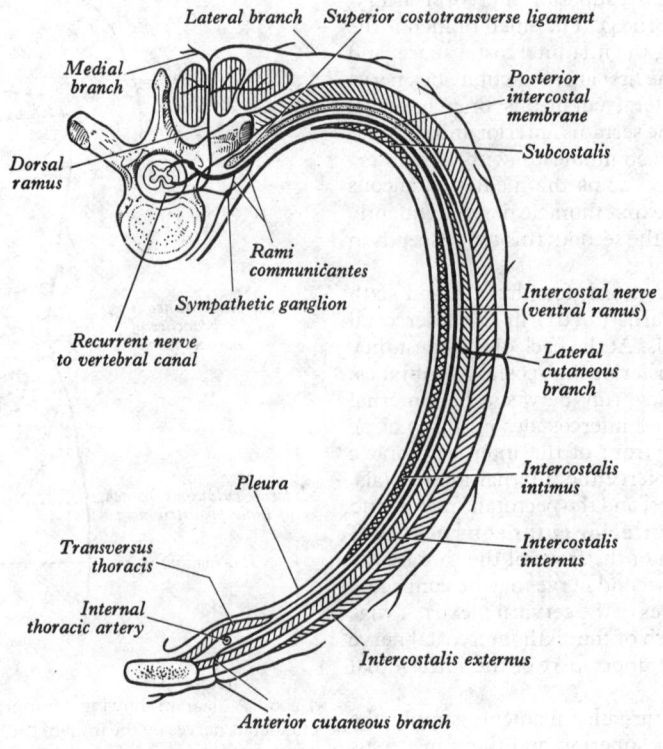

7.207 A diagram of the course of a typical intercostal nerve. The muscular and the collateral branches are not shown.

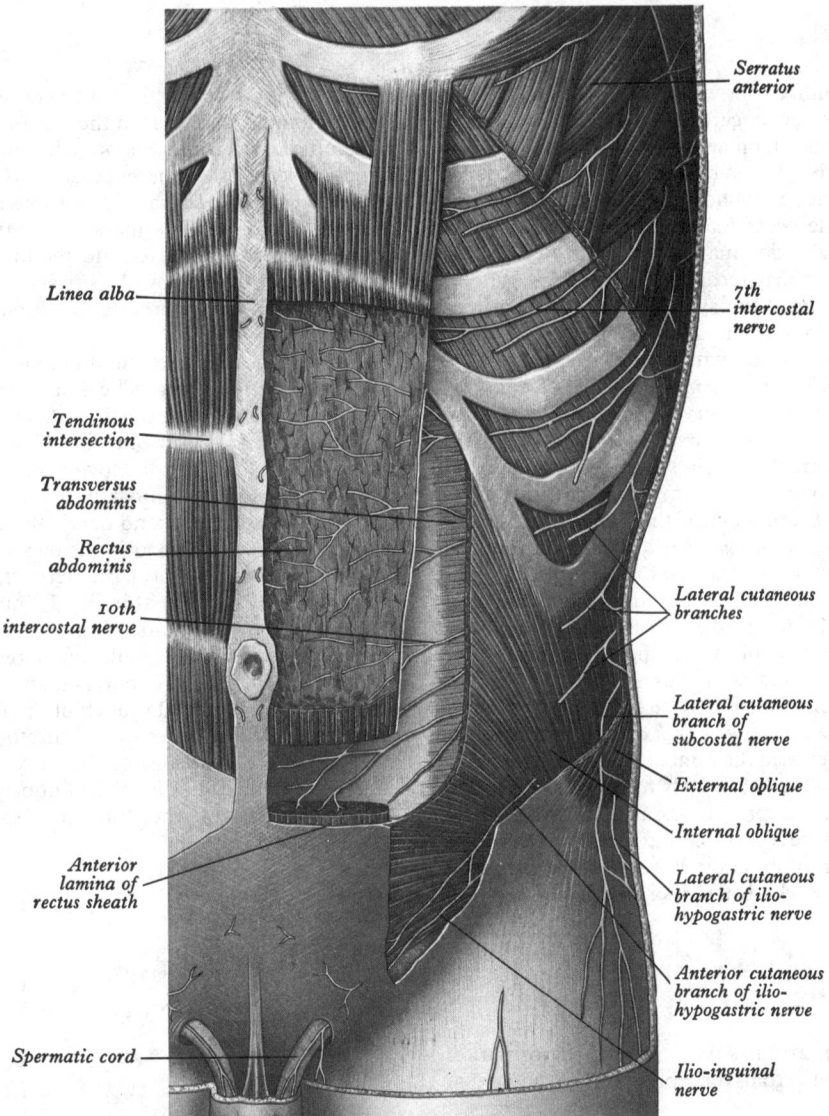

Serratus anterior

Linea alba

7th intercostal nerve

Tendinous intersection

Transversus abdominis

Rectus abdominis

10th intercostal nerve

Lateral cutaneous branches

Lateral cutaneous branch of subcostal nerve

External oblique

Internal oblique

Lateral cutaneous branch of iliohypogastric nerve

Anterior lamina of rectus sheath

Anterior cutaneous branch of iliohypogastric nerve

Spermatic cord

Ilio-inguinal nerve

7.208   Diagram to illustrate the course of the lower intercostal and the cutaneous branches of some lumbar nerves. Portions of the muscles of the anterior abdominal wall have been removed, including most of the anterior layer of the rectus sheath and parts of the rectus abdominis.

and external oblique muscles, gives a twig to the lowest slip of the latter, descends over the iliac crest about 5 cm behind the anterior superior iliac spine (7.215), and is distributed to the skin of the front part of the gluteal region, some filaments reaching as low as the greater trochanter of the femur.

Applied Anatomy. In many diseases affecting the nerve trunks at or near their origins, the pain is referred to their peripheral terminations. Thus, in tuberculosis of thoracic vertebrae, patients often suffer from pain in the abdominal wall. When confined to a single pair of nerves the sensation complained of is often a feeling of constriction, as if a cord were tied round the abdomen, and in these cases the situation of the sense of constriction may serve to localize the disease in the vertebral column. Where the bone disease is more extensive and two or more nerves are involved, a more general, diffused pain in the abdomen is felt.

Occasionally subluxation of an interchondral joint between the lower costal cartilages may trap an intercostal nerve with consequently referred abdominal pain; for a review of the literature of this 'interchondral subluxation' or 'clicking rib syndrome' *see* Abrahams (1976).

Again, it must be borne in mind that the nerves which supply the skin of the abdomen supply also the planes of muscle which

constitute the greater part of the abdominal wall, and this is of importance in protecting the abdominal viscera from injury. A blow on the abdomen, even of considerable force, will do no injury to the viscera if the muscles are in a condition of firm contraction; whereas in cases where the subject has been taken unawares, and the blow has been struck with the abdominal muscles relatively relaxed, an injury insufficient to produce any lesion of the abdominal wall may be accompanied by rupture of some of the abdominal contents. The importance, therefore, of immediate reflex contraction upon the receipt of, or in anticipation of, an injury cannot be overestimated, and the origin of the cutaneous and motor fibres from the same segments of the spinal cord results in a much more rapid response on the part of the muscles to any peripheral stimulation of the cutaneous filaments than would be the case if the two sets of fibres were derived from independent sources.

The nerves supplying the abdominal muscles and skin, derived from the lower intercostal nerves, are intimately connected with the sympathetic nerves supplying the abdominal viscera through the lower thoracic ganglia, from which the splanchnic nerves are derived. In consequence of this, in laceration of the abdominal viscera, and in acute peritonitis, the muscles of the belly wall become firmly contracted, and thus as far as possible preserve the abdominal contents in a condition of rest.

LUMBAR VENTRAL RAMI

The ventral rami of the lumbar nerves increase in size from the first to the last. They are joined, near their origins, by grey rami communicantes from the four lumbar ganglia of the sympathetic trunk. These rami consist of long, slender branches which accompany the lumbar arteries round the sides of the vertebral bodies, under cover of the psoas major. Their arrangement is somewhat irregular: one ganglion may give rami to two lumbar nerves, or one lumbar nerve may receive rami from two ganglia: not infrequently the rami leave the sympathetic trunk between two ganglia. The first and second, and sometimes the third, lumbar nerves are each connected with the lumbar part of the sympathetic trunk by a *white ramus communicans*.

The ventral rami of the lumbar nerves pass downwards and laterally into the psoas major. The first three nerves and the greater part of the fourth form the *lumbar plexus*. The smaller part of the fourth nerve joins with the fifth to form the *lumbosacral trunk*, which assists in the formation of the *sacral* plexus. The fourth nerve is often termed the *nervus furcalis*, from the fact that it is subdivided between the two plexuses. In many cases the fourth lumbar is the nervus furcalis but this arrangement is frequently departed from. The third is occasionally the lowest nerve which enters the lumbar plexus giving at the same time some fibres to the sacral plexus, and thus forming the nervus furcalis; or both the third and fourth may be furcal nerves. When this occurs, the plexus is termed *high* or *prefixed*. More frequently the fifth nerve is divided between the lumbar and sacral plexuses, and constitutes the nervus furcalis; and when this takes place, the plexus is distinguished as a *low* or *postfixed* plexus. These variations necessarily produce corresponding modifications in the sacral plexus. (For further information on variation in primates, including man, consult Piasecka-Kacperska and Gladyskowska-Rzeczycka 1972.)

The Lumbar Plexus

The lumbar plexus (7.209, 210) lies within the posterior part of the psoas major, in front of the transverse processes of the lumbar vertebrae; it is formed by the ventral rami of the first three lumbar nerves and the greater part of the ventral ramus of the fourth; the first lumbar nerve receives a branch from the last thoracic nerve. (It may be noted that the *paravertebral* part of the psoas major muscle is in two planes, a *posterior* mass arising from transverse processes, and an *anterior* mass arising from the lips of vertebral bodies, intervertebral discs and tendinous arches (p. 593); the lumbar plexus is deployed between these planes, and their relationship to the intervertebral foramina should be appreciated.)

Its arrangement varies in different subjects, but the usual condition is the following. The first lumbar nerve, supplemented by a twig from the last thoracic, splits into an upper and a lower branch; the upper, larger branch divides into the iliohypogastric and ilio-inguinal nerves; the lower, smaller branch unites with a branch of the second lumbar to form the genitofemoral nerve. The remainder of the second nerve, the third nerve, and the part of the fourth nerve which joins the plexus, divide into ventral and dorsal branches. The ventral branch of the second unites with the ventral branches of the third and fourth nerves to form the obturator nerve. The dorsal branches of the second and third nerves each divide into a smaller and larger part; the smaller parts unite to form the lateral femoral cutaneous nerve, and the larger parts join with the dorsal branch of the fourth nerve to form the femoral nerve. The accessory obturator, when it exists, arises from the ventral branches of the third and fourth nerves. For details of the blood supply of the lumbar plexus see Day (1964).

The branches of the lumbar plexus may therefore be particularized as follows:

Muscular	T. 12, L. 1, 2, 3, 4.
Iliohypogastric	L. 1.
Ilio-inguinal	L. 1.
Genitofemoral	L. 1, 2.
	Dorsal divisions
Lateral cutaneous, of thigh	L. 2, 3.
Femoral	L. 2, 3, 4.
	Ventral divisions
Obturator	L. 2, 3, 4.
Accessory obturator	L. 3, 4.

Muscular branches are distributed to the quadratus lumborum from the twelfth thoracic and first three or four lumbar nerves; to the psoas minor from the first, to the psoas major from the second, third and, sometimes, from the fourth, and to the iliacus from the second and third lumbar nerves.

The iliohypogastric nerve arises from the first lumbar nerve (7.209). It emerges from the upper part of the lateral border of the psoas major, and crosses obliquely behind the lower part of the kidney, and in front of the quadratus lumborum (7.210; 8.148). Just above the iliac crest it perforates the posterior part of the transversus abdominis, and divides between that muscle and the internal oblique into a lateral and an anterior cutaneous branch; it supplies both.

The *lateral cutaneous branch* pierces the internal and external oblique muscles immediately above the iliac crest at a point a little behind the iliac branch of the twelfth thoracic nerve; it is distributed to the skin of the posterolateral gluteal region.

The *anterior cutaneous branch* (7.206) runs between the internal oblique and transversus, supplying twigs to both muscles. It then pierces the internal oblique at a point about 2 cm on the medial side of the anterior superior iliac spine, perforates the aponeurosis of the external oblique about 3 cm above the superficial inguinal ring, and is distributed to the skin of the abdomen above the pubis.

The iliohypogastric nerve communicates with the subcostal and ilio-inguinal nerves.

The ilio-inguinal nerve, smaller than the iliohypogastric, arises with it from the first lumbar nerve (7.209). It emerges from the lateral border of the psoas major, with or just caudal to the iliohypogastric nerve, and, passing obliquely across the quadratus lumborum and the upper part of the iliacus, perforates the

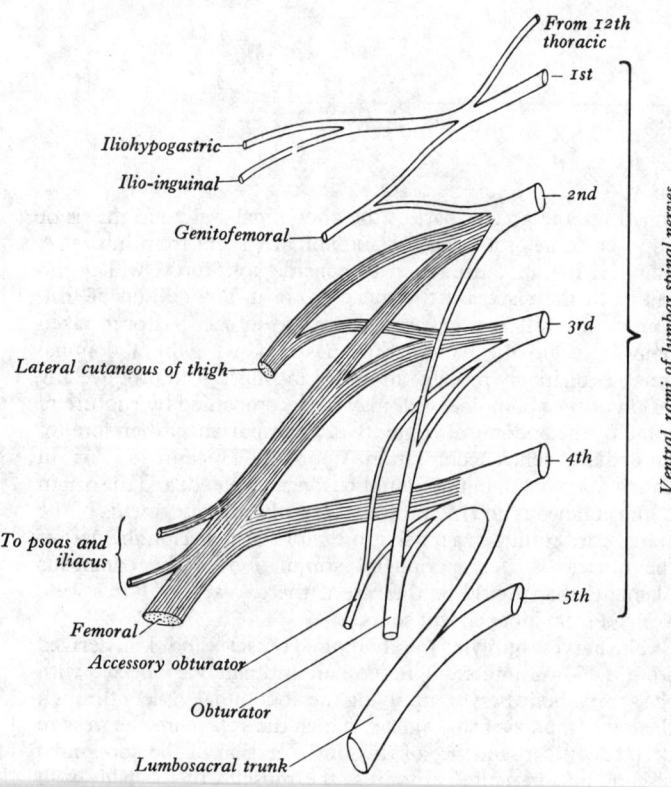

7.209 A plan of the lumbar plexus. The *dorsal divisions* of the second, third and fourth lumbar nerves are shaded.

Labels in figure:
From 12th thoracic
1st
Iliohypogastric
Ilio-inguinal
2nd
Genitofemoral
3rd
Lateral cutaneous of thigh
4th
To psoas and iliacus
5th
Femoral
Accessory obturator
Obturator
Lumbosacral trunk
Ventral rami of lumbar spinal nerves

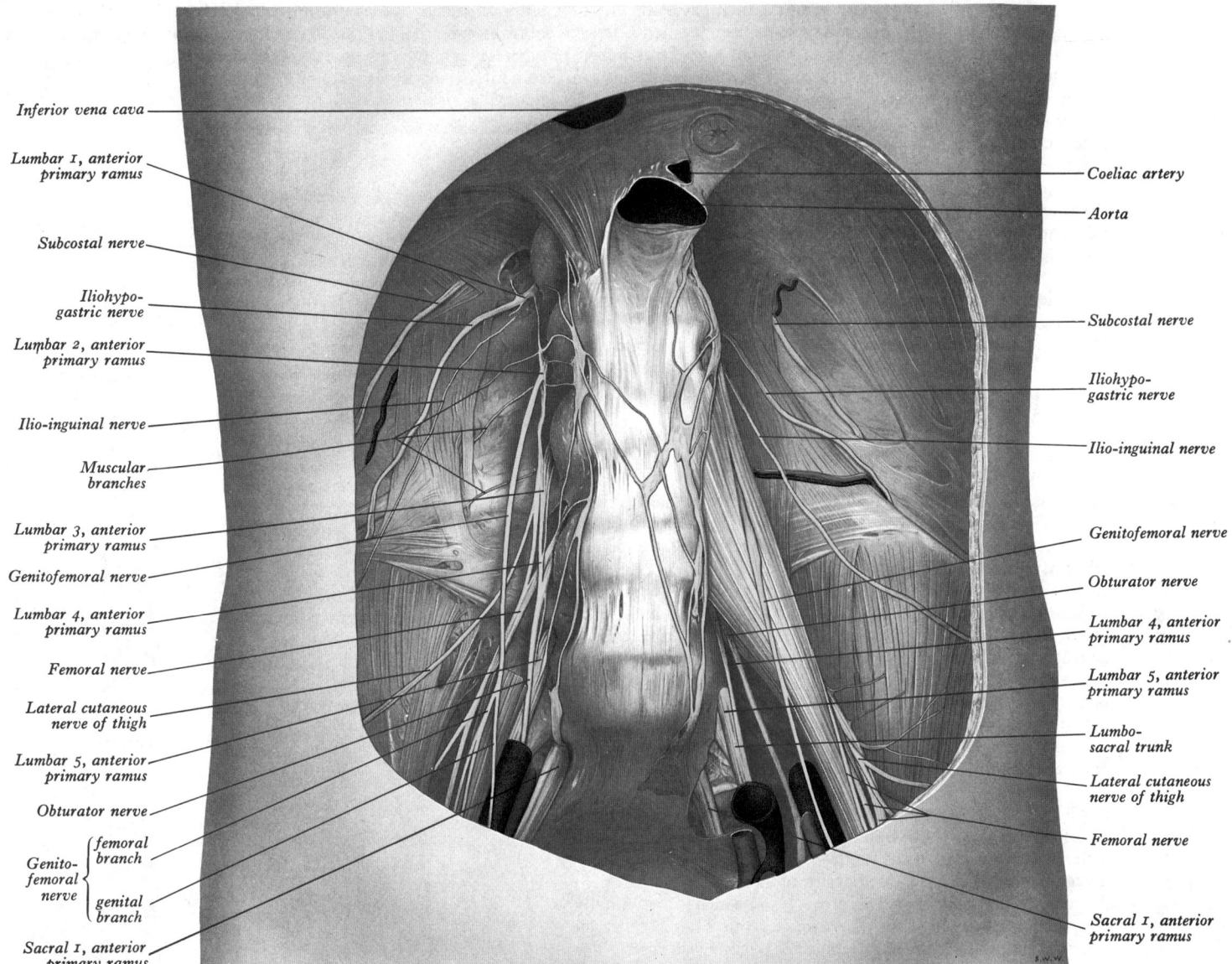

Inferior vena cava

Lumbar 1, anterior primary ramus

Subcostal nerve

Iliohypo-gastric nerve

Lumbar 2, anterior primary ramus

Ilio-inguinal nerve

Muscular branches

Lumbar 3, anterior primary ramus

Genitofemoral nerve

Lumbar 4, anterior primary ramus

Femoral nerve

Lateral cutaneous nerve of thigh

Lumbar 5, anterior primary ramus

Obturator nerve

Genito-femoral nerve { femoral branch / genital branch }

Sacral 1, anterior primary ramus

Coeliac artery

Aorta

Subcostal nerve

Iliohypo-gastric nerve

Ilio-inguinal nerve

Genitofemoral nerve

Obturator nerve

Lumbar 4, anterior primary ramus

Lumbar 5, anterior primary ramus

Lumbo-sacral trunk

Lateral cutaneous nerve of thigh

Femoral nerve

Sacral 1, anterior primary ramus

7.210   A dissection of the posterior abdominal wall to show the lumbar plexus and sympathetic trunks. The right psoas major has been removed.

transversus abdominis, near the anterior part of the iliac crest, and sometimes communicates with the iliohypogastric nerve. It then pierces the internal oblique, distributing filaments to it, lies below the spermatic cord in the inguinal canal and accompanies it through the superficial inguinal ring. It is distributed to the skin of the superomedial area of the thigh, to the skin over the root of the penis and upper part of the scrotum in the male (7.206), and to the skin covering the mons pubis and adjoining part of the labium majus in the female.

The size of the ilio-inguinal nerve is in inverse proportion to that of the iliohypogastric. Occasionally it is very small, and ends by joining the iliohypogastric nerve; in such cases, a branch from the iliohypogastric takes the place of the ilio-inguinal, or the latter nerve may at times be altogether absent. On the analogy of the intercostal nerves the ilio-inguinal nerve may be regarded as the collateral branch of the first lumbar nerve (Davies 1935), and the iliohypogastric as the main trunk, which gives off the lateral cutaneous branch.

The genitofemoral nerve arises from the first and second lumbar nerves (7.209). It passes obliquely forwards and downwards through the substance of the psoas major, and emerges on the abdominal surface of the muscle near its medial border, opposite the third or fourth lumbar vertebra; it then descends on the surface of the psoas major, under cover of the peritoneum, and, crossing obliquely behind the ureter, divides at a variable distance above the inguinal ligament into the genital

and femoral branches. The genitofemoral nerve frequently divides close to its origin, and its two branches then emerge separately through the psoas major.

The *genital branch* crosses the lower end of the external iliac artery, and enters the inguinal canal through the deep inguinal ring; it supplies the cremaster, and gives a few filaments to the skin of the scrotum. In the female, it accompanies the round ligament of the uterus and ends in the skin of the mons pubis and labium majus.

The *femoral branch* descends on the lateral side of the external iliac artery, and sends a few filaments round it; it then crosses the deep circumflex iliac artery, and passing behind the inguinal ligament, enters the femoral sheath, lying lateral to the femoral artery. It pierces the anterior layer of the femoral sheath and the fascia lata, and supplies the skin over the upper part of the femoral triangle (7.211). It communicates with the intermediate cutaneous nerve of the thigh, and gives a few twigs to the femoral artery.

The lateral cutaneous nerve of the thigh arises from the dorsal branches of the ventral rami of the second and third lumbar nerves (7.209). It emerges from the lateral border of the psoas major, and crosses the iliacus obliquely, running towards the anterior superior iliac spine. It supplies branches to the parietal peritoneum of the iliac fossa. On the right side the nerve passes behind and lateral to the caecum, from which it is separated by the fascia iliaca and the peritoneum; on the left side, it passes behind the lower part of the descending colon. It then passes behind or through the inguinal ligament, at a variable distance medial to the anterior superior iliac spine (commonly about 1 cm), and in front of or through the sartorius into the thigh, where it divides into an anterior and a posterior branch (7.211).

The *anterior branch* becomes superficial about 10 cm below the anterior superior iliac spine, and is distributed to the skin of the anterior and lateral parts of the thigh, as far as the knee. Its terminal filaments communicate with the cutaneous branches of the anterior division of the femoral nerve and with the infrapatellar branch of the saphenous nerve, forming with them the *patellar plexus*.

The *posterior branch* pierces the fascia lata at a higher level than the anterior branch, and subdivides into filaments which pass backwards to supply the skin on the lateral surface of the limb, from the level of the greater trochanter to about the middle of the thigh. It may also supply twigs to the skin of the gluteal region.

THE OBTURATOR NERVE

The obturator nerve arises from the ventral branches of the ventral rami of the second, third and fourth lumbar nerves (7.209, 210); the branch from the third is the largest, while that from the second is often very small. It descends through the psoas major, and emerges from its medial border at the brim of the pelvis, where it passes behind the common iliac vessels, and on the lateral side of the internal iliac vessels. It then runs downwards and forwards along the lateral wall of the lesser pelvis lying on the obturator internus, above and in front of the obturator vessels, to gain the upper part of the obturator foramen, through which it enters the thigh. Near the foramen it divides into an anterior and a posterior branch, which are separated at first by a few fibres of the obturator externus, and lower down by the adductor brevis.

The *anterior branch* (7.212) leaves the pelvis anterior to the obturator externus and descends in front of the adductor brevis, and behind the pectineus and adductor longus; at the lower border of the latter muscle it communicates with the medial cutaneous and saphenous branches of the femoral nerve, forming a plexus (termed the *subsartorial plexus*) from which branches are given off to the skin on the medial side of the thigh (7.211). It then descends upon the femoral artery, to which it is finally distributed. Near the obturator foramen this branch gives an articular twig to the hip joint. Behind the pectineus, it distributes branches to the adductor longus, gracilis, usually to the adductor brevis, and often to the pectineus; it receives a filament from the accessory obturator nerve when that nerve is present.

Occasionally the communicating branch to the medial cutaneous and saphenous branches of the femoral nerve is continued down, as a cutaneous branch, to the thigh and leg. When this is so, it emerges from behind the lower border of the adductor longus, descends along the posterior margin of the sartorius to the medial side of the knee, where it pierces the deep fascia, communicates with saphenous nerve, and is distributed to the skin halfway down the medial side of the leg.

The *posterior branch* pierces the anterior part of the obturator externus, and supplies this muscle; it then passes behind the adductor brevis on the front of the adductor magnus, and divides into branches which are distributed to the adductor magnus, and to the adductor brevis when this muscle does not receive a branch from the anterior division of the nerve. It generally gives a slender *articular branch* to the knee joint; this branch perforates the lower part of the adductor magnus or passes through the opening which transmits the femoral artery, and enters the popliteal fossa. Here it descends upon the popliteal artery, to the back of the knee joint, where it pierces the oblique posterior ligament of the knee, and is distributed to the articular capsule. It gives filaments to the popliteal artery.

The accessory obturator nerve (7.209) is occasionally present. It is of small size, and arises from the ventral branches of the ventral rami of the third and fourth lumbar nerves. It descends along the medial border of the psoas major, crosses the superior ramus of the pubis behind the pectineus, and divides into branches. One branch enters the deep surface of the pectineus; another goes to the hip joint; a third communicates with the

anterior branch of the obturator nerve. On occasion the accessory obturator nerve is very small and supplies only the pectineus. Any of these rami may be absent, and occasionally others occur, one sometimes supplying adductor longus. An accessory obturator nerve was present in 69 of 800 instances (p. 599).

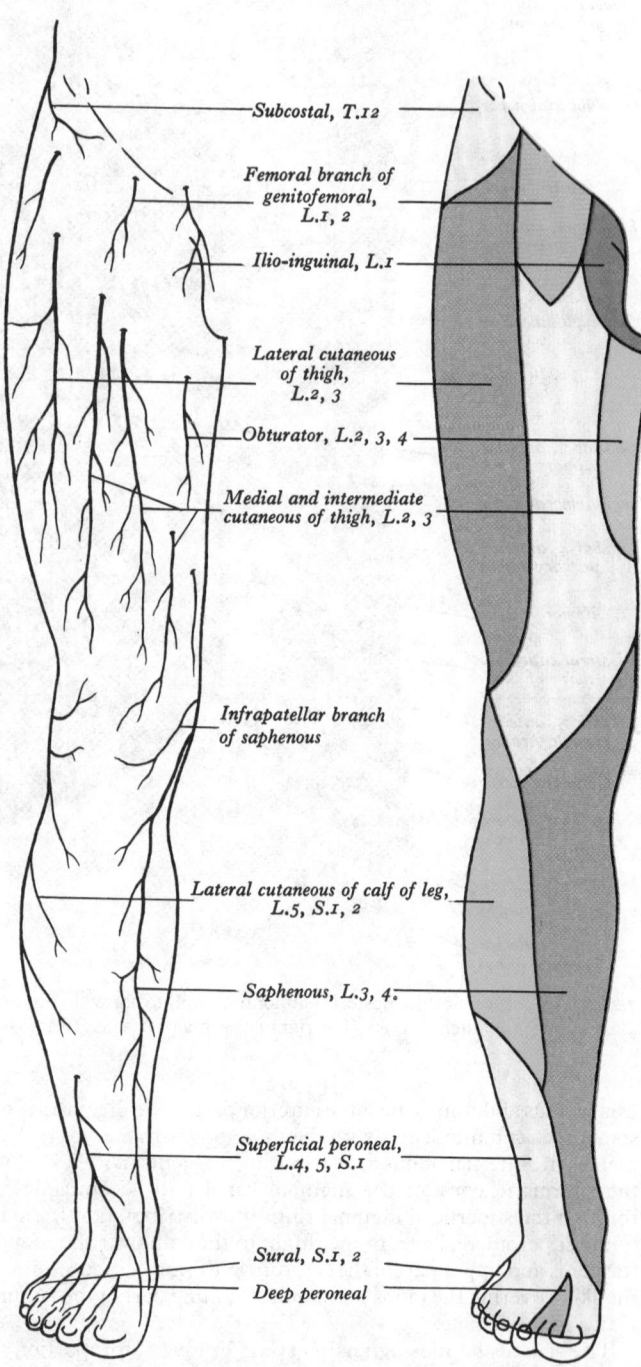

7.211 The cutaneous nerves of the right lower limb, their areas of distribution and segmental origins, viewed from the anterior aspect.

THE FEMORAL NERVE

The femoral nerve (7.209, 210), the largest branch of the lumbar plexus, arises from the dorsal branches of the ventral rami of the second, third and fourth lumbar nerves. It descends through the substance of psoas major, emerging from the muscle at the lower part of its lateral border, and passes down between it and the iliacus, deep to the iliac fascia; it then passes behind the inguinal ligament to enter the thigh, and splits into an anterior and a posterior division. Behind the inguinal ligament it is separated from the femoral artery by part of the psoas major.

Within the abdomen the femoral nerve gives off small branches to the iliacus, the nerve to the pectineus, and a branch which is distributed upon the upper part of the femoral artery; the latter branch may arise in the thigh.

The nerve to the pectineus arises from the medial side of the femoral nerve near the inguinal ligament, passes behind the femoral sheath and enters the anterior surface of the muscle.

The anterior division of the femoral nerve gives off the intermediate and medial cutaneous nerves of the thigh (7.211, 212), and muscular branches to the sartorius.

The intermediate cutaneous nerve of the thigh pierces the fascia lata about 8 cm below the inguinal ligament, either as two branches, or as a single trunk which quickly divides into two branches; these branches descend vertically on the front of the thigh, and supply the skin as low as the knee. They end in the patellar plexus (p. 1108). The lateral branch of the intermediate cutaneous communicates with the femoral branch of the genitofemoral nerve, and frequently pierces the sartorius, to which it may give a branch of supply.

The medial cutaneous nerve of the thigh lies at first on the lateral side of the femoral artery, but at the apex of the femoral triangle it crosses ventral to the artery and divides into an anterior and a posterior branch. Before dividing, the nerve gives off a few filaments which pierce the fascia lata to supply the skin of the medial side of the thigh, in the neighbourhood of the long saphenous vein; one of these filaments emerges through the saphenous opening, and a second becomes subcutaneous about the middle of the thigh. The *anterior branch* runs downwards on the sartorius, perforates the fascia lata at the junction of the middle with the lower one-third of the thigh, and divides into two branches: one supplies the skin as low as the medial side of the knee; the other crosses to the lateral side of the patella, communicating in its course with the infrapatellar branch of the saphenous nerve. The *posterior branch* descends along the posterior border of the sartorius to the knee, where it pierces the fascia lata, communicates with the saphenous nerve, and gives off several cutaneous branches. It then descends to supply the skin of the medial side of the leg. Beneath the fascia lata, at the lower border of the adductor longus, it joins to form a plexiform network (*subsartorial plexus*) with branches of the saphenous and obturator nerves. When the communicating branch from the obturator nerve is large and continued to the skin of the leg, the posterior branch of the medial cutaneous is small, and terminates in the plexus, occasionally giving off a few cutaneous filaments.

The nerve to the sartorius arises in common with the intermediate cutaneous nerve of the thigh.

The posterior division of the femoral nerve gives off the saphenous nerve, and supplies muscular branches to the quadriceps femoris, and articular branches to the knee joint.

The saphenous nerve (7.212) is the largest cutaneous branch of the femoral nerve. It descends on the lateral side of the femoral artery and enters the adductor canal (p. 725) where it crosses in front of the artery obliquely from its lateral to its medial side. At the lower end of the canal it quits the artery, and emerges through the aponeurotic covering of the canal, accompanied by the saphenous branch of the descending genicular artery. It descends vertically along the medial side of the knee behind the sartorius, pierces the fascia lata between the tendons of the sartorius and gracilis, and becomes subcutaneous. It then passes down the medial side of the leg accompanied by the long saphenous vein, descends along the medial border of the tibia, and, at the lower third of the leg, divides into two branches: one continues its course along the medial border of the tibia, and ends at the ankle; the other passes in front of the ankle, and is distributed to the skin on the medial side of the foot, often reaching as far as the metatarsophalangeal joint of the great toe and communicating with the medial branch of the superficial peroneal nerve.

About the mid level of the thigh, the saphenous nerve gives a branch to join the subsartorial plexus.

After leaving the adductor canal it gives off an *infrapatellar branch* (7.211), which pierces the sartorius and fascia lata, and is distributed to the skin in front of the patella. Proximal to the knee it unites with the medial and intermediate cutaneous nerves of the thigh; below the knee, with other branches of the saphenous

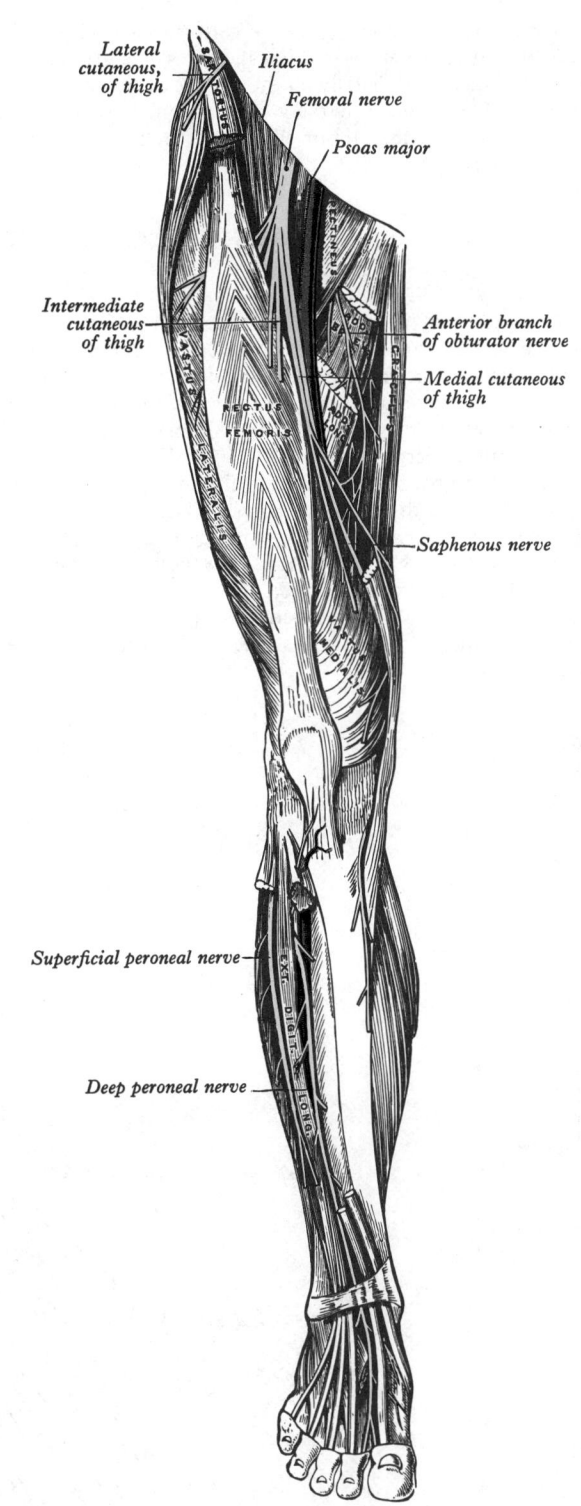

7.212 The nerves of the right lower limb displayed from the anterior aspect.

nerve; on the lateral side of the joint, with branches of the lateral cutaneous nerve of the thigh, forming a *patellar plexus*.

The muscular branches of the posterior division of the femoral nerve supply the quadriceps femoris. The branch to the rectus femoris enters the upper part of the deep surface of the muscle, and supplies a filament to the hip joint. The larger branch to the vastus lateralis forms a prominent neurovascular bundle with the descending branch of the lateral circumflex femoral artery to the lower part of the muscle, and sends an articular filament to the knee joint. The branch to the vastus medialis descends through the upper part of the adductor canal, on the lateral side of the saphenous nerve and the femoral vessels. It enters the muscle about its middle, and gives off a filament which can usually be traced distally on the surface of the muscle to the knee joint. The branches to the vastus intermedius, two or three in number, enter the anterior surface of the muscle about the middle of the thigh; a filament from one of these descends through the muscle to the articularis genus and the knee joint.

Vascular branches are given off from the femoral nerve to the femoral artery and its branches (p. 1131).

SACRAL AND COCCYGEAL VENTRAL RAMI

The ventral rami of the sacral and coccygeal nerves form the sacral and coccygeal plexuses. Those of the upper four sacral nerves enter the pelvis through the pelvic sacral foramina, that of the fifth between the sacrum and coccyx, while that of the coccygeal nerve curves forwards below the rudimentary transverse process of the first piece of the coccyx. The first and second sacral spinal nerves are large; the third, fourth and fifth diminish progressively; the coccygeal nerve is the smallest. Each of these nerves receives a *grey ramus communicans* from the corresponding ganglion of the sympathetic trunk. *Visceral efferent rami* arise from the second, third and fourth sacral nerves; they are named the *pelvic splanchnic nerves* (7.231), and consist of parasympathetic fibres which pass directly to minute ganglia on the walls of the pelvic viscera (p. 1123).

The Sacral Plexus

The sacral plexus (7.213) is formed by the lumbosacral trunk, the ventral rami of the first, second and third sacral nerves, and part of the ventral ramus of the fourth sacral nerve, the remainder of which joins the coccygeal plexus.

The lumbosacral trunk comprises a part of the ventral ramus of the fourth lumbar nerve, and the whole ventral ramus of the fifth lumbar nerve; it appears at the medial margin of the psoas major and descends over the pelvic brim anterior to the sacro-iliac joint to join the first sacral nerve.

The nerves entering into the sacral plexus converge towards the greater sciatic foramen and unite without much interlacement to form an upper large and a lower small band. The upper band is

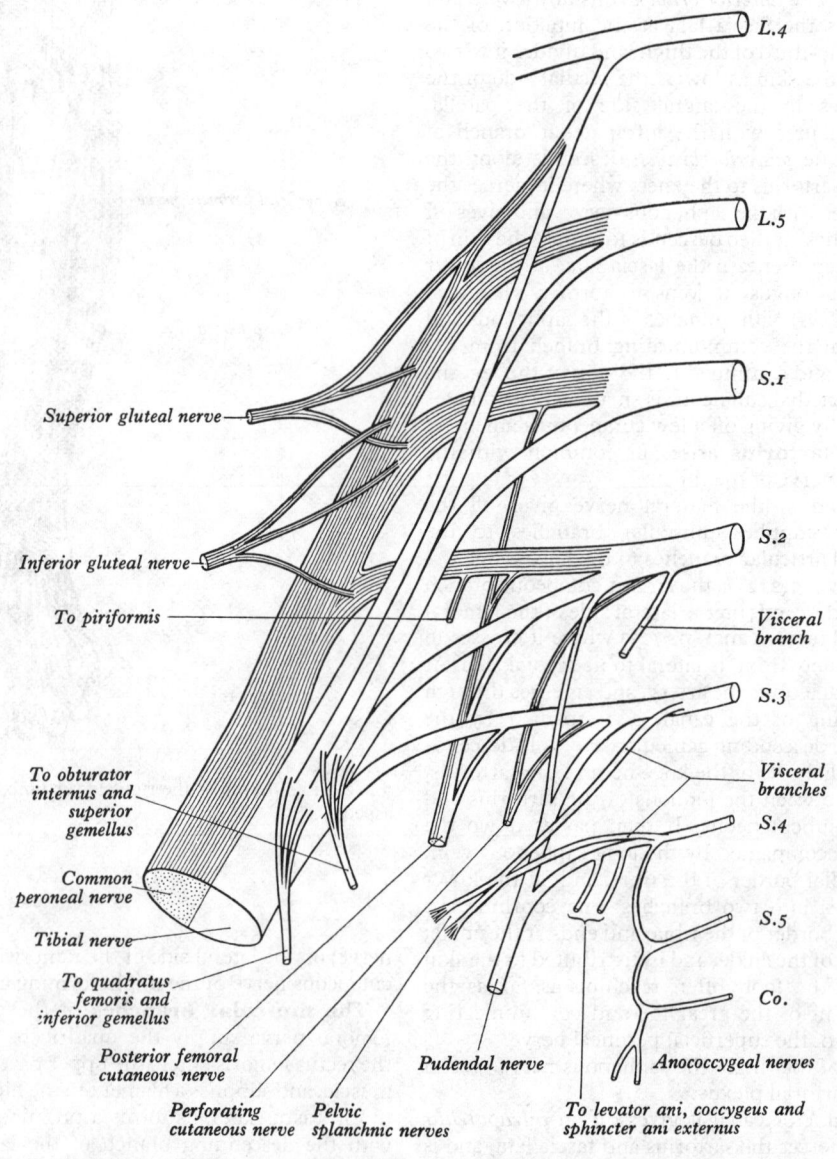

Superior gluteal nerve

Inferior gluteal nerve

To piriformis

To obturator internus and superior gemellus

Common peroneal nerve

Tibial nerve

To quadratus femoris and inferior gemellus

Posterior femoral cutaneous nerve

Perforating cutaneous nerve

Pelvic splanchnic nerves

L.4

L.5

S.1

S.2

Visceral branch

S.3

Visceral branches

S.4

S.5

Co.

Pudendal nerve

Anococcygeal nerves

To levator ani, coccygeus and sphincter ani externus

7.213 A plan of the sacral and coccygeal plexuses. L.4 5, S.1 5 and Co., indicate the *ventral rami* of these lumbar, sacral and coccygeal spinal nerves. The *ventral divisions* of these rami are unshaded, the *dorsal divisions*, and nerves derived from them, are shaded. The contribution from S.2 to the pelvic splanchnic nerves is shown before joining those from 3 and 4.

formed by the union of the lumbosacral trunk with the first, second and the greater part of the third sacral nerves and is continued into the sciatic nerve. The lower band, which has a more plexiform arrangement, results mainly from the junction of the smaller part of the third sacral nerve with the portion of the fourth nerve belonging to the plexus, and is prolonged into the pudendal nerve; it also receives a small contribution from the second sacral nerve. The sciatic nerve is composed of tibial and common peroneal nerves which usually become distinct in the thigh. Above this level, however, these two nerves can be separated to their roots of origin. It is then found that the tibial nerve is formed by the union of ventral divisions of the lumbosacral trunk and first three sacral nerves, while the common peroneal receives the dorsal divisions of the lumbosacral trunk and first two sacral nerves. The component nerves of the sciatic may, however, diverge anywhere along its course from the pelvis. When the division occurs at the plexus the common peroneal usually pierces the piriformis in the greater sciatic foramen. (For details of the blood supply of the sacral plexus *see* Day 1964.)

Relations of the Sacral Plexus

The sacral plexus lies on the posterior wall of the pelvic cavity in front of the piriformis (7.214), and behind the internal iliac vessels, the ureter and the sigmoid colon, on the left side, and the terminal coils of the ileum, on the right side. The superior gluteal vessels run between the lumbosacral trunk and the first sacral nerve, or between the first and second sacral nerves, and the inferior gluteal vessels between the ventral rami of the first and second, or second and third, sacral nerves.

The Branches of the Sacral Plexus

These may be summarized as follows:

	Ventral divisions	Dorsal divisions
To quadratus femoris and gemellus inferior	L. 4, 5, S. 1	
To obturator internus and gemellus superior	L. 5, S. 1, 2	
To piriformis		S. (1) 2
Superior gluteal		L. 4, 5, S. 1
Inferior gluteal		L. 5, S. 1, 2
Posterior femoral cutaneous	S. 2, 3	S. 1, 2
Sciatic—tibial	L. 4, 5, S. 1, 2, 3	
—common peroneal		L. 4, 5 S. 1, 2
Perforating cutaneous		S. 2, 3
Pudendal	S. 2, 3, 4	
To levator ani, coccygeus and sphincter ani externus	S. 4	
Pelvic splanchnic	S. 2, 3, (4)	

The nerve to the quadratus femoris and gemellus inferior arises from the ventral branches of the ventral rami of the fourth and fifth lumbar and first sacral nerves (7.213); it leaves the pelvis through the greater sciatic foramen below the piriformis and descends on the ischium deep to the sciatic nerve, the gemelli and the tendon of the obturator internus. It supplies the gemellus inferior, and enters the anterior surface of the quadratus femoris; it also supplies the hip joint.

The nerve to the obturator internus and gemellus superior arises from the ventral branches of the ventral rami of the fifth lumbar and first and second sacral nerves (7.213). It leaves the pelvis through the greater sciatic foramen below the piriformis and gives a branch which enters the upper part of the posterior surface of the gemellus superior. It then crosses the ischial spine lateral to the internal pudendal vessels, re-enters the pelvis through the lesser sciatic foramen, and pierces the pelvic surface of the obturator internus.

The nerve to the piriformis arises usually from the dorsal branches of the ventral rami of the first and second sacral nerves, the contribution from the first sacral sometimes being absent; it enters the anterior surface of the muscle.

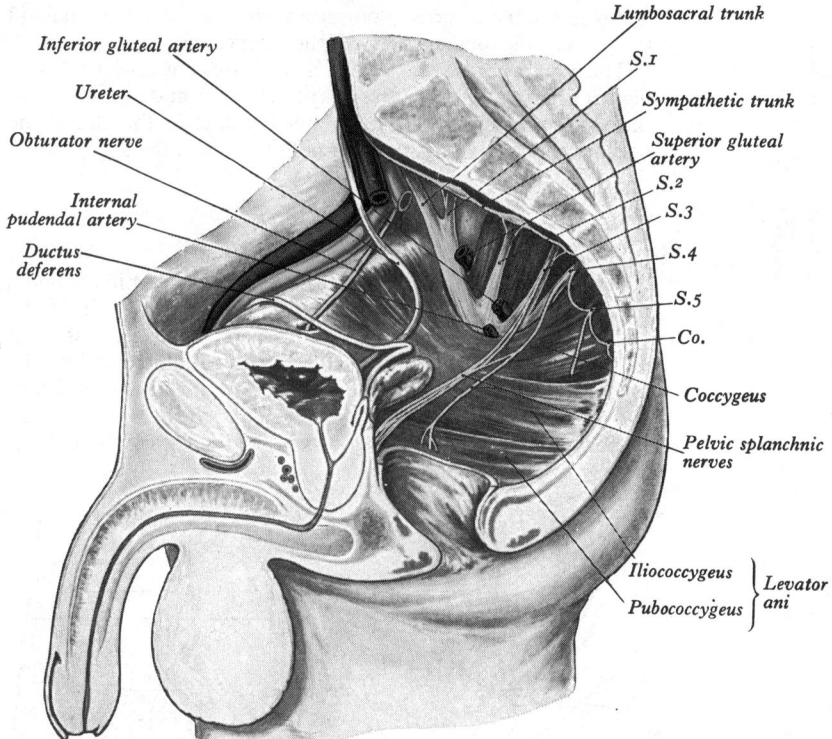

7.214 A dissection of the side wall of the pelvis, showing the sacral and coccygeal plexuses. S.1–5 indicate the anterior rami of the sacral spinal nerves; Co. indicates the ventral ramus of the coccygeal spinal nerve.

The superior gluteal nerve arises from the dorsal branches of the ventral rami of the fourth and fifth lumbar and first sacral nerves (7.213): it leaves the pelvis through the greater sciatic foramen above the piriformis, accompanied by the superior gluteal vessels, and divides into a superior and an inferior branch. The *superior* branch accompanies the upper branch of the deep division of the superior gluteal artery and supplies branches to the gluteus medius and occasionally also to the gluteus minimus. The *inferior* branch runs with the lower branch of the deep division of the superior gluteal artery across the gluteus minimus; it gives twigs to the gluteus medius and gluteus minimus, and ends in the tensor fasciae latae.

The inferior gluteal nerve arises from the dorsal branches of the ventral rami of the fifth lumbar and first and second sacral nerves: it leaves the pelvis through the greater sciatic foramen, below the piriformis, and divides into branches which enter the deep surface of the gluteus maximus.

The posterior femoral cutaneous nerve arises from the dorsal branches of the ventral rami of the first and second, and from the ventral branches of the ventral rami of the second and third, sacral nerves (7.213), and issues from the pelvis through the greater sciatic foramen distal to piriformis. It then descends under cover of the gluteus maximus with the inferior gluteal artery, lying posterior or medial to the sciatic nerve. It descends the back of the thigh superficial to the long head of the biceps femoris, and deep to the fascia lata: at the back of the knee it pierces the deep fascia and accompanies the short saphenous vein as far as the middle of the calf of the leg, its terminal twigs communicating with the sural nerve.

Its branches are all cutaneous, and are distributed to the gluteal region, the perineum, and the flexor aspect of the thigh and leg.

The *gluteal branches*, three or four in number, turn upwards round the lower border of the gluteus maximus, and supply the skin covering the lower and lateral part of that muscle.

The *perineal branch* distributes twigs to the skin at the upper and medial side of the thigh, and then curves forwards across the origin of the hamstrings, below the ischial tuberosity; it pierces the fascia lata, and runs beneath the superficial fascia of the perineum to the skin of the scrotum in the male, and of the labium majus in the female, communicating with the inferior rectal and the posterior scrotal branches of the perineal nerve.

The *branches to the back of the thigh and leg* consist of numerous filaments derived from both sides of the nerve, and distributed to the skin covering the back and medial side of the thigh, the popliteal fossa and the upper part of the back of the leg (7.215).

The Sciatic Nerve

The sciatic nerve (7.213, 216), the largest in diameter in the body, measures at its commencement 2 cm in breadth. It is the continuation of the upper band of the sacral plexus. It passes out of the pelvis through the greater sciatic foramen, below the piriformis, descends between the greater trochanter of the femur and the tuberosity of the ischium, and along the back of the thigh to about its lower one-third, where it divides into two large

Iliohypogastric, L.1
Subcostal, T.12
Dorsal rami, L.1, 2, 3
Dorsal rami, S.1, 2, 3
Gluteal branches Perineal branches | *Post. cutaneous of thigh, S.1, 2, 3*
Lateral cutaneous of thigh, L.2, 3
Obturator, L.2, 3, 4
Medial cutaneous of thigh, L.2, 3
Posterior cutaneous of thigh, S.1, 2, 3
Lateral cutaneous of calf of leg, L.4, 5, S.1
Saphenous, L.3, 4
Sural communicating branch of common peroneal
Sural, L.5, S.1, 2
Medial calcaneal branches of tibial, S.1, 2

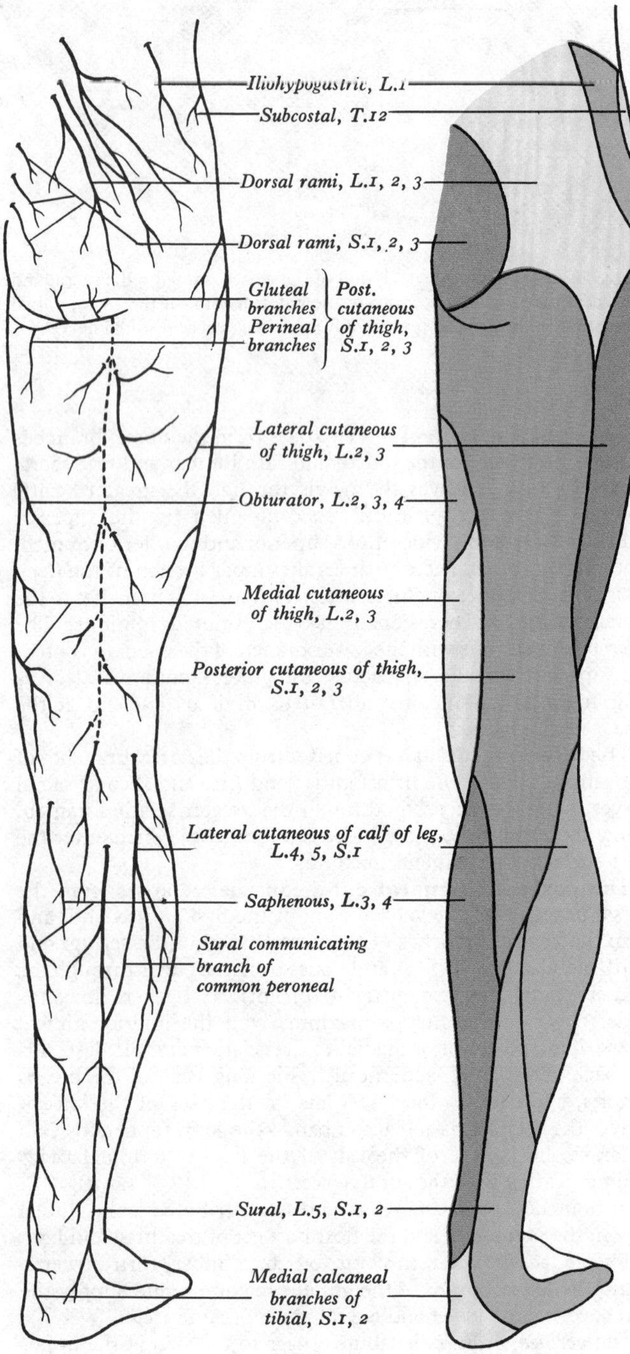

7.215 The cutaneous nerves of the right lower limb and their areas of distribution and segmental origins, viewed from the posterior aspect. The major part of the trunk of the posterior cutaneous nerve of the thigh lies deep to the deep fascia and is therefore shown by an interrupted line.

branches, named the tibial and common peroneal nerves. The nerve also gives off articular and muscular branches.

In the upper part of its course the nerve is situated deep to the gluteus maximus, and rests first upon the posterior surface of the ischium, the nerve to the quadratus femoris intervening; it then crosses the obturator internus and gemelli, and passes on to the quadratus femoris, by which it is separated from the obturator externus and the hip joint; it is accompanied on its medial side by the posterior cutaneous nerve of the thigh and the inferior gluteal artery. More distally it lies upon the adductor magnus, and is crossed obliquely by the long head of the biceps femoris. It can be represented on the back of the thigh by a broad line drawn distally to the apex of the popliteal fossa from just medial to the midpoint of the line joining the ischial tuberosity to the apex of the greater trochanter.

The *articular branches* of the sciatic nerve arise from the upper part of the nerve, and supply the hip joint by perforating the posterior part of its capsule; they are sometimes derived directly from the sacral plexus.

The *muscular branches* of the sciatic nerve are distributed to the biceps femoris, semitendinosus, semimembranosus and to the ischial head of the adductor magnus (p. 599); the branches to the latter two arise by a common trunk. The nerve to the short head of the biceps femoris comes from the common peroneal division, the other muscular branches from the tibial division of the sciatic nerve.

THE TIBIAL NERVE

The tibial (medial popliteal) nerve (7.216), the larger terminal division of the sciatic, arises from the ventral branches of the ventral rami of the fourth and fifth lumbar and first, second and third sacral nerves. It descends along the back of the thigh and through the middle of the popliteal fossa, to the distal border of the popliteus, where it passes with the popliteal artery deep to the arch of the soleus. Thereafter it is continued into the leg. In the thigh it is overlapped by the hamstring muscles above, but it becomes more superficial in the popliteal fossa, where it lies lateral to, and some distance from, the popliteal vessels; it is superficial to these vessels opposite the knee joint and then crosses to the medial side of the popliteal artery (7.217). In the lower part of the popliteal fossa, the nerve is covered by the contiguous margins of the two heads of the gastrocnemius. It can be represented by a line drawn downwards in the midline of the limb from the apex of the popliteal fossa to the level of the neck of the fibula. Continued downwards to a point midway between the medial malleolus and the tendo calcaneus, the line maps out the whole course of the tibial nerve.

In the leg it descends with the posterior tibial vessels to the interval between heel and medial malleolus, ending under the flexor retinaculum by dividing into the medial and lateral plantar nerves. Proximally it is deep to the soleus and gastrocnemius, but in the distal third of the leg it is covered only by skin and fasciae, although it is overlapped sometimes by the medial edge of the flexor hallucis longus. Proximally it is medial to the posterior tibial vessels, but it soon crosses behind them and descends lateral to them until it bifurcates. In most of its course it lies on the tibialis posterior, but in the lower part of the leg it comes into relation with the posterior surface of the tibia.

The branches of this nerve are articular, muscular, sural, medial calcanean and medial and lateral plantar.

Articular branches, usually three, supply the knee joint; one accompanies the superior, and another the inferior medial genicular artery; the third runs with the middle genicular artery. These branches form a plexus with the articular branch of the obturator nerve, and from the plexus branches are distributed to the oblique posterior ligament of the joint and those accompanying the superior and inferior medial genicular arteries supply the medial parts of the capsular ligament. Another articular branch arises from the nerve just above its terminal bifurcation, and supplies the ankle joint. For further details of this nerve supply consult Champetier and Déscours (1968). Gardner and Lenn (1977) have assessed the ratios of myelinated to non-myelinated fibres in genual articular nerves of the monkey as approximately

4:1; they suggest that most of the fibres subserve pain sensation.

The muscular branches arise from the nerve as it lies between the two heads of the gastrocnemius; they supply that muscle, as well as the plantaris, soleus and popliteus. The nerve to

the soleus enters the superficial surface of the muscle. The branch for the popliteus descends, crossing the popliteal vessels obliquely, and turns round the distal border of the muscle to be distributed to its deep surface; it supplies small branches to the tibialis posterior, an articular twig to the upper tibiofibular joint, a medullary branch to the tibia, and an interosseous branch, which descends close to the fibula and reaches the inferior tibiofibular joint.

In the leg, the muscular branches arise either independently or by a common trunk. They supply the soleus (on its deep surface), the tibialis posterior, the flexor digitorum longus and the flexor hallucis longus; the branch to the last muscle accompanies the peroneal vessels.

The sural nerve descends between the two heads of the gastrocnemius, and, piercing the deep fascia in the middle or upper part of the back of the leg, is joined by the sural communicating branch of the common peroneal nerve (7.215). It

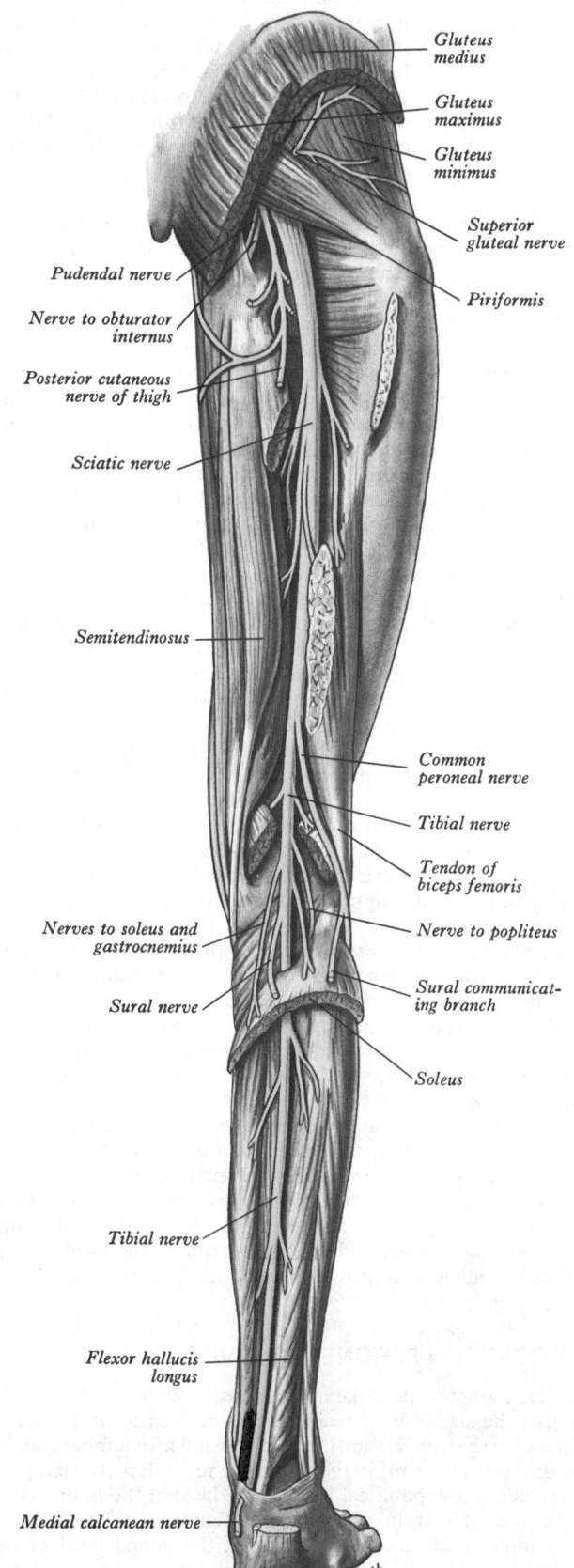

7.216 The nerves of the right lower limb. Posterior aspect. In this figure the gluteus maximus, the gluteus medius and the superficial muscles of the calf of the leg have been removed and the middle part of the long head of the biceps femoris has been excised.

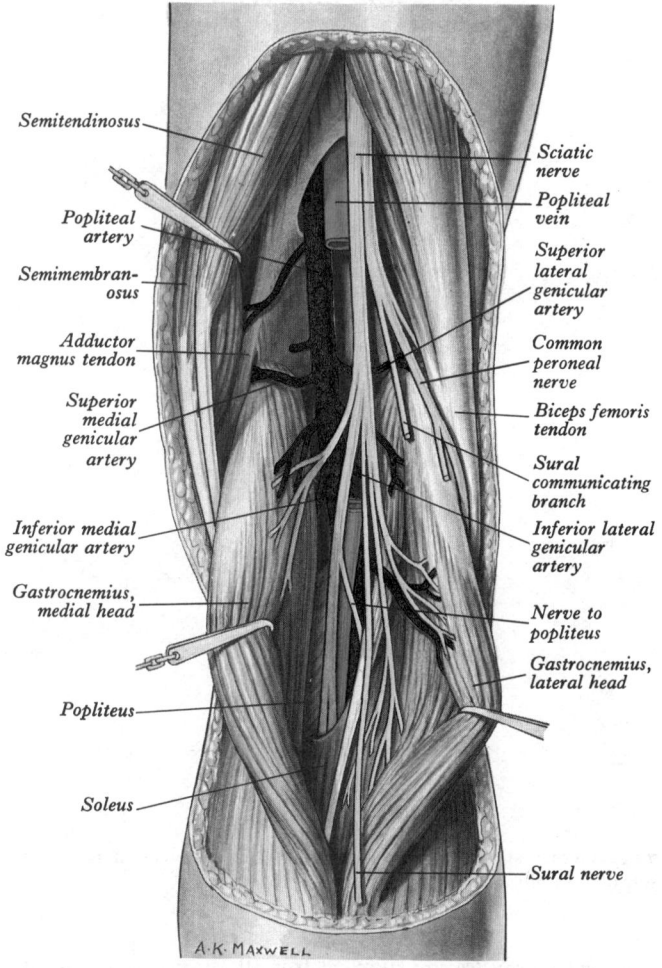

7.217 A dissection of the right popliteal fossa. The two heads of gastrocnemius and the semitendinosus and semimembranosus have been retracted in order to expose the contents of the fossa more fully.

then passes downwards near the lateral margin of the tendo calcaneus, and close to the small saphenous vein, to the interval between the lateral malleolus and the calcaneus; it supplies the skin of the lateral and posterior part of the lower one-third of the leg. It runs forwards below the lateral malleolus, and is continued along the lateral side of the foot and little toe, communicating on the dorsum of the foot with the superficial peroneal nerve. In the leg, its branches communicate with those of the posterior cutaneous nerve of the thigh. For a study of the fibre spectrum and ultrastructure of the fetal sural nerve consult Ochoa (1971).

The *medial calcanean branches* perforate the flexor retinaculum, and supply the skin of the heel and medial side of the sole of the foot.

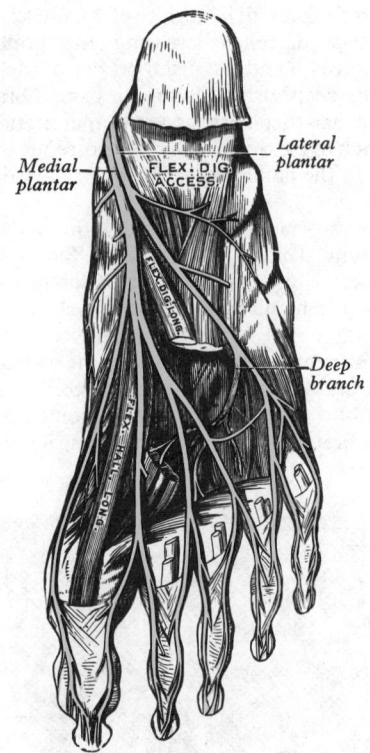

7.218   The plantar nerves of the right foot.

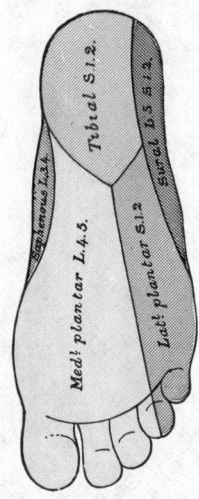

7.219   A diagram showing the distribution of the cutaneous nerves in the sole of the right foot.

Vascular branches are supplied from the tibial nerve and its branches to the arteries which they accompany in the leg and foot (p. 1132).

The medial plantar nerve (7.218), the larger of the two terminal divisions of the tibial nerve, accompanies the medial plantar artery and lies on the lateral side of the vessel. From its origin under cover of the flexor retinaculum it passes deep to the abductor hallucis, and, appearing between this muscle and the flexor digitorum brevis, gives off a digital nerve to the medial side of the great toe and finally divides opposite the bases of the metatarsal bones into three plantar digital nerves.

Branches. Cutaneous branches pierce the plantar aponeurosis between the abductor hallucis and the flexor digitorum brevis and are distributed to the skin of the sole of the foot.

Muscular branches supply the abductor hallucis, the flexor digitorum brevis, the flexor hallucis brevis and the first lumbrical; those for the abductor hallucis and flexor digitorum brevis arise from the trunk of the nerve near its origin and enter the deep surfaces of the muscles; the branch for the flexor hallucis brevis

springs from the digital nerve to the medial side of the great toe, and that for the first lumbrical from the first plantar digital nerve.

Articular branches supply the articulations of the tarsus and metatarsus.

The *proper digital nerve of the great toe* supplies the flexor hallucis brevis and the skin on the medial side of the great toe.

The *three common plantar digital nerves* pass between the divisions of the plantar aponeurosis, and each splits into two proper digital branches. Those of the first plantar digital nerve supply the adjacent sides of the great and second toes; those of the second, the adjacent sides of the second and third toes; and those of the third, the adjacent sides of the third and fourth toes. The third plantar digital nerve receives a communicating branch from the lateral plantar nerve; the first gives a twig to the first lumbrical. Each digital branch gives off cutaneous and articular filaments and opposite the distal phalanx sends upwards a dorsal branch, which supplies the structures around the nail, the continuation of the nerve being distributed to the ball of the toe. It will be observed that the digital branches of the medial plantar nerve *are similar in their distribution to those of the median nerve in the hand*. The muscles supplied by the two nerves also correspond closely. In the hand, the median nerve supplies the abductor and the flexor pollicis brevis, the opponens pollicis and the first and second lumbricals. The opponens is absent in the foot, but the abductor hallucis, flexor hallucis brevis and the first lumbrical are all supplied by the medial plantar nerve. As the flexor digitorum brevis corresponds to the flexor digitorum superficialis (median nerve) of the upper limb, the only difference exists in the innervation of the second lumbrical.

The lateral plantar nerve (7.218) supplies the skin of the fifth toe and lateral half of the fourth, as well as most of the deep muscles, *its distribution being similar to that of the ulnar nerve in the hand*. It passes obliquely forwards in company with the lateral plantar artery, which lies on the lateral side of the nerve, and reaches the lateral side of the foot near the tubercle of the fifth metatarsal bone. It passes between the flexor digitorum brevis and the flexor digitorum accessorius, and ends in the interval between the former muscle and the abductor digiti minimi by dividing into a superficial and a deep branch. Before its division, it supplies the flexor digitorum accessorius and abductor digiti minimi and gives off some small cutaneous branches which pierce the plantar fascia and supply the skin of the lateral part of the sole of the foot (7.219).

The *superficial branch* splits into two plantar digital nerves; of these the lateral supplies the lateral side of the fifth toe, the flexor digiti minimi brevis, and the two interosseous muscles of the fourth intermetatarsal space; the medial communicates with the third plantar digital branch of the medial plantar nerve and divides into two branches which supply the adjoining sides of the fourth and fifth toes.

The *deep branch* accompanies the lateral plantar artery on the deep surface of the tendons of the flexor muscles and the adductor hallucis, and supplies the second, third and fourth lumbricals, the adductor hallucis and all the interossei (except those of the fourth intermetatarsal space). The nerves to the second and third lumbricals pass distally deep to the transverse head of the adductor hallucis, and then pass round its distal border to reach the muscles (7.220).

THE COMMON PERONEAL NERVE

The common peroneal (lateral popliteal) nerve (7.217), about one-half the size of the tibial nerve, is derived from the dorsal branches of the ventral rami of the fourth and fifth lumbar and the first and second sacral nerves. It descends obliquely along the lateral side of the popliteal fossa to the head of the fibula, close to the medial margin of the biceps femoris. It lies between the tendon of the biceps femoris and the lateral head of the gastrocnemius, winds round the lateral surface of the neck of the fibula deep to the peroneus longus, and divides into the superficial and deep peroneal nerves. Its course can be indicated by a line drawn from the apex of the popliteal fossa distally and laterally, along the medial side of the biceps tendon, to the back of the head of the fibula, where the nerve can be rolled against the bone.

Previous to its division it gives off articular and cutaneous branches.

The *articular branches* are three in number; of these one accompanies the superior and another the inferior lateral genicular atery. Both may arise by a common trunk. The third, named the recurrent articular nerve, is given off at or near the point of division of the common peroneal nerve; it ascends with the anterior recurrent tibial artery through the tibialis anterior to supply the anterolateral part of the capsular ligament of the knee joint and also supplies branches to the superior tibiofibular joint.

The *cutaneous branches* (7.215), two in number, frequently spring from a common trunk; they are the lateral cutaneous nerve of the calf and the sural communicating branch.

The *lateral cutaneous nerve of the calf* supplies twigs to the skin on the anterior, posterior and lateral surfaces of the proximal part of the leg. The *sural communicating branch* arises near the head of the fibula, runs obliquely across the lateral head of the gastrocnemius to the middle of the leg, and joins with the sural nerve (p. 1113). It may, however, descend as a separate branch as far as the heel.

The deep peroneal (anterior tibial) nerve (7.212) begins at the bifurcation of the common peroneal nerve, between the fibula and the proximal part of the peroneus longus, passes obliquely forwards deep to the extensor digitorum longus to the front of the interosseous membrane, where it comes into relation with the anterior tibial artery in the proximal third of the leg; it then descends with the artery to the front of the ankle joint, where it divides into lateral and medial terminal branches. It lies at first on the lateral side of the anterior tibial artery, then in front of it, and again on its lateral side at the ankle joint.

In the leg, the deep peroneal nerve supplies *muscular branches* to the tibialis anterior, extensor hallucis longus, extensor digitorum longus and peroneus tertius, and an *articular branch* to the ankle joint.

The *lateral terminal branch* of the deep peroneal nerve passes across the tarsus, deep to the extensor digitorum brevis and, becoming enlarged as a pseudoganglion (c.f. other sites pp. 1097) and then supplies the extensor digitorum brevis. From the enlargement three minute *interosseous branches* are given off which supply the tarsal and metatarsophalangeal joints of the second, third and fourth toes. The first of these sends a filament to the second dorsal interosseous muscle.

The *medial terminal branch* of the deep peroneal nerve runs forwards on the dorsum of the foot, and lies on the lateral side of the arteria dorsalis pedis. At the first interosseous space it communicates with the medial branch of the superficial peroneal nerve, and divides into two dorsal digital nerves which supply the adjacent sides of the great and second toes. Before it divides it gives off an *interosseous branch* which supplies the metatarsophalangeal joint of the hallux and sends a filament to the first dorsal interosseous. The deep peroneal nerve may end by dividing into *three* terminal branches (Geller and Barbato 1970).

The superficial peroneal (musculocutaneous) nerve (7.212) begins at the bifurcation of the common peroneal nerve and lies at first deep to the peroneus longus. It then passes forwards and downwards between the peronei and the extensor digitorum longus, pierces the deep fascia in the distal third of the leg, and divides into a medial and a lateral branch. In its course between the muscles, it gives off muscular branches to the peroneus longus and peroneus brevis, and filaments to the skin of the lower part of the leg.

The *medial branch* passes in front of the ankle joint, and divides into two dorsal digital nerves, one of which supplies the medial side of the great toe, the other, the adjacent sides of the second and third toes. It communicates with the saphenous nerve and with the deep peroneal nerve (7.211).

The *lateral branch*, the smaller, passes along the lateral part of the dorsum of the foot, and divides into dorsal digital branches, which supply the contiguous sides of the third and fourth, and of the fourth and fifth toes. It also supplies the skin of the lateral side of the ankle, and communicates with the sural nerve (7.211).

The branches of the superficial peroneal nerve supply the skin of the dorsal surfaces of all the toes excepting the lateral side of the little toe and the adjoining sides of the first and second toes, the former being supplied by the sural nerve, and the latter by the medial terminal branch of the deep peroneal nerve. Frequently some of the lateral branches of the superficial peroneal are absent, and their places are then taken by branches of the sural nerve.

The perforating cutaneous nerve usually arises from the posterior aspects of the second and third sacral nerves. It pierces the lower part of the sacrotuberous ligament, and, winding round the inferior border of the gluteus maximus, supplies the skin covering the medial and lower parts of that muscle.

The perforating cutaneous nerve may arise from the pudendal nerve or it may be absent; in the latter case its place may be taken by a branch from the posterior cutaneous nerve of the thigh or by a branch from the third and fourth, or fourth and fifth, sacral nerves.

The pudendal nerve (6.94) derives its fibres from the second, third and fourth sacral spinal nerves (7.213). It leaves the pelvis between piriformis and coccygeus through the lower part of the greater sciatic foramen and enters the gluteal region, crossing the sacrospinous ligament close to its attachment to the ischial spine, being situated on the medial side of the internal pudendal vessels which lie on the ischial spine itself. It accompanies the internal pudendal artery through the lesser sciatic foramen into the pudendal canal (p. 562) on the lateral wall of the ischiorectal fossa;

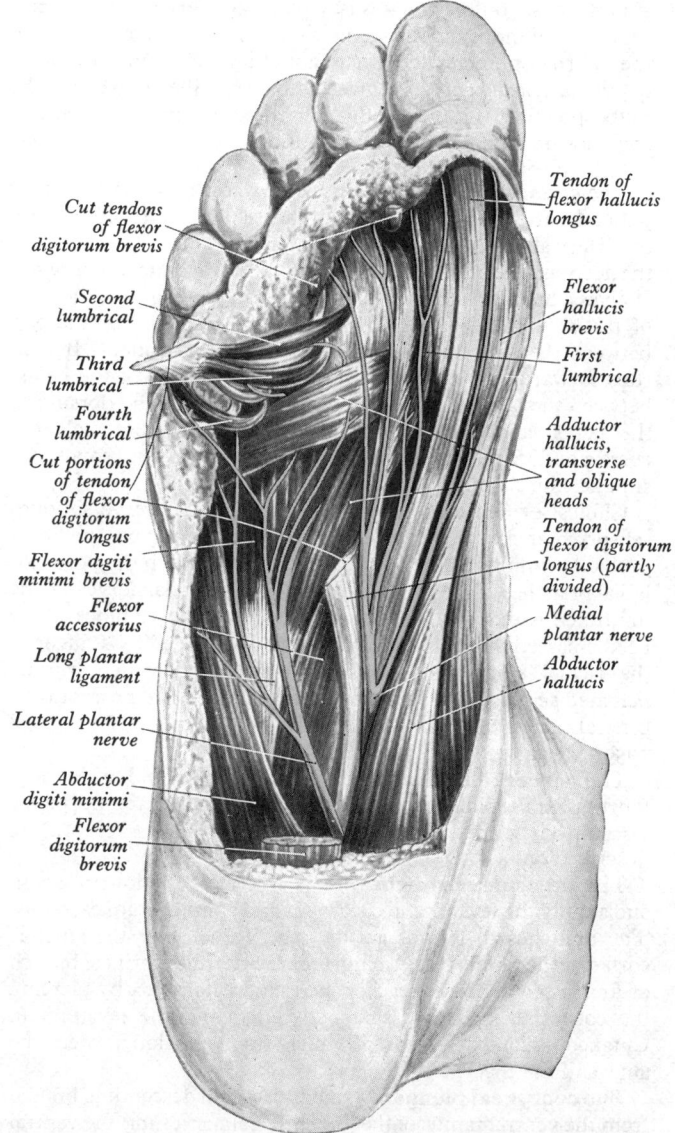

Cut tendons of flexor digitorum brevis

Second lumbrical

Third lumbrical

Fourth lumbrical

Cut portions of tendon of flexor digitorum longus

Flexor digiti minimi brevis

Flexor accessorius

Long plantar ligament

Lateral plantar nerve

Abductor digiti minimi

Flexor digitorum brevis

Tendon of flexor hallucis longus

Flexor hallucis brevis

First lumbrical

Adductor hallucis, transverse and oblique heads

Tendon of flexor digitorum longus (partly divided)

Medial plantar nerve

Abductor hallucis

7.220 A dissection of the lateral and medial plantar nerves of the right foot. Most of the flexor digitorum brevis has been removed. The flexor digitorum longus has been partially divided and its distal end has been displaced together with the second, third and fourth lumbricals.

in the posterior part of this canal it gives off the inferior rectal nerve, and then divides into the perineal nerve and the dorsal nerve of the penis (or clitoris).

The *inferior rectal nerve* pierces the medial wall of the pudendal canal, crosses the ischiorectal fossa with the inferior rectal vessels, and is distributed to the sphincter ani externus, to the lining of the lower part of the anal canal and to the skin round the anus. Branches of this nerve communicate with the perineal branch of the posterior cutaneous nerve of the thigh and with the scrotal nerves. The inferior rectal nerve occasionally arises directly from the sacral plexus. It may traverse the sacrospinous ligament (in 8 out of 40 dissections according to Roberts and Taylor 1973) and it occasionally communicates with the pudendal nerve.

The *perineal nerve*, the inferior and larger terminal branch of the pudendal nerve, runs forwards below the internal pudendal artery. It accompanies the perineal artery and divides into posterior scrotal (or labial) and muscular branches.

The *posterior scrotal branches* number two, medial and lateral. They pierce, or pass superficial to, the inferior fascia of the urogenital diaphragm, and run forwards along the lateral part of the urethral triangle in company with the scrotal branches of the perineal artery; they are distributed to the skin of the scrotum, and communicate with the perineal branch of the posterior cutaneous nerve of the thigh and with the inferior rectal nerve. In the female the corresponding nerves (*posterior labial branches*) supply the labium majus.

The *muscular branches* are distributed to the transversus perinei superficialis, bulbospongiosus, ischiocavernosus, transversus perinei profundus, sphincter urethrae and the anterior parts of the external sphincter and levator ani. A branch, termed the *nerve to the urethral bulb*, is given off from the nerve to the bulbospongiosus; it pierces this muscle, and supplies the corpus spongiosum penis, its terminal fibres ending in the mucous membrane of the urethra.

The *dorsal nerve of the penis* runs forwards above the internal pudendal artery along the ramus of the ischium, and accompanies the artery along the margin of the inferior ramus of the pubis, on the deep surface of the inferior fascia of the urogenital diaphragm. It gives a branch to the corpus cavernosum penis and, at the apex of the membrane, passes through the lateral part of the gap between the structure and the inferior pubic ligament. It then runs forwards, in company with the dorsal artery of the penis, between the layers of the suspensory ligament, to the dorsum of the penis, and ends in the glans penis. In the female the corresponding nerve (*dorsal nerve of the clitoris*) is very small, and supplies the clitoris.

Clinical evidence indicates that the pudendal nerve supplies sensory branches to the lower inch or so of the vagina, the fibres probably running in the inferior rectal nerve and in the posterior labial branches of the perineal nerve. The pudendal nerves can be infiltrated with a local anaesthetic ('pudendal nerve block') by a needle passed through the vaginal wall and guided by a finger to the ischial spine and sacrospinous ligament, which can be palpated *per vaginam*. For various vaginal operative procedures a general anaesthetic can thus usually be avoided (Huntingford 1959; Nakanishi 1967).

The visceral branches arise from the second, third and fourth sacral spinal nerves, and are distributed to the pelvic viscera. They are termed the *pelvic splanchnic nerves* and are described on p. 1123.

The muscular branches are derived from the fourth sacral, and supply the levator ani, coccygeus and sphincter ani externus. The branches to levator ani and coccygeus enter their pelvic surfaces; the ramus to the sphincter ani externus (perineal branch of fourth sacral nerve) reaches the ischiorectal fossa by piercing the coccygeus or by passing between it and the levator ani. Cutaneous filaments from this branch supply the skin between the anus and the coccyx.

The coccygeal plexus is formed by a small descending branch from the ventral ramus of the fourth sacral nerve, and the ventral rami of the fifth sacral and coccygeal nerves. The ventral ramus of the fifth sacral nerve emerges from the sacral hiatus and turns forwards round the lateral margin of the sacrum below the cornu. It pierces the coccygeus to gain its pelvic surface and is then joined by a descending filament from the fourth sacral nerve. The small trunk so formed descends on the pelvic surface of the coccygeus and unites with the minute ventral ramus of the coccygeal nerve, which descends from the sacral hiatus, turns round the lateral margin of the coccyx and pierces the coccygeus to gain the pelvis. This small trunk constitutes the *coccygeal plexus*. The *anococcygeal nerves* arise from the plexus, and consist of a few fine filaments which pierce the sacrotuberous ligament and supply the skin in the region of the coccyx.

Applied Anatomy. The *iliohypogastric nerve* may be cut, as it lies between the muscles inferiorly in the anterior abdominal wall (p. 1106), by an incision ('McBurney's gridiron') through which the vermiform appendix is sometimes approached; the consequent weakness of the muscles in the region of the inguinal canal may predispose to the development of a direct inguinal hernia.

The *lateral cutaneous nerve of the thigh* may be compressed and irritated as it passes through the inguinal ligament (p. 552) or as it pierces the dense fascia lata, and this is said to be one of the causes of a rare condition of pain on the lateral side of the thigh (*meralgia paraesthetica*).

The *femoral nerve* is occasionally injured by wounds in the groin or thigh; the result is paralysis of the quadriceps femoris and diminished cutaneous sensibility on the anterior and medial aspects of the thigh.

Surgical division of the *obturator nerve* is sometimes done for relief of spasm of the adductors of the thigh in certain cases of

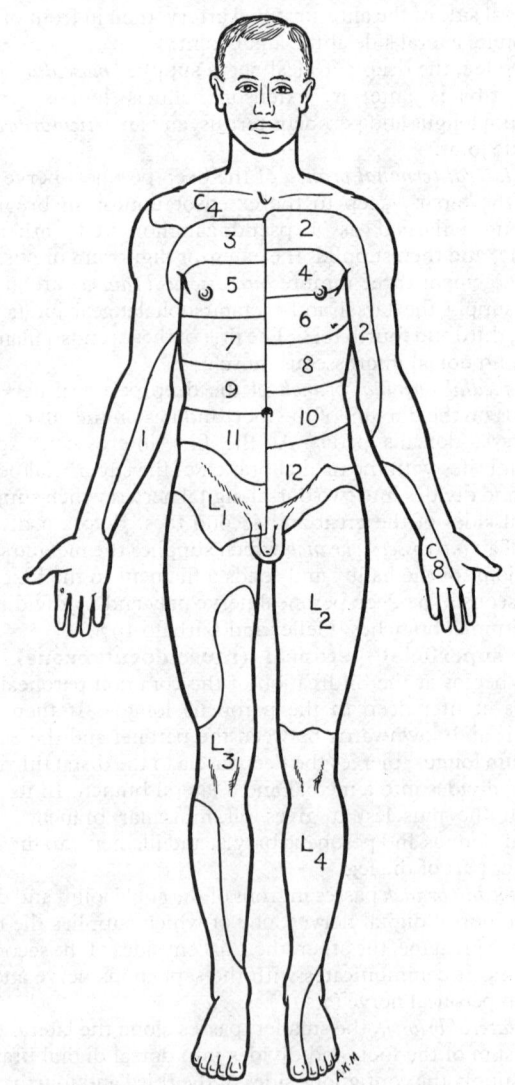

7.221 The cutaneous areas supplied by the ventral rami of the thoracic and upper four lumbar nerves. (After Foerster 1933.) By comparing both sides the degree of overlapping and the area of exclusive supply of any individual nerve may be estimated. See text for the area supplied by T.1 on the trunk.

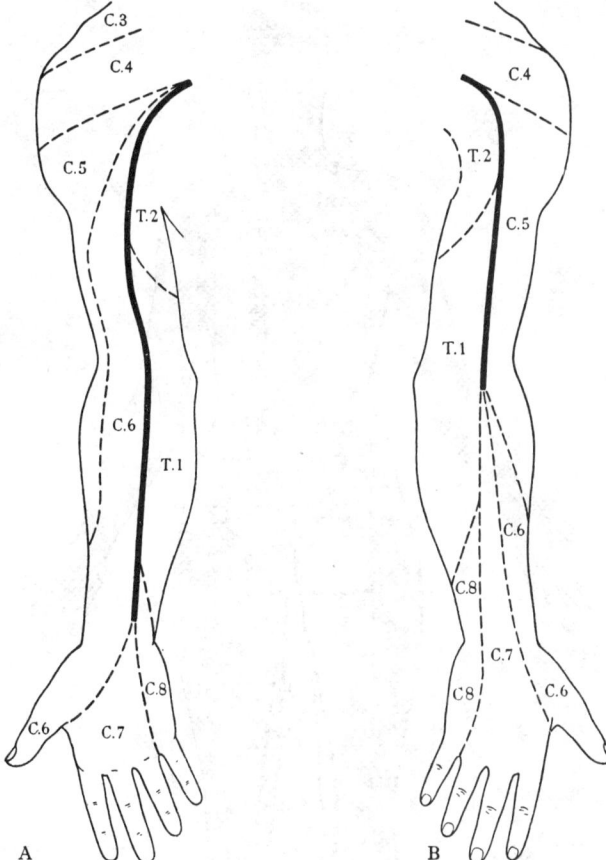

7.222 A *Left:* The arrangement of the dermatomes on the anterior aspect of the upper limb. The *heavy black line* represents the *ventral axial line* and the overlap across it is *minimal*. Across the interrupted lines, the overlap is considerable. *Right:* The arrangement of the dermatomes on the posterior aspect of the upper limb. The *heavy black line* represents the *dorsal axial line* and the overlap across it is *minimal*. Across the interrupted lines the overlap may be, and often is considerable.

spastic paralysis in children, in paraplegia, or in multiple sclerosis. Because of its branches to the hip joint, knee joint and medial side of the thigh, in cases of disease of the hip joint pain may be referred to the medial side of the thigh or to the knee joint.

The *sacral plexus* and *lumbosacral trunk* may be compressed by pelvic tumours, or by the fetal head in pregnancy, and result in pain in the lower limbs which, in the case of malignant growths, may be extremely severe.

The *sciatic nerve* may be injured by posterior dislocations or fracture dislocations of the hip joint; if the lesion of the nerve is complete, which is rare, all muscles below the knee are paralysed and all cutaneous sensibility there is lost, except for the area supplied by the saphenous nerve. In traumatic lesions of the sciatic nerve in middle levels of the thigh, the flexor muscles generally escape because of the high origin of the nerves to these muscles. The surface marking of the sciatic nerve for purposes of injection is given on p. 1112.

The *common peroneal nerve* is the most commonly injured nerve in the lower limb, chiefly because of its exposed position as it winds round the neck of the fibula. Injury here will result in paralysis of all the dorsiflexor and evertor muscles of the foot (tibialis anterior, extensor hallucis longus, extensor digitorum longus, extensor digitorum brevis, peroneus longus and peroneus brevis) producing a 'drop foot'. There is a variable loss of cutaneous sensibility on the anterolateral aspect of the leg and on the dorsum of the foot.

Owing to its deep and protected position, the tibial nerve is rarely injured. Wounds in the popliteal fossa or posterior dislocation of the knee joint may damage the tibial nerve and produce paralysis of the flexor muscles in the leg and the intrinsic muscles of the sole of the foot, resulting in considerable disability.

Furthermore, loss of sensation in the sole of the foot renders its skin liable to 'pressure sores'.

Morphology of Spinal Nerves and Limb Plexuses

The spinal nerves which conform in their behaviour to the more primitive arrangement are the nerves of those segments which have retained to a large extent their *metameric* (segmental) characters, viz. T.2–L.1. These typical spinal nerves are distributed according to a definite plan. The dorsal epaxial ramus passes backwards and downwards lateral to the articular processes and divides into medial and lateral branches which penetrate the deep muscles of the back. Both branches innervate the muscles amongst which they lie, and either one or the other becomes superficial and supplies a band of skin extending from the posterior median line approximately to the scapular line (*see* below).

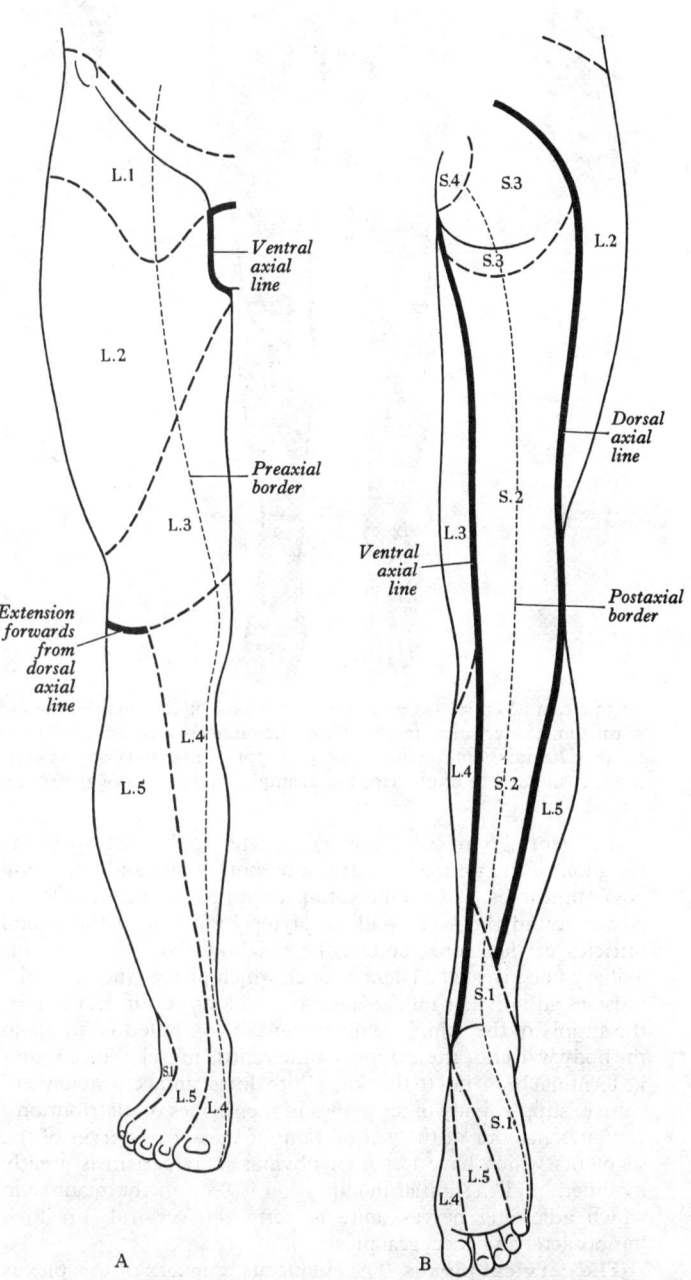

7.222 B *Left:* The segmental distribution of the nerves of the lumbar and sacral plexuses to the skin of the anterior aspect of the lower limb. *Right:* The segmental distribution of the nerves of the lumbar and sacral plexuses to the skin of the posterior aspect of the lower limb. For the significance of the markings see **7.222** caption.

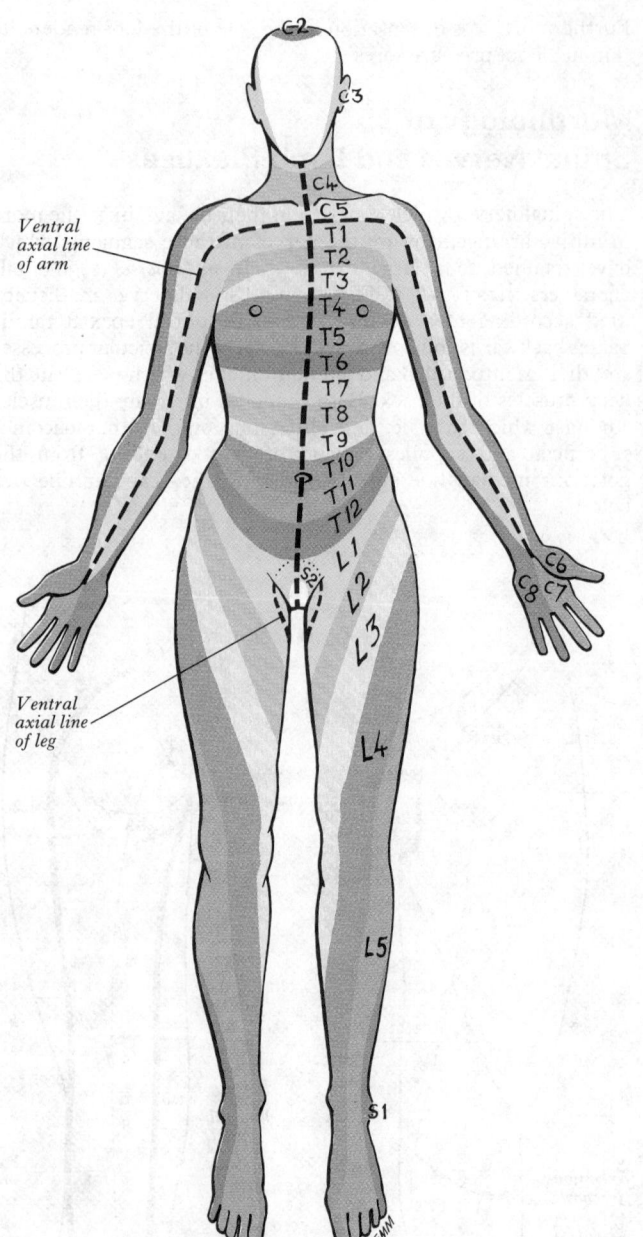

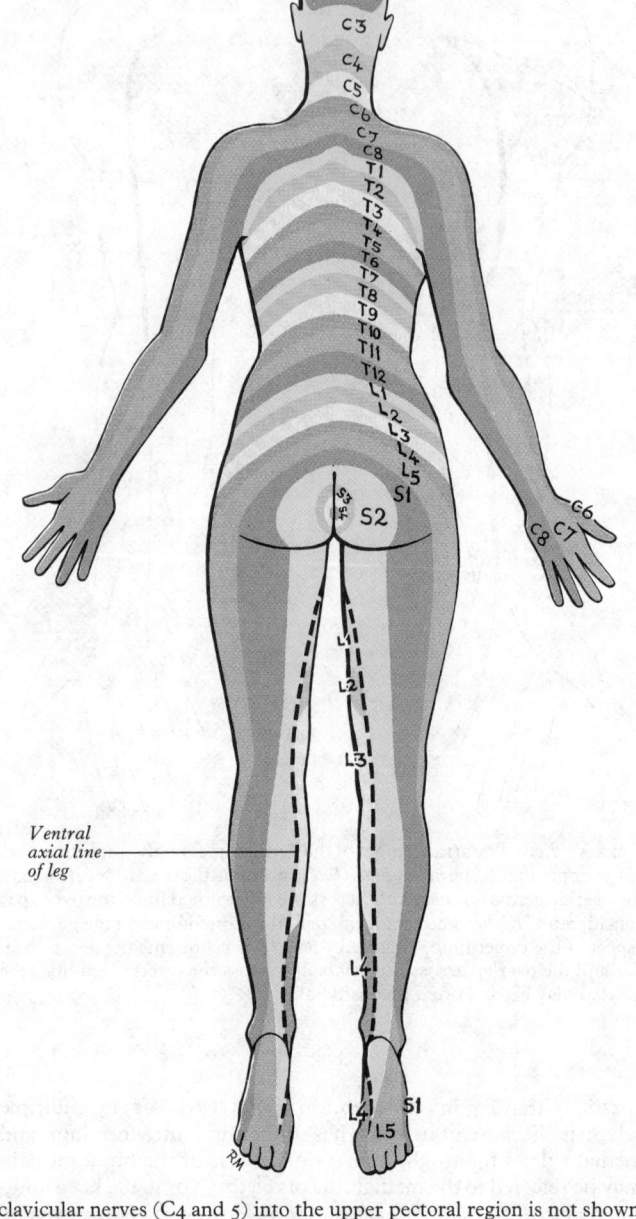

7.233A & B Ventral (A) and dorsal (B) views of the distributions of dermatomic innervation, modified from the studies of Keegan and Garrett (1948). Diagrams of this kind can only be approximate and cannot clearly indicate the degree of overlap. For example, the overlap of the supra-clavicular nerves (C4 and 5) into the upper pectoral region is not shown, and it is possible that the pectoral area assigned to the first thoracic spinal nerve is here exaggerated.

The ventral hypaxial ramus is connected to the corresponding ganglion of the sympathetic trunk by both white and grey rami communicantes. After innervating the prevertebral muscles, it passes round the body wall supplying branches to the lateral muscles of the trunk, and in the neighbourhood of the mid-axillary line gives off a lateral branch which pierces the overlying muscles and divides into an anterior and a posterior division for the supply of the skin. The main trunk is continued forwards in the body wall and, after supplying the ventral muscles, distributes its terminal branches to the skin. (Both dorsal and ventral rami, of course, supply many other tissues in their zones of distribution.)

The behaviour of the ventral rami of the spinal nerves of the segments which have lost their obvious metamerism is greatly modified, and the initial modification is seen in the manner in which adjoining nerves unite to form the cervical, brachial, lumbosacral and coccygeal plexuses.

The cervical plexus. The cutaneous branches of this plexus are homologous with the anterior terminal and the lateral branches of the ventral rami of the typical spinal nerves. The transverse cutaneous nerve of the neck and the medial supraclavicular nerves represent the anterior terminal branches; the lesser occipital and the lateral supraclavicular represent the

lateral branches, while the great auricular and the intermediate supraclavicular probably represent elements of both branches.

The brachial plexus. In the formation of the brachial and lumbosacral plexuses the division of the constituent nerves of the plexus into anterior (ventral or flexor) and posterior (dorsal or extensor) branches is characteristic. In the brachial plexus the division affects the three trunks of the plexus (p. 1094) and, to a remarkable extent, it conforms to the differentiation of the primitive musculature of the limb into a flexor (ventral) and extensor (dorsal) group. So far as the cutaneous innervation is concerned, branches of the ventral divisions of the trunks take a large part in the supply of the skin of the dorsal surface of the limb. This problem has been tentatively explained (Harris 1939) by assuming that each constituent root of the plexus originally divided into ventral and dorsal branches and that, in the evolution of the human type of plexus, inherently dorsal fibres enter the ventral branches of the trunks. As a result of this rearrangement, fibres of the median and ulnar nerves have a wide area of supply on the dorsal surface of the hand.

The position of the developing limb bud on the ventrolateral aspect of the trunk, and the behaviour of the first and second thoracic nerves, provide support for the view that the constituent nerves of the great limb plexuses represent only the lateral branches of the ventral rami of the typical spinal nerves. The second thoracic nerve sends its lateral cutaneous branch into the upper limb as the intercostobrachial nerve, and the size of this nerve varies inversely with the size of the direct contribution which the second thoracic nerve makes to the brachial plexus. Otherwise the second thoracic behaves like a typical spinal nerve. The first thoracic nerve sends a large contribution to the brachial plexus, and this could be homologous with the lateral branch. The remainder of the nerve, despite its small size, behaves in a typical manner, although its fine anterior cutaneous branch is often absent, and, when present, only supplies a limited area of skin.

The lumbosacral plexus. The division of the constituent nerves of the lumbar and sacral plexuses into anterior (ventral or flexor) and posterior (dorsal or extensor) divisions is not so obvious as the corresponding pattern in the brachial plexus, but it can be demonstrated anatomically that the obturator and the tibial nerves arise from ventral and the femoral and peroneal nerves from dorsal divisions. The lateral branches of the twelfth thoracic and first lumbar spinal nerves are drawn over the iliac crest to assist in the innervation of the gluteal skin, but otherwise these nerves behave as typical spinal nerves. The second lumbar nerve behaves in a manner which renders its interpretation difficult, since it not only makes a substantial contribution to the lumbar plexus but also possesses both an anterior terminal branch, the genital branch of the genitofemoral nerve, and a lateral cutaneous branch, represented by the lateral cutaneous nerve of the thigh and the femoral branch of the genitofemoral nerve. The anterior terminal portions of the third, fourth and fifth lumbar and first sacral spinal nerves are suppressed, but the corresponding parts of the second and third sacral nerves supply the skin, etc., of the perineum. The posterior femoral cutaneous nerve provides an interesting example of the principles involved here: it is derived from anterior and posterior divisions and supplies 'flexor' skin in the thigh, but 'extensor' skin through its gluteal rami.

Segmental Innervation of the Skin

The area of skin supplied by any one spinal nerve, through both its rami, constitutes a **dermatome** and, typically, the dermatomes extend round the body from the posterior to the anterior median line (**7.206, 221, 223**). The dermatomes of *consecutive spinal nerves* overlap markedly, and this is seen most clearly in those segments of the body which have been least affected by the development of the limbs, i.e. second thoracic to first lumbar (**7.221, 223**).

In some situations, e.g. the upper part of the anterior thoracic wall, the cutaneous nerves supplying two adjoining areas are *not* derived from consecutive spinal nerves and the overlap between, above, and below, is minimal. When the second thoracic spinal nerve is severed, the line of anaesthesia is sharply demarcated, although there may be some overlap of the painful and thermal elements. Likewise the results found after section of a peripheral nerve (e.g. the ulnar nerve at the wrist) show that the area of tactile loss is always greater than the area of loss of painful and thermal sensibilities, for the degree of overlap of fibres conveying these types of sensibility is always more extensive than the overlap of fibres conveying tactile sensibility. As a result the area of total anaesthesia and analgesia following section of a peripheral nerve is considerably less than might be anticipated from a knowledge of its anatomical distribution (Foerster 1933; Wollard *et al.* 1940).

Cutaneous Innervation of the Neck and Upper Limb

The first cervical spinal nerve has no cutaneous branches. The second cervical usually supplies the skin of the head, from the vertex backwards to the neighbourhood of the superior nuchal line, the cranial surface and most, if not the whole, of the lateral surface of the auricle, the skin over the angle of the mandible and below the chin (**7.182**). The third cervical spinal nerve supplies a very oblique band of skin, commencing behind over the back of the scalp and the upper part of the back of the neck and passing forwards and downwards across the side of the neck. The area increases in extent as it is traced forwards, and in the ventral

median line extends from the hyoid bone down to the level of the first rib. The fourth cervical spinal nerve supplies the upper half or more of the back of the neck, and the area widens as it is traced downwards and forwards round the side of the neck to the anterior aspect of the trunk. It supplies the skin over the clavicle and first intercostal space, as well as over the acromion and the upper part of the deltoid.

Each of these three areas is overlapped by the succeeding area, but the amount of overlapping is slight and is greater for the dorsal rami than for the ventral rami.

The cutaneous distribution of those spinal nerves which contribute to the brachial plexus becomes intelligible only when reference is made to an early stage in the development of the upper limb. In a human embryo of the fourth week the upper limb is represented by a small, somewhat flattened elevation on the ventrolateral aspect of the trunk opposite to the lower four cervical and the first thoracic segments. The ectoderm covering it is directly continuous with the ectoderm of the trunk and draws its nerve supply from the nerves of the corresponding segments. Similarly, its contained mesoderm is also continuous with the mesoderm of the same segments. The lower limb bud appears at a slightly later stage and always lags behind the upper limb bud in its development until after birth.

The limb buds possess ventral and dorsal surfaces and cranial or *preaxial*, and caudal or *postaxial* borders. In the upper limb the *fifth cervical* ventral ramus supplies a strip of skin on both ventral and dorsal surfaces along the preaxial border, and the *first thoracic nerve* has a similar distribution along the postaxial border. The intervening nerves supply approximately parallel strips of skin on both the ventral and dorsal surfaces. As the limb elongates the central nerves of the plexus (C. 6, 7 and 8) become buried proximally and reach the skin only in its more distal part, while the nerves of the adjoining segments (C.4 and T.2 and 3) become drawn in to supply the skin at the root of the limb. In the process of growth, the lengthening limb becomes rotated laterally through roughly 90° and adducted to the trunk (p. 350). Later, therefore, the *preaxial border* runs distally along the lateral aspect of the (supinated) limb to the thumb, which is the *preaxial digit*, while the *postaxial border* runs distally along the medial aspect to the little finger, which is the *postaxial digit*. Accordingly, the cutaneous nerve supply of the lateral aspect of the adult limb is derived from C. 4, which has been drawn in at the root of the limb, C. 5 and C. 6, and its medial aspect from T. 2, T. 1 and C. 8 (**7.222**). On the *front of the limb* the areas supplied by C. 5 and C. 6 adjoin the areas supplied by T. 2, T. 1 and C. 8 but at the dividing line between them, which is termed the *ventral axial line*, the overlap is minimal, for C. 7 is buried proximally and only reaches the skin a little proximal to the wrist (**7.222A**). On the *back of the limb* the condition is very similar, but C.7 (in the posterior cutaneous nerve of the forearm) reaches the skin at, or a little proximal to the elbow so that the *dorsal axial line* ends at a more proximal level (**7.222A**). (The reader must appreciate the effect of pronation of the forearm on the disposition of these lines—*see* p. 350.)

Cutaneous Innervation of the Trunk

The skin of the trunk is supplied by the spinal nerves T. 1 to L. 1 inclusive (**7.206, 221, 223**), by the sacral nerves, except the first, and by the coccygeal nerve. These nerves supply consecutive curved bands of skin, of which the upper are almost horizontal while the lower are disposed obliquely. The upper half of each band receives additional supply from the nerve above and the lower half from the nerve below, so that no appreciable loss of sensibility follows the section of any individual spinal nerve. It is convenient to remember that the band which includes the subcostal angle is supplied by the seventh thoracic nerve and that the umbilicus lies in the upper part of the band supplied by the tenth thoracic nerve.

The area supplied by the dorsal rami of these nerves are limited laterally by the dorsolateral line, which commences above on the back of the head and runs downwards and laterally to the medial end of the acromion. It is then continued downwards to the posterior aspect of the greater trochanter of the femur where it curves medially to the coccyx (**7.222B**). The cutaneous strips supplied by the dorsal rami do not correspond exactly to the strips supplied by the ventral rami, for they differ both in their breadth and in their position.

On the upper part of the ventral aspect of the thorax the third and fourth cervical areas adjoin the first and second thoracic areas (7.206), owing to the fact that the muscles and skin areas supplied by intervening nerves have grown into the upper limb, and a similar but less extensive gap is found on the posterior aspect of the trunk.

A corresponding arrangement is found in the lower part of the trunk, but it is not so obvious owing to the approximation of the lower limbs to one another, but is still apparent in the gluteal region. The first lumbar area adjoins the second sacral area at the root of the penis and scrotum (see p. 1418), for the intervening nerves have been drawn away to supply the lower limb.

Cutaneous Innervation of the Lower Limb

The skin of the lower limb is innervated by the nerves of the segments from which it is derived, viz. T. 12–S. 3 (7.222B, 223). The arrangement originally is precisely similar to that in the upper limb, but its identification in the adult has been rendered difficult on account of the rotation of the lower limb in the early stages of its development (p. 153). Originally the *preaxial border* follows the cephalic border of the limb bud to the hallux, which is the *preaxial digit*, while the *postaxial border* follows its caudal margin to the little toe, which is the *postaxial digit*. As development proceeds, the limb undergoes rotation in a *medial* direction so that the hallux comes to lie on the medial side of the adult foot and the little toe on its lateral side. The tibia, although homologous with the radius, lies on the medial side of the leg. Since the torsion occurs at the hip joint, the gluteal region retains its dorsal (extensor) situation.

The preaxial border commences above on the middle of the front of the thigh and runs down to the knee. It then curves medially as it descends to the medial malleolus to gain the medial side of the foot and the hallux. The post-axial border commences above in the gluteal region and descends to the popliteal fossa. It then declines laterally as it descends to the lateral malleolus to gain the lateral side of the foot. The *ventral* and *dorsal axial lines* necessarily exhibit a corresponding obliquity. The ventral axial line commences proximally at the medial end of the inguinal ligament and descends the posteromedial side of the thigh and leg to end proximal to the heel. The dorsal axial line commences in the lateral part of the gluteal region and descends on the posterolateral aspect of the thigh to the knee. It then inclines medially and ends before it reaches the ankle (7.222B).

The segmental cutaneous distribution of the nerves to the lower limb is shown in 7.222B.

Our knowledge of the extent of the individual dermatomes, especially of the limbs, is necessarily based on clinical evidence, and different authorities have mapped out areas which are far from being identical for the same dermatomes. This is due partly to their failure to adopt a common method in the neurological examination of patients and partly to individual differences between patients suffering from similar lesions. There is more disagreement with regard to the dermatomes in the leg, perhaps in part due to the greater frequency of injuries to the brachial plexus, affording more numerous opportunities of correlation between the areas of cutaneous analgesia or anaesthesia and the exact site of damage to the nerve. The figures of the limb dermatomes here inserted in 7.222 A, B are based on those of the Committee appointed by the Medical Research Council, and published in their *Report on Peripheral Nerve Lesions*, 1942.

When studying these figures it must be clearly understood that the *broken lines* indicate that the nerves on each side of them extend considerably beyond them, the amount of such over-lapping being often difficult to define. But, along the *ventral* and *dorsal axial lines*, shown in *heavy black*, overlap is minimal, for the nerves on each side of the line are *not* derived from consecutive spinal nerves, and the intervening nerve or nerves are buried in the substance of the limb in this situation and only reach the skin at a more distal point.

Some observers (Keegan and Garrett 1948) maintain that, in the embryonic development of the dermatomes of the limbs, the sensory nerves grow spirally from the dorsal surface of the limb buds around their preaxial and postaxial borders to meet on their ventral surface along the ventral axial line, and they deny the existence of a dorsal axial line. By plotting the areas of hyposensitivity, particularly hypalgesia (diminished sensibility to painful stimuli), following damage to *individual* nerve roots they have constructed charts of the limb dermatomes that differ considerably from those shown in 7.222. The views of Keegan and Garrett (7.223) were based on observations of a large number of patients with hypalgesia due to herniation of intervertebral discs. These included 165 cases in the upper limb (47 verified surgically) and 1,264 in the lower limb (707 thus verified). The main difference between these observations is that the derma-tomes are shown as more continuous strips extending from the trunk into and along the limbs.

Segmental Innervation of Muscles

Each spinal nerve originally supplies the musculature derived from the myotome of the same segment. In cases where the derivatives of any one myotome persist as separate entities, they retain their original nerve supply, but when derivatives of adjoining myotomes fuse, the resultant muscle does not necessarily retain its supply from each of the corresponding spinal nerves, although it may and frequently does so. Since the limb muscles develop *in situ* in the mesodermal core of the developing limb, it is impossible to identify the individual segments from which any muscle is derived by the study of its mode of development. The union of the individual spinal nerves and their branches in the brachial and lumbosacral plexuses renders impossible the identification by dissection of the segmental value of the individual motor nerves.

The notion of immutability of nerve-muscle relationship, originally promoted by Furbinger (1873), has been criticized, and exceptions have been recorded. (Consult Haines 1935, and Minkoff 1974.)

Segmental innervation of the Muscles of the Limbs

Most muscles of the limbs are innervated from more than one segment of the spinal cord and the segments involved for individual muscles are indicated in the section on Myology (p. 529). In the list given below, for a given muscle the *predominant segmental origin* of its nerve supply is recorded; damage to these segments or to the motor nerve roots arising therefrom result in maximum paralysis of the appropriate muscles. The information is based chiefly on clinical evidence (Bumke and Foerster 1936; Villiger 1946; Sharrard 1955), but it must be admitted that there is a difference of opinion in the case of a number of muscles, and not all the muscles of the limbs are included in this list. Moreover, though the evidence for some muscles is incontrovertible, it is scanty and uncertain in many other instances. An important example is provided by the intrinsic muscles of the hand, which are often regarded in clinical teaching as innervated solely through the first thoracic spinal nerve. As indicated by the following table the eighth cervical spinal nerve is also involved.

Upper Limb Muscles

C. 3, 4 Trapezius; levator scapulae.

C. 5 Rhomboids; deltoids; supraspinatus; infraspinatus; teres minor; biceps.

C. 6 Serratus anterior; latissimus dorsi; subscapularis; teres major; pectoralis major (clavicular head); biceps; coraco-brachialis; brachialis; brachioradialis; supinator; extensor carpi radialis longus.

C. 7 Serratus anterior; latissimus dorsi; pectoralis major (sternal head); pectoralis minor; triceps; pronator teres; flexor carpi radialis; flexor digitorum superficialis; extensor carpi radialis longus; extensor carpi radialis brevis; extensor digitorum; extensor digiti minimi.

C. 8 Pectoralis major (sternal head); pectoralis minor; triceps; flexor digitorum superficialis; flexor digitorum profundus; flexor pollicis longus; pronator quadratus; flexor carpi ulnaris; extensor carpi ulnaris; abductor pollicis longus;

extensor pollicis longus; extensor pollicis brevis; extensor indicis: abductor pollicis brevis; flexor pollicis brevis; opponens pollicis.

T. 1 Flexor digitorum profundus; intrinsic muscles of the hand (except abductor pollicis brevis; flexor pollicis brevis; opponens pollicis).

Lower Limb Muscles

L. 1 Psoas major; psoas minor.

L. 2 Psoas major; iliacus; sartorius; gracilis; pectineus; adductor longus; adductor brevis.

L. 3 Quadriceps; adductors (magnus, longus, brevis).

L. 4 Quadriceps; tensor fasciae latae; adductor magnus; obturator externus; tibialis anterior; tibialis posterior.

L. 5 Gluteus medius; gluteus minimus; obturator internus; semimembranosus; semitendinosus; extensor hallucis longus; extensor digitorum longus and peroneus tertius; popliteus.

S. 1 Gluteus maximus; obturator internus; piriformis; biceps femoris; semitendinosus; popliteus; gastrocnemius; soleus; peronei (longus and brevis); extensor digitorum brevis.

S. 2 Piriformis; biceps femoris; gastrocnemius; soleus; flexor digitorum longus; flexor hallucis longus; some intrinsic foot muscles.

S. 3 Some intrinsic foot muscles (except abductor hallucis; flexor hallucis brevis; flexor digitorum brevis; extensor digitorum brevis).

(See also p. 873 for a table of lower limb innervation.)

Joint Movements

In terms of movements of joints, the segmental innervation of the limb muscles may be expressed in general as follows:

Shoulder	Abductors and lateral rotators.	C. 5
	Adductors and medial rotators.	C. 6, 7, 8
Elbow	Flexors.	C. 5, 6
	Extensors.	C. 7, 8
Forearm	Supinators.	C. 6
	Pronators.	C. 7, 8
Wrist	Flexors and extensors.	C. 6, 7
Digits	Long flexors and extensors.	C. 7, 8
Hand	Intrinsic muscles.	C. 8, T. 1
Hip	Flexors, adductors, medial rotators.	L. 1, 2, 3
	Extensors, abductors, lateral rotators.	L. 5, S. 1
Knee	Extensors.	L. 3, 4
	Flexors.	L. 5, S. 1
Ankle	Dorsiflexors.	L. 4, 5
	Plantar flexors.	S. 1, 2
Foot	Invertors.	L. 4, 5
	Evertors.	L. 5, S. 1
	Intrinsic muscles	S. 2, 3

THE AUTONOMIC NERVOUS SYSTEM

The autonomic nervous system includes parts of the central and peripheral nervous systems, the latter being concerned with the innervation of viscera, glands, blood vessels and nonstriated muscle. It is the visceral (splanchnic) component of the nervous system. The term 'autonomic' is convenient rather than appropriate. The *autonomy* of this part of the nervous system is illusory. It is intimately responsive to changes in the somatic activities of the body, and while its connexions with somatic elements are not always clear in anatomical terms, the physiological evidence of visceral reflex activities stimulated by somatic events is abundant. (For general information on the anatomy and physiology of the autonomic nervous system consult Sheehan 1936; White *et al.* 1952; Kuntz 1953; Mitchell 1953, 1956; Pick 1970.)

Visceral efferent pathways differ from somatic equivalents in being interrupted by peripheral synapses, at least two neurons being interposed between the central nervous system and the visceral effector organ (7.224). The cells of origin of the primary neurons are sited in the visceral efferent components of cranial nerve nuclei and in the lateral grey columns of the spinal cord. Their axons, which are variably, but usually finely myelinated, traverse the corresponding cranial and spinal nerves to enter ganglia, where they synapse with the dendrites or somata of the secondary neurons. The axons of the secondary or effector neurons are usually nonmyelinated and are distributed to nonstriated muscle or gland cells. There are therefore in the peripheral efferent pathway principal *preganglionic neurons* and *postganglionic neurons*. The latter are more numerous, and one preganglionic neuron may synapse with up to 15 to 20 postganglionic neurons, a circumstance which is associated with the wide diffusion of many autonomic effects. The disproportion between preganglionic and postganglionic neurons is said to be greater in the sympathetic than in the parasympathetic parts of the autonomic nervous system. (Indeed, in an investigation into the human superior cervical ganglion, a ratio of preganglionic to postganglionic fibres of 1 to 196 was claimed—Ebbesson 1968.) The terminations of postganglionic neurons are described on p. 847. The detailed structure of sympathetic ganglia and views concerning further neuronal types, including interneurons is treated on p. 1124.

The visceral afferent paths resemble somatic ones and the cells of origin of the peripheral fibres are unipolar cells in cranial and spinal nerve ganglia. Their peripheral processes are distributed through the autonomic ganglia or plexuses or possibly through somatic nerves without further synapse. Their central processes (axons) accompany the somatic afferent fibres through dorsal spinal nerve roots to the central nervous system (p. 1136).

The autonomic nervous system consists of two complementary parts, the *parasympathetic* and the *sympathetic systems* which differ structurally and in their functions. The preganglionic efferent fibres of the parasympathetic nervous system emerge through certain cranial and sacral spinal nerves and constitute the *craniosacral outflow*. On the other hand the preganglionic efferent fibres of the sympathetic nervous system emerge through the thoracic and upper lumbar spinal nerves and constitute the *thoracolumbar outflow*. The cell bodies of the postganglionic neurons in the parasympathetic system are situated peripherally, either as discrete collections forming ganglia nearer to the structures innervated than to the central nervous system, or sometimes dispersed in the walls of the viscera themselves. The cell bodies of the postganglionic neurons in the sympathetic system are generally situated in ganglia of the sympathetic trunk or as ganglia in more peripheral plexuses, almost always nearer to the spinal cord than to the effectors which they innervate.

Physiologically, parasympathetic reactions are generally localized, whereas sympathetic reactions are mass responses. Thus parasympathetic activity results, for example, in slowing of the heart and increase in the glandular and peristaltic activities of the gut; these may be considered as conservation of body energies. Sympathetic activities result, for example, in general constriction of the cutaneous arteries (with consequent increase in the blood supply to the heart, muscles and brain), acceleration of the heart and increase of the blood pressure, contraction of the sphincters and lessening of the peristalsis of the gut, all of which activities mobilize body energies for dealing with increased activity.

Whereas the passage of nervous impulses along all preganglionic fibres, parasympathetic postganglionic fibres or along somatic efferent fibres is associated with the liberation of *acetylcholine* in the region of the terminals, in the case of the sympathetic postganglionic fibres the substance liberated is *noradrenalin* or *adrenalin*. For this reason the above types of nerves are called *cholinergic* and *adrenergic* respectively. As an exception to this sweat glands are supplied only by postganglionic sympathetic nerves, but these are cholinergic. (*See* also brief references to peripheral *purinergic* fibres on p. 833.)

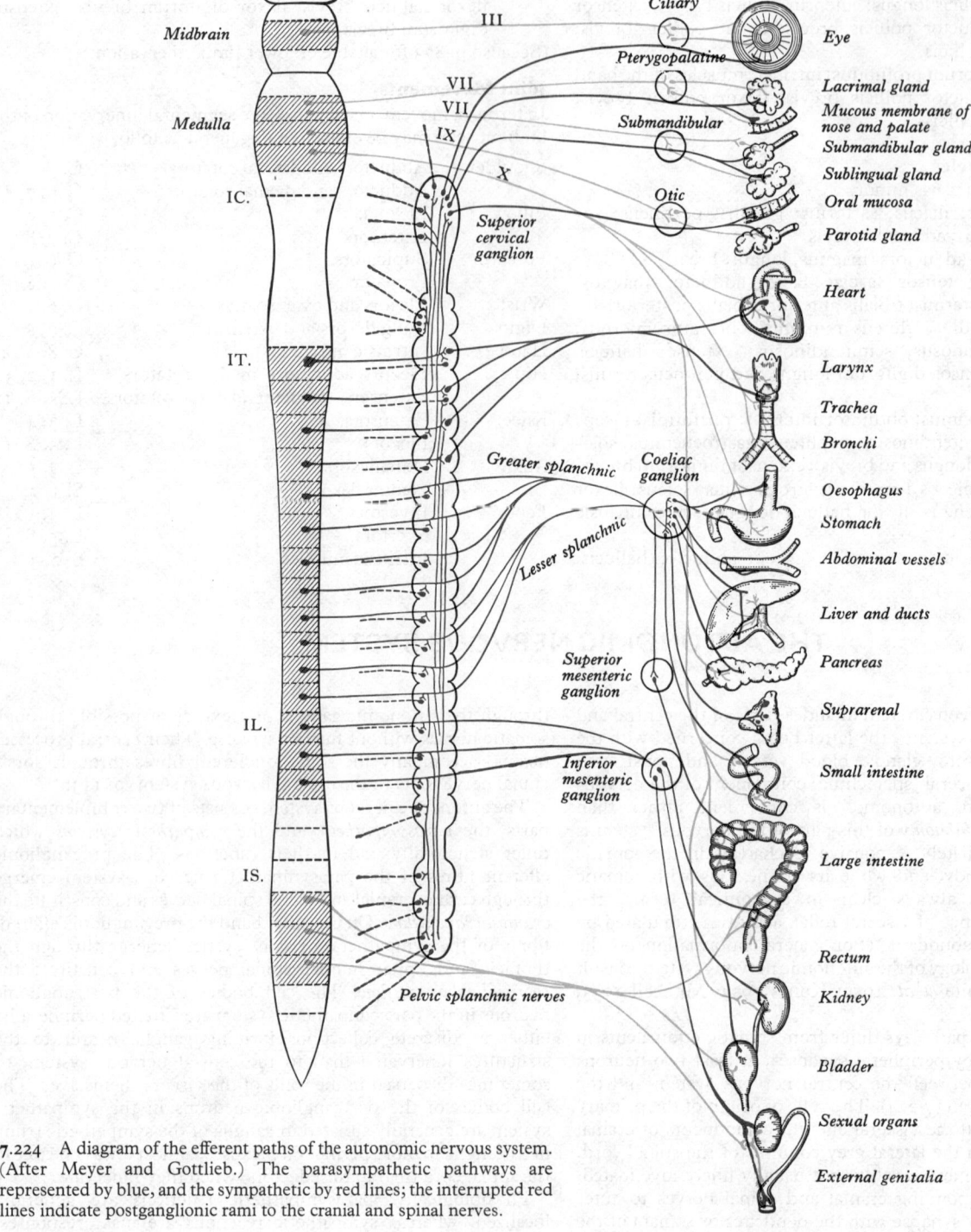

7.224 A diagram of the efferent paths of the autonomic nervous system. (After Meyer and Gottlieb.) The parasympathetic pathways are represented by blue, and the sympathetic by red lines; the interrupted red lines indicate postganglionic rami to the cranial and spinal nerves.

THE PARASYMPATHETIC NERVOUS SYSTEM

Efferent Pathways

The preganglionic parasympathetic fibres are myelinated and occur in (1) the oculomotor, (2) the facial, (3) the glossopharyngeal, (4) the vagus and accessory, and (5) the second, third and fourth sacral spinal nerves. In the cranial part of the parasympathetic system there are four peripheral ganglia which, though small, are readily identified with the naked eye. They are the *ciliary* (p. 1058), *pterygopalatine* (p. 1064), *submandibular* (p. 1072) and *otic* (p. 1076) ganglia, all of which have been described in detail with the cranial nerves. These ganglia are concerned solely with efferent parasympathetic pathways, unlike the trigeminal, facial, glossopharyngeal and vagal ganglia, which are all concerned with afferent impulses and contain the cells of origin of sensory fibres only. The cranial parasympathetic ganglia

are *traversed* by afferent fibres, by postganglionic sympathetic fibres and, in the case of the otic ganglion, even by branchial efferent fibres; but in none are the fibres interrupted during their passage through the ganglia. The postganglionic parasympathetic fibres are usually nonmyelinated and shorter than those of the sympathetic system, since the ganglia where their synapses occur are situated in or near the viscera they supply. Baumann and Gajisin (1975) have emphasized the occurrence of small subsidiary ganglia in the vicinity of those mentioned above, as has been reported by other workers. They claim, in addition, that minute ganglia occur in many other sites in fetal material—along the middle meningeal artery, for example, and in association with certain of the petrosal nerves. These findings, if confirmed, are of physiological importance.

(1) The *oculomotor* parasympathetic fibres commence in the midbrain, and are derived from the accessory oculomotor (Edinger-Westphal) nuclei (p. 1057). The preganglionic fibres travel in the nerve and leave by the branch which it supplies to the inferior oblique to enter the *ciliary ganglion*. There they are relayed, and the postganglionic fibres leave the ganglion in the short ciliary nerves, which pierce the sclera and run forwards in the perichoroidal space, to be distributed to the ciliary muscle (p. 1159) and the sphincter pupillae (p. 1161). These postganglionic fibres are thinly myelinated.

(2) The *facial nerve* contains efferent parasympathetic fibres which are axons of cells in the superior salivatory nucleus (p. 1070) and emerge from the brain in the nervus intermedius. They travel in the facial nerve, leaving it a little above the stylomastoid foramen in the *chorda tympani*, which traverses the tympanic cavity and ultimately reaches the *lingual nerve*. In this way they are conveyed to the submandibular region, where they enter the *submandibular ganglion*, in which the postganglionic secretomotor fibres for the submandibular salivary gland arise. Some preganglionic fibres may synapse around cells in the hilum of the gland (*see also* pp. 1274 and 1275). The secretomotor fibres for the sublingual gland are continued forwards in the lingual nerve after they have arisen in the submandibular ganglion (*see also* pp. 1073 and 1275). Electrical stimulation of the chorda tympani produces dilatation of the arterioles of both these salivary glands in addition to a secretomotor effect. In addition, the facial nerve has usually been said to contain efferent parasympathetic fibres which are secretomotor to the lacrimal gland, travelling by its greater petrosal ramus and nerve of the pterygoid canal and relaying in the pterygopalatine ganglion. The postganglionic branches are said to travel by the zygomatic nerve to the lacrimal gland (p. 1063) and by branches from the ganglion to glands of the nose and palate. Evidence refuting the former route has been reported (Ruskell 1971). This indicated that direct *lacrimal rami* pass to the gland from a *retro-orbital plexus* composed of direct parasympathetic branches of the pterygopalatine ganglion. There is clinical evidence that some of the facial parasympathetic fibres reach the parotid gland (Diamant and Wiberg, 1965: *see also* p. 1070).

(3) The *glossopharyngeal nerve* contains efferent parasympathetic fibres which are secretomotor to the parotid gland. They start in the inferior salivatory nucleus (p. 1074) and travel in the glossopharyngeal nerve and its tympanic branch. After traversing the tympanic plexus, they enter the lesser petrosal nerve and so reach the otic ganglion. There they are relayed and the postganglionic fibres pass by communicating branches to the auriculotemporal nerve, by which they are conveyed to the parotid gland. Electrical stimulation of the lesser petrosal nerve produces a vasodilator as well as a secretomotor effect.

(4) The *vagus nerve* contains efferent parasympathetic fibres which arise in its dorsal nucleus (p. 1076) and travel in the nerve trunk and in its pulmonary, cardiac, oesophageal, gastric, intestinal and other branches. Some cardiac parasympathetic fibres may originate from neurons in or near the nucleus ambiguus. *See* p. 905. The proportion of efferent parasympathetic fibres in the vagus varies at different levels, but is small in relation to its sensory component. These fibres are relayed in minute ganglia which lie in the walls of the individual viscera. The disproportion in the numbers of preganglionic to postganglionic fibres is greater in the vagus than in the efferent parasympathetic components of other cranial nerves and this discrepancy cannot as yet be explained. The cardiac branches are concerned with slowing of the cardiac cycle. They take part in the formation of the cardiac plexuses (p. 1132) and are then relayed in ganglia which are distributed freely over the surfaces of both atria in the subepicardial tissue. The terminal fibres are distributed to the atria and the atrioventricular bundle, and it has been claimed in the past that only through the latter can the vagus exert any control over the ventricular muscle (Cullis and Tribe 1913). The smaller branches of the coronary arteries are innervated mainly by the vagus, whereas their larger branches, though possessing a double innervation, obtain their chief source of supply from the sympathetic system (Wollard 1926). The *pulmonary branches* are motor to the circular nonstriated muscle fibres in the bronchi, and are therefore bronchoconstrictor. The synaptic relays are in the ganglia of the pulmonary plexuses. The *gastric branches* are secretomotor to the glands and motor to the muscular coats of the stomach, but they inhibit the action of the pyloric sphincter. The *intestinal branches* have a corresponding action on the small intestine, caecum, vermiform appendix, ascending colon, right colic flexure and most of the transverse colon, being secretomotor to the glands and motor to the muscular coats of the gut, but inhibitory to the ileocaecal sphincter. The synaptic relays, in this case, are situated in the myenteric (Auerbach's) plexus and the plexus of the submucosa (Meissner's plexus), which are described with the structure of the intestines.

(5) The anterior rami of the *second, third* and often the *fourth sacral spinal nerves* emit visceral branches passing directly to the pelvic viscera. They constitute the *pelvic splanchnic nerves* (7.231), and they unite with branches of the sympathetic pelvic plexuses. Minute ganglia are sited at these points of union and in the walls of the individual viscera. In these ganglia the sacral preganglionic parasympathetic fibres are relayed.

The pelvic splanchnic nerves supply the rectum with motor fibres, the bladder wall with motor and its sphincter with inhibitory fibres, the erectile tissue of the penis or clitoris with vasodilator fibres, the testes or ovaries probably with vasodilator fibres, and the uterine tubes and uterus with vasodilator and possibly inhibitory fibres. In addition, filaments from the pelvic splanchnic nerves pass upwards through the hypogastric plexus to supply the sigmoid colon, descending colon, left colic flexure and terminal part of the transverse colon with visceromotor fibres (Telford and Stopford 1934; Mitchell 1935).

The nervus terminalis (p. 1055) is also considered to contain an as yet unquantified population of visceral efferent fibres by some authorities.

THE SYMPATHETIC NERVOUS SYSTEM

The sympathetic nervous system, which is the larger division of the autonomic, includes the two ganglionated sympathetic trunks, their branches, plexuses and subsidiary ganglia. It has a much wider distribution than the parasympathetic system, for it innervates all the sweat glands of the skin, the arrector muscles of the hairs, the muscular walls of many blood vessels, the heart, lungs and abdomino-pelvic viscera.

Efferent Sympathetic Pathways

The *preganglionic fibres* are the axons of nerve cells in the lateral column of the grey matter of all the thoracic and upper two or three lumbar segments of the spinal cord where they form the intermediomedial and intermediolateral cell groups (p. 872). These fibres are myelinated and have diameters of $1\cdot5$ to $4\cdot0$ μm. They emerge from the spinal cord through the ventral roots of the corresponding spinal nerves and pass into the spinal nerve trunks and the commencement of their ventral rami, which they leave in the *white rami communicantes*, to join either the corresponding ganglia on the sympathetic trunks or their interganglionic parts. Since this outflow is confined to the thoracolumbar region, typical white rami communicantes are also restricted to the fourteen spinal nerves noted above. However, the possibility of a limited outflow of preganglionic fibres in other spinal nerves has been suggested. It is certain that nerve cells of the same type as those in the lateral grey column also exist at other levels, above and below the thoracolumbar outflow (Mitchell 1953), and that small numbers of their fibres issue in corresponding ventral roots. Dorsal spinal nerve roots may also contain vasodilator fibres. Having reached the sympathetic trunk the preganglionic fibres may behave in a number of different ways (7.225). (*a*) They may end in the corresponding ganglion by arborizing with the dendrites of ganglion cells. (*b*) They may pass through the corresponding ganglion and either ascend to a ganglion at a higher level or descend to one at a lower level before terminating in a similar manner; it is believed that preganglionic fibres do not divide into ascending and descending branches on entering the sympathetic trunk. A single preganglionic fibre may, through its collateral and terminal branches, synapse with nerve cells in several of the ganglia which it traverses; other preganglionic fibres distribute branches to one ganglion only. (*c*) They may pass through the corresponding ganglion and may ascend or descend without being interrupted and then emerge in one of the medially directed branches of the sympathetic trunk to enter a plexus of the autonomic system, where they terminate in relation to the ganglion cells therein. Occasionally the interruption of preganglionic fibres occurs in ganglia situated proximal to the sympathetic trunks; these are known as '*intermediate ganglia*' and are most numerous on the grey rami communicantes (*vide infra*) in the cervical and lower lumbar regions (Boyd and Munro 1940). They may be of microscopic size and are sometimes situated in the ventral roots or trunks of the spinal nerves. Branches from more than one preganglionic fibre may synapse with a single postganglionic neuron (*vide infra*).

The sympathetic ganglia include collections of cells of the sympathetic trunks, nerve ganglia in the autonomic plexuses and the 'intermediate' ganglia; in addition some ganglion cells are dispersed through the plexuses. Originally the ganglia on the sympathetic trunks correspond numerically to the ganglia on the dorsal roots of the spinal nerves (p. 1086); but fusion of adjoining ganglia has occurred and in man there are rarely more than twenty-two or twenty-three and there may be fewer discrete ganglia. The subsidiary ganglia in the great autonomic nerve plexuses (e.g. coeliac ganglion, superior mesenteric ganglion, etc.) are derivatives of the ganglia of the sympathetic trunks.

The structure of sympathetic ganglia and, of course, their functional characteristics have been subject to experiment and observation with increasing tempo for many decades. For earlier literature consult the general authorities on the autonomic nervous system (p. 1121), and Gabella (1976) and Eränkö (1978) for more recent contributions. The classical studies of Langley and his successors led to an early view of autonomic ganglia as relay stations, a concept largely corroborated by anatomical observations, though it was quickly recognized that a minor fraction of efferent nerve fibres traverse many ganglia without interruption, as do afferent fibres from viscera and glands. This remains a substantially true representation of ganglia in autonomic pathways, but it has been modified somewhat and markedly extended by application of the techniques of electron microscopy, neurohistochemistry, and electrophysiology. For example, the ratio between nerve fibres entering a ganglion and those leaving it is not on a simple 1:1 basis. The superior cervical sympathetic ganglion, which has been most extensively studied in this and other regards, shows ratios varying from 1:28 to 1:176 in different mammalian species (*see* Billingsley and Ranson 1918; Samuel 1953; Ebbeson 1968). Such results demonstrate that a preganglionic neuron makes connexions with a large number of postganglionic nerve cells, thus providing a structural basis for wide *dissemination* or perhaps *amplification* of sympathetic activity, a well-established characteristic, which is not shared, to the same degree at least, by parasympathetic ganglia. These effects might be achieved in several ways: (*a*) by widespread terminal arborizations of preganglionic nerve fibres, (*b*) by the mediation of interneurons, (*c*) by local diffusion within the ganglion of transmitter substances locally produced (*paracrine* effect) or local response to substances produced elsewhere (*endocrine* effect). There is evidence that all of these mechanisms are involved.

The connective tissue capsule of each ganglion, which is continuous with the epineurium of its connecting rami (roots and branches), also extends as septa into the ganglion, surrounding groups of nerve cells and nerve fibres. Finer subdivisions of this stroma spread amongst these cells, each of which is surrounded by collagen fibres and intercellular matrix. In this stroma are a few fibroblasts and many small blood vessels and capillaries. Satellite cells (amphicytes) encapsulate the somata of the ganglionic neurons and their processes. The external aspect of this thin sheath of satellite cells bears a continuous basal lamina, and these two elements together screen the neurons from direct contact with the stroma of the ganglion and its extracellular matrix. Neuronal elements have direct access only to the internal, juxta-neuronal aspect of the satellite cells; but between the two is a narrow interval of about 15–20 nm. These peri-neuronal spaces are, moreover, linked to the stromal compartment of the ganglion by minute channels situated between the processes of satellite cells. The arrangements just described are of great importance in providing anatomical evidence of possible routes for movements of neurotransmitter and hormonal substances between neurons and also between these and the vascular compartment of the ganglion.

Regarding the characteristics of both the neuronal population of sympathetic (and parasympathetic) ganglia a great volume of observation has been recorded, some aspects only of which can be touched upon here. Many attempts to classify the range of neurons in ganglia have been made, but since the criteria employed were arbitrary and divergent, and in their excesses sometimes confusing, they will not be detailed here. The great majority of the nerve cells are multipolar, with somatic dimensions ranging from about 25 to 50 μm in mankind; a smaller type of cell, of about 15 to 20 μm in 'diameter' and less pronounced multipolar shape, also occurs in much smaller numbers, often clustered together in small groups (De Castro and Herneros 1945), and these latter probably correspond to SIF cells (*vide infra*). While varying in size the multipolar neurons display a more marked variation in the distribution and form of their dendritic arrays. According to McLachland (1974) multipolar ganglion cells in the guinea-pig display a mean of 13 dendrites per cell. The complexity of these dendrites, and particularly of small dendrites which ramify within the capsular, perikaryal space, is greater in human ganglion cells. Dendritic glomeruli have been observed in many ganglia. In general ultrastructural details these neurons resemble others (*see* p. 825). Clusters of small, granular vesicles, adrenergic in type, are dispersed superficially in the perikaryon, and these are also prominent in dendrites, probably to be equated with storage of catecholamines. Ganglionic neurons receive large numbers of axodendritic synapses from preganglionic nerve fibres, axosomatic synapses being much less numerous. Each preganglionic fibre forms several synapses with a number of separate dendrites, which may represent a mechanism for dissemination or amplification, or both. The postganglionic fibres (*vide infra*) commonly branch off from the initial stem of a large dendrite and produce few or no collateral rami.

The existence of *interneurons* in sympathetic ganglia has been strongly denied in the past (*see* Samuel 1953); but according to Williams (1967), Williams and Palay (1969), Libet and Crowman (1974), an interneuronal role can be assigned to some, at least, of the 'small intensely fluorescent' (SIF) cells, which have been identified in many sympathetic ganglia and in a variety of mammals, including mankind (Eränkö and Harkonen 1965; Jacobowitz 1970). Small chromaffin cells have been known to occur in sympathetic ganglia since 1898, or even earlier, where they were recognized by Kohn. Coupland (1965), amongst modern workers, has described them as present in all the ganglia of the sympathetic trunk, and in other sites, in neonatal human material. The distinction between SIF cells and chromaffin cells appears to be a matter of uncertainty or controversy in many accounts. The supposed two types of cells, as far as ganglia are concerned, are identified by separate techniques (the chromaffin reaction and formalin-induced fluorescence) which cannot be applied together to any single cell. However, in a recent study (Santer *et al.* 1975) upon the distribution of SIF and chromaffin cells, in sympathetic ganglia of rats, the former were more numerous and their modes of distribution overlapped but showed some differences. Both types of cell contain catecholamines, and it is possible that some cells contain only enough to be revealed by the more sensitive formaldehyde-induced fluorescence technique (Falck-Hillarp), whereas others have a sufficiently high concentration of catecholamines to produce a positive chromaffin reaction (*see* Gabella 1976 for discussion). Both kinds of cells may

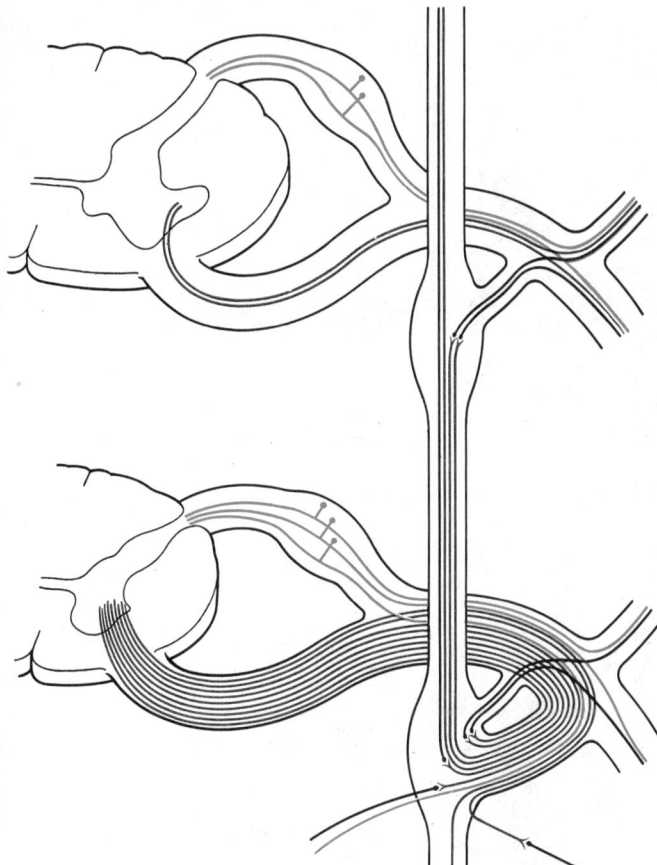

7.225 A scheme showing the constitution of a typical spinal nerve. In the upper part of the diagram the spinal nerve roots show the somatic components; in the lower part of the diagram the spinal roots show the visceral components. *Red:* somatic and preganglionic visceral nerve fibres. *Blue:* somatic and visceral afferent fibres. *Black:* postganglionic afferent visceral fibres.

act as interneurons (Santer *et al.* 1975; Gabella 1976). The putative mode of action of interneuronal SIF cells has been hypothecated by Greengard and Kebabian (1974): dopamine is released by such cells and is united to surface receptors of one or more ganglionic neurons, activating adenylate cyclase and thus increasing the production of cyclic AMP (cyclic adenosine monophosphate), which results in hyperpolarization. In the ganglia of some species two types of SIF cell have been described (Williams 1975): a minority, which have long processes ending near ganglionic neurons and which may therefore be regarded as interneurons (Type I), and a more numerous Type II cell with shorter processes ending in proximity to blood vessels (Chiba *et al.* 1975). In bovine superior cervical sympathetic ganglia 24 per cent of the SIF cells were thus described as belonging to Type I, while in the same ganglion in cats only 20 per cent were ascribed to this group. Although the secretory granules of these Type I cells are regarded as able to act directly upon ganglionic neurons, there is evidence that some SIF cells, presumably of Type II, pass their secretion into the local blood vessels (Polonyi *et al.* 1977), thus exerting more diffuse and distant effects. The role of SIF cells in neurotransmission in sympathetic ganglia has been reviewed by Eränkö (1978).

Quantitative studies of the neuronal populations, in terms of numbers, dimensions, and density of packing, have been recorded for many autonomic ganglia and especially in the superior cervical sympathetic ganglion. Gabella (1976) has assembled and evaluated such data in detail, and the reader is referred to his review for further information.

The axons of the principal ganglion cells are usually fine, non-myelinated fibres and constitute the *postganglionic fibres*. They are distributed to the effector organ in a variety of ways. Postganglionic fibres arising from a ganglion on the sympathetic

trunk may (*a*) pass back to the corresponding spinal nerve through a *grey ramus communicans*; this usually joins the spinal nerve trunk just proximal to the white ramus communicans. Its fibres are distributed through the ventral and dorsal rami of the spinal nerves and their branches to the blood vessels, sweat glands and hairs, etc. in their zone of supply. The extent of the segmental area innervated is variable and the territories supplied through adjacent nerves overlap to a considerable degree; the extent of innervation of different effector systems, e.g., sudomotor and vasomotor, by a particular nerve are not necessarily the same. (*b*) They may pass in a medial branch of a ganglion to be distributed to some particular areas or viscera. (*c*) They may pass to blood vessels in the neighbourhood of the sympathetic trunk and supply these or may be carried along the vessels and their branches towards their peripheral distribution. (*d*) They may ascend to a higher level or descend to a lower level before leaving the sympathetic trunk either in one of its medial branches, in a grey ramus communicans or along adjacent blood vessels.

In addition to white and grey rami mixed types are found. Some of these in the thoracic region represent fusion of white and grey rami but some found in the cervical region contain bundles of thick myelinated fibres, which are somatic efferent in character and are utilizing the grey ramus as a convenient route to reach prevertebral muscles (p. 1127). For a detailed description of rami communicantes and their variations from purely preganglionic to purely postganglionic types consult Winckler (1961).

After diffusing through the plexuses the postganglionic fibres which arise in or join the plexuses are distributed mainly along blood vessels and some ducts.

Functional significance. The efferent postganglionic fibres which pass in the grey rami communicantes to the spinal nerve supply vasoconstrictor fibres to the blood vessels, secretomotor fibres to the sweat glands and motor fibres to the arrectores pilorum muscles in the areas supplied by the corresponding spinal nerve. Those which accompany the motor nerves to voluntary muscles are probably distributed only to the blood vessels supplying the muscles. Thus most, if not all, peripheral branches derived from the spinal nerves contain postganglionic sympathetic fibres. Those which pass to the viscera and other structures are concerned with vasoconstriction, dilatation of the pupils, dilatation of the bronchioles, glandular secretion, movements of the alimentary tract and the urinary bladder (relaxation of the muscle walls and contraction of the sphincters), etc. It is believed that usually, but not invariably, a single preganglionic fibre synapses with the postganglionic neurons innervating one effector system only; therefore a dissociation of sympathetic effects, such as sudomotor and vasomotor activities, can occur. The same is not necessarily true of visceral afferent fibres (p. 890). While in general the sympathetic and para-sympathetic systems exert antagonistic influences on the viscera they supply, this is not always so. In the case of the urinary bladder, for instance, the normal emptying and filling of the viscus are controlled only by the parasympathetic system, the sympathetic being concerned with the supply of the blood vessels of the organ.

Higher autonomic control. The peripheral autonomic nervous system is controlled by the activities of higher levels in the brainstem and cerebral hemispheres. The parts of the brain especially concerned have been described in the section on the Central Nervous System and include the brainstem reticular formation, various thalamic and hypothalamic nuclei, the limbic lobe and the prefrontal neocortex, and a variety of ascending and descending pathways which interconnect these regions.

The sympathetic trunks are two ganglionated nerve cords which extend from the base of the skull to the coccyx. In the neck the trunk is posterior to the carotid sheath and anterior to the transverse processes of the cervical vertebrae; in the thorax it is anterior to the heads of the ribs; in the abdomen it is anterolateral to the bodies of the lumbar vertebrae, and in the pelvis, anterior to the sacrum, medial to the anterior sacral foramina. Anterior to the coccyx the two trunks meet each other in the unpaired terminal *ganglion impar*.

The cervical ganglia are usually reduced to three by fusion of adjoining units, and from the cranial pole of the superior ganglion

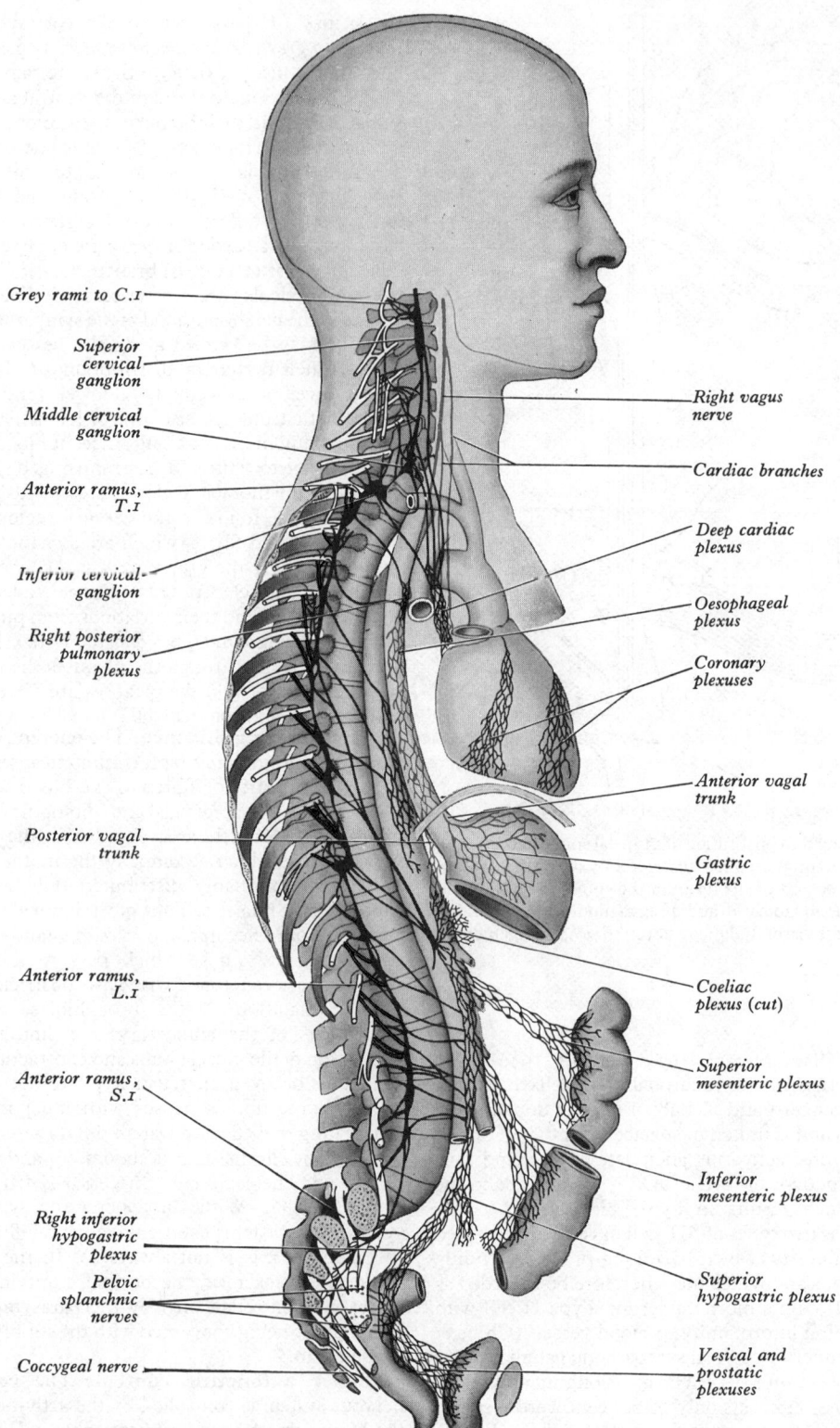

Grey rami to C.I

Superior
cervical
ganglion

Middle cervical
ganglion

Anterior ramus,
T.I

Inferior cervical
ganglion

Right posterior
pulmonary
plexus

Posterior vagal
trunk

Anterior ramus,
L.I

Anterior ramus,
S.I

Right inferior
hypogastric
plexus

Pelvic
splanchnic
nerves

Coccygeal nerve

Right vagus
nerve

Cardiac branches

Deep cardiac
plexus

Oesophageal
plexus

Coronary
plexuses

Anterior vagal
trunk

Gastric
plexus

Coeliac
plexus (cut)

Superior
mesenteric plexus

Inferior
mesenteric plexus

Superior
hypogastric plexus

Vesical and
prostatic
plexuses

7.226 A The right sympathetic trunk and its connexions with the thoracic, abdominal and pelvic plexuses. *Blue:* parasympathetic fibres. *Black:* sympathetic trunk and branches. *Red:* white rami communicantes.

the internal carotid nerve commences. This nerve constitutes an ascending continuation of the sympathetic trunk, and it accompanies the internal carotid artery through its canal into the cranial cavity. In the thorax there are usually eleven ganglia, but the number may be ten or twelve. There are usually four ganglia in the lumbar and four or five in the sacral regions.

CRANIAL PART OF THE SYMPATHETIC SYSTEM

The cranial part of the sympathetic system begins on each side as the **internal carotid nerve**, which is continued up from the

superior cervical ganglion of the sympathetic trunk and contains postganglionic fibres derived from its cells. It ascends behind the internal carotid artery, and, entering the carotid canal in the temporal bone, divides into two branches, one of which is lateral and the other medial to the artery.

The *lateral branch*, the larger, gives filaments to the internal carotid artery, and forms the lateral part of the internal carotid plexus.

The *medial branch* also supplies filaments to the internal carotid artery, and, continuing onwards, forms the medial part of the internal carotid plexus.

The internal carotid plexus surrounds its artery, and occasionally contains a small swelling on the interior aspect of the vessel, the *carotid ganglion*. In addition to this small ganglion, the rest of the plexus also contains some scattered sympathetic nerve cells. The lateral part of the plexus communicates with the trigeminal and pterygopalatine ganglia, with the abducent nerve and with the tympanic branch of the glossopharyngeal nerve; it distributes filaments to the wall of the internal carotid artery.

The branches communicating with the abducent nerve consist of one or two filaments which join that nerve as it lies upon the lateral side of the internal carotid artery. The communication with the pterygopalatine ganglion is effected by a branch named the *deep petrosal*; this branch perforates the cartilage filling the foramen lacerum, and joins the greater petrosal nerve to form the *nerve of the pterygoid canal*, which passes through the pterygoid canal to the pterygopalatine ganglion. The communications with the tympanic branch of the glossopharyngeal nerve are effected by the *superior* and *inferior caroticotympanic nerves*, which traverse the posterior wall of the carotid canal.

The medial part of the internal carotid plexus is inferomedial to the part of the internal carotid artery which is lateral to the sella turcica, in the cavernous sinus. It gives branches to the internal carotid artery, and communicates with the oculomotor, trochlear, ophthalmic and abducent nerves, and with the ciliary ganglion. It also sends vasomotor twigs along the branches of the internal carotid artery which supply the hypophysis cerebri (p. 1443).

The branch to the oculomotor nerve joins that nerve at its point of division; the branch to the trochlear nerve joins the latter as it lies in the lateral wall of the cavernous sinus; filaments also connect with the medial side of the ophthalmic nerve and one joins the abducent nerve. The filament to the ciliary ganglion arises from the anterior part of the plexus and enters the orbit through the superior orbital fissure; it may join the ganglion directly; it may unite with the communicating branch from the nasociliary nerve to the ganglion (p. 1057), or it may travel via the ophthalmic nerve and its nasociliary branch. Its fibres pass through the ciliary ganglion without being interrupted and run in the short ciliary nerves to be distributed to the blood vessels of the eyeball. The fibres which supply the dilatator pupillae usually travel by the ophthalmic, nasociliary and long ciliary nerves, but occasionally some are carried by the short ciliary nerves. Some fibres may also innervate the ciliaris. The preganglionic fibres concerned leave the spinal cord predominantly in T. 1, and pass to the cervicothoracic ganglion, through which they pass uninterruptedly. They then ascend in the cervical section of the sympathetic trunk to reach the superior cervical ganglion, where they are relayed.

The terminal filaments from the internal carotid plexus are prolonged as plexuses around the anterior and middle cerebral arteries and the ophthalmic artery: along the anterior and middle cerebral arteries they may be traced to the pia mater; along the ophthalmic artery they pass into the orbit where they accompany each of the branches of that vessel. The filaments prolonged on the anterior communicating artery connect the sympathetic nerves of the right and left sides and a small ganglion may be found associated with these filaments. Much of the above detail depends upon comparatively old observations, and disagreement and discrepancy still obtains in regard to such details. Mitchell (1953) and Purves (1972) have reviewed the extensive literature. It can, however, be said that the old controversy regarding the autonomic innervation of the cerebral arterial tree is now settled. Moreover, electron microscopy has permitted clear demonstration that this innervation is substantially like that of other vascular systems (*see*, for example, Iwayama *et al.* 1970). The terminals are adrenergic, a finding confirmed histochemically in various mammals, including man (Iwayama 1970). The source of these sympathetic vasoconstrictor fibres is from the internal carotid and vertebral plexuses, but the precise distribution from each is still unresolved. (Although evidence of cholinergic terminals in the walls of cerebral arteries has been recorded, the existence of vasodilator mechanisms has not been clearly demonstrated.) It has also been claimed (Falck *et al.* 1965, and Edvinsson *et al.* 1973) that adrenergic nerve fibres accompany pial and intracerebral arteries which are *not* derived from known

sympathetic sources, but from nerve cells at present merely presumed to be intracranial in location.

CERVICAL PART OF THE SYMPATHETIC SYSTEM

The cervical part of each sympathetic trunk consists of three ganglia distinguished, according to their positions, as the superior, middle and cervicothoracic, connected by intervening cords (7.226). This part sends grey rami communicantes to all the cervical spinal nerves, but receives no white rami communicantes from them; its spinal fibres are derived from the white rami communicantes of the upper thoracic nerves, which enter the corresponding thoracic ganglia of the sympathetic trunk, through which they ascend into the neck. In their course the grey rami communicantes may pierce the longus capitis or the scalenus anterior. For details of the cervical grey rami *see* Potts (1925), Oxford (1928), Pick and Sheehan (1946), Sunderland and Bedbrook (1949).

The superior cervical ganglion, the largest of the three, adjoins the second and third cervical vertebrae and is believed to be formed by the coalescence of four ganglia, corresponding with the upper four cervical nerves. It is in relation, in front, with the sheath of the internal carotid artery behind, with the longus capitis. The internal carotid nerve (p. 1126) ascends from the upper end of the ganglion into the cranial cavity; the lower end of the ganglion is united by the connecting trunk with the middle cervical ganglion. The branches of the ganglion may be divided into lateral, medial and anterior groups.

The *lateral branches* of the superior cervical ganglion consist of grey rami communicantes to the upper four cervical nerves and to certain of the cranial nerves. Delicate filaments run to the inferior ganglion of the vagus, and to the hypoglossal nerve; a branch, named the *jugular nerve*, ascends to the base of the skull and divides into two twigs, one of which joins the inferior ganglion of the glossopharyngeal, and the other the superior ganglion of the

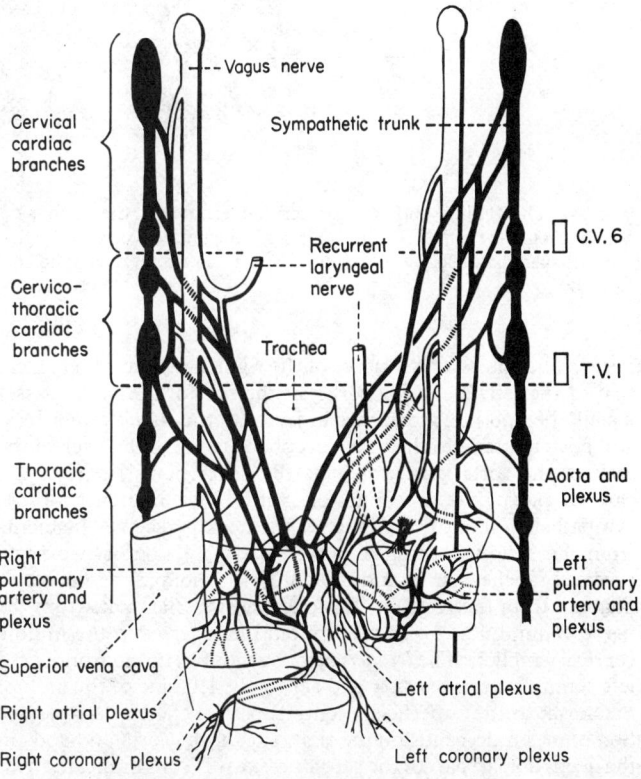

7.226B The human cardiac plexus—a semi-diagrammatic representation of its source from the cervical parts of the vagus nerves and sympathetic trunks and of its extensions—the pulmonary, atrial, and coronary plexuses. Note the numerous junctions between the sympathetic and parasympathetic (vagal) rami which form the plexus. (For further information, particularly concerning the frequent variations, consult N. J. Mizeres, *Amer. J. Anat.* **112**, 1963, to both of whom we are indebted for permission to use this diagram.)

vagus; other twigs pass to the superior jugular bulb, including its associated jugular glomus or glomera, and some are distributed to the meninges in the posterior cranial fossa.

The *medial branches* of the superior cervical ganglion are the laryngopharyngeal and cardiac branches.

The *laryngopharyngeal branches* supply the carotid body, and pass to the side of the pharynx, where they join with branches from the glossopharyngeal and vagus nerves to form the *pharyngeal plexus* (p. 1313).

The *cardiac branch* arises by two or more filaments from the lower part of the superior cervical ganglion, and occasionally receives a twig from the trunk connecting the superior with the middle cervical ganglion. It is believed to contain only efferent fibres, the preganglionic outflow being from the upper thoracic segments of the spinal cord, and to be devoid of any visceral pain fibres from the heart (p. 1132). It runs down the neck behind the common carotid artery, and in front of the longus colli; it crosses in front of the inferior thyroid artery and recurrent laryngeal

surrounding the facial artery supplies a filament to the submandibular ganglion, and the plexus on the middle meningeal artery sends one ramus to the otic ganglion, and another, termed the *external petrosal nerve*, to the ganglion of the facial nerve. Many of the fibres coursing along the external carotid artery and its branches to supply the sweat glands on the face ultimately leave the blood vessels to be distributed through terminal branches of the trigeminal nerve.

The middle cervical ganglion (7.227), the smallest of the three cervical ganglia, is occasionally absent as such, being replaced by minute ganglia in the sympathetic trunk in this region; it may be fused with the superior cervical ganglion. It is usually at the level of the sixth cervical vertebra, anterior or just superior to, the inferior thyroid artery, or it may lie near the cervicothoracic ganglion (p. 1129). It is probably formed by the coalescence of two ganglia corresponding with the fifth and sixth cervical segments, judging by its postganglionic rami, which pass to the fifth and sixth cervical nerves, but also, sometimes to the

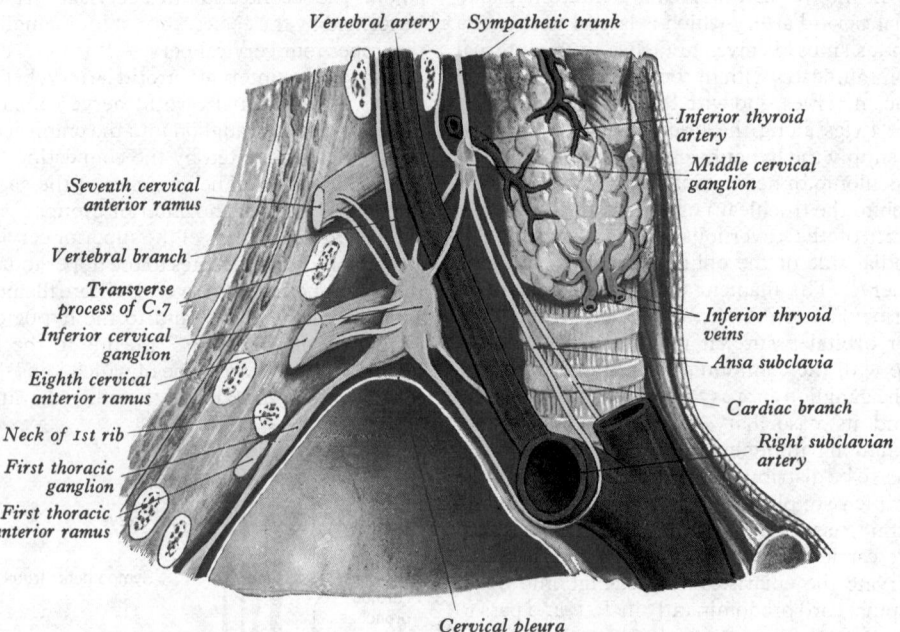

Vertebral artery *Sympathetic trunk*

Inferior thyroid artery

Middle cervical ganglion

Seventh cervical anterior ramus

Vertebral branch

Transverse process of C.7

Inferior cervical ganglion

Eighth cervical anterior ramus

Neck of 1st rib

First thoracic ganglion

First thoracic anterior ramus

Inferior thryoid veins

Ansa subclavia

Cardiac branch

Right subclavian artery

Cervical pleura

7.227 A The middle and inferior cervical ganglia of the right side. Viewed from the right. Note the proximity of the inferior cervical and first thoracic ganglia, usually fused to form a cervicothoracic (stellate) ganglion.

nerve. The course of the nerve of the right side then differs from that of the left. The *right nerve*, at the root of the neck, passes usually behind but sometimes in front of, the subclavian artery, and posterolateral to the brachiocephalic trunk to the back of the arch of the aorta, where it joins the deep (dorsal) part of the cardiac plexus. It is connected with other branches of the sympathetic; about the middle of the neck it receives filaments from the external laryngeal nerve; inferiorly, one or two vagal cardiac branches join it; and as it enters the thorax it is joined by a filament from the recurrent laryngeal nerve. Filaments from the nerve communicate with the thyroid branches from the middle cervical ganglion. The *left nerve*, in the thorax, runs in front of the left common carotid artery and across the left side of the arch of the aorta, to the superficial (ventral) part of the cardiac plexus. Sometimes it descends on the right side of the aorta and ends in the deep (dorsal) part of the cardiac plexus. It communicates with the cardiac branches of the middle cervical and cervicothoracic sympathetic ganglia, and sometimes with the inferior cervical cardiac branches of the left vagus, and branches from these mixed nerves pass down to form a plexus on the ascending aorta.

The *anterior branches* of the superior cervical ganglion ramify upon the common carotid artery, and upon the external carotid artery and its branches, forming around each a delicate plexus in which small ganglia are occasionally found. The plexus

fourth and seventh. The ganglion also gives off thyroid and cardiac branches. It is connected to the cervicothoracic ganglion by two or more cords, which are very variable in their disposition. The posterior cord usually splits to enclose the vertebral artery. The more anterior cord loops down in front of and then below the first part of the subclavian artery, medial to the origin of its internal thoracic branch, and supplies rami to it. This loop is the *ansa subclavia*. It is intimately related to the cervical pleura, frequently consists of more than one filament, and generally communicates with the phrenic nerve. It is uncertain whether this last connexion indicates a contribution to or from the phrenic nerve. Similarly, a connexion between the ansa subclavia and the vagus nerve, usually described, is of uncertain significance.

The *thyroid branches* run along the inferior thyroid artery to the thyroid gland; they communicate with the superior cardiac, external laryngeal and recurrent laryngeal nerves and send branches to the parathyroid glands. The supplies to the thyroid and parathyroid glands are in part vasomotor, but some fibres reach the secretory cells (Raybuck 1952).

The *cardiac branch*, the largest of the sympathetic cardiac branches, is derived from the ganglion itself or, more frequently, from the sympathetic trunk cranial or caudal to it. On the *right side* it descends behind the common carotid artery, and at the root of the neck runs either in front of or behind the subclavian artery; it then descends on the trachea, receives a few filaments from the recurrent laryngeal nerve, and joins the right half of the deep (dorsal) part of the cardiac plexus. In the neck, it communicates with the superior cardiac and recurrent laryngeal nerves. On the *left side*, the nerve enters the thorax between the left common carotid and subclavian arteries, and joins the left half of the deep (dorsal) part of the cardiac plexus.

Fine branches from the middle cervical ganglion also pass to the trachea and oesophagus.

The cervicothoracic (stellate) ganglion is irregularly shaped, and much larger than the middle cervical ganglion, being probably formed by the coalescence of the lower two cervical segmental ganglia with the first thoracic. Sometimes the second (and even the third and fourth) thoracic ganglia are fused with the mass; in other instances the first thoracic ganglion is separate and the upper mass then constituting an *inferior cervical ganglion* (7.226, 227). Owing to the marked change in direction of the sympathetic trunk at the junction of the head and thorax, the long axis of the cervicothoracic ganglion is almost anteroposterior. The ganglion lies on or just lateral to the lateral border of the longus colli and between the base of the transverse process of the seventh cervical vertebra and the neck of the first rib, which are posterior to it, and the vertebral artery and its associated veins which are anterior. Below it is separated from the posterior aspect of the cervical pleura by the suprapleural membrane; the costocervical trunk branches near its lower pole. On its lateral side is the superior intercostal artery.

A small ganglion, the *vertebral ganglion*, may be found on the sympathetic trunk anterior or anteromedial to the commencement of the vertebral artery and directly above the subclavian artery. When present it may give rise to the ansa subclavia and is joined also to the cervicothoracic ganglion by fibres which pass both in front of and behind the vertebral artery. The vertebral ganglion is usually regarded as a detached portion of the middle cervical or cervicothoracic ganglion. Like the middle cervical ganglion it may supply grey rami communicantes to the fourth and fifth cervical spinal nerves.

The cervicothoracic ganglion sends grey rami communicantes to the seventh and eighth cervical and first thoracic nerves, gives off a cardiac branch, supplies branches to neighbouring vessels, and not infrequently sends a branch to join the vagus nerve.

The *grey rami communicantes* to the seventh cervical spinal nerve vary from one to five in number. Two, which is the usual number, are shown in 7.227. Another often ascends medial to the vertebral artery and in front of the transverse process of the seventh cervical vertebra and, after communicating here with the seventh cervical nerve, sends a small branch upwards through the foramen transversarium of the sixth cervical vertebra in company with the vertebral vessels to join the sixth cervical nerve as it emerges from the intervertebral foramen. Another inconstant branch may pass through the foramen transversarium of the seventh vertebra. The grey rami to the eighth cervical spinal nerve are also multiple and vary from three to six in number.

A *cardiac branch* arises from the cervicothoracic ganglion. It descends behind the subclavian artery and along the front of the trachea, to join the deep part of the cardiac plexus. Behind the subclavian artery it communicates with the recurrent laryngeal nerve and the cardiac branch of the middle cervical ganglion. It is often replaced by a variable number of fine branches derived from the cervicothoracic ganglion and the ansa subclavia.

The *branches to blood vessels* form plexuses on the subclavian artery and its branches. The plexus around the subclavian artery is derived from the cervicothoracic ganglion and the ansa subclavia; it extends to the first part of the axillary artery, but fibres may extend further, though not in large numbers. According to Pearson and Sauter (1971) the extension of the subclavian plexus which passes on to the internal thoracic artery is joined by a branch from the phrenic nerve (see also p. 1094). The plexus on the vertebral artery is derived mainly from a thick branch of the cervicothoracic ganglion which ascends behind the vertebral artery to the foramen transversarium of the sixth cervical vertebra, reinforced by branches from the vertebral ganglion or the cervical sympathetic trunk which pass cranially on

the ventral aspect of the artery. From the plexus, branches (*deep rami communicantes*) pass to the anterior rami of the upper five or six cervical spinal nerves. The plexus contains a number of nerve cells. The plexus is continued into the skull along the vertebral and basilar arteries and their branches as far as the posterior cerebral artery, where it meets the plexus derived from that on the internal carotid artery. Some authorities consider that the vertebral plexus represents the main intracranial extension of the sympathetic system (Lazorthes 1949; Mitchell 1952). The plexus on the inferior thyroid artery accompanies the artery to the thyroid gland, and communicates with the recurrent and external

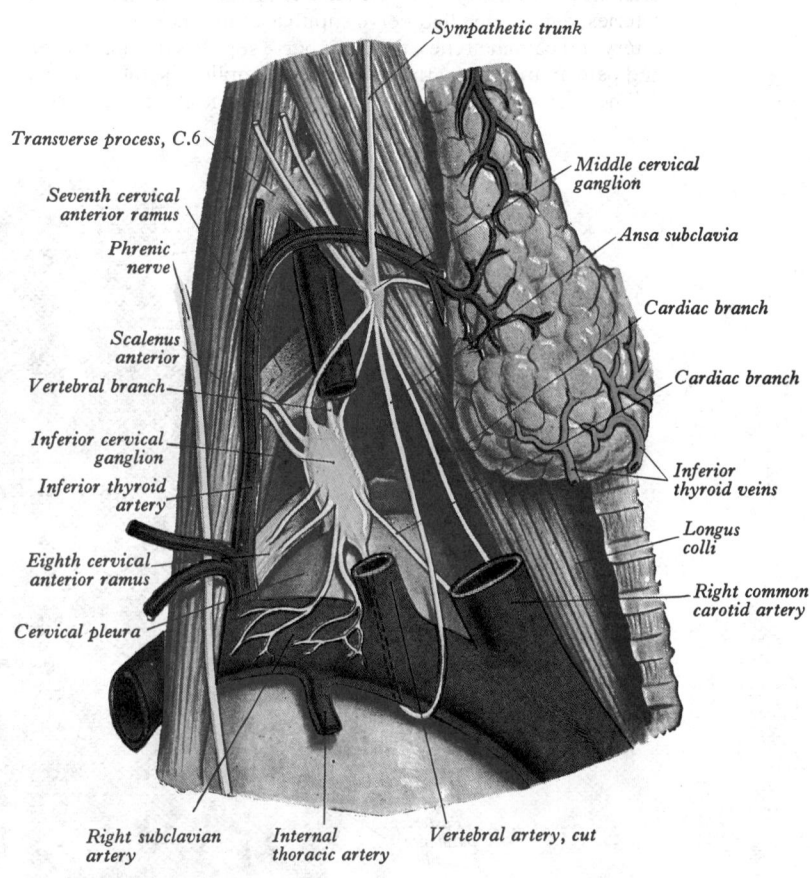

7.227 B   Anterior view of the same structures as in 7.227 A. Part of the vertebral artery has been excised to show the inferior cervical ganglion.

laryngeal nerves, with the cardiac branch of the superior cervical ganglion, and with the plexus on the common carotid artery.

The preganglionic fibres for the head and neck leave the spinal cord through the upper five thoracic nerves (mainly the upper three); they pass up the sympathetic trunk to synapse about cells in the cervical ganglia, whence postganglionic fibres are distributed as indicated above.

The preganglionic fibres concerned with supplying the upper limb are derived from the upper thoracic segments of the spinal cord, probably T. 2–6 (or 7). These fibres ascend the sympathetic trunk to synapse with cells mainly in the cervicothoracic ganglion, whence postganglionic fibres pass to the brachial plexus, mainly the lower trunk. Most of the vasoconstrictor fibres supplying the arteries of the upper limb emerge from the spinal cord in the ventral roots of the second and third thoracic nerves. These arteries can thus be denervated surgically by cutting the sympathetic trunk below the third thoracic ganglion, severing the rami communicantes connected with the second and third thoracic ganglia, or cutting (intradurally) the ventral roots of the second and third thoracic spinal nerves. The white ramus to the cervicothoracic ganglion is not cut, partly because it does not convey many vasomotor or sudomotor fibres to the upper limb, but mainly because it contains most of the preganglionic fibres which pass up the sympathetic trunk to the superior cervical ganglion, from which postganglionic branches pass to supply vasoconstrictor and sudomotor nerves to the face and neck, secretory fibres to the salivary glands, the dilatator pupillae (and probably ciliaris oculi), the non-striated muscle in the upper and lower eyelids and the orbitalis. Destruction of this nerve would result in constriction of the pupil, drooping of the upper eyelid (ptosis), enophthalmos and absence of sweating on the face and neck (*Horner's syndrome*), and possibly some disturbance of accommodation. For a review of such procedures consult Haxton (1954).

The blood vessels of the upper limb beyond the first part of the axillary artery receive their sympathetic nerve supply by means of branches from the brachial plexus through nerves adjacent to the arteries, e.g. the median nerve supplies branches to the brachial artery and palmar arches, the ulnar nerve supplies the ulnar artery and palmar arches and the radial nerve supplies the radial artery.

The first and second (and occasionally the third) intercostal nerves are sometimes connected together in front of the necks of the ribs by filaments which contain postganglionic fibres derived from the grey rami associated with these nerves; these fibres provide another pathway by which postganglionic nerves from the upper thoracic ganglia may pass to the brachial plexus.

THORACIC PART OF THE SYMPATHETIC SYSTEM

The thoracic part of each sympathetic trunk (7.226, 228) contains a series of ganglia, which usually correspond approximately in number to that of the thoracic spinal nerves, but their number is variable. (The number is 11 in more than 70 per cent, occasionally 12, rarely 10 or 13.) The first thoracic ganglion is usually fused with the inferior cervical to form the cervicothoracic ganglion. Jit and Mukerjee (1960) found a fused ganglion in 80 of 100 dissections. Rarely, the middle cervical or second thoracic ganglion may be included.) The succeeding ganglion is called the

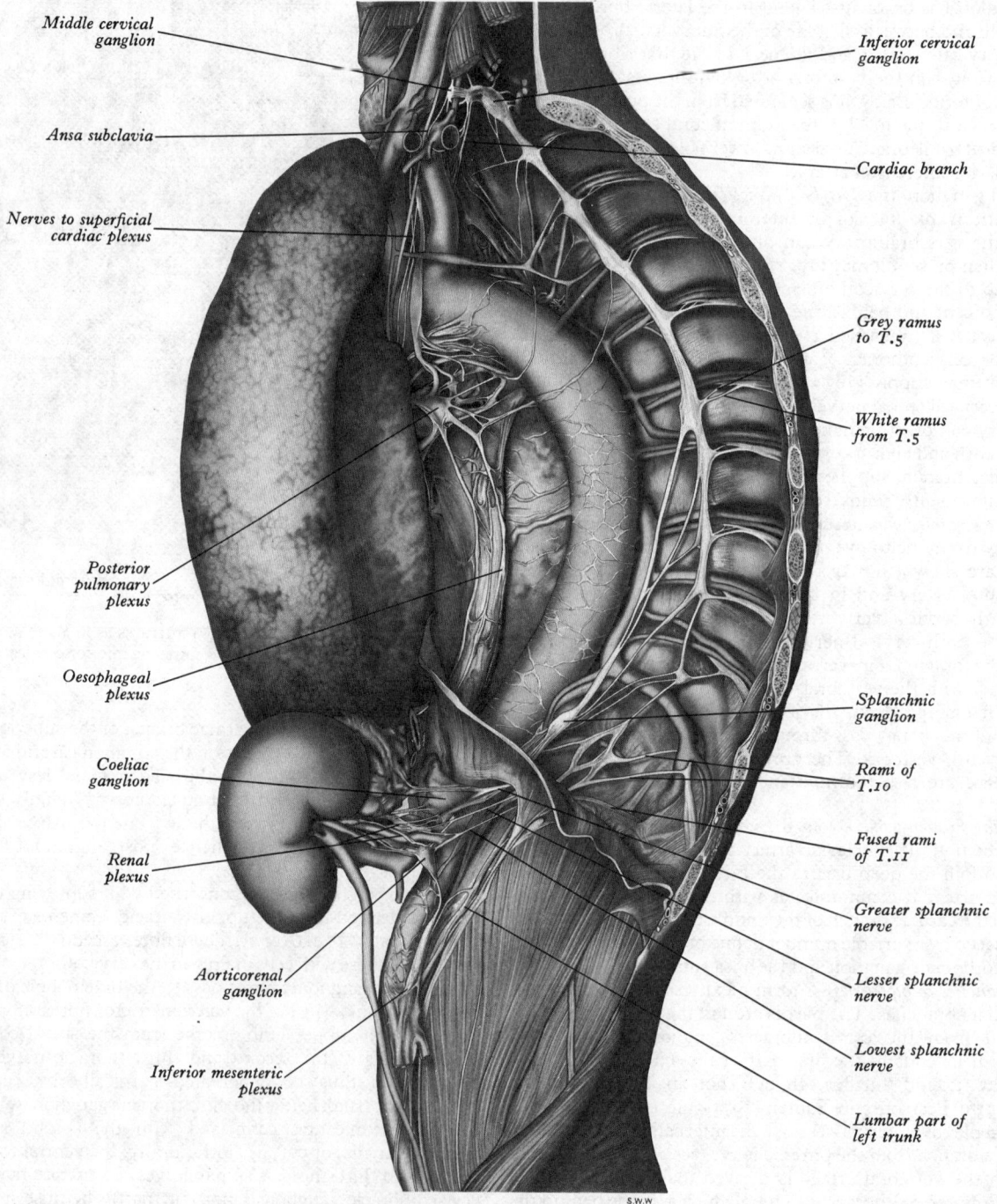

Middle cervical ganglion

Ansa subclavia

Nerves to superficial cardiac plexus

Posterior pulmonary plexus

Oesophageal plexus

Coeliac ganglion

Renal plexus

Aorticorenal ganglion

Inferior mesenteric plexus

Inferior cervical ganglion

Cardiac branch

Grey ramus to T.5

White ramus from T.5

Splanchnic ganglion

Rami of T.10

Fused rami of T.11

Greater splanchnic nerve

Lesser splanchnic nerve

Lowest splanchnic nerve

Lumbar part of left trunk

S.W.W

7.228 The thoracic part of the sympathetic system of the left side. (Drawn from a dissection by the late Dr. G. D. Channell.) Note that the diaphragm has been divided close to its posterior attachment, and the left lung and the left kidney have been drawn forwards and rotated to the right, so as to expose the posterior surface of the left kidney and suprarenal gland.

second thoracic ganglion in order that each thoracic ganglion should correspond numerically with the other segmental structures. With the exception of the last two or three, the thoracic ganglia rest against the heads of the ribs, and are posterior to the costal pleura; the last two or three are placed on the sides of the bodies of the corresponding vertebrae. Inferiorly, the thoracic sympathetic trunk passes dorsal to the medial arcuate ligament (or through the crus of the diaphragm) to become continuous with the lumbar sympathetic trunk. The ganglia are small and are connected together by the intervening portions of the trunk.

Two or more rami communicantes, white and grey, connect each ganglion with its corresponding spinal nerve, the white rami joining the spinal nerve farther distally than the grey. Sometimes a grey and white ramus may be fused to form a single 'mixed' ramus (p. 1123).

The *medial branches from the upper five ganglia* are very small; they supply filaments to the thoracic aorta and its branches. On the aorta they form a delicate plexus (*thoracic aortic plexus*) together with filaments from the greater splanchnic nerve. Twigs from the second to fifth or sixth ganglia enter the posterior pulmonary plexus; others, from the second, third, fourth and fifth ganglia, pass to the deep (dorsal) part of the cardiac plexus. Small branches from these pulmonary and cardiac nerves pass to the oesophagus and trachea.

The *medial branches from the lower seven ganglia* are large; they distribute filaments to the aorta, and unite to form the greater, the lesser and the lowest splanchnic nerves, the last of which is not always identifiable.

The *greater splanchnic nerve* consists mainly of myelinated, preganglionic and visceral afferent fibres; it is formed by branches from the fifth to the ninth or tenth thoracic ganglia, but the fibres in the higher branches may be traced upwards in the sympathetic trunk as far as the first or second thoracic ganglion. Its roots of origin may vary from 1 to 8, 4 being most usual. It descends obliquely on the bodies of the vertebrae, supplies fine branches to the descending thoracic aorta, perforates the crus of the diaphragm, and ends mainly in the coeliac ganglion, but partly in the aorticorenal ganglion and the suprarenal gland. A *splanchnic ganglion* exists on this nerve opposite the eleventh or twelfth thoracic vertebra in 17 to 68 per cent of *dissections* (*see* Jit and Mukerjee 1960 for details); but Mitchell (1953) reported that *microscopic* evidence showed it to be always present.

The *lesser splanchnic nerve* is formed by filaments from the ninth and tenth, sometimes the tenth and eleventh, thoracic ganglia, and from the trunk between the ganglia. It pierces the diaphragm with the preceding nerve, and joins the aorticorenal ganglion.

The *lowest splanchnic nerve* (or renal nerve) arises from the last thoracic ganglion. It gains the abdomen with the sympathetic trunk, and ends in the renal plexus.

Jit and Mukerjee (1960) have described in great detail dissections of the thoracic sympathetic system in 50 cadavers and have surveyed previous findings. For example, the incidence of the splanchnic nerves, according to seven observers is as follows: greater—always present, lesser—94 per cent (86–100 per cent), least—56 per cent (16–98 per cent). A fourth (accessory) splanchnic nerve has been described by de Sousa (1955), but its existence has not been confirmed.

LUMBAR PART OF THE SYMPATHETIC SYSTEM

The lumbar part of each sympathetic trunk (7.228, 229) usually consists of four lumbar ganglia, connected together by the intervening portions of the trunk. It is in the extraperitoneal connective tissue anterior to the vertebral column, along the medial margin of the psoas major. Posterior to the medial arcuate ligament it continues as the thoracic part of the trunk and inferiorly, by passing posterior to the common iliac artery, with the pelvic part. On the right side it is overlapped by the inferior vena cava; on the left by the lateral aortic lymph nodes. It lies in front of the lumbar vessels, but some lumbar veins may pass anterior to it.

The first and second, and sometimes the third, lumbar ventral

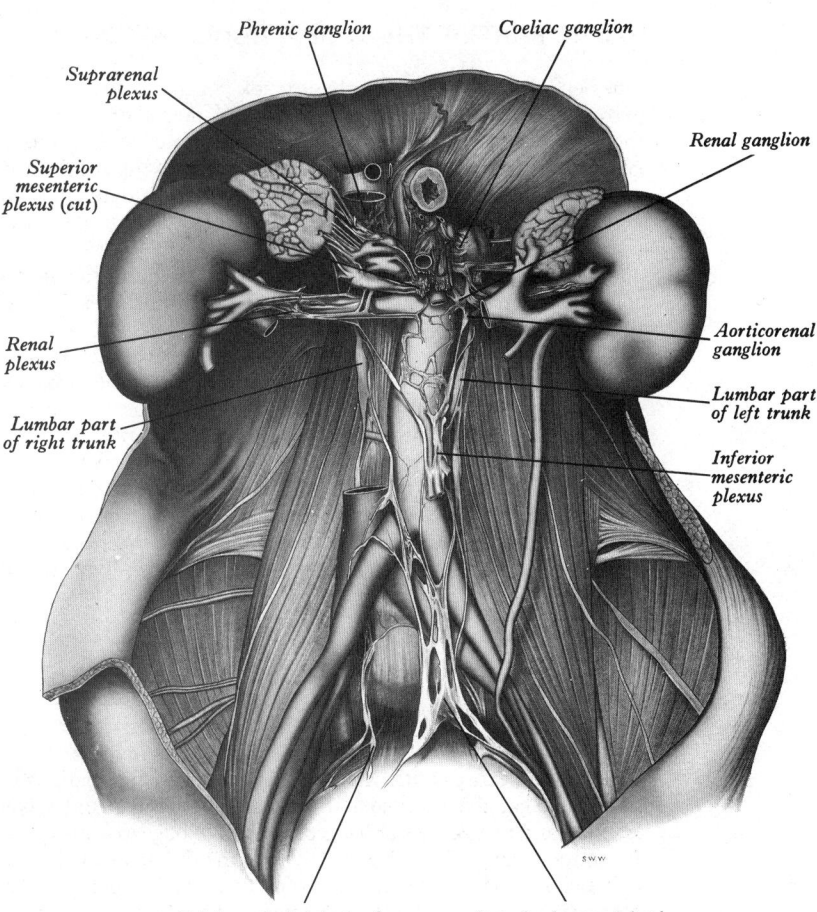

Phrenic ganglion   Coeliac ganglion

Suprarenal plexus

Superior mesenteric plexus (cut)

Renal ganglion

Renal plexus

Aorticorenal ganglion

Lumbar part of right trunk

Lumbar part of left trunk

Inferior mesenteric plexus

Pelvic part of right trunk   Superior hypogastric plexus

7.229 The abdominal part of the sympathetic system. (Drawn from a dissection by the late Dr. G. D. Channell, Guy's Hospital Medical School.)

rami send *white rami communicantes* to the corresponding ganglia.

Grey rami communicantes pass from all the ganglia to the lumbar spinal nerves. These rami are of considerable length and accompany the lumbar arteries round the sides of the bodies of the vertebrae, medial to the fibrous arches to which the psoas major is attached.

Generally four *lumbar splanchnic nerves* pass from the ganglia to join the coeliac, intermesenteric (abdominal aortic) and superior hypogastric plexuses. The first lumbar splanchnic nerve arises from the first ganglion and joins the coeliac, renal and intermesenteric plexuses. The second nerve arises from the second (and sometimes also the third) ganglion and joins the lower part of the intermesenteric plexus. The third nerve issues from the third or fourth ganglion and passes in front of the common iliac vessels to join the superior hypogastric plexus. The fourth lumbar splanchnic, from the lowest ganglion, goes dorsal to the common iliac vessels to join the lower part of the superior hypogastric plexus or the hypogastric 'nerve'.

Vascular branches from all the lumbar ganglia pass to the intermesenteric (aortic) plexus. From the lower lumbar splanchnic nerves, fibres pass to the common iliac arteries, around which they form a plexus continued thence along the internal iliac artery and around the external iliac artery, in the latter case as far as the proximal part of the femoral artery. Many of the postganglionic fibres in the grey rami, connecting the lumbar ganglia to the lumbar spinal nerves, travel in the femoral nerve, and thence in its muscular, cutaneous and saphenous branches, to supply vasoconstrictor nerves to the femoral artery and its branches in the thigh. Other postganglionic fibres travel in the obturator nerve to the obturator artery. Considerable uncertainties persist with regard to the sympathetic supply to the lower limb (*see* Wilde 1951; Wyburn 1956; Pick 1970).

PELVIC PART OF THE SYMPATHETIC SYSTEM

The pelvic part of each sympathetic trunk (7.229) is situated in the extraperitoneal tissue in front of the sacrum, medial or anterior to the anterior sacral foramina. It comprises four or five sacral ganglia, connected by the intervening sections of the trunk. It is continuous cranially with the lumbar part, while caudally, the two pelvic sympathetic trunks converge, and unite on the front of the coccyx in the small *ganglion impar*.

Grey rami communicantes pass from the ganglia to the sacral and coccygeal spinal nerves. No white rami communicantes pass to this part of the sympathetic trunk.

The *medial branches of distribution* communicate on the front of the sacrum with the corresponding branches from the opposite side; twigs from the first two ganglia join the inferior hypogastric plexus (pelvis plexus) or the hypogastric nerve, and others form a plexus on the median sacral artery. Filaments are distributed to the glomus coccygeum from the loop uniting the two trunks. The 'hypogastric nerve', which is itself usually plexiform, is a somewhat redundant term for the connexions, right and left, which exist between the superior and inferior hypogastric plexuses. (*See* p. 1125.)

Vascular branches. Through the grey rami many postganglionic fibres pass to the roots of the sacral plexus, particularly those forming the tibial nerve, to be conveyed to the popliteal artery and its branches in the leg and foot. Others are conveyed by the pudendal and superior and inferior gluteal nerves to the accompanying arteries. Branches to lymph nodes are also described (Woźniak 1967).

The preganglionic fibres concerned with supplying the lower limb are derived from the lower three thoracic and upper two or three lumbar segments of the spinal cord. They reach the lower thoracic and upper lumbar ganglia through the white rami and some pass down the sympathetic trunk to synapse about cells in the lumbar ganglia, whence postganglionic fibres pass to the femoral nerve to be distributed to the femoral artery and its branches in the thigh; other fibres pass down the sympathetic trunk to synapse with cells in the upper two or three sacral ganglia, whence postganglionic axons pass to the tibial nerve to supply the popliteal artery and its branches in the leg and foot. Sympathetic denervation of the vessels of the lower limb can thus be produced by removing the upper three lumbar ganglia and the intervening parts of the sympathetic trunk, all the preganglionic fibres to the lower limb thus being divided.

PLEXUSES OF THE AUTONOMIC NERVOUS SYSTEM

The larger plexuses of the autonomic system are aggregations of nerves and ganglia, situated in the thoracic, abdominal and pelvic cavities, and named the cardiac, coeliac and hypogastric plexuses. From the plexuses branches are given to the thoracic, abdominal and pelvic viscera. Extensions from these major perivascular plexuses pass along most of the branches of the large vessels with which they are associated. Such extensions are usually named after the branch artery along which they are distributed. This leads to a plethora of named plexuses, receiving separate description, the details of which may overshadow the essential continuity of the vascular plexuses in the thorax, abdomen and pelvis.

THE CARDIAC PLEXUSES

The cardiac plexus (7.226, 228) is situated at the base of the heart, and is divided into a *superficial* and a *deep* (dorsal) *part*, which are closely connected. Several small ganglia are found in the plexus, the largest and most constant being the *cardiac ganglion* described below. Mizeres (1963) has emphasized the unity of the cardiac plexus, division of which into superficial and deep parts he considered an artefact of dissection; he was, however, prepared to accord regional names to the coronary, pulmonary, atrial and aortic extensions of the plexus. Since, however, major regions of the plexus are situated as described here, the terms superficial and deep have been retained.

The superficial (ventral) part of the cardiac plexus lies below the arch of the aorta, anterior to the right pulmonary artery. It is formed by the cardiac branch of the superior cervical ganglion of the left sympathetic trunk, and the lower of the two cervical cardiac branches of the left vagus. A small ganglion, termed the *cardiac ganglion*, is usually present in this plexus, and is situated immediately below the arch of the aorta, on the right of the ligamentum arteriosum. The superficial part of the cardiac plexus gives branches (*a*) to the deep part of the plexus, (*b*) to the right coronary plexus, (*c*) to the left anterior pulmonary plexus.

The deep (dorsal) part of the cardiac plexus is situated in front of the bifurcation of the trachea, above the point of division of the pulmonary trunk, and posterior to the aortic arch. It is formed by the cardiac nerves derived from the cervical and upper thoracic ganglia of the sympathetic trunk, and the cardiac branches of the vagus and recurrent laryngeal nerves. The only cardiac nerves which do not join the deep part of the cardiac plexus are those already noted as joining the superficial part of the plexus.

The branches from the *right half* of the deep part of the cardiac plexus pass, some in front of, and others behind, the right pulmonary artery; the former, the more numerous, transmit a few filaments to the right anterior pulmonary plexus, and are then continued onwards to form part of the right coronary plexus; those behind the pulmonary artery distribute a few filaments to the right atrium, and are then continued onwards to form part of the left coronary plexus.

The *left half* of the deep part of the cardiac plexus is connected with the superficial part of the plexus, and gives filaments to the left atrium, and to the left anterior pulmonary plexus, and is then continued to form the greater part of the left coronary plexus.

The left coronary plexus is larger than the right, and accompanies the left coronary artery; it is formed chiefly by filaments prolonged from the left half of the deep part of the cardiac plexus, and by a few from the right half. It gives branches to the left atrium and ventricle.

The right coronary plexus is formed partly from the superficial and partly from the deep parts of the cardiac plexus. It accompanies the right coronary artery, and gives branches to the right atrium and ventricle.

The atrial plexuses, as described by Mizeres (1963), are derivatives of the right and left continuations of the cardiac plexus which extend along the pulmonary arteries. Their fibres are distributed to the corresponding atria, overlapping with those of the coronary plexuses.

All the cardiac branches of the vagus and the sympathetic contain both afferent and efferent fibres, except the cardiac branch of the superior cervical sympathetic ganglion which contains efferent (postganglionic) fibres only.

The *efferent* preganglionic sympathetic fibres arise in the upper four or five thoracic segments of the spinal cord; they pass by white rami communicantes to synapse about cells in the upper thoracic ganglia on the sympathetic trunk, though many travel up the trunk to synapse in the cervical ganglia. From the thoracic and cervical ganglia, postganglionic fibres emerge to form the sympathetic cardiac nerves, the functions of which are cardiac acceleration and dilatation of the coronary arteries. Of the sympathetic fibres arising from the first four or five segments of the spinal cord, the upper ones pass to the ascending aorta, pulmonary trunk and ventricles, while the lower ones supply the atria.

The *efferent* cardiac parasympathetic fibres are derived from the dorsal nucleus of the vagus and from cells near the nucleus ambiguus, and run in the cardiac branches of the vagus to synapse about cells in the cardiac plexuses and in the walls of the atria. These vagal fibres are concerned with slowing of the heart and

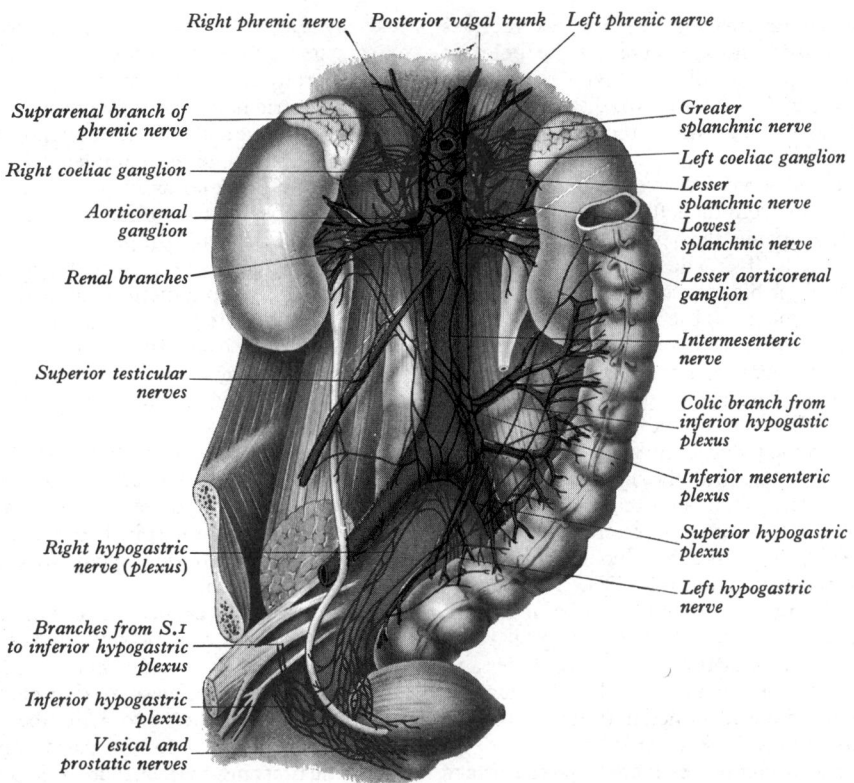

Right phrenic nerve   Posterior vagal trunk   Left phrenic nerve

Suprarenal branch of phrenic nerve

Right coeliac ganglion

Aorticorenal ganglion

Renal branches

Superior testicular nerves

Right hypogastric nerve (plexus)

Branches from S.1 to inferior hypogastric plexus

Inferior hypogastric plexus

Vesical and prostatic nerves

Greater splanchnic nerve

Left coeliac ganglion

Lesser splanchnic nerve

Lowest splanchnic nerve

Lesser aorticorenal ganglion

Intermesenteric nerve

Colic branch from inferior hypogastric plexus

Inferior mesenteric plexus

Superior hypogastric plexus

Left hypogastric nerve

7.230 Autonomic nerves and plexuses in the abdomen and pelvis. (After G. A. G. Mitchell, *Anatomy of the Autonomic Nervous System*, by courtesy of the author and the publishers.) Note the ascending branches of the inferior hypogastric plexus passing up to supply the descending colon. The sympathetic trunks are not shown on the right side; note the upper, middle and lower ureteric nerves.

with constriction of the coronary arteries (see also p. 1121). In man (and most mammals) the intrinsic cardiac nerve cells are limited to the atria and the interatrial septum (Davies *et al.* 1952; King and Coakley 1958); they are most numerous in the subepicardial connective tissue and near the sinuatrial and atrioventricular nodes. For an exposition of variations in the human sympathetic cardiac innervation, consult Ellison and Williams (1969).

THE PULMONARY PLEXUSES

The pulmonary plexuses lie on the anterior and posterior aspects of the bronchial and vascular structures in the hila of the lungs, the anterior pulmonary plexus being much smaller than the posterior. (According to Mizeres 1963, these plexuses are continuous with extensions from the cardiac plexus along the right and left pulmonary arteries.) The plexuses are formed by branches from the vagus and the sympathetic. The efferent parasympathetic fibres arise from the dorsal nucleus of the vagus. The efferent sympathetic fibres are postganglionic branches of the second to fifth thoracic ganglia of the sympathetic trunk.

The *anterior part of the pulmonary plexus* is formed by branches from the cardiac branches of the vagus and cervical sympathetic trunk as well as direct branches from both these sources. The *posterior part of the pulmonary plexus* is formed by rami of the cardiac branches of the vagus, from the cardiac plexus and from the second to fifth or sixth thoracic sympathetic ganglia, the left plexus receiving additional branches from the left recurrent laryngeal nerve.

From the plexuses, nerves pass into the lung to form networks around the branches of the bronchi and the pulmonary and bronchial vessels, extending as far as the visceral pleura. On these nerves, near the hila of the lungs, there are minute collections of nerve cells with which the efferent preganglionic vagal fibres synapse. (In a number of organs, notably the small intestine, *interstitial cells* have been described in the terminal autonomic network which is characteristic in many sites. The presence of these in thoracic organs, apart perhaps from the oesophagus, has not been substantiated—Dijkstra 1969.) The efferent vagal fibres are bronchoconstrictor, secretomotor to the mucous bronchial glands and vasodilator in function. The efferent sympathetic fibres are bronchodilator and vasoconstrictor.

THE COELIAC PLEXUS

The coeliac plexus (7.226, 229), the largest of the three great autonomic plexuses, is situated at the level of the last thoracic and the upper part of the first lumbar vertebra, and is a dense network of nerve fibres which unite together two large *coeliac ganglia*. It surrounds the coeliac artery and the root of the superior mesenteric artery. It lies posterior to the stomach and the omental bursa, anterior to the crura of the diaphragm and the commencement of the abdominal aorta, and between the suprarenal glands. The plexus and the ganglia are joined by the greater and lesser splanchnic nerves of both sides and some filaments from the vagus and phrenic nerves and extend as numerous secondary plexuses along the neighbouring arteries.

The *coeliac ganglia* are two irregularly shaped masses placed, one on each side of the median plane, between the suprarenal gland and the origin of the coeliac artery, and in front of the crura of the diaphragm, that on the right side being placed behind the inferior vena cava and that on the left side behind the splenic vessels. The upper part of each ganglion is joined by the greater splanchnic nerve, while the lower part, which is more or less detached and is named the *aorticorenal* ganglion, receives the lesser splanchnic nerve and gives off the greater part of the renal plexus. The position of the ganglion is very variable, being anywhere in the general vicinity of the origin of the renal artery from the aorta. For a detailed description and discussion of its connexions, and the distribution of its fibres to tubules and glomeruli, in the kidney, consult Norvell (1968).

The secondary plexuses springing from or connected with the coeliac plexus are the phrenic, splenic, hepatic, left gastric, intermesenteric, suprarenal, renal, testicular or ovarian, superior mesenteric, and inferior mesenteric.

The phrenic plexus accompanies the corresponding inferior phrenic artery to the diaphragm, some filaments passing to the suprarenal gland. It arises from the upper part of the coeliac ganglion, and is larger on the right than on the left side. It receives one or two branches from the phrenic nerve. At the point of junction of the right phrenic plexus with the phrenic nerve there is a small mass, the *phrenic ganglion*. This plexus distributes some branches to the inferior vena cava, and to the suprarenal and hepatic plexuses.

The hepatic plexus, the largest derivative of the coeliac plexus, also receives filaments from the left and right vagus and right phrenic nerves. It accompanies the hepatic artery and portal vein and their branches into the liver, and in the liver the nerves are confined to the vicinity of the blood vessels. Branches from the plexus accompany all the branches of the hepatic artery. Those passing to the gall bladder form a scanty *cystic plexus*; branches also pass to the bile ducts. The branches accompanying the right gastric artery supply the pylorus. A considerable plexus accompanies the gastroduodenal artery and its branches. From this plexus branches pass to the pylorus and superior part of the duodenum. Many of the nerves pass with the right gastro-epiploic artery to supply the right part of the stomach and the greater curvature. Others pass with the superior pancreaticoduodenal artery and supply the descending part of the duodenum, head of the pancreas and the lower part of the bile duct. The hepatic plexus contains both afferent and efferent sympathetic and parasympathetic fibres, and it is believed that the vagal constituents are motor to the musculature of the gall bladder and bile ducts and inhibitory to the sphincter of the bile duct. A distinct nerve to the sphincter was identified in twenty-three out of twenty-five human dissections (Petkov 1968).

The left gastric plexus accompanies the left gastric artery along the lesser curvature of the stomach, and joins with the gastric branches of the vagus nerves. The gastric sympathetic nerves are motor to the pyloric sphincter but inhibitory to the muscular coats of the stomach.

The splenic plexus is formed by branches from the coeliac plexus, left coeliac ganglion and right vagus nerve. It accompanies the splenic artery to the spleen, giving off, in its course, subsidiary plexuses along the various branches of the artery. The fibres are principally, if not wholly, sympathetic in origin and terminate on the blood vessels and unstriped muscle of the splenic capsule and trabeculae.

The suprarenal plexus is formed by branches from the coeliac ganglion, coeliac plexus and the greater splanchnic nerve. Relative to its size, the suprarenal gland has a larger autonomic supply than any other organ. The nerves have hitherto been described as myelinated and preganglionic in nature. In the rat, however, the non-myelinated fibres are ten times as numerous as the myelinated; these non-myelinated fibres are regarded as preganglionic. They terminate in synapses, often deeply invaginated, in contiguity with large chromaffin cells, the phaeochromocytes, which are hence homologous with postganglionic sympathetic neurons (p. 1458). A space of 150–200 nm separates the contiguous plasma membranes which often exhibit electron-dense zones. Small vesicles and large vesicles with electron-dense granular contents are present in the endings. Only non-myelinated fibres have been seen innervating chromaffin cells, all of which are related to one or more nerve terminal. Multipolar nerve cell bodies are also found in the adrenal medulla and some of the preganglionic non-myelinated nerve fibres form axodendritic synapses with these cells. The destination of the axons of these nerve cells is not known (Coupland 1965*a*). A preponderance of non-myelinated fibres in the suprarenal plexus has also been described in man (Coupland 1965 *a* and *b*; Grottel 1968).

The renal plexus is a rich plexus formed by filaments from the coeliac ganglion, coeliac plexus, aorticorenal ganglion, lowest thoracic splanchnic nerve, first lumbar splanchnic nerve and aortic plexus. Small collections of nerve cells are found in the plexus, the largest usually lying behind the commencement of the renal artery. The plexus is continued into the kidney around the branches of the renal artery to supply the vessels and the renal glomeruli and tubules, particularly the tubules in the cortex of the kidney (Norvell 1968). For the main part the renal nerves are vasomotor in function. From the renal plexus branches are given to the ureteric and the testicular (or ovarian) plexuses.

The *ureteric plexus* receives fibres from three sources, the upper part by branches from the renal and aortic plexuses, the middle part by branches from the superior hypogastric plexus and the hypogastric nerve, and the lower part by branches from the hypogastric nerve and the inferior hypogastric plexus. The nerves to the ureter are believed to influence its inherent motility.

The testicular plexus accompanies the gonadal artery to the testis. Its upper part is formed by branches from the renal and aortic plexuses. Distally the plexus is reinforced by branches from the superior and inferior hypogastric plexuses. Branches from the plexus pass to the epididymis and the ductus deferens.

The ovarian plexus accompanies the ovarian artery and is distributed to the ovary and uterine tube. The upper part of the plexus is formed by branches from the renal and aortic plexuses; lower down it is reinforced from the superior and inferior hypogastric plexuses.

The nerves in the testicular and ovarian plexuses contain efferent and afferent sympathetic fibres; the efferent fibres are vasomotor in nature and are derived from the tenth and eleventh thoracic segments of the spinal cord; the parasympathetic fibres, derived from the inferior hypogastric plexuses, are probably vasodilator in nature.

The superior mesenteric plexus is a continuation of the lower part of the coeliac plexus, and receives a branch from the junction of the right vagus nerve with the latter plexus. It surrounds the superior mesenteric artery, accompanies it into the mesentery, and divides into a number of secondary plexuses which are distributed to the parts supplied by the artery— pancreatic branches to the pancreas; jejunal and ileal branches to the small intestine; ileocolic, right colic, and middle colic branches, which supply the corresponding parts of the large intestine. The *superior mesenteric ganglion* is situated in the upper part of the plexus, usually immediately above the origin of the superior mesenteric artery.

The sympathetic nerves to the intestine are motor to the ileocaecal sphincter but inhibitory to the muscle of the gut. In addition, they convey vasoconstrictor fibres.

The abdominal aortic plexus (intermesenteric plexus) is formed by branches from the coeliac plexus and ganglia, and receives filaments from the first and second lumbar splanchnic nerves. It is situated upon the sides and front of the aorta, between the origins of the superior and inferior mesenteric arteries. It is not a dense plexus but consists of four to twelve nerves (intermesenteric nerves) connected by obliquely arranged branches. It is continuous above with the coeliac plexus and the coeliac and aorticorenal ganglia, and below with the superior hypogastric plexus. From this plexus parts of the testicular, the inferior mesenteric, the iliac and the superior hypogastric plexuses arise; it also distributes filaments to the inferior vena cava.

The inferior mesenteric plexus is derived chiefly from the aortic plexus, but also receives branches from the second and third lumbar splanchnic nerves. It surrounds the inferior mesenteric artery and is distributed along its branches; thus the left colic plexus supplies the left part of the transverse colon, the descending colon and sigmoid colon, and the superior rectal plexus supplies the rectum. Just above, or below, the origin of the inferior mesenteric artery, a ganglion (the *inferior mesenteric ganglion*) may sometimes be found, but more often small discrete ganglia are scattered about the commencement of the artery in the proximal part of the plexus. In one study (Southam 1959) an inferior mesenteric ganglion occurred in every one of twenty-two human stillborn infants examined. The colic sympathetic nerves are inhibitory to the muscular coats of the colon and rectum. Branches from the parasympathetic pelvic splanchnic nerves run up occasionally through but usually near the superior hypogastric and the inferior mesenteric plexuses to supply the large intestine from the left part of the transverse colon down as far as the rectum (p. 1123 and *vide infra*); impulses along these nerves cause contraction of the musculature of the gut. It is to be emphasized that the parasympathetic supply to the distal colon is largely by these direct branches of the pelvic splanchnic nerves and not *via* the hypogastric and inferior mesenteric plexuses (Mitchell 1935; Woodburne 1956).

THE SUPERIOR HYPOGASTRIC PLEXUS

The superior hypogastric plexus (7.229–231) is situated in front of the bifurcation of the abdominal aorta, the left common iliac vein, the median sacral vessels, the body of the last lumbar vertebra and the promontory of the sacrum, and between the two common iliac arteries. It is often referred to as the *presacral nerve*, but the plexus is seldom sufficiently condensed to resemble a single nerve and moreover the plexus is prelumbar rather than presacral in position. It lies in the extraperitoneal connective tissue, and the parietal peritoneum can easily be stripped off its anterior surface. The plexus varies in breadth and in the degree of condensation of its constituent nerves, and it often lies a little to one side of the median plane (more often to the left side); the root of the sigmoid mesocolon, containing the superior rectal vessels, lies to the left side of the lower part of the plexus. Scattered nerve cells are found in the plexus.

Above, the plexus is formed by the union of branches from the aortic plexus with the third and fourth lumbar splanchnic nerves. Below, the plexus divides into the right and left hypogastric nerves which descend to the two inferior hypogastric plexuses. The superior hypogastric plexus gives off branches to the ureteric and testicular (or ovarian) plexuses and to those on the common iliac arteries. In addition to the sympathetic fibres which descend to form the superior hypogastric plexus, it may contain parasympathetic fibres (from the pelvic splanchnic nerves) which ascend from the inferior hypogastric plexus, though usually these parasympathetic fibres run up to the left of the superior hypogastric plexus and across the sigmoid vessels and the branches of the left colic vessels. These parasympathetic fibres are distributed partly along the branches of the inferior mesenteric artery, but also as independent retroperitoneal nerves, to supply the left part of the transverse colon, the left colic flexure, the descending colon and the sigmoid colon (*see* p. 1123 and 7.230).

THE INFERIOR HYPOGASTRIC PLEXUSES

The superior hypogastric plexus divides below into the right and left *hypogastric 'nerves'*, each of which runs down in the extraperitoneal connective tissue into the pelvis, medial to each internal iliac artery and its branches, to become the inferior hypogastric plexus (7.231). Each nerve may be single or may form an elongated narrow plexus consisting of two or three longitudinal nerves connected by anastomosing filaments. (The hypogastric nerves can scarcely be distinguished from their continuations, the inferior hypogastric plexuses. The latter are joined by the pelvic splanchnic nerves, but this distinction is minimized by the fact that both the nerves and the plexuses contain sympathetic and parasympathetic fibres. Some authorities prefer to describe the superior hypogastric plexus as dividing into two inferior hypogastric plexuses.) From each hypogastric nerve branches may pass to the testicular or ovarian plexus, the ureteric plexus, the plexus on the internal iliac artery and to the sigmoid colon, and each nerve may be joined near its commencement by the lowest lumbar splanchnic nerve, i.e. from the last lumbar sympathetic ganglion.

The inferior hypogastric (or pelvic) plexus lies in the extraperitoneal connective tissue. In the male it is situated on the side of the rectum, seminal vesicle, prostate and the posterior part of the urinary bladder; in the female, each plexus is placed on the side of the rectum, uterine cervix, vaginal fornix and posterior part of the urinary bladder, and extends into the base of the broad ligament of the uterus. Laterally are the internal iliac vessels and their branches and tributaries, levator ani, coccygeus and obturator internus. Behind are the sacral and coccygeal plexuses and above are the superior vesical and obliterated umbilical arteries. The plexuses contain numerous small ganglia. Each plexus is formed by the hypogastric nerve, which conveys most of the sympathetic fibres to the plexus, and by branches from the ganglia, which convey only a few fibres to the plexus; the parasympathetic fibres in the plexus are derived from the pelvic splanchnic nerves. The preganglionic efferent sympathetic fibres originate in the lower three thoracic and upper two lumbar segments of the spinal cord; some of these relay in the ganglia in

the lumbar and sacral parts of the sympathetic trunk, while others synapse about cells in the lower part of the aortic plexus and in the superior and inferior hypogastric plexuses. The preganglionic parasympathetic fibres originate in the second, third and fourth sacral segments of the spinal cord, reach the plexus in the pelvic splanchnic nerves and synapse with cells in the plexus or in the walls of the viscera supplied by the branches of the plexus. From the plexus numerous branches are distributed to the pelvic (and some abdominal) viscera, either directly or by accompanying the branches of the internal iliac artery.

Parasympathetic fibres pass back into the superior hypogastric plexus, or as separate filaments accompanying it, to reach the inferior mesenteric plexus, through the medium of the aortic plexus. By this route the descending and sigmoid parts of the colon receive a parasympathetic innervation.

The middle rectal plexus arises from the upper part of the inferior hypogastric plexus; the fibres pass to the rectum either directly or along the middle rectal artery. The plexus communicates above with branches of the superior rectal plexus and extends inferiorly as far as the internal anal sphincter. The nerve supply of the rectum and anal canal is derived from (*a*) the superior rectal plexus, (*b*) the middle rectal plexus, and (*c*) the inferior rectal (haemorrhoidal) nerves, which are branches of the pudendal nerve. The parasympathetic preganglionic fibres from the superior and middle rectal plexuses synapse with postganglionic neurons in the myenteric plexus, which is well developed in this region. The sympathetic afferents pass through this plexus without interruption. The efferent sympathetic fibres in the rectal plexuses are concerned with inhibition of the expulsive musculature and contraction of the sphincter. Afferent pain impulses pass along both the sympathetic and parasympathetic nerves, but the parasympathetic afferent and efferent fibres are more active in the normal process of defaecation. The inferior rectal nerves supply motor fibres to the external anal sphincter and sensory (somatic) fibres to the lower (ectodermal) part of the anal canal (pp. 202, 1364).

The vesical plexus arises from the anterior part of the inferior hypogastric plexus. It is composed of numerous nerves which accompany the vesical arteries to the bladder. Branches from the plexus pass to the seminal vesicles and deferent ducts. Many small collections of nerve cells are present among the nerve fibres in the muscular wall of the bladder. The sympathetic preganglionic efferent fibres in the plexus arise from the lower two thoracic and upper two lumbar segments of the spinal cord; the cells about which they synapse are scattered in the superior and inferior hypogastric plexuses and in the wall of the bladder. The parasympathetic preganglionic efferent fibres arise from the second, third and fourth sacral segments of the spinal cord and synapse about cells near to or in the walls of the bladder. These neurons convey motor fibres to the muscular coats of the bladder and inhibitory fibres to the sphincter. The efferent sympathetic nerves convey motor fibres to the sphincter and inhibitory fibres to the muscular coats, though some observers maintain that the sympathetic fibres are mainly vasomotor in function and that filling and emptying of the bladder are normally controlled exclusively by the parasympathetic nerves.

The prostatic plexus arises from the lower part of the inferior hypogastric plexus and is composed of relatively large nerves which enter the base and sides of the prostate and contain collections of nerve cells. The nerves are distributed to the prostate, seminal vesicles, prostatic urethra, ejaculatory ducts, corpora cavernosa, corpus spongiosum, membranous and penile parts of the urethra and the bulbo-urethral glands. The nerves supplying the corpora cavernosa form two sets, the lesser and the greater cavernous nerves of the penis; these arise from the front part of the prostatic plexus, join with branches from the pudendal nerve and then pass forwards below the pubic arch. The filaments of the *lesser cavernous nerves* pierce the fibrous covering of the penis near its root and supply the erectile tissue of the corpus spongiosum and the penile urethra. The *greater cavernous nerves* run forwards on the dorsum of the penis, communicate with the dorsal nerve of the penis and are distributed to the erectile tissue; some of the filaments pass to the erectile tissue of the corpus spongiosum. The sympathetic nerves supplying the male genital

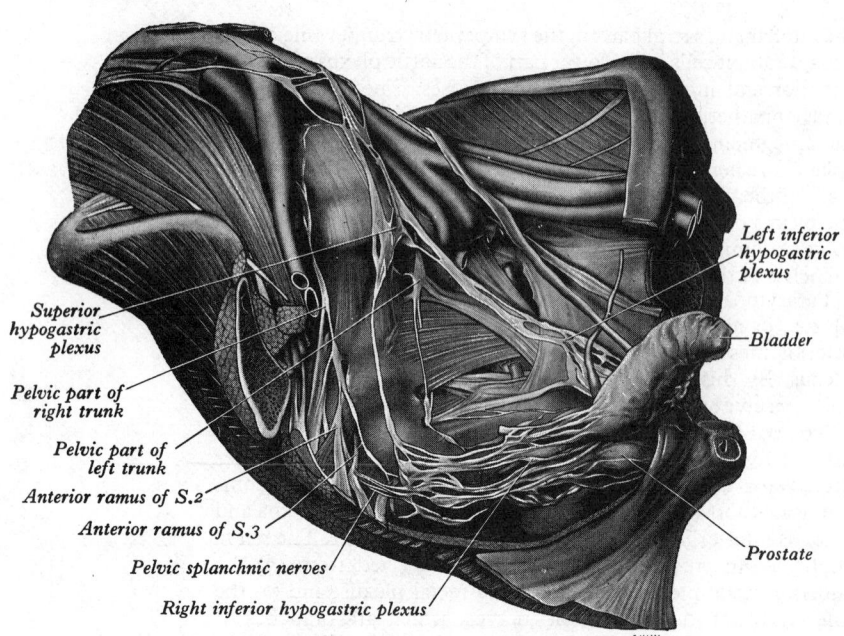

Superior hypogastric plexus

Pelvic part of right trunk

Pelvic part of left trunk

Anterior ramus of S.2

Anterior ramus of S.3

Pelvic splanchnic nerves

Right inferior hypogastric plexus

Left inferior hypogastric plexus

Bladder

Prostate

S.W.W

7.231 The pelvic part of the sympathetic system, viewed from in front and from the right side, a large part of the right hip bone having been removed. The superior hypogastric plexus is seen to divide below into the right and left hypogastric nerves (which are not labelled), which run down to the inferior hypogastric plexuses. (Drawn from a dissection by the late Dr. G. D. Channell, Guy's Hospital Medical School.)

organs produce vasoconstriction, while the parasympathetic produce vasodilatation.

The seminal vesicles are supplied by nerves derived from the vesical plexus, the prostatic plexus and the lower part of the inferior hypogastric nerves. From these nerves filaments pass to the ejaculatory ducts and to the deferent duct. It is generally believed that constriction of the seminal vesicles and seminal ejaculation are brought about by the sympathetic supply, which also produces inhibition of the bladder musculature and contraction of its sphincter during ejaculation, thus preventing reflex of the seminal fluid into the bladder. It has been, however, suggested that contraction of the seminal vesicles is due to parasympathetic impulses (Matthews and Raisman 1969).

The *uterine nerves* arise from the inferior hypogastric plexus, predominantly from the part of the plexus lying in the base of the broad ligament and known as the **uterovaginal plexus**. From the plexus some nerves pass down with the vaginal arteries, while others pass directly to the cervix uteri, or upwards with or near the uterine arteries in the broad ligament. The nerves passing to the cervix form a plexus in which small *paracervical* ganglia are found, one ganglion sometimes being large and called the *uterine cervical ganglion*. The uterine nerves passing upwards with the uterine arteries supply branches to the body of the uterus and, in the upper part of the broad ligament, they supply branches to the uterine tube and communicate with the tubal nerves from the inferior hypogastric plexus and with the nerves of the ovarian plexus. The branches of the uterine nerves ramify in the myometrium and endometrium; most of these nerves accompany the vessels. The efferent preganglionic sympathetic fibres supplying the uterus are derived from the last thoracic and first lumbar segments of the spinal cord; the sites of the cells about which they synapse are not known. The preganglionic parasympathetic fibres arise in the second, third and fourth sacral segments of the cord and relay in the paracervical ganglia. While activity of the sympathetic nerves may produce uterine contraction and vasoconstriction, and that of the parasympathetic nerves produce uterine inhibition and vasodilatation, the results of the activities of these two systems are complicated by the pronounced hormonal control of uterine functions.

The *vaginal nerves* arise from the lower parts of the inferior hypogastric and uterovaginal plexuses, and follow the vaginal

arteries and their branches to be distributed to the walls of the vagina, the erectile tissue of the vestibular bulbs and the clitoris (cavernous nerves of the clitoris), the urethra and the greater vestibular glands. The nerves contain numerous parasympathetic fibres which have a vasodilator effect on the erectile tissue.

Afferent Autonomic Pathways

The efferent autonomic fibres to the viscera and blood vessels are accompanied by others which may be regarded as their sensory counterparts and designated *general visceral afferent* fibres, or perhaps more appropriately as *autonomic afferents*. These nerve fibres are the peripheral processes of unipolar cells in some of the cranial and spinal nerve ganglia. Their central processes (axons) pass in the corresponding nerves to the central nervous system. Their peripheral process (dendrites), which may be myelinated fibres of various sizes or non-myelinated fibres, are distributed with the pre- and post-ganglionic fibres of the parasympathetic and sympathetic subdivisions of the autonomic nervous system, but are not interrupted in autonomic ganglia. Their terminals are variously described as knobs, loops, rings, tendril-like endings and sometimes more elaborate encapsulated endings in the walls of the viscera, including their epithelial linings and serous coverings, and in the walls of blood vessels.

Afferent impulses conducted along these neurons initiate visceral reflexes, but usually they do not reach the level of consciousness. They are also believed to be concerned with organic visceral sensations such as hunger, nausea, sexual sensation, rectal distension, etc. It is also probable that visceral pain fibres follow these peripheral pathways. Although the viscera are insensitive to cutting, crushing or burning, excessive tension and contraction of nonstriated muscle and certain pathological conditions produce visceral pain. It is not always easy to draw a line between what is acceptably pathological and that which is little more than exaggeration of normal activity. 'Abdominal' pain due to excessive intestinal contractions is extremely commonplace. In pathological conditions affecting a viscus vague pain may be felt in the region of the viscus itself (true visceral pain) or in a region of skin or other somatic tissue, the sensory nerve fibres from which enter the same segments of the spinal cord as those which receive pain fibres from the viscus. The latter phenomenon is known as *referred pain*. In addition, if inflammation spreads from a diseased viscus to the parietal serous membrane (e.g. peritoneum) related to it, the somatic pain fibres will be stimulated and cause local somatic pain in this region of the body wall. True visceral pain is poorly localized and dull or heavy; it is also commonly colic in type. Referred pain is often associated with tenderness of the skin surface at the site of reference.

General visceral afferent fibres occur in the vagus and glossopharyngeal nerves and possibly other cranial nerves, in the second, third and fourth sacral nerves, from which they are distributed with the pelvic splanchnic nerves, and in the thoracic and upper lumbar spinal nerves, where they are distributed through the rami communicantes and along the pathways for efferent sympathetic innervation of the viscera and blood vessels.

The vagus is believed to have a large general visceral afferent component. The fibres have their cells of origin in the superior and inferior ganglia of the nerve, which appear to be predominantly sensory ganglia, despite contrary views in respect of the inferior ganglion (p. 1076). Their central processes terminate in the dorsal nucleus of the vagus in the medulla oblongata, or, according to some authorities, in the nucleus of the tractus solitarius. Their peripheral processes have a wide distribution. Their terminals in the walls of the pharynx and oesophagus, together with the terminals of the general visceral afferent fibres in the glossopharyngeal nerve in the pharynx, are concerned with swallowing reflexes. Afferent fibres are also ascribed to the thyroid and parathyroid glands. In the thorax the general visceral afferent fibres are widely distributed. Some terminate in the heart, walls of the great vessels and aortic bodies; some end in pressor receptors in the walls of the great vessels and are stimulated by distension of the vessel walls resulting from raised intravascular pressure. Fibres from the vagus reach the lungs through the pulmonary plexuses and are distributed to (a) the bronchial muscosa; these are probably involved in cough reflexes; (b) the bronchial muscle where they encircle the muscle cells and end in tendril-like arrangements which have been regarded as 'muscle spindles' which are believed to be stimulated by alteration in the length of the muscle cells; (c) the interalveolar connective tissue where they end in knob-like swellings; these, together with the endings on the smooth muscle cells, may initiate the Herring-Breuer reflex; (d) the adventitia of the pulmonary arteries where they may be pressor receptors and the intima of the pulmonary veins where they may function as chemoreceptors. Afferent fibres from the visceral pleura and air passages are also believed to travel with the sympathetic supply to the lungs and to mediate nociceptive responses.

General visceral afferent fibres in the vagus also terminate in the various coats of the stomach and intestines, the digestive glands and in the kidney. The fibres which terminate in the gut and ducts leading to it are said to be stimulated by distension (stretch) or muscle contraction. Impulses from the stomach may be responsible for the sensations of hunger and nausea.

The general visceral afferent element of the glossopharyngeal nerve includes fibres which innervate the posterior region of the tongue, the tonsil and pharynx; the epithelia of all these parts are derived from endoderm. The innervation of the taste buds in these regions is not included, for these are more properly regarded as *special* visceral afferents, and are treated elsewhere (p. 1140). In addition the glossopharyngeal nerve innervates the carotid sinus and carotid body, the receptors there in being sensitive to tension in the vessel wall and to changes in the chemical composition of the blood respectively. Impulses from these receptors play an essential part in circulatory and respiratory reflexes. The cells of origin of the general visceral afferent fibres lie in the ganglia on the glossopharyngeal nerve, and their terminations in the medulla oblongata are probably similar to those of the general visceral afferent fibres in the vagus nerve.

Sensory fibres coursing in the pelvic splanchnic nerves innervate the pelvic viscera and the distal part of the colon. In the urinary bladder, sensory receptors are described at all levels in its wall. Those in muscle strata are connected with heavily myelinated fibres which reach the spinal cord through the pelvic splanchnic nerves; they are believed to be stretch receptors but may be activated by contraction of the muscle. Pain fibres from the bladder and proximal part of the urethra pass in both the pelvic splanchnic nerves and through the inferior hypogastric plexus, the hypogastric nerves, superior hypogastric plexus and lumbar splanchnic nerves to reach their cells of origin in the ganglia on the dorsal roots of the lower thoracic and upper lumbar nerves. The significance of this double sensory pathway is uncertain; lesions of the cauda equina abolish pain resulting from over-distension of the bladder, but section of the hypogastric nerve is ineffective in relieving pain from the organ.

Though fibres from the pelvic splanchnic nerves, some of which are believed to be afferent, are described in the ovary, no supply from this source to the testis has been demonstrated.

Pain fibres from the body of the uterus pass with sympathetic nerves through the hypogastric plexus and the lumbar splanchnic nerves to cells on the dorsal roots of the lowest thoracic and upper lumbar nerves. Thus, surgical section of the hypogastric nerves has been employed for the relief of dysmenorrhoea. Afferent fibres from the cervix, however, pass in the pelvic splanchnic nerves to their cells of origin in the dorsal roots of the upper sacral nerves. Dilatation (stretch) of the cervix uteri causes pain, but cauterization and removal of small portions of mucosa for biopsy do not.

In general the afferent fibres which accompany the pre- and post-ganglionic fibres of the sympathetic system have a segmental arrangement. They end in the same spinal cord segments as the preganglionic fibres of the efferent pathway to the region or viscus (*vide infra*). The general visceral afferent fibres entering the thoracic and upper lumbar spinal segments are, in the main, concerned with the conduction of pain impulses. Painful sensations from the pharynx, oesophagus, stomach, intestines, kidney, ureter, gall bladder and bile ducts seem to be carried along sympathetic pathways. Such impulses from the heart enter the spinal cord through the first to fifth thoracic spinal nerves. They are carried mainly in the middle and inferior cardiac nerves; a few pass directly into the second to fifth thoracic nerves. There are no general visceral afferent fibres in the superior cardiac nerves. Peripherally the fibres pass through the cardiac plexuses and along the coronary arteries. Anoxia of the heart muscle may give rise to the characteristic symptoms of angina pectoris in which there is presternal pain, referred pain over a large part of the left side of the chest, radiating to the left shoulder and inner side of the left arm, upwards along the left side of the neck to the jaw and occiput and downwards to the epigastrium. Occasionally the pain is felt on both sides or confined to the right side. Afferent fibres from the heart are also carried in the cardiac branches of the vagus nerve. These nerves are concerned with cardiac reflex depressing the activity of the heart. In some animals (e.g. rabbit) a separate depressor cardiac nerve is present as a branch of the vagus or of the superior laryngeal nerve. In man, the depressor fibres do not form a separate nerve but run in branches of the superior or internal laryngeal nerves which join in a variable manner, cardiac branches of the vagus or sympathetic.

The pain fibres from the ureter, which also accompany sympathetic fibres, are probably concerned in the painful reflex of renal colic when this duct is obstructed by a calculus.

The afferent (pain) fibres from the testis and ovary run through the corresponding plexuses; their cells of origin are in the ganglia in the dorsal roots of the tenth and eleventh thoracic nerves.

It must be realized that reflex activities in the autonomic nervous system are not initiated solely by impulses conducted through general visceral afferent pathways, nor do impulses travelling in these pathways necessarily activate the general visceral efferent pathways. Indeed, in most situations calling for general sympathetic activity in preparation for effort, the afferent element is almost always somatic, involving either the special senses or skin sensibility. Rises in blood pressure and dilatation of the pupil may result from stimulation of somatic afferent nerves, the receptors of which are located in the skin and other tissues. Conversely, contraction of the rectus abdominis, a somatic structure may result from irritation of abdominal viscera. There is also evidence that reflexes (axon reflexes) may be evoked at the terminals of autonomic postganglionic fibres.

Denervation often has no appreciable effect on the effector organs, nonstriated muscle or glands, innervated by the autonomic system. The contraction of such muscle may be uninfluenced by denervation and no structural changes ensue. This has been variously attributed to the continued activity of ground plexuses or to the intrinsic activity of the visceral muscle cells. In some instances severance of the preganglionic efferent fibres results in hypersensitivity of the postganglionic neurons. In other instances denervation results in cessation of activity, as in the sweat glands, pilomotor muscle, nonstriated muscle of the orbit and in the adrenal medulla (Pick 1970).

DEGENERATION AND REGENERATION IN THE AUTONOMIC NERVOUS SYSTEM

Though these processes have still to be systematically studied, degeneration in the autonomic nervous system is believed to be similar to that in the cerebrospinal nervous system. There is some evidence that the rate of degeneration differs in different regions or with different types of nerve fibre. Regeneration of preganglionic fibres may be influenced by the site of the lesion and, in the case of postganglionic neurons, regeneration may be followed by re-innervation from neighbouring intact nerve fibres.

As far as available experimental evidence goes, the integrity of Schwann cell sheaths is as important a factor in the regeneration of autonomic nerve fibres, whether myelinated or not, as in the case of somatic fibres (Evans and Murray 1954; Kapeller and Mayor 1967; Williams 1971; King and Thomas 1971; Landon 1976). In some studies observations have suggested that the presence of myelinated fibres in proximity to regenerating non-myelinated fibres is necessary to the latter process (Evans and Murray 1954; Williams 1971). With these as yet incomplete studies it is pertinent to mention earlier experiments in which relatively large defects in the sympathetic trunk, in monkeys and other mammals, have apparently been made good by growth of fibres (which might be pre- or post- ganglionic) across a complete gap in the trunk (Tower and Richter 1931; Haxton 1954). Conflicting evidence has been derived from human sympathectomies. The functional recoveries sometimes observed may be explained in other ways, such as incomplete interruption of a sympathetic supply, alternative routes of fibres being overlooked. On the whole the experience of surgeons corroborates the experimental findings stated above (Pick 1970).

SEGMENTAL SYMPATHETIC SUPPLIES

Head and neck	T. 1–5
Upper limb	T. 2–5
Lower limb	T. 10–L. 2
Heart	T. 1–5
Bronchi and lung	T. 2–4
Oesophagus (caudal part)	T. 5–6
Stomach	T. 6–10
Small intestine	T. 9–10
Large intestine as far as splenic flexure	T. 11–L. 1
Splenic flexure to rectum	L. 1–2
Liver and gall bladder	T. 7–9
Spleen	T. 6–10
Pancreas	T. 6–10
Kidney	T. 10–L. 1
Ureter	T. 11–L. 2
Suprarenal	T. 8–L. 1
Testis and ovary	T. 10–11
Epididymis, ductus deferens and seminal vesicles	T. 11–12
Urinary bladder	T. 11–L. 2
Prostate and prostatic urethra	T. 11–L. 1
Uterus	T. 12–L. 1
Uterine tube	T. 10–L. 1

Applied Anatomy. Various parts of the sympathetic nervous system are removed surgically in the treatment of a number of clinical conditions. In operations directed towards the efferent side of the sympathetic, ganglia on the sympathetic trunk are removed, or preganglionic fibres cut, rather than postganglionic fibres severed, since the latter procedure may be followed by regeneration of the nerves. For example, the arteries of the limbs may be denervated in conditions of vascular spasm (Raynaud's disease) or in organic arterial disease where spasm is also present; the parts of the system removed are described above (pp. 1129, 1131). In the treatment of essential hypertension, much more extensive sympathectomy has been performed, involving bilateral removal of the sympathetic trunks (from the eighth thoracic to the first lumbar ganglia) and the greater and lesser thoracic splanchnic nerves.

Sympathectomy is also performed for the relief of pain, for example in cases of severe angina pectoris (*see* p. 1137). Division of the superior hypogastric plexus (presacral neurectomy) does not completely relieve pain associated with disease of the pelvic organs, since, as noted above, many of the pain fibres pass in the pelvic splanchnic nerves. The pain fibres from the body of the uterus, however, pass in the sympathetic nerves via the superior hypogastric plexus, so that this operation is successful in cases of intractable painful menstruation (dysmenorrhoea).

In the male, resection of the superior hypogastric plexus is followed by loss of the power of ejaculation and consequent sterility, owing to the interruption of the sympathetic pathway to the seminal vesicles, deferent ducts and prostate. Knowledge of the pathways pursued by these nerves between the ganglia on the sympathetic trunk and the superior hypogastric plexus is less exact and the pathways may vary in different cases, but in certain individuals the outflow from the first lumbar, and possibly the twelfth thoracic, ganglion is of major importance, while in others the fibres from the third lumbar ganglion are chiefly concerned (White *et al.* 1952). For an extensive text on surgical anatomy of the autonomic nervous system consult Pick (1970).

THE PERIPHERAL APPARATUS OF THE SPECIAL SENSES

In the present section the detailed anatomy of the taste buds, the olfactory epithelium and nasal cavity, the eye and the ear are considered. It should also be noted that the development of these structures is treated in the section devoted to Embryology, and their neurons and nervous connexions are further detailed in the parts of this volume concerned with the Central Nervous System and with the Cranial Nerves.

The Gustatory Apparatus

The peripheral gustatory organs are the *taste buds* (*gustatory caliculi*), which are composed of modified epithelial cells arranged in piriform groups (7.323–235) in the epithelium covering the tongue, the inferior surface of the soft palate, the palatoglossal arches, the posterior surface of the epiglottis and the posterior wall of the oral part of the pharynx. They are most numerous on the sides of the vallate papillae of the tongue (7.233), less so on the walls surrounding these papillae; they are plentiful over the folia linguae and the posterior third of the tongue, but are distributed sparingly on the fungiform papillae of the tongue, the soft palate, epiglottis and pharynx. They are more numerous in the infant than the adult and their atrophy increases with age; those in the extreme posterior part of the tongue and in the epiglottis disappear early in life. There are no taste buds in the mid-dorsal region of the oral part of the tongue.

MICROSCOPIC STRUCTURE OF TASTE BUDS

Each taste bud (7.234) is separated by a basement membrane from the underlying dermis, and opens on to the surface of the epithelium by an aperture termed the *gustatory pore*. In longitudinal section the cells of the taste bud are crescentic in profile, their pointed apices converging on a small cavity which connects with the gustatory pore (7.234, 235). Some of these cells bear 'gustatory hairs' which, ultrastructurally, are seen to be groups of fine microvilli. The base of each taste bud is penetrated by a group of afferent gustatory nerve fibres which give off branches to ramify and spiral around some of the epithelial cells within the taste bud.

It has long been known that the gustatory sensory cells are modified epithelial elements which establish contacts like synapses with the terminal branches of the afferent gustatory fibres. The identification of the gustatory cells amongst the other types of cell in the taste bud has, however, been the subject of some debate. Investigations involving three-dimensional reconstruction of taste buds from electron micrographs of serial sections (Murray and Murray 1970) suggest that there are at least five distinct epithelial cell types, two of which appear to be receptor cells and others supportive, ensheathing or generative in their functions (7.235). The receptor cells are typified by the presence of synaptic contacts with afferent nerve terminals at various points on their surfaces; these show the typical asymmetrical thickening and presynaptic aggregates of small (50 nm) clear synaptic vesicles within the receptor cell, although these appear to be less prominent than in other comparable synapses (p. 828). Also, one of these types of epithelial cell contains dense-cored vesicles about 70 nm in diameter within its cytoplasm, similar to those which have been identified in other sites of the nervous system as catecholamine-containing vesicles (p. 830), whereas in the other type these are absent. It is possible that the two types of cell represent different stages in the maturation of receptors, which are continually being lost and replaced (*vide infra*). It is interesting that at the apices of the receptor cells few microvilli are usually present, and the 'gustatory hairs' of light microscopists are mainly to be found on the supporting cells which surround and enclose the receptors except at their apical surfaces (Miller and Chaudry 1976). One of these non-sensory cell types appears to form a peripheral sheath for the whole taste bud, separating it from the surrounding epithelium; another type is basal in position where it probably forms a blastemal cell capable of giving rise to new cells of the other types by mitosis. The third type is a true supporting cell which insulates each receptor from its neighbours and also provides sheaths at its basal extremities for the afferent nerve fibres which lose their Schwann sheaths on entering the taste bud. Ultrastructurally the true supporting cells contain, together with the usual complement of cell organelles, dense bodies of a secretory nature. These are probably the origin of the material rich in polysaccharides which occupies the apical cavity of the taste bud and into which the sensory terminals project. Gustatory molecules must pass through this material before the sensory surface can be reached, and it may play an important, although undetermined, role in the gustatory process. Occluding junctions are present between the apices of the various cells of the taste bud, preventing ready access of chemical stimuli to non-apical parts (Akisaka and Oda 1978).

The nerve fibres which reach the taste buds from the dermal plexus are complex in their distribution within the tongue, as deduced by electrophysiological recordings from individual nerve fibres within the more proximal fasciculi of the main gustatory nerves (chorda tympani and glossopharyngeal). Each fibre may possess many terminal branches which spread to innervate widely distant taste buds, and, within these structures, to end in relation to more than one sensory cell. Conversely, each taste bud and each of its contained receptor cells, may be innervated by the terminal branches of several different nerve fibres. This cross innervation of taste buds may be of great physiological importance (*vide infra*, and Beidler 1970). No evidence has yet emerged concerning the presence of efferent terminals or of inhibitory phenomena in taste buds.

THE REPLACEMENT OF TASTE BUD CELLS

The cells of the taste bud resemble the cells of the surrounding epithelium in undergoing continual renewal (Beidler and Smallman 1965). Isotopic labelling shows that none of the elements of the taste bud live for more than a few days, except for the basal cells, from which the other types of cell are presumably replaced by mitosis. Newly formed sensory cells must therefore make new synaptic contacts with nerve fibre terminals, and experiments indicate that the precise sensitivity characteristics of the sensory cells are determined by the trophic influence of the nerve fibre and not the reverse (*see* p. 862).

The continual degeneration and replacement of cells in taste buds means that, within a single bud, cells of different types show a spectrum of morphological appearances which adds to the difficulties of precise identification of cells on structural criteria alone.

THE PHYSIOLOGY OF GUSTATORY RECEPTORS

In many aquatic vertebrates chemoreceptors with all the characteristics of gustatory endings are widely distributed over the surface of the body; in terrestrial groups this sense is confined to the buccal cavity and adjoining regions of the pharynx. The sense of taste in man, as commonly understood, is largely a result of the activity of lingual receptors, although others which are palatal and laryngopharyngeal in position may also play a significant part. Classically, four groups of subjective taste qualities have been distinguished, each pertaining to zones of special sensitivity on the surface of the tongue, comprising sweet and salt at the tip of the tongue, sour (acid) at the sides, and bitter at the pharyngeal end. Although these qualities represent subjective sensations, it seemed at one time possible that their special effectiveness in particular areas of the tongue might reflect the presence of receptor cells specific for these general groups. However, the electrophysiological evidence suggests that this is

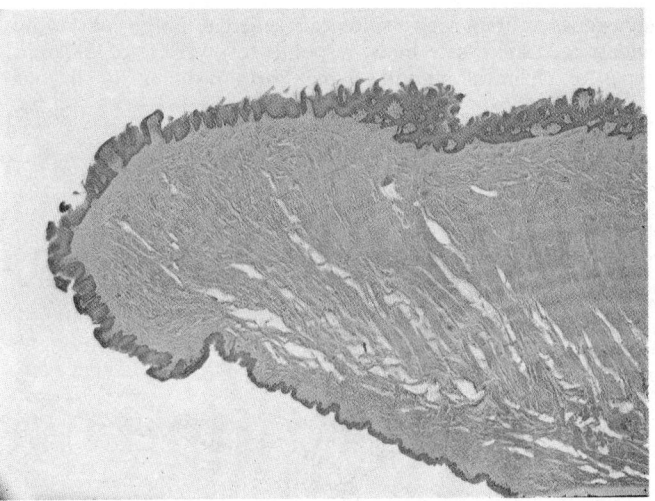

7.232 A low-power light micrograph of a sagittal section through the tip and anterior part of a human tongue, showing: muscle fibres orientated in three different directions; a delicate non-keratinized stratified squamous epithelium on the ventral surface; and a partly keratinized epithelium on the dorsum. The latter is convoluted to produce filiform and fungiform papillae. Dermal papillae project into the deep surface of the epidermal irregularities. Haematoxylin and eosin. Magnication × 10

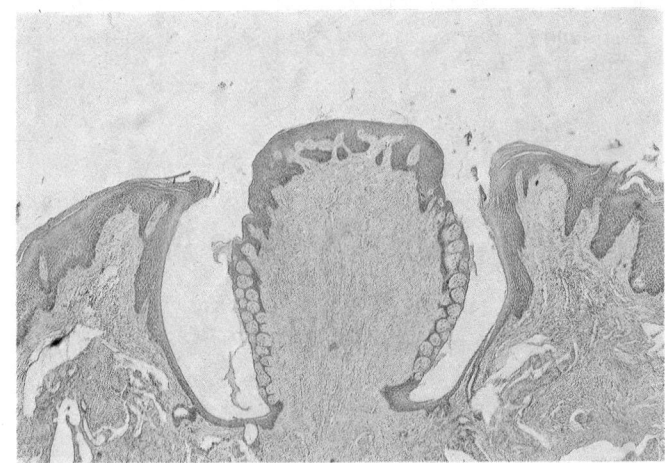

7.233 A light micrograph of a human vallate lingual papilla showing the numerous taste buds clustered along its lateral surfaces. A connective tissue papilla forms the core of the whole structure, and serous glands, situated between the muscle blocks deep to the corium, open into the lateral recesses of the papilla. Haematoxylin and eosin. Magnification × 150

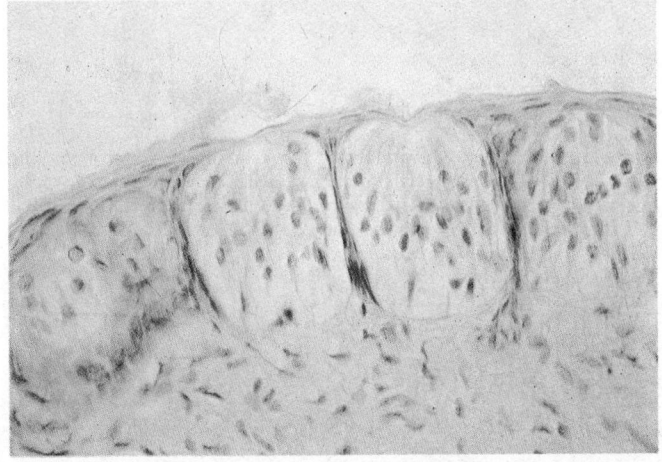

7.234 A light micrograph of a group of taste buds in a vallate lingual papilla, showing the apical cavity and various fusiform cells within the epithelial capsule. Haematoxylin and eosin. Magnification × 400

an over-simplified view, since each afferent fibre, subserving widely separated taste buds, responds to a variety of different types of chemical stimuli. Some fibres respond to all four

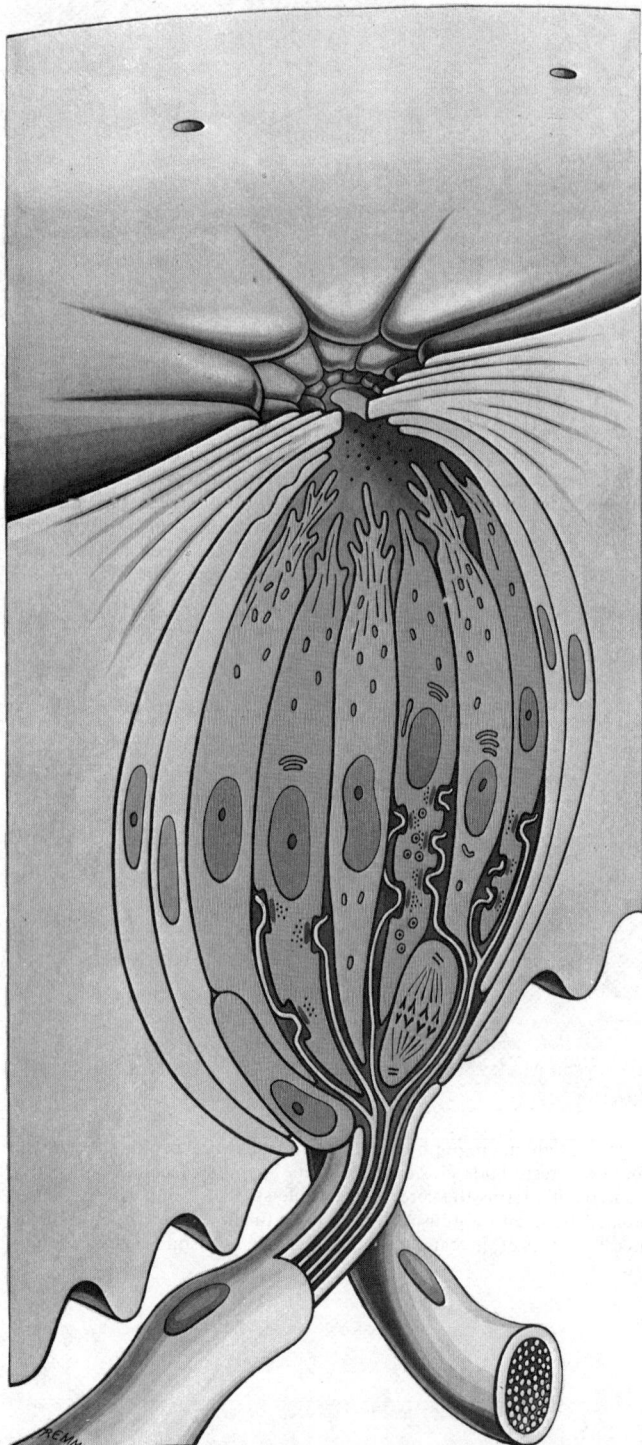

7.235 A schematic reconstruction of the structure of a taste bud, cut away to expose the various cell types. Presumed sensory cells of two types, one with dense-cored vesicles, the other without, are indicated in purple, and their innervation in yellow. The supporting cells are indicated in magenta, and basal cells in brick red. A dense mucosubstance is present in the apical cavity beneath the apical aperture.

qualities, others to less or only to one. Within a particular class of taste qualities receptors are also differentially sensitive to a wide range of similar chemicals. It would therefore seem likely that the sensation of taste represents the results of the analysis of a complex pattern of differential responses over particular areas of the tongue, so that even if there are only a relatively few specific types of taste receptor cells, much more information may be gained about the precise nature of the taste, and its intensity, than if the innervation were of a simpler kind (Beidler 1970; Beidler and Smallman 1965; Pfaffman 1970).

THE GUSTATORY NERVES

The nerve fibres of taste are the peripheral processes of unipolar nerve cells in the genicular ganglion of the facial nerve, the inferior ganglion of the glossopharyngeal nerve and the inferior ganglion of the vagus. The central processes of these cells form the tractus solitarius (p. 903), and their terminals synapse with the neurons which form the nucleus of that tract. The axons of the latter cross the midline, and many then ascend through the brainstem in the dorsomedial part of the medial lemniscus to approach and synapse with the most medially situated neurons of the *nucleus ventralis posterior medialis* (VPM) of the thalamus (in a region sometimes termed the *accessory arcuate nucleus*). From the VPM axons radiate through the internal capsule to reach the antero-inferior part of the sensorimotor cortex and the region of the limen insulae. Broadly, electrophysiological recording from the VPM and the cerebral cortex confirms these anatomical pathways; some units react to one type of stimulus only, whilst others react to a number of different stimuli. Other ascending pathways have been described as ending in a number of the hypothalamic nuclei, through which gustatory information may reach the limbic system (p. 968), and which allow appropriate readjustments of the autonomic nervous system to be made. (*See* also p. 1085.)

The nerve of taste for the anterior part of the tongue, excluding the vallate papillae, is the chorda tympani (through the lingual nerve); these taste fibres in most individuals pass in the chorda tympani to the ganglion of the facial nerve. In a few individuals, they leave the chorda tympani by anastomotic branches connecting it to the otic ganglion and proceed thence in the greater petrosal nerve to the ganglion of the facial nerve (Schwartz and Weddell 1938). This part of the tongue, its *oral* region, is derived from the mandibular branchial arch, and is limited posteriorly by the *sulcus terminalis*. The taste buds in the inferior surface of the soft palate are also supplied mainly by the facial nerve, through the greater petrosal nerve, the nerve of the pterygoid canal and the middle and posterior palatine nerves; the glossopharyngeal also contributes to their supply. The taste buds in the vallate papillae and the pharyngeal part of the tongue, and in the palatoglossal arches and the oral part of the pharynx are innervated by the glossopharyngeal, while those in the extreme back part of the tongue and in the epiglottis are supplied by the internal laryngeal part of the superior laryngeal branch of the vagus. (For discussion of the literature concerning gustatory pathways consult Rollin 1979.) These nerve supplies accord with the embryological development of the tongue (p. 197). It is customary to describe the mandibular and glossopharyngeal territories as the 'anterior two-thirds' and 'posterior third' of the tongue. But since there is an easily recognizable boundary between them—the sulcus terminalis—it is uninformative and misleading to divide so variable an organ in this arbitrary and inevitably inexact manner. *Pre-* and *post-sulcal* regions are better terms.

THE OLFACTORY APPARATUS

The peripheral olfactory organs are said to include the *external nose*, and the *nasal cavity*, which are both divided by a septum into right and left parts. However, the essential olfactory structures are the olfactory epithelium and its nervous connexions.

THE EXTERNAL NOSE

The external nose is pyramidal, and its upper angle, or *root*, is continuous with the forehead; its free angle or tip is termed the *apex*. Its inferior aspect carries two roughly elliptical apertures, the external *nares* or *nostrils*, which are separated from each other by the septum. The lateral surfaces of the nose form, by their union in the median plane, the *dorsum nasi*, the shape and direction of which varies considerably in different individuals; the upper part of the external nose is kept patent by the nasal bones and the frontal processes of the maxillae. The lateral surfaces end below in the rounded *alae nasi*.

The framework of the external nose (7.236) is composed of bones and hyaline cartilages. The *bony framework*, which supports its upper part, consists of the nasal bones, the frontal processes of the maxillae, and the nasal part of the frontal bone. The *cartilaginous framework* consists of the septal, lateral, and major and minor alar nasal cartilages (7.236, 237). These are connected with one another and with the bones by the continuity of the perichondrium and the periosteum. These junctions are of considerable interest, although they have attracted little attention. The cartilages and their intervening connective tissue regions, which surround the nares, have been regarded as a form of valve, controlling the intake of air (Cottle 1960; Galindo *et al.* 1977).

The *septal cartilage* (7.236B, 237), somewhat quadrilateral in form, and thicker at its margins than at its centre, forms almost the whole of the septum between the anterior parts of the nasal cavity. The upper part of its anterosuperior margin is connected to the posterior border of the internasal suture; the middle part is continuous with the lateral cartilages; the lower part is attached to these cartilages by the perichondrium. Its antero-inferior border is connected on each side to the septal process of the major alar cartilage. (Because of this fusion Galindo *et al.* 1977 have suggested that the two lateral and the septal cartilages should be regarded as a unit—the *septodorsal cartilage*. This represents resuscitation of an older view, the term having been used in the Jena Nomina Antomica.) Its posterosuperior border is joined to the perpendicular plate of the ethmoid bone, and its postero-inferior border is attached to the vomer and to the nasal crest of the maxillae and the anterior nasal spine. The cartilage of the septum may extend backwards (especially in children) as a narrow process, termed the *sphenoidal process*, for some distance between the vomer and the perpendicular plate of the ethmoid bone. The antero-inferior part of the nasal septum between the two nostrils is freely movable, and hence is named the *septum mobile nasi*; it is not formed by the cartilage of the septum, but by the septal processes of the major alar cartilages and by the skin.

The *lateral* (upper) *nasal cartilage* (7.236A) is triangular in shape. Its anterior margin is thicker than the posterior, and the upper part is continuous with the cartilage of the septum, but the lower part is separated from this cartilage by a narrow fissure; its superior margin is attached to the nasal bone and the frontal process of the maxilla; its inferior margin is connected by fibrous tissue with the major alar cartilage.

The *major alar* (lower) *cartilage* (7.236A and B) is a thin flexible plate which is situated below the lateral nasal cartilage, and is bent acutely around the anterior part of the naris. The medial part of the plate is narrow, and is termed the *medial crus* (septal process). The latter is loosely connected by fibrous tissue with that of the opposite cartilage, and to the antero-inferior part of the septal cartilage, thus helping to form the septum mobile nasi. The *lateral crus*, which is lateral to the narial orifice, is little more than the inferior free border of the main part of the major alar cartilage. The upper border of the lateral part of the major alar cartilage is attached by fibrous tissue to the lower border of the lateral nasal

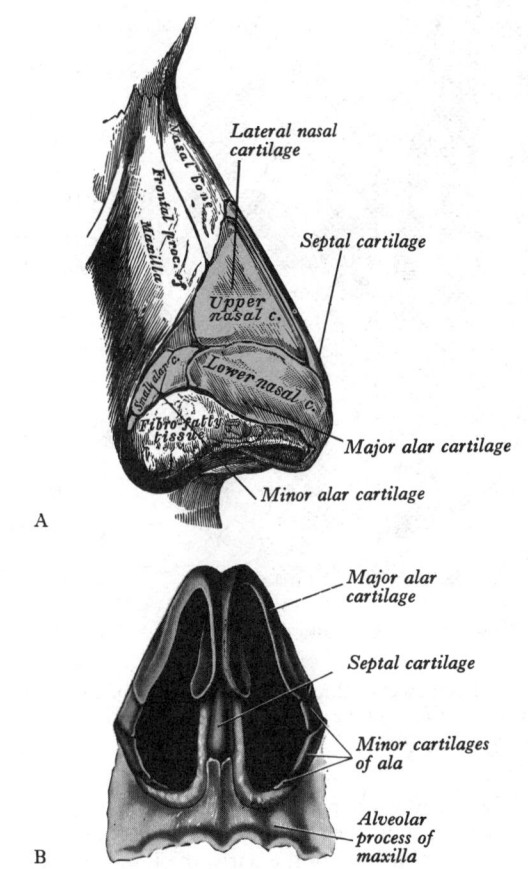

7.236 A   The cartilages of the right side of the nose. Lateral aspect.
B   The cartilages of the nose. Inferior aspect. (Note changes in terminology.)

cartilage. Its posterior, narrow end is connected with the frontal process of the maxilla by a tough fibrous membrane, in which three or four small cartilaginous plates, termed the *minor cartilages of the ala* (7.236A), are found. According to Drumheller (1969) the junction between the adjoining borders of the major alar and lateral cartilages is variable; the two edges may abut each

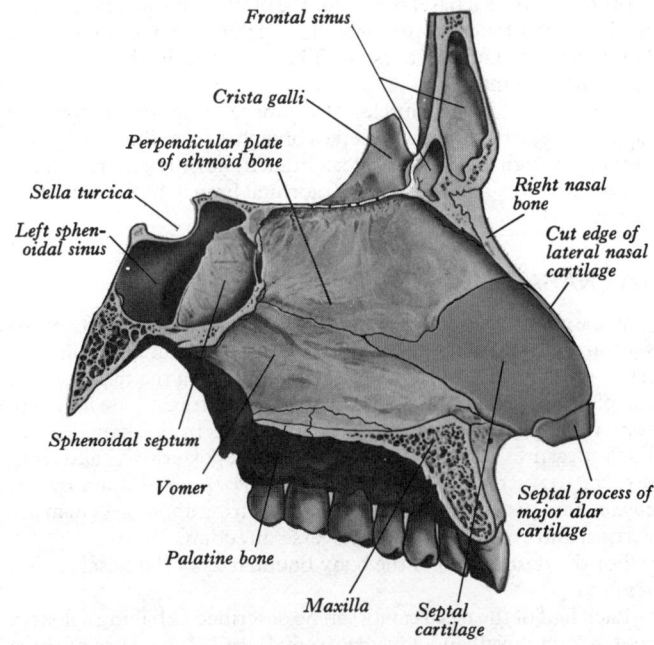

7.237   The right side of the septum of the nose, showing its constituent bones and cartilages. (Note changes in terminology.)

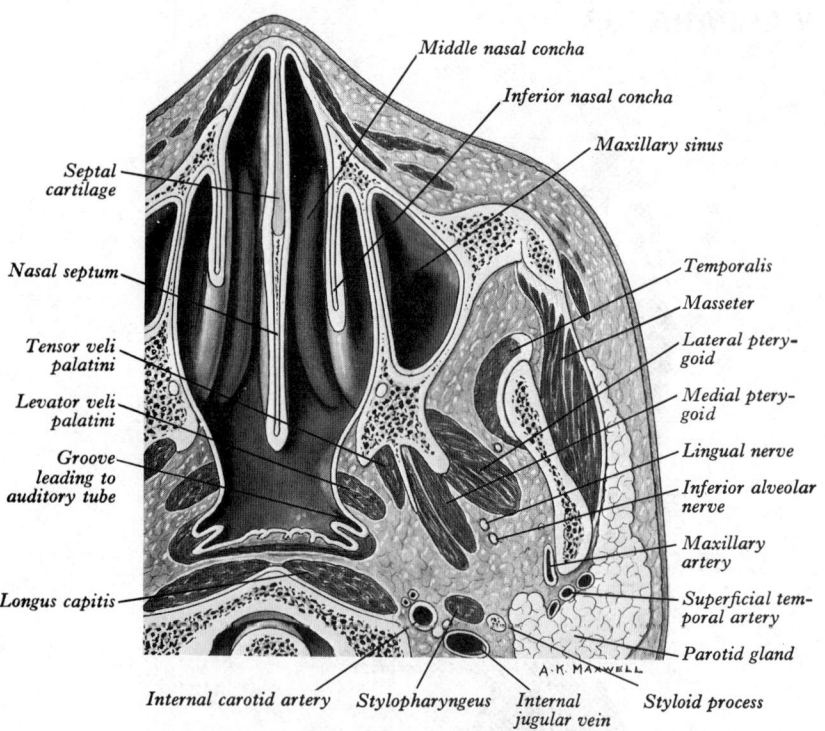

Middle nasal concha

Inferior nasal concha

Maxillary sinus

Septal cartilage

Nasal septum

Tensor veli palatini

Levator veli palatini

Groove leading to auditory tube

Longus capitis

Temporalis

Masseter

Lateral ptery- goid

Medial ptery- goid

Lingual nerve

Inferior alveolar nerve

Maxillary artery

Superficial tem- poral artery

Parotid gland

Internal carotid artery *Stylopharyngeus* *Internal jugular vein* *Styloid process*

A·K·MAXWELL

7.238   A transverse section through the anterior part of the head at a level just inferior to the apex of the odontoid process. Superior aspect.

other or they may overlap, the lateral cartilage being then the more lateral at the junction. Its lower free edge (lateral crus) falls short of the lateral margin of the naris, the lower part of the ala nasi being formed by fatty and fibrous tissue covered with skin. In front, the major alar cartilages are separated by a notch which can be felt at the apex of the nose.

The *muscles* acting on the external nose have been described on p. 531.

The *skin* of the dorsum and sides of the nose is thin, and loosely connected with the subjacent parts; but over the apex and alae it is thicker and more firmly adherent, and is furnished with numerous large sebaceous glands, the orifices of which are usually very distinct.

The *arteries* of the external nose are the alar and septal branches of the facial artery, which supply the ala and lower part of the septum, the dorsal nasal branch of the ophthalmic artery, and the infraorbital branch of the maxillary artery, which supply the lateral aspects and the dorsum. The *veins* end in the facial and ophthalmic veins.

The *nerves* for the muscles of the nose are derived from the facial nerve, while the skin receives branches from the ophthalmic nerve, through its infratrochlear branch and the external nasal nerve (p. 1061), and from the infraorbital branch of the maxillary nerve.

THE NASAL CAVITY

The nasal cavity is divided into right and left halves by the nasal septum (7.238). These two halves open on the face through the nares or nostrils, and communicate behind with the nasal part of the pharynx through the posterior nasal apertures. The *nares* are somewhat piriform apertures, narrower in front than behind. Each measures from 1·5 cm to 2 cm anteroposteriorly, and from 0·5 cm to 1 cm transversely. The *posterior nasal apertures* or *choanae* are two oval openings, each measuring about 2·5 cm in the vertical and 1·25 cm in the transverse direction.

For the description of the bony boundaries of the nasal cavity, *see* p. 313.

Each half of the nasal cavity can be described as having a floor, a roof, a lateral wall and a medial (septal) wall. It consists of three regions—vestibular, olfactory, and respiratory.

The *nasal vestibule* is a slight dilatation just inside the aperture

of the nostril (7.239), bounded laterally by the ala and the lateral part of the lower nasal cartilage, and medially by the septal process of the same cartilage; it extends as a small recess towards the apex of the nose. The vestibule is lined with skin, and coarse hairs and sebaceous and sweat glands are found in its lower part; the hairs (vibrissae) curve downwards and forwards to the naris, and tend to arrest the passage of foreign substances carried with the current of inspired air. In the male, after middle age, they increase considerably in size. The vestibule is limited above and behind by a curved elevation, named the *limen nasi*, which corresponds to the upper margin of the lower nasal cartilage, and along which the skin of the vestibule is continuous with the mucous membrane of the nasal cavity.

The *olfactory region* is limited to the superior nasal concha, the opposed part of the septum and the intervening roof.

The *respiratory region* comprises the rest of the cavity.

The lateral wall of the nasal cavity (7.239, 240) is marked by three elevations, the *superior*, *middle* and *inferior nasal conchae*, and below and lateral to each concha is the corresponding nasal passage or *meatus*. Above the superior concha a triangular fossa, the *spheno-ethmoidal recess*, includes the opening of the sphenoidal sinus. Sometimes a fourth or *highest nasal concha* occurs on the lateral wall of the spheno-ethmoidal recess (7.239); the highest or *supreme nasal meatus* related to it may contain the opening of a posterior ethmoidal sinus. The *superior meatus* is a short oblique passage extending about half-way along the upper border of the middle concha; the posterior ethmoidal sinuses open, usually by one aperture, into the front part of this meatus. The *middle meatus*, deeper in front than behind, is below and lateral to the middle concha, and is continued anteriorly into a shallow depression situated above the vestibule and named the *atrium* of the middle meatus. Above the atrium an ill-defined curved ridge, termed the *agger nasi* (p. 340), runs forwards and downwards from the upper end of the anterior free border of the middle concha; it is better developed in the newborn child than in the adult. When the middle concha is raised or removed the lateral wall of this meatus is displayed fully. A rounded elevation, termed the bulla ethmoidalis and, below, and extending upwards in front of it, a curved cleft, termed the hiatus semilunaris, form the principal features of this wall. The *bulla ethmoidalis* is caused by the bulging of the middle ethmoidal sinuses, which open on or immediately above it, and the size of the bulla varies with that of its contained sinuses. The *hiatus semilunaris*, which is bounded inferiorly by a sharp concave ridge produced by the uncinate process of the ethmoid bone, leads forwards and upwards into a curved channel, which is named the *ethmoidal infundibulum*. The anterior ethmoidal sinuses open into the infundibulum, which in rather more than 50 per cent of subjects is continuous with the frontonasal duct or passage leading from the frontal sinus. In other cases the ethmoidal infundibulum ends blindly in front by forming one or more of the anterior ethmoidal sinuses (*infundibular sinuses*), and the frontonasal duct opens directly into the anterior end of the middle meatus. The opening of the maxillary sinus is situated below the bulla ethmoidalis, and is usually hidden by the flange-like lower edge of the hiatus semilunaris; in a coronal section of the nose this opening is seen to be placed near the roof of the sinus (7.244). An accessory opening of the maxillary sinus is frequently present below and behind the hiatus semilunaris. The *inferior meatus* is below and lateral to the inferior nasal concha; the nasolacrimal duct opens into this meatus under cover of the anterior part of the inferior concha.

The medial wall or nasal septum (7.237) is often deflected from the median plane, thus lessening the size of one half of the nasal cavity and increasing that of the other; ridges or spurs of bone sometimes project from the septum on either side. Immediately superior to the incisive canal at the lower edge of the cartilage of the septum a depression is sometimes seen; it points downwards and forwards, and occupies the position of a canal which connected the nasal with the buccal cavity in early fetal life. On each side of the septum close to this recess a minute orifice may be discerned; it leads backwards into a blind tubular pouch, 2 to 6 mm long, the vestigial *vomeronasal organ*, which is supported by a strip of cartilage named the *vomeronasal cartilage*; it is lined by epithelium consisting mainly of a single layer of tall columnar cells and contains many glands. This organ is particularly prominent in amphibia and reptiles, and moderately well developed in most mammals, but is vestigial in adult primates including mankind (Graziadei 1974). When present it apparently plays a part in the sense of smell, since it is supplied by twigs of the olfactory nerve and is lined with epithelium similar to that in the olfactory region of the nose. (See also p. 994.)

The roof of the nasal cavity is narrow transversely, except at its posterior part, and may be divided, from behind forwards, into sphenoidal, ethmoidal and frontonasal parts, corresponding to the bones which enter into its formation (pp. 141 and 313). The ethmoidal part is almost horizontal, but the frontonasal and sphenoidal parts slope downwards and forwards, and downwards and backwards, respectively. The cavity is therefore deepest where its roof is formed by the cribriform plate of the ethmoid bone.

The floor is transversely concave, anteroposteriorly flat and almost horizontal; its anterior three-fourths are formed by the palatine process of the maxilla, its posterior one-fourth by the horizontal part of the palatine bone. About 2 cm behind the anterior end of the floor a slight depression in the mucous membrane overlies the incisive canals (p. 341).

The nasal conchae add greatly to the surface area of the nasal cavity, especially in macrosmatic mammals such as the Carnivora (Negus 1958). This may not only serve to augment the olfactory area but also to increase turbulence and perhaps improve olfaction by somewhat delaying passage of air through the olfactory part of the cavity. Humidification and warming of the inhaled air is also favoured by the increased area of mucous membrane and by turbulence, even in a microsmatic mammal such as man (Cole 1954). Swirling currents also aid in the trapping of particulate material by the mucous secretion.

The nasal mucous membrane lines the nasal cavities with the exception of the vestibules, and is intimately adherent to the periosteum or perichondrium. It is continuous with the mucous membrane of the nasal part of the pharynx through the posterior nasal apertures, with the conjunctiva, through the nasolacrimal duct and lacrimal canaliculi, and with the mucous membranes of the sphenoidal, ethmoidal, frontal and maxillary sinuses, through their openings.

The mucous membrane is thickest and most vascular over the nasal conchae, especially at their extremities. It is also thick over the nasal septum, but very thin in the meatuses, on the floor of the nasal cavity, and in the various sinuses. The thickness of the membrane reduces materially the size of the bony cavity and the apertures communicating with it. The epithelium of the mucous membrane differs in its characteristics according to the functions of the part of the nose in which it is found. In the *olfactory region*, which extends over the upper 10 mm or so of the septum and over the superior concha and the lateral walls above it, the mucous membrane is yellowish in colour and the epithelium is of a distinctive type.

STRUCTURE OF THE OLFACTORY EPITHELIUM

The olfactory epithelium is considerably thicker than the surrounding non-sensory epithelium of the nasal cavity, and is composed of three chief types of cell (7.241, 242), namely the olfactory receptor cells, supporting (sustentacular) cells and basal cells (Moulton and Beidler 1967), together constituting a pseudostratified epithelium up to 100 μm thick.

The receptor cells are of particular phylogenetic interest since they are *primary sensory neurons*, the cell bodies of which lie close to the sensory surface, a common feature of invertebrates but unique in the vertebrate classes. These cells are *bipolar* neurons, positioned vertically within the olfactory epithelium, the cell bodies being restricted to the basal two-thirds of the epithelial thickness. Basally, a non-myelinated axon stems from each receptor cell body, to run with other axons in small bundles within the epithelium amongst the processes of supporting and basal cells and, finally, to penetrate the basal lamina, where each bundle becomes ensheathed by Schwann cells (Frisch 1967). Such ensheathed bundles (*fila olfactoria*) join with others to form the *fasciculi* of the *olfactory nerve*, and eventually these enter the

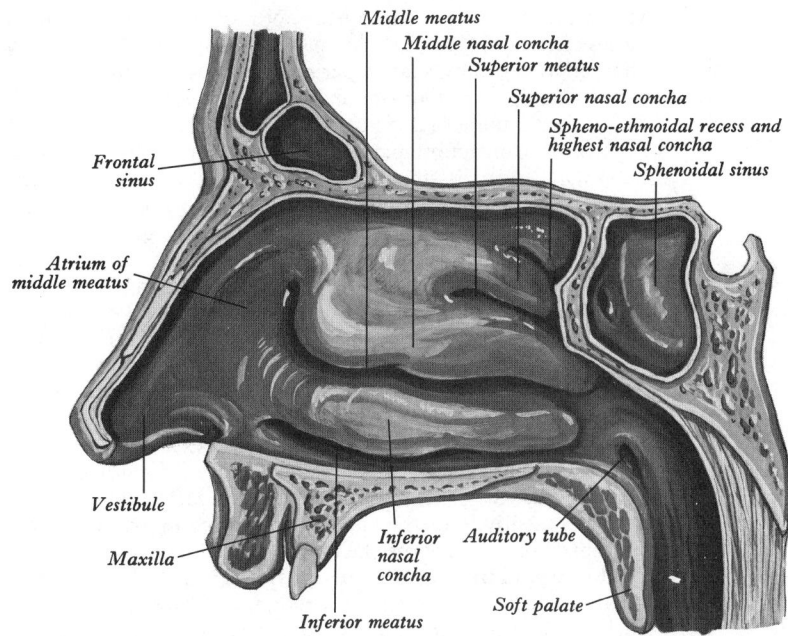

7.239 The lateral wall of the right half of the nasal cavity. Internal aspect.

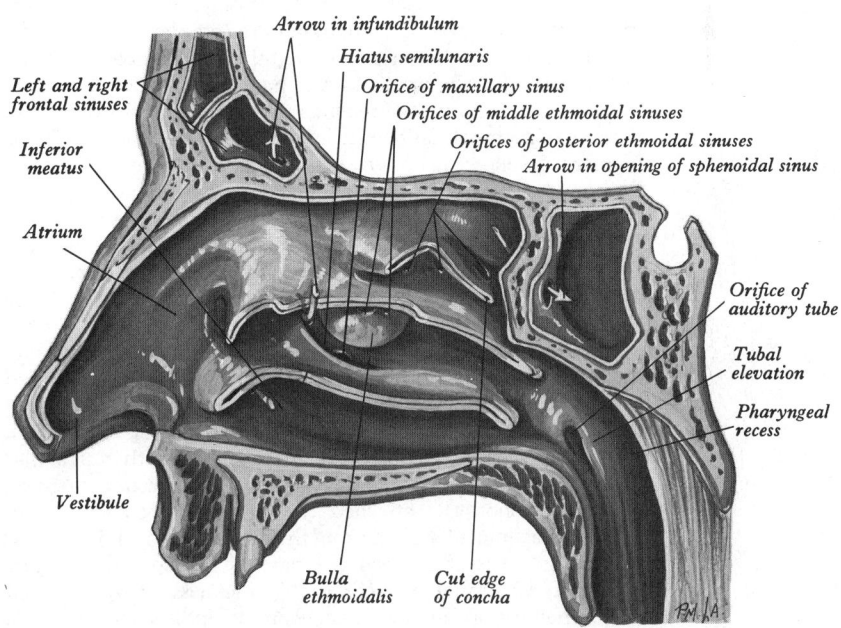

7.240 The lateral wall of the right half of the nasal cavity; the three nasal conchae have been partially removed.

olfactory bulb, where they synapse with second-order sensory neurons (mitral cells, basket cells, periglomerular cells, see p. 992). The olfactory axons are amongst the most slender of nerve fibres, being about 0·2 μm in diameter. Within the fila of the olfactory nerve, the axons are typically grouped together in invaginations of the Schwann cell membrane without any intervening Schwann cell processes (Gasser 1956), perhaps providing possibilities of electrical interaction between adjacent axons.

At the peripheral aspect of the receptor perikaryon (7.243), a single unbranched dendrite extends to the free surface of the epithelium where it expands slightly to form a bulbous olfactory ending (*rod, knob,* or *vesicle*), projecting above the general level of the epithelial surface (Frisch 1967; Reese 1965; Graziadei 1971, 1974; Naessen 1971). Scattered over the surface of the ending itself are numerous *olfactory cilia* which project into the layer of liquid overlying the epithelium. The cilia possess a '9 plus 2' pattern of microtubules internally, similar to that of motile non-sensory cilia elsewhere (p. 18). Amongst the different groups of vertebrates the length and precise form of the olfactory cilium varies considerably, but in mammals it usually takes the form of a short relatively thick 'shaft' tapering after a few microns to a long, thin, distal tip with reduced numbers of microtubules. Deeper in the dendrite, numerous mitochondria are present, indicative of high energy consumption, along with other organelles typical of neuronal dendrites—microtubules, smooth and rough coated vesicles, agranular endoplasmic reticulum and ribosomes; centrioles are also often present, especially in young animals. The *perikaryon* is rich in granular and agranular endoplasmic reticulum, Golgi bodies, and lysosomes, all indicating a high level of metabolism. The nucleus is elliptical and relatively heterochromatic. The exposed regions of the olfactory dendrite and its cilia are covered with a plasma membrane containing a high density of intramembranous particles demonstrable with freeze-fracturing methods (Kerjaschki and Hörandner 1976; Menco *et al.* 1976). It is likely that these particles represent sites of odour reception and of ionic movements associated with sensory activity, and their large numbers can be correlated with the extreme sensitivity of the olfactory receptors to extremely low concentrations of airborne chemicals (Menco 1976).

The supporting cells are irregular columnar elements which separate and partially enclose the receptors. Their large elongate euchromatic nuclei form a distinct layer above the level of the receptor perikarya. The surface of supporting cells is marked by the presence of numerous, long, irregular microvilli which, with the olfactory cilia, form a complex meshwork in the layer of fluid upon the general epithelial surface. Within the cytoplasm are many mitochondria, granular and agranular endoplasmic reticulum and, at the base of the epithelium, lamellated dense bodies which may be similar to the lipofuscin granules of neurons (p. 825), representing the remains of autophagic lysosomes and because of their coloration causing the whole olfactory area to be pigmented yellowish-brown. A prominent Golgi apparatus and lysosomes are also found in the apical parts of supporting cells. Near the epithelial surface, fine microfilaments attaching to desmosomes probably give mechanical coherence to the epithelium. Tight junctions are also reported to occur between supporting cells and olfactory receptors at the level of the epithelial surface (Reese and Brightman 1970).

The basal cells are of two types, one probably supportive and the other capable of continued cell division and so often termed *blastemal cells* (Andres 1966). The first of these classes is confined to the immediate vicinity of the basal lamina of the epithelium, and consists of irregular polygonal cells with rather irregular heterochromatic nuclei and a thin layer of cytoplasm containing numerous tonofibrils attached to desmosomal contacts with adjacent sustentacular cells. From their appearance it is likely that these basal cells strengthen the base of the epithelium. The *blastemal cells* form a layer of variable thickness near the base of the epithelium, and are particularly prominent in young animals. Mitotic figures are often seen amongst these cells, and their nuclei can be labelled with tritiated thymidine and examined with autoradiographic methods. Structurally the blastemal cells have many of the characteristics of embryonic neuroblasts, being rounded, with a large euchromatic nucleus and a layer of basophilic cytoplasm which under the electron microscope is seen to contain numerous free ribosomes with scattered mitochondria and occasional clusters of centrioles. Often some of the basal cells show signs of differentiating into olfactory receptors by sprouting dendrites and other receptor organelles.

Various studies have shown that the blastemal cells constitute a pool of immature stem cells which are constantly contributing new receptor cells by migration into the more apical parts of the epithelium. During early development including the postnatal period (at least in those species which have been examined), this type of recruitment is needed for the expansion of the epithelium, both in its thickness and its surface area, but a slower and continuous addition to the receptor population also occurs throughout life, balanced by the degeneration and death of existing receptor cells which (in rodents) may only last for 2–3 weeks (Moulton, Celebi and Fink 1970; Graziadei 1974). The degenerating receptors are phagocytosed by the neighbouring supporting cells, and their remnants after hydrolysis in

lysosomes, contribute to the olfactory pigment seen at the base of the epithelium (Bannister and Dodson 1975).

This continuous turnover of the olfactory receptor population is no doubt an advantage because of the exposed position of the olfactory endings and their liability to damage from inhaled noxious agents, and also provides possibilities of regeneration if the olfactory epithelium is more seriously damaged, for example by virus infection, although in the latter case functional regeneration appears to depend on the absence of scar tissue which may prevent the regrowth of axons (Douek, Bannister and Dodson 1975). However, it must be remarked that this ability to regenerate primary sensory neurons is unique in the nervous system of mammals, and represents a phenomenon of great neurobiological interest (*see* Graziadei 1978).

The olfactory glands (of Bowman) (7.241, 242) are branched tubular structure beneath the olfactory epithelium, on to the surface of which they pour their secretions, through narrow, vertical ducts. The secretory acini of these glands are composed of cells with basally placed nuclei and granular cytoplasm. Ultrastructurally, two gland cell types have been reported, both containing dense secretory vesicles, but one showing a denser cytoplasm than the other. Their secretions have been reported in some mammals to contain enzymes (acid phosphatase, esterase) and in mice sulphated acid mucosubstances have been demonstrated as a major component. Since this fluid forms the immediate environment of the receptor endings after it has been secreted, it may play an important part in the diffusion of odours from the air to the olfactory receptors and in regulating the passage of ions to and from the sensory cell during electrical activity. It may also possess bactericidal properties.

In addition to the structures described above, a variable population of lymphocytes is often present, particularly in the basal region of the epithelium, as elsewhere in the upper respiratory tract. During life there is a gradual reduction in the number of olfactory receptors and their axons, up to 1 per cent being lost each year; the sensory epithelium may be replaced by ciliated columnar epithelium similar to that of the non-sensory respiratory epithelium lining the adjacent regions of the nasal cavity. These changes parallel the gradual loss of the olfactory sense which usually accompanies aging.

Development of the olfactory epithelium

The olfactory tissue develops from the embryonic olfactory placode in a manner resembling the early ontogeny of the neural tube (p. 157). The cuboidal ectodermal cells of the olfactory placode gradually become taller but remain as a single layer throughout development. Initially, mitosis occurs only near the apical surface of the placode, the dividing cells rounding up then afterwards re-extending to the base of the epithelium, but later, mitoses are seen deeper in the epithelium then finally are confined to the basal layers where cell division persists throughout postnatal life (Smart 1971; Cuschieri and Bannister 1975a). Electron microscopic studies have shown that receptor cells begin to differentiate quite early in development, sprouting apical dendrites and basal axons with large growth cones, growing back to the olfactory bulb to eventually make synaptic contact with second order neurons (Hinds and Hinds 1972; Cuschieri and Bannister 1975b; Farbman 1975). After they reach the apical surface, the dendrites expand at their tips to form spherical bulbs into which centrioles migrate from the receptor cell body, eventually sprouting cilia at the receptor ending. Later, the supporting cells, glands and supportive basal cells differentiate so that the olfactory apparatus is in full functional order before birth.

In the **vomeronasal organ** (of Jacobson), present in most mammals but rudimentary in man (p. 1142), the cellular components are similar to those of the olfactory epithelium except for the reduction or absence of a basal cell layer, the larger size of the receptors, and the presence of microvilli instead of cilia at the sensory endings (Kolnberger 1971). The absence of cilia is particularly interesting since these have been claimed to be the site of chemoreception in the olfactory epithelium; it seems probable that the whole exposed plasma membrane is the chemoreceptive area and that the cilia or microvilli merely serve to increase the surface at which reception of stimuli may take

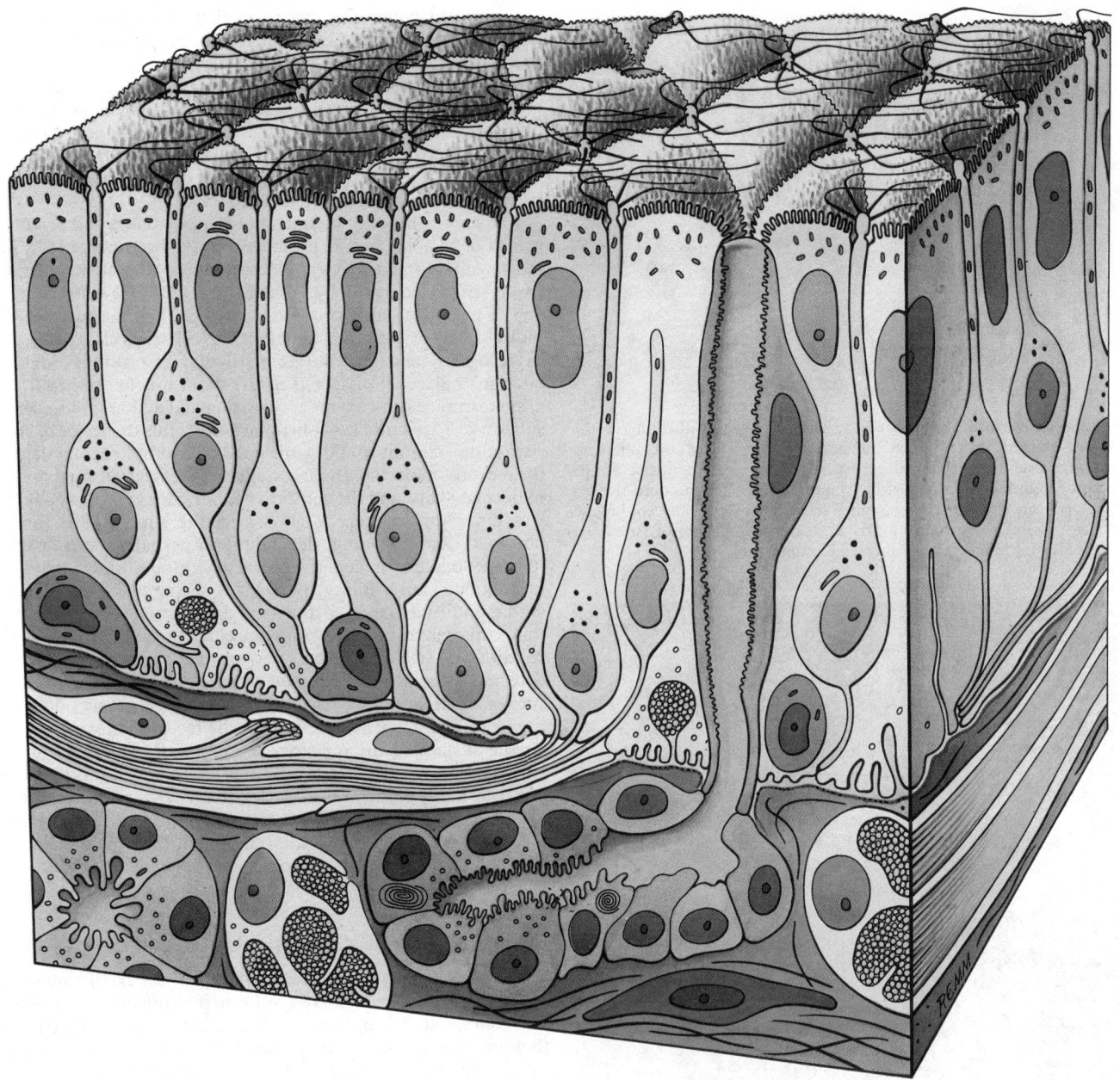

7.241 A schematic reconstruction showing the chief cytological features of the olfactory epithelium. Receptor cells (yellow) are situated among columnar supporting cells. The axons of the receptor cells emerge from the epithelium in groups ensheathed in Schwann cells. Observe the rounded basal epithelial cells and the sub-epithelial glands of Bowman with their intra-epithelial ducts. One of the basal cells is in process of differentiating into a receptor cell. At the surface are cilia of the receptor cells and microvilli of the supporting cells.

place. Electrical recordings show that vomeronasal receptors respond to odours in much the same way as olfactory receptors, although possibly to a narrower range of chemical substances.

Other chemoreceptors in the nasal cavity include various terminals capable of signalling the presence of strong concentrations of odours and of noxious chemicals in aerial solution, for example, ammonia and sulphur dioxide, even in the absence of the olfactory pathway, which may, for instance be interrupted after fracture of the cribriform plate or severance of the olfactory tract. The major nerves supplying the epithelium of the nasal chamber with possible chemosensory endings include trigeminal fibres from ethmoidal branches of the ophthalmic, and various maxillary nerve branches (p. 1148) which give off myelinated fibres; small bundles of such fibres have been detected by Graziadei and Gagne (1973) at the base of the olfactory epithelium, losing their sheaths to penetrate amongst the supporting cells. Another possible chemosensory nerve is the *nervus terminalis*, the functions of which are as yet unknown but which terminates widely in the nasal cavity in those species which have been studied (*see* Graziadei 1974 and p. 1055).

THE PHYSIOLOGY OF OLFACTORY RECEPTORS

The chemical sense subserved by the olfactory epithelium differs from that of the gustatory and other chemical senses in detecting and discriminating between a wide range of *airborne* molecules at very low concentrations. Although in man the sense of smell is only moderately well developed compared with that of macrosmatic species, in which odours play an important, even dominant role, there is evidence that olfaction may be more significant than previously imagined. In mammals, including primates, the production of specific scents ('pheromones') which affect the social or sexual behaviour of others of the same species is becoming recognized as a common phenomenon, and the close relationship between the olfactory pathways and areas of the brain associated with emotional behaviour in man may be important clues to the significance of odours in general. (*See* the limbic system, p. 990.)

Turning to more specific aspects of olfaction, the mechanism of odour reception and signalling to the brain have been approached chiefly by the technique of recording electrical changes in or at the

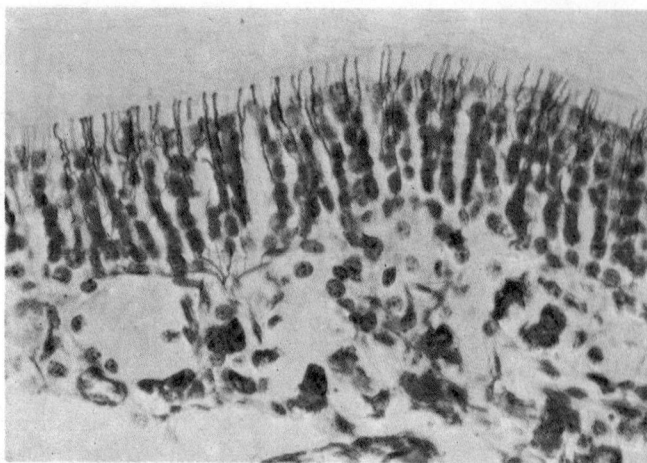

7.242 A A vertical section through the olfactory epithelium of the mouse, stained with Holmes' silver method to show the olfactory dendrites and their terminal expansions (above). The nuclei of the receptor neurons are arranged in columns in this preparation; the fila olfactoria can also be seen emerging from the base of the epithelium (below). (Kindly provided by Dr. A. Cuschieri, Depts. of Anatomy, Guy's Hosp. Med. Sch. and Univ. of Kuwait.)

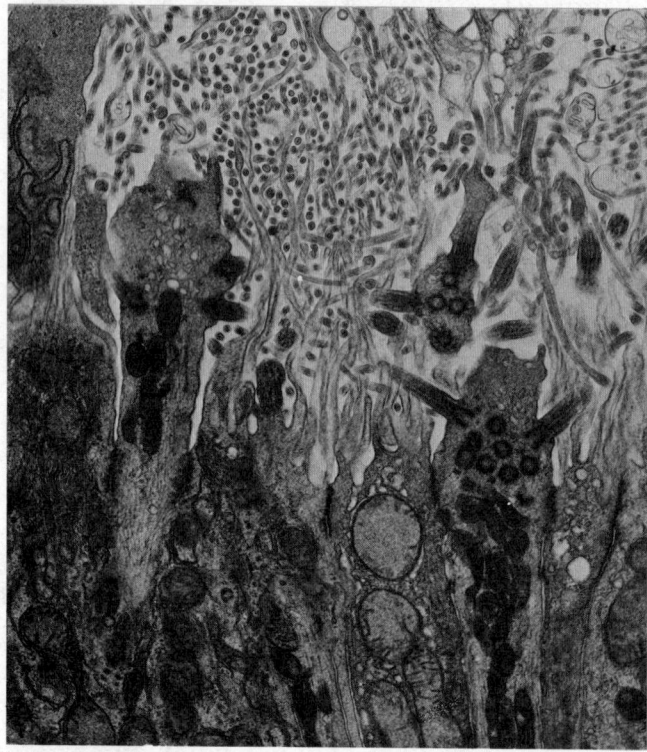

7.242 B An electron micrograph showing receptor endings in the olfactory mucosa of a mouse; note the presence of numerous cilia on the sensory terminals, interspersed by numerous microvilli on the surfaces of the supporting cells. (Kindly provided by Dr. H. C. Dodson.)

surface of the epithelium when it is exposed to odorous stimuli (Moulton and Beidler 1967). Two types of electrical changes can be recorded, namely a slow negative wave (the *electro-olfactogram*) found at the surface and a short distance below it, representing summated ionic fluxes stemming from many individual receptor dendrites; deeper in the epithelium, and in the olfactory nerve bundles, *action potentials* representing all-or-none activity in the axons can be detected instead.

Precisely how an array of receptors provides information about the stimulus type and intensity is the subject of much controversy. It might be imagined that individual receptors could be classified into specific types, each responsive to a particular odour. However, it is possible to distinguish between a large variety of different odours, and it would be necessary to postulate a tremendous range of receptor specificity, each group capable of detecting slight differences in molecular properties, if there is a one-to-one relationship between receptor type and odour. Electrophysiological recording indicates that single receptors can respond in different ways to a wide variety of odours, and that different receptors exhibit diverse patterns of response to single odours. It seems likely that discrimination in the olfactory system results from the analysis of a pattern of differential activity in a wide range of receptors (Gesteland *et al.* 1965).

The precise mode of interaction between odours and receptors is also debatable, and many theories have been proposed. The most widely accepted view is provided by the stereochemical theory first elaborated by Amoore (*see* the review by Amoore 1971); in outline, this proposes that there are various receptor molecules on or in the olfactory receptor cell, each class being able to recognize and respond to particular chemical groups on odorant molecules dissolved in the nasal mucus, rather in the same manner as an enzyme recognizing and acting on its specific substrate (a 'lock and key' phenomenon). Thus the shape of some part of the odorant would correspond to a complementary site in the receptor molecule. In early studies of the relationship between molecular shape and the subjective experience of odour quality it was thought that there might be a quite simple basis for the observed quality coding, depending on perhaps seven or eight 'primary odours' analogous to primary colours in colour vision (e.g. floral, fruity, minty, pungent, musky, etc.). More recent studies indicate that, unfortunately, the position is more complex than this; investigations into specific anosmia, that is, the condition in which one or a few classes of odours cannot be detected although the sense of smell is in general unaffected, suggests that there are perhaps thirty or more 'primary odours' in humans. There is also evidence that the receptor sites are present on the surface of the olfactory receptor cell, embedded in the exposed plasma membrane, and that the odorant binds loosely and transiently to each site to cause a change in the permeability of the membrane to ions. However, further studies at the cellular level are necessary for an understanding of the general nature of the sense of smell which in the past has been rather neglected by scientists.

THE RESPIRATORY EPITHELIUM

Those regions of the mucous membrane of the nasal cavity and associated sinuses which are not formed of olfactory epithelium are composed of columnar or pseudo-stratified ciliated epithelium interspersed with goblet cells, non-ciliated columnar cells with microvilli, and basal cells (Mygind 1975, 1978), collectively termed the *respiratory epithelium* (Negus 1958). In some areas these cells may be low columnar to cuboidal. Beneath the basal lamina are groups of serous and mucous glands which show much variation in cytological detail, opening by branched ducts on to the epithelial surface. The *cavernous tissue* lying beneath the respiratory mucosa in these areas is extensive (*vide infra*), and vascular disturbances cause alterations in the contours of the epithelial surface visible as swelling or shrinkage of the nasal lining.

The endothelium of these cavernous sinuses is particularly interesting since fenestrations have been demonstrated; it is possible that the muscular coats which are associated with changes in blood pressure in their lumina are under endocrine rather than direct autonomic control (Cauna and Hinderer 1969; Cauna *et al.* 1972). Immediately basal to the epithelium and its basement membranes there is a fibrous layer infiltrated with lymphocytes, forming in many parts a diffuse lymphoid tissue, and under this is a nearly continuous layer of mucous and serous glands, the ducts of which pass through the lymphoid layer before opening upon the surface. The abundant amount of mucus secreted by the glands and goblet cells makes the surface of the mucosa moist and sticky. Because of this the dust in the inspired air is deposited on the surface and the air is humidified. The rich vascularity of the membrane ensures warming of the inspired air and accounts for its pink colour. The contaminated mucous film covering the membrane is moved by ciliary action downward and backward, away from the olfactory region and into the nasopharynx at a rate of about 6 mm per minute (Proctor *et al.* 1978). Palate movements then transfer it to the oral pharynx and it is swallowed. Some, however, is passed anteriorly into the vestibule of the nasal cavity.

VASCULATURE AND NERVES OF THE NASAL CAVITY

The *arteries of the nasal cavity* are the anterior and posterior ethmoidal branches of the ophthalmic artery, which supply

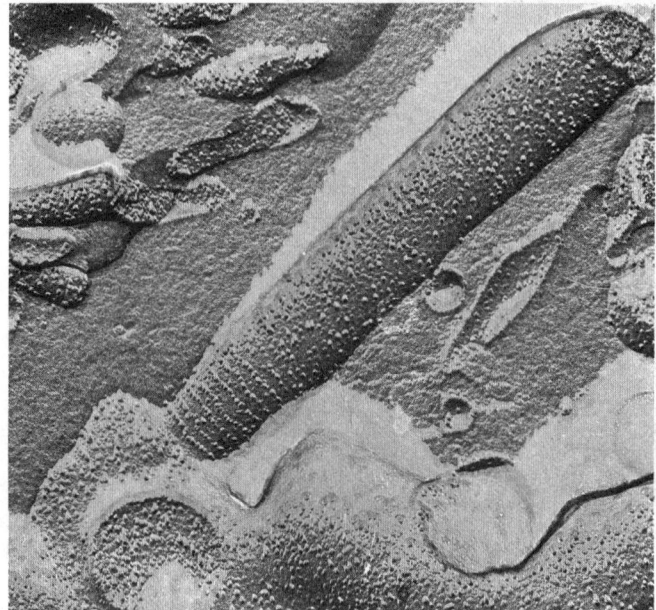

the ethmoidal and frontal sinuses and the roof of the nose; the sphenopalatine branch of the maxillary artery, which supplies the mucous membrane covering the conchae, the meatuses, and septum; the terminal part of the greater palatine artery which ascends through the incisive canal (p. 341); the septal ramus of the superior labial branch of the facial artery, which supplies the part of the septum in the region of the vestibule, anastomosing with the sphenopalatine artery, and is a common site of bleeding from the nose (epistaxis); the infraorbital and superior (anterior and posterior) alveolar branches of the maxillary artery which supply the lining membrane of the maxillary sinus; the pharyngeal branch of the same artery, which is distributed to the sphenoidal sinus. The ramifications of these vessels form a close plexiform network, beneath and in the substance of the mucous membrane.

The *veins of the nasal cavity* form a close cavernous plexus beneath the mucous membrane. Arteriovenous communications are present (Harper 1947). The plexus is especially marked over the lower part of the septum and over the middle and inferior conchae. Some of the veins open into the sphenopalatine vein; others join the facial vein; some accompany the ethmoidal arteries, and end in the ophthalmic veins; a few communicate with the veins on the orbital surface of the frontal lobe of the brain, through the foramina in the cribriform plate of the ethmoid bone. When the foramen caecum is patent it transmits a vein from the nasal cavity to the superior sagittal sinus.

The *nerves of the nasal cavity* include branches of the trigeminal and olfactory cranial nerves.

7.243A   An electron micrograph of an olfactory cilium (bovine) attached at its basal end to an olfactory dendrite terminal, freeze-fractured to expose the numerous intramembranous particles which are thought to be sites of odour reception. An extensive ciliary necklace is present at the base of the cilium, and microvilli of a supporting cell are visible on the left. Magnification × 60,000. (Kindly supplied by Dr. B. Menco, University of Utrecht, Netherlands.)

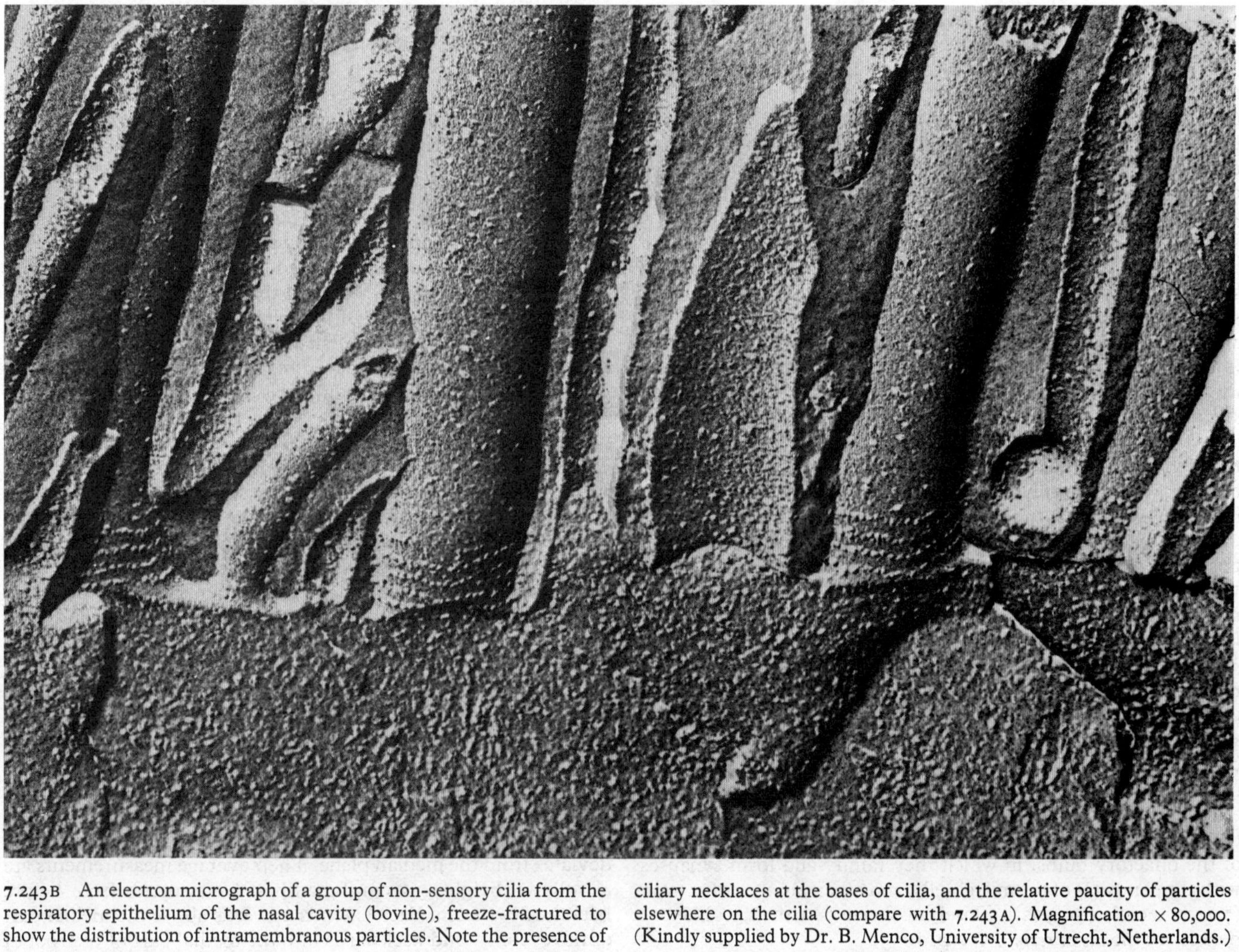

7.243B   An electron micrograph of a group of non-sensory cilia from the respiratory epithelium of the nasal cavity (bovine), freeze-fractured to show the distribution of intramembranous particles. Note the presence of ciliary necklaces at the bases of cilia, and the relative paucity of particles elsewhere on the cilia (compare with 7.243A). Magnification × 80,000. (Kindly supplied by Dr. B. Menco, University of Utrecht, Netherlands.)

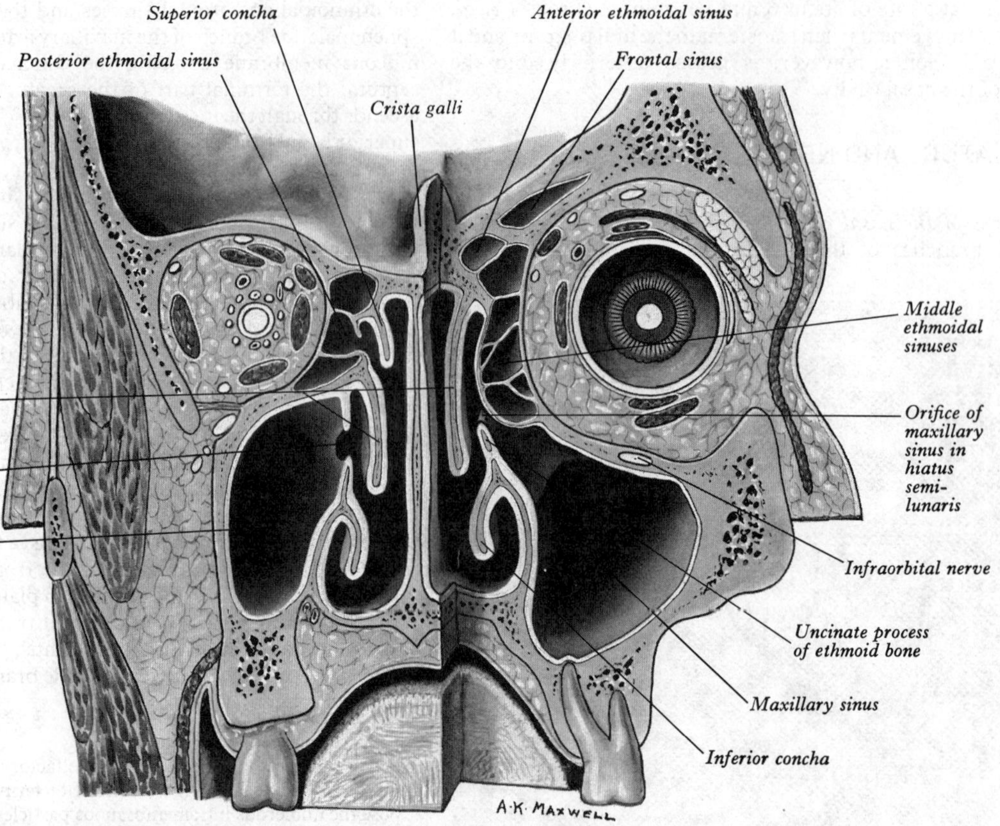

7.244 A coronal section through the nasal cavity, viewed from the posterior aspect. On the right side the plane of the section is more anterior. The normal orifice of the maxillary sinus is shown on the right side, and the not uncommon accessory orifice on the left side.

The *lymph vessels* are described on p. 787.

The *nerves of ordinary sensation* (7.171, 178) supplying the nasal cavity are as follows: the anterior ethmoidal branch of the nasociliary nerve, which supplies the anterior and upper part of the septum, the anterior part of the roof and the anterior parts of the middle and inferior conchae with the lateral wall in front of these; the infraorbital nerve, which supplies the vestibule; the anterior superior alveolar nerve, which supplies the part of the septum and floor near the anterior nasal spine and the anterior part of the lateral wall as high as the opening of the maxillary sinus; the lateral posterior superior nasal and the medial posterior superior nasal nerves (including the nasopalatine nerve), which are branches of the pterygopalatine ganglion, and the posterior inferior nasal branches of the anterior palatine nerve supplying the posterior three-quarters of the lateral wall, roof, floor and septum; branches from the nerve of the pterygoid canal which supply the upper and back part of the roof and septum. It is to be noted that, with the exception of the nasociliary nerve, all the non-olfactory nerves supplying the nasal cavity are derived from the maxillary division of the trigeminal nerve.

Accompanying the sensory fibres in these nerves are postganglionic vasomotor sympathetic fibres to the nasal blood vessels, whilst running with the branches from the pterygopalatine ganglion are postganglionic parasympathetic fibres from the latter which are secretomotor to the nasal glands.

The *olfactory nerves* are, of course, distributed to the olfactory region of the mucosa, and their fibres arise from the bipolar olfactory cells, described above, and are destitute of myelin sheaths. They unite in fasciculi which cross one another in various directions, thus giving rise to the appearance of a plexus in the mucous membrane, and then ascend in grooves or canals in the ethmoid bone; they pass into the skull through the foramina in the cribriform plate of the ethmoid and enter the inferior surface of the olfactory bulbs, in which they ramify and form synapses with the dendrites of the mitral cells, and other varieties of neuron (7.118, see also p. 992). Closely associated with the olfactory nerves are the *nervi terminales* (p. 1055).

THE PARANASAL SINUSES

The frontal, ethmoidal, sphenoidal and maxillary paranasal sinuses (7.238, 239, 240, 244, 245, 246) vary in size and form in different individuals, and are lined with mucous membrane continuous with that of the nasal cavity, an important fact in connexion with the spread of infections. The mucous membrane resembles that of the respiratory region of the nasal cavity, but is thinner, less vascular and more loosely adherent to the bony walls of the sinuses. The mucus secreted by the glands in the mucous membrane is swept into the nose through the apertures of the sinuses by the movement of the cilia covering the surface. The cilia are not uniformly distributed in the lining mucous membrane but are always present near the opening into the nasal cavity. The function of the sinuses is doubtful. They lighten the skull and add resonance to the voice, but the saving in weight would be trivial, for absence of sinuses does not entail an equivalent volume of *solid* bone, and the weight of trabecular bone which would occupy such a volume is small. It is more probable that some, at least, of the sinuses are manifestations of the exceptional growth pattern of the bones in which they occur. They vary considerably in size in different individuals. Most are rudimentary, or even absent, at birth; they enlarge appreciably during the time of eruption of the permanent teeth and after puberty, and this growth is a factor in the alteration in the size and shape of the face at these times.

Two frontal sinuses are posterior to the superciliary arches, between the outer and inner tables of the frontal bone. When of average size, each underlies a triangular area on the surface, the angles of which are formed by the nasion, a point about 3 cm above the nasion and the junction of the medial third with the rest of the supraorbital margin (3.116, 7.245, 246). However, they are rarely symmetrical, because the septum between them frequently deviates from the median plane. Their average measurements are as follows: height, 3·2 cm, breadth, 2·6 cm, depth from before backwards, 1·8 cm. Each extends upwards above the medial part of the eyebrow and backwards into the medial part of the roof of the orbit. The frontal sinus is sometimes divided into a number of intercommunicating recesses by incomplete bony partitions. Rarely, one or both sinuses may be absent, and the degree of prominence of the superciliary arches is no indication of the presence or size of the frontal sinuses. The part of the sinus extending upwards in the frontal bone may be small and the orbital part large, or vice versa. Sometimes one sinus may overlap in front of the other. The sinus may extend posteriorly as far as the sphenoid (lesser wing), but does not invade it. Each opens into the anterior part of the corresponding middle meatus of the nose, either through the *ethmoidal infundibulum* or through the *frontonasal duct*, which traverses the anterior part of the labyrinth of the ethmoid. Rudimentary or absent at birth, they are generally fairly well developed between the seventh and eighth years, but reach their full size only after puberty (see also p. 334). They are usually more prominently developed in males, giving the profile of the forehead an obliquity which contrasts with the vertical or convex outline usual in children and females. The arterial blood supply of the sinus is from the supraorbital and anterior ethmoidal arteries, and the venous drainage is into the anastomotic vein in the supraorbital notch connecting the supraorbital and superior ophthalmic veins. The lymph drainage is to the submandibular nodes. The nerve supply is derived from the supraorbital nerve.

The ethmoidal sinuses (*see* pp. 335, 336) consist of thin-walled cavities in the ethmoidal labyrinth, completed by the frontal, maxillary, lacrimal, sphenoidal and palatine bones. They vary in number and size from three large to eighteen small sinuses on each side, and their openings into the nasal cavity are very variable. They lie between the upper part of the nasal cavity and the orbits, and are separated from the latter by the extremely thin orbital plates of the ethmoid; infection may spread from the sinuses into the orbit and produce orbital cellulitis. On each side they are arranged in three groups—anterior, middle, and posterior, though some anatomists divide them into two groups, anterior and posterior, the anterior group including those described below as the anterior *and* middle groups. The three groups are not sharply delimited from each other and one group may encroach on the territory generally occupied by another. The groups are really only distinguishable on the basis of their sites of communication with the nasal cavity. In each group the sinuses are partially separated by incomplete bony septa. The *anterior group* vary up to eleven in number and open into the ethmoidal infundibulum or the frontonasal duct by one or more orifices; one sinus frequently lies in the agger nasi and the most anterior sinuses may encroach upon the frontal sinus. The *middle group* (bulbar sinuses) generally comprise three cavities which open into the middle meatus by one or more orifices on or above the ethmoidal bulla. The *posterior group* vary from one to seven in number and usually open by one orifice into the superior meatus inferior to the superior concha, though one may open into the highest meatus (when present), and one or more sometimes open into the sphenoidal sinus. The posterior group are very closely related to the optic canal and optic nerve. The ethmoidal sinuses are small, but of clinical importance, at birth; they grow rapidly between the sixth and eighth years and after puberty. They derive their arterial blood supply from the sphenopalatine (p. 684) and the anterior ethmoidal and posterior ethmoidal arteries and are drained by the corresponding veins. The lymphatics of the anterior and middle groups drain into the submandibular nodes and those of the posterior group into the retropharyngeal nodes. The ethmoidal sinuses are supplied by the anterior and posterior ethmoidal nerves and the orbital branches of the pterygopalatine ganglion.

The two sphenoidal sinuses (p. 323, **3**.77, 78), are sited posterior to the upper part of the nasal cavity. Contained within the body of the sphenoid bone, they are, therefore, related above, to the optic chiasma and the hypophysis cerebri, on each side, to the internal carotid artery and the cavernous sinus. If the sinuses are small, they lie in front of the hypophysis cerebri. They vary in size and shape, and, owing to the lateral displacement of the intervening septum, are rarely symmetrical. Frequently one sinus is much the larger of the two and extends across the median plane behind the sinus of the opposite side; occasionally one sinus may

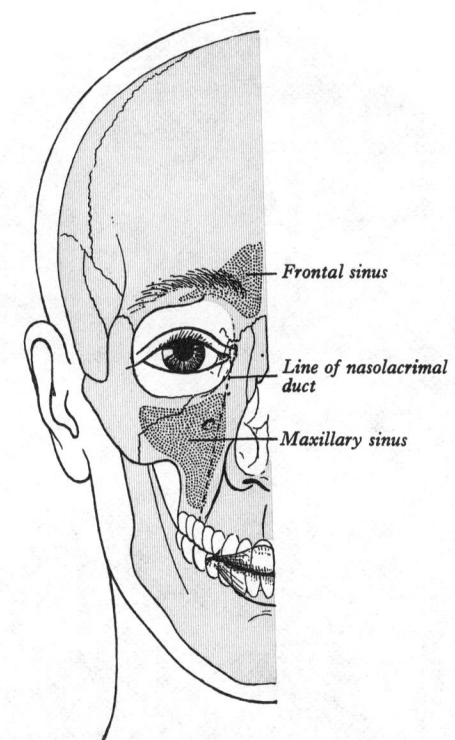

Frontal sinus

Line of nasolacrimal duct

Maxillary sinus

7.245 An outline of the bones of the face, showing the positions of the frontal and maxillary sinuses.

overlap above the other, and rarely there is a communication between the two sinuses. Occasionally one or both sinuses may extend close to and even partially encircle the optic canal on its own side. The following are their average measurements: vertical height, 2 cm; transverse breadth, 1·8 cm; anteroposterior depth, 2·1 cm. When exceptionally large they may extend into the roots of the pterygoid processes or greater wings of the sphenoid, and may invade the basilar part of the occipital bone. Occasionally there are gaps in the bony walls and the mucous membrane may lie directly against the dura mater. Bony ridges, produced by the internal carotid artery and the pterygoid canal, may project into the sinuses from the lateral wall and floor respectively. A posterior ethmoidal sinus may extend into the body of the sphenoid and largely replace a sphenoidal sinus. Each sinus communicates with the spheno-ethmoidal recess by an aperture in the upper part of its anterior wall. They are present as minute cavities at birth, but their main development takes place after puberty. Their blood supply is by means of the posterior ethmoidal vessels and the lymph drainage is to the retropharyngeal nodes. Their nerve supply is from the posterior ethmoidal nerves and the orbital branches of the pterygopalatine ganglion.

The two maxillary sinuses, which are the largest accessory air sinuses of the nose, are pyramidal cavities in the bodies of the maxillae (**3**.117; **7**.244, 245, 246). The base of each is formed by the lateral wall of the nasal cavity; the apex extends into the zygomatic process of the maxilla. The roof, which is the orbital floor, is frequently ridged by the infraorbital canal, while the floor is formed by the alveolar process and is usually about 1·25 cm below the level of the floor of the nose, on a line drawn laterally from the lower border of the ala. Several conical elevations corresponding with the roots of the first and second molar teeth project into the floor, which is sometimes perforated by one or more of these roots. Sometimes the roots of the first and second premolars and the third molar, and occasionally the root of the canine, also project into the sinus (*see* p. 1302). The size of the maxillary sinus varies in different skulls, and even on the two sides of the same skull; when large, its apex may invade the zygomatic bone. The following measurements are those of an average-sized air sinus: vertical height opposite the first molar tooth, 3·5 cm; transverse breadth, 2·5 cm; anteroposterior depth, 3·2 cm. The sinus communicates with the lower part of the hiatus semilunaris through an opening in the anterosuperior part of its base (**7**.244); a

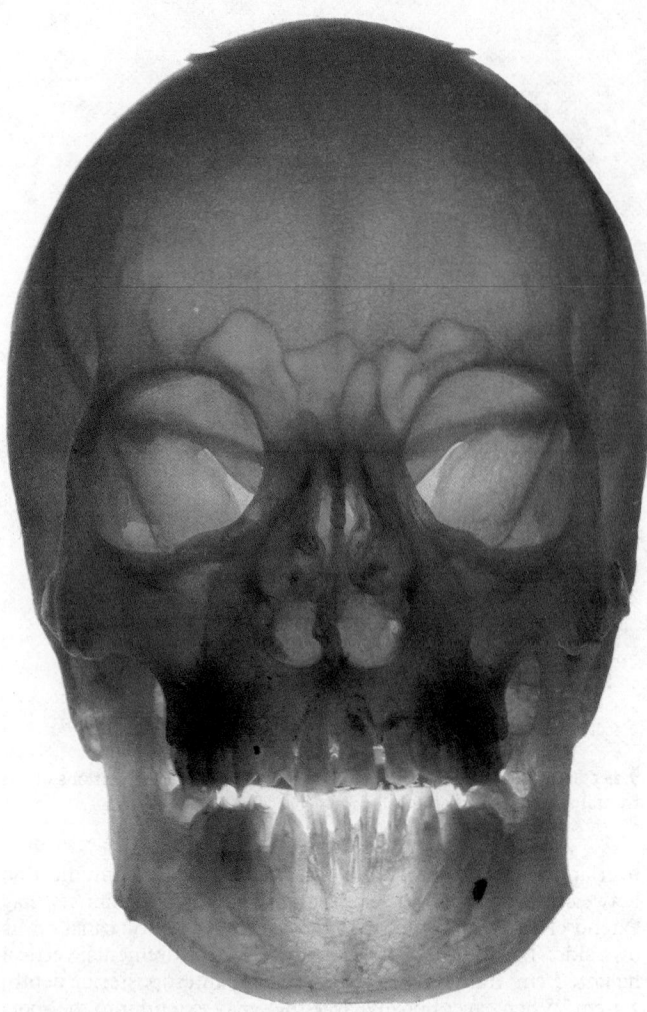

7.246 The skull of an adult woman which has been decalcified and then cleared in methyl salicylate; the specimen was transilluminated and then photographed from the ventral aspect. Note particularly the profiles of the frontal and maxillary paranasal air sinuses, the orbits and superior orbital fissures, and the nasal cavities and conchae. (The specimen was prepared by Dr. D. H. Tompsett of the Royal College of Surgeons of England.)

RADIOLOGICAL APPEARANCES

Normal sinuses are radiolucent, whereas diseased sinuses show varying degrees of opacity. Radiographs also reveal the extent of development of the sinuses. In anteroposterior view (3.62, 68), the sinuses appear as follows. The frontal sinuses are seen above the nasal cavity and the medial part of the orbits and their asymmetry, vertical extent and the presence of bony septa can be assessed. The ethmoidal sinuses are superimposed on each other as well as on the sphenoidal sinus in the radiograph; they lie between the orbits, below the shadow of the cribriform plate. The sphenoidal sinus is not clear in this view. The maxillary sinus forms a pyramidal-shaped translucent area below the orbit and lateral to the lower part of the nasal cavity; inferiorly it extends into the alveolar process of the maxilla. In lateral view, the extent of the frontal sinus both upwards into the frontal bone and backwards into the orbital roof can be seen. The ethmoidal sinuses are seen extending from the shadow of the frontal process of the maxilla as far back as the sphenoidal sinus, the latter being clear and distinct below and in front of the fossa for the hypophysis, though of course the areas of the two sphenoidal sinuses are superimposed, and the individual sphenoidal sinuses are best seen in a superior view. The maxillary sinus is well seen in a lateral view; it lies below the orbit and its extent in relation to the roots of the teeth can be clearly seen.

The maxillary and frontal sinuses can also be examined by the method of trans-illumination. In a dark room an electric torch is placed in the mouth, in the case of the maxillary sinus, or against the superomedial angle of the orbital opening, for the frontal sinus. Normally, a red glow is seen in the region of these sinuses, which may be absent where they are diseased. (For a view of the frontal and maxillary sinuses in a cleared preparation see 7.246.)

Applied Anatomy. Congenital deformities of the nose occur occasionally, such as complete absence of the external nose, an aperture only being present, or perfect development on one side, and suppression or malformation on the other.

The septum of the nose may be displaced or may deviate from the median plane as a result of an injury or of some congenital defect. Sometimes the deviation may be so great that the septum may come into contact with the lateral wall of the nasal cavity, producing complete unilateral obstruction.

Suppuration in the paranasal sinuses is of frequent occurrence, and it is important to note that the middle meatus is of such a form that pus running down from the frontal sinus or the anterior ethmoidal sinuses is directed by the hiatus semilunaris into the opening of the maxillary sinus, so that the latter sinus may, in some cases, act as a secondary reservoir for pus discharged from these sinuses. All the paranasal sinuses can be infected from the nasal cavity, but it should be noted that in the case of the maxillary sinus, the infection may originate from the vicinity of the teeth (p. 1302). This sinus is the one most frequently the seat of chronic suppuration, which may result in loss of cilia from the surface of the mucosa and hence impairment of mucus flow. The normal opening of the maxillary sinus is high above the floor and is poorly placed for natural drainage. Surgical drainage may be effected by puncturing the lateral wall of the inferior nasal meatus, or the canine fossa on the anterior surface of the maxilla.

second orifice is frequently seen in, or immediately below, the hiatus. Both are nearer the roof than the floor of the sinus. The maxillary sinus appears as a shallow groove on the medial surface of the bone about the fourth month of intrauterine life, but does not reach its full size until after the eruption of all the permanent teeth. The blood supply of the sinus is by means of the facial, infraorbital and greater palatine vessels; the lymph drainage is to the submandibular nodes. The nerve supply is derived from the infraorbital and the anterior, middle and posterior superior alveolar nerves.

For further details of the human and comparative anatomy of the paranasal sinuses consult Negus (1958).

THE VISUAL APPARATUS

Introduction

Considering the all-pervading medium of sunlight, in which almost all animal and botanical life is immersed, it is inevitable that responses to this form of electromagnetic radiation should occur. The photosynthetic processes widespread in plant forms have a parallel in the photochemical receptors which occur almost universally in the animal kingdom. Although the range of frequencies in solar electromagnetic radiation is much wider, the spectrum of visible light (400–760 nm) is the range within which

most animal 'light' receptors or *photoreceptors* function. There is evidence that some animal forms have receptors which react outside this range, either in the ultraviolet or infra-red frequencies; in vertebrates, some vipers and boas (Walls 1963) have facial pit organs which respond to infra-red radiation. In general, however, the *visual pigments*, which provide the basis of the photochemical response, display absorption maxima at various points within the visible spectrum. For example, human *rhodopsin* (visual purple) has a maximum at 497 nm. It is a rod pigment; *iodopsin*, a cone pigment, has been identified in some avian species, and several such pigments are present in human cones, where their differing absorption maxima probably account for colour vision, but full details are not yet available (Rushton and Henry 1968). The basis of the response is a light-induced 'bleaching' process, during which the pigment changes to another form, with an associated electrical change, which is propagated through the photoreceptors to the first-order neurons. A rapid restoration process is, of course, a *sine qua non*. A multiplicity of visual pigments have been identified and studied; further details may be obtained in a number of monographs (e.g. Rushton 1962; Pirenne 1967; Davson 1976).

The next elaboration in visual organs is the introduction of a lens, to concentrate light energy upon the photoreceptor and to impart an element of directional sensibility. Many adaptations of this kind have occurred in invertebrates, mostly in two directions: eyes—as we may now call them—with a large number of separate lens-photoreceptor units, the *compound eyes* familiar in most insects and crustaceans, and the *single lens*, focusing light on to an array of photoreceptors, as present in snails and squids. It is the latter type of eye which is universal in vertebrates.

True eyes are able not only to respond to variations in *luminance*—a simple function of photoreceptors; by projecting a focused image upon an array of receptors, each with neural pathways of some degree of specificity, a new modality of vision is introduced—sensitivity to *form*. In both cases movement, which is of great biological significance, may be detected, but the vertebrate type of focusing eye has potentialities for much greater precision in this respect.

Primitively, vision appears to be employed as a form of *distance reception* capable of activating warning systems, and of orientating the animal advantageously with regard to light and shade. The paired eyes of most vertebrates are set in a lateral position in the head, permitting an almost panoramic view of the environment. Such *panoramic vision*, coupled with a system of muscles, by which the eye is reflexly rotated towards any object of significance—such as the movement of a predator—is characteristic of most mammals, and clearly provides a valuable 'early-warning' system. In limited groups of mammals, and some raptorial or predatory birds, the position of the orbits has changed, so that the two uniocular fields subtended by each eye overlap to a greater or lesser extent. This entails that the part of the environment in front of the animal is focused by both eyes. By a gradual refinement of the neural control of the ocular muscles, constantly provided with a feedback from the retinae, the eye movements become sufficiently concerted to 'fuse' the two

slightly dissimilar images falling on the retinae, leading to the establishment of *binocular vision*. This advance is characteristic of carnivorous mammals, who may track down their prey in part by smell, but must rely upon the much more accurately directional nature of vision to carry out the final attack. Primates also possess forward-looking eyes and hence binocular vision; but in their evolution the effect of an arboreal phase of existence is generally regarded as the operative factor in the elaboration of not merely binocular, but *stereoscopic vision*, which is characterized by a more highly evolved motor 'understanding' of the three-dimensional nature of space. Olfaction is less useful in trees, and the acquisition of great skill in not merely climbing but in swinging or leaping from branch to branch could only occur with the development of stereoscopic vision. The same habitat undoubtedly favoured the retention and elaboration of pentadactyl and grasping extremities; and although man, and perhaps his immediate ancestors, is not an arboreal primate, the terrestrial specialization of his feet has not occurred in his hands. Left free of locomotor influences by the adoption of bipedal gait, the hands have in man formed a highly significant partnership. This, coupled with the development of a particularly large brain, able to process with increasing intricacy the highly detailed information from the eyes, and to control both the eyes and hands in increasingly skilful and subtle tasks, can be regarded as the major factors in the extraordinary evolution of human abilities.

The eye, therefore, is not to be regarded in isolation. Its array of modalities—sensitivity to very small changes in luminosity, particularly in dark-adapted or *scotopic vision*, high discriminativeness as to form and movement and to colour in light-adapted or *photopic vision*—do not merely provide interesting information. The information is vital; it is doubtful whether a blind individual could long survive outside human society. The eyes provide a continuous monitoring of all we do, especially in manual tasks. Visual means of communication have proved more valuable and lasting even than auditory. The gradual evolution of visual signs, reacting with auditory communication, has led to the formation of language in all its permutations. Through language, with all its potentialities for communication of increasingly precise information and conceptual influences, it becomes possible for generation after generation to profit from recorded knowledge and skill. The results of this in human culture have become the mainstream of man's evolution. It is against this background that the structure of the visual apparatus should be studied, as we must now proceed to do.

THE PERIPHERAL VISUAL APPARATUS

The eyeball, the peripheral organ of sight, is situated in the cavity of the orbit, the walls of which serve to protect it from injury. The protection is perhaps adventitious, for other considerations, more basic to the function of vision, are more pertinent. It is difficult to imagine how the ocular movements could be concerted in the absence of a socket; nor would the spatial relationship between the two eyes be so precisely preserved in the absence of rigid sockets, a factor of prime importance in animals with binocular vision. Certain accessory structures—the muscles, fasciae, eyebrows, eyelids, conjunctiva and lacrimal apparatus—are intimately associated with the eyeball and will be described in this section.

The eyeball is embedded in the fat of the orbit, but is separated from it by a thin membranous sac, termed the *fascial sheath of the eyeball* (capsule of Tenon) (p. 1181). It is composed of segments of two spheres of different sizes. The anterior segment is one of a small sphere; it is transparent, and it forms about one-sixth of the eyeball. It is more prominent than the posterior segment, which is one of a larger sphere, and is opaque, and it forms about five-sixths of the whole circumference of the eyeball. The *anterior segment* is bounded by the cornea and lens, and is further divided into *anterior* and *posterior chambers*, which are incompletely separated by the iris, being continuous through the pupil. Anteriorly, the periphery of the anterior chamber is

overlapped by the sclera; the angle between the iris and cornea (p. 1152) forming an annulus which is of greater diameter than that of the *limbus*, the junction between sclera and cornea. The difference varies from 1 to 2 mm, the 'angle' being deeper above and below than at the sides. The posterior chamber, between the posterior surface of the iris and the anterior aspect of the lens and its supporting ligament, the zonule (p. 1176), is triangular in section, the apex of the triangle being where the iris touches the lens. Its base, the zonular region, is not, strictly speaking, the zonule itself, since the posterior chamber is regarded as extending among the collagenous bundles of the zonule and even into a *retrozonular space*, the canal of Petit, between the zonule and the vitreous humor in the *posterior segment*, which comprises the parts of the eyeball posterior to the zonule and lens. The term *anterior pole* is applied to the central point of the anterior curvature of the eyeball, and that of *posterior pole* to the central point of its posterior curvature; a line joining the two poles forms the *optic axis*. (By the same convention, the eyeball is considered to have an *equator*, equidistant between the poles; any circumferential line joining the poles is a *meridian*.) The primary axes of the two eyeballs are nearly parallel, and therefore do *not* correspond with the axes of the orbits, which are directed forwards and laterally. The optic nerves follow the direction of the axes of the orbits, and therefore are not parallel; each nerve enters its eyeball about 3 mm to the nasal side of the posterior pole. The vertical diameter (23·5 mm) of the eyeball is rather less than the transverse and anteroposterior diameters (24 mm); the anteroposterior diameter at birth is about 17·5 mm and at puberty from 20 to 21 mm; it may vary considerably from this in *myopia* (29 mm) and *hypermetropia* (20 mm). In the female all three diameters are rather less than in the male. (*See* Stenström 1946, Sorsby and Sheridan 1960.)

The eyeball comprises three tunics, and the contents enclosed by them. From without inwards the three tunics are: (1) the fibrous tunic, consisting of the *sclera* behind and the *cornea* in front; (2) the vascular, pigmented tunic, comprising, from behind forwards, the *choroid, ciliary body* and *iris*, forming together the *uveal tract*; and (3) the nervous layer, termed the *retina*.

The Ocular Fibrous Tunic

The fibrous layer of the eyeball (7.247) consists of an opaque, posterior part, the *tunica sclera*, and a transparent, anterior part, the *tunica cornea*.

THE SCLERA (TUNICA SCLERA)

The sclera, so named from its density and hardness, is a firm membrane which, when distended by the intraocular pressure, serves to maintain the form of the eyeball. It is thickest (about 1 mm) behind, near the entrance of the optic nerve, and thinnest (0·4 mm) at a distance of about 6 mm behind the sclerocorneal junction, in the region of attachment of the recti (p. 1178). Its *external surface* is white, and is in contact with the inner surface of the fascial sheath of the eyeball (p. 1181)— it is smooth, except where the tendons of the orbital muscles are attached to it; its anterior part is covered by the conjunctival epithelium, reflected on to it from the deep surfaces of the eyelids and continuous anteriorly with that covering the cornea. Its *internal surface* is brown, and is marked by grooves in which the ciliary nerves and vessels ramify; it is separated from the external surface of the choroid by an extensive *perichoroidal space*, which is traversed by an exceedingly delicate cellular tissue, termed the *suprachoroid lamina* (or *lamina fusca of the sclera*). Posteriorly, the sclera is pierced by the optic nerve, and is continuous through the fibrous sheath of this nerve with the dura mater. Where the optic nerve pierces the sclera, the latter has the appearance of a cribriform plate and is named the *lamina cribrosa sclerae* (7.258); the minute orifices in this lamina transmit the nerve bundles. One opening, larger than the rest, and occupying the centre of the lamina, transmits the central artery and vein of the retina. The lamina cribrosa is the weakest part of the sclera; if the intraocular pressure be raised for some time, as in cases of chronic glaucoma, the lamina cribrosa becomes bulged outwards producing the condition of 'cupped disc'. Around the lamina cribrosa numerous small apertures are present which transmit the ciliary vessels and nerves, and about midway between these and the sclerocorneal junction there are four or five large apertures for the transmission of veins (*venae vorticosae*). In front, the sclera is directly continuous with the cornea, the line of union being termed the *sclerocorneal junction* (or *limbus*). In the substance of the sclera close to this junction is a canal lined with endothelium running circularly, termed the *sinus venosus sclerae*, which in section presents the appearance of an oval cleft. The outer wall of the cleft is formed by a groove in the sclera. Posteriorly the cleft extends as far as a projecting rim of scleral tissue termed the *scleral spur*, which in section is triangular with the apex directed forwards and inwards. The sinus may be double in parts of its course. The inner wall of the scleral sinus, that is, the aspect of the sinus adjoining the aqueous chamber, is formed by a loose trabecular tissue continuous anteriorly with the posterior limiting lamina of the cornea. Between the fibres of this tissue are spaces through which the aqueous humour in the anterior chamber filters into the sinus (7.250), from which it then passes into the bloodstream, since the scleral sinus drains externally into the anterior ciliary veins. Normally the sinus contains no blood; although the communicating channels between the sinus and the anterior ciliary veins

contain no valves, these channels are oblique and flattened and may prevent reflux of blood into the sinus. Such a valvular mechanism is dubious, pressure gradients being more likely to prevent reflux of blood, since under conditions of venous congestion blood may pass into the sinus. The anterior and outer side of the scleral spur gives attachment to most of the fibres of the trabecular tissue mentioned above and the posterior and inner side to the meridional fibres of the ciliaris. The *iridocorneal angle* (7.248) of the anterior chamber lies between the trabecular tissue and scleral spur anteriorly and outwards and the periphery of the iris posteriorly and inwards.

Structurally, the sclera is formed of white fibrous tissue intermixed with fine elastic fibres; flattened connective tissue cells, some of which are pigmented, are contained in lacunae between the fibres. The fibres are aggregated into bundles arranged in characteristic patterns in different parts of the sclera. Thus they are circumferential with respect to the optic papilla, but become reticular in arrangement anterior to this, and markedly meridional near the attachments of the four recti (Kokott 1934). The individual fibrils vary in diameter from 28 to 280 nm with periodicities of 80 and 21 μm. Collagen accounts for 75 per cent of the dry weight of the sclera. The sclera acts as a viscoelastic structure—an important factor in relation to intraocular circulation and pressure (Gloster *et al.* 1957; Helen and McEwen 1961). Its *vessels* are scanty; its capillaries are small, and unite at wide intervals. Its *nerves* are derived from the ciliary nerves.

THE CORNEA

The cornea (7.247) is the anterior, projecting and transparent part of the external tunic, to which is due the major part of the refraction of the rays of light entering the eye. It is convex anteriorly, and projects as a flattened dome in front of the sclera. Its degree of curvature varies in different individuals, and in the same individual at different periods of life, being more pronounced in youth than in old age. As the curvature of the cornea is greater than that of the rest of the eyeball, a slight furrow, called the *sulcus sclerae*, marks the junction of the cornea and sclera. The cornea is dense and about 1·2 mm thick round its periphery and 0·5–0·6 mm at its centre. Its anterior surface is somewhat elliptical, the transverse diameter being slightly greater than the vertical. Its posterior surface is circular and, because the corneoscleral junction is slightly oblique superiorly and inferiorly, is more extensive than the anterior surface in the vertical axis. The diameter of the cornea is about 11·7 mm on its

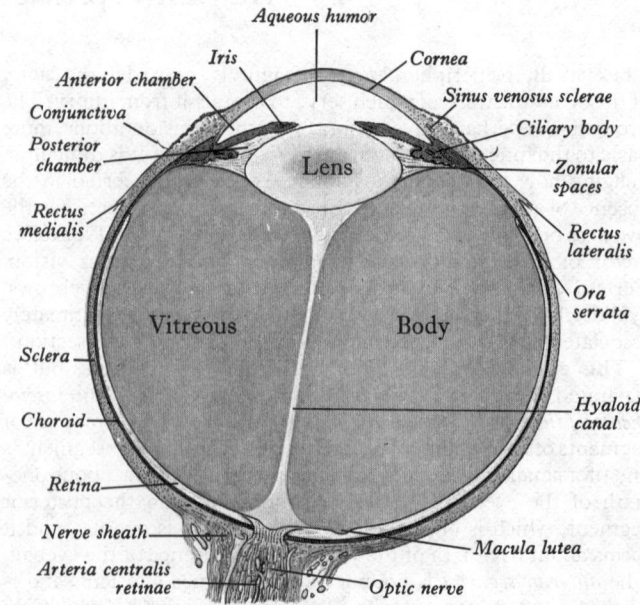

7.247 A horizontal section through a right human eyeball. Superior aspect.

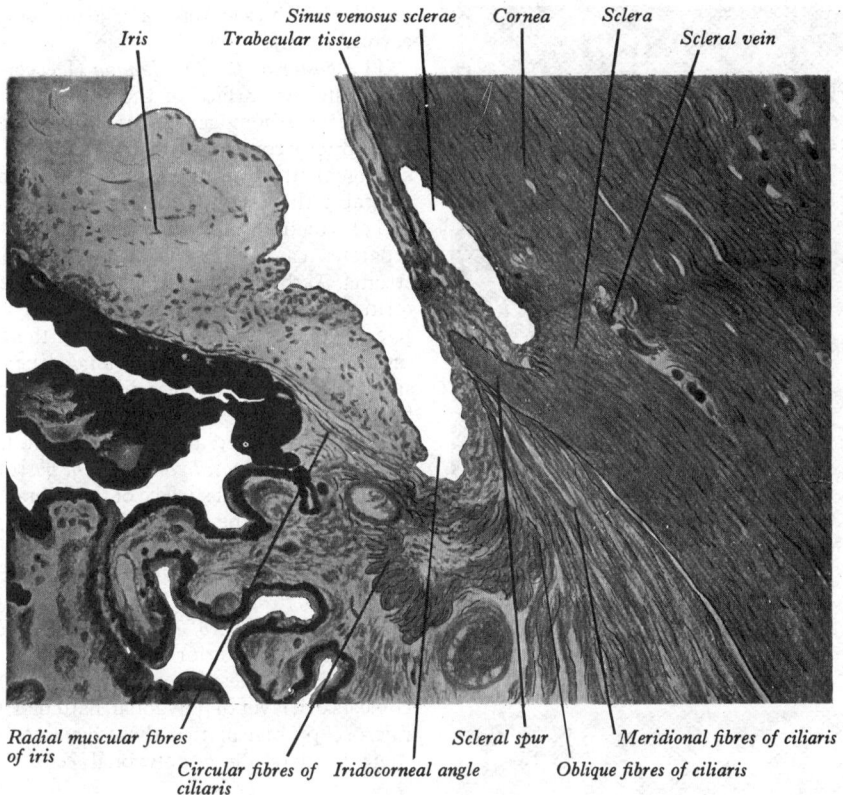

Iris — Sinus venosus sclerae — Trabecular tissue — Cornea — Sclera — Scleral vein

Radial muscular fibres of iris — Circular fibres of ciliaris — Iridocorneal angle — Scleral spur — Oblique fibres of ciliaris — Meridional fibres of ciliaris

7.248   A general view of a meridional section through the iridocorneal angle.

posterior aspect; anteriorly it is 11·7 mm horizontally but only 10·6 mm in the vertical.

Structurally (7.249), the cornea consists from before backwards of five layers—(1) the corneal epithelium, continuous with that of the conjunctiva; (2) the anterior limiting membrane (of Bowman); (3) the substantia propria; (4) the posterior limiting lamina (of Descemet); (5) the endothelium of the anterior chamber.

The *corneal epithelium* covers the front of the cornea and generally consists of five layers of cells. The deepest cells are columnar; their basal surfaces are flat and their outer surfaces rounded, and they contain large round or oval nuclei. The cells of the second layer are polyhedral, with oval nuclei. In the superficial layers the cells become progressively flattened but, unlike the superficial cells of the epidermis, they contain flattened nuclei and they do not normally become keratinized. Most of the cells of the corneal epithelium are 'prickle' cells, similar to those of the germinative zone of the epidermis (p. 1218). At the sclerocorneal junction (or limbus), the epithelium becomes thicker (up to ten or more layers of cells) and is continuous with the conjunctiva covering the sclera. The surface cells of the corneal epithelium display, as do those of the conjunctiva elsewhere, microvilli and microplicae (Dohlman 1971). The latter are considered (Andrews 1975) to be concerned in the retention of an unbroken film of surface secretion, which is essential to what is, of course, the primary refractive surface of the eye.

Scanning electron microscopy of the external surfaces of the corneal epithelial cells has confirmed these interesting features (Pfister 1973; Pfister and Burstein 1977). As is shown in illustration 7.249B, C a pattern of fine irregularities (microvilli) is intermixed with more common, fine, sinuous, intercommunicating ridges. The two types of ultrastructural detail appear to predominate in different cells, giving them a light or dark appearance, the first displaying microvilli, the second microplicae. Some epithelial cells display both features in varying proportions, and a few cells present microprojections merely in the central region of their exposed surfaces. All these observers agree in regarding these projections as helping to maintain the superficial film of tear fluid.

The *substantia propria* is fibrous, compact, unyielding and perfectly transparent. It is composed of about 200–250 flattened,

superimposed lamellae which are made up of bundles of modified connective tissue, the fibres of which are continuous with those of the sclera. Each lamella is about 2 μm thick and of very variable width (10–250 μm). The fibres of each lamella are mostly parallel, but at obtuse angles to those of adjacent lamellae. Fibres frequently pass from one lamella to the next. The dimensions of

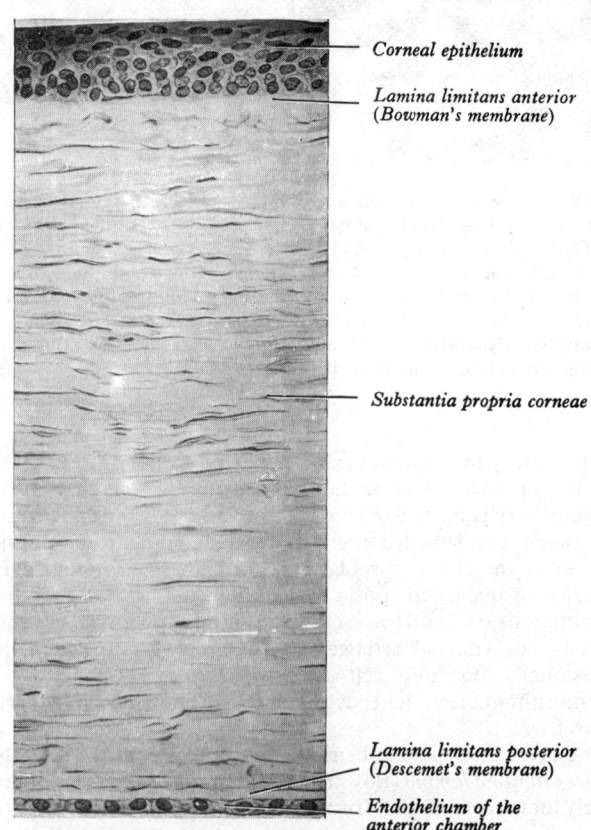

Corneal epithelium

Lamina limitans anterior (Bowman's membrane)

Substantia propria corneae

Lamina limitans posterior (Descemet's membrane)

Endothelium of the anterior chamber

7.249 A   Radial section through the human cornea. × 128.

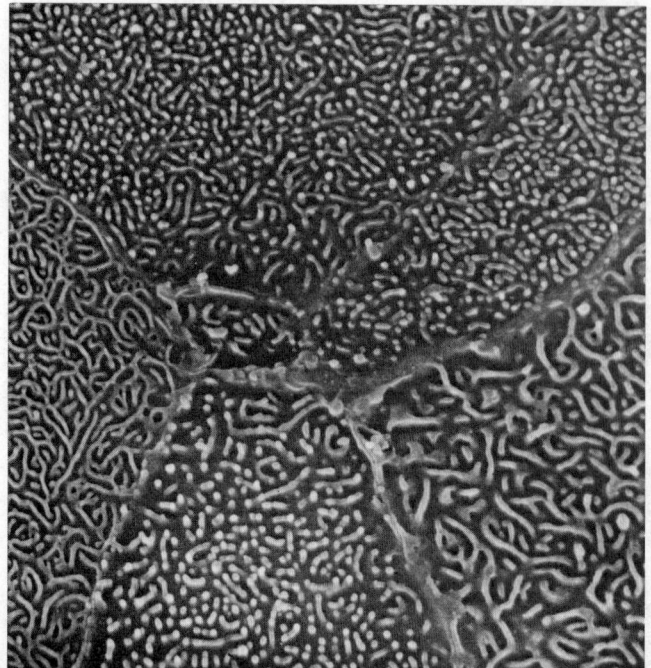

B

C

7.249B, C Scanning electron micrographs of normal human corneal epithelial cells. In B (magnification × 5000) parts of the outlines of several cells are visible; in the upper and lower cells, microvilli predominate, but some microplicae are seen. The cells at the right and left of the field display predominantly microplicae, with only a few microvilli. In C (magnification × 30,000) a small number of microvilli are scattered amongst abundant microplicae. (By kind permission of Drs. R. R. Pfister and N. L. Burstein, *Investigative Ophthalmology* and *Visual Science*.)

the fibres vary in different parts of the cornea, being larger in the vicinity of the posterior limiting membrane. Estimates of diameter vary from 21 to 65 nm with periodicities of 63 and 6 nm.

Between the lamellae there is a small amount of ground substance, in which fibroblasts occur. These are stellate or dendritic in shape and appear to communicate with one another by numerous offsets. However, electron microscopy shows that there is no syncytial arrangement amongst the corneal cells. Occasionally nomadic cells, macrophages, lymphocytes, or polymorphonuclear leucocytes may invade the substantia propria.

The layer immediately beneath the corneal epithelium is the *anterior limiting lamina* (Bowman's membrane). It consists of fine closely interwoven fibrils structurally similar to those found in the substantia propria, but contains no fibroblasts. It is about 8 μm in thickness and its collagen fibres are random in arrangement; views

on their width vary from 14 to 36 nm—much less than those of the corneal stroma.

The *posterior limiting lamina* (Descemet's membrane) covers the posterior surface of the substantia propria, and is a thin, transparent, homogenous membrane. Ultraviolet and polarization microscopy show that it has a stratified, probably fibrillar arrangement, confirmed by electron microscopy. It is considerably thicker than the endothelium to which it is subjacent and of which it is now regarded to be the basement membrane. It separates easily from both the endothelium and the corneal stroma. At the margin of the cornea it breaks up into fibres which form the trabecular tissue on the inner wall of the sinus venosus sclerae (7.250), the spaces between the trabeculae being termed the *spaces of the iridocorneal angle*; they communicate with the sinus venosus sclerae and with the anterior chamber. Some of the fibres of this trabecular tissue pass on to the internal surface of the scleral spur and are continued into the substance of the iris, forming the *pectinate ligament of the iris*; others are connected with the sloping external surface of the scleral spur and a few reach the anterior part of the choroid. The relationships of the trabecular spaces, anterior chamber, and scleral venous sinus, both in structural and functional terms, are still topics of controversy. *See* Davson (1976) for recent views and discussion.

The *endothelium of the anterior chamber* covers the posterior surface of the posterior elastic lamina, is reflected on to the front of the iris, and also lines the spaces of the iridocorneal angle; it consists of a layer of polygonal, flattened, nucleated cells. Electron microscopy reveals that these have no basement membrane. At their borders these polygonal cells display highly complex,

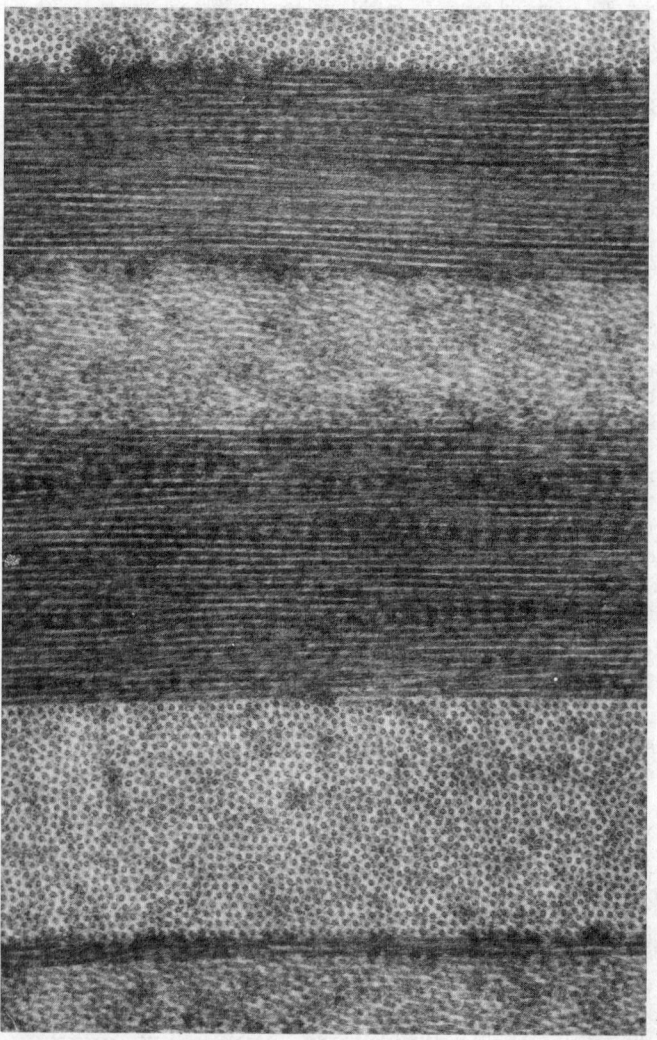

D

7.249D Transmission electron micrograph of the substantia propria of the human cornea; note the geometric precision of the alternation in direction of adjacent layers of collagen fibres. (Preparation by Dr. John Marshall—*see* 7.251B.) Magnification × 48,000.

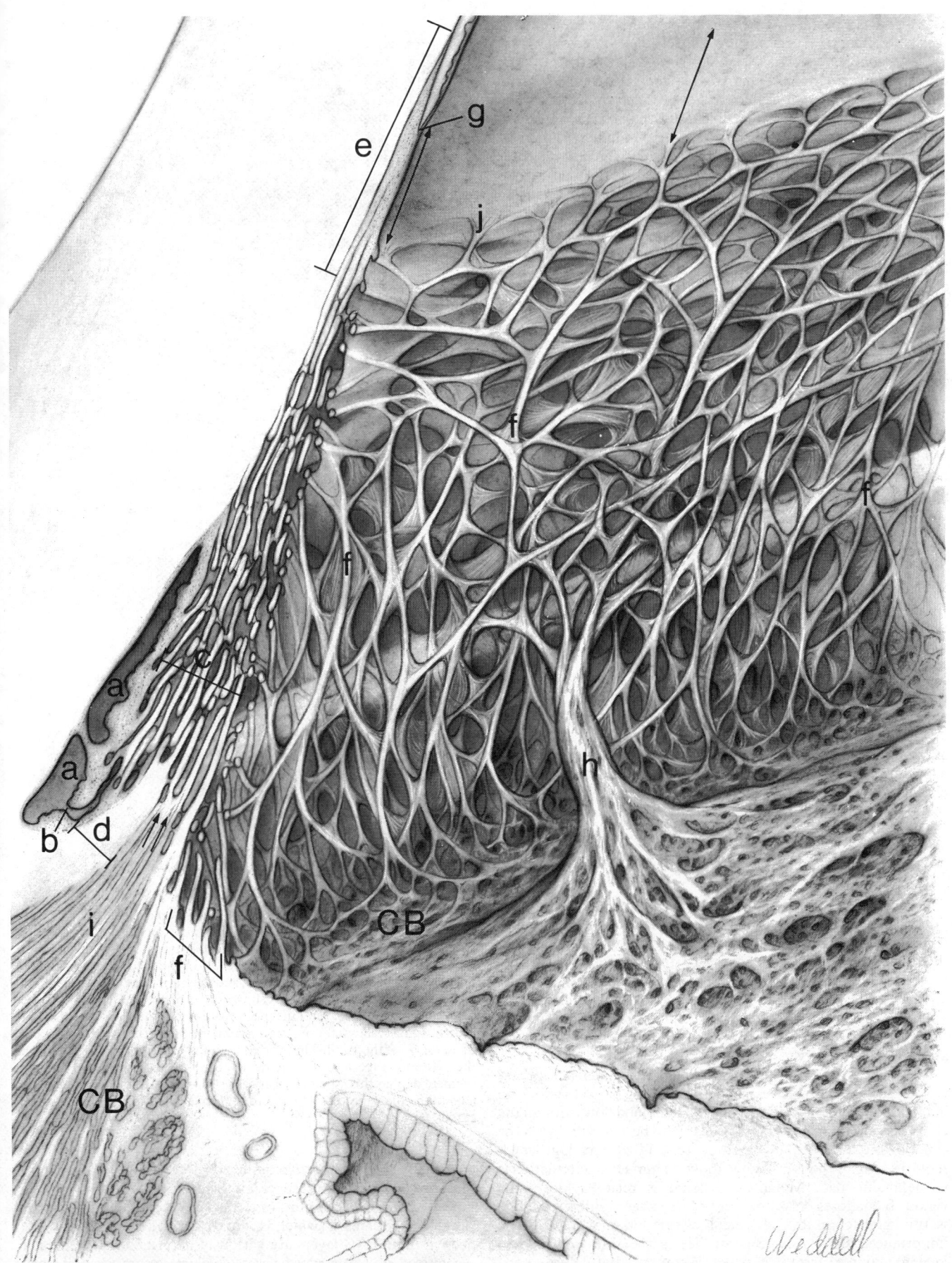

7.250 The iridocorneal angle and adjoining structures, showing the proximity of the scleral venous sinus (*aa*) to the pectinate ligament (*ff*). The trabecular meshwork of the latter is partly uveal, being continuous with the iris (*h*) and ciliary body (CB) and muscle (*i*). Anterior to the scleral spur (*d*) scleral trabecular tissue (*c*) is even closer to the scleral venous sinus. Aqueous fluid percolates through this trabecular region, reaching the lumen of the sinus through small apertures (*b*). The pectinate ligament diminishes as it approaches the corneal limbus (*e*), and in this junctional zone the posterior limiting membrane (of Descemet) also terminates (*g*). The endothelium of the anterior chamber (posterior corneal epithelium) is continuous with the endothelium of the trabeculae (*j*) at the limbus. (From *Histology of the Human Eye*, by Drs. Michael J. Hogan and Jorge A. Alvarado, illustrated by Mrs. Joan E. Weddell, published by W. B. Saunders, Philadelphia, 1971—by kind permission of the authors, artist and publishers. **7.251**, 253, 254, 255, 262, and 265 are from the same source.)

interdigitating profiles when seen by scanning electron microscopy, and their free surfaces show sparsely distributed microvilli.

Vessels and Nerves. The cornea is a non-vascular structure, the
capillary vessels of the conjunctiva and sclera ending in loops at its
circumference. Lymph vessels do not occur in the cornea. The
nerves are numerous and are branches of the ophthalmic nerve,
and particularly the long ciliary nerves. Around the periphery of
the cornea they form an *annular plexus*, from which fibres enter
the substantia propria in a radial pattern. They lose their myelin
sheaths and ramify throughout the substantia propria in a delicate
network, and their terminal filaments form an intricate plexus
beneath the corneal epithelium. This is termed the *subepithelial
plexus*, and from it fine, varicose fibrils are given off which pass
through the anterior limiting membrane to ramify between the
epithelial cells, forming an *intra-epithelial plexus*. Ultrastructural
studies (Matsuda 1968) of human corneal nerve fibres show that
the endoneurium and myelin sheaths are gradually lost as the
fibres enter the cornea. There are no specialized end organs, and
in their epithelial course the nerve fibres are devoid of Schwann
cells; they do not arborize. Electron microscopic studies of the
simian corneal nerves suggest that nerves entering the stroma do
not pass to the corneal epithelium, whose nerves enter it directly
from the limbus (Lim and Ruskell 1978).

The fibrous tunic of the eye is usually dismissed as being merely
protective, but its common and continuous functions are to
provide a smooth translatable external surface (*vide infra*), a
suitable attachment for muscles and a resistance to the intraocular
pressure. In this manner the optical shape and dimensions of the
eyeball are maintained. For further details of the physiology of the
cornea and sclera *see* Jakus (1964), Laugham (1969), Davson and
Graham (1974).

The Vascular Tunic

The vascular tunic or *uveal tract* of the eye consists of the choroid,
the ciliary body and the iris (7.247), forming, of course, a
continuous structure.

The choroid covers the inner surface of the sclera, and extends
as far forwards as the ora serrata of the retina. The ciliary body
continues from the anterior edge of the choroid into the
circumference of the iris. This iris is a circular diaphragm behind
the cornea, and presents near its centre the rounded aperture of
the *pupil*.

THE CHOROID

The choroid is a thin, highly vascular membrane, of a dark brown
or chocolate colour, lining somewhat less than the posterior five-
sixths of the eyeball; it is pierced behind by the optic nerve, and in
this situation is firmly adherent to the sclera. It is thicker
posteriorly. Its external surface is loosely connected with the
sclera by the *suprachoroid lamina* (lamina fusca); its internal
surface is firmly attached to the pigmented layer of the retina. At
the optic disc it is continuous with pial and arachnoid tissues
around the optic nerve.

Structurally, the choroid consists mainly of a dense capillary
plexus, and of small arteries and veins carrying blood to and from
it. (In accord with its high vascularity, the blood flow through the
choroid is also high, but this is probably to be associated with the
effect of the intra-ocular pressure, some 15–20 mm Hg, which
requires a high venous pressure, above 20 mm Hg, if circulation is
to be maintained. Metabolic demand is relatively low—the
venous blood loses only 3 per cent of its oxygen (Bill 1970). The
warming effect of the choroidal circulation may have some
importance.) On its external surface is the thin *suprachoroid
lamina*, about 30 μm thick, which is composed of delicate non-
vascular lamellae, each consisting of a network of fine collagen and
elastic fibres, among which are branched cells containing dark-
brown pigment granules. Ganglionic neurons and plexuses of
nerve fibres are enmeshed in the connective tissue. The
mesothelium-lined interstices, sometimes described, are not
supported by modern observations; pathological accumulations
of fluid may split the lamellae of the suprachoroid. All such

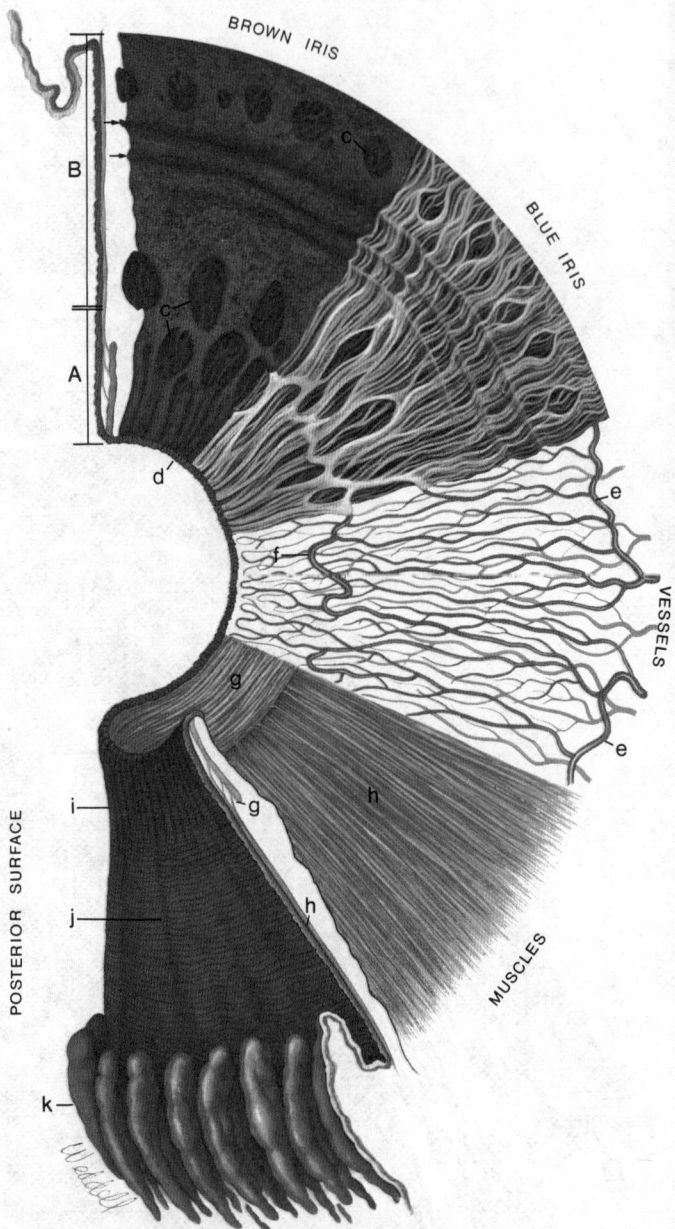

7.251 A Composite view of the surfaces and internal strata of the iris. In
a clockwise direction from above the pupillary (A) and ciliary (B) zones are
shown in successive segments. The first (brown iris) shows the anterior
border layer and the openings of crypts (*c*). In the second segment (blue
iris), the layer is much less prominent, and the trabeculae of the stroma are
more visible. The third segment shows the iridial vessels, including the
major arterial circle (*ee*) and the incomplete minor arterial circle (*f*). The
fourth segment shows the muscle stratum, including the sphincter (*g*) and
dilatator (*h*) of the pupil. The everted 'pupillary ruff' of the epithelium on
the posterior aspect of the iris (*d*) appears in all segments. The final
segment depicts this aspect of the iris, showing radial folds (*i* and *j*) and the
adjoining ciliary processes (*k*). (From *Histology of the Human Eye*—see
7.250.)

potentially weak connective tissue zones are a great attraction to
those in search of lymph spaces in the eye.

The choroid proper (7.251B, 225) is internal to the
suprachoroid lamina (which is in part derived from scleral connective tissue). Its layers are variously described, but generally
recognized are (*a*) an external *vascular lamina* composed of small
arteries and veins and loose supporting connective tissue, in
which are scattered pigment cells, (*b*) an intermediate *capillary
lamina* (choroidocapillaris), and (*c*) a thin, apparently structureless, *basal lamina* (the membrane of Bruch). The vascular lamina
is itself sometimes divided on the basis of the calibre of its vessels,
which naturally decreases towards the capillary layer. In general,
however, these vessels are relatively small and the capillaries

large, and the latter, in any case, dominate the scene. The vascular lamina contains the branches of the short posterior ciliary arteries (p. 689), extending anteriorly in a meridional direction from their entry through the sclera near the optic disc. The veins of the vascular lamina are much larger, and they converge in whorls or vortices upon four or five principal *vorticose veins*, which pass through the sclera to drain into the ophthalmic veins of the orbit. The capillary or choroidocapillary layer, separated from the retina only by the choroidal basal lamina, is almost certainly responsible for the maintenance of the outer layers of the retina, at least in part. It forms a finely meshed network, especially in the posterior region of the eyeball; but the meshes become larger as they approach the ciliary body, where they link up with the capillaries of the ciliary processes. The basal lamina was until recently considered to be a glassy, homogenous layer (lamina vitrea); it is only 2–4 μm in thickness, but electron microscopy has revealed much intimate detail. It consists essentially of a middle stratum of elastic tissue between internal and external collagenous layers, united externally to the basement membrane of the choroidocapillary lamina and internally to the basement membrane of the pigment cells of the retina (Lerche 1965; Nakaizumi 1964). Its exact functional significance is not certain, but it is obviously related to the passage of fluid and solutes from the choroidal capillaries to the retina, and it is said to provide a smooth surface for the precise orientation of the pigment cells and receptors of the retina, a factor important to precise vision. In its development (Takei and Ozanics 1955) the basal lamina is derived from retina and choroid, the former contributing its most internal layer which is in fact the basement membrane of the retinal pigment epithelium. Its most external stratum is closely associated with, and perhaps identical with, the basement membrane of the choroidal capillary endothelium. The functions of the pigment cells of the choroid are speculative. Clearly it is plausible to assume that they prevent the passage of light from the exterior of the sclera; but it seems more probable that they absorb light which has passed through the retina, thus preventing reflexion. In many animals, especially in those of nocturnal habit, specialized cells in the choroid form a reflecting structure, the *tapetum*, which is responsible for the greenish glare visible in the eyes of such animals at night (Walls 1963). The significance of this arrangment is uncertain; it may be a mechanism of aggression or it may bring about increased stimulation of the retinal receptors.

Although there is some clinical evidence of a nerve supply to the choroid, direct anatomical evidence is sparse. In a recent study of the results of crush lesions of the ophthalmic nerves in monkeys (Bergmanson 1977), no evidence of fibres from the ciliary nerves,

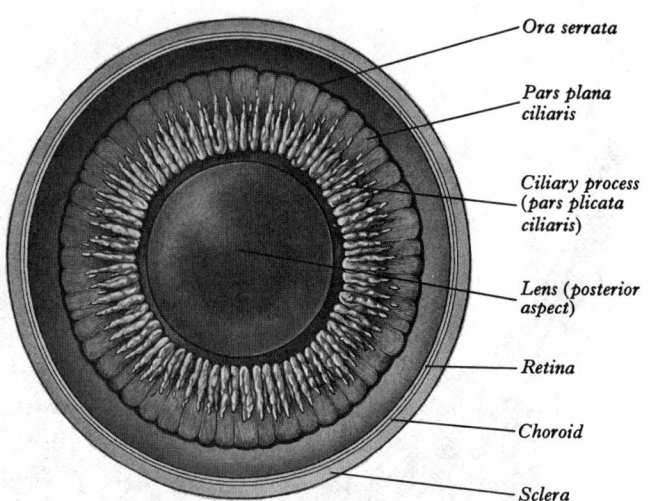

7.252 The interior aspect of the anterior half of the eyeball.

in the suprachoroidal part of their courses, was found; none of their degenerating fibres terminated in the choroid.

THE CILIARY BODY

The ciliary body is a direct anterior continuation of the choroid, and the iris is a further extension of the ciliary body itself. All three regions of the uveal tract evince certain common features and regional differences dependent on variations in function. The ciliary body is specifically concerned with the suspension of the lens and with the mechanism of accommodation, and this accounts for the accumulation of muscle which causes it to bulge towards the interior of the eyeball (7.253). It is also involved as the major producer of aqueous fluid entering the anterior segment of the eye, with which its anterior aspect is related. More posteriorly it is directly contiguous with the vitreous humor, and it is probable that it secretes some of the vitreous glycosaminoglycans. The anterior and the long and short posterior ciliary arteries all meet in the ciliary body (7.255), and it is hence a highly vascular region. This rich circulation is concerned not only with the secretory and muscular activities of the body, but also with the supply of the iris and the limbal region. The ciliary body is also traversed by the major nerves supplying all the anterior ocular tissues.

Externally, the ciliary body extends from a point about 1·5 mm posterior to the corneal limbus (which corresponds also to the scleral spur) to a point 7·5 to 8 mm posterior to this on the temporal side of the eyeball and 6·5 to 7 mm on the medial or nasal side. The body is hence a slightly eccentric structure, extending posteriorly from the scleral spur, to which it is attached, with a meridional width varying from about 5·5 to 6·5 mm. As seen from the interior of the eyeball it presents a posterior periphery, where it is continuous with the choroid, which is crenated or scalloped—the *ora serrata*. Its anterior extremity is confluent with the periphery of the iris, and lateral to this it bounds the iridocorneal angle of the anterior chamber. The internal aspect of the ciliary body is grey in colour, due to the pigment in the deeper layer of its epithelium. It is divisible into an anterior ridged or plicated part, the *corona ciliaris* (pars plicata), which surrounds the base of the iris in an annular manner, and posterior to this a relatively smooth annular strip, the *orbiculus ciliaris* (pars plana, ciliary ring). The orbiculus accounts for more than half of the meridional width of the ciliary body, being 3·5 to 4 mm across. The peripheral rim of the orbiculus is the ora serrata, a dentate limit at which the fully developed *optical* or sensory part of the retina is suddenly reduced to two layers of epithelial cells, prolonged over the whole of the ciliary body as the *pars ciliaris retinae* and beyond this on to the posterior surface of the iris. The corona ciliaris, forming a smaller annular region within the orbiculus, is ridged by seventy to eighty *ciliary processes* which radiate in a meridional direction from the base of the iris towards the orbiculus ciliaris (7.252). Branching

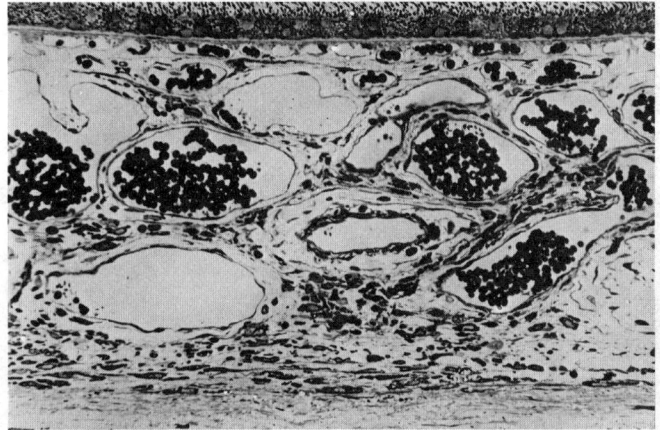

7.251B Light micrograph of a section of human eyeball showing full thickness of the choroid coat and adjacent parts of the retina (above) and sclera (below). Note, from above downwards, rod and cone processes projecting among pigment cells, the basal lamina (membrane of Bruch), layer of capillaries (choriocapillaris), layer of larger vessels, and loose connective tissue merging into the sclera. Numerous pigment cells are scattered throughout the choroid. Magnification × 350. (Kindly provided by Dr. John Marshall, Institute of Ophthalmology, London.)

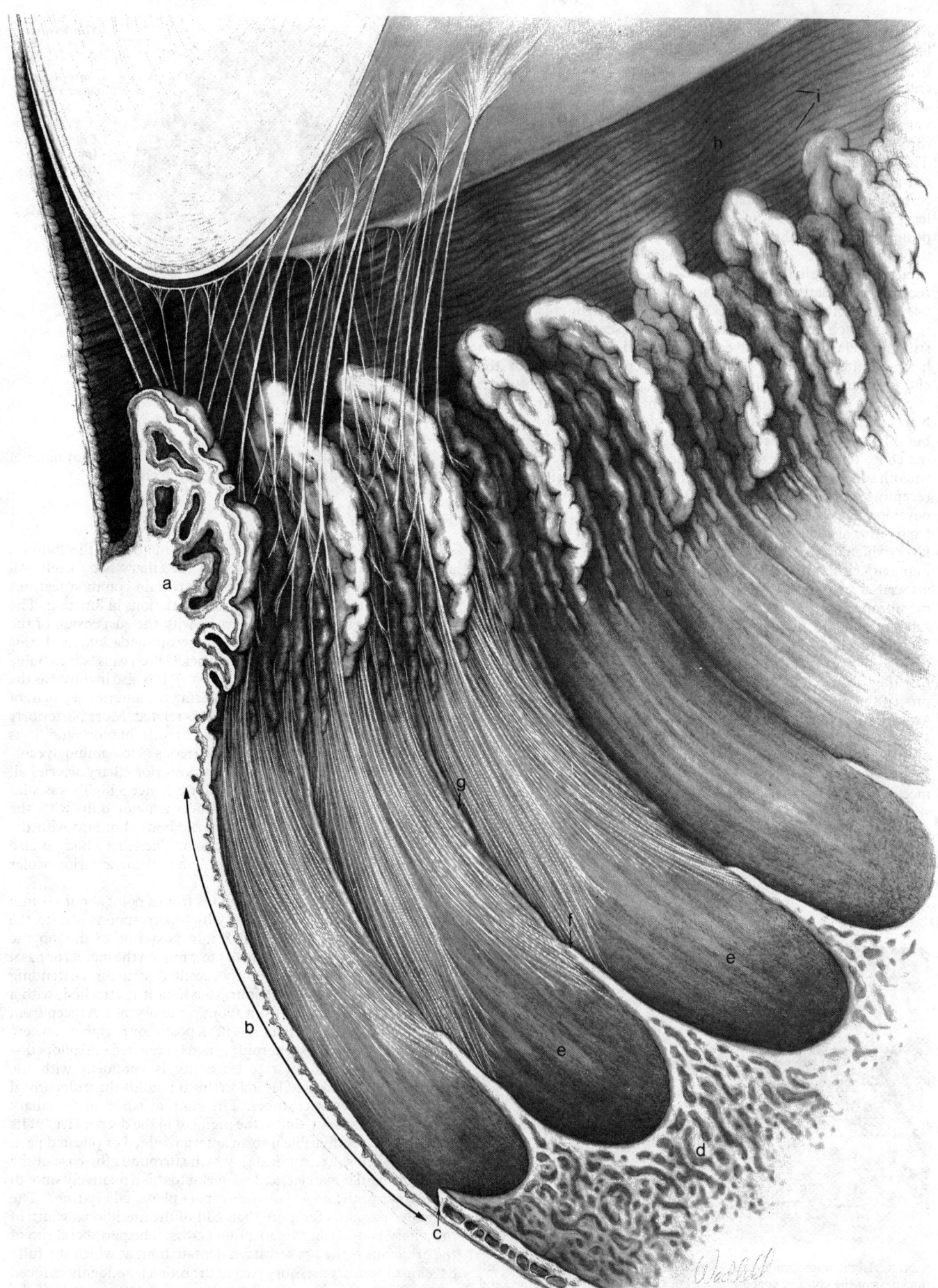

7.253 A magnified view of the ciliary region seen from the ocular interior. Above is the periphery of the lens, attached by the fibres of the *zonule* (suspensory ligament) to the processes of the *corona ciliaris* (pars plicata) of the ciliary body (a). The *orbiculus ciliaris* or pars plana ciliaris (b) has a scalloped boundary, the *ora serrata* (c), which separates it from the retina (d). Flanking the 'bays' (e) of this are the *dentate* processes (f), with which linear ridges or *striae* (g) are continuous. These striae extend forwards between the main ciliary processes, providing an attachment for the longer zonular fibres. The posterior aspect of the iris shows radial (h) and circumferential (i) sulci. (See **7.250** for acknowledgement.)

from the sides of these processes into the valleys between them are numerous minor ridges, the *ciliary plicae*, forming a complex pattern which, in microscopic preparations, presents highly intricate profiles (7.253). Into the valleys between the ciliary processes groups of fibres of the *zonule* (suspensory ligament) of the lens extend, passing beyond the processes to establish continuity with the basement mebrane of the superficial layer of epithelial cells covering the orbiculus ciliaris. The sites of these attachments are marked by striae which pass posteriorly from the valleys of the corona across the orbiculus almost to the apices of the dentate processes of the ora serrata (7.253).

For description it is convenient, if arbitrary, to treat the intimate structure of the ciliary body under three headings: (1) the ciliary epithelium; (2) the ciliary connective tissue and vessels; and (3) the ciliary muscle.

The ciliary epithelium is bilaminar, consisting of two layers of simple epithelium superimposed one upon another and representing the two layers of the optic cup. The *superficial lamina* is formed of cells which are columnar over the orbiculus and cuboidal where they cover the ciliary processes, becoming irregular and more flattened in the intervals between the processes. They contain little or no pigment and are the sole anterior continuation of the neural layers of the retina; this excludes the pigment epithelium of the retina, which is itself continuous with the *deeper layer* of the ciliary epithelium. The cells of the latter are also approximately cuboidal and are loaded with pigment granules. The two layers are firmly united, but pathological accumulations of fluid may separate them, just as the retina detaches from its own pigment epithelium. A basement membrane intervenes between the two epithelia, and the basal aspects of the superficial cells are much infolded, as in the case of other secretory epithelia. These superficial cells exhibit junctions of the desmosomal type, and in their cytoplasm mitochondria are numerous and the endoplasmic reticulum is well developed, the latter often forming stacked arrays in the peri-nuclear zone. The Golgi apparatus is not well developed. Lipid and melanin granules are often present but not prominent.

The pigment epithelium is united to the stroma of the ciliary body by its own basement membrane, which is continued posteriorly into the basal lamina of the choroid tunic (p. 1156). The cytoplasm of these cells contains very numerous round or oval granules containing abundant melanin and measuring about 0·6 to 0·8 μm in diameter. The cells are linked laterally to each other by relatively few desmosomes, these being more numerous between cells in the two strata of the epithelium, despite the intervention of a basement membrane.

The ciliary stroma is composed largely of loosely arranged fasciculi of collagen fibres, and these are aggregated into a considerable mass between the ciliary muscle and the overlying ciliary processes, into both of which the connective tissue extends. In this inner stratum of connective tissue are numerous larger branches of the ciliary arteries and veins with a rich interconnecting network of capillaries of comparatively large calibre; the majority of these are adjacent to the epithelium, and are especially concentrated in the ciliary processes. In these sites capillaries are chiefly of fenestrated type; numerous vessels also enter the ciliary muscle, but the capillaries there evince much less frequent fenestration. Anteriorly, near to the periphery of the iris, is the major arterial circle (7.251, 255), formed chiefly by the long posterior ciliary arteries, branches of the ophthalmic (p. 689), which enter the eye well posterior to the equator and pass anteriorly between the choroid and the sclera to reach the ciliary body. The veins of the body, into which those of the iris drain, pass posteriorly to join the vorticose veins of the choroid.

The ciliary muscle has been variously described by different authorities, their main divergencies being in the number of divisions or parts recognized in this small annular mass of nonstriated muscle. In most descriptions three parts are usually named—*meridional*, *radial* or oblique, and *circular* or sphincteric—but other views have occasionally been stated (*see* Calasans 1953; Rohen 1964). Most and perhaps almost all of the ciliary muscle fibres are attached to the scleral spur (7.254), from which they pass in a variety of directions. It is upon these variations that a somewhat arbitrary division of the whole muscle

into parts is dependent. The most external fibres extend in a meridional or longitudinal direction, passing posteriorly into the stroma of the choroid, where many exhibit terminal branchings or *epichoroidal stars*. The most internal fibres swerve acutely as they leave the scleral spur (7.254) and run circumferentially to form a circular or sphincteric element in the muscle, in close proximity to the periphery of the lens. Between these two muscular strata are fibres which cross obliquely from one to the other, often crossing each other in a lattice of interweaving fibres. This part of the muscle is often referred to as radial in disposition. In its ultramicroscopic features the ciliary muscle exhibits some differences from other nonstriated muscle masses. The fibres show distinct cell walls and possess basement membranes, but they contain an unusual abundance of mitochondria and endoplasmic reticulum. A small bundle of fibres is usually surrounded by a common fibroblastic sheath, forming units not encountered in other nonstriated muscles. Junctions between the fibres within a bundle are described, but their precise nature is still uncertain; they are said to resemble intercalated discs. Three types of nerve ending have been observed, the most common being an indirect contact of synaptic membranes with an interposed basement membrane; contact without a basement membrane also occurs, and most rarely larger and more intimate contacts in depressions in the muscle fibre substance.

Both myelinated and nonmyelinated nerve fibres abound in the ciliary muscle and elsewhere in the ciliary body. The latter are postganglionic fibres derived from the ciliary ganglion, where they link with the parasympathetic outflow of the oculomotor nerve; but there is considerable evidence that some of these fibres are sympathetic. While it is clear that the former supply stimulates the fibres of the ciliary muscle to contract, the role of the sympathetic supply is still unsettled. Cervical sympathetic stimulation in some experimental animals leads to flattening of the lens, which is tantamount to relaxation of accommodation, but the mechanism of this is uncertain. It may be due to an inhibitory effect on the ciliary muscle; but it has also been suggested that the volume of the ciliary body may be reduced by vasoconstriction, thus resulting in tension on the zonule and through this on the periphery of the lens—the reverse of the slackening effect on the zonule of ciliary contraction (Morgan 1944). For a recent critique of this topic consult Alpern (1969). An electron microscopic study of the autonomic plexus in the ciliary muscles of rhesus monkeys has revealed, according to Townes-Anderson and Raviola (1979), a cycle of degeneration and regeneration which becomes more marked in older animals.

THE IRIS

The iris is the delicate and adjustable diaphragm which surrounds the *pupil*, its central orifice (slightly medial to the true centre), which exerts a considerable control over the amount of light entering the eye. The pupil may vary in diameter over a range of at least 1 to 8 mm, and even more under the influence of drugs. This represents an effective aperture range in excess of f 20 to f 2·5, and a ratio of 32:1 in the amount of light permitted to enter the eye. While this is obviously not enough to save the retina from the effects of very intense illumination, it is a factor in smoothing out the wide range of luminosities encountered in ordinary use and in thus preserving useful vision in highly variable conditions. The pupillary diameters noted above and an average diameter of the iris of about 12 mm are, of course, taken as measured through the cornea, whose dioptric power introduces a magnification factor of about 12 per cent. *Constriction* and *dilatation* of the pupil are self-explanatory terms, for which meiosis and mydriasis are also used clinically, though these are more properly reserved for the extreme limits of contraction and dilation. For a most erudite discussion of the immense literature on the pupil and its responses, and an account of iris activity *see* Loewenfeld (1958) and Loewenstein and Loewenfeld (1970).

Though the iris is named after the rainbow, its range of colour is somewhat less, extending from light blue to a very dark brown. The hue may vary in the two eyes and through the same iris. The colour is due to the combined effects of the iridial connective tissue and pigment cells in absorbing or reflecting different

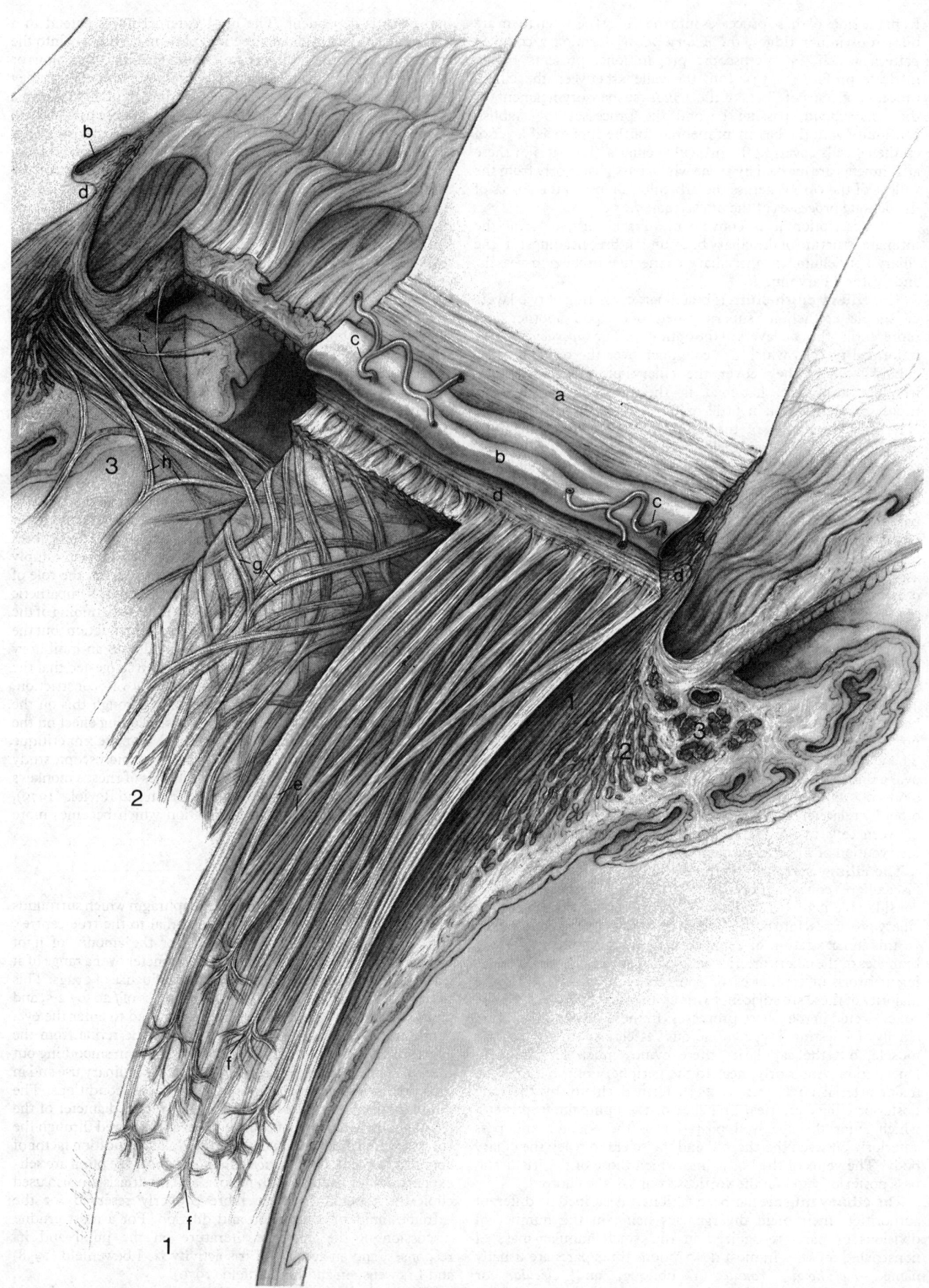

7.254 The ciliary muscle and its components. The meridional or longitudinal (1), radial or oblique (2), and circular or sphincteric (3) layers of muscle fibres are displayed by successive removal towards the ocular interior. The cornea and sclera have been removed, leaving the pectinate ligament (a), the scleral venous sinus (b), collecting venules (c) and scleral spur (d). The meridional fibres often display acutely angled junctions (e) and terminate in epichoroidal stars (f). The radial fibres meet at obtuse angles (g), and similar junctions, at even wider angles (h), occur in the circular stratum of the ciliaris. (See 7.250 for acknowledgement.)

frequencies of light energy in a selective manner. In the absence of significant amounts of pigment, as in the iris of the newborn, the colour is light blue; some degree of pigmentation is necessary to confine light transmission to the pupil and to the centre of the lens, where optical aberrations are least. The concentration of melanocytes is the predominant factor in iris hue, but their distribution is often irregular and may produce in this manner a flecked or maculated appearance.

In shape the iris is not a discoid diaphragm; the anterior convexity of the lens bulges it a little, so that it is more accurately described as a very shallow cone, truncated by the pupillary aperture. It is sited between the cornea and the lens (7.247), immersed in the *aqueous fluid* or humor, and it partially divides the *anterior segment* of the eye into an *anterior chamber*, enclosed by the cornea and iris (which meet at the *iridocorneal angle*) and an unfortunately termed *posterior chamber*, between the iris and the lens. Peripherally, in the latter cavity, the ciliary processes protrude a little between the divisions of the zonular ligament of the lens; and it is here that most of the aqueous fluid is produced, finding its way through the pupil into the anterior chamber and finally to its exit into the scleral venous sinus (p. 1152) at the iridocorneal angle—the 'filtration angle' of clinical parlance.

The microscopic structure of the iris displays a number of unusual features (7.251). Its anterior surface, forming the posterior boundary of the anterior chamber, is not covered by a distinct epithelium, despite frequent statements to the contrary; this surface is merely a modified 'anterior border layer' of the general *stroma*, which forms the bulk of the iris. The stroma contains the vessels and nerves of the region and, near the periphery of the pupil, an aggregation of nonstriated muscle fibres forming an annular contractile structure, the *sphincter pupillae*. The posterior aspect of the iris consists of a continuation of the same two layers of epithelium which cover the ciliary body and which represent the internal and external strata of the optic cup. The pupil, through which this epithelium turns round for a short distance on to the *anterior* surface of the iris as the pigment ruff, or 'border', therefore corresponds with the opening of the optic cup. The deeper, and hence in the iris the more anterior of these epithelial layers, is commonly termed, a little confusingly, the *anterior epithelium* of the iris. It should be emphasized to avoid confusion, that this anterior epithelium is immediately *posterior* to the stroma of the iris. Its cells are pigmented, like those of the same layer in the ciliary epithelium; closely associated with them are the radially arranged nonstriated fibres of the dilatator pupillae, which like the sphincter has a most unusual embryological origin in being derived from the neural ectoderm of the optic cup. Superficial and posterior to this layer of cells is a stratum of heavily pigmented cells, forming the so-called *posterior epithelium*. The layer is continuous with the *non-pigmented*, retinal layer of the ciliary epithelium.

The *anterior border layer* or anterior surface of the iris has been much studied at low magnification by slit-lamp microscopy, with which it is seen to display a somewhat fluffy appearance, except in heavily pigmented irides. Depressions or *crypts*, through which vessels may be visible in the stroma, and various radial and circular folds and striae can be observed, but details of this kind should be sought in a suitable authority (e.g. Vogt 1942). The constituents of the anterior border lamina are chiefly much branched fibroblasts and melanocytes, with no vestige of the endothelium which covers it at birth, and to a rapidly decreasing extent during the early postnatal years (Vrabec 1952). This is confirmed by electron microscopy (Tousimis and Fine 1959). The fibroblasts form an approximately single continuous layer on the surface, with branching processes which form no actual junctions (Smelser and Ishikawa 1966). At the peripheral base of the iris they blend with the connective tissue of the trabecular meshwork (pectinate ligament) at the iridocorneal angle. At the pupillary border they come into contact, but again without specialized junctions, with the pigment epithelium of the posterior surface of the iris. The melanocytes also exhibit intricately branched processes, and again no special junctions have been observed between them. Some capillaries invade the border layer. Macular junctions have recently been reported in the anterior border layer of the iris in monkey and rabbits,

occurring chiefly between fibroblasts. Naked axons were also observed in close apposition to melanocytes and fibroblasts. The nerve fibres were presumed to be sympathetic and it is hypothecated that the cells and axons may release transmitter substances into the anterior chamber (Ringvold 1975).

The *stroma of the iris*, which is derived like the anterior border layer from the mesoderm between the developing lens and optic cup, is also formed by fibroblasts and melanocytes; but in this region there is, in addition, a considerable amount of loose collagenous tissue, whose spaces are filled with fluid and a mucopolysaccharide ground substance. These spaces are apparently in fairly free communication with the fluid in the anterior chamber of the eyeball, and the interchange of fluid between the chamber and the iridial stroma may explain the large changes of volume which appear to accompany contraction and relaxation of the iris diaphragm. The mesodermal stroma also contains not only an abundance of blood vessels and nerves, but also the ectodermal sphincter and dilatator muscles. There is no elastic tissue, and any of the elastic recoil which has been attributed to the iris, and sometimes suggested as a dilatation force when the sphincter is relaxed, must reside in other structures, if it in fact exists. The collagen fibrils, which have a diameter of about 60 nm, and a periodicity of 50 to 60 nm, are very loosely arranged, many describing incomplete circumferential loops around the pupil as centre. 'Clump' cells, mast cells, macrophages and lymphocytes are said to occur in the stroma, but the reader is referred to larger monographs quoted at the end of this account for fuller details.

The sphincter pupillae is an annular flattened band of nonstriated muscle about 0·75 mm in width and 0·15 mm in thickness. Its fusiform cells are closely packed and are often arranged in small groups, as in the ciliary muscle; in accordance with the effect of the muscle on the pupil these fibres are orientated parallel to its margin. Stromal collagen tissue encloses the muscle anteriorly and posteriorly and is particularly dense in the latter situation, where it binds the sphincter pupillae to the pupillary extremity of the dilatator muscle. Ultramicroscopy shows that the muscle fibre groups noted above are not entered by nerve fibres, only one fibre of the group usually being innervated. It is presumed that the depolarizing current of contraction spreads to the rest of the fibres through gap junctions. In other details of cytoplasmic organelles, densities resembling Z-bands, and basement membranes, the muscle fibres are like those of nonstriated muscles elsewhere. Small nerves ramify in the connective tissue between the fibre bundles; most of their fibres are nonmyelinated and they are often enclosed as groups of several axons in the same Schwann cell sheath. They do not approach the surfaces of muscle cells more closely than 0·1 μm.

The dilatator pupillae, a muscle whose existence has been the subject of a prolonged, vexed and at times almost ridiculous controversy, is now a well-established entity, on microscopic, physiological and pharmacological grounds (Alphen 1963; Loewenfeld 1958). It is a thin stratum immediately anterior to the deeper, anterior layer of the epithelium of the posterior aspect of the iris. Its fibres are indeed muscular processes of this anterior layer, whose cells are therefore myoepithelial in character. Their apical processes form the epithelium itself (*supra*). Myofilaments appear in both parts of these cells, but are much more numerous in their basal, muscular processes. The latter are about 4 μm thick, 7 μm wide, and 60 μm in length. They are fusiform and form a stratum 3 to 5 elements thick through most of the iris, from its periphery to a point near the outer perimeter of the sphincters, which it overlaps a little. Towards this perimeter the dilatator rapidly peters out, sending spurs of muscle processes to blend with it. These processes, unlike the apical parts of the myoepithelial cells of the anterior epithelium, show a clear basement membrane, and they are joined by gap junctions like those in the fibres of the sphincter which probably also serve as points of electrical coupling. Their myofilaments are about 3 nm in diameter and numerous densities resembling Z-discs are evident. Nonmyelinated nerve fibres have been described in relation to the muscular processes or 'fibres', and these terminate very close to the cell membrane, the interval being of the order of 20 nm.

The arteries of the iris (7.251, 255, 256) are derived from the long posterior and the anterior ciliary arteries, and from the vessels of the ciliary processes. Each of the two long ciliary arteries, on reaching the attached margin of the iris, divides into an upper and a lower branch; these anastomose with corresponding branches of the artery from the opposite side and with the anterior ciliary arteries, and form a vascular circle (*circulus arteriosus major*). From this circle vessels converge to the free margin of the iris, and there communicate to form a second circle (*circulus arteriosus minor*). This minor circle is incomplete, and some observers regard these vessels as venous. The smaller arteries and veins are very similar in the structure of their walls, and they also share certain peculiarities. Thus, they are often slightly helical—perhaps an adaptation to the great changes in the shape of the iris which occur as the pupil varies in size. Perhaps also to be ascribed to this is the peculiar structure of the vascular wall. All the vessels, including the capillaries, have a non-fenestrated endothelium, with a well-marked and often thick basement membrane. Outside this, in the arteries and veins, there is no elastic lamina, and nonstriated muscle fibres are few, especially in the veins. The connective tissue of the media is loose, and external to this is a remarkably dense collagenous adventitia, which appears to form almost a separate tube outside the

endothelium. The loose stratum of the media has been regarded as a lymph space, but this is improbable; it is about 7 μm in width, and contains a matrix probably derived from the basement membrane of the endothelium (Hogan *et al.* 1971).

The nerves of the iris, like those of the choroid, are chiefly derived from the branches of the long ciliary rami of the nasociliary nerve and from the short ciliary branches of the ciliary ganglion. The latter contain postganglionic but thinly myelinated fibres which innervate the sphincter pupillae. The dilatator pupillae is supplied by postganglionic nonmyelinated fibres derived from the superior cervical ganglion of the sympathetic trunk, but the routes by which these reach the muscle are not precisely established, and they may vary in different species of mammals and may be multiple in man. The sympathetic plexus around the internal carotid artery is said to send a branch through the ciliary ganglion, and these postganglionic fibres reach the eyeball through the short ciliary nerves; but some sympathetic fibres may travel to the eye through the long ciliary nerves. (For details of other routes in monkeys, see p. 1064.) The innervation of the muscles of the iris, as that of the ciliaris, is probably more complex, and both the sphincter and dilatator may possess a double autonomic innervation. Histochemical stains for acetylcholine-esterase and fluorescent techniques have de-

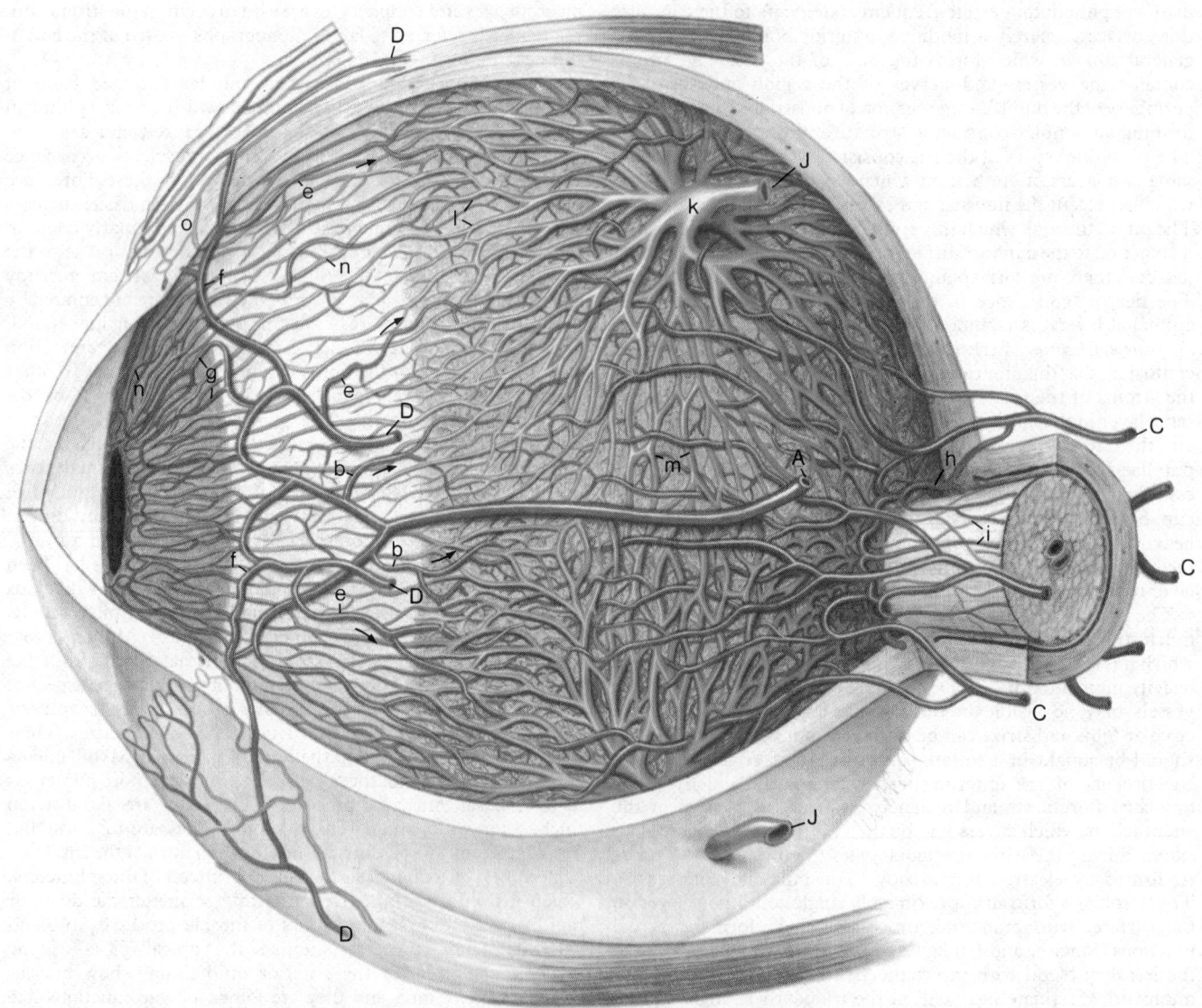

7.255 The vascular arrangements of the uveal tract. The long posterior ciliary arteries, one of which is visible (A), branch at the ora serrata (*bb*) and feed the capillaries of the anterior part of the choroid. Short posteriorly ciliary arteries (CC) divide rapidly to form the posterior part of the choriocapillaris. Anterior ciliary arteries (DD) send recurrent branches to the choriocapillaris (*ee*) and anterior rami to the major arterial circle (*ff*). Branches from the circle extend into the iris (*g*) and to the limbus. Branches of the short posterior ciliary arteries (CC) form an anastomotic circle (*h*) (of Zinn) round the optic disc, and twigs (*i*) from this join an arterial network on the optic nerve. The vorticose veins (*jj*) are formed by the junctions (*k*) of suprachoroidal tributaries (*l*). Smaller tributaries are also shown (*m, n*). The veins draining the scleral venous sinus (*o*) join anterior ciliary veins and vorticose tributaries. (From *Histology of the Human Eye*—see 7.250.)

monstrated both cholinergic and adrenergic activity in both iridial muscles (Ehinger and Falck 1966; Lowenstein and Loewenfeld 1970). Although ganglion cells have occasionally been reported in the iris, it is most likely that all, or almost all the fibres are postganglionic in type, and almost all are also nonmyelinated. They form a plexus at the base of the iris, and from this small nerves and individual fibres extend not only to the two muscles, but to the vessels and to the anterior border layer and the anterior epithelium (but not the pigment layer) of the posterior surface of the iris. Some fibres may be afferent and some are vasomotor, but little is known of either. For details of the occurrrence and distribution of the nonmyelinated sympathetic and parasympathetic nerve fibres in the choroid consult Ruskell (1971). For a discussion of the difficulties in identifying 'free' afferent nerve endings in the iris and ciliary body (and of the uncertainty in distinguishing them from efferent autonomic endings), consult Bergmanson (1978), who has described such endings in the monkey, using the accumulation of mitochondria usually regarded as a criterion of sensory terminals (Cauna 1966; Macintosh 1974). He found that not only did trigeminal (ophthalmic division) terminals display this characteristic, but that it was also observable in autonomic endings in these sites.

Pupillary membrane. In the fetus, the pupil is closed by a delicate, vascular membrane, termed the *pupillary membrane* (p. 177). The vessels of this membrane are partly derived from those of the margin of the iris and partly from those of the capsule of the lens; they end in loops a short distance from the centre of the membrane, which is thus left free from blood vessels. About the sixth month of intrauterine life the membrane begins to disappear by absorption from the centre towards the circumference, and at birth only scattered fragments are present; in exceptional cases it persists and may interfere with vision.

The Retina

The retina (7.247) is the neural, sensory stratum of the eyeball. It is very thin, varying from 0·56 mm near the optic disc to 0·1 mm anterior to the equator of the eyeball, continuing at this thickness to the ora serrata. It is, of course, much thinner at the optic disc and the fovea of the macula (Spence *et al*. 1969). Its external surface is in contact with the choroid, its internal with the hyaloid membrane of the vitreous body. Posteriorly it is continuous with the optic nerve; it gradually diminishes in thickness from the optic disc to the ciliary body; it presents a crenated margin named the *ora serrata* (7.252, 253). Here the nervous tissues of the retina end, but a thin prolongation of the membrane extends forwards over the back of the ciliary processes and iris, forming the *ciliary and iridial parts of the retina*. This forward prolongation consists of the pigmented layer of the retina together with a deeper stratum of columnar epithelium; in the iridial part of the retina both layers of epithelium are cubical and pigmented. The part of the retina extending from the optic disc to the ora serrata is known as the *optic part of the retina*. The retina is soft, translucent, and of a purple tint in the fresh, unbleached state, owing to the presence of a colouring material, named *rhodopsin*, or *visual purple*; but it soon becomes clouded, opaque, and bleached when exposed to light. (It is, in fact, difficult to prepare an eye in such a way as to demonstrate the purple pigment. In the preserved eyes usually dissected, the *fixed* retina has a cloudy white colour.) Near the centre of the posterior part of the retina there is an oval, yellowish area, named the *macula lutea* (7.257); it shows a central depression, termed the *fovea centralis*, where visual resolution is highest. At the fovea centralis the retina is exceedingly thin, some of its layers being absent here, and the dark colour of the choroid is distinctly seen through it. About 3 mm to the nasal side of the macula lutea the optic nerve pierces the retina at the *optic disc*, which has a diameter of about 1·5 mm. The circumference of the disc is slightly raised, while the central part presents a slight depression. The centre of the disc is pierced by the central artery and vein of the retina (7.257, 258). The optic disc is insensitive to light, and is termed the 'blind spot'. On ophthalmoscopic examination the normal disc is seen to be pink; it is, however, much paler than the retina and may be grey or almost white. In

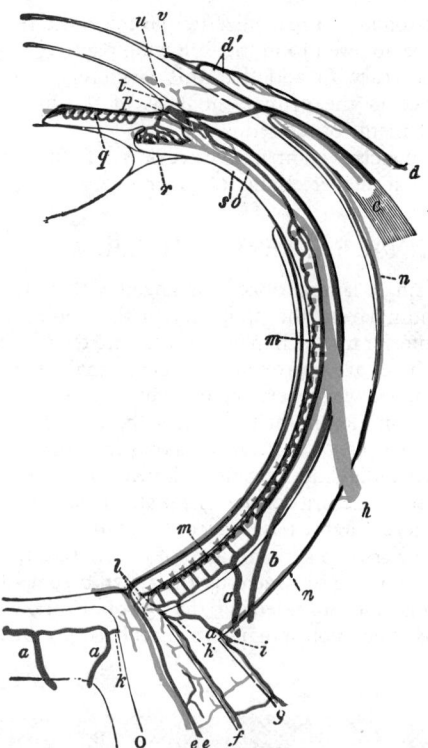

7.256 A diagrammatic representation of the course of the vessels of the eye. Horizontal meridional section. Arteries and capillaries—red; veins—blue. O. Entrance of optic nerve. *a*. Short posterior ciliary arteries. *b* Long posterior ciliary arteries. *c*. Anterior ciliary vessels. *d*. Posterior conjunctival vessels. *d*. Anterior conjunctival vessels. *e*. Central retinal vessels. *f*. Pial vessels. *g*. Dural vessels. *h*. Vorticose veins. *i*. Short posterior ciliary vein. *k*. Branches of the short posterior ciliary arteries to the optic nerve. *l*. Anastomosis of choroidal vessels with those of optic nerve. *m*. Choroidocapillary lamina. *n*. Episcleral vessels. *o*. Recurrent artery of the choroid. *p*. Circulus arteriosus major (in section). *q*. Vessels of iris. *r*. Vessels of ciliary process. *s*. Branch from ciliary muscle to vorticose vein. *t*. Branch from ciliary muscle to anterior ciliary vein. *u*. Sinus venosus sclerae. *v*. Capillary loop at the margin of cornea.

cases of optic atrophy the capillary vessels disappear and the disc appears white. The name 'optic papilla' applied to the disc is misleading since almost all of the normal disc lies in the same plane as the surrounding retina.

The Structure of the Retina

As has already been stated (pp. 169 and 176), the retina is derived from the two layers of the invaginated optic vesicle, the outer of which becomes the *pigment cell* lamina, the inner developing into the far more complex multilaminar structure containing the *photoreceptors*·(rod and cone cells), their *first order neurons* (bipolar cells), the somata and beginnings of the axons of their *second order neurons* (ganglion cells), and *interneurons* arranged across these centripetal pathways (horizontal and amacrine cells). In addition, the retina contains *neuroglial elements* (Müller cells, or sustentacular gliocytes) and a *vascular system* composed chiefly of capillaries. In some descriptions the region of the retina posterior to the ora serrata (*pars optica retinae*)—the region which functions as a sense organ—is separated from its pigment epithelium as 'the retina proper'; but since the epithelium is functionally integrated with the rest of the retina, this arbitrary division into a *stratum pigmentosum* and a *stratum nervosum* will not be followed here. The description will, however, be limited to the sensory region of the retina, whose ciliary and iridial parts have already been considered (pp. 1157, 1159. (The potential space between pigment cells and photoreceptors has been equated with the ventricular cavities of the brain since it is developmentally a hollow outgrowth—*see* p. 176.)

It is customary to recognize ten layers in the retina, and this plan will be adhered to in the following pages (7.259–262). It is hardly necessary to add that this is merely a morphological convenience in the primary analysis of the highly integrated nervous elements of the retina.

For a concise comparative review of the anatomy and physiology of the vertebrate retina *see* Boycott (1974).

THE RETINAL PIGMENT EPITHELIUM

This is a single lamina of cells arranged with marked regularity and extending from the periphery of the optic disc to the ora serrata, anterior to which they continue into the ciliary epithelium (p. 1157). They are approximately flat rectangles in radial sections and hexagonal when seen in tangential aspect. They number about 4 to 6 million in the human eye, becoming more numerous in more aged eyes. Near the macula they measure about 10 and 14 μm in their radial and tangential dimensions, but become much flattened near the ora serrata (Ts'o and Friedman 1967, 1968). Their nuclei are in the basal part of the cytoplasm, adjacent to the basal lamina (Bruch's membrane) of the choroid (p. 1156), from which the cells are separated by their own basement membrane. The latter is much infolded into the cytoplasm of the basal aspects of the cells. Their apical regions project between the rod and cone

processes as microvilli of 5 to 7 μm in length. The intermediate cytoplasmic region contains numerous mitochondria and pigment granules. The pigment is melanin, and various other organelles, associated with the formation of this, have been identified in retinal epithelium, including well-developed granular and agranular endoplasmic reticulum in stacked arrays, premelanosomes and melanosomes (Breathnach and Wylie 1966; Seÿi 1967). Lipofuscin also occurs (Feeney *et al.* 1965), probably as a residue representing the end of a phagosomal process. Additional autoradiographic and ultramicroscopic studies (Young and Bok 1969; Spitznas and Hogan 1970), suggest that the microvilli of the pigment epithelium cells are concerned in a continuous erosion of the external ends of the rod and cone processes. Lamellar inclusions in the cytoplasm of the microvilli closely resemble the lamellar structures in the outer segments of the photoreceptors; they have been termed phagosomes, and they evince progressive disintegration as they pass deeper into the cytoplasm of the cell (Marshall and Ansell 1971).

A typical cell membrane of 'unit' structure encloses the pigment cells. Basally, as stated earlier, this is much infolded by the basement membrane. The latter has a fibrillar structure, and some of its fibrils join the basal membrane of the choroid, a junction which may explain the adherence of the pigment lamina to the choroid rather than to the rest of the retina. Laterally the

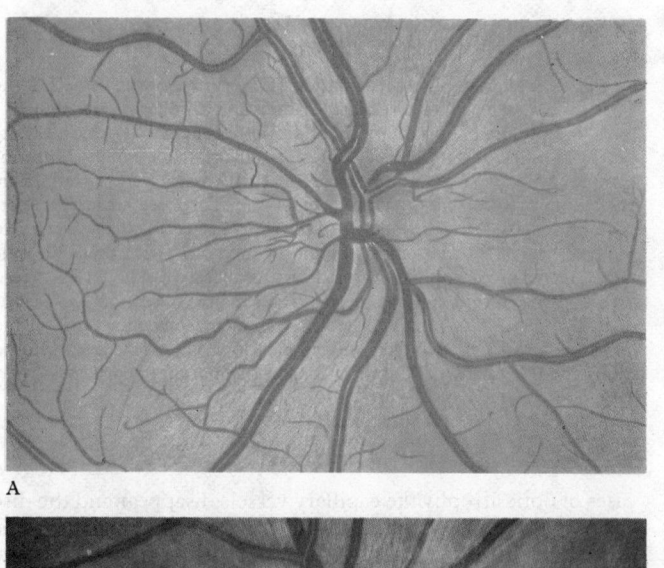

A

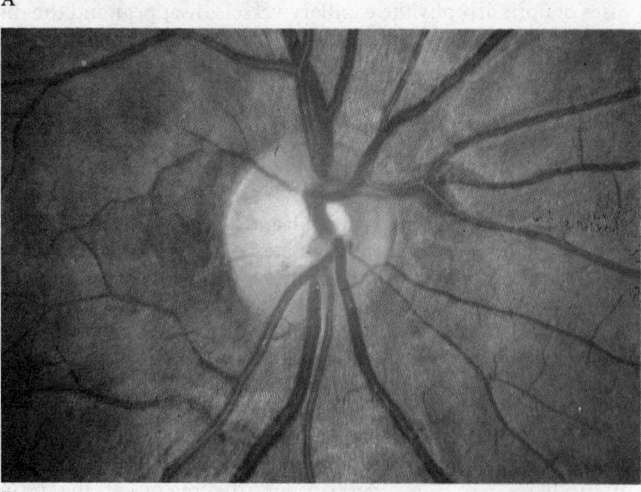

B

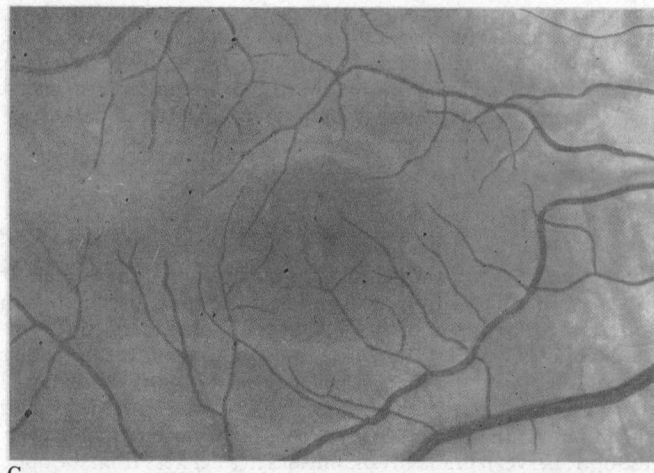

C

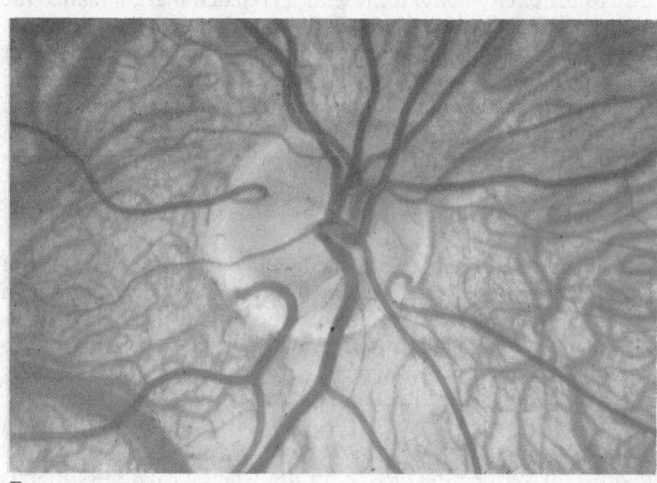

D

7.257A–D Ophthalmoscopic photographs of the right human retina.
A Note dichotomous branching of vessels, arteries being brighter red and showing a more pronounced 'reflex' to light, as a pale stria along their length. The veins are also larger in calibre; more of them cross arteries superficially than is usual. The optic disc, around the entry of the vessels, is a light pink, with a surrounding zone of heavier pigmentation. Compare with 7.257B, from the same Caucasian adult.
B Appearances in a heavily pigmented individual, with a paler optic disc than in 7.257A. Note accentuation of the edge of the disc by retinal and choroidal pigmentation. The arteries cross the veins superficially in this retina. Negroid adult.

C Normal macula of a young Caucasian subject. Note the fovea, showing as a central, paler, circular area. The macular branches of the central retinal artery are approaching from the right. The macula is largely free of vessels of macroscopic size, but the capillaries here form a particularly close network, except at the fovea.
D The region of the optic disc in an eye with poorly developed pigmentation. Three cilioretinal arteries are curving round the edge of the disc, two on the left, one on the right. Between the two cilioretinal arteries a single macular artery is apparent. Due to the depressed pigmentation choroidal vessels are also visible, especially veins; and on the left of the photograph two large vorticose venous tributaries can be seen.

membranes of the pigment cells do not interdigitate markedly, and the variable space between them is sealed off from the apical space (between the microvilli and the photoreceptors) by zonulae occludentes. Other forms of junction also occur. Miki *et al.* (1975) have described retino-choroidal attachment sites in various mammals, including mankind. These include reciprocal inter-digitations, as noted above, and also junctions resembling desmosomes. A viscous, mucopolysaccharide substance occupies the space, which is embryologically derived from the cerebral ventricles.

Although the functions of the pigment epithelium of the retina are still far from elucidated, their ultramicroscopic features strongly suggest a phagocytic activity, possibly a nutritive role, to which may be added a contribution to the spacing and mechanical support of the photoreceptors, and an optical function in absorbing light and preventing back reflexion.

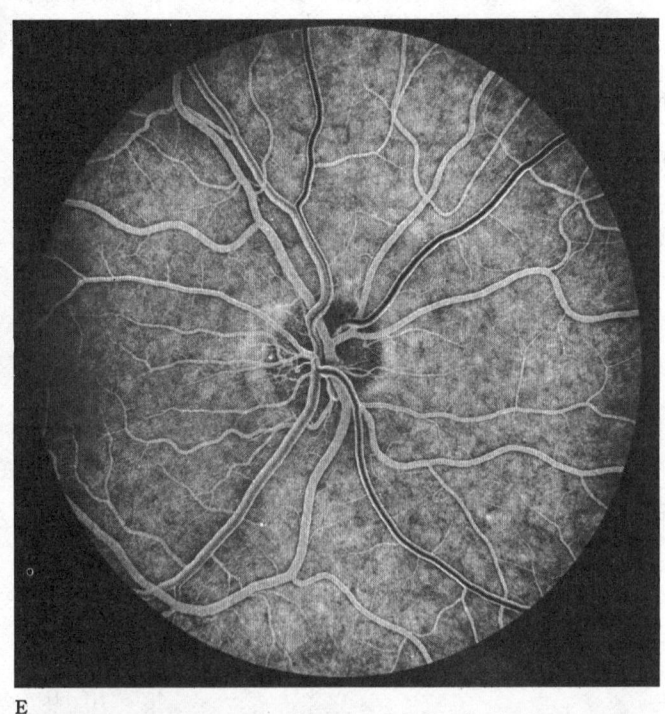

E

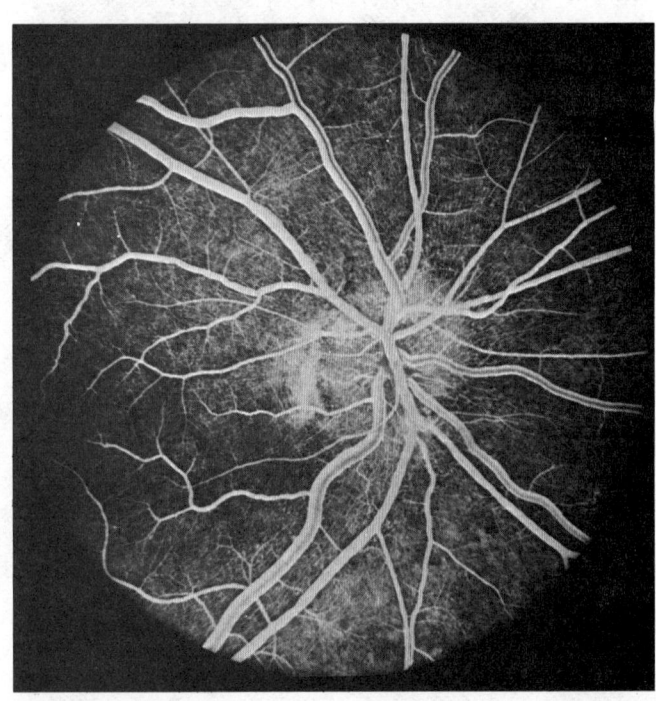

F

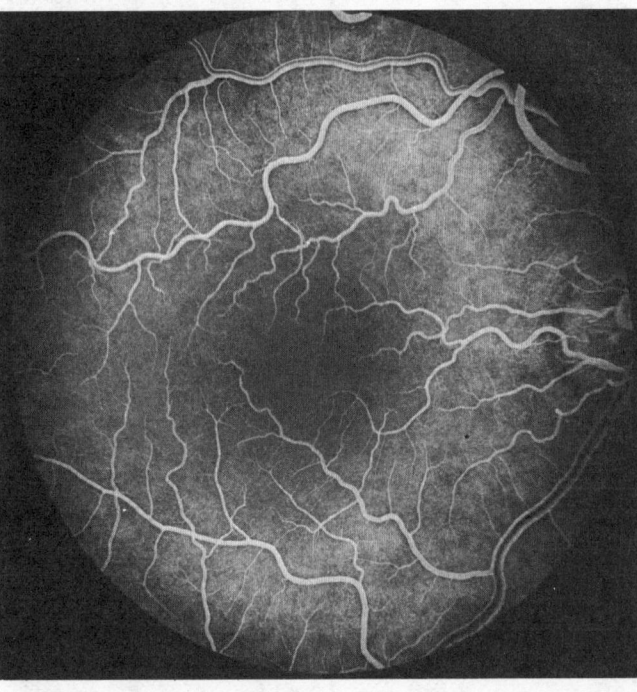

G

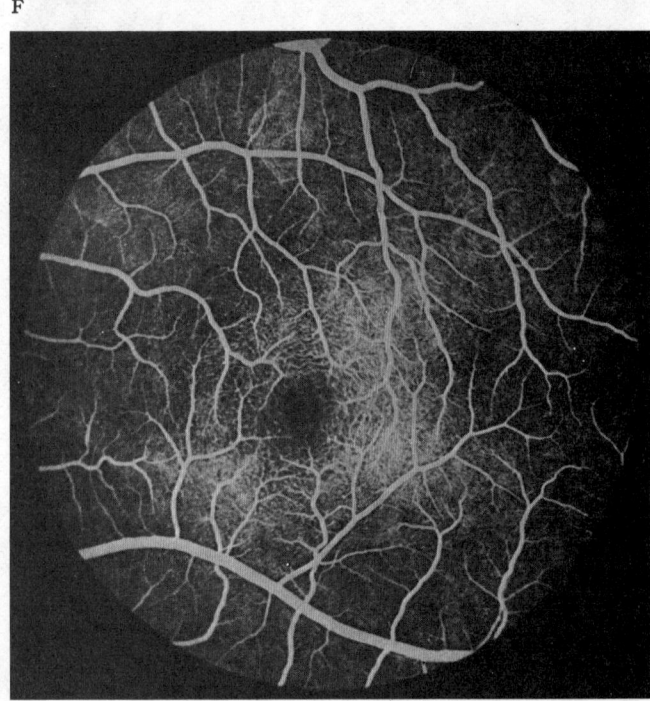

H

7.257E–H   Fluorescence angiograms of the retina. These are produced by photography with a fundus camera at known periods of time following introduction of fluorescein into the circulation. (For details of the technique consult *Fluorescence Photography of the Eye*, by Emanuel S. Rosen, Butterworths, London, 1969, to whom we are indebted for all the colour photographs and angiograms in this illustration.)

E   Angiogram of the same retina as that appearing in 7.257A, taken in 'mid-venous' phase. The arteries display an even fluorescence, but the veins appear striped, due to laminar flow. This appearance is the reverse of, and not to be compared with the arterial 'reflex' seen in 7.257A. The background mottling is due to fluorescence from the choroidal vessels.

F   Angiogram of the left optic disc, showing the major arteries and veins and also their smaller branches. Note particularly the radial pattern in the retinal capillaries. The laminar flow in the veins is less obvious than in 7.257E.

G   Angiogram showing the macular region of a right eye. The main macular vessels are approaching from the right. The subject was an elderly person with considerable macular pigmentation, which masks fluorescence from the choroidal circulation. Compare with 7.257H.

H   Angiogram of the macula of a young subject showing the macular capillaries in detail. Note the central avascular fovea. Left eye. Compare with 7.257G.

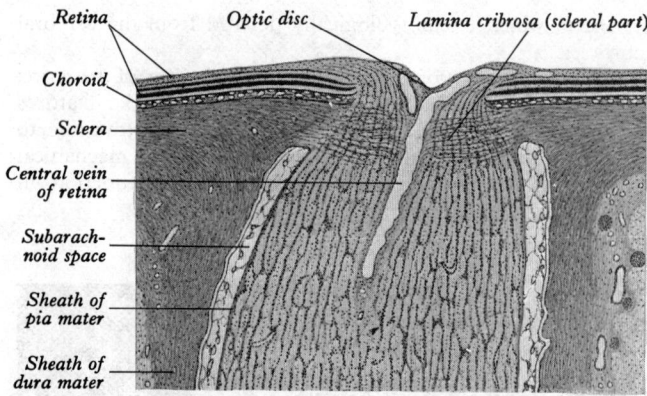

Retina   *Optic disc*   *Lamina cribrosa (scleral part)*

Choroid

Sclera

Central vein of retina

Subarach-noid space

Sheath of pia mater

Sheath of dura mater

7.258A   A horizontal section through the optic nerve at its point of exit from the human eyeball.

THE PROCESSES OF THE ROD AND CONE CELLS

The retinal photoreceptor cells (7.259, 260) consist of a cell body containing the nucleus, which will be described in lamina 4 (*vide infra*), an axonal centripetal process which forms synapses with the retinal neurons in lamina 5, and a photosensitive centrifugal or external process which will be the subject of this section. Although intermediate forms exist in various vertebrates, the great majority of these processes, and hence the cells of which they are a part, fall into two categories, *rods* and *cones*, differentiated principally by the cylindrical and conical form of their processes, but also by the existence of a constricted *outer fibre* connecting the rod process to the soma of its cell (7.260B), a feature absent from cone cells, though in these a small constriction or waist may be apparent. At the junction of the rod processes with their outer fibres and of the cone processes with their cell bodies is the external limiting membrane, through which the photoreceptor process appear to be thrust (7.260B), as if

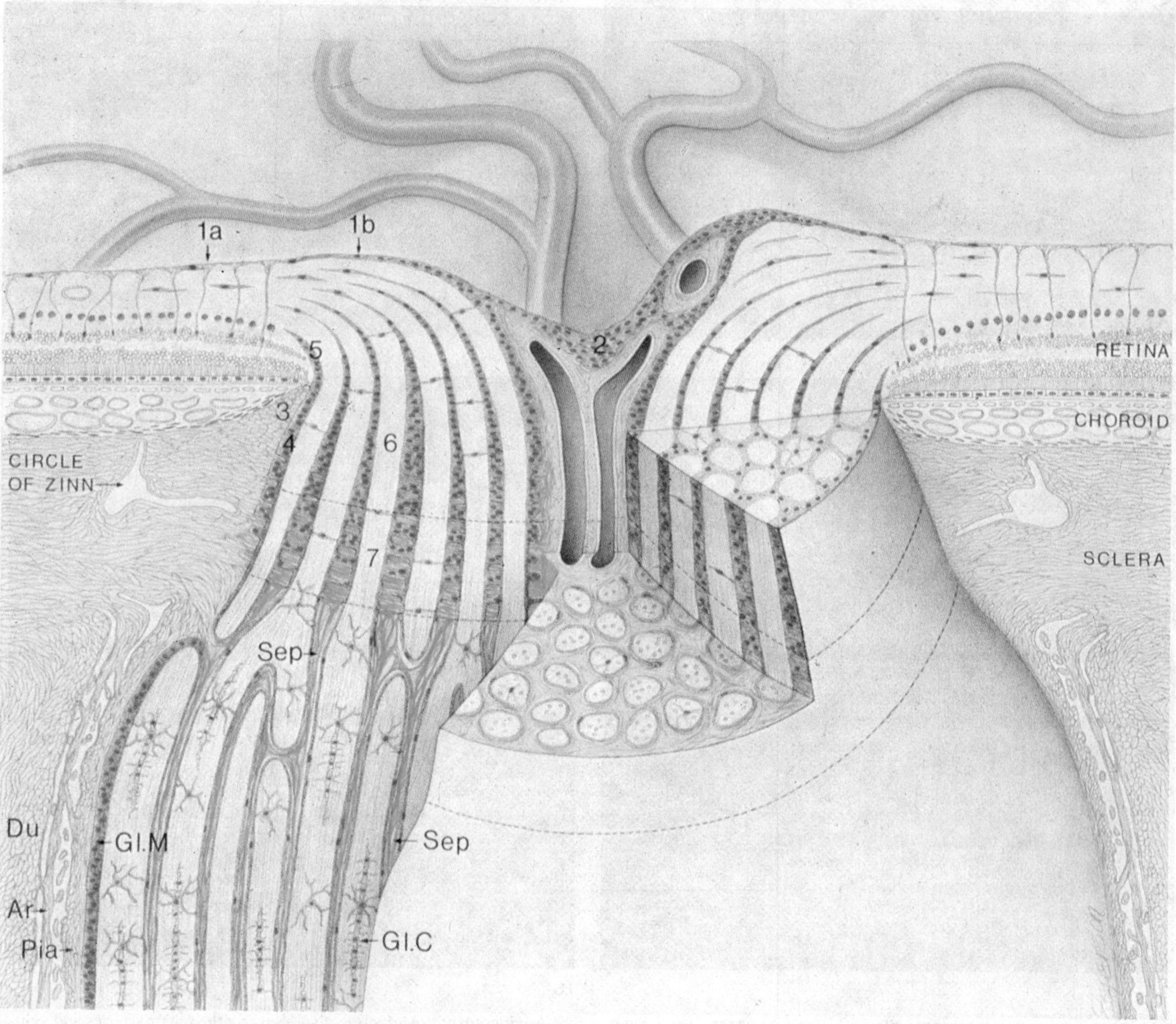

7.258B   Schematic representation of the exit of the human optic nerve from the eyeball, showing the distribution of collagenous (blue) and neuroglial (magenta) tissues. Sep=septa of collagenous connective tissue carried into the nerve from the pia mater and dividing the nerve fibres into numerous fascicles. Gl.M=astroglial membrane separating nerve fibres from connective tissue. Gl.C=astrocytes and oligodendrocytes among the fibres in their fascicles. Du, Ar, Pia=dura, arachnoid and pia maters respectively. 1*a* is the internal limiting lamina of the retina, which is continuous with an astroglial membrane (of Elschnig) covering the optic disc (1*b*). An accumulation of astrocytes forms a central meniscus (of

Kuhnt) in the centre of the disc (2). The anterior or so-called 'choroidal part' of the lamina cribrosa (6) is separated from the choroid by a spur of collagenous tissue (3). The 'border tissue of Jacoby' (4), which is largely astroglia, frequently extends beyond the choroid (5) to separate much of the retina from the 'retinal part' of the optic nerve head. The posterior part of the lamina cribrosa (7) contains collagenous tissue derived from the optic nerve septa and fenestrated sheets of collagen fibres continuous with those of the sclera. (Reproduced by kind permission of Dr. Douglas K. Anderson and Dr. W. F. Hoyt, *Arch. Ophthal.*, **82**(4), 506–530, 1969. Copyright 1969, American Medical Association.)

through a sieve; the nature of this structure will be discussed below (see the external limiting lamina).

The rod and cone processes are closely packed in a highly ordered array, but the density of this diminishes throughout the neuroretina to the ora serrata, where they abruptly cease. In the human retina they are most numerous in and near the *macula*, the region at the optical centre of the retina where vision is most discriminative as to form and colour, though least adapted to functioning in low luminosities. They are entirely absent over the whole of the *optic disc*, where the centrifugal fibres of the retina leave the eyeball to form the optic nerve. The *central area* of the retina is a region 5 to 6 mm in diameter, which contains the *macula lutea*, measuring about 2 mm horizontally and 1 mm vertically, its yellow colour being due either to the presence of xanthophyll, a great reduction in the capillary bed, or perhaps to cell inclusions (other than xanthophyll) in the bipolar and ganglion cells. Approximately at the centre of the macula is the *fovea centralis*, a deep conical depression in the retina, where almost all elements

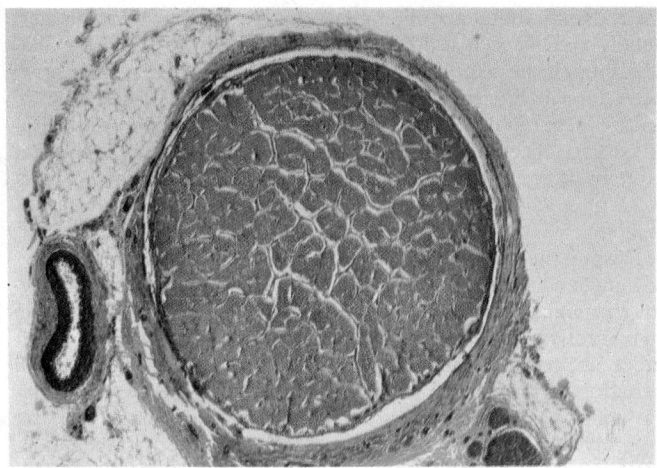

7.258C A transverse section of the optic nerve and its meningeal coverings posterior or proximal to the entry of the central retinal artery, which is visible at the side of the nerve. The dural and pial sheaths are stained green, the subarachnoid pink. Note the fasciculation of the nerve. (Material stained by Masson's trichrome technique and kindly provided by Dr. N. A. Locket, Institute of Ophthalmology, University of London.)

except cones are absent, on its floor at least, the diameter of this being said to be no more than 0·4 mm. This *foveola*, as it is sometimes called, is about 4 mm lateral and 1 mm inferior to the centre of the optic disc; the latter corresponds to the 'blind spot' in the uniocular visual field. The extremely small size of the foveola accounts for the accuracy with which the visual axis must be directed to achieve the most discriminative vision. The macula has been further divided into peri- and para-foveal areas, but for such details recent monographs should be consulted. (*See* also Yamada 1969.) The macula and central fovea will be considered further when the general structure of the retina has been described (p. 1174).

The total number of rods in the human retina has been estimated at 110 to 125 million and of the cones at 6·3 to 6·8 million (Østerberg 1935). Other similar figures have been stated. The distribution of the two types of receptors differs; the cones, which have their densest arrangement at the rod-free foveola (about 147,000 per square millimetre), fall off in numbers very rapidly from this point to a 10-degree circle round the macula to a density of about 5,000 per square millimetre, maintaining this to the ora serrata. The rods, on the other hand, exhibit an almost opposite density, rising from zero at the foveola to an even greater figure than cones at the 10-degree circle (160,000 per square millimetre), and then slowly diminishing in frequency to the periphery of the retina, where there are still estimated to be approximately 30,000 per square millimetre—rods being thus six to thirty times more numerous in the peripheral part of the retina outside the 10-degree circle. This distribution accords well with

the phenomena of photopic (cone) and scotopic (rod) vision. Even with light microscopy it is clear that the neurons in the retina are much less numerous than the rod and cone cells; the ganglion cells (*vide infra*, p. 1172), whose axons form the optic nerve, are probably in the region of a million in number in the entire human retina. It is hence obvious that large numbers of rod and cone cells must activate a single axonal pathway in the optic nerve and beyond. This correlation of rod and cone distribution with photopic and scotopic vision is well illustrated by the retina of cats, which are nocturnal. In the area centralis the concentration of cones, though greater than elsewhere in the feline retina, is about one sixth of the human figure. In contrast the rods, at 275,000 per mm² in the area centralis, reach a maximum of 460,000 per mm² peripheral to this, both concentrations being much greater than anywhere in the human retina.

The rod and cone processes exhibit greater differences with light microscopy than with the electron microscope. Even with the former technique both forms of process show an external or peripheral segment, which with ordinary stains is refractile, positively birefringent and PAS-positive, and an internal segment which stains deeply and has a fibrillar structure. The combined segments of the rod process measure about 100 to 120 μm in the freshly fixed human retina, the cone processes being about 65 to 75 μm (Eichner 1957). These dimensions diminish towards the ora serrata, especially in the cones, which are in addition much narrower at the fovea, where they closely resemble rods in their dimensions. The outer segment of rods contains the photo-sensitive pigments named *rhodopsins* (visual purple), and in the cones (Rushton 1962, Pirenne 1967, Dowthall and Knowles 1977) similar pigments have been detected. These substances display different absorption characteristics, and these account for the different behaviour of rods and cones in conditions of high and low luminosity. Different absorption maxima have also been observed amongst cones themselves in the primate retina, and there is at least some degree of accord between these findings and the trichromatic theory of colour vision. (For a survey of vertebrate photoreceptors and their pigments consult Crescitelli 1972.)

In their *ultrastructural details* the rods and cones are very similar and will be considered together; the rod processes have been more extensively studied (Sjöstrand 1953, 1961; Villegas 1964; Misotten 1962; Dowling 1965; Cohen 1965; Boycott 1974). The rod *outer segment* consists of a remarkably regular series of discoid structures, stacked like thin coins and surrounded by a cell membrane, to the external aspect of which are applied the microvilli of the pigment epithelial cells (*vide supra*, p. 1164). The electron microscope shows that these *discs*, which number from 600 to 1,000 in various species of vertebrate, are flattened sac-like structures, consisting of two unit membranes continuous at the periphery of the disc and separated by a less electron-dense *intradisc space*, except near the periphery, where the space becomes an annular interval of greater width (7.260B). In some vertebrates the discs are infoldings of the cell's membrane, but in the human retina this continuity is lost. The dimensions of these discs have been reported in detail in many species, including primates, and details may be sought in the papers quoted. The discontinuity of the saccular 'discs' in rod outer processes (more recently supported by Cohen 1970, also by Laties and Liebman 1970) provides a distinct *qualitative* difference between these and the discs in cone cells (see also Ripps and Weale 1976), for in the latter the discs do preserve, at part of their periphery, a continuity with the plasma membrane of the outer process. Moreover, as seen in transverse section the cone discs are circular, whereas those of the rod cells display a scalloped profile and are totally separated from the plasma membrane (Cohen 1972). The discoid saccules of cone cells are more flattened than those in rod processes, but are separated by a slightly greater interval. Consequently the number of discs in a unit length is about the same in both types of cell, there being about 30 discs per μm. There is now considerable evidence that the discs of both types of receptor, which develop originally in the vicinity of the cilium (*vide infra*), are progressively pushed away from the basal part of the cell, so that the oldest discs are at the distal end of a rod or cone process and closely related to the pigment epithelial cells. As has

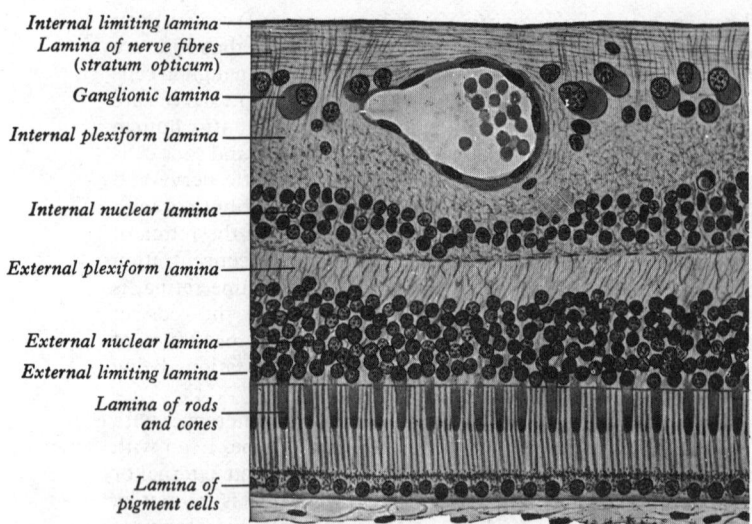

Internal limiting lamina
Lamina of nerve fibres (stratum opticum)
Ganglionic lamina
Internal plexiform lamina
Internal nuclear lamina
External plexiform lamina
External nuclear lamina
External limiting lamina
Lamina of rods and cones
Lamina of pigment cells

7.259A Section through the primate retina, from its vitreous aspect (above) to the choroid tunic (below), showing its layered structure some little distance from the macular region. Diagram to illustrate the customarily recognized strata.

already been mentioned (p. 1164) the epithelial cells are concerned with the phagocytosis of degenerating discs. Young (1971) has shown that in the rhesus monkey almost 1,000 discs are thus removed peripherally and generated proximally in a rod outer segment in the surprisingly short period of 11 days. Since each pigment cell may interdigitate with 24–45 rod segments, the voracious 'appetite' of the former may lead to the destruction of 2,000–4,000 discs *per diem*! These renewal processes have been established by the observed progression of radioactive tracers, introduced into mature animals (rat, mouse, and monkey) by a technique introduced by Droz (1968). There is no such clear evidence for a renewal progress in the outer processes of cones, though Hogan (1972) has described a similar phagocytic process in connexion with the cones of the human fovea. The significance of such a high turnover of a complex intracellular structure as photoreceptor discs has not yet been clarified. The molecular structure of the disc membrane is now considered to consist of a regular arrangement of rhodopsin molecules orientated transversely to the photoreceptor process and embedded in a lipid bilayer, whose molecules extend in the long axis of the process (Schmidt 1938; Weale 1970). Cone (1972) has shown that the orientation of the rhodopsin molecules is labile. Since proteins act as ionic gates for the transference of substances across phospholipid bilayers, the behaviour of rhodopsin molecules

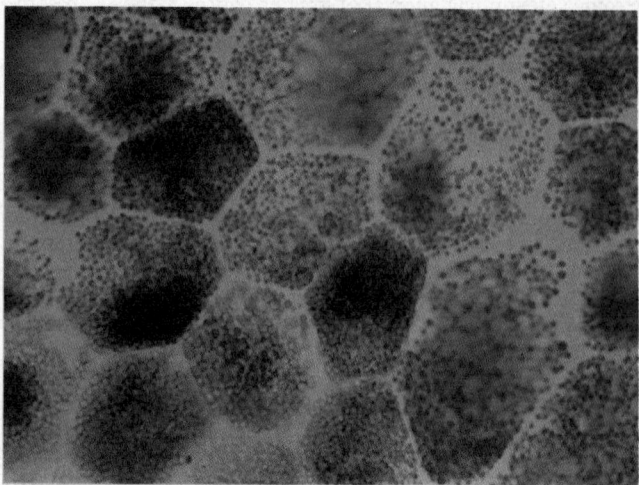

7.259B Colour photograph of unstained retinal pigment epithelium seen in surface view. (Human, aged 40 years.) Magnification ×3,500. (Preparation kindly loaned by Dr. John Marshall, Institute of Ophthalmology, London.)

during photochemical activity is a problem urgently requiring solution. Despite much hypothecation in the form of molecular models, the problem remains (*see* Ripps and Weale 1976, for a discussion).

The rod *inner segments* are longer than the outer and somewhat greater in diameter, both differences being even more accentuated in the inner segments of the cones. The inner segment is itself divisible into two regions—an outer *ellipsoid*, which is acidophilic, contains some glycogen, and displays a large number of mitochondria, and an inner *myoid*, adjacent to the soma of the cell, and containing much randomly arranged agranular endoplasmic reticulum and free ribosomes. The myoid is basophilic, and it contains much more glycogen than the ellipsoid. Extending from the inner to the outer segment are bands of the cytoplasm of the ellipsoid, covered with cell membrane, which are closely applied to the outer segment. Another cytoplasmic process connects the ellipsoid to the outer segment, but in this case there is cytoplasmic continuity. This process, the *cilium*, originates in a basal body and has a similar internal structure to a motile cilium (De Robertis 1960). The cone processes have a very similar structure to that of the rods, the differences being largely dimensional. (For further details *see* Dietrich and Rohen 1970.)

Although the structure of the rod and cone processes has now been described in most extensive detail, it is not yet possible to equate these highly remarkable features to functional studies with any degree of confidence. Their stacked discs have been likened to photomultiplier tubes, intensifying the electrical energy derived from photochemical processes, but this is merely an attractive analogy.

Despite the physiological demonstration of at least three types of human cone cell, each with a characteristic range of sensitivity to the light spectrum, no means of distinguishing these structurally has yet been found. Variations in colour sensitivity of different parts of the retina have not been correlated with regional variations in cone shape or dimensions; and the known sensitivity variations (e.g. achromatism of the extreme periphery) are possibly due to differences in neuronal connexions rather than dimensional or distributional variations. For the same reason the duplicity theory of vision may itself be in part explicable to such factors, as well as the widely accepted equation of photopic and scotopic vision with cones and rods.

THE EXTERNAL LIMITING LAMINA

At the level of the junctions between rod and cone processes with the outer fibres of the rods and the somata of the cone cells light microscopy shows what appears to be a thin membrane, which is commonly regarded as being fenestrated by the above continuities. It extends throughout the neural retina and has been considered for many decades to consist of the fused terminal expansions of the 'fibres of Müller'. These are elaborate neuroglial elements, with cell bodies in the internal nuclear layer (lamina 6, *vide infra*), and from them long processes stretch radially through almost the whole thickness of the retina, from its vitreal surface almost to the pigment epithelium. From these vertical fibres large numbers of subsidiary processes spread out horizontally in a dendriform manner into the plexiform layers (laminae 5 and 7) and form meshworks embracing the somata of cells in the nuclear and ganglion cell layers (laminae 4, 6 and 8). Similar 'fibre baskets' also surround the proximal segments of the photoreceptors. This structural arrangement has naturally suggested that these *retinal gliocytes* or *Müller cells* supply a physical support to the retina, like neuroglia elsewhere in the central nervous system. Moreover, the processes of these cells form a similar *internal limiting membrane* (lamina 10) on the vitreal aspect of the retina, and hence it is not unreasonable to ascribe a stabilizing role to these glial elements. As long ago as 1932 the external limiting membrane was described by Arey as a series of unions between the cell membranes of the rods and cones on the one hand and the 'fibres of Müller' on the other. Ultrastructurally this view has been corroborated (*see* Fine and Zimmerman 1962, for a summary of the earlier literature on this topic), but there is much variation in the types of junction described. They occur between the glial processes and the rod and cone inner segments; a

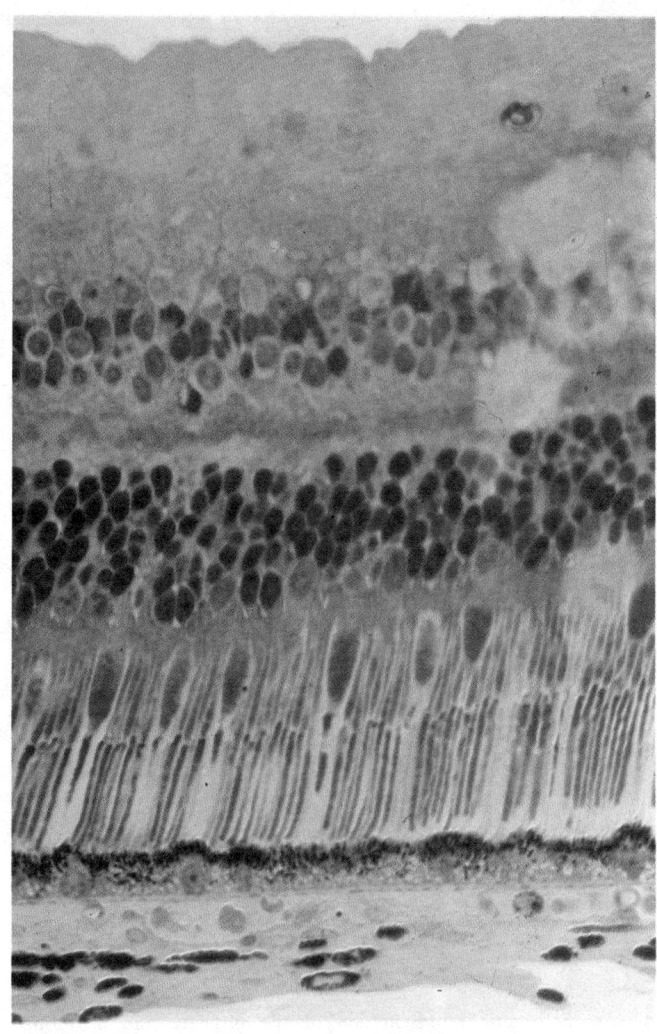

7.259C Thin section of simian retina in araldite-embedded material, stained by toluidine blue. (Kindly provided by Dr. N. A. Locket, Institute of Ophthalmology, University of London.) Compare with 7.259A.

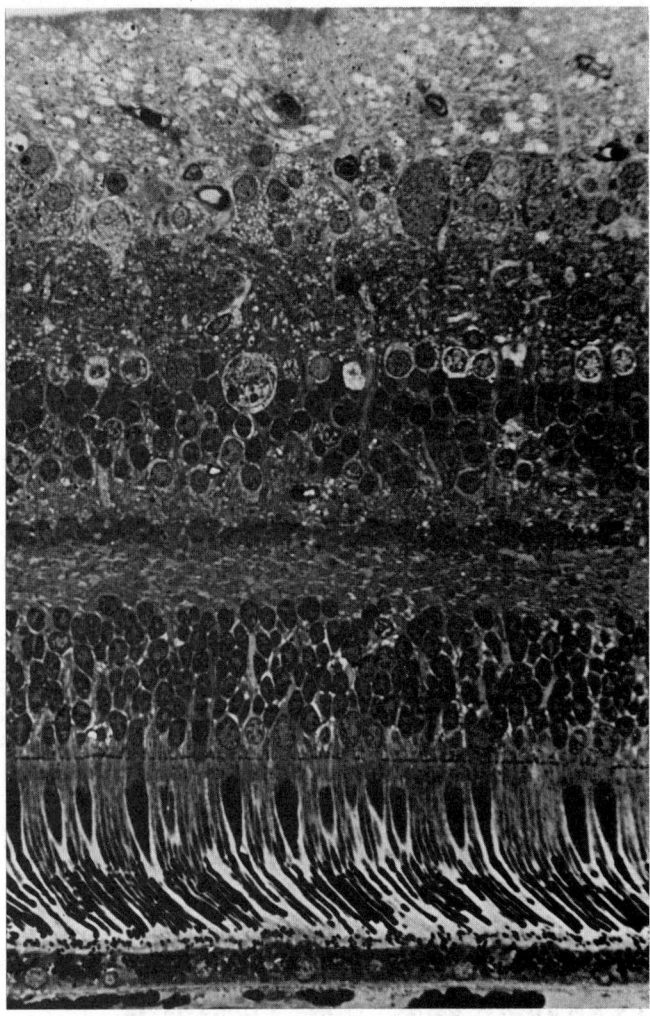

7.259D Light micrograph of the retina of a 19-year-old male. Toluidine blue stain, resin-embedded. (Preparation kindly provided by Dr. John Marshall, Institute of Ophthalmology, London.) Magnification × 1,750.

more recent study of them in the human retina describes them as zonulae adherentes, which also sometimes unite adjoining glial fibres (the classical view) and adjacent photoreceptor processes to each other (Spitznas 1970).

In the plexiform layers the horizontal branches of Müller's cells are closely related to the dendrites and axons of the retinal neurons, and may at times form helical lamellae around individual neurites, without, however, the production of myelin. Like other neuroglial cells they also make extensive contacts with blood vessels, especially capillaries; their basement membranes are at these sites fused with those of nonstriated muscle cells in the media of the vessel or with the basement membrane of the endothelium in the case of capillaries. With their very extensive ramifications and widespread contacts with other retinal elements, these gliocytes take up much of the total volume of the retina, also reducing the extracellular space to exiguous proportions. All the glial processes are, of course, cytoplasmic extensions of the cell body, and it is not surprising, therefore, that other than supportive functions have been ascribed to the retinal gliocytes. There is physiological evidence, for example, that they are concerned in the transport of glucose to retinal neurons, and they are able to synthesize and store glycogen (Cogan and Kuwabara 1967; Radnot and Lovas 1968).

THE EXTERNAL NUCLEAR LAMINA

This retinal lamina contains the parts of the rod and cone cells which are not external to the external limiting membrane or in the external plexiform layer, and this implies in particular the somata of these cells. As they traverse the membrane the photoreceptor processes become the narrower 'outer fibres' of the rods and cones, those of the rods being much more slender and elongated (7.260A). The outer fibres of the cone cells are not much more than a short 'waist' between the process and soma of the cell. The cytoplasm of both types of fibre contains long mitochondria, vesicular agranular endoplasmic reticulum, and many free ribosomes; as it merges into the cell body and spreads round its nucleus, aggregations of microtubules are apparent. The thickness of the external nuclear lamina—and the number of rows of cells in it—varies from 27 μm in the peripheral retina to 50 μm in the fovea (p. 1174), representing in the former a single row of cones with four of rod nuclei, and in the latter about ten rows of cone nuclei.

THE EXTERNAL PLEXIFORM LAMINA

The 'inner fibres' of the rod and cone cells pass centrally (towards the vitreous) and form a most intricate zone of synapses (7.260B) with the dendrites and axons of the bipolar and horizontal neurons of the internal nuclear layer (lamina 6). The inner fibres of the photoreceptors resemble axons, containing a few mitochondria and vesicles, free ribosomes, microfilaments and microtubules. The rod axons are 15 to 25 nm in diameter and 1 μm or even more in length, those of the cones being much thicker and containing more microtubules. Rod axons end in an oval, invaginated *rod spherule*, cone axons as conical or pyramidal *cone pedicles*. These terminations form complex multiple junctions with the bipolar and horizontal neurons, the neurites of which approach them from the internal nuclear layer. The external plexiform layer is sometimes regarded as displaying *three sublaminae*—an outer one of rod and cone inner fibres or axons, an intermediate one of spherules and pedicles, and an internal sublamina of bipolar and horizontal cell processes. (For reviews of this complex subject consult Dowling and Boycott 1966, Misotten and Van den Dooring 1966, Boycott and Dowling 1969, Boycott 1974.)

The rod spherule (7.261) is part of a synaptic complex consisting of three elements—*presynaptic* (the spherule itself), *synaptic* (contacts of the spherular membrane with those of bipolar dendrites and the neurites of horizontal cells), and *postsynaptic* (bipolar and horizontal cell neurites). The spherule is infolded to form a double-walled structure enclosing in its hollow the terminations of two to seven dendrites or processes. The dendrites are derived from *rod bipolar cells* (*vide infra*) and the processes from horizontal neurons; the latter are probably not polarized, conducting impulses in either direction, and hence their neurites cannot properly be termed axons or dendrites. The cytoplasm of the rod spherule contains many presynaptic vesicles and a peculiar osmiophilic lamellar structure, the *synaptic ribbon* (7.261). Microtubules have been described in close association with synaptic ribbons and vesicles in the photoreceptor cells of amphibia (Gray 1976) and primates (Glees and Spoerri 1977). The presynaptic (rod) and postsynaptic (bipolar and horizontal cell) membranes are not thickened as in typical synapses, nor does the 15 nm synaptic cleft contain fibrils or other features. Postsynaptic vesicles may also be present.

The cone pedicle also displays an invaginated pattern, but of a

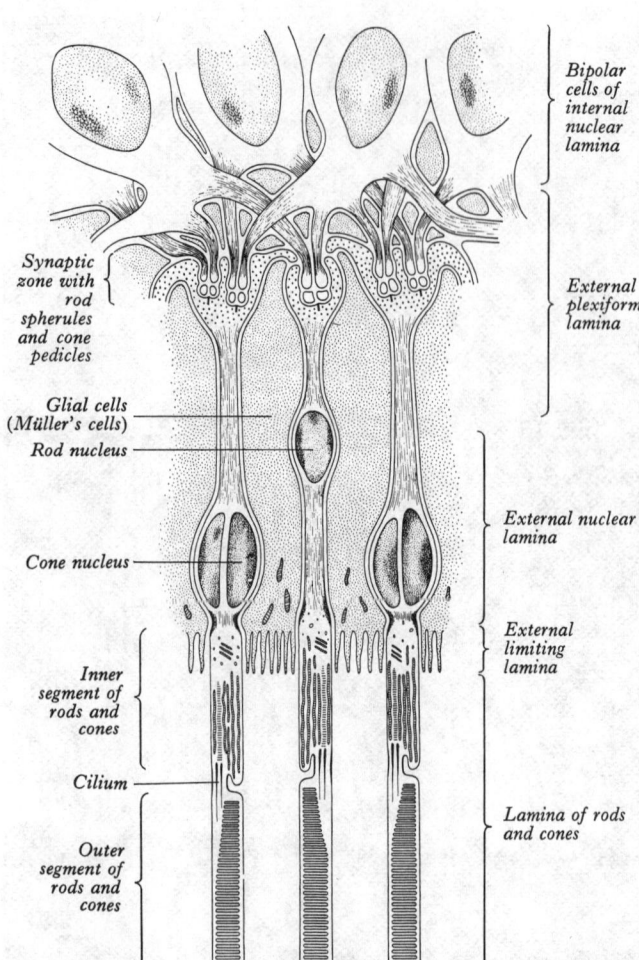

7.260B   Schematic representation of the ultrastructure of retinal photoreceptors and of their connexions with bipolar nerve cells. Note the stacked discs in the outer segments, and refer to 7.261 and 262 and to the text for further details of the synaptic zone. (Reproduced from F. S. Sjöstrand, in: *The Structure of the Eye*, ed. G. K. Smelser, Academic Press, New York, 1961.)

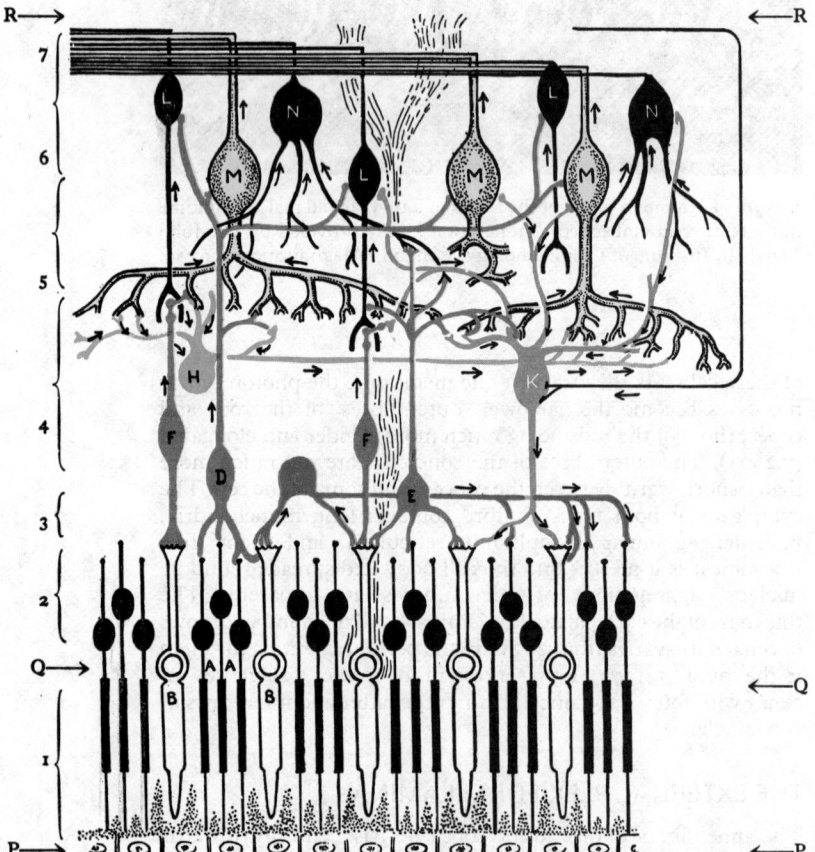

7.260A   Scheme of the retinal neurons. (Modified from Polyak.) 1. Rods and cones. 2. External nuclear lamina. 3. External plexiform lamina. 4. Internal nuclear lamina. 5. Internal plexiform lamina. 6. Lamina of ganglion cells. 7. Lamina of nerve fibres, passing to optic disc. P. Pigment cells. Q. Membrana limitans externa. R. Membrana limitans interna. One sustentacular fibre is indicated in the centre of the diagram. AA. Outer rod fibres. BB. Inner cone segments. C. Horizontal cell. D and E. Bipolar cells forming synapses with rods and cones. FF. 'Midget' bipolar cells synapsing only with cones. HK. Amacrine cells. LLL. Midget ganglion cells. MN. Other ganglion cells. Arrows indicate the probable direction of nerve impulses.

more complex design (7.261). Its cytoplasmic organelles are like those of the rod spherule. A much larger number of neurites make contacts with it. These contacts are of three types: (1) deeply infolded synapses, containing three neurite terminals, two more deeply situated than the other, form *triads*, there being about twenty-five such groups to each cone pedicle; (2) slightly depressed contacts, also on the basal aspect of the pedicle, number as many as 500; and (3) interreceptor contacts connect the periphery of the cone pedicle to adjoining pedicles or rod spherules. The triads contain two 'axon' terminals from horizontal cells and one dendritic terminal from a *midget bipolar neuron* (*vide infra*) or from a horizontal process. (As noted above, some observers consider that the horizontal cell processes cannot be distinguished as axonal or dendritic.) The surface contacts of cone pedicles are synapses with the dendrites of 'flat' bipolar cells (*vide infra*). The interreceptor contacts are devoid of synaptic vesicles, in contrast with other cone pedicle contacts, but they do contain neurotubules. Each pedicle displays six to twelve such contacts. Interreceptor contacts are particularly frequent among the cones of the fovea; elsewhere the contacts are usually between rod and cone cells. In view of the high resolution mediated at the fovea, such contacts have been accepted with reluctance as synapses. However, high resolution microscopy and freeze-fracture techniques (Raviola and Gilula 1973; Witkovsky *et al.* 1974) indicate that they are gap junctions, and that they may provide low resistance routes for the propagation of electrotonic excitation. If this is so, there is at photoreceptor level a separate mechanism of interactions, in parallel with that exercised by horizontal and amacrine cells elsewhere in the retina (*vide infra*).

From this brief description it is clear that each rod cell has direct connexions with two bipolar and perhaps several cone cells, and also with a horizontal neuron. Cone cells display much more complex and numerous contacts, probably between 575 and 600 in number, and involving midget and flat bipolar cells, horizontal, rod and cone cells. The possible pathways for convergence of activities and interactions between the photoreceptors are obviously numerous, but the precise physiological significance of these complex pathways of intercommunication is uncertain. Ultrastructural studies of freeze-fractured material (Raviola 1967) have added somewhat to the structural details of intercellular junctions. The interphotoreceptor contacts have been confirmed as gap junctions, but these and other details do not yet explain the retinal *modus operandi*.

THE INTERNAL NUCLEAR LAMINA

This layer is perhaps unfortunately named. In light microscopic preparations the retina contains *three* tiers of 'nuclei' or cell somata; for the ganglion cells provide the most internal, the 'internal' nuclear layer being thus *intermediate* between this and the layer of rod and cone cells. The layer contains the cell bodies of the *retinal gliocytes* (fibres of Müller) and those of the bipolar, horizontal and amacrine neurons. These cellular components of the inner nuclear layer are arranged in orderly strata. Most external are the cell bodies of the *horizontal cells*, whose processes extend into the adjacent external plexiform layer to form synapses with rod spherules and cone pedicles, as already stated above. Internal to this is a stratum of *bipolar neurons*, which are the primary sensory neurons of the retino-geniculate pathway (p. 977); their dendrites connect with the rod and cone cells, and their axons pass centrally into the internal plexiform layer (lamina 7) to form synapses with the ganglion cells (lamina 8). Internal again to the bipolar cells are the somata of the retinal gliocytes, which have been considered earlier (p. 1168). The most internal stratum of the internal nuclear layer consists of the cell bodies of the *amacrine neurons*, with neurites which spread into the adjacent internal plexiform layer to interconnect with the dendrites of ganglion and the axons of bipolar neurons.

Each of the three categories of neurons in the internal nuclear layer—horizontal, bipolar and amacrine—consists of several types, which are distinguished more by the pattern of their connexions than by their cytological differences, which are not easily defined by light or electron microscopy.

The horizontal neurons are usually divided into two types— *rod* and *cone horizontal cells*. They have multipolar somata, from the angles of which extend a single long process and several short ones (seven in cone and ten to twelve in rod horizontals). Their cytoplasm is like that of bipolars (*vide infra*), except that they contain an organelle rich in ribosomes (Kolmer's crystalloid), which is peculiar to them. The long processes may be up to 1 mm in length—a very long neurite by retinal standards—and their branches make contacts with both rod spherules and cone pedicles. The short neurites of the *cone* horizontal neurons form synaptic junctions with seven cone pedicles, taking part in the formation of several of the triads of each (p. 1170). The short processes of the *rod* horizontal cells establish synapses with ten to twelve rod spherules. The long and short neurites of horizontal neurons are not usually classed as either axons or dendrites, and probably transmit impulses in both directions (7.262).

The bipolar neurons may be classified as *rod bipolar cells*, forming synapses with a variable number of rod spherules, and *midget* or *cone bipolar cells*, which take part in the triad synapses of cone pedicles (7.261). The axon of each rod bipolar connects with up to four ganglion cells while the midget bipolars connect each with but a single midget ganglion cell (*vide infra*). A third type, the *flat bipolar neuron*, connects by its dendrites with many cone pedicles, and through its axon with all types of ganglion cell. Even with electron microscopy the somata of these three kinds of bipolar neuron are difficult to identify and distinguish. They vary somewhat in size and shape, but all are very similar in their organelles—abundant mitochondria, free ribosomes, agranular endoplasmic reticulum, and microtubules.

The amacrine neurons (7.262) are so named because of their

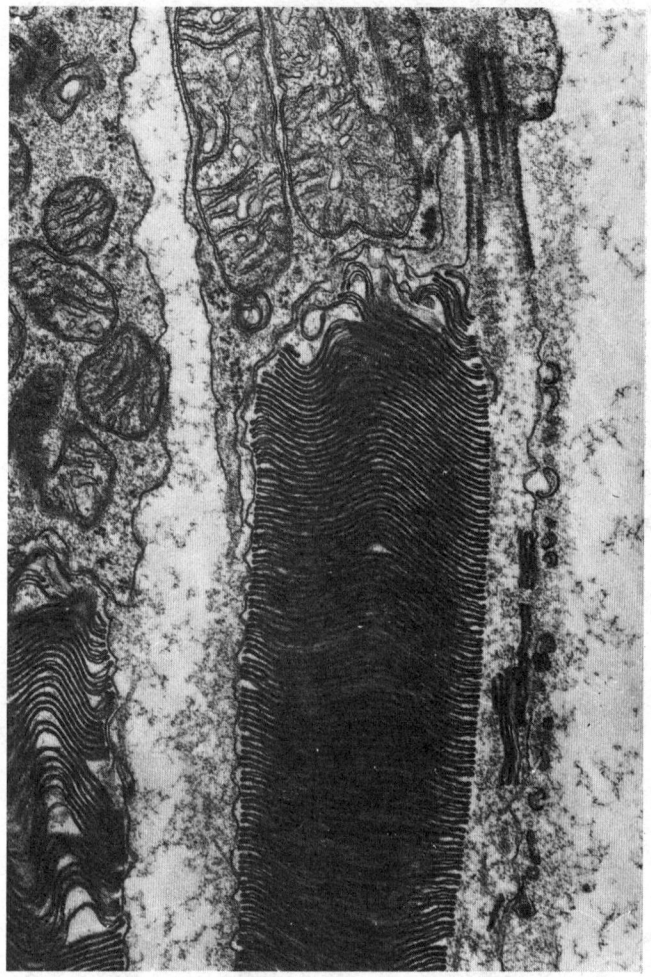

7.260C An electron micrograph of a section of a human retinal rod, showing the junction between the outer and inner segments. The outer segment is made up largely of photoreceptive lamellae (or discs), connected by a short cilium (top right) to the inner segment containing large mitochondria. (Kindly provided by Dr. N. A. Locket, Institute of Ophthalmology, London.) Magnification × 50,000.

lack of a large axonal neurite. All their processes resemble dendrites, but are said to show the cytoplasmic features of both axons and dendrites; the direction of conduction in any process at a particular time will be determined by the polarization of the synapses which are active. Each cell has one or two thick processes from which dendritic trees of variable complexity branch out. It is upon the basis of these patterns that five types of amacrine neuron have been differentiated, though their somata cannot as yet be distinguished by ultrastructural features. They contain an indented nucleus, many cisternae of granular endoplasmic reticulum, free ribosomes, microtubules and occasionally cytoplasmic crystalline bodies, and surface cilia. Subsurface cisterns have been described recently in the amacrine, and also bipolar and ganglionic, neurons of the monkey's retina (Spoerri and Glees 1977). Their significance is uncertain, but some resemblances to synaptic specializations suggests a role in intercellular communication. The processes of amacrine neurons display aggregations of synaptic vesicles at scattered sites, which are presumably points of synaptic contact with bipolar axons and ganglion cell dendrites. *Stratified* and *diffuse* amacrine cell types have been defined on the basis of their dendritic deployment as observed in silver-stained sections; both these types have been further subdivided. Recently Kolb and West (1977) have confirmed the occurrence of a type of amacrine cell (first demonstrated in the teleost retina by Ehinger *et al.* 1969) which has processes extending into both internal and external plexiform layers, in the cat. In the internal layer it makes presynaptic connexions with bipolar neurons and pre- and post-synaptic junctions with other amacrine neurons. Since it also has

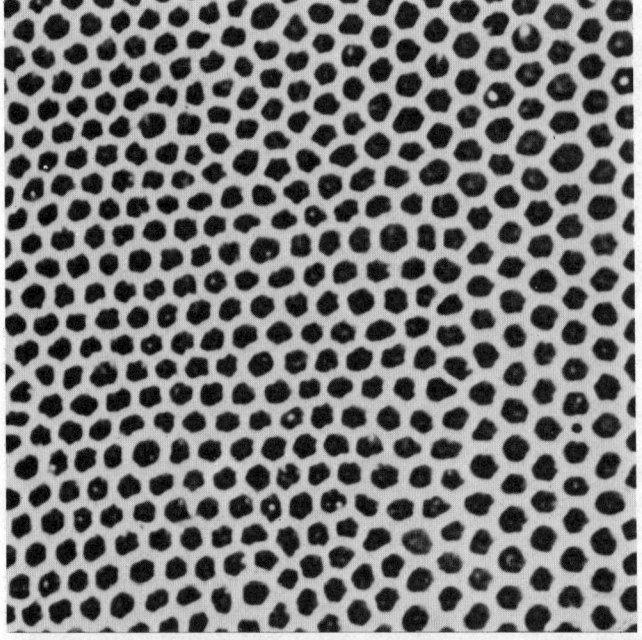

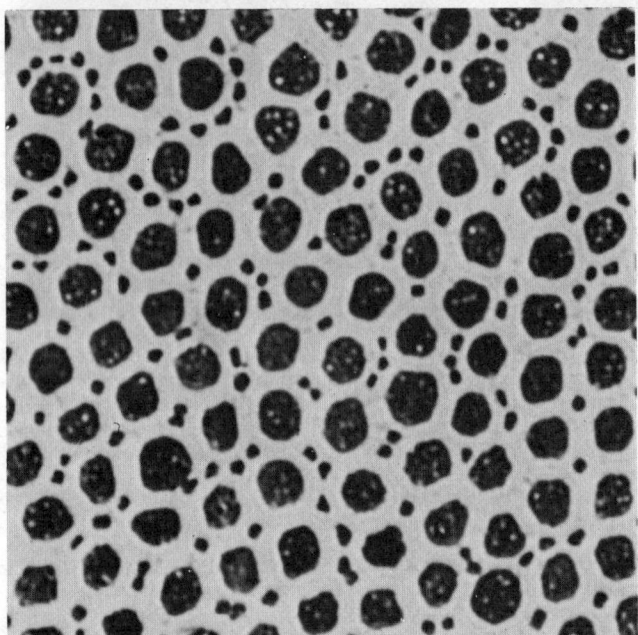

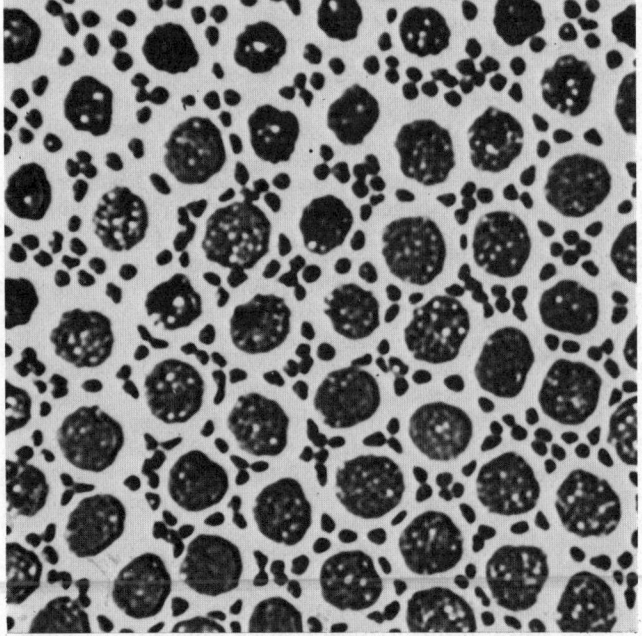

presynaptic connexions with rod and cone bipolar neurons, it apparently mediates feedback between the two plexiform layers, but whether this is excitatory or inhibitory is not yet clear. Similar neurons have been described in the retina of the goldfish by Dowling *et al.* (1976), in which it is considered to be excitatory.

THE INTERNAL PLEXIFORM LAMINA

Between the internal nuclear layer and the lamina of ganglion cells there is a dense neuropil composed of the interconnecting neurites of bipolar, amacrine and ganglionic neurons; it may also contain occasional displaced somata of the latter two categories. It is also traversed, like most of the retinal laminae, by the processes of the retinal gliocytes, which fill the spaces between the neurites, coming into close apposition with them. Even in electron microscrope studies (Allen 1969) the details of the layer are difficult to interpret, but the bipolar axons, amacrine processes, and ganglion cell dendrites can be differentiated, and a wide variety of synaptic contacts can be identified (**7.262**). The bipolar axons form axodendritic and axosomatic synapses with ganglion cells, and axosomatic contacts with amacrine cells. Amacrine cell neurites make synaptic contacts with bipolar cell axons and with the somata of ganglionic cells. These synapses are extremely numerous; in the region of the fovea they are as dense as 2 million per square millimetre.

THE GANGLIONIC CELL LAMINA

The ganglion cells are the second neurons in the visual pathway. Their dendrites are connected with the processes of bipolar and amacrine cells in the internal plexiform layer, while their axons pass into the layer of nerve fibres (lamina 9), which they form. Here they turn tangentially to approach the optic disc, through which they leave the eyeball as the constituent fibres of the optic nerve. Ganglion cells are arranged in a single stratum through most of the retina, but they are progressively more densely packed from its periphery to the macula. In the vicinity of this they are ranked in about ten rows, diminishing again towards the fovea, in which they are largely absent. They are multipolar nerve cells and vary from 10 to 30 μm in diameter; they contain a relatively large nucleus. Ganglion cell dendrites are variable in number and in their patterns of spread and branching; they usually emerge at the opposite pole of the cell with respect to its axon. On the basis of their dendritic patterns ganglion cells have been classified into at least six types (midget, stratified, diffuse, etc.). *See* Cajal (1911); Polyak (1941); Boycott and Dowling (1969); Boycott (1974).

The midget ganglion cells are monosynaptic in their connexions, and have relatively simple dendrites. They are the commonest form in the central area of the retina. They make synaptic contact with the axon terminals of the midget bipolar neurons (**7.260, 261**). Since the latter are usually each connected to a single midget ganglion cell centrally, and to a single cone pedicle by their dendrites—especially in the foveal area—these arrangements provide relatively specific pathways for individual cone processes. However, as reference to the connexions of cone cells will show (p. 1170 and **7.261**), this does not entail that a cone cell discharges exclusively through the midget bipolar and midget ganglion cell associated with it. **Polysynaptic ganglion cells** (rod and flat ganglion cells) include all the other types of ganglionic neuron, for details of which the authorities quoted above should be consulted. They have one or two large dendrites which spread much more widely than those of the midget ganglion cells; and they synapse with the axon terminals of many bipolar neurons—probably with hundreds as a rule. Since it is clear, by light microscopy, that the bipolar neurons are more numerous than ganglion cells (though not by two orders of

7.260D Tangential sections through the primate retina to show the variations in distribution of rod and cone processes in the foveola (top), fovea (middle), and peripheral macula (below). Note the *small* cone processes in the foveola, from which rods are absent. The cones, which are elsewhere larger than the rods, predominate in the fovea, rods becoming more numerous towards the periphery. (Preparations kindly supplied by Dr. John Marshall, Institute of Ophthalmology, London.) $\times$ *c.*5,000.

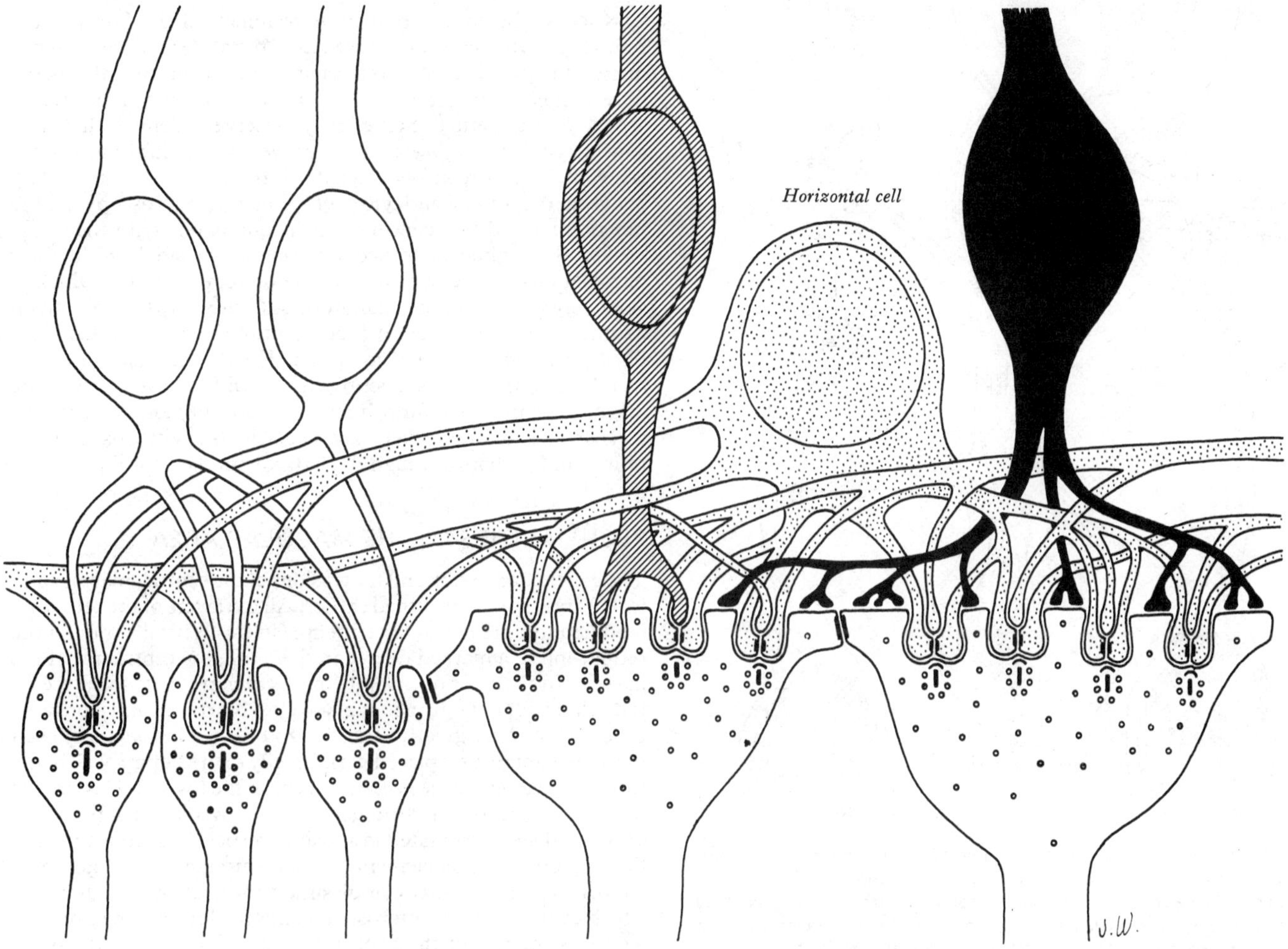

Rod bipolar cell　　　　　*Midget bipolar cell*　　　　*Flat bipolar cell*

Horizontal cell

Rods
(spherules)

Cones
(pedicles)

7.261A A scheme of the synaptic arrangements involving the rod spherules and cone pedicles of the retina. For details consult text.

magnitude), it is obvious that many bipolars must transmit to the same ganglion cell; there is also evidence that the reverse relationship obtains. Such intricate interconnexions at least suggest a structural basis for the summation and lateral inhibition long established in general physiological terms, and more recently explored in detail by unit recording techniques.

The cytoplasm of the ganglion cells contains many fibrils and chromatin granules, the latter appearing as aggregations of cisternae of granular endoplasmic reticulum under the electron microscope. Subsurface cisternae also occur (*see* p. 1171). Free ribosomes and agranular reticulum are also abundant, as are microtubules. It is not possible, nevertheless, to distinguish the types of ganglion cell by their ultrastructure. Their axons are the main component of the nerve fibre lamina which will now be considered.

THE NERVE FIBRE LAMINA

The axons of the ganglion cells converge towards the optic disc from all parts of the retina, forming a lamina of nerve fibres (stratum opticum) which is consequently thickest (20–30 μm) at the periphery of the disc. The axons converge in a simple radial pattern from the medial (nasal) half of the retina, whereas the position of the macular area, inferolateral to the optic disc, somewhat complicates the course of the lateral (temporal) axons. Those from the macula form a *papillomacular fasciculus* which passes directly to the disc, while the more peripheral temporal fibres swerve circumferentially above and below the macula to reach the disc.

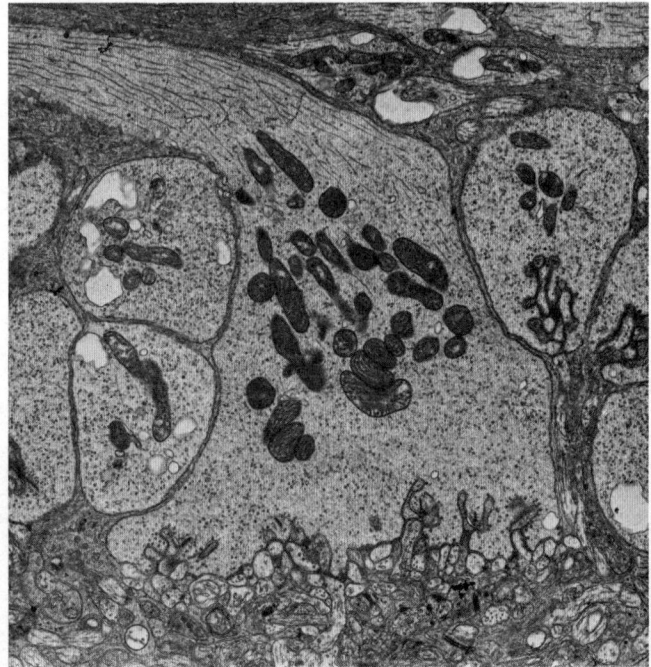

7.261B An electron micrograph of a part of the outer plexiform layer of the retina (monkey), showing synapses at the pedicle base of a cone (centre) and adjacent rod spherules. Note the presence of mitochondria and of synaptic ribbons surrounded by synaptic vesicles in the cone pedicle. The neuropil composed of interweaving dendritic processes of bipolar, horizontal and other cells is shown below. Kindly supplied by Dr. N. A. Locket, Institute of Ophthalmology, London.)

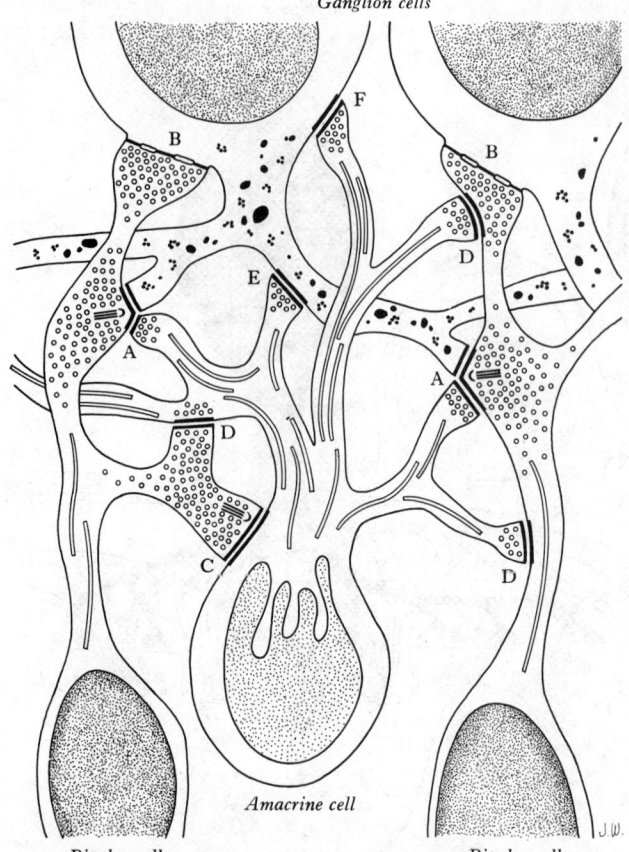

Ganglion cells

Amacrine cell

Bipolar cell   *Bipolar cell*

7.262 A scheme of the synaptic arrangements in the internal plexiform lamina of the retina. Note that bipolar axonal terminals of three types are shown—*axodendritic* (A) in *dyads* involving neurites of amacrine and ganglion cells, *axosomatic* involving ganglion cells (B) and amacrine cells (C). Similarly the neurites of amacrine cells also make three types of contact—with the axons of bipolar neurons (D), and with the dendrites (E) and somata (F) of ganglion cells. (Both **7.261** and **7.262** are from *Histology of the Human Eye*—see **7.250**. Modified from J. E. Dowling and B. B. Boycott, *Proc. R. Soc. B.*, **166**, 1966.)

Ganglion cell retinal axons are non-myelinated—an obvious optical advantage—myelin is highly refractile. The myelin sheaths commence as the axons pass into the optic disc to form the optic nerve. Occasionally, small myelinated fibres have been observed in the human retina; myelination is usual in parts of the retina in many other mammals. The axons in the nerve fibre layers are surrounded by processes of retinal gliocytes and of other neuroglial elements, such as astrocytes and microglial cells (Wolter 1959). Like the somata of the ganglion cells, their axons vary in diameter, ranging from 0·6 to 2·0 μm. They have the ultrastructural appearances typical of axons elsewhere (p. 844).

Centrifugal axons have often been described as present in the retina, but this has almost as often been denied. A considerable literature exists on this topic (Bowin 1895; Mukai 1970), but the existence of such efferent terminals in the human retina is still an open issue. Various sources for efferent retinal fibres have been suggested, including the lateral geniculate nucleus (p. 977), superior colliculus, hypothalamus, and others. They may be vasomotor in nature, but since efferent terminals have been described in relation to amacrine cells by Cajal and others, it is tempting to assume in the visual pathway an efferent analogue to the cortico-olivo-cochlear connexions of the auditory apparatus (pp. 910 and 1204). The majority of studies of this problem have, however, been carried out in birds (Cajal), or mammals remote from man, though this is rarely made plain in textbooks and monographs. There is, nevertheless, good evidence of a corticogeniculate pathway which exerts a modifying influence on the afferent visual pathway (p. 1174), and it may be that a so far ill-defined connexion of this kind extends into the retina itself.

THE INTERNAL LIMITING LAMINA

Classically, the conical branching terminals of the fibres of the retinal gliocytes are said to coalesce at the surface of the vitreous humor to form a continuous membrane, separating the nerve fibre lamina from the vitreous gel, thus 'limiting' the internal aspect of the retina. Some early observers denied this, and considered that the '*membrana hyaloidea*'—the limiting surface layer of the vitreous—in reality formed the sole boundary between the retina and vitreous. Electron microscopy of the junction has led to a concept which compounds both the above views. The gliocyte processes have associated with their terminations a basement membrane of about 0·5 μm, which is sinuously adapted to the glial processes externally but smooth on its internal aspect, where it is adjacent to the vitreous. Collagen fibrils derived from the latter blend with the glial basement membrane, which consequently can be said to have a composite origin. The internal limiting membrane is an obvious factor in the mechanisms of fluid exchange between the vitreous and the retina, and perhaps through the latter, with the choroid.

PECULIARITIES OF THE MACULAR AREA

The general formation and dimensions of the central, macular and foveal areas have been noted earlier. All the layers of the retina are modified to a greater or lesser degree in this central region of the retina, and to a marked degree in the *fovea*, the central pit of the macula. In the floor of this pit (the *foveola*, p. 1167) there are no rod cells at all, but only about 2,500 closely arrayed and elongated cones, which here greatly resemble rod cells. The *somata* of even these elements are displaced peripherally to the sloping wall of the fovea, so that only cone *processes* occur in the foveola. Despite the distorting effect of this arrangement, the foveolar cone processes are nevertheless orientated in a strictly vertical and radial manner, the only other retinal element present (through which light must pass) being the fibrous cytoplasmic processes of the gliocytes, which even here form internal and external limiting membranes. Towards the rim of the conical wall of the foveal pit the other layers begin to appear, and other modifications are also characteristic. Most of the fovea is devoid of rod cells or processes, which only reach its periphery. The rod-free central part of the fovea contains approximately 35,000 cones, and in the whole foveal area (about 1·75 square millimetres) there are in the region of 100,000. Hence, in the fovea, where the cones are most slender and hence ·most densely aggregated—all other layers being absent—there exist the most favourable conditions for photopic vision. Moreover, the cones in this locality possess the most specific connexions (with individual midget bipolar and ganglion cells), and this accords with the highly discriminative nature of foveal vision.

Because of the general displacement of the outer nuclear lamina towards the periphery of the fovea, the internal fibres or 'axons' of the photoreceptors are stretched out in a tangential direction in the external plexiform layer, and hence no cone pedicles or rod spherules are apparent in the central fovea and foveola. The inner nuclear layer is displaced to the edge of the foveal depression, and the internal plexiform, ganglion cell, and nerve fibre layers are almost absent from the whole fovea. Therefore, even on the wall of the fovea the retina is thinner and more transparent. Capillaries approach as far as the foveal margin, invading only the ganglion cell layer at this circumference. The central fovea is normally devoid of all blood vessels.

The *parafoveal region*, extending for about 0·5 mm around the fovea (diameter ·1·5 mm), is the thickest part of the retina, in part due to heaping up of displaced bipolar and ganglion cells. Around the circumference of the parafoveal area another, *perifoveal region* is described. This is the zone in which the density of cones begins to diminish rapidly, and where the incidence of rods shows an opposite tendency.

The fovea is the last part of the primate retina to attain full development, which is not completed until about the fourth postnatal month in mankind. Prenatal development of the primate fovea has been chiefly studied in rhesus monkeys. Details are beyond the scope of this volume; for a résumé of the literature consult Hollenberg and Spira (1973); Hendrickson and Kupfer (1976); *see* also p. 176.

THE OPTIC DISC AND RETINAL BLOOD VESSELS

The retina is interposed between two sets of arteries and veins, the ciliary vessels of the choroid and the branches of the central artery and vein. It is dependent upon both circulations, neither of which is alone sufficient to maintain full visual activity in the retina. The choroidal circulation (p. 1157) and the orbital and intraneural parts of the central retinal vessels have been described elsewhere (pp. 688 and 748).

The **central retinal artery** enters the optic nerve and travels in it to arrive at the 'head' of the nerve, where its constituent fascicles (about 1,000 in man) are passing through the lamina cribrosa—the representative of the sclera in this situation. In fact this region, where the neural tissues of the retina meet the neural elements of the optic nerve (including astrocytes and other glial cells) and also the connective tissues of the sclera and meninges, is highly complex. It is also the entry and exit for the retinal circulation and the only locality in which anastomoses with other arteries (the posterior ciliary arteries, *vide infra*) occur. It can be seen by ophthalmoscopy as the **optic disc** (7.257) or papilla, a region of much clinical interest, since it is here that the central vessels enter and leave the eye and can be inspected directly—the only vessels accessible in this way in the whole body. Oedema of the disc (papilloedema) may be the earliest sign of raised intracranial pressure, which is reflected in the subarachnoid space around the optic nerve and may hence compress and obstruct the central retinal vein where it crosses the space. The optic disc is somewhat medial and superior to the posterior pole of the eyeball and hence not on the visual axis. It is round or oval, being usually about 1·6 mm transversely and 1·8 mm in the vertical. Its appearance is very variable and details should be sought in suitable monographs (Duke-Elder and Wybar 1961). In light-skinned races the general retinal hue is a bright terracotta-red, and the pale pink of the optic disc contrasts sharply with this; its central part is usually even paler and may be a light grey colour. These differences are due to the degree of vascularization of the two regions—this being much less at the optic disc—and the total absence of choroidal and retinal pigment cells, these two ocular tunics being represented by little more than the internal limiting membrane of the retina. Even this does not pass far on to the disc, for the retinal gliocytes are here replaced by astrocytes which belong to the optic nerve (Anderson and Hoyt 1969). In individuals with considerable skin pigmentation both the retina and the disc display darker hues (7.257). The optic disc does not project at all in many eyes, and rarely enough to justify the term *papilla*. It is usually a little more elevated where the papillomacular fibres turn into the optic nerve on the lateral side; and where the retinal vessels pass through its centre there is usually a slight depression.

The *central retinal vessels* pass through individual apertures in the lamina cribrosa. While still at this level, usually just beyond the reach of the ophthalmoscope, the *central retinal artery* divides into two equal branches, superior and inferior; and these divide dichotomously again, after a course of a few millimetres, into superior and inferior nasal and temporal branches. Each of these four vessels supplies its own 'quadrant' of the retina, their territories being in fact rather more than quadrants, since they ramify beyond the equator as far as the ora serrata. A corresponding system of veins unites to form the central retinal vein, but the venous and arterial vessels do not correspond exactly, and the latter often cross the veins, usually lying superficial to them. At such crossings, in severe hypertension, the arteries may impress the veins and cause visible dilation of them, peripheral to the crossings. Pulsation is not visible in the arteries with a monocular ophthalmoscope but may be with instruments showing higher magnification. The branching of the central artery is usually dichotomous—two equal rami diverging from each other at an angle of 45–60°, but smaller branches may leave a larger one singly and at right angles to it. The arteries and veins ramify in the nerve fibre layer, close to the internal limiting membrane, which accounts for the clarity with which they can be seen by opthalmoscopy (7.257). Arteriolar branches pass deeper into the retina and may penetrate as far as the internal nuclear layer, from which venules return to the larger superficial retinal veins. A dense capillary bed extends between such vessels, and this is diffusely arranged and displays no laminar features. In general the structure of the vascular walls is like that of typical vessels of the same calibre, but an internal elastic lamina is lacking in the case of retinal arteries, and muscle cells may appear in their adventitious coat. The capillaries consist of a non-fenestrated endothelium and numerous mural cells or *pericytes*, which extend in the axis of the capillary external to the endothelium, sharing the same basement membrane. The cytoplasm of the two types of cells is similar, but the pericytes do not contain myofilaments; their function remains uncertain. For ultrastructural details of the retinal capillaries *see* Tominaga and Ikui (1964). Microcirculatory studies of the human retina in flat preparations stained after trypsin digestion have revealed many details of capillary arrangement. This resembles the deployment of capillaries in renal glomeruli, a network of capillaries connecting individual arterioles and venules with little or no interconnexion with neighbouring vessels. The distinction between capillaries is not, however, as exclusive as it is in the vessels which feed them. The whole of the retinal arterial and arteriolar tree is for practical purposes devoid of anastomoses, nor do arteriovenous shunts occur. The territories of the quadrantic arteries, for example, do not overlap, nor do the branches within a given quadrant show any anastomoses. For this reason, any blockage of a retinal artery is followed by loss of vision in the corresponding part of the visual field of that eye; the only exception to the endarterial nature of the retinal arterial system is limited to the vicinity of the optic disc. The posterior ciliary arteries enter the eyeball close to the entry of the optic nerve (7.255), and in addition to branches which supply the adjoining region of the choroid, other rami form an anastomotic circle (of Zinn) in the sclera around the 'head' of the optic nerve. Branches from the circle join the pial arteries supplying the nerve (Hayreh 1969), and from any of the arteries in this region small rami may pass into the eyeball to anastomose with a retinal artery (7.257). Such a connexion is called a *cilio-retinal artery*; similarly, small retino-ciliary venous anastomotic channels may sometimes be detected. For other details consult Singh and Dass (1960), in which anastomoses between the central retinal artery and pial branches of the ophthalmic artery within the optic nerve are discussed.

The retinal capillaries, which do not pass beyond the external boundary of the internal nuclear lamina, show regional differences in their density. They are especially numerous in the macula, but absent from the central part of the fovea; they become less numerous in the peripheral retina and are absent from a zone about 1·5 mm wide adjoining the ora serrata.

The central artery is innervated by sympathetic fibres and probably also has a parasympathetic supply (Ruskell 1970).

The Ocular Refractive Media

The contents of the eyeball are the aqueous humor, the vitreous body and the lens. All play a part in refracting the rays entering the eye, and the refracting power of the lens can be varied for near or far vision.

THE AQUEOUS HUMOR

The aqueous humor fills the anterior and posterior chambers. It is small in quantity and is formed by active transport and diffusion from the capillaries of the ciliary processes, from which it passes into the posterior chamber. Thence it passes into the anterior chamber through the pupillary aperture and escapes from the iridocorneal angle into the anterior ciliary veins, through the spaces of the angle and the sinus venosus sclerae. Interference with the resorption of the aqueous humor into the sinus venosus sclerae results in increase of the intraocular pressure—the condition known as *glaucoma*. The optic disc becomes cupped and resultant degenerative changes in the nervous and vascular elements of the retina produced by pressure lead to blindness.

The operation of iridectomy may re-establish the flow of the aqueous humor from the posterior chamber to the anterior in cases where the disease is due to adhesions between the iris and the lens. The aqueous humor is not only a metabolic avenue for the avascular tissues, the lens and cornea, but is also chiefly responsible for maintenance of the intraocular pressure, and hence the constancy of the optical dimensions of the eyeball. It

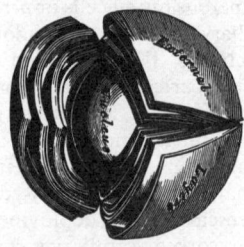

7.263   The human lens, hardened and divided. Enlarged view.

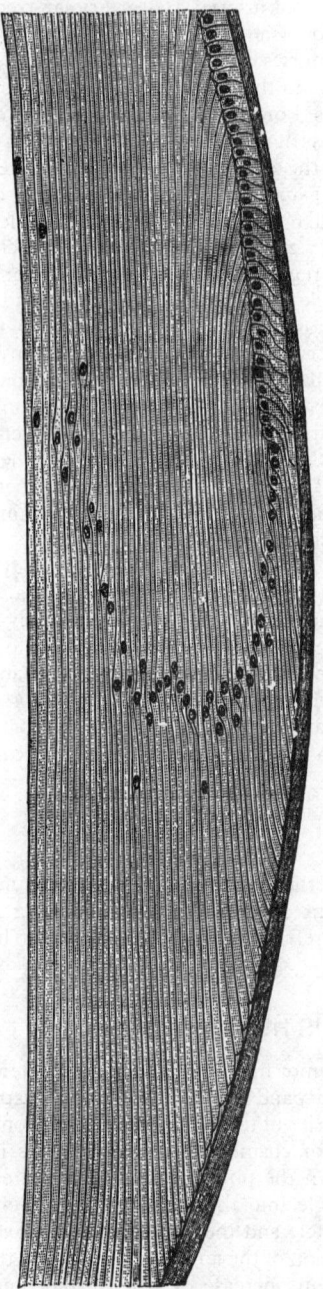

7.264   A section through the margin of the lens, showing the transition of the columnar epithelium into the lens fibres. Note that the lens cells retain their nuclei long after they have assumed the form of a fibre.

carries glucose and amino acids and mediates the exchange of respiratory gases. It contains a high concentration of ascorbic acid.

THE VITREOUS BODY

The vitreous body occupies the vitreous chamber, which constitutes about four-fifths of the eyeball. It fills the concavity of the retina, and is hollowed in front, forming a deep concavity, the *hyaloid fossa*, adjacent to the lens. A colourless, apparently structureless, transparent gel, consisting of about 99 per cent of water, some salts and a little glycoprotein and hyaluronate (but *v. infra*). Fibrils of about 16 nm diameter, and a periodicity of 22 nm, and an interfibrillary substance (*vitreous humor*) may be distinguished by electron microscopy (Schwarz 1961). At its periphery the gel is condensed to form the so-called *vitreous (hyaloid) membrane*. A narrow canal, called the *hyaloid canal*, runs from the optic disc to the centre of the posterior surface of the lens. In the fetus the canal is occupied by the hyaloid artery (p. 176), which normally disappears about six weeks before birth. The vitreous membrane is attached to the ciliary epithelium and processes, and to the edge of the optic disc. Anterior to the ora serrata it is thickened by the accession of radial fibres and is termed the *ciliary zonule*. Here the membrane presents a series of radially arranged furrows, in which the ciliary processes are accommodated and to which they adhere, as is shown by the fact that when they are removed some of their pigment remains attached to the zonule. The ciliary zonule splits into two layers, one of which is thin and lines the hyaloid fossa of the vitreous body; the other, forming a system of *zonular fibres* which collectively comprises the *suspensory ligament of the lens*, is thicker and passes over the ciliary body to be attached to the capsule of the lens a short distance in front of its equator; some of the fibres of the suspensory ligament are attached behind the equator of the lens (7.253). Scattered and delicate fibres are also attached to the region of the equator itself. This ligament retains the lens in position, and is relaxed by the contraction of the meridional fibres of the ciliaris, so that the lens is allowed to become more convex (p. 1184).

The structure of the vitreous body has been studied in much detail, both by light and electron microscopy, and by X-ray diffraction (Suguira 1957; Fine and Tousimis 1961; Brini *et al.* 1968). It contains collagen fibrils of 6–15 nm diameter, the arrangement of which shows regional differences, in particular a cortical condensation. Rounded cells can be identified by phase contrast microscopy; for details of such *hyalocytes*, consult (Hogan 1963). No blood vessels penetrate the vitreous body; so that its nutrition must be mediated by the vessels of the retina and ciliary processes, situated upon its exterior.

THE LENS

The lens (7.263–265), enclosed in its capsule, is situated immediately behind the iris, in front of the vitreous body, and is encircled by the ciliary processes, which slightly overlap its margin or equator.

The *capsule of the lens* is a transparent, elastic membrane which closely surrounds the lens, and is thicker in front than behind. It contains no true elastic tissue, and its elasticity may reside in the disposition of the fine fibrils of which it is formed. It has been described as a very thick basement membrane, which is PAS positive. Its fibrils have a periodicity of 60 nm, which differ from those of the zonule, with which they are continuous at the periphery of the lens. The lens rests, posteriorly, in the hyaloid fossa of the vitreous body; anteriorly, it is in contact with the free border of the iris, but recedes from it at the circumference, thus forming the posterior chamber of the eye; it is retained in its position chiefly by the suspensory ligament already described.

The *lens* is a transparent, biconvex body, the convexity of its anterior being less than that of its posterior surface. the central points of these surfaces are termed respectively the *anterior* and *posterior poles*; a line connecting the poles constitutes the *axis* of the lens, while the marginal circumference is termed the *equator*. Its dioptric power is much less than that of the cornea. All the optical media of the eye have a refractive index not much removed from that of water (1·33), but the corneal surface is in contact with air, and the majority of the 58 dioptres of which the eye is capable is effective here. The advantage of the lens is its potentiality of varying its dioptric power, and this is partly dependent upon a variation in its refractive index from 1·386 in its periphery to 1·406 at its core. The lens contributes about 15 dioptres to the total dioptric power of the eye. The *range* in dioptric power of the lens does not quite reach this, even at birth; most young children show minor refractive errors (Sorsby *et al.* 1961), and available dioptric range decreases with age, being halved by the age of forty and reduced to 1 or 2 dioptres by sixty. For further information on physiological optics consult Bennett and Francis (1962); Duke-Elder (1969).

THE STRUCTURE OF THE LENS

The lens is made up of soft cortical substance and a firm, central part, the so-called nucleus (7.263). Faint sutural lines (radii lentis) radiate from the poles to the equator. In the adult there may be six or more of these lines, but in the fetus there are only three, and these diverge in a Y-shaped manner at angles of 120°; on the anterior surface the Y is upright; on the posterior surface it is inverted (Mann 1924). These lines correspond with the free edges of septa—composed of an amorphous substance, which dip into the substance of the lens. When the lens has been hardened it is seen to consist of a series of concentrically arranged laminae, each of which is interrupted at the septa. Each lamina is built up of a number of ribbon-like lens fibres, the edges of which are more or less serrated—the serrations interdigitating with those of adjacent fibres, which are in part connected by desmosomes (Wanko and Gavin 1961), while the ends of the fibres come into apposition at the sutures. The ultrastructural characteristics of specialized contacts between lental fibres have been described (in the mouse) by Rafferty and Esson (1974); they regard these as gap junctions, a finding confirmed by Philipson *et al.* (1975) in human and bovine lenses. Kuwabara (1975), on the other hand, considers that the cortical cells of the lens are held together by reciprocal 'knob and socket' junctions and by fine ridges in the deeper parts of the lens. The fibres run in a curved manner from the sutures on the anterior surface to those on the posterior surface (7.265). No

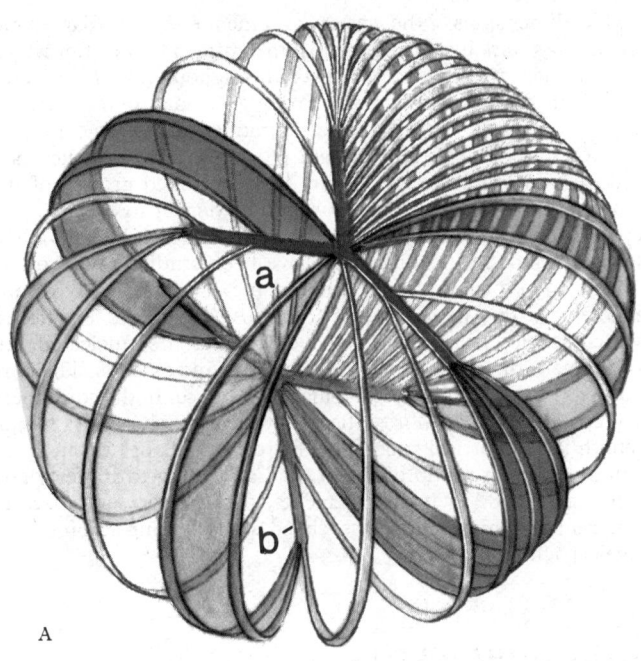

A

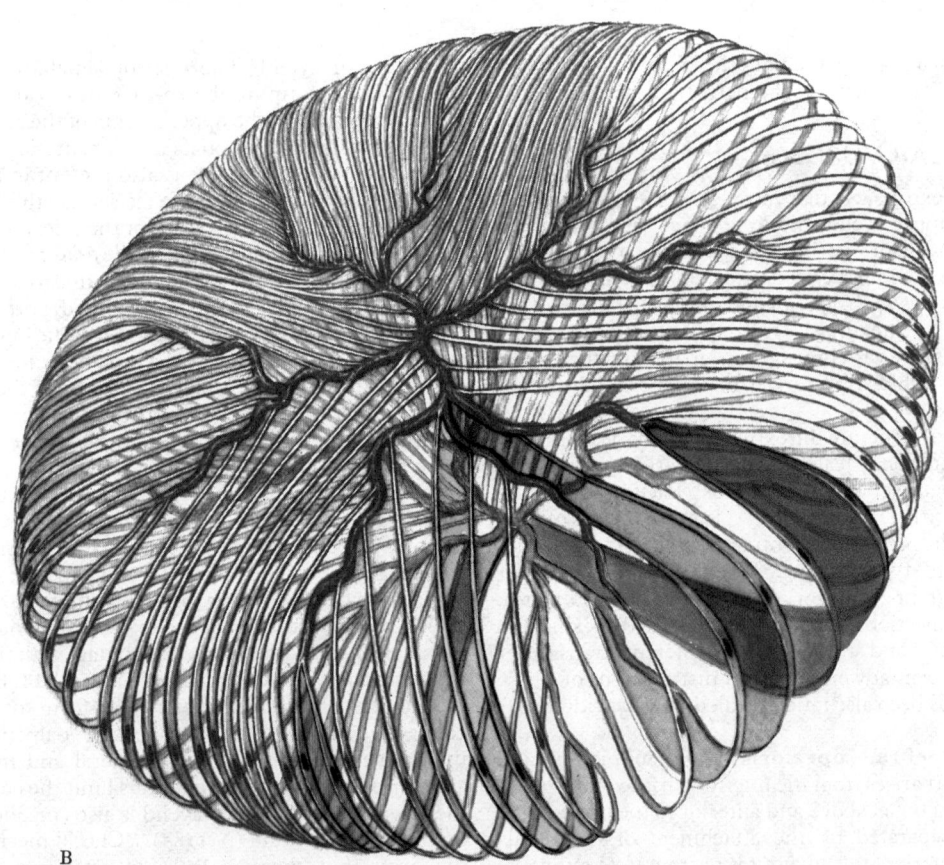

B

7.265A and B   The structure of the fetal (A) and adult (B) human lens, showing the major details of arrangement of the lens cells or fibres. The anterior (*a*) and posterior (*b*) triradiate sutures are shown in the fetal lens, and it is clear that fibres pass from the apex of an arm of one suture to the angle between two arms at the opposite pole, as shown in the coloured segments. Intermediate fibres show the same reciprocal behaviour, ending nearer to one pole, where they start further from the other, and so on. The suture pattern becomes much more complex as successive strata are added to the exterior of the growing lens, the original arms of each triradiate suture showing secondary and tertiary dichotomous branchings. (From *Histology of the Human Eye*—see 7.250.)

fibres pass from pole to pole; they are arranged in such a way that those which begin near the pole on one surface of the lens end near the peripheral extremity on the other, and vice versa. The fibres of the outer layers of the lens are nucleated, and together form a nuclear layer, most distinct towards the equator. The deeper fibres, which have lost their nuclei, are less flexible, and their contours are more deeply serrated. The superficial fibres, the more recently formed, are more flexible; in cross-section they are hexagonal, with two opposite sides longer than the rest. Adhesion along these sides is considered to be less firm, which may provide a basis for the moulding of the lens by its capsule in accommodation. The total number of lens cells or fibres is estimated to be about 2,000 in the adult lens. The anterior surface of the lens is covered by a layer of transparent, nucleated cuboidal epithelium. At the equator the cells become elongated, and their gradual transition into lens fibres can be traced (7.264). In tangential section or surface view these cells are polygonal, with a diameter of about 15 μm. In the central area of the lens they may be flattened and a mere 6 μm in depth. More peripherally they diminish in width but increase greatly in length, and mitoses are more frequent. At the equator of the lens mitotic activity is maximal. The lens 'fibres' retain their nuclei for a long period, but the oldest (and hence most deeply situated) fibres lose their nuclei. Despite the very slow growth of the lens, the epithelial cells are relatively active; experimental injuries in the mouse (Rafferty 1976) heal completely within 3 days.

In the fetus, the lens is nearly spherical and has a slightly reddish tint; it is soft and breaks down readily on the slightest pressure. A small branch (hyaloid artery) from the central artery of the retina runs forwards through the vitreous body to the posterior part of the capsule of the lens, where its branches radiate, forming a plexiform network which covers the posterior surface of the capsule, and is continuous round the margin of the capsule with the vessels of the pupillary membrane and with those of the iris. *In the adult*, the lens is colourless, transparent, firm in texture and devoid of vessels. *In old age*, it is more flattened on both surfaces, slightly opaque, of an amber tint and increased in density. In the condition termed cataract the lens gradually becomes opaque and blindness ensues. In such cases sight may be restored by extraction of the lens and the provision of suitable spectacles.

The dimensions of the lens are of some interest; its diameter at birth is 6·5 mm, increasing to 9·0 mm at fifteen years, after which it continues to grow very slowly throughout life. Its antero-posterior dimension increases from 3·5–4·0 mm at birth to 4·75–5·0 mm at 95(!). Its anterior radius (about 10·0 mm) is greater than the posterior (about 6·0 mm), but both these are reduced during accommodation. The continued growth of the lens is due to the continual slow production of new cells in its epithelium at the lental equator. The ultrastructural details of the lens capsule and its epithelium have been studied (Wanko and Gavin 1960; Cohen 1965). The capsule consists largely of fine filaments with a periodicity of 60 nm. As noted, the lens fibres, which develop from the epithelial cells, are at first nucleated, but their nuclei disappear as the fibre develops and extends. The fibre consists essentially of a fine fibrillary substance. In the adult there are 2,100–2,300 lens fibres; these may be as much as 8–12 mm in length in the cortical zone. The microfilaments of human lental cells are of 5–9 nm in diameter, and according to Rafferty and Goosens (1978) their arrangements, constitution, and sizes in different mammals may be associated with differing accommodative capacities.

THE ACCESSORY VISUAL APPARATUS

The accessory visual apparatus includes the extraocular muscles, the fasciae, the eyebrows, eyelids, conjunctiva and lacrimal gland.

THE EXTRAOCULAR MUSCLES

The extraocular or extrinsic ocular muscles (7.266–270) include an elevator of the upper eyelid, the *levator palpebrae superioris*, and six muscles capable of rotating the eyeball in any direction. The latter comprise the four *recti* (*superior, inferior, medialis* and *lateralis*), and the two *obliqui* (*superior* and *inferior*). This is an old pattern of muscles, extending—with slight modifications—through almost the entire vertebrate group, apart from the levator palpebrae superioris, which is a later delamination from the rectus superior.

In general details of structure the extraocular muscles resemble others, but differ in some features, particularly in details of innervation and pharmacological reactions (Bach-y-Rita *et al.* 1971; Hogan *et al.* 1971; Davson 1976). At least two forms of fibre occur: a slender (9–11 μm), slow-contracting type, and a thicker (11–15 μm), 'twitch' type of fibre. The former show motor endings of the 'en-grappe' category the latter motor end plates. More complex classifications, based on histological data, single or multiple innervation, and responses to different transmitter substances, have been advanced. The distribution of these categories of fibre has been also much studied. (*See* Davidowitz *et al.* 1977, for details.)

The levator palpebrae superioris (7.266–269) is thin and triangular in shape. It arises from the inferior surface of the lesser wing of the sphenoid bone, above and anterior to the optic canal, from which it is separated by the attachment of the rectus superior. At this posterior attachment it is narrow and tendinous, but soon becomes broad and fleshy, the medial margin of the muscle being almost straight, while the lateral is concave. The muscle ends anteriorly in a wide aponeurosis which splits into two lamellae. Some of the fibres of the superior lamella are attached to the anterior surface of the superior tarsus (p. 1185), while others radiate and pass through the overlying orbicularis oculi to the skin of the upper eyelid. The inferior lamella contains nonstriated muscle fibres forming the so-called *superior tarsal muscle*; it is attached directly to the upper margin of the superior tarsus and is covered by conjunctiva on its inferior surface. A less well-marked layer of nonstriated muscle is also present in the lower eyelid; it unites the inferior tarsus to the fascial sheath of the inferior rectus and its expansion to the sheath of the inferior oblique. A further thin layer of nonstriated muscle, the *orbitalis*, bridges the inferior orbital fissure; its precise function is uncertain. The *superior* and *inferior tarsal muscles*, just described, presumably assist in elevation of the upper and depression of the lower eyelid. All three nonstriated muscles receive a sympathetic innervation.

The connective tissue on the surfaces of the levator palpebrae superioris and rectus superior fuse. Where the two muscles separate to reach their insertions, the fascia between them forms a thick mass to which is attached the superior conjunctival fornix, and this is described as an additional insertion of the levator palpabrae superioris. When traced laterally the aponeurosis of the levator palpebrae superioris passes between the orbital and palpebral parts of the lacrimal gland and is fixed to a tubercle on the zygomatic bone, just within the orbital margin (p. 344). When traced medially the aponeurosis loses its tendinous nature as it passes over and comes into close contact with the reflected tendon of the obliquus superior, whence it can be followed towards the medial palpebral ligament in the form of loose strands of connective tissue. When the levator palpebrae contracts, the upper eyelid is raised, but the lateral and medial parts of the aponeurosis are stretched and thus limit the action of the muscle; the elevation of the upper eyelid is also considered to be checked by the orbital septum (p. 1185). ('Check' mechanisms abound in the orbit, but there is little or no direct evidence that the connective tissue structures thus implicated do in fact function in this manner. See also p. 1183.)

The four recti (7.267–270) are attached posteriorly to a fibrous ring which surrounds the superior, medial and inferior margins of the optic canal (7.268), and is termed the *common annular tendon*; this fibrous annulus is continued across the lower and medial part of the superior orbital fissure and is attached to a tubercle on the margin of the greater wing of the sphenoid bone. The tendon is closely adherent to the dural sheath of the optic nerve and to the surrounding periosteum; within it are (1) the anterior aperture of the optic canal transmitting the optic nerve and ophthalmic artery, and (2) the medial part of the superior orbital fissure which transmits the two divisions of the oculomotor nerve, the nasociliary nerve and the abducent nerve. The superior ophthalmic vein may pass through, or above, the annular tendon, the inferior ophthalmic vein through or below it. Two specialized parts of this fibrous ring may be discerned: a lower, which gives origin to the rectus inferior, a part of the rectus medialis and the lower fibres of the rectus lateralis; and an upper, which gives origin to the rectus superior, the other part of the rectus medialis and the upper fibres of the rectus lateralis; a second small tendinous head of origin of the rectus lateralis arises from the orbital surface of the greater wing of the sphenoid bone, lateral to the tendinous ring. Each muscle passes forward in the position implied by its name, to be attached by a tendinous expansion into the sclera, posterior to the margin of the cornea. The average distances of the insertions of the recti from the margin of the cornea are: medialis, 5·5 mm; inferior, 6·5 mm; lateralis, 6·9 mm; superior, 7·7 mm.

The obliquus superior (7.267) is fusiform and occupies a superomedial position in the orbit. It arises from the body of the sphenoid superomedial to the optic canal and to the tendinous attachment of the rectus superior, and, passing forwards, ends in a round tendon, which plays in a fibrocartilaginous loop, the *trochlea*, attached to the trochlear fossa of the frontal bone. The contiguous surfaces of the tendon and the trochlea are separated by a delicate synovial sheath. After traversing the trochlea the tendon passes posterolaterally and inferior to the rectus superior to the lateral part of the eyeball, and is inserted into the sclera, *behind* the equator of the eyeball, in its superolateral posterior quadrant between the rectus superior and rectus lateralis

The obliquus inferior (7.267) is a thin, narrow muscle, near the anterior margin of the floor of the orbit. It arises from the orbital surface of the maxilla lateral to the nasolacrimal groove. Passing laterally, backwards and upwards, at first between the rectus inferior and the floor of the orbit, and then between the eyeball and the rectus lateralis, it is inserted into the lateral part of the sclera, *behind* the equator of the eyeball, in its inferolateral posterior quadrant between the rectus superior and rectus lateralis, near to, but somewhat behind, the insertion of the obliquus superior.

Nerve supply. The levator palpebrae superioris, the obliquus inferior, and the recti superior, inferior and medialis are supplied by the oculomotor nerve, the obliquus superior by the trochlear nerve, the rectus lateralis, by the abducent nerve.

ACTIONS OF THE EXTRAOCULAR MUSCLES

The *levator palpebrae superioris* elevates the upper eyelid, its antagonist being the orbicularis oculi. The degree of elevation—which, apart from blinking, is maintained for long periods of time during waking hours—is a compromise between adequate exposure of the optical media and control of light entering the eye. The amount of light, especially in conditions of very bright sunshine, can be reduced by lowering the upper eyelid, thus reducing glare. Although much is known of the physiology of blinking (McEwen and Goodner 1962), and its significance in the distribution of lacrimal secretions, little information is available concerning the continuous activity of the levator to keep the eyelid raised. The respective roles of the main, striated, voluntary part of the muscle and of the small, inferior stratum of nonstriated tissue (the superior palpebral, tarsal, or Müller's muscle) which is innervated by sympathetic neurons, have not been clarified.

The six *extraocular muscles* all *rotate* the eyeball in directions dependent upon the geometrical relationship between their osseous and global attachments (7.270), which are, of course, influenced by the ocular movements themselves. Before considering these activities it is important to recognize that the *human* extraocular muscles are not accessible to inspection, and that consequently much opinionation regarding them depends on

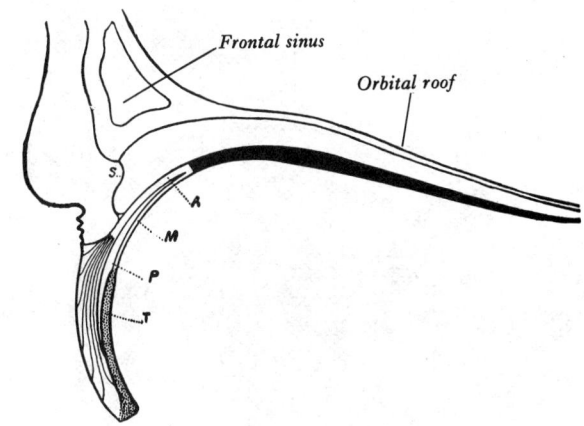

7.266 A diagram of the levator palpebrae superioris, showing its connexions. *A.* Superficial lamella of aponeurosis; *M.* Deep lamella, superior tarsal muscle (of Müller); *P.* Interval between superficial and deep lamellae of aponeurosis; *T.* Tarsus; *S.* Orbital septum.

deductions from disturbances due to lesions of their nerve supplies. A complete assessment of the exact nature of the nerve injury is rarely possible; but this is not to say that, in the vast clinical literature available on this topic, there are not at least some entirely valid observations. It is also essential to note that in any movement of an eyeball, changes in tension and/or length may occur in *all six* of the extraocular muscles, although the direct observation of these has rarely been carried out, even in experimental animals (Sherrington 1905; Szentágothai 1950). It is at least likely that all six muscles are continuously involved, and it is therefore merely a preliminary but necessary exercise to consider each muscle in isolation. Because they form more obvious groupings as antagonists or synergists, it is appropriate to consider the four recti and the two obliques as separate groups, remembering always that they act in concert.

Of the four *recti*, the *medial* and *lateral* muscles exert comparatively straightforward forces on the eyeball. Being approximately horizontal (at least, when the visual axis is in its primary position, directed to the horizon), they rotate the eyeball medially (adduction) or laterally (abduction) about an imaginary vertical axis (7.270). They are antagonists, and by reciprocal adjustment of their lengths the visual axis can be swept through a horizontal arc. When both eyes are involved, as is usual, the four medial and lateral recti can either adjust both visual axes in a *conjugate* movement from point to point at infinity, the axes remaining parallel, or they can *converge* or *diverge* the two axes to and from nearer or more distant objects of interest in the visual field. However, since they do not rotate the eyeball around its transverse axis, the medial and lateral recti cannot make the extremely commonplace action of elevating or depressing the visual axes as the gaze is transferred from nearer to more distant objects or the reverse. This is the contribution of the *superior* and *inferior recti* (aided, as will become apparent, by the two oblique muscles). However, the geometry of these muscles is a little more complex, and the key to the rotational effects which they impart to the eyeball is the obliquity of the orbit (7.270), whose axis does not correspond with the visual axis in its primary position, but is inclined to it by an angle of approximately 23°. The latter value varies somewhat in different individuals being dependent upon the angle between the two orbital axes (7.270A). Hence, the simple rotation caused by an isolated *superior rectus*, when analysed with reference to the three hypothetical axes of the eyeball, appears to be complex—being primarily *elevation* (transverse axis), and secondarily a less powerful *medial rotation* (vertical axis), and a slight *intorsion* (anteroposterior axis) in which the upper rim of the cornea (often referred to as 12 o'clock) is rotated medially towards the nose. These actions—which are in fact a simple, single rotation—are easily appreciated as long as it is realized that the direction of traction of the superior rectus runs *posteromedially* from its attachment *anterior* to the equator and *superior* to the cornea, to its osseous attachment near the apex of the orbit (7.270).

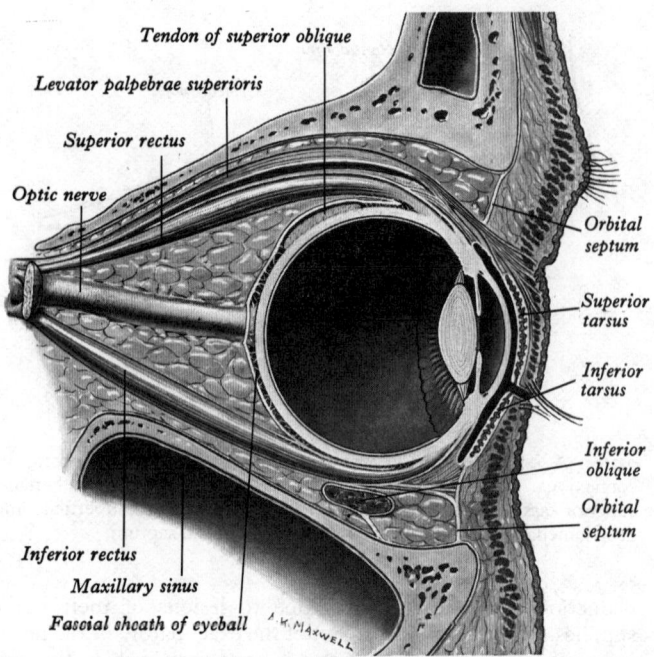

7.267A A sagittal section through the right orbital cavity.

The *inferior rectus* pulls in the same direction, but naturally rotates the visual axis downwards about the transverse axis. It is also clear, that if reference is made to the comparable geometry of the situation, that this muscle will also rotate the eye medially about the vertical axis, but that its action around the anteroposterior axis will *extort* the eye, i.e. rotate it so that the 12 o'clock point on the cornea turns laterally. The superior and inferior recti, therefore, both rotate the eyeball medially, and since their turning movements around the transverse and anteroposterior axes are opposed, their combined contraction could rotate the eye medially. In binocular movements they could thus assist the medial recti in converging the visual axes. By reciprocal adjustment the same two muscles can elevate or depress the visual axes. It may be added that, as the eyeball is rotated laterally, the line of traction of the superior and inferior recti more nearly approaches the plane of the anteroposterior axis (7.270), and hence their rotational effects about this and the vertical axis diminish. In abduction to about 20° or so, these two muscles become almost purely an elevator and a depressor of the visual axis.

The *superior oblique muscle* acts on the eyeball from the trochlea, and since the attachment of the *inferior oblique* is for practical purposes vertically inferior to this, both muscles approach the eyeball at the same angle, being attached in approximately similar positions in the *superior* and *inferior posterolateral* quadrants of the eyeball (7.270). From these attachments it is easy to understand that the superior oblique elevates the *posterior* aspect of the eyeball, the inferior muscle depressing it. Hence, the former rotates the visual axis *downwards*, the latter *upwards*, both movements being around the transverse axis. But the obliquity of both muscles is such that their traction, when the eye is in the primary position, is in a direction posterior to the vertical axis, and thus both rotate the eyeball *laterally* around this axis. In regard to the anteroposterior axis it is not difficult to conclude that in isolation the superior oblique would *intort* the eyeball, the inferior oblique *extorting* it. Like the superior and inferior recti, therefore, the two obliques have a turning movement in common around the vertical axis, but opposed forces in respect of the other two. Acting in concert they could therefore assist the lateral rectus in abduction of the visual axis, as in *divergence* movements of the eyes in transferring attention from near objects to those further away. Just as in the case of the superior and inferior recti, the actions of the oblique muscles also vary with the position of the eyeball; they are more nearly a pure elevator and a depressor as the eye is adducted.

In a short analysis of this kind much must be omitted, and

nothing can be said of the defects of ocular movement. For those desiring further information references Whitnall (1932); Cooper *et al.* (1955); Cogan (1956); Schlossman and Priestley (1966); Alpern (1969); Davson (1972), will prove useful; for those requiring the barest data, the actions of the extraocular muscles may be summarized as follows:

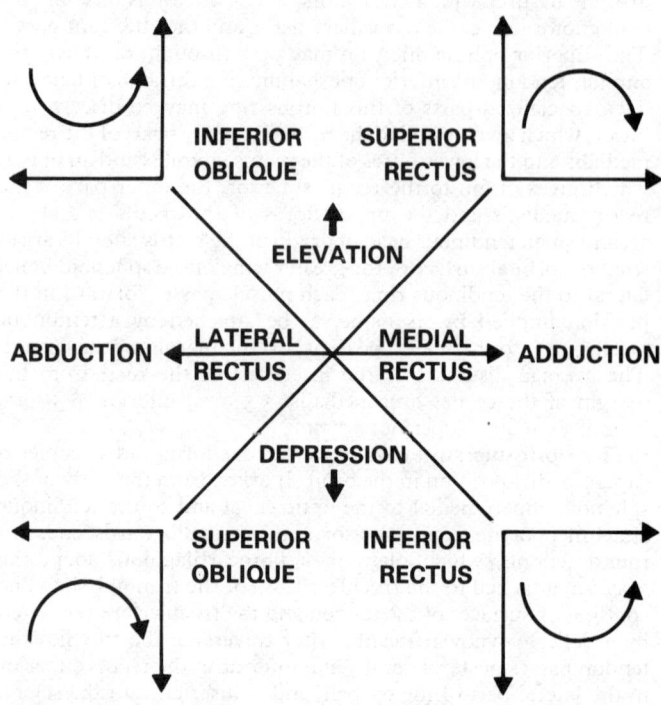

There is a misleading custom of linking extraocular muscles together, usually into pairs—such as the right superior oblique and left inferior rectus—which are supposed thus to deviate both visual axes downwards and to the right, as doubtless they do. Limited views of this kind may have a mnemonic value, but they ignore the inescapable fact that in any ocular rotation *all* six muscles must change in length. Since we do not know enough of the reciprocal innervation circuits of the extraocular muscles, it is impossible to dogmatize as to whether every muscle is contracted precisely in step with the progressive inhibition of an antagonist, although the few experiments previously quoted, in fact, support such a view. However that may be, it would be more useful to link the superior, inferior and medial recti together, as adductors, or *convergence* muscles, and the two obliques with the lateral rectus as abductors, or *divergence* muscles. This is a useful concept in view of the very frequent convergent and divergent movements of

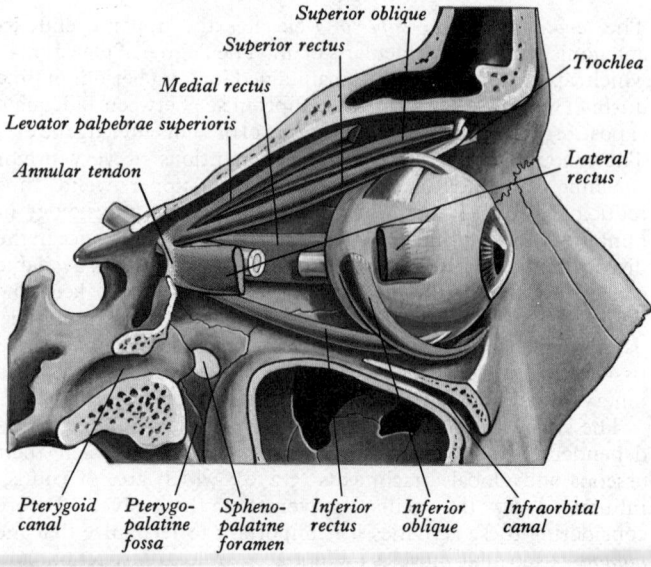

7.267B The muscles of the right orbit. Lateral aspect.

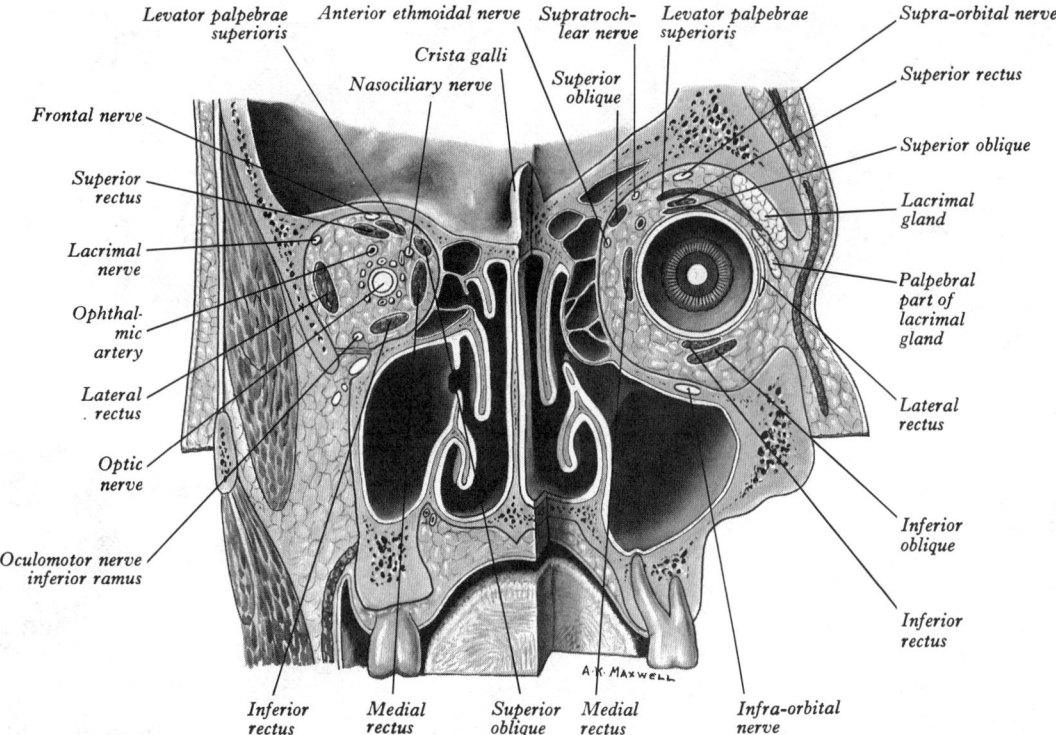

Levator palpebrae superioris — Anterior ethmoidal nerve — Supratrochlear nerve — Levator palpebrae superioris — Supra-orbital nerve

Crista galli — Superior oblique — Superior rectus

Nasociliary nerve — Superior oblique

Frontal nerve — Lacrimal gland

Superior rectus — Lacrimal gland

Lacrimal nerve — Palpebral part of lacrimal gland

Ophthalmic artery — Lateral rectus

Lateral rectus

Optic nerve — Inferior oblique

Oculomotor nerve inferior ramus — Inferior rectus

Inferior rectus — Medial rectus — Superior oblique — Medial rectus — Infra-orbital nerve

7.268 Coronal sections through the two orbits. Posterior aspect. On the left side the plane of the section is more posterior and passes behind the eyeball.

the visual axes required in common activities. It is, nevertheless, an over-simplified view; convergence is commonly accompanied by depression of the visual axes towards objects nearby, and divergence by elevation. In these much more complexly concerted activities, the torsional effects of the muscles become much more important. Analysis shows that in the simple, *cardinal* movements (adduction-abduction, elevation-depression), the torsional effects of the muscles cancel out, and they must also do so in all intermediate positioning of the visual axes to preserve the relationship of corresponding retinal loci and hence binocular (single) vision. So far, no consideration has been given to head movements, but it is common observation that ocular movements are frequently and perhaps usually accompanied by movements of the head, which might be likened to the *coarse* adjustment of an optical instrument such as a microscope, the *finer* adjustments being made by the ocular musculature.

It is interesting to note that while the eye movements are clearly under voluntary control, *torsional* movements cannot be voluntarily incepted. But when the head is tilted relative to the body, reflex torsion movements become apparent, and are in fact necessary to preserve retinal relationships. Any small lapse in the concerted adjustment of the two retinae entails diplopia; it is indeed surprising that such a complex organization of extra-ocular, neck and other muscles is learnt so effectively early in life that diplopia is so rarely experienced by the great majority.

Since the four recti exert a *posterior* traction, and the two obliques pull the globe to some degree *anteriorly*, it is sometimes suggested that they collectively suspend it in the orbital cavity, and that they thus prevent anteroposterior movement of the globe of the eye, assisted perhaps by various 'check ligaments' (*vide infra*).

THE FASCIAL SHEATH OF THE EYEBALL

A thin fascial membrane envelops the eyeball from the optic nerve to the sclerocorneal junction, separating it from the orbital fat, and forming a socket in which it plays (**7.271**). Its inner surface is smooth, and is separated from the outer surface of the sclera by the *episcleral space*; this 'space' is traversed by delicate bands of connective tissue which extend between the fascia and the sclera.

The fascia is perforated behind by the ciliary vessels and nerves, and fuses with the sheath of the optic nerve and with the sclera around the entrance of the optic nerve. In front it blends with the sclera just behind the sclerocorneal junction. It is perforated by the tendons of the orbital muscles, and is reflected on each as a tubular sheath. The sheath of the obliquus superior is carried as far as the fibrous trochlea of that muscle; that on the obliquus inferior reaches as far as the floor of the orbit, to which it gives off a slip. The sheaths on the recti are gradually lost in the perimysium, but they give off important expansions. The expansion from the rectus superior blends with the tendon of the levator palpebrae superioris; that of the rectus inferior is attached to the inferior tarsus and to the sheath of the obliquus inferior. The expansions from the sheaths of the recti medialis et lateralis are strong and

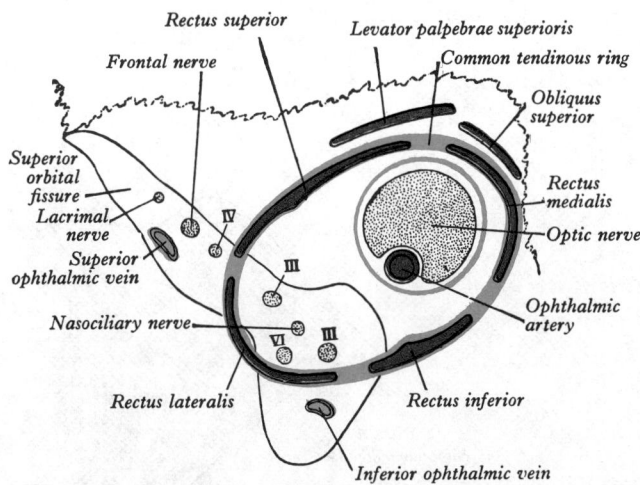

Rectus superior — Levator palpebrae superioris

Frontal nerve — Common tendinous ring

Obliquus superior

Superior orbital fissure — Rectus medialis

Lacrimal nerve — Optic nerve

Superior ophthalmic vein

Nasociliary nerve — Ophthalmic artery

Rectus lateralis — Rectus inferior

Inferior ophthalmic vein

7.269 Scheme to show the common tendinous ring, the origins of the recti, and the relative positions of the nerves entering the orbital cavity through the superior orbital fissure. (Modified from a figure in Whitnall's *Anatomy of the Human Orbit*, 2nd ed., Oxford, 1932.) Note that the ophthalmic veins frequently pass through the common tendinous ring.

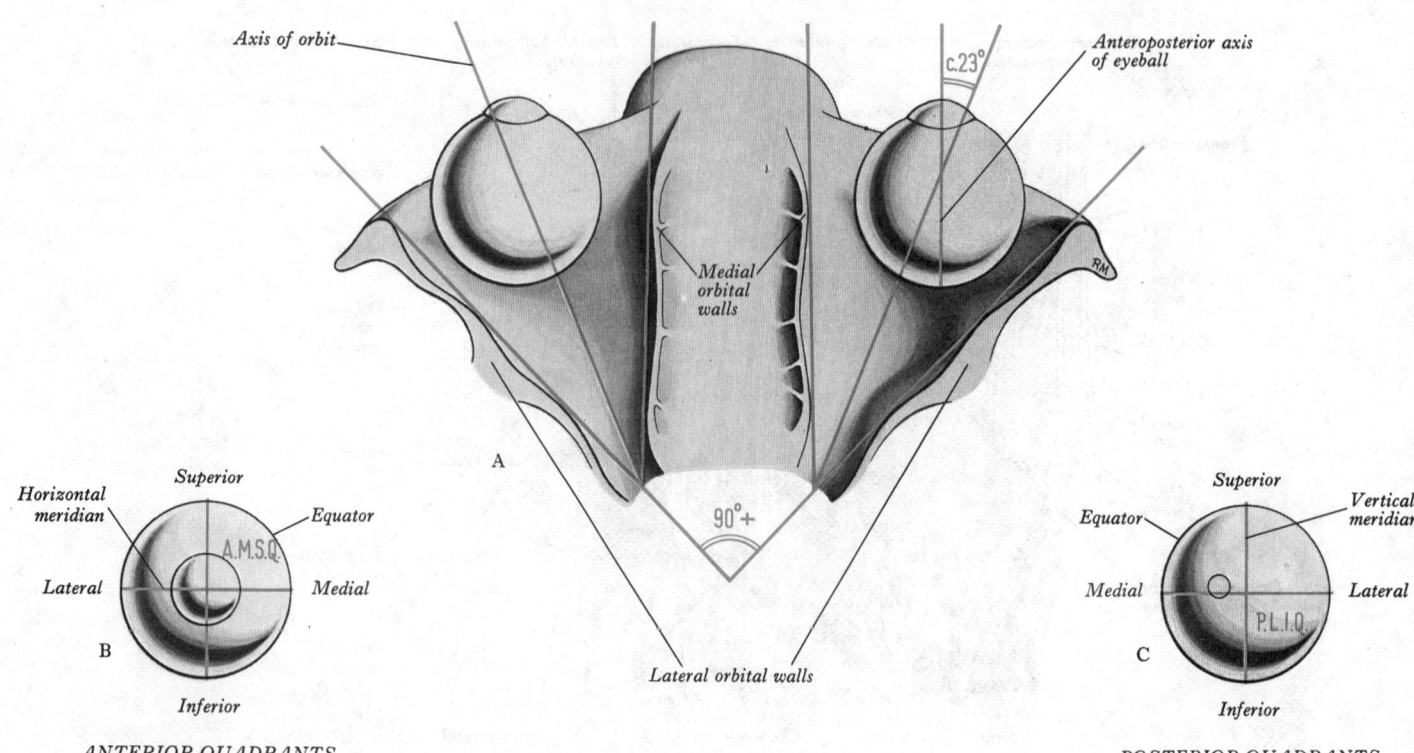

Axis of orbit

Anteroposterior axis
of eyeball

c.23°

Medial
orbital
walls

90°+

Lateral orbital walls

A

Horizontal
meridian

Superior

Equator

A.M.S.Q.

Lateral

Medial

B

Inferior

ANTERIOR QUADRANTS

Superior

Equator

Vertical
meridian

Medial

Lateral

P.L.I.Q.

C

Inferior

POSTERIOR QUADRANTS

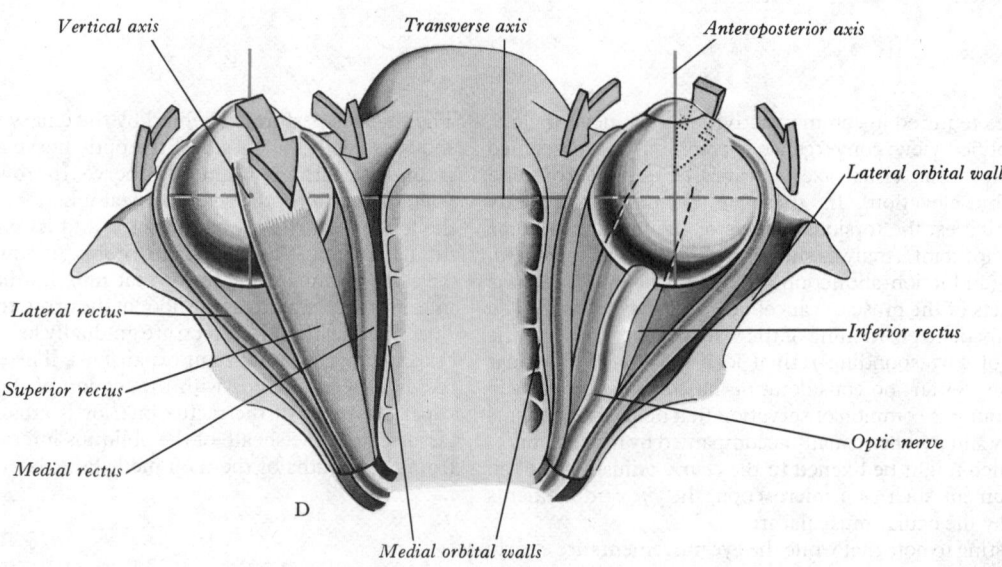

Vertical axis

Transverse axis

Anteroposterior axis

Lateral orbital wall

Lateral rectus

Superior rectus

Inferior rectus

Medial rectus

Optic nerve

D

Medial orbital walls

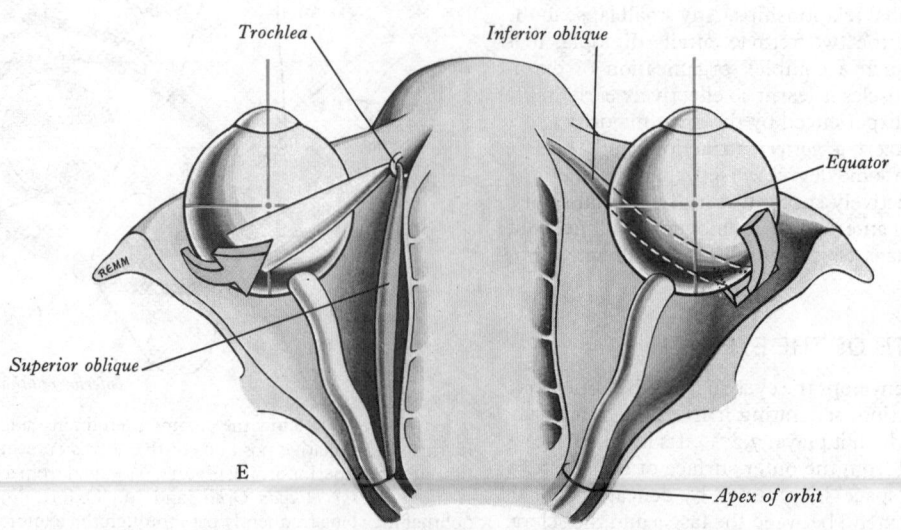

Trochlea

Inferior oblique

Equator

Superior oblique

E

Apex of orbit

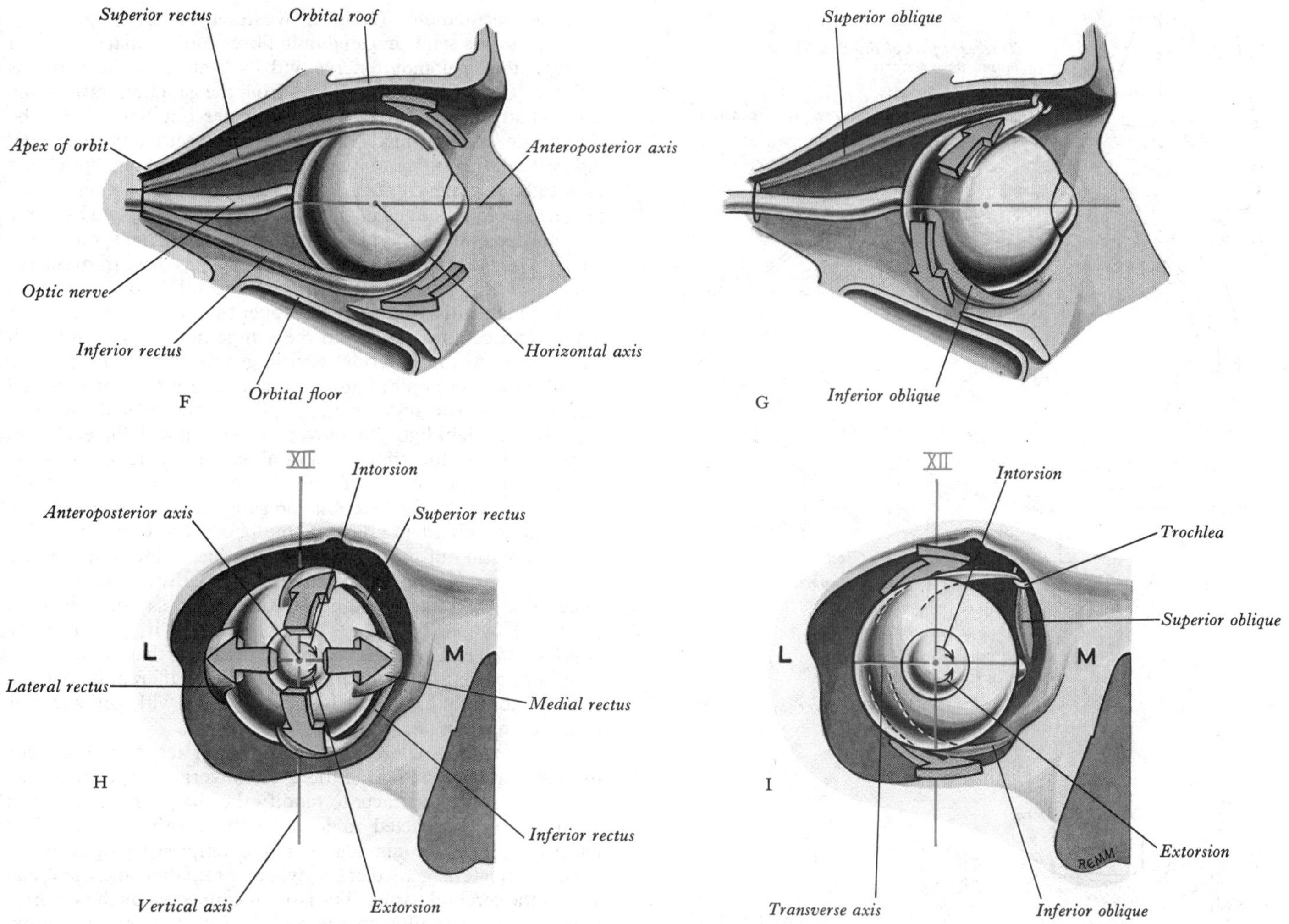

Superior rectus — Orbital roof

Apex of orbit

Optic nerve

Inferior rectus — Orbital floor

Anteroposterior axis

Horizontal axis

F

Superior oblique

Inferior oblique

G

XII — Intorsion

Anteroposterior axis — Superior rectus

Lateral rectus

L — M

Medial rectus

H

Inferior rectus

Vertical axis — Extorsion

XII — Intorsion

Trochlea

Superior oblique

L — M

I

Extorsion

Transverse axis — Inferior oblique

7.270A–I The geometrical basis of ocular movements.

A The relationship between the orbital and ocular axes, with the eyes in the primary position of parallel visual axes.

B and C The ocular globe in anterior and posterior views to show conventional geometry—meridia, equator, etc. A.M.S.Q=anterior medial superior quadrant; P.L.I.Q=posterior lateral inferior quadrant.

D The orbits from above showing the medial and lateral recti, and the superior rectus (left) and inferior rectus (right), indicating turning moments primarily around the vertical axis.

E Superior (left) and inferior (right) oblique muscles showing turning moments primarily around the vertical and also anteroposterior axes.

F Lateral view to show the actions of the superior and inferior recti around the transverse axis.

G Lateral view to show the action of the superior and inferior oblique muscles around the anteroposterior axis.

H Anterior view to show the medial rotational moment of the superior and inferior recti around the vertical axis. Conventionally the 12 o'clock position indicated is said to be *intorted* (superior rectus) or *extorted* (inferior rectus) as indicated by the small arrows on the cornea.

I Anterior view to show the torsional effects of the superior oblique (intorsion) and inferior oblique (extorsion) around the anteroposterior axis, as indicated by the small arrows on the cornea.

triangular in shape, and are attached to the lacrimal and zygomatic bones respectively; since they may check the actions of these two recti they have been named the *medial* and *lateral check ligaments*. A thickening of the lower part of the fascial sheath of the eyeball, is named the *suspensory ligament of the eye* (Lockwood 1886); it is slung like a hammock below the eyeball, being expanded in the centre and narrow at its extremities; it is formed by the union of the margins of the sheath of the rectus inferior with the medial and lateral check ligaments. Anomalies of the various parts of the above fascial arrangement may interfere with the normal movements of the eyes and cause various types of squint (Nutt 1955). It must be added that, apart from the suspensory ligament, there is little real evidence that any of the so-called check ligaments actually limit eye movements. The connective tissue link between the fasciae covering the superior rectus and levator palpebrae superioris is probably of little functional significance. It certainly does not impede independent movement of the lid and eyeball. It is, however, an almost unavoidable deduction that the orbital connective tissue must be so arranged as to locate the eyeball in its orbit, though in such a manner as to offer no

obstruction to the activities of the extrinsic muscles. Moreover, its connexions must be such as to prevent gross displacement of orbital fat, which would interfere with the accurate relative positioning of the two eyes in binocular vision. An interesting re-appraisal of the disposition of the orbital adipose and connective tissues has been made by Koornneef (1977), who emphasizes a series of radial septa which he describes as extending from the fascial sheath of the eyeball to the periorbita. These would provide an excellent suspensory system. His dissections support the concept of separate muscular sheaths, as noted above.

The orbital fascia forms the periosteum of the orbit, but is loosely connected to the bones. Behind, it is united with the dura mater and with the sheath of the optic nerve. In front it is connected with the periosteum at the margin of the orbit, and sends off a stratum which assists in forming the orbital septum. From it two processes are given off: one holds the trochlea of the obliquus superior in position, the other, named the *lacrimal fascia*, forms the roof and lateral wall of the sulcus in which the lacrimal sac is lodged (p. 300). For details of the vascularization of the orbital periosteum and fascia consult Lang (1975).

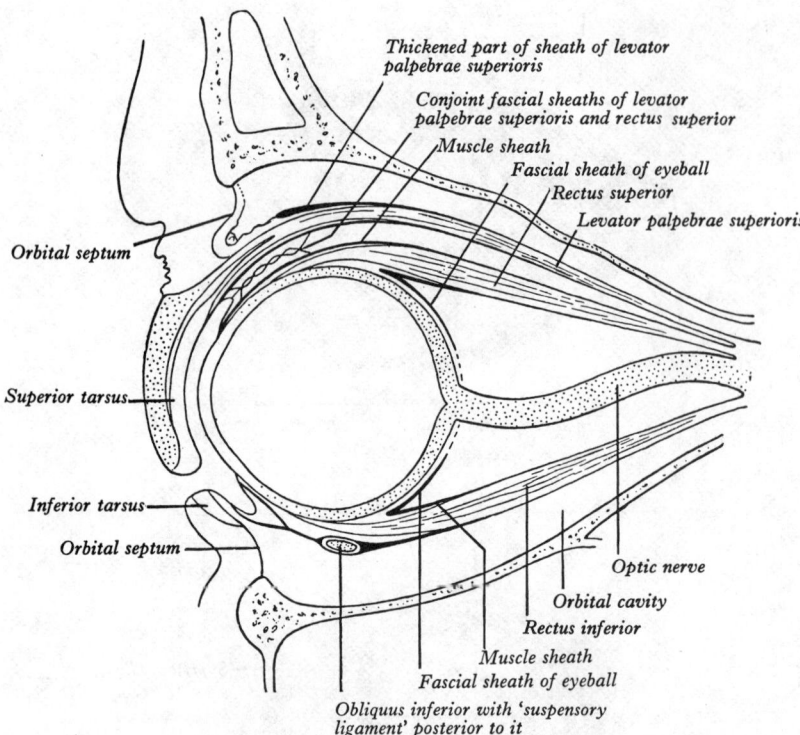

7.271A A scheme of the orbital fascia in sagittal section. (After Whitnall, 1932.)

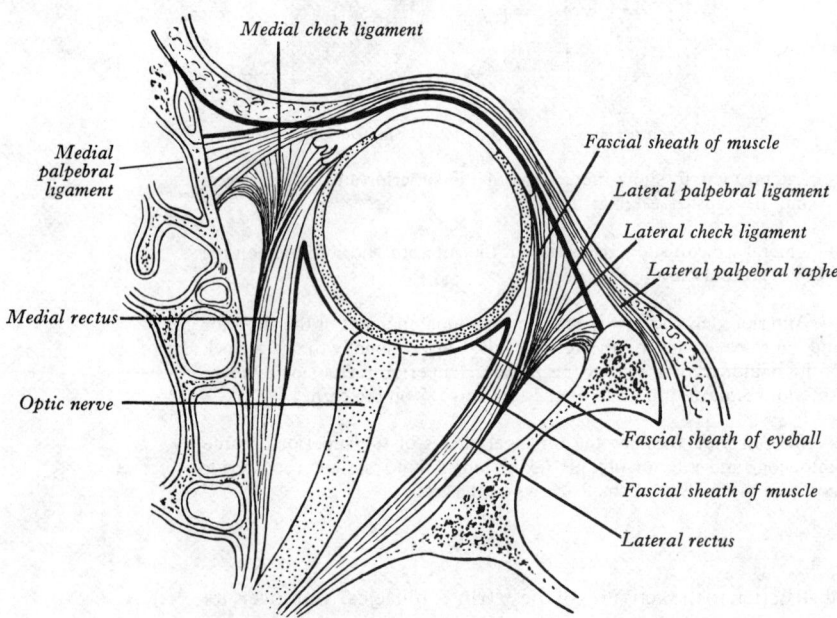

7.271B A scheme of the orbital fascia in horizontal section. (After Whitnall, 1932.)

VISUAL REFLEXES

The impulses concerned with *visual reflexes*, which bring about movements of the eyes, head and neck in response to visual stimuli, follow the pathway provided by the optic nerves and tracts to the superior colliculi. After traversing synapses there, the impulses travel along the tectospinal and tectobulbar tracts to the neurons of the motor columns associated with the spinal and cranial nerves (p. 941).

Pupil light reflex. On exposure of the eye to bright light the pupils contract reflexly. The impulses concerned travel by the optic nerves and tracts to the pretectal nucleus (p. 942) where secondary neuron fibres arise. These, which are very short and run close to the central grey matter, convey the impulses to the

accessory oculomotor (Edinger-Westphal) nucleus (p. 1057), whose neurons send preganglionic fibres to the ciliary ganglion through the oculomotor nerve and its branch to the obliquus inferior. The postganglionic fibres from the ganglion traverse the short ciliary nerves to reach the sphincter pupillae. If light be shone into one eye only, both pupils contract (*consensual pupil light reflex*); this is due to the fact that fibres from one optic tract pass to both pretectal nuclei, the crossing fibres passing via the posterior commissure. The dilatator pupillae is supplied by fibres which arise in the superior cervical ganglion of the sympathetic trunk. The preganglionic fibres of this pathway arise from cells of the lateral grey column in the first and second thoracic segments of the spinal cord and pass by the upper thoracic nerves and their white rami communicantes to the sympathetic trunk, in which they ascend to the superior cervical ganglion (p. 1127). As the condition of the pupil at any time is the result of the *balanced action* of these two systems, the pupil becomes dilated when the stimulus of bright light is removed. The pupil will dilate, also, in response to painful stimulation of almost any part of the body. Presumably the fibres of the sensory pathways establish connexions with the efferent preganglionic neurons of the sympathetic described above. Some believe, however, that this reflex dilatation of the pupil is largely due to inhibition of the accessory oculomotor nucleus, though the pathways involved are uncertain. One manifestation of this reflex is the dilatation produced by pinching the skin of the neck; it is termed the *pupillary skin reflex*, and it provides a reminder that to the above, simplified account must be added many other afferent influences, such as connexions of the reticular system with the superior collicular nucleus (*see* p. 942, 947).

Accommodation reflexes. In the process of accommodation for the viewing of near objects the eyes converge and, at the same time, the ciliaris contracts to modify the shape of the lens, and the pupil is constricted to increase the depth of focus. The pathways for the accommodation reflex comprise the optic nerve, optic tract, lateral geniculate body, optic radiation and the visual area of the cerebral cortex. The latter is connected by the superior longitudinal fasciculus to the eye field of the frontal cortex, whence fibres descend through the internal capsule to the nuclei of the oculomotor nerves in the midbrain. From the accessory oculomotor nuclei fibres pass to the ciliaris and sphincter pupillae (relaying in the ciliary ganglion), and from the ventral part of the oculomotor nucleus (*see* p. 1057) fibres supply the medial recti for the action of convergence of the eyes (*see* p. 1180). These pathways have not been so definitely established as those for the pupil light reflex, and it has been suggested (Wilkinson 1927) that the contraction of the pupil in the accommodation reflex is secondary to the convergence of the eyes and that the afferent impulses arise in the proprioceptor nerve endings in the orbital muscles and travel in the oculomotor nerve direct to the accessory nuclei. In certain diseases of the central nervous system (e.g. tabes dorsalis due to syphilis) the pupil light reflex may be lost while the constriction of the pupil as part of the accommodation reflex is unaffected (Argyll Robertson pupil). The site of the lesion that could produce such an effect is probably located between the accessory oculomotor nucleus and the lateral geniculate body, where the pathways for the two reflexes diverge from each other.

Conjunctival and corneal reflexes. If the conjunctiva or cornea be touched lightly, blinking occurs. Afferent impulses travel via the ophthalmic part of the trigeminal nerve and efferent impulses in the branches of the facial nerve to the orbicularis oculi.

THE EYEBROWS AND EYELIDS

The eyebrows are two arched eminences of skin, which surmount the orbits and support numerous short, thick hairs directed obliquely on the surface. Fibres of the orbicularis oculi, corrugator and frontal belly of the occipitofrontalis are inserted into the skin of the eyebrows.

The eyelids or **palpebrae** are thin, movable folds, adapted to the front of the eye, and protecting it, by their closure, from injury. The upper eyelid is the larger and more movable, and is furnished with an elevator muscle, the levator palpebrae superioris (p. 1178); the two eyelids are united to each other at their ends, and when the eye is open an elliptical space, termed the *palpebral fissure*, is left between their margins; the extremities of the fissure are called the angles of the eye.

The *lateral angle of the eye* (lateral canthus) is more acute than the medial, and lies in close contact with the eyeball. The *medial angle* (medial canthus) is prolonged for a short distance towards the nose, and is about 6 mm away from the eyeball; the two eyelids are here separated by a triangular space, named the *lacus lacrimalis*, in which a small reddish body, termed the *caruncula lacrimalis*, is situated (7.272). On the margin of each eyelid, at the basal angles of the lacus lacrimalis, there is a small conical elevation, termed the *lacrimal papilla*, the apex of which is pierced by the commencement of the lacrimal canaliculus. This minute orifice (7.272) is known as the *punctum lacrimale*.

The *eyelashes* are attached in the free edges of the eyelids from the lateral angle of the eye to the lacrimal papillae. They are short, thick, curved hairs, arranged in double or triple rows: those of the upper eyelid, more numerous and longer than those of the lower, curve upwards; those of the lower eyelid curve downwards so that the upper and lower eyelashes do not interlace when the lids are closed. A number of enlarged and modified sudoriferous glands, termed *ciliary glands*, are arranged in several rows close to the free margin of each lid and open near the attachments of the eyelashes. (*See* also p. 1223.)

Structure of the Eyelids

From without inwards, each eyelid consists of: skin, subcutaneous areolar tissue, fibres of the orbicularis oculi, tarsus and orbital septum, tarsal glands and conjunctiva. The upper eyelid contains, in addition, the aponeurosis of the levator palpebrae superioris (7.273).

The *skin* is extremely thin, and continuous at the margins of the eyelids with the conjunctiva.

The *subcutaneous areolar tissue* is very lax and delicate, and seldom contains any adipose tissue.

The *palpebral fibres of the orbicularis oculi* are thin, pale in colour and parallel with the palpebral fissure. Deep to the muscle there is a layer of loose areolar tissue, which, in the case of the upper eyelid, is continuous with the subaponeurotic layer of the scalp (p. 530), so that effusions of fluid (blood or pus) in this layer of the scalp can pass down into the upper eyelid. It is in this layer of the eyelids that the main nerves lie, so that local anaesthetics have to be injected deep to the orbicularis oculi.

The *tarsi* (7.274) are two thin elongated plates of dense fibrous tissue, about 2·5 cm long; one is placed in each eyelid and contributes to its form and support. The *tarsus of the upper eyelid*, the larger, is of a semi-oval form, about 10 mm in height at the centre, and gradually narrowing towards its extremities. The lowest fibres of the superficial lamella of the aponeurosis of the levator palpebrae superioris are attached to its anterior surface, and the deep lamella of the same aponeurosis is inserted into its upper margin (7.266). The *tarsus of the lower eyelid*, the smaller, is a narrower plate, the vertical diameter of which is about 5 mm. The free or ciliary margins of the tarsi are thick and straight. The attached or orbital margins are connected to the circumference of the orbit by the orbital septum. The lateral ends of the tarsi are attached by a band, named the *lateral palpebral ligament*, to a tubercle on the zygomatic bone, just within the orbital margin; this ligament is separated from the more superficially placed *lateral palpebral raphe* (p. 531) by a few lobules of the lacrimal gland. The medial ends of the tarsi are attached by a strong tendinous band, named the *medial palpebral ligament*, to the upper part of the lacrimal crest, and to the adjoining part of the frontal process of the maxilla in front of this crest; the lower edge of this ligament is separated from the lacrimal sac by some fibres of the orbicularis oculi, since the latter is attached to the ligament.

The *orbital septum* is a weak membranous sheet, attached to the edge of the orbit, where it is continuous with the periosteum. In the upper eyelid it blends with the superficial lamella of the aponeurosis of the levator palpebrae superioris, and in the lower eyelid with the anterior surface of the tarsus. It is perforated by the vessels and nerves which pass from the orbital cavity to the face and scalp, by the aponeurosis of the levator palpebrae superioris, and by the palpebral part of the lacrimal gland.

The *tarsal glands* (7.275) are embedded in the thickness of the tarsi, and may be visible through the conjuctiva on everting the eyelids; they present an appearance like parallel strings of pearls. They are yellow in colour, arranged in a single row, and number about thirty in the upper eyelid, and somewhat fewer in the lower. They are embedded in grooves on the deep surfaces of the tarsi and correspond in length with the breadth of these plates; they are, consequently, longer in the upper than in the lower eyelid. Their ducts open on the free margins of the lids by minute foramina.

The tarsal glands are modified sebaceous glands, each consisting of a straight tube with numerous small lateral diverticula. The tubes are supported by a basement membrane and are lined at their mouths by stratified epithelium; the deeper parts of the tubes and the lateral offshoots are lined by a layer of polyhedral cells. The secretion of the glands spreads over the margin of the eyelid and tends to prevent the tears from overflowing on to the cheek. It has also been suggested that it spreads over the external surface of the tear film and reduces evaporation.

THE CONJUNCTIVA

The conjunctiva is the transparent mucous membrane which passes, in a modified form, over the inner surfaces of the eyelids, and is reflected over the front part of the sclera and the cornea.

The *palpebral conjunctiva* is highly vascular, and has numerous subepithelial connective tissue papillae, its deeper part containing a considerable amount of lymphoid tissue, especially near the fornices. It is intimately adherent to the tarsi. At the margins of the lids it is continuous with the skin, with the lining epithelium of the ducts of the tarsal glands, and, through the lacrimal canaliculi, with the lining membrane of the lacrimal sac and nasolacrimal duct and thence with that lining the nasal cavity. The line of reflexion of the conjunctiva from the eyelids on to the eyeball is named the *conjunctival fornix*, and its different parts are known as the superior and inferior fornices; the ducts of the lacrimal gland open into the lateral part of the superior fornix. Over the sclera the *ocular conjunctiva* is loosely connected to the eyeball; it is thin, transparent, destitute of papillae, and only slightly vascular. Upon reaching the cornea, the ocular conjunctiva continues as the corneal epithelium (p. 1153). The epithelium of the palpebral conjunctiva near the margins of the eyelids is stratified squamous; about 2 mm from the edge of each eyelid there is a groove in which foreign bodies frequently lodge and at which the epithelium comes to consist of two layers, a superficial one of columnar cells and a deeper one of flattened cells. This structure persists throughout most of the palpebral conjunctiva, but as the fornices are approached an intermediate layer of polygonal cells appears and this trilaminar arrangement comprises the structure of the conjunctival epithelium over the sclera. Near the sclerocorneal junction the epithelium changes to the stratified type characteristic of the corneal epithelium (p. 1153). Scattered throughout the conjunctival epithelium there are mucus-secreting goblet cells, but they are few in the palpebral and circumcorneal regions of the epithelium.

The *lacrimal caruncle* (7.272) is a small, reddish, conical body situated in the lacus lacrimalis at the medial angle of the eye; it consists of a small island of skin, and contains sebaceous and sudoriferous glands; a few slender hairs are attached to its surface. Lateral to, and partly obscured by the caruncle there is a semilunar fold of conjunctiva, the *plica semilunaris*, the concave free lateral edge of which is directed towards the cornea. Its epithelium resembles that of the conjunctiva on the sclera but it contains numerous goblet cells. Beneath the epithelium there is some fat and a little nonstriated muscle. The *nictitating membrane*—present as a conjunctival specialization in some amphibians, reptiles and mammals—may be represented by the semilunar fold; but the homologies of these structures are uncertain.

Vessels and nerves. The eyelids receive their blood supply from the medial palpebral branches of the ophthalmic artery and from the lateral palpebral branches of the lacrimal artery (p. 688).

The ocular conjunctiva is supplied by the ophthalmic division

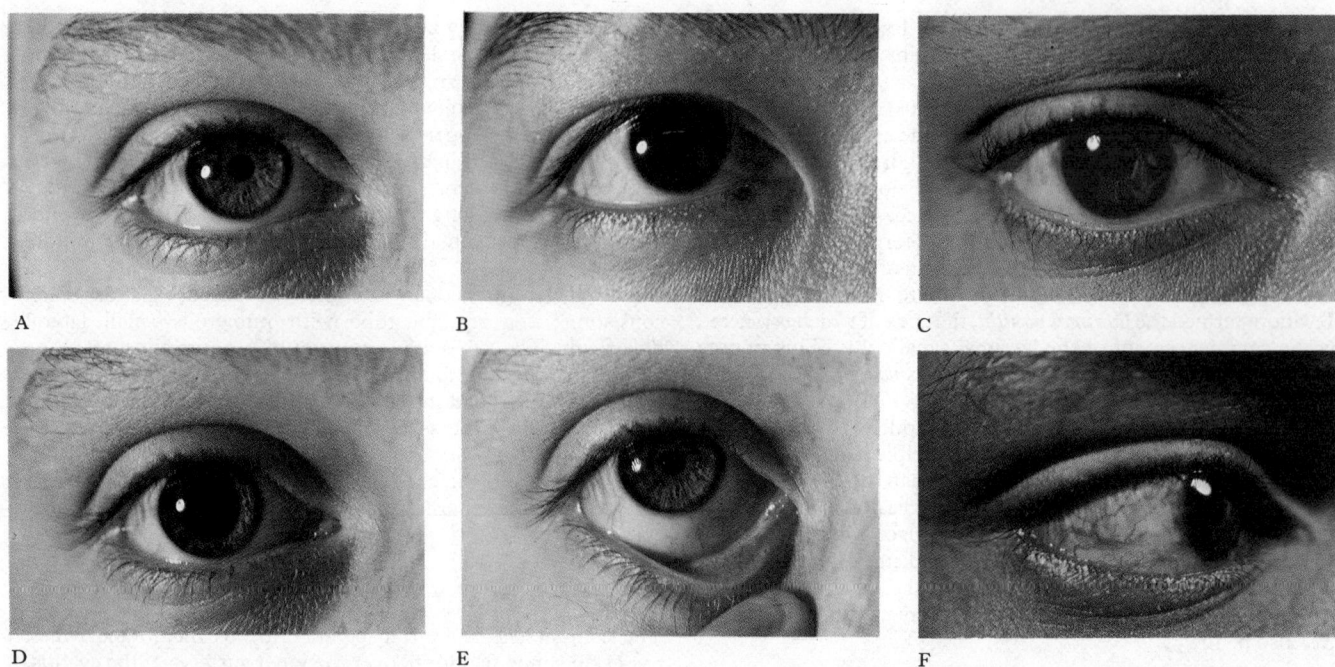

7.272A–F In the top row of photographs the typical appearances of the living eye are compared in a Caucasian female (A), a Mongoloid male (B) and a Negroid male (C). All subjects are about 20 years of age. Note the pale sclera and grey-blue iris in A, the epicanthus overlapping the medial end of the lower eyelid in B, and the dark brown pigmentation of the iris in both B and C, rendering the pupil almost invisible. Compare the size of the pupil in the same eye, photographed under steady bright light in A and by sudden exposure to the same illumination after a period of dark adaptation in D. In E the lower eyelid had been everted somewhat to exhibit the lacrimal punctum and the rich subepithelial network of blood vessels. In F note the circumcorneal pigmentation and the conjunctival blood vessels, deep to which can be seen some details of the episcleral vessels. All photographs are of the right eye.

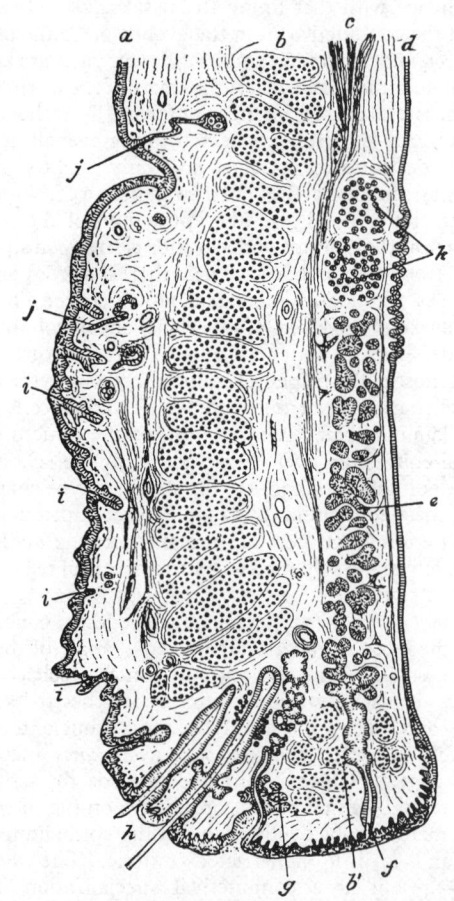

7.273 A sagittal section through the upper eyelid. (After Waldeyer.) *a*. Skin. *b*. Orbicularis oculi. *b′*. Ciliary fasciculi of the orbicularis oculi. *c*. Levator palpebrae superioris. *d*. Conjunctiva. *e*. Tarsal glands embedded in the tarsal plate. *f*. Opening of a tarsal gland. *g*. Ciliary gland. *h*. Eyelashes. *i*. Small hairs of the skin. *j*. Sweat glands. *k*. Posterior tarsal glands.

of the trigeminal nerve. The conjunctiva of the upper eyelid is supplied by the ophthalmic nerve, that of the lower eyelid by the maxillary nerve. Many of the nerves to the conjuctiva end in bulbous corpuscles (p. 855).

The lymph vessels of the eyelids and the conjunctiva are described on p. 786.

Movements of the eyelids. The position of the lids at any particular time depends on the reciprocal tone of the orbicularis oculi and levator palpebrae superioris and on the degree of protrusion of the eyeball. The usual position when the eyes are open is that the margin of the lower lid crosses the eyeball at the lower edge of the circumference of the iris whilst the upper lid covers about half the width of the uppermost portion of the iris. The eyes are closed by movement of *both* upper and lower lids produced by contraction of the orbicularis oculi and at the same time the levator palpebrae superioris relaxes. On looking upwards, the levator palpebrae superioris contracts and the upper lid follows the movement of the eyeball. At the same time the eyebrows are raised by the contraction of the frontal bellies of the occipitofrontalis so as to diminish the degree to which the eyebrows jut beyond the eye. The lower lid lags behind the movement of the eye so that much more of the sclera is exposed below the iris and the lid is bulged forwards to some extent by the pressure exerted against its deep surface by the lower part of the eyeball. When the eye is depressed both lids move, the upper retaining its normal relationship to the eyeball and still covering about half the width of the upper region of the iris. The lower lid is probably dragged downwards by the pull exerted on it by the conjunctiva reflected on to its deep surface from the sclera. In conditions of fear the palpebral fissure is widened by the contraction of the fibres of the nonstriated superior and inferior tarsal muscles, in response to the increased activity of the sympathetic nervous system. For further comments on eyelid movement see p. 1178.

THE LACRIMAL APPARATUS

The lacrimal apparatus (7.276, 279) consists of the lacrimal gland, which secretes a complex fluid, known as the tears, and its excretory ducts which convey the fluid to the surface of the eye;

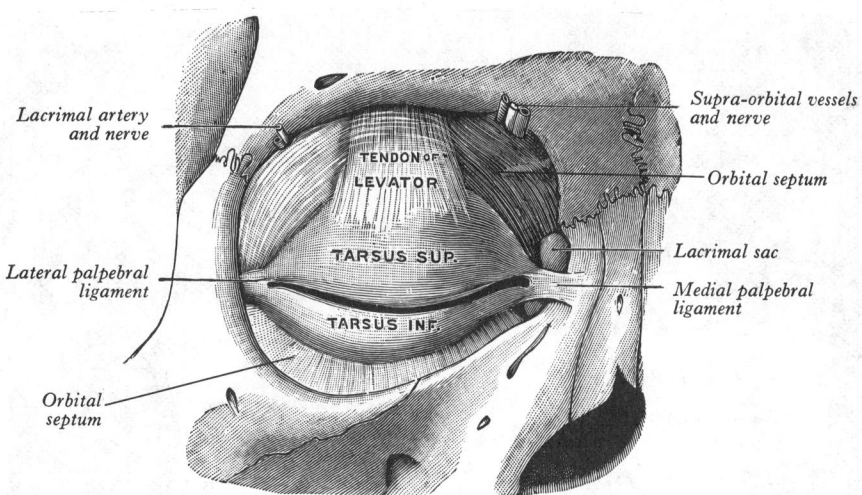

7.274 The tarsi and their ligaments. Anterior aspect.

the lacrimal canaliculi, lacrimal sac, and nasolacrimal duct, by which the fluid is conveyed into the nasal cavity.

The lacrimal gland (7.276) is probably homologous with the *Harderian gland* of lower mammals, and is derived from a serous secreting element and a gland secreting an oily material. In primates the lacrimal (serous) element has migrated from its original position in the lower lid to the upper. The human lacrimal gland consists of a larger upper *orbital part* and a lower smaller *palpebral part*, the two parts being continuous with each other posterolaterally around the lateral concave edge of the aponeurosis of the levator palpebrae superioris. The orbital part is about the size and shape of an almond and is lodged in the lacrimal fossa on the medial side of the zygomatic process of the frontal bone, just within the margin of the orbit. It lies above the levator (and, further laterally, above the lateral rectus); its lower surface is connected to the sheath of the levator, its upper surface is connected to the orbital periosteum, its anterior border is in contact with the orbital septum and its posterior border is attached to the orbital fat. The palpebral part, which is about one-third of the size of the orbital part, is subdivided into two or three lobules and extends below the aponeurosis of the levator into the lateral part of the upper eyelid, where it is attached to the superior fornix of the conjunctiva, through which it can be seen when the eyelid is everted. The ducts of the gland, about twelve in number, open into the superior conjunctival fornix. Those from the orbital part (four or five in number) pass through the palpebral part and are joined by some of the ducts form this latter part, while other ducts of the palpebral part (six to eight in number) open independently. Thus all the ducts pass through the palpebral part, so that excision of this part of the gland is functionally equivalent to removal of the entire gland.

Many small accessory lacrimal glands are present in and near the conjunctival fornices; they are more numerous in the upper lid than in the lower. Their existence may explain why the conjunctiva does not dry up after extirpation of the lacrimal gland proper.

Structure of the lacrimal gland (7.277)

The lacrimal gland is a lobulated tubulo-acinar structure (p. 43), the secretory units or 'endpieces' of which consist of glandular cells resembling those of salivary glands (p. 1275). The secretion produced by the gland is a watery fluid (tears) with an electrolyte content similar to that of plasma, and containing an enzyme, lysozyme, which has bacteriocidal properties. The primary secretion is produced by the secretory units and is modified by the ducts into which they lead.

Although many of the earlier published accounts of the ultrastructure of the lacrimal gland were concerned with rodents and lagomorphs (Obayashi 1959; Scott and Pease 1959; Ichikawa and Nakajima 1962), that of the primate lacrimal gland has now received attention (Ruskell 1968, 1969; Egeberg and Jensen 1969;

Orzalesi *et al.* 1971; Ruskell 1975; Hirsch-Hoffmann 1976, 1978).

The glandular cells of the secretory endpieces are divisible into two categories in the rhesus monkey (Ruskell 1968, 1969; Hirsch-Hoffmann 1976), and into either two (Ito and Shibaski 1964; Kühnel 1968), three (Ruskell 1975) or four (Hirsch-Hoffmann 1978) in man, depending on which ultrastructural and histochemical criteria are considered significant. In the rhesus monkey, glandular cells with uniformly electron-dense secretory granules have been described as *serous*, and those with paler heterogeneous granules as *mucous* (Ruskell 1968). In man, the histochemical and ultrastructural investigations of Ito and Shibasaki (1964) and of Kühnel (1968) led them to propose that there are two distinct types of glandular cell: one, designated the K cell, contains small, electron-lucent granules and has the staining characteristics of a mucous cell, while the other, termed the G cell, contains large, electron-dense granules and has the staining characteristics of a serous cell. In the opinion of Allen *et al.* (1972) and Ruskell (1975) the majority of the glandular cells are of the mucous type, a surprising finding in view of the patently serous nature of tears. Ruskell (1975) distinguishes three categories of glandular cell in man, arbitrarily grouping them into 'light', 'medium' and 'dark' classes according to the electron density of their secretory granules; acini contain either two or all three categories of glandular cell, together with lymphocytes and myoepitheliocytes. In contrast, Hirsch-Hoffmann (1978) groups the glandular cells of the human lacrimal gland into four categories, distinguishable by the number and electron density of their granules, together with the electron density of the surrounding cytoplasm; using this classification, the largest group

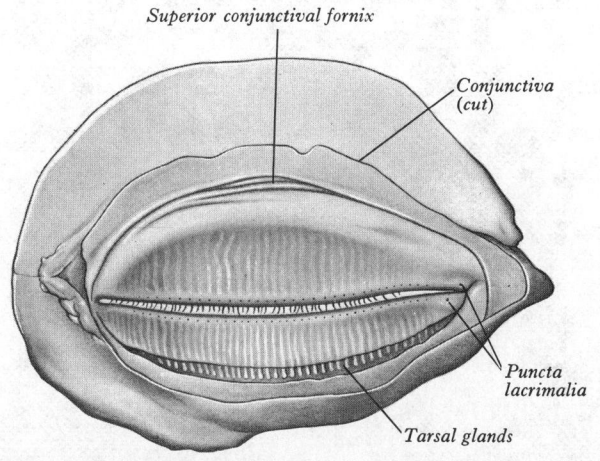

7.275 The posterior surfaces of the upper and lower eyelids of the left side. The orifices of the tarsal glands can be seen on the free margins of the lids.

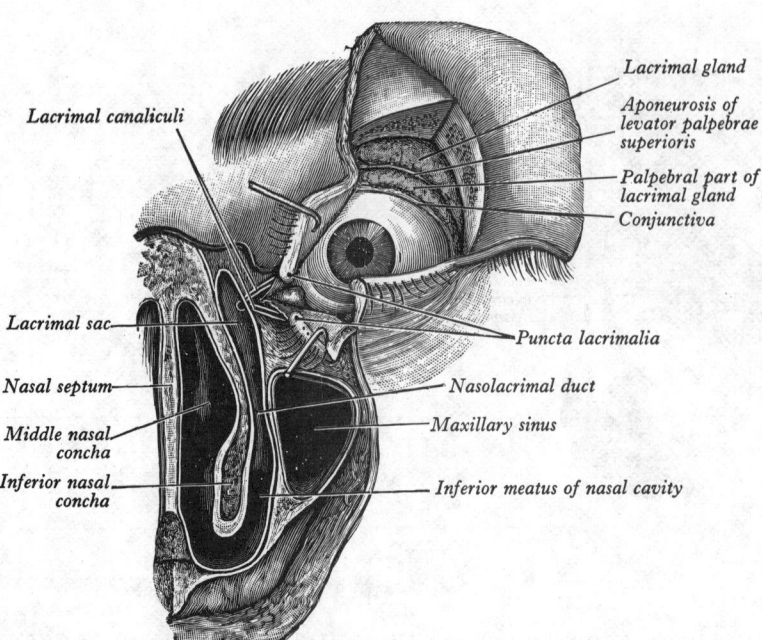

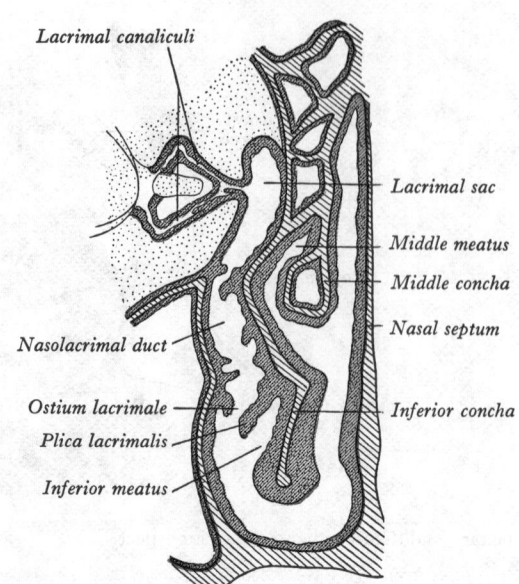

7.276 The left lacrimal apparatus dissected from the anterior aspect.

7.278 Sketch from a coronal section through the right half of the nasal cavity, anterior aspect, to show the relation of the lacrimal passages to the maxillary and ethmoidal sinuses and the inferior nasal concha. The mucous membrane is coloured. (After Whitnall, 1932.)

have pale cytoplasm and numerous granules of variable electron density, a second group have pale cytoplasm and fewer such granules, a third have darker cytoplasm and numerous electron-dense granules, while the remainder have dark cytoplasm and fewer uniformly electron-dense granules. Hirsch-Hoffmann has suggested that at least some of these categories may represent different stages in secretory activity of one or two distinct cell populations, an interpretation which has the support of other investigators (Egebert and Jensen 1969; Orzalesi *et al.* 1971).

The human lacrimal gland contains both interstitital (epi-lemmal, p. 1280) and parenchymal (hypolemmal, **8**.39B) nerve terminals (Ruskell 1975). The ultrastructure of most interstitial and all parenchymal nerve terminals is consistent with their being cholinergic (parasympathetic). A minority of the interstitial terminals contain small granular vesicles, suggesting that they are adrenergic (sympathetic). Of the glandular cells, only the 'dark' or 'serous' types have a parasympathetic, hypolemmal, innervation (Ruskell 1975), a situation similar to that found in the rhesus monkey (Ruskell 1969), where ultrastructural changes following parasympathectomy have led to the proposal that the serous cells

are under direct parasympathetic control, mediated by the hypolemmal terminals, while the mucous cells may function autonomously. Parasympathetic hypolemmal terminals are also associated with the ducts and terminal tubules of the human lacrimal gland. Terminal tubules lie between the secretory endpieces and the ducts, and contain in their walls cells which are intermediate in form between glandular cells and duct cells (Ruskell 1975); these cells of the terminal tubules are smaller than the glandular cells and contain disproportionately fewer granules. Most of the nerve terminals of the ducts and terminal tubules lie adjacent to myoepitheliocytes, suggesting that in these regions their main function is to induce myoepitheliocyte contraction and thus assist in the flow of secretion. Curiously, the myo-epitheliocytes associated with the secretory endpieces do not appear to have a direct, hypolemmal, innervation; Ruskell (1975) has suggested that they may be controlled by nearby interstitial parasympathetic terminals, which may only release transmitter in sufficient quantity to induce their contraction at times of hypersecretion. The role of the interstitial sympathetic terminals in the control of the human lacrimal gland remains to be determined, as do the ultrastructure and functions of its duct system.

The lacrimal canaliculi, one in each eyelid, are about 10 mm long; they commence at the *puncta lacrimalia* (**7**.272, 276, 278). The *superior canaliculus*, smaller and shorter than the inferior, at first ascends, and then bends at an acute angle, and passes medially and downwards to the lacrimal sac. The *inferior canaliculus* at first descends, and then runs almost horizontally to the lacrimal sac. At the angles they are dilated into *ampullae*. The mucous lining of the ducts is covered with stratified squamous epithelium, placed on a basement membrane; outside the latter there is a corium rich in elastic fibres (rendering the ducts easily dilatable during the passage of a probe) and a layer of striped muscular fibres which is continuous with the lacrimal part of the orbicularis oculi. At the base of each lacrimal papilla the muscular fibres are circularly arranged and form a kind of sphincter.

The lacrimal sac (**7**.274, 276, 278) is the upper blind end of the nasolacrimal duct, and is lodged in a fossa formed by the lacrimal bone, the frontal process of the maxilla and the lacrimal fascia. It measures about 12 mm in length, its upper, closed end is flattened from side to side, but its lower part is rounded and is continued into the nasolacrimal duct; the openings of the lacrimal canaliculi are situated in its lateral wall slightly below its upper end.

A layer of fascia, continuous with the periosteum of the orbit and named the *lacrimal fascia*, passes from the lacrimal crest of the

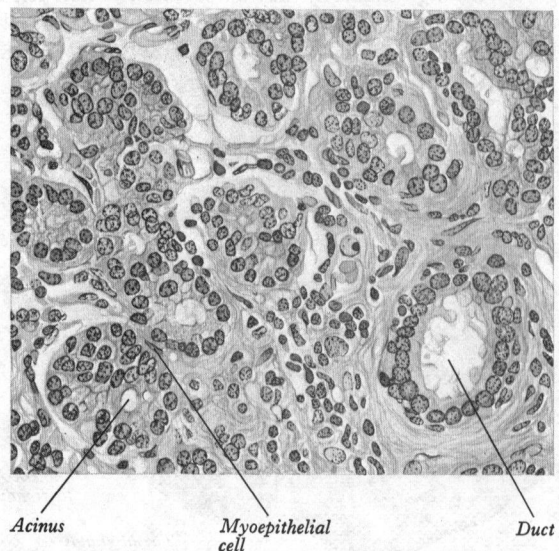

Acinus   *Myoepithelial cell*   *Duct*

7.277 A section through a part of the lacrimal gland, showing the acini lined by a layer of columnar cells; a few myoepithelial cells can be seen outside the basement membrane. Magnification about × 320.

maxilla to the crest of the lacrimal bone, and forms the roof and lateral wall of the fossa in which the lacrimal sac is sited; between the fascia and the lacrimal sac there is a plexus of minute veins. The lacrimal fascia separates the sac from the medial palpebral ligament in front, and from the lacrimal part of the orbicularis oculi behind. The lower half of the fossa which lodges the lacrimal sac is related medially to the anterior part of the middle meatus of the nasal cavity; the upper half is related to the anterior ethmoidal sinuses. (In an examination of 100 skulls, Whitnall 1911, it was found that in 14 the anterior ethmoidal sinuses came into relation only in the posterior wall of the fossa; in 32 they reached as far as the suture between the lacrimal bone and the maxilla; while in 54 one large irregular sinus extended as far as the anterior lacrimal crest.)

The lacrimal sac consists of a fibro-elastic coat, lined internally by mucous membrane; the latter is continuous through the lacrimal canaliculi, with the conjunctiva, and through the nasolacrimal duct with the mucous membrane of the nasal cavity.

The nasolacrimal duct (7.278, 279) is a membranous canal about 18 mm long, which extends from the lower part of the lacrimal sac to the anterior part of the inferior meatus of the nose, where it ends in a somewhat expanded orifice. A fold of the mucous membrane forms an imperfect valve just above the opening and is known as the *lacrimal fold*. The duct is contained in an osseous canal, formed by the maxilla, the lacrimal bone and the inferior nasal concha; it is narrower in the middle than at either end, and is directed downwards, backwards and a little laterally. The mucous lining of the lacrimal sac and nasolacrimal duct is covered with two layers of columnar epithelium which in places is ciliated. Around the duct there is a rich plexus of veins, forming an erectile tissue, engorgement of which may obstruct the duct.

Tear fluid secreted by the lacrimal gland enters the conjunctival sac at its superolateral angle and, under the influence of capillarity aided by blinking movements of the eyelids, is carried across the sac to the lacus lacrimalis mainly along the groove between the lower lid margin and the eyeball. From the lacus it passes into the lacrimal canaliculi. Contraction of the orbicularis oculi tends to press the puncta lacrimalia more firmly into the lacus, and

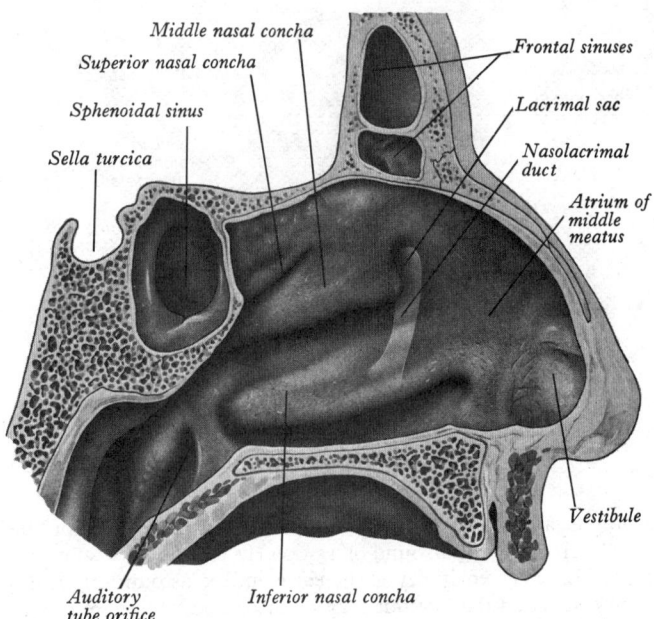

7.279   The left lateral wall of the nasal cavity viewed from the medial side. The lacrimal sac and the nasolacrimal duct of the left side have been projected on to the lateral wall of the nasal cavity to show their positions relative to the middle nasal concha, the middle meatus and the inferior concha.

capillary attraction serves to draw the lacrimal secretion into the lacrimal sac. The sudden dilatation of the lacrimal sac produced by the lacrimal part of the orbicularis oculi during blinking movements (p. 531) probably assists the process. Under normal conditions the secretion of the tarsal glands prevents the tears from overflowing the lid margins and also covers the capillary film of fluid on the front of the eyeball with a film of oil which delays evaporation (Mishima and Maurice 1961; Wolff 1976).

THE AUDITORY AND VESTIBULAR APPARATUS

The peripheral auditory apparatus consists of the various parts of the ear, but including in particular, the *cochlear part* of its *membranous labyrinth*. Essentially, each ear is a *distance receptor* concerned with the collection, conduction, modification, amplification and parametric analysis of the complex sound waveforms which impinge on the head. The latter are converted into coded spatio-temporal patterns of nerve impulses in the afferent fibres of the cochlear part of the vestibulocochlear nerve, for onward transmission to the auditory pathways in the central nervous system (pp. 909, 911, 941, 976, 1021).

The molecular vibrations of the air which constitute the sound waves approaching the head vary according to: (1) the *direction* and *distance*, or *location* of the source of the waves; (2) the *intensity* or *energy content* of the waves; and (3) the relative purity or admixture of different *frequencies* which make up the wave train. The morphological and functional design of the ears is such that they are, within particular ranges of values, extremely sensitive to differences in frequency and intensity of the sound waves, and together, they are also very effective range and direction finders. Further, they are highly responsive to the *rate of change* of all of these sound wave parameters. The *frequency* of a sound wave is expressed as *cycles per second* or *Hertz* (c/s or Hz); it is subjectively appreciated as the *pitch* of the sound, and young adult ears are responsive to frequencies of about 20–20,000 Hz, although higher or lower values are not uncommon in very young ears. The *intensity* of a sound is expressed as the *quantity of energy* which is transmitted in *unit time* through a *unit area* which is perpendicular to the direction of wave propagation. The subjective appreciation of the intensity of a sound is related to the logarithm of its absolute intensity as defined above, but is also dependent upon the

frequency of the sound. The human ear is most sensitive (i.e. has the lowest threshold) for sounds in the 1,500–3,000 Hz range, and above and below these levels the threshold rises sharply. For

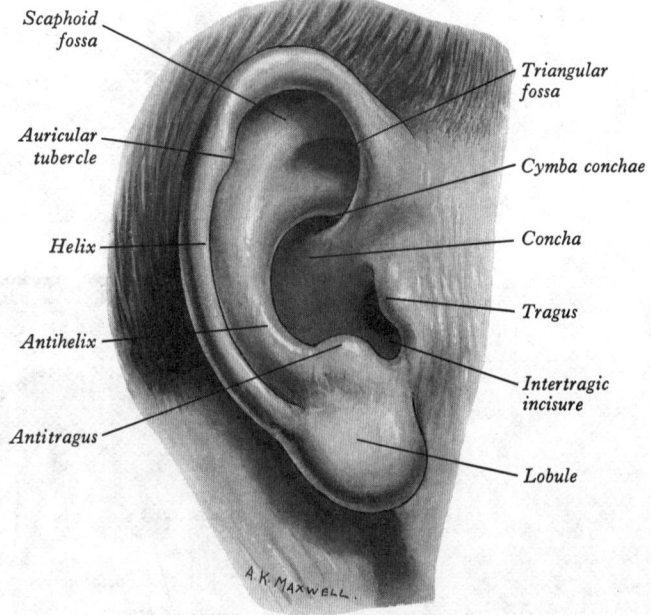

7.280   The right auricle. Lateral aspect.

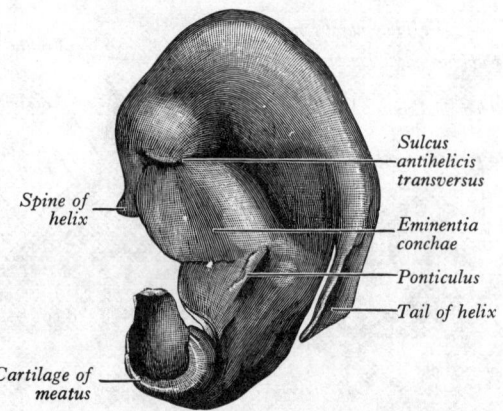

7.281 The medial surface of the right auricular cartilage.

Spine of helix

Sulcus antihelicis transversus

Eminentia conchae

Ponticulus

Tail of helix

Cartilage of meatus

example, about 10,000 times more energy is necessary for the equal perception of a sound of 15,000 Hz than is for a sound of 2,000 Hz. The sensitivity of the ear is indeed astounding; it has been calculated that sounds may be discerned which are due to pressure changes as small as 10^{-10} atmospheres a change equivalent to ascending or descending 1/30,000 of an inch! Because of the great variation in the intensity level of sounds commonly encountered, the concept of the *decibel* has been introduced for convenience. A decibel is defined as 10 times the logarithm of the ratio of the intensity of the sound in question, to the intensity of an accepted reference level. A difference in intensity level of about one decibel is usually just perceptible by the human auditory system.

The *quality* of a sound depends upon the admixture of frequencies it contains. Musical sounds consist of one or more fundamental frequencies each with its mathematically related series of *overtones* or *harmonics*. Mixtures of unrelated or *irregular* frequencies are perceived simply as 'noise'. It is interesting to note that human speech has a considerable noise content. Moreover, some consonants involve somewhat higher frequencies than do vowels. It is due to these facts that speech is easily obscured by concomitant noise, and that, with increasing age and consequent reduction at the high frequency range of the human ear, the elderly find speech less easy to interpret.

The elegant researches into the manner in which the ear operates as a peripheral analyser of frequency and intensity, and how intensity and phase differences impinging on the two ears are related to range and direction finding, can only receive the briefest mention in the present volume. For such details the reader should consult Rasmussen and Windle (1960); Whitfield (1967); De Reuck and Knight (1968); Wolstenholme and Knight (1970).

What follows is a structural account of the human ear, which on phylogenetic, developmental, structural and functional grounds, may be divided into *external*, *middle* and *internal* parts. Intimately associated structurally with the auditory cochlear part of the membranous labyrinth are the sensory receptors in specialized regions of the walls of the utricle and saccule, and in the ampullae of the semicircular ducts. The latter parts of the membranous labyrinth, their contained and surrounding fluids, the bony cavities in which they lie, and the vestibular part of the vestibulocochlear nerve, together constitute **the peripheral vestibular apparatus**. Its essential function is to provide the central nervous system with a constant flow of information concerning the static position of the head in space, or of its state of linear or angular acceleration or deceleration.

(The evolution of the auditory apparatus in land vertebrates has attracted an enormous literature, which cannot even be considered here. For a recent survey and a speculative re-interpretation of the evidence consult Lombard and Bolt 1977.)

The External Ear

The external ear consists of the *auricle*, or *pinna*, and the *external acoustic meatus*. The former projects from the side of the head and serves to collect the air vibrations which constitute the sound waves; the meatus leads inwards from the bottom of the auricle conducting the vibrations which are thereby transmitted to the tympanic membrane.

THE AURICLE

The lateral surface of the auricle (7.280) is irregularly concave, looks slightly anteriorly, and presents numerous eminences and depressions. The curved prominent rim of the auricle is called the *helix*; where it turns postero-inferiorly, a small *auricular tubercle* (of Darwin) is frequently seen; this tubercle is very evident about the sixth month of intrauterine life, when the whole auricle closely resembles that of some adult monkeys. Another curved prominence, parallel with and anterior to the posterior part of the helix, is the *antihelix*; this divides above into two crura, between which is a depressed *triangular fossa*. The narrow curved depression between the helix and the antihelix is the *scaphoid*

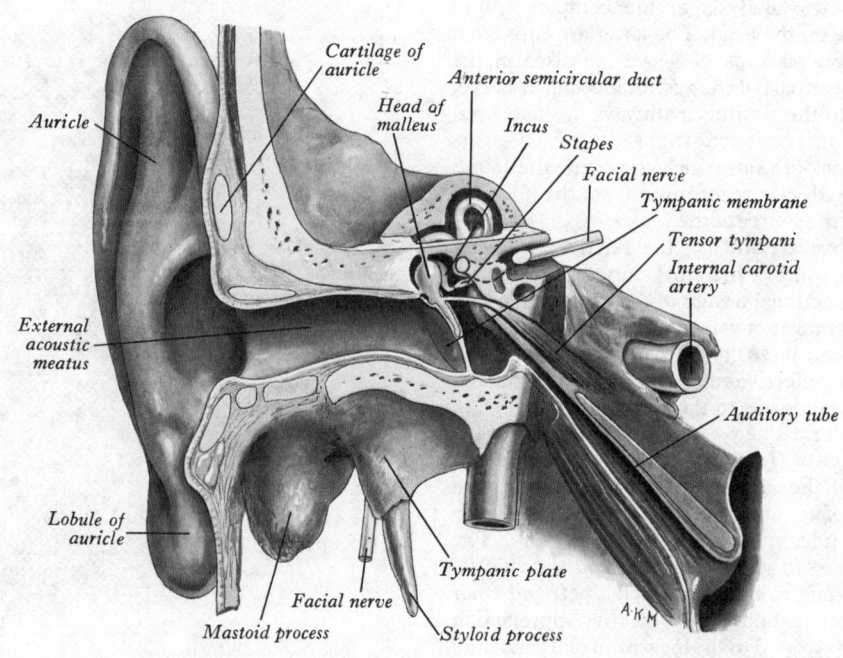

Cartilage of auricle

Anterior semicircular duct

Head of malleus

Incus

Stapes

Facial nerve

Tympanic membrane

Tensor tympani

Internal carotid artery

Auricle

External acoustic meatus

Auditory tube

Lobule of auricle

Facial nerve

Tympanic plate

Mastoid process

Styloid process

7.282 The external and middle regions of the right ear. Anterior aspect.

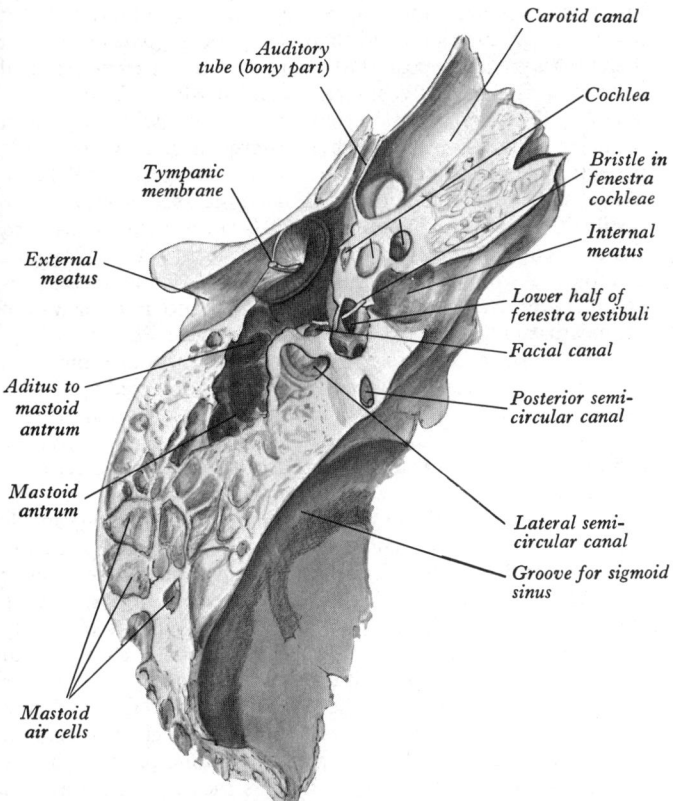

7.283 An oblique section through the left temporal bone viewed from above. (From a section prepared by Mr. P. F. Milling.) Compare with 7.285 and 7.288.

fossa; the antihelix partly encircles a deep, capacious cavity, the *concha of the auricle*, which is incompletely divided into two by the *crus* or anterior end of the helix. The part of the concha above the crus of the helix is the *cymba conchae*; it overlies, and through it can be felt, the supermeatal triangle of the temporal bone (pp. 302, 326), deep to which lies the mastoid antrum. Below the crus of the helix and in front of the concha, a small, curved flap, the *tragus*, projects posteriorly, partly overlapping the orifice of the meatus. Opposite the tragus, separated from it by the *intertragic incisure*, is a small tubercle, the *antitragus*. The *lobule* lies below the antitragus, and being composed of fibrous and adipose tissues it is soft, unlike the rest of the auricle which is firm and elastic.

The cranial surface of the auricle presents elevations which correspond to the depressions on its lateral surface, and after which they are named, e.g. eminentia conchae, eminentia fossae triangularis, etc.

In addition to acting as a collecting 'trumpet' for sound waves and channelling them into the relatively narrow meatus, the asymmetry of the pinnae and their variations in thickness, probably introduce *variable delay paths* in sound transmission which may be important in the efficient binaural (and also the cruder monaural) localization of sound sources.

In its **structure** the auricle is composed of a thin plate of elastic fibrocartilage, covered with skin, and connected with the surrounding parts by ligaments and muscles; it is continuous with the cartilaginous portion of the external acoustic meatus, and the latter is joined to the margins of the bony meatus by fibrous tissue.

The *skin of the auricle* is thin, closely adherent to the cartilage, and covered with fine hairs which are furnished with sebaceous glands; these glands are most numerous in the concha and scaphoid fossa. On the tragus and antitragus, and intertragic incisure the hairs are strong and numerous, especially in the male in old age. The skin of the auricle is continuous with that lining the external acoustic meatus.

The cartilage of the auricle (7.281) consists of a single piece of elastic fibrocartilage; upon its surface the eminences and depressions described above are found. It is absent from the

lobule; it is deficient, also, between the tragus and beginning of the helix, the gap being filled up by dense fibrous tissue. Anteriorly, where the helix bends upwards, there is a small cartilaginous projection, the *spine of the helix*, while at its other extremity the cartilage is prolonged inferiorly as the *tail of the helix*; the latter is separated from the antihelix by the *fissura antitragohelicina*. The cranial surface of the cartilage shows the *eminentia conchae* and the *eminentia triangularis* which correspond to the depressions on the lateral surface. A transverse furrow, the *sulcus antihelicis transversus*, corresponding with the inferior crus of the antihelix on the lateral surface, separates the eminentia conchae from the eminentia triangularis. The eminentia conchae is crossed by an oblique ridge, the *ponticulus*, which gives attachment to the auricularis posterior. There are two fissures in the auricular cartilage, one behind the crus helicis and another in the tragus.

The ligaments of the auricle consist of two sets: (1) extrinsic, connecting it to the temporal bone; (2) intrinsic, connecting various parts of its cartilage together.

The *extrinsic ligaments* are two in number, anterior and posterior. The *anterior ligament* extends from the tragus and spine of the helix to the root of the zygomatic process of the temporal bone; the *posterior ligament* passes from the posterior surface of the concha to the lateral surface of the mastoid process.

The chief *intrinsic ligaments* are: (a) a strong fibrous band, stretching from the tragus to the helix, completing the meatus in front, and forming part of the boundary of the concha; and (b) a band between the antihelix and the tail of the helix. Other less prominent bands are found on the cranial surface of the auricle.

THE AURICULAR MUSCLES

These consist of two sets: *extrinsic*, which connect it with the skull and scalp and move the auricle as a whole; and *intrinsic*, which extend from one part of the auricle to another.

The extrinsic muscles are the auriculares anterior, superior et posterior.

The *auricularis anterior*, the smallest of the three, is thin, fan-shaped, and its fibres are pale and indistinct. It arises from the lateral edge of the epicranial aponeurosis, and its fibres converge to be inserted into the spine of the helix.

The *auricularis superior*, the largest of the three, is thin and fan-shaped. Its fibres arise from the epicranial aponeurosis, and converge to be inserted by a thin, flattened tendon into the upper part of the cranial surface of the auricle.

The *auricularis posterior* consists of two or three fleshy fasciculi, which arise by short aponeurotic fibres from the mastoid portion of the temporal bone, and are inserted into the ponticulus on the eminentia conchae.

Nerve supply. The auriculares anterior et superior are

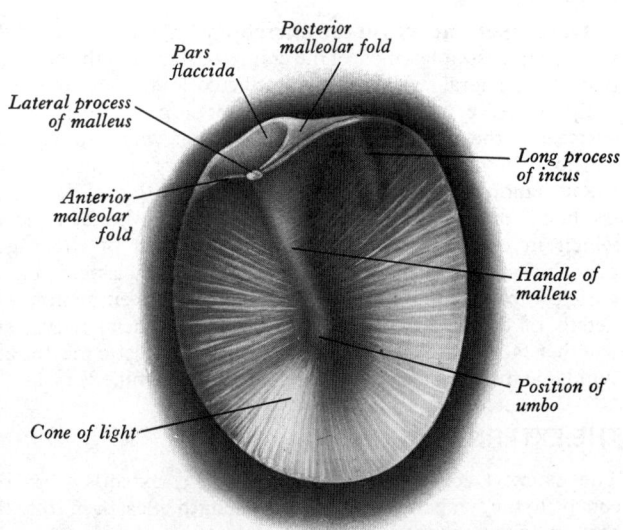

7.284 The left tympanic membrane, external aspect as seen through a speculum.

supplied by the temporal branches, and the auricularis posterior by the posterior auricular branch of the facial nerve.

Actions. In man, these muscles have very little obvious action; the auricularis anterior draws the auricle forwards and upwards; the auricularis superior raises it slightly; and the auricularis posterior draws it backwards. Nevertheless, despite the absence of marked auricular movement, auditory stimuli evoke patterned responses from these small muscles; electromyographic recording of the latter form the basis of the 'crossed acoustic response' of investigative clinical neurology.

The intrinsic muscles are helicis major and minor, tragicus, antitragicus, transversus auriculae and the obliquus auriculae. Their effect in modifying the shape of the pinna is minimal or absent in most human ears. Occasional individuals can, however, modify the shape and position of their external ears to a much greater extent.

The *helicis major* is a narrow vertical band situated upon the anterior margin of the helix. It arises from the spine of the helix, and is inserted into the anterior border of the helix, where the latter is about to curve backwards.

The *helicis minor* is an oblique fasciculus, covering the crus helicis.

The *tragicus* is a short, flattened vertical band on the lateral surface of the tragus.

The *antitragicus* arises from the outer part of the antitragus, and is inserted into the tail of the helix and antihelix.

The *transversus auriculae* is placed on the cranial surface of the auricle. It consists of scattered fibres, partly tendinous and partly muscular, extending from the eminentia conchae to the eminentia scaphae.

The *obliquus auriculae*, also on the cranial surface, consists of a few fibres extending from the upper and posterior parts of the eminentia conchae to the eminentia triangularis.

The nerve supply to the intrinsic muscles on the lateral surface is provided by the temporal branches of the facial nerve, to the intrinsic muscles on the cranial surface by the posterior auricular branch of the same nerve.

The arteries of the auricle are: (*a*) the posterior auricular branch of the external carotid artery, which supplies three or four branches to its cranial surface; twigs from these reach the lateral surface, some by passing through the fissures of the auricular cartilage, and others by turning round the margin of the helix; (*b*) the anterior auricular branches of the superficial temporal artery, which are distributed to the lateral surface; (*c*) a branch from the occipital artery.

The veins of the auricle accompany their corresponding arteries. Arteriovenous anastomoses are numerous in the skin of the auricle.

The lymphatics of the auricle drain into (*a*) the parotid lymph nodes, especially the node in front of the tragus; (*b*) the upper deep cervical lymph nodes; and (*c*) the mastoid lymph nodes.

The sensory nerves of the auricle are: (*a*) the great auricular nerve, which supplies most of the cranial surface and the posterior part of the lateral surface (helix, antihelix, lobule); (*b*) the lesser occipital nerve, which supplies the upper part of the cranial surface; (*c*) the auricular branch of the vagus, which supplies the concavity of the concha and the posterior part of the eminentia; (*d*) the auriculotemporal nerve, which supplies the tragus, the crus of the helix and the adjacent part of the helix; (*e*) the facial nerve, which in company with the auricular branch of the vagus, probably supplies small areas of skin on both aspects of the auricle—in the conchal depression and over its eminence. The details of this cutaneous innervation by the facial nerve, and whether facial fibres also reach the external acoustic meatus and tympanic membrane, however, remain undetermined.

THE EXTERNAL ACOUSTIC MEATUS

The external acoustic meatus (7.282, 283) extends from the concha to the tympanic membrane. Its length, measured from the bottom of the concha, is approximately 2·5 cm (measured from the tragus it is about 4 cm long). It consists of two structurally different parts: the lateral third is *cartilaginous*, and the medial two-thirds *osseous*. It forms an S-shaped curve, and is directed at first medially, anteriorly and slightly superiorly (*pars externa*); it then passes medially, posteriorly and superiorly (*pars media*), and lastly is carried medially, anteriorly, and slightly inferiorly (*pars interna*). The canal is oval in transverse section, with its greatest diameter obliquely placed with a postero-inferior inclination at the external orifice, but it is nearly horizontal at the medial end. There are two constrictions, one near the medial end of the cartilaginous part, and another, the *isthmus*, in the osseous part, about 2 cm from the bottom of the concha. The tympanic membrane, which closes the medial end of the meatus, is obliquely directed; in consequence the floor and anterior wall of the meatus are longer than its roof and posterior wall.

The lateral *cartilaginous part* of the meatus is about 8 mm long; it is continuous with the auricular cartilage and is fixed by fibrous tissue to the circumference of the medial osseous part of the meatus. The meatal cartilage is deficient posterosuperiorly, the deficiency being occupied by a sheet of collagen; two or three deep fissures are present in the anterior part of the cartilage.

The *osseous part* of the meatus is about 16 mm long, and is narrower than the cartilaginous part. It is directed medially, anteriorly and slightly inferiorly, forming in its course a slight curve the convexity of which is posterosuperior in position. Its medial end is smaller than the lateral end, and it is obliquely placed, with the anterior wall projecting medially about 4 mm beyond the posterior wall; this end is marked, except at its upper part, by a narrow groove, the *tympanic sulcus*, to which the circumference of the tympanic membrane is attached. Its lateral end is dilated, and rough in the greater part of its circumference for the attachment of the cartilaginous meatus. The anterior, inferior and most of the posterior parts of the osseous meatus are formed by the tympanic element of the temporal bone, which, in the fetus, exists only as a *tympanic ring* (p. 330). The postero-superior region of the osseous part is formed by the squama of the temporal bone.

The skin which envelops the auricle is continued into the external acoustic meatus and covers the outer surface of the tympanic membrane. It is thin, shows no dermal papillae on section, and is closely adherent to the cartilaginous and osseous parts of the tube; hence inflammatory conditions are extremely painful owing to the increased tension in these tissues. In the thick subcutaneous tissue of the cartilaginous part of the meatus there are numerous *ceruminous glands*, which secrete the ear wax or *cerumen*; their coiled tubular structure resembles that of the sweat glands (p. 1225). When active the cells of the secretory part are columnar, but when quiescent they are cuboidal; they are clothed externally by myoepithelial cells. The ducts of the ceruminous glands open either on to the general epithelial surface or into a neighbouring sebaceous gland of a hair follicle. The cerumen prevents maceration of the meatal lining by trapped water, and may discourage epithelial attacks by insects. Over production, or prolonged retention of ear wax may, however, completely block the meatus, or, when in contact with the tympanic membrane, embarrass its vibratory responses. The ceruminous glands, as well as hair follicles, are largely limited to the cartilaginous part of the meatus, but a few small glands and fine hairs are found in the roof of the lateral part of the osseous meatus.

In addition to the protective value of the cerumen, and the opposition to the entry of insects and foreign bodies afforded by the meatal hairs, the warm humid environment provided by the relatively enclosed meatal air is essential for the effective mechanical responses of the tympanic membrane.

Relations of the Meatus

The condyloid process of the mandible is anterior to the meatus, partially separated from its cartilaginous part by a small part of the parotid gland. A blow on the chin may cause the condyle to break into the meatus. The movements of the mandible influence to some extent the lumen of the cartilaginous part. Superior to the osseous part is the middle cranial fossa; behind it are the mastoid air cells, separated from the meatus by a thin layer of bone. The deepest part of the meatus is related superiorly to the epitympanic recess of the tympanic cavity, and posterosuperiorly to the mastoid antrum, the bone separating the antrum being only 1–2 mm thick, so that it can be opened surgically by this 'transmeatal approach'.

The meatal *arteries* are the posterior auricular branch of the external carotid, the deep auricular branch of the maxillary, and auricular branches of the superficial temporal. The *veins* drain into the external jugular and maxillary veins and the pterygoid plexus. The *lymphatics* drain with those of the auricle (p. 786).

The *nerves* supplying the meatus are derived from the auriculotemporal branch of the mandibular nerve, which supplies the anterior and upper walls of the meatus, and the auricular branch of the vagus nerve which supplies the posterior and inferior walls (*see also* below).

Clinical Examination of the Meatus

The external acoustic meatus can be examined most satisfactorily by light reflected down a funnel-shaped speculum, when the greater part of the canal and tympanic membrane can be viewed. In using this instrument, the auricle should be drawn upwards, backwards and a little laterally, to render the meatus as straight as possible.

At the point of junction of the osseous and cartilaginous parts of the meatus an obtuse angle projects into the tube antero-inferiorly; this produces a constriction which is important when attempting to remove foreign bodies lodged in the meatus. The shortness of the meatus in children should be remembered when an aural speculum is used, because of the risk of injuring the tympanic membrane; indeed, even in the adult, the speculum should not be introduced beyond the constriction which marks the junction of the osseous and cartilaginous parts. Immediately anterolateral to the membrane there is a marked depression, on the floor of the meatus, bounded laterally by a prominent ridge; here foreign bodies may become impacted. By means of the speculum, combined with traction of the auricle upwards and backwards, the greater part of the tympanic membrane is rendered visible (7.284). It is a pearly-grey, slightly glistening membrane, which in the adult is placed obliquely, forming an acute angle of about 55° with the floor of the meatus, while with the roof it forms an obtuse angle. At birth it is more horizontal, situated in almost the same plane as the base of the skull.

A reddish-yellow streak can be seen about midway between the anterior and posterior margins of the membrane which extends from the centre obliquely upwards and forwards; this is due to the handle of the malleus, attached internally to the membrane. At the upper part of this streak, close to the roof of the meatus, a little white, round prominence is clearly seen; this is the lateral or short process of the malleus, projecting against the membrane. The tympanic membrane does not present a plane surface; on the contrary, its centre is drawn inwards, reflecting its connexion with the handle of the malleus, the centre of the concavity corresponding to the *umbo* (p. 1194) on the deep surface of the membrane. A bright reflected 'cone of light', is seen in the antero-inferior quadrant of the membrane. Anterior and posterior to the short process of the malleus, the variably prominent *anterior* and *posterior malleolar folds* are seen, with the flaccid part of the tympanic membrane (p. 1194) between them. Posterior and parallel to the upper part of the handle of the malleus, the long process of the incus is often seen as a whitish streak; sometimes it can be seen to end below near a round spot which is the head of the stapes.

Applied Anatomy. Malformations such as imperfect development of the external parts, supernumerary auricle, preauricular cysts, fistulae and sinuses, or absence of the meatus, are occasionally met with. In the child up to the age of four or five years there is a gap in the antero-inferior wall of the osseous part of the meatus, the *foramen of Huschke* (pp. 331, 348), which is filled by membrane; it may persist in the adult.

The connexions of the nerves of the meatus explain the occurrence of reflex coughing and sneezing, from implication of the vagus, when there exists any source of irritation in the meatus, and the vomiting which may follow syringing the ears of children, and the occasional heart failure similarly induced in elderly people. Probably the association of earache with toothache or with cancer of the tongue is due to involvement of the mandibular branch of the trigeminal nerve, which supplies the teeth and the

tongue also. The upper half of the tympanic membrane is much more vascular than the lower half: for this reason, and also to avoid the chorda tympani nerve and ossicles, incisions through the membrane should be postero-inferior.

The Tympanic Cavity

The middle ear or tympanic cavity (7.283, 285, 287, 288) is an irregular, laterally compressed space within the temporal bone. It is lined with mucoperiosteum (p. 1199) and filled with air, which is conveyed to it from the nasal part of the pharynx through the auditory tube. It contains a chain of movable bones, which connect its lateral to its medial wall and transmit the vibrations of the tympanic membrane across the cavity to the internal ear.

The essential functional significance of the tympanic cavity with its tympanic membrane and associated chain of ossicles, is the efficient transfer of energy from the relatively weak vibrations in the elastic, compressible medium *air* in the external acoustic meatus, to overcome the inertia in the virtually incompressible aqueous *fluids* which surround the delicate membrane-supported receptors of the internal ear. Thus the mechanical coupling between the two systems must be such that their resistances to deformation or 'flow', that is, their *impedances*, are matched as closely as possible. To this end the high-amplitude, low-force per unit area, vibrations of the air are communicated to the *tympanic membrane*, which, in surface area exceeds that of the *footplate of the stapes* (in contact with the perilymph) by 15 to 20 times. In this manner, the *force per unit area* generated by the footplate is increased by a similar amount, whilst its *amplitude* of vibration is little changed.

Protective mechanisms are also incorporated in the design of the tympanic cavity. These include the connexion via the auditory (pharyngotympanic) tube, whereby pressure is equalized on the two sides of the delicate tympanic membrane, and the protection afforded by the shape of the articulations between the ossicles, and the reflex contractions of the stapedius and tensor tympani muscles, which prevent damage due to sudden and potentially excessive excursions of the ossicles.

The tympanic cavity consists of two parts: the *tympanic cavity proper*, opposite the tympanic membrane, and the *epitympanic recess*, above the level of the membrane; the latter contains the upper half of the malleus and the greater part of the incus. Including the epitympanic recess, the vertical and antero-posterior diameters of the cavity are each about 15 mm. The transverse diameter measures about 6 mm above and 4 mm below; opposite the centre of the tympanic membrane it is only about 2 mm. The tympanic cavity is bounded *laterally* by the tympanic membrane; *medially*, by the lateral wall of the internal ear; it communicates, posteriorly, with the mastoid antrum and through it with the mastoid air cells, and anteriorly with the auditory tube (7.285).

THE BOUNDARIES OF THE TYMPANIC CAVITY

The roof of the tympanic cavity (7.286) is a thin plate of compact bone, the *tegmen tympani*, which separates the cranial and tympanic cavities, and forms the greater part of the anterior surface of the petrous part of the temporal bone; it is prolonged posteriorly as the roof of the mastoid antrum, and anteriorly to cover the canal for the tensor tympani. In the young, the unossified petrosquamosal suture (p. 329) may allow direct spread of infection from the tympanic cavity to the cerebral meninges. In the adult, veins from the tympanic cavity pass through this suture to the superior petrosal sinus, or the petrosquamous sinus if present (p. 746), and may transmit infection to the intracranial sinuses.

The floor of the tympanic cavity is narrow, and consists of a thin, convex plate of bone which separates the cavity from the superior bulb of the internal jugular vein (7.288); in places the bone may be deficient, and then the cavity is separated from the vein by mucous membrane and fibrous tissue only. In the floor of the tympanic cavity, near the medial wall, there is a small aperture for the passage of the tympanic branch of the glossopharyngeal

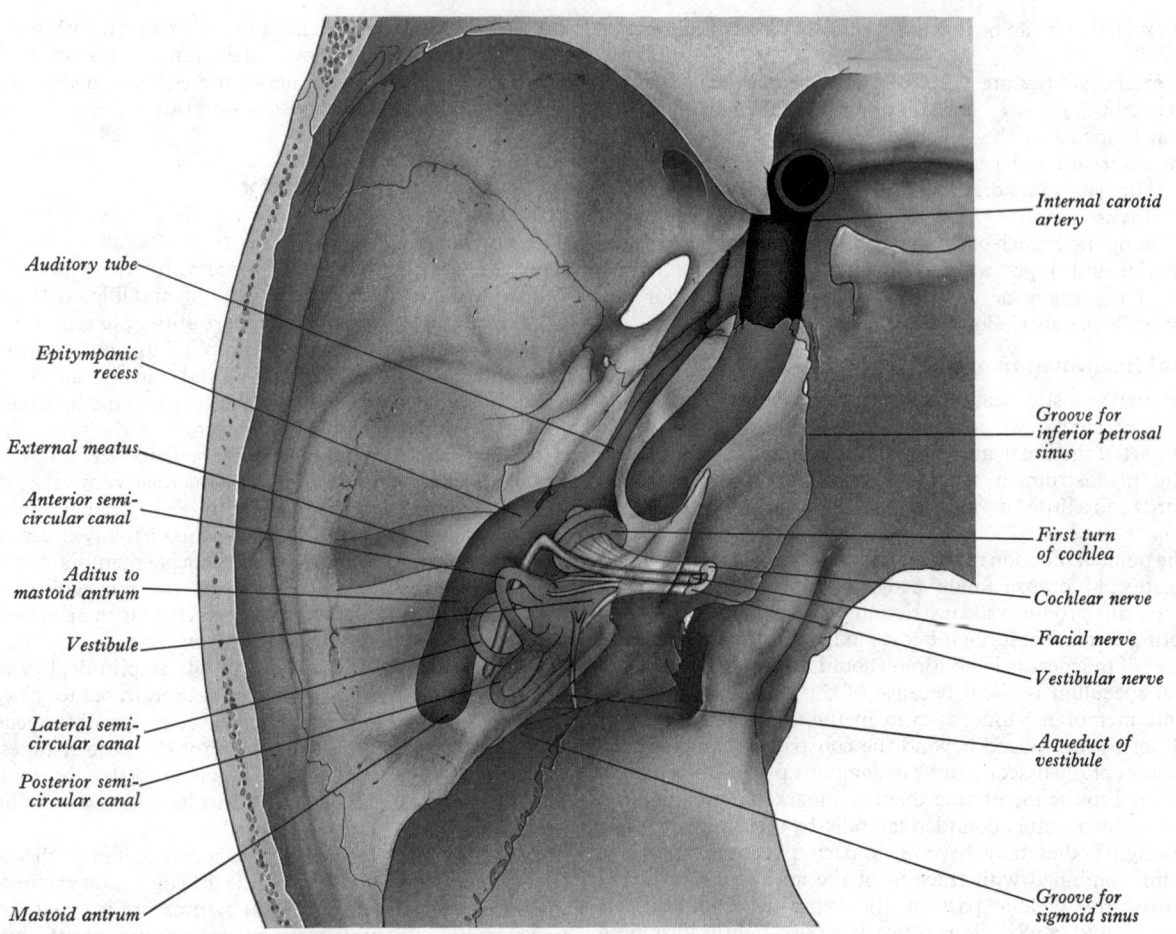

Auditory tube

Epitympanic recess

External meatus

Anterior semi-circular canal

Aditus to mastoid antrum

Vestibule

Lateral semi-circular canal

Posterior semi-circular canal

Mastoid antrum

Internal carotid artery

Groove for inferior petrosal sinus

First turn of cochlea

Cochlear nerve

Facial nerve

Vestibular nerve

Aqueduct of vestibule

Groove for sigmoid sinus

7.285 Scheme showing the parts of the left auditory apparatus as if viewed through a semi-transparent temporal bone. Compare with 7.283 and 7.288.

nerve. The floor is sometimes thick and may contain some accessory mastoid air cells.

The lateral wall of the tympanic cavity (7.286, 287) is formed mainly by the tympanic membrane, but partly also by the ring of bone to which this membrane is attached. There is a deficiency or notch in the upper part of the ring, close to which are three small apertures—the anterior and posterior canaliculi for the chorda tympani nerve, and the petrotympanic fissure.

The *posterior canaliculus for the chorda tympani nerve* is situated in the angle of junction between the posterior and lateral walls of the tympanic cavity immediately behind the tympanic membrane and on a level with the upper end of the handle of the malleus; it leads into a minute canal, which descends in front of the canal for the facial nerve, and ends in that canal about 6 mm above the stylomastoid foramen. Through it the chorda tympani nerve and a branch of the stylomastoid artery enter the tympanic cavity.

The *petrotympanic fissure* opens just above and in front of the ring of bone into which the tympanic membrane is inserted; in this situation it is a mere slit about 2 mm in length. It lodges the anterior process and anterior ligament of the malleus, and transmits to the tympanic cavity the anterior tympanic branch of the maxillary artery.

The *anterior canaliculus for the chorda tympani nerve* is placed at the medial end of the petrotympanic fissure; through it the chorda tympani nerve leaves the tympanic cavity.

The tympanic membrane (7.286, 287) separates the tympanic cavity from the external acoustic meatus. It is thin and semi-transparent, nearly oval in form, somewhat broader above than below, and very obliquely placed, forming an angle of about 55° with the floor of the meatus. Its longest diameter is downwards and forwards, and measures from 9 to 10 mm; its shortest diameter from 8 to 9 mm. The greater part of its circumference is thickened, and forms a *fibrocartilaginous ring* which is attached to the *tympanic sulcus* at the medial end of the

meatus. This sulcus is deficient superiorly, and from the ends of the notch two bands, the *anterior* and *posterior malleolar folds*, are prolonged to the lateral process of the malleus. The small, somewhat triangular part of the membrane situated above these folds is lax and thin, and is named the *pars flaccida*; a small perforation is sometimes present. The remainder of the membrane is taut and is the *pars tensa*. The handle of the malleus is firmly attached to the inner surface of the tympanic membrane as far as its centre, which projects towards the tympanic cavity; the inner surface of the membrane is thus convex, and the point of greatest convexity is named the *umbo*. Although the membrane as a whole is convex on its inner surface, its radiating fibres (*vide infra*) are curved with their concavities directed inwards.

Structurally, the tympanic membrane is composed of three strata: an outer (cuticular), an intermediate (fibrous), and an inner (mucous). The *cuticular layer* is derived from the thin skin which lines the external acoustic meatus, and consists of stratified epithelium. It is hairless, and the subepithelial connective tissue, which carries a number of small blood vessels, does not develop dermal papillae except for a few rudimentary ones around the periphery of the membrane. The *fibrous stratum* consists of two layers: a superficial layer of radiate fibres which diverge from the handle of the malleus, and a deep layer of circular fibres, which are plentiful around the circumference, but sparse and scattered near the centre of the membrane. Near the centre and around the margins of the membrane fine meshes of elastic fibres are said to be interspersed between the collagen fibres (but see below). The *mucous layer* is a part of the mucous membrane of the tympanic cavity; it is thickest towards the upper part of the membrane, and may be covered, it has been claimed, by a layer of ciliated columnar epithelial cells. However, the ciliated epithelium may be present only in patches, or entirely absent, when it is replaced by a low cuboidal or simple squamous epithelium.

The foregoing traditional account of tympanic membrane

structure, based on light microscopy, may well have needed a substantial revision, if the detailed ultrastructural and chemical studies carried out on the membrane of the guinea-pig (Johnson *et al.* 1968) also applied to the human membrane. In this mammal the external epithelium is approximately ten cells thick and consists of two zones, the superficial one consisting of non-nucleated squames, the deep zone resembling, in many respects, the stratum spinosum of the skin with numerous desmosomes and tight junctions between adjacent cells; the deepest cells lie on a continuous basal lamina, but lack epithelial pegs and hemidesmosomes. The internal (mucous) epithelium consists of a single layer of extremely flattened cells, with overlapping or interdigitating boundaries, which carry desmosomes and tight junctions between adjacent cells. Their cytoplasm contains only a sparse population of the usual organelles, micropinocytotic vesicles are few, and the free surfaces of these apparently metabolically inert cells bear occasional irregular microvilli and are coated with an amorphous electron-dense material. Ciliated columnar cells are not present. Most interestingly, the intermediate (fibrous) layer consists of filaments about 10μm in diameter, with what are apparently cross-bridges between adjacent filaments at 25 nm intervals. The filaments are disposed in outer radial, and inner non-radial zones, the former more profuse, and in neither situation do their ultrastructural appearance resemble collagen or elastin. In their amino acid composition, also, the filaments are quite distinctive, and it is proposed that they consist of a protein specialized for the unique function of the tympanic membrane. Recent studies (McMinn and Taylor 1978), however, on the development of the fibrillar component of the tympanic membrane in both guinea-pig and human embryos and fetuses have shed further light on this matter. The findings on the guinea-pig mentioned above were confirmed. In human fetuses, in contrast, small groups of true collagen fibrils were apparent by eleven weeks *in utero*, together with small bundles of elastic microfibrils. Later specimens showed increasing numbers of typically cross-banded collagen fibrils and the development of an amorphous elastin component in association with the microfibrils.

Large fibroblasts lie between the outer radial fibres and the basal lamina of the external epithelium, whilst blood capillaries and their basement membranes lie immediately deep to the basal lamina of the internal epithelium. In the flaccid part of the tympanic membrane the fibrous stratum is replaced by loose connective tissue.

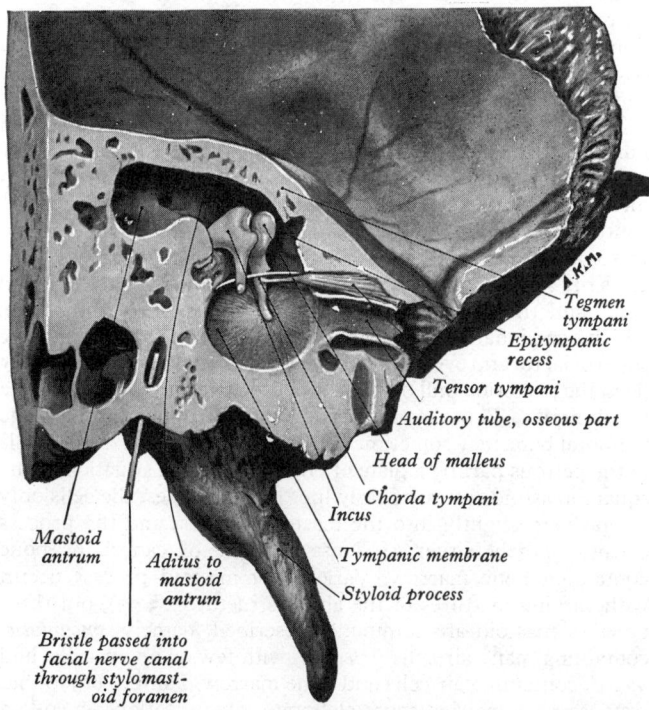

7.286 An oblique vertical section through the left temporal bone, to show the lateral wall of the middle ear and the mastoid antrum.

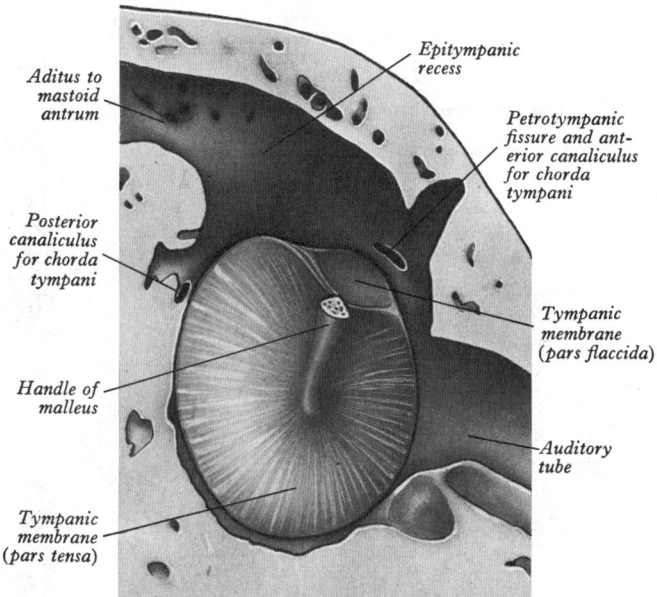

7.287 The lateral wall of the left tympanic cavity.

The *arteries* of the tympanic membrane are derived from the deep auricular branch of the maxillary artery, beneath the cuticular stratum, and from the stylomastoid branch of the occipital or posterior auricular artery, and tympanic branch of the maxillary artery, which are distributed in the mucous surface. The *superficial veins* open into the external jugular vein; those in the *deep surface* drain partly into the transverse sinus and veins of the dura mater and partly into the plexus of veins of the auditory tube. The *nerve supply* of the tympanic membrane is from the auriculotemporal branch of the mandibular nerve, the auricular branch of the vagus nerve, from the tympanic branch of the glossopharyngeal nerve, and possibly from the facial nerve.

The medial wall of the tympanic cavity (7.288, 289) is also the lateral wall of the internal ear. Its structural features are the promontory, the fenestra vestibuli, the fenestra cochleae and the prominence caused by the underlying facial nerve canal.

The *promontory* is a rounded elevation furrowed by small grooves which contain the nerves of the tympanic plexus. It overlies the lateral projection of the basal turn of the cochlea. A minute spicule of bone frequently connects the promontory to the pyramidal eminence on the posterior wall. In front of the promontory the apex of the cochlea is closely related to the medial wall of the tympanum (7.282, 288). A depression behind the promontory, the *sinus tympani*, indicates the position of the ampulla of the posterior semicircular canal.

The *fenestra vestibuli* (*f. ovalis*) is a reniform opening, posterosuperior to the promontory, which connects the tympanic cavity to the vestibule of the internal ear; its long diameter is horizontal, and its convex border is directed superiorly. In life it is occupied by the base of the stapes, the circumference of which is fixed to the margin of the fenestra by an annular ligament.

The *fenestra cochleae* (*f. rotunda*) is below and a little behind the fenestra vestibuli, from which it is separated by the posterior part of the promontory. It lies completely under cover of the overhanging edge of the promontory in a deep hollow or niche. It is placed very obliquely, and, in the macerated bone, opens upwards and forwards from the tympanic cavity into the scala tympani of the cochlea. In life it is closed by the *secondary tympanic membrane*, which is somewhat concave towards the tympanic cavity and convex towards the cochlea, the membrane being bent so that its posterosuperior one-third forms an angle with its antero-inferior two-thirds. This membrane consists of three layers: an external, derived from the mucous lining of the tympanic cavity; an internal, from the lining membrane of the cochlea; and an intermediate, fibrous, layer.

The *prominence of the facial nerve canal* indicates the position of the upper part of the bony canal in which the facial nerve is

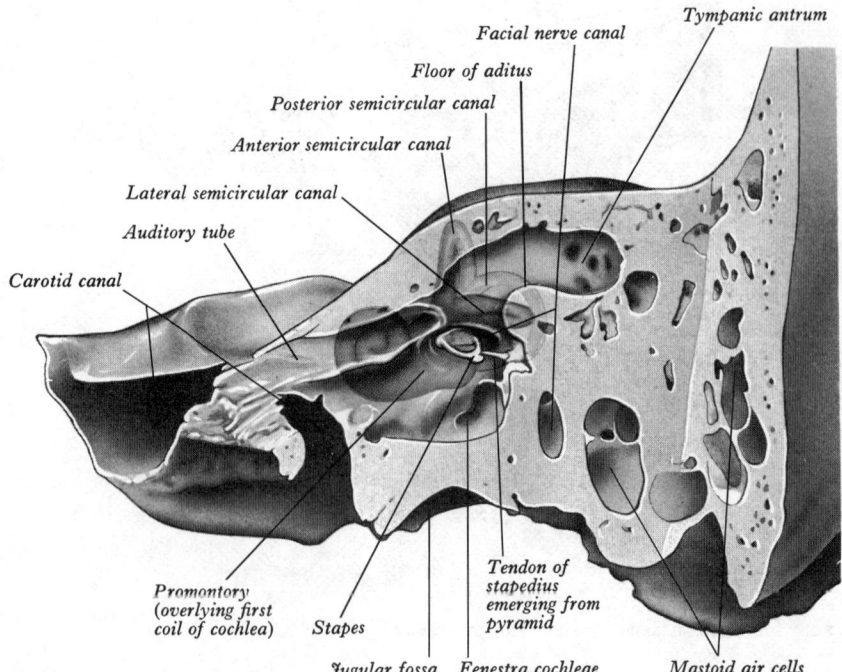

Tympanic antrum
Facial nerve canal
Floor of aditus
Posterior semicircular canal
Anterior semicircular canal
Lateral semicircular canal
Auditory tube
Carotid canal
Promontory (overlying first coil of cochlea)
Stapes
Jugular fossa
Fenestra cochleae
Tendon of stapedius emerging from pyramid
Mastoid air cells

7.288 An oblique section through the left temporal bone, to show the medial wall of the middle ear. The cochlea and the semicircular canals are in blue. Note the relationship of the first coil of the cochlea to the promontory, and the closeness of the facial nerve canal and the lateral semicircular canal to the medial wall of the aditus.

contained; this canal, the lateral wall of which may be partly deficient, traverses the medial wall of the tympanic cavity from before backwards, immediately above the fenestra vestibuli, and then curves downwards in the posterior wall of the cavity.

The posterior wall of the tympanic cavity (7.288) is wider above than below, and its main structural features are the entrance to the mastoid antrum, the pyramid, and the fossa incudis.

The *aditus to the mastoid antrum* is a large irregular aperture, which leads backwards from the epitympanic recess into the upper part of an air sinus, the *mastoid antrum*. On the medial wall of the aditus to the antrum there is a rounded eminence, situated above and behind the prominence of the facial nerve canal; it corresponds with the position of the underlying lateral semicircular canal.

The *pyramidal eminence* is situated immediately behind the

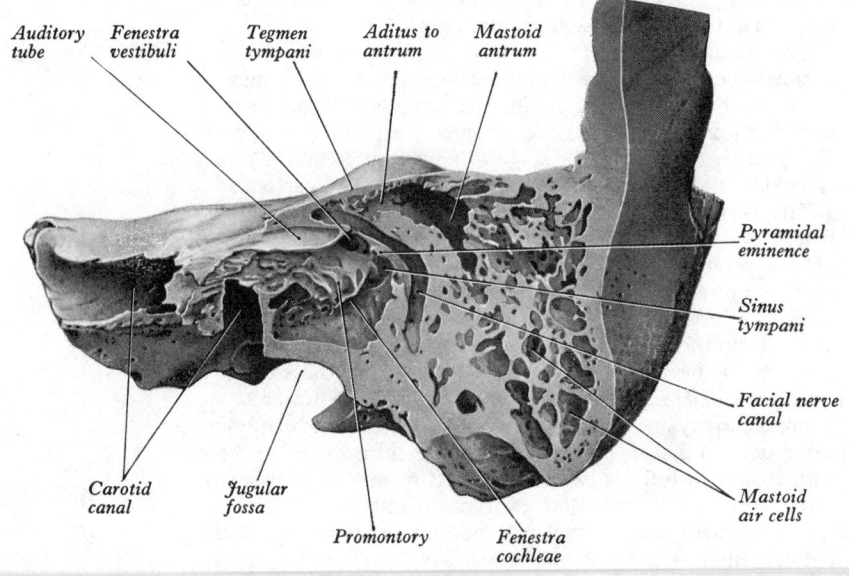

Auditory tube
Fenestra vestibuli
Tegmen tympani
Aditus to antrum
Mastoid antrum
Pyramidal eminence
Sinus tympani
Facial nerve canal
Mastoid air cells
Carotid canal
Jugular fossa
Promontory
Fenestra cochleae

7.289 An oblique section through the left temporal bone, showing the medial wall of the middle ear. Compare with 7.288.

fenestra vestibuli, and in front of the vertical portion of the facial nerve canal; it is hollow, and contains the stapedius; its summit projects forwards towards the fenestra vestibuli, and is pierced by a small aperture which transmits the tendon of the muscle. The cavity in the pyramidal eminence is prolonged downwards and backwards in front of the facial nerve canal, and communicates with the latter by an aperture which transmits a twig from the facial nerve to the stapedius. (See also p. 1072.)

The *fossa incudis* is a small depression in the lower and posterior part of the epitympanic recess; it contains the short process of the incus, which is fixed to the fossa by ligamentous fibres.

The mastoid antrum (7.283, 285, 286, 287, 288, 289) is an air sinus in the petrous part of the temporal bone, and its topographical relations are of considerable surgical importance. In the upper part of its *anterior wall* is an opening, the aditus to the mastoid antrum, which leads forwards into the epitympanic recess; medially, the aditus is related to the lateral semicircular canal. The *medial wall* of the antrum itself is related to the posterior semicircular canal (7.285). *Posteriorly* the antrum is closely related to the sigmoid sinus; some of the mastoid air cells may intervene between them. The *roof* of the antrum, formed by the tegmen tympani, is related to the middle cranial fossa and the temporal lobe of the brain. The floor has a number of apertures through which the antrum communicates with the mastoid air cells. *Antero-inferiorly*, the antrum is related to the descending part of the canal for the facial nerve. The *lateral wall* of the antrum, the usual surgical approach to the cavity, is formed by the postmeatal process of the squamous part of the temporal bone. This wall is only 2 mm thick at birth, but increases in thickness at the rate of approximately 1 mm a year, attaining a final thickness of 12–15 mm. The lateral wall of the antrum in the adult corresponds to the *suprameatal triangle* on the outer surface of the skull (pp. 302, 326); it lies beneath and can be felt through the cymba conchae (p. 1191). The superior side of the triangle, formed by the supramastoid crest, is on a level with the floor of the middle cranial fossa; the antero-inferior side, formed by the posterosuperior margin of the orifice of the external acoustic meatus, lies approximately along the course of the descending part of the canal for the facial nerve; and the posterior side, formed by a vertical tangent to the posterior margin of the meatal orifice, lies just in front of the course of the sigmoid sinus. In the adult, the mastoid antrum has a capacity of about 1 ml, each of its diameters being about 10 mm. At birth, unlike the other cranial air sinuses, it is well developed and almost the same size as in the adult; it lies at a higher level in relation to the external acoustic meatus than in the adult. In the very young child, owing to the thinness of the lateral wall of the antrum and the absence or feeble development of the mastoid process, the stylomastoid foramen and the emerging facial nerve are very superficially situated.

The mastoid air cells (7.283, 288, 289) vary considerably in number, form and size in different individuals. In general, they are a series of intercommunicating cavities, lined by mucous membrane, with a flattened squamous non-ciliated epithelium, continuous with that of the mastoid antrum and tympanic cavity. In some cases they extend throughout the mastoid process, even to its tip, and some of the cells may be separated from the sigmoid sinus and the posterior cranial fossa by extremely thin bone, which occasionally shows deficiencies. Some of the cells may lie superficial to, and even behind the sigmoid sinus, and others may lie in the posterior wall of the descending part of the canal for the facial nerve. Those contained in the squamous part of the temporal bone may sometimes be separated from the deeper cells in the petrous part by a plate of bone lying in the situation of the squamomastoid suture of early life. In other cases, the cells only extend very slightly into the mastoid process, and the process consists largely either of dense bone or of cancellous bone containing bone marrow. Varieties of mastoid process occur, with varying mixtures of the above structures (3.95), but three types of mastoid are commonly described, namely, *pneumatic*, containing many air cells, *sclerotic*, with few or no air cells, and *mixed*, containing air cells and bone marrow. Pannier (1970) has proposed a somewhat more elaborate categorization dependent upon the degree of pneumatization and the relation of air cells to the lateral sinus. In his series (100 mastoid processes) the cells most distant from the antrum were the largest. The cells may extend beyond the confines of the mastoid process. They may extend for some distance into the squamous part of the temporal bone above the supramastoid crest, and also into the posterior root of the zygoma. Others may extend forwards into the roof of the osseous part of the external acoustic meatus, lying immediately below the middle cranial fossa. Some may extend into the floor of the tympanic cavity, lying in very close relation to the superior jugular bulb. Rarely, a few cells may excavate the jugular process of the occipital bone. An important group of cells may extend medially into the petrous part of the temporal bone, even reaching as far as its tip, and they are related to the auditory tube, the carotid canal, the labyrinth and the abducent nerve. Some investigators maintain that these petrous cells are not directly continuous with the mastoid cells proper, but are independent outgrowths from the tympanic cavity. The extensions of the mastoid air cells described above are of considerable importance to the clinician. Infection of the cells may spread to the structures mentioned above as related to them. Whereas the mastoid antrum is well developed at birth, the mastoid air cells, at this time, are only just beginning to develop as tiny diverticula from the antrum. As the mastoid process begins to develop in the second year, the cells gradually extend into it, and by the fourth year they are well formed, though their greatest growth occurs at about the age of puberty. In about 20 per cent of skulls the mastoid process is not excavated by air cells.

The anterior wall of the tympanic cavity is constricted owing to the approximation of the medial and lateral walls of the cavity. Its inferior and larger part consists of a thin lamina of bone which forms the posterior wall of the carotid canal, and is perforated by the superior and inferior caroticotympanic nerves, and the tympanic branch or branches of the internal carotid artery. At the superior part of the anterior wall there are two parallel canals, placed one above the other; superiorly placed is the *canal for the tensor tympani*, inferiorly, the *osseous part of the auditory tube*. These canals incline anteriorly, inferiorly and medially to open in the angle between the squamous and petrous parts of the temporal bone; they are separated by a thin, bony septum. The canal for the tensor tympani and the septum run posterolaterally on the medial wall of the tympanic cavity, and end immediately above the fenestra vestibuli, where the posterior end of the septum is curved laterally to form a pulley, the *processus trochleariformis (cochleariformis)*, over which the tendon of the tensor tympani bends in a lateral direction to reach its attachment to the upper part of the handle of the malleus.

THE AUDITORY TUBE

The *auditory* or *pharyngotympanic* tube (7.282, 285, 286) is the channel through which the tympanic cavity communicates with the nasal part of the pharynx. Through it air passes between pharynx and tympanic cavity thus equalizing the air pressure on the medial and lateral surfaces of the tympanic membrane. Its length is about 36 mm, and its direction is downwards, forwards and medially, forming an angle of some 45° with the sagittal plane and one of about 30° with the horizontal plane. It is formed partly of bone, partly of cartilage and fibrous tissue.

The *bony part of the tube* is about 12 mm long. It begins in the anterior wall of the tympanic cavity, and, gradually narrowing, ends at the angle of junction of the squamous and petrous portions of the temporal bone, its extremity presenting a jagged margin which serves for the attachment of the cartilaginous part; the carotid canal lies on its medial side. It is oblong in transverse section with its greater dimension lying horizontally.

The *cartilaginous part of the tube*, about 24 mm long, is formed of a triangular plate of cartilage, the greater part of which is situated in the posteromedial wall of the tube. The apex of the fibrocartilage is attached by fibrous tissue to the circumference of the medial end of the bony part of the tube, while its base lies directly under the mucous membrane of the lateral wall of the nasal part of the pharynx, where it forms the *tubal elevation*, behind the pharyngeal orifice of the tube. The superior part of the cartilage is bent laterally and downwards, and the cartilage therefore consists of a broad *medial lamina*, and a narrow *lateral lamina*. On transverse section the cartilage has the appearance of a hook; the groove or furrow produced by the bend in the cartilage is open below and laterally, and this part of the wall of the canal is completed by fibrous membrane. The cartilage is fixed to the base of the skull in the groove between the petrous part of the temporal bone and the greater wing of the sphenoid bone; this groove ends near the root of the medial pterygoid plate. The cartilaginous and bony parts of the tube are not in the same plane, the former inclining downwards a little more than the latter. The diameter of the tube is greatest at the pharyngeal orifice, least at the junction of the bony and cartilaginous parts, and again increased towards the tympanic cavity; the narrowest part of the tube is termed the *isthmus*.

The mucous membrane of the tube is continuous medially with that of the pharynx, and laterally with that of the tympanic cavity; it is covered with ciliated columnar epithelium and is thin in the bony part, while in the cartilaginous part it contains many mucous glands, and near its pharyngeal orifice a variable, sometimes considerable aggregation of lymphoid tissue, the *tubal tonsil*.

Relations of the auditory tube. Anterolaterally the tensor veli palatini separates the tube from the otic ganglion, the mandibular nerve and its branches, the chorda tympani nerve and the middle meningeal artery. This muscle receives some fibres from the lateral lamina of the cartilage and from the membranous part of the tube; these fibres constitute the *dilatator tubae*. The salpingopharyngeus (p. 1308) is attached to the inferior part of the cartilage of the tube near its pharyngeal opening. *Posteromedially* the tube is related to the petrous part of the temporal bone and to the levator veli palatini, which arises partly from its medial lamina. The position and relations of the pharyngeal orifice are described with the nasal part of the pharynx (p. 1308).

The tube is opened during deglutition, but the mechanism is uncertain. Some claim that the dilatator tubae, possibly aided by the salpingopharyngeus, is responsible, though others deny the existence of the dilatator tubae. It is also claimed that the levator veli palatini, by elevating the cartilaginous part of the tube, allows the tube to open passively by releasing tension on the cartilage.

In the newborn child the auditory tube is about half as long as that of the adult. Its direction is more horizontal and its bony part is relatively shorter, but much wider than in the adult. Its pharyngeal orifice is a narrow slit, which is on a level with the palate and is devoid of a tubal elevation.

Vessels and nerves. The *arteries* of the auditory tube are derived from the ascending pharyngeal branch of the external carotid artery and from two branches of the maxillary artery, namely, the middle meningeal artery and the artery of the pterygoid canal. The *veins* open into the pterygoid venous plexus. The *nerves* of the tube spring from the tympanic plexus (p. 1200) and from the pharyngeal branch of the pterygopalatine ganglion. The precise contribution from the nerves which form the plexus i.e. the glossopharyngeal, the cervical sympathetic and possibly the facial, remains uncertain in man.

THE AUDITORY OSSICLES

The tympanic cavity contains a chain of three movable ossicles: the *malleus, incus* and *stapes*. The malleus is attached to the tympanic membrane and the base of the stapes to the circumference of the fenestra vestibuli, while the incus is placed between, and articulates with both malleus and stapes.

The malleus (7.290), so named from its fancied resemblance to a mallet, is from 8 to 9 mm long, and is the largest of the auditory ossicles. It consists of a head, neck and three processes— the manubrium or handle, and the anterior and lateral processes.

The *head*, which is the large upper end of the bone, is situated within the epitympanic recess; it is ovoid in shape, and articulates posteriorly with the incus, being clothed with mucous membrane over the rest of its surface. The cartilage-covered facet for articulation with the incus is constricted near the middle, and consists of an upper larger and a lower smaller part, situated nearly at right angles to each other. Opposite the constriction the lower margin of the facet projects in the form of a process which is named the *cogtooth* or *spur* of the malleus.

The *neck* is the narrow part just beneath the head; inferior to the

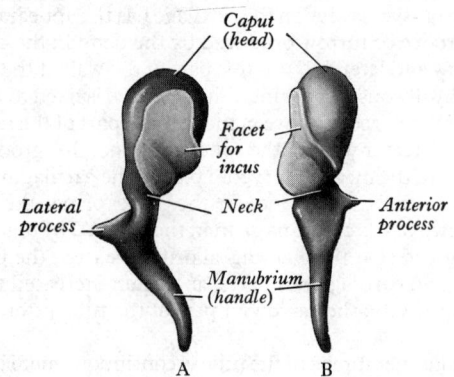

7.290 The left malleus. A Posterior aspect. B Medial aspect.

neck there is an enlargement to which the various processes are attached.

The *manubrium mallei* is connected by its lateral margin with the tympanic membrane. It is directed downwards, medially and backwards; it decreases in size towards its free end, which is curved slightly forwards and flattened transversely. Near the upper end of its medial surface there is a slight projection, into which the tendon of the tensor tympani is inserted.

The *anterior process* is a delicate bony spicule, directed forwards from the enlargement below the neck; it is connected to the petrotympanic fissure by ligamentous fibres (*vide infra*). In the fetus this is the longest process of the malleus, and it is continuous in front with the cartilage of Meckel (p. 150).

The *lateral process* is a conical projection which projects from the root of the handle of the malleus; it is directed laterally, and is attached to the upper part of the tympanic membrane and, by means of the anterior and posterior malleolar folds, to the extremities of the notch at the upper part of the tympanic sulcus.

Ossification. The cartilaginous precursor of the malleus is derived from near the dorsal end of the embryonic Meckel's cartilage. With the exception of its anterior process, it is ossified from a single endochondral centre, which appears near the future neck of the bone in the fourth month of fetal life. The anterior process is ossified separately, in dense connective tissue, and joins the main part of the bone about the sixth month of fetal life.

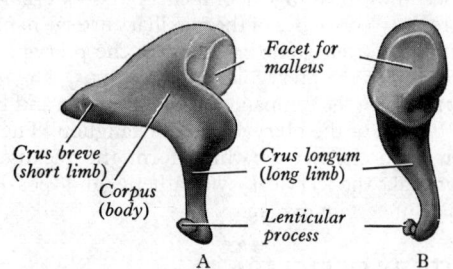

7.291 The left incus. A Medial aspect. B Anterior aspect.

The incus (7.291) received its name from its supposed resemblance to an anvil, but its shape is more like that of a premolar tooth, with two widely diverging roots. It consists of a body and two processes.

The *body* is somewhat cubical, but compressed laterally. On its anterior surface there is a cartilage-covered saddle-shaped facet, for articulation with the head of the malleus.

The *long process*, rather more than half the length of the handle of the malleus, descends nearly vertically, behind and parallel to that process; its lower end bends medially, and terminates in a rounded projection, the *lenticular process*, the medial surface of which is covered with cartilage, and articulates with the head of the stapes.

The *short process*, somewhat conical in shape, projects posteriorly, and is attached by ligamentous fibres to the fossa incudis, in the postero-inferior part of the epitympanic recess.

Ossification. The incus has a cartilaginous precursor which is continuous with the dorsal extremity of the embryonic Meckel's cartilage (*see* p. 145). Its ossification often spreads from a single endochondral centre, which appears in the upper part of its long process in the fourth month of fetal life; the lenticular process, however, may have a separate centre of ossification.

The stapes (7.292), so called from its resemblance to a stirrup, consists of a head, neck, two limbs and a base.

The *head* (*caput*) is directed laterally, and has a small cartilage-covered depression for articulation with the lenticular process of the incus.

The *neck* is the constricted part adjoining the head; the tendon of the stapedius is inserted into its posterior surface.

The *limbs* (*crura*) diverge from the neck and merge at their ends with a roughly oval or reniform plate, the *base*, which forms the footplate of the stirrup, and is fixed to the margin of the fenestra vestibuli by a ring of ligamentous fibres (the annular ligament). The anterior limb is shorter and less curved than the posterior.

Ossification. The stapes is preformed in the perforated dorsal moiety of the hyoid arch cartilage of the embryo. Its ossification starts from a single endochondral centre, which appears in the base of the bone in the fourth month of fetal life, and then gradually spreads through the stapedial limbs to coalesce in its body.

At birth the auditory ossicles have reached an advanced state of maturity.

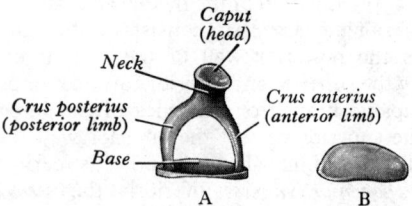

7.292 The left stapes. A Superior aspect. B Basal aspect.

The Articulations of the Auditory Ossicles

These are typical synovial joints. The incudomalleolar joint is a saddle-shaped articulation. The incudostapedial joint is a 'ball and socket' articulation. Their articular surfaces are covered with articular cartilage, and each is enveloped by an articular capsule containing a considerable amount of elastic tissue, and lined by synovial membrane.

The Ligaments of the Auditory Ossicles

The ossicles are connected to the walls of the tympanic cavity by ligaments: three for the malleus, and one each for the incus and stapes. Some of these 'ligaments' are mere folds of mucous membrane which carry blood vessels and nerves to and from the ossicles and their articulations. Others contain a central, strong band of collagen fibres.

The *anterior ligament of the malleus* is attached by one end to the neck of the malleus, just above the anterior process, and by the other to the anterior wall of the tympanic cavity, close to the petrotympanic fissure, some of its collagen fibres being prolonged through the fissure to reach the spine of the sphenoid bone; some fibres are continued into the sphenomandibular ligament. Both the latter ligament and the anterior ligament of the malleus are derived from the fibrous perichondral sheath of the cartilage of Meckel (p. 150). The ligament may contain muscle fibres (the *laxator tympani* or *musculus externus mallei*).

The *lateral ligament of the malleus* is a triangular band passing from the posterior part of the border of the tympanic incisure to the head of the malleus.

The *superior ligament of the malleus* connects the head of the malleus to the roof of the epitympanic recess.

The *posterior ligament of the incus* connects the end of the short process of the incus to the fossa incudis.

The *superior ligament of the incus* is little more than a fold of mucous membrane, passing from the body of the incus to the roof of the epitympanic recess.

The vestibular surface and the circumference of the base of the stapes are covered with hyaline cartilage; that encircling the base is attached to the margin of the fenestra vestibuli by a ring·of elastic fibres, termed the *annular ligament of the base of the stapes*. Certain parts are much narrower, and act as a kind of hinge on which the base of the stapes moves when the stapedius contracts and during acoustic oscillation (*see* below).

THE MUSCLES OF THE TYMPANIC CAVITY

The muscles of the tympanic cavity are the tensor tympani and stapedius.

The tensor tympani (7.282, 286), a long slender muscle, is contained in the bony canal superior to the bony part of the auditory tube, from which it is separated by a thin bony septum. It arises from the cartilaginous part of the auditory tube and the adjoining part of the greater wing of the sphenoid, as well as from the bony canal in which it is contained. Passing backwards through the canal, it ends in a slim tendon which bends laterally round the pulley-like processus trochleariformis, and is attached to the handle of the malleus, near its root.

Nerve supply. The tensor tympani is supplied by a branch of the nerve to the medial pterygoid, which in its turn is a branch of the mandibular nerve and traverses the otic ganglion without interruption (p. 1076). Through this it receives both a motor and a proprioceptive innervation (Candiollo 1965).

The stapedius arises from the wall of a conical cavity in the pyramidal eminence and from the continuation of this cavity which passes down in front of the descending part of the facial canal (p. 1072); its minute tendon emerges from the orifice at the apex of the pyramid, and, passing forwards, is inserted into the posterior surface of the neck of the stapes (7.288). This small muscle is of an asymmetric bipennate form, and consists of numerous small motor units, each containing only six to nine muscle fibres. A few neuromuscular spindles are present near the myotendinous junction.

Nerve supply. The stapedius is supplied by a branch of the facial nerve. (For details of the stapedius muscle and its innervation *see* Blevins 1967.)

Actions. Under normal conditions, the tensor tympani and the stapedius contract simultaneously and reflexly in response to sounds of fairly high intensity, exerting 'a protective damping effect upon sound vibrations reaching the internal ear' (Hallpike 1935). The tensor on contraction pulls inwards the tympanic membrane and renders it more tense; its action also results in the base of the stapes being pushed more tightly into the fenestra vestibuli. The stapedius opposes the latter action of the tensor. Paralysis of the stapedius muscle results in hyperacusis.

MOVEMENTS OF THE AUDITORY OSSICLES

The manubrium faithfully follows all movements of the tympanic membrane, while the malleus and incus rotate together around an axis which runs through the short process and posterior ligament of the incus and the anterior ligament of the malleus. When the tympanic membrane and the manubrium are displaced medially, the long process of the incus also moves in the same direction and pushes the base of the stapes towards the labyrinth. This motion is communicated to the perilymph (p. 1200), the movement of which causes an outward bulging of the secondary tympanic membrane which occupies the fenestra cochleae. The conditions are reversed when the tympanic membrane moves in an outward direction, but if this movement of the membrane is exaggerated the incus does not follow the full outward excursion of the malleus but merely glides on this bone at the incudomalleolar joint, thus avoiding the danger of pulling the base of the stapes out of the fenestra vestibuli. When the manubrium is carried in a medial direction, the tooth or *cog* at the lower margin of the head of the malleus locks the incudomalleolar joint, and this necessitates a medial movement of the long process of the incus; the joint is unlocked when the handle of the malleus is carried outwards. The three bones collectively act as a bent lever, so that the base of the stapes does not move in and out of the fenestra vestibuli like a piston, but rocks on a fulcrum which is situated on the

anteroinferior border of the fenestra, and at this site the annular ligament is thickened. More complex movements of the stapes have also been described (Békésy 1960). The rocking movement around a vertical axis, which has been likened to a door opening and closing, occurs only at moderate sound intensities. With loud, low-pitched sounds, the axis becomes horizontal, the upper and lower margins of the stapedial base oscillating in opposite directions around this central axis, thus immediately and automatically preventing excessive displacement of the perilymph.

THE TYMPANIC MUCOSA

The mucous membrane of the tympanic cavity is continuous with that of the pharynx, through the auditory tube. It invests the auditory ossicles and the muscles and nerves contained in the tympanic cavity, forms the inner layer of the tympanic membrane and the outer layer of the secondary tympanic membrane, and lines the mastoid antrum and mastoid air cells. It forms several vascular folds which extend from the walls of the tympanic cavity to the ossicles; of these, one descends from the roof of the cavity to the head of the malleus and upper margin of the body of the incus, and a second invests the stapedius; other folds invest the chorda tympani nerve and the tensor tympani. These folds separate off saccular recesses, and give the interior of the tympanic cavity a somewhat honeycombed appearance. One of these pouches, termed the *superior recess of the tympanic membrane*, lies between the neck of the malleus and the pars flaccida. Two other recesses, termed the *anterior* and *posterior recesses of the tympanic membrane*, may be mentioned: they are formed by the mucous membrane which envelops the chorda tympani nerve, and are situated, one in front of, and the other behind, the handle of the malleus. In the tympanic cavity the mucous membrane is pale, thin and slightly vascular. It is covered with ciliated columnar epithelium except over the posterior part of the medial wall, the posterior wall, often parts of the tympanic membrane, and the auditory ossicles where the cells are flatter and non-ciliated. Near the orifice of the auditory tube numerous goblet cells are present, but apart from this there are no mucous glands. The mastoid antrum and the mastoid air cells are lined by a flattened non-ciliated epithelium. Undoubtedly, there are considerable variations in the regions of the tympanic cavity and associated structures, which are lined by squamous, cuboidal, columnar, or ciliated columnar epithelium. As yet there has been no systematic exhaustive ultrastructural study of this problem. (For an account of the ultrastructure of restricted regions consult Kawabata and Paparella 1969.)

It is to be noted that the tympanic cavity and mastoid antrum, the auditory ossicles and the structures comprising the internal ear are more or less fully developed by birth and undergo little subsequent alteration. In the fetus the tympanic cavity contains a jelly-like tissue, which has practically disappeared by birth, at which time the cavity is filled with a fluid that is absorbed after birth when air enters the cavity through the auditory tube.

THE VASCULATURE OF THE TYMPANIC CAVITY

The *arteries* supplying the walls and contents of the tympanic cavity are six in number. Two of them are larger than the others—the *anterior tympanic* branch of the *maxillary artery*, which supplies the tympanic membrane, and the *stylomastoid* branch of the *occipital* or *posterior auricular arteries* supplying the posterior tympanic cavity and mastoid air cells. The smaller arteries are—the *petrosal* branch of the *middle meningeal artery*, which enters through the hiatus for the greater petrosal nerve; the *superior tympanic* branch of the *middle meningeal artery*, which traverses the canal for the tensor tympani; a branch from the *ascending pharyngeal artery*, and another from the *artery of the pterygoid canal*, which accompany the auditory tube, and the *tympanic branch*, or branches from the *internal carotid artery*, given off in the carotid canal and perforating the thin anterior wall of the tympanic cavity. In early fetal life the *stapedial artery* passes through the ring of the stapes (p. 189). The *veins* terminate in the pterygoid venous plexus and in the superior petrosal sinus. From the mucous membrane of the mastoid antrum a small group of veins runs medially through the arch formed by the anterior semicircular canal. They emerge on the posterior surface of the petrous part of the temporal bone through the subarcuate fossa, and open into the superior petrosal sinus. These small veins are the remains of the large *subarcuate veins* of the child, and constitute a pathway of infection from the mastoid antrum to the meninges of the brain. The *lymph vessels* are described on p. 786.

THE NERVES OF THE TYMPANIC CAVITY

The *nerves* constitute the *tympanic plexus*, which ramifies upon the surface of the promontory. The plexus is formed by (1) the tympanic branch of the glossopharyngeal nerve, and (2) the caroticotympanic nerves. (For details consult Arslan 1960.) The *tympanic branch of the glossopharyngeal* enters the tympanic cavity by the *canaliculus for the tympanic nerve*, and divides into branches which ramify on the promontory and enter into the formation of the tympanic plexus. The *superior and inferior caroticotympanic nerves*, from the carotid plexus of the sympathetic, pass through the wall of the carotid canal, and join the plexus. The tympanic plexus supplies: (*a*) branches to the mucous lining of the tympanic cavity, auditory tube and mastoid air cells; (*b*) a branch which goes through an opening in front of the fenestra vestibuli and joins the greater petrosal nerve; (*c*) the *lesser petrosal nerve*, which may be looked upon as the continuation of the tympanic branch of the glossopharyngeal nerve through the tympanic plexus. The lesser petrosal nerve traverses a small canal below the canal for the tensor tympani, runs past, and receives a connecting branch from the genicular ganglion of the facial nerve; it emerges from the anterior surface of the temporal bone through a small opening on the lateral side of the hiatus for the greater petrosal nerve. It then passes through the foramen ovale, or through the small canaliculus innominatus (p. 323), and joins the otic ganglion. Postganglionic fibres pass from the otic ganglion, via the auriculotemporal nerve, to provide the secretomotor supply for the parotid gland.

The *chorda tympani* is a nerve derived from the facial nerve, about 6 mm above the stylomastoid foramen. It runs anterosuperiorly in a canal, and enters the tympanic cavity through the *posterior canaliculus*. It then curves anteriorly in the substance of the tympanic membrane lying between its mucous and fibrous layers (p. 1194). After crossing the medial aspect of the upper part of the handle of the malleus it reaches the anterior wall, and enters the *anterior canaliculus*. (For its further course see p. 1072.) The other nerves which are closely related, topographically, to the tympanic cavity include the facial nerve with its genicular ganglion and stapedial and greater petrosal branches; the auricular branch of the vagus; the afferent and efferent terminals of the vestibulocochlear nerve; and the internal carotid sympathetic plexus. These are all described in greater detail elsewhere in the present section.

The *meningeal branch* (p. 1066), of the mandibular nerve, supplies branches to the mastoid air cells.

Applied Anatomy. Fractures of the middle fossa of the base of the skull almost invariably involve the tympanic roof, and are accompanied by a rupture of the tympanic membrane or fracture through the roof of the bony part of the external acoustic meatus. Such injuries are associated with prolonged bleeding from the ear, and, if the dura mater has also been torn, with discharge of cerebrospinal fluid.

The tympanic cavity is frequently the seat of disease, both suppurative and non-suppurative, and in practically every case the inflammation spreads upwards from the nasal cavity and nasal part of the pharynx along the auditory tube. Acute inflammation spreading up to the tympanic cavity is usually associated with much swelling of the mucous membrane of the tube, thus occluding it, and the products of inflammation, confined in the tympanic cavity, may spread directly to the mastoid antrum. In such circumstances the only means of escape for the products is by rupture of the tympanic membrane, which may occur spontaneously or be induced surgically and is followed by a free discharge of pus. Should the swelling of the walls of the auditory tube then subside, the normal drainage of the cavity will be established and the perforation in the drum will heal, but if not—as is often the case because the opening of the tube may be occluded by enlarged lymphatic aggregates in the nasal part of the pharynx or other cause—the pus may continue to accumulate in the middle ear and overflow through the perforation as a chronic otorrhoea. Several intracranial complications may result from purulent material being retained; thus an abscess may form between the bone and dura mater, (*a*) above the roof of the tympanic cavity, and immediately beneath the dura covering the temporal lobe of the brain, or (*b*) between the deep aspect of the mastoid process and the sigmoid sinus, possibly extending widely and surrounding the sinus. In this latter case thrombosis of the sinus may occur, and the infected clot tends to disintegrate and be carried into the general circulation, particles becoming lodged in the capillaries of the lungs and causing abscesses. In addition, bone disease of the tympanic cavity or mastoid antrum may be associated with severe and fatal septic meningitis, or with the formation of abscess in the brain, the most common sites being the temporal lobe of the cerebrum and the hemisphere of the cerebellum.

In some cases of chronic bone disease in the tympanic cavity, the facial nerve becomes exposed as it lies in its canal and an inflammatory process is set up in the nerve, leading to facial paralysis of the infranuclear or peripheral type (p. 1073).

THE INTERNAL EAR

The internal ear consists of two parts—the *osseous labyrinth*, a series of cavities within the petrous part of the temporal bone, and the *membranous labyrinth*, a series of communicating membranous sacs and ducts, contained within the bony cavities.

The Osseous Labyrinth

The osseous labyrinth (7.285, 293) consists of three parts—the *vestibule*, the *semicircular canals* and the *cochlea*. These are cavities hollowed out of the substance of the bone, and lined by periosteum; they contain a clear fluid, known as the *perilymph*, in which the membranous labyrinth is placed. The osseous labyrinth consists of harder, denser bone than the surrounding parts of the petrous portion of the temporal bone, so that it is possible, particularly in the very young skull, to separate the labyrinth from the petrous temporal by artificial dissection.

THE VESTIBULE

The vestibule is the central part of the bony labyrinth, and is situated medial to the tympanic cavity, posterior to the cochlea and anterior to the semicircular canals. It is somewhat ovoid in shape, but flattened transversely; it measures about 5 mm from before backwards, the same from above downwards, and about 3 mm across. In its *lateral wall* there is the opening of the fenestra vestibuli, closed in life by the base of the stapes and its annular ligament. On the anterior part of the *medial wall* there is a small *spherical recess*, containing the saccule and perforated by several minute holes, the *macula cribrosa media*. The recess corresponds to the inferior vestibular area in the bottom of the internal acoustic meatus (7.302), and the foramina transmit filaments of the vestibulocochlear nerve to the saccule. Behind this recess there is an oblique ridge, the *vestibular crest*, the anterior end of which is the *pyramid of the vestibule*; this ridge divides below to enclose a small depression, the *cochlear recess*, which is perforated by a number of holes for the passage of fibre bundles of the vestibulocochlear nerve to the vestibular end of the duct of the cochlea. Posterosuperior to the vestibular crest, and situated in the roof and medial wall of the vestibule there is an *elliptical recess* which contains the utricle. The pyramid and adjoining part of the elliptical recess are perforated by a number of holes, the *macula cribrosa superior*; the holes in the pyramid transmit the nerves to the utricle, and those in the elliptical recess, the nerves to the ampullae of the superior and lateral semicircular ducts. The pyramid and the adjoining part of the elliptical recess correspond to the superior vestibular area at the bottom of the internal acoustic meatus (7.302). The orifice of the *aqueduct of the vestibule* lies below the elliptical recess. This aqueduct extends to the posterior surface of the petrous part of the temporal bone; it transmits one or more small veins, and contains a tubular prolongation of the membranous labyrinth which is termed the *endolymphatic duct*. In the posterior part of the vestibule there are the *five orifices of the semicircular canals*, in the anterior wall, an elliptical opening leading into the *scala vestibuli* of the cochlea.

THE SEMICIRCULAR CANALS

The semicircular canals are three in number, anterior (superior), posterior and lateral, and are situated posterosuperior to the vestibule. They are compressed from side to side, and each occupies about two-thirds of a circle. They are unequal in length, but are all about 0·8 mm in diameter; each presents a dilatation at one end, called the *ampulla*, the diameter of which is nearly twice that of the canal. They open into the vestibule by five orifices, one of which is common to two of the canals.

The *anterior (superior) semicircular canal*, 15 to 20 mm in length, is vertical in direction, and is placed transversely to the long axis of the petrous part of the temporal bone, on the anterior surface of which its arch underlies the arcuate eminence (p. 329). Some maintain, however, that the arcuate eminence does not accurately coincide with the anterior semicircular canal but is adapted to the occipitotemporal sulcus on the inferior surface of the temporal lobe of the cerebral hemisphere. The anterolateral end of the anterior semicircular canal is ampullated, and opens into the upper and lateral part of the vestibule; the opposite end unites with the upper end of the posterior canal to form the *crus commune*, which is about 4 mm long, and opens into the medial part of the vestibule.

The *posterior semicircular canal*, also vertical, is directed backwards, nearly parallel with the posterior surface of the petrous bone; it is from 18 mm to 22 mm long; its ampullated end opens into the lower part of the vestibule, where there are several small holes, the *macula cribrosa inferior*, for the transmission of the nerves to this ampulla, their position corresponding to the *foramen singulare* in the bottom of the internal acoustic meatus (7.302). Its upper end opens into the crus commune.

The *lateral* or *horizontal canal* is from 12 mm to 15 mm long, and its arch is directed horizontally backwards and laterally. Its anterior or ampullated end opens into the upper and lateral angle of the vestibule, just above the fenestra vestibuli and immediately below the ampullated end of the superior canal; its posterior end opens below the orifice of the crus commune.

The lateral semicircular canal of one ear is in the same plane as that of the other ear; while the anterior canal of one ear is in a plane nearly parallel with that of the posterior canal of the other.

Blanks *et al.* (1975) have measured the angular relations of the planes of the semicircular osseous canals in 10 human skulls. Although the planes of the three ipsilateral canals were approximately perpendicular to each other, some variation was apparent, the measured angles being as follows: horizontal/anterior—$111·76 \pm 7·55°$, anterior/posterior—$86·16 \pm 4·72°$, and posterior/horizontal—$95·75 \pm 4·66°$. The planes of similarly orientated canals of the two sides showed marked departure from parallelism: left anterior/right posterior—$24·50 \pm 7·19°$, left posterior/right anterior—$23·73 \pm 6·71°$, and left horizontal/right horizontal—$19·82 \pm 14·93°$. The same observers (Curthoys *et al.* 1977) have also measured the dimensions and radii of curvature of the canals; means for the radii of the osseous canals were as

follows: horizontal—3·25 mm, anterior—3·74 mm, and posterior—3·79 mm. The functional implications of these data are still a matter of speculation. In diameter the osseous canals measure about 1 mm (minor axis) and 1·4 mm (major axis). The membranous ducts are much smaller, also ellipsoid in transverse section, with major and minor axes of 0·23 and 0·46 mm. Representative means for ampullary dimensions are are follows: length—1·94 mm, height—1·55 mm. Curthoys and his collaborators have recorded many other dimensions of the labyrinth both from their own observations and those of others. They attempt to equate these metrical data with theories of labyrinthine mechanics.

THE COCHLEA

The cochlea (7.285, 293, 294, 298) resembles the shell of the common snail; it forms the anterior part of the labyrinth, is conical in form, and placed anterior to the vestibule; it measures about 5 mm from base to apex, and its breadth across the base is about 9 mm. Its apex, or *cupula*, is directed laterally towards the upper and front part of the medial wall of the tympanic cavity (7.285); its base is directed towards the bottom of the internal acoustic meatus, and is perforated by numerous apertures for the

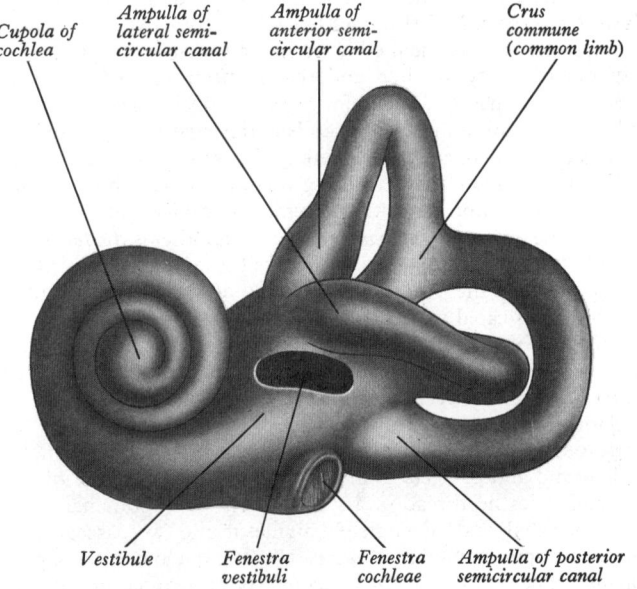

Cupola of cochlea   Ampulla of lateral semi-circular canal   Ampulla of anterior semicircular canal   Crus commune (common limb)

Vestibule   Fenestra vestibuli   Fenestra cochleae   Ampulla of posterior semicircular canal

7.293A The left osseous labyrinth. Lateral aspect.

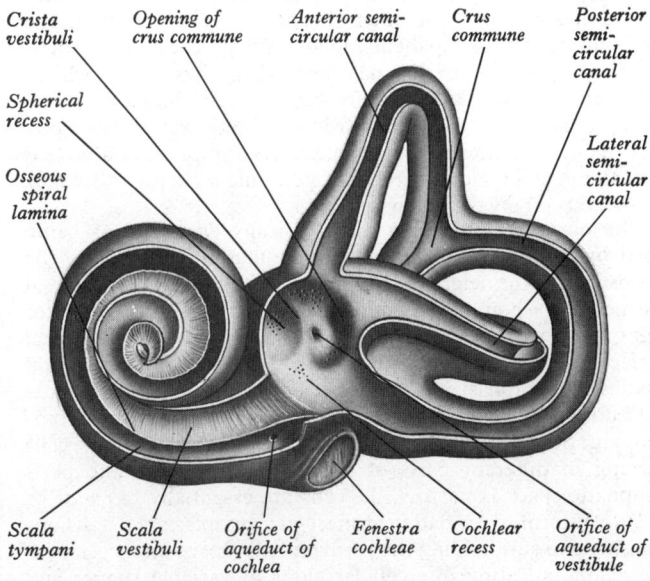

Crista vestibuli   Opening of crus commune   Anterior semicircular canal   Crus commune   Posterior semicircular canal

Spherical recess

Osseous spiral lamina

Lateral semicircular canal

Scala tympani   Scala vestibuli   Orifice of aqueduct of cochlea   Fenestra cochleae   Cochlear recess   Orifice of aqueduct of vestibule

7.293B The interior of the left osseous labyrinth.

passage of the cochlear nerve. The cochlea consists of a cone-shaped central axis, the *modiolus*; of a canal, wound spirally around the central axis for roughly two turns and three-quarters; and of a delicate *osseous spiral lamina*, which projects from the modiolus into the canal, and partially divides it. In the recent state the division of the canal is completed by the *basilar membrane*, which stretches from the free border of the osseous spiral lamina to the outer wall of the bony cochlea; the two passages into which the cochlear canal is thus divided communicate with each other at the apex of the modiolus by a small opening, the *helicotrema*.

The modiolus is the conical, osseous, central pillar of the cochlea. Its base is broad, and appears at the lateral end of the internal acoustic meatus, where it corresponds with the *tractus spiralis foraminosus*, which is perforated by numerous orifices for the transmission of the branches of the cochlear nerve; the nerves for the first turn and a half of the cochlea pass through the foramina of the tractus spiralis; those for the apical turn, through the *foramen centrale*, in the centre of this tract. The canals of the tractus spiralis foraminosus pass through the modiolus and successively bend outwards to reach the attached margin of the osseous spiral lamina. Here they become enlarged, and by their apposition form the *spiral canal of the modiolus*, which follows the course of the attached margin of the osseous spiral lamina and lodges the *spiral ganglion*. The foramen centrale is continued into a canal which runs through the middle of the modiolus to its apex.

The bony canal of the cochlea takes about two turns and three-quarters round the modiolus; the first turn bulges towards the tympanic cavity and there underlies the promontory (p. 1195). It is about 35 mm long, and diminishes gradually in diameter from the base to the summit, where it ends in the *cupula*, which forms the apex of the cochlea. The beginning of this canal is about 3 mm in diameter, and in it there are three openings. One—the *fenestra cochleae*—communicates with the tympanic cavity and in life is closed by the *secondary tympanic membrane*; another is the *fenestra vestibuli* (p. 1195) occupied by the base of the stapes. The third is the aperture of the *cochlear canaliculus*, leading to a minute funnel-shaped canal which opens on the inferior surface of the petrous part of the temporal bone (p. 329). It transmits a small vein to join the inferior petrosal sinus, and establishes a communication between the subarachnoid space and the scala tympani (*vide infra*).

The osseous spiral lamina is a bony shelf or ledge which winds round and projects from the modiolus into the interior of the canal, like the thread of a screw. It reaches about halfway across the canal, and incompletely divides it into two passages or scalae: an upper, named the *scala vestibuli*, and a lower, the *scala tympani*. The width of the osseous spiral lamina gradually decreases from the basal to the apical coil of the cochlea, and near the summit of the cochlea the lamina ends in a hook-shaped process, the *hamulus of the spiral lamina*; this assists in forming the boundary of the *helicotrema*, through which the two scalae communicate with each other. From the spiral canal of the modiolus numerous canaliculi radiate through the osseous spiral lamina as far as its free edge and transmit branches of the cochlear nerve. In the lower part of the first turn of the cochlea a *secondary spiral lamina* projects inwards from the *outer* wall of the bony tube; it does not, however, reach the osseous spiral lamina, so that if the laminae be viewed from the vestibule a narrow *vestibular fissure* is seen between them.

The osseous labyrinth was classically described as being lined by a thin fibroserous membrane closely adherent to the periosteum of the neighbouring bone. The flattened epithelium of the membrane bounds the extensive *perilymphatic space*, the latter being filled with the fluid *perilymph* which bathes the external surface of the membranous labyrinth. However, ultrastructural studies have emphasized that the perilymphatic fluid-filled spaces are bounded by fibrocyte-like *perilymphatic cells*, with accompanying strands of extracellular fibres, the morphology of the cells varying in different parts of the labyrinth. Where the perilymphatic space is narrow, the cells are essentially *reticular* or *stellate* in form, their flattened sheet-like cytoplasmic extensions crossing and subdividing the perilymphatic space into a series of intercommunicating intercellular clefts of variable shapes and dimensions. Such tissue, and its accompanying spaces, occupies

the cochlear canaliculus. Elsewhere, in regions where the perilymphatic space is much wider, as in the scala vestibuli and scala tympani, and throughout much of the vestibule, the perilymphatic cells which line the periosteum and cover the external surface of the membranous labyrinth are extremely flattened, with rather featureless cytoplasm. Despite occasional cytoplasmic projections into the neighbouring perilymph, in such situations the cellular arrangement approaches that of a true squamous epithelium. Over parts of the perilymphatic surface of the basilar membrane, the cells assume a cuboidal form. Closely related to the periosteal or membranous labyrinth aspect of the perilymphatic cells are bundles of collagen fibres which may, in part, be synthesized by these cells. However, in some species, fibres with a helical substructure, which differ from collagen in their ultrastructure, have been described in these situations. Their status in man remains to be determined.

The perilymph which occupies the perilymphatic spaces resembles cerebrospinal fluid fairly closely in its composition, although minor differences have been described (Ormerod 1960). Many regard it simply as an ultrafiltrate of plasma, with perhaps some addition from the cerebrospinal fluid. Its precise source, rate of production, circulation and absorption, cannot yet be regarded as settled. The status of the connexions which exist between the perilymphatic spaces and the general subarachnoid space through the *cochlear canaliculus* has been the subject of some debate. Originally, the canaliculus was often described as containing a simple patent, epithelium-lined duct, which connected the two. Later, this suggestion was rejected by a number of workers (Waltner 1948; Mygind 1948; Young 1952, 1953) who proposed that connective tissue barriers blocked the canaliculus and separated the two fluid compartments. It seems likely, however, that extracellular crevices persist between the perilymphatic cells which occupy the canal; certainly, large-moleculed electron-dense tracers such as thorotrast, when introduced into the craniovertebral subarachnoid space have a ready access to the perilymphatic spaces (Duvall and Quick 1969). Other investigators (Silverstein *et al.* 1969) point out that in cats, even particulates such as india ink or avian erthrocytes, will pass into the perilymphatic spaces via the cochlear canaliculus within twenty-four hours after their introduction into the subarachnoid space of the posterial cranial fossa. The latter authors regard perilymph as probably originating from three sources: (1) a transudate from the blood vessels surrounding the space; (2) from the fluid spaces surrounding the sheaths of the vestibulocochlear nerve fibres; and (3) from a slow continuous flow of cerebrospinal fluid along the cochlear canaliculus. The site of removal of perilymph is uncertain.

The part of the petrous bone which immediately surrounds the labyrinth is developed from the cartilaginous otic capsule; it is denser than the rest of the petrous bone, and exhibits interglobular spaces, which contain cartilage cells (7.295). The modiolus of the cochlea, on the other hand, is formed of trabecular membrane bone (Fraser and Dickie 1914).

The perilymphatic space of the vestibule communicates behind with that of the semicircular canals, and opens anteriorly into the scala vestibuli of the cochlea, which in turn opens into the scala tympani through the helicotrema, at the apex of the cochlea. The scala tympani is separated from the tympanic cavity by the secondary tympanic membrane, but is continuous with the subarachnoid space through the cochlear canaliculus (*vide supra*).

The Membranous Labyrinth

The membranous labyrinth (7.294), while contained within, is much smaller than the bony labyrinth; it is filled with fluid unique in composition named *endolymph*, and in its walls the branches of the vestibulocochlear nerve are distributed. It includes: (*a*) the *utricle* and *saccule*, two small sacs, occupying the vestibule; (*b*) three *semicircular ducts*, enclosed within the semicircular canals; (*c*) the *duct of the cochlea*, contained within the osseous cochlea. The various parts of the membranous labyrinth form a closed system of channels which, however, communicate freely with one another; the semicircular ducts open into the utricle and this into the saccule through the ductus utriculosaccularis which also joins the ductus endolymphaticus, and the saccule opens into the duct of the cochlea through the ductus reuniens.

The membranous labyrinth is fixed at certain points to the wall of the bony labyrinth, but is separated from the greater part of the bony labyrinth by a perilymphatic space (*vide supra*). For details of the fine structure of the membranous labyrinth consult Wersäll (1956); Engström and Wersäll (1958); Iurato (1967); Kimura (1969); Babel *et al.* (1970).

THE UTRICLE

The utricle, the larger of the two vestibular sacs, is irregularly oblong in shape, and occupies the posterosuperior region of the vestibule, lying in contact with the elliptical recess and also the area inferior to it. The part of the utricle in the elliptical recess forms a pouch or cul-de-sac; the lateral half of the floor and the adjoining lower part of the lateral wall of this pouch is thickened over an area measuring about 3 mm by 2 mm to form the *macula of the utricle* (p. 1204), which receives the utricular fibres of the vestibular nerve. The ampullae of the anterior and lateral semicircular ducts open into the lateral part of the utricle, while the ampulla of the posterior duct, the crus commune and the posterior end of the lateral duct open into the medial part of the utricle. The posterior end of the lateral duct widens into a flattened cone which joins the medial end of the utricle at a right angle. From the anteromedial part of the utricle a fine canal, named the *ductus utriculosaccularis*, is given off and opens into the ductus endolymphaticus.

THE SACCULE

The saccule lies in the spherical recess near the opening of the scala vestibuli of the cochlea. When seen from the anterior aspect it presents a nearly globular form, but it is prolonged postero-inferiorly in the form of a cone, part of the upper surface of which is in contact with the under surface of the utricle, and the utricle and saccule have here a common wall. In its anterior wall there is an oval thickening, termed the *macula of the saccule* (pp. 1204, 1205), which lies in a plane at right angles to the macula of the utricle, and to which the saccular fibres of the vestibulocochlear nerve are distributed. Its cavity communicates indirectly through a Y-shaped tube with that of the utricle. From its posterior part the *ductus endolymphaticus* is given off and it is joined by the ductus utriculosaccularis; the ductus endolymphaticus passes

medially and then inferiorly along the aqueduct of the vestibule to end in a blind pouch, the *saccus endolymphaticus*, under the dura mater on the posterior surface of the petrous portion of the temporal bone. From the lower part of the saccule a short tube, the *ductus reuniens*, passes inferiorly and gradually widens into the vestibular or basal end of the duct of the cochlea (7.294).

THE SEMICIRCULAR DUCTS

The semicircular ducts (7.294, 295) are about one-fourth of the diameter of the semicircular canals, but are similar to them in shape and general form. Each has an ampulla at one end, which, of course, lies within the ampulla of the corresponding bony canal. The semicircular ducts open by five orifices into the utricle, one opening being common to the medial end of the anterior, and the upper end of the posterior duct.

In each of the ampullae the wall is thickened, and projects into the cavity as a transverse elevation shaped somewhat like the figure 8, and named the *septum transversum*; the most prominent part of this septum being the *ampullary crest*, which projects from the wall of each ampulla that is most distant from the centre of the circle of which the semicircular duct forms an arc, a situation in which 'any movement of the endolymph would be caught by the crista to the greatest advantage' (Dickie 1920).

The utricle, saccule and semicircular ducts are held in position by fibrocellular bands which stretch across the perilymphatic space to the bony walls.

STRUCTURE OF THE UTRICLE, SACCULE AND SEMICIRCULAR DUCTS

The walls of the utricle, saccule and semicircular ducts are generally described as consisting of three layers. The *outer* layer is composed largely of fibrous tissue containing some blood vessels. Its superficial fibres are, in many places, clothed by flattened perilymphatic cells, and, at some points this outer surface blends with the endosteum of the bony labyrinth. The *middle* layer, a more delicate vascular connective tissue, presents on its internal surface, especially in the semicircular ducts, a number of papilliform projections. The *inner* layer consists in general of a single layer of epithelial cells, which vary from squamous to cubical or polygonal in shape, resting on a basement membrane; they exhibit a *specialized* arrangement in the ampullary crests of the semicircular ducts and in the maculae of the utricle and saccule (7.294, 296). In these special sites, also, the middle coat is

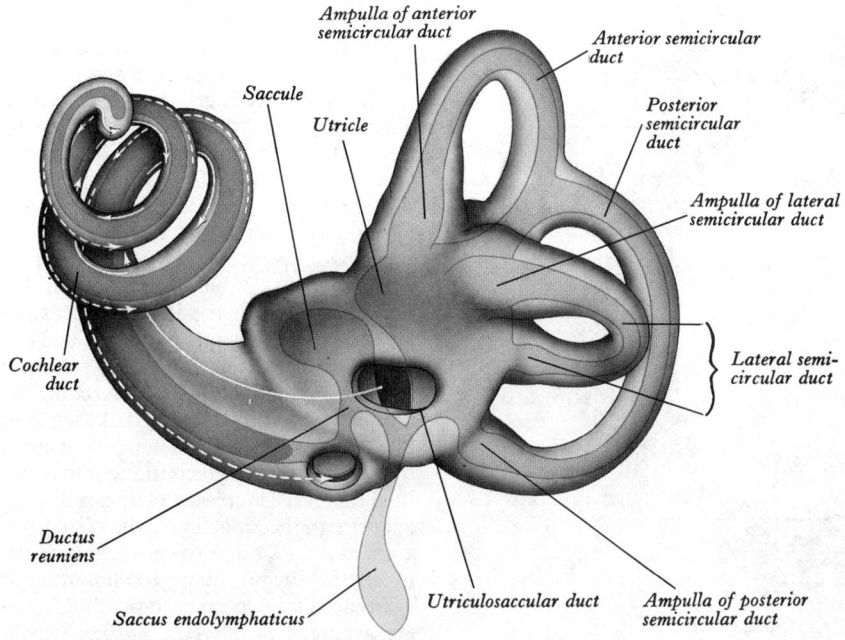

Ampulla of anterior semicircular duct

Saccule

Utricle

Anterior semicircular duct

Posterior semicircular duct

Ampulla of lateral semicircular duct

Lateral semi-circular duct

Cochlear duct

Ductus reuniens

Saccus endolymphaticus

Utriculosaccular duct

Ampulla of posterior semicircular duct

7.294 Scheme of the membranous labyrinth (blue) projected on to the osseous labyrinth. The arrows indicate the direction of sound waves in the cochlea.

thickened. Flanking each crest, between this and the side wall of the ampulla is an area of tall epithelium which in cross-section appears crescentic and is termed the *planum semilunatum*.

Although the so-called 'non-specialized' parts of the membranous labyrinth so far described in fact have epithelia varying from squamous to tall columnar in different regions, ultrastructural studies (Kimura 1969) have also shown that they consist of either *light* or *dark* epithelial cells which are structurally quite dissimilar. In many regions the epithelium differs little from non-secretory simple epithelia elsewhere. The *light cells* comprising these regions contain elliptical or crenated heterochromatic nuclei, their luminal surfaces carry a few microvilli, and relatively few mitochondria, occasional ribosomes and some micropino-cytotic vesicles occur in the cytoplasm. Adjacent cells show junctional complexes near their luminal borders, with also some interdigitation of neighbouring cell surfaces. Infolding of the basal plasmalemma is minimal. In contrast, some patches or strips of epithelium in the utricle and in the general and ampullated walls of the semicircular ducts consist of *dark cells*. These resemble light cells in a few respects only, their luminal surfaces carrying occasional microvilli and pinocytotic invaginations, whilst junctional complexes and interdigitations occur between adjacent cells. Their irregular nuclei are either centrally placed or situated rather near the luminal surface. The dense supranuclear cytoplasm contains numerous small coated vesicles, a profusion of larger smooth-walled vesicles, and many mitochondria, both free and membrane-attached polysomes, lipid droplets, lysosomes, lipofuscin granules, microfilaments and microtubules, and a prominent Golgi apparatus. The infranuclear part of the cell consists of numerous long cytoplasmic processes projecting towards the underlying basal lamina. Each process shows an elongate, fusiform dilatation of its contour, completely occupied by a long narrow mitochondrion. The plasmalemma of the processes is clothed externally by electron-opaque extensions from the basal lamina. Clearly, such cells are highly active, and on general morphological grounds, and their structural similarity to cells in other ion-transporting epithelia such as those in parts of the renal tubules (p. 1391), the ciliary body, the parotid duct, and the salt-secreting glands of various sub-mammalian forms, it has been suggested that the dark cells may be involved in controlling the ionic composition of the endolymph (see also below).

In the *ampullary crests* the epithelium consists of *hair cells* and *supporting cells*. **The hair cells** are the sensory cells and are of two types. *Type I* is piriform with a rounded base and a short neck. Except for its free end it is surrounded by a large goblet-shaped nerve terminal or *calix*; the apposed plasma membranes are separated by an interval of about 20–30 nm in width, but at a number of points the membranes approach each other more closely and the interval is reduced to about 5 nm. The nucleus of the hair cell is basally placed, and is surrounded by numerous mitochondria, and there is also a concentration of these organelles near the free surface of the cell. Scattered in the cytoplasm are occasional cisternae of granular endoplasmic reticulum, free polysomes, microfilaments and microtubules, numerous smooth vesicles about 20 nm in diameter, and a supranuclear Golgi apparatus is present.

The *type II* hair cell is cylindrical with its nucleus placed at varying levels, but usually more centrally placed than in the type I cell. Its cytoplasm contains similar organelles, but its population of smooth-walled vesicles is more abundant, and its supranuclear Golgi apparatus is more prominent. The basal part of the type II hair cell is not surrounded by a nerve terminal calix, but instead is in contact with a number of bud-like synaptic *boutons*. The latter are of two varieties; both contain mitochondria and numerous small membranous vesicles, but in the *non-granulated terminals* the vesicles are clear, whereas in the *granulated terminals* many of the vesicles contain electron-dense cores. It is now generally accepted that the non-granulated terminals are those of *afferent* nerve fibres conducting sensory information towards the central nervous system. The granulated terminals are regarded as derived from *efferent* fibres which innervate the type II hair cells, activity in which probably modifies the effective threshold of the hair cell to sensory stimuli. Thus, the granulated and non-granulated terminals are both considered to be sites of neurochemical transmission, but of course, polarized in opposite directions; however, the details of this transmission remain uncertain. The calix of the type I hair cell is also regarded as the terminal of an afferent vestibular fibre, but whether it operates mainly by neurochemical transmission, or whether low-resistance paths of the 'electrical synapse' type (p. 829) are extensively involved, is still to be determined. A number of granulated boutons are often present applied at points to the external aspect of the calix, and probably these are *efferent* terminals which modify the transmission characteristics of the calix.

In general, the type I hair cells may be regarded as the more 'discriminative' variety. The calices are derived from the larger-diameter, faster-conducting vestibular nerve fibres, and each fibre innervates a small localized group of type I cells. In contrast, the type II cells contact boutons from a number of relatively small-diameter vestibular nerve fibres, and each of the latter innervates a large number of type II cells which are distributed over a substantial area of membrane.

The apical, free surface of both types of hair cell are similar. They carry 40–100 'hairs' or stereocilia which are modified microvilli, of varying length and arranged in a regular hexagonal array when viewed from the surface. The array is, however, polarized with respect to a single long *kinocilium* which is attached to one border of the cell. The kinocilium possesses a typical basal body in the apical cytoplasm of the hair cell, and its shaft carries a ring of nine double microtubules, but the central pair of microtubules, which additionally characterize most other cilia, are sometimes less well developed and have even been claimed to be absent on occasion. Despite its name, any motile activity of this modified cilium remains uncertain. The stereocilia are non-motile microvilli, constricted at their point of attachment to the hair cell, and each contains a complement of longitudinal microfilaments which continue into a well-defined terminal web in the apical cytoplasm of the cell. On the border of the cell opposite to the kinocilium, the stereocilia are short, being about 1 μm in length, but those progressively nearer the cilium increase in length, reaching about 100 μm in the vicinity of the cilium.

In each of the specialized vestibular receptor sites, the hair cells are arranged in *a precise pattern* which is of great functional significance (*vide infra*). In the ampullae of the lateral semicircular ducts, the sides of the hair cells which bear a kinocilium are all directed towards the neighbouring utricular

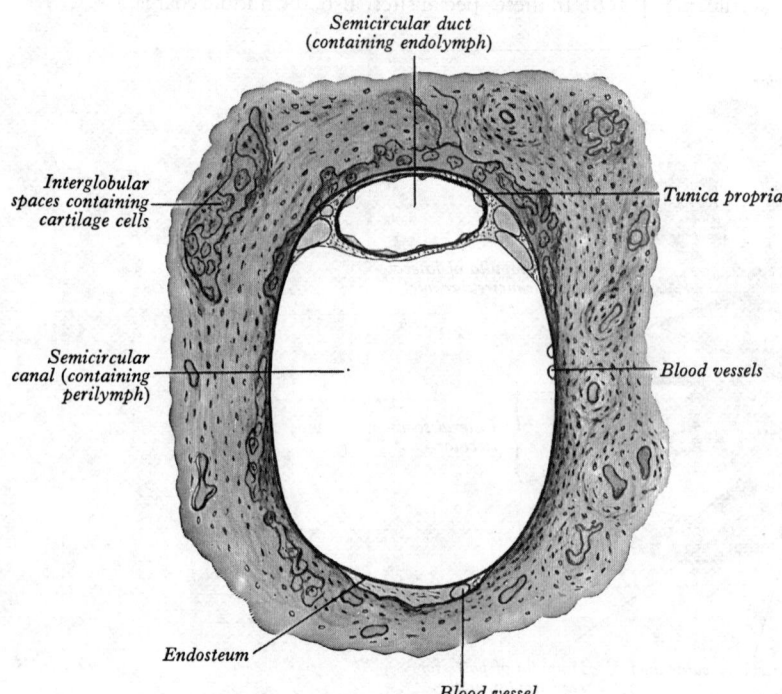

Semicircular duct
(containing endolymph)

Interglobular
spaces containing
cartilage cells

Tunica propria

Semicircular
canal (containing
perilymph)

Blood vessels

Endosteum

Blood vessel

7.295 A transverse section through the left posterior semicircular canal and duct of an adult man. Magnification about ×50. (After Dr. J. K. Milne Dickie.)

cavity, whilst in the ampullae of the anterior and posterior ducts they are directed away from that cavity. In the maculae of the utricle and saccule, there is, in each, a sinuous 'parting line' which crosses the central region of the macula. The polarization of the hair cells is reversed on the opposite sides of this line. In the utricle the hair cells are arranged in curved contours with their kinocilia nearer the parting line, whilst in the saccule they are directed away from this line. For details of the complex relationship of these arrays of hair cells, to the three planes of space, consult Iurato (1967); Babel *et al.* (1970); Ades and Engström (1974). **The supporting cells** are elongated and of variable diameter along their length. They rest on a basal lamina, and the nucleus is usually basal in position. The free surfaces are provided with microvilli, and the cytoplasm contains large osmiophilic granules which may be secretory in nature, in addition to a well-developed Golgi apparatus, numerous vertically running microtubules and microfilaments which enter a prominent subapical terminal web; mitochondria are plentiful. Whether the supporting cells are mainly involved in nutritive support of the hair cells, or in modifying the composition of the endolymph remains uncertain. Junctional complexes are established around the subapical parts of the hair cells and their neighbouring supporting cells. The processes of the hair cells and supporting cells project into a thick, dome-shaped gelatinous protein-polysaccharide containing mass called the cupula. The precise chemical composition of the cupula is, however, undetermined. It possesses a free apical border which almost reaches the opposite ampullary wall. The whole cupula can swing from side to side in response to currents in the endolymph. After displacement in one direction, when the endolymph current ceases, an elastic recoil causes the return of the cupula towards the intermediate vertical position, but some overshoot and oscillation occurs, before it comes to rest.

The epithelial cells of the *planum semilunatum* (Iurato 1967) lie adjacent to the crista on the one hand and gradually change to cuboidal epithelium of the side walls of the ampulla on the other. Their free surfaces are provided with a few microvilli. Basally they contain an abundance of parallel smooth-walled double membranes, arranged perpendicular to the basement membrane. They are infoldings of the basal plasma membrane; they end in vesicular enlargements in the cytoplasm. Between the membrane pairs are vesicles and mitochondria arranged in a linear manner. The nucleus is placed towards the apical part of the cell where there are abundant mitochondria, endoplasmic reticulum, a well-developed Golgi apparatus, polyribosomes and vesicles, some of which contain granules. It will be recalled that this description of the planum cells resembles that of the *dark cells* (*vide supra*) described by other investigators elsewhere in the labyrinth. The planum semilunatum is possibly concerned with the secretion of the endolymph. The *supporting* and *hair cells* of the *maculae* of the *utricle* and *saccule* are generally similar to those of the ampullary crests, but the gelatinous mass into which the cilia project is flatter and is termed an *otolithic membrane* (*membrana statoconiorum*), because it contains numerous minute crystalline bodies called *otoliths, otoconia,* or *statoconia* which consist of calcite (Carlström *et al.* 1953) and associated protein, and give the maculae, when fresh, an opaque white appearance.

The ampullary crests and the maculae of the utricle and saccule are the special end organs concerned with equilibratory vestibular reflexes influencing the position of the eyes in relation to movements of the head through the connexions of the vestibular nerves and their nuclei, via the medial longitudinal fasciculus (p. 839), with the nuclei of the third, fourth and sixth cranial nerves and influencing the general body musculature through the vestibulospinal tracts (p. 875). (The central connexions of the vestibular division of the vestibulocochlear nerve and its intimate relationship with the vestibulocerebellum are considered on pp. 908, 916, 918, 928.) Muscle activity is also influenced by the *position* of the head, and in this case the maculae of the utricle and saccule are the end organs concerned in that the otoliths, under the influence of gravity, exert traction on the cilia of the hair cells in varying positions of the head. The maculae are therefore often referred to as organs of *static balance* (*statotonic reflexes*), whereas the ampullary crests are called organs of *kinetic balance* in that

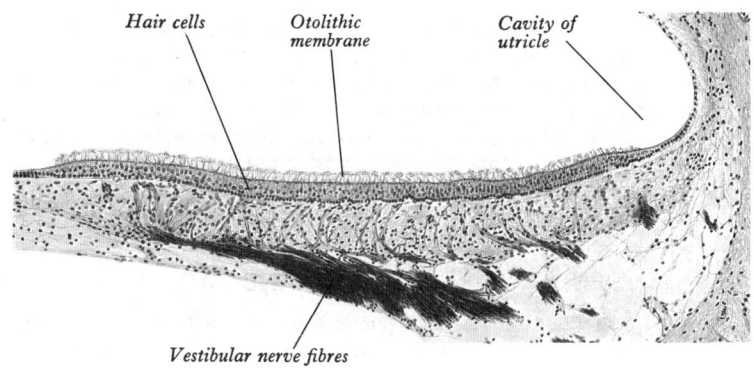

7.296A Section of the macula of the utricle of the cat. Stain—Weigert-Pal and iron haematoxylin. Magnification about ×112. (For source see 7.296B.)

they are stimulated by movement of, or pressure changes in, the endolymph caused by *angular acceleration* of the head (*statokinetic reflexes*) producing deviation of the cupulae. The maculae, however, may also be involved in signalling *linear acceleration* of the head. The terms static and kinetic as used above, therefore, are an oversimplification. The macula of the saccule, although it has the same histological structure as the macula of the utricle, is believed by some not to be concerned in vestibular reflexes but to be associated with the cochlea and concerned with the reception of slow vibrational (auditory) stimuli.

The manner in which deformation of the stereocilia, consequent upon movement of the cupulae or otolithic membrane, results in alteration of the ionic conductances of the hair cell membrane is not understood. However, it has been proposed, on the basis of electrophysiological recordings, that the majority of the vestibular nerve fibres have a steady, continuous basal discharge of afferent nerve impulses when the hair cells are receiving no mechanical stimulation. Bending of the stereocilia towards the kinocilium raises the frequency of nerve impulses, whilst bending in the opposite direction lowers the frequency. Thus, the position of the head, or its state of linear or angular acceleration, is reflected by the state of balance or relative imbalance in the impulse discharge patterns from mutually cooperative pairs of receptor sites in the right and left membranous labyrinths. For example, a horizontal swing of the head to the right results in an increased discharge of impulses from the ampulla of the right lateral semicircular duct, and a

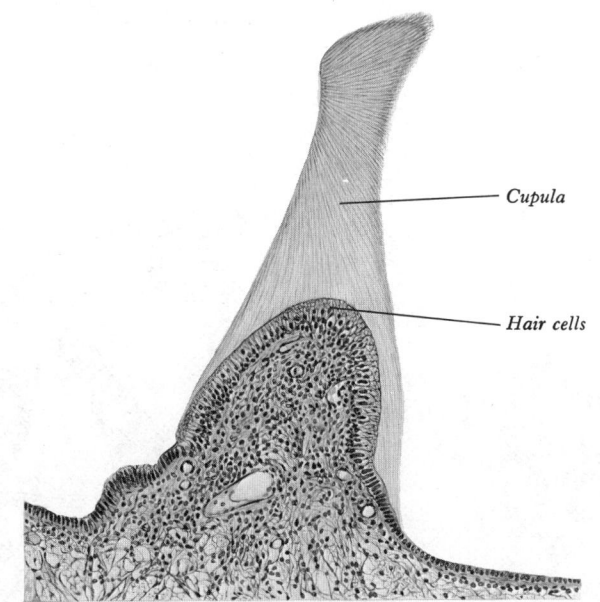

7.296B Section of an ampullary crest of a six month old human fetus. Stained with haematoxylin and eosin. Magnification about ×75. (From a section kindly lent by E. W. Walls, Professor Emeritus, Middlesex Hospital Medical School, University of London.)

decreased discharge from the left. This occurs because the inertia of the endolymph causes a relative movement of this fluid to the left in both these canals. The positioning of the kinocilia and stereocilia in the ampullae is such that it results in their compression in the right ampulla, and their decompression in the left. (For further discussion of post-rotational effects, the effects of more complex movements in other planes, and methods of testing vestibular functional efficiency, *see* Fischer 1956.)

THE ENDOLYMPHATIC DUCT AND SAC

Throughout the endolymphatic duct, its surface cells closely resemble those lining the other non-specialized parts of the membranous labyrinth. As mentioned earlier, the duct continues into a blind-ended saccus endolymphaticus which expands under the dura mater on the posterior surface of the petrous temporal bone. Here the saccus is surrounded by a well-vascularized connective tissue, and its epithelium changes to tall columnar cells of two main types. The one variety has fairly dense cytoplasm, but is otherwise unspecialized. The other type is less dense, its luminal surface bears a profusion of long microvilli, and its cytoplasm carries numerous mitochondria, pinocytotic invaginations and vesicles, and larger smooth-walled vacuoles.

The endolymph which fills the different parts of the membranous labyrinth contrasts sharply in its composition with the perilymph which surrounds it externally. Whilst the latter is roughly comparable with extracellular tissue fluid or cerebro-spinal fluid, the endolymph resembles intracellular fluid in its ionic composition, being rich in potassium ions, but poor in sodium ions. It is widely accepted that endolymph is a form of secretion, but its precise source is still an open question. The various structures considered to be involved in its production include the *dark cells* of the utricle and semicircular ducts, the columnar cells of the *planum semilunatum*, and the specialized epithelial cells and related blood vessels of the *stria vascularis* of the cochlear duct (*vide infra*). Whatever the relative contributions from these different sources, it is thought that the endolymph circulates and then enters the ductus endolymphaticus to be

removed by the specialized epithelial cells of the saccus into the surrounding vascular plexus. Pinocytotic removal of fluid in other parts of the labyrinth is, however, not excluded.

A unique positive electrical potential exists in the endolymphatic spaces. This varies from $+77$ millivolts in the cochlear duct near the stria vascularis, to about $+4$ millivolts in the utricle, whilst it is absent or even negative in the ampullae of the semicircular ducts. Thus, a very large difference of potential of some 150 millivolts exists across the cell membrane of the cochlear hair cells, between the cochlear endolymph and the cell interior. This may account in part for the extreme sensitivity to mechanical deformation shown by these *auditory* hair cells (*vide infra*).

THE COCHLEAR DUCT

The duct of the cochlea (**7.297, 298**) consists of a spirally arranged tube within the bony canal of the cochlea and lying along its outer wall.

As already stated (p. 1202), the osseous spiral lamina extends only part of the distance between the modiolus and the outer wall of the cochlea, while the *basilar membrane* stretches from the free edge of the lamina to the outer wall of the cochlea, and completes the roof of the scala tympani. The endosteum of the outer wall of the cochlea is thickened to form the *spiral ligament of the cochlea*; it projects inwards and to it is attached the outer edge of the basilar membrane. A second and more delicate *vestibular membrane*, extends from the thickened endosteum covering the osseous spiral lamina to the outer wall of the cochlea, where it is attached at some distance above the outer edge of the basilar membrane. A canal is thus shut off between the scala tympani below and the scala vestibuli above; this is the *duct of the cochlea* (**7.298**). It is triangular on transverse section, its roof being formed by the vestibular membrane, its outer wall by the endosteum lining the bony canal, and its floor by the basilar membrane and the outer part of the osseous spiral lamina. The upper extremity of the duct of the cochlea is closed, and is named the *lagaena*; it is attached to the cupola (p. 1202). The lower end turns medially, and narrows

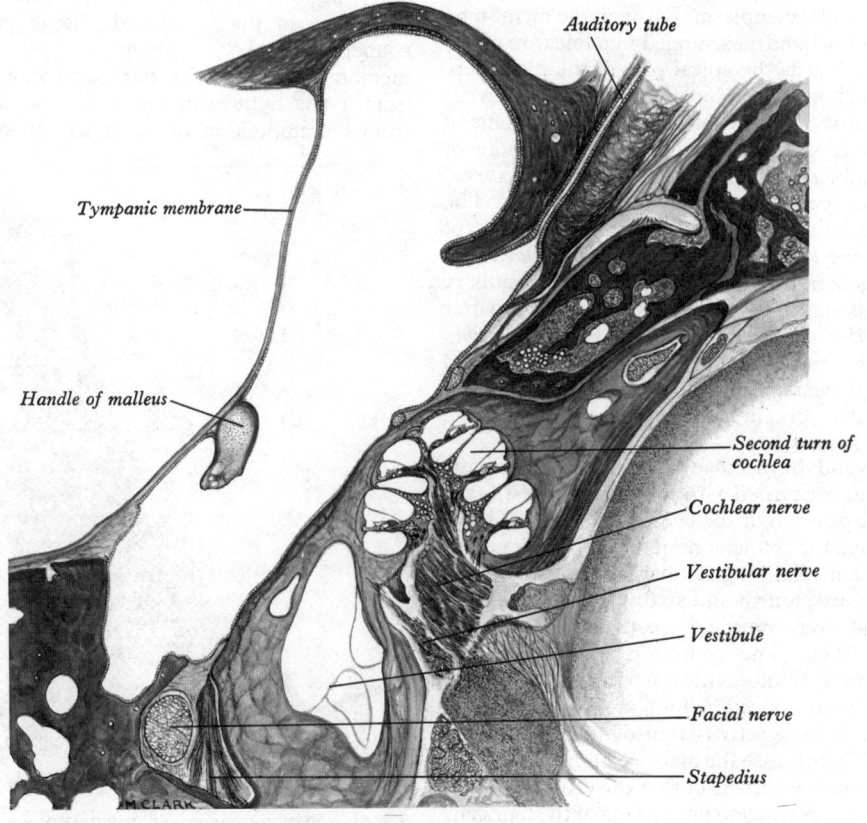

7.297   A horizontal section through the left temporal bone. (Drawn from a section prepared at the Ferens Institute and kindly lent by the late Professor J. Kirk.)

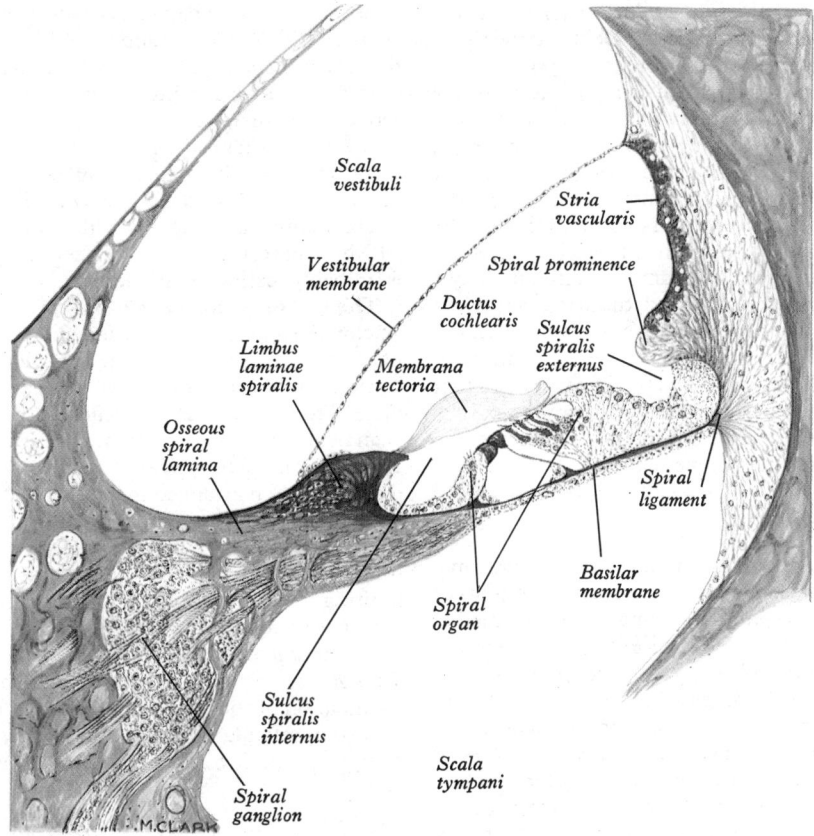

7.298A A section through the second turn of the cochlea indicated in the previous figure. The modiolus is to the left. (Mallory's stain.)

into the *ductus reuniens*, through which it communicates with the saccule (7.294). The spiral organ is situated on the basilar membrane. The vestibular membrane is thin and is covered on its two surfaces by a layer of flattened epithelium. The endosteum forming the outer wall of the duct of the cochlea is greatly thickened and forms the spiral ligament. It projects inwards, inferiorly, as a triangular prominence, termed the *crista basilaris*, to which the outer edge of the basilar membrane is fixed; immediately above this there is a concavity (the *sulcus spiralis externus*), above which the periosteum is thickened, highly vascular and forms a surface projection, the *spiral prominence*, which above this again continues into a specialized periosteal zone, the *stria vascularis*.

Ultrastructural studies have added many details concerning these different regions, including many features of the spiral organ. These can only be touched on briefly here, and the interested reader should consult Babel *et al.* (1970), who include an extensive bibliography.

THE VESTIBULAR MEMBRANE

The vestibular membrane (of Reissner) consists of two layers of flattened epithelial cells with an intervening basal lamina. The aspect facing the scala vestibuli is clothed with perilymphatic cells which are fairly thick in their central perinuclear zone, but elsewhere the cytoplasm is extremely attenuated; adjacent cells establish zonulae occludentes between them. The endolymphatic surface is covered with typical squamous epithelial cells, again joined by zonulae occludentes. Their cytoplasm contains a number of mitochondria and numerous vesicles; the basal surface is sometimes smooth but often complexly invaginated into the cell; the free surface carries numerous, short, irregular microvilli. These cells may be involved in fluid transport.

THE STRIA VASCULARIS

The stria vascularis, as noted above, lies on the outer wall of the cochlear duct immediately above the spiral eminence. It is unique in possessing a specialized stratified epithelium which carries a rich plexus of *intra-epithelial capillaries*. The epithelium consists of three cell types: (1) superficially placed *marginal, dark* or *chromophil cells*; (2) *intermediate, light*, or *chromophobe cells*; and (3) *basal cells*.

The endolymphatic surface is formed exclusively by the dark cells. The intermediate and basal cells are cytologically similar, differing merely in position. Their pale cytoplasm contains scattered mitochondria, numerous pinocytotic vesicles and a number of melanin granules. These cells send cytoplasmic processes towards the surface, where they are insinuated between and around the deeper parts of the marginal cells.

The marginal dark cells are highly specialized, with a dense granular cytoplasm containing many mitochondria and pinocytotic vesicles. The deep part of the cell consists of many long cytoplasmic processes separated by deep invaginations of the plasmalemma; each process contains a series of mitochondria. The intra-epithelial capillaries are closely enveloped both by descending processes from the dark cells and ascending processes from the intermediate and basal cells.

The stria vascularis is considered to be an ion-transporting mechanism which maintains the unique ionic composition of the endolymph. As indicated above, however, other regions of the membranous labyrinth may also be involved in this activity. It has been established by exploration of the stria with microelectrodes that it is the source of the large positive endocochlear electrical potential. The maintenance of this potential is directly and immediately dependent upon an adequate oxygenation level of the epithelial cells, provided by the blood circulation through its intra-epithelial capillaries.

THE OSSEOUS SPIRAL LAMINA

The osseous spiral lamina consists of two plates of bone, and between these are the canals for the transmission of the filaments of the cochlear nerve. On the upper plate of that part of the lamina which is contained within the duct of the cochlea the periosteum is thickened to form the *limbus laminae spiralis* (7.298, 299); this ends externally in a concavity, the *sulcus spiralis internus*, which presents, on section, the form of the letter C; the upper part formed by the overhanging edge of the limbus is the *vestibular lip* (*labium*); the lower part, prolonged and tapering, is the *tympanic lip* (*labium*), and is perforated by numerous foramina for the passage of the branches of the cochlear nerve. The upper surface of the vestibular lip is intersected at right angles by a number of furrows, separated by numerous elevations; these present the appearance of teeth on the free surface and margin of the lip, the *auditory teeth* (*dentes acoustici*) (7.299). The limbus is covered by a layer appearing superficially as a squamous epithelium, but only the cells covering the teeth are flattened, those in the furrows (7.300) being columnar, and occupying the intervals between the elevations. This epithelium is continuous on the one hand with that lining the sulcus spiralis internus, and on the other with that covering the under surface of the vestibular membrane. It is considered by some observers that the interdental cells secrete the material forming the tectorial membrane (*vide infra*).

THE BASILAR MEMBRANE

The basilar membrane (7.298, 299) stretches from the tympanic lip of the osseous spiral lamina to the crista basilaris. It consists of two zones, a thin *zona arcuata*, stretching from its medial attachment, the limbus spiralis, to the bases of the outer rods and supporting the organ of Corti, and an outer thicker part, the *zona pectinata*, commencing beneath the bases of the outer rods and attached laterally to the crista basilaris. The zona arcuata is seen by electron microscopy to be composed of compact bundles of small collagen-like filaments 8–10 nm in diameter, mainly disposed radially. In the zona pectinata the membrane is three-layered with an upper layer composed of a homogenous network of transverse fibres, a lower layer of compact bundles of longitudinal fibres and an intermediate structureless layer containing a few cells. At its attachment to the crista basilaris, the upper and lower layers fuse and the membrane consists of one layer. The length of the basilar membrane is about 35 mm; its width gradually increases from 0·21 mm in the basal turn to 0·36 mm in the apical turn of the cochlea, and this increase is accompanied by a corresponding narrowing of the osseous spiral lamina, and a decrease in the thickness of the crista basilaris. The under surface of the membrane is covered by a layer of vascular connective tissue and elongate perilymphatic cells; one of the vessels in the connective tissue is larger than the rest, and is named the *vas spirale*; it lies below Corti's tunnel (*cuniculum internum*).

THE SPIRAL ORGAN OF CORTI

The spiral organ (Smith and Dempsey 1957; Engström and Wersäll 1958; Babel *et al.* 1970; Ades and Engström 1974) is composed of a series of epithelial structures placed upon the zona arcuata of the basilar membrane (7.298–301). The more central of these structures are two rows of rod-like bodies, the *internal and external rod cells of Corti* or *pillar cells*. The bases or *foot plates* (*crura*) of the rod cells are expanded, and rest on the basilar membrane, touching each other, but apically the neighbouring rods are widely separated; the two rows incline towards each other and, coming into contact above, enclose between them and the basilar membrane the *tunnel of Corti* (*cuniculum internum*) (7.300A), which is triangular in cross-section. On the medial side of the inner rods there is a single row in *internal (inner) hair cells*, and on the lateral side of the outer rods, three or four rows of *external (outer) hair cells*, together with certain supporting cells, the *phalangeal cells of Deiters*, and the *cells of Hensen* (*cellulae limitans externae*). The free ends of the outer hair cells and the apical processes of the cells of Deiters form a highly regular mosaic of cell apices arranged collectively into the *reticular lamina or membrane* (Engström *et al.* 1966). The entire organ is covered by the tectorial membrane, a narrow gap separating this structure from the reticular lamina except where the apical stereocilia of the outer hair cells project to make contact with the tectorial membrane. In addition to the tunnel of Corti, sometimes termed the *inner tunnel* (*cuniculum internum*), described above, other spaces exist in

relation to the outer hair cells which connect through intercellular crevices with each other and with the inner tunnel. These include the *outer tunnel* (*cuniculum externum*) situated between the outermost hair cells and the inner cells of Hensen, beneath the reticular lamina, and the *space of Nuel* (*cuniculum medium*) between the outer rods of Corti and the outer hair cells. The latter is continuous with the extracellular spaces which surrounded the upper two-thirds of the outer hair cells. This complex system of intercommunicating spaces is filled with a fluid termed *cortilymph* which is not continuous with either perilymph or endolymph, and is probably distinct in its chemical constitution (Engström 1960).

The rod or pillar cells of Corti each consist of a base or *crus*, an elongated part or *scapus*, and an upper end or *caput*; each foot plate and head is closely applied to it neighbour but the bodies are separate from each other. The nucleus lies in the triangular foot plate. The body of each rod is finely striated, but in the head there is an oval nonstriated portion which stains deeply with carmine. Electron microscopy shows many microtubules 13–15 nm in diameter, often arranged in parallel bundles, running lengthwise in the body of the rods and then diverging above to terminate in a superficial layer of dense granular cytoplasm, the *cuticle*, in the head. Below, the microtubules arise over a wide area of the limiting membrane of the basal crus from the cytoplasmic densities of an array of hemidesmosomes. Within the body of the rod, transverse sections show that the microtubules are often arranged in a regular square lattice. Detailed analysis of the upper termination of the microtubules shows that many of them curve into the reticular lamina (*vide infra*) to end in junctional complexes (p. 6; **1.4**) established with either similar expansions

7.298B   A transverse section of the spiral organ of Corti (cat), stained with the Mallory trichome method to show the inner and outer hair cells (orange), the basilar membrane (dark blue) and tectorial membrane (light blue), and various supporting cells, including those surrounding the tunnel of Corti. Magnification × 400.

from the supporting cells of Deiters or with the subapical lateral surfaces of the hair cells. The nucleated cytoplasmic zones which partly envelop the rods and extend on to the floor of Corti's tunnel, occupy the angles between the rods and the basilar membrane; these may be looked upon as the less differentiated parts of the cells from which the rods have been formed.

The *inner rods* number nearly 6,000, and their bases rest on the basilar membrane close to the tympanic lip of the sulcus spiralis internus. The body of each is sinuously curved and forms an angle of about 60° with the basilar membrane. The head resembles the proximal end of the ulna, and presents a deep concavity which accommodates a convexity on the head of the outer rod. The head plate, or portion overhanging the concavity, overlaps the head plate of the outer rod.

The *outer rods*, almost 4,000 in number, are longer and more obliquely set than the inner, forming with the basilar membrane an angle of about 40°. Their heads are convex internally; they fit into the concavities on the heads of the inner rods, and are continued outwards as thin flattened plates, the *phalangeal processes*, which unite with the phalangeal processes of Deiters' cells to form the *reticular lamina or membrane* (*vide infra*).

The distances between the bases of the inner and outer rods increase from the base to the apex of the cochlea, while the angles between the rods and the basilar membrane diminish.

The hair cells are short columnar or piriform cells depending upon their site in the spiral organ; their free ends are on a level with the heads of the rods of Corti, and each is surmounted by about 50–100 hair-like stereocilia. The deep ends of the cells reach about halfway along the rods of Corti, and each contains a large open-faced euchromatic nucleus. The plasma membrane at the basal pole of the cell forms synaptic contacts with cochlear nerve fibres. The cytology and supporting structures vary with the position of the hair cells, which are arranged as inner and outer groups set at an angle to each other. The *inner hair cells*, about 3,500 in number, are arranged in a single row on the inner side of the inner rods of Corti, and since their diameters are greater than those of the rods, each is related to more than one rod. The free ends of the inner hair cells are encircled by a cuticular membrane which is fixed to the heads of the inner rods (*vide infra*). Adjoining the inner hair cells there are one or two rows of columnar supporting cells, which, in turn, are continuous with the cubical cells lining the sulcus spiralis internus. The columnar supporting cells which lie adjacent to the inner rods of Corti and surround the inner hair cells are the *inner phalangeal cells*. These are succeeded medially by slender columnar *border cells* of decreasing height which gradually merge medially into the cuboidal epithelium of the sulcus mentioned above. The *outer hair cells* number about 12,000 and are nearly twice as tall as the inner. In the basal coil of the cochlea they are arranged in three regular rows; in the apical coil, in four or five less regular rows.

Electron microscopy has added much about the detailed construction of the cochlear hair cells and their supporting structures (see for example Babel *et al.* 1970). Briefly, the **inner hair cells** resemble in some respects the type I vestibular hair cells described previously (p. 1204). Each possesses a relatively short piriform body, the expanded basal region containing a large euchromatic nucleus (7.300, 301); this is surmounted by a constricted 'neck' which on its free apical surface carries 50–60 hairs or stereocilia, but no kinocilium. The cytoplasm contains an abundance of organelles, indicative of a relatively high metabolic rate. These include numerous mitochondria which are particularly abundant in the apical parts of the cell, free polyribosome groups, agranular endoplasmic reticulum, various vesicles and a few lysosomes. Microtubules and microfilaments are also present, notably in the basal regions of the cytoplasm. Immediately beneath the apical surface of the cell there is a thick layer of filamentous and granular material, the *cuticular plate*, forming a continuous cap except at a cuticle-free gap which is present on the region of the apex nearest to the tunnel of Corti. At the lateral margins of the cell the cuticular plate is attached to the intercellular junctions with the neighbouring phalangeal and pillar cells.

The stereocilia resemble in structure those of vestibular hair cells, being very narrow at their base and expanding to a cylindrical process about 6 μm long at a maximum, and 0·2 μm wide. These structures resemble microvilli in that they are clothed by an extension of the apical plasma membrane of the hair cell, and contain numerous longitudinally directed microfilaments composed of actin. Near their bases an additional dense axial rootlet is present centrally, and this penetrates deep into the cuticular plate to act as an anchoring device. Because of their shape, the stereocilia are able to bend at their narrow basal ends when stimulated by fluid movements or other mechanical disturbances resulting from auditory stimuli arriving at the cochlea (*vide infra*). Unlike the vestibular hair cells, a kinocilium with the typical microtubular interior is not present in mature inner hair cells of the cochlea, although it is found during early development. Its place is taken within the cell by a centriole positioned beneath the apical membrane within the cuticle-free gap, probably as a non-functional vestige.

When viewed from the surface the stereocilia are seen to be arranged in the form of a shallow U with its base pointing towards the centriole, and thus also towards the tunnel of Corti. Each

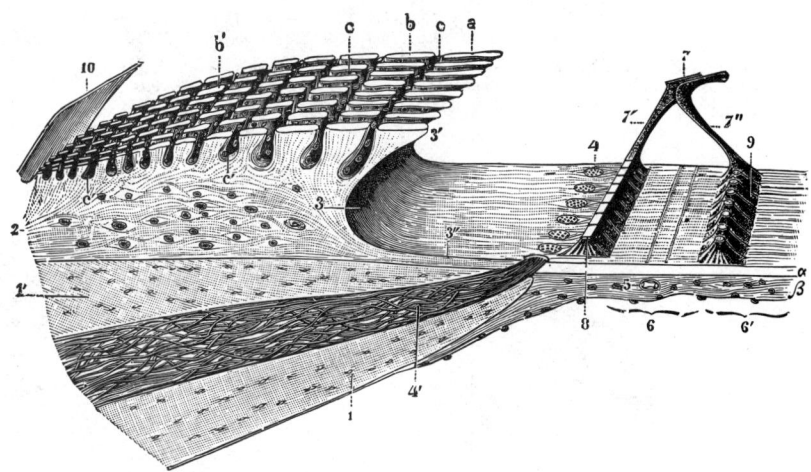

7.299 The limbus laminae spiralis and the basilar membrane. Schematic, after Testut. 1, 1′. Lower and upper lamellae of the lamina spiralis ossea. 2. Limbus laminae spiralis, with *a*, the auditory teeth of the first row; *b*, *b′*, the teeth of the other rows; *c*, *c′*, the grooves between the auditory teeth and the cells which are lodged in them. 3. Sulcus spiralis internus, with 3′, its labium vestibulare, and 3″, its labium tympanicum. 4. Foramina nervosa, giving passage to the nerves from the spiral ganglion. 5. Vas spirale. 6. Zona arcuata, and 6′, zona pectinata of the basilar membrane, with *a*, its hyaline layer, *β*, its connective tissue layer. 7. Summit of the tunnel of Corti, with 7′, its inner rod, and 7″, its outer rod. 8. Bases of the inner rods, from which the cells are removed. 9. Bases of the outer rod. 10. Part of the vestibular membrane.

cluster of stereocilia is made up of three or four rows progressively taller towards the tunnel side of the cell; it is not certain if the tips of the longest touch the tectorial membrane, there being no direct evidence for such a contact. The plasma membrane at the basal pole of the cell is in synaptic contact with two types of synaptic boutons derived from cochlear nerve fibres. The terminals of *afferent* fibres contain relatively few mitochondria and microtubules, and a number of clear vesicles of varying diameter. Both the presynaptic (hair cell) and postsynaptic (bouton) membranes are thickened, the latter more markedly, and the presynaptic cytoplasm contains aggregates of both clear and dense-cored synaptic vesicles. *Accessory synaptic structures* which take the form of electron-dense rods, rings or lamellae around which synaptic vesicles are clustered, are commonly present. The terminal boutons of *efferent* fibres, in contrast, contain a number of mitochondria, microtubules and a profusion of synaptic vesicles, the majority of which are small, spherical and clear, but some are larger with dense cores. Although often described as making synaptic contact with the base of the inner hair cells, it has now been shown that the efferent terminals never make such contacts in the cat, and only infrequently in the guinea-pig and man. Much more numerous are contacts between the efferent terminal and the lateral aspect of an *afferent fibre bouton*. Presumably activity in the efferent fibre modulates the transmission characteristics of the afferent fibre terminal. The inner hair cells are, apart from their areas of synaptic contact just described, wholly surrounded by the cytoplasm of the inner phalangeal supporting cells. The superficial rim of the latter, where it encircles the lateral margin of the apical surface of the hair cell, is joined to it by a circular zone which, in section, is seen to consist of a junctional complex (p. 7; **1**.5). Similar specialized junctions exist between the outer aspect of the rim of the phalangeal cell and expansions from the inner rod of Corti. These surface specializations make up the 'cuticular membrane' of light microscopy mentioned above.

The outer (external) hair cells are more highly specialized than the internal, though with a number of features in common. They are considerably longer cells, cylindrical in form, with a large basally placed euchromatic nucleus. Numerous mitochondria are found along the lateral walls of the cell close to the plasma membrane, and in the basal cytoplasm below the nucleus. Otherwise, the interior of the cell has far fewer organelles in comparison with the inner hair cell; the most notable structures

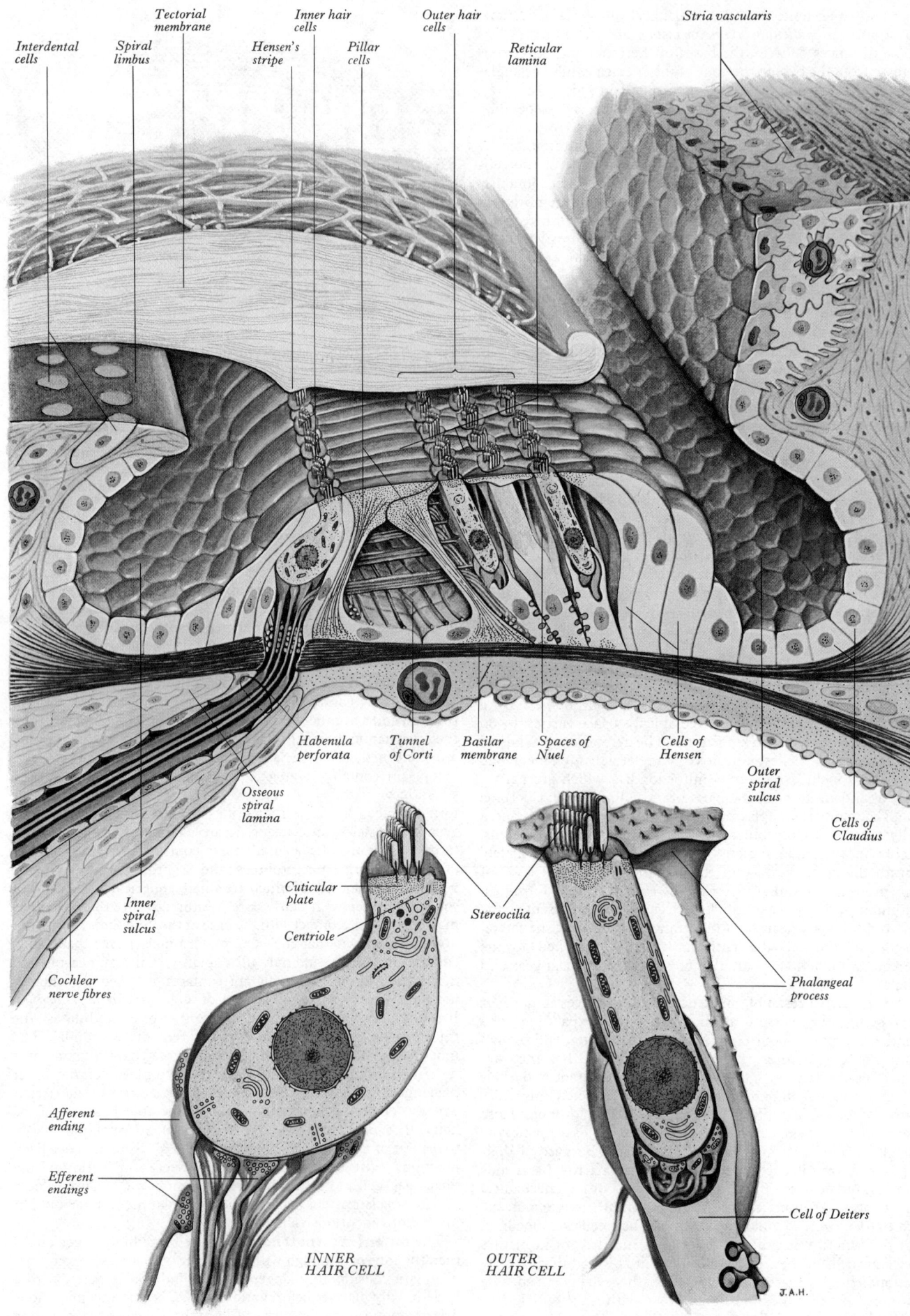

Interdental cells

Spiral limbus

Tectorial membrane

Hensen's stripe

Inner hair cells

Pillar cells

Outer hair cells

Reticular lamina

Stria vascularis

Habenula perforata

Tunnel of Corti

Basilar membrane

Spaces of Nuel

Cells of Hensen

Outer spiral sulcus

Cells of Claudius

Osseous spiral lamina

Inner spiral sulcus

Cochlear nerve fibres

Cuticular plate

Centriole

Stereocilia

Phalangeal process

Afferent ending

Efferent endings

Cell of Deiters

INNER
HAIR CELL

OUTER
HAIR CELL

J.A.H.

7.300A Three-dimensional schema of the structure of the cochlear spiral organ and stria vascularis, showing the arrangement of the various types of cell, and their overall innervation. The organization of the inner and outer hair cells, and their synaptic connexions are depicted below. Sensory nerve terminals are coloured green, and efferent fibres purple. (See text for variant terminology.)

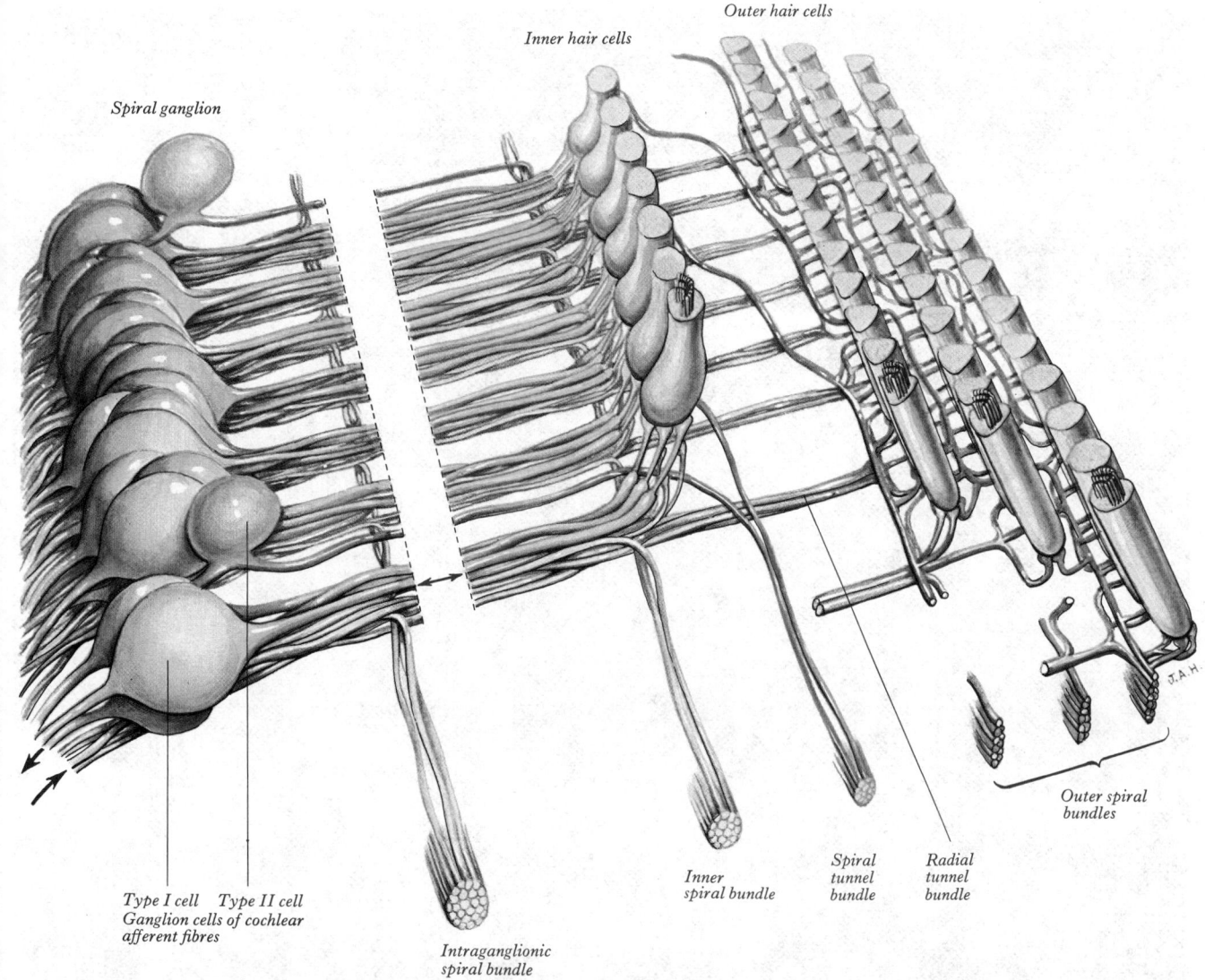

Spiral ganglion

Inner hair cells

Outer hair cells

J.A.H.

Type I cell   Type II cell
Ganglion cells of cochlear
afferent fibres

Intraganglionic
spiral bundle

Inner
spiral bundle

Spiral
tunnel
bundle

Radial
tunnel
bundle

Outer spiral
bundles

7.300B Diagram of the innervation of the spiral organ, showing the distribution of afferent and efferent fibres. The ganglion cells of the sensory nerve fibres include those related to the inner hair cells (dark green) and others innervating the outer hair cells (light green). Efferent fibres are depicted in purple. Note the great contrast between the multiple, convergent afferent innervation of the inner hair cells (about ten fibres to each cell) and the divergent supply of the outer hair cells (one afferent fibre to about ten cells).

are various complexes of flattened membranous cisternae which are attached by thin peg-like filaments to the lateral walls of the cell, or are present in the apical cytoplasm as concentric whorls (Hensen bodies) which are probably precursors of the lateral cisternae. These membranes are likely to act as strengthening elements helping to retain the cylindrical shape of the cell, although some metabolic functions cannot be ruled out. Other structures within the cell include microtubules and microfilaments, polyribosomes, vesicles, lysosomes and glycogen; however, the great majority of the cytoplasm is filled with a pale matrix containing fine filaments and granules of an unknown composition. A cuticular plate similar to that found in the inner hairs cells forms the apical cap to the cytoplasm, and serves as anchorage to the rootlets of the stereocilia; the latter structures although resembling the inner hair cell stereocilia, are slightly narrower and are arranged in the form of a V or W, the base of the letter pointing away from the tunnel of Corti and towards a centriole in the cuticle-free gap in the apex of the cell. Up to 100 stereocilia have been counted in a single group, arranged in ranks of graded length (7.301 A–C). The groups in the apical portion of the cochlea are taller than those in the base, but in all groups the longest stereocilia are embedded in the lower surface of the tectorial membrane (Hunter-Duvar 1976). Both the inner and outer hair cells are so arranged on the basilar membrane that their centrioles are on the side of the cell which is most distant from the central modiolus. This observation is of considerable functional significance and suggests that transverse shearing forces between the hair cells and tectorial membrane would be the most effective in stimulating the sensory receptors.

The basal pole of each outer hair cell is received into a cup-like depression on the upper end of an outer phalangeal cell of Deiters (*vide infra*), except at the synaptic contacts with cochlear nerve fibres. In this case, however, both the afferent and efferent terminals, which are structurally similar to those of the inner cells, are functionally related directly to the plasma membrane of the hair cell itself (Engström *et al.* 1965).

The external phalangeal cells of Deiters (7.300), are between the rows of the outer hair cells; their expanded bases are planted on the basilar membrane, while the opposite end of each presents, as mentioned above, an asymmetrical cup-like region, partially enveloping the base of a hair cell, and a finger-like *phalangeal process*, which extends up between the hair cells to the reticular membrane to form a plate-like expansion (*vide infra*). The cytoplasm of Deiters' cells contain bundles of microtubules, which arise from basally placed hemidesmosomes, and which continue upwards into the phalangeal process. Immediately to the outer side of Deiters' cells there are five or six rows of columnar cells, the *supporting cells of Hensen* or *external limiting cells* (7.300), and even farther on the outer edge, the cells of *Claudius* or *external supporting cells*. Their free surfaces are also beset with microvilli. Near the lagaena these cells contain fat globules which decrease in number and size as the duct of the cochlea is traced towards the

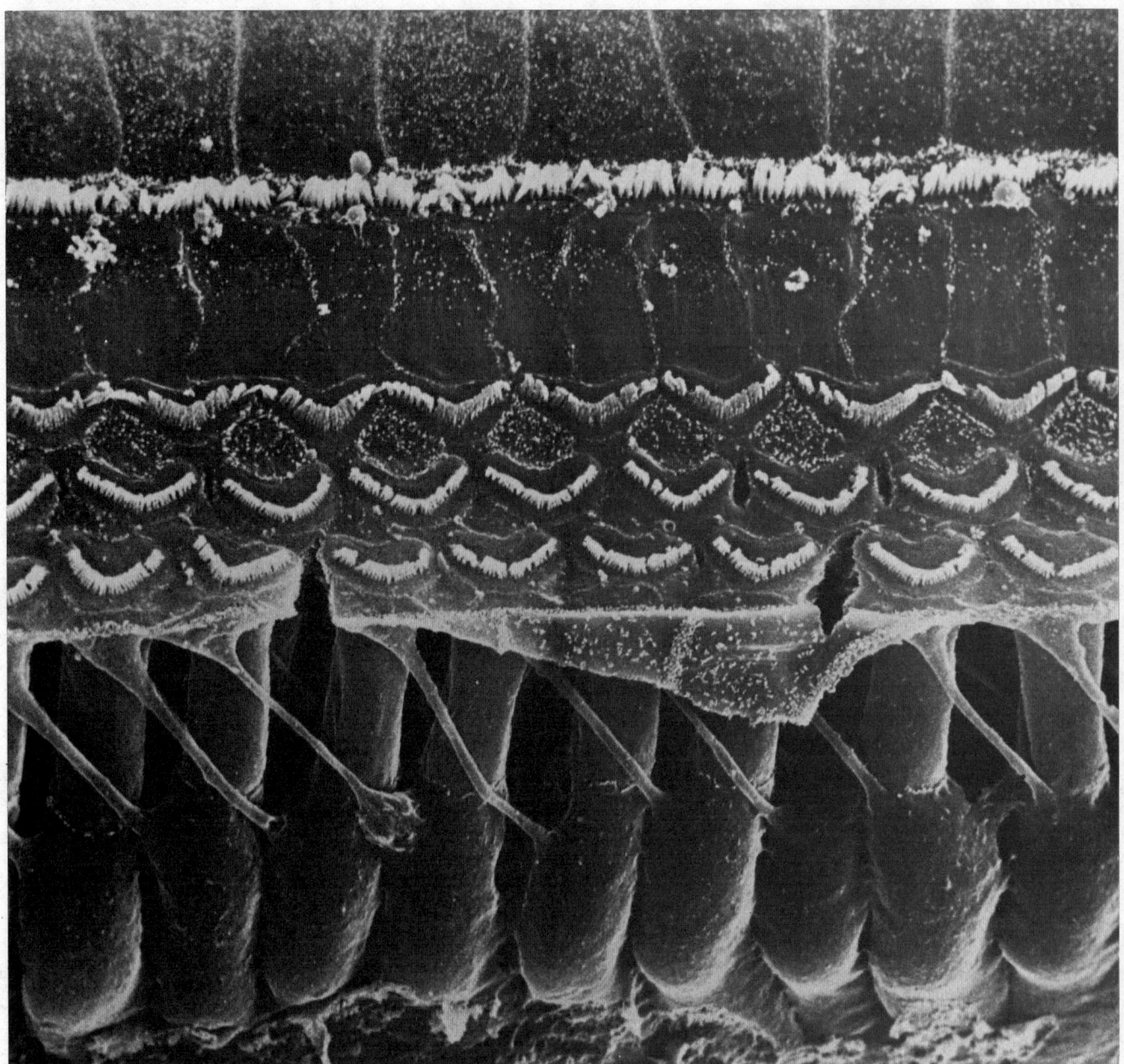

7.301A A scanning electron micrograph of a portion of the spiral organ of Corti (guinea pig) dissected to expose the outer row of outer hair cells and their attendant Deiters' cells with narrow phalangeal processes. The apices of three rows of outer hair cells with their stereocilia are visible in the foreground, and behind them are the apices of the rod cells and a row of inner hair cells and their stereocilia. Magnification × 2,500.

basal coil. It has been suggested (Hallpike 1931) that these globules provide a graduated loading mechanism, which tunes the region of the lagaena to low tones.

The reticular lamina (7.301), when viewed with a light microscope, appears as a delicate framework perforated by circular holes which are occupied by the free ends of the outer hair cells. It extends from the heads of the outer rods of Corti to the external row of the outer hair cells, and is formed by several rows of minute violin-shaped cuticular structures, called *phalanges*, between which are circular apertures containing the free ends of the hair cells. The innermost row of phalanges consists of the phalangeal processes of the outer rods of Corti; the outer rows are formed by the modified free ends of Deiters' cells. Electron microscopy has now shown that the reticular lamina consists of these horizontal expansions from the outer rods and phalangeal cells which carry bundles of microtubules and tonofibrils in the attenuated veil of dense cytoplasm which they contain. The expansions encircle the uppermost rim of the hair cells, where junctional complexes (occluding zones and desmosomes) are formed, and in which the microtubules end. Beneath this delicate supporting system, the upper two-thirds of the lateral surfaces of the outer hair cells are not in contact with supporting cells, but are bathed by the fluid termed cortilymph (*vide supra*).

The reticular lamina is of great significance in auditory stimulation of the sensory cells because it forms a rigid plate capable of transmitting lateral shearing forces generated by the movements of the basal lamina to the hair cells (*vide infra*). It is also interesting that if any hair cells are lost because of acoustic or drug-induced trauma, the phalangeal processes close together to fill the gap, leaving a disturbance in the regular pattern of the lamina (a *phalangeal scar*).

The membrana tectoria (7.300) overlies the sulcus spiralis internus and the spiral organ of Corti. It consists, in fixed preparations, of delicate fibres embedded in a jelly-like matrix (Iurato 1967; Steel 1978). In electron microscopic preparations filaments of 4 mm diameter are seen; they consist of protein which resembles keratin, associated with mucopolysaccharides. The membrana tectoria is wider and thicker in the apical than in the basal part of the cochlea. Its inner part is thin and is attached to the vestibular lip of the limbus laminae spiralis, the attachment reaching as far as the vestibular membrane. The outer part is thick and padlike, the thickness being greatest over, or slightly to the

7.301B A scanning electron micrograph of a group of outer hair cells arranged in three rows, showing the arrangement of their stereocilia and related phalangeal processes of the Deiters' cells, collectively forming the reticular lamina (see text). Short microvilli are visible on the surfaces of the non-sensory cells. Magnification × 3,000.

inner side of, the upper ends of the rods of Corti. Scanning electron microscopy has shown that a network of fine ridges is present on the upper surface, but the lower surface is relatively smooth, except where the tips of the outer hair cell cilia leave a pattern of W- or V-shaped impressions. Another feature seen in a number of species is a longitudinal ridge running along the under surface near the tunnel of Corti; this structure is visible by light microscopy as *Hensen's stripe*. The *interdental cells* of the vestibular lip to which the tectorial membrane is attached have a

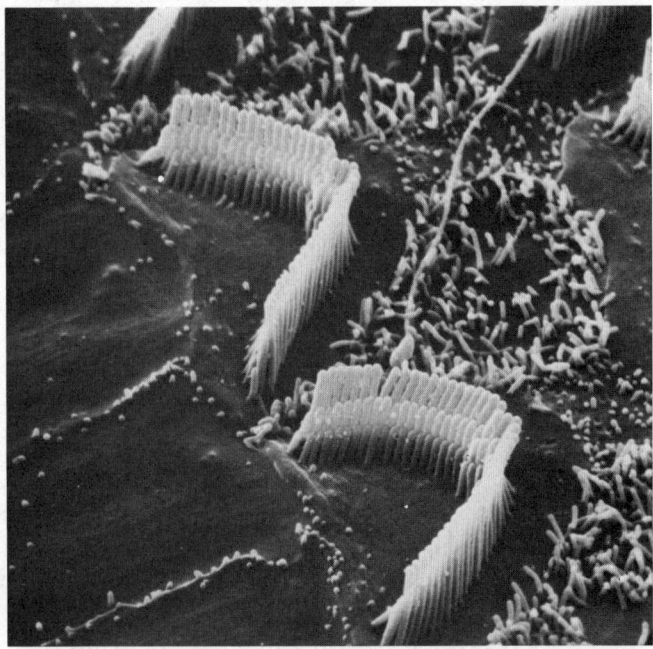

7.301C A scanning electron micrograph of the apices of two outer hair cells showing the different lengths of the stereocilia in the three ranks which constitute each group. Microvilli are also seen on the surface of the Deiters' cells (right). The innermost row of outer hair cells is shown, the tunnel of Corti, roofed by rod cell processes, is on the left. Magnification × 5,000. (Figs 7.301A–F) kindly supplied by Dr. Hilary C. Dodson and photographed by Mr. Michael Crowder, Guy's Hospital Medical School, London.)

well-developed Golgi apparatus, numerous mitochondria, and free polysomes; it is thought that they secrete the membrane. The tips of the outer hair cell stereocilia are embedded in the tectorial membrane, but this attachment is often broken during histological preparation (7.300).

The vestibulocochlear nerve divides near the lateral end of the internal acoustic meatus into an anterior or *cochlear*, and a posterior or *vestibular* component. (The central connexions and proximal parts of these nerves are described on pp. 1073, 1074.)

The vestibular nerve supplies the utricle, the saccule and the ampullae of the semicircular ducts. The *vestibular ganglion*, from the bipolar nerve cells of which the fibres of the nerve take origin, is situated in the trunk of the nerve within the internal acoustic meatus. On the distal side of the ganglion the nerve divides into superior, inferior and posterior vestibular branches. (The nerve sometimes divides on the proximal side of the ganglion, which is then also divided into three parts, one in each branch of the nerve. When this occurs, the ganglion of the posterior division is placed in the foramen singulare.) The filaments of the *superior branch* are transmitted through the foramina in the superior vestibular area, and end in the macula of the utricle and in the ampullary crests of the anterior and lateral semicircular ducts; those of the *inferior branch* traverse the foramina in the inferior vestibular area, and

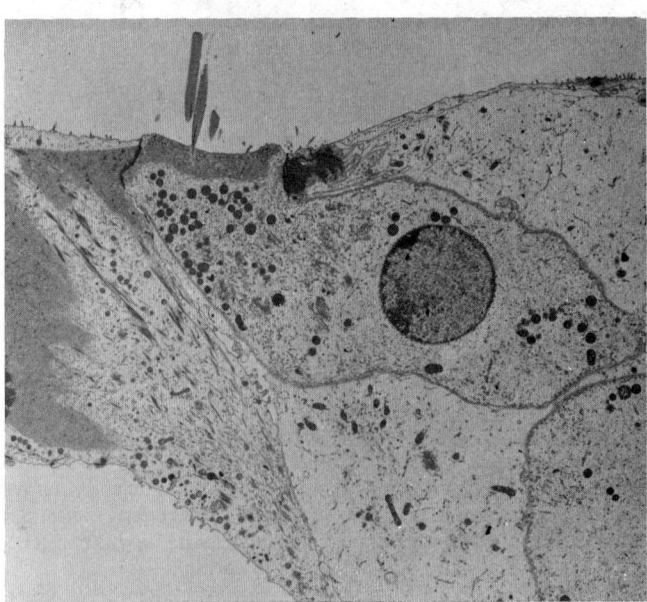

7.301D An electron micrograph of an inner hair cell (guinea pig) showing the apical stereocilia with bases embedded in the dense cuticular plate. Note the centrally placed nucleus and numerous cytoplasmic organelles. The apex of a rod cell is visible on the left, and below it the cavity of the tunnel of Corti. Magnification × 3,000.

end in the macula of the saccule. The *posterior branch* runs through the foramen singulare at the postero-inferior part of the bottom of the meatus (7.302) and divides into filaments for the supply of the ampullary crest of the posterior semicircular duct.

The cochlear nerve, the nerve of hearing, divides into numerous filaments at the base of the modiolus; those for the basal and middle coils pass through the foramina in the tractus spiralis foraminosus, those for the apical coil through the central canal, and the nerves bend outwards and pass between the lamellae of the osseous spiral lamina. The *spiral ganglion* (7.298), consisting of bipolar nerve cells from which the fibres of the nerve take origin, occupies the spiral canal of the modiolus. Reaching the outer edge of the osseous spiral lamina, the nerve fibres lose their myelin sheaths and pass in bundles through the foramina in the tympanic lip (the *habenulae perforatae*); some end by arborizing around the deep ends of the inner hair cells, while others pass between the rods of Corti and across the tunnel of Corti, and end in relation to the outer hair cells. The latter pass between the cells of Deiters and are often enfolded by them. The hair cells in the basal and middle coils are more richly supplied with nerves than

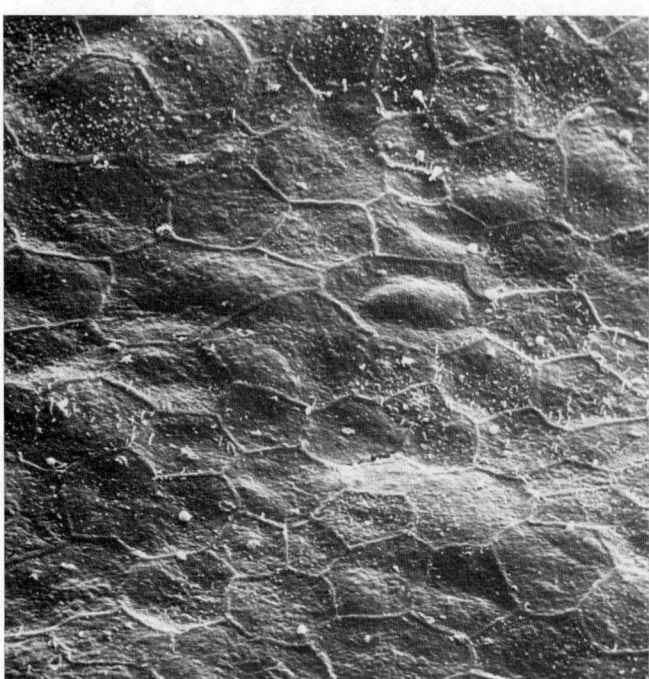

7.301E A scanning electron micrograph of the surface of Reissner's membrane, viewed from the cochlear duct aspect, showing the regular pattern of simple squamous epithelial cells. Magnification ×2,500.

those in the apical coil. The cochlear nerve gives off a vestibular branch to supply the vestibular end of the duct of the cochlea; the filaments of this branch traverse the foramina in the cochlear recess (p. 1201). The cells of the spiral ganglion are also of considerable interest. Two classes (Types I and II) have been described (Spoendlin 1974), the Type I cells being large bipolar elements ensheathed by myelin and giving off large myelinated axons both centrally and peripherally. The Type II cells are smaller and non-myelinated and, oddly, appear to possess only a peripherally directed, non-myelinated process. Various experimental studies have shown that the Type I cells provide the afferent innervation of the inner hair cells, ten ganglion cells being connected to each sensory cell, whereas the Type II ganglion cells are afferent to the outer hair cells, each supplying more than ten sensory cells by extensive branching (Spoendlin 1978).

The structure of the terminal boutons of the efferent and afferent nerve fibres has already been described. The efferent nerve fibres belong to the olivocochlear system first described by Rasmussen (1946). It is now established that this bundle contains both crossed and uncrossed components; the former start from the opposite retrolateral olivary group of neurons, whilst the latter arise from the so-called S-shaped segment of the ipsilateral olivary complex (Whitfield 1967; Iurato 1974). The pharmacology of the pathway is obscure, but it is apparently purely inhibitory in its actions, and alters the threshold level of the outer hair cells, and modifies the transmission through the afferent fibres from the inner hair cells.

The precise distribution of nerve fibres to the inner and outer rows of hair cells is now known in considerable detail. Briefly, the vast majority of the cochlear nerve fibres are distributed to the inner hair cells. The latter each receive the terminals of a number of radially disposed afferent fibres, which are themselves the sites of synaptic terminals from collateral branches of radially disposed inhibitory efferent fibres. The outer hair cells receive the minority of the cochlear nerve fibres. Their efferent fibres are radial in disposition and each fibre establishes inhibitory synapses with a large number of hair cells. The afferent fibres curve into a geometrically organized spiral system of fibres, each of which innervates numerous hair cells. These various fibres form distinct groups with specific orientations and positions within the cochlea, running either radially, or spirally, i.e. parallel to the organ of Corti (Spoendlin 1974). Amongst the different cellular structures of the cochlea lie, firstly, the spiral intraganglionic bundle within

the modiolus consisting of non-myelinated fibres thought to be efferent axons; a similar group of efferent *inner spiral fibres* runs immediately beneath the row of inner hair cells, this bundle being pierced by the groups of axons emerging from the habenulae. More peripherally, the *spiral tunnel fibres* lie in the tunnel of Corti running along this space in bundles or as scattered single axons. Individual fibres also run longitudinally in the spaces between the outer pillar cells and the Deiters' cells, forming three distinct rows flanked and partially invaginated by these different supporting cells; these ranks of fibres collectively form the *outer spiral fibres*, and they consist of both afferent and efferent connexions of the outer hair cells, as do the spiral tunnel fibres already mentioned. Autonomic endings have also been observed, but these appear to be restricted to regions of the spiral organ on the modiolar side of the tunnel of Corti. Superimposed on this spiral pattern are the radial fibres afferent to the inner hair cells, and, crossing the space within the tunnel of Corti, to the outer hair cells. A more recent and detailed analysis of cochlear innervation is summarized in illustration 7.300.

The arteries of the labyrinth are: (1) the labyrinthine artery (p. 696), which may arise from the basilar artery, but is more often derived from the anterior inferior cerebellar artery; and (2) the stylomastoid branch of the occipital or posterior auricular artery. The labyrinthine artery divides at the bottom of the internal meatus into cochlear and vestibular rami. The cochlear branch subdivides into twelve to fourteen twigs, which traverse the canals in the modiolus, and are distributed, in the form of a capillary network, in the lamina spiralis and basilar membrane. The vestibular branches are distributed to the utricle, saccule and semicircular ducts.

The veins of the vestibule and semicircular canals accompany the arteries, and, receiving the veins of the cochlea at the base of the modiolus, unite to form the labyrinthine vein, which ends in the posterior part of the superior petrosal sinus or in the transverse sinus. A small vein, from the basal turn of the cochlea, traverses the cochlear canaliculus and joins the internal jugular vein.

THE MECHANISM OF THE AUDITORY RECEPTORS

Many elegant researches have been directed towards an understanding of the role of the different components of the ears, as analysers of the intensity and frequency patterns, and source location, of the sound wave trains which impinge upon them.

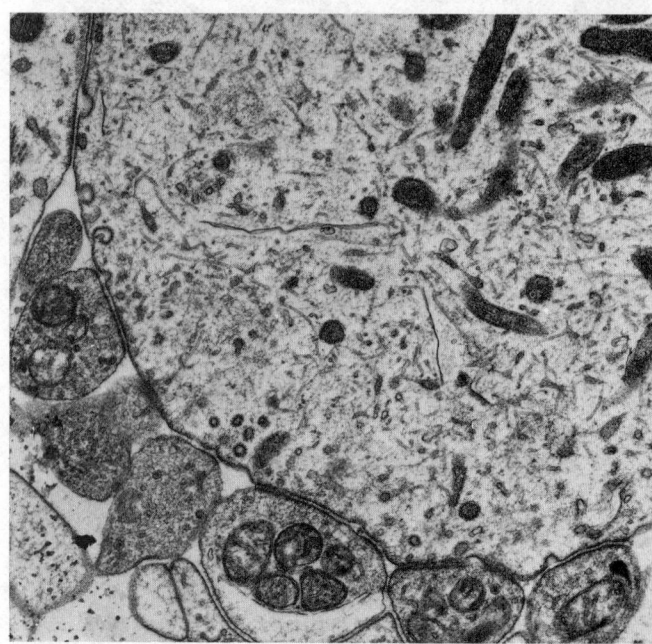

7.301F An electron micrograph of a group of efferent nerve endings at the base of an outer hair cell (guinea pig). Note the numerous mitochondria and vesicles within these endings, and the cytoplasmic densities on both the pre- and post-synaptic membranes ×10,000.

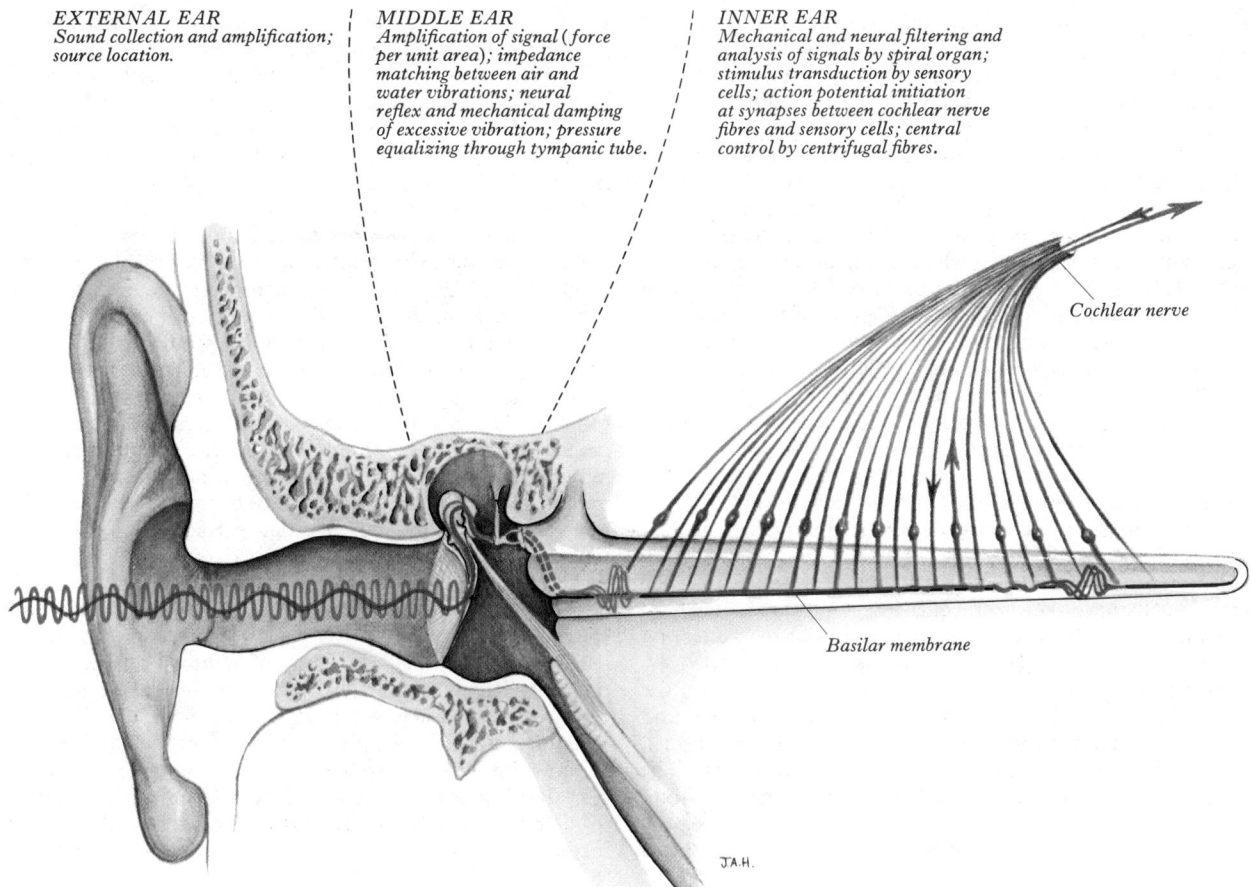

EXTERNAL EAR
Sound collection and amplification; source location.

MIDDLE EAR
Amplification of signal (force per unit area); impedance matching between air and water vibrations; neural reflex and mechanical damping of excessive vibration; pressure equalizing through tympanic tube.

INNER EAR
Mechanical and neural filtering and analysis of signals by spiral organ; stimulus transduction by sensory cells; action potential initiation at synapses between cochlear nerve fibres and sensory cells; central control by centrifugal fibres.

Cochlear nerve

Basilar membrane

7.302 Diagram illustrating the principal activities of the peripheral auditory apparatus. For clarity the cochlea is depicted as though it had been uncoiled. The points of maximal stimulation of the basilar membrane by high frequency (blue) and low frequency (red) vibrations, together with their transmission pathways through the external and middle ear, are also indicated.

These are beyond the scope of a general textbook of anatomy, and monographs and original papers should be consulted.

In outline, sound waves which reach the air column in the external acoustic meatus cause a comparable set of vibrations in the tympanic membrane, and thus in the chain of auditory ossicles. Similar vibrations occur at the foot plate of the stapes, but here the force per unit area of the oscillating surface is increased some twentyfold. These are effective in overcoming the inertia of the perilymph thus producing pressure waves within it, which are conducted almost instantaneously to all parts of the basilar membrane. The latter varies continuously in its width, mass, and stiffness from the basal to the apical end of the cochlea, but its component fibres are *not* under tension. The behaviour of such a mechanical system, when exposed to a periodic oscillating pressure wave in the neighbouring perilymph, varies with the frequency of the oscillations. At very low frequencies, for example 50 Hz, the whole basilar membrane vibrates in phase, and at a similar frequency. As the frequency of the driving fluid pressure waves rises, the different parts of the basilar membrane oscillate less rapidly, and increasingly out of phase, from the basal to the apical end of the cochlea. Consequently, a series of *travelling waves* progresses along the membrane from the base towards the apex. At intermediate frequencies, the *amplitude* of these travelling waves rises slowly as they progress from the basal end, until a maximum amplitude is reached, after which point the amplitude falls rapidly on the apical side of the maximum. With increasing frequency of the driving oscillation, the *position* of the *maximum amplitude* of vibration of the basilar membrane moves progressively from the apical, towards the basal end of the cochlea. The evidence concerning such a distribution of vibration patterns in the basilar membrane stems from three main sources: (1) observation of the excited membrane in cadavers, through drill holes in the bone, using stroboscopic illumination, (2) electrophysiological recording of the *cochlear microphonic potential*, which summates the total electrical activity over short lengths of the hair cell-bearing basilar membrane; and (3) the effects of focal destructive lesions at different points along the membrane.

The vibration pattern of the basilar membrane thus varies with the intensity and frequency of the sound wave train reaching the perilymph. Because of their position and attachment, such oscillations cause a mainly transverse shearing force to be generated between the hair cells carried by the basilar membrane, and the overlying tectorial membrane in which the tips of the stereocilia are embedded. (However, *see* below.) Since at most intermediate frequencies an appreciable length of the basilar membrane is oscillating, it follows that a specific but substantial population of the auditory nerve fibres is active, and it appears that frequency discrimination depends largely upon the topographic limits of the strip of membrane from which these fibres arise. With a just perceptible change in frequency, there is a small shift in these limits, but it must be appreciated that the majority of the nerve fibres active will be common to the two situations. Further, it is considered that at a particular frequency, an increase in the *intensity* of the stimulus is signalled by both an increased *number* of cochlear nerve fibres active, and also by their *rate of discharge* of impulses. For a discussion of these topics consult Tasaki *et al.* (1960); Békésy (1960); Spoendlin (1968); Whitfield (1967). Pye (1979) has studied the dimensional characteristics of the cochlea in most mammalian orders, including primates, and has in particular equated basilar membrane dimensions with the auditory range recorded in many species. However, her results, and others quoted by her, show that there is no simple relation between cochlear spiralling and basilar membrane width, and the frequency range of any particular mammalian group.

The precise roles of the outer and inner hair cells have been much debated in recent years, particularly since it has been realized that the innervation of these two sets of cells is so different. Because of the richness of their afferent supply it has been proposed that the inner hair cells are the major source of auditory sensation, and this view is supported by much experimental and clinical evidence; for example, animals treated with ototoxic antibiotics to cause specific loss of outer hair cells are still able to hear, although their ability to discriminate between two closely similar sounds is impaired (Harrison and Evans 1979). It is therefore feasible that the mechanism of tone discrimination is twofold, a rather broad detection of frequency of the sound waves being first given by the filtering action of the basilar membrane (*vide supra*) which stimulates maximally the inner hair cells of a particular region of the cochlea corresponding to that frequency; and secondly, a sharpening of the 'tuning' of these cells by the activity of the outer hair cells which appear to be able to inhibit the activity of inner hair cells on either side of those responding to the precise wavelength of the auditory stimulus. This type of contrast-sharpening is of course common in sensory systems where lateral inhibition of sensory cells or second order neurons is often achieved by a second set of neural elements. However, the means by which the outer hair cells are able to carry out their functions is by no means clear, particularly as no definite connection between the outer and inner hair cells has yet been demonstrated (Spoendlin 1978). Another mystery is that the outer hair cells have their stereocilia inserted into the tectorial membrane and are thus well arranged to sense lateral shearing between that structure and the reticular lamina caused by movements of the basilar membrane; indeed a large part of the electrical signal detectable with an electrode placed on the round window stems from the outer hair cells (Evans 1974). In contrast, the inner hair cells lack this attachment, and would appear to be stimulated by some other agent, perhaps the movements of fluid flowing between the tectorial membrane and the underlying spiral organ, an arrangement which perhaps allows a wider range of stimulus strength to be detected, but which needs the more precise tuning provided by the outer hair cells.

The electrical responses of the cochlea are still rather obscure in their significance, although many of them are readily recorded with relatively crude extracellular electrode techniques (Evans 1974; Yost and Nielsen 1977). The most significant of these is the *endolymphatic potential*, a steady electrical potential recordable between the cochlear duct and the scala tympani, caused by the different ionic compositions of the fluids in these two compartments. Since the resting potential of the hair cells is about 70 mV (negative within) and the endolymphatic potential is positive within the cochlear duct, the total transmembrane potential across the apices of the hair cells is 150 mV, a higher resting potential than that found anywhere else in the body.

When stimulated by sound vibrations as described above, a rapid oscillatory potential, the *cochlear microphonic* is recorded, exactly matching the frequency of the stimulus and of the movements of the basilar membrane. This appears to depend upon some change in the resistance of the hair cell membranes, which probably induces another extracellular potential, the *summating potential*, a steady direct current shift related to the receptor potentials of the hair cells. The cochlear nerve fibres then begin to respond with *action potentials* also recordable from the cochlea.

Recent intracellular recording of auditory responses of inner hair cells (Russell and Sellick 1977) have shown that these sensory cells resemble other receptors in their activities, their steady receptor potentials being related in size to the amplitude of the sound vibrations. Likewise, the afferent nerve fibres are stimulated by synaptic action at the bases of the inner hair cells, firing more rapidly as the vibration of the basilar membrane increases in amplitude.

THE INTEGUMENT

The integument or skin is an anatomically and physiologically specialized boundary lamina which is of major importance in the life of the individual. It forms the entire external surface and is continuous with the mucosal surfaces of the respiratory, alimentary and urogenital tracts at their respective orifices, where the modified skin of the muco-cutaneous junctions occurs. It also lines the external auditory meatus, covers the lateral aspect of the tympanic membrane, and is continuous with the conjunctiva at the margins of the eyelids and with the lining of the lacrimal canaliculi at the lacrimal puncta.

The skin is adapted to serve many different roles, since it is the major interface between the body and its environment (Montagna 1962). It minimizes, within limits, the potentially injurious effects of mechanical, osmotic, chemical, thermal, and photic environmental stresses; it provides a barrier to invasion by microorganisms, and it limits and regulates the exchange of heat with the environment by special neurovascular mechanisms coupled with thermally insulating properties of layers deep to the skin. The skin is a major sensory surface, containing the receptive fields of a variety of somatic sensory nerve endings; it is capable of limited absorption and excretion, and provides a surface for the conversion of precursor compounds into vitamin D by the action of ultraviolet light; it possesses at its surface good frictional properties, enhancing locomotion and manipulation by its texture and physical structure. Skin also provides a major pathway for social communication, by virtue of vascular responses associated with signalling of emotional states, and muscular responses of expression, creating a complex sign language, and by the equally subtle possibilities of tactile communication. In many primates the skin forms a signalling system related to complicated instinctive behaviour patterns which regulate aggressive, sexual, and other responses of the social group; changes in skin colour and shape in certain areas—the maxillary regions and buttocks for example—often parallel internal changes in hormonal levels, aggressive states and other factors affecting the behaviour of the individual. How far the signals mediated in man by facial expression and other alterations in skin appearance are linked to instinctive behaviour patterns rather than to conditioned behaviour is debatable in adults, although studies on children indicate the fusion of these two elements in a complex fashion.

The Structure of the Skin

The skin is composed of two layers of distinctive structure, properties and embryological origin—the *dermis* or *corium*, a connective tissue layer of mesenchymal origin, which is covered by the *epidermis*, an epithelial layer derived from embryonic ectoderm (7.304). Deep to the dermis lies a layer of loose irregular connective tissue, forming the *superficial fascia*, *hypodermis* or *subcutaneous layer* which in turn is bound to the underlying tissues by a dense fibrous *deep fascia* corresponding to the *epimysium* of muscle blocks, or where a bony or cartilaginous surface is adjacent, to the *periosteum* or *perichondrium*.

The interface between the epidermis and the dermis shows a complex topology, being marked by peg-and-socket or ridge/groove interdigitations between the two (7.304). These arrangements, together with the special anchoring structures linking the epidermis and the dermis, prevent the epithelium from being stripped off the surface of the dermis by shearing forces.

Each of the two layers confers special properties on the skin as a whole. The primary barrier to mechanical damage, desiccation and microbial invasion is the epidermis, particularly the outer horny layers which are highly impermeable to water and chemically rather inert. The epidermis, moreover, has a high capacity for regeneration after damage, and is continually replacing its outer dead cells as they are abraded from the surface. The epidermis also generates the *appendages of the skin*, that is, hairs, nails, sudorific and sebaceous glands. The dermis, in contrast, gives the skin considerable mechanical strength by virtue of the high proportion of collagen fibres intermingled with

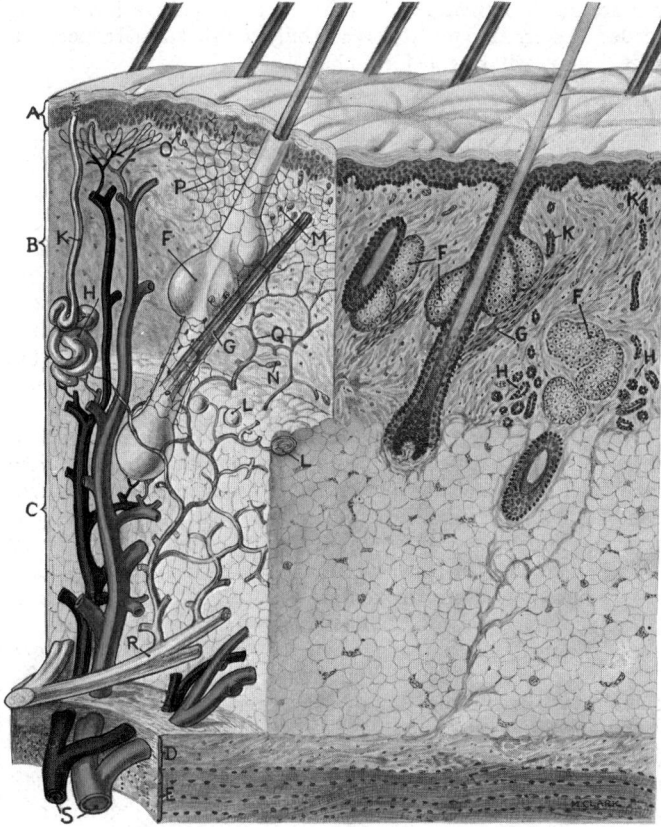

7.303 A diagrammatic scheme showing the structure of the skin. A. Epidermis. B. Dermis. C. Subcutaneous fat. D. Deep fascia. E. Muscle. F. Sebaceous glands in association with a hair follicle. G. Arrector pili muscle. H. Sweat gland. K. Duct of sweat gland. L. Bulbous corpuscle. M. Lamellated end bulb. N. Pressure corpuscle. O. Tactile corpuscle in a papilla. P. Superficial nerve plexus. Q. Deep nerve plexus. R. Cutaneous nerve. S. Cutaneous vessels.

fibres of elastin, and in the various cellular components, provides a reservoir of defensive and regenerative elements capable of combating infection and repairing deep wounds. The vascular supply of the skin is limited entirely to the dermis, and therefore the epidermis relies for its supply of nutrients and metabolic exchange generally on diffusion to and from the capillaries of the most superficial regions of the dermis. The innervation of the skin, however, serves both dermis and epidermis.

Deep to the dermis lie the subcutaneous connective tissue layers of the superficial fascia. In many regions of the body the latter is composed of loosely interwoven irregular connective tissue in the meshes of which lie numerous adipose cells, forming the *panniculus adiposus*. In addition to its mechanical properties as a shock absorber, and its energy-storing capacity, this layer is an important thermal insulator, limiting the flow of heat largely to the channels of the vascular system and thus making thermal regulation by vascular changes possible.

THE EPIDERMIS

The external surface of the epidermis is marked by various furrows, ridges and other irregularities (Montagna and Lobitz 1964). Three principal varieties of surface markings exist (7.305–307): tension and flexure lines and papillary ridges. *Tension lines* form a network of linear furrows of variable size which divide the surface into a large number of polygonal or lozenge-shaped zones. In many areas, such as the back of the hand, they are faint and intersect one another at various angles. Although these lines correspond to variations in the pattern of the fibres in the dermis to some extent, they bear no special relationship to the dermal papillae. *Flexure lines* (*skin joints*) correspond to folds in the dermis associated with habitual joint movements, and to lines of attachment to the underlying deep fascia; they are conspicuous opposite the flexure of joints,

particularly on the surfaces of the palms, soles and digits. *Papillary ridges* (*friction ridges*) are confined to the palmar surface of the hands and to the soles of the feet, including their digits, where they form narrow raised ridges separated by fine but distinct parallel grooves, disposed in curved arrays; the precise forms and positions of these ridges are related to the arrangement and size of the underlying dermal papillae interlocking with the base of the epidermis. In thin, hairy skin, the papillae are simple peg-like structures, scattered randomly over the dermal surface; in thick glabrous (non-hairy) skin, the dermal papillae often show complex forms, are large, and are arranged in rows, a single row occupying the position underneath an epidermal papillary ridge. Sweat glands which open in a row along the midline of each ridge grow down during development into the gaps between papillae, dimpling the basal surface of the epithelium to form secondary epidermal pegs (*rete pegs*) or ridges which protrude into the dermis (7.304A). This complex interdigitation no doubt enhances the stability of the junction between the two layers.

The pattern of papillary ridges, and particularly those of the fingers and thumbs are of considerable interest because of their morphological stability throughout life and also because they are slightly different in each individual, as first recognized by Faulds in 1880. The use of fingerprints for purposes of identification is well known and has led to an intensive study of their variation in human populations (the subject of *dermatoglyphics*; *see* Cummins and Midlo 1961; Cummins 1971). It has become clear that only three major patterns occur in human fingerprints (7.307), consisting respectively of loops, whorls and arches, although various combinations of these, and minor pattern variations in orientation, distortion, ridge width and number, continuity, branching and anastomosing provide the enormous pool of variation which makes their precise form unique to the individual. These three major patterns, and many of the minor features, are determined genetically by multifactorial inheritance along Mendelian lines, although prenatal disturbances of metabolism may also alter their character to some extent. Certain genetic disorders may also be reflected in fingerprint and palmar markings, as, in addition they are in the pattern of flexure lines in some cases, so that these markings may form a useful diagnostic tool in certain circumstances (Valentine 1966; Holt 1968).

THE MICROSCOPIC STRUCTURE OF THE EPIDERMIS

Histologically, the epidermis is composed of keratinizing stratified squamous epithelium (p. 41). In such a tissue, two distinct cellular processes are evident: the replacement of cells continually being lost from the surface, by the mitosis of deeper layers; and the transformation of polygonal living cells to dead, flattened, scale-like structures, filled with increasing amounts of the protein keratin, as they age and are removed towards the epithelial surface by the multiplication of the cells beneath them.

Considerable variation occurs over the body surface in the rate and extent of these two processes, so that in some regions, for example the eyelids, lips and other mucocutaneous junctions, the keratinizing layers and indeed the whole epithelium are extremely thin, whilst in others, particularly the palmar and plantar surfaces of the hands and feet, the keratinized layer may be 2 mm or more thick (7.304). In general the epidermis of flexor surfaces of the limbs (apart from the above exceptions) is more delicate than that of extensor surfaces. Because of important differences in the structure of the epidermal layers in various areas, it is customary to distinguish between the epidermis of *thin, hairy skin*, with hairs and sebaceous glands, which covers most of the body surface, and *thick, glabrous skin*, lacking hairs and sebaceous glands, which is restricted to the palms of the hands, the soles of the feet and the flexor surfaces of the digits.

The degree of keratinization is largely under genetic control, since the overall distribution of heavily and lightly keratinized areas is determined before birth. However, intermittent mechanical pressure or abrasion stimulates the adventitious production of keratinized cells, a response of the epidermis which considerably enhances the stress-resisting properties of skin (Rogers 1969).

When suitably macerated, the outer keratinized zone of the epidermis can be stripped from the underlying layers of the epithelial cells, and it has been customary to divide the layers of the epidermis, albeit in a somewhat arbitrary manner, into a deep *germinative zone* and a more superficial *cornified zone* (7.304). For

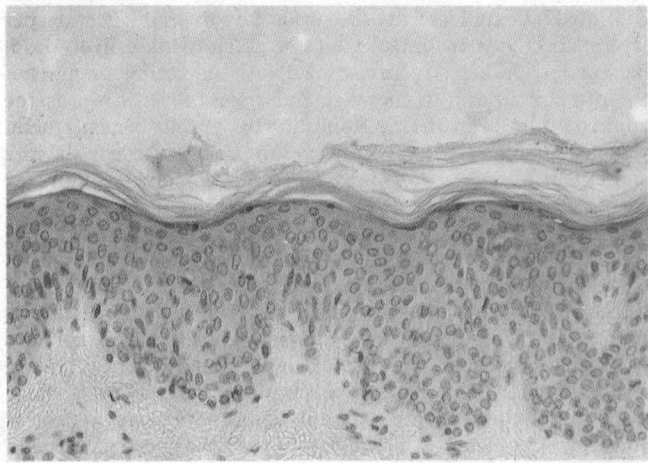

A

B

7.304A and B. A Low-power light micrograph of a vertical section through skin from the sole of the human foot: Mallory's triple stain. The finer collagen fibres of the papillary and reticular layers are stained blue, and the coarser collagen of the reticular layer stains red. The epidermis is differentiated into the stratum corneum and stratum lucidum above (red-brown), and the strata basale, spinosum and granulosum, which stain blue-grey. A spiral duct from a sweat gland is also seen in vertical section. The base of the epidermis is irregularly ridged. B A vertical section through the epidermis of the scalp, between hair follicles. Note the reduced keratinized layer (compare with A). Haematoxylin and eosin.

a more detailed treatment of the structure of these layers, the reader is referred to pp. 40, 41, and only certain special aspects of organization will be considered here.

The Germinative Zone

This consists of living cells with little keratin precursor material visible in their cytoplasm. The deeper layers are capable of mitotic division, chiefly in a vertical direction to generate 'stacks' of epidermal cells which are continuous to the epithelial surface. This zone includes a *basal cell layer* (*stratum basale* or *stratum malpighii*) situated adjacent to a basal lamina, where mitotic activity is particularly intense. Superficial to this is a region several cells thick, the *stratum spinosum* (prickle cell layer), marked by the convoluted, interdigitating profiles of the cells, the *keratinocytes* (prickle cells) which are linked by numerous desmosomes attached to internal tonofibrils of the cytoplasm (7.308). This layer probably provides much of the mechanical coherence of the epithelium. Within the germinative zone three other cell types may be present, two of them with extended dendritic processes ramifying among the surrounding keratinocytes, and one with a more rounded profile. The first two types are the epithelial *melanoblast* (*melanocyte, dendritic cell,* or *clear cell*), which synthesizes melanin (*vide infra*), and the *Langerhans cell* (Zelickson 1967), the functions of which are the subject of some argument (Breathnach and Wylie 1967). The *melanoblasts* (7.306) are structurally distinct from the surrounding keratinocytes by virtue of their lack of desmosomes, their few internal microfilaments, and the presence of specific *melanin bodies* (*melanosomes*), ellipsoidal structures up to 0.7×0.2 μm in their major and minor diameters which show, internally, a regular array of fine (5 nm) granules or fibrils, the latter with regular 5 nm striations. the *Langerhans cells*, in contrast, are devoid of melanin, but contain instead characteristic rod-like *Langerhans granules*, about 0.1 μm long by 0.01 μm wide, having a regular granular interior (Breathnach 1971). Some of these granules are in continuity with the external membrane of the cell. At one time such cells were thought to be degenerating melanoblasts, but the profusion of intracellular organelles and the active DNA synthesis in their nuclei argues against this view; similar cells are known to be present in the dermis, and it is thought by some workers that they are a type of phagocytic cell, similar perhaps to the macrophages of the dermis (Breathnach and Wylie 1967; Zelickson 1971).

There is good evidence that the melanoblasts are of neural crest origin, and that the Langerhans cells come from a different embryonic source.

The other epithelial cell type, found chiefly in hairy skin, is the *Merkel cell* (Zelickson 1971), an element which lies in close association with the tactile sensory nerve endings terminating in the basal parts of the epidermis (*see* p. 852). The cell is marked by an indented nucleus, prominent Golgi complex and numerous dense-cored vesicles each about 80 nm in diameter. These cells appear to play an important role in the sensory process, but the significance of their internal contents is unclear; there is no evidence that the dense-cored vesicles contain catecholamines.

The Zone of Keratinization

In this zone occur the chief cell transformations associated with keratinization (7.304, 308). The *granular layer* (*stratum granulosum*) at its base consists of flattened cells with pyknotic nuclei, surrounded by numerous basophilic granules of the keratin precursor *keratohyalin* (*see* pp. 40, 41). The *clear layer* (*stratum lucidum*) is situated superficially to the granular layer, and appears in sections as a homogenous, slightly striated layer composed of closely packed cells in which traces of flattened nuclei may be seen. This layer is prominent, or indeed visible, only in heavily keratinized areas such as glabrous skin, and contains the keratin precursor *eleidin*. The *cornified layer* (*stratum corneum*) consists of many thicknesses of flattened *squames*, the remnants of cells which have become completely filled with keratin and have lost all other internal structures, including nuclei. At the basal aspect of this layer the cells are closely compacted and adhere to one another strongly, but more superficially they become loosely packed and eventually flake away at the surface. The cells are held together by

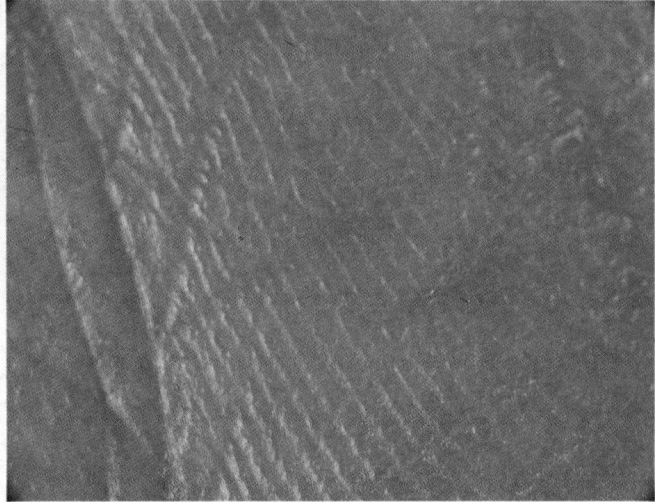

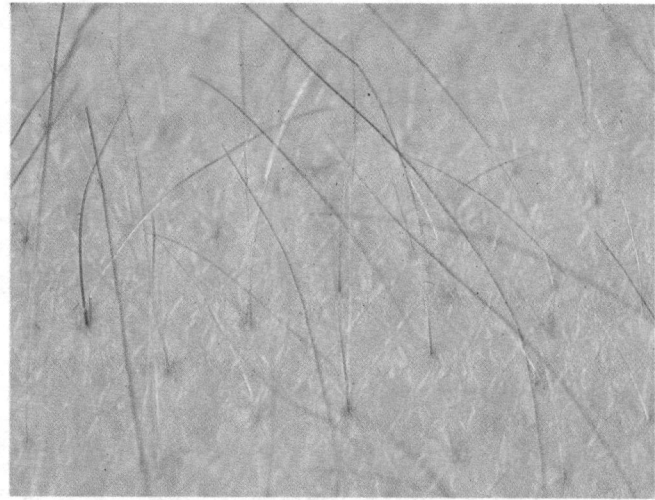

7.305 A low-power light micrograph of hairless skin, in surface view, from the palm of the hand, showing epidermal friction (papillary) ridges and larger flexure lines (left). Mag. ×6.

7.306 A similar light micrograph to that shown in **7.305**, but taken from hairy skin on the extensor aspect of the forearm; note the pattern of surface grooves (tension lines) and hairs. The oblique direction of the emerging hair shafts points away from the pre-axial border of the limb. Mag. ×6.

an extracellular layer of lipid material which confers strong waterproofing properties on the keratinized layer. This material is synthesized in the stratum spinosum in *lamellated granules*, small vacuoles containing rows of lipid lamellae, which finally discharge when the cells reach the stratum granulosum (Strauss and Matolsty 1977).

The Pigmentation of the Skin

The final colour of the skin is determined by the presence of at least five pigments at various levels and places in the integument. These are: *melanin*, a brown pigment, situated chiefly in the germinative zone of the epidermis; *melanoid*, a substance similar to melanin, present diffusely throughout the epidermis; yellow to orange *carotene* in the stratum corneum and the adipose cells of the dermis and superficial fascia; and *haemoglobin* (purple) and *oxyhaemoglobin* (red), contained in the vascular supply of the skin, particularly the superficial venous plexuses (*vide infra*). The amounts of the first three of these pigments vary topographically throughout the body, chronologically with the age of the individual, and genetically between individuals. Their relative contributions determine the characteristic racial pigmentation, although considerable genetically-determined differences may occur within a single ethnic group.

Melanin, the brown pigment, is chemically a protein-like polymer of the amino acid *tyrosin*, and possibly of related

catecholamines also (Duchon *et al.* 1968; Breathnach 1969; Szabó 1969; Fitzpatrick and Quevedo 1971). Tyrosin is converted by the epidermal melanoblasts into dihydroxyphenylalanine (DOPA) by oxidative enzymes amongst which *tyrosinase* is important. DOPA is further converted to DOPA-quinone, commencing a series of reactions during which polymerization takes place to form the final melanoprotein (melanin). Parts of this enzyme system, and hence the identity of the dendritic melanoblasts which contain them, can be demonstrated histochemically by incubating thin pieces or sections of fresh skin with DOPA, which is converted to melanin within the cytoplasm (*the DOPA reaction*).

Complete absence of skin pigments, excepting those of the vascular system (*albinism*), is a recessive Mendelian character and may occur sporadically in any race; one or more enzymes of the DOPA pathway may be absent in this trait, which may also be associated with other disturbances of tyrosin metabolism.

Each epithelial melanoblast possesses many slender branches terminating in flattened expansions applied to the surfaces of neighbouring keratinocytes (**7.313**). The melanin granules formed in the perikaryal region of the melanoblast pass along the dendritic branches, and are either secreted at their tips, being subsequently engulfed by keratinocyte cell processes and incorporated into their cytoplasm, or else phagocytosed within dendritic fragments by these cells (Cruickshank and Harcourt 1964; Cohen and Szabó 1968). Similar melanoblasts sometimes

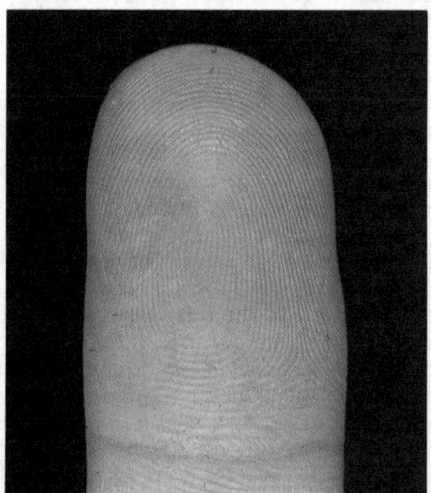
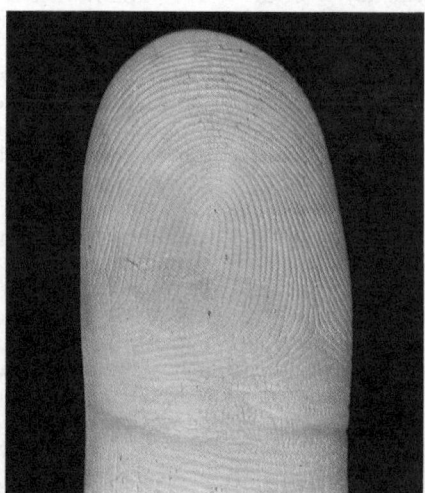
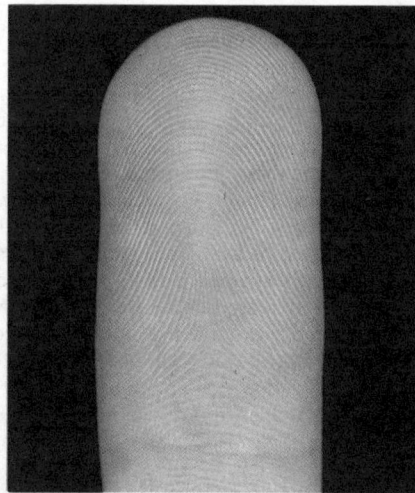

7.307A–C Photographs of the palmar aspect of a terminal phalanx in three different individuals, to show the major types of pattern of the fingerprint ridges. The pattern in A is commonly termed a whorl; B is composed of loops; C is composed of arches. Note interphalangeal flexure lines.

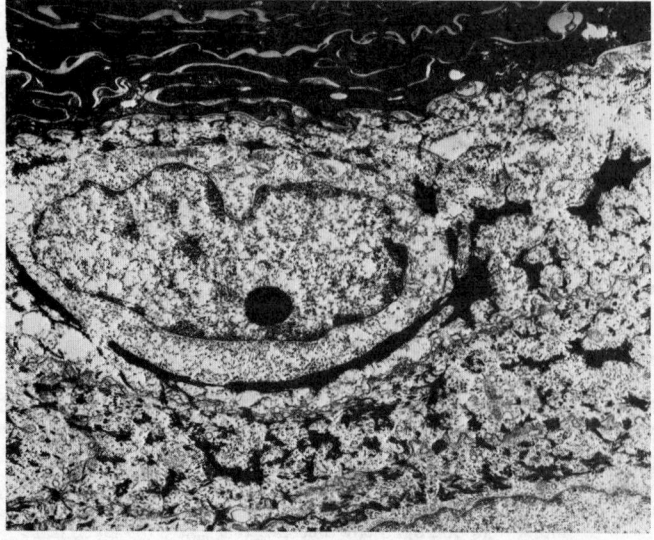

A

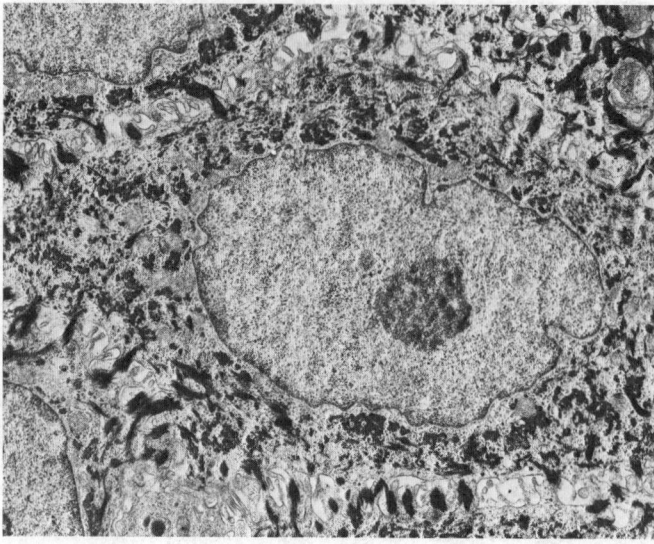

B

7.308A–C Electron micrographs of human skin in vertical section. A This shows the transition between the stratum spinosum (below), stratum granulosum (middle), and stratum corneum (the dark laminae above). The microfilamentous bundles of the keratinocytes, below, become denser and more compact in the stratum granulosum; finally the cells flatten, becoming scale-like and electron dense. B This shows keratinocytes in the stratum spinosum. The cell outlines are highly convoluted and demonstrate numerous desmosomes; many cytoplasmic bundles of microfilaments are visible. C This shows the bases of the cells of the stratum basale; their cytoplasm contains many ribosomes, but few

occur in the dermis and may pass on their melanin to macrophages.

The degree of melanization of any particular body region may depend either upon the number of melanoblasts or on the varying activity of individual melanoblasts. Under normal conditions, melanoblasts are often more numerous at the openings of the mucous membranes, on the surface of the penis, face, and limbs than over the trunk and abdomen, varying between 800 and 2,000/mm³. In newborn infants, up to the age of five months, dermal melanoblasts in the sacral region are responsible for the bluish 'Mongolian Spot' often seen in this vicinity. Melanoblasts are also intimately associated with the bases of hair follicles and provide melanin for incorporation into the cells of hairs (7.313).

In heavily melanized races, the numbers of melanoblasts are about the same as in those with paler skins, so that the difference must lie in the rate of melanin synthesis. However, the melanin content of individual melanosomes is also higher under normal conditions of lighting. In any individual, the rate of melanin synthesis is controlled locally by the incidence of ultraviolet radiation, and systemically by the action of melanocyte stimulating hormone (MSH) from the anterior pituitary gland, amongst other normal and pathological influences. The importance of melanin in protecting the deep, mitosing layers of the epidermis against chromosomal damage by ultraviolet light is seen in the higher incidence of carcinoma of the skin found in Caucasians living in tropical and subtropical regions compared with the indigenous melanized populations. However, melanin also protects the underlying dermis against ultraviolet damage of a more direct nature, that is the inflammatory changes which typically accompany sunburn.

The response of the melanoblasts to varying doses of ultraviolet light are interesting, since at least three distinct phases occur. With brief exposures, the melanin already present in the epidermis darkens appreciably, but later returns to its original colour; with increased exposure, melanin synthesis increases in each melanoblast, and in long exposures the melanocytes themselves increase in number. The extent to which these changes occur depends upon the genetic constitution of the individual as well as upon the dosage.

THE DERMIS

The dermis or corium is tough, flexible and highly elastic. It is very thick in the palms of the hands and soles of the feet; thicker on the posterior than on the anterior aspect of the body, and on the lateral than on the medial sides of the limbs. It is exceedingly thin and delicate in the eyelids, scrotum, and penis.

MICROSCOPIC STRUCTURE OF THE DERMIS

The dermis consists of felted connective tissue, with a varying number of elastic fibres and numerous blood vessels, lymphatic vessels and nerves. The connective tissue is arranged in two layers: a deeper or *reticular*, and a superficial or *papillary*. Nonstriated muscular fibres occur in the superficial layers of the dermis wherever hairs are present; they are also present in the subcutaneous areolar tissue of the scrotum, penis, labia majora and nipples.

The reticular layer consists of strong interlacing bands, composed chiefly of white fibrous tissue, but containing some yellow elastic fibres, which vary in number in different parts. In the deeper part the fasciculi are large, and the wide intervals left by their interlacement are occupied by adipose tissue and sweat glands. Below the reticular layer is the subcutaneous areolar tissue which, except in a few situations, contains fat. The connective tissue bands in the reticular layer lie for the main part in parallel bundles, so that if a *conical* object is stabbed through the skin and then withdrawn it leaves a *linear* wound since the fibres are forced apart without much rupture. The directions taken by the parallel bundles vary in different parts of the body and constitute what are termed the *cleavage lines* of the skin. Surgical incisions made along the direction of cleavage lines heal with minimal formation of scar tissue, whereas incisions across these lines, owing to retraction of the severed fibres, lead to the formation of a broad scar. In general, the cleavage lines are arranged longitudinally in the skin of the limbs and more or less horizontally in the trunk and neck.

With increasing age the yellow elastic fibres atrophy and the skin loses much of its elasticity and becomes wrinkled. If the skin becomes much stretched (as by rapidly growing tumours, fat deposition or pregnancy) the fibres in the reticular layer may undergo partial rupture, followed by scar formation; these areas may show on the surface as white streaks. These are commonly seen on the anterior abdominal wall after pregnancy and are known as *lineae gravidarum*. In many regions the skin is separated from the deep fascia or other structures by loose areolar tissue and in these sites the skin is freely movable over the deeper structures. Elsewhere, however, the skin may be firmly anchored to structures like the periosteum over 'subcutaneous' parts of bones, or to the deep fascia in regions related to movements of underlying joints. In the latter case there may be permanent

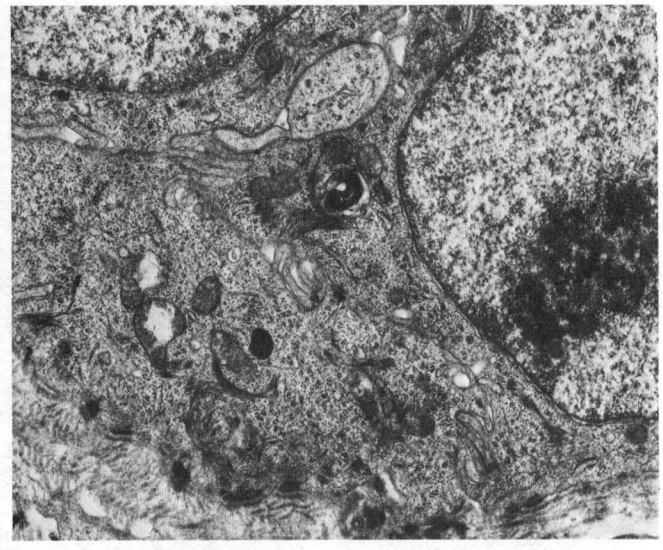

c

microfilaments. Their bases are attached to the basal lamina of the epithelium by punctate dense hemidesmosomes; note dermal collagen (below). Magnifications: A and B ×8,000; C, ×20,000.

creases in the skin known as *flexure lines*; they are particularly evident on the palm of the hand and flexor surfaces of the digits, where they are arranged in relation to the movements of the digits (*see* pp. 473 and 1217).

The papillary layer consists of numerous highly sensitive and vascular eminences, termed the *papillae*, which project perpendicularly (7.304, 310). The papillae are minute conical projections, having round or blunted apices, which may be divided into two or more parts, and are received into corresponding pits on the under surface of the epidermis. On the general surface of the body, and especially in areas endowed with slight sensibility, they are few in number and exceedingly minute; but in some situations, as upon the palmar surfaces of the hands and fingers, and upon the plantar surfaces of the feet and toes, they are large, closely aggregated together, and arranged in parallel curved lines, forming the elevated ridges seen on the surface of the epidermis. Each ridge contains two rows of papillae and between the rows the ducts of the sweat glands pass outwards to open on the summits of the ridges. Each papilla consists of very small and closely interlacing bundles of finely fibrillar tissue, with a few elastic fibres; within this tissue there is a capillary loop, and in some papillae, especially in the palms and the fingers, there are tactile corpuscles (7.31).

For a more detailed description of the cutaneous nerve terminals, and patterns of innervation, consult pp. 851–858.

THE VASCULARIZATION OF SKIN

Within the skin, the blood supply and drainage lie along well-determined pathways (7.309), the precise form of which is related to the metabolic requirements of the various cellular components (Montagna and Parakkal 1974). Amongst the highly active areas of the skin are the epidermis and its cellular extensions—hair follicles, sebaceous and sudorific glands—and also the dermal papillae which contain active fibroblasts and the sensory endings of cutaneous nerves; all of these structures are closely related to rich capillary beds in the dermis. The deeper reticular layer of the dermis, in contrast, is highly fibrous and contains few active cells, and the metabolic requirements are low. Although many vessels pass through the reticular layer, they give off few capillaries and are mostly non-nutritive to this layer.

Blood enters the skin through small arteries which penetrate the superficial fascia from its deep aspect and initially ramify in a sheet-like plexus, the *rete cutaneum*, at the interface of the dermis and superficial fascia. From this plexus, some blood vessels run deeply to supply the subcutaneous adipose tissue and, in those regions where they are present at this depth, the bases of hair follicles and sweat glands. Other vessels arising from the rete cutaneum curve in a superficial direction, giving off capillaries around hair follicles, sebaceous glands and sudorific glands as they pass to the junction of the reticular and papillary layers of the dermis; here they form another flat plexus, the *rete subpapillare* or *superficial plexus*. Capillaries extend from this layer towards the base of the epithelium, within dermal papillae, and loop back to a flat venous plexus which lies immediately beneath the rete subpapillare; this in turn drains into a flat intermediate plexus in the middle of the reticular layer of the dermis which further connects with a deep laminar venous plexus at the dermis—superficial fascia junction. Various capillary beds situated in the dermis around glands and hair follicles drain into these three venous plexuses at appropriate levels within the skin.

In the deeper layers of the dermis, arteriovenous anastomoses are common; in glabrous skin some of these are surrounded by thick sphincter-like groups of smooth muscle and pursue a convoluted course. These vessels are termed *glomera*. Since the nonstriated muscle elements are under autonomic control, heat exchange at the epithelial surface can be regulated by the vasoconstriction of the afferent arterioles of the general cutaneous supply, whilst the arteriovenous anastomoses provide for a maintained deep circulation in the skin under thermal conditions which might otherwise reduce the blood supply of the integument to dangerous levels (see also p. 638).

THE LYMPHATIC DRAINAGE OF THE SKIN

Numerous blind-ending lymphatic vessels terminate in the dermis near the base of the epidermis and drain deeply first into a dermal network in the papillary layer, then into another network at the mid-level of the reticular layer and finally into a network at the junction of the dermis and superficial fascia. Deep to this zone, the lymph flows through wider channels provided with valves, into the main lymphatic drainage of the area. The lymphatic drainage of the skin is quite profuse, and free anastomosis appears to occur between vessels at all levels so that there is free interchange of lymph between areas of the skin which are adjacent to each other (Forbes 1938).

THE INNERVATION OF THE SKIN

Skin is richly innervated (7.29) by myelinated and non-myelinated sensory fibres of the cerebrospinal and autonomic nerves, and via non-myelinated efferent autonomic fibres supplying blood vessels, sweat glands and smooth muscle fibres associated with the

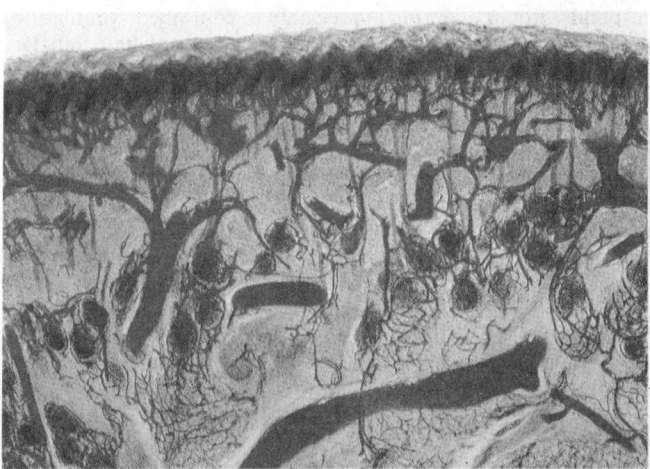

7.309 A thick vertical section through palmar skin, the arteries, arterioles and capillaries of which have been injected with red gelatin to demonstrate the pattern of dermal vascularization. At the base of the dermis a broad flat arterial plexus supplies a more superficial papillary plexus, which in turn gives off capillary loops which enter the dermal papillae. Sweat glands and their ducts are numerous in this specimen; they extend basally into the subcutaneous tissues. Magnification ×200.

hairs (pilomotor nerves). The sensory apparatus has already been described in some cytological detail (p. 851), and will be discussed only in relation to the sensory functions of the whole skin (Sinclair 1967; Kenshalo 1968; Sinclair 1973).

The nerves associated with both efferent and afferent endings penetrate the superficial fascia and ramify through the reticular and papillary layers of the dermis. Conspicuous nerve plexuses are formed around hair follicles and in the papillary layer of the dermis beneath the epithelium where the term *dermal* or *corial plexus* is often applied. Efferent nerves ramify around blood vessels, cutaneous muscle fibres and the bases of sudorific glands. Afferent nerves usually branch to supply a limited area of the skin, or in one type of ending, the Pacinian corpuscle, they may remain unbranched; the area of innervation by a single fibre is often related to the sensitivity threshold of the fibre and inversely to the degree of spatial localization of which it is capable. Each receptive field may overlap with those of other afferent fibres, giving a complex mosaic which creates possibilities of pattern analysis (p. 857).

Since the skin forms the chief interface between the internal and external environments, the elaborate sensory apparatus which it possesses makes it the chief means of comparing physical changes in the two environments and of monitoring their interaction. Information concerning the rate of change of various effective stimuli, their duration, energy content and their spatial and temporal patterning are all being constantly provided by the array of cutaneous receptors. The interpretation and correlation of the various messages arriving in the central nervous system appear to involve the continual analysis of patterns of activity from a wide range of receptor types (p. 851), some of them particularly responsive to certain modalities (e.g. pressure, cold), others responding in a manner determined by their position within the skin, their intrinsic thresholds, and the mechanical characteristics of the non-nervous cells associated with their sensory terminals.

In both hirsute and glabrous skin, the highly branched non-myelinated and myelinated 'free' terminals, which end within the dermis and penetrate the lower layers of the epidermis, form an important sensory component. Some of these fibres subserve mechanoreceptors, others thermoreceptors, whilst yet others are involved in nociception. In hirsute skin, such fibres lie in close association with hair follicles; another type of ending, the Merkel disc is also present in this variety of skin, particularly around the openings of hair follicles, whereas in glabrous skin they are found at the base of the intermediate 'sweat' ridges beneath the papillary ridges of the epidermis. A single myelinated afferent fibre may give rise to several groups of such disc endings, each group corresponding to a 'touch spot' on the surface of the epithelium. Shearing forces applied to the skin surface cause such endings to respond without adapting appreciably to continued stimulation, and it seems likely that the epidermal structures—hair follicles, ridges—to which these endings are attached act as a series of levers, imparting a magnified stress to the associated expanded nerve endings. In hairless skin there are also numerous complex corpuscles, the endings of Meissner, which are situated chiefly within the dermal papillae beneath the papillary ridges (Quilliam 1971). Each Meissner corpuscle receives multiple innervation from the branches of as many as nine separate axons, each of which may also innervate other Meissner corpuscles, so that considerable cross innervation occurs. These endings are some of those responsible for rapidly adapting mechanoreception, and because of their high density on the flexor surfaces of the digits for example, they may be associated with the sensing of rapidly changing, patterned tactile stimuli. Lamellated endings of a somewhat simpler construction are found in other areas such as the mucocutaneous junctions, and mucous membranes (mucocutaneous end organs, genital corpuscles, bulboid endings, lingual end-organs, etc.), where they may have similar functions. Pacinian corpuscles are present chiefly in the deeper layers of the skin, and in the superficial fascia where they are particularly prominent in the flexor surfaces of the digits. Each corpuscle is innervated only by a single axon, and may serve an entire unbranched nerve fibre. The peculiar lamellated structure of the corpuscle appears to possess important properties, making the

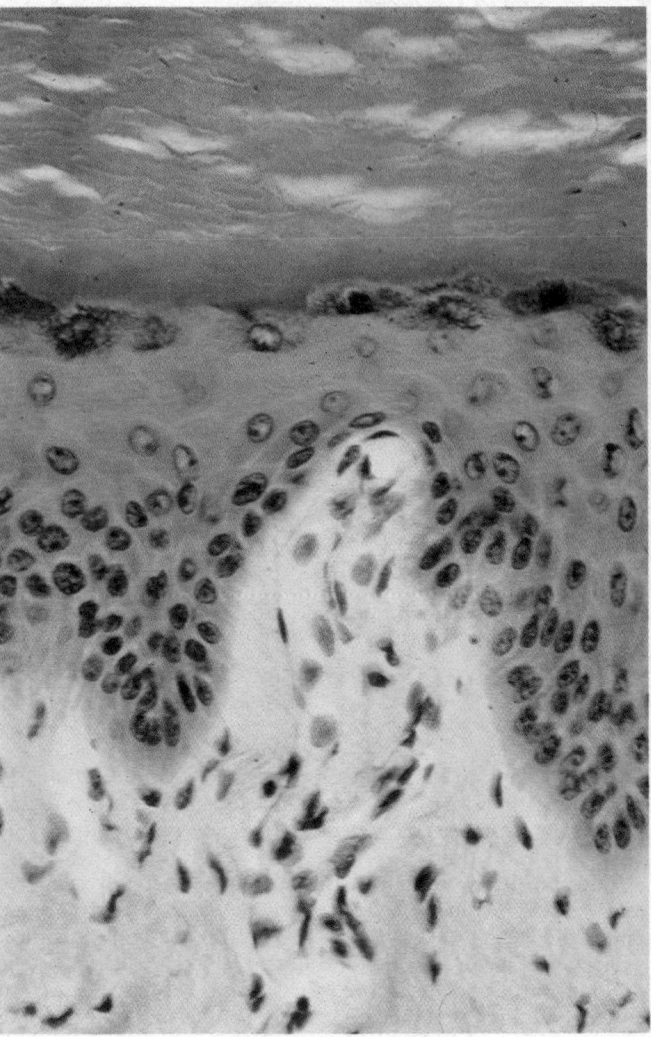

7.310 A vertical section through a dermal papilla and adjacent epidermis, showing a capillary loop. Notice the closeness of the vessel to the basal layer of the epidermis. Mallory's triple stain. Mag. × 800.

end-organ responsive to rapidly changing forces such as vibrations. As with the other types of receptor endings, the position within the skin and the mechanical characteristics of adjacent non-nervous cells may create possibilities of stimulus analysis otherwise not feasible, by virtue of the action of the epidermis and dermis as filters, attenuating mechnical and thermal stimuli in different ways as they are transmitted through their successive layers. Sensory endings situated at different levels may therefore form part of a highly ordered sensory system with organization in both horizontal and vertical planes.

In addition to this minute structuring of the sensory apparatus of the skin, large-scale variation also occurs over the whole body surface. Physiologically, such variations are detectable as changes in thresholds to mechanical and thermal stimuli, in point-to-point discrimination of tactile stimuli, and in the ability to identify the precise locality of a stimulus applied to the body surface. In general, both fine discrimination and location of stimuli follow similar gradients, being of a high order at the extremities of the limbs and becoming progressively cruder towards abdomen, thorax and head, although certain areas such as the lips and tongue, for example, create local reversals in this pattern. Discrimination is not entirely bilaterally symmetrical, the right side usually being more discriminative than the left, at least in dextral subjects. Other gradients, for example, of pressure thresholds, may operate in the opposite directions. Sex differences have also been shown, female persons being generally more sensitive to pressure stimuli than male, a finding possibly related to the slightly thinner epidermis typical of female skin.

The anatomical basis of such variations is still far from being satisfactorily explained, a statement which might also be applied to the whole subject of cutaneous sensation.

APPENDAGES OF THE SKIN

THE NAILS

The nails (7.311) are flattened, elastic structures of a horny texture, placed upon the distal parts of the dorsal surfaces of the fingers and toes. The proximal part of the nail, called the *root*, is implanted into a groove in the skin; the exposed part is the *body* of the nail; the distal end forms the *free border*, and a little proximal to it the skin is attached to the under surface of the nail forming the *hyponychium*. The root of the nail is overlapped by a fold of skin, the *nail fold*, the stratum corneum of which is prolonged distally as a thin cuticular fold, the *eponychium*, to cover completely or partially the white opaque crescentic part of the nail called the *lunule*. The greater part of each collateral border of the nail is overlapped by a fold of skin, termed the *nail wall*. The nail itself is analogous to the horny zone of thick skin, although the keratinized squames are hard and highly coherent (Zaias and Alverez 1968; Zelickson 1971); beneath it lies the germinative zone which, together with the subjacent corium, forms the *nail bed*. Under the greater part of the nail the corium is thick and raised into a series of longitudinal ridges which are very vascular, and this accounts for the pink colour seen through the translucent nail. Near the root of the nail, however, the ridges are smaller, irregularly arranged and less vascular; moreover the tissue of the nail is here more opaque, hence this part of the nail is whiter and constitutes the lunule. The lunule is usually visible in the thumb nail, but in the other digits it becomes progressively more covered by the nail fold towards the little finger, in which it is generally hidden altogether. The germinative zone of the nail bed consists functionally of two parts. The part beneath the root of the nail and the lunule (*germinal matrix*) is thicker and actively proliferative, and is concerned with the growth of the nail, the epidermal cells being gradually converted into the nail substance. On the other hand, the part beneath the rest of the nail (*sterile matrix*) is thinner and is not concerned with nail growth but provides a surface over which the growing nail glides. All growth of the nail therefore occurs at its root; the nail increases in thickness from its root to the distal edge of the lunule and the remainder is of uniform thickness. If a nail be removed without severely damaging the germinal matrix, a new nail will grow from this region. Disturbances of growth of the nails may occur in acute illnesses or local trauma, and transverse grooves may develop on the surface which move gradually distally with growth of the nails. Minute air bubbles, giving rise to white flecks, may develop in the substance of the nail. On an average nails grow about 0·5 mm a week; growth is quicker in summer than in winter, and finger nails grow about four times as fast as toe nails. In the hand, nail growth is most rapid in the longest digit (the middle finger), slowest in the little finger and intermediate in the other digits. Nails act as a rigid background for support of the digital pads of the terminal phalanges, and thus may aid tactile mechanisms. From the evolutionary point of view, nails are derived from the more elaborately structured claws which characterize many other mammals and lower tetrapods.

THE HAIRS

The hairs are found on nearly every part of the surface of the body, but are absent from the palms of the hands, the soles of the feet, the dorsal surfaces of the distal phalanges, the umbilicus, the glans penis, the inner surface of the prepuce, the inner surfaces of the clitoris, labia majora and minora. They vary much in length, thickness and colour in different parts of the body and in different races of mankind. In some parts, as in the skin of the eyelids, they are so short as not to project beyond the follicles containing them; in others, as upon the scalp, they may be remarkably long; the eyelashes, the hairs of the pubic region, and the whiskers and beard are remarkable for their thickness. Straight hairs are stronger than curly hairs and present on transverse section a cylindrical or oval outline: curly hairs, on the other hand, are flat. (Some maintain that the form of the hair does not correspond with its shape in cross-section, however.)

A hair consists of a *root*, the part implanted in the skin, and a *shaft* (*scapus*), the portion projecting from the surface.

The *root* (*radix*) *of the hair* has a proximal enlargement, the *bulb*, which is set in an invagination of the epidermis and superficial portion of the corium, called the *hair follicle* (7.312, 313). When the hair is of considerable length the follicle extends into the subcutaneous tissue. The hair follicle commences on the surface of the skin with a funnel-shaped opening, and passes inwards in an oblique or curved direction—the latter in curly hairs—to become dilated at its deep extremity, where it corresponds with the hair bulb. The ducts of one or more sebaceous glands open into the follicle near the skin surface. At the bottom of each hair follicle there is a small conical vascular eminence or papilla, similar in every respect to those found upon the surface of the skin; it is continuous with the dermal layer of the follicle, and is supplied with myelinated and non-myelinated nerve endings. It is from the capillaries in the papilla that the hair derives its nutrition.

The hair follicle consists of two coats—an outer or dermal, and an inner or epidermal (7.312, 314).

The *outer coat* is formed mainly of fibrous tissue; it is continuous with the dermis, is highly vascular, and is supplied by numerous, minute nerves.

The *inner coat* is closely adherent to the root of the hair, and consists of two strata, named respectively the *outer* and *inner root sheaths*; the outer root sheath corresponds to the stratum spinosum of the epidermis, and resembles it in the rounded form and soft character of its cells; at the bottom of the hair follicle these cells become continuous with those of the root of the hair. The inner root sheath consists of: (1) a delicate cuticle next the hair, composed of a single layer of imbricated scales with atrophied nuclei; (2) one or two layers of horny flattened nucleated cells, known as *stratum epitheliale granuliferum* (*Huxley's layer*); (3) a single layer of cubical cells with clear, flattened nuclei, called *stratum epitheliale pallidum* (*Henle's layer*) (7.314).

The *hair bulb* is moulded over the papilla and composed of polyhedral cells. As they pass upwards into the root of the hair these cells become elongated and spindle-shaped, except those in the centre, which remain polyhedral.

The *shaft of the hair* consists, from within outwards, of the medulla, the cortex and the cuticle. The *medulla* is usually absent from the fine hairs covering the surface of the body, and commonly from those of the head. It is composed of rows of polyhedral cells, with air spaces between, and sometimes within, the cells. The *cortex* constitutes the chief part of the shaft; its cells are elongated and are united to form flattened, fusiform fibres, which contain pigment granules in dark hair, and air in white hair. The *cuticle* consists of a single layer of flat scales which overlap one another from below upwards (Matolsty 1958; Birbeck 1964; Orfanos and Ruska 1968; Zelickson 1971).

Over most parts of the body the hairs are fine and downy and give an appearance of hairlessness. Almost the entire skin of the human at about the middle of fetal life is covered by fine hair, called *lanugo* (primary hairs), and in fact the hairs on the back at

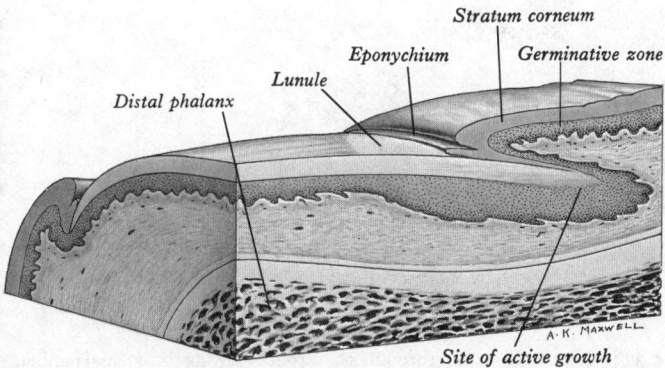

Distal phalanx
Lunule
Eponychium
Stratum corneum
Germinative zone

A·K·MAXWELL

Site of active growth

7.311   A longitudinal section through the root of a nail.

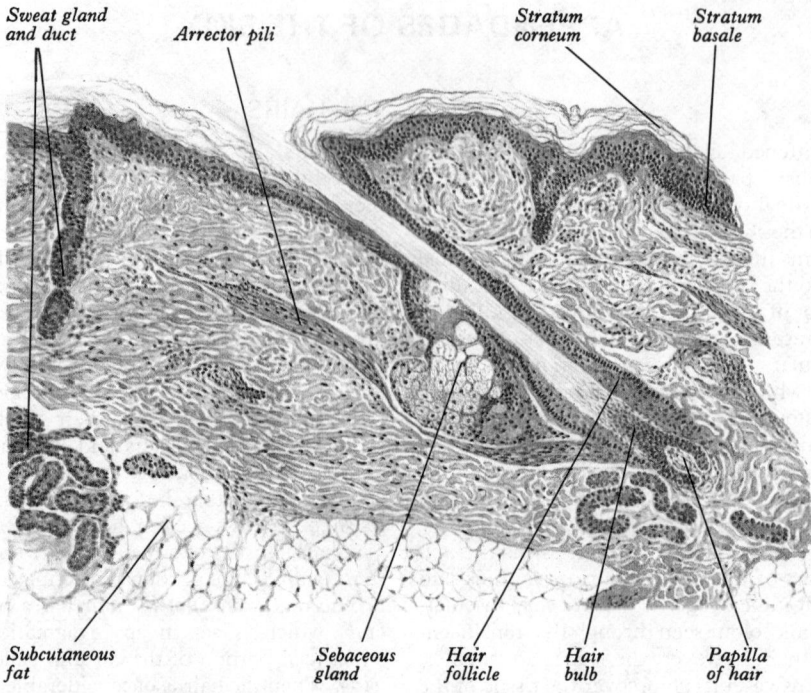

Sweat gland and duct   *Arrector pili*   *Stratum corneum*   *Stratum basale*

Subcutaneous fat   *Sebaceous gland*   *Hair follicle*   *Hair bulb*   *Papilla of hair*

7.312 A section through the skin, showing the epidermis and dermis (corium), a hair in its follicle, an arrector pili muscle, and sebaceous glands opening into the hair follicle. Magnification × 100.

this time are more numerous (per square centimetre) than in the gorilla or chimpanzee of corresponding fetal age. The lanugal hairs are mostly shed by birth and are replaced by fine hairs, termed *vellus* (secondary hairs), in the early months of post-natal life. These are retained in most regions but are replaced by the hairs (*terminal hairs*) of the scalp and eyebrows; also by the axillary and pubic hairs, and those on the face and front of the chest in the male, which appear at puberty, their development and growth being under hormonal control. Hairs on the adult scalp are more numerous (per square centimetre) than in the anthropoid apes. The lanugal and vellous hairs have no medulla. In the male the hairs in the vestibule of the nose and in the external acoustic meatus grow markedly with advancing age.

In furred mammals hair functions in the temperature controlling mechanism by minimizing heat loss; this function in man is served by the subcutaneous fat, and the hairs are concerned largely in the cutaneous sensation of touch (pp. 849, 889). Growth of a hair occurs at the hair bulb, where the cells capping the papilla proliferate and form the *germinal matrix* of the hair. As proliferated cells progressively move towards the surface they become keratinized to form the fibre-like cornified cells of the shaft of the hair. The duration of life of a single hair varies from about four months (eyelashes, axillary hair) to about four years (scalp hair), after which it is shed and is replaced by the sprouting of new cells from the germinal matrix after a period of rest: Rapidly growing hairs are said to be in the *anagen* phase, but when division of the matrix cells ceases, and the hair stops growing, it is termed a *telogen* hair. Growth of a hair varies with its texture, ranging from about 1·5 mm (fine hair) to 2·2 mm (coarse hair) a week. Greying or whitening of hair is due to the collection of minute air bubbles in the cortex (and medulla) of the shaft and to loss of pigment (melanin) formation by cells in the germinal matrix (*see* p. 1218).

Minute bundles of nonstriated muscular fibres, termed the *arrectores pilorum* (7.303, 312), are connected with the hair follicles. They arise from the superficial layer of the corium, and are inserted into the outer coat of the hair follicle, below the entrance of the duct of the sebaceous gland. They are placed on the side towards which the hair slopes, and by their action diminish the obliquity of the follicle and elevate the hair. When they contract, the skin over their origin is depressed, while the skin immediately around the hair is elevated; this results in the appearance of 'goose skin' seen on exposure to cold or in

emotional reactions. The sebaceous gland is situated in the angle which the arrector pili muscle forms with the superficial portion of the hair follicle, and contraction of the muscle thus tends to squeeze the sebaceous secretion out from the duct of the gland. The arrector muscles are supplied by sympathetic nerves.

THE SEBACEOUS GLANDS

The sebaceous glands are small, sacculated, and lodged in the substance of the dermis. They occur in most parts of the dermis, but are especially abundant in the scalp and face; they are also very numerous around the apertures of the ear, nose, mouth and anus, but are absent in the palms of the hands and soles of the feet. Each gland consists of a single duct, relatively capacious, which emerges from a cluster of oval or piriform alveoli, usually from two to five, but in some instances as many as twenty in number. Each alveolus is composed of a basement membrane, enclosing a number of epithelial cells. The outer or marginal cells are small

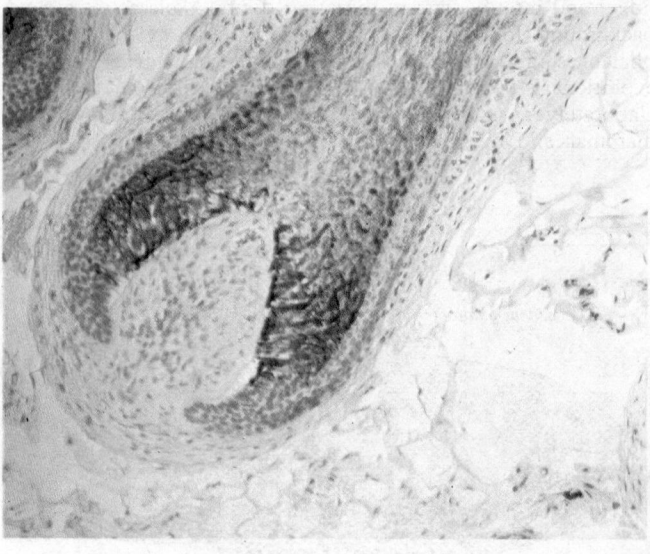

7.313 A vertical section through a hair root, showing the dermal papilla, and numerous melanocyte processes extending into the matrix of the hair. Haematoxylin and eosin. Magnification × 250.

and polyhedral, and are continuous with the cells lining the duct. The remainder of the alveolus is filled with larger cells, containing fat, but in its centre the cells are broken up, leaving a cavity filled with their débris and a mass of fatty matter, which constitutes the *sebum cutaneum* (Charles 1960; Ellis 1967; Zelickson 1971). As the sebaceous glands produce their secretion by complete fatty degeneration of their central cells they are classed as *holocrine* glands. As the central cells disintegrate, they are replaced by proliferation of marginal cells. The ducts open most frequently into hair follicles (the glands being developed as diverticula from the epithelial walls of the follicles themselves) but occasionally upon the general surface, as in the labia minora, glans penis and the free margins of the lips. On the nose and face the glands are of large size, distinctly lobulated, and often become much enlarged from the accumulation of pent-up secretion. The tarsal glands of the eyelids are elongated sebaceous glands with numerous lateral diverticula. Sebum acts as a natural lubricant of the hair and skin protecting skin from the effects of moisture or desiccation and hairs from becoming brittle; it also has some bactericidal action. The secretory activity of the sebaceous glands does not appear to be under nervous control; it is stimulated by hormonal action, particularly androgens.

THE SWEAT GLANDS

Sudoriferous or sweat glands (7.303, 312) occur in almost every part of the skin, and have been classified into two types, *eccrine* and *apocrine* (Biemfica and Montes 1965; Ellis 1968; Zelickson 1971).

The eccrine glands are most numerous and are found in almost every part of the skin. Each consists of a single tube, the

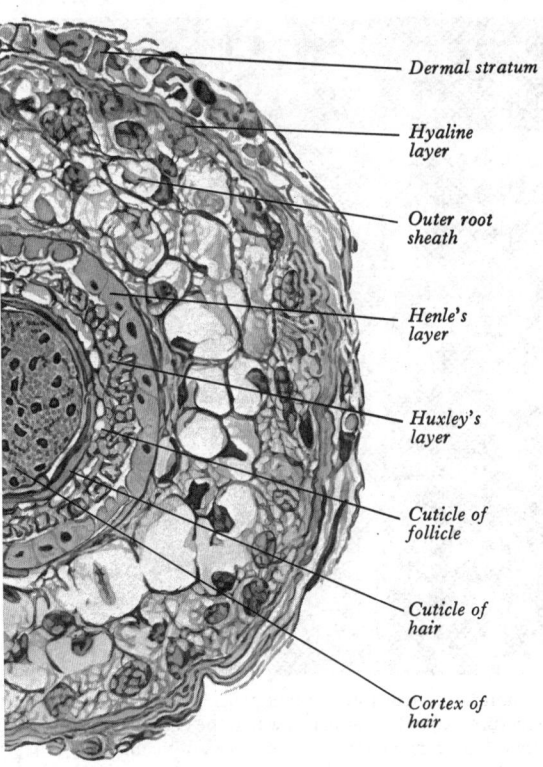

Dermal stratum

Hyaline layer

Outer root sheath

Henle's layer

Huxley's layer

Cuticle of follicle

Cuticle of hair

Cortex of hair

7.314   Transverse section of a hair follicle from the scalp of a newborn infant. Magnification about × 600.

7.315   Scanning electron micrographs of the shafts of hairs from different regions of the body, showing variations in width and in the pattern of cells at their surfaces. Top row (left to right) pubis, eyebrow, axilla, eyelash; bottom row (left to right) moustache, scalp (Negroid), scalp (Caucasian), forearm. Magnification, all × 220.

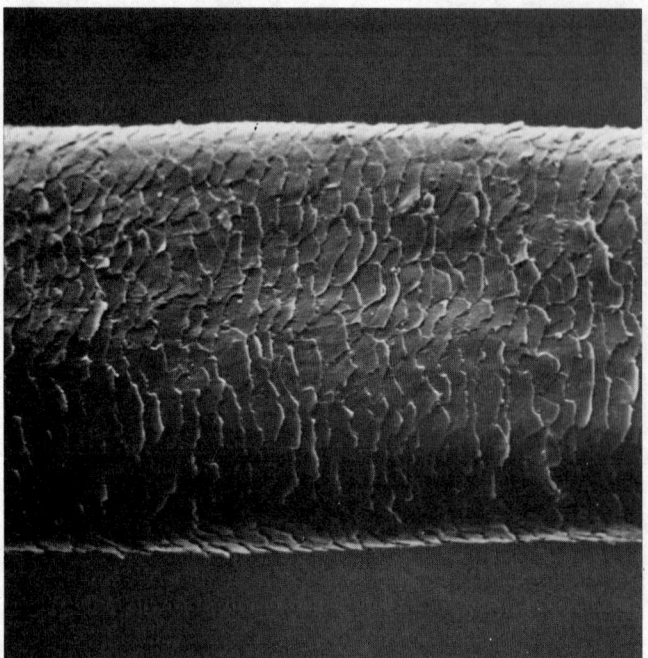

A

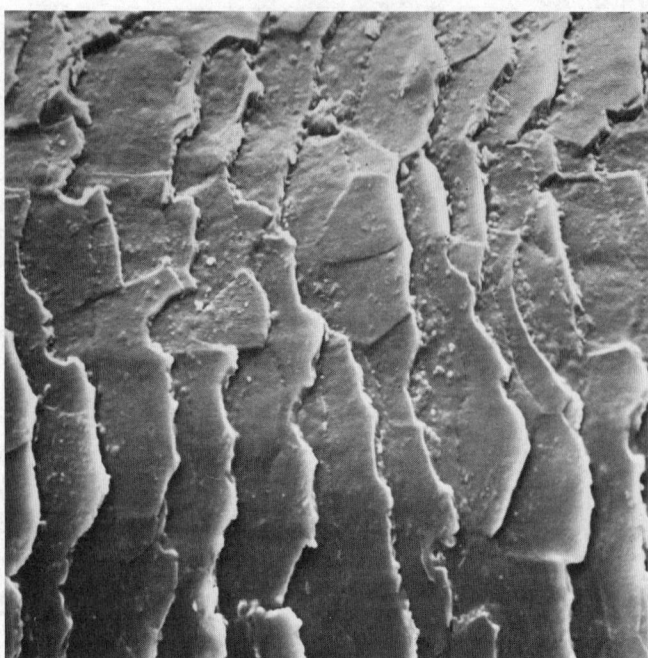

B

7.316A, B   Scanning electron micrographs of a scalp hair showing details of surface structure. Note the manner in which the cuticular cells overlap each other; their free ends point towards the apex of the hair.

Magnification, A × 370; B × 1,850. (Figures 7.315 and 7.316A, B prepared by Mr. Michael Crowder, Guy's Hospital Medical School, London.)

deep part of which is coiled into an oval or spherical ball, which is situated in the deeper layers of the corium or in subcutaneous tissue and is named the *body* of the gland. The superficial part, or duct, traverses the dermis and epidermis and opens on the surface of the skin by an infundibular aperture. In the superficial layers of the dermis the duct is straight, but in the deeper layers it is convoluted or twisted; where the epidermis is thick, as in the palms of the hands and soles of the feet, the part of the duct which passes through it is spirally coiled. The size of the glands varies. They are especially large in those regions where the amount of perspiration is great, as in the axilla and in the groin. Their number varies. They are very plentiful on the palms of the hands and on the soles of the feet, where the orifices of the ducts are exceedingly regular, and open on the curved ridges of the epidermis; they are least numerous in the neck and back. The tube, both in the body of the gland and in the duct, consists of two layers—an outer, of fine areolar tissue, and an inner, of epithelium. The outer layer is thin and is continuous with the superficial stratum of the dermis. In the body of the gland the epithelium consists of cubical or polyhedral cells which may be classified as dark cells and clear cells; the former are rich in ribonucleic acid and mucopolysaccharides and are the main secretory elements. The latter are mostly acidophilic and contain an abundance of glycogen. Electron microscopy reveals canaliculi between the clear cells. Between the basal aspects of these cells and the basement membrane there is an incomplete layer of longitudinally or obliquely arranged elongated, fusiform, myoepitheliocytes. Electron microscopic studies show that their cytoplasm is similar to that of nonstriated muscle cells containing fine (5 nm) filaments and less frequent thicker ones (*see* p. 518). Desmosomes occur where their plasma membranes adjoin those of the secretory cells of the gland. The myoepitheliocytes have been ascribed a contractile role in the expression of secretion from the gland. It is also suggested that they may regulate the flow of metabolites to the secretory cells (Ellis 1965). The ducts are destitute of muscular fibres and are composed of a basement membrane lined by two or three layers of polyhedral cells; the duct passes through the epidermis as a spiral channel which is simply an intercellular cleft between the epidermal cells. When the epidermis is carefully removed from the surface of the dermis, the ducts may be drawn out in the form of short, thread-like processes on its deep surface. The eccrine sweat glands are *merocrine* in nature, i.e., produce their thin watery secretion without demonstrable epithelial cell disintegration.

Apocrine glands occur in the axilla, eyelids, areola and nipple of the breast, in the circumanal region and in association with the external genitalia (Montagna and Parakkal 1974). They are larger than eccrine glands and produce a thicker secretion. In the female they show involution changes related to each menstrual cycle. They are developed in close association with hairs and their ducts typically open into the distal ends of hair follicles.

Apocrine glands are tubular in form, having an extensive coiled secretory portion which may be up to 2 mm across in the larger examples, leading into a narrower straight secretory duct which runs parallel to neighbouring hair follicles. The secretory coils which often anastomose to form a labyrinthine network, are lined with cells being cuboidal to columnar in shape, with pale-staining nuclei, numerous mitochondria and amounts of granular and smooth endoplasmic reticulum varying with their secretory state. Cells often contain large vacuoles of dense material, probably secretory in nature and at the bases of the cells lie numerous longitudinally orientated myoepitheliocytes. The secretions of these glands vary with their anatomical position. In the axilla the glands become distended with a proteinaceous, milky fluid containing various organic compounds including steroids. Bacterial decomposition augments the naturally occurring odour of these secretions to give a characteristic musky, urinous scent. In many mammals similar glands are involved in the production of chemical signals or pheromones which are used in marking territories, in courtship, and in maternal as well as other types of social behaviour. Their existence in humans suggests that these glands may have been important originally in similar types of activity, although their role has been obscured in the exercise of hygiene. The ceruminous glands of the external acoustic meatus are modified apocrine sweat glands. It should be noted that recent work, using the electron microscope, has suggested that the so-called apocrine glands do not lose superficial cytoplasm during secretion and hence all sweat glands may have a merocrine, rather than apocrine, form of secretion.

Eccrine sweat glands are concerned in the temperature control mechanism by surface evaporation of the sweat. They are supplied by sympathetic nerves, though these fibres are cholinergic in nature, and no sweating occurs in a denervated area of skin. Rarely, sweat glands may be congenitally absent, in which case special means have to be adopted to prevent rise of body temperature in hot weather. Apocrine glands are under dual autonomic control, but do not respond markedly to temperature changes, so that strictly speaking they should not be classed as sweat glands at all.

8
SPLANCHNOLOGY

SPLANCHNOLOGY

The grouping together of the organs of respiration, alimentation, urinary excretion, and reproduction, together with the endocrine glands, under the umbrella of 'Splanchnology' is more traditional than rational. There is, perhaps, a certain rough reasonableness in dividing the body's fabric and function into somatic or locomotor and visceral or metabolic moieties, but no effort is customarily exerted to bring osteology, arthrology, myology, angiology and neurology into a corporation. In any case the overlap between all these so-called systems and the miscellany of viscera or *splanchnoi* gathered under the label of 'splanchnology' is most extensive. Again, splanchnology is no sooner formed than it is dismembered into its constituent 'systems', as in this volume. Our only reasons for retaining the term are convenience and tradition.

THE RESPIRATORY SYSTEM

Introduction

The exchange of respiratory gases is a basic essential of the life process, even in its states of suspension, as in seeds or spores, in hibernation or aestivation, when respiration may be greatly diminished but not wholly abolished. In the smallest creatures, such as *Protozoa*, oxygen and carbon dioxide can diffuse directly through the body surface, establishing equilibria along partial pressure gradients with sufficient rapidity to maintain adequate metabolism. In the *Metazoa*, with their greater mass and hence impossibly lengthened diffusion distances, blood transport systems and specific respiratory areas are necessary. The respiratory adaptations assume a dual nature—the transport of respiratory gases and the development, often deep within the animal's body, of specialized areas for gaseous exchange. The breathing tubes or tracheal system of the *Arthropoda* are an example of the former, and it is worthy of note that in the chitinous spirals maintaining patency in such *tracheoles* there is a structural theme which recurs in all vertebrates. In the *Vertebrata*, as in some invertebrate phyla, two major environments, *aqueous* and *terrestrial* (or aerial), have posed different problems of adaptation; but in both habitats there is a basic similarity, in that actual gas exchanges take place between the external environment and circulating blood through a *wet* epithelium, whether in gills, lungs, or through the skin. Cutaneous respiration is limited to some amphibia, and apart from this the respiratory organs of vertebrates are constantly associated with and largely developed from the branchial arch system and the alimentary tube. The first segment of this beyond the mouth is the pharynx, through which water passes to the gills in piscine vertebrates, where the exchanges occur between gases dissolved in water or blood. Diverticula of the pharynx are usual in fishes, and these function as swim-bladders or hydrostatic organs; but in a few extant species they also act as auxiliary respiratory chambers, which can be filled directly from the air. (Such interesting and significant adaptations cannot, unfortunately, detain us here. For these and other aspects of vertebrate life, consult Young 1962; Romer 1970.) It is from such 'lungfish', or allied forms, that the earliest terrestrial vertebrates, the amphibia, have most probably evolved, the acquisition of an air-breathing habit leading to a host of new forms and adaptations.

It is appropriate to note here that some members of all the terrestrial vertebrate classes—amphibia, reptiles, birds and mammals—have returned wholly or partially to an aquatic existence; but only the larvae of amphibians are able to take oxygen directly from water. (There are a few *adult* amphibians, such as *Salamander atra*, which have this ability.) All the others must temporarily suspend breathing while immersed, coming to the surface to expel the contents of their respiratory organs and to refill them with fresh air. To keep out water and to retain inspired air while such air-breathing vertebrates are submerged, some kind of sphincteric mechanism is obviously necessary at the junction of the lungs, however primitive, and the pharynx from which they develop. Even in fish which merely swallow air to fill their swim-bladders, such a sphincter is necessary; and in the piscine ancestors of the land tetrapods, which were able to 'breathe' from water as well as air, the same need exists, and perhaps also for some form of dilatator mechanism. Hence, a sphincter-dilatator mechanism, which becomes the *larynx*, is probably as basic as the lung-sac itself. In land vertebrates the respiratory cavity, or *lung*, is almost always a bilateral structure; and the interior of these paired pharyngeal diverticula, served by a single air-tube, the *trachea*, becomes highly complicated by folding of its epithelium increasing the area of interchange at which blood is exposed to inspired air. Thus a basic pattern of the respiratory organs is early established, and even in small amphibia the pattern of branching of the trachea—which is typically dichotomous—into smaller divisions, *bronchi* and *bronchioles* and their terminal respiratory air sacs or *alveoli*, has already become quite complex. The specialization into a tracheobronchial 'tree' of air-transporting tubes and the terminal array of thin-walled alveoli where gaseous exchanges occur, is universal in land vertebrates including birds, whose lungs, however, display particular specializations which cannot be entered into here (*see* Portmann 1950). Although the tubes acquire the ability to alter their calibre by contraction or relaxation of nonstriated muscle in their walls, it is in the alveoli that the main volumetric changes of inspiration and expiration occur. As we shall see, the means by which inflation and deflation are effected vary. There is considerable variation—even ignoring avian specialization—in the dimensional characteristics of alveoli and bronchi in different vertebrate groups, including the pattern of branching of the bronchi, lobation of the lungs and so on. Within particular groups, however, this pattern displays considerable constancy in its major features; and recognition of this in man (p. 1246) has proved to be especially valuable in problems of diagnosis and surgical treatment of pulmonary disorders.

Like the lungs the *larynx* also varies greatly in its adaptations in diverse vertebrate groups (Negus 1928, 1949). Although primarily a sphincteric mechanism, it also assumes the functions of *phonation*, even in the first air-breathers, the amphibia, whose pre-nuptial coaxation is all too familiar. As might be expected, the musculature, skeletal elements and neurovascular supplies are derived from the pharynx and branchial apparatus, and numerous allusions to this occur in the appropriate sections of this book (*see*, for example pp. 145, 150, 151, 207).

The amphibia still 'swallow' air, which is drawn in through valvular 'nostrils', opening anteriorly into the roof of the mouth, there being no separate nasal cavity. The floor of the mouth is lowered to inspire, and when it is raised the nostrils are closed, and air is pumped into the pharynx, larynx and lungs. The mouth and pharynx therefore subserve both respiration and feeding. Among reptiles more active forms have evolved, demanding more efficient respiration, which is partly achieved by the development of movable ribs and a change to *aspiration* of air by reduction of pressure in the lungs themselves. This new technique of breathing is finally improved, in the mammals, by the evolution of a muscular diaphragm, leading to the formation of a complete musculo-osseous thoracic cavity, whose elaborated musculature becomes a most efficient means of varying the thoracic and hence pulmonary volume. (It also has an additional advantage in aiding the venous return.) In mammals the oronasal cavity becomes divided into respiratory and alimentary regions by the appearance of a palate; but the former, evolving in continuity with the primitive nares of the upper jaw, is *cranial* or *dorsal* to the mammalian oral cavity—the reverse of the relationship between pharynx and larynx. Hence the mammalian airway *crosses* the alimentary path in the middle level of the pharynx, and a great variety of adaptations have occurred in the pharynx, palate, and especially the larynx to overcome the difficulties of this curious arrangement, not only in terms of structure but also in the reflexes associated with breathing and swallowing.

The mammalian 'airway' thus includes the nasal cavity, nasal and oral levels of the pharynx, the larynx, trachea, bronchi and their ultimate ramifications in the lungs. Breathing through the mouth is, of course, also possible; but this interferes with the olfactory and 'air-conditioning' functions of the nasal cavity. The separation of a respiratory and olfactory chamber from the mouth has considerable advantages: predatory mammals can still breathe with a mouth obstructed by prey, whose struggles may be prolonged; and their herbivorous prey may still sense warning odours while feeding. In aquatic vertebrates, such as crocodiles, dolphins and whales, a spout-like elongation of the larynx projects into the nasopharynx, more or less completely separating the airway, so that respiration can continue at the water surface even with the mouth submerged, open, and ready for prey. Of course, in the marine mammals other associated specializations occur (Andersen 1969).

In man, typically less specialized in most structural features than many other vertebrates, there is, nevertheless, one outstanding functional elaboration in association with the respiratory system—*speech*. The nasal cavity is still respiratory, olfactory, and an organ of heat and water vapour exchanges, with a small contribution to the sounds of speech. But the larynx, though still an essential sphincter—obviating the entry of swallowed food and other foreign bodies, and a blockade to build up pressure for coughing or as an aid to extreme muscular efforts—is so frequently involved in phonation that this is often regarded as its prime function. It is certainly in this that it has reached unique abilities in mankind. The elaboration of its musculature, and of the neural control which integrates the laryngeal sound production with respiratory, pharyngeal, palatal, lingual and labial muscles in the expert articulation of speech, provide man with this unique ability. Speech, which must of course be heard and interpreted, and which is later expressed in visible, written form, has led to the most complex and abstract symbolology of language (with which mathematical and musical notation may be coupled). In these activities, for which there is no true parallel in the life of any other animal known to us, the highest integrations of the nervous system and locomotor structures are elaborated. The basic function of the respiratory system remains, naturally, vital to mere existence; but its primitive phonational potentiality, leading finally to speech, in all its permutations, would seem to be a fundamental factor in the evolution of human intelligence. It is curious that in the other most 'vocal' vertebrate class, birds, the larynx is commonly a sphincteric mechanism alone, sounds in great variety being produced by the *syrinx*, situated at the tracheal bifurcation. Some birds, of course, are notable imitators of human speech; but there is no evidence of development of true, symbolic language in any vertebrate other than mankind, in whom, presumably, high manual dexterity, large cerebral (especially cortical) development, and emancipated vision have alone provided the adequate milieu for linguistic evolution.

The Larynx

The larynx, which is an organ of phonation, an air passage, and a sphincteric mechanism, extends from the root of the tongue to the trachea. (*See* Terracol *et al.* 1965, and for a detailed recent monograph on laryngeal biomechanics consult Fink and Demarest 1978.) It projects ventrally between the great vessels of the neck, and is covered anteriorly by the skin, the fasciae and the depressor muscles of the hyoid bone (**8.**1A). Above, it opens into the laryngeal part of the pharynx, of which it forms the anterior wall; below, it is continuous with the trachea. In the adult male it is situated opposite the third to sixth cervical vertebrae, but it occupies a somewhat higher position in the child and in the adult female. In infants of between six and twelve months the tip of the

epiglottis (the highest part of the larynx) is a little above the level of the cartilaginous disc between the dens and the body of the axis. Its average measurements in the European adult are as follows:

	In males	In females
Length	44 mm	36 mm
Transverse diameter	43 mm	41 mm
Anteroposterior diameter	36 mm	26 mm

Until puberty the larynx of the male differs little in size from that of the female. In the female its increase at puberty is only small. In the male the increase is considerable; all the cartilages enlarge and the thyroid cartilage projects in the anterior median line of the neck, while the anteroposterior diameter is nearly doubled.

The skeletal framework of the larynx is formed of cartilages, which are connected by ligaments and membranes, and are moved by a number of muscles. Its internal lining is a mucous membrane continuous above and behind with that of the pharynx and below with that of the trachea.

LARYNGEAL CARTILAGES

The cartilages of the larynx comprise the thyroid, cricoid and epiglottis, and the paired arytenoid, cuneiform and corniculate cartilages.

The thyroid cartilage (**8.**1, 2, 12) is the largest cartilage of the larynx. It consists of two quadrilateral *laminae*, the caudal parts of the anterior borders of which are fused at an angle in the median plane to form a subcutaneous projection named the *laryngeal prominence*, or 'Adam's apple'. This prominence is most distinct at its upper part, and is well marked in the male but scarcely visible in the female. Immediately above it the laminae are separated by a V-shaped notch, the superior *thyroid notch* or *incisure*. Posteriorly the *laminae* diverge and the posterior border of each is prolonged as two slender processes, the *superior* and *inferior cornua*.

On the *external surface* of each lamina an *oblique line* curves downwards and forwards from the *superior thyroid tubercle*, which is situated a little in front of the root of the superior horn, to the *inferior thyroid tubercle* on the lower border of the lamina. This line provides attachment for the sternothyroid, the thyrohyoid and inferior constrictor of the pharynx. The *internal surface* is smooth: above and behind, it is slightly concave and covered with mucous membrane. In the upper part of the angle formed by the junction of the laminae, the thyro-epiglottic ligament is attached; below this, and on each side of the midline, the vestibular and vocal ligaments and the thyro-arytenoid, thyro-epiglottic and vocal muscles gain attachment.

The *superior border* of each lamina is concave behind and convex in front; it gives attachment to the corresponding half of the thyrohyoid membrane. The *inferior border* is concave behind, and nearly straight in front, the two parts being separated by the inferior thyroid tubercle. In and near the median plane it is connected to the cricoid cartilage by the anterior (median) cricothyroid ligament (*see* p. 1233).

The *anterior* border is fused with that of the opposite lamina, forming with it an angle of about 90° in men, and about 120° in women. In men the greater projection of the laryngeal prominence, the correspondingly greater length of the vocal folds and resultant deeper pitch of the voice are all associated with the smaller thyroid angle. The *posterior border*, thick and rounded, receives the attachments of fibres of the stylopharyngeus and palatopharyngeus. The *superior cornu*, long and narrow, is curved upwards, backwards and medially, and ends in a conical extremity, which gives attachment to the lateral thyrohyoid ligament. The *inferior cornu*, short and thick, curves downwards, with a slight inclination forwards and medially; on the medial surface of its lower end there is a small oval facet for articulation with the side of the cricoid cartilage.

During infancy a narrow, diamond-shaped, flexible strip, named the *intrathyroid cartilage*, occupies the interval between the two laminae anteriorly and is joined to them by connective tissue.

The cricoid cartilage (**8.**3, 5, 6, 7) is smaller, but thicker and

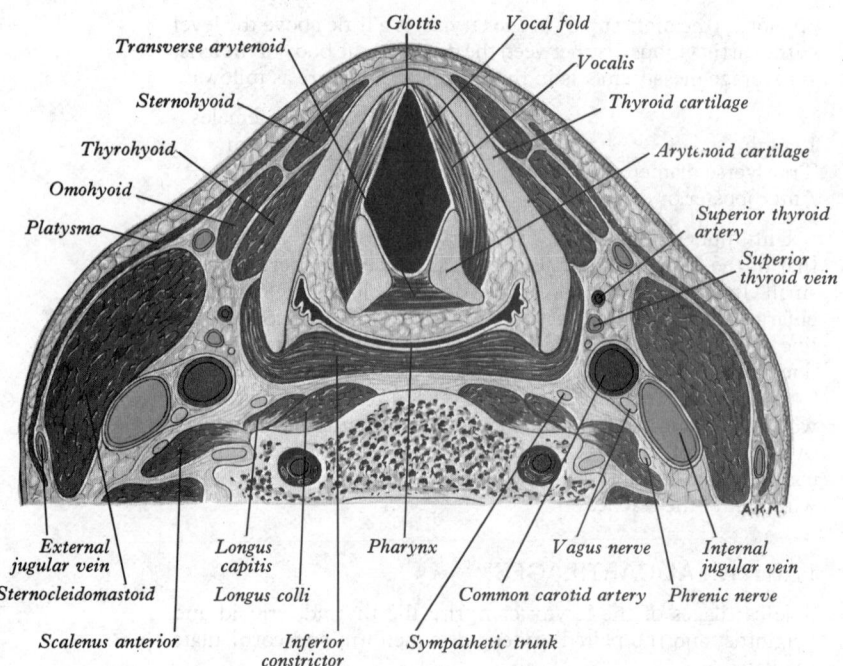

8.1A A transverse section across the ventral region of the neck at the level of the vocal folds. Viewed from the superior aspect.

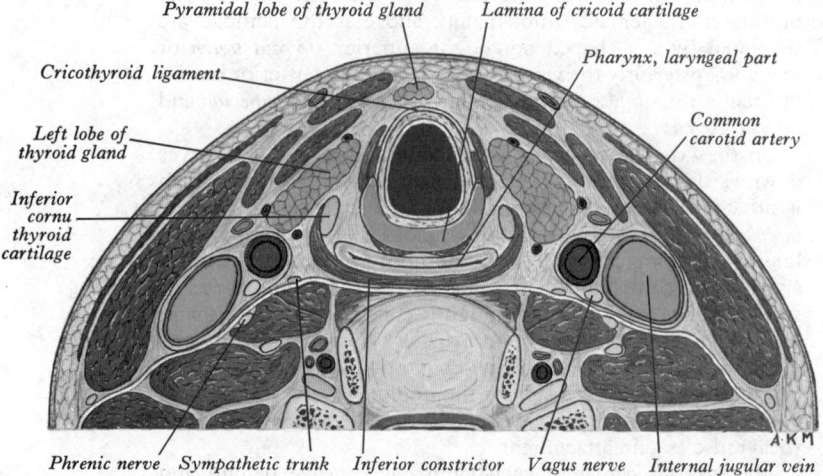

8.1B A transverse section through the ventral region of the neck, between the fifth and sixth cervical vertebrae. Semi-diagrammatic. Viewed from the superior aspect.

stronger than the thyroid cartilage. It is shaped like a signet ring, and forms the caudal parts of the anterior and lateral walls and most of the posterior wall of the larynx. It comprises a quadrate posterior lamina, and a narrow anterior arch.

The *lamina* of the cricoid cartilage is deep and broad, and measures vertically from 2 cm to 3 cm; the posterior surface is marked by a median vertical ridge, to the upper part of which the two fasciculi of the longitudinal fibres of the oesophagus are attached by a tendon (p. 1317). On each side of the ridge there is a shallow depression for the fibres of the posterior crico-arytenoid. The *arch* is narrow in front, measuring vertically from 5 mm to 7 mm, but widens posteriorly as it approaches the lamina. The external surface affords attachment in front and at the side to the cricothyroid, and behind to part of the inferior constrictor of the pharynx. The arch of the cricoid can be felt easily in the living subject below the laryngeal prominence and separated from it by a slight depression in which can be felt the resilient conus elasticus. On each side, at the junction of the lamina with the arch, there is a prominent circular facet, directed laterally and dorsally, for articulation with the inferior horn of the thyroid cartilage. The *inferior border* of the cricoid cartilage is horizontal, and connected to the highest tracheal cartilage by the cricotracheal ligament. The *superior border* runs obliquely upwards and backwards, giving attachment, anteriorly, to the thick median part and, at the sides, to the membranous lateral sheets of the cricothyroid ligament and the lateral crico-arytenoids. Posteriorly, the lamina presents a shallow median notch, and on its superolateral aspect, on each side, there is a smooth, oval, convex surface, directed upwards and laterally, for articulation with the base of the arytenoid cartilage. (*See*, however, Sellars and Keen, 1978). The *internal surface* of the cricoid cartilage is smooth and lined with mucous membrane.

The paired arytenoid cartilages (8.3, 6, 7) are placed on the lateral part of the superior border of the lamina of the cricoid cartilage, at the back of the larynx. Each is pyramidal, and has three surfaces, a base and an apex.

The *posterior surface*, triangular, smooth and concave, is covered by the transverse arytenoid. The *anterolateral surface* is somewhat convex and rough. On it, near the apex of the cartilage, there is an elevation from which a crest curves at first dorsally and then caudally and ventrally to the vocal process. The lower part of this crest intervenes between two depressions (or foveae), an upper, triangular, and a lower, oblong in shape; the upper gives attachment to the vestibular ligament; the lower to the vocalis and lateral crico-arytenoid. The *medial surface* is narrow, smooth and flat; it is covered with mucous membrane, and its lower edge forms the lateral boundary of the intercartilaginous part of the rima glottidis (p. 1233). The *base* is concave, and presents a smooth surface for articulation with the lateral part of the upper border of the lamina of the cricoid cartilage. Its lateral angle or *muscular process*, rounded and prominent, projects backwards and laterally, and gives attachment to the posterior crico-arytenoid

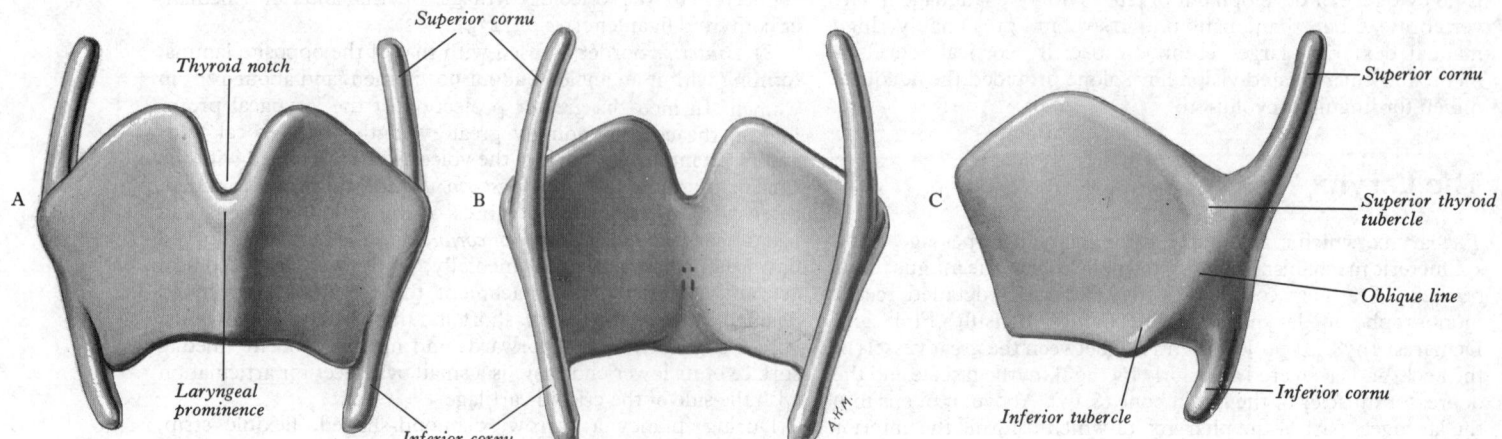

8.2A, B, C The thyroid cartilage. The attachments of the vestibular ligaments (above) and the vocal ligaments (below) are shown in B. A Anterior aspect. B Posterior aspect. C Lateral aspect.

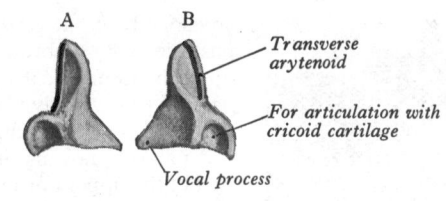

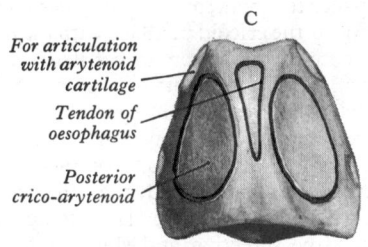

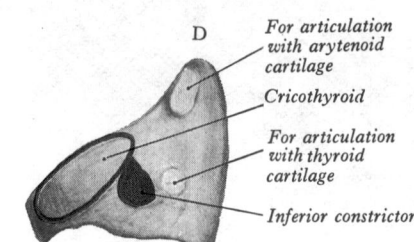

8.3 The arytenoid and cricoid cartilages. A The left arytenoid cartilage, medial aspect. B The right arytenoid cartilage, medial aspect. C The cricoid cartilage, posterior aspect. D The cricoid cartilage, left lateral aspect. Muscle attachments indicated in red.

behind, and to the lateral crico-arytenoid in front. Its anterior angle or *vocal process* is pointed; it projects horizontally forwards and gives attachment to the vocal ligament. The *apex* curves backwards and medially, and articulates with the *corniculate cartilage*.

The **corniculate cartilages** (**8.**6, 7, 11) are two small conical nodules of elastic fibrocartilage which articulate with the summits of the arytenoid cartilages and serve to prolong them backward and medially. They are situated in the posterior parts of the aryepiglottic folds of mucous membrane, and are sometimes fused with the arytenoid cartilages.

The **cuneiform cartilages** (**8.**7, 11) are two small, elongated, club-shaped pieces of elastic fibrocartilage, placed one in each aryepiglottic fold, where they give rise to whitish elevations on the surface of the mucous membrane, anterosuperior to the corniculate cartilages.

The **cartilage of the epiglottis** (**8.**4, 6, 7) is a thin, leaf-like lamella of elastic fibrocartilage, which projects obliquely upwards behind the tongue and the body of the hyoid bone, and in front of the entrance to the larynx. The free extremity, broad and rounded, is directed upwards; the attached part or stalk is long, narrow, and connected by an elastic ligament, named the *thyro-epiglottic ligament*, to the angle formed by the two laminae of the thyroid cartilage, a short distance below the thyroid notch. The sides of the epiglottis are attached to the arytenoid cartilages by the aryepiglottic folds of mucous membrane (p. 1233).

The upper part of the *anterior surface* of the epiglottis is free, and covered with mucous membrane, which is reflected on to the pharyngeal part of the tongue and on to the lateral wall of the pharynx, forming a *median glosso-epiglottic fold* and two *lateral glosso-epiglottic folds*. The depression on each side of the median glosso-epiglottic fold is named the *vallecula*. The lower part of the anterior surface lies behind the hyoid bone and the thyrohyoid membrane, and is connected to the upper border of the hyoid bone by an elastic *hyo-epiglottic ligament*; it is separated from the thyrohyoid membrane by some fatty tissue.

The *posterior surface* of the epiglottis is smooth, concave from side to side, concavoconvex from above downwards, and covered with mucous membrane; its lower part projects backwards as an elevation, known as the *tubercle*. When the mucous membrane is removed, the cartilage is seen to be indented by a number of small pits into which mucous glands project. The cartilage is perforated by branches of the internal laryngeal nerve.

Functions of the Epiglottis

During deglutition (p. 1313) the epiglottis moves upwards and forwards and is squeezed between the base of the tongue and the rest of the larynx, the bolus slipping over its anterior surface as it bends over to close the inlet of the larynx. In man, it is degenerate in function and is separated from the soft palate by a long interval; it is not essential for deglutition, which can take place normally even if the epiglottis is destroyed by disease; neither is it essential for respiration or phonation. Some mammals are keen scented even when the mouth is open for feeding. In these, the epiglottis is large and projects into the nasal part of the pharynx *above* the soft palate; when eating, the epiglottis is drawn downwards and forwards (by the hyo-epiglottic muscle, represented in man by the hyo-epiglottic ligament) against the *upper* surface of the soft palate, so keeping the nasal airway clear and the buccal airway closed. Thus, the function of the epiglottis in these animals is to preserve the integrity of the olfactory sense even when the mouth is open.

Structure of the Laryngeal Cartilages

The corniculate and cuneiform cartilages, the epiglottis, and the apices of the arytenoids consist of *elastic* fibrocartilage, which shows little tendency to calcification. The thyroid, cricoid, and the greater part of the arytenoids consist of hyaline cartilage, and become more or less ossified as age advances. Ossification commences about the twenty-fifth year in the thyroid cartilage, and somewhat later in the cricoid and the arytenoids; by the sixty-fifth year these cartilages commonly show, in radiographs, patchy calcification and may even become ossified.

LARYNGEAL ARTICULATIONS

The joints between the inferior cornua of the thyroid cartilage and the sides of the cricoid cartilage are synovial, and each is

8.4A The epiglottis, posterior aspect. Note its pitted surface.

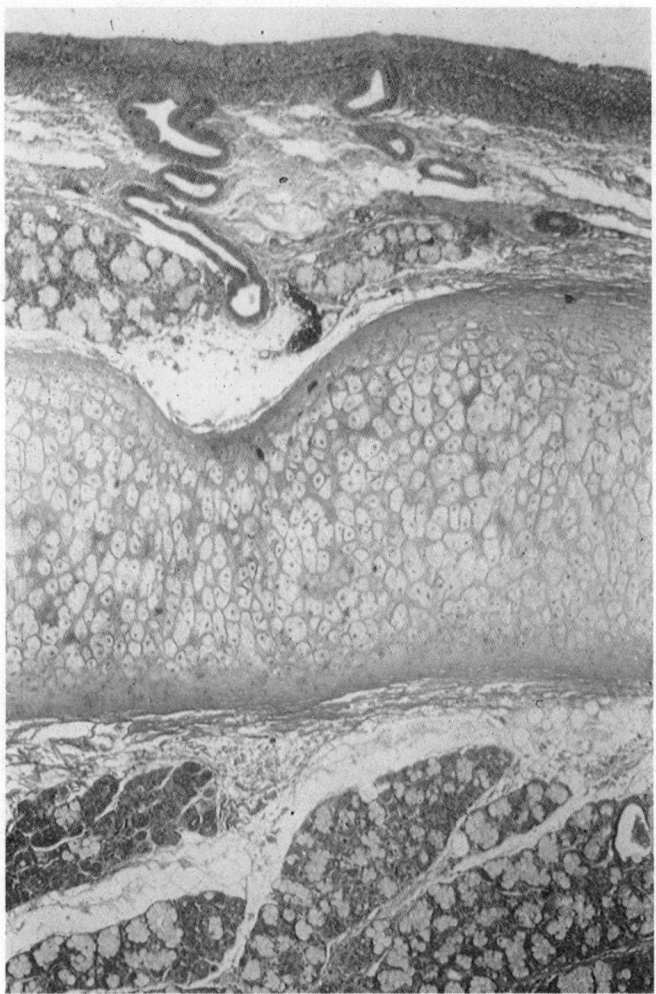

8.4B   A low magnification light micrograph of a transverse section of the epiglottis, showing stratified squamous epithelium covering the anterior surface, mucous glands and fibro-elastic cartilage. Stain—Mallory's Trichrome. (Prepared by Mr. David Ristow; photography Miss Marina Morris, Department of Anatomy, Guy's Hospital Medical School.)

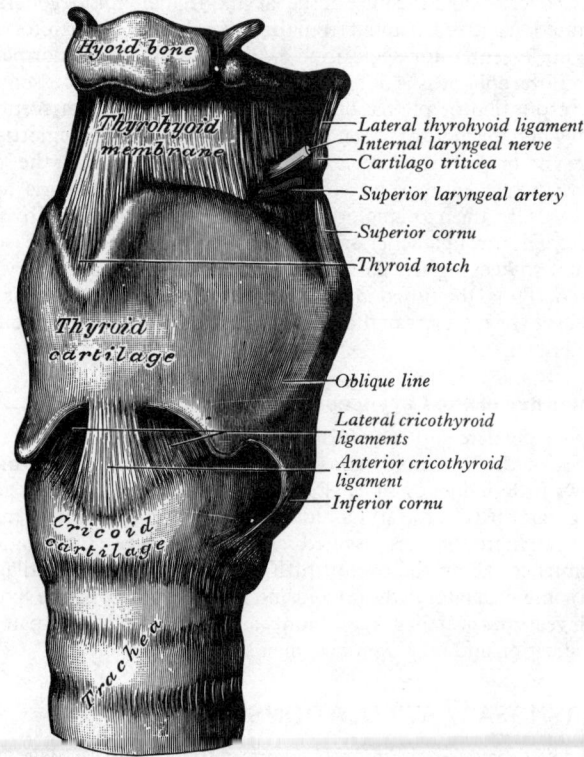

8.5   The ligaments of the larynx. Anterolateral aspect.

enveloped by a capsular ligament, which is strengthened posteriorly by a fibrous band. At these joints the cricoid rotates upon the inferior cornua of the thyroid cartilage around an axis passing transversely through both joints; to a limited extent the cricoid also glides in different directions on the thyroid cornua.

A pair of synovial joints exists between the facets on the lateral part of the upper border of the lamina of the cricoid cartilage and the bases of the arytenoid cartilages; each joint is enclosed by a capsular ligament, and a strong *posterior crico-arytenoid ligament* connects the cricoid cartilage with the medial and posterior part of the base of the arytenoid cartilage. These joints permit of two varieties of movement: one is arytenoid rotation about an axis which is not vertical but runs obliquely from dorsomediocranial to ventrolaterocaudal positions (Frable 1961), whereby the vocal process is moved laterally or medially, and the rima glottidis increased or diminished; the other is a gliding movement, and allows the arytenoid cartilages to approach or recede from each other; the direction and slope of the articular surfaces allows lateral gliding accompanied by a forward and downward movement. The two movements of gliding and rotation are associated, the medial gliding being connected with medial rotation, and the lateral gliding with lateral rotation (von Leden and Moore 1961). The posterior crico-arytenoid ligaments limit the forward movements of the arytenoid cartilages on the cricoid cartilage. (For dimensional details of the crico-arytenoid articulation consult Sonesson 1959 and Sellars and Keen 1978.)

Sometimes a synovial joint exists between the apex of each arytenoid cartilage and the corresponding corniculate cartilage but the junction is usually by cartilage.

Numerous lamellated (Pacinian) corpuscles, a few Ruffini endings and free nerve endings are present in the capsules of the laryngeal joints. In the cat these receptors respond to mechanical stimulation and are involved in the normal coordination of activity in the laryngeal muscles during respiration and phonation (Kirchner and Wyke 1965). The articular supply in man is derived chiefly from the recurrent laryngeal nerves, by independent rami or by branches from the muscular rami of the nerve (Psenicka 1966).

LARYNGEAL LIGAMENTS AND MEMBRANES

Extrinsic Ligaments

The *thyrohyoid membrane* (**8**.5, 6) is a broad, fibroelastic layer, attached below to the cranial border of the thyroid cartilage and to the front of its superior cornua, and above to the *upper* margin of the posterior surface of the body and greater cornua of the hyoid bone. As it ascends, it passes behind the concave posterior surface of the body of the hyoid bone, and is separated from it by a bursa which facilitates the upward movement of the larynx during deglutition. The median and thicker part of the membrane is termed the *median thyrohyoid ligament*; on each side the lateral and thinner part is pierced by the superior laryngeal vessels and the internal laryngeal nerve. Its outer surface is contiguous with the thyrohyoid, sternohyoid and omohyoid, and with the body of the hyoid bone. Its inner surface is related to the epiglottis and the piriform fossa of the pharynx. The *lateral thyrohyoid ligaments* are round elastic cords which form the posterior borders of the thyrohyoid membrane, and connect the tips of the superior cornua of the thyroid cartilage to the posterior ends of the greater cornua of the hyoid bone. A small cartilaginous nodule, termed the *cartilago triticea*, occurs frequently in each ligament.

The epiglottis is attached to the hyoid bone by the hyo-epiglottic ligament and to the thyroid cartilage by the thyro-epiglottic ligament (p. 1231).

The *cricotracheal ligament* unites the lower border of the cricoid cartilage with the first tracheal cartilage. It is continuous below with the fibrous membrane which invests the tracheal cartilages.

Intrinsic Ligaments

Beneath the mucous membrane of the larynx there is a broad sheet of fibrous tissue which contains many elastic fibres, and is termed the *fibroelastic membrane of the larynx*. It is interrupted on each side by the interval between the vestibular and vocal ligaments;

the upper portion, called the *quadrangular membrane*, extends between the arytenoid cartilage and the cartilage of the epiglottis, and is often poorly defined; the lower part is a well-marked membrane forming, with its fellow of the opposite side, the *cricothyroid ligament (cricovocal membrane)*, which connects the thyroid, cricoid and arytenoid cartilages. The joints between the individual cartilages are also provided with ligaments, already described.

The cricothyroid ligament (**8**.5, 7) is the inferior, larger part of the fibro-elastic laryngeal membrane. Composed mainly of elastic tissue, it is described as showing distinctive anterior and lateral parts. The *anterior (median) cricothyroid ligament* is thick, broad below, narrower above, and connects adjacent margins of the cricoid and thyroid cartilages. An anastomosis between the two cricothyroid arteries arches across it and supplies perforating rami to the larynx. The *lateral cricothyroid ligaments* (sometimes termed cricothyroid or cricovocal membranes, are thinner, lined internally by laryngeal mucous membrane and covered externally by the lateral crico-arytenoid and thyro-arytenoid muscles. From the internal rim of the superior border of the cricoid cartilage they ascend and converge, and their free superior edges, which extend from the dorsal aspect of the thyroid angle (just caudal to its midpoint) to the apices of the vocal processes of the arytenoid cartilages, form the basis of the vocal folds (**8**.9). These two thickened edges of the cricothyroid ligamentous complex are distinguished from the rest of it as the *vocal ligaments* (**8**.7). (The picturesque but ill-defined term, *conus elasticus*, is commonly applied to the entire complex sometimes only to its anterior part.)

The cavity of the larynx (**8**.8, 9) extends from the laryngeal inlet, by which it communicates with the pharynx, to the level of the lower border of the cricoid cartilage, where it is continuous with the cavity of the trachea. It is divided into three parts by an upper and a lower pair of folds of mucous membrane which project from the sides of the cavity into its interior. The upper pair are named the *vestibular folds*, and the fissure between them is called the *rima vestibuli*. The lower pair are concerned in the production of the voice, and are therefore named the *vocal folds*, and the fissure between them is called the *rima glottidis*, commonly called the *glottis*.

The *inlet of the larynx* or 'aditus laryngis' (**8**.10) is the aperture through which the laryngeal cavity opens into the pharynx. The plane of the aperture is directed backwards and somewhat upwards, for the anterior wall of the larynx is much longer than its posterior wall and slopes downwards and forwards in its upper part (**8**.8). The opening is bounded anteriorly by the upper edge of the epiglottis, posteriorly by the mucous membrane stretching between the arytenoid cartilages, and on each side by the free edge of a fold of mucous membrane which stretches between the side of the epiglottis and the apex of the arytenoid cartilage and contains ligamentous and muscular fibres; this is the *aryepiglottic fold*, and in the posterior part of its free margin there are two oval elevations, an anterosuperior due to the cuneiform cartilage, and a postero-inferior due to the corniculate cartilage. These are separated by a shallow vertical furrow, which is continuous below with the opening into the sinus of the larynx (*vide infra*).

The *vestibule of the larynx* (**8**.8, 9) is the part between the laryngeal inlet and the level of the vestibular folds; it is wide above, and narrow below. Its anterior wall is much longer than its posterior wall and consists of the posterior surface of the epiglottis, the lower part of which projects backwards as its *tubercle* (p. 1231). Its lateral walls, deep in front and shallow behind, are formed by the medial surfaces of the aryepiglottic folds; its posterior wall consists of the mucous membrane connecting the arytenoid cartilages, above the level of the vestibular folds.

The *middle part* of the laryngeal cavity is the smallest. It reaches from the level of the rima vestibuli to that of the rima glottidis. On each side it opens, through a slit between the vestibular and vocal folds, into a recess which is named the sinus of the larynx.

The *sinus of the larynx* (**8**.8, 9) is a fusiform recess which lies between the vestibular and vocal folds, and ascends for a short distance lateral to the vestibular fold. It is lined with mucous membrane, clothed on the outside by the corresponding thyro-arytenoid. From the anterior part of the sinus a narrow opening leads upwards into the saccule of the larynx.

The *saccule of the larynx* (**8**.9) is a pouch which ascends from the anterior part of the sinus, between the vestibular fold and the inner surface of the thyroid cartilage, occasionally extending as

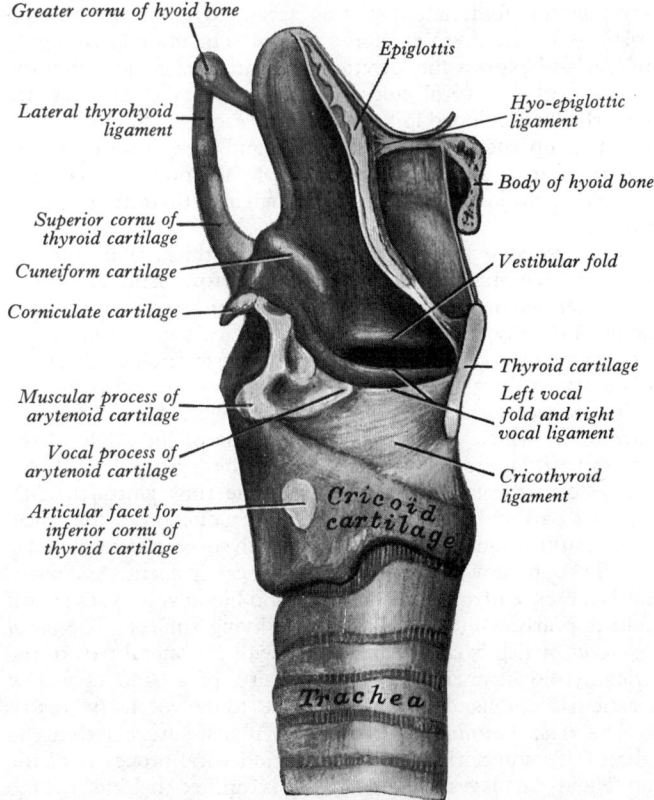

8.6   The ligaments of the larynx. Posterior aspect.

8.7   A dissection to show the right half of the cricothyroid ligament. The right lamina of the thyroid cartilage and the subjacent muscles have been removed.

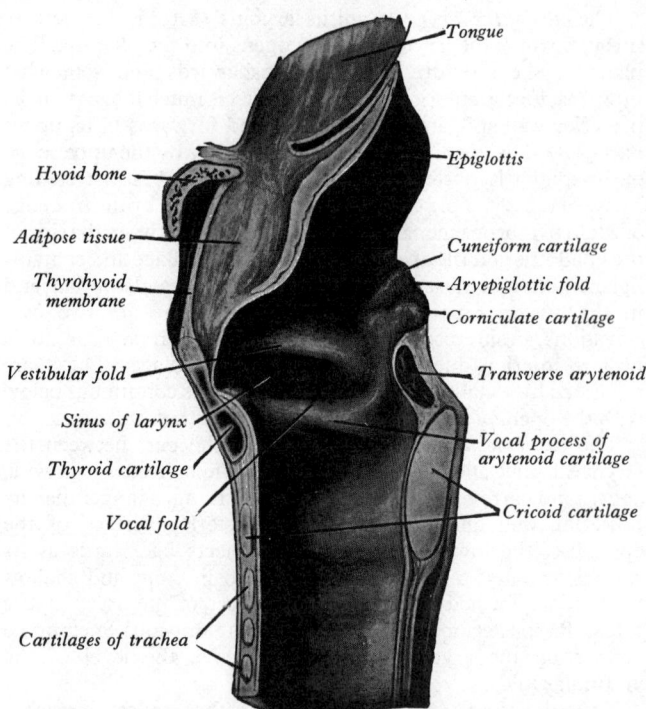

8.8 A sagittal section showing the medial aspect of the right half of the larynx.

high as the upper border of the cartilage; it is conical in form, and curved slightly backwards. On the surface of its mucous membrane there are the openings of sixty to seventy mucous glands, which are lodged in the submucous areolar tissue. The saccule is enclosed in a fibrous capsule continuous below with the vestibular ligament. Its medial surface is covered by a few muscular fasciculi, which arise from the apex of the arytenoid cartilage and, passing forwards between the saccule of the larynx and the mucous membrane of the vestibule, disappear in the aryepiglottic fold; laterally it is separated from the thyroid cartilage by the thyro-epiglottic muscle. The muscles compress the sac and express the secretion of its mucous glands upon the surfaces of the vocal folds. In most apes the saccules are remarkably developed in the form of air sacs, which may extend into the superficial tissues of the neck and even into the axillae; they appear to affect the resonance of the voice. In man, the saccules occasionally protrude through the thyrohyoid membrane.

The *vestibular folds* (**8**.7, 8, 9) are two thick, pink folds of mucous membrane, each enclosing a narrow band of fibrous tissue, termed the *vestibular ligament*, which is fixed in front to the angle of the thyroid cartilage immediately below the attachment of the epiglottic cartilage, and behind to the anterolateral surface of the arytenoid cartilage, a short distance above the vocal process.

The *vocal folds* (**8**.7, 8, 9) are two sharp, white folds of mucous membrane which stretch from the middle of the angle of the thyroid cartilage to the vocal processes of the arytenoid cartilages. They form the lateral boundaries of the rima glottidis in its anterior part and are concerned in the production of the voice. The stratified squamous epithelium which covers the vocal fold is closely bound down to the underlying vocal ligament. As a result of the absence of a submucous layer and blood vessels, the vocal fold is pearly white in colour in the living subject. The *vocal ligament*, which is continuous below with the lateral part of the cricothyroid ligament (p. 1232), consists of a band of yellow elastic tissue, related, on its lateral side, to the vocalis (p. 1235).

The *rima glottidis* or *glottis* (**8**.11) is a fissure between the vocal folds anteriorly, and the bases and vocal processes of the arytenoid cartilages posteriorly; it is limited behind by the mucous membrane passing between the arytenoid cartilages, at the level of the vocal folds. The region between the vocal folds is named the *intermembranous part*, and measures about three-fifths of the length of the entire aperture; that between the arytenoid

cartilages is named the *intercartilaginous part*. The average sagittal diameter of the glottis, in the adult male, is 23 mm; in the adult female, 17 mm. It is the narrowest part of the larynx, but its width and shape vary with the movements of the vocal folds and arytenoid cartilages during respiration and phonation (*see* p. 1236).

The *lower part* of the laryngeal cavity extends from the level of the vocal folds to the lower border of the cricoid cartilage. Its upper part is elliptical in form, but its lower part widens, assumes a circular shape, and is continuous with the cavity of the trachea. It is lined with mucous membrane, and its walls consist of the cricothyroid ligament above, and the inner surface of the cricoid cartilage below.

LARYNGEAL MUSCULATURE

The muscles of the larynx are divisible into two groups: (1) extrinsic and (2) intrinsic.

(1) The extrinsic muscles pass between the larynx and neighbouring structures, and are described in the section on Myology (p. 539).

(2) The intrinsic muscles are: the *cricothyroid, posterior crico-arytenoid, lateral crico-arytenoid, transverse arytenoid, oblique arytenoid* and its subsidiary part the *aryepiglotticus, thyro-arytenoid* and its subsidiary part the *vocalis*, and *thyro-epiglotticus*. With the exception of the transverse arytenoid these muscles are paired.

The cricothyroid (**8**.12), triangular in form, arises from the front and lateral part of the outer surface of the cricoid cartilage; its fibres diverge, and are arranged in two groups. The lower fibres constitute the 'oblique' part and slant backwards and laterally to the anterior border of the inferior cornua, while the superior fibres form the 'straight' part and run upwards and backwards to the posterior part of the lower border of the lamina of the thyroid cartilage.

The medial borders of the two muscles are separated by a triangular interval occupied by the conus elasticus.

The posterior crico-arytenoid (**8**.13) arises from the inferomedial region of the broad depression on the corresponding

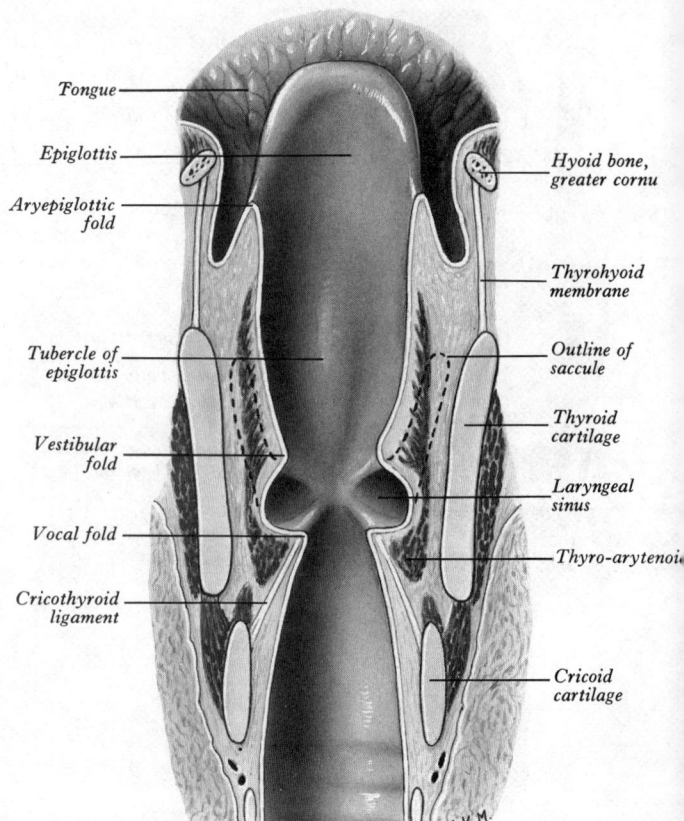

8.9 A coronal section through the larynx and the cranial end of the trachea. Posterior aspect.

half of the posterior surface of the lamina of the cricoid cartilage; its fibres, directed upwards and laterally, converge to be inserted into the back of the muscular process of the arytenoid cartilage. The highest fibres are nearly horizontal, the middle oblique and the lowest almost vertical; some of the latter fibres are inserted into the anterolateral surface of the arytenoid cartilage.

The lateral crico-arytenoid (8.14) is smaller than the preceding muscle; it arises from the upper border of the arch of the cricoid cartilage, and, passing obliquely upwards and backwards, is inserted into the front of the muscular process of the arytenoid cartilage.

The transverse arytenoid (8.13) is a single muscle which bridges the interval between the arytenoid cartilages and fills the posterior concave surfaces of these cartilages. It arises from the back of the muscular process and lateral border of the arytenoid cartilage of one side, and is inserted into the corresponding parts of the cartilage on the opposite side.

The oblique arytenoid (8.13), superficial to the transverse arytenoid, consists of two fasciculi which cross each other like the limbs of the letter X. Each passes from the back of the muscular process of one arytenoid cartilage to the apex of the opposite cartilage. Some of the fibres are continued round the lateral side of the apex of the arytenoid cartilage, and are prolonged into the aryepiglottic fold; they constitute the *aryepiglottic* muscle.

The thyro-arytenoid (8.14) is a broad, thin muscle, which is situated lateral to the vocal fold, the cricothyroid ligament, the sinus and the saccule of the larynx. It arises in front from the lower half of the angle of the thyroid cartilage, and from the crico-thyroid ligament. Its fibres pass backwards, laterally and upwards, to be inserted into the anterolateral surface of the arytenoid cartilage. The lower and deeper fibres of the muscle form a band which, in a coronal section, appears as a triangular bundle, and is attached to the lateral surface of the vocal process and to the inferior impression on the anterolateral surface of the arytenoid cartilage. This muscular bundle is named the **vocalis**, and is parallel with, and just lateral to, the vocal ligament; it is said to be thicker behind than in front, because many deeper fibres start from the vocal ligament, and so do not extend so far forwards as the thyroid cartilage. Other observers do not substantiate this usually accepted description; they consider that all its fibres loop in an arched and plaited manner from the thyroid cartilage to the arytenoid (Tautz and Rohen 1967). A considerable number of the fibres of the thyro-arytenoid are prolonged into the aryepiglottic fold, where some of them cease, while others are continued to the margin of the epiglottis, forming the **thyro-epiglotticus**. A few fibres extend along the wall of the sinus from the lateral margin of the arytenoid cartilage to the side of the epiglottis. The **superior thyro-arytenoid** (8.14), which is not always present, is a small muscle, lying on the lateral surface of the main mass of the thyro-arytenoid and extending obliquely from the angle of the thyroid cartilage to the muscular process of the arytenoid cartilage.

Actions. The muscles of the larynx may be conveniently divided into three groups according to their main actions (Negus 1947): (1) those which vary the glottis—the *posterior* and *lateral crico-arytenoids* and the *oblique* and *transverse arytenoids*; (2) those regulating the tension of the vocal ligaments—the *cricothyroids*, the *posterior crico-arytenoids*, the *thyro-arytenoids* and the *vocales*; (3) those which modify the inlet of the larynx—the *aryepiglottici* and the *thyro-epiglottici* (8.15). It is to be noted that, under normal conditions, the corresponding muscles of the two sides usually act in concert.

The *posterior crico-arytenoids* open the glottis, by rotating the arytenoid cartilage laterally around an axis passing through the crico-arytenoid joints (*see* p. 1232), so that the vocal processes and the attached vocal folds are separated. They also pull back the arytenoids, thus assisting the cricothyroids in making the vocal folds tense. The most lateral fibres of the muscle draw the arytenoids laterally, so that the entire rima glottidis becomes triangular (not rhomboid) when the posterior crico-arytenoids contract.

The *lateral crico-arytenoids* close the glottis, by rotating the arytenoid cartilages medially so as to approximate the vocal processes.

The *transverse arytenoid* approximates the arytenoid cartilages,

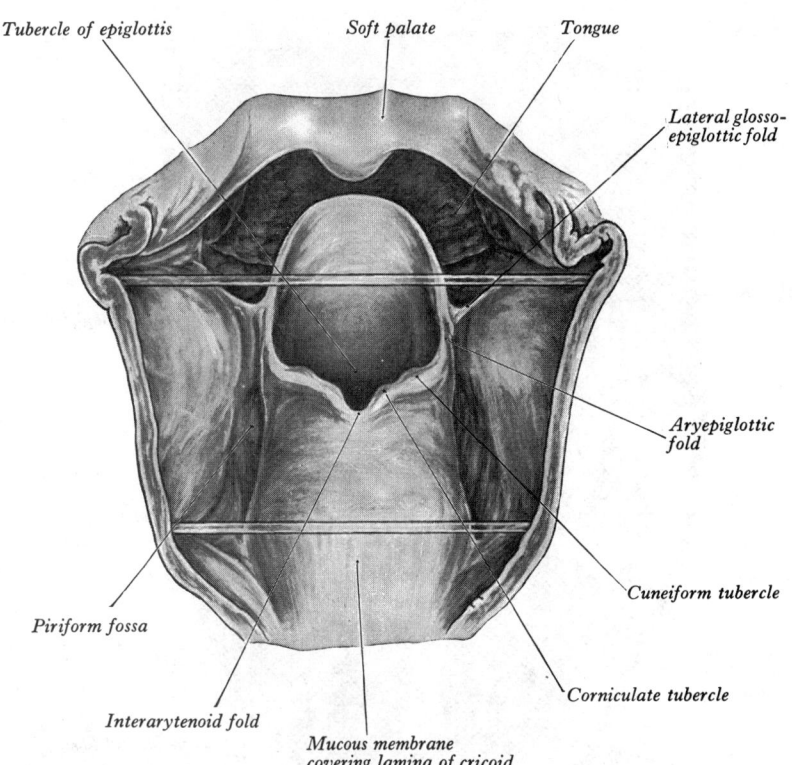

8.10 The inlet of the larynx, viewed from the posterior aspect. The posterior wall of the pharynx has been divided in the median plane and two glass rods have been inserted to keep the cut portions apart.

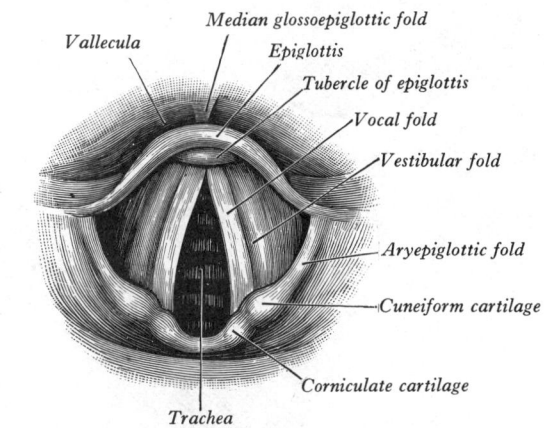

8.11 A laryngoscopic view of the interior of the larynx.

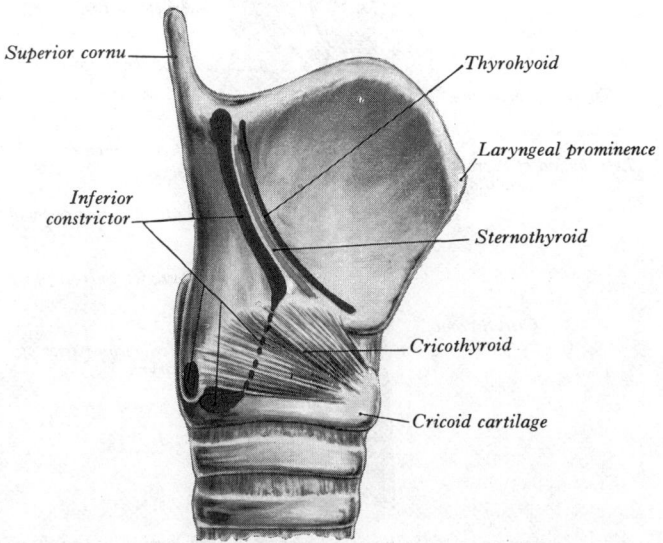

8.12 Lateral view of the larynx, showing the muscular attachments.

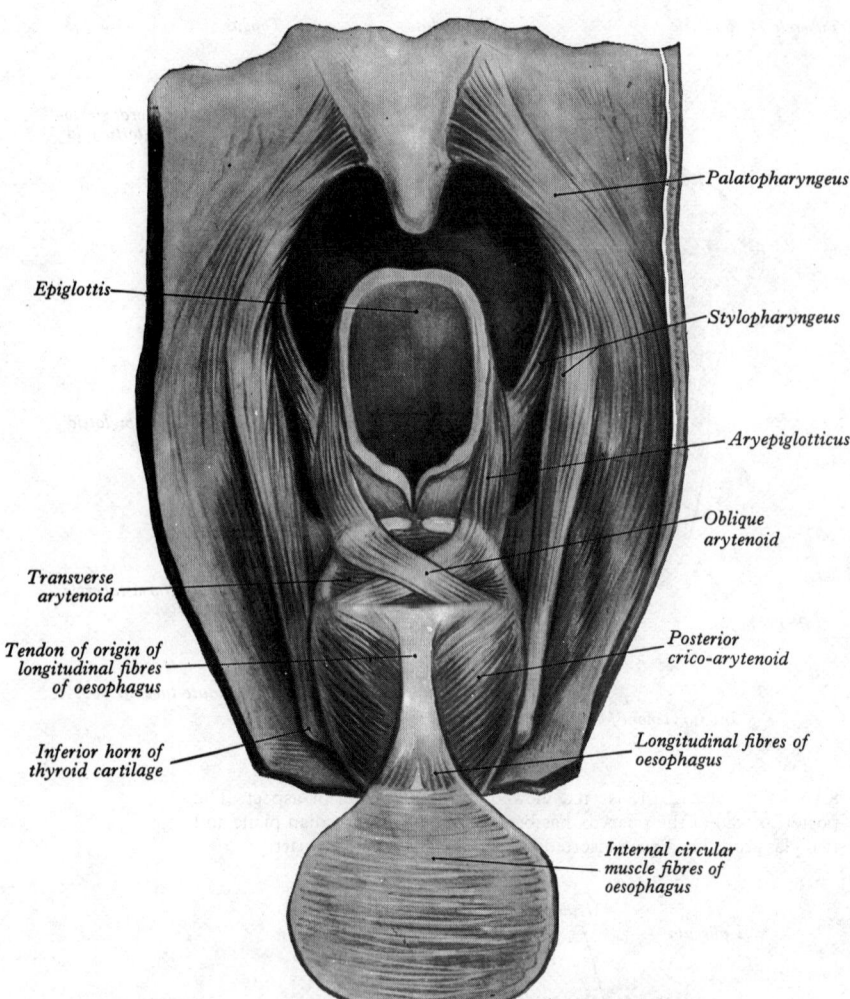

Epiglottis

Transverse
arytenoid

Tendon of origin of
longitudinal fibres
of oesophagus

Inferior horn of
thyroid cartilage

Palatopharyngeus

Stylopharyngeus

Aryepiglotticus

Oblique
arytenoid

Posterior
crico-arytenoid

Longitudinal fibres of
oesophagus

Internal circular
muscle fibres of
oesophagus

8.13   The muscles of the larynx. Posterior aspect.

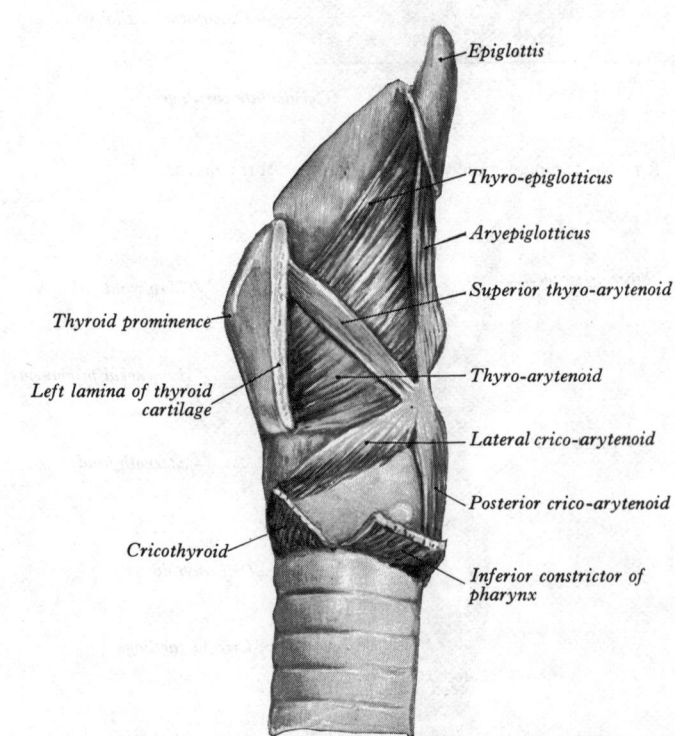

Epiglottis

Thyro-epiglotticus

Aryepiglotticus

Superior thyro-arytenoid

Thyroid prominence

Left lamina of thyroid
cartilage

Thyro-arytenoid

Lateral crico-arytenoid

Posterior crico-arytenoid

Cricothyroid

Inferior constrictor of
pharynx

8.14   The muscles of the larynx. Most of the left lamina of the thyroid
cartilage has been removed. Left lateral aspect.

and thus closes the opening of the glottis, especially in its
posterior part.

The *cricothyroids* produce tension and elongation of the vocal
ligaments by drawing up the arch of the cricoid cartilage and
tilting back the upper border of its lamina; the distance between
the vocal processes and the angle of the thyroid is thus increased,
and the vocal ligaments are consequently put on the stretch. They
also pull the thyroid cartilage forward, increasing the distance
between the angle of the thyroid cartilage and the arytenoids; this
action also renders the vocal folds tense. The latter action of the
muscle is probably the principal one, because during phonation
the lamina of the cricoid is held immovably against the vertebral
column by the action of the cricopharyngeus (p. 1311). During
swallowing, the cricopharyngeus relaxes and allows the cricoid to
be tilted forwards during closure of the inlet of the larynx.

The *thyro-arytenoids* draw the arytenoid cartilages towards the
thyroid cartilage, and thus shorten and relax the vocal ligaments.
At the same time they rotate the arytenoid cartilages medially and
approximate the vocal folds. The deeper fibres, forming the
vocales, produce relaxation of the posterior parts of the vocal
ligaments, while the anterior part is tense, the effect being to raise
the pitch of the voice. For a critique and expansion of the above
simplified views of arytenoid movements, based upon dissection
and electrical stimulation, consult Sellars (1978).

The *oblique arytenoids* and the *aryepiglottici* act as a sphincter of
the inlet of the larynx, by bringing the aryepiglottic folds
together, and by approximating the arytenoid cartilages to the
tubercle of the epiglottis.

The *thyro-epiglottic* muscles widen the inlet by their action on
the aryepiglottic folds.

Neuromuscular spindles are present in all the muscles of the
human larynx (Keene 1961), and their numbers in individual
muscles have been assessed (Voss 1966). The maximal count (23)
was in the transverse arytenoid.

MOVEMENTS OF THE VOCAL FOLDS

In the condition of rest (**8.15**), e.g. in quiet respiration, the
intermembranous part of the rima glottidis is triangular, its apex
being in front and its base behind, the base being represented by a
line (about 8 mm long) connecting the anterior ends of the vocal
processes of the arytenoids; the intercartilaginous part is
rectangular since the medial surfaces of the arytenoids are
parallel.

In forced inspiration, the vocal folds undergo extreme
abduction; the arytenoid cartilages are rotated laterally and their
vocal processes move widely apart. The glottis is thus rhomboid;
both intermembranous and intercartilaginous parts are triangular, the widest part of the aperture being opposite the
attachments of the folds to the vocal processes of the arytenoids.

The movements of the vocal folds during phonation have been
studied by high-speed cinematography (Pressman 1942).
Preparatory to phonation the intermembranous and intercartilaginous parts of the glottis are reduced to a linear chink by
adduction of the vocal folds, and adduction and medial rotation of
the arytenoid cartilages. This is followed by tightening of the
folds, the degree of tension determining the pitch of the sound. As
the pitch rises, the tension of the folds increases and they may
lengthen by as much as 50 per cent in the highest notes. The
photographs suggest that the lengthening affects both extremities
of the folds, indicating that the cricothyroids act not only on the
cricoid cartilage as described above but also tilt the thyroid
cartilage downwards and forwards. In whispering, the intermembranous part of the glottis is closed, but the intercartilaginous part remains widely patent, so that there is free escape of
air during the process.

Fink (1975, 1978) has developed an interpretation of 'laryngeal
biomechanics' which depends particularly upon the behaviour of
the various *folds* which project into the cavity of the larynx to a
highly variable degree dependent upon its activities in respiration, general bodily physical effort, and phonation. In
addition to the well recognized bilateral vocal, vestibular, and
aryepiglottic folds, he identifies two further median folds, which
he designates as the *median thyrohyoid fold* and the *interarytenoid*

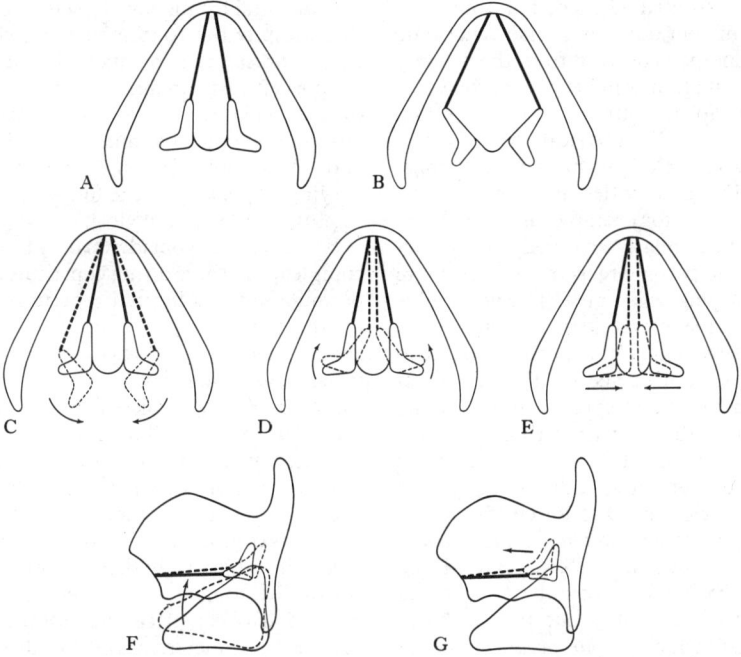

8.15A–G   A series of diagrams to show different positions of the vocal folds and arytenoid cartilages.

A   Position of rest in quiet respiration. The intermembranous part of the rima glottidis is triangular and the intercartilaginous part is rectangular in shape.

B   Forced inspiration. Both parts of the rima glottidis are triangular in shape.

C   Abduction of the vocal folds. The arrows indicate the lines of pull of the posterior crico-arytenoid muscles. The abducted vocal folds and the abducted, retracted and laterally rotated arytenoid cartilages are shown in dotted outline. The entire rima glottidis is triangular.

D   Adduction of the vocal folds. The arrows indicate the lines of pull of the lateral crico-arytenoid muscles. The adducted vocal folds and the medially rotated arytenoid cartilages are shown in dotted outlines.

E   Closure of the rima glottidis. The arrows indicate the line of pull of the transverse arytenoid muscle. Both the vocal folds and the arytenoid cartilages are adducted, but there is no rotation of the latter.

F   Tension of the vocal folds, produced by the action of the cricothyroid muscles which tilt the anterior part of the cricoid cartilage cranially and so carry the arytenoid cartilages dorsally.

G   Relaxation of the vocal folds, produced by the action of the thyro-arytenoid muscles, which draw the arytenoid cartilages ventrally.

fold. The latter consists of the transverse arytenoid muscle, the mucous membrane anterior to which is bulged or *folded* into the larynx when the muscle approximates the two arytenoid cartilages. (It obviously helps to obliterate the intercartilaginous part of the rima glottidis in phonation.) The median thyrohyoid fold is a more complex entity. As can be seen in Fig. **8**.8, the lower part of the epiglottis is attached to the hyoid bone and thyroid cartilage by elastic ligaments, which are separated from the median thyrohyoid ligament by a wedge of adipose tissue; these are the hyo-epiglottic and thyro-epiglottic ligaments (p. 1231). Anterior to the adipose tissue is another elastic ligament, the anterior part of the thyrohyoid membrane including the median thyrohyoid ligament. During swallowing (p. 1313) a marked approximation of the thyroid cartilage and hyoid bone occurs (in addition to the general elevation of pharynx, larynx, and trachea). This approximation causes the structures defined above to bulge posteriorly into the aditus to the larynx as a transverse fold, helping considerably to narrow the entrance to the larynx proper during swallowing. During inspiration the reverse action occurs, and all the folds enumerated above are reduced to a minimum. The intrusion of all the folds, and not just the vocal folds, is also important in phonation since the cavity of the vocal tract (*vide infra*) is thereby altered, thus modifying the resonant properties of the tract. (The *median thyrohyoid fold*, as defined above, contains the so-called '*pre-epiglottic space*' of laryngeal surgery, an important region in the spread of supra-epiglottic tumours. *See* Maguire and Dayal 1974.)

The movements of the larynx, and the manner in which its inlet is closed during deglutition, are described on p. 1313.

THE ANATOMY OF SPEECH

Though the respiratory function of the larynx is basic, its adaptation to speech is paramount to the emergence of human society and culture; the mute, the deaf, the illiterate are at a marked disadvantage. The intelligence, skill, and enterprise of the gregarious primates, especially mankind, is obviously dependent upon clear intercommunication. Language, in all its infinite permutations—firstly in speech, for immediate tribal integrations of effort, and then in recorded ideas and experience, with all the consequent power for accumulation and transmission of knowledge through time and space—is indispensable to human pre-eminence. The acquisition of language is perhaps the most complex sensorimotor development in the individual's life. Large cerebral territories are involved in the sensory, perceptual, and motor aspects of speech (p. 1014), in addition to the auditory system and the intricate assembly of structures producing speech. This assembly includes not only the larynx and other parts of the respiratory system, but also a most extensive array of muscles, from abdominal wall to lips. These muscles, apart from those of the larynx, are described elsewhere. Here we are concerned with the larynx and its associated respiratory spaces.

The larynx is the primary source of the complicated and endlessly varying chains of sounds which are the basis of speech. It is in this context, therefore, that its structure must now be considered, including the associated respiratory spaces—the pharynx, mouth and nasal cavities—which together with it constitute a 'vocal tract'. The physics of this in phonation, the production of sounds and their articulation into *phonemes*, the basic units of speech, has attracted less attention than other aspects of linguistics. These latter have engendered a vast literature, further augmented in recent years by a growing interdisciplinary approximation between workers in such diverse fields as 'hemispheric' anatomy, neurophysiology, and the 'psychological' aspects of speech. Such investigations are outside the compass of this volume; but the acoustics of speech, in terms of the structure of the vocal tract, must be briefly considered.

As in the case of most musical instruments, the speech

mechanism consists of three essential elements: a source of energy, structures capable of periodic and also aperiodic oscillations, and a resonator. Energy is derived from the velocity of the stream of expired air, oscillation primarily from the vocal folds, resonance from the multiform 'column' of air extending from the folds to the lips and nostrils. The physical activities in all three are intricate and rapidly variable, and must require most complex neural integration. These activities are, nevertheless, subject to the same physical laws as, for example, an organ pipe. They can hence, in principle, be similarly analysed, though not with the same metrical simplicity or precision as the more regularly constructed and unvarying organ pipe. The vocal folds vibrate in accordance with the same interactions of length, mass, and tension as a vibrating reed or string. The available increase in length, almost 50 per cent, does not suffice for the octaves of frequencies, on average, characteristic of speech (and certainly not for the range used in singing). Ranges vary with age and sex, the total range for human speech being of the order of 60 to 500 cycles per second, with average central speech frequencies of about 100 for males, 200 for females, and 250 for children. The individual range is achieved by variations in *tension* and *mass*, as well as length. Tensor muscles such as the cricothyroid and posterior crico-arytenoid (p. 1234) affect not only tension, but also length. Changes in mass are probably due to the thyro-arytenoid muscles, a slip of each of which, its *pars vocalis* (p. 1235), is partly attached to its vocal ligament, and may thus play a complex role in adjustments of length, tension and mass. Other intrinsic muscles are of course involved; the geometry of their attachments (p. 1234) is an indication of their effects. Unfortunately, electromyographic and photographic techniques have not provided precise data, and it is as yet unprofitable to attempt detailed analyses of their contribution to phonational mechanics.

The modes of vibration of the vocal folds are not simple (consult Van den Berg 1968, Wyke 1974, Hinchcliffe and Harrison 1976, and Fry 1979, for modern accounts). Not only are numerous harmonics produced above any fundamental tone or frequency, but also the orientation of waves in the vocal folds is complex. High-speed cinematography reveals the occurrence of transverse and longitudinal waves, and the distribution of these over the surface of the folds varies even during a single act of phonation. Commonly, at the start of phonation the glottis closes to a linear chink (rima glottidis) both in its intramembranous and intercartilaginous parts (**8**.15), the latter being approximated by rotation of the arytenoid cartilages. The vocal folds come into complete contact, though this varies in the precise areas apposed, and the parameters of the apposition may vary during production of a single 'sound'. As expiratory air pressure is increased below the approximated folds, these are suddenly forced apart, returning by their own elasticity, because pressure momentarily drops. This cycle of events is repeated at a fundamental frequency (and its harmonics) dependent upon the factors stated above, and upon the velocity and pressure of the air current. Re-apposition of the folds may be aided by muscular effort and also by the 'Bernouilli effect', a kind of suction due to the sudden decrease in pressure as the folds open. The periodicity of these events is measured in milliseconds. They impart pressure waves of like frequencies to the column of air above the vocal folds, which acts as a selective resonator. To describe this as a 'column' is, of course, an oversimplification, for it is not only most variable in its shape and dimensions from level to level, and tortuous in its axis, but also adjustable within wide limits in respect of these same properties and in the tension of its walls. The positioning of the tongue, of the fauces, the soft palate, and changes in other dimensions, all greatly modify the parameters of the vocal tract. The sound impressed upon the resonator, in all the dynamic variation of fundamental tones and their accompanying harmonics, is thus subjected to a second stage of modification, some harmonics being dampened and others enhanced. Techniques are available for the analysis of this ultimate amalgam or 'spectrum' of frequencies in any particular sounds, which may be represented graphically against the coordinates of frequency and amplitude. Such investigations have shown that the vocal tract acts as an intricately selective filter and resonator, which propagates a remarkably similar pattern or 'envelope', irrespective of the fundamental frequency. This is essential to speech, for it ensures that, despite the continuously varying tone of voice, a constant *quality* or *timbre* is maintained. The term *formant* is also used, as in musical instruments; for example, a flute sounds like a flute over a wide register of frequencies, because the formant is characteristic throughout. Each voice similarly has an individual and identifiable formant or quality; if this were constantly varying, especially in the tones given to vowel sounds, linguistic intelligibility would be lost.

So far we have considered only the production of *tones*, however complex, and these do not constitute speech, in which the phonetic permutation of a limited repertoire of vowels is vastly enriched by the imposition of consonants, or *articulation*. Consonants are associated with particular anatomical sites, from which they usually take their designations in the terminology of phonetics. For example, we speak of *labials* (*p* and *b*), *dentals* (*t* and *d*), *nasals* (*m* and *n*) and so on. These sites have two similarities: (*a*) a partial obstruction or constriction at some level of the vocal tract, and (*b*) the production of aperiodic vibration, i.e. noise, which is superimposed upon or interrupts the flow of laryngeal tones. The subject is complex and can only be touched upon here. It is easy to ascertain subjectively that in pronouncing dental consonants the tip of the tongue is apposed to the back of the teeth. This momentarily constricts the passage of escaping air, modifies the resonant parameters of the 'vocal tract', and also generates local noise. The vocal tract is considered to be divided anatomically, in this instance, into a long column stretching from vocal folds to lingual tip and a very short section from teeth to lips. The 'resonator' may similarly be divided by appositional constructions into less unequal moieties by approximation of the tongue to the palate (hard or soft), and at other levels. Each 'setting' of the anatomical resonator, like those of the vocal folds, adds a characteristic to the total acoustic product which must be identifiable as an event of significance in the stream of speech if this is to be intelligible.

It is important to appreciate that these anatomical changes in the larynx and its resonating vocal tract are effected in small fractions of a second, with a speed, adroitness, and subtlety which are difficult to convey in verbal description. The multiplicity of laryngeal, pharyngeal, hyoid, palatal, lingual, and circumoral muscles, combined in rapidly changing combinations in the mechanics of phonation and articulation, reflects the extreme complexity of speech. It is perhaps not surprising that one philospher (Chomsky 1965) has suggested that the cerebrum, and particularly its 'dominant hemisphere', is genetically programmed for language (p. 1024). Quite apart from these wider implications, phonetics is in its own right a larger study, which can only be superficially considered here, even its strictly anatomical connexions. But enough has been said, perhaps, to demonstrate the usefulness of the anatomy of speech to phoneticians and others, and to stimulate the anatomist to a deeper interest in the phonetic functions of the larynx and vocal tract.

LARYNGEAL MUCOUS MEMBRANE

The mucous membrane of the larynx is continuous above with that of the pharynx and mouth, below with that of the trachea. It is loosely attached to the anterior surface of the epiglottis, and to the subjacent tissues in the valleculae. It covers the aryepiglottic folds, which limit the inlet of the larynx; in these folds there is a considerable amount of fibro-areolar tissue. It lines the cavity of the larynx, forms, by its reduplication, the chief parts of the vestibular folds, and is continued into the sinus and saccule of the larynx. It is firmly attached to the posterior surface of the epiglottis and to the laryngeal surfaces of the cuneiform and arytenoid cartilages. The parts covering the vocal ligaments are thin and intimately adherent to them. On the anterior surface and the upper half of the posterior surface of the epiglottis, the upper part of the aryepiglottic folds, and the vocal folds, the epithelium of the mucous membrane is stratified squamous (**8**.4B); other areas of stratified squamous epithelium also occur above the plane of the glottis. The rest of the epithelium of the laryngeal mucous membrane is ciliated columnar in type.

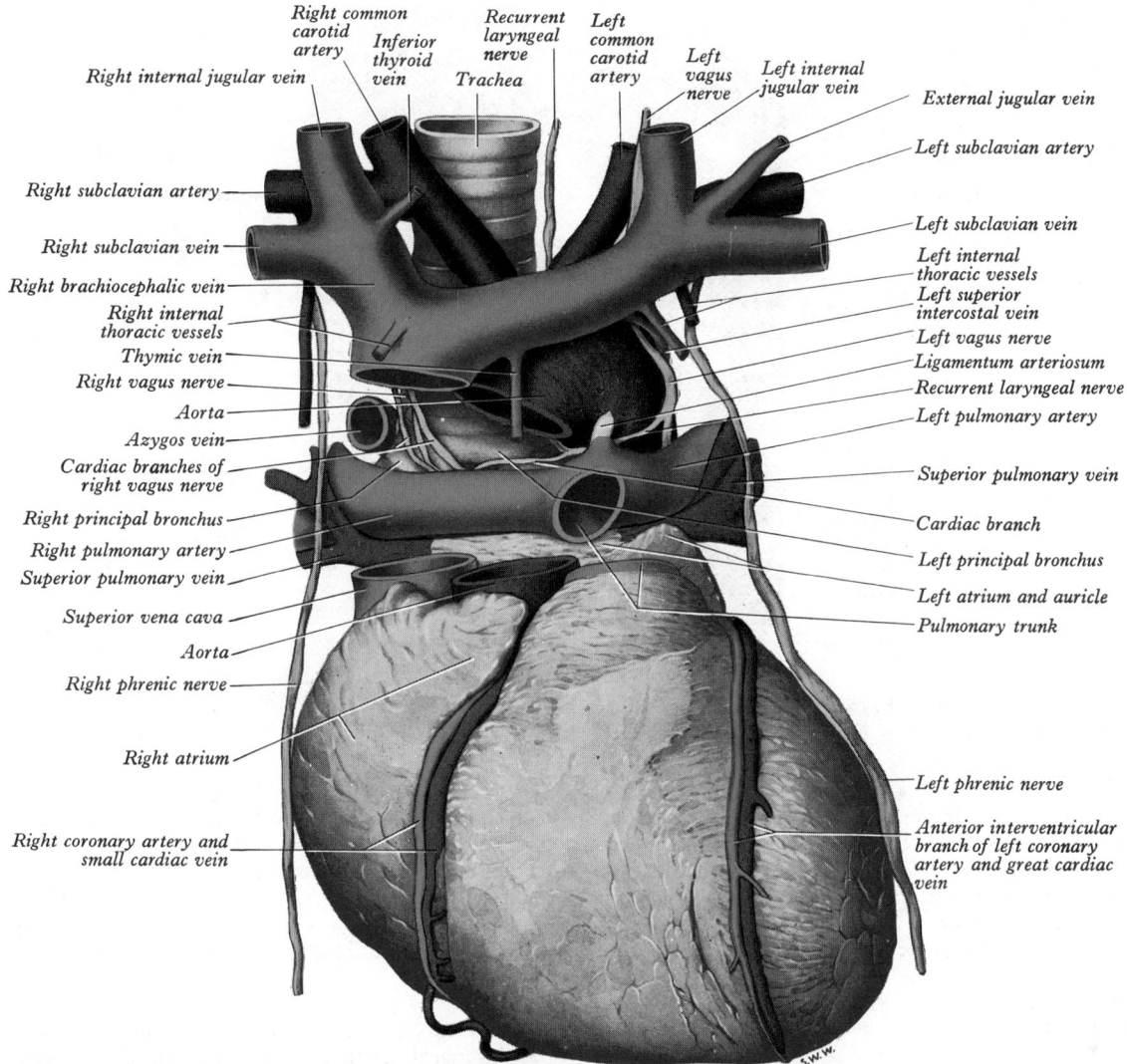

Right common carotid artery
Right internal jugular vein
Inferior thyroid vein
Recurrent laryngeal nerve
Trachea
Left common carotid artery
Left vagus nerve
Left internal jugular vein
External jugular vein
Left subclavian artery
Right subclavian artery
Right subclavian vein
Right brachiocephalic vein
Right internal thoracic vessels
Thymic vein
Right vagus nerve
Aorta
Azygos vein
Cardiac branches of right vagus nerve
Right principal bronchus
Right pulmonary artery
Superior pulmonary vein
Superior vena cava
Aorta
Right phrenic nerve
Right atrium
Right coronary artery and small cardiac vein
Left subclavian vein
Left internal thoracic vessels
Left superior intercostal vein
Left vagus nerve
Ligamentum arteriosum
Recurrent laryngeal nerve
Left pulmonary artery
Superior pulmonary vein
Cardiac branch
Left principal bronchus
Left atrium and auricle
Pulmonary trunk
Left phrenic nerve
Anterior interventricular branch of left coronary artery and great cardiac vein

8.16 A dissection to show the relations of the bifurcation of the trachea and the principal bronchi. Parts of the great vessels have been resected. Vessels containing oxygenated blood are shown in red; those shown in blue contain de-oxygenated blood.

Glands. The mucous membrane of the larynx is furnished with numerous mucous glands; they are very plentiful upon the epiglottis, where they are lodged in little pits; many are present in the margins of the aryepiglottic folds in front of the arytenoid cartilages, where they are termed the *arytenoid glands*. They are large and numerous in the saccules of the larynx and secretion has been observed to flow down over the vocal folds periodically during phonation. The free edges of the vocal folds are devoid of glands. The stratified epithelium covering the vocal folds is, of course, vulnerable to drying, and it is essential that the secretion of neighbouring glands should spread over the folds and be retained. Scanning electron microscope observations have recently demonstrated the existence not only of microvilli, but also of so-called micro-ridges (microplicae) on the surface cells of the epithelium of the folds and elsewhere in the larynx (Andrews 1975; Tillmann *et al.* 1977). Such features have been noted in other epithelia subjected to drying out (e.g. the corneal epithelium), and microplicae are regarded as conducive to retention of surface secretions. Drying of the vocal epithelium interferes with phonation and may lead to superficial necrosis.

Taste buds, similar to those in the tongue, are scattered over the posterior surface of the epiglottis, in the aryepiglottic folds, and less regularly in other parts of the larynx. Their centripetal pathway is through the vagus nerve.

LARYNGEAL VESSELS AND NERVES

The chief *arteries* of the larynx are the laryngeal branches of the superior and inferior thyroid arteries. The *veins* accompanying the superior laryngeal artery join the superior thyroid vein, which opens into the internal jugular vein; those accompanying the inferior laryngeal artery join the inferior thyroid vein, which opens into the left brachiocephalic vein. The *lymph vessels* are divisible into two sets, a superior above the vocal folds, and an inferior below; the superior vessels accompany the superior laryngeal artery, pierce the thyrohyoid membrane, and end in the deep cervical lymph nodes situated near the bifurcation of the common carotid artery; some of the inferior lymph vessels pierce the cricothyroid ligament and open into a lymph node lying in front of that ligament or in front of the upper part of the trachea, while others emerge below the cricoid cartilage and pass to the deep cervical lymph nodes and to the lymph nodes alongside the inferior thyroid artery. The *nerves* are derived from the internal and external branches of the superior laryngeal nerve, from the recurrent laryngeal nerve and from the sympathetic. The internal laryngeal branch is probably entirely sensory and autonomic though some claim that special visceral motor fibres in it innervate the transverse arytenoid (Williams 1951). The nerve enters through the postero-inferior part of the thyrohyoid membrane above the superior laryngeal artery, and divides into branches which supply both surfaces of the epiglottis, the aryepiglottic fold, and the interior of the larynx down as far as the level of the vocal folds. The external laryngeal branch supplies the cricothyroid by entering its lateral surface. The terminal part of the recurrent laryngeal nerve accompanies the laryngeal branch of the inferior thyroid artery, and passes upwards deep to the lower border of the inferior constrictor of the pharynx, immediately behind the cricothyroid joint. It supplies all the

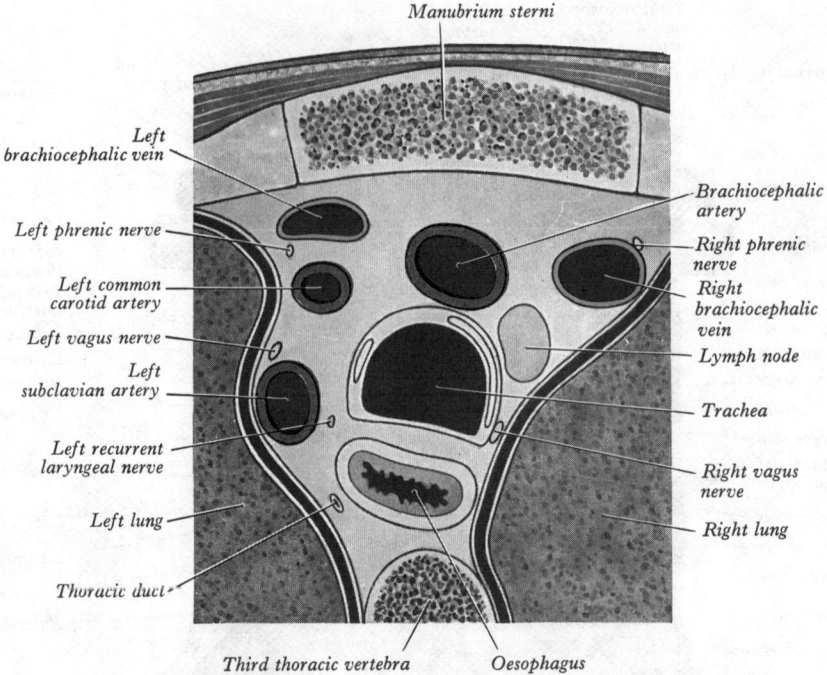

Manubrium sterni

Left brachiocephalic vein

Left phrenic nerve

Left common carotid artery

Left vagus nerve

Left subclavian artery

Left recurrent laryngeal nerve

Left lung

Thoracic duct

Third thoracic vertebra

Brachiocephalic artery

Right phrenic nerve

Right brachiocephalic vein

Lymph node

Trachea

Right vagus nerve

Right lung

Oesophagus

8.17 A transverse section through the mediastinum at the level of the body of the third thoracic vertebra. Viewed from the superior aspect.

intrinsic muscles of the larynx except the cricothyroid, and distributes sensory branches to the laryngeal mucous membrane below the level of the vocal folds. Before entering the larynx it usually divides into a motor and a sensory branch—not into 'adductor' and 'abductor' rami as has sometimes been asserted (Williams 1954). As might be expected in such highly skilled musculature, motor units in the larynx are small; a ratio of 30 muscle fibres to each motor neuron has been estimated (English and Blevins 1969).

Laryngoscopic Examination

The inlet of the larynx, the structures surrounding it, and the cavity of the larynx can be inspected with a laryngoscopic mirror (**8.11**). The epiglottis is seen much foreshortened, but its tubercle can be seen in the median plane. From the margins of the epiglottis the aryepiglottic folds can be traced backwards and medially and, at their posterior extremities, the elevations produced by the cuneiform and the corniculate cartilages can be recognized. The pink vestibular folds and the pearly white vocal folds are visible within the cavity of the larynx and, when the rima glottidis is opened widely, the mucosa and cartilages of the trachea are visible. The piriform fossae can also be inspected.

Radiography

In lateral cervical radiographs, in addition to the ossified parts of the laryngeal cartilages, the epiglottis, aryepiglottic folds, arytenoid, corniculate and sometimes the cuneiform cartilages, and the laryngeal sinus are all visible.

Surface and Applied Anatomy

In or near the midline of the neck the following structures are readily identified (**5.33**). The laryngeal prominence can be felt

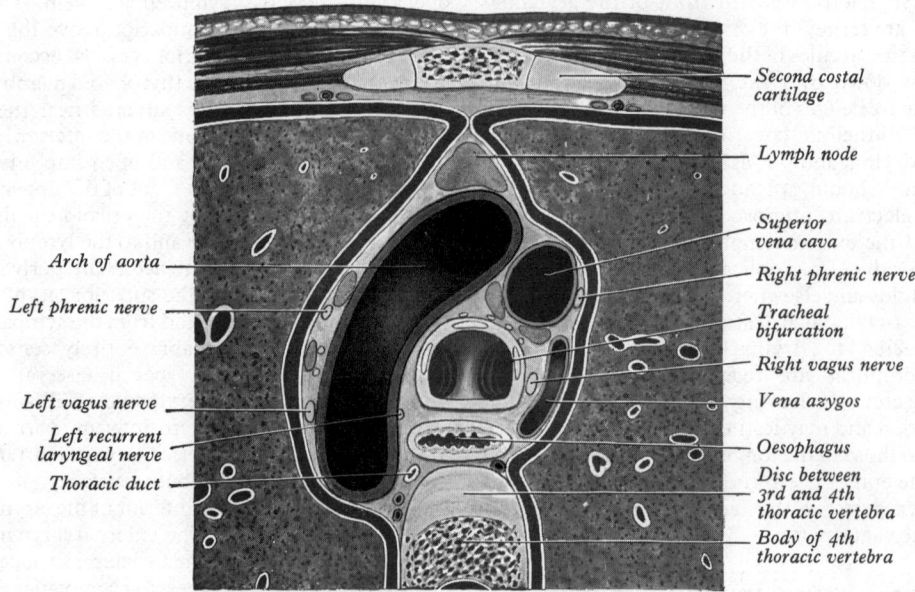

Second costal cartilage

Lymph node

Superior vena cava

Right phrenic nerve

Tracheal bifurcation

Right vagus nerve

Vena azygos

Oesophagus

Disc between 3rd and 4th thoracic vertebra

Body of 4th thoracic vertebra

Arch of aorta

Left phrenic nerve

Left vagus nerve

Left recurrent laryngeal nerve

Thoracic duct

8.18 A transverse section through the mediastinum at the level of the upper part of the body of the fourth thoracic vertebra. Viewed from the superior aspect.

easily; it is visible in men but not always in women. The anterior parts of the upper borders of the laminae of the thyroid cartilage and the thyroid notch are palpable. (The vocal folds are approximately level with the mid point of the anterior border of the thyroid.) Above the thyroid, the body of the hyoid and its greater cornua can be palpated, the latter most readily by gripping the throat at this level between the thumb and finger. The thyrohyoid membrane lies in the depression between the thyroid cartilage and the hyoid. Below the thyroid, the arch of the cricoid can be felt; it is on a level with the lower part of the cricoid lamina and lies opposite the body of the sixth cervical vertebra. The depression between the cricoid and thyroid marks the crico-thyroid ligament. Below the cricoid the first tracheal cartilage can be felt. Foreign bodies may become impacted in the inlet of the larynx or in the rima glottidis and cause suffocation by mechanical obstruction. Small masses may descend into the trachea or bronchi, or lodge in the laryngeal sinus and, by irritating the mucous membrane, cause reflex spasm of the glottis with consequent suffocation. Inflammation of the upper part of the larynx may cause considerable swelling of the mucous membrane through effusion of fluid into the abundant lax submucous tissue; this condition is called 'oedema of the glottis'. The effusion does not involve, or extend below, the vocal folds, since the mucous membrane is closely adherent to the vocal ligaments without the intervention of any submucous tissue. In the above cases, an incision into the larynx below the vocal folds through the cricothyroid ligament (laryngotomy) or into the trachea (tracheotomy) may be necessary to restore a free airway. The mucous membrane of the upper part of the larynx is extremely sensitive, and contact with foreign bodies causes an immediate explosive cough. In suicidal cut-throat the wound is usually through the thyrohyoid membrane damaging the epiglottis, superior thyroid vessels, external and internal carotid arteries and internal jugular veins; less frequently it is above the hyoid with damage to the tongue muscles and the lingual and facial vessels. (For results of damage to the laryngeal nerves, *see* p. 1082.)

The Trachea and Bronchi

THE TRACHEA

The trachea is a cartilaginous and membranous tube, about 10 or 11 cm long continued downwards from the lower part of the larynx (**8.**19, 20). In the cadaver it reaches from the level of the sixth cervical vertebra to that of the upper border of the fifth thoracic vertebra, where it divides into two principal bronchi, one for each lung. The trachea lies mainly in the median plane, though its point of bifurcation usually lies a little to the right of this. In the living subject during deep inspiration the bifurcation descends and may reach the level of the sixth thoracic vertebra. The trachea is very mobile and can extend and shorten very rapidly. It is not quite cylindrical, being flattened posteriorly; its external diameter from side to side is about 2 cm in the adult male, and 1·5 cm in the adult female. In the child the trachea is smaller, more deeply placed and more movable than in the adult. In the living the lumen is smaller than in the cadaver, its diameter in the adult being about 12 mm. In the first year of life the diameter does not

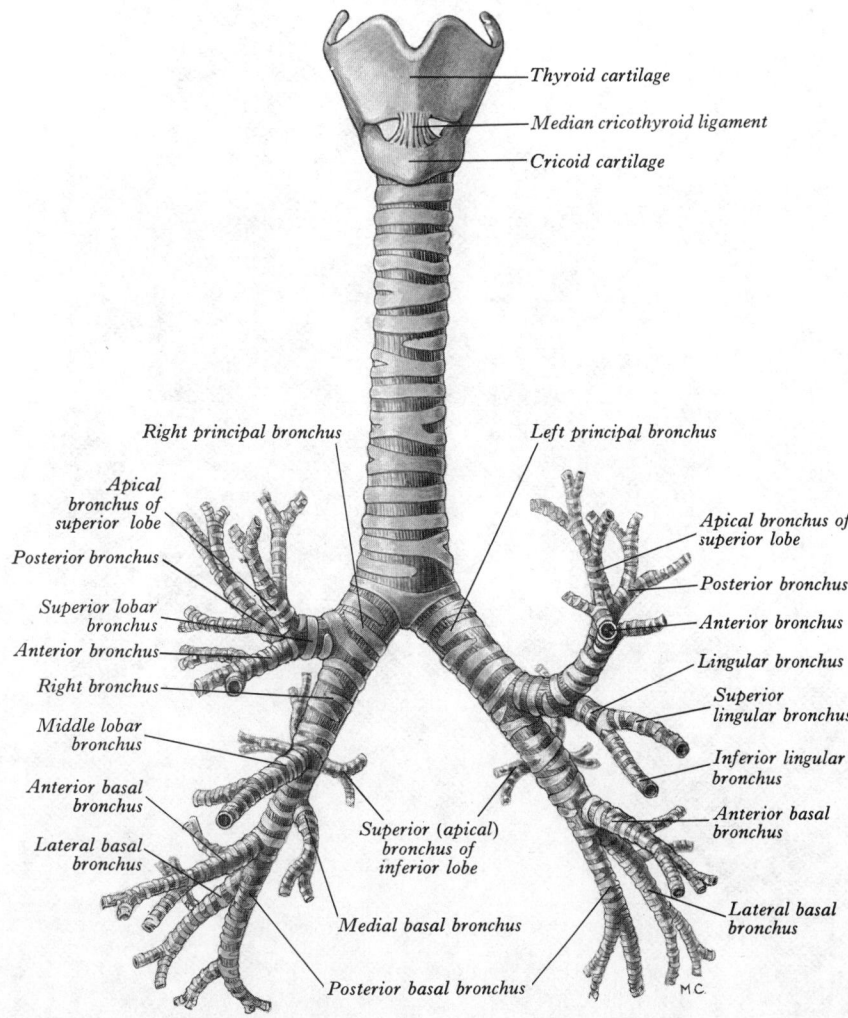

Thyroid cartilage

Median cricothyroid ligament

Cricoid cartilage

Right principal bronchus

Left principal bronchus

Apical bronchus of superior lobe

Posterior bronchus

Superior lobar bronchus

Anterior bronchus

Right bronchus

Middle lobar bronchus

Anterior basal bronchus

Lateral basal bronchus

Apical bronchus of superior lobe

Posterior bronchus

Anterior bronchus

Lingular bronchus

Superior lingular bronchus

Inferior lingular bronchus

Anterior basal bronchus

Lateral basal bronchus

Superior (apical) bronchus of inferior lobe

Medial basal bronchus

Posterior basal bronchus

M.C.

8.19A   The cartilages of the larynx, trachea and bronchi. Anterior aspect. Drawn from a metal cast made by Mr. R. C. (now Lord) Brock, Guy's Hospital, London.

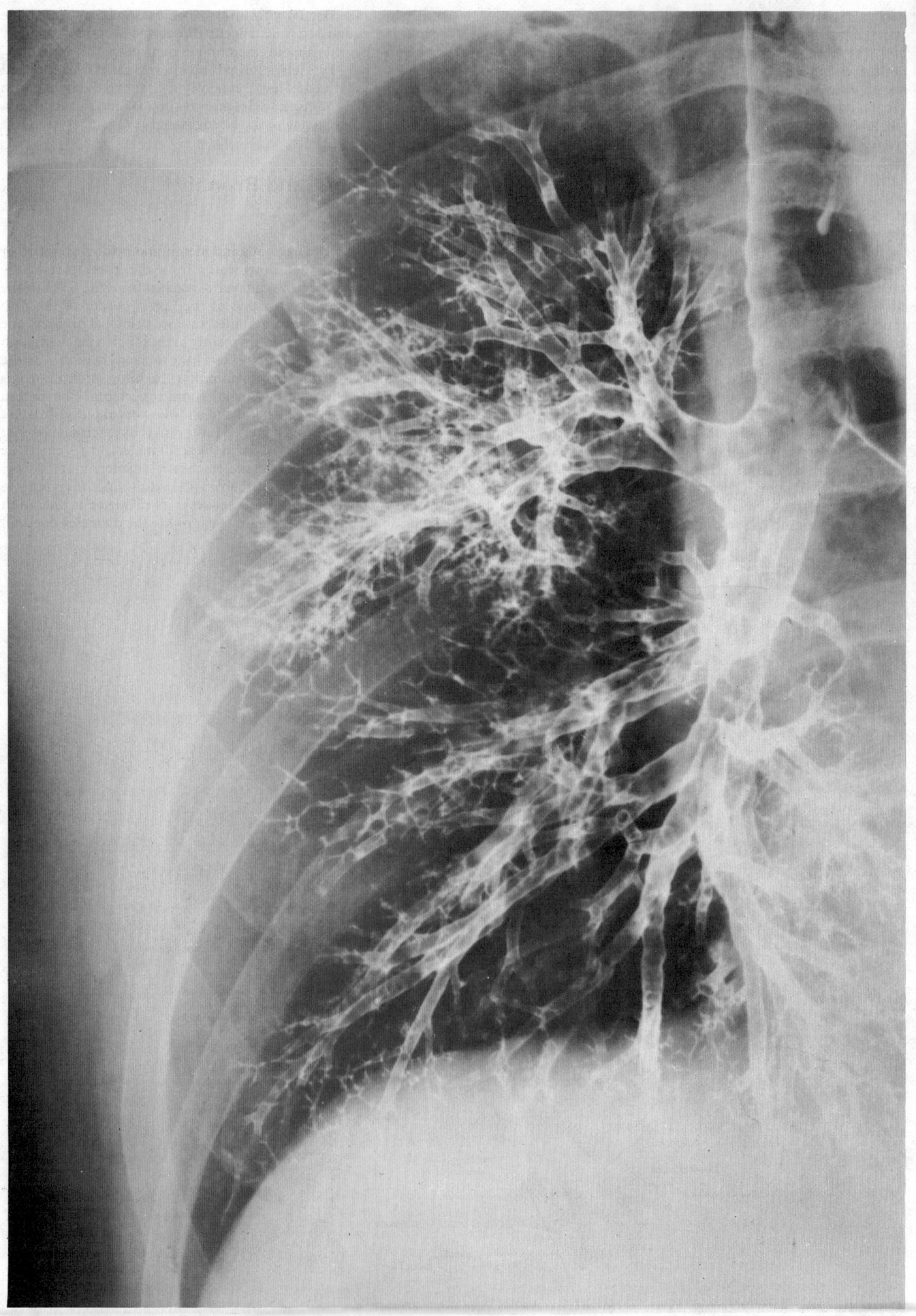

8.19B   A slightly oblique, anteroposterior view of the trachea and right bronchial tree, after insufflation of a radio-opaque contrast medium. For identification of the lobar and segmental bronchi in this bronchogram, compare with illustrations of bronchial tree casts (**8.**19A and **8.**20).

exceed 3 mm, while during childhood the diameter in millimetres corresponds approximately with the age in years. The shape of the lumen in transverse section is very variable, especially in later decades. Thus it may be rounded, lunate, or flattened (Campbell and Liddelow 1967). For an extensive study of tracheobronchial dimensions (in Chinese subjects), *see* Wang and Tai (1965). In this study the lumen of the trachea averaged 16·17 mm, (range 9·5–22·0 mm).

Relations of the Trachea

The *cervical part* of the trachea (**6**.43) is covered anteriorly with the skin and the superficial and deep fasciae. It is crossed by the jugular arch connecting the anterior jugular veins and overlapped by the sternohyoids and sternothyroids. The second, third and fourth tracheal cartilages are crossed by the isthmus of the thyroid gland; immediately above the isthmus an anastomosing vessel connects the two superior thyroid arteries; below the isthmus it is related, in front, to the pretracheal fascia, the inferior thyroid veins, the remains of the thymus, and the arteria thyroidea ima (when that vessel exists). In the child, the brachiocephalic artery crosses obliquely in front of the trachea at, or a little above, the level of the upper border of the manubrium sterni, and the left brachiocephalic vein may extend a little above the upper border of the manubrium. Posteriorly the trachea is related to the oesophagus, which intervenes between it and the vertebral column; the recurrent laryngeal nerves ascend, one on each side, in, or just in front of, the grooves between the sides of the trachea and the oesophagus. *Laterally* the trachea is related to the lobes of the thyroid gland, which descends to the level of the fifth or sixth tracheal cartilage, and to the common carotid and inferior thyroid arteries.

The *thoracic part* of the trachea (**8**.16, 17, 18) descends through the superior mediastinum. *Anteriorly*, it is related to the manubrium sterni, the origins of the sternohyoids and sternothyroids, the remains of the thymus, the inferior thyroid veins, the left brachiocephalic vein, the arch of the aorta, the brachiocephalic and left common carotid arteries, the deep part of the cardiac plexus of nerves, and some lymph nodes. Owing to the divergence of the brachiocephalic and left common carotid arteries as they ascend in the neck, the former vessel comes to lie on the right, and the latter on the left of the trachea. *Posteriorly*, it is related to the oesophagus, by which it is separated from the vertebral column. On the *right* it is related to the right lung and pleura, the right brachiocephalic vein, the superior vena cava, the right vagus nerve, and the azygos vein; on the *left* to the arch of the aorta, the left common carotid and left subclavian arteries. The left recurrent laryngeal nerve, in its upward course, lies at first between the trachea and the arch of the aorta, and then in, or just in front of, the groove between the trachea and the oesophagus.

THE RIGHT PRINCIPAL BRONCHUS

The right principal bronchus (**8**.19, 20), wider, shorter and more vertical than the left, is about 2·5 cm long, and enters the right lung opposite the fifth thoracic vertebra. Its greater width and more vertical course result in a greater frequency of foreign bodies passing into the right than into the left principal bronchus. The azygos vein arches over its superior aspect; the right pulmonary artery lies at first inferior and then anterior to it. After giving off the *superior lobar bronchus* which arises posterosuperior to the pulmonary artery, the right principal bronchus crosses the posterior aspect of the artery to enter the hilum of the lung, postero-inferior to the artery, where it divides into a *middle* and *inferior lobe bronchus*.

The right superior lobar bronchus arises from the lateral aspect of the parent bronchus and runs superolaterally to enter the hilum of the lung. About 1 cm from its origin it divides into three segmental bronchi. One of these, the *apical segmental bronchus*, continues superolaterally towards the apex of the lung, to which it is distributed. Shortly after its origin it divides into an apical and an anterior branch. The second, the *posterior segmental bronchus*, serves the postero-inferior part of the superior lobe of the lung and is directed posterolaterally with a slight superior inclination. Soon after its origin it divides into a lateral and a posterior branch.

The third, the *anterior segmental bronchus*, runs antero-inferiorly to supply the rest of the superior lobe; not far from its origin it divides into a lateral and an anterior branch which are of equal size.

The middle lobar bronchus arises about 2 cm below the origin of the superior lobar bronchus, from the anterior aspect of the parent trunk and descends anterolaterally. It soon divides into a *lateral segmental bronchus* to the lateral part of the middle lobe and a *medial segmental bronchus* to its medial part.

The right inferior lobar bronchus is the continuation of the principal stem beyond the origin of the middle lobar bronchus. At or a little below this point, a large *superior (apical) segmental bronchus* arises from the posterior surface of the inferior lobar bronchus and runs posteriorly to be distributed to the cranial part of the inferior lobe of the right lung. This bronchus subsequently divides into medial, superior and lateral branches, the former two usually arising from a common stem.

Thereafter, the right inferior lobar bronchus continues to descend posterolaterally. The *medial basal segmental bronchus* is given off from its anteromedial aspect, and runs inferomedially to serve a small region below the hilum of the lung. The inferior lobar bronchus continues downwards and then divides into an *anterior basal segmental bronchus* which descends anteriorly, and a trunk which divides almost immediately into a *lateral basal segmental bronchus* which descends laterally, and a *posterior basal segmental bronchus* which descends posteriorly.

In over 50 per cent of right lungs a *subsuperior* (subapical) *segmental bronchus* arises from the posterior surface of the right inferior lobar bronchus between 1 and 3 cm below the superior segmental bronchus. This is distributed to the region of lung between the superior and posterior basal segments.

THE LEFT PRINCIPAL BRONCHUS

The left principal bronchus (**8**.19, 20), narrower, and more transverse than the right, is nearly 5 cm long and enters the root of the left lung opposite the sixth thoracic vertebra. It passes to the left inferior to the aortic arch, and crosses anterior to the oesophagus, thoracic duct and descending aorta; the left pulmonary artery lies at first anterior, and then superior to it. Having entered the hilum of the lung it divides into a superior and an inferior lobal bronchus.

The left superior lobar bronchus arises from the antero-lateral aspect of the parent bronchus. It curves laterally for a short distance and then divides into two bronchi, which correspond to the branches of the right principal bronchus to both the superior and middle lobes of the right lung. They are both distributed to the superior lobe of the left lung, which does not possess a separate middle lobe. The cranial division ascends for about 1 cm before giving off an *anterior segmental bronchus*, and then continues its cranial course for a further 1 cm as the *apicoposterior segmental bronchus*, before dividing into apical and posterior branches. The apical, posterior and anterior segmental bronchi are distributed, for the most part, in a similar manner to the corresponding bronchi of the superior lobe of the right lung. The caudal division of the left superior lobar bronchus descends anterolaterally to be distributed to the antero-inferior part of the superior lobe of the left lung. The latter is termed the *lingula*; unlike the bronchus to the middle lobe of the right lung which divides into medial and lateral segmental bronchi, the *lingular bronchus* divides into *superior lingular* and *inferior lingular segmental bronchi*.

The left inferior lobar bronchus continues to descend posterolaterally for about 1 cm. At this point the *superior* (apical) *segmental bronchus* arises from its posterior surface and is distributed in essentially the same manner as in the right lung. The inferior lobar bronchus continues for a further 1–2 cm before dividing into two stems, an anteromedial and a posterolateral. The *medial basal segmental bronchus* arises in common with the *anterior basal segmental bronchus* from the former; the *lateral basal segmental bronchus* arises in common with the *posterior basal segmental bronchus* from the latter. The territories supplied by these basal segmental bronchi is similar to those on the right. The medial basal segmental bronchus arises independently from the

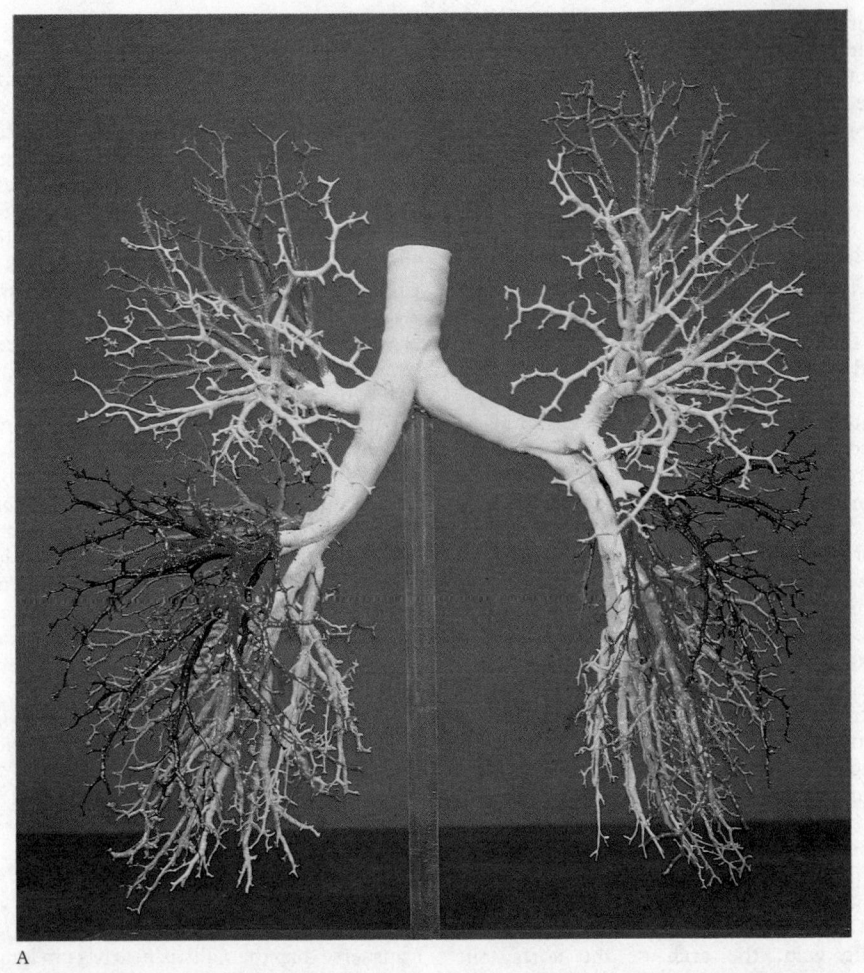

A

B

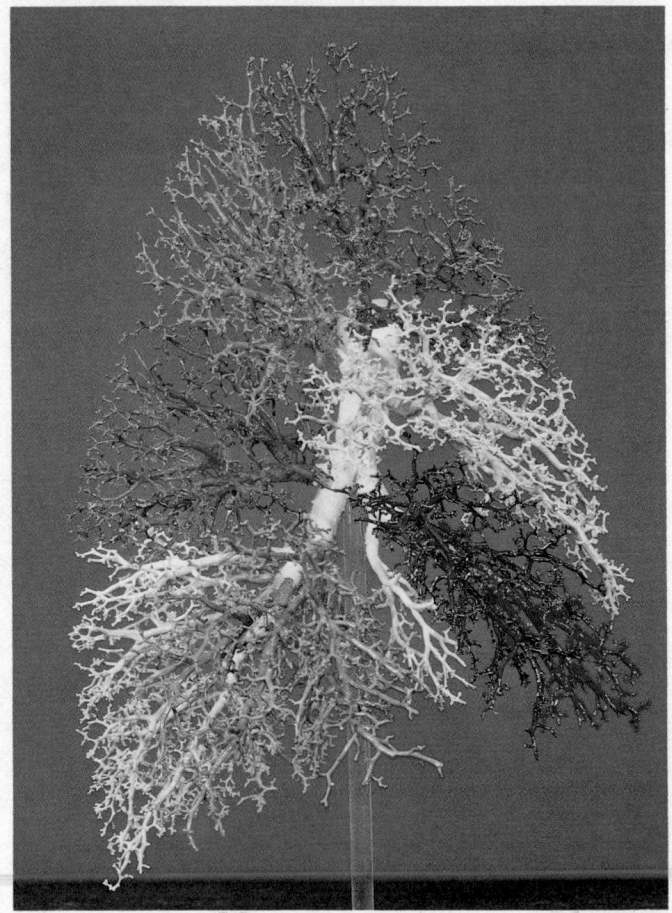

C

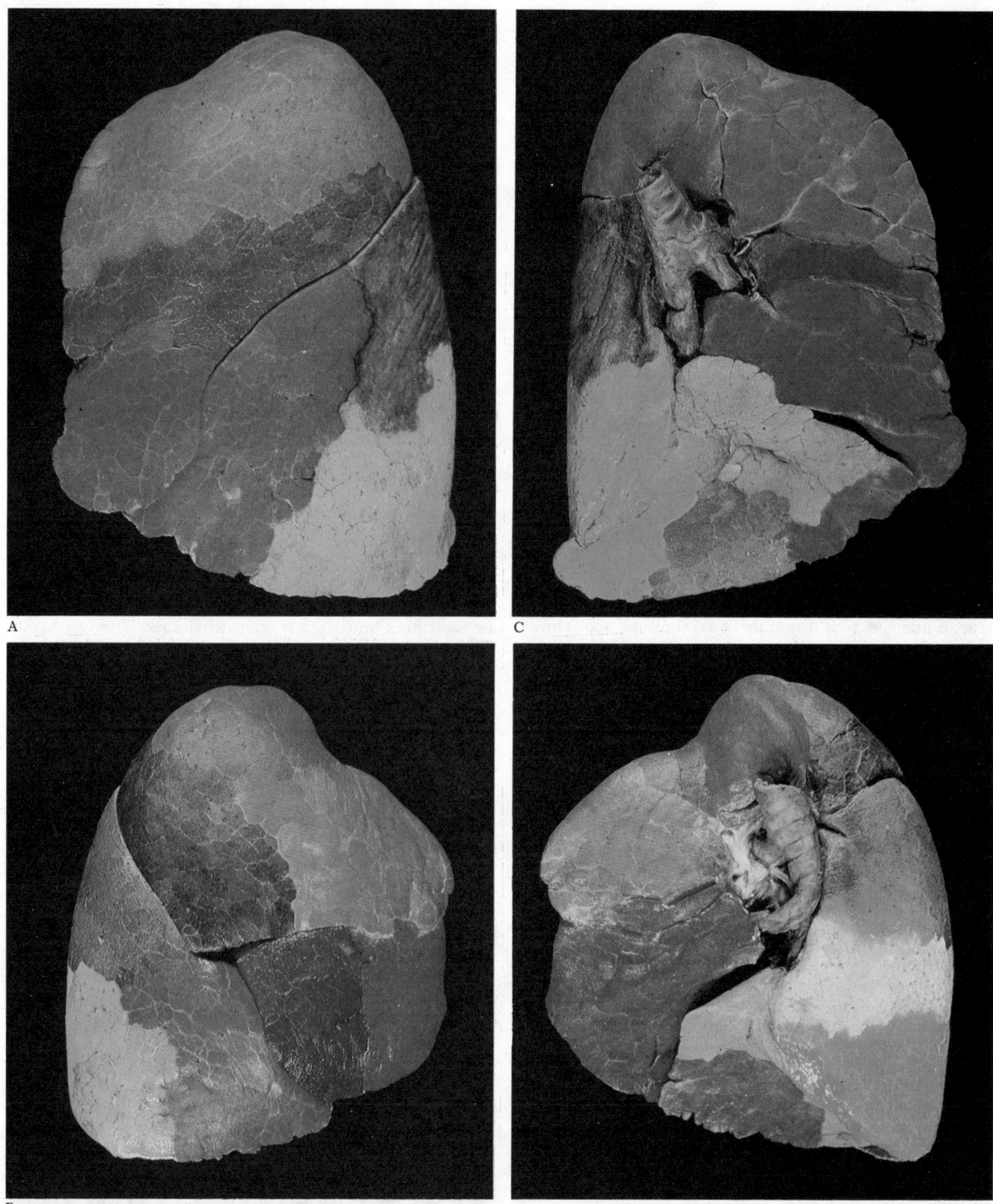

A

B

C

D

8.20A–C. A (opposite) A resin corrosion cast of the adult human lower trachea and bronchial tree photographed from the anterior aspect. The segmental bronchi and their main branches have been coloured: (1) apical—brown; (2) posterior—grey/blue; (3) anterior—pink; (4) lateral (middle lobe) and superior lingular—dark blue; (5) medial (middle lobe) and inferior lingular—red; (6) superior (apical) of inferior lobe—dark green; (7) medial basal—yellow; (8) anterior basal—orange; (9) lateral basal—blue; (10) posterior basal—light green.

B Corrosion cast, left lung, lateral view; colour coding as above.

C Corrosion cast, right lung, lateral view; colour coding as above. (Note that in the final preparation of such specimens large amounts of finer detail have been carefully removed to reveal the major arrangement of the bronchial 'tree'.)

8.21A–D (above) Specimens of human lungs in which the segmental bronchi have been injected with coloured gelatin. Unfixed specimens of the human lungs were obtained at post-morten. The bronchial trees were washed out with water, and individual segmental bronchi, identified by dissection, were cannulated separately and injected with gelatin of contrasting colours. The lungs were subsequently fixed in 10 per cent formaldehyde solution. A costal surface of left lung; B costal surface of right lung; C mediastinal surface of left lung; D mediastinal surface of right lung. Approximately the same colour convention has been used as for the corrosion cast shown in **8.20A**. In these specimens, however, in the right lung a subsuperior segment is present in the inferior lobe, and is coloured white. (All specimens were prepared by Dr. M. C. E. Hutchinson and photographed by Mr. Kevin Fitzpatrick, of the Department of Anatomy, Guy's Hospital Medical School.)

inferior lobe bronchus in up to 10 per cent of lungs, and this, together with the fact that the territory it supplies is similar to that on the right side, lends support to the recognition of this as a separate segmental bronchus. A *subsuperior* (subapical) *segmental bronchus* arises from the posterior surface of the left inferior lobe bronchus in as many as 30 per cent of lungs.

BRONCHOPULMONARY SEGMENTS

Each of the primary branches of the right and left *lobar bronchi* described above are termed *segmental bronchi* because each divides into ramifications that are distributed to self-contained, functionally independent units of lung tissue called *bronchopulmonary segments* (**8**.21). The main segments are *named* and *numbered* as follows:

Right

Superior lobe:	(1) apical, (2) posterior, (3) anterior.
Middle lobe:	(4) lateral, (5) medial.
Inferior lobe:	(6) superior (apical), (7) medial basal, (8) anterior basal, (9) lateral basal, (10) posterior basal.

Left Lung

Superior lobe:	(1) apical, (2) posterior, (3) anterior, (4) superior lingular, (5) inferior lingular.
Inferior lobe:	(6) superior (apical), (7) medial basal, (8) anterior basal, (9) lateral basal, (10) posterior basal.

Each bronchopulmonary segment is surrounded by connective tissue, continuous with that of the visceral pleura, and forms a separate respiratory unit of lung. (For the vascular and lymphatic arrangements of the segments *see* pp. 800, 1265, and for further details of bronchopulmonary segmentation and the bronchial tree consult Brock 1942, 1943, 1944, 1954; Boyden 1955; Bloomer *et al.* 1960; Volpe *et al.* 1969.)

Applied Anatomy. While pathological conditions such as bronchiectasis and certain infective processes may be restricted to one or more bronchopulmonary segments, malignant neoplasms and tuberculosis break through from one segment to adjacent ones.

A knowledge of the mode of branching of the bronchial tree is necessary during bronchoscopy and for the proper interpretation of bronchograms. A similar knowledge is necessary when determining the appropriate postures to be adopted by patients for promoting drainage of infected areas of lung. The superior (apical) segment of the lower lobe is a common site for a lung abscess to occur following aspiration of material when the patient is lying supine. Foreign bodies when inhaled may obstruct a main, lobar, segmental or smaller bronchus according to their size. The interpretation of their effects and the nature of surgical treatment necessarily involve considerations of the pattern of bronchial branching.

When appropriate, surgical resection of a single bronchopulmonary segment can be undertaken; more radical surgical procedures include the removal of a number of segments, of a whole pulmonary lobe (lobectomy), or of the complete lung (pneumonectomy).

STRUCTURE OF TRACHEA AND MAJOR BRONCHI

The trachea and extrapulmonary bronchi consist of a framework of incomplete rings of hyaline cartilage, united by fibrous tissue and nonstriated muscle (**8**.22A). They are lined by mucous membrane (**8**.22B).

The cartilages of the trachea vary from sixteen to twenty in number. Each is an imperfect 'ring' which occupies about the anterior two-thirds of the circumference of the trachea; behind, where the 'rings' are deficient, the tube is flat, and is completed by fibrous and elastic tissue and nonstriated muscular fibres. The cartilages are placed horizontally one above another, and are separated by narrow intervals. They measure about 4 mm vertically and 1 mm in thickness; cranio-caudally their external

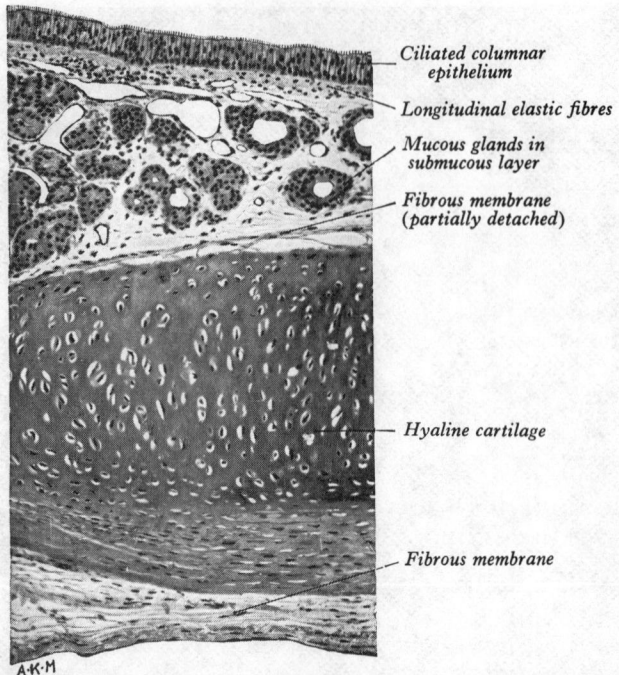

8.22A   A transverse section through a part of the wall of a human trachea.

surfaces are flattened, but their internal surfaces are convex. Two or more of the cartilages often unite, partially or completely, and are sometimes bifurcated at their extremities. They are highly elastic, but may become calcified in advanced life. In the extrapulmonary bronchi the cartilages are shorter, narrower, and rather less regular than those of the trachea, but in general have the same shape and arrangement.

The first and last tracheal cartilages differ from the others (**8**.19). The *first cartilage* is broader than the rest, and often divided at one end; it is connected by the cricotracheal ligament with the lower border of the cricoid cartilage, with which or with the succeeding cartilage it is sometimes blended. The *last cartilage* is thick and broad in the middle, where its lower border is prolonged into a triangular hook-shaped process which curves downwards and backwards between the two bronchi forming a

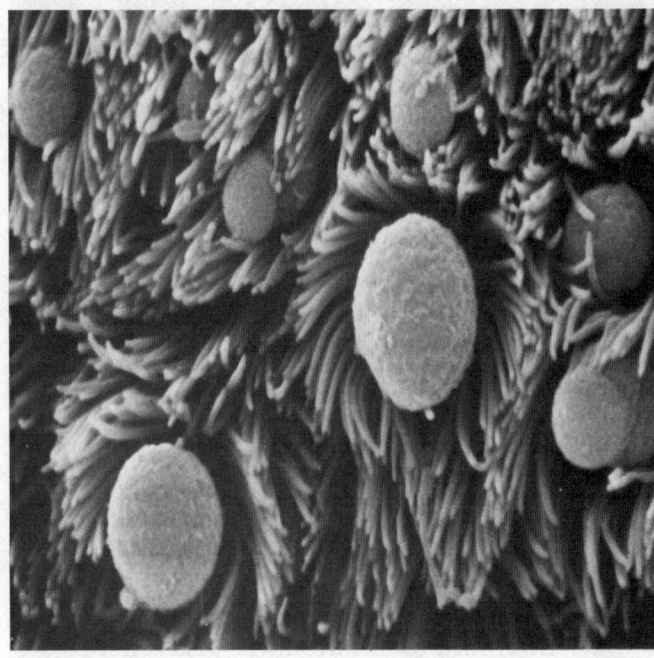

8.22B   A surface view of the ciliated epithelium of the trachea (rat). Mucus-secreting goblet cells occur between the ciliated cells. (Magnification ×5,100.) (Prepared by Mr. Michael Crowder, Department of Anatomy, Guy's Hospital Medical School, London.)

ridge called the *carina*. It forms on each side an imperfect ring which encloses the commencement of the bronchus. The cartilage above the last is somewhat broader at its centre than the others.

Distally, the irregularity of the cartilaginous plates in the extrapulmonary bronchi increases, and as the major bronchi approach their lungs and lobes, the plates encroach on their dorsal aspects, almost, but never completely, encompassing the bifurcations. In the intrapulmonary bronchi, the plates of cartilage become smaller, progressively forming less and less of the bronchial wall, until they disappear at the commencement of bronchioles. (For details *see* **8**.19A, 28A and consult Reid 1976.)

The Fibrous Membrane

Each of the cartilages is enclosed in a perichondrium, which is continuous with a sheet of dense irregular connective tissue forming a fibrous membrane between adjacent hoops of cartilage and at the posterior aspect of the trachea and extrapulmonary bronchi where the cartilage is incomplete. The fibrous layer of the perichondrium and the fibrous membrane are composed mainly of collagen with some elastin fibres intermingled; the fibres cross each other diagonally, allowing changes in diameter of the enclosed airway, and the elastic component provides a measure of elastic recoil when the membrane is stretched.

Nonstriated muscle fibres occur within the fibrous membrane at the posterior aspect of the tube; most of these fibres are transverse, inserting into the perichondrium at the posterior extremities of the cartilages, and forming a transverse sheet in the gaps between these skeletal elements. Contraction of these transverse muscle fibres can therefore alter the cross-sectional area of the trachea and bronchi. A few longitudinal muscle fibres are also present external to the transverse fibres. The nonstriated muscle of the intrapulmonary bronchi is independent of the cartilage plates, and where the latter become less frequent, i.e. in the smaller bronchi, muscular contraction may actually obliterate the lumen (Reid 1976).

The Mucous Membrane

The mucous membrane is continuous with and similar to that of the larynx above and the intrapulmonary bronchi below. It consists of a layer of pseudostratified ciliated epithelium interspersed with goblet cells, both cell types being situated on a basal lamina (**8.22**). Some of the pseudostratified cells possess exceptionally large nuclei and are thought to be polytene in their chromosomal content. Numerous lymphocytes are usually present in the lower parts of the epithelium. The cilia beat the overlying layer of mucus upwards towards the laryngeal aditus (*vide supra*). Deep to the epithelium and its basal lamina are firstly a lamina propria rich in longitudinal elastic fibres, secondly a submucosa of loose irregular connective tissue in which are situated the larger blood vessels, nerve trunks and most of the tubular glands and patches of lymphoid tissue, and thirdly, deep to the submucosa, the perichondrium and fibrous membrane described above. Around the cartilages of the trachea and bronchi, and their fibrous membranes, is the deep fascia merging with the fascial planes of the surrounding muscles, the oesophagus, and other associated structures.

Vessels and Nerves

The trachea is supplied with blood mainly by the inferior thyroid *arteries*. Its thoracic end is supplied by the bronchial arteries, which give off branches ascending to anastomose with the inferior thyroid arteries; all these vessels also supply the oesophagus. The *veins* end in the inferior thyroid venous plexus. The *lymph vessels* pass to the pretracheal and paratracheal lymph nodes. The *nerves* are derived from the vagi and the recurrent laryngeal nerves, and from the sympathetic trunks; they are distributed to the tracheal muscle and to the mucous membrane. The sympathetic nerve endings, when active, cause bronchodilatation by releasing catecholamines, whilst parasympathetic activity causes broncho-constriction. Many small ganglia occur in the autonomic plexuses of both the tracheal and bronchial walls (Feyler 1965). In cats, paraganglia consisting of chromaffin cells and arteriovenous anastomoses have been observed in the bronchial wall (Muratori

1965). For a quantitative technique of assessing the amounts and distribution of the tissues in the bronchial walls, normal or pathological, *see* Hale *et al.* (1968).

Surface and Applied Anatomy. The *trachea*, about 2 cm wide, extends from the cricoid cartilage almost vertically downwards in the median plane as far as the sternal angle; it inclines very slightly to the right as it descends. The *right bronchus* runs from the lower end of the trachea downwards and to the right for 2·5 cm to reach the hilum of the lung opposite the sternal end of the right third costal cartilage. The *left bronchus* runs at a smaller angle from the lower end of the trachea for 5 cm to the left and downwards to reach the hilum of the lung behind the left third costal cartilage, 3·5 cm from the median plane. The trachea may be opened by a median vertical incision, either above the isthmus of the thyroid gland (high tracheotomy) or below it (low tracheotomy). The low operation is more troublesome because the trachea recedes from the surface as it descends and because of the anterior relations of this part of the tube, namely the inferior thyroid veins, the anastomosis between the anterior jugular veins, the arteria thyroidea ima (when present), and (in the child) the brachio-cephalic artery, the left brachiocephalic vein and the thymus. The trachea may be compressed by pathological enlargements of structures related to it, e.g. the thyroid gland, the thymus and the arch of the aorta. (The radiological appearances of the trachea, bronchi and lungs are dealt with on p. 1267.)

Bronchoscopy. By means of a bronchoscope, passed through the mouth and larynx, after the application of a local anaesthetic, the interior of the larynx, trachea and bronchi with the proximal parts of their main branches supplying the bronchopulmonary segments (pp. 1243–1246) may be examined, pathological states studied and biopsies taken for microscopy; foreign bodies may be removed, or accumulations of fluid removed by suction.

The Pleurae

Each lung is invested by a serous membrane arranged in the form of a closed invaginated sac termed the *pleura*. A part of this serous membrane covers the surface of the lung and lines the fissures between its lobes; it is called the *visceral* or *pulmonary pleura*. The rest of the membrane lines the inner surface of the corresponding half of the chest wall, covers a large part of the diaphragm, and is reflected over the structures occupying the middle part of the thorax; this portion is termed the *parietal pleura*. The pulmonary and parietal pleurae are continuous with each other around and below the root of the lung; in health they are in contact in all phases of respiration, the potential space between them being known as the *pleural cavity*. When the lung collapses or when air or fluid collects between the pulmonary and parietal pleurae, the pleural cavity becomes apparent. The right and left pleural sacs are distinct from each other, and come into immediate contact for just a short distance behind the upper half of the body of the sternum (**8.23A**), although they are separated only by a narrow interval behind the oesophagus in the mid-thoracic region. The interval between the two sacs is named the *interpleural space* or *mediastinum*. The right pleural cavity is wider than the left, because the heart extends further to the left than to the right side. The upper and lower limits of the pleural sacs are approximately the same, but the left sac sometimes descends to a lower level in the mid-axillary line.

The pulmonary pleura is inseparably connected with the lung. It is adherent to all the surfaces, including those which bound the fissures between the lobes of the lung; it is absent, however, over an area where the lung root enters, and along a line extending downwards from this and marking the attachment of the so-called pulmonary ligament (**8.27**).

The parietal pleura. Different regions of the parietal pleura have distinctive names; the part lining the internal aspect of the ribs, transversus thoracis (p. 548) and sides of the vertebral bodies is the *costovertebral pleura*; that clothing the thoracic surface of the diaphragm is the *diaphragmatic pleura*; that ascending into the neck over the summit of the lung is the *cervical pleura* (or *dome of the pleura*) and that applied to the structures occupying the interpulmonary region is the *mediastinal pleura*.

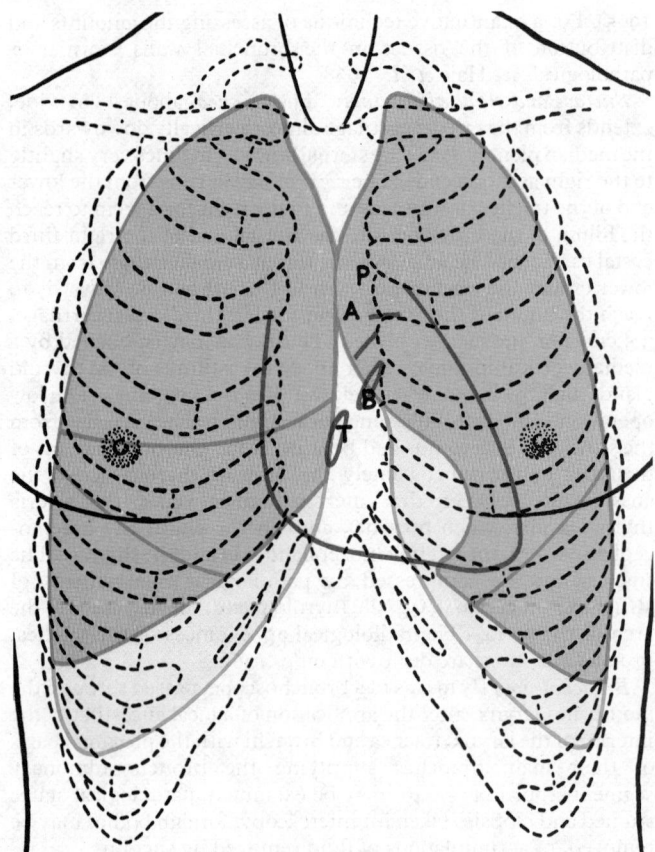

8.23A Ventral aspect of the thorax, showing surface projections. Skeletal, pulmonary (purple), pleural (blue), and cardiac (red outline). A. Orifice of aorta. B. Left atrioventricular (mitral) orifice. P. Orifice of pulmonary trunk. T. Right atrioventricular (tricuspid) orifice.

The *costovertebral* (or '*costal*') *pleura* (**8.23**A–C) lines the sternum, ribs, the constituent parts of the transversus thoracis, and the sides of the bodies of the vertebrae, and is easily separated from them. Outside the costal pleura there is a thin layer of loose areolar tissue, called the *endothoracic fascia*; it corresponds to the transversalis fascia of the abdominal wall. In front, the costal pleura begins behind the sternum where it is continuous with the mediastinal pleura. The line of junction of the mediastinal with the costal pleura extends from behind the sternoclavicular joint downwards and medially to a point near the median plane behind the sternal angle. From here the right and left costal pleurae descend in contact with each other as far as the level of the fourth costal cartilages, below which the line differs on the two sides. On the right side it is continued down to the posterior surface of the xiphisternal joint. On the left it diverges laterally and descends, at a distance of 2 to 25 mm from the margin of the sternum (Woodburne 1947), to the level of the sixth costal cartilage. On each side the costal pleura sweeps laterally, lining the inner surfaces of the costal cartilages, ribs and the constituent parts of the transversus thoracis; at the back of the thorax it passes over the sympathetic trunk and its branches, and on to the side of the bodies of the vertebrae, where it again becomes continuous with the mediastinal pleura. Above, the costal pleura is continuous with the cervical pleura at the inner margin of the first rib. Below, it is continuous with the diaphragmatic pleura along a line which may differ slightly on the two sides. On the right side this *costodiaphragmatic line of reflexion* of the pleura begins dorsal to the xiphoid process, and passes dorsocaudally behind the seventh costal cartilage, reaching the eighth rib in the mid-clavicular line, then reaching the mid-axillary line at the level of the tenth rib; from here the line ascends slightly, and, crossing the twelfth rib, reaches the level of the upper border of the spine of the twelfth thoracic vertebra (**8.23**C). On the left side the line follows at first the ascending part of the sixth costal cartilage, and in the rest of its course may be slightly lower than that on the right side.

The *diaphragmatic pleura* is thin, and covers part of the upper surface of the corresponding side of the diaphragm. The outer

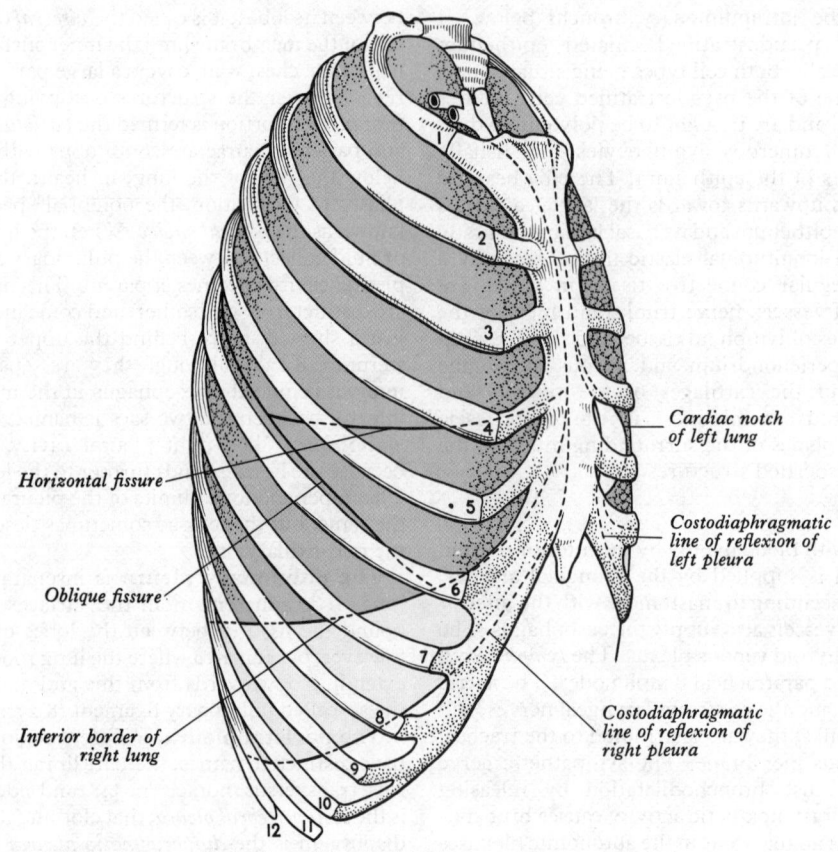

Horizontal fissure

Oblique fissure

Inferior border of right lung

Cardiac notch of left lung

Costodiaphragmatic line of reflexion of left pleura

Costodiaphragmatic line of reflexion of right pleura

8.23B The relations of the pleurae and lungs to the chest wall. Right lateral aspect. Purple—lungs, covered with the pleural sacs. Blue—pleural sac, with no underlying lung.

part of its circumference is the line described above, along which it is continuous with the costal pleura. Medially it is continuous with the mediastinal pleura along the line of attachment of the pericardium to the diaphragm.

The *cervical pleura* or *dome of the pleura* is the continuation of the costal pleura over the apex of the lung (**8.23**B). It extends from the internal border of the first rib medially and upwards to the apex of the lung, its summit reaching as high as the lower edge of the neck of the first rib; it then descends along the side of the trachea to become continuous with the mediastinal pleura. Owing to the obliquity of the first rib, the cervical pleura extends 3 or 4 cm above the first costal cartilage, but does not rise above the level of the neck of the first rib. The cervical pleura is strengthened by a dome-like expansion of fascia, named the *suprapleural membrane*. It is attached in front to the internal border of the first rib and behind to the anterior border of the transverse process of the seventh cervical vertebra; it is covered and strengthened by a few spreading muscular fibres derived from the scaleni. A muscle, frequently present, which is called the *scalenus minimus*, arises from the anterior border of the transverse process of the seventh cervical vertebra and is inserted into the inner border of the first rib, behind the groove for the subclavian artery, and into the dome of the pleura, which, on contraction, it renders tense; some consider that the suprapleural membrane represents the spread-out tendon of this muscle. The cervical pleura (like the apex of the lung) reaches the level of the seventh cervical spine at a distance of 2·5 cm from the median plane. It can be represented by a curved line drawn from the sternoclavicular joint to the junction of the medial and middle thirds of the clavicle, the summit of the curve being 2·5 cm above the clavicle. The subclavian artery, directed upwards and laterally, occupies a furrow a little below the summit of the cervical pleura. the relations of the cervical pleura are similar to those of the apex of the lung (*see* p. 1253, and **8.24**A).

The *mediastinal pleura* forms the lateral boundary of the interpleural space or mediastinum (see also p. 1251). Above the root of the lung it is a continuous sheet between the sternum and the vertebral column. That of the right side is in contact with the

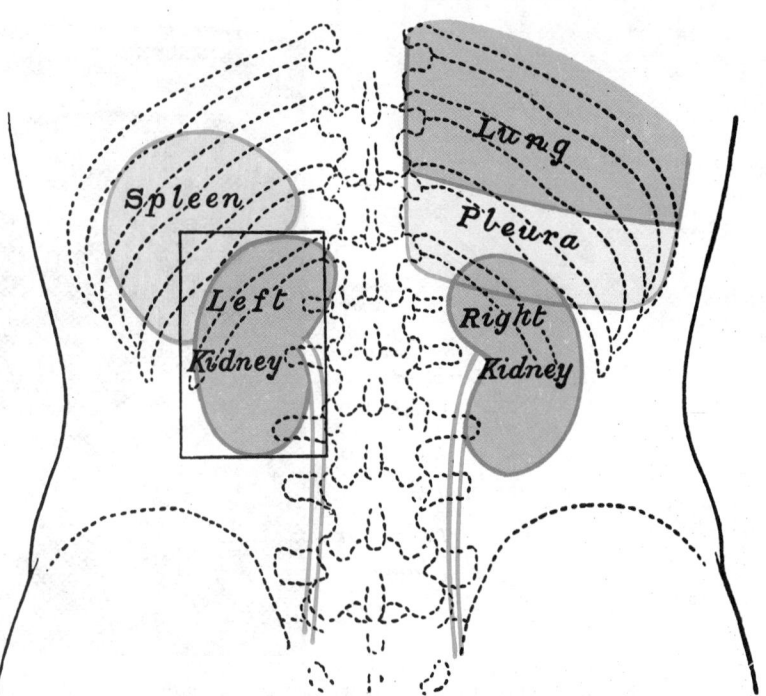

8.23C The lower limits of the lung and pleura, posterior view. The lower portions of the lung and pleura are shown on the right side.

right brachiocephalic vein, the upper part of the superior vena cava, the terminal part of the azygos vein, the right phrenic and right vagus nerves, the trachea and the oesophagus. That of the left side is in relation with the arch of the aorta, the left phrenic and left vagus nerves, the left brachiocephalic and superior intercostal veins, the left common carotid and subclavian arteries, the thoracic duct and the oesophagus. At the root of the lung the

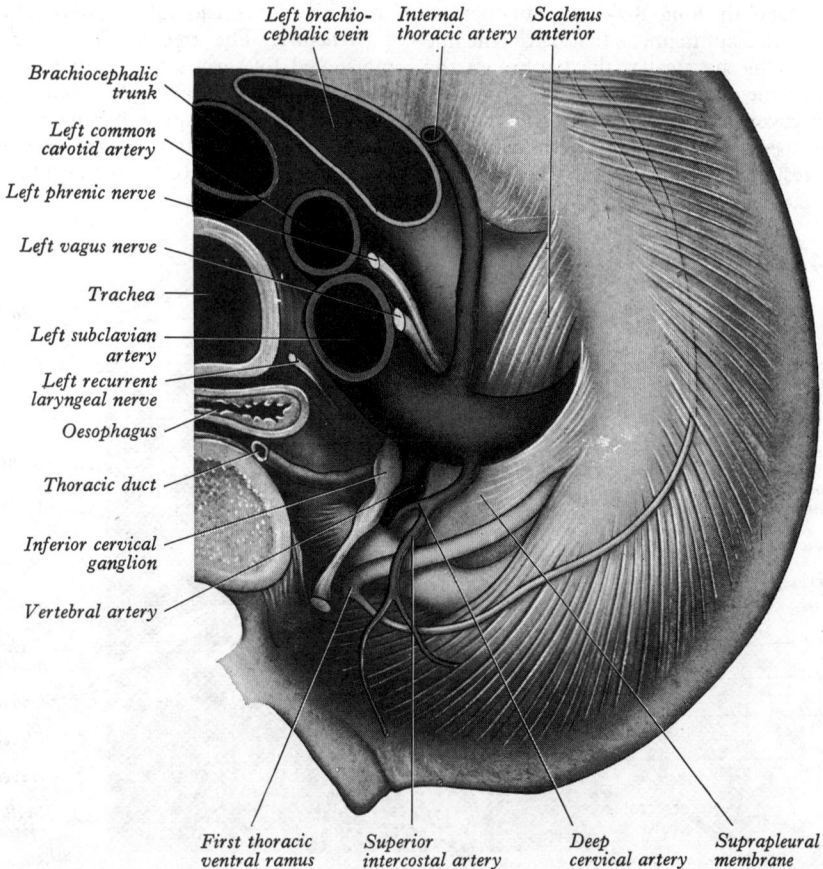

8.24A Structures in relation with the cervical pleura of the left side. Inferior aspect.

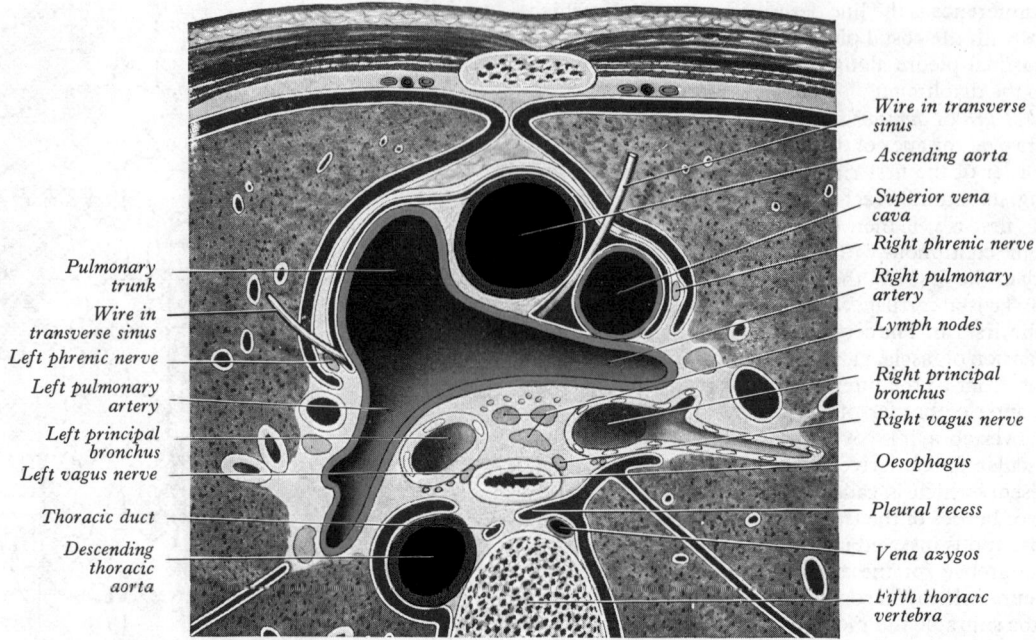

Labels (left, top to bottom): Pulmonary trunk · Wire in transverse sinus · Left phrenic nerve · Left pulmonary artery · Left principal bronchus · Left vagus nerve · Thoracic duct · Descending thoracic aorta

Labels (right, top to bottom): Wire in transverse sinus · Ascending aorta · Superior vena cava · Right phrenic nerve · Right pulmonary artery · Lymph nodes · Right principal bronchus · Right vagus nerve · Oesophagus · Pleural recess · Vena azygos · Fifth thoracic vertebra

8.24B   A transverse section of the mediastinum at the level of the lower part of the body of the fifth thoracic vertebra. Viewed from the superior aspect.

mediastinal pleura is carried laterally as a tube of serous membrane enclosing the structures of the lung root and passing into continuity with the visceral or pulmonary pleura. Below the lung root the mediastinal pleura extends as a double layer from the lateral surface of the oesophagus to the mediastinal surface of the lung, where it is continuous with the pulmonary pleura. This double layer is named the *pulmonary ligament* (**8.**27A, B). It is continuous above with the tube investing the lung root; below it ends in a free falciform border.

The inferior limit of the pleura is on a considerably lower level than the corresponding border of the lung (**8.**23A–C), but does not extend to the attachment of the diaphragm, so that below the line of reflexion of the pleura from the chest wall to the diaphragm, the latter is in direct contact with the rib cartilages and the muscles in the intercostal spaces. Moreover, in ordinary inspiration the thin inferior margin of the lung does not extend as low as the line of the pleural reflexion, with the result that the costal and diaphragmatic

pleurae are here in contact, the intervening narrow slit being termed the *costodiaphragmatic recess*. In quiet respiration the lower limit of the lung is about 5 cm above the lower limit of the pleura. A similar condition exists behind the sternum and rib cartilages, where the anterior thin margin of the lung falls short of the line of pleural reflection, and where the slit-like cavity between the two layers of pleura forms the *costomediastinal recess*. It should be noted that the extent of the costomediastinal recess, the position of the anterior costomediastinal line of pleural reflexion, and the position of the thin anterior margin of the lung are subject to considerable variation in different individuals.

Structure. The free surface of the pleura is smooth, and moistened by serous fluid. Like many serous membranes, its surface is a single layer of flattened cells, which form a meso-thelium, and rest on a basement membrane beneath which are networks of yellow elastic and white fibres embedded in ground substance, containing also fibroblasts, macrophages and

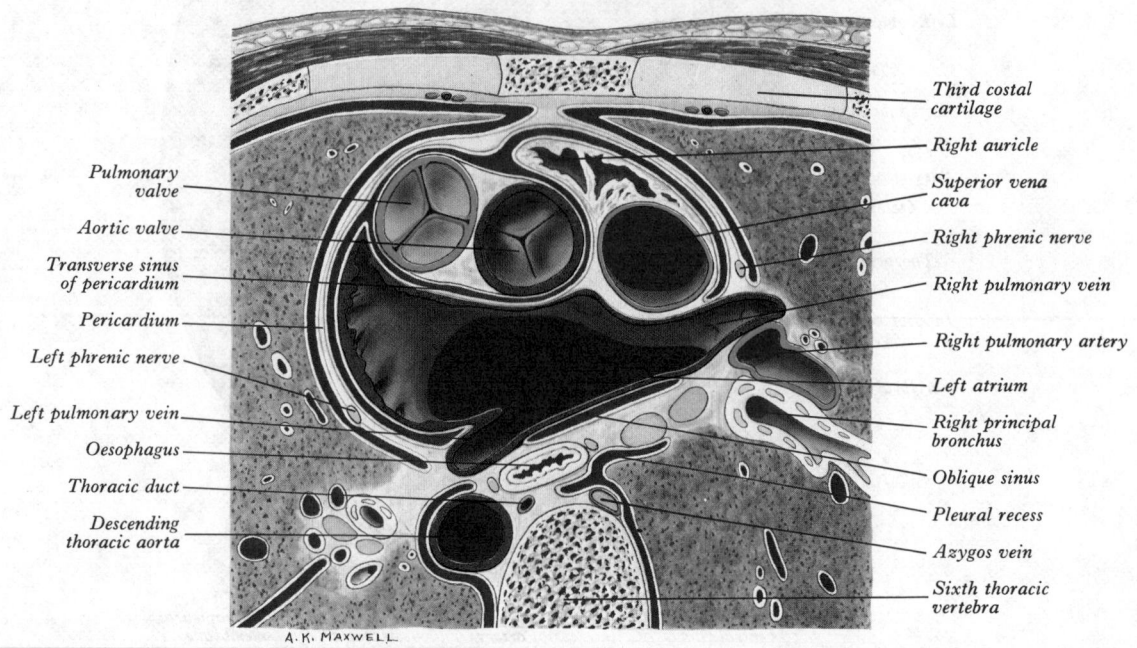

Labels (left, top to bottom): Pulmonary valve · Aortic valve · Transverse sinus of pericardium · Pericardium · Left phrenic nerve · Left pulmonary vein · Oesophagus · Thoracic duct · Descending thoracic aorta

Labels (right, top to bottom): Third costal cartilage · Right auricle · Superior vena cava · Right phrenic nerve · Right pulmonary vein · Right pulmonary artery · Left atrium · Right principal bronchus · Oblique sinus · Pleural recess · Azygos vein · Sixth thoracic vertebra

A. K. MAXWELL

8.24C   A transverse section of the mediastinum at the level of the body of the sixth thoracic vertebra. Viewed from the superior aspect.

the cell types which also characterize loose connective tissue (p. 414). The deeper layers of fibrous tissue in the pulmonary pleura are continuous with the connective tissue around and between the lobules of the lung. Blood vessels, lymph vessels and nerves are distributed in the substance of the pleura. In their ultrastructural details the pleural mesothelial cells are in many respects similar to those of peritoneum (p. 1333). Material studied from human biopsies (de Gasperis and Miani 1969) showed that the squamous cells rest upon a basal lamina 30–40 nm thick, but their highly infolded basal plasma membranes are separated from the lamina by an interval of 20–30 nm. Laterally the surfaces of adjacent cells interdigitate and are joined at intervals by desmosomes. Their free surfaces bear scattered microvilli and numerous cilia (*see* **8.**24D). Micropinocytotic vesicles are common in the cytoplasm. Several structural variants of the squamous cells have been recognized, one type probably being a stem cell.

Vessels and Nerves

The parietal and visceral layers of the pleura are respectively developed from the somatopleural and splanchnopleural layers of the lateral plate mesoderm (pp. 110, 209). Correlated with this origin, the parietal pleura derives its arterial supply from somatic (body wall) arteries (intercostal, internal thoracic and musculo-phrenic); its veins join the systemic veins in the neighbouring parts of the chest wall; its lymphatics also join those in the body wall and drain into the intercostal, parasternal, posterior mediastinal and diaphragmatic nodes, and its nerve supply is derived from the spinal nerves supplying the muscles and skin of the body wall (intercostal and phrenic nerves). The visceral pleura, which is an integral part of the lung itself, derives its vascular supply from the bronchial vessels, its lymphatics join those of the lung, and its nerve supply is derived from the autonomic nerves innervating the lung and accompanying the bronchial vessels. (For details of a special technique for demonstrating the lymphatics of the visceral pleura and of observations thus effected, consult Pennell 1966.) Whereas pain is elicited by the application of tactile or thermal stimuli to the parietal pleura, these do not form adequate stimuli when applied to the visceral pleura (*see* also the peritoneum, p. 1333). The costal pleura and the pleura on the peripheral part of the diaphragm are supplied by the intercostal nerves; the mediastinal pleura and the pleura on the central part of the diaphragm are supplied by the phrenic nerve. Irritation of the former parts of the pleura results in pain referred along the intercostal nerves to the thoracic or abdominal wall, whereas irritation of the latter parts results in pain referred to the lower part of the neck and over the shoulder

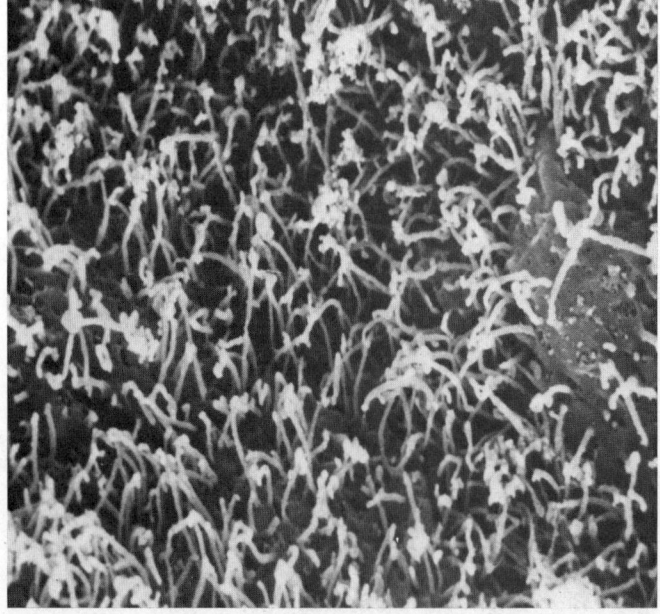

8.24D The surface of the pleura viewed by scanning electron microscopy, showing the numerous cilia of the mesothelial cells lining this structure (murine). (Magnification × 7,000.)

(that is, to the area of skin supplied by the same segments of the spinal cord that give origin to the phrenic nerve, C. 3, 4, 5).

Applied Anatomy. Normally the visceral pleura slides smoothly on the parietal pleura during respiration and does not cause any sound appreciable on auscultation; if, however, the pleura is inflamed, characteristic friction sounds can be heard. If an effusion of fluid occurs into the pleural cavity, the sounds disappear, and as fluid accumulates the lung gradually collapses and the heart and mediastinum are hence displaced towards the opposite side. Entry of air into the pleural cavity (pneumothorax), whether caused accidentally by an external penetrating wound, by rupture of part of the lung, or produced as a therapeutic measure (e.g. to rest the lung in tuberculosis), also results in collapse of the lung, as its elastic tissue recoils. Normally this is prevented by the negative pressure in the pleural cavity and by the cohesion between the opposed parietal and visceral pleura. In operations on the kidney from the back, the relation of the costal pleura to the twelfth rib must be borne in mind. Usually the pleura crosses the rib at the lateral border of the erector spinae, so that its posterior part is superior to the reflexion of the pleura (**8.**23C). If the last rib is too short to project beyond the erector spinae, the eleventh rib may be mistaken for the last when palpated in this position, and an incision which is prolonged up to this level will wound the pleura. It is therefore important to determine whether the lowest palpable rib is the eleventh or twelfth by counting down from the second (at the junction of its costal cartilage with the sternal angle).

The Mediastinum

The mediastinum, strictly speaking, is the *partition* between the two lungs and therefore includes the mediastinal pleura of both sides, but it is generally defined as the *interval* between the two pleural sacs. It extends from the sternum in front to the vertebral column behind (**8.**24B, C), and from the thoracic inlet above to the diaphragm below. For purposes of description it is divided into two parts, an upper, which is named the *superior* mediastinum, and a lower, which is subdivided into the *anterior*, *middle* and *posterior* mediastina. The superior mediastinum is continuous with the lower part at the plane passing through the manubriosternal joint in front and the lower surface of the fourth thoracic vertebra behind.

The superior mediastinum (**8.**17, 18) lies between the manubrium sterni in front, and the upper four thoracic vertebrae behind. It is bounded below by the plane passing through the sternal angle in front, and the lower part of the body of the fourth thoracic vertebra behind; above, by the plane of the thoracic inlet, and laterally by the mediastinal pleurae. It contains the origins of the sternohyoid and sternothyroid and the lower ends of the longus colli muscles; the aortic arch; the brachiocephalic, left common carotid and left subclavian arteries; the brachiocephalic veins and the upper half of the superior vena cava; the left superior intercostal vein; the vagus, cardiac, phrenic and left recurrent laryngeal nerves, and superficial part of the cardiac plexus; the trachea, oesophagus and thoracic duct; the remains of the thymus, and the paratracheal, brachiocephalic and some of the tracheo-bronchial lymph nodes.

The anterior mediastinum lies between the body of the sternum in front and the pericardium behind (**8.**24B); above the level of the fourth costal cartilages, it is exceedingly narrow, owing to the close approximation of the two pleural sacs. It contains some loose areolar tissue, the sternopericardial ligaments, two or three lymph nodes and a few small mediastinal branches of the internal thoracic artery and sometimes also part of the thymus gland or its degenerated remains.

The middle mediastinum (**8.**24 C, 26) is the broadest part of the inferior mediastinum. It contains the heart, pericardium, ascending aorta, lower half of the superior vena cava, terminal part of the azygos vein, the bifurcation of the trachea, the two bronchi, the pulmonary trunk dividing into right and left pulmonary arteries, the right and left pulmonary veins, the phrenic nerves, the deep part of the cardiac plexus and some tracheobronchial lymph nodes.

The posterior mediastinum (8.24B, C; 25, 26) is bounded *in front* by the bifurcation of the trachea, the pulmonary vessels, the pericardium and by the posterior part of the upper surface of the diaphragm; *behind*, by the vertebral column from the lower border of the fourth to the twelfth thoracic vertebrae; *on each side* by the mediastinal pleura. It contains the descending thoracic aorta, the azygos and hemiazygos veins, the vagus and splanchnic nerves, the oesophagus, the thoracic duct, and the posterior mediastinal lymph nodes.

Radiology of the mediastinum (6.27, 28, 29). In anteroposterior radiographs of the chest (6.27), the heart and large blood vessels form an opacity called the mediastinal shadow. Forming the left border of this shadow can be recognized, from above downwards, the left subclavian artery, the arch of the aorta ('the aortic knuckle'), the left auricle and the left ventricle. Immediately below the aortic arch, the infundibulum of the right ventricle or

and the aortic arch and large vessels in the superior mediastinum produce faint shadows.

The Lungs

The lungs are the essential organs of respiration; they are situated one on each side within the thorax, and separated from each other by the heart and the other contents of the mediastinum (6.14). Except for its attachment to the heart and trachea by the root (and the pulmonary ligament), each lung lies freely in the corresponding pleural cavity. The substance of the lung is of a light, porous, spongy texture; it floats in water, and crepitates when handled, owing to the presence of air in its alveoli; it is also highly elastic, hence the retracted state of the lungs when removed from the closed cavity of the thorax. The surface is smooth, shining, and

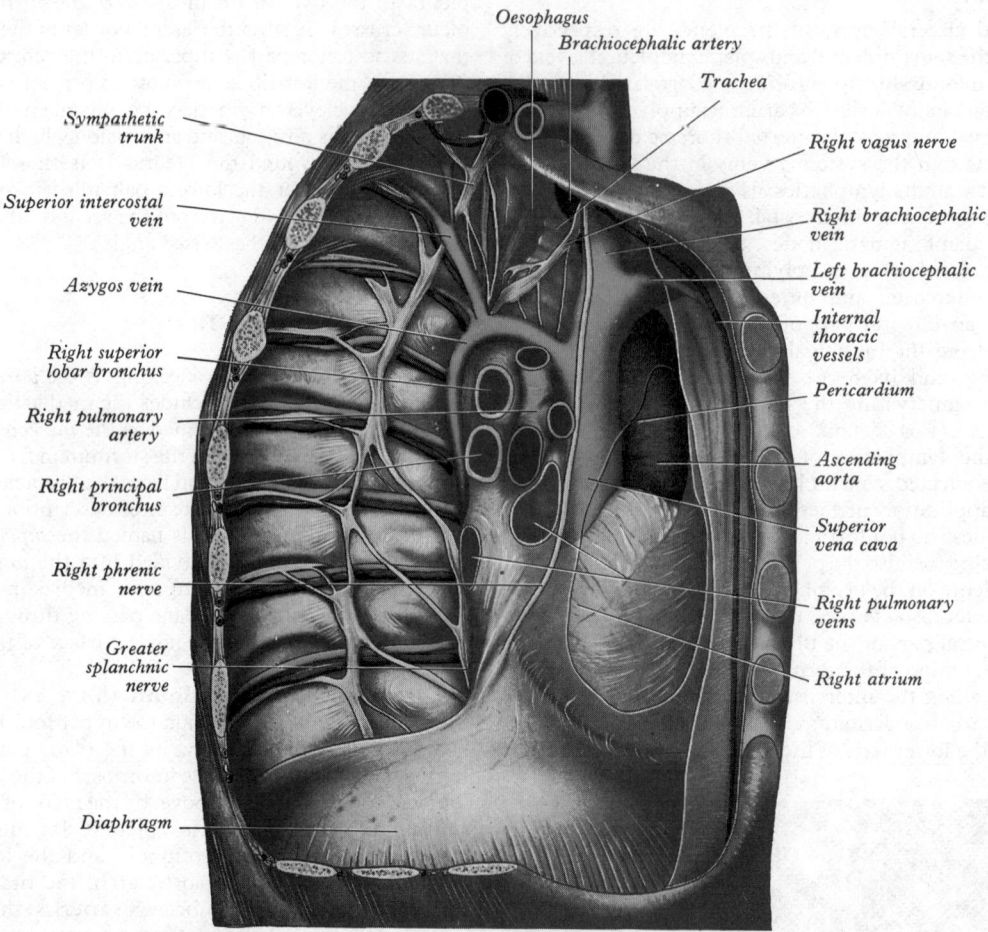

Oesophagus
Brachiocephalic artery
Trachea
Sympathetic trunk
Right vagus nerve
Superior intercostal vein
Right brachiocephalic vein
Azygos vein
Left brachiocephalic vein
Right superior lobar bronchus
Internal thoracic vessels
Right pulmonary artery
Pericardium
Right principal bronchus
Ascending aorta
Right phrenic nerve
Superior vena cava
Greater splanchnic nerve
Right pulmonary veins
Right atrium
Diaphragm

8.25A   The mediastinum, right lateral aspect. A part of the pericardial sac has been removed to expose the lateral surface of the right atrium.

pulmonary trunk may be recognizable upon this border. On the right border of the shadow are seen the right brachiocephalic vein, the superior vena cava, the right atrium and the thoracic part of the inferior vena cava. Enlargements or lateral displacements of any of the above structures accentuate the normal 'bulges' on the borders of the mediastinal shadow. On both sides of the mediastinum, the opacities of the pulmonary vessels entering the lungs constitute the root or hilar shadows. In the upper part of the chest the less dense shadow of the trachea is seen in the median plane.

In lateral or oblique radiographs of the thorax the heart shadow lies above the anterior part of the diaphragm. In front of it is 'the retrosternal space' (the anterior mediastinum), while behind it is 'the retrocardiac space' (posterior mediastinum) containing the oesophagus, rendered visible during the passage of a barium meal (8.90), and the descending thoracic aorta. Above the heart shadow, the 'translucent' trachea and bronchi are recognizable,

marked out by fine, dark, intersecting lines into numerous polyhedral areas, indicating the lobules of the lung; each of these areas is crossed by numerous lighter lines.

At birth the lungs are rose-pink in colour; in adult life the colour is a dark slaty-grey, mottled in patches, and as age advances, this maculation assumes a black colour. The colouring matter consists of granules of inhaled carbonaceous particles deposited in the areolar tissue near the surface of the lung; it increases in quantity as age advances, and is more abundant in men than in women. As a rule, the posterior border of the lung is darker than the anterior. On the upper, less movable parts of the lungs, the surface pigmentation tends to lie opposite the intercostal spaces. The lungs of the fetus, or of the stillborn child who has not respired, differ from those of the child who has done so, in that they are firm to the touch (like the liver), do not crepitate when handled, and, containing no air, sink in water.

The right lung usually weighs about 625 gm, the left 565 gm, but much variation occurs; the weight is also dependent on the amount of blood or serous fluid that they contain. The lungs are heavier in the male than in the female, both absolutely and relative to stature.

Each lung is conical and has an apex, a base, three borders and two surfaces.

The apex, which is rounded, protrudes above the thoracic inlet, in close contact with the cervical pleura. Owing to the obliquity of the inlet (p. 291), this part of the lung reaches from 3 cm to 4 cm above the level of the first costal cartilage, although it does not rise above the level of the neck of the rib. Its summit lies about 2·5 cm above the medial third of the clavicle, and the apex is situated therefore in the root of the neck (**8**.23B). It has been asserted that the apex is in fact intrathoracic, close to the neck of the first rib, and that the anterior surface of this region ascends highest in inspiration (Andreassi 1967). This view requires

Laterally and behind, the base is bounded by a thin, sharp margin which projects for some distance into the costodiaphragmatic recess of the pleura (p. 1250).

The costal surface is smooth, convex, and corresponds to the form of the cavity of the chest, which is deeper behind than in front. It is in contact with the costal pleura, and exhibits, in specimens which have been hardened *in situ*, slight grooves corresponding with the overlying ribs.

The medial surface is divided into a posterior or *vertebral* part, and an anterior or *mediastinal* part. The vertebral part is in contact with the sides of the thoracic vertebrae and intervertebral discs, the posterior intercostal vessels and the splanchnic nerves. The mediastinal part exhibits a deep concavity, which accommodates the pericardium and is termed the *cardiac impression*; this concavity is larger and deeper on the left than on the right lung, because the heart projects more to the left than to the right side of

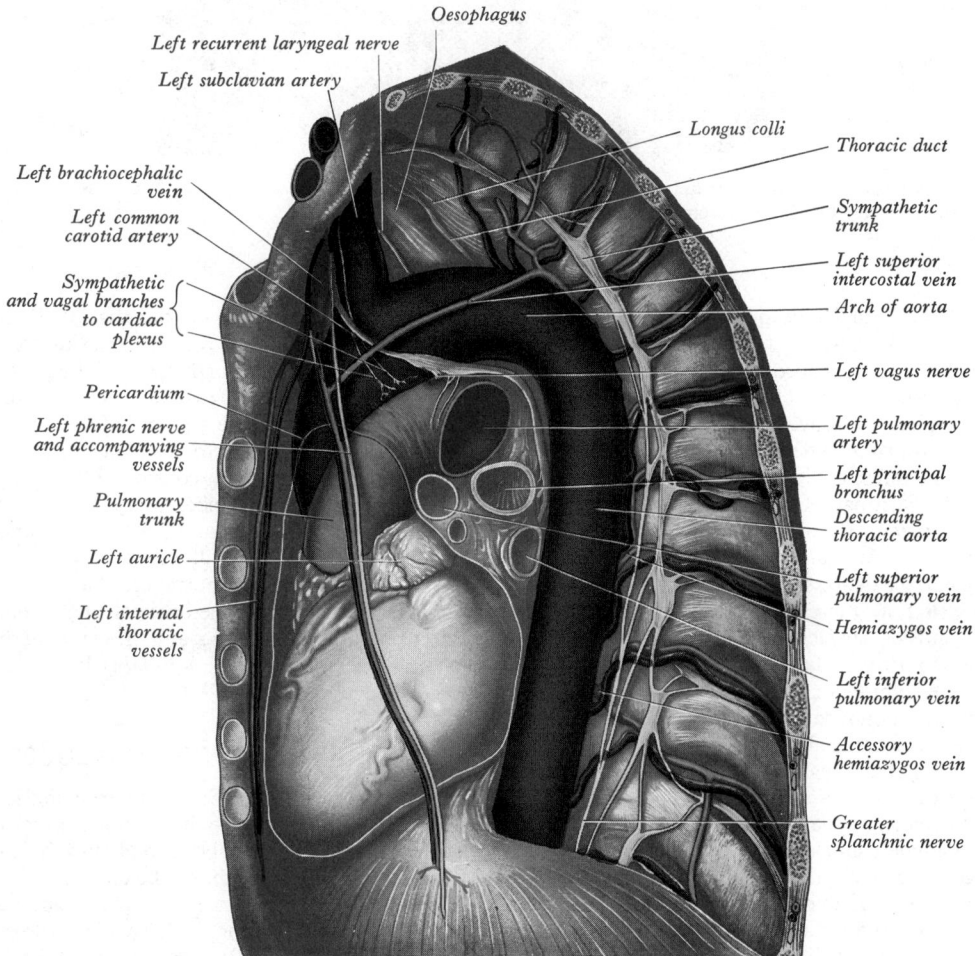

8.25B The mediastinum, left lateral aspect.

confirmation. The cervical pleura intervenes between the apex of the lung and the suprapleural membrane (p. 1249), on which the subclavian artery arches upwards and laterally, producing a groove on the anterior surface of the apex just below its summit and separating it from the scalenus anterior. Posteriorly the apex is related to the cervicothoracic ganglion of the sympathetic trunk, the ventral ramus of the first thoracic nerve and the superior intercostal artery (**8**.24). Laterally, it is related to the scalenus medius; medially, to the brachiocephalic artery, right brachiocephalic vein and trachea on the right side, and to the left subclavian artery and left brachiocephalic vein on the left side.

The base is semilunar in shape, and concave; it rests upon the convex surface of the diaphragm, which separates the right lung from the right lobe of the liver, and the left lung from the left lobe of the liver, the fundus of the stomach and the spleen. Since the diaphragm extends higher on the right side than on the left, the concavity on the base of this is deeper than it is on the left.

the median plane. Above and behind this concavity there is a somewhat triangular depression named the *hilum*, where the structures which form the root of the lung (p. 1255) enter and leave the viscus. These structures are invested by pleura, which extends downwards, below the hilum and behind the cardiac impression, forming the pulmonary ligament.

Apart from these features, which are shared in common by both lungs, the markings on the mediastinal surface seen in specimens hardened *in situ* are different on the two sides (**8**.25A, B; 27A, B). On the *right lung*, the cardiac impression is contiguous with the anterior surface of the right auricle, anterolateral surface of the right atrium and a small part of the anterior surface of the right ventricle. The impression is continued upwards in front of the hilum as a wide groove which lodges the superior vena cava and the lower end of the right brachiocephalic vein (**8**.27A). Posteriorly this groove is joined by a deep, narrow sulcus which arches forwards above the hilum and is caused by the azygos vein.

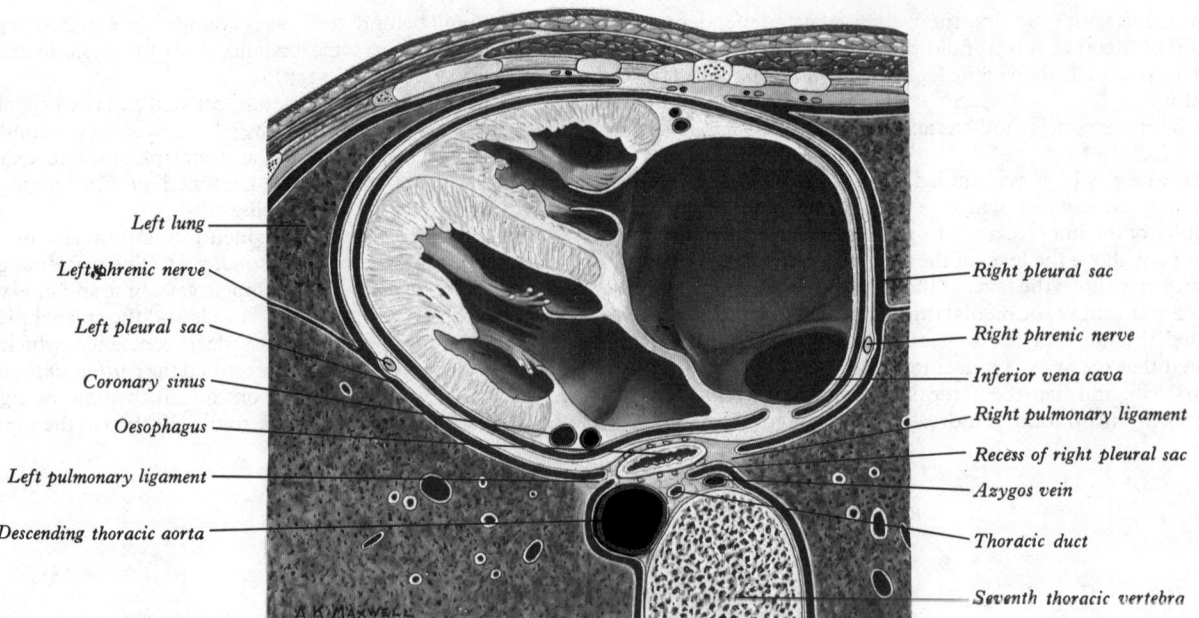

Left lung
Left phrenic nerve
Left pleural sac
Coronary sinus
Oesophagus
Left pulmonary ligament
Descending thoracic aorta

Right pleural sac
Right phrenic nerve
Inferior vena cava
Right pulmonary ligament
Recess of right pleural sac
Azygos vein
Thoracic duct
Seventh thoracic vertebra

8.26 A transverse section through the mediastinum at the level of the body of the seventh thoracic vertebra. Viewed from the superior aspect.

The right edge of the oesophagus produces a shallow groove which runs vertically downwards behind the hilum and the pulmonary ligament. As it approaches the diaphragm, the oesophagus inclines towards the left and passes away from the right lung. The oesophageal groove, therefore, does not extend to the lower limit of this surface. The postero-inferior corner of the cardiac impression is confluent with a short but wide notch which accommodates the thoracic part of the inferior vena cava. Between the apex and the groove for the azygos vein, the trachea and the right vagus nerve are in close relation to the lung, although there is no corresponding surface depression.

On the *left lung* (**8.27**B), the cardiac impression is in relation with the anterior and left surfaces of the left ventricle and left auricle, and the anterior surface of the infundibulum and adjoining part of the right ventricle. It is continued upwards in front of the hilum to accommodate the pulmonary trunk. A wide, deep groove arches backwards above the hilum and downwards behind it and the pulmonary ligament; it lodges the arch and the descending part of the aorta. Near the summit of its curve it is confluent with a narrower groove which ascends towards the apex and is occupied by the left subclavian artery. Behind this groove and above the aortic groove, the lung is in contact with the thoracic duct and the left surface of the oesophagus. In front of the upper part of the groove for the left subclavian artery a faint groove is produced by the left brachiocephalic vein. Inferiorly, the left surface of the oesophagus may make a slight impression in front of the lower end of the pulmonary ligament.

The *inferior border* is thin and sharp where it separates the base from the costal surface and extends into the costodiaphragmatic recess; medially, where it divides the base from the mediastinal surface, it is blunt and rounded. It is represented, during quiet respiration, by a line drawn from the lower end of the anterior border to reach the eighth rib in the mid-axillary line, where it lies nearly 10 cm above the costal margin; the line is then continued medially and slightly upwards across the back to a point 2 cm lateral to the tenth thoracic *spine* (**8.23**C). The *posterior border* separates the costal surface from the vertebral part of the medial surface, and corresponds to the medial margins of the heads of the ribs. It is not marked by any recognizable ridge or line. The thick, rounded posterior part of the lung (the so-called 'posterior border') comprises the adjoining parts of the costal and vertebral surfaces.

The *anterior border* is thin and sharp, and overlaps the front of the pericardium; that of the *right* lung is customarily described as corresponding quite closely to the costomediastinal line of pleural

reflexion, being almost vertical; that of the *left* approaches the costomediastinal line of pleural reflexion in its upper part, but below the level of the fourth costal cartilage it presents a notch of variable size, named the *cardiac notch*. The margin of this notch passes laterally for some 3·5 cm before curving downwards and medially to reach the sixth costal cartilage about 4 cm from the median plane. It thus falls considerably short of the recessed part of the line of pleural reflexion (**8.23**A), leaving the pericardium in this situation covered only with a double layer of pleura. (See Area of Superficial Cardiac Dullness, p. 653.)

It should be noted, however, that cardiothoracic surgeons consider that the costomediastinal line of pleural reflexion, the anterior margin of the lung, and the extent of the costomediastinal pleural recess, show *great variability* between subjects.

PULMONARY FISSURES AND LOBES

The left lung is divided into a superior and an inferior lobe by an *oblique fissure* (**8.27**B), which extends from the costal to the medial surface of the lung both above and below the hilum. As seen on the surface, this fissure begins on the medial surface of the lung at the upper and posterior part of the hilum, and runs backwards and upwards to the posterior border, which it crosses at a point about 6 cm below the apex. It then extends downwards and forwards over the costal surface (**8.23**A), reaching the lower border a little behind its anterior extremity. Its further course can be followed upwards and backwards across the mediastinal surface as far as the lower part of the hilum. On the posterior border of the lung the oblique fissure lies opposite the interval between the spines of the third and fourth thoracic vertebrae (about 2 cm from the median plane), but it may be either above or below this level. Traced downwards and forwards across the costal surface of the lung, the fissure reaches the fifth intercostal space (in or near the mid-axillary line) and follows this until it intersects the inferior border of the lung close to, or just below the sixth costochondral junction (7·5 cm from the median plane). As a rule the oblique fissure of the left lung is more vertical than the corresponding fissure of the right lung. The fissure corresponds roughly with the medial border of the scapula when the arm is abducted above the level of the shoulder, as by placing the hand on the back of the head, the inferior angle of the bone moving outwards and forwards. The *superior lobe* lies above and in front of this fissure, and includes the apex, the anterior border, a considerable part of the costal surface and the greater part of the medial surface of the lung. A small projection is usually present at the lower part of the cardiac notch and is termed the *lingula* of the lung. The *inferior lobe*, the larger of the two, is situated below and behind the fissure, and comprises almost the whole of the *base*, a large portion of the costal surface, and the greater part of the posterior border. (*See* pp. 1243–1246 for details of bronchopulmonary segmentation.)

The right lung is divided into three lobes, superior, middle and inferior, by two fissures (**8.27**A). One of these separates the inferior from the middle and superior lobes, and corresponds closely with the oblique fissure in the left lung. Its direction is, however, less vertical, and cuts the lower border about 7·5 cm behind its anterior end. On the posterior border it lies opposite the spine of the fourth thoracic vertebra or at a slightly lower level. As it descends it crosses the fifth intercostal space and then follows the general line of the sixth rib to the sixth costochondral junction. A short *horizontal fissure* separates the superior from the middle lobe. It begins in the oblique fissure near the mid-axillary line, and, running horizontally forwards, cuts the anterior border on a level with the sternal end of the fourth costal cartilage; on the mediastinal surface it may be traced backwards to the hilum. The *middle lobe* of the right lung is small and cuneiform, and includes a part of the costal surface, the lower part of the anterior border and the anterior part of the base of the lung. Sometimes the medial part of the upper lobe is partially separated from the rest by a fissure of variable depth which contains the terminal part of the azygos vein enclosed in the free margin of a mesentery derived from the mediastinal pleura. The portion of the right lung so defined is sometimes termed the '*lobe of the azygos vein*'. It varies in size and sometimes includes the apex of the lung. It is always supplied by one or more branches of the apical bronchus. Radiographically, a pleural effusion may be limited to the azygos fissure.

Attempts have been made to equate the lobation of the two lungs and in particular to equate the right middle lobe with the lingula. The primary pattern of the development of the lungs does not favour this view (p. 208). For a recent discussion and introduction to the literature consult Yokoh (1977).

Since the diaphragm rises higher on the right side in order to accommodate the liver, the right lung is shorter (by 2·5 cm) than the left, but, owing to the projection of the heart to the left side, it is broader and its total capacity and weight are greater than those of the left lung.

PULMONARY HILA AND ROOTS

The root of the lung (**8.25**) connects the medial surface to the heart and the trachea and is formed by the structures which enter or emerge at the hilum. It comprises the principal bronchus, the pulmonary artery, the two pulmonary veins, the bronchial arteries and veins, the pulmonary plexus of nerves, lymph vessels, bronchopulmonary lymph nodes and areolar tissue, all of which are enveloped by pleura. The roots of the lungs lie opposite the bodies of the fifth, sixth and seventh thoracic vertebrae. That of the right lung lies behind the superior vena cava and part of the right atrium, and below the terminal part of the azygos vein. That of the left lung is below the aortic arch and in front of the descending thoracic aorta. The following relations are common to the two lung roots—*anterior*, the phrenic nerve, the peri-cardiacophrenic artery and vein, and the anterior pulmonary plexus; *posterior*, the vagus nerve and posterior pulmonary plexus; *inferior*, the pulmonary ligament.

The chief structures composing the root of each lung are arranged in a similar manner from before backwards on both sides, viz.: the upper of the two pulmonary veins in front; the pulmonary artery, the principal bronchus behind, with the bronchial vessels on its posterior aspect. Their arrangement differs from above downwards on the two sides; on the right side their position is—superior lobar bronchus, pulmonary artery, right principal bronchus, lower pulmonary vein; but on the left side their position is—pulmonary artery, bronchus, lower

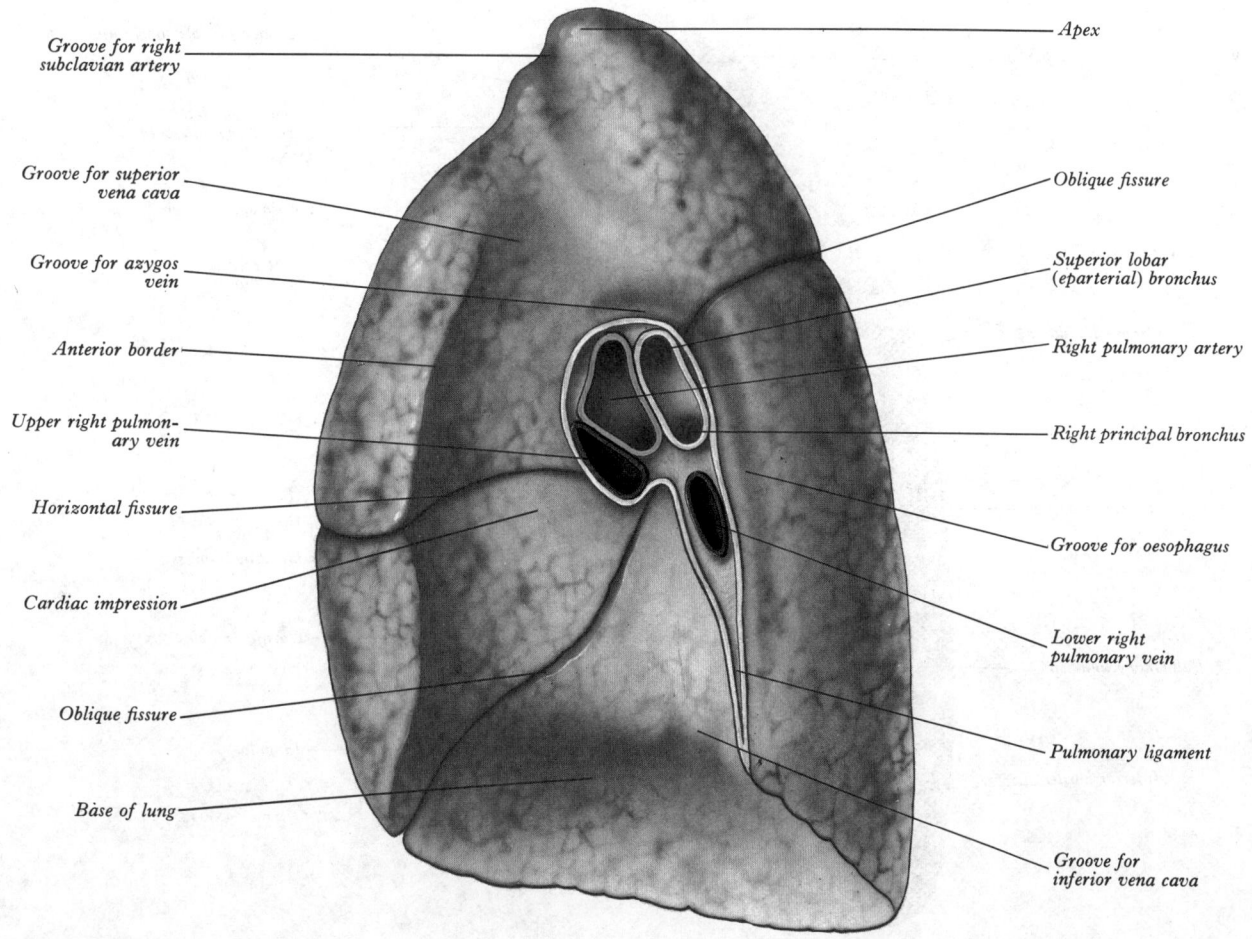

Groove for right subclavian artery

Groove for superior vena cava

Groove for azygos vein

Anterior border

Upper right pulmon-ary vein

Horizontal fissure

Cardiac impression

Oblique fissure

Base of lung

Apex

Oblique fissure

Superior lobar (eparterial) bronchus

Right pulmonary artery

Right principal bronchus

Groove for oesophagus

Lower right pulmonary vein

Pulmonary ligament

Groove for inferior vena cava

8.27A   The medial surface of the right lung.

pulmonary vein. The lower left pulmonary vein is inferior to the principal bronchus, which is lowest in the hilum (8.27A, B).

All parts of the lungs do not move equally in respiration. The region near the root hardly moves at all, while the middle region of the lungs moves very slightly in quiet respiration. The superficial parts of the lungs expand most, but the mediastinal surface, the posterior border and the apical region do not move to the same extent owing to the less movable structures related to them. The diaphragmatic and costomediastinal regions undergo most expansion of all. (*See* Movements of Respiration, p. 550.) Most of the volumetric change in the lungs during respiration occurs in the alveoli. The number of alveoli in the lung of a newborn infant may be over 20 million, increasing to 300 million or more during childhood. Each alveolus varies from 200 to 300 μm in diameter, and a very large number of capillary segments (1,800 is suggested) may make contact with a single alveolus. The area of gas-epithelium-blood interface is relatively enormous even in childhood. Figures of 70 to 100 m² have been estimated (Peters 1969; Fisherman and Pietra, 1974), while the more recent investigations of Gehr et al., (1978) have yielded even higher values.

Structure of the Lung

The lungs arise during development (*see* pp. 199, 207) as epithelial outgrowths of the anterior foregut, forming a system of tubes which branch as they grow, and which become invested externally by mesenchymal tissue rich in blood vessels. This developing mass of tissue protrudes into the anterior coelomic cavity (later the pleural cavities) and becomes lined externally by coelomic epithelium (mesothelium) which eventually forms the lining of the serous coat of the pulmonary pleurae. Within the lung the proximal parts of the tubular system become the *conducting* airways of the bronchial tree and the more distal ones expand to form the cavities across the walls of which *respiratory*

exchange takes place between the atmosphere and the closely applied capillaries. The individual components in this complex will now be considered in some detail.

THE SEROUS COAT AND ASSOCIATED CONNECTIVE TISSUE

The serous coat forms the visceral pleura, and is a thin transparent layer composed of a single layer of mesothelium situated on its underlying lamina propria, inseparably connected with the main mass of lung tissue, which it invests as far as the root. The subserous, loose connective tissue underlies the serous coat and covers the entire surface of the lung, extending inwards along the conducting tubes and blood vessels from the hilum, and delineating the *lung lobules*, which are important functional units of lung structure. Each lobule is a small polyhedral mass of lung tissue which receives a small air passage (a bronchiole) together with the terminal ramifications of the pulmonary arterioles and venules, lymphatics and nerves. Lobules vary in size, those on the surface being large and of a pyramidal form, with their bases turned towards the pleural surface, and are visible as polygonal areas about 5–15 mm across divided from adjacent areas by thin layers of connective tissue. The lobules in the interior are smaller and are of various forms. Because of this lobular structure, the lung is in some respects similar in general organization to a lobulated gland, consisting of terminal lobules and extralobular ducts which are the larger conducting passages of the lung (Miller 1947; von Hayek 1960; Engel 1962; Krahl 1964). The fairly gross pattern of lobulation just described is considered by some investigators to consist of *secondary lobules*; these are delineated on the lung surface by substantial connective tissue septa enclosing areas of 1–2 cm². Close inspection of the latter shows them to be subdivided by delicate septa into small polygonal areas about 1 mm²—the surfaces of the *primary lobules*. A widely accepted terminology for the different suborders of lobulation has yet to be devised.

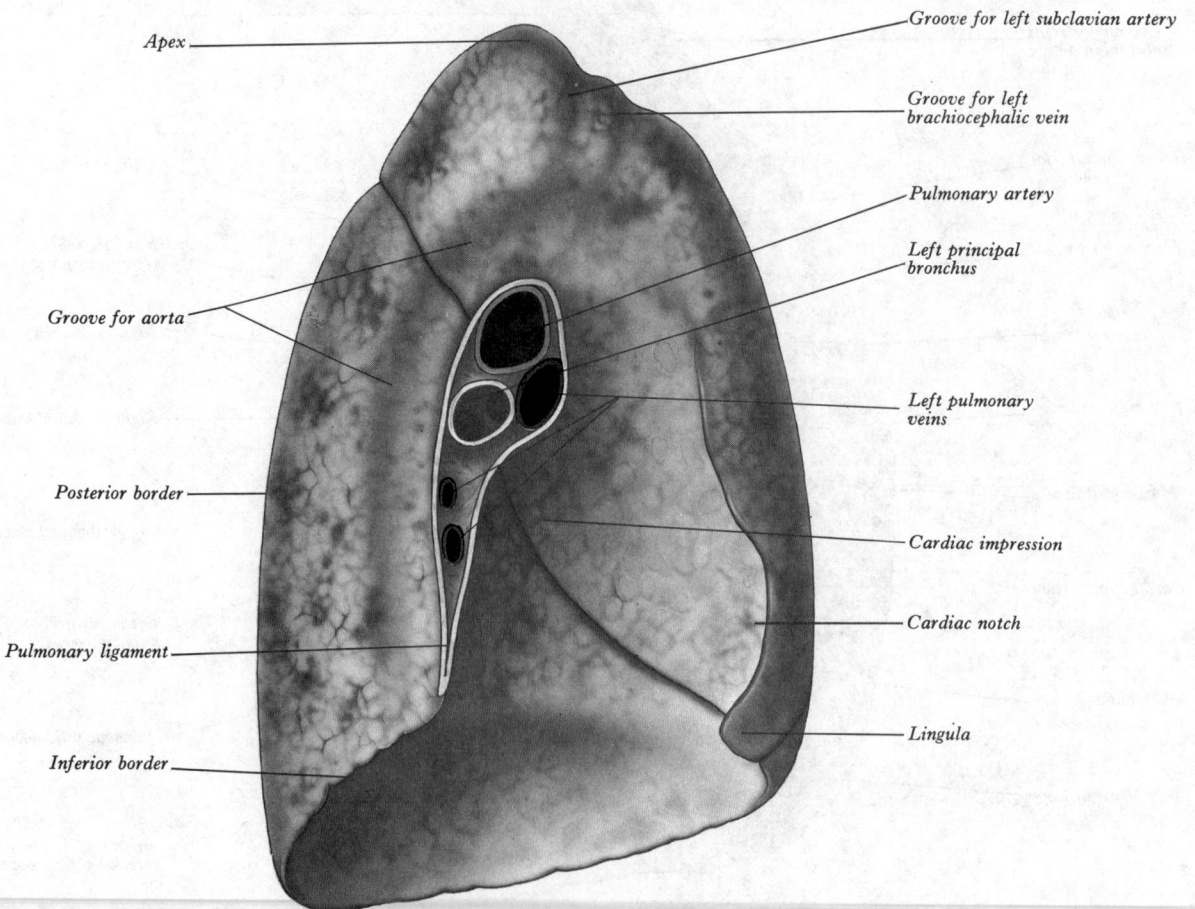

Apex

Groove for left subclavian artery

Groove for left brachiocephalic vein

Pulmonary artery

Left principal bronchus

Groove for aorta

Left pulmonary veins

Posterior border

Cardiac impression

Cardiac notch

Pulmonary ligament

Lingula

Inferior border

8.27B The medial surface of the left lung.

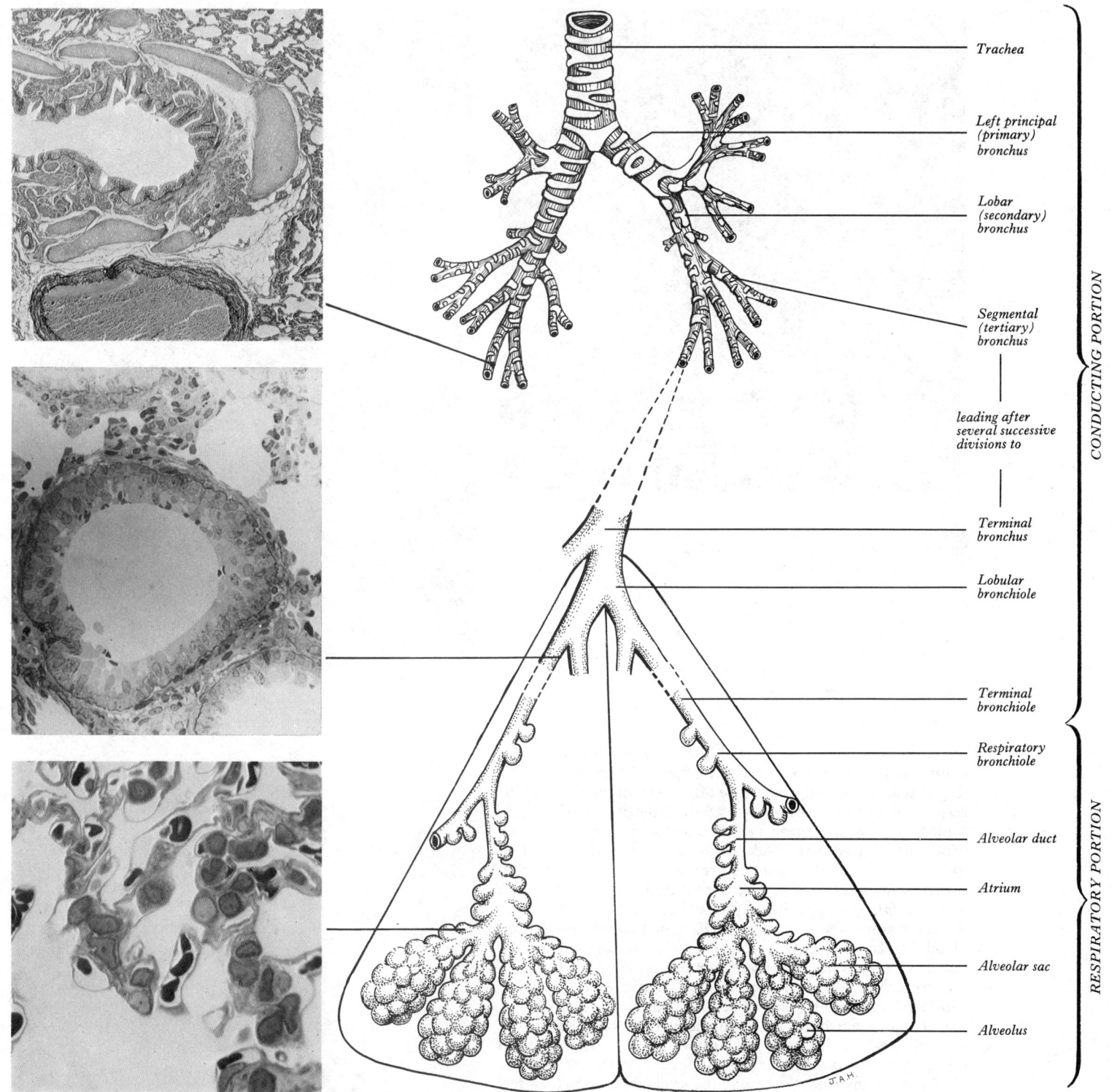

CONDUCTING PORTION

RESPIRATORY PORTION

Trachea

Left principal (primary) bronchus

Lobar (secondary) bronchus

Segmental (tertiary) bronchus

leading after several successive divisions to

Terminal bronchus

Lobular bronchiole

Terminal bronchiole

Respiratory bronchiole

Alveolar duct

Atrium

Alveolar sac

Alveolus

8.28A Bronchopulmonary structure. The diagram on the right shows the gross architecture of the conducting and respiratory parts of the trachea and lungs at segmental levels. On the left are three sections at the levels indicated. *Top:* A small bronchus lined by highly convoluted epithelium, surrounded by plates of hyaline cartilage. A pulmonary arteriole is visible below. *Middle:* A bronchiole. Note absence of cartilage, and conspicuous nonstriated muscle external to the epithelium. (Epoxy resin section prepared by Mrs. Susan Smith; photographed by Miss Marina Morris, Guy's Hospital Medical School.) *Bottom:* An interalveolar septum. Note capillaries containing erythrocytes (dark blue) separated from air spaces by thin alveolar epithelial cells. (Epoxy resin section stained by toluidine blue.)

THE AIR PASSAGES AND THEIR ASSOCIATED RESPIRATORY SPACES

The 'lower' respiratory tract includes larynx, trachea and extrapulmonary bronchi (*vide supra*) and various orders of smaller intrapulmonary tubes which branch repeatedly in a dichotomous manner. It is usual to distinguish between *secondary bronchi*, each supplying a whole lobe, *tertiary bronchi* supplying individual bronchopulmonary segments, and further subdivisions of bronchi which branch repeatedly within bronchopul-monary segments, becoming narrower towards their distal ends. All the intrapulmonary bronchi are supported by irregular cartilaginous plates which become smaller and more widely spaced distally (Reid 1976) until the cartilage is absent, at which point the air passages, now less than 1 mm in diameter, are termed *bronchioles*. After repeated branching, one of the smaller bronchioles enters a lung lobule; this *lobular bronchiole* immediately gives rise to about six *terminal bronchioles* which further subdivide into from one to three generations of *respiratory bronchioles* (Spencer 1977). The terminal bronchioles are the most

1257

B

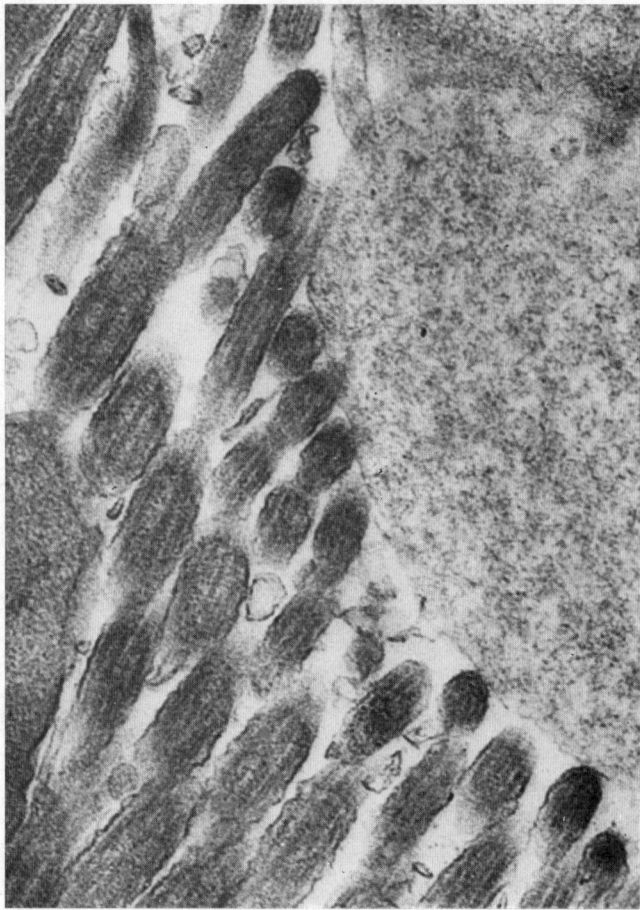

C

8.28B–D. B Electron micrograph showing the epithelium lining the bronchus of a rat. Clustered between the ciliated cells are a variety of non-ciliated cells, including goblet cells and Clara cells. (Magnification ×1,800.) C Electron micrograph showing bronchial cilia with a droplet of the mucus which they transport in contact with their tips. Apical claw-like projections can be seen on one cilium. (Magnification ×35,000.) D Claw-like projections at the apex of a cilium of a bronchial epithelial cell (rat). (Magnification 90,000.) (Preparations B, C, D by Mr. Michael Crowder, Department of Anatomy, Guy's Hospital Medical School, London.)

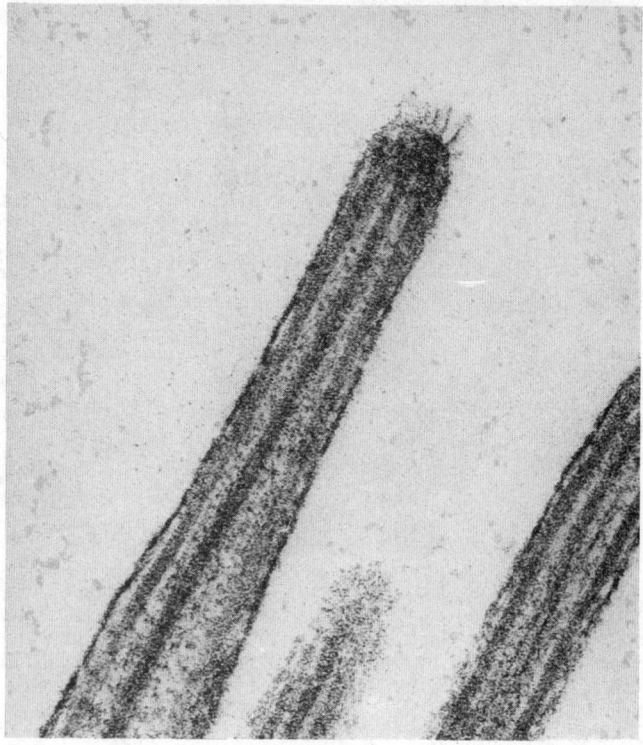

D

distal air passages lined completely by simple columnar epithelium, while the respiratory bronchioles are characterized by the presence of alveoli which arise directly from their walls. The respiratory bronchioles open into two or three *alveolar ducts*, thin walled tubes which terminate in expanded passages, the *atria*, which in turn lead into *alveolar sacs* or *air sacs*. The thin walls of the alveolar ducts, atria and alveolar sacs are studded with small pouches, the *acini* or *alveoli*, separated from each other by thin *interalveolar septa* in which lie the connective tissue elements of the respiratory tissue and the capillary beds of the pulmonary blood supply. The above arrangement, shown diagrammatically in **8.28**A and **8.29**A, B, is normally found in the adult; variations in the pattern occur during periods of active growth during childhood and have been described in detail by Engel (1947).

It should be noted that the terminology for these various orders of branches which has been outlined here is not followed universally and considerable confusion exists in the terminology of all conducting passages beyond the level of the bronchi. Some authors, for example, define the terminal bronchiole as that duct which enters a lung lobule, and others adopt a simpler view of the various passages leading to the alveoli.

Air Passage Epithelium

The epithelial lining of the intrapulmonary bronchi and bronchioles is similar, in many respects, to that of the trachea and extrapulmonary bronchi (Collet *et al.* 1967). The epithelium of the larger intrapulmonary air passages is pseudostratified and, in the main, ciliated, while that of the terminal and respiratory bronchioles consists of a single layer of cells, many of which are non-ciliated.

Only in the rat has there been a systematic survey of the ultrastructure of the epithelial cell types present at different air passage levels (Jeffery and Reid 1975). There are ten distinct cell types within the epithelium: eight of epithelial origin and two

mesenchymal (probably migratory). All the epithelial cells rest on a basal lamina, but in the larger, more proximal bronchi, not all reach as far as the lumen of the air passage. The epithelial cells proper are of the following types: *ciliated, serous, Clara, goblet, intermediate, brush, basal* and *Kulchitsky*. Of these, the serous type has not previously been described in air passage epithelium. The mesenchymal cells are either '*globule*' *leucocytes* or *lymphocytes*. The distinguishing ultrastructural features of these cells are listed in the table below. In the rat, goblet cells occur at all air passage levels, but comprise less than 1 per cent of the cells; in some species they are absent from the smaller bronchioles. Basal cells decrease in number distally and are virtually absent from the smaller bronchioles, where the epithelium is simple in type. Clara cells are represented throughout the system. Even in the most proximal regions over 40 per cent of the cells are non-ciliated (Jeffery and Reid 1975).

The *ciliated cells*, which play an important role in maintaining the ciliary rejection current of the bronchial tree, have electron-lucent cytoplasm, a nucleus with a smooth outline, and cilia which bear apical, claw-like projections (8.28c, d) described by Jeffery and Reid (1975). The cilia are surrounded by a low-viscosity liquid produced, in part, by the serous cells (*vide infra*). Above this periciliary liquid lie flakes and droplets of mucus (Van As and Webster 1972) produced by the goblet cells and by the exocrine cells of the bronchial mucous glands. According to Sleigh (1974) only the tips of the cilia are in contact with this mucus, which they 'claw' along in a disto-proximal direction. The finding of claw-like projections at the tips of the cilia supports this hypothesis. The mucus-secreting *goblet cells* are distended by supranuclear electron-lucent granules which often have electron-dense cores.

The *serous cells* each contain a nucleus of irregular outline, copious rough endoplasmic reticulum and supranuclear homogeneously electron-dense granules. Although most numerous in the trachea and extrapulmonary bronchi, serous cells are occasionally found in the intrapulmonary air passages. Similar cells have been observed in human bronchial submucosal glands (Meyrick and Reid 1970). The serous cells probably contribute to the low-viscosity periciliary fluid mentioned above.

The *Clara cells*, originally thought to be characteristic of the terminal bronchioles (Clara 1937), are also present at more proximal levels of the bronchial tree, and even in the nasal mucosa (Matulionis and Parks 1973). They are non-ciliated cells with blunt luminal projections, irregular homogeneously electron-dense secretory granules, many lysosomes, osmiophilic lamellar bodies, and much smooth endoplasmic reticulum often concentrated in the apical region, which may be connected to the rest of the cell by a narrow 'waist' (Spencer 1977). Niden and Yamada (1966) were the first to suggest that the Clara cell might be a source of surfactant material, a view supported by Etherton and Conning (1971), whose autoradiographic localization of radioactively labelled dipalmitoyl lecithin (a constituent of alveolar surfactant) over the secretory granules of the Clara cells within five minutes of its intraperitoneal injection, cast doubt on the previously held view that type II alveolar cells (p. 1262) were the sole source of surfactant.

Epithelial cells described as '*intermediate*' have been observed in the bronchial epithelium of the rat (Rhodin and Dalhamn 1956; Jeffery and Reid 1975) and in the tracheal epithelium of man (Rhodin 1966). They have long, luminal filopodia, electron-dense cytoplasm but no secretory granules; they do, however, at times contain fibrogranular accumulations, suggestive of ciliogenesis (Sorokin 1968). They may be undifferentiated cells capable of developing into either ciliated or secretory cells (Rhodin 1966).

The *brush cells* are non-ciliated cells with a distinct brush border of microvilli on their luminal aspect. The presence of numerous pinocytotic vesicles near the bases of these microvilli suggests that the brush cells are specialized for absorption. They are most numerous in the tracheal epithelium, where in the rat they constitute about 1 per cent of the epithelial cells, but are also found scattered throughout the rest of the system of air passages (Jeffery and Reid 1975).

The rarest epithelial cell type is the *bronchial Kulchitsky (argentaffin) cell* (Bensch et al. 1965). This has electron-lucent cytoplasm containing dense-cored secretory granules which accumulate between the nucleus and the base of the cell, and discharge their contents in response to hypoxia (Moosavi et al. 1973; Lauweryns and Cokelaere 1973). Termed *Feyrter* cells by Moosavi et al. (1973), and P_a cells by Capella et al. (1978) in a recent report of the ultrastructure of these cells in the human lung, they resemble the D_L cells (p. 1365) of the gastro-entero-pancreatic system, store and secrete amines (Ericson et al. 1972), and should thus be included in the diffuse endocrine system of APUD cells (Pearse and Polak 1971; Hage 1973). Innervated groups of related cells form neuro-epithelial bodies in the human bronchial and bronchiolar mucosa (Lauweryns and Peuskens 1972). Functions suggested for the Kulchitsky cells include an involvement in the regulation of lobular growth (Rosan and Lauweryns 1971) and chemoreception (Lauweryns et al. 1972).

The final group of epithelial cells proper are the undifferentiated *basal cells* found in those parts of the airways lined by pseudostratified epithelium. Division of these cells provides a pool of replacements for the other epithelial cell types, in a manner reminiscent of the activity of the germinal layer of the epidermis (Blenkinsopp 1967). In more distal regions of the bronchial tree, where the epithelium is of a simple type, undifferentiated cells scattered amongst the mature epithelial cells perform this function.

The remaining cells occurring in the normal airway epithelium are either 'globule' leucocytes or lymphocytes, both migratory cells, and not epithelial cells proper. The '*globule*' *leucocytes* have very electron-dense granules and numerous filopodia; they

DISTINGUISHING ULTRASTRUCTURAL FEATURES OF CELLS WITHIN THE EPITHELIUM LINING THE AIR PASSAGES

Cell type	Granules and density	Endoplasmic reticulum	Cytoplasmic density	Surface projections
Epithelial				*Luminal*
Ciliated	−	R	L	Cilia+filiform
Serous	+D	R	D	Filiform
Clara	+D	S	L	Blunt
Goblet	+L	R	D	Filiform
Intermediate	−	R	L or D	Filiform
Brush	−	R	Intermediate	Microvillous
				Intercellular
Basal	−	R (sparse)	D	Blunt
Kulchitsky	Neurosecretory	R	L or D	Elongated
Mesenchymal				
'Globule' leucocyte	+D	R (sparse)	Intermediate	Filiform
Lymphocyte	−	R (sparse)	L	Blunt

+, present	R, rough (granular)	L, electron lucent
−, absent	S, smooth (agranular)	D, electron dense

(Modified from Jeffery and Reid 1975.)

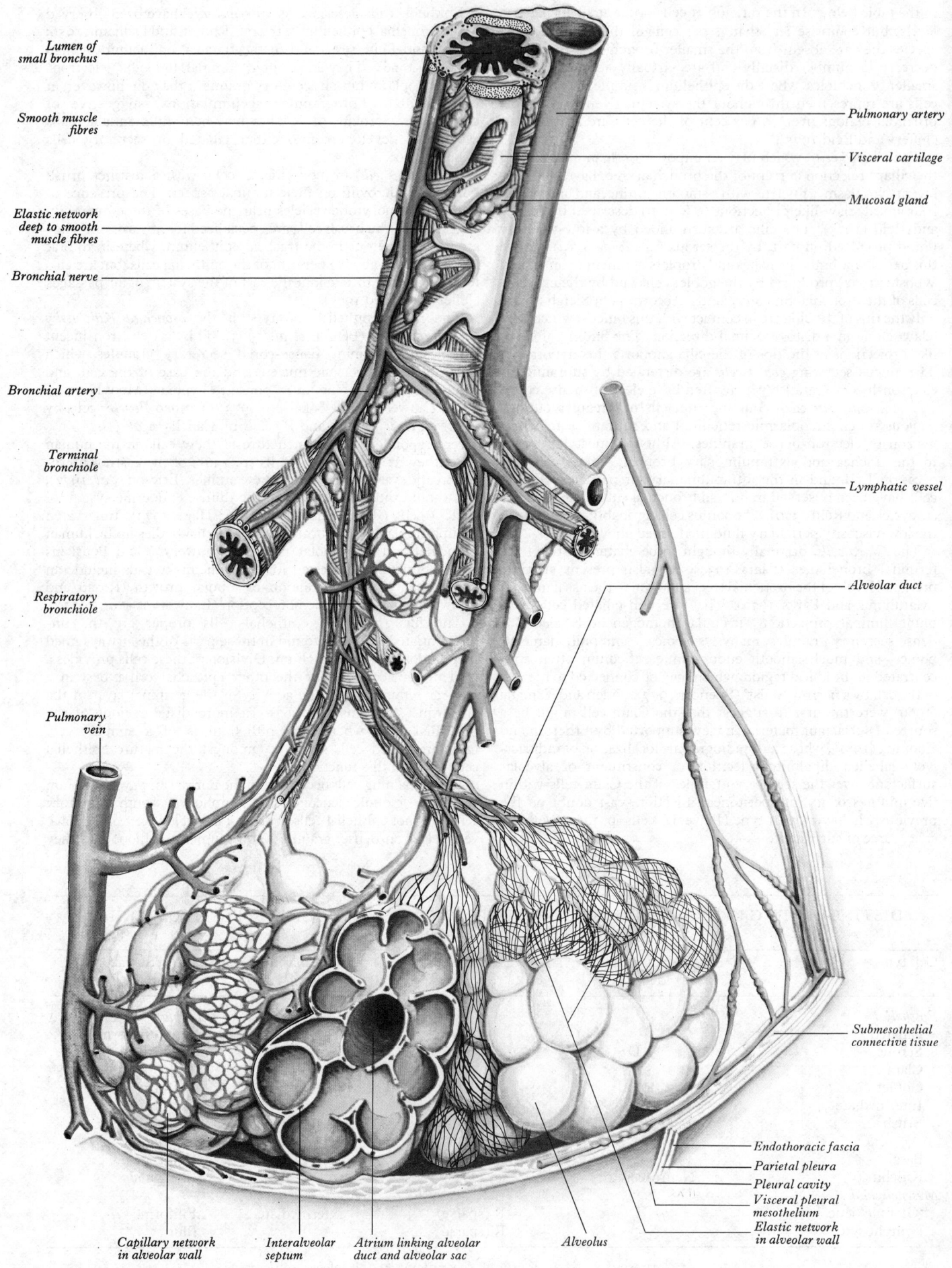

Lumen of small bronchus

Smooth muscle fibres

Elastic network deep to smooth muscle fibres

Bronchial nerve

Bronchial artery

Terminal bronchiole

Respiratory bronchiole

Pulmonary vein

Pulmonary artery

Visceral cartilage

Mucosal gland

Lymphatic vessel

Alveolar duct

Submesothelial connective tissue

Endothoracic fascia

Parietal pleura

Pleural cavity

Visceral pleural mesothelium

Elastic network in alveolar wall

Capillary network in alveolar wall

Interalveolar septum

Atrium linking alveolar duct and alveolar sac

Alveolus

8.29A Diagram of the detailed structure of the respiratory tree and its blood supply and drainage, lymphatic drainage, and nerve supply. Vessels shown in blue contain de-oxygenated blood, those shown in red contain oxygenated blood.

resemble tissue mast cells, from which they may be derived (Murray *et al.* 1968). The *lymphocytes* have sparse electron-lucent cytoplasm and are most numerous in the extrapulmonary airways. Their ultrastructural appearance in human bronchi has been described by Meyrick and Reid (1970). Both cell types may have immunological functions.

The epithelial surface of the smaller bronchi and of the bronchioles is folded into conspicuous longitudinal ridges which allow for the changes in diameter of the passages (8.29A). The epithelium of the respiratory bronchioles is progressively less thick towards the alveolar termination, and is eventually composed of cuboidal, non-ciliated cells; the lateral pouches are lined with simple squamous cells, providing a small accessory respiratory surface.

MICROSCOPICAL STRUCTURE OF THE AIR PASSAGES (8.29A, B, C)

The epithelium of the bronchi and various bronchioles is situated on a basal lamina attached to a lamina propria in which broad, conspicuous, longitudinal bands of elastin follow the course of the bronchial tree, branching at the bifurcations of air passages and finally ramifying amongst the elastin network of the respiratory portions of the lung. The elastic framework which this system creates is an important mechanical element, responsible for much of the elastic recoil of the lung during expiration, although, in the respiratory portions, surface tensional forces may be more important (*vide infra*), and *see* Peters (1969). Nonstriated muscle fibres lie in the submucosal connective tissue, within the confines of the visceral cartilages where these are present, forming two helical tracts which run in opposite directions along the entire intrapulmonary bronchial tree, becoming thinner and finally absent at the bases of the alveoli. These muscle fibres are under nervous and hormonal control, and their contraction narrows the airways, whilst their relaxation permits bronchial dilatation. Normally there is a degree of tone in the muscular bands, which relax slightly during inspiration and contract a little during expiration, thus assisting the tidal flow of air. Abnormal contraction may result if stimulants to nonstriated muscle occur in the circulation or from the local release of such

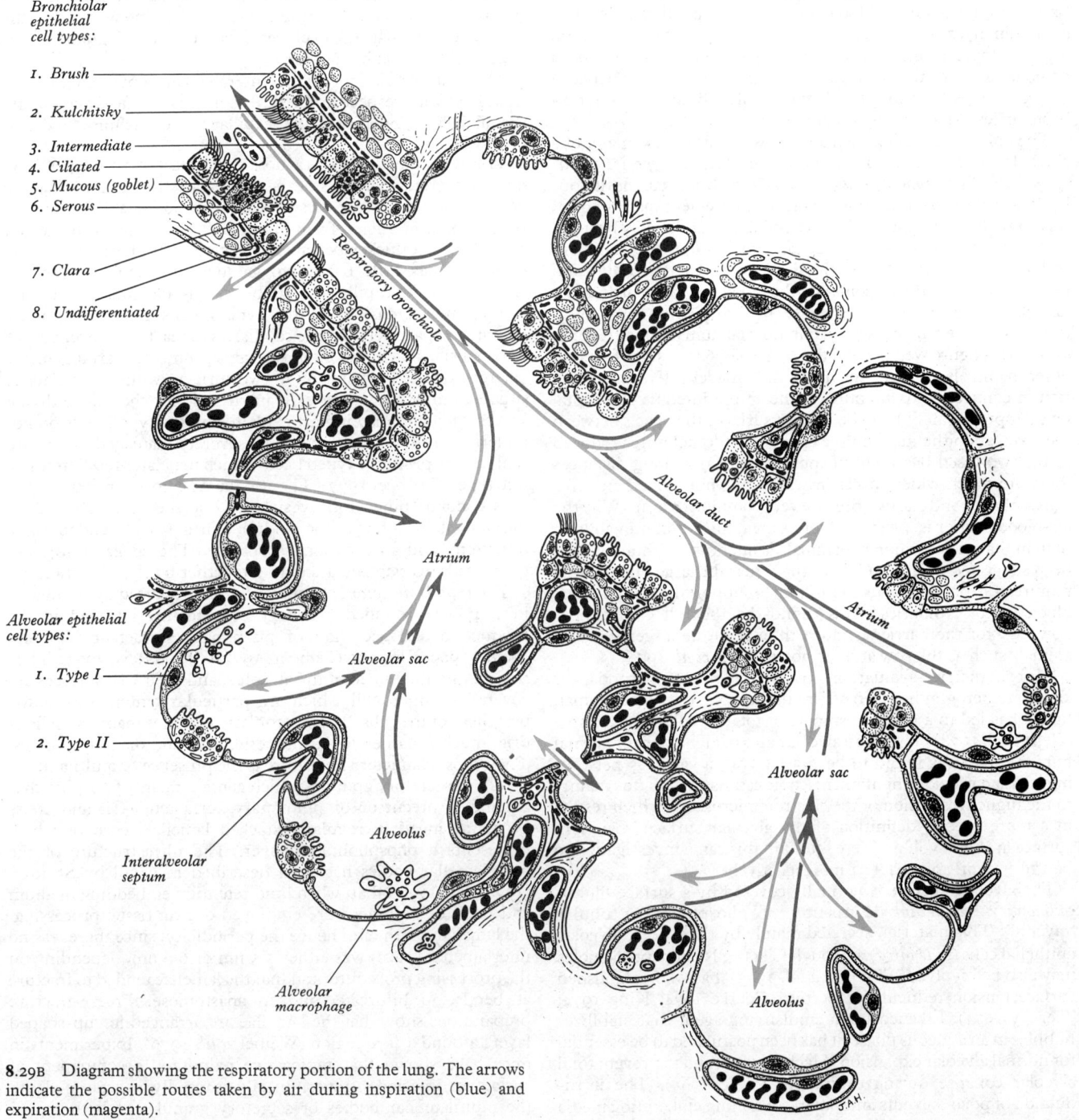

Bronchiolar epithelial cell types:

1. Brush
2. Kulchitsky
3. Intermediate
4. Ciliated
5. Mucous (goblet)
6. Serous
7. Clara
8. Undifferentiated

Respiratory bronchiole

Alveolar duct

Atrium

Atrium

Alveolar epithelial cell types:

1. Type I
2. Type II

Alveolar sac

Alveolar sac

Alveolus

Interalveolar septum

Alveolar macrophage

Alveolus

8.29B Diagram showing the respiratory portion of the lung. The arrows indicate the possible routes taken by air during inspiration (blue) and expiration (magenta).

excitants as serotonin and histamine, to give bronchial spasm and impairment of respiration.

Prolongations of mucous membrane extend beyond the muscle layer in some bronchi and bronchioles. The deep mucosal crypts so formed may develop into canals which provide additional connexions between the peribronchiolar alveoli and the more proximal air passages (Lambert 1955).

Tubular glands are also present within the submucosa of the bronchi and upper parts of the bronchioles; these are *mucous* and *serous* in type, as in the extrapulmonary ducts.

STRUCTURE OF THE ALVEOLUS

The alveoli, consisting of thin-walled pouches, provide collectively a respiratory surface with a total area which is high yet presents only a minimal barrier to the exchange of gases between the atmosphere and the blood of the capillaries lying in close association with the surface lining (Karrer 1956; Collet *et al.* 1967; Klika and Petřick 1965).

The walls of the alveoli are composed of a thin epithelial lining situated on a layer of connective tissue (8.29B, 30). Since the walls are thin and adjoin the walls of adjacent alveoli, the connective tissue and its contained blood vessels are frequently sandwiched as a thin irregular lamina between two layers of epithelium, forming together an *interalveolar septum*. Apertures have been demonstrated in such septa, allowing the passage of air between adjacent alveoli, and also between alveoli and respiratory bronchioles (Macklin 1936; Lambert 1955; Suarez 1979).

The epithelium lining each alveolus consists, in the main, of a single layer of attenuated squamous cells, termed *type I alveolar epithelial cells* or *type I pneumonocytes*, which form a continuous layer as little as 0·05 μm thick. Since the basement membrane of this epithelium and that of the underlying capillaries may be fused together and only about 0·1 μm thick, and the capillary endothelium about 0·05 μm thick, the total barrier to diffusion between the air of the alveolus and the blood may be as little as 0·2 μm (Collet *et al.* 1967). Despite this, the arithmetic mean thickness of the barrier is 2·2 μm in normal human lung (Gehr *et al.* 1978), a figure well beyond the range of 0·6 to 1·5 μm found in other mammals, and due mainly to the broader interstitial space rich in connective tissue fibres found in the interalveolar septa. Over approximately half the barrier surface, the space between the alveolar epithelium and the capillary endothelium is reduced to the two fused basement membranes, but over long distances there are much wider spaces in which free fluid, fixed and free cells, collagen and elastic fibres are very common (8.29B). Why the air-blood barrier is, over much of its area, thicker in human lung than in that of non-human mammals, is unknown, although it has been suggested (Wang and Ying 1977) that the large human lung requires a stronger fibrous continuum to support the weight of the blood in its capillaries. This cannot be the full explanation, however, for the barrier is much thinner in the larger lungs of animals such as the cow and the horse (Gehr *et al.* 1978).

Morphometric evaluation, by stereological methods, of electron micrographs taken of random samples of normal human lungs, has led to a revised estimate of total alveolar surface area (Gehr *et al.* 1978). The mean alveolar area of eight pairs of normal human lungs was found to be 143$\pm$12 m², a value 75 per cent higher than previous light microscopic estimates, and due mainly to the higher resolution of the electron microscope which results in a more precise definition of the alveolar surface. Capillary surface area and volume were found in this same investigation to be 126$\pm$12 m² and 231$\pm$31 ml respectively.

The alveolar surface is normally covered by a surface film of *pulmonary surfactant*, consisting of phospholipid (tubular 'myelin'). This material, secreted mainly by the type II alveolar epithelial cells (*vide infra*), although Clara cells of the bronchioles may also be involved (Etherton *et al.* 1973), has well-established surface tension reducing properties (Goerke 1974; King 1974; Tierney 1974). Likened to an emulsifying agent that stabilizes bubbles in an aqueous phase, it has been postulated to be essential for normal alveolar expansion (Macklin 1954) and to prevent total alveolar collapse during expiration (Pattle 1965). The introduction of polar solvents and water-containing embedding media

in electron microscopy has improved the retention of lipids and carbohydrates in ultrathin sections and has led to a re-assessment of the ultrastructure of pulmonary surfactant (Stratton 1977; Stratton 1978) and of its precursors, the multilamellar bodies of the type II alveolar epithelial cells (Douglas *et al.* 1975; Stratton, 1976). Formerly described as a duplex layer consisting of a basal hypophase in contact with the alveolar epithelial cells and a superficial epiphase (Weibel and Gil 1968), the pulmonary surfactant now appears to be a five-layered material, each layer representing a different stage in the transformation of secreted multilamellar bodies. The layers characteristic of human pulmonary surfactant are as follows: (1) a layer of recently secreted multilamellar bodies lying nearest to the alveolar epithelial cells; (2) a layer of paired lamellae which expand and re-arrange to form tubules; (3) a mature layer of tubular 'myelin' surfactant; (4) a surfactant-air interface layer, usually a single lipid bilayer; and (5) a layer of degraded surfactant, consisting of lipid bilayer spheres which are formed at the interface and degraded in the alveolar space (Stratton 1978). At isolated alveolar sites, degradation of surfactant proceeds until it has been removed completely, leaving the site empty until it can be repopulated by multilamellar bodies which transform into functional surfactant. Localized absence of pulmonary surfactant is part of the natural cycle of secretion and degradation in apparently normal human lung.

The cell population of the interalveolar septa can best be described with respect to the three principal tissue components: the alveolar epithelium, the capillary endothelium, and the interstitial space.

The *alveolar epithelium* is a mosaic of at least two cell types. The most numerous members of the mosaic are the squamous *type I alveolar epithelial cells*; these resemble endothelial cells, having thin cytoplasmic flaps from about 0·05 to 0·2 μm wide which extend from a thicker perinuclear region. The thinness of these flaps provides little barrier to gaseous diffusion between the alveolus and its capillaries. Weibel (1971) has found that deep cytoplasmic extensions of the type I cells pass through the interstitial space between the capillaries to reach and cover part of the opposite side of the interalveolar septum. This arrangement, whereby one cell extends on to two surfaces while its nucleus is situated only on one, makes it impossible for the cell to divide successfully and may in part explain why the type I cells do not undergo proliferative repair. If severely damaged, they are replaced by primitive type II cells which may later redifferentiate into type I (Spencer 1977). The attenuated cytoplasm of the type I cells contains pinocytotic vesicles, while the perinuclear cytoplasm contains a few mitochondria, a little smooth endoplasmic reticulum, and an occasional lysosome. The edges of adjacent type I cells overlap and are bound together by tight junctions.

The *type II alveolar epithelial cells*, first reported by Reinhardt in 1847, are rounded secretory cells mainly engaged in the production and secretion of pulmonary surfactant (Gil and Weibel 1969; King and Clements 1972; Gil and Reiss 1973). They project into the lumen of the alveolus, and their free surfaces are covered by microvilli which are particularly numerous at the periphery of the cells. Their cytoplasm contains many mitochondria, much rough endoplasmic reticulum, and many lysosomes. Their most characteristic feature is the presence of multilamellar bodies or secretory granules which contain mainly phosphatidylcholine, a precursor of pulmonary surfactant (Gil and Reiss 1973), arranged in regularly stacked lamellae, each of which represents a phospholipid bilayer. The ultrastructure of the multilamellar bodies has been described in detail by Stratton (1976) who found that, when lipid retention embedding medium and polar dehydrants were employed during tissue processing, the lamellar width (and hence the periodicity, since there was no interlamellar space) was either 5·5 nm or 6·6 nm, depending on the processing procedure, and that the lamellae tended to fracture at bends, to bifurcate, and to anastomose. Freeze-fracture preparations show that the lamellae are arranged in cup-shaped layers around a core region (Weibel *et al.* 1976). Experiments in mice have shown that radioactive palmitate, a component of surfactant, injected intraperitoneally, is rapidly incorporated into the multilamellar bodies or secretory granules of both type I

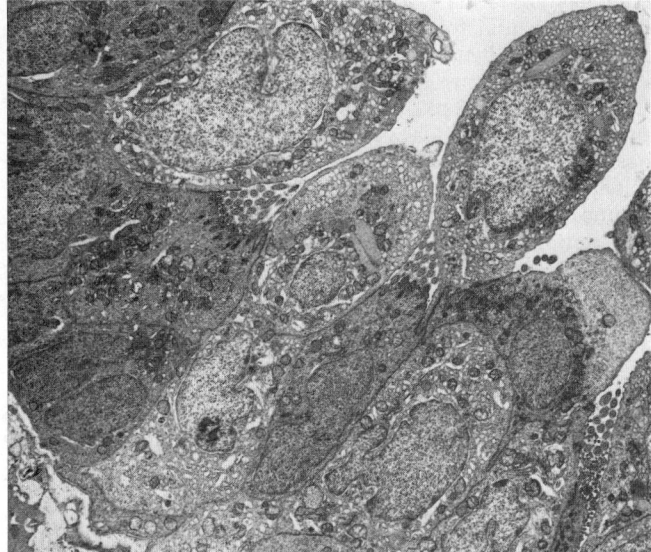

C

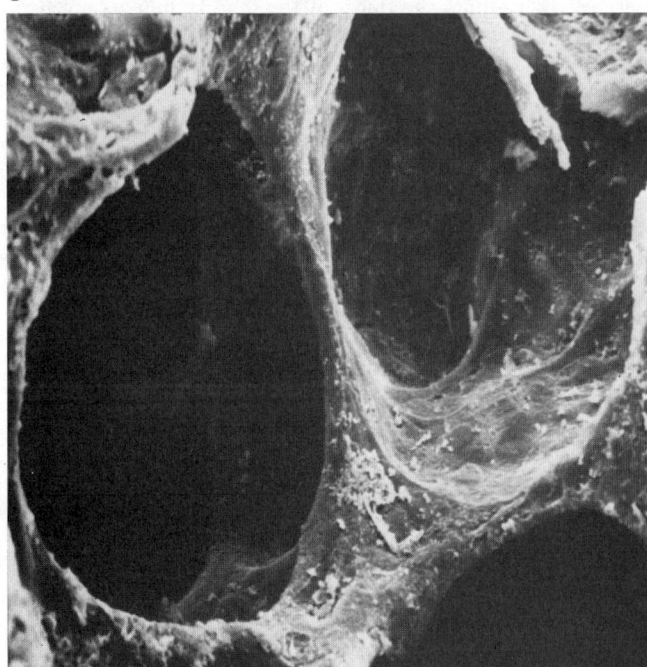

D

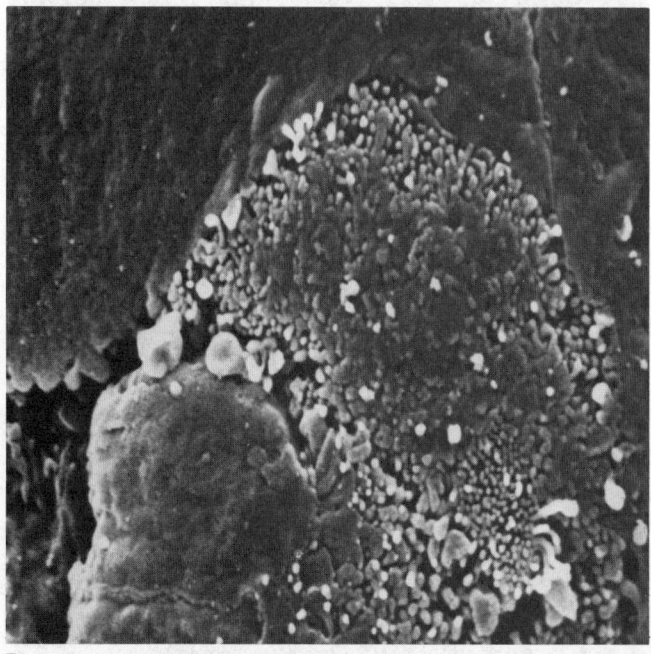

E

8.29C–E. C Transmission electron micrograph of obliquely sectioned bronchiolar epithelium (rat). Many of the cells are non-ciliated, and the majority seen here are mucus-secreting. The basal lamina is at the lower left and the lumen at the upper right corner of the micrograph. (Magnification ×3,500.) D Scanning electron micrograph showing adjacent interalveolar septa in section and the lining epithelium of several alveoli in surface view. (Magnification ×900.) E Scanning electron micrograph showing the surface of a type II alveolar epithelial cell surrounded by several type I alveolar epithelial cells. (Magnification ×7,000.) (Specimens C–E prepared and photographed by Mr. Michael Crowder, Department of Anatomy, Guy's Hospital Medical School, London.)

alveolar epithelial and Clara cells, suggesting that both are involved in pulmonary surfactant production (Niden 1967). Although primarily secretory cells, there is evidence that type II cells have phagocytic properties (Corrin 1970; Suzuki *et al.* 1972). In contrast to the type I cells, type II cells respond to injury by rapid proliferation and it has been suggested that they may replace damaged type I cells (Weibel 1974A).

A third alveolar cell type, the *brush cell* (Meyrick and Reid 1968), is a rare element of the alveolar surface in some species; also called the *type III cell*, it was not observed by Gehr *et al.* (1978) in their exhaustive study of the human lung ultrastructure, although, as they themselves point out, the presence of these rare cells in the human lung cannot be excluded.

In addition to their role in gaseous exchange, the *pulmonary endothelial cells* have many non-respiratory activities: they can clear emboli and thrombi from the blood (Heinemann and Fishman 1969), participate in the metabolism of chylomicrons (Schoefl and French 1968), produce thromboplastin (Zeldis *et al.* 1972), selectively process a variety of hormones and prohormones arriving via the systemic circulation (Vane 1969) and synthesize prostaglandins and related substances (Ryan and Ryan 1977). Their microanatomy and ultrastructure fit them to perform these diverse functions efficiently. The capillary endothelium is a simple layer of squamous cells and of the continuous type (Smith and Ryan 1973). The most remarkable features of its cells are their abundant pinocytotic vesicles (Smith and Ryan 1970), their extreme thinness (less than 0·1 μm in some areas), and their extensive meshwork of luminal projections (Ryan and Ryan 1977). They thus present an enormous surface area to the blood, an area which in man must be many times greater (Ryan and Ryan 1977) than the commonly quoted figure of 70 m² (Fishman and Pietra 1974). According to Ryan and Ryan (1977) the adult human lung may have of the order of 3×10^8 terminal airspaces (alveoli), each of which may have a thousand or more capillary segments in its walls; the capillary bed of the lungs may extend for 1,500 miles or more, with 1·0 ml of blood occupying 10 miles of capillaries. The capillaries lie in the alveolar walls or interalveolar septa. They cross the interstitial space of the septa, interlacing with the connective tissue fibres of this space, and bulge alternately into one or other of the adjacent alveoli which each interalveolar septum separates; thus, in general, on one side of each capillary connective tissue fibres are intercalated between the endothelium and the alveolar epithelium, while on the other the space contains only the fused basement membranes of these two cell layers, providing on approximately half of the capillary surface a minimal barrier to the exchange of gases and the non-volatile solutes of the blood.

The *interstitial spaces* of the interalveolar septa contain, in addition to free fluid, collagen and elastic fibres, a variety of cells of both fixed and free types. The most numerous of these are the fixed cells, which are represented by fibroblasts and pericytes, the latter being demonstrated in human lung by Weibel (1974a). Both fibroblasts and pericytes have contractile properties (Kapanci *et al.* 1974) and are structurally very similar (pp. 46 and 627). They can, however, be distinguished by their relative positions, the fibroblasts generally being associated with the connective tissue fibres whose components they manufacture, while the pericytes lie within the basement membranes of the capillaries. Those fibroblasts situated at the free edges of the interalveolar septa, that is, at the ring-shaped openings of the alveoli into the alveolar sacs resemble smooth muscle cells. Of the free cells of the interstitial space, histiocytes, mast cells and lymphocytes are the most numerous; plasma cells are occasionally present.

Within the cavity of the alveoli another cell type is often seen, the *alveolar phagocyte* (*dust cell*), which is a typical macrophage similar to the connective tissue macrophages beneath the epithelial lining, and to circulatory monocytes. There is evidence that these cells enter the lung in the circulation and migrate out of the capillaries and through the epithelial lining to wander about on the alveolar surface (Collet and Normand-Reuet 1967). Such phagocytic cells are capable of clearing the respiratory portions of the lung of inhaled particles small enough to reach them, subsequently migrating back into the lymphatic drainage of the lung or moving to the base of the bronchial tree where they are taken up by the ciliary rejection current (*vide infra*) and eliminated from the lung by way of the glottis. Whenever extravasation of red blood corpuscles occurs from the lumina of the pulmonary capillaries, the phagocytes become brick-red in colour, by virtue of the erythrocytes which they engulf, and appear in the sputum; such cells are a characteristic feature of congestive heart failure and are often termed '*heart failure cells*' as a result. Other cells of the circulation may also invade the alveoli in pathological conditions, for example neutrophil leucocytes and lymphocytes in inflammatory disturbances, which also give the sputum a characteristic appearance. The presence of these various cell types in sputum is an important diagnostic feature.

PULMONARY DEFENSIVE MECHANISMS

The biological problems associated with the protection and maintenance of the respiratory tract are numerous, since such an exposed structure is subject to damage by desiccation, by microbial invasion, and by the mechanical and chemical action of inhaled particulate matter. Inhaled air is humidified chiefly in the upper respiratory tract where it passes over the mucous membranes of the nasal and bucco-pharyngeal cavities. However, the secretions of the various glands which line the bronchial tree are also of great importance in preventing the desiccation of the conducting passages and of the alveolar surfaces. The precise composition of these secretions is of some interest; the goblet cells secrete sulphated acid mucosubstances, whereas the cells of mucous glands beneath the epithelial surface contain mainly carboxylated mucosubstances, particularly those associated with sialic acid, although sulphated groups are also present (de Haller 1969). The cells of serous glands, in contrast, contain neutral mucosubstances. The goblet cells respond to local irritation and the tubular glands, both mucous and serous, to nervous and hormonal control. Excessive secretion, or an altered composition of the materials secreted, may have serious consequences in causing obstruction to the flow of air in the conducting passages.

In addition to the mucosubstances secreted by the glands of the bronchial tree, antibacterial and antiviral substances are also present within the secreted fluid; these include the bactericidal enzyme lysozyme, antibodies of the IgA type, and also possibly the antiviral material interferon (Havez *et al.* 1966).

Another line of defence against the action of inhaled particles is the ciliary rejection current, already mentioned. The cilia beat the layer of fluid overlying the surface of the bronchioles, bronchi and trachea upwards at a rate of about 1 cm/minute, and those inhaled particles which are trapped in the viscous fluid may be removed from the lung by this route. However, particles smaller than the narrowest respiratory bronchioles may reach the alveoli, and may subsequently be removed by alveolar phagocytes (*vide supra*); if the particles are sufficiently abrasive they may nevertheless cause considerable damage to the respiratory surface, which may be subsequently replaced by fibrous scar tissue with a consequent reduction in respiratory efficiency. Many industrial diseases of the lung, such as pneumoconiosis, caused by inhaling coal dust, are of this pattern. Unless replaced by fibrous connective tissue as a consequence of chronic trauma, the epithelium of the alveoli has limited powers of regeneration and is under normal conditions continually being replaced. The life-span of an alveolar squamous cell is about three weeks; likewise there is a continuous and rapid turnover of alveolar phagocytes, which have a life averaging about four days (Bertalanffy 1966).

Finally, associated with the linings of the various conducting passages are numerous patches of lymphoid tissue which provide foci for the production of lymphocytes, giving local immunological protection against infection.

In addition to all of these cellular mechanisms for protection, the mechanical action of coughing in clearing pulmonary obstructions is, of course, of considerable importance. The initiation of such muscular responses to irritation involves stimulation of sensory endings in the epithelial surface; the identity of such sensory endings is open to argument, except at the laryngeal aditus, where epithelial receptors similar to taste buds are thought to be involved.

PERINATAL PULMONARY DEVELOPMENT

The histology of the lung undergoes considerable changes during development and can, for convenience of description, be divided into three main phases, the glandular, canalicular and alveolar periods (Engel 1947; de Reuck and Porter 1967; Emery 1969; Reid 1976). In the *glandular period*, lasting up to about sixteen weeks of embryonic development, the tubular epithelial outgrowths forming the ducts of the lung rudiments are lined with tall columnar epithelial cells rich in glycogen, which almost obliterate the lumina of the ducts; between these epithelial tubes lie closely packed mesenchymal cells and vascular tissue, so that the whole lung bud is composed of a concentrated mass of cells reminiscent of glandular tissue. At about sixteen weeks, the

A

8.30A–D A Scanning electron micrograph showing part of the capillary plexus of a rat alveolus. The walls of the capillaries are covered by alveolar epithelial cells. (Magnification ×2,500.) B Electron micrograph of a transverse section of a pulmonary capillary, situated in a murine interalveolar septum. A red blood cell (rbc) is present in the capillary

canalicular period commences, being marked by a rapid proliferation of the ducts which branch profusely and become more open as many of the cells transform to a cuboidal type, often ciliated. At this stage some of the cells also become squamous and begin to be associated closely with capillaries, a feature that predominates in the next phase, the *alveolar period*, from about twenty-four weeks and throughout post-natal life, when the alveoli and associated air spaces finally differentiate into their fully formed structure. The expansion of the future air spaces of the respiratory portions of the lung is associated with the passage

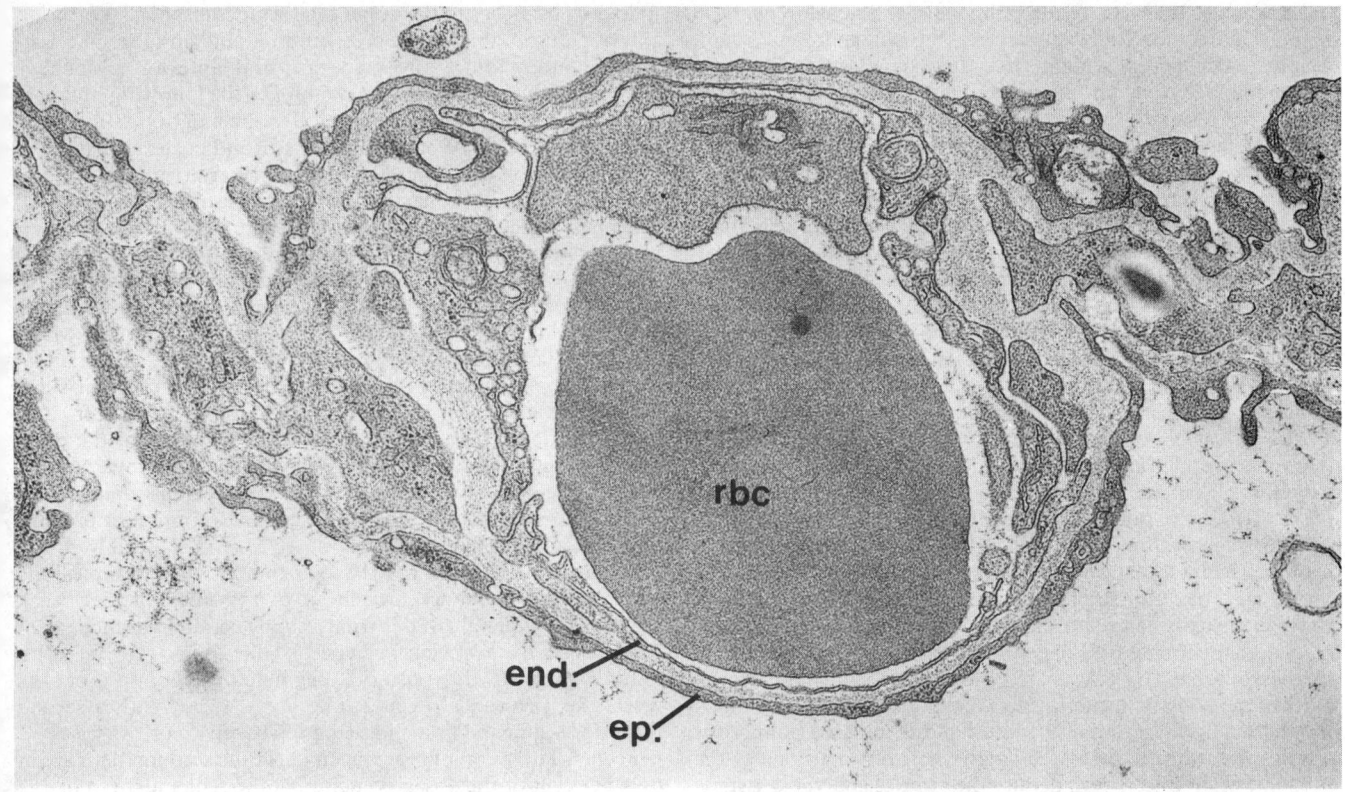

B

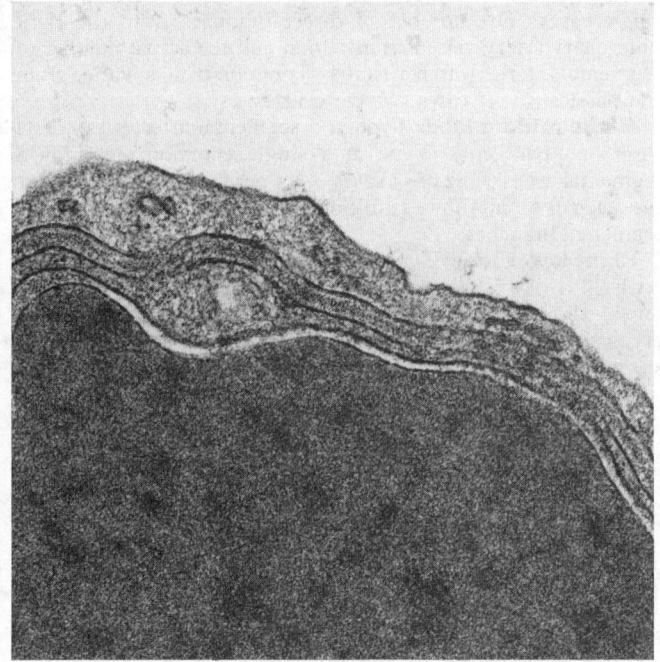

C

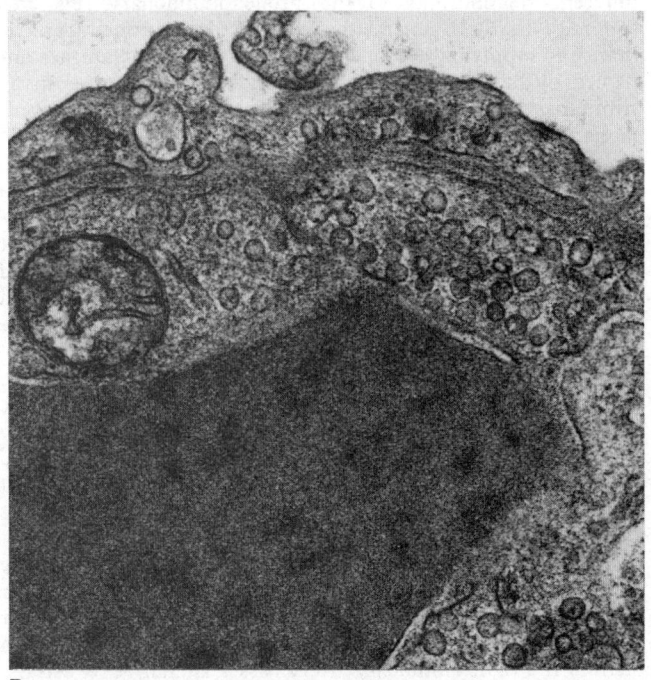

D

lumen; the attenuated endothelium (end.) and the alveolar epithelium (ep.) are indicated. (Magnification × 10,000.) C Transmission electron micrograph of part of an interalveolar septum showing the fused basal laminae separating a type I alveolar epithelial cell and an endothelial cell. (Magnification × 40,000.) D Transmission electron micrograph of part of an interalveolar septum showing many micropinocytotic vesicles in adjacent type I alveolar epithelial and endothelial cells. (Magnification × 35,000.) (Preparations A, C, D by Mr. Michael Crowder, Department of Anatomy, Guy's Hospital Medical School, London.)

of fluid from the lung tissue into their lumina, although it is possible that incipient respiratory movements of the thoracic musculature, causing the inhaling of amniotic fluid, also play a part in the complete expansion of the cavities. By full term, and indeed some time before, the lung is therefore able to support normal respiration once its fluid contents are expelled by expiratory movements in air. After the time of birth, however, minor structural changes also occur, the alveolar surfaces becoming more complex and the alveoli more numerous with the increase in body size.

PULMONARY VASCULATURE AND INNERVATION

The *pulmonary artery* conveys the deoxygenated blood to the lungs; it divides into branches which accompany the segmental and subsegmental bronchi and lie for the main part on their dorsolateral aspects (8.20B). They end in a dense capillary mural network in the alveolar sacs and alveoli (8.29A). The arteries of neighbouring segments are independent of one another. Evidence has been recorded (Elliot and Reid 1965) to show that there is a fairly constant relationship between the luminal and mural dimensions of branches of the pulmonary artery. Moreover, no regional differences can be detected (Simons and Reid 1969), despite evidence suggesting that there is a large difference between blood flow in the upper and lower regions of the lungs.

The *pulmonary capillaries* form plexuses which lie immediately beneath the lining epithelium, in the walls and septa of the alveoli and the alveolar sacs. In the septa between the alveoli the capillary network forms a single layer, the meshes of which are smaller than the vessels themselves; their walls are also exceedingly thin (*see* also above and Suarez 1979).

Arteriovenous shunts have been demonstrated in intact post-mortem human lungs in relation to terminal bronchioles. Experiments showed that such shunts may be capable of passing particles of 500 μm diameter. The functional implications of these findings are not clear (Tobin 1966).

The *pulmonary veins*, two from each lung, arise from the pulmonary capillaries, the radicles coalescing into larger branches which run through the substance of the lung, mostly independently of the pulmonary arteries and bronchi (*vide infra*). After communicating freely with other branches they form large vessels, which ultimately come into relation with the arteries and bronchial tubes, and accompany them to the hilum of the lung, the artery usually being situated dorsolateral, and the vein ventromedial, to the bronchus. Finally, they open into the left atrium of the heart, conveying oxygenated blood to be distributed to all parts of the body by the left ventricle.

Whereas in the region of the hilum of the lung the pulmonary arteries and veins closely accompany the main divisions of the bronchi, when traced peripherally the arteries and veins assume different relationships to the bronchopulmonary segments (pp. 1243–1246). In general, the centrally placed bronchus and its branches supplying a bronchopulmonary segment are accompanied by branches of the pulmonary arteries, whereas many tributaries of the pulmonary veins run between bronchopulmonary segments, so that each venous tributary drains adjacent segments and each segment is drained by more than one vein. Some of the veins also lie beneath the visceral pleura, including that in the interlobar fissures of the lung. Thus a bronchopulmonary segment of lung is not a bronchovascular unit in the sense of possessing its individual bronchus, artery and vein. In surgical removal of bronchopulmonary segments, it should be noted that the planes between adjacent segments are not avascular, but are crossed by pulmonary veins and sometimes by branches of pulmonary arteries. There is considerable variation in the above pattern of the bronchi, arteries and veins, the veins being more variable than the arteries, and the arteries more variable than the bronchi (Brock 1942, 1943, 1944, 1954; Boyden 1955; Cory and Valentine 1959; Bloomer *et al.* 1960; Volpe *et al.* 1969). What follows is a general account of pulmonary vasculature; for details of variations consult the references quoted.

Arteries of the right lung. Before it reaches the hilum of the lung, the right pulmonary artery (p. 667) gives off a large superior branch to the upper lobe, which lies at first in front of the superior lobe bronchus but soon comes to lie posterolateral to it. The main stem of the right pulmonary artery passes laterally between the superior lobar bronchus and the continuation of the principal bronchus and then lies posterolateral to the latter in the oblique fissure of the lung, supplying branches to the middle and lower lobes. The superior branch of the right pulmonary artery gives off the following branches to the bronchopulmonary segments of the upper lobe: apical, anterior descending, anterior ascending, posterior descending and posterior ascending. The branch to the middle lobe arises from the stem of the right pulmonary artery close to (or in common with) the origin of the superior branch to the lower lobe, and it divides into lateral and medial branches which supply the corresponding segments of the middle lobe. The stem of the artery supplies the following branches to the corresponding segments of the lower lobe: superior (apical), subsuperior (subapical), medial basal (cardiac), anterior basal, lateral basal and posterior basal.

Arteries of the left lung. The left pulmonary artery (p. 667), having crossed in front of the left principal bronchus to gain its posterolateral aspect in the hilum of the lung, gives off the

following branches to the upper lobe segments: apical, posterior, anterior descending, anterior ascending and lingular (the last-named subdividing into superior and inferior branches). Thereafter the pulmonary artery supplies the following branches to the lower lobe: superior (apical), subsuperior (subapical), medial basal, anterior basal, lateral basal and posterior basal.

Although the right and left pulmonary arteries rarely depart from their mode of primary branching, the secondary (segmental) branches, which are listed above, display much variation in their precise sites of origin and in their numbers, reduplication being frequent. In a detailed study of the pulmonary hila at 521 partial pneumonectomies Cory and Valentine (1959) observed many variations. For example the right and left upper lobes, whose segmental arteries displayed the greatest range of variation, showed no less than 14 types of pattern in 152 right-sided operations and 29 types in 107 operations on the left. The remaining lobes displayed a lesser scale of variation, the types noted being in the right middle lobe 5, in the right lower lobe 6, and in the left lower lobe only 4. The details of all these observations cannot be given here, but the commonest patterns will be noted briefly.

Right upper lobe: Type 1: 3 segmental arteries, the apical and anterior from a common branch (91 of 152 cases = 60 per cent). Type 2: 3 segmental arteries from a single common branch (26 of 152 cases = 17 per cent). Type 3: like Type 1, but with reduplication of the posterior segmental artery (13 of 152 cases = 8.5 per cent). Type 4: like Type 2, but with a second posterior segmental branch arising separately (9 of 152 cases = 6 per cent). These common arrangements thus accounted for about 91.5 per cent of the observed patterns; the remainder (13 of 152 cases, classified as Types 5 to 14) were mostly represented by single cases. The number of direct branches from the right pulmonary artery varied from 1 to 4, and the actual number of segmental arteries (often arising by common trunks) varied from the classical 3 (118 cases = 77 per cent) to 6.

Right middle lobe: Type 1: 2 segmental arteries (25 of 51 cases = 49 per cent). Type 2: a single common stem for 2 segmental arteries (23 of 51 cases = 45 per cent). Types 3 to 5 (1 case each of 51) all showed multiplicity (3, 4 and 5) of middle lobe segmental branches.

Right lower lobe: Type 1: 3 segmental branches, 1 superior, 2 basal (31 of 43 cases = 72 per cent). Types 3 to 6 (8 of the 43 lobes) showed reduplication of the superior branch and its atypical origin from one of the basal branches. The number of segmental arteries was always either 3 or 4, the individual basal segments usually sharing common trunks.

Left upper lobe: Types 1 to 7 (78 of 107 cases = 70 per cent) showed 4 or 5 segmental branches, and most of these displayed early bifurcation, the resultant vessels being almost always distributed to two segments (with much variation in those supplied). The remaining types (8 to 29) occurred in single instances, showing variation in number (2 to 6 segmental branches) and in their sites of branching from the left pulmonary artery; bifurcation was slightly less common than in Types 1 to 7 and trifurcation occurred in 3 lobes. It appears impossible to define a 'usual' pattern for this lobe.

Left lower lobe: Type 1: showed 1 apical and 2 basal segmental arteries (44 of 57 cases = 77 per cent). Type 2: Type 1 but with 2 apical branches (9 of 57 cases) = 15 per cent). Type 3: like Type 1 but with 3 basal branches (3 of 57 cases). Type 4: showed 2 basal and 3 apical branches.

Veins of the right lung (see also p. 736). In the hilum of the lung the superior right pulmonary vein is formed by the union of three veins from the upper lobe—apical, anterior and posterior—and one vein from the middle lobe (middle lobar vein) which has two principal tributaries, a lateral and medial vein. The inferior right pulmonary vein is formed in the same site by the union of two veins from the lower lobe—superior (apical) and common basal. The common basal vein in its turn is formed by the union of a superior basal and an inferior basal vein.

Veins of the left lung (see also p. 736). In the hilum of the lung the superior left pulmonary vein, which drains the upper lobe, is formed by the union of three veins—apicoposterior (draining the apical and posterior segments), anterior and lingular, the last-named consisting of an upper and a lower part. The inferior left pulmonary vein, which drains the lower lobe, is formed in the hilum by the union of two veins—superior (apical) and common basal, the latter being formed by the junction of a superior basal and an inferior basal vein.

All of the above-named main tributaries of the right and left pulmonary veins receive in their turn smaller tributaries, some of which lie within the bronchopulmonary segments (intrasegmental veins), while others lie between and drain adjacent segments (intersegmental veins, p. 1266).

The bronchial arteries supply blood for the nutrition of the lung; they are derived from the descending thoracic aorta or from the upper posterior (aortic) intercostal arteries, and, accompanying the bronchial tubes, are distributed to the bronchial glands and upon the walls of the larger bronchial tubes and pulmonary vessels. Those supplying the bronchial tubes form, in the muscular coat, a capillary plexus from which branches are given off to form a second plexus in the mucous coat; this plexus communicates with branches of the pulmonary artery, and empties into the pulmonary veins. Others are distributed in the interlobular areolar tissue, and end partly in the deep, partly in the superficial, bronchial veins. Lastly, some ramify upon the surface of the lung, beneath the pleura, where they form a capillary network. The bronchial arteries supply the wall of the air passages only as far as the respiratory bronchioles. They anastomose with branches of the pulmonary arteries in the walls of the smaller bronchi and in the visceral pleura. There is some evidence that such bronchopulmonary anastomoses are more numerous in the newborn and they may later become obliterated to a considerable degree (Wagenvoort and Wagenvoort 1967). In addition to the main bronchial arteries, smaller branches arise from the descending thoracic aorta; one of these may pass in the pulmonary ligament and cause bleeding during surgical removal of the lower lobe.

The bronchial veins form two distinct systems (Marchand *et al.* 1950). *The deep bronchial veins* commence as a network in the intrapulmonary bronchioles and communicate freely with the pulmonary veins; they eventually join to form a single trunk which ends in a main pulmonary vein or in the left atrium. *The superficial bronchial veins* drain the extrapulmonary bronchi, the visceral pleura and the hilar lymph nodes; they also communicate with the pulmonary veins and end, on the right side in the azygos vein, and on the left in the left superior intercostal vein or the accessory hemiazygos vein. The bronchial veins do not receive all the blood conveyed to the lungs by the bronchial arteries, because some enters the pulmonary veins. The main bronchial arteries and veins lie on the dorsal surface of the extrapulmonary bronchi.

The lymph vessels of the lungs are described on p. 800. For reconstructions of the lymphatic networks around bronchi consult Bastianini (1967, 1968). The main ultrastructural features of the pulmonary lymphatics include a rather thin basal lamina, and many vesicles in the cytoplasm of the endothelial cells, together with other evidences of pinocytosis, suggesting a considerable passage of fluid across the lymphatic wall (Lauweryns and Boussaw 1967).

Nerves. The lungs are supplied from the anterior and posterior pulmonary plexuses, formed chiefly by branches from the sympathetic and vagus. The filaments from these plexuses accompany the bronchial tubes, supplying efferent fibres to the bronchial muscle and glands and afferent fibres from the bronchial mucous membrane and from the alveoli of the lung. Small ganglia are found upon these nerves. It is generally accepted that the bronchoconstrictors are supplied by the vagus (pp. 1081, 1123). The sympathetic supply is inhibitory; arrival of impulses along such nerve fibres causes relaxation of the non-striated bronchial musculature, as indeed does the withdrawal or reduction of parasympathetic (vagal) stimulation. In both the actual force which produces bronchodilatation is the pressure of inspired air.

Radiology (3.35, 6.27). The trachea, because of its contained air, is more radio-translucent than the neighbouring tissues and is therefore seen in lateral and anteroposterior radiographs of the neck and upper thorax as a dark area in negatives (or as a light area in positive prints; 3.35). For the same reason the lungs appear as dark areas in the thorax on both sides of the central mediastinal opacity; the lung areas are darker at the end of inspiration and in a diseased condition in which the alveoli are permanently distended (emphysema), but are more radio-opaque in conditions which reduce the amount of air in them (e.g. pneumonia). The dark lung areas are not homogeneous but have superimposed on them the white shadows of the pulmonary blood vessels extending from the hilum of the lung and branching into the lung areas (*see* 6.27 and Lodge 1946). These white branching shadows are sometimes mistaken for bronchi, but the latter (because of their contained air) obviously appear as darker areas. Where a blood vessel is seen end-on, it appears as a homogeneous white circle (6.27); where a bronchus is seen end-on, it appears as a dark circle surrounded by a white line (the latter representing the wall of the bronchus). Lymph nodes at the hilum of the lungs, if enlarged or calcified, appear as mottled areas near the mediastinum. The shape of the lumen of the bronchi and bronchioles can be rendered plainly visible by injecting a suitable radio-opaque oil through the trachea. An effusion of fluid into the pleural cavity appears as an opacity in chest radiographs.

THE ALIMENTARY SYSTEM

Sometimes also described as the *digestive* or *gastro-intestinal* tract, the alimentary system is not merely concerned with the *digestion* of food, nor does it exclude the mouth, pharynx, and oesophagus. It is a muscular canal, lined by mucous membrane, which extends from lips to anal orifice and displays at successive levels a remarkably varied series of accessory structures, dental, glandular and muscular, adapted to the ingestion, mastication, transport, digestion and absorption of foodstuffs of every kind, and also the elimination of the unabsorbed or unabsorbable residues.

Throughout vertebrates, and indeed chordates, the alimentary and respiratory organs have shared a common entry channel from the exterior. In the primary, aquatic era of their evolution the two functions were intimately associated, as they continue to be in their extant representatives. Dissolved oxygen and suspended particulate matter are both absorbed or filtered off from the same stream of water, entering through a single opening, the mouth, and escaping via a branchial apparatus of some kind, as in *Amphioxus*. In this connexion it is pertinent to note that the largest aquatic vertebrates of today, for example, the basking-shark, *Cetorhinus*, and the baleen whales, the *Mysticeti*, still use a filtration technique for food-gathering, though the filter is palatal, rather than branchial, in the latter. In contrast, many fishes, including some teleosts, have developed an ability to gulp in air, either to fill a hydrostatic 'swim-bladder' or a respiratory 'air-bladder', both being pharyngeal diverticula. The teleostean 'swim-bladder' commonly loses its continuity with the pharynx and then derives its contained gas from the associated arterial supply (from the *dorsal* aorta). It is usually accepted that the respiratory system of the terrestrial vertebrates is a development of a ventral 'air-bladder', supplied by branches of the sixth branchial rami of the *ventral* aorta, such as exists in the extant *Dipnoi*, the lung fishes. From the same ancient stock the amphibians are derived, and in them the lungs and bronchial tree, typical of terrestrial vertebrates, have been established, even though many amphibians also depend upon transcutaneous respiration. (Some amphibia of urodelian type, such as the newt, *Necturus*, depend upon skin respiration to such an extent that the lungs are considered to have reverted to a flotation function!) Nevertheless, the buccal cavity and part of the pharynx retain their initially dual function, as they do even after division of the former into definitive buccal and nasal chambers by palatal

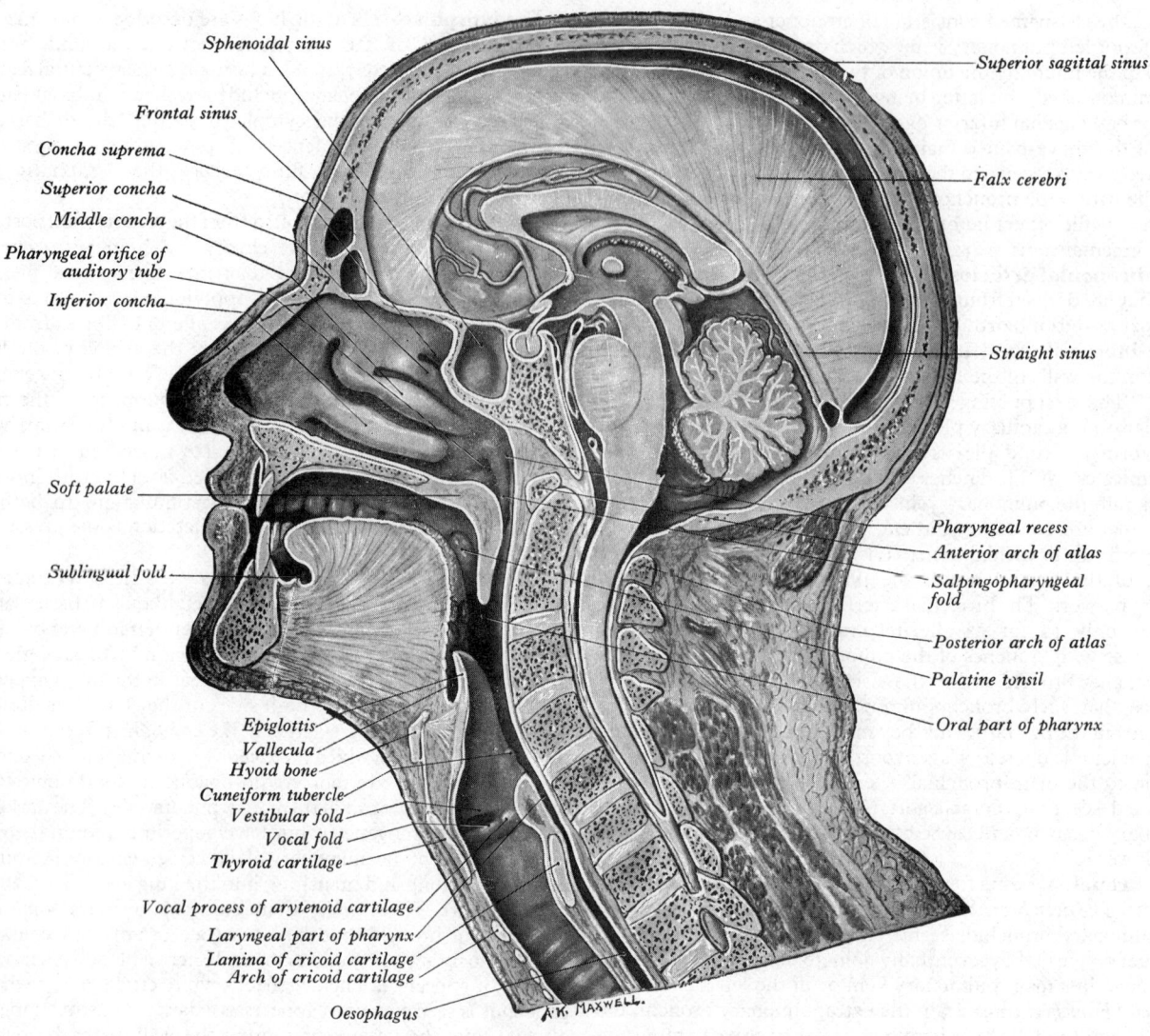

8.31 A median sagittal section through the head and neck. Where it divides the skull and the brain, the section passes slightly to the left of the median plane, but below the base of the skull, it passes slightly to the right of the median plane.

development in reptiles and all subsequent vertebrates. Only where there exist such specializations as that of the crocodilian larynx, which can be elevated into the nasopharynx during breathing, it is possible to achieve a complete separation of the respiratory and alimentary tracts in anatomical terms. In land vertebrates the functional separation is nevertheless complete—into a system which *alternately* inspires and expires air with a variable but relatively regular periodicity and another which is concerned with the unidirectional passage of a much more varied pabulum—solid, fluid, or any intermediate mixture—at intervals varying enormously in different species and often in the individual. The early anatomical association persists, however. Furthermore, the dorsal position of the *respiratory* nasal chamber, relative to the *alimentary* buccal cavity, and the reversal of this relationship, where the larynx diverges *ventrally* from the pharynx, inevitably entail a crossover between the two inlets, for air and food. Since both functions utilize the pharynx, except for its nasal part, mechanisms have evolved to shut off the nasopharynx (by the soft palate) and the respiratory tree (by the larynx) during the act of swallowing. Because of their respiratory function the nasal and oral parts of the pharynx, though possessing muscular walls, are able to preserve their patency. (These matters, which complicate the alimentary function of the pharynx, are considered elsewhere in greater detail. *See* pp. 1228, 1313.)

The alimentary tract, since its traffic is unidirectional, has a

separate exit, caudal in position. This is essentially true even of the saccular urochordate *Ciona*, the sea-squirt, despite its sessile habit; its short gut, already showing distinct mouth, pharynx and intestine, opens into the atriopore, although this is, in fact, near to the mouth. In the cephalochordate *Amphioxus*, the differentiation of the gut has proceeded further, not only in the formation of a caudal anus separate from the atrial chamber into which the pharynx strains its aqueous intake, but also in the development of a diverticulum of the midgut; this has been regarded as 'hepatic' but is more probably a source of digestive enzymes. Both these chordates employ the pharynx, with its multitude of narrow branchial clefts, to strain off food particles from imbibed water, the concentrated solids being passed on into the intestine. Tracts of cilia and mucous glands form a part of this mechanism and of transport in the intestine, a functional cooperation also used in the mammalian respiratory system, though lost in the alimentary canal. These are but small examples of the ability of the lining ectoderm and endoderm (and also associated mesoderm) of the alimentary tract to develop a great variety of adaptations appropriate to the functions of its successive parts. Even in the Agnatha (e.g. *Petromyzon*, the lamprey) the buccal cavity has primitive teeth, a rasping tongue and salivary gland, the pharynx a branchial complex, and the gut is differentiated into oesophagus (foregut) and intestine (midgut), into which a liver delivers its secretion. There is no true stomach; but, while no pancreas is identifiable, some of the cells of the proximal part of the intestine resemble those of the exocrine pancreas of later forms of vertebrate. (Moreover, follicles of intestinal cells have also been described at the same level which do not open into the midgut and may represent the endocrine element of the pancreas.) The lamprey's intestine also shows a surface-increasing specialization, the *typhlosole*, a helical fold of the mucosa like the spiral valve of the elasmobranch intestine. The elasmobranchs display, in addition, a well-defined stomach, pepsin-producing, which is already divisible into cardiac (glandular) and pyloric (muscular) parts. A distinguishable pancreas also appears, secreting into the midgut, together with the large liver. A distinct rectum is also recognized. In teleostean fishes the chief advance is a lengthening and spiralling of the intestine, the degree of this varying with feeding habits. In the amphibian alimentary tract the intestine is clearly separated into intestine proper and colon, a muscular valve intervening.

Except in the 'true', or *Eutherian* mammals, the vertebrate intestine terminates in a cavity, the *cloaca*, which also receives the urinary ducts and usually the genital ducts. Complete separation of the end of the alimentary tube at an anus, though foreshadowed in the Agnatha, was not achieved until the mammals appeared, though even here the monotremes must be excluded. Curiously, a similar peculiarity exists in regard to the *commencement* of the alimentary tract, for salivary glands, which are considered to be present in the Agnatha, are also characteristic of mammals, but are stated to be absent from intervening classes of vertebrate.

It is not possible in a short account to amplify this superficial delineation of the evolutionary history of the alimentary canal, even if restricted to the mammals; but a few further examples must be added to emphasize its apparently endless adaptations to feeding habits. As would be expected, birds display peculiar specializations: the teeth of their Jurassic ancestors (e.g. *Archaeopteryx*) have been lost, and a hard, horny beak of highly variable form provides an alternative feeding tool. The avian oesophagus is commonly dilated into a receptacle, the crop; and the stomach usually consists of a glandular proventriculus and, beyond this, a highly muscular gizzard, with a lining, often horny, and a content of swallowed stones to assist in grinding up hard foodstuffs. Amongst mammals the general pattern of the alimentary tract is alike, and similar to that of mankind; but there is great variation in the size and proportions of its different organs. There is, for example, much variation in the relative size of jaws and their musculature, and even more marked differentiation in the numbers, sizes, and structures of teeth. All such variation reflects the great plasticity of the epithelial tissues and their associated mesodermal derivatives. But it is especially in the mucous membrane lining the alimentary canal that such protean ability is manifest, in the production not only of varied epithelia (columnar, cuboidal, ciliated, stratified, keratinized, and enamel-generating), but also of a wide range of glandular structures, arranged either as single mucous cells or in vast complexes such as the liver and pancreas, capable of secreting a wide spectrum of digestive enzymes and hormones. Apart from exocrine secretion, a range of cells of endocrine function is produced, including not only the thyroid gland and the pancreatic islet tissue, as has long been recognized, but also the more recently identified system of enterochromaffin and other endocrine cells, scattered through the epithelia of much of the alimentary tract (p. 1364).

With the foregoing generalities in mind it is now necessary to consider the more detailed anatomy of the human alimentary tract, considered *seriatim* at its various levels, and including also its accessory organs, the teeth, tongue, salivary glands, liver and pancreas. The microscopic structure of all these is included at appropriate points. It is also convenient, as well as appropriate, to describe the peritoneum together with the subdiaphragmatic parts of the gut with which it is associated.

The alimentary canal, about 9 m long (*see* Underhill 1955, and p. 1342), extends from the mouth to the anus, and is lined almost throughout by mucous membrane. It consists of the following parts: it commences at the *mouth*, where provision is made for the mechanical division of food (*mastication*), and for its admixture with a fluid secreted by the salivary glands (*insalivation*); this is conveyed by the organs of deglutition, the *palate*, the *pharynx* and the *oesophagus*, into the *stomach*, where the first stages of the digestive processes take place; from the stomach it is passed into the *small intestine*, where digestion is continued and many of the resulting products are absorbed into the blood and lymph vessels. Finally the small intestine ends in the *large intestine*, which reaches the surface of the body again at the *anus*. At all levels the alimentary tube is characterized by sphincteric muscles, which will be described *seriatim*. In this connexion consult DiDio and Anderson (1968).

For an account of the general organization of the postpharyngeal alimentary tract *see* p. 1315.

The accessory digestive organs are the *teeth*, which break up and triturate the food in the process of mastication; the three pairs of main *salivary glands*—the *parotid, submandibular* and *sublingual*—the secretion from which mixes with the food in the mouth; the *liver* and the *pancreas*, two large glands in the abdomen, the secretions of which take part in the process of digestion.

The Oral Cavity

The oral cavity, the 'mouth', consists of an outer, smaller part, the vestibule, and an inner, larger part, the oral cavity proper.

The *vestibule* of the mouth is a slit-like space, bounded externally by the lips and cheeks, internally by the gums and teeth (**8**.31). It communicates with the exterior by the *oral fissure*. Above and below, it is limited by the reflexion of the mucous membrane from the lips and cheeks to the gums. When the teeth are apposed it communicates with the oral cavity proper by an aperture behind the third molar teeth on each side, and by narrow clefts between contiguous teeth. On the inner surface of the cheek, opposite the crown of the second upper molar tooth, a small papilla marks the opening of the duct of the parotid salivary gland.

The *oral cavity proper* (**8**.74, 75) is bounded laterally and in front by the alveolar arches, the teeth and the gums; behind, it communicates with the pharynx by the *oropharyngeal isthmus*, between the palatoglossal folds. The roof of the oral cavity consists of the hard palate and soft palate, while the greater part of the floor is formed by the anterior region of the tongue and the remainder by the reflexion of the mucous membrane from the sides and under surface of the tongue to the gum on the inner surface of the mandible. In the median plane a crescentic fold of mucous membrane, the *frenulum linguae*, connects the inferior surface of the anterior part of the tongue to the floor of the mouth. On each side of the lower end of the frenulum there is a small elevation, the *sublingual papilla*, on the surface of which is the orifice of the duct of the submandibular salivary gland. From this papilla a ridge extends laterally and backwards in the mucous membrane of the floor of the mouth; it is produced by the underlying sublingual salivary gland and is hence termed the *sublingual fold*. The minute multiple openings of the ducts of this gland are on the edge of the fold.

The *mucous membrane* of the mouth is continuous with the skin at the free margins of the lips, and with the mucous lining of the pharynx at the oropharyngeal isthmus; it is of a roseate tinge during life, and is very thick where it overlies the hard parts bounding the cavity. It is covered with stratified squamous epithelium, the superficial layers of which, unlike those of the skin, do not become cornified under normal conditions.

The *lymph vessels* of the mouth are described on p. 788.

The lips are two fleshy folds which surround the oral orifice. They are formed externally of skin and internally of mucous membrane, and these two layers enclose the orbicularis oris, the labial vessels and nerves, areolar tissue, and numerous small labial salivary glands. The line of contact between the closed lips (*the oral fissure*) lies opposite the cutting edges of the upper incisor teeth and forms, on each side, the *labial commissure*, which bounds the *angle of the mouth*; the latter usually lies just in front of the first premolar tooth. In the middle of the external surface of the upper lip is a shallow vertical groove named the *philtrum*; it ends below in a slight prominence (the *tubercle*) and is limited on each side by a ridge. The inner surface of each lip is connected in the median plane to the corresponding gum by a fold of mucous membrane, termed the *labial frenulum*—that of the upper lip being the larger.

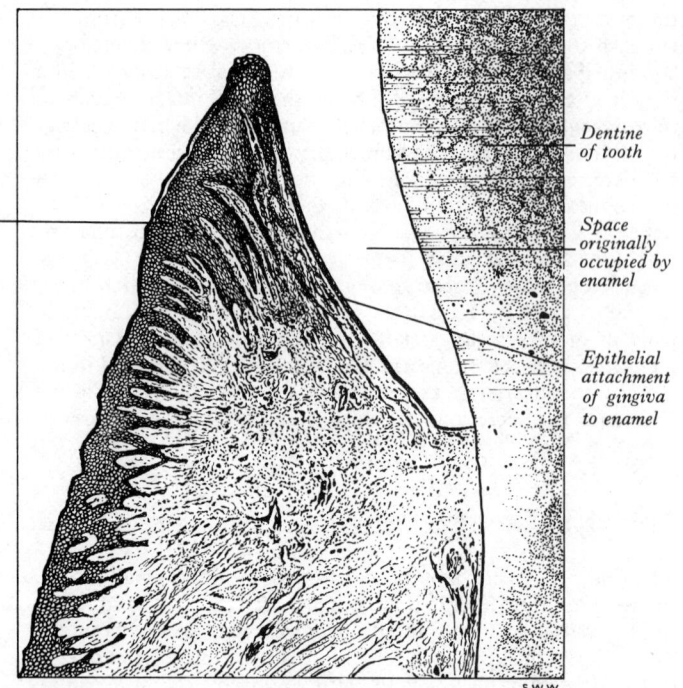

Stratified squamous epithelium of gingiva

Dentine of tooth

Space originally occupied by enamel

Epithelial attachment of gingiva to enamel

S.W.W.

8.32 A section across a tooth and the adjoining part of the gum. The enamel has been decalcified and removed, leaving a space between the dentine and the thin layer of stratified epithelium of the covering gum.

The *labial glands* are situated between the mucous membrane and the orbicularis oris, round the orifice of the mouth. They are about the size of small peas and in structure they resemble the mucous salivary glands (*see* p. 1275). Their ducts open into the vestibule.

The cheeks form a large part of the sides of the face, and are continuous in front with the lips, the junction being indicated externally on each side by a groove, termed *the nasolabial sulcus*, which runs downwards and laterally from the side of the nose to the angle of the mouth. The cheeks are composed of a muscular stratum, and a variable but usually considerable amount of adipose tissue, together with areolar tissue, vessels, nerves and buccal glands, covered with skin externally and with mucous membrane internally.

The *mucous membrane* lining the cheek is reflected above and below on to the outer surfaces of the maxilla and mandible, whence it is continued on to the gums; it is continuous behind with the mucous membrane of the soft palate. Opposite the crown of the second upper molar tooth there is a small papilla, on the summit of which the parotid duct opens. The principal muscle of the cheek is the buccinator; but others enter into its formation, e.g. the zygomaticus major, risorius and platysma.

The *buccal glands* are small mucous glands placed between the mucous membrane and the buccinator. Four or five, larger than the rest, and placed outside the buccinator around the terminal part of the parotid duct, are called *molar glands*; their ducts open in the mouth opposite the last molar tooth by piercing the buccinator.

The *lymph vessels* of the cheeks and lips are described on p. 786.

The gums (*gingivae*) (**8**.32) are composed of dense, vascular fibrous tissue covered with stratified, thinly keratinized squamous epithelium. They consist of two parts: the free part, which surrounds the neck of the tooth like a collar, and the attached part, which is firmly anchored to the alveolar processes of the mandible and maxillae. The fibrous tissue of the gum is continuous with the periosteum lining the alveoli. In young people the stratified squamous epithelium is attached to the surface of the enamel of the teeth (the epithelial attachment), but as age advances the gums recede from the enamel and become attached to the cement. Near the teeth, the mucous membrane on the buccal surface of the gums is thrown into tall papillae, but it becomes smooth where it is related to the enamel (**8**.32). *The nerves* supplying the upper gum are derived from the maxillary nerve through its anterior

palatine, nasopalatine and anterior, middle and posterior superior alveolar branches. The mandibular nerve innervates the lower gum by its inferior alveolar, lingual and buccal branches, the last two supplying the corresponding surfaces of the gum. The buccal nerve supplies the lower gum as far forwards as the mental foramen of the mandible. The vessels for the main part accompany the nerves. (For a highly detailed study of the buccal cavity, including the tongue, consult Combelles 1972.) The lymphatics of the upper gum pass to the submandibular nodes; those from the anterior part of the lower gum pass to the submental nodes, while those from its posterior part enter the submandibular nodes.

The stratified squamous epithelium of the gums, like that of the skin, contains melanocytes, which are obvious in pigmented races. The occurrence of dendritic cells, which are DOPA negative, but probably capable of melanin production, has also been reported. Actual estimates of the incidence of these two types of cells have also been made (Barker 1967). The dendritic cells are located more superficially than the melanocytes, and they are more frequent in the crevices between adjoining teeth, than are melanocytes.

THE PALATE

The palate forms the roof of the mouth: it consists of two regions—the *hard* palate in front, the *soft* palate behind.

The hard palate (**8**.47) is formed by the palatine processes of the maxillae and the horizontal plates of the palatine bones; it is bounded in front and at the sides by the alveolar arches and gums; behind, it is continuous with the soft palate. It is covered with a dense tissue, formed by the periosteum and mucous membrane, which are intimately connected. It presents a median, linear raphe, which ends anteriorly in a small papilla overlying the incisive fossa. On each side of the raphe the mucous membrane of the front part is thick, pale in colour, and corrugated; behind, it is thin, smooth, and of a redder colour: it is covered with keratinized stratified squamous epithelium, and furnished in its posterior half with numerous palatine mucous glands, which lie between the mucous membrane and the periosteum. The upper surface of the hard palate forms part of the floor of the nasal cavity and is largely lined by ciliated epithelium.

The soft palate (**8**.31) is a movable fold, suspended from the posterior border of the hard palate, and extending downwards and backwards between the oral and nasal parts of the pharynx. It consists of a fold of mucous membrane enclosing an aponeurosis, muscular fibres, vessels, nerves, lymphoid tissue and mucous glands. When occupying its usual position (i.e. relaxed and pendant) its anterior (oral) surface is concave, and marked by a median raphe. Its posterior surface is convex, and continuous with the floor of the nasal cavity. Its superior border is attached to the posterior margin of the hard palate, and its sides are blended with the pharynx. Its inferior border is free. The lower region of the soft palate hangs like a curtain between the mouth and the pharynx.

A small conical process, termed the *uvula*, hangs from the middle of its lower border; and two curved folds of mucous membrane, containing muscular fibres, extend laterally and downwards from each side of the base of the uvula (**8**.75). The anterior of the two contains the palatoglossus and is named the *palatoglossal arch*. Below, it reaches the side of the tongue at the junction of the oral and pharyngeal portions and it forms the lateral boundary of the oropharyngeal isthmus. The posterior fold, which is termed the *palatopharyngeal arch*, descends on the lateral wall of the oral part of the pharynx and is described on p. 1309. (The term *isthmus of the fauces* includes both the palatal arches.)

Flanking the median palatine raphe, and near the junction of the hard and soft parts of the palate, there are often to be seen a pair of bilateral depressions, separated by a few millimetres. These *palatine foveae* may extend into pits, a few millimetres deep, and their openings may be anteroposteriorly elongated. It is probable that they are the common superficial orifices of a number of converging ducts from the palatine glands referred to above.

The *mucous membrane* of the soft palate is thin, and consists of stratified squamous epithelium excepting the upper part of its posterior surface and near the pharyngeal orifice of the auditory tube, where it is columnar and ciliated ('respiratory epithelium'), like that of the nasal cavities with which it is continuous. Beneath the mucous membrane on both surfaces there are numerous palatine mucous glands; they are most abundant on the oral surface and round the uvula. Taste buds are present in the epithelium of its oral surface.

Vessels and Nerves

The *arteries* supplying the palate are the greater palatine branch of the maxillary artery, the ascending palatine branch of the facial artery, and the palatine branch of the ascending pharyngeal artery. The *veins* end chiefly in the pterygoid and tonsillar plexuses. The *lymph vessels* pass to the deep cervical lymph nodes. The *sensory nerves* are derived from the greater and lesser palatine, the nasopalatine and the glossopharyngeal nerves. The lesser palatine nerves contain taste fibres from the oral surface of the soft palate.

The Palatine Aponeurosis

A thin, firm, fibrous lamella, termed the *palatine aponeurosis*, which supports the muscles and gives strength to the soft palate, is attached to the posterior border of the hard palate and to the inferior surface of the hard palate behind the palatine crest. It is thick in the anterior two-thirds of the soft palate, but very thin further back. The aponeurosis is actually the expanded tendon of the tensor veli palatini and, near the middle line, it splits to enclose the musculus uvulae. To it, all the other muscles of the soft palate are attached. The anterior (juxta-osseous) part of the soft palate contains little muscle and consists mainly of the palatine aponeurosis, inferior to which are numerous mucous glands; it is less movable and more horizontal than the posterior part and it is upon this part that the tensor veli palatini principally acts.

THE PALATINE MUSCULATURE

The muscles of the palate (**8**.33) include a levator and a tensor of the palate, the muscles underlying the palatoglossal and palatopharyngeal folds and extending into the palate itself, and the muscle of the uvula. For details of their activities in swallowing and speech, see p. 1313.

The levator veli palatini (**8**.33, 84, 85) is a cylindrical muscle situated on the lateral side of the posterior nasal aperture. According to Rohan and Turner (1956) it arises (*a*) by a small tendon from a rough area on the inferior surface of the petrous temporal bone, immediately in front of the lower opening of the carotid canal; (*b*) by fleshy fibres from a sheet of fascia which descends from the vaginal process of the tympanic part of the temporal bone to form the upper part of the carotid sheath; (*c*) by a few fleshy fibres from the inferior surface of the cartilaginous part of the auditory tube. At its origin, the muscle lies inferior rather than medial to the auditory tube and only crosses to the medial side of the tube at the level of the medial pterygoid lamina. After passing within the upper concave margin of the superior constrictor and in front of the salpingopharyngeus, it spreads out in the soft palate between the two strands of the palatopharyngeus, its fibres being inserted into the upper surface of the palatine aponeurosis as far as the median plane, where they blend with those of the opposite muscle.

Action. The levator veli palatini elevates the soft palate.

The tensor veli palatini (**8**.33, 79, 85) is a thin, triangular muscle, which lies lateral to the medial pterygoid plate, the auditory tube and the levator veli palatini. Its lateral surface is in contact with the upper and anterior part of the medial pterygoid, the mandibular, auriculo-temporal, and chorda tympani nerves, the otic ganglion and the middle meningeal artery. It arises from the scaphoid fossa of the pterygoid process, the lateral lamina of the cartilage of the auditory tube and the medial aspect of the spine of the sphenoid bone. As it descends, its fibres converge to form a delicate tendon which turns medially round the pterygoid hamulus, passes through the origin of the buccinator, and is inserted into the palatine aponeurosis and into the surface behind

the palatine crest on the horizontal plate of the palatine bone. Between the tendon and the pterygoid hamulus there is a small bursa.

Actions. Acting singly the tensor veli palatini pulls the soft palate to one side; acting together the two muscles tighten the soft palate (principally its anterior part) and depress it by flattening out its arch.

The musculus uvulae, a bilateral structure, arises from the posterior nasal spine of the palatine bones and from the palatine aponeurosis, between the two laminae of which the two uvular muscles lie; it descends to be inserted into the mucous membrane of the uvula.

Action. The musculus uvulae pulls up and contracts the uvula on its own side.

The palatoglossus is a small, fleshy fasciculus, narrower in the middle than at the ends, forming, with the mucous membrane covering its surface, the palatoglossal arch. It arises from the oral

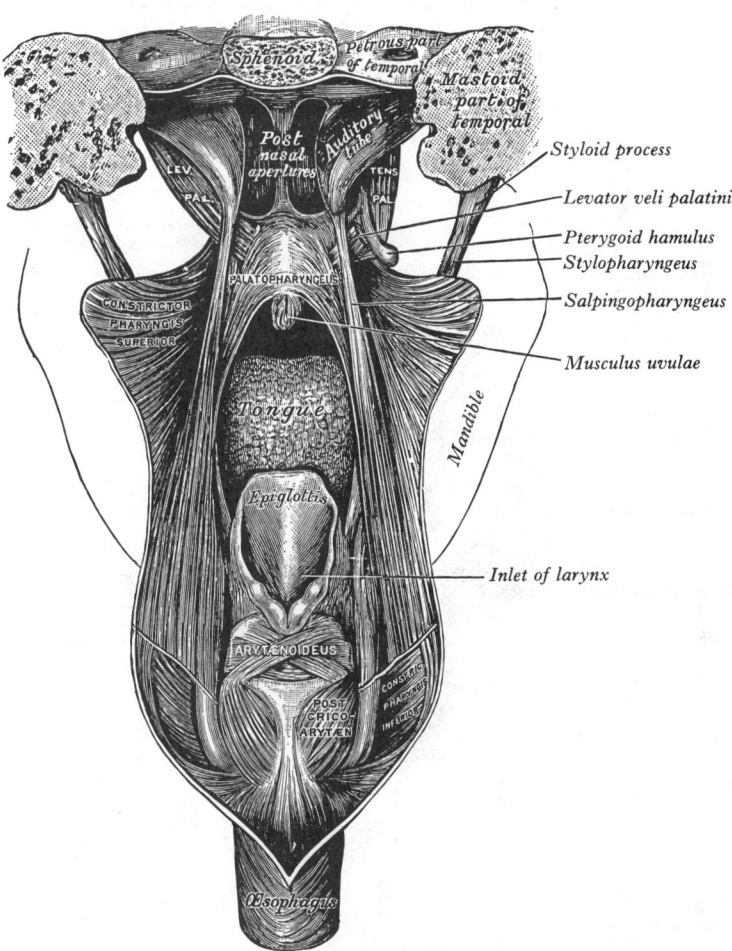

8.33 The muscles of the palate. Exposed from the posterior aspect.

surface of the palatine aponeurosis, where it is continuous with the muscle of the opposite side, and passing antero-inferiorly and laterally in front of the tonsil, is inserted into the side of the tongue, some of its fibres spreading over the dorsum of the tongue, and others passing deeply into its substance to intermingle and blend with the transversus linguae.

Actions. The palatoglossus pulls up the root of the tongue and approximates the palatoglossal arch to the median plane. Both muscles acting together close off the mouth cavity from the oral part of the pharynx.

The palatopharyngeus (**8**.33, 85) forms, with the mucous membrane covering its surface, the palatopharyngeal arch. In the palate it consists of two fasciculi, which are separated by levator veli palatini. The *posterior* fasciculus is in contact with the mucous membrane covering the pharyngeal surface of the palate; it joins with the posterior band of the opposite muscle in the median plane. The *anterior* fasciculus, the thicker, passes between the levator and the tensor veli palatini. It arises from the posterior border of the hard palate and from the palatine aponeurosis, while some of its fibres join in the median plane with the corresponding strand of the opposite muscle. Both fasciculi are attached to the upper surface of the palatine aponeurosis and lie in the same plane in the soft palate. At the posterolateral border of the palate the two layers of the muscle unite and are joined by the fibres of the salpingopharyngeus (p. 1308). Passing laterally and downwards behind the tonsil, the palatopharyngeus descends posteromedial to, and in close contact with, the stylopharyngeus, and is inserted with it into the posterior border of the thyroid cartilage, some of its fibres ending on the side of the pharynx by being attached to the pharyngeal fibrous coat, and others passing across the median plane posteriorly, to decussate with those of the opposite muscle. The palatopharyngeus really forms an internal longitudinal muscular coat for the pharynx.

Actions. The palatopharyngeus pulls the walls of the pharynx, on its own side, upwards, forwards and medially, and so shortens the pharynx during the act of swallowing. Acting together the two muscles approximate the palatopharyngeal arches and draw them forwards.

Nerve supply. With the exception of the tensor veli palatini, which is innervated by the mandibular nerve (p. 1066), all the muscles of the soft palate are supplied by nerve fibres which leave the medulla in the cranial part of the accessory nerve and which reach the pharyngeal plexus via the vagus nerve.

The muscles in the soft palate lie in the following relation to each other: the palatine aponeurosis (tendon of tensor veli palatini) is an intermediate sheet, enclosing the uvular muscles near the median plane; the levator veli palatini and palatopharyngeus are inserted into its upper surface, the two fasciculi of the latter muscle lying in the same plane, respectively in front of and behind the levator veli palatini; the palatoglossus is inserted into the oral surface of the aponeurosis. (For description of *the palatopharyngeal sphincter, see* p. 1309.)

Applied Anatomy. The occurrence of a congenital cleft in the palate has been referred to already as a defect in development p. 149). Paralysis of the soft palate may occur after diphtheria due to the action of the toxin on the nerve cells in the medulla oblongata. It gives rise to a change in the voice, which becomes nasal, and to the regurgitation of fluids into the nose when swallowing is attempted. On inspection, the palate is seen to hang flaccid and motionless when phonation or deglutition is attempted; it is also anaesthetic. Similar results ensue from other pathological processes involving the glossopharyngeal, vagus and accessory nerves, or the nuclei in the medulla oblongata.

The Salivary Glands

In the broadest sense, a salivary gland may be defined as any cell or organ which discharges a secretion into the oral cavity. Conventionally, a distinction is made between the *major salivary glands*, which lie at some distance from the oral mucosa, with which they communicate through one or more extraglandular ducts, and the *minor salivary glands*, which lie in the mucosa or submucosa and open directly or indirectly via many short excretory (collecting) ducts, on to the epithelial surface of the mucosa. In man the major salivary glands comprise three large paired masses—the parotid, submandibular and sublingual glands—while the minor salivary glands comprise the anterior lingual glands, numerous small glands (including von Ebner's glands, p. 1269) in the mucous membrane of the tongue, and small labial, buccal and palatal glands (p. 1281) in relation to the mucous membranes of the lips, cheek and roof of the mouth respectively. Their functions include: the lubrication of food, assisting swallowing; moistening of the buccal mucosa, essential for speech; provision of an aqueous solvent, necessary for taste; provision of a fluid seal, necessary for sucking (including suckling); secretion of digestive enzymes such as salivary amylase; secretion of hormones and others pharmacologically active compounds such as a glucagon-like protein (Lawrence *et al.* 1977) and possibly serotonin (Feyrter 1961).

THE PAROTID GLAND

The parotid gland (**8.34A, B**) is the largest and has an average weight of about 25 gm. It forms an irregular, lobulated, yellowish mass, lying below the external acoustic meatus, between the mandible and the sternocleidomastoid; it projects forwards on to the surface of the masseter, where a small part of it, usually more or less detached, lies between the zygomatic arch above and the parotid duct below; this detached portion is named the *accessory part* of the gland.

The gland is enclosed within a capsule derived from the deep cervical fascia; the part covering its superficial surface is dense, closely adherent to the gland, and attached to the zygomatic arch; the deep part of the capsule is attached to the styloid process, mandible and tympanic plate, and blends with the fibrous sheaths of the muscles related to the gland; a portion of the fascia attached to the styloid process and the angle of the mandible is thickened to form the stylomandibular ligament, which intervenes between the parotid and submandibular glands.

The parotid gland is like an inverted, flattened, three-sided pyramid; it presents a small superior surface, and superficial,

8.34A The right parotid gland. Posteromedial aspect.

Labels (8.34A):
- Area for osseous part of external acoustic meatus
- Facial nerve
- Area for styloid process and its attached muscles
- Posterior auricular artery
- External carotid artery
- Retromandibular vein
- Superior extension behind mandibular condyloid process
- Area for cartilaginous part of external acoustic meatus
- Area for mastoid process
- Area for posterior belly of digastric
- Groove made by anterior border of sternocleidomastoid

Labels (8.34B):
- Area in contact with temporomandibular ligament
- Transverse facial artery
- Grooved area adapted to posterior border of ramus of mandible
- Superficial temporal artery
- Maxillary artery
- External carotid artery
- Retromandibular vein

8.34B The right parotid gland. Anteromedial aspect.

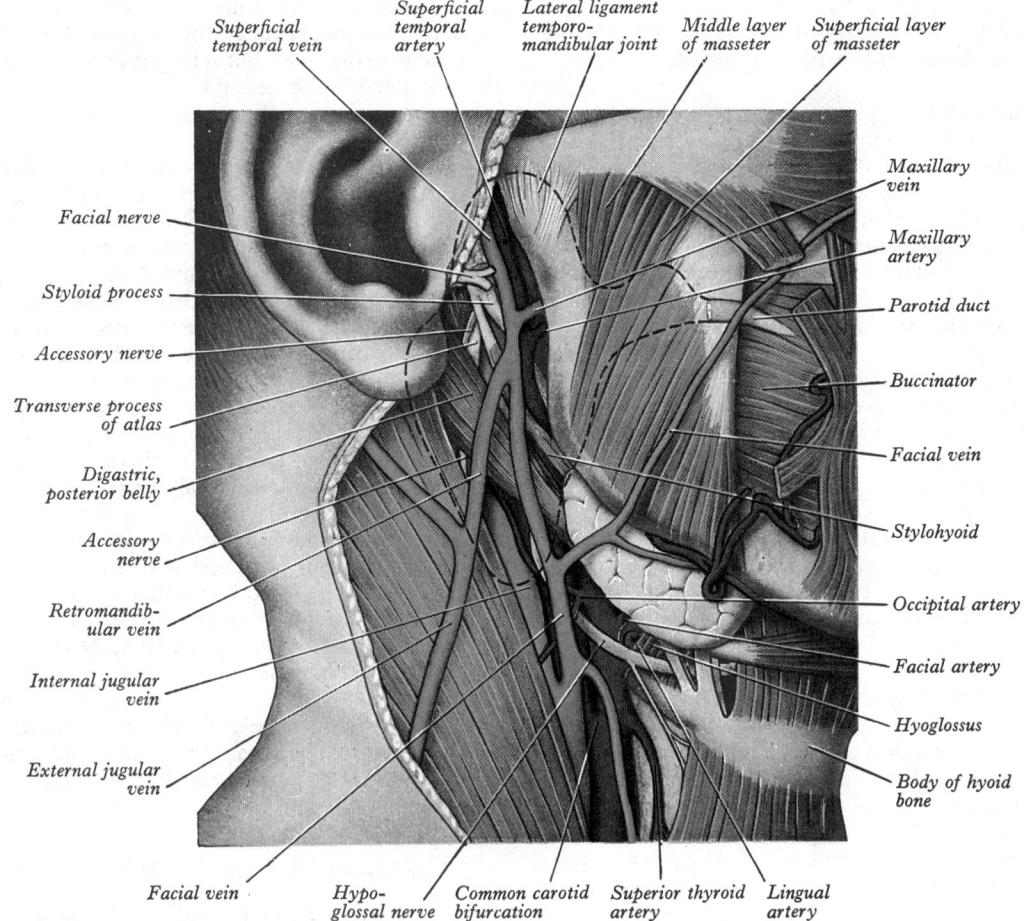

Superficial   Superficial   Lateral ligament
temporal vein   temporal artery temporo-   Middle layer   Superficial layer
                                     mandibular joint of masseter   of masseter

Facial nerve   Maxillary vein

Styloid process   Maxillary artery

Accessory nerve   Parotid duct

Transverse process   Buccinator
of atlas

Digastric,   Facial vein
posterior belly

Accessory   Stylohyoid
nerve

Retromandib-   Occipital artery
ular vein

Internal jugular   Facial artery
vein

                                                                     Hyoglossus

External jugular   Body of hyoid
vein   bone

Facial vein   Hypo-   Common carotid   Superior thyroid Lingual
                 glossal nerve bifurcation   artery   artery

8.35   A drawing of a dissection to show the principal immediate deep relations of the parotid gland. The outline of the parotid gland is indicated by the interrupted black line.

anteromedial and posteromedial surfaces. The lower part of the gland tapers to a blunt apex.

The *superior surface* is concave and is related to the cartilaginous part of the external acoustic meatus, and to the posterior surface of the temporomandibular joint; here the auriculotemporal nerve winds round the neck of the mandible, embedded in the gland or in the capsule around it.

The *apex* of the gland overlaps the posterior belly of the digastric and the carotid triangle to a variable extent.

The *superficial surface* is covered with the skin and the superficial fascia, which contains the facial branches of the great auricular nerve, the superficial parotid lymph nodes (p. 786) and the posterior border of the platysma. It extends upwards to the zygomatic arch, backwards to overlap slightly the anterior border of the sternocleidomastoid, downwards to its apex behind and below the angle of the mandible and forwards across the superficial surfaces of the masseter below the parotid duct (**8.**35).

The *anteromedial surface* is grooved by the posterior border of the ramus of the mandible. It covers the postero-inferior part of the masseter, the lateral aspect of the temporomandibular joint and the adjoining part of the mandibular ramus, and passes forwards on the deep aspect of the ramus to reach the medial pterygoid. The branches of the facial nerve emerge on the face from under cover of the anterior margin of this surface.

The *posteromedial surface* is moulded to the mastoid process and the sternocleidomastoid, and to the posterior belly of the digastric, the styloid process and the styloid group of muscles. The external carotid artery grooves this surface before it enters the substance of the gland. The internal carotid artery and internal jugular vein are separated from the gland by the styloid process and the styloid muscles (**8.**35). The anteromedial and posteromedial surfaces meet along a medial margin which may project so deeply as to be in contact with the side wall of the pharynx.

Within the gland are a number of non-glandular structures, most of which traverse it or even branch within its substance. The external carotid artery enters through the posteromedial surface of the parotid gland, and divides into its terminal branches within its substance. One of these branches—the maxillary artery—emerges from the anteromedial surface, and runs forwards medial to the neck of the mandible, while the other—the superficial temporal artery—gives off its transverse facial branch, and then ascends to exit from the upper limit of the gland (**8.**34B). The posterior auricular artery may start from the external carotid artery within the gland, and it then leaves the latter on its posteromedial surface. The retromandibular vein (p. 740), formed in the upper part of the gland by the union of the maxillary and superficial temporal veins, which enter the gland at the point of exit of the corresponding arteries, is superficial to the intraglandular part of the external carotid artery. It emerges from the gland behind its inferior extremity and joins the posterior auricular vein to form the external jugular vein; before it makes its exit it gives off a communicating branch which leaves the gland in front of its lower extremity and joins the facial vein. On a still more superficial plane the facial nerve traverses the gland. It enters the upper part of the posteromedial surface (**8.**34A), and passes forwards and downwards behind the posterior border of the ramus of the mandible in two main divisions from which its terminal branches arise (**7.**186). These leave the anteromedial surface of the gland above, in front, and below, and pass to their destinations at first deep to its anterior margin.

Developmentally the gland is an outgrowth from the buccal cavity (p. 197) and extends backwards towards the ear. As it does so it covers the facial nerve, but from its deep surface prolongations of the gland penetrate between the branches of the nerve in an irregular manner and constitute its deep portion. The largest of these is found between its main temporal and cervical divisions (Bailey 1947; McKenzie 1948). These outgrowths wrap

1273

themselves round the nerve and its branches, which become buried in its substance. On account of this arrangement the gland may be considered to comprise superficial and deep sections, sometimes termed lobes.

The parotid duct (**8**.36, 40) is about 5 cm long. It begins by the confluence of two main branches within the anterior part of the gland (*see* p. 1281), crosses the masseter, and at the anterior border of this muscle turns inwards nearly at a right angle, passes through the corpus adiposum of the cheek (suctorial pad in the infant) and pierces the buccinator; it then runs for a short distance obliquely forwards between the buccinator and mucous membrane of the mouth, and opens upon a small papilla on the oral surface of the cheek opposite the crown of the second upper molar tooth. While crossing the masseter it receives the duct of the accessory lobe; in this position it lies between the upper and lower buccal branches of the facial nerve; the accessory part of the gland and the transverse facial artery are above it. The buccal branch of the mandibular nerve, as it emerges from beneath the temporalis and masseter, lies just below the duct at the anterior border of the masseter.

Surface Anatomy. The parotid duct can be felt in the living (on the face, or more easily in the vestibule of the mouth), as it dips inwards at the anterior border of the masseter, by pressing the index finger *backwards* on this border of the muscle (with the teeth clenched to make the muscle tense) and moving the finger up and down across the line of the duct. The anterior border of the gland is represented by a line passing downwards and forwards from the upper border of the mandibular condyle to a point just above the middle of the masseter and then downwards and backwards to a point about 2 cm below and behind the angle of the mandible. Its upper border, concave upwards and backwards, corresponds to a curved line drawn from the upper border of the mandibular condyle, across the lobule of the auricle, to the mastoid process. The posterior border corresponds to a straight line joining the ends of the anterior and upper borders. The parotid duct corresponds to about the middle third of a line drawn from the lower border of the tragus to a point midway between the ala of the nose and the red margin of the upper lip.

The wall of the parotid duct is thick, and consists of an external fibrous coat which contains nonstriated muscular fibres, and an internal mucous coat which is lined with short columnar epithelium (see below for further structural details). Its canal is about 3 mm in diameter, but at its orifice on the oral surface of the cheek its lumen is reduced in size.

Vessels and Nerves

The *arteries* supplying the parotid gland are derived from the external carotid artery, and from the branches given off by that vessel in or near the gland. The *veins* empty themselves into the external jugular vein, through some of its tributaries. The *lymph vessels* end in the superficial and deep cervical lymph nodes, interrupted in their course by two or three lymph nodes on the surface and in the substance of the parotid gland. The *efferent innervation* consists of sympathetic and parasympathetic nerves. The sympathetic supply is derived from the plexus on the external carotid artery. It is generally stated that the parasympathetic secretomotor nerves run through the tympanic branch of the glossopharyngeal nerve, are relayed in the otic ganglion, and then travel via the auriculotemporal nerve to the gland. Clinical investigations, however, have led to the conclusion that the human parotid gland also receives secretomotor fibres through the chorda tympani (Reichert and Poth 1933; Diamant and Wiberg 1965). More recently, Holmberg (1972) has shown that in the dog, secretomotor fibres pass to the parotid gland from the surface of the maxillary artery and from the facial nerve as well as from the auriculotemporal nerve. There are no reports, as yet, of these additional routes in man. The exact termination of these nerve supplies is still a topic of controversy. Experimental studies in cats suggest that both parasympathetic and sympathetic fibres end in relation to glandular cells (Genvis-Galvey *et al.* 1966, and *vide infra*).

THE SUBMANDIBULAR GLAND

The submandibular gland (**8**.36) is irregular in form and about the size of a walnut. It consists of a larger superficial part and a smaller deep part, which are continuous with each other around the posterior border of the mylohyoid.

The *superficial part* of the submandibular gland is situated in the digastric triangle, reaching forwards to the anterior belly of

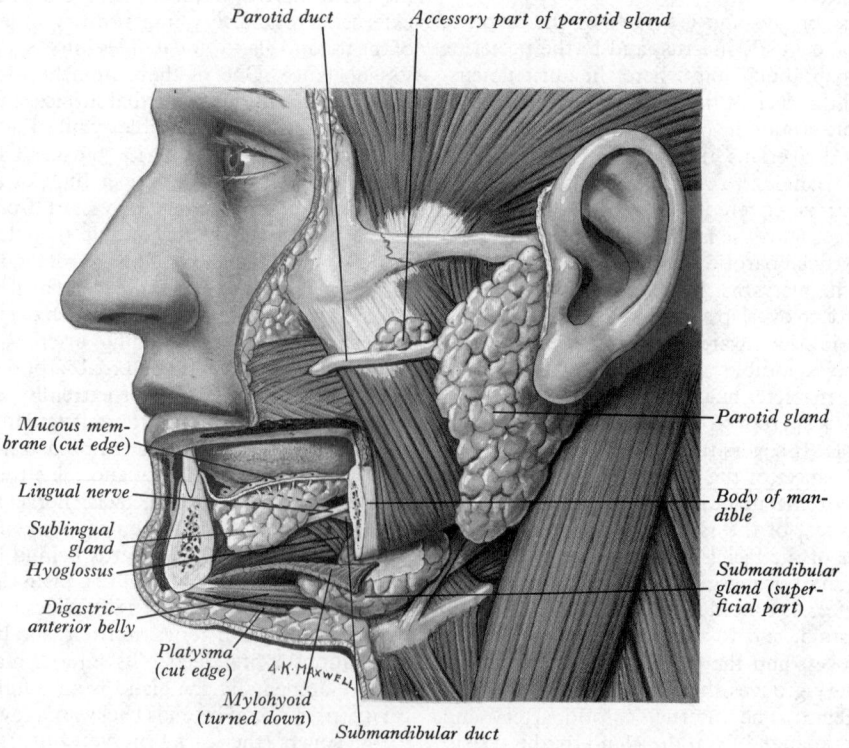

Parotid duct　　*Accessory part of parotid gland*

Mucous membrane (cut edge)

Lingual nerve

Sublingual gland

Hyoglossus

Digastric anterior belly

Platysma (cut edge)

Mylohyoid (turned down)

Submandibular duct

Parotid gland

Body of mandible

Submandibular gland (superficial part)

8.36　A dissection showing the salivary glands of the left side. The cranial region of the superficial part of the submandibular gland has been excised and the cut mylohyoid has been turned down to expose a portion of the deep part of the gland.

the digastric and backwards to the stylomandibular ligament, which intervenes between the submandibular and parotid glands. Above, it extends under the cover of the body of the mandible; below, it usually overlaps the intermediate tendon of the digastric and insertion of the stylohyoid. It has three surfaces, an inferior, a lateral and a medial, and is partially enclosed between two layers of the deep cervical fascia which extend from the greater cornu of the hyoid bone, one layer passing to the lower border of the mandible and covering the inferior surface of the gland, the other passing to the mylohyoid line on the medial surface of the mandible and covering the medial surface of the gland.

The *inferior surface* of the gland is covered by the skin, platysma and deep fascia. It is crossed by the facial vein, and by the cervical branch of the facial nerve; near the mandible the submandibular lymph nodes are in contact with it and a few may actually be embedded in it (p. 787).

The *lateral surface* is in relation with the submandibular fossa on the medial surface of the body of the mandible, and with the insertion of the medial pterygoid. The facial artery is embedded in a groove in the posterosuperior part of the submandibular gland, lying at first deep to it and then emerging between the lateral surface of the gland and the insertion of the medial pterygoid to reach the lower border of the mandible.

The *medial surface* is related, in front, to the mylohyoid, separated from it by the mylohyoid nerve and vessels and by branches of the submental vessels; more posteriorly, the medial surface is related to the styloglossus, stylohyoid ligament and the glossopharyngeal nerve, which separate it from the wall of the pharynx in this region; in its intermediate part the medial surface is related to the hyoglossus, separated from it by the styloglossus, the lingual nerve, the submandibular ganglion, the hypoglossal nerve and the deep lingual vein (in that order from above downwards). Below, the medial surface is related to the stylohyoid and the posterior belly of the digastric.

The *deep part* of the submandibular gland extends forwards as far as the posterior end of the sublingual gland, and lies in the intermuscular interval between the mylohyoid below and laterally, and the hyoglossus and styloglossus medially; above, it is related to the lingual nerve, and below, to the hypoglossal nerve and the deep lingual vein.

In the living, the gland is palpable between the index finger placed on the floor of the mouth and the thumb placed medial to, and just in front of the angle of the mandible, below the floor.

The submandibular duct is about 5 cm long, and its wall is much thinner than that of the parotid duct. It begins as numerous branches in the superficial region of the gland and, emerging from the middle of the deep surface of that part of the gland a little behind the posterior border of the mylohyoid, it runs through the deep part of the gland passing at first upwards and slightly backwards for 4 or 5 mm and then turns forwards to run between the mylohyoid and the hyoglossus; it then passes between the sublingual gland and the genioglossus, and opens by a narrow orifice in the floor of the mouth, on the summit of the sublingual papilla at the side of the frenulum of the tongue (8.74). On the hyoglossus it lies between the lingual and hypoglossal nerves, but at the anterior border of the muscle it is crossed laterally by the lingual nerve; the terminal branches of the lingual nerve ascend on its medial side (7.177). As it passes through the deep part of the gland it receives the small ducts draining this part of the gland. Salivary calculi may occasionally appear in the submandibular duct and are readily detected radiologically.

Vessels and Nerves

The *arteries* supplying the submandibular gland are branches of the facial and lingual arteries. Its *veins* follow the course of the arteries. The *nerves* are derived from the submandibular ganglion, through which it receives fibres from the chorda tympani of the facial nerve, the lingual branch of the mandibular nerve and the sympathetic trunk.

In the dog and cat the submandibular gland receives its nerve supply through a separate, subsidiary ganglion, known as Langley's ganglion (p. 1123), but in the human the cell stations on the parasympathetic nerve supply of the gland (chorda tympani) are mainly in the submandibular ganglion; some sub-ganglia may

occur in the hilum of the gland, and in the branches of the submandibular ganglion passing to it. Garrett and Kemplay (1977), using catecholamine fluorescence and also electron micrography, have re-investigated arrangements in the cat, finding that all adrenergic fibres are derived from the superior cervical sympathetic ganglion.

THE SUBLINGUAL GLAND

The sublingual gland (8.36) is the smallest of the three main salivary glands. It is situated beneath the mucous membrane of the floor of the mouth, in contact with the sublingual fossa on the lingual surface of the mandible, close to the symphysis. It is narrow, flattened, shaped somewhat like an almond, and weighs between 3 and 4 gm. It is in relation, *above*, with the mucous membrane of the mouth, which it raises in the form of the sublingual fold; *below*, with the mylohyoid; *in front*, with its fellow of the opposite side; *behind*, with the deep part of the submandibular gland; *laterally*, with the mandible above the anterior part of the mylohyoid line and *medially*, with the genioglossus, from which it is separated by the lingual nerve and the submandibular duct. Its excretory ducts are from eight to twenty in number. Of the *smaller sublingual ducts*, most open separately into the floor of the mouth on the summit of the sublingual fold; occasionally a few open into the duct of the submandibular gland. From the anterior part of the gland some of the ducts sometimes join to form a *major sublingual duct*, which opens with, or near to, the submandibular duct.

Vessels and Nerves

The sublingual gland is supplied with blood by the sublingual and submental arteries. Its nerves are derived from the lingual and chorda tympani nerves, and from the sympathetic. The parasympathetic cell station is in the submandibular ganglion; occasionally some nerve cells are found on the fibres passing distally from the submandibular ganglion to the lingual nerve and they constitute a sublingual ganglion. The precise terminations of these nerve supplies are not fully established (*vide infra*).

STRUCTURE OF THE MAJOR SALIVARY GLANDS

These are compound racemose glands (8.37A, B; 38A, B, C; 39A, B) consisting of numerous lobes made up of lobules linked by interlobular stroma, a dense areolar tissue containing excretory (collecting) ducts, blood vessels, nerve fibres and small ganglia. Each lobule consists of the ramifications of a single duct, the branches of which end as dilated secretory 'endpieces' which produce the primary secretion. This is secondarily modified as it passes sequentially through the intercalated (intercalary), striated and excretory ducts, the last of which open into one or more main ducts which direct the saliva into the oral cavity.

Secretory 'Endpieces' of Salivary Glands

Variation in the terms used to describe the form and cytology of the secretory endpieces has resulted in considerable confusion. The publications of Garrett (1976) and of Young and van Lennep (1978) include critical appraisals of the problems involved.

In the following description a secretory endpiece is termed an *acinus* if it is approximately spheroidal, and a *tubule* if its long axis is several times greater than its diameter; *tubulo-acini* are intermediate in shape. The glandular cells of acini are pyramidal, having a narrow luminal apex and a broad base, while those of tubules are more cylindrical. These cells, which are the main producers of salivary protein and glycoprotein, are customarily described as *serous*, *seromucous* or *mucous*; unfortunately there is no unanimity in the use of these terms. The observation that some cells produced a serous secretion of low viscosity while others produced a highly viscous mucus was correlated initially with the distinctive appearance of their cells after simple staining procedures; thus serous and mucous cells came to be distinguished. The application of improved histological techniques led to the recognition of a third group of glandular cells, the seromucous variety. It has now been shown that the substances secreted by the glandular cells can be arranged

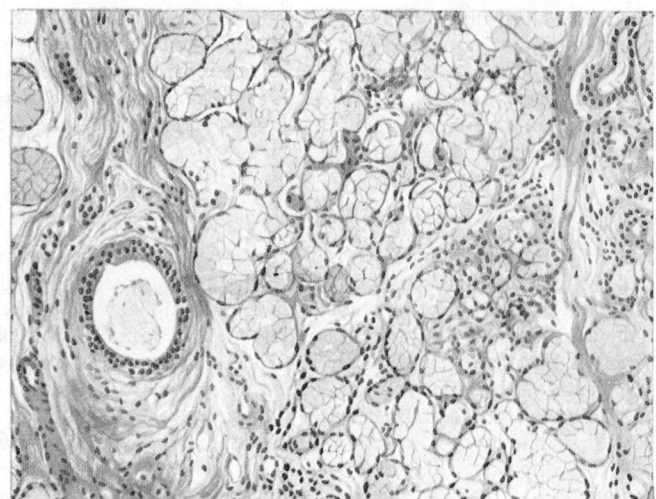

8.37A   A section through the sublingual gland, stained with haematoxylin and eosin. Magnification about × 140. Compare with **8**.38C.

Serous alveolus   *Mucous alveolus*   *Adipose stroma*

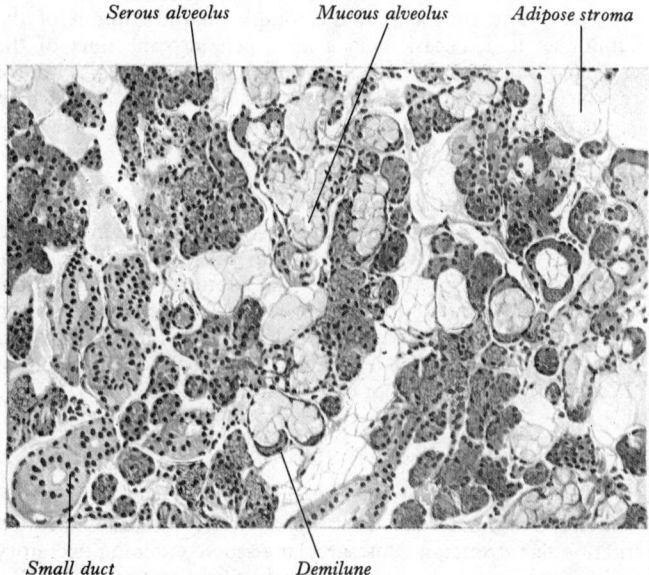

Small duct   *Demilune*

8.37B   A section of the submandibular gland, stained with haematoxylin and eosin. Magnification about × 140. Compare with **8**.38B.

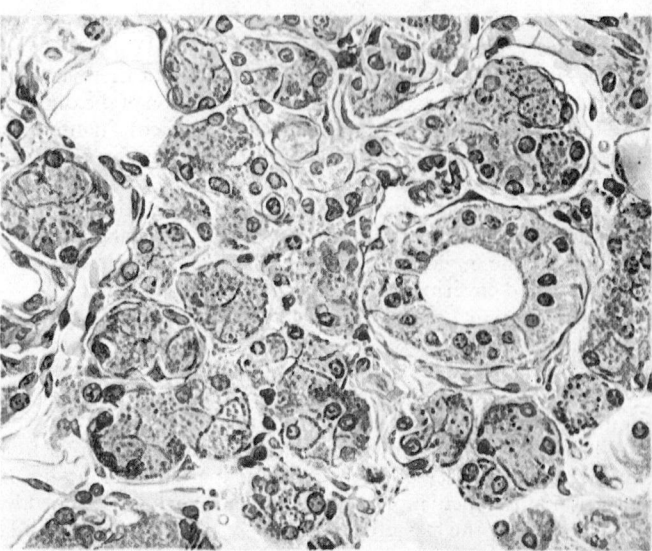

8.38A   A section through the parotid gland, stained with haematoxylin and eosin. Magnification about × 350.

Ductule

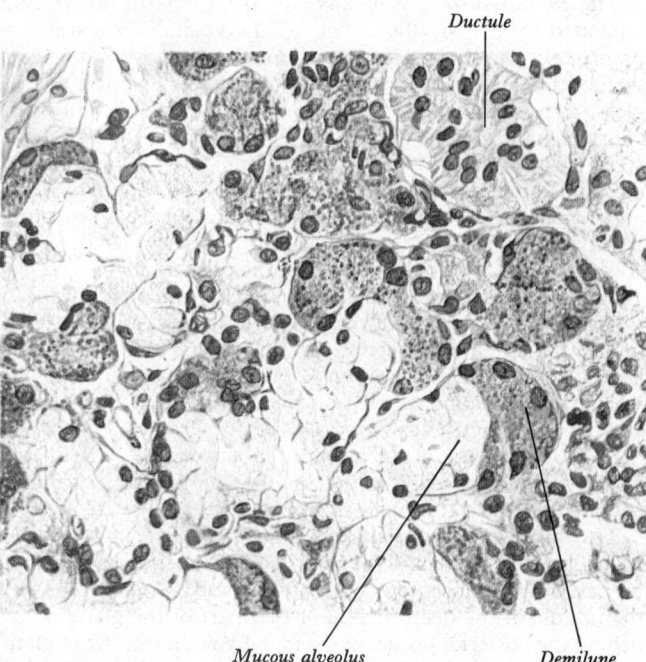

Mucous alveolus   *Demilune*

8.38B   a section through the submandibular gland, stained with haematoxylin and eosin. Magnification about × 350.

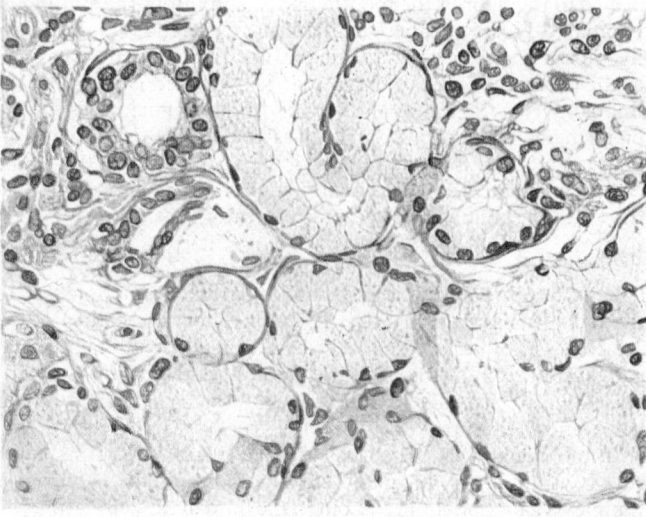

8.38C   A section through the sublingual gland, stained with haematoxylin and eosin. Magnification about × 350.

in an almost continuous series extending from serous secretions containing only a negligible amount of protein-associated acidic carbohydrate to mucous secretions rich in these substances. Attempts have been made to define the descriptive terms serous, seromucous and mucous in strictly histochemical terms (Leblond 1950; Munger 1964; Shackleford and Wilborn 1968) but without general agreement. While recognizing the desirability of a histochemically based definition, it is practically more convenient to make use of readily observable cytological characteristics, such as the appearance of the secretory granules of the cells. Applying the criteria suggested by Young and van Lennep (1978) the glandular cells are here described as *serous* if their granules are small, discrete, homogeneous, generally eosinophilic and electron dense; *mucous* if their granules are larger, closely-packed and ill-defined, with a low affinity for eosin and with a homogeneous, fairly electron-translucent matrix; and *seromucous* if they are intermediate between the serous and mucous types in appearance, having granules which are either closely-packed, eosinophilic and homogeneous, or more discrete, larger than typical serous granules and heterogeneous. A gland or a secretory endpiece containing only one type of glandular cell is termed *homocrine*, while one containing more than one type of such cell is termed *heterocrine*.

The secretory endpieces of the human *parotid gland* consist mainly of seromucous (frequently described as serous) acini;

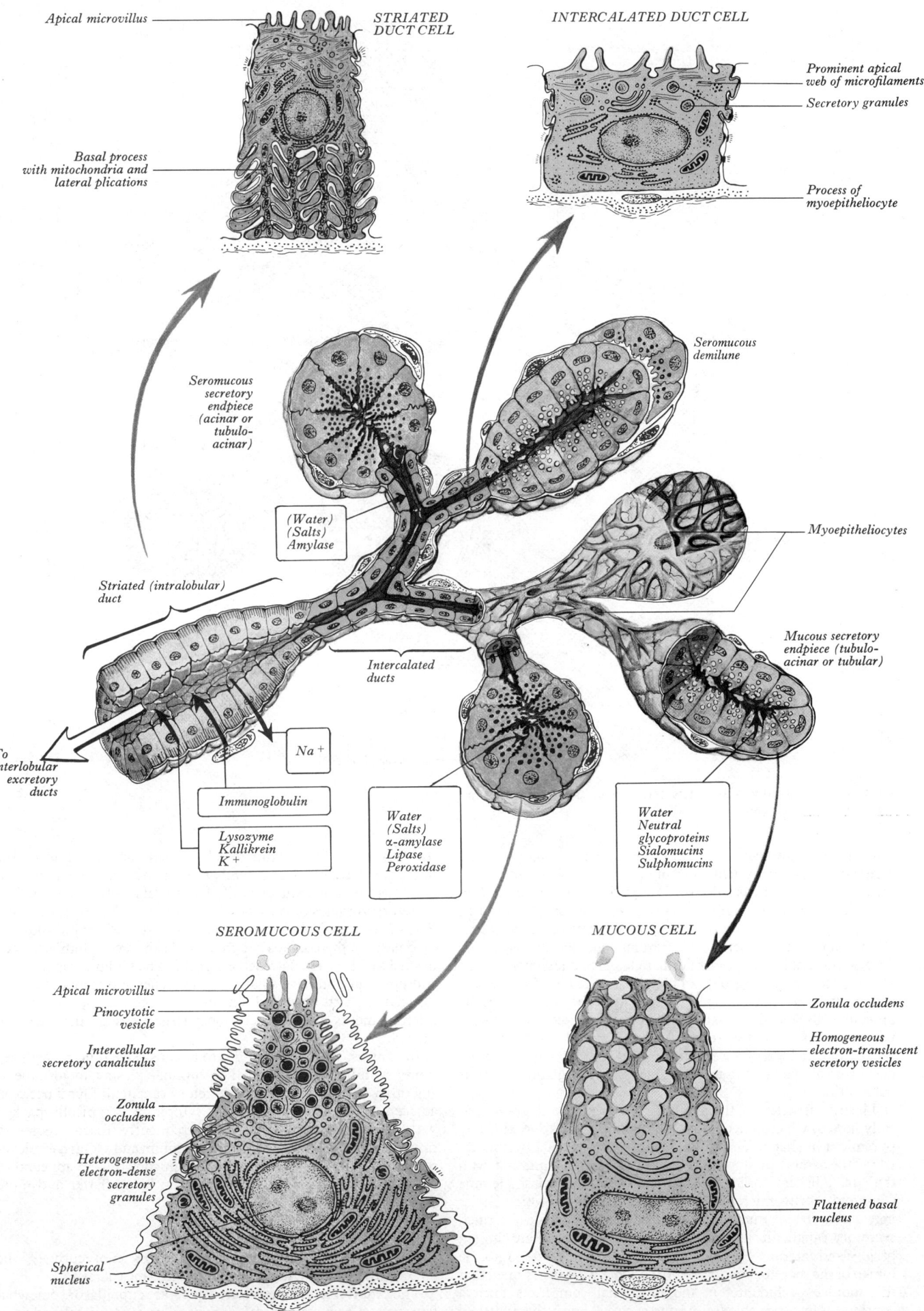

Apical microvillus — STRIATED DUCT CELL

INTERCALATED DUCT CELL

Prominent apical web of microfilaments

Secretory granules

Basal process with mitochondria and lateral plications

Process of myoepitheliocyte

Seromucous secretory endpiece (acinar or tubulo-acinar)

Seromucous demilune

(Water)
(Salts)
Amylase

Myoepitheliocytes

Striated (intralobular) duct

Mucous secretory endpiece (tubulo-acinar or tubular)

To interlobular excretory ducts

Intercalated ducts

Na⁺

Immunoglobulin

Lysozyme
Kallikrein
K⁺

Water
(Salts)
α-amylase
Lipase
Peroxidase

Water
Neutral
glycoproteins
Sialomucins
Sulphomucins

SEROMUCOUS CELL

MUCOUS CELL

Apical microvillus

Zonula occludens

Pinocytotic vesicle

Homogeneous electron-translucent secretory vesicles

Intercellular secretory canaliculus

Zonula occludens

Heterogeneous electron-dense secretory granules

Flattened basal nucleus

Spherical nucleus

8.39A Diagram of the architecture of a generalized salivary gland including ultrastructural details.　　Solid and outlined black arrows indicate direction of transport.

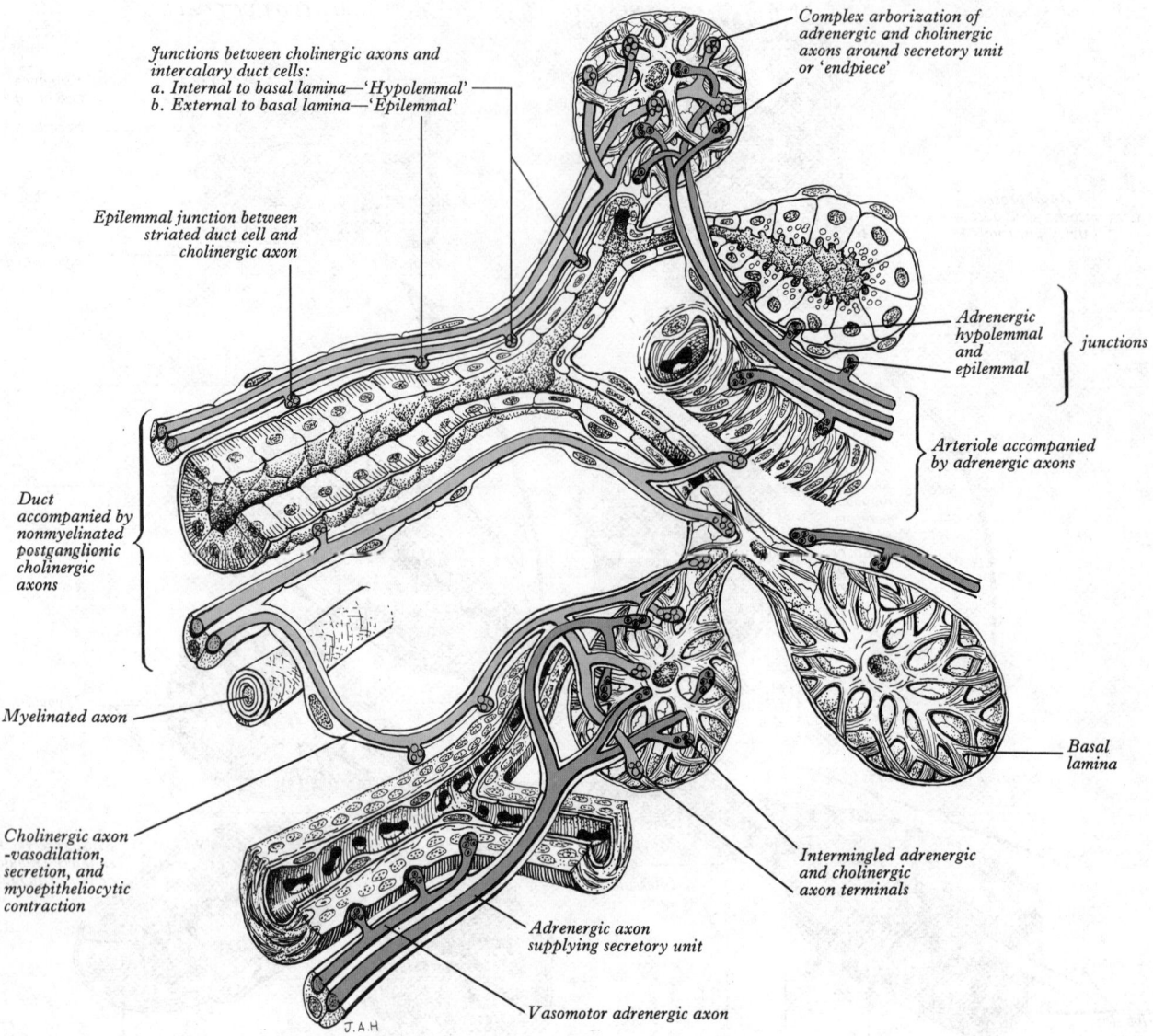

Junctions between cholinergic axons and intercalary duct cells:
a. Internal to basal lamina—'Hypolemmal'
b. External to basal lamina—'Epilemmal'

Complex arborization of adrenergic and cholinergic axons around secretory unit or 'endpiece'

Epilemmal junction between striated duct cell and cholinergic axon

Adrenergic hypolemmal and epilemmal } *junctions*

Duct accompanied by nonmyelinated postganglionic cholinergic axons

Arteriole accompanied by adrenergic axons

Myelinated axon

Basal lamina

Cholinergic axon -vasodilation, secretion, and myoepitheliocytic contraction

Intermingled adrenergic and cholinergic axon terminals

Adrenergic axon supplying secretory unit

Vasomotor adrenergic axon

J.A.H

8.39B   The innervation of the ducts, secretory units, and arterioles in a generalized salivary gland.

mucous acini are rare. Those of the *submandibular gland* are generally seromucous acini, although there are also some of the mucous variety. In the *sublingual gland* the secretory endpieces typically have the form of mucous tubules, but seromucous cells are also present, frequently arranged as acini or as *demilunes* (Young and van Lennep 1978). Seromucous demilunes, which are also present in the submandibular glands of many mammalian species, are crescentic groups of glandular cells found at the base of some mucous secretory endpieces (**8.**39A); they lie between the mucous cells and the basal lamina, apparently communicating with the lumen of the secretory endpiece by means of fine canaliculi which pass between the mucous cells. Using the criteria described above, all the major human salivary glands are thus heterocrine.

The ultrastructure of the glandular cells is shown diagrammatically in **8.**39A. Seromucous (and serous) cells are approximately pyramidal in shape. The basal plasmalemma is usually smooth, while the lateral plasmalemma is plicated, interdigitating with that of adjacent cells. The apical (luminal) plasmalemma commonly forms microvilli, between which pinocytotic vesicles may be seen. A number of discrete, short tubules, termed secretory canaliculi, lie between adjacent cells; they are limited basally and laterally by junctional complexes and open into the lumen of the secretory endpiece. Although it is often assumed that the zonulae occludentes of the junctional complexes form a continuous impermeable seal around each cell, recently published freeze-fracture micrographs of the tight junctions of the secretory endpieces of the rat parotid (de Camilli *et al.* 1976) suggest that

the 'seal' may be incomplete. The shape and position of the nucleus is variable, but it tends to be more spheroidal and less basal than that of a mucous cell. The apical cytoplasm is filled by secretory granules of very variable form (Tandler 1972; Riva and Riva-Testa 1973), while a conspicuous feature of the infranuclear cytoplasm is the copious rough endoplasmic reticulum arranged in stacks of parallel, flattened cisternae. The Golgi complexes are supranuclear and have small coated vesicles and smooth vesicles associated with them. Elongated mitochondria, lysosomes, microfilaments and the occasional large lipid droplet are also present.

In contrast with serous and seromucous cells, mucous cells are more cylindrical in shape. The luminal plasma membrane is smoother, and only rarely are secretory canaliculi found between adjacent cells. The supranuclear cytoplasm is typically packed with large, electron-translucent, frequently fused, secretory droplets. The rough endoplasmic reticulum and Golgi complexes resemble those of the serous and seromucous cells in appearance and position, but the nucleus differs in being flatter and more basal.

Ducts of the Salivary Glands

Leading from the secretory endpieces are, consecutively, the *intercalated*, *striated*, and *excretory* (collecting) *ducts*.

The cells lining the intercalated ducts are cuboidal or somewhat flattened. Their cytoplasm contains elongated mitochondria, a few scattered cisternae of granular endoplasmic reticulum, juxtanuclear Golgi complexes, and some lysosomes and secretory granules. While this unspecialized ultrastructure suggests that they play little part in protein synthesis, it does not preclude an involvement in the addition of water and electrolytes to the saliva. The problem of which cell type of the secretory endpiece-intercalated-duct complex is the main secretor of water and electrolytes is as yet unresolved, and although it is often assumed that the glandular cells have this function, the lining cells of the intercalated ducts could also be responsible for it.

The characteristic feature of the cells lining the striated ducts is their basal striation. This was described by Pflüger (1866) as resembling a thick lawn. In 1909 Regaud and Mawas concluded that the striated appearance was caused by columns of packed, elongated mitochondria, an interpretation confirmed by the ultrastructural studies of Pease (1956). Between the columns of mitochondria the cytoplasm is divided into a small number of basal processes by infolding of the basal plasmalemma. The lateral aspects of these processes may be extensively plicated and interdigitate with neighbouring processes. Desmosomes often link adjacent plications. This folding of the plasmalemma and the local abundance of mitochondria is characteristic of many epithelial cells engaged in electrolyte transport, as these cells certainly are; they transport potassium into the saliva, and by reabsorbing sodium ions in excess of water they render the saliva hypotonic (Thaysen et al. 1954). The lateral plasmalemmae of the lining cells of the striated ducts are linked to those of adjacent cells by slightly pervious junctional complexes (Garrett and Parsons 1974). The luminal plasmalemma forms microvilli, and the cytoplasm often extends into the lumen as apical blebs which are thought to enter the saliva by an apocrine process (Garrett 1976). The supranuclear cytoplasm contains a few scattered cisternae of granular endoplasmic reticulum, mitochondria, lysosomes, vesicles, microtubules and numerous microfilaments, the last being especially noticeable in the terminal web or 'separating zone' (Takano 1969). As well as modifying the electrolyte composition of saliva, the striated ducts also secrete immunoglobulin A (Klaus and Mestecky 1971), lysozyme (Kraus and Mestecky 1971) and kallikrein (Garrett and Kidd 1975). The immunoglobulin A is produced by plasma cells lying deep to the lining epithelium.

There have been few ultrastructural studies of the excretory ducts. They are lined in the rat by a simple columnar or pseudostratified epithelium, which consists mainly of tall columnar cells with perpendicular basal striations containing packed, elongated, mitochondria (Tamarin and Sreebny 1965). The presence of these striations, and the finding of more closely opposed blood capillaries here than in any other part of the duct system, suggests that the excretory ducts are more than merely passive conduits along which saliva is conducted. An involvement in electrolyte transport is suspected (Young and van Lennep 1978).

Myoepitheliocytes of Salivary Glands

These contractile cells are associated with the intercalated ducts, and usually also with the secretory endpieces, lying between the basal lamina and the epithelial cells proper. In a comprehensive review by Garrett and Emmelin (1979) of the activities of salivary myoepitheliocytes (myoepithelial cells) the effects of their contraction are summarized as follows: it speeds up the outflow of saliva, reduces the luminal volume of the intercalated ducts and secretory endpieces, contributes to the secretory pressure, supports the underlying parenchyma, helps salivary flow to overcome increases in peripheral resistance, and may, in certain circumstances, help to expel the contents of the associated secretory cells.

The shape of the salivary myoepitheliocytes depends upon their location: those of the endpieces are stellate, having long, branching, overlapping processes which, sometimes together with those of one or two other such cells, form an encircling basket-shaped network around each endpiece; in contrast, those of the ducts are more fusiform, have fewer branches, and extend along the intercalated ducts in a longitudinal direction. Each endpiece myoepitheliocyte has a central perikaryon from which radiate between four and eight processes, each of which gives rise to two or more generations of branches which cross but do not fuse, and which do not extend on to the ducts. In contrast, the

processes of the myoepitheliocytes of the intercalated ducts seldom branch and often overlap on to the endpieces.

The cytoplasm of each myoepitheliocyte can be divided into two compartments, one filamentous and one non-filamentous. The latter contains the nucleus, juxtanuclear Golgi complexes, a few cisternae of granular endoplasmic reticulum, free ribosomes, lysosome-like bodies, and mitochondria. Globules of neutral fat have been described in human myoepitheliocytes (Garrett 1963). The filaments, which are particularly conspicuous in the processes and their branches, resemble the myofilaments of nonstriated myocytes (p. 518). Both thin (4nm diameter) and thick (10 nm diameter) filaments have been described, the former lying longitudinally in the processes and branches, with the less numerous thick filaments scattered amongst them. Filaments often pass to attachment plaques on the plasmalemma of the basal laminal (stromal) aspect of the cells, causing indentations there. The basal lamina opposite the plaques appears thicker than elsewhere, and is thought to be linked to the cells at these points. Bannerjee et al. (1977) have shown that the basal lamina affords strength to the underlying salivary epithelium. When the myoepitheliocytes contract the basal lamina is probably pulled on at the attachment plaques (Garrett and Emmelin, 1979). Numerous caveolae are commonly associated with the stromal plasmalemma, while that lying adjacent to the epithelial cells has fewer. The myoepitheliocytes are linked to the secretory and ductal cells by desmosomes, and occasionally cilia are found extending from the myoepitheliocytes into indentations on the adjacent epithelial cells. The cilia were first observed by Tandler (1965) in the myoepitheliocytes of the human submandibular gland, and have since been described in other human salivary glands (Tandler et al. 1970), and in glands of other species (Cutler and Chaudhry 1973; Kidd 1978). Garrett and Emmelin (1979) suggest that there may be a cilium on each salivary myoepitheliocyte, as has been proposed by Stirling and Chandler (1976) for the myoepitheliocytes of the human breast (p. 1436). The cilia may have a chemoreceptive and/or a mechanoreceptive role, but as yet their function is obscure.

Salivary myoepitheliocytes appear to have both a sympathetic and a parasympathetic innervation (Garrett 1972; 1976), with several axons of either the same or dissimilar type supplying a single myoepitheliocyte. This is not invariable, however, for in some situations, such as the sublingual gland of the rat, the myoepitheliocytes appear to have only a cholinergic innervation (Templeton and Thulin 1978). Where present, impulses from both sympathetic and parasympathetic nerves can result in myoepitheliocyte contraction (Garrett and Emmelin 1979).

CONTROL OF SALIVARY GLAND ACTIVITY

The wide and rapid variation in the composition of saliva, and in the quantity and rate at which it is produced, in response to environmental stimuli, suggests the existence of an elaborate control mechanism (Emmelin 1972). In some glands saliva is secreted spontaneously, apparently even in the absence of extraneous stimuli; in others secretion can be elicited by the stimulation of a wide variety of extraglandular receptors: gustatory, nociceptive, olfactory and tactile. Secretion may proceed at a continuous but low resting level, partly spontaneously, but mainly as a reflex response to drying of the oral and pharyngeal mucosae. A rapid increase in activity can be superimposed upon the resting level, for example during mastication, in response to the influence of autonomic secretory nerves. Controlled variation in the activity of the many different types of effector cells of the salivary glands—the serous, seromucous and mucous cells of the secretory endpieces, the myoepitheliocytes of the endpieces and intercalary ducts, the lining epithelial cells of all the ductal elements, and the nonstriated myocytes of the blood vessels—affects salivary production quantitatively and qualitatively. Variation in effector cell activity is under hormonal and nervous control.

Hormonal control
The effects of circulating hormones on the secretory activity of the salivary glands have been reviewed by Blair-West et al. (1967).

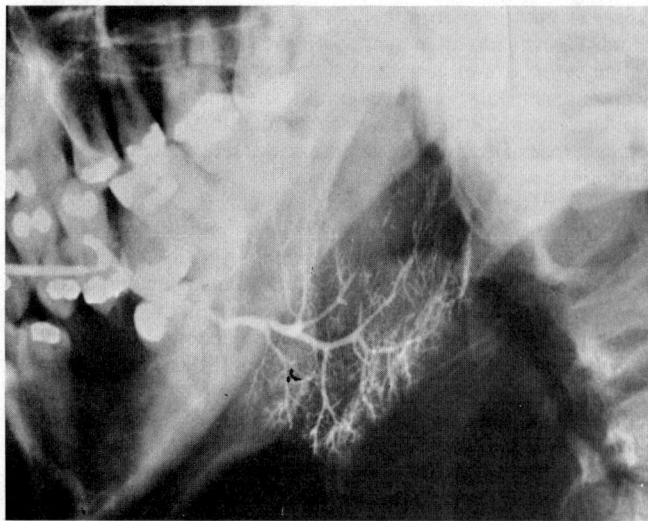

A

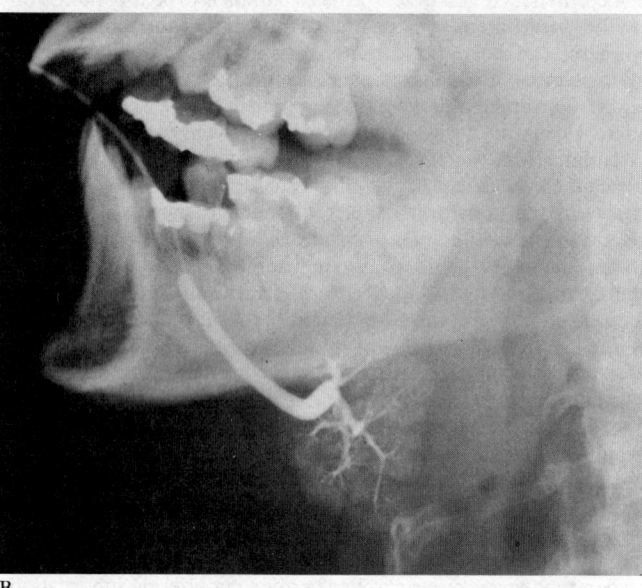

B

8.40A and B A Parotid sialogram; B Submandibular sialogram. In each case the shadow of the cannula used to introduce the radio-opaque medium into the duct of the gland is visible. See text for description.

There is no unequivocal evidence that they can evoke the secretion of saliva directly at physiological levels, although they may alter the response of the glandular cells to neural stimuli. Local hormones, however, have a profound effect on salivary gland activity; it has been shown in the cat submandibular gland, for example, that although vasodilatation is initiated by neural influences, it is maintained by the action of plasma-kinins which are formed locally when kallikrein is released from the secretory cells after stimulation by sympathetic amines (Gautvik *et al.* 1972). Vasodilatation and secretion are functionally related (Hilton and Lewis 1956).

Neural control
Most salivary glands, except those which secrete spontaneously, depend on autonomic nerves to evoke secretion, and in all the greater part of salivary flow is under nervous control. The nerves involved are cholinergic (parasympathetic) and adrenergic (sympathetic); there is no evidence for the involvement of non-cholinergic non-adrenergic nerves in salivary gland control (Garrett 1976).

The general pattern of innervation is shown in Fig. **8.**39B. It should be appreciated that there is considerable variation in different glands and in different species (Garrett 1972), and that the pattern of innervation may vary within a gland with age (Yohro 1971). Only the more constant features are illustrated and

described below. The cholinergic nerves tend to accompany the ducts and form complex arborizations around the secretory endpieces. In contrast, the adrenergic nerves generally enter the glands with the arteries and are distributed with their branches. The main nerve trunks of the glands contain non-myelinated axons predominantly, although a few myelinated axons (presumed to be either those of preganglionic efferent or of afferent neurons) are present. The postganglionic efferent axons, like those of autonomic axons elsewhere (Norberg 1967), have periodic dilatations which contain mitochondria and vesicles, the latter appearing electron-lucent in cholinergic axons and electron-dense in adrenergic axons. Within the glands the nerve fibres intermingle, cholinergic and adrenergic axons commonly lying in adjacent invaginations of the same Schwann cell (Eneroth *et al.* 1969; Garrett 1972, 1976).

At *neuro-effector junctions* (**8.**39B) dilated, vesicle-containing regions of the axons and the effector cells they supply are functionally related. The axonal surface closest to the effector cell is free from Schwann cell cytoplasm. Where the effector cells are epithelial, the neuro-effector junctions may be either *epilemmal* or *hypolemmal* (Garrett 1975), terms introduced by Arnstein (1889, 1895). At epilemmal sites the free surface of the axon and the effector cell are separated by a distance of approximately 100 nm, with the basement membrane intervening. At hypolemmal sites the axon penetrates beneath the basement membrane and is separated from the effector cell by a distance of only 20 nm. A single axon may supply several effector cells directly, and many more indirectly through electrical coupling of adjacent cells (Lowenstein and Kanno 1964); group activity is thus possible. Furthermore, a single effector cell may be supplied by several axons, cholinergic and/or adrenergic. A single axon may act on several different types of effector cell; although there are separate sympathetic axons for secretion and vasoconstriction (Emmelin and Engström 1960), those causing secretion may also induce myoepitheliocyte contraction, and a single parasympathetic axon may, through sequential neuro-effector junctions of the en-passant type (p. 826), induce vasodilatation, secretory cell activity and myoepitheliocyte contraction (Emmelin 1972).

The secretory endpieces generally have the most copious innervation. Cholinergic and adrenergic axons may supply them, with individual secretory cells often having a dual innervation. It has long been accepted that the cholinergic axons are secreto-motor, but it was not until 1974 that evidence was published showing unequivocally that, in the rat parotid gland at least, the sympathetic nerves were also secretomotor (Harrop and Garrett 1974; Hodgson and Spiers 1974). Adrenergically evoked saliva differs from that produced after parasympathetic stimulation both in quantity and in composition. How these differences are brought about is a matter for speculation. In heterocrine glands (p. 1276), adrenergic and cholinergic nerves could activate different types of secretory cell, but in homocrine glands (p. 1276), where all the secretory cells appear to be similar, adrenergic and cholinergic axons presumably evoke different responses from the same cells (Garrett 1972). In some situations sympathetic nerve activity may result in the secondary modification of the saliva produced in response to para-sympathetic stimulation, rather than in the induction of salivary flow directly.

The ductal elements of the salivary glands can produce marked changes in the saliva (p. 1278) and, although they are less well innervated than the secretory endpieces, it appears that their activity is controlled, at least in part, by neural influences. In 1958, Lundberg showed that cells, assumed to be ductal cells, of the submaxillary gland of the cat, responded electrically to both parasympathetic and sympathetic stimulation. Cholinergic (parasympathetic) fibres lie adjacent to the striated ducts of most species, occasionally in a hypolemmal position, although this is more common in the intercalary ducts. In a number of species, including man, adrenergic nerves have also been shown to be associated with the striated ducts (Garrett 1967). The main (excretory) ducts were thought to have only cholinergic nerves associated with them but recently Schneyer (1976) has reported that sympathetic stimulation can alter electrolyte transport across the main duct epithelium in the submaxillary gland of the rat, suggesting that here, at least, there is an adrenergic supply.

The innervation of the myoepitheliocytes which overlie the secretory endpieces and the intercalary ducts has proved difficult to resolve from a physiological standpoint. However, electron microscope observations suggest that both sympathetic and parasympathetic hypolemmal fibres may supply them (Kagayama and Nishiyama 1972). Myoepitheliocytes are stimulated to contract by adrenergic axons; they can respond to a single impulse, suggesting that they have a high sensitivity. The role of cholinergic axons is less certain, but there is experimental evidence that they also may cause contraction (Garrett 1975) and thus assist salivary discharge.

There is considerable morphological evidence that the arterioles of the salivary glands are innervated by both adrenergic and cholinergic axons (see Young and van Lennep 1978). The more numerous adrenergic axons are involved in maintaining a state of vasoconstrictor tone. The cholinergic axons may induce vasodilatation, but this is maintained by locally-produced plasma-kinins (vide supra).

There is surprisingly little published information on the afferent innervation of the salivary glands. The pain associated with obstruction of the salivary ducts and even with sialography suggests that nociceptive endings are present, but these await morphological investigation. Sensory endings associated with the main ducts have been described, and presumably such endings are present elsewhere in the glands. Afferent axons are thought to be present in the main parasympathetic and sympathetic nerve trunks of the glands. It has been reported (see Garrett 1975) that increasing the pressure in the submandibular duct of the dog enhances afferent fibre activity in the chorda tympani; intraglandular baroceptors are presumed to be involved in this response. Detailed cytological studies of the sensory innervation of the salivary glands are clearly necessary.

Accessory glands. Besides the main salivary glands, numerous others occur in the buccal cavity. Some of these occur in the tongue (p. 1306); others lie around and in the tonsil between its crypts, and large numbers are present in the soft palate, the posterior part of the hard palate, the lips and cheeks. These glands are of the same structure as the larger salivary glands and are mainly of the mucous type (see also p. 1272).

Sialography. A cannula can be introduced into the openings of the parotid and submandibular ducts and a substance opaque to X-rays (such as lipiodol) injected into the duct systems of these glands. The normal pattern and calibre of these systems, or their obliteration or dilatation by disease, can then be revealed by radiography. The *parotid duct*, as seen in sialograms (lateral view), is formed about the middle of the posterior border of the ramus of the mandible by the union of two ducts, of slightly smaller calibre, which pass upwards and downwards, respectively, to join the main parotid duct at right angles to its course (8.40A). As it runs across the face, the parotid duct receives from above five or six very small ductules from the accessory parotid gland, and as it bends inwards at the anterior border of the masseter, it is often compressed by that muscle so that the shadow of the lipiodol is considerably attenuated here. The intraglandular part of the main duct receives an alternating series of descending and ascending tributaries. Each of the main tributaries of the duct is formed in turn from an arborization of fine ductules, which are the terminal ducts and acini. The latter normally show no dilatation but are represented in sialograms by the tiny endings of the smallest ducts. The *duct of the submandibular gland* commences from the lowest part of the gland (below the lower border of the body of the mandible as seen in lateral views), passes vertically upwards to just above the lower border of the body of the mandible, and here turns sharply forwards and gradually ascends to its opening. The vertical part of the duct receives tributaries on both its anterior and posterior aspects and, as it turns sharply forwards, it receives a large tributary from the posterior part of the gland (8.40A). Each tributary is formed from an arborization of ductules (with their terminal acini) which have the same appearance as that described in the case of the parotid. Sometimes when a contrast medium for radiography, such as lipiodol, is injected into the submandibular duct, it passes also into the major sublingual duct, and the ductules of the anterior part of the sublingual gland are then revealed in the sialogram.

THE TEETH

Introduction

Teeth are of vital importance to nearly all animals except man. Indeed, the loss of them is incompatible with life, and in many mammals longevity is directly related to the time for which the dentition can withstand the very abrasive process of mastication. In non-mammalian vertebrates the teeth are constantly being replaced throughout life, a condition known as *polyphyodonty*. This is probably related to the fact that many such forms grow throughout life, and larger replacement teeth, more commensurate with the increasing size of the animal, are constantly required. Thus, tooth replacement may be primarily a reflexion of a growth process, and only secondarily related to the maintenance of the dentition against wear and tear.

Limitation of the number of replacement teeth is rare in non-mammalian dentitions; the condition of *diphyodonty* (two dentitions—a *deciduous*, so-called 'milk', and a *permanent* dentition) – is almost diagnostic of a mammal. Some mammals, for example the rat, are *monophyodont*.

The emergence and success of diphyodonty was probably related to the evolution of the condition in which upper and lower teeth meet during mastication. In non-mammalian vertebrates the jaw joint is formed between the *quadrate bone* of the upper jaw and the *articular bone* of the lower jaw, homologous with the incus and malleus of mammals respectively. The lower teeth are set on a curve which lies so far inside the upper teeth that, together with the very restricted lateral movement which can be produced by the jaw muscles, they cannot be moved into a position where they meet the upper teeth. The phylogenetic history of mammals shows that their evolution was closely linked to a posterosuperior growth of the *dentary*, one of several lower jaw bones present in all bony non-mammalian vertebrates, towards the *squamosal bone*, which is homologous with the squamous part of most mammalian temporal bones. At the same time the jaw muscles were rearranged with the result that the mandible (dentary) could be moved more laterally and medially. Together with these trends there was a change in tooth morphology. From the simple conical teeth of their reptilian ancestors, the mammals evolved teeth with complex shearing planes. The wide lateral movements which were possible with the (new) mandible allowed the shearing edges of the lower jaw teeth to grind across those of the upper jaw teeth, with the result that the food could be more effectively comminuted.

It has been argued that in those reptiles ancestral to mammals, the articular and quadrate bones of the jaw joint conducted to the stapes of the ear sound vibrations, which were received by a tympanum which was partly attached to the *angular bone* of the lower jaw (Allin 1976). The backward growth of the dentary had the selective advantage that it reduced the mass of these post-dentary bones which vibrated with the tympanum, thereby enhancing their auditory functioning. Finally, with the evolution of the new jaw joint, the angular bone with its tympanum moved to the cranium, and became the tympanic part of the temporal bone. The quadrate and articular bones remained in contact with each other and with the stapes, and formed the incus and malleus, respectively. The three auditory ossicles then became encompassed within the definitive tympanic cavity. The evolution of the single-boned mammalian lower jaw was thus associated with an increasingly efficient auditory apparatus.

After a few months of active use the newly erupted teeth of a mammal become precisely worn down so that the upper and lower shearing edges are neatly matched. Under such conditions the continued eruption of new teeth would constantly disrupt these matching edges, and therefore, there was selective pressure for reducing the numbers of replacement teeth. However, because there was still the necessity of accommodating teeth in the small jaws of the young, a deciduous dentition was, and is, nearly always maintained in mammals. Thus the deciduous molars, situated in a position of maximum mechanical advantage close to the muscles of mastication, have a shape like that of the permanent molars which at a later age erupt into the mechanically equivalent posterior position in the older, longer jaws.

With the reduction of the number of replacement teeth, tissues

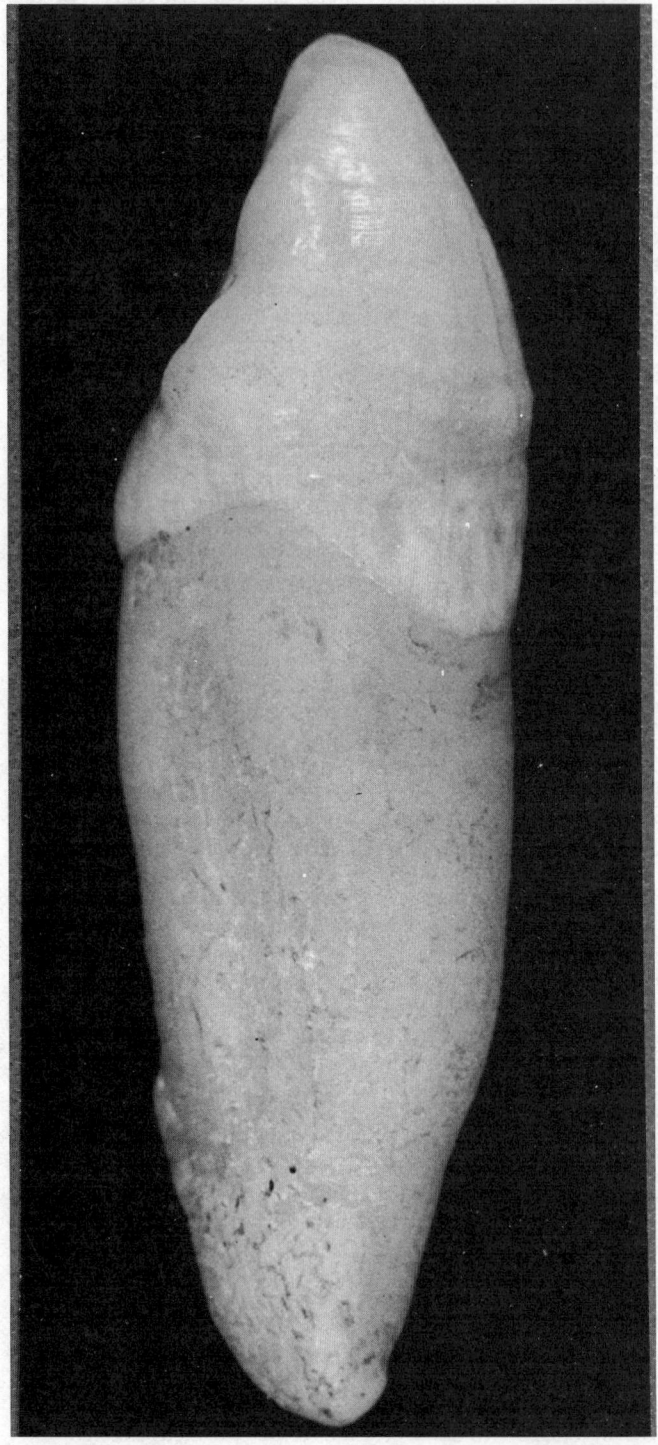

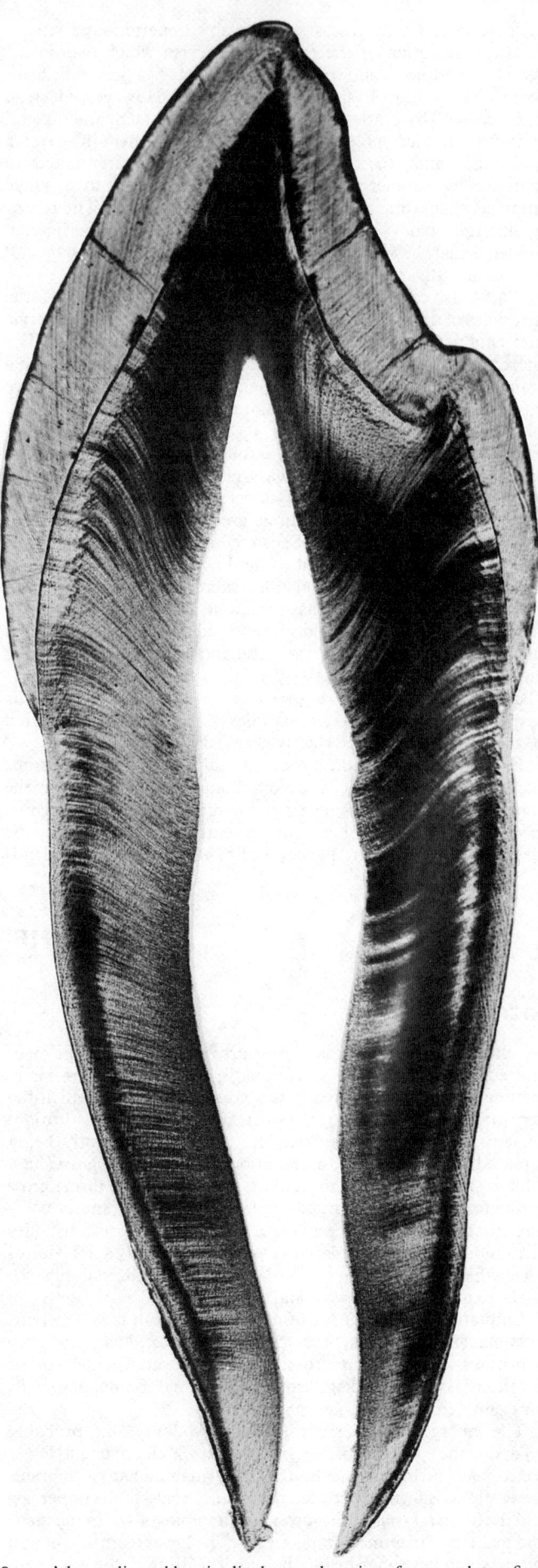

8.41 An extracted upper right canine tooth viewed from its mesial aspect. Note the root covered by cement, and the shiny enamel-covered crown. The sinuous cervical margin projects towards the cusp of the tooth both mesially and distally.

8.42 A bucco-lingual longitudinal ground section of a young lower first premolar tooth, photographed with transmitted light. Note the large buccal and small lingual cusps; the enamel tapers to a knife-edge at the cervical margin. The coarse dark lines perpendicular to the enamel surface are cracks produced during grinding; the fine lines parallel to the

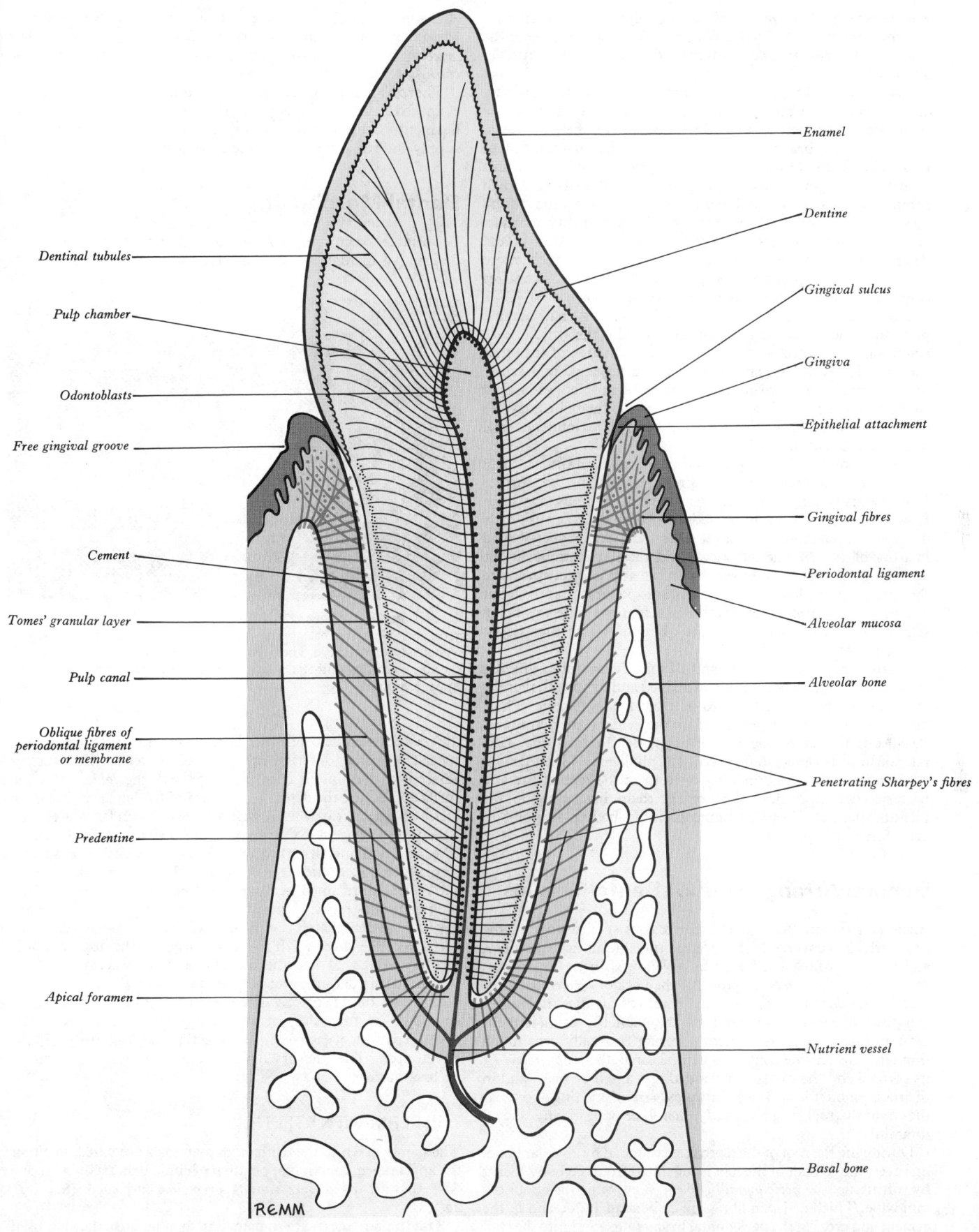

Enamel

Dentine

Gingival sulcus

Gingiva

Epithelial attachment

Gingival fibres

Periodontal ligament

Alveolar mucosa

Alveolar bone

Penetrating Sharpey's fibres

Nutrient vessel

Basal bone

Dentinal tubules

Pulp chamber

Odontoblasts

Free gingival groove

Cement

Tomes' granular layer

Pulp canal

Oblique fibres of
periodontal ligament
or membrane

Predentine

Apical foramen

REMM

cracks indicate the long axes of the enamel prisms; the oblique lines which curve towards the apex of the cusp are the striae of Retzius (the incremental lines of enamel—compare with **8**.55). The primary (S-shaped) curvatures of the coronal dentinal tubules become less pronounced in the root. A thin layer of cement covers the dentine. The

apical foramen is wide open and the pulp cavity is large, in this young tooth.

8.43 (above) Diagram of a longitudinal section of a similar tooth *in situ* and of its environs.

evolved which minimized the effects of tooth wear. Thus, a harder and thicker *prismatic enamel* emerged in mammals to replace the thinner *non-prismatic enamel* covering the teeth of their reptilian ancestors.

The teeth of all reptiles, apart from the crocodiles, are rigidly anchored to the jaw bone by cement, but in mammals the teeth are suspended in sockets by a relatively soft tissue, the *periodontal ligament* or membrane. This allows slight individual movement impossible in the ankylosed teeth of reptiles. The independent suspension permits slight movements of the teeth which compensate for wear, and permit a tooth to be moved by masticatory or other forces into what is presumably its most effective position in the jaw. Appropriate innervation of the deformable periodontal tissues also provides a much more comprehensive flow of proprioceptive data to the nervous centres concerned with the control of masticatory patterns.

With the manufacture of refined carbohydrates, man became prey to caries and periodontal disease. In the wild, these conditions would probably have led to the extinction of any other species. However, by appropriately preparing and, in a sense, predigesting his food, man can overcome the problem of passing suitably comminuted food to the action of the catabolic enzymes of the alimentary canal. The teeth are therefore no longer of vital importance in sustaining life.

It seems probable that, provided civilization continues, there will be only limited selective pressures leading to further evolutionary changes in the dentition of man as a whole. However, there are several variations between the dentitions of different populations. Amongst these are the shovel-shaped incisors of the Mongoloid races, and racial variations in the percentage of upper first molars having an accessory cusp, and in the absence of the lower third molars. It is possible that the latter is a relic of an ancestral population which at one time was being selected for its shortened dentition, but for which there is no longer any selective pressure.

Because they are the hardest and chemically most stable tissues in the body, teeth are selectively preserved and fossilized, thereby providing by far the best record of evolutionary change. Thus, many evolutionary data are based on the evidence of dentitions. Because of this, teeth provide excellent models for the study of the relationship between ontogeny and phylogeny. Further, in modern societies, the durability of teeth in the face of fire and bacterial decomposition often makes them invaluable in the identification of otherwise unrecognizable bodies—a point of great forensic importance.

General Arrangement of Dental Tissues

An extracted tooth (**8.41**) can be seen to consist of two parts—the *crown* which is covered by a very hard translucent tissue, *enamel*, and the root which is covered by a yellowish bone-like tissue, *cement*. The root meets the crown at the neck or *cervical margin*.

*A longitudinal section of a tooth (***8.42***)* reveals that the bulk is composed of *dentine* (ivory) and that the enamel covering is about 1·5 mm thick, while the cement covering is usually very much thinner. The dentine contains a central canal, the *pulp cavity*. At its coronal end the cavity is expanded into a *pulp chamber* and in the root region it is narrowed into a *pulp canal* opening at or near the tip of the root into an *apical foramen*, or occasionally several foramina.

During life the root of the tooth is surrounded by alveolar bone; however, the cement of the root is separated from the bony socket by soft tissue, the *periodontal ligament* (**8.43**) which is about 0·2 mm wide. Thick collagen fibres are embedded at one end in the cement and cross in the periodontal ligament to enter into the wall of the bony socket at their other ends. In most non-mammalian vertebrates the teeth are rigidly cemented directly to the bone providing a rather brittle attachment. Only in mammals (and crocodiles) is there a periodontal ligament which provides an independent and tough suspension for each tooth. Towards the cervical margin (the neck of the tooth), the periodontal ligament is covered by the *gingiva* (or gum) (**8.43**). This covering tissue can be clearly recognized in the healthy mouth because of its pale pink

and stippled appearance (**8.44**): it is continuous with the red, smooth and shiny lining of much of the oral cavity, the *oral mucosa*, and is attached near the cervical margin of the tooth by the *epithelial attachment*.

The *pulp*, from which the pulp cavity takes its name, is a connective tissue, directly continuous with the periodontal ligament via the apical foramen. It carries a vascular nutritive supply and sensory nerves to the dentine.

Dental Morphology

Because the dental arches are curved, the ordinary terms of descriptive anatomy such as anterior and posterior, if applied to

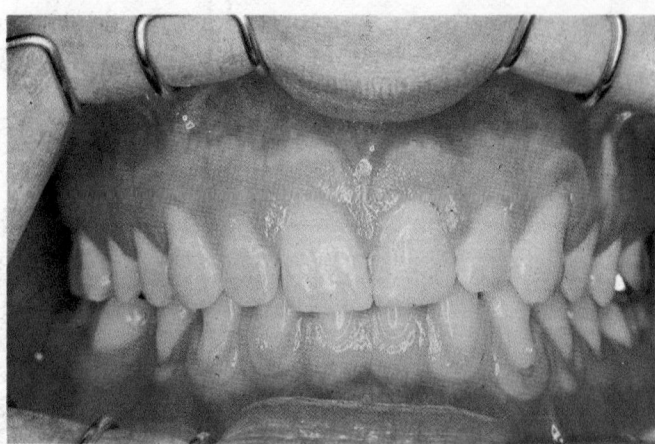

8.44 Anterior view of the dentition in centric occlusion, with the lips retracted. Note the pale pink, stippled gingivae, and the red, shiny, smooth alveolar mucosa. The degree of overbite is rather pronounced, and the gingiva and its epithelial attachment have receded on to the root of the upper left canine.

the teeth, would be confusing. Dental anatomists, therefore, use special terms to describe the surfaces of teeth. The surface adjacent to the lips or cheeks is known as the *labial* or *buccal* surface. (Because the dental arch is curved the labial surface of an incisor faces anteriorly while that of a cheek tooth faces laterally.) The surface adjacent to the tongue is the *lingual* or *palatal* surface. The labial and lingual surfaces of an incisor meet its *mesial* surface medially and its *distal* surface laterally. These two terms, mesial and distal, are retained to describe the equivalent surfaces of the premolar and molar—'cheek' or *postcanine* teeth. However, because the dental arch is curved it will be noted that the mesial surfaces of the cheek teeth face anteriorly and the distal surfaces face posteriorly. Thus, the median point between the central incisors is the datum point for the terms *mesial* (proximal) and *distal*. The biting, or *occlusal*, surfaces of the cheek teeth are composed of tubercles, or *cusps*, separated by fissures. The occlusal pattern of each tooth is characteristic of its position in the dental arch. The incisors have an *incisal edge* rather than an occlusal surface.

THE PERMANENT TEETH

The names given to the teeth of all mammals are based on either the appearance, function or position of equivalent teeth in human jaws: they are the *incisors*, *canines*, *premolars* and *molars* (**8.45**, 46, 47, 48).

The incisor teeth are so named in man because they are used to incise food. The biting or incisal edges originally have three small cusps but these are rapidly smoothed by wear. There are two incisors in each jaw quadrant, a central and a lateral incisor. In buccal view the crowns are roughly trapezoid, the maxillary, particularly the central incisor, being considerably larger than the mandibular. In mesial or distal view, the labial profiles are convex; the lingual surfaces are concavo-convex (ogival) (**8.43**), the convexity towards the cervical margin being provided by

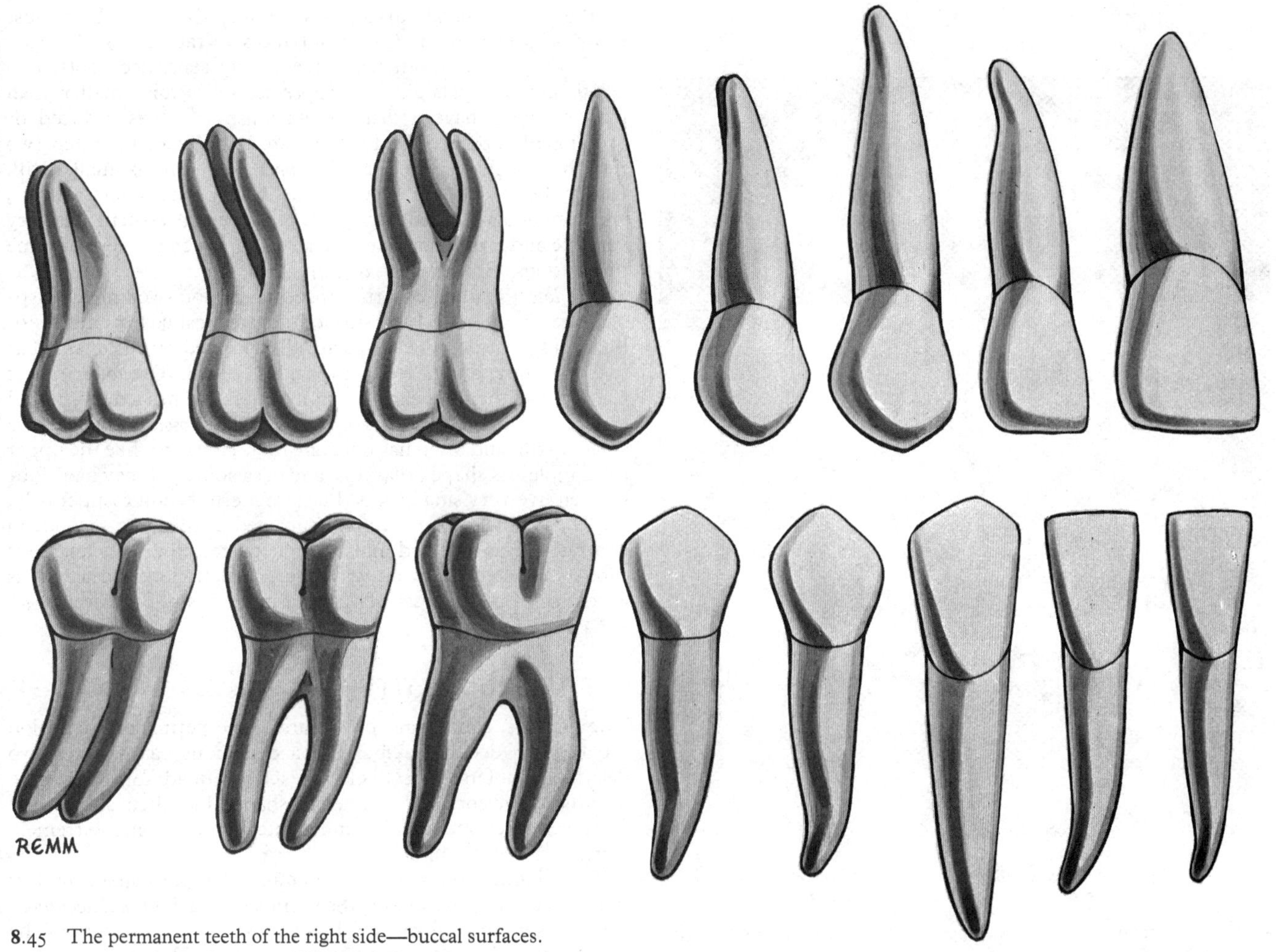

8.45   The permanent teeth of the right side—buccal surfaces.

a low horizontal *ridge* or *cingulum*, which is prominent only on the upper incisors. The roots of the maxillary incisors are rounded while those of the mandibular are flattened mesio-distally. The maxillary lateral incisor may be congenitally absent in about 2 per cent of the population (Brothwell *et al.* 1963), alternatively it may be peg-shaped.

Distal to each lateral incisor is a **canine tooth**, which is rather larger, and each has a single cusp (hence the American term *cuspid*), instead of an incisal edge. Apart from this they are not unlike the incisors, particularly when, in old age, their cusps have been worn flat. The canine roots are the longest in the jaws, and each bulges the associated bone externally, and this can be readily palpated on the face. Occasionally the lower canine may have two roots (Kraus *et al.* 1969).

Distal to each canine are two **premolars**, each with a buccal and a palatal cusp (**8**.42) enclosing an occlusal surface (hence the American term *bicuspid*). The second premolars may be congenitally absent in about 2 per cent of individuals (Garn and Lewis 1962). The occlusal surfaces of the upper premolars are oval (the long axis is bucco-palatal), and a mesiodistal fissure separates the buccal from the palatal cusp. In buccal view the premolars have a shape similar to that of the canine, but they are a little smaller. The *upper first premolar* generally has two roots, one buccal and the other palatal, but may have one root and very occasionally three roots (two buccal and one palatal), the condition in apes. The *upper second premolar* generally has one but may have two roots (one buccal and one palatal).

The occlusal surfaces of the lower premolars are more nearly circular or square than those of the uppers. The buccal cusp of the *lower first premolar* towers over the lingual cusp, to which it is connected by a ridge; the latter separates mesial and distal occlusal pits. In the *lower second premolar* a mesio-distal fissure usually separates the buccal cusp from two poorly separated lingual cusps. Both lower premolars have one root, but very occasionally the root of the first premolar is bifid.

Posterior to each second premolar are three **molars** whose size progressively decreases posteriorly; each has a large rhomboid (upper jaw) or rectangular (lower jaw) occlusal surface, with four or five cusps. The *upper first molar* carries a cusp at each corner of its occlusal surface, the mesio-palatal cusp being joined to the disto-buccal cusp by an *oblique ridge*—a primitive feature which man shares with many lower primates. A fifth cusp may be present

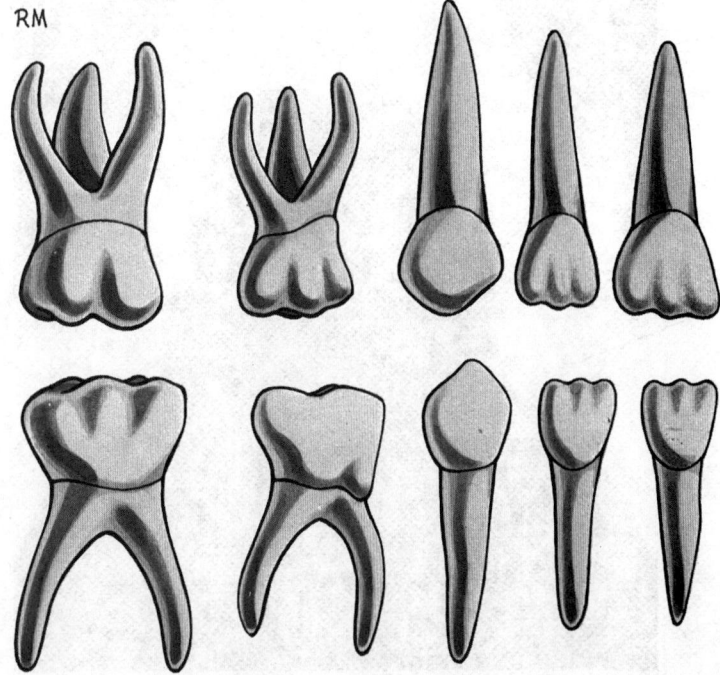

8.46   The deciduous teeth of the right side—buccal surfaces.

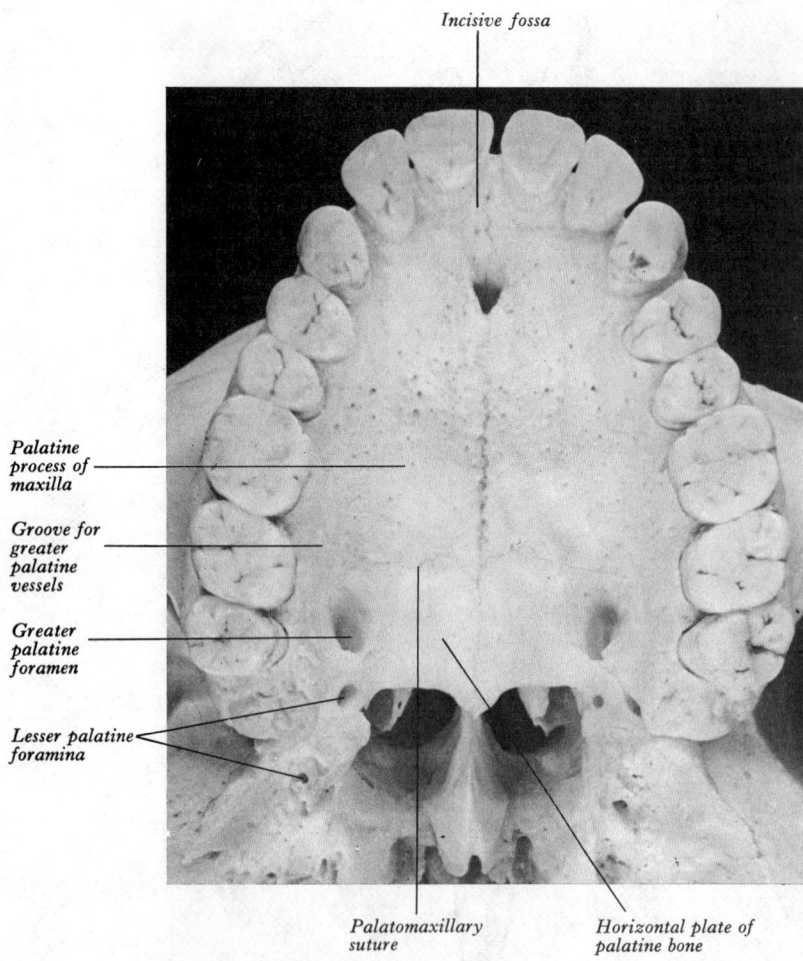

8.47 The permanent teeth of the upper dental arch—inferior aspect.

Incisive fossa

Palatine process of maxilla

Groove for greater palatine vessels

Greater palatine foramen

Lesser palatine foramina

Palatomaxillary suture

Horizontal plate of palatine bone

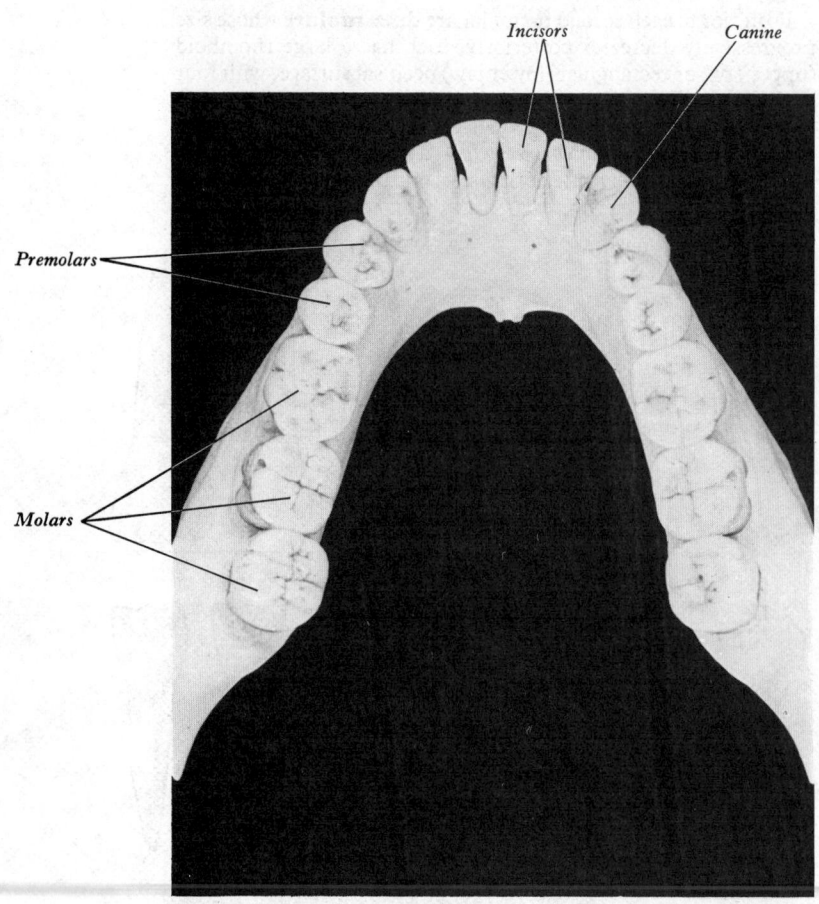

Incisors

Canine

Premolars

Molars

8.48 The permanent teeth of the lower dental arch—superior aspect.

on the mesio-palatal surface—the *cusp of Carabelli*. It is most commonly present in Caucasoid races (Kraus 1959, Alvesalo *et al.* 1975). The tooth has three widely separated roots, two labial and one palatal. The *upper second molar*, smaller than the first, may have a disto-palatal cusp which is reduced or occasionally absent, and its roots are less divergent; often two roots may be fused together. The *upper third molar* is smaller still, usually bears only three cusps (the disto-palatal being absent), and commonly has only one root. However, its structure is very variable and it is congenitally absent in up to 25 per cent of some populations (Brothwell *et al.* 1963).

The *lower first molar* carries three buccal and two lingual cusps on its rectangular occlusal surface, the smallest cusp being disto-buccal in position. The cusps are all separated by fissures. It has two widely separated roots, mesial and distal. The *lower second molar* is like the first, except that it very rarely has a disto-buccal cusp, and its roots are closer together. The *lower third molar* is smaller still, and often has only one root. However, like the upper third molar its shape is variable and occasionally it may have four or even five very small roots. Because it erupts anterosuperiorly, the *lower* third molar is often impacted against the second; in contrast the *upper* third molar erupts posteroinferiorly and so is rarely impacted. Varying with the population concerned, it is absent in from 0·2 to 25 per cent of individuals (Brothwell *et al.* 1963).

THE DECIDUOUS TEETH

The incisors, canine and premolars of the permanent dentition replace two deciduous incisors, a deciduous canine, and two deciduous molars respectively, in each jaw quadrant (**8**.46). The deciduous incisors and canine are shaped like their successors. However, they are smaller and whiter, and become extremely worn in the older child.

The deciduous molars resemble the permanent molars rather than their successors, the premolars. Each *second deciduous molar* has a crown form which is almost identical to that of the posteriorly adjacent first permanent molar. However, the *first upper deciduous molar* has a somewhat triangular occlusal surface (the very rounded 'apex' being lingual), and a fissure separates a double buccal cusp from the palatal cusp. The *lower first deciduous molar* is long and narrow with a central mesio-distal fissure separating long, serrated, buccal and lingual cusps.

Like the permanent molars, the upper deciduous molars have three roots, and the lowers have two roots. However, the roots are more divergent than those of the permanent teeth, each developing premolar being temporarily accommodated directly beneath the crown of its deciduous predecessor.

The roots of the deciduous teeth are progressively resorbed by osteoclasts prior to their being replaced by the permanent teeth. Thus, an extracted deciduous tooth may have very short roots.

DENTAL OCCLUSION

It is possible to bring the jaws together so that the cheek teeth meet, or *occlude*, in innumerable ways (Kraus *et al.* 1969). When the opposing occlusal surfaces meet in such a position that there is maximum 'intercuspation' (i.e. maximum contact between the teeth of opposing arches), the teeth are said to be in *centric occlusion* (**8**.49, 50). In this position the lower teeth normally lie symmetrically just lingual to the upper teeth. The following are a few important features of centric occlusion in a normal dentition. Each lower cheek tooth is slightly in front of the equivalent upper cheek tooth. In the same way the lower canine is in front of the upper canine. The buccal (lateral) cusps of the lower cheek teeth lie between the buccal and palatal cusps of the upper cheek teeth. Thus the lower cheek teeth are slightly lingual and mesial to the equivalent upper teeth. At the front of the jaw the lower incisors bite against the lingual surfaces of the upper incisors, the latter normally obscuring about one-third of the crown length of the lower incisors. The amount by which the upper incisors cover the lower incisors in centric occlusion is called the amount of *overbite*. The extent to which the upper incisors lie anterior to the lower incisors is the extent of *overjet*.

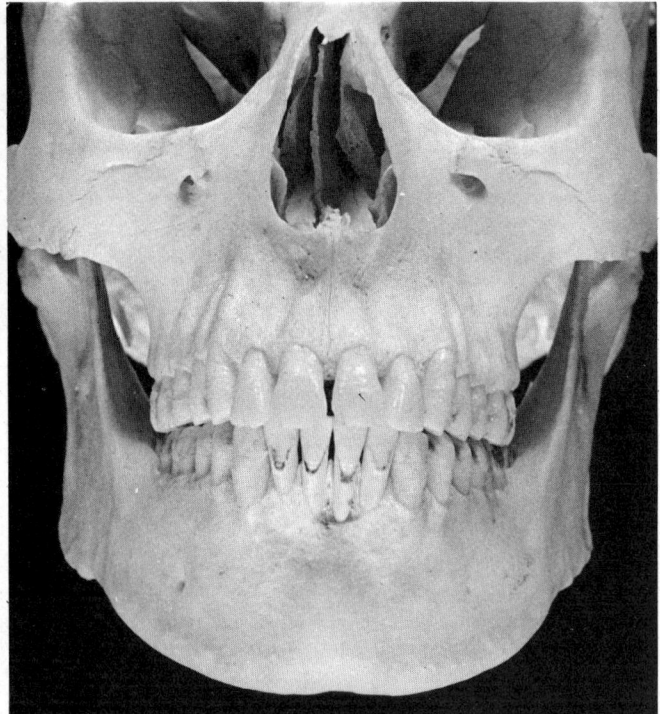

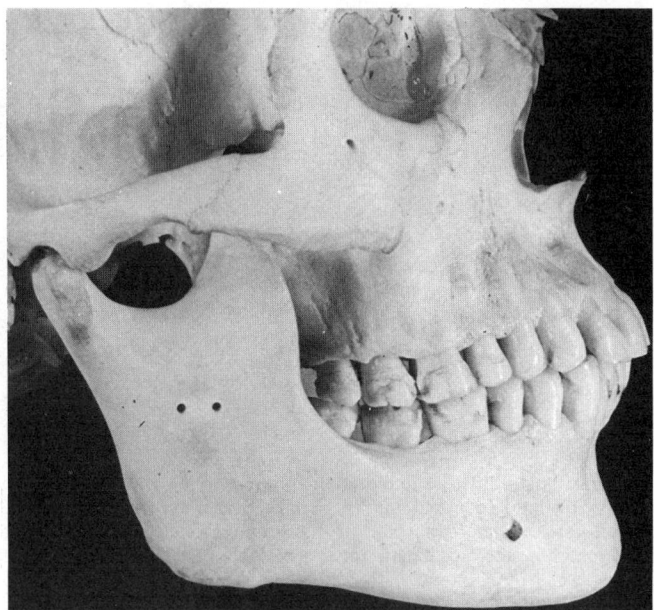

8.50   Lateral view of the dentition in centric occlusion.

8.49   Anterior view of the dentition in centric occlusion. There has been some resorption of bone around the lower incisors.

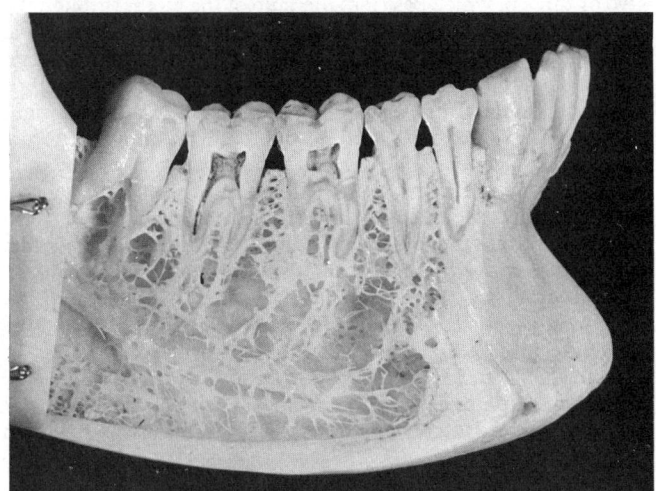

In the most habitual jaw posture, the teeth are normally held slightly apart from that of centric occlusion. The resulting space between the teeth is called the *freeway space* or *interocclusal clearance.*

Each dental arch approximates to a *catenary shape* (i.e. the form assumed by a dependent chain attached at its two extremities)— *see* MacConaill and Scher (1949)—the lower arch being slightly narrower than the upper (**8**.47, 48). When viewed from the side, a line joining the buccal cusps of the lower (or upper) cheek teeth is curved (the *curve of Spee*), the concavity being upwards (Kraus *et al.* 1969). The lower cheek teeth are tilted slightly lingually, so that a line joining the buccal and lingual cusps of the lower left and right first molars (for instance) is curved (*curve of Monson*) with its concavity upwards. The upper cheek teeth have an equivalent tilt outwards. The tilt of each tooth represents to some extent the direction in which the tooth originally erupted. The above curves are of importance in the construction of dentures.

DENTAL BLOOD AND LYMPHATIC VESSELS

The *inferior alveolar artery* is a branch of the maxillary artery. It enters the mandibular foramen and then passes anteriorly in its bony canal, to divide into its *incisive* and *mental* branches, for the supply of the lower teeth, their supporting structures and the body of the mandible, including its cortical bone (Saunders and Röckert 1967). About eight to twelve main channels and a variable number of finer channels supply the alveolar bone together with the teeth (Castelli 1963). There appear to be very few anastomotic channels across the symphyseal region (Howkins 1935). Veins from the alveolar bone and teeth collect either into large veins in the interdental septa or into networks of veins which surround the apex of each tooth, and from thence pass into one of several *inferior alveolar veins*. Some of these drain anteriorly through the mental foramen to join the facial vein, others pass back through the inferior dental foramen to join the pterygoid plexus of veins (Cohen 1959).

The upper jaw is supplied by the *anterior* and *posterior superior alveolar arteries*. The latter branches, from the maxillary artery, course tortuously over the maxillary tuberosity dividing into small stems which supply the alveolar bone, mucosa and teeth of the molar region, together with the adjacent buccal mucosa, where there is an anastomosis with penetrating branches of the

8.51   A vertical section through the right half of the body of the mandible and its dentition. Note: (1) the pulp cavities in the molar teeth; (2) the flat table of bone surmounting the interdental bony septa; (3) the cancellous nature of much of the bone; (4) the cortical plate of compact bone lining the sockets of the teeth (the lamina dura of radiographs—see **8**.71, 72); (5) the compact bone forming the base of the mandible; (6) the inferior dental canal which, in this specimen, is widely separated from the roots of the teeth.

facial artery. Other branches supply the lateral wall of the maxillary sinus. The *anterior superior alveolar artery* is a branch of the infraorbital artery. It takes a curving course through the *canalis sinuosus* (Jones 1939). The canal passes laterally from the infraorbital canal and then downwards and medially below it within the wall of the maxillary sinus. It now follows the curve of the margin of the nasal cavity between the sockets of the canine and incisor teeth and the nasal cavity, to end beside the nasal septum where its terminal branch emerges. The canal may be up to 55 mm long. Occasionally there is a small *middle superior alveolar artery* which forms anastomotic arcades with the anterior and posterior superior dental arteries.

On the palatal side of the upper teeth, the *greater palatine artery* gives branches to the palatal gingivae. Unlike the nasopalatine nerves, the nasopalatine artery does not extend through the incisive canal into the palate.

The veins accompanying the superior alveolar arteries either drain anteriorly to join the facial vein, or posteriorly to the pterygoid venous plexus.

A

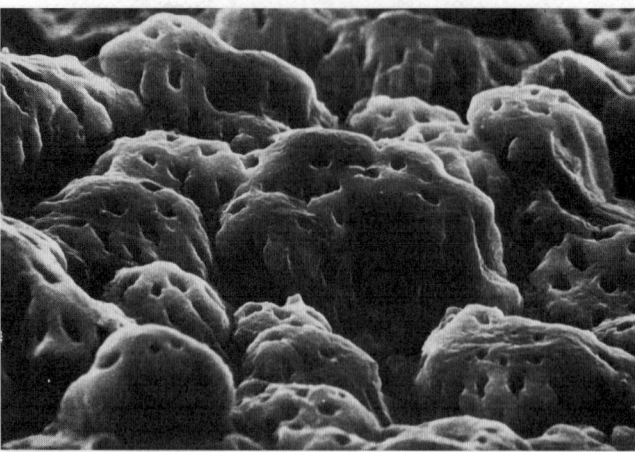

B

8.52A Longitudinal ground section of a tooth viewed by transmitted light showing the dentinal tubules, which appear as dark lines. One of the tubules shows lateral branches. B: A scanning-electron micrograph of the dentine/predentine junction after removal of predentine to expose the

surfaces of calcospherites. The holes (about 1 μm wide) are dentinal tubules. (Kindly provided by Dr. D. Whittaker, The Dental School, University of Wales, Cardiff.)

The periodontal ligaments are supplied by the *dental branches* of the alveolar arteries (Melcher and Bowen 1969). One branch enters the periodontium apically and gives off smaller branches, two or three of which pass into the pulp of the tooth, through the apical foramen, and others which ascend the periodontal ligament. *Interdental arteries* ascend in the interdental septum, ultimately supplying the epithelial attachment and the gingiva. The interdental arteries also give branches which pass at right angles through the bone forming the tooth socket and into the periodontal ligament. Thus, the latter receives its arterial supply from three sets of arteries: one enters from the apical region, a second from the ascending interdental arteries, and a third descending from the crest of the interdental septum. These sets anastomose with each other.

Veins drain the periodontal ligament either into the *interdental veins* or into a *periapical plexus of veins*. It seems probable that the longer vessels seen in the periodontal ligament are veins rather than the anastomosing arteries (Folke and Stallard 1967).

There is little precise knowledge of the lymphatic drainage of the jaws and teeth of man (Saunders and Röckert 1967). Injection

techniques applied to the monkey (MacGregor 1936) have shown that the upper jaw drains mainly towards the submandibular lymph nodes and thence to the supraclavicular nodes. The lower jaw drains mainly towards the submental group and thence to the paratracheal nodes.

Because many dental abscesses in man lead to enlargement of the submandibular and upper deep cervical lymph nodes, this must be considered the common path for lymphatic drainage of both upper and lower teeth. Occasionally there is a buccal lymph node within the cheek and this may be enlarged following infection of the upper cheek teeth. The lower incisors drain to the submental nodes and thence either posteriorly to the submandibular or postero-inferiorly to the lower deep cervical nodes. It is presumed that the alveolar bone, periodontal ligament or membrane, and gingivae share the same lymphatics as the teeth.

DENTAL INNERVATION

The upper teeth are supplied by the *anterior* and *posterior superior alveolar nerves* together with an inconstant *middle superior*

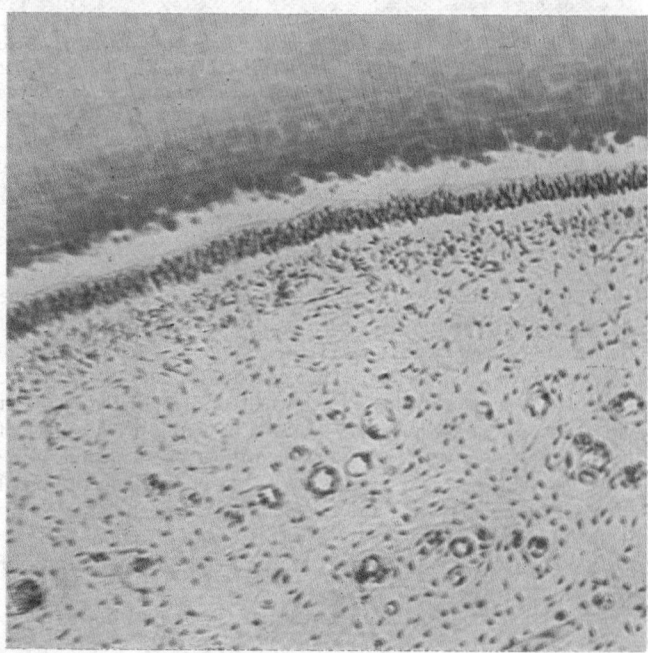

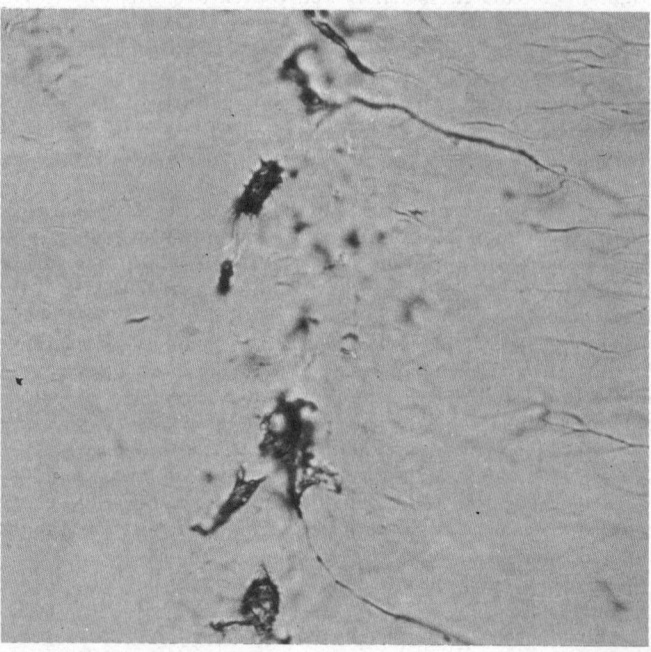

8.53 A transverse section of a demineralized tooth showing pulp below, followed sequentially above by: (1) the layer of odontoblasts, whose stained nuclei show as a purple line; (2) the pale predentine, which adjoins (3) the darkly stained calcospherites which border the dentine at the top of the field.

8.54 Longitudinal ground section of a tooth showing dentine (on the right), the dark spaces of the granular layer of Tomes (centre), and the structureless acellular layer of cement (on the left). c. ×2,000.

alveolar nerve (Fitzgerald and Scott 1958). The teeth are supplied by a plexus which extends from the posterior to the anterior wall of the maxillary antrum, lying partly on the posterior surface of the maxilla and partly in bony canals in the lateral and anterior surfaces of the bone. The *buccal nerve* provides a variable contribution to the buccal molar gingivae, and the *greater palatine* and *nasopalatine* nerves pass to the palatal gingivae, communicating in the region of the canine tooth.

The posterior superior alveolar nerves consist of two or three trunks which arise from the postorbital section of the maxillary trunk. They divide to form a leash of up to eight elements which are invested by the dense periosteum covering the posterior surface of the maxilla. These nerves enter several posterior dental foramina, postero-inferior to the commencement of the infra-orbital groove. The higher branches descend outside the mucous membrane of the antrum to meet the lower branches as they pass forwards above the molar teeth. The variable middle superior alveolar nerve may arise at any point from the infraorbital extension of the maxillary nerve; the anterior superior alveolar nerve runs in the *canalis sinuosus* already described (p. 1287). The bony canals which contain the superior alveolar nerves also contain the corresponding arteries, and it is appropriate therefore to refer to *superior alveolar neurovascular bundles*.

The lower jaw and alveolar bone are largely supplied by branches of the *inferior alveolar nerve* together with branches of the *buccal nerve* to the buccal gingiva of the molar and premolar teeth, and branches of the *lingual nerve* to the lingual gingiva of all the lower teeth. All, of course, are derived from the mandibular division of the trigeminal nerve.

In its simplest and most common form (six out of eight mandibles studied—*see* Carter and Keen 1971) the inferior alveolar nerve is a large trunk travelling through a well-defined osseous mandibular canal close to the roots of the teeth, and giving discrete branches to the teeth and interdental septa. In the region between the premolar teeth, the mental nerve, often multiple, exits via the mental foramen. A small intra-osseous *incisive nerve* continues to supply the first premolar, canine and incisor teeth. From the mental nerve bundle, branches pass anteriorly to form an *incisor plexus* buccal to the teeth. This probably supplies the buccal periodontium and gingiva of these teeth. From the incisor plexus, thick branches curve inferiorly and then lingually to open on to the lingual surface of the mandible opposite the premolar teeth, where they probably communicate with the mylohyoid nerve.

Less commonly (two out of eight mandibles studied), the inferior alveolar nerve lies close to the lower border of the mandible, a considerable distance from the roots of the teeth (**8**.51). From the nerve a variable number of large trunks travel anterosuperiorly towards the roots of the teeth before breaking up into branches to the teeth and interdental septa.

In three out of eight mandibles (in this admittedly small series, which requires amplification) nerves passed from the temporal muscle to enter the mandible through the retromolar fossa and communicated with branches entering the roots of the third to first molar teeth. Buccal to the teeth, fine networks of neurovascular bundles were demonstrated in the molar and incisor regions, the former developed from perforating branches of the bundles to the masseter and temporalis muscles together with the buccal nerve, the latter from branches of the mental nerve (*vide supra*). In the pig (Wedgewood 1966), no fibres of the inferior alveolar nerve have been found supplying teeth on the opposite side of the mandible.

Dental Histology

DENTINE

Dentine (Miles 1967; Symons 1968), which forms the bulk of a tooth, is a hard, elastic, yellowish-white, avascular tissue which is about 70 per cent by weight mineralized (largely by crystalline hydroxyapatite, but some in the form of amorphous calcium phosphates). The remaining 30 per cent is mainly water and collagen, which gives it toughness. Its most characteristic

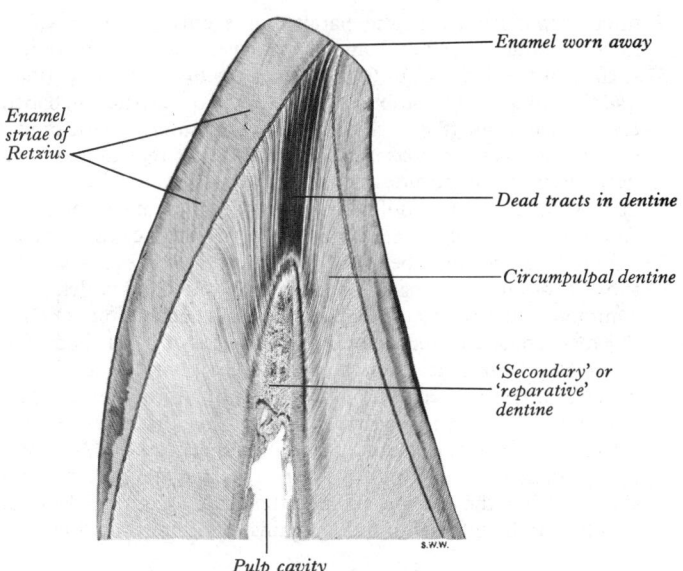

8.55 Longitudinal ground section of an incisor tooth. Compare the brown striae labelled on this section, with those visible on **8**.42.

Enamel worn away

Enamel striae of Retzius

Dead tracts in dentine

Circumpulpal dentine

'Secondary' or 'reparative' dentine

Pulp cavity

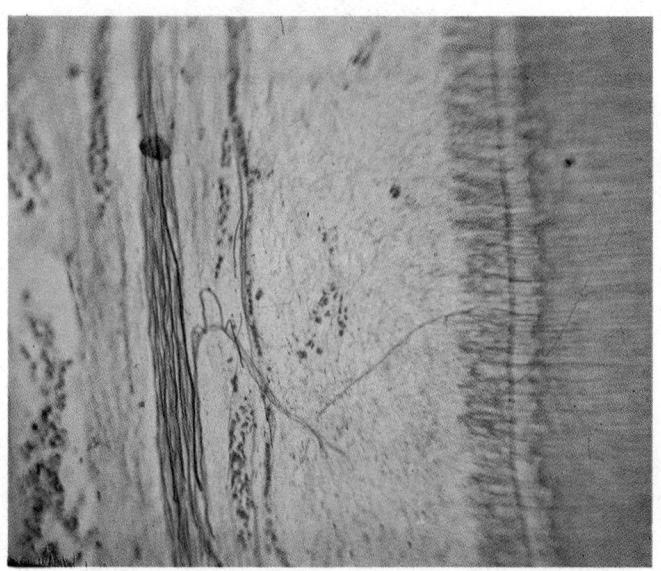

8.56 Longitudinal demineralized section of a tooth stained with a silver impregnation technique. Note the vertical nerve trunk (left of centre) within the pulp, with fine nerve fibres, one of which crosses transversely to pass between the odontoblasts lining the surface of the predentine (the pale-staining vertical layer, right of centre).

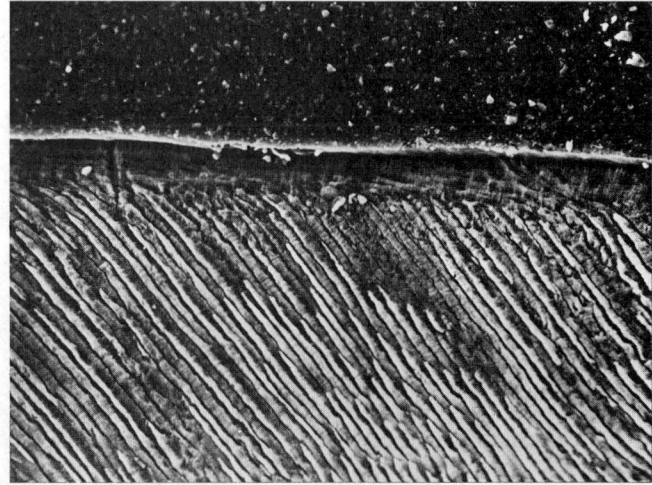

8.57 A scanning electron micrograph of enamel prisms. Each prism is about 5 μm wide and separated from adjacent prisms by interprismatic material which has been removed from this specimen by acid etching. A thick structureless surface layer is present. (Kindly provided by Dr. D. Whittaker, The Dental School, University of Wales, Cardiff.)

microscopic feature is the pattern of evenly spaced *dentinal tubules*, about 1–2 μm in diameter, which extend from the pulpal surface of the dentine out to the *enamel-dentine* (amelo-dentinal) *junction* (**8**.42). The tubules have a characteristic S-shaped *primary curvature* (**8**.42) which is more pronounced in the crown of the tooth. A spiral *secondary curvature* is less regular and has a periodicity and amplitude of a few microns. Towards the enamel-dentine junction the tubules bifurcate and trifurcate, some being continuous with short extensions into the enamel (the *enamel spindles*), particularly beneath the summits of the cusps. The tubules often have many short lateral branches which may communicate with adjacent tubules (**8**.52A). Each tubule contains the membrane-covered cytoplasmic process of a cell body (an *odontoblast*) which is situated in the pulp cavity. The process contains a few mitochondria, lysosomes, microtubules and fine filaments. It is not known whether the processes extend throughout the whole length of a tubule. Recent studies (Holland 1975) suggest that the odontoblast process may only extend a short way into the dentine, and that the remainder of the dentinal tubule may be filled with fluid. Against this interpretation, it is

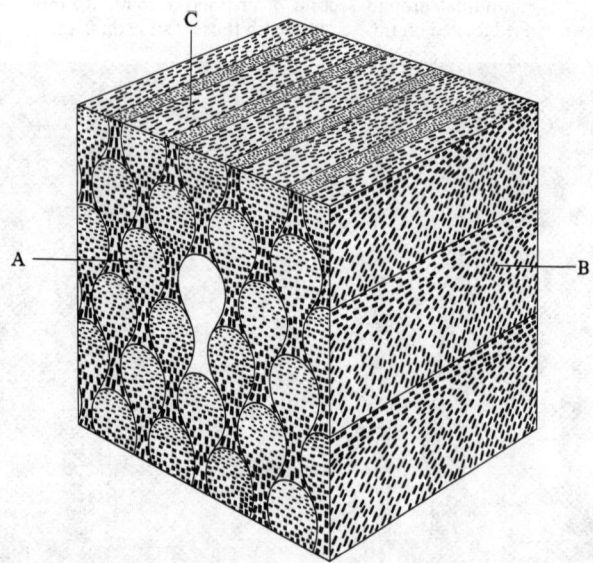

8.58 Diagram showing how the orientation of the crystallites in the enamel determines the appearances when the rods or prisms and the interprismatic substance are cut transversely (face A), longitudinally (face B) and at right angles to both these planes (face C). Note that in this diagram the U-shaped prism sheaths have been arbitrarily extended to meet the subjacent sheaths to produce 'keyhole-shaped' prisms. (Reproduced by kind permission of A. H. Meckel, W. J. Griebstein and R. J. Neal from *Tooth Enamel*, eds., M. V. Stack and R. W. Fearnhead, Bristol, 1965.)

possible that the difficulty of securing adequate histological fixation of the contents of dental tubules may result in their autolytic loss during the preparation of sections for light or electron microscopy. The odontoblasts are arranged as a pseudo-stratified layer of cells lining the pulpal surface of the dentine (**8**.53). Around most tubules there is a very heavily mineralized cylinder of *peritubular dentine* which is devoid of, or has a very sparse, matrix. The remainder constitutes the *intertubular dentine*. The highly mineralized, matrix-deficient, peritubular dentine is lost during the decalcification necessary to produce wax embedded sections. The resultant space seen under the light microscope is, therefore, an artefact.

Throughout life a layer of nonmineralized dentine separates the odontoblasts from the mineralized part; it is called *predentine*. The predentine-dentine border has an irregularly scalloped appearance (**8**.52B). This is because dentine commonly mineralizes in the form of microscopical spheres (*calcospherites*), although mineralization may be linear, as in bone. The enamel-dentine junction appears more regularly scalloped, the convexities facing

the dentine. (This feature is not related to the pattern of mineralization.) The region adjacent to the enamel-dentine junction is less mineralized than the remaining intertubular dentine. It is called *mantle dentine* to distinguish it from the remainder, which is the *circumpulpal dentine*. The collagen fibres in mantle dentine consist predominantly of *Von Korff fibres* arranged parallel to the tubules: those in circumpulpal dentine are arranged at right angles to the tubules.

All the mineralized dental tissues are deposited incrementally and because, unlike bone, dental tissues are rarely remodelled (only cement can be remodelled), each carries a permanent record of its changing shape during development. Dentine has two types of incremental line. In one form the secondary curvatures of adjacent tubules come into phase; this produces a microscopically recognizable *contour line of Owen*. In the other form, junction lines of less mineralized tissue present as the *incremental lines of von Ebner*. The latter are generally much less well marked than the incremental lines in bone. Sometimes there may be slight disturbances during dentine development. These result in the failure of fusion of calcospherites and the production of *interglobular spaces*. They are so common close to the enamel-dentine junction that here they cannot be regarded as pathological.

Immediately internal to the cement, and sometimes indistinguishable from it, is a thin, clear, *hyaline layer* (Owens 1972), followed by the *granular layer of Tomes* (**8**.54). The granularity of the latter layer may either be due to minute interglobular spaces, or it may be due to tiny expansions of the processes of the odontoblasts (Ten Cate 1972).

Throughout life dentine is deposited at an ever-slowing rate. Thus the pulp cavity becomes progressively diminished in size. The dentine produced after a tooth has erupted is called 'regular' *secondary dentine*. It can sometimes be distinguished from the earlier formed, *primary dentine* because of a sudden change in the direction of the dentinal tubules. If the dentine receives a severe stimulus (for example by the attack of caries), an irregular or '*reparative*' *secondary dentine* is laid down (**8**.55). This is a poorly mineralized dentine containing few and irregular tubules, and it may include blood vessels and cell spaces. A less severe stimulus (for example, abrasion of the neck of a tooth due to toothbrush trauma) leads to the sealing of the pulpal ends of the dentinal tubules by a knob of 'reparative' secondary dentine. The region containing the sealed-off tubules is called a *dead tract* (the tubules no longer communicate with odontoblasts).

The peritubular dentine increases in thickness at the expense of the tubule, particularly in the root. Thus in the root region, peritubular dentine gradually obliterates the tubules which now have almost the same refractive index as the rest of the dentine. The result is that in ground sections viewed by transmitted light the dentine appears transparent and is called *translucent dentine*.

THE DENTAL PULP

Within the dentine lies the vascular pulp (Symons 1968), whose surface is covered by odontoblasts; it is continuous with the periodontal ligament and tissue spaces of the alveolar bone via the apical foramen of the tooth. Apart from odontoblasts, the pulp contains all the cells normally found in loose connective tissue (**8**.53).

Several arterioles usually enter the apical foramen and run longitudinally through the tooth giving branches which ultimately form a subodontoblastic capillary plexus in the young tooth (Kramer 1960). The arterioles have remarkably thin walls, possibly related to their inclusion in a small rigid-walled pulp cavity. Several small veins emerge from the pulp. There is a moderate network of lymphatics (Saunders and Röckert 1967; Bernick and Patek 1969), some being perivascular and perineural, others being more isolated trunks.

Non-myelinated postganglionic sympathetic nerve fibres enter the pulp with the arterioles. Myelinated fibres also run longitudinally through the pulp giving branches which ramify in a subodontoblastic *plexus of Raschkow* (Scheinin and Light 1969). Here the fibres lose their myelin sheaths and they continue as beaded axons which pass between the odontoblasts (**8**.56), and

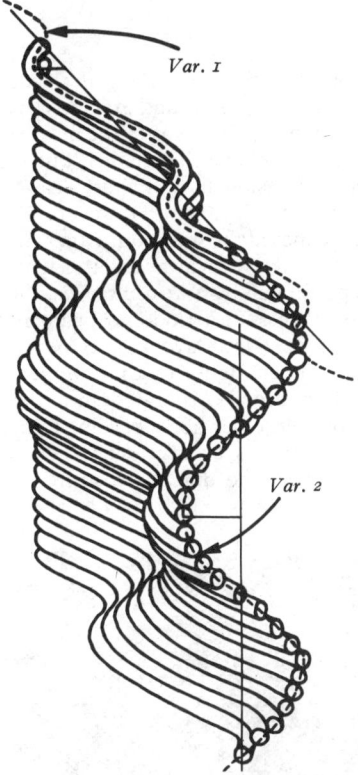

8.59A Diagram illustrating the relationships existing between a vertical stack of enamel prisms. Each prism undulates in the transverse plane of the tooth, but its undulations are out of phase with those of vertically adjacent prisms. Hence, when a section is viewed by reflected light, the undulations are responsible for the characteristic alternation of dark and light bands which cross the prisms obliquely (the Hunter-Schreger bands). Var. 1 and 2 indicate the sine-wave undulations in the transverse and vertical planes, which vary in amplitude and periodicity in the enamel of different species. (From J. W. Osborn in: *Calcified Tissue Research*, **5**, 1970, by courtesy of the author and publishers, Springer.)

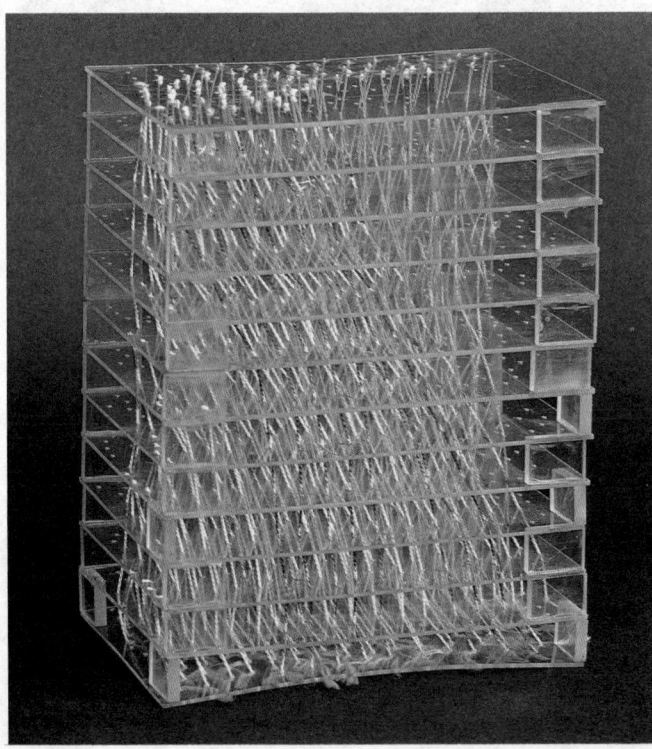

8.59B A reconstruction of enamel prisms in the dog. Each prism is represented by a thread. Note the gradual change in direction between vertically adjacent prisms which are viewed from the cuspal aspect in this photograph. Only a part of one complete prism undulation is shown in each case. Compare with **8.**59A. (Preparation by Dr. J. W. Osborn, Department of Anatomy, Guy's Hospital Medical School.)

may form loops in the predentine (Fearnhead 1967). A few nerve fibres enter the dentinal tubules to lie adjacent to the cytoplasmic processes of odontoblasts for a short distance into the dentine. Although preliminary electron microscopy of this region has been carried out (Symons 1968), the technical difficulties of a comprehensive study of the terminations of such fibres are great. It is assumed that some kind of 'functional connexion' is established with the odontoblast processes; as yet typical synaptic arrangements have not been identified. (For a discussion consult Symons 1968.) It has been implied, therefore, that when dentine is stimulated as, for example, during the cutting of a tooth cavity, some form of impulse may be propagated along the process of the odontoblast and then somehow transmitted on into the nerve fibres; the mechanism is, however, uncertain. Further, this does not explain the sensitivity of newly erupted teeth, which do not yet contain nerve fibres in the dentinal tubules. Also, only about 1 in 2,000 mature tubules contain a nerve fibre, and this appears to be a very sparse innervation for such a sensitive tissue as dentine.

As an alternative to the above hypothesis it is usually suggested that stimuli cause movement of fluid along the dentinal tubules (either intra- or extra-cellular fluid) and that this causes a change in the volume of tissue fluid in the subodontoblastic plexus, which initiates the nervous impulses (Anderson *et al.* 1970). It should be noted that all effective stimuli to either dentine or the pulp are subjectively perceived as pain. (For alternative views of the possible functional roles of fine nerve fibre tissue afferents, *see* p. 891.)

In a recent study (Pimendis and Hinds 1977), labelled proline was injected into the trigeminal ganglion of adult rats, and this was subsequently observed in dentinal tubules close to the enamel—dentine junction, where it had presumably been carried by axoplasmic flow. This observation may re-open the question of dentine sensitivity.

As mentioned above, the pulp cavity diminishes progressively throughout life due to the formation of 'regular' secondary dentine.

THE ENAMEL

Enamel (Fearnhead and Stack 1965, 1971; Miles 1967; Osborn 1973; Boyde 1976), which is the extremely hard translucent tissue covering the crown of the tooth, contains about 96 per cent by weight of inorganic material; this is about 88 per cent by volume. It reaches a thickness of some 2·5 mm over the cusps of teeth and thins down to a knife-edge at the cervical margin.

Enamel is composed of closely packed rods or prisms which are roughly circular in transverse section (**8.**57) extending from the enamel-dentine junction toward the surface of the tooth. Each is separated from the *interprismatic region* by a *prism sheath* which has a U-shape in cross-section, with its convexity towards the tip of the cusp. The concavity of the U-shaped sheath is continuous with the interprismatic region. Prisms are about 3–4 μm wide at the enamel-dentine junction and increase to about 6 μm wide near the tooth surface (Osborn 1968a). For a thickness of about 6 μm the surface enamel is often structureless. The prism sheath is a region of microporosity about 75nm wide. At intervals of about 4 μm along its length each prism is crossed by a dark *striation*. Each 4 μm section is believed to present one day's growth.

The prisms are packed with hydroxyapatite crystallites (**8.**58) which have a preferred orientation depending on their positions in the prisms or interprismatic region. There is a sudden change between crystallite orientation on each side of the prism sheath. However, there is no reason to believe that the interprismatic region and prism body differ in any way, apart from crystallite orientation. Because of this, enamel is sometimes said to be constructed from keyhole-shaped prisms in which the interprismatic region is equivalent to the 'tails' of the keyholes. This difference of opinion is purely one of terminology (Osborn 1968a).

Each prism undulates from side to side in the transverse plane of the tooth (Osborn 1968b and c), the undulations of one prism being slightly out of phase with those above and below it, but being matched by those laterally adjacent to it (**8.**59). Similar regular undulations over the tips of cusps give rise to the appearance of *gnarled enamel* when sections of this region are examined.

The incremental lines in enamel (**8.**55) are termed the *striae of Retzius* (Osborn 1971; Weber *et al.* 1974). They are brown when viewed by transmitted light and blue when seen by incident light. This suggests that light is scattered in the region of striae by some differences in crystallite size or orientation. Where the striae meet the surface of the tooth slight depressions called *perikymata* develop (**8.**60). These can be clearly seen ringing the surface of newly erupted teeth (Weber *et al.* 1974). They become flattened by wear.

A well-marked Retzian stria is seen in the enamel of those teeth whose mineralization spans birth (all deciduous teeth and the first permanent molars). This is the *neonatal line* (Whittaker and Richards 1978). The presence of neonatal lines in the teeth of a dead infant is of forensic importance: it indicates that the infant survived the perinatal period for a few days, because the lines develop during this period.

THE CEMENT

Cement is the bone-like tissue covering the roots of the teeth (Selvig 1965; Furseth 1969; Frank and Steuer 1977). It is about 50 per cent by weight mineral. In terms of the newly erupted tooth, cement generally just overlaps the enamel, but it may also just meet the enamel or occasionally fall short, leaving the dentine exposed to the periodontal ligament. However, in old teeth when the root becomes exposed in the mouth (getting 'long in the tooth'), the cement is often worn away and the dentine is revealed.

The cement is perforated by *Sharpey's fibres*, the extensions of collagen fibres which cross the periodontal ligament to be inserted into the alveolar bone. New layers of cement are deposited throughout life to compensate for tooth movements. In this way new Sharpey's fibres are incorporated into the new cement surface while the deeper, older fibres become mineralized.

The thin first-formed layer of cement generally does not contain cells. However, the apical two-thirds of the root is

8.60 A scanning electron micrograph of the enamel surface showing perikymata. The holes (about 4 μm wide) were occupied by Tomes' processes of ameloblasts when the development of the enamel was completed. (Kindly provided by Dr. D. Whittaker, The Dental School, University of Wales, Cardiff.)

In transverse sections of teeth, tuft-like projections from the enamel-dentine junction can be seen to penetrate between prisms and extend about one-third of the way through the enamel. These are the *enamel tufts*. The 'leaves' of the tufts appear to correspond with thickened prism sheaths and they spread for considerable distances in the longitudinal plane of the tooth (Osborn 1969).

From the outer surface of the enamel, longitudinal sheets of organic or hypocalcified material penetrate the enamel, often as far as, or a little further than, the enamel-dentine junction. They are called the *enamel lamellae*. The extensions into the enamel of the dentinal tubules are *enamel spindles*.

commonly covered by cement which contains *cementocytes*, occupying lacunae joined together by canaliculi which are mainly directed towards the periodontal ligament—their source of nutrition. The *acellular* (**8.**54) and *cellular* forms of cement differ only in the absence or presence of cells; either form may exist over any part of the root (*see* p. 1299).

PERIODONTAL LIGAMENT (MEMBRANE)

The periodontal ligament (Melcher and Bowen 1969) is about 0·2 mm wide. Its main functions are to suspend the tooth in its socket and to provide sensory information by means of pressure receptors (Anderson *et al.* 1970).

The main components of the periodontal ligament are the collagenous *principal fibres*, of which there are several named groups (**8.**43, 61). The *gingival fibres* pass from the more coronal cement and the crest of alveolar bone into the lamina propria of the gingiva, anchoring it firmly against the tooth. They are aided by a *circular group* arranged concentrically around the neck of the tooth. Just apical to these fibres are the *horizontal fibres* which restrict tilting of the tooth. The most numerous elements are the *oblique fibres*. An essential element is the hydrophilic ground substance, providing visco-elastic support (Melcher and Walker 1976).

Each periodontal ligament derives its nerve supply from several sources (*vide supra*). The chief role of these axons seems to be that of subserving mechanoreceptors. They have been described as terminating in 'knob-like' endings, elongated and spindle-shaped endings, Meissner's corpuscle-like endings, and irregularly branched endings. Physiological investigations suggest that such structural variations of these mechanoreceptors are unimportant, but that their spatial arrangement within the ligament determines their response characteristics (Anderson *et al.* 1970; Hannam 1976). The information derived from activity of these nerves probably provides an input to many brainstem centres, including the cerebellum, where the masticatory cycle of jaw movement may, in part, be controlled.

The remaining tissues in the periodontal ligament are, with the following exception, typical of any connective tissue. A network of epithelial cells is scattered through the ligament. These constitute the *epithelial debris of Malassez*, and are remnants of the root sheath (p. 1299). The cells have no obvious function: they give rise to the commonly occurring *dental cysts*.

There is a remarkably rapid turnover of collagen in the periodontal ligaments. Old fibres are removed by fibroblasts acting as fibroclasts (Ten Cate 1974). It is possible to distinguish, ultrastructurally, *fibroblasts* active in fibrogenesis, *fibrocytes* which are largely inactive, and *fibroclasts* which are removing collagen. Alternatively, one cell may be forming and degrading collagen at the same time. The status of the periodontal ligament is determined by a balance between the rate at which fibres are removed and replaced.

THE GINGIVAE (GUMS)

The tissue surrounding the necks of the teeth is the gingiva (Miles 1967; Melcher and Bowen 1969; Squier *et al.* 1976). In a healthy mouth it can be distinguished from the rest of the oral mucosa because it is pale pink and stippled (**8.**44), whereas the adjacent *alveolar mucosa* is red, shiny and smooth; the gingival and palatal epithelia are keratinized (or parakeratinized), the alveolar epithelium is not keratinized (*see* pp. 40, 41).

Around the neck of the tooth, the gingival epithelium is reflected towards the root of the tooth so that its outer surface is attached to the tooth—the *epithelial attachment*. Surrounding the tooth is a shallow *gingival sulcus* which is bordered on one side by the tooth, on the other by the gingiva, its floor being the epithelial attachment. The epithelium attached to the tooth is the *junctional epithelium*, and that lining the gingival sulcus is the *sulcal epithelium*. Junctional epithelium has particularly wide inter-cellular spaces, is non-keratinized, and has a rapid rate of cellular turnover. By means of hemidesmosomes its superficial cells adhere to a basal lamina covering the attached surface of the tooth (Listgarten 1970), and they are shed into the gingival sulcus. It is evident that this basal lamina is produced in its entirety by ectodermal cells. The sulcular epithelium is a non-keratinized stratified squamous epithelium (*see* **8.**32).

Early Dental Development

DECIDUOUS TEETH

At about the 8–9 mm embryonic stage of development the primitive oral epithelium (**8.**63A) begins to bulge into the underlying mesoderm in the region which will later be occupied

by the teeth (**8.**62–66). This horseshoe-shaped thickening in each jaw is at first incomplete across the midline; it is the *dental lamina* (**8.**63B), and from it, and the surrounding cells, are produced the discrete *tooth buds* which later develop into the teeth (Gaunt *et al.* 1971; Provenza and Sisca 1971).

It has been shown that in amphibia, and probably mammals, the dental lamina proliferation is induced by cells of neural crest origin (ectomesenchyme) which have migrated into the developing jaws (Gaunt and Miles 1967; Slavkin 1974). The original site of the embryonic oral membrane has not been traced with certainty in developing oral jaws and therefore it is not known whether the dental lamina is derived from ectoderm or endoderm or from both these germ layers.

At about the 20 mm stage discrete outgrowths of the dental lamina push towards isolated clumps of ectomesenchyme which are found in regions of increased capillary density. An ectodermal outgrowth together with the adjacent ectomesenchyme is known as a *tooth bud*. The ectoderm starts to grow around the

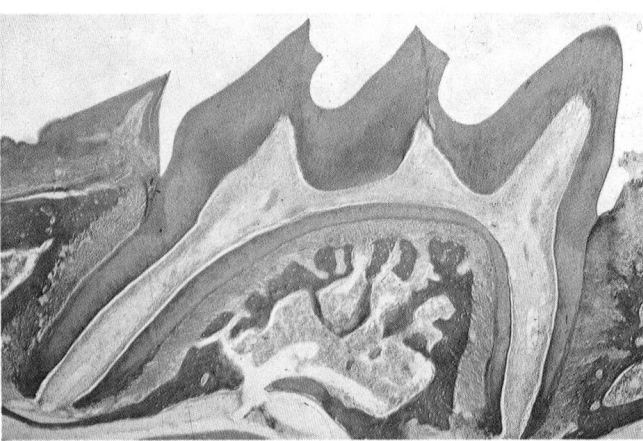

8.61 Demineralized vertical section of a young rat molar tooth *in situ*. Note: (1) the alveolar bone with marrow spaces; (2) the fibres of the periodontal ligament which connect the alveolar bone to the cement of the root; (3) the large pulp cavity typical of a young tooth; (4) the absence of enamel in this demineralized section; and (5) the gingiva adjacent to the enamel space at the neck of the tooth.

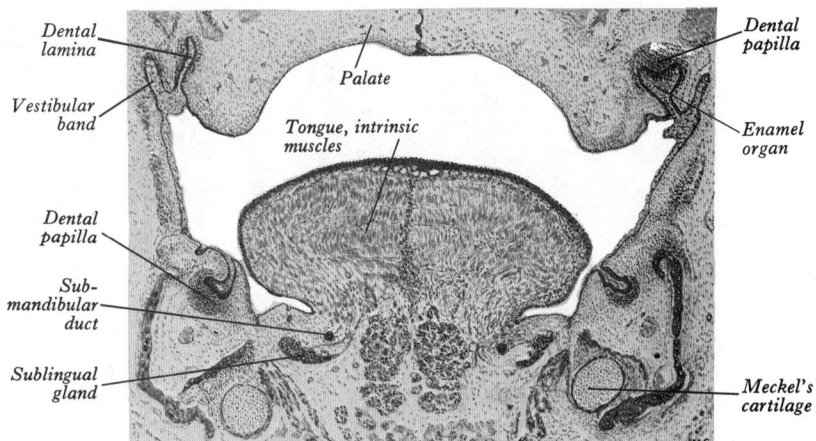

8.62 Coronal section of the head of a human embryo (C.R. length 34 mm), showing developing teeth. The pointer line to Meckel's cartilage passes through the developing mandible.

ectomesenchyme; at this stage the ectodermal part of the tooth bud is called the *enamel organ*, the ectomesenchymal part is the *dental papilla* (**8.**65). Soon, peripheral cuboidal cells of the enamel organ can be distinguished from centrally placed polygonal cells. The encircling edge of the enamel organ is called the *cervical loop* (**8.**63E).

The roughly spherical dental papilla increases in volume, but the cervical loop of the growing enamel organ continues to

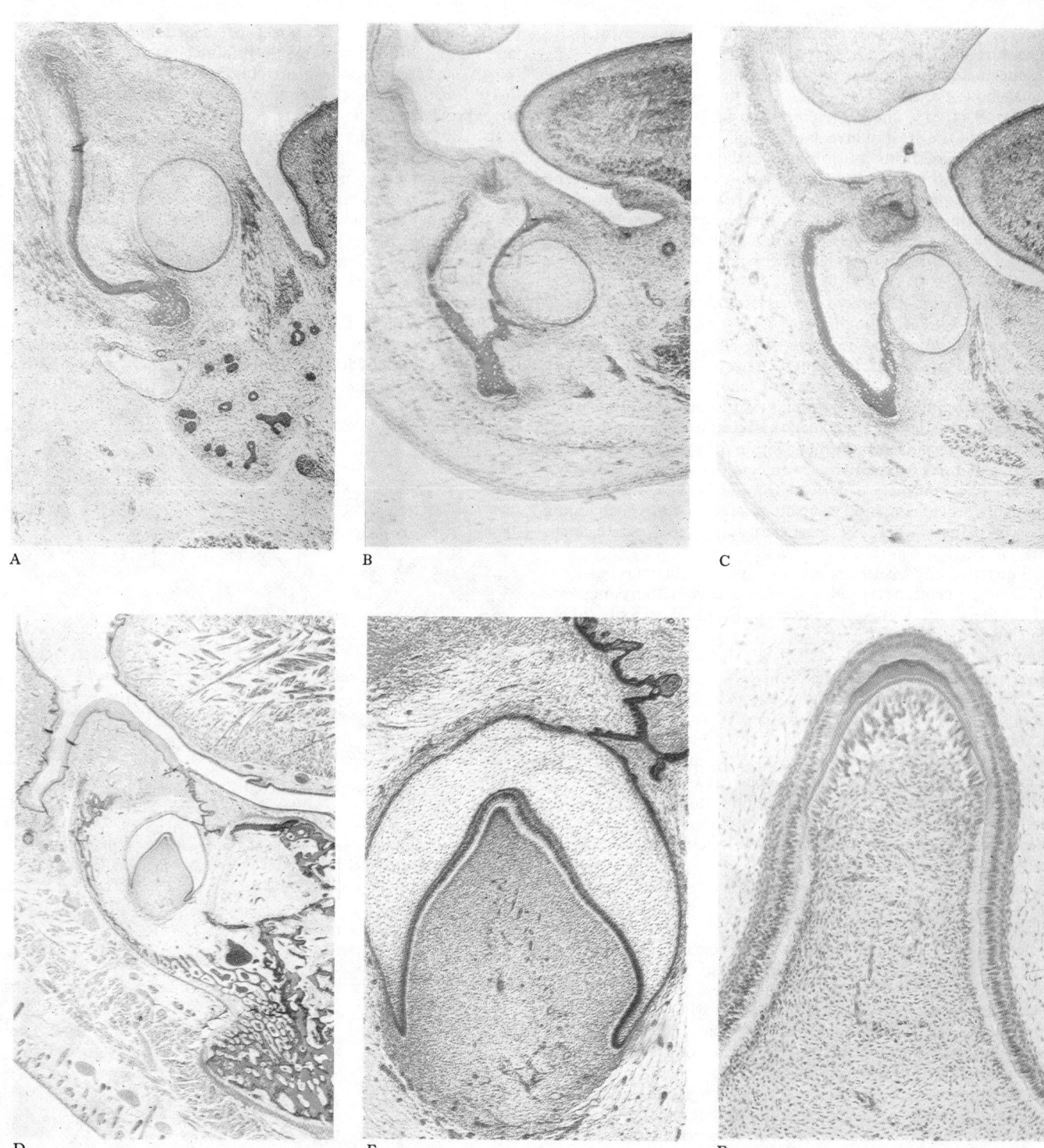

A   B   C

D   E   F

8.63A–C   A series of stages illustrating the early development of teeth. These are all coronal sections through the right half of the body of the mandible, showing the tongue in the top right-hand corner. The mandible is mineralizing to the left of the circular profile of Meckel's cartilage

A   The stage of development before the ingrowth of the dental lamina from the oral epithelium.

B   The dental lamina is growing between the buccal and lingual plates of the ossifying mandible.

C   The cap stage of development. The enamel organ is growing from the dental lamina around the condensation of cells which forms the dental papilla.

8.63D–F   Slightly later stages in tooth development than shown in **8.**63A–C.

D   The bell stage of development. The external enamel epithelium of the enamel organ is connected to the oral mucosa by an irregularly-stranded dental lamina. Lateral to the buccal plate of the mandible the vestibular band has atrophied centrally to initiate the oral vestibule. The tooth germ is separated from the bone by the tooth follicle.

E   A photograph at higher magnification of the bell stage. Note from above downwards: (1) the degenerating dental lamina, top right; (2) the fibrous tooth follicle surrounding the developing tooth; (3) the external enamel epithelium; (4) the delicate stippled appearance produced by the nuclei of the stellate reticulum; (5) the darkly stained, somewhat flattened cells of the stratum intermedium, which is seen more clearly in F; (6) the columnar cells of the internal enamel epithelium; (7) the more closely packed cells of the dental papilla which extend outside the cervical loop; and (8) the capillaries of the pulp and tooth follicle.

F   Dentine formation beginning at the cuspal tip. From above downwards note: (1) loose stellate reticulum; (2) stratum intermedium; (3) layer of columnar ameloblasts; (4) a thin strip of enamel matrix (mauve); (5) mineralized dentine (pink); (6) predentine (pale blue); (7) layer of odontoblasts.

encompass a greater proportion of its periphery until it is sitting on the dental papilla like a cap—the *cap stage* of tooth development (**8**.63C). Meanwhile, the polygonal central cells of the enamel organ have been secreting glycosaminoglycans into their intercellular spaces. This hydrophilic material attracts water into the spaces, which now swell up and compress the cytoplasm of these cells. Because they maintain their desmosomal connexions, the central cells of the enamel organ now look stellate and the tissue is called the *stellate reticulum* (**8**.63E, 64). The original cuboidal cells adjacent to the dental papilla begin to lengthen into columnar cells to form the *internal enamel epithelium* (**8**.63E). The remaining peripheral cells form the *external enamel epithelium* (**8**.63E) which is continuous with the dental lamina.

By its continued growth, the cervical loop now surrounds about three-quarters of the periphery of the enlarging dental papilla, giving the tooth germ the appearance of a bell; *the bell stage* of tooth development (**8**.63D, E).

During the cap and bell stages a clump of ectodermal cells bulges into the stellate reticulum from the internal enamel epithelium. This is the *enamel knot*. From the enamel knot a cord of cells extends through the stellate reticulum towards the connexion between the external enamel epithelium and the dental lamina; this is the *enamel septum* (Berkovitz 1967).

Near the time when the cap stage is merging into the bell stage, a layer of more flattened cells, about three or four cells thick, develops between the inner enamel epithelium and the stellate

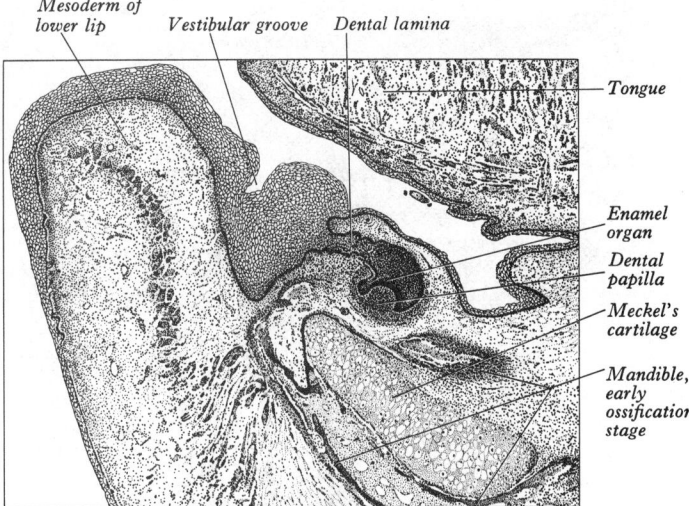

8.65 Part of a sagittal through the head of a human embryo (C.R. 60 mm), passing through the right lower central incisor tooth germ. Stained with haematoxylin and eosin. × 12. (Drawn from a photomicrograph kindly given by Prof. C. H. Tonge, Anatomy Department, King's College, Newcastle-upon-Tyne.)

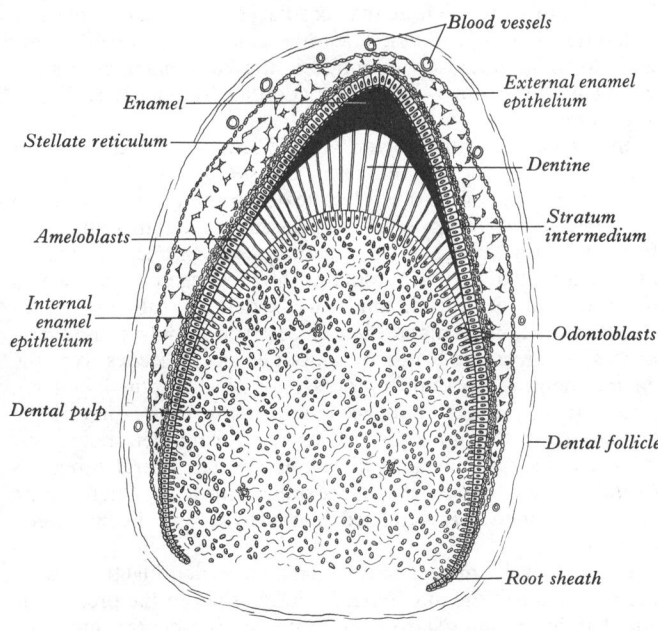

8.64 A simplified diagram of a developing tooth, to show the approximate arrangement of its principal components. Compare with **8**.63F.

reticulum. This is the *stratum intermedium* (**8**.63E, 64); possibly it is derived from the enamel knot or from the *internal enamel epithelium*. Following the bell stage dentine formation starts.

The development of the nerve supply of the deciduous teeth has attracted little attention. Alveolar nerves have grown into the maxillary and mandibular processes at about 5 weeks as the dental laminae are forming. At the 'cap' and 'bell' stages individual bundles of fibres have entered the mesenchymal masses which will produce the dental papillae. (For further details *see* Pearson 1977.)

The Tooth Follicle

A growing enamel organ compresses the surrounding adjacent cells so that they assume a concentric arrangement (Melcher and Bowen 1969). These cells comprise the *tooth follicle* (**8**.63E) which will ultimately be bordered by the developing alveolar bone of the jaw (**8**.63D, 66), the depression in the bone containing the tooth germ and follicle being called a *tooth crypt*.

The growing edge of the cervical loop pushes some

ectomesenchymal cells aside so that they come to lie outside and adjacent to the external enamel epithelium (**8**.63E). This is the *investing layer of the tooth follicle* and from it probably develops the cement of the root. From the outer layer of the tooth follicle develop the osteoblasts and osteoclasts which fashion the bony crypt, and later the socket, of the developing tooth. The middle layer provides further cells and fibres to the periodontal ligament.

The Vestibular Band

At about the same time that the dental lamina appears, there is a similar horseshoe-shaped ingrowth of epithelial cells buccal to it (Tonge 1966). This ingrowth is not associated with demonstrable aggregation of ectomesenchymal cells. It is called the *vestibular band* (**8**.62, 65). (In some accounts of tooth development it is suggested that there is a single primary ingrowth (the *primary epithelial band*), and that this subsequently bifurcates into a buccally placed vestibular band, and the more lingual dental lamina.)

The vestibular band grows deeply into the primitive jaw arches (**8**.65), to separate what will ultimately be the lips and cheeks from the oral mucosa covering the alveolar bone of the jaws; that is to say it will become the site of the vestibule of the mouth. The ectodermal cells divide rapidly to broaden and deepen markedly the vestibular band. A cleft appears within the central cells of the band (**8**.66) at about the time when the tooth germs have reached the cap stage, thereby separating the cheeks and lips from the developing jaws.

THE PERMANENT TEETH

As the jaws grow in length the dental lamina penetrates posteriorly through the jaw mesenchyme as a solid core. From its deep border the buds for the permanent molars develop in mesiodistal sequence, each being initiated within the ramus of the jaw (Scott and Symons 1977); but, with the progressive resorption of the anterior border of the coronoid process, they come to lie within the body of the jaw.

From each of the tooth buds for the deciduous teeth there grows lingually a *successional lamina* out of the region where the external enamel epithelium is in continuity with the dental lamina (**8**.66). Each extends down into the mesenchyme on the lingual side of the deciduous tooth and from its end a tooth bud develops, later to generate the permanent successor.

Each permanent successor to the deciduous incisors and canine, lying lingual to its predecessor, becomes surrounded by its own tooth follicle and bony crypt. However, the follicle maintains a fibrous connexion with the overlying lamina propria

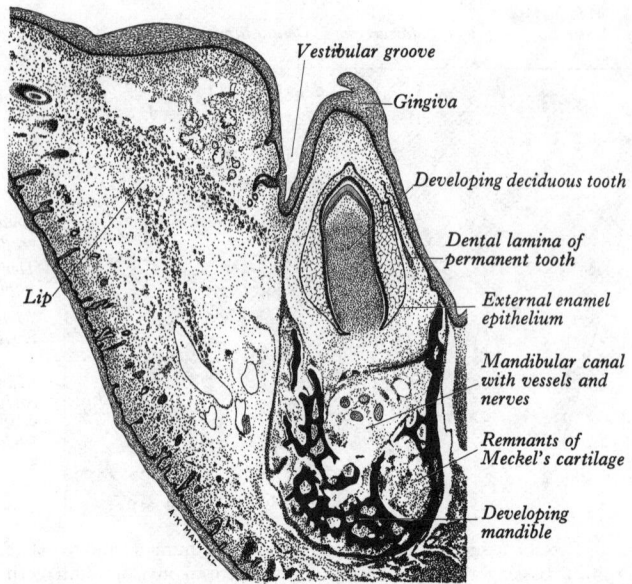

8.66 Developing tooth with mandible and lip *in situ*. (Drawn from a photograph by F. Harrison, Dental Department, University of Sheffield, and reproduced by permission of Messrs. Blackie & Son, Ltd., from *Our Teeth* by R. D. Pedley and F. Harrison.)

of the oral mucosa. These connexions are called *gubernacular cords*. Their original positions, which can easily be seen in young dried skulls, are called *gubernacular canals*. The cords are said to guide the erupting permanent teeth into their correct positions (Scott 1967).

The Fate of the Dental Lamina

About the time dentine and enamel start to develop, the cells of the dental lamina begin to degenerate (**8.63E**), breaking up into discrete clumps, many of which have a whorl-like appearance. These can be found over the developing deciduous teeth, in which position they may occasionally proliferate to become cystic cavities of minor importance known as *eruption cysts*. The latter are recognizable as bluish swellings over the erupting tooth.

In the newborn there can be seen a horseshoe-shaped *dental groove* slightly buccal to the alveolar crest. This is the site of origin of the *dental lamina*. In the upper jaw a more palatally sited *gingival groove* separates the gingival from the palatal epithelium (Scott and Symons 1977).

DEVELOPMENT OF DENTINE

Just prior to dentinogenesis (Miles 1967; Symons 1968), a small space separates the cells of the dental papilla from the, by now, tall columnar cells of the internal enamel epithelium (**8.63F**). The latter, which are still dividing, lie on a basement membrane.

During the late bell stage, the sheet of cells, comprising the inner enamel epithelium, folds in a genetically determined way to take up the definitive outline of the future enamel-dentine junction (Gaunt and Miles 1967; Gaunt *et al.* 1971). Because there are only slight regional variations in the thickness of the enamel, this folding process determines the ultimate shape of the tooth; that is to say, the number and positions of cusps. At the tip of a presumptive dentine cusp the cells of the internal enamel epithelium lengthen towards the undifferentiated cells of the dental papilla and induce the adjacent ectomesenchymal cells to differentiate into presumptive *odontoblasts*, the cells which will lay down the dentine (**8.63F, 67**). Deep to this, a thicker layer of cells differentiates into active fibroblasts; this is the *subodontoblastic layer*, whose cells act as satellites to odontoblasts; they appear to produce many collagen fibres, which penetrate between the odontoblasts and seem to fan out parallel to the presumptive enamel-dentine junction. These are the argyrophylic *von Korff fibres*. It has been demonstrated (Ten Cate *et al.* 1970) that this fibrous appearance may be an artefact produced by the precipitation of silver stains in the intercellular spaces between

the odontoblasts (the penetrating parts of the 'fibres'), and between the odontoblasts and the internal enamel epithelium (the fanned-out ends of the 'fibres').

The newly differentiated odontoblasts, which now have a well-developed endoplasmic reticulum and Golgi apparatus, start secreting the dentine matrix into the region enclosed on one side by the internal enamel epithelium and on the other by their own secreting ends, which are united by desmosomes. At this time each odontoblast sends out several cellular projections towards the basement membrane, some penetrating between the cells of the overlying inner enamel epithelium, ultimately to become the enamel spindles. The accumulation of matrix pushes the odontoblasts back into the papilla and the first layer of unmineralized matrix has been formed. The cellular projections of the odontoblasts lengthen as their perikarya recede, becoming the processes of the odontoblasts enclosed within tubules of matrix. As soon as a few microns of matrix have been laid down, the matrix adjacent to the internal enamel epithelium begins to mineralize (Silva and Kailis 1972), and the basement membrane disappears. *Matrix vesicles* have been described in association with the first layer of dentine to be mineralized (Bernard 1972; see also p. 260).

From this region, at the tip of the presumptive dentine cusp, a wave of differentiation of odontoblasts from dental papilla cells slowly spreads down towards the growing cervical loop. As soon as each odontoblast has differentiated, matrix is laid down, pushing the united layer of cells into the papilla (Osborn 1967). The layer of unmineralized matrix adjacent to the odontoblasts is called the *predentine* (**8.67**). The mineral in dentine is commonly laid down as microscopic spheres, the *calcospherites* (**8.52B**), giving the predentine-dentine junction an irregularly scalloped appearance.

In association with the von Korff fibres, or the intercellular spaces between the odontoblasts (*vide supra*), the first formed, rather thin, layer of dentine is less heavily mineralized. It is called '*mantle dentine*' (Moss 1974) as opposed to the remaining, later formed dentine, which is called *circumpulpal dentine*. The matrix of circumpulpal dentine contains fine collagen fibres which interlace in planes at right angles to the dentinal tubules.

At first, within the mantle dentine, each odontoblast has several processes (Silva and Kailis 1972), but as a cell recedes from the enamel-dentine junction its processes unite so that finally it has a single, main process. This accounts for the bifurcation of the tubules near the junction. However, during dentinogenesis, odontoblasts are constantly developing small lateral processes (Symons 1968), presumably at the predentine level, and these are later surrounded by mineralized dentine to become the fine lateral tubules.

During development it can be seen that the dentinal tubules are much thinner in the mineralized dentine than in the predentine and that this thinning starts near the predentine-dentine junction. It is due to the development of a cylinder of the highly mineralized *peritubular dentine* (**8.52B**) which comes to line the original tubule and progressively reduces its diameter (Miles 1967). Peritubular dentine formation starts at the level where mineralization of the predentine is beginning. Its thickness gradually increases with age. It does not develop in interglobular spaces. (The development of root dentine is considered later in the section on root development.)

DEVELOPMENT OF ENAMEL

Prior to differentiating into ameloblasts the cells of the internal enamel epithelium lengthen and induce the differentiation of odontoblasts from cells of the dental papilla. It is while they are lengthening that they manufacture the intracellular organelles required for producing enamel matrix (Fearnhead and Stack 1965, 1971; Miles 1967; Boyde 1969; Osborn 1970a). At this time the polarity of the cells becomes reversed; the mitochondria which were originally dispersed throughout the cytoplasm become aggregated at the basal (non-secreting) end of the cell where they are joined by the centrioles, whilst the Golgi apparatus migrates to the middle region of the cytoplasm. The cisternae of endoplasmic reticulum increase in amount and together with the

The labels on figure 8.66:
Vestibular groove
Gingiva
Developing deciduous tooth
Dental lamina of permanent tooth
External enamel epithelium
Mandibular canal with vessels and nerves
Remnants of Meckel's cartilage
Developing mandible
Lip

Golgi apparatus become stacked in rows which are parallel to the long axis of the cell; this is now about 40 μm long and 5 μm wide. In cross-section the secretory ends of the cells are regular hexagons, accounting for the classical descriptive term 'honeycomb appearance'. Both the apical and basal margins of adjacent cells are interconnected by junctional complexes (p. 7). Their basal ends are attached by desmosomes to the stratum intermedium, an arrangement which has been suggested as facilitating the transport of materials elaborated by the stratum intermedium to the ameloblasts (Kurahashi and Yoshiki 1972). It is of interest that alkaline phosphatase, an enzyme found in the other cells concerned with the formation of hard tissues (osteoblasts, odontoblasts and cementoblasts), is found in all the cells of the enamel organ apart from the ameloblasts—the cells which actually secrete the enamel precursor.

As soon as mineralized dentine has been formed, the adjacent ameloblasts start secreting enamel precursor material and it is considered that the presence of the dentinal matrix is necessary as an inducing agent to initiate the secretory process. Thus, developing enamel spreads down the sides of the presumptive enamel-dentine junction at the same rate as developing dentine, but lags behind it (8.67). From this it can be visualized that the enamel precursor material is secreted into a region bordered on one side by mineralizing dentine and on the other by the ameloblasts. It seems probable, therefore, that the

rise or potential rise in hydrostatic pressure produced by the accumulation of enamel precursor in this enclosed region, provides the force which pushes the ameloblasts away from the enamel-dentine junction (Osborn 1970b).

At the start of amelogenesis, in each region of the tooth, the adjacent stellate reticulum seems to collapse so that the enamel organ is progressively reduced from four layers (external enamel epithelium, stellate reticulum, stratum intermedium, ameloblasts) to three layers. It is usually suggested that the purpose of this is to bring the ameloblasts, which are on the inside of an avascular enamel organ, closer to their source of nutrition; namely the capillaries which have sprouted prolifically adjacent to the external enamel epithelium. Meanwhile, the cells of the external enamel epithelium, which were originally cuboidal, have become squamous and the surface of the enamel organ is thrown into microscopic folds thereby increasing the area available for diffusion. Apart from the ameloblasts, the remainder of the cells of the enamel organ contain alkaline phosphatase, an enzyme which is probably concerned with transport of materials from the bloodstream to the ameloblasts.

As soon as the ameloblasts have moved a few microns away from the enamel-dentine junction they develop somewhat conical extensions into the developing enamel. These are the *Tomes' processes* of the ameloblasts, whose bases are limited by the apical junctional complex of the cell, and they bear a basal peripheral

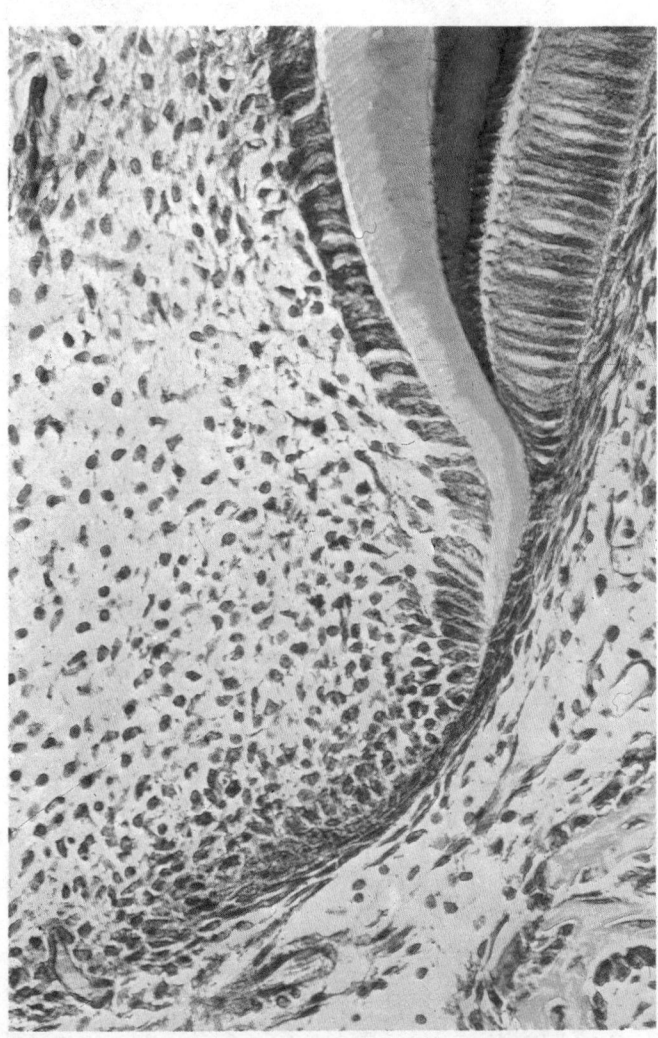

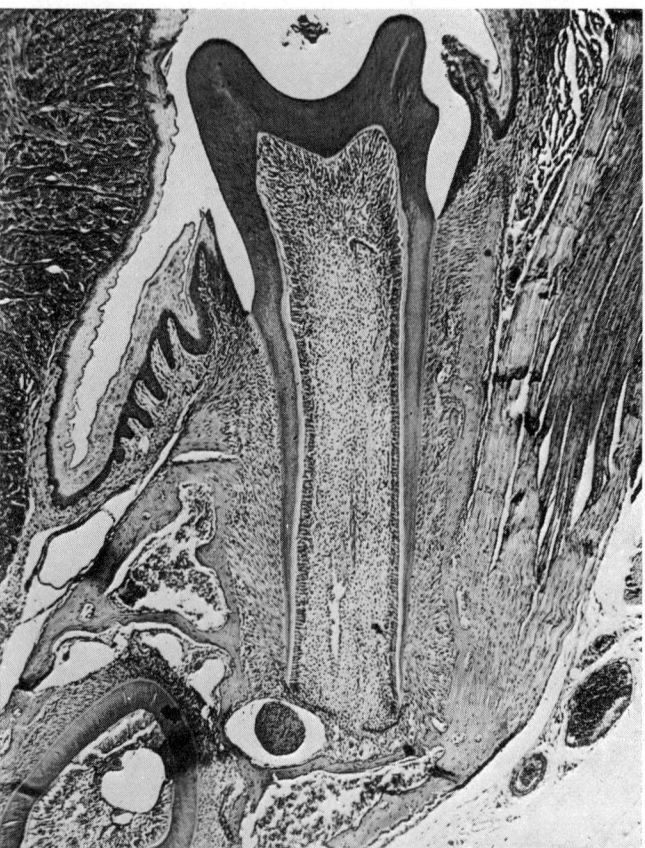

8.68 A longitudinally sectioned developing tooth showing advanced root formation. See text for further details.

8.67 A vertical section through the neck of a developing tooth with the crown above and the developing root below. Note above, from right to left: (1) the columnar ameloblasts; (2) enamel matrix (red); (3) mineralizing dentine (pale mauve); (4) predentine (pale blue); (5) odontoblasts; and (6) the fibroblasts of the pulp. The columnar ameloblasts terminate at the tooth neck where the latter is continuous with the developing root. In this region note from right to left: (7) the fibrous tooth follicle; (8) the layer, two cells thick, of Hertwig's root sheath; (9) the developing dentine; (10) the odontoblasts. Beneath this the root sheath extends to the left beneath the dental papilla.

collar of microvillous projections (for further ultrastructural details of these cells consult Reith 1970). The Tomes' processes project into the mineralizing enamel which, therefore, has a regularly pitted surface (Boyde 1969). Adjacent to each Tomes' process is an unmineralized layer about 50–100 nm thick; this is the enamel precursor material. It is stippled when seen under the electron microscope, and similar material is seen within vesicles in the Tomes' processes. On the enamel side of the stippled material are long thin tape-like crystallites separated by unmineralized material. The first formed enamel contains very irregular prisms and ultrastructurally the enamel (recognized by its long

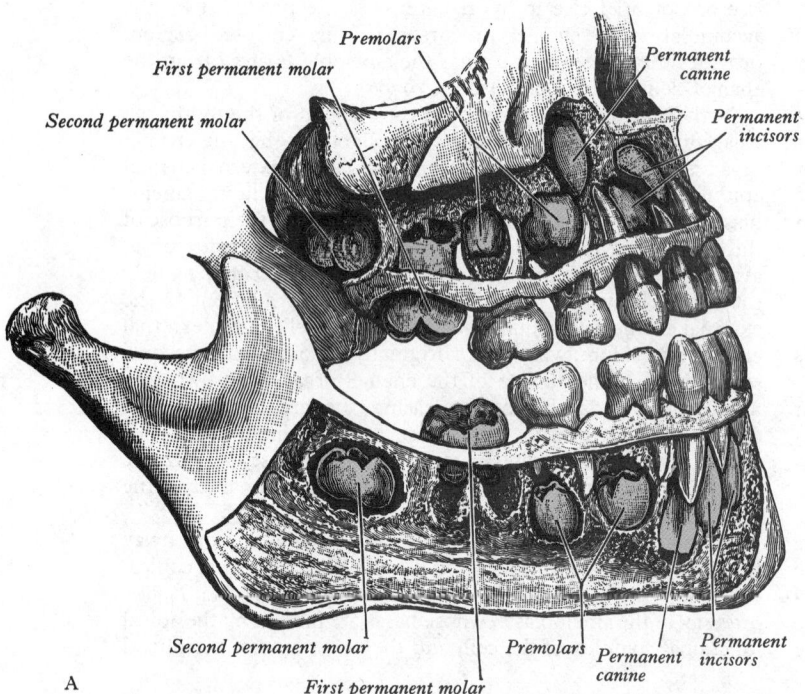

8.69A The teeth of a child aged about six years. The permanent teeth are coloured blue.

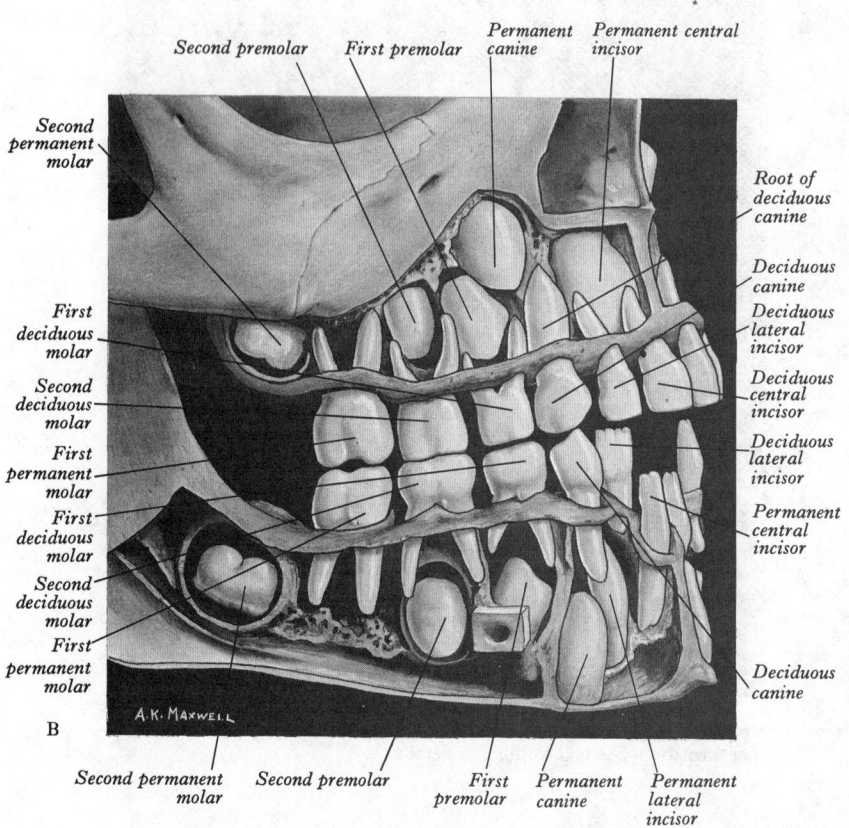

8.69B Normal dentition of a child aged seven years. The lower permanent central incisors have replaced the deciduous central incisors and the first permanent molars have erupted. The upper permanent lateral incisor appears to be congenitally missing from this skull. (Drawn from a photograph by F. Harrison, Dental Department, University of Sheffield, and reproduced by permission of Messrs. Blackie & Son, Ltd., from *Our Teeth* by R. D. Pedley and F. Harrison.)

C, D and E Anterior views of skulls dissected to display the developmental stages of the dentition of children aged: C 4–5 years; D about 8 years; E 9 years.

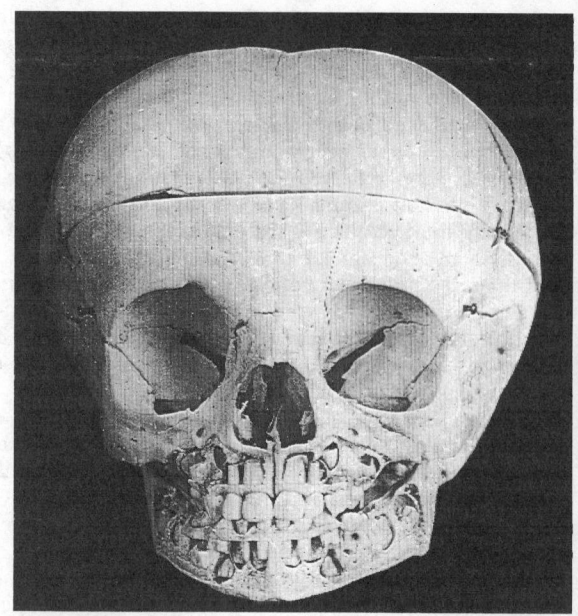

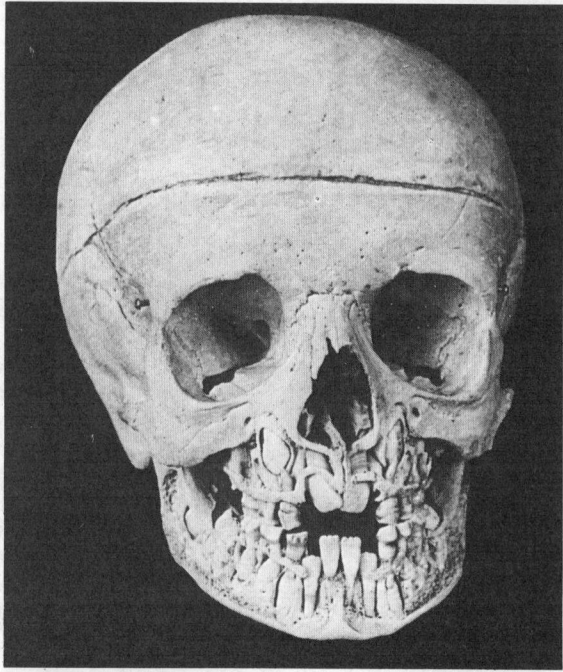

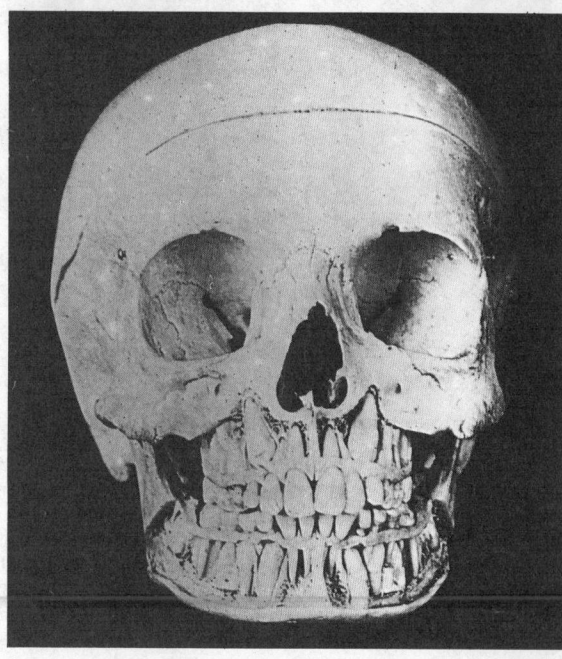

crystallites) appears to merge into the dentine (recognized by the presence of collagen and small crystallites).

Depending upon its position in the enamel prism, the long axis (*c*-axis) of each crystallite has a preferred orientation (**8**.58). Thus, the more cuspal of the crystallites (plane C in **8**.58) are roughly parallel to the long axis of the prism, whereas the crystallites which are more cervical in position are more nearly perpendicular to this axis. The crystallites also have a preferred orientation in plane B which is at right angles to plane C (**8**.58). The central crystallites are roughly parallel to the long axis, whilst the peripheral crystallites are angled to the sides of the prism. There is always a sudden change between crystallite orientation on the two sides of a prism sheath. The hexagonal transverse sectional profiles of the ribbon-like crystallites are randomly orientated with respect to each other.

The mineralizing enamel front appears much the same until amelogenesis is almost complete. As the ameloblasts move outwards the deeper crystallites grow in thickness by the accretion of material from the surrounding matrix. The diminishing volume of calcium-rich matrix is replenished by material secreted by the ameloblasts while water and protein are reabsorbed. The matrix can travel long distances through the developing enamel. Finally, the crystallites which are hexagonal in cross-section can enlarge no more because no further space is available between them. Theoretical analysis suggests that about 12 per cent by volume of unmineralized enamel matrix would thus remain (Carlström 1964), and this accords well with the observation that enamel is 96 per cent by weight mineral (i.e. 88 per cent by volume). No new crystallites appear to be added to the enamel other than at the mineralizing front (Rönnholm 1962).

A large part of the original enamel matrix is removed as the crystallites grow. A proline-rich fraction accounts for much of the protein, but there is also a considerable volume of water (Eastoe 1963; Robinson *et al.* 1975). All this material is removed via the ameloblasts, probably where the interprismatic region develops, because (in the rat), the adjacent region of ameloblasts contains coated vesicles, and their surfaces are beset with microvilli (Reith 1967*a*).

The sequence of mineralization of enamel can be followed using microradiography (Miles1967). It appears that a thin layer of enamel is mineralized rapidly down the presumptive enamel-dentine junction, and from here spreads peripherally more slowly. From this layer, mineralization also spreads rapidly up through the full thickness of the cusp, and from here down the outside of the tooth; it thence passes centrally to fill the less mineralized regions, sandwiched between the fully mineralized inner and outer layers.

When they have finished secreting, the previously conical Tomes' processes of the ameloblasts become flattened and fold into numerous microvilli, which have been shown to reabsorb matrix (Reith 1967*b*). Over some of the enamel surface developed at this time, the crystallites are parallel to each other and there are no prism sheaths. Thus there are non-prismatic regions about 6–12 μm thick (Gwinnett 1967; Osborn 1968 *b* and *c*). Furthermore, the surface layer is more highly mineralized than the rest of the enamel, possibly because of the closer packing of the crystallites permitted by their parallel orientation.

ROOT DEVELOPMENT

During crown development the cervical loop continues to encircle the expanding dental papilla. It now starts to grow away from the developing crown, but still around the lengthening papilla, as a double-layered structure called *Hertwig's root sheath* (Gaunt *et al.* 1971); the two layers of the root sheath are derived from the external and internal enamel epithelia (**8**.67).

The inner layer of the root sheath, like the cells of the inner enamel epithelium from which it is derived, induces the differentiation of odontoblasts from the surface cells of the dental papilla. The odontoblasts now lay down a tube of dentine on the surface of the papilla, under the inner layer of the root sheath (**8**.68). The root sheath continues to grow, outlining the ultimate shape of the roots and inducing the differentiation of more odontoblasts.

As soon as a layer of dentine has been developed in the root region, the adjacent root sheath seems to break up and become fenestrated, probably because the erupting tooth is now pulling the sheath towards the oral cavity (Gaunt *et al.* 1971). This exposes the outside of the thin layer of newly formed dentine to the tooth follicle. From the investing layer of the tooth follicle cementoblasts are differentiated. It appears that these early formed cementoblasts do not elaborate matrix fibres of collagen, but rather that they secrete ground substance which mingles with superficial unmineralized tags of dentine matrix, and with fibres which have developed in the tooth follicle. This composite matrix is now mineralized under the influence of the cementoblasts and, it can be visualized, attaches the tooth to periodontal fibres (Selvig 1965; Owens 1975). The junction between dentine and cement is therefore not easy to define in electron micrographs.

The root sheath continues to outline the presumptive surface of the root. It surrounds the vessels and nerves supplying the dental papilla. These vessels remain and it is because of their presence, particularly those aggregated towards the future apex of the tooth, that foramina open through canals in the dentine into the pulp of the fully developed tooth. However, it will be appreciated that accessory foramina can open anywhere into the root.

Prior to root development capillaries are more densely arranged in regions which will ultimately correspond with the roots of a tooth (Gaunt and Miles 1967). Thus a two-rooted tooth has two capillary aggregations, and a one-rooted tooth has only one. The papilla grows more rapidly in the region of capillaries and less rapidly in the intervening regions. In these intervening regions, for instance in a two-rooted tooth, a growing tip of the root sheath is able to penetrate under the papilla and meet a growing tip from the other side, thereby separating two roots from each other.

After the root sheath breaks up, the epithelial cells of which it is composed seem to move away from the cement and they are ultimately found in the periodontal ligament of the erupted tooth, where they are known as the *epithelial rests of Malassez*.

DEVELOPMENT OF CEMENT

After the breakdown of Hertwig's root sheath around the coronal half of the developing root, the newly differentiated cementoblasts do not manufacture collagen precursors. Unlike the condition in this region, the cementoblasts which differentiate over the remainder of the root (Jande and Bélanger 1970; Boyde and Jones 1972), and in all later formed cement, do contribute collagen fibres, so that the matrix consists of *extrinsic fibres* (of Sharpey) derived from the tooth follicle, and *intrinsic fibres* derived from cementoblasts. The former lie roughly at right angles to, and the latter roughly parallel to the developing surface. The intrinsic fibres are the first to be mineralized. Sharpey's fibres are then mineralized on their surfaces, and if the cement is being formed rapidly, their cores remain unmineralized.

Cementoblasts frequently become trapped in later formed cement. Areas in which they are trapped are known as *cellular cement*, whilst the remainder is *acellular cement*. Cellular cement tends to be formed around the apical two-thirds of the root, and also during rapid cementogenesis. Incremental lines develop parallel to the surface of the root.

DEVELOPMENT OF PERIODONTIUM

It will be recalled that some of the ectomesenchyme cells of the dental papilla are pushed aside by the growing cervical loop of the enamel organ and come to lie adjacent to the external enamel epithelium. It seems probable that the cementoblasts and the inner layers of periodontal fibres are derived from these cells (Melcher and Bowen 1969; Ten Cate 1974). The remainder of the periodontal ligament is developed from the outer layers of the tooth follicle which, since the early developmental stages, has separated the tooth from the bone of its alveolar crypt (**8**.63D). Unlike the inner layer of the follicle, these outer layers have not been associated with the early condensation of mesodermal (ectomesenchymal) cells which contributed to the tooth germ.

Throughout eruption of the tooth, as it appears in tissue sections, the oblique fibres of the periodontal ligament run

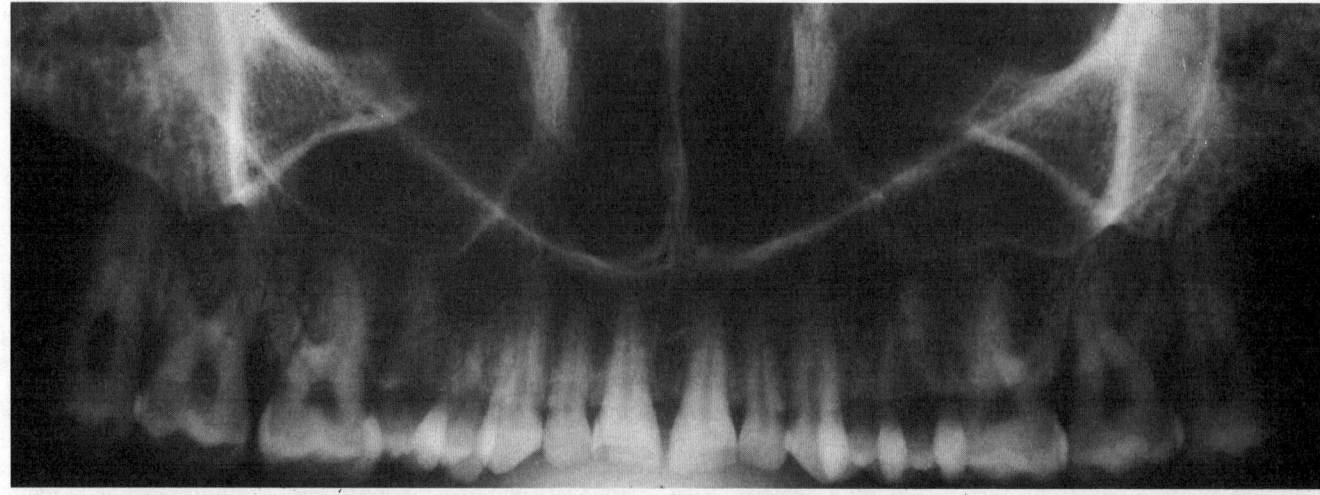

A

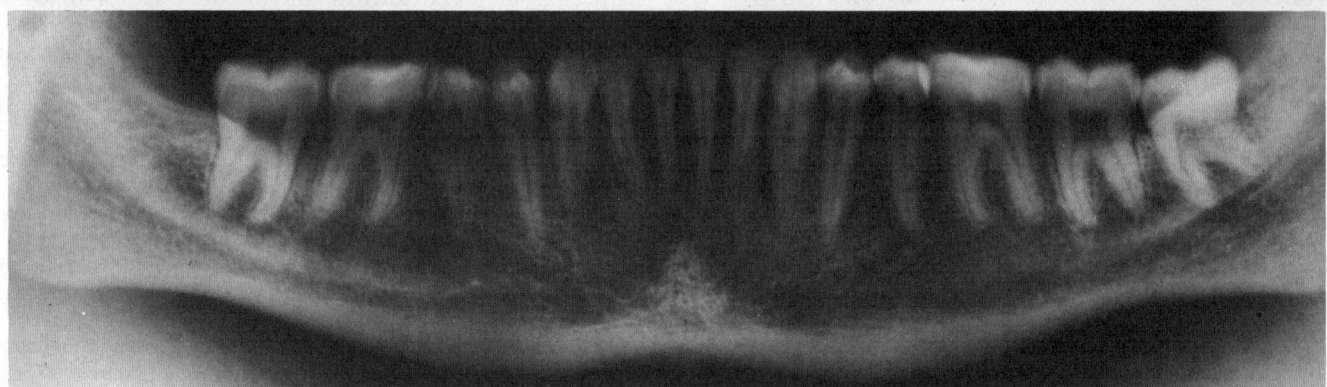

B

8.70A and **B**   Pan-oral radiographs of the whole dentition of the upper and lower jaws of a mature human skull. To achieve these radiographs, an X-ray source is introduced into the buccal cavity and directed towards the palate for the upper jaw, and towards the floor of the cavity for the lower jaw. The X-ray beam is deflected magnetically to disperse anterolaterally, and the film is wrapped around the external aspect of the jaws. The clarity of the radiographs is much greater than that possible during clinical radiography of the living head. (The radiographs were kindly prepared by Mrs. D. White of the X-ray Department of the Royal Dental Hospital, London.) Note: the right lower third molar is missing.

coronally from the cement into the bony wall of the crypt (**8.61**). It will be appreciated that if these fibres remained unbroken with further movement of the tooth towards the oral cavity, they would finally be directed apically from the cement to the bone of the crypt. Based on the appearance of tissue sections, it was at one time thought that fibres joining cement to bone were formed, broken, and reformed in an *intermediate plexus* (Hindle 1967) in the middle of the ligament. Autoradiographic studies now suggest that fibres are formed and ruptured in a plane close to the alveolar bone (Melcher 1976).

DENTAL ERUPTION

When the deciduous dentition is initiated the five tooth buds in each quadrant occupy jaws which are about 1 mm long. By the time the teeth have erupted into the oral cavity they occupy about 3 or 4 cm of jaw; during development these teeth have migrated apart. It is not known how this movement is brought about.

Very soon after root formation starts, the teeth begin to move from their positions in the crypts up towards the oral cavity (Melcher and Bowen 1969). The origin of the forces which move the teeth is not known (Berkovitz 1975). It has been suggested that the lengthening root could induce the movement, or the progressive constriction of the pulp by the growing dentine, or cellular proliferation in the pulp. Such forces involve *pushing* the tooth out of its crypt. Whatever the origin of the force, it has been shown in the dog that the tissue pressure above an erupting tooth is about 10 mm Hg, and within the crown is about 23 mm Hg; in other words an eruptive pressure of 13 mm Hg (van Hassel and McMinn 1972). It has also been suggested that the tooth is progressively *pulled* out of the socket by contractions of the obliquely placed fibres of the developing periodontal ligament (Berkovitz and Thomas 1969).

The permanent incisors and canines develop lingual to their predecessors (**8.63D**, 66). As they develop they migrate buccally but remain slightly lingual to the deciduous teeth. This movement induces the differentiation of osteoclasts from the periodontal ligament of the deciduous tooth, whose root, and ultimately even enamel of the crown, is now intermittently but progressively resorbed (Furseth 1968). In periods of quiescence the resorbed tissues are temporarily repaired by the deposition of thin layers of cement.

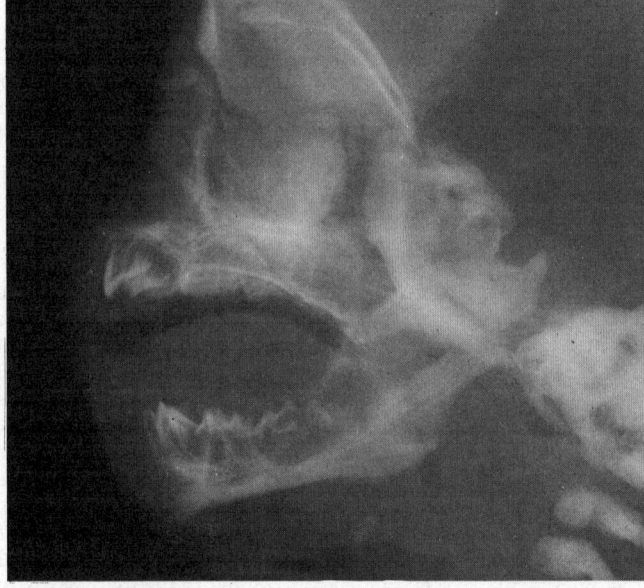

8.71   A lateral radiograph of the jaws of a newborn child. Note the state of development of the jaws and teeth.

Early in their development each premolar moves directly deep to its predecessor to become lodged between widely divergent roots (**8**.46). As the premolar erupts it induces resorption of the deciduous molar. The enamel of the underlying permanent tooth is protected from resorption by its *reduced enamel epithelium* (*see* p. 1297).

Usual times of eruption of the deciduous teeth. (There are considerable variations in these times. *See* **8**.69A–E.)

Central incisors	6–8 months
Lateral incisors	8–10 months
First molars	12–16 months
Canines	16–20 months
Second molars	20–30 months

Calcification of the permanent teeth proceeds in the following order: first molars at birth; incisors (except the upper lateral) and canines about 4 months; upper lateral incisors about 1 year; premolars about 2 years; second molars about 3 years; and third molars about 9 years.

Usual times of eruption of permanent teeth. (The upper teeth erupt a little later than the lower. *See* illustrations **8**.69 A–E.)

First molars	6–7 years
Central incisors	6–8 years
Lateral incisors	7–9 years
Canines	9–12 years
First and second premolars	10–12 years
Second molars	11–13 years
Third molars	17–21 years

ALVEOLAR DEVELOPMENT

The jaws begin to develop at about the time that the dental lamina is being formed. In the lower jaw the growing margin of membrane bone, which remains lateral to Meckel's cartilage, passes backwards caudal to the inferior dental nerve. From this bone, lateral and then medial plates of bone grow cranially (Dixon 1958; Melcher and Bowen 1969). With their apparent downgrowth the developing teeth come to lie between the plates (**8**.66). A septum of bone now grows to divide the teeth, above, from the inferior dental nerve; below, and later still, each tooth becomes separated from adjacent teeth by further upgrowths of bone, thereby isolating each tooth in its own *crypt*. These crypts are wide open on their oral sides; bone does not develop over the deciduous teeth, so that in the skeletonized jaws of the neonate, the teeth are usually lost.

A similar process is involved in the development of crypts in the maxilla.

When the teeth start to move towards the oral cavity, the bone of the socket grows orally, thereby deepening the crypt, increasing the depth of the jaw and carrying the oral mucosa with it. The rate of tooth eruption outstrips the rate of upward bone growth so that the tooth erupts into the oral cavity and finally meets its antagonist. The developing root now starts to penetrate back into the jaw (**8**.68) and some time later the widely open growing apex closes around the nerves and vessels to form the apical foramen.

It is the lengthening of the bony sockets for the accommodation of teeth which leads to the great increase in the depth of the face during and up to puberty (Scott 1967).

CUTICLES AND EPITHELIAL ATTACHMENT

At the end of amelogenesis, the cells of the enamel organ all revert to a squamous shape, becoming a three- or four-cell-thick layer covering the whole surface of the enamel (Gaunt *et al.* 1971). This layer is referred to as the *reduced enamel epithelium*.

When the tooth erupts the reduced enamel epithelium fuses with the oral epithelium so that the junction between the two epithelia cannot usually be recognized. The attachment from tooth to oral epithelium is called the *epithelial attachment* (**8**.32, 68). The oral epithelium grows outside the reduced enamel

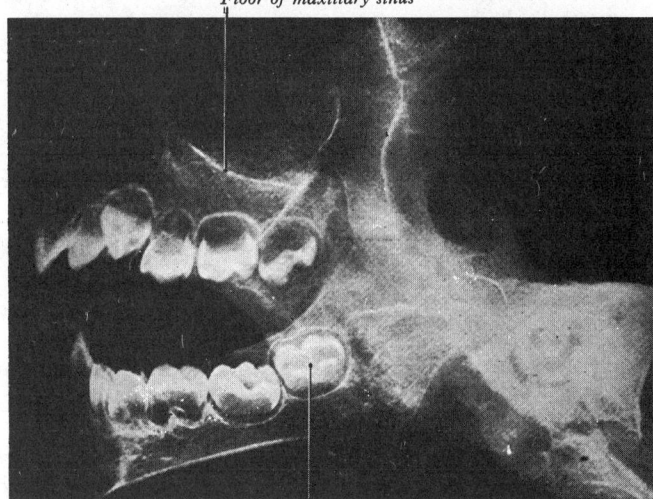

Floor of maxillary sinus

First permanent molar

8.72A   Radiograph of the jaws of an infant, nine months old. (Symington and Rankin's *Atlas of Skiagrams*.) Only the lower central incisor has erupted: the roots of the first lower deciduous molar are just beginning to form; the crown of the first lower permanent molar faces inwards.

epithelium but as the tooth erupts into the oral cavity the reduced enamel epithelium remains attached to the enamel. It is very rapidly worn away from the exposed surface of the tooth. At one time it was thought that the reduced enamel epithelium was separated from the enamel surface by a structureless layer, the *primary enamel cuticle*, about 1 μm thick. This was supposed to be the poorly mineralized, final product of the ameloblasts. Like the reduced enamel epithelium it was rapidly worn from the surfaces of teeth. This cuticle together with the reduced enamel epithelium was called *Nasmyth's membrane*. Because this cuticle cannot be seen under the electron microscope, it is thought to be some form of artefact. However, a form of cement known as *afibrillar cement* has been observed over the enamel around the necks of teeth (Listgarten 1966). Afibrillar cement is only about 100 nm thick, contains no banded collagen, and is probably produced by cementoblasts, following the early disruption of a few cells of the reduced enamel epithelium, at the neck of the tooth.

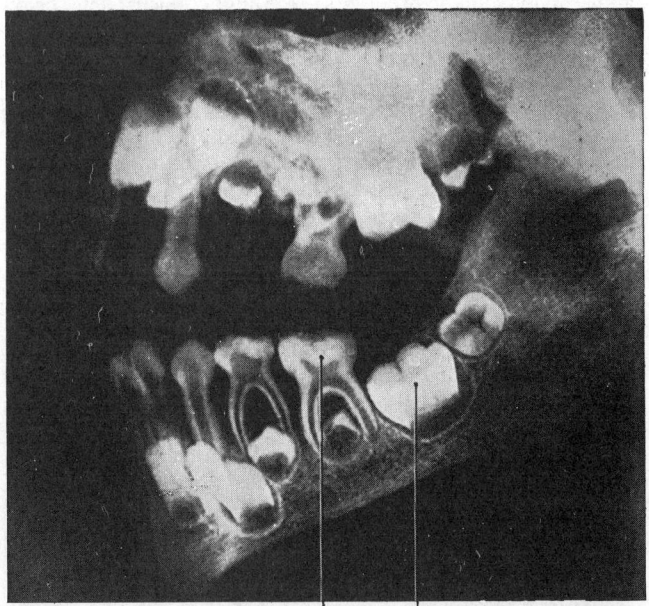

Second deciduous molar   *First permanent molar*

8.72B   Radiograph of the teeth of a boy, aged five years. (Symington and Rankin's *Atlas of Skiagrams*.) In the maxilla the lateral deciduous incisor and the first deciduous molar have been lost, but all the deciduous teeth are present in the mandible. No absorption of the roots of the deciduous teeth has occurred. No permanent teeth have erupted.

Occasionally a cuticle, about 4 μm thick, is found either just beneath the reduced enamel epithelium, or the oral epithelium when it has grown on to the cement. This is the *secondary cuticle*. It may be a product of the epithelial cells, or it may be the remains of blood which has leaked through the epithelial attachment following some slight trauma (Hodson 1966).

Finally, inspissated mucus and debris form a film over the tooth surfaces. This film or cuticle is known as the *acquired cuticle*, and is the foundation of *dental plaque*.

APPLIED ANATOMY

The cortical plate of bone in each jaw is continuous over the alveolar crest with the cortical plate lining each tooth socket (the *lamina dura* of radiographs, **8**.70, 71, 72). Apart from the region of the upper lateral incisor these two cortical plates fuse buccally, with very little trabeculated bone between them (Sicher and Du Brul 1970). It is easier, and obviously more convenient, to extract these teeth by fracturing the buccal plate of bone and removing the tooth buccally than by fracturing the lingual or palatal plate. The apex of the lateral incisor may sometimes be closer to the palate than to the vestibule of the mouth.

In the *lower jaw* both buccal and lingual plates of bone are relatively thin but the buccal plate steadily thickens in the molar region. This thickening is the external oblique line to which the buccinator muscle is attached. Thus, it is mechanically easier to remove an impacted lower third molar through the lingual than the buccal plate of bone. However, great care is necessary to prevent damaging the lingual nerve in this region.

In the *upper jaw* the buccal bone thickens over the molar teeth in the position of the root of the zygomatic arch.

Abscesses developing in relation to the roots of the teeth ultimately perforate either the lingual or buccal plate of bone. The position of the resultant swelling in the soft tissues is largely determined by the relationship of the sinus (the path taken by infected material) in the bone to the muscle attachments. Thus, in the lower incisor region, because the buccal plate of bone is thin, abscesses break through this plate of bone, generally above the level of attachment of mentalis. Therefore, the swelling is usually in the buccal sulcus. However, occasionally the abscess cavity either opens below mentalis and points on to the face, or lingually, when the abscess points in the floor of the mouth. If an abscess on a lower cheek tooth opens out buccally below the attachment of buccinator the swelling is in the neck; if it is above, then the swelling is in the buccal sulcus. If the abscess opens lingually above mylohyoid the swelling is in the lingual sulcus; if it is below, the swelling is in the neck. Because the mylohyoid line rises posteriorly, third molar abscesses tend to discharge into the neck rather than the mouth.

Apart from those of the canine, which has a long root, abscesses on the upper teeth usually open buccally below, rather than above, the attachment of buccinator. Because its root occasionally is nearer the palate, abscesses of the upper lateral incisor may pass back within the submucosa of the palate to 'point' in the soft palate. Abscesses of the upper canine often open into the facial region just below the corner of the eye. Here, the swelling may obstruct drainage from the 'angular' part of the facial vein (p. 739). The angular vein does not possess valves, and it is therefore possible for infected material to travel through the angular and ophthalmic veins, to enter the cavernous sinus. Abscesses on the palatal roots of upper molars generally open on to the palate.

The upper second premolar, and first and second molars are closely related to the maxillary sinus. In individuals with a large antrum the apices of the roots of these teeth may be separated from the sinus solely by its lining mucosa. Sinus infections may stimulate the nerves entering the teeth, simulating toothache. The upper first premolar and third molar may also be closely related to the maxillary sinus.

With the loss of teeth the alveolar bone may be extensively resorbed. Thus, in the lower jaw of the aged edentulous patient, the mental nerve, which originally passes inferior to premolar roots, may come to lie on the crest of the mandible. In the upper jaw, the maxillary sinus may enlarge to approach the oral surface of the remaining alveolar bone. Lingual to the lower premolars or molars, the upper molars, and in the midline of the palate there are occasional marked bony prominences. These are called a *torus mandibularis*, a *torus maxillaris*, and *torus palatinus* respectively (*see* pp. 316, 305, 341). They may need surgical removal before satisfactory dentures can be fitted.

Severe systemic infections during the time the teeth are developing may lead to faults in enamel development. These are often visible as horizontal lines in the incisor teeth (Compare with Harris's growth lines of bone on p. 245).

DENTAL RADIOLOGY

Due to their dense mineralization the enamel and dentine of teeth can readily be seen in radiographs (**8**.70, 71, 72). The pulp appears as a radiolucent region opening into the radiolucent periodontal ligament. Caries, which attacks teeth through the enamel or root surface, can be easily diagnosed by use of intra-oral radiographs because it demineralizes the teeth.

The root of the tooth is separated from the cortical plate of the socket by the radiolucent periodontal ligament. The cortical plate, being formed of compact bone, can be seen as a radio-opaque line, the *lamina dura*. Chronic infections of the pulp of a tooth lead to the spread of infection into the periodontal ligament and then to resorption of the lamina dura around the apex of the tooth. Thus, in a radiograph, continuity of the lamina dura around the apex of a tooth is usually diagnostic of a healthy apical region, except in cases of acute infection in which there has not yet been time for resorption of bone.

Anteriorly the interdental septum forms a sharp crest between teeth; between the molariform teeth it forms a table. With the onset of periodontal disease, which is the eventual outcome of gingival disease, the bony crest or table of the interdental septum is resorbed. Such resorption can be demonstrated by radiographs and the extent of the periodontal involvement judged.

Radiographs are commonly used to analyse the separation between the roots of the upper teeth and the maxillary sinus; it is not uncommon for the roots of the second premolar and the first and second molar teeth to project into the sinus. Lower third molars are commonly prevented from erupting by impaction against the distal sides of the lower second molars. Radiographs are indispensable in assessing the degree of difficulty to be expected when extracting such impacted third molars. Just superior to the mental spines (genial tubercles) many mandibles exhibit a well defined pit terminating in a canal, which presumably contains either a median blood vessel or perhaps some penetrating collagen fibres connected to the lingual frenulum. This pit, together with a radio-opaque ring of compact bone, can nearly always be seen on radiographs of incisors, and can be used to locate the midline in an edentulous mandible.

TONGUE, PHARYNX AND OESOPHAGUS

The Tongue

The tongue (**8**.73–75) is a muscular organ associated with the functions of deglutition, taste and speech; it lies partly in the mouth and partly in the pharynx. By its constituent muscles it is attached to the hyoid bone, the mandible, the styloid processes, the soft palate and the wall of the pharynx. It has a root, a tip, a curved dorsum and an inferior surface. Normally, the mucous membrane on its upper surface is moist and pink.

The *root* of the tongue (**8**.74) is attached to the hyoid bone and the mandible, and between these bones is in contact inferiorly with the geniohyoid and the mylohyoid muscles. The *dorsum* (or posterosuperior surface) is generally convex in all directions in the resting state, and is divided into an anterior part which faces superiorly, and a posterior part which faces posteriorly. These two parts are separated by a V-shaped furrow, termed the *sulcus terminalis*, the limbs of which run laterally and forwards from a median pit, named the *foramen caecum*, to the palatoglossal arches (**8**.73). The foramen caecum marks the site of the upper end of the thyroid diverticulum (p. 198), and the sulcus terminalis serves as the boundary between the oral part (or anterior two-thirds), and the pharyngeal part (or posterior one-third) of the tongue. These two parts differ in the structure of their covering mucous membrane, in their nerve supply and in their development (pp. 197, 1306). (They are more suitably referred to as *presulcal* and *postsulcal* parts.)

The oral part of the tongue (**8**.73, 74) is placed in the cavity and floor of the mouth; its *apex* rests against the incisor teeth; its *margin* is free and in contact with the gums and teeth; its *superior surface* is in relation with the hard and soft palates. On each border, just in front of the palatoglossal arch, there are four or five vertical folds, named the *foliate papillae* (**8**.73). The mucous membrane of the superior surface of the oral part is marked by a median furrow (**8**.73, 74), is intimately adherent to the subjacent muscle, and is covered with papillae. The mucous membrane on the inferior surface is smooth, and of a purplish colour; it is reflected from the tongue to the floor of the mouth and the gums. In the median plane it is connected to the floor of the mouth by the frenulum linguae (**8**.74). Lateral to the frenulum, the deep lingual vein is visible through the mucous membrane, and at the lateral side of the vein there is a fringed fold of mucous membrane, named the *plica fimbriata*, which is directed forwards and medially towards the apex. The oral part of the tongue is developed from the lingual swellings of the mandibular arch, and to a small extent from the tuberculum impar (p. 197). Its nerve of ordinary sensation is the lingual, its nerve of taste, the chorda tympani. (For the origin of the lingual musculature *see* p. 154.)

The pharyngeal part of the tongue (**8**.73) is posterior to palatoglossal arches; its posterior surface (sometimes named the *base* of the tongue) forms the anterior wall of the oral part of the pharynx. The mucous membrane covering it is reflected laterally on to the palatine tonsils and the pharyngeal wall, and posteriorly on to the epiglottic folds. It is devoid of papillae, but exhibits a number of low elevations, due to the presence of underlying nodules of lymphoid tissue, which are embedded in the submucous tissue and collectively constitute the *lingual tonsil*. The pharyngeal part of the tongue is developed from the hypobranchial eminence (p. 197). Its nerves of ordinary sensation and taste are derived from the glossopharyngeal. The branches of the glossopharyngeal nerve extend beyond the sulcus terminalis to supply taste fibres to the vallate papillae, an arrangement which is explained by the forward extension of the part of the hypobranchial eminence derived from the third pharyngeal arch over the posterior part of the lingual swellings (p. 197).

The papillae of the tongue (**8**.73) are projections of the lamina propria which cause elevations of the epithelium above the general level of the lingual surface. They are thickly distributed over the anterior region of the dorsum, giving to this part its characteristic roughness, and are named *papillae vallatae, fungiformes, filiformes* and *simplices*. The papillae are limited to the anterior, presulcal, area of the tongue and are modifications of the mucous membrane visible to the naked eye and designed to increase the area of mucous membrane coming into contact with the fluid which is being tasted. The taste buds (for details, see p. 1138), however, are microscopic specialized cellular arrangements about the endings of the nerves of taste and are much more widespread than the papillae (e.g. entire dorsum and sides of tongue, epiglottis and lingual surface of the soft palate); each of these regions will therefore be innervated by the appropriate nerve of taste (facial, glossopharyngeal, or vagus). The papillae are more readily visible in the living if the tongue is dried.

The *vallate papillae* (**8**.73) are of large size, and vary from eight to twelve in number. They are situated on the dorsum of the tongue, and form a V-shaped row immediately in front of and parallel with the sulcus terminalis. Each papilla is from 1 mm to 2 mm in diameter, and is attached within a circular depression of

the mucous membrane; each depression is surrounded by a wall (vallum) separated from the papilla by a circular sulcus (**8**.76). The papilla is like a truncated cone, the smaller end attached to the tongue; the broader end projects above the general surface of the tongue, and shows numerous small secondary papillae subjacent to the epithelium. The entire papilla and surrounding sulcus are covered with stratified squamous epithelium, and in both walls of the sulcus taste buds abound.

The *fungiform papillae* (**8**.73, 77), more numerous than the

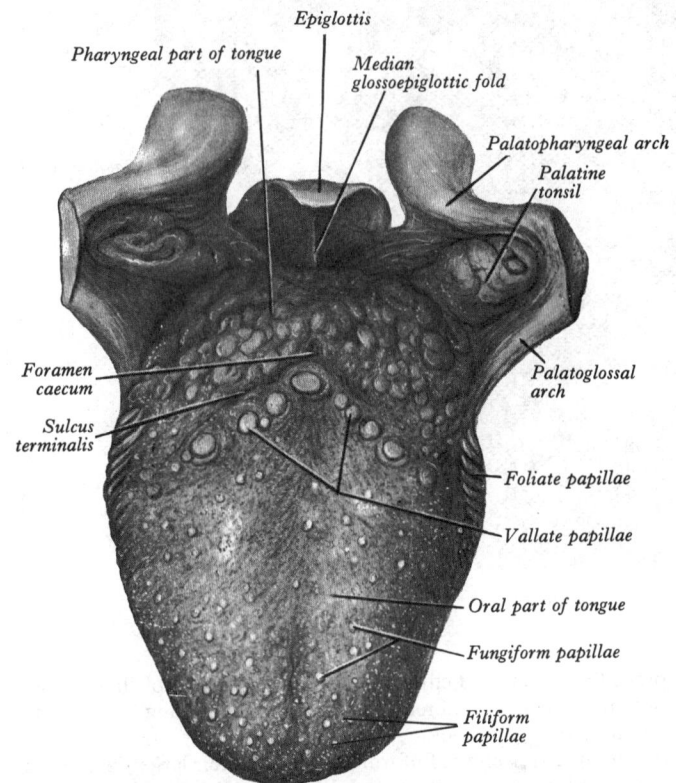

8.73 The dorsum of the tongue.

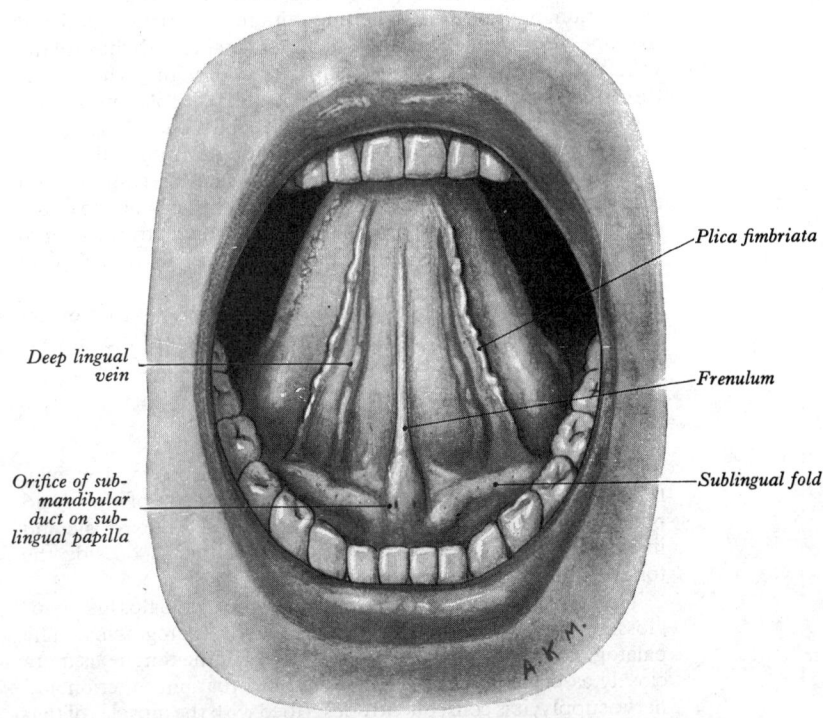

8.74 The cavity of the mouth. The tip of the tongue is turned upwards. In the person from whom the drawing was made the two sublingual papillae formed a single median elevation (*see* p. 1275).

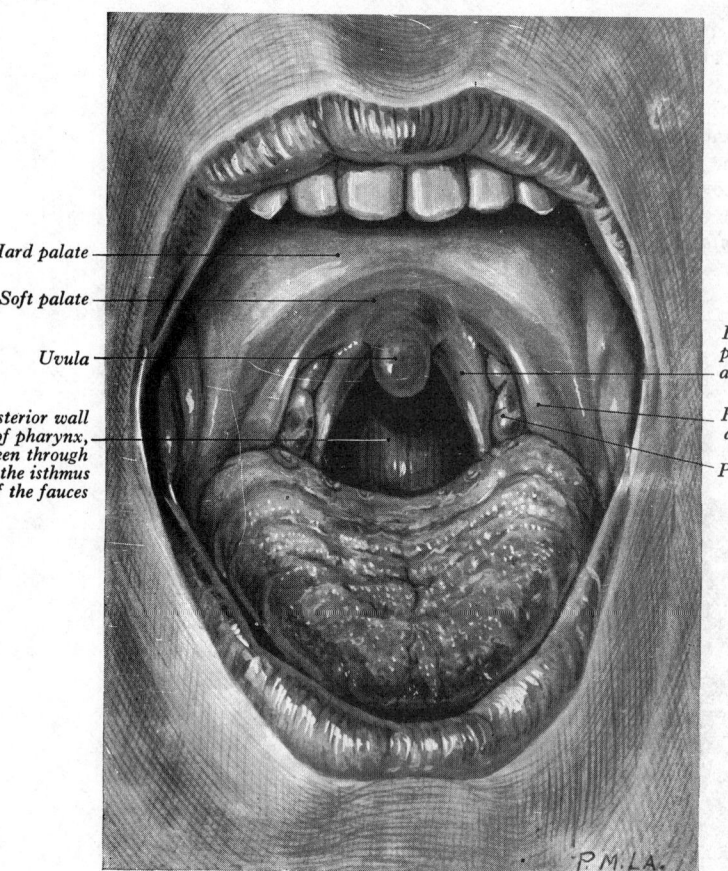

Hard palate

Soft palate

Uvula

Posterior wall
of pharynx,
seen through
the isthmus
of the fauces

Palato
pharyngeal-
arch

Palatoglossal
arch

Palatine tonsil

8.75   The cavity of the mouth.

preceding, are found chiefly at the sides and apex of the tongue, but are usually scattered irregularly and sparingly over the dorsum, though sometimes they may be numerous. They are distinguished from the filiform papillae by their large size, round shape and deep red colour; each has secondary papillae beneath the epithelium and bears taste buds.

The *filiform papillae* (**8**.77), also called conical papillae because of the conical epithelial cap they bear, cover the presulcal area of the dorsum of the tongue. They are minute, conical or cylindrical in shape, and arranged in rows which run parallel with those of the vallate papillae, excepting at the apex of the tongue, where their direction is transverse. The filiform papillae display numerous secondary connective tissue papillae, but these are more pointed than the secondary vallate and fungiform papillae. The epithelium covering the filiform papillae may be split up into filamentous processes, each of which forms the apex of one of the secondary papillae; these processes are of a whitish tint, owing to the thickness and density of the epithelium, the cells of which are elongated and keratinized.

The *papillae simplices* are similar to the papillae of the corium (dermis) of the skin (p. 1217), and cover the whole of the mucous membrane of the tongue, including the larger papillae.

THE LINGUAL MUSCULATURE

The tongue is divided into right and left halves by a median fibrous septum, which is fixed below to the body of the hyoid bone. In each half there are two sets of muscles, extrinsic and intrinsic (**8**.78, 79); the former have attachments outside the tongue, the latter are contained within it.

The *extrinsic muscles* (**8**.79) include the genioglossus, hyoglossus, styloglossus, chondroglossus and palatoglossus. The palatoglossus, although one of the muscles of the tongue, is more closely associated with the soft palate in situation, function and nerve supply; it is consequently described with the muscles of that structure (p. 1271).

The genioglossus is a triangular muscle placed close to and parallel with the median plane. It arises by a short tendon from the

upper genial tubercle on the inner surface of the symphysis of the mandible, just above the origin of the geniohyoid, and spreads out in a fan-like form. The inferior fibres are attached by a thin aponeurosis to the upper part of the anterior surface of the body of the hyoid bone, close to the midline, a few passing between the hyoglossus and chondroglossus to blend with the middle constrictor of the pharynx; the middle fibres pass backwards, and the superior ones upwards and forwards, to enter the whole length of the ventral surface of the tongue, from the root to the apex, intermingling with the intrinsic muscles. The muscles of opposite sides are separated posteriorly by the septum of the tongue (p. 1306); in front, they are more or less blended owing to the decussation of fasciculi in the median plane. A recent study of this muscle in several mammals, including man, suggest that its fibres fall short of the apex of the tongue (Doran and Baggett 1972).

Actions. The genioglossus draws the tongue forwards and protrudes its apex from the mouth. The two muscles acting in their entirety draw the median part of the tongue downwards so as to make the superior surface concave from side to side.

The hyoglossus, thin and quadrilateral, arises from the whole length of the greater cornu and the front of the lateral part of the body of the hyoid bone; it passes almost vertically upwards and enters the side of the tongue, between the styloglossus laterally, and medially the inferior longitudinal muscle. The fibres arising from the body of the hyoid bone overlap those from the greater horn.

Relations. The hyoglossus is in relation by its *superficial surface* with the tendon of the digastric, the stylohyoid, styloglossus and mylohyoid, the lingual nerve and the submandibular ganglion, the sublingual gland, the deep portion of the submandibular gland and the submandibular duct, the hypoglossal nerve and the deep lingual vein. By its *deep surface* it is in relation with the stylohyoid ligament, the genioglossus, the inferior longitudinal muscle, the lingual artery and the glossopharyngeal nerve. In its lower and posterior part, it is separated from the middle constrictor by the lingual artery. This portion of the muscle lies in the lateral wall of the pharynx, a little below the tonsil. The following structures pass deep to the posterior border of the muscle, in order from above downwards: the glossopharyngeal nerve, the stylohyoid ligament and the lingual artery.

Action. The hyoglossus depresses the tongue.

The chondroglossus is sometimes described as a part of the hyoglossus, but it is separated from that muscle by fibres of the genioglossus which pass to the side of the pharynx. It is about 2 cm long, and arises from the medial side and base of the lesser cornu and contiguous part of the body of the hyoid bone; it

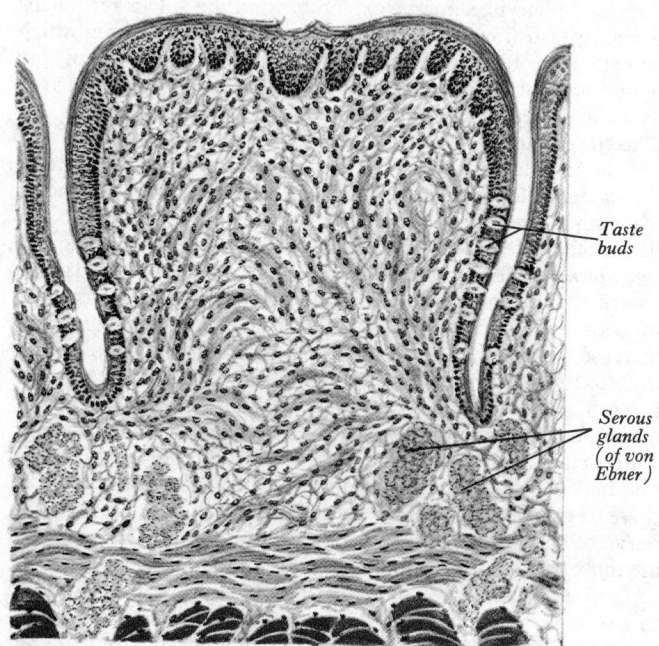

Taste
buds

Serous
glands
(of von
Ebner)

8.76   Section through a vallate papilla. Stained with haematoxylin and eosin. Magnification about × 32. (After Sobotta.)

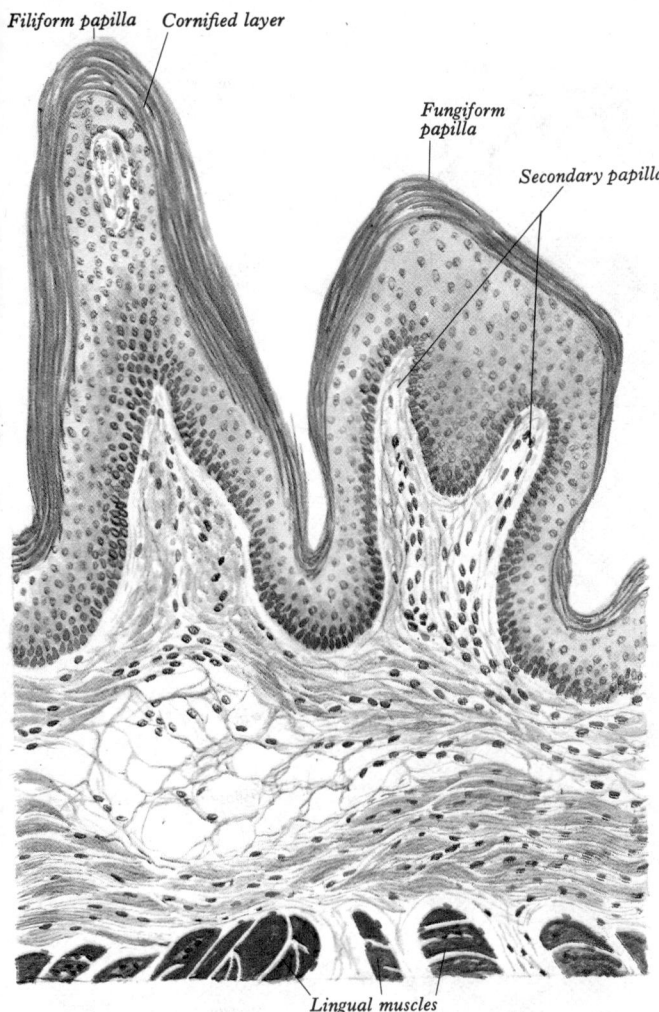

Filiform papilla   *Cornified layer*

Fungiform papilla

Secondary papillae

Lingual muscles

8.77 Section through a filiform papilla and an adjoining fungiform papilla. Stained with haematoxylin and eosin. Magnification about × 30. (Drawn from a preparation kindly lent by Dr. E. E. Hewer.)

ascends and blends with the intrinsic muscular fibres of the tongue, between the hyoglossus and genioglossus.

A small slip arises occasionally from the cartilago triticea in the lateral thyrohyoid ligament and enters the tongue with the posterior fibres of the hyoglossus.

Action. The chondroglossus assists the hyoglossus in depressing the tongue.

The styloglossus, the shortest and smallest of the three styloid muscles, arises from the anterior and lateral surfaces of the styloid process, near its apex, and from the upper end of the stylomandibular ligament. Passing downwards and forwards, it divides upon the side of the tongue into two portions; one, longitudinal, enters the side of the tongue near its dorsal surface, blending with the inferior longitudinal muscle in front of the hyoglossus; the other, oblique, overlaps the hyoglossus and decussates with its fibres.

Action. The styloglossus draws the tongue upwards and backwards.

Nerve supply. With the exception of the palatoglossus (p. 1272) all the extrinsic muscles of the tongue are supplied by the hypoglossal nerve.

The *intrinsic muscles* are the superior and inferior longitudinal, the transverse and the vertical.

The superior longitudinal muscle of the tongue is a thin stratum of oblique and longitudinal fibres immediately subjacent to the mucous membrane on the dorsum of the tongue. It arises from the submucous fibrous layer close to the epiglottis and from the median fibrous septum, and runs forward to the edges of the tongue, some of its fibres being inserted into the mucous membrane.

The inferior longitudinal muscle of the tongue is a narrow

band close to the inferior surface of the tongue between the genioglossus and hyoglossus. It extends from the root to apex of the tongue, some of its posterior fibres being connected with the body of the hyoid bone; in front it blends with the fibres of the styloglossus.

The transverse muscle of the tongue consists of fibres which arise from the median fibrous septum and pass laterally to be inserted into the submucous fibrous tissue at the sides of the tongue, and to blend with the palatopharyngeus (p. 1271).

The vertical muscle of the tongue is found at the borders of the fore part of the tongue. Its fibres extend from the dorsal to ventral surfaces of the organ.

Nerve supply. The intrinsic muscles of the tongue are supplied by the hypoglossal nerve.

Actions. The intrinsic muscles are mainly concerned in altering the shape of the tongue; thus, the superior and inferior longitudinal muscles tend to shorten it, but the former, in addition, turns the tip and sides upwards so as to render the dorsum concave, while the latter pulls the tip downwards making the dorsum convex. The transverse muscle narrows and elongates the tongue, and the vertical muscle renders it flatter and wider. Movements of the tongue are concerned in speech, mastication and deglutition.

LINGUAL STRUCTURE

The tongue consists chiefly of striated muscular tissue, partly invested by mucous membrane.

The *lingual mucous membrane* covering the under surface of the tongue is thin, smooth and identical in structure with that lining the rest of the oral cavity. The mucous membrane of the pharyngeal part of the dorsum of the tongue contains a large number of follicles of lymphoid tissue; each follicle forms a rounded eminence, in the centre of which there is a minute orifice leading into a funnel-shaped cavity or recess; numerous round or oval nodules of lymphoid tissue, each enveloped by a capsule derived from the submucous fibrous layer, are grouped around this recess, which receives the openings of the ducts of some mucous glands in its floor. The mucous membrane on the oral part of the dorsum of the tongue is thin, intimately adherent to the muscular tissue, and covered with numerous *papillae* (p. 1221). It consists of connective tissue (*lamina propria*) covered with stratified squamous epithelium, which also covers each papilla.

The *lamina propria* consists of a dense feltwork of fibrous connective tissue, with numerous elastic fibres, firmly united with the fibrous tissue between the muscular bundles of the tongue. It contains the ramifications of the numerous vessels and nerves from which the papillae are supplied, large plexuses of lymph vessels and the glands of the tongue.

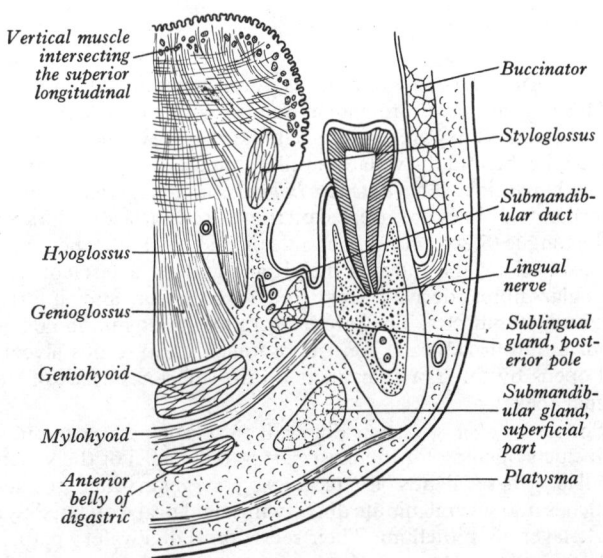

Vertical muscle intersecting the superior longitudinal

Buccinator

Styloglossus

Submandibular duct

Lingual nerve

Hyoglossus

Sublingual gland, posterior pole

Genioglossus

Submandibular gland, superficial part

Geniohyoid

Mylohyoid

Platysma

Anterior belly of digastric

8.78 Diagram of a coronal section through the tongue, the mouth and the body of the mandible opposite the first molar tooth.

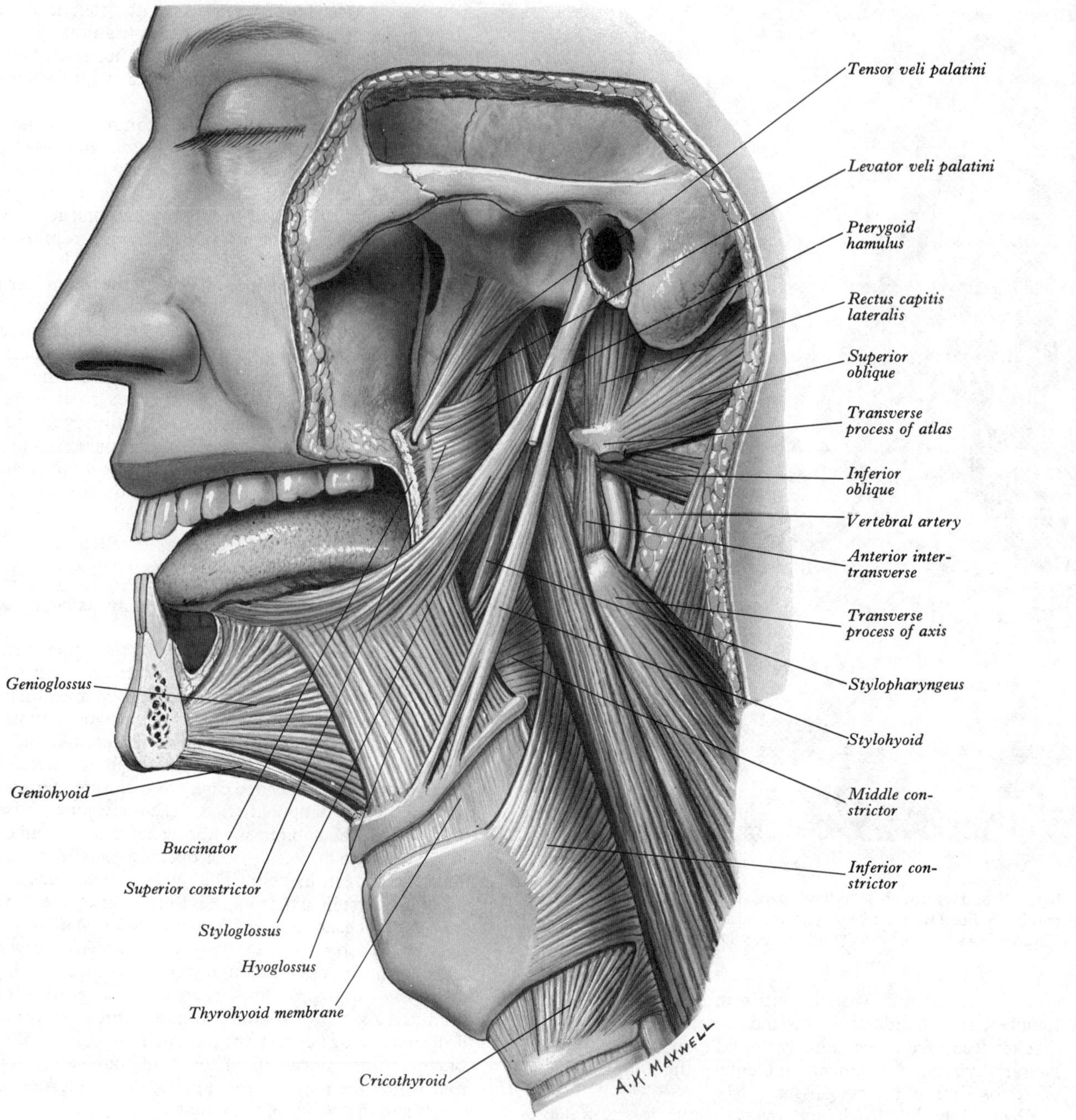

8.79 A dissection showing the muscles of the tongue and pharynx.

The *lingual glands* are of mucous, serous, and mixed varieties.

The *mucous glands* are similar in structure to the labial and buccal glands. They are numerous in the pharyngeal region of the tongue, i.e. behind the vallate papillae, but are also present at the tip and margins. The *anterior lingual salivary glands* require special notice. They are situated on the ventral surface of the apex of the tongue (**8**.80), one on each side of the frenulum, where they are covered by the mucous membrane and by a fasciculus of muscular fibres derived from the styloglossus and inferior longitudinal muscles. They are from 12 mm to 20 mm long, and about 8 mm broad; each consists of mucous and serous alveoli, and opens by three or four ducts on the inferior surface of the lingual apex.

The *serous glands* (of von Ebner, **8**.76) occur near taste buds, their ducts opening for the most part into the sulci of the vallate papillae. These glands are racemose; the main duct of each branches into several minute ducts, which end in alveoli lined by a single layer of epithelium. Their secretion is of a watery nature, and probably assists in distributing the substance to be tasted over the taste area, as well as in washing away the substance after it has been tasted.

The *septum* of the tongue is a median fibrous partition which extends throughout the length of the organ, but does not quite reach the dorsum; it gives origin to the transverse lingual muscle, and is well displayed in a coronal section of the tongue. Posteriorly it expands in a transverse direction and forms what is known as the *hyoglossal membrane*; this membrane connects the root of the tongue to the hyoid bone, and the inferior fibres of the genioglossi are attached to it.

Vessels and Nerves

The main *artery* of the tongue is the lingual branch of the external carotid artery (p. 678), but the tonsillar and ascending palatine branches of the facial and the ascending pharyngeal arteries also give branches to the lingual root. In the vallecula (p. 1231) the epiglottic rami of the superior laryngeal artery anastomose with the inferior dorsal branches of the lingual artery. From this rich anastomotic network the lingual musculature is supplied and a very dense submucosal plexus supplies the mucosa (Combelles 1974). The *veins* are described on p. 742.

The *lymph vessels of the tongue* are described on **6**.159, p. 788.

The *sensory nerves of the tongue* are: (1) the lingual branch of the mandibular nerve, which is the nerve of general sensibility for the presulcal region of the tongue; (2) the chorda tympani branch of the facial nerve (p. 1072), which runs in the sheath of the lingual nerve, and is the nerve of taste for the presulcal region exclusive of the vallate papillae (p. 1138) and is partly derived from the nervus intermedius; (3) the lingual branch of the glossopharyngeal nerve (p. 1076), distributed to the mucous membrane at the base and sides of the tongue, and to the vallate papillae, and is the nerve of taste and of general sensibility for this region; (4) the superior laryngeal nerve (p. 1079), which sends some fine branches to the part immediately in front of the epiglottis.

The existence of a proprioceptor apparatus in the tongue is an old problem, the literature of which has recently been extensively reviewed by Fitzgerald and Sachithanadan (1979). Muscle spindles have been demonstrated in monkeys (Bowman 1968) and also in mankind (e.g. Nakayama 1944; Cooper 1953; Kubota *et al.* 1975), and the workers cited above have confirmed these findings on an extensive scale in simian material. The peripheral route of nerve fibres from these undoubted receptor organs is much less clear, the possible pathways being in the lingual or hypoglossal nerves, and also by cervical spinal nerves which communicate with the latter cranial nerve. In the monkey Fitzgerald and Sachithanadan present strong evidence indicating the hypoglossal nerve as the main vehicle of lingual proprioceptor nerve fibres from the intrinsic and extrinsic musculature, while considerable numbers leave the hypoglossal nerve to enter the second and third cervical anterior primary rami.

In the so-called posterior third of the human tongue, that is the

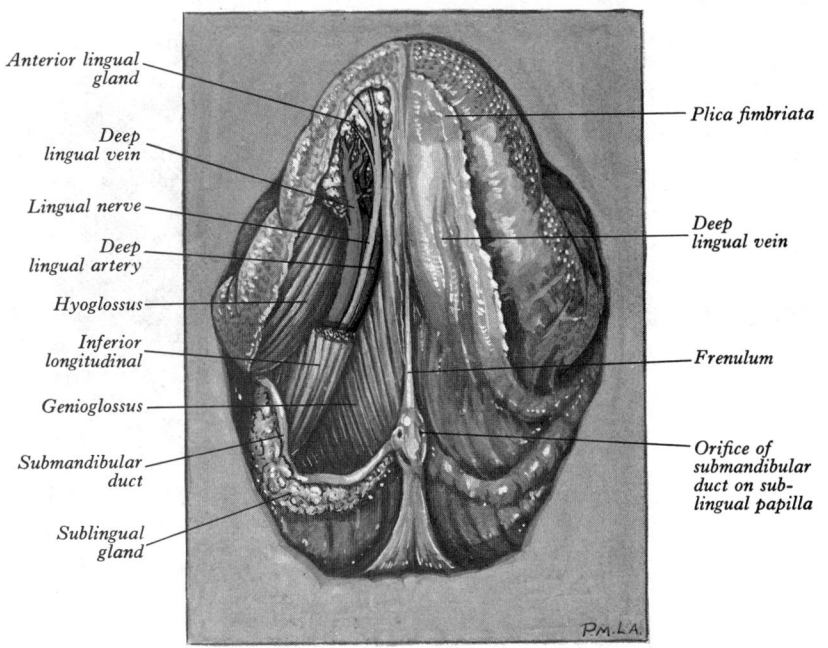

8.80 A dissection of the inferior lingual surface. On the right side (left side of figure) the mucous membrane has been removed, and the inferior longitudinal muscle has been divided and partially resected.

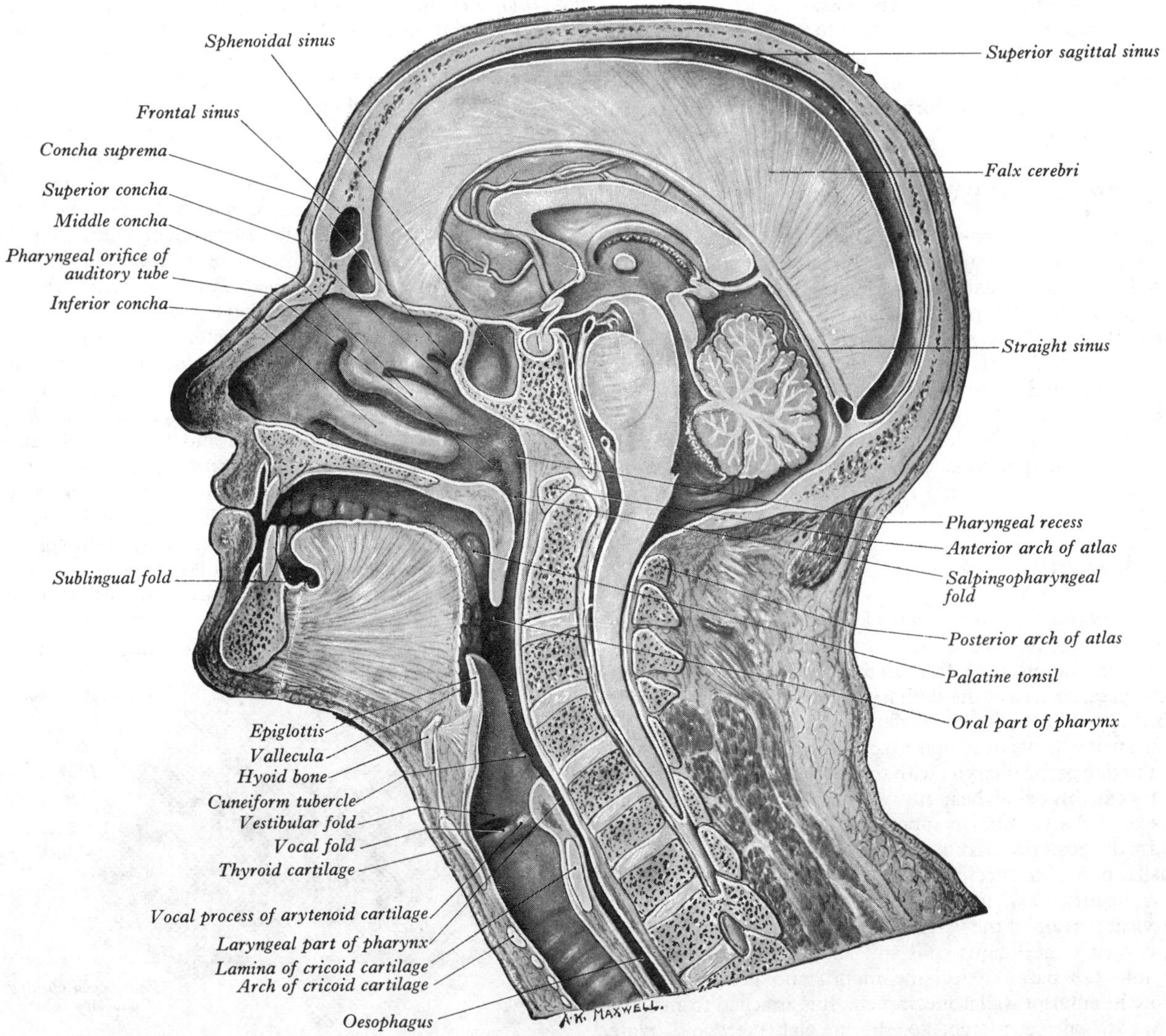

8.81 A sagittal section through the nose, mouth, pharynx and larynx. Where it divides the skull and the brain, the section passes slightly to the left of the median plane, but below the base of the skull, it passes slightly to the right of the median plane.

1307

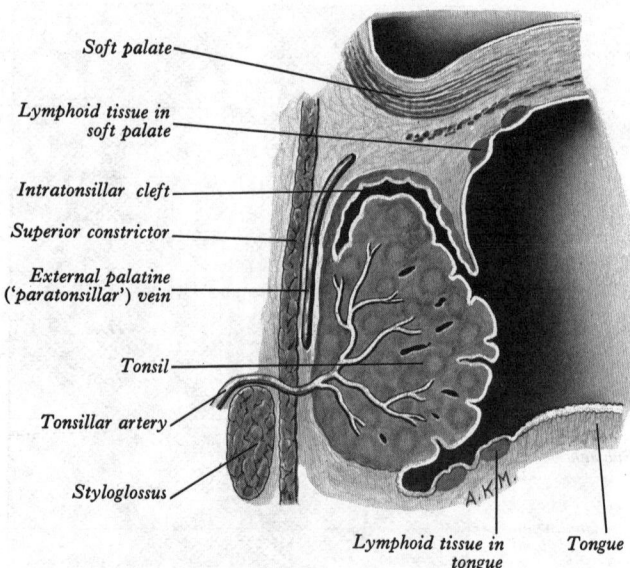

8.82 A coronal section through the palatine tonsil.

Labels (top to bottom): Soft palate; Lymphoid tissue in soft palate; Intratonsillar cleft; Superior constrictor; External palatine ('paratonsillar') vein; Tonsil; Tonsillar artery; Styloglossus; Lymphoid tissue in tongue; Tongue

pharyngeal part, caudal to the sulcus terminalis, scattered isolated nerve cells have been observed. It is hypothecated that they are postganglionic parasympathetic neurons, probably innervating glandular tissue and nonstriated muscle in blood vessels (Chu 1968).

Applied Anatomy. Congenital cysts and fistulae may develop from persistent remains of the thyroglossal duct (p. 198).

It is the attachment of the genioglossi to the genial tubercles on the inner surface of the symphysis of the mandible which prevents the tongue from sinking back and obstructing respiration, and, therefore, anaesthetists pull forward the mandible to get the full benefit of this connexion.

THE OROPHARYNGEAL ISTHMUS

The aperture by which the mouth communicates with the pharynx is called the *oropharyngeal isthmus* (**8.**75). Above, it is bounded by the soft palate, below, by the dorsum of the tongue and, at the sides, by the palatoglossal arches.

The *palatoglossal arch* runs downwards, laterally and forwards on each side from the inferior surface of the soft palate to the side of the tongue, and is formed by the projection of the palatoglossus (p. 1271) with its covering mucous membrane. The approximation of the arches, which helps to shut off the mouth from the oral part of the pharynx, plays an important part in the mechanism of deglutition (p. 1313).

The Pharynx

The pharynx (**8.**82–86) is the part of the digestive tube which is placed behind the nasal cavities, the mouth and the larynx. It is a musculomembranous tube, from 12 to 14 cm long, which extends from the basal surface of the skull to the level of the sixth cervical vertebra opposite the lower border of the cricoid cartilage. Its width is greatest at its uppermost part, where it measures 3·5 cm; at the junction of the pharynx with the oesophagus it is reduced to about 1·5 cm, this being the narrowest part of the alimentary canal (exclusive of the vermiform appendix). The pharynx is limited, *above*, by the posterior part of the body of the sphenoid bone and the basilar part of the occipital bone; *below*, it is continuous with the oesophagus; *behind*, it is separated by loose areolar tissue from the cervical portion of the vertebral column and the prevertebral fascia covering the longus colli and longus capitis; *in front*, it opens into the nasal cavity, the mouth and the larynx, and therefore its anterior wall is incomplete. It is attached from above downwards, on each side, to the medial pterygoid plate, pterygomandibular raphe, mandible, tongue, hyoid bone, and thyroid and cricoid cartilages; *laterally*, it communicates with the

tympanic cavities through the auditory tubes, and is in relation with the styloid processes and their muscles, the common, internal and external carotid arteries, and some of the branches of the latter artery. The pharynx consists of three parts: nasal, oral and laryngeal (**8.**81).

NASAL PART OF THE PHARYNX

The nasal part of the pharynx lies behind the nose and above the level of the soft palate. (From a study of the comparative anatomy of the soft palate, it was suggested by Jones (1940) that the 'nasal part of the pharynx' in man is morphologically an extension of the nasal cavities, and that the soft palate really separates the pharynx from the nasal chambers. This view has been criticized by Leela *et al.* (1974), who consider that the transitional zone between nasal and pharyngeal territories lies just anterior to the orifices of the auditory tubes. They confirm, however, the general concept of nasal and pharyngeal contributions to the region.) With the exception of the soft palate its walls are immovable, and consequently its cavity is never obliterated; in this respect it differs from the oral and laryngeal parts, and resembles the nasal cavities. *Anteriorly* (**8.**81) it communicates with the nasal cavity through the posterior apertures of the nose; these measure about 25 mm vertically, and 12·5 mm transversely, and are separated by the posterior edge of the nasal septum. Between the free edge of the soft palate and the posterior wall of the pharynx the nasal and oral parts of the pharynx communicate through an opening, termed the *pharyngeal isthmus*; in the act of swallowing, this opening is closed by the elevation of the soft palate and the contraction of the palatopharyngeal sphincter (p. 1272). The *lateral wall*, on each side, presents the *pharyngeal opening of the auditory tube*, which lies 10 to 12·5 mm behind and a little below the posterior end of the inferior nasal concha. Somewhat triangular in shape, this opening is bounded above and behind by the *tubal elevation*, a firm prominence which is provided by the underlying pharyngeal end of the cartilage of the auditory tube (p. 1197). The prominent posterior margin facilitates the introduction into the tube of a catheter which has been passed through the nares and along the floor of the nasal cavity. A vertical fold of mucous membrane, termed the *salpingopharyngeal fold*, stretches from the lower part of the tubal elevation downwards in the wall of the pharynx; it contains the *salpingopharyngeus*. A second and smaller fold, termed the *salpingopalatine fold*, stretches from the upper and front part of the elevation to the soft palate. The levator veli palatini, as it enters the soft palate, produces an elevation of the mucous membrane immediately below the pharyngeal opening of the tube (**8.**85). Behind the tubal elevation the mucous membrane lines a recess of variable depth, termed the *pharyngeal recess*. (For literature, radiological and histological details, *see* Khoo *et al.* 1969.) The *roof* and *posterior wall* form a continuous sloping surface which inclines downwards and backwards. It is supported mainly by the basilar part of the occipital bone and, to a lesser extent, by the posterior part of the body of the sphenoid, in front, and the anterior arch of the atlas, below. A collection of lymphoid tissue, best developed in children, lies in the mucous

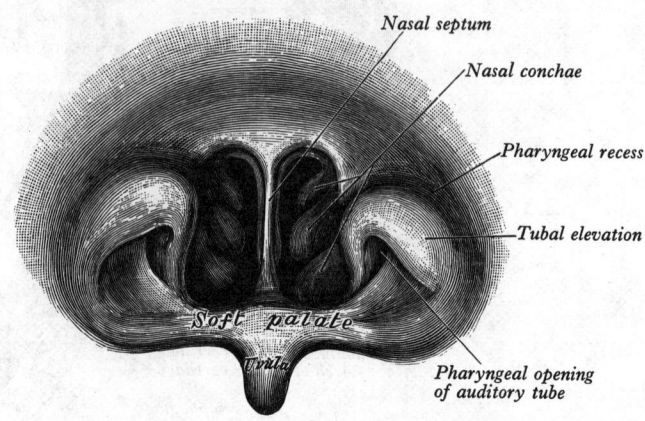

Labels: Nasal septum; Nasal conchae; Pharyngeal recess; Tubal elevation; Soft palate; Uvula; Pharyngeal opening of auditory tube

8.83 Ventral boundary of the nasal part of the pharynx, as seen in posterior rhinoscopy.

membrane of the upper part of this surface and is known as the *pharyngeal tonsil*.

The pharyngeal tonsil is visible to the naked eye during the later months of fetal life and usually increases in size up to the age of six or seven years, after which it not infrequently begins to atrophy. In a child of eighteen months it forms a forwardly projecting pyramidal prominence, the apex of which is near the nasal septum, and the base at the junction of the roof and posterior wall of the nasal part of the pharynx. The prominence consists of a number of folds which radiate forwards and laterally from a median recess, termed the *pharyngeal bursa*, which runs upwards and backwards for some distance into its substance. The folds consist mainly of diffuse lymphoid tissue, but there are also some deeply placed mucous glands. The pharyngeal bursa lies in the base of the pharyngeal tonsil and presents the appearance of a blind recess. In the embryo, the notochord lies for a short distance inferior to the base of the skull, in the region of the developing basilar part of the occipital bone (p. 141); here it is attached to the endoderm forming the roof of the primitive pharynx, and with subsequent growth of this region, the notochordal attachment draws out an angled recess of the endoderm (the *pouch of Luschka*) which forms the pharyngeal bursa. (For bibliographies *see* Cave 1965; Slipka 1972). The lateral prolongation of the pharyngeal tonsil behind the pharyngeal opening of the auditory tube is known as the *tubal tonsil*.

ORAL PART OF THE PHARYNX

The oral part of the pharynx reaches from the soft palate to the upper border of the epiglottis. It opens anteriorly, through the oropharyngeal isthmus, into the mouth and faces the pharyngeal part of the tongue. Its lateral wall presents the palatopharyngeal arch and the palatine tonsil. Posteriorly, it is on a level with the body of the second cervical vertebra and the upper part of the body of the third.

The *palatopharyngeal arch* lies behind and projects further towards the median plane than the palatoglossal arch; it runs downwards, laterally and backwards from the margin of the uvula to the side of the pharynx, and is formed by the projection of the palatopharyngeus (p. 1271), covered with mucous membrane. On each side the palatopharyngeal and palatoglossal arches are separated by a triangular recess, the *tonsillar sinus*, in which the tonsil is contained.

The palatine tonsils (8.75) are two masses of lymphoid tissue, situated in the lateral walls of the oral part of the pharynx. Each tonsil is placed in the triangular recess (*tonsillar sinus*) between the diverging palatoglossal and palatopharyngeal arches. Its medial surface is free and forms a conspicuous projection into the pharynx during childhood, but the size of this projection is not a true indication of the size of the organ. Its deep, or lateral, aspect extends upwards, downwards and forwards beyond the limits of the medial surface and is embedded below the level of the mucous membrane. Inferiorly, it extends into the dorsum of the tongue; superiorly, it invades the soft palate; anteriorly, it may extend for some distance embedded beneath the palatoglossal arch. The tonsil is variable in size and is frequently the seat of inflammatory changes involving hypertrophy. As a result it is difficult to decide which of the many varieties encountered is to be regarded as normal. In late fetal life, a free fold of mucous membrane, which extends backwards from the palatoglossal arch, covers the antero-inferior part of the tonsil and is termed the *plica triangularis*. In the child this fold is usually invaded by lymphoid tissue and becomes incorporated into the tonsil; it is only rarely present, as a small, free fold which extends backwards from the lower part of the palatoglossal arch in adults.

The upper part of the tonsil contains a deep *intratonsillar cleft*, frequently and erroneously termed the *supratonsillar fossa*. This cleft does not lie above the tonsil, but actually in its substance, and its upper wall contains a quantity of lymphoid tissue (8.82) which may reach a large size and extend into the soft palate. The mouth of the cleft is semilunar in shape and is parallel to the curve of the dorsum of the tongue. After puberty the embedded part of the tonsil diminishes considerably in size and the projecting medial surface becomes flattened and much less prominent.

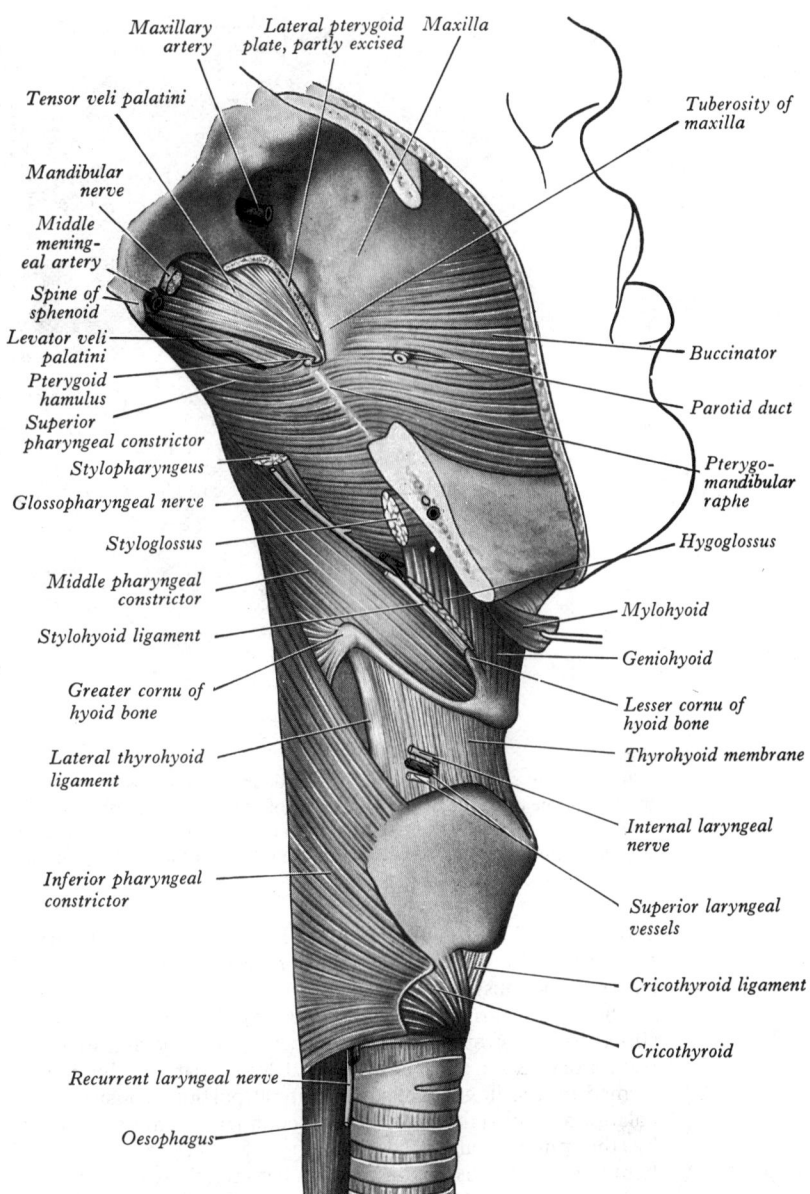

8.84 The buccinator and the muscles of the pharynx.

The *medial surface* of the tonsil presents from twelve to fifteen orifices leading into deep, narrow recesses, termed the *tonsillar crypts*, which penetrate nearly the whole thickness of the tonsil and from which numerous follicles branch out into the tonsillar substance.

The *lateral* or *deep aspect* is covered by a layer of fibrous tissue, termed the *capsule*. In most of its extent the tonsil and its capsule can easily be separated from the muscular wall of the pharynx, which is formed in this situation by the superior constrictor with the styloglossus laterally (8.82, 85). In its antero-inferior part the capsule is firmly connected to the side of the tongue, and behind this point it receives the insertion of some muscular fibres of the palatoglossus and palatopharyngeus. In this situation the tonsillar artery, which is a branch of the facial artery, pierces the superior constrictor and at once enters the tonsil, accompanied by two venae comitantes. An important, and sometimes large, palatine vein (the *external palatine* or 'paratonsillar vein') descends from the soft palate across the lateral aspect of the capsule of the tonsil before piercing the pharyngeal wall (8.82). It is this vessel which is responsible for the excessive venous haemorrhage from the upper angle of the tonsillar sinus sometimes encountered in excision of the tonsil (Browne 1928). The muscular wall of the tonsillar sinus separates the tonsil from the ascending palatine artery and, occasionally, from the facial artery itself (p. 679), which, if very tortuous, may be extremely

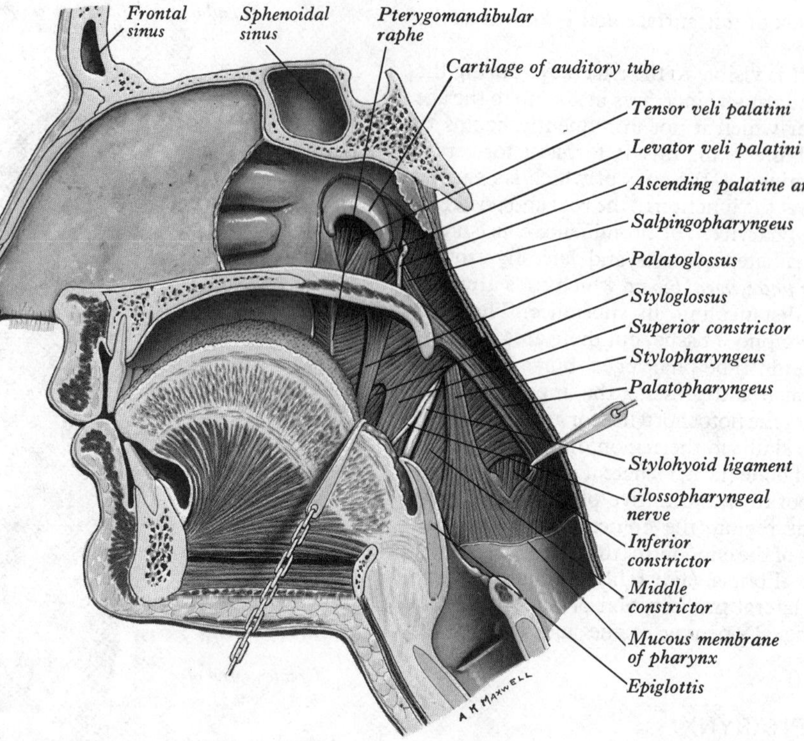

8.85 Median sagittal section of the head, showing a dissection of the interior of the pharynx, after the removal of the mucous membrane. The bodies of the cervical vertebrae have been removed and the cut posterior wall of the pharynx then retracted dorsolaterally. The palatopharyngeus is drawn dorsally to show the cranial fibres of the inferior constrictor, and the dorsum of the tongue is drawn ventrally to display a part of the styloglossus in the angular interval between the mandibular and the lingual fibres of origin of the superior constrictor.

closely related to the pharyngeal wall opposite the lower part of the tonsil. The internal carotid artery lies 25 mm behind and lateral to the tonsil.

The tonsils form part of a circular band of lymphoid tissue (*Waldeyer's ring*) surrounding the opening into the digestive and respiratory tubes. The anterior and lower part of the ring is formed by the lingual tonsil; the lateral portions consist of the palatine and tubal tonsils; the ring is completed behind and above by the pharyngeal tonsil (p. 1309). Smaller collections of lymphoid tissue exist in the intervals between these main masses.

Surface Anatomy. The palatine tonsil lies behind the third lower molar tooth and is represented by an oval area over the lower part of the masseter, a little above and in front of the angle of the mandible.

Structure. The crypts of the tonsil are lined by stratified squamous epithelium, which is continuous with that of the mucous membrane of the pharynx, and is invaded by numerous lymphocytes; some of the latter pass into the mouth and form the so-called salivary corpuscles. The tonsil consists of lymphoid tissue which is arranged in nodules or follicles. The lymphocytes are less closely packed in the centre of each nodule, which is described as a *germinal centre* (*see* p. 768), because multiplication of the lymphocytes goes on in this situation. The crypts may contain cheesy plugs of living and dead lymphocytes, bacteria and desquamated epithelium; these plugs may be gradually eliminated or may remain and become calcified; the bacteria may cause local infection and suppuration or be absorbed and cause general infection. Unlike lymph nodes, there are no lymph sinuses (p. 769) in the tonsil and it has no *afferent* lymph vessels, but a close plexus of vessels surrounds each follicle and from it *efferent* lymph vessels pass to the upper deep cervical lymph nodes, and especially to the jugulodigastric group of lymph nodes (p. 786).

Vessels and nerves. The chief *artery* supplying the tonsil is the tonsillar branch of the facial artery. In addition, it may receive a few twigs from the dorsal lingual branches of the lingual artery, the ascending palatine branch of the facial artery, the ascending pharyngeal artery, and the greater palatine branch of the maxillary artery.

One or more *veins* leave the lower part of the deep aspect of the tonsil and at once pierce the superior constrictor muscle to join the external palatine ('paratonsillar'), pharyngeal or facial veins.

The *nerves* are derived from the pterygopalatine ganglion, through the lesser palatine nerves, and from the glossopharyngeal nerve. The latter, through its tympanic branch, also supplies the mucous membrane of the tympanic cavity; hence tonsillitis may be accompanied by pain referred to the ear.

LARYNGEAL PART OF THE PHARYNX

The laryngeal part of the pharynx reaches from the cranial border of the epiglottis to the caudal border of the cricoid cartilage, where it is continuous with the oesophagus. Its anterior wall presents, from above downwards, the inlet of the larynx (p. 1233), and the posterior surfaces of the arytenoid and of the cricoid cartilages. A small recess, termed the *piriform fossa*, lies on each side of the laryngeal orifice; it is bounded, medially, by the aryepiglottic fold and, laterally, by the thyroid cartilage and the thyrohyoid membrane. Beneath the mucous membrane of the piriform fossa lie the branches of the internal laryngeal nerve after they have pierced the thyrohyoid membrane; foreign bodies may lodge in the piriform fossa and, if the instrument employed to remove them pierces the mucous membrane, the nerve is liable to be damaged, with consequent anaesthesia of the part of the mucous membrane of the larynx supplied by the nerve. Posteriorly the laryngeal part of the pharynx is level with the bodies of the third (lower part), fourth, fifth and sixth (upper part) cervical vertebrae.

Structure

The pharynx is usually described as being composed of three tissue laminae from within outwards—mucous, fibrous and muscular, external to the last being the thin buccopharyngeal fascia, which covers the external surface of the constrictor muscles and extends forwards over the pterygomandibular raphe on to the buccinator.

The *mucous membrane* is continuous with that of the auditory tubes, nasal cavity, mouth and larynx. In the nasal part of the pharynx its epithelium is columnar and ciliated; in the oral and laryngeal regions it is stratified squamous. Between the region covered by ciliated columnar epithelium and that covered by squamous epithelium there is a narrow transitional zone where the epithelium is cubical, and the cilia are imperfect or absent. Superiorly, this zone lies near the nasal septum: laterally it passes over the orifice of the auditory tube and inclines posteriorly at the union of the soft palate with the lateral wall (see also pp. 1270, 1312, 1313). Mucous glands are numerous in the nasal part of the pharynx around the orifices of the auditory tubes.

The *fibrous* intermediate layer lies between the mucous and muscular layers. It is thick above (*pharyngobasilar fascia*) where the muscular fibres are absent, and is firmly connected to the basilar region of the occipital bone and petrous part of the temporal bone medial to the carotid canal, bridging under the auditory tube and extending forwards to be attached to the posterior border of the medial pterygoid plate and to the pterygomandibular raphe. As it descends it diminishes in thickness. It is strengthened posteriorly by a strong fibrous band, which is attached above to the pharyngeal tubercle on the under surface of the basilar portion of the occipital bone, and passes downwards as a median raphe (the *pharyngeal raphe*) which gives attachment to the constrictors. The pharyngeal muscles are usually described as lying external to the fibrous layer. However, the latter is in reality the thickened, deep epimysial covering of the muscles, whilst the thinner external layer of the epimysium constitutes the *buccopharyngeal fascia*.

The *muscular coat* consists of the muscles of the pharynx.

PHARYNGEAL MUSCULATURE

The muscles of the pharynx comprise three *constrictors*, superior, middle and inferior, and a trio of muscles which descend from the styloid process, the cartilaginous torus of the auditory tube and the soft palate. The latter are the *stylo-, salpingo-,* and *palatopharyngei* and they descend obliquely into the muscular wall of the pharynx. The palatopharyngeus has already been described (p. 1271).

The inferior constrictor muscle of the pharynx is the thickest of the constrictors and consists of two parts, the cricopharyngeus and the thyropharyngeus. It arises (**8.**84) from the side of the cricoid cartilage in the interval between the origin of the cricothyroid in front, and the articular facet for the inferior cornu of the thyroid cartilage behind (*cricopharyngeus*). It also arises from the oblique line of the lamina of the thyroid cartilage, from a strip of the surface of the lamina behind this line, from a fine tendinous band, which is thrown across the cricothyroid from the inferior thyroid tubercle to the cricoid cartilage, and, by a small slip, from the inferior cornu (*thyropharyngeus*). The fibres spread backwards and medially, and are inserted with the muscle of the opposite side into a fibrous raphe in the posterior median line of the pharynx. The inferior fibres, which are horizontal, are continuous with the circular fibres of the oesophagus and

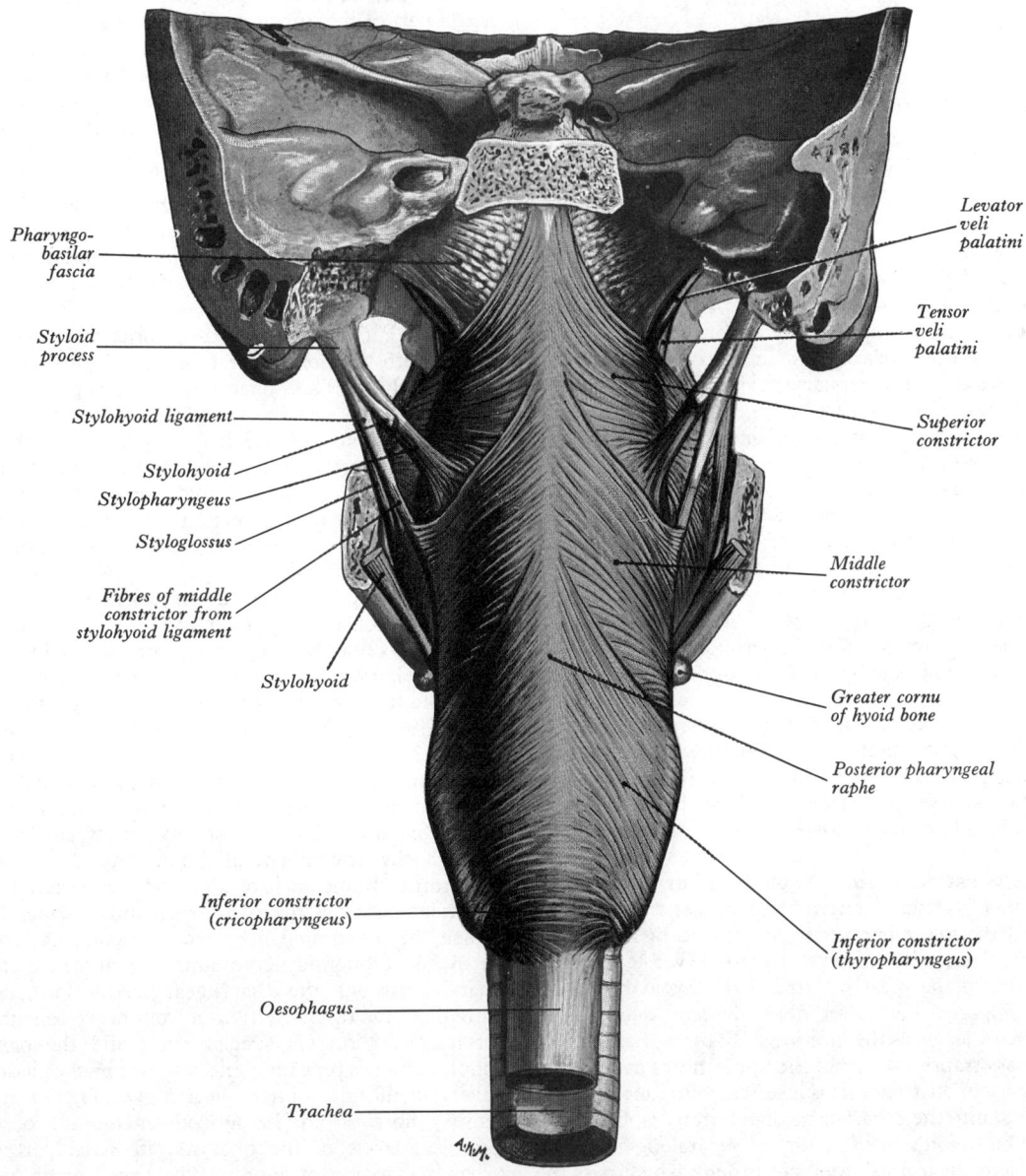

8.86A   The muscles of the pharynx; posterior view. (From Quain's *Anatomy*, 11th ed.)

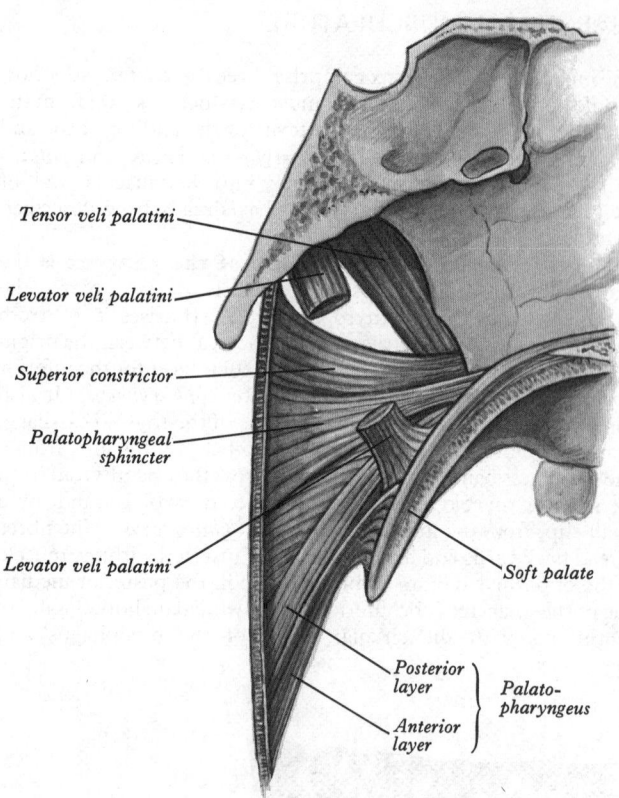

Tensor veli palatini

Levator veli palatini

Superior constrictor

Palatopharyngeal sphincter

Levator veli palatini

Soft palate

Posterior layer } Palato-
Anterior layer } pharyngeus

8.86B The muscles of the left half of the soft palate and adjoining part of the pharyngeal wall. (Dissection by the late Prof. James Whillis, Department of Anatomy, Guy's Hospital Medical School.)

surround the narrowest part of the pharynx; the rest ascend obliquely, and overlap the middle constrictor. During swallowing, the cricopharyngeus acts as the 'sphincteric' part of the muscle (Fuller *et al.* 1959) while the thyropharyngeus acts as the 'propulsive' part; failure of relaxation of the cricopharyngeus may result in a herniation of the pharyngeal mucous membrane posteriorly between the two parts of the muscle (misnamed 'pressure-diverticulum *of the oesophagus*' and sometimes referred to as Killian's dehiscence).

Relations. External to inferior constrictor is the buccopharyngeal fascia. *Behind,* the muscle is in relation with the prevertebral fascia and muscles; *laterally,* with the thyroid gland, the common carotid artery, and the sternothyroid; by its *internal surface,* with the middle constrictor, the stylopharyngeus, the palatopharyngeus and the fibrous coat. The internal laryngeal nerve and the laryngeal branch of the superior thyroid artery run to the thyrohyoid membrane between the upper border of the inferior constrictor and the lower border of the middle constrictor. The external laryngeal nerve runs down on the superficial surface of the muscle, just behind the origin from the thyroid cartilage, and pierces the lower part of the muscle. The recurrent laryngeal nerve and the laryngeal branch of the inferior thyroid artery ascend deep to its lower border, before they enter the larynx.

The middle constrictor muscle of the pharynx is a fan-shaped sheet which is attached anteriorly to the lesser cornu of the hyoid bone and the lower part of the stylohyoid ligament (the *chondropharyngeal* part of the muscle), and to the whole length of the upper border of the greater cornu of the hyoid bone (the *ceratopharyngeal* part). The lower fibres descend deep to the inferior constrictor as far as the inferior end of the pharynx, the middle fibres pass transversely, and the upper fibres ascend and overlap the superior constrictor. It is inserted, with the muscle of the opposite side, into the posterior median fibrous raphe.

Relations. The middle constrictor is separated from the superior constrictor by a small interval, through which pass the glossopharyngeal nerve and the stylopharyngeus, and from the inferior constrictor by the internal laryngeal nerve and

laryngeal branch of the superior thyroid artery. *Posteriorly,* it is related to the prevertebral fascia, the longus colli and the longus capitis. *Laterally,* it is in relation with the carotid vessels, the pharyngeal plexus of nerves and some lymph nodes. Near its origin it is covered with the hyoglossus, from which it is separated by the lingual artery. Its *internal surface* lies upon the superior constrictor, the stylopharyngeus, the palatopharyngeus and the fibrous coat.

The superior constrictor muscle of the pharynx is a quadrilateral sheet, thinner and paler than the other two. It is attached anteriorly to the pterygoid hamulus, and sometimes to the adjoining part of the posterior margin of the medial pterygoid plate, to the pterygomandibular raphe, to the posterior end of the mylohyoid line on the inner surface of the mandible, and by a few fibres to the side of the tongue (**8.**85). According to the different regions of origin of the muscle (as above), the superior constrictor consists of the following parts: *pterygopharyngeal, buccopharyngeal, mylopharyngeal* and *glossopharyngeal.* The fibres curve backwards into the median raphe, being also prolonged by means of an aponeurosis to the pharyngeal tubercle on the basilar part of the occipital bone. The superior fibres arch beneath the levator veli palatini and auditory tube. An interval exists between the cranial border of the muscle and the base of the skull to give passage to the auditory tube. It is bounded anteriorly by the medial pterygoid plate and is closed by the pharyngobasilar fascia (p. 1311).

A constant band of muscle fibres arises from the anterior and lateral part of the upper surface of the palatine aponeurosis and sweeps backwards, lateral to the levator veli palatini, to blend with the internal surface of the superior constrictor near its upper border (**8.**86B). This band has been termed the *palatopharyngeal sphincter,* producing a rounded ridge on the pharyngeal wall (known as the *ridge of Passavant*), which can be seen when the soft palate is elevated in the living (Whillis 1930). These fibres are much hypertrophied in cases of complete cleft palate. The change from columnar, ciliated 'respiratory' epithelium to tall, stratified squamous epithelium on the superior surface of the soft palate takes place where the palatopharyngeal sphincter is attached to the palate.

Relations. The superior constrictor is in relation by its *external surface* with the prevertebral fascia and muscles, the ascending pharyngeal artery and pharyngeal venous plexus, glossopharyngeal and lingual nerves, the styloglossus, middle constrictor and medial pterygoid, the stylohyoid ligament and the stylopharyngeus. The internal carotid artery, the sympathetic trunk, the hypoglossal nerve, the internal jugular vein and the styloid process are more distant relations. By its *internal surface* it is in relation with the palatopharyngeus, the capsule of the tonsil, and the pharyngobasilar fascia. *Superiorly* it is separated from the base of the skull by a crescentic interval in which the levator veli palatini, the tensor veli palatini and the auditory tube are situated. *Inferiorly* its border is separated from the middle constrictor by the stylopharyngeus and glossopharyngeal nerve. *In front* it is separated from the buccinator by the pterygomandibular raphe.

Nerve supply. The constrictors of the pharynx are supplied by the *pharyngeal plexus* (p. 1079). In addition the inferior constrictor receives branches from the external and recurrent laryngeal nerves. Since the pharynx is largely a derivative of branchial structures (p. 199), its motor and sensory supply, through the trigeminal, glossopharyngeal and vagus nerves, which are all branchial in origin, is to be expected. In addition, the glandular tissue in the pharyngeal mucous membrane and, of course, the nonstriated muscle in the blood vessels of the pharynx entail an autonomic nerve supply which in man reaches the pharynx through the pharyngeal plexus. Postganglionic sympathetic fibres reach the plexus from nerve cells in the superior cervical ganglion via special rami; and the parasympathetic supply, which is preganglionic, issues from the medulla oblongata chiefly in the glossopharyngeal nerve. The latter carries back sensory fibres from the mucous membrane of the oral and laryngeal levels of the pharynx, its nasal part being in the trigeminal region of supply. The vagus nerve carries efferent fibres innervating the striated musculature of the pharynx, but most of those probably emerge from the brainstem in the cranial, bulbar part of the accessory nerve.

The pharyngeal rami of the glossopharyngeal and vagus nerves and of the superior cervical ganglion communicate in a plexiform manner in the connective tissue external to the pharyngeal constrictors, particularly the intermediate muscle (Hovelacque 1927). From this plexus, which permits intermingling of the autonomic (sympathetic and parasympathetic) and branchial (efferent and afferent) supplies, mixed rami extend cranially and caudally over the exterior of the superior and inferior constrictors. From these rami branches pass into the muscular layer and others through it to the mucous membrane. This is the pattern of innervation common to most primates (Sprague 1944), including mankind. There is evidence that in some lower primates the plexus is absent or much simplified, and may lack its glossopharyngeal or vagal components. In mammals other than primates arrangements are variable, and in many there is no pharyngeal plexus, but precise information is largely lacking—not only as to the source of the main rami supplying the pharynx, but also, and more significantly, their exact brainstem sources and connexions. The marked development of the pharyngeal plexus in man and some other primates may be associated with the elaboration of phonation, but that is perhaps too facile a view, considering the sparseness of factual evidence.

Actions. The constrictors exercise, in general, a sphincteric or peristaltic action in swallowing. For details, *vide infra*.

The stylopharyngeus (8.79, 85) is a long, slender muscle which is cylindrical above and flattened below. It arises from the medial side of the base of the styloid process of the temporal bone, descends along the side of the pharynx, passes between the superior and the middle constrictors, and spreads out beneath the mucous membrane. Some of its fibres merge into the constrictors and the lateral glosso-epiglottic fold, while others are inserted with the palatopharyngeus into the posterior border of the thyroid cartilage. The glossopharyngeal nerve winds round the posterior border and the lateral side of the stylopharyngeus and passes through the interval between the superior and the middle constrictors to reach the tongue.

Nerve supply. The stylopharyngeus is supplied by a branch from the glossopharyngeal nerve.

Action. The stylopharyngeus is an elevator of the pharynx in swallowing and speech (*vide infra*).

The salpingopharyngeus (8.85) arises from the inferior part of the cartilage of the auditory tube near its pharyngeal opening; it passes downwards and blends with the palatopharyngeus.

Nerve supply. The salpingopharyngeus is supplied by the pharyngeal plexus.

Action. The salpingopharyngeus raises the upper part of the lateral wall of the pharynx, i.e. the part above the attachment of the stylopharyngeus. For its role in swallowing see below.

Vessels and Nerves

The *arteries* supplying the pharynx are derived from the ascending pharyngeal, the ascending palatine and tonsillar branches of the facial artery, branches of the maxillary artery (greater palatine, pharyngeal and the artery of the pterygoid canal) and the dorsal lingual branches of the lingual artery. The *veins* form a plexus which communicates above with the pterygoid plexus and drains into the internal jugular and facial veins.

The *lymph vessels* are described on pp. 787–789.

The *nerve supply* is derived chiefly from the pharyngeal plexus, which is formed by branches from the glossopharyngeal, vagus and sympathetic. (See also p. 1312.) The principal *motor* element is the cranial part of the accessory nerve, which, through branches of the vagus, supplies all the muscles of the pharynx and soft palate, except stylopharyngeus (supplied by the glossopharyngeal) and the tensor veli palatini (supplied by the mandibular nerve). The main *sensory* nerves are the glossopharyngeal and vagus; much of the mucous membrane of the nasal part of the pharynx is supplied by branches of the maxillary nerve (through the pterygopalatine ganglion); the mucous membrane of the soft palate is supplied by the lesser palatine and glossopharyngeal nerves, as is the tonsil. Proprioceptor nerve endings have not been convincingly demonstrated in the pharynx. (*See* Bossy and Vidić 1967 for a discussion.)

PALATINE MOVEMENTS

The movements of the soft palate play an important part in deglutition, in speech and in the act of blowing, and involve a variable degree of closure of the pharyngeal isthmus (p. 1308), necessitated by these activities. Closure is maximal in blowing out through the mouth, when it is essential to prevent entirely any escape of air through the nose. In deglutition closure of the isthmus prevents the food from passing into the nasopharynx, whilst in speech the closure is maximal in the production of the explosive consonants (e.g. *b*, *p*).

Closure of the isthmus is brought about in the following way. The two levatores veli palatini pull the soft palate upwards and backwards towards the posterior pharyngeal wall. Simultaneously the fibres of the palatopharyngeal sphincter raise a rounded ridge on the posterior pharyngeal wall, which meets the nasal surface of the soft palate over a considerable area. (It is at the upper limit of this area of contact that the mucous membrane on the upper surface of the palate changes from the respiratory to the pharyngeal type.)

The tensor veli palatini is active in deglutition rather than in speech, and by producing a localized depression of the anterior part of the palate (p. 1271) squeezes the bolus against the tongue and so helps its descent in the oral pharynx.

MECHANISM OF DEGLUTITION

The *first stage* of swallowing, or deglutition, is *voluntary*. The anterior part of the tongue is raised and pressed against the hard palate, the movement commencing at the tip of the tongue and spreading backwards rapidly. Thus a bolus formed on the tongue behind the tip is pushed to the posterior part of the mouth. At the end of the first stage, the soft palate closes down on to the back of the tongue to help to form the bolus. The movements of the tongue are effected by the intrinsic muscles, especially the superior longitudinal and transverse. At the same time the hyoid bone is raised and moved forwards by contraction of the geniohyoid, mylohyoid, digastric and stylohyoid. By elevation of the posterior part of the tongue posterosuperiorly by the styloglossi, and approximation of the palatoglossal arches, caused by the contraction of the palatoglossi, the bolus is now passed through the oropharyngeal isthmus into the oral pharynx and the second, or *involuntary*, stage of the act of swallowing begins. In swallowing fluids, the intrinsic tongue muscles are used to squirt the fluid back through the mouth; this is succeeded by contraction of the mylohyoid, which bulges the base of the tongue into the oral pharynx. In swallowing solids, only the latter action is used, except when the mouth is being cleansed of saliva and debris after the bolus has been swallowed (Whillis 1946). Hiiemae *et al.* (1978) have made a highly detailed analysis of the activity of the tongue and hyoid apparatus of muscles in swallowing, using the cat and the opossum as their models. By cine-radiographic and electromyographic techniques they have described cyclical activities in the hyoid muscles (see also p. 581) and have also delineated 'envelopes' of such hyoid activity. They regard the lingual movements as essentially a transport of ingested food, the precise combination of movements of the tongue and hyoid bone depending upon the nature of the ingested material. The extension of these studies to primates is to be awaited with great interest.

In the *second stage*, the soft palate is elevated (by the levator muscles) and tightened (by the tensor muscles). In addition it is closely and firmly approximated to the posterior pharyngeal wall by the contraction of the palatopharyngeal sphincter (p. 1272) and the upper fibres of the superior constrictor. The pharyngeal isthmus is tightly closed and the bolus is prevented from passing upwards. At the same time the larynx is drawn upwards behind the hyoid bone and the pharynx ascends with it. This upward displacement is brought about by the stylopharyngeus, salpingopharyngeus, thyrohyoid and palatopharyngeus muscles. Simultaneously the aryepiglottic folds are approximated and the arytenoid cartilages are drawn upwards and forwards by the contraction of the aryepiglottic, oblique arytenoid and thyroarytenoid muscles. This serves to prevent the bolus from entering

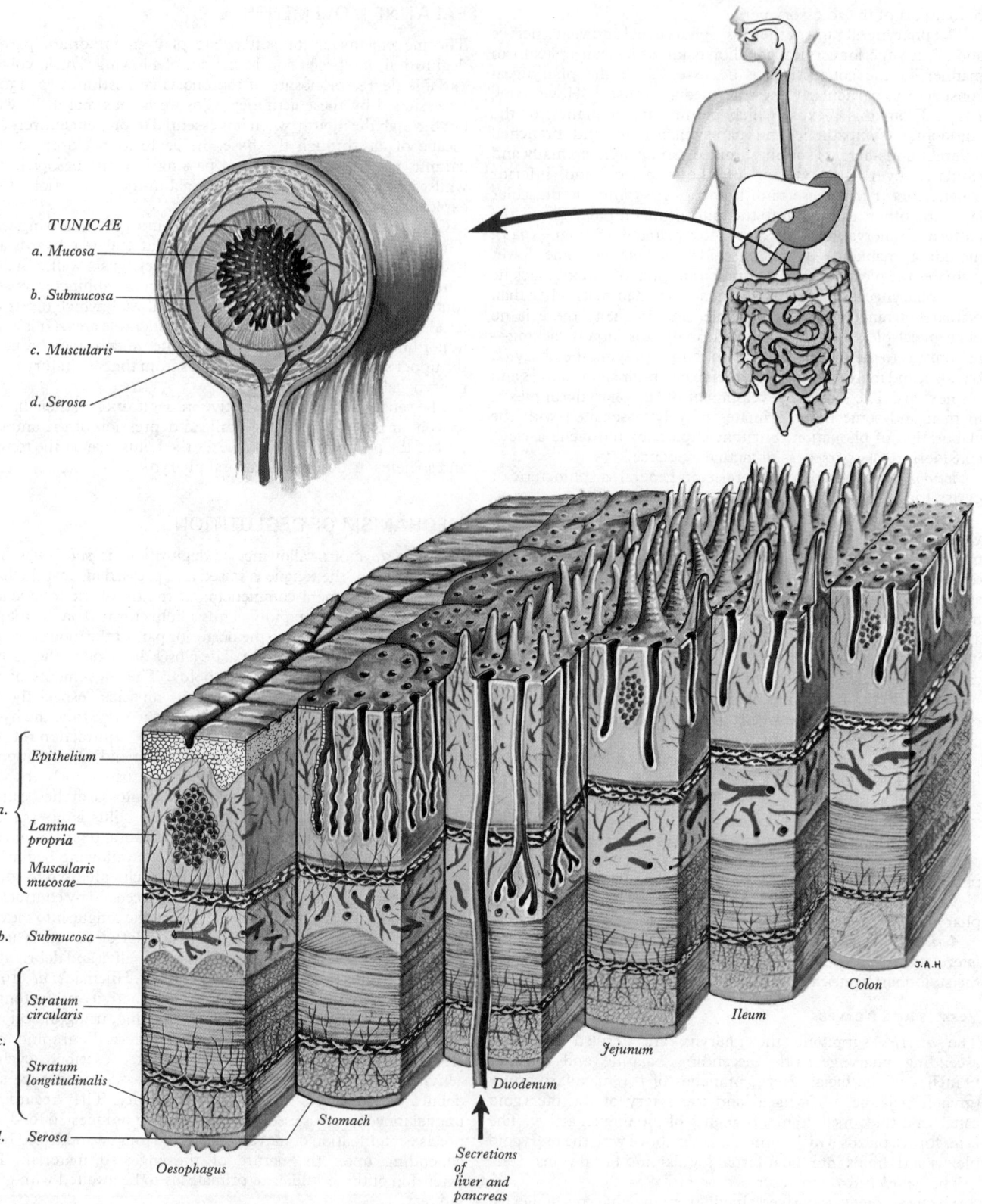

TUNICAE

a. Mucosa

b. Submucosa

c. Muscularis

d. Serosa

Epithelium

a.

Lamina propria

Muscularis mucosae

b.   Submucosa

Stratum circularis

c.

Stratum longitudinalis

d.   Serosa

Oesophagus

Stomach

Duodenum

Secretions of liver and pancreas

Jejunum

Ileum

Colon

8.87   The general arrangement of the alimentary canal, its mural tunicae, and (below) the general histology at the levels indicated (highly diagrammatic). The transverse colon (above right) has been displaced downwards to reveal the duodenum.

the larynx. Partly under the influence of gravity—when the body is in the erect or sitting posture—and partly urged onwards by the successive contractions of the superior and middle constrictors, the bolus slips over the posterior aspect of the epiglottis, the closed inlet of the larynx and the posterior surfaces of the arytenoid cartilages to gain the lowest part of the pharynx. During this stage its passage is facilitated by the action of the palatopharyngei, which shorten the pharynx and pull it upwards. These two muscles, when contracting, convert the surface of the posterior pharyngeal wall into an inclined plane directed

downward and backwards, and on its under surface the bolus descends. The aryepiglottic folds provide lateral food channels leading from the sides of the epiglottis through the piriform fossae to the oesophagus. They are kept more or less tense and upright by the backward pull of the posterior crico-arytenoids on the arytenoid cartilages (to which the folds are attached) and by the muscles in the folds themselves (aryepiglottic and thyroepiglottic), assisted by the cuneiform cartilages, which act as passive props. In paralysis of these muscles (which are supplied by the recurrent laryngeal nerves), the inlet of the larynx is not closed during swallowing, the aryepiglottic folds fall medially, and fluids tend to overflow into the larynx.

The *last stage* in the act is effected by the inferior constrictor, which compresses the bolus onwards into the oesophagus (*see* p. 1312).

These stages follow one another in rapid succession, but it is not difficult for the student to satisfy himself, by palpation of the body of the hyoid bone and the laryngeal prominence during the act, that elevation and forward movement of the hyoid bone precede elevation of the larynx and that the amount of upward movement of the thyroid cartilage is considerable. It is important to emphasize that the thyroid cartilage, and hence the whole larynx, ascends also *relative* to the hyoid bone, an action which shortens the larynx, causes structures between the hyoid bone and thyroid cartilage to bulge posteriorly into the larynx above the vestibular folds, and is also a factor in the increase in curvature of the epiglottis, especially its inferior part, thus aiding the stenosis of the aditus to the larynx during swallowing. Fink and Martin (1977) have drawn special attention to this mechanism.

The actual evidence for the foregoing intimate analysis is, of course, not merely the muscular geometry as a basis for hypothecation. Radiological studies, deductions from the effects of known paralyses, electromyographic observations, and, most recently, cine-radiographic technique have all contributed. Swallowing is, however, a highly complex neuromuscular integration, and considerable disagreement over details, which cannot be reflected in this account, exists among the many observers involved. For full critiques and physiological aspects of the extensive literature on this topic consult Bosma (1957); Doty (1968); Hiiemae (1978).

Applied Anatomy. In young children overgrowth of the lymphoid tissue in the nose and nasal part of the pharynx (adenoids), with or without enlargement of the palatine tonsils, may obstruct nasal respiration, making mouth breathing more or less obligatory. As the child has to keep its mouth open in order to breathe, the hard palate and alveolar arch are habitually out of contact with the dorsum of the tongue; lacking its pressure, they develop with an abnormally high arch and forward projection. Thus the hard palate becomes narrowed laterally, and the projecting alveolar processes afford insufficient room for the permanent teeth, which appear crowded, irregularly set, and overhang those in the lower jaw. The facial surfaces of the maxillae become pinched together, with narrowing of the nasal cavities and maxillary air sinuses. The upper lip is drawn up, still further exposing the projecting front upper teeth. The face is lengthened by dropping of the lower jaw; the whole expression of the child ('adenoid facies') is highly characteristic, suggesting vacuity and inattention, the latter being due to the deafness so often associated with nasal obstruction and caused by blocking of the pharyngeal openings of the auditory tubes.

General Alimentary Organization

Although there is some structural variation along the length of the alimentary tract, a common basic plan of organization exists throughout. This is best appreciated by reference to its embryological development (p. 76). Much of the alimentary canal develops initially as a tube of endoderm enveloped in splanchnopleuric mesoderm, the external surface of which borders the intraembryonic coelom. The endodermal lining forms the epithelium of the tract, and also the secretory and the duct-lining cells of the glands which pour their secretions into the gut lumen, including the pancreas and liver. The main mass of splanchnopleuric mesoderm surrounding the endodermal tube forms the connective tissues, muscle layers, blood vessels and lymphatics of the gut wall, whilst its external surface forms the visceral mesothelium. The latter is, of course, absent throughout the neck and thorax, and also where the hindgut approaches and penetrates the pelvic floor. The disposition and developmental changes in the dorsal and ventral mesenteries of the subdiaphragmatic foregut, and of the dorsal mesentery of the midgut and hindgut have been described elsewhere (p. 200). The nervous elements associated with the gut wall invade it from neighbouring

masses of neural crest tissue (p. 174). Cranially, striated muscle of branchial arch origin and, caudally, striated muscle of less certain origin contribute to the musculature of the extremities of the gut.

In the mature state, therefore, the gut wall is composed of a series of different layers; these are, from the internal to the external surface of the tube: (1) the *epithelium*; (2) the *lamina propria*; (3) the *muscularis mucosae*, which together with the epithelium and the lamina propria, constitutes the *mucosa*; (4) the *submucosa*; (5) the *muscularis externa*, composed of the muscle layers of the main muscle coat; and (6) the *serosa*, or *visceral peritoneum*, throughout most of the subdiaphragmatic gut; throughout the pharynx and most of the oesophagus, however, a serosa is absent and is replaced by a relatively dense connective tissue *adventitia* or fibrous 'coat', which blends with the fascial planes of surrounding structures (8.87). The mesenteries carry blood vessels, autonomic nerves and lymphatic channels to and from the wall of the gut, and at certain points they accompany the ducts of outlying glands.

The epithelium of the gut is the chief region in which interaction between the body and ingested food takes place. Its cytological composition varies in accordance with its particular regional functions; in the pharynx, oesophagus and lower anal canal it is lined with non-keratinizing, stratified, squamous epithelium which protects the underlying tissue and which facilitates the passage of solid material by the lubricant action of its glandular diverticula. In other regions the epithelium is one

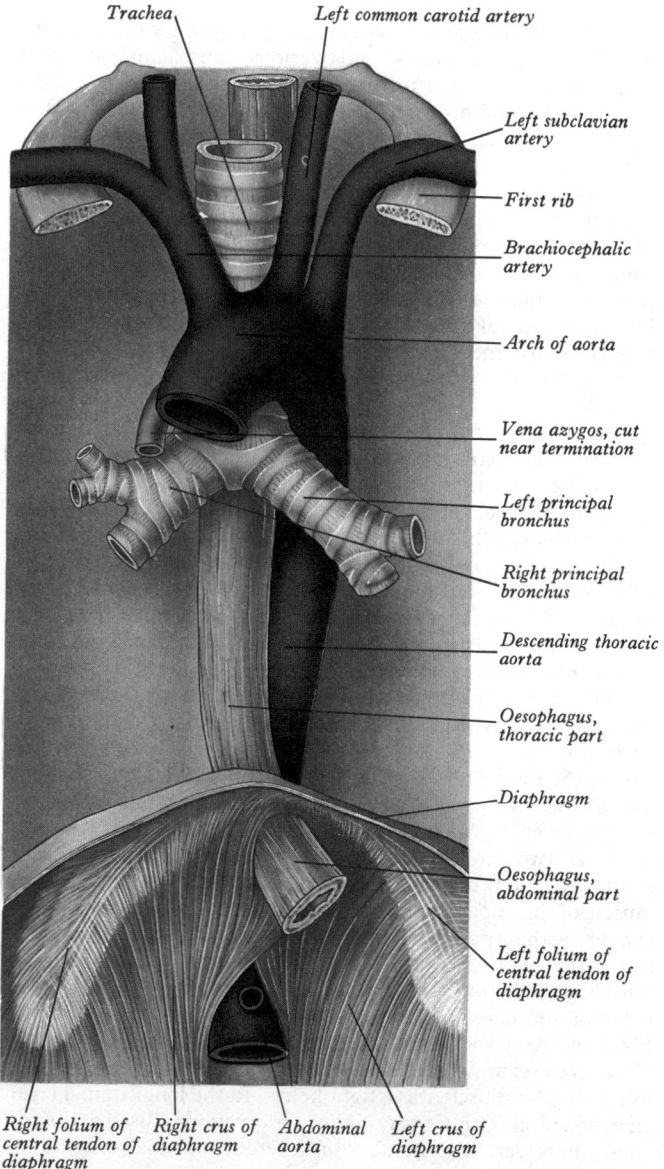

Trachea
Left common carotid artery
Left subclavian artery
First rib
Brachiocephalic artery
Arch of aorta
Vena azygos, cut near termination
Left principal bronchus
Right principal bronchus
Descending thoracic aorta
Oesophagus, thoracic part
Diaphragm
Oesophagus, abdominal part
Left folium of central tendon of diaphragm
Right folium of central tendon of diaphragm
Right crus of diaphragm
Abdominal aorta
Left crus of diaphragm

8.88 A dissection to expose the oesophagus in the posterior mediastinum and in the abdomen.

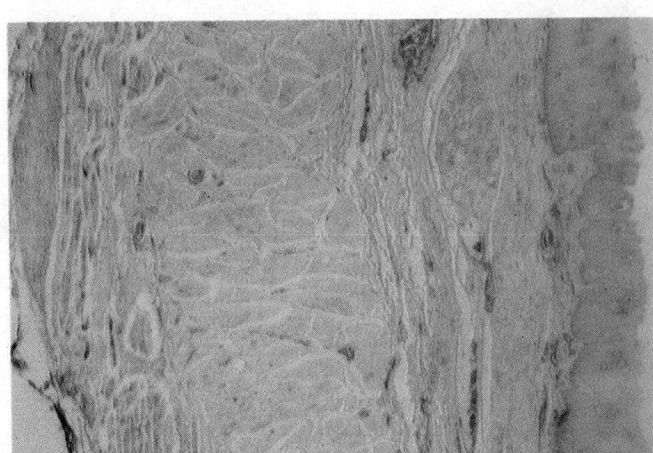

8.89A A low-power micrograph of a vertical section through the wall of a human oesophagus taken in the upper thorax. Mallory's triple stain. Visible in this section are: the epithelium (blue-grey on the right), the lamina propria, muscularis mucosae, submucosa with a group of mucous glands, the external muscle with the circular fibres more deeply placed, and the longitudinal fibres placed externally. See text for further description. Magnification about × 6. (Prepared by Dr. W. Owen, Department of Anatomy, Guy's Hospital Medical School.)

cell thick and is composed predominantly of secretory cells, which provide mucus, enzymes and various ions amongst other materials, and of absorptive cells, which are columnar, with striated or brush borders. The total amount of secretion and of absorption is dependent partly on the surface area of the epithelium, which is maximized by the presence of large folds, finger-like projections (villi), crypts and glands at various points along the tract. The glands are placed at varying positions with reference to the surface of the tract, some in the lamina propria, some in the submucosa, and some completely outside the confines of the alimentary tract's wall, these being the liver and pancreas. The secretions of these glands are important in providing the correct environment for the activity of the different digestive enzymes, and the glands are usually under nervous or hormonal control.

The lamina propria is a layer of compact connective tissue, often rich in elastin fibres, which supports the epithelium; it is bounded at its base by the **muscularis mucosae**, a plexus of nonstriated muscle fibres which can alter the local conformation of the mucosa by the contractile activity of its two layers, the inner of which is circular and the outer longitudinal in direction. **The submucosa** is rich in blood vessels, lymphatics and nerves; variations in its thickness create the macroscopic folds and ridges of the internal aspect of the gut, particularly in the small intestine.

The muscularis externa consists in most regions of two distinct layers, the inner circular and the outer longitudinal. The antagonistic activities of these two components create a series of peristaltic waves which are responsible for the movement of solids within the lumen of the tract. In the stomach, where muscular movements are more complex than in the majority of the tract, an oblique layer is present in some parts of the gastric wall, within the usual two external muscle coats. The muscularis externa is composed chiefly of nonstriated muscle, except where the striated muscle of the upper part of the oesophagus blends with the layers and where the striated fibres of the anal sphincters enter into close association with the tract. The different layers of the external muscle are separated by connective tissue planes in which lie vascular and nervous plexuses. Some controversy (Carey 1921; Elsen and Arey 1966) has existed over the precise direction of the fibres in the external muscle layers, one suggestion being that the inner circular muscle was a tight helix and the longitudinal layer an open spiral (Carey 1921). Other observations (Elsen and Arey 1966), however, on a number of different mammals, including man, indicated that, although some deviations from the precisely circular and longitudinal directions may exist locally within the intestine, fibres do not in general follow spiral pathways. In the

small intestine of man, it appears that adjacent 'rings' of circular muscle are often connected to each other by oblique slips of muscle fibres allowing for some interchange between muscle groups (Schofield 1968).

The nerve plexuses of the alimentary tract, mentioned above, comprise a plexus between the inner and outer layers of external muscle (the *myenteric* or *Auerbach's plexus*), another on the submucosal surface of the external muscle (the *submucosal* or *Meissner's plexus*), and a third at the level of the muscularis mucosae. All three plexuses are connected by fine nerve fibres. The first two plexuses contain scattered ganglion cells of the autonomic system, whereas the mucosal plexus appears to be derived from the submucosal nerve net. The epithelium of the alimentary tract is also innervated from the mucosal plexus, and several types of sensory cells have been postulated to occur on the basis of experimental evidence. The structural counterparts to the physiologically identified receptors have so far evaded observational verification, in most cases, although there is no shortage of suggestions (*see* **8.**125 and p. 1366).

The Oesophagus

The oesophagus (**8.**88) is a muscular tube, about 25 cm (10 in.) long, connecting the pharynx to the stomach. It begins in the neck at the caudal border of the cricoid cartilage, opposite the sixth cervical vertebra, where it is continuous with the pharynx. It descends largely anterior to the vertebral column, through the superior and posterior parts of the mediastinum, pierces the

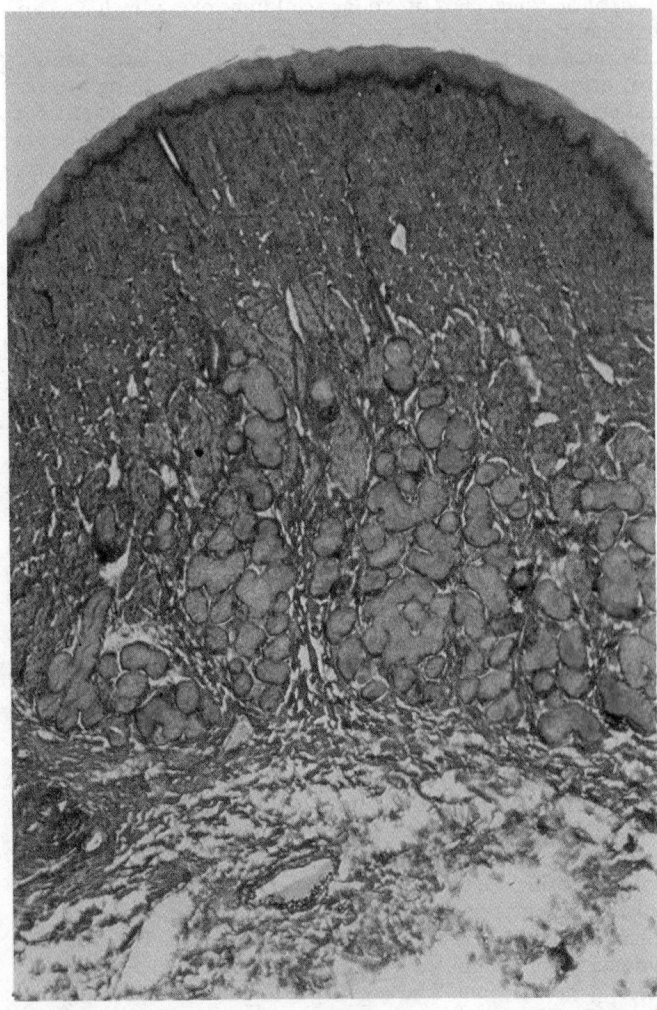

8.89B Low magnification light micrograph of a section of the oesophagus showing the stratified squamous non-keratinized epithelium lining it, and mucous glands (stained turquoise) of the submucosa. Weigert and Van Gieson, with alcian blue. (Prepared by Mr. David Ristow; photographed by Miss Marina Morris, Department of Anatomy, Guy's Hospital Medical School.)

diaphragm level with the tenth thoracic vertebra, and ends at the cardiac orifice of the stomach at the level of the eleventh thoracic vertebra. The general direction of the oesophagus is vertical; but it presents two shallow curves. At its commencement it is median; but it inclines slightly to the left side as far as the root of the neck, gradually passes again to the median plane, at the level of the fifth thoracic vertebra, and again, at the seventh, deviates to the left and then turns anteriorly to the oesophageal opening in the diaphragm. The oesophagus also presents anteroposterior flexures corresponding to the curvatures of the cervical and thoracic parts of the vertebral column. It is the narrowest region of the alimentary tract except for the vermiform appendix, and is constricted (*a*) at its commencement, 15 cm (6 in.) from the incisor teeth, (*b*) where it is crossed by the aortic arch, 22·5 cm (9 in.) from the incisor teeth, (*c*) where it is crossed by the left principal bronchus, 27·5 cm (11 in.) from the incisors, and (*d*) where it pierces the diaphragm, 40 cm (16 in.) from the incisors. The sites of these constrictions are important clinically in connexion with the passage of instruments along the oesophagus.

The cervical part of the oesophagus has the following relations. *Anteriorly* lies the trachea, to the posterior membranous wall of which it is attached by loose connective tissue; the recurrent laryngeal nerves ascend, one on each side, in, or slightly in front of, the groove between the trachea and oesophagus. *Posteriorly*, it adjoins the vertebral column, the longus colli and the prevertebral layer of the deep cervical fascia. *Laterally*, on each side, lie the corresponding common carotid artery and the posterior part of the lobe of the thyroid gland; in the lower part of the neck, where the oesophagus projects to the left side, it has a closer relation to the carotid sheath and the thyroid gland than on the right side. The thoracic duct ascends for a short distance along the left side of the oesophagus.

The thoracic part of the oesophagus (8.17, 18, 24, 88) is at first situated in the superior mediastinum between the trachea and the vertebral column, a little to the left of the median plane. It passes behind and to the right of the aortic arch and descends in the posterior mediastinum along the right side of the descending thoracic aorta. Below, as it inclines to the left, it crosses in front of the aorta, and enters the abdomen through the diaphragm at the level of the tenth thoracic vertebra. It is in relation, *anteriorly* (from above downwards), with the trachea, the right pulmonary artery, the left principal bronchus, the pericardium (separating it from the left atrium), and the diaphragm; *posteriorly* are the vertebral column, the longus colli muscles, the right posterior (aortic) intercostal arteries, the thoracic duct, the azygos vein and the terminal parts of the hemiazygos and accessory hemiazygos veins, and inferiorly, near the diaphragm, the front of the aorta. In the posterior mediastinum an elongated recess of the right pleural sac intervenes between the oesophagus and the vena azygos and vertebral column. On its *left* side, in the superior mediastinum the terminal part of the aortic arch, the left subclavian artery, the thoracic duct, and left pleura are immediate relations, while the left recurrent laryngeal nerve runs upwards in, or just in front of the groove between it and the trachea; in the posterior mediastinum it is in relation with the descending thoracic aorta and the left pleura. On its *right* side it is related to the right pleura, the azygos vein intervening as it arches forwards above the right principal bronchus to join the superior vena cava. Below the roots of the lungs the vagus nerves descend in close contact with it, the right nerve chiefly behind, and the left chiefly in front of it; the two nerves unite to form a plexus around the tube (pp. 1079, 1089).

In the lower part of the posterior mediastinum the thoracic duct lies behind and to the right of the oesophagus; higher up, it is placed behind it, and crossing to the left about the level of the fifth thoracic vertebra, is continued upwards on its left side.

On the right side of the oesophagus, just above the diaphragm, a small serous sac (the *infracardiac bursa*) is sometimes found; it represents the upper detached part of the right pneumato-enteric recess (p. 205).

The abdominal part of the oesophagus, having emerged from the right crus of the diaphragm (p. 549) slightly to the left of the median plane at the level of the tenth thoracic vertebra, lies in the oesophageal groove on the posterior surface of the left lobe of

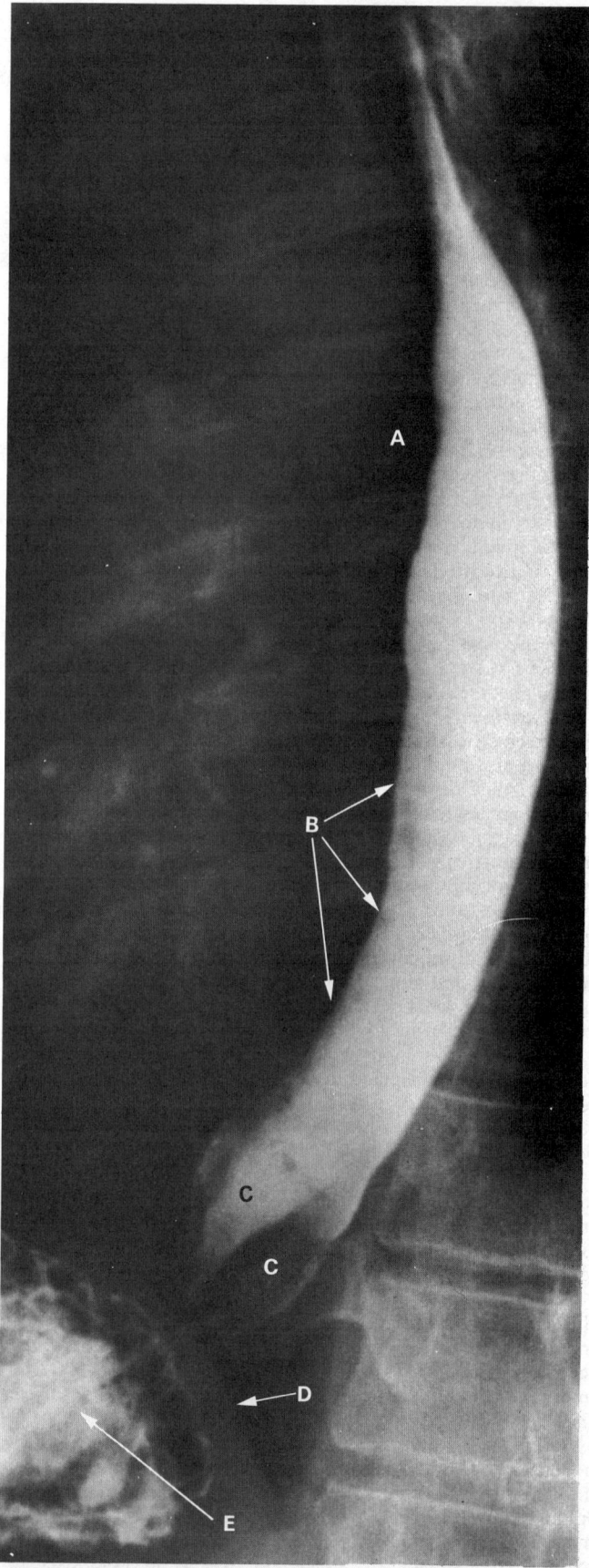

8.90   An oblique radiograph of the thorax during the oesophageal transit of part of a 'meal' of barium sulphate paste. At (A) the translucency of the air-containing right principal bronchus is visible. The concave ventral aspect of the oesophagus (B) is topographically related to the pericardium covering the left atrium of the heart. Longitudinal mucosal folds are visible (C) immediately proximal to the soft tissue shadow of the diaphragm (D). Some barium sulphate is already admixed with the gastric contents (E). The oesophagus in the lower thorax curves ventrally away from the vertebral column to reach the oesophageal orifice in the diaphragm.

the liver. It is shaped like a truncated cone, 1·25 cm long, curving sharply to the left (**8**.88), the base of the cone being continuous with the cardiac orifice of the stomach; its right border continues evenly into the lesser curvature of the stomach while its left border is separated from the fundus of the stomach by the cardiac notch (**8**.105). It is covered by peritoneum in front and on its left side; it is contained in the upper left part of the lesser omentum, the peritoneum reflected from its posterior surface to the diaphragm being part of the gastrophrenic ligament (p. 1326). The oesophageal branches of the left gastric vessels pass to the oesophagus in this peritoneal ligament. Behind the oesophagus are the left crus of the diaphragm and the left inferior phrenic artery. The vagus nerves have variable relations to the oesophagus as it passes through the diaphragm (Doubilet *et al.* 1948). Sometimes one trunk (consisting mainly of left vagal fibres) lies on the front, and one (consisting mainly of right vagal fibres) on the back of the oesophagus, but each vagus may consist of two or three trunks at this level.

Structure (**8**.89). The histological structure of the oesophagus follows the same general pattern as the rest of the alimentary tract, outlined above, being formed of four layers: external or fibrous, muscular, submucous or areolar, and internal or mucous.

The *fibrous layer* consists of an external adventitia of irregular, dense connective tissue, containing many elastin fibres. Its fibres also penetrate and surround the fasciculi of muscle in the deeper layers.

The *muscular layer* (muscularis externa) is composed of the usual two strata, here particularly thick—an outer longitudinal and an inner circular stratum.

The *longitudinal fibres* form a complete investment for nearly the whole of the oesophagus, but at the upper part of the back of the tube, at a point between 3 cm and 4 cm below the cricoid cartilage, they diverge from the median plane and form two longitudinal fasciculi which incline upwards and forwards to the front of the tube. Here they pass deep to the caudal border of the inferior constrictor to end in a tendon attached to the upper part of the ridge on the posterior surface of the lamina of the cricoid cartilage (**8**.13). The V-shaped interval between the diverging longitudinal fasciculi is filled by the circular fibres of the oesophagus, thinly covered below by some decussating longitudinal fibres, and above by the overlapping lower edge of the inferior constrictor. For the main part the longitudinal muscular coat of the oesophagus is thicker than the circular muscle coat.

Accessory slips of non-striated muscle tissue sometimes pass between the oesophagus and the left pleura, or between the oesophagus and the root of the left principal bronchus, trachea, pericardium, or aorta. These muscular slips are sometimes considered to fix the oesophagus to neighbouring structures.

The *circular fibres* are continuous superiorly, on the posterior surface, with the inferior constrictor; anteriorly, the uppermost are inserted into the lateral margins of the tendon of the two longitudinal fasciculi. Inferiorly, the circular muscle fibres are continuous with the oblique fibres of the stomach.

In man, striated muscle is generally limited to the upper two-thirds of the oesophagus; the lower third contains nonstriated muscle only. In the upper quarter, both layers are striated; in the second quarter bundles of nonstriated muscle appear, first on the internal aspect of the muscle layers, and these gradually replace the striated muscle more caudally.

Radiological studies show that swallowed food is momentarily held up in the lower, gastric end of the oesophagus, prior to entry into the stomach (**8**.90). It is hence certain that some form of sphincteric mechanism, capable of contraction and relaxation, must be present at the oesophagogastric junction. The failure of many observers to identify any notable aggregations of muscle to account, with certainty, for the control of entry into or regurgitation from the stomach, has lead to a somewhat meaningless phrase—'a physiological cardiac sphincter'. The frequency of pathological disturbance of this sphincteric control has exacerbated the controversy, for a full discussion of which consult Bombeck *et al.* (1944); Botha (1962); DiDio and Anderson (1968); Code (1968). A wide variety of factors have been adduced to account for the undoubted sphincteric effects at the junction of oesophagus and cardia of the stomach. There are circularly disposed muscle fibres in both organs adjoining the junction, and these are potentially reinforced by the splitting of the right crus of the diaphragm around the oesophagus (p. 549). Some have recorded evidence against the latter view (Atkinson *et al.* 1957). The phrenico-oesophageal ligament, a layer of connective tissue extending from the inferior surface of the diaphragm through the oesophageal orifice to blend with the submucosa and interfascicular septa of the musculature of the termination of the oesophagus, has also been invoked as an accessory factor. The obliquity of the gastro-oesophageal junction, the effects of spiral and longitudinal muscle (Stelzner and Lierse 1967), and various mucosal folds (Creamer 1955), have all been considered to exert a valvular effect. It is hence noticeable that a number of *structures* may possess either alone, or collectively, the potentiality of '*physiological*' sphincteric control at this site of controversy. However, the actual mechanism is still uncertain. (*See* also p. 1336.)

The *submucosa* connects loosely the mucous and muscular coats. It contains the larger blood vessels and nerves, as well as mucous glands.

The *mucous layer* is thick, reddish in colour above, and pale below. It is disposed in longitudinal folds, which disappear on distension of the tube. It consists of: (1) a layer of non-keratinizing stratified squamous epithelium, lining the tube; (2) a layer of connective tissue, papillae from which project into the epithelium; and (3) the muscularis mucosae, a layer of nonstriated muscle. At the commencement of the oesophagus the muscularis mucosae is absent, or only represented by a few scattered bundles; lower down it forms a considerable stratum. At intermediate levels the muscle bundles are mainly longitudinal in direction, but they become more plexiform as the gastro-oesophageal junction is approached. At the gastro-oesophageal junction the stratified squamous epithelium of the oesophagus is abruptly succeeded by the simple columnar epithelium of the stomach. The junction is visible to the naked eye in fresh preparations as a crenated line, the greyish-pink, smooth, oesophageal contrasting with the red, mamillated, gastric mucosa.

The *oesophageal glands* are small, compound racemose glands of the mucous type; they are lodged in the submucous tissue outside the muscularis mucosae, and each opens into the tube by a long duct which pierces the muscularis mucosae. In the abdominal oesophagus, the glands resemble those in the cardiac part of the stomach and are between the muscularis mucosae and lumen (Johns 1952).

Vessels and Nerves

The *arteries* supplying the oesophagus are derived from the inferior thyroid branch of the thyrocervical trunk, from the descending thoracic aorta, from the bronchial arteries, from the left gastric branch of the coeliac artery, and from the left inferior phrenic branch of the abdominal aorta. They have for the most part a longitudinal direction. The *veins* from the cervical part of the oesophagus drain into the inferior thyroid veins and those from the thoracic part into the azygos, hemiazygos and accessory hemiazygos veins. The abdominal part drains partly into the azygos vein and partly into the left gastric vein. The latter vein being a tributary of the portal vein, the abdominal part of the oesophagus is one of the sites where anastomoses between the portal and systemic veins occur (p. 765). In cases of obstruction of the portal circulation (e.g. in cirrhosis of the liver), these veins may become varicose and may burst into the lower part of the oesophagus, causing vomiting of blood and even fatal haemorrhage. The *lymph vessels* are described on pp. 765, 800.

The *nerves* are derived from the vagus and sympathetic. The cervical part of the oesophagus receives branches from the recurrent laryngeal nerves and from the cervical sympathetic trunks (by means of the plexus around the inferior thyroid artery). The thoracic part has branches from the vagal trunks and oesophageal plexus, and from the sympathetic trunks and greater splanchnic nerves. The abdominal part (Mitchell 1938) is supplied by the vagal trunks (anterior and posterior gastric nerves), the thoracic sympathetic trunks, the greater (and occasionally the lesser) splanchnic nerves, and the plexus around the left gastric and inferior phrenic arteries. The nerves form a plexus containing groups of ganglion cells between the two layers of the muscular coat, and a second plexus in the submucous tissue. (See also Semenova 1962.) The literature concerning the 'myenteric' plexus of the oesophagus has been reviewed by Rodrigo *et al.* (1975), who have described intraganglionic nerve endings which they suggest may be afferent in nature.

Radiology (**8**.90). In oblique lateral views after a barium meal the main part of the thoracic portion of the oesophagus is seen in the 'retrocardiac space' behind the heart and diaphragm. The entire thoracic part is seen to be situated a little distance from the vertebral column, its lower part inclining forwards still further from the column. Its anterior wall is indented by the arch of the aorta, the left bronchus, and the left atrium, successively from above downwards. A thin layer of the barium meal may outline the longitudinal grooves of the mucous membrane, producing a characteristically striated shadow.

THE ABDOMEN

The abdomen is the region of the trunk below the diaphragm (**8**.92). It comprises an upper part, the *abdomen proper*, and a lower part, the *lesser pelvis*. These are continuous at the plane of the inlet of the lesser pelvis, which is bounded by the promontory of the sacrum, the arcuate lines of the innominate bones, the pubic crests and the upper border of the symphysis pubis. The abdomen is, to a large extent, bounded by muscles; consequently its shape and size can alter under different conditions, such as varying degrees of distension of the contained hollow organs and the phases of respiration. Further, the tone of these muscles is an important factor in maintaining the abdominal and pelvic viscera in position (pp. 558 and 561).

The *abdomen proper* is bounded *in front* by the rectus abdominis muscles, the pyramidales and the aponeurotic parts of the flat abdominal muscles (obliquus externus abdominis, obliquus internus abdominis and the transversus abdominis); *at the side*, by the fleshy parts of these flat abdominal muscles, the iliacus muscles and the iliac bones; *behind*, by the lumbar part of the vertebral column, the crura of the diaphragm, the psoas and quadratus lumborum muscles and the posterior parts of the iliac bones; *above*, by the diaphragm; while *below*, it is continuous with the pelvis through the superior aperture of the lesser pelvis (p. 385). As the diaphragm, which forms the dome-like roof of the abdominal cavity, arches upwards, a considerable part of the cavity extends superiorly and is internal to the bony framework of the thorax (*see* pp. 550–551 for variations in the position of the diaphragm during respiration, etc.). The abdomen proper contains the greater part of the digestive tube, as well as the liver, pancreas, spleen, kidneys, parts of the ureters, suprarenal glands and numerous blood vessels, lymph vessels, lymph nodes and nerves.

The *lesser pelvis* is roughly infundibular, like an inverted, truncated cone, and extends dorso-inferiorly from the lower end of the abdominal cavity (**8**.95, **3**.181). It is bounded *antero-laterally* by the parts of the hip bones below the pubic crests and arcuate lines, and by the obturator internus muscles; *supero-dorsally*, by the sacrum, coccyx and the piriformis and coccygeus muscles; and *inferiorly*, by the levatores ani, which with their covering fasciae form the pelvic diaphragm (pp. 559–561), and by the transversus perinei profundus and sphincter urethrae, which together with *their* fascial coverings constitute the urogenital diaphragm. The lesser pelvis contains the urinary bladder, terminal parts of the ureters, the sigmoid colon, rectum and a few coils of small intestine, and the internal genitalia, together with blood vessels, lymph vessels, lymph nodes and nerves.

The muscles bounding the abdomen proper and the pelvis are, like muscles in general, ensheathed in fascia; the layer of fascia covering the deep surface of the muscles immediately adjacent to the abdominal and pelvic cavities is named differently in the various regions. For example, that which adjoins the internal surface of the transversus abdominis is the *transversalis fascia* (p. 559), that inferior to the diaphragm is the *diaphragmatic fascia*, that covering the psoas and iliac is the *iliac fascia* (p. 593), that covering the anterior surface of the quadratus lumborum is the *anterior layer of the thoracolumbar fascia* (p. 542), and that covering the muscles in the pelvis is the *pelvic fascia* (p. 560).

Most of the organs in the abdominal and pelvic cavities are largely covered with a serous membrane, termed the *peritoneum* (pp. 1321–1333).

ABDOMINAL REGIONS

For purposes of location of the viscera, especially in clinical usage, the abdomen is divided into nine regions by imaginary planes, two horizontal and two sagittal, passing through the cavity, the edges of the planes being indicated by lines projected upon the surface of the body (**8**.91). The upper horizontal plane, or *transpyloric plane* (of Addison), is indicated by a line encircling the body at a level midway between the suprasternal notch and the symphysis pubis (roughly midway between the umbilicus and the inferior end of the body—not the xiphoid process—of the sternum, or a hand's breadth below the xiphisternal joint); it intersects the front of the body of the first lumbar vertebra near its lower border and

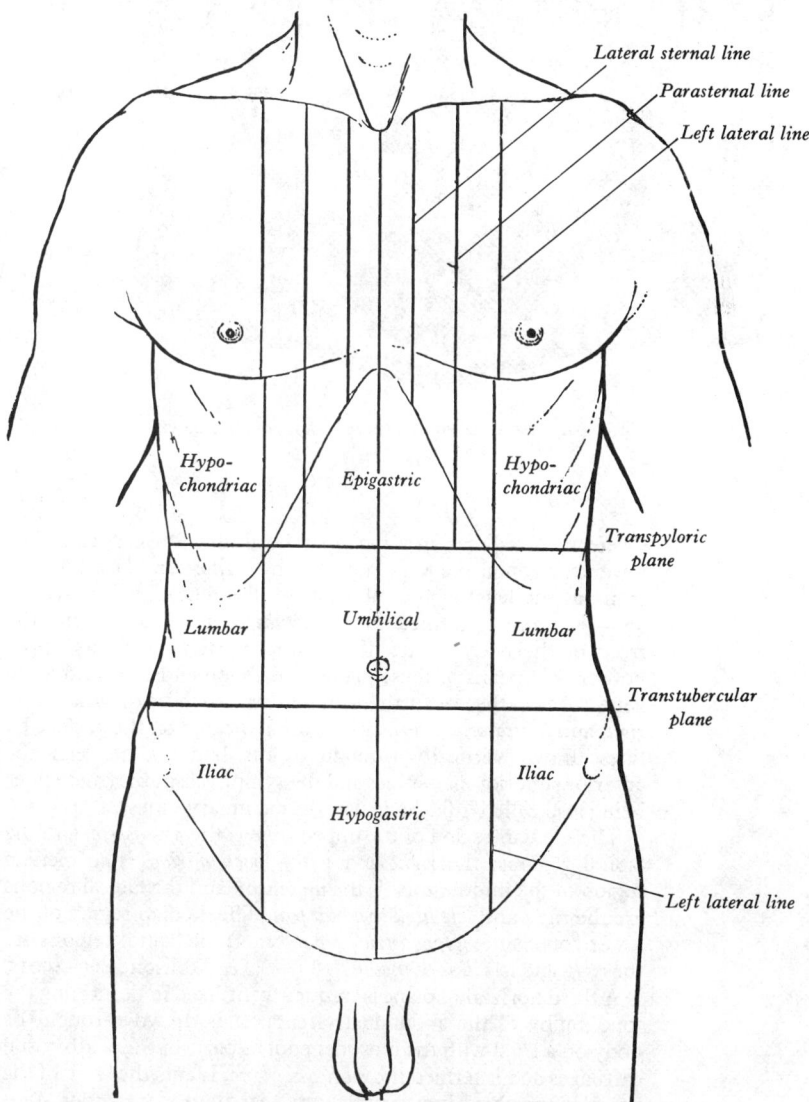

8.91 Reference lines on the anterior aspect of the thorax and abdomen for use in delineating surface projections.

The *umbilicus* is variable in position; in the young adult it usually lies on a level with the disc between the third and fourth lumbar vertebrae. As age advances, and in conditions of deficient tone of the abdominal muscles, it sinks to a lower position. It is also lower in the child because of the under-developed condition of the pelvic region.

On the posterior surface of the body, a transverse line joining the highest parts of the two iliac crests cuts through the spinous process of the fourth lumbar vertebra in the midline; it is known as the *supracristal plane*. This landmark is used for the purpose of identifying individual vertebral spinous processes.

When the anterior abdominal wall is removed (**8**.92), the viscera are partly exposed as follows: above and to the right side the liver is visible, situated chiefly under the shelter of the right ribs and their cartilages, but extending across the median plane, where it reaches down to the transpyloric plane. The stomach is exposed in the angle between the left costal margin and the lower border of the liver. From its lower border an apron-like fold of peritoneum, termed the *greater omentum*, descends for a varying distance, anterior to the other viscera to a greater or lesser extent. Below the greater omentum, however, some of the coils of the small intestine can generally be seen, while in the right iliac region the caecum and, in the left iliac region, the lower portion of the descending colon are partly exposed (*see* p. 1330). The urinary bladder occupies the anterior part of the pelvis, and if distended, projects above the symphysis pubis; the rectum is placed in the concavity of the sacrum, but is usually hidden by coils of the small intestine. The sigmoid colon may be seen lying between the rectum and the bladder.

When the stomach is followed from left to right, it is seen to be continuous with the first part of the small intestine (duodenum), the point of continuity being marked by a thickened ring (of muscle), which indicates the position of the pyloric sphincter. The duodenum passes towards the under surface of the liver, and then, curving downwards, is lost to sight. If, however, the greater omentum, together with the transverse colon which lies beneath it, be turned upwards over the chest, the horizontal part of the duodenum can be traced across the vertebral column towards the left side, where it ascends to the second lumbar vertebra to become continuous with the coils of the jejunum and ileum, which constitute the remainder of the small intestine. These measure about six metres in length (see also p. 1342), and if followed downwards the ileum is seen to end in the right iliac fossa by opening into the large intestine at the junction of the caecum and ascending colon. From the caecum the large intestine takes a looped course, passing at first upwards on the right side, then across the median plane below the liver and stomach, and then downwards on the left side, forming respectively the ascending, transverse, and descending parts of the colon. In the pelvis the colon assumes the form of a loop, termed the sigmoid colon, and ends in the rectum.

The spleen lies behind the stomach in the left hypochondriac region, and may be exposed in part by pulling the stomach over towards the right side.

The glistening appearance of the deep surface of the abdominal wall and of the surfaces of the exposed viscera is due to the fact that the former is lined, and the latter are largely covered, with a serous membrane, termed the *peritoneum*.

The relations of the organs described in the following pages refer to the body in the recumbent position. But it must be noted that the position of the viscera is affected not only by posture but by respiratory movements and by the condition of the hollow organs as regards quantity of contents. In addition there is a wide range of variation associated with bodily build, i.e. the shape of the chest, abdomen and pelvis; also the organs may vary in the same individual at different times depending, for example, on their physiological activity and their degree of mobility. Thus, in the absence of radiological examination, the surface outlines of the viscera, particularly the hollow organs, described in the text and illustrated in the figures, must be regarded as variable within wide limits.

With regard to their different types of physique individuals have been classified into two extreme groups, namely, *hypersthenic* (for pyknic) and *asthenic* (or leptosomatic), with inter-

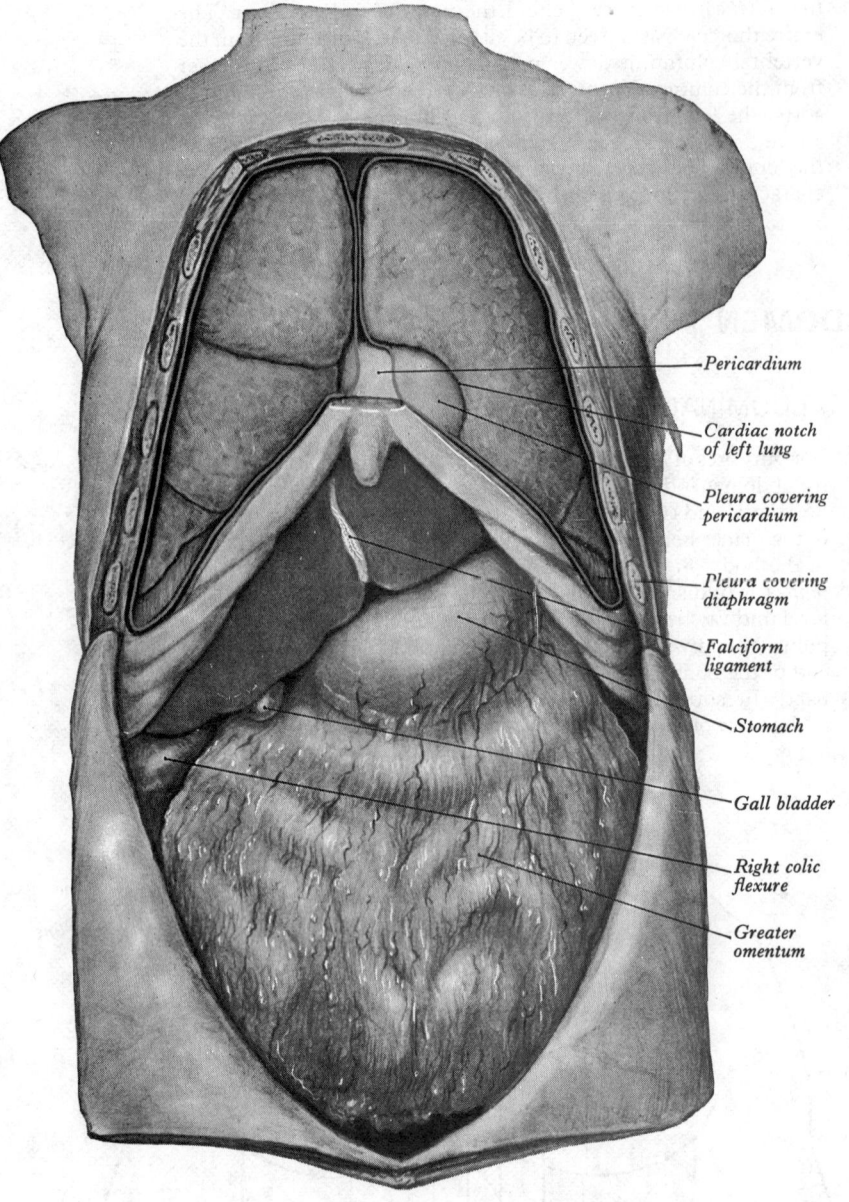

Pericardium

Cardiac notch of left lung

Pleura covering pericardium

Pleura covering diaphragm

Falciform ligament

Stomach

Gall bladder

Right colic flexure

Greater omentum

8.92 Anterior aspect of the thoracic and abdominal viscera.

meets the costal margin at the tip of the ninth costal cartilage. The lower horizontal plane is indicated by a line carried round the trunk at the level of the tubercles on the iliac crests (**8**.91, pp. 378–380) and is termed the *transtubercular plane*; it cuts the front of the body of the fifth lumbar vertebra near its upper border. By means of these planes the abdomen is cut into three zones; each of these is further subdivided into three regions by the *right* and *left lateral planes*, which are indicated on the surface by lines drawn vertically through points halfway between the anterior superior iliac spines and the symphysis pubis (these lines being also called 'midclavicular' or 'mammary' lines).

The median region of the upper zone is the *epigastric*, and the lateral regions, the *right* and *left hypochondriac*. The median region of the middle zone is the *umbilical*, and the lateral regions are the *right* and *left lumbar* or *lateral*. The median region of the lower zone is the *hypogastric* or *pubic*, and the lateral regions are the *right* and *left iliac* or *inguinal* (**8**.91). (*See* Addison 1899–1901.)

A third horizontal plane is frequently utilized in describing the topography of the abdominal viscera. It is drawn through the body on a level with the most dependent parts of the tenth costal cartilages and is termed the *subcostal plane*. It cuts the front of the body of the third lumbar vertebra near its upper border. It is frequently utilized instead of the transpyloric plane for the purpose of dividing the abdomen into the regions named above.

mediate grades, *sthenic* and *hyposthenic* (Mills 1917, 1922). In the hypersthenic type, with massive physique, the thorax is wide and short and the subcostal angle is very obtuse, so that the heart and lungs are wide transversely; the abdomen is widest in its upper part and the stomach is less elongated in a vertical direction with the pylorus relatively high, while the transverse colon is more truly transverse. In the asthenic type, with frail and slender physique, the thorax is long and narrow and the subcostal angle is acute, so that the heart and lungs are long and narrow; the

abdomen is widest in its lower part, the stomach is long with the pylorus relatively low, while the colon is long with the transverse colon descending in a V-shaped manner to the pelvis. The varieties of human physique (somatotypes) have also been classified as endomorphic (massive bodybuild), mesomorphic (intermediate) and ectomorphic (slender), with intermediate grades, each type having predominant psychological characteristics (Sheldon *et al.* 1940; Sheldon and Stevens 1942). (Consult current monographs of physical anthropology for details.)

THE PERITONEUM

The peritoneum (Brizon *et al.* 1956) is the largest and most complexly arranged serous membrane in the body, and consists, in the male, of a closed sac, a part of which lines the abdominal wall, whilst the remainder is reflected over the contained viscera. In the female the lateral ends of the uterine tubes open into the peritoneal cavity. The part which lines the abdominal wall (or parieties) is named the *parietal* portion of the peritoneum; that which is reflected over the contained viscera constitutes the *visceral* portion of the peritoneum. The *free surface* of the membrane is smooth, covered with a layer of flattened *mesothelium* and kept moist and smooth by a thin film of serous fluid. Hence the viscera can glide on the wall of the cavity or on one another with complete freedom—within the limits dictated by their attachments to the walls of the abdominal cavity, or to other structures, such as vascular supplies. Some organs are themselves sessile, but are covered by peritoneum wherever they are in contact with mobile viscera.

The peritoneal cavity is, of course, a *coelom*—a discontinuity in the mesoderm with its own special surface epithelium (mesothelium) which maintains the surface. Loss of this epithelium entails adherence of the underlying tissues and a consequent interference with function which may be serious and even lethal (p. 1332), providing also convincing evidence of at least one essential contribution of such serosal epithelia—the separation of viscera sufficiently to allow them unimpeded activity. As a biological arrangement the coelomic principle is sufficiently basic to divide all animals into the *Coelomata* and the *Acoelomata*.

Many different functions are served by coelomic spaces in various invertebrates and vertebrates (Jones 1913; Romer 1970). Thus, excretory organs such as nephridia drain from a general body coelom, and the nephric systems of vertebrates are in part derived from it. As animals evolve to a greater size, a coiled gut becomes essential. It is difficult to envisage how such coiling could evolve phylogenetically or develop ontogenetically without the emergence of a coelomic cavity. In elementary chordates such as *Amphioxus*, the gametes are extruded from the gonads into a coelom, and in some early vertebrates the same arrangement exists in both male and female, persisting in the female in all forms, including mankind. The special ducts which evolve to connect the kidneys and testes to the cloaca and its later derivatives are in part formed from the coelomic epithelium, which is also closely associated with the formation of the gonads themselves (p. 210).

General Structure

A considerable amount of areolar connective tissue intervenes between the parietal peritoneum and the abdominal walls, with the fascial lining of which it blends. It is known as the *extraperitoneal tissue*. It varies in quantity and contains a varying amount of fat in different regions. While this tissue loosely connects the parietal peritoneum to the abdominal and pelvic walls in general and so allows the peritoneum to be relatively easily stripped from them, it is denser on the inferior surface of the diaphragm and behind the linea alba, so that the parietal peritoneum is more firmly adherent to these parts. It is especially loosely arranged in some places to allow alteration in the size of certain organs; for example, in the front part of the pelvis and lower part of the anterior abdominal wall where it allows the urinary bladder to distend in an upward direction behind the

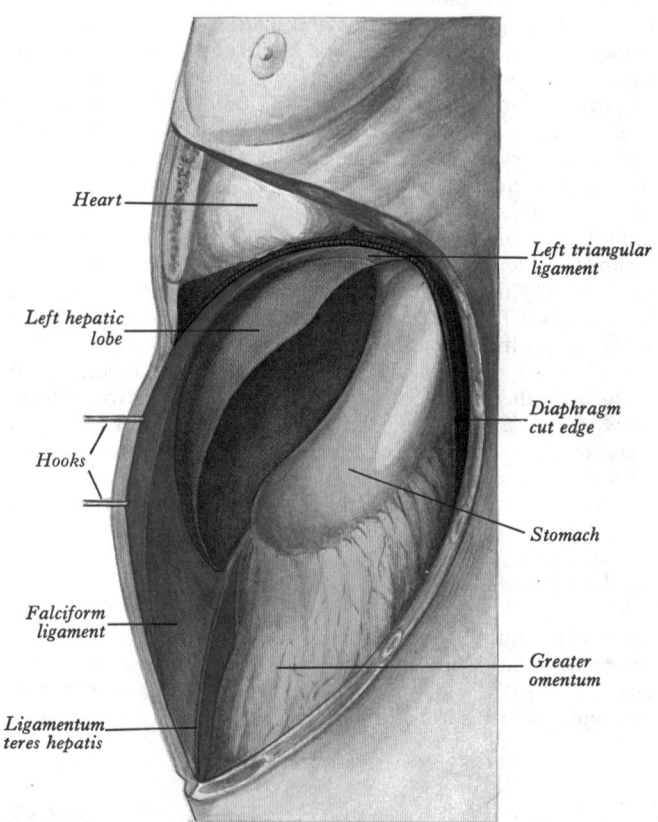

8.93 A dissection to expose the left side of the falciform fold or ligament of the liver.

Labels on figure: Heart — Left triangular ligament — Left hepatic lobe — Diaphragm cut edge — Hooks — Stomach — Falciform ligament — Greater omentum — Ligamentum teres hepatis

anterior abdominal wall, from which it strips off the peritoneum as it ascends. It is usually heavily laden with fat on the posterior abdominal wall in relation to the kidneys. The visceral peritoneum, on the other hand, is firmly united to the viscera which it covers, and cannot be readily stripped off them. In fact, the connective tissue layer (*tela subserosa*) of the visceral peritoneum is directly continuous with the fibrous tissue stroma of the viscera; thus from the point of view of pathological conditions of an organ, the visceral peritoneum must be considered to be part of the viscus itself.

THE PERITONEAL CAVITY

The parietal and visceral layers of the peritoneum are in actual contact; the potential space between them is the *peritoneal cavity*. The latter consists of (1) a main region, termed the *greater sac* (or *cavum peritonei*), and (2) a diverticulum from this, the *omental bursa* or *lesser sac*, which is situated behind the stomach and adjoining structures; the neck or communication between the greater sac and the lesser sac is the *epiploic foramen* (or *aditus to the lesser sac*).

The disposition of the complexly arranged peritoneum is best understood by first studying the development of the alimentary canal (pp. 197–207), followed by examination of the various regions of the peritoneal cavity in a cadaver which has not been made unnaturally rigid by preservative fluids.

To trace the peritoneum from one viscus to another, and from the viscera to the parieties, it is helpful to follow its continuity in the vertical and horizontal directions, and it is simpler to describe the greater sac and the lesser sac separately.

VERTICAL DISPOSITION OF THE PERITONEUM

It is convenient to commence tracing the arrangement of the greater peritoneal sac in the vertical plane (8.97) from the anterior abdominal wall at the level of the umbilicus. A fibrous cord, the *ligamentum teres* or *obliterated left umbilical vein* (p. 665), ascends from the umbilicus to the inferior surface of the liver; it commences at the umbilicus, inclines slightly to the right of the midline and recedes slightly from the anterior abdominal wall as it passes upwards; it raises a triangular fold of parietal peritoneum from the anterior abdominal wall and inferior surface of the diaphragm in this upward course—the *falciform ligament of the liver* (p. 1376 and 8.93). The latter consists of two layers of peritoneum, right and left, with intervening connective tissue (8.99). The inferior, juxta-umbilical region of the falciform ligament has a posterior free border, extending from the umbilicus to the inferior border of the liver and containing the ligamentum teres (obliterated left umbilical vein). Superior to this the ligament extends from the diaphragm to become continuous with the visceral peritoneum on the anterosuperior surface of the liver (8.93). At the site of reflexion from the anterior part of the inferior surface of the diaphragm to the upper surface of the liver, the two layers of the falciform ligament diverge from each other (8.135), the right layer passing more or less transversely to the right and forming the *superior layer of the coronary ligament of the liver* (which thus passes from the diaphragm to the upper surface of the right lobe of the liver), while the left layer passes to the left to form the *anterior layer of the left triangular ligament* of the liver (which passes from the diaphragm to the upper surface of the left lobe).

The visceral peritoneum on the supero-anterior surface of the liver is continued round the sharp lower border to the inferior (visceral) surface where it has the following arrangements. To the right of the gall bladder it covers the inferior surface of the right lobe and is reflected from the posterior part of this lobe to the right

suprarenal gland and the upper end of the right kidney, forming the *inferior layer of the coronary ligament*; the peritoneum frequently passes directly from the inferior surface of the liver to the front of the right kidney, forming a fold termed the *hepatorenal ligament*. From the right kidney the peritoneum passes downwards to the front of the superior part of the duodenum and of the right colic flexure; it also passes medially in front of a short segment of the inferior vena cava (between the duodenum and the liver), where it is continuous with the posterior wall of the omental bursa (8.98). Between the two layers of the coronary ligament there is a large, triangular area on the back of the right lobe devoid of peritoneal covering; this is termed the *bare area* of the liver and here the liver is attached to the diaphragm by areolar tissue.

Towards the right margin of the liver the superior and inferior layers of the coronary ligament gradually approach each other and ultimately fuse to form a small triangular fold, connecting the right lobe of the liver to the diaphragm and called the *right triangular ligament* of the liver (8.135). The latter forms the apex of the bare area, the base being formed by the *groove for the inferior vena cava*.

The visceral peritoneum covers the inferior surface and sides of the gall bladder, the inferior surface of the quadrate lobe of the liver as far back as the anterior margin of the porta hepatis, and the inferior surface of the left lobe; from the posterior surface of the left lobe it is reflected to the diaphragm as the *posterior layer of the left triangular ligament*. The peritoneum along the anterior margin of the porta hepatis is continuous at the right extremity of the porta with the peritoneum of the omental bursa, the latter being reflected from the posterior margin of the porta (8.135). The visceral peritoneum passes into the depth of the fissure for the ligamentum venosum (8.135), between the caudate and left lobes of the liver, in two layers, anterior and posterior. The anterior layer is continuous with the peritoneum reflected from the anterior margin of the porta hepatis (8.135). From this L-shaped line formed by the left margin of the fissure for the ligamentum venosum and the anterior margin of the porta hepatis, the peritoneum is reflected to the lesser curvature of the stomach and about the first 2 cm of the duodenum, forming the anterior layer of the *lesser omentum*.

The part of the lesser omentum connecting the liver to the stomach is called the *hepatogastric ligament*, while the part passing

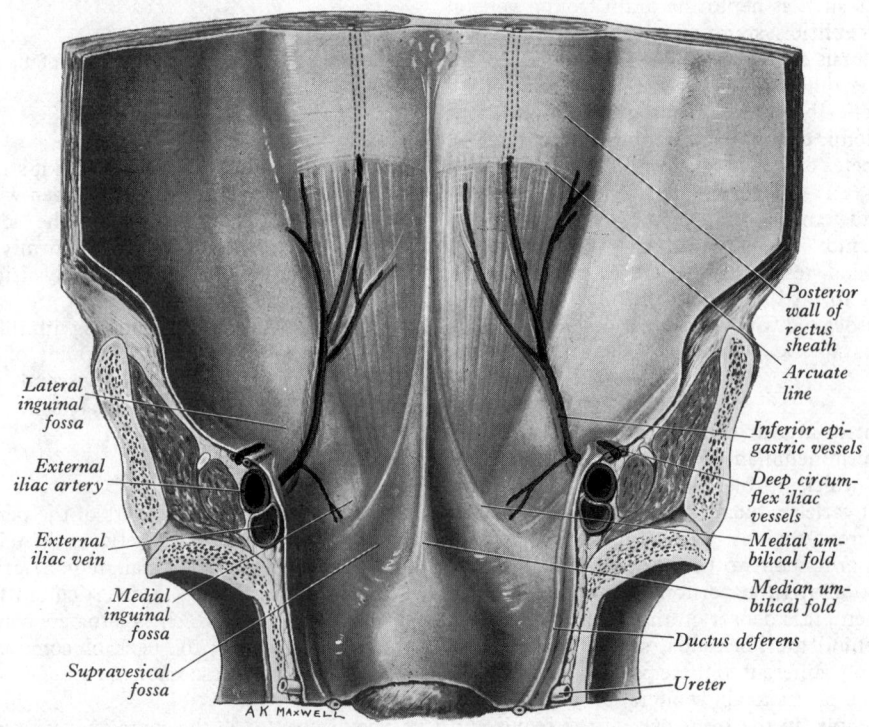

Lateral inguinal fossa

External iliac artery

External iliac vein

Medial inguinal fossa

Supravesical fossa

A K MAXWELL

Posterior wall of rectus sheath

Arcuate line

Inferior epigastric vessels

Deep circumflex iliac vessels

Medial umbilical fold

Median umbilical fold

Ductus deferens

Ureter

8.94 The infra-umbilical part of the anterior abdominal wall of a male subject. Posterior surface, with the peritoneum *in situ*.

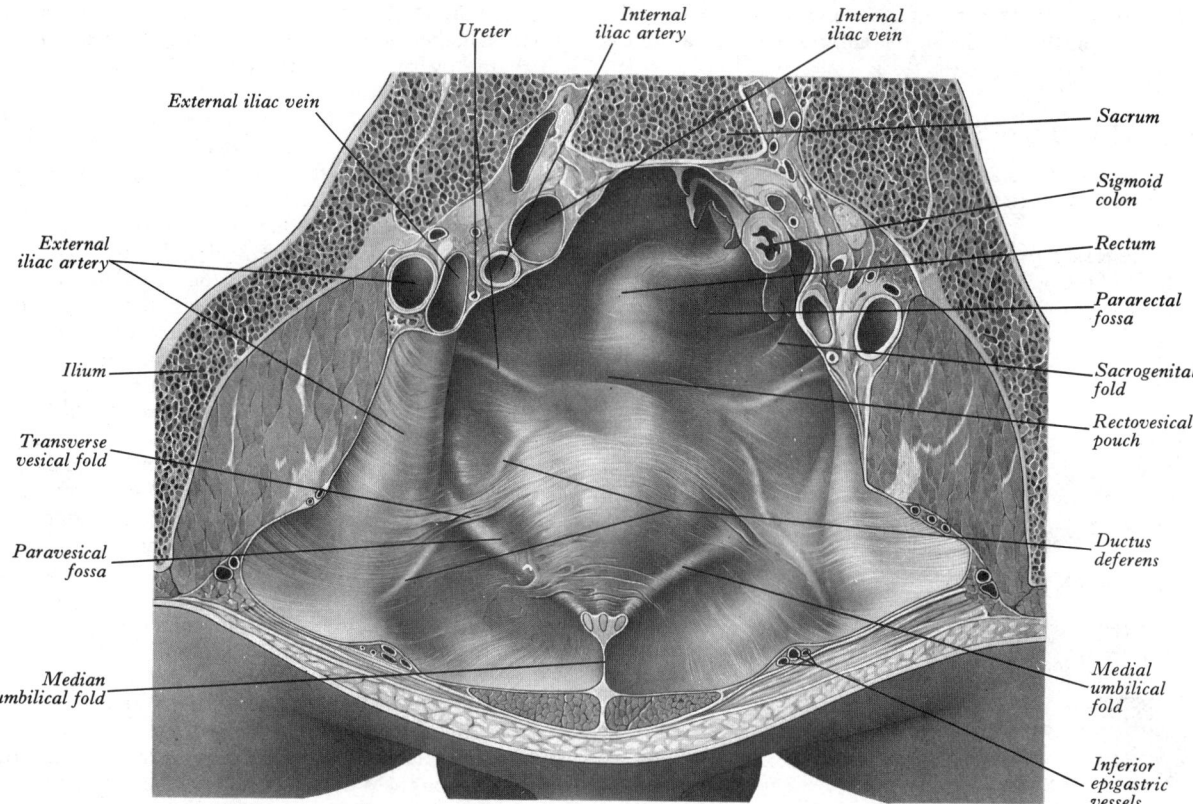

8.95 The peritoneum of the male pelvis. Anterosuperior view. The median umbilical fold contains both the unpaired median and the paired medial umbilical ligaments in the plane of section in this subject.

from the liver to the duodenum is named the *hepatoduodenal ligament*. The anterior layer of the lesser omentum, if traced to the right, is found to pass in front of the hepatic artery proper, bile duct and portal vein, and then to turn round the right side of these structures to become continuous behind them with the posterior layer of the lesser omentum; the latter here forms the anterior wall of the omental bursa. Thus the lesser omentum has a free right border, in which lie the hepatic artery proper, bile duct and portal vein, and behind which is the *epiploic foramen* (**8**.99). The anterior layer of the lesser omentum is continuous below with the visceral peritoneum that covers the front of the stomach and the first 2 cm of the duodenum. This layer of peritoneum then descends from the greater curvature of the stomach and the neighbouring part of the duodenum to form the most anterior layer of a large free fold, called the *greater omentum*. Reaching the free lower margin of this fold, it turns upwards, forming the most posterior layer of the greater omentum; the latter runs to the anterosuperior aspect of the transverse colon (opposite the *taenia omentalis*), to which it is adherent but from which it can be stripped off. It then passes backwards, adherent to but separable from the upper layer of the transverse mesocolon, to the anterior surface of the head and the anterior border of the body of the pancreas; it leaves the latter as the upper layer of the transverse mesocolon (**8**.97). This passes to the posterior surface of the transverse colon (opposite the *taenia mesocolica*), covers the upper, anterior and lower surfaces of that part of the gut, and passes from its posterior surface to the front of the head and the anterior border of the body of the pancreas as the inferior layer of the transverse mesocolon. Thence it is continued over the lower part of the anterior surface of the head and over the inferior surface of the body of the pancreas on to the front of the horizontal and ascending parts of the duodenum. From the latter it passes downwards on the posterior abdominal wall; it is also in part carried forward on the superior mesenteric vessels to the jejunum and ileum as the *right layer of the mesentery*. It invests this part of the gut and then passes to the posterior abdominal wall as the *left layer of the mesentery*; it then descends over structures like the abdominal aorta, inferior vena cava, the ureter and the psoas major into the lesser pelvis. It is reflected from the posterior pelvic

wall, as the anterior layer of the *sigmoid mesocolon*, invests the sigmoid colon and returns to the pelvic wall as the posterior layer of the sigmoid mesocolon. It then descends, covering the front and sides of the upper third of the rectum and the front of the middle third of the rectum.

In the male, it leaves the front of the rectum (at the junction of its middle and lower thirds) and passes forwards on to the upper ends of the seminal vesicles and the upper surface of the urinary bladder. Between the rectum and the bladder it dips slightly downwards forming a recess, the *rectovesical pouch*, the bottom of which is a little below the level of the upper ends of the seminal vesicles and about 7·5 cm from the anal orifice. From the apex of the bladder it is carried along the median and medial umbilical ligaments (**8**.94) to the anterior abdominal wall up to the level of the umbilicus (from which a start was made). When the bladder is distended the peritoneum is stripped away from the lower part of the anterior abdominal wall, so that a considerable part of the anterior surface of the bladder lies directly against the abdominal wall without the intervention of peritoneum (p. 1405). An instrument can therefore be passed through the abdominal wall into the distended bladder without passing through the peritoneal cavity.

In the female, the peritoneum passes from the front of the rectum on to the posterior fornix of the vagina and thence on to the back of the cervix and body of the uterus, forming the *recto-uterine fold*. This fold dips downwards to form the *recto-uterine pouch* (of Douglas), the bottom of which is about 5·5 cm from the anal orifice. The peritoneum continues over the fundus of the uterus and descends on its anterior (vesical) surface as far as the junction of the body and cervix, from which site it is reflected forwards on to the upper surface of the bladder, forming a shallow recess, the *vesico-uterine pouch*. The layers of peritoneum on the anterior and posterior surfaces of the uterus are reflected laterally, from the lateral margins of the uterus, to the side walls of the pelvis, forming on each side an expanded fold, termed the *broad ligament of the uterus*; the latter thus consists of two layers, antero-inferior and posterosuperior, which are continuous above to form the free upper border of the broad ligament; between the two layers at the upper border is the uterine tube. The reflexion of the peritoneum from the bladder to the anterior abdominal wall is similar to that in the male.

HORIZONTAL DISPOSITION OF THE PERITONEUM

Below the transverse colon the arrangement is simple, and it may be considered in three regions: pelvic, lower abdominal, and upper abdominal.

(1) **In the lesser pelvis.** The peritoneum here follows closely the surfaces of the pelvic viscera and the inequalities of the pelvic walls, and presents important differences in the two sexes. (a) **In the male** (8.95) it almost encircles the sigmoid colon, from which

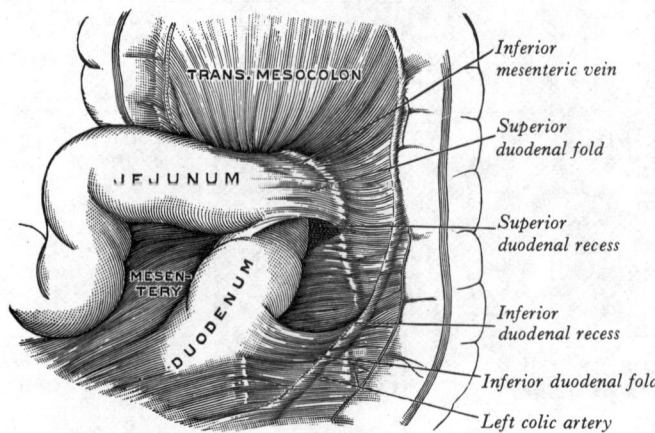

8.96 The superior and inferior duodenal recesses. The transverse colon and jejunum have been displaced. (After Jonnesco. From Poirier and Charpy's *Traité d'Anatomie humaine*. Masson et Cie.)

it is reflected to the posterior wall of the pelvis as the *sigmoid mesocolon*. It leaves the sides and, finally, the front of the rectum, and is continued over the upper parts of the seminal vesicles to the superior surface of the bladder; on each side of the rectum it forms a *pararectal fossa*, which varies in size with the distension of the rectum. Anterior to the rectum the peritoneum forms the *recto-vesical pouch*, which is limited laterally by peritoneal folds extending from the sides of the bladder posteriorly, on either side of the rectum, to the anterior aspect of the sacrum. These folds are known, from their position, as the *sacrogenital folds*; each is lateral to the corresponding pararectal fossa. The peritoneum of the anterior pelvic wall covers the superior surface of the bladder, and on each side of this viscus forms a *paravesical fossa*, which is limited laterally by a ridge of peritoneum elevated by the ductus deferens. The size of this fossa is dependent on the state of distension of the bladder, and when the bladder is empty, a variable *transverse vesical* fold of peritoneum bisects the fossa. Under the same conditions, the anterior ends of the sacrogenital folds may sometimes be joined by a fold which demarcates an anterior part of the recto-vesical pouch, termed the middle fossa (8.95). On the peritoneum between the paravesical and pararectal fossae the only elevations are those produced by the ureters and the internal iliac vessels. (b) **In the female,** pararectal and paravesical fossae similar to those in the male are present; the lateral limit of the paravesical fossa is the peritoneum investing the round ligament of the uterus. The rectovesical pouch is, however, divided by the uterus and vagina into a small, anterior, vesico-uterine and a deep, posterior, recto-uterine pouch (8.183). The folds forming the margins of the latter are the *recto-uterine folds* (p. 1432); they correspond to the sacrogenital folds of the male, and pass backwards from the sides of the cervix uteri, on each side of the rectum, to the front of the sacrum. The *broad ligaments* extend from the sides of the uterus to the lateral walls of the pelvis; the uterine tubes are contained in their free margins, and the ovaries are attached to their posterior layers. Below, the broad ligaments are continuous with the peritoneum on the lateral walls of the pelvis. In the angle between the elevations produced

by the obliterated umbilical artery and the ureter on the lateral pelvic wall, there is a shallow fossa, known as the *ovarian fossa*, in which the ovary lies in the nulliparous female. It is situated behind the lateral attachment of the broad ligament.

(2) **In the lower abdomen.** The peritoneum lining the lower part of the anterior abdominal wall is raised into five ridges or folds which converge as they pass upwards (8.94). One of these is placed in the median plane and extends from the apex of the urinary bladder to the umbilicus. It contains the urachus (p. 1405) and is termed the *median umbilical fold*. To its lateral side the obliterated umbilical artery forms the *medial umbilical fold*, as it ascends from the pelvis towards the umbilicus. The depressions between these folds form the two *supravesical fossae*. Further to the lateral side, the inferior epigastric artery raises a fold, called the *lateral umbilical fold*, below the point at which it enters the sheath of the rectus muscle. The *medial inguinal fossa* is the depression situated between the lateral and medial umbilical folds; the *lateral inguinal fossa*, which overlies the deep inguinal ring, lies to the lateral side of the lateral umbilical fold and indicates the site where the processus vaginalis extended into the anterior abdominal wall during the descent of the testis. A fourth depression is placed below and slightly medial to the lateral inguinal fossa and is separated from it by the medial end of the inguinal ligament. It overlies the femoral ring (p. 725) and is termed the *femoral fossa*.

Traced from the linea alba, below the level of the transverse colon, and followed in a horizontal direction to the right, the peritoneum covers the inner surface of the abdominal wall almost as far as the lateral border of the quadratus lumborum; it is reflected over the sides and front of the ascending colon and encloses the caecum and vermiform appendix; it may then be traced over the duodenum, psoas major, and inferior vena cava towards the median plane, whence it passes along the superior mesenteric vessels to invest the small intestine, and back again to the large vessels in front of the vertebral column, forming the *mesentery* (8.97, 98), the layers of which enclose the jejunum, ileum, the superior mesenteric blood vessels, nerves, lacteals and lymph nodes. It is then continued across the abdominal aorta and the left psoas major; it covers the sides and front of the descending colon, and, reaching the abdominal wall, is carried on it to the median plane.

(3) **In the upper abdomen** (8.98, 99A). Above the transverse colon, the peritoneum of the greater sac is more complexly arranged. Starting in front of the part of the inferior vena cava lying immediately above the superior part of the duodenum, the peritoneum of the greater sac is here continuous to the left, behind the epiploic foramen, with the peritoneum forming the posterior wall of the omental bursa (8.99A). From the front of the inferior vena cava, it passes to the right over the front of the right suprarenal gland and upper part of the right kidney to the anterolateral abdominal wall. From the anterior median line a double fold passes backwards and to the right, to become continuous with the peritoneum investing the liver, and forms the *falciform ligament*. Continuing to the left, the peritoneum lines the anterolateral abdominal wall and covers the lateral part of the front of the left kidney, and is thence reflected to the posterior border of the hilum of the spleen as the posterior or lateral layer of the *lienorenal* (or *phrenicolienal*) *ligament* (8.99A). It can then be traced over the surfaces of the spleen to the front of its hilum, and thence to the cardiac end of the greater curvature of the stomach as the left layer of the *gastrosplenic ligament*. It covers the anterosuperior surface of the stomach and commencement of the duodenum, and ascends from the lesser curvature of the stomach to the liver as the anterior layer of the lesser omentum. The right free border of the latter has been previously described (p. 1323), and at this border the anterior layer of the lesser omentum (formed by peritoneum of the greater sac) becomes continuous with the posterior layer of the lesser omentum (formed by the peritoneum of the omental bursa).

THE OMENTAL BURSA (LESSER SAC)

The omental bursa (or lesser sac of the peritoneum), is a large potential recess of irregular form behind the stomach and extending beyond its limits. Its name is due to the concept that it forms a bursa (p. 522) facilitating movement of the *posterior aspect* of the stomach. It is not, however, closed, and the fact that its communication with the greater sac is so limited is due rather to embryological factors than to functional demands. The stomach expands or contracts and moves about just as freely in vertebrates in which no special narrow-necked diverticulum exists. (It is also perhaps stating the obvious, that movements of the *anterior aspect* of the stomach place similar functional demands on greater sac peritonealized surfaces.) The anterior and posterior walls of the omental bursa are extensive and they are limited by variable inferior, right, left and superior borders. The recess is shut off from the greater sac except in the upper part of its right border,

The epiploic foramen (aditus to the lesser sac) is a short, vertically flattened passage, about 3 cm long, which leads out from the upper part of the right border of the omental bursa into the greater sac. Its *anterior wall* is formed by the right margin of the lesser omentum, which contains between its two layers in this situation the bile duct, the portal vein and the hepatic artery proper (**8**.99A). Traced upwards, the two layers separate and the posterior layer covers the caudate process of the liver, forming the *roof* of the epiploic foramen (**8**.100), and then descends in front of the inferior vena cava, forming the *posterior wall* of the foramen. At, or a little below, the upper border of the superior part of the duodenum, this layer passes forwards from the front of the inferior vena cava and above the head of the pancreas to become

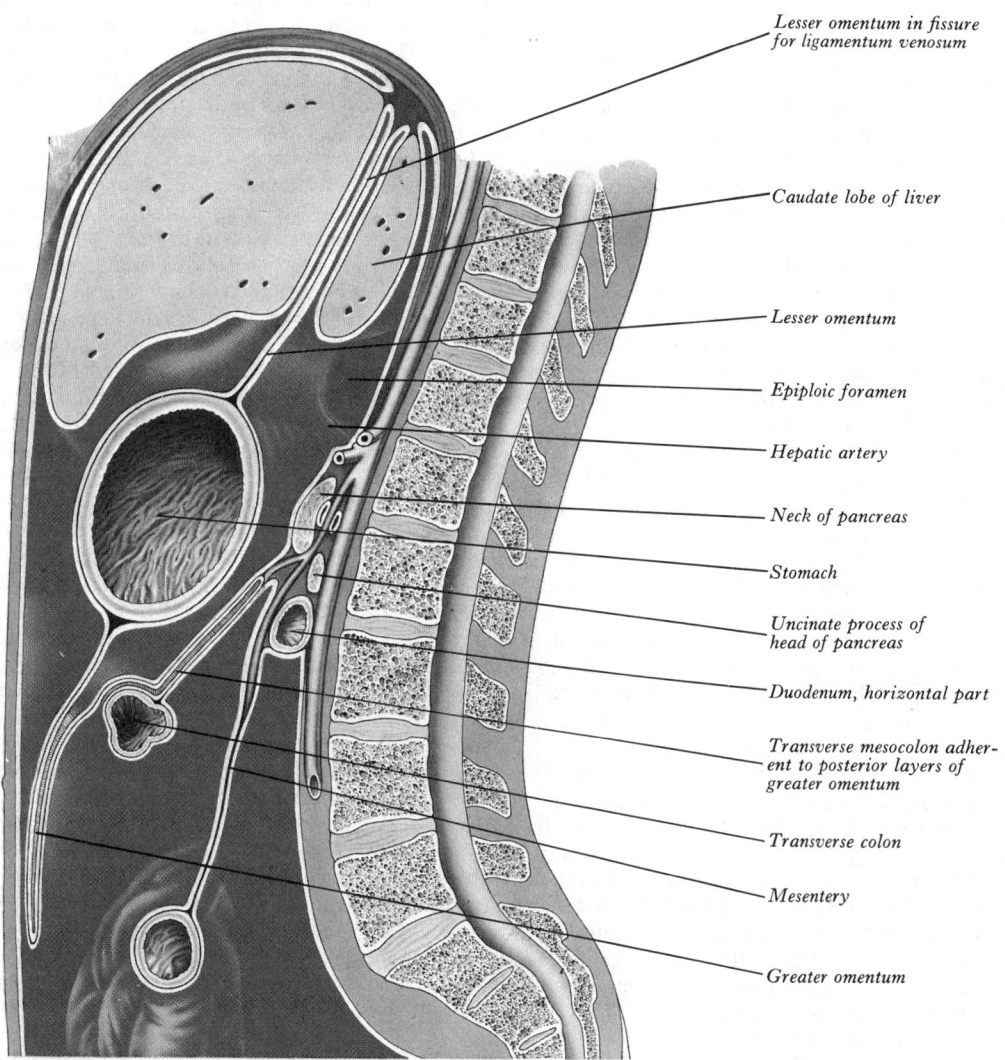

Lesser omentum in fissure for ligamentum venosum

Caudate lobe of liver

Lesser omentum

Epiploic foramen

Hepatic artery

Neck of pancreas

Stomach

Uncinate process of head of pancreas

Duodenum, horizontal part

Transverse mesocolon adherent to posterior layers of greater omentum

Transverse colon

Mesentery

Greater omentum

8.97 A sagittal section through the abdomen, approximately in the median plane. Compare with **8**.98. The section cuts the posterior abdominal wall along the line YY in **8**.98. The peritoneum is shown in blue except along its cut edges, which are left white.

where a communication is established through a vertical, slit-like opening. In its upper part the posterior wall of the lesser sac is formed by a single layer of peritoneum, closely applied to the posterior abdominal wall (**8**.97), but, below the pancreas, the sac is carried into the interior of the greater omentum and its posterior wall is formed by the posterior two layers of that structure, which, above the transverse colon, are blended with the transverse mesocolon (**8**.97). It is usual to state this part of the peritoneum to consist of four layers; but while such layers may be appreciated, it must be understood that the true, epithelial peritoneum is lost, except where a surface persists. Where two originally separate 'folds' of peritoneum adhere and blend into a single structure, the opposed epithelia disappear, although subepithelial layers of supporting connective tissue may still be recognizable.

continuous with the posterior layer of the lesser omentum, and in this situation, it forms the *floor* of the epiploic foramen. The medial end of the floor is continuous inferiorly with the right border of the lower part of the lesser sac (**8**.98), and it is by passing forwards below the medial end of the floor that the hepatic artery is able to insinuate itself between the two layers of the lesser omentum (**8**.99A). The narrow passage lying to the left of the epiploic foramen, below the caudate process of the liver and above the superior part of the duodenum, is called the *vestibule* of the omental bursa. Traced laterally, to the right, all boundaries of the epiploic foramen are directly continuous with the peritoneum of the greater sac. The roof is continuous with the peritoneal covering of the inferior surface of the right hepatic lobe (**8**.135); the posterior wall with the peritoneum on the right suprarenal gland (**8**.98); the anterior wall with the anterior layer of the lesser omentum round the portal vein and the bile duct (**8**.99A); the floor with the peritoneum covering the lower part of the right suprarenal gland and the adjacent parts of the duodenum and right kidney (**8**.98). The anterior and posterior walls of the epiploic foramen are normally in contact with each other.

The omental bursa and its boundaries can now be considered in detail. The *anterior wall* is formed (1) by the peritoneum which covers the postero-inferior aspect of the stomach and about the first 2 cm of the duodenum. Traced downwards this layer becomes the posterior of the anterior two layers of (2) the greater omentum; traced upwards and to the right, it leaves the stomach along the lesser curvature and the duodenum at its upper border, and becomes the posterior layer of (3) the lesser omentum. The omental bursa is usually described as passing upwards *behind* the caudate lobe of the liver, but this description is scarcely accurate, for the caudate lobe projects into the omental bursa from its right border and is covered by peritoneum on its anterior as well as on its posterior surface (**8**.97, 135).

The *posterior wall* is formed by the anterior of the posterior two layers of the greater omentum. Above, the posterior of these two layers is fused with, but separable from, the peritoneum on the upper aspect of the transverse colon and the upper layer of the transverse mesocolon. Surgical separation of the omentum from the transverse colon and its mesocolon thus provides access to the posterior wall of the stomach through the posterior wall (greater omentum) of the omental bursa. Dissection where the omentum meets the transverse colon opens up the embryological plane ('bloodless plane') of adhesion of the omentum to the colon and its mesocolon, and thus passes between the omental vessels (from the gastro-epiploic vessels) in the greater omentum and the middle colic vessels in the transverse mesocolon (Freder 1905; Lardennois and Okinczyc 1913; Grégoire 1922; Ogilvie 1935). There are no anastomoses between the omental and colic vessels across this plane. Above the anterior border of the pancreas, the posterior wall of the omental bursa lines the posterior abdominal wall, covering a small part of the front of the head, and the whole of the front of the neck and body of the pancreas, a small part of the anterior aspect of the left kidney and most of the anterior aspect of the left suprarenal gland, the commencement of the abdominal aorta and the coeliac artery, and a considerable area of the diaphragm. In addition, the inferior phrenic, the splenic, the left gastric and, to a much smaller extent, the hepatic arteries course behind the omental bursa (**8**.98, 99A).

The borders of the omental bursa are formed by the lines along which its peritoneal posterior wall is reflected to become continuous with the peritoneal anterior wall, and they are subject to considerable variation. The *inferior border* is, developmentally (p. 205), the lower border of the greater omentum, but, as a rule, partial fusion of the constituent layers of the greater omentum occurs after birth, so that the cavity of the omental bursa in the adult does not usually extend much below the transverse colon. (As an accompaniment of this fusion, the two internally opposed epithelial surfaces, of course, lose their epithelium.) The *upper border* of the omental bursa is narrow and extends between the right side of the oesophagus and the upper end of the fissure for the ligamentum venosum of the liver (**8**.135). In this interval the peritoneal posterior wall of the omental bursa is reflected forwards from the diaphragm and becomes continuous with the posterior layer of the lesser omentum.

The *right border* of the omental bursa corresponds, below, to the right free border of the greater omentum. Above the upper end of the latter it is formed by the reflexion of the peritoneum from the neck and head of the pancreas on to the inferior aspect of the superior part of the duodenum (**8**.99B). The line of this reflexion passes upwards and to the left along the medial side of the gastroduodenal artery. Near the upper border of the duodenum the right border becomes continuous with the floor of the epiploic foramen round the hepatic artery proper (**8**.98). Above the opening, which interrupts its continuity, the right border is formed by the reflexion of the peritoneum from the diaphragm to the right margin of the caudate lobe of the liver where it follows the left edge of the inferior vena cava (**8**.98).

The *left border* of the omental bursa corresponds, below, to the left margin of the greater omentum. Above the root of the

transverse mesocolon (**8**.98) the left border is broader. It is formed by the *lienorenal* and the *gastrosplenic ligaments* (**8**.99A), which together represent a part of the original dorsal mesogastrium (p. 200). The lienorenal ligament extends from the front of the left kidney to the hilum of the spleen (*lien*) as a bilaminar fold, in which the splenic vessels and tail of the pancreas (**8**.98, 99A) are enclosed. From the hilum of the spleen these two layers are continued forwards to the greater curvature of the stomach as the gastrosplenic ligament. The inner (or right) layer of the lienorenal ligament is directly continued into the inner (or right) layer of the gastrosplenic ligament; but the outer (or left) layer of the lienorenal ligament, on reaching the back of the hilum of the spleen, is continuous with the visceral peritoneum of the spleen. The latter is then reflected from the front of the hilum of the spleen as the outer (or left) layer of the gastrosplenic ligament. The spleen thus projects to the left into the greater sac (**8**.99A). The part of the omental bursa projecting towards the spleen, between the lienorenal and gastrosplenic ligaments, is known as the *splenic recess* of the omental bursa. At their superior ends the lienorenal and the gastrosplenic ligaments merge into a short fold, the *gastrophrenic ligament*, which passes from the diaphragm, behind, to the posterior aspect of the fundus of the stomach, in front. The two layers of this ligament diverge as they approach the oesophagus and a part of the posterior surface of the stomach is devoid of a peritoneal covering. In this situation the upper end of the left border becomes continuous with the left extremity of the roof and the left gastric artery turns forwards to gain the lesser omentum. (A considerable number of peritoneal *folds* are misleadingly termed 'ligaments'. They have little in common either in structure or function with skeletal ligaments, being often the neurovascular pedicles of organs which are inevitably covered by peritoneum. In a few instances they may also have a supportive function, but the actual evidence for this is usually tenuous. The lienorenal 'ligament' might be more suitably termed *splenorenal*.)

The omental bursa is encroached upon by two crescentic folds of peritoneum which are drawn into the sac by the hepatic and left gastric arteries. The *left gastropancreatic fold* is formed by the left gastric artery as it passes from the posterior abdominal wall to reach the lesser curvature of the stomach; the *right gastropancreatic fold*, at a lower level, is formed by the hepatic artery as it passes forwards from the posterior abdominal wall to the lesser omentum (**8**.98). The folds show considerable variation in their depths but, when well marked, they constrict the lesser sac and enclose a *foramen bursae omenti majoris*. The *superior recess* of the omental bursa lies above the foramen and communicates through it with the *inferior recess*, which represents the pancreatico-enteric recess of the embryo (p. 205). The superior recess thus lies behind the lesser omentum and the liver, while the inferior recess lies behind the stomach and in the greater omentum.

During a considerable part of fetal life the transverse colon is suspended from the posterior abdominal wall by a mesentery of its own, the posterior two layers of the greater omentum passing at this stage in front of the colon (**2**.135). This condition occasionally persists throughout life, but as a rule adhesion occurs between the mesentery of the transverse colon and the posterior layer of the greater omentum; even so, these layers of peritoneum are separable in the adult (more readily in the living subject than in the formalinized cadaver) (*see* pp. 1327, 1328), though, of course, the epithelial elements have disappeared from the fused layers, leaving only the subjacent connective tissue. In the adult the omental bursa intervenes between the stomach and the structures on which that viscus lies and which form the 'stomach bed' (p. 1334); it performs therefore the functions of a serous bursa for the stomach, facilitating the movements of the latter over the neighbouring structures.

A number of peritoneal folds extend between various organs or connect them to the abdominal and pelvic walls; they enclose the vessels and nerves proceeding to the viscera and, although they are clearly not designed to sustain any weight, they may help to retain certain of the viscera in contact with one another. They are named ligaments, omenta and mesenteries. (The inappropriate nature of the term 'ligament' in this context has been alluded to above. An 'omentum' is a cover and the word may have been used to denote an apron. Hence, the term is suitable for the greater

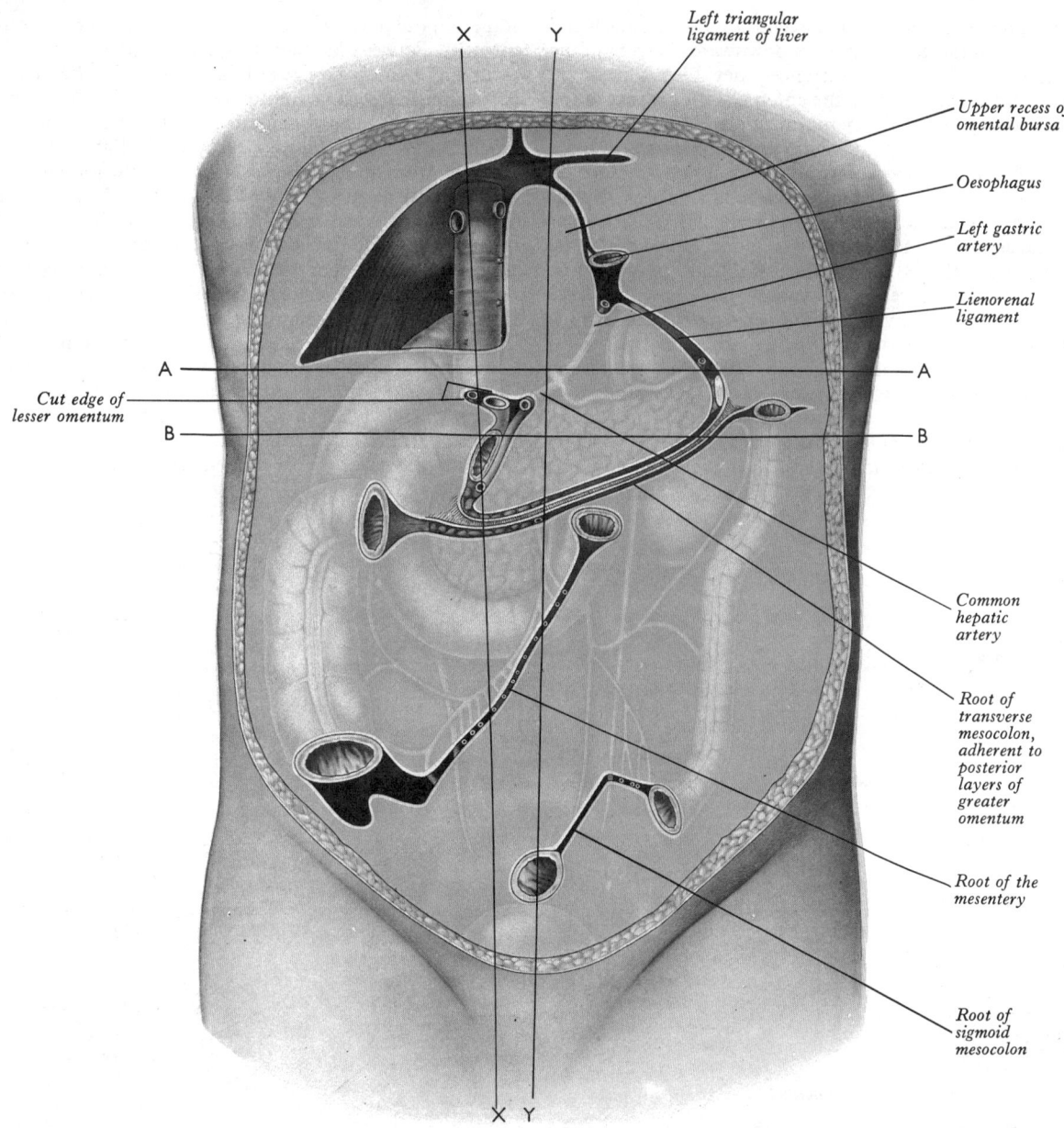

Left triangular ligament of liver

Upper recess of omental bursa

Oesophagus

Left gastric artery

Lienorenal ligament

X Y

A —————————————————————————————— A

Cut edge of lesser omentum

B —————————————————————————————— B

Common hepatic artery

Root of transverse mesocolon, adherent to posterior layers of greater omentum

Root of the mesentery

Root of sigmoid mesocolon

X Y

8.98   The posterior abdominal wall, showing the lines of peritoneal reflexion, after removal of the liver, spleen, stomach, jejunum, ileum, caecum, transverse colon and sigmoid colon. The various sessile (retroperitoneal) organs are seen shining through the posterior parietal peritoneum. Note the ascending and descending colon, duodenum, kidneys, suprarenals, pancreas and inferior vena cava. Line YY represents the plane of **8**.97. Line AA represents the plane of **8**.99 A. Line XX represents the plane of **8**.100. Line BB represents the plane of **8**.99 B.

omentum, but has been used by extension for other peritoneal folds connected with the stomach.) The peritoneal 'ligaments' will be described with their respective organs.

THE OMENTA

There are two omenta, the lesser and the greater.

The lesser omentum is the fold of peritoneum which extends to the liver from the lesser curvature of the stomach and the commencement of the duodenum. It is continuous with the two layers which cover the anterosuperior and postero-inferior surfaces of the stomach and about the first 2 cm of the duodenum. From the lower part of the lesser curvature of the stomach and the upper border of the duodenum, these two layers ascend as a double fold to the porta hepatis; from the upper part of the lesser curvature, the two layers pass to be attached to the bottom of the fissure for the ligamentum venosum. The hepatic attachment of the lesser omentum is, therefore, ⅃-shaped, the horizontal limb corresponding to the margins of the porta hepatis, and the vertical limb to the floor of the fissure for the ligamentum venosum. At the superior limit of the latter, the lesser omentum reaches the diaphragm, where the two layers separate to embrace the abdominal part of the oesophagus. At the right border of the omentum the two layers are continuous, and form a free margin which constitutes the anterior boundary of the epiploic foramen. The portion of the lesser omentum extending between the liver and stomach is named the *hepatogastric ligament*, and that between the liver and duodenum the *hepatoduodenal ligament*. Close to its right free margin the two layers of the lesser omentum enclose the hepatic artery proper, portal vein and bile duct, a few lymph nodes and lymph vessels, and the hepatic plexus of nerves—all these structures being enclosed in a fibrous capsule, termed the *perivascular fibrous capsule*. The right and left gastric arteries, the corresponding veins, branches from the gastric (vagus) nerves (p. 1081), and some of the left gastric lymph nodes and their vessels, lie between the layers of the lesser omentum, where these are attached to the stomach. The left part of the lesser omentum is thinner than the right part and may be fenestrated. This variation in thickness is dependent upon the amount of connective tissue and, especially fat.

The greater omentum is the largest peritoneal fold. It consists of a double sheet, folded on itself so that it is made up of four layers. The two layers which descend from the stomach and commencement of the duodenum pass downwards in front of the small intestine for a variable distance; they then turn upon themselves, and ascend as far as the anterosuperior aspect of the transverse colon (opposite the taenia omentalis). They adhere to, but are separable from, the peritoneum on the upper surface of the transverse colon and the upper layer of the transverse mesocolon (*see* pp. 1326, 1329). The left border of the greater omentum is continuous above with the gastrosplenic ligament; its right border extends as far as the commencement of the duodenum. (It must be emphasized that the greater omentum and gastrosplenic 'ligaments' are not merely continuous; they are the same structure, separated only by descriptive convenience and terminology. This continuity has been further obscured by *changes* in terminology; the gastrosplenic *ligament* was formerly regarded as an *omentum*, but it is now officially the gastro-*lienal* ligament—a somewhat inadvisable use of the stem 'lien', the spleen, in a region where almost all else is 'splenic'.) The greater omentum is usually thin, and presents a cribriform appearance, but it always contains some adipose tissue, which in fat people is present in considerable quantity. Between its anterior two layers, about a finger's breadth from the greater curvature of the stomach, the right and left gastro-epiploic vessels anastomose with each other. Variations in the distribution and anastomoses of the arteries in the greater omentum have been recently recorded by Jiang Dian-fu (1978).

Apart from functioning as a storehouse for fat, the greater omentum may limit spread of infection in the peritoneal cavity. When the abdomen is carefully opened without disturbing the organs, the greater omentum is frequently found wrapped about the organs in the upper part of the abdomen; only occasionally is it evenly dependent anterior to the intestines. It has a lesser capacity for absorption than the peritoneum in general. That it is not a vital physiological organ is indicated by the fact that it is occasionally congenitally absent and that it may be removed without apparent ill effect. The greater omentum contains numerous fixed macrophages, which can be mobilized as free macrophages. These cells may accumulate in places into dense, oval or round patches, visible to the naked eye as '*milky-spots*' on the omentum. Similar

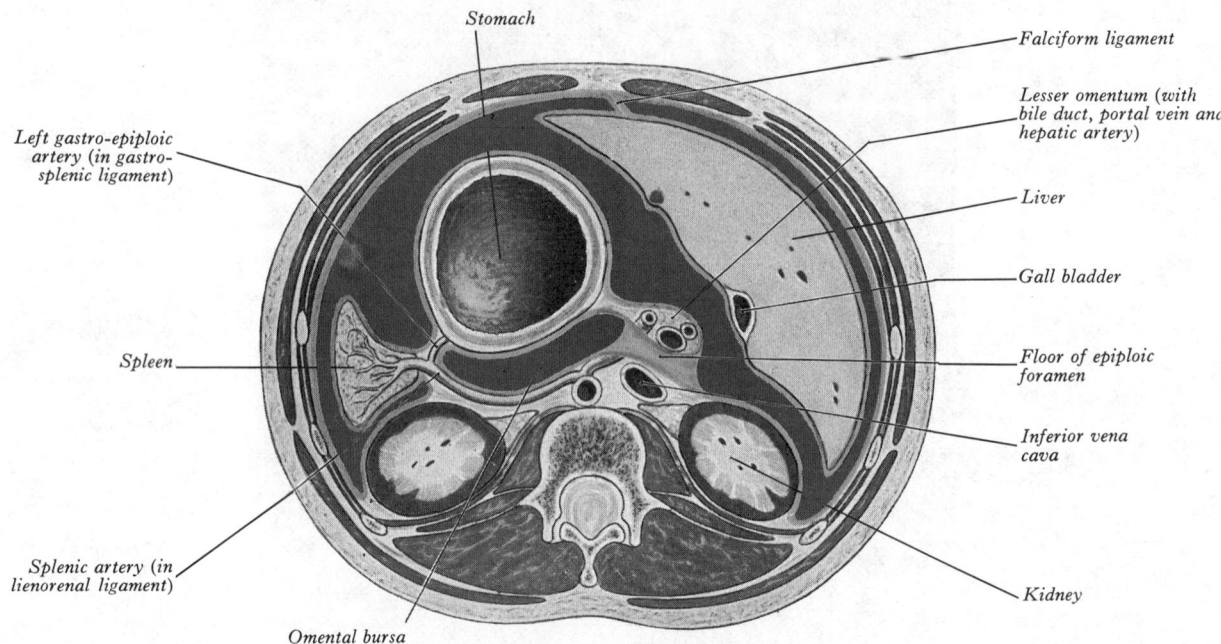

8.99A   A transverse section through the abdomen, at the level of line AA in **8.**98, viewed from above. The peritoneal cavity is shown in dark blue; the peritoneum and its cut edges in lighter blue.

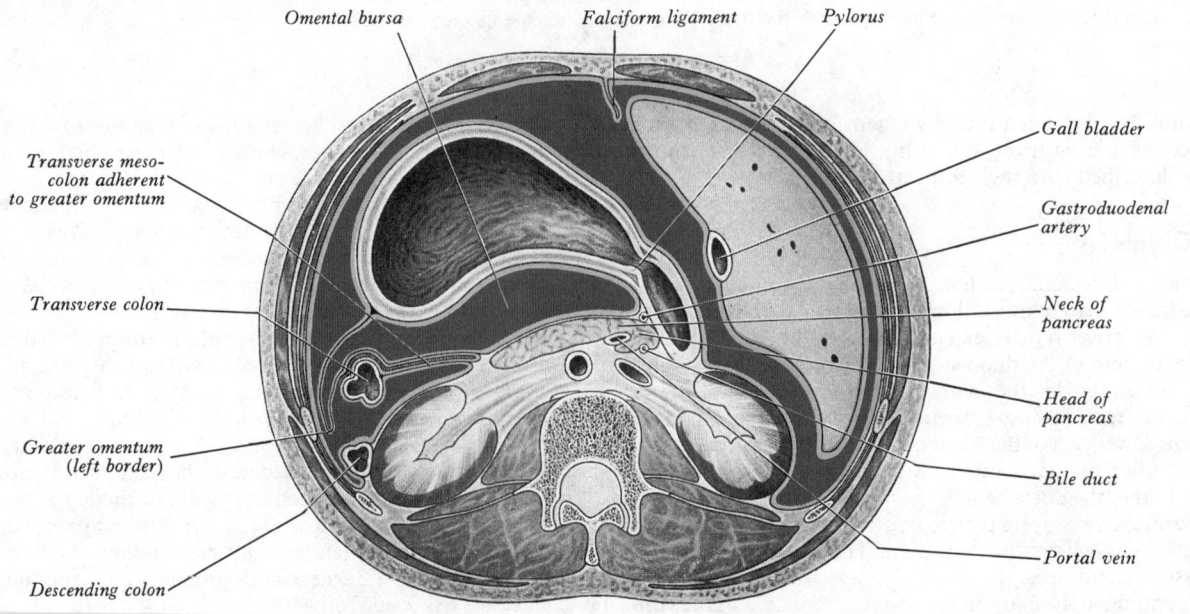

8.99B   A transverse section through the abdomen, at the level of the line BB in **8.**98, viewed from above. Colours as in **8.**99A.

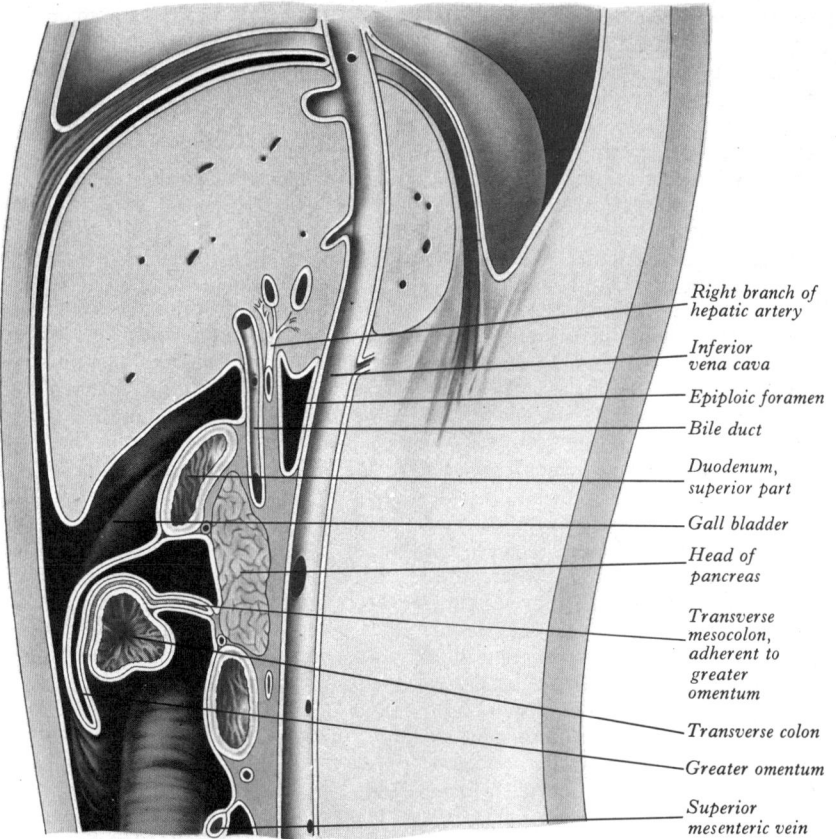

Right branch of
hepatic artery

Inferior
vena cava

Epiploic foramen

Bile duct

Duodenum,
superior part

Gall bladder

Head of
pancreas

Transverse
mesocolon,
adherent to
greater
omentum

Transverse colon

Greater omentum

Superior
mesenteric vein

8.100   A section through the upper part of the abdominal cavity, along the line **XX** in **8**.98. The boundaries of the epiploic foramen are shown, and a small recess of the omental bursa is displayed in front of the head of the pancreas. Note that the transverse colon and its mesocolon are adherent to the posterior two layers of the greater omentum.

THE MESENTERIES

spots may be found on other serous membranes (pleura, pericardium and, on occasion, in association with the leptomeninges).

The peritoneal folds collectively known as *mesenteries* include the mesentery of the small intestine (the mesentery proper), the mesoappendix, the transverse mesocolon and the sigmoid mesocolon. An ascending and a descending mesocolon are sometimes present and the gall-bladder occasionally has a mesentery.

The mesentery (of the small intestine) is a broad, fan-shaped fold of peritoneum connecting the coils of jejunum and ileum to the posterior abdominal wall. The border attached to the posterior wall of the abdomen is called the *root of the mesentery*; it is about 15 cm (6 in.) long and is directed obliquely downwards to the right from the duodenojejunal flexure (at the left of the second lumbar vertebra) to the upper part of the right sacro-iliac joint. (Schmidt, 1974, measured the mesenteric 'root' in 44 cadavers, finding a mean length of 13·9 cm, with extremes of 7·4 and 19·3 cm.) In this course it passes successively in front of the horizontal part of the duodenum (where the superior mesenteric vessels enter the mesentery), the abdominal aorta, the inferior vena cava and the right ureter and right psoas major. The intestinal border of the mesentery is about 6 m (20 ft) long and is thrown into numerous pleats or frills. (Its length is, however, subject to much variation, *see* p. 1342.) The pleating diminishes towards the posterior abdominal wall where the root is attached along almost a straight line. The central part of the mesentery is the longest (measured from its root to its intestinal border) and attains a maximum of about 20 cm (8 in.); it becomes shorter towards each end. The mesentery consists of two layers of the peritoneum of the greater sac—right and left—between which lie the jejunal and ileal branches of the superior mesenteric artery, with their accompanying veins, nerve plexuses and lymph vessels (here called *lacteals*), the mesenteric lymph nodes, connective tissue

and fat. The fat is most abundant in the lower part of the mesentery and here extends from the root to the intestinal border; in the upper part, the mesentery contains less fat and this tends to accumulate near the root and to leave oval or circular fat-free, translucent areas in the mesentery adjoining the upper part of the jejunum. At the intestinal border of the mesentery, the two layers of peritoneum separate to enclose the gut, forming its visceral

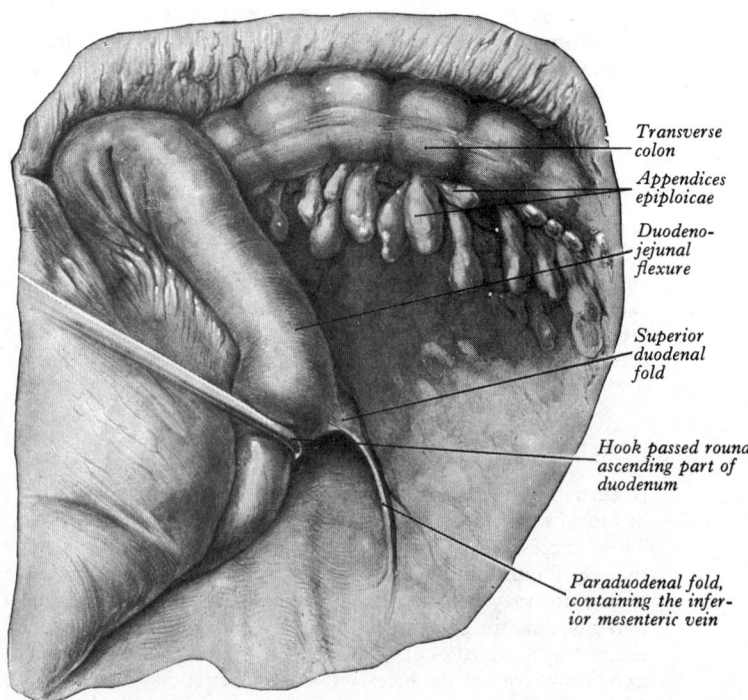

Transverse
colon

Appendices
epiploicae

Duodeno-
jejunal
flexure

Superior
duodenal
fold

Hook passed round
ascending part of
duodenum

Paraduodenal fold,
containing the infer-
ior mesenteric vein

8.101   The paraduodenal recess.

1329

peritoneal coat. At the root of the mesentery, the right layer of the peritoneum is reflected, in its lower part, over the posterior abdominal wall to cover the ascending colon, and in its upper part, to become continuous with the inferior layer of the transverse mesocolon; whereas the left layer passes to the left over the posterior abdominal wall and descending colon. (It is helpful to bear this arrangement in mind when determining which is the proximal and which the distal part of a coil of small intestine while still *in situ.*)

The **mesoappendix** (**8**.102) is a triangular fold of the peritoneum around the vermiform appendix, and is attached to the back of the lower end of the mesentery, close to the ileocaecal junction. It usually extends to the tip of the appendix, but sometimes it fails to reach the distal third or so, and is here represented by a low peritoneal ridge containing fat. Its layers enclose the blood vessels, nerves and lymph vessels of the vermiform appendix, together with a lymph node (p. 795).

The **transverse mesocolon** is a broad fold which connects the transverse colon to the posterior abdominal wall. Its two layers pass from the anterior surface of the head and the anterior border of the body of the pancreas to the posterior surface of the transverse colon (opposite the *taenia mesocolica*), where they separate to surround that part of the gut. The upper layer is adherent to, but separable from, the greater omentum (*see* pp. 1326, 1328 and **8**.97). Posteriorly, the lower layer of the transverse mesocolon covers the inferior surface of the pancreas and passes thence on to the front of the horizontal and ascending parts of the duodenum. Between the layers of the transverse mesocolon are the blood vessels, nerves and lymphatics of the transverse colon. The middle colic artery descends to the right, leaving a large avascular area of the fold to its left, and a similar but smaller area to its right.

The **sigmoid mesocolon** is a fold of peritoneum which attaches the sigmoid colon to the pelvic wall. Its line of attachment has the form of an inverted V, the apex of which is near the division of the left common iliac artery (**8**.98); the left limb descends medial to the left psoas major; the right limb descends into the pelvis and ends in the median plane at the level of the third sacral vertebra. The sigmoid and superior rectal vessels run between the layers of the sigmoid mesocolon and the left ureter descends into the pelvis behind its apex.

Usually the peritoneum covers only the front and sides of the ascending and descending parts of the colon, but sometimes these are surrounded by peritoneum and attached to the posterior abdominal wall by an ascending and a descending mesocolon respectively (p. 1355). A fold of peritoneum, termed the *phrenicocolic ligament,* is continued from the left colic flexure to the diaphragm opposite the tenth and eleventh ribs; it has an anterior free border and passes inferolateral to the lateral end of the spleen; it is sometimes given the misleading name of *sustentaculum lienis* or, more explicitly, 'the splenic shelf'—both implying a supportive role which is as hypothetical as it is in the case of most peritoneal folds or 'ligaments'.

The *appendices epiploicae* are small appendages of the peritoneum filled with adipose connective tissue and situated along the colon; they are best marked on the transverse (**8**.101) and sigmoid parts of the colon, absent from the rectum and rudimentary on the caecum and appendix. Many contain a small arteriole which enters them from the wall of the gut. In the case of the colon, they are most numerous along the line of the *taenia libera* (p. 1362).

PERITONEAL RECESSES

In certain parts of the abdomen, peritoneal folds may sometimes be found which bound fossae or recesses of the peritoneal cavity. These recesses are of surgical importance since they may become the site of 'internal' herniae, that is, a piece of intestine may enter a recess and may be constricted (strangulated) by the peritoneal fold guarding the entrance to the recess. Since the entrance to the recess may need to be cut to relieve strangulation and allow the gut to be drawn out of the recess, it is necessary to note whether the fold is significantly vascularized. From a surgical point of view the omental bursa can be considered to belong to this category, with

its opening at the epiploic foramen, bounded in front by the free border of the lesser omentum. Other recesses, of much smaller size, are sometimes found in relation to the duodenum, caecum and sigmoid mesocolon.

1. Duodenal Recesses

(*a*) The *superior duodenal recess* (**8**.96) is present in about 50 per cent of subjects; it may exist alone but usually occurs together with the inferior duodenal recess. It lies on the left side of the upper portion of the ascending part of the duodenum, opposite the second lumbar vertebra, and is situated behind a crescentic fold of peritoneum, the *superior duodenal fold* (or *duodenojejunal fold*), which has a semilunar free lower margin and which merges, on the left, with the peritoneum covering the front of the left kidney. The inferior mesenteric vein lies behind the point of junction between the left end of the superior duodenal fold and the posterior parietal peritoneum. The recess is about 2 cm deep and admits a fingertip; its orifice looks downwards and is situated in the angle formed by the left renal vein as it crosses in front of the abdominal aorta.

(*b*) The *inferior duodenal recess* (**8**.96) is present in about 75 per cent of subjects, usually in association with a superior duodenal recess, with which it may share a single oval orifice. It lies on the left side of the lower portion of the ascending part of the duodenum, opposite the third lumbar vertebra, and is situated behind a non-vascular, triangular peritoneal fold, the *inferior duodenal fold* (or *duodenomesocolic fold*), which has a sharp upper free margin. The recess is about 3 cm deep and admits the tips of one or two fingers; its orifice looks upwards, facing that of the superior duodenal recess. The inferior duodenal recess may sometimes extend behind the ascending part of the duodenum and to the left in front of the ascending branch of the left colic artery and the inferior mesenteric vein; in these circumstances this large fossa is liable to become the site of an internal hernia.

(*c*) The *paraduodenal recess* (**8**.101) may occur together with the superior and inferior duodenal recesses. It is found more frequently in the fetus and the newborn child than in the adult, in whom it occurs in about 2 per cent of subjects. It lies a little to the left of the ascending part of the duodenum, behind a falciform peritoneal fold (*paraduodenal fold*), the right free edge of which contains the inferior mesenteric vein accompanied by the ascending branch of the left colic artery, the fold forming a mesentery for these vessels. The free margin of the fold lies in front of the wide orifice of the recess, which is directed towards the right.

(*d*) The *retroduodenal recess* is only occasionally present. It is the largest of the duodenal recesses and lies behind the horizontal and ascending parts of the duodenum, in front of the abdominal aorta. It extends upwards nearly as far as the duodenojejunal junction, being about 8 to 10 cm deep, and is bounded on both sides by peritoneal folds (the duodenoparietal folds); its orifice looks downwards and to the left.

(*e*) The *duodenojejunal* or *mesocolic recess* is present in about 20 per cent of subjects and is rarely or never accompanied by any other variety of duodenal recess. It is about 3 cm deep and lies on the left side of the abdominal aorta, between the duodenojejunal junction and the root of the transverse mesocolon; it is bounded above by the pancreas, on the left by the left kidney, and below by the left renal vein. Its orifice is circular, bounded by two peritoneal folds and looks downwards and to the right.

(*f*) The *mesentericoparietal recess* (of Waldeyer) is found more frequently in the fetus and the newborn child than in the adult, in whom it occurs in about 1 per cent of subjects. It lies just below the horizontal part of the duodenum and invaginates the upper part of the mesentery towards the right. Its orifice is large and looks towards the left; it is guarded in front by a fold of the mesentery raised by the superior mesenteric artery.

2. Caecal Recesses

(*a*) The *superior ileocaecal recess* (**8**.102) is usually present, and best developed, in children, but it may become reduced in size and is often absent in the aged, especially in the obese. It is formed by a peritoneal fold (the *vascular fold of the caecum*) which arches over the branch of the ileocolic artery (and its accompanying vein) that supplies the ileocaecal junction on its anterior surface (anterior caecal artery). The recess is a narrow chink and is bounded, in front, by the vascular fold of the caecum, behind, by the mesentery of the ileum, below, by the terminal part of the ileum and, on the right, by the ileocaecal junction. Its orifice opens downwards and to the left.

(*b*) The *inferior ileocaecal recess* (**8**.102) is well marked in the young subject but is frequently obliterated by fat in advanced age. It is produced by a peritoneal fold, the *ileocaecal fold*, which extends from the anterior and inferior surfaces of the terminal part of the ileum to the front of the mesoappendix (or to the appendix or the caecum). The ileocaecal fold is also known as the '*bloodless fold of Treves*', but it sometimes contains blood vessels; if inflamed, and especially if the appendix and its mesentery lie behind the caecum, the fold may be mistaken for the mesoappendix. (Regarding the source of vessels in this fold, *see* Cabanie and Javelle 1966.) The inferior ileocaecal recess is bounded, in front, by the ileocaecal fold; above, by the posterior surface of the ileum and its mesentery; to the right by the caecum, and, behind, by the upper part of the mesoappendix. Its orifice opens towards the left and downwards.

(*c*) The *retrocaecal recess* (**8**.102) lies behind the caecum; it varies much in size and extent and may occasionally extend upwards for some distance behind the ascending colon and be deep enough to admit an entire finger. It is bounded, in front, by the caecum (and sometimes the lower part of the ascending colon); behind, by the parietal peritoneum and, on each side, by the *caecal folds* (parietocolic folds) of peritoneum passing from the caecum to the posterior abdominal wall. The vermiform appendix frequently lies in this recess (pp. 1352, 1353).

3. The Intersigmoid Recess

This is constantly present in the fetus and during infancy, but may disappear as age advances. It lies behind the apex of the V-shaped parietal attachment of the root of the sigmoid mesocolon and forms a funnel-shaped recess which is directed upwards; its orifice opens downwards. The recess varies in size from a mere dimple to a fossa which will admit the little finger, and its posterior wall, formed by the peritoneum on the posterior abdominal wall, covers the left ureter as it crosses the bifurcation of the left common iliac artery. Occasionally the recess lies within the layers of the sigmoid mesocolon nearer the gut than the root of the mesocolon. The presence of the recess is due to imperfect blending of the mesocolon with the posterior parietal peritoneum.

ANOMALOUS PERITONEAL FOLDS

In addition to the folds described above in connexion with the peritoneal recesses, certain other peritoneal folds, bands, or ligaments are sometimes found in the abdomen. They are of interest in that some of them are considered to cause obstruction to the passage of its contents along the gut by exerting traction on, and producing angulation of, sections of the intestine. Others are thought to be of importance in limiting the spread of peritoneal effusions to certain localities in the abdomen. Their exact mode of origin is doubtful and they have been variously attributed to errors in development, to previous inflammation (peritonitis) and to mechanical traction by the gut, possibly associated with the evolution of the upright posture in man, though the latter factor is improbable. These anatomically anomalous folds must be distinguished from pathological adhesions which are definitely due to peritonitis, and it must be borne in mind that, when coils of intestine are pulled out of their normal position by the observer, they may be artificially kinked by the traction thus exerted upon their mesenteries, with resultant simulation of bands, whereas none actually existed with the gut *in situ*. Further, these anomalous folds only become of clinical importance if it be proved, in a given case, that they interfere with the normal function of the gut, and mere discovery of their presence should not cause a search for another—the real—cause of the symptoms to be neglected. The principal anomalous folds which are encountered are as follows:

(*a*) Occasionally the lesser omentum is prolonged to the right of the usual site of the epiploic foramen in the form of a peritoneal

fold which may pass from the gall bladder to the superior part of the duodenum (*cystoduodenal ligament*), or in front of the latter to the greater omentum or right colic flexure, or from the under surface of the right lobe of the liver to the right colic flexure (*hepatocolic ligament*).

(*b*) The duodenojejunal junction is sometimes joined to the transverse mesocolon by a peritoneal band.

(*c*) The greater omentum may be attached to the front of the ascending colon or extend over it to the lateral abdominal wall. A thin sheet of peritoneum (*Jackson's membrane*), containing fine blood vessels, may pass from the front of the ascending colon and caecum to the lateral part of the posterior abdominal wall; it may be continuous, on the left, with the greater omentum. Occasionally a peritoneal band passes from the right side of the ascending colon to the lateral abdominal wall at about the level of the iliac crest; it has been called the 'sustentaculum hepatis', but it is only closely related to the liver in fetal and early post-natal life, when that organ is relatively larger than in the adult. Other folds passing from the ascending colon to the posterolateral abdominal wall may divide the right lateral paracolic gutter (the groove between the right side of the ascending colon and the posterior abdominal wall) into several small recesses.

(*d*) The ascending colon, and less frequently the descending colon, may have a mesentery.

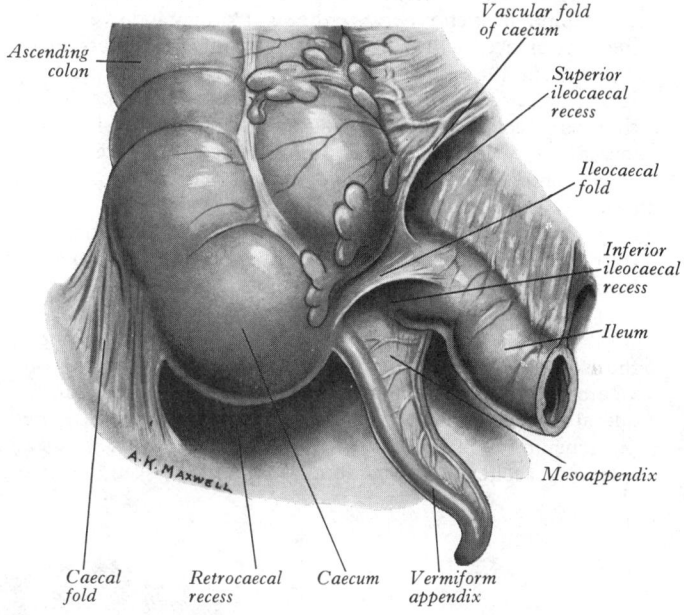

8.102 The peritoneal folds and recesses in the caecal region.

(*e*) The proximal and distal ends of the sigmoid colon may be bound close to each other by a fibrous band.

(*f*) Frequently a fan-shaped peritoneal fold (the *presplenic fold*) extends from the anterior surface of the gastrosplenic ligament (near the greater curvature of the stomach), below the lateral end of the spleen, to blend with the phrenicocolic ligament. It may be adherent to the spleen or to the diaphragm, and it contains branches from the splenic or left gastro-epiploic artery. The omental bursa may be prolonged into the fold. The fold is better marked in the fetus than in the adult, in whom it often appears to be merely a part of the phrenicocolic ligament. It may be of importance in limiting peritoneal effusions in the left supracolic space (*vide infra*), and, if adherent to the spleen or diaphragm, it may form a vascular obstruction in the surgical removal of the spleen.

(*g*) A fibrous band, described as passing from the terminal part of the ileum to the posterior abdominal wall, and a similar one passing from the proximal part of the sigmoid colon to the posterior abdominal wall, were formerly thought to be the cause of partial obstruction by producing kinking of these parts of the gut, but such a view does not receive much support at the present time.

SPECIAL PERITONEAL REGIONS

From the point of view of the spread of pathological collections of fluid, the peritoneal cavity presents a number of potential spaces or recesses which are normally in communication with each other, but which may become sealed off from one another by pathological adhesions between the neighbouring peritoneum and viscera. These spaces are as follows:

(1) The *supracolic space* (or *subphrenic region*) lies between the diaphragm above, and the transverse colon and its mesocolon below. It is subdivided into (*a*) the *right subphrenic space*, which lies between the diaphragm and the anterior, superior and right lateral surfaces of the right lobe of the liver, bounded to the left by the falciform ligament, and behind by the upper layer of the coronary ligament; (*b*) the *left subphrenic space*, which lies between the diaphragm, the anterior and superior surfaces of the left lobe of the liver, the anterosuperior surface of the stomach, and the diaphragmatic surface of the spleen; it is bounded to the right by the falciform ligament, and behind by the anterior layer of the left triangular ligament; (*c*) the *right subhepatic space* (also known as the *hepatorenal recess*, or *Morison's pouch*), which is bounded, above and in front, by the inferior surface of the right lobe of the liver and by the gall bladder; below and behind, by the right suprarenal gland, the upper part of the right kidney, the descending part of the duodenum, the right colic flexure, the transverse mesocolon and part of the head of the pancreas; above and behind, it extends between the right kidney and liver as far as the inferior layer of the coronary ligament and the right triangular ligament; (*d*) the *left subhepatic space*, which is the omental bursa.

(2) The *right infracolic space* lies below and behind the transverse colon and mesocolon and to the right side of the mesentery, owing to the obliquity of which the space is widest above. The vermiform appendix often lies in the lower part of this space.

(3) The *left infracolic space* lies below and behind the transverse colon and mesocolon and to the left of the mesentery; it is widest below and in free communication with the pelvis.

(4) The *pelvic cavity*.

(5) The *paracolic gutters* are the longitudinal channels alongside the ascending and descending colon (which are normally sessile), where their visceral peritoneum dips dorsally on the medial and lateral aspects of the gut, to become continuous with the parietal peritoneum of the dorsolateral abdominal wall. Thus *medial* and *lateral* paracolic gutters exist both on the *right* (ascending colon)

and on the *left* (descending colon) sides of the abdomen. Of particular surgical significance is the *right lateral paracolic gutter*. Superiorly it skirts the superolateral aspect of the hepatic flexure of the colon to become continuous with the hepatorenal pouch (of Morison), and beyond this, through its aditus, with the cavity of the omental bursa and its superior recess. Inferiorly, skirting the lateral margin of the caecum and passing unimpeded over the brim of the lesser pelvis, the right lateral paracolic gutter is continuous with the *recto-vesical* (male) or *recto-uterine* (female) *pouch* (of Douglas). Intimately related to the right lateral gutter, and its superior extension, are the vermiform appendix, the right kidney, the gall bladder, the lesser curvature of the stomach, and the first and second parts of the duodenum—all common sites of acute abdominal disease. Slow percolation of infected fluid along this channel, from one site to another, may result in uncharacteristic symptoms and signs, with consequent errors in diagnosis. Further, with the patient supine, infected fluid in the right lateral gutter, tends to *ascend* and may enter, and accumulate in, the superior recess of the omental bursa, with potentially grave consequences, because of its deep inaccessible site, and its close relation to the pleural and pericardial cavities. Conversely, with the patient nursed in the semi-sitting posture, such fluid *descends* either to the relatively accessible recto-vesical pouch or to the recto-uterine pouch which may be approached surgically through the rectum or vagina.

(6) Two extraperitoneal 'spaces' in the subphrenic region are defined, which may likewise become the site of localized infection. They are: (*a*) the *right extraperitoneal space*, which lies between the two layers of the coronary ligament, the 'bare area' of the liver, and the diaphragm; and (*b*) the *left extraperitoneal space*, which comprises the extraperitoneal connective tissue around the left suprarenal gland and the upper pole of the left kidney.

PERITONEAL ABSORPTION

With regard to the question of the absorption of fluid effusions from the peritoneal cavity, substances in complete solution (solutes) are probably absorbed directly into the blood capillaries, whereas particulate matter in suspension probably passes into the lymph vessels, with the aid of phagocytes (granular leucocytes and monocytes). After abdominal or pelvic operations, it has been customary to prop up the patient in bed so that any inflammatory intraperitoneal effusion will gravitate into the pelvis. One presumed reason for adopting this position was that the

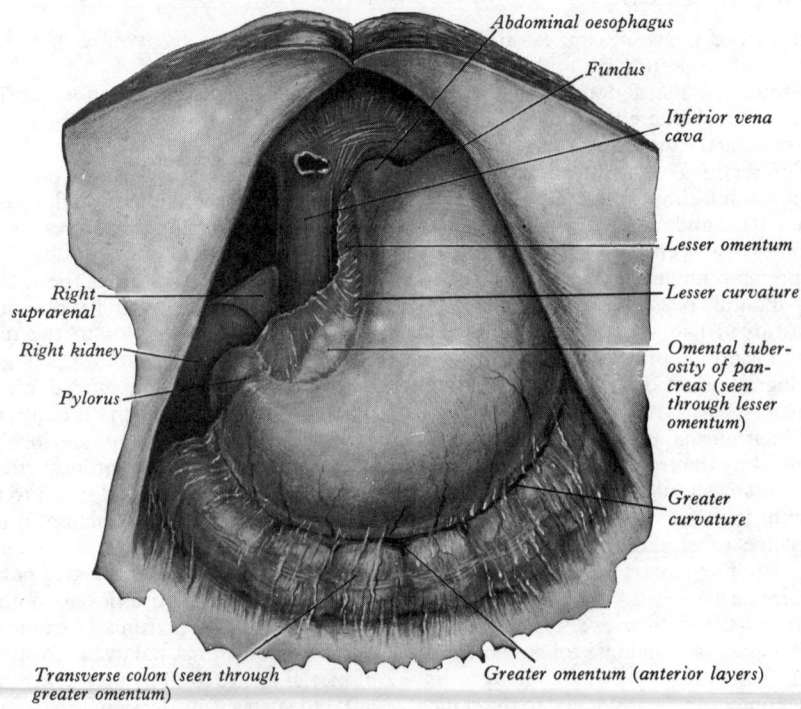

Abdominal oesophagus

Fundus

Inferior vena cava

Lesser omentum

Lesser curvature

Omental tuberosity of pancreas (seen through lesser omentum)

Greater curvature

Right suprarenal

Right kidney

Pylorus

Transverse colon (seen through greater omentum)

Greater omentum (anterior layers)

8.103   The stomach *in situ*, after removal of the liver.

peritoneum in the subphrenic region has a greater absorptive capacity than the other regions; hence inflammatory products, if they gained access to this region, would more rapidly pass into the general circulation. It was held by some that in the subphrenic region there were gaps (peritoneal stomata) between the mesothelial cells lining the peritoneum and similar gaps (endothelial stigmata) between the endothelial cells lining the lymph vessels subjacent to the peritoneum, and that these gaps greatly facilitated absorption. It is now generally believed that these gaps are artefacts produced during the histological technique employed to demonstrate them, that absorption is more or less equally rapid in all parts of the peritoneum, and that the greater absorption in the upper part of the abdomen is to be correlated partly with the larger area of the peritoneal surface in the subphrenic region and partly with the fact that respiratory movements expedite absorption in this zone.

PERITONEAL STRUCTURE

The peritoneum consists of a single layer of flattened mesothelial cells which covers a layer of loose connective tissue. In most areas the mesothelium forms a continuous surface. Adjacent mesothelial cells are joined by junctional complexes, which probably allow the passage of macrophages to and from the underlying connective tissue, in the same manner as endothelial cell junctions allow leucocytes to pass from the bloodstream. In other areas, however, as in the greater omentum, the peritoneum may be discontinuous, presenting a series of fenestrations which may be visible to the unaided eye. At such points the mesothelial surface layer is continuous over the trabeculae of connective tissue which interlace around the margins of the fenestrae.

The sub-mesothelial connective tissue carries the cells usually found in loose connective tissues, but the population of macrophages, lymphocytes, and in some regions adipocytes, are particularly numerous. Aggregations of lymphocytes occur in some regions and form macroscopic 'milky spots' under the mesothelium. It has been claimed that the mesothelial cells possess a phagocytic capacity, and that they may leave the surface to form free macrophages. They may also transform into fibroblasts, and fusion between layers of fibroblasts of mesothelial origin may lead to macroscopic adhesions between the peritoneal surfaces of adjacent structures; if extensive, these may have serious clinical consequences, interfering with intestinal motility, or even leading to complete obstruction of the gut.

The mesothelium is similar in many respects to the endothelial lining of blood vessels, in that it forms a dialysing membrane across which fluids and small molecules of various solutes may pass. Numerous pinocytotic vesicles are present near the cell surfaces, the remaining cytoplasm being relatively poorly provided with organelles, indicating a low level of metabolic activity (Tesi and Forssmann 1970). Normally, small volumes of fluid are transferred across the peritoneal surfaces. Therapeutically, however, considerable volumes of fluid may be administered via the intraperitoneal route, whilst conversely, certain blood-borne substances such as urea can be dialysed from the bloodstream into fluid artificially circulated through the peritoneal cavity.

PERITONEAL FLUID

The fluid layer which covers the peritoneal surfaces, as already stated, contains water, electrolytes and other solutes derived from the interstitial fluid of the neighbouring tissue and from the plasma of adjacent blood vessels. It also contains proteins and a variety of cell types (Carr 1967). The latter vary in their numbers, structure and type in different pathological conditions, and they are hence of diagnostic importance. Normally the cells consist of desquamated flat mesothelial elements derived from the peritoneal surfaces, and of wandering macrophages, mast cells, fibroblasts, lymphocytes and small numbers of other leucocytes. Some of these cells, particularly the macrophages, can migrate freely between the peritoneal cavity and the surrounding connective tissue; particulate material injected intraperitoneally may therefore be ingested by these cells and transported to various other sites in the body. The lymphocytes in the fluid provide both cellular and humoral immunological defence mechanisms.

PERITONEAL VESSELS AND NERVES

The parietal and visceral layers of the peritoneum are respectively developed from the somatopleural and splanchnopleural layers of the lateral plate mesoderm (p. 118). Correlated with their embryological origin is the fact that the parietal peritoneum derives its arterial supply from the somatic (body wall) arteries supplying the abdominal and pelvic walls; its veins join the systemic veins in the neighbouring parts of the body wall, its lymphatics also join those in the body wall and thus drain into parietal lymph nodes, and its nerve supply is derived from the spinal nerves which also supply the muscles and skin of the parietes. The visceral peritoneum, however, which is to be considered as an integral part of the viscera themselves, derives its arterial supply from the arteries supplying the appropriate viscera, its veins and lymphatics join the visceral veins and lymph vessels, and its nerve supply is derived from the autonomic nerves innervating the viscera. The difference in the sensibility of the two layers of the peritoneum is thus to be correlated with their different innervation. Whereas pain is elicited by the application of tactile, thermal or chemical stimuli to the parietal peritoneum (in the conscious patient), these stimuli are ineffectual when applied to the visceral peritoneum (or to the viscera themselves). For example, the liver, stomach or intestine can be cut, pinched, clamped or burned in the conscious subject without evoking pain, the insensibility of the alimentary canal to these forms of stimulation extending from about the middle of the oesophagus down to the junction of the endodermal and ectodermal parts of the anal canal. On the other hand, a stimulus which evokes pain when applied to viscera or visceral peritoneum is tension, such as that accompanying over-distension of the hollow viscera or traction on the mesenteries, which stretches the nerve plexuses in the walls of the organs or the nerves in the mesenteries. Other effective stimuli are spasm of visceral muscle, and ischaemia (deprivation of blood supply). The somatic nerves which supply the parietal peritoneum also supply the corresponding segmental area of skin and trunk muscles, and in cases where the parietal peritoneum is irritated, the muscles are reflexly contracted, thus causing rigidity of the abdominal wall in that region. The parietal peritoneum of the under surface of the diaphragm is supplied centrally by both phrenic nerves and peripherally by the lower six intercostal and the subcostal nerves. Irritation of the peripheral part of the diaphragmatic peritoneum may result in pain, in tenderness and muscular rigidity in the area of distribution of the lower intercostal nerves. On the other hand, irritation of the peritoneum over the central portion of the diaphragm may result in pain in the area of distribution of the cutaneous branches of the third, fourth and fifth cervical nerves over the shoulder region and can lead to diagnostic errors.

The Stomach

The stomach (ventriculus or gaster) is the most dilated part of the alimentary canal, and is situated between the end of the oesophagus and the beginning of the small intestine. It lies in the epigastric, umbilical, and left hypochondriac regions of the abdomen, and occupies a recess bounded by the upper abdominal viscera, and completed in front and on the left side by the anterior abdominal wall and the diaphragm. Its shape and position are modified by changes within itself and the surrounding viscera, and no one form or position is typical. Its mean capacity varies with age, being about 30 ml at birth, increasing gradually to about 1000 ml at puberty, and commonly reaching to about 1500 ml in the adult.

The stomach has two openings, and is described as if it had two borders or curvatures, and two surfaces. In reality, of course, its external surface is a continuum, and it is not divided by any readily perceptible 'borders'. Since, however, the peritoneal surface is interrupted by the attachments of the greater and lesser omenta, along profiles which define the gastric shadow in radiographs, these 'borders' or curvatures may be conveniently regarded as separating the surfaces. (Similar arbitrary borders are assigned to the heart, see p. 639.)

THE GASTRIC ORIFICES

The opening by which the oesophagus communicates with the stomach is the *cardiac orifice*, and is situated on the left of the median plane, behind the seventh costal cartilage 2·5 cm (1 in.) from its junction with the sternum, and at the level of the eleventh thoracic vertebra. It is placed about 10 cm (4 in.) from the anterior abdominal wall and is 40 cm (16 in.) from the incisor teeth. The short abdominal part of the oesophagus is like a truncated cone and curves sharply left, the base of the cone being continuous with the cardiac orifice of the stomach. The right side of the oesophagus is continuous with the lesser curvature of the stomach, while the left side joins the greater curvature at an acute angle, termed the *cardiac notch*. The part of the stomach to the left of and above the cardiac orifice is called the *fundus*—a curiously inappropriate term, but it is the *bottom* of the stomach, if entered surgically from below.

The opening into the duodenum is the *pyloric orifice*, and its position is usually indicated (**8**.103) by a circular groove on the surface of the organ, termed the *pyloric constriction*, which indicates the position of the pyloric sphincter. In the living subject, at operation, it can be identified by the prepyloric vein, which runs vertically across its anterior surface. The pyloric orifice lies about 1·2 cm (0·5 in.) to the right of the median plane near the level of the lower border of the first lumbar vertebra (transpyloric plane), when the body is in the supine position and the stomach is empty.

THE GASTRIC CURVATURES

The lesser curvature, extending between the cardiac and pyloric orifices, forms the right (or posterosuperior border) of the stomach. It descends as a continuation of the right margin of the oesophagus in front of the decussating fibres of the right crus of the diaphragm, and then, turning to the right, it curves below the omental tuberosity of the pancreas and ends at the pylorus (**8**.103, 105). The most dependent part of the curve may form a notch, named the *angular incisure*, which varies somewhat in position with the state of distension of the viscus; it may be used to separate the stomach into right and left parts. The lesser curvature gives attachment to the lesser omentum, the two layers of which contain the right and left gastric vessels, adjacent to the lesser curvature.

The greater curvature is directed antero-inferiorly, and is four or five times as long as the lesser curvature. Starting from the cardiac orifice at the cardiac notch, it forms an arch backwards, upwards, and to the left; the highest point of the convexity (of the *fundus*) is on a level with the left fifth intercostal space and lies just below the left nipple, though this level, like that of the diaphragm, varies with the phases of respiration (*see* pp. 550–551). From this level it may be followed downwards and forwards, with a slight convexity to the left almost as low as the cartilage of the tenth rib, when the body is in the supine position; it then turns to the right, to end at the pylorus. Directly opposite the angular incisure of the lesser curvature the greater curvature presents a bulge, which is the left extremity of the *pyloric part* of the stomach; this is limited on the right by a slight groove, which indicates the subdivision of the pyloric part into a pyloric antrum and a pyloric canal. The latter is only 2 to 3 cm in length and terminates at the pyloric constriction. At its commencement the greater curvature is covered by peritoneum continuous with that on the front of the stomach. On the left side of the fundus and the adjoining part of the body, the greater curvature gives attachment to the gastrosplenic ligament, while to its lower region are attached the two layers of the greater omentum, separated from each other by the gastro-epiploic vessels. The gastrosplenic ligament and the greater omentum (together with the gastrophrenic and lienorenal ligaments, *see* pp. 1324, 1326) are directly continuous, being parts of the original dorsal mesentery of the stomach (dorsal mesogastrium) (p. 205). The separate names merely indicate regions of the same peritoneal fold.

THE GASTRIC SURFACES

When the stomach is empty and its walls contracted, its surfaces are almost superior and inferior, but when it is distended they become anterior and posterior respectively. They may therefore be described as anterosuperior and posterio-inferior.

Anterosuperior surface. The left part of this surface is posterior to the left costal margin. It is in contact with the diaphragm, which separates it from the left pleura, the base of the left lung, the pericardium, and the sixth, seventh, eighth and ninth ribs and intercostal spaces of the left side. It is related to the costal attachment of the upper fibres of origin of the transversus abdominis, which intervene between it and the seventh, eighth and ninth costal cartilages. The upper and left part of this surface becomes posterolateral and is in contact with the gastric surface of the spleen. The right half is in relation with the left and quadrate lobes of the liver and with the anterior abdominal wall. When the stomach is empty, the transverse colon may lie on the front of this surface. The whole surface is covered with peritoneum, and a part of the greater sac of the peritoneum intervenes between it and the above structures.

Postero-inferior surface. This is related to the diaphragm, the left suprarenal gland, the upper part of the front of the left kidney, the splenic artery, the anterior surface of the pancreas, the left colic flexure, and the upper layer of the transverse mesocolon. These structures form the shallow *stomach bed*, (**8**.104), but the stomach is separable from them, and can slide over them, due to the intervening omental bursa (lesser sac). The gastric surface of the spleen is also generally described as part of the stomach bed, but as stated above it is separated from the stomach by a part of the greater sac. Further, the greater omentum and the transverse mesocolon separate the stomach from the duodenojejunal flexure and small intestine. The postero-inferior surface is covered with peritoneum, except near the cardiac orifice, where there is a small, somewhat triangular area, in direct contact with the left crus of the diaphragm, and sometimes with the left suprarenal gland. The left gastric vessels reach the lesser curvature of the stomach at the right extremity of this area (in the left gastropancreatic fold, p. 1326), and from its left side a short peritoneal fold, termed the *gastrophrenic ligament*, which is continuous below with the lienorenal and gastrosplenic ligaments, passes to the inferior surface of the diaphragm.

A plane passing through the angular incisure on the lesser curvature and the left limit of the opposed bulge on the greater curvature divides the stomach into a large, left portion or *body* and a small, right, or *pyloric part*.

Radiology

By means of X-rays the form and position of the stomach can be studied in the living subject after swallowing a suitable 'meal' containing barium sulphate (**8**.108). During the process of digestion, it is divided by a muscular constriction into a large, dilated, left region, and a narrow, contracted, tubular, right portion. The constriction is in the body of the stomach, and does not follow any of the anatomical landmarks; indeed, it shifts gradually towards the left as digestion progresses. The position of the stomach varies with the posture, with the amount of the stomach contents and with the condition of the intestines on which it rests. It is also influenced by the tone of the abdominal muscle and of the musculature of the organ itself, and by the type of body build of the individual. In the commonest type of stomach, the empty organ is somewhat J-shaped and, in the erect posture, the pylorus descends to the level of the second or the upper part of the third lumbar vertebra, and the most dependent part of the stomach is below the level of the umbilicus. The fundus is usually distended with gas. Variation in the amount of its contents affects mainly the body of the stomach, the pyloric portion remaining in a more or less contracted condition during the process of digestion. As the stomach fills it tends to expand forwards and downwards in the direction of least resistance, but when this is interfered with by a distended condition of the colon or intestines the fundus presses upwards on the liver and diaphragm and gives rise to the feelings of oppression and palpitation complained of in such cases. When hardened *in situ*

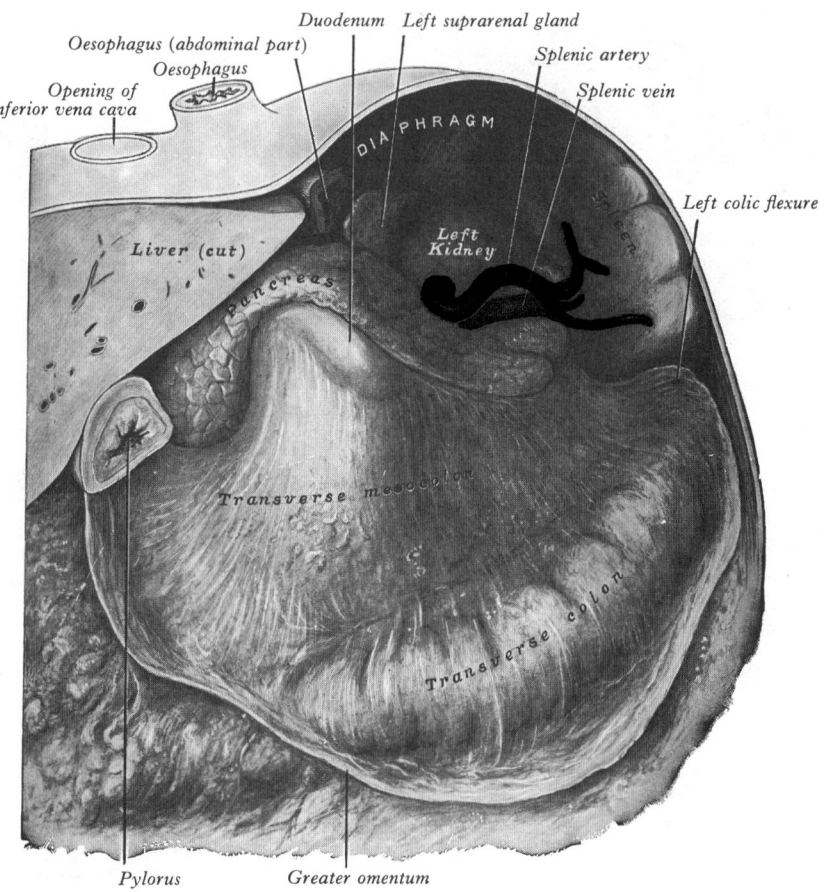

8.104   The stomach bed: a dissection in which the stomach has been removed to show its posterior relations.

the contracted stomach is crescentic, the fundus looking directly backwards. The surfaces are superior and inferior, the upper having, however, a gradual downward slope to the right. The greater curvature is in front of and at a slightly higher level than the lesser.

The position of the full stomach depends, as already indicated, on the state of the intestines: when the latter are empty the fundus expands vertically and also forwards, the pylorus is displaced towards the right, and the whole organ assumes an oblique position, so that its surfaces are directed more forwards and backwards. The lowest part of the stomach is at the pyloric antrum, which reaches below the umbilicus. Where the intestines interfere with the downward expansion of the fundus the stomach retains the horizontal position which is characteristic of the contracted viscus. Less commonly the stomach may lie almost transversely, even in the erect posture; this is known as the 'steer-horn' type of stomach. Intermediate types of stomach, between the J-shaped and 'steer-horn' varieties, also occur (Barclay 1936).

Interior of the Stomach

When examined after death, the stomach is usually fixed at some stage of the digestive process. A common form is that shown in (**8**.109). When the viscus is laid open by a section through the plane of its two curvatures, it is seen to consist of two segments: (*a*) a large globular portion on the left; and (*b*) a narrow tubular part on the right. The transition between the two regions is gradual, and this division is purely arbitrary. The cardiac incisure lies to the left of the abdominal part of the oesophagus: the projection of this notch into the cavity of the stomach increases as the organ distends, and has been supposed to act as a valve preventing regurgitation into the oesophagus. The elevation corresponding to the angular incisure is seen at the beginning, and the circular thickening of the pyloric sphincter at the end, of the pyloric region.

Modelling of the gastric epithelium in the human fetus (Lewis 1912) has shown that a channel (the *gastric canal*) extends along

the lesser curvature from the cardiac orifice to the angular incisure (**8**.105). It was also demonstrated radiologically that such a canal exists in the adult (Jefferson 1915); in most cases examined whilst in the act of swallowing radio-opaque fluid it was found that the fluid was at first confined to the part of the stomach adjacent to the lesser curvature, and concluded that the oblique muscular coat of the stomach is so arranged that by its contraction it will cause a temporary separation of a canal along the lesser curvature.

The pyloric sphincter is a muscular ring formed by a marked thickening of the circular layer of the muscular coat. Some of the longitudinal fibres turn in and interlace with the fibres of the sphincter. (*See* DiDio and Anderson 1968.)

A *cardiac sphincter* is sometimes described at the oesophageal end of the stomach, formed from the circular fibres of the gastric

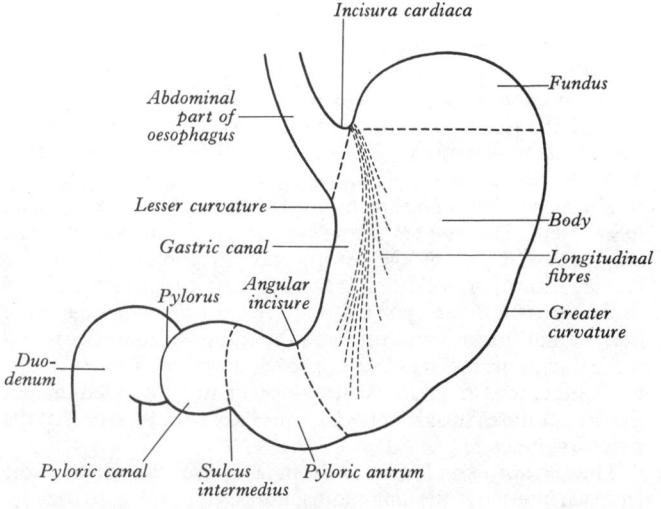

8.105   The parts of the stomach.

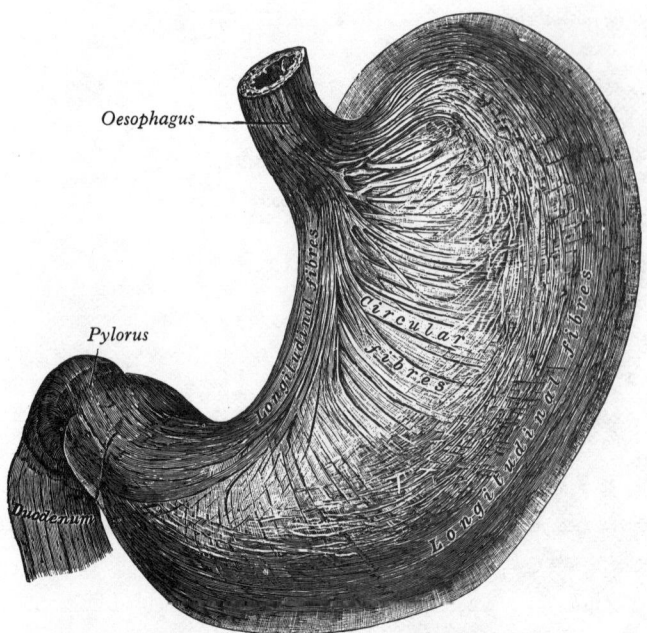

8.106 The longitudinal and circular gastric muscular fibres. Antero-superior aspect. (Spalteholz.)

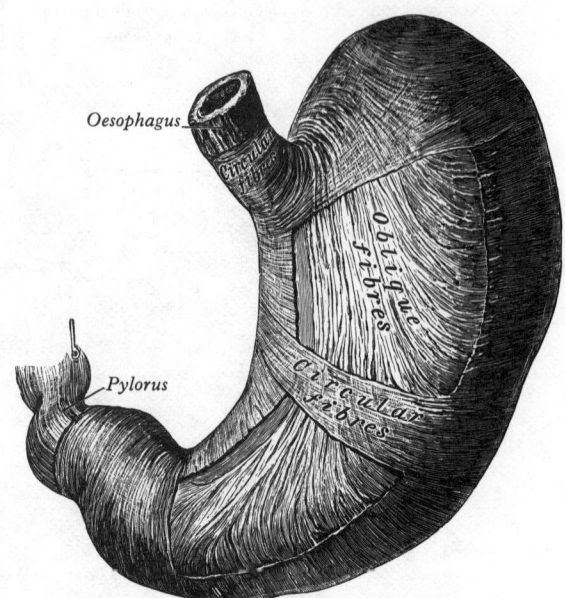

8.107 The oblique muscular fibres of the stomach, shown by partial dissection of its wall. Anterosuperior aspect.

wall. Although radiological observation strongly suggests the presence of some force capable of delaying entry of oesophageal contents into the stomach, the histological evidence for this is uncertain and requires further study. It is often asserted that the muscle fibres of the right crus of the diaphragm, which decussate obliquely around the termination of the oesophagus, may exert a compressive or kinking effect at this level; but this is an anatomical speculation rather than a physiological fact. Bowden and El-Ramli (1967) have described the structure of this part of the diaphragm (see also p. 549), and they concluded that while the right crus invariably embraces the termination of the oesophagus, the relationship does not suggest an effective sphincter, although they agreed that in some positions of the trunk a kinking effect might occur. Others have suggested that the sphincter may be formed by circular muscle of the oesophageal wall (pp. 550, 1318).

GASTRIC STRUCTURE

The wall of the stomach consists of the four usual layers: serous, muscular, submucous and mucous, together with their vessels and nerves.

The serosa, or visceral peritoneum, covers the entire surface of the organ, excepting (*a*) along the greater and lesser curvatures at the lines of attachment of the greater and lesser omenta, where the two layers of peritoneum leave a small space in which vessels and nerves lie, and (*b*) a small area on the postero-inferior surface of the stomach, close to the cardiac orifice, where the stomach is in contact with the inferior surface of the diaphragm at the site of reflexion of the gastrophrenic and left gastropancreatic folds.

The muscularis externa (**8.**106, 107) is situated immediately beneath the serous covering, with which it is closely connected by subserous areolar tissue. It consists of three layers of visceral muscular fibres: longitudinal, circular and oblique.

The *longitudinal fibres* are the most superficial and are arranged in two sets. The first set consists of fibres continuous with the longitudinal fibres of the oesophagus; they radiate from the cardiac orifice, are best developed near the curvatures and end proximal to the pyloric portion. The second set commences in the body of the stomach and passes to the right, its fibres becoming more thickly arranged as they approach the pylorus. Some of the more superficial longitudinal fibres pass on to the duodenum, but the deeper fibres turn inwards and interlace with the fibres of the pyloric sphincter.

The *circular fibres* form a uniform layer over the whole of the stomach internal to the longitudinal fibres. At the pylorus they are most abundant, and are there aggregated into an annular mass, the

pyloric sphincter. The circular fibres of the gastric wall are continuous with those of the oesophagus, but they are sharply marked off from the circular fibres of the duodenum by a connective tissue septum.

The *oblique fibres*, internal to the circular layer, are limited chiefly to the body of the stomach and are most developed near the cardiac orifice. They sweep downwards from the cardiac notch and run more or less parallel with the lesser curvature. On the right they present a free and well-defined margin (**8.**107); on the left they blend with the circular fibres.

The peristaltic contraction of the musculature of the pyloric antrum thoroughly mixes the stomach contents in this region, returning some to the body of the stomach and propelling some into the duodenum. The pyloric sphincter contracts intermittently during contraction of the antral musculature, but, when the stomach is at rest, it is relaxed, leaving the pylorus open (Atkinson *et al.* 1957; Spira 1957). The exact mechanism of control of the transfer of stomach contents through the pylorus into the duodenum is still not fully clarified. For an authoritative summary consult Hunt and Knox (1968).

The submucosa consists of loose, areolar tissue, connecting the mucous and muscular layers.

The mucous membrane is thick and its surface is smooth, soft and velvety. In the fresh state it is of a pinkish tinge at the pyloric end, and of a red or reddish-brown colour over the rest of its surface. During the contracted state of the organ it is thrown into numerous folds or rugae which for the most part have a longitudinal direction, and are best marked towards the pyloric end of the stomach, and along the greater curvature (**8.**109). These folds are obliterated when the organ is distended.

STRUCTURE OF GASTRIC MUCOSA

When examined with a lens, the luminal surface of the mucous membrane (**8.**109) has a honeycomb appearance, because it is covered with small depressions or alveoli, of a polygonal or slit-like form, about 0·2 mm in diameter. These are the *gastric pits* and at the bottom of each are the orifices of the gastric glands. The surface of the mucous membrane including the gastric pits is covered with a single layer of columnar secretory epithelial cells, the *surface mucous cells*, which liberate mucus from their apices on to the surface of the stomach. This acts as a lubricant and also protects the gastric lining against its own secretions of acid and enzymes. This type of epithelium commences very abruptly at the cardiac orifice, where there is a sudden transition from the stratified epithelium of the oesophagus. The *gastric glands* comprise: (1) cardiac glands; (2) main glands of the body and fundus; and (3) pyloric glands.

The cardiac glands (**8**.110A) are infrequent and confined to a small area near the cardiac orifice; some are simple tubular glands whilst others are compound racemose in type. Mucus-secreting cells predominate whilst oxyntic and zymogenic cells are infrequent.

The main gastric glands of the body and fundus, of which from three to seven open into each gastric pit, are cytologically the most highly differentiated of the gastric glands (**8**.110B, C). At least four distinct cell types have been distinguished:

(1) The *chief (peptic* or *zymogenic) cells* are present particularly in the basal parts of the glands. These are cuboidal, typical protein-synthesizing cells, containing prominent secretory bodies, much rough endoplasmic reticulum, a prominent Golgi complex, and by light microscopy they are strongly basophilic because of their contained RNA. These cells are the source of the digestive enzymes of the stomach (**8**.109).

(2) The *oxyntic (parietal) cells*, large, rounded and eosinophilic, are most numerous on the side walls and near the duct of the gland. They occur only at intervals, and with the light microscope appear to be applied to the external surface of the other cell types, or partly intercalated between their external aspects. They bulge into the adjacent lamina propria, producing a moniliform or beaded appearance. They are connected with the lumen of the gland by fine processes that pass between the adjacent cells. Ultrastructural studies have shown that the oxyntic cells possess tortuous intracellular canaliculi, the surfaces of which are covered with microvilli, and which open directly into the lumen of the gland on the apices of the fine processes which pass centrally between the adjacent cells as mentioned above. Each cell possesses a large, round, centrally-placed nucleus, and the surrounding cytoplasm carries only sparse endoplasmic reticulum and no secretory granules, but a very large number of closely packed mitochondria. Minute smooth-walled membranous tubules converge on, and open into, the intracellular canaliculi (**8**.109).

(3) *Mucous 'neck' cells* are disseminated between the other types of cell and are particularly numerous around the necks of the glands. They are typical mucus-secreting cells, but their secretions are distinct histochemically from those of the surface mucous cells.

(4) *Argentaffin cells* occur in all types of gastric gland but more commonly in those of the body and fundus than in the pyloric glands. They are more usual in the deeper parts of the gland, lying between the zymogenic cells and the basal lamina; they rarely reach the glandular lumen. Their irregular nuclei are surrounded by a granular cytoplasm which is impregnated strongly with silver-staining methods. Ultrastructurally, the cytoplasm contains many dense membrane-bound vacuoles of varying size, some of which are large (0·3 μm diameter) and which are responsible for the staining reaction. These cells are part of the gastroenteropancreatic endocrine system, described in detail elsewhere (p. 1364), and are now subdivided into a variety of types on the basis of differences in their detailed ultrastructure and secretions (**8**.109, 129).

(5) *Undifferentiated columnar cells* are also present in smaller numbers, and these appear to be the origin of new cells to replace the existing ones as they are lost. Surface mucous cells last for about three days, and mucous neck cells for about one week. Other cell types appear to live considerably longer.

The pyloric glands each consist of two or three short convoluted tubes opening into a conical pit, which occupies about two-thirds of the depth of the mucous membrane (**8**.110D). The epithelial cells are predominantly mucous in type, oxyntic cells being sparse. The enteric hormone *gastrin* has been isolated from these glands in man. Gastrin is released by mechanical stimuli, and acts to increase stomach motility and the secretory activity of chief and oxyntic cells (see also p. 1365).

Although the oxyntic cells are relatively few in pyloric glands, they are apparently invariably present, both in fetal and postnatal material; in adults they may also appear in the mucous membrane of the duodenum, but only in its proximal part, near the pylorus (Leela and Kanagasuntheram 1968).

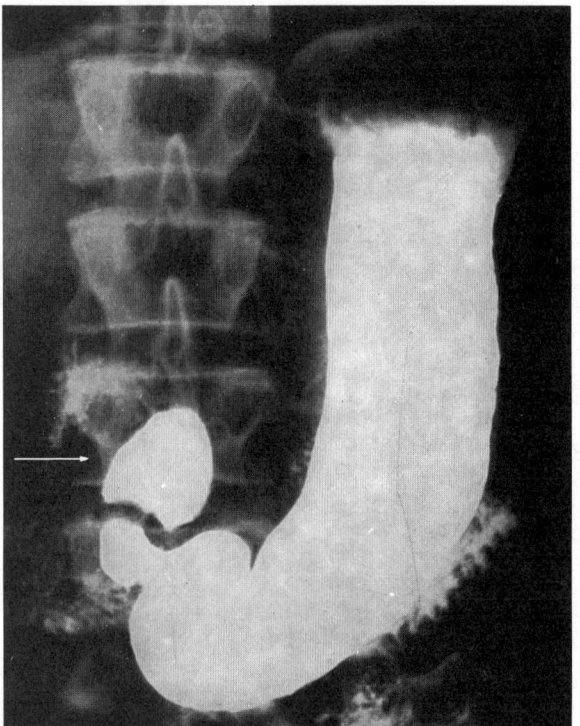

8.108A Radiograph of a normal stomach after a barium meal. The tone of the muscular wall is good and supports the weight of the column in the body of the organ. The arrow points to the duodenal cap, below which a gap in the barium indicates the position of the pylorus.

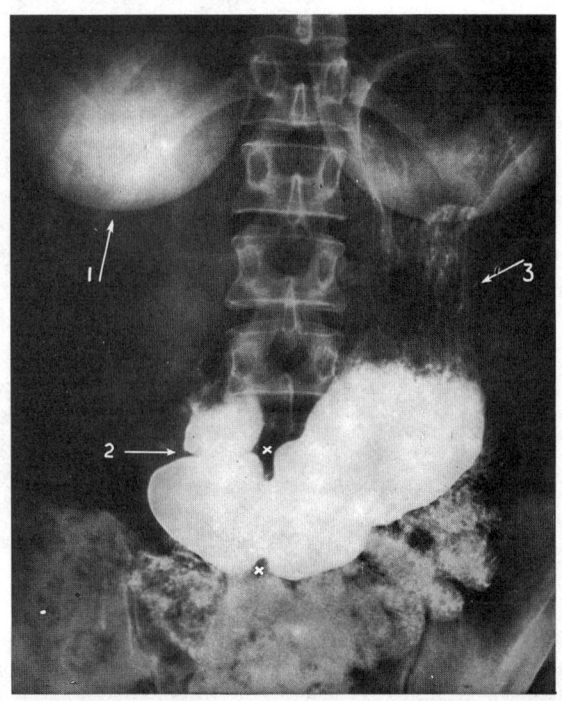

8.108B Radiograph of an atonic stomach after a barium meal. Note that this stomach contains the same amount of barium suspension as the stomach in **8**.108A. Arrow 1 points to the shadow of the right breast, arrow 2, to the pylorus, arrow 3, to the upper part of the body of the stomach, where longitudinal folds can be seen in the mucous membrane. XX marks a wave of peristalsis.

Between the glands the lamina propria consists of a connective tissue framework, and lymphoid tissue. In places, this latter tissue, especially in early life, is collected into little masses which resemble the solitary follicles of the intestine, and are termed the *gastric lymphatic follicles*. In the mucous membrane, deep to the glands, is a thin stratum of nonstriated muscle fibres, the *muscularis mucosae*; it consists of an inner circular and an outer

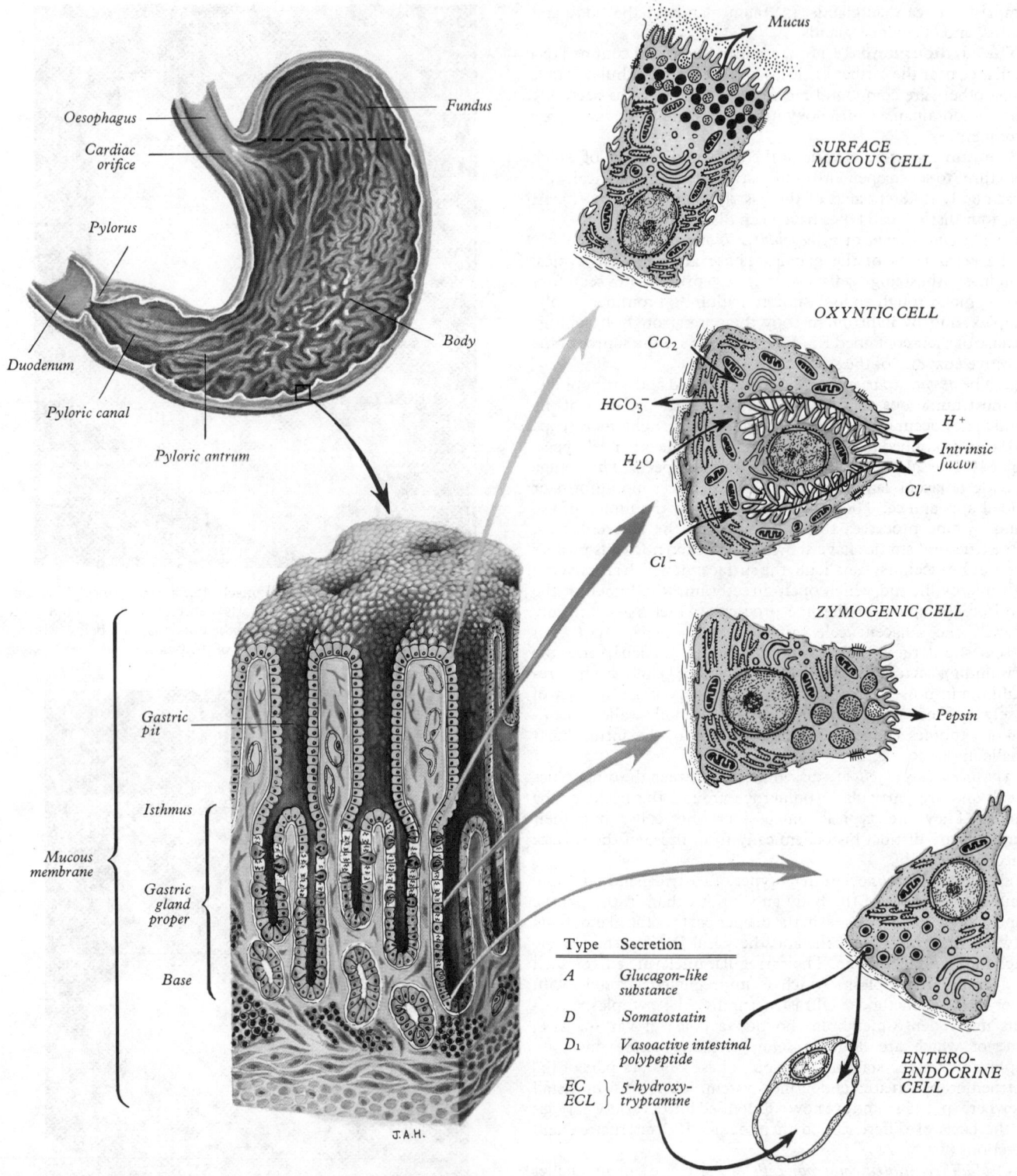

8.109   A diagram showing the principal regions of the interior of the stomach, and the histology and ultrastructure of its mucous membrane. Undifferentiated, dividing, cells are shown in white.

longitudinal layer (with a third, outer circular layer, in places). The inner layer sends strands between the glands, the contraction of which probably aids the emptying of the glands.

Vessels and Nerves

The stomach is supplied by the left gastric artery (from the coeliac artery), the right gastric and right gastro-epiploic arteries (from the common hepatic artery), and the left gastro-epiploic and short gastric arteries (from the splenic artery). These vessels not only anastomose extensively on the serosal surface of the stomach, as described elsewhere (p. 712), but they also form anastomotic

networks in their intramural distribution at intramuscular, submucosal, and mucosal levels; but it is at the submucosal level that a true plexus of small arteries and arterioles is deployed. This *submucosal plexus*, from which the mucosa is supplied, presents considerable regional variations, not only in the gastric wall, but also in the proximal part of the duodenum. In view of the possible implications of a vascular factor in the genesis of peptic ulcer in these regions, the local details of angio-architecture are of considerable interest. It has been claimed that arteriovenous anastomoses occur in the gastroduodenal mucosa (Spanner 1932; De Busscher 1948; Barlow *et al.* 1951; Boulter and Parkes 1963), and that dysfunction in these might produce local ischaemia and hence ulceration. Mucosal end-arteries have also been described, and larger vessels such as the supraduodenal artery (p. 713) have similarly been designed end-arterial. These problems have been most recently discussed, from the anatomical point of view, by Piasecki (1974, 1977), and the results of his observations on fetal, neonatal, and adult human stomachs, injected with India ink, form the basis of this account of the gastric intramural arteries. From the anastomotic arcades formed along the greater and lesser curvatures by the main arteries of supply mentioned above, large numbers of branches pass on to both anterior and posterior aspects of the stomach, in directions approximately transverse to the organ's long axis. In addition to these *anterior* and *posterior gastric arteries* smaller rami, which are often paired, pass directly towards the part of the gastric wall subjacent to attachments of omenta. As these vessels ramify on the external surface and then penetrate the muscular layer of the wall to reach the submucous and mucosal levels of connective tissue, they form plexuses—subserosal, intramuscular, and submucosal, of which the second is the most richly developed (8.111B, c). The muscular plexus is supplied by branches from both the subserous and submucosal plexuses, and the muscular vessels vary in direction in the different laminae of muscle, perhaps in adaptation to their characteristic directions of contraction. The arteries of the submucosa anastomose freely, but the incidence of arterial anastomoses varies in different regions. Counts by Piasecki (1974) showed that while the *number* of anastomoses along, for example,

the lesser curvature increased from its cardiac to its pyloric end, the mean calibre of the anastomosing arteries showed a reverse tendency. The mucosal arteries, which fill the capillary networks supporting the epithelium and its glands, are chiefly derived from the submucous plexus; but along both curvatures a few mucosal arteries are derived directly from extramural (subserosal) sources. These pass through the muscular layers and submucosa, often without lateral communications with submucosal arteries. The incidence of these vessels apparently increases from the cardiac to the pyloric region of the stomach in man. The capillary networks supplied by them are largely independent of those fed by adjacent submucosal arteries, and to that extent the patch of mucosa supplied by such a vessel may be considered more vulnerable to vascular obstruction. According to Piasecki mucosal arteries in general do not form lateral anastomoses, but since their submucosal feeders do, vascular obstruction is to that degree less likely. The same observer examined the pyloric canal and sphincter, and showed a different pattern of supply. Rami of the right gastric and gastro-epiploic arteries ('*pyloric*' *arteries*) pierce the duodenum immediately distal to the pyloric sphincter around its entire circumference, passing through the muscular layer to reach the submucosa, where each divides into two or three branches. These turn into the pyloric canal, internal to the sphincter, to traverse the submucosa as far as the termination of the pyloric antrum (8.111C), supplying the whole of the mucosa of the pyloric canal. Branches of these pyloric submucosal arteries may anastomose at their commencement with duodenal sub-

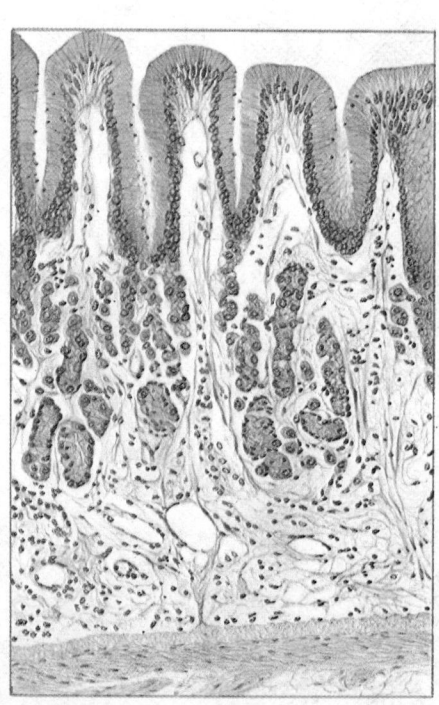
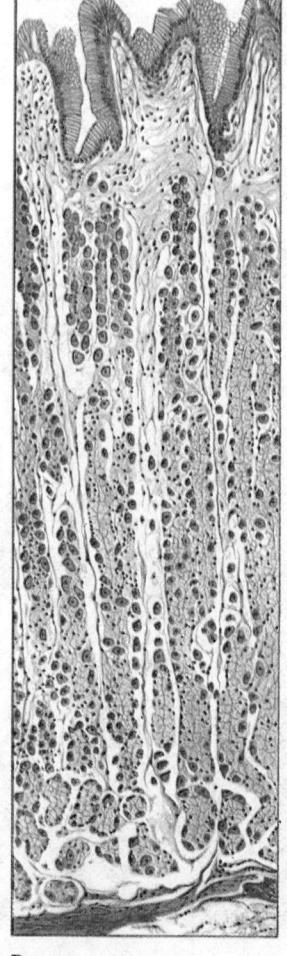
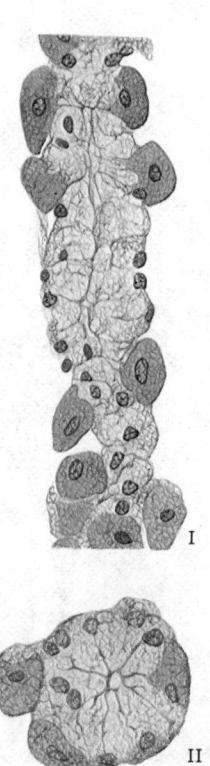
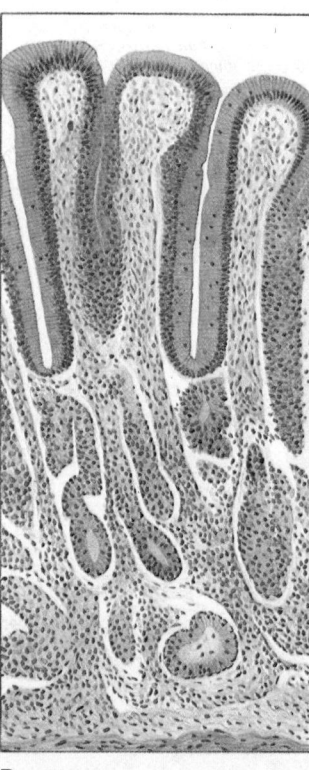

A   B   C   D

8.110A Vertical section through the mucous membrane of the cardiac part of the stomach (human). Stained with haematoxylin and eosin. Magnification about × 150.

8.110B Vertical section through the mucous membrane of the fundus of the stomach (cat). Stained with haematoxylin and eosin. Magnification about × 100. Note the beaded appearance given by the oxyntic cells.

8.110C (I) Gland from fundus of stomach (cat). (II) Lower part of gland cut transversely. Stained with haematoxylin and eosin. Magnification about × 530. The peripherally placed cells staining deeply with eosin are the oxyntic cells.

8.110D Vertical section through the mucous membrane of the pyloric part of the stomach (cat). Stained with haematoxylin and eosin. Magnification about × 75.

mucosal arteries and, by their terminal branches, with the corresponding gastric arteries. The pyloric sphincter itself is supplied by gastric and pyloric arteries, rami of both of which leave their parent vessels in the subserosal and submucosal parts of their courses to penetrate the sphincter.

The gastric veins commence as straight vessels between the glands of the mucosa, and these drain into submucosal veins. Their further arrangement has not received the same attention as the corresponding arteries; but the larger veins accompany the main arteries to their ultimate drainage into the splenic and superior mesenteric veins, while some pass directly to the portal vein. The smaller lymphatic vessels are said to resemble the veins in distribution. The regional lymph nodes and their final drainage are described on p. 793.

The *nerves* are derived from multiple sources. The sympathetic supply is mainly from the coeliac plexus through the plexuses around the gastric and gastro-epiploic arteries. Some branches from the plexus around the hepatic artery proper reach the lesser curvature by passing between the layers of the hepatogastric ligament (p. 1327). Branches from the left phrenic plexus pass to the cardiac end of the stomach, which also receives a twig from the branch of the left phrenic nerve to the right crus of the diaphragm. Inconstant branches are given to the stomach from the left thoracic splanchnic nerves and from the thoracic and lumbar sympathetic trunks.

The parasympathetic supply is derived from the vagus nerves. Usually one or two nerve trunks lie on the anterior and one or two on the posterior aspect of the gastro-oesophageal junction; the anterior nerves comprise for the most part left vagal fibres, and the posterior right vagal fibres, which have emerged from the oesophageal plexus. The anterior nerves supply several filaments to the cardiac orifice and then divide near the upper end of the lesser curvature into branches. Of these: (*a*) *gastric branches* (4–10) radiate on the anterior surface of the body and fundus of the stomach; one is larger than the others and lies in the lesser omentum near the lesser curvature (the *greater anterior gastric nerve*); (*b*) *pyloric branches*, generally two in number, one running almost horizontally to the right in the lesser omentum towards its free edge and then turning down on the left side of the hepatic artery proper to reach the pylorus, the other usually arising from

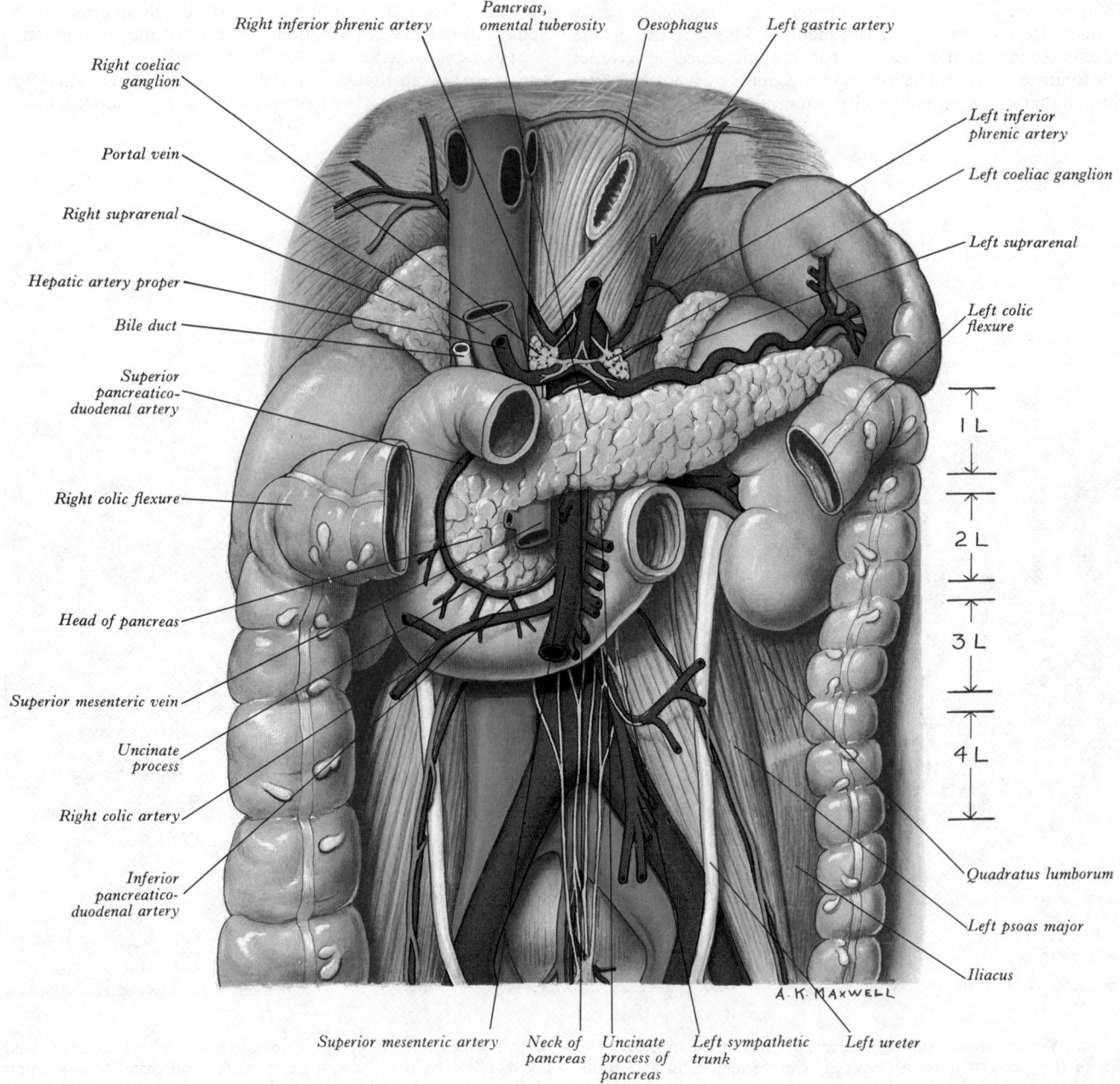

Right inferior phrenic artery
Pancreas, omental tuberosity
Oesophagus
Left gastric artery
Right coeliac ganglion
Portal vein
Right suprarenal
Hepatic artery proper
Bile duct
Superior pancreatico-duodenal artery
Right colic flexure
Head of pancreas
Superior mesenteric vein
Uncinate process
Right colic artery
Inferior pancreatico-duodenal artery
Left inferior phrenic artery
Left coeliac ganglion
Left suprarenal
Left colic flexure
1 L
2 L
3 L
4 L
Quadratus lumborum
Left psoas major
Iliacus
A. K. MAXWELL
Superior mesenteric artery
Neck of pancreas
Uncinate process of pancreas
Left sympathetic trunk
Left ureter

8.111A A dissection to show the duodenum, pancreas, stem arterial rami of the gastro-intestinal tract, and surrounding structures. The right and left hepatic veins have been cut away at their points of entry into the inferior vena cava. The superior hypogastric plexus is shown in front of the sacral promontory and the sympathetic nerves which form it are seen descending across the bifurcation of the aorta, the left common iliac vein and the body of the fifth lumbar vertebra. (In this specimen the left renal artery is situated anterior to the left renal vein at the hilum of the kidney.)

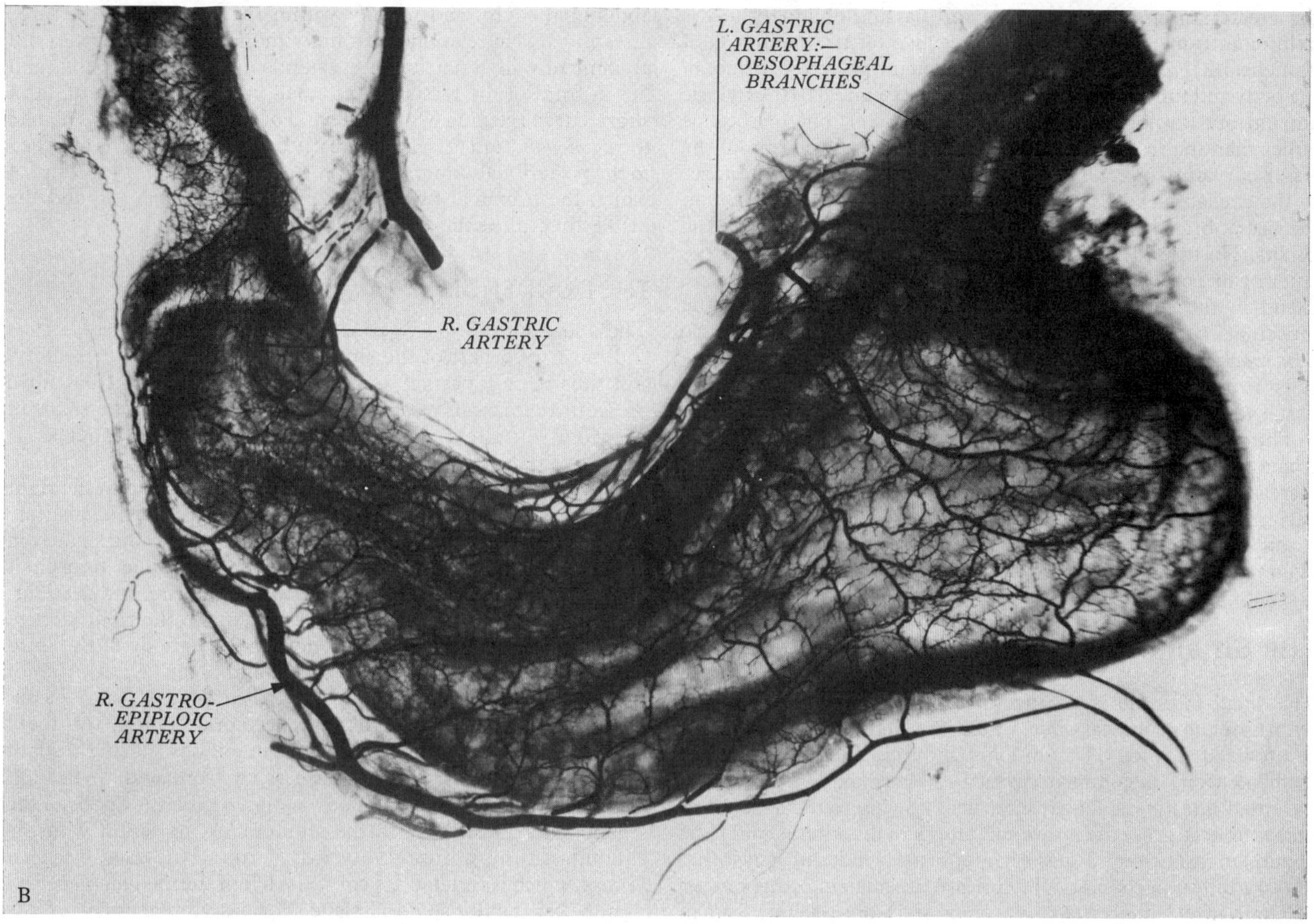

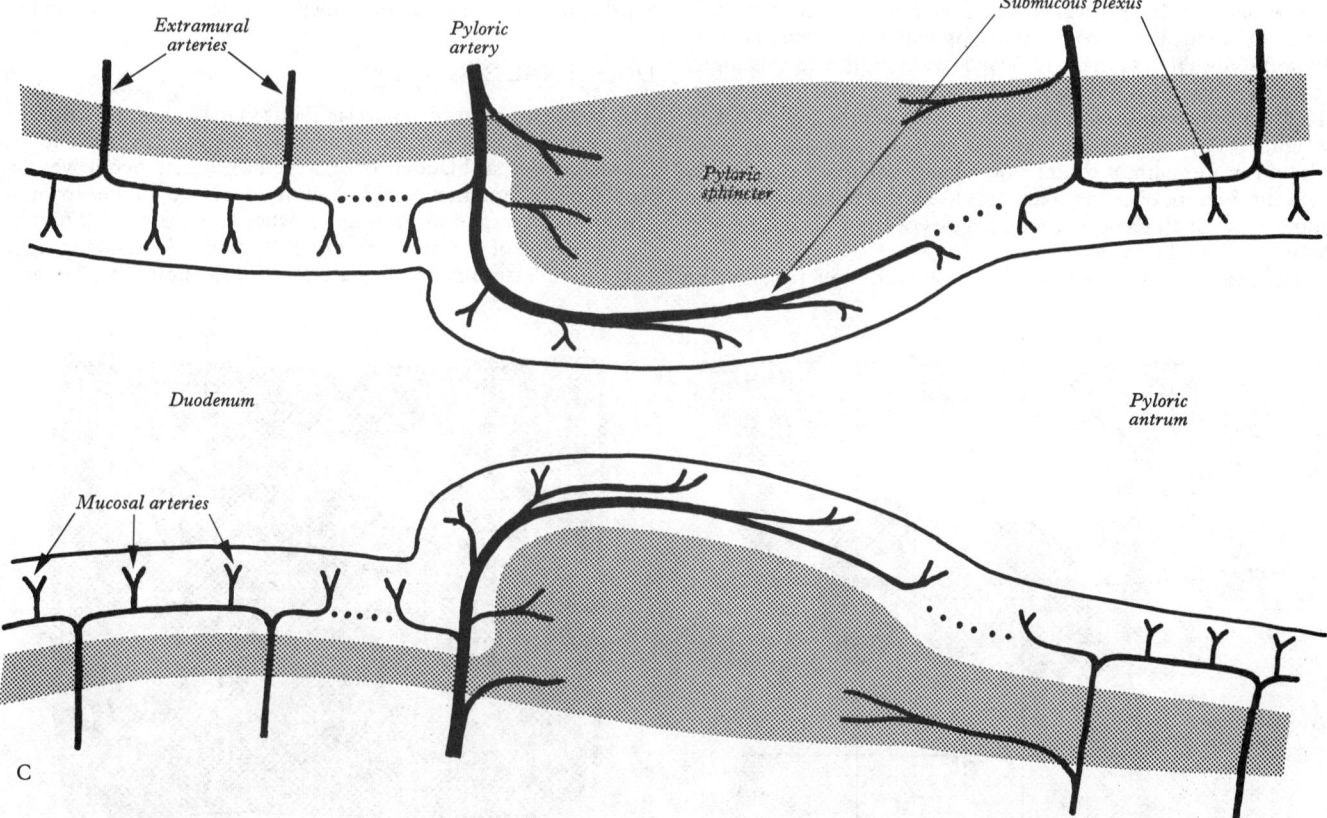

8.111B and c Blood supply of the stomach and the proximal duodenum. B Arterial system in a fetal human stomach. The muscle layer has been removed. Note double arcade along the lesser curvature. The arteries have been injected with a mixture of 2 per cent gelatin and Indian ink and subsequently cleared by the Spalteholz technique. (Magnification approximately ×6.) c A scheme of arterial arrange-ments at the gastroduodenal junction. Dotted lines indicate sites where the submucous plexus may be deficient in continuity. Shaded areas represent the muscular layer of the visceral wall. (By courtesy of Dr. C. Piasecki, Department of Anatomy, Royal Free Hospital School of Medicine, London, and the *Journal of Anatomy*.)

the greater anterior gastric nerve and passing obliquely to the pyloric antrum. The posterior nerves give off two main sets of branches: (*a*) *gastric branches*, radiating on the posterior surface of the body and fundus of the stomach; they extend on to the pyloric antrum but do not reach the pyloric sphincter; one of these is larger than the rest and passes along the posterior margin of the lesser curvature (*greater posterior gastric nerve*), giving branches to the coeliac plexus; (*b*) *coeliac branches*, which are larger than the gastric branches, and pass in the lesser omentum to the coeliac plexus. No true plexiform array of nerves occurs on either the anterior or posterior surface of the stomach. Nerve plexuses are found in the submucosal coat and between the layers of the muscular coat. The latter corresponds to the myenteric (*Auerbach's*) plexus of the intestine and contains numerous nerve cells. From these plexuses fibres are distributed to the muscular tissue and the mucous membrane.

The vagus has both secretory and motor influences on the stomach; stimulation evokes a secretion rich in pepsin and increased gastric motility, while after vagotomy the stomach becomes flaccid and empties slowly. The sympathetic supplies vasomotor fibres to the gastric blood vessels and provides the main pathway for pain fibres from the stomach.

The Small Intestine

The small intestine is a convoluted tube, extending from the pylorus to the ileocaecal valve, where it joins the large intestine. It is often stated to be 6 to 7 m long, and gradually diminishes in diameter from its commencement to its termination. However, the small intestine is longer after death owing to the absence of muscle tone; during life its average length in the adult is said to be about 5 m (*see* below). It has been reported (Underhill 1955) that in 109 adult subjects shortly after death the small intestine ranged in length from 3·35 to 7·16 m in women and from 4·88 to 7·85 m in men, the average length being 5·92 m in women and 6·37 m in men. The length was found to be correlated with the height of the individual, but independent of the age. The large intestine was found to be much more constant in length in this particular study. Jit and Grewal (1975) have reviewed the literature on this topic, reporting their own findings in a series of 137 Indian subjects. They confirmed an association with height, and the lack of it with weight. They observed a variable reduction in length after fixation in formalin, a contraction which sometimes reached 44 per cent. Various observers have passed flexible tubes through the alimentary canal, recording total lengths of from 2·7 m to 4·5 m (cited by Jit and Grewal 1975).

The small intestine lies in the central and lower parts of the abdominal cavity and usually within the confines of the large intestine; it is in relation, in front, with the greater omentum and abdominal wall; a portion of it extends down into the pelvis and lies in front of the rectum. The small intestine consists of: (1) a short, curved section which is devoid of a mesentery and is named the *duodenum*; and (2) a long, greatly coiled part which is attached to the posterior abdominal wall by the mesentery (p. 1329), and of which the proximal two-fifths constitutes the *jejunum*, and the distal three-fifths the *ileum*.

THE DUODENUM

The duodenum (**8**.111A), 20–25 cm long, is the shortest, widest and most fixed part of the small intestine; it has no mesentery, and thus is only partially covered with peritoneum. Its course presents a remarkably constant curve, somewhat of the shape of an incomplete circle, which encloses the head of the pancreas. It lies entirely above the level of the umbilicus.

It begins at the pylorus, passes backwards, up and to the right for about 5 cm, under cover of the posterior part of the quadrate lobe of the liver, to the neck of the gall bladder, varying slightly in direction according to the degree of distension of the stomach; it then makes a sharp curve (*superior duodenal flexure*) and descends for about 7·5 cm in front of the medial part of the right kidney, usually to the level of the lower border of the body of the third lumbar vertebra, just medial to the lateral plane (**8**.112). Here it makes a second bend (*inferior duodenal flexure*), and passes horizontally to the left across the vertebral column for about 5–10 cm, just above the level of the umbilicus, having a slight inclination upwards; it then ascends in front and to the left of the abdominal aorta for about 2·5 cm, and ends opposite the second lumbar vertebra in the jejunum. At its union with the jejunum it turns abruptly forwards, forming the *duodenojejunal flexure*, which is situated 2·5 cm to the left of the median plane and 1 cm below the transpyloric plane. For descriptive purposes it is divided into first, second, third and fourth parts. The first and second parts are respectively superior and descending, while the third and fourth parts are described as horizontal and ascending.

DUODENAL RELATIONS

The superior part (*first part*) is about 5 cm long, and is the most movable of the four sections; it begins at the pylorus, and ends at the neck of the gall bladder. It is covered with peritoneum over the whole of its anterior aspect, but it is devoid of peritoneum posteriorly, *except near the pylorus*, where it takes a small part in the formation of the anterior wall of the omental bursa; the right part of the lesser omentum is attached to the upper border, and

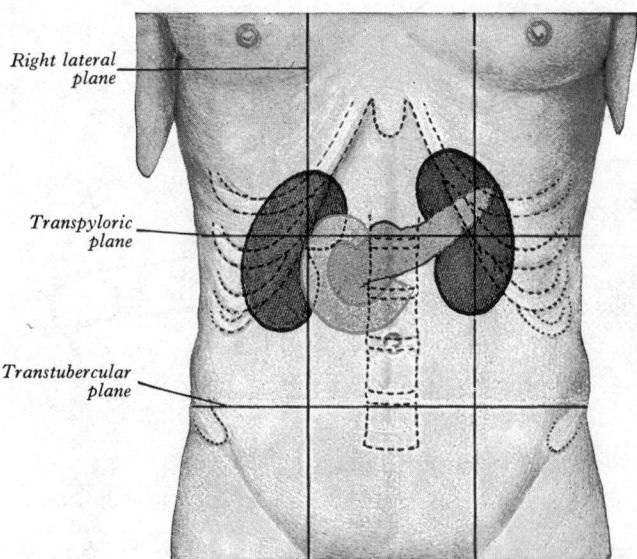

Right lateral plane

Transpyloric plane

Transtubercular plane

8.112   The surface projection of the duodenum, pancreas and kidneys. The vertebra just above the umbilicus is the 3rd lumbar.

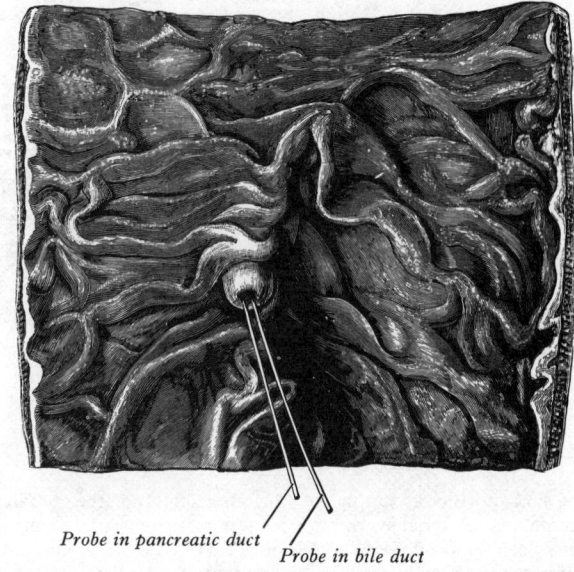

Probe in pancreatic duct   *Probe in bile duct*

8.113   The interior of the descending (second) part of the duodenum, showing the major duodenal papilla.

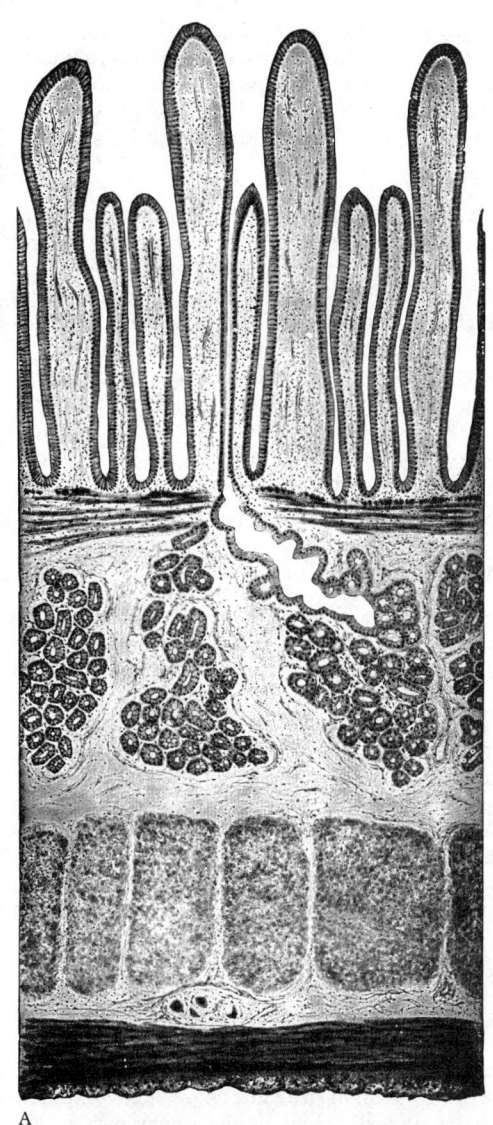

Villi

Intestinal glands

Muscularis mucosae

Duodenal glands in
submucosa

Circular muscular
layer

Longitudinal
muscular layer

Serous layer

A

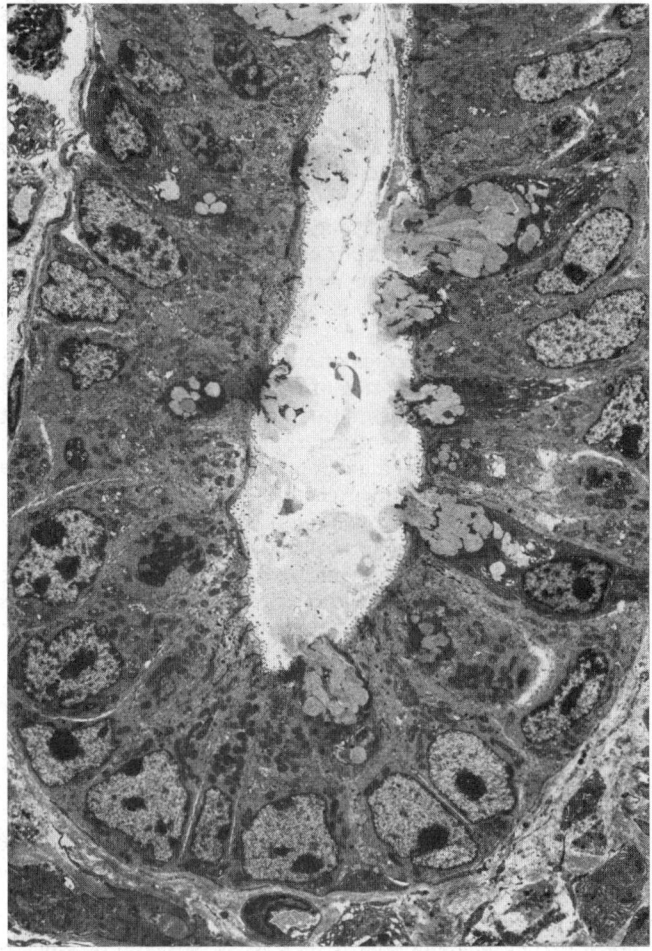

B

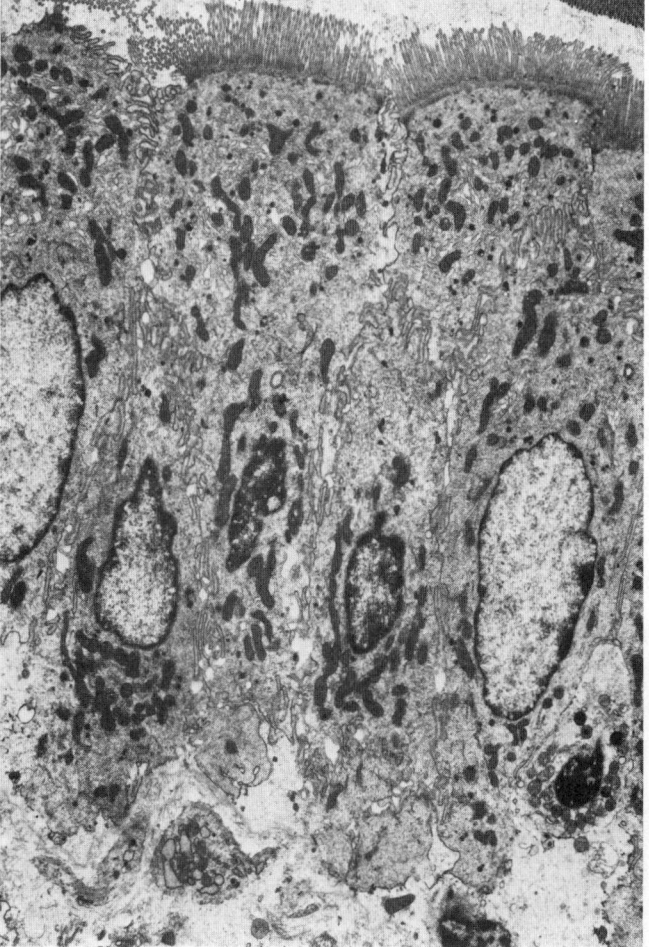

8.114A–C A A longitudinal section of feline duodenal wall. (Magnification about ×60.) B Electron micrograph of the base of a duodenal crypt showing absorptive columnar epithelial cells interspersed with mucus-secreting goblet cells (rat). (Magnification ×1,700.) C Electron micrograph showing absorptive columnar epithelial cells of a duodenal villus (rat). (Magnification ×3,700.) (Specimens B and C prepared and photographed by Mrs. Susan Smith, Department of Anatomy, Guy's Hospital Medical School, London.)

the greater omentum to the lower border of the proximal half. It is in relation above and in front with the quadrate lobe of the liver and the gall bladder; above, and on a more posterior plane, with the epiploic foramen; behind, with the gastroduodenal artery, the bile duct and the portal vein, and below and behind, with the head and neck of the pancreas. It is in such close relation with the gall bladder that it is usually found to be stained by bile after death, especially on its anterior surface.

The descending part (*second part*), from 8 to 10 cm long, descends from the neck of the gall bladder along the right side of the vertebral column as low as the lower border of the body of the third lumbar vertebra. It is crossed by the transverse colon, the posterior surface of which is connected to the duodenum by a small quantity of areolar tissue. The parts above and below the transverse colon are covered in front with peritoneum. It is in relation, in front, from above downwards, with the duodenal impression on the right lobe of the liver, the transverse colon and the root of the transverse mesocolon, and the jejunum; behind, it has a variable relation to the front of the right kidney in the neighbourhood of its hilum, and is connected to it by loose areolar tissue; the right renal vessels, edge of the inferior vena cava, and psoas major are also behind it. Its medial side is related to the head of the pancreas and the bile duct; its lateral side, to the right colic

C

flexure. Sometimes a small part of the head of the pancreas is actually embedded in the wall of the descending part of the duodenum. The bile duct and the pancreatic duct come into contact at the medial side of this part of the duodenum. The two ducts enter the wall of the gut obliquely and there unite to form a short, dilated tube which is named the *hepatopancreatic ampulla* (*see* p. 1383). The narrow, distal end of this ampulla opens on the summit of a *major duodenal papilla*, which is situated within the descending part of the duodenum at the junction of its medial and posterior walls (**8**.113, 132), from 8 cm to 10 cm distal to the pylorus. The accessory pancreatic duct, when present, opens about 2 cm proximal to the major papilla, on a small rounded *minor duodenal papilla*.

The horizontal part (inferior or *third part*), about 10 cm long, begins at the right side of the lower border of the third lumbar vertebra and passes from right to left, with a slight inclination upwards, in front of the inferior vena cava, and ends in the fourth part in front of the abdominal aorta. Its anterior surface is covered with peritoneum, except near the median plane, where it is crossed by the superior mesenteric vessels and the root of the mesentery. Its posterior surface is uncovered by peritoneum, except towards its left extremity, where the left layer of the mesentery sometimes covers it to a variable extent. This surface rests upon the right ureter, the right psoas major, the right testicular (or ovarian) vessels, the inferior vena cava and the abdominal aorta (with the origin of the inferior mesenteric artery). The upper surface is in relation with the head of the pancreas; the lower, with the coils of the jejunum.

The ascending part (*fourth part*), about 2·5 cm long, ascends on or immediately to the left of the aorta, as far as the level of the upper border of the second lumbar vertebra, where it turns ventrally at the *duodenojejunal flexure* and is continuous with the jejunum. It lies in front of the left sympathetic trunk, left psoas major, the left renal and testicular vessels and the inferior mesenteric vein. Along its right border it gives attachment to the upper part of the root of the mesentery, the left layer of which is continued over its anterior surface and left side. To its left there are the left kidney and ureter; above, there is the body of the pancreas; in front, there is the transverse colon and transverse mesocolon (the latter separating the duodenojejunal flexure from the omental bursa and stomach).

The superior part of the duodenum, as stated above, is to a slight degree mobile, but the rest is relatively fixed, and is sessile upon neighbouring viscera and the posterior abdominal wall. Radiologically, after a barium meal, the superior part of the duodenum is seen as a somewhat triangular homogeneous shadow, called the 'duodenal cap' (**8**.108).

The terminal part of the duodenum and the duodenojejunal flexure are usually described as suspended and fixed in position by the '*suspensory muscle of the duodenum*' (suspensory muscle, or ligament, of Treitz). This is often said to be in two parts: (*a*) a slip of *striated* muscle, derived from the diaphragm near its oesophageal opening and ending in the connective tissue adjacent to the coeliac arterial trunk, and (*b*) a fibromuscular band, containing *nonstriated* muscle, which passes from the duodenum (third and fourth parts and duodenojejunal flexure) to blend with the same pericoeliac connective tissue. Treitz (1853) described both entities, naming the former one the *Hilfsmuskel* (accessory muscle). Some subsequent authorities (Lockwood 1886, and Low 1907) regarded these two structures as parts of a digastric muscle, naming the whole as the suspensory muscle of Treitz, a description perpetuated by many textbooks. The confusion was increased by Haley and Peden (1943), who derived the 'suspensory muscle' from the right crus, and by Argème *et al.* (1970), who described an intermediate tendon but regarded this as part of a 'false' digastric muscle. Jit (1952, 1977) has persistently supported the dual nature of the original description of Treitz, on embryological and histological grounds. The diaphragmatic slip (the Hilfsmuskel) has no satisfactory official name. It is supplied, according to Jit, by myelinated nerve fibres, derived probably from the phrenic nerve (pp. 550, 1094). This slip is sometimes considered to be an aberrant part of iliocostalis thoracis. The suspensory muscle proper (nonstriated) is supplied by autonomic fibres from the coeliac and superior mesenteric

plexuses (Jit and Grewal, 1977). Descriptions of the extent of the duodenal attachments of the suspensory muscle vary; none of these accounts contain a convincing view of the function of the muscle, the usual suggestion being that it may kink the duodenojejunal flexure even further, producing a valvular effect.

Vessels and Nerves

The *arteries* supplying the duodenum are derived from the right gastric, supraduodenal, right gastro-epiploic and the superior and inferior pancreaticoduodenal arteries. (They are described on pp. 713, 716; **6**.86.) The superior part of the duodenum receives a leash of small branches from the hepatic artery proper which runs in the right part of the lesser omentum, and a similar leash of vessels from the gastroduodenal artery. These vessels also supply the neighbouring part of the pyloric canal, and there is some anastomosis in the wall of the alimentary canal between these vessels across the pyloroduodenal junction (p. 1338). The *veins* end in the splenic, superior mesenteric and portal veins. The *nerves* are derived from the coeliac plexus.

THE JEJUNUM AND ILEUM

The rest of the small intestine extends from the duodenojejunal flexure to the ileocaecal valve, where it ends in the junction of the caecum and ascending colon of the large intestine; it is arranged in a series of coils or loops which are attached to the posterior abdominal wall by its mesentery. This part of the gut is completely covered with the peritoneum, except for a narrow strip along its mesenteric border, where the two layers of the mesentery diverge from each other to enclose it. It is divided into jejunum and ileum, the former name being given to the first two-fifths and the latter to the distal three-fifths. There is no identifiable level of distinction between these two parts, and the division is arbitrary; but at the same time the character of the intestine gradually undergoes a change from the beginning of the jejunum to the end of the ileum, so that samples from these two situations present characteristic differences.

The jejunum has a diameter of about 4 cm, and is thicker, redder and more vascular than the ileum. The circular folds (p. 1345) of its mucous membrane are large and thickly set, and its villi surpass those of the ileum in size. The aggregated lymphatic follicles (p. 1349) are almost absent in the upper part of the jejunum; in the lower part they are fewer and smaller than in the ileum and tend to assume a circular form. When the jejunum is grasped between the finger and thumb the circular folds can be felt through the wall of the gut; as these folds are absent from the lower part of the ileum, it is possible in this way to distinguish the upper from the lower part of the small intestine.

For the most part the jejunum lies in the umbilical region, but it may extend into any of the surrounding areas. The first coil occupies a recess between the left part of the transverse mesocolon and the anterior surface of the left kidney.

The ileum has a diameter of 3·5 cm, and its wall is thinner than that of the jejunum. A few circular folds are present in the upper part of the ileum, but they are small and disappear almost entirely towards its lower end; the aggregated lymphatic follicles are, however, larger and more numerous than in the jejunum. For the most part the ileum is situated in the hypogastric (pubic) and pelvic regions. The terminal part of the ileum usually lies in the pelvis, from which it ascends over the right psoas major and right iliac vessels to end in the right iliac fossa by opening into the medial side of the junction of the caecum and ascending colon.

The jejunum and ileum are attached to the posterior abdominal wall by an extensive fold of peritoneum, termed the *mesentery*, which allows of very free movement, so that each coil can accommodate itself to changes in form and position.

The mesentery (p. 1329) is fan-shaped; its vertical border or root, about 15 cm long, is attached to the posterior abdominal wall along a line running from the left side of the body of the second lumbar vertebra to the right sacro-iliac joint, and crossing successively the horizontal part of the duodenum, the aorta, the inferior vena cava, the right ureter and right psoas major (**8**.98). Its average breadth from the vertebral to the intestinal border is about 20 cm, but is greater in the middle than at its upper and

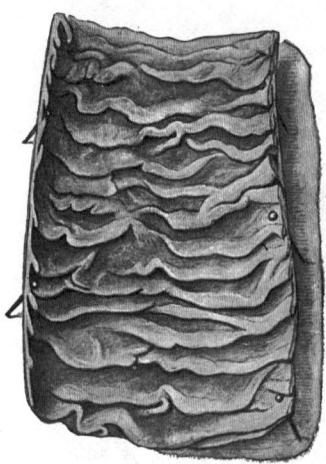

8.115A Internal aspect of representative sample of proximal jejunum, showing circular folds.

lower ends. The two layers of the mesentery contain the jejunum, ileum, the jejunal and ileal branches of the superior mesenteric blood vessels, nerves, lacteals and lymph nodes, together with a variable amount of fat.

The diverticulum ilei (of Meckel) projects from the antimesenteric border of the lower part of the ileum in about 3 per cent of subjects. Its average position is about 1 m above the ileocaecal valve, and its average length about 5 cm. Its calibre is generally similar to that of the ileum, and its blind extremity may be free or may be connected with the abdominal wall or with some other part of the intestine by a fibrous band. It represents the persistent proximal part of the vitelline or yolk duct, which connects the yolk sac and the primitive digestive tube in early fetal life (pp. 120, 200, 205). The mucous membrane of the diverticulum usually has the same structure as that of the neighbouring ileum, but occasionally small regions of the mucous membrane may have a structure similar to that of the body and fundus of the stomach, with oxyntic cells which secrete acid. Sometimes small heterotopic areas of pancreatic or other tissues may be found in the wall of the diverticulum. In a recently published study of 1816 late fetal and neonatal cadavers Miyabara *et al.* (1974) found a diverticulum ilei present in 61 individuals (3·4 per cent). Of these, gastric mucosa was present in 11, jejunal mucosa in 2, colonic mucosa in 2, and pancreatic tissue in only 1.

STRUCTURE OF THE SMALL INTESTINE

The intestinal wall is composed of the series of layers described previously—serous, muscular, submucous and mucous (**8.116**).

The *serous layer* is formed of visceral peritoneum which merges with a subserous stratum of areolar connective tissue.

The *muscularis externa* is thicker in the cranial than in the caudal part of the small intestine; it consists of a thin outer longitudinal and a thick inner circular layer of nonstriated muscle fibres.

The *submucous layer* consists of loose connective tissue carrying blood vessels, lymphatics and nerves.

The *mucous membrane* is thick and highly vascular in the upper part of the small intestine, but thinner and less vascular in the lower part. It is thrown into circular or spiral pleats or plicae, the *circular folds*, and the whole surface is studded with finger-like, filiform or tongue-shaped projections, the intestinal villi.

The circular folds (*plicae circulares*, or '*valves*' *of Kerkring*— **8.**115, 119), are large, crescentic folds of mucous membrane which project into the intestinal lumen transversely to its long axis. Unlike the folds in the stomach they are permanent: they are not obliterated when the intestine is distended. The majority extend round the intestine for about one-half or two-thirds of its circumference, but some form complete circles, some bifurcate and join adjacent folds, and others have a spiral direction; the latter usually extend a little more than once round the lumen, but occasionally two or three times. The larger folds are about 8 mm

in depth at their broadest part; but the greater number are of smaller size. The larger and smaller folds often alternate. Circular folds do not exist in the commencement of the duodenum, but begin to appear about 2·5 to 5 cm beyond the pylorus. Distal to the point where the common bile duct (ductus choledochus) and pancreatic duct enter the duodenum, the folds are very large and closely approximated. In the proximal half of the jejunum they are large and numerous, but from here to a point midway along the ileum, they diminish considerably in size. In the distal ileum they are almost entirely absent; hence the comparative thinness of this portion of the intestine, as compared with the duodenum and jejunum. The circular folds retard the passage of the food and afford an increased surface for absorption (**8.**119).

The intestinal villi are highly vascular processes, just visible to the naked eye; they project from the mucous membrane of the whole of the small intestine, and give to its surface a velvety appearance. They are large and numerous in the duodenum and jejunum, but are smaller and fewer in the ileum. In the first part of the duodenum they are broad, ridge-like structures, changing to tall leaf-like villi in the distal duodenum and proximal jejunum thereafter they gradually transform into shorter finger-like extensions in the distal jejunum and ileum (Verźar and McDougall 1936; McMinn and Mitchell 1954). They vary in density from 10 to 40 per square millimetre, and are from about 0·5 to 1·0 mm in height. The villi increase the surface area, compared with an unfolded surface, about eightfold. (For details of an experimental analysis of the formation of intestinal villi after lesions of the mucosa in the small intestine of the cat, consult McMinn and Mitchell 1954.)

The mucous membrane (**8.**116) of the small intestine has three layers. Next to the submucous coat is the *muscularis mucosae* which consists of an outer longitudinal and an inner circular layer of nonstriated muscular fibres. This layer extends into the circular folds. Internal to it is a quantity of reticular tissue, the lamina propria, in which, in addition to fibroblasts and fibres, there are lymphocytes, eosinophilic leucocytes, macrophages,

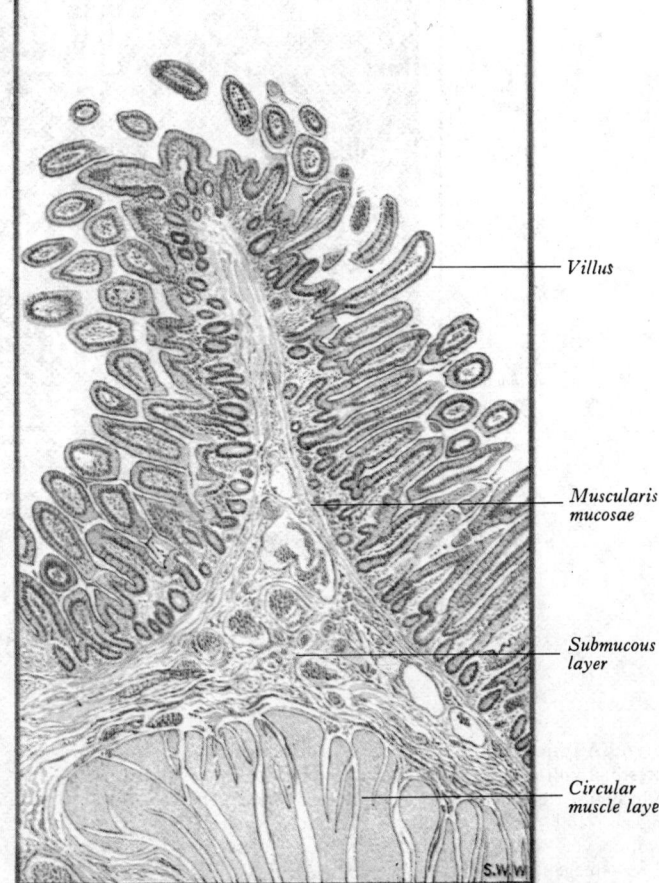

Villus

Muscularis mucosae

Submucous layer

Circular muscle layer

8.115B Section through a circular fold from human small intestine. Stained with haematoxylin and eosin. Magnification about × 19.

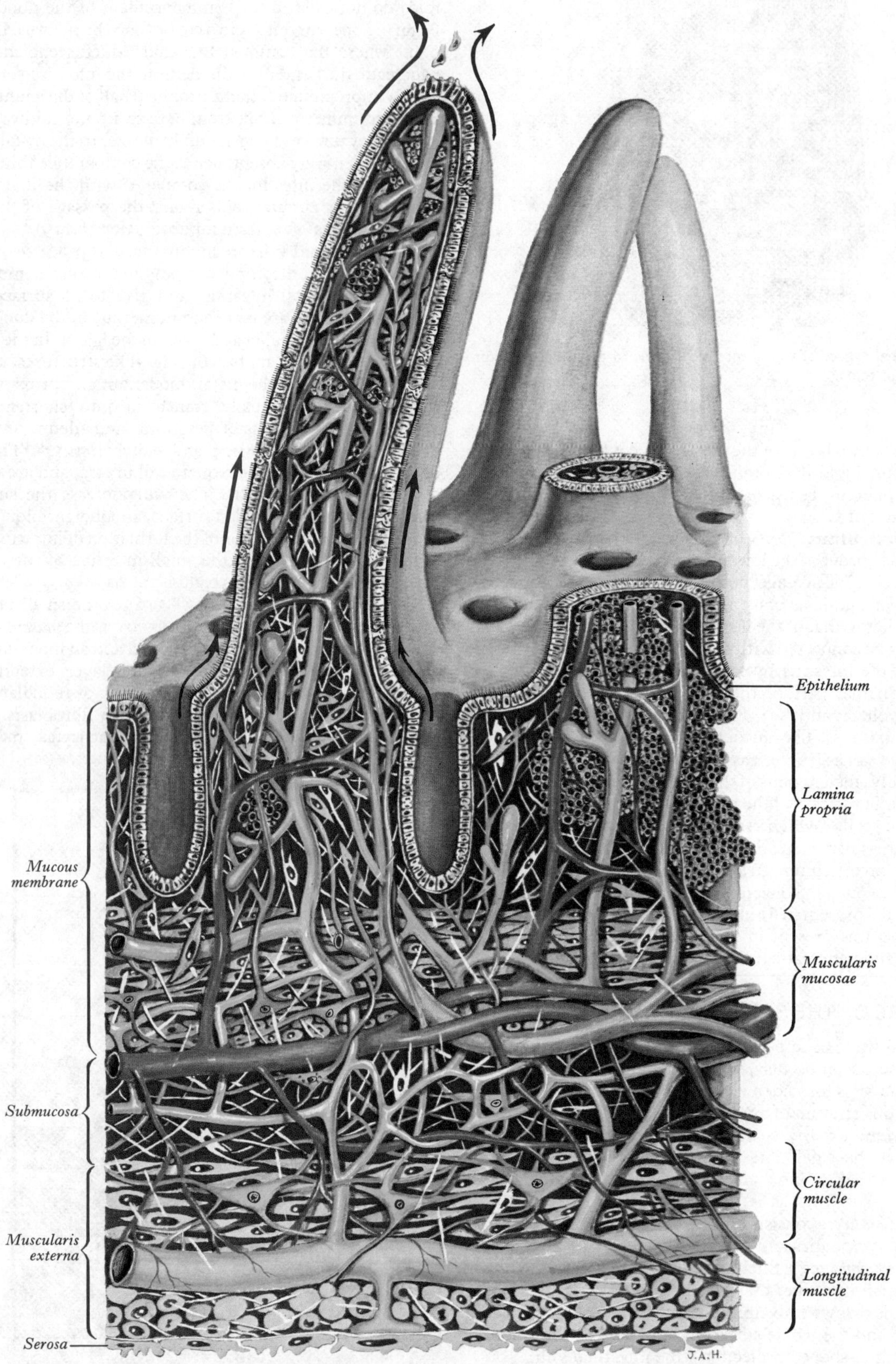

8.116 A three-dimensional reconstruction of the architecture of intestinal villi and subjacent wall; the principal layers of the latter are indicated. Arteries and arterioles—red; veins and venules—blue; central lacteals and other lymphatic channels—orange; aggregations of lymphocytes—yellow; neural elements—green; nonstriated muscle fibres—magenta; fibroblasts—white. Note the orifices of the intestinal crypts (of Lieberkühn). Types of cells in the epithelium include absorptive cells, goblet cells, and entero-endocrine cells. Arrows indicate direction of cell migration. The various layers are not drawn to scale.

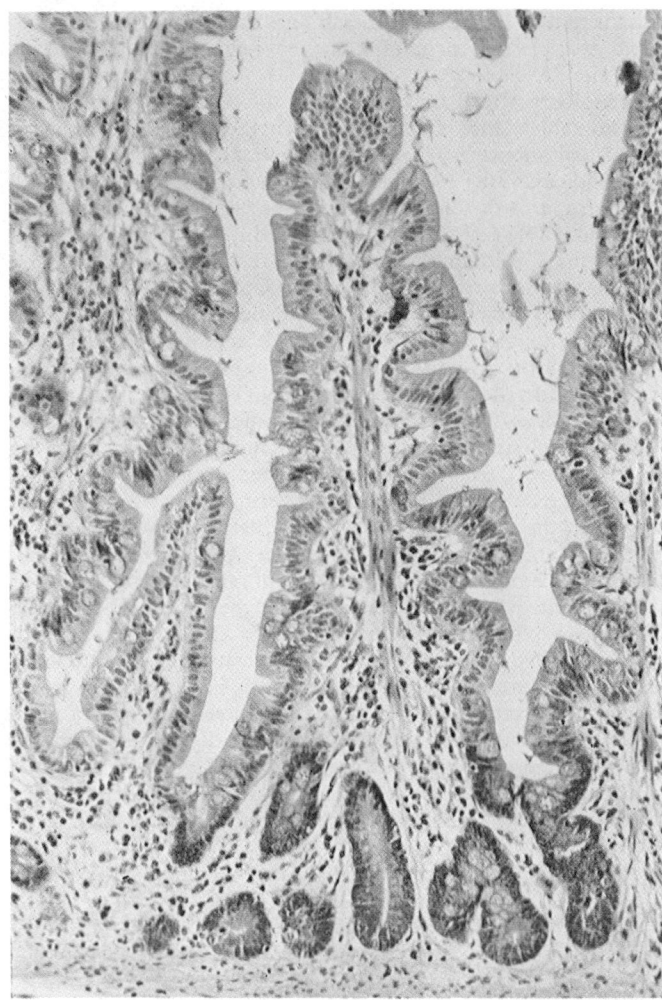

8.117A A light micrograph of part of the mucosa of the murine small intestine, showing a villus in longitudinal section. Nonstriated myocytes (pink) can be seen in the lamina propria of the villus. Stained with haematoxylin, eosin, and periodic acid/Schiff. (Prepared and photographed by Mr. Stephen Sitch, Department of Anatomy, Guy's Hospital Medical School, London.) Magnification about ×250.

mast cells, blood capillaries, lymphatic vessels and nonmyelinated nerves. The plasma cells are numerous and the lymphocytic cells are in many regions collected into solitary and aggregated lymphatic follicles, some of which may extend through the muscularis mucosae into the submucous coat. Internal to the lamina propria is a basement membrane supporting an epithelium composed mainly of tall columnar cells, except over lymphatic follicles where they are partly replaced by *membrane* or *M cells*, specialized for transport of antigens (Owen and Nemamic 1978). Extending into the mucosa from the surface between the intestinal villi are simple, tubular, *intestinal glands* or *crypts* (of *Lieberkühn*). They reach almost to the muscularis mucosae. In the duodenum there are also mucous, tubulo-alveolar *duodenal glands* (of *Brunner*), the ducts of which extend through the muscularis mucosae to expand into the submucous coat; the secretions of these glands contain mucosubstances and bicarbonate ions.

Structure of the Intestinal Villi

Each villus has a core of reticular tissue containing a lymph vessel or lacteal, blood vessels, nerves and some nonstriated muscle fibres covered by a layer of columnar epithelium resting on a basement membrane (**8.**116, 117A–D).

The *lacteal*, usually single but occasionally double, commences in a dilated blind extremity near the summit of the villus and then courses along the axis of the villus and empties into the plexus of lymph vessels in the underlying lamina propria. Its wall is composed of a single layer of endothelial cells.

The *muscle fibres* are derived from the muscularis mucosae,

and are arranged in bundles around the lacteal vessel, extending from the base to the summit of the villus, and giving off, laterally, individual muscle cells, which are enclosed by the reticulum, and by it are attached to the basement membrane and to the lacteal. Contraction of the muscularis mucosae therefore has the effect of 'milking' the lacteals.

The *blood vessels* form a capillary plexus in the lamina propria, and are enclosed in the reticular tissue. The capillaries are lined by a fenestrated endothelium, a modification which may be of considerable importance in ensuring the rapid uptake of nutrients diffusing from the epithelium (Clementi and Palade 1969).

The *epithelium* covering the superficial surface of the mucous membrane and the intestinal villi consists mainly of absorptive columnar cells, interspersed with goblet cells.

The columnar cells (1.1) are granular in appearance; each possesses a clear oval nucleus located in its basal half. At the free surface of the cell there is a highly refractile, vertically striated border about 1 μm in depth, the *striated* or *brush border*. This border is very rich in alkaline phosphatase and is concerned with the process of active absorption; electron microscope studies show that this border is composed of minute parallel cylindrical microvilli, each about 1 μm long and 0·1 μm broad. Applied to the outer surface of the enveloping plasma membrane of the microvilli is a coating of ultramicroscopic fine filaments, rich in mucosubstances, believed to be formed by the epithelial cells and to constitute an integral part of the surface. In the cytoplasmic core of each microvillus are fine filaments which are continuous at its base with a plexiform band of similar filaments in the apical cytoplasm of the cell; this band is the *terminal web* and, except for occasional vesicles, is free of organelles (**8.**118B). The lateral plasma membranes of the columnar cells are often plicated and interdigitated; at their luminal extremities they show typical junctional complexes (p. 7, 1.4). Elsewhere scattered desmosomes occur between adjacent plasma membranes; basally, the lateral plasma membranes may be separated by intervals constituting intercellular canaliculi. The basal plasma membranes are usually smooth and lie adjacent to the basement membrane which is about 20 nm wide. Rod-shaped mitochondria are scattered through the cell cytoplasm together with a moderate amount of rough and smooth surfaced endoplasmic reticulum. The Golgi apparatus is supranuclear in position and the apical part of the cell, beneath the terminal web, contains numerous membrane-bound lysosomes. At the summits of the intestinal villi the cells often show signs of degeneration and their microvilli may be stunted and degenerate. The role of these cells in digestion and absorption of nutrients has attracted much attention. It appears likely that absorption of amino acids and simple carbohydrates occurs by facilitated diffusion across the cell membranes, these materials passing

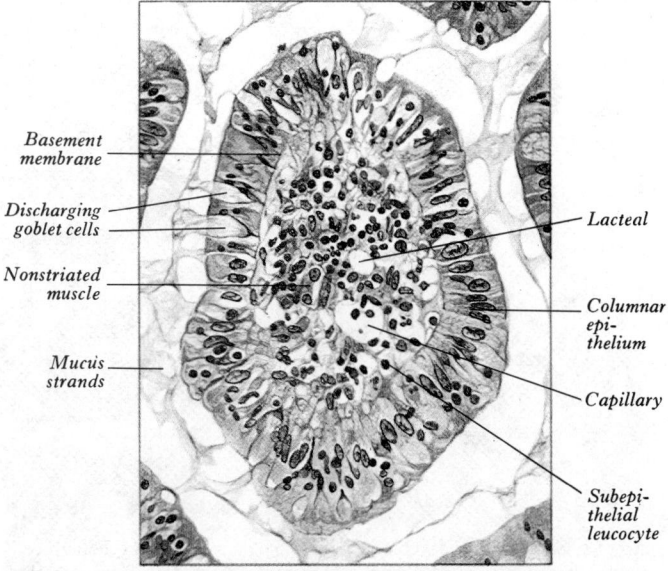

Basement membrane

Discharging goblet cells

Nonstriated muscle

Mucus strands

Lacteal

Columnar epithelium

Capillary

Subepithelial leucocyte

8.117 B Transverse section through a villus in human jejunum. Stained with haematoxylin and eosin. Magnification about ×380.

through the cell to the underlying capillary arrays of the lamina propria. Lipid absorption also appears to occur by the diffusion of small molecules (fatty acids, etc.) through the membrane of the luminal surface, the lipid accumulating in the vacuoles within the cytoplasm in the apical region of the cell, before being passed to the underlying lymphatics. It has also been possible to isolate the striated borders of absorptive cells by cell fractionation and centrifugation; digestive enzymes such as disaccharidases are bound to the surface, and there are indications that much of the intestinal digestion may occur close to the site of absorption, digestive enzymes being absorbed on to the cell surfaces, perhaps in the polysaccharide cell coat.

The goblet cells are intercalated at intervals in the epithelial layer. Their nuclei are elongated and basally situated and their apical parts are distended with membrane-bound mucin granules. The Golgi apparatus is well developed and supranuclear in position and the granular endoplasmic reticulum is abundant infranuclearly. The goblet cells have less frequent and more irregular microvilli at their superficial surfaces and the terminal web is poorly marked. The mucin is believed to be generally

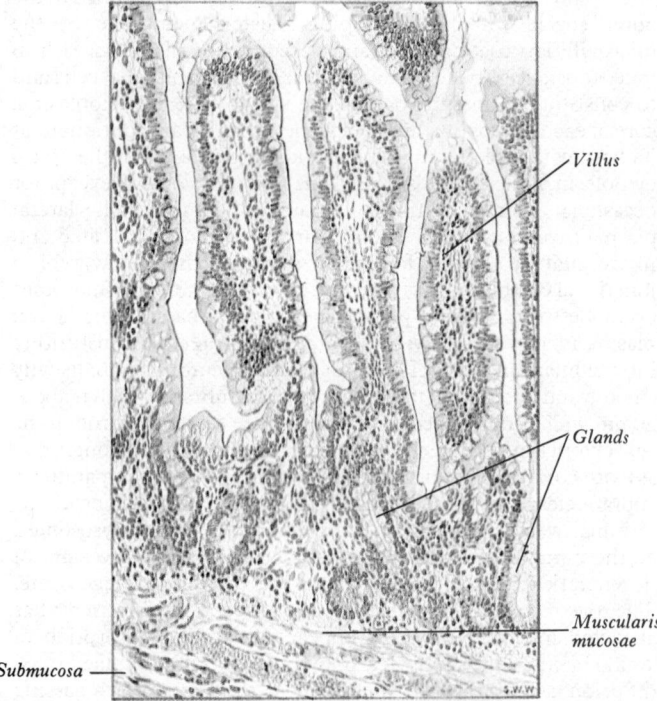

8.117 C   Intestinal glands and villi in human small intestine. Stained with haematoxylin and eosin. Magnification about × 120.

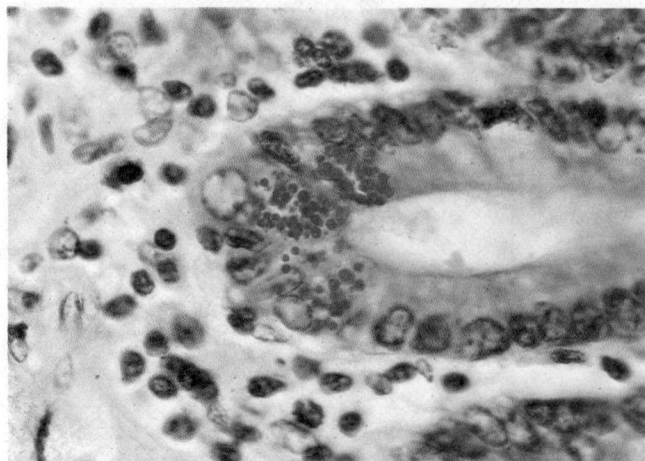

8.117 D   Part of a transverse section of the ileum, showing Paneth cells containing orange-stained zymogen granules at the base of an intestinal gland. 'Undifferentiated' epithelial cells are also visible. Mallory's azan stain. Magnification about × 400.

discharged by fusion of the membranes surrounding the granules with the plasma membrane in the manner of a merocrine secretion (**8.118A**).

The intestinal glands (*crypts of Lieberkühn*) occur in considerable numbers over every part of the mucous membrane of the small intestine. They are simple tubular pits, arranged perpendicularly to the surface, upon which they open by small circular apertures. Their orifices may be seen with the aid of a lens as minute dots scattered between the villi. Their walls are thin, consisting of columnar epithelium on a basement membrane and associated with a rich mucosal capillary plexus (**8.116**).

The cells lining the intestinal glands consist of so-called 'undifferentiated cells', Paneth and argentaffin cells. **The undifferentiated cells** are the most numerous and proliferate by mitotic division, passing upwards out of the intestinal glands along the sides of the villi, where they differentiate into columnar absorptive elements or into goblet cells; eventually they reach the tips of the villi, from which they are shed. In this way there is a continual renewal of the epithelial covering of the villi. When not dividing, the free surfaces of these cells have fewer and more irregular microvilli than the surface cells with occasional pseudopodia; their lateral plasma membranes are straighter but the attachment areas are similar in both types of cell. Their nuclei are basally placed and the terminal webs are poorly developed. Membrane-bound secretory granules occur in their cytoplasm and are believed to be discharged both in the manner of an apocrine secretion through the surface blebs and by the more usual merocrine mechanism commonly described. The undifferentiated epithelial cells in the intestinal glands multiply at the rate of 1 cell per 100 cells per hour and constitute one of the most rapidly proliferating of tissues in the body (Lipkin *et al.* 1963, MacDonald *et al.*, 1964).

The zymogenic cells (of Paneth) are numerous in the deeper parts of the intestinal crypts, particularly those of the duodenum. They are rich in zinc and contain granules (**8.117D**) which stain with phosphotungstic haematoxylin. Electron microscopy shows irregular microvilli to be present at the apices of these cells, and prominent membrane-bound granules to be present in the supranuclear cytoplasm. Scattered mitochondria, lysosomes and a large amount of granular endoplasmic reticulum also exist, especially in the basal parts of the cells. These zymogenic cells are the source of the digestive enzymes produced by the wall of the small intestine.

Between the cells lining the intestinal glands, and less commonly among those covering the villi, are pyramidal or columnar cells, the cytoplasm of which contains granules that have an affinity for silver salts, by which they are stained black, and for chromium salts, by which they are stained brown. They are called **enterochromaffin cells** or **argentaffin cells** and are a variety of APUD cell (*see* p. 1366). The distributions of these and other cells of the gastro-entero-pancreatic endocrine system are shown in **8.129C**.

The duodenal glands (of Brunner) are limited to the duodenum (**8.114**), and are sited *in the submucous areolar tissue*, i.e. they penetrate through the muscularis mucosae. They are largest and most numerous near the pylorus, forming an almost complete layer in the superior part and proximal half of the descending part of the duodenum; beyond this they gradually diminish in number and disappear at the junction of the duodenum and jejunum. They are small, compound, acinotubular glands, each consisting of a number of alveoli lined with short columnar epithelium and opening by a duct on the inner surface of the intestine. They seemingly contain only one kind of exocrine cell in man; this is a typical mucous element. The nuclei of these cells are small and basal, and they show variations according to the secretory cycle. The Golgi apparatus is extensive; mucin droplets are numerous. Many argentaffin cells (*vide supra*) are present amongst the mucinogenic cells.

The solitary lymphatic follicles are seen to be scattered throughout the mucous membrane of the small intestine but are most numerous in the lower part of the ileum. Their free surfaces are covered with rudimentary villi, except at the summits, and each follicle is surrounded by the openings of the intestinal glands. Each consists of a dense, interlacing, reticular tissue closely packed with lymphocytes, and permeated by an abundant, capillary network. The interspaces of the reticular tissue are continuous with larger lymph spaces which surround the follicle, and by this means they are enabled to communicate with the lacteal system. They are situated partly in the submucous tissue and partly in the mucous coat. They are partly covered by 'M' cells (p. 1347).

The aggregated lymphoid follicles (Peyer's patches, **8.**120A, B) form circular or oval patches, individual ones containing from 10 to 260 follicles, and varying in length from 2 to 10 cm. Like the other collections of lymphoid tissue in the body (except the lymph nodes), solitary and aggregated lymphatic follicles are most numerous around puberty and thereafter diminish in number and size but many persist to old age (Cornes 1965). Aggregated follicles are largest and most numerous in the ileum. In the distal part of the jejunum they are small, circular and few in number. They are occasionally seen in the duodenum. They are placed lengthwise in the intestine, and are in the part of its circumference most distant from the attachment of the mesentery. Each patch is formed of a group of solitary lymphoid

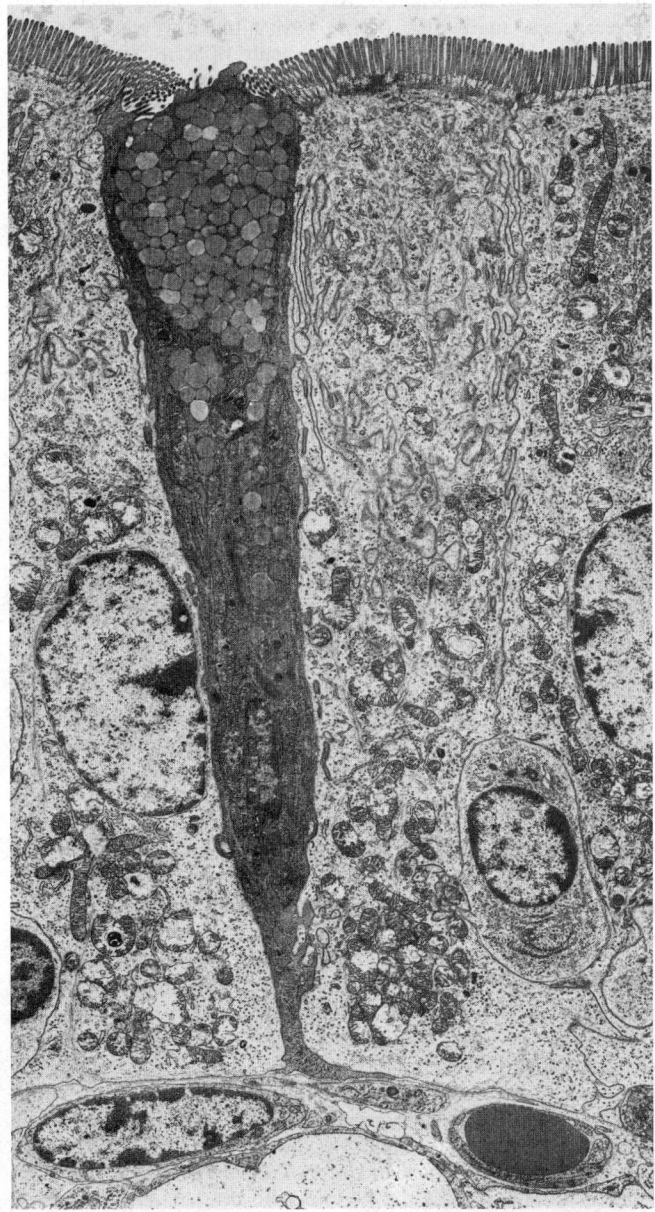

8.118A Transmission electron micrograph of the columnar epithelium lining the murine small intestine, showing a mucus-secreting goblet cell between two absorptive cells which bear microvilli. The cells rest on a delicate basal lamina deep to which is the vascular lamina propria. (Magnification × 4,800.) (Prepared and photographed by Mr. Derrick J. Lovell, Department of Anatomy, Guy's Hospital Medical School, London.)

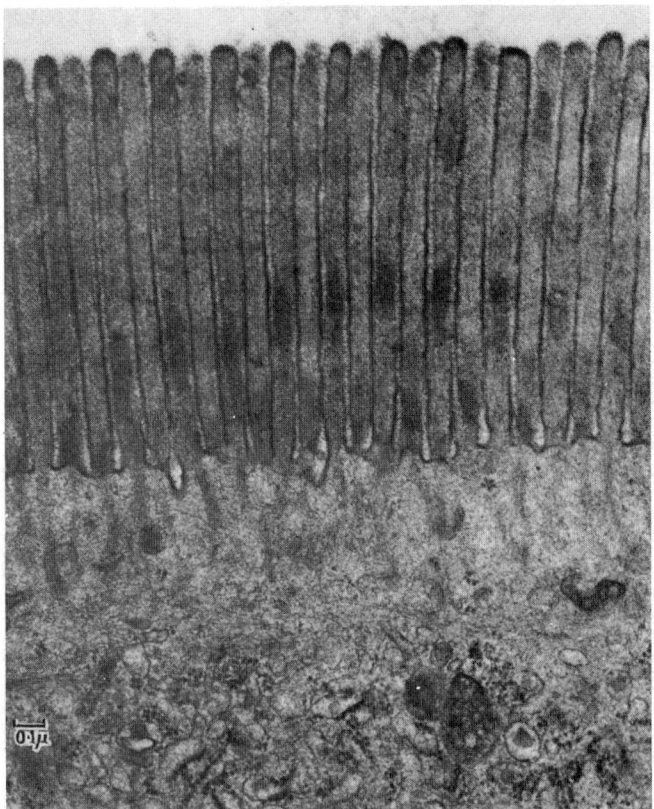

8.118B Electron micrograph of the apical region in a columnar cell from the jejunum showing the regular series of microvilli which constitutes the striated border of light microscopy. Filaments can be seen passing from the microvilli to the terminal web. Rat. Magnification about × 32,500.

follicles covered with columnar epithelial cells and 'M' cells (p. 1347); the patches do not, as a rule, possess villi on their free surfaces. They are freely supplied with blood vessels, which form an abundant plexus around each follicle and give off fine branches to permeate the lymphoid tissue in the interior of the follicle. The plexuses of lymph vessels are especially abundant around these patches. In typhoid fever, ulceration of these aggregated lymphatic follicles may occur; the ulcers are thus oval in shape, their long axes are in the long axis of the bowel (so subsequent fibrosis does not constrict the gut), they are present chiefly in the lower part of the ileum, and are situated on or near its antimesenteric border.

Vessels and nerves. The *arteries* to the jejunum and ileum (**8.**120C, D) stem from the superior mesenteric artery, the jejunal and ileal branches of which, having reached the mesenteric border extend between the serous and muscular coats. From these vessels numerous branches are given off, which pierce the muscular coat, supplying it and forming an intricate plexus in the submucous tissue. From this plexus minute vessels pass to the glands and villi of the mucous membrane (*see* p. 1347). The anastomoses between the terminal intestinal branches are by no means free, and there is a distinct tendency for the alternate vessels to be distributed to opposite sides of the gut. The *veins* have a course and arrangement similar to the arteries. (For an extensive and detailed investigation into the course, distribution and variations in the coeliac and superior mesenteric arteries consult Nesebar *et al.* 1969.) The *lymph vessels* of the small intestine (lacteals) are arranged in two sets, viz. those of the mucous membrane and those of the muscular coat. The lymph vessels of the villi commence in these structures in the manner described on p. 1347. They form a highly intricate plexus in the mucous and submucous tissue, being joined by the lymph vessels from lymph spaces at the bases of the solitary follicles, and from there pass to larger vessels at the mesenteric border of the gut. The lymph vessels of the muscular coat are situated to a great extent between the two layers of muscular fibres, where they form a close plexus; throughout their course they communicate freely with those from the mucous

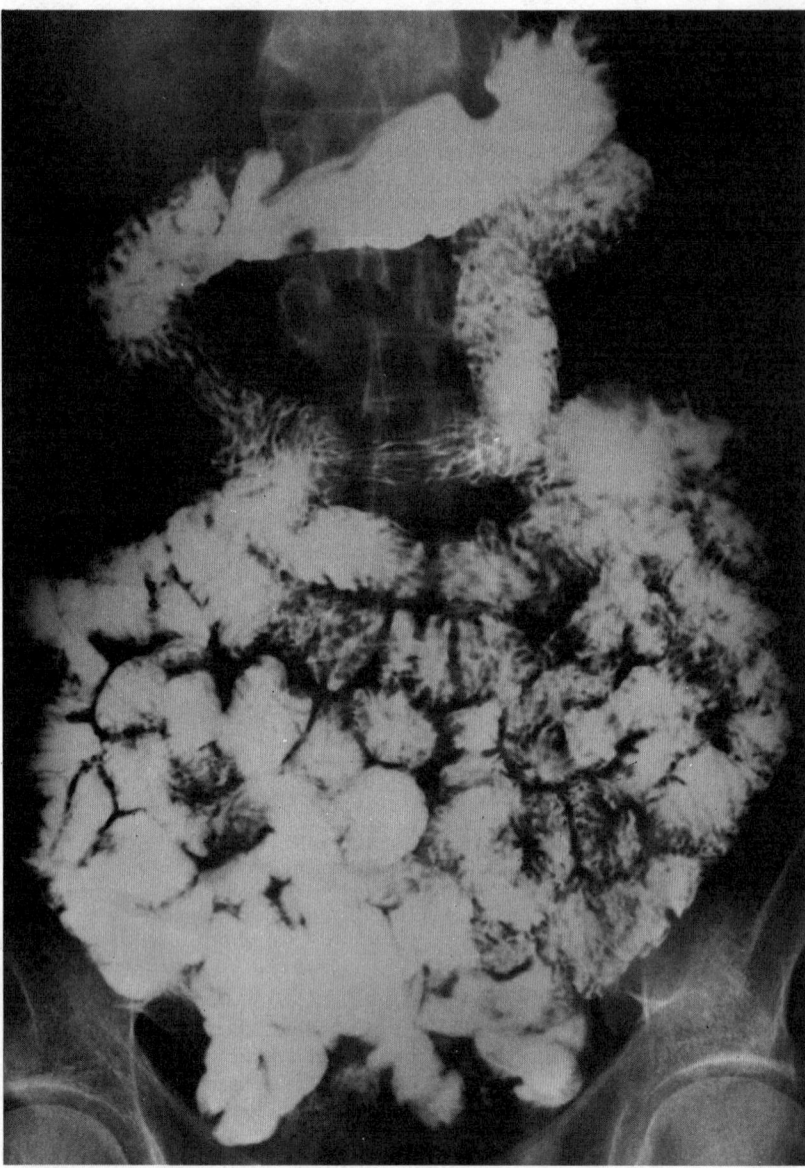

8.119 A radiograph of the pyloric end of the stomach, duodenum, jejunum and ileum, one hour after taking a barium meal. The feathery appearance of the profile of the small intestine is due to the presence of numerous circular mucosal folds.

membrane, and open in the same manner as these into the origins of the lacteal vessels at the attached border of the gut.

The *nerves* of the small intestine are derived from the vagus and splanchnic nerves through the coeliac ganglia and the plexuses around the superior mesenteric artery. They run to the *myenteric plexus* (p. 1316) of nerves and ganglia, situated between the circular and longitudinal layers of the muscularis externa, which they supply. From the myenteric plexus a secondary plexus, termed the *submucous plexus*, is derived, and is formed by branches which have perforated the circular muscular layer. This plexus also contains ganglionic neurons from which the nerve fibres pass to the muscularis mucosae and to the mucous membrane. The nerve bundles of the submucous plexus are finer than those of the myenteric plexus. The neurons in both plexuses are essentially parasympathetic (vagal). An old controversy as to the source of the postganglionic neurons in the enteric ganglia has been renewed latterly (Andrew 1971). Endodermal and mesodermal origins have been suggested, but the evidence (though largely equivocal) indicates a derivation from the neural crest. In general the sympathetic is inhibitory to the peristaltic movements of the alimentary canal, but stimulates the sphincters and also the muscularis mucosae. While the parasympathetic is generally an augmentor of the peristaltic movements and an inhibitor of the sphincters, the result of stimulation of the parasympathetic appears to depend on the state of contraction or relaxation of the organ at the time of stimulation. The parasympathetic is also augmentory to the intestinal secretion. For an extensive evaluation of the status of the 'intestinal' cells (of Cajal) in the intestinal wall, consult Rogers and Burnstock (1966). It seems probable that these are connective tissue cells, and not neurons. There is some evidence that some of the neurons in the ganglia of the intestinal wall may be afferent (Fehér and Vajda 1974).

The Large Intestine

The large intestine extends from the end of the ileum to the anus, and is about 1·5 m long. Its calibre is greatest at its commencement at the caecum, and gradually diminishes as far as the rectum, where there is a dilatation of considerable size just above the anal canal. Its functions being chiefly the absorption of fluid and solutes, it differs considerably in appearance, structure, size and arrangement from the small intestine. (1) It has a greater calibre. (2) For the most part, it is more fixed in position. (3) Its longitudinal muscular fibres, although distributed as a *complete* layer, are particularly concentrated to form three longitudinal bands or *taeniae coli*. (4) Since these taeniae are often held to be 'shorter' than the circular muscular coat, the intervening colonic wall is puckered and thrown into *sacculations* (alternatively termed *haustrations*). It is not certain that the sacculation is adequately explained in this manner (*see* p. 1362, and Hamilton 1946; Pace 1968). (5) Small, peritoneum-covered, adipose projections, termed *appendices epiploicae*, are found scattered over the free surface of the whole of the large intestine, with the exceptions of the caecum, the vermiform appendix and the rectum.

In its course the large intestine curves around and usually encloses the convolutions of the small intestine. It commences in the right iliac region, in a dilated part termed the *caecum* (*intestinum crassum caecum*—the term *caecum*, like rectum, duodenum and others, is an adjective, rapidly becoming by linguistic abbreviation a noun). The colon ascends the right lumbar and hypochondriac regions to the inferior surface of the liver; here it bends (the *right colic flexure*, **8**.111) to the left, and, curving with a downward and a forward convexity, passes, as the *transverse colon*, across the abdomen to the left hypochondriac region; it then bends again (the *left colic flexure*, **8**.111), and descends through the left lumbar and iliac regions to the lesser pelvis, where it forms a sinuous loop, the *sigmoid colon* (**8**.125); from this it is continued along the lower part of the posterior wall of the pelvis as the rectum and anal canal. It is divided into the caecum (including the vermiform appendix), the colon, the rectum and the anal canal.

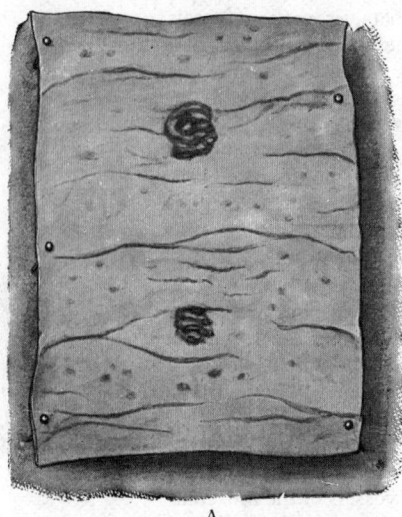

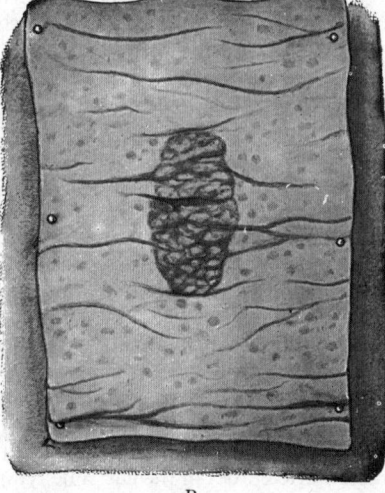

A

B

8.120A and B Aggregated lymphatic follicles, in the proximal (A) and distal (B) parts of the ileum.

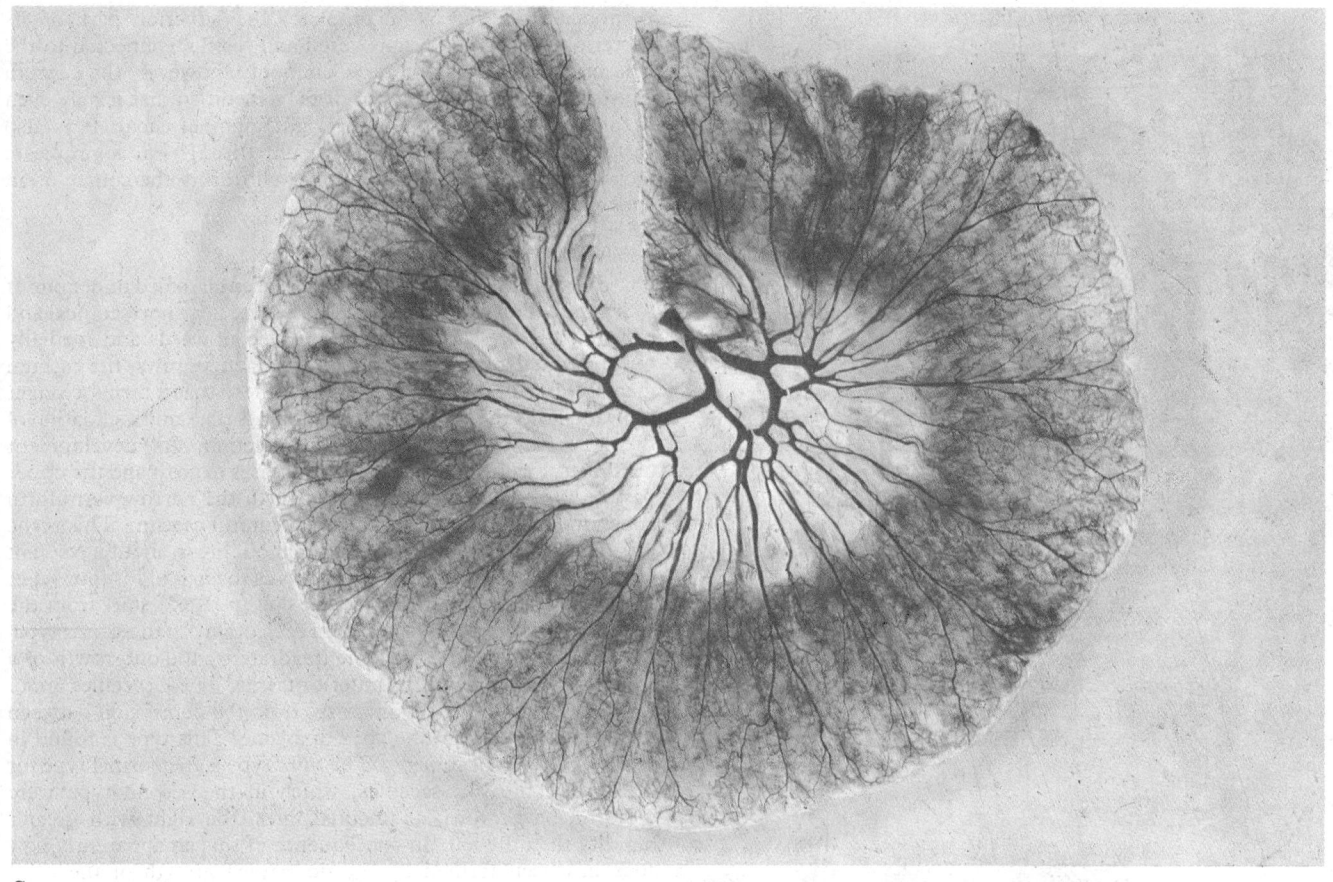

C

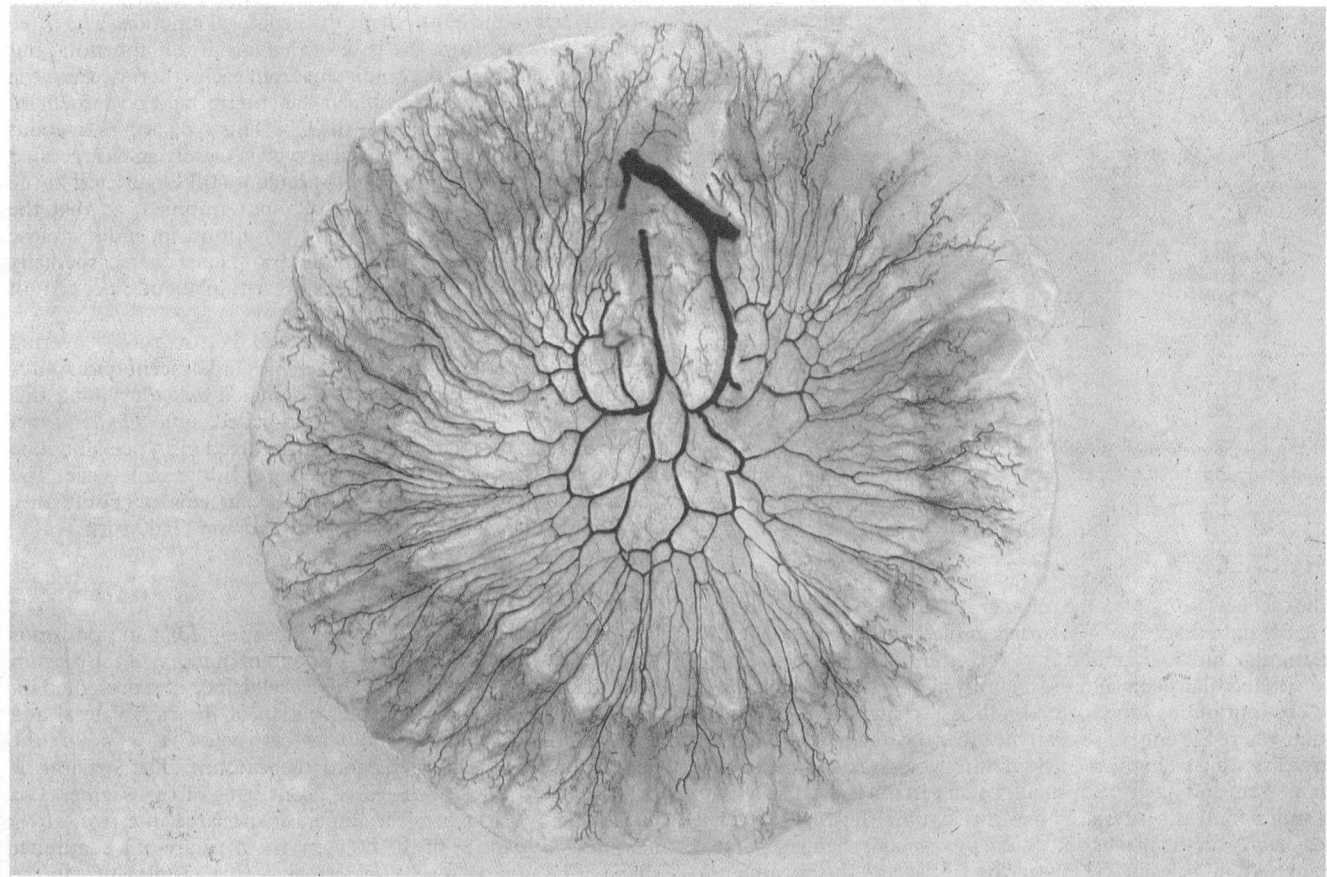

D

8.120C and D   Specimens of the jejunum (C) and ileum (D) from a subject in whom the superior mesenteric artery was injected with a red coloured mass of gelatin before fixation. Subsequently the specimens were dehydrated, and then cleared in benzene followed by methyl salicylate. The largest vessels present are the jejunal and ileal branches of the superior mesenteric artery, and these are succeeded by anastomotic arterial arcades, which are relatively few in number (1–3) in the jejunum, becoming more numerous (5–6) in the ileum. From the arcades, straight arteries pass towards the gut wall; frequently, successive straight arteries are distributed to opposite sides of the gut. Note the denser vacularity of the jejunal wall. (Specimens prepared by Dr. Michael C. E. Hutchinson, Department of Anatomy, Guy's Hospital Medical School.)

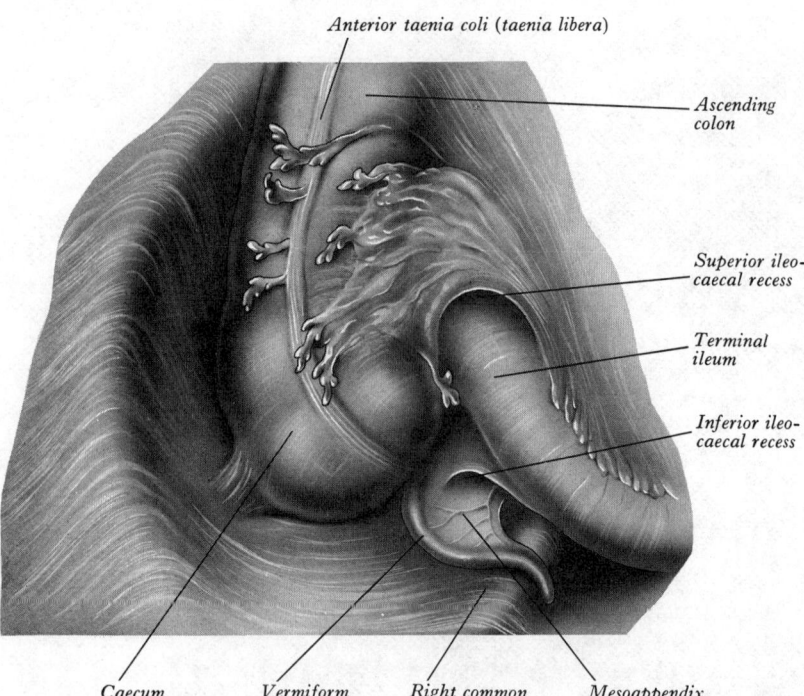

Anterior taenia coli (taenia libera)

Ascending colon

Superior ileo-caecal recess

Terminal ileum

Inferior ileo-caecal recess

Caecum   *Vermiform appendix*   *Right common iliac artery*   *Mesoappendix*

8.121A   The terminal ileum, caecum and vermiform appendix. Anterior aspect.

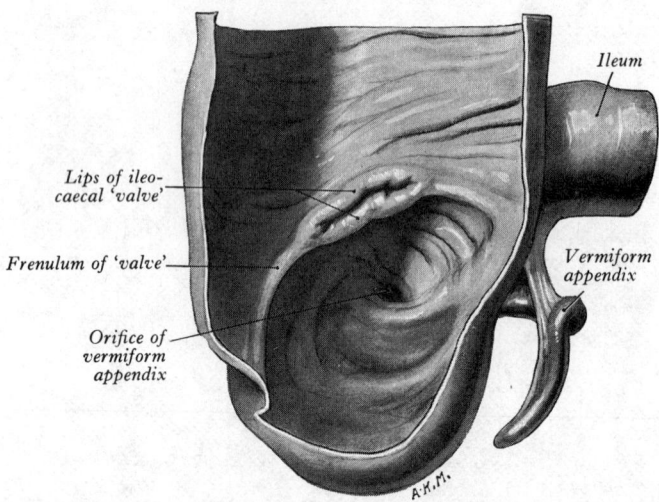

Ileum

Lips of ileo-caecal 'valve'

Frenulum of 'valve'

Vermiform appendix

Orifice of vermiform appendix

8.121B   The interior of the caecum and commencement of the ascending colon, showing the ileocaecal 'valve'. (See text for discussion.)

THE CAECUM

The caecum (**8.**121A, B), the commencement of the large intestine, is in the right iliac fossa—its surface projection hence occupies the triangular area bounded by the right lateral plane, the transtubercular plane and the inguinal ligament. It is a large cul-de-sac continuous superiorly with the ascending colon, and at the point where the one passes into the other the ileum opens into the large intestine from the medial side. Its average axial dimension is about 6 cm and its breadth about 7.5 cm. In the right iliac fossa it is superior to the lateral half of the inguinal ligament: it rests posteriorly on the iliacus and on the psoas major, being separated from both muscles by their covering fasciae and the peritoneum, and posterior to it is the *retrocaecal recess* (p. 1331) which frequently contains the vermiform appendix. In addition, the lateral cutaneous nerve of the thigh intervenes between it and the iliacus. Anteriorly it is usually in contact with the anterior abdominal wall, but the greater omentum, and, if the caecum is empty, some coils of small intestine may be interposed. Usually, it is entirely enveloped by peritoneum, but sometimes the

peritoneal covering is incomplete, the superior part of its posterior surface being uncovered and sessile, connected to the iliac fascia by areolar tissue. Commonly, however, the caecum enjoys a considerable amount of movement, so that it may even become herniated through the right inguinal canal. It is also commonplace to deliver the caecum through an appropriate incision in the anterior abdominal wall during the course of an appendicectomy.

Caecal Variations

The caecum varies in shape, and it has been classified under one of four types (Treves 1885). In early fetal life it is short, conical and broad at the base, with its apex turned upwards and medially towards the ileocaecal junction. As the fetus grows, the caecum increases in length more than in breadth, so that it forms a longer tube and lacks the broad base, but still has the same inclination of the apex towards the ileocaecal junction. As development continues, the lower part of the tube ceases to grow and the upper part becomes increased, so that at birth the narrow vermiform appendix extends from the apex of a conical caecum. This is the *infantile form* and as it persists throughout life in about 2 per cent of subjects, it was regarded by Treves as the *first* of his four types of human caeca. The three taeniae coli (p. 1362) start from the appendix and are equidistant from each other. In the *second* type, the conical caecum has become quadrate by the outgrowth of a saccule on each side of the anterior taenia. These saccules are of equal size, and the appendix arises from the depression between them, instead of from the apex of a cone. This type is found in about 3 per cent of subjects. The *third* type is the normal type for man. Here the two saccules, which in the second type were uniform, have grown at unequal rates, the right with greater rapidity than the left. In consequence of this an apparently new apex has been formed by the downward growth of the right saccule, and the original apex, with the appendix attached, is pushed over to the left towards the ileocaecal junction. The three taeniae still start from the base of the vermiform appendix, but they are now no longer equidistant from each other, because the right saccule has grown between the anterior and posterolateral taeniae, pushing them over to the left. This type occurs in about 90 per cent of subjects. The *fourth* type is merely an exaggerated condition of the third; the right saccule is still larger, and at the same time the left saccule has become atrophied, so that the original apex of the caecum, with the vermiform appendix, is close to the ileocaecal junction, and the anterior taenia courses medially to the same situation. This type is present in about 4 per cent of subjects.

In a more recent study (Pavlov and Pétrov 1968) covering eighty-two males and forty-four females (adolescent and adult), the third type noted above was designated *ampullary*, and this accounted for 78 per cent. A so-called *infundibular* type, approximating to the conical group, occurred in 13 per cent. The remaining 9 per cent were intermediate. In the same series, the caecum was mobile 20 per cent more often in females. (For further developmental analyses consult Balthazar and Gade 1976.)

THE ILEOCAECAL VALVE

The lower end of the ileum opens into the medial and posterior aspect of the large intestine, at the point of junction of the caecum with the colon (**8.**121B). The ileocaecal orifice is represented on the surface at the point of intersection of the right lateral and transtubercular planes; about 2 cm below this point the vermiform appendix opens into the caecum. The opening is provided with a so-called 'valve', consisting of two segments or flaps which project into the lumen of the large intestine. *After inflation and fixation* of the caecum, the flaps are of a semilunar shape. The upper, approximately horizontal, is attached to the line of junction of the ileum with the colon; the lower, which is longer and more concave, is attached to the line of junction of the ileum with the caecum. At the ends of the aperture the two segments of the valve coalesce, and are continued as narrow membranous ridges for a short distance, forming the *frenula* of the valve. The left or anterior end of the aperture is rounded, the right or posterior is narrow and pointed. In the fresh condition, or in specimens which have been hardened *in situ*, the lips of the valve project as thick folds into the lumen of the caecum, and the opening may present the appearance of a slit or may be somewhat oval in shape. The circular and longitudinal muscle coats of the terminal part of the ileum are continued into the valve and form a sphincter. It must be added that direct observation of the living ileocaecal 'valve' does not corroborate this description (Rosenberg and DiDio 1969); in nine cases studied (through a caecostomy) the ileal projection was papillary in shape. Moreover, radiological evidence contradicts the concept of an effective valve at this junction.

It may be added here that accumulations of circular fibres, sometimes described as sphincters, have been observed at various levels in all parts of the colon, and a large literature has developed on this topic (*see* DiDio and Anderson 1968; Rosenberg and DiDio 1969). The functional reality of most of these entities has not been established. It should be noted that all such sphincteric mechanisms must be balanced by an opposite, dilatatory activity.

The margin of the ileocaecal valve is formed by a reduplication of the mucous membrane and of the circular muscular fibres of the intestine. The longtitudinal fibres are partly reduplicated as they extend into the valve (Jit and Singh 1956), though the more superficial longitudinal elements and the peritoneum are continued uninterruptedly from the small to the large intestine.

The surfaces of the valve directed towards the ileum are covered with villi and present the characteristic structure of the mucous membrane of the small intestine; while those turned towards the large intestine are destitute of villi and marked with the orifices of the numerous tubular glands peculiar to the mucous membrane of the large intestine. It is usually considered that the valve not only prevents reflux from the caecum into the ileum, but in all probability it also acts as a sphincter at the end of the ileum and prevents the contents of the ileum from passing too quickly into the caecum; the valve is kept in a condition of tonic contraction by impulses which reach it through the sympathetic nerves. The taking of food into the stomach initiates contraction of the duodenum and the rest of the small intestine, followed by the passage of ileal contents into the large intestine through the ileocaecal opening (the so-called gastro-ileal reflex).

THE VERMIFORM APPENDIX

The vermiform appendix (**8.**121A, B) is a narrow, worm-shaped tube, which springs from the posteromedial wall of the caecum, 2 cm or less below the end of the ileum, and may occupy one of several positions: (*a*) it may lie behind the caecum and the lower part of the ascending colon (*retrocaecal and retrocolic*); (*b*) it may descend over the brim of the lesser pelvis (*pelvic* or *descending*), in which case it lies in close relation to the right uterine tube and ovary in the female; (*c*) it may lie below the caecum (*subcaecal*); (*d*) it may lie in front of the terminal part of the ileum and may then be in contact with the anterior abdominal wall; or (*e*) it may lie behind the terminal part of the ileum. In a study of 10,000 subjects (Wakeley 1933), the vermiform appendix was retrocaecal and retrocolic in 65·28 per cent, pelvic in 31·01 per cent, subcaecal in 2·26 per cent, pre-ileal in 1·0 per cent, and post-ileal in 0·4 per cent. Although these classical figures were based upon a very large series, subsequent literature, both anatomical and surgical, shows much contradiction. Buschard and Kjaeldgaard (1973), reporting upon a short series (234 autopsies), have compared the results of a number of studies (from 1885 to 1973), of which Wakeley's remains by far the largest. They classify all positions into two groups: *Anterior* (=pelvic and ileocaecal) and *Posterior* (retrocaecal and subcaecal). On this basis all but three of the eleven series quoted found the *anterior* positions more frequent. Like Wakeley these observers found *posterior* positions more commonly in their own Danish series, while in their German autopsies, the finding was reversed. Collins (1932) in the second largest series (4680), returned percentages which are the reverse of Wakeley's, the partition between anterior and posterior being 78·5 per cent and 21·5 per cent (Collins) and 32·4 per cent and 67·6 per cent (Wakeley). In view of these and other less contrasted disagreements, it is scarcely useful to continue to quote such figures. Perhaps the different observers involved have used different criteria in their examinations; perhaps, even, there are population divergencies. For the present, however, positional percentages for the vermiform appendix must be regarded as unreliable. The surface marking most used for the base of the appendix is the junction of the lateral and middle thirds of the line joining the right anterior superior iliac spine to the umbilicus (*McBurney's point*). (The latter is a useful surgical approximation, but as noted, considerable variation may occur). The three taeniae coli on the ascending colon and caecum converge on the base of the appendix, where they merge into its longitudinal muscular layer. The anterior taenia of the caecum is generally distinct and can be easily traced to the root of the appendix, thus affording a ready guide to it. The appendix varies from 2 to 20 cm in length, the average being about 9 cm. It is longer in the child than in the adult and may atrophy and become smaller after mid-adult life. It is connected by a short *mesoappendix* to the lower part of the mesentery of the ileum. This fold, in the majority of cases, is more or less triangular, and as a rule extends along the entire length of the tube. The main artery to the appendix, a branch of the lower division of the ileocolic artery (p. 716), runs behind the terminal part of the ileum and enters the mesoappendix a short distance from the base of the appendix. Here it gives off a recurrent branch which anastomoses at the base of the appendix with a branch of the posterior caecal artery, the anastomosis sometimes being of considerable size. The main appendicular artery runs towards the tip of the appendix, lying at first near to and afterwards in the free border of the mesoappendix. The terminal part of the artery, however, lies actually on the wall of the appendix and may become

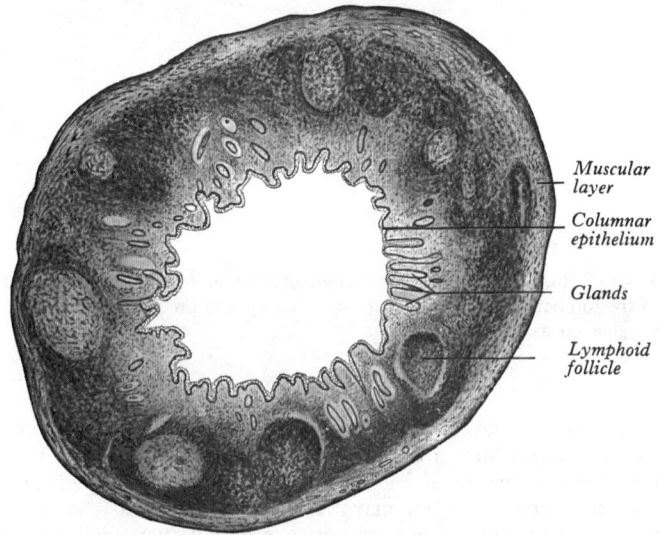

Muscular layer

Columnar epithelium

Glands

Lymphoid follicle

8.122   A transverse section of human vermiform appendix. Magnification about × 20.

thrombosed in inflammation of the appendix, which may result in gangrene or necrosis of its distal part. However, the arterial supply of the vermiform appendix may vary considerably. Numbers of accessory arteries are common; in 80 per cent of individuals there are two or more arteries of supply (Solanke 1968). The canal of the vermiform appendix is small, and communicates with the caecum by an orifice which is placed below and a little behind the ileocaecal opening. The orifice is sometimes guarded by a semilunar valve formed by a fold of mucous membrane. The lumen of the appendix may be partially or completely obliterated after mid-adult life. In view of its rich blood supply and histological differentiation, the vermiform appendix is probably more correctly regarded as a specialized than as a degenerate, vestigial structure. The configuration of the caecum and appendix in man and the anthropoid apes is probably less primitive than in the monkeys.

Structure. The layers of the vermiform appendix are the same as those of the intestine: serous, muscular, sub-mucous and mucous. The *serous* coat forms a complete investment for the tube, except along the narrow line of attachment of its mesentery.

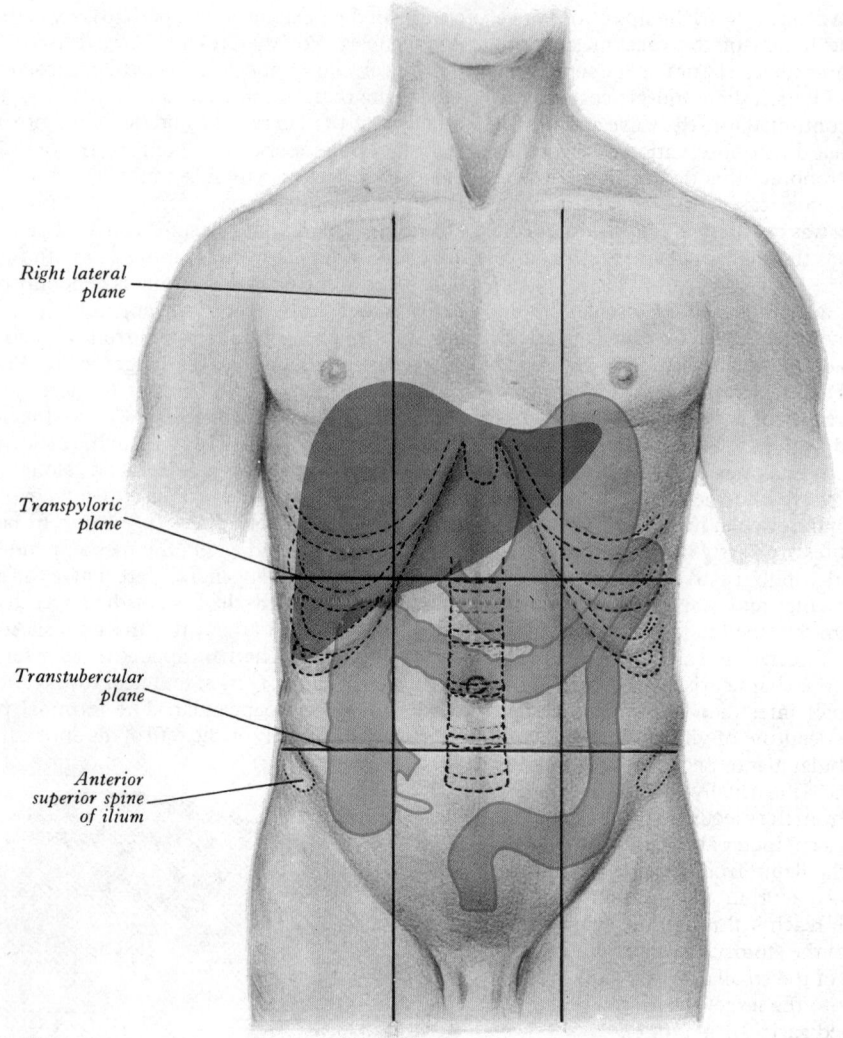

Right lateral
plane

Transpyloric
plane

Transtubercular
plane

Anterior
superior spine
of ilium

8.123A   Surface projection of the stomach, liver and colon. The outlines of the lumbar vertebral bodies, lower ribs, xiphoid process, and parts of the iliac crests are indicated.

Beneath it lies a layer of subserous areolar tissue. The *longitudinal muscular fibres* form a uniformly thick layer which invests the whole organ, except at one or two points where the longitudinal and circular layers may be both deficient, so that the peritoneal and submucous coats are contiguous over small areas. At the base of the appendix, the longitudinal muscle becomes thickened around the perimeter, to form incipient taeniae coli which becomes continuous with those of the caecum and colon. The *circular muscular fibres* form a thicker layer than the longitudinal fibres, and are separated from them by a small amount of connective tissue. The *submucous layer* is well developed, and contains a large number of masses of lymphoid tissue which cause the mucous membrane to bulge into the lumen and so render the latter of small size and irregular shape. The *mucous membrane* is covered by columnar epitheliocytes and attenuated antigen-transporting membrane or M cells (Owen and Nemanic 1978). Glands are few in number and penetrate deeply amongst the lymphoid tissue (**8**.122). In the normal human appendix the lymphoid tissue lies in the lamina propria and in the submucosa, where follicular and parafollicular zones can be distinguished; clusters of lymphocytes or immunoblasts also lie within or between surface epithelial cells where they possibly mature or differentiate into plasma cells (Gorgollón 1978). The lymphoid tissue of the lamina propria contains many plasma cells, together with lymphocytes, acidophilic leucocytes, mast cells and macrophages, all embedded in a fibrocellular reticulum. The submucosal follicles (germinal centres) contain immunoblasts, lymphocytes, macrophages, plasma cells and dendritic reticular cells, the last two being most abundant in the central regions of the

follicles; cells similar to the dendritic reticular cells have been found in the human thymus (Kaiserling *et al.* 1974). The parafollicular zones are distinguished by aggregations of small lymphocytes, a scarcity of plasma cells, and by the presence of post-capillary venules lined by a tall endothelium through which lymphocytes may migrate. These endothelial cells have a surface coat of immunoglobulins which may be involved in the control of lymphocyte recirculation (Sordat *et al.* 1971). The *lymphoid masses* provide a local defence against infection, and have also been suggested as a possible homologue of the *bursa of Fabricius* in birds (p. 61) which is concerned with the acquisition of immunological competence by certain lymphocytes. Experimental evidence is, however, lacking. In many mammals, particularly herbivores, the caecum and associated appendix are large, and form an important site for the digestion of cellulose by symbiotic bacteria.

THE COLON

The colon may be considered in four parts—the ascending, transverse, descending and sigmoid.

The ascending colon, about 15 cm long, is narrower than the caecum. It begins at the caecum, and ascends to the inferior surface of the right lobe of the liver, where it is lodged in a shallow colic depression; here it bends abruptly forwards and to the left, forming the *right colic flexure* (**8**.111A). In surface projection it ascends to the right of the right lateral plane (**8**.123A), from the transtubercular plane to midway between the subcostal and transpyloric planes. It is surrounded by peritoneum except where its posterior surface is connected by areolar tissue to the fascia over the iliacus, iliolumbar ligament, quadratus lumborum and the aponeurotic origin of the transversus abdominis, and to the perirenal fascia in front of the inferolateral part of the right kidney. The lateral cutaneous nerve of the thigh, usually the fourth lumbar artery and, sometimes, the ilio-inguinal and iliohypogastric nerves cross behind it. Sometimes it is almost completely invested with peritoneum, thus possessing a distinct but narrow mesocolon. In a series of 100, 52 per cent had neither an ascending nor a descending mesocolon, 14 per cent had both, 12 per cent an ascending, and 22 per cent a descending mesocolon (Treves 1885). It is in relation, anteriorly, with the convolutions of the ileum, the right edge of the greater omentum and the abdominal wall.

The right colic flexure comprises the terminal part of the ascending colon and the commencement of the transverse colon, which turns downwards, forwards and to the left. Behind, it is in relation with the lower and lateral part of the anterior surface of the right kidney. Above and anterolaterally, it is related to the right lobe of the liver; anteromedially, to the descending part of the duodenum and the fundus of the gall bladder. It is not covered by peritoneum on its posterior surface, so that this surface is in direct contact with the renal fascia. The flexure is not so acute as the left colic flexure.

The transverse colon (8.103), about 50 cm long, begins at the right colic flexure, placed in the right lumbar region, passes across the abdomen into the left hypochondriac region, and here curves sharply on itself, downwards and backwards, beneath the lateral end of the spleen, forming the *left colic flexure*. In its course across the abdomen it describes an arch, the concavity of which is usually directed backwards and upwards; towards its splenic end there is often an abrupt U-shaped curve which may descend lower than the main curve. Its surface projection (8.123) is drawn from a point, situated immediately lateral to the right lateral plane and midway between the subcostal and transpyloric planes, to the umbilicus, and then upwards and to the left to a point a little above and lateral to the intersection of the left lateral and transpyloric planes. The precise position occupied by the transverse colon is difficult to define, for it not only shows individual variation but its position varies in the same individual from time to time. Very commonly it lies in the lower umbilical or upper hypogastric region. It may be higher but frequently descends in a V-shaped manner, the apex of the V reaching well below the level of the iliac crests (*see* p. 1321). In a radiological assessment of the position of the transverse colon in the living (upright position), the level of the lowest part of the tube in 1,000 young adults varied greatly, even descending into the true pelvis. Moreover, the level varied as much as 17 cm in the same individual between the upright and recumbent positions (Moody 1927).

The posterior surface of its right extremity is devoid of peritoneum, and is attached by areolar tissue to the front of the descending part of the duodenum and the head of the pancreas. Between the head of the pancreas and the left colic flexure, the transverse colon is almost completely invested by peritoneum, and is connected to the anterior border of the body of pancreas by the *transverse mesocolon*. It is in relation, by its upper surface, with the liver and gall bladder, the greater curvature of the stomach, and the lateral end of the spleen; by its under surface, with the small intestine; by its anterior surface with the posterior layers of the greater omentum; its posterior surface is in relation with the descending part of the duodenum, the head of the pancreas, the upper end of the mesentery, the duodenojejunal flexure and some of the coils of the jejunum and ileum.

The left colic flexure (8.111) is the junction of the transverse and descending sections of the colon in the left hypochondriac region, and is in relation with the lower part of the spleen and the tail of the pancreas above, and with the anterior aspect of the left kidney medially; the flexure is so acute that the end of the transverse colon usually lies in contact with the front of the descending colon. The left colic flexure lies at a higher level than, and on a plane posterior to, the right colic flexure, and is attached to the diaphragm, opposite the tenth and eleventh ribs, by a peritoneal fold, named the *phrenicocolic ligament,* which lies below the lateral end of the spleen (p. 1330).

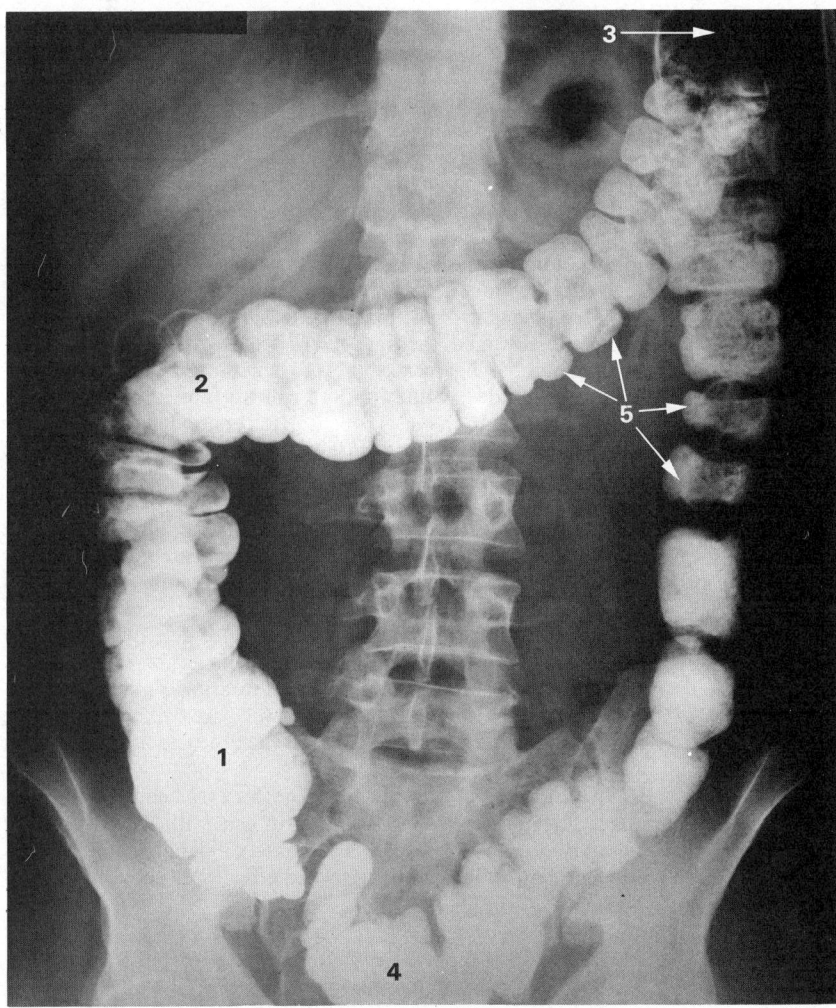

8.123B A radiograph of the abdomen after the administration of a barium enema which has filled the whole of the large intestine as far as the caecum and ileocaecal valve. (1) the caecum; (2) the right or hepatic flexure of the colon, which is much inferior to (3) the left or splenic flexure of the colon; (4) the sigmoid colon; and (5) the sacculations, or haustrations, which are clearly visible throughout most of the colon.

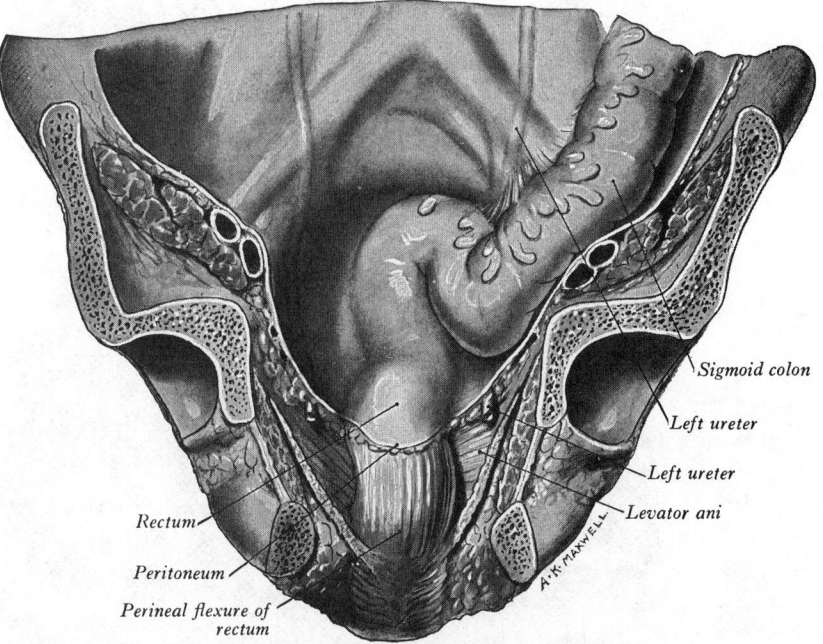

8.123C Oblique coronal section through the pelvis to expose the anterior aspect of the rectum.

The descending colon (8.111A), about 25cm long, passes downwards through the left hypochondriac and lumbar regions. At first it follows the lower part of the lateral border of the left kidney and then, at the lower pole of that organ, it descends, in the angle between psoas major and quadratus lumborum, to the crest of the ilium; it then curves downwards and medially in front of the iliacus and psoas major, and ends in the sigmoid colon at the inlet of the lesser pelvis. (The descending colon is sometimes described as ending at the level of the iliac crest, the part between that level and the inlet of the true pelvis being named the *iliac colon*.) In surface projection (8.123A) it passes downwards, just lateral to the left lateral plane, from a point situated a little above and to the left of the intersection of the transpyloric and left lateral planes, as far as the inguinal ligament. The peritoneum covers its anterior surface and sides, while its posterior surface is connected by areolar tissue with the fascia over the lower and lateral part of the left kidney, the aponeurotic origin of the transversus abdominis, the quadratus lumborum, the iliacus and the psoas major (8.111A). Numerous structures cross behind it. They include: the subcostal vessels and nerve, the iliohypogastric and ilio-inguinal nerves, the fourth lumbar artery (as a rule), the lateral femoral cutaneous, femoral and genitofemoral nerves, the testicular (or ovarian) vessels and the external iliac artery, all of the left side. The descending colon is smaller in calibre, more deeply placed, and more frequently covered with peritoneum on its posterior surface, than the ascending colon (p. 1355). Anteriorly it is related to coils of the jejunum, except in its lower part, which can be felt through the anterior abdominal wall when the abdominal muscles are relaxed.

The sigmoid colon (pelvic colon) (8.123C) begins at the inlet of the lesser pelvis, where it is continuous with the descending colon; it forms a loop which varies greatly in length, but averages about 40 cm, and normally lies within the lesser pelvis. The loop

consists of three parts; the first part descends in contact with the left pelvic wall; the second crosses the pelvic cavity, between rectum and bladder in the male, and rectum and uterus in the female, and may come into contact with the right pelvic wall; the third arches backwards and reaches the median plane at the level of the third piece of the sacrum, where it bends downward and ends in the rectum. The sigmoid colon is closely surrounded by peritoneum, which forms a mesentery, the *sigmoid mesocolon* (p. 1330); this diminishes in length from the centre towards the ends of the loop, where it disappears, so that the loop is fixed at its junctions with the descending colon and rectum, but enjoys a considerable range of movement in its central region. Its relations are therefore subject to considerable variation. *Laterally* it is related to the external iliac vessels, the obturator nerve, the ovary (in the female), the ductus deferens (in the male) and the lateral pelvic wall. *Posteriorly* it is related to the internal iliac vessels, the ureter, the piriformis and the sacral plexus, all of the left side. *Inferiorly* it rests on the bladder, in the male, and on the uterus and bladder, in the female. *Above* and on its *right* side, it is in contact with the terminal coils of the ileum.

The position and shape of the sigmoid colon vary very much, and depend on (*a*) its length; (*b*) the length and freedom of its mesocolon; (*c*) the condition of distension; when distended it rises out of the lesser pelvis into the abdominal cavity, and when empty it sinks again into the pelvis; (*d*) the condition of the rectum and bladder (and the uterus, in the female); when these organs are distended the sigmoid colon tends to rise, and conversely. Racial variation in the size of the sigmoid colon has been noted (Lisowski 1969); in some groups—and particularly in Ethiopians—the incidence of a suprapelvic loop, which may be conducive to volvulus, is particularly high.

THE RECTUM

The rectum (8.123C–124C) is continuous with the sigmoid colon at the level of the third sacral vertebra, the junction being indicated by the lower end of the sigmoid mesocolon. From its origin it descends, following the concavity of the sacrum and coccyx, forming an anteroposterior curve known as the *sacral flexure* of the rectum. It thus passes at first downwards and backwards, then downwards, and finally downwards and forwards to become continuous with the anal canal by passing through the pelvic diaphragm (p. 560). The *anorectal junction* is situated 2 to 3 cm in front of, and slightly below the tip of the coccyx; from this level, which in the male is opposite the apex of the prostate, the anal canal passes downwards and backwards from the lower end of the rectum, the backward bend of the gut at the anorectal junction being termed the *perineal flexure* of the rectum. In addition to its anteroposterior curve, the rectum deviates from the midline in the form of three lateral curves; the upper one is convex to the right, the middle one (which is the most prominent) bulges to the left, and the lower one is convex to the right; the beginning and end of the rectum are in the median plane (8.124A).

The rectum is about 12 cm long and its upper part has the same diameter as the sigmoid colon (about 4 cm in the empty state), but its lower part is dilated to form the *rectal ampulla*. The rectum differs from the sigmoid colon in that it has no sacculations, appendices epiploicae or mesentery, while the taeniae coli blend about 5 cm above the junction of the rectum and sigmoid colon to form two wide muscular bands which descend, one in the anterior and the other in the posterior wall of the rectum. The peritoneum is related only to the upper two-thirds of the rectum, covering at first its front and sides, but lower down its front only; from the latter it is reflected on to the bladder in the male, forming the rectovesical pouch of peritoneum, and on to the posterior wall of the vagina in the female, forming the recto-uterine pouch. The level of peritoneal reflexion is higher in the male, the rectovesical pouch being about 7·5 cm from the anus (the height to which the index finger inserted through the anus can reach); in the female the recto-uterine pouch is about 5·5 cm from the anus. In the male fetus the peritoneum extends on the front of the rectum as far as the lower end of the prostate (*see* p. 1420). On the sigmoid colon, the peritoneum is firmly attached to the muscle coat by fibrous

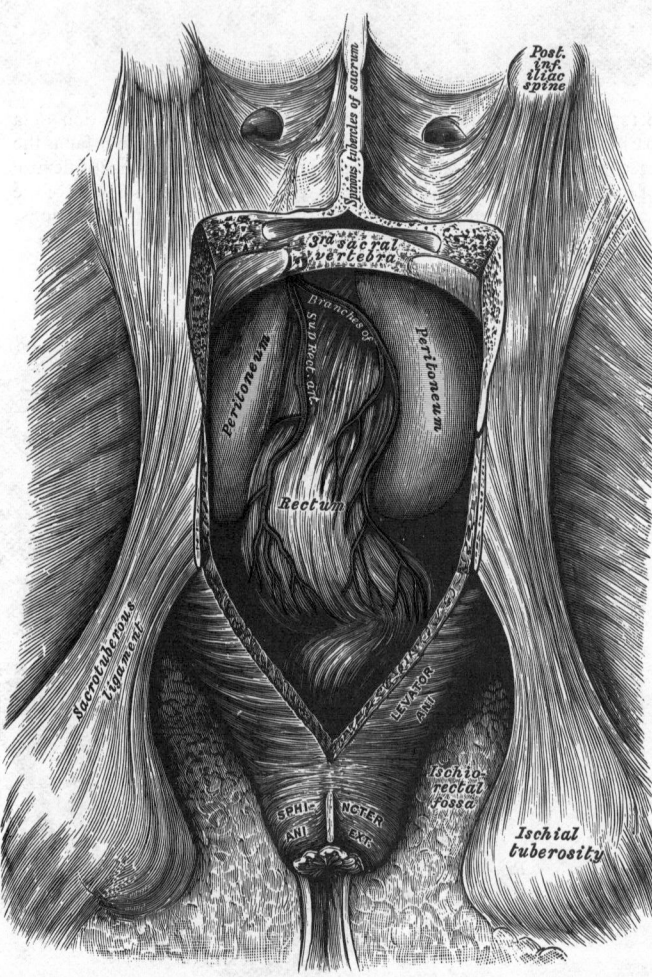

8.124A Posterior aspect of the rectum exposed by removal of the lower part of the sacrum and coccyx. Note superior rectal artery (red) and peritoneum of the pararectal fossae (blue).

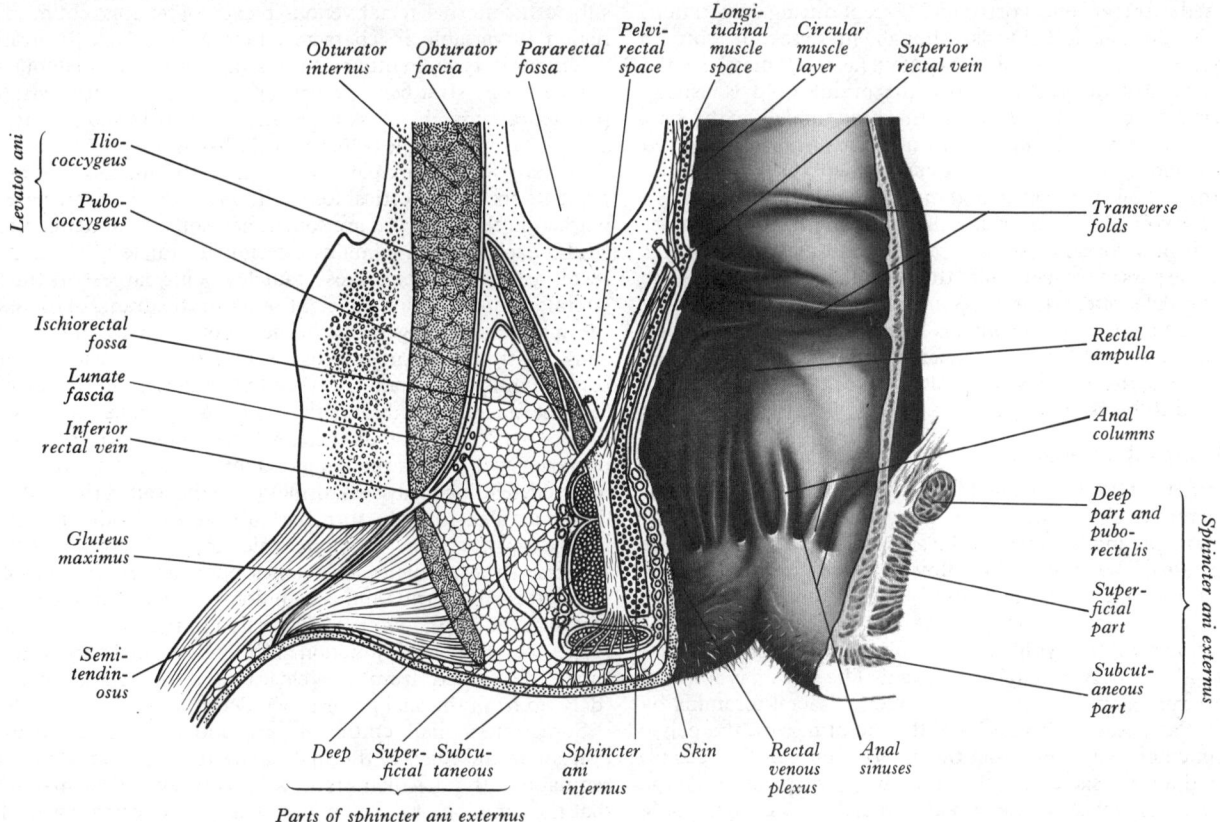

8.124B Diagram of a coronal section of the rectum and anal canal and the adjacent structures. (Adapted from Rauber-Kopsch, *Lehrbuch und Atlas der Anatomie des Menschen*, 1929.) The internal pudendal vessels, the dorsal nerve of the penis and the perineal nerve are shown transected in the lateral wall of the ischiorectal fossa, where they are traversing the 'lunate fascia' (pudendal canal).

connective tissue; but as it descends on the rectum, the peritoneum becomes more loosely attached to the muscle by fatty areolar tissue, thus allowing considerable expansion of this part of the gut.

In the empty state of the rectum, the mucous membrane of its lower part presents a number of longitudinal folds which are effaced by distension of the rectum. Besides these, there are permanent *transverse* or *horizontal folds* of a semilunar shape, which are most marked when the rectum is distended. Two forms of horizontal folds have been recognized (Jit 1961). One, consisting of mucous membrane, the circular muscle coat, and part of the longitudinal muscle coat, is marked on the outer surface of the rectum by an indentation. The other form is devoid of longitudinal muscle coat fibres, and bears no external surface marking. Commonly three folds are present, but their number is variable. The upper one is situated near the commencement of the

rectum and may be on the left or the right side; occasionally it may encircle the gut and the lumen of the gut is then somewhat constricted at this site. The middle fold is the largest and most constant, and is situated immediately above the ampulla of the rectum; it projects from the anterior and right walls of the rectum just below the level at which the peritoneum is reflected from the anterior surface of the rectum; the circular muscle in this fold is more marked than in the others. The lowest fold is inconstant and lies on the left side, about 2·5 cm below the middle fold; sometimes a fourth fold is present on the left side, about 2·5 cm above the middle fold described above.

It has been suggested (Paterson 1912) that the rectum consists functionally of two parts, one above and the other below the middle fold, the upper part containing faeces and being free to distend towards the peritoneal cavity, while the lower part occupies a more confined situation, enclosed in a tube of

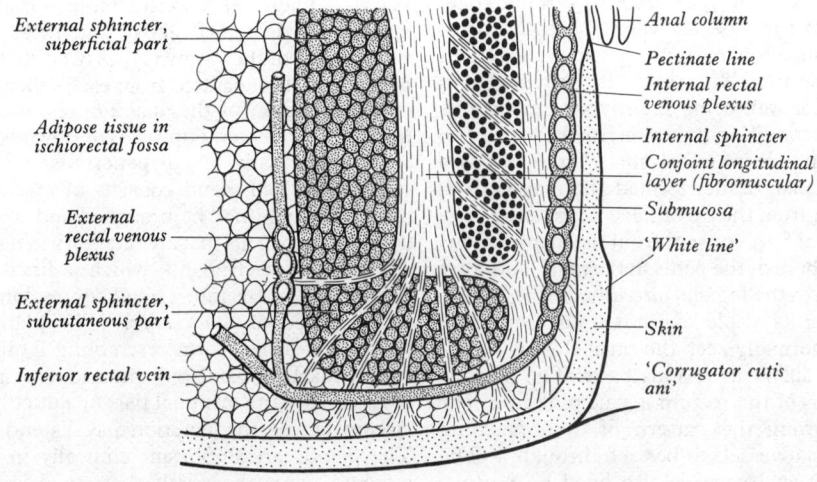

8.124C Part of **8.124B**, enlarged to show greater detail.

condensed extraperitoneal tissue and (except during defaecation) empty in normal individuals, though in cases of chronic constipation, or after death, it may contain faeces. It may be noted that the part of the rectum above the middle fold is usually considered to be developed from the hindgut while the part below, together with the upper part of the anal canal, is derived from the cloaca (post-allantoic gut). Others (O'Beirne 1833; Hurst 1919), however, considered that the sigmoid colon acts as a faecal reservoir, that in normal individuals the rectum is empty (though it may contain faeces in cases of chronic constipation), and that the passage of faeces into the rectum normally excites the desire to defaecate. It has been shown experimentally, by registering the pressure in balloons inserted into the rectum and anal canal, that considerable distension of the rectum results in a desire to defaecate and relaxation of the anal sphincters (Denny-Brown and Robertson 1935).

Relations of the Rectum

Posteriorly the rectum is related in the median plane to the lower three sacral vertebrae, the coccyx, the median sacral vessels, the ganglion impar and branches of the superior rectal vessels, while on each side of the midline the following structures, particularly those of the left side, lie behind the rectum: namely, the piriformis, the anterior rami of the lower three sacral and coccygeal nerves, the sympathetic trunk, the lower lateral sacral vessels and the coccygei and levatores ani. The rectum is attached to the sacrum along the lines of the anterior sacral foramina by fibro-areolar tissue which encloses the sacral nerves, the pelvic splanchnic nerves passing from the anterior rami of the second, third (and fourth) sacral nerves to join the pelvic plexuses on the rectal wall, branches of the superior rectal vessels, lymphatic vessels and lymph nodes, and loose perirectal fat. *Anteriorly*, the relations of the rectum differ in the two sexes. In the male, above the site of reflexion of the peritoneum from the rectum, the upper parts of the base of the bladder and of the seminal vesicles, and the rectovesical pouch of the peritoneum and its contents (terminal coils of the ileum and the sigmoid colon) are in front of the rectum; while below the peritoneal reflexion the rectum is related anteriorly to the lower parts of the base of the bladder and of the seminal vesicles, the deferent ducts, the terminal parts of the ureters and the prostate. In the female, above the peritoneal reflexion, the uterus, the upper part of the vagina and the recto-uterine pouch of the peritoneum and its contents (terminal coils of the ileum and the sigmoid colon) lie in front of the rectum, while below the peritoneal reflexion the rectum is related anteriorly to the lower part of the vagina. *Laterally*, the upper part of the rectum is related to the pararectal fossa of the peritoneum and its contents (sigmoid colon or lower part of the ileum), while below the peritoneal reflexion the pelvic sympathetic plexuses, the coccygei and levatores ani, and branches of the superior rectal vessels constitute its immediate lateral relations.

THE ANAL CANAL

The anal canal (*see* Milligan *et al.* 1937; Gabriel 1945; Wilde 1949; Goligher *et al.* 1955; Fowler 1957) begins where the lower end of the ampulla of the rectum suddenly narrows, passing downwards and backwards to end at the anus (**8**.124B, C). It is about 3.8 cm long in the adult, its anterior wall being slightly shorter than its posterior, and in the empty condition its lumen has the form of an anteroposterior or triradiate longitudinal slit. *Posteriorly*, is a mass of fibrous and muscular tissue, termed the *anococcygeal ligament*, which separates it from the tip of the coccyx; *anteriorly*, it is separated by the *perineal body* (p. 562) from the membranous part of the urethra and the bulb of the penis in the male, and from the lower end of the vagina in the female; *laterally*, it is related to the ischiorectal fossae. Over its whole length it is surrounded by sphincter muscles, which normally keep the canal closed.

The lining of the canal differs in various parts. The mucous membrane of the lower part of the rectum is pale pink in colour and semitransparent, allowing the pattern of the branching radicles of the superior rectal vessels to be seen through it. The upper half (15 mm) of the anal canal is also lined by mucous membrane and is plum-coloured owing to the blood in the subjacent internal rectal venous plexus. The epithelium in this region is variable in character; in some cases it is stratified columnar in type, in others it is mainly stratified squamous with patches of stratified columnar, together with stratified polyhedral cells (the cells nearest the lumen being columnar), and a single layer of simple columnar cells like those lining the rectum (Walls 1958). In this part of the canal the mucous membrane presents six to ten vertical folds, the *anal columns*, which are well marked in the child but are sometimes not so well defined in the adult. Each column contains a terminal radicle of the superior rectal artery and vein, these radicles being largest in the left-lateral, right-posterior and right-anterior quadrants of the wall of the anal canal; enlargements of the venous radicles in these three sites constitute primary internal haemorrhoids. The lower ends of the columns are joined together by small crescentic valve-like folds of mucous membrane, the *anal valves*, above each of which lies a small recess or *anal sinus*. The sinuses, deepest on the posterior wall of the canal, may retain faecal matter and become infected, leading to abscess formation in the wall of the anal canal; the anal valves may be torn by hard faeces, producing an anal fissure (p. 1361). The line along which the anal valves are situated is termed the *pectinate line*; it lies opposite the middle of the sphincter ani internus and is commonly considered to be the site at which the anal membrane is situated in the early fetus; thus it represents the place of junction of the endodermal part of the anal canal (developed from the cloaca) and the ectodermal part (derived from the anal pit or proctodeum).

Sometimes small epithelial projections (anal papillae) are present on the edges of the anal valves; they are considered to be remnants of the anal membrane. It has, however, been maintained that the junction of the ectodermal and endodermal parts of the anal canal is situated lower down, at the lower border of the pecten (Johnson 1914). The succeeding part of the anal canal extends for about 15 mm below the anal valves, and is known as the *transitional zone* or *pecten*. Its epithelium is stratified and is intermediate in thickness between the epithelium lining the upper part of the canal and the skin lining the lower part; unlike the latter, it contains no sweat glands. The transitional zone also overlies part of the internal rectal venous plexus and is shiny and bluish in appearance. Its submucosa contains fairly dense connective tissue, in contrast with the lax connective tissue of the upper half of the anal canal, suggesting a firm support and anchorage of the lining of the pecten to the surrounding muscle coats of this part of the anal canal. The transitional zone ends below at a narrow wavy zone, commonly called the '*white line*' (of Hilton); in the living subject this 'line' is bluish pink in colour and is only rarely recognizable macroscopically (Ewing 1954). Its only interest lies in the fact that it is situated at the level of the interval between the subcutaneous part of the external sphincter and the lower border of the internal sphincter, and on digital examination of the anal canal in the living subject an *anal intersphincteric groove* can be felt at this site. Below the white line, the lower 8 mm or so of the anal canal are lined by true skin which may be dull white or brownish in colour and contains sweat glands and sebaceous glands. There are considerable variations in the zones of epithelium described above and there is frequently an inter-penetration of the various types of epithelia, so that the zones may not be rigidly separated from each other.

In the region of the anal sinuses, *anal glands* (Fowler 1957; McColl 1967) extend upwards or downwards into the submucosa and occasionally even penetrate deeply into the internal sphincter. Each gland consists of one to six spiral or straight tubules, which may be branched and which are lined by two or three layers of cells that are secretory in nature and contain mucin. The duct of each gland, which is lined by stratified columnar epithelium, opens into a small depression of the lining of the anal canal, called an *anal crypt*. The glands are surrounded by lymphocytes in a form resembling lymphatic follicles, and the submucosal nonstriated muscle is thick in their vicinity. Occasionally the terminal part of a duct is not canalized; in these circumstances, the secretion may distend the gland to form a cyst. The glands are important clinically in that they may become infected, with the result that an abscess or a fistula may be produced. The glands vary widely in number and in their depth of penetration, and they may even extend in the submucosa above the anorectal junction. (For a detailed and extensive study of the comparative anatomy and pathology of the anal glands, consult McColl 1967. In this series, fifty normal anal canals were examined, and half of these had anal glands passing right through the internal sphincter; the average number of such extensions was four, but they may number up to sixteen. This author considers that the human glands are not homologous with the *anal scent glands* of some other mammals.)

ANAL MUSCULATURE

The walls of the anal canal are surrounded by a complex of muscular sphincters (**8**.124), divided into internal and external parts. At the anorectal junction the circular muscle coat of the rectum becomes considerably thickened (5 to 8 mm) to form the **sphincter ani internus**. This nonstriated sphincter surrounds the upper three-quarters (30 mm) of the anal canal and ends below at the level of the white line (*vide supra et infra*).

The sphincter ani externus surrounds the whole length of the anal canal; it is usually described as consisting of three parts, each composed of striated muscle. The *subcutaneous part* of the external sphincter is a flat band, about 15 mm broad, which surrounds the lower part of the anal canal and lies horizontally below the lower borders of the internal sphincter and of the superficial part of the external sphincter; it lies beneath the skin at the anal orifice and, centrally, it is subcutaneous below the white line in the canal. Anteriorly a few fibres are attached to the perineal body (or the superficial transverse perineal muscles), and posteriorly some fibres are usually attached to the anococcygeal ligament. The *superficial part* of the external sphincter is elliptical and lies deep to the subcutaneous part. It is the only part of the external sphincter that is attached to bone, arising from the posterior surface of the terminal segment of the coccyx by a median fibrous aponeurosis, the anococcygeal raphe; anteriorly, after surrounding the lower part of the internal sphincter, it is inserted chiefly into the perineal body. The *deep part* of the external sphincter is a thick annular band which surrounds the upper part of the internal sphincter; its deeper fibres are fused with and inseparable from those of the puborectalis (p. 561). In front of the anal canal many of the fibres of the deep part of the external sphincter decussate and become continuous with the superficial transverse perineal muscles, this arrangement being more marked in the female. Posteriorly some fibres are usually attached to the anococcygeal ligament. However, it has been asserted by some that there is no clear separation of the three parts of the external sphincter from each other (Goligher *et al.* 1955). In the female, according to Oh and Kark (1972), the muscle is a single band, at least, in its anterior part. (*See* also Wendell-Smith and Wilson 1977).

The tone of both internal and external anal sphincters keeps the anal canal and anus closed; during defaecation these muscles are relaxed and the lower part of the anal canal is opened out and flattened, so that the mucous membrane of the upper part of the canal appears at the surface. The external sphincter can be voluntarily contracted and thus more firmly occlude the anus. The nerve supply of the external sphincter is derived from the inferior rectal branch of the pudendal nerve (S. 2 and 3) and from the perineal branch of the fourth sacral nerve.

At the anorectal junction the pubococcygeal fibres of the levator ani fuse with the longitudinal nonstriated muscle coat of the rectum to form a *conjoint longitudinal coat* for the anal canal, between the internal and external sphincters (**8**.124). Distally, this conjoint coat becomes increasingly fibro-elastic, and at the level of the white line it breaks up into 9 to 12 fibro-elastic septa which spread out fanwise and pass mainly through the subcutaneous part of the external sphincter to become attached to the corium of the skin around the anus. These septa consist mainly of yellow elastic fibres and the most lateral septum passes between the subcutaneous and superficial parts of the external sphincter and becomes lost in the fat of the ischiorectal fossa. It has been maintained that the most medial septum passes between the internal sphincter and the subcutaneous part of the external sphincter to reach the lining of the anal canal at the white line,

being called the *anal intermuscular septum*, and that it produces the *anal intersphincteric groove* referred to above. Other investigators (Wilde 1949; Goligher *et al.* 1955; Fowler 1957), however, point out that the longitudinal fibres in this position, compared with those penetrating through the subcutaneous part of the external sphincter, are too weak and scanty to warrant the name of anal intermuscular septum, and maintain that the intersphincteric groove is caused by the mass of muscle on each side of it (internal sphincter above, subcutaneous external sphincter below), as well as by contraction of the subcutaneous external sphincter. In the submucosa of the anal canal below the level of the anal sinuses there is a well-marked layer composed of nonstriated muscle, yellow elastic fibres and connective tissue. These fibres are derived mainly from strands of the conjoint longitudinal coat of the anal canal which pass inwards and downwards between bundles of the internal sphincter (**8**.124C). Some of the strands end inferiorly by passing laterally around the lower edge of the internal sphincter to join the main longitudinal coat. The majority, however, continue downwards and laterally, superficial to the subcutaneous part of the external sphincter, to be attached to the corium of the skin from the level of the white line to well beyond the anus. These fibres produce the corrugation of the skin characteristic of this region and constitute the so-called *corrugator cutis ani muscle*. It has been claimed (Wilde 1949) that these fibres are exclusively yellow elastic fibres, although others (Goligher *et al.* 1955) have noted the presence of nonstriated muscle fibres among them. Another investigator (Fowler 1957) who failed to find muscle fibres in this site suggested that the puckering of the perianal skin is due to the combined effects of levator ani pulling on the longitudinal coat and the tone of the subcutaneous external sphincter.

A *muscularis mucosae* has been described in the anal canal immediately above the pectinate line and possibly extending below it (Jit 1974).

Between the subcutaneous external sphincter and the skin of the anal canal lies the inferior part of the internal rectal venous plexus; veins connect the external and internal rectal plexuses and thus establish connexions between the portal and systemic venous systems. The radiating elastic septa end below by breaking up into a network which subdivides the narrow interval between the subcutaneous external sphincter and the skin into a compact honeycomb; this arrangement accounts for the severe pain produced by collections of pus or blood which may occur in this region and for the localization of a haemorrhage following the rupture of a vein of the external rectal plexus. The submucosa above the white line, containing the superior part of the internal rectal venous plexus, is known as the *submucous space*, while that below the white line, containing the inferior part of the internal rectal plexus, is the *perianal space*. These two spaces are separated by the dense submucous layer of nonstriated muscle and connective tissue referred to above, which is especially well marked for a short distance below the anal valves.

It has been concluded (Fowler 1957) that the anal canal proper extends between two readily palpable muscular landmarks, namely, the *anorectal ring* above and the *intersphincteric ring* below. In the relaxed state the lower border of the internal sphincter and the intersphincteric groove lie at the anal orifice, the subcutaneous external sphincter being lateral to the orifice, and it is only when the external sphincter contracts that it becomes withdrawn to surround the lower part of the 'apparent' anal canal.

At the anorectal junction the puborectalis, deep external sphincter and internal sphincter collectively form the *anorectal ring* of muscle, which can be felt by a finger in the anal canal; surgical division of the ring results in rectal incontinence. The anterior part of the ring is not so well marked, since relatively few fibres of the puborectalis pass in front of the anorectal junction, most of the fibres of this muscle forming a sling which loops round the sides and back of the gut at this site, slinging the anorectal junction forwards towards the pubis.

Correlated with the *dual development* of the anal canal, the part above the anal valves being derived from the endodermal cloaca and the part below from the ectodermal proctodeum (p. 202), the following facts may be noted. In the *ectodermal part*, the lining is skin which is supplied by spinal nerves (inferior rectal nerve), the

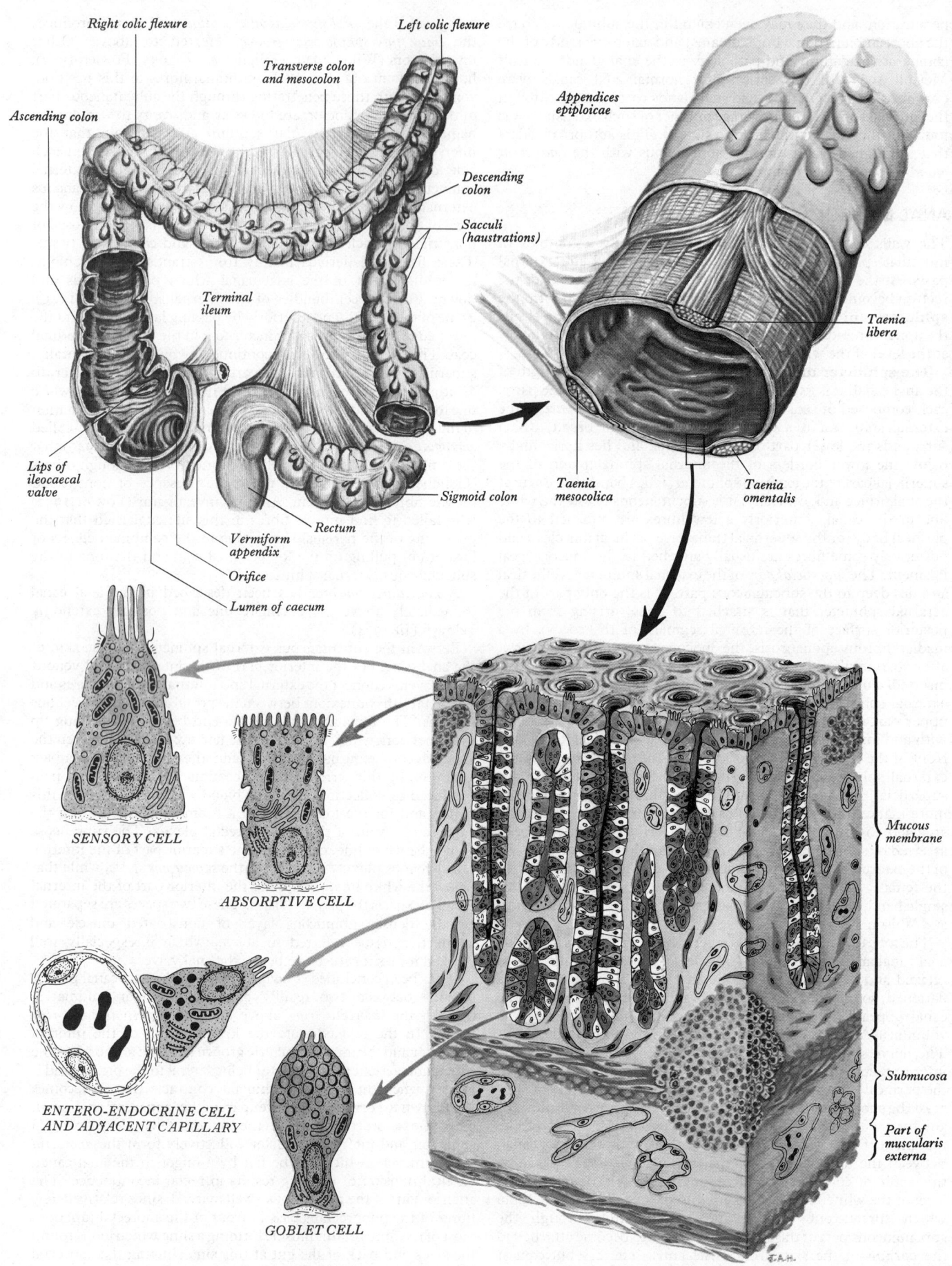

Right colic flexure

Left colic flexure

Transverse colon and mesocolon

Ascending colon

Descending colon

Sacculi (haustrations)

Appendices epiploicae

Terminal ileum

Taenia libera

Lips of ileocaecal valve

Taenia mesocolica

Taenia omentalis

Sigmoid colon

Rectum

Vermiform appendix

Orifice

Lumen of caecum

SENSORY CELL

ABSORPTIVE CELL

Mucous membrane

ENTERO-ENDOCRINE CELL AND ADJACENT CAPILLARY

Submucosa

GOBLET CELL

Part of muscularis externa

8.125 Diagrams of the disposition of the major regions of the large intestine, the micro-architecture and histology of the colonic wall, and the ultrastructure of its epithelial cells. Note aggregations of lymphocytes (shown in yellow), and undifferentiated epithelial cells (shown in white).

arterial blood supply is from the inferior rectal artery, the venous drainage is by the inferior rectal vein which passes to the internal pudendal vein (a systemic vein), and the lymphatics drain with those of the perianal skin into the superficial inguinal lymph nodes. In the *endodermal part*, the mucous membrane is supplied by autonomic nerves, the arterial blood supply (Griffiths 1961) is mainly from the superior rectal artery, the venous drainage is by the superior rectal vein which continues as the inferior mesenteric vein (a tributary of the portal venous system), and the lymphatics drain with those of the rectum (p. 795). In cases of obstruction of the portal venous system, the collateral circulation opened up by the anastomosis between the portal and systemic veins in the anal canal may result in the dilatation of these veins. The different nerve supply of the two parts is correlated with a response to different types of stimuli; the lower part is very sensitive and responds to touch, pain and thermal stimuli like skin in general; the upper part, like the gut, has a high threshold for the above stimuli and responds more readily to increase in tension. The effects of the difference in the innervation of the two parts of the anal canal are seen in cases of haemorrhoids (varicosities in the anal canal) which may be covered with skin in their lower parts and mucous membrane in their upper parts; in injection to cause thrombosis in such distended veins, the needle is inserted into the insensitive upper part and not into the very sensitive ectodermal part. Fissure *in ano* (tearing of the anal valves) is very painful, because it involves the lower sensitive part of the anal canal.

RECTAL EXAMINATION

On inserting the index finger through the anal orifice, it is first grasped by the subcutaneous part of the external sphincter, and then, higher up in the anal canal, by the internal sphincter, the superficial and deep parts of the external sphincter and the puborectalis; beyond this, it may reach the inferior (or even the middle) transverse rectal fold. Many of the structures related to the anal canal and lower part of the rectum may be felt through the walls of these parts of the gut.

In the male, through the anterior rectal wall, the bulb of the penis and (particularly if a catheter is placed in the urethra) the membranous part of the urethra are first identified, and then, about 4 cm above the anus, the prostate can be felt; beyond this the seminal vesicles (if enlarged) and the base of the bladder (especially if the viscus is distended) may be recognized. Posteriorly, the pelvic surfaces of the lower part of the sacrum and the coccyx may be palpated. Laterally, the ischial spine, ischial tuberosity and (if enlarged) the internal iliac lymph nodes may be felt. Pathological thickening of the ureters, swellings in the ischiorectal fossae and abnormal contents of the rectovesical peritoneal pouch may also be detected.

In the female, the uterine cervix can be palpated through the anterior wall of the rectum (and, for example, its degree of dilatation during childbirth determined), and pathological conditions which cause tenderness or changes of shape, size, consistency or position of the ovaries, uterine tubes, broad ligaments and recto-uterine pouch may be detected. In both sexes tenderness of an inflamed vermiform appendix can be elicited, if that organ occupies a pelvic position.

THE RECTAL FASCIAE AND 'SPACES'

Various parts of the pararectal pelvic fascia are composed of loose connective tissue, whilst others are more dense, with particular orientations and attachments. The latter are often considered as mechanical 'supports' for the rectum, and their surgical division is necessary to produce mobilization of the organ. From the anterior surface of the lower part of the sacrum a stout avascular condensation of fascia passes forwards to the posterior aspect of the anorectal junction; it is known as the *fascia of Waldeyer*. The fascia around the middle rectal vessels passes from the posterolateral wall of the lesser pelvis (at the level of the third sacral vertebra) to the rectum, constituting on each side the *lateral ligament of the rectum*. Anteriorly the fascia between the rectum and the seminal vesicles and prostate (the *rectovesical fascia*, p. 1421) is more loosely attached to the latter structures than to

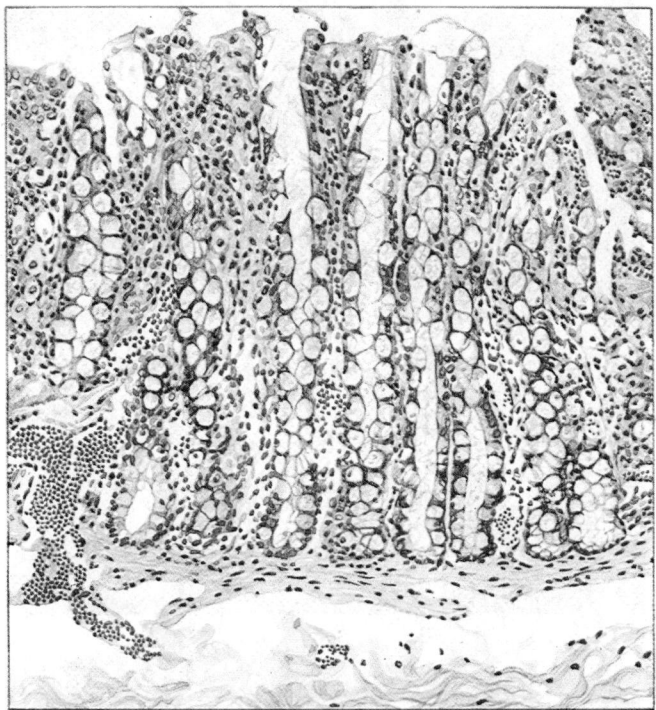

8.126A Section of the mucous membrane of feline large intestine. Stained with haematoxylin and eosin. Magnification about × 100. Note the presence of large numbers of goblet cells and the vascularity of the mucosa.

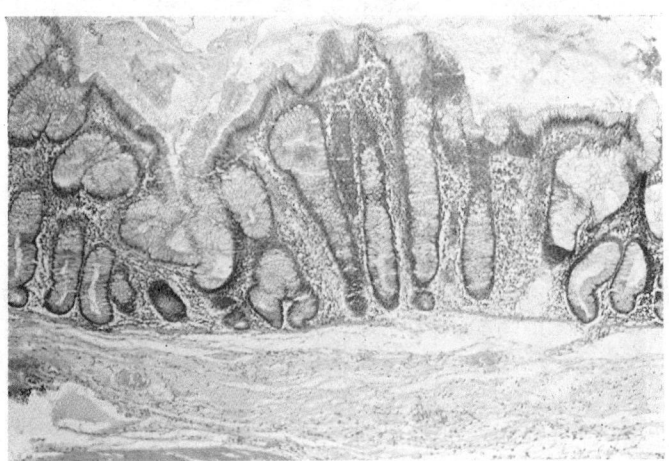

8.126B Medium power light micrograph of human large intestinal mucosa showing crypts containing goblet cells. Stained with haematoxylin and eosin. (Material kindly provided by Dr. J. Heaton; prepared and photographed by Mr. Stephen Sitch, Department of Anatomy, Guy's Hospital Medical School, London.)

the rectum, and in surgical excision of the rectum this fascia is separated along the plane between it and the prostate and seminal vesicles.

In addition to the ischiorectal fossae (p. 561), the following *spaces* of surgical importance are described in relation to the rectum and anal canal. The *pelvirectal space* comprises the loose extraperitoneal connective tissue above the levator ani; it is divided into anterior and posterior regions by the *lateral ligaments of the rectum*. The *submucous space* of the anal canal lies between the mucous membrane of the canal (above the white line) and the internal sphincter, and contains the superior part of the internal rectal venous plexus and lymphatics; above, it is continuous with the rectal submucosa, and below with the *perianal space* (*see* p. 1358); the lateral part of which is bounded above by the most lateral of the radiating elastic septa that pass through the subcutaneous external sphincter; the latter septum divides the ischiorectal fossa into an upper main space containing coarsely

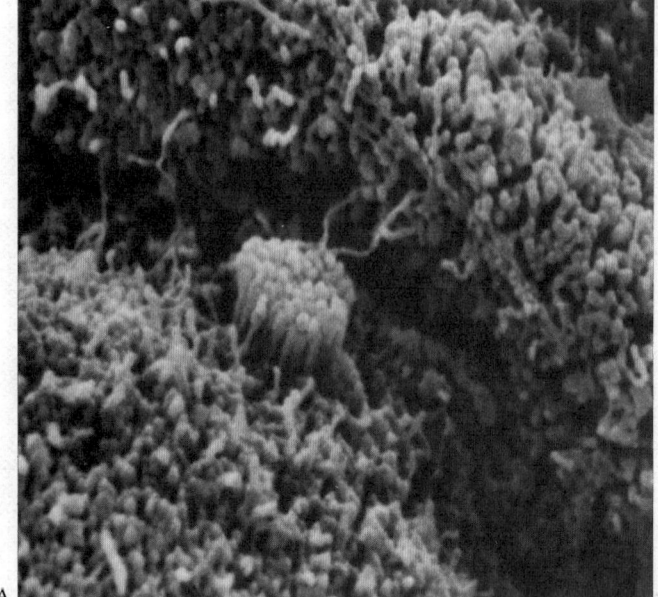

A

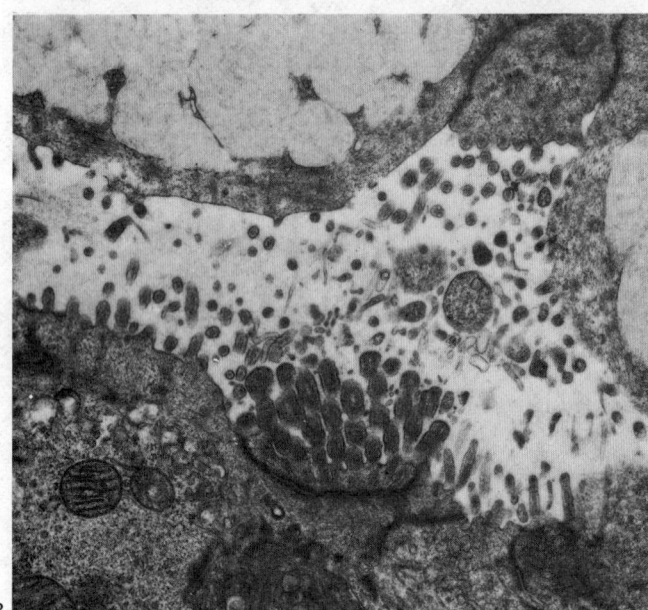

B

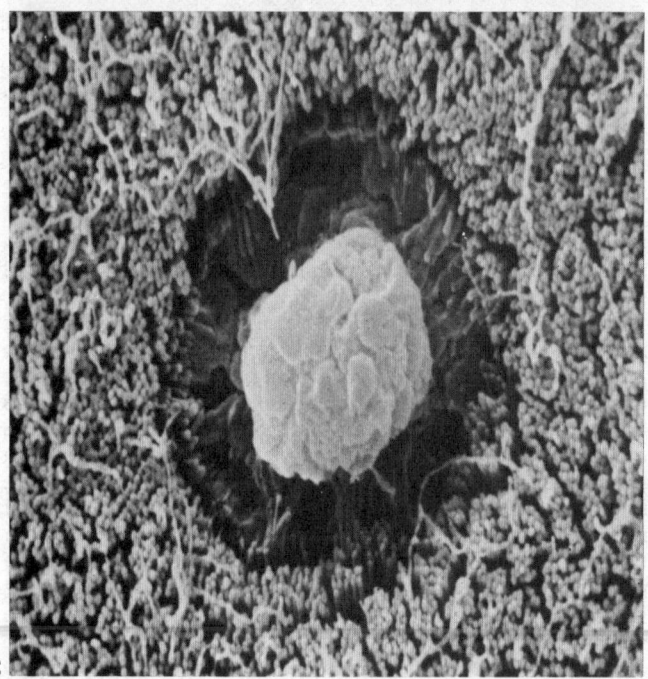

1362 C

lobulated fat and a lower perianal space which contains finer and more compact fat. The perianal space contains the subcutaneous external sphincter, the external rectal venous plexus and the terminal branches of the inferior rectal vessels and nerves. Owing to the arrangement of the radiating septa passing through the subcutaneous external sphincter, pus in the perianal space tends to spread to the anal canal at the white line or to the surface of the perianal skin rather than to the main part of the ischiorectal fossa. Since the perianal space surrounds the lower part of the anal canal, pus on one side may spread round the canal in the space.

STRUCTURE OF THE LARGE INTESTINE

The tissue layers in the wall of the large intestine are the same as in the small intestine.

The *serosa* or *visceral peritoneum* invests the different regions of the large intestine to a variable extent. In the course of the colon the peritoneum is thrown into a number of small pouches filled with fat, the *appendices epiploicae*; they are most numerous on the sigmoid and transverse colon, but absent on the rectum. Beneath the peritoneum lies a subserous layer of loose connective tissue.

The *muscularis externa* consists of an outer longitudinal and an inner circular layer of nonstriated muscle fibres.

The *longitudinal fibres* form a continuous layer over the surface of the large intestine (Hamilton 1946), but in certain situations this layer is thickened to form conspicuous longitudinal bands, termed *taeniae coli*, and in the intervals between them the longitudinal coat is less than half the thickness of the circular coat. In the caecum and colon three taeniae are present, ranging from 6 to 12 mm in width in different individuals. One (the *taenia libera*) is placed anteriorly on the caecum, ascending, descending and sigmoid colon, but is placed inferiorly on the transverse colon; the second (the *taenia mesocolica*) is situated on the posteromedial surface of the caecum, ascending, descending and sigmoid colon, but posteriorly on the transverse colon, at the site of attachment of the transverse mesocolon; the third (the *taenia omentalis*) is placed posterolaterally in the caecum, ascending, descending and sigmoid colon, but is situated on the anterosuperior surface of the transverse colon, at the site where the posterior (ascending) layers of the greater omentum meet this part of the large intestine. These bands are said to be shorter than the other coats of the intestine, and may produce the sacculi or haustrations which are characteristic of the caecum and colon; accordingly, when they are dissected off, the tube can be lengthened, and its sacculated character becomes lost. In the descending colon the taeniae increase in thickness at the expense of the rest of the longitudinal coat, and on the sigmoid colon this coat undergoes a real increase in its total bulk. In the sigmoid colon the longitudinal fibres become more scattered; round the rectum they spread out and form a layer, which completely encircles this portion of the gut, but is thicker on the anterior and posterior surfaces, so that an anterior and a posterior broad band can be recognized. At the rectal ampulla, a few strands of the anterior longitudinal fibres pass forwards to the perineal body (p. 562); they constitute the *recto-urethralis muscle*. In addition, two fasciculi of nonstriated muscle arise from the front of the second and third coccygeal vertebrae, and pass downwards and forwards to blend with the longitudinal muscular fibres on the posterior wall of the anal canal. These are known as the *rectococcygeal muscles* (Wesson 1951).

The *circular fibres* form a thin layer over the caecum and colon, being especially accumulated in the intervals between the sacculi; in the rectum they form a thick layer, and in the anal canal they

8.127A–C   A   Scanning electron micrograph of the epithelium lining the colon (rat), showing a cell of suggested sensory function, bearing an apical tuft of particularly long microvilli. (Magnification ×12,000.) B   Transmission electron micrograph showing part of a colonic epithelial cell bearing sensory microvilli. The smaller absorptive microvilli of adjacent cells can also be seen. (Magnification ×14,000.) c   Scanning electron micrograph of the epithelium lining the colon (rat), showing a goblet cell surrounded by absorptive cells bearing microvilli. (Magnification ×8,000.) (Specimens A–C prepared and photographed by Mr. Michael Crowder, Guy's Hospital Medical School, London.)

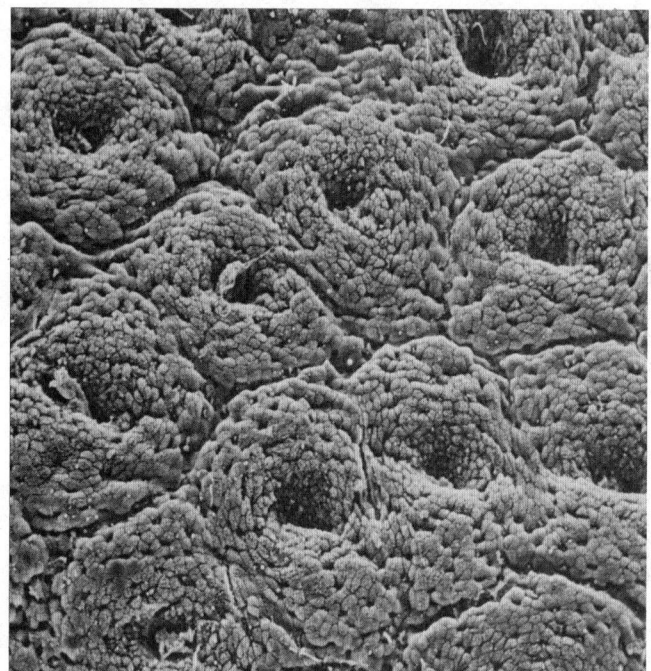

8.128 Scanning electron micrograph of the luminal surface of the human rectal mucosa. The outlines of cells bearing microvilli, and the openings of rectal crypts can be seen. (Magnification ×240.) (Material kindly supplied by Dr. D. S. Rampton. Prepared and photographed by Mr. Michael Crowder, Guy's Hospital Medical School, London.)

become numerous, and constitute the sphincter ani internus. Older observations, which recorded an interchange of fascicles of muscle fibres between the circular and longitudinal layers, have amply been confirmed in a study covering 112 post-mortem cadavers (from early fetal life to 88 years). By microdissection and histological sampling, interchanges of fibres, especially near the taenia coli, were observed to be commonplace. The deviation of longitudinal fibres from the taenia towards the circular layer may, in some instances, account for the haustration of the colon (Pace 1968).

The *submucosa* and *muscularis mucosae* are similar to those of the small intestine.

The *mucous membrane* of the caecum and colon is pale, smooth, destitute of villi, and raised into numerous crescentic folds which correspond with the intervals between the sacculi; that of the rectum is thicker, of a darker colour, more vascular, and connected more loosely with the muscular coat. The *epithelium* of the caecum, colon, and upper rectum consists of scattered mucus-secreting (goblet), columnar absorptive (Pittman and Pittman 1966) and M cells (p. 1354). Epithelium also lines the walls of the intestinal glands which are numerous and provide a large surface area for mucus secretion and absorption of water, salts and other materials; these functions are important in providing lubrication for the passage of faeces, and in resorbing many of the substances secreted into the alimentary canal in the buccal cavity, stomach and small intestine. The large intestine also provides an environment for a large bacterial flora; in many animals intestinal bacteria, in addition to being of digestive importance, metabolize essential organic compounds which supplement the vitamin intake in the normal diet. The columnar cells, which carry microvilli on their luminal surfaces, show junctional complexes at the luminal ends of their lateral borders, and their cytoplasm contains a subapical terminal web and the usual cytoplasmic organelles. Many of them also contain secretory granules in their apical cytoplasm. The secretion provided by such cells appears to be mucoid in nature, but is also rich in antibodies of the IgA group (p. 62) which provide a measure of protection against invasion by micro-organisms (Schofield and Atkins 1970). Other columnar cells posses an apical tuft of particularly long microvilli (**8.**127A, B), and have been suggested as a type of sensory ending (Silva 1966).

The *glands* of the large intestine are minute tubular

prolongations of the mucous membrane arranged perpendicularly to its surface; they are longer, more numerous, and placed in much closer apposition than those of the small intestine; they open by minute orifices upon the surface (**8.**128), giving it a cribriform appearance. As mentioned above, each gland is lined with short columnar epithelium, the majority of the cells being goblet cells (**8.**126), between which lie columnar absorptive cells.

The *solitary lymphoid follicles* of the large intestine are most abundant in the caecum and vermiform appendix, but are irregularly scattered over the rest of the large intestine also. They are similar to those of the small intestine, and they have epithelial M cells, specialized for antigenic transport, associated with them (Owen and Nemanic 1978).

VESSELS AND NERVES OF THE LARGE INTESTINE

The *arteries* which supply the part of the large intestine developed from the mid-gut (caecum, appendix, ascending colon, right two-thirds of transverse colon) are derived from the colic branches of the superior mesenteric artery; those supplying the left part of the transverse colon, descending colon, sigmoid colon, rectum and upper half of the anal canal (hindgut derivatives) are the inferior mesenteric artery (and its terminal branch, the superior rectal) and the middle rectal artery. They give off large branches, which ramify between and supply the muscular coats, and after dividing into small vessels in the submucous tissue pass to the mucous membrane. The arteries of the rectum and anal canal are (*a*) the superior rectal artery (the continuation of the inferior mesenteric), which is the chief artery of the rectum; this divides into two branches that run down, one each side of the rectum, and break up into terminal branches that pierce the muscular coat and pass in the submucosa of the rectum and thence in the anal columns as far as the anal valves, where they form looped anastomoses; (*b*) the middle rectal arteries, which run in the 'lateral ligaments of the rectum' to supply the muscle coats of the lower part of the rectum and anastomose with each other but only form poor anastomoses with the superior and inferior rectal arteries; (*c*) the inferior rectal arteries, which supply the internal and external sphincters, the anal canal below the anal valves, and the perianal skin; (*d*) the median sacral artery, which supplies the posterior wall of the anorectal junction and of the anal canal. The *veins of the large intestine* are the superior and inferior mesenteric veins, which drain the same parts of the large intestine as are supplied by the corresponding arteries. The veins of the rectum and anal canal are (*a*) the superior rectal veins, which commence from the internal rectal plexus in the anal canal and pass up in the rectal submucosa in the form of about six vessels of considerable size, to pierce the muscular wall of the rectum about 7.5 cm above the anus and unite to form a single trunk, the superior rectal vein, which is continued as the inferior mesenteric vein; (*b*) the middle rectal veins, which begin in the submucosa of the rectal ampulla and drain chiefly the muscular walls of this part of the rectum; (*c*) the inferior rectal veins, which begin from the external rectal plexus and drain the lower part of the anal canal.

The *nerve supply of the large intestine* (exclusive of the lower half of the anal canal) is derived from the sympathetic and parasympathetic systems. The caecum, appendix, ascending colon and the right two-thirds of the transverse colon (all derivatives of the midgut) have their sympathetic supply from the coeliac and superior mesenteric ganglia, and their parasympathetic supply from the vagus; in each case the nerves are distributed to the gut in the plexuses around the branches of the superior mesenteric artery. The left third of the transverse colon, descending colon, sigmoid colon, rectum and upper half of the anal canal (derivatives of the hindgut) derive their sympathetic supply from the lumbar part of the trunk and the superior hypogastric plexus by means of the plexuses on the branches of the inferior mesenteric artery. The sympathetic supply of the colon is largely vasomotor in function. The parasympathetic supply to this part of the gut is derived from the pelvic splanchnic nerves (nervi erigentes). From these latter, fibres pass to the inferior hypogastric plexuses to supply the rectum and upper half of the anal canal: in addition, some fibres pass up uninterruptedly through the superior hypogastric plexus to be distributed along

Hypothetical receptor sites sensitive to luminal stimuli

LUMEN OF GUT

Gap junctions allow communication with adjacent cells

Stimulus-secretion coupling

BASAL SECRETION OF GRANULE CONTENTS

'PARACRINE' EFFECTS
Diffusion of hormones to local sites of action

ENDOCRINE EFFECTS
Passage of hormones to distant sites of action via vessels

8.129A Diagram showing the ultrastructure and possible modes of action of an entero-endocrine cell.

the inferior mesenteric artery to the transverse, descending and sigmoid colon (p. 1356). Further, branches from the pelvic splanchnic nerve pass up on the posterior abdominal wall behind the peritoneum, independently of the inferior mesenteric artery, to be distributed directly to the left colic flexure and descending colon (Mitchell 1953). The ultimate distribution in the gut wall is similar to that in the wall of the small intestine (p. 1350). A study of the adrenergic and cholinergic activity in the nerve supply of the taenia coli, coupled with observations on the distribution of nerve fibres, suggests that (in the guinea-pig) few muscle cells are directly innervated, propagation of activity being chiefly through the 'electrical synapse' type of intercellular contact (Bennett and Rogers 1967).

The sympathetic nerves to the rectum and upper part of the anal canal pass mainly along the inferior mesenteric and superior rectal arteries and partly via the superior and inferior hypogastric plexuses, the latter supplying the lower part of the rectum and the internal sphincter. The parasympathetic supply from the pelvic splanchnic nerve passes forwards as long strands (about 3 cm long) from the sacral nerves to join the inferior hypogastric

plexuses on the sides of the rectum, being motor to the musculature of the rectum and inhibitory to the internal sphincter. The external sphincter ani is supplied by the inferior rectal branch of the pudendal nerve (S. 2, 3) and the perineal branch of the fourth sacral nerve (p. 1115). In surgical excision of the rectum, the dissection is kept close to the rectal wall, otherwise these nerves may be damaged, with consequent dysfunction of the bladder and, in the male, impotence resulting from failure of erection of the penis. Afferent impulses underlying sensations of physiological distension are conveyed by the parasympathetic nerves, while pain impulses are conducted by both the sympathetic and parasympathetic nerves supplying the rectum and upper part of the anal canal. In the condition of *aganglionosis* of the colon (*megacolon*) there is commonly a marked reduction or complete absence of (postganglionic) autonomic neurons in the wall of the organ (Bodian *et al.* 1961; Bodian 1966; Soltero-Harrington *et al.* 1969). Garrett *et al.* (1969) have studied the myenteric plexus and its ganglion cells by electron microscopy and histochemical techniques for transmitter substances. They found that in cases of megacolon there is not only a variable diminution, amounting sometimes to absence, of ganglion cells, but that there is also a failure of innervation of the muscle layers, even when ganglionic neurons are present.

The *lymph nodes and vessels* of the large intestine are described on p. 795.

The Gastro-entero-pancreatic Endocrine System

The gastro-entero-pancreatic (GEP) endocrine system (Fujita 1973) consists of the scattered, frequently solitary, hormone-

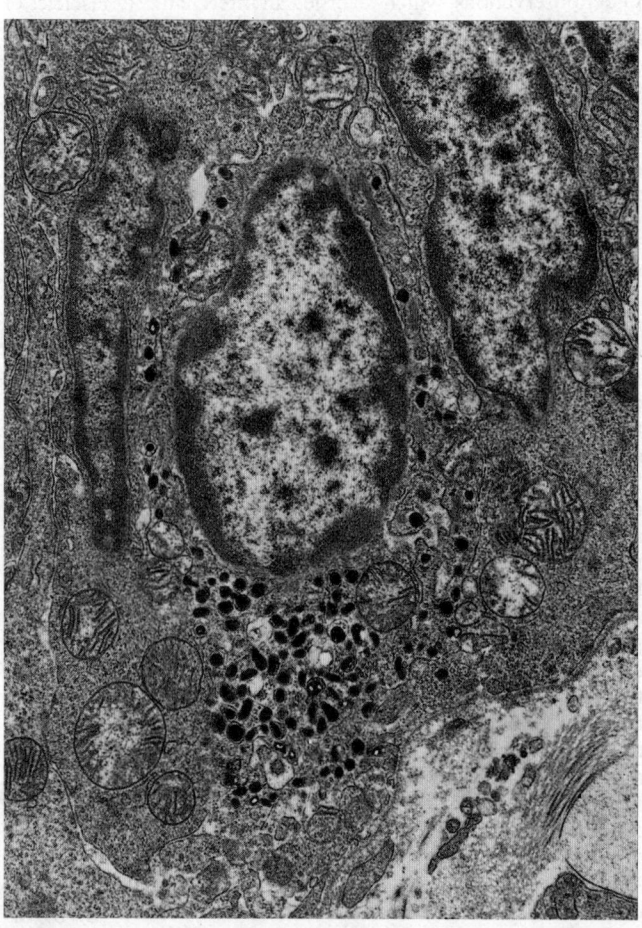

8.129B Transmission electron micrograph of an entero-endocrine (APUD) cell of the epithelium lining the colon (rat). Secretory vesicles can be seen towards the basal aspect of the cell. Absorptive columnar cells lie on either side of the APUD cell. (Magnification × 11,000.) (Prepared and photographed by Mr. Michael Crowder, Guy's Hospital Medical School, London.)

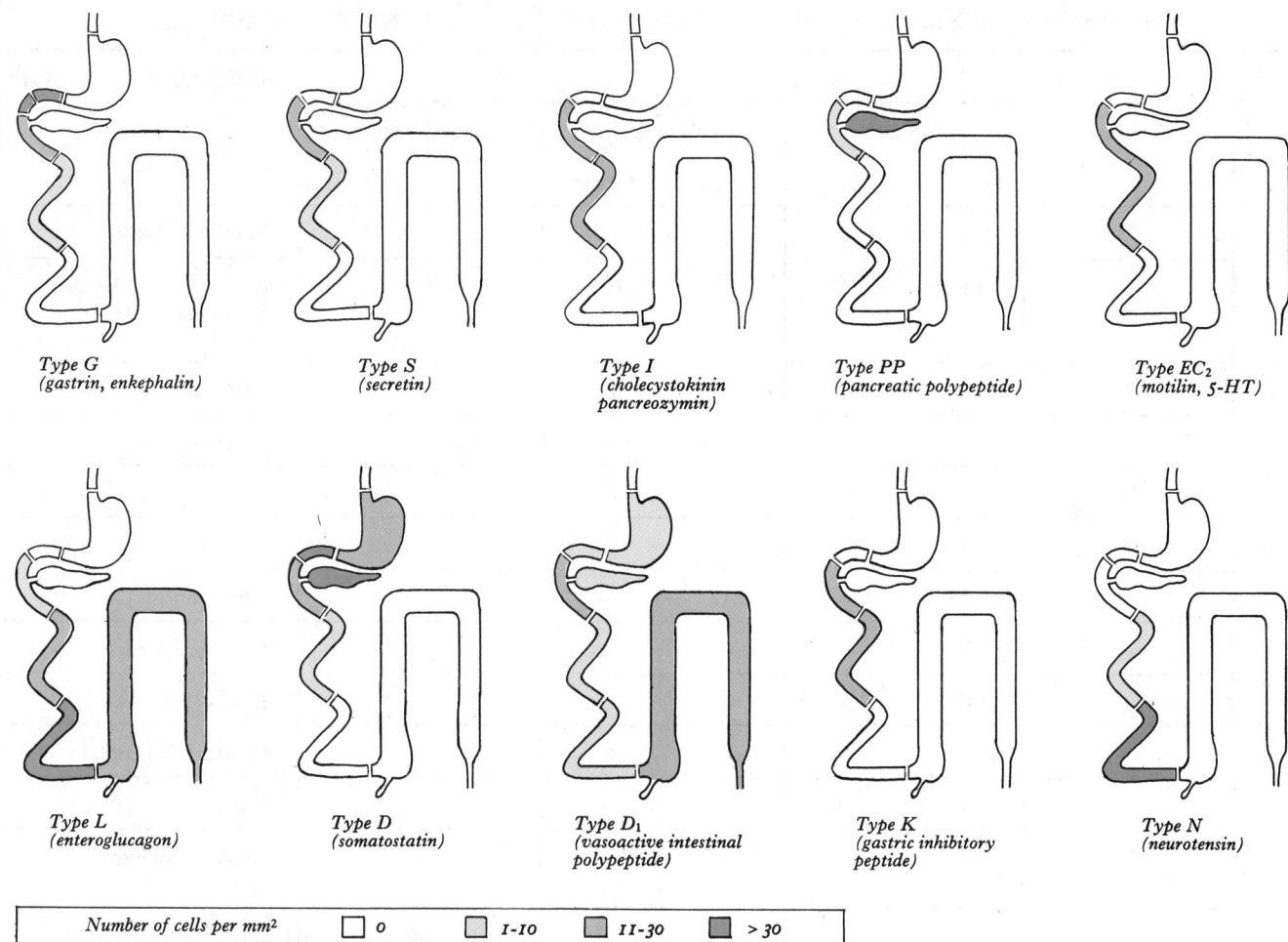

Type G
(gastrin, enkephalin)

Type S
(secretin)

Type I
(cholecystokinin
pancreozymin)

Type PP
(pancreatic polypeptide)

Type EC$_2$
(motilin, 5-HT)

Type L
(enteroglucagon)

Type D
(somatostatin)

Type D$_1$
(vasoactive intestinal
polypeptide)

Type K
(gastric inhibitory
peptide)

Type N
(neurotensin)

Number of cells per mm²	□ 0	1-10	11-30	> 30

8.129C Approximate quantitative distribution of a selection of human gastro-entero-pancreatic (GEP) endocrine cells. Highly diagrammatic. (After Bloom and Polak 1978, by kind permission of Churchill-Livingstone.)

producing cells of the stomach, intestine and pancreas. The application of electron microscopy to ultra-thin (0·07 μm) sections and of immunocytochemical techniques to adjacent semi-thin (1 μm) sections of resin-embedded tissue has led to the identification of a wide range of GEP endocrine cell types and their secretory products (*see* p. 1366). Of the fifteen specific cell types so far recognized in man, only one (type B) is restricted to the pancreatic islets, and at least three are distributed in both the gastro-intestinal mucosa and the pancreas.

The ultrastructure of human GEP endocrine cells has been described in detail by Rubin (1972), Sasagawa *et al.* (1973), Capella *et al.* (1976) and Cavallero *et al.* (1976) and summarized by Solcia *et al.* (1978). The gastric and intestinal endocrine cells are scattered in the epithelium lining the lumen and glands of the gut, lying with their bases resting on the basal lamina. As a rule, those of the oxyntic mucosa do not extend as far as the lumen of the gut and are referred to as 'closed', while most of those of the pyloric and intestinal mucosae do and are referred to as 'open'. Their osmiophilic secretory granules vary characteristically in shape, size and ultrastructure in the different cell types and are generally infranuclear in position, while the Golgi complexes are supranuclear; the luminal aspects of cells of the 'open' type are covered with microvilli of variable number, length and shape. Typical entero-endocrine cells are shown in Fig. **8.**129A, B. Human *P cells* contain very small (100–140 nm mean diameter) secretory granules with a slight reactivity to Grimelius' silver (Capella *et al.* 1977); they are rare in normal adult tissues, and possibly contain a bombesin-like polypeptide (Polak *et al.* 1976). *EC cells*, which contain osmiophilic, argentaffin, Grimelius' silver-reactive granules, are sub-divided into three types, termed EC$_1$, EC$_2$ and EC$_n$; in addition to storing 5-hydroxytryptamine, the EC$_1$ cells store substance P (Heitz *et al.* 1977), the EC$_2$ cells store motilin (Polak *et al.* 1975), and the EC$_n$ cells an as yet unidentified material. The *D$_1$ cells* contain granules which are argyrophilic and about 140–190 nm in diameter (Capella *et al.* 1977); they store a vaso-active intestinal peptide-like material. The *PP cells*, which occur commonly in the pancreatic islets and rarely in the exocrine pancreas, store pancreatic polypeptide in granules of 150–170 nm mean diameter; they are equivalent to the F cells identified in other mammals (Baetens *et al.* 1976). Human *D, B* and *A cells* are described in detail in the section on the endocrine pancreas (p. 1371). *X cells* have been described in the human oxyntic mucosa by Solcia *et al.* (1977); their function is as yet unknown. Human *ECL cells* (Vassallo *et al.* 1971) store an amine with reducing powers, possibly 5-hydroxytryptamine, in granules with an intensely argyrophilic core; histamine many also be present. Human *G cells* (Vassallo *et al.* 1971) manufacture gastrin, and possibly enkephalin (Polak *et al.* 1978), and have slightly argyrophilic granules with floccular contents. The *S cells* (Capella *et al.* 1976) are scattered in the duodenojejunal mocosa and produce secretin (Larsson *et al.* 1977); they are structurally similar to D$_1$ cells but differ in their secretory product. Human *I cells* (Capella *et al.* 1976) are most common in the duodenum and jejunum, occurring only rarely in the ileum; they are the source of cholecystokinin-pancreozymin (more correctly, pancreaticozymin) (Buchan *et al.* 1977). Human *K cells* (Capella *et al.* 1976) contain large granules (approximately 350 nm in diameter) with osmiophilic, argyrophobic cores; like the I cells, they are most common in the duodenum and jejunum. K cells produce gastric inhibitory peptide. *N* and *L* cells are difficult to distinguish cytologically in man; the granules of the N cells are, however, generally homogeneous and about 300 nm in

HUMAN GASTRO-ENTERO-PANCREATIC ENDOCRINE CELLS AND THEIR PRODUCTS

	LOCATION					SECRETORY PRODUCTS
	Pancreas	Stomach		Small Intestine		Large Intestine	
		Oxyntic	Pyloric	Upper	Lower		
C E L L T Y P E S	P*	P	P	P	—	—	Bombesin-like?
	—	EC_n	EC_n	EC_2	EC_1	EC_1	EC_1: 5-HT, Substance P EC_2: 5-HT, Motilin EC_n: 5-HT, Undetermined
	D_1	D_1	D_1	D_1	D_1	D_1	VIP-like?
	PP	—	—	PP?	—	—	Pancreatic Polypeptide
	D	D	D	D	—	—	Somatostatin
	B	—	—	—	—	—	Insulin
	A	A?*	—	A?	—	—	Glucagon
	—	X	—	—	—	—	Undetermined
	—	ECL	—	—	—	—	H or 5-HT
	—	—	G	G	—	—	Gastrin, Enkephalin
	—	—	—	S	S	—	Secretin
	—	—	—	I	I	—	Cholecystokinin-pancreozymin
	—	—	—	K	K	—	GIP
	—	—	—	—	N	—	Neurotensin
	—	—	—	L	L	L	Enteroglucagon or GLI

*Found in the fetus or new-born, rare in adults.

Abbreviations used for secretory products:
5-HT = 5-hydroxytryptamine
VIP = Vaso-active intestinal peptide
H   = Histamine
GIP = Gastric inhibitory peptide
GLI = Glicentin

(Based mainly on the Lausanne 1977 classification of gastro-entero-pancreatic endocrine cells reported by Solcia *et al.* in 1978).

diameter, while those of the L cells sometimes have argyrophilic cores and tend to be somewhat smaller (about 260 nm in diameter). The N cells produce neurotensin (Orci *et al.* 1976) while the L cells produce enteroglucagon or glicentin.

The concentration of endocrine cells in the gastro-intestinal mucosa is very low and generally decreases progressively in an aboral direction. This is illustrated diagrammatically in Fig. 8.129C.

The GEP endocrine cells have certain common features which allow them to be classified in a variety of other ways. They are all able to produce peptides and/or amines active as hormones or as neurotransmitters, and are all derived from neuro-endocrine-programmed cells of ectoblastic origin. As such they belong to the *APUD cell series* (Pearse 1968; Pearse 1977; Pearse and Polak 1978) and modulate not only the actions of the autonomic nervous system but also of each other. The APUD concept is described in more detail elsewhere (p. 1454). An intriguing finding, reported in recent years, of supposedly similar neurohormones and neurotransmitter-like peptides in the neurons of the brain and in some GEP endocrine cells, together with other neuron-like characteristics of the latter cells, has led to their designation as *paraneurons* (Fujita 1976). Peptides common to the brain and the gastro-intestinal mucosa include the following: substance P, somatostatin, vaso-active intestinal peptide, bombesin, neurotensin, cholecystokinin and the opiate-like enkephalin (Bloom and Polak 1978). The GEP endocrine cells are presumed to have receptor sites on their surfaces, adequate stimulation of which by 'secretogogues' triggers stimulus-secretion coupling (Kanno 1973) as in chromaffin cells (Douglas 1968); the GEP endocrine cells can thus also be termed *receptosecretory cells* (Fujita 1976).

There is still some doubt as to the mode of action of the endocrine cells restricted to the gastro-intestinal mucosa. It might be supposed, from their ultrastructure and proximity to blood capillaries, that their secretory products are endocrinal in action, exerting distant, diffuse effects after transportation in the blood. However, of the many products of these cells, only the following have been shown to act as circulating hormones: gastrin, secretin, cholecystokinin-pancreozymin, gastric inhibitory peptide, motilin and enteroglucagon (Bloom and Polak 1978). There are several fundamental differences between the endocrine cells of the gastric and intestinal mucosae and those of the endocrine system proper: they are not aggregated into glands, but are scattered among what may well be their local target cells; most are closely related to the lumen of the gut, allowing specialized regions of their plasmalemmae to detect and respond to luminal stimuli directly; there are no common hyposecretory or hypersecretory syndromes; and the relationship between plasma hormone levels and functional response (modulation of the neural control of gut motility, for example) is not stoichiometric. It has been suggested (Wingate 1976) that the gastro-intestinal hormones may have local 'paracrine' as well as distant 'endocrine' effects; direct actions on gastro-intestinal smooth muscle, on neighbouring endocrine cells, on other enterocytes, and on the neurons of the gut are speculative but feasible. Although the gut can be regarded as the largest endocrine organ of the body (Pearse 1974) it is, perhaps, better to regard it as a region in which neural, paracrine and endocrine control of activity are intimately linked.

Applied Anatomy of the Intestine

The infrequency of rupture of the small intestine by external injury to the abdominal wall is attributable to its elasticity and the ease with which the coils glide over each other; the more fixed duodenum, particularly its horizontal part as it crosses the vertebral column, is more liable to such damage.

HERNIA

In external hernia the ileum is the intestinal part most frequently herniated. When a part of the large intestine is involved it is usually the caecum or the sigmoid colon.

The chief sites at which external hernia may occur are the inguinal region, the femoral canal and the umbilical region.

Inguinal Hernia

In this form, the viscus is protruded through the inguinal region of the abdominal wall. The two principal varieties are oblique, and direct.

In *oblique inguinal hernia* the intestine is protruded through the lateral inguinal fossa (which lies behind the deep inguinal ring). Here the herniated gut pushes before it a pouch of the parietal peritoneum and extraperitoneal areolar tissue. It enters the inguinal canal at the deep inguinal ring and becomes invested by the internal spermatic fascia which encloses the constituents of the spermatic cord. In passing along the canal it displaces upwards the arched fibres of the transversus abdominis and obliquus internus, receives a covering from the cremasteric fascia and muscle, and lies in front of the constituents of the spermatic cord. It emerges from the canal at the superficial inguinal ring and here becomes invested by the external spermatic fascia. Lastly, it descends into the scrotum, here receiving additional coverings from the superficial fascia and the skin. Such a herniated part of the gut may become constricted at the deep inguinal ring, with consequent interference with its blood supply (strangulation). In relieving the strangulation, the deep inguinal ring should be cut in an upward and lateral direction to avoid the inferior epigastric vessels. In most cases, the occurence of oblique inguinal hernia depends upon congenital defects in the processus vaginalis, the peritoneal pouch which precedes the descent of the testis (p. 217). Obliteration of the processus vaginalis may be complete at birth, or it may begin shortly before birth and be completed subsequently; the closure occurs first at the deep inguinal ring and at the top of the epididymis and gradually extends until the whole of the intervening portion is converted into a fibrous cord. Complete or partial failure of closure of the processus may occur, with consequent variations in the relation of the hernial protrusion to the testis and tunica vaginalis; e.g. where the processus is patent throughout, the herniated gut descends in front of the testis into the tunica vaginalis (complete congenital hernia), the processus and tunica vaginalis constituting the sac of the hernia. In incomplete congenital hernia (or hernia into the funicular process), the gut descends as far as the top of the testis, where the processus is sealed off from the tunica vaginalis (8.130). Although the above types of inguinal hernia are called congenital, the actual extrusion of the hernia into the pre-existing peritoneal sac may not take place until adult life and then be produced by such factors as increased intra-abdominal pressure or sudden

muscular strain. (For recent critique of the anatomy of inguinal hernia consult Lytle 1979.)

In *direct inguinal hernia* the protrusion makes its way through some part of the inguinal triangle, which is bounded inferiorly by the medial half of the inguinal ligament, medially by the lower part of the lateral border of the rectus abdominis, and laterally by the inferior epigastric artery. The triangle overlies the medial inguinal fossa and part of the supravesical fossa (p. 1324). A direct hernia may pass either through (a) the medial inguinal fossa where only extraperitoneal tissue and transversalis fascia intervene between the peritoneum and the aponeurosis of the external oblique; or through (b) the supravesical fossa and the falx inguinalis (conjoint tendon) which lies in front of the fossa. In the former the hernial protrusion escapes from the abdomen on the lateral side of the conjoint tendon, pushes before it the peritoneum, extraperitoneal tissue and transversalis fascia, and enters the inguinal canal. It passes along nearly the whole length of the canal and finally emerges from the superficial ring, receiving an investment from the external spermatic fascia. The coverings of this form of hernia are similar to those of the oblique form, except that a portion derived from the general layer of transversalis fascia replaces the internal spermatic fascia, so that the hernia lies between the innermost and the middle covering of the spermatic cord.

In the second form, which is the more frequent, the hernia is either forced through the fibres of the falx inguinalis, or the falx is gradually distended in front of it so as to form a complete investment for it. The intestine then enters the lower end of the inguinal canal, escapes at the superficial ring, lying on the medial side of the cord, and receives additional coverings from the external spermatic fascia, the superficial fascia and the skin. The coverings of this form therefore differ from those of the oblique form in that the conjoint tendon is substituted for the cremaster, and the internal spermatic fascia is replaced by a portion of the general layer of the transversalis fascia. In all the varieties of inguinal hernia the most superficial covering is an investment

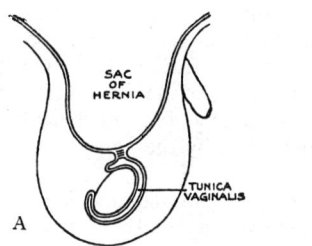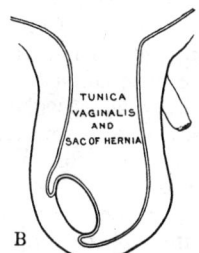

8.130A and B Diagrams representing varieties of oblique inguinal hernia.   A   Incomplete congenital.   B   Complete congenital.

from the external spermatic fascia and is identical with the outermost covering of the spermatic cord. An oblique inguinal hernia lies within the spermatic cord and shares all its coverings, but the covering which a direct hernia acquires from the transversalis fascia is distinct from the covering which the spermatic cord receives from that layer.

Direct inguinal hernia is of much less frequent occurrence than oblique, and occurs more often in men than in women. The main differences in position between it and the oblique form are: (a) it is placed over the body of the pubic bone and not in the course of the inguinal canal; (b) the inferior epigastric artery runs lateral (*not* medial) to the neck of the sac; (c) the spermatic cord lies along its lateral and posterior sides, not directly behind it as in oblique inguinal hernia. A direct hernia is always of the acquired variety.

The seat of stricture in both varieties of direct hernia is usually found either at the neck of the sac or at the superficial ring. In that form which perforates the conjoint tendon it may occur at the edges of the fissure through which the gut passes. In all cases of inguinal hernia, whether direct or oblique, it is proper to divide the stricture upwards: by cutting in this direction the incision is made parallel to the inferior epigastric artery, and all chance of wounding the vessel is thus avoided.

Femoral Hernia

In femoral hernia the protrusion of the intestine takes place through the femoral ring (p. 725). This ring is closed by the femoral septum, a partition of modified extraperitoneal tissue; it is therefore a weak spot in the abdominal wall, especially so in the female, where the ring is larger, and where profound changes are produced in the tissues by pregnancy. Femoral hernia is therefore more common in women than in men.

When a section of the intestine is forced through the femoral ring, it carries before it a pouch of peritoneum which forms the hernial sac. It receives an investment from the extraperitoneal tissue (or femoral septum), and descends along the femoral canal as far as the saphenous opening; at this point it changes its course, being prevented from extending further down the sheath on account of the narrowing of the latter, its close contact with the vessels, and also the close attachment of the superficial fascia and femoral sheath to the lower part of the circumference of the saphenous opening. (See also p. 595.) The hernia is consequently directed forwards, pushing before it the cribriform fascia, and then curves upwards over the inguinal ligament and the lower part of the aponeurosis of the obliquus externus. While the hernia is contained in the femoral canal it is usually of small size, owing to the resisting nature of the surrounding parts, but when it escapes from the saphenous opening into the loose areolar tissue of the groin it becomes considerably enlarged. The direction taken by a femoral hernia is at first downwards, then forwards and upwards; in the application of taxis (manual pressure) for the reduction of a femoral hernia, therefore, pressure should be directed in the reverse order, and the thighs should be passively flexed in order that the greatest degree of relaxation may be obtained.

The coverings of a femoral hernia from within outwards are: peritoneum, femoral septum, femoral sheath, cribriform fascia, superficial fascia and skin. A fibrous investment (*fascia propria*) of a femoral hernia lies immediately external to the peritoneal sac but is frequently separated from it by some adipose tissue. Surgically it is important to remember the frequent existence of this layer on account of the ease with which an inexperienced operator may mistake the fascia for the peritoneal sac and the contained extraperitoneal fat for omentum, as there is often a great excess of subperitoneal fatty tissue enclosed in the 'fascia propria'. In many cases it resembles a fatty tumour, but on further dissection the true hernial sac will be found in the centre of the mass of fat. The fascia propria is merely a modified femoral septum which has been thickened to form a membranous sheet by the pressure of the hernia.

When the intestine descends along the femoral canal only as far as the saphenous opening the condition is known as *incomplete femoral hernia*, in contradistinction to the *complete hernia*, which has passed through the opening. The small size of the protrusion in the incomplete form of hernia renders it an exceedingly dangerous variety of the disease, from the extreme difficulty of detecting the existence of the swelling, especially in corpulent subjects.

The site of strangulation of a femoral hernia varies: it may be at the neck of the hernial sac; more often it is at the point of junction of the falciform margin of the saphenous opening with the free edge of the pectineal part of the inguinal ligament, or it may be at the margin of the saphenous opening. (*See*, however, comments on the 'lacunar ligament' on p. 552.) The stricture should in every case be divided in a direction upwards and medially for a distance of about 4 to 6 mm. All normally positioned vessels and other structures of importance in relation to the neck of the sac will thus be avoided. (However, an abnormal obturator artery may be imperilled by such an incision—see p. 721.)

The pubic tubercle forms an important landmark in serving to differentiate the inguinal from the femoral variety of hernia. The neck of the inguinal protrusion is above and medial to the tubercle, while the neck of the femoral protrusion is below and lateral to it.

Umbilical Hernia

There are three varieties of umbilical hernia.

(*a*) *Congenital umbilical hernia.* This variety is due to a failure of retraction of the umbilical loop of gut (p. 205).

(*b*) *Infantile umbilical hernia.* This is due to stretching of the scar tissue in the umbilical region. It occurs usually within the first three years after birth and is associated with conditions causing increased intra-abdominal pressure.

(*c*) *Acquired umbilical hernia.* This variety really occurs through the linea alba, usually immediately above the umbilicus (para-umbilical hernia), and most frequently occurs in obese multiparous females.

Rarely, hernia may occur at other sites, e.g. through the *lumbar triangle* (p. 565), via the *obturator foramen*, the *greater* or *lesser sciatic foramen*, or the *ischiorectal fossa*. Occasionally *incisional hernia* may occur at the site of a scar following an abdominal operation, particularly if the wound becomes septic.

Both the small intestine and colon are subject to considerable variations, which may be of medical and surgical importance. Consult Goligher (1967), and Kanagasuntheram (1970) for details and literature.

The Pancreas

The pancreas is a soft, lobulated, greyish-pink gland, 12–15 cm long, extending nearly transversely across the posterior abdominal wall, behind the stomach, from the duodenum to the spleen. Its broad, right extremity is called the *head*, and is connected to the main part, or *body*, by a slightly constricted *neck*; its narrow, left extremity forms the *tail*. It passes obliquely to the left and slightly upwards, across the posterior wall of the abdomen, in the epigastric and left hypochondriac regions.

RELATIONS OF THE PANCREAS

The structures in close topographic relationship to the pancreas are considered systematically with respect to the head, neck, body and tail (*see* 8.111, 131, 132).

The head, flattened from before backwards, is sited within the curve of the duodenum. Its upper border is overlapped by the superior part of the duodenum; the other borders are grooved to receive the adjacent margin of the duodenum, which they overlap in front and behind to a variable extent. Sometimes a small part of the head of the pancreas is actually embedded in the wall of the descending part of the duodenum. From the lower and left part of the head there is a prolongation, the *uncinate process*, which projects upwards and to the left behind the superior mesenteric vessels. In or near the groove between the duodenum and the right and lower borders of the head are the anastomosing superior and inferior pancreaticoduodenal arteries (pp. 713, 716; **6**.86).

Anterior surface. From the anterosuperior aspect of the head of

Related to spleen

Related to left kidney

Related to diaphragm

Coeliac artery

Superior mesenteric artery

Portal vein

Splenic vein

Bile duct

Duodenum

Related to inferior vena cava

8.131 Posterior aspect of the pancreas and duodenum.

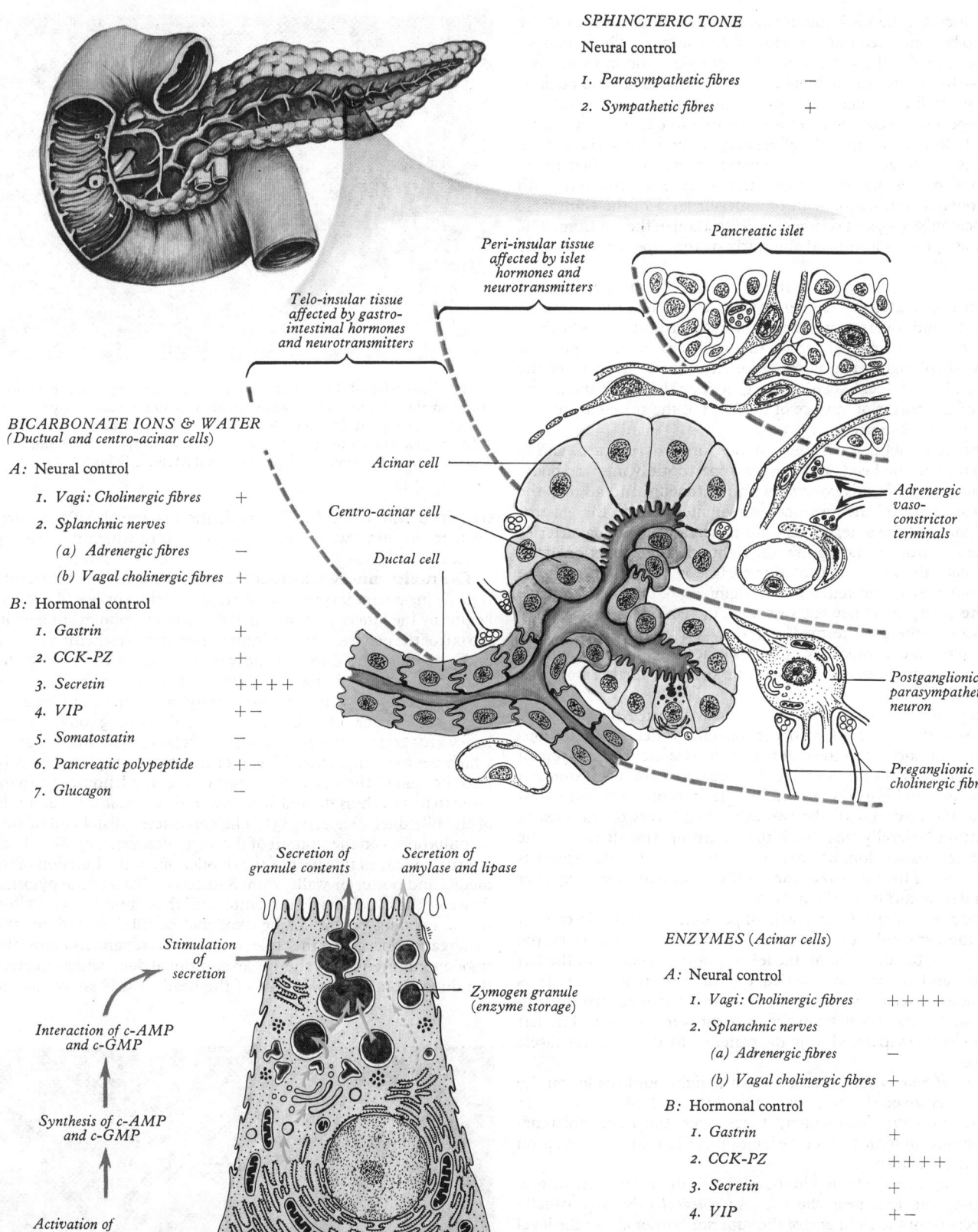

SPHINCTERIC TONE

Neural control

1. *Parasympathetic fibres* —
2. *Sympathetic fibres* +

Pancreatic islet

Peri-insular tissue affected by islet hormones and neurotransmitters

Telo-insular tissue affected by gastro-intestinal hormones and neurotransmitters

BICARBONATE IONS & WATER
(Ductual and centro-acinar cells)

A: Neural control

  1. *Vagi: Cholinergic fibres* +

  2. *Splanchnic nerves*

    (a) Adrenergic fibres —

    (b) Vagal cholinergic fibres +

B: Hormonal control

  1. *Gastrin* +

  2. *CCK-PZ* +

  3. *Secretin* + + +

  4. *VIP* + −

  5. *Somatostatin* —

  6. *Pancreatic polypeptide* + −

  7. *Glucagon* —

Acinar cell

Centro-acinar cell

Ductal cell

Adrenergic vaso-constrictor terminals

Postganglionic parasympathetic neuron

Preganglionic cholinergic fibres

Secretion of granule contents

Secretion of amylase and lipase

Stimulation of secretion

Interaction of c-AMP and c-GMP

Synthesis of c-AMP and c-GMP

Activation of
A
B } *Adenyl cyclase*
C *Guanyl cyclase*

Zymogen granule (enzyme storage)

J.A.H.

Binding sites for
A Acetylcholine
B CCK-PZ and/or gastrin
C Secretin and/or VIP

ENZYMES (Acinar cells)

A: Neural control

  1. *Vagi: Cholinergic fibres* + + + +

  2. *Splanchnic nerves*

    (a) Adrenergic fibres —

    (b) Vagal cholinergic fibres +

B: Hormonal control

  1. *Gastrin* +

  2. *CCK-PZ* + + + +

  3. *Secretin* +

  4. *VIP* + −

  5. *Glucagon* —

8.132 Diagram of the ultrastructure of the exocrine pancreas and the mechanisms by which its secretion is controlled. The hormones referred by acronyms are as follows: CCK-PZ=cholecystokinin-pancreatico-zymin; VIP=vaso-active intestinal polypeptide.

the pancreas, the neck juts forwards, upwards and towards the left, to be continued into the body of the pancreas. The boundary between the head and neck, on the right side (and in front), is a groove for the gastroduodenal artery; on the left side (and behind) a deep notch intervenes between the head and the neck, and in it the superior mesenteric and splenic veins unite to form the portal vein. Below and to the right of the neck the anterior surface of the head is in contact with the transverse colon, only areolar tissue intervening, while still lower the surface is covered with peritoneum continuous with the inferior layer of the transverse mesocolon (8.98), and is in contact with a coil of the jejunum. The uncinate process is crossed anteriorly by the superior mesenteric vessels.

Posterior surface. The posterior surface of the head of the pancreas is related to the inferior vena cava, which runs upwards behind it and covers nearly the whole of this aspect. In addition, it is related to the terminal parts of the renal veins and the right crus of the diaphragm. The uncinate process passes in front of the aorta. The bile duct lies either in a groove on the upper and lateral part of the posterior surface of the head of the pancreas, or in a canal in its substance (p. 1383).

The neck, about 2 cm long, extends forwards, upwards and to the left from the head, and merges imperceptibly into the body. Its anterior surface is covered with peritoneum and adjoins the pylorus, part of the omental bursa intervening; the gastro-duodenal and the anterior superior pancreaticoduodenal arteries descend in front of the gland at the right side of the junction of the neck with the head; its posterior surface is in relation with the superior mesenteric vein and the beginning of the portal vein.

The body is somewhat prismoid in section, and has three surfaces, generally termed anterior, posterior and inferior (although, more precisely, these surfaces are largely antero-superior, posterior and antero-inferior, and furthermore are obliquely set).

The *anterior surface* is concave, and is directed forwards and upwards; it is covered with peritoneum, which is continuous antero-inferiorly with the anterior of the two ascending layers of the greater omentum (8.97), and is separated from the stomach by the omental bursa. On reaching the level of the taenia mesocolica, the posterior of the two ascending layers of the greater omentum generally fuses with the anterosuperior surface of the transverse mesocolon, while the anterior layer continues upwards to the root of the mesocolon, and is then reflected superiorly over the anterior surface of the pancreas.

The *posterior surface* is devoid of peritoneum, and is in contact with the aorta and the origin of the superior mesenteric artery, the left crus of the diaphragm, the left suprarenal gland and the left kidney and its vessels, particularly the left renal vein. It is intimately related to the splenic vein, which courses from left to right and separates it from the structures mentioned. The left kidney is also separated from the pancreas by the perirenal fascia and fat.

The *inferior surface* is narrow on the right but broader on the left, and is covered with peritoneum of the postero-inferior layer of the transverse mesocolon; it lies upon the duodenojejunal flexure and on some coils of the jejunum; its left extremity rests on the left colic flexure.

The *superior border* is blunt and flat to the right; narrow and sharp to the left, near the tail. An *omental tuberosity* usually projects from the right end of the superior border above the level of the lesser curvature of the stomach, and is in contact with the posterior surface of the lesser omentum. It is in relation above with the coeliac artery, from which the common hepatic artery courses to the right just above the gland, while the splenic artery runs towards the left following a wavy course along this border.

The *anterior border* separates the anterior from the inferior surface, and along this border the two layers of the transverse mesocolon diverge from each other: one passing upwards over the anterior surface, the other backwards over the inferior surface.

The *inferior border* separates the posterior from the inferior surface: the superior mesenteric vessels emerge under its right extremity.

The tail of the pancreas is narrow, and usually lies in contact with the inferior part of the gastric surface of the spleen. It is

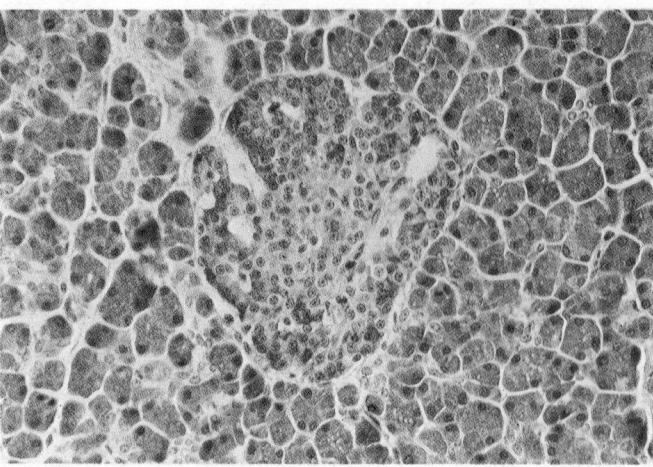

8.133A An islet of Langerhans and surrounding exocrine glandular tissue in the pancreas of a rhesus monkey, stained with orange G and aldehyde fuchsin. Within the islet, the B cells stain purple, whereas the A cells are pale yellow in colour. (Kindly provided by Dr. J. Henderson, Department of Physiology, Guy's Hospital Medical School.)

contained within the two layers of the lienorenal (splenorenal) ligament together with the splenic vessels, to which it is closely related.

The main pancreatic duct traverses the pancreas from left to right, lying nearer its posterior than its anterior surface (8.132). It begins by the junction of the small ducts of the lobules situated in the tail of the pancreas, and, running from left to right through the body, receives the ducts of the various lobules composing the gland, the latter joining the main duct almost at right angles ('herringbone pattern'). Considerably augmented in size, it reaches the neck of the pancreas, and turning downwards, backwards and to the right, comes into relation with the bile duct, which lies to its right side. Together the two ducts pass obliquely into the wall of the descending part of the duodenum, and there unite to form a short dilated *hepatopancreatic ampulla* (or ampulla of the bile duct) (*see* p. 1383). The constricted distal end of this ampulla opens on the summit of the *major duodenal papilla*, which is situated within this part of the duodenum at the junction of its medial and posterior walls, from 8 to 10 cm distal to the pylorus. As a rule the two ducts do not unite until they approach very close to the opening on the major duodenal papilla. Sometimes the pancreatic duct and the bile duct open separately into the duodenum. Frequently there is an additional duct, which receives the ducts from the lower part of the head, and is known as the

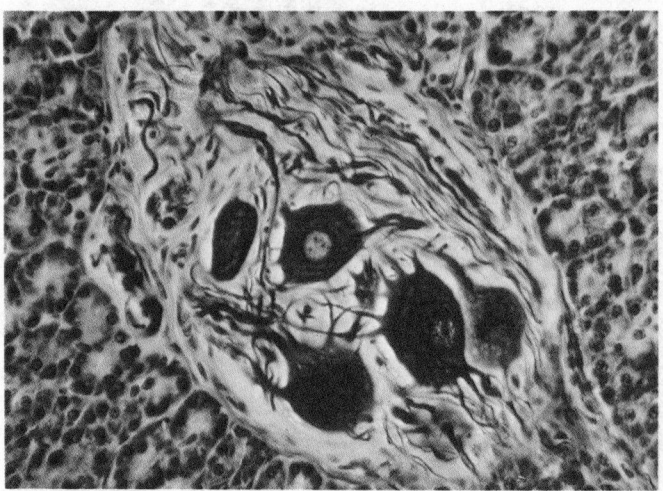

8.133B A low-power micrograph showing a cluster of autonomic ganglionic neurons with dendritic trees and axonal bundles, situated amongst pancreatic acinar cells of the goat. Palmgren silver impregnation. (Kindly provided by Dr. J. Henderson, Department of Physiology, Guy's Hospital Medical School, London.)

accessory pancreatic duct (**8**.132). It runs upwards in front of the main pancreatic duct, to which it is connected by a communicating duct, and opens into the duodenum about 2 cm above and slightly ventral to the major duodenal papilla on a small rounded *minor duodenal papilla*. The terminal part of the accessory duct may fail to expand and the secretion from the lower part of the head of the pancreas is then diverted along the communicating duct into the main pancreatic duct. It is, however, usually fully patent (Dawson and Langmann 1961).

Surface Anatomy. The head of the pancreas lies within the curve of the duodenum. The neck lies in the transpyloric plane, behind the pylorus. The body passes obliquely to the left and slightly upwards for about 10 cm, its left part lying a little above the transpyloric plane. The tail lies a little above and to the left of the intersection of the transpyloric and left lateral planes.

PANCREATIC STRUCTURE

The pancreas is composed of two quite separate types of glandular tissue which are, however, in intimate topographic association with each other. The main mass of tissue is the *exocrine part* of the pancreas, embedded in which are clusters of *endocrine cells* constituting the *pancreatic islets*.

THE EXOCRINE PANCREAS

The exocrine part of the pancreas is a lobulated, branched, acinar gland, surrounded and partially divided into lobules by delicate loose connective tissue (de Reuck and Cameron 1962; Beck and Sinclair 1971). The secretory acinar cells, pyramidal in shape, are arranged in flask-shaped or tubular groups. A narrow intercalated (intralobular) duct is deeply inserted into each secretory mass, the initial part of the duct walls being lined by low cuboidal *centro-acinar cells*, whilst more distally these are replaced by taller cuboidal and eventually columnar cells. The larger ducts are interlobular in position, and are surrounded by areolar tissue which contains nonstriated muscle fibres, and autonomic nerve fibres. Argentaffin cells (p. 1366) are present amongst the undifferentiated columnar cells of the duct walls, and mast cells are numerous in the connective tissue sheaths.

The acinar cells of the exocrine pancreas are typical zymogenic cells, containing a basally situated nucleus, which is surrounded with basophilic cytoplasm, seen ultrastructurally as regular arrays of granular endoplasmic reticulum, interspersed with mitochondria and dense secretory granules. In the apical, supranuclear compartment of the cell, a prominent Golgi complex is present, surrounded by numerous larger secretion granules; the latter are membrane-bound bodies containing the powerful enzymic constituents of the pancreatic secretion. These enzymes are in an inactive form whilst within the granules, and only become activated after their release. Because of the highly ordered nature of their cell contents, acinar cells have been widely used to establish the morphological pathway of secretion synthesis and transport, as an example of typical protein-secreting cells, (see also *Cytology*). In post-mortem specimens the hydrolytic enzymes of the pancreas tend to cause rapid degeneration of cellular detail.

Ganglionic cells and cords of relatively undifferentiated epithelial cells are also present in the exocrine pancreas; the latter may provide stem cells for the replacement of the degenerating exocrine cells, and perhaps, also for islet tissue.

The structure of the exocrine pancreas and the mechanisms by which its activity is controlled are shown diagrammatically in **8**.132 and have been described in detail by Webster *et al.* (1977); Singh and Webster (1978); Case (1979); and Wormsley (1979).

THE ENDOCRINE PANCREAS

This consists of the pancreatic islets (of Langerhans), spheroidal or ellipsoidal clusters of cells randomly embedded in the exocrine part of the pancreas (Laguesse 1906; Lane 1907) together with scattered, often solitary, endocrine cells (Heitz *et al.* 1976).

The islets, which may number more than one million in a normal individual, tend in man to be most numerous in the tail of the pancreas (Findlay and Ashcroft 1975). Each consists of a mass of polyhedral endocrine cells pervaded by a network of fenestrated capillaries (Goldstein and Davies 1968) and with a rich autonomic innervation (Gerich and Lorenzi 1978). Histological staining procedures have been used successfully to distinguish between the major types of secretory cell (Lane 1907; Bensley 1911; Bloom 1931; Kito and Hosoda 1977) and the development of highly specific fluorescent antibody labelling of sections of pancreas (Lacy and Davies 1957), often accompanied by electron microscopic examination of adjacent ultrathin sections (Heitz *et al.* 1976; Baetens *et al.* 1977) has led to confirmation of the identity of their secretory products and to the recognition of other less numerous endocrine cell types (**8**.134).

The most numerous endocrine cells, types A (or alpha) and B (or beta), secrete glucagon (Baum *et al.* 1962) and insulin (Lacy and Davies 1957) respectively. Although there is interspecific variation (Findlay and Ashcroft 1975), in man the A cells tend to be arranged towards the periphery of the islets while the B cells are more central in position (Orci 1976). The cytoplasmic storage granules of the A cells are fixed by alcohol, are generally smaller than those of the B cells, stain brilliant orange or red with the dyes Orange G and Mallory-Azan, and are aldehyde-fuchsin negative. Those of the B cells are alcohol-soluble and, after appropriate fixation, are aldehyde-fuchsin positive. A third type of endocrine cells, the D cell, discovered in human pancreatic islets by Bloom (1931), and generally described as producing gastrin, has recently been shown to contain somatostatin or a closely related peptide (Orci *et al.* 1975). In man the D cells tend to be peripherally located in the islets, as are the A cells. It has been suggested (Orci and Unger 1975) that the islets may consist of two functionally distinct regions: a homocellular medulla containing mainly B cells, where insulin is secreted at a relatively constant rate in response, for example, to the presence of glucose in the surrounding intercellular fluid, and a heterocellular cortex of A, B and D cells, particularly rich in vascular and neural elements, where rapid changes in secretory activity are made in response to various environmental stimuli. In this latter region somatostatin release by the D cells may inhibit the secretory activity of adjacent A or B cells (Orci 1976). Since in many mammalian species including man, D cells are more frequently in contact with A cells than with B cells, it might be expected that the main effect of pancreatic somatostatin would be to inhibit glucagon release. Convincing evidence that this is so has come from the organ culture studies of Barden *et al.* (1977) who found that when antisomatostatin serum was incubated with rat islets there was a ten-fold increase in glucagon release but no significant change in the release of insulin. The mechanism by which somatostatin inhibits glucagon, and possibly also insulin, release is not yet known, but it has been suggested that it may act intracellularly after transmission through gap junctions from adjacent cells (Gerich and Lorenzi 1978). Another suggested hormonal modifier of islet cell activity is gastric inhibitory peptide which appears to potentiate the insulin secretory response to glucose (Dupré *et al.* 1973). The 'neurohormones' of the autonomic nervous system, acetylcholine and noradrenalin, also affect secretion, acetylcholine augmenting insulin and glucagon release, while noradrenalin inhibits glucose-induced insulin release; they may also affect somatostatin and pancreatic polypeptide secretion. At present the relative importance of circulating noradrenalin and adrenalin of adrenal origin, and of neurogenous noradrenalin acting locally on islet cell secretion, remains to be clarified. The control of islet cell activity is shown diagrammatically in **8**.134.

Peptide-secreting cells with smaller cytoplasmic granules than those of the A, B and D cells have been identified recently in the human pancreas. There are at least two types: one, termed the PP cell, has been shown to contain the hormone pancreatic polypeptide, while the other has an ultrastructure similar to that of the D_1 cells of the gastric mucosa (p. 1365). The pancreatic 'D_1' cells differ from the PP cells in that the granules of the former do not react with anti-bovine pancreatic polypeptide serum while those of the latter do (Baetens *et al.* 1977). Although the nature of the secretory product of the gastro-enteric D_1 cells is still uncertain, it has been suggested that it may be related to vaso-

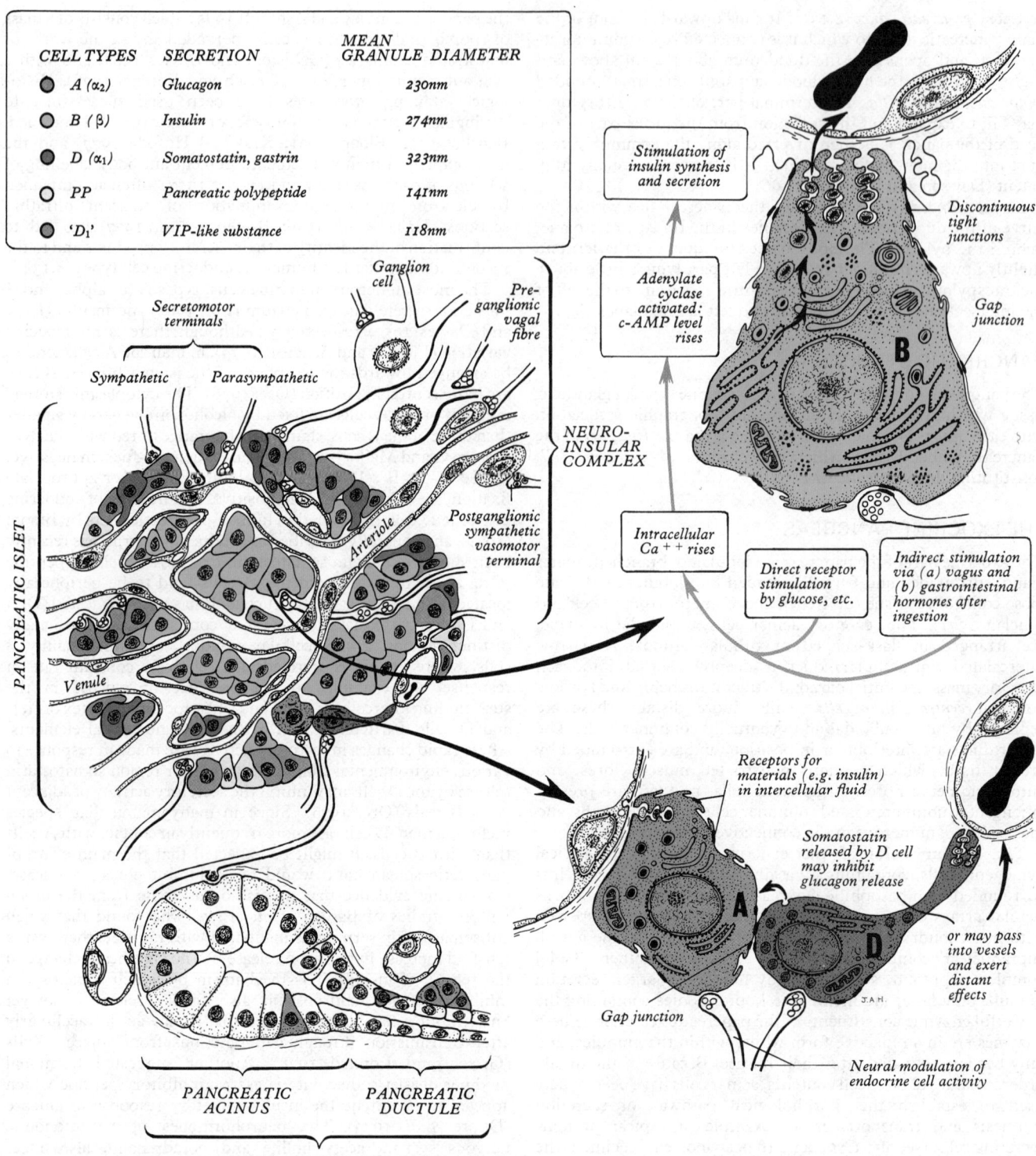

CELL TYPES	SECRETION	MEAN GRANULE DIAMETER
A (α_2)	Glucagon	230nm
B (β)	Insulin	274nm
D (α_1)	Somatostatin, gastrin	323nm
PP	Pancreatic polypeptide	141nm
'D₁'	VIP-like substance	118nm

8.134 Diagram of the histology, ultrastructure, and mode of operation of the endocrine pancreas. (VIP = vaso-active intestinal polypeptide.)

active intestinal polypeptide (Buffa *et al.* 1977); that of the pancreatic 'D₁' cells is still uncertain.

'D₁' and PP cells are not restricted to the islets but are also found scattered throughout the predominantly exocrine compartment of the pancreas (**8**.134).

Islet vessels and nerves

The *arteries* of the pancreas are derived from the splenic artery and the pancreaticoduodenal arteries (pp. 713, 716 and **6**.86). Its *veins* drain into the portal, splenic and superior mesenteric veins. Its *lymph vessels* are described elsewhere (p. 793). The larger blood and lymph vessels travel with ducts and nerves in the

interlobular connective tissue and from them branches supply the lobules of the gland. Bunnag *et al.* (1963) have shown that in mice one to three afferent arterioles arise from the arterial branches to supply each islet, before which they may supply acini. Once within an islet they branch into a network of capillaries so rich that it is reminiscent of a renal glomerulus. The islet capillary bed is drained by from one to six venules which join together before entering an adjacent intralobular vein. McCuskey and Chapman (1969) have reported that flow through individual islet capillaries is intermittent, local interruption being brought about by bulging of the endothelial cells into the capillary lumen. The islet capillaries are fenestrated.

The pancreatic *nerve supply* is derived from the coeliac plexus and enters the organ along with branches of the arteries supplying the pancreas. As well as afferent fibres about which little is known, there is an efferent supply consisting of sympathetic postganglionic fibres from the coeliac ganglion and parasympathetic preganglionic fibres from the right vagus nerve. The fibres, which are mainly nonmyelinated (Benscome 1959) are vasomotor (sympathetic) and parenchymal (sympathetic and parasympathetic) in their distribution. Fine branches ramify among the islet cells from peri-insular plexuses (Findlay and Ashcroft 1975). Fibres frequently make synaptic contact with acinar cells before innervating the islets, suggesting a close linkage between the neural control of the exocrine and endocrine components of the pancreas. Many fibres enter the islets in company with arterioles (Coupland 1958).

Parasympathetic ganglia lie in the inter- and intra-lobular connective tissue, and in the latter case are frequently associated with islet cells, together with which they constitute *neuro-insular complexes*. These intimate associations of islet cells and neural elements were first described by van Campenhout (1925) who named them the 'complexes sympathico-insulaires', the term being revised to 'complexes neuro-insulaires' by Simard (1937). They were classified into two groups by Fujita (1959), one consisting of nerve cells and islet cells (Fig. **8**.134) and the other of nerve fibres and islet cells, the latter being described in detail by Kobayashi and Fujita (1969). The islet cells involved in the neuro-insular complexes include both A and B types. The significance and role of the complexes are still speculative; if, however, the islet cells have a neural crest origin, as suggested by Pearse (1969) and Weichert (1970) then the close anatomical association of the islet cells and neural elements is not surprising, nor is the involvement of the autonomic system in the control of islet secretion.

Three types of nerve terminal have been identified in the islets (Smith and Porte 1976): cholinergic (having 30–50 nm diameter agranular vesicles), adrenergic (having 30–50 nm dense-cored vesicles) and an as yet uncharacterized type (having 60–200 nm dense-cored vesicles). No selective association with any particular islet cell type has been found, and sometimes more than one type of terminal is associated with a single cell type (Esterhuizen *et al.* 1968). Orci *et al.* (1973) have shown that some of the chemical synapses between axon terminal and islet cell show areas of narrowing of the synaptic cleft suggesting the presence of an electrical synapse or gap junction. Similar gap junctions have been found between the islet cells (Orci 1974), and the concept of electrical coupling of the nerve supply to a functional network of islet cells has been introduced. Some nerve terminals appear to end remote from the surface of the islet cells (Kobayashi and Fujita 1969); neurotransmitters released from such endings into the intercellular spaces of the islets could, perhaps, diffuse to and affect numerous islet cells.

Applied Anatomy. Cysts of the pancreas may attain a large size, and cause symptoms by pressing on the stomach, diaphragm or bile duct. They generally push their way forwards between the stomach and transverse colon, and may then be felt in the upper part of the abdomen as a definite tumour in the median plane. The tumour is fixed and does not move with respiration. The pancreas may be the site of cancer; this usually affects the head, and therefore speedily involves the bile duct, leading to persistent jaundice; or it may press upon the portal vein, causing ascites, or involve the stomach, causing pyloric obstruction. The descending part of the duodenum is occasionally encircled by the head of the pancreas (annular pancreas), and should the latter then be the seat of malignant disease or chronic inflammation it may cause obstruction of the duodenum. (For a case report of annular pancreas and a resumé of the literature *see* Kasai *et al.* 1974.) Similarly, if the bile duct lies in a canal in the head of the pancreas (p. 1383), chronic inflammatory disease of the pancreas may obstruct the duct and produce jaundice. Accessory nodules of pancreatic tissue may sometimes be present in the wall of the duodenum (most commonly), the jejunum, the ileum or the ileal diverticulum (p. 1345). Those in the duodenum may be associated with the occasional presence of duodenal diverticula. These take the form of small protrusions which involve all the coats of the

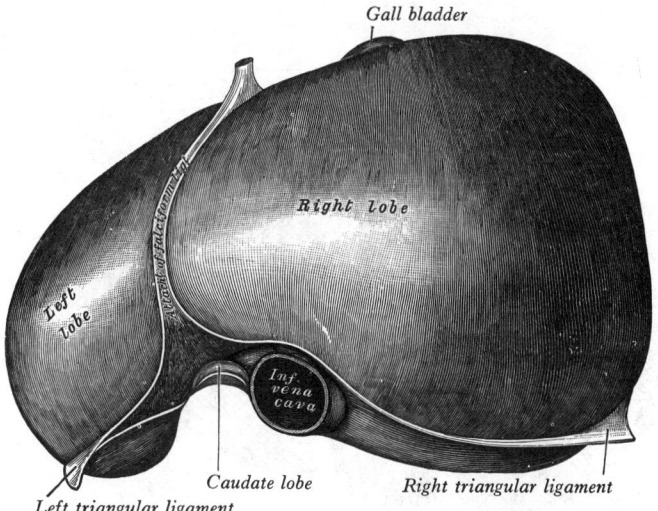

8.135A   The superior, anterior and right lateral surfaces of the liver.

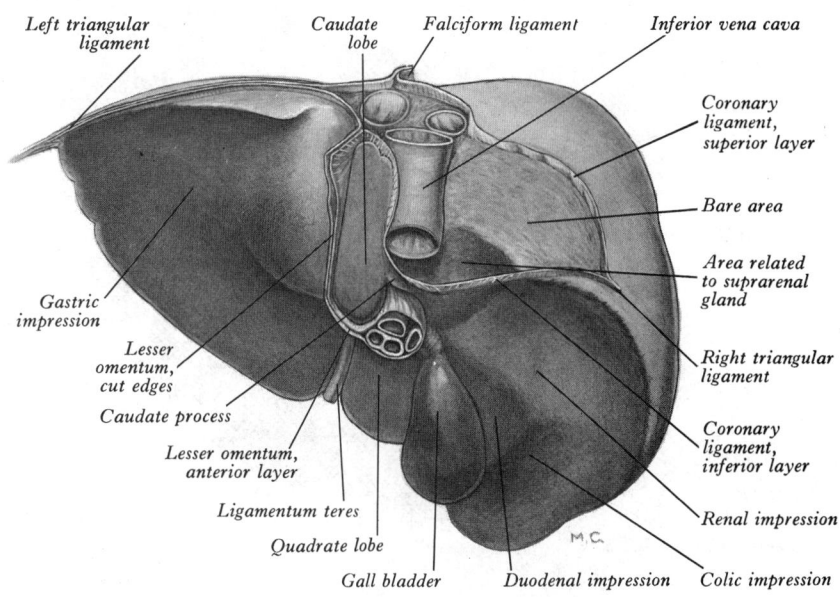

8.135B   Posterior aspect of the liver, showing its peritoneal connexions divided close to its surfaces.

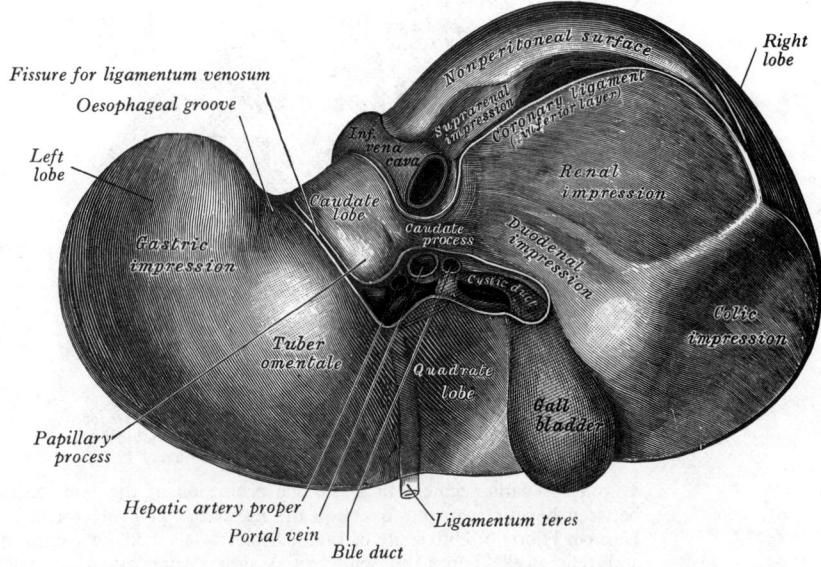

8.135C   The inferior surface of the liver.

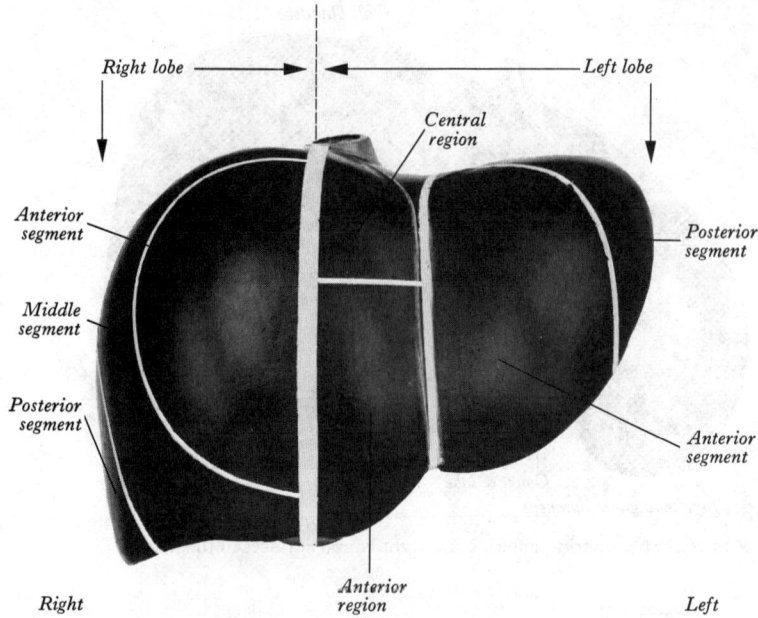

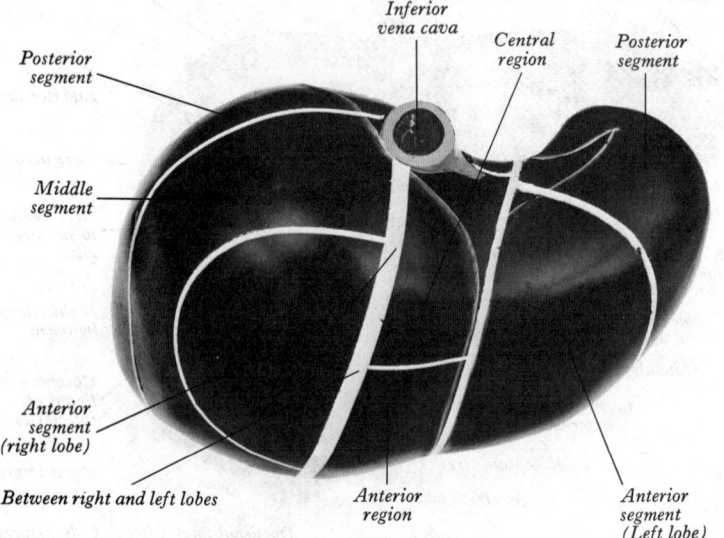

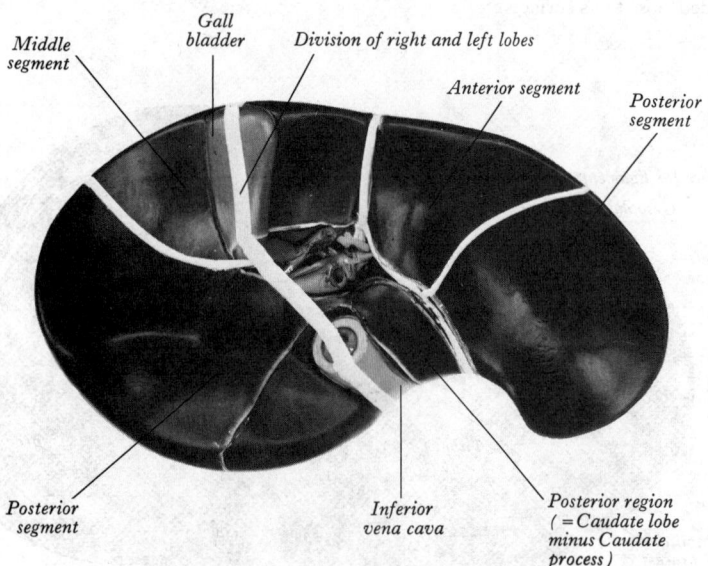

8.136A Hepatic segmentation. Surface projection of the boundaries between hepatic segments based on the researches of Professor Carl-Herman Hjortsjö, University of Lund, Sweden—see text for comment and references. Top: Anterior view; Middle: Anterosuperior view; Bottom: Inferior (visceral) view.

duodenum, or only the mucous and submucous layers, and they are usually situated on the wall of the duodenum adjacent to the pancreas and in close relation to the opening of the bile duct.

The Liver

The liver (hepar), the largest gland in the body, is situated in the upper and right parts of the abdominal cavity, occupying almost the whole of the right hypochondrium, the greater part of the epigastrium, and extending into the left hypochondrium as far as the left lateral line (Rouiller 1964). In the male it commonly weighs from 1·4 to 1·8 kg., in the female from 1·2 to 1·4 kg. with, however, a range of 1·0–2·5 kg. It is relatively much larger in the fetus than in the adult. It is somewhat wedge-shaped, reddish brown and, although firm and pliant to the touch, it is friable and easily lacerated. For this reason wounds of the liver must not be too tightly sutured. Owing to its great vascularity, wounds of the liver cause considerable haemorrhage. In spite of its relatively great weight, it is widely held that the liver, like the other abdominal organs, is maintained in its position, not by its peritoneal folds (p. 1376) or connective tissue attachments, but by the general intra-abdominal pressure due to the tonus of the abdominal muscles. Continuity of the hepatic veins with the inferior vena cava also provides some support. However, it should be noted that rather dogmatic statements such as the foregoing should be treated with caution. In the absence of systematic studies of the intra-abdominal perihepatic pressure gradients and their variations with posture, respiration, gastrointestinal dilatation and so forth, together with studies of the statics and dynamics of the peritoneal folds, connective tissues, adjacent viscera and vascular pedicles, our knowledge of the mechanisms maintaining the position of the liver (and other abdominal organs) remains quite rudimentary.

BORDERS OF THE LIVER

The superior, anterior and right surfaces are continuous at rounded 'borders', but a sharp *inferior border* (**8**.135), separates the right lateral and anterior surfaces from the inferior surface. Somewhat rounded where it intervenes between the right lateral and inferior surfaces, the inferior border is thin and sharp where it forms the lower margin of the anterior surface and is notched by the *ligamentum teres*, just to the right of the median plane. Lateral to the fundus of the gall bladder, which often corresponds to a second notch 4 to 5 cm to the right of the median plane, this border generally corresponds with the costal margin. To the left of the fundus of the gall bladder, it ascends less obliquely than the right costal margin and, crossing the infrasternal angle, passes behind the left costal margin in the neighbourhood of the tip of the eighth costal cartilage. Thereafter it ascends sharply and merges with the thin left margin of the left lobe. As it crosses the infrasternal angle the inferior border is closely related to the deep surface of the anterior abdominal wall and is readily accessible to examination in the living subject by percussion, though normally it is not palpable; in the median plane the inferior border of the liver lies on the transpyloric plane, about a hand's breadth below the xiphisternal joint (**8**.123). In women and children this border usually lies at a slightly lower level, and it tends to project downwards for a short distance below the right costal margin. As will be apparent below, the other 'borders' are indistinct and arbitrary.

THE HEPATIC LOBES

The liver is divisible into a right lobe and a much smaller left lobe. The original basis for the demarcation between these lobes was entirely a concatenation of superficial features (attachment of falciform fold, fissures for ligamenta teres et venosum), and this description persists in most anatomical texts. However, the classical studies of Hjortsjö (1948, 1951, 1956), upon the segmental branching of the bile ducts, hepatic artery, and portal vein within the liver, have emphasized that the primary anatomical and functional lobation is better defined by the distribution of the right and left hepatic ducts. Although there is individual variation and naturally some interdigitation between the smaller tributaries of the right and left hepatic ducts, the zone of division between the two lobes is a plane considerably to the right of the customary demarcation noted above (8.136C). Inferiorly and posteriorly the division is indicated by a line extending from the fossa for the gall bladder towards the inferior vena cava. Thus the quadrate and caudate lobes (*vide infra*) are part of the *left* lobe, *not* the right, except for the processus caudatus, which is served by the right hepatic duct. The plane between the right and left lobes is not quite sagittal in position, being tilted in both the horizontal and vertical planes.

A considerable number of other workers have confirmed Hjortsjö's classification (e.g. Healey and Schroy 1953, Goldsmith and Woodburne 1957, Stucke 1959) during investigations—by corrosion cast techniques, dissection and radiology—of the further subdivision of the lobes into segments (*vide infra*). In such preparations, particularly those produced by casts of the ducts or vessels, so-called 'fissures' are visible (8.136A) at the frontiers between the territories of their right and left branches, and also less clearly between lobar subdivisions (segments). It is important to note that these fissures, even the main interlobar zone described above, do not produce reliable surface indications which could be utilized with confidence in right or left lobectomy.

The *right lobe* of the liver, much greater in volume than the left, contributes to all surfaces of the organ, as described later, including the whole of the right or costal aspect. These somewhat arbitrarily separated surfaces—anterior, superior, inferior and posterior—all pass uninterruptedly on to the *left lobe*, except where shallow grooves partially demarcate the quadrate and caudate 'lobes', which are really parts of the left lobe.

The *quadrate lobe* is placed on the inferior surface, and is somewhat rectangular in outline. It is bounded in front by the inferior border of the liver; on the left by the fissure for the ligamentum teres; behind by the porta hepatis: and on the right by

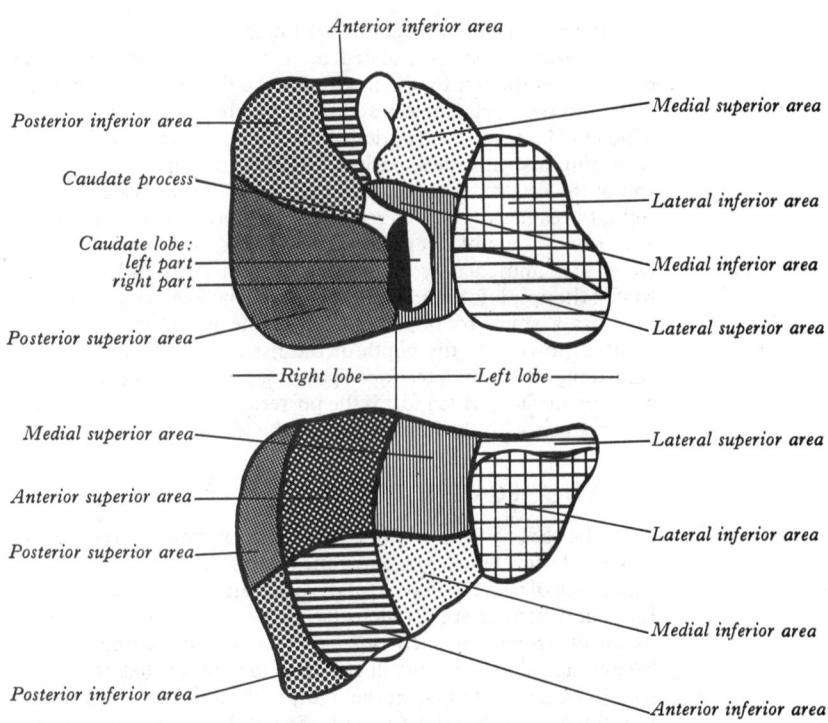

8.136B The segmentation of the liver, based upon the principal divisions of the hepatic artery and accompanying hepatic ducts. The upper drawing is of the visceral surface of the liver, the lower drawing is of the diaphragmatic surface. See text for further description. (After J. E. Healey and P. C. Schroy, from *Celiac and Superior Mesenteric Arteries* by R. A. Nesebar, P. L. Kornblith, J. J. Pollard and N. A. Michels, published by Little, Brown and Company, by courtesy of the authors and publishers.)

8.136C A resin corrosion cast of the blood vessels and duct systems of the liver of a woman, prepared by Dr. D. H. Tompsett of the Royal College of Surgeons of England. Bile duct, cystic duct, gall bladder, and their tributaries—yellow; the hepatic artery and its branches—red; the portal vein and its tributaries—light blue; the inferior vena cava, hepatic veins and their tributaries—dark blue. The photograph is of the visceral surface of the organ, and was taken before the finer blood vessels and ducts were removed by trimming. Posterior aspect is above.

a shallow fossa which lodges the gall bladder.

The *caudate lobe* is situated on the posterior surface. It is bounded on the left by the fissure for the ligamentum venosum, below by the porta hepatis and on the right by the deep groove which lodges the upper portion of the inferior vena cava. Above, it is continuous with the superior surface on the right of the upper end of the fissure for the ligamentum venosum. Below and to the right, the caudate lobe is connected to the mass of the right lobe by a narrow tongue of liver substance, termed the *caudate process*, which lies immediately behind the porta hepatis and forms the roof of the epiploic foramen. Below and to the left the caudate lobe presents a small rounded projection, the *papillary process*. In addition, owing to the depth of the fissure for the ligamentum venosum, the caudate lobe possesses a peritoneal-covered anterior surface, which forms the posterior wall of the fissure and is in contact with the hepatic part of the lesser omentum.

HEPATIC SEGMENTATION

As in the case of other organs with a hilum at which enter or leave single arteries, veins and sometimes ducts, investigations of the branching of these structures in the liver has promoted a system of lobes and further subdivisions (sectors or segments). From the mode of growth of such organs, with the branching and re-branching of vessels and ducts, it is inevitable that territories should be associated, as in the liver, with individual rami of the portal triad of tubes (i.e. fine branches of the portal vein, hepatic artery and hepatic ducts). The size and hence number of such territories, or segments, is dependent upon the size of tube recognized as 'segmental' or, in other words, upon the level within the arborization which is arbitrarily selected. The hepatic artery, portal vein and common bile duct divide and subdivide upon a common pattern, as is implied in the classical observations of Glisson (1654) and confirmed by many subsequent workers. Moreover, no evidence of intrahepatic anastomosis upon a significant scale between the elements of these dendriform systems has been recorded. For example, despite variation in its origin (and the occurrence of accessory vessels), the hepatic artery is a system of end-arteries (Michels 1966), a frequent consequence of the development of any organ from a single vascularized blastema of dichotomizing potentiality. However, the implications of this pattern of growth were not followed up until the late nineteenth century, apart from the recognition of right and left hepatic lobes (consult McIndoe and Counsellor 1927 and Hjortsjö 1948, for earlier literature). The division into two lobes, based upon the primary triadic divisions, and not upon surface features, is generally accepted, although confusing descriptions based on the latter persist in many texts. Inevitably, however, further subdivision has been suggested, if only in the interests of partial hepatectomy; and the classical figure in this field was Hjortsjö (1948, 1951, 1956, 1975), who was the first to propose a complete segmental model of the liver (8.136A) basing his concepts of subdivision of the hepatic lobes on dissections, injections, and radiographic technique, particularly applied to the biliary ducts and portal vein. Many subsequent workers have modified or enlarged upon his scheme of segmentation, subdividing some segments, redefining and renaming others. (*See*, for example, Elias and Petty 1952, Healey and Schroy 1953, Couinaud 1954, Goldsmith and Woodburne 1957, Bilbey and Rappaport 1960.) The main extension of Hjortsjö's scheme has been the division of his segments into superior and inferior parts, chiefly by Healey and Schroy 1953 (8.136B) and Couinaud 1954. Apart from this, and some minor differences in delineating the major segments, there is general agreement amongst these observers upon a division of the right hepatic lobe into approximately 'anterior', intermediate and 'posterior' segments, and of the left lobe into lateral and medial parts. Most of these reports have been based upon injections and casts, and principally upon corrosion casts of one or more of the biliary, hepatic arterial and portal venous ramifications. It is easy to discern in such casts (8.136C) a sagittal zone between the right and left lobes, the *fissura principalis*, which is picturesquely described by Hjortsjö (1956) as being in the same plane as the left tympanic membrane. All other 'fissures', as described with some variation by different investigators, are intersegmental, dividing the right and left lobes into segments. The term '*spatium*' has been suggested for 'fissure', and Hjortsjö recognized a number of these (8.136A). The appearance of such 'spaces' in corrosion casts is, of course, due to an absence of all but the smallest visible rami of the portal triad of structures. However, these demarcations do not correspond to any substantial zones of connective tissue in the living organ, which might produce superficial or internal indications for the purposes of subtotal resection (cf. broncho-pulmonary segments). The surface projections of the segmental fissures are shown in 8.136A and 8.136B, according to the schemes of Hjortsjö (1948) and of Healey and Schroy (1953), the chief difference being that in the latter scheme superior and inferior regions are shown in each of the major segments (as also recognized in the studies of Elias and Petty 1952, Couinaud 1954, and Platzer and Maurer 1966). Most recently Gupta *et al.* (1978) have put forward a similar scheme (9 segments), comparing previous findings with their own. No attempt will be made here to equate the segments described, with differing names and delimitations, by the investigators quoted, because this would not improve the reliability of this information in practical application. As Gupta *et al.* (1978) have pointed out, the segmental pattern (apart from the minor disagreements of different observers) is in itself variable; in their own series of 41 corrosion casts more than half of these showed marked variation in the volume of one segment or another, and this variability has been noted by others. Although most observers regard the segments as functionally independent and uncomplicated by, for example, intrahepatic arterial anastomosis, occasionally this has been noted. Surgical opinion is divided upon the reliability of hepatic segmentation in hepatic resections. Dawson (1974) and others consider resection of less than a lobe as hazardous, while others, such as Ryncki (1974) have recorded successful segmental resections. Even if such operations are to be attempted, individual variation points to the need for portal venography and cholangiography to define segmental patterns before operation, wherever feasible. Finally, the disposition of the hepatic veins and their tributaries is not a favourable factor. These veins, which have also been studied by corrosion casts (Goldsmith and Woodburne 1957), do not follow the pattern of the hepatic triads; they drain parts of adjoining segments. It is therefore most difficult to plan a resection plane which is optimal in respect of the triadic structures and the hepatic veins.

PERITONEAL CONNEXIONS OF THE LIVER

With the exception of an extensive, triangular area on the posterior surface of the right lobe, the liver is almost completely invested with peritoneum. It is connected to the stomach and duodenum, to the diaphragm and to the anterior abdominal wall by a number of peritoneal folds, and the lines along which they meet the organ are also necessarily devoid of a peritoneal covering. These folds include the falciform ligament, the right and left triangular ligaments, the coronary ligament and the lesser omentum.

The *falciform ligament* (8.93) is a sickle-shaped fold, consisting of two closely applied layers of peritoneum, which connects the liver to the diaphragm and the supra-umbilical part of the anterior abdominal wall. Its convex margin is fixed to the inferior surface of the diaphragm and to the posterior surface of the anterior abdominal wall, extending downwards to the umbilicus; as this attachment ascends from the umbilicus it passes slightly to the right of the median plane. The falciform ligament is attached to the notch for the ligamentum teres on the inferior border of the liver and to the anterior and superior surfaces of the liver. Its free edge, which extends from the umbilicus to the notch for the ligamentum teres, contains the latter structure and the small para-umbilical veins, and lies in front of the pyloric region of the stomach. At its upper end the two layers of the falciform ligament separate from each other and expose a small triangular area on the superior surface of the liver which is devoid of peritoneum. The left layer becomes continuous with the anterior layer of the left triangular ligament: the right with the upper layer of the coronary ligament.

The *coronary ligament* (**8.**135) is formed by the reflexion of the peritoneum from the diaphragm to the superior and posterior surfaces of the right lobe. It consists of an upper and a lower layer, continuous at their right extremities with the right triangular ligament of the liver then diverging widely to the left so as to enclose posteriorly a large triangular area of the right lobe which is not covered with peritoneum and is termed the 'bare area'. The upper layer is continuous with the right layer of the falciform ligament, skirts anteriorly the upper end of the groove for the inferior vena cava and then gradually descends from the posterior part of the upper surface to the upper part of the posterior surface. There it is continuous with the anterior layer of the right triangular ligament. The lower layer is continuous with the posterior layer of the right triangular ligament and passes almost horizontally along the lower limit of the posterior surface of the right lobe. In this situation the peritoneum may be reflected on to the upper part of the anterior surface of the right kidney (*hepatorenal ligament*), instead of on to the diaphragm beyond the margin of that organ. At its left extremity the lower layer of the coronary ligament passes in front of the lower end of the groove for the inferior vena cava and becomes continuous with the line of peritoneal reflexion from the right border of the caudate lobe, i.e. the right margin of the upper recess of the omental bursa.

The *left triangular ligament* passes from the upper surface of the left lobe upwards and backwards to the under surface of the diaphragm. It consists of two closely applied layers of peritoneum which become continuous with each other when traced to the left, where the ligament ends in a free margin. Traced to the right the anterior layer becomes continuous with the left layer of the falciform ligament, and the posterior layer with the anterior layer of the lesser omentum at the upper end of the fissure for the ligamentum venosum. The triangular ligament is placed in front of the abdominal part of the oesophagus, the upper end of the lesser omentum and part of the fundus of the stomach. Much variation occurs in this peritoneal fold; it may contain blood vessels of considerable size (Qutrequin *et al.* 1967).

The *right triangular ligament* is a short V-shaped fold which connects the lateral part of the posterior aspect of the right lobe to the diaphragm. The apex of the V forms a free right margin for the ligament, around which its two layers become continuous with each other. The ligament really constitutes the right extremity of the coronary ligament.

The *lesser omentum* has already been described (p. 1327) in detail. At the upper end of the fissure for the ligamentum venosum, its anterior layer becomes continuous with the posterior layer of the left triangular ligament, and its posterior layer with the line of reflexion of the peritoneum from the upper end of the right border of the caudate lobe and so, indirectly, with the lower layer of the coronary ligament (**8.**135B).

HEPATIC SURFACES

The superior surface of the liver (**8.**135) includes portions of the right and left lobes. It fits under the vault of the diaphragm, and is covered with peritoneum, except over a small triangular area where the two layers of the upper part of the falciform ligament diverge. Its right and left portions are convex, but its central part presents a shallow *cardiac impression*, which corresponds with the position of the heart on the upper surface of the diaphragm. It is related to the diaphragmatic pleura of the right side and the base of the right lung, to the pericardium and ventricular part of the heart, and, to a much smaller extent, to the diaphragmatic pleura of the left side and the base of the left lung.

It should be noted here that because of the overall convexity of the superior surfaces of the right and left lobes, the superior surface grades insensibly into the so-called anterior surface and also into the peritonealized part of the posterior surface of the right lobe. No definite border or other anatomical feature separates the superior, anterior, right lateral and right posterior *aspects* of the liver, and in many ways it is more appropriate to group these as the *diaphragmatic surface*. In contrast, most of the latter is separated from the *visceral surface* by a well defined, narrow edge or border.

The anterior surface, which is triangular in shape, is covered with peritoneum except at the line of attachment of the falciform ligament. A large part of this surface is in contact with the diaphragm, which separates it, on the right side, from the pleura and the sixth to the tenth ribs and their cartilages and, on the left side, from the seventh and eighth costal cartilages. The thin anterior margin of the base of the lung is related to the upper part of this surface, but the relationship is much more extensive on the right than it is on the left side. The median part of the anterior surface of the liver lies behind the xiphoid process of the sternum and the anterior abdominal wall in the infracostal angle (**8.**92).

The profile of the liver may be marked out on the anterior aspect of the trunk as follows (**8.**123): its upper border corresponds to a line drawn through the xiphisternal joint and ascending to the right to a point a little below the right nipple (fourth intercostal space) and ascending less sharply to the left to a point a little below and medial to the left nipple; its right border corresponds to a curved line, convex to the right, drawn from the right end of the upper border to a point 1 cm below the right costal margin at the tip of the tenth costal cartilage; its lower border is drawn by joining the ends of the upper and right borders, this line crossing the median plane at the level of the transpyloric plane and showing a slight concavity opposite the right linea semilunaris.

The right surface is covered with peritoneum and is related to the right portion of the diaphragm, which separates it from the right lung and pleura, and the seventh to eleventh ribs. Over its upper third both lung and pleura intervene between the diaphragm and the ribs; over its middle third only the costodiaphragmatic recess of the pleura is interposed; over its lower third the diaphragm is in actual contact with the costal arches.

The posterior surface is wide and convex on the right, but narrow on the left. A deep concavity marks its median region and corresponds with the forward convexity formed by the vertebral column (**8.**135A). A large part of this surface is devoid of peritoneal covering and is attached to the diaphragm by areolar tissue. This non-peritoneal surface constitutes the 'bare area' of the liver. It is triangular and is limited above and below by the superior and inferior layers of the coronary ligament. Its base, to the left, is the deep groove for the inferior vena cava, while its apex, directed downwards and laterally, corresponds with the right triangular ligament. The *groove for the inferior vena cava* is a deep, linear depression, occasionally a complete tunnel, on the posterior surface of the liver and is devoid of peritoneal covering. It lodges the upper part of the vessel and its floor is pierced by the hepatic veins (p. 763). At its lower end the groove is separated from the porta hepatis in front by the caudate process. Immediately lateral to the lower end of the groove, the 'bare area' adjoins the upper part of the right suprarenal gland (*suprarenal impression*). On the left side of the groove for the inferior vena cava the *caudate lobe* occupies part of the posterior surface of the left lobe. It lies in the superior recess of the omental bursa, and the peritoneum covering its posterior aspect is continued round its left border on to its anterior aspect, which forms the posterior wall of the fissure for the ligamentum venosum (**8.**135). The caudate lobe should therefore be regarded as projecting into the superior recess of the omental bursa from its right border. The posterior surface of the caudate lobe is related to the crura of the diaphragm above the aortic opening and to the right inferior phrenic artery, and is separated by them from the descending thoracic aorta. The *papillary process* often projects downwards in front of the origin of the coeliac artery.

The lips of the *fissure for the ligamentum venosum* (**8.**135) separate the posterior aspect of the caudate from the main part of the left lobe. The fissure itself cuts deeply into the liver in front of the caudate lobe and contains the two layers of the lesser omentum. At its lower end it curves laterally in front of the papillary process, and reaches the left extremity of the porta hepatis. The *ligamentum venosum*, which is the fibrous remnant of the ductus venosus (p. 665), is attached below to the posterior border of the left branch of the portal vein. It ascends in the floor of the fissure and passes laterally at the upper end of the caudate lobe to join the left hepatic vein near its point of entry into the inferior vena cava, or sometimes the vena cava itself.

The posterior aspect of the left lobe is marked by a shallow *oesophageal impression* near the upper end of the fissure for the ligamentum venosum, which is occupied by the abdominal part of the oesophagus. To the left of this impression the left lobe is related to a part of the fundus of the stomach.

The inferior or visceral surface (8.135) is directed downwards, backwards and to the left, and, in the formalin-hardened organ, bears the imprints of the neighbouring viscera. It is invested with visceral peritoneum except at the porta hepatis, the fissure for the ligamentum teres and the fossa for the gall bladder. On the inferior surface of the left lobe, in direct continuity with the oesophageal impression, the *gastric impression* is moulded over the stomach. On the right of this impression there is a rounded *omental tuberosity*. This occupies the concavity of the lesser curvature of the stomach and is in contact with the lesser omentum. The *fissure for the ligamentum teres* is a cleft of variable depth which passes upwards and backwards from the corresponding notch on the inferior border of the liver to the left end of the porta hepatis where it meets the lower end of the fissure for the ligamentum venosum. It forms the left boundary of the quadrate lobe and may be, partially or completely, bridged over by a band of liver substance. Its floor lodges the *ligamentum teres of the liver*, which is the obliterated vestige of the left umbilical vein of the fetus (p. 665). Commencing at the umbilicus it ascends in the free margin of the falciform ligament to the inferior border of the liver, traverses the fissure and ends by joining the left branch of the portal vein at the left extremity of the porta hepatis opposite the attachment of the ligamentum venosum.

The gastric impression may be continued on to the anterior part of the *quadrate lobe*, which is hollowed out and moulded over the pyloric part of the stomach and the beginning of the duodenum, when these organs are dilated. The posterior part of the quadrate lobe is in contact with the right free border of the lesser omentum and its contained structures. When the stomach is empty the quadrate lobe is related to the superior part of the duodenum and a segment of the transverse colon.

The *porta hepatis* is placed on the inferior surface of the liver between the quadrate lobe in front and the caudate process behind. It is a deep fissure which runs transversely between the upper ends of the fissure for the ligamentum teres and the fossa for the gall bladder. Through the porta hepatis the portal vein, the hepatic artery proper and the hepatic plexus of nerves enter the liver, and the right and left hepatic ducts and some lymph vessels emerge. The hepatic ducts are situated anteriorly, the portal vein and its right and left branches posteriorly and the hepatic artery proper and its right and left branches are intermediate in position.

The *caudate process* connects the lower and lateral part of the caudate lobe (of the left lobe) to the right lobe. It is placed behind the porta hepatis and in front of the inferior vena cava and forms the roof of the epiploic foramen. The caudate process is commonly considered a part of the right lobe, whereas the caudate lobe is, of course, part of the territory of the left hepatic duct, i.e. it is part of the *left* lobe (*vide supra*).

The *fossa for the gall bladder* forms the right boundary of the quadrate lobe and extends from the inferior border of the liver to the right extremity of the porta hepatis. It is usually shallow and devoid of peritoneal covering, but the breadth of this bare area is subject to individual variation.

To the right of the fossa for the gall bladder, the inferior surface of the right lobe is marked by three impressions, viz. colic, renal and duodenal. The *colic impression* is related to the right colic flexure and is within the anterior part of the area, immediately adjoining the inferior border of the liver. The *renal impression* is usually well marked; it is situated behind the colic impression and is separated from the neck and the adjoining part of the gall bladder by the duodenal impression. It is related to the upper part of the anterior surface of the right kidney, and in its superomedial part to the lower pole of the right suprarenal gland. When the lower layer of the coronary ligament is reflected from the liver on to the right kidney, the renal and suprarenal impressions extend for a short distance on to the lower part of the 'bare area'. The *duodenal impression* lies to the lateral side of the neck and adjoining part of the gall bladder, and is related to the termination of the superior part and the commencement of the descending part of the duodenum.

The relations of the liver show considerable variation, for which posture and the movements of respiration are only partly responsible. Those which have been enumerated refer to the body in the supine recumbent position.

The branches of the portal vein and the tributaries of the hepatic veins are more numerous before birth; after birth a reduction in their number takes place by fusion of vessels or by degeneration. In the fetus the portal vein joins the umbilical vein in a smooth curve to the right; this direction is maintained after

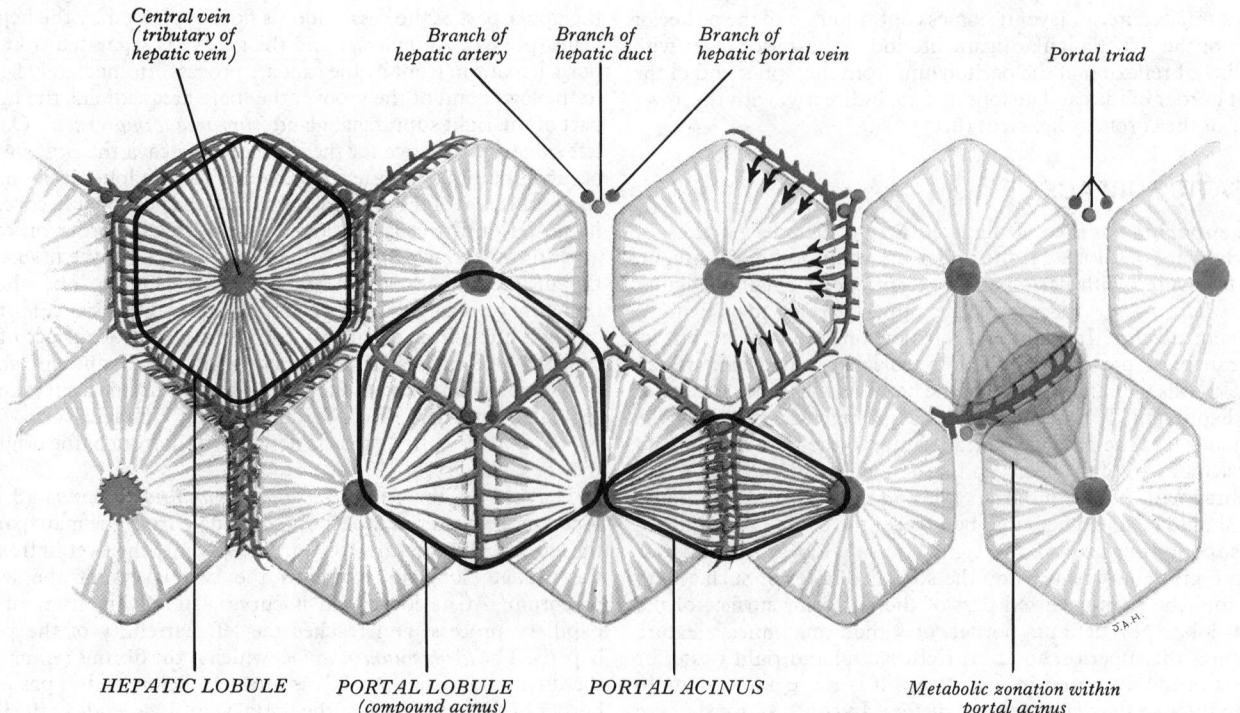

Central vein
(tributary of
hepatic vein)   Branch of   Branch of   Branch of
                       hepatic artery   hepatic duct   hepatic portal vein   Portal triad

HEPATIC LOBULE   PORTAL LOBULE   PORTAL ACINUS   Metabolic zonation within
                        (compound acinus)   portal acinus

8.137 A diagram of the histological organization of the liver, showing the principal types of subdivisions which have been proposed. For purposes of clarity, the territories of the classical hepatic lobules are shown as regular hexagons, unlike their real appearance which is highly variable (see text).

birth and there is a sharp angle between the trunk of the portal vein and its left branch, so that the left (physiological) lobe of the liver would appear to be at a circulatory disadvantage and not able to keep pace with the growth of the right lobe. At the left end of the left lobe in the adult there is sometimes a fibrous band (*the fibrous appendix of the liver*), which represents the atrophied remains of the more extensive left lobe in the young child; it contains atrophied remnants of bile ducts, known as the *vasa aberrantia of the liver*. Similar remnants may be present in the edges of the left lobe and near the inferior vena cava. Occasionally the lower border of the right lobe, a little to the right of the gall bladder, may project downwards for a considerable distance as a broad tongue-like process (*Riedel's lobe*).

HEPATIC VESSELS AND NERVES

The vessels connected with the liver are the portal vein, the hepatic artery proper and the hepatic veins.

The *portal vein* and *hepatic artery proper* ascend between the layers of the lesser omentum to the porta hepatis, where each divides into two branches; the *bile duct* and *lymph vessels* descend from the porta hepatis between the layers of the same omentum. They are all enveloped in a loose areolar tissue, termed the *perivascular fibrous capsule* (or *hepatobiliary capsule of Glisson*), which surrounds the vessels in their course through the portal canals in the interior of the liver and is continuous with the fibrous capsule of the liver.

The hepatic artery and its branches pursue a variable course in the porta hepatis. The smaller branches finally become associated with a specific branch of the portal vein and are distributed to the same territory. There are no anastomoses between hepatic arterial territories and hench each branch is an end-artery (Glauser 1953).

The *hepatic veins* (**8.**136C) convey blood from the liver to the inferior vena cava, and are described on p. 763. They have very little fibrous investment, but what there is binds them closely to the walls of the canals through which they run; so that, on section of the liver, they remain widely open and are solitary, and may be easily distinguished from the branches of the portal vein, which are more or less collapsed, and always accompanied by an artery and duct.

The *lymph vessels* of the liver are described on p. 794. Lymph from the liver has a very rich protein content and obstruction of the venous drainage of the liver results in a considerable increase of lymph in the thoracic duct. The importance of the trans-diaphragmatic lymph drainage of the liver into the internal mammary and diaphragmatic lymph nodes has been emphasized by Nidden *et al.* (1973). They confirm that this drainage reaches the right lymphatic duct, partly via the tracheobronchial group of nodes.

The *nerves* of the liver are derived from the hepatic plexus (p. 1134) and contain both sympathetic and parasympathetic (vagal) fibres. They enter the liver at the porta hepatis and for the most part accompany the blood vessels and bile ducts; very few run amongst the liver cells and the node of termination of these is not known with certainty. Nerve fibres, both myelinated and non-myelinated, also reach the liver parenchyma from nerves in the various peritoneal folds of the organ (Sutherland 1965).

HEPATIC STRUCTURE

In the following remarks, hepatic structure (*see* **8.**137–140) is considered firstly in relation to its lobular architecture, followed by the cytology of the hepatocyte, the microanatomy of the lobules, the vascular channels and the duct system.

The greater part of the liver is invested with peritoneum, which covers a thin capsule of connective tissue (Glisson's capsule). As classically described in the pig and in several other species of mammals, the liver consists of a very large number of polyhedral *hepatic lobules* (which often appear to be hexagonal in histological sections), each being about 1 mm in diameter and having a small *central vein* (a tributary of the hepatic veins) as its central axis, and surrounded at its edges by groups of three tubes, each group being termed a *portal triad* (**8.**137). Each portal triad contains a branch of the portal vein, a branch of the hepatic artery and an

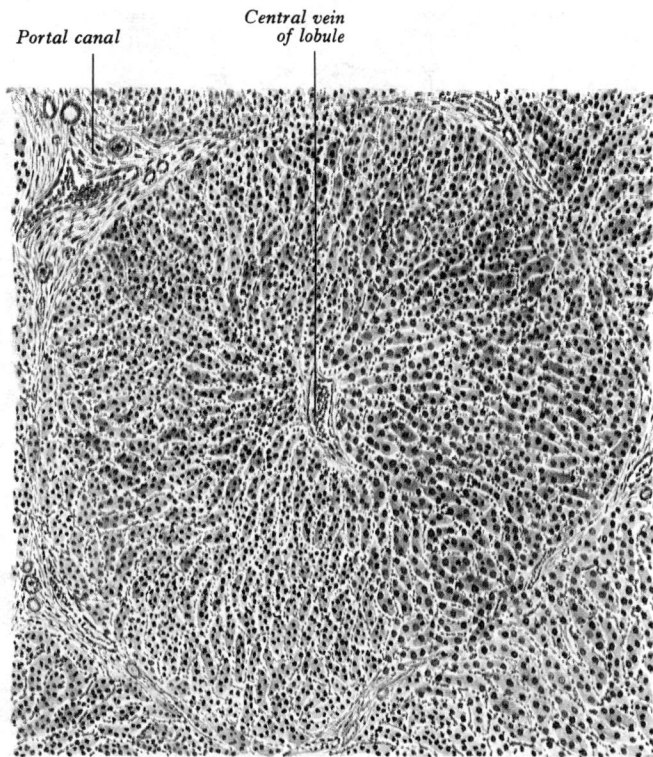

Portal canal   *Central vein of lobule*

8.138A   A section through a hepatic lobule. Human. Stained with haematoxylin and eosin. Magnification about × 70. (After Sobotta.)

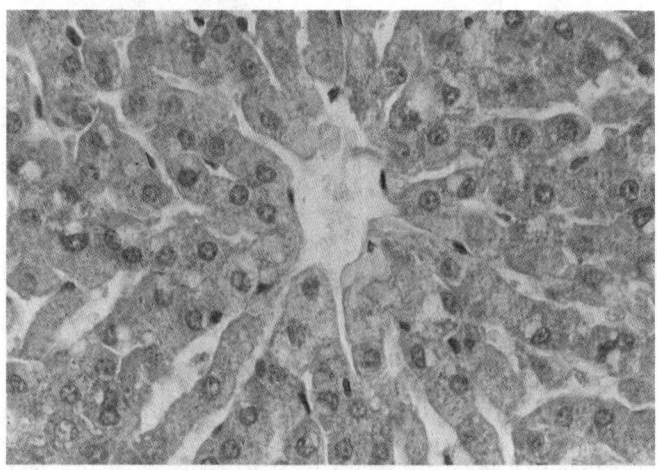

8.138B   A section through a number of hepatic cords radiating from a central hepatic venule of the rabbit. The hepatocytes appear cuboidal, and the Kupffer cells, which line the sinusoids, have flattened, densely staining nuclei. Haematoxylin and eosin.

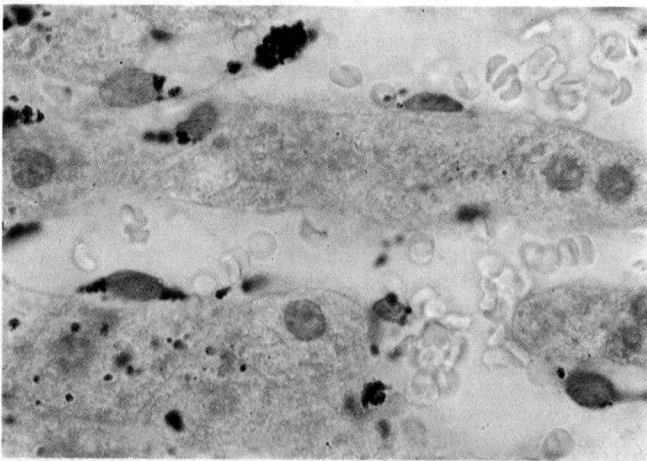

8.138C   A section similar to that shown in B, but taken from the liver of a rabbit previously injected intravenously with carbon particles. The Kupffer cell nuclei are outlined with phagocytosed particles which demonstrate the limits of the cell cytoplasm.

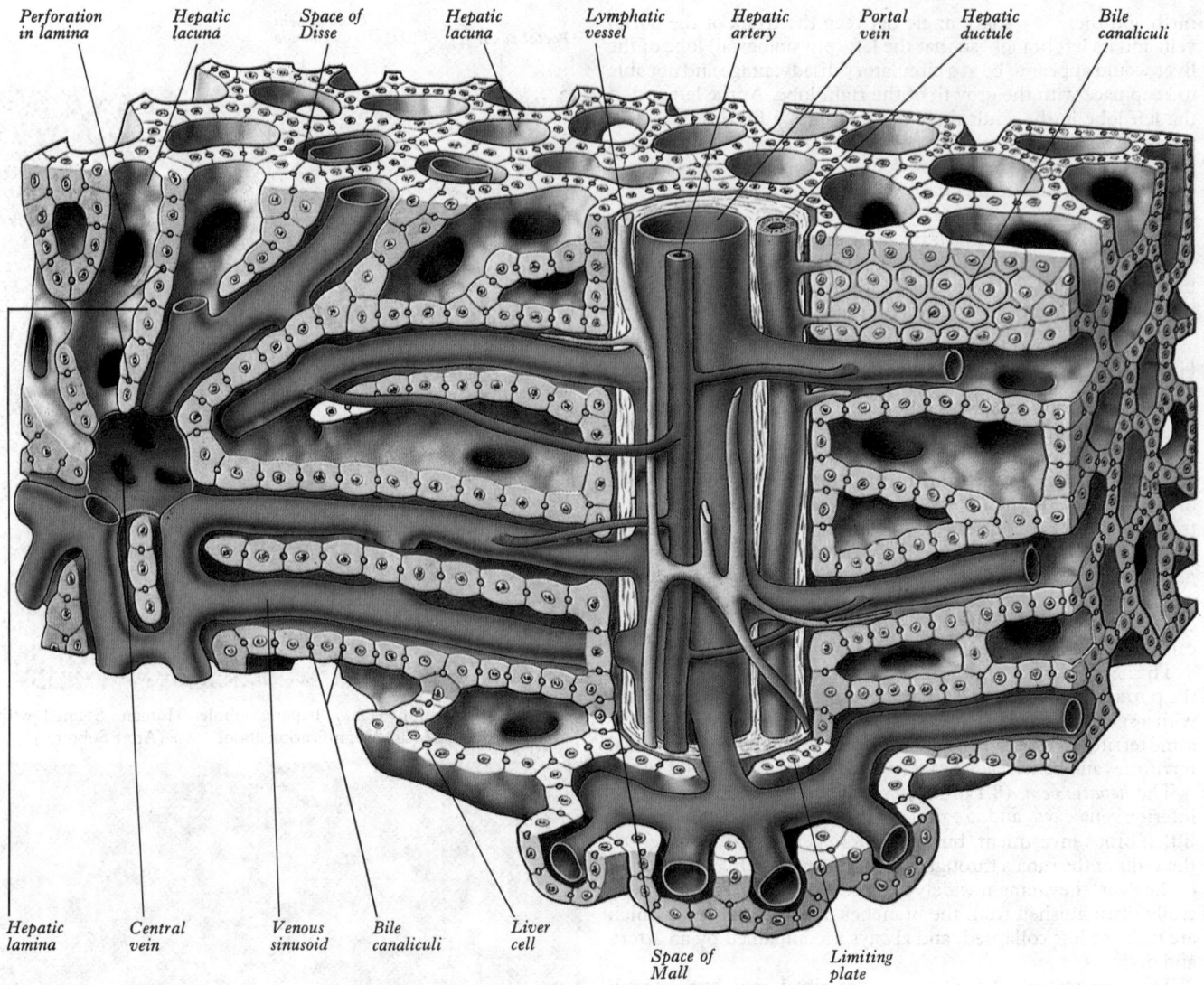

Perforation in lamina — Hepatic lacuna — Space of Disse — Hepatic lacuna — Lymphatic vessel — Hepatic artery — Portal vein — Hepatic ductule — Bile canaliculi

Hepatic lamina — Central vein — Venous sinusoid — Bile canaliculi — Liver cell — Space of Mall — Limiting plate

8.139   Diagram of hepatic structure (after Prof. H. Elias, Department of Anatomy, Chicago Medical School).

interlobular bile ductule; all these structures are in a connective tissue sheath, the *portal canal* or *perivascular fibrous capsule*. In the adult pig, each *hepatic lobule* is sharply marked off from neighbouring lobules by connective tissue septa, but in man (and many other mammals) the hepatic lobules are not normally separated by such septa. A '*portal lobule*', on the other hand, consists of the adjoining parts of three hepatic lobules, the bile from which drains into a bile ductule in the portal canal at the meeting place of the three hepatic lobules; in sections therefore, a portal lobule is the polygonal territory centred on a portal triad, its boundary line passing through adjacent central veins. The concept of a third type of unit structure, the *portal acinus*, is useful in considerations of blood flow, patterns of oxygenation, and pathological degeneration. This unit is centred around a preterminal branch of an hepatic arteriole, and includes the parenchyma served by this vessel, bounded by the territories of other portal acini and by two adjacent central veins (**8.**137). Because several phenomena, for example, zones of anoxic damage, glycogen deposition and removal, and toxic trauma are related to the arterial blood flow patterns, they tend to follow the acinar pattern rather than any other. Rappaport (1969) has also proposed multiple units, (*compound portal acini*) for more complex groups of hepatic units. Neither the hepatic nor the portal lobules are to be considered as fixed anatomical entities; under normal conditions the '*hepatic lobular structure*' is evident, but this can be changed to a '*portal lobular structure*' by alteration in the relative blood pressure in the portal and hepatic veins (e.g. by raising the hepatic venous pressure or lowering the portal pressure), the change being reversible. Such alterations in the

venous pressure gradients are produced by various pathological conditions.

The bulk of the cells within the liver, constituting about 80%, are the *hepatocytes* or *parenchymal cells*, which are derived from the endoderm of the caudal part of the foregut during embryonic development, and with which they retain their connexion through the bile ductules and hepatic ducts. Each hepatocyte is polyhedral, having five to twelve sides, and is from 12 to 25 μm across. The nucleus is spheroidal and euchromatic, often also being polyploid or present in multiples (2 or more) within each cell (Doljansky 1960). The cytoplasm typically displays copious amounts of both granular and agranular endoplasmic reticulum, high concentrations of mitochondria, lysosomes, and a well-developed Golgi complex—all features indicating a high metabolic activity. Glycogen granules and lipid vacuoles are usually prominent storage substances. Peroxisomes and vacuoles containing enzymes such as urease (uricosomes) in distinctive crystalline forms are characteristic features, reflecting the complex metabolic pathways operating in these cells. Their role in iron metabolism is seen in the common presence of iron storage vacuoles containing crystals of ferritin and of haemosiderin. Where hepatocytes adjoin bile canaliculi, microvilli are present at their surface, and numerous membrane-bound vesicles are clustered near the lumen (**8.**140A).

Hepatocytes carry out a multitude of metabolic activities. They synthesize and release into the blood various plasma proteins including albumins, clotting factors and complement components; they deaminate amino acids by means of the urea cycle pathway, liberating urea into the blood for excretion through the urinary system; they convert bilirubin, formed by breakdown of haemoglobin, to the bile pigment biliverdin which is secreted into the biliary tract; they synthesize bile salts which are also secreted into the bile as important agents in the emulsification of fats during digestion; they eliminate many endogenous and exogenous substances from the bloodstream, and are thus important in detoxification of the blood; they convert the thyroid hormone tetra-iodothyronin to the more active tri-iodothyronin. Hepatocytes also store carbohydrates as glycogen, and tri-glycerides as lipid droplets, metabolizing these substances when required and releasing glucose and lipid into the bloodstream. Most of the lipid secreted in this way passes out of the perisinusoidal surface of the cell in secretory vesicles, having been conjugated with protein in the endoplasmic reticulum and Golgi apparatus, and forms the 'very low density lipoprotein' moiety of the blood plasma (Claude 1970). Some cholesterol is also secreted into the bile, apparently as an excretory activity. Iron is stored in the cells as ferritin.

All of these metabolic processes also result in the production of heat by the liver which is therefore important in the maintenance of body temperature.

The large number of chemical activities carried out by the hepatocytes is reflected in their structural complexity, which varies according to metabolic demand. This is seen, for example in the proliferation of the agranular endoplasmic reticulum which occurs during barbiturate detoxification (Jones and Fawcett 1966), and is also observed in the effects of starvation, when glycogen and lipid reserves disappear. Also, because of their involvement in many chemical processes, hepatocytes are particularly vulnerable to anoxia, various poisons and carcinogens which cause characteristic patterns of degeneration within the portal acini.

The arrangement of hepatocytes within the liver has been the subject of some controversy. It has been customary to consider the hepatic lobule as consisting of 'cords' of liver cells radiating from a central vein, each cord consisting of two rows of liver cells, with a bile capillary between them, and radiating venous sinusoids lying between adjacent cords in such a way that the bile capillary is separated from the sinusoids by the thickness of a liver cell. More recent studies, however, indicate that the liver cells are really arranged as plates or sheets (*hepatic laminae*), *one cell thick*, which form a continuous system throughout the liver (**8.**139). These laminae form a 'wall-work' or *muralium* and are irregularly arranged, with interlaminar bridges of liver cells connecting adjacent laminae. Between the laminae lie spaces (*hepatic lacunae*) which contain the venous sinusoids, and the laminae are perforated to allow the passage of anastomoses between the sinusoids. Where the liver cells adjoin portal canals or tributaries of the hepatic vein, they form a sheet, the *limiting plate*, which surrounds the vessels and is perforated by radicles of these vessels as well as by branches of the hepatic artery and bile ductules; a similar limiting plate, consisting of a single layer of liver cells, lies beneath the capsule of the liver.

In histological sections of the liver (**8.**138A), the rows of liver cells seen radiating from the central vein to the periphery of the hepatic lobule really represent sections through the hepatic laminae. These rows of cells, with intervening sinusoids, do not pass regularly and straight like the spokes of a wheel to the periphery of the lobule, but pursue irregular courses, because the hepatic laminae themselves are irregular and branched. Some of the rows of cells (and sinusoids) pass at the periphery of the lobule between adjacent portal canals, while others pursue very indirect courses between the central vein and the portal canals.

The intralobular venous sinusoids are wider than blood capillaries and are lined by thin and highly fenestrated endothelial cells interspersed with *hepatic macrophages* also known as *stellate cells of von Kupffer* (*Kupffer cells*). The hepatic macrophages form a major part of the mononuclear phagocyte system and they are responsible for clearing much of the cellular debris and other particulate material from the circulation and for removing microbial infections, these activities being shared with the spleen and circulating leucocytes (Wisse 1970). The radicles of the portal vein contained in the portal canals give off branches, called *inlet venules*, which pass through holes in the limiting plate

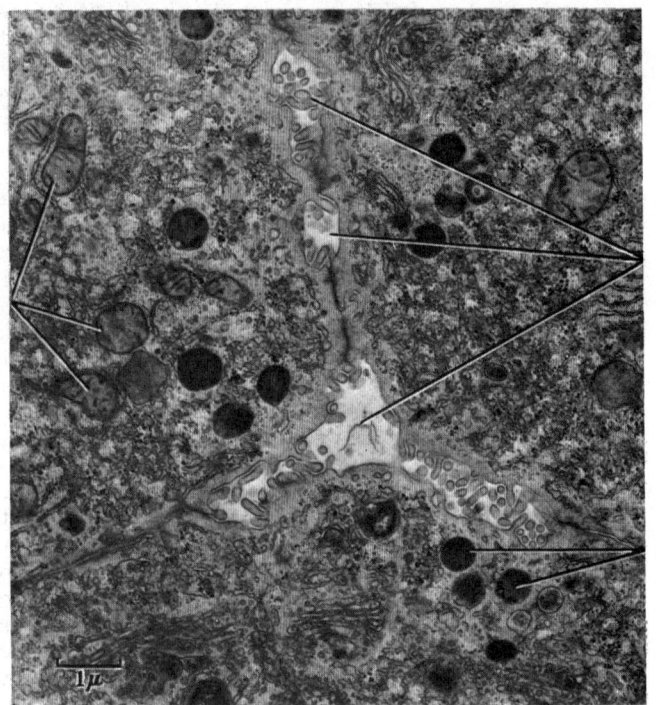

8.140A Electron micrograph showing portions of three adjacent hepatocytes and the intervening bile canaliculi. Magnification × 10,000.

surrounding the portal canal in order to enter the hepatic lobules, where they become continuous with the sinusoids. These convey blood mainly from the interlobular branches of the portal vein (in the portal canals) to the central veins. The sinusoids lie in the hepatic lacunae between the hepatic laminae, and are separated from the liver plates by the *perisinusoidal space of Disse* (**8.**140B), which becomes distended in conditions of anoxia. Reticulin fibres are present here, and irregular microvilli of the phagocytic cells in these walls and of adacent hepatocytes also project into the space of Disse, itself continuous at the periphery of the lobule with a *space of Mall* which surrounds the vessels and bile ductules in the portal canals; and it is in the latter space that the lymph vessels of the liver commence (as in other organs) by blind ends. Very few, extremely small, lymph vessels penetrate into the periphery of the lobule. Adipocytes are also found occasionally in the space of Disse.

The *central veins* from adjacent lobules join to form *interlobular veins* and these in their turn unite to form the *hepatic veins*, which drain the blood from the liver into the inferior vena cava.

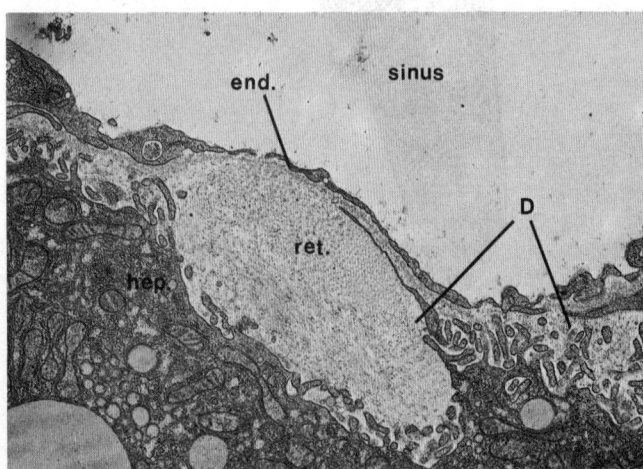

8.140B A transmission electron micrograph of the border of a hepatic sinusoid (*sinus*) showing part of an hepatocyte (*hep.*) and the tenuous fenestrated endothelium (*end.*) separated by the space of Disse (*D*) in part of which lies a reticulin bundle (*ret.*). Magnification × 5,000.

In the hepatic lobules, minute *bile canaliculi* (**8.**140A) form networks of polygonal meshes, each mesh surrounding an individual liver cell except on the surfaces of the cell apposed to the venous sinusoids. Individual bile canaliculi lie between the walls of the liver cells which form their borders. At the periphery of the hepatic lobules the bile canaliculi join to form extremely thin intralobular bile ductules (*terminal ductules* or *canals of Hering*) lined by cuboidal epithelium, penetrating the terminal plates to enter the interlobular hepatic ductules within the portal tracts. The intralobular ductules are distinct from other canals of the biliary tract in their structure and reactions to injury, proliferating greatly, for example, when the flow of bile is interrupted by extrahepatic obstruction (Biava 1964; Jones *et al.* 1975). The hepatic ductules in the portal canals are lined by cuboidal to columnar cells which may contain crystals of cholesterol and droplets of fat.

The relationship between the blood, the hepatocytes and the bile canaliculi is clearly seen in experiments with colloidal tracers, the diffusion barriers to which can be determined by electron microscopy. Horseradish peroxidase, for example, injected into the circulation, diffuses freely from the sinusoids through the gaps between the stellate endothelial cells, into the intercellular spaces between hepatocytes, which are held together largely by communicating junctions (p. 7). Diffusion is, however, halted at the edges of the bile canaliculi by the tight junctions which mark the boundaries of the hepatocytes. It is inferred that blood-borne materials have free access to the hepatocyte, but that substances can only enter the bile canaliculi through the cytoplasm of the hepatocytes, which are therefore in many respects exocrine glandular cells secreting into a duct system.

The preterminal branches of the hepatic arteries, contained in the portal canals, give off branches which convey arterial blood to the liver sinusoids by a variety of routes. The chief pathway is through the system of fine capillaries which form a plexus around the interlobular hepatic ductules and ducts and drain into various branches of the hepatic portal veins, inlet venules and hepatic sinusoids. Some arterial blood also passes directly into the hepatic sinusoids without entering such capillary networks, but this appears to form only a small proportion of the total flow (Burkel 1970; Healey 1970; Jones and Spring-Mills 1977). The sinusoids therefore contain mixed venous and arterial blood which

nourishes the liver cells. The composition, amount and velocity of the blood streaming at any time through any minute area of the liver may change and be adapted to the necessities of the moment by sphincteric arrangements in the inlet venules and in the branches of the hepatic arteries, and by the contractile walls of the sinusoids. Each portal triad (branches of the hepatic arteries, portal vein and bile ducts) supplies its own sharply delimited territory, and normally there are no anastomoses between these territories. On the other hand, the hepatic veins for the main part run independently of the portal veins, hepatic arteries and bile ductules, and cross the boundary lines of the territories of the liver supplied by the latter structures.

During fetal life the liver acts as one of the main haemopoietic organs, both red and white blood corpuscles being developed in the mesenchyme covering the endothelium of the sinusoids (p. 66).

The Biliary Ducts and Gall Bladder

The excretory apparatus of the liver consists of: (1) the *common hepatic duct*, formed by the junction of the *right* and *left hepatic ducts* which leave the liver at the porta hepatis; (2) the *gall bladder* which serves as a reservoir for the bile; (3) the *cystic duct*, or duct of the gall bladder; (4) the *bile duct*, formed by the junction of the common hepatic and cystic ducts.

THE COMMON HEPATIC DUCT

Two main ducts (right and left hepatic) issue from the liver and unite near the right end of the porta hepatis to form the common hepatic duct, which passes downwards for about 3 cm, and is joined on its right side and at an acute angle by the cystic duct; by the union of the common hepatic with the cystic duct, the bile duct is formed (**8.**141). The common hepatic duct is on the right of the hepatic artery proper and in front of the portal vein.

THE GALL BLADDER

The gall bladder (**8.**135, 141, 142) is a slate-blue, piriform sac partly contained in a fossa on the inferior surface of the right hepatic lobe, extending from near the right extremity of the porta hepatis to the inferior border of the liver. Its upper surface is attached to the liver by connective tissue; its under surface and sides are covered with peritoneum continued from the surface of the liver. Occasionally it is completely invested with peritoneum and may be connected to the liver by a short mesentery. It is from 7 to 10 cm long, 3 cm broad at its widest part, and from 30 to 50 ml in capacity. It is divided into a fundus, body and neck.

The *fundus* or expanded end, is directed downwards, forwards and to the right. It projects beyond the inferior border of the liver, and comes into relationship with the posterior surface of the anterior abdominal wall below the ninth right costal cartilage, behind the point where the lateral edge of the right rectus abdominis crosses the costal margin; posteriorly the fundus is in relation with the transverse colon, near its commencement. (These relations are, however, considerably altered when the gall bladder descends lower in the abdomen, as it frequently does, particularly in slender females. *See* Fleischner and Sayegh 1958.) It is entirely covered with peritoneum. The *body* is directed upwards, backwards and to the left; near the right end of the porta hepatis it is continuous with the neck. It is in relation by its upper surface with the liver; by its under surface with the right part of the transverse colon; and further back with the superior part of the duodenum and the upper end of the descending part. The *neck* is narrow; it curves upwards and forwards, and then, turning abruptly backwards and downwards, becomes continuous with the cystic duct; at its point of continuity with the cystic duct there is a constriction. The neck is attached to the liver by areolar tissue in which the cystic artery is embedded. The mucous membrane which lines the neck projects into its lumen in the form of oblique ridges, forming a sort of spiral valve; when the neck is distended, this valve causes the surface of the neck to present a spiral constriction.

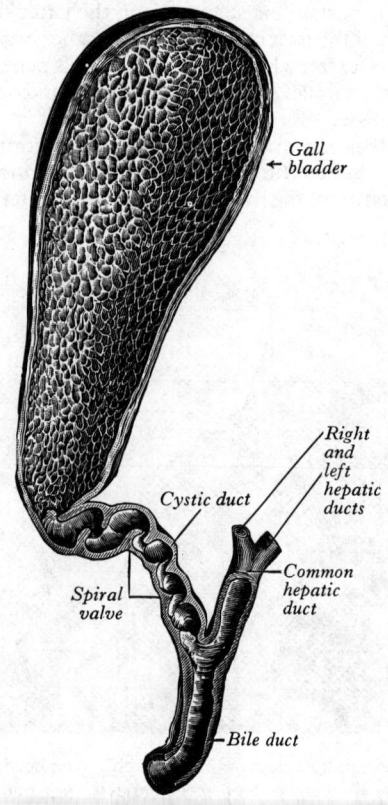

Gall bladder →

Right and left hepatic ducts

Cystic duct

Common hepatic duct

Spiral valve

Bile duct

8.141   Interior of the gall bladder and bile ducts.

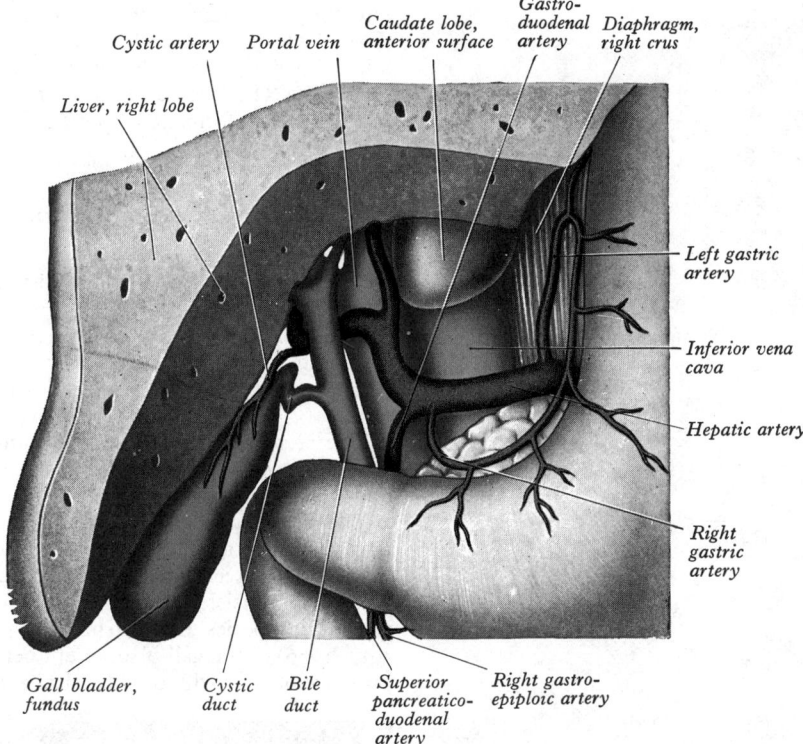

Cystic artery   *Portal vein*   *Caudate lobe, anterior surface*   *Gastro-duodenal artery*   *Diaphragm, right crus*

Liver, right lobe

Left gastric artery

Inferior vena cava

Hepatic artery

Right gastric artery

Gall bladder, fundus   *Cystic duct*   *Bile duct*   *Superior pancreatico-duodenal artery*   *Right gastro-epiploic artery*

8.142   Dissection to show the relations of the hepatic artery, bile duct and portal vein to each other in the lesser omentum. Anterior aspect.

From the right wall of the neck of the gall bladder a small pouch may project downwards and backwards towards the duodenum. This pouch, often termed Hartmann's pouch (although originally described by Broca) has been widely regarded as a constant feature of the normal gall bladder, but investigations (Davies and Harding 1942) have shown that it is always associated with pathological conditions, especially dilatation. When the pouch is well marked, the cystic duct arises from its upper and left wall and not from what appears to be the apex of the gall bladder.

The cystic duct (**8**.141, 142), from 3 to 4 cm long, passes backwards, downwards and to the left from the neck of the gall bladder, and joins the common hepatic duct to form the bile duct; it runs parallel with and adheres to the common hepatic duct for a short distance before joining with it. The junction is usually situated immediately below the porta hepatis, but it may be at a considerably lower level. In the latter event the cystic duct lies in the right free margin of the lesser omentum. The mucous membrane lining its interior is thrown into a series of crescentic folds, from five to twelve in number, similar to those found in the neck of the gall bladder. They project into the duct in regular succession, and are directed obliquely round the tube, presenting much the appearance of a crescentic, *spiral valve* (**8**.141). When the duct is distended, the spaces between the folds are dilated, and the exterior of the duct appears twisted in the same manner as the neck of the gall bladder.

The bile duct is formed near the porta hepatis by the junction of the cystic and common hepatic ducts; it is usually about 7·5 cm long and about 6 mm in diameter (*vide infra*).

It runs at first downwards, backwards and slightly to the left, anterior to the epiploic foramen; here it lies in the right border of the lesser omentum, in front of the right edge of the portal vein, and on the right of the hepatic artery proper (**8**.142). It passes behind the superior part of the duodenum with the gastro-duodenal artery on its left, and then runs in a groove on the upper and lateral part of the posterior surface of the head of the pancreas (**8**.131); here it is situated in front of the inferior vena cava, and is sometimes completely embedded in the pancreatic tissue (pp. 1370, 1373). It has been pointed out (Lytle 1959) that the bile duct may lie close to the left border of the descending part of the duodenum or lie as far away as 2 cm from the duodenal wall, and that, even when the duct is embedded in the pancreas, its position

is indicated by a groove on the back of the head of the pancreas. This groove can be palpated by the fingers of the left hand passed behind the descending part of the duodenum, with the thumb on the front of the head of the pancreas to act as a counter pressure, and through the groove stones in the bile duct may be detected and removed. At the left side of the descending part of the duodenum the bile duct comes into contact with the pancreatic duct and accompanies it into the wall of this part of the gut, and there the two ducts usually unite to form the *hepatopancreatic ampulla* (p. 1370); the distal, constricted end of this ampulla opens into the descending part of the duodenum on the summit of the major duodenal papilla (**8**.132) situated about 8 to 10 cm from the pylorus (p. 1344).

The bile duct may be indicated on the anterior surface of the abdomen by a line which begins 5 cm above the transpyloric plane and 2 cm to the right of the median plane, and runs downwards for 7·5 cm.

Vessels and Nerves

The *cystic artery* and its distribution are described in some detail on p. 714 and the *cystic veins* on p. 765. The lower part of the bile duct receives several branches from the *posterior superior pancreatico-duodenal artery* (p. 714), while the upper part of the bile duct and the hepatic ducts are supplied with branches from the *cystic artery*. The *right hepatic artery* gives branches to the middle part of the bile duct, though these are very small, the main supply being from the *cystic* and *posterior superior pancreaticoduodenal arteries*. There is considerable variation in the arrangement of the above vessels (Shapiro and Robillard 1948; Michels 1962). The posterior superior pancreaticoduodenal artery ends below by anastomosing with the posterior branch of the inferior pancreaticoduodenal artery in the vicinity of the hepatopancreatic ampulla; in cases where this anastomosis is poor, ligation of the posterior superior pancreaticoduodenal artery may result in gangrene or stricture of the bile duct (Henley 1955). The *veins* from the upper part of the bile duct and the hepatic ducts, like those from the gall bladder and cystic duct, generally enter the liver directly, while those from the lower part of the bile duct enter the portal vein. The *lymph vessels* of the gall bladder and bile ducts are described on p. 794. The *nerves* of the gall bladder, which are sympathetic and parasympathetic, are derived from the

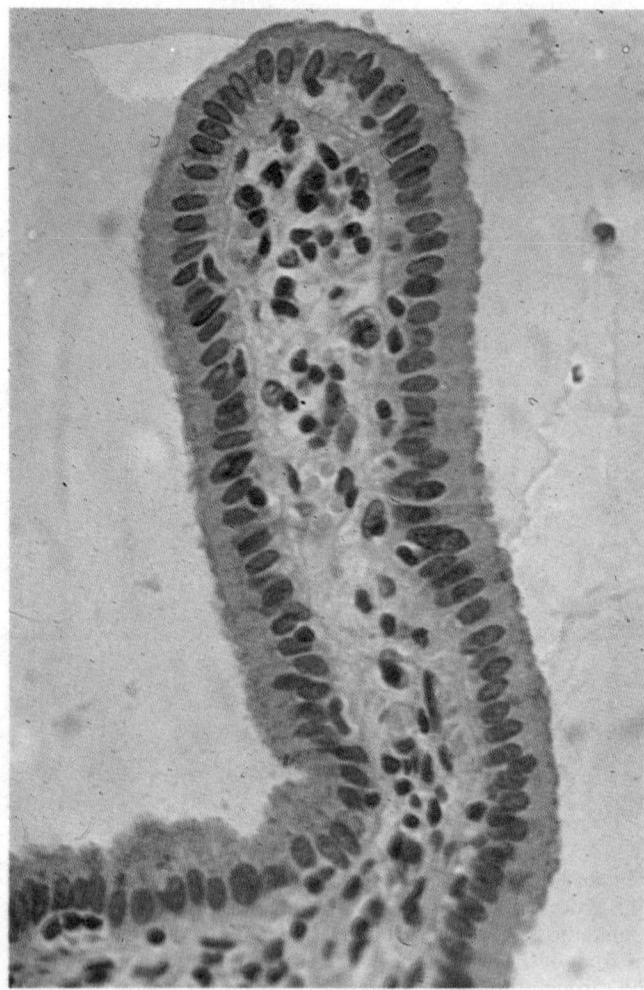

8.143 A section through a surface projection (ruga) of the gall bladder showing the columnar epithelial cells and lamina propria. Haematoxylin and eosin. Magnification × 500.

coeliac plexus and pass along the hepatic artery and its branches. Autonomic plexuses are present in the muscular and submucous layers, and the occurrence of ganglion cells, presumably parasympathetic, in these plexuses has been confirmed in monkeys (Sutherland 1966, 1967). Fibres from the right phrenic nerve, through the communications between the phrenic plexus and the coeliac plexus, also appear to reach the gall bladder in the hepatic plexus, as evidenced by the referred 'shoulder pain' experienced in diseases of the gall bladder.

VARIATIONS IN GALL BLADDER AND BILE DUCTS

The gall bladder frequently varies in size and shape; in rare cases it may be (a) duplicated (Mincsev 1967), the two gall bladders having separate ducts or a common cystic duct, or (b) absent, though the cystic duct may be present in a few of these cases.

The level of junction of the cystic and common hepatic ducts varies from the porta hepatis to behind, or even below, the superior part of the duodenum; when the junction is low the two ducts in their course may be closely connected by fibrous tissue, rendering it very difficult, in removing the gall bladder, to clamp the cystic duct without injuring the common hepatic (or the main bile) duct. Occasionally the cystic duct joins the right hepatic duct. The cystic duct may pass to the left behind, or in front of, the common hepatic duct and then joins the latter on its left surface. Accessory hepatic ducts may emerge from the liver, more frequently from its right lobe, to join the main hepatic ducts, or very rarely the gall bladder itself. Rarely, a failure of canalization of the bile ducts during prenatal development leads to congenital atresia or stenosis of these ducts with fatal results in a few months after birth. The bile duct and pancreatic duct may open separately

into the duodenum or they may join together before passing through the duodenal wall. Variations of the arteries related to the bile ducts are much more common than those of the bile ducts themselves. (For the variations in the hepatic arteries, *see* p. 713.) For further information consult Santulli and Blanc (1961); Boyden *et al.* (1967).

STRUCTURE OF GALL BLADDER AND BILIARY DUCTS

The wall of the gall bladder displays three layers: serous, fibromuscular and mucous.

The *serous layer* is derived from the peritoneum; it completely invests the fundus, but covers only the inferior surfaces and sides of the body and neck. Beneath it lies a subserous layer of areolar tissue.

The *fibromuscular layer* is a stratum of fibrous tissue, mixed with nonstriated myocytes ('fibres'); these are arranged in loose bundles disposed in longitudinal, circular and oblique directions.

The *mucous layer* is loosely connected with the fibrous layer. It is generally of a yellowish-brown colour, and is elevated into minute rugae giving it a honeycomb appearance (8.141, 143). The epithelium consists of a single layer of columnar cells which vary in size with species. Electron microscope investigation (Chapman *et al.* 1966) of the gall bladder epithelium of the dog shows the presence of microvilli on the apical surface; these structures are

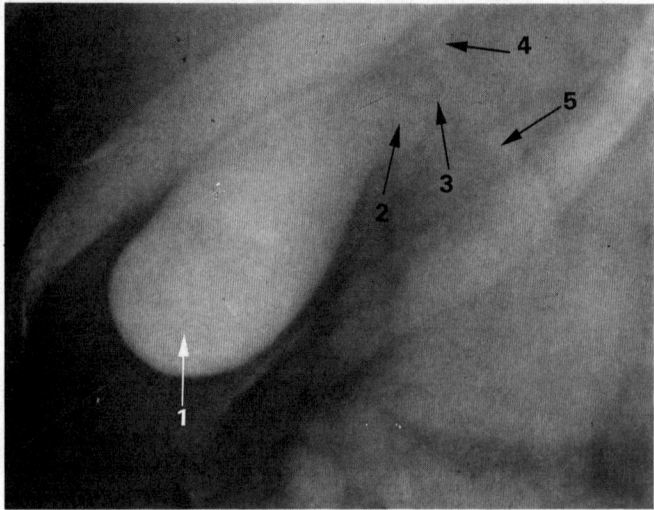

8.144A An anteroposterior radiograph of the gall bladder and biliary ducts after the oral adminstration of sodium tetra-iodophenolphthalein the previous day.

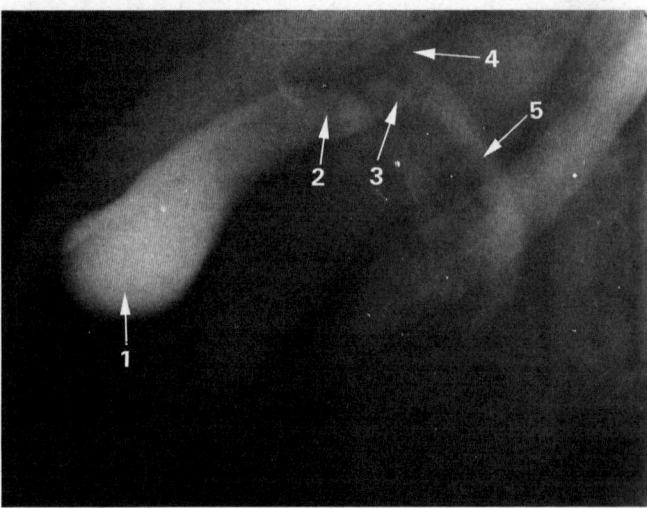

8.144B The same field as **8.144A** but taken twenty minutes after a fatty meal, demonstrating the contraction and partial emptying of the gall bladder. In both A and B; (1) fundus of gall bladder; (2) neck of gall bladder; (3) cystic duct; (4) common hepatic duct; and (5) the bile duct.

regularly arranged and between their bases pinocytotic activity is observed. The basal intercellular spaces show considerable dilatation and many capillaries lie close to the basement membrane. These features indicate active absorption of water and solutes from the bile, rendering it more concentrated. The basal intercellular spaces are particularly large when active water absorption is occurring (Kaye *et al.* 1966). The mechanism of bile concentration appears to involve the active transport of sodium and calcium which creates an osmotic gradient from the lumen of the gall bladder, to the capillaries of the lamina propria. Mucous granules are present in the apical half of some cells, particularly those near the duct; these are secreted into the lumen (Johnson *et al.* 1962; Mueller *et al.* 1972).

The coats of the large biliary ducts are an external or fibrous, and an internal or mucous. The fibrous layer is composed of fibroareolar tissue, intermingled with a few nonstriated muscle fibres that are arranged in a longitudinal, oblique and circular manner. (In one investigation, muscle fibres were detected in only 12 out of 100 human common bile ducts—*see* Mahour *et al.* 1967.) The mucous layer is continuous with the lining membrane of the hepatic ducts and gall bladder, and also with that of the duodenum; like the mucous membrane of these structures, its epithelium is of the columnar variety; many lobulated mucous glands are present. In the bile duct the mucous membrane is provided with numerous tubulo-alveolar glands that are not evenly distributed throughout the mucosa but are arranged in clusters. The glands secrete mucin, at least some of which is of the sulphated variety (McMinn and Kugler 1961). An electron microscope study of the bile duct epithelium of the guinea-pig shows that microvilli are present on the luminal surface of the cells. Secretion granules, some composed of mucus, are found in the apical cytoplasm. The epithelial cells in the rat, which has no gall bladder, can be divided into two types, termed light and dark, owing to a difference in electron density. The light cell has longer and more regular microvilli, but the basal intercellular space between adjacent dark cells is greater. The mucous membrane appears to actively modify the bile and thus, in this species, compensates for the absence of a gall bladder (Riches and Palfrey 1966).

The circular muscle around the lower part of the bile duct, including the ampulla and the terminal part of the main pancreatic duct, is thickened and is called the *sphincter of the hepatopancreatic ampulla* (or the *sphincter of Oddi*). The latter comprises musculature at three levels: (*a*) at the end of the bile duct (*sphincter ductus choledoci*); (*b*) around the terminal part of the pancreatic duct (*sphincter ductus pancreatici*); and (*c*) around the ampulla. Only the choledochal sphincter is constantly present. The mechanism for the emptying of the gall bladder appears to be under hormonal, rather than nervous, control. The presence of fat or acid in the duodenum is believed to cause the liberation of a hormone (cholecystokinin) which stimulates the gall bladder to contract. When the pressure of the stored bile in the gall bladder exceeds 100 mm of bile, the gall bladder contracts, the sphincter of Oddi relaxes and bile is poured into the duodenum. It has been maintained (Kirk 1944) that there is no sphincteric arrangement of the musculature around the opening of the bile and pancreatic ducts into the duodenum, but that the sphincter of Oddi surrounds the bile duct as it passes through the submucosal zone of the duodenal wall and is continuous with the circular muscle coat of the duodenum, which is thickened at this site. However, subsequent studies suggest strongly that in man and other primates there is a common sphincteric apparatus surrounding both ducts, and that the common bile duct has a second sphincter of the type described above (Boyden 1957, 1966). The terminal part of the united bile and pancreatic ducts is packed with villous, valvular folds of the mucous membrane, and the muscle fibres extend into the connective tissue cores of these folds. This arrangement suggests that contraction of the muscle fibres results in retraction and erection or aggregation of the folds, thus preventing reflux of duodenal contents into the ducts and controlling the exit of bile into the duodenum. In the cat, stimulation of branches of the vagus nerves to this region results in relaxation of the opening of the bile duct into the duodenum; in the human, the *myenteric (Auerbach's) plexus* is well developed at the site of termination of the ducts. Inflammatory swelling of the villous folds may obstruct the ducts.

Applied Anatomy. On account of its large size, its fixed position, and its friability, the liver may be accidentally ruptured and haemorrhage may be severe, because the hepatic veins are contained in rigid canals in the liver and are unable to contract. The liver may be torn by the end of a broken rib perforating the diaphragm.

There is clinical evidence that the bloodstreams conveyed to the portal vein by the superior mesenteric vein and the splenic vein remain largely separate from each other in the portal vein, and pass respectively along the right and left branches of the portal vein to the right and left physiological lobes (p. 1374) of the liver, e.g. malignant or infective emboli may be more pronounced in the right lobe than the left, if the primary infection lies in a part of the gut drained by the superior mesenteric vein, and in the left lobe if the primary site is in splenic vein or inferior mesenteric vein territory. Some injection experiments in living animals also indicate the tendency for the two venous streams to remain separate in the portal vein.

The results of studies on the vascular segments of the liver were referred to previously (p. 1376); such information is essential to the successful performance of the modern surgical operation of *partial hepatectomy.* As yet, attempts at *liver transplantation* are still tentative.

The gall bladder may become distended by obstruction of the cystic duct or by gall stones within it, and may form a large swelling which projects downwards and forwards towards the umbilicus. It moves with respiration, since it is attached to the liver.

Obstruction of the bile duct, apart from stone, is often due to occlusion of this canal by pressure of malignant tumours, especially those commencing in the pylorus or pancreas. It is also seen following ulceration of the duct, cicatricial contraction of the scar tissue taking place. Enormous distension, both of the bile duct itself and of its radicles in the liver substance, may occur.

Cholecystography (**8**.144). The gall bladder is not opaque to X-rays. If certain radio-opaque substances are given orally or intravenously, they are excreted by the liver from the blood into the bile. In the gall bladder the substance is concentrated and thus becomes more opaque to X-rays. In this way the form, position and emptying process of the gall bladder can be demonstrated radiographically. By this means it has been shown that the position and form of the gall bladder vary with the general body build (somatotype) of the individual (Davies 1927). In broad (hypersthenic) types the gall bladder is broad and lies high up and far laterally (at the level of the first lumbar vertebra), whereas in narrow (asthenic) types it is narrow, lies more medially and may reach as low as the fourth lumbar vertebra.

THE UROGENITAL SYSTEM

Introduction

So customary has it become to link the organs concerned with urinary excretion and reproduction as the *urogenital system,* that the suitability of this amalgamation is rarely even questioned. Yet functionally the two have nothing in common, and to associate them in description may seem at first absurd (Romer 1970). In fully developed human organs there is little more structural overlap than the use of the male urethra both as a urinary and seminal duct. Of course, in their embryological development, the gonadal ducts have a *nephric* origin; but in the human female even the *oviduct* ('female', paramesonephric or Müllerian duct) is no longer formed from nephric tissue, and there is, indeed, no adequate reason to continue to define a 'urogenital' system in female mammals in general. Even in the human male, as in other mammals, in which the duct system of the testis is derived from nephric tubules and the *mesonephric duct*, these no longer function as excretory structures. In reptiles the functional kidney is no longer a mesonephros; and the intromittent *penis*, developed from cloacal tissues, consists of two bilateral parts, which enclose a purely seminal groove when erected for copulation, urine being voided into the cloaca. Apart from its development, therefore, the genital system has already achieved complete dissociation from the urinary organs in both sexes at the level of reptilian evolution. Birds, like amphibians, have no intromittent organ, though this is probably a secondary loss in the avian group, some more primitive birds possessing a rudimentary penis of reptilian type.

In all vertebrates (as distinct from elementary chordates) the nephric excretory tubules and the gonads develop from the coelomic epithelium, together with their collecting ducts as they evolve. Both develop in a dorsal situation on each side of the major vessels of the body; and it is perhaps this propinquity, combined with the more medial location of the gonads, which has led to the adoption of urinary ducts and tubules to serve the gonads, especially in the male.

In elementary *Chordata*, such as *Amphioxus*, the excretory tubules, or *nephridia*, have *glomeruli* or capillary 'tufts' close to them, and they secrete from the adjoining bloodstream and the coelom directly into a peribranchial space, the atrium. These excretory organs are ectodermal in origin and are unrelated to the mesodermal nephric tubules of vertebrates. The evolutionary relationship of *Amphioxus* to the early vertebrates is, in any case, uncertain; it has even been suggested that this chordate is itself a paedomorphic form of a primitive vertebrate (Grassé 1948; Young 1962) but these are matters which cannot be discussed here.

In vertebrates living in water, whether saline or fresh, the problems of preserving internal osmotic constants in blood and body fluids, despite variable and sometimes very large intakes of water and salts, has led to a continuous evolution of more efficient and usually more elaborate excretory tubules. These are derived from mesoderm intermediate between the somites and the lateral plate in the embryo, forming structures named *nephrotomes* (p. 210). Though these arise segmentally, their large numbers overshadow this, and in most lower vertebrates they form elongated masses or cords of nephrogenic tissue projecting into the coelom in a dorsolateral position. The gonads develop medial to these *nephrogenic cords* (p. 210); and since the *primary excretory ducts* (p. 210), carrying urine to the exterior, evolve lateral to the nephros, gonadal access to them is effected by adaptation of nephric tubules. The degree of development and functional status of different parts of the elongate vertebrate kidney, or *holonephros*, extending through many body segments, has displayed much variation in different forms. There is a tendency for the most cranial part of the kidney to disappear, except where it functions during embryonic life or in a larval form, as in some cyclostomes (e.g. *Petromyzon*, the lamprey). In the anamniotic gnathostomes (fish and amphibia) an intermediate nephrogenic region develops into a functional kidney, and while this is still elongate in shape, it occupies a comparatively small number of body segments, with which its vascular connexions approximately accord. In the reptiles and mammals an even more caudal mass of nephrotomes differentiates into the definitive kidney, which becomes a more localized and rounded organ, of familiar reniform shape, particularly in mammals, losing almost all indications of a plurisegmental origin in its full development.

This progressive caudal shift of the functioning kidney has prompted a division of the nephrogenic mass into pro-, meso-, and meta-nephros, giving rise to somewhat misleading concepts. The amphibian kidney, for example, is often described as a mesonephros, and the mammalian organ as a metanephros; nevertheless, the mode of formation of the nephric tubules is substantially the same at all levels, and it cannot be assumed that these regional terms can be applied over large and diverse vertebrate groups with anything more than approximate segmental uniformity. The terms are perhaps more of a

descriptive convenience than vehicles of biological truth. (For a discussion of these and other problems in the development of the urogenital organs, consult Fraser 1950.)

Distinctions between the three regional levels of the nephrogenic ridge depend more upon their functional status and duct systems than their intrinsic morphology. Primitively, in cyclostomes, the excretory tubules discharge their products, as do the gonads their gametes, into the coelomic cavity, from which they escape to the exterior through the *abdominal pores* or short ducts situated in a caudal position. In gnathostomes, the *nephric tubules* lose continuity with the coelom, developing a saccular extremity, with each of which a *glomerulus* becomes closely related. The tubules establish continuity with a longitudinal *archinephric duct (primary excretory duct)*, which at first serves the pronephros, and is taken over by the succeeding more caudal part of the kidney, the mesonephros. It is through mesonephric tubules that the testis establishes an outlet through the urinary duct, the arrangement typical of amphibians. Both urine and spermatozoa reach the cloaca through such *mesonephric* (Wolffian) *ducts*. The ovary also acquires a duct, which in some forms is derived from the mesonephric duct; but this *paramesonephric, female*, or *Müllerian duct*, becomes a separate entity and does not function as a channel for transmission of urine. Moreover, the oviduct never establishes direct continuity with the ovary, which retains its primitive habit of shedding ova into the coelom, though in very close proximity to the open, coelomic beginnings of the oviduct. (This relationship is constant and occurs in the human female. The oviduct of mammals, and of tetrapods in general, appears to develop independently by evagination from the coelomic lining.)

In reptiles, birds and mammals, the functioning kidney is a *metanephros*, developing caudal to the mesonephros. Its main distinction is that it has no association with genital function, developing in part as an outgrowth of the caudal region of the mesonephric duct, but ultimately forming a separate opening into the bladder by its own duct, the *ureter*. The mesonephros is thus relieved of all excretory function, its tubules and ducts persisting only in so far as they are modified to form the *vasa efferentia* of the testis, the *epididymis* and *ductus deferens*.

To summarize—the pronephros is an embryonic or larval excretory organ, having a duct which is purely excretory. The mesonephros, which takes over this duct and the function of excretion, also provides a pathway for the conduction of spermatozoa to the exterior, and also, in some vertebrates by division of its duct, a separate exit for ova. The metanephros is again a purely excretory organ, the male genital ducts continuing as mesonephric derivatives, and the oviduct as an increasingly independent development.

The complete segregation of urinary and genital organs in the reptiles—apart from their embryonic development and the opening of their separate ducts in a common cavity, the cloaca— becomes modified once more by the evolution of a male intromittent organ to effect the internal fertilization associated with full emancipation from reproduction in water. The production of eggs, or rather embryos, protected from desiccation in a terrestrial environment by shells impervious to water (though not to respiratory gases), necessitates the fusion of gametes in the oviduct proximal to specialization of its wall to produce such coverings. Similarly, the production of offspring which have already reached a sufficient stage of development to live exposed to the external environment naturally also entails internal fertilization. Apart from monotremes, all mammals exhibit this viviparous habit. Although internal fertilization is accomplished without an intromittent organ in birds, an intromittent and hence erectile *penis* is universal in marsupials and eutherian mammals. Since this is traversed by the urethra, used to void both urine and semen, the male mammal, including mankind, exhibits to this extent an association of excretory and reproductive organs. Both systems display many adaptations and specializations, which in mammals are concerned with greatly improved ability to preserve a steady internal environment in terms of total body water volume, and the concentrations of ions and other dissolved substances. This entails an increasing complexity in the structure of glomeruli and renal tubules, forming the *nephrons* or kidney units, capable not only of removing variable and sometimes large volumes of water from the body, but also of selective reabsorption of some substances such as glucose and salts, and the retention of others, such as urea, in the urine.

The kidney ducts open into various forms of reservoir in many vertebrates, and in mammals a cloacal derivative the *bladder*. In the reproductive organs of mammals a number of accessory structures have evolved in connexion with intromission in the male, and the retention and nurture of developing embryos in the female. These include the seminal vesicles and prostate, the

uterus and vagina. All these developments have already been considered in their embryological development in a previous section (p. 210); they must now be studied in their fully developed primate morphology. The customary association of the two groups of organs has been retained, since even in description of the completely differentiated urinary and genital systems reference to their ontogeny and phylogeny will occasionally be necessary. It is, nevertheless, perhaps worthy of repetition that their association in the post-natal human being is limited to the dual function of the penile urethra.

THE URINARY ORGANS

The urinary organs comprise: (1) the *kidneys* (renes), which 'secrete' the urine; (2) the *ureters*, which convey it to (3) the *urinary bladder* (vesica urinaria), where it is stored temporarily; and (4) the *urethra*, through which it is discharged from the urinary bladder. It must be added that while the female urethra is purely a urinary duct, the male urethra serves two functions—urinary and reproductive. Of these the latter is essential, and the former function merely utilizes a tube which is almost entirely genital. Loss of the penile urethra need not greatly affect urination, whereas the genital function is for all practical purposes abolished. It would, therefore, be more logical to include almost the whole of the urethra with the rest of the genitalia in the male.

The Kidneys

The kidneys excrete end products of the metabolic activities of the body and excess water, and are thus essential in controlling the concentration of the constituents of the body fluids, maintaining, for example, the electrolyte and water balance in the tissue fluids. They also have an endocrine function, producing and releasing into the bloodstream *erythropoietin*, which affects blood formation, *renin* which influences blood pressure, and *1,25-hydroxycholecalciferol*, which is involved in the control of calcium metabolism. The last of these hormones is a derivative of vitamin D and may modify the action of parathyroid hormone (O'Riordan 1978). The kidneys are two reddish-brown organs situated in the posterior part of the abdomen, one on each side of the vertebral column, behind the peritoneum; they are surrounded by a mass of adipose connective tissue. Their cranial ends are level with the superior border of the twelfth thoracic vertebra, their caudal limits with the third lumbar vertebra. The right kidney is usually

slightly inferior to the left, probably on account of its relationship to the liver; the left is a little longer and narrower than the right and is a little nearer to the median plane. The long axis of each kidney is directed downwards and laterally, the transverse axis posterolaterally. This entails that the anterior and posterior aspects, as described below, are in fact *anterolateral* and *posteromedial*; but the simpler terms have been retained for brevity. The transpyloric plane (p. 1320) passes through the superior part of the hilum of the right kidney, and through the inferior part of the hilum of the left.

Each kidney is about 11 cm in length, 6 cm in breadth, and about 3 cm in anteroposterior thickness. In the adult male the weight of the kidney averages about 150 g., in the adult female 135 g. In a thin individual with a lax abdominal wall the lower pole of the kidney may just be felt in full inspiration by bimanual examination of the loin. Usually, however, it is impalpable.

RENAL SURFACE PROJECTIONS

In the recumbent position, the outline of each kidney can be projected to the anterior or posterior surface of the abdominal wall as follows, bearing in mind that the right kidney is a little lower (about 1·25 cm) than the left. (*a*) *Anterior surface*. The centre of the hilum is approximately at the transpyloric plane, about 5 cm from the median plane and slightly medial to the tip of the ninth costal cartilage. The hilum of the left kidney is just above the transpyloric plane and that of the right kidney just below it (**8**.145). In relation to this position of the hilum, a kidney-shaped figure is drawn, 11 cm long and 4·5 cm broad, so that the upper pole is about 2·5 cm and the lower pole about 7·5 cm from the midline. (As the kidney lies obliquely, the width of the figure is about 1·5 cm less than the actual width of the organ.)

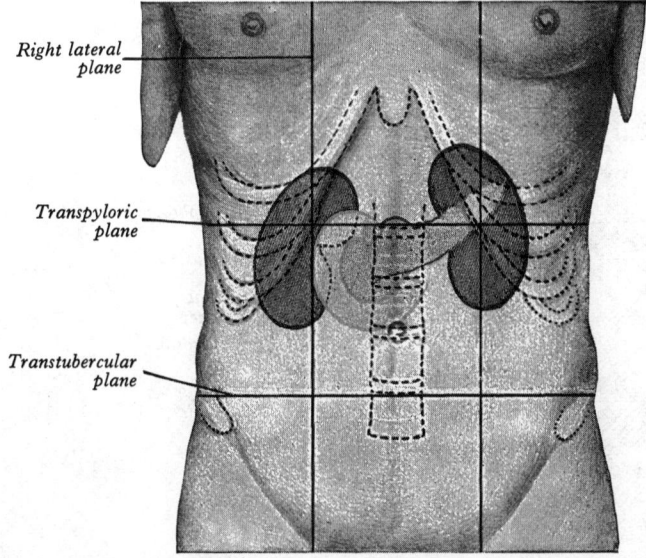

8.145 Surface projections of the duodenum, pancreas and kidneys. The lower ribs and the lumbar vertebrae are also indicated.

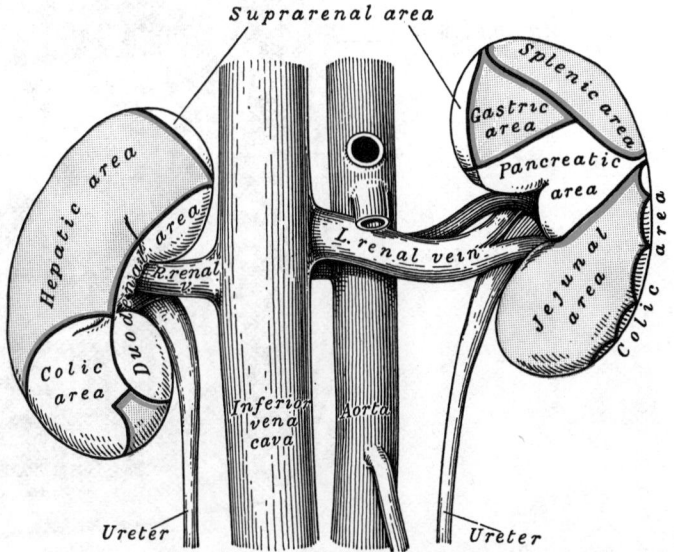

8.146 The anterior surfaces of the kidneys, showing the areas related to neighbouring viscera.

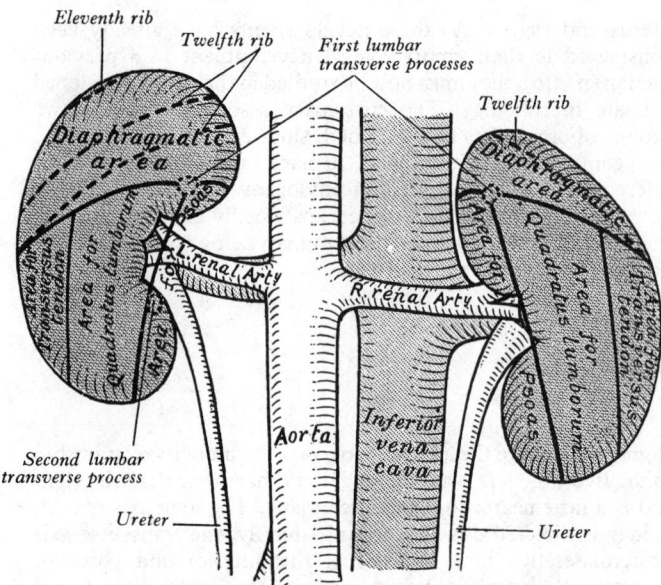

8.147 The posterior surfaces of the kidneys, showing the areas of relation to the posterior abdominal wall.

(b) *Posterior surface.* The centre of the hilum lies opposite the lower border of the spine of the first lumbar vertebra, about 5 cm from the median plane. In relation to this point, a figure is constructed in the same way as that described above for the anterior surface. The lower pole is usually a little (2·5 cm) above the highest part of the iliac crest. The kidneys lie about 2·5 cm lower in the standing than in the recumbent position and they move up and down with respiration.

RENAL RELATIONS

The *anterior surface* (**8.**111A, 146) is convex, and actually faces anterolaterally. Its relations to adjacent viscera differ on the two sides of the body.

(a) *Anterior surface of right kidney.* A small area of the superior pole is in contact with the right suprarenal gland, which may overlap it or the upper part of the medial border. A large area just below this and involving about three-fourths of the surface, lies in

the renal impression on the right lobe of the liver, and a narrow area near the medial border is in contact with the descending part of the duodenum. Inferiorly the anterior surface is in contact laterally with the right colic flexure, and medially with part of the small intestine. The area in relation with the small intestine and almost the whole of the area in contact with the liver are covered with peritoneum (with the intervention of the renal fascia); the suprarenal, duodenal and colic areas are devoid of peritoneum.

(b) *Anterior surface of left kidney.* A small area along the superior pole of the medial border is in relation with the left suprarenal gland, and about the upper two-thirds of the lateral half of the anterior surface are in contact with the renal impression on the spleen. A somewhat quadrilateral field, about the middle of the anterior surface, is in contact with the body of the pancreas and the splenic vessels. Above this there is a small triangular region, between the suprarenal and splenic areas, which is in contact with the stomach. The size of the gastric area of contact is very variable. Below the pancreatic and splenic areas the lateral part is in relation with the left colic flexure and the commencement of the descending colon, and the medial part with the first coils of the jejunum. The jejunal area is extensive but the colic area forms an irregular, narrow strip immediately adjoining the lateral border of the kidney. The area adjacent to the stomach is covered with the peritoneum of the omental bursa, while the areas in relation to the spleen and the jejunum are covered with the peritoneum of the greater sac; behind the peritoneum of the jejunal area some branches of the left colic vessels are related to the kidney. The suprarenal, pancreatic and colic areas are devoid of peritoneum, because these organs are in actual contact with the kidney, without any possibility of mutual movements.

The *posterior surface* (**8.**147, 148, 150) is actually directed posteromedially. It is embedded in fat, and is devoid of peritoneal covering. It lies upon the diaphragm, the medial and lateral arcuate ligaments, the psoas major, the quadratus lumborum, and aponeurotic tendon of the transversus abdominis, the subcostal vessels, and the subcostal, iliohypogastric and ilio-inguinal nerves. The upper pole of the right kidney is level with the twelfth rib, that of the left with the eleventh and twelfth. The diaphragm separates the kidney from the pleura, which descends to form the costodiaphragmatic recess (**8.**148), but sometimes the muscular fibres of the diaphragm are defective or absent over a triangular area immediately above the lateral arcuate ligament; when this is so the perirenal adipose tissue is in contact with the diaphragmatic pleura.

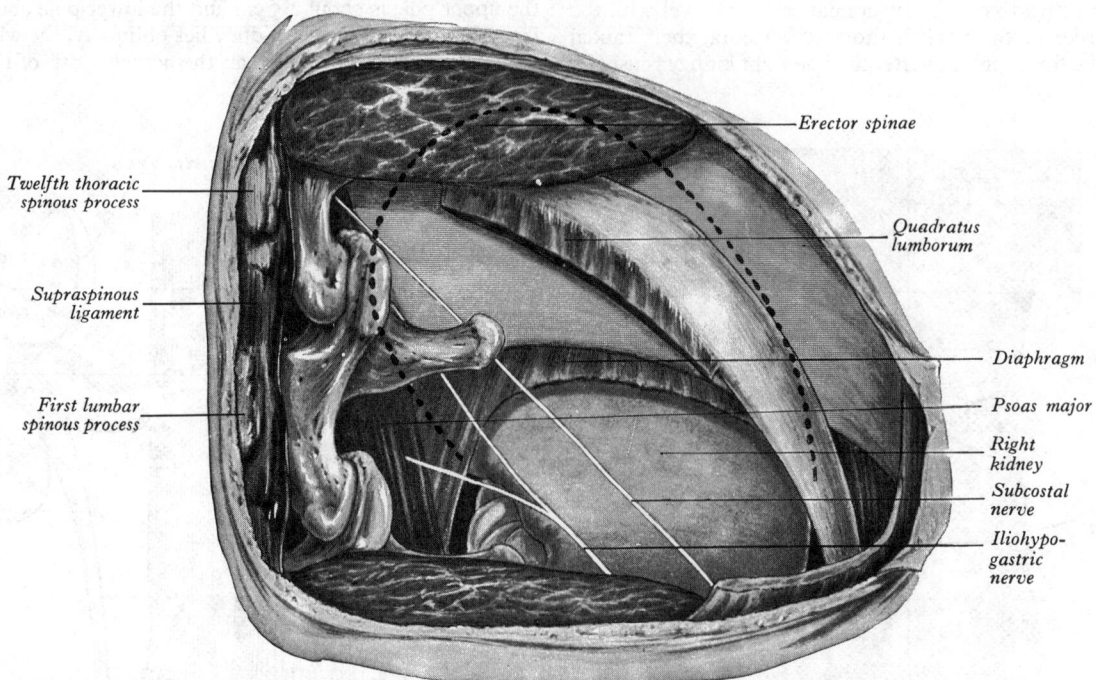

8.148 The right kidney, posterior exposure. The blue area represents the pleura, the broken red line the upper part of the kidney. The subcostal nerve has been displaced downwards. Parts of the diaphragm and the quadratus lumborum have been resected.

The *superior pole* of each kidney is thick and round, and is nearer the median plane than the lower; it is related to the suprarenal gland. The *inferior pole*, smaller and thinner than the upper, extends to within 2·5 cm of the iliac crest.

The *lateral border* is convex; that of the left kidney is covered superiorly with greater sac peritoneum which separates it from the spleen and, below this, it is in contact with the descending colon; the lateral border of the right kidney is separated by peritoneum of the greater sac from the right lobe of the liver.

The *medial border* of each kidney is convex adjacent to the poles and concave between these curvatures; it slopes downwards and laterally. In its central part there is a deep vertical fissure opening anteromedially, termed the *hilum*, which is bounded by an anterior and a posterior lip, and contains the renal vessels and nerves and the funnel-shaped continuation of the upper end of the ureter, the *renal pelvis*. The relative positions of the main hilar structures are as follows: the renal vein is in front, the renal artery in the middle and the pelvis of the kidney behind. Commonly one of the branches of the renal artery enters the hilum behind the renal pelvis, and it is not uncommon to find one of the tributaries of the renal vein issuing from the hilum in the same plane. Above the hilum the medial border is in relation with the suprarenal gland, and below with the commencement of the ureter.

The hilum leads into a central recess named the *renal sinus*, which is lined by a continuation of the capsule of the kidney and is almost entirely filled by the pelvis of the kidney and renal vessels; numerous nipple-like elevations, termed the *renal papillae*, indent the wall of the sinus. The renal pelvis extends outside the hilum to become continuous with the ureter. Within the sinus it divides into two, sometimes three, large branches, which are named the *major calices*, and each of these divides again into several short branches, named the *minor calices* (**8**.149, 152). In all, there are usually from seven to thirteen of these minor calices; each expands as it approaches the wall of the renal sinus, and the expanded end is indented and moulded round from one to three renal papillae (**8**.149). The wall of the expanded end of each calix is firmly adherent to the capsule lining the renal sinus; it is perforated by the collecting tubules which open on the summits of the renal papillae.

The kidney and its vessels are embedded in a mass of adipose connective tissue, termed the *perirenal (perinephric) fat*, which is thickest at the borders of the kidney and is prolonged through the hilum into the renal sinus. The fibro-areolar tissue surrounding the kidney and perirenal fat is condensed to form a sheath termed the *renal fascia* (**8**.150A, B).

THE RENAL FASCIA

At the lateral border of the kidney the two layers of the renal fascia are fused. The anterior layer is carried medially in front of the kidney and its vessels, and at the level of the latter it merges with the connective tissue around the aorta and inferior vena cava. This medial continuation is very thin and does not extend higher than the superior mesenteric artery. The posterior layer extends medially behind the kidney and in front of the fascia on the quadratus lumborum and psoas major, and is attached to that fascia at the lateral and medial borders of the psoas, and to the vertebrae and intervertebral discs. There is also a deeper stratum (not shown in **8**.150) attaching the anterior and posterior layers to one another at the medial border of the kidney and pierced by the renal vessels (Martin 1942). This may account for the fact that a perirenal effusion of fluid does not usually extend across to the opposite perirenal space. However, injected air *does* diffuse along this route (Grossman 1954). Above the suprarenal gland the two layers of the renal fascia fuse, and are connected with the fascia of the diaphragm; it is generally agreed that below the kidney they remain separate, enclosing the ureter, the anterior layer being gradually lost in the extraperitoneal tissue of the iliac fossa, while the posterior layer blends with the fascia over the iliacus; but this has been denied (Mitchell 1950). The renal fascia is connected to the fibrous capsule of the kidney by numerous trabeculae, which traverse the perirenal fat, and are strongest near the lower pole. Behind the renal fascia there is a considerable quantity of fat, termed the *pararenal (paranephric) body*. The kidney is held in

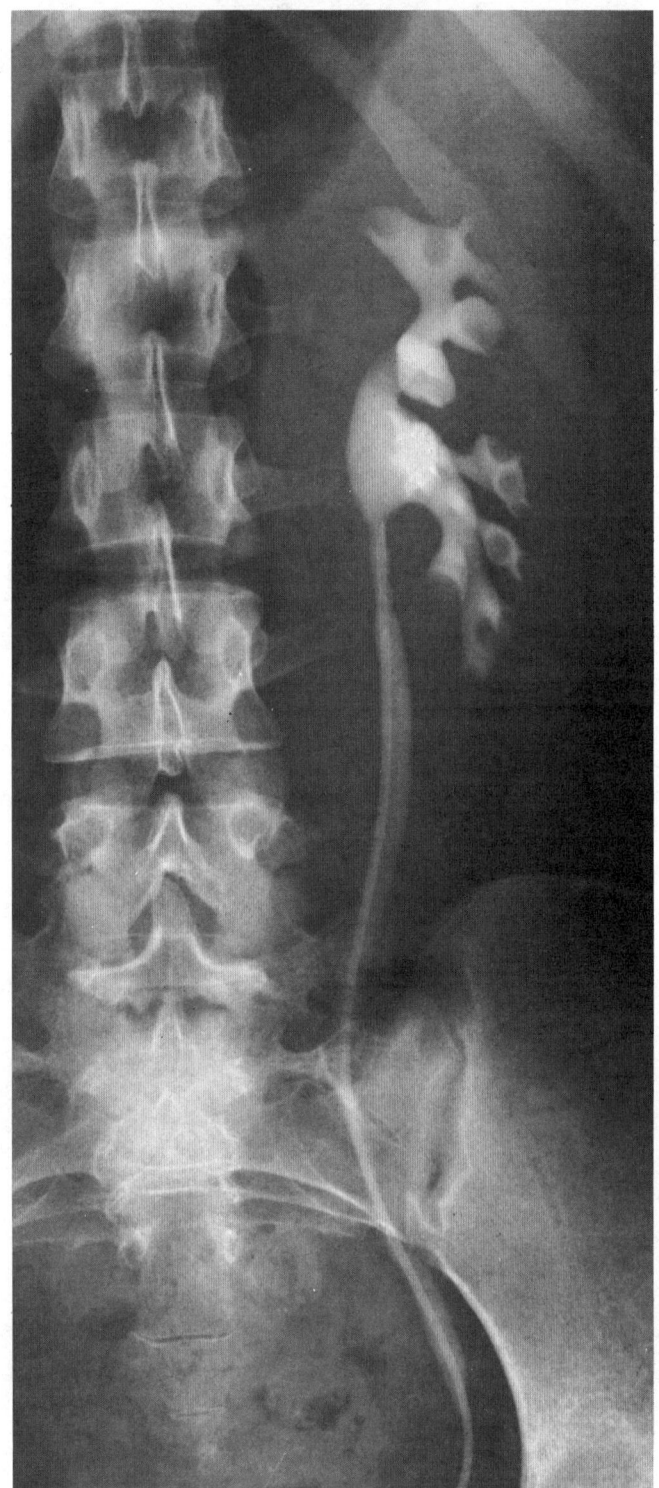

8.149 A left retrograde pyelogram. The contrast medium has been introduced into the calices, pelvis and upper ureter via a ureteric catheter, which is still in position, and can be clearly identified. There is considerably greater density of the shadows of the calices than that achieved by the intravenous method. Note the relation of the ureter to the tips of the lumbar transverse processes, and the characteristic 'cupping' or 'champagne glass' profiles of the tips of the lesser calices where they surround the renal pyramids. 'Calix' means a cup, and such 'cupping' of the minor calices is the normal appearance. (Note that the major calices are not cups and are hence inappropriately named.)

position partly through the attachments of the renal fascia, but principally by the apposition of the neighbouring viscera.

In the fetus the kidney consists of about twelve distinct lobules (**8**.151), but in the adult these are fused and the kidney presents a uniformly smooth surface, though traces of the fetal lobulation may remain.

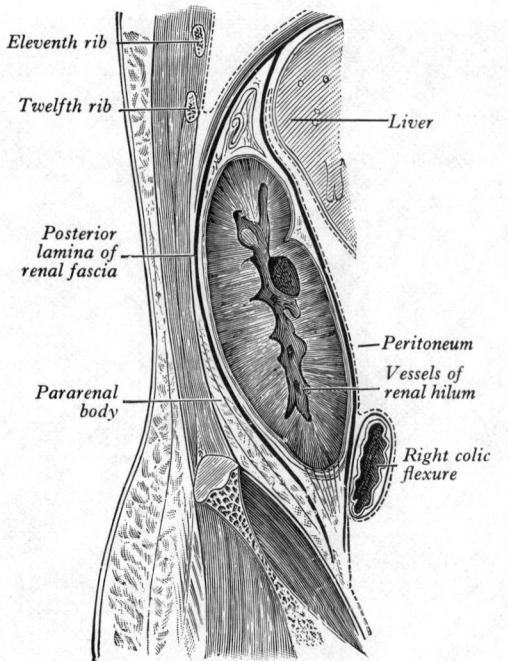

Eleventh rib

Twelfth rib

Liver

Posterior lamina of renal fascia

Peritoneum

Vessels of renal hilum

Pararenal body

Right colic flexure

8.150A   A sagattal section through the posterior abdominal wall showing the relations of the renal fascia of the right kidney.

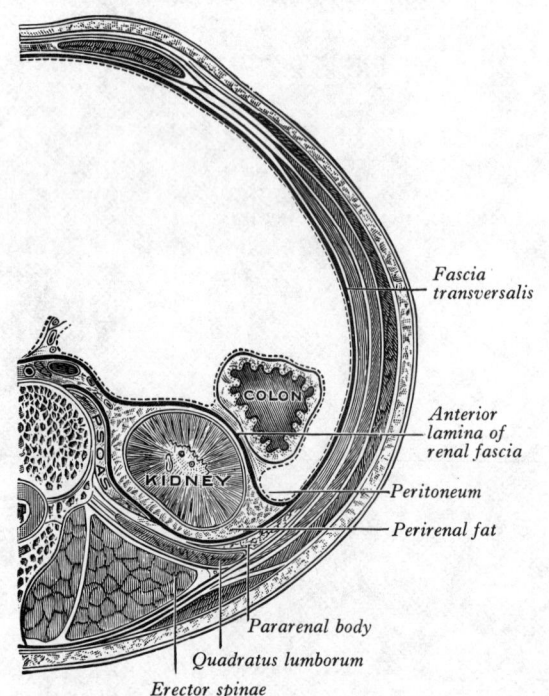

Fascia transversalis

COLON

Anterior lamina of renal fascia

KIDNEY

Peritoneum

Perirenal fat

Pararenal body

Quadratus lumborum

Erector spinae

8.150B   A transverse section, showing the relations of the renal fascia.

GENERAL RENAL STRUCTURE (8.152A–D)

The kidney is invested by a thin capsule that is easily stripped off and consists of white fibrous tissue, with a few yellow elastic and nonstriated muscle fibres. In some diseases of the kidney the capsule becomes adherent to the kidney and cannot be easily stripped away.

The kidney has an internal *medulla* and an external *cortex*.

The *renal medulla* consists of a number of pale, striated, conical masses, termed the *renal pyramids*, the bases of which are directed towards the periphery of the kidney, while their apices converge towards the renal sinus, where they form prominent papillae projecting into the interior of the calices, each minor calix receiving from one to three papillae. Each pyramid, capped by

cortical substance, forms a *'lobe'* of the kidney. Estimates of the number of papillae—and hence of pyramids or renal 'lobes'—are somewhat variable. In counts of them in 375 human kidneys (Lake *et al.* 1966), the range was 5 to 11 in 89 per cent with a most frequent value of 8 papillae (in 26 per cent). Total numbers of terminal uriniferous ducts (of Bellini) opening on the papillae varied (in each of 208 kidneys) from 116 to 776; no marked peak frequency was observed, but 23 per cent of the kidneys displayed in the region of 275 openings of uriniferous ducts. In another series of 54 kidneys (Arvis 1969), the number of papillae was 6 to 14.

The *renal cortex* (**8.**152A) lies immediately beneath the fibrous capsule, arches over the bases of the pyramids, and dips in between adjacent pyramids towards the renal sinus. The parts dipping in between the pyramids are named the *renal columns* (**8.**152C), while the regions which connect the renal columns with each other and intervene between the bases of the pyramids and the fibrous capsule are called the *cortical arches* or *cortical lobules*. If this latter part of the cortex be examined with a lens, it will be seen to be traversed by a series of radially orientated lighter-coloured *medullary rays* (**8.**152C), and a darker-coloured intervening substance, which from the complexity of its structure is named the *convoluted part*. The rays gradually taper towards the capsule of the kidney, and consist of a series of outward prolongations from the base of each renal pyramid.

The cortex is also divisible into an *outer* and *inner zone* according to its histological organization. The inner cortex is divided from the medulla by a line of tangentially running blood vessels (the arcuate arteries and veins) although a thin layer of cortical tissue (the '*subcortex*') is present on the medullary side of this line. The layers of cortex close to the medulla are also termed the *juxtamedullary cortex* by some authors.

RENAL HISTOLOGY

The kidney is composed of a very large number of tortuous, closely packed *uriniferous tubules*, bound together by a little connective tissue in which run the blood vessels, lymphatics and nerves. Each uriniferous tubule consists of two parts, which are embryologically distinct (p. 210), namely (1) the *nephron* or secreting part, which elaborates the urine, and (2) a *collecting tubule*. The *nephron* comprises (*a*) the *renal corpuscle*, which is concerned with filtration of substances from the plasma; (*b*) the *renal tubule*, which is concerned with the selective resorption of substances from the glomerular filtrate until it approaches the composition of urine. The *collecting tubule* carries the fluid from a number of renal tubules to a terminal *papillary duct* or *duct of Bellini*; the latter opens into a minor calix at the apex of a renal papilla. If the surface of one of the papillae be examined with a lens, it will be seen to be studded over with numerous minute openings (*vide supra*), the orifices of the ducts of Bellini, and if pressure be made on a fresh kidney, urine will be seen to exude from these orifices.

THE RENAL CORPUSCLE (8.152D, 153A)

The renal (Malpighian) corpuscles, which are small rounded masses, averaging about 0·2 mm in diameter, are visible in the renal cortex and columns (of Bertin) except a narrow zone around the perimeter of the cortex (the *cortex corticis*). There are one to two million renal corpuscles in each kidney, the numbers falling with age (Dunnil and Halley 1973). Each of these bodies is composed of two parts: a central *glomerulus* of vessels, and a membranous envelope, termed the *glomerular capsule*, which is the small pouch-like commencement of a renal tubule.

The glomerulus is a lobulated tuft of convoluted, capillary blood vessels, held together by scanty connective tissue. This capillary network is derived from a small *afferent arteriole*, which enters the capsule, generally at a point opposite to that at which the latter is connected with the tubule; the *efferent arteriole* emerges from the capsule at the same point, which is thus known as the *vascular pole* of the capsule. The glomeruli have a simple form up to late prenatal life and, while some may maintain this form for about six months after birth, the majority of these mature by six years of age and all are of adult form by twelve years

(Macdonald and Emery 1959). In the fetal rabbit, guinea-pig and sheep the precursor of a glomerulus is a solid sphere of mesodermal cells in which vessels develop by canalization. The glomerulus comes to consist of a compact anastomosing plexus of vessels, and not of independent capillary loops, and the lobules of the glomerulus are not independent (Lewis 1958).

The glomerular capsule (of Bowman) is the blind, expanded end of the renal tubule, deeply invaginated by the glomerulus. Its lining consists of epithelial cells which are of the squamous type in its outer (parietal) wall; in contrast, surrounding the capillaries of the glomerulus (glomerular or visceral wall) are specialized, epithelial *podocytes* (Latta 1973; Spinelli 1974). Thus, between the glomerulus and the outer layer of the capsule there is a flattened cavity, the *urinary space*, which varies in size according to the secretory activity of the glomerulus (*vide infra*), and is continuous with the lumen of the proximal convoluted tubule. A distinct basal lamina underlies the cells of the capsule and, in the glomerulus, this fuses with the basal lamina surrounding the capillary endothelial cells. The podocytes which surround the capillary loops are highly distinctive, consisting of flattened stellate cells, the major dendrite-like processes of which are curved around the capillaries, interdigitating tightly with the processes of other podocytes. Where they approach neighbouring capillaries, many end-feet (pedicels), encroach on the basal lamina, leaving narrow gaps between them. Each podocyte contains numerous mitochondria, microtubules, microfilaments, and vesicles of various types, indicating a metabolically active state. The *endothelium* of the glomerulus is of the finely fenestrated type; the only barriers to the passage of fluid from the capillary lumen to the cavity of the glomerular capsule are therefore the fused basal laminae of the endothelium and podocytes. This layer, the *glomerular basement membrane*, is about 0·33 μm thick in man, and is finely fibrillar. The membrane is composed of three distinct layers, the first and third of which are pale-staining (the *laminae rarae interna* and *externa*) and the middle one dense and fibrous (the *lamina densa*). Unlike many fenestrated endothelia, there does not appear to be a regular dense lamina across the pores of the glomerular endothelial cells, but on the distal margin of the glomerular membrane the area between adjacent podocyte end-feet is covered by a clear dense layer, *the glomerular slit diaphragm*, through which the filtrate must pass before it enters the urinary space. This diaphragm has recently been shown to be made of fine filaments arranged in two highly regular rows looking rather like a zip-fastener (Rodewald and Karnovsky 1974) which may be significant in forming a further sieve in the glomerular filter. Much evidence indicates that the glomerular basement membrane acts as a selective filter, allowing the passage from the blood, under pressure, of water and a variety of the smaller types of molecule present in the circulation.

Molecules such as haemoglobin are able to pass through, but larger molecules are held back. The selectivity of this process appears to be related to the physical and chemical characteristics of the glomerular filter. Tracer experiments show that some of the material withheld from the urine by the glomerulus passes through all but the outer surface of the filter, which therefore acts in a much more complex fashion than would a simple physical filter (Rennke *et al.* 1975).

In addition to the cell types mentioned above, irregular *mesangial cells* with phagocytic and contractile capacities have been demonstrated, particularly around the base of the glomerular capillary tufts. The mesangial cells are important in clearing the glomerular filter of enmeshed substances which may otherwise clog it; for example, immune complexes, cellular debris, etc. (*see* Caulfield and Farquhar 1974).

THE RENAL TUBULE

Each renal tubule (**8.**152D) consists of the following parts in sequence: (1) the *glomerular capsule*, already described; (2) the *proximal* (or *first*) *convoluted tubule*, connected to the glomerular capsule by a short *neck* (which in man is only slightly narrower than the rest of the tubule); the initial part forms a series of loops (the *convoluted portion*), while the terminal part of the proximal tubule becomes straight or slightly spiral (the *straight portion*) and runs towards the medulla to become (3) the *descending limb* of the *loop of Henle* connected by (4) a 'U' turn to (5) the *ascending limb*. The descending and ascending limbs of the loop of Henle may be narrower and thin-walled along much of their length, forming the *thin segment*. The upper part of the ascending limb of the loop of Henle (the *thick segment*), however, has a diameter similar to that of the proximal convoluted tubule. The ascending limb becomes continuous with (6) the *distal* (or *second*) *convoluted tubule*, divisible into a *straight* and a *convoluted section*, between which is a special thickened region, the *macula densa*; the nephron finally straightening out as (7) the *junctional* or *connecting tubule*, which ends by joining (8) a *collecting* or *straight duct*. Between the distal convoluted tubule and the junctional tubule, there is often a short angular segment of the renal tubule that is called the *zigzag* (or *irregular*) *tubule*.

The *collecting* (or *straight*) ducts (**8.**154B) commence in the medullary rays of the cortex; they unite at short intervals with one another and finally open into wider tubes, termed the *papillary ducts* or *ducts of Bellini*, which in turn open on the summit of a papilla, the numerous duct openings giving the tip of the papilla a perforated, cribriform, appearance (*area cribrosa*).

STRUCTURE OF THE RENAL TUBULE (8.152D, 153A)

The renal tubules are lined by a single layer of epithelial cells outside which is a basal lamina of varying thickness (Maunsbach 1973). The precise form of the constituent cells varies from one part of the tubule to another, according to their distinctive roles in the functioning of the kidney. These cells are concerned with the active transport and passive diffusion of various ions and water into and out of the tubules, and with the removal of any proteins and certain other organic substances which may pass the glomerular filter (Bulger 1977).

The proximal tubule is lined by cuboidal epithelial cells bearing a brush border of tall microvilli at their luminal surfaces. The precise shape of these cells depends on the pressure of the fluid within the tubule; in life this distends the lumen and stretches the lining cells to a slightly flattened cuboidal shape, which changes to a taller form when the glomerular blood pressure falls at death or on removal of tissue for histology. The cytoplasm of proximal tubule cells is strongly eosinophilic, and the nucleus euchromatic and centrally positioned. Faint striae are visible with the light microscope at the bases of the cells, which on inspection with the electron microscope are seen to be highly folded in a complex series of pleats between which lie numerous mitochondria orientated perpendicular to the basal lamina of the tubule. Careful reconstruction of these basal infoldings has shown that the complex of cellular processes is really formed by the inter-digitations between the lateral margins of adjacent epithelial

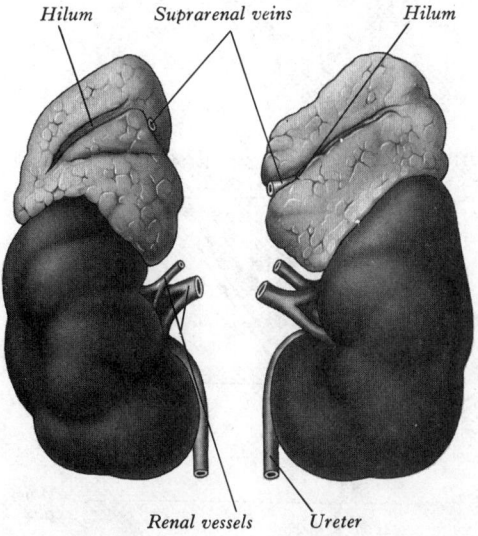

Hilum *Suprarenal veins* *Hilum*

Renal vessels *Ureter*

8.151 The kidneys and suprarenal glands of a newborn infant. Anterior aspect. Note lobulation of renal surface, and relative size of the organs.

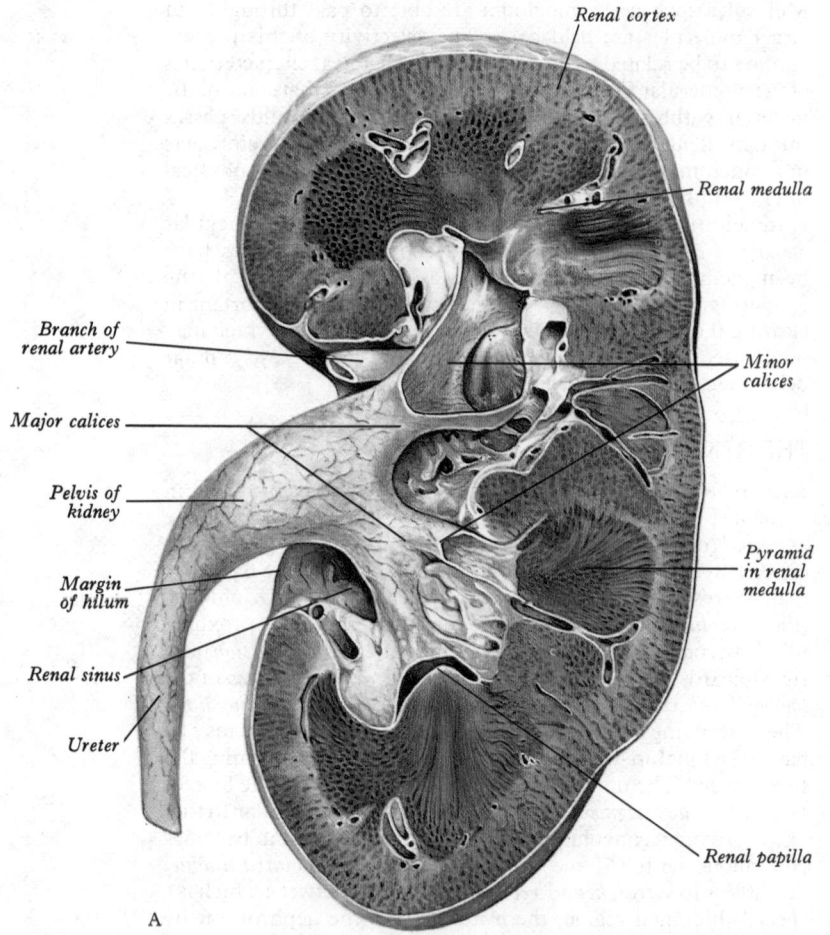

Renal cortex

Renal medulla

Branch of
renal artery

Minor
calices

Major calices

Pelvis of
kidney

Pyramid
in renal
medulla

Margin
of hilum

Renal sinus

Ureter

Renal papilla

A

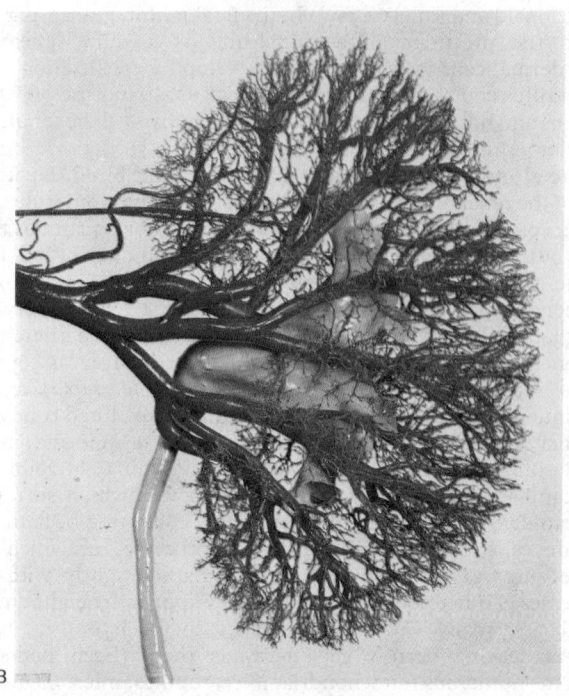

B

8.152A–D The structural and functional organization of the kidney. A Longitudinal section through a kidney; note pelvis of the ureter and its division into calices; and also the macroscopic appearance of the normal kidney. The pelvis and major calices have not been opened. B A corrosion cast of a human kidney, showing minor and major calices, ureteric pelvis and upper ureter (all in yellow), and renal arterial tree (in red). Note also suprarenal branches from the renal artery and direct from the aorta. (Prepared by Dr. M. C. E. Hutchinson; photographed by Mr. Kevin Fitzpatrick, Anatomy Dept., Guy's Hosp. Med. School. C A diagram illustrating the major structures in the kidney cortex and medulla (left), the position of cortical and juxtamedullary nephrons (middle) and the major blood vessels (right). D A schematic diagram of the regional structure and principal activities of a kidney nephron and collecting duct. For clarity, a nephron of the long loop (juxtamedullary) type is depicted.

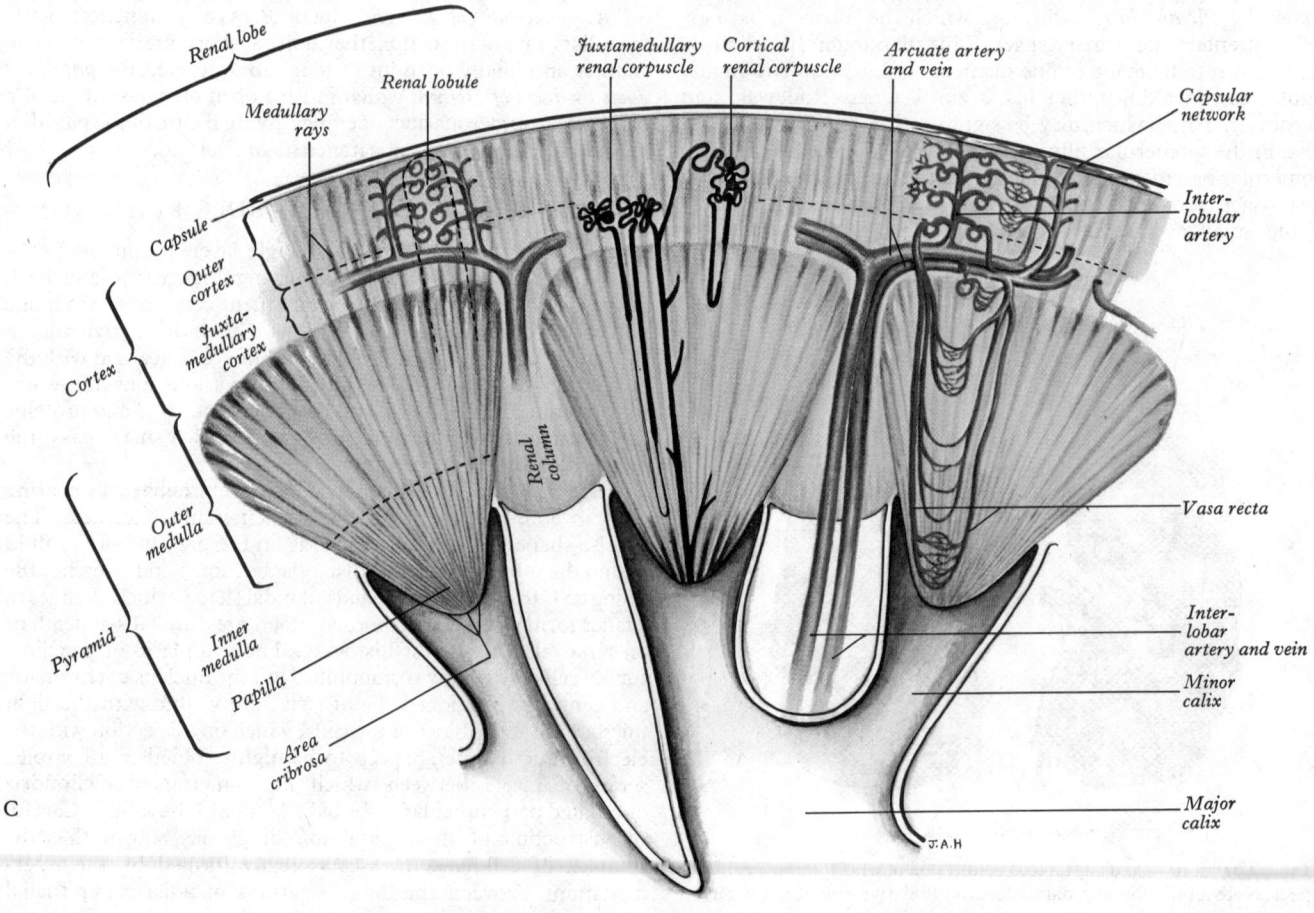

Renal lobe

Juxtamedullary
renal corpuscle

Cortical
renal corpuscle

Arcuate artery
and vein

Renal lobule

Medullary
rays

Capsular
network

Capsule

Inter-
lobular
artery

Outer
cortex

Juxta-
medullary
cortex

Cortex

Renal column

Outer
medulla

Vasa recta

Pyramid

Inner
medulla

Inter-
lobar
artery and vein

Papilla

Minor
calix

Area
cribrosa

Major
calix

C

J.A.H

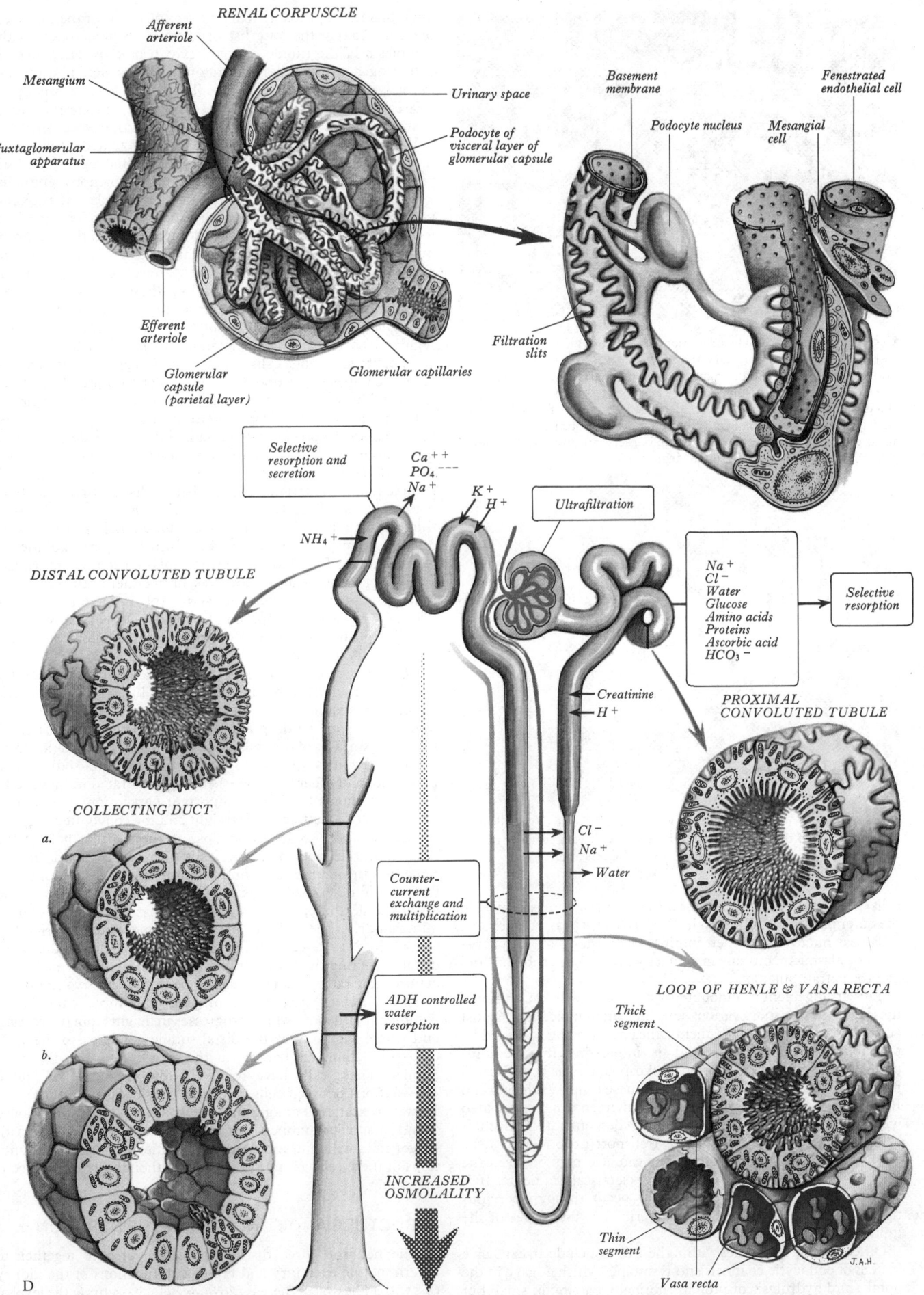

RENAL CORPUSCLE

Afferent arteriole

Mesangium

Juxtaglomerular apparatus

Efferent arteriole

Glomerular capsule (parietal layer)

Glomerular capillaries

Urinary space

Podocyte of visceral layer of glomerular capsule

Basement membrane

Podocyte nucleus

Fenestrated endothelial cell

Mesangial cell

Filtration slits

Selective resorption and secretion

Ca^{++}
PO_4^{---}
Na^+

K^+
H^+

Ultrafiltration

NH_4^+

Na^+
Cl^-
Water
Glucose
Amino acids
Proteins
Ascorbic acid
HCO_3^-

Selective resorption

DISTAL CONVOLUTED TUBULE

PROXIMAL CONVOLUTED TUBULE

Creatinine
H^+

COLLECTING DUCT

a.

Cl^-
Na^+
Water

Counter-current exchange and multiplication

LOOP OF HENLE & VASA RECTA

Thick segment

ADH controlled water resorption

b.

INCREASED OSMOLALITY

Thin segment

Vasa recta

J.A.H.

D

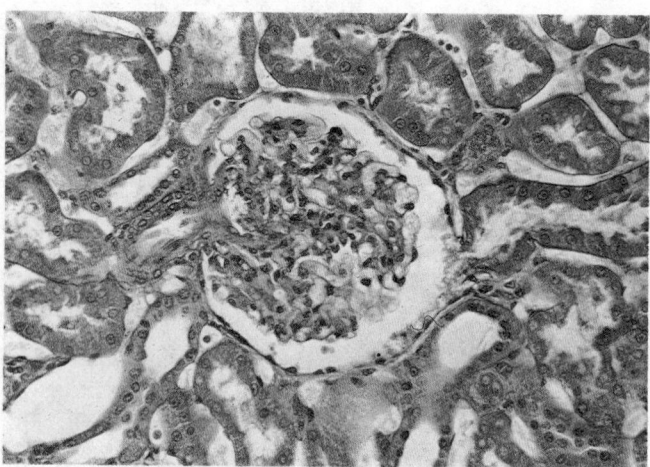

8.153A A renal corpuscle in section, showing a glomerulus (centre) within its capsule, the urinary space of which is continuous with the proximal convoluted tubule (right). A glomerular arteriole is visible at the vascular pole of the capsule, where it is associated with a *macula densa* of a distal convoluted tubule (left), denoted by a group of closely spaced nuclei. Profiles of proximal convoluted tubules (with brush borders) and distal convoluted tubules (lacking brush borders) are also visible. Simian kidney. Masson's trichrome stain. Magnification × 150.

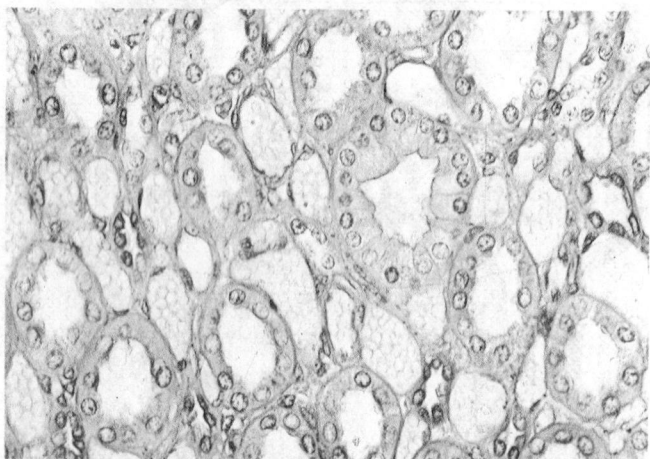

8.153B Part of a medullary ray in cross-section, showing large collecting ducts, and small thin segments, interspersed with vasa recta distended with erythrocytes. Tissue and stain as in A. Magnification × 250.

cells, thus creating a labyrinthine meshwork of cytoplasmic feet in the outer part of the tubule wall (Bulger 1965, 1977). As the same cells bear microvilli on their luminal faces, they have very large areas of plasma membrane in contact with both the tubular fluid and the extratubular space.

This arrangement is undoubtedly connected with the ability of these cells to transport various ions and small molecules against very steep concentration gradients, the energy being provided by the abundant mitochondria. It is interesting that sodium/potassium-stimulated adenosine triphosphatase has been located in both apical and basal membranes of these cells (Wachstein and Bradshaw 1965) and that numerous other enzymes associated with ionic transport have also been demonstrated in their cytoplasm. Apart from the active transport of ions across the epithelial cells, water and other ions can also pass between the cells passively, following osmotic and electrical gradients. It is probable that the passive movements occur through pervious tight junctions which have been described at the apices of the tubule cells (see Frömter 1979).

Other notable organelles within these cells include an extensive system of endocytic channels and lysosomes which engage in the uptake and hydrolysis of proteins normally passing in small but significant amounts into the glomerular filtrate (Caulfield and Farquhar 1975). Peroxisomes and lipid droplets are also frequently seen, and, at the base of each cell, a system of

microfilaments, probably maintaining the shape of the tubule, is present. Among the long list of enzymes demonstrated in the cytoplasm of the tubule cells are cytochrome oxidase, succinic dehydrogenase and other respiratory enzymes, acid phosphatase in the lysosomes, glucose-6-phosphatase and leucine aminopeptidase; the high activities of these enzymes reflects the highly energetic nature of these cells in kidney function (*vide infra*).

Variations occur in cell structure, enzyme activity and functions of the proximal tubule cells. The initial region, or *neck* is very short and consists of the simple squamous epithelial extension of the outer wall of the renal corpuscle; the proximal tubule beyond this has been subdivided into three (or sometimes four) regions, the first typified by long microvilli, the second by microvilli of medium length, the third, again (in man) by long microvilli (*see* Moffat 1975). The patterns of lysosomal and transport activities appear to follow these subdivisions, too, although their functional importance is not understood.

The loop of Henle (renal loop, ansa nephroni) consists of the thin segment, narrow in diameter (about 30 μm) and lined by low cuboidal to squamous cells; and the thick segment (about 60 μm across) composed of cuboidal cells similar to those of the distal tubule with which it is sometimes classified as the *straight portion* of that structure. *The thin segment* forms the major part of the loop in the more deeply placed (juxtamedullary and deep cortical) nephrons which reach far into the medulla. In such nephrons, the descending limb has moderately thick epithelial cells with complex interdigitations and narrow belts of tight junctions between cells, whereas the ascending limb appears to be lined by thin squamous cells with wider junctional zones. These differences may be related to the distinctive permeabilities to water and ions found in these two tubular structures (*see* Bulger 1971). Very few organelles are present in either of the cell types, indicating a passive rather than active role in ionic movements. *The thick segment* is composed of cuboidal epithelium with cells containing numerous mitochondria, deep basolateral folds and short apical microvilli.

The distal tubule is lined by cells resembling those of the proximal tubule, except that the microvilli are few, small and irregularly spaced, and that the basolateral folds containing mitochondria are so deep and extensive that they almost reach the apical surfaces of the cells, as seen in vertical section (**8.152D**). Various enzymes related to active transport of sodium and potassium and other ions have been demonstrated in these cells, which are known to play an important part in ionic regulation. At the junction between the straight and the convoluted regions, the distal tubule comes to lie close to the vascular pole of the renal corpuscle, and the tubule cells form a specialized sensory structure, the *macula densa*, important in the regulation of blood flow and ionic exchange (*vide infra*). At the termination of the distal tubule the epithelium has fewer basal folds and sparser mitochondria, the tubule here sometimes being termed a *connecting duct*, since it has several characteristics of the more distal collecting duct, but unlike the latter is still formed from nephrogenic rather than ductal tissue during embryogenesis.

The collecting ducts are composed of simple cuboidal to columnar epithelium which progresses in height from the cortical end where they receive the distal tubule contents, to the wide ducts of Bellini, discharging at the area cribrosa on the renal papilla. These cells have relatively few organelles or lateral convolutions between cells, and bear only occasional microvilli. However, scattered among these cells are other, *dark cells*, also found in smaller numbers in the distal convoluted tubule, with longer microvilli and more numerous mitochondria. The functions of these cells are not yet known, although their presence is intriguing.

STRUCTURE AND FUNCTION IN THE NEPHRON

In the nephron three distinct processes operate together to determine the excretory and regulatory functions of the kidney (**8.152D**). The first of these is *filtration*, which occurs at the level of the glomerulus; the second is *selective resorption* of various materials from the filtrate as it passes along the nephric tubule; and the third involves *secretion* of various substances into the filtrate by the cells of the tubule. The first two of these processes have been the more intensively studied, and more clearly correlated with the cytological structure of the nephron in its various regions.

Glomerular filtration is the term used to describe the passage of water, containing a wide variety of small molecules, from the blood into the urinary space of the glomerular capsule, the larger molecules such as those of the plasma proteins, polysaccharides and lipids being substantially retained in the circulation as though by a physical filter of minute pore-size placed between the blood and the lumen of the capsule. Such a filter exists in the glomerular basement membrane (see above for details of this, and of the fate of the small quantities of protein which pass through the filter).

Filtration occurs along the considerable pressure gradient existing between the inside of the large capillaries and the urinary space, across the glomerular basement membrane, which is the only structure separating the two. This pressure gradient is far greater than the osmotic gradient related to the colloid osmotic pressure of the blood, which tends to oppose the outward flow of filtrate. In the peripheral parts of the renal cortex, the arteriolar pressure gradient is enhanced by the dimensional inequality of the afferent and efferent arterioles of the glomerulus, the former having a larger diameter than the latter (*vide infra*). Further, in all glomeruli, the rate of filtration can be altered by vasomotor responses of the glomerular arterioles. When first formed, the *glomerular filtrate* is isotonic with the blood of the glomerulus, its composition also being identical in its concentration of ions and small molecules.

Selective resorption of many materials actively from the filtrate occurs mainly in the proximal convoluted tubule, particularly of glucose, amino acids, phosphate, chloride, sodium, calcium and bicarbonate. The proximal tubular cells are freely permeable to water, which moves from the tubule lumen along the osmotic gradient created by the resorption of these various solutes, so that the filtrate in this region remains isotonic with the blood. The numerous microvilli, folded lateral and basal surfaces, and profusion of mitochondria, indicate an active role in absorption by the proximal tubule cells. Further selective absorption, particularly of sodium ions, also occurs in the distal convoluted tubule.

The remainder of the renal tubule appears to provide a mechanism for the resorption of the great majority (95 per cent) of the water content of the glomerular filtrate, so that when it reaches the calices of the kidney, the urine has a greatly diminished volume, and is *hypertonic* to blood. Along the *descending* limb of the loop of Henle, sodium and water pass freely between the lumen of the tubule and the surrounding extratubular spaces of the renal medulla, which many loops occupy. In at least part of the *ascending* limb, chloride ions are actively transported from the lumen to the interstitial spaces and sodium ions follow passively, but in this tubular region the lining cells do not allow water to follow the sodium. Some of this sodium now diffuses back into the descending limb, being added to that already present in the filtrate passing along it; in the ascending limb, sodium is again extracted from this enriched solution, being added to the intercellular ions and again diffusing into the descending limb. Alternatively, sodium may not actually enter the descending limb, but water is merely withdrawn because of the raised tonicity of the extratubular fluid-filled space, thus concentrating the filtrate. In either manner, considerable concentrations of sodium and chloride ions are built up in the renal medulla, the responsible mechanism being termed a *countercurrent multiplier system*. The potentially rapid removal of ions from the renal medulla by the circulatory system is minimized by another looped *countercurrent exchange system*. In the latter, the arterioles entering the medulla pass for long distances parallel to the venules leaving it, before breaking up into capillary beds around the tubules; this arrangement allows the passage of ions by diffusion from the outflowing to the inflowing blood, so that these vessels (the arteriolar and venular *vasa recta*) conserve the high osmotic pressure of the medulla as a whole.

Because of the selective extrusion of sodium and chloride by the cells of the ascending limb and distal tubule (under aldosterone control), the filtrate, when it reaches the distal end of the convoluted tubules, is *hypotonic* to the blood. However, as it passes on into the collecting ducts which descend again through the medulla, it moves into a region of high osmotic pressure and, since the cells composing the duct walls are, under the action of the antidiuretic hormone (ADH) from the neurohypophysis, variably permeable to water, the latter passes along the osmotic gradient into the neighbouring extratubular spaces (Gottschalk and Mylle 1959). Thus, throughout the collecting duct, the tonicity of the urine gradually rises, until at the renal pyramids it

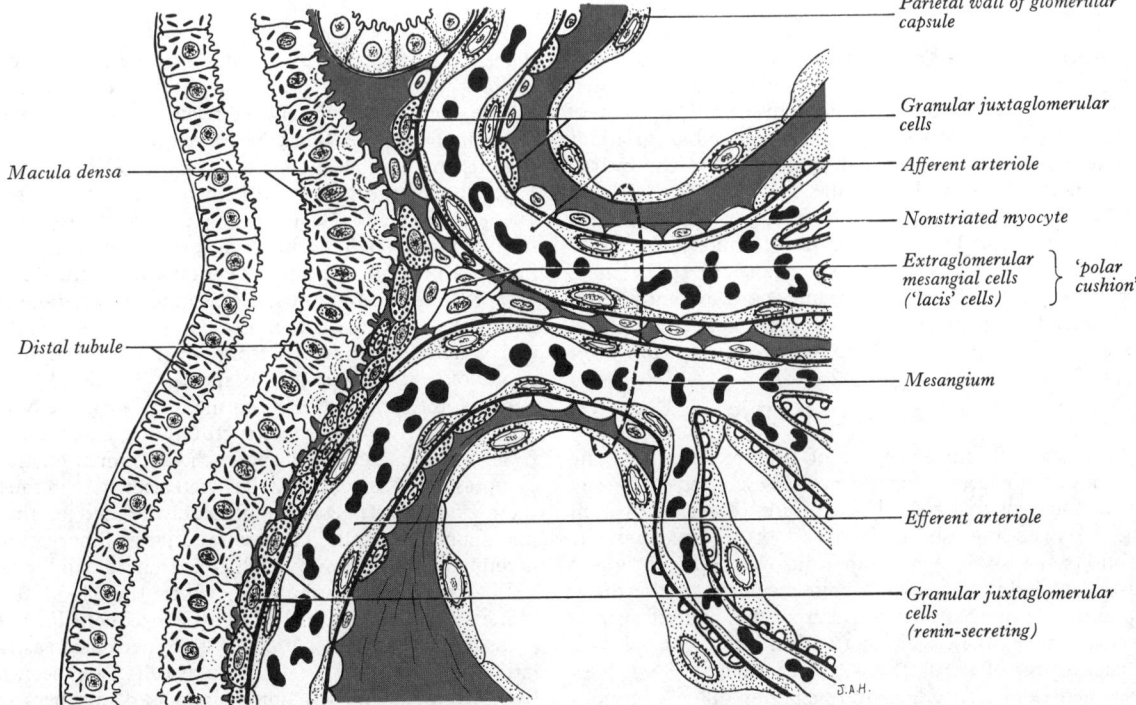

Macula densa

Distal tubule

Parietal wall of glomerular capsule

Granular juxtaglomerular cells

Afferent arteriole

Nonstriated myocyte

Extraglomerular mesangial cells ('lacis' cells) } 'polar cushion'

Mesangium

Efferent arteriole

Granular juxtaglomerular cells (renin-secreting)

8.154 A diagram showing the organization of the juxtaglomerular complex including the macula densa (left), granular juxtaglomerular cells (middle) and the vascular pole of the glomerular capsule (right).

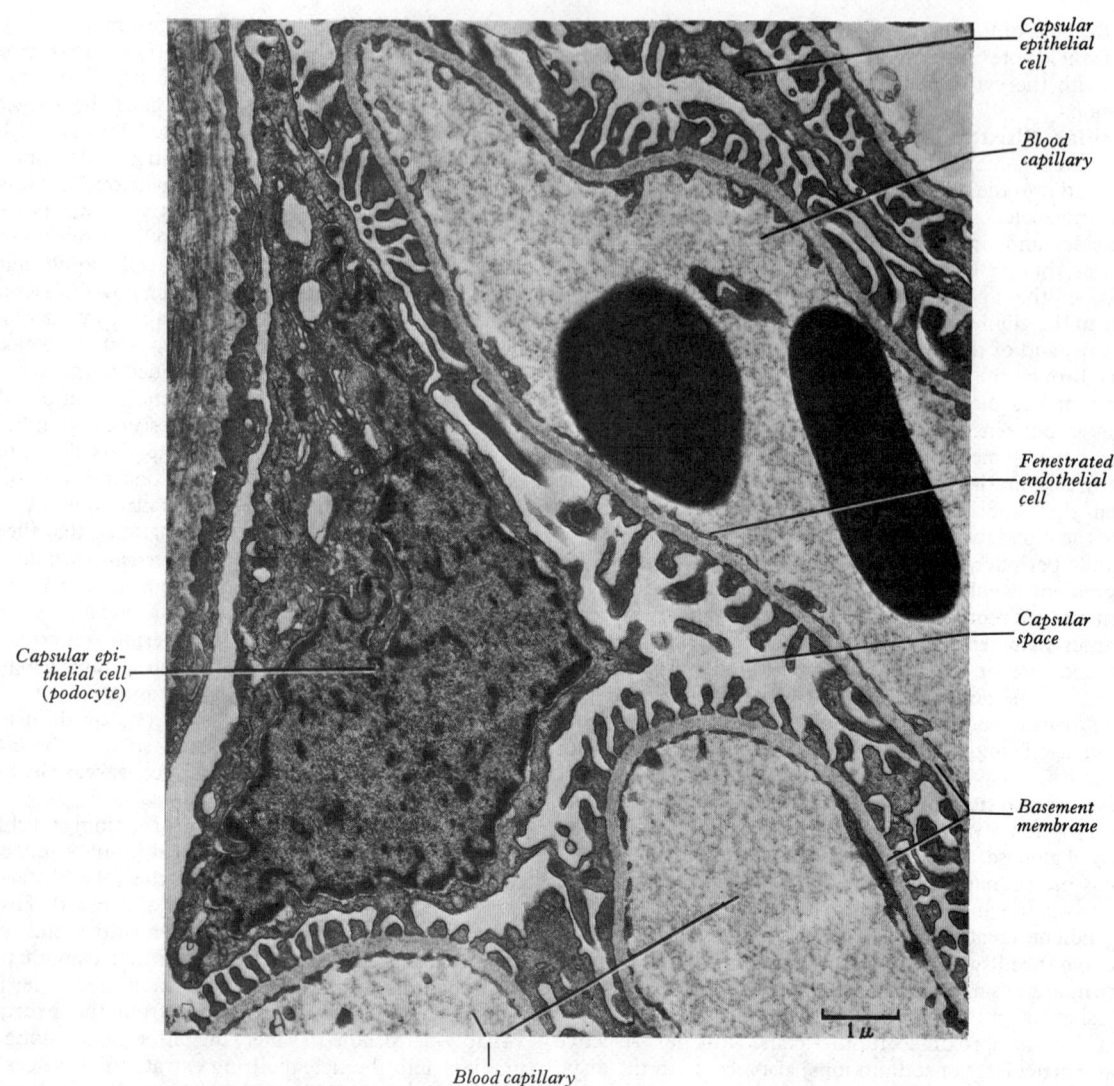

Capsular epithelial cell

Blood capillary

Fenestrated endothelial cell

Capsular space

Basement membrane

Capsular epi-thelial cell (podocyte)

1 μ

Blood capillary

8.155A Electron micrograph showing capillaries and epithelial cells of a renal corpuscle (rat). Note the basement membrane and fenestrated endothelial cells as well as the foot processes of the epithelial cell (podocyte). Magnification about × 9,200.

is finally *hypertonic* to the blood. In this manner, as mentioned above, up to 95 per cent of the water in the original glomerular filtrate is resorbed into the bloodstream. This complex system of controls is a highly flexible one, allowing considerable variations in filtration and absorption rates, depending upon the overall needs of the body at any particular time.

Secretion of various ions occurs at a number of sites, hydrogen ions and ammonium being particularly important in the regulation of acids and bases in the blood. Many other substances are also secreted, including various organic acids, and antibiotics, when administered therapeutically. The proximal and distal tubules are the chief regions carrying out these activities (8.152D).

THE JUXTAGLOMERULAR APPARATUS

The afferent and efferent arterioles of a glomerulus join the vascular pole of a glomerular capsule more or less opposite the site of origin of the proximal convoluted tubule (8.152D). In each nephron, the ascending limb of Henle's loop returns from the medulla and passes towards the glomerulus which gives origin to that particular nephron, and the commencement of the distal convoluted tubule lies between the afferent and efferent vessels at the vascular pole, in close contact with the vascular pole and the external aspect of the walls of the arterioles. The cells of the tunica media of the arterioles in this region differ from non-striated muscle cells of blood vessels in general in that they are large, rounded and 'epithelioid' in type and have large spherical nuclei; as revealed by special stains, their cytoplasm is granular.

These cells are known as *juxtaglomerular cells* and they lie in close contact with the cells of the slightly dilated distal convoluted tubule, which are closely aggregated together in this region, forming the *macula densa*; these two groups of cells, together with various mesangial elements, constitute the *juxtaglomerular apparatus* of the kidney (Edelman and Hartroft 1961; Barajas 1966; Dalton and Hagenau 1967; Hartroft 1968). Ultrastructurally, the juxtaglomerular cells possess bundles of cytoplasmic fibrils, a prominent granular endoplasmic reticulum and Golgi complex, and numerous dense, membrane-bound secretory granules, which mature in association with the Golgi complex, and which vary in shape from spheroid, through ellipsoid, to rhomboid, according to the species (Barajas 1971). In many animals, interspersed between the foregoing, are other so-called '*lacis cells*', which are ultrastructurally similar, but lack granules. Cells of the macula densa approach a columnar form, with a paler cytoplasm and fewer organelles than elsewhere in the tubule, and their external aspects are intimately related to the surface of juxtaglomerular cells. The latter receive an innervation from non-myelinated nerve fibres which are adrenergic in type.

In experimental animals in which the blood supply to the kidney has been diminished, with consequent increase of blood pressure, and in some patients with hypertension associated with kidney disease, the juxtaglomerular cells have been found to be hypertrophied and to contain increased numbers of granules. These granules contain *renin*, an enzyme which converts a polypeptide in the blood, *angiotensinogen*, to *angiotensin I*; this in turn is converted, notably in the lung, by other enzymes into

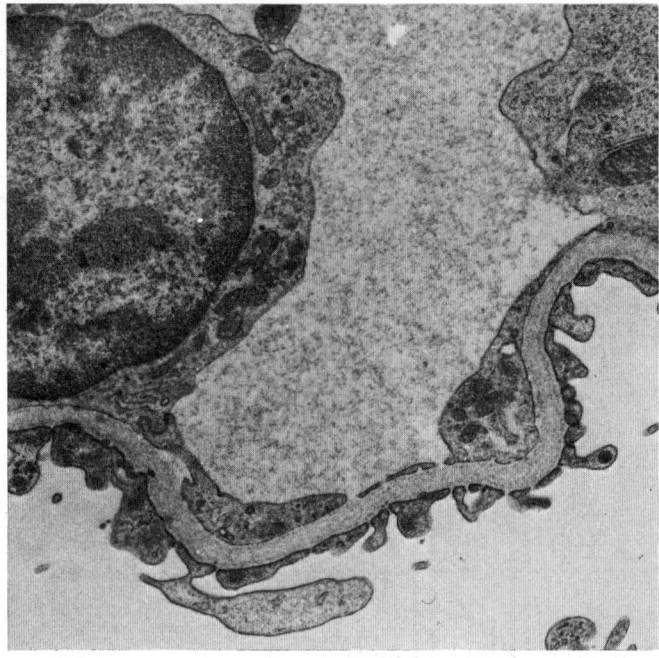

8.155B A high-power electron micrograph of part of a glomerular capillary in a canine renal corpuscle to show, in greater detail, the structures intervening between the blood-vascular and urinary spaces. The capillary lumen (top centre) is bounded to the left by the nucleated part of an endothelial cell, and below by the thin cytoplasmic extension from a neighbouring endothelial cell, in which a few fenestrations are present. The urinary capsular space (below), the podocyte processes, and the compound glomerular basement membrane, which shows zonal variations in electron-opacity, may be identified. Magnification × 18,000.

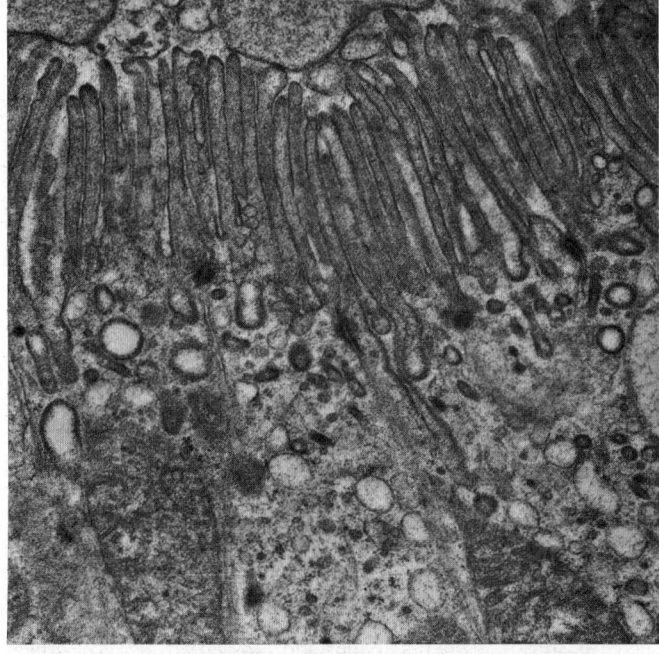

8.155C A high-power electron micrograph of the apical (luminal) parts of four proximal convoluted tubular epithelial cells. The apical array of microvilli, subapical junctional complexes between adjacent cells, and the highly active cytoplasm containing lysosomes, parts of very large mitochondria, and a wide variety of granules, vesicles and vacuoles, are visible. Magnification × 20,000.

angiotensin II, a polypeptide with a wide variety of actions on the body, including the elevation of blood pressure and the stimulation of aldosterone release from the adrenal cortex (leading to more active removal of sodium ions from the distal convoluted tubule). The details of this system of renin secretion and consequent change in tubule activity are not yet clear, but the juxtaglomerular apparatus appears to provide an important feedback control regulating the flow of fluid through the

glomerular filter, ionic resorption in the nephric tubule, and thus the final concentration of urine. There is evidence that renin secretion is controlled by at least three major factors (Davies and Freeman 1976): (1) the influence of the cells of the macula densa which react to changes in fluid passing them through the distal tubule; (2) the pressure of blood in the glomerular arterioles which in some way affects the secretory activity of the granule cells in their walls; (3) stimulation by the numerous sympathetic fibres ending close to the juxtaglomerular cells. These and other agents appear to enable the juxtaglomerular apparatus to correlate various aspects of blood flow, filtration rate and osmoregulation and to mediate appropriate action to maintain homeostasis.

OTHER CELLS IN THE KIDNEY

Between the renal tubules and blood vessels lie various cellular elements important to the structure and activities of the kidney. Amongst these are connective tissue elements, inconspicuous though present in the cortex, but prominent in the medulla, where they secrete the glycoprotein matrix, particularly in the papillary regions. The medullary *interstitial cells*, some of which at least, appear to be modified fibroblasts, form vertical piles of tangentially orientated cells between the more distal collecting ducts, giving a distinctive ladder-like appearance. Various studies have shown that these cells are able to secrete prostaglandins with widespread effects on visceral muscle and other cells in the body (Muirhead *et al.* 1972), although the significance of this finding is not yet clear.

RENAL BLOOD VESSELS

The complicated vascular patterns of the kidney show considerable regional specialization, and are closely moulded to the geometric organization and functional roles of the renal corpuscles, tubules and collecting ducts (8.152, 156, 157). A very large research literature has accumulated concerning not only renal angioarchitecture, but also renal haemodynamics, and the control factors which may operate to secure appropriate functional variations in blood flow. Analysis of this must, however, be omitted here; similarly, no attempt will be made to describe the important species variations in pattern known to exist, or the minor variations in the human pattern. For these considerations, and as an introduction to the human to the literature, the interested reader should consult works devoted exclusively to these topics, for example Trueta *et al.* (1947) published over thirty years ago, and the admirable reviews by Fourman and Moffat (1971), and Moffat (1975).

The renal vasculature may, of course, be studied at various levels of organization, commencing with the origin, course, structure, variations and primary branches of the principal *renal artery*, and the disposition of any *accessory renal arteries* which may also be present. Together with the primary patterns of branching, and their substantial areas of distribution, may be considered the concept of *vascular segmentation* of the kidney. Arising sequentially from these primary branches are *lobar, interlobar, arcuate* and *interlobular* arteries, the *afferent* and *efferent glomerular arterioles*, the cortical *intertubular capillary plexuses*, and the venous radicles which drain these regions; the *vasa recta* and associated capillary plexuses which supply the renal medulla must also be considered. (The disposition of the vessels and their relationship to the uriniferous tubules are summarized in 8.152, 156.)

The main features of the origins, courses and extrarenal distribution of the *renal arteries* have already been described (p. 718). Single arteries on right and left sides occur in about 70 per cent of individuals, but they vary somewhat on the two sides and in different individuals, in terms of their level of origin (the right often being superior), their calibre, obliquity and precise topographic relationships. (For a most extensive review of this clinically important topic in almost 11,000 kidneys, *see* Merklin and Michels 1958.) In its extrarenal course (Schneider *et al.* 1969) each renal artery gives off one or more inferior suprarenal arteries, and rami of supply to the perinephric tissue, renal capsule, pelvis and proximal part of the ureter; and upon or just before reaching

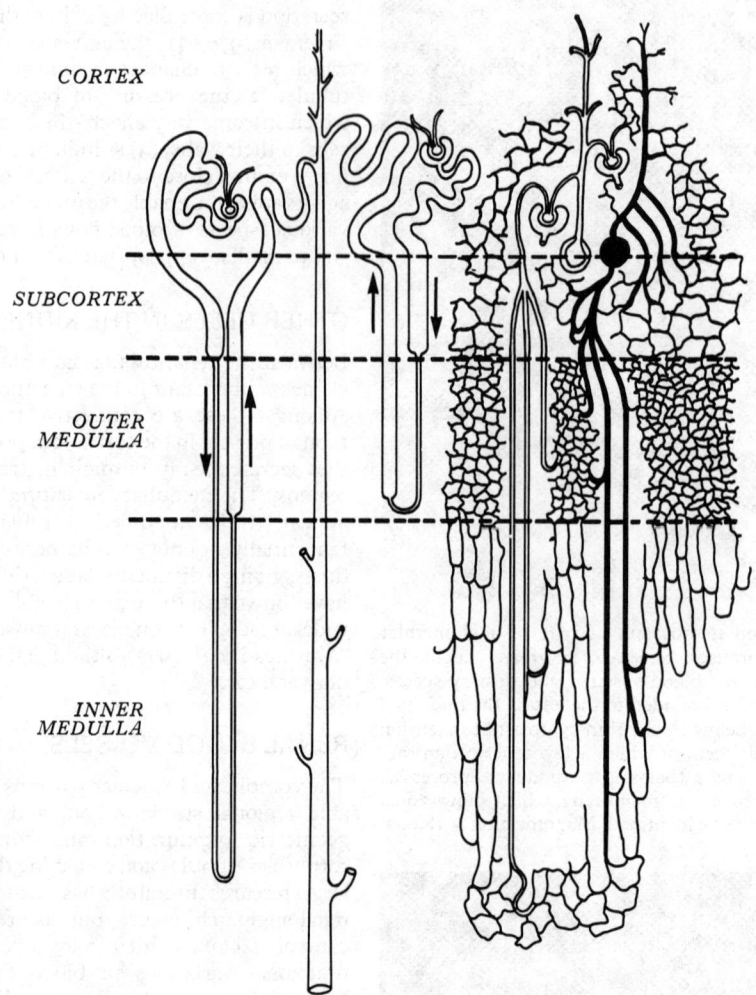

CORTEX

SUBCORTEX

OUTER MEDULLA

INNER MEDULLA

8.156A A diagram to illustrate the arrangement of the tubules (left) and blood vessels (right) in various structural zones of the kidney. Note the variations in the pattern of the tubules with either long or short medullary loops; tubules of intermediate length also occur. Compare the different structural and functional segments of the tubules which occur together in the cortical, subcortical, and outer and inner medullary zones, with their related vascular patterns. Arteries—black outlines; capillaries—single black lines; veins—full black. Compare with **8.156B**. (From *the Blood Vessels of the Kidney*, by Julia Fourman and D. B. Moffat, by courtesy of the authors and publishers, Blackwell Scientific Publications, Oxford, 1971.)

the hilum of the kidney, the renal artery on each side divides into an *anterior* and a *posterior division*. The primary branches of the divisions (the *segmental arteries*) supply the *vascular segments* of the kidney.

Accessory renal arteries are common, occurring in up to 30 per cent of individuals. Usually such arteries arise from the aorta, above or below the renal artery, and pursue a course parallel with it to the renal hilum. Higher or lower origins are, however, not uncommon, the accessory artery or leash of arteries passing to the superior or inferior pole of the kidney. The foregoing are widely regarded as members of the embryonic *lateral splanchnic* series of arteries, which formed precociously, and persisted throughout subsequent development. Accessory vessels to the inferior pole of the kidney cross the ureter, and have been suggested as one cause of hydronephrosis, but this causal relationship cannot be regarded as proven. Very occasionally accessory renal arteries may arise from the coeliac or superior mesenteric arteries, near the bifurcation of the aorta or from the common iliac artery.

The fact of *renal vascular segmentation* was initially recognized by John Hunter in 1794, but the first detailed account of the primary branching pattern was provided by Graves (1954, 1956 *a* and *b*) who studied casts and radiographs of injected kidneys. He described five vascular segments: (1) *apical* occupying the medial side and anterior part of the superior pole; (2) *upper (anterior)* including the rest of the superior pole and the anterosuperior part of the central area; (3) the *lower* which encompasses the whole of the lower pole; (4) the *middle (anterior)* which lies between the upper (anterior) and lower segments; and (5) the *posterior* segment which includes the whole of the posterior aspect of the kidney between the apical and lower segments. Although Graves' terminology has been adopted internationally and used by some researchers in this field (Smith 1963; Sykes 1963, 1964), others have proposed more complex schemes including, for example, three posterior segments (Faller and Ungvary 1962); yet others have emphasized the great variability of the precise areas supplied by the segmental arteries (Fine and Keen 1966). Despite these somewhat divergent views, it is nevertheless important to emphasize that the vascular segments are supplied by arteries which have virtually no anastomoses with their neighbours and are thus end-arteries. In contrast, the larger *intrarenal veins* have no segmental organization, and anastomose freely.

Early this century (Brodel 1901) a relatively avascular longitudinal zone (the *'bloodless' line of Brodel*) was described as existing along the convex lateral border of the kidney, and this was proposed as the most suitable route for renal surgical incisions. However, it is now clear that many vessels cross this zone, which is far from 'bloodless', and that planned radial or intersegmental incisions are preferable.

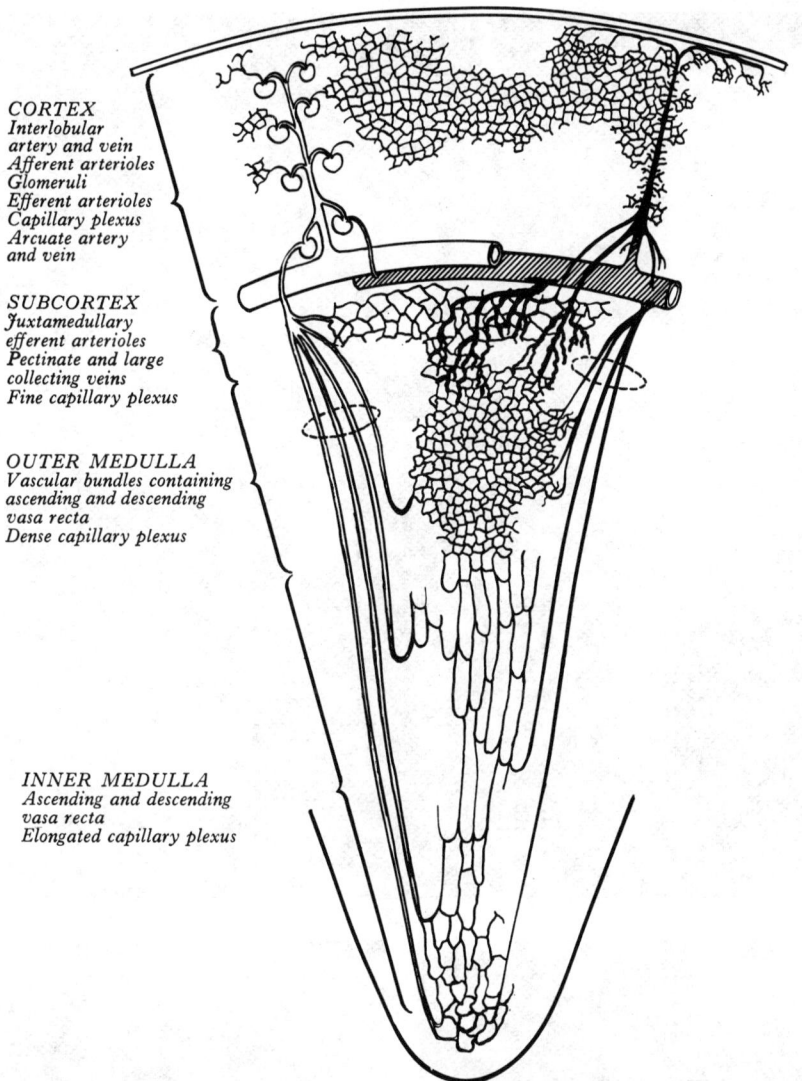

CORTEX
Interlobular
artery and vein
Afferent arterioles
Glomeruli
Efferent arterioles
Capillary plexus
Arcuate artery
and vein

SUBCORTEX
Juxtamedullary
efferent arterioles
Pectinate and large
collecting veins
Fine capillary plexus

OUTER MEDULLA
Vascular bundles containing
ascending and descending
vasa recta
Dense capillary plexus

INNER MEDULLA
Ascending and descending
vasa recta
Elongated capillary plexus

8.156B   A diagram of the basic arrangements of the blood vessels in the mammalian kidney. Arteries—black outlines; capillaries—single black lines; veins—cross-hatched or full black. Note the variations in the pattern of the meshes in the capillary networks. See accompanying text for further description. (From *The Blood Vessels of the Kidney*, by Julia Fourman and D. B. Moffat, by courtesy of the authors and publishers, Blackwell Scientific Publications, Oxford, 1971.)

The initial branches of the segmental arteries provide the *lobar arteries*, usually one for each renal pyramid, which again divide just before entering the kidney substance into two or three *interlobar arteries*, extending towards the cortex on each side of the pyramid (i.e. between adjacent renal lobes). At the junction between the cortex and medulla, the interlobar arteries divide dichotomously into *arcuate arteries* which diverge at right angles from the parent stem. As they pursue their curved courses between cortex and medulla, each arcuate artery undergoes several further divisions, and from each of these branches a series of *interlobular arteries* ascend radially into the cortical substance. The terminations of adjacent arcuate arteries do not anastomose, but finally turn into the cortex as further interlobular arteries. Although the majority of the latter are branches of arcuate arterial branches, some also arise from the stem vessel and even from the terminal part of the interlobar arteries (**8.**156B).

The *interlobular arteries* may follow a simple course towards the superficial part of the cortex, or they may branch a few times *en route*, whilst some pursue a more tortuous course, recurving towards the medulla once or twice before again proceeding towards the renal surface. Some of the interlobular arteries pass straight through the surface of the kidney as *perforating arteries* (Hammersen and Staubesand 1961) to anastomose with the *capsular plexus* of vessels (which also receives *capsular branches*

from the *inferior suprarenal*, *renal* and *testicular* or *ovarian* arteries).

The *afferent glomerular arterioles* are mainly side branches from the interlobular arteries, but a few arise directly from the arcuate and interlobar arteries. Those from the interlobular artery vary in their direction and angle of origin from the parent vessel—the deeper ones are directed obliquely back towards the medulla, the intermediate ones pass out horizontally from the parent stem, whilst the more superficial ones pass obliquely towards the renal surface before ending in a glomerulus (**8.**156A, B).

From the majority of the glomeruli (i.e. with the exception of the juxtamedullary ones, and a small proportion of those at intermediate levels in the cortex), the *efferent glomerular arterioles* soon divide to form a fine-meshed *peritubular capillary plexus*, which runs between and around the proximal and distal convoluted tubules. Thus, in the main renal cortical circulation there are *two* distinct sets of *capillaries* in series, glomerular followed by peritubular, the two being linked by the efferent glomerular arterioles. From the venous ends of the peritubular plexus, fine radicles converge to join *interlobular veins*, one of which accompanies each artery of the same name. Many of the interlobular veins commence immediately beneath the fibrous capsule of the kidney, by the convergence of a number of *stellate veins* which drain the most superficial parts of the renal cortex,

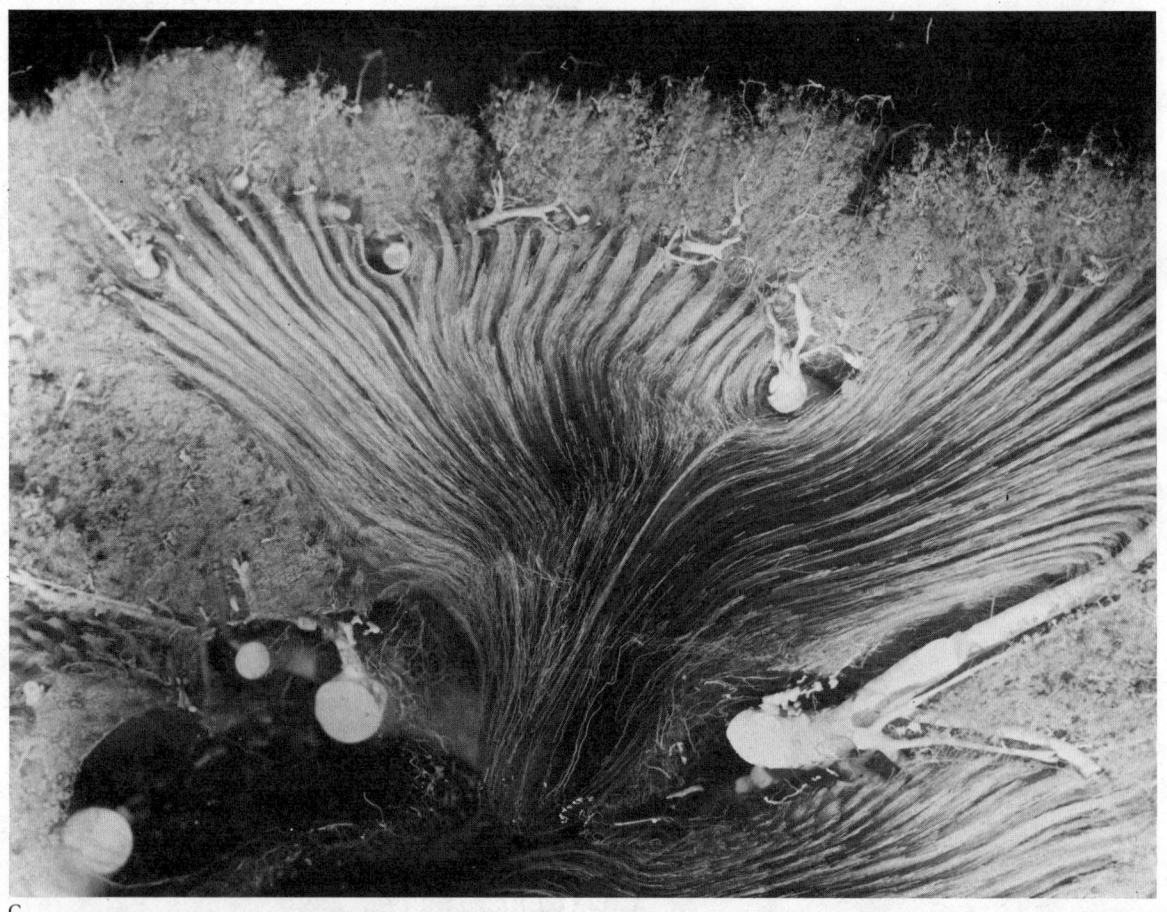

C

D

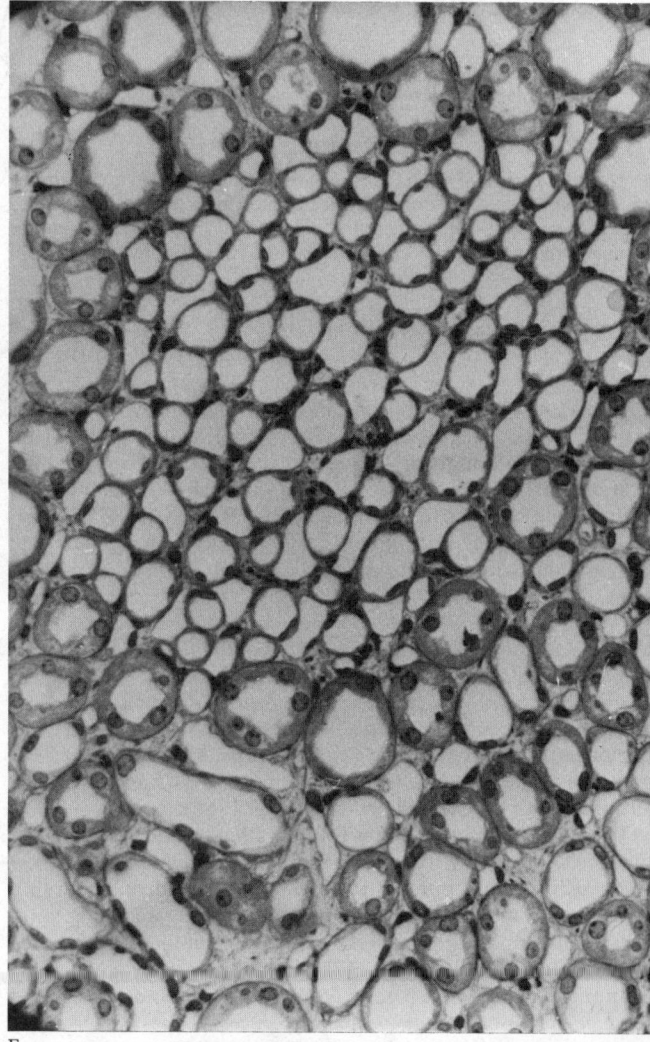

E

8.156C, D, E (opposite page) C A low-power, survey micrograph of a single nephric lobe following 'microfil' injection of the arterial tree of the human kidney. Note the clear distinction between cortex peripherally and the medulla converging on the renal papilla. The lobe is flanked by large interlobar arteries. Note also the arcuate arteries at the cortico-medullary junction, the interlobular arteries ascending into the cortex, where the glomeruli are also visible, and the converging vascular bundles of the medulla. (Compare with 8.156B.)

D A higher magnification of the same preparation at the cortico-medullary junction. Note juxtamedullary efferent arterioles, leaving the glomeruli, to form medullary vascular bundles (descending vasa recta).

E Transverse section through a vascular bundle (consisting of descending and ascending vasa recta). Surrounding the vascular bundle are sections of large collecting ducts, and thin and thick segments of the medullary loops (of Henle). (Preparations C–E kindly provided by Professor D. B. Moffat, Department of Anatomy, University College of Wales, Cardiff.)

and which are so named because of their appearance in surface view. During their course towards the corticomedullary junction, the interlobular veins also receive the terminations of some of the ascending vasa recta (*vide infra*), and then they end in the *arcuate veins* which accompany the arteries of the same name; but, unlike them, they anastomose with neighbouring veins to some extent. In turn, the arcuate veins drain into *interlobar veins* which by their convergence and anastomosis with neighbouring channels, ultimately form the renal vein.

The vascular supply of the *renal medulla* is derived in the main from the efferent arterioles of juxtamedullary glomeruli, but these are supplemented by efferent arterioles from more superficially placed glomeruli, and by a number of 'aglomerular' arterioles (probably the result of previous glomerular degeneration). The efferent arterioles destined for the medulla are relatively long, wide vessels, which contribute side branches to the neighbouring capillary plexuses before entering the medulla, and each then divides into twelve to twenty-five *descending vasa recta*. The latter, as their name suggests, pursue straight courses (to varying depths) in the renal medulla, and contribute side branches to a radially elongated capillary plexus (8.156A) which is closely applied to the descending and ascending limbs of Henle's loop and to the collecting ducts. The venous ends of the capillaries converge to *ascending vasa recta* which return to drain into the arcuate or interlobular veins. An essential structural feature of the vasa recta is that, particularly in the outer medulla, a considerable number of both ascending and descending vessels are grouped into *vascular bundles*. Within the latter the external aspects of the descending and ascending vasa recta are closely apposed to each other, whilst around the margins of the bundles, the vessels are closely applied to the limbs of the loop of Henle and to the collecting ducts. As the vascular bundles are traced centrally into the renal medulla, they come to contain fewer vessels, as some of their numbers terminate at successive levels in the neighbouring capillary plexuses. This close parallel apposition of descending and ascending vessels, with each other, and with the surrounding duct systems mentioned above, provides the structural basis for the countercurrent exchange and multiplier phenomena mentioned on a previous page, which characterize the physiological role of the renal medulla (p. 1395; 8.152D; 156D, E).

In terms of their *ultrastructure*, the renal vessels in their various regions show features that have been described elsewhere in this volume. The renal, interlobar and arcuate arteries are fairly typical 'large muscular arteries' (p. 625); the interlobular vessels are similar to 'small muscular arteries' in other organs, whilst the afferent glomerular vessels show a typical arteriolar structure with a muscular coat which is two to three cells thick. This and the connective tissue strata gradually diminish as the glomerulus is approached, until a point some 30–50 μm proximal to the glomerulus, where the arteriolar cells begin to show the modifications which characterize the *juxtaglomerular apparatus* (p. 1396). The efferent glomerular arterioles from the majority of the cortical glomeruli have slightly thicker walls and a narrower calibre than their corresponding afferent vessels. The peritubular

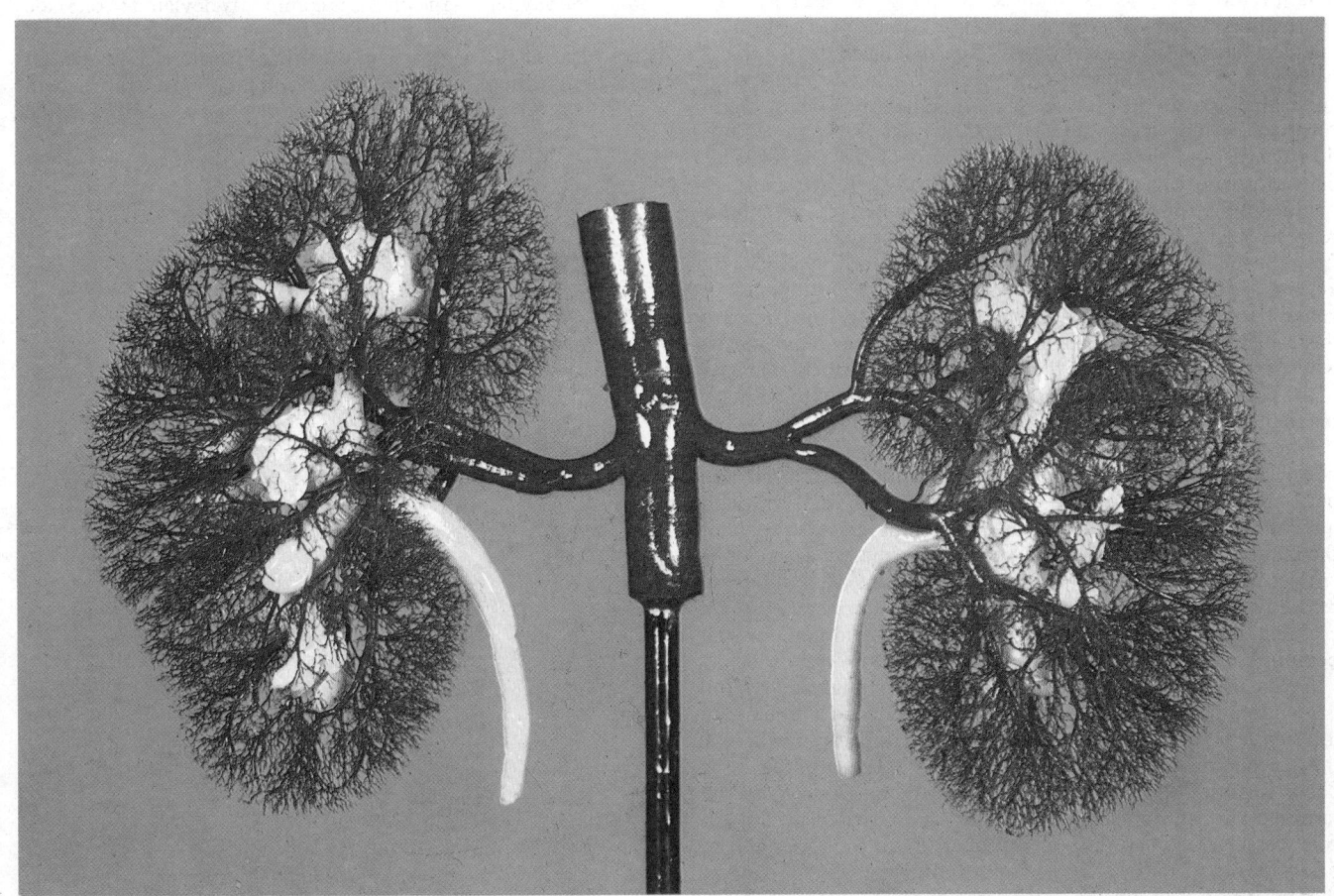

8.157 Resin corrosion cast of human kidneys prepared by Dr. D. H. Tompsett of the Royal College of Surgeons of England. Ureter, pelvis and calices—yellow; aorta, renal arteries and their branches—red. See text for a detailed description of the renal blood vessels.

and medullary capillaries possess a well-defined pericapillary basal lamina, and their endothelial cells have a typically fenestrated cytoplasm (p. 628)—a structure also possessed by the ascending vasa recta, in contrast to the descending vasa, which have a thicker and continuous endothelium. The structure of the glomerular capillaries was considered above (p. 1399). Brief comments concerning the functional associations between the different parts of the nephric tubules and their associated blood vessels were given in the previous pages (p. 1395); for a much more detailed analysis consult Fourman and Moffat (1971).

Nerves of the kidney. The general sources of the nerves to the kidney are described on p. 1134). Nerve fibres passing directly from the plexuses around the arcuate arteries to innervate the juxtamedullary efferent arterioles and vasa recta have been described (Munkacsi and Newstead 1971). It is suggested that these might control blood flow between the cortex and medulla of the kidney, without affecting glomerular circulation.

Lymph vessels of the kidney are described on p. 797.

Renal anomalies. The early pelvic position of the kidney (p. 213) may persist, and in these cases the organ usually derives its blood supply from the common iliac artery, and the hilum lies on its anterior aspect. Occasionally the kidneys may be connected by a transverse bar of renal tissue forming a 'horseshoe kidney'; generally the bar connects the lower poles of the kidneys and only rarely does it connect their upper poles. The ureters curve across the ventral aspect of the connecting bar of renal tissue, and may suffer partial obstruction at this level, with resultant hydronephrosis. Congenital absence or imperfect development of one kidney may occur, and may be compensated by enlargement of the kidney on the opposite side. Rarely the two kidneys may lie on the same side of the body. Single or multiple renal cysts of congenital origin are not uncommon; alternatively, widespread congenital polycystic disease may occur.

The Ureters

The ureters are the two muscular tubes whose peristaltic contractions convey the urine from the kidneys to the urinary bladder. Each measures from 25 to 30 cm in length, and is a thick-walled, narrow, cylindrical tube which is directly continuous superiorly with the funnel-shaped renal pelvis; a slight constriction may mark the site of transition (8.152A). It runs downwards and slightly medially in front of the psoas major, passes into the pelvic cavity, and opens into the base of the urinary bladder. Its position can be projected on the surface, where it is represented by a line from a point on the transpyloric plane 5 cm from the median plane, drawn almost vertically downwards, with a very slightly medial inclination, to the pubic tubercle. In general, its diameter is about 3 mm, but it is slightly constricted in three places: superiorly, at its junction with the renal pelvis; where it crosses the brim of the lesser pelvis at the medial border of the psoas major; and as it passes through the wall of the urinary bladder (this latter part being its narrowest portion).

The *pelvis of the kidney* has already been described (p. 1389).

The *abdominal part* of the ureter descends posterior to the peritoneum on the medial part of the psoas major, which intervenes between it and the tips of the transverse processes of the lumbar vertebrae (8.149). Anterior to the psoas major it crosses in front of the genitofemoral nerve and is itself obliquely crossed by the gonadal vessels. It enters the lesser pelvis by crossing anterior to either the end of the common, or the beginning of the external, iliac vessels.

At its origin the *right* ureter is usually overlapped by the descending part of the duodenum; in its descent it is lateral to the inferior vena cava, and is crossed anteriorly by the right colic and the ileocolic vessels, whilst near the superior aperture of the lesser pelvis it passes behind the lower part of the mesentery and the terminal part of the ileum. The *left ureter* is crossed by the left colic vessels and, near the superior aperture of the lesser pelvis, passes posterior to the sigmoid colon and its mesentery, lying in the posterior wall of the intersigmoid recess. At operation, because of these differences, the abdominal part of the left ureter is easier to expose than the right.

The *pelvic part* of the ureter is about the same length as the abdominal part; in both sexes it lies in the extraperitoneal areolar tissue. At first it descends posterolaterally on the lateral wall of the lesser pelvis, following the anterior border of the greater sciatic notch. Opposite the ischial spine it turns anteromedially to run in the fibrous adipose tissue above the levator ani to reach the base of the bladder. On the posterolateral pelvic wall it is anterior to the internal iliac artery and the commencement of its anterior trunk, dorsal to which are the internal iliac vein, lumbosacral nerve and sacro-iliac joint. Laterally it lies on the fascia covering the obturator internus and it progressively crosses and is medial to the umbilical artery, the obturator nerve, artery and vein, the inferior vesical artery and the middle rectal artery.

In the male, as it descends anteromedially, the ureter is crossed above and in front, and from lateral to medial, by the ductus deferens. Thereafter the ureter passes in front of, and slightly above, the upper end of the seminal vesicle and finally enters the bladder wall obliquely to open into the bladder at the lateral angle of the trigone (8.162). The terminal part of the ureter is surrounded by tributaries of the vesical veins.

In the female, the pelvic part of the ureter at first has the same general relations as in the male, though where it lies anterior to the internal iliac artery it is situated immediately behind the ovary and here forms the posterior boundary of the ovarian fossa (p. 1423). In its later forward and medial course to the bladder it has important relations to the uterine artery, the cervix of the uterus and the fornices of the vagina. It lies in the extraperitoneal connective tissue in the lower and medial part of the broad ligament of the uterus (parametrium; p. 1432); here it is intimately related to the uterine artery, which lies above and in front of the ureter for a distance of 2·5 cm and then crosses to gain the medial side of the ureter to ascend alongside the uterus. The ureter runs forwards slightly above the lateral fornix of the vagina and is here situated commonly about 2 cm lateral to the supravaginal portion of the cervix of the uterus (p. 1428), though this distance may vary from 1 to 4 cm. It then inclines medially to reach the bladder and in doing so it has a variable relation to the front of the vagina. As the uterus is commonly deviated to one side of the median plane, one ureter may be more extensively apposed to the front of the vagina than the other. In most cases the left ureter has the more extensive relation and it may cross the median plane; the reverse may occur and sometimes one ureter may not lie

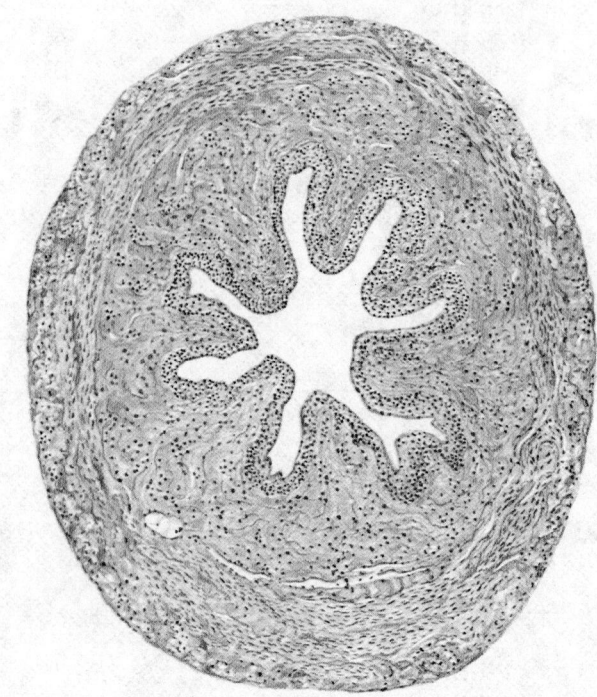

8.158A A transverse section through the lower third of a human ureter. Note great thickness of muscular wall. Stained with haematoxylin and eosin. Magnification about × 30.

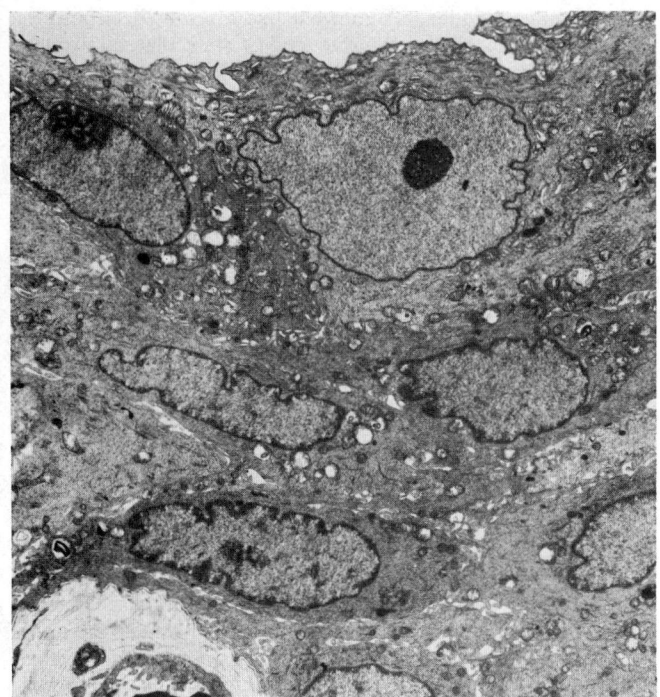

8.158B Transmission electron micrograph showing the transitional epithelium lining the ureter (rat). (Magnification × 4,000.) (Prepared and photographed by Mrs. Susan Smith, Department of Anatomy, Guy's Hospital Medical School, London.)

in front of the vagina, in which case a much longer part of the other ureter than usual lies in front of it.

In the distended bladder, in both sexes, the ureteric openings are about 5 cm apart, this distance being reduced somewhat in the empty viscus. In its oblique course through the vesical wall, the ureter becomes compressed and hence flattened when the bladder is distended, and this may prevent regurgitation of urine from the bladder into the ureter, though peristalsis of the ureteric muscle is also a factor in this effect.

The ureter, the renal pelvis and the calices can be demonstrated in the living subject by radiographs taken (1) after the intravenous injection of a radio-opaque substance excreted in the urine (descending or *excretion pyelography*), or (2) after injection of similar solutions into the ureter by means of a ureteric catheter passed through an operating cystoscope (ascending or *retrograde pyelography*). The resulting radiographs are termed *pyelograms* (*see* e.g. **8.149**). The normal cupping of the minor calices by the projecting renal papillae is clinically important, because it may be obliterated by pathological conditions (such as hydronephrosis) which are associated with chronic distension of the ureter and renal pelvis consequent upon some source of urinary back-pressure.

URETERIC STRUCTURE

The ureter has three coats: fibrous, muscular and mucous.

The *fibrous coat* is continuous at one end with the fibrous capsule of the kidney in the floor of the renal sinus, while at the other it is lost on the wall of the bladder.

The *muscular coat* in the pelvis, calices and upper two-thirds of the ureter consists of two layers of non-striated myocytes, an inner longitudinal and an outer circular. The former begins at the attachment of the minor calices to the renal papillae. The circular muscle in this region becomes prominent and forms rings around the bases of the papillae; the periodic contractions of this muscle may have the effect of 'milking' the papillae and squeezing urine out of the ducts of Bellini. Subsequent observations in the main confirm this, but separated fasciculi of muscle fibres have been shown to extend into the caliceal wall. No pelvi-ureteric sphincter has been detected. Clear definition of circular and longi-tudinal strata of muscle fibres in the upper urinary tract was considered to be obscured by a plexiform arrangement of

fibres (Gosling 1970). In the lower third of the ureter, an additional outer longitudinal muscular layer is added, and the inner longitudinal layer becomes less distinct. In its oblique passage through the wall of the bladder the muscle is arranged entirely in a longitudinal direction, and this, on contraction, has the effect of keeping patent this part of the ureter. The muscle of the ureter is not compactly arranged as in the intestine but is infiltrated by connective tissue derived from the mucosal and fibrous layers. Furthermore, as in the case of the urinary bladder, the muscular strata described above intermingle to such an extent that they cannot be separated into three clearly defined strata. Various accounts of the arrangement of muscle fasciculi at the ureterovesical junction have been recorded. The muscle fibres of the ureter are continued into the superficial strata of the trigone (p. 1406), but disagreement over details has not yet been resolved (Tanagho and Pugh 1963; Tanagho *et al.* 1968). The muscular coat of the ureter undergoes peristaltic contractions progressing downwards from the renal pelvis and calices. The muscular tone and peristaltic activity of the ureter are materially influenced by the rate of urine excretion by the kidney, both being increased if the renal output rises. With average renal output the ureter propels jets of urine into the bladder at the rate of four or five a minute.

The *mucous coat* is smooth, and presents about six longitudinal folds which become effaced by distension. It is continuous with the mucous membrane of the bladder below, while it is prolonged over the papillae of the kidney above. It consists of fibrous tissue containing many elastic fibres, and covered with transitional epithelium (urothelium), four or five cells thick in the ureter and two or three cells thick in the pelvis and calices. No distinct submucosa is present. The mucous membrane at the entrance of the ureter into the bladder is slightly infolded but it is very doubt-ful whether this arrangement has any valvular action.

The *arteries* supplying the ureter (Daniel and Shackman 1952) are derived from the renal, abdominal aortic, testicular (or ovarian), common iliac, internal iliac, vesical and uterine vessels, the branches supplying the different parts of the ureter in its course and being subject to much variation. The longitudinal ana-stomosis between these branches on the wall of the ureter is good. In the treatment of certain clinical conditions affecting the urinary bladder, the lower part of the ureter may be divided, a short length mobilized and implanted into the wall of a neighbouring part of the colon. It has been shown that there is a danger of ischaemic necrosis of the lower part of the divided ureter if the anastomotic vessels in this region are only minute, and that the ureter may be more safely divided 2 cm below the bifurcation of the common iliac artery. The branches from the inferior vesical artery (p. 720) are constant in their occurrence and supply the lower part of the ureter as well as a large part of the trigone of the bladder.

The *lymph vessels* of the ureter are described on p. 797.

The *uretic nerves* (p. 1134; **7**.230) are derived from the renal, aortic, and superior and inferior hypogastric plexuses; through these plexuses fibres are derived from the lower three thoracic and first lumbar and the second to the fourth sacral segments of the spinal cord. Small ganglia and isolated ganglion cells have been reported in the fibrous and muscular coats. They are most numerous in the lower part of the ureter and relatively sparse elsewhere. The ureteric plexuses contain both sympathetic and parasympathetic fibres, but their function is not clear, and it has been maintained that they are largely sensory in nature, since they can be stripped off the ureter without affecting its muscular activities. In rodents (Dixon and Gosling 1971) a combination of ultrastructural observations with the histochemical localization of adrenergic and cholinergic nerve fibres has indicated some regional differences, the former type predominating in the major calices and pelvis of the ureter. Ganglion cells have not been observed in the wall of the superior region of the ureter. Schulman (1975) has studied the development, pre- and peri-natal, of the ureteric nerves in mammals (rabbit) by histochemical demonstration of adrenergic and cholinergic fibres. The latter appear earlier and the former are poorly developed even at birth, a finding which, on phylogenetic grounds, is considered to support the views of Cauna *et al.* (1969), Gosling and Dixon (1974), and others that the cholinergic fibres are afferent.

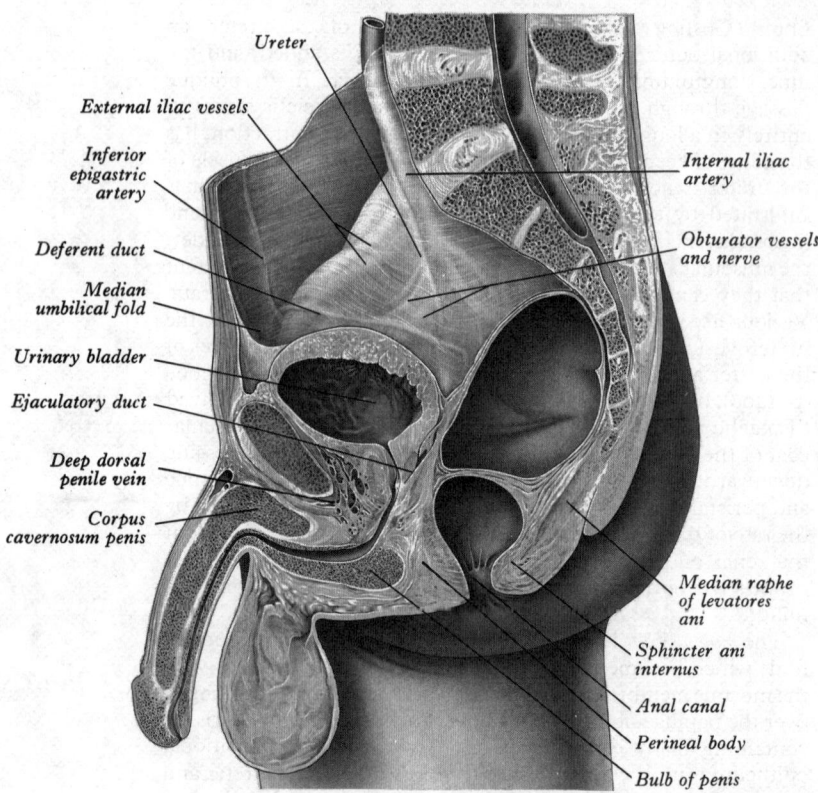

8.159 A median sagittal section to show male internal and external genitalia, etc. A number of structures (e.g. obturator vessels, ureter) are only faintly visible through the overlying peritoneum.

Ultrastructural studies (Notley 1969, 1971) of the intrinsic nerves of the ureter largely confirm the findings of light microscopy. Non-myelinated nerve fibres lie in close proximity to muscle fibres of vessels and the ureteric wall, but no specialized motor endings have been observed. No ganglion cells were noted.

Applied Anatomy. As in the case of the gut (p. 1333), excessive ureteric distension or spasm of its muscle is an adequate stimulus to provoke severe pain (renal colic). This may be produced, for example, by a stone (calculus) causing incomplete and intermittent obstruction in the ureter, particularly if the stone is gradually forced down the ureter by the muscle spasm. The pain, which is spasmodic and agonizing, is referred to cutaneous areas innervated from the same segments of the spinal cord as supply the ureter, mainly T. 11–L. 2 (p. 1138). It commences in the loin and shoots downwards and forwards to the groin and the scrotum or labium majus; it may extend into the upper part of the front of the thigh by radiation along the genitofemoral nerve (L. 1, 2), and the cremaster (innervated by this nerve) may undergo reflex contraction and retract the testis. A ureteric stone is liable to become impacted at one of the constricted parts of the ureter, namely, its cranial end, where it crosses the brim of the lesser pelvis, or as it passes through the bladder wall; radiologically, such an impacted stone would be seen respectively near the tip of the transverse process of the second lumbar vertebra, or overlying the sacro-iliac joint, or slightly medial to the ischial spine. Sometimes the ureter is duplicated on one or both sides and the two tubes may remain distinct as far as the bladder; they rarely open separately into the bladder in such cases.

The Urinary Bladder

The urinary bladder (**8.**159), solely a reservoir, varies in its size, shape, position and relations according to the amount of fluid that it contains, as well as with the state of distension of neighbouring viscera. When empty it is entirely within the lesser pelvis, but as it becomes distended it expands upwards and forwards into the abdominal cavity.

The empty bladder is somewhat like a tetrahedron and has a base, neck, apex, superior surface and two inferolateral surfaces. The *fundus* or *base* is triangular and directed backwards and downwards. In the female (**8.**183) it is closely related to the anterior wall of the vagina. In the male it is related to the rectum, but its upper part is separated from the rectum by the rectovesical pouch of peritoneum, and below that the seminal vesicles and deferent ducts separate the two viscera (**8.**160). In the triangular interval between the two deferent ducts, the bladder and the rectum are separated only by the rectovesical fascia (p. 1421); the inferior part of this triangular interval may often be obliterated by the approximation of the deferent ducts above the prostate. Although the fundus (or base) of the bladder should be, by definition, the lowest region of the bladder, it is the *neck* which is in fact the lowest and also the most fixed part of the bladder; it lies 3 to 4 cm behind the lower part of the symphysis pubis (i.e. a little above the plane of the inferior aperture of the lesser pelvis). It is pierced by the internal urethral orifice and is subject to but little alteration in position with varying conditions of the bladder and rectum. There is no special constriction of the bladder at the site of the neck. In the male the neck rests on, and its wall is in direct continuity with, the base of the prostate; in the female it is related to the pelvic fascia which surrounds the upper part of the urethra. The *apex* in both sexes is directed forwards towards the upper part of the symphysis pubis, and from it the median umbilical ligament (urachus, p. 1405) is continued upwards on the posterior surface of the anterior abdominal wall to the umbilicus; the peritoneum folded over this ligament is the median umbilical fold. The *superior surface*, triangular in shape, is bounded on each side by a lateral border that runs from the apex to the entrance of the ureter into the bladder, and by a posterior border which corresponds with a line joining the entrances of the ureters into the bladder. In the male the superior surface is completely covered with peritoneum, which extends slightly on to the base and is continued behind into the rectovesical pouch, at the sides into the paravesical fossae, and at the front into the median umbilical fold. It is in contact with the sigmoid colon and the terminal coils of the small intestine. In the female the superior surface is almost entirely covered with peritoneum, but near its posterior border the peritoneum is reflected from it to the uterus at the level of the internal os (i.e. the junction of the body and cervix of the uterus), forming the vesico-uterine pouch. The

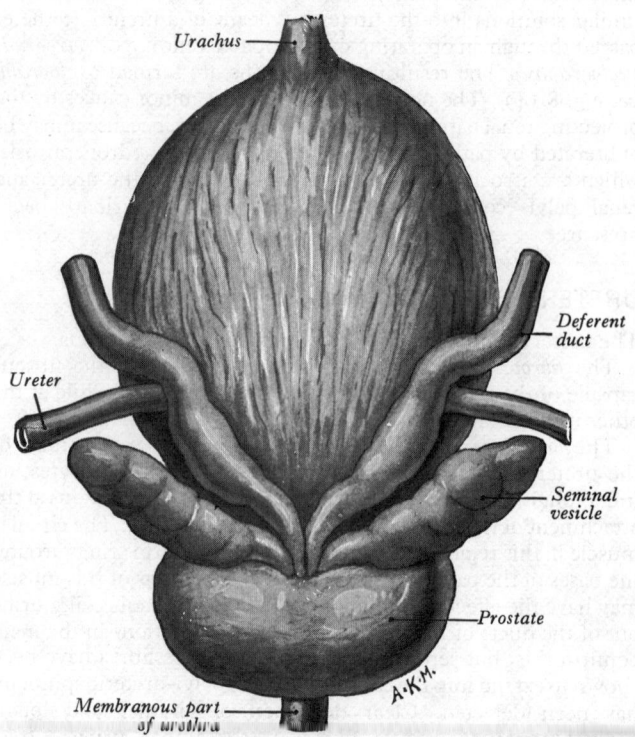

8.160 Posterosuperior aspect of the male internal urogenital organs.

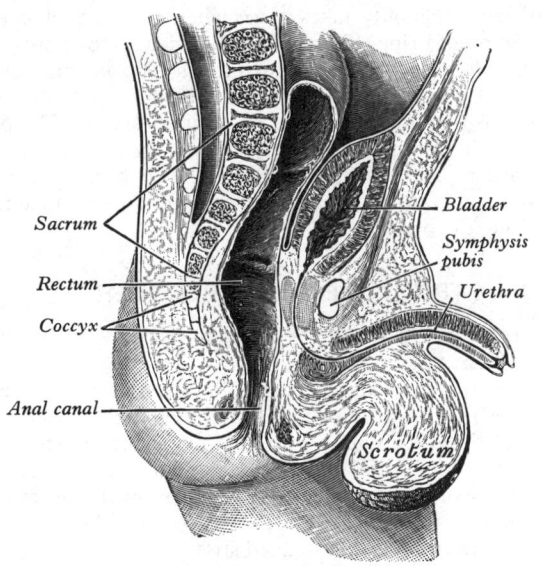

8.161 A sagittal section of the pelvis of a newborn male infant. Note the abdominal position of the urinary bladder.

posterior part of the superior surface, devoid of peritoneum, is separated from the supravaginal part of the uterine cervix by fibro-areolar tissue. Each *inferolateral surface* in the male is separated anteriorly from the pubis and puboprostatic ligaments by an adipose mass termed the retropubic pad; further posteriorly it is separated by fascia from the levator ani and obturator internus. In the female the relations are the same, except that the puboprostatic are replaced by the pubovesical ligaments. The inferolateral surfaces are not covered by peritoneum.

As the bladder fills, its borders become rounded off and it assumes an ovoid form. In front it displaces the parietal peritoneum on the suprapubic region of the abdominal wall, so that the inferolateral surfaces (now becoming the anterior surface of the distended bladder) rest against the anterior abdominal wall without the intervention of peritoneum for some distance above the symphysis pubis; this distance varies with the degree of

distension of the bladder, but commonly, when the bladder is distended to an average degree, it measures about 5 cm, though in excessive distension it may extend as far as the umbilicus or even higher. Thus the full bladder may be approached surgically by incisions or punctures through the anterior abdominal wall a little above the symphysis pubis without traversing the peritoneum. The summit of the full bladder is directed upwards and forwards above the point of attachment of the median umbilical ligament, so that the peritoneum, which follows the ligament, forms a supravesical recess of varying depth between the summit and the anterior abdominal wall; the recess often contains coils of small intestine.

At birth (**8.161**), the bladder lies at a relatively higher level than in the adult; the internal urethral orifice is at the level of the *upper* border of the symphysis pubis and the bladder is an abdominal, rather than a pelvic organ, extending about two-thirds of the distance up to the umbilicus. It progressively descends until it reaches its adult position shortly after puberty.

THE VESICAL LIGAMENTS

Each side of the bladder is connected to the tendinous arch of the pelvic fascia (p. 561) by a condensation of fibro-areolar tissue, termed the *lateral true ligament* of the bladder. Anteriorly the same tissue forms two thickened bands, on each side of the median plane, termed the lateral and medial puboprostatic ligaments in the male. The *lateral puboprostatic ligament* extends from the anterior end of the tendinous arch of the pelvic fascia downwards and medially to blend with the upper part of the sheath of the prostate; the *medial puboprostatic ligament* is attached to the back of the pubic bone near the middle of the symphysis and passes downwards and backwards to the sheath of the prostate, forming the floor of the retropubic space. In the female similar bands are termed *pubovesical ligaments*. The *apex* of the bladder is joined to the umbilicus by the remains of the urachus, which forms the *median umbilical ligament*. The lumen of the lower part of the urachus persists throughout life and is lined by modified, transitional epithelium. It may communicate with the cavity of the bladder (Begg 1930). As the veins of the vesical venous plexus stream backwards from the lateral borders of the base of the bladder to join the internal iliac veins, they are enveloped on each side in a band of fibro-areolar tissue which is sometimes termed

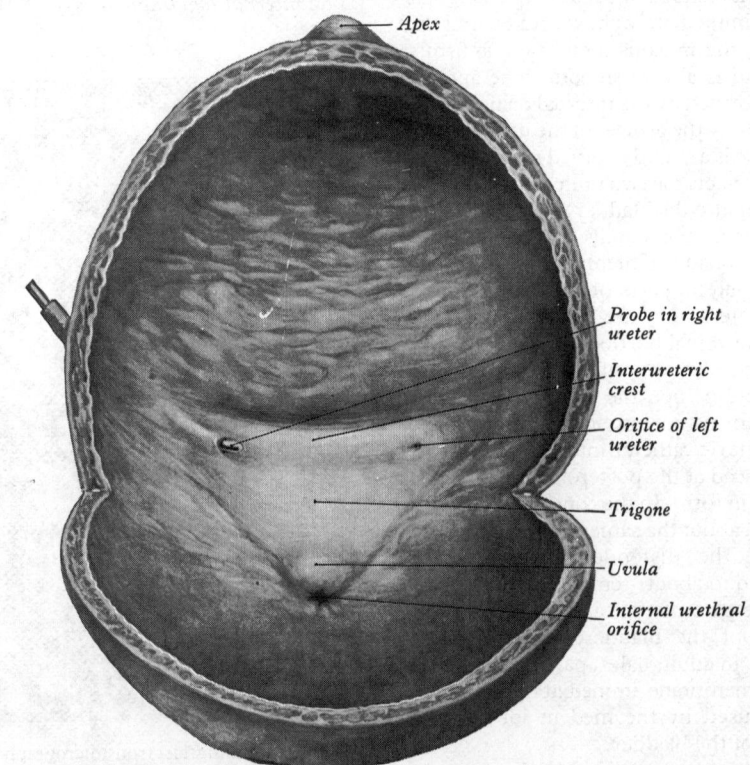

8.162 Anterior aspect of the interior of the urinary bladder.

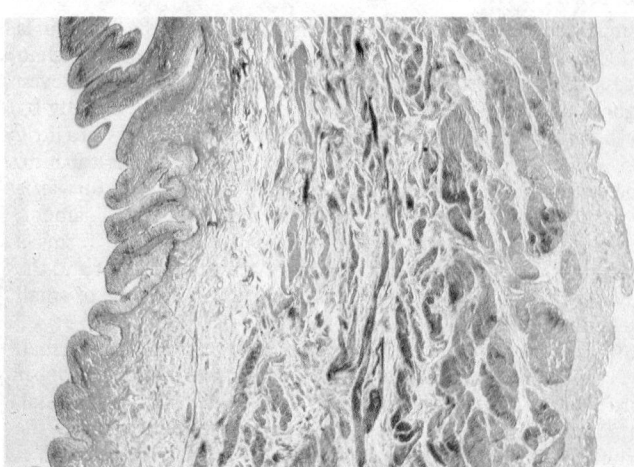

8.163A A transverse section through the urinary bladder wall of a monkey. Masson's trichrome stain. The section shows the folded transitional epithelium (grey-brown on the left), the connective tissue (deep green), and fasciculi of the detrusor muscle (green-brown, to the right), surrounded externally by the connective tissue adventitia. × 20.

the *posterior ligament* of the bladder. The functional role of the various vesical 'ligaments' is *presumed* to be supportive.

From the superior surface of the bladder the peritoneum is carried off in a series of folds which are termed the 'false' ligaments of the bladder, on the dubious distinction that they do not share the supportive function imputed to the (true) 'ligaments' noted above. Anteriorly there are three folds: the *median umbilical fold* over the median umbilical ligament, and two *medial umbilical folds* over the obliterated umbilical arteries (**8.**94). The reflexions of the peritoneum from the bladder to the side walls of the pelvis form the *lateral false ligaments* while the sacrogenital folds (p. 1324) constitute the *posterior false ligaments*.

THE VESICAL INTERIOR

The mucous membrane of the bladder (**8.**163) is, over the greater part of the viscus, only loosely attached to the subjacent muscular layer and develops folds when the bladder is contracted: the folds are effaced when the bladder is distended. Over a small triangular area, the *trigone of the bladder*, immediately above and behind the internal orifice of the urethra, the mucous membrane is firmly bound to the muscle layer, and is always smooth. The antero-inferior angle of the trigone is formed by the internal orifice of the urethra, its posterolateral angles by the orifices of the ureters. The superior boundary of the trigone is a slightly curved ridge, termed the *interureteric crest*, which connects the two ureteric orifices and is produced by the continuation into the bladder wall of the inner longitudinal coats of the ureters. (For details of the musculature of the trigonal region, consult Uhlenhuth *et al.* 1952; Woodburne 1965, 1968.) The lateral parts of this ridge extend beyond the openings of the ureters; they are named the *ureteric folds* and are produced by the terminal portions of the ureters as they run obliquely through the bladder wall. When the living bladder is examined with a cystoscope the interureteric crest appears as a pale band, and forms an important guide to the ureteric orifice in guiding a ureteric catheter into it.

The *orifices of the ureters*, placed at the posterolateral angles of the trigone, are usually slit-like in form. In the contracted bladder they are about 2·5 cm apart and about the same distance from the internal urethral orifice; in the distended bladder these measurements may be increased to about 5 cm.

The *internal urethral orifice* is placed at the apex of the trigone, in the most dependent part of the bladder, and is usually somewhat crescentic in section; in adult males, particularly those past middle age, the mucous membrane immediately behind it exhibits a slight elevation, caused by the median lobe of the prostate and termed the *uvula* of the bladder.

The mean capacity of the living urinary bladder in the male adult is 220 ml, varying from 120 to 320 ml (Thompson 1919).

Micturition commonly takes place when the bladder contains about 280 ml of urine. Filling of the bladder up to about 500 ml may be tolerated, but at levels much above this, pain is experienced due to the tension in the bladder wall, and reflex contractions then begin to make micturition urgent. The pain is referred to the cutaneous areas supplied by the same segments of the spinal cord as supply the bladder (T. 11–L. 2; S. 2–4), namely the lower part of the anterior abdominal wall, the perineum and, in the male, the penis.

VESICAL STRUCTURE OF THE BLADDER

The bladder has a three-layered wall—serous, muscular and mucous (**8.**163A).

The *serous*, peritoneal covering has been described above.

The *muscular* stratum, which constitutes the *detrusor muscle*, consists of three layers of nonstriated myocyte: an external and an internal of longitudinal fibres, and a middle of circular fibres. As in the case of the ureter (p. 1403), there is much intermingling of the muscle fibres in these layers, so that they cannot be separated into three clearly defined strata.

The fibres of the *external longitudinal layer* pass in a more or less longitudinal manner, along the inferolateral surfaces of the bladder, over its apex on to the superior surface, and then descend over the base to blend with the capsule of the prostate or with the anterior region of the vagina. Some of the longitudinal fibres are carried on to the front of the rectum, and are named the *rectovesical muscle*. Others traverse the medial puboprostatic ligaments and are attached to the lower part of the pelvic surface of the pubis on each side. They are termed the *pubovesical muscles*. (It is the presence of these muscular components in the ligaments of the bladder which is the chief evidence of their supportive role.) At the sides of the bladder the fibres are arranged obliquely and intersect one another.

The fibres of the *middle, circular, layer* are very thinly and irregularly scattered on the body of the organ, and, although to some extent placed transverse to its long axis, are for the most part arranged obliquely. Towards the cervical part of the bladder they are disposed in a circular layer, forming the *sphincter vesicae*, which surrounds the internal urethral orifice. Many authorities state that no completely sphincteric fascicles occur; some of these strands are continuous with prostatic muscle fibres which encircle the proximal part of the urethra in the male.

The *internal longitudinal layer* is thin, and its fasciculi have a

8.163B A scanning electron micrograph of the luminal surface of the bladder (murine) showing the plate like organization of the urothelial plasma membranes of its lining cells. (Magnification × 7,000.)

reticular arrangement, but with a tendency to assume for the most part a longitudinal direction.

Two bands of oblique fibres, originating behind the orifices of the ureters, converge to the dorsum of the prostate, and are inserted by means of a fibrous process into the median lobe of the organ. They are the *muscles of the ureters*, described long ago (Bell 1812) as serving to maintain the oblique direction of the ureters, and so prevent the reflux of the urine into them during contraction of the bladder.

There is no muscularis mucosae in the bladder wall.

The *sphincter vesicae* (internal sphincter), composed of nonstriated muscle, and the *sphincter urethrae* (external sphincter), composed of striated muscle, both control the outflow from the bladder; it is usually considered that they can function independently, and the sphincter vesicae prevents urine leaking into the prostatic urethra, where its presence would arouse a desire to micturate. (*See*, however, further discussion on p. 1409.)

The *mucous membrane* is pale pink in colour; it is continuous above with that of the ureters, and below with that of the urethra; the epithelium covering it is of the transitional variety. The loose texture of the lamina propria allows the mucosa to be thrown into folds or *rugae* when the bladder is empty. Over the trigone it is closely attached to the muscular coat, and is not thrown into folds, but is smooth and flat. There are no true glands in the mucous membrane of the bladder, though certain mucous follicles, which exist especially near the neck of the bladder, have been regarded as such. Some believe that there are true mucous glands near the internal urethral orifice, both in the bladder (*subtrigonal glands*) and the prostatic urethra (*subcervical glands*), and maintain that enlargement of these may obstruct the outflow of urine from the bladder. The transitional epithelium of the ureter and bladder contains alkaline phosphatase, though the significance of this is not clear. Transplantation of pieces of bladder epithelium into the sheath of the rectus abdominis induces the formation of bone in that site (Huggins 1931), but it is suggested (Johnson and McMinn 1956) that, apart from alkaline phosphatase, other factors may be involved in the induction of this heterotopic bone formation. Ectopic ossification in sites containing potentially osteogenic cells is a well-known phenomenon, of which this is merely an instance.

Vessels and Nerves

The principal *arteries* of supply to the bladder are the superior and inferior vesical, derived from the anterior trunk of the internal iliac artery. The obturator and inferior gluteal arteries also send small branches to it, and in the female additional branches are derived from the uterine and vaginal arteries.

The *veins* form a complicated plexus on the inferolateral surfaces near the prostate, and pass backwards in the posterior ligaments of the bladder to end in the internal iliac veins.

The *lymph vessels* are described on p. 797.

The *nerves* supplying the bladder form the vesical plexus (*see* pp. 1135, 1137) and consist of both sympathetic and parasympathetic components, each of which contains motor (efferent) and sensory (afferent) fibres. The efferent parasympathetic fibres, arising from the second to the fourth sacral segments of the spinal cord (nervi erigentes), convey motor fibres to the detrusor muscle and inhibitory fibres to the sphincter vesicae; if they are destroyed normal micturition is not possible. The sympathetic efferent fibres from the lower two thoracic and upper two lumbar segments of the spinal cord are said to supply inhibitory fibres to the detrusor and motor fibres to the sphincter vesicae, though it is claimed that they are mainly vasomotor in function and that normal filling and emptying of the bladder are controlled exclusively by the parasympathetic nervous system. The sphincter urethrae (external sphincter), a somatic striated muscle, is also concerned in controlling micturition and is supplied by the pudendal nerve (S. 2–4). As regards the sensory innervation of the bladder, there are fibres concerned with pain and those concerned with conscious awareness of bladder distension. The pain fibres, which are stimulated by distension or by spasm of the bladder wall, or by a stone or inflammation or malignant disease irritating the bladder, run in both the sympathetic and parasympathetic nerves, the latter pathway being predominant. Because of this

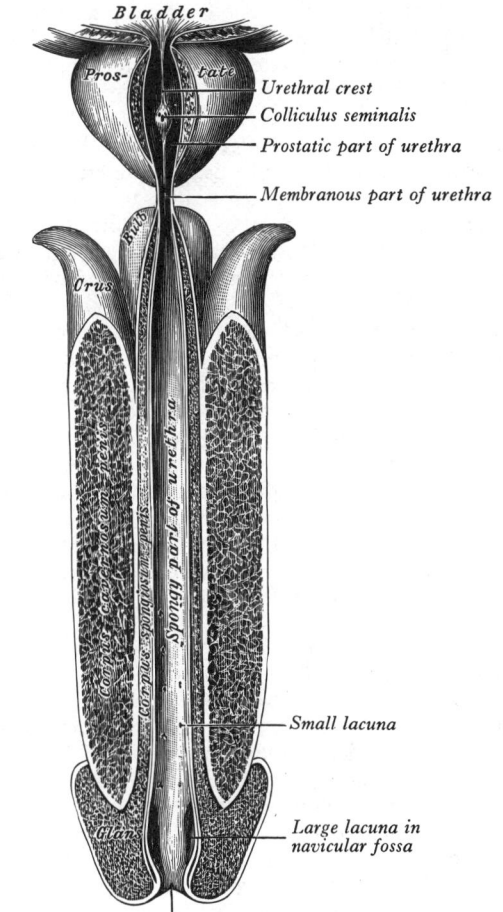

Bladder

Pros-
tate

Urethral crest
Colliculus seminalis
Prostatic part of urethra

Membranous part of urethra

Bulb

Crus

Small lacuna

Large lacuna in navicular fossa

Glans

External urethral orifice

8.164 The whole length of the lumen of the male urethra exposed by an incision extending into it from its dorsal aspect. Note openings of prostatic utricle and ejaculatory ducts on the colliculus seminalis.

double pathway of the pain fibres, simple division of the appropriate sympathetic path (e.g. 'presacral neurectomy') or division of the superior hypogastric plexus (p. 1135) does not materially relieve bladder pain. In the spinal cord the pain fibres lie in the anterolateral white columns, and bladder pain may be considerably relieved by cutting these tracts (bilateral anterolateral cordotomy). On the other hand, the nerve fibres subserving awareness of fullness of the bladder run in the posterior columns of the spinal cord (fasciculus gracilis), so that after anterolateral cordotomy for the relief of pain, the patient is still aware of bladder fullness and of the desire to micturate.

Applied Anatomy. The distended bladder may be ruptured by injuries of the lower half of the anterior abdominal wall or by fracture of the pelvis; the rupture may be extraperitoneal or, if involving the superior surface of the bladder, it is usually complicated by tearing of the peritoneum and consequent escape of vesical contents into the peritoneal cavity. If there is progressive chronic obstruction to the outflow of urine from the bladder, e.g. by enlargement of the prostate (p. 1143) or by stricture of the urethra, the bladder becomes distended and its musculature hypertrophies, its fasciculi increasing in size and interlacing in all directions, giving rise to what is known as 'the trabeculated bladder'. The mucous membrane between these bundles forms 'diverticula' of the bladder which may contain phosphatic concretions. When the bladder outflow is thus obstructed, it is not able to empty itself completely and after micturition some urine is retained and may become infected; the infection may extend along the ureters to the kidneys. Back pressure from the bladder thus distended may cause gradual dilatation of the ureter and its pelvis, and of the renal collecting tubules. Lesions of the fasciculus gracilis in the spinal cord (e.g. tabes dorsalis) cause loss of appreciation of bladder distension and of the desire to micturate; the bladder distends and may empty merely by overflow. In severe lesions of the spinal cord above its sacral segments, the efferent and afferent nerve tracts involved in normal micturition may be interrupted, resulting in 'automatic' emptying of the bladder.

The vesical interior can be examined with the cystoscope, introduced along the urethra after distending the bladder with fluid. A special cystoscope can be used to catheterize the ureter, for the purpose of obtaining a specimen of urine from either kidney, or in order to inject radio-opaque fluid into ureter and renal pelvis for retrograde pyelography (8.149). The outline of the bladder can also be demonstrated when filled with the same fluid.

Puncture of the distended bladder may be performed just above the symphysis pubis without injuring the peritoneum (*suprapubic cystostomy*). When the bladder contains about 300 ml of fluid, its antero-inferior surface is in contact with the anterior abdominal wall, without the interposition of peritoneum, for about 7·5 cm above the pubic crest. Access to the bladder, for the purpose of removing calculi, is almost always effected by this suprapubic route. In the female, owing to the shortness of the urethra and its ready dilatability, calculi and foreign bodies and new growths, when of small size, may be removed by the urethral route. Congenital abnormalities of the bladder are described on p. 221.

The Male Urethra

The male urethra (8.159, 164), from 18 to 20 cm long, extends from an internal orifice in the urinary bladder to an external opening, or meatus, at the end of the penis. It may be considered in three regional parts: *prostatic, membranous* and *spongiose*, and presents a double curve in the ordinary flaccid state of the penis (8.159). Except during the passage of fluid along it, the urethral canal is a mere slit; in the prostatic part the slit is transversely arched; in the membranous portion, stellate; in the spongiose portion, transverse; while at the external orifice it is sagittal.

The prostatic part, which is the widest and most dilatable part of the urethra, is about 3 cm long, and runs almost vertically through the prostate from its base to its apex, but with some angulation between its supra- and infra-collicular sections (Glenister 1962). It lies nearer the anterior than the posterior surface of the prostate. It is widest in the middle and narrowest below, adjoining the membranous part; on section it is crescentic, convex ventrally.

Its shape on section is due to the presence on the posterior wall (or floor) of a narrow median longitudinal ridge formed by an elevation of the mucous membrane and its subjacent tissue, termed the *urethral crest* (8.164). On each side of the crest there is a shallow depression, termed the *prostatic sinus*, the floor of which is perforated by the *orifices of the prostatic ducts*. About the middle of the length of the urethral crest the *colliculus seminalis* forms an elevation on which the slit-like orifice of the prostatic utricle is situated; on each side of, or just within, this orifice there is the small opening of the ejaculatory duct. The *prostatic utricle* is a cul-de-sac about 6 mm long, which runs upwards and backwards in the substance of the prostate behind the median lobe. Its walls are composed of fibrous tissue, muscular fibres, and mucous membrane; the latter presents the openings of numerous small glands. It is developed from the paramesonephric ducts or urogenital sinus, and is thought to be homologous with the vagina of the female (p. 221). The prostatic utricle is therefore called by some 'the *vagina masculina*', but the more usual view is that it is a uterine homologue, and hence the term 'utricle'. The ejaculatory ducts are described on p. 1417.

The membranous part is the shortest, least dilatable, and, with the exception of the external orifice, the narrowest section of the urethra. It descends with a slight ventral concavity from the prostate to the bulb of the penis (8.159), passing through the perineal membrane about 2·5 cm postero-inferior to the pubic symphysis. The hinder part of the bulb of the penis is closely apposed to the inferior aspect of the urogenital diaphragm (perineal membrane), but, anteriorly, it is slightly separated from the latter, so that the wall of the urethra is related anteriorly neither to the perineal membrane nor the penile bulb. If this part

of the anterior wall of the urethra is regarded as 'membranous', it measures about 2 cm, whilst the length of this part of the urethra is posteriorly only 1·25 cm.

The membranous urethra is surrounded by the fibres of the sphincter urethrae (p. 563). In front of it, the deep dorsal vein of the penis enters the pelvis between the transverse perineal ligament and the arcuate ligament of the pubis; one bulbo-urethral gland is sited on each side of this part of the urethra.

The spongiose part is contained in the corpus spongiosum penis (p. 1419). It is about 15 cm long, and extends from the end of the membranous urethra to the external urethral orifice on the glans penis. Commencing below the perineal membrane, it continues the ventrally concave curve of the membranous urethra to a point anterior to the lowest level of the symphysis pubis. From here, when the penis is flaccid, it curves downwards in the 'free' part of the penis. It is narrow, with a uniform diameter of about 6 mm in the penis; it is dilated at its commencement as the *intrabulbar fossa*, and again within the glans penis, where it becomes the *navicular fossa*. The enlargement of the intrabulbar fossa affects the floor and side walls but not the roof of the urethra. The bulbo-urethral glands open into the spongiose section of the urethra about 2·5 cm below the perineal membrane.

The *external urethral orifice* is the narrowest part of the urethra: it is a sagittal slit, about 6 mm long, bounded on each side by a small labium.

The epithelium of the urethra, except in its most anterior part, presents the orifices of numerous small mucous glands and follicles situated in the submucous tissue, and named the *urethral glands*. Besides these there are a number of small pit-like recesses, or *lacunae*, of varying sizes; the orifices of these are directed forwards, and may intercept the point of a catheter in its passage along the canal. One, larger than the rest, the *lacuna magna*, is situated on the roof of the navicular fossa.

Urethral Sphincters

Two sphincters surround the urethra. The internal *sphincter vesicae* (p. 1407) controls the neck of the bladder and the prostatic urethra above the opening of the ejaculatory ducts. It is composed of nonstriated muscle, is not under voluntary control, and is supplied by sympathetic and parasympathetic fibres derived from the vesical plexus (*vide supra* and also, p. 1135). The external sphincter, or *sphincter urethrae* (p. 563), closely surrounds the membranous urethra and consists of striated muscle; it is supplied by the perineal branch of the pudendal nerve (S. 2, 3 and 4) and is under voluntary control after early infancy.

The existence of a definable internal sphincter at the vesical exit, however, cannot be stated dogmatically, for there is much difference of opinion in regard to this. Many observers state that the nonstriated muscle fibres of the vesical neck are arranged in many directions, as in the wall of the bladder, and that no true band of circumferential fibres can be identified (for a short critical review, *see* Woodburne 1961; Angell 1969). However, all authorities appear to agree that a substantial aggregation of muscle tissue, with a considerable admixture of elastic and collagenous fibres, does exist at the vesical outlet (Vincent 1966*a*). The significance of this in the mechanisms of micturition is briefly described on p. 1409.

Male Urethral Structure

The urethra is composed of a mucous membrane, supported by submucous tissue which connects with the various structures through which it passes, and surrounded by muscle.

The *mucous membrane* of the urethra is continuous internally with that of the bladder, and externally with the skin covering the glans penis; it is prolonged into the ducts of the urethral, bulbo-urethral and prostatic glands, and into the deferent ducts and seminal vesicles, through the ejaculatory ducts. (This continuity is an important consideration in the spread of urethral infections.) In the spongiose and membranous urethral regions it is arranged in longitudinal folds when the tube is empty. Small papillae are found upon it near the external urethral orifice; its epithelial lining is of the transitional variety as far as the orifice of the ejaculatory duct; thereafter it is composed of patches of pseudostratified columnar and stratified epithelium, amongst which are situated diverticula of various sizes, some extending into the lamina propria as mucous glands (of Littré). Concretions similar to those found in the prostate may occur in these glands in old age. Near the external urethral orifice, the epithelium is stratified squamous in type.

The *submucous tissue* consists of a vascular erectile layer; outside this there is a layer of nonstriated myocytes, arranged into an inner longitudinal and an outer circular layer and best marked in the prostatic and membranous sections of the urethra.

The Female Urethra

The female urethra is about 4 cm long and 6 mm in diameter. It begins at the internal urethral orifice of the bladder, approximately opposite the middle of the symphysis pubis, and runs antero-inferiorly behind the symphysis pubis, embedded in the anterior wall of the vagina. It traverses the perineal membrane and ends at the external urethral orifice, an anteroposterior slit with rather prominent margins, which is situated directly anterior to the opening of the vagina and about 2·5 cm behind the glans clitoridis. Except during the passage of urine the anterior and posterior walls of the urethra are in apposition, and the epithelium is thrown into longitudinal folds, one of which, on the posterior wall of the canal, is termed the *urethral crest*. Many small mucous *urethral glands* and minute pit-like recesses or *lacunae* open into the urethra. On each side, near the lower end of the urethra, a number are grouped together and open into a duct, named the *para-urethral duct*; each duct runs down in the submucous tissue and ends in a small aperture on the lateral margin of the external urethral orifice. The urethral glands are considered to be the homologue of the male prostate, and the term '*female prostate*' is sometimes applied, but only to the glands of the proximal urethra. The glands, however, do not resemble male prostatic glandular tissue microscopically. Developmentally, the female urethra corresponds to the part of the prostatic urethra in the male that lies proximal to the opening of the prostatic utricle.

Female Urethral Structure

The female urethra contains muscular, erectile and mucous layers in its wall, as in the male.

The *muscular layer* is continuous with that of the bladder; it extends the whole length of the tube, and consists of internal longitudinal and external circular fibres. At the vesical end of the urethra, the circular muscle of the bladder is thickened and forms the sphincter vesicae (internal sphincter). In addition to this, just above the perineal membrane, the female urethra is surrounded by the sphincter urethrae, as in the male. The nerve supply of these two sphincters is the same as that in the male (pp. 563, 1135, 1407).

A thin layer of *spongiose erectile tissue*, containing a plexus of large veins, intermixed with bundles of nonstriated muscle fibres and connective tissue rich in elastin fibres, lies immediately beneath the mucous coat.

The *mucous membrane* is pale; it is continuous externally with that of the vulva, and internally with that of the bladder. Near the bladder it consists of transitional epithelium, grading more distally into non-keratinizing stratified squamous epithelium interspersed with pseudostratified columnar epithelium, and small mucous recesses and glands. Its external orifice is surrounded by a few mucous follicles.

The female urethra is much more readily dilatable than that of the male, and the passage of catheters or cystoscopes is a much easier process.

MICTURITION

Accumulation of urine in the bladder is accompanied by modification of tone of the detrusor muscle, so that the intravesical tension does not rise greatly until a considerable filling of the bladder has occurred. When the tension in the bladder rises, sensory nerves are stimulated, the desire to micturate is felt and, if neglected, a sensation of fullness of the bladder and finally pain supervenes. These sensations are accompanied by rhythmic reflex contractions of the detrusor muscle. Until micturition starts, voluntary restraint is exercised by coincident inhibition of the detrusor and contraction of the sphincter urethrae and the perineal muscles (Denny-Brown and Robertson 1933). There is no evidence that the sphincter vesicae is under voluntary control, but a reciprocal increase in tone of this muscle accompanies relaxation of the detrusor. The first stage of the act of micturition is a relaxation of the perineal muscles, except the sphincter urethrae, and a contraction of the muscles of the abdominal wall. This is followed by firm contraction of the detrusor and relaxation of the sphincter vesicae. The flow of urine begins on subsequent relaxation of the sphincter urethrae, the bladder being then emptied by the contraction of the detrusor assisted by the action of the muscles of the abdominal wall which raise the intra-abdominal pressure. As the act is completed, the bladder muscle relaxes and the sphincter vesicae contracts. Finally the sphincter urethrae is closed and, in the male, the last drops of urine are expelled from the bulbar portion of the urethra by the action of the bulbospongiosus.

This straightforward account of the supposed behaviour of the various muscular forces involved in micturition represents widely held views, but does not take into account certain dissensions and modifications. Much of the controversy centres around the role of the so-called *sphincter vesicae* (*vide supra*), most observers being in agreement on the efficacy of the external sphincter, the voluntarily controlled, striated, *sphincter urethrae*. An interesting theory, supported by considerable experimental evidence (Vincent 1966), suggests that when the pelvic floor is relaxed, the aggregation of nonstriated muscle at the bladder neck, even if contracted, cannot completely oppose the detrusor muscle. Conversely, the upward pressure of the pelvic floor, when it returns to normal tonus, moulds the bladder neck tissues into an efficient plug to the outlet. This theory receives support from observations on the intraluminal pressure in the proximal urethra; this begins to fall several seconds before micturition starts (Tanagho and Miller 1970). Most of the evidence quoted here is, therefore, generally in accord with the concept that micturition is initiated by relaxation of the pelvic floor, including the sphincter urethrae. Other authorities accord the initiating role to the vesical detrusor (Lapides 1969). For further information on this controversial topic, consult Boyarsky (1973).

During ejaculation, the sphincter vesicae, as well as stopping the escape of urine, also prevents the regurgitation of the semen into the bladder.

Applied Anatomy. The urethra may be ruptured, in which case extravasation of urine will complicate micturition. Extravasation most frequently occurs superficial to the perineal membrane but deep to the membranous layer of the superficial fascia. Both these layers of fascia are attached firmly to the ischiopubic rami. It is clear, therefore, that extravasated fluid cannot pass posteriorly, because the two layers are continuous with each other around the superficial transverse perineal muscles; it cannot extend laterally, on account of the connexion of these layers with the rami of the pubis and ischium; it cannot find its way into the lesser pelvis, because the opening into this cavity is closed by the perineal membrane, and, therefore, so long as this layer remains intact, the only direction in which the fluid can make its way is anteriorly into the areolar tissue of the scrotum and the penis, and thence on to the anterior wall of the abdomen. When the lesser pelvis is crushed the urethra may be ruptured between its prostatic and membranous parts; the extravasation of urine then takes place into the extraperitoneal tissue of the lesser pelvis.

(The lower urinary tract is subject to a number of congenital anomalies (*see* p. 221) which are amenable to surgical correction. For a survey of such anomalies from the standpoint of anatomy, consult Paul and Kanagasuntheram 1956.)

REPRODUCTIVE ORGANS OF THE MALE

The male genital organs include the *testes, epididymides, deferent* and *ejaculatory ducts* and *penis*, together with certain accessory glandular structures—the *seminal vesicles, prostate* and *bulbo-urethral glands*.

The Testes

The testes, the *primary reproductive organs* or *gonads* in the male, are suspended in the scrotum by the scrotal tissues (especially the nonstriated dartos muscles—see below), and by the structures forming the spermatic cords, the left testis usually hanging somewhat lower (about 1 cm) than its fellow. The average dimensions of the testis are from 4 to 5 cm in length, 2·5 cm in breadth and 3 cm in the anteroposterior diameter; its weight varies from 10·5 to 14 g. Each testis is of an ellipsoidal form (8.165), compressed laterally, and has an oblique position in the scrotum; the upper extremity is tilted anterolaterally, the lower, posteromedially. The anterior border is convex, the posterior border nearly straight, and the spermatic cord is attached to the latter.

The anterior border, the medial and lateral surfaces, and the extremities of the testis, are convex, smooth and invested by the visceral layer of the serosal tunica vaginalis by which they are separated from the parietal layer and hence scrotal tissues external to this. The posterior border receives only a partial investment from the tunica serosa. The epididymis lies along the lateral part of the posterior border.

The epididymis consists essentially of a tortuous canal which forms the first part of the efferent route from the testis. This canal is folded on itself and tightly packed into the form of a long, narrow, flattened body attached to the posterolateral aspect of the testis. It consists of a central region, or *body*, a superior enlarged *head*, and an inferior pointed end, the *tail*. The head is directly connected with the cranial pole of the testis by its *efferent ductules*;

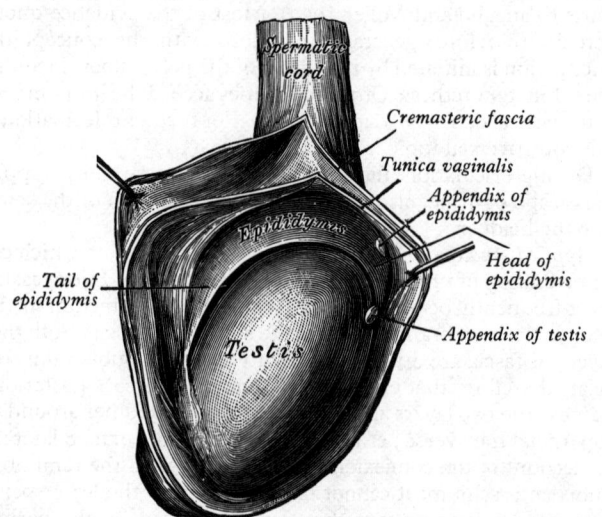

8.165 The right testis, exposed by incising and laying open the cremasteric fascia and parietal layer of the tunica vaginalis on the lateral aspect of the testis.

the tail is linked with the caudal pole by areolar tissue and a reflexion of the tunica vaginalis. The lateral surfaces of the head and tail of the epididymis are covered by the tunica vaginalis and are hence 'free'; the body is also invested by it, except along its posterior border. A recess of the tunica vaginalis, the *sinus of the epididymis*, lies between the body of the epididymis and the lateral surface of the testis.

The testicular and epididymial appendices. At the upper extremity of the testis, just inferior to the head of the epididymis, there is a minute, oval, sessile body, termed the *appendix of the*

testis; it is a remnant of the upper end of the paramesonephric duct. On the head of the epididymis there is a small, stalked appendage (sometimes duplicated), the *appendix of the epididymis*, which is usually considered to be a mesonephric vestige.

The testis is invested by the tunica vaginalis, tunica albuginea and tunica vasculosa.

The tunica vaginalis (8.165, 171) is the inferior extremity of the processus vaginalis of the peritoneum, which, in the fetus, precedes the descent of the testis from the abdomen into the scrotum (p. 218). After the testis has reached the scrotum the cranial part of the processus vaginalis, from the internal inguinal ring to within a short distance of the testis, contracts and undergoes obliteration. The distal part remains as a closed sac, into which the testis is invaginated. It is reflected on to the internal surface of the scrotum and hence it may be described as consisting of visceral and parietal layers.

The *visceral layer* covers the lateral and medial aspects and the anterior border of the testis, but leaves most of the posterior border uncovered. At the medial side of the posterior border it is reflected forwards to become continuous with the parietal layer. At the lateral side of the posterior border it is reflected on to the medial aspect of the epididymis, lining the sinus of the epididymis, and then passes over its lateral aspect as far as its posterior border, where it is reflected forwards to become continuous with the parietal layer. The continuity between the visceral and parietal layers is established also at the upper and lower poles of the testis, but at the upper pole the visceral layer covers the upper surface of the head of the epididymis before being reflected.

The *parietal layer* is more extensive than the visceral; it reaches below the testis and extends upwards for some distance in front and on the medial side of the spermatic cord. The inner surface of the tunica vaginalis is smooth, and covered with a layer of mesothelial cells. The potential space between the visceral and parietal layers constitutes the cavity of the tunica vaginalis.

In the embryo, the gonads (testis or ovary) project into the coelomic cavity and are covered by what for long was called the 'germinal' epithelium; some have considered that even in the adult the testis (like the ovary) is not covered by 'typical' peritoneum, the layer of mesothelium on the surface of the organ being the remnant of the original 'germinal' epithelium. The tunica vaginalis would thus be considered to consist of a parietal layer only and to become continuous with the germinal epithelium at the posterior border of the testis. Structurally, however, the parietal and visceral epithelia of the tunica are not only similar, but also like that of many areas of the general peritoneum. (Doubtless the term germinal epithelium reflects the misconception that generations of gonocytes developed from the specialized mesothelium of the gonadal or genital ridge.)

The obliterated part of the processus vaginalis may frequently be seen as a fibrous thread in the anterior part of the spermatic cord; sometimes this may be traced from the internal end of the inguinal canal, where it is connected with the peritoneum, as far as the tunica vaginalis; sometimes it is lost in the spermatic cord. In some instances the upper part of the processus vaginalis is not obliterated, and the peritoneal cavity then communicates with the tunica vaginalis; in others the proximal processus vaginalis may persist, but its distal end is shut off from the tunica vaginalis (pp. 218, 1367). Occasionally the processus may preserve its original cavity at an intermediate level as a cyst in the spermatic cord.

The tunica albuginea forms a fibrous covering for the testis. It is a dense membrane, of a bluish-white colour, composed of interlacing bundles of white fibrous tissue. It is covered externally by the visceral layer of the tunica vaginalis, except at the head and tail of the epididymis, and along the posterior border of the testis, where the testicular vessels and nerves enter the gland. It is applied to the tunica vasculosa, and, at the posterior border of the testis, is projected into its interior, forming a thick but incomplete, vertical septum, the **mediastinum testis** (8.166). This extends from the upper to near the lower end of the testis,

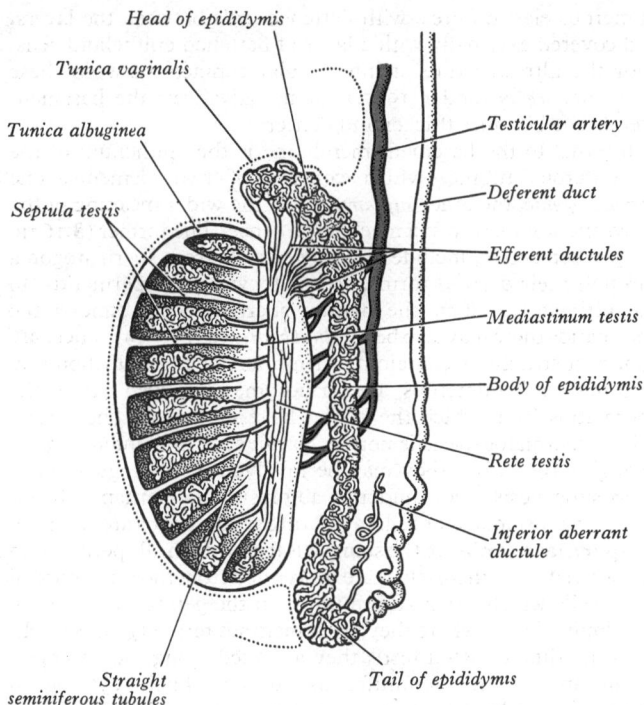

Tunica vaginalis

Tunica albuginea

Septula testis

Straight
seminiferous tubules

Testicular artery

Deferent duct

Efferent ductules

Mediastinum testis

Body of epididymis

Rete testis

Inferior aberrant
ductule

Tail of epididymis

8.166   A vertical section through the testis and epididymis, showing the arrangement of the ducts of the testis and the mode of formation of the deferent duct.

and is wider above than below. From its front and sides numerous incomplete septa (*septula testis*) are given off and radiate towards the surface of the testis, where they are attached to the deep aspect of the tunica albuginea. They divide the organ incompletely into a number of cone-shaped lobules.

The bases of the lobules are at the surface of the testis, and their apices converge upon the mediastinum. Dhingra (1977) has

contributed a detailed comparative account of the mediastinum testis. The arrangement of its connective tissue is very variable, and some species, including man (Holstein 1967), display nonstriated myocytes amongst the collagen fibres (*see* p. 1415). It is traversed by many vessels, and by efferent and retial tubules. Leydig cells are absent from the mediastinum testis in man, but are present in some ungulates.

The tunica vasculosa is the vascular layer of the testis, consisting of a plexus of blood vessels held together by delicate areolar tissue. It extends over the internal aspect of the tunica albuginea and covers the septa, and therefore forms an investment to all the lobules of the testis.

STRUCTURE OF THE TESTIS

The surface covering of the testis is a layer of flattened mesothelial cells similar to those which line the peritoneal cavity; by some these mesothelial cells have been considered to be remnants of a 'germinal' epithelium. The main internal architecture is formed by the *lobules of the testis* (**8.**166). Their number, in a human testis, is estimated to be between 200 and 300. They differ in size according to their position, those in the middle of the testis being larger and longer. Each lobule contains from one to three or more minute *convoluted seminiferous tubules*. When the latter have been unravelled by careful dissection under water, they are seen to commence either by free blind ends or by anastomotic loops. Clermont and Huckins (1961) and Roosen-Runge (1961) have described these tubules (in the rat) as closed loops, opening at both ends into the tubuli recti, and thence into the rete testis, a view which is now generally accepted. They are supported by loose connective tissue which contains here and there groups of *interstitial cells* (**8.**167A) containing yellow pigment granules. The total number of tubules in each testis is estimated to be between 400 and 600, and the length of each is 70 to 80 cm. Their diameter varies from 0·12 to 0·3 mm. The tubules are pale in colour in early life, but in old age they contain much fatty matter and acquire a deep yellow tinge. Each tubule (**8.**167A) consists of a basement membrane formed of laminated connective tissue containing

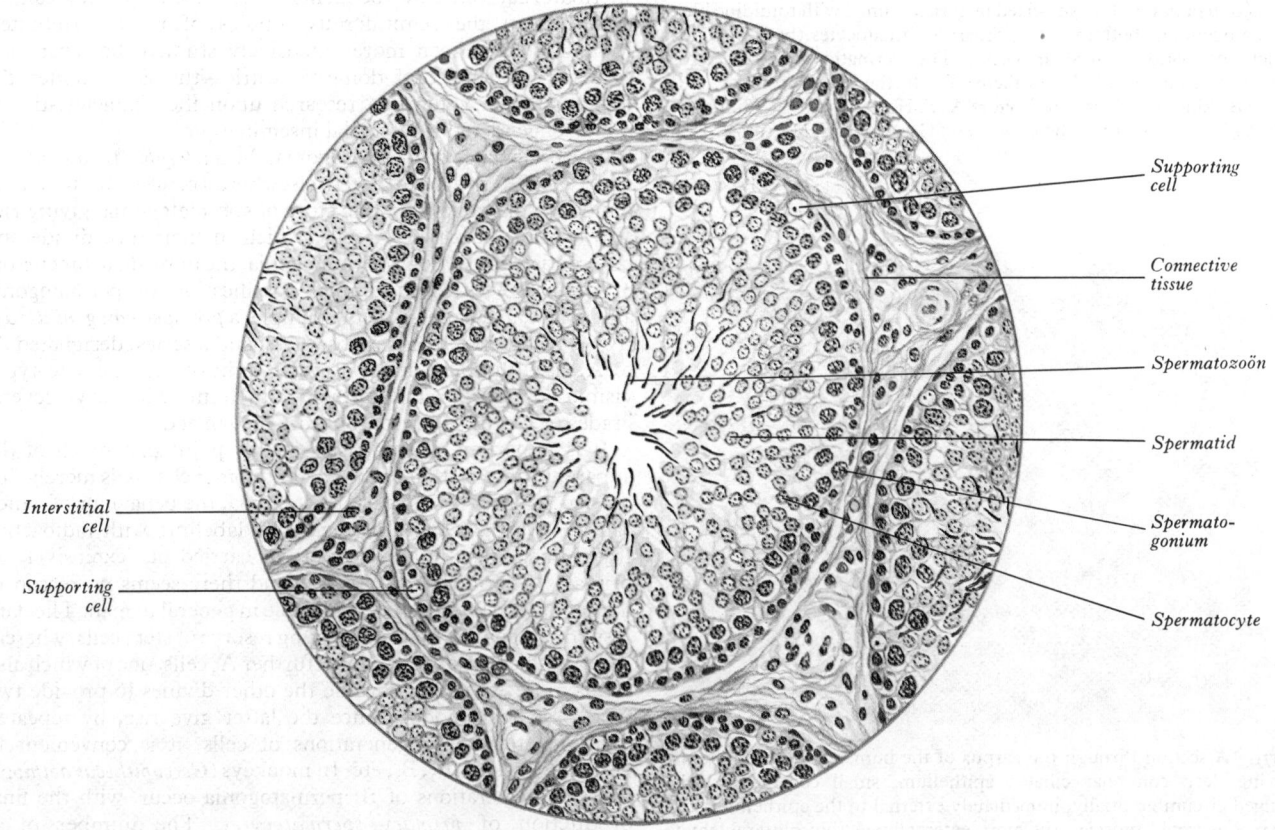

Supporting
cell

Connective
tissue

Spermatozoön

Spermatid

Spermato-
gonium

Spermatocyte

Interstitial
cell

Supporting
cell

8.167A   A tranverse section through a part of a human testis. Stained with iron haematoxylin and Van Gieson's stain. Magnification about ×350.

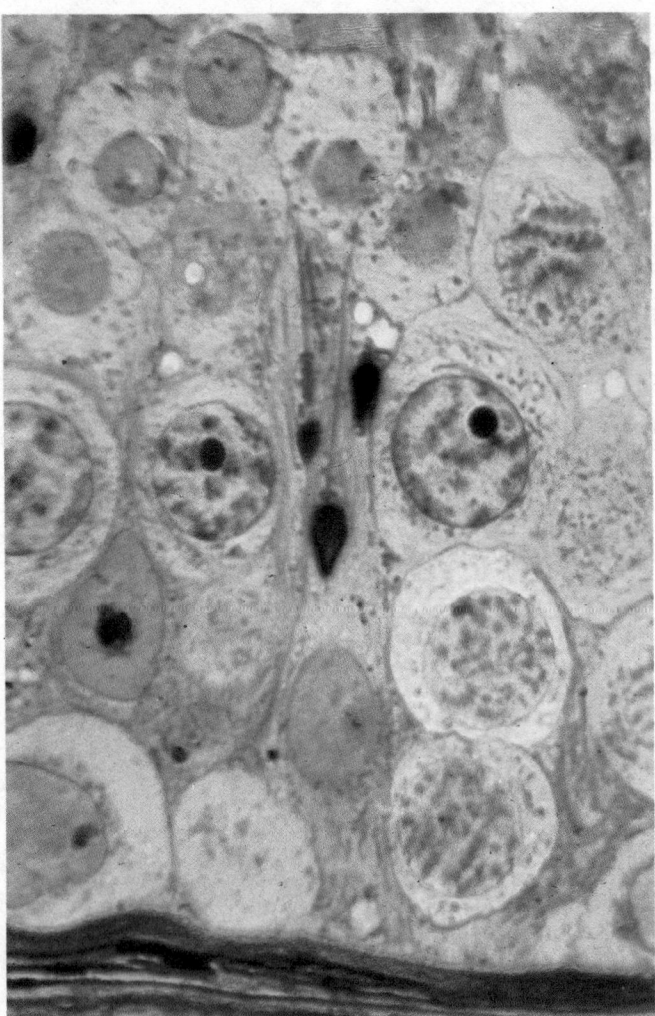

8.167B The epithelium of the seminiferous tubule in an adult man. A thin section of glutaraldehyde-fixed material, stained with toluidine blue. Spermatogonium (bottom left), primary spermatocytes (bottom right), and late spermatids (central) are visible. The spermatids are attached to a supportive cell of Sertoli (below them). For further details compare with **8.168**. (By kind permission of Doctors A. F. Holstein and U. Wulfhekel, *Andrologie*, **3**, 1971 and of the publishers, Grosse Verlag, Berlin.)

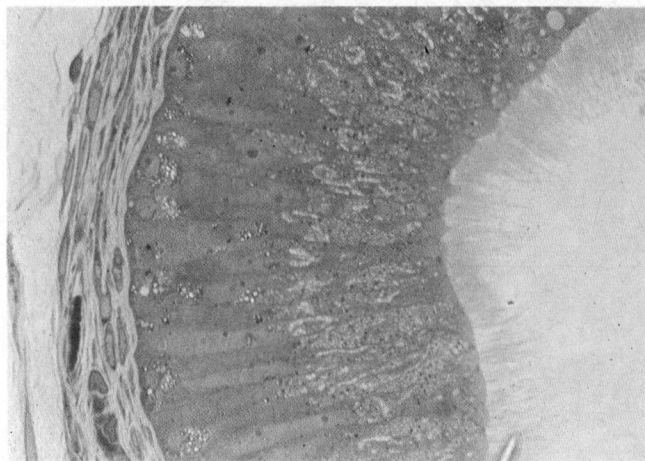

8.167C A section through the corpus of the human adult epididymis, showing deep columnar ciliated epithelium, small contractile cells, arranged circumferentially, immediately external to the epithelium, and larger nonstriated muscle cells more externally. A thin glutaraldehyde fixed section stained with toluidine blue. (Kindly supplied by Dr. A. F. Holstein, University of Hamburg. For details see H. G. Baumgarten, A. F. Holstein and E. Rosengren, *Z. Zellforsch. mikrosk. Anat.*, **120**, 1971.)

numerous elastin fibres, with flattened cells between the layers, and covered externally with a layer of flattened epithelioid cells. (For the ultrastructure of intercellular contacts between these elements, *see* Nicander 1967.) As age advances, the basement membrane becomes thicker and denser.

Internal to the basement membrane is the epithelium of the seminiferous tubule, which consists of two elements, one *spermatogenic*, the other *supportive*—in the wider meaning of the word and not merely in a mechanical sense. The former (**8.167B**, 168), when active, include an array of cells from spermatogonia through their derived forms, spermatocytes and spermatids, to the ultimate product, the mature spermatozoön. Among the spermatids there may also be *residual bodies* (pp. 96, 99). These are spherical structures containing membranous and mitochondrial residues and numerous free ribosomes, derived from the spermatids from which the bodies have separated. Their exact role in spermatogenesis is not yet clarified, despite intense study (Vaughn 1966), but they may be involved in the regulation of spermatogenesis; they undergo autolysis and perhaps phagocytosis by sustentacular cells, as mature spermatozoa are released.

Spermatogonia are the stem cells from which all spermatozoa are derived; they themselves are descended from those primordial germ cells which reach and multiply in the genital cords of the developing testis, where they are sometimes termed *gonocytes*. In the fully differentiated testis they are sited along the basement membrane of the seminiferous tubule. Their cytological characteristics have been noted in a previous section (p. 98), as have those of the generations of spermatocytes and spermatids which are produced from them in serial cell divisions. It has been recognized since the beginning of the twentieth century that several types of spermatogonia can be recognized (Regaud 1901); and on the basis of cell and nuclear dimensions, distribution of nuclear chromatin, and histochemical and ultrastructural data, a considerable number of spermatogonial types have been identified, though there are disagreements over minor details (Courot *et al.* 1970). It may be added here that in the same manner, various stages in the maturation of spermatocytes and spermatids (p. 98) have also been recognized, so that a rather complex lineage of cells, extending from spermatogonia to spermatozoa, can now be defined step by step, and can be identified in the seminiferous tubules of most vertebrates. Mammals have been more extensively studied, however, and especially rodents and domestic cattle—the latter under the impetus of the volume of research upon the characteristics of semen engendered by artificial insemination.

The basic types of spermatogonia, the *dark type A*, *pale type A*, and *type B*, have been noted elsewhere (p. 98); the first type divides to maintain the basic store of spermatogonia, giving rise also to some *pale* type A cells, which in their turn divide and differentiate into type B spermatogonia, the immediate precursors of spermatocytes. More complex classifications of spermatogonia are, in fact, in current use; for example, a *pro-spermatogonial stage* (A_0) is recognized by most observers, and a series, designated A_1 to A_4, is utilized in describing the train of intermediate types visible in active seminiferous epithelium. Similarly, several grades of B type spermatogonia are recognized.

It is important to emphasize at this point that much of the apparent complication of the current nomenclature is merely due to successive divisions of spermatogonia, the behaviour of which can be investigated in this regard by labelling with radioactive thymidine. Such studies have been carried out extensively in primates (Clermont 1963, 1969), and there seems no reason to doubt that the findings are applicable in general to man. The dark A spermatogonia (A_1) form a resting reserve of stem cells, whereas the pale type (A_2) divide to form further A_2 cells, one of which also forms more A_2 elements while the other divides to provide two type B spermatogonia. Since the latter give rise, by repeated divisions, to several generations of cells, it is convenient to designate these as B_1, B_2, etc. In monkeys (*Cercopithecus aethiops*) five such generations of B spermatogonia occur, with the final production of *primary spermatocytes*. The number of B spermatogonial generations in man is four, and since the number of cells (without allowing for degeneration of some) is theoretically doubled at each stage, there is a basis for the production of the very large number of spermatozoa formed during the active reproductive life of male vertebrates. The number of spermatogonial generations varies, being high in fish (six to fourteen), low in birds (one to three) and ranging from four to seven in the mammals so far investigated, (see p. 98 and Holstein 1976).

The primary spermatocytes, with a diploid chromosome content, divide to form **secondary spermatocytes** with a haploid complement of chromosomes; this constitutes the nuclear division designated meiosis I (see p. 98). In turn the secondary spermatocytes undergo the second meiotic division to form spermatids. Both types of spermatocytes and the spermatids pass through a series of changes, recognized chiefly on the basis of nuclear activities, although of course in the latter there are

are also identified. It is, however, generally assumed that almost all the non-germinal cells in the tubules of the adult, functioning mammalian testis are of this limited type. Such sustentacular cells are sited on the limiting membrane of the tubules, occupying it almost to the exclusion of all but occasional spermatogonia. In tangential sections they are polygonal cells, in transverse sections somewhat irregular, but approximately columnar. Their apical profiles are, however, much complicated by recesses into which the spermatids and spermatozoa fit until the latter are mature enough for release; and long cytoplasmic processes also extend among the spermatogonia and spermatocytes. This arrangement strongly suggests that the sustentacular cells maintain structural cohesion between the varied elements of the epithelium. Electron microscopy (Fawcett 1975) has disproved the once strongly held

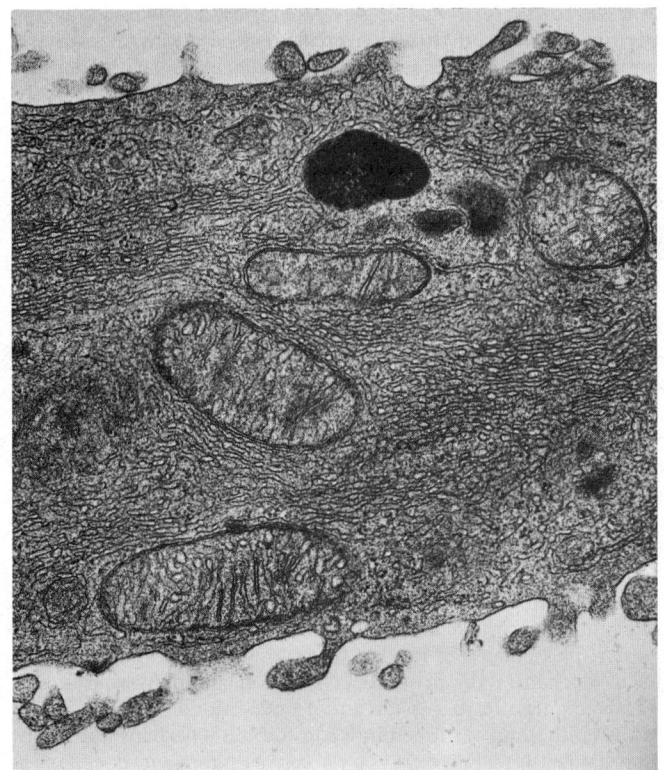

8.167D A transmission electron micrograph of part of an interstitial cell of the testis (Leydig cell) showing the profuse agranular endoplasmic reticulum and prominent mitochondria (murine). (Magnification × 20,000.)

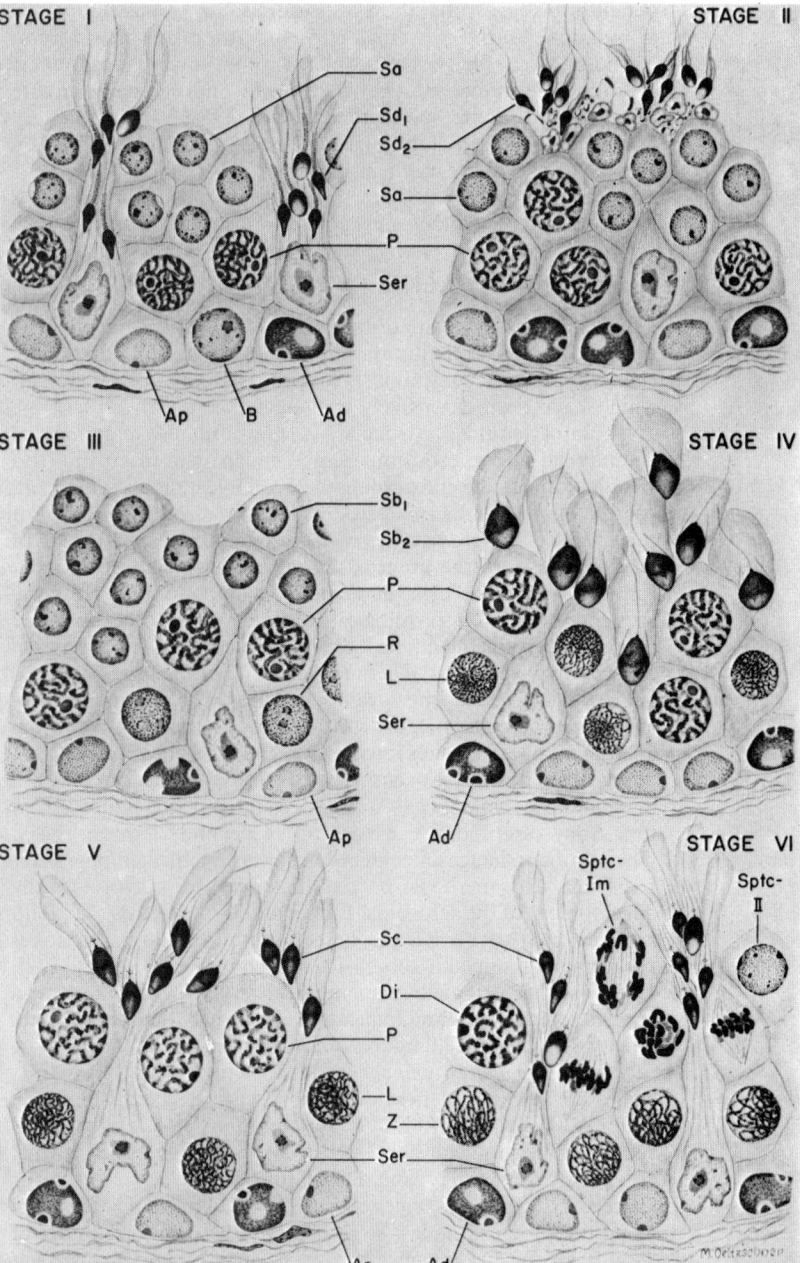

8.168 The cycle of spermatogenesis in the human seminiferous tubule. For details of the six stages see accompanying text. Ser = sustentacular cells of Sertoli. Ad, Ap and B = A type (dark and pale) and B type of spermatogonia. R = resting primary spermatocyte. L, Z, P, and Di = primary spermatocytes in leptotene, zygotene, pachytene, and diplotene stages. Sptc-Im = dividing primary spermatocyte. Sptc-II = secondary spermatocyte in interphase. Sa, b, etc. = generations of spermatids. (From Prof. Y. Clermont, *Amer. J. Anat.*, **112**, 1963, with acknowledgements to the author and publishers—the Wistar Institute of Anatomy and Biology.)

profound accompanying modifications in the soma of each spermatid to reach the mature flagellated spermatozoön.

The primary spermatocyte passes through a resting or *proleptotene* stage, followed by identifiable leptotene, zygotene, pachytene and diplotene stages (see p. 98), and the completion of cell division. The secondary spermatocytes divide much more rapidly into spermatids.

The spermatids gradually mature into spermatozoa through a series of nuclear and cytoplasmic modifications (see p. 99), and several stages have been identified on the basis of nuclear shape, 'round' and 'elongating' spermatids being generally recognized. With their formation, spermatogenic multiplication is usually regarded as coming to an end. Thus, it will be appreciated that the spectrum of identifiable cell types in the seminiferous epithelium is of considerable complexity, and to this must be added the supportive element in the epithelium.

The sustentacular or supporting cells (of Sertoli), or *sustentocytes*, are the only non-germinal individuals in the complex cell population of the seminiferous tubules. They are somewhat polymorphic cells, both in nuclear and somatic characteristics. They have been the subject of much controversy, and in the developing seminiferous tubules a variety of types of supporting cells, which do not fall within the 'Sertolian' category,

view that these cells formed a syncytium, revealing that their contiguous surfaces are in fact held together by a variety of specialized contacts superficially similar to 'tight' junctions (*see* p. 7), but of at least three types (Nicander 1967; Flickinger and Fawcett 1967). (Sertoli, himself, demonstrated the individual nature of these cells in 1865.) The nucleus is irregular, often indented, and weakly positive to Feulgen staining. It contains one or two pyroninophilic nucleoli, and these may be tripartite, consisting of a nucleolonema (*see* p. 27) and two associated electron-dense bodies (see also the general description of nucleolar structure on p. 27). The cytoplasm contains abundant organelles, particularly mitochondria, endoplasmic reticulum, secretion granules, Golgi complexes, ribosomes, microfilaments and microtubules. The last named form, presumably, a kind of cytoskeleton and may also be a factor in the cohesive effect of the sustentacular cells on other epithelial units. These and other details also suggest that these cells exercise a metabolic influence in relation to the germinal elements; they are in addition considered to be phagocytic (Carr *et al.* 1968) and perhaps to exert an endocrine influence. They themselves undergo considerable changes during the spermatogenic cycle (*vide infra*) and are influenced by the hypophysial hormones, LH and FSH. They also play a part in the mechanisms of the '*blood-testis barrier*'. For a recent general review of these and other aspects of testicular structure and physiology consult Johnson and Gomes (1977).

THE SPERMATOGENIC CYCLE

The numerous elements of the seminiferous epithelium briefly described above constitute an intimately related population of cells which is in a continuously dynamic state, the parameters of which have been extensively investigated, not only in a range of laboratory animals (particularly the rat), but also in animals of commercial or 'domestic' importance, and in man. In addition to a wealth of information regarding the cytological progression from spermatogonia to spermatozoa, various quantitative aspects of this cycle of changes have been assessed, including the timing of various phases of the process, the relative numbers of different forms of germinal cells at definable points in this temporal sequence, and so on. Cytological studies have clearly established that at any specified locus in a seminiferous tubule the activities in the several generations of germinal cells (identifiable as described above), and hence the overall qualitative and quantitative associations of cell types, pass through a *cycle*, the length of which and the number of stages involved vary in different species (Clermont 1972). The cycle occupies 12 to 13 days in rats, but only 8·6 in the mouse; in monkeys (Arsenieva *et al.* 1961) the figure is 10·5 to 11·6 days, and about 16 in mankind (Heller and Clermont 1964). The number of stages recognized has been established on a surer basis by two methods of classification, one involving the development of the acrosomal system of the spermatids (Leblond and Clermont 1952), the other based upon morphological changes in germ cell nuclei (Roosen-Runge and Giesel 1950). Both methods were originally elaborated in rats, but have also been applied to many other mammals and some birds, and also to man. The two methods yield figures of fourteen and eight stages in the rat; the *acrosomalic technique* leads to identification of only six stages in man, and these will shortly be detailed. In many mammals all parts of the circumference of a short length of seminiferous tubule are observed to be in the same stage, while adjoining cylindrical regions display the numerically preceding or succeeding stages. All stages can thus be recognized in succession along the tubules, the last stage being followed by a cylinder of activity in the first stage. Of course, each annular locus itself passes through all stages and then begins again, so that there exists not only a *local* succession of stages, spread over the characteristic period of the cycle in the species involved, but also a procession of stages, which appears to move in an undulatory manner along the tubule. Such a '*spermatogenic wave*' has been measured in some species (for example, the rat and bull—*see* Mochereau 1963), but the 'wave-length' has been found to be irregular and subject to modifications. In man, not even all quadrants of a local cylinder of seminiferous epithelium are in the same cyclic phase at any one time. These stages are as follows (**8**.168).

Stage I is typified by the presence of two generations of spermatids—a newly formed one with spherical nuclei, and an older one with elongating nuclei. These latter cells are deeply invaginated in groups into the cytoplasm of sustentacular cells. Associated with them are primary spermatocytes at the start of the lengthy pachytene stage of the prophase of the first meiotic division; A (dark and pale) and B type spermatogonia are arranged close to the basement membrane.

Stage II also contains two stages of spermatid maturation, but these are further advanced, and the older spermatids are close to final modification into spermatozoa, and hence the presence of residual bodies. The primary spermatocytes are still in the pachytene stage, but the B type spermatogonia have now appeared in greater numbers amongst the dark and pale A types.

Stage III. The older spermatids have reached the end of spermatogenesis and are released as spermatozoa (spermiation). Nearest to these is a single generation of well-advanced spermatids, external to which are two generations of primary spermatocytes, the older still in the pachytene stage, with younger cells in the proleptotene or resting stage of meiosis. Type A spermatogonia predominate, the type B having become spermatocytes.

Stage IV. Spermatid nuclei are commencing to elongate, but are irregular in shape. The older primary spermatocytes are in the pachytene stage and the younger ones are moving into the leptotene stage. Type A spermatogonia are gathered external to the above cells.

Stage V. Groups of spermatids display elongated nuclei orientated centrifugally towards the periphery of the tubule. The primary spermatocytes are now in the *late* pachytene condition, and some in the leptotene stage.

Stage VI. Groups of maturing spermatids are visible between primary spermatocytes undergoing division into secondary spermatocytes. Some of the latter are seen in the interphase stage, and external to them are primary spermatocytes in the zygotene stage. Type A spermatogonia (pale and dark) now predominate again close to the basement membrane.

The details of these stages may seem a little complex, but they fit into the succession of divisions and differentiations known to occur in the germinal cell series. The process appears much less regular, particularly in human biopsy material, perhaps in part due to mechanical disturbance. In addition, an unknown number of cells undergo retrogression and degeneration.

The epithelium of the seminiferous tubule is perhaps unique in its organization (Fawcett 1975), consisting, as it does, of a permanent population of non-proliferative Sertoli or sustentacular cells, sessile upon its basement membrane, and, interspersed amongst these, the proliferating spermatogonia and their derivatives. The latter are related to the basal parts of the Sertolian cells, the lateral junctions of which (**8**.167B) are now considered to create a '*basal compartment*', in which are the preleptotene spermatocytes. These junctions must at some point break down to allow the spermatocytes to move up towards the tubular lumen and into a so-called '*adluminal compartment*'. It is suggested that in primates the 'blood-testis barrier' resides in these modifiable junctions, which may be under hormonal control. This implies that, whereas the spermatogonia and early spermatocytes are directly accessible to all substances transported across the capillary wall, the later spermatocyte stages and spermatids are screened from substances incapable of passing the 'barrier'. As the spermatids mature they move up towards the lumen from the lateral aspects of Sertolian cells, at first embedded largely in the cytoplasm of the latter and attached by a junctional specialization which itself undergoes a series of changes (Ross 1976) until the spermatozoa are released into the lumen of the seminiferous tubule.

THE EFFERENT DUCTULES AND EPIDIDYMIS

The events described in the preceding section occur in the highly coiled part of the seminiferous tubules. As they reach the apices of their lobules, they become less convoluted, assume an almost straight course, and unite to form twenty to thirty larger straight ducts, about 0·5 mm in diameter (**8**.166).

The straight seminiferous tubules enter the fibrous tissue of the mediastinum, and pass upwards and backwards, forming, in their ascent, a close network of anastomosing tubes lined by flattened epithelium; this network is named the *rete testis*. At the upper end of the mediastinum these tubes terminate in from twelve to twenty ducts, termed the *efferent ductules*; they perforate the tunica albuginea, and pass from the testis to the epididymis. Their course is at first straight; then they become enlarged and exceedingly convoluted, and form a series of conical masses, known as the *lobules of the epididymis*, which together form the head (caput) of the epididymis. Each lobule consists of a single convoluted duct, from 15 to 20 cm in length. Opposite the bases of the lobules the ducts open into a single canal, the *duct of the epididymis*, whose complex convolutions form the body (corpus) and tail (cauda) of the epididymis. When the convolutions are unravelled, this tube measures upwards of 6 m in length; it increases in diameter and thickness as it approaches the tail of the epididymis where it becomes the deferent duct. The convolutions are held together by fine areolar tissue, and by bands of fibrous tissue. It is to be noted that the body and tail of the epididymis consist of a single tube.

The efferent ductules are lined by ciliated columnar epithelium and have a thin circular layer of nonstriated myocytes in their walls. In the duct of the epididymis the muscle coat becomes thicker and the epithelium is columnar pseudostratified, the superficial cells having long ($15\ \mu$m) regular microvilli with a diameter of about $0.2\ \mu$m. Because of their superficial resemblance to cilia these structures are often termed 'stereocilia', but they lack the characteristic ultrastructural details of true cilia. Tracer experiments have demonstrated the uptake of proteins by the cells of the efferent ductules by phagocytic vesicle formation. It seems likely that these cells modify the composition of seminal fluid. Electron microscopic studies have revealed inclusion bodies in the nuclei of the epithelial cells of the human epididymis (Horstmann 1962, 1966). In man fragments of ingested spermatozoa have also been observed in small round cells, with their cytoplasmic processes projecting into the duct of the epididymis (Holstein 1967). In several experimental animal species the localization of lysosomal enzymes supports the hypothesis of an absorptive function in the epididymal epithelium (Moniem and Glover 1972).

Further studies (Baumgarten *et al.* 1970) suggest that three ultrastructurally distinguishable types of contractile cell occur in both the ductules and the epididymis. These cells receive only a sparse adrenergic innervation, and few of the nerve fibres penetrate between the nonstriated myocytes (in contrast to the more profuse innervation of the nonstriated muscle of the caput epididymis and ductus deferens). These morphological differences accord with the slow, spontaneous and local contractions of the ductules and proximal epididymis, and the rapid, reflex contractions of the caput and ductus during emission.

THE TESTICULAR INTERSTITIAL TISSUE

The interstitial cells of the testis, usually equated with the *cells of Leydig* (1850), also include various other cells, some of connective tissue type, together with the vessels and nerves in the tissue between the seminiferous tubules. The origin of the specific interstitial cells (of Leydig) is probably mesenchyme, but it is also suggested that they arise from the mesonephric blastema (Witschi 1951). These cells, isolated or clustered, occur in the intertubular tissue of most vertebrates, including man. They are large, polyhedral cells with an eccentric nucleus containing one to three nucleoli, and they possess scanty, poorly staining cytoplasm (Hooker 1971). They contain (**8.**167D) much agranular endoplasmic reticulum (rich in ascorbic acid), vacuoles containing fats, phospholipids and cholesterol, resembling in such features the interstitial cells of the ovary (p. 1424), the luteal cells of the corpus luteum (p. 1424), and cells in the adrenal cortex (p. 1457). Their masculinizing effect was demonstrated almost eighty years ago (Bouin and Ancel 1903), but doubt as to whether they are the sole source of testicular androgens is not yet entirely removed. Most of the experimental evidence demonstrates the lack of endocrine function of this kind in the tubules, rather than proving that the interstitial cells are the sole source of androgens. However, it is generally accepted that they do secrete androgens, and that no other function can at present be attributed to them (Christensen 1975). They are themselves stimulated into activity by the interstitial cell stimulating hormone (ICSH—identical with LH) and possibly follicle stimulating hormone (FSH) of the anterior lobe of the hypophysis cerebri (p. 1439). In cryptorchidism, where the testes are retained in the inguinal canal and hence subjected to a warmer environment, the rate of production of androgens, as well as of spermatozoa, is depressed, without, however, any recognized cytological changes in the interstitial cells. Androgens stimulate the growth and activity of the male accessory reproductive glands (prostate, seminal vesicles and bulbo-urethral glands) and also the secondary sexual changes at puberty—growth of facial, axillary and pubic hair, enlargement of larynx and paranasal air sinuses, and the additional skeletal growth characteristic of the male. Most of these changes are inhibited by oestrogens (in sufficient amounts), especially when androgenic output is depressed, as occurs in some eunuchoid conditions. (*See* also p. 1439.)

The proportion of interstitial cells (of Leydig) to the other components of the intertubular tissue displays variations at different ages, and in certain clinical conditions. After an initial period of multiplication in early fetal life, they appear to atrophy around the time of birth, though there is some disagreement as to whether they actually disappear or are merely depressed in number in the immediate post-natal period (Cooper 1929; Sohval 1954). It is generally agreed that they reappear in the prepubertal period, perhaps developing from mesenchyme cells (Hisakayu and Harrison 1971). There is no reduction in the numbers of Leydig cells in the elderly (55–65 years) according to Kothari and Gupta (1974). In cryptorchid testes these cells are absent, while the total amount of intertubular tissue is increased. For techniques of quantitative assessment of Leydig cells in human and other mammalian testes consult papers by Dykes (1969); Heller *et al.* (1971); Christensen (1975); and Kothari *et al.* (1978).

THE TESTICULAR CAPSULE

The coverings of the testis have already been noted (p. 1410); they may be briefly summarized, proceeding from the surface internally, as the skin, dartos muscle, superficial perineal fascia, external and internal spermatic fasciae with the cremasteric fascia interposed, the parietal layer of the tunica vaginalis, the capillary interval between this and the visceral layer, the tunica albuginea and tunica vasculosa. The last three strata form the so-called 'testicular capsule', a concept which has become of considerable significance as a result of the reported demonstration of a contractile element in what is usually regarded as an inert and largely fibrous tissue structure (Davis and Langford 1969). In rabbits and rats isolated preparations of the testicular capsule can be made, and these show spontaneous contractions and also react to both cholinergic and adrenergic agents. Abundant autonomic nerve endings in relation to blood vessels have been described in the tunica albuginea (Norberg *et al.* 1967), and nonstriated muscle cells have been demonstrated in the tunica in various rodents and in man (Holstein 1967). The distribution of such muscle fibres varies, and while they are widely scattered in the human tunica albuginea, they are concentrated chiefly in a posterior position, adjacent to the epididymis. The role of the contractile testicular capsule is most probably to massage or pump the duct system of the testis, and thus to impel spermatozoa in an onward direction. For a review of the pertinent literature, consult Davis *et al.* (1971).

TEMPERATURE REGULATION AND THE TESTIS

The testes are external or *scrotal* in most marsupials and eutherians, but *abdominal* in fish, amphibians, reptiles, birds and monotremes. There are some exceptions amongst mammals: the testes are abdominal in location in sloths, elephants and hyraces. In aquatic mammals they are either abdominal (whales and dolphins) or close to the body in the inguinal region (seals). In some mammals they are scrotal in position during the breeding season, but inguinal or even abdominal in position at other times (rodents, bats and insectivores). The variations in mammals are puzzling, but there is a general assumption that the predominance of scrotally suspended testes in mammals is associated with their homeothermic specialization, though this does not explain the abdominal testes of birds. The regulation of the internal body temperature (Hammel 1968) appears to create, in general, too warm an environment for spermatogenesis; and experimental evidence, reviewed by Bishop and Walton (1960), amply confirms this general concept. The scrotal temperature is usually several degrees below that of the abdomen (three degrees in man), but obviously the optimum range for spermatogenesis—though lower—must also be maintained. The mechanisms involved in this are numerous (Waites 1970), and the factors concerned in testicular thermoregulation can merely be mentioned here. The scrotal skin is usually hairy, but this potential heat-conserving covering is most variable; it is scant in man, profuse in marsupials. Sweat glands are well developed and usually numerous, but subcutaneous fat is generally absent. The radiant area of the scrotal skin, which is well vascularized, can be much reduced in colder conditions by contraction of the dartos muscle, probably by direct stimulation, or perhaps by local reflexes. The primate scrotal skin contains many nerve endings which respond to chilling or to warming, the latter being more numerous (Iggo 1969). The cremaster muscles, by elevating the testes closer to the perineum, may assist in conserving heat. A form of *countercurrent heat exchange* is considered to occur between the arteries and veins of the spermatic cord, where they are both considerably coiled and in close apposition (Harrison and Weiner 1949). In dogs experiment has shown that the blood delivered to the testis is pre-cooled by 3 °C. by this mechanism (Waites and Moule 1961). There is increasing evidence that in addition to heat, respiratory gases and a number of other substances (including labelled testosterone and prostaglandin-F), can be transferred between these arteries and veins, in several species, including man. (For discussion of the literature *see* Free 1977.)

There thus exists a surprising array of factors concerned in testicular thermoregulation and the maintenance of fertility; and while these appear to be jointly involved in this particular function, there are suggestions, at least, that the scrotum may also act as an agent in the reciprocal control of general body temperature.

Vessels and Nerves

The *testicular artery* is described on p. 718 and the *testicular veins* on p. 762. The intratesticular course of the arterial supply has received much attention in man (Hundeiker and Keller 1963). In a recent study (Kormano and Suoranta 1971), marked coiling of the arteries was noted; the capillaries adjoining the seminiferous tubules penetrate the layers of the interstitial tissues. These capillaries have acquired a particular interest as a part of the anatomical site of the 'blood-testis barrier' (Setchell and Waites 1975; Neaves 1977). The capillaries run either parallel to the tubules or transversely across them; they do not enter the walls of the seminiferous tubules, being separated from the germinal and supporting cells by a basement membrane, and by variable amounts of interstitial tissue containing Leydig cells. It is at this level that highly interesting selective exchange phenomena occur in connexion with androgens and immune substances.

The *lymph vessels* of the testis end in the lateral and pre-aortic lymph nodes (p. 797).

The *nerves* accompany the testicular vessels, and are derived from the tenth and eleventh thoracic segments of the spinal cord, through renal and aortic autonomic plexuses (pp. 1131, 1134). Catecholamine-containing nerve fibres have been identified in the human testis and epididymis. Such fibres form plexuses around smaller blood vessels and amongst the interstitial cells (Baumgarten and Holstein 1967). For a review of testicular innervation, consult Hodson (1970).

Applied Anatomy. At an early period of fetal life the testes are located in the posterior part of the abdominal cavity. Their descent into the scrotum is described on p. 217. The descent appears to be under hormonal control (gonadotropins and androgens). In the scrotum the testes are in a cooler environment

than that in the abdomen, and it is believed that the lower temperature favours spermatogenesis (*vide supra*).

The descent of the testis may be arrested. It may be retained in the abdomen; or it may be arrested at the deep inguinal ring, or in the inguinal canal; alternatively it may just emerge from the superficial inguinal ring without reaching the bottom of the scrotum. The retained testis is probably reproductively useless; so that a man in whom both testes are retained (*anorchism*) is sterile, though he may not be impotent. The absence of one testis is termed *monorchism*. When a testis is retained in the inguinal canal the situation is often complicated by a co-existent congenital hernia, the processus vaginalis of the peritoneum remaining patent. The testis may descend through the inguinal canal, but may miss the scrotum and assume some abnormal position (*ectopia testis*, p. 218).

The testis may be inverted within the scrotum so that its posterior or attached border is anterior and the tunica vaginalis is posterior.

Fluid collections of a serous character frequently occur in the

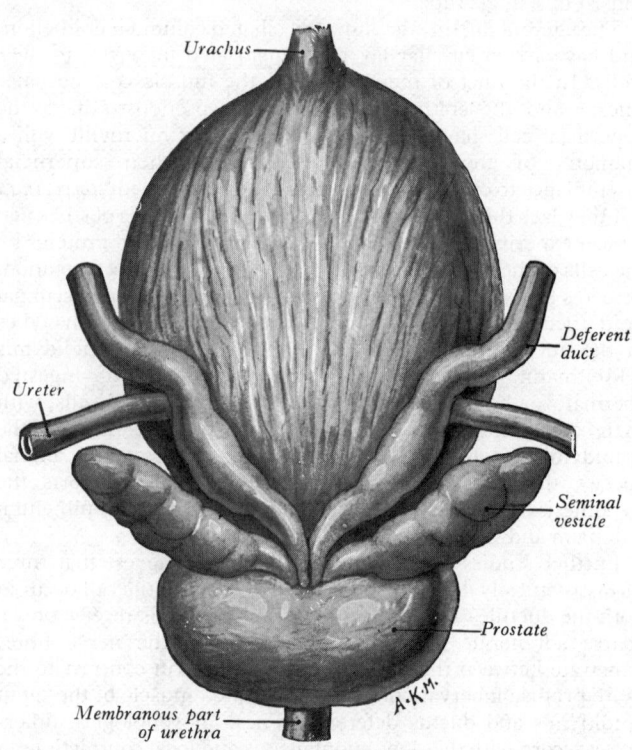

8.169   Posterosuperior aspect of the male internal urogenital organs.

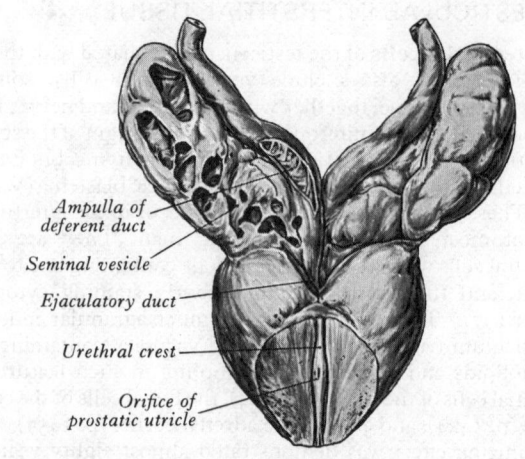

8.170   Anterior aspect of the seminal vesicles, terminal parts of the deferent ducts, and the prostate. The lamina of the right seminal vesicle, the ampulla of the right deferent duct, and of the prostatic part of the urethra have been exposed by appropriate removal of tissues.

scrotum. To these the term *hydrocele* is applied. The most common form is the ordinary *vaginal hydrocele*, in which the fluid is contained in the sac of the tunica vaginalis. In another form, the *congenital hydrocele*, the fluid is in the sac of the tunica vaginalis, but this sac communicates with the general peritoneal cavity, owing to non-obliteration of the proximal part of the processus vaginalis. A third variety, known as an *infantile hydrocele*, occurs in those cases where the processus vaginalis is obliterated only at or near the deep inguinal ring. It resembles the vaginal hydrocele, except as regards its shape, the collection of fluid extending up the cord into the inguinal canal. Fourthly, the processus vaginalis may be obliterated both at the deep inguinal ring and above the epididymis, leaving a central unobliterated portion, which may become distended with fluid, giving rise to a condition known as *encysted hydrocele of the cord.*

Encysted hydrocele of the epididymis, or *spermatocele*, is the name given to a cyst found in connexion with the head of the epididymis. Among its contents may be found a varying number of spermatozoa, and it is probably a retention cyst of one of the tubules.

The Ductus Deferens (Vas Deferens)

The deferent duct (ductus or vas deferens) is the continuation of the duct of the epididymis (8.166). Commencing at the tail of the epididymis, it is at first very tortuous, but, becoming gradually straighter, it ascends along the posterior border of the testis and the medial side of the epididymis. From the superior pole of the testis it runs upwards in the posterior part of the spermatic cord, and traverses the inguinal canal to the deep inguinal ring. Here it leaves other structures of the spermatic cord, curves round the lateral side of the inferior epigastric artery, and ascends for about 2·5 cm anterior to the external iliac artery. It is next directed backwards and slightly downwards, and, crossing the external iliac vessels obliquely, enters the lesser pelvis, where it continues posteriorly between the peritoneum and the lateral wall of the pelvis, and on the medial side of the obliterated umbilical artery, the obturator nerve and vessels, and the vesical vessels (8.159). It then crosses the ureter (8.169), and, reaching the medial side of this tube, bends at an acute angle, and runs medially and slightly forwards between the posterior surface of the bladder and the upper end of the seminal vesicle. Reaching the medial side of the seminal vesicle, it is directed downwards and medially in contact with it, and gradually approaches the opposite duct. Here it lies between the base of the bladder and the rectum, from which it is separated by the rectovesical fascia. Lastly, it passes downwards to the base of the prostate, and is joined at an acute angle by the duct of the seminal vesicle to form the ejaculatory duct (8.170). Owing to the thickness of its wall relative to the small size of its lumen, the deferent duct feels hard and cord-like when grasped by the finger and thumb. Its lumen in the greater part of its extent is of very small calibre, but posterior to the bladder it becomes dilated and tortuous, as the *ampulla*; its terminal part, which joins the duct of the seminal vesicle, is again greatly diminished in calibre (8.169).

Aberrant ductules. A narrow blind tubule, the *caudal aberrant ductule*, frequently occurs, and is connected with the caudal part of the duct of the epididymis, or with the commencement of the deferent duct. Its length, when it is uncoiled, varies from 5 to 35 cm, and it may be dilated towards its blind extremity, or may be of uniform diameter throughout. Its structure is similar to that of the deferent duct. Occasionally it is found unconnected with the epididymis. A second tubule, termed the *rostral aberrant ductule*, occurs in the head of the epididymis, and is connected with the rete testis. These aberrant ductules are derived from mesonephric tubules (p. 213).

Paradidymis. This term is applied to a small collection of convoluted tubules, situated anteriorly in the inferior region of the spermatic cord superior to the head of the epididymis. These tubes are lined with ciliated columnar epithelium, and probably represent the remains of a part of the mesonephros (p. 213).

Structure. The wall of the deferent duct has the usual layers of such tubes—external areolar, intermediate muscular, and

internal mucosal. The muscular tunic is thick and of nonstriated fibres, arranged as an external longitudinal stratum and an internal circular one. There is also an internal longitudinal layer at the commencement of the duct. All the muscle strata intermingle to a marked extent. The internal mucous tunic is arranged in longitudinal folds. The epithelium is columnar and non-ciliated through most of the duct; a variable section towards the distal end of the duct has a bilaminar epithelium of columnar cells, the superficial tier displaying non-motile stereocilia. Many of the columnar epithelial cells are secretory in nature, and for histochemical studies on their secretory activies consult Wendler (1968).

The deferent duct is supplied by a specific vessel, the artery of the ductus deferens (p. 720), usually derived from the superior vesical artery. It anastomoses with the testicular artery, and thus forms part of the vascular supply to the epididymis and testis. The anastomosis is of special interest in connexion with the toxic effects of cadmium on the mammalian testis (*see* Gunn and Gould 1975, and Johnson 1977, for literature). These effects appear to be due to interference with vascularization, but in this respect, the artery to the ductus deferens (in rats) is apparently immune.

The Seminal Vesicles and Ejaculatory Ducts

The seminal vesicles (8.169) are two sacculated and contorted tubes, placed between the posterior surface of the bladder and the rectum. Each vesicle is about 5 cm long, and is somewhat pyramidal in form, the base being directed backwards, upwards and laterally. It consists of a single tube, coiled upon itself, and giving off several irregular diverticula (8.170); the separate coils, as well as the diverticula, are connected together by fibrous tissue. The tube has a diameter of 3–4 mm and its length when uncoiled varies from 10 to 15 cm; it ends above in a cul-de-sac; its lower extremity becomes constricted into a narrow straight duct, which joins with the corresponding deferent duct to form the ejaculatory duct. The *anterior surface* is in contact with the posterior surface of the bladder, extending from near the termination of the ureter to the base of the prostate. The *posterior surface* is in relation with the rectum, from which it is separated by the rectovesical fascia. The vesicles diverge from each other superiorly, and are related to the deferent ducts and the terminations of the ureters, and are partly covered with peritoneum; each is enveloped in a dense, fibromuscular sheath. Along the medial margin of each vesicle runs the ampulla of a deferent duct. Lateral to the vesicle, the veins of the prostatic venous plexus pass posteriorly to join the internal iliac vein.

Structure. The seminal vesicles have three tunics: an external *areolar*; a middle *muscular*, thinner than that of the deferent duct and arranged in two layers, an external longitudinal and an internal circular; and an internal *mucosal* tunic, which has a reticular structure. The epithelium is columnar, and in the diverticula occur goblet cells, the secretion of which forms a large part of the seminal fluid. The secretion is lightly alkaline, and contains fructose (which supplies the energy requirements of the sperms) and a coagulating enzyme (vesiculase). The seminal vesicles do not form a reservoir for spermatozoa. The latter, passing from the testis, are stored in the epididymis (and possibly in the ampulla of the deferent duct). The vesicles contract during ejaculation and their secretion forms the bulk of the ejaculated fluid.

Electron microscopy of human seminal vesicle mucosa has revealed a second type of epithelial cell amongst the columnar cells. This is a small stellate cell, usually sited between the basal parts of the columnar cells, which contains few cytoplasmic organelles. The columnar cells carry microvilli on their luminal surfaces, contain numerous mitochondria, well-developed granular endoplasmic reticulum, and a Golgi apparatus characterized by numerous secretory vacuoles (Riva 1967).

Vessels and Nerves

The *arteries* supplying the seminal vesicles are derived from the inferior vesical and middle rectal arteries. The *veins* and *lymph vessels* accompany the arteries. The *nerves* are derived from the pelvic plexuses.

Applied Anatomy. The seminal vesicles can be palpated from the rectum. Abscesses of the seminal vesicle may rupture into the peritoneal cavity.

The ejaculatory ducts (8.159, 170), one on each side of the median plane, are formed by the union of the duct of a seminal vesicle with the terminal part of a deferent duct, and are nearly 2 cm long. They commence at the base of the prostate, run antero-inferiorly between the median and right (or left) lobes, pass along the sides of the prostatic utricle, and end on the colliculus seminalis in slit-like orifices on, or just within, the margins of the opening of the utricle (p. 1408). The ducts diminish in size, and also converge, towards their terminations.

Structure. The walls of the ejaculatory ducts are thin. They contain: an *outer fibrous layer*, which almost entirely disappears beyond the entrance of the ducts into the prostate, then non-striated myocytes consisting of a thin outer circular, and an inner longitudinal layer, a *mucous membrane* covered with columnar epithelium. For a detailed description of the ejaculatory musculature, *see* Schlager (1967).

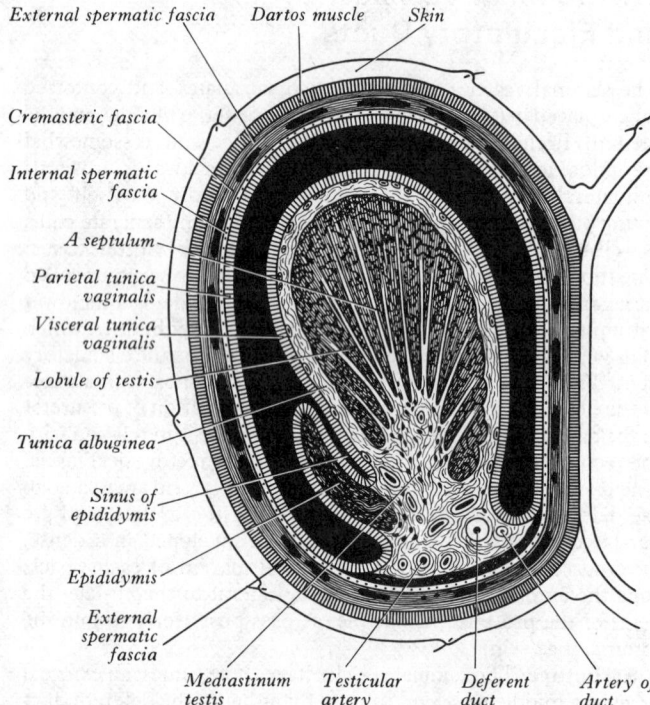

External spermatic fascia   *Dartos muscle*   *Skin*

Cremasteric fascia

Internal spermatic fascia

A septulum

Parietal tunica vaginalis

Visceral tunica vaginalis

Lobule of testis

Tunica albuginea

Sinus of epididymis

Epididymis

External spermatic fascia

Mediastinum testis   *Testicular artery*   *Deferent duct*   *Artery of duct*

8.171 A diagrammatic transverse section through the left half of the scrotum and the left testis. The tunica vaginalis is represented as artificially distended to show its visceral and parietal layers.

The Spermatic Cord and its Coverings

When the testis descends through the abdominal wall into the scrotum, it carries its vessels and nerves and the deferent duct with it. These structures meet at the deep inguinal ring and together form the *spermatic cord*, which suspends the testis in the scrotum, and extends from the deep inguinal ring to the posterior border of the testis; the left spermatic cord is a little longer than the right. Between the superficial inguinal ring and the testis the cord is anterior to the rounded tendon of the adductor longus, and is here crossed anteriorly by the superficial external pudendal artery and posteriorly by the deep external pudendal artery.

The spermatic cord traverses the inguinal canal (p. 559) having the walls of the canal as its relations, with the ilio-inguinal nerve inferior to it. In passing through the canal it acquires coverings from the layers of the abdominal wall. These coverings extend downwards into the wall of the scrotum and are named the

internal spermatic, cremasteric and external spermatic fasciae.

The *internal spermatic fascia* is a thin layer which loosely invests the spermatic cord, and is derived from the transversalis fascia (p. 559).

The *cremasteric fascia* encloses some striated muscular fasciculi, united to one another by areolar tissue; the muscular fasciculi constitute the cremaster and are continuous with the obliquus internus abdominis (p. 555).

The *external spermatic fascia* is a thin fibrous stratum continuous superiorly with the aponeurosis of the obliquus externus abdominis, and descending from the crura of the superficial ring (p. 553).

Structure of the spermatic cord. The spermatic cord is composed of arteries, veins, lymph vessels, nerves and the deferent duct, connected together by areolar tissue.

The *arteries* of the spermatic cord are the testicular (p. 718), cremasteric (p. 724) and deferential (p. 720).

The *testicular veins* are described on p. 762.

The *lymph vessels* of the testis are described on p. 797.

The *nerves* are the genital branch of the genitofemoral nerve (p. 1107), the cremasteric nerve and the testicular plexus of the sympathetic (p. 1134), joined by filaments from the pelvic plexus which accompany the deferential artery.

The Scrotum

The *scrotum* is a cutaneous and fibromuscular sac containing the testes and the lower parts of the spermatic cords, and dependent below the pubic symphysis in front of the upper parts of the thighs. It is divided on its surface into right and left halves by a cutaneous ridge, or *raphe*, which is continued ventrally to the inferior surface of the penis, and dorsally along the middle line of the perineum to the anus; the left usually descends a little more than the right, in correspondence with the greater length of the left spermatic cord. The raphe indicates the bilateral origin of the scrotum from the genital swellings (p. 220). The external appearance varies in different circumstances: thus, under the influence of warmth, and in old and debilitated persons, the scrotum is smooth, elongated and flaccid; but, under the influence of cold, and in the young and robust, it is short, corrugated, and closely applied to the testes. It consists of the skin and the dartos muscle, together with the external spermatic, cremasteric and internal spermatic fasciae, already described in connexion with the spermatic cord. The inner surface of the internal spermatic fascia is loosely attached to the parietal layer of the tunica vaginalis (8.171).

The *scrotal skin* is very thin, of a brownish colour, and often thrown into folds or rugae. It is beset with thinly scattered, crisp hairs, the roots of which are visible through the skin; it is provided with sebaceous glands, the secretion of which has a peculiar odour. It also contains numerous sweat glands, pigment cells, and nerve endings responding to mechanical stimulation of the hairs and skin, and to variations in the circumambient temperature. Subcutaneous adipose tissue is lacking.

The *dartos muscle* is a thin layer of nonstriated muscular fibres, continuous around the base of the scrotum with the superficial fascia of the groin and of the perineum. It sends inwards a sagittal *septum of the scrotum*, which connects the raphe to the inferior surface of the radix of the penis, and divides the scrotum into two cavities for the testes. The scrotal septum is composed of all the layers of the scrotal wall, except the skin, which forms one continuous investment to the entire scrotum. The dartos muscle is closely united to the skin, but is connected with the subjacent parts by delicate areolar tissue, by means of which it is able to move with great independence. However, Shafik (1977) has described a fibromuscular band, which he calls the '*scrotal ligament*', extending from the dartos sheet to the inferior pole of the testis. He regards this as part of the thermoregulatory mechanism (p. 1415).

Vessels and Nerves

The *arteries* supplying the scrotum are the external pudendal branches of the femoral artery (p. 726), the scrotal branches of the internal pudental artery (p. 722), and the cremasteric branch from the inferior epigastric artery (p. 724). The subcutaneous plexuses of the scrotal vessels are dense and carry a substantial blood flow to effect loss of heat (Esser 1932). Arteriovenous anastomoses of a simple but large-calibre type are prominent (Molyneux 1965). (See also a short discussion on temperature regulation of the testis on p. 1415.) The *veins* follow the course of the corresponding arteries. The *lymph vessels* end in the inguinal lymph nodes (p. 791). The *nerves* are the ilio-inguinal and genital branch of the genitofemoral (p. 1107), the two posterior scrotal branches of the perineal nerve (p. 1116), and the perineal branch of the posterior

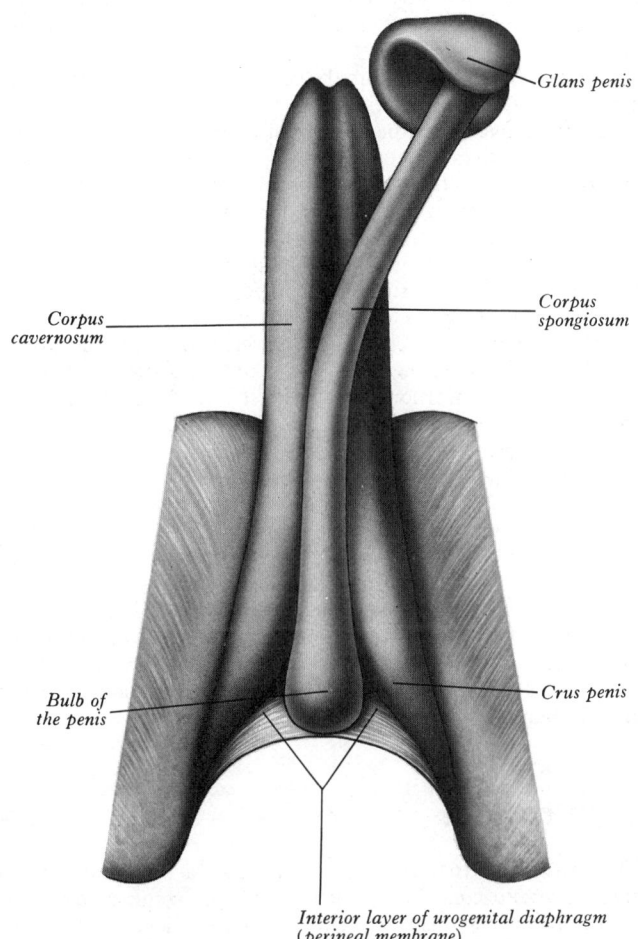

8.172 Ventral aspect of the constituent erectile masses of the penis in erect position. The glans penis and the distal part of the corpus spongiosum are shown detached from the corpora cavernosa penis and turned to the left.

femoral cutaneous nerve (p. 1111). It is to be noted that the anterior third of the scrotum is supplied mainly from the first lumbar segment of the spinal cord (through the ilio-inguinal and genitofemoral nerves), whereas the posterior two-thirds are supplied mainly from the third sacral segment (through the perineal and posterior femoral cutaneous nerves). The ventral axial line of the lower limb (p. 1117) passes between these two differently innervated parts of the scrotum. A spinal anesthetic consequently needs to be injected much higher up to anaesthetize the anterior rather than the posterior part of the scrotum.

The Penis

The penis is the male organ of copulation. It comprises an attached portion, termed the *radix* or *root*, which is situated in the perineum, and a free, normally pendulous part, termed the *corpus* or *body*, which is completely enveloped in skin.

The radix of the penis comprises the three masses of erectile

tissue which lie in the urogenital triangle of the perineum. They include the two crura and the bulb of the penis, which are firmly attached to the margins of the pubic arch and the perineal membrane respectively. The crura and bulb of the penis are, in fact, no more than the posterior regions of the corpora cavernosa and the corpus spongiosum; and the junctions between the former (in the radix penis) and the latter (in the 'free' corpus penis) are imperceptible and impossible of exact definition.

The *crus penis* (8.172) commences posteriorly as a blunt, rounded process, which is attached to bone immediately anterior to the ischial tuberosity. It is an elongated structure, closely applied and firmly adherent to the everted border of the conjoined pubic and ischial rami, and covered by the ischiocavernosus (p. 563). Anteriorly it converges towards its fellow and the median plane, presenting a slight enlargement just posterior to this point. Near the inferior border of the pubic symphysis the two crura bend sharply downwards and forwards to become continuous with the corpora cavernosa of the body of the penis.

The *bulb of the penis* (8.159, 172) occupies the space between the two crura and is firmly connected to the inferior aspect of the fascia forming the perineal membrane, from which it receives a fibrous investment. Oval in section, it narrows anteriorly to become continuous with the corpus spongiosum of the body of the penis, bending sharply downwards and forwards as it does so. Its convex superficial surface is completely overlapped by the bulbospongiosus; its flattened deep surface is pierced above its centre by the urethra, which traverses its substance to reach the corpus spongiosum. It is this part of the urethra which exhibits the intrabulbar fossa (p. 1408).

The corpus of the penis is composed of three elongated masses of erectile tissue which are capable of considerable enlargement when they are engorged with blood during erection of the organ. When flaccid, the body of the penis is cylindrical, but when erect, it approaches the form of a triangular prism with rounded angles. The surface which is posterosuperior during erection is spoken of as the *dorsum* of the penis, and the opposite aspect is the *urethral surface*. The masses of erectile tissue are termed the right and left corpora cavernosa and the corpus spongiosum penis, and it is to be emphasized again that they are the continuations of the crura and bulbus penis.

The *corpora cavernosa penis* form the greater part of the substance of the body of the penis. Throughout their length they are in close apposition with each other, being surrounded by a common fibrous envelope and separated only by a median fibrous septum. On the urethral surface the combined mass shows a wide median groove, adjoining the corpus spongiosum (8.173), and on the dorsal surface, a similar but narrower groove contains the deep dorsal vein of the penis. The two corpora cavernosa do not reach the end of the penis but terminate, within the hollow internal (proximal) aspect of the glans penis, in a rounded conical extremity, on which each forms a small projection (8.172). Proximally, each is continuous with the corresponding crus penis.

The corpora cavernosa penis are surrounded by a strong, fibrous envelope (tunica albuginea) consisting of superficial and deep strata. The superficial fibres are longitudinal, and form a single tube which encloses both corpora; the deep fibres are arranged circularly and surround each corpus separately, and form by their junction in the median plane the *septum of the penis*. This septum is thick and complete proximally, but imperfect in the more distal region of the penis, where it consists of a series of bands arranged like the teeth of a comb; it is therefore sometimes named the *pectiniform septum*.

The *corpus spongiosum penis*, which is traversed throughout its whole length by the spongiose part of the urethra, lies in the median groove on the urethral surface of the conjoined corpora cavernosa. It is cylindrical and tapers slightly towards its distal end. It is surrounded by a fibrous sheath (tunica albuginea). Near the extremity of the penis it suddenly expands to form a somewhat conical enlargement, more aptly described as being like an acorn, whence its name, the glans penis (8.172); traced towards the perineum, it becomes the bulbus penis.

The *glans penis* projects dorsally over the free extremities of the corpora cavernosa, presenting a shallow concave surface to which they are attached. The base of the glans has a projecting margin,

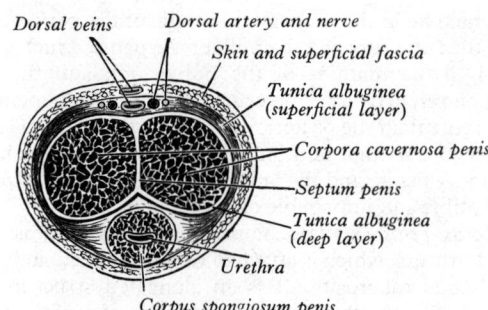

Dorsal veins   Dorsal artery and nerve

Skin and superficial fascia

Tunica albuginea
(superficial layer)

Corpora cavernosa penis

Septum penis

Tunica albuginea
(deep layer)

Urethra

Corpus spongiosum penis

8.173 A transverse section of human penis.

the *corona glandis*, which overhangs an obliquely grooved constriction, the *neck of the penis*. The navicular fossa (p. 1408) of the urethra lies within the glans and opens by a sagittal slit on or near its apex.

The skin covering the penis is remarkable for its thinness, its dark colour and its looseness of connexion with the fascial sheath of the organ. At the neck of the penis it is folded upon itself to form the *prepuce* or *foreskin*, which overlaps the glans for a variable distance. The internal layer of the prepuce is confluent along the line of the neck with the thin skin which covers and adheres firmly to the glans and is continuous with the mucous membrane of the urethra at the external urethral orifice. On the urethral surface of the glans penis a small median fold passes from the deep surface of the prepuce to a point on the glans immediately proximal to the external urethral orifice; this is the *frenulum* of the prepuce. The high degree of cutaneous sensitivity possessed by the general surface of the glans penis is even more marked near the frenulum. The prepuce is separated from the glans penis by a potential space or cleft—the *preputial sac*—which presents two shallow fossae, one on each side of the frenulum. On the corona of the glans and on the neck of the penis there are numerous small *preputial glands*; these secrete a sebaceous material, the *smegma*.

The *superficial fascia of the penis* is devoid of fat and consists of very loosely arranged areolar tissue, into which a few fibres of the dartos muscle extend from the scrotum (p. 1418). As in the suprapubic region of the abdominal wall, the deepest layer is somewhat condensed; it forms the *fascia penis*, which surrounds both the corpora cavernosa and the corpus spongiosum and separates the superficial from the deep dorsal vein. It does not extend beyond the neck of the penis, where it blends with the fibrous envelopes of the corpora cavernosa and corpus spongiosum. Proximally, it is continuous with the dartos muscle and the fascia covering the urogenital region of the perineum (p. 1418).

The weight of the body of the penis is supported by two ligaments, both continuous with the fascia of the penis and consisting very largely of elastin fibres. The *fundiform ligament* (5.45) springs from the lower part of the linea alba and splits into two lamellae which pass one on each side of the penis and unite below with the septum of the scrotum. The *suspensory ligament* is inferior or deep to the fundiform ligament. It is triangular in shape and is attached, above, to the front of the pubic symphysis. Below, it blends with the fascia penis on each side of the organ.

Structure of the penis. From the internal surface of the fibrous envelope of the corpora cavernosa penis, as well as from the sides of the septum, numerous *trabeculae* arise, and cross the corpora cavernosa in all directions, thus dividing them into a number of *cavernous spaces*, and giving the entire structure a spongy form (8.173). These trabeculae are composed of white fibrous tissue, elastin fibres and nonstriated muscular fibres, and they contain numerous arteries and nerves. The cavernous spaces are filled with blood during erection, but many are empty in the flaccid penis. They are lined with a continuous layer of flattened endothelial cells without fenestrae (Leeson and Leeson 1965).

The fibrous envelope (tunica albuginea) of the corpus spongiosum penis is thinner, whiter in colour, and more elastic than that of the corpora cavernosa penis. It is formed partly of nonstriated muscular fibres, and a layer of the same tissue surrounds the epithelium of the urethra (p. 1408).

Vessels and Nerves

The *arteries* bringing the blood to the cavernous spaces are the deep arteries of the penis (p. 722), and branches from the dorsal arteries of the penis, which perforate the fibrous capsule along the upper surface, especially near the extremity of the organ. On entering the cavernous structure the arteries divide into branches which are supported and enclosed by the trabeculae. Some of these arteries end in a capillary network, which opens directly into the cavernous spaces; others assume a tendril-like appearance, and form convoluted and somewhat dilated vessels, named *helicine arteries*. They open into the cavernous spaces, and from them small capillary branches go to supply the trabecular structure. They are most abundant in the posterior parts of the corpora cavernosa. For a detailed description *see* Alvarez-Morujo (1968).

The *veins* drain blood from the cavernous spaces by means of a series of vessels, some of which emerge from the base of the glans penis and converge on the dorsum of the penis to form the deep dorsal vein; others pass out on the upper surface of the corpora cavernosa and join the same vein; some emerge from the inferior surface of the corpora cavernosa and, receiving branches from the corpus spongiosum, wind round the side of the penis to end in the deep dorsal vein; but many pass out at the root of the penis and join the prostatic plexus. (See also p. 761.)

Erection of the penis is a purely vascular phenomenon, independent of muscular compression exerted by the ischiocavernosi and the bulbospongiosus. Rapid inflow from the helicine arteries fills the cavernous spaces and the resulting distension of the corpora cavernosa acts as a contributory factory by pressing on the veins which drain the erectile tissue.

The *lymph vessels* are described on p. 797.

The *nerves* are from the second, third and fourth sacral spinal nerves, through the pudendal nerve and the pelvic plexuses (p. 1115). On the glans and on the bulb of the penis some filaments of the cutaneous nerves have lamellated corpuscles connected with them, and many of them end in peculiar end bulbs (p. 853). Stimulation of the cutaneous receptors of the glans and frenulum of the penis is, of course, of great importance in the maintenance of the erect condition of the penis and in the initiation of orgasm and ejaculation of semen.

The Prostate

The prostate (8.164, 169, 170) is a firm, partly glandular and partly fibromuscular body, surrounding the commencement of the urethra in the male. It is situated at a low level in the lesser pelvis, behind the inferior border of the symphysis pubis and the pubic arch, and anterior to the ampulla of the rectum, through the wall of which it may be palpated. It is somewhat conical in shape, and thus presents for examination a base or vesical aspect, an apex, posterior, anterior and two inferolateral surfaces.

The *base* is for the greater part of its extent directly contiguous with the neck of the urinary bladder superior to it: the urethra enters this surface nearer to its anterior border.

The *apex* is inferior and is in contact with the fascia on the superior aspects of the sphincter urethrae and the transversus perinei profundus (p. 536).

The *posterior surface* is transversely flat and vertically convex; it is separated from the rectum by the prostatic sheath and loose connective tissue external to the sheath. Near to its superior (juxtavesical) border is a depression where the two ejaculatory ducts penetrate the prostate, dividing this surface into a superior and an inferior, larger, part. The superior part is variable in size; it is usually regarded as the external aspect of the *median lobe*. The inferior area displays a shallow, median sulcus, which is usually considered to mark a partial separation of *right and left lateral lobes*, which account for the main mass of the prostate and are really continuous posterior to the urethra. A band of fibromuscular tissue, ventral to the urethra, joins the lateral lobes together, and is often referred to as the anterior lobe. This contains fewer glandular structures than are found in the rest of the gland. This simplified view of the lobation of the prostate gland, based primarily upon the classical study of Lowsley (1912) is presented here, despite the large number of modifications introduced by subsequent observers, among whom there persist confusions and inconsistencies which have not yet been completely resolved. Some investigators deny the existence of lobes which can be distinguished by dissection; and those who favour lobation differ among themselves in their boundaries and terminology. McNeal (1975) has analysed these disagreements. Lobation has assumed an importance through the conviction among some investigators that malignant tumours are derived from particular regions of the prostate. This pathological and surgical interest prompted the study of Tisell and Salander (1975), who claim to have confirmed a recognizable lobar structure (by dissection of more than 100 human prostate glands). From their extensive observations they recognize two large *lateral lobes*, but they consider that these do not form any part of the dorsal (rectal) aspect of the prostate. They describe this dorsal region as paired *dorsal lobes*, which extend to the lateral aspects of the gland and also form all of its apex. They recognize *median lobes*, but define these as deeply placed in the prostate, extending around the urethra (except at the apex of the prostate) deep to the dorsal and lateral lobes, thus forming a deeper layer, as if patterned like an onion. Their median lobes may be equated with the *internal zone* described below. All three pairs of lobes are, they affirm, separable by dissection. At present it is difficult to reconcile the differences between this and other descriptions of prostatic lobation, and the question must be left *sub judice*. (*See*, however, Goland 1975.)

The *anterior surface* is transversely narrow and convex, and extends from apex to base. It lies about 2 cm behind the pubic symphysis, separated by a plexus of veins and some loose adipose tissue. Near its superior limit it is connected to the pubic bones by the puboprostatic ligaments. The urethra emerges from this surface a little anterosuperior to the apex of the prostate.

The *inferolateral surfaces* are prominent, and are related to the anterior parts of the levatores ani, which are, however, separated from the gland by a plexus of veins embedded in fibrous tissue which forms the lateral part of the sheath of the organ.

The prostate measures about 4 cm transversely at the base, about 2 cm in its anteroposterior, and 3 cm in its vertical diameters. Its weight is about 8 gm. It has a fibrous sheath, partly vascular and partly avascular. On each side the sheath consists of fibrous tissue in which are embedded the veins of the prostatic venous plexus (6.138). Anteriorly, it is continuous with the puboprostatic ligaments (p. 1405) and, inferiorly, it blends with the fascia on the deep surfaces of the sphincter urethrae and each transversus perinei profundus, and with the perineal body (5.56). The posterior wall of the sheath has a different constitution and is avascular. In the male fetus, at the fourth month, the rectovesical peritoneal pouch extends downwards to the pelvic floor and separates the prostate from the rectum. The lower part of this recess becomes obliterated and the fused peritoneal layers form the posterior wall of the prostatic sheath (Smith 1908). This fibrous layer has been termed the *rectovesical fascia*. Traces of its origin from two separate layers are evident, for there exists a central plane of cleavage. Above, it extends upwards over the posterior aspects of the seminal vesicles and the deferent ducts and is connected to the peritoneal floor of the rectovesical pouch (5.54). On each side, it is connected with the posterior ligament of the bladder (p. 1406) and, below, where it becomes closely adherent to the prostate, it merges into the perineal body. Others have denied such peritoneal fusion and maintain that the rectovesical fascia is formed simply by condensation of loose areolar tissue (Silver 1956). The anterior portions of the levatores ani pass posteriorly from the pubis and embrace the sides of the prostate; from the support thus afforded to the organs these parts of the muscles are named the *levatores prostatae*.

The prostate is traversed by the urethra and the ejaculatory ducts and contains the prostatic utricle. The urethra usually lies along the junction of its anterior with its middle one-third. The ejaculatory ducts pass obliquely antero-inferiorly through the posterior part of the prostate, and open into the prostatic region of the urethra (p. 1408).

Structure (8.174, 175). The prostate gland varies in colour from grey through pale pink to reddish according to its activity, and is of great density. It is enveloped in a thin but firm fibrous capsule, distinct from the sheath derived from the pelvic fascia, the latter containing a plexus of veins. The capsule is firmly adherent to the prostate and is structurally continuous with a median septum in the urethral crest which separates the lateral masses below the level of the colliculus seminalis. The capsule is also continuous with numerous fibromuscular septa which enmesh the glandular tissue.

The *muscular tissue* is mainly nonstriated (Hutch and Rambo 1970); ventral to the urethral lumen is a distinct layer of nonstriated muscle which curves to merge with the great mass of such muscle found in the fibromuscular septa of the gland. Superiorly the former is continuous with the nonstriated muscle of the bladder. However, a striated muscle, crescentic in sections transverse to the urethra, is present anterior to the nonstriated muscle ventral to the urethral lumen, and is continuous inferiorly with the sphincter urethrae in the deep perineal pouch. Its fibres lie transversely internal to the fibrous capsule, into which they are inserted laterally by diffuse collagen bundles; other collagen bundles pass posteromedially through the gland, merging with its

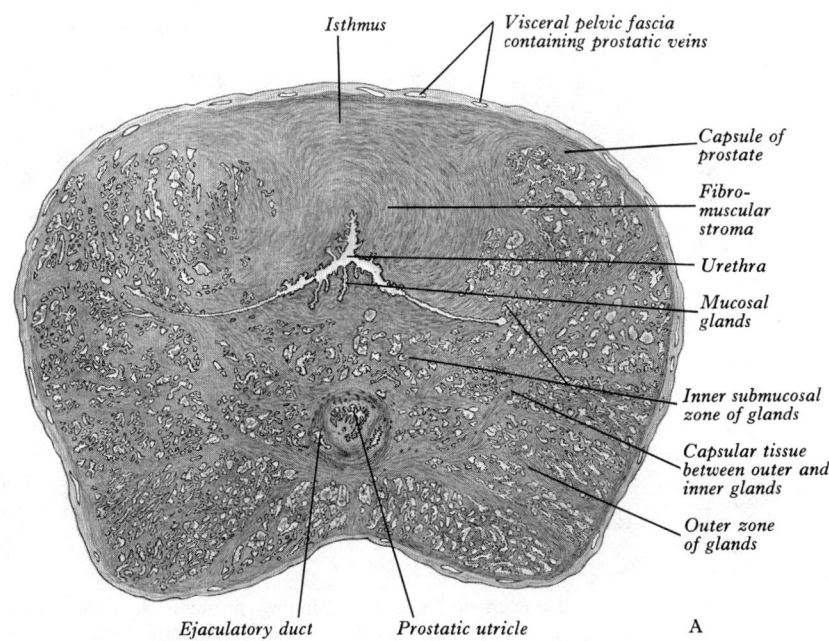

Isthmus — Visceral pelvic fascia containing prostatic veins

Capsule of prostate

Fibro-muscular stroma

Urethra

Mucosal glands

Inner submucosal zone of glands

Capsular tissue between outer and inner glands

Outer zone of glands

Ejaculatory duct   Prostatic utricle   A

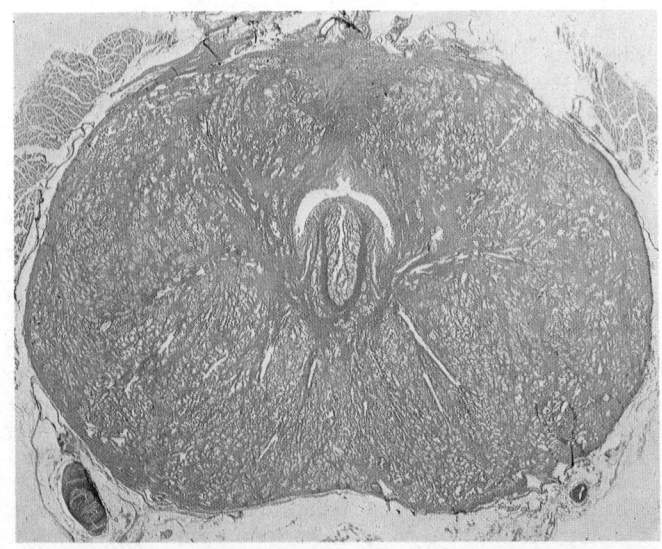

B

8.174A and B Human prostate. A Transverse section at a level to show the prostatic utricle and ejaculatory ducts. B Transverse section to show the urethral crest (stained by haematoxylin and eosin and kindly lent by Dr. L. M. Franks, Imperial Cancer Research Institute, London). Compare with A, and see text for further details. Both A and B are slightly more than twice natural size. Ventral is to the top in both figures.

fibromuscular septa and the septum of the urethral crest. This muscle, supplied by the pudendal nerve, is generally held to be a compressor urethrae (Haines 1969). From the above description however, it is suggested by Brookes (personal communication) that it is possibly a 'dilator' urethrae, pulling the urethral crest backwards and the prostatic sinuses forwards. Simultaneously, the glands are compressed and their contents expelled into the urethral lumen, now expanded into a reservoir adequate to contain the seminal fluid (3–5 ml) during the variable period of sexual excitement prior to ejaculation.

The *glandular substance* is composed of numerous follicles with frequent internal papillary elevations. The follicles open into elongated canals which join to form from twelve to twenty small excretory ducts. They are connected together by areolar tissue, supported by prolongations from the fibrous capsule and muscular stroma, and enclosed in a delicate capillary plexus. The epithelium which lines the canals and the follicles is of the columnar variety. Electron microscopic studies reveal microvilli at the luminal surfaces of the cells. There are relatively few short lengths of granular endoplasmic reticulum and some free

the peripheral zone and the 'submucosal' glands are deficient. The peripheral and internal zones are separated by an ill-defined, irregular 'capsule'. Carcinoma affects almost exclusively the former zone, while the internal zone is particularly prone to benign hypertrophy (hyperplasia); this latter growth projects into the bladder and displaces the peripheral zone postero-inferiorly, producing thereby a more distinct 'capsule' between the outer and inner zones, which provides a 'cleavage plane' during surgical enucleation of the hypertrophic growth.

Vessels and nerves. The *arteries* (Clegg 1955, 1956) supplying the prostate are derived from the internal pudendal, inferior vesical and middle rectal arteries. Its *veins* form a plexus around the sides and base of the gland (p. 761); they receive in front the deep dorsal vein of the penis, and end in the internal iliac veins. The *lymph vessels* are described on p. 797. The *nerves* are derived from the inferior hypogastric (pelvic) plexus (p. 1135).

Age Changes in the Prostate

At birth (Swyer 1944) the prostate consists of a duct system embedded in a stroma which forms a large part of the bulk of the

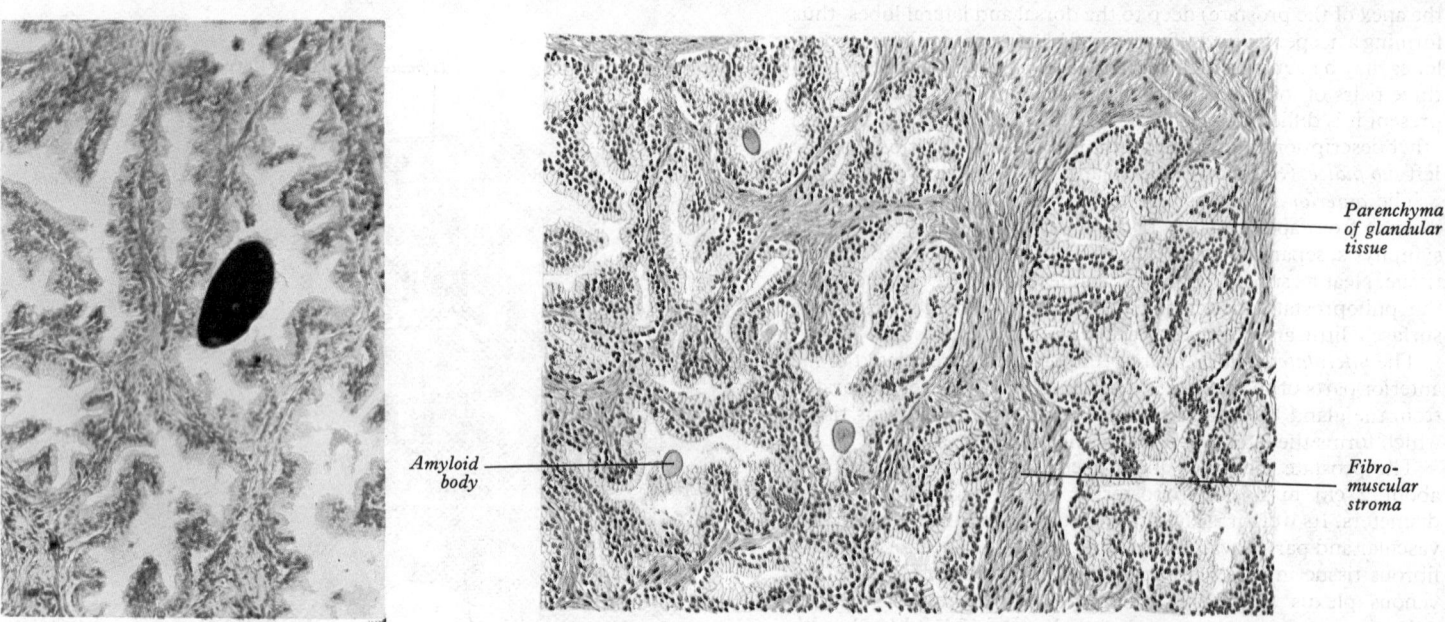

Parenchyma of glandular tissue

Amyloid body

Fibro-muscular stroma

8.175 Characteristic fields of the human prostate. Note the highly convoluted columnar epithelium, forming papillary projections into the lumen of a follicle, and abundant interfollicular fibromuscular tissue.

(Specimen on left provided by Dr. Abla el Nasser, Department of Anatomy, Guy's Hospital Medical School, from a University of London Ph.D. Thesis entitled *The Embryology of the Prostate Gland*, 1980.)

ribosomes in the cytoplasm. Ultramicroscopic vesicles, the contents of some of which give a positive reaction for acid phosphatase, are also present. The apposed borders of the cells are mainly straight and the plasma membranes are united by desmosomes (Fisher and Jeffrey 1965). The *prostatic ducts* open mainly into the prostatic sinuses in the floor of the prostatic part of the urethra, and have a bilaminar epithelium, the luminal layer consisting of columnar and the external of small cubical cells. Small colloid masses, known as *amyloid bodies*, are often found in the glandular tubes. The prostatic secretion and the secretion of the seminal vesicle together form the bulk of the seminal fluid. The prostatic secretion is slightly acid and contains acid phosphatase and fibrinolysin.

Histological sections of the prostate (8.174) do not show a lobar pattern, but two well-defined, concentric zones of glandular tissue, partially surrounding the prostatic urethra, are recognizable (Le Duc 1939; Franks 1954, Fergusson and Gibson 1956). The larger, *peripheral zone* is composed of long, branched glands, the ducts of which curve posteriorly to open mainly into the prostatic sinuses, though some open into the lateral walls of the urethra. The *internal zone* consists of a set of 'submucosal' glands, the ducts of which open into the floor of the prostatic sinuses, and partly on the colliculus seminalis, and an *innermost* group of short, simple 'mucosal' glands, which surround the upper part of the prostatic urethra. Anteriorly, in the isthmus of the prostate,

gland. Follicles are represented by small end buds on the ducts. There is a hyperplasia and squamous metaplasia of the epithelium of the ducts, the colliculus seminalis, and the prostatic utricle which are possibly due to the action of maternal oestrogens circulating in the fetal blood. These changes subside in about six or seven weeks and then the prostate undergoes little structural change until about the ninth year, when there occurs a hyperplasia of the duct epithelium and the formation of side buds leading to an elaboration of the duct system. During the period up to puberty there is a slow and continuous increase in the size of the prostate.

At puberty changes occur very rapidly over a period of about six months to one year. There is rapid increase to more than twice the size of the pre-pubertal gland, due almost entirely to the development of follicles, partly from the end buds on the ducts, and partly from modification of branches of the ducts. This change is associated with some condensation of the stroma, which becomes reduced in volume relative to the amount of glandular tissue. These changes are probably due to the secretion of testosterone into the bloodstream by the testis.

During the third decade the glandular epithelium is increased by irregular complication of the infolding of epithelium into the lumen of the follicles.

After the third decade the size remains fairly constant until the age of forty-five to fifty. Infoldings of the epithelium tend to disappear so that the outlines of the follicles are more regular and amyloid bodies increase in number. All these changes indicate the beginning of prostatic involution.

After the age of forty-five to fifty the prostate either may undergo benign hypertrophy so that its size increases gradually until death, or it may undergo progessive atrophy.

Applied Anatomy. In advanced life the prostate often becomes considerably enlarged and projects into the bladder so as to impede the passage of urine by elongating and distorting the prostatic urethra. In some cases the median lobe enlarges most, and even a small enlargement of this lobe may act injuriously, by forming a sort of valve over the internal urethral orifice, preventing the passage of urine; and the more the patient strains, the more completely will it block the opening into the urethra. The hypertrophied part of the gland projecting into the bladder may be removed by the operation of prostatectomy.

Valveless venous communications between the prostatic plexus and the extradural venous plexuses have been shown to be a normal feature, and to be probably an important factor in the metastasis of prostatic neoplasms to the vertebral bodies (Batson 1940; Franks 1953).

THE BULBO-URETHRAL GLANDS

The bulbo-urethral glands (6.97) are two small, round and somewhat lobulated bodies, of a yellow colour and about a centimetre in diameter. Each is placed lateral to the membranous urethra, superior to the perineal membrane. They lie above the bulb of the penis, and are enclosed by the fibres of the sphincter urethrae. They gradually diminish in size as age advances.

The excretory duct of each gland is nearly 3 cm long; it passes obliquely forwards external to the mucous membrane of the membranous part of the urethra, penetrates the inferior fascia of the urogenital diaphragm, and opens by a minute orifice on the floor of the spongiose part of the urethra about 2·5 cm below the inferior fascia of the urogenital diaphragm (perineal membrane).

Structure. Each gland is made up of several lobules which are held together by a fibrous investment. Each lobule consists of a number of acini of columnar epithelial cells. The secretion of the bulbo-urethral glands is an additional constituent of the seminal fluid, but the glands are very small in man as compared with many animals, and the part which they play is uncertain.

REPRODUCTIVE ORGANS OF THE FEMALE

The female's reproductive system may be considered to consist of the internal and the external genitalia. The *internal organs* are situated within the lesser pelvis, and are the ovaries, the uterine tubes, the uterus and the vagina. The *external organs* are antero-inferior to the pubic arch and they include the mons pubis, the labia majora et minora pudendi, the clitoris, the bulb of the vestibule, the greater vestibular glands, and the vestibule itself.

The Ovaries

The ovaries (8.176, 183) are homologous with the testes in the male, developing also from the genital ridge (p. 217). They are situated one on each side of the uterus close to the lateral wall of the lesser pelvis, and attached to the posterosuperior aspect of the broad ligament of the uterus, postero-inferior to the uterine tube (8.176). They are of a greyish-pink colour, and present a smooth surface before regular ovulation begins, but thereafter the surface is distorted by the cicatrization which follows degeneration of successive corpora lutea. Each ovary is almond-shaped (amyg-daloid) and is about 3 cm long, 1·5 cm wide, and about 1 cm thick. The position of the ovary is subject to a wide range of variation in

women who have borne children, because it is displaced in the first pregnancy and probably never returns to its original location. It is also variably mobile and may change its position to some degree, according to the state of surrounding organs, such as the intestines. The description given here applies to that of the nulliparous woman. In the upright position the long axis of the ovary is vertical; the gonad presents lateral and medial surfaces, tubal and uterine extremities, and mesovarian and free borders. The ovary occupies a depression, the *ovarian fossa*, on the lateral pelvic wall; this fossa is bounded anteriorly by the obliterated umbilical artery, and posteriorly are the ureter and the internal iliac artery. The *tubal* (superior) *extremity* is near the external iliac vein; to it are attached the ovarian fimbria of the uterine tube and a fold of peritoneum, the *suspensory ligament of the ovary*, which contains the ovarian vessels and nerves, and passes superiorly over the external iliac vessels (8.183) to become continuous with the peritoneum on the psoas major, posterior to the caecum or descending colon. The *uterine* (inferior) *extremity* is directed downwards towards the pelvic floor; it is usually narrower than the tubal extremity, and is attached to the lateral angle of the uterus, postero-inferior to the uterine tube, by a rounded cord termed the *ligament of the ovary*, which lies within the broad

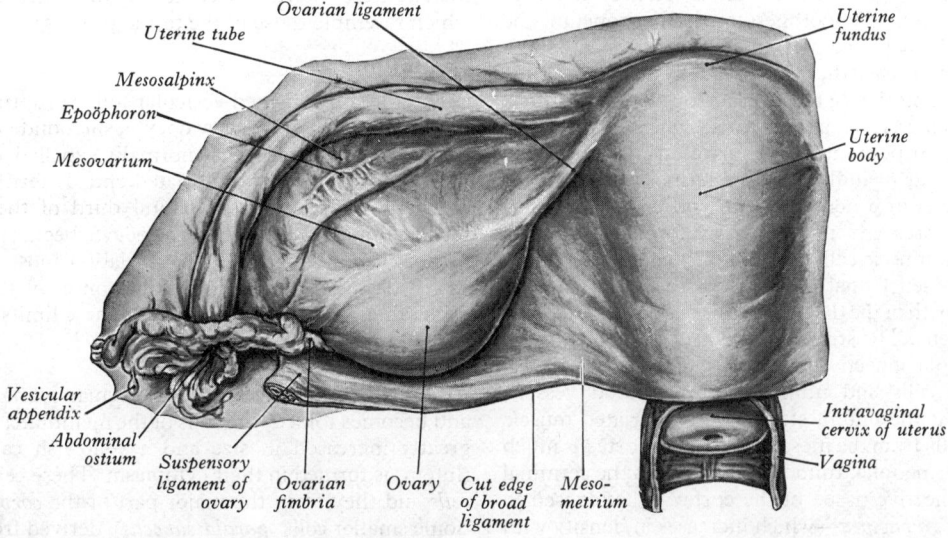

8.176 Posterosuperior aspect of the uterus and the left broad ligament. The 'ligament' has been spread out and the ovary is displaced downwards.

ligament and contains some nonstriated muscular fibres. The *lateral surface* is in contact with the parietal peritoneum which lines the ovarian fossa; it separates the ovary from the extraperitoneal tissue and the obturator vessels and nerve. The *medial surface* is to a large extent covered by the uterine tube, and the peritoneal recess between this aspect of the gland and the mesosalpinx which overlaps it is usually termed the *ovarian bursa*. The *mesovarian border* is straight and is directed towards the obliterated umbilical artery; it is attached to the back of the broad ligament by a short fold named the *mesovarium*. Between the two layers of this fold the blood vessels and nerves pass to the hilum of the ovary. The *free border* is convex, and is directed towards the ureter. The uterine tube arches over the ovary, running upwards in relation to its mesovarian border, curving over its tubal extremity, and then passing downwards on its free border and medial surface (**8**.183).

In embryonic and early fetal life the ovaries are situated, like the testes, in the lumbar region near the kidneys, but they gradually descend into the lesser pelvis (p. 385). Accessory ovaries may occur, either in the mesovarium or in the adjacent part of the broad ligament.

Structure of the Ovary

The surface of the ovary is covered, in the young female, with a layer of cuboidal cells which become flattened later in life. This so-

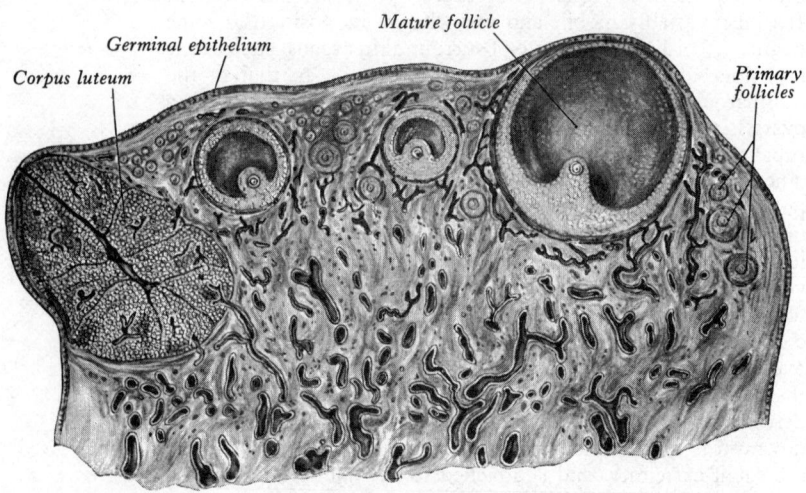

Germinal epithelium

Mature follicle

Corpus luteum

Primary follicles

8.177 Semi-diagrammatic section of an ovary.

called *germinal epithelium* gives to the ovary a dull grey colour as compared with the shining smoothness of the peritoneum; the transition between the flattened mesothelium of the peritoneum and the cuboidal cells covering the ovary is usually marked by a fine white line around the anterior, or mesovarian, border of the ovary.

The ovary, after puberty, has a thick cortex which contains the ovarian follicles and corpora lutea and surrounds, except at the hilum, a richly vascular medulla. The interstitial framework or *stroma* of the *cortex* is of a dense texture and consists of networks of reticular fibres and numerous fusiform cells, which resemble nonstriated muscle cells, though they contain no fibrils and are possibly mesenchymal elements. These stromal cells contribute to the growth of the theca folliculi (*vide infra*), and they may secrete oestrogens. The stroma of the *medulla* is of a looser texture and consists of connective tissue, with many elastin fibres, nonstriated muscle cells and numerous large blood vessels, particularly veins. At the hilum, strands of nonstriated muscle cells enter the medulla from the mesovarium. The cortex is much less vascular than the medulla. Immediately beneath the germinal epithelium, the connective tissue of the cortex is condensed to form a delicate *tunica albuginea*—which increases in density with advancing years. It is a collagenous layer, in contradistinction to the general stroma of the ovary. During prenatal life the stroma of

the cortex contains small groups of *interstitial cells*, but after puberty these cells are only present in the thecae of atretic ovarian follicles.

Ovarian follicles

At birth, the cortex of the ovary contains very numerous *primary ovarian follicles*. Each of these consists of a large central cell, the *oögonium*, surrounded by a single layer of small cuboidal or flattened elements, the *follicular cells*. The prenatal development of these follicles is described in the Embryology Section (p. 217). Many of them degenerate both during childhood and after puberty. After puberty some develop each month to form *vesicular ovarian (Graafian) follicles*, one of which usually matures and ruptures (ovulation). After puberty, and during the 'child-bearing period' of life (that is, up to the time of the menopause), the cortex contains ovarian follicles, corpora lutea in all stages of development, and atretic follicles (**8**.177).

The primary follicle develops into a vesicular follicle in the following manner. The follicular cells multiply and become many layered; a cavity (the *antrum folliculi*) containing fluid (the *liquor folliculi*) appears between the cells and separates them into two collections. The outer set forms the *membrana granulosa* and the inner set, which surrounds the ovum and attaches it to one pole of the follicle, forms the *cumulus ovaricus* (**8**.178). The stromal cells of the cortex of the ovary form a sheath (the *theca folliculi*) around the follicle and this sheath is differentiated into an inner part, the *tunica interna*, which is cellular in character and permeated by a capillary plexus, and an outer part, the *tunica externa*, which is fibrous in character. The tunica interna is separated from the membrana granulosa by a delicate basement membrane, and as the follicle is approaching its full development, the tunica interna becomes well defined and constitutes the '*thecal gland*'. The cells of the tunica interna and those of the membrana granulosa produce oestrogenic hormones (principally oestradiol), and the development of the follicle itself is stimulated by the gonadotropic hormone of the hypophysis cerebri (follicle stimulating hormone: FSH). Meanwhile the oögonium in the primary follicle transforms into a *primary oöcyte*, and the latter divides into a *secondary oöcyte* and a first polar cell. When the fully developed vesicular ovarian follicle ruptures (ovulation), it is the secondary oöcyte which is extruded, and the second polar cell is not formed unless fertilization takes place. During this maturation process the ovum also grows in size, from about 30 μm in the primordial follicle to 110 μm in the fully developed vesicular follicle. Much of this is due to the accumulation of substances to form the *yolk* or *deutoplasm*.

A fully developed follicle is 10 mm or more in diameter. As a rule, only one follicle matures and ruptures in each monthly cycle; *anovulatory* cycles may occur. However, several other primary follicles also develop to varying extents to form small vesicular follicles; but these degenerate into atretic follicles, and the cells of their tunicae internae become the interstitial cells of the ovary, which resemble those of the testis (p. 1415).

Ovulation

When the fully ripened vesicular follicle ruptures on the surface of the ovary, the secondary oöcyte, surrounded by follicular cells of the cumulus ovaricus, is normally expelled and passes into the uterine tube along its fimbriated end. If fertilization follows, it normally happens in the lateral third of the uterine tube. If fertilization does not occur, the oöcyte begins to degenerate after a short time (24–48 hours). Ovulation most frequently occurs 12–16 days before the anticipated onset of the next menstrual cycle, though it may occur outside these limits.

Corpus luteum

After ovulation, the wall of the vesicular ovarian follicle collapses and becomes folded, the cells of the membrana granulosa become greatly increased in size and a yellowish carotenoid pigment (lutein) is formed in their cytoplasm. These cells are called *luteal cells* and they form the major part of the *corpus luteum* (**8**.177). Some smaller cells (*paraluteal cells*), derived from the cells of the tunica interna, also lie on and between the superficial luteal cells. The basement membrane external to the granulosa cells vanishes, and capillaries grow in from the vessels in the tunica interna and lie between the luteal cells. A little blood clot occupies the interior of the corpus luteum. If fertilization does not occur, the corpus luteum has a functional life of about 12–14 days, after which it shows progressive degenerative changes. Such a corpus luteum is known as a *corpus luteum of menstruation*. The degenerative changes include fatty degeneration of the luteal cells and their gradual replacement by fibrous tissue so that eventually, after a period of about two months, only a small fibrous cicatrix remains (*corpus albicans*). The synergic activities of the hypophysial gonadotropic hormones—follicle stimulating hormone (FSH), luteinizing hormone (LH) and luteotropic hormone (LTH, prolactin, or lactogen)—are responsible for the conversion of the granulosal into luteal cells. The luteal cells produce the hormone progesterone, and both the luteal and paraluteal cells produce some oestradiol. The action of these ovarian hormones (progesterone and oestradiol) on the endometrium of the uterus is described in the Embryology Section (pp. 126–129).

If pregnancy occurs, implantation of the blastocyst in the

The *veins* emerge from the hilum in the form of a leash, named the *pampiniform plexus*; the ovarian vein is formed from this plexus, and leaves the lesser pelvis in company with the artery (p. 762). The *lymph vessels* are described on p. 798. The *nerves* are derived from the ovarian plexus (p. 1134). Apart from the facts that postganglionic sympathetic and parasympathetic fibres and autonomic afferent fibres innervate the ovary, little is known of their actual distribution or their function within the ovary, particularly in humans. Bulmer (1965) and Owman *et al.* (1967) have demonstrated nerve fibres in the stroma by cholinesterase and catecholamine staining techniques, amongst others. (*See* Neilson *et al.* 1970, for a review of this literature.) Balboni (1971) has described the distribution of adrenergic fibres. However, the structures innervated, apart from vessels, and the functions of such innervation remain uncertain.

Vestigial Structures

The *epoöphoron* (**8**.176) lies in the lateral part of the mesosalpinx (p. 213) between the ovary and the uterine tube, and consists of

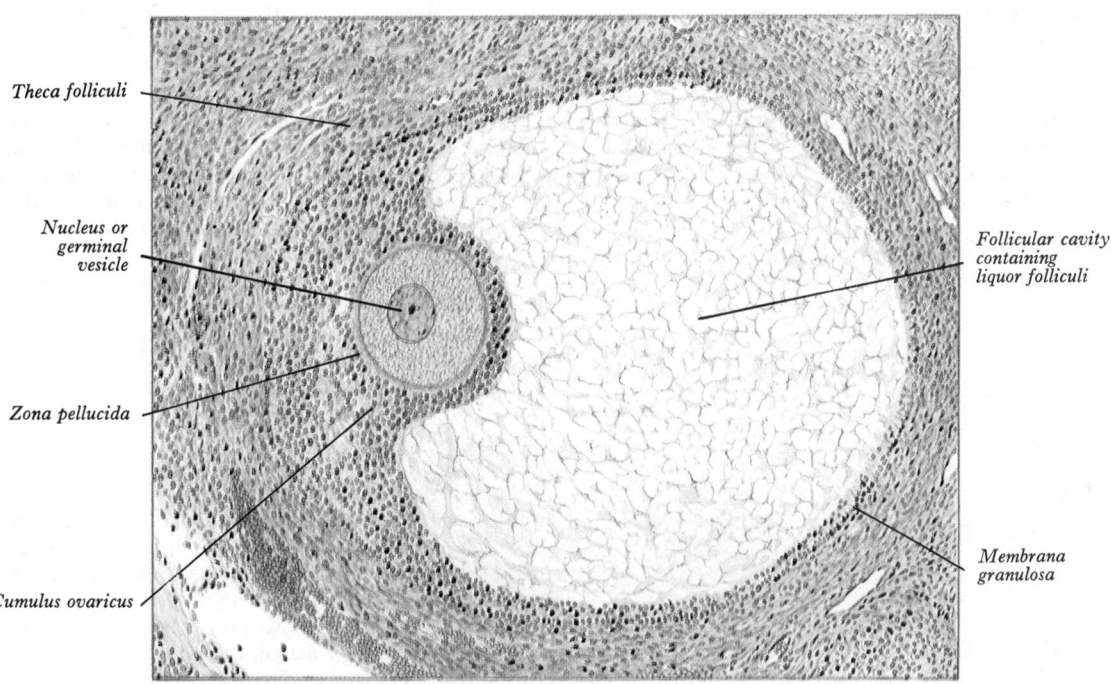

Theca folliculi

Nucleus or germinal vesicle

Zona pellucida

Cumulus ovaricus

Follicular cavity containing liquor folliculi

Membrana granulosa

8.178 Section through an ovarian follicle from a woman aged twenty-eight years. Stained with haematoxylin and eosin. Magnification about ×90).

uterine endometrium normally begins on the seventh day following fertilization, and the trophoblast produces hormones (follicle stimulating hormone, luteinizing hormone, progesterone and oestradiol). These follicle-stimulating and luteinizing hormones are known as chorionic gonadotropins and they stimulate the corpus luteum of menstruation to increase in size and prolong its activity. Such a corpus luteum is known as a *corpus luteum of pregnancy*. Whereas the corpus luteum of menstruation is active for only about 12–14 days and usually is about 1 cm in diameter, the corpus luteum of pregnancy increases in size to reach about 2·5 cm in diameter about the middle of gestation and remains active until late. By the end of pregnancy it is reduced to about 1 cm in diameter and gradually, over a period of some months, undergoes the same degenerative changes as occur in the corpus luteum of menstruation.

As the female becomes older, the ovary becomes more and more fibrotic due to the formation of corpora albicantia and, after the menopause, which occurs usually near the age of fifty years, the above changes in the ovary, involving the formation of follicles and corpora lutea, come to an end.

Vessels and nerves. The *arteries* of the ovaries and uterine tubes are the ovarian arteries from the abdominal aorta (p. 718).

ten to fifteen short tubules, the *transverse ductules* of the epoöphoron, which converge towards the ovary and end blindly, while their other ends open into a rudimentary duct, the *longitudinal duct of the epoöphoron*, which runs medially in the broad ligament of the uterus, parallel with the lateral part of the uterine tube. Frequently, between the epoöphoron and the fimbriated end of the uterine tube, one or more small cysts are present (the *vesicular appendices*).

In a small proportion of subjects the longitudinal duct of the epoöphoron (*duct of Gartner*) can be followed alongside the uterus to near the level of the internal os. Here it pierces the muscular wall of the uterus and descends in the wall of the cervix uteri, gradually approaching the mucous membrane, without, however, quite reaching it. The duct then runs downwards in the lateral wall of the vagina and ends at, or close to, the free margin of the hymen.

The *paroöphoron* consists of a few scattered rudimentary tubules, best seen in the child, situated in the broad ligament between the epoöphoron and the uterus.

The tubules of the epoöphoron and of the paroöphoron are remnants of mesonephric tubules; the duct of the epoöphoron is a persistent portion of the mesonephric duct (p. 213).

The Uterine Tubes

The two **uterine (Fallopian) tubes** (8.176, 182) transmit ova from the ovaries to the cavity of the uterus, and are situated in the upper margins of the broad ligaments of the uterus. Each tube is about 10 cm long, its medial end opening into the superior angle of the cavity of the uterus, the lateral end into the peritoneal cavity close to the ovary. The uterine ostium or opening is very small, and admits only a fine bristle; the opening into the peritoneal cavity is named the *abdominal ostium*, and when its muscular wall is relaxed has a diameter of about 3 mm. The abdominal opening is situated at the bottom of a trumpet-shaped expansion of the uterine tube, the *infundibulum*, the circumference of which is prolonged by a varying number of irregular processes, called *fimbriae*, and therefore this extremity of the tube is sometimes called the *fimbriated end*. The inner surfaces of the fimbriae are lined by mucous membrane, and in the larger fimbriae this exhibits longitudinal folds which are continuous with similar folds in the mucous lining of the infundibulum. One fimbria, longer and more deeply grooved than the others, is closely applied to the tubal extremity of the ovary, and is named the *ovarian fimbria*. The infundibulum opens into the *ampulla* of the tube, which is thin-walled and tortuous and forms rather more than one-half of its entire length. The ampulla is succeeded by the *isthmus*, which is round and cord-like and constitutes approximately the medial one-third. The part continued from the isthmus through the wall of the uterus is about 1 cm long, and is named the *uterine* (intramural) *part* of the tube.

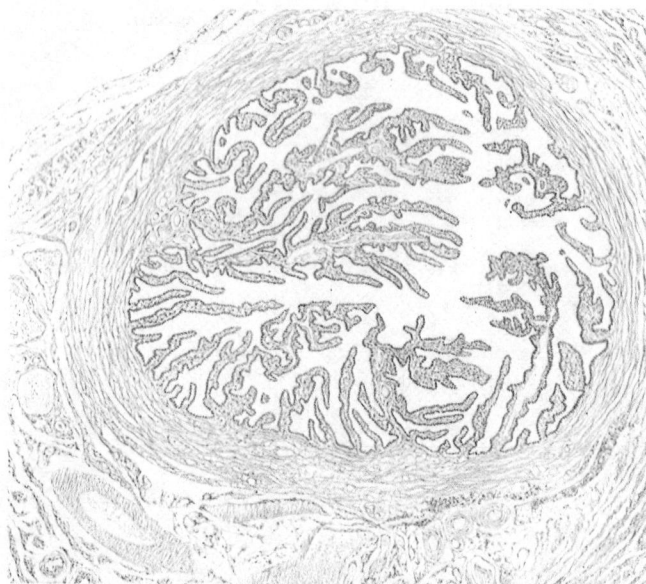

8.179   Transverse section through the ampulla of a human uterine tube. Stained with haematoxylin and eosin. Magnification about × 15.

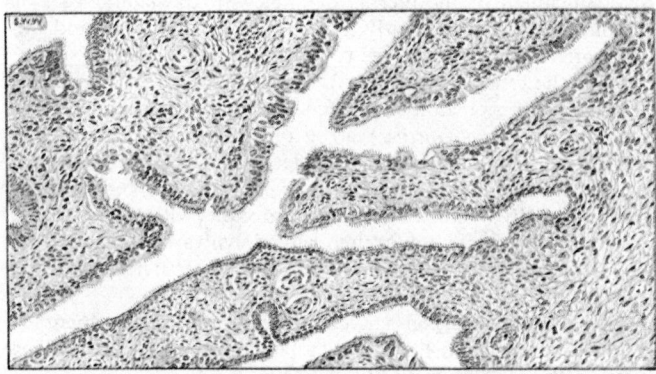

8.180   Section showing the plicated mucous membrane in the ampullary part of human uterine tube. Note the columnar epithelium. Stained with haematoxylin and eosin. Magnification about × 75.

The uterine tube extends laterally as far as the medial (uterine) pole of the ovary, and then ascends along the mesovarian border to the tubal pole, over which it arches; finally it turns downwards and ends in relation to the free border and medial surface of the ovary. In connexion with the fimbriae of the uterine tube, or with the broad ligament close to them, there are frequently one or more small pedunculated vesicles; these are termed the *vesicular appendices* (see above and **8**.176).

Structure (**8**.179, 180). The uterine tube resembles other hollow viscera in the main components of its wall, which consists of an external *serosal* layer, an intermediate stratum of *muscle*, and an internal *mucosal* layer. The *serosa* is a layer of peritoneum with subjacent areolar tissue. The *muscular layer* consists of external longitudinal and internal circular strata of nonstriated muscle; in addition an internal longitudinal group of fibres appears in some parts of the tube. The uterine or intramural region has attracted much attention (Lisa *et al.* 1954; Sweeney 1962), because of the idea that a sphincter might exist to cut off the uterine cavity from the tube and peritoneum as a guard against infection. The two extensive studies quoted revealed no intramural sphincter, but the occurrence of internal longitudinal fibres was confirmed. The isthmus, however, has the thickest muscular layer of any part of the uterine tube, there being a most marked reduction in this as the isthmus continues into the ampulla. The thickness of the muscle here largely affects the circular stratum; and it is here that the lumen of the tube is usually smallest, varying according to different observations from 0·1 to 1·0 mm. The mucosa is less complexly folded in the isthmus, there usually being three to six primary folds. There is considerable evidence that this stretch of the tube acts as a sphincteric mechanism which delays the progress of the segmenting zygote towards the uterus, perhaps to ensure that the former has reached a sufficient stage of development to implant in the endometrium. (The interesting suggestion has been made that a pharmacological agent capable of overruling this sphincteric activity might be valuable as a post-coital contraceptive—Mastroianni 1962.) In the ampullary part of the uterine tube the internal longitudinal muscle is absent, and the external longitudinal and circular fibres are considerably intermingled. The musculature of the infundibulum is similar in arrangement, but attempts to explain movements to appose the fimbriae to the ovary have entailed a variety of descriptions, none of which are as yet convincing. A sphincter at the abdominal ostium of the tube has been described (Whitelaw 1933; Woodruff and Pauerstein 1969), on the basis of experimental distension, but no histological confirmation of this has been clearly established.

The *mucous membrane* of the uterine tube consists of an epithelium and an underlying connective tissue stratum containing the mucosal blood and lymph vessels and nerve fibres. The epithelium is classically described as columnar and ciliated, but at least three kinds of cells can be identified, of which only one is ciliated, the others being secretory and 'intercalary'. At their basal aspects these cells adjoin an incomplete basement membrane. The intercalary cells, and another undifferentiated type with a darkly staining nucleus (Pauerstein and Woodruff 1967), may be precursors of the secretory cells, or exhausted remnants of them. The ultrastructural details of the ciliated cells present no peculiarities (pp. 18–20); the secretory cells show variations during the menstrual cycle, observable by both light (Snyder 1924) and electron (Hashimoto *et al.* 1960) microscopy. Briefly, these indicate increased secretory preparation during the follicular phase, and augmented secretory discharge during the luteal phase of the menstrual cycle.

The epithelium of the uterine tube is invaginated into the lumen of the tube in a series of major plicae or folds, each of which displays secondary and even tertiary projections. Thus, in transverse sections (**8**.180) the lumen is extensively invaded by plicae, each with its core of connective tissue, capillaries and supporting vessels. The functional role of the tube is, of course, the transport of ova and spermatozoa, and their joint product, the zygote, and the pre-implantation morula and blastocyst (p. 103) into which it develops. Moreover, both spermatozoa and ova, but especially the former, must be maintained in a viable state prior to fertilization; and similarly, the developing blastocyst, although perhaps more dependent upon its own nutritional resources, must

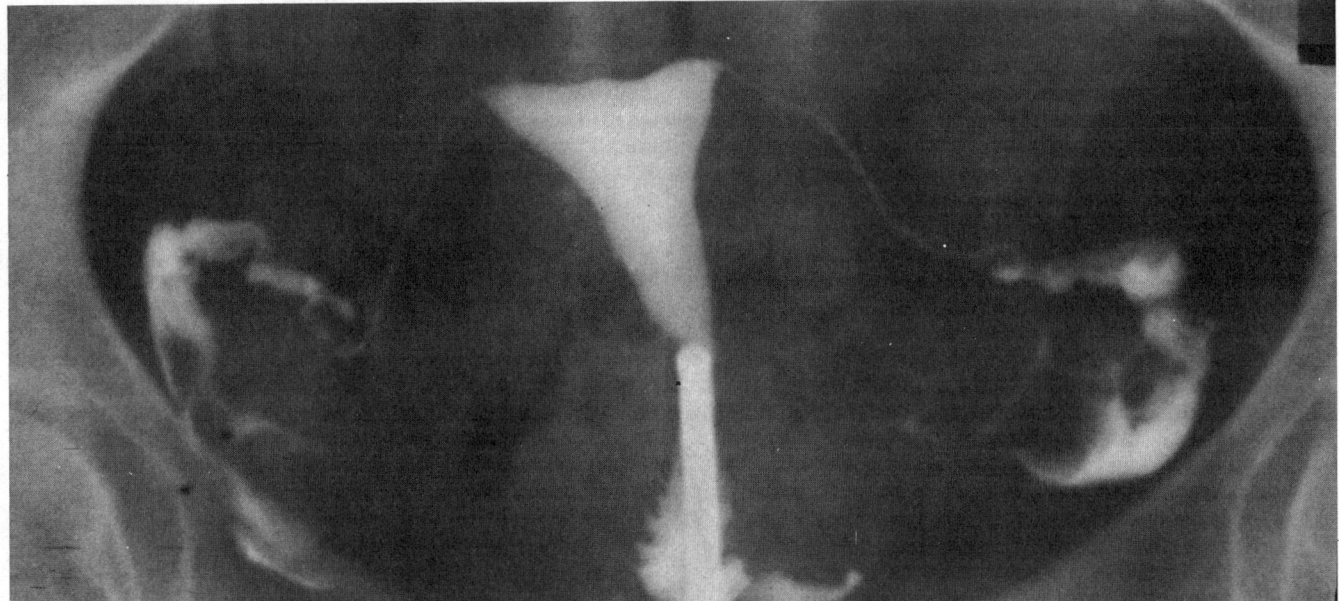

8.181   A radiograph of the uterus and uterine tubes after the introduction of a radio-opaque contrast medium through a cannula passed through the vagina into the cervix uteri. The shadow of the cannula is visible; above this is the triangular cavity of the uterine body and fundus. From the superior angles the lumina of the narrow (less than 1·0 mm diameter) intramural and isthmic parts of the uterine tubes may be traced inferolaterally, where they expand into the wider (2–4 mm diameter) ampullary parts of the tubes. Some contrast medium has escaped into the pelvic cavity from the abdominal ostia. (Radiograph kindly provided by Dr. J. Hilliger Smitham, Chelsea Hospital for Women, London.)

be maintained in a fluid environment entirely adapted to its needs. A great amount of research has been applied to the dynamics of these activities, details of which are beyond the scope of this text. No explanations are yet forthcoming as to why such large numbers of spermatozoa are required, or why relatively so few actually reach the ampulla, where they await the ovum and fertilize it. The fluid in the tube acts as an intermediary for respiratory gases in the case of spermatozoa, whose demands are high, and doubtless for its other passengers. Capacitation (p. 97) also occurs in the tube. The details of the entry of ova into the tube are still uncertain; direct conveyance by fimbriae, suction due to ciliary action, and so on, have been suggested, but convincing evidence is scant. The highly plicated nature of the tubal epithelium, the isthmic sphincter, and other details, obviously accord with the necessities of interchange of gases, nutrients and hormones, and with the delay necessary to ensure blastocyst development for implantation; but the immediate mechanisms are by no means clear. The contractile patterns, and the effects on them of nerve supplies, hormones (oestrogens and progesterone), and transmitter substances, have been widely investigated in experimental animals; details must be sought in appropriate papers and monographs.

The arrest of the zygote at the isthmo-ampullary junction, the so-called 'isthmic block', has attracted special attention. It seems probable that the isthmic sphincter is adrenergic, like many other sphincters in other organs. Histochemical and experimental findings suggest that there is no cholinergic control in some parts, at least, of the human uterine tube (Nakanishi *et al.* 1967). Prostaglandins (pp. 97, 1397), which occur in the plasma of human semen, stimulate the muscle in the proximal part of the tube and inhibit it in the more distal (ampullary) region. All these studies are continuing, and doubtless more precise information will soon become available.

Vessels and nerves.   The arterial and venous vessels of the uterine tube are basically derived from the *ovarian* and *uterine* stems. Some disagreements as to how much is supplied by the two arteries are apparent in the reports and results of different observers, and it has even been claimed that the intramural part of the tube is supplied by the ovarian artery. There is, however, convincing arteriographic evidence (Borell and Fernstrom 1953) that the uterine artery usually supplies about the medial two-thirds of the tube, the ovarian artery the remainder, though the partition between the two sources (which in any case anastomose) is somewhat variable. (For a detailed study of the arteries of the

uterine tube consult Koritke *et al.* 1967.) The *veins* are arranged similarly; intrinsic networks in the mucosa, muscular layer and subserosa have been described (Gatsalov 1963; Koritké *et al.* 1967). The *lymphatic drainage* follows the veins (p. 798); its intrinsic details have been extensively described (Sampson 1937; Gatsalov 1963). The *nerve supply*, which, of course, largely reaches the uterine tube along the ovarian and uterine arteries, shows a correspondingly similar pattern of distribution. Most parts of the tube appear to receive both sympathetic and parasympathetic supplies; in the case of the latter, vagal fibres reach the lateral half of the tube, pelvic splanchnic fibres the medial moiety. The sympathetic supply is from the tenth thoracic to second lumbar segments of the spinal cord. The afferent fibres travel along the same path as the sympathetic motor supply and enter the cord through the eleventh thoracic to second lumbar dorsal nerve roots. Modified Pacinian corpuscles have been noted in the submucous connective tissue of the ampulla of the tube (Chiara 1959). The details of the tubal intrinsic innervation have been studied by metallic impregnations (Chiara 1959; Damiani *et al.* 1961), and more recently by fluorescence microscopy (Owman and Capodacqua 1967; Kubo 1970). The sympathetic nerves associated with the uterine tube are composed of postganglionic fibres, as well as the visceral afferents referred to above. The parasympathetic supply must form synapses, but ganglia have not been reliably reported in the wall or vicinity of the tube; the locus of relay may be in the paracervical ganglia. It is said that afferent autonomic fibres also accompany the parasympathetic nerves of the uterine tube.

Applied Anatomy.   Pelvic peritonitis is said to occur more frequently in the female than in the male, because infective conditions of the vagina, uterus or uterine tube may involve the peritoneum by direct spread, owing to the communication which exists between the peritoneal cavity and the lumen of the tube through the abdominal ostium. Pus in the recto-uterine pouch may be palpated through the posterior fornix of the vagina.

Tubal inflammation (*salpingitis*) is usually due to infections which have spread upwards by way of the vagina and uterus. In some cases the fimbriated end of the tube may become closed by adhesions and a collection of pus forms in the tube (*pyosalpinx*).

Fertilization of the ovum (p. 100) usually occurs in the ampulla of the uterine tube, and normally the segmenting zygote is then passed on into the uterus; however, it may adhere to and undergo development in the tube, giving rise to the commonest variety of *ectopic gestation* (p. 110). In such cases the amnion and chorion are formed, but a true decidua never develops; the gestation usually ends by extrusion of the ovum through the abdominal ostium, although it is not uncommon for the tube to rupture into the peritoneal cavity, this being accompanied by severe haemorrhage, necessitating surgical interference. For further information *see* Pauerstein and Woodruff 1967; Woodruff and Pauerstein 1969.

The Uterus

The uterus (**8**.176, 182, 183) is a hollow, thick-walled, muscular organ normally located entirely in the lesser pelvis between the urinary bladder in front and the rectum behind. Into its upper part the uterine tubes open, one on each side, while below, its cavity communicates with that of the vagina. When the ova are discharged from the ovaries they are carried to the uterine cavity through the uterine tubes. If an ovum be fertilized it embeds itself in the uterine wall and is normally retained in the uterus until prenatal development is completed, the uterus undergoing changes in size and structure to accommodate itself to the needs of the growing embryo and fetus. After parturition the uterus returns almost to its former condition, though it is somewhat larger than in the nulliparous state.

In the *adult nulliparous state* the uterus is flattened from before backwards and is piriform, with the narrow end being postero-inferior. It lies between the bladder antero-inferiorly, and the sigmoid colon and rectum posterosuperiorly, and is completely below the level of the pelvic inlet.

The long axis of the uterus usually lies approximately in the axis

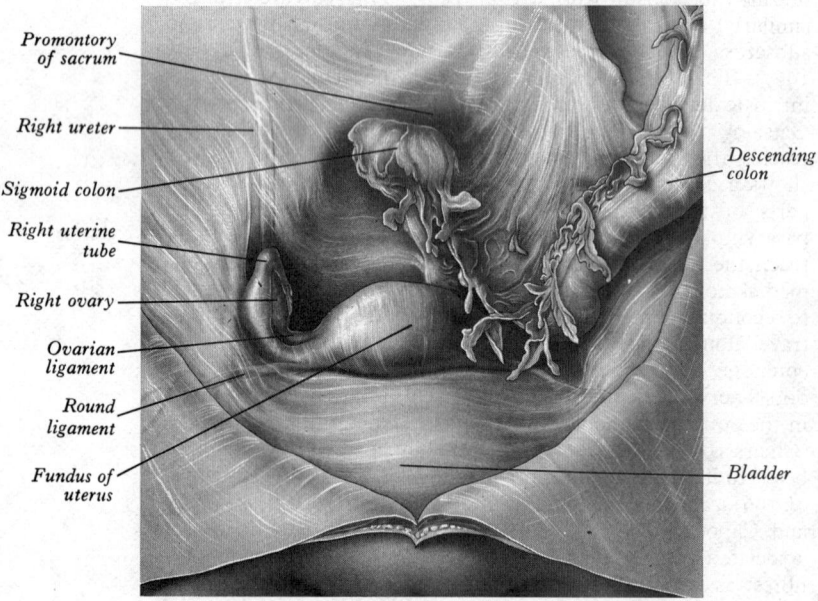

Promontory of sacrum
Right ureter
Sigmoid colon
Right uterine tube
Right ovary
Ovarian ligament
Round ligament
Fundus of uterus
Descending colon
Bladder

8.182 A female pelvis and its contents. Anterosuperior view.

of the pelvic *inlet* (p. 384), but as the organ is freely movable its position necessarily varies with the state of distension of the bladder and rectum. Except when much displaced by a distended bladder, it forms almost a right angle with the vagina, since the axis of the vagina corresponds to the axes of the cavity and *outlet* of the lesser pelvis (p. 385).

The uterus measures about 7·5 cm in length, 5 cm in breadth at its widest, superior level, and nearly 2·5 cm in thickness; it weighs from 30 to 40 gm. It is divisible into two regions. On the surface, a little below the middle, there is a slight constriction, which corresponds to a narrowing of the uterine cavity, named the *internal os* of the uterus. The part above the internal os is termed the *corpus* or *body*, and that below, the *cervix*. The part of the body which lies above a plane passing through the points of entrance of the uterine tubes is known as the *fundus* (despite the fact that it is the *highest* part of the uterus. It is, however, the *deepest* region when approached through the cervix.)

The corpus uteri. The body gradually narrows from the

fundus to the internal os. The *vesical*, or *anterior*, *surface* is in apposition with the urinary bladder. It is flattened and covered with peritoneum, which is reflected on to the bladder as the uterovesical fold at the level of the internal os. The recess between the bladder and the uterus is named the *vesico-uterine pouch*. It is usually empty, but may be occupied by a part of the intestine (p. 1323).

The *intestinal*, or *posterior*, *surface* is convex transversely, and is covered with peritoneum, which is continued downwards on the cervix uteri and the upper part of the vagina before being reflected backwards on to the rectum (**8**.183). It is related to the sigmoid colon, from which it is usually separated by the terminal coil of the ileum.

The *fundus* is convex like a dome, and covered with peritoneum continuous with that on the vesical and intestinal surfaces. Some coils of small intestine, and occasionally the distended sigmoid colon, are in contact with it.

The *lateral margins* or sides of the uterus are transversely convex. At the upper end of each the uterine tube passes through the uterine wall. Antero-inferior to this the round ligament of the uterus is attached; postero-inferior to it the ligament of the ovary is attached. These three structures lie within the broad ligament, which stretches from the margin of the uterus to the lateral wall of the pelvis (p. 1432).

The cervix uteri is about 2·5 cm in length; it is narrower and hence more cylindrical than the body, and is a little wider in the middle than above or below. Owing to its relationships it is less freely movable than the body, so that its long axis and that of the body are seldom in the same straight line. The long axis of the uterus as a whole presents the form of a curved line with its concavity forward, and the organ is described as being *anteflexed*. In extreme cases there may be an angular bend at the region of the internal os—acute anteflexion. When the bladder is empty the long axis of the cervix meets the long axis of the vagina at an angle which faces antero-inferiorly, and the whole uterus is therefore turned anteriorly on the vagina, or *anteverted*.

The cervix projects through the anterior wall of the vagina, which divides it into upper, supravaginal and lower, vaginal regions (**8**.183).

The *supravaginal part* of the cervix is separated *in front* from the bladder by cellular connective tissue, the *parametrium*, which extends also on to the sides of the cervix, and laterally between the layers of the broad ligaments. The uterine arteries reach the lateral aspects of the cervix in this tissue, whilst on each side the ureter runs downwards and forwards in it at a distance of about 2 cm from the cervix (*see* p. 1402). The relationship of the arteries and the ureters is not always symmetrical, and in particular one or other of the ureters may be somewhat anterior to the cervix. *Posteriorly* the supravaginal cervix is covered with peritoneum, which is prolonged below on to the posterior vaginal wall, whence it is reflected to the rectum, forming the *recto-uterine recess* (**8**.183). It is in relation with the rectum, from which it may be separated by the terminal coil of the ileum.

The *vaginal part* of the cervix projects into the anterior wall of the vagina forming the vaginal fornices (p. 1433). On its projecting rounded extremity there is a small, depressed, circular aperture, the *external os of the uterus*, through which the cavity of the cervix communicates with that of the vagina. In women who have borne children the external os is bounded by two lips, anterior and posterior, of which the anterior is the shorter and thicker, although, because of the slope of the cervix, it projects lower than the posterior. Normally both lips are in contact with the posterior vaginal wall.

The cavity of the uterus (**8**.181, 184) is small in comparison with the size of the organ, partly due to its thick wall.

The *cavity of the body* is a mere slit in sagittal section because the anterior and posterior walls are almost in contact. In coronal section it is seen to be triangular in shape, the base being formed by the internal surface of the fundus between the orifices of the uterine tubes, the apex by the internal os of the uterus; through this orifice the cavity of the body communicates with the canal of the cervix (**8**.183).

The *canal of the cervix* is somewhat fusiform, flattened from before backwards, and broader in the middle than at the ends. It communicates above, through the *internal os*, with the cavity of the body, and below, through the *external os*, with the vaginal cavity. A longitudinal ridge is a feature of both the anterior and posterior walls of the canal and from each a number of small oblique *palmate folds* ascend laterally, giving the appearance of branches from the stem of a tree (*arbor vitae uteri*). The folds on the two walls are not opposed, but fit between one another so as to close the cervical canal.

The total length of the uterine cavity from the external os to the fundus is about 6 cm.

Approximately the upper third of the cervix has been termed the *isthmus*, because it presents certain features which differentiate it from the rest (Stieve 1927; Frankl 1933). Although it is

The fundus is just below the level of the superior pelvic aperture. The palmate folds are distinct, and extend to the upper part of the uterine cavity.

In the adult the position of the uterus is liable to considerable variation, depending chiefly on the condition of the bladder and rectum. When the bladder is empty the entire uterus is inclined anteriorly, and is at the same time curved at the junction of body and cervix, so that the body lies upon the bladder. As the latter fills, the uterus gradually becomes more and more erect, until with a fully distended bladder the fundus may be inclined towards the sacrum.

During menstruation the organ is slightly enlarged, and more vascular, and its surfaces are rounder; the external os is rounded,

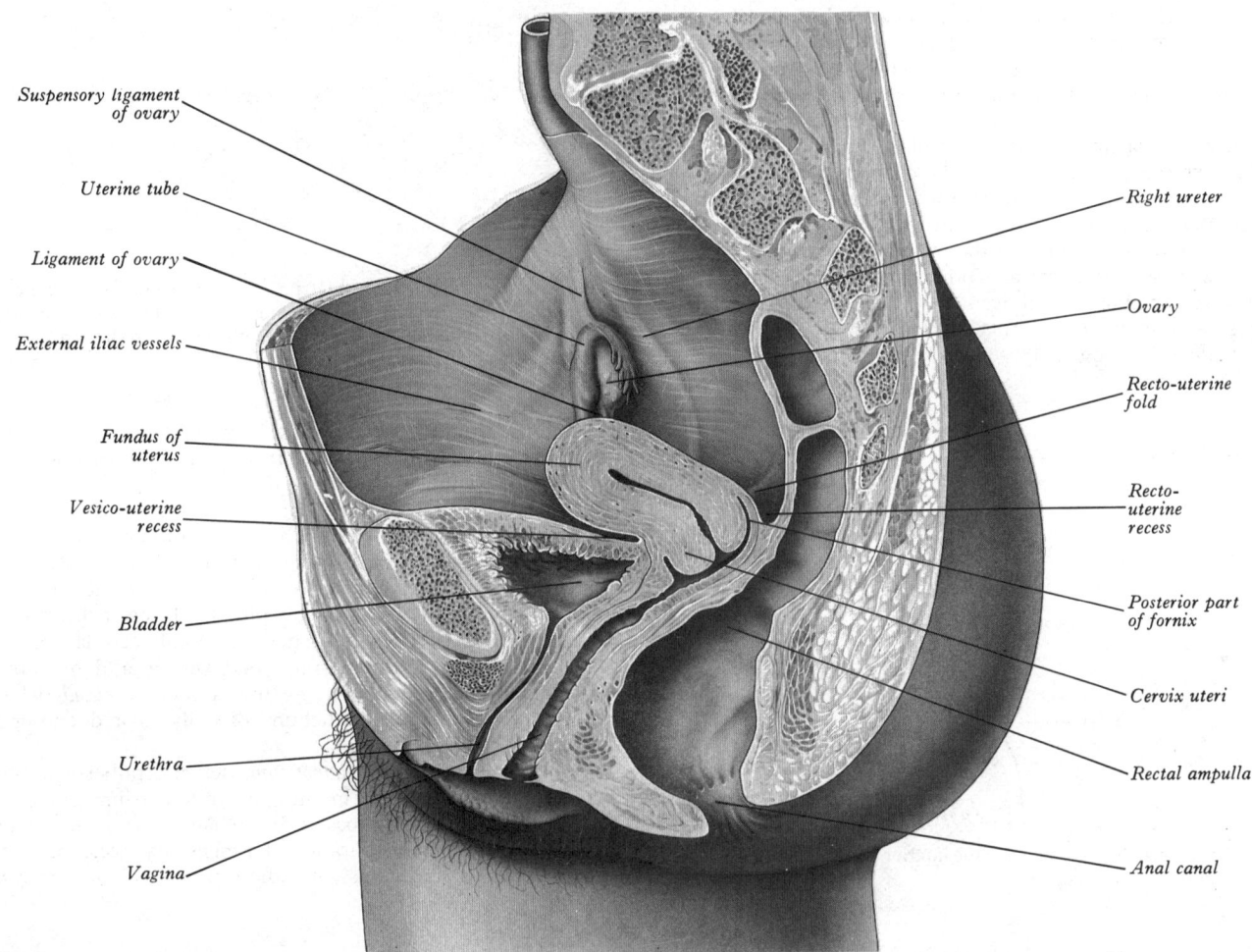

8.183 A median sagittal section through a human female pelvis. The peritoneum is shown in blue.

unaffected in the first month of pregnancy, it is gradually taken up into the body of the uterus during the second month and forms the 'lower uterine segment' of British obstetricians. The fetal membranes, though firmly blended with the rest of the uterine mucosa, are not attached to the lower uterine segment. In the non-pregnant uterus the isthmus undergoes changes associated with menstruation similar to, but less pronounced than those which occur in the body of the organ. Histologically the isthmus resembles the body more than it resembles the cervix; its lining epithelium is low cylindrical in type and is ciliated; its mucous coat is thinner and the glands are fewer in number than they are in the cervix.

Age and reproductive changes. The form, size and situation of the uterus vary at different periods of life and in different circumstances.

In fetal life the uterus projects above the superior aperture of the lesser pelvis. The cervix is considerably larger than the body.

At puberty the uterus is piriform, and weighs from 14 to 17 gm.

its lips swollen, and the lining membrane of the body is a darker colour.

During pregnancy the uterus becomes enormously enlarged, and in the eighth month reaches the epigastric region. The increase in size is mainly due to hypertrophy of pre-existing muscular fibres, and partly to development of new fibres. As pregnancy proceeds, the wall of the uterus becomes progressively thinner.

After parturition the uterus nearly regains its usual size, weighing about 42 gm; but its cavity is larger than in the nulliparous state, its vessels are tortuous, and its muscular layers are a little thicker and more defined; the external os is more prominent, and its edges present one or more fissures.

In old age the uterus becomes atrophied, paler, and denser in texture; a more distinct constriction separates the body and cervix. The internal os is frequently, and the external os occasionally, obliterated, whilst the lips of the external os almost entirely disappear.

Structure. The uterine wall consists of three tissue strata, an external serous, a middle muscular, and an internal mucous layer.

The *serous layer (perimetrium)* is the peritoneum which posteriorly covers the body and supravaginal part of the cervix, but in front covers the body only. In the lower quarter of the posterior surface the peritoneum is not closely connected with the uterus, being separated from it by a layer of loose cellular tissue and some large veins. Beneath the serous covering is a subserous layer of areolar tissue.

The *muscular layer (or myometrium)* forms most of the substance of the uterus. In nulliparae it is dense, firm, of a greyish colour, and (in the fixed state) cuts almost like cartilage. It is about 1·25 cm thick at the middle of the body and fundus, and thin at the orifices of the uterine tubes. It consists of bundles of nonstriated muscular fibres, intermixed with areolar tissue, blood vessels, lymph vessels and nerves. During pregnancy the muscular tissue becomes more prominently developed, the fibres being greatly enlarged. Although the muscular fibres interlace in all directions, they are arranged in three more or less distinct layers: external, middle and internal. The muscle of the cervix contains more collagen and elastin fibres than that of the body.

The external layer consists chiefly of longitudinal fibres, which pass over the fundus and converge at the lateral angle on each side of the uterus, to be continued on the uterine tube, the round ligament and the ligament of the ovary; some pass on each side into the broad ligament, and others run backwards from the cervix into the uterosacral ligaments. The intermediate layer of fibres is the thickest, but presents no regularity in its arrangement, being disposed longitudinally, obliquely and

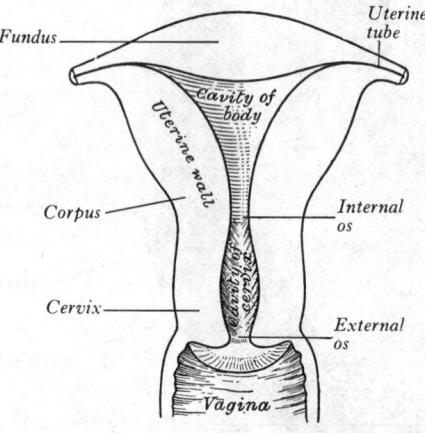

8.184 Sectional diagram showing the interior divisions of the uterus and its continuity with the vagina.

transversely; it contains the larger blood vessels. The internal layer consists of longitudinal and circular fibres. The deep ends of the uterine glands come into close relation with the fibres of the internal layer. An infundibular arrangement of elastin and nonstriated muscle fibres has been described in the cervix, with its narrow extremity adjacent to the external os (Hamperl 1970). It is considered that this may be able to exert a valvular effect in keeping the os normally closed. Toth (1977) has described bilateral longitudinal fascicles of nonstriated muscle embedded in the wall of the uterus, one on each side in the lateral region and extending from the fundal angle to the cervix. They are said to be largely submucosal. Each fascicle consists, in its juxtafundal part, of a number of somewhat dispersed sub-fascicles; but the bundle is more compact as it approaches the cervix. In places, epithelial rests were observed in the axis of the bundle, and it is suggested that the origin of the structure may be mesonephric. The muscle cells of the bundle ('*fasciculus cervico-angularis*') are structurally different from those of the myometrium proper, and this has prompted the speculation that the fasciculus may function as a relatively fast conducting pathway in the integration of uterine muscular activity. Confirmation and expansion of these findings will be awaited with interest.

The *mucous membrane* or *endometrium* (8.185) of the uterus is

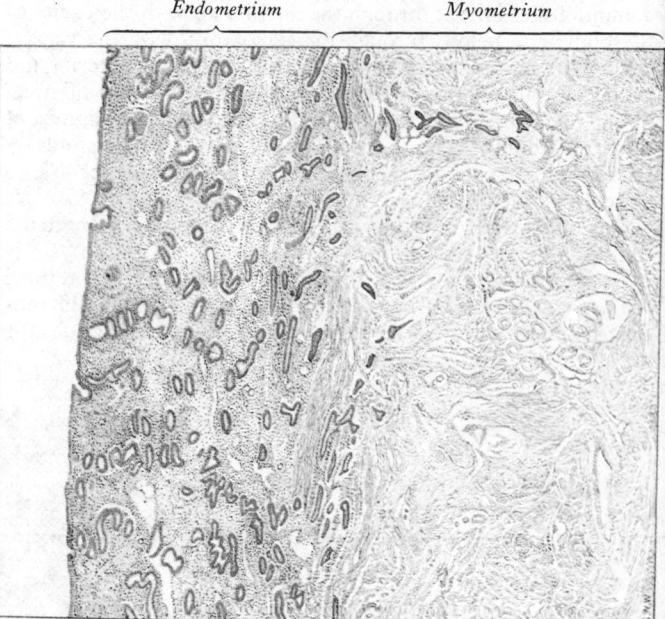

Endometrium   *Myometrium*

8.185 Section of human endometrium and underlying musculature in the interval phase. Note that some glands extend into the more internal layers of the myometrium. Stained with haematoxylin and eosin. Magnification about × 20.

continuous, through the fimbriated extremities of the uterine tubes, with the peritoneum, and, through the external os of the uterus, with the mucosa of the vagina.

In the body of the uterus the mucous membrane is covered with columnar epithelium. Prior to puberty the epithelium is ciliated, but owing to its periodic destruction in the process of menstruation and pregnancy it is usually non-ciliated over large areas in the adult uterus. The subepithelial layer of the mucosa consists of an embryonic, nucleated and highly cellular form of connective tissue in which run blood vessels and numerous lymphatic spaces. It contains many tube-like *uterine glands*, which are lined with columnar epithelium, patchily ciliated, and open into the cavity of the uterus.

Cyclical changes in the endometrium. Between puberty and the menopause, in each lunar month, the endometrium lining the body of the uterus undergoes cyclical changes which constitute the *menstrual cycle*. The duration of a menstrual cycle may vary from individual to individual, or in the same individual from cycle

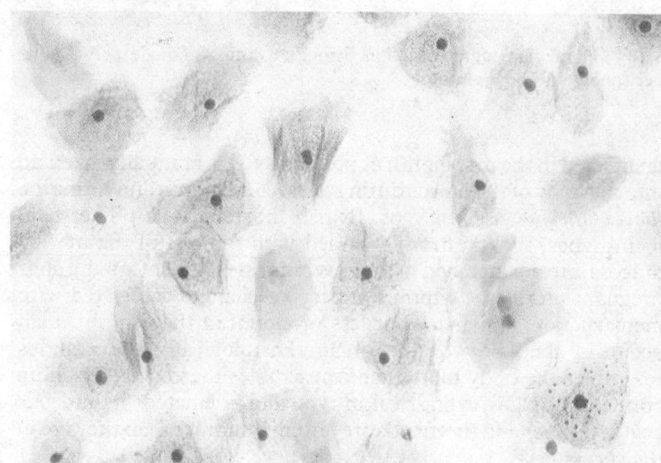

8.186 Cells of the mature vaginal epithelium in a diagnostic smear preparation stained by Papanicolaou's technique (Shorr modification). All the cells shown are superficial keratinizing squames, some of which show actual keratin granules. The pink cells are the oldest and most superficial. (Preparation kindly provided by Dr. Max Levene, St. Helier Hospital, Carshalton.)

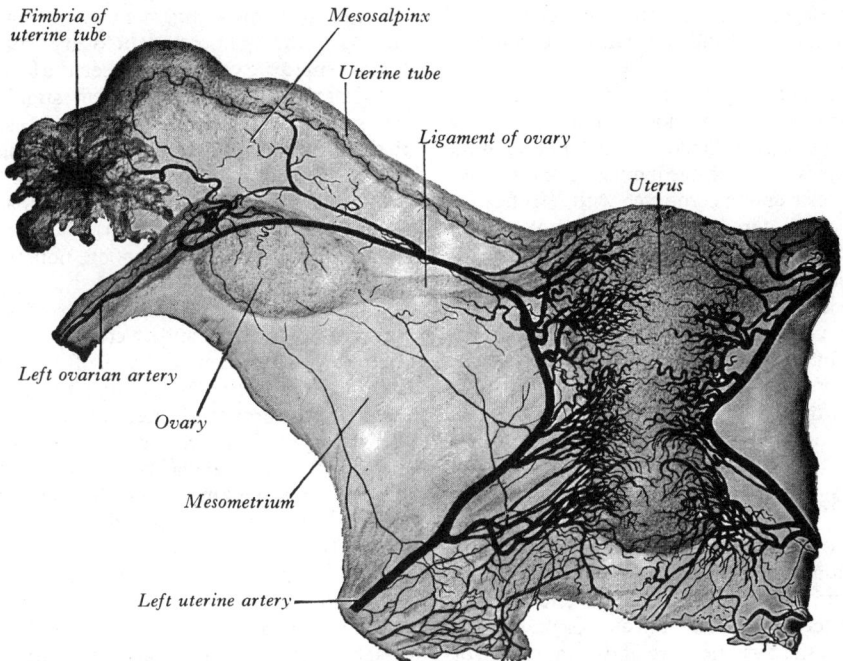

8.187 Posterior aspect of a cleared injected specimen (prepared by Dr. Hamilton Drummond) to show the distribution of the left uterine and ovarian arteries of a female aged seventeen and a half years.

to cycle. The details of the histological changes, the hormonal factors influencing them and the variations that may occur in the cycles are described in the section of this volume devoted to Embryology (pp. 126–130).

In the upper two-thirds of the cervix, the mucous membrane is provided with numerous, deep, glandular follicles which secrete a clear, viscid, alkaline mucus; in addition, extending through the whole length of the canal, a variable number of little cysts are found, presumably follicles which have become occluded and distended with retained secretion. They are called the *ovula Nabothi*. The mucous membrane covering the lower one-half of the cervical canal presents numerous papillae. The epithelium of the upper two-thirds is cylindrical and ciliated, but below this it loses its cilia, and, close to the external os, changes to stratified squamous epithelium. On the vaginal surface of the cervix the epithelium is similar to that lining the vagina, viz. stratified squamous. The mucous membrane lining the lower two-thirds of the cervix does not undergo the (menstrual) cyclical changes such as occur in the body of the uterus (see also pp. 126–130).

Vessels and nerves. *The main arterial supply* to the whole of the uterus is by way of the two uterine branches of the internal iliac arteries (p. 720). These vessels anastomose superiorly with the ovarian and inferiorly with the vaginal arteries; and while this provides alternative channels, the predominance of the uterine arteries is emphasized by the fact that whilst they undergo marked hypertrophy during pregnancy, the other vessels mentioned do not. The two uterine arteries anastomose extensively with each other across the midline by small branches, and one or other stem artery can be ligated without serious interference with the blood supply (Siegel and Mengert 1961), and even more extensive ligations have been safely undertaken. The tortuosity of the uterine arteries, as they ascend in the broad ligaments, is repeated in their branches in the uterine wall; all these sinuosities straighten out as the uterus expands in pregnancy, but it is interesting to note that they are present in the nulliparous state. Each uterine artery produces large numbers of branches which pass at once into the uterine wall, dividing there into groups of *anterior* and *posterior arcuate arteries*. These pass circumferentially in the myometrium, and their terminal branches anastomose across the midline with similar vessels from the opposite side (**8**.187). This entails that both the anterior and posterior aspects of the uterus show considerable zones in the median region which are free of larger vessels; these zones are sometimes referred to as

relatively avascular—a misleading and erroneous concept. The arcuate arteries supply large numbers of tortuous *radial branches* which travel centripetally through the deeper myometrial layers, supplying these *en route*, to reach the endometrium. As microradiographic and injection techniques show, these vessels provide the basis of a very rich series of capillary plexuses in the myometrium and endometrium (Farrer-Brown *et al*. 1970). In the myometrium numerous *spiral* or helical arteriolar branches of the radial arteries pass towards and supply the endometrium, their form being affected considerably by the endometrial cycle. During the proliferative phase the spiral arterioles are much less prominent; in the secretory phase they appear to grow in length and calibre and become even more tortuous (Ramsey 1955).

The *uterine veins* are arranged upon a plan similar to that of the arteries (Farrer-Brown *et al*. 1970). Volumetrically they are even more prominent than the arteries in the uterine wall. The occurrence of minute venous sinuses or lakes in the endometrium, where they are a constant feature in the pregnant uterus, has been recorded in the resting uterus (Schlegel 1946) but not confirmed. During pregnancy, the *ovarian veins*, unlike the accompanying arteries, may be noticeably enlarged (O'Leary and O'Leary 1966).

The *lymphatic vessels* and drainage of the uterus are described on p. 798. The intrinsic lymphatic plexuses of the uterine wall have been little studied in the human female. A most detailed injection study of primate arrangements concerns the macaque monkey (Wislocki and Dempsey 1939).

The *nerves* supplying the uterus have been described on p. 1136. The autonomic supply is derived directly from the ovarian and hypogastric plexuses; the sympathetic preganglionic fibres are axons of nerve cells in the twelfth thoracic and first lumbar spinal segments; the parasympathetic preganglionic axons issue in the second, third and fourth ventral spinal roots of the sacral cord. Whilst there is general agreement as to the pathways by which autonomic fibres reach the uterus, the details of their distribution and physiological effects are still ill-defined. Both cholinergic and adrenergic fibres have been identified in the muscular and submucous strata of the cervix, the former predominating. A number of observers consider that the corpus uteri is innervated only by sympathetic nerves. In a histochemical study (Owman *et al*. 1967), only adrenergic nerve terminals were considered to be present in most parts of the female internal genitalia, including the uterus. The effects of the uterine innervation are also complicated by hormonal influences and by the evocation of different responses by the same active agent (for example, adrenalin) under differing prevailing conditions (Wansbrough *et al.* 1967).

Applied Anatomy. A certain degree of *anteversion* or *retroversion* of the uterus can take place without being regarded as pathological; but when the degree of flexion at the junction of the body with the cervix becomes considerable it must be so regarded. This is especially true of retroversion combined with retroflexion. Retroversion alone is a posterior inclination of the whole uterus, so that the cervix points forwards towards the os pubis; retroflexion is a posterior curvature of the body only, at its junction with the cervix. The two conditions are usually combined. Prolapse of the uterus is another common condition. The organ sinks to an abnormally low level in the pelvis, and sometimes protrudes beyond the vulva. This condition is usually due to imperfect repair of the pelvic floor following injury of the perineum during parturition (p. 561).

THE UTERINE LIGAMENTS

The uterus is connected to the bladder, the rectum and the walls of the lesser pelvis by a number of 'ligaments'; some of these are merely peritoneal folds and have little ligamentous effect, whilst others consist of nonstriated muscle and fibrous tissue and can function as real ties, with also some measure of dynamic control.

The *anterior ligament* consists of the *uterovesical fold* of peritoneum, which is reflected on to the upper surface of the bladder from the front of the uterus, at the junction of the cervix and body.

The *posterior ligament* consists of the *rectovaginal fold* of peritoneum, which is reflected from the back of the posterior fornix of the vagina on to the front of the rectum. It forms the bottom of the deep *recto-uterine pouch* which is bounded anteriorly by the posterior wall of the body of the uterus, the supravaginal portion of the cervix uteri and the posterior fornix of the vagina; posteriorly, by the rectum; laterally, by two crescentic folds of peritoneum which pass backwards from the cervix uteri, one on each side of the rectum, to the posterior wall of the lesser pelvis. These are the *recto-uterine folds* and they contain a considerable amount of fibrous tissue and nonstriated muscular fibres, which are attached to the front of the sacrum and constitute the *uterosacral ligaments.* On rectal examination the uterosacral ligaments can be palpated as they pass backwards at the sides of the rectum.

The two **broad ligaments** (8.176) pass from the sides of the uterus to the lateral walls of the pelvis. Together with the uterus they form a septum across the cavity of the lesser pelvis dividing it into two parts. The anterior part contains the bladder; the posterior part, the rectum, and usually, the terminal coil of the ileum and a part of the sigmoid colon.

When the bladder is empty or only slightly distended, the surfaces of the broad ligament are directed superiorly and inferiorly and it has a free anterior and an attached posterior border. As the bladder fills, the plane of the ligament alters and its free border becomes superior in position. In this condition of the bladder, the broad ligament consists of anterior and posterior layers, which are continuous with each other at its upper, free border, and diverge from each other below, where they approach the superior surface of the levator ani. The uterine tube is contained in the free border, and the part of the ligament between the tube and the ligament of the ovary and mesovarium is termed the *mesosalpinx.* The infundibulum of the tube projects from the free border near its lateral extremity. The ovary is attached to the posterior layer by the *mesovarium.* The part of the broad ligament which extends from the infundibulum of the tube and the upper pole of the ovary to the lateral wall of the lesser pelvis contains the ovarian blood vessels, nerves and lymph vessels and is named the *suspensory ligament of the ovary* (in some respects a less appropriate term than the older, but unofficial, *infundibulopelvic ligament*). It is continued laterally over the external iliac vessels as a distinct fold. Between the ovary and the uterine tube, the mesosalpinx contains the epoöphoron (p. 1425) and, at its medial end, the paroöphoron (p. 1425) and anastomosing branches of the uterine and ovarian vessels. The term *mesometrium* is applied to the part of the broad ligament extending from the pelvic floor to the ovary, the ligament of the ovary and the body of the uterus. The uterine artery passes between the layers of the broad ligament at its inferior border, about 1·5 cm lateral to the cervix and after it has crossed the ureter (p. 720). It then ascends in the medial part of the ligament and turns laterally below the uterine tube to anastomose with the ovarian artery. In addition to all the structures already enumerated, the broad ligament encloses the ligament of the ovary (p. 1423), the proximal part of the round ligament of the uterus and some nonstriated muscle and fibroareolar tissue.

The *round ligaments* of the uterus (8.182) are two narrow, flat bands between 10 cm and 12 cm long, situated between the layers of the broad ligament in front of and below the uterine tubes. Commencing at the lateral angle of the uterus each ligament is directed forwards and laterally across the vesical vessels, the obturator vessels and nerve, and the obliterated umbilical artery and over the external iliac vessels. It then passes through the deep inguinal ring, hooking round the commencement of the inferior epigastric artery, to traverse the inguinal canal. It finally breaks up into strands which merge with the areolar tissue in the labium majus. Near the uterus the round ligament contains much nonstriated muscle; this becomes progressively less in amount and the terminal part is purely fibrous. It is accompanied by blood vessels, nerves and lymphatics; the latter drain from the parts of the uterus near the entry of the uterine tube and pass to the superficial inguinal lymph nodes (p. 791), and malignant disease of the uterus may spread along this route. In the fetus a tubular process of the peritoneum (*processus vaginalis*) is carried with it for a short distance into the inguinal canal. The processus vaginalis is generally obliterated in the adult, but sometimes remains patent even in advanced life. It is homologous with the peritoneal processus which precedes the descent of the testis.

In its passage through the inguinal canal, the round ligament receives coverings corresponding to those related to the spermatic cord in the male (p. 1418), though they are thinner and blend with the ligament itself; the ligament may even fail to reach the labium majus and end by fusing with these coverings.

The round ligament and the ligament of the ovary are together homologous with the gubernaculum testis in the male (p. 218).

The *transverse cervical ligament* (of Mackenrodt) is attached to the side of the cervix uteri and to the vault and lateral fornix of the vagina, and is continuous with the fibrous tissue which surrounds the pelvic blood vessels, probably playing a considerable part in maintaining the position of the uterus. It is of importance in gynaecological surgery. Other dense parts of the pelvic fascia connect the cervix of the uterus and the upper part of the vagina to the posterior aspect of the pubis.

While the above ligaments (as well as the vagina) act in varying (and undetermined) measure as mechanical supports for the uterus, maintaining it in its normal position, the levatores ani and coccygei, the muscles of the urogenital diaphragm and the perineal body appear to be of particular importance in this respect. For the role of the musculature of the pelvic floor consult a review and discussion by Wendell-Smith and Wilson (1977).

The Vagina

The vagina (8.183), the female organ of copulation, is a fibromuscular tube lined with a stratified epithelium, which extends from the vestibule, or cleft between the labia minora, to the uterus, and is situated between the bladder and urethra anteriorly, and the rectum and anal canal posteriorly; it is inclined posterosuperiorly, its axis forming with that of the uterus an angle of over ninety degrees, opening anteriorly, but which varies with the conditions of the bladder and rectum. Its walls are ordinarily in contact, and the usual shape of its lower part on transverse section is that of an H, but the transverse limb being slightly curved forwards or backwards, while the lateral limbs are somewhat convex towards the median plane; its middle part has the appearance of a transverse slit. Its length is 7·5 cm along its anterior wall, and 9 cm along its posterior wall; its width gradually increases from below upwards. Its upper end surrounds the vaginal portion of the cervix uteri a short distance from the external os of the uterus, its attachment extending higher on the posterior than on the anterior wall of the cervix. To the arched recess between the cervix uteri and vagina the term *fornix* is applied; anterior, posterior and lateral parts of this annular groove are often so designated, but the groove is essentially continuous.

The *anterior* part of the vaginal wall is in relation with the base of the bladder, and with the urethra, which is actually embedded in the anterior wall of the vagina. The *posterior part* of the vaginal wall, which is covered with peritoneum in its upper quarter, is separated from the rectum by the recto-uterine pouch above, and by some loose fibro-areolar tissue in its middle half; the lower quarter is separated from the anal canal by a mass of muscular and fibrous tissue, the *perineal body*. At the sides are the levatores ani (p. 561) and pelvic fascia. As the terminations of the ureters pass forwards and medially to reach the fundus of the bladder, they are close to the lateral fornices of the vagina, and as they enter the bladder are usually placed in front of the vagina (p. 1402). The ureter is crossed transversely in this situation by the uterine artery (p. 720).

Structure. The vagina consists of a mucous membrane and a muscular stratum, the lamina propria of the former containing in its deepest layer a large number of thin-walled veins.

The *mucous membrane* is firmly fixed to the muscular layer; on its epithelial surface there are two median longitudinal ridges, one on the anterior and the other on the posterior wall of the vagina. These ridges are called the *vaginal columns*, and from them numerous transverse ridges or rugae extend laterally on each side. These rugae are divided by furrows of variable depth, giving to the mucous membrane the appearance of being studded with conical projections or papillae; they are most numerous on the posterior wall and near the orifice of the vagina, especially before parturition. The epithelium of the mucous membrane is of the non-keratinized, stratified squamous variety. After puberty it becomes thick and is rich in glycogen. Unlike the condition in many mammals, the vaginal epithelium does not undergo very marked changes during the menstrual cycle; its glycogen increases in the post-ovulatory phase and diminishes towards the end of the cycle. The fermentative action of certain bacteria (Döderlein's bacillus) on the glycogen renders the fluid in the vagina acid. There are no glands in the vaginal mucous membrane, which is thus lubricated by mucus derived from the glands of the cervix of the uterus.

Cyclical variations in the vaginal epithelium have been studied extensively in a number of animals, and in man, by taking specimens or 'smears' to examine the cell types and their relative frequencies (Stockard and Papanicolaou 1917). The cells observed are mostly epithelial cells and leucocytes (**8**.186). This technique has proved most fruitful in reproductive studies in experimental animals, and also in clinical medicine. The effects of hormonal therapy are clearly reflected in changes in the vaginal cellular débris. The method has also been extended to serve as an early warning of uterine carcinoma, especially of the cervix (Papanicolaou and Traut 1943). For developments in this technique consult Koss (1968).

The *muscular layers* are nonstriated: an external longitudinal, which is by far the stronger, and an internal circular layer. The longitudinal fibres are continuous with the superficial muscular fibres of the uterus. The strongest fasciculi are those attached to the rectovesical fascia on each side. The two layers are not distinctly separable from one another, but are connected by oblique decussating fasciculi. In addition to this, the lower end of the vagina is surrounded by a band of striated muscular fibres, termed the *bulbospongiosus* (p. 563).

External to the muscular coat there is a layer of areolar tissue, containing a large plexus of blood vessels.

Vessels and nerves. The *arteries* of the vagina are derived from the vaginal, uterine, internal pudendal, and middle rectal branches of the internal iliac arteries (p. 720; **6**.94). The *veins* form plexuses at the sides of the vagina, and these plexuses are drained through the vaginal veins into the internal iliac veins. The *lymph vessels* are described on p. 798. The *nerves* are derived from the vaginal plexuses, and from the pelvic splanchnic nerves (p. 1135). The lower part of the vagina is supplied by the pudendal nerve

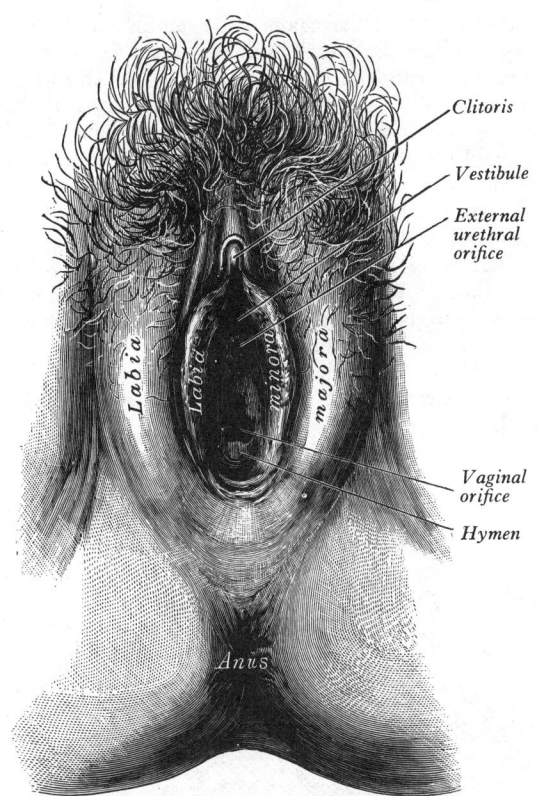

Clitoris

Vestibule

External urethral orifice

Labia

Labia minora

majora

Vaginal orifice

Hymen

Anus

8.188 Female external genitalia, with the labia majora et minora separated.

(p. 1115). Many nerve fibres in the lamina propria and muscle coat react strongly to tests for cholinesterase and are probably of the cholinergic type.

The External Genital Organs of the Female

The female external genitalia (**8**.188) include the mons pubis, the labia majora et minora pudendi, the clitoris, the vestibule of the vagina, the bulb of the vestibule and the greater vestibular glands. The term *pudendum* or *vulva*, as generally applied, includes all these parts.

The mons pubis, the rounded eminence in front of the pubis symphysis, is formed by a collection of subcutaneous adipose connective tissue. It becomes covered with coarse pubic hair at the time of puberty over an area which often has an approximately horizontal upper limit. In the male the *pubic* hair proper has a similar upper limit and its apparent continuation upwards to the umbilicus consists of ordinary body hair.

The labia majora are two prominent, longitudinal, cutaneous folds which extend from the mons pubis into the perineum and form the lateral boundaries of the *pudendal cleft*, into which the vagina and urethra open. Each labium has two surfaces—external, pigmented and covered with crisp hairs, and internal, smooth but studded with large sebaceous follicles. Between the two surfaces there is a considerable quantity of areolar and adipose tissue, and this is intermixed with nonstriated muscle fibres resembling the dartos muscle of the scrotum, together with vessels, nerves and glands. The round ligament of the uterus ends in the adipose tissue and skin of the front part of the labium. A peristent processus vaginalis and a congenital inguinal hernia may reach the labium. The labia are thicker in front, where they form by their meeting the *anterior commissure*. Posteriorly they are not really joined, but appear to become lost in the neighbouring skin, ending close to, and nearly parallel with, each other; together with the connecting skin between them, they form the *posterior commissure*, or posterior boundary of the pudendal fissure. The interval between the posterior commissure and the anus, from 2·5 to 3 cm in length, constitutes the '*gynaecological*' *perineum*.

The labia minora are two small cutaneous folds, devoid of fat, situated between the labia majora, and extending from the clitoris obliquely downwards, laterally and backwards for about 4 cm on each side of the orifice of the vagina, between which and the labia majora they end; in the virgin the posterior ends of the labia minora are usually joined across the median plane by a fold of skin, named the *frenulum of the labia minora*. Anteriorly, each labium minus divides into two; the upper division passes above the clitoris to meet its fellow, forming a fold which overhangs the glans clitoridis and is named the *prepuce* of the clitoris; the lower division passes below the clitoris and is united to its under surface, forming, with its fellow of the opposite side, the *frenulum* of the clitoris. Numerous sebaceous follicles are placed on the opposed surfaces of the labia minora.

The vestibule is the cleft between the labia minora: in it the vaginal and external urethral orifices are situated, and, between

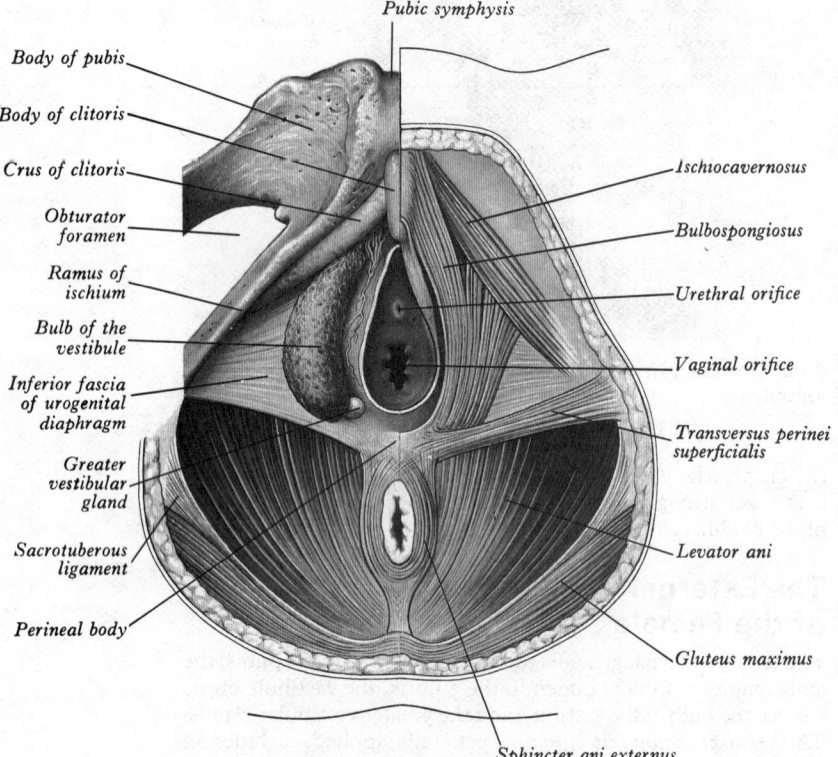

8.189 Dissection of the female perineum to show the bulb of the vestibule and greater vestibular gland on the right; on the left the muscles superficial to these structures have been left *in situ*.

them, numerous small mucous *lesser vestibular glands* open on the surface of the vestibule. The part of the vestibule between the vaginal orifice and the frenulum of the labia minora consists of a shallow depression named the *vestibular fossa*.

The clitoris is an erectile structure, homologous with the penis. It is postero-inferior to the anterior commissure, partially enclosed between the anterior ends of the labia minora. The *body of the clitoris* consists of two corpora cavernosa, composed of erectile tissue enclosed in a dense layer of fibrous tissue, and separated along their medial surfaces by an incomplete fibrous *pectiniform septum*; each corpus cavernosum is connected to the pubic and ischial rami by a *crus*. The free extremity, or *glans clitoridis*, is a small rounded tubercle, consisting of spongy erectile tissue, and its epithelium has a highly developed cutaneous sensitivity, of considerable importance in sexual responses. The clitoris is provided, like the penis, with a 'suspensory' ligament, and with two small muscles, named the ischiocavernosi (p. 563), which are inserted into the crura of the clitoris. The clitoris is thus in many details a smaller version of the penis, but it differs basically in being entirely separate from the urethra.

The vaginal orifice (or *introitus*) is usually a median slit below and behind the opening of the urethra; its size varies inversely with that of the *hymen*. This orifice is, of course, like all parts of

the vagina capable of great distension during parturition, and to a much lesser degree, during coitus.

The hymen vaginae is a thin fold of mucous membrane situated at the orifice of the vagina; the internal surfaces of the fold are normally in contact with each other, and the vaginal orifice appears as a cleft between them. The hymen varies much in shape and extent. When stretched, its commonest form is that of a ring, generally broadest posteriorly; sometimes it is represented by a semilunar fold, with its concave margin turned towards the pubes. Occasionally it is cribriform or its free margin forms a membranous fringe. It may be entirely absent, or may form a complete septum across the lower end of the vagina; the latter condition is known as an imperforate hymen. When the hymen has been ruptured, small rounded elevations known as the *carunculae hymenales* are found as its remains. The hymen has no known function.

The external urethral orifice is about 2·5 cm postero-inferior to the glans clitoridis and immediately anterior to the orifice of the vagina: it usually assumes the form of a short, sagittal cleft with slightly raised margins. It is markedly distensible.

The bulbs of the vestibule (8.189) are the homologue of the single bulb of the penis and corpus spongiosum of the male. They are two elongated masses of erectile tissue, one on each side of the vaginal orifice and united to each other in front by a narrow median band termed the *commissura bulborum* (pars intermedia). Each lateral mass measures about 3 cm in length. Their posterior ends are expanded and are in contact with the greater vestibular glands; their anterior ends are tapered and joined to one another by the commissure and to the glans of the clitoris by two slender bands of erectile tissue; their deep surfaces are in contact with the inferior fascia of the urogenital diaphragm; superficially they are covered with the bulbospongiosus. Thus, in the female, the corpus spongiosum is, as it were, cleft into two bilateral masses, through all but its most anterior region, by the vestibule and the vaginal and urethral orifices.

The greater vestibular glands (8.189) are the homologues of the bulbo-urethral glands in the male. They consist of two small, round, or oval bodies of a reddish-yellow colour, situated one on each side of the vaginal orifice, in contact with, and often overlapped by, the posterior end of the lateral mass of the bulb of the vestibule. Each gland opens by means of a duct, about 2 cm long, immediately lateral to the hymen, in the groove between its attached border and the labium minus.

Vessels and nerves. The arterial blood supply, venous and lymph drainage and the nerve supply of the structures comprising the external genital organs of the female are similar to those relating to the homologous structures in the male. The arterial supply, from two external and one internal pudendal artery on each side, is potentially massive. Hence, haemorrhage from injuries of the vulva may be very severe. The different sensory innervation of the anterior and posterior parts of the labium majus resembles that of the scrotum (*see* p. 1418).

The Mammae or Mammary Glands

The mammae exist in both sexes. In the male they are rudimentary throughout life; in the female they are undeveloped before puberty, but undergo considerable growth and elaboration at and after puberty. They attain their greatest development during the later months of pregnancy and especially for some time after parturition, during lactation. The mamma consists of glandular tissue (the *mammary gland* proper), which secretes milk, and fibrous and adipose tissue in between the lobes and lobules of the glandular tissue, together with blood vessels, lymph vessels and nerves.

The female mamma or breast. In the young adult female, each breast forms a rounded eminence lying within the superficial fascia chiefly anterior to the thorax, but spreading variably on to its lateral aspect. Its shape varies markedly in different individuals, and races, and in the same individual at different ages. It may be hemispherical, conical, and is variably pendulous, becoming either piriform or thin and flattened. Most of its bulk is adipose tissue, except during lactation; and hence its shape and consistency are primarily dependent upon this. In the lateral plane (**8**.91) its base extends vertically from the second to the sixth rib, and at the level of the fourth costal cartilage it extends transversely from the side of the sternum to near the mid-axillary line. The superolateral part of the breast is prolonged upwards and laterally towards the axilla, forming the *axillary tail*, which extends along the lower border of the pectoralis major and may pass through the deep fascia to lie in close relationship to the pectoral group of axillary lymph nodes. The deep aspect of the breast is slightly concave and is related to the pectoralis major, serratus anterior, obliquus externus abdominis and the aponeurosis of the latter muscle as it forms the anterior wall of the sheath of the rectus abdominis. The breast is, however, separated from these muscles by the deep fascia, and between the breast and the deep fascia there is a zone of loose areolar tissue, the *retromammary* or *submammary* 'space', which allows the breast some degree of movement on the deep fascia covering pectoralis major. Advanced mammary carcinoma may, by invasion, fix the breast to the pectoralis major. Occasionally small projections of the mammary glandular tissue may penetrate through the deep fascia into the superficial part of this muscle. The *mammary papilla* or *nipple* is a cylindrical or conical projection from just below the centre of the anterior surface of the breast; it commonly lies at the level of the fourth intercostal space in nulliparous females. It is pink or light brown in colour and is traversed by fifteen to twenty lactiferous ducts, which open by minute orifices on its wrinkled tip. It contains numerous nonstriated muscle fibres; most of these are arranged circularly and their contraction when the papilla is mechanically stimulated (e.g. by suckling) causes erection of the papilla; other fibres are arranged longitudinally and their contraction may retract it. Occasionally the papilla may not evert during its prenatal development (p. 156) and thus remains permanently retracted, a condition that causes difficulty in suckling. The base of the papilla is encircled by a coloured area of skin called the *areola*, which is rose-pink in the nulliparous Caucasian female. During the second month of pregnancy it becomes larger and darker in colour and as pregnancy advances it becomes dark brown, the depth of the colour varying with the female's complexion; this colour diminishes in intensity after the end of lactation, but the areola never returns to its original hue. It contains numerous sebaceous *areolar glands*, which become much enlarged during pregnancy and lactation to form '*tubercles*' beneath the skin; the oily secretion of these glands provides a protective lubricant for the skin of the areola and papilla during lactation. Some of the glands appear to be intermediate in structure between the sebaceous and sweat varieties, whilst others (Montgomery's glands) are of the mammary type. There is no fat immediately beneath the skin of the areola and papilla.

STRUCTURE OF THE BREAST

The mammary gland (**8**.190–193) consists of: (*a*) glandular tissue of the tubulo-alveolar type; (*b*) fibrous tissue, connecting its lobes; (*c*) adipose tissue in the intervals between the lobes (Cowie 1974). The subcutaneous tissue encloses the gland (but does not form a distinct capsule) and sends numerous septa into it to support its various lobules. From the part of the fascia which covers the gland fibrous processes pass forwards to the skin and the papilla; these are better developed over the upper part of the breast and constitute the *suspensory ligaments* (of Cooper). These ligaments may become contracted by fibrosis in cancer of the breast, thus causing the overlying skin to become pitted and retracted. The normal gland tissue is of a pale reddish colour, firm in texture, and forms a lobulated mass which is flattened anteroposteriorly and thicker in the centre than at the circumference. It consists of fifteen to twenty lobes, and these are composed of lobules, connected together by loose connective tissue, blood vessels and ducts. The smallest lobules, when fully developed, consist of a cluster of rounded alveoli which open into the smallest branches of the lactiferous ducts; these branches unite to form larger ducts which end in the terminal or *lactiferous ducts*, each of which drains a lobe of the gland. The lactiferous ducts hence also vary from fifteen to twenty in number; they converge towards the areola,

beneath which they form variable dilatations, or *lactiferous sinuses*, which may serve as reservoirs for milk. (Some authorities, however, consider the lactiferous 'sinus' or dilatation to be a preparative artefact.) At the base of the papilla they contract, and pursue a straight course to its summit, ending as separate orifices considerably narrower than the ducts themselves. The ducts are surrounded by connective tissue containing longitudinal and transverse elastic fibres; they are lined by columnar epithelium and, outside these, a layer of longitudinally orientated myoepitheliocytes resting on a basal lamina. In the larger ducts the epithelium consists of two or more layers of cells, and near the openings on the papilla it becomes keratinizing stratified squamous in type. Mammary carcinoma is usually caused by neoplastic changes in the duct cells which form tumours within the ductular tissues, or infiltrate the surrounding connective tissue.

The mammary gland varies in its structure with *age* and during *pregnancy* and *lactation*. (The prenatal development of the gland is described on p. 156.) At *birth* it consists almost entirely of lactiferous ducts, no alveoli being present. This condition persists until puberty, only very slight branching of the ducts taking place, and the slight enlargement of the breast is due to deposition of fat and growth of the fibrous tissue stroma. After *puberty*, under the

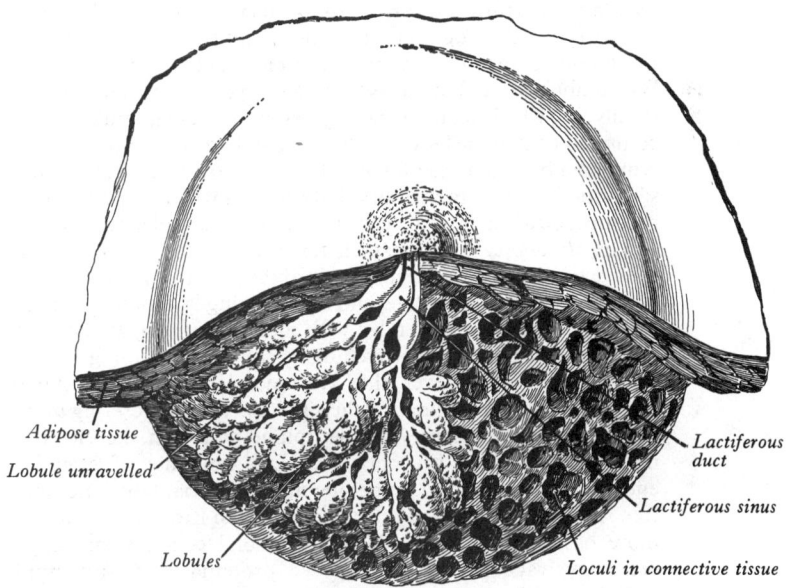

8.190   A dissection showing increased secretory lobules in the breast during lactation.

Adipose tissue

Lobule unravelled

Lobules

Lactiferous duct

Lactiferous sinus

Loculi in connective tissue

stimulating influence of ovarian oestrogenic hormones, the ducts develop branches and their ends form small, solid, spheroidal masses of granular polyhedral cells, which are potential alveoli. In the resting state the glandular epithelium is separated from the neighbouring vascularized stroma by a thin zone of fibroblasts amongst which vessels do not penetrate. This '*epithelio-stromal junction*' may exercise a control over the passage of materials to the secretory cells (Ozzello 1974). True secreting alveoli only appear during pregnancy, when the ducts branch markedly and their terminal parts develop lumina which increase in size as milk is secreted into them. This growth is due to the rising output of oestrogen and progesterone from the placenta. The amount of adipose tissue is also increased and there is a greater content of circulating blood. The secretory activity of the cells lining the alveoli increases progressively during the latter half of pregnancy. The secretion formed in the later stages of pregnancy and for a few days after parturition is known as *colostrum*. It contains cells, called *colostral corpuscles*, which have numerous fat globules in their cytoplasm. The nature of these cells is uncertain; some believing that they are the innermost cells lining the primitive alveoli, which undergo fatty degeneration and are shed into the alveolar lumen, whereas others maintain that they are macrophages which phagocytose the small amount of fat present in

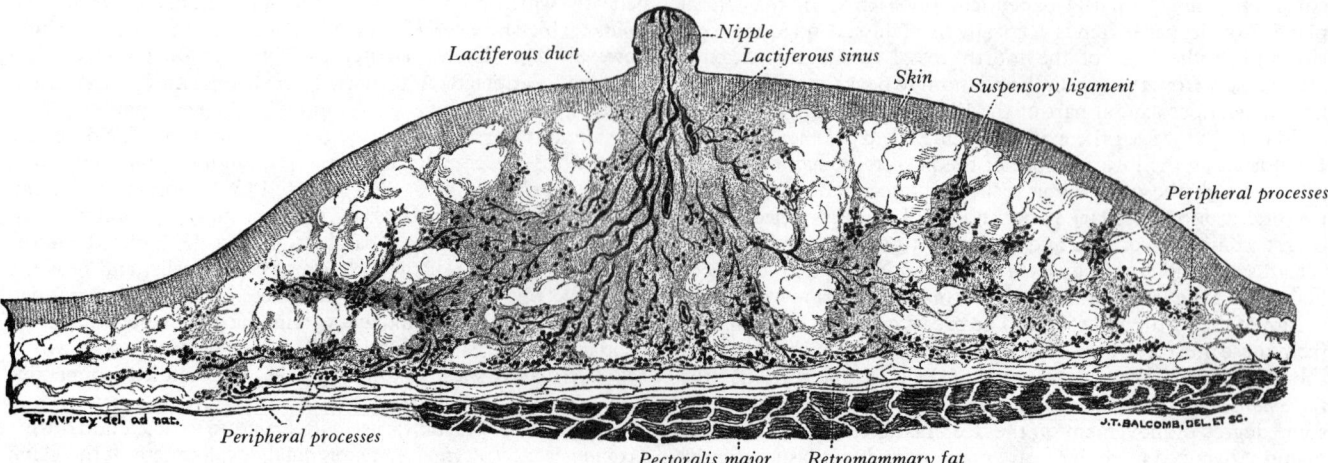

8.191 Horizontal section of the breast at the level of the nipple in a multiparous female, aged forty years. (From Quain's *Anatomy*, vol. iii, part iv.)

the secretion at this time. True milk secretion commences a few days after parturition, due to the reduction in blood levels of oestrogen and progesterone, which appears to trigger off production of prolactin by the anterior lobe of the hypophysis (*see* Wolstenholme and Knight 1972, and p. 1439). The milk distends the alveoli, which are then lined by a single layer of granular, short columnar cells at the bases of which are stellate myoepitheliocytes lying on a basement membrane; the cells become flattened as the secretion increases in quantity. Fat droplets which accumulate in the superficial parts of the cells pass into the alveolar lumen.

The *alveolar cells* vary in their form and contents according to their activities in the secretory cycle of the mammary gland, being cuboidal in the resting state, and taller during lactation, although when the alveoli are distended with milk they may again become cuboidal because of the stretching of the alevolar lining; in the latter condition, however, they are much larger cells with huge secretory vacuoles distending their apices. Whilst secreting milk, their cytoplasm is somewhat basophilic and with the electron microscope is seen to contain copious amounts of granular endoplasmic reticulum, numerous mitochondria, lysosomes and free ribosomes. Apical to the nucleus, which is positioned in the more basal part of the cell, lies the Golgi complex and large secretory vacuoles of two types, proteinaceous and lipid containing. The protein vacuoles contain granules of casein in a micellar form, and other proteins of the milk; these are formed in the granular endoplasmic reticulum and passed to the Golgi

apparatus where they are compounded into larger vacuoles and eventually secreted in a merocrine manner into the lumen of the gland. The lipid vacuoles do not apparently pass through the Golgi apparatus, but accumulate in the apical portion of the cell, often fusing to form large 'milk vacuoles' up to 10 μm across which frequently protrude from the surface of the cell before they are secreted. The lipid vacuoles are eventually discharged from the cell with a thin surrounding rim of cell membrane and cytoplasm (Pitelka 1977; Tobon and Salazar 1975; Saacke and Heald 1974; Hollmann 1974). This type of secretion may be considered as *apocrine* since actual cytoplasm is lost from the secreting cell, although not to the extent envisaged by those who first coined this term.

The alveolar cells are joined to each other apically by occluding tight junctions which prevent the passage of substances from the lumen of the gland into the intercellular spaces and *vice versa*; numerous desmosomes and communicating junctions also occur between cells (Pitelka *et al.* 1973).

In pregnant rats, one or two additional types of protein granule are synthesized, as precursors of the normal type of granule, and these may form part of the colostrum (Murad 1970).

The passage of the milk from the alveoli into and along the ducts is thought to be caused by the contraction of *myoepitheliocytes* (p. 39) sited between the lining cells and the basement membrane. The flow of milk is initiated by suckling, which stimulates the abundant nerves in the papilla and areola; these afferent impulses cause the formation of oxytocin by hypothalamic nuclei and its liberation into the circulation from the neurohypophysis, the oxytocin apparently stimulating contraction of the myoepitheliocytes and of the nonstriated muscle related to the ducts. In the absence of suckling, secretion of milk soon ceases. Oestrogenic hormones stimulate growth of the ducts whereas progesterone (produced by the the corpus luteum) directly or indirectly stimulates the formation of small alveoli at the ends of the ducts. The formation of true secretory alveoli in pregnancy is controlled by the synergic action of oestrogens, progesterone (produced mainly by the placenta), and hypophysial (pituitary) hormones (prolactin and growth hormone, which also maintain milk secretion). Commonly lactation is active for about five or six months after childbirth and then progressively diminishes, so that the infant is weaned at about nine months of age. When lactation ceases, the glandular tissue returns to a 'resting' condition, any milk remaining is absorbed and the alveoli shrink, many losing their lumen. In some cases the glandular tissue of the breast may not develop sufficiently during the whole of pregnancy to produce milk, whereas in others milk secretion may cease within a few weeks of the birth of the child. After the menopause, the mammary gland atrophies, the cellular elements of the alveoli and ducts degenerating and disappearing, though a few ducts may remain (Ozzello 1974). The connective tissue stroma becomes much less cellular and the collagenous fibres

8.192 A sample section of human non-lacting breast from a young adult female.

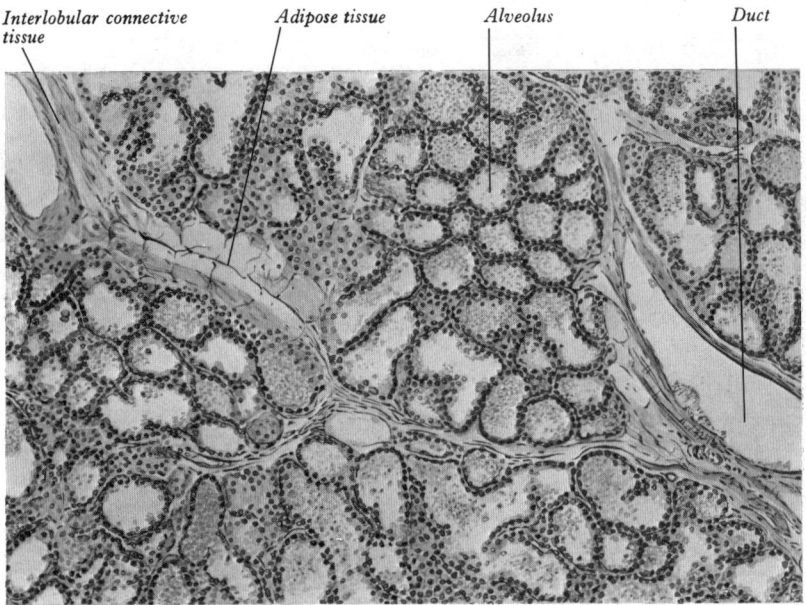

| Interlobular connective tissue | Adipose tissue | Alveolus | Duct |

8.193 A sample section of an actively lactating human breast. Magnification about × 60.

decrease in number. The amount of fat varies considerably.

In either sex, in the first week or two after birth, the cells lining the ducts may, under the influence of maternal oestrogenic hormones that have passed across the placenta, secrete a little fat-free fluid ('witch's milk').

Milk is a complex fluid, composed in the human of about 88 per cent water, 7 per cent lactose, 4 per cent fat and 1 per cent protein, as well as containing various ions, notably calcium, phosphate, sodium, potassium and chloride. Vitamins and antibodies mainly of the IgA (secretory) class are also present, the latter being largely responsible for the sterility of the milk during lactation (Jennesse 1974). The proteins are chiefly caseins and a lactalbumen; these, together with lactose and several of the triglycerides secreted in milk, are synthesized from various precursors entering the gland in the circulation, and are unique to the mammary gland. It should be added that the milk secreted initially, that is *colostral milk*, has a markedly different composition, being poor in nutrients and having an ionic composition rather similar to that of blood plasma. For a recent description of lactation and of the structure of the human breast *see* Larson (1978).

The male mamma is rudimentary throughout life and consists of small ducts (without alveoli), together with a little supporting adipose and fibrous tissue. Sometimes the ducts are largely represented by solid cellular cords. It may undergo a very slight, temporary enlargement at puberty. Generally the duct system does not extend beyond the limits of the areola, which is, however, well developed. The mammary papilla is relatively very small.

Vessels and nerves. The *arteries* supplying the mammary gland are derived from the thoracic branches of the axillary artery, and from the internal thoracic and intercostal arteries. The *veins* describe an anastomotic circle round the base of the nipple, the *circulus venosus*. From this circle and from the glandular tissue branches transmit the blood to the circumference of the gland, and end in the axillary and internal thoracic veins. The *lymph vessels* are described on p. 790. The *nerves* are derived from the anterior and lateral cutaneous branches of the fourth, fifth and sixth thoracic nerves. These nerves convey sympathetic fibres to the breast, but its secretory activities are largely under the control of hormones derived from the ovary and the hypophysis cerebri. The papilla is served by a dense nerve plexus supplying numerous sensory end organs such as Meissner's corpuscles and Merkel's discs, and also terminating in free nerve endings. This innervation is of great importance in signalling suckling (*vide supra*).

Applied Anatomy. The ducts descending from the papilla radiate through the gland, and incisions should hence be radial. A lactiferous duct may become obstructed and distended forming a cyst, or *galactocele*. Abscesses may occur between the septa, in the actual glandular tissue, subcutaneously near the papilla, or between the glandular tissue and the deep fascia on the anterior surface of pectoralis major.

Supernumerary mammae (polymastia) or papillae (polythelia) may occur in male or female, usually somewhere along a line which extends from the axilla to the pubic region. Occasionally the male mamma may undergo hypertrophy after puberty (gynaecomastia), a condition usually due to an imbalance between oestrogenic and androgenic hormones. (*See* pp. 1438, 1440, 1454, and 1461 for mammary endocrine influences.)

THE ENDOCRINE SYSTEM

For a multicellular organism to survive and maintain its integrity in a variable and often adverse environment, regulatory mechanisms are essential. The metazoa have attained a measure of functional freedom from the vagaries of their external environment (over which only the most advanced have much control and this only of a short-term nature) by developing the capacity to regulate, with varying degrees of success, the composition and properties of the immediate surroundings of the majority of their cells. These surroundings comprise the intercellular or tissue fluid, and, in those metazoa with a vascular system, the fluid component of the blood and lymph. The effectiveness with which the stability of these media is maintained is directly related to the success and longevity of the organism.

Tissue fluid and plasma are maintained in a stable state through the co-ordinated regulatory activity of the autonomic division of the nervous system and the endocrine system, the latter being subdivided into the diffuse neuro-endocrine system (p. 1454) and the endocrine system proper. All operate through intercellular communication, but differ in the manner and speed of this communication, and in the degree of localization of the effects

1437

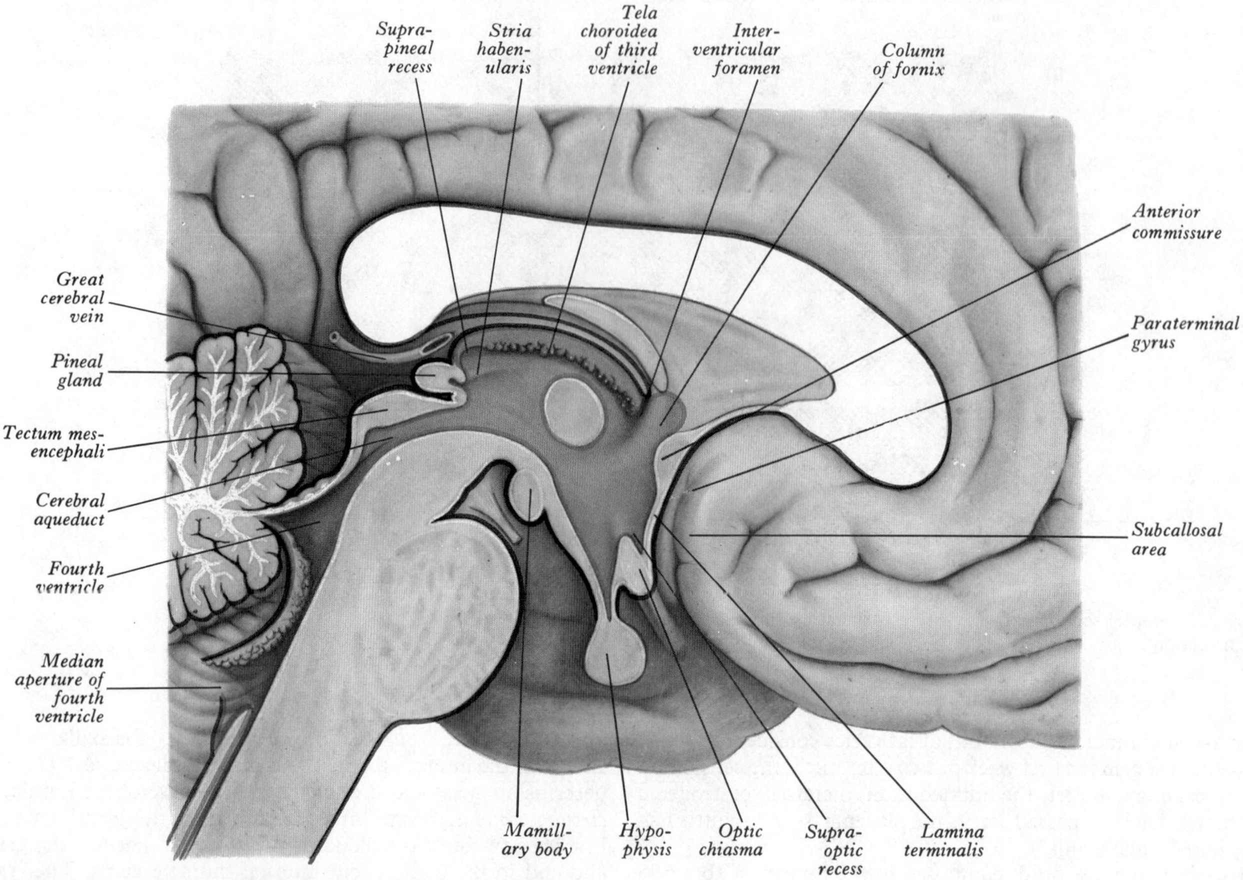

8.194 Part of a median section through the brain to demonstrate the location of the hypophysis and its immediate surroundings, particularly hypothalamic structures. Cut edge of pia mater shown in red, ependyma in blue. Compare with fig. 7.110A and B.

which they produce. The *autonomic system* utilizes neural impulse transmission and neurotransmitter release to transmit information, is swift in action and localized in the responses which it induces; the diffuse *neuro-endocrine system* makes use of secretions only, is slower in action, and the responses which it induces are less localized, for the secretions can act on contiguous cells in the manner of neurotransmitters, on groups of nearby cells which they reach by diffusion in the manner of a paracrine secretion, or on distant cells which they reach by way of the blood in the manner of a hormone; the *endocrine system* proper, comprising those isolated cells, clusters of cells, and discrete ductless glands which produce hormones, organic molecules which are transported in the blood to distant effector cells, is still slower in action and even less localized, though the effects which it produces are specific to certain cell types and of longer duration. The regulatory systems described above are not mutually exclusive but overlap in form and function. There is a gradation from the neural autonomic system, through the intermediate diffuse neuro-endocrine system, to the hormonal endocrine system proper.

That the autonomic and endocrine systems are closely interrelated both structurally and functionally becomes apparent when the hypothalamus is considered (p. 970). This acts as an integrating centre for both systems, and is the major site at which their activities combine. As well as its classic nervous functions, it also operates as an endocrine organ, producing by neurosecretion a wide variety of peptide hormones including releasing and release-inhibiting factors which control the activity of the adenohypophysis, a major endocrine gland itself. Although conventionally considered separately, the autonomic system, diffuse neuro-endocrine and endocrine system proper, collectively represent a single neuro-endocrine regulatory system which operates in a co-ordinated manner to control the metabolic activities and internal environment of the organism, providing

conditions in which it can function in a biologically successful manner.

There are, in addition to the discrete endocrine glands and diffuse neuro-endocrine system described in the following section, other hormone-producing cells forming minor components of other systems of the body and described with those systems. These include the pancreatic islets (p. 1371), the gastro-entero-endocrine cells (p. 1364), certain cells of the thymus (p. 780) and kidney (p. 1396), endocrine cells of the lung (p. 1259), the interstitial cells of the testis (p. 1415), and the interstitial, follicular, ovarian and luteal cells of the ovary (p. 1424). During pregnancy the placenta (p. 135) is also a major endocrine organ.

The Hypophysis Cerebri

The hypophysis cerebri (pituitary gland; 8.194–198) is a reddish-grey and somewhat ovoid body, measuring about 12 mm in transverse and 8 mm in anteroposterior diameter. It weighs about 500 mg. It is continuous with the apex of the infundibulum, a hollow conical projection from the inferior aspect of the tuber cinereum (p. 966). The gland lies in the hypophysial fossa of the sphenoid bone, where it is overlapped by a circular fold of dura mater, the *diaphragma sellae*; this fold has a small central aperture through which the infundibulum passes, and it separates the anterior part of the superior surface of the hypophysis from the optic chiasma (8.195). On each side, the hypophysis is related to the cavernous sinus and the structures which it contains (p. 746). Inferiorly, it is separated from the floor of the fossa by a large, partially loculated, venous sinus, which communicates with the circular sinus (p. 748). The meninges blend with the capsule of the hypophysis and cannot be identified as separate layers in the fossa.

The terminology of the divisions and subdivisions of the hypophysis cerebri is still somewhat confused. There is, however, a basic division into two regions of different embryological, morphological and functional characteristics. One is a downgrowth from the diencephalon connected by neural pathways with the hypothalamus, the other is a derivative of the ectodermal roof of the stomatodeum. These are named respectively the *neurohypophysis* and *adenohypophysis*. Both include parts of the *infundibulum* or hypophysial stalk, and in this differ from the older terms—anterior and posterior lobes, which do not include the stalk. The infundibulum itself consists of a neural core, the *infundibular stem*, which contains the neural connexions of the hypophysis, and this in turn is continuous with the *median eminence* of the tuber cinereum (p. 966). Thus, the *neurohypophysis* is commonly, but not invariably, taken to include the median eminence, the infundibular stem, and the main posterior part of the hypophysis—the *neural lobe* or *pars posterior*. The macroscopic infundibulum also comprises an extension of the adenohypophysis, which largely surrounds most of the neural infundibular stem; this is the *pars tuberalis* or *infundibularis*. The major region of the adenohypophysis is divisible into a *pars anterior* (pars distalis) and a *pars intermedia*, separated in fetal and early post-natal life by the hypophysial cleft, a vestige of the cavity of the ectodermal diverticulum, Rathke's pouch (p. 199), from which the adenohypophysis develops. This cleft usually becomes obliterated in childhood, persisting only as a variable series of cystic cavities scattered near the adeno-neurohypophysial interface and sometimes invading the neural lobe. The pars intermedia is, in any case, rudimentary in mankind, and because it may also be partially displaced into the neural lobe it has been included in both the pars anterior and pars posterior by different authorities. Apart from this terminological equivocation, which is of little significance in the human hypophysis, in view of the exiguous status of the pars intermedia, the pars anterior and pars posterior (or nervosa) may be equated with the terms anterior and posterior lobes. When the associated parts of the infundibulum which are continuous with these two lobes are considered with them, the names adenohypophysis and neurohypophysis become appropriate usage. To recapitulate therefore, in the present account the following terms are used:

(1) The *neurohypophysis* includes the pars posterior (pars nervosa, posterior, or neural lobe), the infundibular stem and the median eminence.

(2) The *adenohypophysis* includes the pars anterior (pars distalis or glandularis) and the pars intermedia, together with the pars tuberalis.

The various subdivisions of the hypophysis cerebri are shown in **8**.195, 196A, 198. They differ considerably in the type of cells, their arrangement, and in the details of their vascular and neural supplies, and these features have engendered a formidable literature, which can only be represented here briefly and simply.

THE ADENOHYPOPHYSIS

The adenohypophysis (**8**.196, 197) is a highly vascular structure, consisting of epithelial cells of varying size and shape arranged in cords, irregular masses or follicles, separated by thin-walled vascular sinusoids and supported by a complex network of reticular tissue.

It is known from numerous experimental and clinical investigations that at least seven distinct hormones are synthesized and released from the adenohypophysis, most of them being *tropic hormones* such as *somatotropin* (STH) which is involved in the control of body growth to maturity; *mammotropin* or *lactogenic hormone* (LTH) which stimulates the growth and secretory activity of the female mammae during pregnancy; *adrenocorticotropin* (ACTH) which governs the secretion of some of the hormones by the adrenal cortex; *thyrotropin* (TSH) which stimulates thyroid activity; *follicle-stimulating hormone* (FSH) which stimulates growth of ovarian follicles and their secretion of oestrogens in the female, and spermatogenesis in the male; *interstitial-cell-stimulating hormone* (ICSH) which activates androgen secretion by the Leydig cells of the testis and progesterone secretion by the corpus luteum (then termed luteinizing hormone, LH); and finally *melanocyte-stimulating hormone* (MSH) which causes an increase in cutaneous pigmentation. Most of these hormones also have many other complex metabolic effects which are, however, beyond the scope of the present account.

CELL TYPES OF THE ADENOHYPOPHYSIS

The problem of the cellular sites of secretion of the adenohypophysial hormones has attracted much study. A wide variety of methods have been developed to distinguish between different cell types by their staining reactions (Harris and Donovan 1966); the earlier techniques used the simple criterion of the differential binding affinities of the cells for mixtures of acidic and basic dyes; an example of the former is orange-G and of the latter, aldehyde fuchsin. Those cells which stained strongly were termed *chromophilic cells*, those showing little affinity for dyes, *chromophobic cells*. Of the chromophilic cells, those which bound acidic dyes were classed as *acidophils* (α-cells), in contrast to the *basophils* (β-cells) which stained strongly with basic dyes.

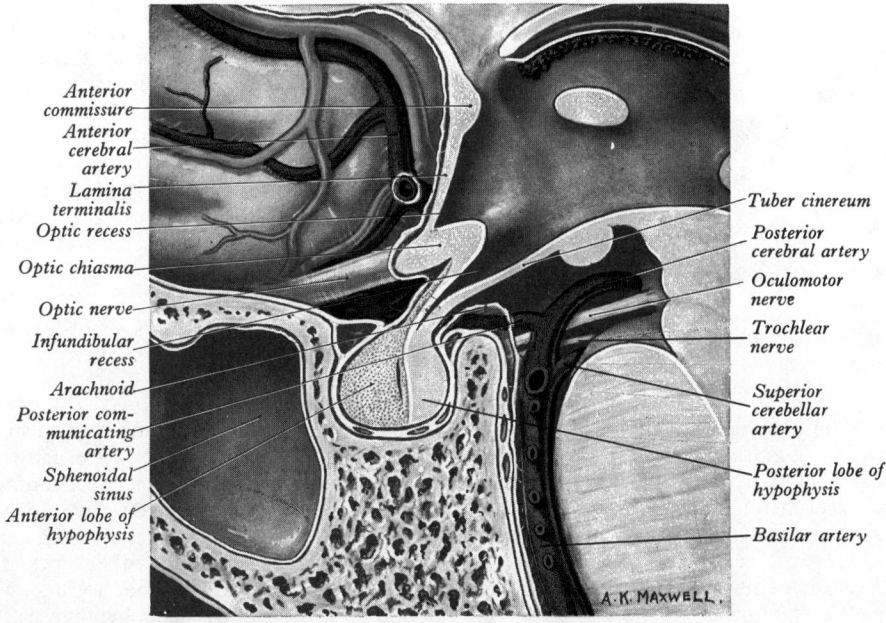

Anterior commissure
Anterior cerebral artery
Lamina terminalis
Optic recess
Optic chiasma
Optic nerve
Infundibular recess
Arachnoid
Posterior communicating artery
Sphenoidal sinus
Anterior lobe of hypophysis

Tuber cinereum
Posterior cerebral artery
Oculomotor nerve
Trochlear nerve
Superior cerebellar artery
Posterior lobe of hypophysis
Basilar artery

A. K. MAXWELL.

8.195 A median section through the hypophysis, *in situ*.

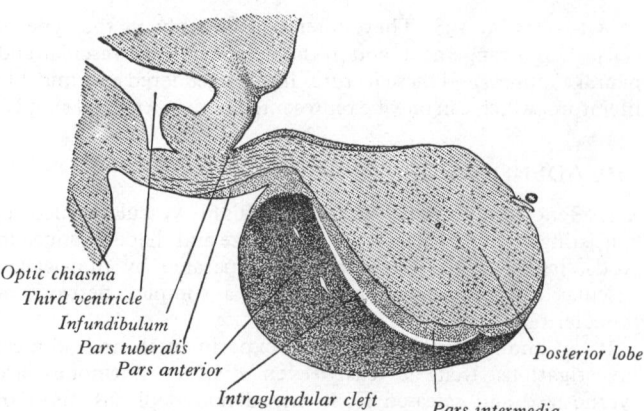

Optic chiasma
Third ventricle
Infundibulum
Pars tuberalis
Pars anterior

Posterior lobe

Intraglandular cleft

Pars intermedia

8.196A Diagram of a median section through the hypophysis cerebri of an adult monkey.

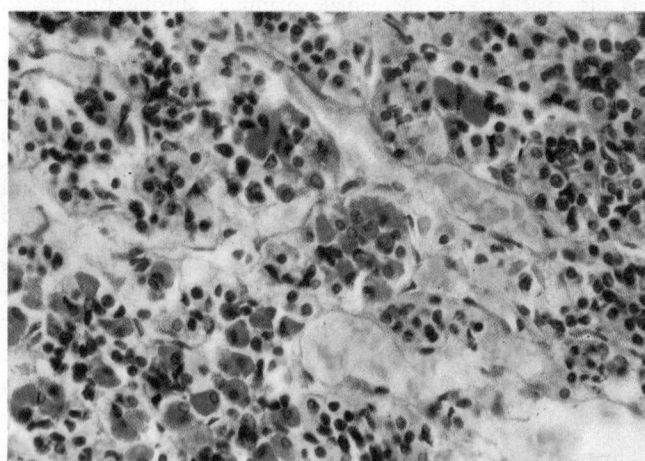

8.196B A medium-power micrograph of adenohypophysial cells stained with the PAS-orange G technique, to distinguish between α-cells (yellow), β-cells (red to brown) and chromophobic cells (pale grey with little cytoplasm). A number of large vascular sinusoids containing yellow-stained erythrocytes are also prominent. Magnification × 400.

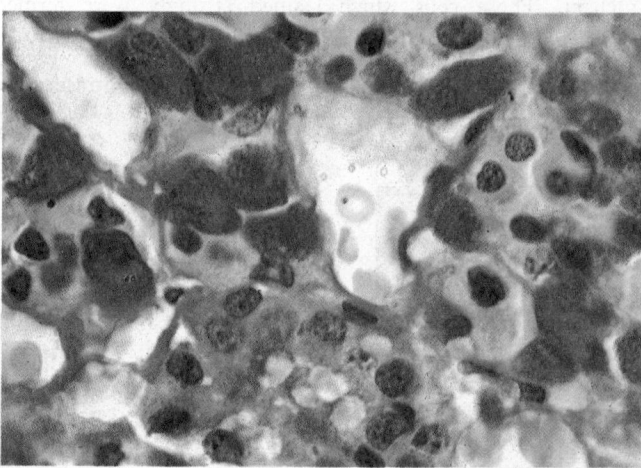

8.196C A high-power micrograph of a group of cells stained as for B, grouped around a vascular sinusoid. Note *a*-cells (yellow) and *β*-cells (red). Magnification × 1,000.

Modifications of such techniques using complex, multi-stage staining methods, further distinguished many sub-categories of cells, each with its own tinctorial nuance. These techniques, applied particularly to experimental or pathological pituitary glands from individuals with known hormonal defects, were used to correlate specific cell types with specific hormones. However, many uncertainties prevailed, because of the difficulty in the standardization of stains, and the further difficulty of classifying the endless minor variations in different individuals and species.

More recently, the methods of immunochemistry have been brought to bear on this problem (Nakane 1970), and it has been possible by immunofluorescence techniques to identify the site of synthesis of individual hormones with some certainty. Such methods, combined with ultrastructural and cell fractionation studies, have clarified the problem considerably although some difficulties still remain. At the present time all of the major hormones have been assigned to particular cell types; the terminology stemming from the older schemes of classification has, however, been generally retained, and the cells will be described briefly here. Ultrastructural studies of the normal human adenohypophysis are rare, and much of the information available on the fine structure of the cells of the adenohypophysis has been obtained, of necessity, from studies on non-human primates. In the following description the cytological details given refer, unless otherwise stated, to the adenohypophysis of the Rhesus monkey (Costoff 1977).

The pars anterior contains the following cell varieties:

A. *The chromophil cells*

1. *Acidophils (a cells)*
(*a*) *Somatotrophs.* Usually arranged in groups along the sinusoids, these ovoid cells are the largest and the most abundant of the adenohypophysial acidophils. They secrete proteinaceous STH (GH) and their distinguishing stain is orange G. They contain large numbers of electron-dense, spherical, secretory granules of 350 to 500 nm diameter, and have a well-developed Golgi complex although the granular endoplasmic reticulum is not particularly extensive. Cells with a single cilium have been observed. Cells having a similar fine structure have been found to characterise human eosinophilic adenomata (Zambrano *et al.* 1968).
(*b*) *Mammotrophs.* These secrete the proteinaceous hormone prolactin, and are the dominant cell type during pregnancy. They hypertrophy during lactation. Their distinguishing stain is erythrosin but they also have an affinity for azocarmine. Their secretory granules, which are the largest of any hypophysial cell type (600 to 1000 nm in diameter), are evenly dense and either ovoid or irregular, the latter type resulting from the fusion of the former. Excess granules fuse with lysosomes to form autophagic vacuoles within which they are degraded by hydrolytic enzymes. In active cells the granular endoplasmic reticulum and Golgi complex are both well-developed. Mammotrophs are present in the adenohypophyses of males as well as females but have more abundant secretory granules in the latter.
(*c*) *Corticotrophs.* These secrete ACTH. They are small cells, grouped in clusters, often arranged as follicles around central masses of glycoprotein (von Lawzewitsch *et al.* 1972) and stain faintly with periodic acid/Schiff. They are irregularly stellate in shape and their sparse secretory granules, some 200 to 250 nm in diameter, are generally located immediately beneath the plasma membrane. The characteristic shape of the cells and the peripheral location of the secretory granules are useful aids to the identification of corticotrophs (Siperstein and Miller 1970). The granular endoplasmic reticulum is of a scattered vesicular or short lamellar type and the Golgi complex is somewhat small. Although classified here as α-cells, some authorities have placed the corticotrophs in the β-cell category (Baker *et al.* 1970). It is possible that their staining characteristics and hence their classification may vary with their metabolic state. The smallness and sparseness of their secretory granules, together with their poor staining properties, have frequently led to the classification of corticotrophs as chromophobes.

2. *Basophils (β cells)*
(*a*) *Thyrotrophs* (Beta-basophils). These secrete TSH. They are elongate polygonal cells arranged in clusters towards the centre of the adenohypophysis. They are usually found deep within the cell cords and are thus not in direct contact with the sinusoids. They are stained selectively by aldehyde fuchsin. Their peripherally arranged granules are irregular in shape and are less electron-dense than those of other types of basophil; being only 100 to 150 nm in diameter, they are among the smallest granules found in the cells of the adenohypophysis.
(*b*) *Gonadotrophs (Delta-basophils).* These are larger than the

thyrotrophs, are round in shape, and usually lie next to the sinusoids. Their secretory granules have an affinity for periodic acid/Schiff. In some cells, usually peripheral in position, the granules stain purple, while in others, more centrally placed, they stain red, and it has been suggested that the former secrete FSH and the latter LH or ICSH (Purves 1961). Ultrastructural studies of the rat adenohypophysis have also distinguished two separate cell types (Kurosumi and Oota 1968). However, immunocyto-chemical studies of Phifer *et al.* (1973) in man and of Moriarity (1975) in the rat suggest that FSH and LH may be present simultaneously in the same cell. Gonadotrophs have pleomorphic nuclei. Their spherical granules are from 200 to 300 nm in diameter, their granular endoplasmic reticulum is vesicular and the Golgi complex well-developed. Gonadotrophs similar to those of the Rhesus monkey have been described in man (von Lawzewitsch *et al.* 1972).

B. *The chromophobe cells*

Chromophobe cells are present as inconspicuous clusters of small elements with scanty cytoplasm, and an ultrastructural appearance suggestive of metabolic inactivity. They may represent a non-secretory phase in the activity of other glandular cell types.

Variations occur in the distribution of these diverse cell types in the adenohypophysis. Generally, similar cells occur in groups, probably reflecting their pattern of development.

The pars intermedia contains numerous β-cells, and chromophobe cell-follicles surrounding PAS-positive colloidal material. Some of these are pouches derived from the embryonic intrahypophysial cleft (i.e. the cavity of the pouch of Rathke). In some species in which this cavity remains large in the fully developed gland, a ciliated lining epithelium has been described.

The secretory cells of the pars intermedia contain granules containing either α-endorphin or β-endorphin scattered uniformly throughout their cytoplasm (Bloom *et al.* 1977; Guilleman 1978). These same cells have also been shown by numerous investigators to contain a variety of peptide hormones including ACTH and α-MSH (melanocyte stimulating hormone). They are considered to belong to the APUD series of cells (p. 1454), as are the other secretory cells of the adenohypophysis. Endorphin has also been found in the pars anterior, but here it is localised to peripherally placed granules in discrete groups of cells.

The pars tuberalis is characterized by the large number of blood vessels traversing it. Between these are cords or balls of undifferentiated cells admixed with some α- and β-cells.

The secretion of adenohypophysial hormones into the circulation appears to occur by the exocytosis of the vesicular contents into the perivascular spaces of the sinusoids, the latter being lined with a fenestrated endothelium which facilitates diffusion into the bloodstream. The signal for secretion is the liberation of chemical releasing factors from neurons (McKelvy 1974) in the median eminence, the nucleus infundibularis (arcuate nucleus), and other hypothalamic centres, into the upper radicles of the portal system of veins which carries them to, and distributes them within, the adenohypophysis. These neurosecretory cells are effectively *neuroendocrine transducers* (Wurtman 1970), in that they receive neural signals and respond to them by secreting hormones; they thus convert one type of signal into another. The neurons producing releasing factors are peptidergic, while the neurons which modulate their activity are largely monoaminergic and make either axosomatic or axo-axonic synapses with them.

Tanycytes (p. 836, **8.197A** and B) may also be involved in secretion control, possibly transporting hormones from the cerebrospinal fluid to the capillaries of the portal system (Knigge 1976) and/or from hypothalamic neurons to the cerebrospinal fluid (Joseph and Knigge 1978). The presence of tight junctions between them and neighbouring ependymal cells (Brightman *et al.* 1975), although of a somewhat rudimentary nature compared, say, with those of the choroid plexus, impedes the movement of peptides and even of some amino acids (Weindl and Joynt 1972),

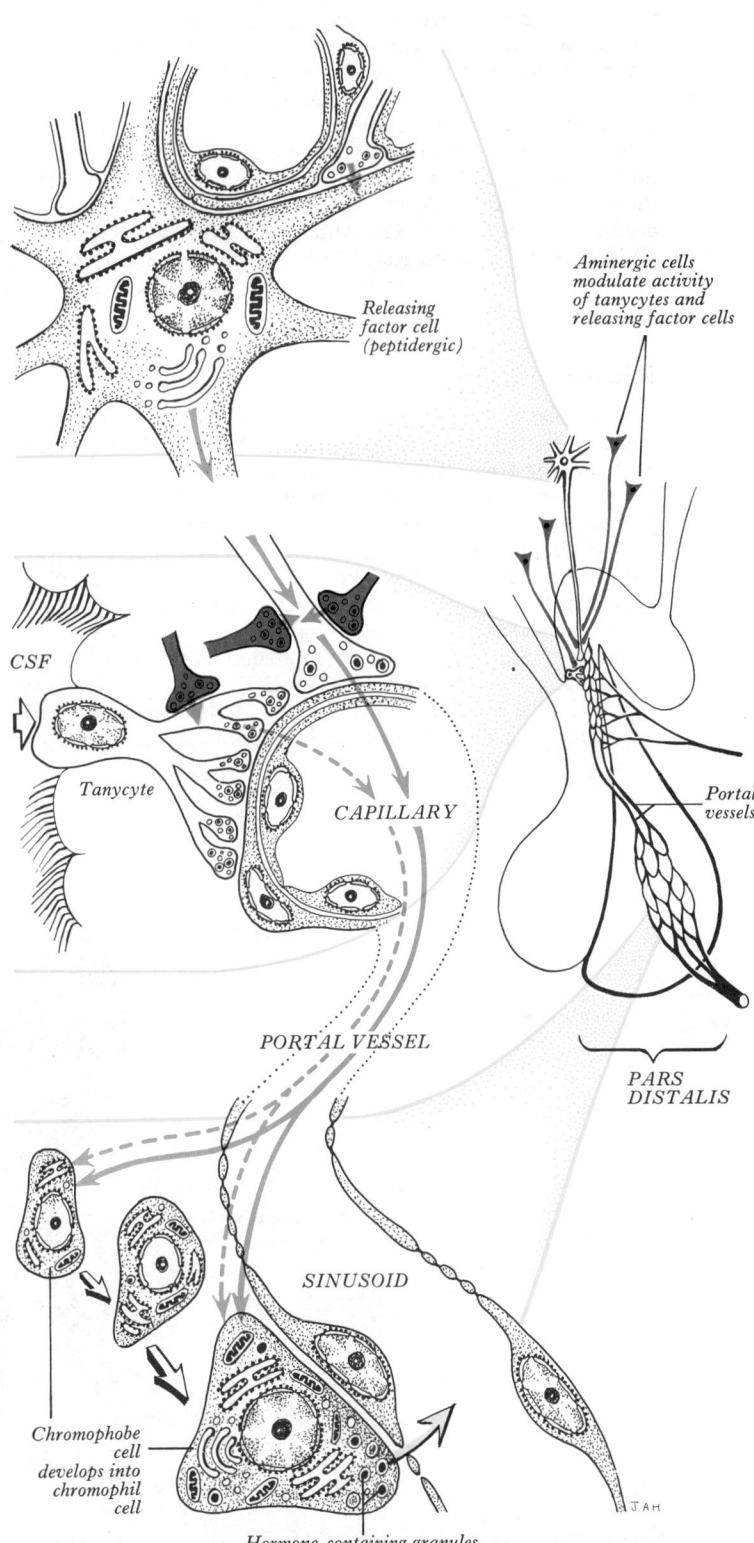

8.197A A diagram illustrating the control systems which influence the hormonal output of the adenohypophysis. The small diagram on the right shows the ventral hypothalamus, median eminence, and adenohypophysis (pars distalis), and their associated neurons and vasculature. On the left three zones are shown in greater detail. The concatenation of events is as follows: under the influence of blood-borne factors and neural stimuli, the hypothalamic neuron shown above liberates specific releasing factors into capillaries of the median eminence. Modulation at this point is mediated by aminergic neurons and tanycytes. Onward transport of these factors through portal vessels is indicated by blue arrows. Stimulation and hormone production and release on the part of the adenohypophysial cell are indicated.

and suggests that materials such as releasing factors would have to pass across the ependyma through, rather than between, the tanycytes. The endocytotic (and presumably also the exocytotic) activity of tanycytes appears to be under monoaminergic neuronal control (Kobayashi 1975; Nozaki *et al.* 1975). The tanycytes of the walls of the ventricular recess appear to be well suited for the transport of releasing factors from neurosecretory cells to the cerebrospinal fluid, for they course through the bed of the arcuate nucleus and are linked to its neurons by junctions which permit the passage of low molecular weight substances such as peptide hormones between the cells which they unite (del Cerro and Knigge 1977). The suggested involvement of the tanycytes in secretion control by transporting releasing factors awaits experimental verification.

THE NEUROHYPOPHYSIS

The posterior lobe of the hypophysis is developed as a downgrowth from the floor of the diencephalon, and during early fetal life contains a cavity continuous with that of the third ventricle. In some animals, for example the cat, this cavity persists throughout life. The posterior lobe, infundibular stem and median eminence are frequently termed collectively the *neurohypophysis (vide supra)*. (For general reviews consult Harris 1955; Heller and Clark 1962; Scharrer and Scharrer 1963; Gabe 1966; Donovan 1970; Knowles 1974.)

Axons stemming from perikarya situated in the hypothalamus (supra-optic and paraventricular nuclei, and other sites) are present in the neurohypophysis; some of the fibres are short, ending in the median eminence and infundibular stem in relation

these hormones has been described (p. 1439); they are passed from the perikarya along the axons of the tract, and are finally released at the nerve terminals. Ligation of the tract causes a proximal damming of the secretion granules, which can be stained, by virtue of their high glycoprotein content, by the PAS technique. The active hormones are simple polypeptides, but they are elaborated and transported within the cell with a glycoprotein, *neurophysin*. When released into the circulation the association between the two is broken, and the hormone is then carried by plasma glycoproteins to its target sites.

Histologically, the neurohypophysis is composed of thin, non-myelinated nerve fibres and associated cells, which are the terminal ramifications of the hypothalamohypophysial tract. Proximally, in the infundibulum, the fibres are ensheathed by typical astrocytes, but towards the posterior lobe another type of cell, the *pituicyte*, makes its appearance. The latter is a dendritic cell with variable appearance, but often possessing long processes which lie parallel to the adjacent axons, and which form the bulk of the non-excitable tissue of the neurohypophysis. The cytoplasmic processes of many pituicytes terminate on or near the surfaces of neighbouring capillaries and sinusoids between terminations of nerve fibres. The latter end in the perivascular spaces, and, although they approach the wall of the sinusoid fairly closely, they remain separated from it by two basal laminae, one applied to the nerve ending, the other to the surface of the endothelial cells. Some fine collagen fibres are also often interposed between the two. In regions other than the perinuclear zone the cytoplasm of the endothelial cells is extremely attenuated, and presents regular fenestrations (**8.**197C).

Three distinct types of nerve terminals have been described in

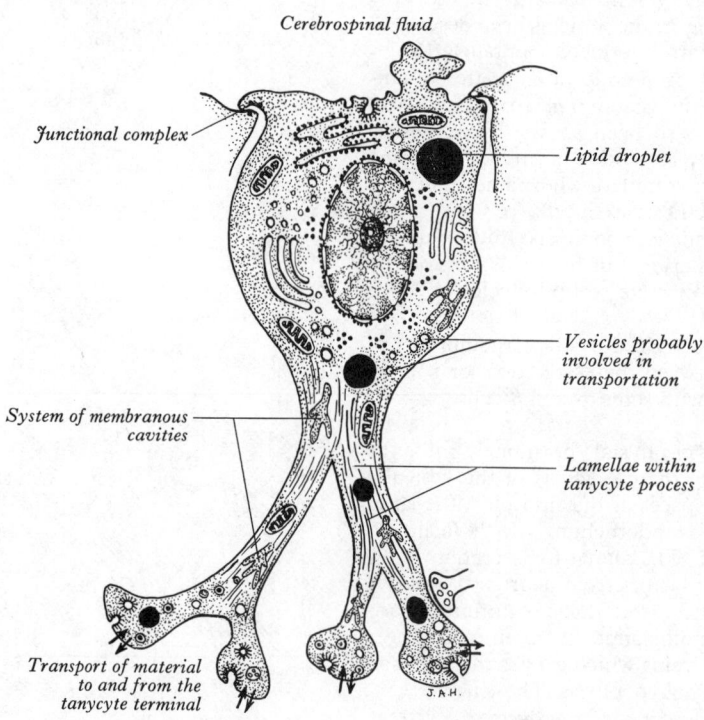

Cerebrospinal fluid

Junctional complex

Lipid droplet

Vesicles probably involved in transportation

System of membranous cavities

Lamellae within tanycyte process

Transport of material to and from the tanycyte terminal

J.A.H.

8.197B The general ultrastructure of a tanycyte. (See text on pp. 836 and 1441 for explanatory details.)

to the superior capillary beds of the venous portal circulation, and thus providing the possibility of neural control of adenohypophysial function (*vide infra*). Longer axons from these nuclei pass to the main mass of the neurohypophysis, where they terminate in relation to the vascular sinusoids. These nerve fibres constitute the neurosecretory hypothalamohypophysial tract, which is considered in greater detail elsewhere (p. 969). The chief hormones concerned are vasopressin (antidiuretic hormone, ADH) which controls the reabsorption of water by the kidney tubules, and oxytocin, which promotes the contraction of uterine and mammary nonstriated muscle. The site of production of

the posterior lobe. These are: (1) terminal axonal swellings lying adjacent to the vascular sinusoids, which contain large (200–300 nm) dense vesicles, characterized biochemically as containing hormones bound to glycoproteins; they also contain clear, small (40–60 nm) spherical vesicles; (2) peri-axonal endings containing small (80 nm) dense-cored vesicles, similar to catecholamine-containing vesicles found in sympathetic nerve fibre endings (p. 775); (3) peri-axonal endings containing small (40–60 nm) clear spherical vesicles; the latter form synapses with the large hormone-containing endings (Bargmann 1966). The interactions between these various nerve terminals is not entirely clear. The hormones appear to be released by exocytosis from the endings containing large dense vesicles. It is possible that excitation or inhibition of release is mediated by the other types of endings, or possibly by the excitation of the hormone-containing endings themselves.

THE VESSELS OF THE HYPOPHYSIS

These have been studied in preparations injected with neoprene latex or with the dye Berlin blue, and also by certain techniques which stain erythrocytes (Popa and Fielding 1930 *a, b*; Green and Harris 1949; Xuereb *et al.* 1954 *a, b*; Stanfield 1960). More recently, scanning electron microscopy has been used to examine corrosion casts in attempts to determine the details of the angio-architecture of the median eminence and the vascular re-lationships within and between the neurohypophysis and the adenohypophysis (Page and Bergland 1977).

The arteries of the pituitary gland stem from branches of the internal carotid artery, a single *inferior* and a series of *superior hypophysial arteries* on each side. The inferior hypophysial artery arises from the cavernous portion of the internal carotid artery, while the superior hypophysial arteries arise from its supraclinoid portion and from the anterior and posterior cerebral arteries. Each inferior hypophysial artery divides into medial and lateral branches which anastomose with their fellows from the opposite side forming an *arterial ring* around the infundibular process of the neurohypophysis. From this ring several small branches pass centrally to supply the neurohypophysis, breaking up into a capillary bed within its substance. The superior hypophysial arteries supply the median eminence, the upper part of the infundibulum and, via branches termed the *arteries of the trabeculae*, the lower infundibulum. A confluent capillary bed extends through the whole of the neurohypophysis (the median eminence, infundibular stem and pars nervosa), and is supplied with arterial blood from two sources, the inferior hypophysial arteries below and the superior hypophysial arteries above. It has been shown that *reversal* of blood flow can occur in cerebral capillary beds lying between two sources of arterial supply (Gillilan 1974), and similar reversal has been suggested in the neurohypophysial capillary bed (Page and Bergland 1977).

The arteries of the median eminence and the upper and lower parts of the infundibulum end in characteristic tufts of capillaries, the most complex being in the upper infundibulum. Two distinct capillary plexuses are present in the median eminence: an *external* or 'mantle' plexus (Green 1951) and an *internal* or 'deep' plexus (Duvernoy 1972). The external plexus is supplied by the superior hypophysial arteries, is continuous with the capillary bed of the infundibular stem, and is drained by *long portal vessels* which descend to the pars anterior. The internal plexus projects into the bowl formed by the external plexus, by whose vessels it is supplied. The internal plexus is continuous posteriorly with the capillary bed of the infundibular stem and, like the external plexus, is drained by *long portal vessels* which descend to the pars anterior. *Short portal vessels* run from the lower part of the infundibulum to the pars anterior. Both varieties of portal vessel open into vascular sinusoids lying between the secretory cell cords and clumps of the glandular part of the pituitary, providing virtually the whole of its vascular supply; there is no direct arterial supply (Wislocki and King 1936). The *portal system* is considered to be of great functional significance, carrying hormone-releasing factors, probably elaborated in the parvocellular groups of hypothalamic neurons and which control the secretory cycles of the cells of the pars anterior. Little is known of the blood supply of the pars intermedia, which appears to be avascular according to corrosion casts of the hypophysis (Page and Bergland 1977).

The venous drainage of the neurohypophysis is by three potential routes: to the adenohypophysis via the long and short portal vessels, to the systemic circulation via the large inferior hypophysial veins which open into the dural venous sinuses, or to the hypothalamus via small capillaries which pass between it and the median eminence. The venous drainage allows the hypo-physial hormones to leave the gland and be carried to their target organs. It also facilitates the feedback control of secretion. In contrast to the rich venous drainage of the neurohypophysis, that

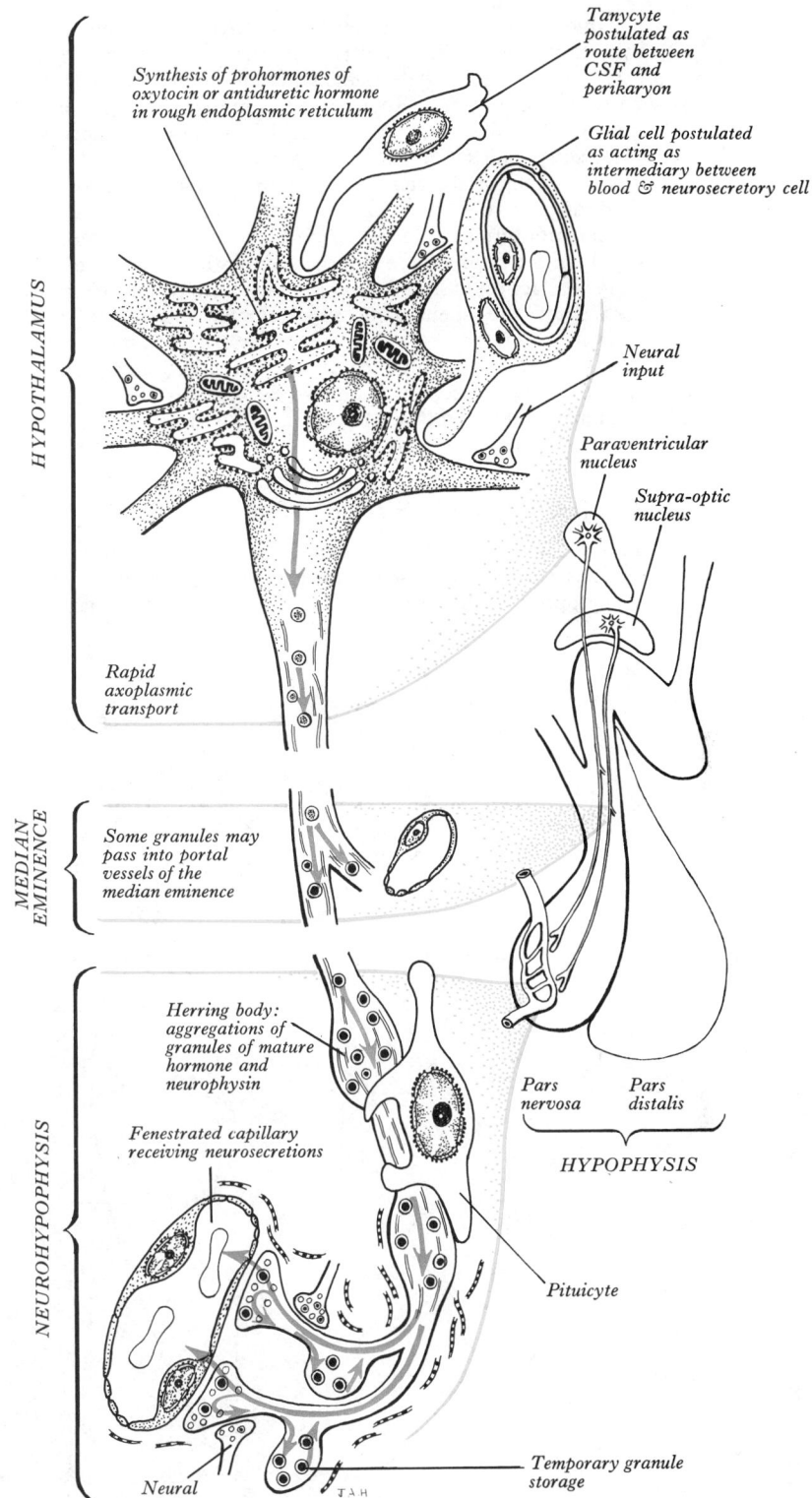

8.197C A diagram summarizing the control systems which influence the production of neurohypophysial hormones.

of the adenohypophysis appears to be somewhat restricted. Few veins connect it with the surrounding systemic venous structures directly, and this raises the question of the route by which blood leaves it. If flow in the short portal system between the adeno- and neurohypophysis were reversible (Adams *et al.* 1969), then these vessels could act as drainage channels. Adenohypophysial hormones would pass into the neurohypophysial capillaries before entering the systemic venous system and this would provide the 'short feedback' loop which has been postulated on the basis of endocrinological investigation. Reversed flow within the neurohypophysial capillary bed (from the neurohypophysis to the hypothalamus (Török 1954) would provide a vascular route by which hormones released in the neurohypophysis could reach the tanycytes of the ventricular surface and hence the cerebrospinal fluid. The demonstration of neurohypophysial hormones in this fluid (Robinson and Zimmerman 1973) and of releasing factors both here and in the tanycytes (Knigge and Joseph 1974) adds weight to this hypothesis. The concept that the neurohypophysis secretes into the third ventricle, proposed by Cushing in 1912, may yet prove to be correct.

The implications of recent advances in determining the blood supply of the pituitary are far-reaching. Rather than the median eminence acting as the final common pathway for the neural control of the adenohypophysis (Harris 1947), the entire neurohypophysis may be involved; its capillary bed may actively and selectively 'determine the destination of both hypothalamic and pituitary secretions, conveying some to the glandular pituitary, others to distant target organs, and yet others to the brain' (Page and Bergland 1977).

THE PHARYNGEAL HYPOPHYSIS

This is a small collection of adenohypophysial glandular tissue which is consistently present in the mucoperiosteum of the roof of the nasopharynx in man (Boyd 1956) and in many other mammals

(McGrath 1974). By 28 weeks of human intrauterine development, it is capable of hormone secretion and is richly vascularized, receiving its blood supply from the systemic vessels of the roof of the nasopharynx. At this stage it is covered posteriorly by fibrous tissue, but in the second half of fetal life this is replaced by numerous venous sinuses and an extended trans-sphenoidal portal venous system develops, bringing it under the same hypothalamic control as the cranial adenohypophysial tissue (McGrath 1978). The peripheral vascularity of the pharyngeal hypophysis persists until about the fifth year, after which it becomes re-invested in fibrous tissue and is presumed to be again under the control of factors carried in the systemic blood.

Although in the male the pharyngeal hypophysis does not change in size significantly after birth, in the female it becomes temporarily smaller, returning to its volume at birth during the fifth decade (McGrath 1971), when once again its control may be through factors carried in the trans-sphenoidal extension of the hypothalamo-hypophysial portal venous system. It has been suggested that the human pharyngeal hypophysis may provide a reserve of potentially functional adenohypophysial tissue which may be stimulated, particularly in the female, to synthesize and secrete adenohypophysial hormones in middle age when the intra-cranial adenohypophysial tissue is beginning to fail.

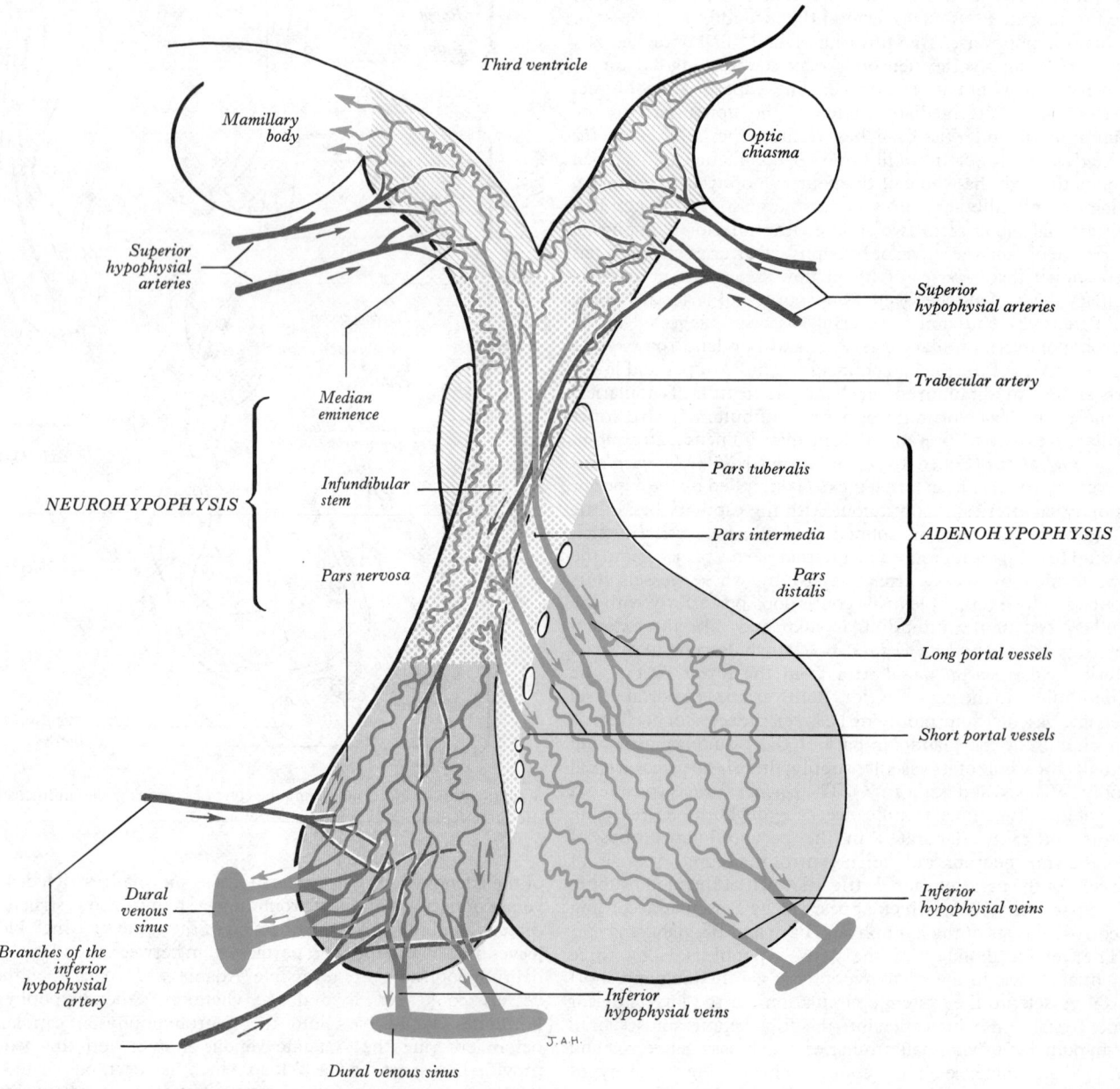

8.198 A diagram summarizing the vasculature of the hypothalamic median eminence, infundibulum, and the rest of the hypophysis cerebri. (See text for a detailed account and possible significances.)

The Pineal Gland

The pineal gland or *epiphysis cerebi* (**8**.194) is a small, piriform, reddish-grey organ which occupies the depression between the superior colliculi. (For classical reviews *see* Gladstone and Wakeley 1940; Kappers 1960; Wolfe *et al.* 1962; Kappers and Schadé 1965; Wurtman *et al.* 1968; Wolstenholme and Knight 1971: and see below for more recent references.) It is inferior to the splenium of the corpus callosum, but is separated from it by the tela choroidea of the third ventricle and the contained cerebral veins. It is enveloped by the lower layer of the tela, which is then reflected over the tectum (**8**.194). The pineal gland measures about 8 mm in length, and its base, directed anteriorly, is attached by a *peduncle* or stalk which divides anteriorly into two laminae, superior and inferior, separated from each other by the *pineal recess* of the third venticle (**8**.194). The inferior lamina contains the posterior commissure (p. 963) and the superior lamina the habenular commissure (p. 962).

Aberrant commissural fibres may loop into the substance of the gland through its stalk, but they do not terminate in relation to parenchymal cells. Nerve fibres enter the dorsal or dorsolateral aspects of the gland from the region of the tentorium cerebelli where they form a single or paired *nervus conarii*. This runs in a subendothelial position in the wall of the straight sinus (p. 745) and its fibres are derived from the cells of the superior cervical ganglia. The nerve fibres are adrenergic sympathetic elements and run in association with blood vessels and parenchymal cells (Kappers 1960; Wolfe *et al.* 1962). Björklund *et al.* (1972) have shown in the rat that postganglionic sympathetic fibres from the *nervus conarii* supply neurons in the habenular nuclei, from which some of the fibres of the *habenulopineal tract* (p. 963) may arise. Møller (1978) has reported that in the human fetus, fibres from this tract extend to the *ganglion conarii* (Pastori ganglion) at the pineal apex (**8**.199A); the site of the perikarya of origin of these fibres needs verification. Møller (1978, 1979) has also described a hitherto undiscovered unpaired nerve containing non-myelinated fibres and bipolar neurons with axosomatic and axodendritic synapses, the *nervus pinealis*, in human fetuses; it lies in the subarachnoid space in or near the median plane just caudal to the pineal gland, connecting it with the posterior commissure; it is an ephemeral structure, presumed to degenerate late in intrauterine life, since it has not been found postnatally. Its functional significance is uncertain, but since it is presumed to be homologous with the sensory pineal nerve of fish and amphibians (Møller 1978) which have pineal complexes containing photoreceptive cells, it has been suggested that it might be able to transmit light-generated impulses. However, although in some mammals, for example neonatal rats, the pinealocytes have a transient photoreceptor-like ultrastructure (Zimmerman and Tso 1975) those of human fetuses do not, so this suggestion is not supported. The nervus pinealis of the human fetus could also be involved in pineal differentiation. In the absence of further information, it can be regarded as a phylogenetic vestige present briefly during development; its loss may be associated with the evolution of the pineal as a secretory (*vide infra*) rather than a photoreceptive organ.

Møller (1978) has also described a new ganglion, presumed to be parasympathetic, lying rostral to the pineal gland and close to the choroid plexus of the third ventricle. An anteriorly located *intrapineal ganglion* has also been found (**8**.199A).

STRUCTURE OF THE PINEAL GLAND

The pineal gland consists of cords and follicles of pinealocytes and neuroglial cells, between which ramify copious blood vessels and nerve fibres. Connective tissue septa extend into the gland from the surrounding pia mater.

The pinealocytes are the characteristic cells of the pineal parenchyma. Extending from the cell body, which contains a spherical, oval or lobulated nucleus, are one or more tortuous basophilic processes within which microtubules lie in parallel array (Knight *et al.* 1973). The processes end in slightly expanded terminal buds close to blood capillaries or, less frequently, near the ependymal cells of the pineal recess. The terminal buds

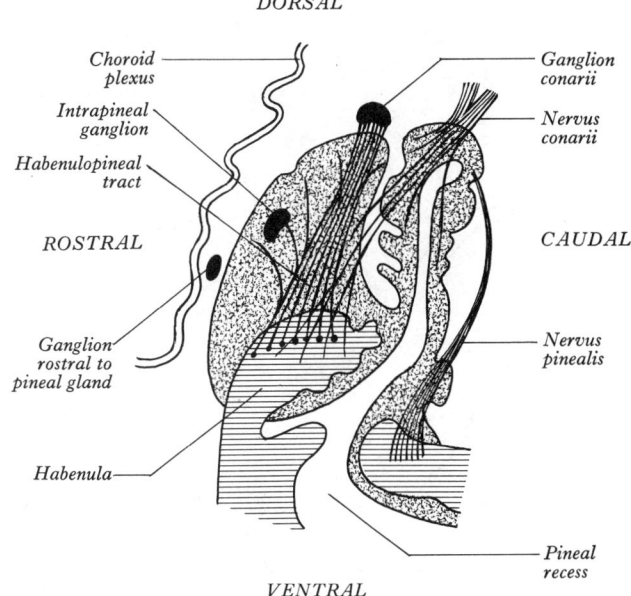

8.199A A diagram showing the principal neural pathways which have been described in connexion with the human fetal pineal gland (epiphysis cerebri).

contain granular endoplasmic reticulum, mitochondria and electron-dense cored vesicles. These vesicles store monoamines and polypeptide hormones (Sheridan and Sladek 1975), the release of which appears to require sympathetic innervation. Lukaszyk and Reiter (1975) have proposed that the polypeptide hormones (which in their opinion could also be produced by the neuroglial cells and neurons of the pineal) form complexes with specific carrier proteins, which they term *neuro-epiphysins* to distinguish them from the neurophysins of the pituitary. They are released by exocytosis, together with fragments of the vesicle membrane which contained them, the latter forming exocytotic debris. Once released, the complex is believed to dissociated, the hormones being exchanged from calcium ions. The calcium-carrier complex so formed is, in the case of the pineal, deposited in concentric layers around the exocytotic debris, forming *corpora arenacea* or 'brain sand' (**8**.199B). No evidence supports the notion, once widely held, that the pineal atrophies with age and that the development of corpora arenacea is a sign of this atrophy; on the contrary, the presence of these multilaminar corpuscles may indicate continued secretory activity. Wildi and Frauchiger (1965) have found no evidence of pineal degeneration in the elderly.

The cell bodies of the pinealocytes contain both granular and agranular endoplasmic reticulum, together with extensive Golgi complexes, lipid droplets and numerous mitochondria, as befits cells with a secretory function. Occasionally an unusual organelle, consisting of groups of microtubules and perforated lamellae, is found close to the granular endoplasmic reticulum and lipid droplets. Termed 'canaliculate lamellar bodies' by Lin (1967) and McNeil (1977), 'annulate lamellae' by Friere and Cardinali (1975) and 'mikrotubuli' by Gusek (1976), it has been suggested that these organelles may be involved in secretion (McNeil, 1977); their function is, however, as yet somewhat enigmatic.

Another feature of the pinealocyte of some mammals is the presence of synaptic ribbons, which may be involved in impulse transmission between adjacent cells. Close to the ribbons are vesicles containing neurotransmitters such as gamma amino butyric acid (Krstic 1976). These structures may arise from microtubular sheaths which in turn arise from centrioles (Karasek 1976). Similar arrangements of organelles have been found in the photoreceptor cells of the mammalian retina and in the simpler photoreceptor cells of a range of submammalian vertebrates, supporting the theory that the mammalian pinealocyte is phylogenetically derived from a photoreceptive element (Kap-

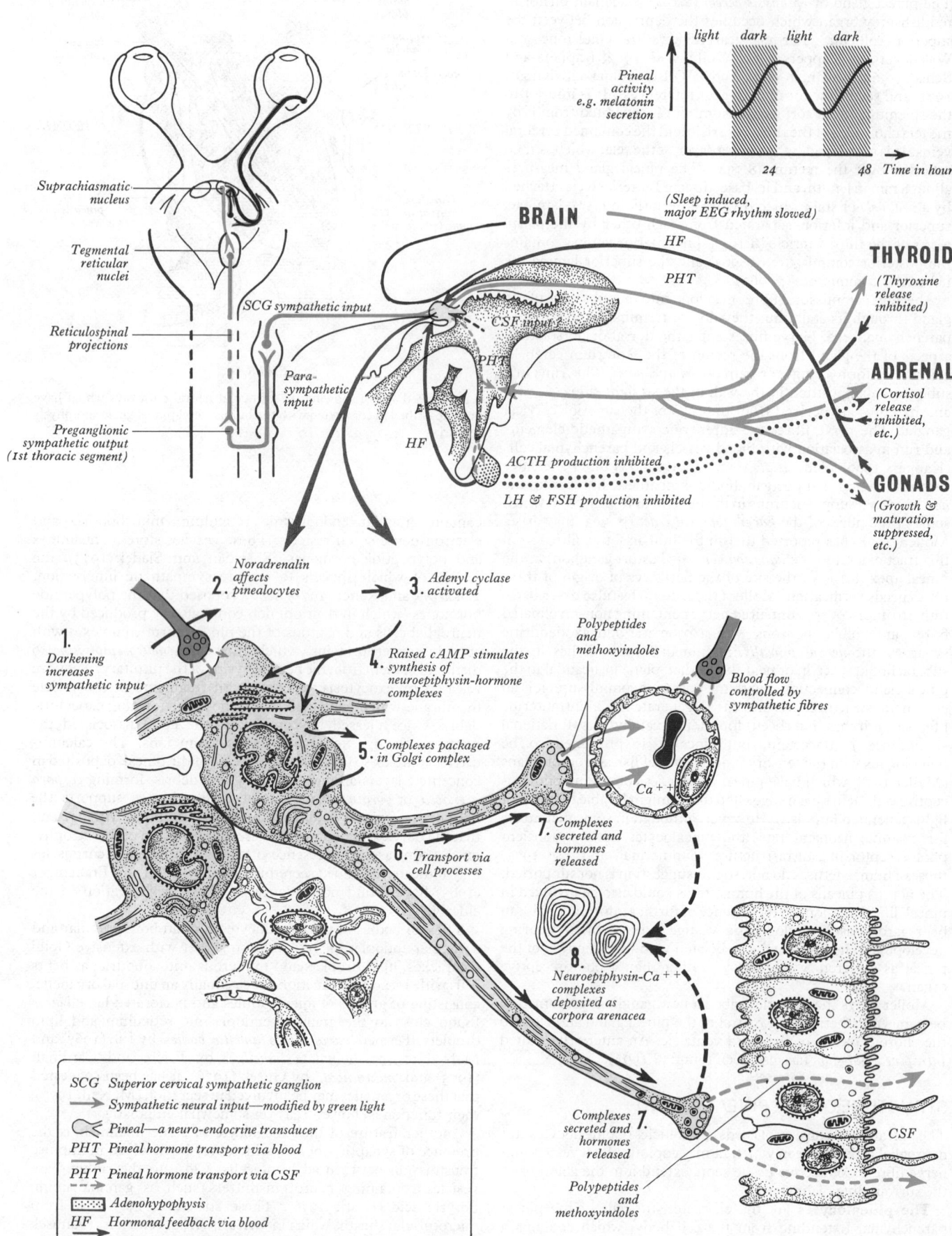

Suprachiasmatic nucleus

Tegmental reticular nuclei

Reticulospinal projections

Preganglionic sympathetic output (1st thoracic segment)

SCG sympathetic input

Parasympathetic input

BRAIN

(Sleep induced, major EEG rhythm slowed)

HF

PHT

CSF input

PHT

HF

ACTH production inhibited

LH & FSH production inhibited

Pineal activity e.g. melatonin secretion

light   dark   light   dark

24   48   Time in hours

THYROID

(Thyroxine release inhibited)

ADRENAL

(Cortisol release inhibited, etc.)

GONADS

(Growth & maturation suppressed, etc.)

1. Darkening increases sympathetic input

2. Noradrenalin affects pinealocytes

3. Adenyl cyclase activated

4. Raised cAMP stimulates synthesis of neuroepiphysin-hormone complexes

5. Complexes packaged in Golgi complex

6. Transport via cell processes

7. Complexes secreted and hormones released

Polypeptides and methoxyindoles

Blood flow controlled by sympathetic fibres

Ca^{++}

8. Neuroepiphysin-Ca^{++} complexes deposited as corpora arenacea

7. Complexes secreted and hormones released

Polypeptides and methoxyindoles

CSF

SCG Superior cervical sympathetic ganglion

Sympathetic neural input—modified by green light

Pineal—a neuro-endocrine transducer

PHT Pineal hormone transport via blood

PHT Pineal hormone transport via CSF

Adenohypophysis

HF Hormonal feedback via blood

8.199B   A diagrammatic review of current hypotheses regarding the control and effects of pineal function. (See p. 1447 for details.)

pers 1976; Relkin 1976). Further support for this comes from the finding by Zimmerman and Tso (1975) of a transient similarity between pinealocytes and retinal photoreceptor cells in the neonatal rat.

There is ultrastructural evidence that human fetal pinealocytes have a secretory function early in intrauterine life (Møller, 1974). As in adults, they contain all the organelles necessary for such activity, together with copious microfilaments, some microtubules, and occasional cilia with a 9+0 microtubular pattern. Cilia of this type are also associated with secretory cells of other endocrine glands, such as the pituitary (Barnes 1961; Andersen *et al.* 1970). Fetal pinealocytes are linked by gap junctions, desmosomes and 'intermediate-like junctions' (Møller 1976); the first of these allows for electrotonic coupling between adjacent pinealocytes and could make group activity possible, while the others may be of an adhesive nature. Synaptic ribbons do not appear to be present in human fetal pinealocytes.

(*vide supra*), or directly through the leptomeningeal surface of the pineal. They terminate either in the perivascular spaces, between the pinealocytes, or occasionally in synaptic relation with these cells. The nerve fibres passing along the pineal stalk from other brain centres are of rather imprecise origin, but the habenular complex of nuclei has become a strong proponent.

FUNCTIONS OF THE PINEAL GLAND

Once considered to be a phylogenetic relic, the vestigial remains of a dorsal third eye, and an organ of little functional significance, the mammalian pineal gland has now been demonstrated to be an endocrine gland of major regulatory importance, modulating the activity of the adenohypophysis, neurohypophysis, endocrine pancreas, parathyroids, adrenal cortex, adrenal medulla and gonads (De Vries and Kappers 1971; Kappers 1976; Relkin 1972; Reiter 1977; Klein 1978).

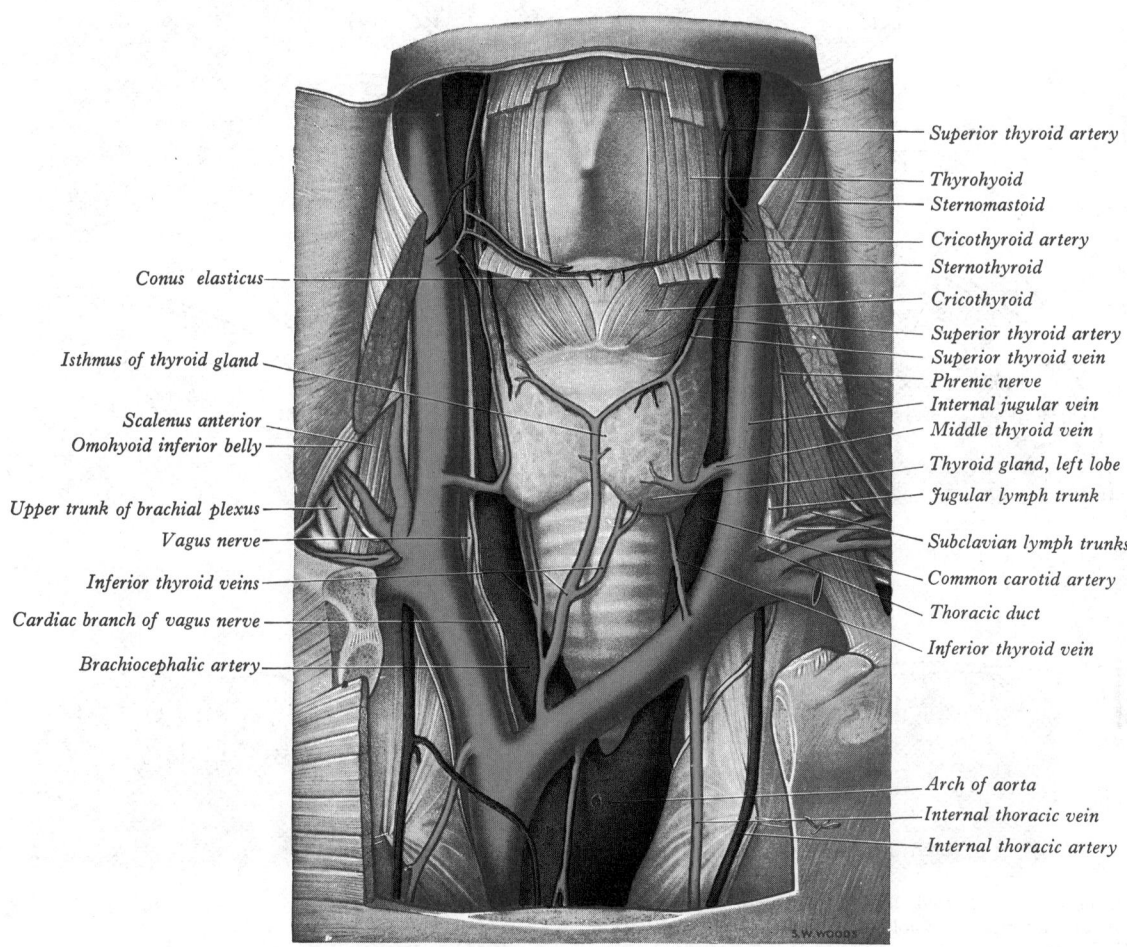

Conus elasticus

Isthmus of thyroid gland

Scalenus anterior
Omohyoid inferior belly

Upper trunk of brachial plexus

Vagus nerve

Inferior thyroid veins

Cardiac branch of vagus nerve

Brachiocephalic artery

Superior thyroid artery
Thyrohyoid
Sternomastoid
Cricothyroid artery
Sternothyroid
Cricothyroid
Superior thyroid artery
Superior thyroid vein
Phrenic nerve
Internal jugular vein
Middle thyroid vein
Thyroid gland, left lobe
Jugular lymph trunk
Subclavian lymph trunks
Common carotid artery
Thoracic duct
Inferior thyroid vein

Arch of aorta
Internal thoracic vein
Internal thoracic artery

8.200 The thyroid gland and its environs. The manubrium sterni and the sternal ends of the clavicles and first costal cartilages have been removed and the pleural sac and lung have been retracted on each side.

The neuroglial cells are situated amongst the pinealocytes which they partially ensheath and separate. They are similar to astrocytes, and many of those situated in the pineal stalk have highly elongated cytoplasmic processes which pass longitudinally, composing most of the substance of the stalk. Ultrastructurally these glial elements are seen to possess numerous filaments which extend to all parts of their cytoplasmic processes.

The *capillaries* of the pineal are lined by thin and sometimes fenestrated endothelial cells, external to which is a tenuous basal lamina, sometimes incomplete. Most non-myelinated autonomic *nerve fibres* are noradrenergic, containing dense-cored vesicles in their terminal and pre-terminal expansions. These nerve fibres, which arise from the superior cervical ganglion, penetrate the pineal either by way of perivascular spaces, or via the nervi conarii

In general, the effects of pineal secretions are inhibitory. The indole-amine and polypeptide hormones secreted by its pinealocytes are believed to reduce the level of synthesis and release of various hormones of the pars anterior, for example, both by direct action on the secretory cells of the gland, and indirectly by inhibiting the production of releasing factors within the hypothalamus. The pineal secretions may reach their target cells either via the cerebrospinal fluid (Sheridan *et al.* 1969; Knight *et al.* 1973) or via the circulatory system.

A number of pineal indole-amines, including melatonin, and the enzymes necessary for their biosynthesis, such as serotonin N-acetyltransferase, show a circadian rhythm in their concentration and activity within the pineal glands of a variety of mammals. In the rat there appears to be an *endogenous circadian oscillator* in the

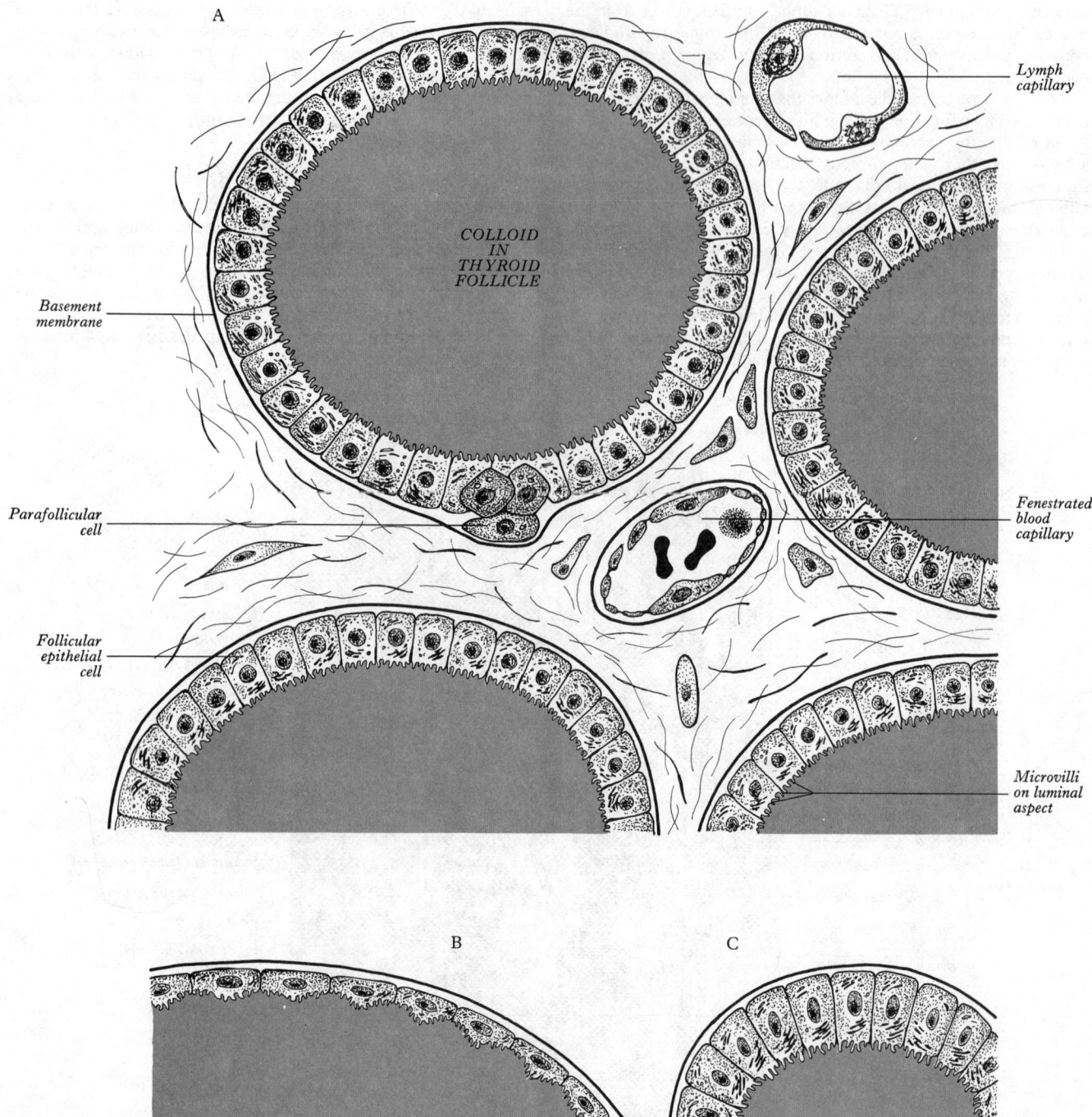

8.201 A–C   Diagrams of thyroid follicular structure.   A   Normal structure under average physiological conditions.   B   'Resting' state.   C   Highly active state. (See text for details.)

suprachiasmatic nucleus of the hypothalamus, whose intrinsic rhythmicity is responsible for the cyclical behaviour of the pineal (Klein 1978). Photic stimuli, in particular changes in the level of yellow-green light acting on a rhodopsin-like retinal photopigment, seem to be involved in regulating this rhythm (Cardinali *et al.* 1971; Minneman *et al.* 1974), with the result that the pineal gland is most active during darkness. If the visual system and the sympathetic nerve supply of the pineal gland remain intact, exposure to light after several hours of darkness depresses pineal activity (Klein and Weller 1972). Neurons of the suprachiasmatic nucleus are supplied directly by axons of retinal nerve cells, allowing response to changes in illumination (Nishino *et al.* 1976). In the rat the neuronal pathway from the suprachiasmatic nucleus to the pineal gland includes the tegmental nuclei and the upper thoracic intermediolateral cell column (Saper *et al.* 1976). Preganglionic sympathetic fibres from the latter pass to the superior cervical ganglia, from which postganglionic fibres can be traced to the pinealocytes. The release of catecholamines from these fibres causes a receptor-mediated increase in the production of cyclic AMP in the pinealocytes, which in turn triggers a 70- to 100-fold increase in the activity of serotonin N-acetyltransferase activity, and is responsible for the daily changes in melatonin production and hence in plasma melatonin levels. So far interest has centred on the sympathetic innervation of the pineal; the gland also has a parasympathetic supply but its role in the control of pineal activity is as yet unknown. (The foregoing general hypotheses are summarized in **8**.199B.)

There is now evidence of the existence of a circadian rhythm in human pineal activity, as demonstrated by changes in plasma melatonin levels (Vaughan *et al.* 1976). As in other mammals, the level rises during darkness and falls during the day. Whether or not control mechanisms similar to those so elegantly demonstrated in other animals also operate in man remains to be resolved.

The Thyroid Gland

The thyroid gland (8.200) is a brownish-red, highly vascular organ, situated anteriorly in the lower part of the neck, at the level of the fifth, sixth and seventh cervical and the first thoracic vertebrae. It is ensheathed by the pretracheal layer of the deep cervical fascia, and consists of right and left lobes connected across the median plane by a narrow region, termed the *isthmus*. Its weight is somewhat variable, but is usually about 25 g. It is slightly heavier in the female, in whom it becomes enlarged during menstruation and pregnancy.

The lobes are approximately conical in shape, the apex of each ascending and diverging laterally to the level of the oblique line of the thyroid cartilage; the base is on a level with the fourth or fifth tracheal 'ring'. Each lobe is about 5 cm long; its greatest transverse and anteroposterior dimensions being about 3 cm and 2 cm. The posteromedial aspect of each lobe is attached to the side of the cricoid cartilage by a ligamentous band, called the *lateral ligament* of the thyroid gland (p. 1081). The *lateral* or *superficial surface* is convex. External to the sheath of pretracheal fascia, this aspect of the gland is closely covered with the sternothyroid and it is the insertion of this muscle into the oblique line on the lamina of the thyroid cartilage which prevents the upper part of the lobe from extending on to the thyrohyoid muscle. More anteriorly still are the sternohyoid and the superior belly of the omohyoid, overlapped below by the anterior border of the sternocleidomastoid. The *medial surface* is adapted to the larynx and trachea. At its superior pole, it is in contact with the inferior pharyngeal constrictor and the posterior part of the cricothyroid, which intervene between the gland and the posterior part of the lamina of the thyroid and the side of the cricoid cartilages. The external laryngeal nerve is medial to this part of the gland on its way to the cricothyroid. Below, it is related to the side of the trachea in front and to the recurrent laryngeal nerve (*see* p. 1081) and (especially on the left side) to the oesophagus posteriorly. The *posterolateral surface* is related to the carotid sheath and overlaps the common carotid artery. The *anterior border*, closely related to the anterior branch of the superior thyroid artery, is thin and descends obliquely and medially. The *posterior border*, blunt and rounded, is between the posterior and the medial surfaces and is closely related below to the inferior thyroid artery and an anastomosing branch which connects it to the posterior branch of the superior thyroid artery. In addition, the parathyroid glands are usually related to the posterior border (p. 1453) The lower end of the posterior border of the left lobe is closely related to the thoracic duct.

The isthmus connects the lower parts of the two lobes; it measures about 1·25 cm transversely, and the same vertically, and usually extends anterior to the second and third rings of the trachea, though it is often placed at a higher, or occasionally lower, level. Its situation and size present, however, many variations. Anteriorly, it is separated by the pretracheal fascia from the sternothyroids. More superficially it is covered by the sternohyoids, the anterior jugular veins, the fascia and the skin. An anastomotic branch uniting the two superior thyroid arteries runs along its upper border; at its lower border the inferior thyroid veins leave the gland. Occasionally the isthmus is absent.

A third, conical *pyramidal lobe* is often present; it ascends towards the hyoid bone from the upper part of the isthmus, or from the adjacent part of either lobe (more commonly the left). It is occasionally quite detached, or may occur as two or more separate parts.

A fibrous or fibro-muscular band sometimes descends from the body of the hyoid bone to the isthmus of the gland, or its pyramidal lobe; when muscular, it is termed the *levator of the thyroid gland*.

Small detached masses of thyroid tissue sometimes occur in the vicinity of the lobes or superior to the isthmus; they are called *accessory thyroid glands*. Vestiges of the thyroglossal duct (p. 198) may persist between the isthmus and the foramen caecum of the tongue, and may give rise to accessory nodules or cysts of thyroid tissue, situated in or near the median plane, and even in the substance of the tongue.

Structure. The thyroid gland is invested by a thin capsule of connective tissue and it is divided into masses of irregular form and size by extensions of this connective tissue. The thyroid parenchyma is mainly derived from the endoderm of the thyroglossal duct (p. 198), a generally ephemeral structure which in the embryo connects it to the tongue. Branching, solid, epithelial cords and sheets grow out from the distal end of the duct, and lumina filled with a yellow, viscid colloid appear within them. The endodermally derived epithelia are usually considered to develop into separate follicles, approximately spherical structures of from about 0·02 to 0·9 mm in diameter, each consisting of a central core of colloid surrounded by a single layer of epithelial cells and enclosed in a basal lamina; however, three-dimensional reconstructions of mature thyroid glands have shown that the follicles are not usually separate units, but typically occur as aggregates, with a shared sheet of epithelial cells bordering on several masses of colloid, and thus linking the follicles together (Isler *et al.* 1968). The colloid, which stains pink with eosin, consists of the iodinated glycoprotein thyroglobulin, a precursor of the thyroid hormones tri-iodothyronine (T_3) and tetra-iodothyronine or thyroxine (T_4), and is a product of the follicular epithelial cells which surround it.

The aggregates of follicles nestle in the delicate connective tissue stroma of the gland, surrounded by close-meshed plexuses of fenestrated blood capillaries. The stroma also contains an extensive network of lymphatic vessels. Sympathetic nerve fibres supply the arterioles and capillaries, and some end in apposition to the follicular epithelial cells (Melander *et al.* 1975; Melander 1977).

In addition to the follicular cells proper, the thyroid parenchyma contains a second cell type, the parafollicular, light, clear or C cells, which produce the peptide hormone thyrocalcitonin. The C cells are a form of APUD cell (p. 1454), and are derived from the ultimobranchial bodies (p. 199). They are a minor component of the parenchyma, although the application of improved detection techniques for these cells (Solcia *et al.* 1968, 1969) suggests that in some species their numbers may be higher than previously estimated; in the dog, for example, Kameda (1971, 1976) has found from 30 to 90 parafollicular cells per 100 follicular cells. Occasionally branched tubules containing desquamated cells are found in the thyroid gland; like the C cells, these are of ultimobranchial origin (Halmi 1978). Other cell types, including ciliated cells, have been described in the thyroid glands of rodents (Wollman and Nève 1971); their functional significance is uncertain.

The *follicular cells proper* vary in shape between squamous and columnar, depending on their level of activity, which is mainly controlled by circulating pituitary thyrotropin (8.201). In the absence of thyrotropin (TSH), the follicular cells are squamous and are referred to as 'resting', while the luminal colloid is copious and dense, reflecting increased storage of iodinated thyroglobulin. The secretion of TSH is followed by the uptake of colloidal droplets from the lumen by endocytosis at the luminal aspect of the follicular cells, and cavities may be seen in the luminal colloid where it adjoins the epithelium (8.202). The prolonged action of high levels of circulating TSH induces follicular cell hypertrophy and even hyperplasia, accompanied by the progressive resorption of luminal colloid and an increase in stromal vascularity.

The follicular cells have a striking ultrastructural and functional polarity. They are concurrently engaged, when activated by TSH or possibly by adrenergic nerve terminals (*vide infra*), in the apically (that is, luminally) directed processes of thyroglobulin synthesis and exocytosis, and in the basally directed processes of thyroglobulin endocytosis, degradation, and liberation of the thyroid hormones T_3 and T_4 into the blood capillaries. This dual polarity of function is reflected in the arrangement of the organelles concerned with these processes. Collectively forming a continuous wall around each mass of colloidal thyroglobulin, the follicular cells are linked at their apices by junctional complexes, and lie with their bases resting against a basal lamina (8.201 and 8.202). Each follicular cell has a basally disposed nucleus, prominent granular endoplasmic reticulum, and a supranuclear Golgi complex, the last being particularly prominent in TSH activated cells. Lying above the Golgi complex, in the apical half of the cell, are numerous Golgi-derived secretory vesicles which transport the glycoprotein

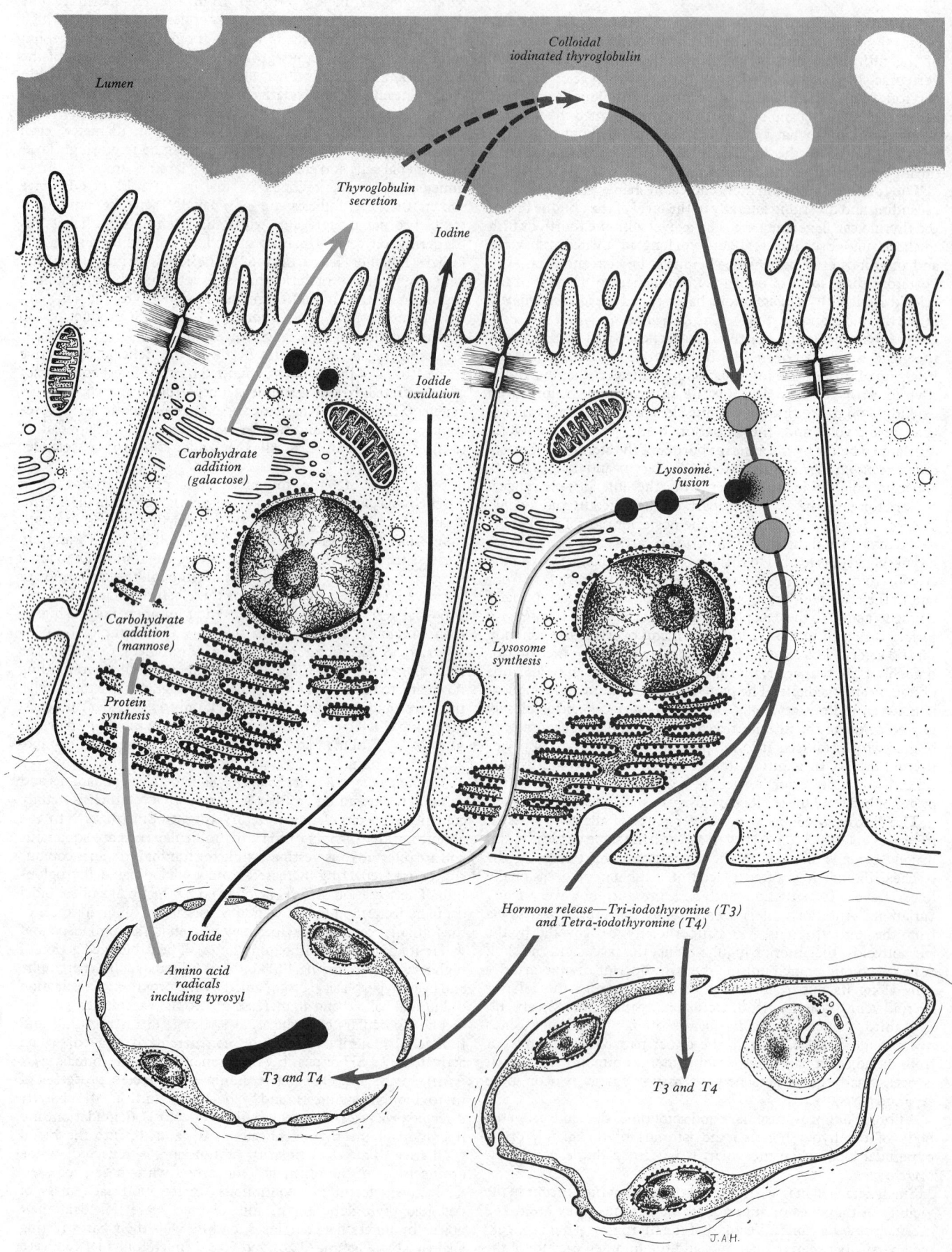

Lumen

Colloidal iodinated thyroglobulin

Thyroglobulin secretion

Iodine

Carbohydrate addition (galactose)

Iodide oxidation

Lysosome fusion

Carbohydrate addition (mannose)

Lysosome synthesis

Protein synthesis

Iodide

Amino acid radicals including tyrosyl

Hormone release—Tri-iodothyronine (T3) and Tetra-iodothyronine (T4)

T3 and T4

T3 and T4

J.A.H.

8.202   The functional architecture of the thyroid follicular cells, showing on one aspect the colloid-containing cavity, and on the other blood and lymphatic capillaries. Arrows indicate metabolic flow pathways.

manufactured by the consecutive activity of the rough endo-plasmic reticulum and the Golgi complex to the apical plasmalemma, where it is released by exocytosis into the lumen of the follicle. The iodine required to complete the formation of thyroglobulin enters the basal aspect of the follicular cells in the form of iodide by active transport across the basal plasmalemma from the blood capillaries, and quickly assumes a predominantly luminal location (Wolff 1964). Here it is rapidly oxidized to iodine, mainly by the activity of thyroid peroxidase in the apical plasmalemma (Taurog 1970). The iodine then becomes attached to the tyrosyl groups of the secreted glycoprotein manufactured by the follicular cells, forming mono- and di-iodotyrosyls. Coupling of these results in the formation of iodothyronyl groups (thyroid hormones in peptide linkage) and completes the formation of iodinated thyroglobulin, the large precursor of the thyroid hormones (Bjorkman and Ekholm 1973; Robbins et al. 1974; Haeberli et al. 1975). Investigation of the distribution of endogenous peroxidase in the follicular cells has shown it to be present in the perinuclear cisternae, granular endoplasmic reticulum, Golgi complex, apical vesicles, and at the apical plasmalemma, particularly over the microvilli which project into the lumen of the follicle (Strum and Karnovsky 1970). It is believed, however, that it is only when incorporated into the apical plasmalemma that it is involved in iodide oxidation, and that the rest of its distribution indicates the route along which it passes to reach this site. The apical microvilli are numerous but short in resting follicular cells (Wetzel et al. 1965), but become longer and often branched after stimulation by TSH. Such stimulation also results in the extension of long cytoplasmic processes into the luminal colloid; these fuse around portions of the colloid with the result that it is taken up into the cell. Shortly after colloid endocytosis, lysosomes, which in the resting state tend to be basal in position, migrate to the apex of the cell and fuse with the intracellular droplets of colloid, forming secondary lysosomes or phagolysosomes. These then migrate back to the base of the cell, and during this period the colloid gradually disappears as the acid proteases and peptidases of the phago-lysosomes degrade the iodinated thyroglobulin releasing the thyroid hormones T_3 and T_4. These pass to the base of the cell where they are released, leaving the gland mainly via the blood capillaries. The more numerous precursor molecules 3-mono-iodotyrosine and 3,5-di-iodotyrosine are de-iodinated by a dehalogenase, and the iodine released migrates apically to be reused in the iodination of newly synthesized thyroglobulin molecules. Both microtubules and microfilaments appear to be actively involved in thyroid hormone secretion (Wolff and Williams 1973), for treatment with colchicine and cytochalasin B, which interfere with microtubule and microfilament activity respectively, inhibit TSH and dibutyryl cAMP-induced secretion. Some iodinated thyroglobulin escapes from the follicles intact and can be demonstrated in the blood by radioimmuno-assay, probably reaching it by way of the lymphatic vessels of the thyroid. The activity of the follicular cells in thyroglobulin formation and thyroid hormone secretion is illustrated diagram-matically in **8**.202. The reviews of Greer and Haibach (1974), DeGroot and Niepomniszcze (1977), and Taurog (1978) pro-vided detailed accounts of these processes.

Of the two thyroid hormones, it is thought that tri-iodo-thyronine in the body cells is the chief agent which stimulates and increases the rate of cellular metabolism, its action being very powerful and immediate, whereas *thyroxine* (tetra-iodothyronine) is powerful but delayed in its similar action. Over-production of these hormones causes the condition of *thyrotoxicosis* (exophthal-mic goitre); hyposecretion in the adult produces *myxoedema*, and in the post-natal period, *cretinism* (p. 822). Their development and activity are mainly controlled by thyrotropic hormone (TSH) secreted by the adenohypophysis. However, in *Graves' disease*, human thyroid-stimulating immunoglobulin (HTSI) antibodies bind to the TSH-receptor sites on the thyroid follicular cells, interfering with this control mechanism and resulting in excessive thyroid hormone production (Werner 1978). The thyroid hor-mones increase the sensitivity of the body tissues to the effects of adrenalin and noradrenalin secreted by the medulla of the suprarenal glands.

Although the activity of the follicular cells is mainly controlled by the level of circulating TSH, there is considerable evidence in a number of mammalian species, including man, of a direct sympathetic influence on the follicular cells, in addition to that induced by sympathetically mediated changes in the microcircu-lation of the gland (Melander 1978). Fluorescence histochemistry and electron autoradiography have been used to demonstrate the presence of adrenergic nerve terminals associated with both blood vessels and follicular cells. In mice where TSH secretion has been eliminated to avoid indirect effects on thyroid secretion, unilateral stimulation of the sympathetic fibres supplying the thyroid induces secretion of thyroid hormones only in those regions of the gland innervated by the stimulated nerve (Melander et al. 1972). Furthermore, studies using normal human thyroid tissue *in vitro* suggest that noradrenalin can directly induce changes within the follicular cells which might be expected to lead to thyroid hormone secretion; the changes noted included intracellular colloid droplet formation and lysosome migration (Melander 1978). Similar *in vitro* investigations using isolated calf thyroid cells have shown that catecholamines can enhance the in-corporation of iodine and synthesis of thyroid hormones (Melander et al. 1973). The stimulation of follicular cell activity induced by catecholamines can be abolished both *in vivo* and *in vitro* by drugs that block adrenergic receptors (Melander 1970, Maayan and Ingbar 1978). Treatment with these agents does not, however, affect the abililty of the cells to respond to TSH. It would thus appear that catecholamines and TSH interact with different receptors on the follicular cell surface, and that they can act independently. The effects which they set in motion are, however, similar; both induce activation of adenyl cyclase, resulting in increased formation of cyclic-AMP, which leads to increased release of thyroid hormones (Melander 1978). Although there can be little doubt from the results of both anatomical and physiological investigations that follicular cells as well as blood vessels have a direct sympathetic innervation, the degree and significance of sympathetic control of thyroid hormone secretion is uncertain. Although TSH may be of greater importance as a long term regulator of thyroid activity, the existence of a link between the sympathetic nervous system and the follicular cells may provide for rapid secretory responses to certain stimuli.

The *parafollicular cells* (C, clear or light cells) are a type of APUD cell (p. 1454). They are situated singly or in small groups in the spaces between the follicles of the thyroid, with the outer borders of which they are in close juxtaposition, lying within the follicular basement membrane. Often they are partly insinuated between the adjacent borders of follicular cells, but they do not reach the lumen of the follicle. They are larger than the follicular cells and are polyhedral or oval in outline. The grouping of the parafollicular cells is sometimes more marked as, for example, in the thyroid gland of dogs that have been perfused with fixative in preparation for electron microscopy. In this case follicle-like groups of such cells are seen, with expanded central extracellular spaces; it is possible that these spaces act as a hormone store, but more evidence of their significance is required. Unlike the follicular cells, they do not appear to have a direct nerve supply.

The most distinctive ultrastructural feature of the para-follicular cells is the presence in their cytoplasm of numerous membrane-bound secretory granules. It is generally assumed that these contain a stored form of the peptide hormone thyrocalci-tonin, an assumption supported by immunocytochemical light microscope studies (Pearse 1966; Wolfe et al. 1974). As befits cells actively engaged in the manufacture of exportable protein, their cytoplasm contains granular endoplasmic reticulum whose form appears to vary with the level of activity of the cells, well-developed Golgi complexes and numerous mitochondria. There are also many free ribosomes. The nucleus is approximately oval, with a smooth or slightly irregular membrane, and is generally eccentric in position.

It is generally stated that the main factor controlling the release of thyrocalcitonin is the concentration of serum calcium: a rise in the concentration of calcium in the blood perfusing the thyroid stimulates thyrocalcitonin secretion, while hypocalcaemia suppresses it, and there is thus a reciprocal relationship between

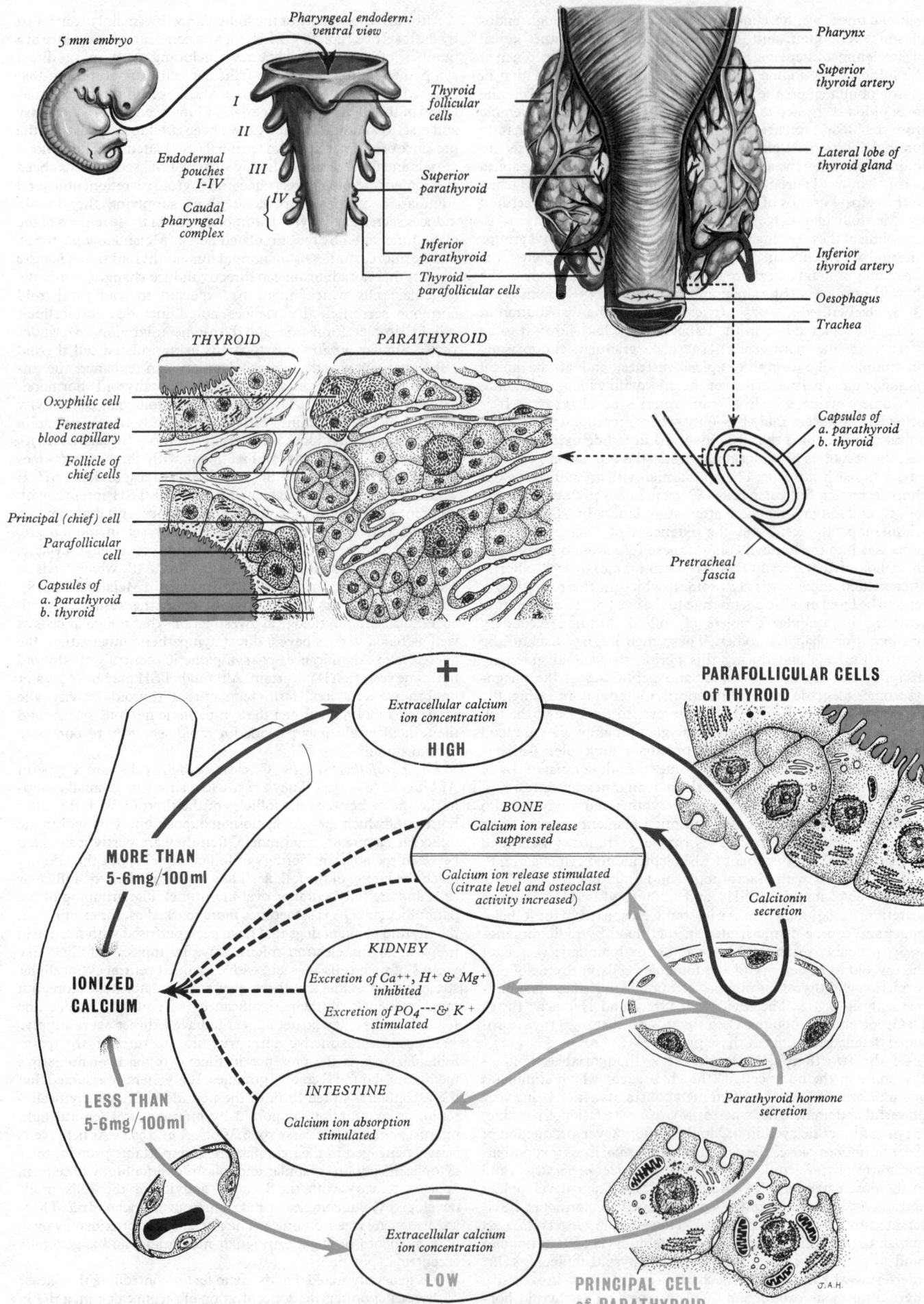

8.203 Diagrammatic representation of the roles of the parathyroid and thyroid glands in the control of calcium metabolism.

the secretion of thyrocalcitonin and the secretion of parathyroid hormone (**8**.203). However, while thyrocalcitonin would appear to be an important regulator of calcium metabolism in many species, acting predominantly by suppressing bone resorption, its status in man in uncertain. This uncertainty stems in part from the difficulties which have been encountered in detecting thyrocalcitonin in normal human plasma even with highly sensitive radioimmunoassays, and in part from the failure as yet to detect a disease associated with thyrocalcitonin deficiency (O'Riordan 1978).

Vessels and nerves. The *arteries* supplying the thyroid gland are the superior (p. 698) and inferior thyroid arteries (p. 678); sometimes there is an additional *lowest* thyroid artery, the *arteria thyroidea ima*, a branch from the brachiocephalic trunk or the aortic arch which ascends upon the front of the trachea. The arteries are remarkable for their large size and frequent anastomoses, not only on the surface of the gland, but also in its substance. The *veins* form a plexus on the surface of the gland and on the front of the trachea; from this plexus the superior, middle and inferior thyroid veins arise; the superior and middle end in the internal jugular vein, the inferior in the left brachiocephalic vein. The capillary blood vessels form a dense plexus in the connective tissue around the follicles, between the epithelium of the follicles and endothelium of the lymph capillaries which surround a greater or smaller part of the circumference of the follicle. The *lymph vessels* run in the interlobular connective tissue, not uncommonly surrounding the arteries which they accompany, and communicate with a network in the capsule of the gland; they may contain colloid material. They end in the thoracic duct and the right lymphatic duct. The *nerves* are derived from the superior, middle and inferior cervical ganglia of the sympathetic.

Applied Anatomy. Apart from variable but often considerable enlargement during menstruation and pregnancy, which is regarded as being within physiological limits, any enlargement of the thyroid gland is called a goitre. Pressure may be exerted by the enlarged gland on any of the structures related to it. Symptoms are most commonly due to pressure on the trachea or recurrent laryngeal nerves. The external laryngeal nerve or the cricothyroid (which is supplied by the nerve and whose action is to render tense the vocal folds) may be damaged by the pressure of a much enlarged thyroid gland or in the operation of thyroidectomy. In such cases the voice becomes monotonous in character, incapable of varying its pitch, and slightly tremulous (Harries 1955).

Partial extirpation of the thyroid may be required in hyperthryoidism and other conditions of thyroid enlargement. Enough of the gland is removed to relieve symptoms, but, except in malignant disease, the whole gland is not removed, for this is followed by the development of myxoedema. In ligating the inferior thyroid artery, the position of the recurrent laryngeal nerve (pp. 698 and 1080) must be remembered, lest it should be also ligated and divided. Temporary aphonia not uncommonly follows from mere bruising of the nerve and, if nothing more serious has occurred, soon passes off. In partial removal of the thyroid gland it is customary to leave behind the posterior part of each lobe, because during its removal there is great risk of coincident removal of the parathyroid glands with resultant serious disturbance of calcium metabolism.

The Parathyroid Glands

The parathyroid glands (**8**.203) are small, yellowish-brown, ovoid or lentiform structures, which usually lie between the posterior borders of the lobes of the thyroid gland and its capsule. They vary in size, but commonly measure about 6 mm longitudinally, 3–4 mm transversely, and 1–2 mm anteroposteriorly; each weighs about 50 mg. Usually there are four, two on each side, and they are called, from their positions, the superior and inferior para-thyroids. The anastomotic artery connecting the superior to the inferior thyroid artery runs along the posterior border of the lobe of the thyroid gland (p. 678) and, as it usually passes very close to the parathyroids, it forms a good guide to them.

The *superior parathyroid gland* is more constant in position than the inferior parathyroid, and is usually situated at the level of the middle of the posterior border of the lobe of the thyroid gland, though it may lie at a higher level. The *inferior parathyroid gland* may lie in various positions (Walton 1931; Gilmour 1938; Murley and Peters 1961), e.g. (1) within the fascial sheath of the thyroid gland, below the inferior thyroid artery and near the inferior pole of the lobe of the thyroid; (2) behind and outside the fascial sheath of the thyroid gland, immediately above the inferior thyroid artery; or (3) within the substance of the lobe of the thyroid gland near the inferior end of its posterior border. These variable positions are important surgically, since a tumour of the inferior parathyroid which occupies position (1) tends to descend along the inferior thyroid veins, in front of the trachea into the superior mediastinum of the thorax, whereas if the gland occupies position (2) the tumour tends to extend postero-inferiorly, behind the oesophagus into the posterior mediastinum. The superior parathyroids are usually dorsal, and the inferior pair ventral, to the recurrent laryngeal nerves (Pyrtek and Painter 1964).

The parathyroid glands are developed from the endoderm of the pharyngeal saccules or pouches (p. 199), the inferior para-thyroids from the third pouch and therefore referred to as parathyroids III, and the superior parathyroids from the fourth pouch and therefore called parathyroids IV. The inferior parathyroid is closely connected in the early stages of its development with the diverticulum from the third pouch, which forms the thymus, and it is drawn down with the thymus in the caudal migration of the latter. Normally the inferior parathyroid migrates only as far as the inferior pole of the lobe of the thyroid gland, but it may descend with the thymus into the thorax, or it may not descend at all and remain above its normal level, near the bifurcation of the common carotid artery.

The parathyroid glands vary in number; there may be many minute islands of parathyroid tissue scattered in the connective tissue in the region of the usual position of the glands, or there may only be three (Hintzsche 1937; Vail and Coller 1967).

Vessels and nerves. The parathyroid glands receive a very rich blood supply from the *inferior thyroid arteries* or from the anastomoses between the superior and inferior thyroid arteries. Their *lymph* vessels are numerous, and are associated with those of the thyroid and thymus glands. Their *nerve supply* is derived from the sympathetic, either directly from the superior or middle cervical ganglia, or indirectly through a plexus in the fascia on the posterior surface of the lobes of the thyroid gland. The nerves are not secretomotor in function, but probably vasomotor, and parathyroid activity appears to be controlled by alterations in the calcium content of the blood, being inhibited by a rise and stimulated by a fall.

Structure (**8**.203). Each parathyroid has a thin connective tissue capsule, from which septa pass into the gland but do not subdivide it into distinct lobules. In the child, the gland consists of wide, irregular anastomosing columns or cords of cells, the *chief cells* (or *principal cells*). These cells are responsible throughout life for the synthesis and secretion of parathyroid hormone (or parathormone). Three types of chief cell, namely *light*, *dark*, and *clear*, can be distinguished according to the depth of staining of their cytoplasm, which by light microscopy appears homogenous. Between the columns of cells lies a rich network of sinusoidal capillaries, via which parathyroid hormone leaves the gland.

The ultrastructure of parathyroid glands was reviewed in detail by Capen in 1975. The appearance of human chief cells differs according to the level of their activity (Munger and Roth 1963). Active chief cells are characterized by large Golgi complexes with which are associated numerous vesicles and small membrane-bound granules, the latter probably representing prosecretory granules; larger secretory granules proper are rare, cytoplasmic glycogen sparse, and much of the cytoplasm is taken up by flattened sacs of granular endoplasmic reticulum aggregated into parallel arrays. In contrast, the cytoplasm of inactive chief cells contains only small Golgi complexes with which are associated only a few vesicles and a small number of membrane-bound secretory granules arranged in groups; there is abundant glycogen and many lipofuschin granules but the sacs of granular endoplasmic reticulum are rare and dispersed. In the normal human parathyroid gland inactive chief cells outnumber the active variety by a ratio of 3–5:1.

The active chief cells synthesize, package and then secrete parathyroid hormone. It is generally assumed that the dense-cored, membrane-bound granules found in the parathyroid chief cells of all mammalian species investigated contain parathyroid hormone, although this remains to be proved (Capen and Roth 1973). There follows a period of gradual involution during which the granules first become peripheral in position and then, under the appropriate stimulus (*vide infra*), their membranes fuse with the plasmalemma, and their contents, presumably including parathyroid hormone, are released. Involution continues with an increase in lysosomal activity and a reduction in size and complexity of the Golgi complexes and the granular endoplasmic reticulum; glycogen reaccumulates, lipofuschin granules form, and the cells enter a temporary 'inactive' phase. In contrast to the thyroid, where the activity of adjacent cells is synchronized, each chief cell of the parathyroid appears to pass through the secretory cycle independently (Roth and Capen 1974).

A second cell type, the *oxyphil* (or *eosinophil*) cell, appears in the parathyroid glands just before puberty and increases in number with age (Roth 1962). Only in man, the macaque monkey and in cattle have such cells been described. The oxyphil cells are larger than the chief cells and contain more cytoplasm, which by light microscopy appears granular and which stains deeply with eosin. The oxyphil nucleus is also smaller and more darkly staining than that of the chief cell. Ultrastructural observations (Munger and Roth 1963; Gaillard *et al.* 1965; Roth and Capen 1974; Capen 1975) have shown that the 'granules' seen by light microscopy are actually mitochondria, which are extremely numerous and tightly packed within the cytoplasm, and which are often bizarre in form. The cytoplasm also contains a few sacs of granular endoplasmic reticulum, some glycogen, and, rarely, small Golgi complexes. No secretory granules have been reported. These findings suggest that the oxyphil cells in normal parathyroid glands are not involved directly in hormone synthesis or secretion, although the copious mitochondria contained in their cytoplasm suggest a high level of metabolic activity. The role of these cells remains to be determined.

The arrangement of the chief and oxyphil cells in the parathyroid gland is shown in **8.203**.

Parathyroid hormone (PTH) is a single-chain polypeptide of 84 amino acid residues (Potts *et al.* 1971), concerned with the control of the level and distribution of calcium and phosphorus in the body (**8.203**). Two other hormones, calcitonin (p. 261) and 1,25-hydroxycholecalciferol, are also involved in the control process. The latter of these is produced by the sequential action of cells of the liver and kidney on vitamin D (O'Riordan 1978). The hormonal control of calcium metabolism has been described in detail by Copp and Talmage (1978). The secretion of PTH is dependent on the concentration of calcium ions in the blood supplying the parathyroid glands. PTH acts directly upon osteocytes and osteoclasts, its rapid initial effect being to increase the rate of release of calcium from bone mineral into the blood, apparently as a result of stimulation of osteocytic osteolysis (Bélanger 1969). It also produces a delayed effect if a high level of secretion is maintained, stimulating internal bone remodelling by promoting osteoclast activity. Changes in the membrane potential of the osteoclasts appears to be involved in the latter effect (Mears 1971). PTH also affects ion transport in the kidney (Puschett 1978), increasing the excretion of phosphate, sodium and potassium, while decreasing that of calcium. It may also affect the intestinal transport of calcium. 1,25-hydroxycholecalciferol, the production of which is regulated by PTH, shares many of these effects and may well be a modulator of PTH action (O'Riordan 1978). The mechanism by which PTH affects its target cells is not yet fully understood; however, adenyl cyclase activation and a consequent rise in intracellular cAMP appear to be involved (Chase and Aurbach 1967; Chabardes *et al.* 1975).

If all the parathyroids are removed, the muscles undergo convulsive spasms (*tetany*) and, as the respiratory muscles (including the laryngeal muscles) are involved, death ensues. These tetanic contractions are due to the fall in blood calcium levels. Excess of parathyroid secretion, such as occurs in tumours

of the glands, results in removal of calcium ions from the bones, so that they become soft, a condition known as *generalized osteitis fibrosa*. The calcium ions pass from the bones into the blood (hypercalcaemia) and are excreted in the urine and may cause calcification in the renal tubules with resultant death from kidney disease.

The Chromaffin System

Chromaffin cells or *phaeochromocytes* are classically defined as elements derived from neuro-ectoderm, innervated by pre-ganglionic sympathetic nerve fibres, capable of synthesizing and secreting catecholamines (dopamine, noradrenalin or adrenalin) and storing them in sufficient quantities to give an intense yellow-brown coloration, termed a positive chromaffin reaction, when treated with aqueous solutions of chromium salts, in particular potassium dichromate (Coupland 1965). Groups of such hormone-secreting cells, associated structurally and functionally with the sympathetic nervous system, comprise the *chromaffin system*. This includes (*a*) the medullae of the suprarenal glands, (*b*) the para-aortic bodies, (*c*) the paraganglia proper, (*d*) certain cells of the carotid bodies, and (*e*) small masses of cells scattered irregularly and variably among the ganglia of the paravertebral sympathetic chains, splanchnic nerves and the great (pre-vertebral) autonomic plexuses, and may, therefore, be closely related to various organs (heart, liver, kidney, ureter, prostate, epididymis, ovary, etc.). The distribution of the main components of the chromaffin system in the newborn infant, is shown diagrammatically in Fig. **8**.204.

There are *three main groups* of cells giving a positive chromaffin reaction (Coupland 1976): (1) *true chromaffin cells*, as defined above, (2) *enterochromaffin cells*, found in the epithelial tissue lining the gastrointestinal and respiratory tracts (p. 1364), and (3) *amine-storing mast cells*, found in the connective tissues of the gut, pancreas and liver. The similarity of the ultrastructure of the first two groups, and the ability of all three groups to take up and decarboxylate amino acids (*see* APUD cells, p. 1364), the observation that other non-chromaffin cells in the walls of the gastrointestinal and respiratory tracts, pancreas and other endocrine glands have similar ultrastructural and amino-acid uptake characteristics, and the discovery of paraneurons having many of the features of chromaffin cells in sympathetic ganglia (p. 1124), raise questions about the advisability of continuing to restrict the term 'chromaffin system' to the 'true' chromaffin cells, as it is traditional to do. It may be more appropriate to consider it as part of a diffuse neuro-endocrine system (*vide infra*).

The Diffuse Neuro-endocrine System

The studies of Feyrter (1938) drew attention to the existence of isolated groups of hormone-secreting cells which were not restricted to specific endocrine glands but which were widely scattered throughout the tissues of the body. Feyrter described these cells as '*helle Zellen*', *clear cells*, and noted that they were particularly prominent in the gut and the pancreas. The clear cells of Feyrter have now been classified as types of 'APUD' cells (Pearse 1966; Pearse 1968; Pearse 1974), cells whose acronymic title is derived from the initial letters of their characteristic amine-handling properties, namely *a*mine *p*recursor *u*ptake and *d*ecarboxylation. Most APUD cells have now been shown to manufacture structurally related peptides which act as *hormones* or as *neurotransmitters*, although in others the main secretion is a similarly acting *amine*. Collectively the APUD cells comprise a 'system' far more extensive than that visualized by Feyrter, including, among others, chromaffin cells (*vide supra*), SIF cells (p. 1124), peptide-producing cells of the hypothalamus (p. 970), pituitary (p. 1440), pineal (p. 1445), parathyroids (*supra*) and placenta, and the Kulchitsky cells of the lung (p. 1259). So far some forty different cell types have been categorized as APUD cells and included in what is now generally described as the *diffuse neuro-endocrine system* (Pearse and Polak 1978). These cells are listed in the accompanying Table.

THE APUD CELLS OF THE DIFFUSE NEURO-ENDOCRINE SYSTEM

I. APUD cells of neural crest origin

Location	Type	Main secretion	
		Peptide	Amine
Thyroid	Parafollicular (C)	Calcitonin	5-HT, Da
Ultimobranchial body	C	Calcitonin	5-HT, Da
Carotid body	Type I Glomus	—	Da, NA
Sympathetic ganglia	SIF	—	NA
Adrenal medulla	Chromaffin	—	Ad
Adrenal medulla	Chromaffin	—	NA
Skin	Melanoblast	—	Promelanin
Urogenital tract	EC	—	5-HT
Urogenital tract	E	—	—

II. APUD cells of placodal or specialized ectodermal origin

Location	Type	Main secretion	
		Peptide	Amine
Hypothalamus	N pv	Oxytocin, CRF	—
	N so	Vasopressin	—
	N sch	—	—
	N dm/vm	TRF	—
	N arc	LHRF	Da
	N ant/post	SRF, CRF	—
	N periv	Somatostatin	—
Pineal gland	P	LHRF	5-HT, MT
Parathyroid	Chief	PTH	—
Pituitary	Somatotroph	Somatotropin	Da
	Mammotroph	Prolactin	Da
	Gonadotroph	Follitropin	Da
	Gonadotroph	Lutropin	Da
	Corticotroph	Corticotropin	—
	M	Melanotropin	T
	Thyrotroph	Thyrotropin	Da

II. (continued)

Location	Type	Main secretion	
		Peptide	Amine
Placenta	Endocrine	Gonadotropin	—
	Endocrine	Somato-mammotropin	—
	Endocrine	Corticotropin	—

III. APUD cells of disputed origin (possibly endodermal)

Location	Type	Main secretion	
		Peptide	Amine
Pancreas	A	Glucagon	5-HT
	B	Insulin	5-HT
	D	Somatostatin	Da
	D$_1$	VIP-like	Da
	P	Bombesin-like	—
	PP	Pancreatic polypeptide	Da
Stomach	A	Glucagon	—
	D	Somatostatin	—
	ECL	—	H?
	EC$_1$	Substance P	5-HT
	G	Gastrin, Enkephalin	—
	X	—	—
Intestine	D	Somatostatin	—
	D$_1$(H)	VIP	—
	EC$_1$	Substance P	5-HT
	EC$_2$	Motilin	5-HT
	EC$_n$	—	5-HT
	I	Cholecystokinin	—
	K	GIP	—
	L	Enteroglucagon	—
	N	Neurotensin	—
	S	Secretin	—
Lung	Kulchitsky (P$_a$)	—	—

ABBREVIATIONS

Ad	Adrenalin
CRF	Corticotropin releasing factor
Da	Dopamine
GIP	Gastric inhibitory peptide
H	Histamine
5-HT	5-hydroxytryptamine
LHRF	Luteotropin releasing factor (Luteinizing hormone releasing factor)
MT	Melatonin
NA	Noradrenalin
N ant/post	Anterior and posterior nuclear 'zones' of hypothalamus
N arc	Nucleus arcuatus (Nucleus infundibularis)
N dm/vm	Nucleus dorsomedialis/ventromedialis
N periv	Nuclei periventriculares
N pv	Nucleus paraventricularis
N sch	Nucleus suprachiasmaticus
N so	Nucleus supraopticus
PTH	Parathyroid hormone
SIF	Small intensely fluorescent
SRF	Somatotropin releasing factor
T	Tryptamine
TRF	Thyrotropin releasing factor
VIP	Vasoactive intestinal peptide
—	Unidentified

In 1966, Pearse proposed that cells manufacturing peptide hormones shared a common set of cytochemical characteristics, the most striking being related to the production of biogenic amines (adrenalin, noradrenalin, dopamine, 5-hydroxytryptamine, etc.) and suggested that the uptake of 5-hydroxytryptophan (5-HTP) and its decarboxylation to 5-hydroxytryptamine (5-HT) could be linked to the process of peptide hormone production in general. It was from this concept that the designation 'APUD' cell arose (Pearse 1968). Pearse (1977) has also suggested that all cells of the APUD series are derived from *neuroendocrine-programmed cells* of the *ectoblast*, a proposition in need of revision, since the investigations of others suggest that some APUD cells, such as the gastro-entero-pancreatic endocrine cells (p. 1364), may be similarly programmed but of *endodermal* origin (Leblond and Cheng 1976; Le Douarin 1978).

In Pearse's opinion, the APUD cells may collectively be considered as a third division of the nervous system, acting as third-line effectors to support, modulate or amplify the actions of neurons in the autonomic and somatic divisions and of each other. They possess activities which are slower in onset and longer in duration than those of cells in the autonomic division, which in turn bear a similar functional relationship to the faster-acting neurons of the somatic division. The various secretions of the APUD cells (the cells of the diffuse neuro-endocrine system) may act upon *contiguous* cells, upon *groups of nearby* cells, or upon *distant* cells after transport in the blood; they may, in this respect, be considered as intermediate between the locally-acting transmitters produced by neurons and the distantly-acting secretions of the discrete endocrine glands. The diffuse neuroendocrine system thus complements and links the nervous and endocrine systems, *all three systems* interacting to provide a sensitive mechanism allowing for homeostatic control.

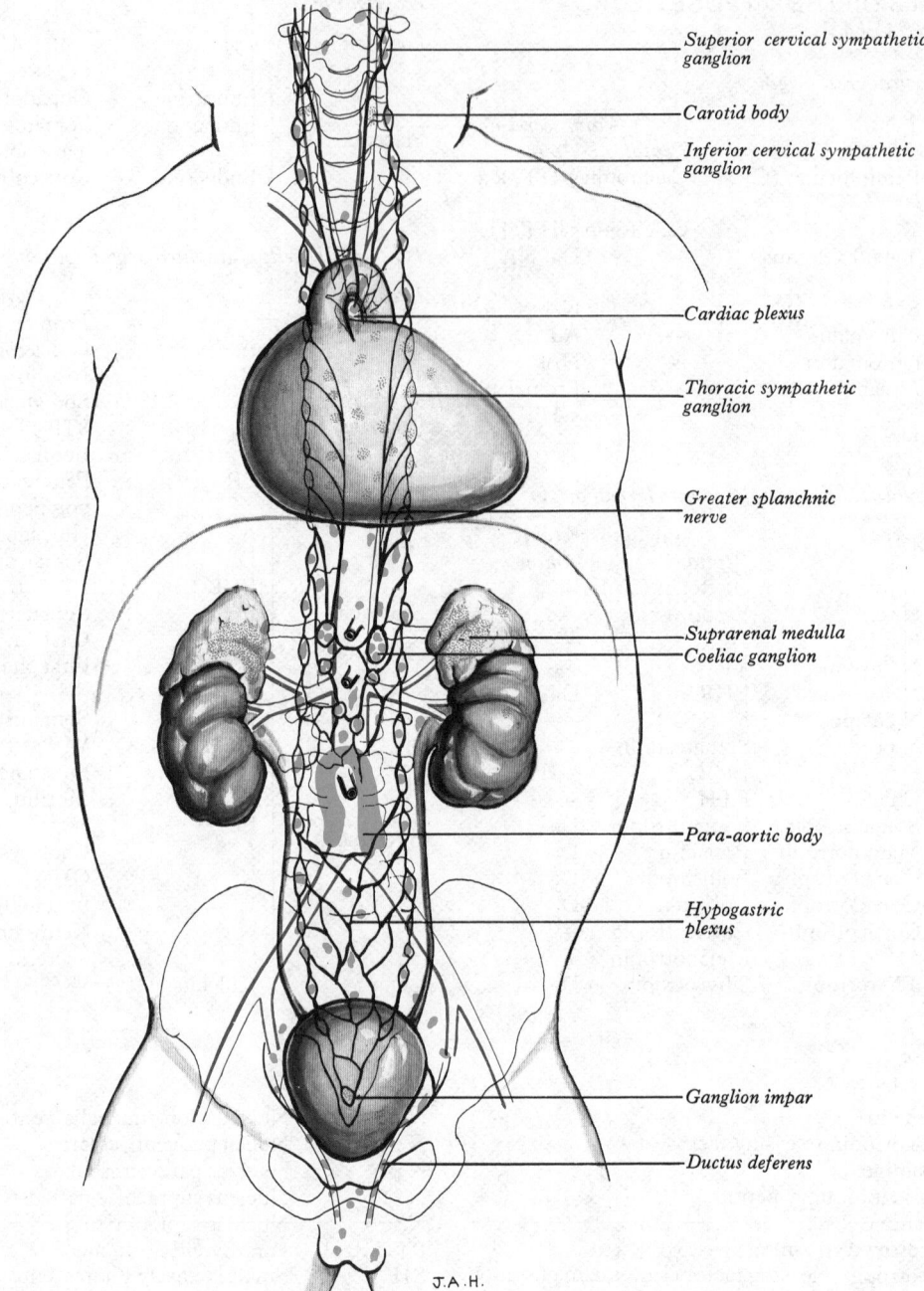

Superior cervical sympathetic ganglion

Carotid body

Inferior cervical sympathetic ganglion

Cardiac plexus

Thoracic sympathetic ganglion

Greater splanchnic nerve

Suprarenal medulla
Coeliac ganglion

Para-aortic body

Hypogastric plexus

Ganglion impar

Ductus deferens

J.A.H.

8.204 The principal aggregations of 'classical' chromaffin tissue in the human neonatal child. The aggregates in stippled blue lie deep to overlying structures.

It has been suggested that deviations in the relative levels of the secretions of the different cell types of the diffuse neuroendocrine system may result in many of the disorders currently described as psychosomatic (Pearse and Polak 1978) or frankly psychotic (Webster 1978). If this proves to be so, then the growing understanding of this system may lead to significant clinical progress in their treatment.

The Suprarenal Glands

The suprarenal (or adrenal) glands (**8**.98, 205, 206) are two small bodies of a yellowish colour, flattened anteropostiorly, and situated one on each side of the median plane, behind the peritoneum, and immediately anterosuperior to the superior pole of each kidney. They are surrounded by areolar tissue containing a considerable amount of perinephric fat (p. 1389). They are enclosed, together with the kidneys, in the renal fascia, but are separated from the kidneys by a little fibro-areolar tissue. Each

gland consists of an external cortical zone, which is rich in lipids and contains no chromaffin tissue, and an internal medulla, which stains deeply with chromic salts. In man it is not uncommon to find small masses of tissue, identical with the suprarenal cortex, in the neighbourhood of the gland or in other situations. They are termed 'cortical bodies'. Ontogenetically, phylogenetically, structurally, and functionally, the cortex and the medulla of the suprarenal gland are distinct from each other, but together they constitute a single topographical entity.

The right gland is somewhat pyramidal, resembling an irregular tetrahedron; the left is semilunar, and is usually larger and extends to a more cranial level than the right. Each gland in the adult measures about 50 mm vertically, 30 mm transversely, and 10 mm in the anteroposterior plane, and weighs about 5 g. (the medulla being about one-tenth of the total weight); but all these measurements are variable. At birth the gland is about one-third of the size of the kidney, whereas in the adult it is only about one-thirtieth. The adult gland is little larger than it is at birth. This change in proportions is not only due to renal growth but also to the fact that after birth the gland begins to diminish in size, due to the involution of the fetal cortex (*see* p. 175). By the end of the second month its weight is only about one-half of that at birth. In the latter half of the second year the gland begins to increase in size and gradually attains its birth weight at or slightly before puberty, after which it only slightly increases in weight in adult life.

RELATIONS OF THE SUPRARENAL GLAND

The right suprarenal gland (8.111) is situated posterior to the inferior vena cava and the right lobe of the liver, and anterior to the diaphragm and superior pole of the right kidney. It is roughly triangular; the base, which is inferior, is in contact with the medial and anterior surfaces of the superior pole of the right kidney. Frequently the base overlaps the upper part of the medial border of the right kidney and not its superior pole. The *anterior surface* faces a little laterally, and presents a medial, narrow, vertical area, not covered by peritoneum, and posterior to the inferior vena cava, and a lateral, somewhat triangular, in contact with the liver. The upper part of the latter surface is devoid of peritoneum, and is in contact with the inferomedial angle of the bare area of the liver, while its inferior part may be covered by peritoneum, reflected on to it from the inferior layer of the coronary ligament; occasionally the duodenum overlaps this area. A little inferior to the apex, and near the anterior border of the gland, there is a short sulcus, forming the hilum, from which the right suprarenal vein emerges to join the inferior vena cava. The *posterior surface* is divided into upper and lower parts by a curved ridge; the superior, slightly convex, rests against the diaphragm; the inferior, concave, is in contact with the superior pole and the adjacent part of the anterior surface of the right kidney. The thin *medial border* of the gland is related to the right coeliac ganglion, which is medial to it inferiorly, and to the right inferior phrenic artery, as the vessel courses superolaterally on the right crus of the diaphragm.

The **left suprarenal gland** (8.111) is crescentic, its concavity being adapted to the medial border of the superior pole of the left kidney. Its medial aspect is convex, its lateral concave; its superior border is sharp, the inferior rounded. Its *anterior surface* has two areas: a superior, covered with the peritoneum of the omental bursa, which separates it from the cardiac end of the stomach and sometimes from the posterior extremity of the spleen; an inferior, which is not covered with peritoneum, but is in direct contact with the pancreas and splenic artery. The hilum, which faces ventrocaudally, is near the lower part of the anterior surface. From it the left suprarenal vein emerges to join the left renal vein. Its *posterior surface* is divided into two areas by a ridge; the lateral adjoins the kidney, the medial and smaller is in contact with the left crus of the diaphragm. The convex *medial border* is related to the left coeliac ganglion, which is inferomedial, and to the left inferior phrenic and left gastric arteries, as they ascend on the left crus of the diaphragm.

Small *accessory suprarenal glands*, which may consist of cortical tissue only, often occur in the areolar tissue around the principal glands; they are sometimes present in the spermatic cord and epididymis, and in the broad ligament of the uterus.

STRUCTURE OF THE SUPRARENAL GLANDS

If a suprarenal gland is cut across (**8.**207), it is seen with the naked eye to consist of an outer part, called the *cortex*, which is yellow in colour and forms the main mass of the gland, and a thin inner part, called the *medulla*, which forms only about one-tenth of the whole gland and is dark red or pearly grey in colour, depending on its content of blood. The medulla is completely enclosed by the cortex, except at the hilum, where the suprarenal vein emerges from the gland. The gland is invested by a thick, collagenous capsule from which trabeculae pass to varying depths into the cortex; the capsule contains a rich plexus of arteries from which branches pass into the gland.

The suprarenal cortex is seen on histological examination (**8.**208, 209) to consist of three zones of cells. The outer is immediately subjacent to the capsule and is called the *zona glomerulosa*; it consists of small polyhedral cells arranged in rounded groups or curved columns, the cells having deeply staining nuclei and a scanty basophilic cytoplasm in which a few lipid droplets may be present. Ultrastructurally (Lever 1955; Long and Jones 1967; Bloodworth and Powers 1968; Shelton and Jones 1971), the cytoplasm contains numerous microtubules, characteristic elongated mitochondria, and abundant agranular endoplasmic reticulum. The latter feature is typical of cells which synthesize steroids elsewhere in the body, and also in the deeper layers of the suprarenal cortex (*vide infra*). The zona glomerulosa is relatively poorly developed in man. Internal to it is the broader *zona fasciculata*; this consists of large polyhedral cells with basophilic cytoplasm, which are arranged in straight columns two cells thick, with fenestrated venous sinusoids coursing parallel with and between the columns. The cells contain numerous lipid

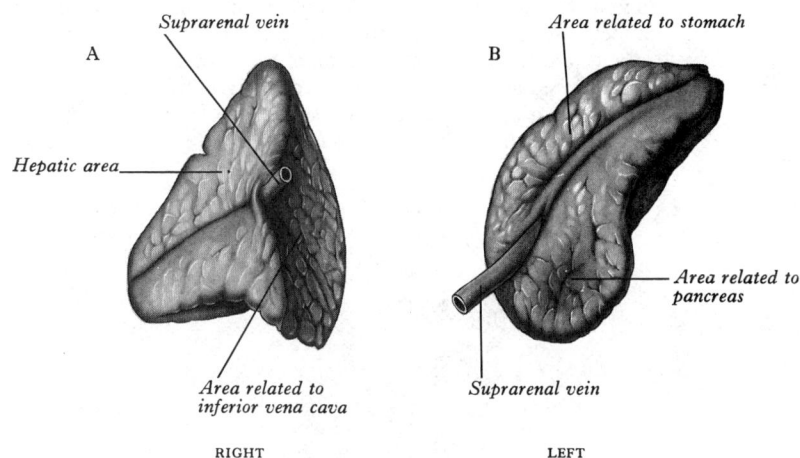

8.205A and B   The suprarenal glands. Anterior aspect.

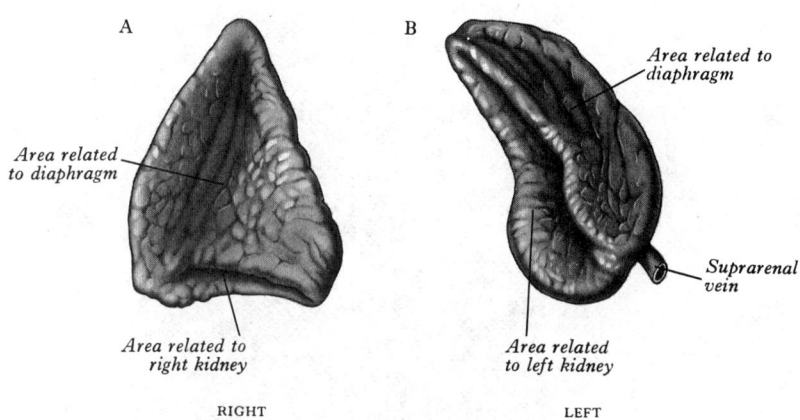

8.206A and B   The suprarenal glands. Posterior aspect.

droplets and large amounts of phospholipids, fats, fatty acids and cholesterol, embedded in a complex arrangement of agranular endoplasmic reticulum. The mitochondria are typically spherical with tubular cristae, and the Golgi complex is extensive.

The innermost zone of cortex is called the *zona reticularis*; it consists of branching and anastomosing columns of rounded cells. Their cytoplasm contains much agranular endoplasmic reticulum, numerous lysosomes, and some pigment bodies which may be evidence of degeneration. It is believed by some workers that the cells in the zona glomerulosa, particularly those in its deeper part, undergo continuous proliferation, some of the newly formed cells migrating into the zona fasciculata and thence to the zona reticularis; in the latter the cells degenerate and are absorbed, but conclusive evidence is lacking. Autoradiographic studies indicate that most of the cell proliferation occurs in the zona glomerulosa and outer reticularis, but mitoses are also found in the other regions of the cortex (Reiter and Hoffman 1967). There is certainly little ultrastructural indication of cell death in the zona reticularis.

The deeper part of the zona fasciculata becomes significantly wider in pregnancy (Whiteley and Stoner 1957) and, in the summer, in women of childbearing age (MacKinnon and MacKinnon 1958). It has also been noted that atrophy of the cortex in old age (in males) is greatest in the deeper part of the zona fasciculata and least in its periphery (MacKinnon and MacKinnon 1960). The cells of the cortex produce various hormones and, particularly those of the zonae fasciculata and reticularis, are very rich in *ascorbic acid* (*vitamin C*). The cells of the zona glomerulosa appear to produce *aldosterone*, which affects electrolyte and water balance in the body tissues; the cells in the zona fasciculata produce hormones that maintain carbohydrate balance, in particular *cortisol* (*hydrocortisone*); the cells of the zona reticularis may produce sex hormones (*progesterone*, *oestrogens* and *androgens*). The cortex is essential to life; complete removal is lethal unless replacement therapy is instituted. The cortex appears to exercise a control over lymphocytes and lymphoid tissue, and an increase in secretion of corticosteroids is associated with a reduction in the number of lymphocytes. In some mammals the cortex undergoes cycles of hypertrophy and regression during the oestrous cycle. Between the cells of the

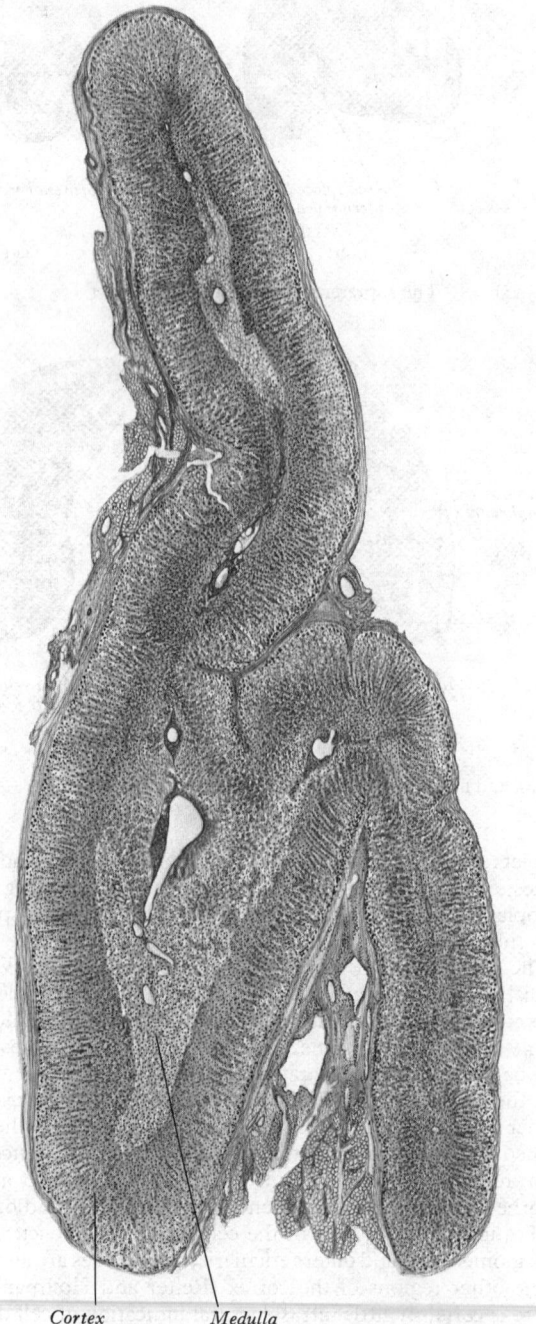

Cortex   *Medulla*

8.207   Vertical section through a whole adult human suprarenal gland.

cortex are sinusoids, into which most of the capsular vascular plexus and cortical arteries open. Some arteries pass straight through the cortex to supply medullary sinusoids. The cortical sinusoids finally discharge blood into the medullary sinusoids. The endothelial cells of the sinusoids are phagocytic and belong to the macrophage (reticulo-endothelial) system (p. 765).

The development of the suprarenal gland is described on p. 175. At birth, the relatively large size of the suprarenal is due to the very thick *fetal cortex* (Johannisson 1968), the *definitive cortex* forming only a thin peripheral zone. About the time of birth the fetal cortex begins to undergo regression and it has largely disappeared after a few weeks. Too rapid involution of this cortex may be complicated by fatal haemorrhage in the suprarenal glands. This transient fetal cortex occurs only in anthropoids. There is no conclusive evidence that it produces androgenic hormones. It is very poorly developed in anencephalic fetuses. It does not represent the X-zone (or androgenic zone) occurring, for instance, in the young mouse as a zone surrounding the medulla (Jones 1957).

The suprarenal medulla is composed of groups and columns of *chromaffin cells* (or *phaeochromocytes*) with wide venous sinusoids permeating between them. Small groups of nerve cells, or even single nerve cells, occur here and there in the medulla.

The chromaffin cells synthesize and secrete noradrenalin and adrenalin into the venous sinusoids, the release being under preganglionic sympathetic control (*see* Coupland 1965a, for review). In several species of mammals these substances have been identified in two distinct cell types (Yates *et al.* 1962), the noradrenalin storing cells usually being situated more peripherally than those which store adrenalin. All the cells, however, are large columnar elements which are arranged in rows one cell thick, along the margins of the venous sinusoids. The bases of the cells, near which their nuclei are situated, point away from the sinusoids, and adjoin expanded extracellular spaces, into which nerve terminals penetrate to synapse with the chromaffin cells. The latter may form follicular groups, although these are not of the same type of organization as the thyroid follicles (Al-Lami 1970). The cytoplasm of the chromaffin cells is basophilic, and ultrastructurally shows a well-developed granular endoplasmic reticulum, mitochondria and Golgi complex, indicating a high metabolic activity (Al-Lami 1970; Coupland 1965b). Numerous secretory vesicles are also present. In noradrenalin-storing cells, these are typically rounded or ellipsoidal bodies which, after treatment with aldehyde and osmium, are highly electron-dense. In adrenalin-storing cells, after similar treatment, the vesicles have a paler appearance (Coupland *et al.* 1964), often with a clear zone between the granular contents and the bounding membrane. In the human suprarenal, cells with mixed adrenalin and noradrenalin vesicles have been reported (Brown *et al.* 1970) so that both may be secreted from the same cell. The vesicles are released at the cell apices into the perivascular space, and gain access to the circulation through the thin endothelial lining of the venous sinusoid, which is fenestrated (Elfvin 1965; Al-Lami 1969).

The sinusoids open into the suprarenal vein at the hilum of the gland. Normally, little adrenalin and noradrenalin are secreted, but under such conditions as fear, anger and stress, the secretion is considerably increased. Noradrenalin produces cardiac acceleration, vaso-constriction, raised blood pressure, etc., whilst adrenalin has a more marked effect on carbohydrate metabolism. Unlike the cortex, the medulla of the suprarenal is not essential to life, and its removal has no apparent effect. The chromaffin cells of the medulla develop and migrate from the neural crests (sympatho-chromaffin tissue, p. 175). The chromaffin reaction (brown staining of the granules in the cells by potassium bichromate due to oxidation of the adrenalin and noradrenalin in the granules) is positive in the fifth month of fetal life, but adrenalin is present as early as the third month (Keene and Hewer 1927).

Interaction between cortex and medulla. It is only in mammals that the chromaffin tissue of the suprarenal becomes almost completely enclosed by cortical tissue. In elasmobranches (e.g. the dogfish) the cortical tissue exists as a pair of structures between the kidneys, the *inter-renal bodies*, whereas the chromaffin tissue is completely separate and forms segmentally

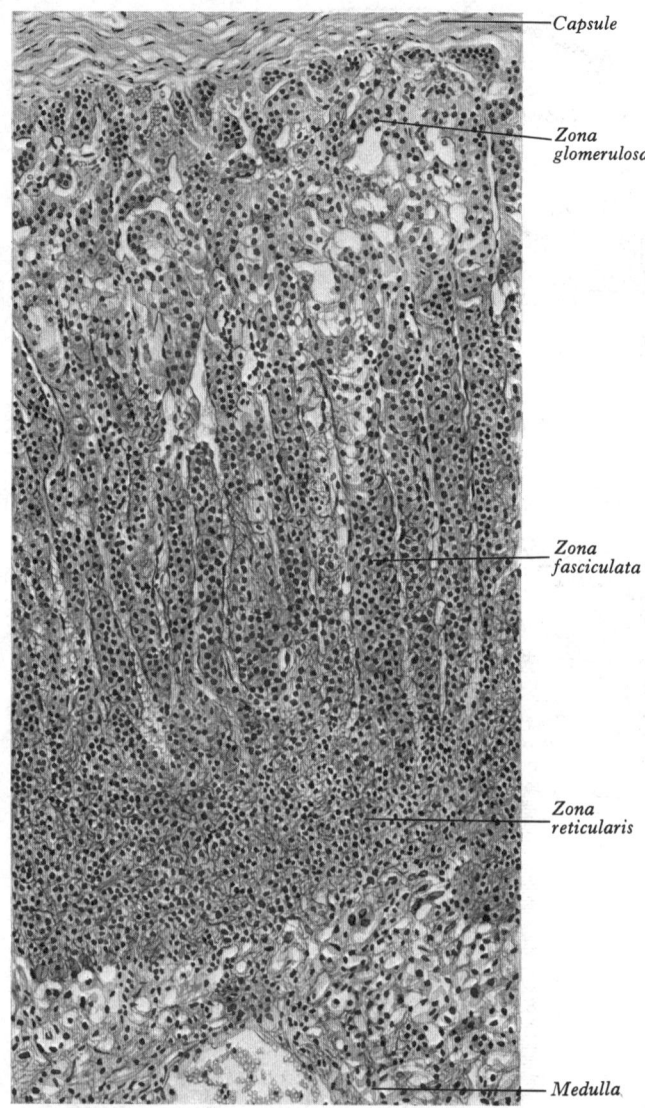

Capsule

Zona
glomerulosa

Zona
fasciculata

Zona
reticularis

Medulla

8.208A Section of adult human suprarenal gland. Magnification about × 200.

8.208B Medium magnification light micrograph of a section of the cortex of a human suprarenal gland. Beneath the capsule (blue) lie the zonae glomerulosa, fasciculata and reticularis. M.S.B. triple stain. (Prepared and photographed by Mr. Stephen Sitch, Department of Anatomy, Guy's Hospital Medical School, London.)

arranged bodies closely related to the ganglia of the sympathetic trunk. In amphibia and birds, cords of chromaffin and cortical cells are intimately associated, but a true medulla is not formed Coupland (1965a). It appears that the presence of cortical tissue in close proximity to the chromaffin tissue may be associated with the appearance of adrenalin, formed by the methylation of the primary amine, noradrenalin (*vide infra*).

Vessels and nerves. The suprarenal gland is exceedingly vascular. It is supplied by three groups of vessels, the superior, middle and inferior suprarenal *arteries*, which arise from the inferior phrenic artery, abdominal aorta, and renal artery respectively (Harrison and Hoey 1960). Most of the branches of the suprarenal arteries ramify over the capsule before entering the gland and dividing within it to form a narrow subcapsular plexus from which fenestrated sinusoids pass around the clusters of cells in the zona glomerulosa and then inwards between the columns of cells in the zona fasciculata to a deep plexus located in the zona reticularis. From this, small venules pass between the chromaffin cells of the medulla to the medullary veins, which they enter by passing between prominent bundles of longitudinally-arranged muscle fibres. It would appear that these bundles of fibres regulate blood flow through a vascular 'dam' at the cortico-medullary junction, that is, at the deep aspect of the zona reticularis (Dobbie and Symington 1966). Since this would in turn control the rate of blood flow through the zona reticularis and zona fasciculata, it could provide a mechanism for control, in part, of the availability of ACTH to the secretory cells of these regions, and hence of its possible uptake (Griffiths and Cameron 1975).

Some major arterial branches bypass the system described above and supply the medulla directly. It thus has a dual blood supply, both indirect and direct (**8**.209); blood reaching it by the indirect route, that is, by way of the cortical sinusoids, is probably sufficiently rich in glucocorticoids to induce and maintain the synthesis of phenylethanolamine-N-methyl-transferase, an enzyme needed for the synthesis of adrenalin from noradrenalin, whereas that reaching it by the direct, non-cortical route is not. Whether a medullary chromaffin cell can make adrenalin or noradrenalin depends on the presence or absence of this enzyme, and may well be determined by its blood supply (Wurtman and Pohorecky 1971); changes in the relative blood flow along the two routes could thus have profound physiological consequences.

The medullary veins drain into the *suprarenal vein* which emerges from the hilum of the gland; the right opens into the inferior vena cava, that on the left into the left renal vein.

The *lymph* vessels end in the lateral aortic nodes (p. 795).

The *nerves* are exceedingly numerous. They are mainly myelinated preganglionic sympathetic fibres and are described on p. 1134. They are distributed to the chromaffin cells in the medulla of the gland. The activities of the suprarenal cortex are largely controlled by adrenocorticotropic hormone (ACTH) secreted by the anterior lobe of the hypophysis.

Applied Anatomy. Various clinical conditions may occur as the result of lesions affecting the cortex or medulla of the suprarenal and they are attributable to the effects of excess or deficiency of the secretions of these parts of the gland.

Atrophy or tuberculosis of the suprarenal cortex, with

8.209 The suprarenal gland, displaying its gross sectional appearance, histology, vasculature, and ultrastructure. Brief functional summaries are appended.

consequent insufficiency of cortical secretion, results in Addison s disease, which is characterized by muscular weakness, low blood pressure, anaemia, brownish pigmentation of the skin, changes in electrolyte and fluid balance of the tissues, and terminal circulatory and renal failure. Excessive cortical secretion due to tumours or hyperplasia of the cortex may produce various effects. (1) In the adult, Cushing's syndrome may result, characterized by obesity, excessive hairiness of the face and trunk, diabetes mellitus, and impotence and hypogonadism in the male or amenorrhoea in the female. (2) In women, masculinization of the secondary characters (virilism) may occur due to excessive secretion of androgenic hormones. (3) In men, feminization, particularly breast enlargement, may occur. (4) In children, there may be precocious body growth and development of the external genital organs, with early menstruation in the female. (5) In the female fetus, hyperplasia of the cortex occurring between the third and fourth months gives rise to the condition of female pseudohermaphroditism, the excessive androgen secretion interfering with the differentiation of the urogenital sinus which occurs at this time, so that the urethra and vagina open into a persistent urogenital sinus; the clitoris also enlarges and the external genital organs thus resemble those of the male. In the male fetus, cortical hyperplasia at this time causes excessive development of the external genital organs.

Bilateral removal of the suprarenal glands (adrenalectomy) is practised in the treatment of some advanced, inoperable cases of disseminated carcinoma of the breast or prostate, which do not respond to radiotherapy and where the malignant changes are considered to be dependent on hormonal control (androgens or oestrogens). The suprarenal glands can be demonstrated radiologically if air is injected into the perirenal fat.

Tumours of the suprarenal medulla and of the para-aortic bodies (phaeochromocytomata) may occur; the consequent excessive secretion of adrenalin and noradrenalin produces attacks of palpitations, excessive sweating, pallor of the skin, hypertension, headaches, and, if the tumours are of long duration, retinitis and vascular changes in the kidneys.

THE PARAGANGLIA

The paraganglia (Zuckerkandl 1901; Köhn 1903) are generally defined as extra-adrenal aggregations of chromaffin tissue (p. 1458), widely distributed along and within the autonomic nervous system (Coupland 1965a; Mascorro and Yates 1971; Hervonen et al. 1978a). Cells similar to those which characterize the paraganglia proper (which, as their name implies, adjoin various autonomic ganglia) are also found within sympathetic ganglia, where they are referred to as *small, intensely fluorescent* (SIF) *cells* (Williams et al. 1975), in the walls of various viscera and in a variety of retroperitoneal and mediastinal locations (Ramsdale et al. 1972; Hervonen et al. 1976; Hervonen et al. 1978b). All share a common neuro-ectodermal origin and an ability to synthesize and store catecholamines; their function, however, differs with their location, the intraneurally disposed acting as interneurons (p. 809) while the remainder are sources of a variety of endocrine secretions, including a tryptophan-containing protein as well as catecholamines (Hervonen et al. 1978a). This diffuse collection of extra-adrenal chromaffin tissue has been referred to as the *paraganglion system* (Mascorro and Yates 1975); it is particularly prominent in the fetus, where it provides the main source of catecholamines while the adrenal medulla is in process of development (West et al. 1955; Kovrishko 1964). Although many paraganglia degenerate soon after birth (Coupland 1965a), the application of specific fluorescent histochemical techniques for the detection of catecholamines (Eränkö 1967) has led to the location of large numbers of persistent, though often minute, paraganglia in adult man and other mammals, and to the need for a revision of the earlier concept of general involution of the paraganglia after birth.

Paraganglia contain two characteristic varieties of cell: *type I* (or granule-containing) and *type II* (or satellite) cells (Mascorro and Yates 1975). The type I cells each have a large nucleus, round or elongated mitochondria, some granular endoplasmic reticulum also well-developed Golgi complexes, glycogen deposits, and numerous, membrane-bound, electron-dense cytoplasmic granules containing catecholamines and, possibly, the tryptophan-containing protein mentioned earlier; these features, together with their neuro-ectodermal origin, place them in the APUD cell category (Pearse 1969; Pearse and Polak 1974). On the basis of their cytoplasmic density, the type I cells can be classified as either 'light' or 'dark'; although the general belief is that 'dark' cells appear so as a fixation artefact (Mugnaini 1965; Benedeczky and Smith 1972), an alternative possibility is that they and the 'light' cells represent different stages of a secretory cycle. The type II cells lack cytoplasmic granules, and have processes which envelop adjacent type I cells either partially or completely.

The type I cells of the paraganglia receive a 'preganglionic' sympathetic innervation (Mascorro and Yates 1974) like that of the chromaffin cells of the adrenal medulla (Cummings 1969). For much of their length the non-myelinated nerve fibres are separated from the type I cells by Schwann cell cytoplasm and the cytoplasmic projections of the type II cells, approaching them closely only at synapses. The presynaptic endings contain mitochondria, glycogen granules and many synaptic vesicles, most of which are electron-lucent, although some have electron-dense cores. There is evidence that in vagal paraganglia such endings are cholinergic and efferent in type (Chen and Yates 1970).

The paraganglia are well vascularized, and the secretory type I cells generally lie next to one or more fenestrated capillaries, often with only basal lamina intervening between them, although fine collagen fibres and the cytoplasmic processes of type II cells may also be present. There is thus little to obstruct the passage of hormones from the type I cells to the blood system (Mascorro and Yates 1975).

Currently available evidence suggests that the paraganglia should be considered as endocrine organs whose secretory cells manufacture catecholamines and proteins, storing them as cytoplasmic granules until stimulated to release them by intrinsic or extrinsic factors. As well as having a remote endocrine effect, a local paracrine action of these secretions on nearby cells cannot be ruled out. The paraganglia collectively comprise a diffuse system which may act throughout life as a source of catecholamines additional to those provided by the adrenal medulla, and as such assume a role of considerable metabolic and clinical significance (*vide infra*).

THE PARA-AORTIC BODIES

The para-aortic bodies progressively develop during fetal life and attain their maximum size in the first three years of post-natal life, when the largest take the form of two elongated, brownish bodies, about 1 cm long, which lie one on each side of the abdominal aorta in the region of origin of the inferior mesenteric artery. They are usually united across the front of the aorta by a horizontal mass, which lies immediately above the origin of the inferior mesenteric artery, so as to form collectively an inverted horseshoe, or H-shaped arrangement. They are intimately related to the intermesenteric and superior hypogastric plexuses. The constituent cells undergo dispersal and some atrophy, and by the age of fourteen years have been said to be completely disintegrated (*see* Coupland 1965a). When well-developed, they consist of masses of polygonal chromaffin cells embedded in a wide-meshed capillary plexus. The chromaffin cells of the para-aortic bodies secrete noradrenalin.

Other small chromaffin bodies are found in the fetus in all parts of the abdominal and pelvic prevertebral sympathetic plexuses. They reach their maximum size between the fifth and eighth months of fetal life, and in the adult they are present as discernible structures mainly in the vicinity of the coeliac and superior mesenteric arteries, whilst only microscopic collections of chromaffin cells persist in association with the lower parts of the intermesenteric plexus.

Although the chromaffin cells of the sympathetic ganglia, as noted above, may act as interneurons, those of the other extra-adrenal chromaffin tissues described above are endocrine in nature and probably subserve the adrenal medulla as sources of catecholamines (Chen et al. 1976), particularly in pre- and early post-natal life when the adrenal medulla and the autonomic nervous system are not fully differentiated (West *et al.* 1953). Support for this proposition has come from the findings of Coupland and Weakley (1970) that the extra-adrenal chromaffin cells are structurally similar to those of the adrenal medulla. There is also ultrastructural evidence in the rat that the chromaffin cells associated with the nodes of the solar plexus have

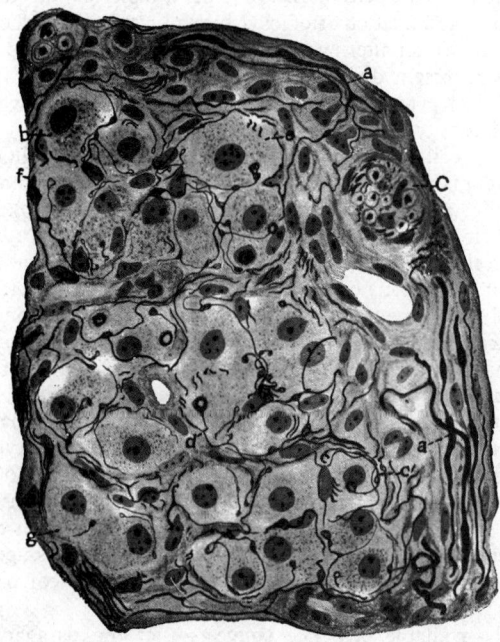

8.210A A section of the carotid body, showing nerve fibres distributed to the cells. (de Castro.) *a*, Myelinated fibre dividing into two fine branches; *b*, cell closely surrounded by nerve fibrils; *c*, section of a small nerve, composed of several myelinated fibres; *f*, a nerve fibril apparently ending within the cytoplasm of a cell; *g*, a nerve fibril ending between the cells.

processes extending beyond the ensheathing glial cells towards blood capillaries into which their catecholamines pass (Levkova and Kakabadze 1977).

The Carotid Bodies (Glomera Carotica)

The two carotid bodies are reddish-brown, ellipsoidal structures, situated one on each side of the neck, in close relation to the carotid sinus (p. 676). Each is about 5 to 7 mm in height and 2·5 to 4 mm in width and varies slightly in position, being either posterior to the bifurcation of the common carotid artery, or completely or partially wedged in between the commencements of the internal and external carotid arteries, and is attached to, and sometimes partially embedded in, the outer fibrous layer of these arteries. Occasionally the organ is in the form of a number of separate nodules. Aberrant spheroidal 'miniglomera', microstructurally similar to the carotid bodies but with a diameter of 600 μm or less, have been identified in the adventitia and in the adipose tissue of the carotid sinus region in some human cadavers (Garfia 1980). The carotid body is innervated by the carotid branch or branches of the glossopharyngeal nerve, including the *carotid sinus nerve*, and by a fine *plexus* containing glossopharyngeal, vagal and sympathetic components (*see* pp. 1075, 1128 for variations in origin and arrangement). Its copious blood supply is derived from neighbouring branches of the external carotid artery.

Structure and Function

The carotid body is an *arterial chemoreceptor*. When stimulated by hypoxia, hypercapnia or elevated hydrogen ion concentration in the blood, it elicits a reflex increase in the rate and volume of ventilation through neural connexions with the array of respiratory centres in the brainstem. Although its main physiological role is known, it is uncertain which of its diverse components function as chemoreceptors (Chen *et al.* 1976). The carotid body may also be an *endocrine gland*; its *glomus cells* (*vide infra*) are included in the APUD series (p. 1454), which has been

extended to include virtually all the specialized cells responsible for peptide hormone production (Pearse 1976), but as yet no specific peptide hormone has been assigned to them. A suggested role in the control of erythropoiesis (Tramezzani *et al.* 1971) has now been refuted (Paulo *et al.* 1973).

The carotid body has a fibrous capsule, whose invading septa divide it into lobules, each containing tightly packed collections of epithelioid '*glomus*' (type I) cells, enveloped by *sheath (sustentacular or type II) cells* (**8.210B**), the latter separating the former from an extensive anastomosing network of fenestrated sinusoids. Between the sheath cells and the sinusoid endothelium are non-myelinated nerve fibres, Schwann cells, the attenuated processes of connective tissue cells, and collagen fibres (Chen *et al.* 1976). Many of the nerve fibres are afferent, passing between the sheath cells and making synaptic contacts with the glomus cells. Preganglionic efferent fibres pass to the sparse population of parasympathetic and sympathetic *ganglion cells*; these fibres are derived from the carotid sinus and sympathetic nerves (McDonald and Mitchell 1975), while the cells are located either separately or in small groups near the surface of the carotid body. Axons proceed from the ganglion cells to the local blood vessels, the parasympathetic efferent fibres probably causing vasodilation (Biscoe *et al.* 1969) and the sympathetic vasoconstriction (Purves 1970).

'*Glomus*' (glomeral) *cells* are more numerous than sheath cells. They are moderately large, with abundant cytoplasm, and carry a few dendritic processes which extend into neighbouring intercellular spaces. Membrane bound electron-dense granules occur near the Golgi apparatus and close to the plasma membrane. Occasional sites of fusion of the limiting membranes of these granules and the plasma membrane have been found (Hansen 1977). The glomus cells store dopamine (Kobayashi 1969), other neurotransmitters and the protein 'glomin' (Pearse 1969), presumably in these granules; they are now accepted as a variety of chromaffin cell (Böck and Gorgas 1976), although a positive chromaffin response can only be detected by electron microscopy (p. 1458). They have also recently been classified as paraneurons (Fujita 1976). Their granular endoplasmic reticulum is not abundant, but an arrangement similar to neuronal Nissl substance has been described in the human carotid body (Böck *et al.* 1970). In the rat two types of glomus cell are recognized: types A and B (MacDonald and Mitchell 1975). Type A cells have larger and more numerous electron-dense granules than type B cells, and while the former usually have a smooth, globular contour with relatively few stubby dendritic processes, the latter are more irregular and bear several long thin processes.

Nerve endings are seldom found in synaptic contact with type B glomus cells, but at least two types of fibres synapse with type A: of these, over 95 per cent are *chemo-afferent axons* which leave the carotid body in the carotid sinus nerve, having their cell bodies in the sensory ganglia of the glossopharyngeal nerve, while less than 5 per cent are *preganglionic efferent axons* from the cervical sympathetic trunk, which enter the carotid body in company with postganglionic axons from the superior sympathetic ganglion. Apparently no efferent axons from the glossopharyngeal nerve synapse with glomus cells, although some are preganglionic to parasympathetic ganglion cells. In the rabbit, efferent fibres of uncertain origin, presumed to be inhibitory, make synaptic contact with the chemo-afferent axons (Verna 1975). Of the 'afferent' nerve endings, some are presynaptic to type A glomus cells, others postsynaptic, and yet others form *reciprocal synapses* with them. Reciprocal synapses have also been found linking adjacent glomus cells (McDonald and Mitchell 1975); they are dendrodendritic in form (Reese and Shepherd 1972). Similar organizations have been found in several parts of the central nervous system (pp. 829, 831) but not hitherto in the *peripheral* nervous system. The various neural circuits claimed to exist in the carotid body of the rat and rabbit (McDonald and Mitchell 1975; Verna 1975) are shown diagrammatically in Fig. **8.210**. Evidence concerning the human carotid body is awaited with much interest.

Ultrastructural and neurophysiological evidence suggests that the glomus cells could have at least three functions. Firstly, they are known to release neurotransmitters in response to hypoxia

(Eyzaguire *et al.* 1972), and may therefore be *sensory cells*. Secondly, they have an ultrastructure which has led to their being classified as APUD cells (Pearse 1969), and may therefore act as *effector cells* modifying, by varying the release of dopamine from their secretory vesicles, the sensitivity of neighbouring chemoreceptive nerve endings (Biscoe *et al.* 1970). Thirdly, since they have a synaptic input and output, are enveloped by gliaform

sheath cells, and have dendritic processes but no axons, they may operate as *interneurons* (McDonald and Mitchell 1975). The simultaneous involvement in multiple functional roles as those postulated above is reminiscent of the possible phylogenetic progenitor neuronal cell.

As stated above, there is no uniform view as to which of the components of the carotid body act as chemoreceptors: *glomus*

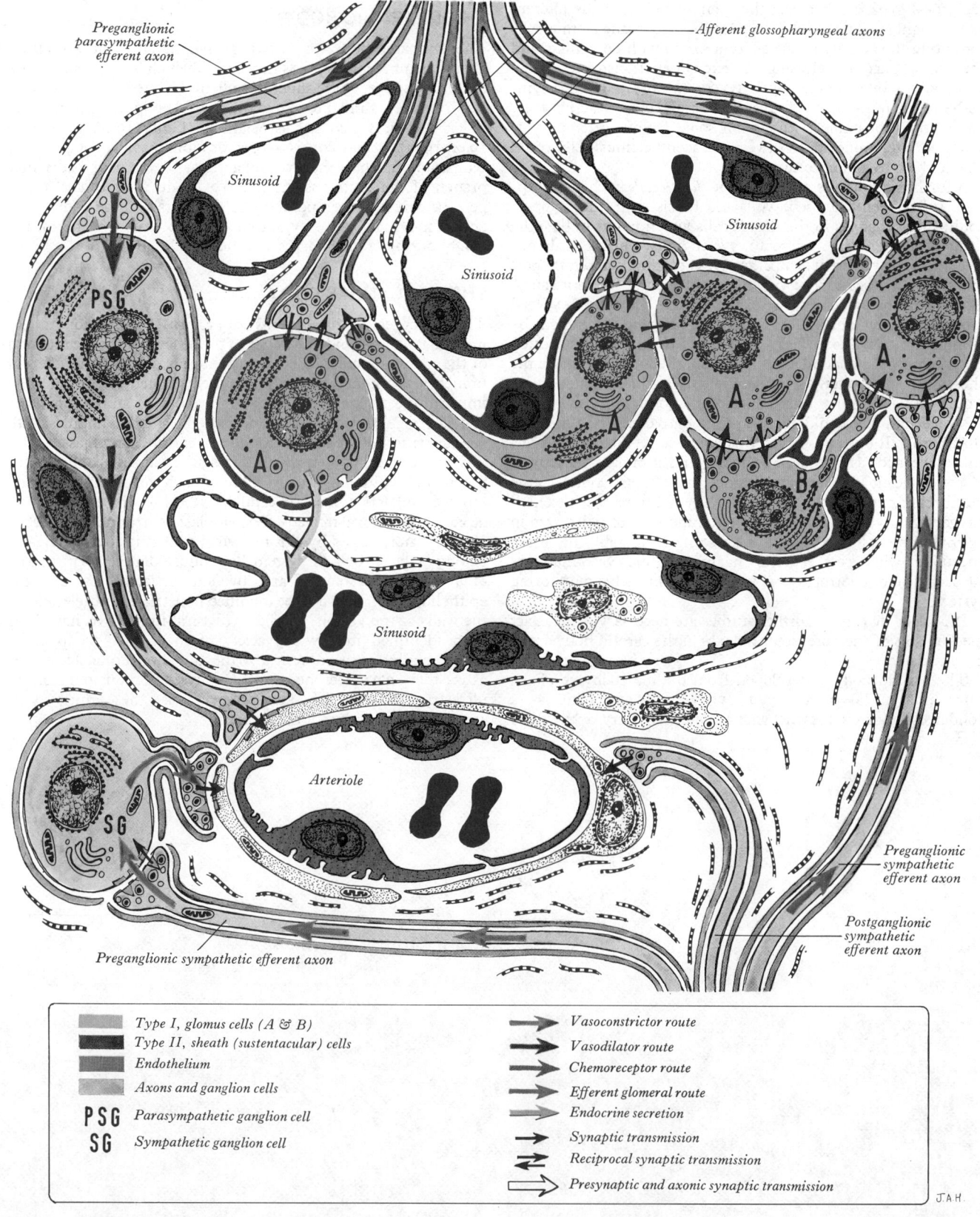

8.210B The cellular, neural and vascular architecture of the carotid body. Functional pathways are indicated. (Consult text for detailed discussion.)

cells (Lever *et al.* 1959; Eyzaguire *et al.* 1972), *afferent nerve terminals* (McDonald and Mitchell 1975), and *sheath cells* (Mills and Jöbsis 1972) have all been suggested as possibilities. A recent hypothesis is that the afferent axons and glomus cells co-operate as chemoreceptive units (McDonald and Mitchell 1975). It is proposed that: (1) the afferent nerve endings, interconnected with glomus cells by reciprocal synapses, are the true chemoreceptors; (2) the glomus cells are dopaminergic interneurons which modify the sensitivity of the chemoreceptive endings; (3) the reciprocal synapses between the glomus cells and the afferent axons might mediate an inhibitory feedback loop, the glomus cells inhibiting the activity of the afferent axons with dopamine, and the afferent axons releasing an excitatory transmitter when stimulated, for example, by hypoxia; (4) preganglionic sympathetic neurons may decrease chemoreceptive activity by stimulating dopamine release from some glomus cells; and (5) synaptic interconnexions between adjacent glomus cells could enable them to influence each other.

A further hypothesis implicates the *sheath cells* in chemoreception. It has been proposed (Mills and Jöbsis 1972) that hypoxia could cause changes in the sheath cells which might initiate the firing of adjacent 'free' afferent nerve endings, whilst the glomus cells, with their efferent sympathetic innervation, could form part of a feedback circuit modifying the activity of the afferent axons. About half the cytochrome a_3 of the carotid body is of a special type with a low affinity for oxygen. It has been concluded, by elimination, that this special type of cytochrome a_3 may be located in sheath cell mitochondria. If this is so, then it would make the sheath cells particularly sensitive to hypoxia, for the cytochrome would remain in its reduced state in conditions of oxygen shortage, and this would depress the oxidative metabolism of the sheath cells. It is proposed that this might cause them to release a substance, such as potassium, which would stimulate the activity of the adjacent afferent axons. Since these axons contain large numbers of mitochondria containing a cytochrome a_3 with a high oxygen affinity, they would be able to function normally even in an environment low in oxygen. It is also of considerable interest to speculate what role a local haemodynamic vasodilatator/vasoconstrictor control system would serve in such a monitoring system.

It is clear that many further studies are necessary if the exact mechanisms of chemoreceptor reflexes in the carotid body are to be elucidated.

The carotid body is developed from the mesenchyme of the third pharyngeal arch (Boyd 1937), first appearing as a condensation of the mesenchyme around the artery of the third pharyngeal arch, and its nerve supply is mainly from the nerve of that arch, the glossopharyngeal.

Other small bodies, with a structure similar to that of the carotid bodies, are found near the arteries of the fourth and sixth pharyngeal arches; they hence are close to the arch of the aorta, the ductus arteriosus and the right subclavian arteries, and are supplied by branches of the vagus nerves. They are also believed to function as chemoreceptors.

THE TYMPANIC BODY

The tympanic body (or *jugular glomus*) is an ovoid structure, about 0·5 mm long and 0·25 mm broad, which is in the adventitia of the upper part of the superior bulb of the internal jugular vein. Its structure is similar to that of the carotid body, and presumably it has similar functional associations (Guild 1941; Kjaegaard 1944, 1973). It may consist of two or more masses related to the tympanic branch of the glossopharyngeal nerve or the auricular branch of the vagus, as these nerves lie in their canals in the petrous part of the temporal bone. Tumours of these bodies may occur and may cause symptoms due to involvement of the neighbouring cranial nerves and the middle ear.

THE COCCYGEAL BODY

The coccygeal body (*glomus coccygeum*) is anterior to, or immediately inferior to, the apex of the coccyx, at the termination of the median sacral vessels, which supply afferent and efferent branches to the organ, and it is closely related to the ganglion impar of the sympathetic trunks. It is about 2·5 mm in diameter and is irregularly oval in shape; several smaller nodules with similar structure are found around or near the main mass.

The coccygeal body consists of irregular masses of spherical or polyhedral 'epithelioid' cells, those of each mass being grouped around a dilated, sinusoidal capillary vessel. Each cell contains a large, round or oval nucleus, the cytoplasm surrounding which is clear and not stained by chromic salts, so that it does not clearly belong to the chromaffin system. The sinusoidal vessels form part of a complex system of arteriovenous anastomoses, and the epithelioid cells appear to be modified nonstriated muscle cells of the walls of the vessels. Similar structural formations have long been identified in a wide variety of mammals, occupying a corresponding position, ventral to the caudal vertebrae. However, as yet there have been no detailed analyses of their innervation, ultrastructure and functional associations; these must await anatomical researches of the future.

BIBLIOGRAPHY

BIBLIOGRAPHY

A

Abbie, A. A. (1933). The blood supply of the lateral geniculate body, with a note on the morphology of the choroidal arteries, *J. Anat.*, **67**, 491–521.

Abbie, A. A. (1934). Projection of forebrain on pons and cerebellum, *Proc. R. Soc. B.*, **115**, 504–22.

Abbie, A. A. (1934). The morphology of the fore-brain arteries, with especial reference to the evolution of the basal ganglia, *J. Anat.*, **68**, 433–70.

Abbie, A. A. (1950). Closure of cranial articulations in the skull of the Australian aborigine, *J. Anat.*, **84**, 1–12.

Abbie, A. A. (1952). A new approach to the problem of human evolution, *Trans. R. Soc. S. Aust.*, **75**, 70–88.

Abbot, M. E. (1936). *Atlas of Congenital Heart Disease*, American Heart Association: New York.

Acheson, R. M. (1960). Effects of nutrition and disease on human growth. In: *Human Growth* (Tanner, J. M. ed.), pp. 73–92, Pergamon: London, Paris, New York.

Adachi, B. & Hasehe, K. (1928). *Das Arteriensystem der Japaner*, Volume 1, Maruzen Company: Kyoto.

Adams, C. W. M. (1965). *Neurohistochemistry* (editor), Elsevier: Amsterdam.

Adams, J., Daniel, P. M. and Prichard, M. M. (1969). The blood supply of the pituitary gland of the ferret with special reference to infarction after stalk section, *J. Anat.*, **104**, 209–25.

Adams, W. E. (1957). On the possible homologies of the occipital artery in mammals, with some remarks on the phylogeny and certain anomalies of the subclavian and carotid systems, *Acta anat.*, **29**, 90–113.

Adams, W. E. (1958). *The Comparative Morphology of the Carotid Body and Carotid Sinus*, Thomas: Springfield, Ill.

Ades, H. W. (1944). Midbrain auditory mechanisms in cats, *J. Neurophysiol.*, **7**, 415–24.

Addison, C. (1899–1901). On the topographical anatomy of abdominal viscera in man, especially the gastro-intestinal canal, I. *J. Anat.*, **33**, 565–86; II. *J. Anat.*, **34**, 427–50; III. *J. Anat.*, **35**, 166–204 and 277–304.

Adinolfi, A. M. and Pappas, G. D. (1968). The fine structure of the caudate nucleus of the cat, *J. comp. Neurol.*, **133**, 167–84.

Adrian, E. D. (1941). Afferent discharges to the cerebral cortex from peripheral sense organs, *J. Physiol., Lond.*, **100**, 159–91.

Afifi, A. and Kaelber, W. W. (1965). Efferent connections of the substantia nigra in the cat, *Expl Neurol.*, **11**, 474–82.

Aguayo, A. J., Peyronnard, J. M. and Bray, G. M. (1973). A quantitative ultrastructural study of regeneration from isolated proximal stumps of transected unmyelinated nerves, *J. Neuropath. exp. Neurol.*, **32**, 256–69.

Aguayo, A., Perkins, S., Duncan, I. and Bray, G. (1978). Human and animal neuropathies studied in experimental nerve transplants. In: *Peripheral Neuropathies* (Canal, N. and Pozza, G. eds.), pp. 37–48, Elsevier/North-Holland Biomedical Press: Amsterdam.

Aho, A. (1950). On the venous network of the human heart and its arteriovenous anastomoses. *Annls Med. exp. Biol. Fenn., Suppl.* 1, **29**, 1–90.

Aitken, J. T. and Bridger, J. E. (1961). Neuron size and population density in lumbosacral region of the cat's spinal cord, *J. Anat.*, **95**, 38–53.

Ajmone Marsan, C. (1965). The thalamus. Data on its functional anatomy and on some aspects of thalamocortical integration, *Archs ital. Biol.*, **103**, 847–82.

Akelaitis, A. J. (1942). Studies on corpus callosum; orientation (temporalspatial gnosis) following section of corpus callosum, *Archs Neurol. Psychiat., Chicago*, **48**, 914–37.

Akerblom, B. (1948). *Standing and Sitting Posture* (Synge, A. trans.), Nordiska Bokhandeln: Stockholm.

Akert, K., Pfenninger, K., Sandri, C. and Moor, H. (1972). Freeze-etching and cytochemistry of vesicles and membrane complexes in synapses of the central nervous system. In: *Structure and Function of Synapses* (Pappas, G. D. and Purpura, D. P. eds.), pp. 67–86, Raven Press: New York.

Albe-Fessard, D. and Liebeskind, J. (1966). Origine des messages somato-sensitifs activant les cellules du cortex moteur chez le singe, *Expl Brain Res.*, **1**, 127–246.

Albert, M. L. and Obler, L. K. (1978). *The Bilingual Brain*, Academic Press: New York, London.

Aleksandrowicz, R., Gosek, M. and Prorok, M. (1974). Normal and pathologic dimensions of the abdominal aorta, *Folia morphol. (Warsz.).*, **33**, 309–15.

Alexander, R. McN. and Vernon, A. (1975). The mechanics of hopping by kangaroos, *J. zool, Lond.*, **177**, 265–303.

Alexander, R. McN. and Goldspink, G. (1977). *Mechanics and Energetics of Animal Locomotion* (editors), Chapman & Hall: London.

Alexander, R. McN. and Bennet-Clark, H. C. (1977). Storage of elastic strain energy in muscle and other tissues, *Nature, Lond.*, **265**, 114–17.

Al-Lami, F. (1969). Light and electron microscopy of the adrenal medulla of Macaca mulatta monkey, *Anat. Rec.*, **164**, 317–32.

Al-Lami, F. (1970). Follicular arrangements in hamster adrenomedullary cells: light and electron microscopic studies, *Anat. Rec.*, **168**, 161–78.

Allanson, J. T. and Whitfield, I. C. (1955). *Third London Symposium on Information Theory*, Butterworth: London.

Allen, E., Pratt, J. P., Newell, Q. U. and Bland, L. J. (1930). Human tubal ova; related early corpora lutea and uterine tubes, *Contr. Embryol.*, **22**, 45–75.

Allen, L. (1967). Lymphatics and lymphoid tissues, *A. Rev. Physiol.*, **29**, 197–224.

Allen, M., Wright, P. and Reid, L. (1972). The human lacrimal gland. A histochemical and organ culture study of the secreting cells, *Archs Ophthal.*, **88**, 493–7.

Allen, R. M. (1969). The mental age-visual perception issue assessed. *Exceptional Child.*, **35**, 748–9.

Allison, A. C. (1954). The secondary olfactory areas in the human brain, *J. Anat.*, **88**, 481–88.

Allison, A. C. (1973). The role of microfilaments and microtubules in cell movements, endocytosis and exocytosis. In: *Locomotion of Tissue Cells, Ciba Symposium 14* (Porter, R. & Fitzsimons, D. W. eds.), pp. 109–43, Associated Scientific Publishers: Amsterdam.

Allison, P. R. (1951). Reflux esophagitis, sliding hiatal hernia and anatomy of repair, *Surgery Gynec. Obstet.*, **92**, 419–31.

Alpern, M. (1969). Movements of the eyes. In: *The Eye* (Davson, H. ed.), Volume 3, Second edition, Part I, Academic Press: New York, London.

Altman, J. and Carpenter, M. B. (1961). Fiber projections of the superior colliculus in the cat, *J. comp. Neurol.*, **116**, 157–78.

Altman, J. (1972a). Postnatal development of the cerebellar cortex in the rat. I. The external germinal layer and the transitional molecular layer, *J. comp. Neurol.*, **145**, 353–98.

BIBLIOGRAPHY

Altman, J. (1972b). Postnatal development of the cerebellar cortex in the rat. II. Phases in the maturation of Purkinje cells and of the molecular layer, *J. comp. Neurol.*, **145**, 399–464.

Altman, J. (1972c). Postnatal development of the cerebellar cortex in the rat. III. Maturation of the components of the granular layer, *J. comp. Neurol.*, **145**, 465–514.

Altman, J. and Bayer, S. (1975). Postnatal development of the hippocampal dentate gyrus under normal and experimental conditions. In: *The Hippocampus* (Isaacson, R. L. and Pribram, K. H. eds.), Volume 1, Plenum Press: New York, London.

Alvarez-Morujo, A. (1968). Terminal arteries of the penis, *Acta anat.*, **67**, 387–98.

Amoroso, E. C. (1952). Placentation. In: *Marshall's Physiology of Reproduction* (Parkes, A. S. ed.), Volume 2, 127–311, Longmans, Green: London.

Amos, L. A. and Klug, A. (1974). Arrangement of subunits in flagellar microtubules, *J. Cell Sci.*, **14**, 523–37.

Amprino, R. (1948). Recherches et considérations sur la structure du cartilage hyalin, *Acta anat.*, **5**, 123–46.

Amprino, R. (1968). Bone histophysiology, *Guy's Hosp. Rep.*, **116**, 51–69.

Amprino, R. and Engström, A. (1952). Studies on X-ray absorption and diffraction of bone tissue, *Acta anat.*, **15**, 1–22.

Amprino, R. and Camosso, M. (1955). Richerche sperimentali sulla morfogenesi degli arti nel pollo, *J. exp. Zool.*, **129**, 453–94.

Anand, B. K. (1970). Regulation of visceral activities by the central nervous system. In: *Control Processes in Multicellular Organisms* (Wolstenholme, G. E. W. and Knight, J. eds.), pp. 356–80, Ciba Foundation Symposium, Churchill: London.

Ånberg, A. (1957). The ultrastructure of the human spermatozoon, *Acta Obstet. Gynec. Scand.*, **36**, Suppl. 2, 1–33.

Andersen, H. T. (1969). *The Biology of Marine Mammals* (editor), Academic Press: New York, London.

Andersen, P., Eccles, J. C., Schmidt, R. F. and Yokota, T. (1964). Identification of relay cells and interneurons in the cuneate nucleus, *J. Neurophysiol.*, **27**, 1080–95.

Andersen, P., Eccles, J. C. and Sears, T. A. (1964). The ventro-basal complex of the thalamus: Types of cells, their responses and their functional organisation, *J. Physiol., Lond.*, **174**, 370–99.

Andersen, P., Brooks, C. McC., Eccles, J. C. and Sears T. A. (1964). The ventrobasal nucleus of the thalamus: Potential fields, synaptic transmission and excitability of both presynaptic and postsynaptic components, *J. Physiol., Lond.*, **174**, 348–69.

Andersen, P., Eccles, J. C. and Løyning, Y. (1964a). Location of postsynaptic inhibitory synapses on hippocampal pyramids, *J. Neurophysiol.*, **27**, 592–607.

Andersen, P., Eccles, J. C. and Løyning, Y. (1964b). Pathway of postsynaptic inhibition in the hippocampus, *J. Neurophysiol.*, **27**, 608–19.

Andersen, P., Blackstad, T. W. and Lømo, T. (1966). Location and identification of excitatory synapses on hippocampal pyramidal cells, *Expl Brain Res.*, **1**, 236–48.

Anderson, C. E. (1962). The structure and function of cartilage, *J. Bone Jt Surg.*, **44A**, 777–86.

Anderson, C. E. and Parker, J. (1966). Invasion and resorption in endochondral ossification. An electron microscopic study, *J. Bone Jt Surg.*, **48A**, 899–914.

Anderson, D. J., Hannam, A. G. and Matthews, B. (1970). Sensory mechanisms in mammalian teeth and their supporting structures, *Physiol. Rev.*, **50**, 171–95.

Anderson, D. R. and Hoyt, W. (1969). Ultrastructure of intraorbital portion of human and monkey optic nerve, *Archs Ophthal., N.Y.*, **82**, 506–30.

Anderson, H. C. (1967). Electron microscopic studies of induced cartilage development and calcification, *J. Cell Biol.*, **35**, 81–101.

Anderson, H. C. and Reynolds, J. J. (1973). Pyrophosphate stimulation of calcium uptake into cultured embryonic bones. Fine structure of matrix vesicles and their role in calcification, *Dev. Biol.*, **34**, 211–27.

Anderson, P., Röhlich, P. and Slorach, S. A. (1974). Morphological and storage properties of rat mast cell granules isolated by different methods, *Acta physiol. Scand.*, **91**, 145–53.

Anderson, P. and Uvnäs, B. (1975). Selective localization of histamine to electron dense granules in antigen-challenged sensitised rat mast cells and to similar granules isolated from sonicated mast cells. An electron microscope autoradiographic study, *Acta physiol. Scand.*, **94**, 63–73.

Anderson, R. H. and Becker, A. E. (1980). *Cardiac Anatomy*. Gower Medical Publishing Co.: London.

Andreassi, G. (1967). Sur la topographie de l'apex pulmonaire chez l'homme, *C. r. Ass. Anat.*, **137**, 141–7.

Andres, K. H. (1966). Der Feinbau der Regio olfactoria von Makrosmatikern, *Z. Zellforsch. mikrosk. Anat.*, **69**, 140–54.

Andrew, A. (1971). The origin of intramural ganglia. IV. The origin of enteric ganglia: a critical review, *J. Anat.*, **108**, 169–84.

Andrews, J. (1971). Streak gonads and the Y chromosome, *J. Obstet. Gynec.*, **78**, 448–57.

Andrews, P. M. (1975). Microplicae, *J. Cell Biol.*, **67**, 11a.

Andy, O. J. and Stephan, H. (1968). The septum in the human brain, *J. comp. Neurol.*, **133**, 383–410.

Angaut, P. and Brodal, A. (1967). The projection of the 'vestibulo-cerebellum' onto the vestibular nuclei in the cat, *Archs ital Biol.*, **105**, 441–79.

Angaut, P. and Repérant, J. (1976). Fine structure of the optic fibre termination layers in the pigeon optic tectum, *Neuroscience*, **1**, 93–105.

Angell, J. C. (1969). Treatment of benign prostatic hyperplasia by phenol injection, *Br. J. Urol.*, **41**, 735–8.

Angevine, J. B. Jr. (1975). Development of the hippocampal region. In: *The Hippocampus* (Isaacson, R. L. and Pribram, K. H. eds.), Volume 1, Plenum Press: New York, London.

Annett, M. (1964). A model of the inheritance of handedness and cerebral dominance, *Nature, Lond.*, **204**, 59–60.

Annis, D. (1962). A study of the regenerative ability of the epithelial lining of the urinary bladder, *Ann. R. Coll. Surg.*, **31**, 23–45.

Anson, B. J., Morgan, E. H. and McVay, C. B. (1960). Surgical anatomy of the inguinal region based upon a study of 500 body-halves, *Surgery Gynec. Obstet.*, **111**, 707–25.

Anson, B. J., Jamieson, R. W., O'Connor, V. J. and Beaton, L. E. (1953). The pectoral muscles, *Q. Bull. North West Univ. Med. Sch.*, **27**, 211–18.

Appenzeller, O. (1964). Electron microscope study of the innervation of the auricular artery in the rat, *J. Anat.*, **98**, 87–91.

Applebaum, A. E., Beall, J. E., Foreman, R. D. and Willis, W. D. (1975). Organization and receptive fields of primate spinothalamic tract neurons. *J. Neurophysiol.*, **38**, 572–86.

Appleton, A. B. (1922). On the hypotrochanteric fossa and accessory adductor groove of the primate femur, *J. Anat.*, **56**, 295–306.

Appleton, A. B. (1934). Postural deformities and bone growth; experimental study, *Lancet*, **1**, 451–4.

Arbib, M. (1964). *Brains, Machines and Mathematics*, McGraw-Hill: New York.

Arey, L. B. (1932). Retina, choroid and sclera. In: *Cytology and Cellular Pathology of the Nervous System* (Penfield, W. ed.), pp. 743–836, Hoeber: New York.

Arey, L. B. (1949). The craniopharyngeal canal re-interpreted on the basis of its development, *Anat. Rec.*, **103**, 420.

Arey, L. B. (1974). *Developmental Anatomy*, Seventh edition, Saunders: London.

Ariens Kappers, C. U. (1920). *Die vergleichende Anatomie des Nervensystems der Wirbeltiere und des Menchen*, Boln: Haarlem.

Ariens Kappers, C. U. (1921). On structural laws in the nervous system: The principles of neurobiotaxis, *Brain*, **44**, 125–49.

Ariens Kappers, C. U. (1934). Differences in the effect of various impulses on the structure of the central nervous system, *Ir. J. med. Sci.*, **105**, 495–519.

Ariens Kappers, C. U., Huber, G. C. and Crosby, E. C. (1936). *The Comparative Anatomy of the Nervous System of Vertebrates, Including Man*, Macmillan: New York.

Armstrong, J., Richardson, K. C. and Young, J. Z. (1956). Staining neural end feet and mitochondria after postchroming and carbowax embedding, *Stain Technol.*, **31**, 263–70.

Arnold, F. (1851). *Handbuch der anatomie des Menschen*, Volume 2, Emmerling and Herder: Freiburg.

Arnold, G. (1974). Festigkeit und Kraft-Längenänderungs-Verhalten der Strecksehnen des menschlichen Fusses, *Res. exp. Med.*, **164**, 123–36.

Arnstein, K. (1889). Über die Nerven der Schweissdrüsen, *Anat. Anz.*, **4**, 378–83.

Arnstein, K. (1895). Zur Morphologie der sekretorischen Nervendapparate, *Anat. Anz.*, **10**, 410–19.

Arsenieva, N. A., Dubinin, N. P., Orlova, N. N. and Bakulina, E. D. (1961). A radiation analysis of the duration of meiosis phases in the spermatogenesis of Macaca mulatta, *Dokl. Akad. Nauk. S.S.R.*, **141**, 1486–90.

Arslan, M. (1960). The innervation of the middle ear, *Proc. R. Soc. Med.*, **53**, 1068–74.

Arvis, G. (1969). Considérations anatomiques sur le hile et le sinus du rein, *Annls Radiol.*, **12**, 75–106.

Asanuma, H. and Sakata, H. (1967). Functional organisation of a cortical efferent system examined with focal depth stimulation in cats, *J. Neurophysiol.*, **30**, 35–54.

Ascenzi, A. and Bonucci, E. (1968). The compressive properties of single osteons, *Anat. Rec.*, **161**, 377–92.

Ascenzi, A. and Bell, G. H. (1972). Bone as a mechanical and engineering problem. In: *The Biochemistry and Physiology of Bone* (Bourne, G. H. ed.), Second edition, Academic Press: London, New York.

Ashby, W. R. (1960). *Design for a Brain*, Wiley: New York.

Ashley, G. T. (1952). The manner of insertion of the pectoralis major muscle in man, *Anat. Rec.*, **113**, 301–8.

Ashley, G. T. (1954). Morphological and pathological significance of synostoses at the manubrio-sternal joint, *Thorax*, **9**, 159–66.

Ashley, G. T. (1956). The relationship between the pattern of ossification and the definitive shape of the mesosternum in man, *J. Anat.*, **90**, 87–105.

Ashton, E. H. and Zuckerman, S. (1956). The base of the skull in immature hominids, *Am. J. phys. Anthrop.*, **14**, 611–24.

Ashworth, M. A. and Leach, F. N. (1973). Development of insulin secretion in the human fetus, *Arch. Dis. Child.*, **48**, 151–2.

Askansas, V. and Engel, W. K. (1975). A technique of fiber selection from human muscle tissue culture for histochemical-electron microscopic studies, *J. Histochem. Cytochem.*, **23**, 144–6.

Assali, N. S., Rauramo, L. and Peltonen, T. (1960). Measurement of uterine blood flow and uterine metabolism. VIII. Uterine and fetal blood flow and oxygen consumption in early human pregnancy, *Am. J. Obstet. Gynec.*, **79**, 86–98.

Athias, M. (1897). Recherches sur l'histogenese de l'ecorce du cervelet, *J. de l'Anatomie*, **33**, 372–404.

Atkinson, M., Edwards, D. A., Honour, A. J. and Rowlands, E. N. (1957). Comparison of cardiac and pyloric sphincters: a manometric study, *Lancet*, **2**, 918–22.

Auerbach, R. (1960). Morphogenetic interactions in the development of the mouse thymus gland, *Devl Biol.*, **2**, 271–84.

Auerbach, R. (1961). Experimental analysis of the origin of cell types in the development of the mouse thymus, *Devl Biol.*, **3**, 336–54.

Austin, C. R. (1951). Observations on the penetration of the sperm into the mammalian egg, *Aust. J. sci. Res.*, Ser. B, **4**, 581–96.

Austin, C. R. (1961). *The Mammalian Egg*, Blackwell Scientific Publications: Oxford.

Austin, C. R. (1963). Fertilisation and transport of the ovum. In: *Mechanisms Concerned with Conception* (Hartman, C. G. ed.), pp. 285–320, Pergamon Press: Oxford, London, New York, Paris.

Austin, C. R. (1965). *Fertilisation*, Foundation of developmental biology series, Prentice-Hall Inc.: New Jersey.

Austin, C. R. (1968). *Ultrastructure of Fertilisation*, Holt, Rinehart and Winston: New York, Chicago, San Francisco, Toronto, London.

Austin, C. R. and Braden, A. W. H. (1956). Early reactions of the rodent egg to spermatozoon penetration, *J. exp. Biol.*, **33**, 358–65.

Austin, C. R. and Bishop, M. W. H. (1958). Some features of the acrosome and perforatorium in mammalian spermatozoa, *Proc. R. Soc. B.*, **149**, 234–40.

Austin, C. R. and Walton, A. (1960). Fertilisation. In: *Marshall's Physiology of Reproduction* (Parkes, A. S. ed.), Third edition, Volume I, Part 2, pp. 310–416, Longmans, Green: London.

Avenando, S., Croxatto, H. D., Pereda, J. and Croxatto, H. B. (1975). A seven-cell human egg recovered from the oviduct, *Fertil. Steril.*, **26**, 1167–72.

B

Babel, J., Bischoff, A. and Spoendlin, H. (1970). *Ultrastructure of the Peripheral Nervous System and Sense Organs*, Churchill: London.

Bachmann, G. (1916). The inter-auricular time interval, *Am. J. Physiol.*, **41**, 309.

Bach-y-Rita, P. and Collins, C. C. (1971). *The Control of Eye Movements*, Academic Press: New York, London.

Backhouse, K. M. (1964). The gubernaculum testis Hunteri: testicular descent and maldescent, *Ann. R. Coll. Surg.*, **35**, 15–33.

Backhouse, K. M. and Butler, H. (1958). The development of the coverings of the testis and cord, *J. Anat.*, **92**, 645P.

Baetens, D., De Mey, J. and Gepts, W. (1977). Immunohistochemical and ultrastructural identification of the pancreatic polypeptide-producing (PP-cell) in the human pancreas, *Cell Tissue Res.*, **185**, 239–46.

Bagnall, K. M., Harris, P. F. and Jones, P. R. M. (1977). A radiographic study of the human fetal spine. 2. The sequence of development of ossification centres in the vertebral column, *J. Anat.*, **124**, 791–802.

Bailey, H. (1947). Parotidectomy; indications and results, *Br. med. J.*, **1**, 404–7.

Bailey, P. and Bonin G. von (1951). *The Isocortex of Man*, University of Illinois Press: Urbana.

Baillarger, J. G. F. (1840). Recherches sur la structure de la couche corticale des circonvolutions du cerveaux, *Mém. Acad. roy. Méd., Paris*, **8**, 149–83.

Bainton, D. F., Ullyot, J. L. and Farquhar, M. G. (1971). The development of neutrophilic polymorphonuclear leukocytes in human bone marrow. Origin and content of azurophil and specific granules, *J. exp. Med.*, **134**, 907–34.

Baker, B. L., Pek, S., Midgley, A. R. and Gerstein, B. E. (1970). Identification of the corticotrophin cell in rat hypophyses with peroxidase-labelled antibody, *Anat. Rec.*, **166**, 537–68.

Baker, J. R. (1960). *Cytological Techniques*, Fourth edition, Methuen: London.

Baker, T. G. (1963). A quantitative and cytological study of germ cells in human ovaries, *Proc. R. Soc. B.*, **158**, 417–33.

Baker, T. G. (1966). A quantitative and cytological study of oogenesis in the rhesus monkey, *J. Anat.*, **100**, 761–76.

Baker, T. G. and Neal, P. (1974). Oogenesis in human fetal ovaries maintained in organ culture. *J. Anat.*, **117**, 591–604.

Balinsky, B. I. (1975). *Introduction to Embryology*, Fourth edition, Holt-Saunders: Eastbourne.

Baló, J. (1950). The dural venous sinuses. *Anat. Rec.*, **106**, 319–26.

Balthasar, K. (1952). Morphologie der spinalen Tibialis- und Peronaeus-Kerne bei der Katze: Topographie, Architektonik, Axon- und Dendritenverlauf der Motonerurone und Zwischenneurone in den Segmenten L₆–S₂, *Arch. Psychiat. NervenKrankh.*, **188**, 345–78.

Balthazar, E. J. and Gade, M. (1976). The normal and abnormal development of the appendix. *Radiology*, **121**, 599–604.

Bannerjee, S. D., Cohn, R. H. and Bernfield, M. R. (1977). Basal lamina of embryonic salivary epithelia. Production by the epithelium and role in maintaining lobular morphology, *J. Cell Biol.*, **73**, 445–63.

Bannister, L. H. (1976). Sensory terminals of peripheral nerves. In: *The Peripheral Nerve* (Landon, D. N. ed.), pp. 396–463, Chapman & Hall: London.

Bannister, L. H. and Dodson, H. C. (1975). Proliferative units in the olfactory epithelium of the mouse, *J. Anat.*, **119**, 407–8.

Barajas, L. (1966). The development and ultrastructure of the juxta-glomerular cell granule, *J. Ultrastruct. Res.*, **15**, 400–13.

Barbenel, J. C. (1972). The biomechanics of the temporo-mandibular joint, *J. Biomech.*, **5**, 251–6.

Barbizet, J. (1963). Defect of memorizing of hippocampal-mamillary origin: a review, *J. Neurol. Neurosurg. Psychiat.*, **26**, 127–35.

Barclay, A. E. (1936). *The Digestive Tract*, Cambridge University Press: London.

Barclay, A. E. and Franklin, K. J. (1938). Time of functional closure of foramen ovale in lamb, *J. Physiol., Lond.*, **94**, 256–8.

Barclay, A. E., Barcroft, J., Barron, D. H. and Franklin, K. J. (1939). A radiographic demonstration of the circulation through the heart in the adult and in the fetus and the identification of the ductus arteriosus, *Br. J. Radiol.*, **12**, 505–17.

Barclay, A. E., Franklin, K. J. and Prichard, M. M. L. (1942). Further data about circulation and about the cardiovascular system before and just after birth, *Br. J. Radiol.*, **15**, 249–56.

Barclay, A. E., Franklin, K. J. and Prichard, M. M. L. (1944). *The Foetal Circulation*, Blackwell: Oxford.

Barclay-Smith, E. (1896). The astragalo-calcaneo-navicular joint, *J. Anat*, **30**, 390–412.

Barcroft, J. (1941). Four phases of birth, *Lancet*, **2**, 91–5.

Bard, Ph. and Rioch, D. McK. (1937). A study of four cats deprived of neocortex and additional portions of the forebrain, *Bull. Johns Hopkins Hosp.*, **60**, 73–147.

Barden, H. (1969). The histochemical relationship of neuromelanin and lipofuscin, *J. Neuropath. exp. Neurol.*, **28**, 419–41.

Bargmann, W. (1966). Neurosecretion, *Int. Rev. Cytol.*, **19**, 183–201.

Bargmann, W. and Schadé, J. P. (1963). *The Rhinencephalon and Related Structures* (editors), *Prog. Brain Res.*, **3**.

Bargmann, W. and Scharrer, B. (1970). *Aspects of Neuroendocrinology* (editors), Springer-Verlag: Berlin, N.Y.

Barker, D. S. (1967). The dendritic cell system in human gingival epithelium, *Archs oral Biol.*, **12**, 203–8.

Barker, D. (1974). The motor innervation of muscle spindles. In: *Essays on the Nervous System* (Bellairs, R. and Gray, E. G. eds.), pp. 131–54, Clarendon Press: Oxford.

Barker, D. (1974). Morphology of muscle receptors. In: *Handbook of Sensory Physiology* (Hunt, C. C. ed.), Volume 3, Part 2, pp. 191–234, Springer: Berlin.

Barlow, T. E. Bentley, F. H. and Walder, D. N. (1951). Arteries, veins and arteriovenous anastomoses in the human stomach, *Surgery Gynec, Obstet.*, **93**, 657–71.

Barlow, T. E., Haigh, A. L. and Walder, D. N. (1961). Evidence for two vascular pathways in skeletal muscle, *Clin. Sci.*, **20**, 367–85.

Barnard, J. W. (1940). The hypoglossal complex of vertebrates, *J. comp. Neurol.*, **72**, 489–524.

Barnard, J. W. and Woolsey, C. N. (1956). A study of localisation in the cortico-spinal tracts of monkey and rat, *J. comp. Neurol.*, **105**, 25–50.

Barnes, R. D. (1976). In-vitro fertilisation, *Lancet*, **i**, 1016–17.

BIBLIOGRAPHY

Barnett, C. H. (1952). Locking at the knee joint, *J. Anat.*, **86**, 485P.

Barnett, C. H. (1956). Phases of human gait, *Lancet*, **2**, 617–21.

Barnett, C. H. and Napier, J. R. (1952). The axis of rotation at the ankle joint in man. Its influence upon the form of the talus and the mobility of the fibula, *J. Anat.*, **86**, 1–9.

Barnett, C. H. and Lewis, O. J. (1958). The evolution of some traction epiphyses in birds and mammals, *J. Anat.*, **92**, 593–607.

Barnett, C. H., Davies, D. V. and MacConaill, M. A. (1961). *Synovial Joints, Their Structure and Mechanics,* Longmans: London.

Baroldi, G. and Scomazzoni, G. (1967). *Coronary Circulation in the Normal and the Pathologic Heart,* Office of the Surgeon General: Washington.

Barondes, S. H. (1970). Cerebral protein synthesis inhibitors block long-term memory, *Int. Rev. Neurobiol.*, **12**, 177–205.

Barr, M. L. and Bertram, E. G. (1949). A morphological distinction between neurones of the male and female, and the behaviour of the nucleolar satellite during accelerated nucleoprotein synthesis, *Nature, Lond.*, **163**, 676–8.

Barrett, W. C. (1951). A note of the internal cremaster muscle, *Anat. Rec.*, **109**, 392(a).

Barrow, M. V. (1971). A brief history of teratology to the early 20th century, *Teratology*, **4**, 119–30.

Barry, A. (1951). The aortic arch derivatives in the human adult, *Anat. Rec.*, **111**, 221–38.

Barson, A. J. (1970). The vertebral level of termination of the spinal cord during normal and abnormal development, *J. Anat.*, **106**, 489–97.

Bartelmez, G. W. and Dekaban, A. S. (1962). The early development of the human brain, *Contr. Embryol.*, **37**, 13–32.

Barth, M., Tongio, J. and Warter, P. (1976). Interprétation embryologique de l'anatomie des artères digestives, *Ann. Radiol.*, **19**, 305–13.

Basmajian, J. V. (1959). 'Spurt' and 'shunt' muscles: An electromyographic confirmation, *J. Anat.*, **93**, 551–3.

Basmajian, J. V. (1967). *Muscles Alive,* Second edition, Williams and Wilkins: Baltimore.

Basmajian, J. V. and Spring, W. B. (1955). Electromyography of the male (voluntary) sphincter urethrae, *Anat. Rec.*, **121**, 388(a).

Basmajian, J. V. and Latif, A. (1957). Integrated actions and functions of the chief flexors of the elbow. A detailed electromyographic analysis, *J. Bone Jt Surg.*, **39A**, 1106–18.

Basmajian, J. V. and Travill, A. A. (1961). Electromyography of the pronator muscles in the forearm, *Anat. Rec.*, **139**, 45–9.

Basmajian, J. V. and Griffin, W. R. (1972). Function of anconeus muscle. An electromyographic study. *J. Bone Jt Surg.*, **54A**, 1712–14.

Bassett, C. A. L. (1962). Current concepts of bone formation, *J. Bone Jt Surg.*, **44A**, 1217–44.

Bassett, C. A. L. (1965). Electrical effects in bone, *Scient. Am.*, **213**, 18–25.

Bast, T. H. and Anson, B. J. (1949). *The Temporal Bone and the Ear,* Thomas: Springfield, Ill.

Bastianini, A. (1967). Aspetti microscopici della trama linfatica polmonare in condizioni sperimentali. I. Anossia intrauterina, *Atti Acad. fisiocr.*, **16**, 1304–8.

Bastianini A. (1967). Osservazioni sulla morfologia microscopica e l'istotopografia dei vasi linfatici del polmone umano, *Boll. Soc ital. Biol. sper.*, **43**, 1567–70.

Bates, M. W. (1948). The early development of the hypoglossal musculature in the cat, *Am. J. Anat.*, **83**, 329–56.

Batson, O. V. (1940). Function of vertebral veins and their role in the spread of metastases, *Ann. Surg.*, **112**, 138–49.

Batson, O. V. (1957). The vertebral vein system, *Am. J. Roentg.*, **78**, 195–212.

Bauer, W., Ropes, M. W. and Waine, H. (1940). Physiology of articular structures, *Physiol. Rev.*, **20**, 272–312.

Baum, J., Simons, B. E., Unger, R. H. and Madison, L. L. (1962). Localisation of glucagon in the alpha cells in the pancreatic islet by immunofluorescent technics, *Diabetes*, **11**, 371–4.

Baumann, J. A and Gajisin, S. (1975). Sur la multiplicité et la dispersion des ganglions parasympathiques de la tète, *Bull. Ass. Anat.*, **59**, 329–32.

Baumel, J. J. and Beard, D. Y. (1961). The accessory meningeal artery of man, *J. Anat.*, **95**, 386–402.

Baumel, J. J. (1974). Trigeminal-facial nerve communications. Their function in facial muscle innervation and reinnervation, *Arch. Otolaryngol.*, **99**, 34–44.

Baumgarten, H. G. and Holstein, A. F. (1967). Catecholaminhaltige Nervenfasern im Hoden des Menschen, *Z. Zellforsch. mikrosk. Anat.*, **79**, 389–95.

Baumgarten, H. G., Holstein, A. F. and Owman, C. (1970). Auerbach's plexus of mammals and man: electron microscopic identification of three different types of neuronal processes in myenteric ganglia from rhesus monkeys, guinea pigs and man, *Z. Zellforsch. mikrosk. Anat.*, **106**, 376–97.

Baxter, J. S. (1953). *Frazer's Manual of Embryology*, Third edition, Baillière, Tindall and Cox: London.

Beard, J. (1896). The history of a transient nervous apparatus in certain Ichthyopsida. An account of the development and degeneration of ganglion cells and nerve fibres, Part I, *Raja batis, Zoll. Jb.*, **9**, 319.

Bearn, J. G. (1961). An electromyographic study of the trapezius, deltoid, pectoralis major, biceps and triceps muscles, during static loading of the upper limb, *Anat. Rec.*, **140**, 103–8.

Bearn, J. G. (1967). Direct observation on the function of the capsule of the sterno-clavicular joint in clavicular support, *J. Anat.*, **101**, 159–70.

Beatty, R. A. (1957). *Parthenogenesis and Polyploidy in Mammalian Development*, University Press: Cambridge.

Beatty, R. A. (1967). Parthenogenesis in vertebrates. In: *Fertilisation* (Metz, C. B. and Monroy, A. eds.), Volume 1, pp. 413–40, Academic Press: New York, London.

Beaudet, A. and Descarries, L. (1979). Radioautographic characterisation of a serotonin-accumulating nerve cell group in adult rat hypothalamus, *Brain Res.*, **160**, 231–43.

Beck, I. T. and Sinclair, D. G. (1971). *The Exocrine Pancreas* (editors), Churchill: London.

Becks, H., Asling, C. W., Collins, D. A., Simpson M. E. and Evans, H. M. (1948). Changes with increasing age in the ossification of the third metacarpal of the female rat, *Anat. Rec.*, **100**, 577–92.

Bedford, J. M. (1977). Sperm/egg interaction: the specificity of human spermatozoa, *Anat. Rec.*, **186**, 477–88.

Bedford, J. M., Calvin, H. and Cooper, G. W. (1973). The maturation of spermatozoa in the human epididymis, *J. Reprod. Fert., Suppl.* **18**, 199–213.

Bedford, J. M. and Calvin, H. I. (1974). Changes in -S-S- linked structures of the sperm tail during epididymal maturation, with comparative observations in sub-mammalian species, *J. exp. Zool.*, **187**, 181–204.

Begg, R. C. (1930). The urachus: its anatomy, histology and development, *J. Anat.*, **64**, 170–83.

Behrens, J. C., Walker, P. S. and Shoji, H. (1974). Variations in strength and structure of cancellous bone at the knee, *J. Biomech.*, **7**, 201–7.

Beidler, L. M. (1970). Physiological properties of mammalian taste receptors. In: *Taste and Smell in Vertebrates* (Wolstenholme, G. E. W. and Knight, J. eds.), pp. 51–67, Ciba Foundation Symposium, Churchill: London.

Beidler, L. M. and Smallman, R. L. (1965). Renewal of cells within taste buds, *J. Cell Biol.*, **27**, 263–72.

Bélanger, L. F. (1956). Autoradiographic studies of the formation of the organic matrix of cartilage, bones and the tissue of teeth. In: *Bone Structure and Metabolism* (Wolstenholme, G. E. W. and O'Connor, C. M. eds.), pp. 75–87, Ciba Foundation Symposium, Churchill: London.

Bélanger, L. F. (1969). Osteocyte osteolysis, *Calc. Tiss. Res.*, **4**, 1–12.

Bélanger, L. F., Robichon, J., Migicovsky, B. B., Copp, D. H. and Vincent, J. (1963). Resorption without osteoclasts (osteolysis). In: *Mechanisms of Hard Tissue Destruction* (Sognnaes, R. F. ed.), pp. 531–56, Am. Ass. Adv. Sci.: Washington.

Bell, E. T. and Loraine, J. A. (1965). Time of ovulation in relation to cycle length, *Lancet*, **i**, 1029–30.

Bell, G. H., Cuthbertson, D. P. and Orr, J. (1941). Strength and size of bone in relation to calcium intake, *J. Physiol., Lond.*, **100**, 299–317.

Bellairs, A. d'A. (1969). *The Life of Reptiles*. Chapter 7, p. 1088, Weidenfeld & Nicolson: London.

Bellairs, R. (1959). The development of the nervous system in chick embryos, studied by electron microscopy, *J. Embryol. exp. Morph*, **7**, 94–115.

Belmonte, N. (1968). Estudios anatomicos sobre la vascularizacion del nervio optico, *Archos Soc. oftal, hisp.-am.*, **28**, 801–10.

Belt, E. (1965). *Leonardo the Anatomist*, University of Kansas Press: Kansas.

Beltrani, F. (1946). Considérations biologiques sur le mandibule chez l'homme, *Rev. Stomat.*, **47**, 1–9.

Bender, M. B. (1964). *The Oculomotor System* (editor), Hoeber: New York.

Bender, M. B. and Weinstein, E. A. (1943). Functional representation in oculomotor and trochlear nuclei, *Archs. Neurol. Psychiat., Chicago*, **49**, 98–106.

Benedeczky, I. and Smith, A. D. (1972). Ultrastructural studies on the adrenal medulla of the golden hamster: origin and fate of secretory granules. *Z. Zellforsch.*, **124**, 367–86.

Bennett, A. G. and Francis, J. L. (1962). Refraction at plane and spherical surfaces. In: *The Eye* (Davson, H. ed.), Volume 4, pp. 19–34, Academic Press: New York, London.

Bennett, H. S., Luft, J. H. and Hampton, J. C. (1957). Morphological classification of vertebrate blood capillaries, *Am. J. Physiol.*, **196**, 381–90.

Bennett, M. R. and Rogers, D. C. (1967). A study of the innervation of the taenia coli, *J. Cell Biol.*, **37**, 573–96.

Benninghoff, A. (1925). Spaltlinien am Knocken. *Anat. Anz.*, **60**, 189–206.

Bensch, K. G., Gordon, G. B. and Miller, L. R. (1965). Studies on the bronchial counterpart of the Kultschitzky (Argentaffin) cell and innervation of bronchial glands, *J. Ultrastruct. Res.*, **12**, 668–86.

Benscome, S. (1959). Studies on the terminal autonomic nervous system with special reference to the pancreatic islets, *Lab. Invest.*, **8**, 629–46.

Bensley, R. R. (1911). Studies on the pancreas of the guinea pig, *Am. J. Anat.*, **12**, 297–388.

Benson, R. L., Sacktor, B. and Greenawalt, J. W. (1971). Studies on the ultrastructural localisation of intestinal disaccharides, *J. Cell Biol.*, **48**, 711–16.

Berard, A. (1835). *Arch. générale de Médicine, Series* II, 176–83.

Bergamson, J. P. G. (1978). Ophthalmic terminals in the iris and the ciliary body of monkeys, *Alb. von Graefes Arch. Klin. exp. Ophthal.*, **206**, 39–47.

Bergland, R. and Ray, B. S. (1969). The arterial supply of the human optic chiasm, *J. Neurosurg.*, **31**, 327–34.

Berkovitz, B. K. B. (1967). An account of the enamel cord in *Setonix braychurus* (Marsupiala) and on the presence of an enamel knot in *Trichosurus vulpecula*, *Archs oral Biol.*, **12**, 49–59.

Berkovitz, B. K. B. and Thomas, R. N. (1969). Unimpeded eruption in the root-transected lower incisor of the rat with preliminary note on root transection, *Archs oral Biol.*, **14**, 771–80.

Berlin, N. I., Waldmann, T. A. and Weissman, S. M. (1959). Life span of red blood cell, *Physiol. Rev.*, **39**, 577–616.

Berman, N. and Jones, E. G. (1977). A retino-pulvinar projection in the cat, *Brain Res.*, **134**, 237–48.

Bernard, G. W. and Pease, D. C. (1969). An electron microscopic study of initial intramembranous osteogenesis, *Am. J. Anat.*, **125**, 271–90.

Bernick, S. and Patek, R. (1969). Lymphatic vessels of the dental pulp in dogs, *J. dent. Res.*, **48**, 959–64.

Berquist, H. (1932). Zur Morphologie des Zwischen-hirns bei neideren Wirbeltieren, *Acta Zool.*, *Stockholm*, **13**, 57–304.

Berquist, H. (1952). Studies on the cerebral tube in vertebrates. The neuromeres, *Acta zool. Stockh.*, **33**, 117–91.

Berquist, H. (1968). Uber die Differenzierung der Neuraloknes besounders des Stratum Zonale, *Z. Zellforsch. mikrosk. Anat.*, **86**, 401–21.

Berry, A. C. (1975). Factors affecting the incidence of non-metrical skeletal variants, *J. Anat.*, **120**, 519–35.

Berry, A. C. and Berry, R. J. (1967). Epigenetic variation in the human cranium, *J. Anat.*, **101**, 361–80.

Berry, M. (1974). Development of the cerebral neocortex of the rat. In: *Aspects of Neurogenesis: Studies on the Development of Behaviour and the Nervous System* (Gottlieb, G. ed.), Volume 2, Academic Press: New York, London.

Berry, M. and Rogers, A. W. (1965). The migration of neuroblasts in the developing cerebral cortex, *J. Anat.*, **99**, 691–709.

Berry, M. and Eayrs, J. T. (1966). The effects of X-irradiation on the development of the cerebral cortex, *J. Anat.*, **100**, 707–22.

Berthold, C.-H. (1968). Ultrastructure of the node-paranode region of mature feline ventral lumbar spinal-root fibres, *Acta Soc. Med. upsal.*, **73**, Suppl. 9.

Berthold, C.-H. and Carlstedt, T. (1977). Observations on the morphology at the transition between the peripheral and the central nervous system in the cat, *Acta physiol. scand.*, Suppl. **446**, 5–85.

Bickers, W. (1960). Sperm migration and uterine contractions, *Fert. Steril.*, **11**, 286–90.

Bilbey, D. L. J. and Rappaport, A. M. (1960). The segmental anatomy of human liver, *Anat. Rec.*, **136**, 330.

Billingham, R. E. and Silvers, W. K. (1960). The melanocytes of mammals, *Q. Rev. Biol.*, **35**, 1–40.

Billings-Gagliardi, S., Chan-Palay, V. and Palay, S. L. (1974). A review of lamination in Area 17 of the visual cortex of *Macaca mulatta*, *J. Neurocytol.*, **3**, 619–29.

Billingsley, P. R. and Ransom, S. W. (1918). On the number of nerve cells in the ganglion cervicale superioris, *J. comp. Neurol.*, **29**, 359–66.

Bischoff, R. (1975). Regeneration of single skeletal muscle fibres in vitro, *Anat. Rec.*, **182**, 215–35.

Biscoe, T. J., Bradley, G. W. and Purves, M. J. (1969). The relation between carotid body chemoreceptor activity and carotid sinus pressure in the cat, *J. Physiol., Lond.*, **203**, 40P.

Biscoe, T. J., Lall, A. and Samson, J. R. (1970). Electron microscopical and electrophysiological studies on the carotid body following intracranial section of the glossopharyngeal nerve, *J. Physiol., Lond.*, **208**, 133–52.

Bishop, A. (1964). The use of the hand in lower primates. In: *Evolutionary and Genetic Biology of Primates* (Buettner-Janusch, J. ed.), Academic Press: New York, London.

Bishop, D. W. (1962). Sperm motility, *Physiol. Rev.*, **42**, 1–59.

Bishop, D. W. and Tyler, A. (1956). Fertilizins of mammalian eggs, *J. exp. Zool.*, **132**, 575–601.

Bishop, M. W. H. and Walton, A. (1960). Spermatogenesis and the structure of mammalian spermatozoa. In: *Marshall's Physiology of Reproduction* (Parkes, A. S. ed.), Third edition, Volume I, Part II, pp. 1–129, Longmans: New York, London.

Bizarro, A. H. (1921). On sesamoid and supernumerary bones of the limbs, *J. Anat.*, **55**, 256–68.

Björklund, A., Owman, Ch. and West, K. A. (1972). Peripheral sympathetic innervation of serotonin cells in the habenular region of the rat brain, *Z. Zellforsch. mikrosk. Anat.*, **127**, 570–9.

Björklund, A. and Nobin, A. (1973). Fluorescence histochemical and microspectrofluorometric mapping of dopamine, and noradrenaline cell groups in the rat diencephalon, *Brain Res.*, **51**, 193–205.

Björkman, A. and Wohlfart, G. (1936). Faseranalyse der Nn. oculomotoris, trochlearis und abducens des Menschen, *Z. mikrosk.-anat. Forsch.*, **39**, 631–47.

Björkman, U. and Ekholm, R. (1973). Thyroglobulin synthesis and intracellular transport studied in bovine thyroid slices, *J. Ultrastruct. Res.*, **45**, 231–53.

Black, D. (1917). The motor nuclei of the cerebral nerves in phylogeny; a study of the phenomena of neurobiotaxis. I. Cyclostomi and pisces, *J. comp. Neurol.*, **27**, 117–26.

Blackhall-Morison, A. (1922). The musculature of the aorta and of the cardiac valves, anatomically, physiologically and clinically considered, *Trans. Med-Chir. Soc. Edinb.*, **102**, 149–63 (in *Edinb. med. J.*, N.S. **30**, 1923).

Blackstad, T. W. (1956). Commissural connections of the hippocampal region in the rat, with special reference to their mode of termination, *J. comp. Neurol.*, **105**, 417–538.

Blackstad, T. W. (1958). On the termination of some afferents to the hippocampus and fascia dentata, *Acta anat.*, **35**, 202–14.

Blackstad, T. W. (1967). Cortical grey matter. A correlation of light and electron microscopic data. In: *The Neuron* (Hydén, H. ed.), pp. 49–118, Elsevier: Amsterdam.

Blackstad, T. W. and Flood, P. R. (1963). Ultrastructure of hippocampal axo-somatic synapses, *Nature, Lond.*, **198**, 542–3.

Blackstad, T. W. and Kjaerheim, A. (1967). Special axo-dendritic synapses in the hippocampal cortex: electron and light microscopic studies on the layer of mossy fibres, *J. comp. Neurol.*, **117**, 133–60.

Blackwood, H. J. J. (1959). The development, growth and pathology of the mandibular condyle, M.D. Thesis, Queen's University, Belfast.

Blackwood, H. J. J. (1965). Vascularisation of the condylar cartilage of the human mandible, *J. Anat.*, **99**, 551–63.

Blair-West, J. R., Coghlan, J. P., Denton, D. A. and Wright, R. D. (1967). Effect of endocrines on salivary glands. In: *Handbook of Physiology*, Section 6, *Alimentary Canal*, Volume II (Code, C. F. and Heidel, W. eds.), pp. 633–64, American Physiological Society: Washington, DC.

Blakemore, C. B. and Falconer, M. A. (1967). Long-term effects of anterior temporal lobectomy on certain cognitive functions, *J. Neurol. Neurosurg. Psychiat.*, **30**, 364–7.

Blalock, A. (1947). Use of shunt or by-pass operations in the treatment of certain circulatory disorders including portal hypertension and pulmonic stenosis, *Ann. Surg.*, **125**, 129–41.

Blandau, R. J. and Rumery, R. E. (1964). The relationship of swimming movements of epididymal spermatozoa to their fertilising capacity, *Fert. Steril.*, **15**, 571–9.

Bleier, R. (1977). Ultrastructure of supraependymal cells and ependyma of hypothalamic third ventricle of mouse, *J. comp. Neurol.*, **174**, 359–76.

Blenkinsopp, W. K. (1967). Proliferation of respiratory tract epithelium in the rat, *Exp. Cell Res.*, **46**, 144–54.

Blevins, C. E. (1967). Innervation patterns of the human stapedius muscle, *Archs otolar.*, **86**, 136–42.

Blinkov, S. M. and Glezer, I. I. (1968). *The Human Brain in Figures and Tables*, Basic Books: New York.

Blix, M. (1882–1883). Experimentala bidrag till lösning af frågan om hudnervenas specifika energi, *Uppsala läkareförhandlingar*, **18**, 87–102.

Block, E. (1953). Quantitative morphological investigation of follicular system in newborn female infants, *Acta anat.*, **17**, 201–6.

Bloodworth, J. M. B. and Powers, K. L. (1968). The ultrastructure of the normal dog adrenal, *J. Anat.*, **102**, 457–76.

Bloom, G. and Nicander, L. (1961). On the ultrastructure of the protoplasmic droplet of spermatozoa, *Z. Zellforsch. mikrosk. Anat.*, **55**, 833–44.

Bloom, S. R. and Polak, J. M. (1978). Gut hormone overview. In: *Gut Hormones* (Bloom, S. R. ed.), pp. 3–18. Churchill Livingstone: Edinburgh, London.

Bloom, W. (1931). New types of granular cell in islets of Langerhans of man, *Anat. Rec.*, **49**, 363–71.

Bloom, W. (1938). Tissue cultures of blood and blood-forming tissues. In: *Handbook of Haematology* (Downey, H. ed.), pp. 1471–1585, Hamish Hamilton: London.

BIBLIOGRAPHY

Bloom, W. and Bartelmez, G. W. (1940). Hematopoiesis in young human embryos, *Am. J. Anat.*, **67**, 21–54.

Bloom, W. and Fawcett, D. W. (1968). *A Textbook of Histology*, Ninth edition, Saunders: Philadelphia, London.

Bloor, C. M. and Lowman, R. M. (1963). Myocardial bridges in coronary atherosclerosis, *Am. Heart J.*, **65**, 195–9.

Blose, S. H. and Chako, S. (1976). Rings of intermediate (100Å) filament bundles in the perinuclear region of vascular endothelial cells: their mobilization by colcemid and mitosis, *J. Cell Biol.*, **70**, 459–66.

Blum, J. S., Chow, K. C. and Pribram, K. H. (1950). Behavioral analysis of the organisation of the parieto-temporo-preoccipital cortex, *J. comp. Neurol.* **93**, 53–100.

Blunt, M. J. (1951). Posterior wall of inguinal canal, *Br. J. Surg.*, **39**, 230–3.

Blunt, M. J. (1954). The blood supply of the facial nerve, *J. Anat.*, **88**, 520–6.

Blunt, M. J. (1959). The vascular anatomy of the median nerve in the forearm and hand, *J. Anat.*, **93**, 15–22.

Böck, P., Stockinger, L. and Vyslonzil, E. (1970). Die Feinstrucktur des Glomus caroticum beim Menschen, *Z. Zellforsch. mikrosk. Anat.*, **105**, 543–68.

Böck, P. and Gorgas, K. (1976). Catecholamine and granule content of carotid body type I-cells. In: *Chromaffin, Enterochromaffin and Related Cells* (Coupland, R. E. and Fujita, T. eds.), pp. 355–74, Elsevier: New York, Amsterdam.

Bockman, D. E. and Cooper, M. D. (1973). Pinocytosis by epithelium associated with lymphoid follicles in the bursa of Fabricius, appendix and Peyer's patches. An electron microscopic study, *Am. J. Anat.*, **136**, 455–77.

Bodian, D. (1936). A new method for staining nerve fibers and nerve endings in mounted paraffin sections, *Anat. Rec.*, **65**, 89–97.

Bodian, D. (1962). Discussion in: *Interhemispheric Relations and Cerebral Dominance* (Mountcastle, V. B. ed.), Johns Hopkins University Press: Baltimore.

Bodian, D. (1970). A model of synaptic and behavioral ontogeny. In: *The Neurosciences, a Second Study Program* (Schmitt, F. O., Quarton, G. C., Melnechuk, T. and Adelman, G. eds.), pp. 129–40, Rockefeller University Press: New York.

Bodian, D. (1970). An electron microscopic characterisation of classes of synaptic vesicles by means of controlled aldehyde fixation, *J. Cell Biol.*, **44**, 115–24.

Bodian, D. and Mellors, R. C. (1945). Regenerative cycle of motoneurons, with special reference to phosphatase activity, *J. exp. Med.*, **81**, 469–88.

Bodmer, W. F. and Cavelli-Sforza, L. L. (1976). *Genetics, Evolution and Man*, Freeman: San Francisco.

Bogen, J. E. and Gazzinga, M. S. (1965). Cerebral commissurotomy in man, *J. Neurosurg.*, **23**, 394–9.

Boggon, R. P. and Palfrey, A. J. (1973). The microscopic anatomy of human lymphatic trunks, *J. Anat.*, **114**, 389–405.

Boggs, D. R. (1967). The kinetics of neutrophilic leukocytes in health and disease, *Seminars Hemat.*, **4**, 359–86.

Böhme, C. C. (1962). The fine structure of Clarke's nucleus of the spinal cord, Thesis, University of Pennsylvania.

Boime, I. and Boguslawski, S. (1974). The synthesis of human placental lactogen by ribosomes derived from human placenta, *Proc. natn. Acad. Sci. U.S.A.*, **71**, 1322–5.

Bois, E., Feingold, J., Benmaiz, H. and Briard, M. L. (1975). Congenital urinary tract malformations: epidemiologic and genetic aspects, *Clin. Genetics*, **8**, 37–47.

Bojsen-Møller, F. (1975). Demonstration of terminalis, olfactory, trigeminal and perivascular nerves in the rat nasal septum, *J. comp. Neurol.*, **159**, 245–56.

Bojsen-Møller, F. and Tranum-Jensen, J. (1971). Whole-mount demonstration of cholinesterase containing nerves in the right atrial wall, nodal tissue and atrioventricular bundle of the pig heart, *J. Anat.*, **108**, 375–86.

Bojsen-Møller, F. and Tranum-Jensen, J. (1971). On nerves and nerve-endings in the conducting system of the moderator band (septomarginal trabecula), *J. Anat.*, **108**, 387–95.

Bojsen-Møller, F. and Schmidt, L. (1974). The palmar aponeurosis and the central spaces of the hand, *J. Anat.*, **117**, 55–68.

Bok, H. E. (1966). *De Foetale Transformatie van het Middenoorgebied*, Elsevier: Amsterdam.

Bolk, L., Goppert, G., Kallius, E. and Lubosch, W. (1931–9). *Handbuch der vergleichenden Anatomie der Wirbeltiere*, Urban: Berlin.

Bollobás, B. and Hajdu, I. (1975). The development of the tympanic sinus, *ORL*, **37**, 97–102.

Bonucci, E. (1967). Fine structure of early cartilage calcification, *J. Ultrastruct. Res.*, **20**, 33–50.

Bonucci, E. (1970). Fine structure and histochemistry of calcifying globules in epiphyseal cartilage, *Z. Zellforsch. mikrosk. Anat.*, **103**, 192–217.

Böök, J. A., Lejeune, J., Levan, A., Chu, E. H. Y., Ford, C. E., Fraccaro, M., Harnden, D. G., Hsu, T. C., Hungerford, D. A., Jacobs, P. A., Makino, S., Puck, T. T., Robinson, A. and Tjio, J. H. (1960). A proposed standard system of nomenclature of human mitotic chromosomes, *Lancet*, **1**, 1063–5.

Borell, U. and Fernstrom, I. (1953). Adnexal branches of uterine artery; arteriographic study in human subjects, *Acta radiol.*, **40**, 561–82.

Borell, U. and Fernstrom, I. (1960). Radiologic pelvimetry, *Acta radiol.*, Suppl. **191**, 3–97.

Borg, E. (1973). A neuroanatomical study of the brainstem auditory system of the rabbit. *Acta morph. neerl.-scand.*, **11**, 49–62.

Born, G. V. R., Dawes, G. S., Mott, J. C. and Widdicombe, J. G. (1954). Changes in the heart and lungs at birth, *Cold Spring Harb. Symp. quant. Biol.*, **19**, 102–8.

Bors, J. (1926). Uber das Zahlenverhaltnis zwischen Nerven und Muskelfasern, *Anat. Anz.*, **60**, 415–20.

Bortolami, R., Veggetti, A., Callegari, E., Lucchi, M. L. and Palmieri, G. (1977). Afferent fibres and sensory ganglion cells within the oculomotor nerve in some mammals and man. I. Anatomical investigations. *Arch. ital. Biol.*, **115**, 355–85.

Borun, T. W. and Stein, G. S. (1972). The synthesis of acidic chromosome proteins during the cell cycle of HeLa S3 cells. II. The kinetics of residual protein synthesis and transport, *J. Cell Biol.*, **52**, 308–15.

Borun, T. W., Gabrielli, F. and Ajiro, K. (1975). Further evidence of transcriptional and translational control of histone messenger RNA during the HeLa S3 cycle, *Cell*, **4**, 59–67.

Bosma, J. F. (1957). Deglutition: pharyngeal stage, *Physiol. Rev.*, **37**, 275–300.

Bossy, J. (1968). Sur la présence d'un noyau sensitif du annexe au tractus spinal du V, *Acta anat.*, **70**, 332–40.

Bossy, J. and Vidić, B. (1967). Existe-il une innervation proprioceptive des muscles du pharynx chez l'homme? *Archs Anat. Histol. Embryol.*, **50**, 273–84.

Botha, G. S. M. (1962). *The Gastro-Oesophageal Junction*, Churchill: London.

Boucher, B. J. (1957). Sex differences in the foetal pelvis, *Am. J. phys. Anthrop.*, N.S. **15**, 581–600.

Bouillard, J. P. (1825). Recherches cliniques propres à démontrer que la perte de la parole correspond à la lésion des lobules antérieurs du cerveau. *Arch. gén. Méd.*, **8**, 24–45.

Bouin, P. and Ancel, P. (1903). Le gland interstitielle du testicule chez le cheval, *Archs Zool. exp. Gén.*, **3**, 391.

Boulter, P. S. and Parkes, A. G. (1963). Submucosal vascular patterns of the alimentary tract and their significance, *Br. J. Surg.*, **47**, 546–50.

Bounoure, L. (1939). *L'Origine des Cellules Réproductrices et le Problème de la Lignée Germinale*, Gauthiers-Villars: Paris.

Bourgeois, S. and Monod, J. (1970). *In vitro* studies of the *lac* operon regulatory system. In: *Control Processes in Multicellular Organisms* (Wolstenholme, G. E. W. and Knight, J. eds.), pp. 3–21, Ciba Foundation Symposium, Churchill: London.

Bourne, G. H. (1968). *The Structure and Function of Nervous Tissue* (editor). Academic Press: New York, London.

Bøving, B. G. (1959). Implantation, *Ann. N.Y. Acad. Sci.*, **75**, 700–25.

Bøving, B. G. (1959). The biology of trophoblast, *Ann. N.Y. Acad. Sci.*, **80**, 21–43.

Bøving, B. G. (1963). Implantation mechanisms. In: *Mechanisms Concerned with Conception* (Hartman, C. G. ed.), pp. 321–96, Pergamon Press: Oxford, London, New York, Paris.

Bowden, D. M., German, D. C. and Poynter, W. D. (1978). An autoradiographic, semistereotaxic mapping of major projections from the locus coeruleus and adjacent nuclei, *Brain Res.*, **145**, 257–76.

Bowden, R. E. M. (1955). Surgical anatomy of the recurrent laryngeal nerve, *Br. J. Surg.*, **43**, 153–63.

Bowden, R. E. M. (1966). The functional anatomy of striated muscle, *Ann. R. Coll. Surg.*, **38**, 41–59.

Bowman, J. P. (1968). Muscle spindles in the intrinsic and extrinsic muscles of the rhesus monkey's tongue, *Anat. Rec.*, **161**, 483–8.

Boyarsky, S. (1973). Management of the neurogenic bladder: current status and recent developments, *Clin. Neurosurg.*, **20**, 409–23.

Boycott, B. B. and Dowling, J. E. (1969). Origin of the primate retina: light microscopy, *Phil. Trans. R. Soc.*, Ser. B, **255**, 109–84.

Boycott, B. B. (1974). Aspects of the comparative anatomy and physiology of the vertebrate retina. In: *Essays on the Nervous System* (Bellairs, R. and Gray, E. G. eds.), pp. 223–57, Clarendon Press: Oxford.

Boycott, B. B. and Wässle, H. (1974). The morphological types of ganglion cells of the domestic cat's retina, *J. Physiol., Lond.*, **240**, 397–419.

Boyd, G. I. (1930). The emissary foramina of the cranium in man and the anthropoids, *J. Anat.*, **65**, 108–21.

Boyd, I. A. (1954). The histological structure of the receptors in the knee joint of the cat correlated with their physiological response, *J. Physiol., Lond.*, **124**, 476–88.

Boyd, I. A. (1962). The structure and innervation of the nuclear bag muscle fibre system and nuclear chain muscle fibre system in mammalian muscle spindles. *Phil. Trans. R. Soc.*, Ser. B, **245**, 81–136.

Boyd, I. A. and Roberts, T. D. M. (1953). Proprioceptive discharges from stretch-receptors in the knee joint of the cat, *J. Physiol., Lond.*, **122**, 38–58.

Boyd, I. A. and Davey, M. R. (1968). *Composition of Peripheral Nerves*, Livingstone: Edinburgh.

Boyd, J. D. (1933). The classification of the upper lip in mammals, *J. Anat.*, **67**, 409–16.

Boyd, J. D. (1937). The development of the human carotid body, *Contr. Embryol.*, **26**, 1–31.

Boyd, J. D. (1956). Observations on the human pharyngeal hypophysis, *J. Endocrinol.*, **14**, 66–77.

Boyd, J. D. (1961). The inferior aortico-pulmonary glomus, *Br. med. Bull.*, **17**, 127–31.

Boyd, J. D. (1964). Development of the human thyroid gland. In: *The Thyroid Gland* (Pitt-Rivers, R. and Trotter, W. R. eds.), pp. 9–31, Butterworths: London.

Boyd, J. D. and Trevor, J. C. (1953). Race, sex, age and stature from skeletal material. In: *Modern Trends in Forensic Medicine* (Simpson, K. ed.), Part 1, Chapter 7, Butterworths: London.

Boyd, J. D. and Hamilton, W. J. (1966). Electron microscopic observations on the cytotrophoblastic contribution to the syncytium in the human placenta, *J. Anat.*, **100**, 535–48.

Boyd, J. D., Hamilton, W. J. and Boyd, C. A. (1968). The surface of the syncytium of the human chorionic villus, *J. Anat.*, **102**, 553–63.

Boyd, J. D. and Hamilton, W. J. (1970). *The Human Placenta*, Heffer: Cambridge.

Boyd, W., Blincoe, H. and Hayner, J. C. (1965). Sequence of action of the diaphragm and quadratus lumborum during quiet breathing, *Anat. Rec.*, **151**, 579–82.

Boyde, A. (1969). Electron microscopic observations relating to the nature and development of prism decussation in mammalian dental enamel, *Bull. Grp. int. Rech. sci. Stomat.*, **12**, 151–207.

Boyden, E. A. (1955). *Segmental Anatomy of the Lungs*, McGraw-Hill: New York.

Boyden, E. A. (1957). The anatomy of the choledochoduodenal junction in man, *Surgery Gynec. Obstet.*, **104**, 641–52.

Boyden, E. A. (1966). The pancreatic sphincters of the baboon as revealed by serial sections of the choledochoduodenal junction, *Surgery, St. Louis*, **60**, 1187–94.

Boyden, E. A. (1972). Development of the human lung. In: *Brennemann's Practice of Pediatrics*, Volume 4, Chapter 64, Harper & Row: New York.

Boyden, E. A., Cope, J. G. and Bill, A. H. (1967). A new look at the etiology of duodenal atresia, *Anat. Rec.*, **157**, 218P.

Bozler, E. (1948). Conduction, automaticity and tonus of visceral muscles, *Experientia*, **4**, 213–18.

Braak, H. (1975). On the fine structure of the external glial layer in the isocortex of man, *Cell Tiss. Res.*, **157**, 367–90.

Braak, H. (1977). The pigment architecture of the human occipital lobe, *Anat. Embryol.*, **150**, 229–50.

Braak, H. (1978). On magnopyramidal temporal fields in the human brain, *Anat. Embryol.*, **152**, 141–69.

Brachet, J. (1944). *Chemical Embryology*, Interscience Publishers: New York.

Brachet, J. (1960). *The Biochemistry of Development*, Pergamon Press: Oxford, New York, Toronto, Sydney.

Brachet, J. (1965). *The Biochemistry of Animal Development*, Academic Press: New York, London.

Brachet, J. (1967). Biochemical changes during fertilisation and early development. In: *Cell Differentiation* (De Reuck, A. V. S. and Knight, J. eds.), pp. 39–61, Ciba Foundation Symposium, Churchill: London.

Brachet, J. (1969). Acides nucléiques et différenciation embryonnaire, *Annales d'Embryologie et de Morphogenèse*, Suppl. I, 21–37.

Brachet, J. and Mirsky, A. E. (1960). *The Cell*, Volume 1 (editors), Academic Press: New York, London.

Brachet, J. and Quertier, J. (1963). Cytochemical detection of cytoplasmic deoxyribonucleic acid (DNA) in amphibian oocytes, *Expl Cell Res.*, **32**, 410–13.

Braden, A. W. H. (1955). The reactions of isolated mucopolysaccharides to several histochemical tests, *Stain Technol.*, **30**, 19–26.

Bradley, R. L. (1973). Surgical anatomy of the gastro-duodenal artery, *Int. Surg.*, **58**, 393–6.

Brain, R. (1961). The neurology of language, *Brain*, **84**, 145–66.

Braitenberg, V. (1967). Is the cerebellar cortex a biological clock in the millisecond range? *Prog. Brain Res.*, **25**, 334–46.

Braitenberg, V. (1977). Cortical architectonics, general and areal. In: *Architectonics of the Cerebral Cortex* (Brazier, M. A. B. and Petsche, H. eds.), Raven Press: New York.

Braitenberg, V. and Atwood, R. P. (1958). Morphological observations on the cerebellar cortex, *J. comp. Neurol.*, **109**, 1–33.

Braithwaite, F., Channel, G. D., Moore, F. T. and Whillis, J. (1948). The applied anatomy of the lumbrical and interosseous muscles of the hand, *Guy's Hosp. Rep.*, **97**, 185–95.

Braithwaite, J. L. (1951). The arterial supply of the urinary bladder, *J. Anat.*, **85**, 413P.

Braithwaite, J. L. (1952). Variations in origin of the parietal branches of the internal iliac artery, *J. Anat.*, **86**, 423–30.

Braithwaite, J. L. and Adams, D. J. (1957). The venous drainage of the rat spleen, *J. Anat.*, **91**, 352–7.

Brantigan, O. C. and Voshell, A. F. (1943). Tibial collateral ligament; its function, its bursae and its relation to medial meniscus, *J. Bone Jt Surg.*, **25**, 121–31.

Branton, D. (1971). Freeze-etching studies of membrane structure, *Phil Trans. Roy. Soc. Lond. B.*, **261**, 133–8.

Brash, J. C. (1924). The growth of the jaws and the palate. In: *The Growth of the Jaws, Normal and Abnormal, in Health and Disease*, Dental Board of the United Kingdom: London.

Brash, J. C. (1934). Some problems in the growth and developmental mechanisms of bone, *Edinb. med. J.*, N.S. **41**, 305–87.

Brauer, K. and Winkelmann, E. (1976). Zur Rolle der Interneuronen im corpus geniculatum laterale der Säugetiere, *Z. mikrosk. anat. Forsch.*, **90**, 334–51.

Brazier, M. A. B. (1959). The historical development of neurophysiology. In: *Handbook of Physiology* (Field, E. J. and Magoun, H. W. eds.), Section 1, Volume 1, pp. 1–58, American Physiological Society: Washington.

Brazier, M. A. B. (1969). *The Interneuron* (editor), University of California Press: Berkeley.

Breathnach, A. S. (1969). Normal and abnormal melanin pigmentation of the skin. In: *Pigments in Pathology* (Wolman, M. ed.), pp. 353–94, Academic Press: New York, London.

Breathnach, A. S. (1971). *An Atlas of the Ultrastructure of Human Skin*, Churchill: London.

Breathnach, A. S. and Wylie, L. M. A. (1966). Ultrastructure of retinal pigment epithelium of the human fetus, *J. Ultrastruct. Res.*, **16**, 584–97.

Brekelmans, W. A. M., Poort, H. W. and Slooff, T. J. (1972). A new method to analyse the mechanical behaviour of skeletal parts, *Acta orthop. scand.*, **43**, 301–17.

Bretscher, M. S. and Raff, M. C. (1975). Mammalian plasma membranes, *Nature, Lond.*, **258**, 43–9.

Brightman, M. W. and Palay, S. L. (1963). The fine structure of ependyma in the brain of the rat, *J. Cell Biol.*, **19**, 415–39.

Brightman, M. W. and Reese, T. S. (1969). Junctions between intimately apposed cell membranes in the vertebrate brain, *J. Cell Biol.*, **40**, 648–77.

Brightman, M. W., Klatzo, I., Olsson, Y. and Reese, T. (1970). The blood brain barrier to proteins under normal and pathological conditions, *J. Neurol. Sci.*, **10**, 215–39.

Brini, A., Porte, A. and Stoeckel, M. E. (1968). *Biology and Surgery of the Vitreous Body*, Masson: Paris.

Brizon, J., Castaing, J. and Hourtaille, F. C. (1956). *Le Péritoine*, Maloine: Paris.

Brizzi, E., Serantoni, C., Ciani, P. A., Orlandini, A. and Pernice, L. (1973). The distribution of the vagus nerves in the stomach, *Chir. Gastroent. (Engl. Ed.)*, **7**, 17–34.

Broca, P. P. (1861). Perte de la parole, ramollissement chronique et destruction partielle du lobe antérieur gauche du cerveau, *Bull. Soc. Anthrop.*, **2**, 235–8; 301–21.

Broca, P. P. (1875). Instructions craniologiques et craniométriques de la Société d'Anthropologie de Paris, *Bull. Soc. Anthrop.*, **6**, 534–6.

Broca, P. (1878). Anatomie comparée des circonvolutions cérébrales. Le grand lobe limbique et la scissure dans la série des mammifères, *Revue Anthrop.*, **1**, 385–498.

Brock, R. C. (1942). The use of silver nitrate in the production of aseptic obliterative pleuritis, *Guy's Hosp. Rep.*, **91**, 111–30.

Brock, R. C. (1943). Observations on the anatomy of the bronchial tree, with special reference to the surgery of lung abscess, *Guy's Hosp. Rep.*, **92**, 26–37.

Brock, R. C. (1944). A note on secondary diphtheritic infection of empyema and thoracotomy wounds, *Guy's Hosp. Rep.*, **93**, 62–6.

Brock, R. C. (1949). Surgery of pulmonary stenosis, *Br. med. J.*, **ii**, 399–406.

Brock, R. C. (1952). Congenital pulmonary stenosis, *Am. J. Med.*, **12**, 706–19.

Brock, R. C. (1954). *The Anatomy of the Bronchial Tree*, Oxford University Press: London.

Brock, R. C. (1955). Control mechanisms in the outflow tract of the right ventricle in health and disease, *Guy's Hosp. Rep.*, **104**, 356–77.

BIBLIOGRAPHY

Brodal, A. (1940). Modification of the Gudden method for the study of cerebral localisation, *Archs Neurol. Psychiat., Chicago*, **43**, 46–58.

Brodal, A. (1947). Central course of afferent fibers for pain in facial, glossopharyngeal, and vagus nerves. Clinical observations, *Archs Neurol. Psychiat., Chicago*, **57**, 292–306.

Brodal, A. (1949). Spinal afferents to the lateral reticular nucleus of the medulla oblongata in the cat. An experimental study, *J. comp. Neurol.*, **91**, 259–95.

Brodal, A. (1957). *The Reticular Formation of the Brain Stem. Anatomical Aspects and Functional Correlations*, Oliver and Boyd: Edinburgh.

Brodal, A. (1968). The termination of spinovestibular fibers in the cat, *Brain Res.*, **5**, 494–500.

Brodal, A. (1969). *Neurological Anatomy*, Oxford University Press: London.

Brodal, A. and Walberg, F. (1952). Ascending fibres in the pyramidal tract of the cat, *A.M.A. Archs Neurol. Psychiat.*, **68**, 755–75.

Brodal, A. and Pompeiano, O. (1957). The vestibular nuclei in the cat, *J. Anat.*, **91**, 438–54.

Brodal, A., Pompeiano, O. and Walberg, F. (1962). *The Vestibular Nuclei and Their Connexions, Anatomical and Functional Correlations*, Oliver and Boyd: Edinburgh.

Brodal, A. and Angaut, P. (1967). The termination of spino-vestibular fibres in the cat, *Brain Res.*, **5**, 494–500.

Brodal, P. (1968). The corticopontine projection in the cat. I. Demonstration of a somatotopically organized projection from the primary sensorimotor cortex, *Expl Brain Res.*, **5**, 212–37.

Brodal, P. (1968). The corticopontine projection in the cat. II. Demonstration of a somatotopically organized projection from the second somatosensory cortex, *Archs ital. Biol.*, **106**, 310–32.

Brodie, A. G. (1941). On the growth pattern of the human head. From the third month to the eighth year of life, *Am. J. Anat.*, **68**, 209–62.

Brodmann, K. (1909). *Vergleichende Lokaliastionslehre der Grosshirnrinde in ihren Prinzipien dargestellt auf Grund des Zellenbaues*, Barth: Leipzig.

Brokaw, C. J. (1975). Molecular mechanism for oscillation in flagella and muscle, *Proc. natn. Acad. Sci. U.S.A.*, **72**, 102–6.

Bronk, D. W. and Ferguson, L. K. (1935). Nervous control of intercostal respiration, *Am. J. Physiol.*, **110**, 700–7.

Brookes, M. (1958). The vascularisation of long bones in the human fetus, *J. Anat.*, **92**, 261–7.

Brookes, M. (1963). Cortical vascularisation and growth in foetal tubular bones, *J. Anat.*, **97**, 597–609.

Brookes, M. (1964). The blood supply of bones. In: *Modern Trends in Orthopaedics* (Clark, J. M. P. ed.), Volume 4, *Science of Fractures*, Butterworths: London.

Brookes, M. (1967). The osseous circulation, *Bio-Med. Eng.*, **2**, 294–9.

Brookes, M. (1971). *The Blood Supply of Bone*, Butterworths: London.

Brookes, M. and Harrison, R. G. (1957). The vascularisation of the rabbit femur and tibio-fibula, *J. Anat.*, **91**, 61–72.

Brookes, M. and Landon, D. N. (1963). The juxta-epiphysial vessels in the long bones of foetal rats, *J. Bone Jt Surg.*, **46B**, 336–45.

Brooks, S. T. (1955). Skeletal age at death: reliability of cranial and pubic age indicators, *Am. J. phys. Anthrop.*, N.S. **13**, 567–97.

Broom, R. (1901). On the structure and affinities of *Udenodon, Proc. zool. Soc. Lond.*, **2**, 162–90.

Broom, R. (1930). *The Origin of the Human Skeleton: An Introduction to Human Osteology*, Witherby: London.

Brothwell, D. R. (1968). *The Skeletal Biology of Earlier Human Populations* (editor), Symp. Soc. Study Hum. Biol., Volume 8, Pergamon: Oxford, London, New York, Paris.

Brouwer, B. (1918). Klinisch-anatomische Untersuchung über den oculometoriuskern, *Zentbl. ges. Neurol. Psychiat.*, **40**, 152–89.

Brouwer, B. and Zeeman, W. P. C. (1926). The projection of the retina in the primary optic neuron in monkeys, *Brain*, **49**, 1–35.

Browder, J. and Kaplan, H. A. (1976). *Cerebral Dural Sinuses and their Tributaries*, Thomas: Springfield.

Brown, A. G. (1973). Ascending and long spinal pathways: dorsal columns, spinocervical tract and spinothalamic tract. In: *Handbook of Sensory Physiology* (Iggo, A. ed.), Volume II, pp. 315–38, Springer: Berlin.

Brown, D. D. (1966). The nucleolus and synthesis of ribosomal RNA during oogenesis and embryogenesis of *Xenopus laevis, Nat. Cancer Inst. Monogr.*, **23**, 297–309.

Brown, D. D. and Gurdon, J. B. (1964). Absence of ribosomal RNA synthesis in the anucleolate mutant of *Xenopus laevis, Proc. natn. Acad. Sci. U.S.A.*, **51**, 139–46.

Brown, D. D. and Gurdon, J. B. (1965). Absence of ribosomal RNA in the anucleolate mutant of *Xenopus laevis*. In: *Molecular and Cellular Aspects of Development* (Bell, E. ed.), pp. 3–39, Harper and Row: New York, Evanston, London.

Brown, J. R. (1949). Localising cerebellar syndromes, *J. Am. med. Ass.*, **141**, 518–21.

Brown, J. W. (1950). *Congenital Heart Disease*, Second edition, Staples Press: London.

Brown, L. T. (1974). Coticorubral projections in the rat, *J. comp. Neurol.*, **154**, 149–68.

Brown, R. L. (1944). Rate of transport of spermia in human uterus and tubes, *Am. J. Obstet. Gynec.*, **47**, 407–11.

Brown, W. J., Barajas, L. and Latta, H. (1970). The ultrastructure of the human adrenal medulla: with comparative studies of the white rat, *Anat. Rec.*, **169**, 173–83.

Brown, W. V. and Bertke, E. M. (1969). *Textbook of Cytology*, C. V. Mosby: St. Louis.

Browne, D. (1928). The surgical anatomy of the tonsil, *J. Anat.*, **63**, 82–6.

Browne, D. (1938). Diagnosis of undescended testicle, *Br. med. J.*, **2**, 168–71.

Bruesch, S. R. and Arey, L. B. (1942). The number of myelinated and unmyelinated fibres in the optic nerve of vertebrates, *J. comp. Neurol.*, **77**, 631–65.

Brun, A. (1965). The subpial granular layer of the foetal cerebral cortex in man, *Acta path. microbiol. scand. Suppl.*, **179**, 1–98.

Bruni, C. and Porter, K. R. (1965). The fine structure of the parenchymal cell of the normal rat liver. I. General considerations, *Am. J. Path.*, **46**, 69–75.

Bruns, R. R. and Palade, G. E. (1968). Studies on blood capillaries. I. General organization of blood capillaries in muscles, *J. Cell Biol.*, **37**, 244–76.

Bryce, T. H. and Teacher, J. H. (1908). *Contributions to the Study of the Early Development and Imbedding of the Human Ovum*, Part I, Maclehose: Glasgow.

Buchan, A. M. J., Polak, J. M., Facer, P., Bloom, S. R., Szelke, M., Hudson, D. and Pearse, A. G. E. (1977). Abstract: Use of synthetic fragments for specific immunostaining of CCK cells. Presented at the British Society of Gastroenterology, April 1977.

Buchanan, A. R. (1937). The course of the secondary vestibular fibers in the cat, *J. comp. Neurol.*, **67**, 183–204.

Buckland-Wright, J. C. (1977). Microradiographic and histological examination of the split-line formation in bone, *J. Anat.*, **124**, 193–203.

Buckland-Wright, J. C. (1978). Bone structure and the patterns of force transmission in the cat skull (*Felis cattus*), *J. Morphol.*, **155**, 35–62.

Buckley, I. K. and Porter, K. R. (1967). Cytoplasmic fibrils in living cultured cells. A light and electron microscope study. *Protoplasma*, **64**, 349–80.

Bucy, P. C. (1944). *The Precentral Motor Cortex* (editor), University of Illinois Press: Urbana.

Buffa, R., Capella, C., Solcia, E., Frigerio, B. and Saio, S. I. (1977). Vasoactive intestinal peptide (VIP) cells in the pancreas and gastrointestinal mucosa. An immunohistochemical and ultrastructural study, *Histochemistry*, **50**, 217–27.

Bulger, R. E. (1965). The shape of rat kidney tubule cells, *Am. J. Anat.*, **116**, 237–55.

Buller, A. J. (1969). The physiology of the motor unit. In: *Disorders of Voluntary Muscle* (Walton, J. N. ed.), pp. 17–28, Churchill: London.

Bullough, P. and Goodfellow, J. (1968). The significance of the fine structure of articular cartilage, *J. Bone Jt Surg.*, **50B**, 852–7.

Bullough, W. S. (1967). *The Evolution of Differentiation*, Academic Press: New York, London.

Bullough, W. S., Laurence, E. B., Iverson, O. H. and Elgjo, K. (1967). The vertebrate epidermal chalone, *Nature, Lond.*, **214**, 578–80.

Bulmer, D. (1957). Observations on the development of the vaginal wall, *J. Anat.*, **91**, 599P.

Bulmer, D. (1959). Histochemical observations on the foetal vaginal epithelium, *J. Anat.*, **93**, 36–42.

Bulmer, M. G. (1970). *The Biology of Twinning in Man*, Clarendon Press: Oxford.

Bumke, O. and Foerster, O. (1936). *Handbuch der Neurologie* (editors), Springer-Verlag: Berlin.

Bunge, R. P. (1968). Glial cells and the central myelin sheath, *Physiol. Rev.*, **48**, 197–251.

Bunnag, S. C., Bunnag, S. and Warner, N. E. (1963). Microcirculation in the islets of Langerhans of the mouse, *Anat. Rec.*, **146**, 117–23.

Bunnell, S. (1949). *Surgery of the Hand*, Second edition, Saunders: Philadelphia.

Bunning, P. S. C. and Barnett, C. H. (1963). Variations in the talocalcaneal articulation, *J. Anat.*, **97**, 643P.

Bunning, P. S. C. and Barnett, C. H. (1965). A comparison of adult and foetal talocalcaneal articulations, *J. Anat.*, **99**, 71–6.

Buntine, J. A. (1970). The omohyoid muscle and fascia: morphology and anomalies, *Aust., N.Z. J. Surg.*, **40**, 86–8.

Burdi, A. R. and Faist, K. (1967). Morphogenesis of the palate in normal human embryos with special emphasis on the mechanisms involved, *Am. J. Anat.*, **120**, 149–60.

Burgess, P. R. and Perl, E. R. (1973). Cutaneous mechanoreceptors and nociceptors. In: *Handbook of Sensory Physiology*, Volume 2, *Somatosensory System* (Iggo, A. ed.), pp. 39–78, Springer: Berlin.

Burke, R. E. and Tsairis, P. (1973). Anatomy and innervation ratios in motor units of cat gastrocnemius, *J. Physiol., Lond.*, **234**, 749–65.

Burke, R. E., Levine, D. and Tsairis, P. (1973). Physiological types and histochemical profiles in motor units of the cat gastrocnemius. *J. Physiol., Lond.*, **234**, 723–48.

Burkitt, A. N. and Lightoller, G. S. (1926). The facial musculature of the Australian aboriginal, *J. Anat.*, **61**, 14–39 (Part 1); *J. Anat.*, **62**, 33–57 (Part 2).

Burnard, E. D. (1959). The cardiac murmur in relation to symptoms in the newborn, *Br. med. J.*, **1**, 134–8.

Burnet, F. M. (1962). The immunological significance of the thymus: an extension of the clonal selection theory of immunity, *Australas. Ann. Med.*, **11**, 79–91.

Burns, R. K. (1955). Urogenital system. In: *Analysis of Development* (Willier, B. H., Weiss, P. A. and Hamburger, V. eds.), pp. 462–91, Saunders: Philadelphia, London.

Burnside, M. B. (1971). Microtubules and microfilaments in newt neurulation, *Dev. Biol.*, **26**, 416–41.

Burnstock, G. (1970). Structure of smooth muscle and its innervation. In: *Smooth Muscle* (Bülbring, E., Brading, A. F., Jones, A. W. and Tomita, T. eds.), pp. 1–69, Arnold: London.

Burnstock, G. (1975). Control of smooth muscle activity in vessels by adrenergic nerves and circulating catecholamines, Smooth Muscle Pharmacology and Physiology, *INSERM (Paris)*, **50**, 251–64.

Burton, H. and Loewy, A. D. (1977). Projections to the spinal cord from medullary somatosensory relay nuclei, *J. comp. Neurol.*, **173**, 773–92.

Buschard, K. and Kjaeldgaard, A. (1973). Investigations and analysis of the position, fixation, length and embryology of the vermiform appendix, *Acta chir. scand.*, **139**, 293–8.

Bussolati, G. and Pearse, A. G. (1967). Immunofluorescent localisation of calcitonin in the 'C' cells of pig and dog thyroid, *J. Endocr.*, **37**, 205–9.

Butler, H. (1957). The development of certain human dural venous sinuses, *J. Anat.*, **91**, 510–26.

Butler, H. (1967). The development of mammalian dural venous sinuses with special reference to the post-glenoid vein, *J. Anat.*, **102**, 33–6.

Butterfield, D. A. (1977). Electron spin resonance study of membrane protein alterations in erythrocytes in Huntington's disease, *Nature, Lond.*, **267**, 453–5.

Büttner-Ennever, J. A. (1977). Pathways from the pontine reticular formation to structures controlling horizontal and vertical eye movements in the monkey. In: *Developments in Neurosciences*, Volume I, Elsevier: New York.

C

Cabanie, M. H. and Javelle, J. (1966). La frange péritonéale sous-iléo-terminale, *C. r. hebd, Séanc. Acad. Sci., Paris*, **131**, 248–52.

Cajal, S. R. y (1890). Origen y terminación de las fibras nerviosas olfatorias, *Gazz, sanit., Barcelona*, pp. 1–21.

Cajal, S. R. y (1900). *Textura del Sistema Nervioso del Hombre y de los Vertebrados*, Volume 1, Moya: Madrid.

Cajal, S. R. y (1908). Structure et connexions des neurons. In: *Les Prix Nobel en 1906*, pp. 1–25, Norstedt and Söner: Stockholm.

Cajal, S. R. y (1911). *Histologie du système nerveux de l'homme et des vertébrés*, Maloine: Paris.

Cajal, S. R. y (1919). Acción neurotrópica de los epitelios. Algunos detalles sobre el mecanismo genético de las ramificaciones nerviosas intraepiteliales sensitivas y sensoriales, *Trab. del Lab. de Investig. biol.*, **17**, 65–86.

Cajal, S. R. y (1928). *Degeneration and Regeneration of the Nervous System*, Oxford University Press: London.

Cajal, S. R. y (1955). *Studies on the Cerebral Cortex* (Kraft, L. M. trans.), Lloyd-Luke: London.

Calasans, O. M. (1953). Arquitetura do músculo ciliar no homen, *An. Fac. Med. Univ. S. Paolo*, **27**, 3–98.

Caldwell, W. E. and Moloy, H. C. (1933). Anatomical variations in the female pelvis and their effect in labor with suggested classification, *Am. J. Obstet, Gynec.*, **26**, 479–505.

Caldwell, W. E. Moloy, H. C. and D'Esopo, D. A. (1940). The more recent conceptions of the pelvic architecture. *Am. J. Obstet, Gynec.*, **40**, 558–65.

Calne, D. B. (1970). *Parkinsonism*, Arnold: London.

Cameron, D. A. (1961). Erosion of the epiphysis of the rat tibia by capillaries, *J. Bone Jt Surg.*, **43B**, 590–4.

Campain, R. and Minckler, J. (1976). A note on the gross configurations of the human auditory cortex, *Brain and Language*, **3**, 318–23.

Campbell, A. H. and Liddelow, A. G. (1967). Significant variations in the shape of the trachea and large bronchi, *Med. J. Aust.*, **54**, 1017–20.

Campbell, A. W. (1905). *Histological Studies on the Localisation of Cerebral Function*, Cambridge University Press: Cambridge.

Campbell, E. J. M. (1955). The role of the scalene and sternomastoid muscles in breathing in normal subjects. An electromyographic study, *J. Anat.*, **89**, 378–86.

Campbell, M. (1965). Causes of malformations of the heart, *Br. med. J.*, **2**, 895–904.

Campos-Ortega, J. A., Glees, P. and Neuhoff, V. (1968). Ultrastructural analysis of individual layers in the lateral geniculate body of the monkey, *Z. Zellforsch. mikrosk. Anat.*, **87**, 82–100.

Campos-Ortega, J. A., Hayhow, W. R. and de V. Clüver, P. F. (1972). A note on the problem of retinal projections to the inferior pulvinar of primates; *Brain Res.*, **22**, 126–30.

Candiollo, L. (1965). Richerche anatomo-comparative sul musculo tensore del timpano, con riferimento alla innergazione propriocettiva, *Z. Zellforsch. mikrosk. Anat.*, **67**, 34–56.

Candiollo, L. and Levi, A. C. (1969). Studies on the morphogenesis of the middle ear muscles in man, *Arch. Ohr.-, Nas.-u. KehlkHeilk.*, **195**, 55–67.

Cannieu, A. (1886). Recherche sur l'innervation de l'éminence thénar par le cubital, *J. Méd. Bordeaux*, 377–9.

Cannon, W. B. and Britton, S. W. (1925). Studies on the conditions of activity in endocrine glands. XV. Pseudaffective medulliadrenal secretion, *Am. J. Physiol.*, **72**, 283–94.

Capella, C., Solcia, E., Frigerio, B. and Buffa, R. (1976). Endocrine cells of the human intestine. An ultrastructural study. In: *Endocrine Gut and Pancreas* (Fujita, T. ed.), pp. 43–60, Elsevier: New York, Amsterdam.

Capella, C., Hage, E., Solcia, E. and Usellini, L. (1978). Ultrastructural similarity of endocrine-like cells of the human lung and some related cells of the gut, *Cell Tiss. Res.*, **186**, 25–37.

Capen, C. C. (1975). Functional and fine structural relationships of parathyroid glands, *Adv. Vet. Sci. Comp. Med.*, **19**, 249–86.

Capen, C. C. and Roth, S. I. (1973). Ultrastructural and functional relationships of normal and pathologic parathyroid cells. In: *Pathobiology Annual 1973* (Ioachim, H. L. ed.), Volume 3, pp. 129–75, Appleton-Century-Crofts: New York.

Cardinali, D. P., Larin, F. and Wurtman, R. J. (1971). Action spectra for effects of light on hydroxyindole-o-methyltransferases in rat pineal, retina and harderian gland, *Endocrinology*, **91**, 877–86.

Carey, E. J. (1921). Studies on the structure and function of the small intestine, *Anat. Rec.*, **21**, 189–216.

Carey, E. J., Zeit, W. and McGrath, B. F. (1927). Studies in the dynamics of histogenesis. XII. The regeneration of the patellae of dogs, *Am. J. Anat.*, **40**, 127–58.

Carlon, N. and Stahl, A. (1973). Les premiers stades du développement des gonades chez l'Homme et les Vertébrés supérieurs, *Pathol.-Biol.*, **21**, 903–14.

Carlson, B. M. (1973). The regeneration of skeletal muscles. A review, *Am. J. Anat.*, **137**, 119–49.

Carlstedt, T. and Berthold, C.-H. (1977). Observations on the morphology at the transition between the peripheral and the central nervous system in the cat, *Acta physiol. Scand., Suppl.* **446**.

Carlström, D. (1964). Polarisation microscopy of dental enamel with reference to incipient carious lesions. In: *Advances in Oral Biology*, pp. 255–96. Academic Press: New York, London.

Carman, J. B., Cowan, W. M. and Powell, T. P. S. (1963). The organisation of the cortico-striate connexions in the rabbit, *Brain*, **86**, 525–62.

Carman, J. B., Cowan, W. M. and Powell, T. P. S. (1964). Cortical connexions of the thalamic reticular nucleus, *J. Anat.*, **98**, 587–98.

Carman, J. B., Cowan, W. M., Powell, T. P .S. and Webster, K. E. (1965). A bilateral cortico-striate projection, *J. Neurol. Neurosurg. Psychiat.*, **28**, 71–7.

Carmel, P. and Starr, A. (1963). Acoustic and non-acoustic factors modifying middle-ear muscle activity in waking cats, *J. Neurophysiol.*, **26**, 598–616.

Caro, A. J. (1977). *A genetic problem in East Anglia: Huntington's Chorea*. Ph.D. Thesis, University of East Anglia.

Carpenter, M. B. (1950). Athetosis and basal ganglia; review of the literature and a study of forty-two cases, *Archs Neurol. Psychiat., Chicago*, **63**, 875–901.

Carpenter, M. B., Whittier, J. R. and Mettler, F. A. (1950). Analysis of choreoid hyperkinesia in the rhesus monkey. Surgical and pharmacological analysis of hyperkinesia resulting from lesions in the subthalamic nucleus of Luys, *J. comp. Neurol.*, **92**, 293–332.

Carpenter, M. B. and Carpenter, C. S. (1951). Analysis of somatotopic relations of the corpus Luysi in man and monkey. Relation between the

BIBLIOGRAPHY

site of dyskinesia and distribution of lesions within the subthalamic nucleus, *J. comp. Neurol.*, **95**, 349–70.

Carpenter, M. B. and Whittier, J. R. (1952). Study of methods for producing experimental lesions of the central nervous system with special reference to stereotaxic technique. *J. comp. Neurol.*, **97**, 73–132.

Carpenter, M. B. and Britten, G. M. (1958). Subthalamic hyperkinesia in the rhesus monkey: effects of secondary lesions in the red nucleus and brachium conjunctivum, *J. Neurophysiol.*, **21**, 400–13.

Carpenter, M. B., Correll, J. W. and Hinman, A. (1960). Spinal tracts mediating subthalamic hyperkinesia. Physiological effects of selective partial cordotomies upon dyskinesia in the rhesus monkey, *J. Neurophysiol.*, **23**, 288–304.

Carpenter, M. B., Stein, B. M. and Shriver, J. E. (1968). Central projections of spinal dorsal roots in the monkey. II. Lower thoracic, lumbosacral and coccygeal dorsal roots, *Am. J. Anat.*, **123**, 75–118.

Carpenter, M. B. and Peter, P. (1971). Accessory oculomotor nuclei in the monkey, *J. Hirnforsch.*, **12**, 405–18.

Carpenter, M. B. and Peter, P. (1972). Nigrostriatal and nigrothalamic fibres in the rhesus monkey, *J. comp. Neurol.*, **144**, 93–116.

Carr, I. (1967). Nuclear membranous whorls, *Z. Zellforsch. mikrosk. Anat.*, **80**, 140–4.

Carr, I. (1970). The fine structure of the mammalian lymphoreticular system, *Int. Rev. Cytol.*, **27**, 283–348.

Carr, I., Clegg, E. J. and Meek, G. A. (1968). Sertoli cells as phagocytes: an electron microscopic study, *J. Anat.*, **102**, 501–10.

Carrow, R. E., Brown, R. E. and van Huss, W. D. (1967). Fiber sizes and capillary to fiber ratios in skeletal muscle of exercised rats, *Anat. Rec.*, **159**, 33–40.

Carter, R. B. and Keen, E. N. (1971). The intramandibular course of the inferior alveolar nerve, *J. Anat.*, **108**, 433–40.

Case, R. M. (1979). Pancreatic secretion: cellular aspects. In: *Scientific Basis of Gastroenterology* (Duthi, H. L. and Wormsley, K. G. eds.), pp. 163–98, Churchill Livingstone: Edinburgh, London.

Caspersson, T., Farber, S., Foley, G. E., Kudynowski, J., Modest, E. J., Simonsson, E., Wagh, V. and Zech, L. (1968). Chemical differentiation along metaphase chromosomes, *Exp. Cell Res.*, **49**, 219–22.

Caspersson, T. and Zech, L. (1973). *Chromosome Identification*, Academic Press: New York.

Castelli, W. A. (1963). Vascular architecture of the human adult mandible, *J. dent Res.*, **42**, 786–92.

Castleman, B. (1966). The pathology of the thymus gland in myasthenia gravis, *Ann. N.Y. Acad. Sci.*, **135**, 496–505.

Catton, W. T. (1970). Mechanoreceptor function, *Physiol. Rev.*, **50**, 297–318.

Catton, W. T. and Gray, J. E. (1951). Electromyographic study of the action of the serratus anterior muscle in respiration, *J. Anat.*, **85**, 412P.

Cauldwell, E. W., Siekert, R. G., Lininger, R. E. and Anson, B. J. (1948). Bronchial arteries. Anatomic study of 150 human cadavers, *Surg. Gynec. Obstet.*, **86**, 395–412.

Cauna, N. (1966). Fine structure of the receptor organs and its probable functional significance. In: *Touch, Heat and Pain* (de Reuck, A. V. S. and Knight, J. eds.), pp. 117–27, Ciba Foundation Symposium, Churchill: London.

Cauna, N. and Mannan, G. (1959). Developmental and post-natal changes of digital Pacinian corpuscles (*corpuscula lamellosa*) in the human hand, *J. Anat.*, **93**, 271–86.

Cauna, N. and Hinderer, K. H. (1969). Fine structure of blood vessels of the human nasal respiratory epithelium, *Ann. Otol. Rhinol. Lar.*, **78**, 865–79.

Cavallero, C., Spagnoli, L. G. and Villaschi, S. (1976). An electron microscopic study of human pancreatic islets. In: *Endocrine Gut and Pancreas* (Fujita, T. ed.), pp. 61–72, Elsevier: New York, Amsterdam.

Cavatorti, P. (1908). Il tipo normale e le variazioni delle arterie della base dell'encefalo nell'uomo, *Monitore zool. ital., Firenze*, **19**, 248–58.

Cave, A. J. E. (1937). The innervation and morphology of the cervical intertransverse muscles, *J. Anat.*, **71**, 497–515.

Cave, A. J. E. (1975). The morphology of the mammalian cervical pleurapophysis, *J. Zool., Lond.*, **177**, 377–93.

Cave, A. J. E. and Brown, R. W. (1952). On the tendon of the subclavius muscle, *J. Bone Jt Surg.* **34B**, 466–9.

Cave, A. J. E., Griffiths, J. D. and Whiteley, M. M. (1955). Osteo-arthritis deformans of Luschka joints, *Lancet*, **1**, 176–9.

Cave, A. J. E. and Porteous, C. J. (1958). The attachments of m. semimembranosus, *J. Anat.*, **92**, 638P.

Cawley, J. C. and Hayhoe, F. G. J. (1973). *Ultrastructure of Haemic Cells*, Saunders: London.

Celtis, A. and Porter, A. J. (1952). Lymphatics of the thorax, *Acta radiol.*, **38**, 461–70.

Chabardes, D., Imbert, M., Clique, A., Montegut, M. and Morel, F. (1975). PTH sensitive adenyl cyclase activity in different segments of the rabbit nephron, *Pflügers Archiv.*, **354**, 229–39.

Challice, C. E. and Viragh, S. (1973). *Ultrastructure of the Mammalian Heart*, Academic Press: New York, London.

Chamberlain, D. W., Nopajaroonskri, C. and Simon, G. T. (1973). Ultrastructure of the pulmonary lymphoid tissue, *Am. Rev. Respir. Dis.*, **108**, 621–31.

Chamberlain, H. D. (1928). The inheritance of left-handedness, *J. Hered.*, **19**, 557–9.

Chambers, M. R., Andres, K. H., Duering, M. von and Iggo, A. (1972). The structure and function of the slowly adapting type II receptor in hairy skin, *Q. Jl exp. Physiol.*, **57**, 417–45.

Chambers, W. W. and Sprague, J. M. (1955). Functional localisation in the cerebellum. II. Somatotopic organisation in cortex and nuclei, *Archs Neurol. Psychiat., Chicago*, **74**, 653–80.

Chambers, W. W. and Liu, C.-N. (1957). Cortico-spinal tract of the cat. An attempt to correlate the pattern of degeneration with deficits in reflex activity following neocortical lesions, *J. comp. Neurol.*, **108**, 23–56.

Champetier, J. and Descours, C. (1968). The branches of the posterior tibial nerve in the tibiotarsal joint, *C. r. Ass. Anat.*, **141**, 677–85.

Chang, H. T. (1951). Caudal extension of Clarke's column in the spider monkey, *J. comp. Neurol.*, **95**, 43–51.

Chang, I. W. (1969). Tympanosclerosis. Electron microscopic study, *Acta Otolaryngol.*, **68**, 62–72.

Chang, M. C. (1951). Fertilizing capacity of spermatozoa deposited into fallopian tubes, *Nature, Lond.*, **168**, 697–8.

Chang, M. C. and Hunter, R. H. F. (1975). Capacitation of mammalian sperm: biological and experimental aspects. In: *Handbook of Physiology* (Hamilton, D. W. and Greep, R. O. eds.), Section 7, Volume V, pp. 339–52, American Physiological Society: Washington.

Chaplin, D. M. and Greenlee, T. K. (1975). The development of human digital tendons, *J. Anat.*, **120**, 253–74.

Chapman, G. B., Chiardo, A. J., Coffey, R. J. and Weineke, K. (1966). The fine structure of mucosal epithelial cells of a pathological human gall bladder. *Anat. Rec.*, **154**, 579–616.

Charcot, J. M. (1883). *Lecture on Localization of Cerebral and Spinal Disease*, Edited and translated by M. A. Hodden, New Sydenham Society: London.

Charnley, J. (1959). The lubrication of animal joints. In: *Proceedings of a Symposium on Biomechanics*, Institution of Mechanical Engineers: London.

Chase, L. R. and Aurbach, G. D. (1967). Parathyroid function and the renal excretion of 3′5′-adenylic acid, *Proc. natn. Acad. Sci. U.S.A.*, **58**, 518–25.

Chau-Pham, T. T. (1978). The opiate receptors and the discovery of opioid-like peptides, *Drug Metab. Rev.*, **7**, 255–94.

Chayen, D. and Nathan, H. (1974). Anatomical observations on the sub-galeotic fascia, *Acta anat.*, **87**, 427–32.

Chen, I.-Li. and Yates, R. D. (1970). Ultrastructural studies of vagal paraganglia in Syrian hamsters, *Z. Zellforsch.*, **108**, 309–23.

Chen, I.-Li., Mascorro, J. A. and Yates, R. D. (1976). Morphology and functional considerations of the carotid body and paraganglia. In: *Chromaffin, Enterochromaffin and Related Cells* (Coupland, R. E. and Fujita, T. eds.), pp. 333–53, Elsevier: New York, Amsterdam.

Chen, J. M. (1952). Studies on the morphogenesis of the mouse sternum. I. Normal embryonic development, *J. Anat.*, **86**, 373–86.

Chen, L. and Weiss, L. (1972). Electron microscopy of the red pulp of the human spleen, *Am. J. Anat.*, **134**, 425–58.

Chen, L. and Weiss, L. (1973). The role of the sinus wall in the passage of erythrocytes through the spleen, *Blood*, **41**, 529–37.

Chiara, F. (1959). Study of the fine innervation of the female genitalia, I. Uterus, *Annali Ostet. Ginec.*, **81**, 553–76.

Chiba, T. and Yamauchi, A. (1970). On the fine structure of the nerve terminals in the human myocardium, *Z. Zellforsch. mikrosk. Anat.*, **108**, 324–38.

Chiechi, M. A., Lees, W. M. and Thompson, R. (1956). Functional anatomy of normal mitral valve, *J. Thor. Surg.*, **32**, 378–98.

Chiquoine, A. D. (1959). Electron microscopic observations on the developmental cytology of the mammalian ovum, *Anat. Rec.*, **133**, 258–9.

Chomsky, N. (1965). *Aspects of the Theory of Syntax*, M.I.T. Press: Cambridge, Mass.

Christ, J. F. (1969). Derivation and boundaries of the hypothalamus, with atlas of hypothalamic grisea. In: *The Hypothalamus* (Haymaker, W., Anderson, E. and Nauta, W. J. H. eds.), pp. 13–60. Thomas: Springfield, Ill.

Christie, G. A. (1963). The development of the limbus fossae ovalis in the human heart—a new septum, *J. Anat.*, **97**, 45–54.

Chrzanowska, G. and Beben, A. (1973). Weight of the brain and body weight in man between the ages of 20 and 89 years, *Folia morphol., Warsz.*, **32**, 391–406.

Chrzanowska, G. and Krechowiecki, A. (1975). Hängt das Gehirngewicht von der Körperlänge ab? *Gegenb. morph. Jahrb. Leipzig*, **121**, 192–208.

Church, R. B. and Schultz, G. A. (1974). Differential gene activity in the pre- and postimplantation mammalian embryo, *Curr. Top. Dev. Biol.*, **8**, 179–202.

Chu-Wu, T. and Wen-Kuei, W. (1965). Further observations on the development and connections of the mesencephalic nucleus of the trigeminal nerve in the human brain, *Acta anat. sin.*, **8**, 352–5.

Chvapil, M. (1967). *The Physiology of Connective Tissue*, Butterworths: London.

Čihák, R. (1970). Variations of lumbosacral joints and their morphogenesis, *Acta univ. Carolin. Med.*, **16**, 145–65.

Čihák, R. (1972). Ontogeny of the skeleton and intrinsic muscles of the human hand and foot, *Ergeb. Anat. Entwicklungsgesch.*, **46**, 5–194.

Čihák, R. and Popelka, S. (1961). Částečné defekty velkého svalu prsního. Morfologická a kliniká studie, *Acta chir. orthop. traum., Cech.*, **28**, 185–94.

Clara, M. (1937). Zur Histobiologie des Bronchalepithels, *Z. mikrosk.-anat. Forsch.*, **41**, 321–47.

Clark, E. R. (1938). Arterio-venous anastomoses, *Physiol. Rev.*, **18**, 229–47.

Clark, S. L. (1934). Innervation of the choroid plexuses and the blood vessels within the central nervous system, *J. comp. Neurol.*, **60**, 21–36.

Clark, W. E. L. G. (1920). On the Pacchionian bodies, *J. Anat.*, **55**, 40–8.

Clark, W. E. L. G. (1926). The mammalian oculomotor nucleus, *J. Anat.*, **60**, 426–48.

Clark, W. E. L. G. (1932). A morphological study of the lateral geniculate body, *Br. J. Ophthalmol.*, **16**, 264–84.

Clark, W. E. L. G. (1933). The medial geniculate body and the nucleus isthmi, *J. Anat.*, **67**, 536–48.

Clark, W. E. L. G. (1936). The topography and homologies of the hypothalamic nuclei in man, *J. Anat.*, **70**, 203–14.

Clark, W. E. L. G. (1936). The thalamic connections of the temporal lobe of the brain in the monkey, *J. Anat.*, **70**, 447–64.

Clark, W. E. L. G. (1937). The termination of ascending tracts in the thalamus of the macaque monkey, *J. Anat.*, **71**, 7–40.

Clark, W. E. L. G. (1940). Vascular mechanism related to great vein of Galen, *Br. med. J.*, **1**, 476.

Clark, W. E. L. G. (1941). The laminar organisation and cell content of the lateral geniculate body in the monkey, *J. Anat.*, **75**, 419–33.

Clark, W. E. L. G. (1945). Deformation patterns in the cerebral cortex. In: *Essays on Growth and Form* (Clark, W. E. L. G. and Medawar, P. B. eds.), pp. 1–22, Clarendon Press: Oxford.

Clark, W. E. L. G. (1947). *The Anatomical Patterns as the Essential Basis of Sensory Discrimination*, Clarendon Press: Oxford.

Clark, W. E. L. G. (1949). Laminar pattern of lateral geniculate nucleus considered in relation to colour vision, *Documenta ophth.*, **3**, 57–64.

Clark, W. E. L. G. (1951). The projection of the olfactory epithelium on the olfactory bulb in the rabbit, *J. Neurol. Neurosurg. Psychiat.*, **14**, 1–10.

Clark, W. E. L. G. (1957). Inquiries into the anatomical basis of olfactory discrimination, *Proc. R. Soc. B.*, **146**, 299–319.

Clark, W. E. L. G. (1971). *The Tissues of the Body*. Sixth edition, Clarendon Press: Oxford.

Clark, W. E. L. G. and Penman, G. C. (1934). The projection of the retina in the lateral geniculate body, *Proc. R. Soc. B.*, **114**, 291–313.

Clark, W. E. L. G., Beattie, J., Riddoch, G. and Dott, N. M. (1938). *The Hypothalamus. Morphological, Functional, Clinical and Surgical Aspects*, Oliver and Boyd: Edinburgh.

Clark, W. E. L. G. and Russell, W. R. (1939). Observations on the efferent connexions of the centre median nucleus, *J. Anat.*, **73**, 255–62.

Clark, W. E. L. G. and Meyer, M. (1947). The terminal connexions of the olfactory tract in the rabbit, *Brain*, **70**, 304–28.

Clark, W. E. L. G. and Powell, T. P. S. (1953). On thalamocortical connexions of general sensory cortex of Macaca, *Proc. R. Soc. B.*, **141**, 467–87.

Clarke, C. A. (1970). *Human Genetics and Medicine*, Arnold: London.

Clarke, E. and O'Malley, C. D. (1968). *The Human Brain and Spinal Cord*, University of California Press: Los Angeles.

Clarke, I. C. (1973a). Correlation of scanning electron microscopy replication and light microscopy studies of the bearing surfaces in human joints. In: *Scanning Electron Microscopy Workshop Proceedings: Part 3* (Johari, O. and Corvin, I. eds.), Illinois Research Institute: Chicago.

Clarke, I. C. (1973b). Quantitative measurement of human articular surface topography *'in vitro'* by profile recorder and stereomicroscopy techniques, *J. microsc.*, **97**, 309–14.

Clarke, I. C. (1974). Articular cartilage: a review and scanning electron microscope study, *J. Anat.*, **118**, 261–80.

Clarke, J. A. (1963). An X-ray microscopic investigation of the vasa vasorum of the human ascending aorta, *J. Anat.*, **97**, 630–631P.

Clawson, R. C. and Domm, L. V. (1969). Origin and early migration of primordial germ cells in the chick embryo. A study of the stages definitive primitive streak through eight somites, *Am. J. Anat.*, **125**, 87–112.

Clay, R. S. and Court, T. H. (1932). *The History of the Microscope*, Griffin: London.

Clegg, E. J. (1955). The arterial supply of the human prostate and seminal vesicles, *J. Anat.*, **89**, 209–16.

Clegg, E. J. (1956). The vascular arrangements within the human prostate gland, *Br. J. Urol.*, **28**, 428–35.

Clements, J. A. (1968). Surface active materials in the lung. In: *The Lung* (Liebow, A. A. and Smith, D. E. eds.), Williams and Wilkins: Baltimore.

Clementi, F. and Palade, G. E. (1969). Intestinal capillaries. I. Permeability to peroxidase and ferritin, *J. Cell Biol.*, **41**, 33–58.

Clermont, Y. (1963). The cycle of the seminiferous epithelium in man, *Am. J. Anat.*, **112**, 35–52.

Clermont, Y. (1966). Renewal of spermatogonia in man, *Am. J. Anat.*, **118**, 509–24.

Clermont, Y. (1969). Two classes of spermatogonial stem cells in the monkey (*Cercopithecus aethiops*), *Am. J. Anat.*, **126**, 57–72.

Clermont, Y. and Leblond, C. P. (1955). Spermiogenesis of man, monkey, ram and other mammals as shown by the 'periodic acid-Schiff' technique, *Am. J. Anat.*, **96**, 229–54.

Close, R. I. (1972). Dynamic properties of mammalian skeletal muscles, *Physiol Rev.*, **52**, 129–97.

Cnockaert, J. C. and Pertuzon, E. (1974). Sur la géometrie musculo-squelettique du triceps brachii. Application à la détermination dynamique de sa compliance, *Europ. J. appl. Physiol.*, **32**, 149–58.

Coakley, J. B. and King, T. S. (1959). Cardiac muscle relation of the coronary sinus, the oblique vein of the left atrium and the left precaval vein in mammals, *J. Anat.*, **93**, 30–5.

Cobb, J. L. S. and Benett, T. (1969). A study of nexuses in visceral smooth muscle, *J. Cell Biol.*, **41**, 287–97.

Cochran, G. V., Pawluk, R. J. and Bassett, C. A. (1968). Electromechanical characteristics of bone under physiologic moisture conditions, *Clin. Orthop.*, **58**, 249–70.

Cockett, F. B. (1956). Diagnosis and surgery of high-pressure venous leaks in the leg, *Br. med. J.*, **2**, 1399–403.

Code, C. F. (1968). *Handbook of Physiology*, Section 6, *Alimentary Canal*, Volume 4 (editor), American Physiological Society: Washington.

Coërs, C. and Woolf, A. L. (1959). *The Innervation of Muscle*, Blackwell: Oxford.

Cogan, D. G. (1956). *Neurology of the Ocular Muscles*, Second edition, Thomas: Springfield, Ill.

Cogan, D. G. and Kuwabara, T. (1967). The sphingolipidoses and the eye, *Archs Ophthal., N.Y.*, **79**, 437–52.

Coggeshall, R. E., Coulter, J. D. and Willis, W. D. (1973). Unmyelinated fibres in the ventral root, *Brain Res.*, **57**, 229–33.

Coghill, G. E. (1929). *Anatomy and the Problem of Behaviour*, University Press: Cambridge.

Cohen, A. I. (1965). New details of the ultrastructure of the outer segments and ciliary connections of the rods of human and macaque retinas, *Anat. Rec.*, **152**, 63–80.

Cohen, A. I. (1965). The electron microscopy of the normal human lens, *Invest. Ophthal.*, **4**, 433–46.

Cohen, A. I. (1970). Further studies on the question of the patency of saccules in outer segments of vertebrate photoreceptors, *Vision Res.*, **10**, 445–53.

Cohen, A. I. (1972). Rods and cones. In: *Handbook of Sensory Physiology* (Fuortes, M. G. F. ed.), Volume VII/2, pp. 63–110, Springer-Verlag: Berlin.

Cohen, A. S., McNeill, M., Calkins, E., Sharp, J. T. and Schubart, A. (1967). The 'normal' sacroiliac joint: analysis of 88 sacroiliac roentgenograms, *Am. J. Roentg.*, **100**, 559–63.

Cohen, D., Chambers, W. W. and Sprague, J. M. (1958). Experimental study of the efferent projections from the cerebellar nuclei to the brain stem of the cat, *J. comp. Neurol.*, **109**, 233–59.

Cohen, J. and Harris, W. H. (1958). The three dimensional anatomy of Haversian systems, *J. Bone Jt Surg.*, **40A**, 419–34.

Cohen, J. and Szabó, G. (1968). Study of pigment donation *in vitro*, *Expl Cell Res.*, **50**, 418–34.

Cohen, L. (1959). Venous drainage of the mandible. *Oral Surg.*, **12**, 1447–9.

Cohen, S. (1958). A nerve growth-promoting protein. In: *The Chemical Basis of Development* (McElroy, W. D. and Glass, B. eds.), pp. 665–76, Johns Hopkins: Baltimore.

Cohen, S. and Levi-Montalcini, R. (1956). A nerve growth-stimulating factor isolated from snake venom, *Proc. natn. Acad. Sci. U.S.A.*, **42**, 571–4.

Cohn, Z. A., Hirsch, J. G. and Weiner, E. (1963). The cytoplasmic granules of phagocytic cells and the degradation of bacteria. In: *Lysosomes* (de Reuck, A. V. S. and Cameron, M. P. eds.), pp. 126–44, Ciba Foundation Symposium, Churchill: London.

Cokkinis, A. J. (1930). Observations on the mesenteric circulation, *J. Anat.*, **64**, 200–5.

BIBLIOGRAPHY

Cole, F. J. (1944). *A History of Comparative Anatomy. From Aristotle to the Eighteenth Century*, Macmillan: London.

Cole, P. (1953). Some aspects of temperature, moisture and heat relationships in the upper respiratory tract, *J. Lar. Otol.*, **67**, 449–56.

Cole, P. (1954). Recordings of respiratory air temperature, *J. Lar. Otol.*, **68**, 295–307.

Coleman, S. S. and Anson, B. J. (1961). Arterial patterns in the hand based upon a study of 650 specimens, *Surgery Gynec. Obstet.*, **113**, 409–24.

Collet, A., Basset, F. and Normand-Reuet, C. (1967). Etude au microscope électronique du poumon humain normal et pathologique, *Poumon Cœur*, **23**, 747–85.

Collet, A. and Normand-Reuet, C. (1967). Aspects infrastructureaux de la traversée de la paroi alvéolaire du poumon par des cellules migratrices, *Sém. Hôp. Paris*, **43**, 1928–37.

Collier, J. and Buzzard, E. F. (1903). The degenerations resulting from lesion of posterior nerve roots and from transverse lesions of the spinal cord in man. A study of 20 cases, *Brain*, **26**, 559–91.

Collis, J. L., Satchwell, L. M. and Abrams, L. D. (1954). Nerve supply to the diaphragm, *Thorax*, **9**, 22–5.

Colonnier, M. (1964). The tangential organization of the visual cortex, *J. Anat.*, **98**, 327–44.

Colonnier, M. L. (1966). The structural design of the neocortex. In: *Brain and Conscious Experience* (Eccles, J. C. ed.), pp. 1–18, Springer-Verlag: Berlin.

Colonnier, M. (1967). The tangential organization of the visual cortex, *J. Anat.*, **98**, 327–44.

Colonnier, M. (1967). The fine structural arrangement of the cortex, *Archs Neurol. Psychiat., Chicago*, **16**, 651–7.

Colonnier, M. (1968). Synaptic patterns on different cell types in the different laminae of the cat visual cortex, *Brain Res.*, **9**, 268–87.

Colonnier, M. (1974). Spatial inter-relationships as physiological mechanisms in the central nervous system. In: *Essays on the Nervous System* (Bellairs, R. and Gray, E. G. eds.), pp. 344–66, Clarendon Press: Oxford.

Colonnier, M. and Guillery, R. W. (1964). Synaptic organisation in the lateral geniculate nucleus of the monkey, *Z. Zellforsch. mikrosk. Anat.*, **62**, 333–5.

Combelles, R. (1972). Vascularization de la cavité buccale, *Arch. Anat. Histol. Embryol.*, **55**, 179–208.

Comings, D. E. and Okada, T. (1972). The architecture of meiotic cells and mechanisms of chromosome pairing, *Adv. Cell molec. Biol.*, **2**, 309–84.

Condé, F. and Condé, H. (1973). Etude de la morphologie des cellules du noyau rouge du chat par la méthode de Golgi-Cox, *Brain Res.*, **53**, 249–71.

Cone, R. A. (1972). Rotational diffusion of rhodopsin in the visual receptor membrane, *Nature, New Biol.*, **236**, 39–43.

Conel, J. L. (1939–59). *The Post-natal Development of the Human Cerebral Cortex*, Volumes I–VI, Harvard University Press: Cambridge, Massachusetts.

Conel, J. L. (1942). The origin of the neural crest, *J. comp. Neurol.*, **76**, 191–216.

Congdon, E. D. (1922). Transformation of the aortic-arch system during the development of the human embryo, *Contr. Embryol.*, **14**, 47–110.

Conklin, J. L. (1962). Cytogenesis of the human fetal pancreas, *Am. J. Anat.*, **111**, 181–93.

Conklin, J. L. (1968). The development of the human fetal adenohypophysis, *Anat. Rec.*, **160**, 79–92.

Conn, H. J. (1948). *The History of Staining*, Biotechnical Publications: Geneva, New York.

Connor, J. D. (1968). Caudate unit responses to nigral stimuli: evidence for a possible nigro-neostriatal pathway, *Science, N.Y.*, **160**, 899–900.

Connors, T. A. (1975). Cytotoxic agents in teratogenic research. In: *Teratology* (Berry, C. L. and Poswillo, D. E. eds.), Springer-Verlag: Berlin.

Cook, G. M. W. (1968). Glycoproteins in membranes, *Biol. Rev.*, **43**, 363–92.

Cook, P. J., Robson, E. B. and Buckton, K. E. (1973). Segregation of genetic markers in families with chromosomal polymorphism and structural rearrangements involving chromosome 1, *Ann. Hum. Genet.*, **37**, 261–74.

Coombs, J. S., Curtis, D. R. and Landgren, S. (1956). Spinal cord potentials generated by impulses in muscle and cutaneous afferent fibres, *J. Neurophysiol.*, **19**, 452–67.

Cooper, E. R. A. (1929). The histology of the retained testis in the human subject at different ages, and its comparison with the scrotal testis, *J. Anat.*, **64**, 5–27.

Cooper, E. R. A. (1945). The development of the human lateral geniculate body, *Brain*, **68**, 222–42.

Cooper, E. R. A. (1946). Development of substantia nigra, *Brain*, **69**, 22–33.

Cooper, E. R. A. (1946). Development of human red nucleus and corpus striatum, *Brain*, **69**, 34–43.

Cooper, E. R. A. (1946). Accessory optic tracts in the human foetus, *Brain*, **69**, 45–9.

Cooper, E. R. A. (1946). Development of nuclei of oculomotor and trochlear nerves (somatic efferent column), *Brain*, **69**, 50–7.

Cooper, E. R. A. (1950). The development of the thalamus, *Acta anat.*, **9**, 201–26.

Cooper, E. R. A. (1958). Nerves of the meninges and choroid plexus. In: *The Cerebrospinal Fluid* (Wolstenholme, G. E. W. and O'Connor, C. M. eds.), pp. 80–91, Ciba Foundation Symposium, Churchill: London.

Cooper, G. W. and Prockop, D. J. (1968). Intracellular accumulation of protocollagen and extrusion of collagen by embryonic cartilage cells, *J. Cell Biol.*, **38**, 523–37.

Cooper, P. R., Milgram, J. W. and Robinson, R. A. (1966). Morphology of the osteon, *J. Bone Jt Surg.*, **48A**, 1239–71.

Cooper, S. (1953). Muscle spindles in the intrinsic muscles of the human tongue, *J. Physiol., Lond.*, **122**, 193–202.

Cooper, S., Daniel, P. D. and Whitteridge, D. (1955). Muscle spindles and other sensory endings in the extrinsic eye muscles; the physiology and anatomy of these receptors and of their connexions with the brain stem, *Brain*, **78**, 564–83.

Cope, V. Z. (1917). The internal structure of the sphenoidal sinus, *J. Anat.*, **51**, 127–36.

Copp, D. H., Cockcroft, D. W. and Kueh, Y. (1967). Calcitonin from ultimobranchial glands of dogfish and chickens, *Science, N.Y.*, **158**, 924–5.

Copp, D. H. and Talmage, R. V. (1978). *Endocrinology of Calcium Metabolism* (editors), International Congress Series No. 421, Elsevier: New York, Amsterdam.

Corazza, R., Fadiga, E. and Parmeggiani, P. L. (1963). Patterns of pyramidal activation of cat's motoneurons, *Archs ital. Biol.*, **101**, 337–64.

Corbin, K. B. and Harrison, F. (1940). Function of mesencephalic root of the fifth cranial nerve, *J. Neurophysiol.*, **3**, 423–35.

Corner, G. W. (1938). Quantitative studies of experimental menstruation-like bleeding due to hormone deprivation, *Am. J. Physiol.*, **124**, 1–12.

Cornes, J. S. (1965). Number, size and distribution of Peyer's patches in the human small intestine. II. The development of Peyer's patches, *Gut*, **6**, 225–9.

Costoff, A. (1977). Ultrastructure of the pituitary gland. In: *The Pituitary, a Current Review* (Allen, M. B. Jr. and Mahesh, V. B. eds.), pp. 59–76, Academic Press: New York, London.

Cottle, H. M. (1966). Concepts of nasal physiology, *Archs Otolar.*, **72**, 11–20.

Couinaud, C. (1954). Les enveloppes vasculo-biliaires du foie ou capsule de Glisson, *Lyon Chir.*, **49**, 489–607.

Coulombre, A. J. (1964). Problems in corneal morphogenesis. In: *Advances in Morphogenesis* (Abercrombie, M. and Brachet, J. eds.), Volume 4, pp. 81–110, Academic Press: New York, London.

Coupland, R. E. (1952). The pre-natal development of the abdominal para-aortic bodies in man, *J. Anat.*, **86**, 357–72.

Coupland, R. E. (1954). Post-natal fate of the abdominal para-aortic bodies in man, *J. Anat.*, **88**, 455–64.

Coupland, R. E. (1958). The innervation of the pancreas of the rat, cat and rabbit as revealed by the cholinesterase technique, *J. Anat.*, **92**, 143–9.

Coupland, R. E. (1965). Electron microscope observations on the structure of the rat adrenal medulla. I. The ultrastructure and organisation of chromaffin cells in the normal adrenal medulla. II. Normal innervation, *J. Anat.*, **99**, 231–54, 255–72.

Coupland, R. E. (1965). *The Natural History of the Chromaffin Cell*, Longmans: London.

Coupland, R. E., Pyper, A. S. and Hopwood, D. (1964). A method for differentiating between noradrenaline- and adrenaline-storing cells in the light and electron microscope, *Nature, Lond.*, **201**, 1240–2.

Coupland, R. E. and Weakley, B. S. (1970). Electron microscopic observation on the adrenal medulla and extra-adrenal chromaffin tissue of the post-natal rabbit, *J. Anat.*, **106**, 213–31.

Courot, M., Hochereau de Reviers, M.-T. and Ortevant, R. (1970). Spermatogenesis. In: *The Testis* (Johnson, A. D., Gomes, W. R. and Vandemark, N. L. eds.), Volume I, pp. 339–432, Academic Press: New York, London.

Courtney, H. (1949). Posterior subsphincteric space; its relation to posterior horseshoe fistula, *Surgery Gynec. Obstet.*, **89**, 222–6.

Courville, J. (1966). Rubrobulbar fibres to the facial nucleus and the lateral reticular nucleus (nucleus of the lateral funiculus). An experimental study in the cat with silver impregnation methods, *Brain Res.*, **1**, 317–37.

Courville, J. (1966). The nucleus of the facial nerve; the relation between cellular groups and peripheral branches of the nerve, *Brain Res.*, **1**, 338–54.

1478

Courville, J. (1966). Somatotopical organisation of the projection from the nucleus interpositus anterior of the cerebellum to the red nucleus. An experimental study in the cat with silver impregnation, *Expl Brain Res.*, **2**, 191–215.

Courville, J. and Brodal, A. (1966). Rubrocerebellar connections in the cat. An experimental study with silver impregnation methods, *J. comp. Neurol.*, **126**, 471–85.

Cowan, W. M. and Clarke, P. G. H. (1976). The development of the isthmo-optic nucleus, *Brain Behav. Evol.*, **13**, 354–75.

Cox, R. P. (1974). *Cell Communication* (editor), Wiley: New York.

Cox, R. W. and Peacock, M. A. (1977). The fine structure of developing elastic cartilage, *J. Anat.*, **123**, 283–96.

Cragg, B. G. (1962). Centrifugal fibres to the retina and olfactory bulb, and composition of the supra optic commissures in the rabbit, *Expl Neurol.*, **5**, 406–27.

Cragg, B. G. (1967). The density of synapses and neurones in the motor and visual areas of the cerebral cortex, *J. Anat.*, **101**, 639–54.

Cragg, B. G. (1969). The topography of the afferent projections in the circumstriate visual cortex of the monkey studied by the Nauta method, *Vision Res.*, **9**, 733–47.

Cragg, B. R. (1970). What is the signal for chromatolysis? *Brain Res.*, **23**, 1–21.

Cragg, B. G. (1976). Ultrastructural features of human cerebral cortex, *J. Anat.*, **121**, 331–62.

Cralley, J., Fitch, K. and McGonagle, W. (1975). Lumbrical muscles and contracted toes, *Anat. Anz.*, **138**, 348–53.

Crandall, W. R. (1938). A quantitative study of the influence of the ovarian hormones on hyperplasia by mitosis in the rabbit uterus in early pregnancy, *Anat. Rec.*, **72**, 195–210.

Creamer, B. (1955). Oesophageal reflux, *Lancet*, **1**, 279–81.

Creemers, J. and Jacques, P. J. (1971). Endocytic uptake and vesicular transport of injected horseradish peroxidase in the vacuolar apparatus of rat liver cells, *J. Ultrastruct. Res.*, **67**, 188–203.

Crescitelli, F. (1972). The visual cells and visual pigments of the vertebrate eye. In: *Handbook of Sensory Physiology* (Dartnall, H. J. A. ed.), Volume 8, Part 1, pp. 245–95, Springer: Berlin.

Critchley, M. (1953). *The Parietal Lobes*, Arnold: London.

Crock, H. V. (1965). A revision of the anatomy of the arteries supplying the upper end of the human femur, *J. Anat.*, **99**, 77–88.

Crock, H. V. (1967). *The Blood Supply of the Lower Limb Bones in Man*, Livingstone: London.

Crompton, A. W. and Hiiemae, K. (1969). How mammalian molar teeth work, *Discovery*, **5**, 23–34.

Crosby, E. C. (1953). Relations of brain centers to normal and abnormal eye movements in the horizontal plane, *J. comp. Neurol.*, **99**, 437–80.

Crosby, E. C. and Woodburne, R. T. (1940). Comparative anatomy of the pre-optic area and hypothalamus, *Res. Publs Ass. Res. nerv. ment. Dis. Proc.*, **20**, 52–169.

Crosby, E. C. and Humphrey, T. (1941). Studies of vertebrate telencephalon; nuclear pattern of anterior olfactory nucleus, tuberculum olfactorium and amygdaloid complex in adult man, *J. comp. Neurol.*, **74**, 309–52.

Crosby, E. C. and Woodburne, R. T. (1943). Nuclear pattern of non-tectal portions of midbrain and isthmus in Primates, *J. comp. Neurol.*, **78**, 441–82.

Crosby, E. C. and Henderson, J. W. (1948). The mammalian midbrain and isthmus regions. Part II. Fiber connections of the superior colliculus. B. Pathways concerned in automatic eye movements, *J. comp. Neurol.*, **88**, 53–92.

Crosby, E. C., Yoss, R. E. and Henderson, J. W. (1952). The mammalian midbrain and isthmus regions. II. The fiber connections. D. The pattern for eye movements on the frontal eyefield and the discharge of specific portions of this field to and through midbrain levels, *J. comp. Neurol.*, **97**, 357–84.

Crosby, E. C., Humphrey, T. and Lauer, E. W. (1962). *Correlative Anatomy of the Nervous System*, Macmillan: New York.

Crosby, E. C. and Dejonge, B. R. (1963). Experimental and clinical studies of the central connections and central relations of the facial nerve, *Ann. Otol. Rhinol. Lar.*, **72**, 735–55.

Crowder, R. E. (1957). The development of the adrenal gland in man, with special reference to origin and ultimate location of cell types and evidence in favor of the 'cell migration' theory, *Contr. Embryol.*, **36**, 193–210.

Cruickshank, C. N. D. and Harcourt, S. A. (1964). Pigment donation *in vitro*, J. invest. Derm., **42**, 183–4.

Csapo, A. (1962). Smooth muscle as a contractile unit, *Physiol. Rev.*, **42**, Suppl. 5, 7–33.

Cserr, H. F. (1971). Physiology of the choroid plexus. *Physiol. Rev.*, **51**, 273–311.

Cuajunco, F. (1940). Development of the neuromuscular spindle in human fetuses, *Contr. Embryol.*, **28**, 95–128.

Cuajunco, F. (1942). Development of the human motor endplate, *Contr. Embryol.*, **30**, 127–52.

Cullis, W. and Tribe, E. (1913). Distribution of nerves in the heart, *J. Physiol.*, *Lond.*, **46**, 141–50.

Cummins, H. and Midlo, C. (1961). *Finger Prints, Palms and Soles. An Introduction to Dermatoglyphics*, Dover Publishing Company: New York.

Cunningham, B. A. (1976). Structure and significance of beta 2-microglobulin, *Fedn. Proc.*, **35**, 1171–6.

Curtis, A. S. G. (1967). *The Cell Surface: its Molecular Role in Morphogenesis*, Logos Press: London.

Curtis, A. S. G. (1973). Cell adhesion, *Prog. Biophys. molec. Biol.*, **27**, 316–86.

Cuschieri, A. and Bannister, L. H. (1975a). The embryonic development of the olfactory organ of the mouse. 1. Light microscopy, *J. Anat.*, **119**, 277–86.

Cuschieri, A. and Bannister, L. H. (1975b). The embryonic development of the olfactory organ of the mouse. 2. Electron microscopy, *J. Anat.*, **119**, 471–98.

Cushing, H. (1912). *The Pituitary Body and its Disorders*, Lippincott: Philadelphia.

Czarnetzki, A. (1971). Epigenetische Skeletlmerkmale im Populationsvergleich. I. Rechts-links-Unterschiede bilateral angelegter Merkmale, *Z. morph. anthrop.*, **63**, 238–54.

D

Daems, W. Th. (1968). On the fine structure of human neutrophilic leukocyte granules, *J. Ultrastruct. Res.*, **24**, 343–8.

Dahlström, A. and Fuxe, K. (1964). Evidence for the existence of monoamine neurons in the central nervous system, *Acta physiol. scand.*, *Suppl.* **232**, 1–55.

Dail, W. G. and Evan, A. P. (1974). Neural and vascular development in the human phallus, *Investig. Urol.*, **11**, 427–38.

Dalcq, A. M. (1954). Nouvelles données structurales et cytochimiques sur l'œuf des mammifères, *Rev. gén. Sci. pur. appl.*, **61**, 19–41.

Dale, H. H., Feldburg, W. W. and Vogt, M. (1936). Release of acetylcholine at voluntary motor nerve endings, *J. Physiol.*, *Lond.*, **86**, 353–80.

Dallenbach-Hellweg, G. and Nette, G. (1964). Morphological and histochemical observations on trophoblast and decidua of the basal plate of the human placenta at term, *Am. J. Anat.*, **115**, 309–26.

Dal Pont, G. (1960). Contribution à l'étude de la structure fonctionelle du maxillaire, *Ann. di stomat.*, **9**, 921–32.

Dalton, A. J. and Hagenau, F. (1967). *Ultrastructure of the Kidney* (editors), Academic Press: New York, London.

Dalton, A. J. and Hagenau, F. (1968). *The Membranes* (editors), Academic Press: New York, London.

Dalton, A. J. and Hagenau, F. (1968). *The Nucleus* (editors), Academic Press: New York, London.

Damiani, N. and Capodacqua, A. (1961). On the intrinsic innervation of the fallopian tube, *Annali Ostet. Ginec.*, **83**, 436–46.

Dancis, J. (1959). The placenta, *J. Pediat.*, **55**, 85–101.

Daniel, J. C. Jr. and Olson, J. D. (1966). Cell movement, proliferation and death in the formation of the embryonic axis of the rabbit, *Anat. Rec.*, **156**, 123–7.

Daniel, O. and Shackman, R. (1952). Blood supply of human ureter in relation to utero-colic anastomoses, *Br. J. Urol.*, **24**, 334–43.

Daoust, R. and Clermont, Y. (1955). Distribution of nucleic acids in germ cells during the cycle of the seminiferous epithelium in the rat, *Am. J. Anat.*, **96**, 255–84.

Dart, A. M. (1971). Cells of the dorsal column nuclei projecting down into the spinal cord, *J. Physiol.*, *Lond.*, **219**, 29–30.

Daseler, E. H. and Anson, B. J. (1943). The plantaris muscle. An anatomical study of 780 specimens, *J. Bone Jt Surg.*, **25**, 822–7.

Davenport, H. A. and Ranson, S. W. (1930). The red nucleus and adjacent cell groups. A topographical study in the cat and rabbit, *Archs Neurol. Psychiat.*, *Chicago*, **24**, 257–66.

Davidowitz, J., Philips, G. and Breinin, G. M. (1977). Organization of the orbital surface layer in rabbit superior rectus, *Invest. ophthal. vis. sci.*, **16**, 711–29.

Davidson, R. (1969). Regulation of melanin synthesis in mammalian cells, as studied by somatic hybridization. 3. A method of increasing the frequency of cell fusion, *Exp. Cell Res.*, **55**, 421–6.

Davidson, W. M. and Smith, D. R. (1954). A morphological sex difference in the polymorphonuclear neutrophil leukocytes, *Br. med. J.*, **2**, 6–7.

Davies, D. V. (1950). Structure and function of synovial membrane, *Br. med. J.*, **1**, 92–5.

Davies, D. V. (1951). Blood supply of the tendon of extensor pollicis longus, *Br. med. J.*, **2**, 56.

Davies, D. V. and Edwards, D. A. W. (1948). Blood supply of synovial membrane and intra-articular structures, *Ann. R. Coll. Surg.*, **2**, 142–56.

Davies, D. V. and Young, L. (1954). The distribution of radioactive sulphur (^{35}S) in the fibrous tissues, cartilages and bones of the rat following its administration in the form of inorganic sulphate. *J. Anat.*, **88**, 174–83.

Davies, D. V., Barnett, C. H., Cochrane, W. and Palfrey, A. J. (1962). Electron microscopy of articular cartilage in the young adult rabbit, *Ann. rheum. Dis.*, **21**, 11–22.

Davies, F. (1927). Normal cholecystography, *Br. med. J.*, **1**, 1138–40.

Davies, F. (1935). A note on the first lumbar nerve (anterior ramus), *J. Anat.*, **70**, 177–8.

Davies, F. (1944). A previllous human ovum, aged nine to ten days (the Davies-Harding ovum), *Trans. Roy. Soc. Edinb.*, **61**, 315–28.

Davies, F., Gladstone, R. J. and Stibbe, E. P. (1932). The anatomy of the intercostal nerves, *J. Anat.*, **66**, 323–33.

Davies, F. and Harding, H. E. (1942). Pouch of Hartmann, *Lancet*, **1**, 193–5.

Davies, F. and Francis, E. T. B. (1946). The conducting system of the vertebrate heart, *Biol.-Rev.*, **21**, 173–88.

Davies, F. and Francis, E. T. B. (1952). The conduction of the impulse for cardiac contraction, *J. Anat.*, **86**, 302–9.

Davies, F., Francis, E. T. B. and King, T. S. (1952). Neurological studies of the cardiac ventricles of mammals. *J. Anat.*, **86**, 130–43.

Davies, J. (1960). *Survey of Research in Gestation and the Developmental Sciences*, Williams and Wilkins: Baltimore.

Davies, J. and Routh, J. I. (1957). Comparison of the foetal fluids of the rabbit, *J. Embryol. exp. Morph.*, **5**, 32–9.

Davis, C. L. (1923). Description of a human embryo having twenty paired somites, *Contr. Embryol.*, **15**, 1–51.

Davis, C. L. (1927). Development of the human heart from its first appearance to the stage found in embryos of twenty paired somites, *Contr. Embryol.*, **19**, 245–84.

Davis, J. R. and Langford, G. A. (1969). Response of the testicular capsule to acetylcholine and noradrenaline, *Nature, Lond.*, **222**, 386–7.

Davis, J. R., Langford, G. A. and Kirby, P. J. (1970). The testicular capsule. In: *The Testis* (Johnson, A. D., Gomes, W. R. and Vandemark, N. L. eds.), Volume I, pp. 282–338, Academic Press: New York, London.

Davis, P. R. (1955). The thoracolumbar mortice joint, *J. Anat.*, **89**, 370–7.

Davis, P. R. (1959). The medial inclination of the human thoracic intervertebral articular facets, *J. Anat.*, **93**, 68–74.

Davis, P. R. (1961). Human lower lumbar vertebrae: some mechanical and osteological considerations, *J. Anat.*, **95**, 337–44.

Davis, P. R. (1963). Some effects of lifting, pulling and pushing on the human trunk, *Ergonomics*, **6**, 303–4.

Davis, P. R., Troup, J. D. G. and Burnard, J. H. (1965). Movements of the thoracic and lumbar spine when lifting: a chrono-cyclophotographic study, *J. Anat.*, **99**, 13–26.

Davson, H. (1962). *The Eye* (editor), Volume 2, Academic Press: New York, London.

Davson, H. (1970). *Physiology of the Cerebro-spinal Fluid*, Churchill: London.

Davson, H. (1972). *Physiology of the Eye*, Third edition, Churchill-Livingstone: London.

Davson, H. and Danielli, J. F. (1952). *The Permeability of Natural Membranes*, Cambridge University Press: Cambridge.

Davson, H. and Graham, L. T. (1975). *The Eye*, Volume 5, *Comparative Physiology* (editors), Academic Press: New York, London.

Dawes, G. S. (1961). Changes in the circulation at birth, *Br. med. Bull.*, **17**, 148–53.

Dawes, G. S. (1969). *Foetal and Neonatal Physiology*, Year Book Medical Publishers: Chicago.

Dawson, J. L. (1974). Tumours of the liver. In: *Surgical Forum—the Liver* (Smith, R. ed.), Butterworths: London.

Dax, M. (1865). Lésions de la moitié gauche de l'encéphale coincident avec l'oubli désignés de la pensée, *Gaz. Hebdom.*, **11**, 259–60.

Day, H. J., Hohnsen, H. and Hovig, T. (1969). Subcellular particles of human platelets, *Scand. J. Haematol.*, *Suppl.* **7**, 3–35.

Day, M. A. (1964). Postural reflex patterns, *Nurs. Res.*, **13**, 139–47.

Day, M. H. and Napier, J. R. (1961). The two heads of flexor pollicis brevis, *J. Anat.*, **95**, 123–30.

de Barros, H. F. (1959). Considérations sur la structure anatomique de la mandibule et du canal dentaire inférieure, *Rev. brasil. Odontol.*, **17**, 171–8.

de Beer, G. R. (1937). *The Development of the Human Skull*, Oxford University Press: London.

de Busscher, G. (1948). Les anastomoses artériveneuses, *Acta neerl. morphol.*, **6**, 87–105.

de Camilli, P., Peluchetti, D. and Meldolesi, J. (1976). Dynamic changes in the luminal plasmalemma in stimulated parotid acinar cells. A freeze-fracture study, *J. Cell Biol.*, **70**, 59–74.

de Castro, F. (1932). Sensory ganglia of the cranial and spinal nerves. In: *Cytology and Cellular Pathology of the Nervous System* (Penfield, W. G. ed.), Hoeber: New York.

de Castro, F. and Herreros, M. L. (1945). Actividad del ganglio cervical superior, *Trab. Inst. Cajal Invest. Biol.*, **37**, 287–342.

de Duve, C. (1963). The lysosome, *Scient. Am.*, **208**, 64–72.

De Feúdis, F. V. (1974). *Central Cholinergic Systems and Behaviour*, Academic Press: London, New York.

de Gasperis, C. and Miani, A. (1969). Observations sur l'ultrastruture du mesothelium pleural de l'homme, *Bull. Ass. Anat., Paris*, **145**, 188–202.

Degroot, L. J. and Niepomniszcze, H. (1977). Biosynthesis of thyroid hormone, basic and clinical aspects, *Metab. Clin. Exp.*, **26**, 665–718.

De Gruchy, G. C. (1970). *Clinical Haematology in Medical Practice*, Third edition, Blackwell: Oxford.

de Haan, R. L. (1958). Cell migration and morphogenetic movements. In: *The Chemical Basis of Development* (McElroy, W. D. and Glass, B. eds.), Johns Hopkins Press: Baltimore.

Dejean, C., Hervouët, F. and Leplat, G. (1958). *L'Embryologie de l'Oeil et sa Tératologie*, Masson: Paris.

Déjerine, J. and Déjerine-Klumpke, H. (1901). *Anatomie des Centres Nerveux*, Tôme 2, Rueff: Paris.

Dekaban, A. (1953). Human thalamus: An anatomical, developmental and pathological study, *J. comp. Neurol.*, **99**, 639–83.

DeKock, L. L. and Dunn, A. E. G. (1968). Electron microscopic investigation of the nerve endings in carotid body. In: *Arterial Chemoreceptors* (Torrance, R. W. ed.), pp. 179–88, Blackwell: Oxford.

Del Galindo, S. E. C. De G. and Ramirez, L. C. (1977). Anatomical and functional account on the lateral nasal cartilages, *Acta anat.*, **97**, 393–9.

del Rio Hortega, P. (1924). Le névroglie et le troisième élément des centres nerveux, *Bull. Soc. Sci. méd. biol., Montpellier*, **5**.

de Marsh, Q. B., Windle, W. F. and Alt, H. L. (1942). Blood volume of newborn infants in relation to early and late clamping of the umbilical cord, *Am. J. Dis. Child*, **63**, 1123–9.

De Meyer, W. (1959). Number of axons and myelin sheaths in adult human medullary pyramids. Study with silver impregnation and iron hematoxylin staining methods. *Neurology, Minneap.*, **9**, 42–7.

Dennison, M. (1971). Electron stereoscopy as a means of classifying synaptic vesicles, *J. Cell Sci.*, **8**, 525–40.

Denny-Brown, D. and Robertson, E. G. (1933). On the physiology of micturition, *Brain*, **56**, 149–90.

Denny-Brown, D. and Robertson, E. G. (1935). Investigation of nervous control of defaecation, *Brain*, **58**, 256–310.

de Palma, A. F. (1957). *Degenerative Changes in the Sternoclavicular and Acromioclavicular Joints in Various Decades*, Thomas: Springfield, Ill.

De Pierre, J. W. and Karnovsky, M. L. (1973). Plasma membranes of mammalian cells, *J. Cell Biol.*, **56**, 275–303.

de Reuck, A. V. S. and Cameron, M. P. (1962). *The Exocrine Pancreas: Normal and Abnormal Functions* (editors), Ciba Foundation Symposium, Churchill: London.

de Reuck, A. V. S. and Porter, R. (1967). *Development of the Lungs* (editors), Ciba Foundation Symposium, Churchill: London.

de Reuck, A. V. S. and Knight, J. (1968). *Hearing Mechanisms in Vertebrates* (editors), Ciba Foundation Symposium, Churchill: London.

de Reuck, J. (1972). The cortico-subcortical arterial angio-architecture in the human brain, *Acta neurol. belg.*, **72**, 232–9.

De Robertis, E. (1960). Some observations on the ultrastructure and morphogenesis of photoreceptors, *J. gen. Physiol.*, **43**, 1–13.

De Robertis, E. D. P. and Bennett, H. S. (1955). Some features of the sub-microscopic morphology of synapses in frog and earthworm, *J. biophys. biochem. Cytol.*, **1**, 47–58.

De Robertis, E. D. P. and Carrera, R. (1965). Biology of neuroglia (editors), *Prog. Br. Res.*, **15**.

De Robertis, E. D. P., Saez, F. A. and De Robertis, E. M. F. (1975), *Cell Biology*, Saunders: Philadelphia.

Derry, D. E. (1923). On the sexual and racial characters of the human ilium, *J. Anat.*, **58**, 71–83.

de Sousa, E. (1955). Alcoolização do espaço extradural: tratamento em certas algias rebeldes, *Rev. brasil cir.*, **30**, 353–60.

de Sousa, O. M. (1964). Estudo electromiogràfico do m. platysma, *Folia clin. et Biol.*, **33**, 42–52.

de Sousa, O. M. and Vitti, M. (1966). Estudio electromiográfico de los músculos adductores largo y mayor, *Arch. Mex. de Anat.*, **7**, 52–3 (abstract).

Detwiler, S. M. (1936). *Neuroembryology, an Experimental Study*, Macmillan: New York.

Detwiler, S. R. (1955). Experiments on the origin of the ventrolateral trunk musculature in the urodele (*Amblystoma*), *J. exp. Zool.*, **129**, 45–76.

Deuchar, E. M. (1958). Regional differences in cathepetic activity in *Xenopus laevis* embryos, *J. Embryol. exp. Morph.*, **6**, 223–37.

de Vries, P. A. and Saunders, J. B. de C. M. (1962). Development of the ventricles and spiral outflow tract in the human heart, *Contr. Embryol.*, **37**, 87–114.

de Vries, R. A. C. and Kappers, J. A. (1971). Influence of the pineal gland on the neurosecretory activity of the supraoptic hypothalamic nucleus in the male rat, *Neuroendocrinol.*, **8**, 359–66.

Diamant, H. and Wiberg, A. (1965). Does the chorda tympani in man contain secretory fibres for the parotid gland?, *Acta oto-laryngol.*, **60**, 255–64.

Diamond, I. T., Jones, E. G. and Powell, T. P. S. (1969). The projection of the auditory cortex upon the diencephalon and brain stem in the cat, *Brain Res.*, **15**, 305–40.

Dible, J. H. and West, C. M. (1940). A human ovum at the previllous stage, *J. Anat.*, **75**, 269–81.

Di Chiro, G. (1962). Angiographic patterns of cerebral convexity veins, *Am. J. Roentg.*, **87**, 308–21.

Dickie, J. K. M. (1920). Note on the anatomy of the membranous labyrinth, *J. Laryngol.*, **35**, 76–81.

Dickmann, Z. (1965). Sperm penetration into and through the zona pellucida of the mammalian egg. In: *Preimplantation Stages of Pregnancy* (Wolstenholme, G. E. W. and O'Connor, M. eds.), pp. 169–78, Ciba Foundation Symposium, Churchill: London.

Dickson, A. D. (1957). The development of the ductus venosus in man and the goat, *J. Anat.*, **91**, 358–68.

DiDio, L. J., Zappalá, A. and Carney, W. P. (1967). Anatomico-functional aspects of the musculus articularis genus in man, *Acta anat*, **67**, 1–23.

DiDio, L. J. and Anderson, M. C. (1968). *The 'Sphincters' of the Digestive System*, Williams and Wilkins: Baltimore.

Diem, K. and Lentner, C. (1970). *Scientific Tables: Documenta Geigy*, Seventh edition, Geigy: Basle.

Dieters, O. F. K. (1865). *Untersuchungen über Gehirn und Rückenmark des Menschen und der Säugethiere*, Vieweg und Sohn: Braunschweig.

Dijkstra, C. (1969). Structure of the autonomic terminal network of the thoracic organs after visualisation with the osmium zinc iodide method, *Mikroskopie*, **24**, 161–71.

Dilly, P. N., Wall, P. D. and Webster, K. E. (1968). Cells of origin of the spinothalamic tract in the cat and the rat, *Expl Neurol.*, **21**, 550–62.

Dimond, S. J. and Beaumont, J. G. (1973). *Hemisphere Function in the Human Brain* (editors), Wiley: New York.

Dingle, J. T. (1962). Aetiological factors in the collagen diseases. Lysosomal enzymes and the degradation of cartilage matrix, *Proc. R. Soc. Med.*, **55**, 109–11

Dingle, J, T., Fell, H. B. and Dean, R. T. (1969–1976). *Lysosomes in Biology and Pathology*, Volumes 1–3 (editors), North Holland: Amsterdam.

Dintenfass, L. (1963). Lubrication in synovial joints: a theoretical analysis, *J. Bone Jt Surg.*, **45A**, 1241–56.

Dixon, A. D. (1958). Development of the jaws, *Dent. Pract.*, **9**, 10–12.

Dixon, A. F. (1920). Note on the vertebral epiphyseal discs, *J. Anat.*, **55**, 38–9.

Dixon, J. S. and Gosling, J. A. (1971). Histochemical and electron microscopic observations on the innervation of the upper segment of the mammalian ureter, *J. Anat.*, **110**, 57–66.

Dobbie, J. W. and Symington, T. (1966). The human adrenal gland with special reference to the vasculature, *J. Endocrinol.*, **34**, 479–89.

Dodd, H. (1959). The varicose tributaries of the superficial femoral vein passing into Hunter's canal, *Postgrad. med. J.*, **35**, 18–23.

Dodd, H. and Cockett, F. B. (1956). *The Pathology and Surgery of the Veins of the Lower Limb*, Livingstone: Edinburgh.

Dogiel, A. S. (1908). *Der Bau der Spinalganglien des Menschen und der Säugethiere*, Fischer: Jena.

Dogson, M. C. H. (1962). *The Growing Brain*, Wright: Bristol.

Dohlman, C. H. (1971). The function of the corneal epithelium in health and disease, *Invest. Ophthal.*, **10**, 303–407.

Domnić-Stošić, T. and Jeličić, N. (1974). Morphological differences between meningeal arteries and the arteries of the scalp in the fetus and neonate, *Srpski Arkhiv*, **102**, 175–80.

Donovan, B. T. (1970). *Mammalian Neuroendocrinology*, McGraw-Hill: London, New York.

Doran, G. A. and Baggett, H. (1972). The genioglossus muscle: a reassessment of its anatomy in some mammals, including man, *Acta anat.*, **83**, 403–10.

Doty, R. (1968). Neural organisation of deglutition. In: *Handbook of Physiology*, Section 6, Alimentary Canal, Volume 4 (Code, C. F. ed.), pp. 1861–1902, American Physiological Society: Washington.

Doubilet, H., Shafiroff, B. G. P. and Mulholland, J. H. (1948). Anatomy of periesophageal vagi, *Ann. Surg.*, **127**, 128–35.

Douek, E. E., Bannister, L. H. and Dodson, H. C. (1975). Olfaction and its disorders, *Proc. R. Soc. Med.*, **68**, 467–70.

Douglas, W. H. J., Redding, R. A. and Stein, M. (1975). The lamellar substructure of osmiophilic inclusion bodies in rat type II alveolar pneumocytes, *Tissue & Cell*, **7**, 137–42.

Douglas, W. W. (1968). Stimulus-secretion coupling: the concept and clues from chromaffin and other cells, *Br. J. Pharmacol.*, **34**, 451–74.

Døving, K. B. (1967). Problems in the physiology of olfaction. In: *Chemistry and Physiology of Flavors* (Schultze, H. W., Day, E. A. and Libbey, L. M. eds.), pp. 52–94, Avi: Connecticut.

Dow, R. S. (1938). The electrical activity of the cerebellum and its functional significance, *J. Physiol., Lond.*, **94**, 67–86.

Dow, R. S. (1939). Cerebellar action potentials in response to stimulation of various afferent connections, *J. Neurophysiol.*, **2**, 543–55.

Dow, R. S. (1942). The evolution and anatomy of the cerebellum, *Biol. Rev.*, **17**, 179–220.

Dow, R. S. (1949). Action potentials of cerebellar cortex in response to local electrical stimulation, *J. Neurophysiol.*, **12**, 245–56.

Dow, R. S. (1969). Cerebellar syndromes including vermis and hemispheric syndromes. In: *Handbook of Clinical Neurology* (Vinken, P. J. and Bruyn, G. W. eds.), Volume II, North Holland Publishing Co.: Amsterdam.

Dow, R. S. and Moruzzi, G. (1958). *The Physiology and Pathology of the Cerebellum*, University of Minnesota Press: Minneapolis.

Dowling, J. E. (1965). Foveal receptors of the monkey retina: fine structure, *Science, N.Y.*, **147**, 57–9.

Dowling, J. E. and Boycott, B. B. (1966). Organisation of the primate retina: electron microscopy, *Proc. R. Soc. B.*, **166**, 80–111.

Dowson, D., Wright, V. and Longfield, M. D. (1969). Human joint lubrication, *Bio-Med. Eng.*, **4**, 160–5.

Doyle, J. F. (1970). The perforating veins of the gluteus maximus, *I. J. Med. Sci.*, **3**, 285–8.

Doyle, J. L., Watkins, H. O. and Halbert, D. S. (1967). Undescended laryngeal nerve, *Tex. med. J.*, **63**, 53–6.

Draper, M. H. and Mya-Tu, M. (1959). A comparison of the conduction velocity in cardiac tissues of various mammals, *Q. Jl exp. Physiol.*, **44**, 91–109.

Draper, M. H., Ladefoged, P. and Whitteridge, D. (1960). Expiratory pressures and air flow during speech, *Br. med. J.*, **1**, 1837–43.

Droz, B. (1963). Dynamic conditions of proteins in the visual cells of rats and mice as shown by radio-autography with labelled amino-acids, *Anat. Rec.*, **145**, 157–68.

Druckman, R. (1952). A critique of 'suppression', with additional observations in the cat, *Brain*, **75**, 226–43.

Drumheller, G. (1969). Anatomical observations of the lower lateral nasal cartilages, *Archs Otolar.*, **89**, 599–601.

Drummond, H. (1914). The arterial supply of the rectum and pelvic colon, *Br. J. Surg.*, **1**, 677–85.

Drury, R. A. B. and Wallington, E. A. (1967). *Carlton's Histological Technique*, Fourth edition, Oxford University Press: London.

Dubois, P. (1967). Etude au microscope électronique de la pars distalis de l'hypophyse de l'embryon humain, *C. R. d'Ass. Anat., Nancy*, **138**, 429–33.

Dubowitz, V. (1969). Histochemical aspects of muscle disease. In: *Disorders of Voluntary Muscle* (Walton, J. N. ed.), pp. 239–76, Churchill: London.

Duce, I. R. and Keen, P. (1977). An ultrastructural classification of the neuronal cell bodies of the rat dorsal root ganglion using zinc iodide-osmium impregnation, *Cell Tiss. Res.*, **185**, 263–77.

Duchon, J., Fitzpatrick, T. B. and Seiji, M. (1968). Melanin 1968: some definitions and problems. In: 1967–1968 *Yearbook of Dermatology* (Kopf, A. W. and Andrade, R. eds.), pp. 1–33, Chicago Year Book Publishing Co.: Chicago.

Duckworth, W. L. H. (1947). *Some Complexities of Human Structure*, Oxford University Press: London.

Dudley, H. R. and Spiro, D. (1961). The fine structure of bone cells, *J. biophys. biochem. Cytol.*, **11**, 627–49.

Duenhoelter, J. H. and Pritchard, J. A. (1973). Human fetal respiration, *Obstet. Gynecol.*, **42**, 746–50.

Duke-Elder, J. S. (1969). *The Practice of Refraction*, Eighth edition, Churchill: London.

Duke-Elder, W. S. (1932). *Textbook of Ophthalmology*, Volume I, Kimpton: London.

Duke-Elder, W. S. (1963). *System of Ophthalmology*, Volume 3, Part I, *Embryology*, Kimpton: London.

Duke-Elder, J. S. and Wybar, K. C. (1961). *A System of Ophthalmology*, Volume 2, Kimpton: London.

Du Praw, E. J. (1968). *Cell and Molecular Biology*, Academic Press: New York, London.

Dupré, J., Ross, S. A., Watson, D. and Brown, J. C. (1973). Stimulation of insulin secretion by gastric inhibitory polypeptide in man, *J. clin. Endocr. Metab.*, **37**, 826–8.

Durcan, D. J., Shea, J. J. and Sleeckx, J. P. (1967). Bifurcation of the facial nerve, *Archs Otolar.*, **86**, 619–31.

Dusser de Barenne, J. G. (1933). 'Corticalization' of function and functional localisation in cerebral cortex, *Archs Neurol. Psychiat., Chicago*, **30**, 884–901.

Dusser de Barenne, J. G., Garol, H. W. and McCulloch, W. S. (1942). Physiological neuronography of the cortico-striatal connections, *Res. Publs Ass. Res. nerv. ment. Dis.*, **21**, 246–66.

Duvall, A. J. and Quick, C. A. (1969). Tracers and endogenous debris in delineating cochlear barriers and pathways. An experimental study, *Ann. Otol. Rhinol. Lar.*, **78**, 1041–57.

Duvernoy, H. (1972). The vascular architecture of the median eminence. In: *Brain-Endocrine Interaction. Median Eminence: Structure and Function* (Knigge, K. M., Scott, D. E. and Weindl, A. eds.), pp. 79–108, Karger: Basel.

Duvernoy, H. M. (1975). *The Superficial Veins of the Human Brain*, Springer-Verlag: Berlin.

Duvernoy, H., Koritké, J. G. and Monnier, G. (1969). On the vascularisation of the lamina terminalis in the human, *Z. Zellforsch. mikrosk. Anat.*, **102**, 49–77.

Dykes, J. R. W. (1969). Histometric assessment of human testicular biopsies, *J. Path.*, **97**, 429–40.

Dziewiatkowski, D. D. (1962). Intracellular synthesis of chondroitin sulfate, *J. Cell Biol.*, **13**, 359–64.

E

Eager, R. P. (1963). Efferent cortico-nuclear pathways in the cerebellum of the cat, *J. comp. Neurol.*, **120**, 81–103.

Eager, R. P. (1966). Patterns and mode of termination of cerebellar cortico-nuclear pathways in the monkey (*Macaca mulatta*), *J. comp. Neurol.*, **126**, 551–65.

Earle, K. M. (1952). The tract of Lissauer and its possible relation to the pain pathway, *J. comp. Neurol.*, **96**, 93–111.

Eastoe, J. E. (1963). The amino acid composition of proteins from the oral tissues. II. The matrix proteins in dentine and enamel from deciduous teeth, *Archs oral Biol.*, **8**, 633–52.

Eastoe, J. E. and Eastoe, B. (1954). Organic constituents of mammalian compact bone, *Biochem. J.*, **57**, 453–9.

Eayrs, J. T. (1955). Cerebral cortex of normal and hypothyroid rats, *Acta anat.*, **25**, 160–83.

Ebbesson, S. O. E. (1968). Quantitative studies of superior cervical sympathetic ganglia in a variety of primates including man. II. Neuronal packing density, *J. Morph.*, **124**, 181–6.

Eccles, J. C. (1957). *The Physiology of the Nerve Cell*, Johns Hopkins Press: Baltimore.

Eccles, J. C. (1964). *The Physiology of Synapses*, Springer-Verlag: Berlin.

Eccles, J. C. (1964). The excitatory responses of spinal neurons, *Prog. Brain Res.*, **12**, 65–91.

Eccles, J. C. (1965). Functional meaning of the patterns of synaptic connections in the cerebellum. *Perspect. Biol. Med.*, **8**, 289–310.

Eccles, J. C. (1970). Neurogenesis and morphogenesis in the cerebellar cortex, *Proc. natn. Acad. Sci. U.S.A.*, **66**, 294–301.

Eccles, J. C. (1973). *The Understanding of the Brain*, McGraw-Hill: New York, London.

Eccles, J. C., Fatt, P., Landgren, S. and Winsbury, G. J. (1954). Spinal cord potentials generated by volleys in large muscle afferents, *J. Physiol., Lond.*, **125**, 590–606.

Eccles, J. C., Eccles, R. M., Iggo, I. and Lundberg, A. (1960). Electrophysiological studies on gamma motoneurons, *Acta physiol. scand.*, **50**, 32–40.

Eccles, J. C. and Schadé, J. P. (1964). Organisation of the spinal cord (editors), *Prog. Brain Res.*, **11**.

Eccles, J. C. and Schadé, J. P. (1964). Physiology of spinal neurons (editors), *Prog. Brain Res.*, **12**.

Eccles, J. C., Llinas, R. and Sasaki, K. (1964). Excitation of cerebellar Purkinje cells by the climbing fibres, *Nature, Lond.*, **203**, 245–6.

Eccles, J. C., Llinas, R. and Sasaki, K. (1966). The inhibitory interneurone within the cerebellar cortex, *Expl Brain Res.*, **1**, 1–16.

Eccles, J. C., Llinas, R. and Sasaki, K. (1966). Parallel fibre stimulation and the responses induced thereby in the Purkinje cells of the cerebellum, *Expl Brain Res.*, **1**, 17–39.

Eccles, J. C., Llinas, R. and Sasaki, K. (1966). The excitatory synaptic action of climbing fibres on the Purkinje cells of the cerebellum, *J. Physiol., Lond.*, **182**, 268–96.

Eccles, J. C., Llinas, R. and Sasaki, K. (1966). Interaction experiments on the responses evoked in Purkinje cells by climbing fibres, *J. Physiol., Lond.*, **182**, 297–315.

Eccles, J. C., Llinas, R. and Sasaki, K. (1966). The action of antidromic impulses on the cerebellar Purkinje cells, *J. Physiol., Lond.*, **182**, 316–45.

Eccles, J. C., Llinas, R. and Sasaki, K. (1966). Intracellularly recorded responses of the cerebellar Purkinje cells, *Expl Brain Res.*, **1**, 161–83.

Eccles, J. C., Sasaki, K. and Strata, P. (1966). The profiles of physiological events produced by a parallel fibre volley in the cerebellar cortex, *Expl Brain Res.*, **2**, 18–34.

Eccles, J. C., Sasaki, K. and Strata, P. (1967). Interpretation of the potential fields generated in the cerebellar cortex by a mossy fibre volley, *Expl Brain Res.*, **3**, 58–80.

Eccles, J. C., Sasaki, K. and Strata, P. (1967). A comparison of the inhibitory action of Golgi cells and of basket cells, *Expl Brain Res.*, **3**, 81–94.

Eccles, J. C., Ito, M. and Szentágothai, J. (1967). *The Cerebellum as a Neuronal Machine*, Springer-Verlag: Berlin.

Edelman, G. M. (1974). Antibody structure and cellular specificity in the immune response, *Harvey Lectures*, **68**, 149–84.

Edelman, R. and Hartcroft, P. M. (1961). Localisation of renin in juxtaglomerular cells of rabbit and dog through the use of the fluorescent-antibody technique, *Circulation Res.*, **9**, 1069–77.

Edvinson, L., Lindvall, M., Nielson, K. C. and Owman, C. (1973). Are brain vessels innervated also by central (non-sympathetic) adrenergic neurons?, *Brain Res.*, **63**, 496–9.

Edwards, D. A. W. (1946). The blood supply and lymphatic drainage of tendons, *J. Anat.*, **80**, 147–52.

Edwards, J. H. (1958). Congenital malformations of the central nervous system in Scotland, *B. J. prev. soc. Med.*, **12**, 115–30.

Edwards, R. G. (1965). Maturation *in vitro* of mouse, sheep, cow, pig, rhesus monkey and human ovarian oocytes, *Nature, Lond.*, **208**, 349–51.

Edwards, R. G., Bavister, B. D. and Steptoe, P. C. (1969). Early stages of fertilisation *in vitro* of human oocytes matured *in vitro*, *Nature, Lond.*, **221**, 632–5.

Egeberg, J. and Jensen, O. A. (1969). The ultrastructure of the acini of the human lacrimal gland, *Acta ophthal.*, **47**, 400–10.

Ehinger, B. and Falck, B. (1966). Concomitant adrenergic and parasympathetic fibres in the rat iris, *Acta physiol. scand.*, **67**, 201–7.

Ehinger, B., Falck, B. and Laties, A. M. (1969). Adrenergic neurons in teleost retina, *Z. Zellforsch. mikrosk. Anat.*, **97**, 285–97.

Ehrenberg, C. G. (1833). Notwendigkeit einer feineren mechanischen Zerlegung des Gehirns und der Nerven, *Poggendorff's Ann. phys. Chem.*, **22**, 449–65.

Ehrlich, P. (1886). Über die Methylenblaureaction der lebenden Nervensubstanz, *Dt. med. Wschr.*, **12**, 49–52.

Eichner, D. (1957). Über Histologie und Topochemie der sehschicht in der Netzhaut des Menschen, *Z. mikrosk.-anat. Forsch.*, **63**, 82–93.

Einarson, L. (1932). Method for progressive selective staining of Nissl and nuclear substances in nerve cells, *Am. J. Path.*, **8**, 295–308.

Eisenberg, B. R., Kuda, A. M. and Peter, J. B. (1974). Stereological analysis of mammalian skeletal muscle. I. Soleus muscle of the adult guinea pig, *J. Cell Biol.*, **60**, 732–54.

Eisler, P. (1912). *Die Muskeln des Stammes*, Jena.

Elftman, H. (1966). Biomechanics of muscle with particular application to studies of gait, *J. Bone Jt Surg.*, **48A**, 363–77.

Elfvin, L. G. (1965). The fine structure of the cell surface of chromaffin cells in the rat adrenal medulla, *J. Ultrastruct. Res.*, **12**, 263–86.

Elias, H. (1955). Liver morphology, *Biol. Rev.*, **30**, 263–310.

Elias, H. (1967). Recruitment in human bile duct formation, *Acta hepatosplenol. (Stuttg.)*, **14**, 253–60.

Elias, H. and Petty, D. (1952). Gross anatomy of the blood vessels and ducts within the human liver, *Am. J. Anat.*, **90**, 59–111.

Elišková, M. (1973). Blood vessels of the ciliary ganglion in man, *Br. J. Ophthal.*, **57**, 766–72.

Elliot, F. M. and Reid, L. (1965). Some new facts about the pulmonary artery and its branching pattern, *Clin. Radiol*, **16**, 193–9.

Elliott, H. C. (1942). Studies on the motor cells of the spinal cord. I. Distribution in the normal human cord, *Am. J. Anat.*, **70**, 95–117.

Elliott, H. C. (1943). Studies on the motor cells of the spinal cord. II. Distribution in the normal human fetal cord, *Am. J. Anat.*, **72**, 29–38.

Ellis, L. C. and Hargrove, J. L. (1977). Prostaglandins. In: *The Testis* (Johnson, A. D. and Gomes, W. R. eds.), Academic Press: New York, London.

Ellis, R. A. (1965). Fine structure of the myoepithelium of the eccrine sweat glands of man, *J. Cell Biol.*, **27**, 551–64.

Ellison, J. P. and Williams, T. H. (1969). Sympathetic nerve pathways to the human heart and their variations, *Am. J. Anat.*, **124**, 149–62.

El-Najjar, M. Y. and Dawson, G. L. (1977). The effect of artificial cranial deformation on the incidence of Wormian bones in the lambdoidal suture, *Am. J. phys. Anthrop.*, **46**, 155–60.

Elsen, J. and Arey, L. B. (1966). On spirality in the intestinal wall, *Am. J. Anat.*, **118**, 11–20.

Emery, J. (1969). *The Anatomy of the Developing Lung* (editor), Spastics International Medical Publications: New York.

Emmelin, N. (1972). Control of salivary glands. In: *Oral Physiology* (Emmelin, N. and Zotterman, Y. eds.), pp. 1–16, Pergamon: Oxford.

Emmelin, N. and Engström, J. (1960). On the existence of specific secretory sympathetic fibres for the cat's submaxillary gland, *J. Physiol., Lond.*, **153**, 1–8.

Enders, A. C. and Schlafke, S. J. (1965). The fine structure of the blastocyst: some comparative studies. In: *Preimplantation Stages of Pregnancy* (Wolstenholme, G. E. W. and O'Connor, M. eds.), pp. 29–54, Ciba Foundation Symposium, Churchill: London.

Eneroth, C.-M., Hökfelt, T. and Norberg, K.-A. (1969). The role of the parasympathetic and sympathetic innervation for the secretion of human parotid and submandibular glands, *Acta oto-laryngol.*, **68**, 369–75.

Engel, S. (1947). *The Child's Lung*, Arnold: London.

Engel, S. (1962). *Lung Structure*, Thomas: Springfield, Ill.

English, D. T. and Blevins, C. E. (1969). Motor units of laryngeal muscles, *Archs Otolar*, **89**, 778–84.

English, K. B. (1977). The ultrastructure of cutaneous Type I mechanoreceptors (Haarsheiben) in cats following denervation, *J. comp. Neurol.*, **172**, 137–64.

Engström, H. and Wersäll, J. (1958). Myelin sheath structure in nerve fibre demyelinization and branching regions, *Expl Cell Res.*, **14**, 414–25.

Enlow, D. H. and Harris, D. B. (1964). Study of the post-natal growth of the human mandible, *Am. J. Orth.*, **50**, 25–50.

Ennabli, E. and Niveiro, M. (1967). Etude embryonnaire des artères intercostales. Reconstruction par la méthode de Born de deux embryons humains (14 et 17 mm.), *Path. Biol., Paris*, **15**, 92–8.

Enoch, D. M. and Kerr, F. W. L. (1967). Hypothalamic vasopressor and vesicopressor pathways. I. Functional studies, *Archs Neurol. Psychiat., Chicago*, **16**, 290–306.

Enoch, D. M. and Kerr, F. W. L. (1967). Hypothalamic vasopressor and vesicopressor pathways. II. Anatomic study of their course and connections, *Archs Neurol. Psychiat., Chicago*, **16**, 307–20.

Epling, G. E. (1966). Electron microscope observations of pericytes of small blood vessels in the lungs and hearts of normal cattle and swine, *Anat. Rec.*, **155**, 513–30.

Epstein, B. S. (1966). An anatomic, myelographic and cinemyelographic study of the dentate ligaments, *Am. J. Roentg.*, **98**, 704–12.

Eränkö, O. (1967). The practical histochemical demonstration of catecholamines by formaldehyde-induced fluorescence, *Jl R. microsc. Soc.*, **87**, 259–76.

Eränkö, O. (1978). Small intensely fluorescent (SIF) cells and neurotransmission in sympathetic ganglia, *Ann. Rev. Pharmacol. Toxicol.*, **18**, 417–30.

Eränkö, O. and Härkönen, M. (1965). Monoamine-containing cells in the superior cervical ganglion of the rat, *Acta physiol. scand.*, **63**, 511–12.

Ericson, E., Hakanson, R., Larson, B., Owman, Ch. and Sundler, F. (1972). Fluorescence and electron microscopy of amine-staining enterochromaffin-like cells in tracheal epithelium of mouse, *Z. Zellforsch. mikrosk. Anat.*, **124**, 532–45.

Erlanger, J. and Gasser, H. S. (1937). *Electrical Signs of Nervous Activity*, University of Pennsylvania Press: Philadelphia.

Esser, P. H. (1932). Über die Funktion und den Bau des Scrotums, *Z. Zellforsch. mikrosk. Anat.*, **31**, 108–74.

Estable, C., Acosta-Ferreira, W. and Sotelo, J. R. (1957). An electron microscope study of the regenerating nerve fibres, *Z. Zellforsch. mikrosk. Anat.*, **46**, 387–99.

Esterhuizen, A., Spriggs, T. and Lever, J. (1968). Nature of islet cell innervation in the cat pancreas, *Diabetes*, **17**, 33–6.

Etherton, J. E. and Conning, D. M. (1971). Early incorporation of labelled palmitate into mouse lung, *Experientia*, **27**, 554–5.

Etherton, J. E., Conning, D. M. and Corrin, B. (1973). Autoradiographical and morphological evidence for apocrine secretion of dipalmitoyl lecithin in the terminal bronchiole of mouse lung, *Am. J. Anat.*, **138**, 11–35.

Evans, D. H. L. and Murray, J. G. (1954). Histological and functional studies on the fibre compostion of the vagus nerve of the rabbit, *J. Anat.*, **88**, 320–37.

Evans, D. H. L. and Murray, J. G. (1954). Regeneration of non-medullated nerve fibres, *J. Anat.*, **88**, 465–80.

Evans, E. (1949). Congenital heart disease, *J. Florida Med. Assoc.*, **35**, 487–91.

Evans, F. G. (1973). *Mechanical Properties of Bone*, Thomas: Springfield.

Everett, N. B. (1943). Observational and experimental evidences relating to origin and differentiation of definitive germ cells in mice, *J. exp. Zool.*, **92**, 49–91.

Ewing, M. R. (1954). Lingual thyroid, *Proc. R. Soc. Med.*, **47**, 510–12.

Eyzaguirre, C., Nishi, K. and Fidone, S. (1972). Chemoreceptor synapses in the carotid body, *Fedn Proc. Fedn Am. Socs exp. Biol.*, **31**, 1385–93.

Falck, B. (1962). Observations on the possibilities of the cellular localisation of monoamines by a fluorescence method, *Acta physiol. Scand., Suppl.* **197**(56), 1–25.

Falck, B., Hillarp, N. A., Thieme, G. and Torp, A. (1962). Fluorescence of catecholamines and related compounds condensed with formaldehyde, *J. Histochem. Cytochem.*, **10**, 348–54.

Falck, B. and Owman, C. (1965). A detailed methodological description of the fluorescence method for the cellular demonstration of biogenic monoamines, *Acta Univ. lund.*, Sect. II, **7**, 1–23.

Falconer, M. A. (1949). Intramedullary trigeminal tractotomy and its place in the treatment of facial pain, *J. Neurol. Neurosurg. Psychiat.*, **12**, 297–311.

Falconer, M. (1967). Brain mechanisms suggested by neurophysiologic studies. In: *Brain Mechanisms Underlying Speech and Language* (Darley, F. L. ed.), pp. 185–90. Grune and Stratton: New York.

Falconer, M. A. and Wilson, J. L. (1958). Visual field changes following anterior temporal lobectomy, their significance in relation to 'Meyer's loop' of the optic radiation, *Brain*, **81**, 1–14.

Farbman, A. I. and Gesteland, R. C. (1975). Developmental and electrophysiological studies of olfactory mucosa in organ culture. In: *Olfaction and Taste V* (Denton, D. and Coghlan, J. P. eds.), pp. 107–10, Academic Press: New York, London.

Farnarier, G., Planche, D. and Rohner, J. J. (1977). Blocage des afférences nociceptives par stimulation périphérique percutanee chez le chat, *C.R. Soc. Biol.*, **171**, 1054–8.

Farquhar, M. G., Wissig, S. L. and Palade, G. E. (1961). Glomerular permeability. I. Ferritin transfer across the normal glomerular capillary wall, *J. exp. Med.*, **113**, 47–66.

Farrer-Brown, G., Beilby, J. O. W. and Tarbit, M. M. (1970). The blood supply of the uterus. I. Arterial vasculature, *J. Obstet. Gynaec. Br. Commonw.*, **77**, 673–81.

Farrer-Brown, G., Beilby, J. O. W. and Tarbit, M. M. (1970). The blood supply of the uterus. II. Venous pattern, *J. Obstet. Gynaec. Br. Commonw.*, **77**, 682–9.

Faucett, B. (1895). The structure of the inferior maxilla with special reference to the position of the inferior dental canal, *J. Anat. Lond.*, **29**, 355.

Fawcett, D. W. (1961). Cilia and flagella. In: *The Cell*, Volume II (Brachet, J. and Mirsky, A. E. eds.), pp. 217–98. Academic Press: New York, London.

Fawcett, D. W. (1961). Intercellular bridges, *Expl Cell Res.*, Suppl. 8, 174–87.

Fawcett, D. W. (1965). The anatomy of the mammalian spermatozoon with particular reference to the guinea pig, *Z. Zellforsch. mikrosk. Anat.*, **67**, 279–96.

Fawcett, D. W. (1966). *The Cell, An Atlas Of Fine Structure*, Saunders: Philadelphia, London.

Fawcett, D. W. (1968). The topographical relationship between the plane of the central pair of flagellar fibrils and the transverse axis of the head in guinea-pig spermatozoa, *J. Cell Sci.*, **3**, 187–98.

Fawcett, D. W. (1975). Ultrastructure and function of the Sertoli cell. In: *Handbook of Physiology*, Section 7, *Endocrinology*, V, pp. 21–55, American Physiological Society: Washington, DC.

Fawcett, D. W. and Burgos, M. H. (1956). Observations on the cytomorphosis of the germinal and interstitial cells of the human testis. In: *Colloquia on Ageing*, Volume 2, *Ageing in Transient Tissues* (Wolstenholme, G. E. W. and Millar, E. C. P. eds.), pp. 86–96. Ciba Foundation Symposium. Churchill: London.

Fawcett, D. W. and Burgos, M. H. (1956). A comparison of the structural organisation of mammalian and amphibian sperm tails, *Anat. Rec.*, **124**, 289P.

Fawcett, D. W., Ito, S. and Slautterback, D. (1959). The occurrence of intercellular bridges in groups of cells exhibiting synchronous differentiation, *J. biophys. biochem., Cytol.*, **5**, 453–60.

Fawcett, D. W. and Ito, S. (1965). The fine structure of bat spermatozoa, *Am. J. Anat.*, **116**, 567–609.

Fawcett, D. W. and McNutt, N. S. (1969). The ultrastructure of the cat myocardium. I. Ventricular papillary muscle, *J. Cell Biol.*, **42**, 1–45.

Fawcett, E. (1905). On the early stages in the ossification of the pterygoid plates of the sphenoid bone of man, *Anat. Anz.*, **26**, 280–6.

Fawcett, E. (1907). On the completion of ossification of the human sacrum, *Anat. Anz.*, **30**, 414–21.

Fawcett, E. (1911). Some notes on the epiphyses of the ribs, *J. Anat.*, **45**, 172–8.

Fawcett, E. (1911). The development of the human maxilla, vomer and paraseptal cartilage, *J. Anat.*, **45**, 378–406.

Fawcett, E. (1917). The primordial cranium of *Microtus amphibius* (water

BIBLIOGRAPHY

rat) as determined by sections and a model of the 25 mm stage, with comparative remarks, *J. Anat.*, **51**, 309–59.

Fawcett, E. (1918). The primordial cranium of *Erinaceus europaeus*, *J. Anat.*, **52**, 211–50.

Fawcett, E., Brash, J. C., Northcroft, G. and Keith, A. (1924). *The Growth of the Jaws*, Dental Board of the United Kingdom: London.

Fawcett, E. and Blachford, J. V. (1933). The circle of Willis: an examination of 700 specimens, *J. Anat.*, **40**, 65–9.

Fearnhead, R. W. (1967). The innervation of the dental tissues. In: *Structural and Chemical Organisation of Teeth* (Miles, A. E. W. ed.), Volume 2, pp. 247–81, Academic Press: New York, London.

Fearnhead, R. W. and Stack, M. V. (1965). *Tooth Enamel*, I (editors), Wright: Bristol.

Fearnhead, R. W. and Stack, M. V. (1971). *Tooth Enamel*, II, Wright: Bristol.

Fedorko, M. E. and Hirsch, J. G. (1970). Structure of monocytes and macrophages, *Seminars Haemat.*, **7**, 109–24.

Feldman, M. and Globerson, A. (1974). Reception of immunogenic signals by lymphocytes, *Curr. Top. Dev. Biol.*, **8**, 1–40.

Fell, H. B. (1939). The origin and developmental mechanics of the avian sternum, *Phil. Trans. R. Soc.*, Ser. B, **229**, 407–63.

Fell, H. B. and Canti, R. G. (1934). Observations on the early development of the knee joint *in vivo* and *in vitro*, *Proc. R. Soc. B.*, **116**, 316–51.

Feltz, P. and Mackenzie, J. S. (1969). Properties of caudate unitary responses to repetitive nigral stimuli, *Brain Res.*, **13**, 612–16.

Ferguson, M. W. (1977). The mechanism of palatal shelf elevation and the pathogenesis of cleft palate, *Virchow's Arch. Pathol. Anat.*, **375**, 97–113.

Fergusson, J. D. and Gibson, E. C. (1956). Prostatic smear diagnosis, *Br. med. J.*, **1**, 822–5.

Ferraro, A. and Barrera, S. E. (1935). Posterior column fibers and their termination in *Macacus rhesus*, *J. comp. Neurol.*, **62**, 507–30.

Ferraro, A. and Barrera, S. E. (1935). The nuclei of the posterior funiculi in *Macacus rhesus*. An anatomic and experimental investigation, *Archs Neurol. Psychiat.*, *Chicago*, **33**, 262–75.

Ferraro, A. and Barrera, S. E. (1936). Lamination of the medial lemniscus in *Macacus rhesus*, *J. comp. Neurol.*, **64**, 313–24.

Ferraro, J. A. and Minckler, J. (1977). The brachium of the inferior colliculus. The human auditory pathways: a quantitative study, *Brain and Language*, **4**, 156–64.

Ferraro, J. A. and Minckler, J. (1977). The human lateral lemniscus and its nuclei, *Brain and Language*, **4**, 277–94.

Ferrier, D. (1874). The localisation of function in the brain, *Proc. R. Soc. B*, **22**, 229–32.

Feyler, K. P. (1965). Quantitative Untersuchungen über die vegetativen Ganglien im Paries membranaceus tracheae des Menschen, *Anat. Anz.*, **117**, 371–9.

Feyrter, F. (1938). Über diffuse endokrine epitheliale organe, *Zentbl. Innere Med.*, **545**, 31–41.

Field, E. J. (1951). The development of the conducting system in the heart of sheep, *Br. Heart J.*, **13**, 129–47.

Field, W. S., Bruetman, M. E. and Weibel, J. (1965). *Collateral Circulation of the Brain*, Williams and Wilkins: Baltimore.

Findlay, J. A. and Lever, J. D. (1971). A preliminary study of the effects of the diabetogenic agent Streptozotocin on the islets of Langerhans in the genetically obese mouse (Bar Harbor strain), *J. Anat.*, **108**, 583P.

Findlay, J. A. and Ashcroft, J. J. H. (1975). Cells of the islets of Langerhans. In: *The Cell in Medical Science* (Beck, F. and Lloyd, J. B. eds.), Volume 3, pp. 243–306, Academic Press: New York, London.

Fine, B. S. and Tousimis, A. J. (1961). The structure of the vitreous body and the suspensory ligaments of the lens, *Archs Ophthal.*, N.Y., **65**, 95–110.

Fine, B. S. and Zimmerman, L. E. (1962). Miller's cells and the 'middle limiting membrane' of the human retina. An electron microscopic study, *Invest. Ophthal.*, **1**, 304–26.

Fine, H. and Keen, E. N. (1966). The arteries of the human kidney, *J. Anat.*, **100**, 881–94.

Finnegan, M. (1972). *Population definition on the North West coast by analysis of discrete character variation*, Ph. D. Dissertation, University of Colorado, Boulder.

Finnegan, M. (1978). Non-metric variation of the infracranial skeleton, *J. Anat.*, **125**, 23–37.

Finnegan, M. and Faust, M. A. (1974). *Bibliography of Human and Non-human Non-metric Variation*. Research report, no. 14, Dept Anthropology, University of Massachusetts.

Fischer, A. (1977). Autonomy for a specific gene product in oocytes: experimental evidence in the polychaetous annelid Platynereis dumerilii, *Dev. Biol.*, **55**, 46–58.

Fischer, J. (1956). *The Labyrinth: Physiology and Functional Tests*, Grune and Stratton: New York.

Fischer, L., Machenaud, A. and Morin, A. (1974). Contribution à l'étude de la vascularisation du cubitus, *Arch. Anat. Path.*, **22**, 261–5.

Fisher, E. R. and Jeffrey, W. (1965). Ultrastructure of human normal and neoplastic prostate; with comments related to prostatic effects of hormonal stimulation in the rabbit, *Am. J. clin. Path.*, **44**, 119–34.

Fisher, S. K. and Linberg, K. A. (1975). Intercellular junctions in the early human embryonic retina, *J. Ultrastruct. Res.*, **5**, 69–78.

Fishman, A. P. and Hecht, H. H. (1969). *The Pulmonary Circulation and Interstitial Space* (editors), University of Chicago Press: Chicago.

Fishman, A. P. and Pietra, G. G. (1974). Handling of bioactive materials by the lung, *New Engl. J. Med.*, **291**, 884–9.

Fitzgerald, M. J. T. (1956). The occurrence of a middle superior alveolar nerve in man, *J. Anat.*, **90**, 520–2.

Fitzgerald, M. J. T. and Scott, J. H. (1958). Observations on the anatomy of the superior dental nerves, *Br. dent. J.*, **104**, 205–8.

Fitzgerald, M. J. T. and Sachithanadan, S. R. (1979). The structure and source of lingual proprioceptors in the monkey, *J. Anat.*, **128**, 523–52.

Fitzpatrick, T. B. and Quevedo, W. C. Jr, (1971). Biological processes underlying melanin pigmentation and pigmentary disorders. In: *Modern Trends in Dermatology* (Borrie, P. ed.), Volume 4, pp. 122–49, Butterworth: London.

Flechsig, P. (1876). *Die Leitungsbahnen im Gehirn und Rückenmark des Menschen auf Grund entwicklungsgeschichtlicher Untersuchungen*, Engelmann: Leipzig.

Flecker, H. (1929). Röntgenographic study of movements of abduction at normal shoulder joint, *Med. J. Aust.*, **2**, 123–4.

Fleischner, F. G. and Sayegh, V. (1958). Assessment of the size of the liver: roentgenologic considerations, *New Engl. J. Med.*, **259**, 271–4.

Flickinger, C. and Fawcett, D. W. (1967). The junctional specialisations of Sertoli cells in the seminiferous epithelium, *Anat. Rec.*, **158**, 207–22.

Flood, S. and Jansen, J. (1966). The efferent fibres of the cerebellar nuclei and their distribution on the cerebellar peduncles in the cat, *Acta anat.*, **63**, 137–66.

Florey, L. (1966). The endothelial cell, *Br. med. J.*, **2**, 487–90.

Florian, J. (1930). The formation of the connecting stalk and the extension of the amniotic cavity towards the tissue of the connecting stalk in young human embryos, *J. Anat.*, **64**, 454–76.

Floyd, W. F. and Silver, P. H. S. (1950). Electromyographic study of patterns of activity of the anterior abdominal wall muscles in man, *J. Anat.*, **84**, 132–45.

Floyd, W. F. and Silver, P. H. S. (1955). The function of the erectores spinae muscles in certain movements and postures in man. *J. Physiol.*, *Lond.*, **129**, 184–203.

Flyger, G. and Hjelmquist, U. (1957). Normal variations in the caliber of the human cerebral aqueduct, *Anat. Rec.*, **127**, 151–62.

Foerster, O. (1929). Beiträge zur Pathophysiologïe der Sehbahn und der Sehsphäre, *J. Psychol. Neurol.*, *Lpz.*, **39**, 463–85.

Foerster, O. (1933). The dermatomes in man, *Brain*, **56**, 1–39.

Foerster, O. (1936). Motorische Felder und Bahnen. In: *Handbuch der Neurologie* (Bumke, O. and Foerster, O. eds.), Volume 6, pp. 1–357, Springer-Verlag: Berlin.

Foley, J. M. and Baxter, D. (1958). On the nature of pigment granules in the cells of the locus coerulus and substantia nigra, *J. Neuropath. exp. Neurol.*, **17**, 586–98.

Foley, J. O. and DuBois, F. S. (1937). Quantitative studies of the vagus nerve in the cat, *J. comp. Neurol.*, **67**, 49–67.

Folke, L. E. A. and Stallard, R. E. (1967). Periodontal microcirculation as revealed by plastic microspheres, *J. periodont. Res.*, **2**, 53–63.

Foltz, E. L. and White, L. E. Jr. (1962). Pain 'relief' by frontal cingulumotomy, *J. Neurosurg.* **19**, 89–100.

Forbes, G. (1938). Lymphatic drainage of the skin, with a note on lymphatic watershed areas, *J. Anat.*, **72**, 399–410.

Ford, E. B. (1974). *Genetics for Medical Students*, Seventh edition, Methuen: London.

Ford, E. H. R. (1956). The growth of the foetal skull, *J. Anat.*, **90**, 63–72.

Forrest, W. J. (1967). Motor innervation of human thenar and hypothenar muscles in 25 hands: a study combining E. M. G. and percutaneous nerve stimulation, *Canad. J. Surg.*, **10**, 196–9.

Forsberg, J. G. (1963). *The Derivation and Differentiation of the Vaginal Epithelium*, Dissertation, Lund.

Forssner, H. (1928). Über den Deszensus der Geschlechtsdrüsen beim Menschen. *Acta obstet. gyn. Scand.*, **7**, 379–406.

Fourman, J. and Moffat, D. B. (1971). *The Blood Vessels of the Kidney*, Blackwell: Oxford.

Fountain, F. P., Minear, W. L. and Allison, R. D. (1966). Function of longus colli and longissimus cervicis muscles in man, *Archs phys. Med.*, **47**, 665–9.

Fowler, R. Jr. (1957). Primary peritonitis, *Aust. N.Z. J. Surg.*, **26**, 204–13.

Fox, C. A., Ubeda-Purkiss, M., Ihrig, H. K. and Biagioli, D. (1951). Zinc-chromate modification of Golgi technic, *Stain Technol.*, **26**, 109–14.

Fox, C. A. and Barnard, J. W. (1957). A quantitative study of the Purkinje cell dendritic branchlets and their relationship to afferent fibres, *J. Anat.*, **91**, 299–313.

Fox, C. A., Hillman, D. E., Siegesmund, K. A. and Sether, L. A. (1966). The primate globus pallidus and its feline and avian homologues: a Golgi and electron microscopic study. In: *Evolution of the Forebrain* (Hassler, R. and Stephan, H. eds.), pp. 237–48, Thieme: Stuttgart.

Fox, C. A. and Snider, R. S. (1967). *The Cerebellum* (editors), *Prog. Brain Res.*, **25**.

Fox, C.A., Hillman, D.E., Siegesmund, K. A. and Dutta, C. R. (1967). The primate cerebellar cortex: a Golgi and electron microscopic study, *Prog. Brain Res.*, **25**, 174–225.

Foxon, G. E. H. (1955). Problems of the double circulation in the vertebrates, *Biol Rev.*, **30**, 196–226.

Foxon, G. E. H. and Bannister, L. H. (1974). Circulation and circulatory systems, *Encyclopaedia Britannica*, fifteenth edition.

Frable, M. A. (1961). Computation of motion at the crico-arytenoid joint, *Archs Otolar.*, **73**, 551–6.

Fraenkel, L. and Papanicolaou, G. N. (1938). Growth, desquamation and involution of the vaginal epithelium of fetuses and children with a consideration of the related hormonal factors, *Am. J. Anat.*, **62**, 427–52.

Fraley, E. E. and Weiss, L. (1961). An electron microscopic study of the lymphatic vessels in the penile skin of the rat, *Am. J. Anat.*, **109**, 85–102.

François, J. and Neetens, A. (1969). Physioanatomy of the axial vascularisation of the optic nerve, *Documenta ophth.*, **26**, 38–49.

François, R. J. and Dhem, A. (1974). Microradiographic study of the normal human vertebral body, *Acta anat.*, **89**, 251–65.

Frank, J. S. and Langer, G. A. (1974). The myocardial interstitium: its structure and its role in ionic exchange, *J. Cell Biol.*, **60**, 586–601.

Frankfurter, A., Weber, J. T., Royce, G. J., Strominger, N. L. and Harting, J. K. (1976). An autoradiographic analysis of the tecto-olivary projection in primates, *Brain Res.*, **118**, 245–57.

Frankl, O. (1933). On the physiology and pathology of the isthmus uteri. *J. Obstet. Gynaec. Br. Commonw.*, **40**, 397–422.

Franklin, K. J. (1937). *A Monograph on Veins*, Williams and Wilkins: Baltimore.

Franklin, K. J. (1939). Radiographic demonstration of circulation through the heart in the adult and in the foetus, and identity of the ductus arteriosus, *Br. J. Radiol.*, **12**, 505–17.

Franks, L. M. (1953). Spread of prostatic cancer to bone, *J. Path. Bact.*, **66**, 91–3.

Franks, L. M. (1954). Benign nodular hyperplasia of prostate: review, *Ann R. Coll. Surg.*, **14**, 92–106.

Franzini-Armstrong, C. (1973). The structure of a simple Z-line, *J. Cell Biol.*, **58**, 630–42.

Fraser, F. C. (1959). Causes of congenital malformations in human beings, *J. chron. Dis.*, **10**, 97–110.

Fraser, G. and Mayo, O. (1975). *Textbook of Human Genetics*, Blackwell: Oxford.

Fraser, H. M., Gunn, A. and Jeffcoate, S. L. (1975). Effect of active immunisation to luteinizing hormone releasing hormone on serum and pituitary gonadotrophins, testes and accessory sex organs in the male rat, *J. Endocrinol.*, **63**, 399–406.

Fraser, J. S. and Milne Dickie, J. K. (1914). A reconstruction model of the right middle and inner ear, *J. Anat.*, **49**, 119–35.

Fraser, L. R., Zanellotti, H. M., Paton, G. R. and Drury, L. M. (1976). Increased incidence of triploidy in embryos derived from mouse eggs fertilised *in vitro*, *Nature, Lond.*, **260**, 39–40.

Fraser Roberts, J. A. (1970). *An Introduction to Medical Genetics*, Fifth edition, Oxford University Press: London.

Frazer, J. E. (1914). The second visceral arch and groove in the tubo-tympanic region, *J. Anat.*, **48**, 391–408.

Frazer, J. E. (1926). The disappearance of the pre-cervical sinus, *J. Anat.*, **61**, 132–43.

Frazer, J. E. (1928). Development of the region of the isthmus rhombencephali, *J. Anat.*, **63**, 7–18.

Frazer, J. E. and Robbins, R. H. (1915). On the factors concerned in causing rotation of the intestine in man, *J. Anat.*, **50**, 75–110.

Fredrickson, J. M., Figge, U., Scheid, P. and Kornhuber, H. H. (1966). Vestibular nerve projection to the cerebral cortex of the rhesus monkey, *Expl Brain Res.*, **2**, 318–27.

Freeman, M. A. R. (1973). *Adult Articular Cartilage* (editor), Pitman Medical: London.

Freeman, M. A R. and Wyke, B. (1967). The innervation of the knee joint, *J. Anat.*, **101**, 505–32.

Freitag, P. and Engel, M. B. (1970). Autonomic innervation in rabbit salivary glands, *Anat. Rec.*, **167**, 87–106.

Friend, D. S. and Gilula, N. B. (1972). Variations in tight and gap junctions in mammalian tissues, *J. Cell Biol.*, **53**, 758–76.

Frisch, D. (1967). Ultrastructure of mouse olfactory mucosa, *Am. J. Anat.*, **121**, 87–120.

Fritsch, G. and Hitzig, E. (1870). Ueber die elektrische Erregbarkeit des Grosshirns, *Arch. Anat. Physiol.*, **37**, 300–32.

Frutiger, P. (1969). Zur Frühentwicklung der Ductus paramesonephrici und des Müllerrerschen Hugels beim menschen, *Acta anat.*, **72**, 233–45.

Fry, G. N., Devine, C. E. and Burnstock, G. (1977). Freeze-fracture studies of nexuses between smooth muscle cells. Close relationship to sarcoplasmic reticulum, *J. Cell Biol.*, **72**, 26–34.

Fujita, H. and Fujita, S. (1963). Electron microscopic studies on neuroblast differentiation in the central nervous system of domestic fowl, *Z. Zellforsch. mikrosk. Anat.*, **60**, 463–78.

Fujita, S. (1963). The matrix cell and cytogenesis in the developing central nervous system, *J. comp. Neurol.*, **120**, 37–42.

Fujita, S. (1967). Quantitative analysis of cell proliferation and differentiation in the cortex of the postnatal mouse cerebellum, *J. Cell Biol.*, **32**, 277–87.

Fujita, S., Shimada, M. and Nakamura, T. (1966). ³H-thymidine autoradiographic studies on the cell proliferation and differentiation in the external and internal granular layers of the mouse cerebellum, *J. comp. Neurol.*, **128**, 191–208.

Fujita, T. (1959). Histological studies on the neuro-insular complex in the pancreas of some mammals, *Z. Zellforsch.*, **50**, 94–109.

Fujita, T. (1973). *Gastro-Entero-Pancreatic System. A Cell-Biological Approach* (editor), Igaku Shoiu: Tokyo.

Fujita, T. (1976). The gastro-enteric endocrine cell and its paraneuronic nature. In: *Chromaffin, Enterochromaffin and Related Cells* (Coupland, R. E. and Fujita, T. eds.), pp. 191–208, Elsevier: New York, Amsterdam.

Fujita, T., Miyoshi, M. and Tokunaga, J. (1970). Scanning and transmission electron microscopy of human ejaculate spermatozoa with special reference to their abnormal forms, *Z. Zellforsch. mikrosk. Anat.*, **105**, 483–97.

Fukuda, T. (1976). Ultrastructure of primordial germ cells in human embryo, *Virchows Arch. B. cell Path.*, **20**, 85–9.

Fukuda, T. and Hedinger, C. (1975). Ultrastructure of developing germ cells in the fetal human testis, *Cell Tissue Res.*, **161**, 55–70.

Fuld, H. and Irwin, D. T. (1954). Clinical application of portal venography, *Br. med. J.*, **1**, 312–13.

Fulton, C. (1971). Centrioles. In: *Origin and Continuity of Cell Organelles* (Reinert, J. and Ursprung, H. eds.), pp. 170–221, Springer: Berlin.

Fulton, J. F. and Sheenan, D. (1935). The uncrossed lateral pyramidal tract in higher primates, *J. Anat.*, **69**, 181–7.

Fuller, A. P., Fozzard, J. A. and Wright, G. H. (1959). Sphincteric action of crico-pharyngeus: radiographic demonstration, *Br. J. Radiol.*, **32**, 32–5.

Fürbinger, M. (1873). Zur vergleichenden Anatomie der Schulter-muskeln, *Jena. Z. Naturw.*, **7**, 237–320.

Furseth, R. (1968). The resorption process of human deciduous teeth studied by light microscopy, microradiography and electron microscopy, *Archs oral. Biol.*, **13**, 417–31.

Furseth, R. (1969). The fine structure of the cellular cement of young human teeth, *Archs oral Biol.*, **14**, 1147–58.

Furschpan, E. J. and Potter, D. D. (1968). Low resistance junctions between cells in embryos and tissue culture. In: *Current Topics in Developmental Biology* (Moscona, A. and Monroy, A. eds.), Volume 3, pp. 95–128, Academic Press: New York, London.

Fuṣijawa, H., Morioka, H., Watanabe, K. and Nakamura, H. (1976). A decay of gap junctions in association with cell differentiation of neural retina in chick embryonic development, *J. Cell Sci.*, **22**, 585–96.

Fuxe, K., Hökfelt, T., Jonsson, G. and Ungerstedt, U. (1970). Fluorescence microscopy in neuroanatomy. In: *Contemporary Research Methods in Neuroanatomy* (Nauta, W. J. H. and Ebbesson, S. O. E. eds.), pp. 275–314. Springer-Verlag: Berlin.

G

Gabe, M. (1966). *Neurosecretion*, Pergamon Press: Oxford, New York.

Gabella, G. (1973). Cellular structures and electrophysiological behaviour. Fine structure of smooth muscle, *Phil. Trans. R. Soc. Lond.*, **265B**, 7–16.

Gabella, G. (1976). *Structure of the Autonomic Nervous System*, Chapman & Hall: London.

Gabriel, A. C. (1958). Some anatomical features of the mandible. *J. Anat.*, **92**, 580–6.

Gabriel, W. B. (1945). *The Principles and Practice of Rectal Surgery*, Lewis: London.

BIBLIOGRAPHY

Gacek, R. R. (1974). Transection of the post-ampullary nerve for the relief of benign paroxysmal postural vertigo, *Ann. Otol. Rhinol. Laryngol.*, **83**, 596–605.

Gaddum, P. (1968). Sperm maturation in the male reproductive tract: Development of motility, *Anat. Rec.*, **161**, 471–82.

Gadow, H. F. (1933). *The Evolution of the Vertebral Column* (Gaskell, J. F. and Green, H. L. H. H. eds.), Cambridge University Press: Cambridge.

Gaillard, P. J., Talmage, R. V. and Budy, A. M. (1965). *The Parathyroid Glands: Ultrastructure, Secretion and Function* (editors), University of Chicago Press: Chicago.

Gamble, H. J. and Eames, R. A. J. (1964). An electron microscope study of the connective tissue of human peripheral nerve, *J. Anat.*, **98**, 655–63.

Gamble, H. J., Fenton, J. and Allsopp, G. (1978). Electron microscope observations on the changing relationships between unmyelinated axons and Schwann cells in human fetal nerves, *J. Anat.*, **127**, 363–78.

Ganguly, D. N. and Roy, K. K. (1964). A study on the cranio-vertebral joint in the man, *Anat. Anz.*, **114**, 433–52.

Gardner, D. L. (1972). The influence of microscopic technology on knowledge of cartilage surface structure, *Ann. Rheum. Dis.*, **31**, 235–58.

Gardner, D. L. and Woodward, D. (1969). Scanning electron microscopy and replica studies of articular surfaces of guinea pig synovial joints, *Ann. Rheum. Dis.*, **28**, 379–91.

Gardner, E. D. (1948). The innervation of the knee joint, *Anat. Rec.*, **101**, 109–30.

Gardner, E. D. (1948). The innervation of the shoulder joint, *Anat. Rec.*, **102**, 1–18.

Gardner, E. D. (1950). Physiology of movable joints, *Physiol. Rev.*, **30**, 127–76.

Gardner, E. (1967). Spinal cord and brain stem pathways for afferents from joints. In: *Myotatic, Kinesthetic and Vestibular Mechanisms* (de Reuck, A. V. S. and Knight, J. eds.), pp. 56–76, Ciba Foundation Symposium, Churchill: London.

Gardner, E. and Gray, D. J. (1950). Prenatal development of the human hip joint, *Am. J. Anat.*, **87**, 162–212.

Gardner, E. and Gray, D. J. (1953). Prenatal development of the human shoulder and acromioclavicular joints, *Am. J. Anat.*, **92**, 219–76.

Gardner, E. and O'Rahilly, R. (1968). The early development of the knee joint in staged human embryos, *J. Anat.*, **102**, 289–99.

Gardner, E. and O'Rahilly, R. (1976). The nerve supply and conducting system of the human heart at the end of the embryonic period proper, *J. Anat.*, **121**, 571–87.

Gardner, E. and Lenn, N. J. (1977). Fibres in monkey posterior articular nerves, *Anat. Rec.*, **187**, 99–106.

Garey, L. J. and Powell, T. P. S. (1967). The projection of the lateral geniculate nucleus upon the cortex in the cat, *Proc. R. Soc. B.*, **169**, 107–26.

Garey, L. J. and Powell, T. P. S. (1968). The projection of the retina in the cat, *J. Anat.*, **102**, 189–222.

Garey, L. J., Fisken, R. A. and Powell, T. P. S. (1976). Cellular changes in the lateral geniculate nucleus of the cat and monkey after section of the optic tract, *J. Anat.*, **121**, 15–27.

Garfia, A. (1980). Glomus tissue in the vicinity of the human carotid sinus, *J. Anat.*, **130**, 1–12.

Garn, S. M. and Rohmann, C. G. (1960). Variability in the order of ossification of the bony centers of the hand and wrist, *Am. J. phys. Anthrop.*, N.S. **18**, 219–30.

Garn, S. M. and Lewis, A. B. (1962). The relationship between third molar agenesis and reduction in tooth number, *Angle orthod.*, **33**, 14–18.

Garrett, F. D. (1948). Development of the cervical vesicles in man, *Anat. Rec.*, **100**, 101–14.

Garrett, J. R. (1966). The innervation of salivary glands. I. Cholinesterase-positive nerves in normal glands of the cat, *Jl R. microsc. Soc.*, **85**, 135–48.

Garrett, J. R. (1966). The innervation of salivary glands. II. The ultrastructure of nerves in normal glands of the cat, *Jl R. microsc. Soc.*, **85**, 149–62.

Garrett, J. R. (1967). The innervation of normal human submandibular and parotid salivary glands demonstrated by cholinesterase histochemistry, catecholamine fluorescence and electron microscopy. *Archs oral Biol.*, **12**, 1417–36.

Garrett, J. R. (1972). Neuro-effector sites in salivary glands. In: *Oral Physiology* (Emmelin, N. and Zotterman, Y. eds.), pp. 83–97, Pergamon: Oxford.

Garrett, J. R. (1975). Recent advances in physiology of salivary glands, *Br. med. Bull.*, **31**, 152–5.

Garrett, J. R. (1976). Structure and innervation of salivary glands. In: *Scientific Foundations of Dentistry* (Cohen, B. and Kramer, J. R. H. eds.), pp. 499–516, Heineman: London.

Garrett, J. R. and Parsons, P. A. (1974). Movement of horseradish peroxidase in submandibular glands of rabbits after arterial injection, *J. Physiol., Lond.*, **237**, 3–4P.

Garrett, J. R. and Kidd, A. (1975). Effects of nerve stimulation and denervation on secretory material in submandibular striated duct cells of cats, and the possible role of these cells in the secretion of salivary kallikrein, *Cell Tissue Res.*, **161**, 71–84.

Garrett, J. R. and Emmelin, N. (1979). Activities of salivary myoepithelial cells: a review, *Med. Biol.*, **57**, 1–28.

Garrod, A. E. (1963). *Inborn Errors of Metabolism*, Oxford University Press: Oxford.

Gasic, G. and Gasic, T. (1962). Removal and regeneration of the cell coating in tumour cells, *Nature, Lond.*, **196**, 170.

Gasser, H. S. (1956). Olfactory nerve fibres, *J. gen. Physiol.*, **39**, 473–98.

Gastaut, H. and Lammers, H. J. (1960). *Anatomie du rhinencéphale. Les grandes activités du rhinencéphale*, Masson: Paris.

Gatsalov, M. D. (1963). Intra-organic venous circulation of the human uterine tube, *Arkh. Anat.*, **44**, 87–92.

Gatsalov, M. D. (1963). Limfaticheskaia sistema slizistoĭ obolochki fallopievoĭ truby cheloveka, *Akush. Ginek.*, **39**, 85–90.

Gauer, O. H. (1968). Osmocontrol versus volume control, *Fedn Proc. Fedn Am. Socs exp. Biol.*, **27**, 1132–6.

Gaughran, G. R. L. (1963). Mylohyoid boutonnière and sublingual bouton, *J. Anat.*, **97**, 565–8.

Gaunt, W. A. (1971). *Microreconstruction*, Pitman Medical: London.

Gaunt, W. A. and Miles, A. E. W. (1967). Fundamental aspects of tooth morphogenesis. In: *Structural and Chemical Organisation of Teeth* (Miles, A. E. W. ed.), Volume I, pp. 151–97, Academic Press: New York, London.

Gaunt, A. W., Osborn, J. W. and Ten Cate, A. R. (1971). *Advanced Dental Histology*, Second edition, Wright: Bristol.

Gauthier, G. F. and Schaeffer, S. F. (1974). Ultrastructural and cytochemical manifestations of protein synthesis in the peripheral sarcoplasm of denervated and newborn skeletal muscle fibres, *J. Cell Sci.*, **143**, 113–37.

Gautvik, K. M., Kriz, M. and Lund-Larsen, K. (1972). Adrenergic vasodilation in the cat submandibular salivary gland. In: *Oral Physiology* (Emmelin, N. and Zotterman, Y. eds.), pp. 161–2, Pergamon: Oxford.

Gaze, R. M. (1970). *The Formation of Nerve Connections*, Academic Press: London, New York.

Gazzaniga, M. S. (1970). *The Bisected Brain*, Appleton-Century-Crofts: New York.

Gazzaniga, M. S. and Sperry, R. W. (1967). Language after section of the cerebral commissures, *Brain*, **90**, 131–48.

Gebhardt, W. (1901). Über funktionell wichtige Anordnungsweisen der grösseren und feineren Bauelemente des Wirbeltierknochens, *Arch. EntwMech. Org.*, **11**, 383–498; **20**, 187–334.

Gehr, P. Bachofen, M. and Weibel, E. R. (1978). The normal human lung: ultrastructure and morphometric estimation of diffusion capacity, *Resp. Physiol.*, **32**, 121–40.

Gelfan, S. and Rapisarda, A. F. (1964). Synaptic density on spinal neurons of normal dogs and dogs with experimental hind-limb rigidity, *J. comp. Neurol.*, **123**, 73–96.

Gelfan, S., Kao, G. and Ruchkin, D. S. (1970). The dendritic tree of spinal neurons, *J. comp. Neurol.*, **139**, 355–412.

Geller, M. and Barbato, D. (1970). Nervus peronaeus profundus. A study of the terminal branches and their variations, *Hospital, Rio de J.*, **77**, 679–98.

Gemzell, C. A. and Roos, P. (1966). Pregnancies following treatment with human gonadotrophin, with special references to the problem of multiple births, *Am. J. Obstet. Gynec.*, **94**, 490–6.

Geniec, P. and Morest, D. K. (1971). The neuronal architecture of the human posterior colliculus—a study with the Golgi method. *Acta otolaryng., Stockh., Suppl.* **295**.

Génis-Gálvez, J. M. (1957). Innervation of the ciliary muscle, *Anat. Rec.*, **127**, 219–30.

Génis-Gálvez, J. M., Santos Gutierrez, L. and Martin Lopez, M. (1966). On the double innervation of the parotid gland. An experimental study, *Acta anat.*, **63**, 398–403.

Genovese, S. T. (1959). *Diferencias sexuales en el Huesco Coxal*, Universidad Nacional Autonoma de Mexico, No. 49, Mexico.

Genovese, S. T. and Messmacher, M. (1959). Valor de los patrones tradicionales para la determinacion de la edad par medio de las suturas en craneos Mexicanos (Indigenas y Mestizos), *Cuad. Inst. Hist., Méx.*, No. 7.

Geren, B. B. (1954). The formation from the Schwann cell surface of myelin in the peripheral nerves of chick embryos, *Expl Cell Res.*, **7**, 558–62.

Gerich, J. E. and Lorenzi, M. (1978). The role of the autonomic nervous system and somatostatin in the control of insulin and glucagon secretion. In: *Frontiers in Neuroendocrinology* (Ganong, W. F. and Martini, L. eds.), Volume 5, Raven Press: New York.

Gernandt, B. E., Iranyi, M. and Livingston, R. B. (1959). Vestibular influences on spinal mechanisms, *Expl Neurol.*, **1**, 248–73.

Geschwind, N. and Levitsky, W. (1968). Human brain: left-right asymmetries in temporal speech region, *Science, N.Y.*, **161**, 186–7.

Gesteland, R. C., Lettvin, J. Y., Pitts, W. H. and Rojas, A. (1963). In: *Olfaction and Taste* (Zotterman, Y. ed.), pp. 19–34, Pergamon Press: Oxford.

Getz, B. and Sirnes, T. (1949). The localisation within the dorsal motor vagal nucleus. An experimental investigation, *J. comp. Neurol.*, **90**, 95–110.

Ghadially, F. N. and Roy, S. (1969). *Ultrastructure of Synovial Joints in Health and Disease*, Butterworth: London.

Ghadially, F. N., Ailsby, R. L. and Oryschak, A. F. (1974). Scanning electron microscopy of superficial defects in articular cartilage, *Ann. Rheum. Dis.*, **44**, 327–32.

Ghadially, F. N., Ghadially, J. A., Oryschak, A. F. and Yong, N. K. (1976). Experimental production of ridges on rabbit articular cartilage, *J. Anat.*, **121**, 119–32.

Ghadially, F. N., Ghadially, J. A., Oryschak, A. F. and Yong, N. K. (1977). The surface of dog articular cartilage: a scanning electron microscopic study, *J. Anat.*, **123**, 527–36.

Ghadially, F. N., Moshurchak, E. M. and Thomas, I. (1977). Humps on young human and rabbit articular cartilage, *J. Anat.*, **124**, 425–35.

Ghadially, J. A. and Ghadially, F. N. (1975). Evidence of cartilage flow in deep defects in articular cartilage, *Virchows Arch. B. Cell Path.*, **18**, 193–204.

Gianelli, F. (1970). *Human Chromosomes and DNA Synthesis*, Monographs in Human Genetics, Volume 5, Karger: Basel, New York.

Gibbons, I. R. and Grimstone, A. V. (1960). On flagellar structures in certain flagellates, *J. biophys. biochem. Cytol.*, **7**, 697–716.

Gibbons, I. R. and Rowe, A. J. (1965). Dynein: a protein with adenosine triphosphatase activity from cilia. *Science, N.Y.*, **149**, 424–6.

Giblett, E. R. (1969). *Genetic Markers in Human Blood*, Davis: Philadelphia.

Gil, J. and Weibel, E. R. (1969). Improvements in demonstration of lining layer of lung alveoli by electron microscopy, *Resp. Physiol.*, **8**, 13–36.

Gil, J. and Reis, O. K. (1973). Isolation and characterisation of lamellar bodies and tubular myelin from rat lung homogenates, *J. Cell Biol.*, **58**, 152–71.

Gilbert, P. W. (1952). The origin and development of the head cavities in the human embryo, *J. Morph.*, **90**, 149–87.

Gilbert, P. W. (1957). The origin and development of the human extrinsic ocular muscles, *Contr. Embryol.*, **36**, 59–78.

Giles, E. and Elliot, O. (1960). Negro-white identity from the skull, *Proc. 6th Internat. Anthrop. Congr.*, Paris. Quoted in: *The Human Skeleton in Forensic Medicine* (Krogman, W. M. ed.), Thomas: Springfield, Ill.

Gillies, C. B. (1973). Synaptonemal complex and chromosome structure, *Ann. Rev. Genet.*, **9**, 91–109.

Gillilan, L. A. (1941). The connexions of the basal optic root (posterior accessory optic tract) and its nucleus in various mammals, *J. comp. Neurol.*, **74**, 367–408.

Gillilan, L. (1958). The arterial blood supply of the human spinal cord, *J. comp. Neurol.*, **110**, 75–104.

Gillilan, L. A. (1972). Anatomy and embryology of the arterial system of the brain stem and cerebellum. In: *Handbook of Clinical Neurology* (Vinken, I. J. and Bruyn, G. W. eds.), Volume 11, pp. 24–44, North Holland Publishing Company: Amsterdam.

Gillilan, L. A. (1974). Potential collateral circulation to the human cerebral cortex, *Neurology, Minneap.*, **24**, 941–8.

Gillman, J. (1948). The development of the gonads in man, with a consideration of the role of fetal endocrines and the histogenesis of ovarian tumors, *Contr. Embryol.*, **32**, 81–131.

Gilmour, J. R. (1938). Gross anatomy of the parathyroid glands, *J. Path. Bact.*, **46**, 133–49.

Gilula, N. B. and Satir, P. (1972). The ciliary necklace: A ciliary membrane specialisation, *J. Cell Biol.*, **53**, 494–509.

Gingerich, P. D. (1971). Functional significance of mandibular translation in vertebrate jaw mechanics, *Postilla*, **152**, 1–10.

Giok, S. P. (1956). *Localisation of Fibre Systems Within the White Matter of the Medulla Oblongata and the Cervical Cord in Man*, Ijido: Leiden.

Girgis, F. G., Marshall, J. L. and Al Monajem, A. R. S. (1975). The cruciate ligaments of the knee joint, *Clin. Orthop.*, **106**, 216–31.

Gladstone, R. J. (1929). Development of the inferior vena cava in the light of recent research, with especial reference to certain abnormalities, and current descriptions of the ascending lumbar and azygos veins, *J. Anat.*, **64**, 70–93.

Gladstone, R. J. and Wakeley, C. P. G. (1940). *The Pineal Organ*, Baillière, Tindall and Cox: London.

Glaister, J. and Brash, J. C. (1937). *Medicolegal Aspects of the Buck Ruxton Case*, Livingstone: Edinburgh.

Glees, P. (1941). The termination of optic fibres in the lateral geniculate body of the cat, *J. Anat.*, **75**, 434–40.

Glees, P. (1944). The anatomical basis of cortico-striate connexions, *J. Anat.*, **78**, 47–51.

Glees, P. (1946). Terminal degeneration within the central nervous system as studied by a new silver method, *J. Neuropath. exp. Neurol.*, **5**, 54–9.

Glees, P. (1961). Terminal degeneration and trans-synaptic atrophy in the lateral geniculate body of the monkey. In: *The Visual System* (Jung, R. and Körnmüller, H. eds.), pp. 104–10, Springer-Verlag: Berlin.

Glees, P. (1961). *Experimental Neurology*, Clarendon Press: Oxford.

Glees, P. (1963). *Neuroglia, Morphology and Function*, Blackwell: Oxford.

Glees, P., Cole, J., Whitty, C. W. M. and Cairns, H. (1950). The effects of lesions in the cingular gyrus and adjacent areas in monkeys, *J. Neurol. Neurosurg. Psychiat.*, **13**, 178–90.

Glees, P. and Spoerri, P. E. (1977). Microtubule-vesicle-ribbon associations in the monkey retina, *J. Neurocytol.*, **6**, 353–4.

Glenister, T. W. (1954). The origin and fate of the urethral plate in man, *J. Anat.*, **88**, 413–25.

Glenister, T. W. (1962). The development of the utricle and of the so-called 'middle' or 'median' lobe of the human prostate, *J. Anat.*, **96**, 443–55.

Glenister, T. W. (1976). An embryological view of cartilage, *J. Anat.*, **122**, 323–30.

Glickstein, M., King, R. A., Miller J. and Berkley, M. (1967). Cortical projections from the dorsal lateral geniculate nucleus of cats, *J. comp. Neurol.*, **130**, 55–76.

Glisson, F. (1654). *De Hepate*.

Gloor, P. (1960). Amygdala. In: *Handbook of Physiology*, Volume II, Section I, *Neurophysiology* (Field, J., Magoun, H. W. and Hall, V. E. eds.), pp. 1395–1420, American Physiological Society: Washington.

Gloster, J., Perkins, E. S. and Pommier, M. (1957). Extensibility of strips of sclera and cornea, *Br. J. Ophthal.*, **41**, 103–10.

Goddard, V. (1964). Functions of the amygdala, *Psychol. Bull.*, **62**, 89–109.

Godlewski, C., Hedon, B. and Castel, C. (1975). Contribution à l'étude de l'origine de l'artère hépatique chez l'embryon et le foetus, *Bull. Ass. Anat.*, **59**, 411–18.

Godwin-Austen, R. B. (1967). The identification of mechanoreceptors in the costo-vertebral joint excited by displacement of the ribs, *J. Physiol., Lond.*, **202**, 35P–36P.

Goerke, J. (1974). Lung surfactant. *Biochim. biophys. Acta.*, **344**, 241–61.

Goldberg, J. M. and Moore, R. Y. (1967). Ascending projections of the lateral lemniscus in the cat and monkey, *J. comp. Neurol.*, **129**, 143–56.

Goldberg, S. (1970). The origin of the lumbrical muscles in the hand of the South African native, *The Hand*, **2**, 168–71.

Goldman, R. and Knipe, D. M. (1973). Functions of cytoplasmic fibres in non-muscle cell motility, *Cold Spring Harbor Symp. Quant. Biol.*, **37**, 523–34.

Goldsmith, N. A. and Woodburne, R. T. (1957). The surgical anatomy pertaining to liver resection, *Surgery Gynec. Obstet.*, **105**, 310–18.

Goldspink, G. (1968). Sarcomere length during post-natal growth of mammalian muscle fibres, *J. Cell Sci.*, **3**, 539–48.

Goldstein, M. B. and Davies, E. A. Jr. (1968). The three-dimensional architecture of the islets of Langerhans, *Acta anat.*, **71**, 161–71.

Goligher, J. C. (1967). *Surgery of the Anus, Rectum and Colon*, Baillière, Tindall and Cox: London.

Goligher, J. C., Leacock, A. G. and Brossy, J.-J. (1955). Surgical anatomy of anal canal, *Br. J. Surg.*, **43**, 51–61.

Goltz, F. (1891–2). Der Hund ohne Grosshirn; siebente Abhandlung über die Verrichtungen des Grosshirns. *Pflügers Arch. ges. Physiol.*, **51**, 570–614.

Goodenough, D. A. and Revel, J. P. (1970). A fine structural analysis of intercellular junctions in the mouse liver, *J. Cell Biol.*, **45**, 272–90.

Goodman, D. C., Hallett, R. E. and Welch, R. B. (1963). Patterns of localisation in the cerebellar corticonuclear projections of the albino rat, *J. comp. Neurol.*, **121**, 51–67.

Goodman, L. S. and Gilman, A. (1970). *The Pharmacological Basis of Therapeutics* (editors), Fourth edition, Macmillan: London.

Goodrich, E. S. (1958). *Studies on the Structure and Development of Vertebrates*, Macmillan: London (republication of 1930 edition).

Gordon, K. C. D. (1967). A comparative anatomical study of the distribution of the cystic artery in man and other species, *J. Anat.*, **101**, 351–9.

Gorp, P. E. V. and Kennedy, W. R. (1974). Localization of muscle spindles in the human extensor indicis muscle for biopsy purposes, *Anat. Rec.*, **179**, 447–52.

Gorter, E. and Grendel, F. J. (1925). On bimolecular layers of lipoids on the chromocytes of the blood, *J. exp. Med.*, **41**, 439–43.

Gosling, J. A. (1970). The musculature of the upper urinary tract, *Acta anat.*, **75**, 408–22.

Gosling, J. A. and Dixon, J. S. (1971). The fine structure of the vasa recta and associated nerves in the rabbit kidney, *Anat. Rec.*, **165**, 503–13.

Goto, F. (1959). Histological and histochemical studies on human fetal membranes, *Acta Med. Okoyama*, **13**, 276–300.

Gotsev, T. (1939). Blood volume in lambs, *J. Physiol., Lond.*, **94**, 539–49.

BIBLIOGRAPHY

Gottlieb, G. (1972, 1974). *Studies on the Development of Behaviour and the Nervous System* (editor), Academic Press: New York, London.

Gottschaldt, K. M., Iggo, A. and Young, D. W. (1973). Functional characteristics of mechanoreceptors in sinus hair follicles of the cat, *J. Physiol., Lond.*, **235**, 287–315.

Gottschalk, C. W. and Mylle, M. (1959). Micropuncture studies of the mammalian urinary concentrating mechanism: evidence for the counter current hypothesis, *Am. J. Physiol.*, **196**, 927–36.

Gould, B. S. (1968). *Treatise on Collagen* (editor), Volume 2, Academic Press: New York, London.

Gowans, J. L. (1966). Lifespan, recirculation and transformation of lymphocytes, *Int. Rev. exp. Path.*, **5**, 1–24.

Gowans, J. L. and Knight, E. J. (1964). The route of recirculation of lymphocytes in the rat, *Proc. R. Soc. B.*, **159**, 257–82.

Graham, R. C. and Karnovsky, M. J. (1966). Glomerular permeability. Ultrastructural and cytochemical studies using peroxidase as protein tracer, *J. exp. Med.*, **124**, 1123–34.

Graham, R. C. and Karnovsky, M. J. (1966). The early stages of absorption of injected horseradish peroxidase in the proximal tubules of mouse kidney: ultrastructural cytochemistry by a new technique, *J. Histochem. Cytochem.*, **14**, 291–302.

Grainger, F., James, D. W. and Tresman, R. L. (1968). An electron microscopic study of the early outgrowth from chick spinal cord *in vitro*, *Z. Zellforsch. mikrosk. Anat.*, **90**, 53–67.

Grainger, F. and James, D. W. (1970). Association of glial cells with the terminal parts of neurite bundles extending from the chick spinal cord in vitro, *Z. Zellforsch. mikrosk. Anat.*, **108**, 93–104.

Granit, R. (1962). *Receptors and Sensory Perception*, Yale University Press: New Haven.

Granit, R. (1970). *The Basis of Motor Control*, Academic Press: New York, London.

Granit, R., Henatsch, H. D. and Steg, G. (1956). Tonic and phasic ventral horn cells differentiated by post-tetanic potentiation in cat extensors, *Acta Physiol. scand.*, **37**, 114–26.

Grant, G. and Oscarrson, O. (1966). Functional organisation of the spinoreticulocerebellar path with identification of its spinal component, *Expl Brain Res.*, **1**, 306–19.

Grant, G. and Oscarsson, O. (1966). Mass discharges evoked in the olivocerebellar tract on stimulation of muscle and skin nerves, *Expl Brain Res.*, **1**, 329–37.

Grant, P. G. (1973). Lateral pterygoid: two muscles?, *Am. J. Anat.*, **138**, 1–10.

Grant, R. T. (1930). Observations on local arterial reactions in the rabbit's ear, *Heart*, **15**, 257–80.

Grant, R. T. (1930). Observations on direct connections between arteries and veins in the rabbit's ear, *Heart*, **15**, 281–303.

Grant, R. T. and Regnier, M. (1926). The comparative anatomy of the cardiac coronary vessels, *Heart*, **13**, 285–317.

Grant, R. T. and Bland, F. F. (1931). Observations on arterio-venous anastomoses in human skin and in the bird's foot with special reference to reaction to cold, *Heart*, **15**, 385–407.

Grant, R. T. and Payling Wright, H. (1968). Further observations on the blood vessels of skeletal muscle (rat cremaster), *J. Anat.*, **103**, 553–65.

Grant, R. T. and Payling Wright, H. (1970). Anatomical basis for non-nutritive circulation in skeletal muscle exemplified by blood vessels of rat biceps femoris tendon, *J. Anat.*, **106**, 125–34.

Grant, R. T. and Payling Wright, H. P. (1971). The peculiar vasculature of the external spermatic fascia in the rat: possibilities subserving thermoregulation, *J. Anat.*, **109**, 293–305.

Grassé, P. P. (1948). *Traité de Zoologie*, Volume XI (editor), Masson: Paris.

Grassé, P. P. (1954). *Traité de Zoologie*, Volume XII (editor), Masson: Paris.

Graves, F. T. (1954). Anatomy of intrarenal arteries and its application to segmental resection of the kidney, *Br. J. Surg.*, **42**, 132–9.

Graves, F. T. (1956). Ganglion in muscle belly of peroneus longus, *Br. J. Surg.*, **43**, 438–9.

Graves, F. T. (1956). The aberrant renal artery, *J. Anat.*, **90**, 553–8.

Gray, D. F. (1970). *Immunology*, Second edition, Arnold: London.

Gray, D. J. and Gardner, E. D. (1943). The human sternochondral joints, *Anat. Rec.*, **87**, 235–54.

Gray, D. J. and Gardner, E. (1950). Prenatal development of the human knee and superior tibiofibular joints, *Am. J. Anat.*, **86**, 235–88.

Gray, D. J. and Gardner, E. (1951). Prenatal development of the human elbow joint, *Am. J. Anat.*, **88**, 429–69.

Gray, D. J., Gardner, E. and O'Rahilly, R. (1957). The prenatal development of the skeleton and joints of the human hand, *Am. J. Anat*, **101**, 169–224.

Gray, E. G. (1959). Axo-somatic and axo-dendritic synapses of the cerebral cortex: an electron microscope study, *J. Anat.*, **93**, 420–33.

Gray, E. G. (1961). The granule cells, mossy synapses and Purkinje spine synapses of the cerebellum: light and electron microscope observations, *J. Anat.*, **95**, 345–56.

Gray, E. G. (1969). Electron microscopy of excitatory and inhibitory synapses: a brief review, *Prog. Brain Res.*, **31**, 141–55.

Gray, E. G. (1974). Synaptic morphology with special reference to microneurons. In: *Essays on the Nervous System* (Bellairs, R. and Gray, E. G. eds.), Clarendon Press: Oxford.

Gray, E. G. and Guillery, R. W. (1966). Synaptic morphology in the normal and degenerating nervous system, *Int. Rev. Cytol.*, **19**, 111–82.

Gray, E. G. and Basmajian, J. V. (1968). Electromyography and cinematography of leg and foot ('normal' and flat) during walking, *Anat. Rec.*, **161**, 1–16.

Gray, J. (1958). The movement of the spermatozoa of the bull, *J. exp. Biol.*, **35**, 96–108.

Gray, J. A. B. and Sato, M. (1953). Properties of receptor potentials in Pacinian corpuscles, *J. Physiol., Lond.*, **122**, 610–36.

Graybiel, A. M. (1975). Anatomical organization of retinotectal afferents in the cat: an autoradiographic study, *Brain Res.*, **96**, 1–23.

Graybiel, A. M. and Harting, E. A. (1974). Some afferent connexions of the oculomotor complex in the cat, *Brain Res.*, **81**, 543–51.

Graziadei, P. P. C. (1971). The olfactory mucosa of vertebrates. In: *Handbook of Sensory Physiology*, Volume 4, *Chemical Senses* (Beidler, L. M. ed.), Springer-Verlag: Berlin.

Graziadei, P. P. C. and Monti-Graziadei, G. A. (1978). Continuous nerve cell renewal in the olfactory system. In: *Handbook of Sensory Physiology* (Jacobson, M. ed.), Volume 9, pp. 55–83, Springer: Berlin.

Green, J. D. (1951). The comparative anatomy of the hypophysis with special references to its local blood supply and innervation, *Am. J. Anat.*, **88**, 225–311.

Green, J. D. and Harris, G. W. (1949). Observations of hypophysioportal vessels of living rat, *J. Physiol., Lond.*, **108**, 359–61.

Green, N. A., Griffiths, J. D. and Lavy, G. A. D. (1958). Venous drainage of anterior tibio-fibular compartment of leg, with reference to varicose veins, *Br. med. J.*, **1**, 1209–10.

Greengard, P. and Kebabian, J. W. (1974). Role of cyclic AMP in synaptic transmission in the mammalian peripheral nervous system, *Fed. Proc.*, **33**, 1059–67.

Greenwald, A. S. and Haynes, D. W. (1972). Weight-bearing areas in the human hip joint, *J. Bone Jt Surg.*, **54B**, 157–63.

Greep, R. O., Dyke, H. B. van and Chow, B. F. (1940). The effect of pituitary gonadotrophins on the testicles of hypophysectomised immature rats, *Anat. Rec.*, **78**, 88.

Greer, M. and Haibach, H. (1974). Thyroid secretion. In: *Handbook of Physiology*, Volume III, *Endocrinology*, p. 135, American Physiological Society: Washington, DC.

Grégoire, R. (1922). *Anatomo médico-chirugicale de l'abdomen. Région sous-thoracique*, Baillière: Paris.

Gregson, N. A. (1975). The chemistry of myelin. In: *The Peripheral Nerve* (Landon, D. N. ed.), pp. 512–604, Chapman & Hall: London.

Greulich, W. W. (1951). The growth and development of Guamarian schoolchildren, *Am. J. phys. Anthrop.*, N.S. **9**, 55–70.

Greulich, W. W. and Thoms, H. (1938). The dimensions of the pelvic inlet of 789 white females, *Anat. Rec.*, **72**, 45–52.

Greulich, W. W. and Pyle, S. I. (1959). *Radiographic Atlas of Skeletal Development of Hand and Wrist*, Second edition, Stanford University Press: Stanford.

Grieve, D. W. and Cavanagh, P. R. (1974). The validity of quantitative statements about surface electromyograms recorded during locomotion, *Scand. J. Rehab. Med., Suppl.* **3**, 19–25.

Griffiths, H. E. (1943). Treatment of injured workmen (Hunterian lecture, abridged), *Lancet*, **1**, 729–33.

Griffiths, J. D. (1961). Extramural and intramural blood supply of the colon, *Br. med. J.*, **1**, 323–6.

Grimstone, A. V. and Klug, A. (1966). Observations on the substructure of flagella fibres, *J. Cell Sci.*, **1**, 351–62.

Grobstein, C. (1967). The problem of the chemical nature of embryonic inducers. In: *Cell Differentiation* (de Reuck, A. V. S. and Knight, J. eds.) pp. 131–8, Ciba Foundation Symposium, Churchill: London.

Gross, L. (1921). *The Blood Supply to the Heart*, Oxford University Press: London.

Grosser, O. (1936). Ueber hergleichende Anatomie und Phylogenese der Placenta. Stufenfolge der Placenten, Amnionbildung, Misch-placenta (Katze), Alterung der Placenta, *Verh. anat. Ges., Jena*, **81**, 15–33.

Grossman, J. (1954). A note on the radiological demonstration of perirenal space. *J. Anat.*, **88**, 407–9.

Grotte, G. (1956). Passage of dextran molecules across blood-lymph barrier, *Acta chir. scand., Suppl.* **211**, 1–84.

Grottel, K. (1968). The innervation of the suprarenal gland of man, dog, cat and rabbit, *Pr. Tow. Przyjac. Nauk poznań.*, **37**, 41–79.

Grüneberg, H. (1963). *The Pathology of Development*, Blackwell Scientific Publications: Oxford.

Grüneberg, H. (1973). A ganglion probably belonging to the N. terminalis system in the nasal mucosa of the mouse, *Z. Anat. EntwGesch.*, **140**, 39–52.

Grzybiak, M., Szostakiewicz-Sawicka, H. and Treder, A. (1975). Remarks on pathways of venous drainage from the left upper intercostal spaces in man, *Folia morph., Warsz.*, **34**, 301–13.

Guasp, F. T. (1957). *Anatomica Functional del Corazon*, Paz Montalvo, Madrid.

Guasp, F. T. (1970). *The Electrical Circulation*, Guasp, Denia.

Gudden, B. A. (1870). Experimentaluntersuchungen über das peripherische und centrale Nervensystem, *Arch. Psychiat. NervKrankh.*, **2**, 693–723.

Guerrier, Y. (1975). Le nerf facial. Quelques points d'anatomie topographique, *Ann. Oto-Laryng.*, **92**, 161–71.

Guffarth, A. and Graumann, W. (1975). Über die Lagebezielung der Arteria carotis externa zur Glandula parotis, *Archs Oto-Rhino-Laryngol.*, **211**, 17–23.

Guild, S. R. (1941). A hitherto unrecognised structure, the Glomus Jugularis, in man, *Anat. Rec.*, **79**, 28P.

Guillemin, R. (1978). Peptides in the brain: the new endocrinology of the neuron, *Science, N.Y.*, **202**, 390–402.

Guillery, R. W. (1972). Experiments to determine whether retino-geniculate axons can form translaminar sprouts in the dorsal lateral geniculate nucleus of the cat, *J. comp. Neurol.*, **146**, 407–20.

Guillery, R. W. (1973). Quantitative studies of transneuronal atrophy in the lateral geniculate nucleus of cats and kittens, *J. comp. Neurol.*, **149**, 423–38.

Guillery, R. W. and Colonnier, M. (1966). A study of Golgi preparations from the dorsal lateral geniculate body of the adult cat, *J. comp. Neurol.*, **128**, 21–50.

Guillery, R. W. and Colonnier, M. (1970). Synaptic patterns in the dorsal lateral geniculate nucleus of the monkey, *Z. Zellforsch. mikrosk. Anat.*, **103**, 90–108.

Guiot, G., Hardy, J. and Albe-Fessard, D. (1962). Precise delimitation of the subcortical structures and identification of thalamic nuclei in man by stereotatic electrophysiology, *Neurochirurgia*, **5**, 1–18.

Guiot, G., Albe-Fessard, D., Arfel, G. and Derome, P. (1964). Dérivations d'activités limitaires en cours d'interventions stéréotaxiques, *Neurochirurgia*, **10**, 427–35.

Gupta, C. D., Gupta, S. C., Arora, A. K. and Singh, J. P. (1976). Vascular segments in the human spleen, *J. Anat.*, **121**, 613–16.

Gupta, S. C., Gupta, C. D. and Arora, A. K. (1977). Subsegmentation of the human liver, *J. Anat.*, **124**, 413–23.

Gupta, K. K. and Knoell, A. C. (1973). Mathematical modelling and structural analysis of the mandible, *Biomat. Med. Dev. Art. Org.*, **1**, 469–79.

Guraya, S. S. (1963). Histochemistry of the cytoplasmic droplet in the mammalian spermatozoon, *Experientia*, **19**, 94–5.

Gurdjian, F. S. and Lissner, H. R. (1945). Deformations of the skull in head injury studied by the 'stresscoat' technique, *Surgery Gynec. Obstet.*, **83**, 219–33.

Gurdon, J. B. (1967). Nuclear transplantation and cell differentiation. In: *Cell Differentiation* (de Reuck, A. V. S. and Knight, J. eds.), pp. 65–74, Ciba Foundation Symposium, Churchill: London.

Gurdon, J. B. (1968). Nucleic acid synthesis in embryos and its bearing on cell differentiation, *Essays in Biochemistry*, **4**, 25–68.

Gurney, C. W. (1968). Erythropoietin. In: *Advances in Metabolic Disorders* (Levine, R. and Luft, R. eds.), Volume 3, pp. 279–304, Academic Press: New York, London.

Gurr, G. (1965). *The Rational Use of Dyes in Biology*, Hill: London.

Gustafson, G. and Gustafson, A.-G. (1967). Microanatomy and histochemistry of enamel. In: *Structural and Chemical Organisation of Teeth* (Miles, A. E. W. ed.), Volume 2, pp. 75–134, Academic Press: New York, London.

Gustafson, T. and Wolpert, L. (1962). Cellular mechanisms in the morphogenesis of the sea urchin larva. Changes in shape of cell sheets, *Exp. Cell Res.*, **27**, 260–79.

Gwinnett, A. S. (1967). The ultrastructure of the 'prismless' enamel of permanent human teeth, *Archs oral Biol.*, **12**, 381–7.

Gwyn, D. G. and Waldron, H. A. (1968). A nucleus in the dorsolateral funiculus of the spinal cord of the rat, *Brain Res.*, **10**, 342–51.

Gwyn, D. G., Nicholson, G. P. and Flumerfelt, B. A. (1977). The inferior olivary nucleus of the rat: a light and electron microscopic study, *J. comp. Neurol.*, **174**, 489–520.

H

Ha, H. and Liu, C. N. (1966). Organisation of the spino-cervico-thalamic system, *J. comp. Neurol.*, **127**, 445–69.

Ha, H. and Liu, C. N. (1968). Cell origin of the ventral spinocerebellar tract, *J. comp. Neurol.*, **133**, 185–206.

Haas, L. L. (1952). Roentgenological skull measurements and their diagnostic application, *Am. J. Roentg.*, **67**, 197–209.

Haberich, F. J. (1968). Osmoreception in the portal circulation, *Fedn Proc. Fedn Am. Socs exp. Bio.*, **27**, 1137–41.

Habermann, E. (1961). Similar behaviour of reticulin and collagen against collagenase, *Hoppe-Seyler's Z. physiol. Chem.*, **324**, 232–42.

Hackenbrock, C. R. (1972). States of activity and structure in mitochondrial membranes, *Ann. N.Y. Acad. Sci.*, **195**, 492–505.

Hage, E. (1973). Electron microscopic identification of several types of endocrine cells in branchial epithelium of human foetuses, *Z. Zellforsch. mikrosk. Anat.*, **141**, 401–12.

Hagen, E., Knoche, H., Sinclair, D. C. and Weddell, G. (1953). The role of specialised nerve terminals in cutaneous sensibility, *Proc. R. Soc. B.*, **141**, 279–88.

Häggqvist, G. (1936). Analyse der Faserverteilung in einem Rückenmarkquerschnitt (Th. 3), *Z. Zellforsch. mikrosk. Anat.*, **39**, 1–34.

Haines, R. W. (1933). Cartilage canals, *J. Anat.*, **68**, 45–64.

Haines, R. W. (1935). A consideration of the constancy of muscular nerve supply, *J. Anat.*, **70**, 33–55.

Haines, R. W. (1937). The primitive form of epiphysis in the long bones of tetrapods, *J. Anat.*, **72**, 323–43.

Haines, R. W. (1942). The tetrapod knee joint, *J. Anat.*, **76**, 270–301.

Haines, R. W. (1942). Eudiarthrodial joints in fishes, *J. Anat.*, **77**, 12–19.

Haines, R. W. (1944). The mechanism of rotation at the first carpometacarpal joint, *J. Anat.*, **78**, 44–6.

Haines, R. W. (1946). Movements of the first rib, *J. Anat.*, **80**, 94–100.

Haines, R. W. (1947). The development of joints, *J. Anat.*, **81**, 33–55.

Haines, R. W. (1969). The striped compressor of the prostatic urethra, *Br. J. Urol.*, **41**, 481–493.

Haines, R. W. (1974). The pseudoepiphysis of the first metacarpal of man, *J. Anat.*, **117**, 145–58.

Haines, R. W. (1975). The histology of epiphyseal union in mammals, *J. Anat.*, **120**, 1–25.

Haines, R. W. and Mohuiddin, A. (1968). Metaplastic bone, *J. Anat.*, **103**, 527–38.

Hajniš, K. (1974). A contribution to the problem of the biphalangy of the thumb, *Folia morph.*, **22**, 291–5.

Hale, A. R. (1952). Morphogenesis of volar skin in the human fetus, *Am. J. Anat.*, **91**, 147–81.

Hale, F. C., Olsen, C. R. and Mickey, M. R. (1968). The measurement of bronchial wall components, *Am. Rev. resp. Dis.*, **98**, 978–87.

Haley, J. C. and Peden, J. K. (1943). Suspensory muscle of the duodenum, *Am. J. Surg.*, **59**, 546–50.

Hall, D. A. and Jackson, D. S. (1968). *International Review of Connective Tissue Research* (editors), Volume 4.

Hall, M. C. (1965). *The Locomotor System*, Thomas: Springfield, Ill.

Hall, M. C. (1966). *The Architecture of Bone*, Thomas: Springfield, Ill.

Haller, R. de (1969). Development of mucus-secreting elements. In: *The Anatomy of the Developing Lung* (Emery, J. ed.), pp. 94–115, Spastics International Medical Publications: London.

Hallermann, H. (1934). Die Beziehungen der Werkstoffmechanik und Werstofforschung zur allgemeinen Knocken-Mechanik, *Verh. Deutsch. orthop. Gesh.*, **62**, 347–60.

Hallpike, C. S. (1931). The precise anatomy of the Hensen's cells in the cochlea of the guinea pig and its physiological significance, *J. Physiol., Lond.*, **73**, 8P.

Hallpike, C. S. (1935). Function of the tympanic membrane, *Proc. R. Soc. Med.*, **28**, 226–31.

Hally, A. D. (1959). The fine structure of the gastric parietal cell in the mouse, *J. Anat.*, **93**, 217–25.

Halmi, N. J. (1978). The normal thyroid, anatomy and histochemistry. In: *The Thyroid—a Fundamental and Clinical Text* (Werner, S. C. and Ingbar, S. H. eds.), pp. 9–21, Harper & Row: New York, London.

Ham, A. W. (1969). *Histology*, Sixth edition, Pitman: London; Lippincott: Philadelphia.

Hamashima, Y., Harter, J. G. and Coons, A. H. (1964). The localisation of albumin and fibrinogen in human liver cells, *J. Cell Biol.*, **20**, 271–80.

Hamburger, V. (1952). Development of the nervous system, *Ann. N.Y. Acad. Sci.*, **55**, 117–32.

Hamburger, V. (1960). *A Manual of Experimental Embryology*, Second impression, University Press: Chicago.

Hamburger, V. (1975). Cell death in the development of the lateral motor column of the chick embryo, *J. comp. Neurol.*, **160**, 535–46.

Hamilton, G. F. (1946). The longitudinal muscle coat of the human colon, *J. Anat.*, **80**, 230 P.

Hamilton, W. J. (1944). Phases of maturation and fertilisation in the human ova, *J. Anat.*, **78**, 1–4.

Hamilton, W. J. (1949). Early stages of human development, *Ann. R. Coll. Surg.*, **4**, 281–94.

Hamilton, W. J. and Boyd, J. D. (1951). Observations on the human placenta, *Proc. R. Soc. Med.*, **44**, 489–96.

BIBLIOGRAPHY

Hamilton, W. J. and Boyd, J. D. (1960). Development of the human placenta in the first three months of gestation, *J. Anat.*, **94**, 297–328.

Hamilton, W. J., Boyd, J. D. and Mossman, H. W. (1962). *Human Embryology*, Heffer: Cambridge.

Hamlyn, L. H. (1962). The fine structure of the mossy fibre endings in the hippocampus of the rabbit, *J. Anat.*, **96**, 112–20.

Hammel, H. T. (1968). Regulation of internal body temperature, *Ann. Rev. Physiol.*, **30**, 641–710.

Hammer, G. and Rådberg, C. (1961). The sphenoidal sinus. An anatomical and roentgenologic study with reference to transsphenoidal hypophysectomy, *Acta radiologica.*, **56**, 401–22.

Hammersen, F. and Staubesand, J. (1961). Arteries and capillaries of the human renal pelvis with special reference to the so-called spiral arteries. I. Angio-architectural studies on the kidneys, *Z. Anat. EntwGesch.*, **122**, 314–47.

Hammond, B. T. and Charnley, J. (1967). The sphericity of the femoral head, *Med. Biol. Eng.*, **5**, 445–53.

Hammond, J. (1966). Fertility. In: *Marshall's Physiology of Reproduction* (Parkes, A. S. ed.), Volume 2, pp. 648–740, Longmans, Green: London.

Hammond, W. S. (1941). The development of the aortic arch bodies in the cat, *Am. J. Anat.*, **69**, 265–94.

Hamori, J. (1972). Developmental morphology of dendritic postsynaptic specialisations. In: *Recent Development of Neurobiology in Hungary* (Lissák, K. ed.), Kiado: Budapest.

Hámori, J. and Szentágothai, J. (1965). Purkinje cell baskets: ultrastructure of an inhibitory synapse, *Acta biol. hung.*, **15**, 465–79.

Hámori, J. and Szentágothai, J. (1966). Identification under the electron microscope of climbing fibers and their synaptic contacts, *Expl Brain Res.*, **1**, 65–81.

Hámori, J. and Szentágothai, J. (1966). Participation of Golgi neuron processes in the cerebellar glomeruli: an electron microscope study, *Expl Brain Res.*, **2**, 35–48.

Hampson, J. L., Harrison, C. R. and Woolsey, C. N. (1952). Cerebro-cerebellar projections and the somatotopic localization of motor function in the cerebellum, *Res. Publs Ass. Res. nerv. ment. Dis.*, **30**, 299–316.

Hanak, H. and Böck, P. (1971). Die Feinstruktur der Muskel-sehnenverbindung von Skelett-und Herzmuskel, *J. Ultrastruct. Res.*, **36**, 68–75.

Hanaway, J. and Young, R. R. (1977). Localization of the pyramidal tract in the internal capsule of man, *J. neurol. Sci.*, **34**, 63–70.

Hancox, N. M. and Boothroyd, B. (1963). Structure-function relationships in the osteoclast. In: *Mechanisms of Hard Tissue Destruction* (Sognnaes, R. F. ed.), pp. 497–514, Am. Ass. Adv. Sci.: Washington.

Hand, P. J. (1966). Lumbosacral dorsal root terminations in the nucleus gracilis of the cat. Some observations on terminal degeneration in other medullary sensory nuclei, *J. comp. Neurol.*, **126**, 137–56.

Hand, P. J. and Liu, C. N. (1966). Efferent projections of the nucleus gracilis, *Anat Rec.*, **154**, 353–4.

Hanna, R. E. and Washburn, S. L. (1953). Determination of sex of skeletons, as illustrated by a study of the Eskimo pelvis, *Hum. Biol.*, **25**, 21–7.

Hansen, J. T. (1977). Freeze-fracture study of the carotid body, *Am. J. Anat.*, **148**, 295–300.

Hanson, J. and Lowy, J. (1963). The structure of F-actin and of actin filaments isolated from muscle, *J. molec. Biol.*, **6**, 46–60.

Hardisty, M. W. (1967). The numbers of vertebrate primordial germ cells, *Biol. Rev.*, **42**, 265–87.

Harker, D. W. (1972). The structure and innervation of sheep superior rectus and levator palpebrae muscles. I. Extrafusal muscle fibres, *Invest. Ophthal.*, **11**, 956–69.

Harkmark, W. (1954). Cell migrations from the rhombic lip to the inferior olive, the nucleus raphe and the pons. A morphological and experimental investigation on chick embryos, *J. comp. Neurol.*, **100**, 115–210.

Harness, D. and Sekeles, E. (1971). The double anastomotic innervation of thenar muscles, *J. Anat.*, **109**, 461–6.

Harness, D., Sekeles, E. and Chaco, J. (1974). The double motor innervation of the opponens pollicis muscles: an electromyographic study, *J. Anat.*, **117**, 329–31.

Harper, W. F. (1947). Observations on the blood vasculature of the turbinate mucosa in man and other mammals, *J. Anat.*, **81**, 392P.

Harpman, J. A. and Wollard, H. H. (1938). The tendon of the lateral pterygoid muscle, *J. Anat.*, **73**, 112–15.

Harries, D. J. (1955). Thyroid enlargement and cricothyroid muscle, *Br. med. J.*, **1**, 1012–13.

Harris, A. E. (1965). Differentiation and degeneration in the motor horn of the foetal mouse, Ph.D. Thesis, University of Cambridge.

Harris, G. W. (1955). *Neural Control of the Pituitary Gland*, Arnold: London.

Harris, G. W. and Donovan, B. T. (1966). *The Pituitary Gland* (editors), Butterworths: London.

Harris, H. A. (1933). *Bone Growth in Health and Disease*, Oxford University Press: London.

Harris, H. A. (1939). Anatomical and physiological basis of physical training, *Br. med. J.*, **2**, 939–43.

Harris, H. (1970). *Cell Fusion*, Clarendon Press: Oxford.

Harris, H. (1970). *Nucleus and Cytoplasm*, Clarendon Press: Oxford.

Harris, J. W. (1963). *The Red Cell*, Harvard University Press: Cambridge, Massachusetts.

Harris, P. F. and Jones, P. R. M. (1976). A radiological study of morphology and growth in the human fetal colon, *Br. J. Radiol.*, **49**, 316–20.

Harris, W. (1939). *The Morphology of the Brachial Plexus*, Oxford University Press: London.

Harris, W. (1952). Fifth and seventh cranial nerves in relation to nervous mechanism of taste sensation; new approach, *Br. med. J.*, **1**, 831–6.

Harrison, R. G. (1906). Further experiments on the development of peripheral nerves, *Am. J. Anat.*, **5**, 121–32.

Harrison, R. G. (1907). Experiments in transplanting limbs and their bearing upon the problem of the development of nerves, *Anat. Rec.*, **1**, 58–9.

Harrison, R. G. (1907). Experiments in transplanting limbs and their bearing upon the problems of the development of nerves, *J. exp. Zool.*, **4**, 239–82.

Harrison, R. G. (1907). Observations on the living developing nerve fiber, *Anat. Rec.*, **1**, 116–18.

Harrison, R. G. (1910). The outgrowth of the nerve fiber as a mode of protoplasmic movement, *J. exp. Zool.*, **9**, 787–848.

Harrison, R. G. (1914). The reaction of embryonic cells to solid structures, *J. exp. Zool.*, **17**, 521–44.

Harrison, R. G. (1924). Neuroblast versus sheath cell in the development of peripheral nerves, *J. comp. Neurol.*, **37**, 123–206.

Harrison, R. G. and Barclay, A. E. (1948). Distribution of testicular artery (internal spermatic artery) to human testis, *Br. J. Urol.*, **20**, 57–66.

Harrison, R. G. and Weiner, J. S. (1949). Vascular patterns of the mammalian testis and their functional significance, *J. exp. Biol.*, **26**, 304–16.

Harrison, R. G., Connolly, R. C. and Abdalla, A. (1969). Kinship of Smenkhkare and Tutankamen demonstrated serologically, *Nature, Lond.*, **224**, 325–6.

Harrison, T. J. (1957). Pelvic growth, Ph.D. Thesis, Queen's University, Belfast.

Harrop, T. J. and Garrett, J. R. (1974). Effects of preganglionic sympathectomy on secretory changes in parotid acinar cells of rats on eating, *Cell Tissue Res.*, **154**, 135–50.

Harvey, S. C. and Burr, H. S. (1926). The development of the meninges, *Archs Neurol. Psychiat., Chicago*, **15**, 545–67.

Hashimoto, M., Komori, A., Kosaka, M., Mori, Y., Shimoyama, T. and Akashi, H. (1960). Electron microscopic studies on the smooth muscle of the human uterus, *J. Jap. Obstet. Gynec. Soc.*, **7**, 115–21.

Halsewood, G. A. D. (1967). *Bile Salts*, Methuen: London.

Hassler, R. (1950). Projections of the cerebellum to the midbrain and the thalamus, *Dt. Z. Nervenheilk.*, **163**, 629–71.

Hassler, R. (1959). Anatomy of the thalamus. In: *Introduction to Stereotaxis with an Atlas of the Human Brain* (Schaltenbrand, G. and Bailey, P. eds.), Volume I, pp. 230–90. Grune and Stratton: New York.

Hast, M. H., Fischer, J. M., Wetzel, A. B. and Thompson, V. E. (1974). Cortical motor representation of the laryngeal muscles in *Macaca mulatta*, *Brain Res.*, **73**, 229–40.

Haug, H. (1956). Remarks on the determination and significance of the grey cell coefficient, *J. comp. Neurol.*, **104**, 473–92.

Havel, R. J. (1965). Autonomic nervous system and adipose tissue. In: *Handbook of Physiology* (Renold, A. E. and Cahill, G. F. Jr. eds.), Section 5, pp. 575–82, Williams and Wilkins: Baltimore.

Havers, Cl. (1691). *Osteologia Nova*, London.

Havez, R., Decaud, P., Roussel, P., Voisin, C., Biserte, C. and Gernez-Rieux, Ch. (1966). Identification des gamma-globulines, de la kallikreine, de la transferrine dans la muquesque bronchique humaine, *C. r. hebd. Séanc. Acad. Sci., Paris*, **262d**, 1777.

Haxton, H. A. (1954). Sympathetic nerve supply of the upper limb in relation to sympathectomy, *Ann. R. Coll. Surg.*, **14**, 247–66.

Haymaker, W., Anderson, E. and Nauta, W. J. H. (1959). *The Hypothalamus* (editors), Thomas: Springfield, Ill.

Hayreh, S. S. (1963). Arteries of the orbit in the human being, *Br. J. Surg.*, **50**, 938–53.

Hayreh, S. S. (1969). Blood supply of the optic nerve head and its role in optic atrophy, glaucoma and oedema of the optic disc, *Br. J. Ophthal.*, **53**, 721–48.

Hayward, J. (1961). The lower end of the oesophagus, *Thorax*, **16**, 36–41.

Head, H. and Rivers, W. H. R. (1920). *Studies in Neurology*, Froude: London.

Headon, M. P. and Powell, T. P. S. (1973). Cellular changes in the lateral geniculate nucleus of infant monkeys after suture of the eyelids, *J. Anat.*, **116**, 135–45.

Headon, M. P. and Powell, T. P. S. (1978). The effect of bilateral eye closure upon the lateral geniculate nucleus in infantile monkeys, *Brain Res.*, **143**, 147–54.

Healey, J. E. and Schroy, P. C. (1953). Anatomy of biliary ducts within the human liver; analysis of prevailing pattern of branchings and major variations of biliary ducts, *A.M.A. Archs Surg.*, **66**, 599–616.

Hécaen, H. and de Ajuriaguerra, J. (1964). *Left-Handedness: Manual Superiority and Cerebral Dominance*, Grune & Stratton: New York.

Heimann, P. (1966). Ultrastructure of the human thyroid: a study of normal thyroid, untreated and treated diffuse toxic goitre, *Acta endocr., Copenh.*, **53**, Suppl. 110.

Heimer, L. (1968). Synaptic distribution of centripetal and centrifugal nerve fibres in the olfactory system of the rat. An experimental anatomical study, *J. Anat.*, **103**, 413–32.

Heimer, L. and Wall, P. D. (1968). The dorsal root distribution to the substantia gelatinosa of the rat with a note on the distribution in the cat, *Expl Brain Res.*, **6**, 89–99.

Heinemann, H. O. and Fishman, A. P. (1969). Nonrespiratory functions of mammalian lung, *Physiol Rev.*, **49**, 1–47.

Heintzberger, C. F. M. (1974). The development of the sinu-atrial node in the mouse, *Acta morph. neerl.-scand.*, **12**, 317–30.

Heitz, Ph., Polak, J. M., Bloom, S. K. and Pearse, A. G. E. (1976). Identification of the D_1-cell as the source of human pancreatic polypeptide (HPP), *Gut*, **17**, 755–8.

Held, H. (1893). Die centrale Gehörleitung, *Arch. mikrosk. Anat. EntwMech.*, 201–48.

Heller, C. H. and Clermont, Y. (1964). Kinetics of the germinal epithelium in man, *Recent Prog. Horm. Res.*, **20**, 545–75.

Heller, H. and Clark, R. B. (1962). *Neurosecretion*, Academic Press: New York, London.

Heller, J. H. (1960). (Editor). *Reticuloendothelial Structure and Function*, Ronald Press: New York.

Heller, L. and Langman, J. (1964). The menisco-femoral ligaments of the human knee, *J. Bone Jt Surg.*, **46B**, 307–13.

Heller-Steinberg, M. (1951). Ground substance, bone salts and cellular activity in bone formation and destruction, *Am. J. Anat.*, **89**, 347–79.

Henderson, J. R. (1969). Serum-insulin or plasma-insulin? *Lancet*, **2**, 545–7.

Henderson, R. G. (1978). The position of the nutrient foramen in the growing tibia and femur of the rat, *J. Anat.*, **125**, 593–9.

Hendrickson, A. and Kupfer, C. (1976). The histogenesis of the fovea in the macaque monkey, *Invest. ophthal.*, **15**, 746–56.

Hendry, N. G. C. (1958). The hydration of the nucleus pulposus and its relation to intervertebral disc derangement, *J. Bone Jt Surg.*, **40B**, 132–44.

Henkind, P. and Levitsky, M. (1969). Angioarchitecture of the optic nerve. I. The papilla, *Am. J. Opthal.*, **68**, 979–86.

Henle, J. (1876). *Handbuch der Gefasslebre des Menschen*, Braunschweig.

Henle, W., Henle, G. and Chambers, L. A. (1938). Studies on ontogenic structure of some mammalian spermatozoa, *J. exp. Med.*, **68**, 335–52.

Henley, F. A. (1955). Blood supply of common bile duct and its relationship to the duodenum, *Br. J. Surg.*, **43**, 75–80.

Herbert, M. C. and Graham, C. F. (1974). Cell determination and biochemical differentiation of the early mammalian embryo, *Curr. Top. Dev. Biol.*, **8**, 151–78.

Herlant, M. (1964). The cells of the adenohypophysis and their functional significance, *Int. Rev. Cytol.*, **17**, 299–382.

Herman, P. G., Yamamoto, I. and Mellins, H. Z. (1973). Microcirculation of the aortic wall in experimental atheromatosis, *Radiology*, **107**, 265–71.

Herndon, R. M. (1963). The fine structure of the Purkinje cell, *J. Cell Biol.*, **18**, 167–80.

Herrick, C. J. (1922). *Introduction to Neurology*, Saunders: Philadelphia.

Hertig, A. T. (1935). Angiogenesis in the early human chorion and in the primary placenta of the macaque monkey, *Contr. Embryol.*, **25**, 37–82.

Hertig, A. T. and Rock, J. (1941). Two human ova of pre-villous stage, having ovulation age of about 11 and 12 days respectively, *Contr. Embryol.*, **29**, 127–56.

Hertig, A. T. and Rock, J. (1945). Two human ova of the pre-villous stage, having a developmental age of about seven and nine days respectively, *Contr. Embryol.*, **31**, 65–84.

Hertig, A. T., Rock, J., Adams, E. C. and Mulligan, W. J. (1954). On the pre-implantation stages of the human ovum: a description of four normal and four abnormal specimens ranging from the second to the fifth day of development, *Contr. Embryol.*, **35**, 199–220.

Hertig, A. T., Rock, J. and Adams, E. C. (1956). A description of 34 human ova within the first 17 days of development, *Am. J. Anat.*, **98**, 435–93.

Hertwig, W. A. O. (1881). *Die Coelomtheorie*, Jena.

Hervonen, A., Vaalasti, A., Vaalasti, T., Partenen, M. and Kanerva, L.
(1976). Paraganglia in the urogenital tract of man, *Histochemistry*, **48**, 307–13.

Hervonen, A., Vaalasti, A., Partanen, M. and Kanerva, L. (1978a). The endocrine nature of the paraganglia of man, *Experientia*, **34**, 111–12.

Hervonen, A., Vaalasti, A. and Partanen, M. (1978b). Paraganglia of the bladder, *J. Urol.*, **119**, 335–7.

Hess, A. (1970). Vertebrate slow muscle fibers, *Physiol. Rev.*, **50**, 40–62.

Hess, A. and Young, J. Z. (1952). The nodes of Ranvier, *Proc. R. Soc. B.*, **140**, 301–20.

Hess, W. R., Brügger, M. and Bucher, V. (1946). Zur Physiologie von Hypothalamus, Area praeoptica und Septum, sowie angrenzender Balken- und Stirnhirnberache, *Mschr. Psychiat. Neurol.*, **III**, 17–59.

Heuser, C. H. and Streeter, G. L. (1941). Development of macaque embryo, *Contr. Embryol.*, **29**, 15–56.

Heuser, C. H. and Corner, G. W. (1957). Developmental horizons in human embryos. Description of age group X, 4 to 12 somites, *Contr. Embryol.*, **36**, 29–39.

Hewitt, A. B. (1977). An investigation using holographic interferometry of surface strain in bone induced by orthodontic forcés, *Br. J. Orthodont.*, **4**, 39–41.

Hewitt, W. (1958). The development of the human caudate and amygdaloid nuclei, *J. Anat.*, **92**, 377–82.

Hewitt, W. (1960). The median aperture of the fourth ventricle, *J. Anat.*, **94**, 549–57.

Hewitt, W. (1961). The development of the human internal capsule and lentiform nucleus, *J. Anat.*, **95**, 191–9.

Hewitt, W. (1962). The development of the human corpus callosum, *J. Anat.*, **96**, 355–8.

Heylings, D. J. A. (1978). Supraspinous and interspinous ligaments of the human lumbar spine, *J. Anat.*, **125**, 127–31.

Heyssel, R. M. (1961). Determination of human platelet survival utilising C-14 labelled serotonin, *J. clin. Invest.*, **40**, 2134–42.

Hicks, J. H. (1953). The mechanics of the foot. I. The joints, *J. Anat.*, **87**, 345–57.

Hicks, J. H. (1953). The mechanics of the foot. II. The plantar aponeurosis and the arch, *J. Anat.*, **88**, 25–30.

Hicks, J. H. (1955). The foot as a support, *Acta anat.*, **25**, 34–45.

Hicks, R. M. (1965). Permeability barriers in the rat ureter transitional epithelium, *J. Anat.*, **99**, 932P.

Hicks, R. M., Ketterer, B. and Warren, R. C. (1974). The ultrastructure and chemistry of the lumenal plasma membrane of the mammalian urinary bladder: a structure with low permeability to water and ions, *Phil. Trans. R. Soc. Lond.*, **268B**, 23–38.

Hicks, S. P., Amato, C. J. D'. and Lowe, M. J. (1959). The development of the mammalian nervous system. I. Malformations of the brain, especially the cerebral cortex, induced in rats by radiation, *J. comp. Neurol.*, **113**, 435–69.

Highstein, S. M. (1977). Abducens to medial rectus pathway in the MLF. In: *Eye Movements* (Brooks, B. A. and Bajandas, F. J. eds.), Plenum Press: New York.

Hiiemae, K. M. (1978). Mammalian mastication: a review of the activity of the jaw muscles and the movements they produce in chewing. In: *Development, Function and Evolution of Teeth* (Butler, P. M. and Joysey, K. A. eds.), pp. 359–98, Academic Press: New York, London.

Hiiemae, K., Thexton, A. J. and Crompton, A. W. (1978). Intra-oral food transport: the fundamental mechanism of feeding. In: *Muscle Adaptation in the Craniofacial Region* (Carlson, D. S. and McNamara, J. A. eds.), Monograph 8, Craniofacial Growth Series, University of Michigan Center for Human Growth and Development: Michigan.

Hill, A. V. (1970). *First and Last Experiments in Muscle Mechanics*, Cambridge University Press: Cambridge.

Hill, J. P. (1932). The developmental history of the primates, *Phil. Trans. R. Soc.*, Ser. B, **221**, 45–178.

Hillman, P. and Wall, P. D. (1969). Inhibitory and excitatory factors influencing the receptive fields of lamina 5 spinal cord cells, *Exp. Brain Res.*, **9**, 284–306.

Hilton, S. M. and Lewis, G. P. (1956). The relationship between glandular activity, bradykinin formation and functional vasodilatation in the submandibular salivary gland, *J. Physiol., London.*, **134**, 471–83.

Hindle, M. O. (1967). The intermediate plexus of the periodontal membrane. In: *Mechanisms of Tooth Support*, pp. 66–71, Wright: Bristol.

Hinds, J. W. (1971). Early neuroblast differentiation in the mouse olfactory bulb, *Anat. Rec.*, **169**, 340–1.

Hinds, J. W. and Hinds, P. L. (1972). Reconstruction of dendritic growth cones in neonatal mouse olfactory bulb, *J. Neurocytol.*, **1**, 169–87.

Hinrichsen, C. F. L. and Larramendi, L. M. H. (1969). Features of trigeminal mesencephalic nucleus structure and organization, *Am. J. Anat.*, **126**, 497–506.

Hintzsche, E. (1937). Uber den Einfluss der Schilddrüsengrosse auf die Lage der Epithelkörperchen, *Anat. Anz.*, **84**, 18–25.

Hirsch, E. F. (1963). The innervation of the human heart. III. The conductive system, *Archs Path.*, **74**, 427–39.

Hirsch, E. F. (1963). The innervation of the human heart. IV. (1) The fiber connections of the nerves with the perimyseal plexus (Gerlach-Hoffman). (2) The role of nerve tissues in the repair of infarcts, *Archs Path.*, **75**, 378–401.

Hirsch, E. F. and Borghard-Erdle, A. M. (1962). The innervation of the human heart. II. The papillary muscles, *Archs Path.*, **73**, 101–17.

Hirsch, S. (1960). *Morphologie et Physiologie des Anastomoses Arterioveneuses*, Acta Tertii Europaei de Cordis Scientia Conventus, Romae. *Excerpta Medica*, **i**, 61–4.

Hirsch-Hoffman, H. U. (1976). Licht- und elektronenmikroskopische Untersuchungen an der Tränendrüse des Affen (Macac mulatta). *Z. mikrosk.-anat. Forsch.*, **90**, 369–84.

Hirsch-Hoffman, H. U. (1978). Die Ultrastruktur von Drüsenendstücken der menschlichen Tränendrüse, *Klin. Mbl. Augenheilk.*, **172**, 80–7.

His, W. (1879). Ueber die Anfänge des peripherischen Nervensystems, *Arch. Anat. Physiol., Lpz., Anat. Abt.*, 455–82.

His, W. (1883). Ueber das Auftreten der weissen Substanz und der Wurzelfasern am Rückenmark menschlicher Embryonen, *Arch. Anat. Physiol., Lpz., Anat. Abt.*, 163–70.

His, W. (1887). Zur Geschichte des menslichen Rückenmarks und der Nervenwurzeln, *Abh. Gesch. Math.*, **13**, 477–14.

His, W. (1887). On the development of the roots of the nerves and on their propagation to the central organs and to the periphery, *Brit. Ass. Rep.*, 773–5.

His, W. (1890). Histogenese und Zusammenhang der Nervenelemente, *Arch. Anat. Physiol., Lpz., Anat. Abt.*, Suppl., 95–117.

Hjarnø, J., Jørgensen, J. B. and Vesely, M. (1974). *Archeological and Anthropological Investigations of Late Heathen Graves in Upernarvik District*, Reitzels Forlag: København.

Hjortsjö, C.-H. (1948). Die Anatomie der Intrahepatischen Gallengänge beim Menschen; Mittels Röntgen- und Injections-Technik stadiert, *Lunds Univ. Årssk. N.F. Avd. 2*, **44**, 1–112.

Hjortsjö, C.-H. (1951). The topography of the intrahepatic duct systems, *Acta anat.*, **11**, Suppl. 14–15.

Hjortsjö, C.-H. (1956). The intrahepatic ramification of the portal vein, *Lunds Univ. Årssk, N.F. Avd. 2*, **52**, 1–30.

Hjortsjö, C.-H. (1975). Den Segmentella Indelingen av Levern, *Comm. Dept Anat. Univ. Lund*, 4.

Hodge, A. J. (1967). Structure at the electron microscopic level. In: *Treatise on Collagen*, Volume 1, *Chemistry of Collagen* (Ramachandran, G. N. ed.), pp. 185–206. Academic Press: New York, London.

Hodgkin, A. L. (1964). *The Conduction of the Nervous Impulse*, Thomas: Springfield, Ill.

Hodgson, C. and Spiers, R. L. (1974). The effect of preganglionic cervical sympathectomy on the amylase content of parotid glands in fasted and fed rats, *J. Physiol., Lond.*, **237**, 56–57P.

Hodson, J. J. (1966). The distribution, structure, origin and nature of the dental cuticle of Gottlieb. I and II. *Periodontics*, **5**, 237–50, 296–302.

Hodson, N. (1970). The nerves of the testis, epididymus and scrotum. In: *The Testis* (Johnson, A. D., Gomes, W. R. and Vandemark, N. L. eds.), Volume I, pp. 47–100, Academic Press: New York, London.

Hoerr, N. L., Pyle, S. J. and Francis, C. C. (1962). *Radiographic Atlas of Skeletal Development of Foot and Ankle*, Thomas: Springfield, Ill.

Hoff, E. C. and Hoff, H. E. (1934). Spinal termination of the projection fibers from the motor cortex of primates, *Brain*, **57**, 454–74.

Hoffman, B. F. and Cranefield, P. F. (1960). *Electrophysiology of the Heart*, McGraw-Hill: New York.

Hoffman, E. and Thiel, W. (1956). Untersuchungen vermeintlicher und wirkincher Abflusswege aus dem Subdural- und Subarachnoideatraum, *Z. Anat. EntwGesch*, **119**, 283–310.

Hoffman, H. H. and Kuntze, A. (1957). Vagus nerve components, *Anat. Rec.*, **127**, 551–68.

Hogan, M. J. (1963). The vitreous, its structure and relation to the ciliary body and retina, *Invest. Ophthal*, **2**, 418–45.

Hogan, M. J. (1972). Role of the retinal pigment epithelium in macular disease, *Trans. Am. Acad. Ophthal. Otol.*, **76**, 64–80.

Hogan, M. J., Alvarado, J. A. and Weddell, J. E. (1971). *Histology of the Human Eye*, Saunders: Philadelphia.

Hökfelt, T. (1968). *In vitro* studies on central and peripheral monoamine neurons at the ultrastructural level, *Z. Zellforsch. mikrosk. Anat.*, **91**, 1–74.

Hökfeldt, T., Elde, R., Johannson, O., Luft, R., Nilsson, G. and Arimura, A. (1976). Immunohistochemical evidence for separate populations of somatostatin-containing and substance P-containing primary afferent neurons in rat, *Neuroscience*, **1**, 131–6.

Hollenberg, M. J. and Spira, A. W. (1973). Human retinal development, *Am. J. Anat.*, **137**, 357–86.

Holliday, R. (1964). A mechanism for gene conversion in fungi, *Genet. Res.*, **5**, 282–304.

Hollinshead, W. H. (1971). *Anatomy for Surgeons*, Harper & Row: New York.

Hollmann, K. H. (1966). Sur des aspects particuliers des protéines élaborées dans la glande mammaire. Etude au microscope électronique chez la lapine en lactation, *Z. Zellforsch. mikrosk. Anat.*, **69**, 395–402.

Holloway, R. L. (1968). The evolution of the primate brain: some aspects of quantitative relations, *Brain Res.*, **7**, 121–72.

Holmberg, J. (1972). On the nerves of the parotid gland. In: *Oral Physiology* (Emmelin, N. and Zotterman, Y. eds.), pp. 17–19, Pergamon: Oxford.

Holmdahl, D. E. and Ingelmark, B. E. (1950). The contact between the articular cartilage and the medullary cavities of the bone, *Acta orthopaed. scand.*, **20**, 156–65.

Holmes, G. (1939). The cerebellum of man, *Brain*, **62**, 1–30.

Holmes, G. and Lister, W. T. (1916). Disturbances of vision, from cerebral lesions, with special reference to the cortical representation of the macula, *Brain*, **39**, 34–73.

Holstein, A. F. (1967). Zur Nervenzellverteilung in glattmuskeligen Sphinkteren, *Verh. anat. Ges., Jena*, **61**, 269–75.

Holstein, A. F. (1967). Spermiophagen im Neenhoden des Menschen, *Naturwissenschaften*, **54**, 98–9.

Holtzer, H. (1972). The cell cycle, myogenesis and psoriasis, *J. Invest. Dermatol.*, **59**, 33–4.

Holtzer, H., Strahs, K. and Biehl, J. (1975). Thick and thin filaments in postmitotic mononucleated myoblasts, *Science, N.Y.*, **188**, 943–5.

Holtzman, E. (1976). *Lysosomes: A Survey*, Cell Biology Monographs, Springer: Vienna.

Hooker, C. W. (1971). The intertubular tissue of the testis. In: *The Testis* (Johnson, A. D., Gomes, W. R. and Vandemark, N. L. eds.), Volume I, pp. 483–550, Academic Press: New York, London.

Hope, J. (1965). The fine structure of the developing follicle of the rhesus ovary, *J. Ultrastruct. Res.*, **12**, 592–610.

Horowitz, A. L. and Dorfman, A. (1968). Subcellular sites for synthesis of chondromucoprotein of cartilage, *J. Cell Biol.*, **38**, 358–68.

Horridge, G. A. (1968). *Interneurons*, Freeman: London, San Francisco.

Horsley, V. A. H. and Clarke, R. H. (1908). The structure and functions of the cerebellum examined by a new method, *Brain*, **31**, 45–124.

Hörstadius, S. (1950). *The Neural Crest*, Oxford University Press: London.

Horstmann, E. (1966). Uber das Endothel der Zottenkapillaren im Dünndarm des Meerschwanchens und des Menschen, *Z. Zellforsch. mikrosk. Anat.*, **72**, 364–9.

Horton, R. C. (1952). The gastro-epiploic arteries, *Guy's Hosp. Rep.*, **101**, 108–10.

Hoshiko, M. (1962). Electromyographic investigation of the intercostal muscles during speech, *Archs phys. Med.*, **43**, 115–19.

Hovelacque, A. (1927). *Anatomie des nerfs craniens et rachidiens et du système grand sympathétique chez l'homme*, Doin: Paris.

Howarth, F. and Cooper, E. R. A. (1949). Departure of substances from spinal theca, *Lancet*, **2**, 937–40.

Howkins, C. H. (1935). Blood supply of the lower jaw, *Proc. R. Soc. Med.*, **29**, 506–7.

Hoyte, D. A. N. (1960). Alizarin as an indicator of bone growth, *J. Anat.*, **94**, 432–42.

Hoyte, D. A. N. (1966). Experimental investigations of skull morphology and growth, *Int. Rev. gen. exp. Zool.*, **2**, 345–408.

Hoyte, D. A. N. (1975). A critical analysis of the growth in length of the cranial base, *Birth Defects*, **11**, 255–82.

Hrdlička, A. (1939). *Practical Anthropometry*, Wistar Institute: Philadelphia.

Hruban, Z. and Swift, H. (1964). Uricase: localisation in hepatic microbodies, *Science, N.Y.*, **146**, 1316–18.

Hubel, D. H. and Wiesel, T. N. (1961). Integrative action in the cat's lateral geniculate body, *J. Physiol., Lond.*, **155**, 385–98.

Hubel, D. H. and Wiesel, T. N. (1962). Receptive fields, binocular interaction and functional architecture in the cat's visual cortex, *J. Physiol., Lond.*, **160**, 106–54.

Hubel, D. H. and Wiesel, T. N. (1963). Receptive fields of cells in striate cortex of very young, visually inexperienced kittens, *J. Neurophysiol.*, **26**, 994–1002.

Hubel, D. H. and Wiesel, T. N. (1965). Binocular interaction in striate cortex of kittens reared with artificial squint, *J. Neurophysiol.*, **28**, 1041–59.

Hubel, D. H. and Wiesel, T. N. (1968). Receptive fields and functional architecture of monkey striate cortex, *J. Physiol., Lond.*, **195**, 215–43.

Hubel, D. H. and Wiesel, T. N. (1969). Anatomical demonstration of columns in the monkey striate cortex, *Nature, Lond.*, **221**, 747–50.

Hubel, D. H. and Wiesel, T. N. (1971). Aberrant visual projections in the Siamese cat, *J. Physiol., Lond.*, **218**, 33–62.

Hubel, D. H., Wiesel, T. N. and LeVay, S. (1977). Plasticity of ocular dominance columns in monkey striate cortex, *Phil. Trans. R. Soc. Ser. B.*, **278**, 377–409.

Huber, G. C. (1909). The development of the albino rat, from the end of the first to the tenth day after insemination, *Am. J. Anat.*, **9**, 84–8.

Huber, A. C. and Crosby, E. C. (1933). The reptilian optic tectum, *J. comp. Neurol.*, **57**, 57–164.

Hudson, C. L., Moritz, A. R. and Wearn, J. T. (1932). The extracardiac anastomoses of the coronary arteries, *J. exp. Med.*, **56**, 919–25.

Hudson, R. E. B. (1965). *Cardiovascular Pathology*, Arnold: London.

Hudson, R. E. B. and Wendell-Smith, C. P. (1966). Congenital abnormalities of the heart and great vessels. In: *Systemic Pathology* (Payling Wright, G. and Symmers, W. St. C. eds.), pp. 59–90, Longmans: London.

Huelke, D. F. (1962). The dorsal scapular artery—a proposed term for the artery to the rhomboid muscles, *Anat. Rec.*, **142**, 57–62.

Hugget, A. S. (1927). Foetal blood-gas tensions and gas transfusion through the placenta of the goat, *J. Physiol., Lon.*, **62**, 373–84.

Huggins, C. B. (1931). Formation of bone under the influence of epithelium of the urinary tract, *Archs Surg., Chicago*, **22**, 377–408.

Hughes, A. F. W. (1953). The growth of embryonic neurites. A study on cultures of chick neural tissue, *J. Anat.*, **87**, 150–62.

Hughes, A. F. W. (1968). *Aspects of Neural Ontogeny*, Academic Press: New York, London.

Hughes, A. (1974). Endocrines, neural development and behaviour. In: *Aspects of Neurogenesis* (Gottlieb, G. ed.), Academic Press: New York, London.

Hughes, A. (1976). The development of the dorsal funiculus in the human spinal cord, *J. Anat.*, **122**, 169–75.

Hughes, H. (1952). The factors determining the direction of the canal for the nutrient artery in the long bones of mammals and birds, *Acta anat.*, **15**, 261–80.

Humphrey, J. H. and White, R. G. (1970). *Immunology for Students of Medicine*, Third edition, Blackwell Scientific Publications: Oxford.

Humphrey, T. (1944). Primitive neurons in the embryonic central nervous system, *J. comp. Neurol.*, **81**, 1–45.

Humphrey, T. (1947). Sensory ganglion cells within the central canal of the embryonic human spinal cord, *J. comp. Neurol.*, **86**, 1–36.

Humphrey, T. (1960). The development of the pyramidal tracts in human fetuses correlated with cortical differentiation. In: *Structure and Function of the Cerebral Cortex* (Tower, D. B. and Schadé, J. P. eds.), pp. 93–103, Elsevier Press: Amsterdam.

Humphrey T. (1964). Some observations on the development of the human hippocampal formation, *Trans. Am. neurol. Ass.*, **89**, 207–9.

Humphrey, T. (1967). The development of the human hippocampal fissure, *J. Anat.*, **101**, 655–76.

Humphrey, T. (1969). The central relations of the trigeminal nerve. In: *The Surgery of Pain* (Kahn, E. A. ed.), Second edition, Thomas: Springfield, Ill.

Hundeiker, M. and Keller, L. (1963). Die Gefässarchitektur des menschlichen Hodens, *Morph. Jb.*, **105**, 26–73.

Hunt, C. C. (1974). The physiology of muscle receptors. In: *Handbook of Sensory Physiology* (Hunt, C. C. ed.), Volume 3, Part 2, pp. 191–234, Springer: Berlin.

Hunt, J. N. and Knox, M. T, (1968). Regulation of gastric emptying. In: *Handbook of Physiology*, Section 6, *Alimentary Canal* (Code, C. F. ed.), Volume 4, pp. 1917–36, American Physiological Society: Washington.

Hunt, R. K. and Jacobson, M. (1974). Neuronal specificity, revisited. In: *Current Topics in Developmental Biology* (Moscona, A. A. and Monroy, A. eds.), Volume 8, Academic Press: New York, London.

Hunter, J. (1762). Observations on the state of the testis in the foetus, and on the hernia congenita. In: *Medical Commentaries* (Hunter, W.), Part 1, London.

Hunter, R. F. (1974). Chronological and cytological details of fertilization and early embryonic development in the domestic pig, *Sus scrofa*, *Anat. Rec.*, **178**, 169–86.

Huntingford, P. J. (1959). Pudendal nerve block; the results of its routine use, with special reference to the trans-vaginal technique, *J. Obstet. Gynaec. Br. Commonw.*, **66**, 26–31.

Huntington, G. S. (1908). The genetic interpretation of the development of the mammalian lymphatic system, *Am. J. Anat.*, **2**, 19–45.

Huntington, G. S. (1920). The morphology of the pulmonary artery in the Mammalia, *Anat. Rec.*, **17**, 165–202.

Hurrell, D. J. (1934). The vascularisation of cartilage, *J. Anat.*, **69**, 47–61.

Hurst, A. F. (1919). *Chronic Constipation*, Oxford University Press: London.

Hutch, J. A. and Rambo, O. S. Jr. (1970). A study of the anatomy of the prostate, prostatic urethra and urinary sphincter system, *J. Urol.*, **104**, 443–52.

Hutchinson, M. C. E. (1978). A study of the atrial arteries in man, *J. Anat.*, **125**, 39–54.

Huxley, A. F. and Niedergerke, R. (1954). Structural changes in muscle during contraction, *Nature, Lond.*, **173**, 971–3.

Huxley, A. F. and Taylor, R. E. (1958). Local activation of striated muscle, *J. Physiol., Lond.*, **144**, 426–41.

Huxley, H. E. (1963). Electron microscope studies on the natural and synthetic protein filaments from striated muscle, *J. molec. Biol.*, **7**, 281–308.

Huxley, H. E. (1964). Evidence for continuity between the central elements of the triads and extracellular space in frog sartorius muscle, *Nature, Lond.*, **202**, 1067–71.

Huxley, H. E. (1966). The fine structure of striated muscle and its functional significance, *Harvey Lect.*, **60**, 85–118.

Huxley, H. E. (1969). The mechanism of muscular contraction, *Science, N.Y.*, **164**, 1356–66.

Huxley, H. E. and Hanson, J. (1954). Changes in the cross striations of muscle during contraction and stretch and their interpretation, *Nature, Lond.*, **173**, 973–6.

Huxley, H. E. and Brown, W. (1967). The low angle X-ray diagram of vertebrate striated muscle and its behaviour during contraction and rigor, *J. molec. Biol.*, **30**, 383–434.

Huxley, H. E. and Klug, A. (1971). *New Developments in Electron Microscopy*, The Royal Society: London.

Huxley, J. S. (1942). *Evolution. The Modern Synthesis*, Harper: New York.

Hyde, J. B., Akeson, E. J. and Berinstein, E. (1973). Asymmetrical growth of superior temporal gyri in man, *Experientia*, **29**, 1131.

Hydén, H. (1960). The neuron. In: *The Cell* (Brachet, J. and Mirsky, A. E. eds.), Volume IV, pp. 215–324, Academic Press: New York, London.

Hydén, H. (1967). *The Neuron* (editor), Elsevier: Amsterdam.

Hyndman, O. R. and Wolkin, J. (1943). Anterior cordotomy; further observations on physiologic results and optimum manner of performance, *Archs Neurol. Psychiat., Chicago*, **50**, 129–48.

I

Ibuka, N. and Kawamura, H. (1975). Loss of circadian rhythm in sleep-wakefulness cycle of the rat by suprachiasmatic nucleus lesions, *Brain Res.*, **96**, 76–81.

Ichikawa, A. and Nakajima, Y. (1962). Electron microscope study on the lacrimal gland of the rat, *Tohuku J. exp. Med.*, **77**, 136–49.

Iggo, A. (1969). Cutaneous thermoreceptors in primates and sub-primates, *J. Physiol., Lond.*, **200**, 403–30.

Iggo, A. (1974). Activation of cutaneous receptors and their actions on dorsal horn neurons, *Adv. Neurol.*, **4**, 1–9.

Iklé, F. A. (1961). Trophoblast cells in the circulating blood, *Schweiz. med. Wschr.*, **91**, 943–5.

Iklé, F. A. (1964). Dissemination von Syncytiotrophoblastzellen im mütterlichen Blut während der Gravidität, *Bull. schweiz. Akad. med. Wiss.*, **20**, 62–72.

Illis, L. (1964). Spinal cord synapses in the cat: the normal appearances by the light microscope, *Brain*, **87**, 543–54.

Imamoto, K. and Leblond, C. P. (1977). Presence of labelled monocytes, macrophages and microglia in a stab wound of the brain following an injection of bone marrow cells labelled with ³H-uridine into rats, *J. comp. Neurol.*, **174**, 255–80.

Imamoto, K. and Leblond, C. P. (1978). Radioautographic investigation of gliogenesis in the corpus callosum of young rats. II. Origin of microglial cells, *J. comp. Neurol.*, **180**, 139–64.

Ince, H. and Young, M. (1940). The bony pelvis and its influence on labour: a radiological and clinical study of 500 women, *J. Obstet. Gynaec. Br. Commonw.*, **47**, 130–90.

Ingle, D. and Schneider, G. E. (1969). *Subcortical Visual Systems* (editors), Karger: Basel.

Ingram, T. T. (1969). The new approach to early diagnosis of handicaps in childhood, *Dev. Med. Child Neurol.*, **11**, 279–90.

Ingram, V. M. (1963). *The Haemoglobins in Genetics and Evolution*, Columbia University Press: New York.

Ingram, W. R. (1940). Nuclear organisation and chief connections of the primate hypothalamus, *Res. Publs Ass. Res. nerv. ment. Dis.*, **20**, 195–244.

Inman, V. T. (1969). The influence of the foot-ankle complex on the proximal skeletal structure, *Artif. Limbs*, **13**, 59–65.

Inman, V. T. and Saunders, J. B. de C. M. (1937). The ossification of the human frontal bone, with special reference to its presumed pre- and post-frontal elements, *J. Anat.*, **71**, 383–94.

Inman, V. T., Saunders, J. B. de C. M. and Abbot, L. C. (1944). Observations on the function of the shoulder joint, *J. Bone Jt. Surg.*, **26**, 1–30.

Inoue, H. (1973). Three-dimensional observation of collagen framework of intervertebral discs, *Arch. hist jap.*, **36**, 39–56.

Inoué, S. and Sato, H. (1967). Cell motility by labile association of molecules: the nature of mitotic spindle fibres and their role in chromosome movement, *J. gen. Physiol.*, **50**, 259–92.

Ip, M. C. and Chang, K. S. F. (1968). A study on the radial supply of the human brachialis muscle, *Anat. Rec.*, **162**, 363–72.

Irving, M. H. (1964). The blood supply of the growth cartilage in young rats, *J. Anat.*, **98**, 931–9.

Irving, R. and Harrison, J. M. (1967). The superior olivary complex and audition: A comparative study, *J. comp. Neurol.*, **130**, 77–86.

Isaacson, R. L. and Pribram, K. H. (1975). *The Hippocampus* (editors), Plenum Press: New York.

Isler, H., Sarkar, S. K., Thompson, B. and Tonkin, R. (1968). The architecture of the thyroid gland: a 3-dimensional investigation, *Anat. Rec.*, **161**, 325–35.

Isotupa, K. (1972). Alizarin trajectories in experimental studies of skull growth, *Proc. Finn. dent. Soc.*, **68**, *Suppl. II*, 1–49.

Ito, M., Yoshida, M. and Obata, K. (1964). Monosynaptic inhibition of the intracerebellar nuclei induced from the cerebellar cortex, *Experienta*, **20**, 575–6.

Ito, M. and Yoshida, M. (1966). The origin of cerebellar-induced inhibition of Deiter's neurones. I. Monosynaptic initiation of the inhibitory postsynaptic potentials, *Expl Brain Res.*, **2**, 330–49.

Ito, S. (1967). Anatomic structure of the gastric mucosa. In: *Handbook of Physiology*, Section 6, *Alimentary Canal* (Code, D. F. ed.), Volume 2, pp. 705–42, American Physiological Society. Washington.

Ito, T. and Shibasaki, S. (1964). Lichtmikroskopische Untersuchungen über die Glandula lacrimalis des Menschen, *Arch. hist. jap.*, **25**, 117–43.

Iurato, S. (1967). *Submicroscopic Structure of the Inner Ear*, Pergamon Press: Oxford, London, Toronto.

Iwayama, T. (1970). Ultrastructural changes in the nerves innervating the cerebral artery after sympathectomy, *Z. Zellforsch. mikrosk. Anat.*, **109**, 465–80.

Iwayama, T., Furness, J. B. and Burnstock, G. (1970). Dual adrenergic and cholinergic innervation of the cerebral arteries of the rat: an ultrastructural study, *Circ. Res.*, **26**, 635–46.

J

Jabbur, S. J. and Towe, A. L. (1961). Cortical excitation of neurons in dorsal column nuclei of cat, including an analysis of pathways, *J. Neurophysiol.*, **24**, 499–509.

Jackson, I. T. and Saint Onge, R. A. (1977). The use of palmaris longus tendon to stabilise trapezium implants. A preliminary report, *Hand*, **9**, 42–4.

Jackson, K. M., Joseph, J. and Wyard, S. J. (1977). Sequential muscular contraction, *J. Biomechanics*, **10**, 97–106.

Jacob, F. and Monod, J. (1961). Genetic regulatory mechanisms in the synthesis of proteins, *J. molec. Biol.*, **3**, 318–56.

Jacob, F. and Monod, J. (1963). Genetic repression, allosteric inhibition and cellular differentiation. In: *Cytodifferentiation and Macromolecular Synthesis* (Locke, M. ed.), pp. 30–64, Academic Press: New York, London.

Jacobowitz, D. M. and Palkovits, M. (1974). Topographic atlas of catecholamine and acetylcholinesterase-containing neurons in the rat brain. I. Forebrain (telencephalon, diencephalon), *J. comp. Neurol.*, **157**, 13–28.

Jacobs, L. and Comroe, H. J. (1971). Reflex apnoea, bradycardia and hypotension produced by serotonin on the nodose ganglia of the cat, *Circ. Res.*, **29**, 145–55.

Jacobsen, K. (1974). Area intercondylaris tibiae: osseous surface structure and its relation to soft tissue structures and applications to radiography, *J. Anat.*, **117**, 605–18.

Jacobson, M. (1969). Development of specific neuronal connections, *Science, N.Y.*, **163**, 543–7.

Jacobson, M. (1970). *Developmental Neurobiology*, Holt, Rhinehart & Wilson: New York.

Jacobson, M. (1970). Development, specification and diversification of neuronal connections. In: *The Neurosciences: Second Study Program* (Schmitt, F.O. ed.), Rockefeller University Press: New York.

Jacobson, M. (1974). A plentitude of neurons. In: *Aspects of Neurogenesis* (Gottlieb, G. ed.), Volume 2, Academic Press: New York, London.

Jain, K. K., Bhandari, G. J. and Koranne, S. P. (1973). Histogenesis of human eye lid, *East. Arch. Ophthal.*, **3**, 8–15.

Jakus, M. A. (1964). *Ocular Fine Structure*, Churchill: London.

James, D. W. (1974). Growth cones and synaptic connections in tissue culture. In: *Essays on the Nervous System* (Bellairs, R. and Gray, E. G. eds.), Clarendon Press: Oxford.

James, D. W. and Tresman, R. L. (1969). An electron microscopic study of the *de novo* formation of neuromuscular junctions in tissue culture, *Z. Zellforsch. mikrosk. Anat.*, **100**, 126–40.

James, D. W. and Tresman, R. L. (1969). Synaptic profiles in the outgrowth from chick spinal cord *in vitro*, *Z. Zellforsch. mikrosk. Anat.*, **101**, 598–606.

James, T. N. (1960). The arteries of the free ventricular walls in man, *Anat. Rec.*, **136**, 371–84.

James, T. N. (1974). Anatomy of the coronary arteries and veins. In: *The Heart* (Hurst, J. W. ed.), Third edition, McGraw-Hill: New York.

James, T. N. (1978). Anatomy of the conducting system of the heart. In: *The Heart, Arteries and Veins* (Hurst, J. W. editor-in-chief), Fourth edition, McGraw-Hill: New York.

James, T. N., Sherf, L. and Urthaler, F. (1974). Fine structure of the bundle-branches, *Br. Heart J.*, **36**, 1–18.

Jamieson, J. D. and Palade, G. E. (1967). Intracellular transport of secretory proteins in the pancreatic exocrine cell. I. Role of the peripheral elements of the Golgi complex, *J. Cell Biol.*, **34**, 577–96.

Jamieson, J. K. and Dobson, J. F. (1907). Lectures on the lymphatic system of the caecum and appendix, *Lancet*, **1**, 1061–6.

Jamieson, J. K. and Dobson, J. F. (1908). The lymphatics of the colon, *Proc. R. Soc. Med.*, **2**, 149–74.

Jamieson, J. K. and Dobson, J. F. (1910). The lymphatics of the testicle, *Lancet*, **1**, 493–5.

Jamieson, J. K. and Dobson, J. F. (1910). On the injection of lymphatics by Prussian blue, *J. Anat.*, **45**, 7–10.

Jamieson, J. K. and Dobson, J. F. (1920). The lymphatics of the tongue; with particular reference to the removal of lymphatic glands in cancer of the tongue, *Br. J. Surg.*, **8**, 80–7.

Jancsó, G., Kiraly, E. and Jancsó-Gábor, A. (1977). Pharmacologically-induced selective degeneration of chemosensitive primary sensory neurones, *Nature, Lond.*, **270**, 741–3.

Janda, V. and Stará, V. (1965). The role of thigh adductors in movement patterns of the hip and knee joint, *Courrier, Centre Internat. de l'Enfance*, **15**, 1–3.

Jande, S. S. and Belanger, L. F. (1970). Fine structural study of rat molar cementum, *Anat. Rec.*, **167**, 439–63.

Jansen, J. and Brodal, A. (1954). *Aspects of Cerebellar Anatomy* (editors), Grundt Tanum: Oslo.

Jansen, J. and Jansen, J. Jr. (1955). On the efferent fibers of the cerebellar nuclei in the cat, *J. comp. Neurol.*, **102**, 607–32.

Jansen, J. and Korneluissen, H. (1977). Morphogenesis and morphology of the brainstem nuclei of Cetacea. The hypoglossal nucleus, *J. Hirnforsch.*, **18**, 253–69.

Jasper, H. H. (1966). Recording from microelectrodes in stereotatic surgery for Parkinson's disease, *J. Neurosurg.*, **24**, Suppl. 11, 219–21.

Jasper, H., Proctor, L. D., Knighton, R. S. and Costello, R. T. (1958). *Reticular Formation of the Brain* (editors), Henry Ford Hospital International Symposium, Little, Brown and Co.: Boston.

Jaworek, T. E. (1973). The intrinsic vascular supply to the first metatarsal. Surgical considerations, *J. Am. Podiatry Ass.*, **63**, 189–97.

Jayaraman, A., Batton, R. R. and Carpenter, M. B. (1977). Nigrotectal projections in the monkey: an autoradiographic study, *Brain Res.*, **135**, 147–52.

Jayatilika, A. D. P. (1965). Arachnoid granulations in sheep, *J. Anat.*, **99**, 315–27.

Jayatilika, A. D. P. (1965). An electron microscopic study of sheep arachnoid granulations, *J. Anat.*, **99**, 635–49.

Jefferson, G. (1915). The human stomach and the canalis gastricus (Lewis), *J. Anat.*, **49**, 165–81.

Jefferson, N. C., Ogawa, T., Syleos, C., Zambetoglou, A. and Necheles, H. (1960). Restoration of respiration by nerve anastomosis, *Am. J. Physiol.*, **198**, 931–3.

Jeffrey, P. K. and Reid, L. (1975). New features of rat airway epithelium: a quantitative and electron microscopic study, *J. Anat.*, **120**, 295–300.

Jenkins, F. A. (1969). The evolution and development of the dens of the mammalian axis, *Anat. Rec.*, **164**, 173–84.

Jit, I. (1952). The development and the structure of the suspensory muscle of the duodenum, *Anat. Rec.*, **113**, 395–407.

Jit, I. (1956). Estimation of stature from clavicles, *Ind. J. med. Res.*, **44**, 137–55.

Jit, I. (1961). The structure and development of the valves of Houston, *Ind. J. med. Res.*, **49**, 635–47.

Jit, I. and Charnalia, J. (1959). The vertebral level of the termination of the spinal cord, *J. Anat. Soc. Ind.*, **8**, 93–101.

Jit, I. and Gandhi, O. P. (1966). The value of pre-auricular sulcus in sexing bony pelves, *J. Anat. Soc. India*, **15**, 104–7.

Jit, I. and Singh, S. (1966). The sexing of the adult clavicles, *Ind. J. Med. Res.*, **54**, 551–71.

Jit, I. and Grewal, S. S. (1975). Lengths of the small and large intestines in North Indian subjects, *J. Anat., India*, **24**, 89–100.

Jit, I. and Kulkaria, M. (1976). Times of appearance and fusion of epiphysis at the medial end of the clavicle, *Ind. J. med. Res.*, **64**, 773–82.

Johannisson, E. (1968). The foetal adrenal cortex in the human. Its ultrastructure at different stages of development and in different functional states, *Acta endocr., Copnh.*, **58**, Suppl. 130.

Johansen, K. and Martin, A. W. (1965). Comparative aspects of cardiovascular function in vertebrates. In: *Handbook of Physiology* (Hamilton, W. F. and Dow, P. eds.), Section 2, Circulation, Volume III, pp. 2583–614. American Physiological Society: Washington.

John, B. (1976). Myths and mechanisms of meiosis, *Chromosoma (Berlin)*, **54**, 295–325.

Johns, B. A. (1952). Developmental changes in the oesophageal epithelium in man, *J. Anat.*, **86**, 431–42.

Johnson, A. D., Gomes, W. R. and Vandemark, N. L. (1970). *The Testis* (editors), Volume I, Academic Press: New York, London.

Johnson, D. A., Roth, G. M. and Craig, W. M. (1952). Autonomic pathways in the spinal cord, *J. Neurosurg.*, **9**, 599–605.

Johnson, F. P. (1914). The development of the rectum in the human embryo, *Am. J. Anat.*, **16**, 1–58.

Johnson, F. R. and McMinn, R. M. H. (1956). Transitional epithelium and osteogenesis, *J. Anat.*, **90**, 106–16.

Johnson, F. R., McMinn, R. M. H. and Birchenough, R. F. (1962). The ultrastructure of the gall bladder epithelium of the dog, *J. Anat.*, **96**, 477–87.

Johnson, F. R., McMinn, R. M. H. and Atfield, G. N. (1968). Ultrastructural and biochemical observations on the tympanic membrane, *J. Anat.*, **103**, 297–310.

Johnston, J. B. (1909). The morphology of the forebrain vesicle in vertebrates, *J. comp. Neurol.*, **19**, 458–539.

Johnston, J. and Parkinson, D. (1974). Intracranial sympathetic pathways associated with the sixth nerve, *J. Neurosurg.*, **40**, 236–43.

Johnston, M. C. and Platt, K. M. (1975). A developmental approach to teratology. In: *Teratology* (Berry, C. L. and Poswillo, D. E. eds.), Springer-Verlag: Berlin.

Jollie, M. (1962). *Chordate Morphology*, Reinhold: Pittsburgh.

Jones, D. S., Beargie, R. J. and Pauly, J. E. (1953). An electromyographic study of some muscles of costal respiration in man, *Anat. Rec.*, **117**, 17–24.

Jones, E. G. and Powell, T. P. S. (1968). The ipsilateral cortical connexions of the somatic sensory areas in the cat, *Brain Res.*, **9**, 71–94.

Jones, E. G. and Powell, T. P. S. (1969). Connexions of the somatic sensory cortex of the rhesus monkey. I. Ipsilateral cortical connexions, *Brain*, **92**, 477–502.

Jones, E. G. and Powell, T. P. S. (1970). An anatomical study of converging sensory pathways within the cerebral cortex of the monkey, *Brain*, **93**, 793–820.

Jones, E. G., Burton, H. and Porter, R. (1975). Commissural and cortico-cortical 'columns' in the somatic sensory cortex of primates, *Science, N.Y.*, **190**, 572–4.

Jones, E. G., Coulter, J. D., Burton, H. and Porter, R. (1977). Cells of origin and terminal distribution of corticostriatal fibres arising in the sensori-motor cortex of monkeys, *J. comp. Neurol.*, **173**, 53–80.

Jones, E. G. and Hartman, B. K. (1978). Recent advances in neuroanatomical methodology, *Annual Review of Neuroscience*, **1**, 215–96.

Jones, F. W. (1911). On the grooves upon the ossa parietalia commonly said to be caused by the arteria meningea media, *J. Anat.*, **46**, 228–38.

Jones, F. W. (1912). Some nerve markings on lumbar vertebrae, *J. Anat.*, **47**, 118–20.

Jones, F. W. (1913). The function of the coelom and the diaphragm, *J. Anat.*, **47**, 282–318.

Jones, F. W. (1931). The non-metrical morphological characters of the skull as criteria for racial diagnosis. I. General discussion of the morphological characters employed in racial diagnosis. II. The non-metrical morphological characters of the Hawaiian skull. III. The non-metrical morphological characters of the prehistoric inhabitants of Guam, *J. Anat.*, **65**, 179–95, 368–78, 438–45.

Jones, F. W. (1939). The anterior superior alveolar nerve and vessels, *J. Anat.*, **73**, 583–91.

Jones, F. W. (1939). The so-called maxillary antrum of the gorilla, *J. Anat.*, **73**, 116–19.

Jones, F. W. (1940). The nature of the soft palate, *J. Anat.*, **74**, 147–70.

Jones, F. W. (1941). *The Principles of Anatomy as Seen in the Hand*, Baillière, Tindall and Cox: London.

Jones, F. W. (1949). *Structure and Function as Seen in the Foot*, Baillière, Tindall and Cox: London.

Jones, I. C. (1957). *The Adrenal Cortex*, Cambridge University Press: Cambridge.

Jones, J. P. and Fox, H. (1977). Syncytial knots and intervillous bridges in the human placenta: an ultrastructural study, *J. Anat.*, **124**, 275–86.

Jones, R. L. (1937). Cell fibre ratios in the vagus nerve, *J. comp. Neurol.*, **67**, 469–82.

Jones, R. L. (1941). The human foot. An experimental study of its mechanics, and the role of its muscles and ligaments in the support of the arch, *Am. J. Anat.*, **68**, 1–40.

Jones-Seaton, A. (1950). Etude de l'organisation cytoplasmique de l'œuf rongeurs, principalement quant à la basophilie ribonucléique, *Archs Biol., Paris*, **61**, 291–444.

Jonsson, B. (1974). Function of the erector spinae muscle on different working levels, *Acta morph. neerl. scand.*, **12**, 211–14.

Jonsson, B. and Steen, B. (1962). Function of the hip and thigh muscles in Romberg's test and "standing at ease", *Acta morph. neerl. Scand.*, **5**, 267–76.

Jonsson, B. and Hagberg, M. (1974). The effect of different working heights on the deltoid muscle, *Scand. J. rehab. Med., Suppl.* **3**, 26–32.

Jordan, R. K., McFarlane, B. and Scothorne, R. J. (1973). An electron microscopic study of the histogenesis of the ultimobranchial body and of the C-cell system in the sheep, *J. Anat.*, **114**, 115–36.

Jørgensen, F. and Bentzon, M. W. (1968). The ultrastructure of the normal human glomerulus. Thickness of glomerular basement membrane, *Lab. Invest.*, **18**, 42–8.

Joseph, J. (1951). Further studies of the metacarpophalangeal and interphalangeal joints of the thumb, *J. Anat.*, **85**, 221–9.

Joseph, J. (1951). The sesamoid bones of the hand and the time of fusion of the epiphyses of the thumb, *J. Anat.*, **85**, 230–41.

Joseph, J. (1960). *Man's Posture: Electromyographic Studies*, Thomas: Springfield, Ill.

Joseph, J. (1973). Sequential contraction of muscles producing the same movement at a joint. In: *New Developments in EMG and Clinical Neurophysiology* (Desmedt, J. E. ed.), pp. 665–74, Karger: Basel.

Joseph, J. (1975). Movements at the hip joint, *Ann. R. Coll.Surg.*, **56**, 192–201.

Joseph, J., Nightingale, A. and Williams, P. L. (1955). Detailed study of electric potentials recorded over some postural muscles while relaxed and standing, *J. Physiol., Lond.*, **127**, 617–25.

Joseph, J. and Williams, P. L. (1957). Electromyography of certain hip muscles, *J. Anat.*, **91**, 286–94.

Jost, A. (1961). The role of fetal hormones in pre-natal development, *Harvey Lect.*, **55**, 201–26.

Joubert, D. M. (1955). Growth of muscle fibre in the foetal sheep, *Nature, Lond.*, **175**, 936–7.

Jouvet, M. (1962). Recherches sur les structures nerveuses et les mécanismes responsables des différentes phases du sommeil physiologique, *Archs ital. Biol.*, **100**, 125–206.

Jouvet, M. (1964). Etude neurophysiologique unique des troubles de la conscience, *Acta neurochir.*, **12**, 258–69.

Jouvet, M. (1965). Paradoxical sleep—a study of its nature and mechanisms, *Prog. Brain Res.*, **18**, 20–62.

Jovanović, S. and Živanović, S. (1965). The establishment of the sex by the great sciatic notch, *Acta anat.*, **61**, 101–7.

K

Kaada, B. (1960). Cingulate, posterior orbital, anterior insular and temporal pole cortex. In: *Handbook of Physiology*, Section I, Volume II (Field, J., Magoun, H. and Hall, V. E. eds.), pp. 1345–72, American Physiological Society: Washington.

Kaada, B. (1967). Brain mechanisms related to aggressive behaviour. In: *Aggression and Defense. Neural Mechanisms and Social Patterns* (Clemente, C. D. and Lindsley, D. B. eds), pp. 95–216, University of California Press: Berkeley.

Kagayama, M. and Nishiyama, A. (1972). Comparative aspect on the innervation of submandibular glands in cat and rabbit; an electron microscopic study, *Tohuku J. exp. Med.*, **108**, 179–93.

Kaiser, O. (1891). *Die Funktionen der Ganglionzellen des Halsmarkes*, Nijhoff: Haag.

Kameda, Y. (1971). The occurrence and distribution of the parafollicular cells in the thyroid, parathyroid IV and thymus IV in some mammals, *Arch. histol. jap.*, **33**, 283–99.

Kameda, Y. (1976). Fine structural and endocrinological aspects of thyroid parafollicular cells. In: *Chromaffin, Enterochromaffin and Related Cells* (Coupland, R. E. and Fujita, T. eds.), pp. 155–70, Elsevier: New York, Amsterdam.

Kampmeier, O. F. (1969). *Evolution and Comparative Morphology of the Lymphatic System*, Thomas: Springfield, Ill.

Kanagasuntheram, R. (1957). Development of the human lesser sac, *J. Anat.*, **91**, 188–206.

Kanagasuntheram, R. (1960). Some observations on the development of the human duodenum, *J. Anat.*, **94**, 231–40.

BIBLIOGRAPHY

Kanagasuntheram, R. (1967). A note on the development of the tubotympanic recess in the human embryo, *J. Anat.*, **101**, 731–42.

Kanagasuntheram, R. (1970). Some unresolved mysteries in the anatomy of the visual system, *Singapore Med. J.*, **11**, 63–70.

Kanaseki, T. and Kadota, K. (1969). The 'vesicle in a basket', *J. Cell Biol.*, **42**, 202–20.

Kanno, T. (1973). Unidirectional cellular processes in stimulus-secretion coupling in cells of the GEP system. In: *Gastro-Entero-Pancreatic Endocrine System. A Cell-Biological Approach* (Fujita, T. ed.), pp. 64–70, Igaku Shoiu: Tokyo.

Kapanci, Y., Assimacopoulos, A., Irle, C., Zwahlen, A. and Gabbiani, G. (1974). 'Contractile interstitial cells' in pulmonary alveolar septa: a possible regulator of ventilation/perfusion ratio? Ultrastructure, immunofluorescence and 'in vitro' studies, *J. Cell Biol.*, **60**, 375–92.

Kapandji, I. A. (1963). *Physiologie Articulaire, Fascicule 1. Membre Supérieur*, Maloine: Paris.

Kapandji, I. A. (1970). *The Physiology of the Joints*, Livingstone: London.

Kapandji, I. A. (1972). La rotation du ponce sur axe longitudinal lors l'opposition, *Revue chir. Ortho.*, **58**, 273–89.

Kapeller, K. and Mayor, D. (1967). Ultrastructural appearance of degeneration in constricted sympathetic nerves, *J. Anat.*, **101**, 602P.

Kapeller, K. and Mayor, D. (1967). The accumulation of noradrenaline in constricted sympathetic nerves as studied by fluorescence and electron microscopy, *Proc. R. Soc. B.*, **167**, 282–92.

Kaplan, E. B. (1958). The iliotibial tract. Clinical and morphological significance, *J. Bone Jt Surg.* **40A**, 817–31.

Kaplan, E. B. (1965). *Functional and Surgical Anatomy of the Hand*, Lippincott: Philadelphia.

Kaplan, H. A. (1956). Arteries of the brain; anatomic study, *Acta radiol.*, **46**, 364–470.

Kaplan, H. A. and Ford, D. H. (1966). *The Brain Vascular System*, Elsevier: Amsterdam.

Kaplan, H. A., Browder, A. and Browder, J. (1973). Nasal venous drainage and the foramen caecum, *Laryngoscope*, **83**, 327–9.

Kappers, C. U. A. (1947). *Anatomie Comparée du Systeme Nerveux*, Masson: Paris.

Kappers. J. A. (1960). The development, topographical relations and innervation of the epiphysis cerebri in the albino rat, *Z. Zellforsch. mikrosk. Anat.*, **52**, 163–215.

Kappers, J. A. (1976). The mammalian pineal gland, a survey, *Acta neurochir.*, **34**, 109–49.

Kappers, J. A. and Schadé, J. P. (1965) (editors), *Prog. Brain Res.*, **10**.

Karfunkel, P. (1971). The role of microtubules and microfilaments in neurulation in Xenopus. *Dev. Biol.*, **25**, 30–56.

Karnovsky, M. L. (1967). Energetics of transport, *Protoplasma*, **63**, 76–85.

Karnovsky, M. L. (1968). The metabolism of leukocytes, *Seminars Hemat.*, **5**, 156–65.

Karrer, H. E. (1956). The ultrastructure of mouse lung. General architecture of capillary and alveolar walls, *J. biophys. biochem. Cytol.*, **2**, 241–52.

Kasai, T., Takahashi, G. and Aiyama, S. (1974). A case report of annular pancreas, *Acta anat., Nippon*, **49**, 103–19.

Kashef, R. (1966). The Node of Ranvier, Ph.D. Thesis, University of London.

Katchburian, E. (1973). Membrane-bound bodies as initiators of mineralization of dentine, *J. Anat.*, **116**, 285–302.

Kate, B. R. (1968). The torsion of the humerus in central India, *J. Ind. anthrop. Soc.*, **3**, 17–30.

Kate, B. R. and Robert, S. L. (1965). Some observations on the upper end of the tibia in squatters, *J. Anat.*, **99**, 137–42.

Katz, B. (1966). *Nerve, Muscle and Synapse*, McGraw-Hill: New York.

Katz, B. and Miledi, R. (1965). Propagation of electrical activity in motor nerve terminals, *Proc. R. Soc. B.*, **161**, 453–82.

Katz, B. and Miledi, R. (1965). The measurement of synaptic delay, and the time course of acetylcholine release at the neuromuscular junction, *Proc. R. Soc. B.*, **161**, 483–95.

Katz, B. and Miledi, R. (1965). The effect of calcium on acetylcholine release from motor nerve terminals, *Proc. R. Soc. Ser. B.*, **161**, 496–503.

Kauer, J. M. G. (1974). The interdependence of carpal articulation chains, *Acta anat.*, **88**, 481–501.

Kawabata, I. and Paparella, M. M. (1969). Ultrastructure of normal human middle ear mucosa. Preliminary report, *Ann. Otol. Rhinol. Lar.*, **78**, 125–38.

Kaye, G. I., Wheeler, H. O., Whitlock, R. T. and Lane, N. (1966). Fluid transport in the rabbit gall bladder. A combined physiological and electron microscopic study, *J. Cell Biol.*, **30**, 237–68.

Keagy, R. D., Brumlik, J. and Bergan, J. L. (1966). Direct electro-myography of the psoas major muscle in man, *J. Bone Jt Surg.*, **48A**, 1377–82.

Keech, M. K. (1960). Electron microscope study of the normal rat aorta, *J. biophys. biochem. Cytol.*, **7**, 533–8.

Keegan, J. J. and Garrett, F. D. (1948). The segmental distribution of the cutaneous nerves in the limbs of man, *Anat. Rec.*, **102**, 409–37.

Keen, J. A. (1950). Study of differences between male and female skulls, *Am. J. phys. Anthrop.*, **8**, 65–79.

Keene, M. F. L. (1961). Muscle spindles in human laryngeal muscles, *J. Anat.*, **95**, 25–9.

Keene, M. F. L. and Hewer, E. E. (1927). Observations on the development of the human suprarenal gland, *J. Anat.*, **61**, 302–24.

Keene, M. F. L. and Hewer, E. E. (1931). Some observations on myelination in the human central nervous system, *J. Anat.*, **66**, 1–13.

Keene, M. F. L. and Hewer, E. E. (1935). The sub-commissural organ and the mesocoelic recess in the human brain, with a note on Reissner's fibre, *J. Anat.*, **69**, 501–7.

Kefalides, N. A. (1973). Structure and biosynthesis of basement membranes, *Int. Rev. Connect. Tis. Res.*, **6**, 63–104.

Keith, A. (1948). *Human Embryology and Morphology*, Sixth edition, Arnold: London.

Kelley, R. O. (1973). Fine structure of the apical rim-mesenchyme complex during limb morphogenesis in man, *J. Embryol. exp. Morph.*, **29**, 117–31.

Kellgren, J. H. and Samuel, E. P. (1950). Sensitivity and innervation of the articular capsule, *J. Bone Jt Surg.*, **32B**, 84–92.

Kelly, A. M. and Zacks, S. I. (1969). The fine structure of motor end plate morphogenesis, *J. Cell Biol.*, **42**, 154–69.

Kelly, D. E. and Cahill, M. A. (1972). Filamentous and matrix components of skeletal muscle Z-disks, *Anat. Rec.*, **172**, 623–42.

Kelly, P. J. (1968). Anatomy, physiology and pathology of the blood supply of bones, *J. Bone Jt Surg.*, **50A**, 766–83.

Kemp, J. M. (1968). An electron microscopic study of the termination of afferent fibres in the caudate nucleus, *Brain Res.*, **11**, 484–7.

Kemp, J. M. (1968). Observations on the caudate nucleus of the cat impregnated with the Golgi method, *Brain Res.*, **11**, 467–70.

Kemp, J. M. (1970). The termination of strio-pallidal and strio-nigral fibres, *Brain Res.*, **17**, 125–8.

Kemp, J. M. and Powell, T. P. S. (1970). The cortico-striate projections in the monkey, *Brain*, **93**, 525–46.

Kennedy, W. P. (1967). Epidemiological aspects of the problem of congenital malformations, *Birth Defects*, **3**, 1–18.

Kenny, M. (1944). The clinically suspect pelvis and its radiographical investigation in 1,000 cases, *J. Obstet. Gynaec. Br. Commonw.*, **51**, 277–92.

Kenshalo, D. R. (1968). *The Skin Senses*, Thomas: Springfield, Ill.

Kerjaschki, D. and Hörandner, H. (1976). The development of mouse olfactory vesicles and their cell contacts. A freeze-etching study, *J. Ultrastruct. Res.*, **54**, 420–44.

Kerr, F. W. L. (1962). Facial, vagal and glossopharyngeal nerves in the cat. Afferent connexions, *Archs Neurol. Psychiat., Chicago*, **6**, 264–81.

Kerr, F. W. L. (1969). Preserved vagal visceromotor function following destruction of the dorsal motor nucleus, *J. Physiol., Lond.*, **202**, 755–69.

Kessel, R. G. (1968). Annulate lamellae, *J. Ultrastruct. Res.*, **Suppl. 10**, 1–82.

Keswani, N. H. and Hollinshead, W. H. (1956). Localisation of the phrenic nucleus in the spinal cord of man, *Anat. Rec.*, **125**, 683–700.

Kettlekamp, D. B. and Jacobs, W. (1972). Tibio-femoral contact area: determination and implications, *J. Bone Jt Surg.*, **54A**, 349–56.

Keynes, G. (1954). The physiology of the thymus gland, *Br. med. J.*, **2**, 659–63.

Khoo, F. Y., Kanagasuntheram, R. and Chia, K. B. (1969). Variations of the lateral recesses of the naso-pharynx, *Archs Otolar.*, **88**, 456–62.

Kielbasinski, G. (1976). Arteries of the inferior part of the vermis cerebelli in man, *Folia morphol., Warsz.*, **25**, 149–57.

Kier, E. L. (1966). Embryology of the normal optic canal and its anomalies, *Investigative Radiology*, **1**, 346–62.

Kier, E. L. (1977). The cerebral ventricles: a phylogenetic and ontogenetic study. In: *Radiology of the Skull and Brain* (Newton and Potts, eds.), Volume 3, Mosby: New York.

Kimmel, D. L. (1961). The nerves of the cranial dura mater and their significance in dural headache and referred pain, *Chicago med. Sch. Q.*, **22**, 16–26.

Kimmel, D. L. (1961). Innervation of spinal dura mater and dura mater of the posterior cranial fossa, *Neurology, Minneap.*, **11**, 800–9.

Kimura, R. S. (1969). Distribution, structure and function of dark cells in the vestibular labyrinth, *Ann. Otol. Rhinol. Lar.*, **78**, 542–61.

King, R. H. M. and Thomas, P. K. (1971). Aberrant regeneration of unmyelinated axons in the vagus nerve of the rabbit, *J. Anat.*, **108**, 596P.

King, R. J. (1974). The surfactant system of the lung, *Fedn Proc. Fedn Am. Socs exp. Biol.*, **33**, 2238–47.

King, R. J. and Clements, J. A. (1972). Surface active materials from dog lung. I. Composition and physiological correlations, *Am. J. Physiol.*, **223**, 715–26.

King, T. J. and Briggs, R. (1965). Serial transplantation of embryonic nuclei. In: *Molecular and Cellular Aspects of Development* (Bell, E. ed.) pp. 171–93, Harper and Row: New York, Evanston, London.

King, T. S. (1954). The anatomy of hare-lip in man, *J. Anat.*, **88**, 1–12.

King, T. S. and Coakley, J. B. (1958). The intrinsic nerve cells of the cardiac atria of mammals and man, *J. Anat.*, **92**, 353–76.

Kinmonth, J. B. (1964). Some general aspects of the investigation and surgery of the lymphatic system, *J. cardiovasc. Surg.*, **5**, 680–2.

Kinmonth, J. B. and Taylor, G. W. (1964). Chylous reflux, *Br. med. J.*, **1**, 529–32.

Kinnaert, P. (1973). Anatomical variations of the cervical part of the thoracic duct in man, *J. Anat.*, **115**, 45–52.

Kinman, J. (1977). Surgical aspects of the anatomy of the sphenoidal sinuses and the sella turcica, *J. Anat.*, **124**, 541–53.

Kirchner, J. A. and Wyke, B. D. (1965). Articular reflex mechanisms in the larynx, *Ann. Otol. Rhinol. Lar.*, **74**, 749–68.

Kirk, J. (1944). Observations on the histology of the choledochduodenal junction and papilla duodeni, with particular reference to the ampulla of Vater and sphincter of Oddi, *J. Anat.*, **78**, 118–20.

Kisch, B. (1958). New investigations on cardiac nerves. An electron microscopic study, *Exp. Med. Surg.*, **16**, 81–95.

Kiss, F. (1932). Sympathetic elements in the cranial and spinal ganglia, *J. Anat.*, **66**, 488–98.

Kito, H. and Hosoda, S. (1977). Triple staining for simultaneous visualization of cell types in islets of Langerhans of pancreas. Successive application of argyrophil, aldehyde-fuchsin and lead-hematoxylin stains in a single tissue section, *J. Histochem. Cytochem.*, **25**, 1019–20.

Klein, D. C. (1978). The pineal gland: a model of neuroendocrine regulation, *Res. Publs Ass. Res. nerv. ment. Dis.*, **56**, 303–27.

Klein, D. C. and Weller, J. L. (1972). A rapid light-induced decrease in pineal serotonin N-acetyltransferase activity, *Science, N.Y.*, **177**, 532–3.

Klika, E. and Petřík, P. (1965). A study of the structure of the lung alveolar and bronchiolar epithelium (a histological and histochemical study using the method of membranous preparations), *Acta histochem.*, **20**, 331–42.

Klintworth, G. K. (1967). The ontogeny and growth of the human tentorium cerebelli, *Anat. Rec.*, **158**, 433–42.

Klintworth, G. K. (1968). The comparative anatomy and phylogeny of the tentorium cerebelli, *Anat. Rec.*, **160**, 635–42.

Klosovskii, B. N. (1963). *The Development of the Brain and its Disturbance by Harmful Factors* (Haigh, B. trans.), Pergamon Press: Oxford, New York, London, Paris.

Klüver, H. and Bucy, P. C. (1937). Psychic blindness and other symptoms following temporal lobectomy in rhesus monkeys, *Am. J. Physiol.*, **119**, 352–3.

Klüver, H. and Barrera, E. (1953). Method for combined staining of cells and fibers in the nervous system, *J. Neuropath. exp. Neurol.*, **12**, 400–3.

Knigge, K. M. and Joseph, S. A. (1974). Thyrotropin releasing factor (TRF) in cerebrospinal fluid of the third ventricle of rat, *Acta endocrinol.*, **76**, 209–13.

Knigge, K. M., Scott, D. E., Kobayashi, H. and Ishi, S. (1975). *Brain-Endocrine Interaction II* (editors), Karger: Basel, London, New York.

Knight, B. K., Hayes, M. M. M. and Symington, R. B. (1973). The pineal gland—a synopsis of present knowledge with particular emphasis on its possible role in control of gonadotrophin function, *S. Afr. J. Anim. Sci.*, **3**, 143–6.

Knisely, M. H. (1936). Spleen studies. I. Microscopic observations of the circulatory system of living unstimulated mammalian spleens, *Anat. Rec.*, **65**, 23–50.

Kobayasi, H. (1975). Absorption of cerebrospinal fluid by ependymal cells of the median eminence. In: *Brain-Endocrine Interaction. II. 2nd International Symposium* (Knigge, K. M., Scott, D. E., Kobayashi, H. and Ishi, S. eds.), pp. 109–22. Karger: Basel, London, New York.

Kobayashi, S. and Fujita, T. (1969). Fine structure of mammalian and avian pancreatic islets with special reference to D cells and nervous elements, *Z. Zellforsch. mikrosk. Anat.*, **100**, 340–63.

Koch, J. C. (1917). The laws of bone architecture, *Am. J. Anat.*, **21**, 177–298.

Koella, W. P. (1969). Control of skeletal motor activity, with emphasis on the role of diencephalic mechanisms. In: *The Hypothalamus* (Haymaker, W., Anderson, E. and Nauta, W. J. H. eds.), pp. 645–58, Thomas: Springfield, Ill.

Koguerman-Lepp, E. P. (1968). Hepatic veins and venous blood outflow from liver segments in man, *Arkh. Anat. Gistol. Embriol.*, **55**, 105–10.

Köhn, A. (1903). Die paraganglien, *Ark. Mikrosk. Anat.*, **62**, 263.

Kohnstamm, O. (1898). Zur Anatomie und Physiologie des Phrenicuskernes, *Fortschr. Med.*, **16**, 643–53.

Kokott, W. (1934). Das Spaltlinienbild der Sklera (Ein Beitrag zum funktionellen Bau der Sklera), *Klin. Mbl. Augenheilk.*, **92**, 177–85.

Kolb, H. and West, R. W. (1977). Synaptic connection of the interplexiform cell in the retina of the cat, *J. Neurocytol.*, **6.**, 155–70.

Kolliker, A. (1896). *Handbuch der Gewebelehre des Menschen*, Sixth edition, Engelmann: Leipzig.

Kolmodin, G. M. (1957). Integrative processes in single spinal interneurones with proprioceptive connections, *Acta physiol. scand.*, **40**, Suppl. 139, 1–89.

Kolmodin, G. M. and Skoglund, C. R. (1960). Analysis of spinal interneurons activated by tactile and nociceptive stimuli, *Acta physiol. scand.*, **50**, 337–55.

Komai, T. and Fukoka, G. (1934). A study of the frequency of left-handedness and left-footedness among Japanese school children, *Hum. Biol.*, **6**, 33–42.

Konigsmark, B. W. (1970). Methods for the counting of neurons. In: *Contemporary Research Methods in Neuroanatomy* (Nauta, W. J. H. and Ebbesson, S. O. E. eds.), pp. 315–39, Springer-Verlag: Berlin.

Konigsmark, B. W., Kalyanarama, U. P., Corey, P. and Murphy, E. A. (1969). An evaluation of techniques in neuronal population estimates: the sixth nerve nucleus, *Johns Hopkins Hosp. Bull.*, **125**, 146–58.

Koornneef, L. (1977). New insights in the human orbital connective tissue, *Arch. Ophthal.*, **95**, 1269–73.

Koritké, J. G., Gillet, J. Y. and Pietri, J. (1967). Les artères de la trompe uterine chez la femme, *Archs Anat. Histol. Embryol.*, **50**, 47–70.

Kormano, M. and Suoranta, H. (1971). Microvascular organisation of the adult human testis, *Anat. Rec.*, **170**, 31–40.

Korneliussen, A. K. (1967). Cerebellar corticogenesis in *Cetacea* with special reference to regional variations, *J. Hirnforsch.*, **9**, 151–85.

Korneliussen, A. K. (1968). On the ontogenetic development of the cerebellum (nuclei, fissures and cortex) of the rat, with special reference to regional variations in corticogenesis, *J. Hirnforsch.*, **10**, 379–412.

Kornguth, S. E., Anderson, J. W. and Scott, G. (1966). Observations on the ultrastructure of the developing cerebellum of the *Macaca mulatta*, *J. comp. Neurol.*, **130**, 1–23.

Kosinski, C. (1926). Observations on the superficial venous system of the lower extremity, *J. Anat.*, **60**, 131–42.

Koskinen, L., Isotupa, K. and Koski, K. (1976). A note on craniofacial sutural growth, *Am. J. phys. Anthrop.*, **45**, 511–16.

Koss, L. G. (1968). *Diagnostic Cytology*, Second edition, Pitman: London.

Kostick, E. L. (1963). Facets and imprints on the upper and lower extremities of femora from a Western Nigerian population, *J. Anat.*, **97**, 393–402.

Kovrishko, N. M. (1964). Postnatal development and structural characteristics of the principal paraganglia in man, *Fed. Proc. (Trans. Suppl.)*, **22**, 740.

Kozielec, T. and Józwa, H. (1977). Variation in the course of the facial artery in the prenatal period in man, *Folia morph., Warsz.*, **36**, 55–61.

Krahl, V. E. (1944). An apparatus for measuring the torsion angle in long bones, *Science, N.Y.*, **99**, 498.

Krahl, V. E. (1964). Anatomy of the mammalian lung. In: *Handbook of Physiology* (Fenn, W. O. and Rahn, H. eds.), Section 3, Volume 1, pp. 213–84, American Physiological Society: Washington.

Krahl, V. E. (1976). The phylogeny and ontogeny of humeral torsion, *Am. J. phys. Anthrop.*, **45**, 595–9.

Kramer, I. R. H. (1960). The vascular architecture of the human dental pulp, *Archs oral Biol.*, **2**, 177–89.

Kraus, B. (1959). Occurrence of the Carabelli trait in Southwest ethnic groups, *Am. J. phys. Anthrop.*, **17**, 117–24.

Kraus, F. W. and Mestecky, J. (1971). Immunohistochemical localization of amylase, lysozyme and immunoglobulins in the human parotid gland, *Archs oral Biol.*, **16**, 781–9.

Kraus, K. S., Jordan, R. E. and Abrams, L. A. (1969). *Dental Anatomy and Occlusion*, Williams and Wilkins: Baltimore.

Krause, W. J. and Leeson, C. R. (1967). The origin, development and differentiation of Brunner's glands in the rat, *J. Anat.*, **101**, 309–20.

Krieckhaus, E. E. (1967). The mamillary bodies: their function and anatomical connections, *Acta Biol. exp., Vars.*, **27**, 319–37.

Krieckhaus, E. E. and Randall, D. (1968). Lesions of the mamillothalamic tract in the rat produce no decrements in recent memory, *Brain*, **91**, 369–78.

Krmpotić-Nemanić, J. and Keros, P. (1973). Funktionale Bedeutung der Adaptation des Dens axis beim Menschen, *Verh. Anat. Ges.*, **67, S**, 393–7.

Krogman, W. M. (1941). *Bibliography of Human Morphology*, Chicago University Press: Chicago.

Krogman, W. M. (1962). *The Human Skeleton in Forensic Medicine*, Thomas: Springfield, Ill.

Krompecher, S. (1967). Local tissue metabolism and the quality of callus, *Symp. Biol. Hung.*, **7**, 275–81.

Kubik, S. (1967). The efferent lymph vessels and the regional lymph nodes of the female genital organs, *Prog. Lymphology*, 196–7.

Kubik, S. (1970). Lung lymphatics, *Prog. Lymphology*, **II**, 29–31.

Kubik, S. and Müntener, M. (1969). Zur Topographie der spinalen Nervenwurzeln. II. Der Einfuss des Wachstums des Duralsackes, sowie der Krümmagen und der Bewegungen der spinalen Nerven-wurzeln, *Acta anat.*, **74**, 149–68.

Kubo, J. (1970). Some observations on the autonomic innervation of the human oviduct, *Int. J. Fert.*, **15**, 30–5.

Kubota, K., Negishi, T. and Nasegi, T. (1975). Topological distribution of muscle spindles in the human tongue, *Bull. Tokyo med. dent. Univ.*, **22**, 235–42.

Kuczynski, K. (1974). Carpometacarpal joint of the human thumb, *J. Anat.*, **118**, 119–26.

Kudo, H. and Nori, S. (1974). Topography of the facial nerve in the human temporal bone, *Acta anat.*, **90**, 467–80.

Kuffler, S. W. and Nicholls, J. G. (1976). *From Neuron to Brain*, Sinauer Associates Inc.: Sunderland, Mass.

Kugel, M. A. (1927). Anatomical studies on the coronary arteries and their branches. I. Arteria anastomotica auricularis magna, *Am. Heart J.*, **3**, 260–70.

Kuhlenbeck, H. (1954). Human diencephalon; summary of development, structure, function and pathology, *Confinia neurol.*, Suppl. 14, 1–230.

Kuhlenbeck, H. and Miller, R. N. (1949). The pretectal region of the human brain, *J. comp. Neurol.*, **91**, 369–408.

Kühnel, W. (1968). Vergleichende histologische histochemische und elektronenmikroskopische Untersuchungen an Tränendrüsen. VI. Menschliche Tränendrüsen, *Z. Zellforsch.*, **89**, 550–72.

Kummer, B. K. F. (1966). Photoelastic studies on the functional structure of bone, *Fol. biotheoret.*, **6**, 31–40.

Kummer, B. K. F. (1972). Biomechanics of bone. In: *Biomechanics* (Fung, Y. C., Perrone, N. and Anliker, M. eds.), Chapter 10, Prentice-Hall: New Jersey.

Kuntscher, G. (1934). Die Darstellung des Kraftflusses im Knocken, *Z. beit. Chir.*, **61**, 2130–6.

Kuntz, A. (1953). *The Autonomic Nervous System*, Fourth edition, Lea and Febiger: Philadelphia.

Kupfer, C., Chumbley, L. and Downer, J. de C. (1967). Quantitative histology of optic nerve, optic tract and lateral geniculate nucleus of man, *J. Anat.*, **101**, 393–402.

Kurahashi, Y. and Yoshiki, S. (1972). Electron microscopic localisation of alkaline phosphatase in the enamel organ of the young rat, *Archs oral Biol.*, **17**, 155–63.

Kuré, K., Saégusa, G., Kawaguchi, K. and Shiraishi, K. (1930). On the parasympathetic (spinal parasympathetic) fibres in the dorsal or posterior roots of the lumbar region of the spinal cord, *Q. Jl exp. Physiol.*, **20**, 333–44.

Kuré, K., Murakami, S. and Okinaka, S. (1934). Die Spinalpara-sympathetischen Ganglionzellen im den Spinalganglien und der Spinalparasympathetiens des Halssegmentes, *Z. Zellforsch. mikrosk. Anat.*, **22**, 54–79.

Kurosumi, K. and Oota, Y. (1968). Electron microscopy of two types of gonadotrophs in the anterior pituitary glands of persistent estrus and diestrus rats, *Z. Zellforsch. mikrosk. Anat.*, **85**, 34–46.

Kurrat, H. J. and Oberländer, W. (1978). The thickness of the cartilage in the hip joint, *J. Anat.*, **126**, 145–55.

Kuru, Y. (1967). Meningeal branches of the ophthalmic artery, *Acta radiol.*, **6**, 241–51.

Kuwabara, T. (1975). The maturation of the lens cell, *Exp. Eye Res.*, **20**, 427–43.

Kuypers, H. G. J. M. (1958). Cortico-bulbar connexions to the pons and lower brain stem in man. An anatomical study, *Brain*, **81**, 364–88.

Kuypers, H. G. J. M. (1960). Central cortical projections to motor and somato-sensory cell groups, *Brain*, **83**, 161–84.

Kuypers, H. G. J. M. and Lawrence, D. G. (1967). Cortical projections to the red nucleus and the brain stem in the rhesus monkey, *Brain Res.*, **4**, 151–88.

Kuypers, H. G. J. M. and Maisky, V. A. (1975). Retrograde axonal transport of horseradish peroxidase from spinal cord to brainstem cell groups in the cat, *Neurosci. Lett.*, **1**, 9–14.

Kvinnsland, S. and Kvinnsland, S. (1975). Growth in craniofacial cartilages studied by ³H-thymidine incorporation, *Growth*, **39**, 305–14.

Kyber, E. (1870). Uber die Milz des Menschen und einiger Saugetiere, *Arch. mikrosk. Anat. EntwMech.*, **6**, 540–70.

L

Lacroix, P. (1951). *The Organisation of Bones*, Churchill: London.

Lacy, D. (1960). Light and electron microscopy and its use in the study of factors influencing spermatogenesis in the rat, *Jl R. microsc. Soc.*, **79**, 209–25.

Lacey, P. E. (1957). Electron microscopic identification of different cell types in the islets of Langerhans of the guinea pig, rat, rabbit and dog, *Anat. Rec.*, **128**, 255–68.

Lacey, P. E. (1967). The pancreatic beta cell, *New Engl. J. Med.*, **276**, 187–95.

Lacy, P. E. and Davies, J. (1957). Preliminary studies on the demonstration of insulin in the islets by the fluorescent antibody technic, *Diabetes*, **6**, 354–7.

Laguesse, E. (1906). La glande nouvelle ou endocrine (Ilots de Langerhans), *Revue gén. Hist.*, **2**, 1–286.

Laitio, M., Lev, R. and Orlic, D. (1974). The developing human fetal pancreas: an ultrastructural and histochemical study with special reference to exocrine cells, *J. Anat.*, **117**, 619–34.

Lallemand, R. C. and Newman, D. L. (1973). Role of the bifurcation in atheromatosis of the abdominal aorta, *Surgery Gynec. Obstet.*, **137**, 987–90.

Lam, J. H., Ranganathan, N. and Wigle, E. D. (1970). Morphology of the human mitral valve. I. Chordae tendineae: a new classification, *Circulation*, **41**, 449–58.

Lambert, M. W. (1955). Accessory bronchio-alveolar communications, *J. Path. Bact.*, **70**, 311–14.

Lamberty, B. G. H. and Živanovic, S. (1973). The retro-articular vertebral artery ring of the atlas and its significance, *Acta anat.*, **85**, 113–22.

Landgren, S., Phillips, C. G. and Porter, R. (1962). Cortical fields of origin of the monosynaptic pyramidal pathways to some alpha motoneurons of the baboon's hand and forearm, *J. Physiol., Lond.*, **161**, 112–25.

Landis, D. M. D. and Reese, T. S. (1947). Differences in membrane structure between excitatory and inhibitory synapses in the cerebellar cortex, *J. comp. Neurol.*, **155**, 93–126.

Landon, D. N. (1966). Electron microscopy of muscle spindles. In: *Control and Innervation of Skeletal Muscle* (Andrew, B. L. ed.), pp. 96–111, Thompson: Dundee.

Landon, D. N. (1970). The influence of fixation upon the fine structure of the Z disk of rat striated muscle, *J. Cell Sci.*, **6**, 257–76.

Landon, D. N. (1970). Observations on the morphogenesis of rat skeletal muscle, *J. Anat.*, **107**, 385P.

Landon, D. N. (1972). The fine structure of the equatorial regions of developing muscle spindles in the rat, *J. Neurocytol.*, **1**, 189–210.

Landon, D. N. and Williams, P. L. (1963). Ultrastructure of the node of Ranvier, *Nature, Lond.*, **199**, 575–7.

Landon, D. N. and Langley, O. K. (1971). The local chemical environment of nodes of Ranvier: a study of cation binding, *J. Anat.*, **108**, 419–32.

Landsmeer, J. M. F. (1949). The anatomy of the dorsal aponeurosis of the human finger and its functional significance, *Anat. Rec.*, **104**, 31–44.

Landsmeer, J. M. F. (1955). Anatomical and functional investigations on the articulation of the fingers, *Acta anat.*, Suppl. 24.

Landsmeer, J. M. F. (1976). *Atlas of Anatomy of the Hand*, Churchill Livingstone: London.

Lane, M. A. (1907). The cytological characters of the areas of Langerhans, *Am. J. Anat.*, **7**, 709–22.

Lang, J. and Schäfer, K. (1977). Über Form, Grösse und Variabilität des Plexus choroideus ventriculi III, *Gegenb. morph. Jahrb. Leipzig*, **123**, 727–41.

Lang, J. and Brunner, F. X. (1978). Über die Rami centrales der Aa. cerebri anterior und media, *Gegenb. morph. Jahrb. Leipzig*, **124**, 364–74.

Lanyon, L. E. (1973). Analysis of surface bone strain in the calcaneus of sheep during normal locomotion, *J. Biomech.*, **6**, 41–9.

Langham, M. E. (1969). *The Cornea* (editor), Johns Hopkins Press: Baltimore.

Lapides, J. (1969). Urinary diversion by anterior transposition of the male urethra, *J. Urol.*, **101**, 338–42.

Lardennois, G. and Okinczyc, J. (1913). La typhlosigmoïdostomie en Y dans la traitement des coutes rebelles et de la stase du gros intestin, *Bull. et Mém. Soc. Anat. de Paris*, **39**, 858–72.

Larsell, O. (1934). Morphogenesis and evolution of the cerebellum, *Archs Neurol. Psychiat., Chicago*, **31**, 373–95.

Larsell, O. (1937). The cerebellum. A review and interpretation, *Archs Neurol. Psychiat., Chicago*, **38**, 580–607.

Larsell, O. (1947). The development of the cerebellum in man in relation to its comparative anatomy, *J. comp. Neurol.*, **87**, 85–130.

Larsell, O. (1948). The development and subdivisions of the cerebellum of birds, *J. comp. Neurol.*, **89**, 123–90.

Larsell, O. (1952). The morphogenesis and adult pattern of the lobules and fissures of the cerebellum of the white rat, *J. comp. Neurol.*, **97**, 281–356.

Larsell, O. (1953). The cerebellum of the cat and monkey, *J. comp. Neurol.*, **99**, 135–200.

Larsson, L.-I., Sundler, F., Alumets, G., Håkanson, R., Schaffalitzky de Muckadell, O. and Fahrenkrug, J. (1977). Distribution, ontogeny and ultrastructure of the mammalian secretin cell, *Cell Tissue Res.*, **181**, 361–8.

Laruelle, L. and Reumont, M. (1933). Etude de l'anatomie microscopique de la moelle épinière par la méthode des coupes longitudinales plurisegmentalés, *Rev. Neurol.*, **44**, 1130–41.

Lasek, R. J. and Hoffman, P. N. (1976). The neuronal cytoskeleton, axonal transport and axonal growth. In: *Cell Motility* (Goldman, R., Pollard, T. and Rosenbaum, J. eds.), Cold Spring Harbor Laboratories: New York.

Lashley, K. S. and Clark, G. (1946). The cytoarchitecture of the cerebral cortex of Ateles: A critical examination of architectonic studies, *J. comp. Neurol.*, **85**, 223–306.

Lasi, G. N. (1959). Pre- and post-natal changes in the thymus gland, Ph.D. Thesis, University of London.

Lassek, A. M. (1940). The human pyramidal tract. II. A numerical investigation of the Beta cells of the motor area, *Archs Neurol. Psychiat., Chicago*, **44**, 718–24.

Lassek, A. M. (1942). The human pyramidal tract. IV. A study of the mature, myelinated fibers of the pyramid, *J. comp. Neurol*, **76**, 217–25.

Lassek, A. M. (1942). The pyramidal tract. The effects of pre- and post-central cortical lesions on the fiber components of the pyramids in the monkey, *J. nerv. ment. Dis.*, **95**, 721–9.

Lassek, A. M. (1954). *The Pyramidal Tract*, Thomas: Springfield, Ill.

Lassek, A. M. and Rasmussen, G. L. (1939). The human pyramidal tract. A fiber and numerical analysis, *Archs Neurol. Psychiat., Chicago*, **42**, 872–6.

Last, R. J. (1948). Some anatomical details of knee joint, *J. Bone Jt Surg.*, **30B**, 683–8.

Last, R. J. (1950). The popliteus muscle and the lateral meniscus, *J. Bone Jt Surg.*, **32B**, 93–9.

Last, R. J. (1951). Specimens from Hunterian collection: synovial cavity of knee joint (specimen S 110A); ligaments of knee (specimen S 95A), *J. Bone Jt Surg.*, **33B**, 442–5.

Last, R. J. (1968). *Wolff's Anatomy of the Eye and Orbit* (editor), Lewis: London.

Latarjet, M., Neidhart, J. H., Morrin, A. and Autissier, J.-M. (1967). L'entrée du nerf musculo-cutané dans le muscle coraco-brachial, *C.r. Ass. Anat.*, **138**, 755–65.

Latham, R. A. (1966). Observations on the growth of the cranial base in the human skull, *J. Anat.*, **100**, 435P.

Latham, R. A. (1973). Development and structure of the premaxillary deformity in bilateral cleft lip and palate, *Br. J. Plast. Surg.*, **26**, 1–11.

Latham, R. A. and Deaton, T. G. (1976). The structural basis of the philtrum and the contour of the vermilion border: a study of the musculature of the upper lip, *J. Anat.*, **121**, 151–60.

Laties, A. M. and Liebman, P. A. (1970). Cones of living amphibian eyes: selective staining, *Science, N.Y.*, **165**, 1475–7.

Laurie, W. and Woods, J. D. (1958). Anastomoses of the coronary circulation, *Lancet*, **ii**, 812–16.

Lauweryns, J.-M. and Boussaw, L. (1967). L'ultrastructure des vaisseaux lymphatiques pulmonaires. *C. r. Ass. Anat.*, **138**, 766–75.

Lauweryns, J. M. and Peuskens, J. C. (1972). Neuro-epithelial bodies (neuroreceptor or secretory organs?) in human infant bronchial and bronchiolar epithelium, *Anat. Rec.*, **172**, 471–82.

Lauweryns, J. M., Cokelaere, M. and Theunynck, P. (1972). Neuro-epithelial bodies in the respiratory mucosa of various mammals, *Z. Zellforsch. mikrosk. Anat.*, **135**, 569–92.

Lauweryns, J. M. and Cokelaere, M. (1973). Hypoxia-sensitive neuro-epithelial bodies, intrapulmonary secretory neuro-receptors modulated by the CNS, *Z. Zellforsch. mikrosk. Anat.*, **145**, 521–46.

Lavelle, C. L. (1974). The effect of age on human third molar and rat molar teeth, *Acta anat.*, **87**, 110–18.

Lawn, A. M. (1966). The localisation, in the nucleus ambiguus of the rabbit, of the cells of origin of motor nerve fibers in the glosso-pharyngeal nerve and various branches of the vagus nerve by means of retrograde degeneration, *J. comp. Neurol.*, **127**, 293–305.

Lazarides, E. (1976). Actin, alpha-actinin and tropomyosin interaction in the structural organisation of actin filaments in non-muscle cells, *J. Cell Biol.*, **68**, 202–19.

Lazorthes, G. (1949). *Le Système Neurovasculaire*, Masson: Paris.

Lazorthes, G., Espagno, J., Lazorthes, Y. and Zadeh, J. O. (1968). The vascular architecture of the cortex and cortical blood flow, *Prog. Brain Res.*, **30**, 27–32.

Lazorthes, G., Gouaze, A., Zadeh, J. O., Santini, J. J., Lazorthes, Y. and Burdin, P. (1971). Arterial vascularization of the spinal cord, *J. Neurosurg.*, **35**, 253–62.

Leak, L. V. and Burke, J. F. (1968). Ultrastructural studies on the lymphatic anchoring filaments, *J. Cell Biol.*, **36**, 129–49.

Leao, A. A. P. (1944). Spreading depression of activity in cerebral cortex, *J. Neurophysiol.*, **7**, 359–90.

Leblond, C. P. (1950). Distribution of periodic acid-reactive carbo-hydrates in the adult rat, *Am. J. Anat.*, **86**, 1–50.

Leblond, C. P. and Gross, J. (1948). Thyroglobulin formation in thyroid follicle visualized by 'coated autograph' technique, *Endocrinology*, **43**, 306–24.

Leblond, C. P. and Clermont, Y. (1952). Definition of the stages of the cycle of the seminiferous epithelium in the rat, *Ann. N.Y. Acad. Sci.*, **55**, 548–73.

Leblond, C. P. and Clermont, Y. (1952). Spermiogenesis of rat, mouse, hamster and guinea pig as revealed by the 'periodic acid-fuchsin sulfurous acid' technique, *Am. J. Anat.*, **90**, 167–216.

Leborgne, J., Letenneur, J., Pannier, M., Visset, J., Bainvel, J.-V. and Barbin, J.-Y. (1973). Considérations sur la vascularisation artérielle du muscle carré crural, *Archs Anat. Path.*, **21**, 359–63.

Lebourg, L. and Champagne, G. (1951). A propos du dévelopment mandibulaire post-natal. Precisions sur la chronologie de la suture symphysaire, *Rev. Stomat.*, **52**, 891–7.

Lecco, V. and Balli, R. (1968). Alcune considerazioni etiopatogenetiche e clinche su due casi di diverticulosi multipla dell'esofago, *Archo ital. Otol. Rinol. Lar.*, **79**, 497–506.

Le Douarin, N. M. (1978). The embryological origin of the endocrine cells associated with the digestive tract. Experimental analysis based on the use of a stable cell marking technique. In: *Gut Hormones* (Bloom, S. R. ed.), pp. 49–56, Churchill Livingstone: Edinburgh, London.

Le Douarin, N. M. and Teillet, M. A. (1974). Experimental analysis of the migration and differentiation of neuroblasts of the autonomic nervous system and of neuroectodermal mesenchymal derivatives, using a biological marker technique, *Dev. Biol.*, **41**, 162–84.

LeDuc, I. E. (1939). Anatomy of the prostate and pathology of early benign hyperplasia, *J. Urol.*, **42**, 1217–41.

Lee, C.-S. and Tsai, T.-L. (1974). The relation of the sciatic nerve to the piriformis muscle, *J. Formosan Med. Ass.*, **73**, 75–80.

Lee, W. R. (1964). Appositional bone formation in canine bone: a quantitative microscopic study using tetracycline markers, *J. Anat.*, **98**, 655–77.

Leela, K. and Kanagasuntheram, R. (1968). A microscopic study of the human pyloro-duodenal junction and proximal duodenum, *Acta anat.*, **71**, 1–12.

Leela, K., Kanagasuntheram, R. and Khoo, F. Y. (1974). Morphology of the primate nasopharynx, *J. Anat.*, **117**, 333–40.

Leeson, T. S. (1957). The fine structure of the mesonephros of the 17-day rabbit embryo, *Expl Cell Res.*, **12**, 670–2.

Leeson, T. and Leeson, C. R. (1968). The fine structure of Brunner's glands in man, *J. Anat.*, **103**, 264–76.

Leeson, T. S. and Leeson, C. R. (1964). A light and electron microscope study of developing respiratory tissues in the rat, *J. Anat.*, **98**, 183–93.

Leeson, T. S. and Leeson, C. R. (1965). The fine structure of cavernous tissue in the adult rat penis, *Invest. Urol.*, **3**, 144–54.

Legait, H., Contet-Audonneau, J. L., Burlet, C. and Floquet, J. (1973). Etude des rapports des volumes du cerveau de l'hypothalamus et des lobes hypophysaires chez divers mammifères, *Bull. Ass. Anat.*, **57**, No. 158.

Lehninger, A. L. (1964). *The Mitochondrion*, Benjamin: New York.

Lehninger, A. L. (1965). *Bioenergetics*, Benjamin: New York, Amsterdam.

Leikola, A. (1976). The neural crest: migrating cells in embryonic development, *Folia morphol.*, **24**, 155–72.

Leithner, C., Sinzinger, H., Hohenecker, J., Wicke, L., Olbert, F. and Feigl, W. (1975). Radiologic anatomy of the abdominal aorta and their large branches, *Okajimas Folia anat. jap.*, **52**, 119–50.

Lele, P. P. and Weddell, G. (1956). The relationship between neuro-histology and corneal sensibility, *Brain*, **79**, 119–54.

LeMay, M. and Culebras, A. (1972). Human brain—morphologic differences in the hemispheres demonstrable by carotid arteriography, *New Eng. J. Med.*, **287**, 168–70.

Lenneberg, E. H. (1964). *New Directions in the Study of Language*, Massachusetts Institute of Technology Press: Cambridge.

Leppi, T. J. and Spicer, S. S. (1966). The histochemistry of mucins in certain primate salivary glands, *Am. J. Anat.*, **118**, 833–60.

Lerche, W. (1965). Elektronenmikroskopische Beobachtungen über die Histogenese der bruchschen Membran des Menschen, *Z. Zellforsch. mikrosk. Anat.*, **65**, 163–75.

Leslie, D. R. (1954). The tendons on the dorsum of the hand, *Aust. N.Z. J. Surg.*, **23**, 253–6.

Leuchtenberger, C. and Schrader, F. (1950). The chemical nature of the acrosome in the male germ cells, *Proc. natn. Acad. Sci. U.S.A.*, **36**, 677–83.

Lev, R. and Orlic, D. (1973). Uptake of protein in swallowed amniotic fluid by monkey fetal intestine *in utero*, *Gastroenterol.*, **65**, 60–8.

Lev, R. and Orlic, D. (1974). Histochemical and radioautographic studies of normal human fetal colon, *Histochemistry*, **39**, 301–11.

LeVay, S. (1971). On the neurons and synapses of the lateral geniculate nucleus of the monkey, *Z. Zellforsch. mikrosk. Anat.*, **113**, 396–419.

LeVay, S., Stryker, M. P. and Shatz, C. J. (1978). Ocular dominance

BIBLIOGRAPHY

columns and their development in layer IV of the cat's visual cortex, *J. comp. Neurol.*, **179**, 223–44.

Levene, C. (1964). The patterns of cartilage canals, *J. Anat.*, **98**, 515–38.

Lever, J. D. (1955). Electron microscopic observations on the adrenal cortex, *Am. J. Anat.*, **97**, 409–30.

Lever, J. D., Lewis, P. R. and Boyd, J. D. (1959). Observations on the fine structure and histochemistry of the carotid body in the cat and rabbit, *J. Anat.*, **93**, 478–90.

Lever, J. D., Irvine, G. and Chick, W. J. (1965). The vesiculated axons in relation to arteriolar smooth muscle in the pancreas. A fine structural and quantitative study, *J. Anat.*, **99**, 299–313.

Lever, J. D., Spriggs, T. L. B. and Graham, J. D. P. (1968). A formol-fluorescence, fine-structural and autoradiographic study of the adrenergic innervation of the vascular tree in the intact and sympathectomised pancreas of the cat, *J. Anat.*, **103**, 15–34.

Levi-Montalcini, R. (1950). The origin and development of the visceral system in the spinal cord of the chick embryo, *J. Morph.*, **86**, 253–83.

Levi-Montalcini, R. (1952). Effects of mouse tumor transplants on the nervous system, *Ann. N.Y. Acad. Sci.*, **55**, 330–43.

Levi-Montalcini, R. (1960). Destruction of the sympathetic ganglia in mammals by an antiserum to the nerve growth-promoting factor, *Proc. natn. Acad. Sci. U.S.A.*, **46**, 384–91.

Levi-Montalcini, R. (1967). Differentiation and growth control mechanisms in the nervous system. In: *Morphological and Biochemical Aspects of Cytodifferentiation* (*Exp. Biol. Med.*, **1**, 170–82), Karger: Basel, New York.

Levi-Montalcini, R. and Levi, G. (1942). Les conséquences de la destruction d'un territoire d'innervation périphérique sur le développement des centres nerveux correspondants dans l'embryon de poulet, *Archs Biol., Liège*, **53**, 537–45.

Levi-Montalcini, R. and Hamburger, V. (1951). Selective growth-stimulating effects of mouse sarcoma on the sensory and sympathetic nervous system of the chick embryo, *J. exp. Zool.*, **116**, 321–62.

Levi-Montalcini, R. and Booker, B. (1960). Excessive growth of the sympathetic ganglia evoked by a protein isolated from mouse salivary glands, *Proc. natn. Acad. Sci. U.S.A.*, **46**, 373–83.

Levi-Montalcini, R. and Angeletti, P. U. (1963). Essential role of the nerve growth factor in the survival and maintenance of dissociated sensory and sympathetic embryonic nerve cells *in vitro*, *Devl Biol.*, **7**, 653–9.

Levin, P. M. and Bradford, F. K. (1938). The exact origin of the cortico-spinal tract in the monkey, *J. comp. Neurol.*, **68**, 411–22.

Levinger, I. M. and Kedem, J. (1974). A method for the evaluation of the surface area of cerebral ventricles in animals, *J. Anat.*, **117**, 481–5.

Levkova, N. A. and Kakabadze, S. A. (1977). Ultrastructural organisation of chromaffin paraganglia in the nodes of the solar plexus, *Arkh. Anat. Gistol. Embriol.*, **72**, 53–8.

Levy, J. and Nagylaki, T. (1972). A model for the genetics of handedness, *Genetics*, **72**, 117–28.

Lewin, B. M. (1970). *The Molecular Basis of Gene Expression*, Wiley: London, New York, Sydney, Toronto.

Lewin, M. L. (1976). Spheno-pharyngeal meningocoele and cleft palate. Case report with 12 year follow up, *Cleft Palate J.*, **13**, 61–73.

Lewin, W. (1961). Observations on selective leucotomy, *J. Neurol. Neurosurg. Psychiat.*, **24**, 37–44.

Lewis, F. T. (1912). The form of the stomach in human embryos with notes upon the nomenclature of the stomach, *Am. J. Anat.*, **13**, 477–503.

Lewis, K. R. and John, B. (1970). *The Organisation of Heredity*, Arnold: London.

Lewis, O. J. (1957). The blood vessels of the adult mammalian spleen, *J. Anat.*, **91**, 245–50.

Lewis, O. J. (1957). The formation and development of the blood vessels of the mammalian cerebral cortex, *J. Anat.*, **91**, 40–6.

Lewis, O. J. (1958). The development of the blood vessels of the metanephros, *J. Anat.*, **92**, 84–97.

Lewis, O. J. (1958). The tubercle of the tibia, *J. Anat.*, **92**, 587–92.

Lewis, O. J. (1958). The vascular arrangement of the mammalian renal glomerulus as revealed by a study of its development, *J. Anat.*, **92**, 433–40.

Lewis, O. J. (1959). The coraco-clavicular joint, *J. Anat.*, **93**, 296–303.

Lewis, O. J. (1962). The comparative morphology of M. flexor accessorius and the associated long flexor tendons, *J. Anat.*, **96**, 321–33.

Lewis, O. J. (1964). The homologies of the mammalian tarsal bones, *J. Anat.*, **98**, 195–208.

Lewis, O. J. (1964). The tibialis posterior tendon in the primate foot, *J. Anat.*, **98**, 209–18.

Lewis, O. J. (1965). The evolution of the Mm. Interossei in the primate hand, *Anat. Rec.*, **153**, 275–88.

Lewis, O. J. (1977). Joint remodelling and the evolution of the human hand, *J. Anat.*, **123**, 157–201.

Lewis, O. J., Hamshere, R. J. and Bucknill, T. M. (1970). The anatomy of the wrist joint, *J. Anat.*, **106**, 539–52.

Lewis, P. R. and Shute, C. C. D. (1959). Selective staining of visceral efferents in the rat brain stem by a modified Koelle technique, *Nature, Lond.*, **183**, 1743–4.

Lewis, P. R. and Lever, J. D. (1960). The association of certain chemical activities with intracellular structure, *Jl R. microsc. Soc.*, **78**, 104–10.

Lewis, P. R. and Shute, C. C. D. (1966). The distribution of cholinesterase in cholinergic neurons demonstrated with the electron microscope, *J. Cell Sci.*, **1**, 381–90.

Lewis, W. B. (1878). On the comparative structure of the cortex cerebri, *Brain*, **1**, 79–96.

Lewis, W. H. (1945). Axon growth and regeneration, *Anat. Rec.*, **91**, 287 (abstract).

Lewis, W. H. and Hartman, C. G. (1933). Early cleavage stages of the egg in the monkey (*Macacus rhesus*), *Contr. Embryol.*, **24**, 187–201.

Lewy, F. H. and Kobrak, H. (1936). The neural projection of the cochlear spirals on the primary acoustic centers, *Archs Neurol. Psychiat., Chicago*, **35**, 839–52.

Leydig, F. (1850). Zur Anatomie der mannlichen Geschelechtsorgane und Analdrusen der Saugethiere, *Z. wiss. Zool.*, **2**, 1–10.

Leyton, A. S. and Sherrington, C. S. (1917). Observations on the excitable cortex of the chimpanzee, orang utan and gorilla, *Q. Jl exp. Physiol.*, **11**, 135–222.

Libet, B. and Owman, C. (1974). Concomitant changes in formaldehyde-induced fluorescence of dopamine interneurons and in slow inhibitory post-synaptic potentials, *J. Physiol., Lond.*, **237**, 635–62.

Liddell, E. G. T. and Phillips, C. G. (1950). Thresholds of cortical representation, *Brain*, **73**, 125–40.

Lieb, F. J. and Perry, J. (1968). Quadriceps function. An anatomical and mechanical study using amputated limbs, *J. Bone Jt Surg.*, **50A**, 1535–48.

Lieberman, A. R. (1968). An investigation by light and electron microscopy of chromatolytic and other phenomena induced in mammalian nerve cells by experimental lesions, Ph.D. Thesis, University of London.

Lieberman, A. R. (1973). Neurons with presynaptic perikarya and presynaptic dendrites in the rat lateral geniculate nucleus, *Brain Res.*, **59**, 35–59.

Lieberman, A. R. (1974). Some factors affecting retrograde neuronal responses to axonal lesions. In: *Essays on the Nervous System* (Bellairs, R. and Gray, E. G. eds.), pp. 71–104, Clarendon Press: Oxford.

Lieberman, A. R. (1974). Comments on the fine structural organization of the dorsal lateral geniculate nucleus of the mouse, *Z. Anat. EntwGesch.*, **145**, 261–7.

Lieberman, A. R. (1976). Sensory ganglia. In: *The Peripheral Nerve* (Landon, D. N. ed.), pp. 188–278, Chapman & Hall: London.

Lieberman, A. R. and Webster, K. E. (1974). Aspects of the synaptic organisation of intrinsic neurons in the dorsal lateral geniculate nucleus, *J. Neurocytol.*, **3**, 677–710.

Lightoller, G. H. S. (1925). Facial muscles. The modiolus and muscles surrounding the rima oris with some remarks about the panniculus adiposus, *J. Anat.*, **60**, 1–85.

Like, A. A. (1967). The ultrastructure of the secretory cells of the islets of Langerhans in man, *Lab. Invest.*, **16**, 937–52.

Lillie, F. R. (1913). The mechanism of fertilisation, *Science, N.Y.*, **38**, 524–8.

Lillie, F. R. (1917). The free-martin; a study of the action of sex hormones in the foetal life of cattle, *J. exp. Zool.*, **23**, 371–452.

Lillie, R. D. (1944). Factors influencing the Romanowsky staining of blood films and the role of methylene violet, *J. Lab. clin. Med.*, **29**, 1181–97.

Lim, C. H. and Ruskell, G. L. (1978). Corneal nerve access in monkeys, *Alb. von Graefes Arch. Klin. exp. Ophthal.*, **208**, 15–23.

Limbrick, A. R. and Finean, J. B. (1970). X-ray diffraction and electron microscope studies of the brush border membrane of guinea pig intestinal epithelial cells, *J. Cell Sci.*, **7**, 373–86.

Lind, J. and Wegelius, C. (1954). Human fetal circulation: Change in the cardiovascular system at birth and disturbances in the post-natal closure of the foramen ovale and ductus arteriosus, *Cold Spring Harb. Symp. quant. Biol.*, **19**, 109–25.

Lindahl, P. E. and Drevius, L. O. (1964). Observations on bull spermatozoa in a hypotonic medium related to sperm mobility mechanisms, *Expl Cell Res.*, **36**, 632–46.

Lindahl, U. and Rodén, L. (1966). The chondroitin 4-sulfate-protein linkage, *J. biol. Chem.*, **241**, 2113–19.

Lindsay, R. D. and Scheibel, A. B. (1974). Quantitative analysis of the dendrite branching pattern of small pyramidal cells from adult rat somesthetic and visual cortex, *Exp. Neurol.*, **45**, 424–35.

Lindvall, M., Edvinsson, L. and Owman, C. (1977). Histochemical study on regional differences in the cholinergic nerve supply of the choroid plexus from various laboratory animals, *Exp. Neurol.*, **55**, 152–9.

Lindvall, O. and Björklund, A. (1974). The organization of the ascending catecholamine neuron systems in the rat brain as revealed by the glyoxylic acid fluorescence method, *Acta physiol. Scand.*, **412**, 1–48.

Ling, E. A. (1978). Evidence for a haematogenous origin of some of the macrophages appearing in the spinal cord of the rat after dorsal rhizotomy, *J. Anat.*, **128**, 43–154.

Ling, E. A., Paterson, J. A., Privat, A., Mori, S. and Leblond, C. P. (1973). Investigation of glial cells in semithin sections. I. Identification of glial cells in the brain of young rats, *J. comp. Neurol.*, **149**, 43–72.

Linkevich, V. R. (1969). Embryogenesis of female internal genitalia, *Akush. Gimek., Mosk.*, **7**, 43–7.

Lipkin, M., Sherlock, P. and Bell, B. (1963). Cell proliferation kinetics in the gastro-intestinal tract of man. II. Cell renewal in stomach, ileum, colon and rectum, *Gastroenterology*, **45**, 721–9.

Lippert, H. and Käfer, H. (1974). Biomechanik des Schädeldachs. 2. Dicken der Knochenschichten, *Mschr. Unfallheilk.*, **77**, 329–39.

Lisa, J. R., Gioia, J. D. and Rubin, I. C. (1954). Observations on interstitial portion of the fallopian tube, *Surgery Gynec. Obstet.*, **99**, 159–69.

Lissák, K. (1967). *Recent Developments of Neurobiology in Hungary* (editor), Akadémiai Kiadó: Budapest.

Lissák, K. (1967). *Results in Neuroanatomy, Neurochemistry, Neuropharmacology and Neurophysiology* (editor), Akadémiai Kiadó: Budapest.

Lissmann, H. W. and Machin, K. E. (1958). The mechanism of object location in *Gymnarchus niloticus* and similar fish, *J. exp. Biol.*, **35**, 451–86.

Lister, U. M. (1968). Ultrastructure of the human amnion, chorion and fetal skin, *J. Obstet. Gynaec. Br. Commonw.*, **75**, 327–41.

Listgarten, M. A. (1966). Phase contrast and electron microscopic study of the junction between reduced enamel epithelium and enamel in unerupted human teeth, *Archs oral Biol.*, **11**, 999–1016.

Listgarten, M. A. (1970). Changing concepts about the dentoepithelial junction, *J. Can. dent. Ass.*, **36**, 70–5.

Litt, M. (1964). Eosinophils and antigen-antibody reactions, *Ann. N.Y. Acad. Sci.*, **116**, 964–85.

Liu, C. N. and Chambers, W. W. (1964). An experimental study of the cortico-spinal system in the monkey (*Macaca mulatta*). The spinal pathways and preterminal distribution of degenerating fibers following discrete lesions of the pre- and postcentral gyri and bulbar pyramid, *J. comp. Neurol.*, **123**, 257–84.

Livingston, R. B. (1970). Some general integrative aspects of brain function. In: *Control Processes in Multicellular Organisms* (Wolstenholme, G. E. W. and Knight, J. eds.), pp. 384–400, Ciba Foundation Symposium, Churchill: London.

Locke, S. (1961). The projection of the magnocellular medial geniculate body, *J. comp. Neurol.*, **116**, 179–93.

Locke, S. (1967). Thalamic connections to insular and opercular cortex of monkey, *J. comp. Neurol.*, **129**, 219–40.

Lockwood, C. B. (1886). The anatomy of the muscles, ligaments and fasciae of the orbit, including an account of the capsule of tenon, the check ligaments of the recti and of the suspensory ligaments of the eye, *J. Anat.*, **20**, 1–25.

Lodge, T. (1946). Anatomy of blood vessels of the human lung as applied to chest radiology, *Br. J. Radiol.*, **19**, 1–7.

Loewenfeld, I. E. (1958). Mechanisms of reflex dilation of the pupil. Historical review and experimental analysis, *Documenta ophth.*, **12**, 185–448.

Loewenstein, W. R. (1971). Mechano-electric transduction in the Pacinian corpuscle. Initiation of sensory impulse in mechanoreceptors. In: *Handbook of Sensory Physiology* (Loewenstein, W. R. ed.), Volume 1, pp. 269–90, Springer: Berlin.

Loewy, A. G. and Siekevitz, P. (1969). *Cell Structure and Function*, Second edition, Holt, Rinehart and Winston: New York.

Loken, A. C. and Brodal, A. (1970). A somatotopical pattern in the human lateral vestibular nuclei, *Archs Neurol. Psychiat., Chicago*, **23**, 350–7.

Lombard, R. E. and Bolt, J. R. (1979). Evolution of the tetrapod ear: an analysis and reinterpretation, *Biol. J. Linn. Soc.*, **11**, 19–76.

Lømo, T. and Westgaard, R. H. (1974). Contractile properties of muscle: control by pattern of muscle activity in the rat, *Proc. R. Soc.*, **187B**, 99–103.

Lømo, T. and Westgaard, R. H. (1975). Control of ACh sensitivity in rat muscle fibers, *Cold Spring Harb. Symp. quant. Biol.*, **40**, 263–74.

Long, C. (1968). Intrinsic-extrinsic muscle control of the fingers. Electromyographic studies, *J. Bone Jt Surg.*, **50A**, 973–84.

Long, C. and Brown, M. E. (1964). Electromyographic kinesiology of the hand: muscles moving the long finger, *J. Bone Jt Surg.*, **46A**, 1683–1706.

Long, J. A. and Jones, A. L. (1967). Observations on the fine structure of the adrenal cortex of man, *Lab. Invest.*, **17**, 355–70.

Longfield, M. D., Dowson, D., Walker, P. S. and Wright, V. (1969). 'Boosted lubrication' of human joints by fluid enrichment and entrapment, *Bio-Med. Eng.*, **4**, 517–22.

Longmore, R. B. (1976). Reflected light interference microscopy (RLIM) of load-bearing human articular cartilage, *Proc. R. microsc. Soc. Lond.*, **11**, 60–1.

Longmore, R. B. and Gardner, D. L. (1978). The surface structure of ageing human articular cartilage: a study by reflected light interference microscopy (RLIM), *J. Anat.*, **126**, 353–65.

Longo, D. (1972). Placental transfer mechanisms—an overview, *Obstet. Gynec. Ann.*, **1**, 103–38.

Lopashov, G. V. and Stroeva, O. G. (1961). Morphogenesis of the vertebrate eye. In: *Advances in Morphogenesis* (Abercrombie, M. and Brachet, J. eds.), **1**, pp. 331–78, Academic Press: New York, London.

Lorente de Nó, R. (1933). Anatomy of the eighth nerve. I. The central projections of the nerve endings of the internal ear, *Laryngoscope, St. Louis*, **43**, 1–38.

Lorente de Nó, R. (1934). Studies on the striation of the cerebral cortex. II. Continuation of the study of the ammonic system, *J. Psychol. Neurol., Lpz.*, **46**, 113–77.

Lorente de Nó, R. (1949). Cerebral cortex: architecture, intracortical connections, motor projections. In: Fulton's *Physiology of the Nervous System*, Third edition, pp. 288–312, Oxford University Press: London.

Louw, J. H. (1959). Congenital intestinal atresia and stenosis in the newborn. Observations on its pathogenesis and treatment, *Ann. R. Coll. Surg.*, **25**, 209–34.

Low, A. (1907). A note on the crura of the diaphragm and the muscle of Treitz, *J. Anat.*, **42**, 93–6.

Lowenstein, O. and Loewenfeld, I. E. (1969). The pupil. In: *The Eye* (Davson, H. ed.), Volume 3, Second edition, pp. 256–337, Academic Press: New York, London.

Lowenstein, W. R. and Kanno, Y. (1964). Studies on an epithelial (gland) cell junction. I. Modification of surface membrane permeability, *J. Cell Biol.*, **22**, 565–86.

Lowy, J. and Small, J. V. (1970). The organisation of myosin and actin in vertebrate smooth muscle, *Nature, Lond.*, **227**, 46–51.

Lubińska, L. (1964). Axoplasmic streaming in regenerating and in normal nerve fibres, *Prog. Brain Res.*, **13**, 1–71.

Luck, D. J. and Reich, E. (1964). DNA in mitochondria of *Neurospora crassa*, *Proc. natn. Acad. Sci. U.S.A.*, **52**, 931–8.

Luckett, W. P. (1975). The development of primordial and definitive amniotic cavities in early rhesus monkey and human embryos, *Am. J. Anat.*, **144**, 149–67.

Lucy, J. A. (1968). Ultrastructure of membranes: micellar organisation, *Br. med. Bull.*, **24**, 127–9.

Luk, S. C., Nopajaroonskri, C. and Simon, G. T. (1973). The architecture of the normal lymph node and hemolymph node. A scanning and transmission electron microscopic study, *Lab. Invest.*, **29**, 258–65.

Lundberg, A. (1958). Electrophysiology of salivary glands, *Physiol. Rev.*, **38**, 21–40.

Lutfi, A. M. (1970). Mode of growth, fate and functions of cartilage canals, *J. Anat.*, **106**, 135–46.

Lutfi, A. M. (1974). The role of cartilage in long bone growth: a reappraisal, *J. Anat.*, **117**, 413–17.

Lyon, M. F. (1962). Sex chromatin and gene action in the mammalian X-chromosome, *Am. J. hum. Genet.*, **14**, 135–48.

Lytle, W. J. (1959). The common bile-duct groove in the pancreas, *Br. J. Surg.*, **47**, 209–12.

Lytle, W. J. (1970). The deep inguinal ring, development, function and repair, *Br. J. Surg.*, **57**, 531–6.

M

Maayan, M. L. and Ingbar, S. H. (1970). Effects of epinephrine on iodine and intermediary metabolism in isolated thyroid cells, *Endocrinology*, **87**, 588–95.

MacCabe, J. A., Errick, J. and Saunders, J. W. Jr. (1974). Ectodermal control of the dorsoventral axis in the leg bud of the chick embryo, *Dev. Biol.*, **39**, 69–82.

McCall, J. G. (1968). Scanning electron microscopy of articular surfaces, *Lancet*, **ii**, 1194.

McClure, C. F. W. and Butler, E. G. (1925). The development of the vena cava inferior in man, *Am. J. Anat.*, **35**, 331–84.

McColl, I. (1967). The comparative anatomy and pathology of anal glands, *Ann. R. Coll. Surg.*, **40**, 36–67.

MacConaill, M. A. (1932). The function of intra-articular fibrocartilages, with special reference to the knee and inferior radio-ulnar joints, *J. Anat.*, **66**, 210–27.

MacConaill, M. A. (1941). The mechanical anatomy of the carpus and its bearings on some surgical problems, *J. Anat.*, **775**, 166–75.

BIBLIOGRAPHY

MacConaill, M. A. (1945). The postural mechanism of the human foot, *Proc. R. Ir. Acad. L. Sect B.*, **14**, 265–78.

MacConaill, M. A. (1946). Some anatomical factors affecting stabilising functions of muscles, *Ir. J. med. Sci.*, 6th series, 160–4.

MacConaill, M. A. (1949). Movements of bones and joints; function of musculature, *J. Bone Jt Surg.*, **31B**, 100–4.

MacConaill, M. A. (1950). Rotary movements and functional décalage, with some references to rehabilitation, *Br. J. phys. Med. ind. Hyg.*, **13**, 50–6.

MacConaill, M. A. (1953). The movements of bones and joints. V. The significance of shape, *J. Bone Jt Surg.*, **35B**, 290–7.

MacConaill, M. A. (1964). Joint movements, *Physiotherapy*, November, 359–67.

MacConaill, M. A. (1966). The geometry and algebra of articular kinematics, *Bio-Med. Eng.*, **1**, 205–12.

MacConaill, M. A. (1966). The compound polarizer, *Lab. Pract.*, **15**, 659–63.

MacConnaill, M. A. (1975). The muscular slings of the mandible, *J. Ir. dent. Ass.*, **21**, 22–4.

MacConaill, M. A. and Basmajian, J. V. (1977). *Muscles and Movement*, Krieger Publishing Company: New York.

McCotter, R. E. (1912). The connections of the vomeronasal nerves with the accessory olfactory bulb in the oppossum and other mammals, *Anat. Rec.*, **6**, 299–318.

McCotter, R. E. (1915). A note on the course and distribution of the nervus terminalis in man, *Anat. Rec.*, **9**, 243–6.

McCuskey, R. S. and Chapman, T. M. (1969). Microscopy of the living pancreas *in situ*, *Am. J. Anat.*, **126**, 395–407.

McCutchen, C. W. (1959). Sponge-hydrostatic and weeping bearings, *Nature, Lond.*, **184**, 1284–5.

McDonald, D. A. (1960). *Blood Flow in Arteries*, Arnold, London.

McDonald, D. M. and Mitchell, R. A. (1975). The innervation of glomus cells, ganglion cells and blood vessels in the rat carotid body: a quantitative ultrastructural analysis, *J. Neurocytol.*, **4**, 177–230.

MacDonald, M. S. and Emery, J. L. (1959). The late intrauterine and postnatal development of human renal glomeruli, *J. Anat.*, **93**, 331–40.

MacDonald, W. C., Trier, J. S. and Everett, N. B. (1964). Cell proliferation and migration in the stomach, duodenum and rectum of man, radioautographic studies, *Gastroenterology*, **46**, 405–17.

MacDougall, J. D. B. (1955). The attachments of the masseter muscle, *Br. dent. J.*, **98**, 193–9.

McEwen, W. K. and Goodner, E. T. (1962). Secretion of tears and blinking. In: *The Eye* (Davson, H. ed.), Volume 3, pp. 341–78, Academic Press: New York, London.

McGrath, P. (1971). The volume of the human pharyngeal hypophysis in relation to age and sex, *J. Anat.*, **110**, 275–82.

McGrath, P. (1974). The pharyngeal hypophysis in some laboratory animals, *J. Anat.*, **117**, 95–115.

McGrath, P. (1977). The cavernous sinus: an anatomical survey, *Aust. N.Z. J. Surg.*, **47**, 601–13.

McGrath, P. (1978). Aspects of the human pharyngeal hypophysis in normal and anencephalic fetuses and neonates and their possible significance in the mechanism of its control, *J. Anat.*, **127**, 65–81.

MacGregor, A. (1936). An experimental investigation of the lymphatic system of the teeth and jaws, *Proc. R. Soc. Med.*, **29**, 1237–72.

McGuigan, J. E. (1968). Antibodies to the C-terminal tetrapeptide amide of gastrin, *Gastroenterology*, **54**, 1012–17.

McFarland, G. B. Jr., Krusen, U. L. and Weathersby, H. T. (1962). Kinesiology of selected muscles acting on the wrist: electromyographic study, *Archs phys. Med.*, **43**, 165–71.

Machado, A. B. and DiDio, L. J. (1967). Frequency of the musculus palmaris longus studied *in vivo* in some Amazon indians, *Am. J. phys. Anthrop.*, **27**, 11–20.

Maciewicz, R. J., Kaneko, C. R. and Highstein, S. M. (1975). Morphophysiological identification of interneurons in the oculomotor nuclei that project to the abducens nucleus in the cat, *Brain Res.*, **96**, 60–5.

Maciewicz, R. J., Eagen, K., Kaneko, C. R. S. and Highstein, S. M. (1977). Vestibular and medullary brain stem afferents to the abducent nucleus in the cat, *Brain Res.*, **123**, 229–40.

McIndoe, A. H. and Counsellor, V. S. (1927). The bilaterality of the liver, *Am. med. Assoc. Arch. Surg.*, **13**, 589–612.

McIntosh, R. (1959). The problem of congenital malformations; general consideration, *J. chron. Dis.*, **10**, 139–51.

McIntosh, J. R., Hepler, P. K. and Van Wie, D. G. (1969). Model for mitosis, *Nature, Lond.*, **224**, 659–63.

MacIntosh, S. R. (1974). The innervation of the conjunctiva in monkeys, *V. Graefes Arch. Ophthal.*, **192**, 105–16.

McIntyre, A. K., Holman, M. E. and Veale, J. L. (1967). Cortical responses to impulses from single Pacinian corpuscles in the cat's hind limb, *Expl Brain Res.*, **4**, 243–55.

McKay, D. G., Hertig, A. T., Bardawil, W. A. and Velardo, J. T. (1956). Histochemical observations on the endometrium. I. Normal endometrium, *Obstet. Gynec., N.Y.*, **8**, 22–39.

MacKenzie, D. W. Jr., Whipple, A. O. and Winterstiener, M. P. (1941). Studies on the microscopic anatomy and physiology of living transilluminated mammalian spleens, *Am. J. Anat.*, **68**, 397–456.

McKenzie, J. (1948). The parotid gland in relation to the facial nerve, *J. Anat.*, **82**, 183–6.

McKenzie, J. (1955). A Bronze Age burial near Stonehaven, Kincardineshire, *J. Anat.*, **89**, 579P.

McKenzie, J. (1955). The foot as a half-dome, *Br. med. J.*, **1**, 1068–70.

McKern, T. W. and Stewart, T. D. (1957). *Skeletal Age changes in Young American Males, Analysed from the Standpoint of Identification*, H.Q.M. Res. and Dev. Command. Tech. Rep. EP-45: Natick, Mass.

McKibbin, B. (1968). The action of the iliopsoas muscle in the newborn, *J. Bone Jt Surg.*, **50B**, 161–5.

MacKinnon, I. L. and MacKinnon, P.C.B. (1958). Seasonal rhythm in the morphology of the suprarenal cortex in women of child-bearing age, *J. Endocr.*, **17**, 456–67.

MacKinnon, P. C. B. and MacKinnon, I. L. (1960). Morphologic features of the human suprarenal cortex in men aged 20–86 years, *J. Anat.* **94**, 183–91.

Macklin, C. C. (1936). Alveolar pores and their significance in the human lung, *Archs Path.*, **21**, 202–16.

Macklin, C. C. (1954). The pulmonary alveolar mucoid film and the pneumonocyte, *Lancet*, **i**, 1099–1104.

McKusick, V. A. (1969). *Human Genetics*, Second edition, Prentice-Hall: New Jersey.

McKusick, V. A. (1975). *Mendelian Inheritance in Man*, Fourth edition, Johns Hopkins Press: Baltimore.

McLachlan, E. M. (1974). The formation of synapses in mammalian sympathetic ganglia, *J. Physiol., Lond.*, **237**, 217–42.

McLean, F. C. and Urist, M. R. (1969). *Bone: an Introduction to the Physiology of Skeletal Tissue*, Chicago University Press: Chicago.

MacLean, P. D. (1958). The limbic system with respect to self-preservation and the preservation of the species, *J. nerv. ment. Dis.*, **127**, 1–11.

MacLean, P. D. (1969). The hypothalamus and emotional behavior. In: *The Hypothalamus* (Haymaker, W., Anderson, E. and Nauta, W. J. H. eds.), pp. 659–78, Thomas: Springfield, Ill.

McMahon, T. A. (1977). Scaling quadrupedal galloping: frequencies, stresses and joint angles. In: *Scale Effects in Animal Locomotion* (Pedley, T. J. ed.), pp. 143–51, Academic Press: London, New York.

McMasters, R. E., Weiss, A. H. and Carpenter, M. B. (1966). Vestibular projections to the nuclei of the extraocular muscles. Degeneration resulting from discrete partial lesions of the vestibular nuclei in the monkey, *Am. J. Anat.*, **118**, 163–94.

McMinn, R. M. H. (1969). *Tissue Repair*, Academic Press: New York, London.

McMinn, R. M. H. and Mitchell, J. E. (1954). The formation of villi following artificial lesions of the mucosa in the small intestine of the cat, *J. Anat.*, **88**, 99–107.

McMinn, R. M. H. and Kugler, J. H. (1961). The glands of the bile and pancreatic ducts: autoradiographic and histochemical studies, *J. Anat.*, **95**, 1–11.

McNamara, J. A. (1972). Dual functions of the lateral pterygoid muscle: a study of *Macaca mulatta*, *Anat. Rec.*, **172**, 360.

McNutt, N. S. and Weinstein, R. S. (1970). The ultrastructure of the nexus. A correlated thin-section and freeze-cleave study, *J. Cell Biol.*, **47**, 666–88.

McVay, C. B. and Anson, B. J. (1940). Aponeurotic and fascial continuities in the abdomen, pelvis and thigh, *Anat. Rec.*, **76**, 213–32.

McVay, C. B. and Anson, B. J. (1940). Composition of the rectus sheath, *Anat. Rec.*, **77**, 213–25.

Mabuchi, M. and Kusama, T. (1966). The corticorubral projection in the cat, *Brain Res.*, **2**, 254–73.

Maggio, E. (1965). *Microhemocirculation*, Thomas: Springfield, Ill.

Mahour, G. H., Wakim, K. G., Soule, F. H. and Ferris, D. O. (1967). The common bile duct after cholecystectomy: comparison of common bile ducts in patients who have intact biliary systems with those in patients who have undergone cholecystectomy, *Ann. Surg.*, **166**, 964–7.

Maillot, C., Koritke, J. G. and Laude, M. (1976). La vascularisation de la toile choroidienne inférieure chez l'homme, *Archs anat. Histol. Embryol.*, **59**, 33–70.

Malhotra, S. K. (1970). Organisation of the cellular membranes, *Prog. Biophys. mol. Biol.*, **20**, 67–131.

Malis, L. I., Loevinger, R., Kruger, L. and Rose, J. E. (1957). Production of laminar lesions in the cerebral cortex by heavy ionising particles, *Science, N.Y.*, **126**, 302–3.

Mall, F. P. (1911). On the muscular architecture of the ventricles of the human heart, *Am. J. Anat.*, **11**, 211–78.

Mall, F. P. (1912). On the development of the human heart, *Am. J. Anat.*, **13**, 249–98.

Mankin, H. J. and Lippiello, L. (1969). The turnover of adult rabbit articular cartilage, *J. Bone Jt Surg.*, **51A**, 1591–1600.

Mann, I. C. (1924). Notes on the anatomy of the living eye, as revealed by the Gullstrand slitlamp, *J. Anat.*, **59**, 155–65.

Mann, I. C. (1927). The relations of the hyaloid canal in the foetus and in the adult, *J. Anat.*, **62**, 290–6.

Mann, I. C. (1964). *The Development of the Human Eye*, Third edition, Cambridge University Press: London.

Mann, T. (1949). Metabolism of semen, *Adv. enzymol.*, **9**, 329–90.

Mann, T. (1967). Sperm metabolism. In: *Fertilisation* (Metz, C. B. and Monroy, A. eds.), Volume 1, pp. 99–116, Academic Press: New York, London.

Manni, E., Bortolami, R. and Deriu, P. L. (1970). Presence of cell bodies of the afferents from the eye muscles in the semilunar ganglion, *Arch. ital. Biol.*, **108**, 106–20.

Manotaya, T. and Potter, E. L. (1963). Oocytes in prophase of meiosis from squash preparations of human fetal ovaries, *Fert. Steril.*, **14**, 378–92.

Marchand, P., Gilroy, J. C. and Wilson, V. H. (1950). Anatomical study of bronchial vascular system and its variations in disease, *Thorax*, **5**, 207–21.

Marchesi, V. T. (1962). The passage of colloidal carbon through inflamed endothelium, *Proc. R. Soc.*, **156B**, 550–2.

Marchesi, V. T. and Gowans, J. L. (1964). The migration of lymphocytes through the endothelium of venules in lymph nodes, *Proc. R. Soc. B.*, **159**, 283–90.

Marchi, V. and Algeri, G. (1815–16). Sulle degenerazioni discendenti consecutive a lesioni sperimentale in diverse zone delle corteccia cerebrale, *Riv. sper. Freniatria Med. legal.*, **11**, 492–4; **12**, 208–52.

Marcus, A. J. and Zucker, M. B. (1965). *The Physiology of Blood Platelets*, Grune and Stratton: New York.

Marikovsky, Y. and Danon, D. (1969). Electron microscope analysis of young and old red blood cells stained with colloidal iron for surface charge evaluation, *J. Cell. Biol.*, **43**, 1–7.

Marinesco, G. (1909). *La Cellule Nerveuse*, Dom: Paris.

Mark, R. (1974). *Memory and Nerve Cell Connections*, Clarendon Press: Oxford.

Markee, J. E., Logue, J. T. Jr., Williams, M., Stanton, W. B., Wrenn, R. N. and Walker, L. B. (1955). Two joint muscles of the thigh, *J. Bone Jt Surg.*, **37A**, 125–42.

Markowski, J. (1911). Über die Entwicklung der Sinus durae matris und der Hirnvenen bei menschlichen Embryonen von 15·5–49 mm Scheitel-Steisslänge, *Bull. int. Acad. Sci. Lett. Cracovie B*, 590–611.

Markowski, J. (1922). Entwicklung der Sinus durae matris und der Hirnvenen des Menschen, *Bull. int. acad. pol. sci. Lett. B*, 1–269.

Marneffe, R. de (1951). *Recherches morphologiques et expérimentales sur la vascularisation osseuse*, Acta medica Belgica: Brussels.

Maroudas, A. (1976). Transport of solutes through cartilage: permeability to large molecules, *J. Anat.*, **122**, 335–47.

Maroudas, A., Stockwell, R., Nachemson, A. and Urban, J. (1975). Factors involved in the nutrition of the human lumbar intervertebral disc: cellularity and diffusion of glucose *in vitro*, *J. Anat.*, **120**, 113–30.

Marsden, C. D. (1961). Pigmentation in the nucleus substantiae nigrae of mammals, *J. Anat.*, **95**, 256–61.

Marshall, J. and Ansell, P. L. (1971). Membranous inclusions in the retinal pigmented epithelium: phagosomes and myeloid bodies, *J. Anat.*, **110**, 91–104.

Martin, B. F. (1958). The annular ligament of the superior radio-ulnar joint, *J. Anat.*, **92**, 473–82.

Martin, C. B. Jr. (1965). Uterine blood flow and placental circulation, *Anaesthesiology*, **26**, 447–59.

Martin, C. P. (1932). The cause of torsion of the humerus and of the notch on the anterior edge of the glenoid cavity of the scapula, *J. Anat.*, **67**, 573–82.

Martin, C. P. (1942). A note on the renal fascia, *J. Anat.*, **77**, 101–3.

Martin, R. (1928). *Lehrbuch der anthropologie*, Second edition, Fischer: Jena.

Martin, R. and Saller, K. (1961). *Lehrbuch der Anthropologie*, Third edition, Urban und Schwarzenburg: Stuttgart.

Martins, R. (1961). Lateral cervical and pre-auricular sinuses: their transmission as dominant characters, *Br. med. J.* **1**, 255–6.

Mascorro, J. A. and Yates, R. D. (1971). Ultrastructural studies of the effects of reserpine on mouse abdominal sympathetic paraganglia, *Anat. Rec.*, **170**, 269–80.

Mascorro, J. A. and Yates, R. D. (1974). Innervation of abdominal paraganglia. An ultrastructural study, *J. Morphol.*, **142**, 153–64.

Mascorro, J. A. and Yates, R. D. (1975). A review of abdominal paraganglia. Ultrastructure, mitotic cells, catecholamine release, innervation, light and dark cells, vascularity. In: *Electron Microscopic Concepts of Secretion. Ultrastructure of Endocrinal and Reproductive Organs* (Hess, M. ed.), pp. 435–52, Wiley: New York.

Mason, C. A. and Lincoln, D. W. (1976). Visualization of the retinohypothalamic projection in the rat by cobalt precipitation, *Cell Tiss. Res.*, **168**, 117–31.

Massopust, L. C. Jr. and Daigle, H. J. (1960). Cortical projection of the medial and spinal vestibular nuclei in the cat, *Exp. Neurol.*, **2**, 179–85.

Masterton, R. B., Jane, J. A. and Diamond, I. T. (1967). Role of brainstem auditory structures in sound localisation. I. Trapezoid body superior olive, and lateral lemniscus, *J. Neurophysiol*, **30**, 341–59.

Mastroianni, L. (1962). The structure and function of the fallopian tube. A correlative review, *Clin. Obstet. Gynec.*, **5**, 781–90.

Mathiasen, M. S. (1973). Determination of bone age and recording of minor skeletal hand anomalies in normal children, *Dan. med. Bull.*, **20**, 80–5.

Matoltsy, A. G. (1958). The chemistry of keratinisation. In: *The Biology of Hair Growth* (Montagna, W. and Ellis, R. A. eds.), pp. 135–69, Academic Press: New York, London.

Matusda, H. (1968). Electron microscopic study on the corneal nerve with special reference to its endings, *Acta soc. ophthal. jap.*, **12**, 163–73.

Matthews, M. A. (1968). An electron microscopic study of the relationship between axon diameter and the initiation of myelin production in the peripheral nervous system, *Anat. Rec.*, **161**, 337–52.

Matthews, M. R. and Raisman, G. (1969). The ultrastructure and somatic efferent synapses of small granule-containing cells in the superior cervical ganglion, *J. Anat.*, **105**, 255–82.

Matthews, P. B. C. (1971). Recent advances in the understanding of the muscle spindle. In: *Scientific Basis of Medicine. Annual Review* (Gilliland, I. and Francis, J. eds.), pp. 99–128, Athlone Press: London.

Matthews, P. B. C. (1972). *Mammalian Muscle Receptors and Their Central Actions*, Arnold: London.

Matulionis, D. H. and Parks, H. F. (1973). Ultrastructural morphology of the normal nasal respiratory epithelium of the mouse, *Anat. Rec.*, **175**, 68–84.

Maul, G. G. (1971). Structure and formation of pores in fenestrated capillaries, *J. Ultrastruct. Res.*, **36**, 768–82.

Maxwell, D. S. and Pease, D. C. (1956). The electron microscopy of the choroid plexus, *J. biophys. biochem. Cytol.*, **2**, 467–74.

Mayfield, J. K., Johnson, R. P. and Kilcoyne, R. F. (1976). The ligaments of the human wrist and their functional significance, *Anat. Rec.*, **186**, 417–28.

Mayhall, J. T., Dahlberg, A. A. and Owen, D. G. (1970). Torus mandibularis in an Alaskan Eskimo population, *Am. J. phys. Anthrop.*, **33**, 57–60.

Mechanik, N. (1934). Das Venensystem der Herzwände, *Z. Anat. EntwGesch.*, **103**, 813–43.

Meckel, J. F. (1832). *Manual of Anatomy*, Volume 3, Carey & Lea: Philadelphia.

Meessen, H. and Olszewski, J. (1949). *A Cytoarchitectonic Atlas of the Rhombencephalon of the Rabbit*, Karger: Basel.

Mehler, W. R. (1962). The anatomy of the so-called 'pain tract' in man: An analysis of the course and distribution of the ascending fibers of the fasciculus anterolateralis. In: *Basic Research in Paraplegia* (French, J. D. and Porter, R. W. eds.), pp. 26–55, Thomas: Springfield, Ill.

Mehler, W. R. (1966). Further notes on the centre médian nucleus of Luys. In: *The Thalamus* (Purpura, D. P. and Yahr, M. D. eds.), pp. 109–27, Columbia University Press: New York.

Mehler, W. R., Feferman, M. E. and Nauta, W. J. H. (1960). Ascending axon degeneration following anterolateral cordotomy. An experimental study in the monkey, *Brain*, **83**, 718–50.

Mehta, H. J. and Gardner, W. U. (1961). A study of lumbrical muscles in the human hand, *Am. J. Anat.*, **109**, 227–38.

Mei, N. and Dussardier, M. (1966). Etudes des lésions pulmonaires produites par la section des fibres sensitives vagales, *J. Physiol., Paris*, **58**, 427–31.

Meikle, T. H. and Sprague, J. M. (1964). The neural organisation of the visual pathways in the cat, *Int. Rev. Neurobiol.*, **6**, 150–91.

Meininger, V. and Baudrimont, M. (1977). The cytoarchitecture of the inferior colliculus in the cat, *J. neurol. Sci.*, **34**, 25–36.

Melander, A. (1970). Amines and mouse thyroid activity: release of thyroid hormone by catecholamines and indoleamines and its inhibition by adrenergic blocking drugs, *Acta endocrinol. (Kbh.)*, **65**, 371–84.

Melander, A. (1977). Aminergic regulation of thyroid activity: importance of the sympathetic innervation and of the mast cells of the thyroid gland, *Acta med. scand.*, **201**, 257–62.

Melander, A. (1978). Sympathetic nervous-adrenal medullary system. In: *The Thyroid—a Fundamental and Clinical Text* (Werner, S. C. and Ingbar, S. H. eds.), pp. 216–21, Harper & Row: New York, London.

Melander, A., Nilsson, E. and Sundler, F. (1972). Sympathetic activation of thyroid hormone secretion in mice, *Endocrinology*, **90**, 194–9.

BIBLIOGRAPHY

Melander, A., Sundler, F. and Westgren, U. (1973). Intrathyroidal amines and the synthesis of thyroid hormone, *Endocrinology*, **93**, 193–200.

Melander, A., Ericson, L. E., Sundler, F. and Westgren, U. (1975). Intrathyroidal amines in the regulation of thyroid activity, *Rev. Physiol.*, **73**, 39–71.

Melcher, A. M. and Bowen, W. H. (1969). *The Biology of the Periodontium* (editors), Academic Press: New York, London.

Melcher, A. M. and Walker, T. W. (1978). The periodontal ligament in attachment and as a shock absorber. In: *The Eruption and Occlusion of Teeth* (Poole, D. F. G. and Stack, M. V. eds.), pp. 183–92, Butterworths: London.

Meller, K. and Glees, P. (1969). The development of the mouse cerebellum. In: *Neurobiology of Cerebellar Evolution and Development* (Llinas, R. ed.), pp. 783–801, AMAERF: Chicago.

Melzack, R. and Wall, P. D. (1965). Pain mechanisms: A new theory, *Science, N.Y.*, **150**, 971–9.

Menco, B. P. M., Dodd, G. H., Davey, M. and Bannister, L. H. (1976). Presence of membrane particles in freeze-etched bovine olfactory cilia, *Nature, Lond.*, **263**, 597–9.

Mendel, G. (1886). Versuche über Pflanzenhybriden, *Verh. naturf. Vere Brünn.*, **4**, 3–44.

Mendell, L. M. (1966). Physiological properties of unmyelinated fiber projections to the spinal cord, *Expl Neurol.*, **16**, 316–32.

Mendell, L. M. and Wall, P. D. (1964). Pre-synaptic hyperpolarisation: a role for fine afferent fibres, *J. Physiol., Lond.*, **172**, 274–94.

Menning, von A., Schumacher, G.-H., Lau, H., Schultz, M. and Himstedt, H. W. (1974). Zur Topographie der muskulären Nervenausbreitungen. 6. Untere Extremität. Glutealmuskeln, *Anat. Anz.*, **135**, 302–14.

Merendino, K. A., Johnson, R. J., Skinner, H. H. and Maguire, R. X. (1956). Intradiaphragmatic distribution of phrenic nerve with particular reference to placement of diaphragmatic incisions and controlled segmental paralysis, *Surgery Gynec. Obstet.*, **39**, 189–98.

Merklin, R. J. and Michels, N. A. (1958). The variant renal and suprarenal blood supply with data on the inferior phrenic, ureteral and gonadal arteries: a statistical analysis based on 185 dissections and a review of the literature, *J. int. Coll. Surg.*, **29**, 41–76.

Metz, C. B. and Monroy, A. (1969). *Fertilisation* (editors), Academic Press: New York, London.

Meyer, A. W. (1917). Studies on hemal nodes. VII. The development and function of hemal nodes, *Am. J. Anat.*, **21**, 375–406.

Meyer, A. (1971). *Historical Aspects of Cerebral Anatomy*, Oxford University Press: London.

Meyer, R. (1938). Zur Frage der Entwicklung der menschlichen Vagina; Vagina infima septa und andere besonderheiten, *Arch. Gynaek.*, **167**, 306–38.

Meynert, T. (1867–8). Der Bau der Gross-Hirnrinde und seine ortlichen Verschiedenheiten nebst einem pathologisch-anatomischen corollarium, *Vjschr. Psychiat., Vienna*, **1**, 77–93, 198–217; **2**, 88–113.

Meyrick, B. and Reid, L. (1968). The alveolar brush cell in rat lung—a third pneumonocyte, *J. Ultrastruct. Res.*, **23**, 71–80.

Meyrick, B. and Reid, L. (1970). Ultrastructure of cells in the human bronchial submucosal glands, *J. Anat.*, **107**, 281–99.

Miale, I. L. and Sidman, R. L. (1961). An autoradiographic analysis of the histogenesis of the mouse cerebellum, *Exp. neurol.*, **4**, 227–96.

Michels, N. A. (1962). The anatomic variations of the arterial pancreaticoduodenal arcades: their import in regional resection involving the gall bladder, bile ducts, liver, pancreas and parts of the small and large intestines, *J. int. Coll. Surg.*, **37**, 13–40.

Michels, N. A. (1966). Newer anatomy of the liver and its variant blood supply and collateral circulation, *Am. J. Surg.*, **112**, 337–47.

Middleton, W. S. (1923). Costodiaphragmatic adhesions and their influence on the respiratory function, *Am. J. med. Sci.*, **166**, 222–8.

Midgley, A. R., Pierce, G. B. Jr., Deneau, G. A. and Gosling, J. R. (1963). Morphogenesis of syncitiotrophoblast *in vivo*: an autoradiographic demonstration, *Science, N.Y.*, **141**, 349–50.

Mierzwa, J. and Kozielec, T. (1975). Variation of the anterior cardiac veins, *Folia morph., Warsz.*, **34**, 125–33.

Miki, H., Bellhorn, M. B. and Henkind, P. (1975). Specializations of the retinochoroid juncture, *Invest. Ophthal.*, **14**, 701–7.

Mikić, Ž. Dj. (1978). Age changes in the triangular fibrocartilage of the wrist joint, *J. Anat.*, **126**, 367–84.

Milaire, J. (1957). Contribution à la connaissance morphologique et cytochimique des bourgeons de membres chez quelques Reptiles, *Archs Biol., Liège*, **68**, 429–512.

Milaire, J. (1965). Aspects of limb morphogenesis in mammals. In: *Organogenesis* (DeHaan, R. L. and Ursprung, H. eds.), pp. 283–300, Holt: New York.

Milburn, A. (1973). The early development of muscle spindles in the rat, *J. Cell Sci.*, **12**, 175–95.

Miles, A. E. W. (1967). *Structural and Chemical Organisation of Teeth* (editor), Academic Press: New York, London.

Millen, J. W. (1959). Some aspects of the relationship between environment and congenital malformations, *Ir. J. med. Sci.*, **397**, 23–9.

Millen, J. W. and Woollam, D. H. M. (1953). Vascular patterns in the choroid plexus, *J. Anat.*, **87**, 114–23.

Millen, J. W. and Woollam, D. H. M. (1961). On the nature of the pia mater, *Brain*, **84**, 514–20.

Millen, J. W. and Wollam, D. H. M. (1962). *The Anatomy of the Cerebrospinal Fluid*, Oxford University Press: London.

Miller, E. J. (1971a). Isolation and characterisation of a collagen from chick cartilage containing three identical a-chains, *Biochemistry*, **10**, 1652.

Miller, E. J. (1971b). Collagen cross-linking: identification of two cyanogen bromide peptides containing sites of intermolecular cross-link formation in cartilage collagen, *Biochem. biophys. Res. Commun.*, **45**, 444.

Miller, F., De Harven, E. and Palade, G. E. (1966). The structure of eosinophil leukocyte granules in rodents and in man, *J. Cell Biol.*, **31**, 349–62.

Miller, J. F. A. P. and Osoba, D. (1967). Current concepts of the immunological function of the thymus, *Physiol. Rev.*, **47**, 437–520.

Miller, M. R., Ralston, H. J. and Kasahara, M. (1958). The pattern of cutaneous innervation of the human hand, *Am. J. Anat.*, **102**, 183–218.

Miller, R. L. and Chaudhry, A. P. (1976). Comparative ultrastructure of vallate, foliate and fungiform taste buds of golden Syrian hamster, *Acta anat.*, **95**, 75–92.

Miller, W. S. (1947). *The Lung*, Second edition, Thomas: Springfield, Ill.

Milligan, E. T. C., Morgan, C. N., Jones, L. E. and Officer, R. (1937). Surgical anatomy of anal canal and operative treatment of haemorrhoids, *Lancet*, **2**, 1119–24.

Millikan, C. H. and Darley, F. L. (1967). *Brain Mechanisms Underlying Speech and Language* (editors), Grune and Stratton: New York.

Mills, E. and Jöbsis, F. F. (1972). Mitochondrial respiratory chain of carotid body and chemoreceptor response to changes in oxygen tension, *J. Neurophysiol.*, **35**, 405–28.

Mills, R. W. (1917). The relation of bodily habits to visceral form, position, tonus and motility, *Am. J. Roentg.*, **4**, 155–69.

Mills, R. W. (1922). X-ray evidence of abdominal small intestinal states embodying an hypothesis of the transmission of gastro-intestinal tension, *Am. J. Roentg.*, **9**, 199–225.

Milner, B. (1958). Psychological defects produced by temporal lobe excision, *Res. Publs. Ass. Res. nerv. ment. Dis.*, **36**, 244–57.

Milner, B. (1967). Brain mechanisms suggested by studies of temporal lobes. In: *Brain Mechanisms Underlying Speech and Language* (Millikan, C. H. and Darley, F. L. eds.), pp. 122–31, Grune and Stratton: New York.

Milner, B. (1974). Hemispheric specialization: scope and limits. In: *The Neurosciences, Third Study Program* (Schmitt, F. O. and Worden, F. G. eds.), MIT Press: Cambridge, U.S.A.

Milner, B., Branch, C. and Rasmussen, T. (1964). Observations on cerebral dominance. In: *Disorders of Language* (Wolstenholme, D. W. and O'Connor, M. eds.), Churchill: London.

Milner, B. and Teuber, H. L. (1968). Alteration of perception and memory in man. In: *Analysis of Behavioural Change* (Weiskrantz, G. L. ed.), Harper & Row: New York.

Mincsev, M. (1967). Bilocular gall bladder, *Orvoskepzes*, **42**, 286–98.

Minkoff, E. C. (1974). The Fürbinger hypothesis of nerve-muscle specificity re-examined, *Can. J. Zool.*, **52**, 525–32.

Minkowski, M. (1913). Experimentelle Untersuchungen über die Beziehungen der Grosshirnrinde und der Netzhaut zu den primären optischen Zentren, besonders zum Corpus geniculatum externum, *Arb. hirnanat. Inst., Zürich*, **7**, 259–362.

Minns, R. J. and Stevens, F. S. (1977). The collagen fibre organisation in human articular cartilage, *J. Anat.*, **123**, 437–57.

Minsky, M. (1965). *Matter, Mind and Models*, Spartan Books: Washington.

Mintz, B. (1960). Embryological phases of mammalian gametogenesis, *J. cell. comp. Physiol.*, **56**, Suppl. 1, 31–47.

Mintz, B. (1962). Experimental study of the developing mammalian egg: removal of the zona pellucida, *Science, N.Y.*, **138**, 594–5.

Mintz, B. (1964). Synthetic processes and early development in the mammalian egg, *J. exp. Zool.*, **157**, 267–72.

Mishima, S. and Maurice, D. M. (1961). The effect of normal evaporation on the eye, *Expl Eye Res.*, **1**, 46–52.

Misotten, L. (1962). L'ultrastructure des cônes de la rétine humaine, *Bull. soc. belge Ophtal.*, **130**, 472–502.

Mitchell, G. A. G. (1935). Innervation of distal colon, *Edinb. med. J.*, **42**, 11–20.

Mitchell, G. A. G. (1938). Nerve supply of the gastro-oesophageal junction, *Br. J. Surg.*, **26**, 333–45.

Mitchell, G. A. G. (1939). A macroscopic study of the nerve supply of the stomach, *J. Anat.*, **75**, 50–63.

Mitchell, G. A. G. (1950). Renal fascia, *Br. J. Surg.*, **37**, 257–66.

Mitchell, G. A. G. (1952). Rostral extremities of sympathetic trunks, *Nature, Lond.*, **129**, 533–4.

Mitchell, G. A. G. (1953). *Anatomy of the Autonomic Nervous System*, Livingstone: Edinburgh.

Mitchell, G. A. G. (1956). *Cardiovascular Innervation*, Livingstone: Edinburgh, London.

Mitchell, G. A. G. and Warwick, R. (1955). The dorsal vagal nucleus, *Acta anat.*, **25**, 371–95.

Mittwoch, U. (1967). *Sex Chromosomes*, Academic Press: New York, London.

Miyabara, S., Okamoto, N., Akimoto, N., Satow, Y. and Hidaka, N. (1974). Meckel's diverticulum found at autopsy, *Hiroshima J. med. Sci.*, **23**, 179–90.

Mizeres, N. J. (1963). The cardiac plexus in man, *Am. J. Anat.*, **112**, 141–51.

Mizuno, N. (1966). An experimental study of the spino-olivary fibers in the rabbit and the cat, *J. comp. Neurol.*, **127**, 267–91.

Modi, J. P. (1957). *Jurisprudence and Toxicology*, Tripathi Private: Bombay.

Moe, G. K., Preston, J. B. and Burlington, H. (1956). Physiologic evidence for a dual A V transmission system, *Circ. Res.*, **4**, 357.

Moens, P. (1974). Quantitative electron microscopy of chromosome organization at meiotic prophase, *Cold Spring Harb. Symp. quant. Biol.*, **28**, 99–107.

Moffat, D. B. (1959). Developmental changes in the aortic arch system of the rat, *Am. J. Anat.*, **105**, 1–36.

Moffat, D. B. (1961). The development of the ophthalmic artery in the rat, *Anat. Rec.*, **140**, 217–22.

Moffat, D. B. (1961). The development of the anterior cerebral artery and its related vessels in the rat, *Am. J. Anat.*, **108**, 17–30.

Moffet, B. C. Jr. (1957). The prenatal development of the human temporomandibular joint, *Contr. Embryol.*, **36**, 19–28.

Moffet, B. C. Jr., Johnson, L. C., McCabe, J. B. and Askew, H. C. (1964). Articular remodelling in the adult human temporomandibular joint, *Am. J. Anat.*, **115**, 119–42.

Mohuiddin, A. (1953). Vagal preganglionic fibres to the alimentary canal, *J. comp. Neurol.*, **99**, 289–318.

Møller, M. (1978). Presence of a pineal nerve (nervus pinealis) in the human fetus; a light and electron microscopical study of the innervation of the pineal gland, *Brain Res.*, **154**, 1–12.

Møller, M. (1979). Presence of a pineal nerve (nervus pinealis) in fetal mammals, *Prog. Brain Res.*, **52**, 103–6.

Møllgård, K., Møller, M. and Kimble, J. (1973). Histochemical investigations on the human fetal sub-commissural organ, *Histochemie*, **37**, 61–74.

Monro, A. (1746). *The Anatomy of the Human Bones and Nerves*, Hamilton & Balfour: Edinburgh.

Monroy, A. and Tyler, A. (1967). The activation of the egg. In: *Fertilisation* (Metz, C. B. and Monroy, A. eds.), Volume 1, pp. 369–412, Academic Press: New York, London.

Montagna, W. (1968). *The Structure and Function of Skin*, Second edition, Academic Press: New York, London.

Montagna, W. and Ellis, R. A. (1961). *Advances in the Biology of the Skin* (editors), Pergamon Press: Oxford, London, New York, Toronto, Sydney.

Montagna, W. and Lobitz, W. C. (1964). *The Epidermis* (editors), Academic Press: New York, London.

Montagna, W. and Parakkal, P. F. (1974). *The Structure and Function of Skin*, Third edition, Academic Press: New York.

Montagu, M. F. A. (1935). The premaxilla in the primates, *Q. Rev. Biol.*, **10**, 32–59, 181–208.

Montagu, M. F. A. (1951), Wallbrook frontal bone, *Am. J. phys. Anthrop.*, **9**, 5–14.

Montagu, M. F. A. (1960). *An Introduction to Physical Anthropology*, Thomas: Springfield, Ill.

Moody, R. O. (1927). The position of the abdominal viscera in healthy young British and American adults, *J. Anat.*, **61**, 223–31.

Moore, C. R. (1947). *Embryonic Sex Hormones and Sex Differentiation*, Thomas: Springfield, Ill.

Moore, M. A. and Owen, J. J. (1967). Experimental studies on the development of the thymus, *J. exp. Med.*, **126**, 715–26.

Moore, R. Y. (1973). Retinohypothalamic projection in mammals: a comparative study, *Brain Res.*, **49**, 403–9.

Moore, R. Y. and Goldberg, J. M. (1963). Ascending projections of the inferior colliculus in the cat, *J. comp. Neurol.*, **121**, 109–36.

Moore, R. Y. and Goldberg, J. M. (1966). Projections of the inferior colliculus in the monkey, *Expl Neurol.*, **14**, 429–38.

Moore, R. Y. and Lenn, N. J. (1972). A retino-hypothalamic projection in the rat, *J. comp. Neurol.*, **146**, 1–14.

Moore, R. Y. and Eichler, V. B. (1972). Loss of a circadian adrenal corticosterone rhythm following suprachiasmatic lesions in the rat, *Brain Res.*, **42**, 201–6.

Moore, R. Y. and Klein, D. C. (1974). Visual pathways and the central neural control of a circadian rhythm in pineal serotonin n-acetyltransferase activity, *Brain Res.*, **71**, 17–34.

Moores, G. R. and Partridge, T. A. (1974). In: *The Cell in Medical Science* (Beck, F. and Lloyd, J. B. eds.), Volume 1, pp. 75–104, Academic Press: London.

Moorrees, C. F., Osborne, R. and Wilde, E. (1952). Torus mandibularis. Its occurrence in Aleut children and its genetic determinants, *Am. J. phys. Anthrop.*, **10**, 319–30.

Moosavi, H., Smith, P. and Heath, D. (1973). The Feyrter cell in hypoxia, *Thorax*, **28**, 729–41.

Mooseker, M. S. and Tilney, L. G. (1976). Organization of an actin filament-membrane complex. Filament polarity and membrane attachment in the microvilli of intestinal epithelial cells, *J. Cell Biol.*, **67**, 725–43.

Morant, G. M. (1936). A biometric study of the human mandible, *Biometrics*, **28**, 84–122.

Morest, D. K. (1964). The neuronal architecture of the medial geniculate body of the cat, *J. Anat.*, **98**, 611–30.

Morest, D. K. (1965). The lateral geniculate system of the midbrain and the medial geniculate body: a study with Nauta and Golgi methods in the cat, *J. Anat.*, **99**, 611–34.

Morest, D. K. (1967). Experimental study of the projections of the nucleus of the tractus solitarius and the area postrema in the cat, *J. comp. Neurol.*, **130**, 277–300.

Morest, D. K. (1971). Dendrodendritic synapses of cells that have axons: the fine structure of the Golgi type II cell in the medial geniculate body of the cat, *Z. Anat. EntwGesch.*, **133**, 216–46.

Morgan, J. D. (1959). Blood supply of the growing rabbit's tibia, *J. Bone Jt Surg.*, **41B**, 185–203.

Morgan, M. W. (1944). Accommodation and its relation to convergence, *Am. J. Optom.*, **21**, 183–95.

Mori, S. and Leblond, C. P. (1969). Electron microscopic features and proliferation of astrocytes in the corpus callosum of the rat, *J. comp. Neurol.*, **137**, 197–226.

Mori, S. and Leblond, C. P. (1970). Electron microscopic identification of three classes of oligodendrocytes and a preliminary study of their proliferative activity in the corpus callosum of young rats. *J. comp. Neurol.*, **139**, 1–60.

Moriarity, G. C. (1975). Electron microscopic-immunocytohisto-chemical studies of rat pituitary gonadotrophs: a sex difference in morphology and cytochemistry of LH cells, *Endocrinology*, **97**, 1215–25.

Morin, F., Schwartz, H. G. and O'Leary, J. L. (1951). Experimental study of the spino-thalamic and related tracts, *Acta psychiat. neurol. scand.*, **26**, 371–96.

Morin, F. and Catalano, J. V. (1955). Central connections of a cervical nucleus (nucleus cervicalis lateralis of the cat), *J. comp. Neurol.*, **103**, 17–32.

Morison, A. B. (1954). The levatores costarum and their nerve supply, *J. Anat.*, **88**, 19–24.

Morris, E. W. T. (1976). Observations on the source of embryonic myoblasts, *J. Anat.*, **121**, 47–64.

Moruzzi, G. and Magoun, H. W. (1949). Brain stem reticular formation and activation of the EEG, *Electroenceph. clin. Neurophysiol.*, **1**, 455–73.

Moss-Salentijn, L. (1975). Cartilage canals in the human spheno-occipital synchondrosis during fetal life, *Acta anat.*, **92**, 595–606.

Mossman, H. W. (1937). Comparative morphogenesis of the fetal membranes and accessory uterine structures, *Contr. Embryol.*, **26**, 133–247.

Moulton, D. G. and Beidler, L. M. (1967). Structure and function of the peripheral olfactory system, *Physiol. Rev.*, **47**, 1–52.

Moulton, D. G., Celebi, G. and Fink, R. P. (1970). Olfaction in mammals—two aspects: proliferation of cells in the olfactory epithelium and sensitivity to odours. In: *Taste and Smell in Vertebrates* (Wolstenholme, G. E. W. and Knight, J. eds.), pp. 227–45, Ciba Foundation Symposium, Churchill: London.

Mountcastle, V. B. (1957). Modality and topographic properties of single neurons of cat's somatic sensory cortex, *J. Neurophysiol.*, **20**, 408–34.

Mountcastle, V. B. (1968). *Medical Physiology*, Twelfth edition, Mosby: St. Louis.

Mountcastle, V. B. and Powell, T. P. S. (1959). Central nervous mechanisms subserving position sense and kinesthesis, *Bull. Johns Hopkins Hosp.*, **105**, 173–200.

Mountcastle, V. B. and Powell, T. P. S. (1959). Neural mechanisms subserving cutaneous sensibility, with special reference to the role of afferent inhibition in sensory perception and discrimination, *Bull. Johns Hopkins Hosp.*, **105**, 201–32.

Mountcastle, V. B., Lynch, C. J., Georgopoulos, A., Sakata, H. and Acuna, A. (1975). Posterior parietal association cortex of the monkey, *J. Neurophysiol.*, **38**, 871–908.

BIBLIOGRAPHY

Mourant, A. E. (1954). *The Distribution of the Human Blood Groups*, Blackwell: Oxford.

Movat, H. Z. and Fernanado, N. V. (1963). Acute inflammation: The earliest fine structural changes at the blood-tissue barrier, *Lab. Invest.*, **12**, 895–910.

Moyers, R. E. (1950). Electromyographic analysis of certain muscles involved in temporomandibular movement, *Am. J. Orthodont.*, **36**, 481–515.

Mueller, K. H., Trias, A. and Ray, R. D. (1966). Bone density and composition. Age-related and pathological changes in water and mineral content, *J. Bone Jt Surg.*, **48A**, 140–8.

Mugnaini, E. (1965). "Dark cells" in electron micrographs from the central nervous system of vertebrates, *J. Ultrastruct. Res.*, **12**, 235–6.

Mugnaini, E. (1970). The relationship between cytogenesis and the formation of different types of synaptic contact, *Brain Res.*, **17**, 169–79.

Mugnaini, E. and Forstrønen, P. F. (1967). Ultrastructural studies on cerebellar histogenesis. I. Differentiation of granule cells and development of glomeruli in the chick embryo, *Z. Zellforsch. mikrosk. Anat.*, **77**, 115–43.

Muir, A. R. (1954). The development of the ventricular part of the conducting tissue in the heart of the sheep, *J. Anat.*, **88**, 381–91.

Muir, A. R. (1957). The cellular structure of rabbit cardiac muscle, as observed with the electron microscope, *J. Anat.*, **91**, 570P.

Muir, A. R. (1957). Observations on the fine structure of the Purkinje fibres in the ventricles of the sheep's heart, *J. Anat.*, **91**, 251–8.

Muir, A. R. (1957). An electron microscope study of the intercalated disc in the heart of the rabbit, *J. biophys. biochem. Cytol.*, **3**, 193–202.

Mukai, N. (1970). Axonal reaction of the optic nerve following heat coagulation. Histochemical evidence for antidromic conduction, *Can. J. Ophthal.*, **5**, 78–90.

Müller, F. (1977). The development of the anterior falcate and lacrimal arteries in the human, *Anat. Embryol.*, **150**, 207–27.

Muller, T. (1959). Variations in the abductor pollicis longus and extensor pollicis brevis in the South African Bantu, *S. Afr. J. Lab. clin. Med.*, **5**, 56–62.

Mulnard, J. G. (1964). Obtention in vitro du développement continue de l'œuf de souris du stade II au stade du blastocyste, *C. r. hebd. Séanc. Acad. Sci., Paris*, **258**, 6228–9.

Mulnard, J. G. (1965). Studies of regulation of mouse ova *in vitro*. In: *Preimplantation Stages of Pregnancy* (Wolstenholme, G. E. W. and O'Connor, M. eds.), pp. 123–38, Ciba Foundation Symposium, Churchill: London.

Munger, B. L. (1964). Histochemical studies on seromucous and mucous-secreting cells of human salivary glands, *Am. J. Anat.*, **115**, 411–29.

Munger, B. L. (1965). The intraepidermal innervation of the snout skin of the oppossum. A light and electron microscope study with observations on the nature of Meckel's *Tastzellen*, *J. Cell Biol.*, **26**, 79–98.

Munger, B. L. and Roth, S. I. (1963). The cytology of the normal parathyroid glands of man and Virginia deer; a light and electron microscopic study with morphologic evidence of secretory activity, *J. Cell Biol.*, **16**, 379–400.

Münzer, E. and Wiener, M. (1902). Das Zwischen- und Mittel- hirn des Kaninchens, *Mschr. Psychiat. Neurol.*, **12**, 241–79.

Murad, T. M. (1970). Ultrastructural study of rat mammary gland during pregnancy, *Anat. Rec.*, **167**, 17–36.

Muratori, G. (1965). Struttura microscopica del seno carotideo nel gatto, cane e conglio, *Boll. Soc. ital. Biol. sper.*, **42**, 301–3.

Murley, R. S. and Peters, P. M. (1961). Inadvertent parathyroidectomy, *Proc. R. Soc. Med.*, **54**, 487–9.

Murray, M. R. (1965). Nervous tissues *in vitro*. In: *Cells and Tissues in Culture* (Willmer, E. B. ed.), Volume II, pp. 373–455, Academic Press: New York, London.

Murray, M., Miller, H. R. P. and Jarrett, W. F. H. (1968). The globule leucocyte and its derivation from the subepithelial mast cell, *Lab. Invest.*, **19**, 222–34.

Murray, P. D. F. (1936). *Bones*, Cambridge University Press: Cambridge.

Murray, P. D. F. and Huxley, J. S. (1924). Self differentiation of the grafted limb-bud of the chick, *J. Anat.*, **59**, 379–84.

Murray, R. G. and Murray, A. (1970). The anatomy and ultrastructure of taste endings. In: *Taste and Smell in Vertebrates* (Wolstenholme, G. and Knight, J. eds.), pp. 3–24, Ciba Foundation Symposium, Churchill: London.

Muskens, L. J. J. (1914). An anatomico-physiological study of the posterior longitudinal bundle in its relation to forced movement, *Brain*, **36**, 352–426.

Mustafa, G. Y. and Gamble, H. J. (1979). Changes in axonal numbers in developing human trochlear nerve, *J. Anat.*, **128**, 323–30.

Myers, D. B., Highton, T. C. and Rayns, D. G. (1969). Acid mucopolysaccharides closely associated with collagen fibrils in normal human synovium, *J. Ultrastruct. Res.*, **28**, 203–13.

Myers, R. D. (1969). Temperature regulation: neurochemical systems in the hypothalamus. In: *The Hypothalamus* (Haymaker, W., Anderson, E. and Nauta, W. J. H. eds.), pp. 506–23, Thomas: Springfield, Ill.

Myers, R. E. (1959). Localisation of function in the corpus callosum. Visual gnostic transfer, *Archs Neurol. Psychiat., Chicago.*, **1**, 74–7.

Myers, R. E. and Henson, C. O. (1960). Role of corpus callosum in transfer of tactuokinesthetic learning in chimpanzee, *Archs Neurol. Psychiat., Chicago*, **3**, 404–9.

Mygind, N. (1975). Scanning electron microscopy of the human nasal mucosa, *Rhinology*, **13**, 57–75.

Mygind, S. H. (1948). Further labyrinthine studies; on labyrinthine transformation of acoustic vibrations to pitch-differentiated nervous impulses, *Acta otolar.*, Suppl. 68, 53–80.

Mysorekar, V. R. (1967). Diaphysial nutrient foramina in human long bones, *J. Anat.*, **101**, 813–22.

N

Nachemson, A. (1960). Lumbar intradiscal pressure. Experimental studies on post-mortem material, *Acta orthopaed. scand.*, Suppl. 43, 1–104.

Nadler, N. J., Young, B. A., Leblond, C. P. and Mitmaker, B. (1964). Elaboration of thyroglobulin in the thyroid follicle, *Endocrinology*, **74**, 333–54.

Naessen, R. (1971). The 'receptor surface' of the olfactory organ (epithelium) of man and guinea pig. A descriptive and experimental study, *Acta otolaryngol.*, **71**, 335–48.

Nagylaki, T. and Levy, J. (1973). "The sound of one paw clapping" is not sound, *Behav. Genet.*, **3**, 279–92.

Nakai, J. (1960). Studies on the mechanism determining the course of nerve fibers in tissue culture. II. The mechanism of fasciculation, *Z. Zellforsch. mikrosk. Anat.*, **52**, 427–49.

Nakai, J. and Kawasaki, Y. (1959). Studies on the mechanism determining the course of nerve fibers in tissue culture. I. The reaction of the growth cone to various obstructions, *Z. Zellforsch. mikrosk. Anat.*, **51**, 108–22.

Nakaizumi, Y. (1964). The ultrastructure of Bruch's membrane. I. The human, monkey, rabbit, guinea pig and rat eyes, *Archs Ophthal., N.Y.*, **72**, 380–7.

Nakane, P. K. (1970). Classifications of anterior pituitary cell types with immunoenzyme histochemistry, *J. Histochem. Cytochem.*, **18**, 9–20.

Nakanishi, H., Wansbrough, H. and Wood, C. (1967). Postganglionic sympathetic nerve innervating the human fallopian tube, *Am. J. Physiol.*, **213**, 613–19.

Nakanishi, T. (1967). Studies on the pudendal nerve. I. Macroscopic observations on the pudendal nerve in humans, *Acta anat. nippon.*, **42**, 223–39.

Nakayama, M. (1944). Nerve terminations in the muscle spindle of the human lingual muscles, *Tohuku med. J.*, **34**, 367–77.

Namba, T., Nakamura, T. and Grob, D. (1968). Motor nerve endings in human extraocular muscles, *Neurology, Minneap.*, **18**, 403–7.

Nanninga, N. (1973). Structural aspects of ribosomes, *Int. Rev. Cytol.*, **35**, 135–88.

Napier, J. R. (1955). The form and function of the carpo-metacarpal joint of the thumb, *J. Anat.*, **89**, 362–9.

Napier, J. R. (1956). The prehensile movements of the human hand, *J. Bone Jt Surg.*, **38B**, 902–13.

Napier, J. R. (1961). Prehensility and opposability in the hands of primates, *Symp. Zool. Soc. Lond.*, **5**, 115.

Napier, J. R. (1966). Functional aspects of the anatomy of the hand. In: *The Hand* (Pulvertaft, R. G. ed.), pp. 1–31, Butterworths: London.

Napolitano, L. (1965). The fine structure of adipose tissue. In: *Handbook of Physiology* (Renold, A. E. and Cahill, G. F. Jr. eds.), Section 5, pp. 109–24, Williams and Wilkins: Baltimore.

Napolitano, L. M. and Scallen, T. J. (1969). Observations on the fine structure of peripheral nerve myelin, *Anat. Rec.*, **163**, 1–6.

Nass, M. M. K. (1968). Mitochondrial DNA: advances, problems and goals, *Science, N.Y.*, **165**, 25–35.

Nass, M. M. K. (1969). Uptake of isolated chloroplasts by mammalian cells, *Science, N.Y.*, **165**, 1128–31.

Nathan, H. and Gloobe, H. (1974). Flexor digitorum brevis—anatomical variations, *Anat. Anz.*, **135**, 295–301.

Nathan, P. W. (1963). Results of antero-lateral cordotomy for pain in cancer, *J. Neurol. Neurosurg. Psychiat.*, **26**, 353–62.

Nathan, P. W. (1976). The gate-control theory of pain. A critical review, *Brain*, **99**, 123–58.

Nathan, P. W. and Smith, M. C. (1955). The Babinski response. A review and new observations, *J. Neurol. Neurosurg. Psychiat.*, **18**, 250–9.

Nathan, P. W. and Smith, M. C. (1955). Long descending tracts in man. I. Review of present knowledge, *Brain*, **78**, 248–303.

Nathan, P. W. and Smith, M. C. (1959). Fasciculi proprii of the spinal cord in man. Review of present knowledge, *Brain*, **82**, 610–68.

Nauta, W. J. H. and Mehler, W. R. (1966). Projections of the lentiform nucleus in the monkey, *Brain Res.*, **1**, 3–42.

Nauta, W. J. H. and Haymaker, W. (1969). Hypothalamic nuclei and fiber connections. In: *The Hypothalamus* (Haymaker, W., Anderson, E. and Nauta, W. J. H. eds.), pp. 136–209, Thomas: Springfield, Ill.

Nauta, W. J. H. and Ebesson, O. E. (1970). *Contemporary Research Methods in Neuroanatomy* (editors), Springer-Verlag: Berlin.

Navaratnam, V. (1963). Observations on the right pulmonary arch artery and its nerve supply in human embryos, *J. Anat.*, **97**, 569–73.

Navaratnam, V. (1965). Development of the nerve supply to the human heart, *Br. Heart J.*, **27**, 640–50.

Neal, H. V. (1918). The history of the eye muscles, *J. Morph*, **30**, 433–53.

Needham, J. (1942). *Biochemistry and Morphogenesis*, Cambridge University Press: Cambridge.

Needham, J. (1959). *A History of Embryology*, Second edition, Cambridge University Press: Cambridge.

Negami, S. (1964). Dynamic Mechanical Properties of Synovial Fluid, M.Sc. Thesis, Lehigh University, Bethlehem, Pennsylvania.

Negus, V. E. (1928). *The Mechanism of the Larynx*, Heinemann: London.

Negus, V. E. (1947). (1947). Intrinsic carcinoma of larynx; a review of a series of cases, *Proc. R. Soc. Med.*, **40**, 515–24.

Negus, V. E. (1949). *The Comparative Anatomy and Physiology of the Larynx*, Heinemann: London.

Negus, V. E. (1958). *The Comparative Anatomy and Physiology of the Nose and Paranasal Sinuses*, Livingstone: Edinburgh.

Nelson, D. S. (1968). *Macrophages and Immunity*, North Holland Publishing Company: Amsterdam.

Nelson, E., Blinziger, K. and Hager, H. (1961). Electron microscopic observations on subarachnoid and perivascular spaces of the Syrian hamster brain, *Neurology, Minneap.*, **11**, 285–95.

Nelson, E. and Rennels, M. (1970). Innervation of intracranial arteries, *Brain*, **93**, 475–90.

Nelson, L. (1967). Sperm motility. In: *Fertilisation* (Metz, C. B. and Monroy, A. eds.), Volume 1, pp. 27–98, Academic Press: New York.

Nesebar, R. A., Kornblith, P. L., Pollard, J. J. and Michels, N. A. (1969). *Celiac and Superior Mesenteric Arteries: A Correlation of Angiograms and Dissections*, Little, Brown & Co.: Boston.

Neuman, W. F. and Neuman, M. W. (1958). *The Chemical Dynamics of Bone Mineral*, Chicago University Press: Chicago.

Newman, D. L., Gosling, R. G. and Bowden. R. (1971). Changes in aortic distensibility and area ratio with the development of atherosclerosis, *Atherosclerosis*, **14**, 231–40.

Neyfakh, A. A. and Kostomarova, A. A. (1971). Migration of newly synthesised RNA during mitosis. I. Embryonic cells of the loach (Misgurnus fossilis L), *Exp. Cell Res.*, **65**, 340–4.

Nicander, L. (1967). An electron microscopical study of cell contacts in the seminiferous tubules of some mammals, *Z. Zellforsch. mikrosk. Anat.*, **83**, 375–97.

Nicholls, J. G. and Kuffler, S. W. (1964). Extracellular space as a pathway for exchange between blood and neurons in the central nervous system of the leech: ionic composition of glial cells and neurons, *J. Neurophysiol.*, **27**, 645–71.

Nicolson, G. L. (1976). Transmembrane control of the receptors on normal and tumour cells. I. Cytoplasmic influence over cell surface components, *Biochem. Biophys. Acta.*, **457**, 57–108.

Nicolson, G. L. and Poste, G. (1976). The cancer cell: dynamic aspects and modifications in cell-surface organisation, *New Engl J. Med.*, **295**, 197–203.

Niden, A. H. (1967). Bronchiolar and large alveolar cell in pulmonary phospholipid metabolism, *Science, N.Y.* **158**, 1323–4.

Niden, A. H. and Yamada, E. (1966). Some observations on the fine structure and function of the non-ciliated bronchiolar cells. In: *VIth International Congress for Electron Microscopy*, p. 599. Maruzen.

Nielsen, K. C. and Owman, C. (1968). Difference in cardiac adrenergic innervation between hibernators and non-hibernating mammals, *Acta physiol. scand., Suppl.* **316**, 1–30.

Niemineva, K. (1950). Observations on the development of the hypophysial-portal system, *Acta paediat., Stockh.*, **39**, 366–77.

Nieuwenhuys, R. (1967). Comparative anatomy of the cerebellum, *Prog. Brain Res.*, **25**, 1–93.

Niewenkoop, P. D. (1967). Problems of embryonic induction and pattern formation in amphibians and birds. In: *Morphological and Biochemical Aspects of Cytodifferentiation* (Hagen, E., Wechsler, W. and Zilliken, P. eds.), pp. 22–36 (*Exp. Biol. Med.*, **1**), Karger: Basel, New York.

Niimi, K., Fujiwara, N., Takimoto, T. and Matsungi, S. (1962). The course and termination of the ascending fibers of the brachium conjunctivum in the cat as studied by the Nauta method, *Tokushima J. exp. Med.*, **8**, 269–84.

Niimi, K., Kanaseki, T. and Takimoto, T. (1963). The comparative anatomy of the ventral nucleus of the lateral geniculate body in mammals, *J. comp. Neurol.*, **121**, 313–24.

Niimi, K. and Naito, F. (1974). Cortical projections of the medial geniculate body in the cat, *Exp. Brain Res.*, **19**, 326–42.

Nilsson, O. (1962). Electron microscopy of the glandular epithelium in the human uterus. I. Follicular phase, *J. Ultrastruct. Res.*, **6**, 413–21.

Nilsson, O. (1962). Electron microscopy of the glandular epithelium in the human uterus. II. Early and late luteal phase, *J. Ultrastruct. Res.*, **6**, 422–31.

Nilsson, O. (1967). Attachment of rat and mouse blastocysts onto uterine epithelium, *Int. J. Fert.*, **12**, 5–13.

Nishino, N. Koizumi, K. and Brooks, C. Mc.. (1976). The role of the suprachiasmatic nuclei of the hypothalamus in the production of circadian rhythm, *Brain Res.*, **112**, 45–59.

Nissl, F. (1892). Ueber experimentell erzengte Veränderungen an den Vorderhornzellen des Rückenmarks bei Kaninchen mit Demonstrationen mikroskopischer Präparate, *Allg. Zt. Psychiat.*, **48**, 675–82.

Noback, C. R. and Moss, M. L. (1953). The topology of the human premaxillary bone, *Am. J. phys. Anthrop.*, **11**, 181–7.

Nopajaroonskri, C., Luk, S. C. and Simon, G. T. (1971). Ultrastructure of the normal lymph node, *Am. J. Path.*, **65**, 1–24.

Norberg, K.-A. (1967). Transmitter histochemistry of the sympathetic adrenergic nervous system, *Brain Res.* **5**, 125–70.

Norberg, K.-A., Risley, P. L. and Ungerstedt, U. (1967). Adrenergic innervation of the male reproductive ducts in some mammals. I. The distribution of adrenergic nerves, *Z. Zellforsch. mikrosk. Anat.*, **76**, 278–86.

Norris, E. H. (1916). The morphogenesis of the follicles in the human thyroid gland, *Am. J. Anat.*, **20**, 411–48.

Norris, E. H. (1938). The morphogenesis and histogenesis of the thymus gland in man: in which the origin of the Hassal's corpuscles of the human thymus is discovered, *Contr. Embryol.*, **27**, 191–207.

Norton, A. C. (1968). *U.C.L.A. Brain Information Service, Updated Review Project—Cutaneous Sensory Pathways: Dorsal Column—Medial Lemniscus System*. University of California Press: Berkeley.

Norvell, J. E. (1968). The aorticorenal ganglion and its role in renal innervation, *J. comp. Neurol.*, **133**, 101–12.

Nossal, G. J. V. (1969). *Antibodies and Immunity*, Nelson: London.

Notley, R. G. (1969). The innervation of the upper ureter in man and in the rat: an ultrastructural study, *J. Anat.*, **105**, 393–402.

Notley, R.G. (1971). The structural basis for normal and abnormal ureteric motility. The innervation and musculature of the human ureter, *Ann. R. Coll. Swg. Engl.*, **49**, 250–67.

Nussbaum, A. (1912). Über das gefässystem des Herzens, *Arch. mikrosk. Anat. EntwMech.*, **80**, 450–77.

Nutt, A. B. (1955). Significance and surgical treatment of congenital ocular palsies, *Ann. R. Coll. Surg.*, **16**, 30–59.

Nyberg-Hansen, R. (1964). Origin and termination of fibers from the vestibular nuclei descending in the medial longitudinal fasciculus. An experimental study with silver impregnation methods in the cat, *J. comp. Neurol.*, **122**, 355–67.

Nyberg-Hansen, R. (1964). The location and termination of tectospinal fibers in the cat, *Expl Neurol.*, **9**, 212–27.

Nyberg-Hansen, R. (1965). Anatomical demonstration of gamma motoneurons in the cat's spinal cord, *Expl Neurol.*, **13**, 71–81.

Nyberg-Hansen, R. (1965). Sites and mode of termination of reticulospinal fibers in the cat. An experimental study with silver impregnation methods, *J. comp. Neurol.*, **124**, 71–100.

Nyberg-Hansen, R. (1966). Sites of termination of interstitiospinal fibers in the cat. An experimental study with silver impregnation methods, *Archs ital. Biol.*, **104**, 98–111.

Nyberg-Hansen, R. (1966). Functional organisation of descending supraspinal fibre systems to the spinal cord. Anatomical observations and physiological correlations, *Ergebn. Anat. EntwGesch.*, **39**, Heft 2, 1–48.

Nyberg-Hansen, R. (1969). Cortico-spinal fibres from the medial aspect of the cerebral hemisphere in the cat. An experimental study with the Nauta method, *Exp. Brain Res.*, **7**, 120–32.

Nyberg-Hansen, R. and Brodal, A. (1963). Sites of termination of cortico-spinal fibers in the cat. An experimental study with silver impregnation methods, *J. comp. Neurol.*, **120**, 369–91.

Nyberg-Hansen, R. and Rinvik, E. (1963). Some comments on the pyramidal tract, with special reference to its individual variations in man, *Acta neurol. scand.*, **39**, 1–30.

Nyberg-Hansen, R. and Brodal, A. (1964). Sites and mode of termination of rubrospinal fibres in the cat. An experimental study with silver impregnation methods, *J. Anat.*, **98**, 235–53.

1507

Nyberg-Hansen, R. and Mascitti, T. (1964). Sites and mode of termination of fibers of the vestibulospinal tract in the cat. An experimental study with silver impregnation methods, *J. comp. Neurol.*, **122**, 369–87.

Nyby, O. and Jansen, J. (1951). An experimental investigation of the cortico-pontine projection in *Macaca mulatta*, *Skr. Norske Vidensk.-Akad.*, **I**, Mat.-nat. Kl. No. 3, 1–47.

O

Oakes, B. W. and Bailkower, B. (1977). Biomechanical and ultrastructural studies on the elastic wing tendon from the domestic fowl, *J. Anat.*, **123**, 369–87.

Obayashi, T. (1959). Electron microscope study on lacrimal gland of normal rabbit, *Acta Soc. ophthal. jap.*, **63**, 2631–45.

Obersteiner, H. (1883). Der feinere Bau der Kleinhirnrinde beim Menschen und bei Tieren, *Biol. Zentralbl.*, **3**, 145–55.

Obletz, B. E. and Halbstein, B. M. (1938). Non-union of fractures of carpal navicular, *J. Bone Jt Surg.*, **20**, 424–8.

Ochoa, J. (1971). The sural nerve of the human foetus: electron microscope observations and counts of axons, *J. Anat.*, **108**, 213–45.

Ochoa, J. and Mair, W. E. P. (1969). The normal sural nerve in man. I. Ultrastructure and numbers of fibres and cells, *Acta neuropath.*, **13**, 197–216.

Ochoa, J. and Mair, W. E. P. (1969). The normal sural nerve in man. II. Changes in the axons and Schwann cells due to ageing, *Acta neuropath.*, **13**, 217–39.

O'Connell, J. E. A. (1934). Some observations on cerebral veins, *Brain*, **57**, 484–503.

Odgers, P. N. B. (1930). Some observations on the development of the ventral pancreas in man, *J. Anat.*, **65**, 1–7.

Odgers, P. N. B. (1934). The formation of the venous valves, the foramen secundum and the septum secundum in the human heart, *J. Anat.*, **69**, 412–22.

Odgers, P. N. B. (1937). An early human ovum (Thomson) *in situ*, *J. Anat.*, **71**, 161–8.

Odgers, P. N. B. (1938). The development of the pars membranacea septi in the human heart, *J. Anat.*, **72**, 247–59.

Odor, D. L. (1960). Electron microscope studies on ovarian oöcytes and unfertilised tubal ova in the rat, *J. biophys. biochem. Cytol.*, **7**, 567–74.

Offer, G., Moos, C. and Starr, R. (1973). A new protein of the thick filaments of vertebrate skeletal myofibrils. Extraction, purification and characterisation, *J. mol. Biol.*, **74**, 653–76.

Ogawa, T., Jefferson, N. C., Toman, J. E., Chiles, T., Zambetoglou, A. and Necheles, H. (1960). Action potentials of accessory respiratory muscles in dogs, *Am. J. Physiol.*, **199**, 569–72.

Ogilvie, H. (1952). First part of the duodenum, *Lancet*, **2**, 1077–81.

Ogura, J. H. and Bello, J. A. (1952). Laryngectomy and radical neck dissection for carcinoma of the larynx, *Laryngoscope, St. Louis*, **62**, 1–52.

Ohno, S. and Smith, J. B. (1964). Role of fetal follicular cells in meiosis of mammalian oöcytes, *Cytogenetics*, **3**, 324–33.

Ohta, M., Offord, K. and Dyck, P. J. (1974). Morphometric evaluation of first sacral ganglia of man, *J. Neurol. Sci.*, **22**, 73–82.

Ojemann, G. A., Fedio, P. and van Buren, J. M. (1968). Anomia from pulvinar and subcortical parietal stimulation, *Brain*, **91**, 99–116.

Okamoto, E. and Ueda, T. (1967). Embryogenesis of intramural ganglia of the gut and its relation to Hirschsprung's disease, *J. pediat. Surg.*, **2**, 437–43.

O'Leary, J. L. and O'Leary, J. A. (1966). Uterine artery ligature in the control of intractable post-partum hemorrhage, *Am. J. Obstet. Gynec.*, **94**, 920–4.

Olivier, G. (1975). Biometry of the human occipital bone, *J. Anat.*, **120**, 507–18.

Olsson, Y. (1971). Studies on vascular permeability in peripheral nerves, *Acta neuropath.*, **17**, 114–26.

Olszewski, J. (1954). The cytoarchitecture of the human reticular formation. In: *Brain Mechanisms and Consciousness* (Adrian, E. D., Bremer, F. and Jasper, H. H. eds.), Blackwell: Oxford.

Olszewski, J. and Baxter, D. (1954). *Cytoarchitecture of the Human Brain Stem*, Karger: Basel.

O'Malley, C. D. (1964). *Andreas Vesalius of Brussels, 1514–1564*, University of California Press: Berkeley.

Oppel, O. (1963). Microscopic investigations of the number and caliber of the medullated nerve fibers of the optic fasciculus in man, *Albrecht v. Graefes Arch. Ophthal.*, **166**, 19–27.

O'Rahilly, R. (1956). Developmental deviations in the carpus and tarsus, *Clin. Orthop.*, **10**, 9–18.

O'Rahilly, R. (1966). The early development of the eye in staged human embryos, *Contr. Embryol.*, **38**, 1–42.

O'Rahilly, R. (1975). The prenatal development of the human eye, *Exp. Eye Res.*, **21**, 93–112.

O'Rahilly, R., Gardner, E. and Gray, D. J. (1956). The ectodermal thickening and ridge in the limbs of staged human embryos, *J. Embryol. exp. Morph.*, **4**, 254–64.

O'Rahilly, R. and Meyer, D. B. (1959). The early development of the eye in the chick *Gallus domesticus* (stages 8 to 25), *Acta anat.*, **36**, 20–58.

O'Rahilly, R. and Gardner, E. (1971). The timing and sequence of events in the development of the human nervous system during the embryonic period proper, *Z. Anat. EntwGesch.*, **134**, 1–12.

O'Rahilly, R. and Gardner, E. (1972). The initial appearance of ossification in staged human embryos, *Am. J. Anat.*, **134**, 291–301.

O'Rahilly, R. and Muecke, E. C. (1972). The timing and sequence of events in the development of the human urinary system during the embryonic period proper, *Z. Anat. EntwGesch.*, **138**, 99–109.

O'Rahilly, R. and Boyden, E. A. (1973). The timing and sequence of events in the development of the human respiratory system during the embryonic period proper, *Z. Anat. EntwGesch.*, **141**, 237–50.

O'Rahilly, R. and Tucker, J. A. (1973). The early development of the larynx in staged human embryos. Part 1: Embryos of the first five weeks (to stage 15), *Ann. Otol. Rhinol. Lar.*, **82**, Suppl. 7.

O'Rahilly, R. and Gardner, E. (1975). The timing and sequence of events in the development of the limbs in the human embryo, *Anat. Embryol.*, **148**, 1–23.

Orci, L. (1974). A portrait of a pancreatic B-cell, *Diabetologia*, **10**, 163–87.

Orci, L. (1976). Morphofunctional aspects of the islets of Langerhans. The microanatomy of the islets of Langerhans, *Metabolism*, **25**, Suppl. 1, 1303–13.

Orci, L. and Unger, R. H. (1975). Functional subdivision of islets of Langerhans and possible role of D-cells, *Lancet*, **ii**, 1243–4.

Orci, L., Baetens, D., Dubois, M. P. and Rufener, C. (1975). Evidence for the D-cell of the pancreas secreting somatostatin, *Horm. Metab. Res.*, **7**, 400–2.

Orci, L., Baetens, D., Ravazzola, M., Malaisse-Lagae, F., Amherdt, M. and Rufener, C. (1976). Somatostatin in the pancreas and gastrointestinal tract. In: *Endocrine Gut and Pancreas* (Fujita, T. ed.), pp. 73–88, Elsevier: New York, Amsterdam.

Orfanos, C. and Ruska, H. (1968). Die Feinstruktur des menschlichen Haars. I. Die Haar-Cuticula, *Arch. klin. exp. Derm.*, **231**, 97–110.

O'Riordan, J. L. H. (1978). Hormonal control of mineral metabolism. In: *Recent Advances in Endocrinology and Metabolism* (O'Riordan, J. L H. ed.), pp. 189–217, Churchill Livingstone: Edinburgh, London, New York.

Ormerod, F. C. (1960). The physiology of the endolymph, *J. Lar. Otol.*, **74**, 659–67.

Orofino, C., Sherman, M. S. and Schechter, D, (1960). Luschka's joint—a degenerative phenomenon, *J. Bone Jt Surg.*, **42A**, 853–8.

Orzalesi, N., Riva, A. and Testa, F. (1971). Fine structure of human lacrimal gland. I. The normal gland, *J. submicrosc. Cytol.*, **3**, 283–98.

Osborn, J. W. (1967). A mechanistic view of dentinogenesis and its relation to the curvatures of the processes of the odontoblasts, *Archs oral Biol.*, **12**, 275–80.

Osborn, J. W. (1968). Directions and interrelationships of enamel prisms from the sides of human teeth, *J. dent. Res.*, **47**, 223–32.

Osborn, J. W. (1968). Directions and interrelationships of prisms in cuspal and cervical enamel in human teeth, *J. dent. Res.*, **47**, 395–402.

Osborn, J. W. (1969). The three-dimensional morphology of the tufts in human enamel, *Acta anat.*, **73**, 481–95.

Osborn, J. W. (1970a). The mechanism of prism formation in teeth: a hypothesis, *Calc. Tiss. Res.*, **5**, 115–32.

Osborn, J. W. (1970b). The mechanism of ameloblast movement: a hypothesis, *Calc. Tiss. Res.*, **5**, 344–59.

Osborn, J. W. (1971). A relationship between the striae of Retzius and prism directions in the transverse plane of the tooth, *Arch oral Biol.*, **16**, 1061–70.

Oscarsson, O. (1965). Functional organisation of the spino- and cuneo-cerebellar tracts, *Physiol. Rev.*, **45**, 495–522.

Oscarsson, O. (1967). Termination and functional organisation of a dorsal spino-olivocerebellar path, *Brain Res.*, **5**, 531–4.

Oscarsson, O. and Uddenberg, N. (1964). Identification of a spino-cerebellar tract activated from forelimb afferents in the cat, *Acta physiol. scand.*, **62**, 125–36.

Osen, K. K. (1969). The intrinsic organisation of the cochlear nuclei, *Acta otolar.*, **67**, 352–9.

Osen, K. K. (1970). Course and termination of the primary afferents in the cochlear nuclei of the cat. An experimental anatomical study, *Archs ital. Biol.*, **108**, 21–51.

Osmond, D. G. (1966). The origin of peritoneal macrophages from the bone marrow, *Anat. Rec.*, **154**, 397(a).

Osterberg, G. A. (1935). Topography of the layers of the rods and cones in the human retina, *Acta ophthal.*, Suppl. 6.

Otsuka, R. and Hassler, R. (1962). On the striation and segmentation of the cortical center of vision in the cat, *Arch. Psychiat. NervKrankh.*, **203**, 212–34.

Outrequin, G., Caix, M. and Casanova, G. (1967). Variations du ligament triangulaire gauche du foie en fonction du type morphologique, *C.r. Ass. Anat.*, **136**, 756–62.

Owen, M. (1976). Cell differentiation in bone, *J. Anat.*, **122**, 189.

Owen, R. (1868). *On the Anatomy of Vertebrates*, Longmans, Green: London.

Owen, R. L. and Jones, A. L. (1974). Epithelial cell specialization within human Peyer's patches. An ultrastructural study of intestinal lymphoid follicles, *Gastroenterol.*, **66**, 189–203.

Owen, R. L. and Nemanic, P. (1978). Antigen processing structures of the mammalian intestinal tract: an SEM study of lymphoepithelial organs. *Scanning Electron Microscopy*, Volume 11, SEM Inc. AMF O'Hare, Il. 60666 USA.

Owman, C., Rosenbren, E. and Sjoberg, N. O. (1967). Adrenergic innervation of the human female reproductive organs: a histochemical and chemical investigation, *Obstet. Gynec.*, **30**, 763–73.

P

Pace, J. L. (1968). Stereoscopic micro-anatomy of human colonic mucosa and its blood vessels, *J. Anat.*, **103**, 602P.

Pacini, A. and Gremiger, D. (1975). Alcune modalità nella distribuzione del nervo mascellare, *Arch. Ital. Anat. Embriol.*, **80**, 29–35.

Padgett, D. H. (1948). The development of the cranial arteries in the human embryo, *Contr. Embryol.*, **32**, 205–61.

Padgett, D. H. (1957). The development of the cranial venous system in man, from the viewpoint of comparative anatomy, *Contr. Embryol.*, **36**, 79–140.

Padykula, H. A. and Gauthier, G. F. (1970). The ultrastructure of the neuromuscular junctions of mammalian red, white and intermediate skeletal muscle fibers, *J. Cell Biol.*, **46**, 27–41.

Page, R. B. and Bergland, R. M. (1977). Pituitary vasculature. In: *The Pituitary. A Current Review* (Allen, M. B. and Mahesh, V. B. eds.), pp. 9–17, Academic Press: New York, London.

Page, S. G. (1965). A comparison of the fine structures of frog slow and twitch muscle fibers, *J. Cell Biol.*, **26**, 477–97.

Pakkenberg, H. (1966). The number of nerve cells in the cerebral cortex of man, *J. comp. Neurol.*, **128**, 17–20.

Palade, G. E. (1968). Small pore and large pore systems in capillary permeability. In: *Hemorheology* (Copley, A. L. ed.), pp. 703–20, Pergamon: Paris, New York, London.

Palade, G. E. (1975). Intracellular aspects of the process of protein secretion. Nobel lecture, *Science, N.Y.*, **189**, 347–58.

Palay, S. L. and Palade, G. E. (1955). The fine structure of neurons, *J. biophys. biochem. Cytol.*, **1**, 69–88.

Palay, S. L. and Chan Palay, V. (1977). General morphology of neurons and neuroglia. In: *Handbook of Physiology* (Kandel, E. R. ed.), Volume 1, Part 2, pp. 803–53, Physiological Society: Bethesda.

Palfrey, A. J. and Davies, D. V. (1966). The fine structure of chondrocytes, *J. Anat.*, **100**, 213–26.

Pallie, W. and Manuel, J. K. (1968). Intersegmental anastomoses between dorsal spinal rootlets in some vertebrates, *Acta anat.*, **70**, 341–51.

Palma, A. F. de (1957). *Degenerative Changes in the Sternoclavicular and Acromioclavicular Joints in Various Decades*, Thomas: Springfield, Ill.

Pandya, P. N. and Kuypers, H. G. J. M. (1969). Cortico-cortical connections in the rhesus monkey, *Brain Res.*, **13**, 13–36.

Pandya, P. N. and Vignolo, L. A. (1969). Interhemispheric projection of the parietal lobe in the rhesus monkey, *Brain Res.*, **106**, 365–70.

Panner, B. J. and Honig, C. R. (1967). Filament ultrastructure and organisation in vertebrate smooth muscle, *J. Cell Biol.*, **35**, 303–21.

Papanicolau, G. N. and Traut, H. F. (1943). *Diagnosis of Uterine Cancer by the Vaginal Smear*, Commonwealth Fund: London, Oxford.

Papathanassion, B. T. (1968). A variant of the motor branch of the median nerve in the hand, *J. Bone Jt Surg.*, **50B**, 156–7.

Papez, J. W. (1927). Subdivisions of the facial nucleus, *J. comp. Neurol.*, **43**, 159–91.

Papez, J. W. (1937). A proposed mechanism of emotion, *Archs Neurol. Psychiat., Chicago*, **38**, 725–43.

Pappas, G. D. and Tennyson, V. M. (1962). An electron microscopic study of the passage of colloidal particles from the blood vessels of the ciliary processes and choroid plexus of the rabbit, *J. Cell Biol.*, **15**, 227–39.

Parkes, A. S. (1952–66). *Marshall's Physiology of Reproduction* (editor), volume 1, Part 1, 1956; Volume 2, 1960; Volume 3, 1966, Longmans, Green: London.

Parkinson, D. (1973). Carotid cavernous fistula: direct repair with preservation of the carotid artery. Technical note, *J. Neurosurg.*, **38**, 99–106.

Parkinson, D., Johnston, J. and Chaudhuri, A. (1978). Sympathetic connections to the fifth and sixth cranial nerves, *Anat. Rec.*, **191**, 221–6.

Parry, E. W. (1970). Some electron microscope observations on the mesenchymal structures of full-term umbilical cord, *J. Anat.*, **107**, 505–18.

Parsons, F. G. (1903). On the meaning of some of the epiphyses, *J. Anat.*, **37**, 315–23.

Parsons, F. G. (1904). Observations on traction epiphyses, *J. Anat.*, **38**, 248–58.

Parsons, F. G. (1905). On pressure epiphyses, *J. Anat.*, **39**, 402–12.

Partlow, G. D., Colonnier, M. and Szabo, J. (1977). Thalamic projections of the superior colliculus in the rhesus monkey (*Macaca mulatta*). A light and electron microscope study, *J. comp. Neurol.*, **171**, 285–318.

Partridge, S. M. (1970). Isolation and characteristics of elastin. In: *Chemistry and Molecular Biology of the Intercellular Matrix* (Balazs, E. A. ed.), Volume I, Chapter 4, pp. 593–616, Academic Press: New York, London.

Patake, S. M. and Mysorekar, V. R. (1977). Diaphyseal nutrient foramina in human metacarpals and metatarsals, *J. Anat.*, **124**, 299–304.

Paterson, A. M. (1904). *The Human Sternum*, Williams and Norgate: London.

Paterson, A. M. (1912). The form of the human stomach, *J. Anat.*, **47**, 356–9.

Patten, B. M. (1956). The development of the sinuventricular conduction system, *Univ. Mich. Med. Bull.*, **22**, 1–21.

Patten, B. M. (1968). *Human Embryology*, Third edition, McGraw-Hill: New York, Toronto, Sydney, London.

Patten, B. M. and Kramer, T. C. (1933). Initiation of contraction in embryonic chick heart, *Am. J. Anat.*, **53**, 349–75.

Pattle, R. E. (1965). Surface lining of lung alveoli, *Physiol. Rev.*, **45**, 48–79.

Pauerstein, C. J. and Woodruff, J. D. (1967). The role of the 'indifferent' cell of the tubal epithelium, *Am. J. Obstet. Gynec.*, **98**, 121–5.

Paul, J. and Gilmour, R. S. (1968). Organ-specific restriction of transcription in mammalian chromatin, *J. molec. Biol.*, **34**, 305–16.

Paul, M. and Kanagasuntheram, R. (1956). Congenital anomalies of lower urinary tract, *Br. J. Urol.*, **28**, 64–74.

Paula-Barbosa, M. M. and Sousa-Pinto, A. (1973). Auditory cortical projections to the superior colliculus in the cat, *Brain Res.*, **50**, 47–61.

Paulo, L. G., Fink, G. D., Roh, B. L. and Fischer, J. W. (1973). Influence of carotid body ablation on erythropoietin production in rabbits, *Am. J. Physiol.*, **224**, 442–4.

Pauwels, F. (1965). *Gesammelte Abhandlungen zur funktionellen Anatomie des Bewegungsapparatus*, Springer-Verlag: Berlin.

Pavlov, S. and Pétrov, V. (1968). Sur l'anse sous-clavière de l'artère sous-clavière droite rétro-oesophagienne, *Folia med., Plovdiv.*, **10**, 73–8.

Payton, C. G. (1934). The position of the nutrient foramen and direction of the nutrient canal in the long bones of the madder-fed pig, *J. Anat.*, **68**, 500–10.

Peacock, A. (1951). Observations on the pre-natal development of the intervertebral disc in man, *J. Anat.*, **85**, 260–74.

Peacock, A. (1952). Observations on the postnatal structure of the intervertebral disc in man, *J. Anat.*, **86**, 162–79.

Pearee, G. W. (1960). Some cortical projections to the midbrain reticular formation. In: *Structure and Function of the Cerebral Cortex* (Tower, D. B. and Schadé, J. P. eds.), pp. 131–7, Elsevier: Amsterdam.

Pearse, A. G. E. (1966). The cytochemistry of the thyroid C cells and their relationship to calcitonin, *Proc. R. Soc. B.*, **164**, 478–87.

Pearse, A. G. E. (1966). 5-Hydroxytryptophan uptake by dog thyroid C cells and its possible significance in polypeptide hormone production, *Nature, Lond.*, **211**, 598–600.

Pearse, A. G. E. (1968). *Histochemistry*, Third edition, Churchill: London.

Pearse, A. G. E. (1969). The cytochemistry and ultrastructure of polypeptide hormone-producing cells (the APUD series) and the embryologic, physiologic and pathologic implications of the concept, *J. Histochem. Cytochem.*, **17**, 303–13.

Pearse, A. G. E. (1974). The gut as an endocrine organ, *Br. J. hosp. Med.*, **11**, 697–704.

Pearse, A. G. E. (1976). Neurotransmission and the APUD concept. In: *Chromaffin, Enterochromaffin and Related Cells* (Coupland, R. E. and Fujita, T. eds.), pp. 147–54, Elsevier: New York, Amsterdam.

Pearse, A. G. E. (1977). The apudomas; with particular reference to those of gastroenteropancreatic origin. In: *The Gastrointestinal Tract* (Yardley, J. H., Morson, B. C. and Abell, M. R. eds.), pp. 206–18, Williams & Wilkins: Baltimore.

BIBLIOGRAPHY

Pearse, A. G. E. (1977). The diffuse neuroendocrine system and the APUD concept: related 'endocrine' peptides in brain, intestine, pituitary, placenta and anuran cutaneous glands, *Med. Biol.*, **55**, 115–25.

Pearse, A. G. E. and Polak, J. M. (1971). Cytochemical evidence for the neural crest origin of mammalian C cells, *Histochemie*, **37**, 96–102.

Pearse, A. G. E. and Polak, J. M. (1974). Endocrine tumours of neural crest origin: neurolophomas, apudomas and the APUD concept, *Med. Biol.*, **52**, 3–18.

Pearse, A. G. E. and Polak, J. M. (1978). The diffuse neuroendocrine system and the APUD concept. In: *Gut Hormones* (Bloom, S. R. ed.), Churchill Livingstone: Edinburgh, London.

Pearson, A. A. (1941). The development of the nervus terminalis in man, *J. comp. Neurol.*, **75**, 39–66.

Pearson, A. A. (1943). Trochlear nerve in human fetuses, *J. comp. Neurol.*, **78**, 29–43.

Pearson, A. A. (1944). The oculomotor nucleus in the human fetus, *J. comp. Neurol.*, **80**, 47–68.

Pearson, A. A. (1949). The development and connections of the mesencephalic root of the trigeminal nerve in man, *J. comp. Neurol.*, **90**, 1–46.

Pearson, A. A. and Sauter, R. W. (1971). Observations on the caudal end of the spinal cord, *Am. J. Anat.*, **131**, 463–70.

Pearson, A. A. and Sauter, R. W. (1971). The internal thoracic (mammary) nerve, *Thorax*, **26**, 354–6.

Pearson, R. C. A. and Powell, T. P. S. (1975). The cortico-cortical connexions to area 5 of the parietal lobe from the primary somatic sensory cortex of the monkey, *Proc. R. Soc.*, **200B**, 103–8.

Pease, D. C. (1956). An electron microscopic study of red bone marrow, *J. Hematol.*, **11**, 501–26.

Pease, D. C. and Quilliam, T. A. (1957). Electron microscopy of the Pacinian corpuscle, *J. biophys. biochem. Cytol.*, **3**, 331–42.

Peck, H. M. and Hoerr, N. L. (1951). The intermediary circulation in the red pulp of the mouse spleen, *Anat. Rec.*, **109**, 447–78.

Peck, H. M. and Hoerr, N. L. (1951). The effect of environmental temperature changes on the circulation of the mouse spleen, *Anat. Rec.*, **109**, 479–94.

Pedersen, H. (1969). Ultrastructure of the ejaculated human spermatozoa, *Z. Zellforsch. mikrosk. Anat.*, **94**, 542–54.

Penfield, W. (1957). Vestibular sensation and the cerebral cortex, *Ann. Otol. Rhinol. Lar.*, **66**, 691–8.

Penfield, W. (1966). Speech, perception and the uncommitted cortex. In: *Brain and Conscious Experience* (Eccles, J. C. ed.), pp. 217–37, Springer-Verlag: Berlin.

Penfield, W. and McNaughton, F. (1940). Dural headache and innervation of the dura mater, *Archs Neurol. Psychiat., Chicago*, **44**, 43–75.

Penfield, W. and Rasmussen, T. (1950). *The Cerebral Cortex of Man*, Macmillan: New York.

Penfield, W. and Welch, K. (1951). The supplementary motor area of the cerebral cortex. A clinical and experimental study, *Archs Neurol. Psychiat., Chicago*, **66**, 289–317.

Penfield, W. and Jasper, H. (1954). *Epilepsy and the Functional Anatomy of the Human Brain*, Little, Brown & Co.: Boston.

Penfield, W. and Roberts, L. (1959). *Speech and Brain Mechanisms*, Princeton University Press: Princeton.

Pennel, T. C. (1966). Anatomical study of the peripheral pulmonary lymphatics, *J. thorac. cardiovasc. Surg.*, **52**, 629–34.

Percheron, G. (1977). Les artères du thalamus humain. Les artères choroïdiennes. I–V, *Rev. Neurol.*, **133**, 533–58.

Perera, H. and Edwards, F. R. (1957). Intradiaphragmatic course of the left phrenic nerve in relation to diaphragmatic incisions, *Lancet*, **2**, 75–7.

Perloff, W. H. and Steinberger, E. (1964). *In vivo* survival of spermatozoa in cervical mucus, *Am. J. Obstet. Gynec.*, **88**, 439–42.

Pernkopf, E. (1963). *Atlas of Topographical and Applied Human Anatomy*, Saunders: Philadelphia.

Perry, S. V. (1968). The role of myosin in muscular contraction. In: *Aspects of Cell Motility*, S.E.B. Symposia, **22**, 1–16.

Persaud, T. V. N. (1977). *Problems of Birth Defects*, MTP Press: Lancaster, England.

Person, R. S. and Roshchina, N. A. (1958). Elektromiograficheskow issledovanie koordinatsii deiatel'nostimyshtsantagonistov pri dvizhenii pal'tsev ruti cheloveka, *Fiziol. Zh. SSSR*, **44**, 455–62.

Perutz, M. F., Rossmann, M. G., Cullis, A. F., Muirhead, H., Will, G. and North, A. C. T. (1960). Structure of haemoglobin. A three-dimensional Fourier synthesis at 5·5 Å resolution, obtained by X-ray analysis, *Nature, Lond.*, **185**, 416–22.

Peters, A. (1966). The node of Ranvier in the central nervous system, *Q. Jl exp. Physiol.*, **51**, 229–36.

Peters, A. and Palay, S. L. (1966). The morphology of the laminae A and A₁ of the dorsal nucleus of the lateral geniculate body of the cat, *J. Anat.*, **100**, 451–86.

Peters, A., Palay, S. L. and Webster, H. de F. (1970). *The Fine Structure of the Nervous System*, Harper and Row: New York.

Peters, H. (1899). *Ueber die einbettung des menschlichen eies und das frueheste bisher bekannte menschliche Placentationsstadium*, Deutcke: Leipzig, Vienna.

Peters, R. M. (1969). *The Mechanical Basis of Respiration*, Churchill: London.

Peterson, M. R. and Leblond, C. P. (1964). Uptake by the Golgi region of glucose labelled with tritium in the 1 or 6 position, as an indicator of synthesis of complex carbohydrates, *Expl Cell Res.*, **34**, 420–3.

Peterson, M. and Leblond, C. P. (1964). Synthesis of complex carbohydrates in the Golgi region, as shown by radioautography after injection of labelled glucose, *J. Cell Biol.*, **21**, 143–8.

Petkov, P. (1968). Anatomotopographic investigations on the anterior hepatic nerve plexus, *Scripta Scient. med. Varna*, **7**, 95–7.

Petras, J. M. and Cummings, J. F. (1977). The origin of spinocerebellar pathways: the nucleus centrobasalis of the cervical enlargement and the nucleus dorsalis of the thoraco-lumbar spinal cord, *J. comp. Neurol.*, **173**, 693–716.

Petrovic, A. G. (1972). Mechanisms and regulation of mandibular condylar growth, *Acta morph. neerl.-scand.*, **10**, 25–34.

Pfaffman, C. (1970). Physiological and behavioural processes of the sense of taste. In: *Taste and Smell in Vertebrates* (Wolstenholme, G. E. W. and Knights, J. eds.), pp. 31–44, Ciba Foundation Symposium, Churchill: London.

Pfeifer, R. A. (1940). *Die angioarchitecktonische areale Gliederung der Grosshirnrinde*, Thieme: Leipzig.

Pfenninger, K., Sandri, C., Akert, K. and Eugster, G. H. (1969). Contribution to the problem of structural organisation of the presynaptic area, *Brain. Res.*, **12**, 10–18.

Pfenninger, K. H. and Rees, R. P. (1976). From the growth cone to the synapse: properties of membranes involved in synapse formation. In: *Neuronal Recognition* (Barondes, S. H. ed.), pp. 131–78, Plenum: New York.

Pfister, R. R. (1973). The normal surface of corneal epithelium: a scanning electron microscopic study, *Invest. Ophthal.*, **12**, 654–68.

Pfister, R. R. and Burstein, N. (1977). The normal and abnormal human corneal epithelial surface: a scanning electron microscope study, *Invest. Ophthal.*, **16**, 614–22.

Pflüger, E. (1866). Ueber die Epithelien der Glandula submaxillaris, *Zbl. med. Wiss.*, **4**, 193–5.

Phifer, R. F., Midgley, A. R. and Spicer, S. S. (1973). Immunologic and histologic evidence that follicle stimulating and luteinizing hormones are present in the same cell types in the human pars distalis, *J. clin. Endocr. Metab.*, **36**, 125–41.

Philipson, T., Hanninen, L. and Balazs, E. A. (1975). Cell contacts in human and bovine lenses. *Expl Eye Res.*, **21**, 205–19.

Phillips, D. M. (1975). Mammalian sperm structure. In: *Handbook of Physiology* (Hamilton, D. W. and Greep, R. O. eds.), Section 7, Volume V, pp. 405–20, American Physiological Society: Washington.

Phillips, D. M. and Olson, G. (1974). Cinematographic analysis of mammalian spermatozoa. In: *Functional Anatomy of the Spermatozoon* (Afzelius, B. ed.), Pergamon: London.

Piasecka-Kacperska, K. and Gladyskowska-Rzeczycka, J. (1972). Splot krzyżowy u naczelnych, *Folia morphol. (Warsz.)*, **31**, 21–33.

Piasecki, C. (1974). Blood supply to the human gastro-duodenal mucosa, *J. Anat.*, **118**, 295–335.

Piasecki, C. (1977). Role of ischaemia in the initiation of peptic ulcer, *Ann. R. Coll. Surg.*, **59**, 476–8.

Pick, J. W., Anson, B. J. and Ashley, F. L. (1942). The origin of the obturator artery. A study of 640 body halves, *Am. J. Anat.*, **70**, 317–44.

Pick, J. and Sheehan, D. (1946). Sympathetic rami in man, *J. Anat.*, **80**, 12–20.

Pick, J. (1970). *The Autonomic Nervous System*, Lippincott: Philadelphia.

Pickel, V. M., Segal, M. and Bloom, F. E. (1979). A radioautographic study of the efferent pathways of the nucleus locus coeruleus, *J. comp. Neurol.*, **155**, 15–42.

Pickel, V. M., Joh, T. H., Reis, D. J., Leeman, S. E. and Miller, R. J. (1979). Electron microscope localization of substance P and enkephalin in axon terminals related to dendrites of catecholaminergic neurons, *Brain Res.*, **160**, 387–400.

Pickem, L. (1960). *The Organisation of Cells*, Clarendon Press: Oxford.

Pickett-Heaps, J. D. (1975). Aspects of spindle evolution, *Ann. N.Y. Acad. Sci.*, **253**, 352–61.

Pieron, A. P. (1973). The mechanism of the first carpo-metacarpal joint, *Acta orthop. Scand., Suppl.*, **148**.

Pierson, R. and Carpenter, M. B. (1974). Anatomical analysis of pupillary reflex pathways in the rhesus monkey, *J. comp. Neurol.*, **158**, 121–44.

Pikó, L. (1967). Gamete structure and sperm entry in mammals. In: *Fertilisation* (Metz, C. B. and Monroy, A. eds.), Volume 2, Chapter 8, Academic Press: New York, London.

Pikó, L. and Tyler, A. (1964). Fine structural studies of sperm penetration in the rat, *Proc. Vth Congr. Internazionale per la Riproduzione Animale e la Fecundazione Artificale, Trento*, Section I, Volume II, 372–7.

Pimendis, M. Z. and Hinds, J. W. (1977). An autoradiographic study of the sensory innervation of teeth. I. Dentin, II. Dental pulp and periodontium, *J. Dent. Res.*, **56**, 827–34; 835–40.

Pinching, A. J. and Powell, T. P. S. (1971). The neuron types of the glomerular layer of the olfactory bulb, *J. Cell Sci.*, **9**, 305–46.

Pinching, A. J. and Powell, T. P. S. (1971). The neuropil of the glomeruli of the olfactory bulb, *J. Cell Sci.*, **9**, 347–78.

Pinching, A. J. and Powell, T. P. S. (1971). The neuropil of the periglomerular region of the olfactory bulb, *J. Cell Sci.*, **9**, 379–409.

Pinkerton, J. H. M., McKay, D. G., Adams, E. C. and Hertig, A. T. (1961). Development of the human ovary—a study using histochemical technics, *Obstet. gynec., N.Y.*, **18**, 152–81.

Pinkus, H. (1958). Embryology of hair. In: *The Biology of Hair Growth* (Montagna, W. and Ellis, R. A. eds.), pp. 1–32, Academic Press: New York, London.

Pirenne, M. (1967). *Vision and the Eye*, Second edition, Chapman and Hall: London.

Pittman, F. E. and Pittman, J. C. (1966). An electron microscopic study of the epithelium of normal human sigmoid colonic mucosa, *Gut*, **7**, 644–61.

Platz, F. and Adelmann, G. (1976). Zur Anatomie der 'Vena arcuata cruris posterior', und ihrer Tiefen-anastomosen (Vv. communicantes sive perforantes), *Verh. anat. ges., Jena.*, **70**, 709–14.

Platzer, W. and Maurer, H. (1966). Zur Segmenteinteilung der Leber, *Acta anat.*, **63**, 8–31.

Plentl, A. A. (1958). The origin of amniotic fluid. In: *Gestation* (Villee, C. A. ed.), pp. 71–114, Macy Foundation: New York.

Plets, C. (1969). The arterial blood supply and angioarchitecture of the posterior wall of the third ventricle, *Acta neurochir.*, **21**, 309–17.

Poggio, G. F. and Mountcastle, V. B. (1960). A study of the functional contributions of the lemniscal and spinothalamic systems to somatic sensibility, *Bull. Johns Hopkins Hosp.*, **106**, 266–316.

Poggio, G. F. and Mountcastle, V. B. (1963). The functional properties of ventrobasal thalamic neurons studied in unanaesthetised monkeys, *J. Neurophysiol.*, **26**, 775–806.

Poirier, L. J. and Bertrand, C. (1955). Experimental and anatomical investigations of lateral spinothalamic and spinotectal tracts, *J. comp. Neurol.*, **102**, 745–57.

Poisel, S. and Golth, D. (1974). Zur Variabilität der grossen Arterien im Trigonum caroticum, *Wien. Med. Wochenschr.*, **124**, 229–32.

Polacek, P. (1961). Relation of myocardial bridges and loops on the coronary arteries to coronary occlusion, *Am. Heart J.*, **61**, 44–62.

Polak, J. M. and Pearse, A. G. E. (1976). Hypothalamic peptides in the endocrine (APUD) cells of the gut and pancreas. In: *Endocrine Gut and Pancreas* (Fujita, T. ed.), pp. 103–12, Elsevier: New York, Amsterdam.

Polani, P. (1973). The incidence of developmental and other genetic abnormalities, *Guy's Hosp. Rep.*, **122**, 53–63.

Polge, C. (1957). Low-temperature storage of mammalian spermatozoa, *Proc. R. Soc. B.*, **147**, 498–508.

Policard, A. (1950). Sur quelques caractères histophysiologiques des formations lymphoïdes bronchiques, *Bull. Hist. Appl.*, **27**, 118.

Politzer, G. (1952). Zur normalen und abnormen Entwicklung des menschlichen Gesichtes, *Z. Anat. EntwGesch.*, **116**, 332–47.

Pollack, R., Osborn, M. and Weber, K. (1975). Patterns of organisation of actin and myosin in normal and transformed cultured cells, *Proc. natn Acad. Sci., U.S.A.*, **72**, 994–8.

Pollard, T. D., Shelton, E., Weihing, R. R. and Korn, E. D. (1970). Ultrastructural characterisation of F-actin isolated from *Acanthamoeba castellanii* and identification of cytoplasmic filaments as F-actin by reaction with rabbit heavy meromyosin, *J. molec. Biol.*, **50**, 91–8.

Polónyi, J., Kapeller, K. and Mráz, P. (1977). SGG (SIF) cells in autonomic ganglion: evidence for a possible secretion of their contents into the blood vessels, *Z. mikrosk. Anat. Forsch.*, **91**, 581–9.

Polyak, S. (1941). *The Retina*, University of Chicago Press: Chicago.

Polyak, S. (1957). *The Vertebrate Visual System*, University of Chicago Press: Chicago.

Pomeranz, B., Wall, P. D. and Weber, W. V. (1968). Cord cells responding to fine myelinated afferents from viscera, muscle and skin, *J. Physiol., Lond.*, **199**, 511–32.

Pomerat, C. M., Hendelman, W. J., Raiborn, C. W. Jr. and Massey, J. F. (1967). Dynamic activities of nervous tissue *in vitro*. In: *The Neuron* (Hydén, H. ed.), pp. 119–78, Elsevier: London, Amsterdam, New York.

Pompeiano, O. and Brodal, A. (1957). The origin of vestibulospinal fibres in the cat. An experimental-anatomical study, with comments on the descending medial longitudinal fasciculus, *Archs ital. Biol.*, **95**, 166–95.

Pompeiano, O. and Brodal, A. (1957). Experimental demonstration of a somato-topical origin of rubro-spinal fibers in the cat, *J. comp. Neurol.*, **108**, 225–52.

Pompeiano, O. and Brodal, A. (1957). Spino-vestibular fibers in the cat, *J. comp. Neurol.*, **108**, 353–82.

Pontes, C., Reis, F. F. and Sousa-Pinto, A. (1975). The auditory cortical projections on to the medial geniculate body in the cat, *Brain Res.*, **91**, 43–63.

Poole, B., Higashi, T. and de Duve, C. (1970). The synthesis and turnover of rat liver peroxisomes. III. The size distribution of peroxisomes and the incorporation of new catalase, *J. Cell Biol.*, **45**, 408–15.

Popa, G. T. and Fielding, U. (1930). Vascular link between pituitary and hypothalamus, *Lancet*, **2**, 238–40.

Popa, G. T. and Fielding, U. (1930). A portal circulation from the pituitary to the hypothalamic region, *J. Anat.*, **65**, 88–91.

Pope, A. (1967). Microchemical architecture of human isocortex, *Archs Neurol. Psychiat., Chicago*, **16**, 351–6.

Popescu, L. M. and Diculescu, I. (1975). Calcium in smooth muscle sarcoplasmic reticulum in situ. Conventional and X-ray analytical electron microscopy, *J. Cell Biol.*, **67**, 911–18.

Popoff, N. W. (1934). Digital vascular system, with reference to the state of the glomus in inflammation, arteriosclerotic gangrene, diabetic gangrene, thrombo-angitis obliterans and supernumerary digits in man, *Archs Path.*, **18**, 295–330.

Popper, K. R. and Eccles, J. C. (1977). *The Self and its Brain*, Springer International: Berlin.

Porteous, C. J. (1960). The olecranon epiphyses, *J. Anat.*, **94**, 286P.

Posselt, U. (1952). Studies in mobility of human mandible, *Acta odont. scand.*, **10**, Suppl. 10, 3–160.

Potenza, A. D. (1963). Critical evaluation of flexor-tendon healing and adhesion formation within artificial digital sheaths. An experimental study, *J. Bone Jt Surg.*, **45A**, 1217–33.

Potts, T. K. (1925). The main peripheral connections of the human sympathetic nervous system, *J. Anat.*, **59**, 129–35.

Potts, T. R. Jr., Murray, T., Peacock, M., Niall, H. D., Tregear, G. W., Keutmann, H. T., Powell, D. and Deftos, L. J. (1971). Parathyroid hormone: sequence, synthesis, immunoassay studies, *Am. J. Med.*, **50**, 639–49.

Powell, T.P.S. and Cowan, W. M. (1956). A study of thalamostriate relations in the monkey, *Brain*, **79**, 364–90.

Powell, T. P. S. and Mountcastle, V. (1959). The cytoarchitecture of the postcentral gyrus of the monkey *Macaca mulatta*, *Bull. Johns Hopkins Hosp.*, **105**, 108–31.

Powell, T. P. S. and Mountcastle, V. B. (1959). Some aspects of the functional organisation of the cortex of the postcentral gyrus of the monkey. A correlation of findings obtained in a single unit analysis with cytoarchitectonics, *Bull. Johns Hopkins Hosp.*, **105**, 173–200.

Powell, T. P. S. and Cowan, W. M. (1963). Centrifugal fibres in the lateral olfactory tract, *Nature, Lond.*, **199**, 1296–7.

Powell, T. P. S., Cowan, W. M. and Raisman, G. (1965). The central olfactory connexions, *J. Anat.*, **99**, 791–813.

Präder, A. (1947). Die frühembryonale Entwicklung der menschlichen Zwischenwirbelscheibe, *Acta anat.*, **3**, 68–83.

Prakash, S., Chopra, S. R. K. and Jit, I. (1979). Ossification of the human patella, *J. Anat., Ind.*, **28**, 78–83.

Pressman, J. J. (1942). Physiology of vocal cords in phonation and respiration, *Archs Otolar.*, **35**, 355–98.

Prestige, M. C. (1965). Cell turnover in the spinal ganglia of *Xenopus laevis* tadpoles, *J. Embryol. exp. Morph.*, **13**, 63–72.

Prestige, M. C. (1967). The control of cell number in the lumbar spinal ganglia during the development of *Xenopus laevis* tadpoles, *J. Embryol. exp. Morph.*, **17**, 453–71.

Prestige, M. C. (1970). Differentiation, degeneration, and the role of the periphery: Quantitative considerations. In: *The Neurosciences, a Second Study Program* (Schmitt, F. O., Quarton, G. C., Melnechuk, T. and Adelman, G. eds.), pp. 73–82, Rockefeller University Press: New York.

Preston, J. B. and Whitlock, D. G. (1961). Intracellular potentials recorded from motoneurons following precentral gyrus stimulation in primates, *J. Neurophysiol.*, **24**, 91–100.

Price, H. M. (1969). Ultrastructure of the skeletal muscle fibre, In: *Disorders of Voluntary Muscle* (Walton, J. N. ed.), pp. 29–56, Churchill: London.

Price, J. L. (1968). The termination of centrifugal fibres in the olfactory bulb, *Brain Res.*, **7**, 483–6.

Price, J. L. and Powell, T. P. S. (1970). The morphology of the granule cells of the olfactory bulb, *J. Cell Sci.*, **7**, 91–123.

Price, J. L. and Powell, T. P. S. (1970). The synaptology of the granule cells of the olfactory bulb, *J. Cell Sci.*, **7**, 125–56.

Primrose, W. B. (1952). Chest movements and intercostal muscles, *Br. J. Anaesth.*, **24**, 3–24.

Printz, R. H. and Hall, J. L. (1974). Evidence for a retinohypothalamic pathway in the golden hamster, *Anat. Rec.*, **179**, 57–66.

Prinzmetal, M., Simkin, B., Bergman, H. C. and Kruger, H. E. (1947). Studies on the coronary circulation, *Am. Heart J.*, **33**, 420–42.

BIBLIOGRAPHY

Pritchard, J. J., Scott, J. H. and Girgis, F. G. (1956). The structure and development of cranial and facial sutures, *J. Anat.*, **90**, 73–87.

Privat, A. (1975). Postnatal gliogenesis in the mammalian brain, *Int. Rev. Cytol.*, **40**, 281–323.

Prokop, O. and Uhlenbruck, G. (1969). *Human Blood and Serum Groups*, Maclaren: London.

Provenza, D. V. and Sisca, R. F. (1971). Electron microscopic study of human dental primordia, *Archs oral Biol.*, **16**, 121–33.

Pšenicka, P. (1966). Beitrag zur Kenntnis der Innervation der Kehlkopfgelenke, *Anat. Anz.*, **118**, 1–6.

Puchades-Orts, A., Nombela-Gomez, M. and Ortuño-Pacheco, G. (1976). Variation in the form of the circle of Willis, *Anat. Rec.*, **185**, 119–24.

Pulec, J. L., Kamio, T. and Graham, M. D. (1975). Eustachian tube lymphatics, *Ann. Otol. Rhinol. Otolaryngol.*, **84**, 483–92.

Purpura, D. P. and Schadé, J. P. (1964). *Growth and Maturation of the Brain*, Elsevier: Amsterdam.

Purpura, D. P. and Yahr, M. D. (1966). *The Thalamus*, Columbia University Press: New York, London.

Purves, H. D. (1961). Morphology of the hypophysis related to its function. In: *Sex and Internal Secretion* (Young, W. C. ed.), pp. 161–238, William & Wilkins: Baltimore.

Purves, M. J. (1970). The role of the cervical sympathetic nerve in the regulation of oxygen consumption of the carotid body of the cat, *J. Physiol., Lond.*, **209**, 417–31.

Purves, M. J. (1972). *The Physiology of the Cerebral Circulation*, Cambridge University Press: Cambridge.

Püschel, J. (1930). Wassergehalt normaler und degenerierter Zwischenwirbelscheiben, *Beitr. path. Anat.*, **84**, 123–30.

Puschett, J. B. (1978). Renal tubular effects of parathyroid hormone. An update, *Clin. Orthop.*, **135**, 249–59.

Putschar, W. (1931). *Entwicklung, Wachstum und Pathologie der Beckenverbindungen des Menschen mit besonderer Berücksichtigung von Schwangerschaft, Geburt und ihren Folgen. Aus dem Pathologischen Institute der Universität Göttingen*, Fischer Verlag: Stuttgart.

Pyle, S. I. and Hoerr, N. L. (1955). *Radiographic Atlas of Skeletal Development of the Knee*, Thomas: Springfield, Ill.

Pyrtek, L. J. and Painter, R. L. (1964). An anatomic study of the relationship of the parathyroid glands to the recurrent laryngeal nerve, *Surgery Gynec. Obstet.*, **119**, 509–12.

Q

Quilliam, T. A. (1966). Unit design and array patterns in receptor organs. In: *Touch, Heat and Pain* (de Reuck, A. V. S. and Knight, J. eds.), pp. 86–112, Ciba Foundation Symposium, Churchill: London.

Quilliam, T. A. (1971). Some extrinsic and intrinsic determinants of sensory acuity in the human finger tip, *J. Anat.*, **109**, 337P.

Qvist, G. (1977). The course and relations of the left phrenic nerve in the neck, *J. Anat.*, **124**, 803–5.

R

Rabey, G. P. (1968). *Morphanalysis*, Lewis: London.

Rabey, G. P. (1971). Craniofacial morphanalysis, *Proc. R. Soc. Med.*, **64**, 103–11.

Rabinowicz, T. (1964). The cerebral cortex of the premature infant of the 8th month, *Prog. Brain Res.*, **4**, 39–92.

Rabinowicz, Th. (1967). Quantitative appraisal of the cerebral cortex of the premature infant of 8 months. In: *Regional Development of the Brain in Early Life* (Minkowski, A. ed.), Blackwell: London.

Race, R. R. and Sanger, R. (1975). *Blood Groups in Man*, Sixth edition, Blackwell Scientific Publications: Oxford.

Racher, E. (1975). Reconstruction, mechanism of action and control of ion pumps, *Biochem. Soc. Trans.*, **3**, 785–802.

Radnót, M. and Lovas, B. (1968). Data on the ultrastructure of Miller's cells of the retina, *Archs Soc. Am. Opthal. Optom.*, **6**, 393–404.

Rafferty, N. S. and Esson, E. A. (1974). An electron microscope study of adult mouse lens, *J. Ultrastruct. Res.*, **46**, 239–53.

Rafferty, N. S. and Goosens, W. (1978). Cytoplasmic filaments in the crystalline lens of various species, *Expl Eye Res.*, **26**, 177–90.

Raisman, G. (1966). Neural connections of the hypothalamus, *Br. med. Bull.*, **22**, 197–201.

Raisman, G., Cowan, W. M. and Powell, T. P. S. (1965). The extrinsic afferent, commissural and association fibers of the hippocampus, *Brain*, **88**, 963–96.

Rakic, P. (1971a). Guidance of neurons migrating to the foetal monkey neocortex, *Brain Res.*, **33**, 471–6.

Rakic, P. (1971b). Neuron-glia relationship during granule cell migration in developing cerebellar cortex. A Golgi and EM study in *Macacus rhesus*, *J. comp. Neurol.*, **141**, 283–312.

Rakic, P. (1975). Editor. *Bull. Neurosci. Res. Prog.*, **13**, 291–446.

Rakic, P. (1976). Prenatal genesis of connexions subserving ocular dominance in the rhesus monkey, *Nature, Lond.*, **261**, 467–71.

Rakic, P. (1977). Genesis of the dorsal lateral geniculate nucleus in the rhesus monkey, *J. comp. Neurol.*, **176**, 23–52.

Rakic, P. and Yakovlev, P. I. (1968). Development of the corpus callosum and cavum septi in man, *J. comp. Neurol.*, **132**, 45–72.

Rall, W. (1977). Core conductor theory and cable properties of neurons. In: *Handbook of Physiology* (Kandel, E. R. ed.), Volume 1, Part 1, pp. 39–97, Physiological Society: Bethesda.

Rall, W., Shepherd, G. M., Reese, T. and Brightman, M. W. (1966). Dendrodendritic synaptic pathway for inhibition in the olfactory bulb, *Expl Neurol.*, **14**, 44–56.

Ralston, H. J. (1965). The organization of the substantia gelatinosa Rolandi in the cat lumbrosacral cord, *Z. Zellforsch. mikrosk. Anat.*, **67**, 1–23.

Ralston, H. J. (1974). On the neuronal organization of the spinal cord. In: *Essays on the Nervous System* (Bellairs, R. and Gray, E. G. eds.), Clarendon Press: Oxford.

Rambourg, A. and Leblond, C. P. (1967). Electron microscope observations on the carbohydrate-rich cell coat present at the surface of cells in the rat, *J. Cell Biol.*, **32**, 27–53.

Ramón-Moliner, E. and Nauta, W. J. H. (1966). The iso-dendritic core of the brain stem, *J. comp. Neurol.*, **126**, 311–35.

Ramón-Moliner, E. and Dansereau, J. A. (1974). The peribrachial region of the cat, *Cell Tiss. Res.*, **149**, 173–90; 191–204.

Ramsdale, D. R., Dixon, J. S. and Gosling, J. A. (1972). Chromaffin cells in the mammalian urethra, *J. Anat.*, **113**, 290–1.

Ramsey, E. M. (1955). Vascular patterns in endometrium and placenta, *Angiology*, **6**, 321–39.

Ramsey, G. M., Corner, G. W. Jr. and Donner, M. W. (1963). Serial and cineradioangiographic visualisation of maternal circulation in the primate (hemochorial) placenta, *Am. J. Obstet. Gynec.*, **86**, 213–25.

Rang, M. (1969). *The Growth Plate and its Disorders*, Livingstone: Edinburgh, London.

Ranganathan, N., Lam, J. H. and Wigle, E. D. (1970). Morphology of the human mitral valve. II. The valve leaflets, *Circulation*, **41**, 459–67.

Ranson, S. W. and Magoun, W. W. (1933). The central path of the pupilloconstriction reflex in response to light, *Archs Neurol. Psychiat., Chicago*, **30**, 1193–1202.

Ranvier, L. A. (1871). Contributions à l'histologie et à la physiologie des nerfs périphériques, *C. r. hebd. Séanc. Acad. Sci., Paris*, **73**, 1168–71.

Rappaport, A. M. (1958). The structural and functional unit in the human liver, *Anat. Rec.*, **130**, 673–90.

Rappaport, A. M., Black, R. G., Lucas, C. C., Ridout, J. H. and Best, C. H. (1966). Normal and pathologic microcirculation of the living mammalian liver, *Rev. Int. Hepat.*, **16**, 813–28.

Rao, G. S. and Sahu, S. (1974). The localization within the dorsal motor nucleus of the vagus in the buffalo (*Bubalus bubalis*), *Acta anat.*, **90**, 388–93.

Rascol, M. M. and Izard, J. Y. (1974). The subdural neurothelium of the cranial meninges in man, *Anat. Rec.*, **186**, 429–36.

Rasmussen, A. T. (1932). Secondary vestibular tracts in the cat, *J. comp. Neurol.*, **54**, 143–72.

Rasmussen, A. T. and Peyton, W. T. (1946). Origin of ventral external arcuate fibres and their continuity with the straie medullares of the fourth ventricle in man, *J. comp. Neurol.*, **84**, 325–37.

Rasmussen, G. L. (1942). An efferent cochlear bundle, *Anat. Rec.*, **82**, 441P.

Rasmussen, G. L. (1946). The olivary peduncle and other fiber projections of the superior olivary complex, *J. comp. Neurol.*, **84**, 141–220.

Rasmussen, G. L. (1957). Selective silver impregnation of synaptic endings. In: *New Research Techniques of Neuroanatomy* (Windle, W. F. ed.), pp. 27–39, Thomas: Springfield, Ill.

Rasmussen, G. L. (1960). Efferent fibers of the cochlear nerve and cochlear nucleus. In: *Neural Mechanisms of the Auditory and Vestibular Systems* (Rasmussen, G. L. and Windle, W. F. eds.), pp. 105–15, Thomas: Springfield, Ill.

Rasmussen, G. L. (1967). Efferent connections of the cochlear nerve. In: *Sensori-neural Hearing Processes and Disorders* (Graham, A. B. ed.), pp. 61–75, Henry Ford Hospital International Symposium, Little, Brown & Co.: Boston.

Rasmussen, G. L. and Windle, W. F. (1960). *Neural Mechanisms of the Auditory and Vestibular Systems* (editors), Thomas: Springfield, Ill.

Raven, C. P. (1958). *Morphogenesis: The Analysis of Molluscan Development*, Pergamon Press: New York, Oxford, London.

Raven, C. P. (1961). *Oogenesis*, Pergamon Press: Oxford, New York, Toronto, Sydney.

Raven, C. P. (1963). Differentiation in mollusc eggs. In: *Cell Differentiation (Symposium Soc. Exp. Biol.,* **17**, 274–84), Cambridge University Press: Cambridge.

Raven, C. P. (1966). *Analysis of Molluscan Development*, Pergamon Press: Oxford, New York, Toronto, Sydney.

Raven, C. P. (1966). *An Outline of Developmental Physiology*, Pergamon Press, Oxford, New York, Toronto, Sydney.

Raviola, G. and Raviola, E. (1967). Light and electron microscopical observations on the inner plexiform layer of the rabbit retina, *Am. J. Anat.,* **120**, 403–25.

Raviola, E. and Karnovsky, M. J. (1972). Evidence for a blood-thymus barrier using electron opaque tracers, *J. exp. Med.,* **136**, 466–98.

Raviola, E. and Gilula, N. B. (1973). Gap junctions between photoreceptor cells in the Vertebrate retina, *Proc. natn Acad. Sci., U.S.A.,* **70**, 1677–81.

Rawles, M. E. (1948). Origin of melanophores and their role in development of color patterns in vertebrates, *Physiol Rev.,* **28**, 383–408.

Ray, R. D., Johnson, R. J. and Jameson, R. M. (1951). Rotation of the forearm, *J. Bone Jt Surg.,* **33A**, 993–6.

Raybuck, H. E. (1952). The innervation of the parathyroid glands, *Anat. Rec.,* **112**, 117–24.

Redler, I. (1974). A scanning electron microscopic study of human normal and osteoarthritic articular cartilage, *Clin. Orthop. rel. Res.,* **103**, 262–8.

Redler, I. and Zimny, M. L. (1970). Scanning electron microscopy of normal and abnormal articular cartilage and synovium, *J. Bone Jt Surg.,* **52A**, 1395–1404.

Reed, C. I. and Reed, B. P. (1948). Comparative study of human and bovine sperm by electron microscopy, *Anat. Rec.,* **100**, 1–8.

Reedy, M. K. (1968). Ultrastructure of insect flight muscle. I. Screw sense and structural grouping in the rigor cross-bridge lattice, *J. molec. Biol.,* **31**, 155–76.

Rees, L. A. (1954). The structure and function of the mandibular joint, *Br. dent. J.,* **96**, 125–33.

Rees, S. (1976). A quantitative electron microscopic study of the aging human cerebral cortex, *Acta neuropath.,* **36**, 347–62.

Reese, T. (1965). Olfactory cilia in the frog, *J. Cell Biol.,* **25**, 209–30.

Reese, T. and Brightman, M. W. (1970). Olfactory surface and central olfactory connexions in some vertebrates. In: *Taste and Smell in Vertebrates* (Wolstenholme, G. E. W. and Knight, J. eds.), pp. 115–43, Ciba Foundation Symposium, Churchill: London.

Reese, T. S. and Shepherd, G. M. (1972). Dendrodendritic synapses in the central nervous system. In: *Structure and Function of Synapses* (Pappas, G. D. and Purpura, D. P. eds.), pp. 121–36, Raven Press: New York.

Regaud, C. and Mawas, J. (1909). Sur les mitochondries des glandes salivaires chez les mammifères, *C. r. Séanc. Soc. Biol.,* **66**, 97–100.

Reichert, F. L. and Poth, E. J. (1933). Pathways for the secretory fibres of the salivary glands in man, *Proc. Soc. exp. Biol. Med.,* **30**, 973–7.

Reid, L. (1976). Visceral cartilage, *J. Anat.,* **122**, 349–55.

Reil, L. C. (1807). Fragments über die Bildung des Kleinen Gehirns im Menschen, *Arch. Physiol. Halle,* **8**, 1–58.

Reilly, D. T. and Burstein, A. H. (1974). The mechanical properties of cortical bone, *J. Bone Jt Surg.,* **56**, 1001–22.

Reimann, A. F., Daseler, E. H., Anson, B. J. and Beaton, L. E. (1944). The palmaris longus muscle and tendon, *Anat. Rec.,* **89**, 495–505.

Reinius, S. (1967). Ultrastructure of blastocyst attachment in the mouse, *Z. Zellforsch. mikrosk. Anat.,* **77**, 257–66.

Reiter, R. J. and Hoffman, R. A. (1967). Adrenocortical cytogenesis in the adult male golden hamster. A radioautographic study using tritiated thymidine, *J. Anat.,* **101**, 723–30.

Reiter, R. J. (1977). *The Pineal.—1977*, Annual Research Reviews, Eden Press: Montreal.

Reith, E. J. (1967). The absorptive activity of ameloblasts during the maturation of enamel, *Anat. Rec.,* **157**, 577–88.

Reith, E. J. (1967). The early stages of amelogenesis as observed in molar teeth of young rats, *J. Ultrastruct. Res.,* **17**, 503–26.

Reith, E. J. (1970). The stages of amelogenesis as observed in the molar teeth of young rats, *J. Ultrastruct. Res.,* **30**, 111–51.

Relkin, R. (1972). Effect of pinealectomy on adrenal secretion of testosterone in castrate rats, *Acta Endocrinol. Panam.,* **3**, 129–33.

Remak, R. (1836). Vorläufige Mitteilung mikroscopischen Beobachtungen über den innern Bau der Cerebrospinalnerven und über die Entwicklung ihrer Formelemente, *Arch. Anat. Physiol.,* pp. 145–61.

Renfrew, S. and Melville, I. D. (1960). The somatic sense of space (choraesthesia) and its threshold, *Brain,* **83**, 93–112.

Renshaw, B. (1941). Influence of discharge of motoneurons upon excitation of neighbouring motoneurons, *J. Neurophysiol.,* **4**, 167–83.

Renshaw, B. (1946). Central effects of centripetal impulses in axons of spinal ventral roots, *J. Neurophysiol.,* **9**, 191–204.

Réthelyi, M. and Szentágothai, J. (1969). The large synaptic complexes of the substantia gelatinosa, *Expl Brain Res.,* **7**, 258–74.

Revel, J. P. (1962). The sarcoplasmic reticulum of the bat cricothyroid muscle, *J. Cell Biol.,* **12**, 571–88.

Revel, J. P. and Hay, E. D. (1963). An autoradiographic and electron microscopic study of collagen synthesis in differentiating cartilage, *Z. Zellforsch. mikrosk. Anat.,* **61**, 110–44.

Rexed, B. (1952). The cytoarchitectonic organization of the spinal cord in the cat, *J. comp. Neurol.,* **96**, 415–95.

Rexed, B. (1954). A cytoarchitectonic atlas of the spinal cord in the cat, *J. comp. Neurol.,* **100**, 297–379.

Rexed, B. (1964). Some aspects of the cytoarchitectonics and synaptology of the spinal cord, *Prog. Brain Res.,* **11**, 58–90.

Reynolds, E. L. (1945). Bony pelvic girdle in early infancy; roentgenometric study, *Am. J. phys. Anthrop.,* **3**, 321–54.

Reynolds, E. L. (1947). The bony pelvis in prepuberal childhood, *Am. J. phys. Anthrop.,* N.S. **5**, 165–200.

Reynolds, J. J. (1974). The role of 1,25-Dihydroxy-cholecalciferol in bone metabolism, *Biochem. Soc. Spec. Publ.,* **3**, 91–102.

Reynolds, J. J. (1976). Calcification of bone, *J. Anat.,* **122**, 92.

Reynolds, S. R. M. and Zweifach, B. W. (1959). *The Microcirculation*, University of Illinois Press: Urbana.

Rhinelander, F. W. (1968). The normal microcirculation of diaphysial cortex and its response to fracture, *J. Bone Jt Surg.,* **50A**, 784–800.

Rhodin, J. A. G. (1962). The diaphragm of capillary endothelial fenestrations, *J. Ultrastruct. Res.,* **6**, 171–85.

Rhodin, J. A. G. (1962). Fine structure of vascular walls in mammals, with special reference to the smooth muscle component, *Physiol Rev.,* **42**, Suppl. 5, 48–81.

Rhodin, J. A. G. (1966). Ultrastructure and function of the human tracheal mucosa, *Ann. Rev. Respir. Dis.,* **93**, 1–15.

Rhodin, J. A. G. (1967). The ultrastructure of mammalian arterioles and precapillary sphincters, *J. Ultrastruct. Res.,* **18**, 181–223.

Rhodin, J. A. G. (1968). Ultrastructure of mammalian venous capillaries, venules and small collecting veins, *J. Ultrastruct. Res.,* **25**, 452–500.

Rhodin, J. A. G., Del Missier, P. and Reid, L. C. (1961). The structure of the specialised impulse-conducting system of the steer heart, *Circulation,* **24**, 348–67.

Rhodin, J. A. G. and Terzakis, J. (1962). The ultrastructure of the human full-term placenta, *J. Ultrastruct. Res.,* **6**, 88–106.

Rhodin, J. and Dalhamn, T. (1956). Electron microscopy of the tracheal ciliated mucosa in rat, *Z. Zellforsch. mikrosk. Anat.,* **44**, 345–412.

Rhoton, A. L. Jr., Kobayashi, S. and Hollinshead, W. H. (1968). Nervus intermedius, *J. Neurosurg.,* **29**, 609–18.

Richards, B. M. and Pardon, J. F. (1970). The molecular structure of nucleohistone (DNH), *Expl Cell Res.,* **62**, 184–96.

Richardson, A. P. and Hinsey, J. C. (1933). Functional study of the nodose ganglion of the vagus with degeneration methods, *Proc. soc. exp. Biol Med.,* **30**, 1141–3.

Richardson, K. C. (1949). Contractile tissues in the mammary gland with special reference to myoepithelium in the goat, *Proc. R. Soc. B.,* **136**, 30–45.

Richardson, K. C. (1962). The fine structure of autonomic nerve endings in smooth muscle of the rat vas deferens, *J. Anat.,* **96**, 427–42.

Richardson, P. J. Quoted in Taylor, S. (1965). Surgical treatment of cancer of the thyroid, *Br. J. Surg.,* **52**, 740–2.

Riches, D. J. and Palfrey, A. J. (1966). The ultrastructure of the bile duct epithelium of the rat, *J. Anat.,* **100**, 429–30P.

Rieck, N. W. (1959). Motor responses from the macaque occipital lobe, *J. comp. Neurol.,* **112**, 203–30.

Riggs, H. E. and Rupp, C. (1963). Variation in form of circle of Willis, *Archs Neurol. Psychiat., Chicago,* **8**, 8–14.

Riisager, M. and Weddell, G. (1962). Nerve terminations in the human carotid sinus, *J. Anat.,* **96**, 25–38.

Riley, D. A. and Allin, E. F. (1973). The effects of inactivity, programmed stimulation and denervation on the histochemistry of skeletal muscle fibre types, *Exp. Neurol.,* **40**, 391–413.

Riley, H. A. (1930). Lobules of mammalian cerebellum and cerebellar nomenclature, *Archs Neurol. Psychiat., Chicago,* **24**, 227–56.

Ringo, D. L. (1967). The arrangement of subunits in flagellar fibers, *J. Ultrastruct. Res.,* **17**, 266–77.

Ringvist, M. (1974). Fibre types in human masticatory muscles. Relation to function, *Scand. J. dent. Res.,* **82**, 333–55.

Ringvold, A. (1975). Distribution of ascorbic acid in the ciliary body of albino rabbit, guinea pig and rat, *Acta ophthal.,* **53**, 751–9.

Rinvik, E. (1968). The corticothalamic projection from the second somatosensory cortical area in the cat. An experimental study with silver-impregnation methods, *Expl Brain Res.,* **5**, 153–72.

Rinvik, E., (1968). The corticothalamic projection from the pericruciate and coronal gyri in the cat. An experimental study with silver-impregnation methods, *Brain Res.*, **10**, 79–119.

Ripps, H. and Weale, R. A. (1976). The visual photoreceptors. In: *The Eye* (Davson, H. ed.), Volume 2A, Second edition, Academic Press: New York, London.

Ritchie, J. M. and Rogart, R. B. (1977). The density of sodium channels in mammalian myelinated nerve fibers and the nature of the axonal membrane under the myelin sheath, *Proc. natn Acad. Sci. U.S.A.*, **74**, 211–15.

Riva, A. (1967). Fine structure of human seminal vesicular epithelium, *J. Anat.*, **102**, 71–86.

Riva, D. and Riva-Testa, F. (1973). Fine structure of acinar cells of human parotid gland, *Anat. Rec.*, **176**, 149–65.

Robb, G. P. and Steinberg, I. (1939). Visualisation of the chambers of the heart, pulmonary circulation, and great blood vessels in man; practical method, *Am. J. Roentg.*, **41**, 1–17.

Roberts, W. and Taylor, W. H. (1973). Inferior rectal nerve variation as it relates to pudendal block, *Anat. Rec.*, **177**, 461–3.

Robertson, J. D. (1955). The ultrastructure of adult vertebrate peripheral myelinated nerve fibers in relation to myelinogenesis, *J. biophys. biochem. Cytol.*, **1**, 271–8.

Robinson, A. G. and Zimmerman, E. A. (1973). Cerebrospinal fluid and ependymal neurophysin, *Clin. Invest.*, **52**, 1260–7.

Robinson, R. A. (1964). Observations regarding compartments for tracer calcium in the body. In: *Bone Biodynamics* (Frost, H. M. ed.), pp. 423–9, Churchill: London.

Robson, J. A. and Hall, W. C. (1977). The organization of the pulvinar in the grey squirrel (Sciurus carolinensis). II. Synaptic organization and comparisons with the dorsal lateral geniculate nucleus, *J. comp. Neurol.*, **173**, 389–416.

Roche, A. F. and Lewis, A. B. (1974). Sex differences in the elongation of the cranial base during pubescence, *Angle Orthodont.*, **44**, 279–94.

Rock, J. and Hertig, A. T. (1948). The human conceptus during the first two weeks of gestation, *Am. J. Obstet. Gynec.*, **55**, 6–17.

Rockell, A. J. and Jones, E. G. (1973). The neuronal organization of the inferior colliculus of the adult cat, *J. comp. Neurol.*, **147**, 11–60.

Rodrigo, J., Hernández, C. J., Vidal, M. A. and Pedrosa, J. A. (1975). Vegetative innervation of the oesophagus, *Acta anat.*, **92**, 79–100.

Rogers, A. W. (1973). *Techniques of Autoradiography*, Second edition, Elsevier: Amsterdam.

Rogers, D. C. (1965). The development of the rat carotid body, *J. Anat.*, **99**, 89–101.

Rogers, D. C. and Burnstock, G. (1966). Multiaxonal autonomic junctions in intestinal smooth muscle of the toad *(Bufo marinus)*, *J. comp. Neurol.*, **126**, 625–52.

Rohan, R. F. and Turner, L. (1956). The levator palati muscle, *J. Anat.*, **90**, 153–4.

Rohon, J. V. (1884). Zur Histiogenese des Rückenmarkes der Forelle, *Sitz. ber. mathemat.-phys. Klasse Königl. bayr. Akad, Wiss.*, **14**, 301–56.

Roitt, I. M. (1977). *Essential Immunology*, Third edition, Blackwell: Oxford.

Rollet, F. (1899). *De la mensuration de os longs des membres*, Thèse pour le doc. en méd., 1st ser., **43**, 1–128.

Rollin, H. (1977). Course of the peripheral gustatory nerves, *Ann. Otol. Rhinol. Laryngol.*, **86**, 251–8.

Romanes, G. J. (1941). Cell columns in the spinal cord of a human foetus of fourteen weeks, *J. Anat.*, **75**, 145–52.

Romanes, G. J. (1941). The development and significance of the cell columns in the ventral horn of the cervical and upper thoracic spinal cord of the rabbit, *J. Anat.*, **76**, 112–30.

Romanes, G. J. (1942). The spinal cord in a case of congenital absence of the right limb below the knee, *J. Anat.*, **77**, 1–5.

Romanes, G. J. (1946). Motor localisation and the effects of nerve injury on the ventral horn cells of the spinal cord, *J. Anat.*, **80**, 117–31.

Romanes, G. J. (1951). The motor cell columns of the lumbosacral cord of the cat, *J. comp. Neurol.*, **94**, 313–63.

Romanes, G. J. (1953). The motor cell columns of the lumbosacral spinal cord of the cat. In: *The Spinal Cord* (Wolstenholme, G. E. W. ed.), Ciba Foundation Symposium, pp. 24–38, Churchill: London.

Romanes, G. J. (1964). The motor pools of the spinal cord, *Prog. Brain Res.*, **11**, 93–119.

Romer, A. S. (1942). Cartilage—an embryonic adaptation, *Am. Nat.*, **76**, 394–404.

Romer, A. S. (1970). *The Vertebrate Body*, Fourth edition, Saunders: Philadelphia.

Rönnholm, E. (1962). The amelogenesis of human teeth as revealed by electron microscopy. II. The development of the enamel crystallites, *J. Ultrastruct. Res.*, **6**, 249–303.

Roosen-Runge, E. C. (1952). The third maturation division in mammalian spermatogenesis, *Anat. Rec.*, **112**, 463 (D50).

Roosen-Runge, E. C. and Giesel, L. O. (1950). Quantitative studies on spermatogenesis in the albino rat, *Am. J. Anat.*, **87**, 1–30.

Rosa, M. and Borzone, M. (1973). The venous system of the corpus striatum. Angiographic study, *Neuroradiol.*, **6**, 219–30.

Rosan, R. C. and Lauweryns, J. (1971). Secretory cells in the premature human lung lobule, *Nature, Lond.*, **232**, 60–1.

Rose, G. C. (1963). *Cinematography in Biology* (editor), Academic Press: New York, London.

Rose, J. E. (1952). The cortical connections of the reticular complex of the thalamus, *Res. Publs. Ass. Res. nerv. ment. Dis.*, **30**, 454–79.

Rose, J. E. and Woolsey, C. N. (1949). The relations of thalamic connections, cellular structure and evocable electrical activity in the auditory region of the cat, *J. comp. Neurol.*, **91**, 441–66.

Rose, J. E. and Woolsey, C. N. (1958). *Cortical Connexions and Functional Organisation of the Thalamic and Auditory System of the Cat*, University of Wisconsin Press: Madison.

Rose, J. E. and Malis, L. I. (1965). Geniculo-striate connections in the rabbit. I. Retrograde changes in the dorsal lateral geniculate body after destruction of cells in various layers of the striate region, *J. comp. Neurol.*, **125**, 95–120.

Rose, J. E. and Malis, L. I. (1965). Geniculo-striate connections in the rabbit. II. Cytoarchitectonic structure of the striate region and of the dorsal lateral geniculate body; organisation of the geniculo-striate projections, *J. comp. Neurol.*, **125**, 121–40.

Rose, M. (1926). Die sog. Riechrinde beim Menschen und beim Affen, *J. Psychol. Neurol., Lpz.*, **34**, 261–401.

Rose, M. D. (1975). Functional proportions of primate lumbar vertebral bodies, *J. Hum. Evol.*, **4**, 21–38.

Rose, V., Izukawa, T. and Moës, C. A. F. (1975). Syndromes of asplenia and polysplenia. A review of cardiac and non-cardiac malformations in 60 cases with special reference to diagnosis and prognosis, *Br. Heart J.*, **37**, 840–52.

Rosenberg, J. C. and DiDio, L. J. A. (1969). *In vivo* appearance and function of the termination of the ileum as observed directly through a cecostomy, *Am. J. gastroent.*, **52**, 411–19.

Rosenthal, J. (1977). Trophic interactions of neurons. In: *Handbook of Physiology* (Kandel, E. R. ed.), Volume 1, Part 2, pp. 775–801, Physiological Society: Bethesda.

Ross, K. F. A. (1967). *Phase Contrast and Interference Microscopy for Cell Biologists*, Arnold: London.

Ross, M. A. (1971). Functional anatomy of the tensor palati. Its relevance in cleft palate surgery, *Arch. Otolaryngol.*, **93**, 1–3.

Ross, M. D. (1969). The general visceral efferent component of the eighth cranial nerve, *J. comp. Neurol.*, **135**, 453–78.

Ross, R. (1968). The fibroblast and wound repair, *Biol Rev.*, **43**, 57–96.

Ross, R. and Klebanoff, S. J. (1967). Fine structural changes in uterine smooth muscle and fibroblasts in response to estrogen, *J. Cell Biol.*, **32**, 155–67.

Ross, R. and Klebanoff, S. J. (1971). The smooth muscle cell. I. *In vivo* synthesis of connective tissue proteins, *J. Cell Biol.*, **50**, 159–71.

Rossi, G. F. (1964). A hypothesis on the neural basis of consciousness. Considerations based upon some experimental work, *Acta neurochir.*, **12**, 187–97.

Roth, S. I. and Capen, C. C. (1974). Ultrastructural and functional correlations of the parathyroid gland, *Int. Rev. exp. Path.*, **13**, 161–221.

Rothblat, L. A. and Schwarz, M. L. (1979). The effect of monocular deprivation on dendritic spines in visual cortex of young and adult albino rats: evidence for a sensitive period, *Brain Res.*, **161**, 156–61.

Rothschild, L. (1957). The fertilising spermatozoon, *Discovery, Lond.*, **18**, No. 2.

Rouiller, C. (1964). *The Liver: Morphology, Biochemistry, Physiology* (editor), two volumes, Academic Press: New York, London.

Rouviére, H. (1933). *Anatomie des Lymphatiques de l'Homme*, Masson: Paris.

Rowbotham, G. F. and Little, E. (1962). The circulation and reservoir of the brain, *Br. J. Surg.*, **50**, 244–50.

Rowe, R. W. D. (1973). The ultrastructure of Z-discs from white, intermediate and red fibres of mammalian striated muscles, *J. Cell Biol.*, **57**, 261–77.

Rowe, R. W. D. and Goldspink, G. (1969). Muscle fibre growth in five different muscles of both sexes of mice. I. Normal mice, *J. Anat.*, **104**, 519–30.

Rowntree, T. (1949). Anomalous innervation of the hand muscles, *J. Bone Jt Surg.*, **31B**, 505–10.

Rozendal, R. H. and Molen, N. H. (1972). The relevancy of the concept of 'shunt' and 'spurt' muscles in functional anatomy, *Acta Morphol. neerl. scand.*, **10**, 347–50.

Rubin, W. (1972). An unusual intimate relationship between endocrine cells and other types of epithelial cells in the human stomach, *J. Cell Biol.*, **52**, 219–30.

Rubin, W. (1972). Endocrine cells in the human stomach. A fine structural study, *Gastro-enterol.*, **63**, 784–800.

Rudolph, A. M., Drorbaugh, J. E., Auld, P. H., Rudolph, A. J., Nades, A. S., Smith, C. A. and Aubbell, J. P. (1961). Studies on the circulation in the neo-natal period. The circulation in respiratory distress syndrome, *Pediatrics, Springfield*, **27**, 551–66.

Rushmer, R. F. (1976). *Structure and Function of the Cardiovascular System*, Second edition, Saunders: London, Philadelphia.

Rushton, W. A. H. (1962). *Visual Pigments in Man*, Liverpool University Press: Liverpool.

Rushton, W. A. H. and Henry G. H. (1968). Bleaching and regeneration of cone pigments in man, *Vision Res.*, **8**, 617–31.

Ruskell, G. L. (1968). The fine structure of nerve terminations in the lacrimal glands of monkeys, *J. Anat.*, **103**, 65–76.

Ruskell, G. L. (1969). Changes in nerve terminals and acini of the lacrimal gland and changes in secretion induced by autonomic denervation, *Z. Zellforsch.*, **94**, 261–81.

Ruskell, G. L. (1970). An ocular parasympathetic nerve pathway of facial nerve origin and its influence on intraocular pressure, *Expl Eye Res.*, **10**, 319–30.

Ruskell, G. L. (1971). Ocular autonomic nerves of faeial nerve origin, *J. Anat.*, **107**, 374P.

Ruskell, G. L. (1971). The distribution of autonomic postganglionic nerve fibres in the lacrimal gland in the rat, *J. Anat.*, **109**, 229–42.

Ruskell, G. L. (1974). Form of the choroidocapillaris, *Expl Eye Res.*, **18**, 411–12.

Ruskell, G. L. (1975). Nerve terminals and epithelial cell variety in the human lacrimal gland, *Cell Tiss. Res.*, **158**, 121–36.

Russel, G. V. (1955). Schematic presentation of thalamic morphology and connections, *Tex. Rep. Biol. Med.*, **13**, 989–92.

Russell, J. G. (1969). Radiology in the diagnosis of fetal abnormalities, *J. Obstet. Gynec.*, **76**, 345–50.

Rustioni, A. and Dekker, J. J. (1974). Non-primary afferents to the dorsal column nuclei of cat: distribution pattern and cells of origin, *Anat. Rec.*, **178**, 454–5.

Rusznyák, I., Földi, M. and Szabó, G. (1960). *Lymphatics and Lymph Circulation; Physiology and Pathology*, Pergamon Press: New York, London.

Ryan, J. W. and Ryan, U. S. (1977). Pulmonary endothelial cells, *Fedn Proc. Fedn. Am. Socs exp. Biol.*, **36**, 2683–91.

Rybicki, E. F., Smonen, F. A. and Weis, E. B. (1972). On the mathematical analysis of stress in the human femur, *J. Biomech.*, **5**, 203–15.

Ryncki, P. V. (1974). Anatomie chirugicale du foie, *Helv. Chir. Acta*, **41**, 543–74.

S

Saavedra, J. P., Mascitti, T. A. and Vaccarezza, O. L. (1969). Lamination of the *Cebus* lateral geniculate nucleus, *Z. Zellforsch.*, **94**, 346–51.

Sabin, F. R. (1907). A model of the medullated fiber paths in the thalamus of a new-born brain, *Am. J. Anat.*, **1**, 54–68.

Sabin, F. R. (1912). On the origin of the abdominal lymphatics in mammals from the vena cava and the renal veins, *Anat. Rec.*, **6**, 335–42.

Sager, R. (1965). Genes outside the chromosomes, *Scient. Am.*, **212**, 70–9.

Saha, K. (1961). *Theory of Shoulder Mechanism*, Thomas: Springfield, Ill.

St. Helen, R. and McEwen, W. K. (1961). Rheology of the human sclera. I. Anelastic behavior, *Am. J. Ophthal.*, **52**, 539–48.

Salpeter, M. M., McHenry, F. A. and Feng, H. H. (1974). Myoneural junctions in the extraocular muscles of the mouse, *Anat. Rec.*, **179**, 201–24.

Salsbury, C. R. (1937). The interosseous muscles of the hand, *J. Anat.*, **71**, 395–403.

Salter, N. (1955). Methods of measurement of muscle and joint function, *J. Bone Jt Surg.*, **37B**, 474–91.

Salzmann, J. A. (1961). *Roentgenographic Cephalometrics*, Saunders: Philadelphia.

Samarasinghe, D. D. (1965). The innervation of the cerebral arteries in the rat: an electron microscope study, *J. Anat.*, **99**, 815–28.

Samols, E., Tyler, J. and Megyesi, C. (1966). Immunochemical glucagon in human pancreas, gut and plasma, *Lancet*, **2**, 727–9.

Sampson, J. A. (1937). Lymphatics of mucosa of fimbriae of the fallopian tube, *Am. J. Obstet. Gynec.*, **33**, 911–30.

Samuel, E. P. (1953). Chromidial studies on the superior cervical ganglion of the rabbit; (a) caudally projected postganglionic axons; (b) intercalary 'commissural' neurons, *J. comp. Neurol.*, **98**, 93–112.

Sandborn, E. B., LeBuis, J. J. and Bois, P. (1966). Cytoplasmic microtubules in blood platelets, *Blood*, **27**, 247–52.

Sanides, F. (1964). The cyto-myeloarchitecture of the human frontal lobe and its relation to phylogenetic differentiation of the cerebral cortex, *J. Hirnforsch.*, **6**, 269–82.

Santer, R. M., Lu, K.-S., Lever, J. D. and Presley, R. (1975). A study of the distribution of chromaffin-positive (CH+) and small intensely fluorescent (SIF) cells in sympathetic ganglia of the rat at various ages, *J. Anat.*, **119**, 589–99.

Santulli, T. V. and Blanc, W. A. (1961). Congenital atresia of the intestine: pathogenesis and treatment, *Ann. Surg.*, **154**, 939–48.

Saper, C. B., Loewi, A. D., Swanson, L. W. and Cowan, W. M. (1976). Direct hypothalamoautonomic connections, *Brain Res.*, **117**, 305–12.

Sapin, M. R. and Borziak, E. I. (1974). Anatomie des ganglions lymphatiques du médiastin, *Acta anat.*, **90**, 200–25.

Sargent, P. (1910). Some points in the anatomy of the intra-cranial blood-sinuses, *J. Anat.*, **45**, 69–72.

Sarnat, B. G. (1951). *The Temporomandibular Joint*, Second edition, Thomas: Springfield, Ill.

Sarton, G. (1954). *Galen of Pergamon*, University of Kansas Press: Kansas.

Sasagawa, T., Kobayashi, S. and Fujita, T. (1973). Electron microscope studies on the endocrine cells of the human gut and pancreas. In: *Gastro-Entero-Pancreatic Endocrine System. A Cell-Biological Approach* (Fujita, T. ed.), pp. 17–38, Igaku Shoiu: Tokyo.

Satir, P. (1965). Studies on cilia. II. Examination of the distal region of the ciliary shaft and the role of the filaments in motility, *J. Cell Biol.*, **26**, 805–34.

Satir, P. (1974). The present status of the sliding microtubule model of ciliary locomotion. In: *Cilia and Flagella* (Sleigh, M. A. ed.), pp. 131–42, Academic Press: London.

Satiukova, G. S. and Rassokhina-Volkova, L. J. (1972). Compensatory-adaptive changes in the lymphatic system of organs in experiment and disease, *Biblphie anat.*, **11**, 481–7.

Sato, T. (1973a). A new classification of the transverso-spinalis system, *Proc. Jap. Acad.*, **49**, 51–56.

Sato, T. (1973b). Innervation and morphology of the musculi levatores costarum longi, *Proc. Jap. Acad.*, **49**, 555–8.

Satoh, J. (1974). The Mm. subcostales in Man and monkeys, *Okajimas Fol. anat. jap.*, **50**, 345–58.

Sauer, F. C. (1935a). The cellular structure of the neural tube, *J. comp. Neurol.*, **63**, 13–23.

Sauer, F. C. (1935b). Mitosis in the neural tube, *J. comp. Neurol.*, **62**, 337–405.

Sauer, F. C. (1936). The interkinetic migration of embryonic epithelial nuclei, *J. morphol.*, **60**, 1–11.

Sauer, M. E. and Chittenden, A. C. (1959). Deoxyribonucleic acid content of cell nuclei in the neural tube of the chick embryo: evidence for intermitotic migration of nuclei, *Exp. cell Res.*, **16**, 1–6.

Saunders, J. W. Jr. (1948). The proximo-distal sequence of origin of the parts of the chick wing and the role of the ectoderm, *J. exp. Zool.*, **108**, 363–404.

Saunders, R. L. de C. H. and Röckert, H. Ö. E. (1967). Vascular supply of dental tissues, including lymphatics. In: *Structural and Chemical Organisation of Teeth* (Miles, A. E. W. ed.), pp. 199–245, Academic Press: New York, London.

Sawaki, Y. (1971). Retinohypothalamic projection: electrophysiological evidence for the existence in female rats, *Brain Res.*, **120**, 336–41.

Saxén, L. (1970). Defective regulatory mechanisms in teratogenesis, *Int. J. Gynaec. Obstet.*, **8**, 798–804.

Sayfi, Y. (1967). Note sur l'innervation du dos de la main, *Archs Anat. path.*, **15**, 139–40.

Scapinelli, R. (1968). Studies on the vasculature of the human knee joint, *Acta anat.*, **70**, 305–31.

Schachner, M., Hedley-White, E. T., Hsu, D. W., Schoonmaker, G. and Bignami, A. (1977). Ultrastructural localization of glial fibrillary acidic protein in mouse cerebellum following immunoperoxidase labelling, *J. Cell Biol.*, **75**, 67–73.

Schadé, J. P. (1964). On the volume and surface area of spinal neurons, *Prog. Brain Res.*, **11**, 261–77.

Schaefer, K. P. and Schneider, H. (1968). Reizversuche im Tectum opticum des Kaninchens. Ein experimenteller Beitrag zur seno-mororischen Koordination des Hirnstammes, *Arch. Psychiat. Nerv-Krankh.*, **211**, 118–37.

Schaper, A. (1897). Die Frühesten Differenzirungsvorgange in Central-nervensystem, *Arch. Entwicklungsmechanik*, **5**, 81–132.

Scharf, J. H. (1958). Sensible ganglien. In: *Handbuch der mikroskopischen Anatomie des Menschen*, Bd 4/3, Springer: Berlin.

Scharrer, E. and Scharrer, B. (1963). *Neuroendocrinology*, Columbia University Press: New York.

Scheibel, M. E. and Scheibel, A. B. (1966). Spinal motoneurons, interneurons and Renshaw cells: A Golgi study, *Archs ital. Biol.*, **104**, 328–53.

BIBLIOGRAPHY

Scheibel, M. E. and Scheibel, A. B. (1966). The organisation of the nucleus reticularis thalami: A Golgi study, *Brain Res.*, **1**, 43–62.

Scheibel, M. E. and Scheibel, A. B. (1966). Patterns of organisation in specific and non-specific thalamic fields. In: *The Thalamus* (Purpura, D. P. and Yahr, M. D. eds.), pp. 12–46, Columbia University Press: New York.

Scheibel, M. E. and Scheibel, A. B. (1966). Terminal axonal patterns in cat spinal cord. I. The lateral corticospinal tract, *Brain Res.*, **2**, 333–50.

Scheibel, M. E. and Scheibel, A. B. (1967). Structural organisation of nonspecific thalamic nuclei and their projection toward cortex, *Brain Res.*, **6**, 60–94.

Scheibel, M. E. and Scheibel, A. B. (1968). Terminal axonal patterns in cat spinal cord. II. The dorsal horn, *Brain Res.*, **9**, 32–58.

Scheibel, M. E. and Scheibel, A. B. (1970). Elementary processes in selected thalamic and cortical subsystems—the structural substrates. In: *The Neurosciences, a Second Study Program* (Schmitt, F. O., Quarton, G. C., Melnechuk, T. and Adelman, G. eds.), pp. 443–57. Rockefeller University Press: New York.

Scheinin, A. and Light, E. I. (1969). Innervation of the dental pulp, *Acta odnt. scand.*, **27**, 313–19.

Scheuer, J. L. (1964). Fibre size frequency distribution in normal human laryngeal nerves, *J. Anat.*, **98**, 99–104.

Schippel, K. and Reissig, D. (1968). Zur Feinstruktur des Muskel-Sehnenüberganges, *Z. mikrosk.-anat. Forsch.*, **78**, 235–55.

Schlager, F. (1967). Uber die Muskulatur der Ductus ejaculatorii beim Menschen, *Z. Zellforsch. mikrosh. Anat.*, **76**, 268–76.

Schlegel, J. V. (1946). Arterio-venous anastomoses in the endometrium in man, *Acta anat.*, **1**, 284–325.

Schlossman, A. and Priestley, B. S. (1966). *Strabismus*, Little, Brown & Co.: Boston.

Schmalbruch, H. and Kamieniecka, Z. (1974). Fibre types in the human brachial biceps muscle, *Exp. Neurol.*, **44**, 313–28.

Schmalbruch, H. and Hellhammer, U. (1976). The number of satellite cells in normal human muscle, *Anat. Rec.*, **185**, 279–87.

Schmidt, H.-M. (1974). Über den Verlauf der Radix mesenterii beim Menschen, *Z. Anat. EntwGesch.*, **144**, 187–93.

Schmidt, R. F. (1973). Control of the access of afferent activity to somatosensory pathways. In: *Handbook of Sensory Physiology* (Iggo, A. ed.), Volume II, pp. 151–206, Springer: Berlin.

Schmidt, W. J. (1938). Polarisationsoptische Analyse eines Eiweiss-Lipoid Systems, *Kolloidzschr.*, **85**, 137–48.

Schmitt, F. O., Quarton, G. C., Melnechuck, T. and Adelman, G. (1970). *The Neurosciences, a Second Study Program* (editors), Rockefeller University Press: New York.

Schmitt, F. O., Parvati, D. and Smith, B. H. (1976). Electronic processing of information by brain cells, *Science, N.Y.*, **193**, 114–20.

Schneider, U., Inke, G. and Schneider, I. G. (1969). Zahle, Abstand der Verzweigungsstellen vom Rand des sinus renalis und Kaliber der extrarenalen Nierengefässe des Menschen, *Anat. Anz.*, **124**, 278–91.

Schneyer, L. H. (1976). Sympathetic control of Na+, K+ transport in perfused submaxillary main duct of rat, *Am. J. Physiol.*, **230**, 341–5.

Schneyer, L. H. and Schneyer, C. A. (1967). *Secretory Mechanisms of Salivary Glands* (editors), Academic Press: New York, London.

Schnitzlein, H. N., Rowe, L. C. and Hoffman, H. H. (1958). The myelinated component of the vagus nerves in man, *Anat. Rec.*, **131**, 649–67.

Schoefl, G. T. (1972). The migration of lymphocytes across the vascular endothelium in lymphoid tissue. A re-examination, *J. exp. Med.*, **136**, 568–88.

Schoefl, G. T. and French, J. E. (1968). Vascular permeability to particulate fat: morphological observations on vessels of lactating mammary glands and of lung, *Proc. R. Soc. Lond.*, **169B**, 153–65.

Schoenberg, M. D., Mumaw, V. R., Moore, R. D. and Weisberger, A. S. (1964). Cytoplasmic interaction between macrophages and lymphocytic cells in antibody synthesis, *Science, N.Y.*, **143**, 964–5.

Schofield, G. C. (1968). Anatomy of muscular and neural tissue in the alimentary canal. In: *Handbook of Physiology* (Code, C. F. ed.), Section 6, *Alimentary Canal*, Volume 4, pp. 1579–1628, American Physiological Society: Washington.

Schofield, G. C. and Silva, D. G. (1968). The ultrastructure of enterochromaffin cells in the mouse colon, *J. Anat.*, **103**, 1–14.

Schofield, G. C. and Atkins, A. M. (1970). Secretory immunoglobulin in columnar epithelial cells of the large intestine, *J. Anat.*, **107**, 491–504.

Schoultze, T. W. and Swett, J. E. (1972). The fine structure of the Golgi tendon organ, *J. Neurocytol.*, **1**, 1–26.

Schoultze, T. W. and Swett, J. E. (1974). Ultrastructural organization of the sensory fibres innervating the Golgi tendon organ, *Anat. Rec.*, **179**, 147–62.

Schramm, L. P., Stribling, J. M. and Adair, J. R. (1976). Developmental reorientation of sympathetic preganglionic neurons in the rat, *Brain Res.*, **106**, 166–71.

Schrödinger, E. (1967). *What is Life?: Mind and Matter*, Cambridge University Press: Cambridge.

Schuknecht, H. F. (1960). Neuroanatomical correlates of auditory sensitivity and pitch discrimination in the cat. In: *Neural Mechanisms of the Auditory and Vestibular systems* (Rasmussen, G. L. and Windle, W. F. eds.), pp. 76–90. Thomas: Springfield, Ill.

Schulter, F. P. (1976). Studies of the basicranial axis: a brief review, *Am. J. phys. Anthrop.*, **45**, 545–52.

Schultze, A. H. (1969). *The Life of Primates*, Weidenfeld and Nicolson: London.

Schumacher, G. H. (1961). *Funktionelle Morphologie der Kaumuskulatur*, Fischer: Jena.

Schumacher, G. H., Lau, H., Freund, E., Schultz, M., Himstedt, H. W. and Menning, A. (1976). Zur Topographie der muskulären Nervenausbreitungen. 9. Kaumuskeln M. pterygoideus medialis und lateralis, *Anat. Anz.*, **139**, 71–87.

Schunke, G. B. (1938). The anatomy and development of the sacro-iliac joint in man, *Anat. Rec.*, **72**, 313–31.

Schwalbe, G. (1876). *Z. Anat. EntwGesch.*, **1**, 307–52.

Schwartz, H. and Weddell, G. (1938). Observations on pathways transmitting the sensation of taste, *Brain*, **61**, 99–115.

Schwartz, H. G., Roulhac, G. E., Lam, R. L. and O'Leary, J. (1951). Organisation of the fasciculus solitarius in man, *J. comp. Neurol.*, **94**, 221–37.

Schwarz, W. (1961). Electron microscopic observations on the human vitreous body. In: *The Structure of the Eye* (Smelser, G. K. ed.), pp. 283–91, Academic Press: New York, London.

Scollo-Lavizzari, G. and Akert, K. (1963). Cortical area 8 and its thalamic projection in *Macaca mulatta*, *J. comp. Neurol.*, **121**, 259–70.

Scott, B. L. and Pease, D. C. (1959). Electron microscopy of the salivary and lacrimal glands of the rat, *Am. J. Anat.*, **104**, 115–61.

Scott, J. H. (1957). The shape of the dental arches, *J. dent. Res.*, **36**, 996–1003.

Scott, J. H. (1967). *Dento-facial Development and Growth*, Pergamon Press: London.

Scott, J. H. and Symons, N. B. B. (1971). *Introduction to Dental Anatomy*, Sixth edition, Livingstone: Edinburgh, London.

Sebuwufu, P. H. (1968). Ultrastructure of human fetal thymic cilia, *J. Ultrastruct. Res.*, **24**, 171–80.

Sedzmir, C. B. (1959). An angiographic test of collateral circulation through the anterior segment of the circle of Willis, *J. Neurol. Neurosurg. Psychiat.*, **22**, 64–8.

Seeman, P. M. and Palade, G. E. (1967). Acid phosphatase localisation in rabbit eosinophils, *J. Cell Biol.*, **34**, 745–56.

Seipel, M. (1948). Studies on the structure of the mandible, *Acta odont. Scand.*, **8**, 81–191.

Selvig, K. A. (1965). The fine structure of human cementum, *Acta odont. Scand.*, **23**, 423–41.

Sem-Jacobsen, C. W., Petersen, M. C., Dodge, H. W., Jacks, Q. D., Lazarte, J. A. and Holman, C. B. (1956). Electric activity of the olfactory bulb in man, *Am. J. med. Sci.*, **232**, 243–51.

Semenova, G. S. (1962). Innervation of the pharynx and the cervical portion of the oesophagus in man and certain animals, *Trudȳ Smolensk. Med. Inst.*, **15**, 18–27.

Senior, H. D. (1919). The development of the arteries of the human lower extremity, *Am. J. Anat.*, **25**, 55–96.

Senior, H. D. (1920). The development of the human femoral artery, a correction, *Anat. Rec.*, **17**, 271–80.

Sensenig, E. C. (1949). The early development of the human vertebral column, *Contr. Embryol.*, **33**, 21–41.

Sensenig, E. C. (1951). The early development of the meninges of the spinal cord in human embryos, *Contr. Embryol.*, **34**, 145–57.

Sensenig, E. C. (1957). The development of the occipital and cervical segments and their associated structures in human embryos, *Contr. Embryol.*, **36**, 141–52.

Serafini-Fracassini, A. and Smith, J. W. (1966). Observations on the morphology of the protein polysaccharide complex of bovine nasal cartilage and its relation to collagen, *Proc. R. Soc. B.*, **165**, 440–9.

Serafini-Fracassini, A. and Smith, J. W. (1974). *The Structure and Biochemistry of Cartilage*, Churchill-Livingstone: Edinburgh, London.

Severn, C. B. and Holyoke, E. A. (1973). Human acardiac anomalies, *Am. J. Obstet. Gynec.*, **116**, 358–65.

Shackleford, J. M. and Wilborn, W. H. (1968). Structural and histochemical diversity in mammalian salivary glands, *Ala. J. med. Sci.*, **5**, 180–203.

Shackleford, J. M. and Schneyer, L. H. (1971). Ultrastructural aspects of the main excretory duct of rat submandibular gland, *Anat. Rec.*, **169**, 679–98.

Shafik, A. (1977). The cremasteric muscle. In: *The Testis* (Johnson, A. D. and Gomes, W. R. eds.), Academic Press: New York, London.

Shah, P. M., Scarton, H. A. and Tsapogas, M. J. (1978). Geometric anatomy of the aortic-common iliac bifurcation, *J. Anat.*, **126**, 451–8.

Shakir, A. and Zaini, S. (1974). Skeletal maturation of the hand and wrist of young children in Baghdad, *Ann. hum. Biol.*, **1**, 189–99.

Shaner, R. F. (1921). The development of the pharynx and aortic arches of the turtle, with a note on the fifth and pulmonary arches of mammals, *Am. J. Anat.*, **29**, 407–30.

Shaner, R. F. (1929). The development of the atrioventricular node, bundle of His and sinoatrial node in the calf; with a description of a third embryonic node-like structure, *Anat. Rec.*, **44**, 85–100.

Shaner, R. F. (1962). Anomalies of the heart bulbus, *J. Pediat.*, **61**, 233–41.

Shanks, M. F., Rockel, A. J. and Powell, T. P. S. (1975). The commissural fibre connexions of the primary somatic sensory cortex, *Brain Res.*, **98**, 166–71.

Shanks, M. F., Pearson, R. C. A. and Powell, T. P. S. (1978). The intrinsic connexions of the primary somatic sensory cortex of the monkey, *Proc. R. Soc.*, **200B**, 95–101.

Shapiro, A. L. and Robillard, G. L. (1948). Arterial blood supply of common and hepatic bile ducts with references to problems of common duct injury and repair: based on a series of 23 dissections, *Surgery, St. Louis*, **23**, 1–11.

Shariff, G. A. (1953). Cell counts in the primate cerebral cortex, *J. comp. Neurol.*, **98**, 381–400.

Sharrard, W. J. W. (1955). The distribution of the permanent paralysis in the lower limb in poliomyelitis, *J. Bone Jt Surg.*, **37B**, 540–58.

Sharrard, W. J. W. (1956). Poliomyelitis: The distribution of the paralysis. In: *British Surgical Progress*, p. 83.

Sheehan, D. (1933). On unmyelinated fibres in the spinal nerves, *Anat. Rec.*, **55**, 111–16.

Sheehan, D. (1936). Discovery of the autonomic nervous system, *Archs Neurol. Psychiat., Chicago*, **35**, 1081–1115.

Sheehan, D., Mulholland, J. H. and Shariroff, B. (1941). Surgical anatomy of the carotid sinus nerve, *Anat. Rec.*, **80**, 431–42.

Shehata, R. (1966). The crura of the diaphragm and their nerve supply, *Acta anat.*, **63**, 49–54.

Shelanski, M. L., Gaskin, R. and Cantor, C. R. (1971). Isolation of filaments from brain, *Science, N.Y.*, **174**, 1242–5.

Sheldon, H. and Robinson, R. A. (1958). Studies on cartilage: electron microscopic observations on normal rabbit ear cartilage, *J. biophys. biochem. Cytol.*, **4**, 401–6.

Sheldon, W. H., Stevens, S. S. and Tucker, W. B. (1940). *The Varieties of Human Physique*, Harper: London.

Sheldon, W. H. and Stevens, S. S. (1942). *The Variety of Temperament*, Harper: New York.

Shellshear, J. L. (1922). Blood supply of the dentate nucleus of the cerebellum, *Lancet*, **1**, 1046.

Shellshear, J. L. (1927). The blood supply of the hypoglossal nucleus, *J. Anat.*, **61**, 279–82.

Shelton, J. H. and Jones, A. L. (1971). The fine structure of the mouse adrenal cortex and the ultrastructural changes in the zona glomerulosa with low and high sodium diets, *Anat. Rec.*, **170**, 147–82.

Shepard, E. (1951). Tarsal movements, *J. Bone Jt Surg.*, **33B**, 258–63.

Shepherd, G. M. (1974). *The Synaptic Organization of the Brain*, Oxford University Press: New York.

Shepherd, G. M. (1978). Microcircuits in the nervous system, *Sci. Amer.*, **238**, 92–103.

Sheps, J. G. (1945). Nuclear configuration and cortical connections of human thalamus, *J. comp. Neurol.*, **83**, 1–56.

Sherrington, C. S. (1905). On reciprocal innervation of anatagonistic muscles, *Proc. R. Soc. B.*, **76**, 160–3.

Sherrington, C. S. (1906). *The Integrative Action of the Nervous System*, Scribner: New York, reprinted by Yale University Press in 1947.

Shettles, L. B. (1953). Observations on human follicular and tubal ova, *Am. J. Obstet. Gynec.*, **66**, 235–47.

Shettles, L. B. (1955). A morula stage of human ovum developed *in vitro*, *Fert. Steril.*, **6**, 287–9.

Shettles, L. B. (1955). Chorionepithelioma following a full-term pregnancy, *Am. J. Obstet. Gynec.*, **69**, 869–73.

Shettles, L. B. (1957). Parthenogenetic cleavage of the human ovum, *Bull. Sloane Hosp. for Women*, **3**, 59–61.

Shettles, L. B. (1958). The living human ovum, *Am. J. Obstet. Gynec.*, **76**, 398–406.

Shettles, L. B. (1960). *Ovum Humanum: Growth, Maturation, Nourishment, Fertilisation and Early Development*, Hafner: New York.

Shimizu, N. and Imamoto, K. (1970). Fine structure of the locus coeruleus in the rat, *Archs histol. jap.*, **31**, 229–46.

Sholl, D. A. (1953). Dendritic organization in the neurons of the visual and motor cortices of the cat, *J. Anat.*, **87**, 387–406.

Sholl, D. A. (1955). The organisation of the visual cortex in the cat, *J. Anat.*, **89**, 33–46.

Sholl, D. A. (1956). *The Organisation of the Cerebral Cortex*, Methuen: London.

Showers, M. J. C. (1959). The cingulate gyrus: additional motor area and cortical autonomic regulator, *J. comp. Neurol.*, **112**, 231–310.

Showers, M. J. C. and Crosby, E. C. (1958). Somatic and visceral responses from the cingulate gyrus, *Neurology, Minneap.*, **8**, 561–5.

Showers, M. J. C. and Lauer, E. W. (1961). Somatovisceral motor pathways in the insula, *J. comp. Neurol.*, **117**, 107–16.

Shrewsbury, M. M. and Kuzynski, K. (1974). Flexor digitorum superficialis tendon in the fingers of the human hand, *The Hand*, **6**, 121–33.

Shriver, J. E., Stein, B. M. and Carpenter, M. B. (1968). Central projections of spinal dorsal roots in the monkey. I. Cervical and upper thoracic dorsal roots, *Am. J. Anat.*, **123**, 27–74.

Shultze, A. H. (1956). Postembryonic age changes, *Primatologia*, **1**, 887–964.

Shute, C. C. D. (1956). The evolution of the mammalian ear drum and the tympanic cavity, *J. Anat.*, **90**, 261–81.

Sicher, H. and Du Brul, E. L. (1970). *Oral Anatomy*, Mosby: St Louis.

Sidman, R. L. (1970). Cell proliferation, migration and interaction in the developing mammalian central nervous system. In: *The Neurosciences, a Second Study Program* (Schmitt, F. O., Quarton, G. C., Melnechuk, T. and Adelman, G. eds.), pp. 100–7, Rockefeller University Press: New York.

Sidman, R. L., Miale, I. L. and Feder, N. (1959). Cell proliferation and migration in the primitive ependymal zone, *Exp. Neurol.*, **1**, 322–33.

Siegel, P. and Mengert, W. F. (1961). Internal iliac artery ligation in obstetrics and gynecology, *J. Am. med. Ass.*, **178**, 1059–62.

Sierociński, W. (1975). Arteries supplying the left colic flexure in man, *Folia morph., Warsz.*, **34**, 117–24.

Sierociński, W. (1976). Anastomoses of the arteries supplying the descending and sigmoid colon in man, *Folia morph., Warsz.*, **35**, 467–79.

Siffert, R. S. (1956). The effect of staples and longitudinal wires on epiphysial growth: an experimental study, *J. Bone Jt Surg.*, **38A**, 1077–88.

Silberberg, R., Silberberg, M. and Feir, D. (1964). Cycle of articular cartilage cells: an electron microscope study of the hip joint of the mouse, *Am. J. Anat.*, **114**, 17–48.

Silva, D. G. (1966). The fine structure of multivesicular cells and large microvilli in the epithelium of the mouse colon, *J. Ultrastruct. Res.*, **16**, 693–705.

Silva, D. G. and Hart, J. A. L. (1967). Ultrastructural observations on the mandibular condyle of the guinea pig, *J. Ultrastruct. Res.*, **20**, 227–43.

Silva, D. G. and Kailis, D. G. (1972). Ultrastructural studies on the cervical loop and the development of the amelo-dentinal junction in the cat, *Archs oral. Biol.*, **17**, 279–90.

Silver, M. D., Lam, J. H. C., Ranganathan, N. and Wigle, E. D. (1971). Morphology of the human tricuspid valve, *Circulation*, **43**, 333–48.

Silver, P. H. (1956). The role of the peritoneum in the formation of the septum recto-vesicale, *J. Anat.*, **90**, 538–46.

Silverstein, H., Davies, D. G. and Griffin, W. L. Jr. (1969). Cochlear aqueduct obstruction: changes in perilymph biochemistry, *Ann. Otol. Rhinol. Lar.*, **78**, 532–41.

Simard, L. C. (1937). Les complexes neuro-insulaires du pancréas humain, *Arch. Anat. microsc.*, **33**, 49–64.

Simionescu, N., Simionescu, M. and Palade, G. E. (1973). Permeability of muscle capillaries to exogenous myoglobin, *J. Cell Biol.*, **57**, 424–52.

Simionescu, N., Simionescu, M. and Palade, G. E. (1975). Permeability of muscle capillaries to small hemepeptides. Evidence for the existence of patent transendothelial channels, *J. Cell Biol.*, **64**, 586–607.

Simons, P. and Reid, L. (1969). Muscularity of pulmonary arterial branches in the upper and lower lobes of the normal young and aged lung, *Br. J. dis. Chest.*, **63**, 38–44.

Sinclair, D. (1967). *Cutaneous Sensation*, Oxford University Press: London.

Sindou, M., Quoex, O. and Baleydier, C. (1974). Fibre organization at the posterior spinal cord-rootlet junction in man, *J. comp. Neurol.*, **153**, 15–26.

Singer, C. (1931). *A Short history of Biology*, Clarendon Press: Oxford.

Singer, C. (1956). *Galen on Anatomical Proceedings* (translation of the surviving books and introduction and notes), Oxford University Press: London.

Singer, C. and Underwood, A. E. (1962). *A Short History of Medicine*, Clarendon Press: Oxford.

Singer, R. (1953). Estimation of age from cranial suture closure: report on its unreliability, *J. For. Med.*, **1**, 52–9.

Singer, S. J. and Nicolson, G. L. (1972). The fluid mosaic model of the structure of cell membranes, *Science, N.Y.*, **175**, 720–31.

Singh, I. (1959). Variations in the metacarpal bones, *J. Anat.*, **93**, 262–7.

Singh, I. (1960). Variations in the metatarsal bones, *J. Anat.*, **94**, 345–50.

Singh, I. (1963). The prenatal development of enterochromaffin cells in the human gastro-intestinal tract, *J. Anat.*, **97**, 377–87.

Singh, I. (1964). On argyrophile and argentaffin reactions in individual granules of enterochromaffin cells of the human gastro-intestinal tract, *J. Anat.*, **98**, 497–500.

Singh, M. and Webster, P. D. (1978). Neurohormonal control of pancreatic secretion. A review, *Gastroenterol.*, **74**, 294–309.

Singh, S. (1965). Variations of the superior articular facets of atlas vertebrae, *J. Anat.*, **99**, 565–71.

Singh, S. and Dass, R. (1960). The central artery of the retina. I. Origin and course. II. A study of its distribution and anastomoses, *Br. J. Ophthal.*, **44**, 193–212, 280–99.

Singh, S. and Singh, S. P. (1974). Weight of the femur—a useful measurement for identification of sex, *Acta anat.*, **87**, 141–5.

Singh, S. and Singh, S. P. (1975). Identification of sex from tarsal bones, *Acta anat.*, **93**, 568–73.

Singh, S. and Potturi, B. R. (1978). Greater sciatic notch in sex determination, *J. Anat.*, **125**, 619–24.

Siperstein, E. R. and Miller, K. J. (1970). Further cytophysiologic evidence for the identity of the cells that produce adrenocorticotrophic hormone, *Endocrinology*, **86**, 451–86.

Sirang, H. (1973). Ein Canalis alae ossis ilii und seine Bedeutung, *Anat. Anz.*, **133**, 225–38.

Sisca, R. F. and Provenza, D. V. (1972). Initial dentin formation in human deciduous teeth. An electron microscopic study, *Calcif. Tiss. Res.*, **9**, 1–16.

Sissons, H. A. (1969). Anatomy of the motor unit. In: *Disorders of Voluntary Muscle* (Walton, J. N. ed.), pp. 1–16, Churchill: London.

Siwe, S. A. (1931). The cervical part of the ganglionated cord, with special reference to its connections with the spinal nerves and certain cerebral nerves, *Am. J. Anat.*, **48**, 479–97.

Sjöstrand, F. S. (1953). Ultrastructure of outer segments of rods and cones of the eye as revealed by the electron microscope, *J. cell comp. Physiol.*, **42**, 15–44.

Sjöstrand, F. S. (1953). Ultrastructure of inner segments of retinal rods of the guinea pig eye as revealed by the electron microscope, *J. cell comp. Physiol.*, **42**, 45–70.

Sjöstrand, F. S. (1961). Electron microscopy of the retina. In: *The Structure of the Eye* (Smelser, G. K. ed.), pp. 1–20, Academic Press: New York, London.

Sjöstrand, F. S. (1967). *Electron Microscopy of Cells and Tissues*, Academic Press: New York, London.

Skoglund, S. (1956). Anatomical and physiological studies of knee joint innervation in the cat, *Acta physiol. scand.*, **36**, Suppl. 124, 1–101.

Skoglund, S. (1973). Joint receptors and kinaesthesis. In: *Handbook of Sensory Physiology* (Iggo, A. ed.), Volume II, pp. 110–36, Springer: Berlin.

Skornicki, R., Zieniánski, A. and Orebbwski, A. (1968). Galáz zewnetrzna nerwu krtaniowego górnego u człowieka i psa, *Folia morph.*, **27**, 79–87.

Slater, E. C. (1972). *Mitochondria: Biogenesis and Bioenergetics*, North Holland: Amsterdam.

Śledziński, Z. and Tyszkiewicz, T. (1975). Hepatic veins of the right part of the liver in man, *Folia morph., Warsz.*, **34**, 315–22.

Sleigh, M. A. (1962). *The Biology of Cilia and Flagella*, Pergamon Press: Oxford, London, Sydney, Paris.

Sleigh, M. A. (1974). *Cilia and Flagella*. Editor. Academic Press: London.

Slipka, J. (1972). Early development of the bursar pharyngea, *Folia Morphol.*, **20/2**, 138–40.

Small, J. V. (1974). Contractile units in vertebrate smooth muscle cells, *Nature, Lond.*, **249**, 324–7.

Smart, I. (1971). Location and orientation of mitotic figures in the developing mouse olfactory epithelium, *J. Anat.*, **109**, 243–51.

Smelser, G. K. (1966). Electron microscopy of a typical epithelial cell and of the normal human ciliary process, *Trans. Am. Acad. Ophthal. Otolar.*, **70**, 738–54.

Smith, A. D. (1972). Storage and secretion of hormones, *Scientific Basis of Medicine, Annual Reviews*, 74–102.

Smith, C. A. (1959). *The Physiology of the Newborn Infant*, Third edition, Blackwell: Oxford.

Smith, C. A. and Dempsey, E. W. (1957). Electron microscopy of the organ of Corti, *Am. J. Anat.*, **100**, 337–68.

Smith, C. G. (1941). Incidence of atrophy of olfactory nerves in man, *Archs Otolar.*, **34**, 533–9.

Smith, C. G. and Richardson, W. F. G. (1966). The course and distribution of the arteries supplying the visual (striate) cortex, *Am. J. Ophthal.*, **61**, 1391–6.

Smith, D. S. (1972). *Muscle*, Academic Press: New York, London.

Smith, F. B. and Blair, H. C. (1954). Tibial collateral ligament strain due to occult derangements of medial meniscus: confirmed by operation in 30 cases, *J. Bone Jt Surg.*, **36A**, 88–93.

Smith, G. E. (1903). The primary subdivisions of the mammalian cerebellum, *J. Anat.*, **36**, 381–5.

Smith, G. E. (1903). Notes on the morphology of the cerebellum, *J. Anat.*, **37**, 329–32.

Smith, G. E. (1907). New studies on the folding of the visual cortex and the significance of the occipital sulci in the human brain, *J. Anat.*, **41**, 198–207.

Smith, G. E. (1908). Studies in the anatomy of the pelvis, with special reference to the fasciae and visceral supports. I and II, *J. Anat.*, **42**, 191–218, 252–70.

Smith, G. E. (1908). The cerebral cortex in *Lepidosiren* with comparative notes on the interpretation of certain features of the forebrain with other vertebrates, *Anat. Anz.*, **33**, 513–50.

Smith, G. E. (1930). The cortical representation of the macula, *J. Anat.*, **64**, 477–8.

Smith, G. T. (1963). The renal vascular patterns in man, *J. Urol.*, **89**, 275–88.

Smith, J. M. and Savage, R. J. G. (1959). The mechanics of mammalian jaws, *School Sci. Rev.*, **141**, 289–301.

Smith, J. W. (1953). The act of standing, *Acta orthopaed. scand.*, **23**, 159–68.

Smith, J. W. (1954). Muscular control of the arches of the foot in standing: an electromyographic assessment, *J. Anat.*, **88**, 152–63.

Smith, J. W. (1956). Observations on the postural mechanism of the human knee joint, *J. Anat.*, **90**, 236–60.

Smith, J. W. (1958). The ligamentous structures in the canalis and sinus tarsi, *J. Anat.*, **92**, 616–20.

Smith, J. W. (1960). The arrangement of collagen fibres in human secondary osteones, *J. Bone Jt Surg.*, **42B**, 588–605.

Smith, J. W. (1962). The relationship of epiphysial plates to stress in some bones of the lower limb, *J. Anat.*, **96**, 58–78.

Smith, J. W. (1962). The structure and stress relation of fibrous epiphysial plates, *J. Anat.*, **96**, 209–25.

Smith, J. W. (1968). Molecular pattern in native collagen, *Nature, Lond.*, **219**, 157–8.

Smith, J. W. and Walmsley, R. (1951). Experimental incision of the intervertebral disc, *J. Bone Jt Surg.*, **33B**, 612–25.

Smith, J. W. and Walmsley, R. (1959). Factors affecting the elasticity of bone, *J. Anat.*, **93**, 505–23.

Smith, J. W. and Serafini-Fracassini, A. (1967). The relationship of hyaluronate and collagen in the bovine vitreous body, *J. Anat.*, **101**, 99–112.

Smith, J. W., Peters, T. J. and Serafini-Fracassini, A. (1967). Observations on the distribution of the proteinpolysaccharide complex and collagen in bovine articular cartilage, *J. Cell Sci.*, **2**, 129–36.

Smith, M. C. (1951). Use of Marchi staining in late stages of human tract degeneration, *J. Neurol. Neurosurg. Psychiat.*, **14**, 222–5.

Smith, M. C. (1957). The anatomy of the spino-cerebellar fibers in man. I. The course of the fibers in the spinal cord and brain stem, *J. comp. Neurol.*, **108**, 285–352.

Smith, M. C. (1967). Stereotatic operations for Parkinson's disease— Anatomical considerations. In: *Modern Trends in Neurology* (Williams, D. ed.), pp. 21–52, Butterworths: London.

Smith, P. and Porte, D. (1976). Neuropharmacology of the pancreatic islets, *Ann. Rev. Pharmacol. Toxicol.*, **16**, 269–85.

Smith, R. B. (1970). The development of the intrinsic innervation of the human heart between the 10 and 70 mm stages, *J. Anat.*, **107**, 271–80.

Smith, R. B. (1971). Intrinsic innervation of the human heart in foetuses between 70 mm and 420 mm crown–rump length, *Acta anat.*, **78**, 200–9.

Smith, U. and Ryan, J. W. (1970). An electron microscopic study of the vascular endothelium as a site for bradykinin and adenosine-5′-triphosphate inactivation in the rat lung, *Adv. exp. med. Biol.*, **8**, 249–61.

Smith, U. and Ryan, J. W. (1973). Electron microscopy of endothelial cells collected on cellulose acetate paper, *Tissue and Cell*, **5**, 333–6.

Smout, C. F. V., Jacoby, F. and Lillie, E. W. (1969). *Gynaecological and Obstetrical Anatomy and Functional Histology*, Arnold: London.

Smyth, G. E. (1939). Systemisation and central connections of spinal tract and nucleus of trigeminal; clinical and pathological study, *Brain*, **62**, 41–87.

Sneath, R. S. (1955). The insertion of the biceps femoris, *J. Anat.*, **89**, 550–3.

Snider, R. S. (1936). Alterations which occur in mossy terminals of the cerebellum, following transection of the brachium pontis, *J. comp. Neurol.*, **64**, 417–31.

Snider, R. S. (1940). Morphology of the cerebellar nuclei in the rabbit and the cat, *J. comp. Neurol.*, **72**, 399–415.

Snider, R. S. (1945). Electro-anatomical studies on a tecto-cerebellar pathway, *Anat. Rec.*, **91**, 299.

Snider, R. S. (1950). Recent contributions to the anatomy and physiology of the cerebellum, *Archs Neurol. Psychiat., Chicago*, **64**, 196–219.

Snider, R. S. (1952). Inter-relations of cerebellum and brain stem, *Res. Publs Ass. Res. nerv. ment. Dis.*, **30**, 267–81.

Snider, R. S. (1967). Functional alterations of cerebral sensory areas by the cerebellum, *Prog. Brain Res.*, **25**, 322–33.

Snider, R. S. and Stowell, A. (1942). Evidence of a representation of tactile sensibility in the cerebellum of the cat, *Fedn. Proc. Fedn. Am. Socs exp. Biol.*, **1**, 82.

Snider, R. S. and Stowell, A. (1944). Receiving areas of the tactile, auditory and visual systems in the cerebellum, *J. Neurophysiol.*, **7**, 331–57.

Snider, R. S. and Eldred, E. (1951). Electro-anatomical studies on cerebrocerebellar connections in the cat, *J. comp. Neurol.*, **95**, 1–16.

Snider, R. S. and Eldred, E. (1952). Cerebro-cerebellar relationships in the monkey, *J. Neurophysiol.*, **15**, 27–40.

Snook, T. (1950). A comparative study of the vascular arrangements in mammalian spleens, *Am. J. Anat.*, **87**, 31–78.

Snyder, F. F. (1924). Changes in the human oviduct during the menstrual cycle and pregnancy, *Bull. Johns Hopkins Hosp.*, **35**, 141–6.

Sohval, A. R. (1954). Histopathology of cryptorchidism; studies based upon comparative histology of retained and scrotal testes from birth to maturity, *Am. J. Med.*, **16**, 346–62.

Solanke, T. F. (1968). The blood supply of the vermiform appendix in Nigerians, *J. Anat.*, **102**, 353–62.

Solcia, E., Vassalo, G. and Capella, C. (1968). Selective staining of endocrine cells by basic dyes after acid hydrolysis, *Stain Technol.*, **43**, 257–63.

Solter, M. and Paljan, D. (1973). Variations in shape and dimensions of sigmoid groove, venous portion of jugular foramen, jugular fossa, condylar and mastoid foramina classified by age, sex and body size, *Z. Anat. EntwGesch.*, **140**, 319–35.

Soltero-Harrington, L. R., Garcia-Rinaldi, R. and Albe, L. W. (1969). Total aganglionosis of the colon: recognition and management, *J. Pediat. Surg.*, **4**, 330–8.

Somlyo, A. V. and Somlyo, A. P. (1968). Electromechanical and pharmacomechanical coupling in vascular smooth muscle, *J. Pharmacol. exp. Ther.*, **159**, 129–45.

Sommer, J. R. and Johnson, E. A. (1968). Cardiac muscle. A comparative study of Purkinje fibers and ventricular fibres, *J. Cell Biol.*, **36**, 497–526.

Somogyi, B., Undi, F. and Kausz, M. (1973). Blood supply of the spinal ganglia, *Morph. és. Ig. Arv. Szemle.*, **13**, 191–5.

Sorenson, G. D. (1960). An electron microscopic study of popliteal lymph nodes from rabbits, *Am. J. Anat.*, **107**, 73–96.

Sorenson, G. D. (1960). An electron microscopic study of hematopoiesis in the liver of the fetal rabbit, *Am. J. Anat.*, **106**, 27–40.

Sorokin, S. P. (1968). Reconstructions of centriole formation and ciliogenesis in mammalian lungs, *J. Cell Sci.*, **3**, 207–30.

Sorsby, A. and Sheridan, M. (1960). The eye at birth: measurements of the principal diameters in forty-eight cadavers, *J. Anat.*, **94**, 192–7.

Sorsby, A., Benjamin, B., Sheridan, M., Stone, J. and Leary, G. A. (1961). Refraction and its components during the growth of the eye from the age of three, *M.R.C. Sp. Rep. Lond.*, **301**, 1–67.

Sotelo, J. R. and Porter, K. R. (1959). An electron microscope study of the rat ovum, *J. biophys. biochem. Cytol.*, **5**, 327–42.

Soupart, P. and Noyes, R. W. (1964). Sialic acid as a component of the zona pellucida of the mammalian ovum, *J. Reprod. Fertil.*, **8**, 251–3.

Soupart, T. P. and Clewe, T. H. (1965). Sperm penetration of rabbit zona pellucida inhibited by treatment of ova with neuraminidase, *Fert. Steril.*, **16**, 677–89.

Soupart, P. and Strong, P. A. (1974). Ultrastructural observations on human oöcytes fertilized in vitro, *Fert. Steril.*, **25**, 11–44.

Soupart, P. and Strong, P. A. (1975). Ultrastructural observations on polyspermic penetration of zona pellucida-free human oöcytes inseminated in vitro, *Fert. Steril.*, **26**, 523–37.

Sousa-Pinto, A. (1973). Cortical projections of the medial geniculate body in the cat, *Adv. Anat. Embryol. Cell Biol.*, **48**, 1–42.

Sousa-Pinto, A. and Brodal, A. (1969). Demonstration of a somatotopical pattern in the cortico-olivary projection in the cat. An experimental-anatomical study, *Expl Brain Res.*, **8**, 364–86.

Southam, J. A. (1959). The inferior mesenteric ganglion, *J. Anat.*, **93**, 304–8.

Sow, M. L., Dintimille, H., Padonov, N., Sylla, S. and Argenson, C. (1975). La vascularisation veineuse du pancreas, *Bull. Ass. Anat.*, **59**, 255–64.

Spalteholtz, W. (1924). *Die Arterien der Herzwand. Anatomische Untersuchungen an Menschen und Tieren*, Hirzel: Leipzig.

Spanner, R. (1932). Neue Befunde über die Blutwege der Darmwand und ihre funktionelle Bedentung, *Gegn. morph. Jbach.* **1**. Ab. 69, 394–454.

Speidel, C. C. (1932). Studies of living nerves. I. The movements of individual sheath cells and nerve sprouts correlated with the process of myelin sheath formation in amphibian larvae, *J. exp. Zool.*, **61**, 279–331.

Speidel, C. C. (1933). Studies of living nerves. II. Activities of ameboid growth cones, sheath cells and myelin segments, as revealed by prolonged observations of individual nerve fibers in frog tadpoles, *Am. J. Anat.*, **52**, 1–80.

Speidel, C. C. (1935). Studies of living nerves. III. Phenomena of nerve irritation and recovery, degeneration and repair, *J. comp. Neurol.*, **61**, 1–82.

Speller, A. M. and Moffat, D. B. (1977). Tubulo-vascular relationships in the developing kidney, *J. Anat.*, **123**, 487–500.

Spemann, H. (1938). *Embryonic Development and Induction*, Yale University Press: New Haven.

Spencer, H. (1977). *Pathology of the Lung*, Volume 1, Third edition, Pergamon Press: Oxford.

Spencer, L. M., Foos, R. Y. and Straatsma, B. R. (1969). Meridional folds and meridional complexes in the peripheral retina, *Trans. Am. Acad. Ophthal. Oto-Lar.*, **73**, 204–21.

Sperry, R. W. (1951a). Mechanisms of neural maturation. In: *Handbook of Experimental Psychology* (Stevens, S. ed.), pp. 236–80, Wiley: New York.

Sperry, R. W. (1951b). Regulative factors in the orderly growth of neural circuits, *Growth Symp.*, **10**, 63–87.

Sperry, R. W. (1958). Developmental basis of behavior. In: *Behavior and Evolution* (Roe, A. and Simpson, G. T. eds.), Yale University Press: New Haven.

Sperry, R. W. (1963). Chemoaffinity in the orderly growth of nerve fiber patterns and connections, *Proc. natn. Acad. Sci. U.S.A.*, **50**, 703–10.

Sperry, R. W. (1965). Embryogenesis of behavioral nerve nets. In: *Organogenesis* (DeHann, R. L. and Ursprung, H. eds.), pp. 161–86, Holt, Rinehart & Winston: New York.

Sperry, R. W. (1966). Brain bisection and mechanisms of consciousness. In: *Brain and Conscious Experience* (Eccles, J. C. ed.), pp. 298–308, Springer-Verlag: Berlin.

Sperry, R. W. (1970). Perception in the absence of the neocortical commissures, *Percept. Disord. (A.R.N.M.D.)*, **48**, 123–38.

Sperry, R. W. (1971). How a developing brain gets itself properly wired for adaptive function. In: *Biopsychology of Development* (Tobach, ed.), Academic Press: New York, London.

Sperry, R. W. (1974). Lateral specialization in the surgically separated hemispheres. In: *The Neurosciences, Third Study Program* (Schmitt, F. O. and Worden, F. G. eds.), MIT Press: Cambridge, Mass.

Sperry, R. W. (1977). Problems outstanding in the evolution of brain function. In: *The Encyclopaedia of Ignorance* (Duncan, R. and Weston-Smith, M. eds.), Pergamon Press: Oxford.

Sperry, R. W., Gazzaniga, M. S. and Bogen, J. E. (1969). Interhemispheric relationships. In: *Clinical Neurology* (Vinken, P. J. and Bruyn, G. W. eds.), Volume 4, North Holland Publishing Co.: Amsterdam.

Spira, A. (1962). Die Lymphknotengruppen (Lymphocentra) bei der Säugernein ein Homologisierungsversuch, *Anat. Anz.*, **111**, 294–364.

Spira, A. W. and Hollenberg, M. J. (1973). Human retinal development: ultrastructure of the inner retinal layers, *Dev. Biol.*, **31**, 1–21.

Spira, J. J. (1957). Comparison of cardiac and pyloric sphincters, *Lancet*, **2**, 1008.

Spirin, A. S. and Gavrilova, L. P. (1969). *The Ribosome*, Springer-Verlag: Berlin.

Spitznas, M. (1970). Zur feinstruktur der sog. Membrana limitans externa der menschlichen Retina, *Albrecht v. Graefes Arch. Ophthal.*, **180**, 44–56.

Spitznas, M. and Hogan, M. J. (1970). Outer segments of photo-receptors and the pigmented epithelium interrelationships in the human eye, *Archs Ophthal., N.Y.*, **84**, 810–19.

Spoerri, P. E. and Glees, P. (1977). Subsurface cisterns in the *Cynomolgus* retina, *Cell Tiss. Res.*, **182**, 33–8.

Sprague, J. M. (1944). The innervation of the pharynx in the rhesus monkey, and the formation of the pharyngeal plexus in primates, *Anat. Rec.*, **90**, 197–208.

Sprague, J. M. (1948). A study of motor cell localization in the spinal cord of the rhesus monkey, *Am. J. Anat.*, **82**, 1–26.

Sprague, J. M. (1958). The distribution of dorsal root fibres on motor cells in the lumbosacral spinal cord of the cat, and the site of excitatory and inhibitory terminals in monosynaptic pathways, *Proc. R. Soc. B.*, **149**, 534–56.

Sprague, J. M. (1963). Corticofugal projection to the superior colliculus in the cat, *Anat. Rec.*, **145**, 288 (abstract).

Sprague, J. M. and Ha, H. (1964). The terminal fields of dorsal root fibers in the lumbosacral spinal cord of the cat, and the dendritic organisation of the motor nuclei, *Prog. Brain Res.*, **11**, 120–52.

Spudich, J. A. and Lin, S. (1972). Cytochalasin B, its interaction with actin and actomyosin from muscle, *Proc. natn. Acad. Sci. U.S.A.*, **69**, 442–6.

BIBLIOGRAPHY

Srivastava, H. C. (1977). Development of ossification centres in the squamous portion of the occipital bone in man, *J. Anat.*, **124**, 643–9.

Srivastava, P. N., Adams, C. E. and Hartree, E. F. (1965). Enzymic action of acrososmal preparations on the rabbit ovum *in vitro*, *J. Reprod. Fert.*, **10**, 61–7.

Stack, H. G. (1962). Muscle function in the fingers, *J. Bone Jt Surg.*, **44B**, 899–1022.

Stack, H. G. (1973). *The Palmar Fascia*, Churchill Livingstone: London.

Staehelin, L. A. (1974). Structure and function of intercellular junctions, *Int. Rev. Cytol.*, **39**, 191–283.

Staehelin, L. A. (1975). A new occludens-like junction linking endothelial cells of small capillaries (probably venules) of rat jejunum, *J. Cell Sci.*, **18**, 545–51.

Stalsberg, H. and de Hann, R. L. (1968). Endodermal movements during foregut formation in the chick embryo, *Dev. Biol.*, **18**, 198–215.

Stamm, T. T. (1931). The constitution of the ligamentum cruciatum cruris, *J. Anat.*, **66**, 80–3.

Stämpfli, R. (1954). Saltatory conduction in nerve, *Physiol Rev.*, **34**, 101–12.

Stanfield, J. P. (1960). The blood supply of the human pituitary gland, *J. Anat.*, **94**, 257–73.

Stanier, M. W. (1977). The function of muscles around a simple joint, *J. Anat.*, **123**, 827–30.

Straprans, I. and Dirksen, E. R. (1974). Microtubule protein during ciliogenesis in the mouse oviduct, *J. Cell Biol.*, **62**, 164–74.

Stark, D. (1965). *Embryologie*, Second edition, Thieme: Stuttgart.

Steel, F. L. D. and Tomlinson, J. D. W. (1958). The 'carrying angle' in man, *J. Anat.*, **92**, 315–17.

Steele, D. G. (1970). Estimation of stature from fragments of long limb bones. In: *Personal Identification in Mass Disasters* (Stewart, T. D. ed.), Smithsonian Institute: Washington DC.

Steele, E. J. and Blunt, M. J. (1956). The blood supply of the optic nerve and chiasma in man, *J. Anat.*, **90**, 486–93.

Stein, B. M. and Carpenter, M. B. (1967). Central projections of portions of the vestibular ganglia innervating specific parts of the labyrinth in the rhesus monkey, *Am. J. Anat.*, **120**, 281–318.

Steiger, H.-J. and Büttner-Ennever, J. (1978). Relationship between motor neurons and internuclear neurons in the abducens nucleus, *Brain Res.*, **148**, 181–8.

Stein, G. S. and Borun, T. W. (1972). The synthesis of acidic chromosomal proteins during the cell cycle of HeLa S3 cells. I. The accelerated accumulation of acidic residual nuclear protein before initiation of DNA replication, *J. Cell Biol.*, **52**, 292–307.

Steinberger, E. and Steinberger, A. (1975). Spermatogenic function of the testis. In: *Handbook of Physiology* (Hamilton, D. W. and Greep, R. O. eds), Section 7, Volume V, pp. 1–19, American Physiological Society: Washington.

Steindler, A. (1955). *Kinesiology of the Human Body*, Thomas: Springfield, Ill.

Steinman, R. M., Lustig, D. S. and Cohn, Z. A. (1974). Identification of a novel cell type in peripheral lymphoid organs of mice. II. Functional properties *in vitro*, *J. exp. Med.*, **139**, 380–97.

Stelzner, F. and Lierse, W. (1967). Über das Verschluss-system der terminalen Speiseröhre, *Thoraxchirurgie*, **15**, 676–9.

Strenström, S. (1946). Untersuchungen über die Variation und Kovariation der optischen Elemente des menschlichen Auges, *Acta ophthal.*, Suppl. 26, 1–103.

Stephan, F. K. and Zucker, I. (1972). Circadian rhythms in drinking behavior and locomotor activity of rats are eliminated by hypothalamic lesions, *Proc. natn. Acad. Sci., U.S.A.*, **69**, 1583–6.

Stephan, H. and Andy, O. J. (1962). The septum (a comparative study in its size in insectivores and primates), *J. Hirnforsch.*, **5**, 229–44.

Stephens, R. E. and Edds, K. T. (1976). Microtubule structure, chemistry and functions, *Physiol Rev.*, **56**, 709–77.

Steptoe, P. C. and Edwards, R. G. (1970). Laparoscopic recovery of pre-ovulatory human oöcytes after priming of ovaries with gonadotrophins, *Lancet*, **1**, 683–9.

Steptoe, P. C. and Edwards, R. G. (1976). Reimplantation of a human embryo with subsequent tubal pregnancy, *Lancet*, **1**, 880–2.

Sterling, P. and Kuypers, H. G. J. M. (1967). Anatomical organisation of the brachial spinal cord of the cat. I. The distribution of dorsal root fibers, *Brain Res.*, **4**, 1–15.

Sterling, P. and Kuypers, H. G. J. M. (1967). Anatomical organisation of the brachial spinal cord of the cat. II. The motoneuron plexus, *Brain Res.*, **4**, 16–32.

Stern, J. T. Jr. (1971). Investigations concerning the theory of 'spurt' and 'shunt' muscles, *J. Biomech.*, **4**, 437–53.

Stevenson, P. H. (1924). Age order of epiphysial union in man, *Am. J. phys. Anthrop.*, **7**, 53–93.

Stewart, T. D. (1954). Evaluation of evidence from the skeleton. In:

Legal Medicine (Gradwohl, R. E. H. ed.), pp. 407–50, Mosby: St Louis.

Stieve, H. (1926). Ein 13½ Tage altes, in der Gebärmutter ergaltenes und durch Eingriff gewonnenes menschliches Ei, *Arch. mikrosk. Anat. EntwMech.*, **7**, 295–402.

Stieve, H. (1927). *Der Halsteil der menschlichen Gebärmutter sein Bau und seine Aufgaben während der Schwangerschaft, der Geburt und des Wochenbeltes*, Akadem. Verlags-gesellschaft: Leipzig.

Stieve, H. (1936). Ein ganz junges, in der Gebärmutter erhaltenes menschliches Ei (Keimling Werner), *Z. Zellforsch. mikrosk. Anat.*, **40**, 281–322.

Steive, H. (1948). Der Bau der Primatenplacenta, *Anat. Anz.*, **96**, 299–328.

Stillwell, D. L. Jr. (1957). The innervation of tendons and aponeuroses, *Am. J. Anat.*, **100**, 289–318.

Stirling, J. W. and Chandler, J. A. (1976). Ultrastructural studies of the female breast. I. 9+0 cilia in myoepithelial cells, *Anat. Rec.*, **186**, 413–16.

Stockard, C. R. and Papanicolau, G. N. (1917). The existence of a typical oestrus cycle in guinea pigs and its histology, *Anat. Rec.*, **11**, 411P.

Stockwell, R. A. (1967). Lipid content of human costal and articular cartilages, *J. Anat.*, **101**, 607P.

Stolinsky, C. and Breathnach, A. S. (1975). *Freeze-Fracture Replication of Biological Tissues*, Academic Press: London.

Stopford, J. S. B. (1915). The arteries of the pons and medulla oblongata. Part I, *J. Anat.*, **50**, 131–64.

Stopford, J. S. B. (1916). The arteries of the pons and medulla oblongata. Part II, *J. Anat.*, **50**, 255–80.

Stopford, J. S. B. (1921). The nerve supply of the interphalangeal and metacarpo-phalangeal joints, *J. Anat.*, **56**, 1–11.

Stratton, C. J. (1976). The high resolution ultrastructure of the periodicity and architecture of lipid-retained and extracted lung multilamellar body laminations, *Tissue and Cell*, **8**, 713–28.

Stratton, C. J. (1977). The three dimensional aspect of the mammalian lung surfactant myelin figure, *Tissue and Cell*, **9**, 285–300.

Stratton, C. J. (1978). The ultrastructure of multilamellar bodies and surfactant in the human lung. *Cell Tiss. Res.*, **193**, 219–29.

Straus, W. L. (1927). Human ilium: sex and stock, *Am. J. phys. Anthrop.*, **11**, 1–28.

Straus, W. L. Jr. and Rawles, M. E. (1953). Effects of fluorine on calcium metabolism and bone growth in pigs, *Am. J. Anat.*, **92**, 361–90.

Streeter, G. L. (1918). The developmental alterations in the vascular system of the brain of the human embryo, *Contr. Embryol.*, **8**, 5–38.

Streeter, G. L. (1919). Factors involved in the formation of the filum terminale, *Am. J. Anat.*, **25**, 1–12.

Streeter, G. L. (1922). Development of the auricle in the human embryo, *Contr. Embryol.*, **14**, 111–38.

Streeter, G. L. (1942). Developmental horizons in human embryos. Descriptions of age group XI, 13 to 20 somites, and age group XII, 21 to 29 somites, *Contr. Embryol.*, **30**, 211–45.

Streeter, G. L. (1945). Developmental horizons in human embryos. Description of age group XIII, embryos of about 4 or 5 millimeters long, and age group XIV, period of indentation of the lens vesicle, *Contr. Embryol.*, **31**, 27–63.

Streeter, G. L. (1948). Developmental horizons in human embryos. Description of age groups XV, XVI, XVII, and XVIII, being the third issue of a survey of the Carnegie collection, *Contr. Embryol.*, **32**, 133–203.

Streeter, G. L. (1949). Developmental horizons in human embryos. Description of age group XI, 13 to 20 somites, and age group XII, 21 to 29 somites, *Contr. Embryol.*, **30**, 211–45.

Streeter, G. L. (1949). Developmental horizons in human embryos (fourth issue): A review of the histiogenesis of cartilage and bone, *Contr. Embryol.*, **33**, 149–67.

Strum, J. M. and Karnovsky, M. J. (1970). Cytochemical localization of endogenous peroxidase in thyroid follicular cells, *J. Cell Biol.*, **44**, 655–66.

Stuurman, F. J. (1916). Die Lokalisation der Zungenmuskeln im Nucleus hypoglossi, *Anat. Anz.*, **48**, 593–610.

Sudeck, P. (1907). Uber die Gefässversorgung des Mastdarmes in Hinsicht auf die operative Gangrän, *Münch. med. Wchnschr.*, **54**, 1314–17.

Sullivan, F. M. (1976). Effects of drugs on fetal development. In: *Fetal Physiology and Medicine* (Beard, R. W. and Nathanielsz, P. W. eds.), Saunders: London.

Sunderland, S. (1938). The production of cortical lesions by de-vascularisation of cortical areas, *J. Anat.*, **73**, 120–9.

Sunderland, S. (1945). Arterial relations of the internal auditory meatus, *Brain*, **68**, 56–72.

Sunderland, S. (1945). The actions of the extensor digitorum communis, interosseous and lumbrical muscles, *Am. J. Anat.*, **77**, 189–209.

Sunderland, S. (1946). The innervation of the first dorsal interosseous muscle of the hand, *Anat. Rec.*, **95**, 7–10.

Sunderland, S. (1974). Meningeal-neural relations in the intervertebral foramen, *J. Neurosurg.*, **40**, 756–63.

Sunderland, S. and Hughes, E. S. R. (1946). The pupilloconstrictor pathway and the nerves to the ocular muscles in man, *Brain*, **69**, 301–9.

Sunderland, S. and Bedbrook, G. M. (1949). Relative sympathetic contribution to individual roots of the brachial plexus in man, *Brain*, **72**, 297–310.

Sutherland, S. D. (1965). The intrinsic innervation of the liver, *Rev. Int. Hepat.*, **15**, 569–78.

Sutherland, S. D. (1966). The intrinsic innervation of the gall bladder in *Macaca rhesus* and *Cavia porcellus*, *J. Anat.*, **100**, 261–8.

Sutherland, S. D. (1967). The neurons of the gall bladder and gut, *J. Anat.*, **101**, 701–10.

Sutton, R. N. (1974). The practical significance of mandibular accessory foramina, *Aust. Dent. J.*, **19**, 167–73.

Suzuki, N. (1972). An electromyographic study of the role of the muscles in arch support of the normal and flat foot, *Nagoya Med. J.*, **17**, 57–79.

Suzuki, Y., Churg, J. and Ono, T. (1972). Phagocytic activity of the alveolar epithelial cells in pulmonary asbestosis, *Am. J. Path.*, **69**, 373–88.

Swarz, J. R. (1976). The presence of Bergmann fibres in prenatal mouse cerebellum and its implications in cerebellar histogenesis, *Anat. Rec.*, **184**, 543.

Swarz, J. R. and del Cerro, M. P. (1975). Lack of evidence for glial cells originating from the external granular layer in the mouse cerebellum, *Neurosci. Abs.*, **1**, 760.

Swarz, J. R. and del Cerro, M. (1977). Lack of evidence for glial cells originating from the external granular layer in mouse cerebellum, *J. Neurocytol.*, **6**, 241–50.

Sweeney, W. J. (1962). The interstitial portion of the uterine tube—its gross anatomy, course and length, *Obstet. Gynec.*, **19**, 1–8.

Swyer, G. I. M. (1944). Post-natal growth changes in the human prostate, *J. Anat.*, **78**, 130–45.

Sykes, D. (1963). The arterial supply of the human kidney with special reference to accessory renal arteries, *Br. J. Surg.*, **50**, 368–74.

Sykes, D. (1964). The correlation between renal vascularisation and lobulation of the kidney, *Br. J. Urol.*, **36**, 549–55.

Sylvén, B. (1951). On the biology of the nucleus pulposus, *Acta orthopaed. scand.*, **20**, 275–9.

Symons, N. B. B. (1968). *Symposium on Dentine and Pulp* (editor), Livingstone: London.

Szabo, T. and Dussardier, M. (1964). Les noyaux d'origine du nerf vague chez le mouton, *Z. Zellforsch. mikrosk. Anat.*, **63**, 247–76.

Székely, G. (1974). Problem of neuronal specificity in the development of some behavioural patterns in amphibia. In: *Aspects of Neurogenesis* (Gottlieb, G. ed.), Volume 2, Academic Press: New York.

Szentágothai, J. (1942). Die innere Gliederung des Oculomotoriuskernes, *Arch Psychiat. NervKrankh.*, **115**, 127–35.

Szentágothai, J. (1943). Die Lokalisation der Kehlkopfmuskulatur in den Vaguskernen, *Z. Anat. EntwGesch.*, **112**, 704–10.

Szentágothai, J. (1948). Anatomical considerations of monosynaptic reflex arcs, *J. Neurophysiol.*, **11**, 445–54.

Szentágothai, J. (1948). Representation of facial and scalp muscles in facial nucleus, *J. comp. Neurol.*, **88**, 207–20.

Szentágothai, J. (1949). Functional representation in the motor trigeminal nucleus, *J. comp. Neurol.*, **90**, 111–20.

Szentágothai, J. (1950). Recherches expérimentales sur les voies oculogyres, *Sem. Hôp. Paris*, **26**, 2989–95.

Szentágothai, J. (1963). The structure of the synapse in the lateral geniculate body, *Acta anat.*, **55**, 166–85.

Szentágothai, J. (1964). Neuronal and synaptic arrangement in the substantia gelatinosa Rolandi, *J. comp. Neurol.*, **122**, 219–39.

Szentágothai, J. (1970). Glomerular synapses, complex synaptic arrangements and their operational significance. In: *The Neurosciences, a Second Study Program* (Schmitt, F. O., Quarton, G. C., Melnechuck, T. and Adelman, G. eds.), pp. 427–43, Rockefeller University Press: New York.

Szentágothai, J. (1975). The 'module-concept' in cerebral cortex architecture, *Brain Res.*, **95**, 475–96.

Szentágothai, J., Flerkó, B., Mess, B. and Halász, B. (1962). *Hypothalamic Control of the Anterior Pituitary*, Akadémiai Kiadó: Budapest.

Szollosi, D. (1967). Modification of the endoplasmic reticulum in some mammalian oöcytes, *Anat. Rec.*, **158**, 59–73.

Szollosi, D. (1970). Cortical cytoplasmic filaments of cleaving eggs: A structural element corresponding to the contractile ring, *J. Cell Biol.*, **44**, 192–209.

Szollosi, D. G. and Ris, H. (1961). Observations on sperm penetration in the rat, *J. biophys. biochem. Cytol.*, **10**, 275–83.

T

Taber, E. (1961). The cytoarchitecture of the brain stem of the cat. I. Brain stem nuclei of the cat, *J. comp. Neurol.*, **116**, 27–70.

Taboada, R. P. (1927). Note sur la structure du corps genouille externe, *Trab. Lab. Invest. biol. Univ. Madr.*, **25**, 319–29.

Tachi, S., Tachi, C. and Lindner, H. R. (1970). Ultrastructural features of blastocyst attachment and trophoblastic invasion in the rat, *J. Reprod. Fert.*, **21**, 37–56.

Takahashi, D. (1913). Zur vergleichenden Anatamoie des Seitenharns im Rückenmark der Vertebraten, *Arb. neurol. Inst. Univ. Wien.*, **20**, 62–83.

Takahashi, Y. (1975). Anthropological studies on the humerus of the recent Japanese, *J. Anthrop. Soc. Nippon*, **83**, 219–32.

Takahashi, Y. (1976). Anthropological studies on the humerus of the recent Japanese, *Acta anat. Nippon*, **51**, 79–88.

Takano, K. (1969). Electron microscopic study of the so-called "separating zone" in the striated duct cell of the parotid gland, *Okajimas Folia anat. jap.*, **46**, 201–29.

Takei, Y. and Ozanics, V. (1975). Origin and development of Bruch's membrane in monkey fetuses, *Invest. Ophthal.*, **14**, 903–16.

Tamarin, A. and Sreebny, L. M. (1965). The rat submaxillary salivary gland. A correlative study by light and electron microscopy, *J. Morphol.*, **117**, 296–352.

Tanagho, E. A. and Pugh, R. C. B. (1963). The anatomy and function of the uterovesical junction, *Br. J. Urol.*, **35**, 151–65.

Tanagho, E. A., Meyers, F. H. and Smith, D. R. (1968). The trigone: anatomical and physiological considerations. I. In relation to the uterovesical junction, *J. Urol.*, **100**, 623–32.

Tanagho, E. A. and Miller, F. R. (1970). Initiation of voiding, *Br. J. Urol.*, **42**, 175–83.

Tandler, B. (1965). Ultrastructure of the human submaxillary gland. III. Myoepithelium, *Z. Zellforsch.*, **65**, 852–63.

Tandler, B. (1972). Microstructure of salivary glands. In: *Proceedings of Symposium on Salivary Glands and their Secretions* (Rowe, N. H. ed.), pp. 8–21, University of Michigan: Ann Arbor.

Tandler, B., Denning, C. R., Mandel, J. D. and Kutscher, A. H. (1970). Ultrastructure of human labial salivary glands. III. Myoepithelium and ducts, *J. Morphol.*, **130**, 227–46.

Tandler, J. (1912). *Keibel and Mall's Manual of Embryology*, Lippincott: London, Philadelphia.

Tanner, J. M. (1962). *Growth at Adolescence*, Blackwell Scientific Publications: Oxford.

Tappen, N. C. (1954). A comparative functional analysis of primate skulls by the split-line technique, *Human Biol.*, **26**, 220–38.

Tappen, N. C. (1970). Main patterns and individual differences in baboon split-lines, *Am. J. phys. Anthrop.*, **33**, 61–72.

Tarkowski, A. K. (1961). Mouse chimaeras developed from fused eggs, *Nature, Lond.*, **190**, 857–60.

Tarkowski, A. K. (1963). Studies on mouse chimeras developed from eggs fused *in vitro*, *Nat. Cancer Inst. Monogr.*, **11**, 51–71.

Tarkowski, A. K. (1965). Embryonic and postnatal development of mouse chimeras. In: *Pre-implantation Stages of Pregnancy* (Wolstenholme, G. E. W. and O'Connor, M. eds.), pp. 183–93, Ciba Foundation Symposium, Churchill: London.

Tarlov, E. (1972). Anatomy of the two vestibulo-oculomotor projection systems, *Prog. Brain Res.*, **37**, 489–91.

Tarlov, E. (1975). Synopsis of current knowledge about association projections from the vestibular nuclei. In: *The Vestibular System* (Naunton, R. F. ed.), pp. 55–69, Academic Press: New York, London.

Taton, R. (1966). *A General History of the Sciences*, Thames & Hudson: London.

Taurog, A. (1970). Thyroid peroxidase and thyroxine biosynthesis, *Recent Prog. Horm. Res.*, **26**, 189–247.

Taurog, A. (1978). Thyroid hormone synthesis and release. In: *The Thyroid. A Fundamental and Clinical Text* (Werner, S. C. and Ingbar, S. H. eds.), pp. 31–61, Harper & Row: New York.

Taussig, H. B. (1961). *Congenital Malformations of the Heart*, Second edition, Volumes I and II, Harvard University Press: Cambridge, Mass.

Tautz, C. and Rohen, H. W. (1967). Ueber den konstruktiven Bau des M. vocalis beim Menschen, *Anat. Anz.*, **120**, 409–29.

Taylor, A. (1960). The contribution of the intercostal muscles to the effort of respiration in man, *J. Physiol., Lond.*, **151**, 390–402.

Taylor, E. W. (1965). Control of DNA synthesis in mammalian cells in culture, *Exp. Cell Res.*, **40**, 316–32.

Taylor, J. R. (1975). Growth of human intervertebral discs and vertebral bodies, *J. Anat.*, **120**, 49–68.

Taylor, K. J. W. (1978). *Atlas of Grey Scale Ultrasonography*, Churchill Livingstone: Edinburgh.

BIBLIOGRAPHY

Taylor, R. B. (1965). Pluripotential stem cells in mouse embryo and liver, *Br. J. exp. Path.*, **46**, 376–83.

Taylor, S. (1968). *Calcitonin: Proceedings of a Symposium on Thyrocalcitonin and the 'C' cells* (editor), Heinemann: London.

Teir, H. and Rytömaa, T. (1967). *Control of Cellular Growth in Adult Organisms*, Academic Press: New York, London.

Teitelbaum, S. L., Moore, K. E. and Shieber, W. (1970). C cell follicles in the dog thyroid: demonstrated by *in vitro* perfusion, *Anat. Rec.*, **168**, 69–78.

Telford, D. and Stopford, J. S. B. (1934). Autonomic nerve supply of distal colon; anatomical and clinical study, *Br. med. J.*, **1**, 572–4.

Templeton, D. and Thulin, A. (1978). Secretory, motor and vascular effects in the sublingual gland of the rat caused by autonomic nerve stimulation, *Q. Jl exp. Physiol.*, **63**, 59–66.

Ten Cate, A. R. (1972). An analysis of Tomes' granular layer, *Anat. Rec.*, **172**, 137–47.

Ten Cate, A. R., Melcher, A. H., Pudy, G. and Wagner, D. (1970). The non-fibrous nature of the von Korff fibres in developing dentine. A light and electron microscope study, *Anat. Rec.*, **168**, 491–524.

Tench, E. N. (1936). Development of the anus in the human embryo, *Am. J. Anat.*, **59**, 333–46.

Tennyson, V. M. (1969). The fine structure of the developing nervous system. In: *Developmental Neurobiology* (Himwich, H. ed.), Part 2, Chapter 3, Thomas: Springfield, Ill.

Tennyson, V. M. (1970). The fine structure of the axon and growth cone of the dorsal root neuroblast of the rabbit embryo, *J. Cell Biol.*, **44**, 62–79.

Terni, T. (1922). Ricerche sulla struttura e sull'evoluzione del simpatico dell'uomo, *Monitore zool. ital.*, **33**, 63–72.

Terracol, J., Calvet, J., Granel, F., Ardouin, P. and Fabre, L. (1965). L'anatomie fonctionelle du larynx, *Biologie méd.*, **54**, 180–255.

Terzakis, J. A. (1963). The ultrastruture of normal human first trimester placenta, *J. Ultrastruct. Res.*, **9**, 268–84.

Terzian, H. and Ore, G. D. (1955). Syndrome of Klüver-Bucy reproduced in man by bilateral removal of the temporal lobes, *Neurology, Minneap.*, **5**, 373–80.

Tesi, D. and Forssmann, W. G. (1970). Untersuchungen am mesenterium der Ratte, *Anat. Anz.*, **126**, 365–73.

Teubér, H.-L., Battersby, W. S. and Bender, M. B. (1960). *Visual Field Defects After Penetrating Missile Wounds of the Brain*, Harvard University Press: Cambridge, Massachusetts.

Teuber, H. L. (1974). Why two brains? In: *The Neurosciences, Third Study Program* (Schmitt, F. O. and Worden, F. G. eds.), MIT Press: Cambridge, Massachusetts.

Thaemert, J. C. (1969). Fine structure of neuromuscular relations in mouse heart, *Anat. Rec.*, **163**, 575–86.

Thaysen, J. H., Thorn, N. A. and Schwartz, I. L. (1954). Excretion of sodium, potassium, chloride and carbon dioxide in human parotid saliva, *Am. J. Physiol.*, **178**, 155–9.

Thebesius, A. C. (1708). *Dissertatio de Circulo Sanguinis in Corde*, Lugdunum: Batavorum.

Theiler, K. (1957). Uber die Differenezierung der Rumpfmyotome beim Menschen und die Herkunft der Bauchwandmuskeln, *Acta anat.*, **30**, 842–64.

Thiery, J. P. and Bader, J. P. (1966). Ultrastructure des îlots de Langerhans du pancréas humain normal et pathologique, *Annls Endocr.*, **27**, 625–47.

Thom, R. (1969). A mathematical approach to morphogenesis: archetypal morphologies, *Wistar Inst. Symp. Mongr.*, **9**, 165–74.

Thom, R. (1975). *Structural Stability and Morphogenesis*, Benjamin Inc.: Reading, Mass.

Thomas, C. E. (1965). The ultrastructure of human amnion epithelium, *J. Ultrastruct. Res.*, **13**, 65–83.

Thomas, C. E. (1967). An electron- and light-microscope study of sinus structure in perfused rabbit and dog spleens, *Am. J. Anat.*, **120**, 527–52.

Thomas, P. K. (1963). The connective tissue of peripheral nerve: an electron microscope study, *J. Anat.*, **97**, 35–44.

Thompson, A. R. (1919). The maturation of the human ovum, *J. Anat.*, **53**, 172–208.

Thompson, J. S. and Thompson, M. W. (1973). *Genetics in Medicine*, Second edition, Saunders: Philadelphia.

Thompson, R. H. S. and King, E. J. (1969). *Biochemical Disorders in Human Disease*, Third edition, Churchill: London.

Thompson, W. D'Arcy (1942). *Growth and Form*, Cambridge University Press: Cambridge.

Thomas, H. (1940). Roentgen pelvimetry as a routine prenatal procedure, *Am. J. Obstet. Gynec.*, **40**, 891–905.

Thorel, C. (1909). Vorläufige Mitteilung über eine besondere Muskeln verbindung zwischen der Cava superior und dem Hisschen Bündel, *Munch. Med. Wochenschr.*, **56**, 2159.

Thorel, C. (1910). Über den Aufbaum des Sinusknotens und seine Verbindung mit der Cava superior und den Wenckebachschen Bündeln, *Munch. Med. Wochenschr.*, **57**, 183.

Thornton, M. W. and Schweisthal, M. R. (1969). The phrenic nerve: its terminal divisions and supply to the crura of the diaphragm, *Anat. Rec.*, **164**, 283–90.

Thuma, B. D. (1928). Studies on the diencephalon of the cat. I. The cytoarchitecture of the corpus geniculatum laterale, *J. comp. Neurol.*, **46**, 173–200.

Tiedemann, F. (1860). *Anatomie und Bildungsgeschichte des Gehirns im Foetus des Menschen*, Steinishen: Nuremburg.

Tierney, D. F. (1974). Lung metabolism and biochemistry, *Ann. Rev. Physiol.*, **36**, 209–31.

Tillmann, B., Pietzsch-Rohrschneider, I. and Hoenges, H. L. (1977). The human vocal cord surface, *Cell Tiss. Res.*, **185**, 279–83.

Tilney, F. (1933). Behavior in its relation to the development of the brain. Part II. Correlation between the development of the brain and behavior in the albino rat from embryonic states to maturity, *Bull. Neurol. Inst. N.Y.*, **3**, 252–358.

Tilney, F. and Riley, H. A. (1928). *The Brain From Ape to Man*, Hoeber: New York.

Tobias, P. V. (1970). Brain size, grey matter and race—fact or fiction? *Am. J. phys. Anthrop.*, **32**, 3–25.

Tobias, P. V. (1971). *The Brain in Hominid Evolution*, Columbia University Press: New York.

Tobin, C. E. (1966). Arteriovenous shunts in the peripheral pulmonary circulation in the human lung, *Thorax*, **21**, 197–204.

Tobon, H. and Salazar, H. (1974). Ultrastructure of the human mammary gland. I. Development of the fetal gland throughout gestation, *J. Clin. Endocr. Metab.*, **39**, 443–56.

Todd, T. W. (1920). Age changes in the pubic bone. I. The male white pubis, *Am. J. phys. Anthrop.*, **3**, 285–334.

Todd, T. W. (1920). Age changes in the pubic bone. II. The pubis of the male Negro-white hybrid. III. The pubis of the white female. IV. The pubis of the female Negro-white hybrid, *Am. J. phys. Anthrop.*, **4**, 1–70.

Todd, T. W. (1921). Age changes in the pubic bone. V. Mammalian pubic metamorphosis, *Am. J. phys. Anthrop.*, **4**, 333–406.

Todd, T. W. (1921). Age changes in the pubic bone. VI. The interpretation of variations in the symphyseal area, *Am. J. phys. Anthrop.*, **4**, 407–24.

Todd, T. W. (1931). Differential skeletal maturation in relation to sex, race, variability and disease, *Child Dev.*, **2**, 49–56.

Todd, T. W. (1937). *Atlas of Skeletal Maturation of the Wrist*, Mosby: St. Louis.

Todd, T. W. and Lyon, D. W. Jr. (1924). Endocranial suture closure, its progress and age relationship. I. Adult males of white stock, *Am. J. phys. Anthrop.*, **7**, 325–84.

Todd, T. W. and Lyon, D. W. Jr. (1925). Endocranial suture closure, its progress and age relationship. II. Ectocranial closure in adult males of white stock, *Am. J. phys. Anthrop.*, **8**, 23–45.

Todd, T. W. and Lyon, D. W. Jr. (1925). Endocranial suture closure, its progress and age relationship. III. Endocranial closure in adult males of Negro stock, *Am. J. phys. Anthrop.*, **8**, 47–71.

Todd, T. W. and Lyon, D. W. Jr. (1925). Endocranial suture closure, its progress and age relationship. IV. Ectocranial closure in adult males of Negro stock, *Am. J. phys. Anthrop.*, **8**, 149–68.

Todd, T. W. and D'Erico, J. Jr. (1928). The clavicular epiphyses, *Am. J. Anat.*, **41**, 25–50.

Todd, T. W. and Lindàla, A. (1928). Dimensions of the body; Whites and American Negros of both sexes, *Am. J. phys. Anthrop.*, **12**, 35–119.

Todd, T. W. and Tracy, B. (1930). Racial features in American negro cranium, *Am. J. phys. Anthrop.*, **15**, 53–110.

Toivonen, S. (1967). Mechanism of primary induction. In: *Morphological and Biochemical Aspects of Cytodifferentiation* (Hagen, E., Wechsler, W. and Zilliken, P. eds.), pp. 1–7 (*Exp. Biol. Med.*, **1**), Karger: Basel, New York.

Tomasch, J. and Malpass, A. J. (1958). The human motor trigeminal nucleus, *Anat. Rec.*, **130**, 91–102.

Tömböl, T. (1967). Short neurons and their synaptic relations in the specific thalamic nuclei, *Brain Res.*, **3**, 307–26.

Tominaga, Y. L. and Ikui, H. (1964). The fine structure of the arteriovenous crossing parts in the human retina, *Acta Soc. Ophthal. jap.*, **68**, 148–50.

Toncray, J. E. and Kreig, N. J. S. (1946). Nuclei of the human thalamus: comparative approach, *J. comp. Neurol.*, **85**, 421–59.

Töndury, G. (1943). Zur Anatomie der Halswirbelsäule. Gibt es Uncovertebralgelenke? *Z. Anat. EntwGesch.*, **112**, 448–59.

Tondury, G. (1958). *Entwicklungsgeschichte und Fehlbildungen der wirbesaule*, Thieme: Stuttgart.

Tonge, C. H. (1953). The early development of teeth, *Proc. R. Soc. Med.*, **46**, 313–18.

Tongerson, J. (1951). Developmental, genetic and evolutionary meaning of metopic suture, *Am. J. phys. Anthrop.*, **9**, 193–210.

Tonna, E. A. and Cronkite, E. P. (1962). Use of tritiated thymidine for the study of the origin of the osteoclast, *Nature, London.*, **190**, 495–6.

Törnqvist, G. (1967). The relative importance of the parasympathetic and sympathetic nervous systems for accommodation in monkeys, *Invest. ophthal.*, **6**, 612–17.

Török, B. (1954). Lebeudbeobachtung des hypophysenkreislaufes an hunden, *Acta Morph. Acad. Sci. Hung.*, **4**, 83–9.

Torr, J. B. D. (1957). *The Blood Supply of the Human Cord*, M.D. Thesis, University of Manchester.

Torrey, T. W. (1954). The early development of the human nephros, *Contr. Embryol.*, **35**, 175–97.

Torvik, A. (1957). The spinal projections from the nucleus of the solitary tract. An experimental study in the cat, *J. Anat.*, **91**, 314–22.

Torvik, A. and Brodal, A. (1954). The cerebellar projection of the perihypoglossal nuclei (nucleus intercalatus, nucleus praepositus hypoglossi and nucleus of Roller) in the cat, *J. Neuropath. exp. Neurol.*, **13**, 515–27.

Torvik, A. and Brodal, A. (1957). The origin of reticulospinal fibers in the cat. An experimental study, *Anat. Rec.*, **128**, 113–37.

Tournade, A., Maillot, C. and Koritke, J. G. (1972). Les veines superficielles du tronc cérébral chez l'homme, *Archs Anat. Histol. Embryol.*, **55**, 233–81.

Tournay, A. and Paillard, J. (1953). Electromyographie des muscles radiaux a l'état normal, *Revue neurol.*, **89**, 277–9.

Tousimis, A. J. and Fine, B. S. (1959). Ultrastructure of the iris: intercellular stromal components, *Archs Ophthal., N.Y.*, **62**, 974–6.

Tow, P. M. and Whitty, C. W. M. (1953). Personality changes after operations on the cingulate gyrus in man, *J. Neurol. Neurosurg. Psychiat.*, **16**, 186–93.

Tower, D. B. (1960). Chemical architecture of the central nervous system. In: *Handbook of Physiology*, Section 1, Volume 3 (Field, J. ed.), pp. 1793–1813, American Physiological Society: Washington.

Tower, S. S. and Richter, C. P. (1931). Injury and repair within the sympathetic nervous system; preganglionic neurons, *Archs Neurol. Psychiat., Chicago*, **26**, 485–95.

Townes, P. L. and Holtfreter, J. (1965). Directed movements and selective adhesion of embryonic amphibian cells. In: *Molecular and Cellular Aspects of Development* (Bell, E. ed.), pp. 3–39, Harper and Row: New York, Evanston, London.

Townes-Anderson, E. and Raviola, G. (1978). Degeneration and regeneration of autonomic nerve endings in the anterior part of rhesus monkey ciliary muscle, *J. Neurocytol.*, **7**, 583–600.

Toyoda, Y. and Chang, M. C. (1974). Fertilisation of rat eggs *in vitro* by epididymal spermatozoa and the development of eggs following transfer, *J. Reprod. Fertil.*, **36**, 9–22.

Tozer, F. M. (1911). On the presence of ganglion cells in the roots of III, IV and VI cranial nerves, *Proc. Physiol. Soc. Lond.*, 1910–1911 p. xv.

Tramezzani, J. H., Morita, E. and Chiocchio, S. R. (1971). The carotid body as a neuroendocrine organ involved in the control of erythropoiesis, *Proc. natn Acad. Sci. U.S.A.*, **68**, 52–5.

Traquair, H. M. (1948). *An Introduction to Clinical Perimetry*, Fifth edition, Kimpton: London.

Travill, A. A. (1964). Transmission of pressures across the elbow joint, *Anat. Rec.*, **150**, 243–7.

Travis, A. M. (1955). Neurological deficiencies following supplementary motor area lesions in *Macaca mulatta*, *Brain*, **78**, 174–98.

Trier, J. S. (1968). Morphology of the epithelium of the small intestine. In: *Handbook of Physiology* (Code, C. F. ed.), Section 6, Volume 3, pp. 1125–77. American Physiological Society: Washington.

Trier, J. S. and Rubin, C. E. (1965). Electron microscopy of the gut: a word of caution, *Gastroenterology*, **47**, 313–15.

Trier, J. S., Lorenzsonn, V. and Groehler, K. (1967). Pattern of secretion of Paneth cells of the small intestine of mice, *Gastroenterology*, **53**, 240–9.

Tripathi, R. C. (1970). Mechanism of the aqueous outflow across the trabecular wall of Schlemm's canal, *Expl Eye Res.*, **10**, 111–16.

Trolle, D. (1947). *Accessory Bones of the Human Foot*, MunsKgaard: Copenhagen.

Trotter, M. (1937). Accessory sacro-iliac articulations, *Am. J. phys. Anthrop.*, **22**, 247–61.

Trotter, M. and Gleser, G. C. (1958). A re-evaluation of estimation of stature based on measurements of stature taken during life and of long bones after death, *Am. J. phys. Anthrop.*, N.S. **16**, 79–123.

Trueta, J. (1957). The normal vascular anatomy of the femoral head during growth, *J. Bone Jt Surg.*, **39**B, 353–8.

Trueta, J., Barclay, A. E., Daniel, P. M., Franklin, K. J. and Prichard, M. M. L. (1947). *Studies of the Renal Circulation*, Blackwell: Oxford.

Trueta, J. and Morgan, J. D. (1960). The vascular contribution to osteogenesis. I. Studies by the injection method, *J. Bone Jt Surg.*, **42B**, 97–109.

Truex, R. C. and Bishof, J. K. (1958). Conducting system in human hearts with interventricular septal defect, *J. thorac. Surg.*, **35**, 421–39.

Truex, R. C., Smythe, M. Q. and Taylor, M. J. (1967). Reconstruction of the human sinoatrial node, *Anat. Rec.*, **159**, 371–8.

Truex, R. C. and Smythe, M. Q. (1967). Reconstruction of the human atrioventricular node, *Anat. Rec.*, **158**, 11–20.

Truex, R. C., Taylor, M. J., Smythe, M. Q. and Gildenberg, P. (1970). The lateral cervical nucleus of cat, dog and man, *J. comp. Neurol.*, **139**, 93–104.

Tsanev, R. (1975). Cell differentiation and the structure of chromatin. In: *Biochemistry of the Cell Nucleus: Mechanism and Regulation of Gene Expression* (Hidvegi, E. J. ed.), pp. 409–17, North Holland Press: Amsterdam.

Tsanev, R. and Sendov, B. (1971). Possible molecular mechanism for cell differentiation in multicellular organisms, *J. theoret. Biol.*, **30**, 337–93.

Tschumi, P. (1957). The growth of the hindlimb bud of *Xenopus laevis* and its dependence upon the epidermis, *J. Anat.*, **91**, 149–73.

Ts'o, M. O. and Friedman, E. (1967). The retinal pigment epithelium. I. Comparative histology, *Archs Ophthal., N.Y.*, **78**, 641–9.

Ts'o, M. O. and Friedman, E. (1968). The retinal pigment epithelium. III. Growth and development, *Archs Ophthal., N.Y.*, **80**, 214–16.

Tsuchida, U. (1906). Ueber die Ursprungskerne der Augenbewegungsnerven und über die mit diesen in Beziehung stehenden bahnen im Mittel- und Zwishenhirn; normal-anatomische, embryologische, pathologisch-anatomische und vergleichend-anatomische Untersuchungen, *Arb. hirnanat. Inst. Zürich*, **2**, 1–205.

Turnbull, I. M., Brieg, A. and Hassler, O. (1966). Blood supply of cervical spinal cord in man. A microangiographic cadaver study, *J. Neurosurg.*, **24**, 951–65.

Turnbull, W. D. (1970). Mammalian masticatory apparatus. Fieldiana: *Geology*, **18**, 149–356.

Turner, D. C., Wallimann, T. and Eppenberger, H. M. (1973). A protein that binds specifically to the M-line of skeletal muscle is identified as the muscle form of creatine kinase, *Proc. natn Acad. Sci. U.S.A.*, **70**, 702–5.

Turner, D. R. (1969). The vascular tree of the haemal node in the rat, *J. Anat.*, **104**, 481–94.

Turner, R. S. (1943). Chromatolysis and recovery of efferent neurons, *J. comp. Neurol.*, **79**, 73–8.

Turner, W. (1886). The index of the pelvic brim as a basis of classification, *J. Anat.*, **20**, 125–43.

Turner-Warwick, R. T. (1959). The lymphatics of the breast, *Br. J. Surg.*, **46**, 574–82.

Tyler, A. (1967). Problems and procedures of comparative gametology and syngamy. In: *Fertilisation* (Metz, C. B. and Monroy, A. eds.), Volume 1, pp. 2–26, Academic Press: New York, London.

Tyler, A. (1967). Masked messenger RNA and cytoplasmic DNA in relation to protein synthesis and processes of fertilisation and determination in embryonic development, *Devl Biol.*, Suppl. 1, 170–226.

Tyler, A. and Bishop, D. W. (1963). Immunological phenomena. In: *Mechanisms Concerned with Conception* (Hartman, C. G. ed.), pp. 397–481, Pergamon Press: Oxford, London, New York, Paris.

U

Uchizono, K. (1965). Characterisation of excitatory and inhibitory synapses in the central nervous system of the cat, *Nature, Lond.*, **207**, 642–3.

Uddenburg, N. (1968). Functional organisation of long, second-order afferents in the dorsal funiculus, *Expl Brain Res.*, **4**, 377–82.

Uhlenhuth, E., Hunter, D. W. T. and Loechel, W. E. (1952). *Problems in the Anatomy of the Pelvis*, Lippincott: Philadelphia.

Uitto, J. and Lichtenstein, J. R. (1976). Defects in the biochemistry of collagen in diseases of connective tissue, *J. Invest. Dermatol.*, **66**, 59–79.

Ullah, M. (1978). Localization of the phrenic nucleus in the spinal cord of the rabbit, *J. Anat.*, **125**, 377–86.

Underhill, B. M. L. (1955). Intestinal length in man, *Br. med. J.*, **2**, 1243–6.

Undi, F., Somogyi, B. and Kausz, M. (1973). Data on blood supply of spinal nerve roots, *Acta morph. Acad. Sci., Hung.*, **21**, 311–18.

Ungerstedt, U. (1971). Stereotaxic mapping of the monoamine pathways in the rat brain, *Acta physiol. Scand.*, **367**, 1–48.

Unterharnscheidt, F., Jachnik, D. and Gött, H. (1968). *Der Balkenmangel. Monographien aus dem Gesamtgebiete der Neurologie und Psychaitrie*, Heft **128**, pp. 1–232, Springer-Verlag: Berlin.

BIBLIOGRAPHY

Updyke, B. V. (1977). Topographic organization of the projections from cortical areas 17, 18 and 19 on to the thalamus, pretectum and superior colliculus in the cat, *J. comp. Neurol.*, **173**, 81–122.

Urist, M. R. (1966). Origins of current ideas about calcification, *Clin. Orthop. rel. Res.*, **44**, 13–39.

V

Vail, A. D. and Coller, F. C. (1967). The parathyroid glands: clinico-pathologic correlation of parathyroid disease as found in 200 unselected autopsies, *Missouri Med.*, **64**, 234–8.

Valentine, G. H. (1966). *The Chromosome Disorders*, Heinemann: London.

Valverde, F. (1961). Reticular formation of the pons and medulla oblongata: A Golgi study, *J. comp. Neurol.*, **116**, 71–100.

Valverde, F. (1961). A new type of cell in the lateral reticular formation of the brain stem, *J. comp. Neurol.*, **117**, 189–95.

Valverde, F. (1962). Reticular formation of the albino rat's brain stem; cytoarchitecture and corticofugal connections, *J. comp. Neurol.*, **119**, 25–53.

Valverde, F. (1965). *Studies on the Piriform Lobe*, Harvard University Press: Massachusetts.

van As, A. and Webster, I. (1972). The organisation of ciliary activity and mucus transport in pulmonary airways, *S. Afr. med. J.*, **46**, 347–50.

van Beusekom, G. T. (1955). *Fibre analysis of the anterior and lateral funiculi of the cord in the cat*, Thesis, University of Leiden.

van Buskirk, C. (1945). The seventh nerve complex, *J. comp. Neurol.*, **82**, 303–33.

van Campenhout, E. (1925). Etude sur le développement et la signification morphologique des ilots endocrine du pancréas chez l'embryon de mouton, *Archs Biol., Liège*, **35**, 45–88.

van Campenhout, E. (1956). Le développement embryonnaire comparé des nerfs olfactif et audatif, *Acta Oto-Rhin-Laryng Belg.*, **11**, 279–87.

Vane, J. R. (1969). The release and fate of vaso-active hormones in the circulation, *Br. J. Pharmacol.*, **35**, 209–42.

Van Gehuchten (1892). Contribution à l'étude des ganglions cérébro-spinaux, *La Cellule*, **8**, 209–31; 233–54.

van Hassel, H. J. and McMinn, R. G. (1972). Pressure differential favouring tooth eruption in the dog, *Archs oral Biol.*, **17**, 183–90.

van Mow, C., Lai, W. M. and Redler, I. (1974). Some surface characteristics of articular cartilage. I. A scanning electron microscopic study and a theoretical model for the dynamic interaction of synovial fluid and articular cartilage, *J. Biomech.*, **7**, 449–56.

Van Valen, L. (1974). Brain size and intelligence in man, *Am. J. phys. Anthrop.*, **40**, 417–24.

Varon, S. and Bunge, R. P. (1978). Trophic mechanisms in the peripheral nervous system, *Ann. Rev. Neurosci.*, **1**, 327–61.

Vassallo, G., Capella, C. and Solcia, E. (1971). Endocrine cells of the human gastric mucosa, *Z. Zellforsch.*, **118**, 49–67.

Vastesaeger, M. M., van der Straeten, P. P., Friart, J., Candaele, G., Ghys, A. and Bernard, R. M. (1957). Les anastomoses inter-coromariennes telles qu'elles apparaissent à la coronographie postmortem, *Acta cardiol.*, **12**, 365–401.

Vaughan, D. W. and Peters, A. (1974). Neuroglial cells in the cerebral cortex of rats from young adulthood to old age: an electron microscope study, *J. Neurocytol.*, **3**, 405–29.

Vaughan, G. M., Pelham, R. W., Pang, S. F., Loughlin, L. L., Wilson, K. M., Sandock, K. L., Vaughan, M. K., Koslow, S. H. and Reiter, R. J. (1976). Nocturnal elevation of plasma melatonin and urinary 5-hydroxyindoleacetic acid in young men: attempts at modification by brief changes in environmental lighting and sleep and by autonomic drugs, *J. Clin. Endocr. Metab.*, **42**, 752–64.

Vaughn, J. C. (1966). The relationship of the 'sphere chromatophile' to the fate of displaced histones following histone transition in rat spermiogenesis, *J. Cell Biol.*, **31**, 257–78.

Venning, P. (1956). Radiological studies of variations in the segmentation and ossification of the digits of the human foot. I. Variations in the number of phalanges and centers of ossification of the toes, *Am. J. phys. Anthrop.*, **14**, 1–34.

Venning, P. (1956). Radiological studies of variations in the segmentation and ossification of the digits of the human foot. II. Variations in length of the digit segments correlated with differences of segmentation and ossification of the toes, *Am. J. phys. Anthrop.*, **14**, 129–51.

Verbout, A. J. (1976). A critical review of the 'Neugliederung' concept in relation to the development of the vertebral column, *Acta biotheoret.*, **25**, 219–58.

Verhaart, W. J. C. (1953). The fibre structure of the cord in the cat, *Acta anat.*, **18**, 88–100.

Verhaart, W. J. C. and Mechelse, K. (1954). The Pedunculus Cerebri and the Capsula Interna, *Mschr. Psychiat. Neurol.*, **127**, 65–80.

Verhaart, W. J. C. and Beusekom, G. T. van (1958). Fibre tracts in the cord in the cat, *Acta physiol. neurol. scand.*, **33**, 359–76.

Verna, A. (1975). Observations on the innervation of the carotid body of the rabbit. In: *Peripheral Arterial Chemoreceptors* (Purves, M. J. ed.), pp. 75–99, Cambridge University Press: Cambridge.

Verney, E. B. (1947). The antidiuretic hormone and the factors which determine its release, *Proc. R. Soc. B.*, **135**, 25–106.

Verzár, F. and McDougall, E. J. (1936). *Absorption from the Small Intestine*, Longmans, Green: London.

Vidal, F. (1940). Pallidohypothalamic tract, or x bundle of Meynert, in the rhesus monkey, *Archs Neurol. Psychiat., Chicago*, **44**, 1219–23.

Vidić, B. (1968). The origin and course of the communicating branch of the facial nerve in the lesser petrosal nerve in man, *Anat. Rec.*, **162**, 511–16.

Vidić, B. and Young, P. A. (1967). Gross and microscopic observations on the communicating branch of the facial nerve to the lesser petrosal nerve, *Anat. Rec.*, **158**, 257–61.

Vilas, E. (1932). Über die Entwicklung der menschlichen Scheide, *Z. Anat. EntwGesch.*, **98**, 263–92.

Vilas, E. (1933). Über die Entwicklung des Utriculus prostaticus beim Menschen, *Z. Anat. EntwGesch.*, **99**, 399–421.

Villegas, G. M. (1964). Ultrastructure of the human retina, *J. Anat.*, **98**, 501–13.

Villiger, E. (1946). *Die Periphere Innervation*, Tenth edition, Schwabe: Basel.

Vincent, S. A. (1966). Postural control of urinary incontinence. The curtsey sign, *Lancet*, **2**, 631–2.

Vitti, M., Basmajian, J. V., Ouellette, P. L., Mitchell, D. L., Eastman, W. P. and Seaborn, R. D. (1975). Electromyographic investigations of the tongue and circumoral muscular sling, *J. Dent. Res.*, **54**, 844–9.

Vitti, M. and Basmajian, J. V. (1977). Integrated actions of masticatory muscles, *Anat. Rec.*, **187**, 173–90.

Vizoso, A. D. and Young, J. Z. (1948). Internode length and fibre diameter in developing and regenerating nerves, *J. Anat.*, **82**, 110–34.

Vlahovitch, B., Fuentes, J. M. and Verger, A. C. (1973). Angioarchitecture insulaire chez l'homme et chez les primates, *Arch. Anat. Path.*, **21**, 395–9.

Voetmann, E. (1949). On striations and surface area of human choroid plexuses, quantitative anatomical study, *Acta anat.*, **8**, 1–116.

Vogelberg, K. (1957). Die Lichtungsweite der Koronarostein an normalen und hypertrophen Herzen, *Z. Kreislaufforsch.*, **46**, 101–15.

Vogt, A. (1942). *Lehrbuch und Atlas der Spaltlampenmikroskopie des lebenden Auges. Teil 3. Iris, Glaskörper, Bindehaut*, Enke: Stuttgart.

Vogt, C. and Vogt, O. (1926). Die vergleichendarchitektonische und die vergleichendreizphysiologische Felderung der Grosshirnrinde unter besonderer Berücksichtigung der menschlichen, *Naturwissenschaften*, **14**, 1190–4.

Volkman, A. (1966). The production of monocytes and related cells (abstract). In: *Congr. Internat. Soc. Hematol.*, 11th, p. 28, N.S.W. Government Press: Sydney.

Volkman, A. and Gowans, J. L. (1965). The production of macrophages in the rat, *Br. J. exp. Path.*, **46**, 50–61.

von Bonin, G. (1950). *Essay on the Cerebral Cortex*, Thomas: Springfield, Ill.

von Bonin, G. (1962). Anatomical asymmetries of the cerebral hemispheres. In: *Interhemispheric Relations and Cerebral Dominance* (Mountcastle, V. B. ed.), Johns Hopkins University Press: Baltimore.

von Economo, C. and Koskinas, G. N. (1925). *Die Cytoarchitecktonik der Hirnrinde*, Springer-Verlag: Berlin.

Voneida, T. J. (1965). Visual loss following midline section through the mesencephalic tegmentum in cats, *Anat. Rec.*, **151**, 429 (abstract).

von Euler, U. S. (1936). On specific vasodilating and plain muscle stimulating substances from accessory genital glands in man and certain animals (prostaglandin and vesiglandin), *J. Physiol., London.*, **88**, 213–34.

von Hayek, H. (1960). *The Human Lung*, Hafner: New York.

von Lawzewitsch, I., Dickmann, G. H., Amezua, L. and Pardal, C. (1972). Cytobiological and ultrastructural characterisation of the human pituitary, *Acta anat.*, **81**, 286–316.

von Leden, H. and Moore, P. (1961). Vibratory pattern of the vocal cords in unilateral laryngeal paralysis, *Acta otolar.*, **53**, 493–506.

von Luschka, H. (1850). *Die Nerven in der harten Hirnhaut*, Laupp: Tubingen.

von Luschka, H. (1860). *Icones Nervorum Capitis*, Mohr: Heidelberg.

von Monakow, C. (1882). Weitere Mittheilungen über durch Exstirpation circumscripter Hirnrinden-regionen bedingte Entwicklungshemmungen des Kaninchengehirns, *Arch. Psychiat. NervKrankh.*, **12**, 141–56, 535–49.

von Monakow, C. (1905). *Gehirnpathologie*, Second edition, Hölder: Vienna.

Von Nooren, G. K. (1973). Histological studies of the visual system in monkeys with experimental amblyopia, *Invest. Ophthal.*, **12**, 727–38.

Voogd, J. (1964). *The Cerebellum of the Cat. Structure and Fibre Connexions*, Proefschr. Van Gorcum & Co. N.V.: Assen.

Voss, H. (1966). Untersuchungen über Vorkommen, Zahne und individuelle Variation der Muskelspindeln in den Muskeln des menschlichen Kehlkopfes, *Anat. Anz.*, **118**, 306–9.

Vrabec, F. (1952). Sur la question de l'endothélium de la surface antérieure de l'iris humain, *Ophthalmologica*, **123**, 20–30.

Vraa-Jensen, G. F. (1942). *The Motor Nucleus of the Facial Nerve*, Munksgaard: Copenhagen.

W

Wachstein, M. and Bradshaw, M. (1965). Histochemical localisation of enzymes acting in the kidneys of three mammalian species during their postnatal development, *J. Histochem. Cytochem.*, **13**, 44–56.

Wada, J. (1949). A new method for the determination of the side of cerebral speech dominance, *Igakuto Seibutsugaku*, **14**, 221–2.

Waddington, C. H. (1956). *Principles of Embryology*, Allen & Unwin: London.

Waddington, C. H. (1962). *New Patterns in Genetics and Development*, Columbia University Press: New York, London.

Waddington, C. H. (1966). The nucleolus—retrospect and prospect, *Nat. Cancer Inst. Monogr.*, **23**, 563–72.

Wagenvoort, C. A. and Wagenvoort, N. (1967). Arterial anastomoses, bronchopulmonary arteries and pulmobronchial arteries in perinatal lungs, *Lab. Invest.*, **16**, 13–24.

Waggener, J. D. and Beggs, J. (1967). The membranous coverings of neural tissues: an electron microscopy study, *J. Neuropath. exp. Neurol.*, **26**, 412–26.

Wagner, B. M. and Smith, D. E. (1967). *The Connective Tissue* (editors), Williams and Wilkins: Baltimore.

Waite, P. M. E. (1977). Normal nerve fibres in the barrel region of developing and adult mouse cortex, *J. comp. Neurol.*, **173**, 165–74.

Waites, G. M. H. (1970). Temperature regulation and the testis. In: *The Testis* (Johnson, A. D., Gomes, W. R. and Vandemark, N. L. eds.), Volume I, pp. 241–81, Academic Press: New York, London.

Waites, G. M. H. and Moule, G. R. (1961). Relation of vascular heat exchange to temperature regulation in the testis of the ram, *J. Reprod. Fert.*, **2**, 213–20.

Wakeley, C. P. C. (1929). A note on the architecture of the ilium, *J. Anat.*, **64**, 109–10P.

Wakeley, C. P. C. (1933). The position of the vermiform appendix as ascertained by an analysis of 10,000 cases, *J. Anat.*, **67**, 277–83.

Walberg, F. (1960). Further studies on the descending connections to the inferior olive. Reticulo-olivary fibers: an experimental study in the cat, *J. comp. Neurol.*, **114**, 79–87.

Walberg, F. and Brodal, A. (1953). Spino-pontine fibers in the cat. An experimental study, *J. comp. Neurol.*, **99**, 251–88.

Walberg, F. and Brodal, A. (1953). Pyramidal tract fibers from temporal and occipital lobes. An experimental study in the cat, *Brain*, **76**, 491–508.

Walberg, F., Bowsher, D. and Brodal, A. (1958). The termination of primary vestibular fibers in the vestibular nuclei in the cat. An experimental study with silver methods, *J. comp. Neurol.*, **110**, 391–419.

Waldeyer, H. (1888). *Das Gorilla-Rückenmark*, Akademie der Wissenschaften: Berlin.

Waldeyer, W. (1891). Ueber einige neuere Forschungen im Gebiete des Anatomie des Centralnervensystems, *Deutsche med. Wschr.*, **17**, 1213–18, 1244–6, 1267–9, 1287–9, 1331–2, 1352–6.

Walker, A. E. (1934). The thalamic projection to the central gyri in *Macacus rhesus*, *J. comp. Neurol.*, **60**, 161–84.

Walker, A. E. (1937). Experimental anatomical studies of the topical localisation within the thalamus of the chimpanzee, *Proc. K. ned. Akad. Wet.*, **40**, 198–206.

Walker, A. E. (1938). *The Primate Thalamus*, Chicago University Press: Chicago.

Walker, A. E. (1942). Somatotopic localisation of spinothalamic and secondary trigeminal tracts in mesencephalon, *Archs Neurol. Psychiat., Chicago*, **48**, 884–9.

Walker, A. E. (1943). Central representation of pain. In: *Pain* (Wolff, H. G., Gasser, H. S. and Hinsey, J. C. eds.) (*Res. Publs Ass. Res. nerv. ment. Dis.*, **23**, 63–85).

Walker, A. E. and Fulton, J. F. (1938). The thalamus of the chimpanzee. III. Metathalamus. Normal structure and cortical connections, *Brain*, **61**, 250–68.

Walker, P. S., Sikorski, J., Dowson, D., Longfield, M. D., Wright, V. and Buckley, T. (1969). Behaviour of synovial fluid on surfaces of articular cartilage. A scanning electron microscope study, *Ann. Rheum. Dis.*, **28**, 1–14.

Wall, P. D. (1962). The origin of a spinal cord slow potential, *J. Physiol., Lond.*, **164**, 508–26.

Wall, P. D. (1967). The laminar organization of dorsal horn and effects of descending impulses, *J. Physiol., Lond.*, **188**, 403–23.

Wall, P. D. (1970). The sensory and motor role of impulses travelling in the dorsal columns towards cerebral cortex, *Brain*, **93**, 505–24.

Wall, P. D. (1978). The gate control theory of pain mechanisms. A re-examination and re-statement, *Brain*, **101**, 1–18.

Waller, A. V. (1850). Experiments on the section of the glossopharyngeal and hypoglossal nerves of the frog, and observations of the alterations produced thereby in the structure of their primitive fibres, *Phil. Trans. R. Soc.*, Ser. B, **140**, 423–9.

Walls, E. W. (1947). The development of the specialised conducting tissue of the human heart, *J. Anat.*, **81**, 93–110.

Walls, E. W. (1958). Observations on the microscopic anatomy of the human anal canal, *Br. J. Surg.*, **45**, 504–12.

Walls, G. L. (1963). *The Verebrate Eye*, Hafner: New York, London.

Walmsley, R. (1937). The sheath of the rectus abdominis, *J. Anat.*, **71**, 404–14.

Walmsley, R. (1953). The development and growth of the intervertebral disc, *Edinb. med. J.*, **60**, 341–64.

Walmsley, R. (1958). The orientation of the heart and the appearance of its chambers in the adult cadaver, *Br. Heart J.*, **20**, 441–58.

Walmsley, T. (1915). The costal musculature, *J. Anat.*, **50**, 165–71.

Walmsley, T. (1928). Articular mechanism of diarthroses, *J. Bone J Surg.*, **10**, 40–5.

Walmsley, T. (1929). The heart. In: *Quain's Elements of Anatomy*, Volume IV, Part III, Longmans, Green & Co.: London.

Walter, W. G. (1953). *The Living Brain*, Duckworth: London.

Walther, J. B. and Rasmussen, G. L. (1960). Descending connections of auditory cortex and thalamus in the cat, *Fedn. Proc.*, **19**, 291.

Waltner, J. G. (1948). Barrier membrane of cochlear aqueduct; histologic studies on the patency of the cochlear aqueduct, *Archs otolar.*, **47**, 656–69.

Walton, A. J. (1931). Surgical treatment of parathyroid tumours, *Br. J. Surg.*, **19**, 285–91.

Wang, K. P. and Thai, H. P. (1965). An analysis of variations of the segmental vessels of the right lower lobe in 50 Chinese lungs, *Acta anat. sin.*, **8**, 408–23.

Wang, N. S. and Ying, W. L. (1977). A scanning electron microscopic study of alkali-digested human and rabbit alveoli, *Am. Rev. resp. Dis.*, **115**, 449–60.

Wanko, T. and Gavin, M. A. (1960). Electron microscope study of lens fibers, *J. biophys. biochem. Cytol.*, **6**, 97–102.

Wanko, T. and Gavin, M. A. (1961). Cell surfaces in the crystalline lens. In: *The Structure of the Eye* (Smelser, G. K. ed.), pp. 221–34, Academic Press: New York, London.

Wansbrough, H., Nakanishi, H. and Wood, C. (1967). Effect of epinephrine on human uterine activity *in vitro* and *in vivo*, *Obstet. Gynec.*, **30**, 779–89.

Warbrick, J. C. (1960). The early development of the nasal cavity and upper lip in the human embryo, *J. Anat.*, **94**, 351–62.

Ward, F. O. (1838). Outlines of Human Osteology, Renshaw: London.

Warner, F. D. and Satir, (1974). The structural basis of ciliary bend formation. Radial spoke positional changes accompanying microtubule sliding, *J. Cell Biol.*, **63**, 35–63.

Warr, W. B. (1966). Fiber degeneration following lesions in the anterior ventral cochlear nucleus of the cat, *Expl Neurol.*, **14**, 453–74.

Warren, J. M. and Akert, K. (1964). *The Frontal Granular Cortex and Behaviour* (editors), McGraw-Hill: New York.

Warshawsky, H., Leblond, C. P. and Droz, B. (1963). Synthesis and migration of proteins in the cells of the exocrine pancreas as revealed by specific activity determined from radioautographs, *J. Cell Biol.*, **16**, 1–28.

Wartenberg, H. and Stegner, H. E. (1960). Ueber die elektronenmikroskopische Feinstrukter des menschlichen Ovarialeies, *Z. Zellforsch. mikrosk. Anat.*, **52**, 450–74.

Wartenberg, H. and Holstein, A. F. (1975). Morphology of the spindle shaped body in the developing tail of human spermatids, *Cell Tiss. Res.*, **159**, 435–43.

Warwick, R. (1950). Study of retrograde degeneration in oculomotor nucleus of rhesus monkey, with a note on method of recording its distribution, *Brain*, **73**, 532–43.

Warwick, R. (1950). The relation of the direction of the mental foramen to the growth of the human mandible, *J. Anat.*, **84**, 116–20.

Warwick, R. (1951). A juvenile skull exhibiting duplication of the optic canals, *J. Anat.*, **85**, 289–91.

BIBLIOGRAPHY

Warwick, R. (1953). Observations upon certain reputed accessory nuclei of the oculomotor complex, *J. Anat.*, **87**, 46–53.

Warwick, R. (1953). Representation of the extra-ocular muscles in the oculomotor nuclei of the monkey, *J. comp. Neurol.*, **98**, 449–503.

Warwick, R. (1954). The ocular parasympathetic nerve supply and its mesencephalic sources, *J. Anat.*, **88**, 71–93.

Warwick, R. (1954). The peculiarities of ciliary ganglion neurons, *J. Anat.*, **88**, 555P.

Warwick, R. (1955). The so-called nucleus of convergence, *Brain*, **78**, 92–114.

Warwick, R. (1964). Oculomotor organization. In: *The Oculomotor System* (Bender, M. ed.), pp. 173–202, Harper & Row: New York.

Warwick, R. (1968). The skeletal remains. In: *The Romano-British Cemetery at Trentholme Drive, York* (Wenham, L. P. ed.), pp. 113–78, Ministry of Public Building and Works, Archaeological Report 5, H.M.S.O.: London.

Warwick, R. and Mitchell, G. A. G. (1956). Localization of the phrenic nucleus in the spinal cord of man, *J. comp. Neurol.*, **105**, 683–700.

Warwick, R. and Pond, J. B. (1968). Trackless lesions in nervous tissues produced by high intensity focused ultrasound (high frequency mechanical waves), *J. Anat.*, **102**, 387–406.

Washburn, S. L. (1947). The relation of the temporal muscle to the form of the skull, *Anat. Rec.*, **99**, 239–48.

Washburn, S. L. (1949). Sex differences in the pubic bone of Bantu and Bushman, *Am. J. phys. Anthrop.*, **7**, 425–32.

Wasserman, G. D. (1973). Molecular genetics and developmental biology, *Nature, New Biol.*, **245**, 163–5.

Watanabe, H. and Yamamoto, T. Y. (1974). Freeze-etch study of smooth muscle cells from vas deferens and taeniae coli, *J. Anat.*, **117**, 553–64.

Watanabe, I., Donahue, S. and Hoggatt, N. (1967). Method for electron microscopic studies of circulating human leukocytes and observations on their fine structure, *J. Ultrastruct. Res.*, **20**, 366–82.

Watanabe, S., Mikata, A. and Niki, R. (1971). Cell dynamic of the lymphatic system under the influence of antigen stimulation: difference between conventional and germ-free mice, *Saishin Igaku*, **26**, 1425–36.

Watanabe, Y. (1960). An experimental study on the coronary luminal communicating channels in coronary circulation, *Jap. Circ. J.*, **24**, 11–26.

Watson, D. M. S. (1917). The evolution of tetrapod shoulder girdle and forelimb, *J. Anat.*, **52**, 1–63.

Watson, J. D. (1976). *The Molecular Biology of the Gene*, Third edition, Benjamin: London.

Watson, J. D. and Crick, F. H. C. (1953). Genetical implications of the structure of deoxyribonucleic acid, *Nature, Lond.*, **171**, 964–7.

Watson, J. D. and Crick, F. H. C. (1953). Molecular structure of nucleic acids. A structure for deoxyribose nucleic acid, *Nature, Lond.*, **171**, 737–8.

Watterson, R. L. (1965). Structure and mitotic behaviour of the early neural tube. In: *Organogenesis* (DeHaan, R. L. and Hirsprung, H. eds.), Holt, Rinehart & Wilson: New York.

Watterson, R. L., Veneziano, P. and Bartha, A. (1956). Absence of a true germinal zone in neural tubes of young chick embryos as demonstrated by the colchicine technique, *Anat Rec.*, **124**, 379.

Watzka, M. (1955). Die Leydigschen Zwischenzellen im Funiculus spermaticus des Menschen, *Z. Zellforsch. mikrosk. Anat.*, **43**, 206–13.

Weale, R. A. (1970). Optical properties of photoreceptors, *Br. med. bull.*, **26**, 134–7.

Wearn, J. T. (1941). Morphological and functional alterations of coronary circulation, *Harvey Lect.*, **17**, 754–77.

Wearn, J. T., Mettier, S. R., Klumpp, T. G. and Zschiesche, L. J. (1933). The nature of the vascular communications between the coronary arteries and the chambers of the heart, *Am. Heart J.*, **9**, 143–64.

Weber, E. H. (1834). *De Pulsu, Resorptione, Auditu et Tactu*, Koehler: Leipzig.

Webster, H. de F. (1964). Some ultrastructural features of segmental demyelination and myelin regeneration in peripheral nerve, *Prog. Brain Res.*, **13**, 151–72.

Webster, H. de F. (1971). The geometry of peripheral myelin sheaths during their formation and growth in rat sciatic nerves, *J. Cell Biol.*, **48**, 348–67.

Webster, K. E. (1961). Cortico-striate interrelations in the albino rat, *J. Anat.*, **95**, 532–44.

Webster, K. E. (1965). The cortico-striate projection in the cat, *J. Anat.*, **99**, 329–37.

Webster, K. E. (1974). Changing concepts of the organization of the central visual pathways in birds. In: *Essays on the Nervous System* (Bellairs, R. and Gray, E. G. eds.), pp. 258–98, Clarendon Press: Oxford.

Webster, K. E. (1975). Structure and function of the basal ganglia. A non-clinical view, *Proc. R. Soc. Med.*, **68**, 203–10.

Webster, K. E. (1977). Somaesthetic pathways, *Br. med. Bull.*, **33**, 113–19.

Webster, K. E. (1978). The brainstem reticular formation. In: *The Biological Basis of Schizophrenia* (Hennings, G. and Hemmings, W. A. eds.), M.T.P. Press: Lancaster.

Webster, P. D., Black, U., Mainz, D. L. and Singh, M. (1977). Pancreatic acinar cell metabolism and function, *Gastroenterol.*, **73**, 1434–49.

Weddell, G. (1941). The pattern of cutaneous innervation in relation to cutaneous sensibility, *J. Anat.*, **75**, 346–67.

Weddell, G., Palmer, E. and Pallie, W. (1955). Nerve endings in mammalian skin, *Biol Rev.*, **30**, 159–95.

Wedgwood, M. (1966). The peripheral course of the inferior dental nerve, *J. Anat.*, **100**, 639–50.

Wee, E. L., Wolfson, L. G. and Zimmerman, E. F. (1976). Palate shelf movement in mouse embryo culture: evidence for skeletal and smooth muscle contractility, *Devl Biol.*, **48**, 91–103.

Weed, L. H. (1920). The experimental production of an internal hydrocephalus, *Carnegie Inst. Contrib. Embryol.*, **9**, no. 272, 425–46.

Weed, L. H. (1938). Meninges and cerebrospinal fluid, *J. Anat.*, **72**, 181–215.

Weibel, E. R. (1971). The mystery of 'non-nucleated plates' in the alveolar epithelium of the lung explained, *Acta anat.*, **78**, 425–43.

Weibel, E. R. (1974a). A note on differentiation and divisibility of alveolar epithelial cells, *Chest*, **65**, *Suppl. 4*, 19S–21S.

Weibel, E. R. (1974b). On pericytes, particularly their existence on lung capillaries, *Microvasc. Res.*, **8**, 218–35.

Weibel, E. R. and Gil, J. (1968). Electron microscopic demonstration of an extra-cellular duplex lining layer of alveoli, *Respir. Physiol.*, **4**, 42–57.

Weibel, E. R. and Elias, H. (1969). *Quantitative Methods in Morphology*, Springer-Verlag: Berlin.

Weibel, E. R., Gehr, P., Haies, D., Gil, J. and Bachofen, M. (1976). The cell population of the normal lung. In: *Lung Cells in Disease* (Bouhuys, A. ed.), pp. 3–16, North-Holland Publishing Co.: Amsterdam.

Weichert, R. F. (1970). The neural ectodermal origin of the peptide-secreting endocrine glands, *Am. J. Med.*, **49**, 232–41.

Weigert, C. (1882). Über eine neue Untersuchungs–methode des Zentral-vervensystems. In: *Gesammelte Abhandlungen* (1906), Volume 2, pp. 533–8.

Weil, A. J. (1965). The spermatozoa-coating antigen (SCA) of the seminal vesicle, *Ann. N.Y. Acad. Sci.*, **124**, 267–9.

Weiner, J., Spiro, D. and Loewenstein, W. R. (1965). Ultrastructure and permeability of nuclear membranes, *J. Cell Biol.*, **27**, 107–18.

Weiner, N. and Schadé, J. P. (1963). *Nerve, Brain and Memory Models* (editors), Elsevier: Amsterdam.

Weinmann, J. P. and Sicher, H. (1955). *Bone and Bones—Fundamentals of Bone Biology*, Second edition, Kimpton: London.

Weinstock, A. and Albright, J. T. (1967). The fine structure of mast cells in normal human gingiva, *J. Ultrastruct. Res.*, **17**, 245–56.

Weintraub, W. (1953). *Thèse de l'Université de Genève*, Paris.

Weisberg, J. A. and Rustioni, A. (1977). Cortical cells projecting to the dorsal column nuclei of Rhesus monkey, *Exp. Brain Res.*, **28**, 521–8.

Weisengreen, H. H. (1975). Observation of the articular disc, *Oral Surg.*, **40**, 113–21.

Weisl, H. (1953). *The relation of movement to structure in the sacroiliac joint*, Ph. D. Thesis, University of Manchester.

Weisl, H. (1954). The ligaments of the sacro-iliac joint examined with particular reference to their function, *Acta anat.*, **20**, 201–13.

Weisl, H. (1954). The articular surfaces of the sacro-iliac joint and their relation to the movements of the sacrum, *Acta anat.*, **22**, 1–14.

Weisl, H. (1955). Movements of the sacro-iliac joint, *Acta anat.*, **23**, 80–91.

Weiss, C., Rosenberg, L. and Helfert, A. J. (1968). An ultrastructural study of normal young adult human articular cartilage, *J. Bone Jt Surg.*, **50A**, 663–74.

Weiss, L. (1957). A study of the structure of splenic sinuses in man and the albino rat with the light microscope and the electron microscope, *J. biophys. biochem. Cytol.*, **3**, 599–610.

Weiss, L. (1967). *The Cell Periphery, Metastasis and other Contact Phenomena*, North Holland Publishing Company: Amsterdam.

Weiss, P. (1941). Nerve patterns: The mechanics of nerve growth, *Growth*, Suppl. 5, 163–203.

Weiss, P. (1950). An introduction to genetic neurology. In: *Genetic Neurology*, pp. 1–39, Chicago University Press: Chicago.

Weiss, P. (1961). Guiding principles in cell locomotion and cell aggregation, *Expl Cell Res.*, Suppl. 8, 260–81.

Weiss, P. (1970). Neural development in biological perspective. In: *The Neurosciences, a Second Study Program* (Schmitt, F. O., Quarton, G. C., Melnechuk, T. and Adelman, G. eds.), pp. 53–61, Rockefeller University Press: New York.

Weiss, P. and Hiscoe, H. B. (1948). Experiments on the mechanism of nerve growth, *J. exp. Zool.*, **107**, 315–96.

Weller, G. L. Jr. (1933). Development of the thyroid, parathyroid and thymus glands in man, *Contr. Embryol.*, **24**, 93–139.

Wellings, S. R. and Philip, J. R. (1964). The function of the Golgi apparatus in lactating cells of the BALB/cCrgl mouse. An electron microscopic and autoradiographic study, *Z. Zellforsch. mikrosk. Anat.*, **61**, 871–82.

Wells, L J. (1954). Development of the human diaphragm and pleural sacs, *Contr. Embryol.*, **35**, 107–34.

Wen, C. Y., Tan, C. K. and Wong, W. C. (1977). Presynaptic dendrites in the cuneate nucleus of the monkey (Macaca fascicularis), *Neurosci. Lett.*, **5**, 129–32.

Wenckebach, K. F. (1908). Beitrage zur Kenntnis der menschlichen Herztätigkeit, *Arch. Anat. Physiol.*, **3**, 53.

Wende, S., Nakayama, N. and Schwerdtfeger, P. (1975). The internal auditory artery: (embryology, anatomy, angiography, pathology), *J. Neurol.*, **210**, 21–31.

Wendell-Smith, C. P. (1967). Studies on the morphology of the pelvic floor, Ph.D. Thesis, University of London.

Wendell-Smith, C. P. and Wilson, P. M. (1977). Musculature of the pelvic floor. In: *Scientific Foundations of Obstetrics and Gynaecology* (Philipp, E. E., Barnes, J. and Newton, M. eds.), Second edition, pp. 78–84, Heinemann Medical: London.

Wendler, D. (1968). Histologisch-histochemische Befunde an der Schleimhaut des Ductus deferens (pars funicularis) beim geschlectsreifen Mann, *Acta histochem.*, **31**, 48–69.

Wendt, R. and Albe-Fessard, D. (1962). Sensory responses of the amygdala with special reference to somatic afferent pathways. In: *Physiologie de l'hippocampe*, pp. 171–200, Ed. Centre National de la Recherche Scientifique: Paris.

Wenisch, H. J. (1976). Retino-hypothalamic projections in the mouse: electron microscopic and iontophoretic investigations of hypothalamic and optic centres, *Cell Tiss. Res.*, **167**, 547–61.

Wennberg, E. and Weiss, L. (1968). Splenic erythroclasia: an electron microscopic study of hemoglobin H disease, *Blood*, **31**, 778–90.

Wentink, G. H. (1976). The action of the hind limb musculature of the dog in walking, *Acta anat.*, **96**, 70–80.

Werner, S. (1969). *The Thyroid* (editor), Third edition, Harper and Row: New York.

Werner, S. (1978). Immune system. III. Role in thyroid disease. In: *The Thyroid* (Werner, S. and Ingbar, S. H. eds.), pp. 615–23, Harper and Row: New York.

Wertheim, M. G. (1847). Mémoire sur l'élasticité et la cohésion des principaux tissus du corps humain, *Ann. Chim. Phys.*, **21**, 385.

Wessels, N. K., Spooner, B. S., Ash, J. F., Bradley, M. O., Luduena, M. A., Taylor, L. E., Wrenn, J. T. and Yamada, K. M. (1971). Microfilaments in cellular and developmental processes, *Science, N. Y.*, **171**, 135–43.

Wesson, M. B. (1951). Rationale of prostatectomy, *Am. J. Surg.*, **82**, 714–19.

West, G. B., Shepherd, D. M., Hunter, R. B. and MacGregor, A. R. (1955). The function of the organs of Zuckerkandl, *Clin. Sci.*, **12**, 317–26.

Weston, J. A. (1970). The migration and differentiation of neural crest cells. In: *Advances in Morphogenesis* (Abercrombie, M. and Brachet, J. eds.), Volume 8, pp. 41–114, Academic Press: New York, London.

Wetzel, B. K., Spicer, S. S. and Wollman, S. H. (1965). Changes in fine structure and acid phosphatase localization in rat thyroid cells following thyrotropin administration, *J. Cell Biol.*, **25**, 593–618.

Wharton, L. R. (1932). Innervation of ureter, with respect to denervation, *J. Urol.*, **28**, 639–73.

Wharton Young, M. (1952). The termination of the perilymphatic duct, *Anat. Rec.*, **112**, 404–5.

Wharton Young, M. (1953). The perilymphatic sac, *Anat. Rec.*, **115**, 419–20.

Whillis, J. (1930). A note on the muscles of the palate and the superior constrictor, *J. Anat.*, **65**, 92–5.

Whillis, J. (1931). Lower end of the oesophagus, *J. Anat.*, **66**, 132–3P.

Whillis, J. (1946). Movements of the tongue in swallowing, *J. Anat.*, **80**, 115–16.

White, E. G. (1935). Die Struktur des Glomus caroticum, *Beitr. path. Anat.*, **96**, 177–227.

White, J. C., Smithwick, R. H. and Simeone, F. A. (1952). *The Autonomic Nervous System*, Third edition, Kimpton: London.

White, J. W. (1943). Torsion of achilles tendon; its surgical significance, *Archs Surg.*, **46**, 784–7.

White, L. E. (1959). Ipsilateral afferents to the hippocampal formation in the albino rat. I. Cingulum projections, *J. comp. Neurol.*, **113**, 1–42.

White, L. E. Jr. (1965). A morphological concept of the limbic lobe, *Int. Rev. Neurobiol.*, **8**, 1–34.

White, M. J. D. (1973). *Animal Cytology and Evolution*, Third edition, Cambridge University Press: Cambridge.

Whitehouse, H. L. K. (1973). *Towards an Understanding of the Mechanism of Heredity*, Third edition, Arnold: London.

Whitehouse, H. L. K. and Hastings, P. J. (1965). The analysis of genetic recombination on the polaron hybrid DNA model, *Genet. Res.*, **6**, 27–92.

Whitehouse, W. J. (1975). Scanning electron micrographs of cancellous bone from the human sternum, *J. Pathol.*, **116**, 213–24.

Whitehouse, W. J. (1977). Cancellous bone in the anterior part of the iliac crest, *Calc. Tiss. Res.*, **23**, 67–76.

Whitehouse, W. J. and Dyson, E. D. (1974). Scanning electron microscope studies of trabecular bone in the proximal end of the human femur, *J. Anat.*, **118**, 417–44.

Whitelaw, M. J. (1933). Tubal contractions in relation to estrus cycle as determined by uterotubal insufflation, *Am. J. Obstet. Gynec.*, **25**, 475–84.

Whiteley, H. J. and Stoner, H. B. (1957). The effect of pregnancy on the human adrenal cortex, *J. Endocr.*, **14**, 325–34.

Whitfield, I. C. (1967). *The Auditory Pathway*, Arnold: London.

Whitfield, I. C. and Evans, E. F. (1965). Responses of auditory cortical neurons to changing frequency, *J. Neurophysiol.*, **28**, 655–72.

Whitlock, D. G. and Nauta, W. J. H. (1956). Subcortical projections from the temporal neocortex in *Macaca mulatta*, *J. comp. Neurol.*, **106**, 183–212.

Whitnall, S. E. (1911). The relation of the lacrimal fossa to the ethmoidal cells, *Ophthal. Rev.*, **30**, 321–5.

Whitnall, S. E. (1932). *Anatomy of the Human Orbit*, Second edition, Oxford University Press: London.

Whittaker, D. K. and Adams, D. (1971). The surface layer of human foetal skin and oral mucosa: a study by scanning and transmission electron microscopy, *J. Anat.*, **108**, 453–64.

Whittier, J. R. and Mettler, F. A. (1949). Studies on the subthalamus of the rhesus monkey. I. Anatomy and fibre connections of the subthalamic nucleus of Luys, *J. comp. Neurol.*, **90**, 281–318.

Widén, L. and Ajmone Marsan, C. (1961). Action of afferent and corticofugal impulses on single elements in the dorsal lateral geniculate nucleus. In: *Neurophysiologie und Psychophysik des visuellen Systems* (Jung, R. and Kornhuber, H. eds.), pp. 125–32, Springer-Verlag: Berlin.

Wiesel, T. N. and Hubel, D. H. (1963a). Effects of visual deprivation on morphology and physiology of cells in the cat's lateral geniculate body, *J. Neurophysiol*, **26**, 978–93.

Wiesel, T. N. and Hubel, D. H. (1963b). Single cell responses in striate cortex of kittens deprived of vision in one eye, *J. Neurophysiol.*, **26**, 1003–17.

Wiesenhaan, P. F. (1972). Fetography, *Am. J. Obstet. Gynec.*, **113**, 819–22.

Wilde, C. E. Jr. (1961). The differentiation of vertebrate pigment cells. In: *Advances in Morphogenesis* (Abercrombie, M. and Brachet, J. eds.), Volume 1, pp. 267–300, Academic Press: New York, London.

Wilde, F. R. (1949). Anal intermuscular spasm, *Br. J. Surg.*, **36**, 279–85.

Wilde, F. R. (1951). Perivascular neural pattern of femoral region, *Br. J. Surg.*, **39**, 97–105.

Wiles, P. (1935). Movements of lumbar vertebrae during flexion and extension, *Proc. R. Soc. Med.*, **28**, 647–51.

Wilkin, P. and Bursztein, M. (1958). Etude quantitative de l'évolution au cours de la grossesse, de la superficie de la membrane d'échange du placenta humain. In: *Le Placenta Humain* (Snoek, J. ed.), pp. 211–48, Masson: Paris.

Wilkinson, H. J. (1927). Argyll-Robertson pupil: contribution toward its understanding, *Med. J. Aust.*, **1**, 267–72.

Wilkinson, J. L. (1953). The insertions of the flexores pollicis longus et digitorum profundus, *J. Anat.*, **87**, 75–88.

Williams, A. F. (1951). Nerve supply of laryngeal muscles, *J. Lar. Otol.*, **65**, 343–8.

Williams, A. F. (1954). Recurrent laryngeal nerve and the thyroid gland, *J. Lar. Otol.*, **68**, 719–25.

Williams, E. E. (1959). Gadow's arcualia and the development of tetrapod vertebrae, *Q. Rev. Biol.*, **34**, 1–32.

Williams, J. F. and Svensson, N. L. (1968). A force analysis of the hip joint, *Bio-Med. Eng.*, **3**, 365–70.

Williams, P. L. and Kashef, R. (1968). Asymmetry of the node of Ranvier, *J. Cell Sci.*, **3**, 341–56.

Williams, P. L. and Hall, S. M. (1970). *In vivo* observations on mature, myelinated nerve fibres of the mouse, *J. Anat.*, **107**, 31–8.

Williams, P. L. and Hall, S. M. (1971). Prolonged *in vivo* observations of normal peripheral nerve fibres and their acute reactions to crush and local trauma, *J. Anat.*, **108**, 397–408.

Williams, P. L. and Wendell-Smith, C. P. (1971). Some parametric variations between peripheral nerve fibre populations, *J. Anat.*, **109**, 505–26.

Williams, T. H. (1967). Electron microscopic evidence for an autonomic interneuron, *Nature, Lond.*, **214**, 309–10.

Williams, T. H. and Palay, S. L. (1969). Ultrastructure of the small neurones in the superior cervical ganglion, *Brain Res.*, **15**, 17–34.

BIBLIOGRAPHY

Williams, T. H., Black, A. C. Jr., Chiba, T. and Bhalla, R. C. (1975). Morphology and biochemistry of small intensely fluorescent cells of sympathetic ganglia, *Nature, Lond.*, **256**, 315–17.

Williamson, A. R. (1976). The biological origin of antibody diversity, *Ann. Rev. Biochem.*, **45**, 467–500.

Willier, B. H., Weiss, P. A. and Hamburger, V. (1955). *Analysis of Development* (editors), Saunders: Philadelphia, London.

Willis, A. G. and Tange, J. D. (1959). Studies on the innervation of the carotid sinus of man, *Am. J. Anat.*, **104**, 87–114.

Willis, A. G. and Tange, J. D. (1959). The argentophil cells of the human carotid body, *Am. J. Anat.*, **105**, 141–64.

Willis, R. A. (1936). Growth of embryo bones transplanted whole in rat's brain, *Proc. R. Soc. B.*, **120**, 496–8.

Willis, T. (1664). *Cerebri Anatome*, Martyn and Allestry: London.

Willis, T. A. (1949). Nutrient arteries of the vertebral bodies, *J. Bone Jt Surg.*, **31A**, 538–40.

Willis, W. D. and Willis, J. C. (1966). Properties of interneurons in the ventral spinal cord, *Archs ital. biol.*, **104**, 354–86.

Wilsman, N. J. and Van Sickle, D. C. (1972). Cartilage canals, their morphology and distribution, *Anat. Rec.*, **173**, 79–93.

Wilson, A., Obrist, A. R. and Wilson, H. (1953). Some effects of extracts of thymus glands removed from patients with myasthenia gravis, *Lancet*, **2**, 368–71.

Wilson, G. H. (1920). *A Manual of Dentral Prosthetics*, Kimpton: London.

Wilson, H. G. (1928). Postnatal development of the lung, *Am. J. Anat.*, **41**, 97–122.

Wilson, M. E. and Cragg, B. G. (1967). Projections from the lateral geniculate nucleus in the cat and monkey, *J. Anat.*, **101**, 677–92.

Winckler, G. (1960). Remarks on the histological structure of the leptomeninx in man, *Archs Anat. Histol. Embryol.*, **43**, 259–77.

Winckler, G. (1972). Remarques sur la structure de l'artère vertébrale, *Quad. Anat. Pract.*, **28**, 105–15.

Winer, J. A., Diamond, I. T. and Raczkowski, D. (1977). Subdivisions of the auditory cortex of the cat: the retrograde transport of horseradish peroxidase to the medial geniculate body and posterior thalamic nuclei, *J. comp. Neurol.*, **176**, 387–418.

Winer, J. W. (1977). A review of the status of the horseradish peroxidase method in neuroanatomy, *Behavioral Rev.*, **1**, 45–54.

Wingate, D. (1976). The eupeptide system: a general theory of gastrointestinal hormones, *Lancet*, **i**, 529–32.

Wingerd, J., Peritz, E. and Sproul, A. (1974). Race and stature differences in the skeletal maturation of the hand and wrist, *Ann. hum. Biol.*, **1**, 201–9.

Wintrobe, M. M. (1974). *Clinical Hematology*, Seventh edition, Lea & Febiger: Philadelphia.

Wischnitzer, S. (1973). The submicroscopic morphology of the interphase nucleus, *Int. Rev. Cytol.*, **34**, 1–48.

Wislocki, G. B. (1929). On the placentation of primates, with a consideration of the phylogeny of the placenta, *Contr. Embryol.*, **20**, 51–80.

Wislocki, G. B. (1937). The meningeal relations of the hypophysis cerebri. II. An embryological study of the meninges and blood vessels of the human hypophysis, *Am. J. Anat.*, **61**, 95–130.

Wislocki, G. B. and King, L. S. (1936). The permeability of the hypophysis and hypothalamus to vital dyes, with a study of the hypophyseal vascular supply, *Am. J. Anat.*, **58**, 421–72.

Wislocki, G. B. and Streeter, G. L. (1938). On the placentation of the macaque (*Macaca mulatta*), from the time of implantation until the formation of the definitive placenta, *Contr. Embryol.*, **27**, 1–66.

Wislocki, G. B. and Dempsey, E. W. (1939). Remarks on the lymphatics of the reproductive tract of female rhesus monkey (*Macaca mulatta*), *Anat. Rec.*, **75**, 341–63.

Wislocki, G. B. and Leduc, E. (1953). The cytology and histochemistry of the subcommissural organ and Reissner's fiber in rodents, *J. comp. Neurol.*, **97**, 515–44.

Wislocki, G. B. and Ladman, A. J. (1958). The fine structure of the mammalian choroid plexus. In: *The Cerebrospinal Fluid* (Wolstenholme, G. E. W. and O'Connor, C. M. eds.), pp. 55–74, Ciba Foundation Symposium, Churchill: London.

Witkovsky, P., Shakib, M. and Ripps, H. (1974). Inter-receptoral junctions in the teleost retina, *Invest. ophthal.*, **13**, 996–1009.

Witschi, E. (1948). Migration of the germ cells of human embryos from the yolk sac to the primitive gonadal folds, *Contr. Embryol.*, **32**, 67–80.

Witschi, E. (1951). Embryogenesis of the adrenal and reproductive glands, *Recent Prog. Horm. Res.*, **6**.

Wittmann, H. G. (1976). Structure, function and evolution of ribosomes, *Europ. J. Biochem.*, **61**, 1–13.

Wladmirow, B. (1968). Arterial sources of blood supply of the knee joint in man, *Acta med.*, **47**, 1–10.

Wolfe, D. E., Potter, L. T., Richardson, K. C. and Axelrod, J. (1962). Localising tritiated norepinephrine in sympathetic axons by electron microscope autoradiography, *Science, N.Y.*, **138**, 440–2.

Wolfe, H. J., Voelkel, E. F. and Tashjian, A. H. Jr. (1974). Distribution of calcitonin-containing cells in the normal adult human thyroid gland: a correlation of morphology with peptide content, *J. Clin. Endocrinol. Metab.*, **38**, 688–94.

Wolff, C. F. (1781). *Acta acad. scient. Petropol*, Part 1, p. 211.

Wolff, J. (1964). Transport of iodide and other anions in the thyroid gland, *Physiol Rev.*, **44**, 45–90.

Wolff, J. R. (1976). An ontogentically defined angioarchitecture of the neocortex. (Proceedings), *Arzneim Forsch.*, **26**, 1239.

Wolff, J. and Williams, J. A. (1973). The role of microtubules and microfilaments in thyroid secretion, *Rec. Prog. Horm. Res.*, **29**, 229–85.

Wolffson, D. M. (1950). Scapula shape and muscle function, with special reference to vertebral border, *Am. J. phys. Anthrop.*, **8**, 331–42.

Wolinsky, H. and Glagov, S. (1964). Structural basis for the static mechanical properties of the aortic media, *Circulation Research*, **14**, 400.

Wolinsky, H. and Glagov, S. (1967a). A lamellar unit of aortic medial structure and function in mammals, *Circ. Res.*, **20**, 99–111.

Wolinsky, H. and Glagov, S. (1967b). Nature of species differences in the medial distribution of aortic vasa vasorum in mammals, *Circ. Res.*, **20**, 409–21.

Wollman, S. H. and Nève, P. (1971). Ultimobranchial follicles in the thyroid glands of rats and mice, *Rec. Prog. Horm. Res.*, **27**, 213–34.

Wolpert, L. (1971). Cell movement and cell contact, *Sci. Basis Med. Ann. Rev.*, 81–98.

Wolstenholme, G. E. W. and O'Connor, M. (1960). *Haemopoiesis* (editors), Ciba Foundation Symposium, Churchill: London.

Wolstenholme, G. E. W. and Knight, J. (1970). *Sensorineural Hearing Loss* (editors), Ciba Foundation Symposium, Churchill: London.

Wolstenholme, G. E. W. and Knight, J. (1971). *The Pineal Gland* (editors), Ciba Foundation Symposium, Churchill-Livingstone: London.

Wolter, J. R. (1959). Glia of the human retina, *Am. J. Ophthal.*, **48**, 370–93.

Wolter, J. R. and Liss, L. (1956). Zentrifugale (antidrome) Nervenfasern im menschlichen Sehnerven, *Albrecht v. Graefes Arch. Ophthal.*, **158**, 1–7.

Woo, J.-K. (1949). Ossification and growth of the human maxilla, premaxilla and palate bone, *Anat. Rec.*, **105**, 737–62.

Wood, J. G. (1967). The relationship of nucleotidase activity to catecholamine storage sites in adrenomedullary tissue, *Am. J. Anat.*, **121**, 671–704.

Wood, N. K., Wragg, L. E. and Stuteville, O. H. (1967). The premaxilla. Embryological evidence that it does not exist in man, *Anat. Rec.*, **158**, 485–90.

Wood, N. K., Wragg, L. E., Stuteville, O. H. and Oglesby, R. J. (1969). Osteogenesis of the human upper jaw. Proof of the non-existence of a separate pre-maxillary centre, *Archs oral Biol.*, **14**, 1331–9.

Wood, P. and Bunge, R. (1975). Evidence that sensory axons are mitogenic for Schwann cells, *Nature, Lond.*, **256**, 662–4.

Wood, W. B. (1973). Post natal ossification timing in two population groups from Papua, New Guinea, *J. Anat.*, **116**, 482P.

Woodburne, R. T. (1956). The sacral parasympathetic innervation of the colon, *Anat. Rec.*, **124**, 67–76.

Woodburne, R. T. (1962). Segmental anatomy of the liver: blood supply and collateral circulation, *Univ. Mich. Med. Bull.*, **28**, 189–99.

Woodburne, R. T., Crosby, E. C. and McCotter, R. E. (1946). The mammalian hindbrain and isthmus region. II. The fiber connections. A. The relations of the tegmentum of the midbrain with the basal ganglia in *Macaca mulatta*, *J. comp. Neurol.*, **85**, 67–92.

Woodburne, R. T. and Olsen, L. L. (1951). Arteries of the pancreas, *Anat. Rec.*, **111**, 255–70.

Woodruff, J. D. and Pauerstein, C. J. (1969). *The Fallopian Tube*, Williams and Wilkins Co.: Baltimore.

Woollam, D. H. M. and Millen, J. W. (1960). The modification of the activity of certain agents exerting a deleterious effect on the development of the mammalian embryo. In: *Congenital Malformations* (Wolstenholme, G. E. W. and O'Connor, C. M. eds.), pp. 158–72, Ciba Foundation Symposium, Churchill: London.

Woollard, H. H. (1926). The innervation of the heart, *J. Anat.*, **60**, 345–73.

Woollard, H. H., Weddell, G. and Harpman, J. A. (1940). Observations on the neurohistological basis of cutaneous pain, *J. Anat.*, **74**, 413–40.

Woolsey, C. N. (1964). Cortical localization as defined by evoked potential and electrical stimulation studies. In: *Cerebral Localization and Organization* (Schaltenbrand, G. and Woolsey, C. N. eds.), pp. 17–32, University of Wisconsin Press: Madison.

Woolsey, C. N. and Walzl, E. M. (1942). Topical projection of nerve fibers from local regions of the cochlea to the cerebral cortex of the cat, *Bull. Johns Hopkins Hosp.*, **71**, 315–44.

Woolsey, C. N., Settlage, B. H., Meyer, D. R., Spencer, W., Hamuy, T.

P. and Travis, A. M. (1952). Patterns of localisation in precentral and 'supplementary' motor areas and their relation to the concept of a premotor area, *Res. Publs Ass. Res. nerv. ment. Dis.*, **30**, 238–64.

Woolsey, T. A. and van der Loos, H. (1970). The structural organisation of layer IV in the somatosensory region (SI) of mouse cerebral cortex. The description of a cortical field composed of discrete cytoarchitectonic units, *Brain Res.*, **17**, 205–42.

World Health Organisation (1972). *Genetic Disorders*, WHO Technical Report Series 497, Geneva.

Wormsley, K. G. (1979). Pancreatic secretion: physiological control. In: *Scientific Basis of Gastroenterology* (Duthie, H. L. and Wormsley, K. G. eds.), pp. 199–248, Churchill Livingstone: Edinburgh, London.

Woźniak, W. (1966). Odcinki krzyzowe pni wspolczulnych (pies, kot, czlowiek), *Folia morph.*, **25**, 433–40.

Woźniak, W. and Skowrońska, U. (1967). Comparative anatomy of the pelvic plexus in the cat, dog, rabbit, macaque and man, *Anat. Anz.*, **120**, 457–73.

Wright, D. M. and Moffett, B. C. (1974). The postnatal development of the human temporomandibular joint, *Am. J. Anat.*, **141**, 235–50.

Wright, N. L. (1969). Dissection study and mensuration of the human aortic arch, *J. Anat.*, **104**, 377–85.

Wright, R. H. (1964). Odor and molecular vibration: the far infra-red spectra of some perfume chemicals, *Ann. N. Y. Acad. Sci.*, **116**, 552–8.

Wulle, K. G. and Lerche, W. (1967). Zur Feinstruktur der embryonalen menschlichen Linsenblase, *Albrecht v. Graefes Arch. Ophthal.*, **173**, 141–52.

Wurtman, R. J., Axelrod, J. and Kelly, D. E. (1968). *The Pineal*, Academic Press: New York, London.

Wurtman, R. J. and Pohorecky, L. A. (1971). Adrenocortical control of epinephrine synthesis in health and disease, *Adv. Metab. Disord.*, **5**, 53–76.

Wyburn, G. M. (1937). The development of the infra-umbilical portion of the abdominal wall, and remarks on the aetiology of ectopia vesicae, *J. Anat.*, **71**, 201–31.

Wyburn, G. M. (1956). Uncertainties of anatomy of vascular innervation of lower limb, *Scot. med. J.*, **1**, 201–5.

Wyckoff, R. W. G. and Young, J. Z. (1956). The motoneurone surface, *Proc. R. Soc. B.*, **144**, 440–50.

Wyke, B. D. (1947). Clinical physiology of the cerebellum, *Med. J. Aust.*, **2**, 533–40.

Wyke, B. D. (1967). The neurology of joints, *Ann. R. Coll. Surg.*, **41**, 25–50.

Wynn, R. M. and French, G. L. (1968). Comparative ultrastructure of the mammalian amnion, *Obstet. Gynec., N.Y.*, **31**, 759–74.

Yates, R. D., Wood, J. G. and Duncan, D. (1962). Phase and electron microscope observations on two cell types in the adrenal medulla of the Syrian hamster, *Tex. Rep. Biol. Med.*, **20**, 494–502.

Yoffey, J. M. (1962). The present status of the lymphocyte problem, *Lancet*, **1**, 206–11.

Yoffey, J. M. (1962). A note on the thick-walled and thin-walled arteries of bone marrow, *J. Anat.*, **96**, 425.

Yoffey, J. M. (1964). The lymphocyte, *A. Rev. Med.*, **15**, 125–48.

Yoffey, J. M. and Courtice, F. C. (1970). *Lymphatics, Lymph and the Lymphomyeloid Complex*, Academic Press: New York, London.

Yohro, T. (1971). Nerve terminals and cellular junctions in young and adult mouse submandibular glands, *J. Anat.*, **108**, 409–17.

Yoshikawa, T. and Suzuki, T. (1969). Comparative anatomical study of the masseter of the mammal, *Anat. Anz.*, **125**, 363–87.

Yoss, R. E. (1952). Studies of the spinal cord. I. Topographic localization within the dorsal spinocerebellar tract in *Macaca mulatta*, *J. comp. Neurol.*, **97**, 5–20.

Yoss, R. E. (1953). Studies of the spinal cord. II. Topographic localization within the ventral spinocerebellar tract in the macaque, *J. comp. Neurol.*, **99**, 613–38.

Young, J. A. (1968). Microperfusion investigation of chloride fluxes across the epithelium of the main excretory duct of the rat submaxillary gland, *Pflügers Arch. ges. Physiol.*, **303**, 366–74.

Young, J. A. and van Lennep, E. E. (1978). *The Morphology of Salivary Glands*, Academic Press: London, New York.

Young, J. (1940). Relaxation of pelvic joints in pregnancy: pelvic arthropathy of pregnancy, *J. Obstet. Gynaec. Br. Commonw.*, **47**, 493–524.

Young, J. Z. (1958). Anatomical considerations, *Electroenceph. clin. Neurophysiol.*, Suppl. 10, 9–11.

Young, J. Z. (1962). *The Life of Vertebrates*, Oxford University Press: London.

Young, J. Z. (1964). *A Model of the Brain*, Clarendon Press: Oxford.

Young, J. Z. (1971). *An Introduction to the Study of Man*, Clarendon Press: Oxford.

Young, J. Z. (1978). *Programs of the Brain*, Oxford University Press: London.

Young, J. Z. and Zuckerman, S. (1936). The course of fibres in the dorsal roots of *Macaca mulatta*, the rhesus monkey, *J. Anat.*, **71**, 447–57.

Young, M. and Turnbull, H. M. (1931). Analysis of data collected by status lymphaticus investigating committee, *J. Path. Bact.*, **34**, 213–58.

Young, R. W. and Bok, D. (1969). Participation of the retinal pigment epithelium in the rod outer segment renewal process, *J. Cell Biol.*, **42**, 392–403.

Yuris, J. J. (1977). *New Chromosomal Syndromes*, Academic Press: London, New York.

X

Xuereb, G. P., Prichard, M. M. L. and Daniel, P. M. (1954(a)). The arterial supply and venous drainage of the human hypophysis cerebri, *Q. Jl exp. Physiol.*, **39**, 199–218.

Xuereb, G. P., Prichard, M. M. L. and Daniel, P. M. (1954(b)). The hypophyseal portal system of vessels in man, *Q. Jl exp. Physiol.*, **39**, 219–27.

Y

Yagita, K. (1910). Experimentelle Untersuchungen über den Ursprung des Nervus facialis, *Anat. Anz.*, **37**, 195–218.

Yamada, E. (1969). Some structural features of the fovea centralis in the human retina, *Archs Ophthal., N.Y.*, **82**, 151–9.

Yamamoto, M., Shimoyama, I. and Highstein, S. M. (1978). Vestibular nucleus neurons relaying excitation from the anterior canal to the oculomotor nucleus, *Brain Res.*, **148**, 31–42.

Yamauchi, A. (1973). Ultrastructure of the innervation of the mammalian heart. In: *Ultrastructure of the Mammalian Heart* (Challice, C. E. and Viragh, S. eds.), Academic Press: New York, London.

Yamauchi, A. and Burnstock, G. (1969). Postnatal development of smooth muscle cells in the mouse vas deferens. A fine structural study, *J. Anat.*, **104**, 1–15.

Yasuda, Y. (1973). Differentiation of human limb buds *in vitro*, *Anat. Rec.*, **175**, 561–78.

Z

Zaias, N. and Alvarez, J. (1968). The formation of the primate nail plate. An autoradiographic study in the squirrel monkey, *J. invest. Derm.*, **51**, 120–36.

Zaki, W. (1960). The trochlear nerve in man. Study relative to its origin, its intracerebral traject and its structure, *Archs Anat. Histol. Embryol.*, **43**, 105–20.

Zaki, W. (1973). Aspect morphologique et fonctionnel de l'annulus fibrosus du disque intervertébral de la colonne dorsale, *Archs Anat. Path.*, **21**, 401–3.

Zamboni, L. (1971). Acrosome loss in fertilising mammalian spermatozoa: a clarification, *J. Ultrastruct. Res.*, **34**, 401–5.

Zamboni, L. and Pease, D. C. (1961). The vascular bed of red bone marrow, *J. Ultrastruct. Res.*, **5**, 65–85.

Zamboni, L. and De Martino, C. (1968). Embryogenesis of the human renal glomerulus. I. A histologic study, *Arch. Path.*, **86**, 279–91.

Zamboni, L., Thompson, R. S. and Moore-Smith, D. (1972). Fine morphology of human oocyte maturation *in vitro*, *Biol. Reprod.*, **7**, 425–57.

Zambrano, D., Amezuo, L., Dickmann, G. and Franke, E. (1968). Ultrastructure of human pituitary adenomata, *Acta neurochir.*, **18**, 78–94.

Zangwill, O. L. (1960). *Cerebral Dominance and its Relations to Psychological Function*, Oliver & Boyd: Edinburgh.

Zangwill, O. L. (1967). Speech and the minor hemisphere, *Acta neurol. psychiat. Belg.*, **67**, 1013–20.

Zeeman, E. C. (1976). Catastrophe theory, *Sci. Am.*, **234**, 65–83.

Zeki, S. M. (1969). Representation of central visual fields in prestriate cortex of monkey, *Brain Res.*, **14**, 271–91.

BIBLIOGRAPHY

Zeki, S. M. (1970). Interhemispheric connections of prestriate cortex in monkey, *Brain Res.*, **19**, 63–75.

Zeki, S. M. (1974). The mosaic organization of the visual cortex in the monkey. In: *Essays on the Nervous System* (Bellairs, R. and Gray, E. G. eds.), pp. 327–43, Clarendon Press: Oxford.

Zeki, S. M. (1977). Simultaneous anatomical demonstration of the representation of the vertical and horizontal meridians in areas V2 and V3 of rhesus monkey visual cortex, *Proc. R. Soc.*, **195B**, 517–23.

Zeldis, S. M., Nemerson, Y., Pitlick, F. A. and Lentz, T. L. (1972). Tissue factor (thromboplastin): localization to plasma membranes by peroxidase-conjugated antibodies, *Science, N.Y.*, **175**, 766–8.

Zelickson, A. S. (1967). *Ultrastructure of Normal and Abnormal Skin* (editor), Kimpton: London.

Zelickson, A. S. (1971). Ultrastructure of the human epidermis. In: *Modern Trends in Dermatology* (Borrie, P. ed.), Volume 4, pp. 31–52. Butterworths: London.

Zimmerman, B. L. and Tso, M. O. M. (1975). Morphological evidence of photoreceptor differentiation of pinealocytes in the neonatal rat, *J. Cell Biol.*, **66**, 60–75.

Zimmerman, J. (1966). The functional and surgical anatomy of the heart, *Ann. R. Coll. Surg. Eng.*, **39**, 348–66.

Živanović, S. (1973). The menisco-fibular ligament of the knee joint, *Acta vet., Beograd.*, **23**, 89–94.

Živanović, S. (1974). Menisco-meniscal ligaments of the human knee joint, *Anat. Anz.*, **135**, 35–42.

Zucker-Franklin, D. (1967). Electron microscopic study of human basophils, *Blood*, **29**, 878–90.

Zucker-Franklin, D. (1968). Electron microscopic studies of human granulocytes: structural variations related to function, *Seminars Hemat.*, **5**, 109–33.

Zucker-Franklin, D. (1969). Microfibrils of blood platelets: their relationship to microtubules and the contractile protein, *J. clin. Invest.*, **48**, 165–75.

Zucker-Franklin, D. (1969). The ultrastructure of lymphocytes, *Seminars Hemat.*, **6**, 4–27.

Zuckerkandl, E. (1901). Ueber Nebenorgane des Sympatheticus in Retroperitoneaolraum des Menschen, *Anat. Anz.*, **15**, 97.

Zuckerman, S. (1940). The histogenesis of tissues sensitive to oestrogens, *Biol. Rev.*, **15**, 231–72.

Zuckerman, S. (1951). The number of oöcytes in the mature ovary. In: *Recent Progress in Hormone Research* (Pincus, G. ed.), Volume 6, pp. 63–110, Academic Press: New York, London.

Zuckerman, S. (1956). The regenerative capacity of ovarian tissue. In: *Colloquia on Ageing*, Volume 2, *Ageing in Transient Tissues* (Wolstenholme, G. E. W. and Millar, E. C. P. eds.), pp. 31–52, Ciba Foundation Symposium, Churchill: London.

Zuckerman, S., Ashton, E. H., Flinn, R. M., Oxnard, C. E. and Spence, T. E. (1973). Some locomotor features of the pelvic girdle in Primates, *Symp. zool. Soc. Lond.*, **33**, 71–166.

Zweifach, B. W. (1959). The microcirculation of the blood, *Scient. Am.*, **200**, 54–60.

Zweifach, B. W. (1961). The structural basis of the microcirculation. In: *Development and Structure of the Cardiovascular System* (Luisada, A. A. ed.), pp. 198–205, McGraw-Hill: New York, London.

Zweifach, B. W. and Kossmann, C. E. (1937). Micromanipulation of small blood vessels in mouse, *Am. J. Physiol.*, **120**, 23–35.

Zweifach, B. W. and Metz, D. B. (1955). Selective distribution of blood through the terminal vascular bed of mesenteric structures and skeletal muscle, *Angiology*, **6**, 282–9.

Zwilling, E. (1961). Limb morphogenesis. In: *Advances in Morphogenesis* (Abercrombie, M. and Brachet, J. eds.), **1**, pp. 301–30, Academic Press: New York, London.

INDEX

INDEX

Main references are given in bold figures

INDEX

1536

G

INDEX

M

INDEX

INDEX

X

Y

Z

W